CHILTON®

Asian
SERVICE MANUAL
2010 EDITION
VOLUME IV
MAZDA
MITSUBISHI
SUBARU
SUZUKI

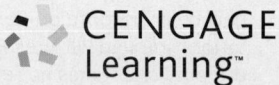

CENGAGE
Learning™

Australia • Brazil • Japan • Korea • Mexico • Singapore • Spain • United Kingdom • United States

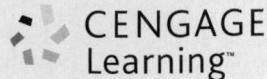
CENGAGE
Learning™

CHILTON®
Asian Service Manual
2010 Edition
Volume IV
Mazda, Mitsubishi, Subaru, Suzuki

Vice President,
Technology Professional
Business Unit:
  Gregory L. Clayton

Publisher,
Technology Professional
Business Unit:
  David Koontz

Director of Marketing:
  Beth A. Lutz

Director Education Production:
  Carolyn Miller

Marketing Manager:
  Jennifer Barbic

Marketing Coordinator:
  Rachael Torres

Chilton Content Specialist:
  Paula Baillie

Graphical Designer:
  Melinda Possinger

Art Director:
  Benjamin Gleeksman

Sr. Content Project Manager:
  Elizabeth C. Hough

Senior Editor:
  Christine L. Sheeky

Editors:
  Jim Bailey

  Nick D'Andrea

  Eugene F. Hannon Jr., A.S.E.

  John Howard

  Maureen Lazarz

  David G. Olson

  Jon Wallace

© 2011 Chilton, a part of Cengage Learning

For product information and technology assistance, contact us at
Professional & Career Group customer Support, 1-800-648-7450.
For permission to use material from this text or product,
submit all requests online at
www.cengage.com/permissions.
Further permissions questions can be e-mailed to
permissionrequest@cengage.com

ISBN-13: 978-1-1110-3767-3
ISBN-10: 1-1110-3767-1
ISSN: 1939-621X

Chilton
5 Maxwell Drive
Clifton Park, NY 12065-2919
USA

Chilton products are represented in Canada by Nelson Education, Ltd.

NOTICE TO THE READER

Printed in the United States of America
1 2 3 4 5 6 7 15 14 13 12 11

Contents

Model Index

USING THIS INFORMATION

Organization

To find where a particular model section or procedure is located, look in the Table of Contents. Main topics are listed with the page number on which they may be found. Following the main topics is an alphabetical listing of all of the procedures within the section and their page numbers.

Manufacturer and Model Coverage

This product covers 2009–2010 Asian models that are produced in sufficient quantities to warrant coverage, and which have technical content available from the vehicle manufacturers before our publication date. Although this information is as complete as possible at the time of publication, some manufacturers may make changes which cannot be included here. While striving for total accuracy, the publisher cannot assume responsibility for any errors, changes, or omissions that may occur in the compilation of this data.

Part Numbers and Special Tools

Part numbers and special tools are recommended by the publisher and vehicle manufacturer to perform specific jobs. Before substituting any part or tool for the one recommended, you must be completely satisfied that neither your personal safety, nor the performance of the vehicle will be endangered.

ACKNOWLEDGEMENT

The publisher would like to express appreciation to the following vehicle manufacturers for their assistance in producing this manual: Mazda Motor Corporation, Mitsubishi Motors North America, Inc., Suzuki Motor Corporation. No further reproduction or distribution of the material in this manual is allowed without the expressed written permission of the vehicle manufacturers and the publisher.

PRECAUTIONS

Before servicing any vehicle, please be sure to read all of the following precautions, which deal with personal safety, prevention of component damage, and important points to take into consideration when servicing a motor vehicle:

• Always wear safety glasses or goggles when drilling, cutting, grinding or prying.

• Steel-toed work shoes should be worn when working with heavy parts. Pockets should not be used for carrying tools. A slip or fall can drive a screwdriver into your body.

• Work surfaces, including tools and the floor should be kept clean of grease, oil or other slippery material.

• When working around moving parts, don't wear loose clothing. Long hair should be tied back under a hat or cap, or in a hair net.

• Always use tools only for the purpose for which they were designed. Never pry with a screwdriver.

• Keep a fire extinguisher and first aid kit handy.

• Always properly support the vehicle with approved stands or lift.

• Always have adequate ventilation when working with chemicals or hazardous material.

• Carbon monoxide is colorless, odorless and dangerous. If it is necessary to operate the engine with vehicle in a closed area such as a garage, always use an exhaust collector to vent the exhaust gases outside the closed area.

• When draining coolant, keep in mind that small children and some pets are attracted by ethylene glycol antifreeze, and are quite likely to drink any left in an open container, or in puddles on the ground. This will prove fatal in sufficient quantity. Always drain the coolant into a sealable container.

• To avoid personal injury, do not remove the coolant pressure relief cap while the engine is operating or hot. The cooling system is under pressure; steam and hot liquid can come out forcefully when the cap is loosened slightly. Failure to follow these instructions may result in personal injury. The coolant must be recovered in a suitable, clean container for reuse. If the coolant is contaminated it must be recycled or disposed of correctly.

• When carrying out maintenance on the starting system be aware that heavy gauge leads are connected directly to the battery. Make sure the protective caps are in place when maintenance is completed. Failure to follow these instructions may result in personal injury.

• Do not remove any part of the engine emission control system. Operating the engine without the engine emission control system will reduce fuel economy and engine ventilation. This will weaken engine performance and shorten engine life. It is also a violation of Federal law.

• Due to environmental concerns, when the air conditioning system is drained, the refrigerant must be collected using refrigerant recovery/recycling equipment. Federal law requires that refrigerant be recovered into appropriate recovery equipment and the process be conducted by qualified technicians who have been certified by an approved organization, such as MACS, ASI, etc. Use of a recovery machine dedicated to the appropriate refrigerant is necessary to reduce the possibility of oil and refrigerant incompatibility concerns. Refer to the instructions provided by the equipment manufacturer when removing refrigerant from or charging the air conditioning system.

• Always disconnect the battery ground when working on or around the electrical system.

• Batteries contain sulfuric acid. Avoid contact with skin, eyes, or clothing. Also, shield your eyes when working near batteries to protect against possible splashing of the acid solution. In case of acid contact with skin or eyes, flush immediately with water for a minimum of 15 minutes and get prompt medical attention. If acid is swallowed, call a physician immediately. Failure to follow these instructions may result in personal injury.

• Batteries normally produce explosive gases. Therefore, do not allow flames, sparks or lighted substances to come near the battery. When charging or working near a battery, always shield your face and protect your eyes. Always provide ventilation. Failure to follow these instructions may result in personal injury.

- When lifting a battery, excessive pressure on the end walls could cause acid to spew through the vent caps, resulting in personal injury, damage to the vehicle or battery. Lift with a battery carrier or with your hands on opposite corners. Failure to follow these instructions may result in personal injury.

- Observe all applicable safety precautions when working around fuel. Whenever servicing the fuel system, always work in a well-ventilated area. Do not allow fuel spray or vapors to come in contact with a spark, open flame, or excessive heat (a hot drop light, for example). Keep a dry chemical fire extinguisher near the work area. Always keep fuel in a container specifically designed for fuel storage; also, always properly seal fuel containers to avoid the possibility of fire or explosion. Do not smoke or carry lighted tobacco or open flame of any type when working on or near any fuel-related components.

- Fuel injection systems often remain pressurized, even after the engine has been turned OFF. The fuel system pressure must be relieved before disconnecting any fuel lines. Failure to do so may result in fire and/or personal injury.

- The evaporative emissions system contains fuel vapor and condensed fuel vapor. Although not present in large quantities, it still presents the danger of explosion or fire. Disconnect the battery ground cable from the battery to minimize the possibility of an electrical spark occurring, possibly causing a fire or explosion if fuel vapor or liquid fuel is present in the area. Failure to follow these instructions can result in personal injury.

- The EPA warns that prolonged contact with used engine oil may cause a number of skin disorders, including cancer! You should make every effort to minimize your exposure to used engine oil. Protective gloves should be worn when changing oil. Wash your hands and any other exposed skin areas as soon as possible after exposure to used engine oil. Soap and water, or waterless hand cleaner should be used.

- Some vehicles are equipped with an air bag system, often referred to as a Supplemental Restraint System (SRS) or Supplemental Inflatable Restraint (SIR) system. The system must be disabled before performing service on or around system components, steering column, instrument panel components, wiring and sensors. Failure to follow safety and disabling procedures could result in accidental air bag deployment, possible personal injury and unnecessary system repairs.

- Always wear safety goggles when working with, or around, the air bag system. When carrying a non-deployed air bag, be sure the bag and trim cover are pointed away from your body. When placing a non-deployed air bag on a work surface, always face the bag and trim cover upward, away from the surface. This will reduce the motion of the module if it is accidentally deployed.

- Electronic modules are sensitive to electrical charges. The ABS module can be damaged if exposed to these charges.

- Brake pads and shoes may contain asbestos, which has been determined to be a cancer-causing agent. Never clean brake surfaces with compressed air. Avoid inhaling brake dust. Clean all brake surfaces with a commercially available brake cleaning fluid.

- When replacing brake pads, shoes, discs or drums, replace them as complete axle sets.

- When servicing drum brakes, disassemble and assemble one side at a time, leaving the remaining side intact for reference.

- Brake fluid often contains polyglycol ethers and polyglycols. Avoid contact with the eyes and wash your hands thoroughly after handling brake fluid. If you do get brake fluid in your eyes, flush your eyes with clean, running water for 15 minutes. If eye irritation persists, or if you have taken brake fluid internally, immediately seek medical assistance.

- Clean, high quality brake fluid from a sealed container is essential to the safe and proper operation of the brake system. You should always buy the correct type of brake fluid for your vehicle. If the brake fluid becomes contaminated, completely flush the system with new fluid. Never reuse any brake fluid. Any brake fluid that is removed from the system should be discarded. Also, do not allow any brake fluid to come in contact with a painted or plastic surface; it will damage the paint.

- Never operate the engine without the proper amount and type of engine oil; doing so will result in severe engine damage.

- Timing belt maintenance is extremely important! Many models utilize an interference- type, non freewheeling engine. If the timing belt breaks, the valves in the cylinder head may strike the pistons, causing potentially serious (also time-consuming and expensive) engine damage.

- Disconnecting the negative battery cable on some vehicles may interfere with the functions of the on-board computer system(s) and may require the computer to undergo a relearning process once the negative battery cable is reconnected.

- Steering and suspension fasteners are critical parts because they affect performance of vital components and systems and their failure can result in major service expense. They must be replaced with the same grade or part number or an equivalent part if replacement is necessary. Do not use a replacement part of lesser quality or substitute design. Torque values must be used as specified during reassembly.

MAZDA

1

B2300 • B4000

SPECIFICATIONS AND MAINTENANCE CHARTS

ENGINE AND VEHICLE IDENTIFICATION

| | | Engine | | | | | Model Year | |
Code ①	Liters (cc)	Cu. In.	Cyl.	Fuel Sys.	Type	Eng. Mfg.	Code ②	Year
D	2.3 (2261)	138	4	EFI	DOHC	Ford	9	2009
E	4.0 (3998)	244	6	EFI	DOHC	Ford		

EFI: Electronic fuel injection

DOHC: Dual Overhead Camshafts

① 8th digit of the Vehicle Identification Number (VIN)

② 10th digit of the Vehicle Identification Number (VIN)

37671_BPUP_C0001

GENERAL ENGINE SPECIFICATIONS

Year	Model	Engine Displacement Liters	Engine (VIN)	Net Horsepower @ rpm	Net Torque @ rpm (ft. lbs.)	Bore x Stroke (in.)	Compression Ratio	Oil Pressure @ rpm
2009	B2300	2.3	D	143@5250	154@3750	3.45X3.70	9.7:1	40-60@2000
	B4000	4.0	E	207@5250	238@3000	3.95x3.35	9.0:1	40-60@2000

37671_BPUP_C0002

GASOLINE ENGINE TUNE-UP SPECIFICATIONS

Year	Engine Displacement Liters	Engine VIN	Spark Plug Gap (in.)	Ignition Timing (deg.) ① MT	Ignition Timing (deg.) ① AT	Fuel Pump (psi)	Idle Speed (rpm) MT	Idle Speed (rpm) AT	Valve Clearance In.	Valve Clearance Ex.
2009	2.3	D	0.041-0.045	10B	10B	56-72	①	①	HYD	HYD
	4.0	E	0.061-0.068	10B	10B	35-45	①	①	HYD	HYD

NOTE: The Vehicle Emission Control Information label often reflects specification changes changes made during production. The label figures must be used if they differ from those in this chart.

B: Before top dead center

HYD: Hydraulic

① Electronically controlled and cannot be adjusted

37671_BPUP_C0003

CAPACITIES

Year	Model	Engine Displacement Liters	Engine VIN	Engine Oil with Filter (qts.)	Transmission (pts.) 5-Spd	Transmission (pts.) Auto.	Transfer Case (pts.)	Drive Axle Front (pts.)	Drive Axle Rear (pts.)	Fuel Tank (gal.)	Cooling System (qts.)
2009	B2300	2.3	D	4.0	5.6	19.8	NA	NA	5.0	①	②
	B4000	4.0	E	5.0	5.6	③	2.50	3.25	5.0	①	④

NOTE: All capacities are approximate. Add fluid gradually and check to be sure a proper fluid level is obtained.

NA: not available

① Regular Cab short wheel base: 17 gal.

   Regular Cab long wheel base: 20.3 gal.

   Super Cab: 19.5 gal.

② w/MT: 11.2

   w/AT: 10.9

③ 2WD: 20.0

   4WD: 20.6

④ w/MT: 15.2

   w/AT: 14.8

37671_BPUP_C0004

FLUID SPECIFICATIONS

Year	Model	Engine Displacement Liters	Engine ID/VIN	Engine Oil	Auto. Trans.	Drive Axle	Power Steering Fluid	Brake Master Cylinder
2009	B-Series	2.3	D	5W-20 ①	Mercon® V	80W-90 ② ③	Mercon®	DOT 3
		4.0	E	5W-20 ①	Mercon® V	80W-90 ② ③	Mercon®	DOT 3

DOT: Department Of Transportation

® Registered trademark

① Synthetic motor oil is recommended

② 8.8 inch. ring gear high torque 75W-140 synthetic lubricant

③ Friction modifier XL-3

37671_BPUP_C0015

VALVE SPECIFICATIONS

Year	Engine Displacement Liters	Engine VIN	Seat Angle (deg.)	Face Angle (deg.)	Spring Test Pressure (lbs. @ in.)	Spring Installed Height (in.)	Stem-to-Guide Clearance (in.) Intake	Stem-to-Guide Clearance (in.) Exhaust	Stem Diameter (in.) Intake	Stem Diameter (in.) Exhaust
2009	2.3	D	45	45	①	1.129	0.0009-0.0027	0.0012-0.0029	0.2154-0.2159	0.2152-0.2157
	4.0	E	45	45	202-225@ 1.413-1.445	1.569-1.608	0.0010-0.0020	0.0010-0.0030	0.2740-0.2750	0.2740

NA: not available

① Intake: 97.0@1.201

   Exhaust: 93.3@1.201

37671_BPUP_C0005

CAMSHAFT SPECIFICATIONS CHART

All measurements are given in inches.

Year	Engine Displ. Liters	Engine VIN	Journal Dia.	Brg. Oil Clearance	Shaft End-play	Runout	Lobe Height Intake	Exhaust
2009	2.3	D	0.9827-0.9834	0.0014-0.0031	0.0035-0.0094	0.0020	1.6710 0.1569	1.6210 0.1569
	4.0	E	1.0990-1.1010	0.0020-0.0040	0.0003-0.0070	0.0012	0.2590	0.2590

NA - Not available

37671_BPUP_C0007

CRANKSHAFT AND CONNECTING ROD SPECIFICATIONS

All measurements are given in inches.

Year	Engine Displacement Liters	Engine VIN	Crankshaft Main Brg. Journal Dia.	Main Brg. Oil Clearance	Shaft End-play	Thrust on No.	Connecting Rod Journal Diameter	Oil Clearance	Side Clearance
2009	2.3	D	2.0460-2.0470	0.0007-0.0013	0.0080-0.0160	3	1.9670-1.9680	0.0010-0.0020	0.0760-0.1200
	4.0	E	2.2430-2.2440	0.0008-0.0015	0.0020-0.0126	3	2.1250-2.1260	0.0003-0.0024	0.0036-0.0106

37671_BPUP_C0008

PISTON AND RING SPECIFICATIONS

All measurements are given in inches.

Year	Engine Displacement Liters	Engine VIN	Piston Clearance	Ring Gap Top Compression	Bottom Compression	Oil Control	Ring Side Clearance Top Compression	Bottom Compression	Oil Control
2009	2.3	D	0.0009-0.0017	0.0060-0.0012	0.0120-0.0180	0.0070-0.0270	0.0008-0.0013	0.0004-0.0011	0.0025-0.0054
	4.0	E	0.0008-0.0019	0.015-0.023	0.015-0.023	0.015-0.055	0.0010-0.0030	0.0010-0.0030	SNUG

37671_BPUP_C0006

TORQUE SPECIFICATIONS
All readings in ft. lbs.

	Engine Displacement Liters	Engine VIN	Cylinder Head Bolts	Main Bearing Bolts	Rod Bearing Bolts	Crankshaft Damper Bolts	Flywheel Bolts	Manifold		Spark Plugs	Oil Pan Drain Plug
								Intake *	Exhaust		
2009	2.3	D	①	⑫	⑬	②	③	13	40	9	21
	4.0	E	⑪	⑦	⑧	⑨	⑩	7	16	13	18

NA: Information not available

* NOTE: Applies to Lower Manifold only.

① Step 1: 44 inch lbs.

  Step 2: 11 ft. lbs.

  Step 3: 33 ft. lbs.

  Step 4: +90 degrees

  Step 5: +90 degrees

② Step 1: 74 ft. lbs.

  Step 2: +90 degrees

③ Step 1: 37 ft. lbs.

  Step 2: 59 ft. lbs.

  Step 3: 83 ft. lbs.

④ Step 1: 59 ft. lbs.

  Step 2: Back off 1 full turn

  Step 3: 34-40 ft. lbs.

  Step 4: 63-73 ft. lbs.

⑤ Step 1: 89 inch lbs.

  Step 2: 17 ft. lbs.

⑥ 8mm bolts: 24 ft. lbs.

  12mm bolts: 24 ft. lbs. +80 degrees, +80 degrees

⑦ Step 1: 26 ft. lbs.

  Step 2: +57 degrees

⑧ Step 1: 15 ft. lbs.

  Step 2: +90 degrees

⑨ Step 1: 37 ft. lbs.

  Step 2: +90 degrees

⑩ Step 1: 10 ft.lbs.

  Step 2: 52 ft. lbs.

⑪ Step 1: 12mm bolts 9 ft. lbs.

  Step 2: 12mm bolts 18 ft. lbs.

  Step 3: 8mm bolts 24 ft. lbs.

  Step 4: 12 mm bolts +90 degrees

  Step 5: 12mm bolts +90 degrees

⑪ Step 1: 26 ft. lbs.

  Step 2: +57 degrees

37671_BPUP_C0009

WHEEL ALIGNMENT

Year	Model		Caster		Camber		Toe-in (in.)
			Range (+/-Deg.)	Preferred Setting (Deg.)	Range (+/-Deg.)	Preferred Setting (Deg.)	
2009	B Series	2WD	1.0	①	0.70	-0.50	0.06+/-0.25
		4WD	1.0	②	0.70	-0.50	0.12+/-0.25
		Rear	—	—	0.75	0	0

① Left: +3.5

  Right: +3.9

② Left: +2.8

  Right: +3.5

37671_BPUP_C0012

TIRE, WHEEL AND BALL JOINT SPECIFICATIONS

Year	Model	OEM Tires		Tire Pressures (psi)		Wheel Size	Ball Joint Inspection	Lug Nut Torque (ft. lbs.)
		Standard	Optional	Front	Rear			
2009	B2300	P225/70R15	none	①	①	7-JJ	②	100
	B4000	P235/75R15	none	①	①	7-JJ	②	100

OEM: Original Equipment Manufacturer

PSI: Pounds Per Square Inch

① See placard on door post

② Upper 0 - 0.032
  Lower 0 - 0.008

37671_BPUP_C0011

BRAKE SPECIFICATIONS

All measurements in inches unless noted

Year	Model	Brake Disc			Brake Drum Diameter			Minimum Lining Thickness	Brake Caliper	
		Original Thickness	Minimum Thickness	Maximum Runout	Original Inside Diameter	Max. Wear Limit	Maximum Machine Diameter		Bracket Bolts (ft. lbs.)	Mounting Bolts (ft. lbs.)
2009	B-Series	NA	0.988	0.0030	10.00	NA	①	0.0030	85	27

NOTE: Due to changes made during production, refer to manufacturer's specifications if they differ from those in this chart

NA: Not Available

① Molded into the drum

37671_BPUP_C0010

SCHEDULED MAINTENANCE INTERVALS
MAZDA B SERIES

TO BE SERVICED	TYPE OF SERVICE	VEHICLE MILEAGE INTERVAL (x1000)												
		5	10	15	20	25	30	35	40	45	50	55	60	65
Engine oil & filter	R	✓	✓	✓	✓	✓	✓	✓	✓	✓	✓	✓	✓	✓
Tires	Rotate	✓	✓	✓	✓	✓	✓	✓	✓	✓	✓	✓	✓	✓
Auto trans. fluid	I									✓				
Brake pads/shoes	I			✓			✓			✓			✓	
Wheel ends ①	I			✓			✓			✓			✓	
Coolant hoses	S/I			✓			✓			✓			✓	
Steering linkage	I/L			✓			✓			✓			✓	
Suspension and driveshaft	I			✓			✓			✓			✓	
Cabin air filter	R			✓			✓			✓			✓	
Ball joints (2WD)	I/L			✓			✓			✓			✓	
Exhaust system	I						✓						✓	
Engine air filter	R						✓						✓	
Fuel filter	R						✓						✓	
Manual trans fluid	R												✓	
Auto trans fluid	R						✓						✓	
Wheel bearings (2WD) ②	I/L												✓	
Premium Gold coolant	I												✓	
Spark plugs	R	every 100,000 miles												
PCV valve ③	R	every 60,000 miles												
Rear axle fluid	R	every 150,000 miles												
Accessory drive belts	I	every 100,000 miles												
Accessory drive belts	R	every 150,000 miles, if not previously replaced												

R: Replace   S: Service   I: Inspect   L: Lubricate

① Check for play and noise
② Replace after 150,000 miles
③ At 60,000 miles (96,000 km), the dealer will replace the PCV valve at no cost, except Canada and CA

Special Operating Condition Requirements

When towing a trailer or using a camper or car-top carrier:

Change engine oil and install a new oil filter every 4,800 km (3,000 miles), 3 months or 200 hours of engine operation (whichever occurs first).

Change transfer case fluid every 96,000 km (60,000 miles).

Change manual transmission fluid as required.

Change automatic transmission fluid, lubricate 4x2 wheel bearings, install new grease seals and adjust bearings every

During extensive idling and/or low speed driving for long distances, as in heavy commercial use such as delivery, taxi or livery:

Change engine oil and install a new oil filter every 4,800 km (3,000 miles), 3 months or 200 hours of engine operation (whichever occurs first).

Lube front lower control arm and steering linkage ball joints with zerk fittings (if equipped) every 4,800 km (3,000 miles) or 3 months.

Inspect brake system and check battery electrolyte level (Patrol cars) every 8,000 km (5,000 miles).

Install a new fuel filter every 24,000 km (15,000 miles).

Change automatic transmission fluid, lubricate 4x2 wheel bearings, install new grease seals and adjust bearings every

48,000 km (30,000 miles). If equipped, change the in-line service installed transmission fluid filter.

Install new spark plugs and change transfer case fluid every 96,000 km (60,000 miles).

Change transfer case fluid every 96,000 km (60,000 miles).

Install a new cabin air filter as required.

When operating in dusty conditions such as unpaved or dusty roads:

Change engine oil and install a new oil filter every 3,000 miles (4,800 km) or 3 months.

Install a new fuel filter every 24,000 km (15,000 miles).

Change automatic transmission fluid every 48,000 km (30,000 miles). If equipped, change the in-line service installed transmission fluid filter.

Install a new engine air filter as required.

Install a new cabin air filter as required.

When operating in off-road conditions:

Change automatic transmission fluid every 48,000 km (30,000 miles). If equipped, change the in-line service installed transmission fluid filter.

Change transfer case fluid every 96,000 km (60,000 miles).

Install a new cabin air filter as required.

Inspect and lubricate U-joints.

37671_BPUP_C0013

BRAKES   INFORMATION AND PRECAUTIONS

ANTI-LOCK SYSTEMS

• Certain components within the ABS system are not intended to be serviced or repaired individually.

• Do not use rubber hoses or other parts not specifically specified for and ABS system. When using repair kits, replace all parts included in the kit. Partial or incorrect repair may lead to functional problems and require the replacement of components.

• Lubricate rubber parts with clean, fresh brake fluid to ease assembly. Do not use shop air to clean parts; damage to rubber components may result.

• Use only DOT 3 brake fluid from an unopened container.

• If any hydraulic component or line is removed or replaced, it may be necessary to bleed the entire system.

• A clean repair area is essential. Always clean the reservoir and cap thoroughly before removing the cap. The slightest amount of dirt in the fluid may plug an orifice and impair the

system function. Perform repairs after components have been thoroughly cleaned; use only denatured alcohol to clean components. Do not allow ABS components to come into contact with any substance containing mineral oil; this includes used shop rags.

• The Anti-Lock control unit is a microprocessor similar to other computer units in the vehicle. Ensure that the ignition switch is **OFF** before removing or installing controller harnesses. Avoid static electricity discharge at or near the controller.

• If any arc welding is to be done on the vehicle, the control unit should be unplugged before welding operations begin.

DISC AND DRUM SYSTEMS

> #### ❋❋ CAUTION
> Dust and dirt accumulating on brake parts during normal use may contain asbestos fibers from production or aftermarket brake linings. Breathing

excessive concentrations of asbestos fibers can cause serious bodily harm. Exercise care when servicing brake parts. Do not sand or grind brake lining unless equipment used is designed to contain the dust residue. Do not clean brake parts with compressed air or by dry brushing. Cleaning should be done by dampening the brake components with a fine mist of water, then wiping the brake components clean with a dampened cloth. Dispose of cloth and all residue containing asbestos fibers in an impermeable container with the appropriate label. Follow practices prescribed by the Occupational Safety and Health Administration (OSHA) and the Environmental Protection Agency (EPA) for the handling, processing, and disposing of dust or debris that may contain asbestos fibers.

BRAKES   BLEEDING THE BRAKE SYSTEM

BLEEDING PROCEDURE

BLEEDING PROCEDURE

> #### ❋❋ WARNING
> Use of any other than the approved DOT 3 brake fluid will cause permanent damage to brake components and will render the brakes inoperative. Failure to follow these instructions may result in personal injury.

> #### ❋❋ WARNING
> Brake fluid contains polyglycol ethers and polyglycols. Avoid contact with eyes. Wash hands thoroughly after handling. If brake fluid contacts eyes, flush eyes with running water for 15 minutes. Get medical attention if irritation persists. If taken internally, drink water and induce vomiting. Get medical attention immediately. Failure to follow these instructions may result in personal injury.

> #### ❋❋ CAUTION
> Do not allow the brake master cylinder reservoir to run dry during the bleeding operation. Keep the master

cylinder reservoir filled with the specified brake fluid. Never reuse the brake fluid that has been drained from the hydraulic system.

> #### ❋❋ CAUTION
> Brake fluid is harmful to painted and plastic surfaces. If brake fluid is spilled onto a painted or plastic surface, immediately wash it with water.

➡ When any part of the hydraulic system has been disconnected or a new component is installed, air may enter the system, causing spongy brake pedal action. This requires the bleeding of the hydraulic system after it has been correctly connected.

> #### ❋❋ CAUTION
> Be sure to check the brake fluid level in the brake master cylinder reservoir often. Do not let it run dry.

1. Fill the brake master cylinder reservoir with the specified brake fluid.
2. Begin bleeding the system, going in order from the RH rear wheel, to the LH rear wheel, then to the RH front wheel and ending with the LH front wheel.
3. Attach a rubber drain hose to the rear bleeder screw and submerge the free end in

a container partially filled with clean brake fluid.
4. Have and assistant pump the brake pedal 10 times and then hold firm pressure on the brake pedal.
5. Loosen the bleeder screw until the fluid flow stops. Maintain pressure on the brake pedal and tighten the bleeder screw.
6. RepeatRepeat steps 4 and 5 until clear, bubble-free brake fluid flows.
7. Tighten the bleeder screw and refill the brake master cylinder reservoir as necessary.
8. Continue bleeding the brake hydraulic system at each wheel.
9. Fill the brake master cylinder reservoir with the specified brake fluid.

BLEEDING THE ABS SYSTEM

PRECAUTIONS

➡ Bleeding the Hydraulic Control Unit (HCU) is required only when removing or installing the HCU or master cylinder, or opening the lines to the HCU.

➡ Carrying out the System Bleed function drives trapped air from the HCU. Subsequent bleeding removes the air from the brake hydraulic system through the bleeder screws.

➡Adequate voltage to the HCU module is required during the anti-lock control portion of the system bleed.

1. Connect the WDS or equivalent Tester.

2. Access the SYSTEM BLEED FUNCTION. Go to the Tool Tab-Chassis-Braking-ABS Service Bleed and follow the directions on the diagnostic tool.

3. Manually bleed the brake hydraulic system.

4. Repeat the procedure carrying out a total of two diagnostic tool cycles and two manual bleed cycles.

MANUAL BLEEDING PROCEDURE

❋❋ CAUTION

Use of any other than approved DOT 3 brake fluid will cause permanent damage to brake components and will render the brakes inoperative. Failure to follow these instructions may result in personal injury.

❋❋ WARNING

Brake fluid contains polyglycol ethers and polyglycols. Avoid contact with eyes. Wash hands thoroughly after handling. If brake fluid contacts eyes, flush eyes with running water for 15 minutes. Get medical attention if irritation persists. If taken internally, drink water and induce vomiting. Get medical attention immediately. Failure to follow these instructions may result in personal injury.

❋❋ CAUTION

Brake fluid is harmful to painted and plastic surfaces. If brake fluid is spilled onto a painted or plastic surface, immediately wash it with water.

❋❋ CAUTION

Do not allow the brake master cylinder reservoir to run dry during the bleeding operation. Keep the brake master cylinder reservoir filled with the specified brake fluid. Never reuse the brake fluid that has been drained from the hydraulic system.

➡When any part of the hydraulic system has been disconnected for repair or replacement, air can get into the system and cause spongy brake pedal action. This requires bleeding of the hydraulic system after it has been properly connected. The hydraulic system can be bled manually or with pressure bleeding equipment.

➡Bleeding the Hydraulic Control Unit (HCU) is required only when removing or installing the HCU or master cylinder, or opening the lines to the HCU.

➡Carrying out the System Bleed function drives trapped air from the HCU. Subsequent bleeding removes the air from the brake hydraulic system through the bleeder screws.

1. Clean all dirt from and remove the brake master cylinder filler cap and fill the brake master cylinder reservoir with the specified brake fluid.

2. Place a box end wrench on the RH rear bleeder screw. Attach a rubber drain tube to the RH rear bleeder screw and submerge the free end of the tube in a container partially filled with clean brake fluid.

3. Have an assistant pump, and then hold firm pressure on, the brake pedal.

4. Loosen the RH rear bleeder screw until a stream of brake fluid comes out. While an assistant maintains pressure on the brake pedal, tighten the RH rear bleeder screw.

5. Repeat until clear, bubble-free fluid comes out.

6. Refill the brake master cylinder reservoir as necessary.

7. Tighten the RH rear bleeder screw.

8. Tighten to 71 inch lbs. (8 Nm).

9. Repeat Steps 2-5 for the LH rear bleeder screw.

10. Place a box end wrench on the RH front disc brake caliper bleeder screw. Attach a rubber drain tube to the RH front disc brake caliper bleeder screw, and submerge the free end of the tube in a container partially filled with clean brake fluid.

11. Have an assistant pump, and then hold firm pressure on, the brake pedal.

12. Loosen the RH front disc brake caliper bleeder screw until a stream of brake fluid comes out. While an assistant maintains pressure on the brake pedal, tighten the RH front disc brake caliper bleeder screw.

13. Repeat until clear, bubble-free fluid comes out.

14. Refill the brake master cylinder reservoir as necessary.

15. Tighten the RH front disc brake caliper bleeder screw.

16. Tighten to 11 ft. lbs. (15 Nm).

17. Repeat Steps 7-10 for the LH front disc brake caliper bleeder screw.

PRESSURIZED BLEEDING PROCEEDURE

➡Master cylinder pressure bleeder adapter tools are available from various manufacturers of pressure bleeding equipment. Follow the instructions of the manufacturer when installing the adapter. Install the bleeder adapter to the brake master cylinder reservoir, and attach the bleeder tank hose to the fitting on the adapter.

1. Bleed the longest line first. Make sure the bleeder tank contains enough specified brake fluid to complete the bleeding operation. Place a box-end wrench on the RH rear bleeder screw. Attach a rubber drain tube to the RH rear bleeder screw, and submerge the free end of the tube in a container partially filled with clean brake fluid.

2. Open the valve on the bleeder tank.

3. Loosen the RH rear bleeder screw. Leave open until clear, bubble-free brake fluid flows, then tighten the RH rear bleeder screw and remove the rubber hose.

4. Tighten to 71 inch lbs. (6 Nm).

5. Continue bleeding the rest of the system, going in order from the LH rear bleeder screw to the RH front disc brake caliper bleeder screw, ending with the LH front disc brake caliper bleeder screw.

6. Tighten the brake caliper bleeder screws to 11 ft. lbs. (15 Nm).

7. Close the bleeder tank valve. Remove the tank hose from the adapter, and remove the adapter.

BRAKES

ANTI-LOCK BRAKE SYSTEM (ABS)

WHEEL SPEED SENSORS

REMOVAL & INSTALLATION

Front

See Figure 1.

1. Before servicing the vehicle, refer to the Precautions Section.
2. Disconnect the negative battery cable.
3. Remove the front wheel and tire assembly.

➡**The harness connector is located in the engine compartment.**

4. Disconnect the harness electrical connector.
5. Remove the brake disc (4WD models)
6. Remove the 3 brake disc shield bolts and the brake disc shield. (4WD models)

7. Release the grommet from the body and pull the connector through.
8. Release the sensor harness retainer.

➡**Clean the area surrounding the wheel speed sensor and bolt before removal.**

9. Remove the wheel speed sensor bolt.

➡**Thoroughly clean the wheel speed sensor and wheel speed sensor mounting surface. Apply wheel bearing grease to the mounting surface.**

To install:

10. Install the wheel speed sensor bolt.
11. Tighten wheel speed sensor bolt to 9 ft. lbs. (12 Nm).
12. Install the sensor harness retainer.

13. Tighten wheel speed sensor bolt to 11 ft. lbs. (15 Nm).
14. Connect the harness electrical connector.
15. Install the 3 brake disc shield bolts and the brake disc shield. (4WD models).
16. Tighten the 3 brake disc shield bolts to 9 ft. lbs. (12 Nm).
17. Install the brake disc (4WD models)
18. Connect the negative battery cable.

Rear

➡**On 4WD vehicles, the rear sensor ring is integral to the rear axle half-shaft and cannot be repaired separately.**

1. Before servicing the vehicle, refer to the Precautions Section.
2. Disconnect the negative battery cable.
3. Raise the vehicle on a hoist.
4. Release the grommet and pull the connector through the body.
5. Disconnect the wheel speed sensor harness electrical connector.
6. Remove the wheel speed sensor harness bolts.

➡**Clean the area surrounding the wheel speed sensor and bolt before removal.**

7. Remove the wheel speed sensor bolt and wheel speed sensor.
8. On FWD vehicles, remove the rear ABS sensor nut and ring.

➡**Thoroughly clean the wheel speed sensor and mounting surface before installation.**

9. To install, reverse the removal procedure. Tighten ABS sensor and ring to 214 ft. lbs. (290 Nm). Tighten all remaining fasteners to 80 inch lbs. (9 Nm).

WHEEL SPEED SENSOR RINGS

REMOVAL & INSTALLATION

The front tone ring for 2WD models is part of the front brake rotor. Refer to the Brake Rotor removal and installation. On 4WD models, the tone ring is incorporated into the hub and bearing assembly. Refer to the Front Hub and Bearing replacement. The rear (exciter) tone ring is inside the rear axle housing affixed to the ring gear.

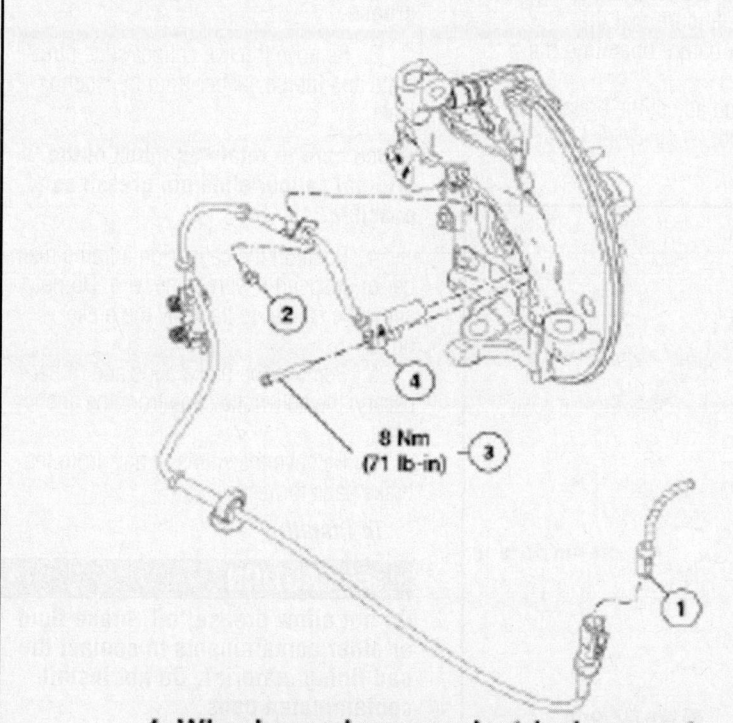

8 Nm
(71 lb-in)

1. **Wheel speed sensor electrical connector**
2. **Wheel speed sensor harness bolt**
3. **Wheel speed sensor bolt**
4. **Wheel speed sensor**

22140_BPUP_G0026

Fig. 1 Front wheel sensor and wiring exploded view

BRAKES

FRONT DISC BRAKES

BRAKE CALIPER

REMOVAL & INSTALLATION

See Figure 2.

1. Before servicing the vehicle, refer to the Precautions Section.
2. Loosen the wheel lug nuts.
3. Raise and safely support the front of the vehicle. Remove the wheel.
4. Remove the brake caliper flow bolt and position the brake hose aside.
5. Discard the 2 copper washers.

➡**Plug the brake hose.**

6. Remove the two caliper slide pin bolts and lift the caliper from the anchor plate.

➡**Use care to retain as much of the original caliper slide pin grease as possible.**

7. Position the caliper on a frame member or suspend it with some wire. Do not allow the caliper to hang by the brake hose.
8. Disconnect and plug the brake hose at the caliper. Remove the caliper from the rotor.

To install:

9. Position the caliper over the brake pads and align the slide pin mounting holes.
10. Install the slide pin bolts.
11. Tighten the bolts to 27 ft. lbs. (36 Nm)
12. Install the caliper brake hose using new washers. Tighten the bolt to 25 ft. lbs. (34 Nm).
13. Bleed the brake caliper.
14. Install the wheel and snug the lug nuts.
15. Lower the vehicle and tighten the lug nuts to 100 ft. lbs. (135 Nm).

➡**The first couple of times you apply the brakes, the pedal may go to the floor. Continue to pump the brake pedal until it feels firm.**

16. Start the engine and apply the brakes several times to readjust the caliper pistons. Ensure that the pedal feels firm before operating the vehicle.
17. Check and adjust the brake fluid level, as required.

DISC BRAKE PADS

REMOVAL & INSTALLATION

See Figure 3.

1. Before servicing the vehicle, refer to the Precautions Section.
2. Raise and safely support the front of the vehicle. Remove the wheels.

✳✳ CAUTION

Do not pry in the caliper sight hole to retract the pistons, as this can damage the pistons and boots

➡**Do not remove the anchor pins unless installing new pins. Inspect the anchor plate guide pins. They should slide in and out of the anchor plates with no binding. Check the boots for cracks or tears. If the pins do not slide easily or the boots are cracked or torn, install new boots and pins. Lubricate the pins using silicone brake caliper grease.**

3. Remove the two caliper slide pin bolts and lift the caliper from the anchor plate.

➡**Use care to retain as much of the original caliper slide pin grease as possible.**

4. Position the caliper on a frame member or suspend it with some wire. Do not allow the caliper to hang by the brake hose.
5. Remove the brake pads and, if necessary, the anti-rattle clips from the anchor plate.
6. Remove the shims, if any, from the brake pads for re-use.

To install:

✳✳ CAUTION

Do not allow grease, oil, brake fluid or other contaminants to contact the pad lining material. Do not install contaminated pads.

7. If removed, install new anti-rattle clips.
8. Install the brake pads to the anchor plate.
9. Position the caliper over the brake pads and align the slide pin mounting holes.
10. Tighten the bottom caliper bolt before tightening the top caliper bolt.

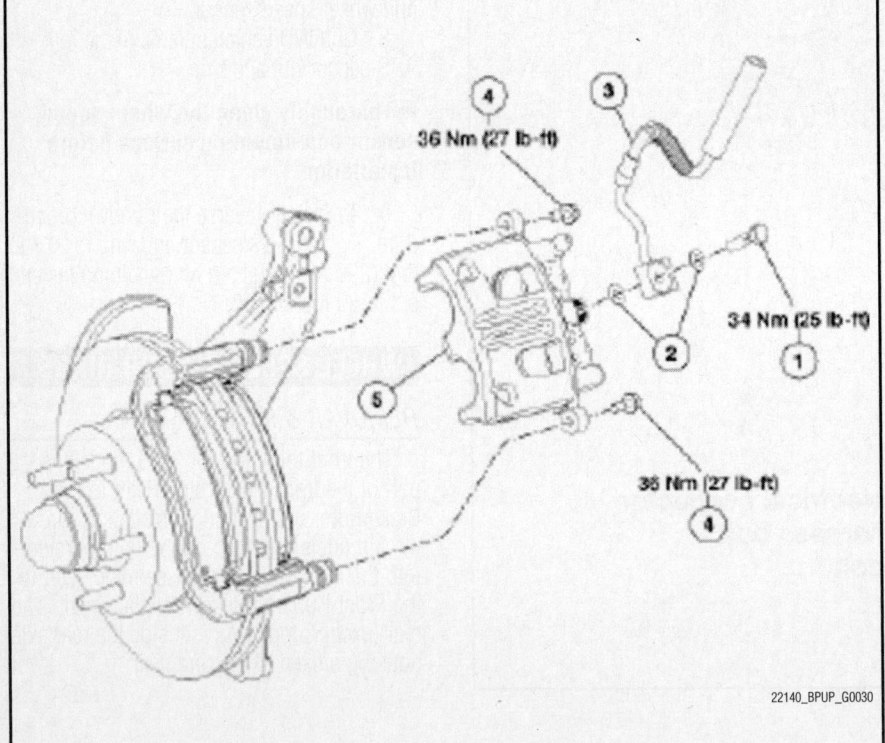

36 Nm (27 lb-ft)

34 Nm (25 lb-ft)

36 Nm (27 lb-ft)

22140_BPUP_G0030

Fig. 2 Brake caliper removal and installation

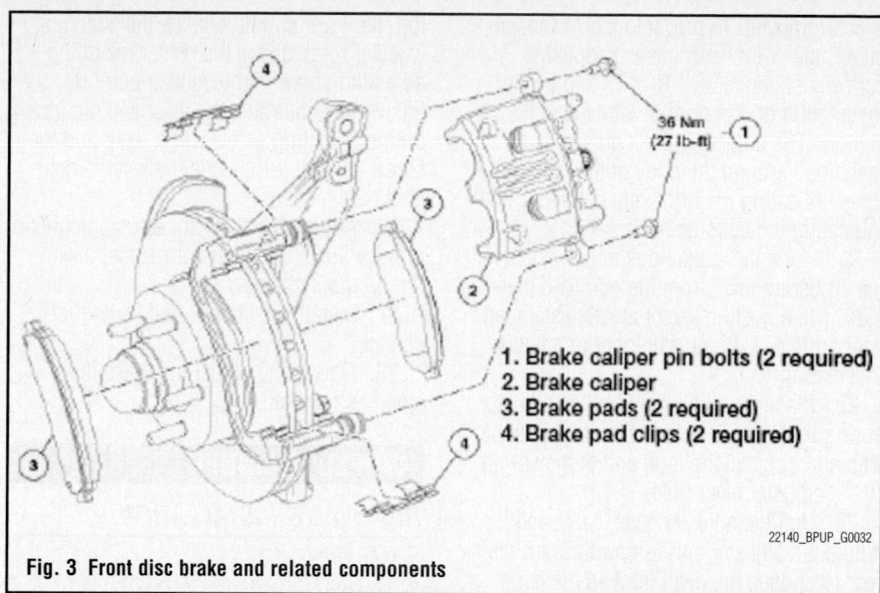

1. Brake caliper pin bolts (2 required)
2. Brake caliper
3. Brake pads (2 required)
4. Brake pad clips (2 required)

22140_BPUP_G0032

Fig. 3 Front disc brake and related components

11. Install the slide pin bolts and tighten them to 27 ft. lbs. (36 Nm).

12. Install the wheel and snug the lug nuts.

13. Lower the vehicle and tighten the lug nuts to 100 ft. lbs. (135 Nm).

➡**The first couple of times you apply the brakes, the pedal may go to the floor. Continue to pump the brake pedal until it feels firm.**

14. Start the engine and apply the brakes several times to readjust the caliper pistons. Ensure that the pedal feels firm before operating the vehicle.

15. Check and adjust the brake fluid level, as required.

BRAKES

REAR DRUM BRAKES

BRAKE DRUM

REMOVAL & INSTALLATION

1. Before servicing the vehicle, refer to the Precautions Section.

2. Raise and safely support the vehicle. Remove the wheel and tire assembly.

3. Remove the retaining nuts, if equipped, and remove the brake drum.

4. Inspect the brake drum surface for wear, scoring and runout. Machine or replace, as necessary.

To install:

5. Install the brake drum and secure in place with the retainer nuts, if equipped.

6. Adjust the rear brakes.

7. Install the wheel. Lower the vehicle.

BRAKE SHOES

REMOVAL & INSTALLATION

See Figure 4.

1. Before servicing the vehicle, refer to the Precautions Section.

2. Raise and safely support the vehicle. Remove the wheel and tire assembly and the brake drum.

3. Pull backward on the adjusting lever cable to disengage the adjusting lever from the adjusting screw. Move the outboard side of the adjusting screw upward and back off the pivot nut as far as it will go.

4. Pull the adjusting lever, cable and automatic adjuster spring down and toward the rear to unhook the pivot hook from the

large hole in the secondary shoe web. Do not pry the pivot hook from the hole.

5. Remove the automatic adjuster spring and adjusting lever.

6. Remove the secondary shoe-to-anchor spring using a suitable brake spring

removal/installation tool. Using the tool, remove the primary shoe-to-anchor spring and unhook the cable anchor. Remove the anchor pin plate, if equipped.

7. Remove the cable guide from the secondary shoe.

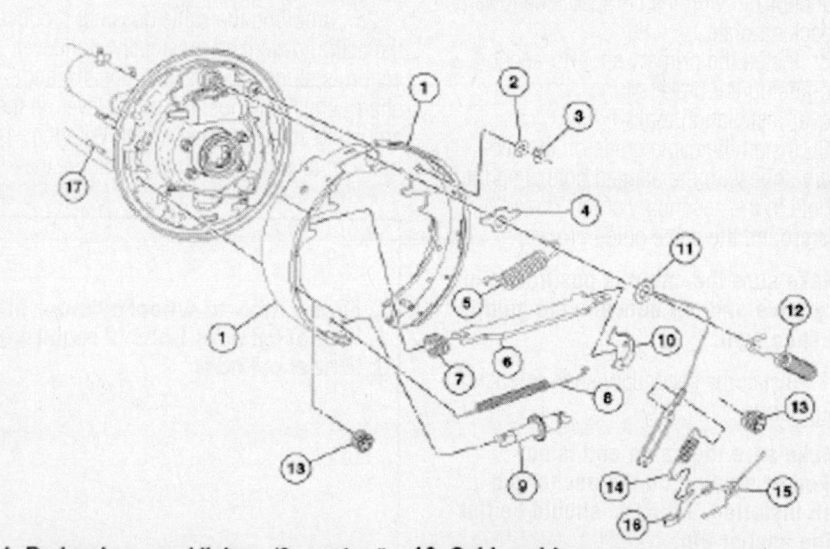

1. Brake shoes and linings (2 required)
2. Washer
3. Parking brake lever pin retainer
4. Brake shoe guide plate
5. Primary brake shoe retracting spring
6. Parking brake strut
7. Parking brake strut spring
8. Brake shoe adjusting screw spring
9. Brake shoe adjusting screw
10. Cable guide
11. Brake shoe adjusting lever cable
12. Secondary brake shoe retracting spring
13. Brake shoe hold-down spring
14. Brake shoe adjusting lever cable spring
15. Adjusting lever return spring
16. Brake shoe adjusting lever
17. Brake shoe hold-down spring pin

22140_BPUP_G0031

Fig. 4 Rear brake assembly and related components

8. Remove the shoe hold-down springs, shoes, adjusting screw, pivot nut and socket. Note the color and position of each hold-down spring so they can be reassembled in the same position.

9. Remove the parking brake link and spring. Disconnect the parking brake cable from the parking brake lever.

10. Remove the secondary brake shoe. On 9 inch rear brakes, remove the parking brake lever from the shoe. On 10 inch rear brakes, remove the retainer clip and spring washer and remove the parking brake lever.

To install:

11. Clean the backing plate ledge pads and sand lightly. Apply a light coating of high temperature lithium grease to the points where the brake shoes touch the backing plate. Lubricate the adjusting cable eye and the anchor pin area.

12. Install the parking brake lever on the secondary shoe. On 10 inch brakes, secure with the spring washer and retaining clip.

13. Position the brake shoes on the backing plate and install the hold-down spring pins, springs and cups. Install the parking brake link, spring and washer. Connect the parking brake cable to the parking brake lever.

14. Install the anchor pin plate, if equipped, and place the cable anchor over the anchor pin with the crimped side toward the backing plate.

15. Install the primary shoe-to-anchor spring using the brake spring removal/installation tool.

16. Install the cable guide on the secondary shoe with the flanged hole fitted into the hole in the secondary shoe. Thread the cable around the cable guide groove.

➡ Make sure the cable is positioned in the groove and not between the guide and shoe web.

17. Install the secondary shoe-to-anchor (long) spring.

➡ Make sure the cable end is not cocked or binding on the anchor pin when installed. All parts should be flat on the anchor pin.

18. Apply high temperature lithium grease to the threads and the socket end of the adjusting screw. Turn the adjusting screw into the adjusting pivot nut to the end of the threads and then loosen, ½ turn.

19. Place the adjusting socket on the screw and install the assembly between the shoe ends with the adjusting screw nearest the secondary shoe.

20. Be sure to install the adjusting screw on the same side of the vehicle from which it was removed. To prevent incorrect installation, the socket end of each adjusting screw is stamped with **R** or **L**, to indicate installation on the right or left side of the vehicle. The adjusting pivot nuts have lines machined around the body of the nut, 2 lines indicating the right side nut and 1 line indicating the left side nut.

21. Hook the cable hook into the hole in the adjusting lever from the outboard plate side. The adjusting levers are also stamped with an **R** or **L** to indicate right or left side installation.

22. Place the hooked end of the adjuster spring in the large hole in the primary shoe web and connect the loop end of the spring to the adjuster lever hole.

23. Pull the adjuster lever, cable and automatic adjuster spring down toward the rear to engage the pivot hook in the large hole in the secondary shoe web.

24. After installation, check the action of the adjuster by pulling the section of the cable between the cable guide and the adjusting lever toward the secondary shoe web far enough to lift the lever past a tooth on the adjusting screw wheel. The lever should snap into position behind the next tooth and releasing the cable should cause the adjuster spring to return the lever to its original position. This return action will turn the adjusting screw 1 tooth.

25. If pulling the cable does not produce the action described previously, or if lever action is sluggish instead of positive and sharp, check the position of the lever on the adjusting screw toothed wheel. With the brake in a vertical position, anchor at the top, the lever should contact the adjusting wheel 1 tooth above the centerline of the adjusting screw. If the contact point is below the centerline, the lever will not lock on the adjusting screw wheel teeth and the screw will not turn, since the lever is actuated by the cable.

26. Adjust the brake shoes using either a brake adjustment gauge or manually with the drums installed.

27. Install the wheels, and lower the vehicle.

28. Check and adjust the brake fluid level, as required.

WHEEL CYLINDER

REMOVAL & INSTALLATION

See Figure 5.

1. Before servicing the vehicle, refer to the Precautions Section.

2. Raise and support the vehicle safely.

3. Remove the wheel and tire assembly.

4. Remove the brakes shoes.

5. Disconnect the brake line at the wheel cylinder.

6. Remove the wheel cylinder bolts.

To install:

7. Installation is the reverse of the removal procedure.

8. Torque the brake tube-to-wheel cylinder fitting to 13 ft. lbs. (18 Nm).

9. Torque the retaining bolts to 11 ft. lbs. (15 Nm).

10. Bleed the brake system.

11. Correct and adjust the brake fluid level.

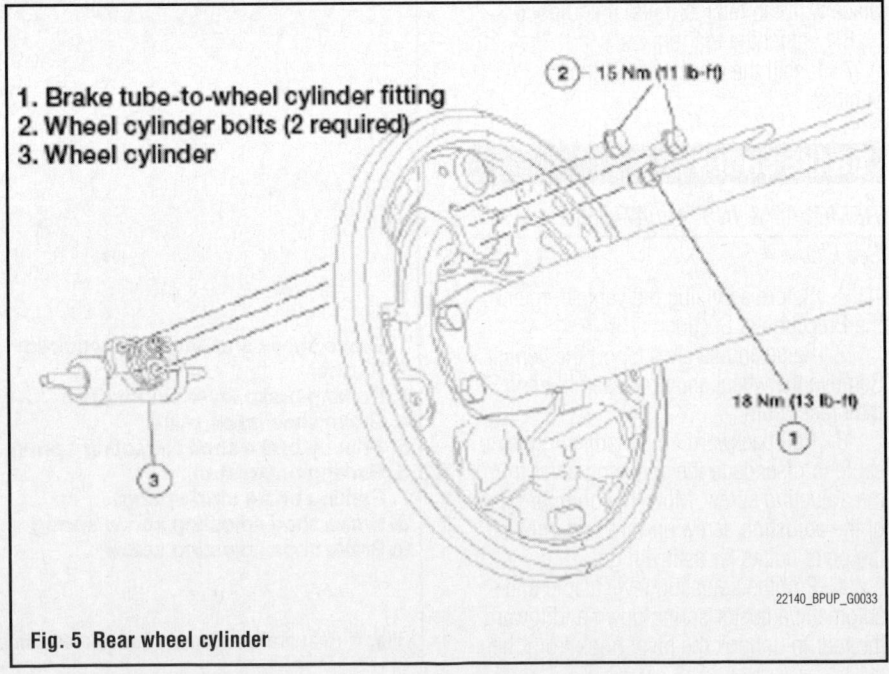

1. Brake tube-to-wheel cylinder fitting
2. Wheel cylinder bolts (2 required)
3. Wheel cylinder

2 – 15 Nm (11 lb-ft)

18 Nm (13 lb-ft)

22140_BPUP_G0033

Fig. 5 Rear wheel cylinder

BRAKES

PARKING BRAKE

PARKING BRAKE CABLES

ADJUSTMENT

1. After installation of the rear brake shoes, check the action of the adjuster by pulling the section of the cable between the cable guide and the adjusting lever toward the secondary shoe web far enough to lift the lever past a tooth on the adjusting screw wheel. The lever should snap into position behind the next tooth and releasing the cable should cause the adjuster spring to return the lever to its original position. This return action will turn the adjusting screw 1 tooth.

2. If pulling the cable does not produce the action described previously, or if lever action is sluggish instead of positive and sharp, check the position of the lever on the adjusting screw toothed wheel. With the brake in a vertical position, anchor at the top, the lever should contact the adjusting wheel 1 tooth above the centerline of the adjusting screw. If the contact point is below the centerline, the lever will not lock on the adjusting screw wheel teeth and the screw will not turn, since the lever is actuated by the cable.

3. Adjust the brake shoes using either a brake adjustment gauge or manually with the drums installed.

4. Install the wheels, and lower the vehicle.

PARKING BRAKE SHOES

REMOVAL & INSTALLATION

The rear drum brake shoes serve as the parking brakes. Refer to the procedures under Rear Drum Brakes.

CHASSIS ELETRICAL

AIR BAG (SUPPLEMENTAL RESTRAINT SYSTEM)

GENERAL INFORMATION

✳✳ CAUTION

Vehicles equipped with an air bag system must be disarmed before performing service on, or around, system components, the steering column, instrument panel components, wiring and sensors. Failure to follow the safety precautions and the disarming procedure could result in accidental air bag deployment, possible injury and unnecessary system repairs.

SERVICE PRECAUTIONS

Disconnect and isolate the battery negative cable before beginning any airbag system component diagnosis, testing, removal, or installation procedures. Allow system capacitor to discharge for two minutes before beginning any component service. This will disable the airbag system. Failure to disable the airbag system may result in accidental airbag deployment, personal injury, or death.

DISARMING THE SYSTEM

✳✳ CAUTION

All vehicles are equipped with an air bag system. The system MUST BE disabled before performing service on or around system components, steering column, instrument panel components, wiring and sensors. Failure to follow safety and disabling procedures could result in accidental air bag deployment, possible per-sonal injury and unnecessary system repairs.

➡**This procedure may also be referred to as Depowering the SRS.**

✳✳ CAUTION

The Supplemental Inflatable Restraint (SIR) system must be disarmed before performing service around SIR system components or SIR system wiring. Failure to do so may cause accidental deployment of the air bag, resulting in unnecessary SIR system repairs and/or personal injury.

The positive battery cable must be disconnected for a minimum of 1 minute before beginning any air bag work to de-energize the back-up power supply. It is a good idea to disengage both the positive and negative battery cables to ensure that the Air Bag system is definitely discharged.

1. Before servicing the vehicle, refer to the Precautions Section.
2. Turn all vehicle accessories. **OFF.**
3. Turn the ignition switch to **OFF**.
4. At the central junction box (SJB), located below the left side of the instrument panel, remove the cover and the restraints control module (RCM) fuse(s) from the SJB. See the Owner's Manual.
5. Turn the ignition **ON** and visually monitor the air bag indicator for at least 30 seconds. The air bag indicator will remain lit continuously (no flashing) if the correct RCM fuse has been removed. If the air bag indicator does not remain lit continuously, remove the correct RCM fuse before proceeding.
6. Turn the ignition **OFF**.

✳✳ CAUTION

To avoid accidental deployment and possible personal injury, the backup power supply must be depleted before repairing or replacing any front or side air bag supplemental restraint system (SRS) components and before servicing, replacing, adjusting or striking components near the front or side air bag sensors, such as doors, instrument panel, console, door latches, strikers, seats and hood latches.

The front impact severity sensor is located on the radiator support bracket.

The first row side impact sensors (if equipped) are located at or near the base of the B-pillars.

The second row side impact sensors (if equipped) are located on the C-pillar.

➡**To deplete the backup power supply energy, disconnect the battery ground cable and wait at least one minute. Be sure to disconnect auxiliary batteries and power supplies (if equipped).**

7. Disconnect the battery ground cable and wait at least one minute.

ARMING THE SYSTEM

➡**This procedure may also be referred to as Repowering the SRS.**

1. Before servicing the vehicle, refer to the Precautions Section.

⊹⊹ **CAUTION**

The restraint system diagnostic tool is for restraint system service only. Remove from vehicle prior to road use. Failure to remove could result in injury and possible violation of vehicle safety standards.

2. Make sure all restraint system diagnostic tool(s) that may have been installed during the repair have been removed from the vehicle and all SRS components are connected.

3. Turn the ignition switch from **OFF** to **ON**.

4. Install RCM fuse(s) to the SJB and close the cover.

⊹⊹ **CAUTION**

Be sure that nobody is in the vehicle and that there is nothing blocking or set in front of any air bag module when the battery ground cable is connected.

5. Connect the battery ground cable.

6. Prove out the supplemental restraint system (SRS) as follows:

7. Turn the ignition key from **ON** to **OFF**. Wait 10 seconds, and then turn the key back to ON and visually monitor the air bag indicator with the air bag modules installed. The air bag indicator will light continuously for approximately six seconds and then turn off. If an air bag supplemental restraint system (SRS) fault is present, the air bag indicator will either:

- Fail to light
- Remain lit continuously
- Flash

8. The flashing might not occur until approximately 30 seconds after the ignition switch has been turned from the **OFF** to the **ON** position. This is the time required for the restraints control module (RCM) to complete the testing of the SRS. If the air bag indicator is inoperative and a SRS fault exists, a chime will sound in a pattern of five sets of five beeps. If this occurs, the air bag indicator and any SRS fault discovered must be diagnosed and repaired.

9. Clear all continuous DTC's from the restraints control module using a scan tool.

DRIVETRAIN

CLUTCH DRIVEN DISC & PRESSURE PLATE

REMOVAL & INSTALLATION

See Figures 6 and 7.

1. Before servicing the vehicle, refer to the precautions in the beginning of this section.

2. Remove or disconnect the following:
   - Negative battery cable
   - Transmission see manual transmission removal in this section.

➡️**If the clutch disc and pressure plate are to be reinstalled, bolts must be** removed evenly or permanent damage to the diaphragm spring will occur resulting in complete clutch release.

- Bolts, clutch pressure plate and the clutch disc.

➡️**If the parts are to be reused, index-mark the clutch pressure plate to the flywheel.**

To install:

3. Lubricate the transmission input shaft pilot bearing with front axle grease.

4. Using a suitable press, press downward on the pressure plate fingers until the adjusting ring moves freely.

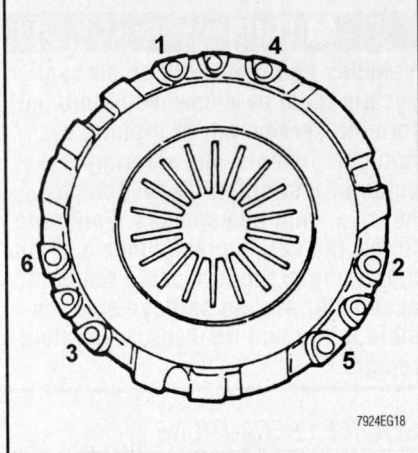

Fig. 7 Tighten the bolts gradually in the correct sequence to avoid warping the pressure plate

5. Rotate the adjusting ring counterclockwise to compress the tension springs. Hold the adjusting ring in this position.

6. Release the pressure on the fingers, the adjusting ring will stay in the reset position.

7. Position the clutch disc on the flywheel.

➡️**If reusing the clutch pressure plate and flywheel, align the marks made during removal.**

8. Align the clutch disc and the clutch pressure plate. Install the bolts and tighten in a star pattern sequence to 24 ft. lbs. (35 Nm).

- Install the transmission.

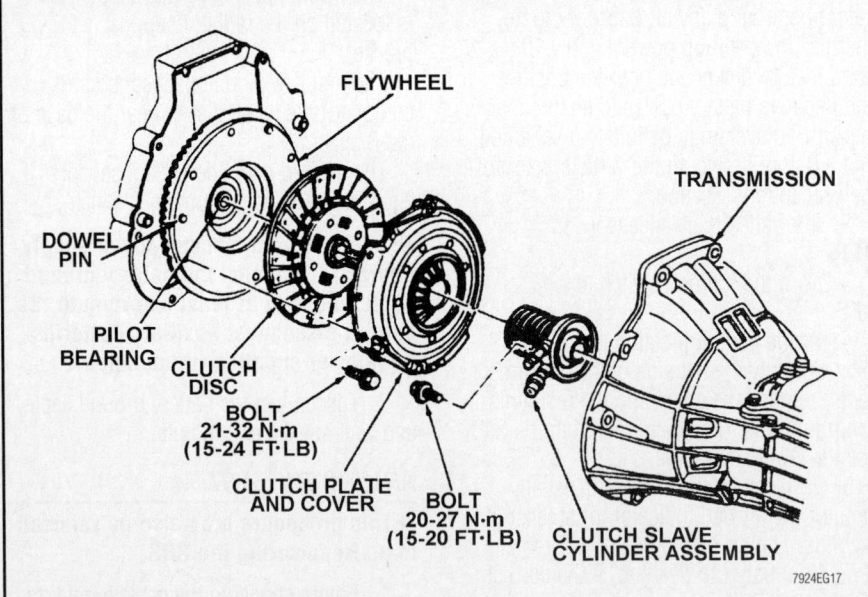

Fig. 6 Clutch disc, pressure plate and bearing assembly

ADJUSTMENTS

Because the clutch is hydraulically driven, there is no adjustment required.

In the event the clutch pedal develops a squeak or uneven feel when depressing, spray the pedal bushing assembly with penetrating oil and work the pedal back-and-forth.

CLUTCH MASTER CYLINDER

REMOVAL & INSTALLATION

1. Before servicing the vehicle, refer to the Precautions Section.
2. Disconnect the negative battery cable.
3. Disconnect the clutch master cylinder rod from the clutch pedal.
4. Remove the clutch pedal bracket nut.

✳✳ CAUTION

Brake fluid contains polyglycol ethers and polyglycols. Avoid contact with eyes. Wash hands thoroughly after handling. If brake fluid contacts eyes, flush eyes with running water for 15 minutes. Get medical attention if irritation persists. If taken internally, drink water and induce vomiting. Get medical attention immediately. Failure to follow these instructions may result in personal injury.

✳✳ WARNING

Brake fluid is harmful to painted and plastic surfaces. If brake fluid is spilled onto a painted or plastic surface, wash it immediately with water.

5. Disconnect the clutch master cylinder hose from the clutch master cylinder and plug it to prevent excess fluid loss.
6. Disconnect the clutch slave cylinder tube from the clutch master cylinder.
7. Remove the clutch master cylinder nut.
8. Remove the clutch master cylinder.
9. To install, reverse the removal procedure. Torque all fasteners to specification.
10. Bleed the air from the system.

BENCH BLEEDING PROCEDURE

1. Position the clutch master cylinder reservoir above the clutch master cylinder and the quick connect fitting below the clutch master cylinder.
2. Press the internal mechanism of the quick connect fitting to open the valve.

3. Compress and hold the clutch master cylinder push rod.
4. Release the quick connect fitting valve.
5. Release the clutch master cylinder push rod.
6. Fill the clutch master cylinder reservoir.
7. Repeat steps 2 and 3 four times.

CLUTCH SLAVE CYLINDER

REMOVAL & INSTALLATION

✳✳ CAUTION

Brake fluid contains polyglycol ethers and polyglycols. Avoid contact with eyes. Wash hands thoroughly after handling. If brake fluid contacts eyes, flush eyes with running water for 15 minutes. Get medical attention if irritation persists. If taken internally, drink water and induce vomiting. Get medical attention immediately. Failure to follow these instructions may result in personal injury.

✳✳ WARNING

Brake fluid is harmful to painted and plastic surfaces. If brake fluid is spilled onto a painted or plastic surface, wash it immediately with water. Remove the clutch slave cylinder to clutch line adapter.

1. Before servicing the vehicle, refer to the Precautions Section.
2. Disconnect the negative battery cable.
3. Remove the transmission.
4. Remove the clutch slave cylinder bolts.
5. Remove the clutch slave cylinder.
6. To install, reverse the removal procedure. Torque slave cylinder bolts to 15 ft. lbs. (20 Nm).
7. Bleed the air from the system.

CLUTCH HYDRAULIC SYSTEM BLEEDING

BLEEDING PROCEDURE

The following procedure is recommended for bleeding the clutch hydraulic system installed on the vehicle. It is recommended that the original clutch tube, with quick-connect fitting be replaced when servicing the hydraulic system, because air can be trapped in the quick-connect fitting and prevent complete bleeding of the system. The replacement tube does not include a quick-connect fitting.

1. Before servicing the vehicle, refer to the precautions in the beginning of this section.
2. Clean the dirt and grease from the dust cap.
3. Remove the cap and diaphragm and fill the reservoir to the top with approved brake fluid C6AZ-19542-AA or BA, (ESA-M6C25-A).

➡**To keep brake fluid from entering the clutch housing, route a suitable rubber tube of appropriate inside diameter from the bleed screw to a container.**

4. Loosen the bleed screw, located in the slave cylinder body, next to the inlet connection. Fluid will now begin to move from the master cylinder down the tube to the slave cylinder.

➡**The reservoir must be kept full at all times during the bleeding operation, to ensure no additional air enters the system.**

5. Observe the bleed screw outlet. When the slave cylinder is full, a steady stream of fluid will flow from the outlet port. Tighten the bleed screw.
6. Depress the clutch pedal to the floor and hold for 1–2 seconds. Release the pedal as rapidly as possible. The pedal must be released completely. Pause for 1–2 seconds. Repeat 10 times.
7. Check the fluid level in the reservoir. The fluid should be level with the step when the diaphragm is removed.
8. Hold the pedal to the floor, slightly open the bleed screw to allow any additional air to escape. Close the bleed screw, then release the pedal.
9. Check the fluid in the reservoir. The hydraulic system should now be fully bled, and should actuate the clutch.
10. Check the vehicle by starting, pushing the clutch pedal to the floor and selecting reverse gear. There should be no grating of gears. If there is, and the hydraulic system still contains air; repeat the bleeding procedure.

TRANSFER CASE ASSEMBLY

REMOVAL & INSTALLATION

See Figures 8 through 10.

1. Before servicing the vehicle, refer to the Precautions Section.

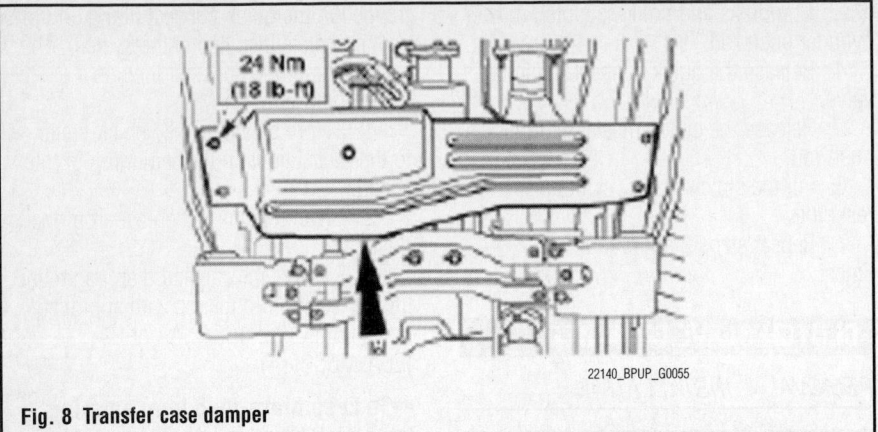

Fig. 8 Transfer case damper

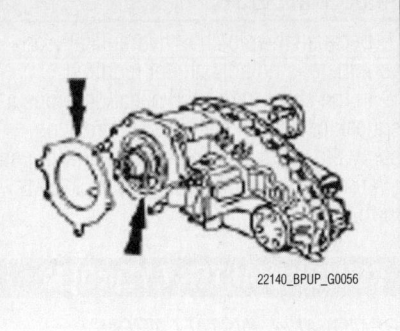

Fig. 9 New gasket on the front of the transfer case.

2. Disconnect the negative battery cable.

➡**On some vehicles, when the battery cable is disconnected and reconnected, some abnormal drive symptoms may occur while the vehicle relearns its adaptive strategy. The vehicle may need to be driven to relearn its strategy.**

3. With the vehicle in **NEUTRAL**, raise and support the vehicle.

4. Remove the skid plate.

5. Remove the damper.

6. Disconnect the transfer case harness connector and position it aside.

7. If transfer case disassembly is necessary, remove the drain plug and drain the fluid. Install the drain plug when all of the fluid has drained.

➡**Index-mark the front output shaft assembly and the front driveshaft constant velocity (CV) joint.**

※※ WARNING

Always disconnect the front driveshaft from the transfer case first. Otherwise, the weight of the driveshaft can pinch the boot between the shaft and the boot can and cause the boot to tear.

8. Index-mark the front output shaft assembly and the front driveshaft constant velocity (CV) joint.

9. Remove and discard the bolts and washers.

10. Disconnect the front driveshaft from the transfer case and position the driveshaft aside.

➡**Index-mark the front flange on the rear driveshaft and the flange on the transfer case.**

11. Remove the rear driveshaft.

※※ CAUTION

Secure the transfer case to the jack with safety straps.

12. Position a high lift jack under the transfer case.

13. Remove the five bolts retaining the transfer case to the extension housing.

14. Slide the transfer case rearward and off of the transmission output shaft.

15. Remove and discard the front extension housing gasket, and clean the mating surfaces.

To install:

16. Install the transfer case with a new gasket.

17. Tighten the bolts that retain the transfer case to the extension housing in a clockwise direction beginning with the upper LH bolt. Torque to 40 ft. lbs. (54 Nm).

18. Install the front driveshaft with new

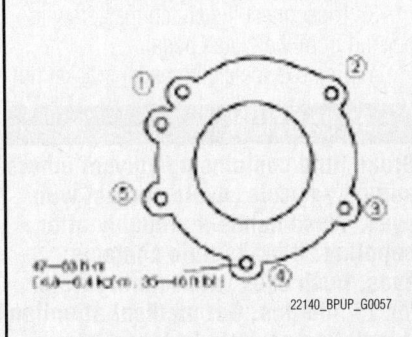

Fig. 10 Install procedure of the axle to the transfer case

bolts and washers and the rear driveshaft with new bolts. If new bolts are not available, coat the threads of the original bolts with Threadlock and Sealer E0AZ-19554-AA, or equivalent.

➡**When installing the front driveshaft, always connect it to the axle first and then connect it to the transfer case.**

➡**Align the index marks when installing the front and rear driveshafts.**

19. The remainder of installation is the reverse of the removal procedure.

20. Check and, if necessary, fill the transfer case with the specified type and quantity of fluid.

ENGINE COOLING

✳✳ CAUTION

Never remove the pressure relief cap or coolant recovery reservoir cap while the engine is operating or when the cooling system is hot. Failure to follow these instructions can result in damage to the cooling system or engine or personal injury. To avoid having scalding hot coolant or steam blow out of the coolant reservoir when removing the pressure relief cap, wait until the engine has cooled, then wrap a thick cloth around the pressure relief cap and turn it slowly. Step back while the pressure is released from the cooling system. When you are sure all the pressure has been released, (still with a cloth) turn and remove the pressure relief cap. Failure to follow these instructions may result in personal injury.

✳ WARNING

Engine coolant provides freeze protection, boil protection, cooling efficiency and corrosion protection to the engine and cooling components. In order to obtain these protections, the engine coolant must be maintained at the correct concentration and fluid level. When adding engine coolant, use a 50/50 mixture of engine coolant and clean, drinkable water.

✳✳ WARNING

Do not add Mazda Specialty Orange Engine Coolant, VC-2. Mixing coolants may degrade the coolant's corrosion protection.

THERMOSTAT

REMOVAL & INSTALLATION

1. See all precautions in this section.
2. Disconnect the negative battery cable.
3. Drain and recycle the engine coolant.
4. Remove the upper and lower coolant hoses on 2.3L engine.
5. Remove the upper coolant hose on 4.0L engine.
6. Remove the thermostat housing.
7. Remove the bolts and separate the water outlet adapter from the thermostat housing.

➡The thermostat is indexed and must be installed correctly.

8. Remove the thermostat and the O-ring seal.

➡Clean and inspect the sealing surfaces.

9. Fill the cooling system.
10. See Filling and Bleeding Procedure in this section.

WATER PUMP

REMOVAL & INSTALLATION

2.3L Engine
See Figure 11.

1. See all precautions in this section.
2. Disconnect the negative battery cable.

➡On some vehicles, when the battery cable is disconnected and reconnected, some abnormal drive symptoms may occur while the vehicle relearns its adaptive strategy. The vehicle may need to be driven to relearn its strategy.

3. Remove the air cleaner outlet pipe.
4. Drain the cooling system.
5. Remove the drive belt.
6. Remove the fan if equipped with a clutch driven fan, see cooling fan removal in this section.
7. Remove the water pump pulley.
8. Remove the water pump.

To install:
9. Clean the mating surfaces where the water pump attaches to the engine.

➡Lubricate the water pump O-ring.

10. To install, reverse the removal procedure. Torque the water pump mount bolts to 89 inch lbs. (10 Nm). Torque the pulley bolts to 18 ft. lbs. (25 Nm).
11. See all Filling and Bleeding Procedure in this section.
12. Maintain the engine speed at 2,500 RPM for an additional three minutes.
13. Increase engine speed to 4,000 RPM and hold for five seconds.
14. Return the engine speed to 2,500 RPM and hold for an additional three minutes.
15. Repeat the previous two steps.
16. Stop the engine and check for leaks.
17. Verify the correct fluid level after the engine cools for 20 minutes. Top off the degas bottle to the maximum fill level.

4.0L Engine
See Figure 12.

1. See all precautions in this section.
2. Disconnect the negative battery cable.

➡On some vehicles, when the battery cable is disconnected and reconnected, some abnormal drive symptoms may occur while the vehicle relearns its adaptive strategy. The vehicle may need to be driven to relearn its strategy.

3. Drain the cooling system.
4. Remove or disconnect the following:
   • Fan shroud, see cooling fan removal in this section

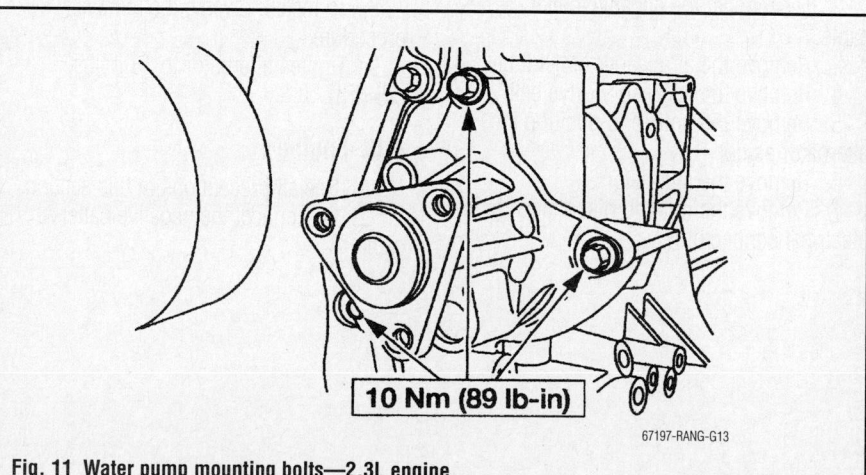

10 Nm (89 lb-in)

67197-RANG-G13

Fig. 11 Water pump mounting bolts—2.3L engine

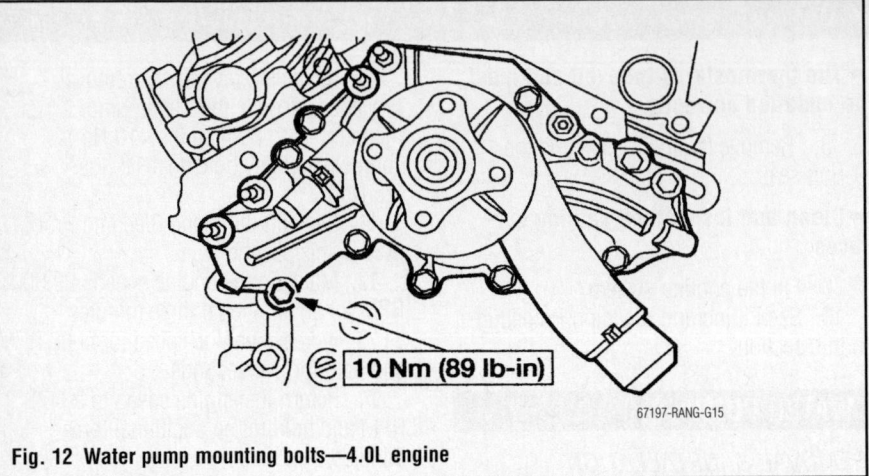

Fig. 12 Water pump mounting bolts—4.0L engine

- Accessory drive belt
- Idler pulley
- Water bypass hose
- Heater hose
- Lower radiator hose
- Water pump pulley
- Water pump

To install:

5. Clean the mating surfaces where the water pump attaches to the engine.

✳✳ WARNING

Use care when scraping the water pump-to-engine block mating surfaces. Gouges in the aluminum could form leak paths.

6. Clean all the sealing surfaces.
7. To install, reverse the removal procedure. Torque the water pump bolts to 89 inch lbs. (10 Nm). Torque the pulley bolts to 18 ft. lbs. (25 Nm).
8. Fill the cooling system.
9. See all Filling and Bleeding Procedures in this section.
10. Maintain the engine speed at 2,500 RPM for an additional three minutes.
11. Increase engine speed to 4,000 RPM and hold for five seconds.
12. Return the engine speed to 2,500 RPM and hold for an additional three minutes.
13. Repeat the previous two steps.
14. Stop the engine and check for leaks.
15. Verify the correct fluid level after the engine cools for 20 minutes. Top off the degas bottle to the maximum fill level.

ENGINE ELECTRICAL

✳✳ WARNING

Do not allow any metal object to come in contact with the housing and the internal diode cooling fins with key on or off. A short circuit will result and burn out the diodes.

ALTERNATOR

REMOVAL & INSTALLATION

2.3L Engine

See Figures 13.

1. See all precautions in this section.
2. Disconnect the negative battery cable.
3. Remove the air cleaner outlet tube.
4. Remove the accessory drive belt.
5. Remove the bolts and position the alternator aside.
6. Remove the alternator.
7. Remove the nut and disconnect the electrical connectors.

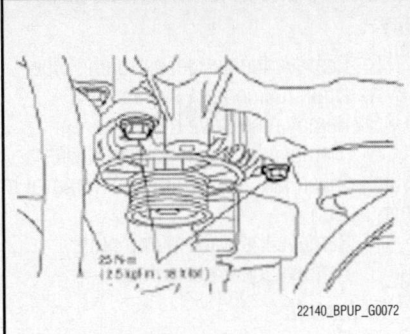

Fig. 13 Alternator and system components—2.3L engine

8. To install, reverse the removal procedure.
9. Tighten alternator to 18 ft. lbs. (25 Nm).

4.0L Engine

1. See all precautions in this section.
2. Disconnect the negative battery cable.

CHARGING SYSTEM

3. Remove the air cleaner outlet tube.
4. Rotate the front end accessory drive belt tensioner counterclockwise and position the accessory drive belt aside.
5. Disconnect the 2 electrical connectors from the generator.
6. Position the protective cover aside, remove the nut and position the generator B+ terminal aside.
7. Remove the 2 bolts from the generator.
8. Remove the stud bolt, position the generator harness bracket aside and remove the generator.
9. To install, reverse the removal procedure.
10. Tightening the B+ terminal to 71 inch lbs. (8 Nm).
11. Tightening the alternator bolts to 35 ft. lbs. (47 Nm).
12. Tightening the stud bolt to 35 ft. lbs. (47 Nm).

ADJUSTMENT

The ignition timing is preset to 10 degrees Before Top Dead Center (BTDC) and is not adjustable.

IGNITION COIL PACK

REMOVAL & INSTALLATION

2.3L Engine

1. Remove or disconnect the following:
   - Negative battery cable.
   - Ignition wires

➡ **Do not pull on the spark plug wire as it may separate from the spark plug wire connector inside the spark plug wire boot. Remove the spark plug wires by slightly twisting while pulling upwards.**

   - Ignition coil electrical connectors
   - Remove the bolts and remove the ignition coil.

❈❈ CAUTION

Correct installation of the ignition wires is critical to engine operation. If one spark plug wire is not correctly installed at either the spark plugs or the ignition coil, both spark plugs connected to the ignition coil may not fire under load.

2. To install, reverse the removal procedure.

3. Wipe the coil towers with a clean cloth dampened with soap and water. Remove any soap film and dry with compressed air. Inspect for cracks, carbon tracking and dirt. To install, reverse the removal procedure.

4. Apply silicone dielectric compound to the inside of the spark plug wire boots.

5. Torque coil bolts to 53 inch lbs. (6 Nm).

4.0L Engine

1. Remove or disconnect the following:
   - Negative battery cable.
   - Ignition wires

➡ **Do not pull on the spark plug wire as it may separate from the spark plug wire connector inside the spark plug wire boot. Remove the spark plug wires by slightly twisting while pulling upwards.**

   - Ignition coil electrical connectors
   - Remove the bolts and remove the ignition coil.

➡ **Correct installation of the ignition wires is critical to engine operation. If one spark plug wire is not correctly installed at either the spark plugs or the ignition coil, both spark plugs connected to the ignition coil may not fire under load.**

2. To install, reverse the removal procedure.

3. Wipe the coil towers with a clean cloth dampened with soap and water. Remove any soap film and dry with compressed air. Inspect for cracks, carbon tracking and dirt. To install, reverse the removal procedure.

4. Apply silicone dielectric compound to the inside of the spark plug wire boots.

5. Torque coil bolts to 53 inch lbs. (6 Nm).

FIRING ORDER

The 2.3L engine firing order is: 1–3–4–2. The 4.0L engine firing order is :1–4–2–5–6.

IGNITION TIMING

The ignition timing is controlled by the Powertrain Control Module and is not adjustable.

SPARK PLUGS

REMOVAL & INSTALLATION

2.3L Engine

1. Disconnect the negative battery cable.
2. Remove or disconnect the following:

   - 4 ignition coil-on-plug electrical connectors.
   - 4 ignition coil-to-valve cover bolts.
   - 4 ignition coils.

➡ **Use compressed air to remove any foreign material in the spark plug well before removing the spark plugs.**

   - Remove the 4 spark plugs.

3. Inspect the spark plugs. Replace as necessary.

4. Apply a small amount of dielectric grease to the inside of the ignition coil boots before attaching to the spark plugs.

5. To install, reverse removal procedure.

6. Torque spark plugs to 144 inch lbs. (15 Nm).

4.0L Engine

➡ **All B-Series trucks with 4.0L engine have factory-installed, color-coded spark plugs. These spark plugs are not interchangeable between the right and left bank.**

1. Disconnect the negative battery cable.
2. Remove the right front tire and wheel.
3. Remove the right front fender splash shield.
4. Remove or disconnect the following:
   - 6 ignition coil-on-plug electrical connectors
   - 6 ignition coil-to-valve cover bolts
   - 6 ignition coils

➡ **Use compressed air to remove any foreign material in the spark plug well before removing the spark plugs.**

   - Remove the 6 spark plugs.

5. Inspect the spark plugs. Replace as necessary.

➡ **Platinum-enhanced spark plugs should not require adjustment throughout their service life.**

6. Apply a small amount of dielectric grease to the inside of the ignition coil boots before attaching to the spark plugs.

7. To install, reverse removal procedure.

8. Torque spark plugs to 144 inch lbs. (15 Nm).

ENGINE ELECTRICAL

❋ CAUTION

When carrying out maintenance on the starting system be aware that heavy gauge leads are connected directly to the battery. Make sure protective caps are in place when maintenance is completed. Failure to follow these instructions may result in personal injury.

STARTER

REMOVAL & INSTALLATION

2.3L Engine

1. Before servicing the vehicle, refer to the Precautions Section.

2. Disconnect the negative battery cable.

➡ **On some vehicles, when the battery cable is disconnected and reconnected, some abnormal drive symptoms may occur while the vehicle relearns its adaptive strategy. The vehicle may need to be driven to relearn its strategy.**

3. Raise and support the vehicle safely.
4. Disconnect the electrical connections from the starter.
5. Disconnect the positive battery cable.
6. Remove the starter retaining bolts. Remove the starter from the engine.

To install:

7. Installation is the reverse of the removal procedure.

STARTING SYSTEM

8. Torque the starter retaining bolts to 9 ft. lbs. (12 Nm).

4.0L Engine

1. Before servicing the vehicle, refer to the precautions in the beginning of this section.
2. Remove or disconnect the following:
   - Negative battery cable
   - Starter harness connectors
   - Starter motor

To install:

3. Install or connect the following:
   - Starter motor. Tighten the bolts to 24–33 ft. lbs. (32–46 Nm).
   - Starter harness connectors. Tighten the battery cable nut to 87–104 inch lbs. (10–12 Nm).
   - Negative battery cable

ENGINE MECHANICAL

ACCESSORY DRIVE BELTS

INSPECTION

Inspect the drive belt for signs of glazing or cracking. A glazed belt will be perfectly smooth from slippage, while a good belt will have a slight texture of fabric visible. Cracks will usually start at the inner edge of the belt and run outward. All worn or damaged drive belts should be replaced immediately.

ADJUSTMENT

Belt tensioners are preloaded and no adjustments can be made.

REMOVAL & INSTALLATION

2.3L Engine

See Figure 14.

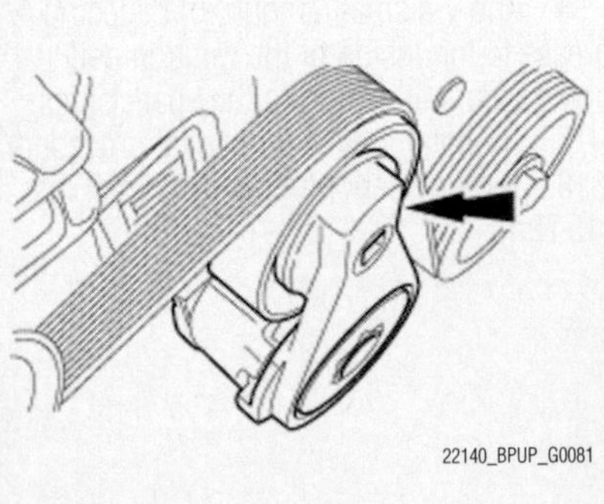

1. Disconnect the negative battery cable.
2. Rotate the accessory drive belt tensioner clockwise and remove the accessory drive belt.
3. To install, reverse the removal procedure.

4.0L Engine

See Figure 15.

1. Disconnect the negative battery cable.
2. Remove the air cleaner outlet tube.
3. Remove the drive belt.
4. Rotate the accessory drive belt tensioner clockwise counterclockwise for

5. Remove the belt.
6. To install, reverse the removal procedure.

BALANCE SHAFT

REMOVAL & INSTALLATION

2.3L Engine

➡ **Due to the precision interior construction of the balancer unit, it cannot be disassembled.**

1. Install the service tool 303—507 as shown in the figure.
2. Turn the crankshaft clockwise the crankshaft is in the No.1 cylinder TDC position (until the balance weight is attached to the Special Service Tool 303—507.
3. Install the adjustment shim to the seat face of the balancer unit.
4. With the balancer unit marks at the exact top center, assemble the unit to the cylinder block. Set the SST as shown, then measure the gear backlash using a dial gauge.

➡ **For an accurate measurement of gear backlash, insert a screwdriver into the crankshaft No. 1 balance weight area and set both the rotation and the thrust direction with the screwdriver, using a prying action, as shown in the figure.**

➡ **If the backlash exceeds the specified range, re-measure the backlash and,**

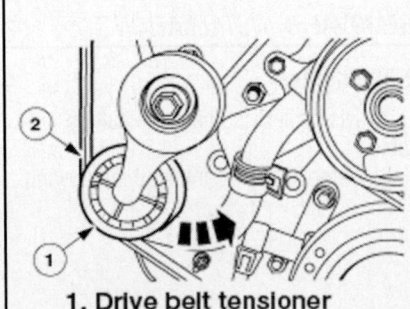

Fig. 14 Drive belt removal & installation—2.3L engine

22140_BPUP_G0081

1. Drive belt tensioner
2. Drive belt.

22140_BPUP_G0082

Fig. 15 Drive belt removal & installation—4.0L engine

using the adjustment shim selection table, select the proper shim, according to the following procedure.

✳✳ CAUTION

When measuring the backlash, rotate the crankshaft one full rotation and verify that it is within the specified range at all of the following six positions: 10°, 30°, 100°, 190°, 210°, 280° ATDC.

5. Value range is 0.005–0.101 mm {0.00019–0.0039 in}

6. Using master adjustment shim (No.50), assemble the balancer unit to the cylinder block, and then measure the backlash.

7. Select the proper adjustment shim according to the measured value.

8. Install the selected adjustment shim to the balancer unit, then assemble the balancer unit to the cylinder

CAMSHAFT AND VALVE LIFTERS

REMOVAL & INSTALLATION

➡Although Mazda suggests that this component is removable while the engine is installed in the vehicle, depending on the particular options with which your truck is equipped, working clearance may be extremely tight and this procedure may be much easier to perform with the engine removed. Before commencing, read through this procedure and make certain enough clearance, or working room, exists with the engine in the vehicle; if there is not enough space, the engine should be removed.

2.3L Engine

See Figures 16 through 25.

➡On some engines the crankshaft, crankshaft sprocket and the pulley are fitted together by friction, with diamond washers between the flange on each part. For that reason, the crankshaft sprocket is also unfastened if you loosen the pulley. Therefore, the engine must be retimed each time the damper is removed. Otherwise severe damage can occur.

1. Before servicing the vehicle, refer to the Precautions Section.
2. Relieve the fuel system pressure.
3. Disconnect the negative battery cable.

➡On some vehicles, when the battery cable is disconnected and reconnected, some abnormal drive symptoms may occur while the vehicle relearns its adaptive strategy. The vehicle may need to be driven to relearn its strategy.

4. Drain the cooling system.
5. Properly discharge the air conditioning system.
6. Remove or disconnect the following:
   - Drive belt
   - Accessory drive belt tensioner
   - Cooling fan drive pulley
   - Coolant pump pulley
   - Power steering pump
   - Engine oil level indicator assembly
   - Engine oil level indicator.
   - Engine oil level indicator tube
   - Water outlet tube
   - Water outlet tube.
   - Air conditioning compressor.

➡The alternator will be removed with the accessory bracket.

   - Accessory bracket
   - Right motor mount
   - Coolant hose from the thermostat
   - Coolant hose from the EGR valve
   - Coolant tube assembly
   - Exhaust manifold and gasket
   - Block heater (if so equipped)
   - Water outlet
   - EGR valve
   - Power steering pump and reservoir as an assembly
   - Idle Air Control (IAC) valve
   - Throttle Position Sensor (TPS)
   - Manifold Absolute Pressure (MAP) sensor
   - Swirl control valve monitor electrical connector
   - Crankshaft Position (CKP) sensor and the wiring harness pin-type retainers
   - Knock Sensor (KS)
   - Electric thermostat
   - Swirl control valve
   - Camshaft Position (CMP) sensor electrical connector and disconnect the PCV hose from the intake manifold
   - Engine wiring harness pin-type retainers from the intake manifold
   - Engine wiring harness connector bracket. Position the engine wiring harness aside
   - EGR tube
   - Fuel supply line clip from the front of the intake manifold. Disconnect the vacuum hose from the intake manifold

   - Intake manifold assembly
   - Fuel injector electrical connectors. Detach the wiring harness pin-type retainers
   - Ignition coil and the Cylinder Head Temperature (CHT) sensor electrical connectors
   - Engine wiring harness anchors from the valve cover studs. Remove the engine wiring harness
   - Ignition coil
   - Bypass hose
   - Thermostat housing
   - Knock sensor and the engine vent cover
   - Left motor mount
   - Fuel injector supply manifold with the injectors and the ground strap
   - Water pump pulley
   - Water pump
   - CMP sensor
   - CHT sensor
   - Spark plugs
   - Valve cover
   - CKP sensor
   - Crankshaft vibration damper

➡There is one front cover bolt behind the cooling fan drive pulley. To remove this bolt, align one of the cooling fan drive pulley access holes with the bolt head to access the bolt.

   - Front cover
   - Timing chain tensioner
   - Timing chain guides
   - Timing chain assembly

➡Use a wrench on the flats between cylinders No. 1 and No. 2 to hold the camshaft in place.

   - Camshaft drive sprockets
   - Oil pump chain tensioner and guide

➡The oil pump chain sprocket must be held in place

   - Oil pump chain and sprockets

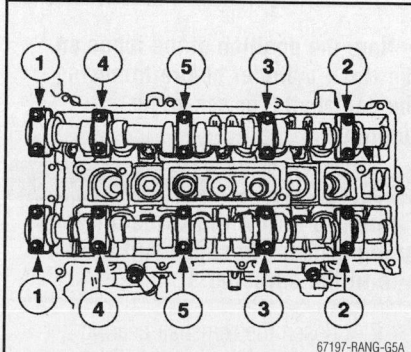

67197-RANG-G5A

Fig. 16 Camshaft cap loosening sequence—2.3L engine

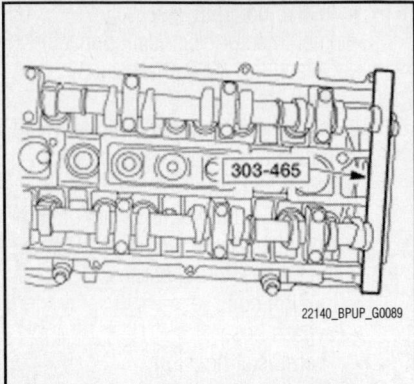

Fig. 17 Camshaft alignment plate 303-465

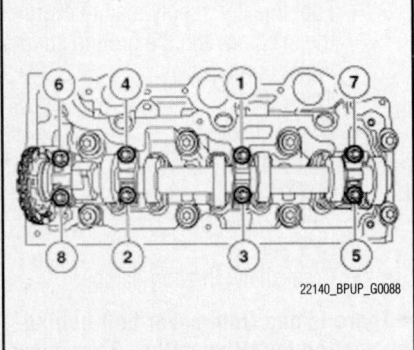

Fig. 18 Position of the camshaft bearing caps tightening sequence

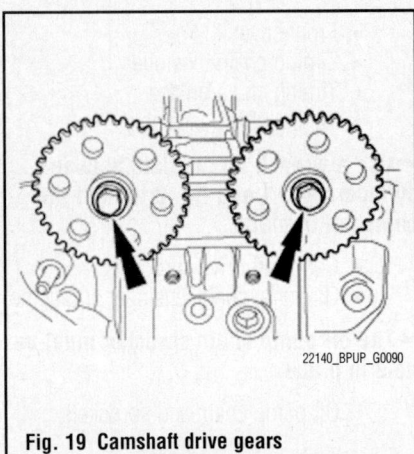

Fig. 19 Camshaft drive gears

→Note the position of the lobes on the No. 1 cylinder before removing the camshafts for assembly reference.

✳✳ WARNING

Failure to follow the camshaft loosening procedure can result in damage to the camshafts.

7. Loosen the camshaft bearing cap bolts in sequence, one turn at a time. Repeat the first step until all tension is released from the camshaft bearing

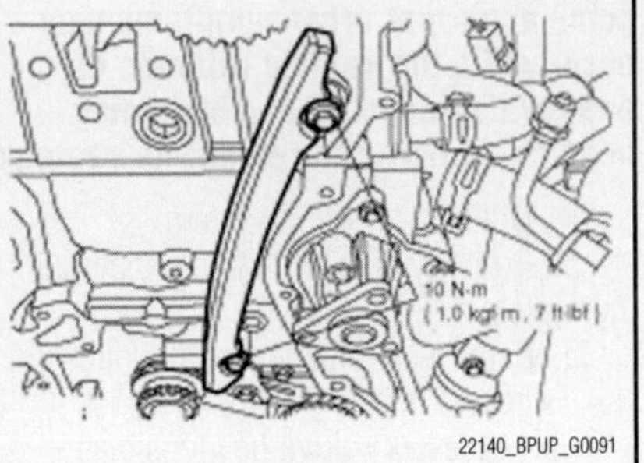

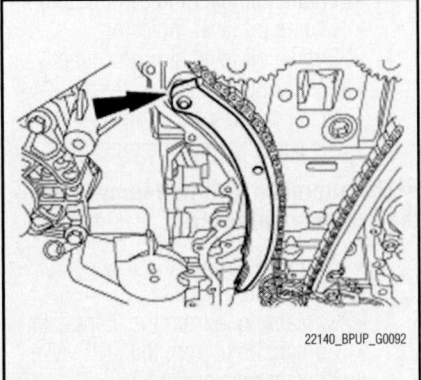

Fig. 20 LH timing chain guide and bolts

10 N·m (1.0 kgf·m, 7 ft·lbf)

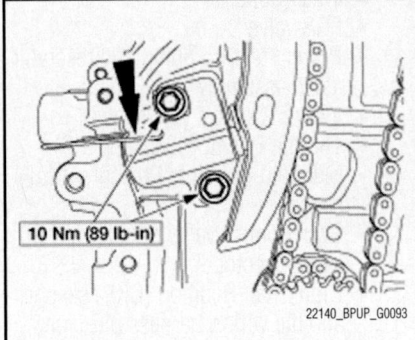

Fig. 21 RH timing chain guide and bolts

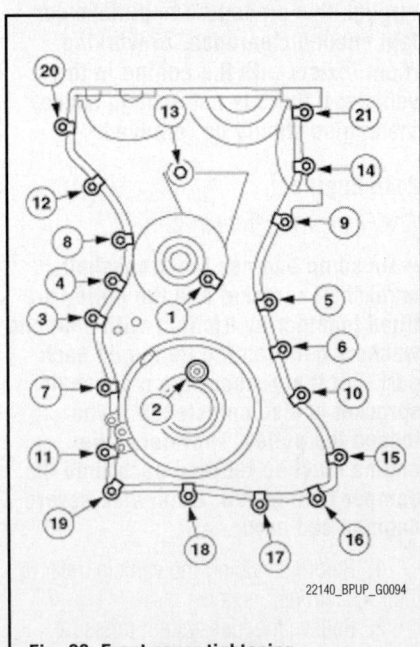

10 Nm (89 lb-in)

Fig. 22 Timing chain tensioner and the bolts—2.3L engine

caps. Remove the camshaft bearing caps.

8. Remove the camshaft alignment plate 303-465

9. Remove the camshafts.

To install:

→Install the camshafts with the alignment notches in the camshaft lined up so the camshaft alignment plate can be installed without rotating the camshafts. Make sure the lobes on the No. 1 cylinder are in the same position as noted in the disassembly procedure. Rotating the camshafts, or installing

the camshafts 180 degrees out of position can cause severe damage to the valves and pistons.

→Lubricate the camshaft journals and bearing caps with clean engine oil. Install the camshafts and bearing caps.

✳✳ WARNING

Do not rely on the camshaft alignment plate to prevent camshaft rotation. Damage to the tool or the camshaft can occur. If necessary, remove the bolts and the camshaft sprockets. Use the flats on the camshaft to prevent camshaft rotation.

10. Remove the camshaft alignment plate 303-465

→Tighten the bolts in the sequence shown in three stages.

- Step 1: Tighten the camshaft bearing caps one turn at a time until tight.
- Step 2: Tighten the bolts to 7 Nm (62 inch lbs.)
- Step 3: Tighten the bolts to 16 Nm (12 ft. lbs.)

11. Install the camshaft drive gears and hand-tighten the bolts.

→Do not tighten the bolts at this time.

12. Install the LH timing chain guide and bolts.

13. Tighten the LH timing chain guide bolts to 89 inch lbs. (10 Nm)

Fig. 23 Front cover tightening sequence—2.3L engine

14. Install the timing chain.

15. Install the RH timing chain guide.

16. Tighten the RH timing chain guide bolts to 89 inch lbs. (10 Nm)

17. Install the timing chain tensioner and the bolts.

18. Tighten the timing chain tensioner bolts to 89 inch lbs. (10 Nm).

19. Remove the paper clip to release the piston.

20. Install the camshaft alignment plate.

☀ WARNING

Do not rely on the camshaft alignment plate to prevent camshaft rotation. Damage to the tool or the camshafts can result. Using the flats on the camshafts to prevent camshaft rotation.

21. Tighten the camshaft sprockets bolts to 53 ft. lbs. (72 Nm)

22. Install the front cover

23. Tighten the bolts in the sequence shown, to the following specifications:

- 8 mm bolts—7 ft. lbs. (10 Nm).
- 10 mm bolts—18 ft. lbs. (25 Nm).
- 13 mm bolts—35 ft. lbs. (48 Nm).

☀ WARNING

There is one bolt behind the cooling fan drive pulley. This bolt can be accessed by lining up one of the holes in the pulley with the bolt.

24. Position the power steering pump and install the lower retaining bolt.

25. Tighten the power steering pump bolts to 18 ft. lbs. (25 Nm).

26. Connect the PSP switch electrical connector.

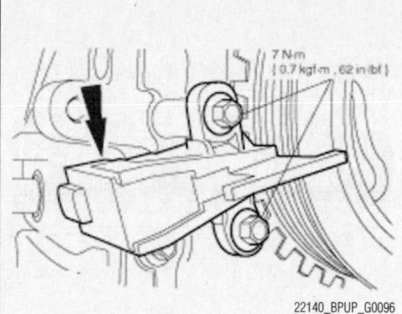

Fig. 25 CKP sensor with the alignment tool—2.3L engine

27. Install the water pump pulley.

28. Tighten the water pump pulley bolts to 15 ft. lbs. (20 Nm).

29. Install the belt tensioner.

30. Tighten the belt tensioner bolts to 36 ft. lbs. (50 Nm).

☀☀ CAUTION

The crankshaft, the crankshaft sprocket and the pulley are fitted together by friction, with diamond washers between the flange faces on each part. For that reason, the crankshaft sprocket is also unfastened if you loosen the pulley. Therefore, the engine must be retimed each time the damper is removed. Otherwise severe engine damage can occur.

31. Install the crankshaft vibration damper:

➡**Do not reuse the crankshaft pulley bolt. Tighten the bolt in two stages.**

32. Tighten the water pump pulley bolts to 15 ft. lbs. (20 Nm).

☀☀ CAUTION

A new CKP sensor must be installed whenever the old sensor is removed. Install a new CKP sensor, do not tighten the bolts at this time

33. Tighten the CKP sensor bolts to 82 inch. lbs. (7 Nm).

34. Connect the CKP sensor electrical connector and the wiring harness pin-type retainers.

35. Connect the battery ground cable.

4.0L Engine—Right Side

See Figures 26 through 29.

1. Before servicing the vehicle, refer to the Precautions Section.

2. Properly relieve the fuel system pressure.

3. Disconnect the negative battery cable.

➡**On some vehicles, when the battery cable is disconnected and reconnected, some abnormal drive symptoms may occur while the vehicle relearns its adaptive strategy. The vehicle may need to be driven to relearn its strategy.**

4. Remove the valve covers. Remove the fuel supply manifold.

➡**Mark each camshaft roller follower to ensure its original position during reassembly.**

5. Using tool 303-581, remove the roller followers.

6. Remove the RH hydraulic chain tensioner.

7. Install the camshaft sprocket holding tool and the adaptor 303-575 on the rear of the RH cylinder head.

➡**The camshaft sprocket is a left hand threaded bolt pattern.**

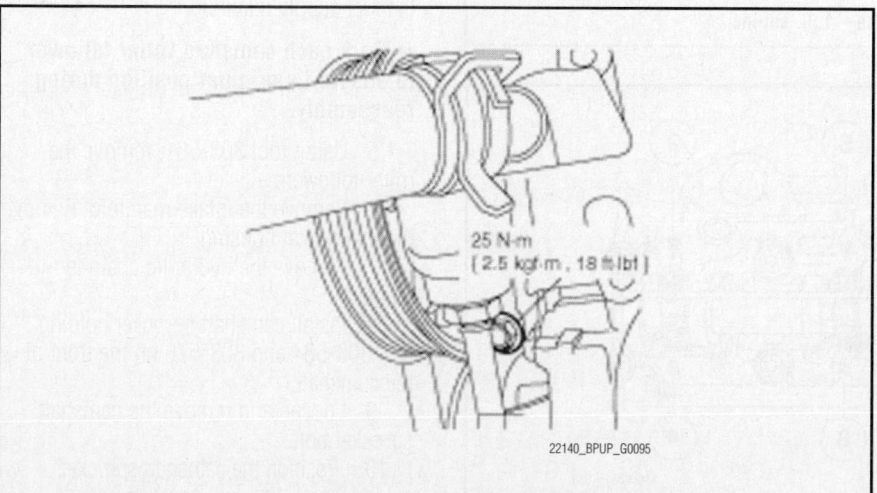

Fig. 24 Power steering pump and install the lower retaining bolt—2.3L engine

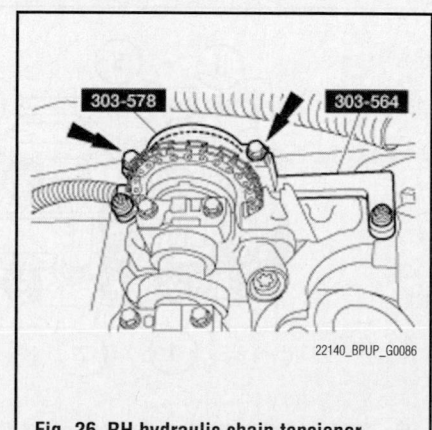

Fig. 26 RH hydraulic chain tensioner.

8. Loosen and remove the camshaft sprocket bolt. Position the camshaft sprocket aside.

➡ **Mark the position of the camshaft bearing caps so that they can be reinstalled in their original positions.**

9. Remove the bolts in the proper sequence and remove the bearing caps.

10. If equipped, remove the oil supply tube.

11. Remove the camshaft.

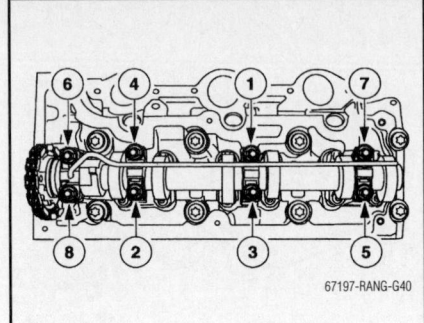

Fig. 29 Camshaft bolt torque sequence—4.0L engine

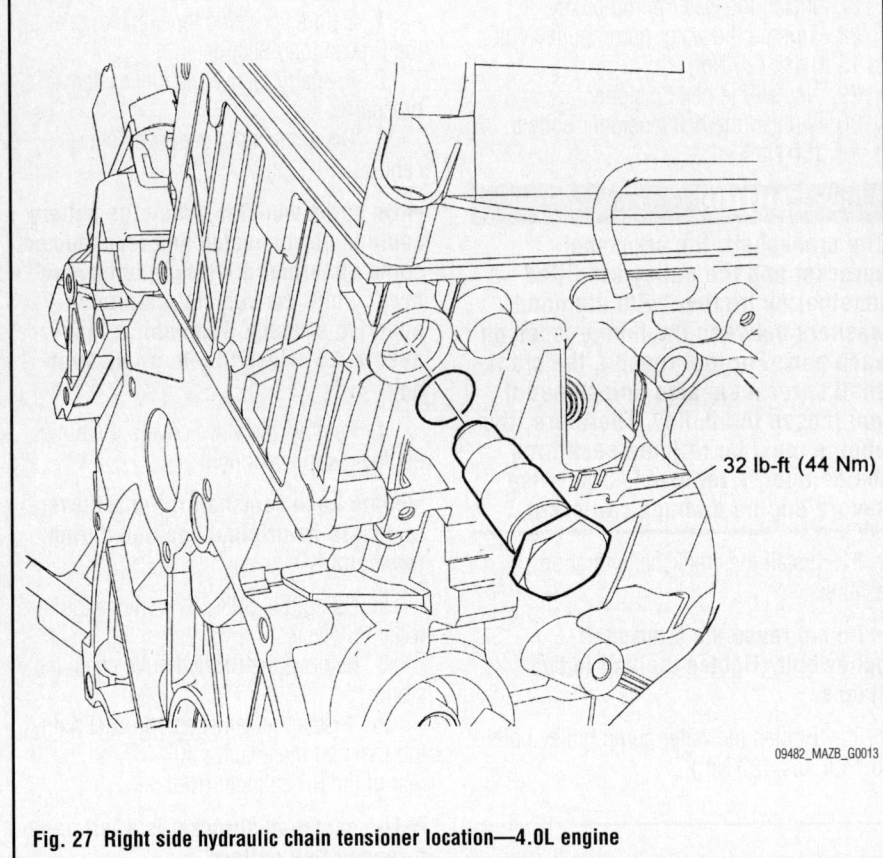

Fig. 27 Right side hydraulic chain tensioner location—4.0L engine

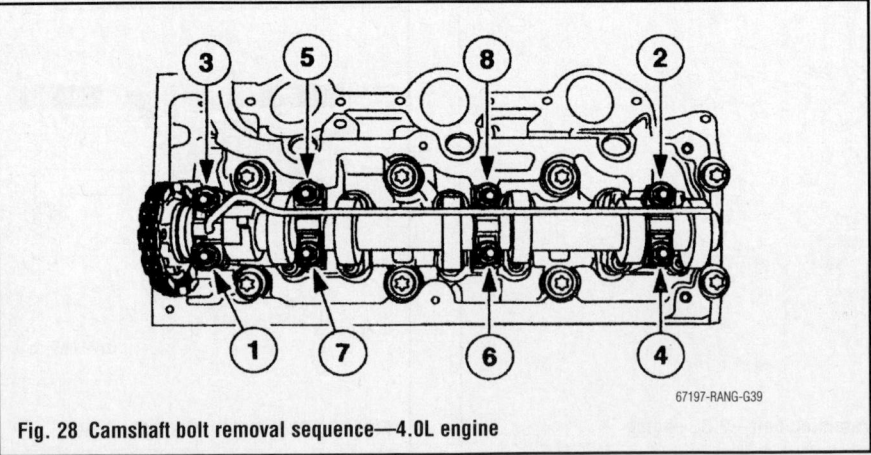

Fig. 28 Camshaft bolt removal sequence—4.0L engine

To install:

12. Lubricate all of the moving parts with SAE 50W engine oil.

13. Install camshaft onto the cylinder head.

14. Position the oil supply tube, if equipped, the camshaft bearing caps and the bolts.

15. Position bearing caps and bolts.

16. Torque the bolts in 2 steps:

17. Step 1—53 inch lbs. (6 Nm).

18. Step 2—12 ft. lbs. (16 Nm).

➡ **The camshaft Gear must turn freely on the camshaft. Do not tighten the bolt at this time.**

19. Install the camshaft sprocket and loosely install the bolt.

➡ **The camshafts must be retimed or engine damage may occur.**

20. Retime the camshafts. See timing chain installation section.

21. Continue the installation in the reverse order of the removal procedure.

4.0L Engine–Left Side

See Figures 30 through 32.

1. Before servicing the vehicle, refer to the Precautions Section.

2. Properly relieve the fuel system pressure.

3. Disconnect the negative battery cable.

➡ **On some vehicles, when the battery cable is disconnected and reconnected, some abnormal drive symptoms may occur while the vehicle relearns its adaptive strategy. The vehicle may need to be driven to relearn its strategy.**

4. Remove the valve covers. Remove the fuel supply manifold.

➡ **Mark each camshaft roller follower to ensure its original position during reassembly.**

5. Using tool 303-581, remove the roller followers.

6. Remove the intake manifold. Remove the thermostat housing.

7. Remove the hydraulic chain tensioner.

8. Install camshaft sprocket holding tool 303-564 and 303-578 on the front of the camshaft.

9. Loosen and remove the camshaft sprocket bolt.

10. Position the camshaft sprocket aside.

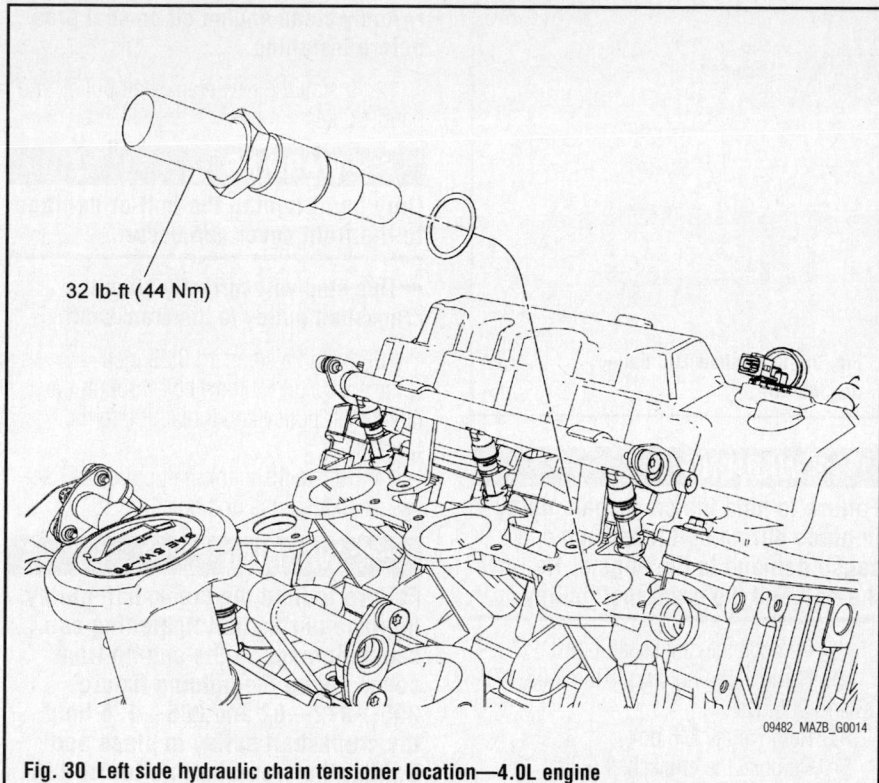

32 lb-ft (44 Nm)

09482_MAZB_G0014

Fig. 30 Left side hydraulic chain tensioner location—4.0L engine

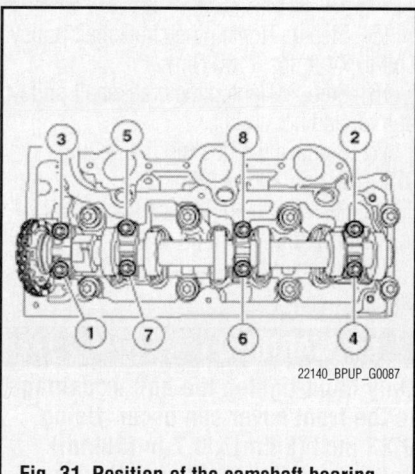

22140_BPUP_G0087

Fig. 31 Position of the camshaft bearing caps

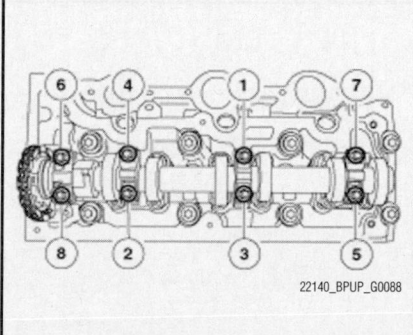

22140_BPUP_G0088

Fig. 32 Position of the camshaft bearing caps tightening sequence

➡**Mark the position of the camshaft bearing caps so that they can be reinstalled in their original positions.**

11. Remove the bolts in the proper sequence and remove the bearing caps.
12. If equipped, remove the oil supply tube.
13. Remove the camshaft.

To install:

14. Lubricate all of the moving parts with SAE 50W engine oil.

15. Install camshaft onto the cylinder head.
16. Position the oil rail, if equipped and install the bearing caps and bolts. Torque the bolts in 2 steps:
17. Step 1—53.5 inch lbs. (6 Nm).
18. Step 2—12 ft. lbs. (16 Nm).

➡**The camshaft Gear must turn freely on the camshaft. Do not tighten the bolt at this time.**

19. Install the camshaft sprocket and loosely install the bolt.

➡**The camshafts must be retimed or engine damage may occur.**

20. Retime the camshafts. See timing chain installation section.
21. Continue the installation in the reverse order of the removal procedure.

CRANKSHAFT DAMPER

REMOVAL & INSTALLATION

2.3L Engine

See Figures 33 through 39.

✳✳ CAUTION

The crankshaft, the crankshaft sprocket and the pulley are fitted together by friction, with diamond washers between the flange faces on each part. For that reason, the crankshaft sprocket is also unfastened if you loosen the pulley. Therefore, the engine must be retimed each time the damper is removed. Otherwise severe damage can occur.

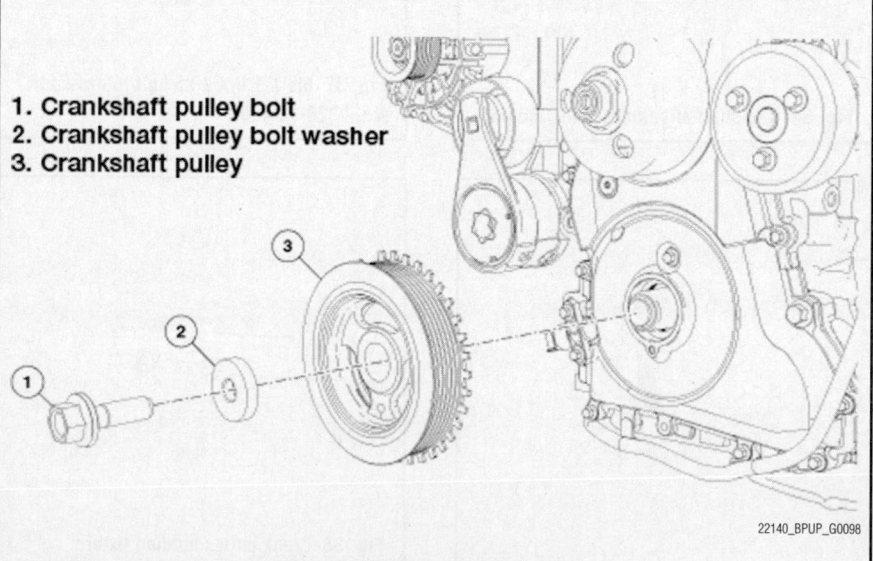

1. Crankshaft pulley bolt
2. Crankshaft pulley bolt washer
3. Crankshaft pulley

22140_BPUP_G0098

Fig. 33 Crankshaft pulley removal and installation exploded view

1. Before servicing the vehicle, refer to the Precautions Section.

2. Disconnect the negative battery cable.

➡ **On some vehicles, when the battery cable is disconnected and reconnected, some abnormal drive symptoms may occur while the vehicle relearns its adaptive strategy. The vehicle may need to be driven to relearn its strategy.**

3. Install the camshaft alignment plate 303-465 in the slots on the rear of both camshafts.

4. Remove the plug bolt.

➡ **Only turn the engine in the normal direction of rotation. Installing the special tool in this step will prevent the engine from being rotated in the clockwise direction. However, the engine can still be rotated in the counterclockwise direction.**

5. Install the crankshaft TDC timing peg 303-507—2.3L engine.

6. Install a M6 X 1.0 X 2.25 bolt (Mazda part No. 1F20-12-428) to verify ignition timing.

7. Install the holding fixture 205-072-02 and 205-126.

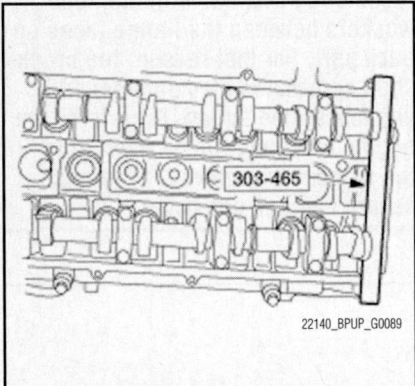

Fig. 34 Camshaft alignment plate 303-465

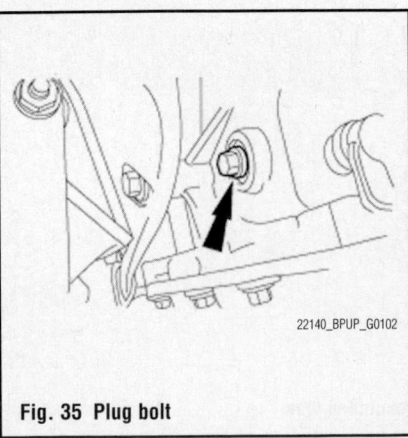

Fig. 35 Plug bolt

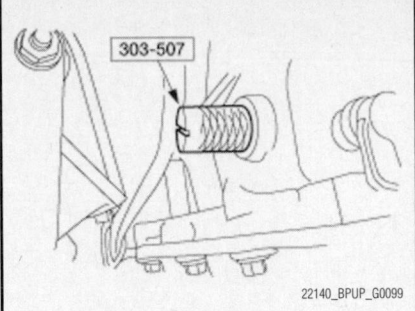

Fig. 36 Crankshaft TDC timing peg303-507

✳✳ WARNING

Failure to hold the crankshaft pulley in place during bolt loosening can cause damage to the engine. Remove the bolt and the crankshaft pulley.

8. Remove the crankshaft bolt.

9. Remove the holding fixture 205-072-02 and 205-126.

10. Remove the M6 bolt.

11. Remove the crankshaft pulley.

To install:

➡ **Do not reuse the crankshaft damper bolt or washer.**

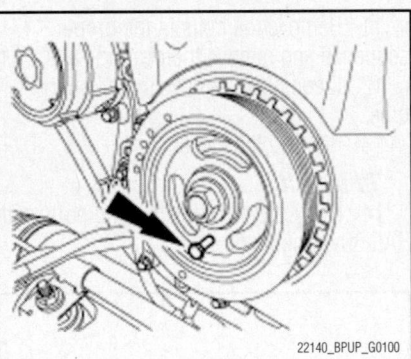

Fig. 37 M6 X 1.0 X 2.25 bolt (Mazda part No. 1F20-12-428)

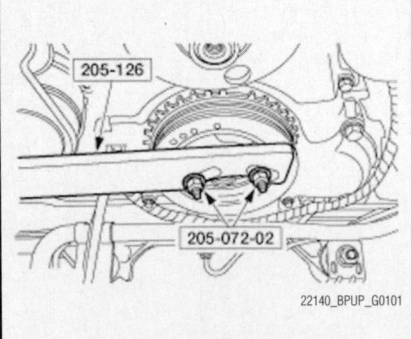

Fig. 38 Crank pulley holding fixture 205-072-02 and 205-126

➡ **Apply clean engine oil on seal area before installing.**

12. Install the new crankshaft pulley and hand-tighten the bolt.

✳✳ WARNING

Only hand-tighten the bolt or damage to the front cover can occur.

➡ **This step will correctly align the crankshaft pulley to the crankshaft.**

13. Install a standard 0.23 inch (6mm) x 0.7 in (18mm) bolt through the crankshaft pulley and thread it into the front cover.

14. Rotate the crankshaft pulley as necessary to align the bolt holes.

✳✳ WARNING

Failure to hold the crankshaft pulley in place during bolt tightening can cause damage to the engine front cover. Using the holding fixture 205—072—02 and 205—126 hold the crankshaft pulley in place and tighten the crankshaft pulley bolt in two stages.

15. Step 1: Tighten the crankshaft pulley bolt to 74 ft. lbs. (100 Nm).

16. Step 2: Tighten the crankshaft pulley bolt an additional 90°.

17. Remove the holding fixture205—072—02 and 205—126

18. Remove the 0.23 inch (6mm) x 0.7 in (18mm) bolt.

19. Remove the crankshaft TDC timing peg.

✳✳ WARNING

Only hand-tighten the bolt or damage to the front cover can occur. Using 0.23 inch (6mm) x 0.7 in (18mm) bolt, check the position of the crankshaft pulley.

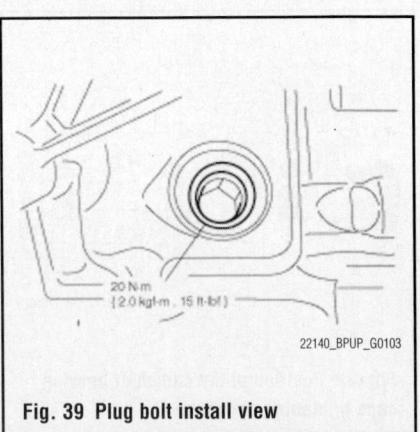

Fig. 39 Plug bolt install view

20. If it is not possible to install the 0.23 inch (6mm) x 0.7 in (18mm) bolt, correct the engine timing.

21. Using the camshaft alignment plate 303—465, check the position of the camshafts.

22. If it is not possible to install the camshaft alignment plate 303—465, correct the engine timing.

23. Remove the camshaft alignment plate.

24. Remove the 6 mm 0.23 inch (6mm) x 0.7 in (18mm) bolt.

25. Remove the crankshaft TDC timing peg.

26. Install the plug bolt.

27. Tighten the Plug Bolt to 15 ft. lbs. (20 Nm).

28. Install the valve cover.

29. Install the accessory drive belt.

30. Install the fan and shroud.

4.0L Engine

See Figures 40 through 43.

1. Before servicing the vehicle, refer to the Precautions Section.

2. Disconnect the negative battery cable.

➡**On some vehicles, when the battery** cable is disconnected and reconnected, **some abnormal drive symptoms may occur while the vehicle relearns its adaptive strategy. The vehicle may need to be driven to relearn its strategy.**

3. Remove the fan shroud. See removal and installation of the cooling fan.

4. Remove the accessory drive belt. See removal and installation of drive belts.

✳✳ CAUTION

This bolt is a torque-to-yield design and cannot be reused.

5. Using the strap wrench 303—D055, remove the crankshaft pulley bolt.

➡**If the crankshaft pulley bolt is not installed, the crankshaft vibration damper remover 303—773 will bottom out before the pulley is off. Install the crankshaft pulley bolt 2-3 turns. Using the crankshaft vibration damper remover 303—773 and the 8 mm x 1.25 x 100 mm bolts, remove the pulley.**

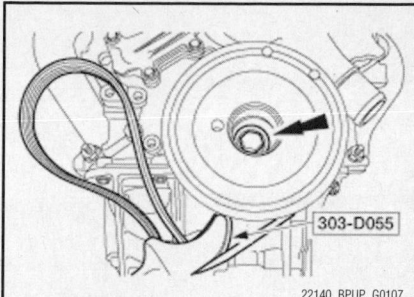

22140_BPUP_G0107

Fig. 41 Strap wrench 303—D055, and crankshaft pulley bolt to be removed and installed.

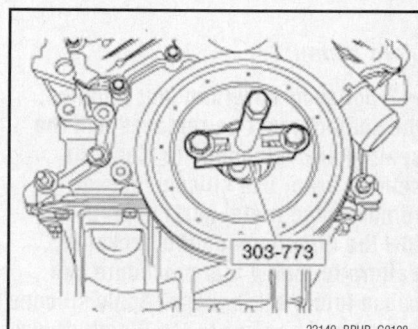

22140_BPUP_G0108

Fig. 42 Crankshaft vibration damper remover 303—773 tool

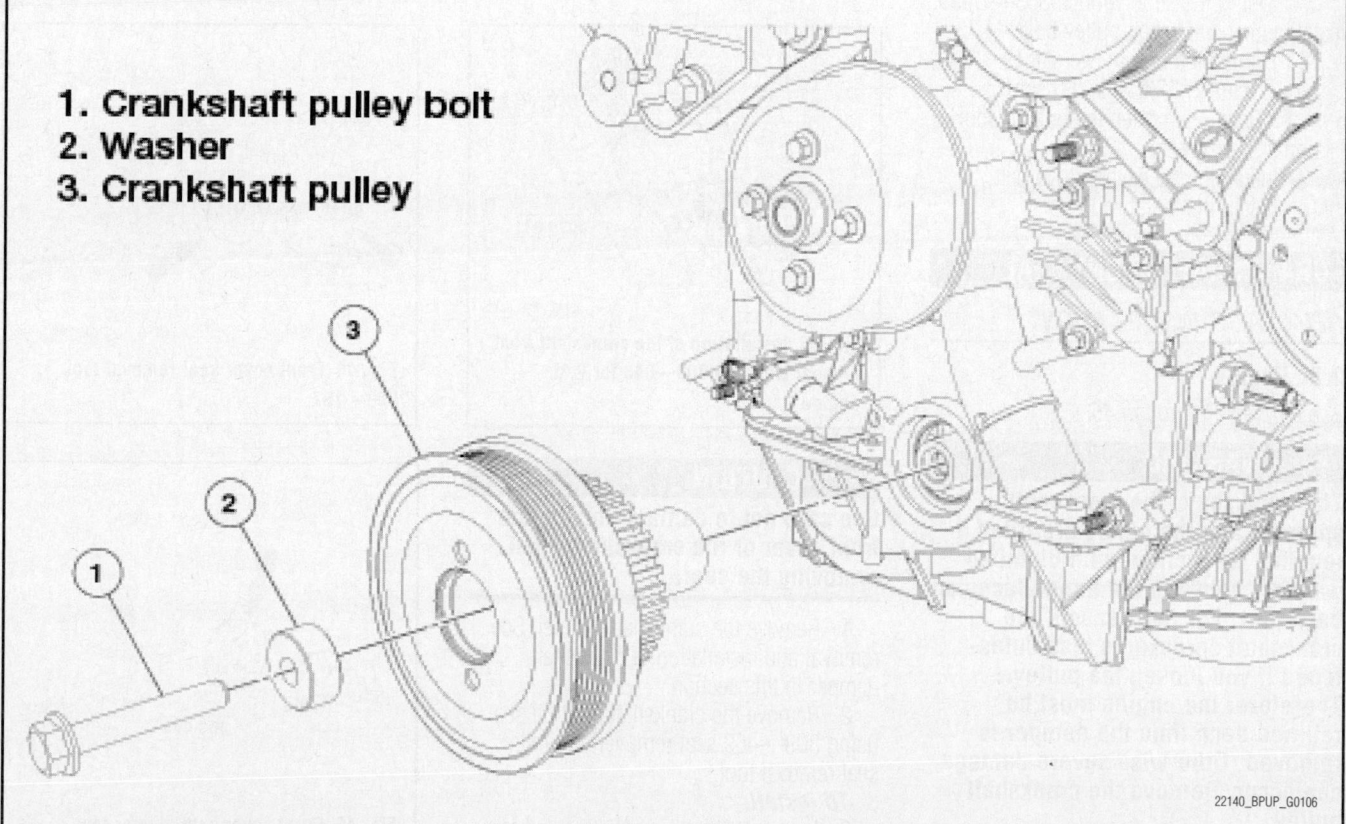

1. Crankshaft pulley bolt
2. Washer
3. Crankshaft pulley

22140_BPUP_G0106

Fig. 40 Crankshaft damper exploded view

Fig. 43 Crankshaft vibration damper installer 303—102 tool

To install:

➡️If not secured within four minutes, the sealant must be removed and the sealing area cleaned. To clean the sealing area, use silicone gasket remover and metal surface prep. Follow the directions on the packaging. Failure to follow this procedure can cause future oil leakage. Apply silicone gasket and sealant to the Woodruff key slot on the crankshaft pulley.

6. Using the crankshaft vibration damper installer 303—102, install the crankshaft pulley.

7. Using the strap wrench 303—D055, install a new crankshaft pulley bolt.

8. Tighten the bolt in two stages:
   a. Stage 1: 37 ft. lbs. (50 Nm).
   b. Stage 2: Tighten the bolt an additional 90 degrees.

9. Install the accessory drive belt.

10. Install the fan shroud.

CRANKSHAFT FRONT SEAL

REMOVAL & INSTALLATION

2.3L Engine

See Figures 44 through 46.

✳️✳️ WARNING

The crankshaft, the crankshaft sprocket and the pulley are fitted together by friction, with diamond washers between the flange faces on each part. For that reason, the crankshaft sprocket is also unfastened if you loosen the pulley. Therefore, the engine must be retimed each time the damper is removed. Otherwise severe damage can occur. Remove the crankshaft pulley.

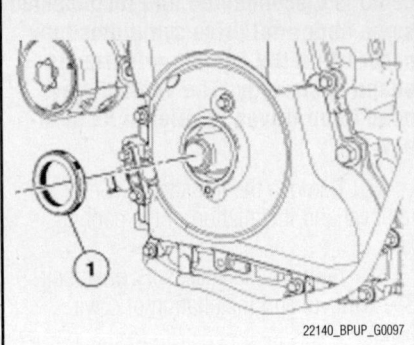

Fig. 44 Crankshaft front seal removal / installation exploded view—2.3L engine

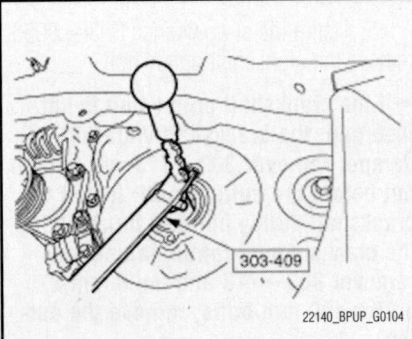

Fig. 45 Removal of the crankshaft front oil seal with tool 303—409 for 2.3L engine

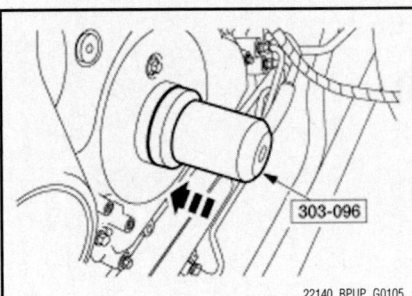

Fig. 46 Installation of the crankshaft front oil seal with tool 303—096 for 2.3L engine

✳️✳️ CAUTION

Use care not to damage the engine front cover or the crankshaft when removing the seal.

1. Remove the crankshaft damper. See removal and installation of crankshaft damper in this section.

2. Remove the crankshaft front oil seal using 303—409 seal remover or similar seal removal tool.

To install:

3. Lubricate the oil seal with clean engine oil.

4. Using the crankshaft front oil seal installer 303—096, install the crankshaft front oil seal.

5. Install the crank shaft pulley.

4.0L Engine

See Figures 47 through 49.

1. Remove the crankshaft pulley. See crankshaft pulley removal and installation in this section.

2. Using the front cover seal remover 303—107, remove the crankshaft front oil seal.

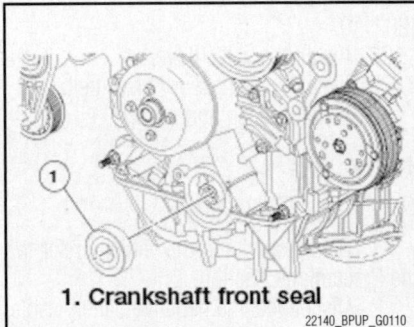

1. Crankshaft front seal

Fig. 47 Crankshaft front oil seal removal and installation exploded view

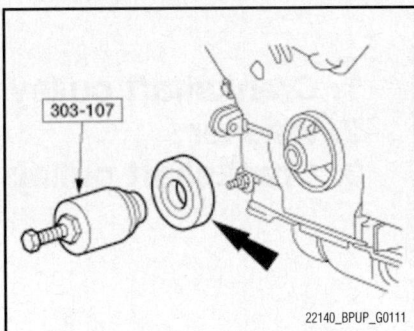

Fig. 48 Front cover seal removal tool 303—107

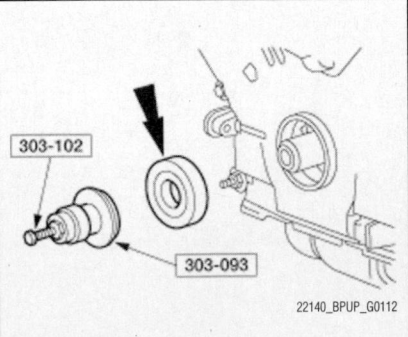

Fig. 49 Front cover seal aligner tool 303—093 and 303—102

To install:

➡**Lubricate the seal lip with clean engine oil. Using the front cover seal remover 303-107 and front cover aligner replacer 303-093 and 303-102, install the crankshaft front oil seal.**

3. Install the crankshaft pulley.

CYLINDER HEAD

REMOVAL & INSTALLATION

2.3L Engine

See Figures 50 through 52.

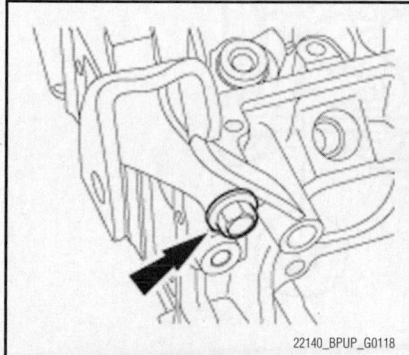

22140_BPUP_G0118

Fig. 50 Rear lifting eye and the front bracket mounting location

1. Before servicing the vehicle, refer to the precautions in the beginning of this section.
2. Relieve the fuel system pressure.
3. Drain the cooling system.
4. Properly discharge the A/C system.
5. Remove the rear lifting eye and the front bracket.
6. Remove the Cylinder Head Temperature (CHT) sensor.
7. Remove or disconnect the following:
   - Negative battery cable
   - Drive belt
   - Engine oil level indicator assembly
   - Engine oil level indicator
   - Engine oil level indicator tube
   - Water outlet tube
   - Water outlet tube
   - A/C compressor

➡**The alternator will be removed with the accessory bracket.**

   - Accessory bracket
   - Right motor mount
   - Coolant hose from the thermostat
   - Coolant hose from the Exhaust Gas Recirculation (EGR) valve
   - Coolant tube assembly
   - Exhaust manifold and gasket

   - Block heater (if so equipped)
   - EGR valve
   - Power steering pump and reservoir as an assembly
   - Idle Air Control (IAC) valve
   - Throttle Position (TP) sensor
   - Manifold Absolute Pressure (MAP) sensor
   - Swirl control valve monitor electrical connector
   - Crankshaft Position (CKP) sensor and the wiring harness pin-type retainers
   - Knock Sensor (KS)
   - Electric thermostat
   - Swirl control valve
   - Camshaft Position (CMP) sensor electrical connector and disconnect the Positive Crankcase Ventilation (PCV) valve hose from the intake manifold
   - Engine wiring harness pin-type retainers from the intake manifold
   - Engine wiring harness connector bracket.

➡**Position the engine wiring harness aside.**

   - EGR tube
   - Fuel supply line clip from the front of the intake manifold

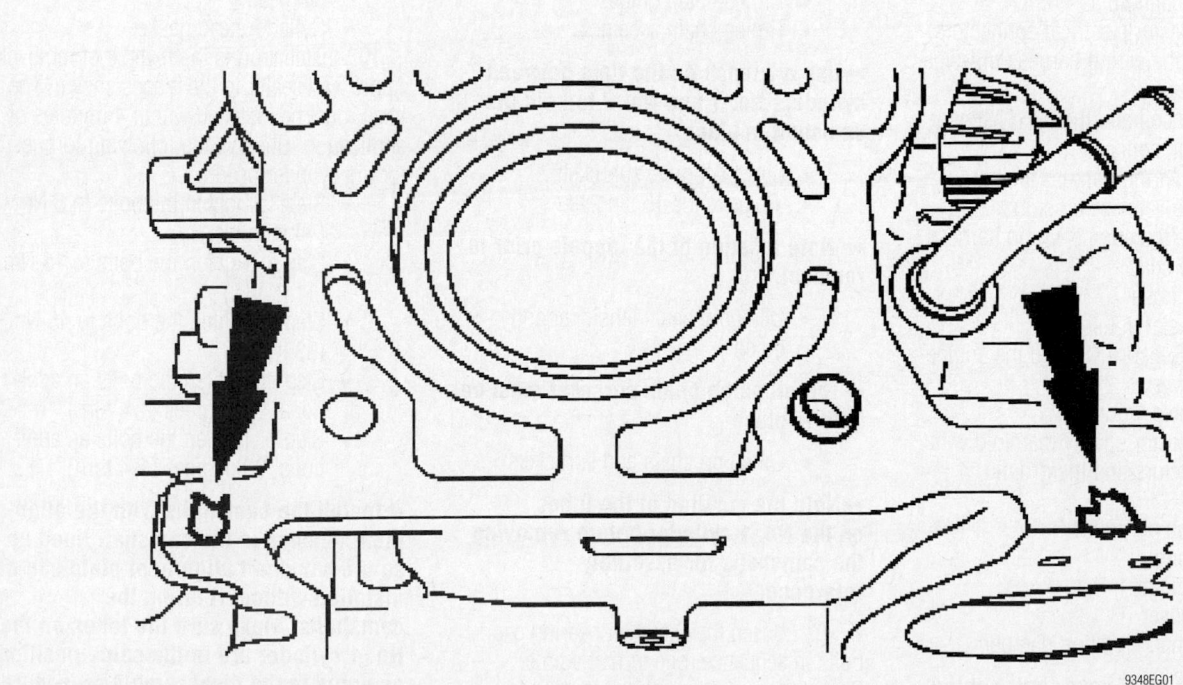

9348EG01

Fig. 51 RTV sealer application

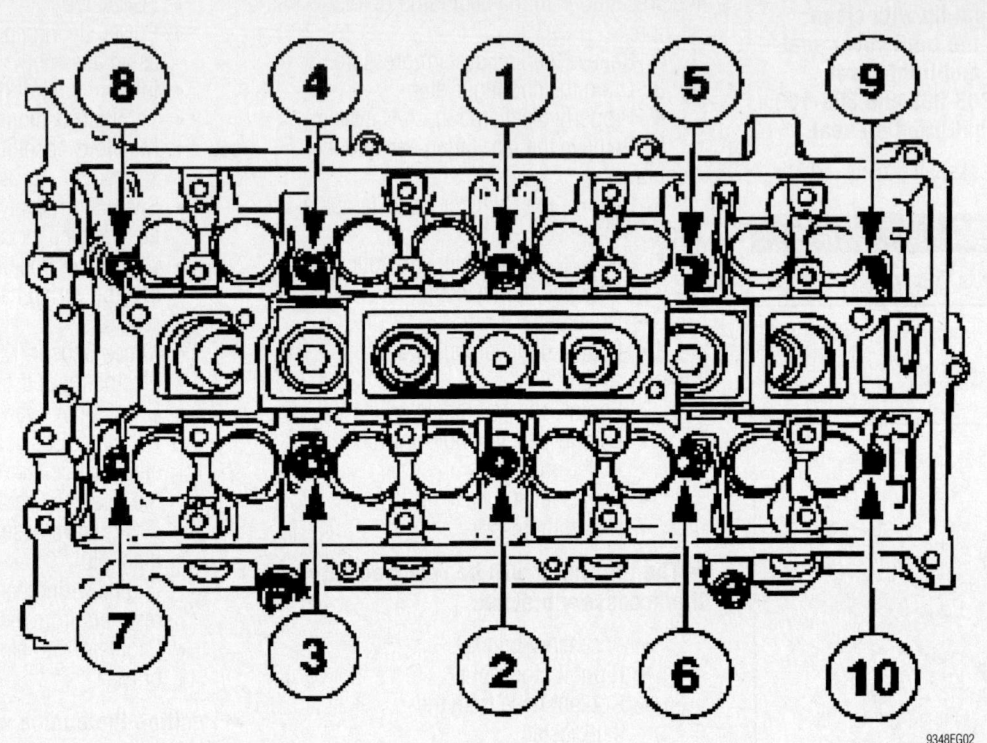

Fig. 52 Head bolt torque sequence

- Disconnect the vacuum hose from the intake manifold
- Intake manifold assembly
- Fuel injector electrical connectors. Detach the wiring harness pin-type retainers
- Ignition coil and the (CHT) sensor electrical connectors
- Engine wiring harness anchors from the valve cover studs
- Remove the engine wiring harness
- Ignition coil
- Bypass hose
- Thermostat housing
- Knock Sensor (KS) and the engine vent cover
- Left motor mount
- Fuel injector supply manifold with the injectors and the ground strap
- Water pump pulley
- Water pump
- Spark plugs
- Valve cover
- Crankshaft vibration damper

➡There is one front cover bolt behind the cooling fan drive pulley. To remove this bolt, align one of the cooling fan drive pulley access holes with the bolt head to access the bolt.

- Front cover
- Timing chain tensioner
- Timing chain guides
- Timing chain assembly

➡Use a wrench on the flats between cylinders No. 1 and No. 2 to hold the camshaft in place

- Camshaft drive sprockets
- Remove the 16 tappets

➡ Note location of the tappets prior to removal.

- Oil pump chain tensioner and guide

➡The oil pump chain sprocket must be held in place

- Oil pump chain and sprockets

➡Note the position of the lobes on the No. 1 cylinder before removing the camshafts for assembly reference.

8. Loosen the camshaft bearing cap bolts in sequence, one turn at a time. Repeat the first step until all tension is released from the camshaft bearing caps. Remove the camshaft bearing caps.

9. Remove or disconnect the following:

- Camshafts
- Cylinder head bolts and the cylinder head
- Cylinder head gasket

10. Installation is the reverse of removal. Apply RTV sealer to the places shown. The head must be installed within 4 minutes of application. Observe the following torques for the cylinder heads:

- Step 1: Tighten the bolts to 5 Nm (44 inch lbs.)
- Step 2: Tighten the bolts to 15 Nm (11 ft. lbs.)
- Step 3: tighten the bolts to 45 Nm (33 ft. lbs.)
- Step 4: Tighten the bolts an additional 90 degrees (1/4 turn)
- Step 5: Tighten the bolts an additional 90 degrees (1/4 turn)

➡Install the camshafts with the alignment notches in the camshaft lined up so the camshaft alignment plate can be installed without rotating the camshafts. Make sure the lobes on the No. 1 cylinder are in the same position as noted in the disassembly procedure. Rotating the camshafts, or installing the camshafts 180 degrees out of position can cause severe damage to the valves and pistons. Lubricate the camshaft journals and bearing caps with clean engine oil. Install the camshafts and bearing caps. Tighten the bolts in the sequence shown in three stages.

11. Observe the following torques for the camshafts.
- Step 1: Tighten the camshaft bearing caps one turn at a time until tight
- Step 2: Tighten the bolts to 7 Nm (62 inch lbs.)
- Step 3: Tighten the bolts to 16 Nm (12 ft. lbs.)

Observe the following torques for the crankshaft vibration damper.

➡**Do not reuse the crankshaft pulley bolt. Tighten the bolt in two stages.**

- Step 1: Tighten the bolt to 40 Nm (30 ft. lbs.)
- Step 2: Tighten the bolt and additional 90 degrees (1/4 turn).

4.0L Engine

See Figures 53 through 55.

➡**If only one cylinder head is to be removed, only follow the procedures that apply. The following tools, or their equivalents are absolutely necessary to properly perform this procedure:**

- Cam Chain Tensioner tool T97T-6K254-A
- Cam Gear Removal tool T97T-6256-F
- Cam Gear Torque adapter T97T-6256-G
- Camshaft Gear Positioning/Holding tool T97T-6256-B
- Camshaft Gear Positioning/Holding tool adapter T97T-6256-A
- Camshaft holding tool T97T-6256-C
- Crankshaft holding tool T97T-6303-A
- Camshaft holding tool adapter T97T-6256-D

1. Before servicing the vehicle, refer to the Precautions Section.
2. Properly relieve the fuel system pressure.
3. Disconnect the negative battery cable.

➡**On some vehicles, when the battery cable is disconnected and reconnected, some abnormal drive symptoms may occur while the vehicle relearns its adaptive strategy. The vehicle may need to be driven to relearn its strategy.**

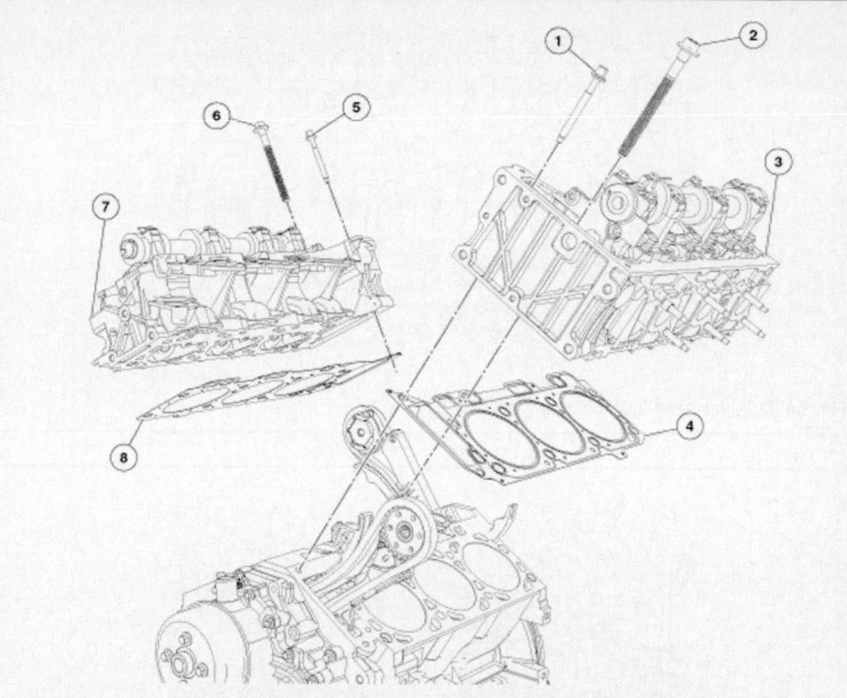

1. LH cylinder head bolts (M8) (2 required)
2. LH cylinder head bolts (M12) (8 required)
3. LH cylinder head
4. LH cylinder head gasket
5. RH cylinder head bolts (M8) (2 required)
6. RH cylinder head bolts (M12) (8 required)
7. RH cylinder head
8. RH cylinder head gasket

22140_BPUP_G0117

Fig. 53 Cylinder head removal exploded view

4. Drain the cooling system.
5. Remove or disconnect the following:
- Lower intake manifold
- Fan blade and shroud
- Valve cover
- Roller followers, if equipped
- Drive belt
- Upper radiator hose and tube
- Alternator electrical connectors
- Alternator mounting bracket
- Engine accessory bracket and move it aside
- Camshaft Position (CMP) electrical connector
- Crankshaft Position (CKP) sensor electrical connector
- Engine Coolant Temperature (ECT) sensor electrical connector
- Coil pack electrical connector
- Exhaust Gas Recirculation (EGR) valve electrical connector
- EGR valve bracket and move it aside
- Heater hoses
- Fuel injector electrical connectors
- Water bypass hose
- Thermostat housing
- Spark plug wires
- Fuel injection supply manifold
- Fuel injectors
- Crankcase vent separator spring

- Oil dipstick housing
- Exhaust manifold
- Hydraulic chain tensioner
- Cassette retaining bolt
- Camshaft sprocket

✳✳ WARNING

Remove the camshaft sprocket to prevent breaking the cassette and to gain clearance to remove the cylinder head.

➡**Hold the chain and cassette with a rubber band to aid in removal and prevent the chain from falling into the cylinder block.**

- Cylinder head and discard the gasket

➡**Discard the head bolts.**

To install:

6. Thoroughly clean all gasket mating surfaces. Remove all traces of old gasket material, oil, grease or dirt.
7. Insure that the rubber band is holding the right-hand chain to the cassette.
8. Install a new head gasket and the cylinder head. Be sure to use new cylinder head bolts.

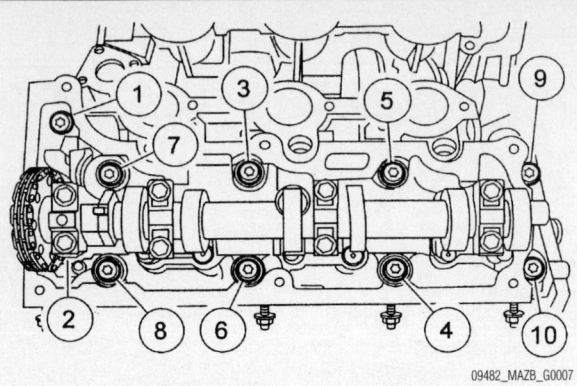

Fig. 54 Cylinder head bolt loosening sequence

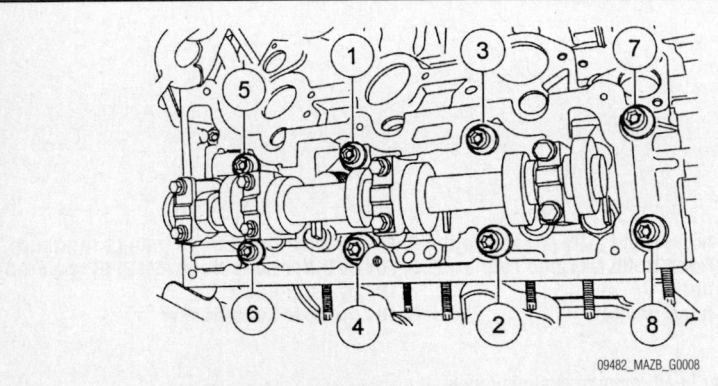

Fig. 55 Cylinder head bolt torque sequence

✳✳ WARNING

Cylinder head bolts are a torque-to-yield design. New bolts must be installed.

9. Torque the new cylinder head bolts in sequence as follows:

10. Install bolts 8 (12mm) and torque, in sequence to:
- Step 1: 9 ft. lbs. (12 Nm)
- Step 2: 18 ft. lbs. (25 Nm)
- Step 3: Install bolts 5 and 6 (8mm), and torque to 24 ft. lbs. (32 Nm)
- Step 4: 8 (12mm) bolts plus 90 degrees
- Step 5: 8 (12mm) bolts plus an additional 90 degrees

11. Install or connect the following:
- Camshaft sprocket in the cassette and make certain that the camshaft sprocket turns freely on the camshaft
- Cassette retaining bolt. Torque the bolt to 89 inch lbs. (10 Nm)
- Exhaust manifold
- Oil level indicator tube. Torque the bolt to 18 ft. lbs. (25 Nm)
- Crankcase vent separator and spring

- Thermostat housing. Torque the bolts to 8 ft. lbs. (11 Nm)
- Water bypass hose
- Heater hoses
- EGR bracket. Torque the bolt to 89 inch lbs. (10 Nm)
- EGR tube. Torque the nut to 30 ft. lbs. (40 Nm)
- ECT sensor electrical connector
- Electrical harness retainer. Torque the bolt to 89 inch lbs. (10 Nm)
- CKP and CMP electrical connectors
- Accessory bracket. Torque the bolts to 31 ft. lbs. (42 Nm)
- Alternator mounting bracket. Torque the bolts to 31 ft. lbs. (42 Nm)
- Alternator and electrical connectors
- Drive belt
- Fan shroud
- Roller followers
- Valve cover
- Lower intake manifold
- Negative battery cable

12. Change the engine oil and filter.
13. Refill the cooling system.
14. Start the engine and check for leaks, repair if necessary.

EXHAUST MANIFOLD

REMOVAL & INSTALLATION

2.3L Engine

1. Before servicing the vehicle, refer to the Precautions Section.
2. Disconnect the negative battery cable.

➡ **On some vehicles, when the battery cable is disconnected and reconnected, some abnormal drive symptoms may occur while the vehicle relearns its adaptive strategy. The vehicle may need to be driven to relearn its strategy.**

3. Remove or disconnect the following:
- Exhaust flange nuts
- Drive belt
- Coolant
- Upper radiator hose and the engine reservoir hose
- Air conditioning compressor
- Heater hose
- Oil indicator and the upper bolt for the tube assembly
- Lower bolt and remove the oil indicator tube assembly
- Front radiator tube

4. Remove the pushpins and position the right inner fender splash shield out of the way.

5. Remove or disconnect the following:
- Alternator electrical connectors
- Lower Front End Accessory Drive (FEAD) mounting bolts
- Upper mounting bolt and the FEAD assembly
- Two nuts and position the coolant tube out of the way
- Exhaust manifold
- Exhaust manifold gasket

To install:

6. Install or connect the following:
- Exhaust manifold gasket
- Exhaust manifold and the nuts
- Coolant tube and the nuts
- FEAD assembly and the upper mounting bolts, finger tight.
- Lower FEAD mounting bolts, finger tight. Then, torque all FEAD bolts, in the sequence shown, to 35 ft. lbs. (47 Nm).
- Alternator electrical connectors
- Right inner splash shield and pushpins
- Upper radiator tube and install the bolts

- Oil indicator tube assembly and the lower bolt
- Oil indicator tube upper bolt and the oil indicator
- Heater water hose
- Air conditioning compressor
- Upper radiator hose and the engine reservoir hose

7. Fill the cooling system.
8. Install the serpentine drive belt.
9. Install the exhaust flange nuts.
10. Connect the battery ground cable.

4.0L Engine—Left Side

See Figure 56.

1. Before servicing the vehicle, refer to the Precautions Section.
2. Disconnect the negative battery cable.

➡**On some vehicles, when the battery cable is disconnected and reconnected, some abnormal drive symptoms may occur while the vehicle relearns its adaptive strategy. The vehicle may need to be driven to relearn its strategy.**

3. Raise and support the vehicle safely. Remove the exhaust pipe flange retaining bolts.
4. Lower the vehicle.
5. Disconnect the Differential Pressure Feedback Exhaust Gas Recirculation (DPFE) transducer hoses.
6. Disconnect the Exhaust Gas Recirculation (EGR) valve to exhaust manifold tube.
7. Disconnect the EGR valve to exhaust manifold tube from the EGR valve and remove the EGR valve to exhaust manifold tube.
8. Remove the exhaust manifold retaining nuts. Remove the exhaust manifold from the engine.

To install:

9. Clean the mating surfaces for the exhaust manifold and cylinder head
10. Install a new gasket and the exhaust manifold. Torque the exhaust flange nuts to 33 ft. lbs. (46 Nm).
11. Continue the installation in the reverse order of the removal procedure.

4.0L Engine—Right Side

See Figure 57.

1. Before servicing the vehicle, refer to the Precautions Section.
2. Disconnect the negative battery cable.

➡**On some vehicles, when the battery cable is disconnected and reconnected, some abnormal drive symptoms may occur while the vehicle relearns its adaptive strategy. The vehicle may need to be driven to relearn its strategy.**

3. Raise and support the vehicle safely. Remove the exhaust pipe flange retaining bolts.

4. Lower the vehicle.
5. Remove the exhaust manifold retaining nuts. Remove the exhaust manifold from the engine.

To install:

6. Clean the mating surfaces for the exhaust manifold and cylinder head
7. Install a new gasket and the exhaust manifold. Torque the bolts to specification.
8. Continue the installation in the reverse order of the removal procedure.

FLYWHEEL

REMOVAL & INSTALLATION

2.3L Engine

➡**Engine balancing is not required. Balance weights should not be installed on the new flywheel.**

➡**Special bolts are used for installation. Do not use standard bolts. Inspect the pilot bearing. Install a new pilot bearing as necessary**

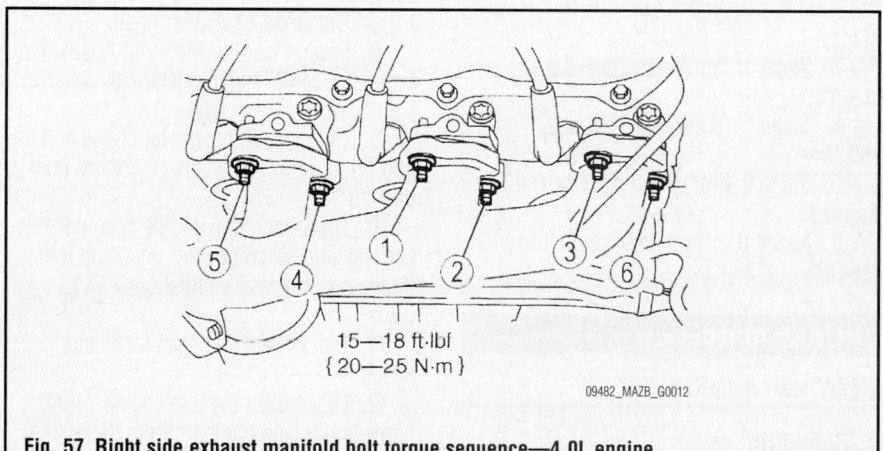

15—18 ft·lbf
{20—25 N·m}

09482_MAZB_G0012

Fig. 57 Right side exhaust manifold bolt torque sequence—4.0L engine

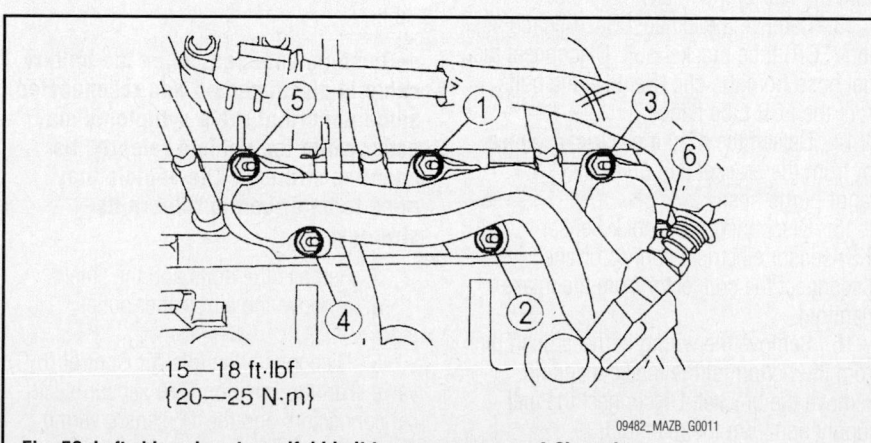

15—18 ft·lbf
{20—25 N·m}

09482_MAZB_G0011

Fig. 56 Left side exhaust manifold bolt torque sequence—4.0L engine

1. Before servicing the vehicle, refer to the Precautions Section.
2. Disconnect the negative battery cable.
3. Remove transmission See transmission installation and removal in this section.
4. Remove the clutch disc and the clutch cover.
5. Remove bolts and the flywheel.
6. To install, reverse removal procedure.
7. Using the special tool, install the bolts in the sequence shown, in 3 stages.—2.3L engine

8. Stage 1: Tighten to 37 ft. lbs. (50 Nm).

9. Stage 2: Tighten to 59 ft. lbs. (80 Nm).

10. Stage 3: Tighten to 83 ft. lbs. (112 Nm).

4.0L Engine

➡Engine balancing is not required. Balance weights should not be installed on the new flywheel.

➡Special bolts are used for installation. Do not use standard bolts. Inspect the pilot bearing. Install a new pilot bearing as necessary

1. Disconnect the negative battery cable.

2. Remove transmission See transmission installation and removal in this section.

3. Remove the clutch disc and the clutch cover.

4. Remove bolts and the flywheel.

5. To install, reverse removal procedure.

6. Using the special tool, install the bolts in the sequence shown, in 3 stages.

7. Stage 1: Tighten to 10 ft. lbs. (13 Nm).

8. Stage 2: Tighten to 37 ft. lbs. (50 Nm).

9. Stage 3: Tighten an additional 90 degrees.

10. Install the clutch disc and clutch cover.

INTAKE MANIFOLD

REMOVAL & INSTALLATION

2.3L Engine

See Figure 58.

1. Before servicing the vehicle, refer to the Precautions Section.

2. Relieve the fuel system pressure.

3. Disconnect the negative battery cable.

➡On some vehicles, when the battery cable is disconnected and reconnected, some abnormal drive symptoms may occur while the vehicle relearns its adaptive strategy. The vehicle may need to be driven to relearn its strategy.

4. Remove the accelerator control splash shield. Remove the air cleaner outlet tube.

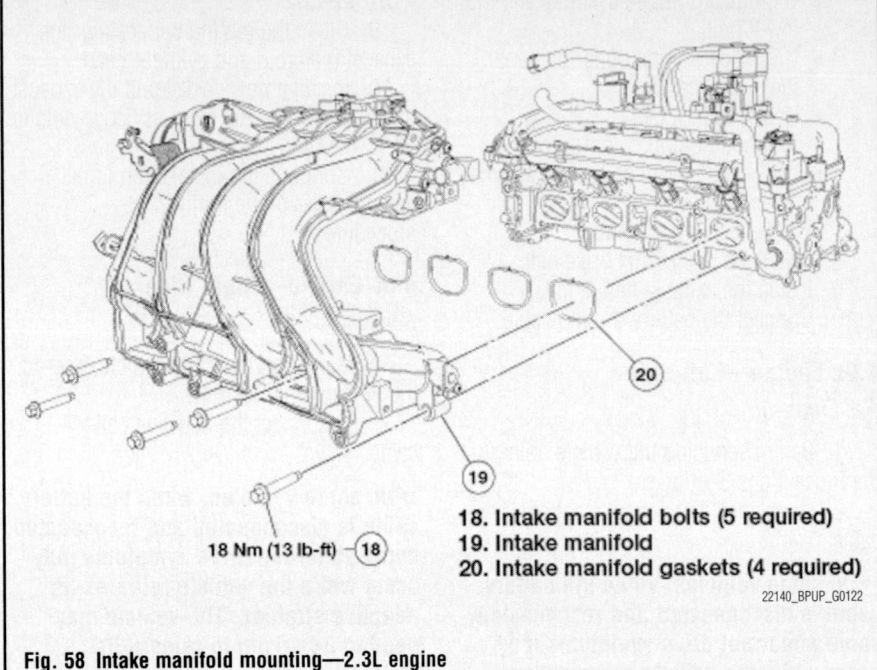

18 Nm (13 lb-ft) —(18)

19 —

(20)

18. Intake manifold bolts (5 required)
19. Intake manifold
20. Intake manifold gaskets (4 required)

22140_BPUP_G0122

Fig. 58 Intake manifold mounting—2.3L engine

5. Disconnect the throttle cables from the intake manifold.

6. Disconnect the Throttle Position (TP) sensor.

7. Disconnect breather hose.

8. Disconnect the Manifold Absolute Pressure (MAP) sensor electrical connector.

9. Disconnect the vacuum hose and the Idle Air Control (IAC) valve electrical connector.

10. Disconnect the engine vacuum harness and brake booster hose and the Positive Crankcase Ventilation (PCV) valve.

11. Disconnect and plug the heater hoses.

12. Disconnect the two engine wiring harness pin type retainers from the intake manifold. Disconnect the pin type retainer from the rear of the intake manifold.

13. Remove the Exhaust Gas Recirculation (EGR) tube bracket bolt. Disconnect the fuel hose from the clip. Remove the bolts from the EGR tube flange.

14. Detach the 42 pin electrical connector from the bracket and disconnect the vapor purge hose.

15. Disconnect the Knock Sensor (KS) sensor electrical connector and disconnect the connector from the intake manifold.

16. Remove the wiring harness push pin from the bottom of the intake manifold. Remove the bracket. Disconnect the fuel supply line from the manifold.

17. Remove the intake manifold retaining bolts. Remove the intake manifold from the engine.

To install:

18. Be sure to clean all old gasket material from the sealing surfaces, using the proper tools.

19. Installation is the reverse of the removal procedure.

20. Be sure to use new gaskets.

21. Torque the bolts to 13 ft. lbs. (18 Nm).

22. Be sure to fill the cooling system with the proper grade and type engine coolant.

23. Start the engine and check for leaks, correct as required.

4.0L Engine

See Figures 59 through 60.

1. Before servicing the vehicle, refer to the Precautions Section.

2. Disconnect the negative battery cable.

➡On some vehicles, when the battery cable is disconnected and reconnected, some abnormal drive symptoms may occur while the vehicle relearns its adaptive strategy. The vehicle may need to be driven to relearn its strategy.

3. Remove the bolts and the shield.

4. Remove the air cleaner outlet pipe.

5. Disconnect the Idle Air Control (IAC) valve, Throttle Position (TP) sensor electrical connectors and the TP sensor wiring pin-type retainer.

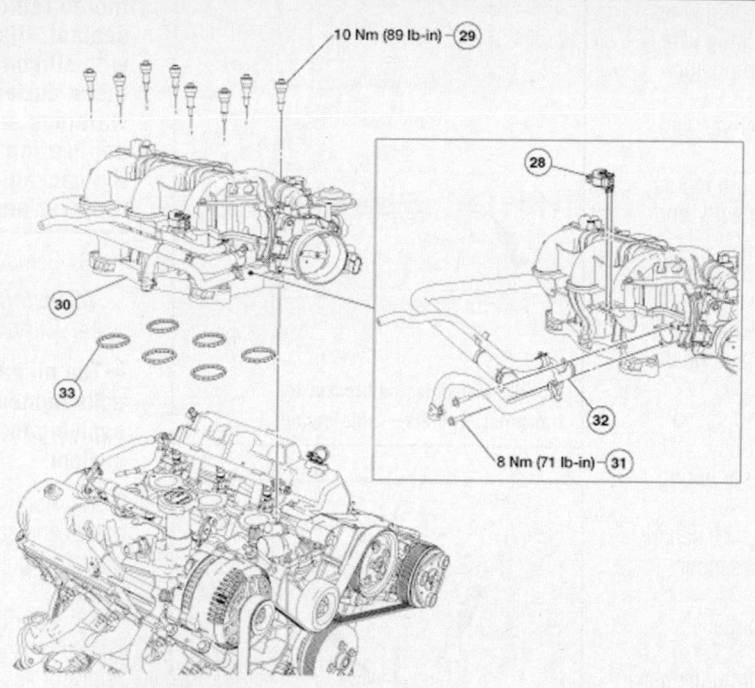

28. Knock sensor (KS) electrical connector
29. Intake manifold bolt and grommet assembly (8 required)
30. Intake manifold
31. Heated positive crankcase ventilation (PCV) fitting bolts (2 required)
32. Heated PCV fitting
33. Intake manifold gaskets (6 required)

22140_BPUP_G0135

Fig. 59 Intake manifold component identification part 5—4.0L engine

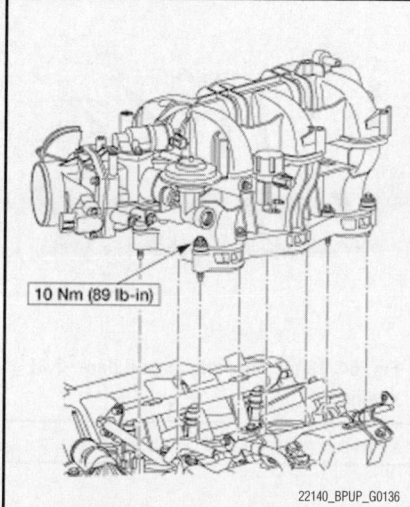

22140_BPUP_G0136

Fig. 60 Intake manifold installation—4.0L engine

6. Disconnect the Mass Air Flow (MAF) sensor wiring pin-type retainer.

7. Detach the accelerator and speed control cables from the throttle body.

8. Detach the accelerator and speed control cables from the bracket, and position the cables aside.

9. Disconnect the Exhaust Gas Recircu-lation (EGR) valve vacuum hose and tube fitting.

10. Disconnect the Exhaust Gas Recircu-lation (EGR) vacuum regulator solenoid valve electrical connector and vacuum hose.

11. Loosen the clamp and disconnect the brake booster vacuum hose.

➡It is important to twist the spark plug wire boots while pulling upward to avoid possible damage to the spark plug wire.

➡Mark the spark plug wire locations before removing them.

12. Disconnect the RH spark plug wires from the coil. Remove the spark plug wire routing clip pin-type retainer and position the wires aside.

13. Remove the wiring harness bracket retainer, then position the wiring harness aside.

14. Remove the accelerator cable routing clip pin-type retainer and the wiring harness pin-type retainer.

15. Remove the bolts.

16. Remove the bolts and position the coil and bracket aside.

17. Disconnect the vacuum hoses.

18. Remove the nut.

19. Disconnect the Powertrain Control Module (PCM) electrical connector.

20. Remove the retainer and position the ground wires aside.

21. Detach the electrical connector retainer.

22. Remove the intake manifold bolts and lift up the intake manifold.

23. Remove the heated Positive Crankcase Ventilation (PCV) hose retainers and remove the heated PCV fitting.

24. Remove the intake manifold retaining bolts. Remove the intake manifold from the engine.

To install:

25. Be sure to clean all old gasket material from the sealing surfaces, using the proper tools.

26. Installation is the reverse of the removal procedure.

27. Be sure to use new gaskets.

28. Remove the bolts.

29. Remove the accelerator cable routing clip pin-type retainer and the wiring harness pin-type retainer.

30. Remove the wiring harness bracket retainer, then position the wiring harness aside.

31. Disconnect the RH spark plug wires from the coil. Remove the spark plug wire routing clip pin-type retainer and position the wires aside.

32. Connect the brake booster vacuum hose.

33. Disconnect the EGR vacuum regulator solenoid valve electrical connector and vacuum hose.

34. Disconnect the EGR valve vacuum hose and tube fitting.

35. Detach the accelerator and speed control cables from the bracket, and position the cables aside.

36. Detach the accelerator and speed control cables from the throttle body.

37. Disconnect the MAF sensor wiring pin-type retainer.

38. Disconnect the IAC valve, TP sensor electrical connectors and the TP sensor wiring pin.

39. Remove the air cleaner outlet pipe

40. Torque the intake manifold bolts to 89 inch lbs. (10 Nm).

OIL PAN

REMOVAL & INSTALLATION

2.3L Engine

See Figures 61 through 66.

1. Before servicing the vehicle, refer to the precautions in the beginning of this section.

2. Drain the engine oil.

3. Only the LH side of the engine will be raised. Remove the LH engine support insulator-to-engine bracket nuts.

4. Install the lifting eyes 301-D030 or similar hoist ring.

5. Install the 3–Bar Engine Support Kit 301-F072 and raise the LH side of the engine approximately 25 mm (1 inch).

6. Detach the wiring harness retainers from the engine front cover.

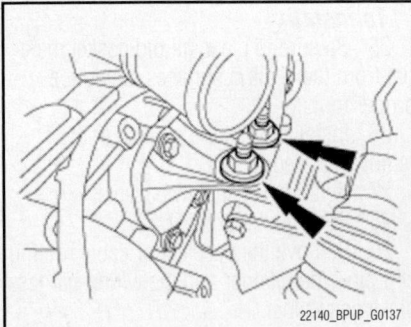

Fig. 61 LH engine support insulator-to-engine bracket nuts—2.3L engine

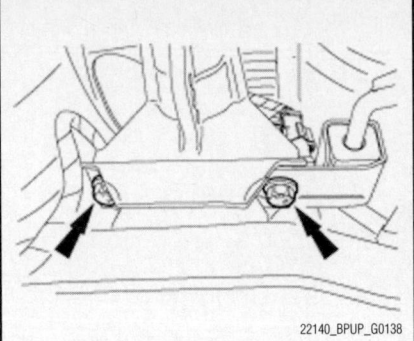

Fig. 62 Transmission bracket-to-transmission bolts—2.3L engine

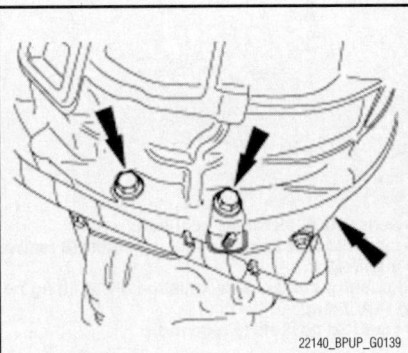

Fig. 63 4 transmission-to-oil pan bolts—2.3L engine

7. Remove the 4 engine front cover-to-oil pan bolts.

8. Remove the transmission bracket-to-transmission bolts.

9. Remove the 4 transmission-to-oil pan bolts.

❋❋ WARNING

To prevent damage to the transmission, do not loosen the transmission-to-engine bolts more than 5 mm (0.19 inch).

10. Loosen the 7 transmission-to-engine bolts 5 mm (0.19 inch).

11. Slide the transmission rearward 5 mm (0.19 inch).

➡ **If necessary, lift the rear of the transmission to remove the oil pan.**

12. Remove the 13 oil pan bolts and oil pan.

❋❋ WARNING

Do not use metal scrapers, wire brushes, power abrasive discs or other abrasive means to clean the sealing surfaces. These tools cause scratches and gouges, which make leak paths. Use a plastic scraping

tool to remove all traces of old sealant. Clean the sealing surfaces with silicone gasket remover and metal surface prep. Observe all warnings and cautions and follow all application directions contained on the packaging of the silicone gasket remover and metal surface prep.

13. Remove the engine oil pan.

To install:

14. Clean and inspect all mating surfaces.

➡ **The oil pan must be installed and the bolts tightened with four minutes of applying the silicone gasket and sealant.**

15. Apply a 2.5 mm bead of silicone gasket and sealant to the oil pan. Install the oil pan. Tighten the oil pan in the sequence shown.

16. Lubricate the O-ring with clean engine oil and install the engine oil level indicator assembly.

17. Install the engine front cover-to-oil pan bolts.

18. Torque to 89 inch lbs. (10 Nm).

19. Tighten the oil pan bolts in the sequence shown.

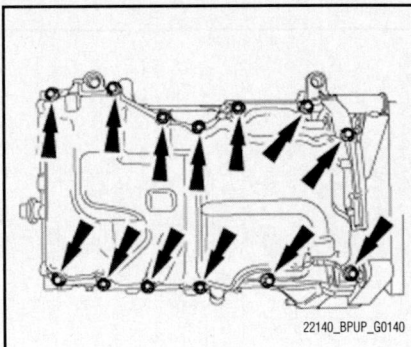

Fig. 64 13 oil pan bolts and oil pan—2.3L engine

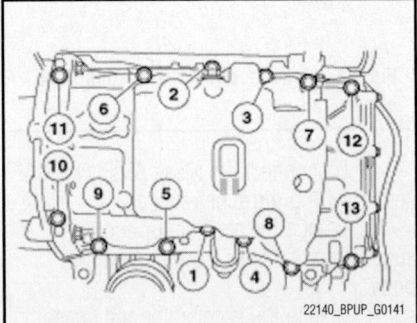

Fig. 65 Oil pan bolt tightening sequence—2.3L engine

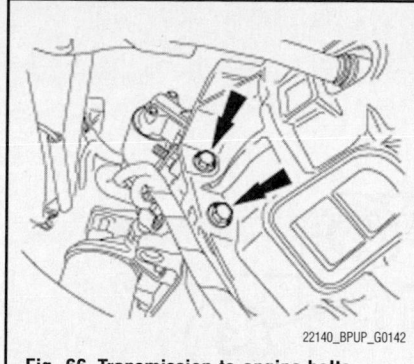

Fig. 66 Transmission-to-engine bolts location—2.3L engine

20. Torque to 18 ft. lbs. (25 Nm).
21. Attach the wiring harness retainers to the engine front cover.
22. Slide the transmission forward and tighten the 7 transmission-to-engine bolts.
23. Torque to 35 ft. lbs. (48 Nm).
24. Install the 4 transmission-to-oil pan bolts. Torque to 35 ft. lbs. (48 Nm).
25. Install the transmission bracket-to-transmission bolts. Torque to 66 ft. lbs. (90 Nm).
26. Lower the engine and remove the 3 bar engine support kit and the lifting eyes.
27. Install the engine support insulator nuts and torque to 75 ft. lbs. (102 Nm).
28. Install the oil level tube assembly.
29. Connect the battery cable.

4.0L Engine

See Figure 67.

1. Before servicing the vehicle, refer to the Precautions Section.
2. Disconnect the negative battery cable.

➡**On some vehicles, when the battery cable is disconnected and reconnected, some abnormal drive symptoms may occur while the vehicle relearns its adaptive strategy. The vehicle may need to be driven to relearn its strategy.**

3. With the vehicle in neutral, position. raise and support the vehicle safely.
4. Drain the engine oil.
5. Remove the oil pan retaining bolts.
6. Remove the oil pan and discard the gasket

To install:

7. Torque the bolts to 19 ft. lbs. (26 Nm).
8. Install the negative battery cable.
9. Fill the engine with clean oil.

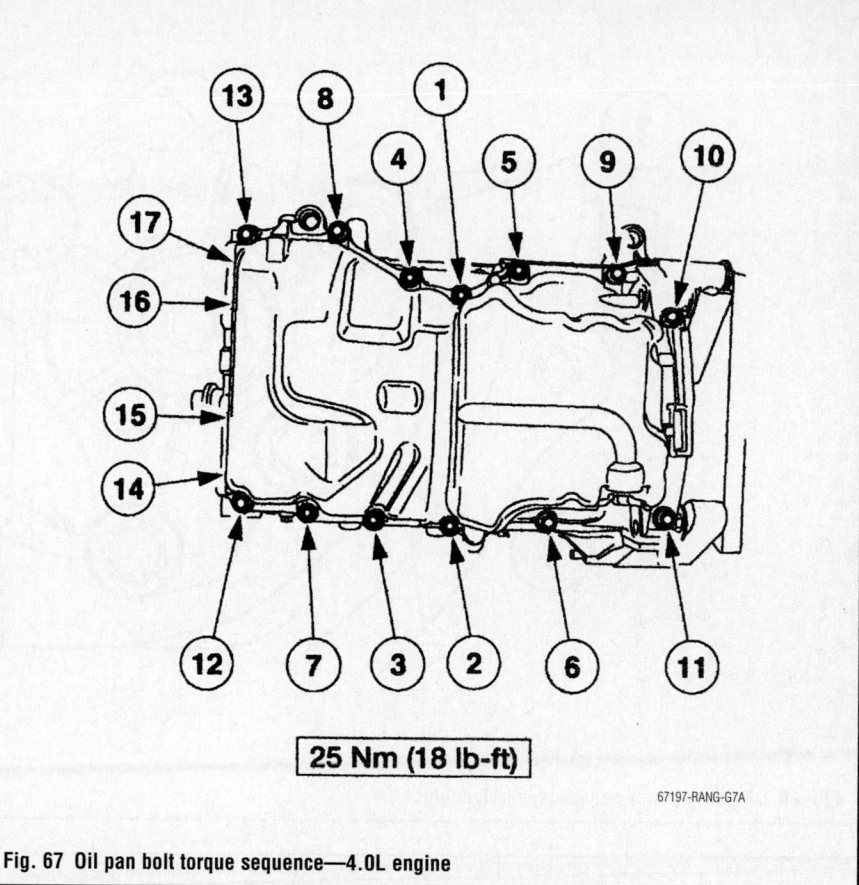

25 Nm (18 lb-ft)

Fig. 67 Oil pan bolt torque sequence—4.0L engine

10. Start the vehicle and check for leaks, repair if necessary.

OIL PUMP

REMOVAL & INSTALLATION

2.3L Engine

See Figures 68 and 69.

➡**The oil pump is located on the front of the engine and is turned by the timing belt.**

1. Before servicing the vehicle, refer to the precautions in the beginning of this section.
2. Remove or disconnect the following:
   • Negative battery cable
   • Timing chain
   • Oil pump chain and sprockets
   • Oil pan
   • Oil pump pickup tube and gasket
   • Oil pump assembly and gasket

To install:

3. Turn the crankshaft clockwise to position the No. 1 piston.
4. Remove the plug bolt.

5. Install the engine timing peg 303-507.

➡**Clean the gasket surface with metal surface cleaner.**

6. Install a new gasket and the oil pump assembly. Tighten the bolts in the sequence shown in two stages.
   • Step 1: Tighten the bolts to 80 inch lbs. (10 Nm)
   • Step 2: Tight the bolts to 17 ft. lbs. (23 Nm)

7. Install a new oil pump pickup tube gasket and the pickup tube. Tighten the bolts in the sequence shown

4.0L Engine

See Figure 70.

➡**The oil pump cannot be removed with the engine in the vehicle.**

1. Before servicing the vehicle, refer to the Precautions Section.
2. Disconnect the negative battery cable.

➡**On some vehicles, when the battery cable is disconnected and reconnected, some abnormal drive symptoms may occur while the vehicle relearns its adaptive strategy. The vehicle may**

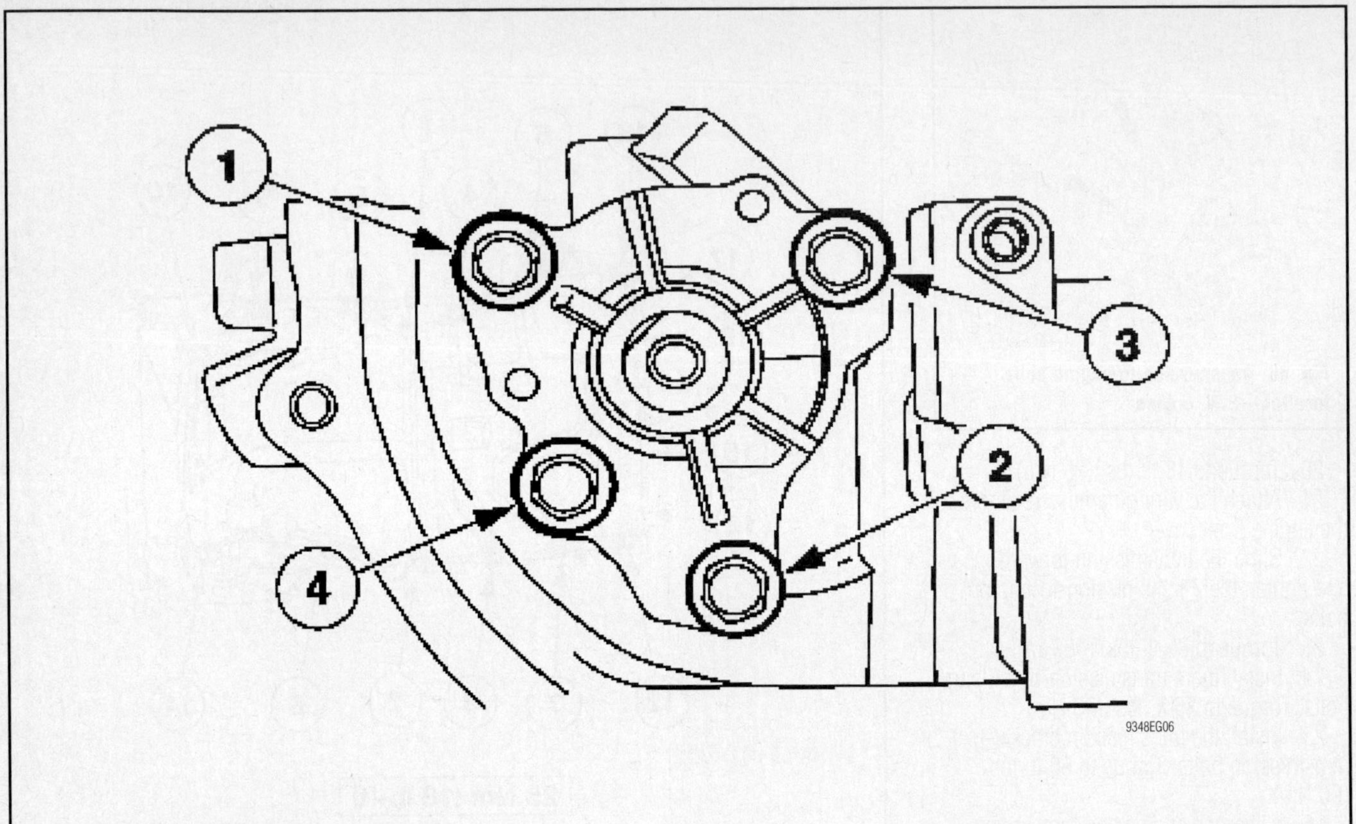

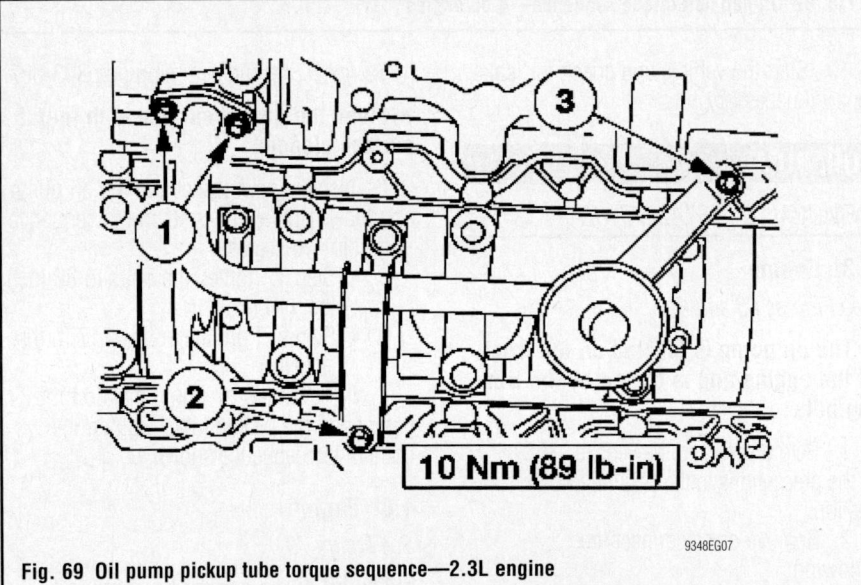

Fig. 68 Oil pump torque sequence—2.3L engine

10 Nm (89 lb-in)

Fig. 69 Oil pump pickup tube torque sequence—2.3L engine

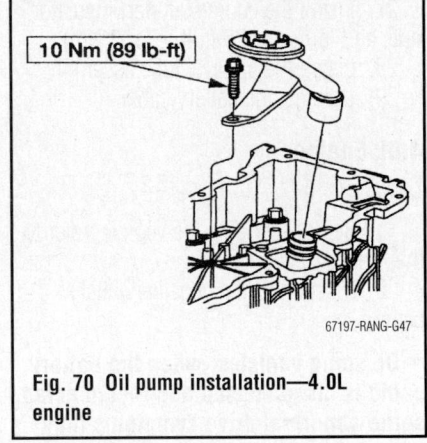

10 Nm (89 lb-ft)

Fig. 70 Oil pump installation—4.0L engine

need to be driven to relearn its strategy.

3. Raise and support the vehicle safely.

4. Drain the engine oil.

5. Remove or disconnect the following:
- Engine from the vehicle
- Oil pan
- Unbolt the oil pick-up tube

- The 8 ladder frame bolts that were under the oil pan
- The 2 rear outer ladder frame bolts
- The 7 left-hand and the 8 right-hand ladder frame bolts
- The ladder frame from the engine
- The 2 oil pump attaching bolts and the pump

To install:

6. Submerge the pump in clean engine oil to prime it.

7. Install or connect the following:
- The ladder frame on the engine
- The 8 right-hand and 7 left-hand ladder frame bolts
- The 2 rear outer and the 8 frame bolts under the pan
- The oil pump. Torque the bolts to 14 ft. lbs. (19 Nm).
- Oil pick-up tube
- Oil pan
- Engine to the vehicle

• Negative battery cable
8. Fill the engine with clean oil.
9. Start the vehicle and check for leaks, repair if necessary.

PISTON AND RING

POSITIONING

See Figures 71 through 73.

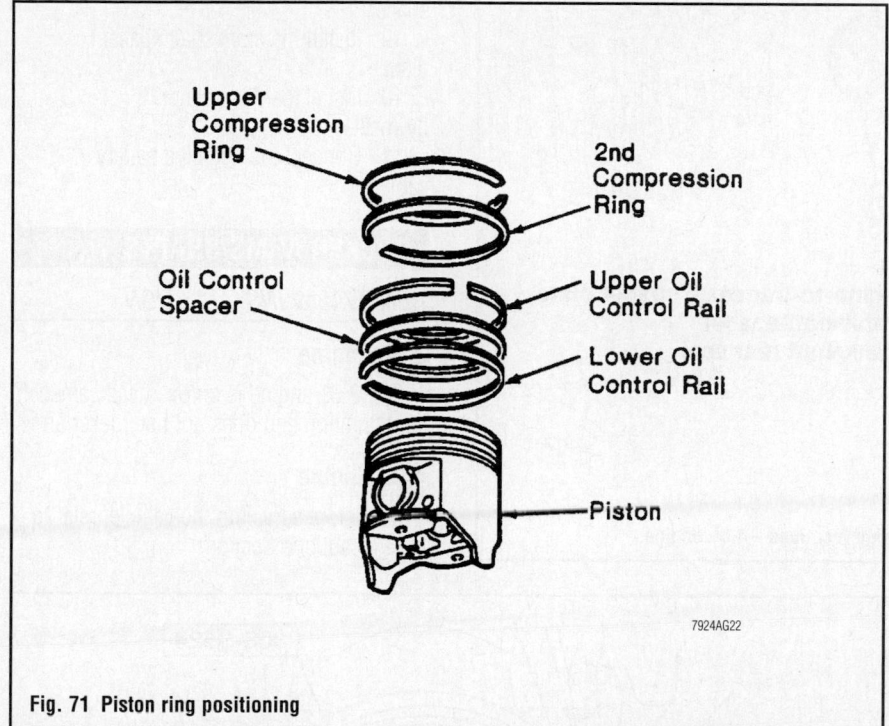

Fig. 71 Piston ring positioning

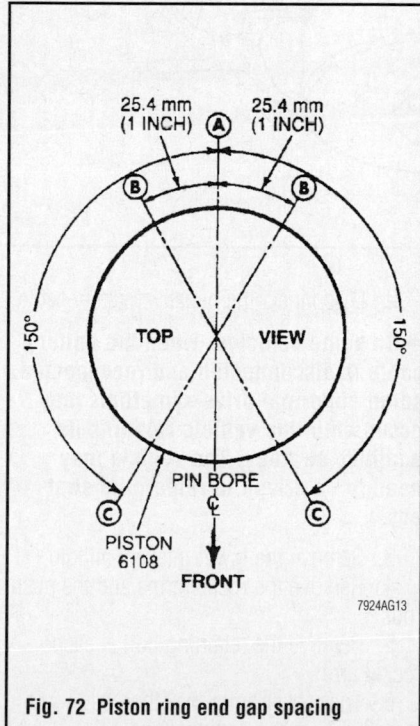

Fig. 72 Piston ring end gap spacing

REAR MAIN SEAL

REMOVAL & INSTALLATION

2.3L Engine

See Figure 74.

1. Before servicing the vehicle, refer to the Precautions Section.

Fig. 73 Piston and connecting rod positioning on 4.0L engine

2. Disconnect the negative battery cable.

➡On some vehicles, when the battery cable is disconnected and

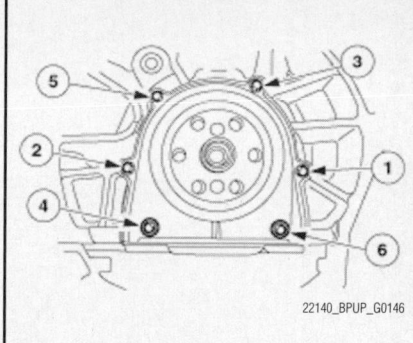

Fig. 74 Rear main seal tightening sequence—2.3L engine

reconnected, some abnormal drive symptoms may occur while the vehicle relearns its adaptive strategy. The vehicle may need to be driven to relearn its strategy.

3. Raise and support the vehicle safely.
4. Remove the flywheel or flexplate.
5. Remove the oil pan.
6. Remove the bolts and the crankshaft rear oil seal.

To install:

7. Install the crankshaft rear main oil seal.
8. Tighten the bolts in the sequence shown to 7 ft. lbs. (10 Nm).
9. Install the oil pan.
10. Install the flywheel or flexplate.
11. Connect the negative battery cable.

4.0L Engine

See Figures 75 through 78.

1. Before servicing the vehicle, refer to the Precautions Section.
2. Remove the flexplate or flywheel.
3. Disconnect the negative battery cable.

✳✳ WARNING

Avoid scratching or damaging the oil crankshaft seal running surface during removal of the crankshaft rear oil seal.

4. Using the special tool 303-409 remove the crankshaft rear oil seal.

To install:

➡**Be sure the crankshaft rear sealing surface is clean and free of any rust or corrosion. To clean the crankshaft rear sealing surface, use extra-fine emery cloth or extra-fine 0000 steel wool with metal surface cleaner.**

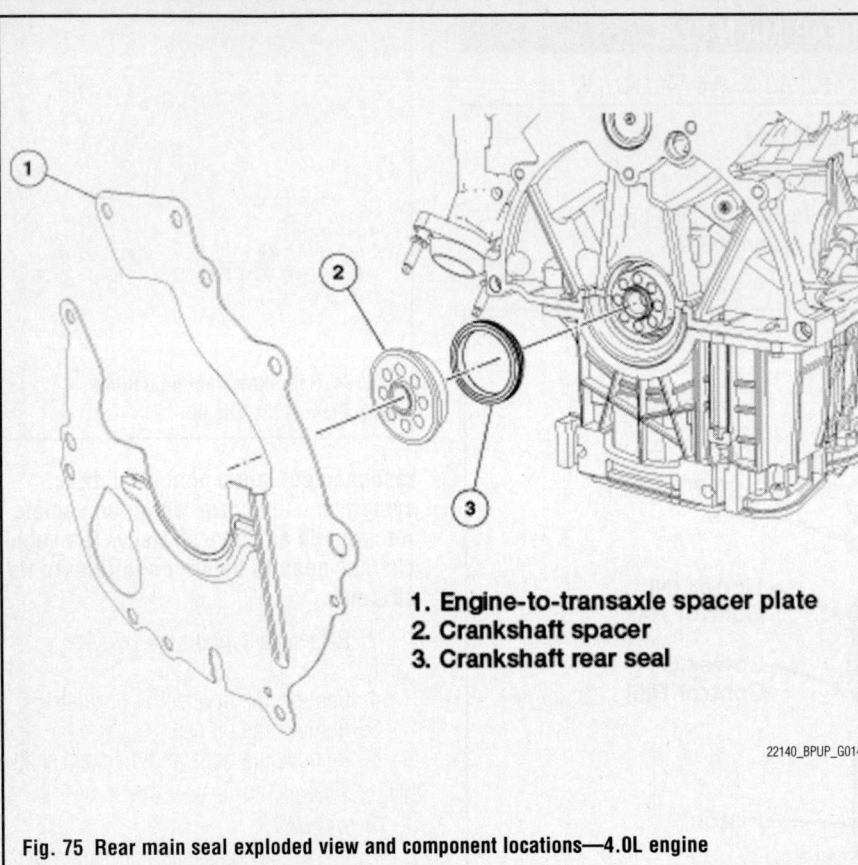

1. Engine-to-transaxle spacer plate
2. Crankshaft spacer
3. Crankshaft rear seal

22140_BPUP_G0147

Fig. 75 Rear main seal exploded view and component locations—4.0L engine

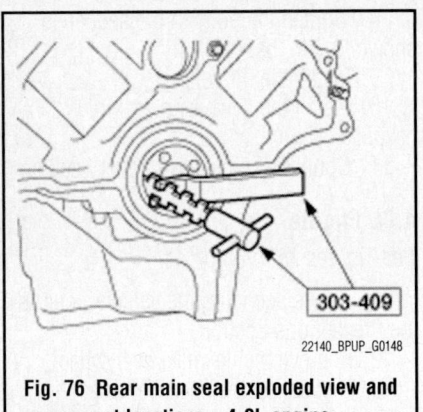

303-409

22140_BPUP_G0148

Fig. 76 Rear main seal exploded view and component locations—4.0L engine

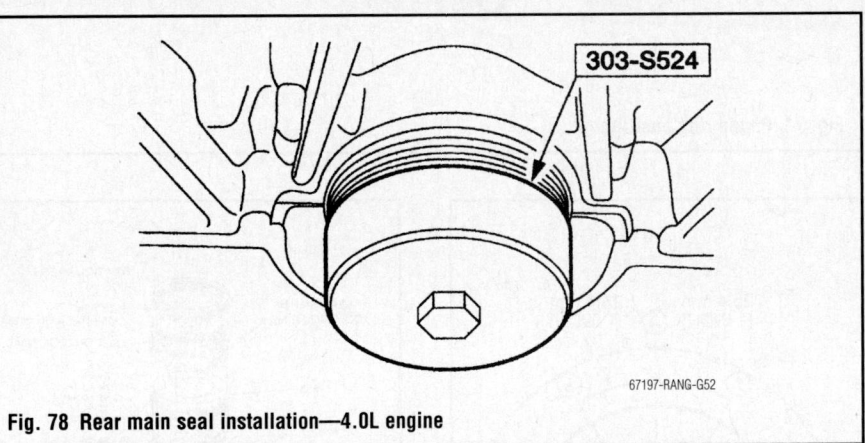

303-S524

67197-RANG-G52

Fig. 78 Rear main seal installation—4.0L engine

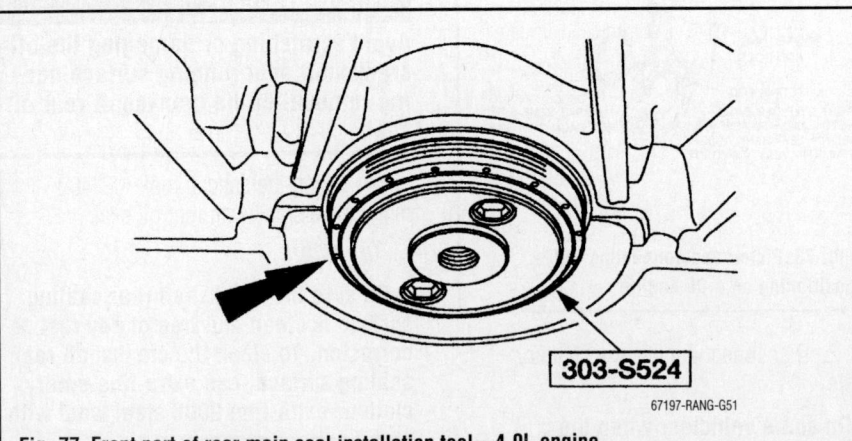

303-S524

67197-RANG-G51

Fig. 77 Front part of rear main seal installation tool—4.0L engine

5. Lubricate the crankshaft rear oil seal with clean engine oil and install on the rear main seal protector, which is part of the crankshaft rear oil seal service set 49 UN-01-1360.

6. Using the special tool 49 UN-01-1360, install the crankshaft rear oil seal.

7. Remove all the special tools.

8. Install the crankshaft spacer.

9. Install the crankshaft spacer plate.

10. Install the flexplate or flywheel.

11. Connect the negative battery cable.

ROCKER ARMS/SHAFTS

REMOVAL & INSTALLATION

2.3L Engine

The 2.3L engine is an overhead camshaft configuration and does not use rocker arms.

4.0L Engine

1. Before servicing the vehicle, refer to the Precautions Section.

2. Disconnect the negative battery cable.

➡**On some vehicles, when the battery cable is disconnected and reconnected, some abnormal drive symptoms may occur while the vehicle relearns its adaptive strategy. The vehicle may need to be driven to relearn its strategy.**

3. Remove the lower intake manifold.

4. Remove the rocker arms and the push rods.

5. Remove the retaining bolt at each rocker arm.

6. The rocker arms may then be removed from the engine. Keep all

rocker arms and pushrods in order so they may be installed in their original locations.

To install:

7. Be sure to clean all old gasket material from the sealing surfaces, using the proper tools. Do not use metal scrappers, wire brushes, power abrasive discs or other abrasive means to clean the sealing areas.

8. Lubricate the rocker arm assemblies with SAE 50W engine oil.

9. Ensure that the fulcrums are properly seated into the cylinder head. Torque the rocker arm fulcrum bolts to 19 ft. lbs. (26 Nm).

TIMING CHAIN COVER AND SEAL

REMOVAL & INSTALLATION

2.3L Engine

See Figures 79 through 81.

➡On some engines the crankshaft, crankshaft sprocket and the pulley are fitted together by friction, with diamond washers between the flange on each part. For that reason, the crankshaft sprocket is also unfastened if you loosen the pulley. Therefore, the engine must be retimed each time the damper is removed. Otherwise severe damage can occur.

1. Before servicing the vehicle, refer to the Precautions Section.

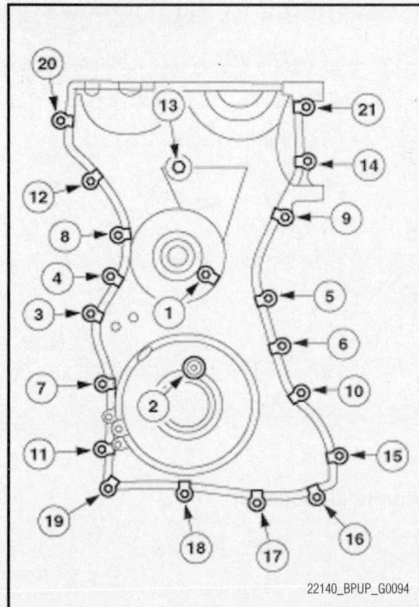

Fig. 79 Front cover tightening sequence— 2.3L engine

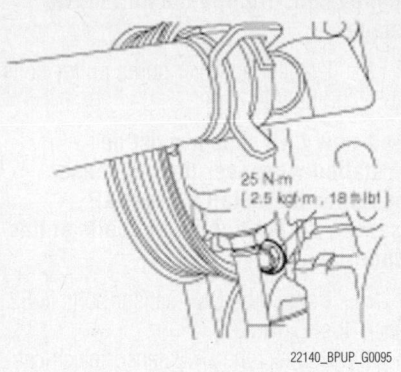

Fig. 80 Power steering pump and install the lower retaining bolt. 2.3L engine

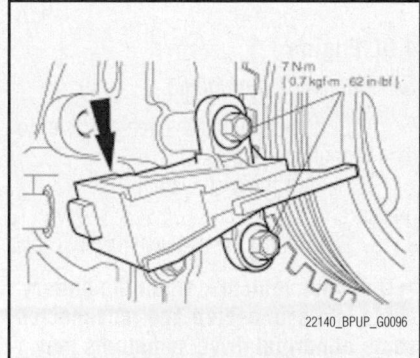

Fig. 81 CKP sensor with the alignment tool—2.3L engine

2. Relieve the fuel system pressure.

3. Disconnect the negative battery cable.

➡On some vehicles, when the battery cable is disconnected and reconnected, some abnormal drive symptoms may occur while the vehicle relearns its adaptive strategy. The vehicle may need to be driven to relearn its strategy.

4. Drain the cooling system.

5. Properly discharge the air conditioning system.

6. Remove or disconnect the following:
   - Drive belt
   - Accessory drive belt tensioner
   - Cooling fan drive pulley
   - Coolant pump pulley
   - Power steering pump
   - Engine oil level indicator assembly
   - Engine oil level indicator
   - Engine oil level indicator tube
   - Water outlet tube
   - Water outlet tube
   - Air conditioning compressor

➡The alternator will be removed with the accessory bracket.

   - Accessory bracket
   - Right motor mount.
   - Coolant hose from the thermostat
   - Coolant hose from the Exhaust Gas Recirculation (EGR) valve
   - Coolant tube assembly
   - Exhaust manifold and gasket
   - Block heater (if so equipped)
   - Water outlet
   - EGR valve
   - Power steering pump and reservoir as an assembly
   - Idle Air Control (IAC) valve
   - Throttle Position (TP) sensor
   - Manifold Absolute Pressure (MAP) sensor
   - Swirl control valve monitor electrical connector
   - Crankshaft Position (CKP) sensor and the wiring harness pin-type retainers
   - Knock Sensor (KS)
   - Electric thermostat
   - Swirl control valve
   - Camshaft Position (CMP) sensor electrical connector and disconnect the Positive Crankcase Ventilation (PCV) hose from the intake manifold
   - Engine wiring harness pin-type retainers from the intake manifold
   - Engine wiring harness connector bracket. Position the engine wiring harness aside
   - Exhaust Gas Recirculation (EGR) tube
   - Fuel supply line clip from the front of the intake manifold. Disconnect the vacuum hose from the intake manifold
   - Intake manifold assembly
   - Fuel injector electrical connectors. Detach the wiring harness pin-type retainers
   - Ignition coil and the Cylinder Head Temperature (CHT) sensor electrical connectors
   - Engine wiring harness anchors from the valve cover studs. Remove the engine wiring harness
   - Ignition coil
   - Bypass hose
   - Thermostat housing
   - Knock sensor and the engine vent cover
   - Left motor mount
   - Fuel injector supply manifold with the injectors and the ground strap

- Water pump pulley
- Water pump
- CMP sensor
- CHT sensor
- Spark plugs
- Valve cover
- CKP sensor
- Crankshaft vibration damper

➥There is one front cover bolt behind the cooling fan drive pulley. To remove this bolt, align one of the cooling fan drive pulley access holes with the bolt head to access the bolt.

7. Remove the front cover.

To install:

8. Install the front cover

9. Tighten the bolts in the sequence shown, to the following specifications:
- 8 mm bolts—7 ft. lbs. (10 Nm).
- 10 mm bolts—18 ft. lbs. (25 Nm).
- 13 mm bolts—35 ft. lbs. (48 Nm).

❄ WARNING

There is one bolt behind the cooling fan drive pulley. This bolt can be accessed by lining up one of the holes in the pulley with the bolt.

10. Position the power steering pump and install the lower retaining bolt.

11. Tighten the power steering pump bolts to 18 ft. lbs. (25 Nm).

12. Connect the power steering pump switch electrical connector.

13. Install the water pump pulley.

14. Tighten the water pump pulley bolts to 15 ft. lbs. (20 Nm).

15. Install the belt tensioner.

16. Tighten the belt tensioner bolts to 36 ft. lbs. (50 Nm).

❄ WARNING

The crankshaft, the crankshaft sprocket and the pulley are fitted together by friction, with diamond washers between the flange faces on each part. For that reason, the crankshaft sprocket is also unfastened if you loosen the pulley. Therefore, the engine must be retimed each time the damper is removed. Otherwise severe engine damage can occur.

17. Install the crankshaft vibration damper:

➥Do not reuse the crankshaft

pulley bolt. Tighten the bolt in two stages.

18. Tighten the water pump pulley bolts to 15 ft. lbs. (20 Nm).

➥A new CKP sensor must be installed whenever the old sensor is removed. Install a new CKP sensor, do not tighten the bolts at this time.

19. Tighten the CKP sensor bolts to 82 inch. lbs. (7 Nm).

20. Connect the CKP sensor electrical connector and the wiring harness pin-type retainers.

21. Connect the battery ground cable.

4.0L Engine

See Figures 82 through 85.

1. Before servicing the vehicle, refer to the Precautions Section.

2. Relieve the fuel system pressure.

3. Disconnect the negative battery cable.

➥On some vehicles, when the battery cable is disconnected and reconnected, some abnormal drive symptoms may occur while the vehicle relearns its adaptive strategy. The vehicle may need to be driven to relearn its strategy.

4. With the vehicle in **NEUTRAL**, position it on a hoist.

5. Drain the cooling system.

6. Remove the nut and detach the A/C hose bracket from the front cover.

7. Remove the block cradle to front cover bolts.

8. Remove or disconnect the following:
- A/C compressor
- Power steering bracket bolts
- Alternator wiring
- Accessory drive belt tensioner
- Generator mounting bracket
- Crankshaft Position (CKP) sensor electrical connector
- CKP wiring harness retainer.
- Wiring harness anchors
- Bypass hose clamp
- Heater hose from the coolant pump
- Lower radiator hose from the coolant pump

➥This bolt is a torque-to-yield design and cannot be reused.

9. Using the strap wrench 303—D055, remove the crankshaft pulley bolt.

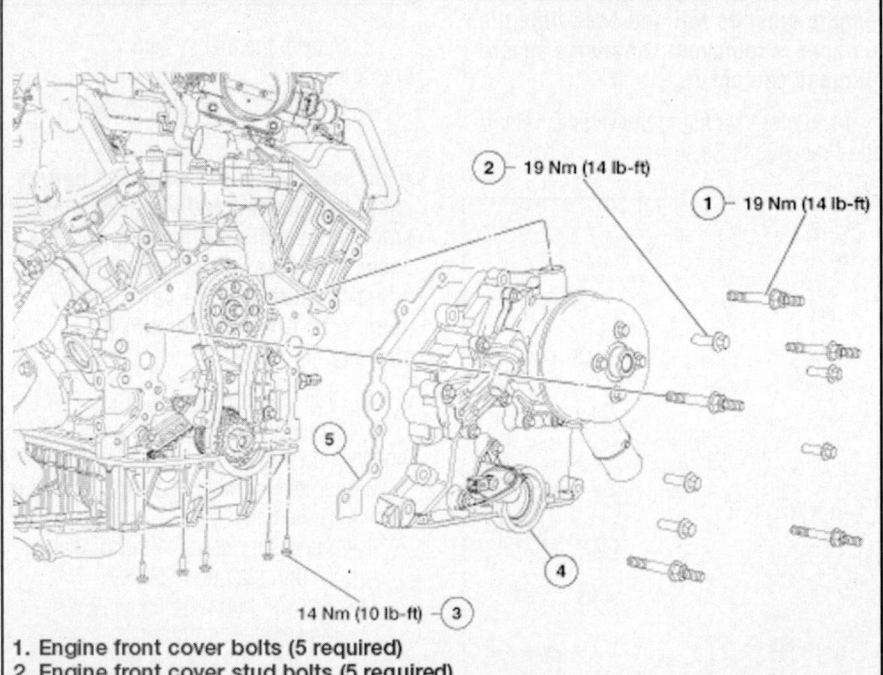

1. Engine front cover bolts (5 required)
2. Engine front cover stud bolts (5 required)
3. Cylinder block cradle-to-engine front cover bolts (5 required)
4. Engine front cover
5. Engine front cover gasket

22140_BPUP_G0157

Fig. 82 Engine front cover removal and installation exploded view of accessories—4.0L engine

➡️**If the crankshaft pulley bolt is not installed, the crankshaft vibration damper remover 303-773 will bottom out before the pulley is off. Install the crankshaft pulley bolt 2-3 turns. Using the crankshaft vibration damper remover 303-773 and the 8 mm x 1.25 x 100 mm bolts, remove the pulley.**

10. Remove the crankshaft front seal.

➡️**Note the positions of the stud bolts for installation reference.**

11. Remove the 5 bolts, the 5 stud bolts, the engine front cover and the gasket.

To install:

➡️**If not secured within four minutes, the sealant must be removed and the sealing area cleaned. To clean the sealing area, use silicone gasket remover and metal surface prep. Follow the directions on the packaging. Failure to follow this procedure can cause future oil leakage. Apply silicone gasket and sealant to the Woodruff key slot on the crankshaft pulley.**

12. Make sure the stud bolts are installed in their original positions.

13. Position the engine front

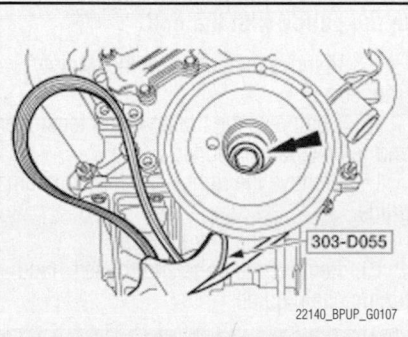

Fig. 83 Strap wrench 303—D055, and crankshaft pulley bolt to be removed and installed

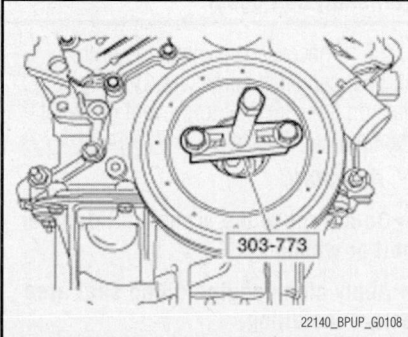

Fig. 84 Crankshaft vibration damper remover 303-773 tool

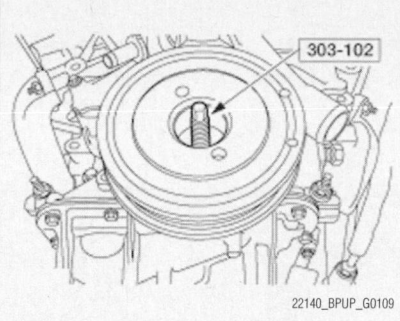

Fig. 85 Crankshaft vibration damper installer 303—102 tool

cover and install the 5 bolts and 5 stud bolts.

14. Tighten to 14 ft lbs. (19 Nm).

➡️**The 5 bolts may or may not have washers. All the bolts should be of the same type. Install the cylinder block cradle-to-front cover bolts.**

15. Tighten to 89 inch lbs. (10 Nm).With washers.

16. Tighten to 10 ft lbs. (10 Nm).Without washers.

17. Using the crankshaft vibration damper installer 303-102, install the crankshaft pulley.

18. Using the strap wrench 303-D055, install a new crankshaft pulley bolt.

19. Tighten the bolt in two stages.

20. Torque Stage 1: 37 ft. lbs. (50 Nm).

21. Torque Stage 2: Tighten the bolt an additional 90 degrees

22. Position the bypass hose clamp.

23. Connect the lower radiator hose to the coolant pump.

24. Connect the heater hose to the coolant pump.

25. Position the wiring harness and install the bolt.

26. Tighten the wiring harness to 89 inch lbs. (10 Nm).

27. Install new wiring harness anchors and position the CKP sensor wiring.

28. Attach the wiring harness retainer.

29. Connect the CKP sensor electrical connector.

30. Position the A/C compressor and power steering bracket and install the 4 bolts.

31. Tighten to 31 ft. lbs. (42 Nm).

32. Position the generator bracket and install the 3 bolts.

33. Tighten to 35 ft. lbs. (47 Nm).

34. Connect the generator electrical connectors and the wiring harness retainer.

35. Attach the generator terminal and install the nut.

36. Tighten to 71 inch lbs. (8 Nm).

37. Position the A/C tube bracket and install the nut.

38. Tighten to 15 ft. lbs. (20 Nm).

39. Drain the engine oil. Install the drain plug when finished.

40. Tighten to 19 ft. lbs. (26 Nm).

41. Fill the engine with clean engine oil.

42. Connect the battery.

43. Fill the engine cooling system.

TIMING CHAIN AND SPROCKETS

REMOVAL & INSTALLATION

2.3L Engine

See Figures 86 through 97.

➡️**On some engines the crankshaft, crankshaft sprocket and the pulley are fitted together by friction, with diamond washers between the flange on each part. For that reason, the crankshaft sprocket is also unfastened if you loosen the pulley. Therefore, the engine must be retimed each time the damper is removed. Otherwise severe damage can occur.**

1. Before servicing the vehicle, refer to the Precautions Section.

2. Disconnect the negative battery cable.

➡️**On some vehicles, when the battery cable is disconnected and reconnected, some abnormal drive symptoms may occur while the vehicle relearns its adaptive strategy. The vehicle may need to be driven to relearn its strategy.**

3. Remove or disconnect the following:
   - Fan and shroud
   - Drive belt
   - Valve cover

4. Install the camshaft alignment plate 303-465 in the slots on the rear of both camshafts.

5. Remove the plug bolt.

➡️ **Only turn the engine in the normal direction of rotation. Installing the special tool in this step will prevent the engine from being rotated in the clockwise direction. However, the engine can still be rotated in the counterclockwise direction.**

6. Install the crankshaft TDC timing peg (2.3L engine) 303-507.

7. Install a M6 X 1.0 X 2.25 bolt (Mazda part No. 1F20-12-428) to verify ignition timing.

8. Install the holding fixture 205-072-02 and 205-126.

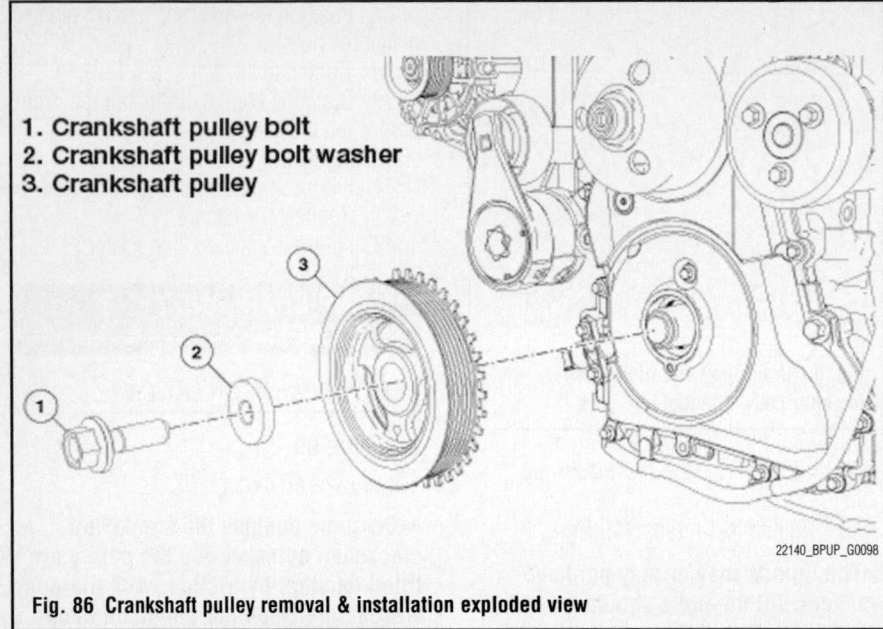

1. Crankshaft pulley bolt
2. Crankshaft pulley bolt washer
3. Crankshaft pulley

22140_BPUP_G0098

Fig. 86 Crankshaft pulley removal & installation exploded view

✳✳ WARNING

Failure to hold the crankshaft pulley in place during bolt loosening can cause damage to the engine. Remove the bolt and the crankshaft pulley.

9. Remove the crankshaft bolt.

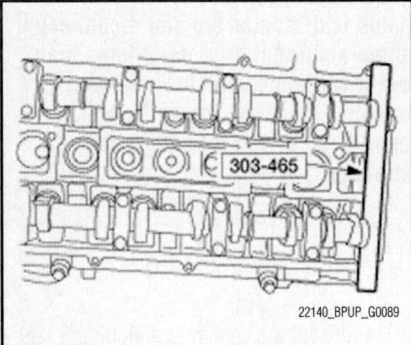

22140_BPUP_G0089

Fig. 87 Camshaft alignment plate 303-465

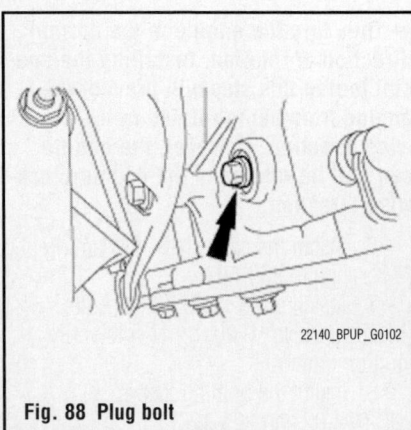

22140_BPUP_G0102

Fig. 88 Plug bolt

10. Remove the holding fixture 205-072-02 and 205-126.
11. Remove the M6 bolt.
12. Remove the crankshaft pulley.
13. Remove or disconnect the following:
   • Camshaft pulley
   • Crankshaft position sensor

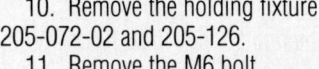

303-507

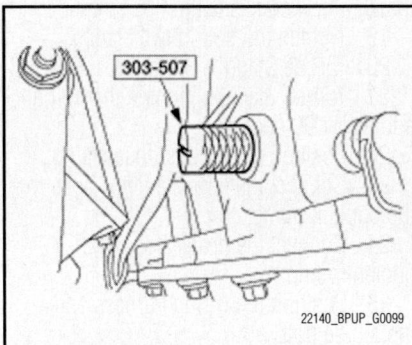

22140_BPUP_G0099

Fig. 89 Crankshaft TDC timing peg303—507

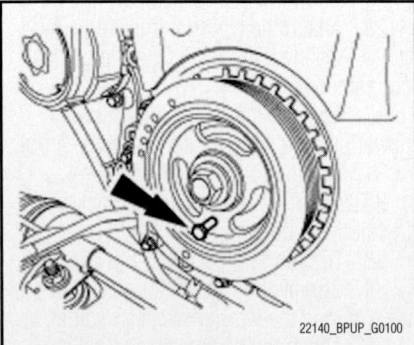

22140_BPUP_G0100

Fig. 90 M6 X 1.0 X 2.25 bolt (Mazda part No. 1F20-12-428)

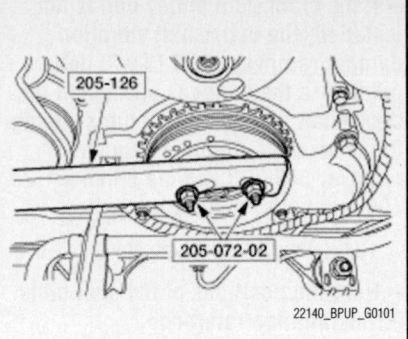

205-126

205-072-02

22140_BPUP_G0101

Fig. 91 Crank pulley holding fixture 205—072—02—and 205—126

   • Belt tensioner
   • Water pump pulley
   • Power steering high pressure hose. Remove the nylon O-ring.
   • Power steering return hose
   • Power steering pump

➡This step is needed only if a new front cover is being installed.

14. Using a three-jaw puller, remove the fan drive pulley.

➡There is one bolt behind the cooling fan drive pulley. This bolt can be accessed by lining up one of the holes in the pulley with the bolt.

15. Remove the bolts and the engine front cover.
16. Compress the timing chain tensioner and remove the tensioner.
17. Remove the right-hand timing chain guide.
18. Remove the timing chain.
19. Remove the bolts and the left-hand timing chain guide.

✳✳ WARNING

Do not rely on the Camshaft Alignment Plate to prevent camshaft rotation. Damage to the tool or the camshaft can occur.

20. If necessary, remove the bolts and the camshaft sprockets. Use the flats on the camshaft to prevent camshaft rotation.

To install:

➡Do not reuse the crankshaft damper bolt or washer.

➡Apply clean engine oil on seal area before installing.

21. Install the new crankshaft pulley and hand-tighten the bolt.

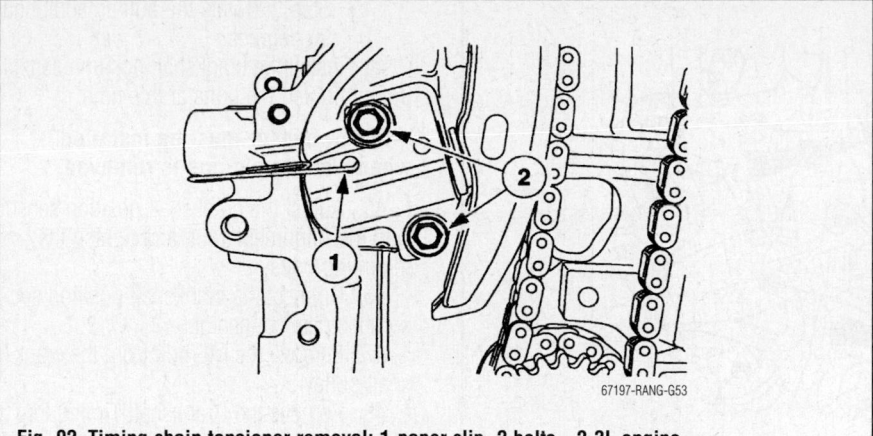

Fig. 92 Timing chain tensioner removal; 1-paper clip, 2 bolts—2.3L engine

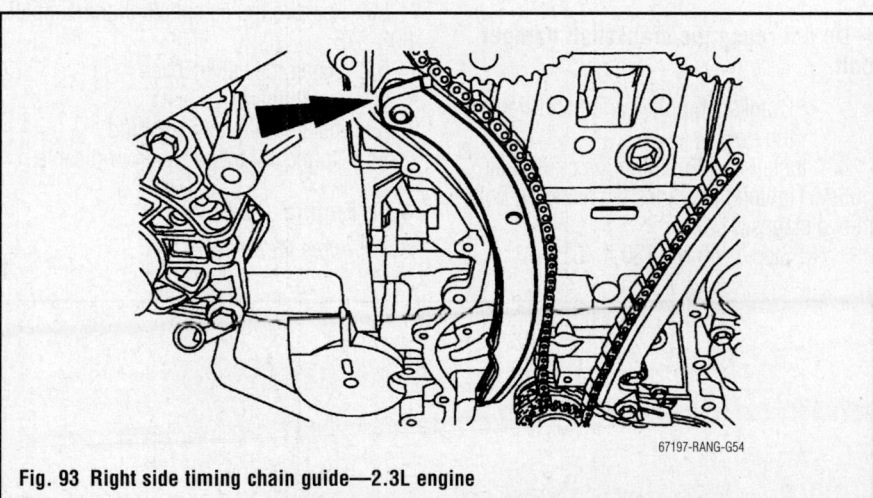

Fig. 93 Right side timing chain guide—2.3L engine

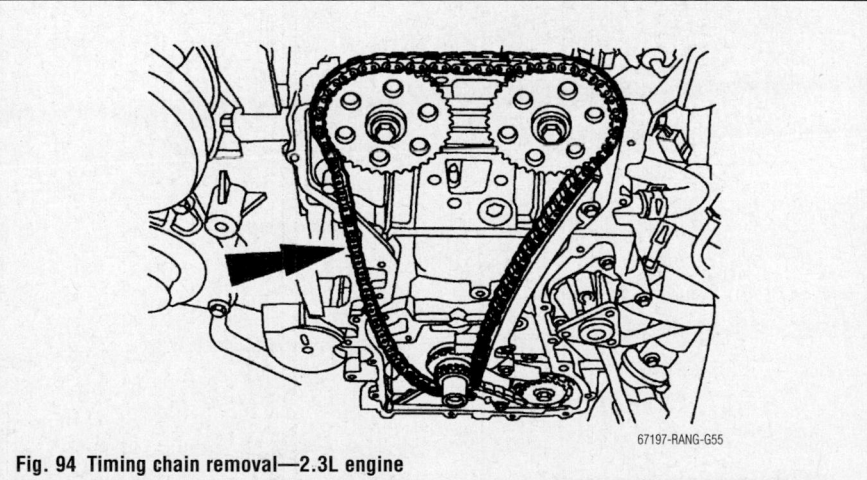

Fig. 94 Timing chain removal—2.3L engine

✴✴ WARNING

Only hand-tighten the bolt or damage to the front cover can occur.

➡**This step will correctly align the crankshaft pulley to the crankshaft.**

22. Install a standard 0.23 inch (6mm) x 0.7 in (18mm) bolt through the crankshaft pulley and thread it into the front cover.

23. Rotate the crankshaft pulley as necessary to align the bolt holes.

✴✴ WARNING

Failure to hold the crankshaft pulley in place during bolt tightening can cause damage to the engine front cover. Using the holding fixture 205—072—02 and 205—126 hold the crankshaft pulley in place and tighten the crankshaft pulley bolt in two stages.

24. Step 1: Tighten the crankshaft pulley bolt to 74 ft. lbs. (100 Nm).

25. Step 2: Tighten the crankshaft pulley bolt an additional 90°.

26. Remove the holding fixture 205-072-02 and 205-126

27. Remove the 0.23 inch (6mm) x 0.7 in (18mm) bolt.

28. Remove the crankshaft TDC timing peg

➡**Only hand-tighten the bolt or damage to the front cover can occur. Using 0.23 inch (6mm) x 0.7 in (18mm) bolt, check the position of the crankshaft pulley.**

29. If it is not possible to install the 0.23 inch (6mm) x 0.7 in (18mm) bolt, correct the engine timing.

30. Using the camshaft alignment plate 303-50/, check the position of the camshafts.

31. If it is not possible to install the camshaft alignment plate 303-507, correct the engine timing.

32. Remove the camshaft alignment plate.

33. Remove the 6mm 0.23 inch (6mm) x 0.7 in (18mm) bolt.

34. Remove the crankshaft TDC timing peg.

35. Install the plug bolt.

36. Tighten the Plug Bolt to 15 ft. lbs. (20 Nm).

37. Install the fan drive pulley using a nut and bolt with flat washers.

38. Clean and inspect the mounting surfaces of the engine and the front cover.

➡**The engine front cover must be installed and the bolts tightened within four minutes of applying the silicone gasket and sealant.**

39. Apply a 2.5 mm bead of silicone gasket and sealant to the cylinder head and oil pan joint areas.

40. Apply a 2.5 mm bead of silicone gasket and sealant to the front cover.

41. Install the front cover. Tighten the bolts in the sequence shown, to the following specifications:
- Step 1: 8 mm bolts to 10 Nm (89 inch lbs.)

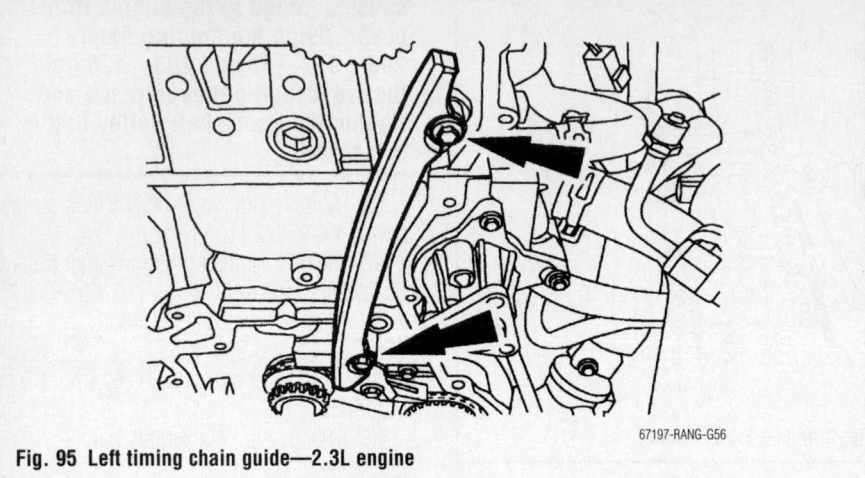

Fig. 95 Left timing chain guide—2.3L engine

67197-RANG-G56

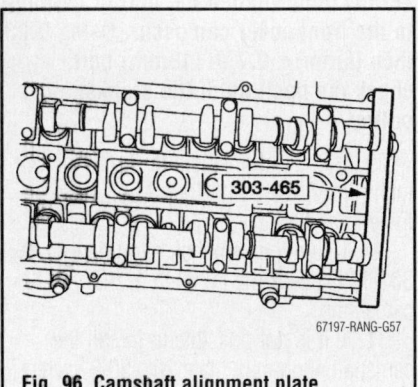

303-465

67197-RANG-G57

Fig. 96 Camshaft alignment plate installed—2.3L engine

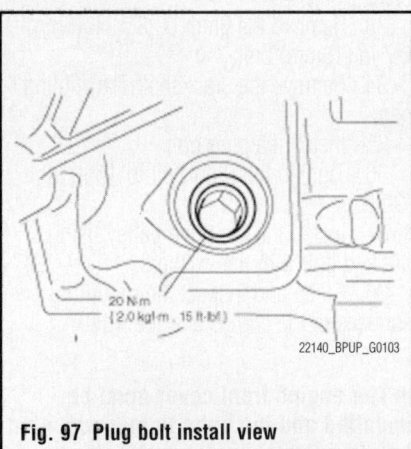

20 Nm
(2.0 kgf-m, 15 ft-lbf)

22140_BPUP_G0103

Fig. 97 Plug bolt install view

- Step 2: 10 mm bolts to 25 Nm (18 ft. lbs.)
- Step 3: 13 mm bolts to 48 Nm (35 ft. lbs.)

42. Install or connect the following:
- Power steering pump and lower retaining bolt
- Power steering return hose
- New nylon O-ring and install the high pressure line.
- Water pump pulley
- Belt tensioner

➡**Do not reuse the crankshaft damper bolt.**

- Crankshaft pulley and hand-tighten the bolt

43. Install an M6 bolt in the crankshaft pulley. Tighten the crankshaft retaining bolt in two stages.
- Step 1: 40 Nm (30 ft. lbs.)
- Step 2: Rotate the bolt an additional 90 degrees.

44. Install the crankshaft position sensor, do not tighten the bolts at this time.

➡**A new sensor must be installed whenever the old one is removed.**

45. Adjust the crankshaft position sensor with the Alignment Tool, and tighten the mounting bolts.

46. Connect the crankshaft position sensor electrical connector.

47. Remove the M6 bolt from the crankshaft pulley.

48. Remove the Crankshaft Timing Peg 303-507.

49. Install the plug.

50. Remove the camshaft alignment plate 303-376.

51. Install the valve cover.

52. Install the drive belt.

53. Install the fan and shroud.

54. Connect the battery ground cable.

4.0L Engine

See Figures 98 through 110.

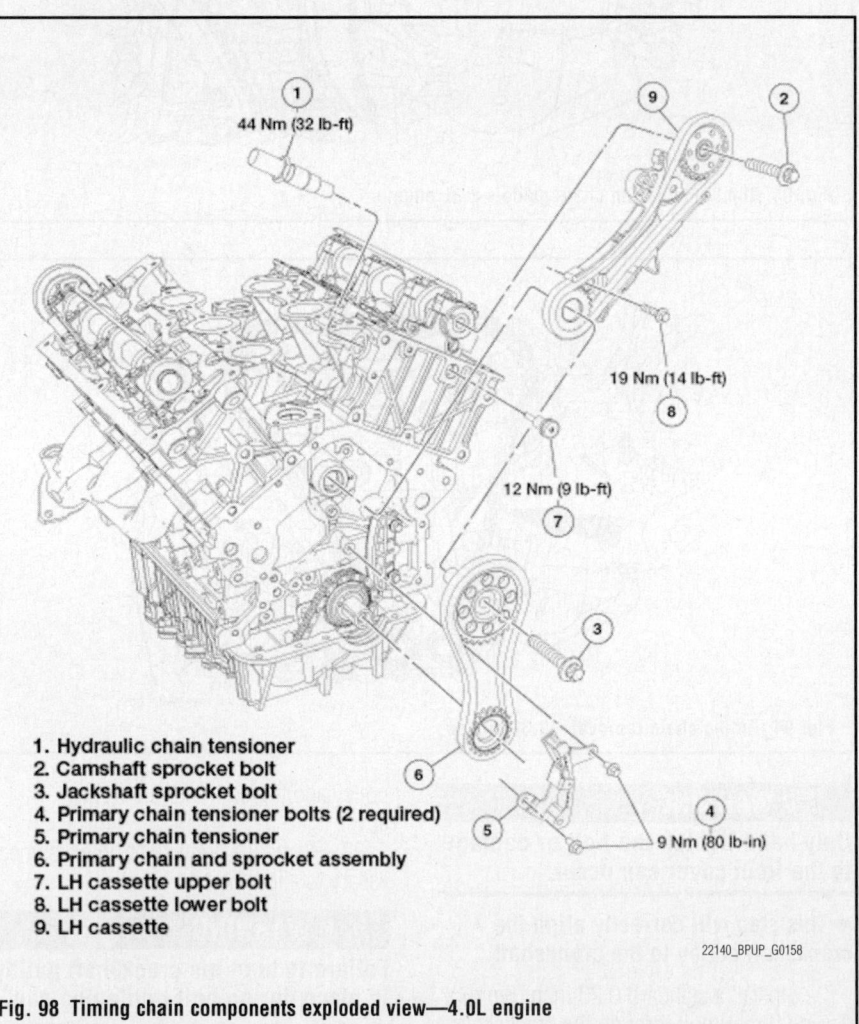

44 Nm (32 lb-ft)

19 Nm (14 lb-ft)

12 Nm (9 lb-ft)

9 Nm (80 lb-in)

1. Hydraulic chain tensioner
2. Camshaft sprocket bolt
3. Jackshaft sprocket bolt
4. Primary chain tensioner bolts (2 required)
5. Primary chain tensioner
6. Primary chain and sprocket assembly
7. LH cassette upper bolt
8. LH cassette lower bolt
9. LH cassette

22140_BPUP_G0158

Fig. 98 Timing chain components exploded view—4.0L engine

1. Before servicing the vehicle, refer to the Precautions Section.

2. Disconnect the negative battery cable.

➡**On some vehicles, when the battery cable is disconnected and reconnected, some abnormal drive symptoms may occur while the vehicle relearns its adaptive strategy. The vehicle may need to be driven to relearn its strategy.**

3. With the vehicle in neutral, position it on a hoist.

4. Remove the intake manifold.

5. Remove the fuel supply manifold.

6. Remove the accessory drive belt.

7. Remove the thermostat housing.

8. Remove the roller followers.

9. The crankshaft timing tool 303—573 must be installed on the damper and should contact the engine block, this positions the engine at TDC.

❋❋ WARNING

Do not rotate the engine counterclockwise. Rotating the engine counterclockwise will result in incorrect timing of the engine.

➡**You must retime the LH and RH camshafts when either camshaft is disturbed. Turn the crankshaft clockwise to position the number 1 cylinder at top dead center (TDC).**

10. Remove the LH hydraulic chain tensioner.

11. Remove the LH camshaft sprocket bolt.

12. Install the camshaft sprocket holding tool 303—564 and the adapter

13. Remove the LH camshaft sprocket bolt.

14. Using the crankshaft holding tool 303-674 to prevent the crankshaft from turning, remove the jackshaft sprocket bolt.

15. Remove the bolts and the primary chain tensioner.

16. Remove the primary chain and sprockets as an assembly.

17. Remove the LH cassette upper bolt

18. Remove the LH cassette lower bolt and the LH cassette.

To install:

19. The camshaft chain sprockets must be oriented correctly. Position the LH cassette.

20. Install the LH cassette lower bolt.

21. Torque the LH cassette lower bolt to 14 ft. lbs. (19 Nm).

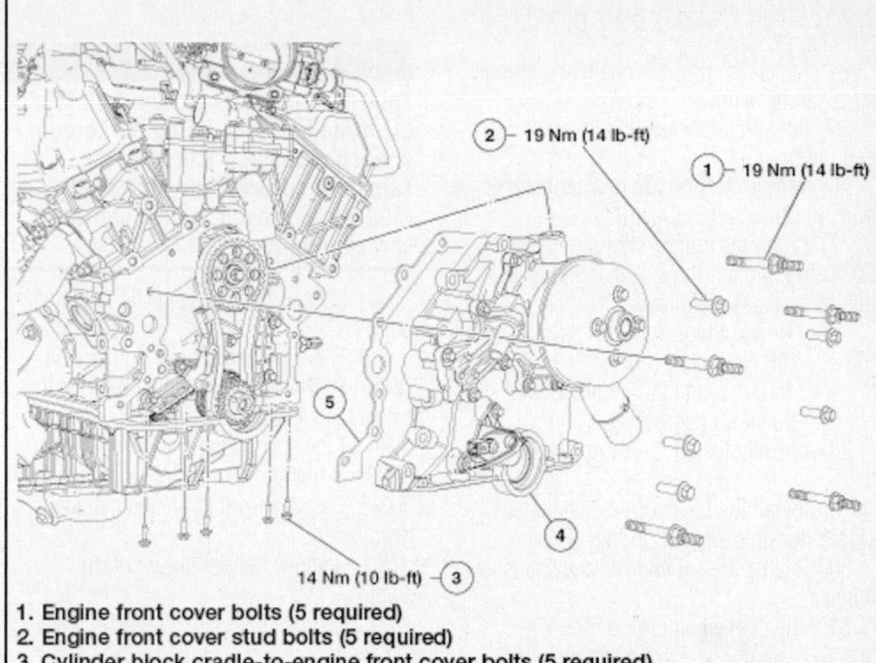

1. Engine front cover bolts (5 required)
2. Engine front cover stud bolts (5 required)
3. Cylinder block cradle-to-engine front cover bolts (5 required)
4. Engine front cover
5. Engine front cover gasket

22140_BPUP_G0157

Fig. 99 Engine front cover—4.0L engine

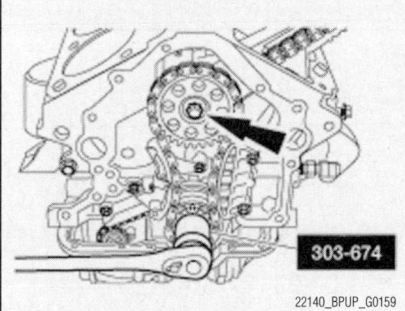

22140_BPUP_G0159

Fig. 100 Crankshaft holding tool 303—674 to prevent the crankshaft from turning—4.0L engine

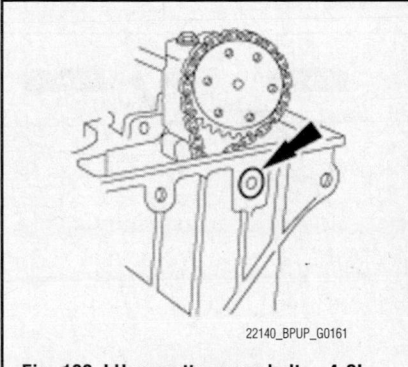

22140_BPUP_G0161

Fig. 102 LH cassette upper bolt— 4.0L engine

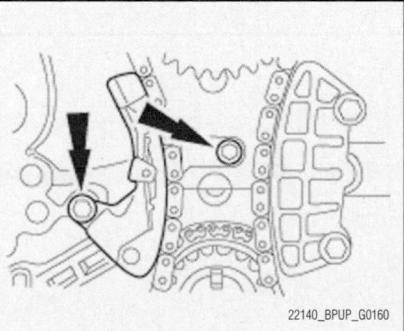

22140_BPUP_G0160

Fig. 101 Primary chain tensioner location and bolt locations— 4.0L engine

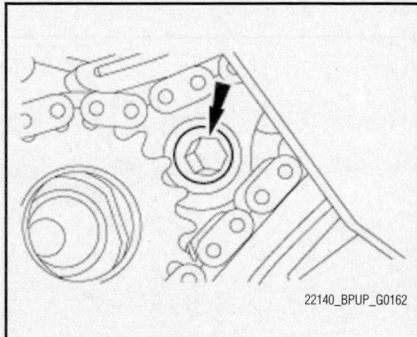

22140_BPUP_G0162

Fig. 103 LH cassette lower bolt and the LH cassette.— 4.0L engine

22. Install the LH cassette upper bolt.

23. Torque the LH cassette upper bolt to 8 ft. lbs. (12 Nm).

24. Install the primary chain and sprockets as an assembly.

25. Install the primary chain tensioner and the bolts.

26. Torque the primary chain tensioner bolts to 80 inch lbs. (9 Nm).

27. Using the crankshaft holding tool 303–ning, tighten the jackshaft sprocket bolt in 2 stages as follows:
- Torque Stage 1: 33 ft. lbs. (45 Nm)
- Torque Stage 2: Tighten an additional 90 degrees

28. Loosely install the camshaft sprocket bolt.

29. Install the LH hydraulic chain tensioner during camshaft timing.

30. Retime the LH and RH camshafts as follows.

31. The crankshaft timing tool 303-578 and 303-564 must be installed on the damper and should contact the engine block, this positions the engine at TDC.

32. Install the crankshaft timing tool.

33. Install the camshaft sprocket holding tool 303-75 and the adaptor to the RH cylinder head and tighten the 2 top clamp bolts.

34. Loosen the top 2 special tool clamp bolts.

35. The camshaft timing slots are off-center. Position the camshaft timing slots below the centerline of the camshaft to correctly fit the special tools and install the camshaft holding tool 303-576 and adaptor 303-577 on the front of the RH cylinder head.

36. Remove the RH lower splash shield.

37. Remove the RH hydraulic camshaft tensioner.

38. Install the timing chain tensioner 303-571.

39. Tighten the special tool top 2 clamp bolts to 89 inch lbs. (10 Nm).

40. Using the torque wrench extension service tool with the camshaft sprocket nut socket 303-565, tighten the camshaft bolt.

41. Tighten camshaft bolts to 45 ft. lbs. (61 Nm).

42. Remove the timing chain tensioner 303-571

43. Install a new O-ring seal on the tensioner and lubricate with clean engine oil. Install the RH hydraulic camshaft tensioner.

44. Tighten camshaft bolts to 32 ft. lbs. (44 Nm).

45. Install the RH lower splash shield.

46. Remove the special tools from the RH cylinder head.

47. Install the camshaft sprocket holding tool 303-564 and-578 the adaptor SST on the front of the LH cylinder head and tighten the top 2 clamp bolts.

48. Tighten camshaft sprocket bolts to 89 inch lbs. (10 Nm).

49. Loosen the LH camshaft sprocket bolt.

50. Loosen the top 2 clamp bolts on the camshaft sprocket holding tool to allow the camshaft sprocket to rotate freely.

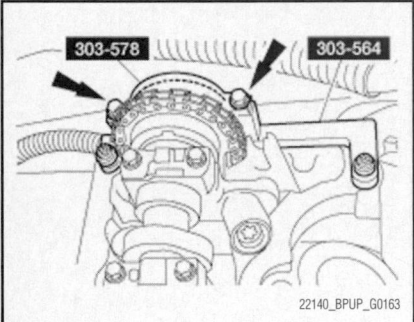

Fig. 104 Crankshaft holding tool 303-578 and 303-564 to prevent the crankshaft from turning—4.0L engine

22140_BPUP_G0163

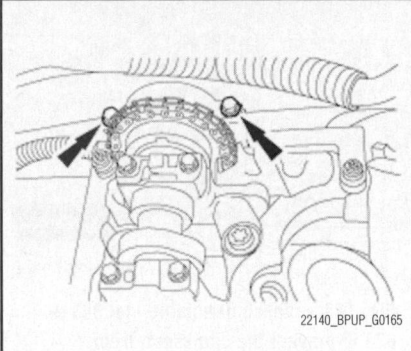

Fig. 106 RH camshaft sprocket bolt.—4.0L engine

22140_BPUP_G0165

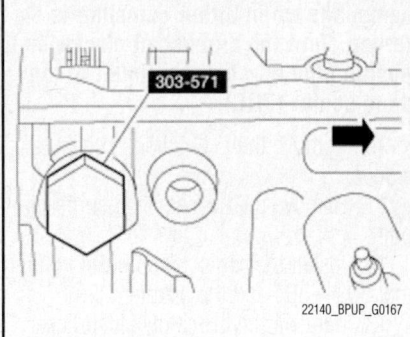

Fig. 108 Timing chain tensioner service tool303—571—4.0L engine

22140_BPUP_G0167

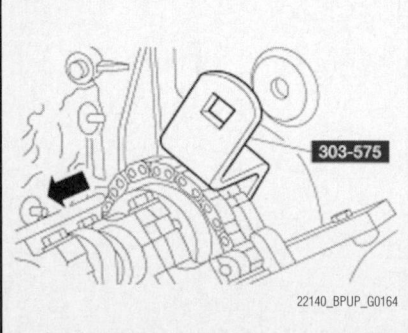

Fig. 105 Camshaft holding tool 303—575 and adaptor—4.0L engine

22140_BPUP_G0164

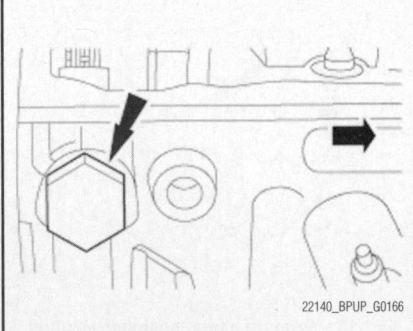

Fig. 107 RH hydraulic camshaft tensioner—4.0L engine

22140_BPUP_G0166

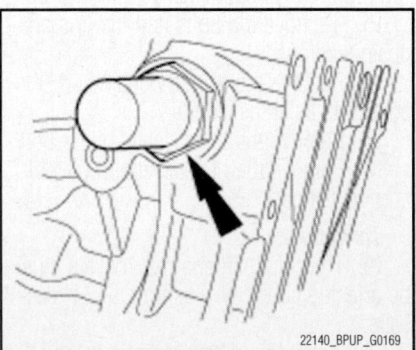

Fig. 109 LH hydraulic camshaft tensioner—4.0L engine

22140_BPUP_G0169

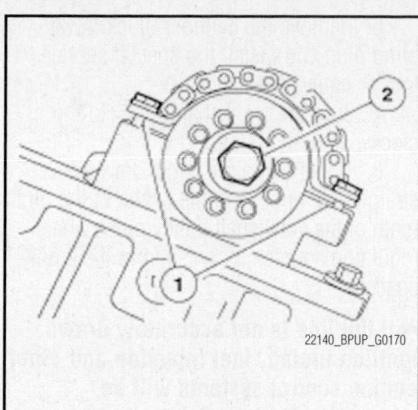

Fig. 110 Top 2 clamp bolts and the LH camshaft bolt.—4.0L engine

51. The camshaft timing slots are off-center. Position the camshaft timing slots below the centerline of the camshaft to correctly fit the special tools and install the camshaft holding tool 303-576 and adaptor 303-577 on the rear of the LH cylinder head.

52. Remove the LH hydraulic camshaft tensioner.

53. Install the timing chain tensioner 303-571.

54. Tighten camshaft top 2 clamp bolts to 89 inch lbs. (10 Nm).

55. Tighten LH camshaft bolt to 63 ft. lbs. (85 Nm).

56. Remove the timing chain tensioner 303-571.

57. Install a new O-ring seal on the tensioner and lubricate with clean engine oil. Install the LH hydraulic camshaft tensioner.

58. Tighten camshaft sprocket bolts to 32 ft. lbs. (44 Nm).

59. Remove the crankshaft timing tool 303—573.

60. Position the A/C hose and install the nut.

61. Install the camshaft roller followers.

62. Install the thermostat housing.

63. Install the engine front cover.

64. Install the fuel rail.

65. Install the valve covers.

66. Connect the battery cable.

ENGINE PERFORMANCE & EMISSION CONTROL

CAMSHAFT POSITION (CMP) SENSOR

LOCATION

See Figures 111 and 112.

Refer to the accompanying illustrations for CMP sensor location.

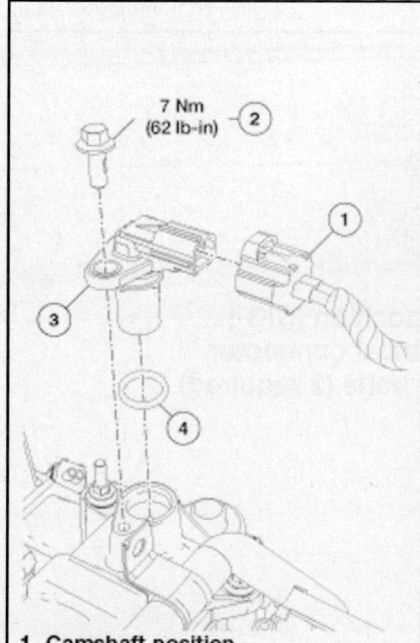

1. Camshaft position (CMP) sensor electrical connector
2. CMP sensor bolt
3. CMP sensor
4. CMP sensor O-ring seal

Fig. 111 Camshaft Position Sensor (CMP)—2.3L engine

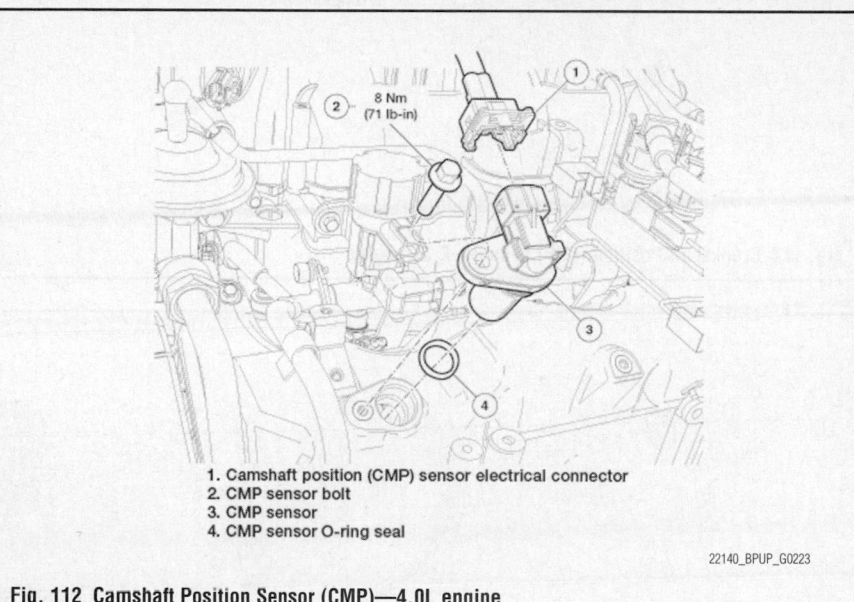

1. Camshaft position (CMP) sensor electrical connector
2. CMP sensor bolt
3. CMP sensor
4. CMP sensor O-ring seal

Fig. 112 Camshaft Position Sensor (CMP)—4.0L engine

REMOVAL & INSTALLATION

See Figure 116.

1. Disconnect the negative battery cable.

2. Disconnect the CMP sensor connector.

3. Remove the CMP sensor installation bolt.

4. Remove the CMP sensor.

5. Install the CMP sensor in the reverse order of removal. Tighten to 71 inch lbs. (8 Nm).

CRANKSHAFT POSITION (CKP) SENSOR

LOCATION

See Figures 113 and 114.

Refer to the accompanying illustrations for CKP sensor location.

REMOVAL & INSTALLATION

See Figures 115 through 117.

1. Remove the right front wheel.

2. Remove the splash shield.

3. Disconnect the CKP sensor connector.

4. Remove the installation bolts to remove the CKP sensor.

To install:

➡**Perform the following procedure so that piston No.1 is at the top dead center.**

5. Remove the right side driveshaft.

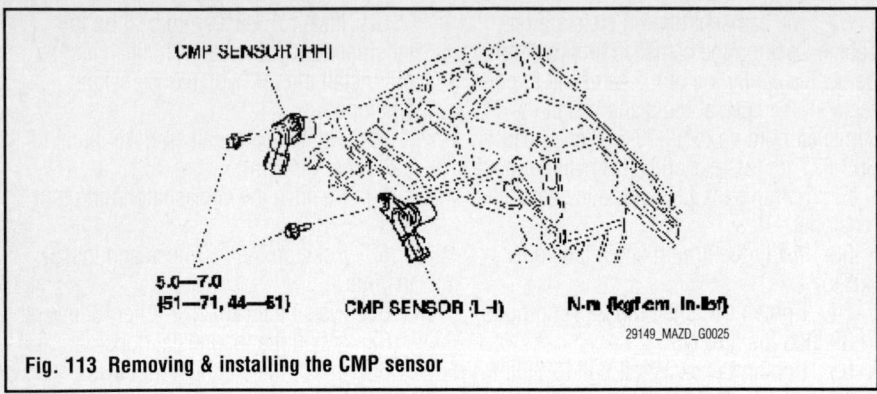

Fig. 113 Removing & installing the CMP sensor

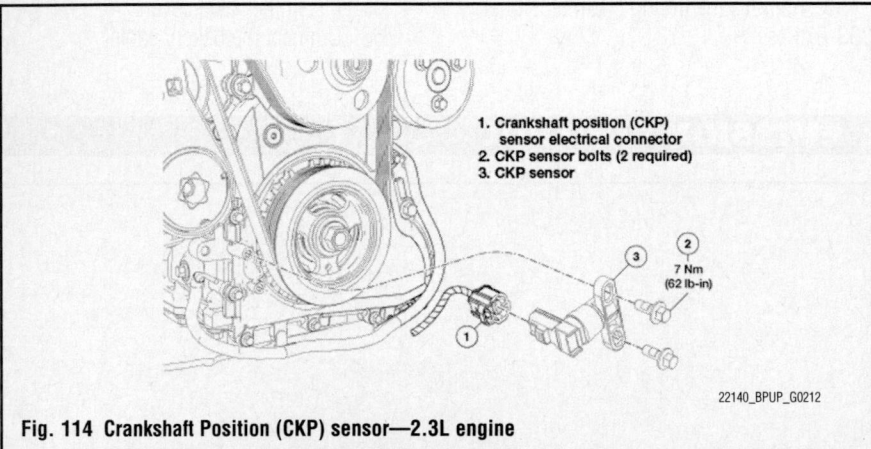

1. Crankshaft position (CKP) sensor electrical connector
2. CKP sensor bolts (2 required)
3. CKP sensor

Fig. 114 Crankshaft Position (CKP) sensor—2.3L engine

6. Remove the cylinder block lower blind plug and install the special service tool or equivalent.

7. Turn the crankshaft pulley to the clockwise until it stops.

8. Using a straight edge, draw a straight line directly in the center of the ninth tooth of the crankshaft pulley pulse wheel (counting counterclockwise from the empty space).

➡**If the line is not accurately drawn, ignition timing, fuel injection and other engine control systems will be adversely affected. Drawn the straight line carefully using a straight edge.**

9. Align the centerline of the crankshaft position sensor and the line drawn, then install the sensor.

10. Install the CKP sensor fitting bolts. Tightening torque 4.1–5.5 ft lbs.

11. Remove the special service tool or equivalent then install the cylinder block lower blind plug.

12. Tightening torque 15 ft. lbs. (20 Nm).

13. Install the right side driveshaft.

14. Install the splash shield.

15. Install the right front wheel

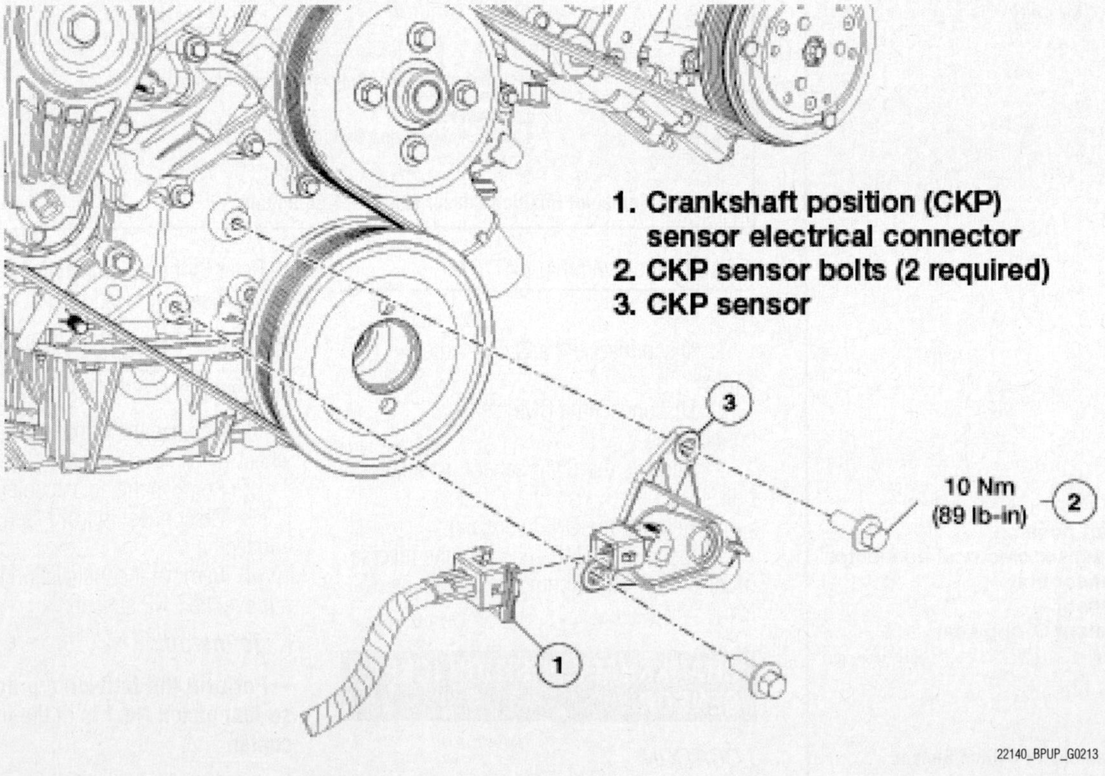

1. **Crankshaft position (CKP) sensor electrical connector**
2. **CKP sensor bolts (2 required)**
3. **CKP sensor**

10 Nm (89 lb-in)

Fig. 115 Crankshaft Position (CKP) sensor—4.0L engine

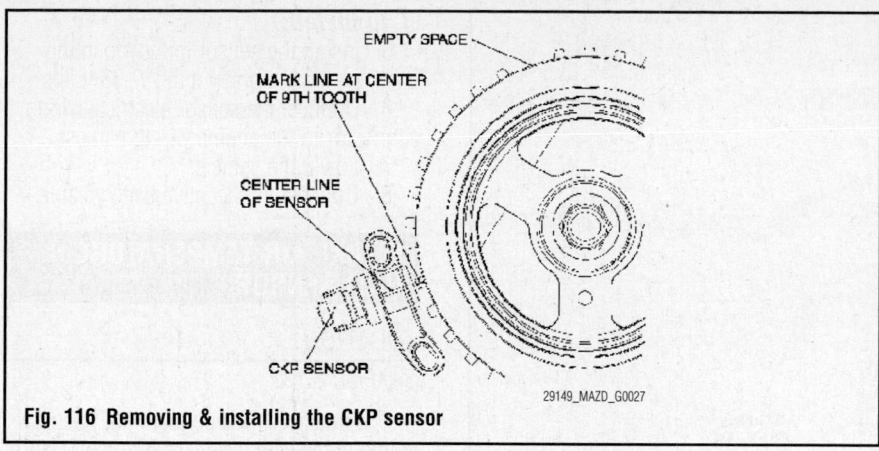

Fig. 116 Removing & installing the CKP sensor

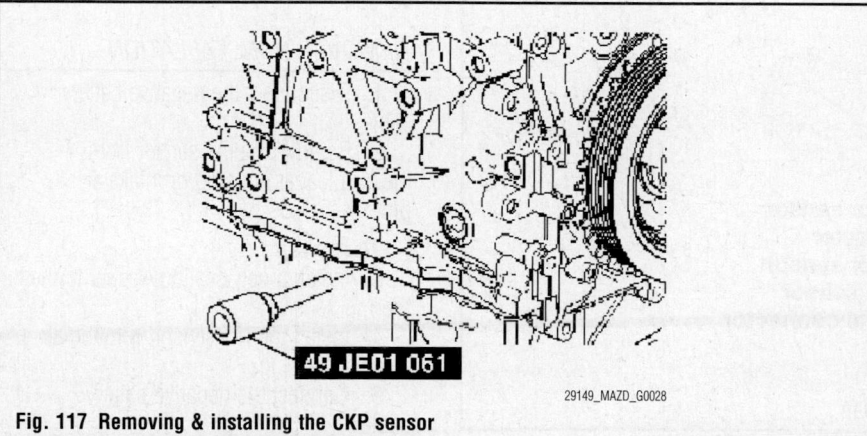

49 JE01 061

Fig. 117 Removing & installing the CKP sensor

ENGINE COOLANT TEMPERATURE (ECT) SENSOR

LOCATION

See Figures 118 and 119.

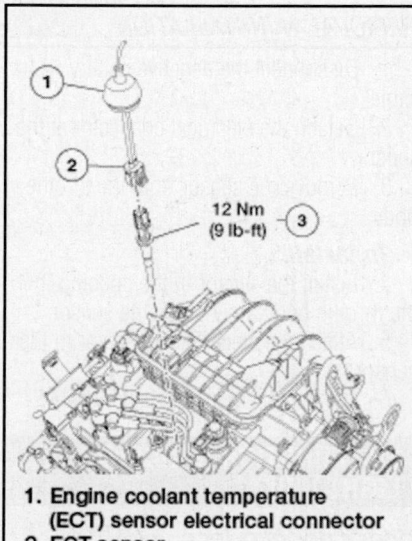

1. Engine coolant temperature (ECT) sensor electrical connector
2. ECT sensor

Fig. 118 Engine Coolant Temperature (ECT) sensor—2.3L engine

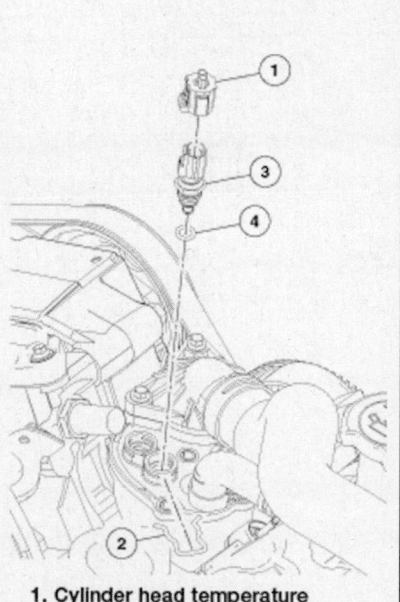

1. Cylinder head temperature (CHT) sensor cover
2. CHT sensor electrical connecor
3. CHT sensor

Fig. 119 Engine Coolant Temperature (ECT) sensor—4.0L engine

The Engine Coolant Temperature (ECT) sensor is located in the cylinder head.

REMOVAL & INSTALLATION

1. Partially drain the engine cooling system until the coolant level is below the ECT sensor-mounting hole.
2. Disconnect the negative battery cable.
3. Detach the wiring harness connector from the ECT sensor.
4. Using an open-end wrench, remove the coolant temperature sensor from the intake manifold or thermostat housing.

To install:

5. Thread the sensor into the intake manifold, or thermostat housing, by hand, then tighten it securely.
6. Connect the negative battery cable.
7. Refill the engine cooling system.
8. Start the engine, check for coolant leaks and top off the cooling system.

HEATED OXYGEN (HO2S) SENSOR

LOCATION

See Figures 120 and 121.

The oxygen sensors are located in the exhaust manifold and downstream from the catalytic converters.

REMOVAL & INSTALLATION

✲✲ CAUTION

The temperature of the exhaust system is extremely high after the engine has been run. To prevent personal injury, allow the exhaust system to cool completely before removing sensor from the exhaust system.

1. Disconnect the negative battery cable.
2. Raise and safely support the vehicle on jack stands.
3. Disconnect the HO2S from the engine control sensor wiring.

➡**If excessive force is needed to remove the sensors lubricate the sensor with penetrating oil prior to removal.**

4. Remove the sensors with a sensor removal tool, such as Ford Tool T94P-9472-A or equivalent.

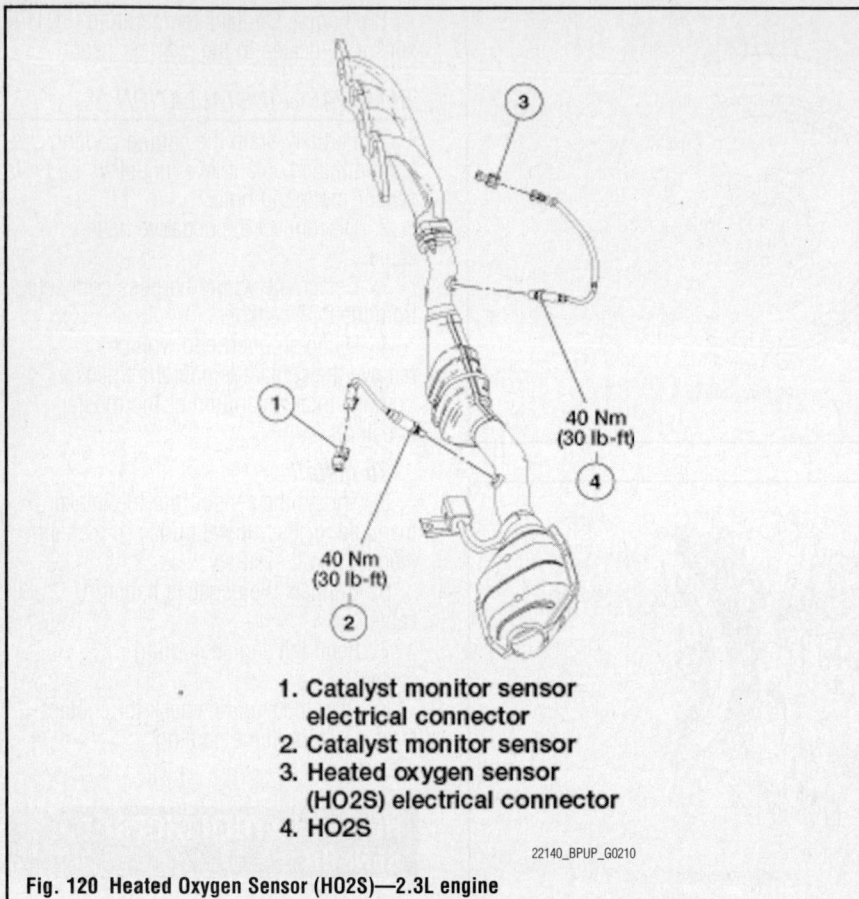

1. Catalyst monitor sensor electrical connector
2. Catalyst monitor sensor
3. Heated oxygen sensor (HO2S) electrical connector
4. HO2S

22140_BPUP_G0210

Fig. 120 Heated Oxygen Sensor (HO2S)—2.3L engine

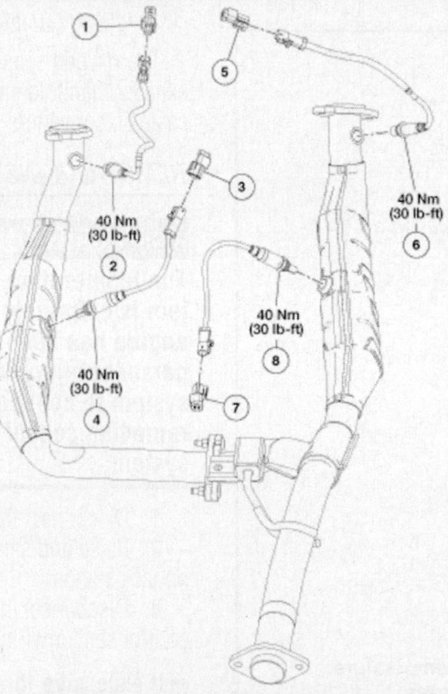

1. Heated oxygen sensor (HO2S) electrical connector
2. HO2S
3. Catalyst monitor sensor electrical connector
4. Catalyst monitor sensor
5. HO2S electrical connector
6. HO2S
7. Catalyst monitor sensor electrical connector
8. Catalyst monitor sensor

22140_BPUP_G0211

Fig. 121 Heated Oxygen Sensor (HO2S)—4.0L engine

To install:

5. Install the sensor in the mounting boss, and then tighten it to 27–33 ft. lbs.

6. Connect the sensor electrical wiring connector to the engine wiring harness.

7. Lower the vehicle.

8. Connect the negative battery cable.

INTAKE AIR TEMPERATURE (IAT) SENSOR

LOCATION

See Figure 122.

The Intake Air Temperature (IAT) sensor is integrated with the Mass Air Flow (MAF) sensor.

REMOVAL & INSTALLATION

1. Disconnect the negative battery cable.

2. Remove the IAT sensor from the air cleaner housing or intake air pipe.

To install:

3. Install a new sealing washer to the thermo sensor.

4. Install the IAT sensor from the air cleaner housing.

5. Connect the negative battery cable.

KNOCK SENSOR (KS)

LOCATION

See Figures 123 and 124.

Refer to the accompanying illustrations for KS location.

REMOVAL & INSTALLATION

1. Disconnect the negative battery cable.

2. Detach the electrical connector at the sensor.

3. Remove the sensor from the throttle body.

To install:

4. Install the sensor in the opening in the throttle body and tighten the sensor.

5. Attach the electrical connector to the sensor.

6. Connect the negative battery cable.

MALFUNCTION INDICATOR LIGHT (MIL)

RESET PROCEDURES

To turn off the Malfunction Indictor Light(MIL) after a repair, a reset command from the IDS or equivalent tester must be

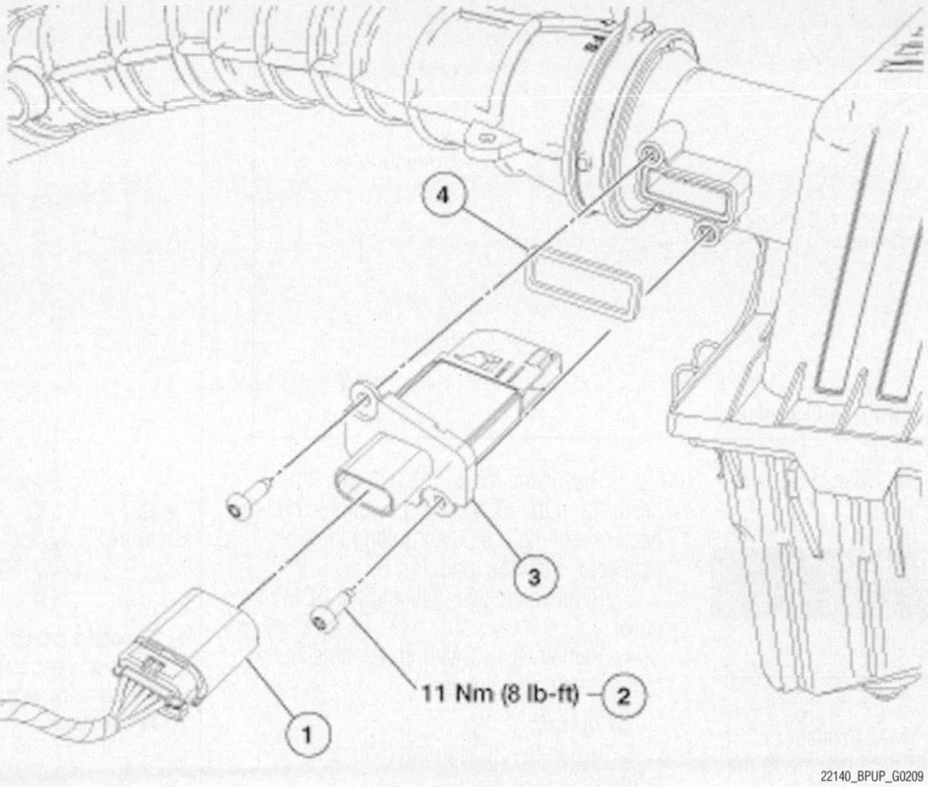
4
3
11 Nm (8 lb-ft)
2
1
22140_BPUP_G0209

Fig. 122 Mass Air Flow and Intake Air Temperature (MAF/IAT) sensor (3)

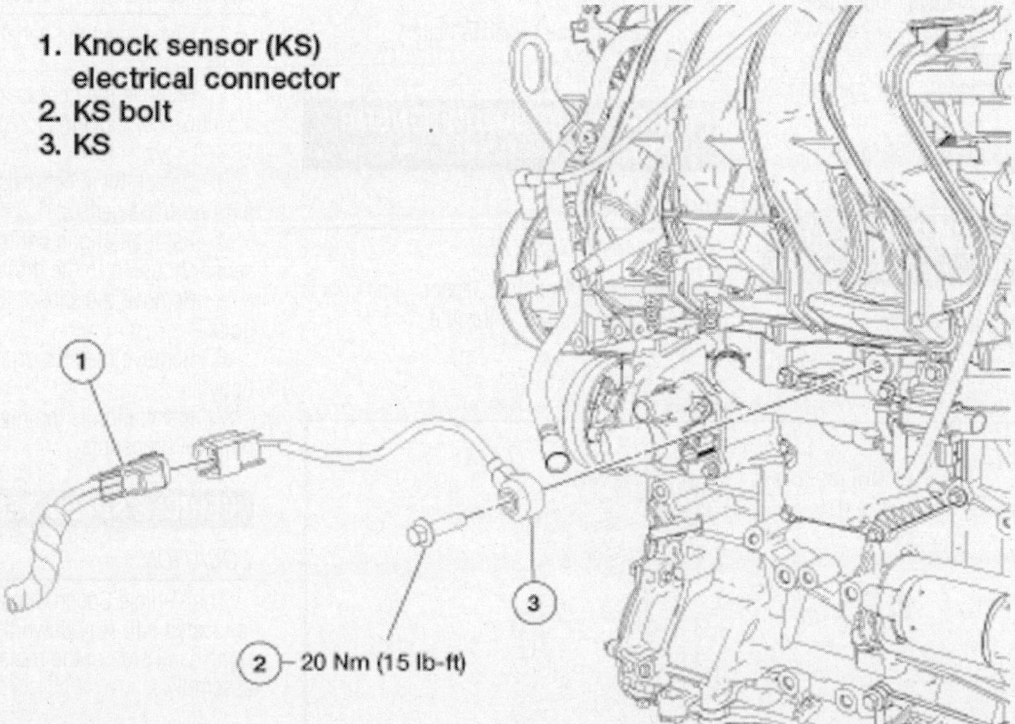
1. Knock sensor (KS) electrical connector
2. KS bolt
3. KS
1
3
2
20 Nm (15 lb-ft)
22140_BPUP_G0220

Fig. 123 Knock Sensor (KS)—2.3L engine

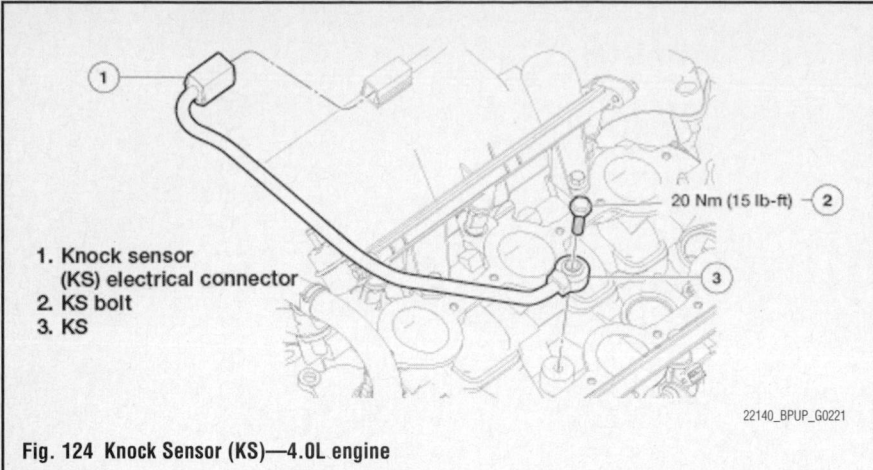

1. Knock sensor
   (KS) electrical connector
2. KS bolt
3. KS

20 Nm (15 lb-ft)

22140_BPUP_G0221

Fig. 124 Knock Sensor (KS)—4.0L engine

sent, or three consecutive drive cycles must be completed without a fault.

MASS AIR FLOW (MAF) SENSOR

LOCATION

See Figure 122.

The Mass Air Flow (MAF) sensor is located in the intake air duct leading to the throttle housing.

REMOVAL & INSTALLATION

See Figure 125.

1. Remove the air cleaner outlet tube.
2. Disconnect the Mass Air Flow (MAF) sensor electrical connector.
3. Remove the attaching four nuts and the MAF sensor.
4. Installation is the reverse of the removal procedure.

POWERTRAIN CONTROL MODULE (PCM)

REMOVAL & INSTALLATION

1. Disconnect the negative battery cable.

2. Disengage the wiring harness connector from the PCM by loosening the connector retaining bolt, then pulling the connector from the module.
3. Remove the two nuts and the PCM cover.
4. Remove the PCM from the bracket by pulling the unit outward.

To install:

5. Install the PCM in the mounting bracket.
6. Install the PCM cover and tighten the two nuts.
7. Attach the wiring harness connector to the module, then tighten the connector-retaining bolt.
8. Connect the negative battery cable.

THROTTLE POSITION SENSOR (TPS)

LOCATION

See Figures 126 and 127.

The Throttle Position Sensor (TPS) is located on the throttle hosing.

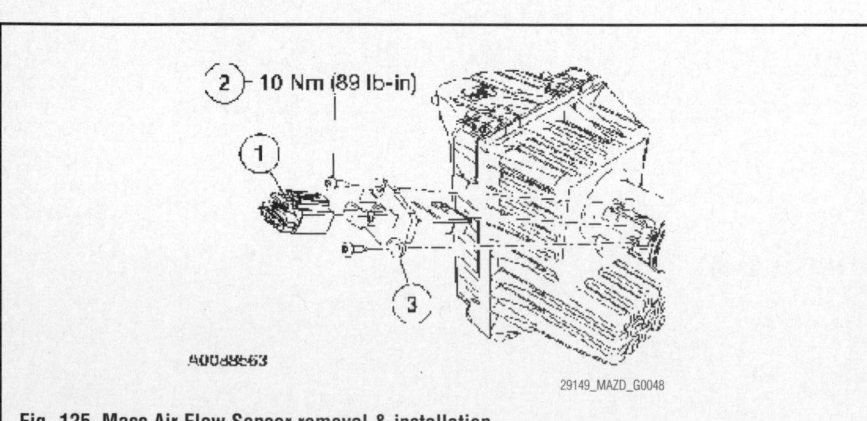

2 — 10 Nm (89 lb-in)

A0048863

29149_MAZD_G0048

Fig. 125 Mass Air Flow Sensor removal & installation

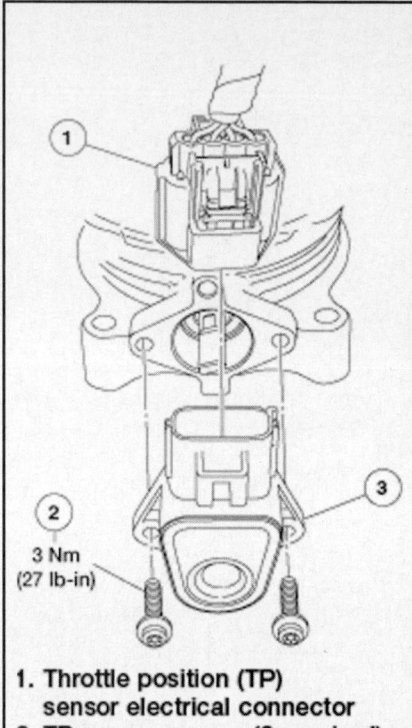

1. Throttle position (TP)
   sensor electrical connector
2. TP sensor screws (2 required)
3. TP sensor

3 Nm
(27 lb-in)

22140_BPUP_G0218

Fig. 126 Throttle Position Sensor (TPS)—2.3L engine

REMOVAL & INSTALLATION

1. Disconnect the negative battery cable.
2. Remove the necessary air intake components to access the TPS.
3. Detach the electrical wire harness plug from the sensor.
4. Paint an alignment mark on the sensor housing to the throttle body.
5. Remove the sensor attaching bolts.
6. Remove the sensor from the throttle body.
7. Installation is the reverse of the removal procedure.

VEHICLE SPEED SENSOR (VSS)

LOCATION

The Vehicle Speed Sensor (VSS) is located half-way down the right-hand side of the transmission assembly.

REMOVAL & INSTALLATION

The VSS is located half-way down the right-hand side of the transmission assembly.

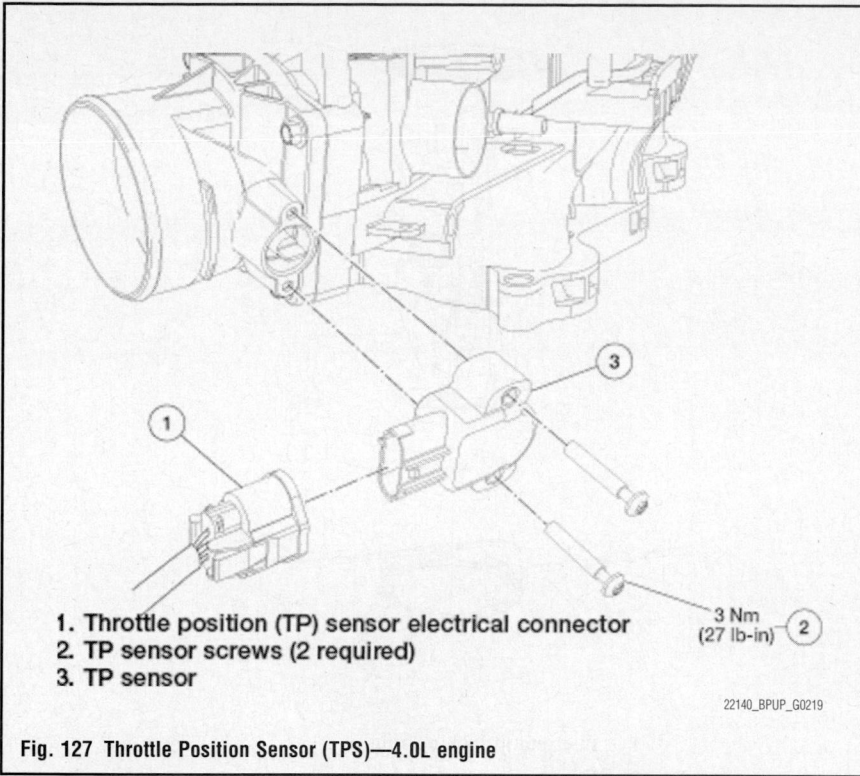

1. Throttle position (TP) sensor electrical connector
2. TP sensor screws (2 required)
3. TP sensor

3 Nm
(27 lb-in) 2

22140_BPUP_G0219

Fig. 127 Throttle Position Sensor (TPS)—4.0L engine

1. Apply parking brake, block the rear wheels, then raise and safely support the front of the vehicle on jack stands.

2. From under the right-hand side of the vehicle, disengage the wiring harness connector from the VSS.

3. Loosen the VSS hold-down bolt, then pull the VSS out of the transmission housing.

To install:

4. If a new sensor is being installed, transfer the driven gear retainer and gear to the new sensor.

5. Ensure that the O-ring is properly seated in the VSS housing.

6. For ease of assembly, engage the wiring harness connector to the VSS, then insert the VSS into the transmission assembly.

7. Install and tighten the VSS hold-down bolt to 62–88 inch lbs.

8. Lower the vehicle and remove the wheel blocks.

FUEL SYSTEM

GASOLINE FUEL INJECTION SYSTEM

FUEL SYSTEM SERVICE PRECAUTIONS

Safety is the most important factor when performing not only fuel system maintenance but any type of maintenance. Failure to conduct maintenance and repairs in a safe manner may result in serious personal injury or death. Maintenance and testing of the vehicle's fuel system components can be accomplished safely and effectively by adhering to the following rules and guidelines.

• To avoid the possibility of fire and personal injury, always disconnect the negative battery cable unless the repair or test procedure requires that battery voltage be applied.

• Always relieve the fuel system pressure prior to disconnecting any fuel system component (injector, fuel rail, pressure regulator, etc.), fitting or fuel line connection. Exercise extreme caution whenever relieving fuel system pressure to avoid exposing skin, face and eyes to fuel spray. Please be advised that fuel under pressure may penetrate the skin or any part of the body that it contacts.

• Always place a shop towel or cloth around the fitting or connection prior to loosening to absorb any excess fuel due to

spillage. Ensure that all fuel spillage (should it occur) is quickly removed from engine surfaces. Ensure that all fuel soaked cloths or towels are deposited into a suitable waste container.

• Always keep a dry chemical (Class B) fire extinguisher near the work area.

• Do not allow fuel spray or fuel vapors to come into contact with a spark or open flame.

• Always use a back-up wrench when loosening and tightening fuel line connection fittings. This will prevent unnecessary stress and torsion to fuel line piping.

• Always replace worn fuel fitting O-rings with new. Do not substitute fuel hose or equivalent where fuel pipe is installed.

Before servicing the vehicle, make sure to also refer to the precautions in the beginning of this section as well.

RELIEVING FUEL SYSTEM PRESSURE

All Sequential Fuel Injection (SFI) fuel injected engines are equipped with a pressure relief valve located on the fuel supply manifold. Remove the fuel tank cap and attach fuel pressure gauge T80L-9974-B, to the valve to release the fuel pressure. Be

sure to drain the fuel into a suitable container and to avoid gasoline spillage. If a pressure gauge is not available, disconnect the vacuum hose from the fuel pressure regulator and attach a hand-held vacuum pump. Apply about 25 in. Hg (84 kPa) of vacuum to the regulator to vent the fuel system pressure into the fuel tank through the fuel return hose. Note that this procedure will remove the fuel pressure from the lines, but not the fuel. Take precautions to avoid the risk of fire and use clean rags to soak up any spilled fuel when the lines are disconnected.

An alternate method of relieving the fuel system pressure involves disconnecting the inertia switch.

FUEL FILTER

REMOVAL & INSTALLATION
See Figure 128.

✳✳ CAUTION

Do not smoke or carry lighted tobacco or open flame of any type when working on or near any fuel-related component. Highly flammable mixtures are always present and can be ignited, resulting in possible personal injury.

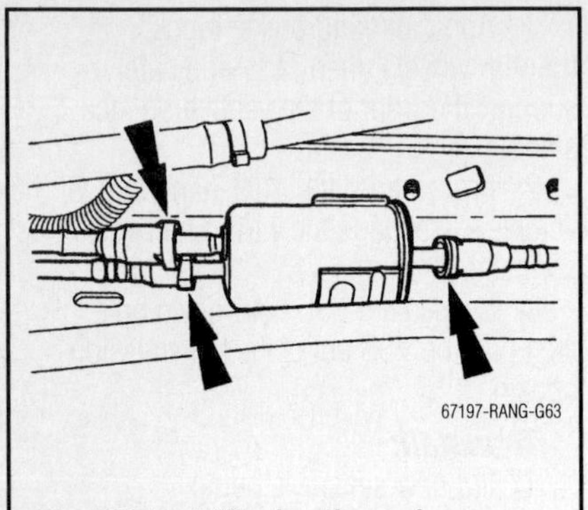
67197-RANG-G63

Fig. 128 Fuel filter location and connections

⁂ CAUTION

Fuel in the fuel system remains under high pressure even when the engine is not running. Before repairing or disconnecting any of the fuel lines or fuel system components, the fuel system pressure must be relieved to prevent accidental spraying of fuel, causing personal injury or a fire hazard.

1. Before servicing the vehicle, refer to the precautions in the beginning of this section.
2. Properly relieve the fuel system pressure.
3. Remove or disconnect the following:
   - Negative battery cable
   - Push connect and R-clip fittings from the fuel filter
   - Fuel filter

To install:

4. Install or connect the following:
   - Fuel filter. Torque the nut to 17 ft. lbs. (23 Nm).
   - R-clip and push connect fittings
   - Negative battery cable
5. Start the vehicle, check for leaks and repair if necessary.

FUEL LEVEL SENDING UNIT

REMOVAL & INSTALLATION

See Figure 129.

1. Disconnect the negative battery cable.

➡On some vehicles, when the battery cable is disconnected and reconnected, some abnormal drive symptoms may occur while the vehicle relearns its adaptive strategy. The vehicle may need to be driven to relearn its strategy.

The fuel pump must be handled carefully to avoid damage to the float arm and filter. Remove the fuel pump assembly.

2. Properly relieve the fuel system pressure
3. Clean the area around the fuel pump mounting flange.
4. Using the fuel tank sender unit wrench 310-123, remove the fuel tank sending unit assembly locking retainer ring.
5. Remove and discard the fuel pump mounting gasket.

To install:

6. Clean the fuel pump mounting flange and the fuel tank mounting surface.
7. Install a new fuel pump mounting gasket
8. Install the fuel tank sending unit assembly with the float toward the rear of the fuel tank. Align the arrows molded into the fuel tank flange.
9. Install the locking retainer ring while compressing the fuel tank sending unit assembly into the tank.
10. Using the fuel tank sender unit wrench, tighten the fuel tank sending unit assembly locking retainer ring until it locks in place.
11. Install the fuel tank.

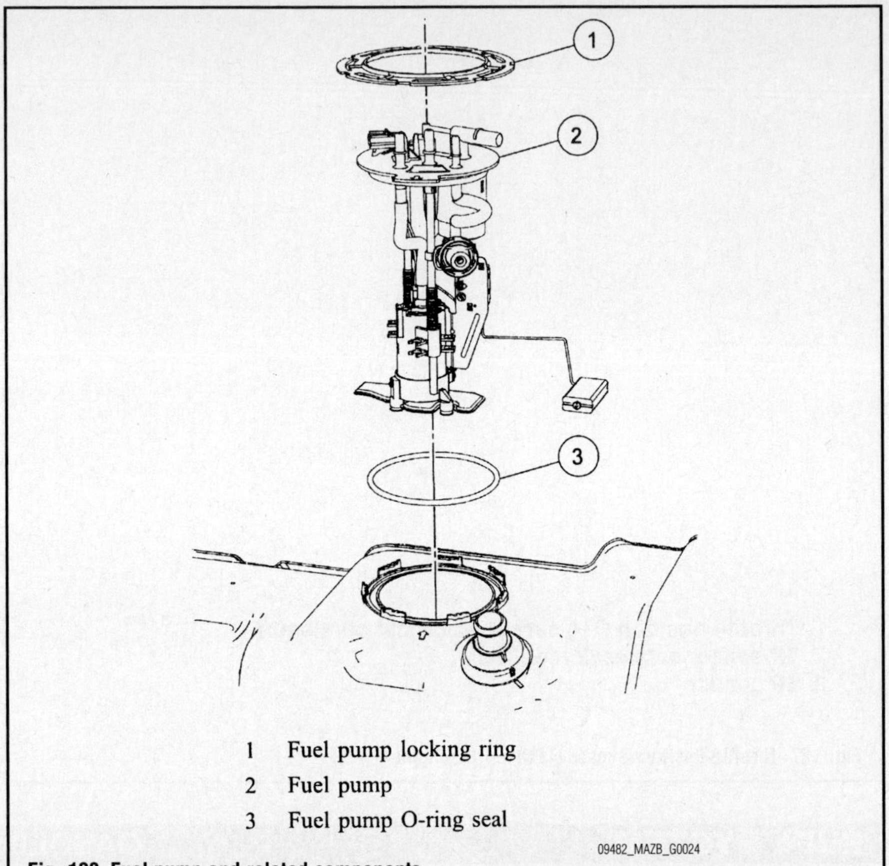
1
2
3
1 Fuel pump locking ring
2 Fuel pump
3 Fuel pump O-ring seal
09482_MAZB_G0024

Fig. 129 Fuel pump and related components

FUEL PUMP

REMOVAL & INSTALLATION

See Figures 129.

1. Before servicing the vehicle, refer to the Precautions Section.
2. Properly relieve the fuel system pressure.
3. Disconnect the negative battery cable.

➡On some vehicles, when the battery cable is disconnected and reconnected, some abnormal drive symptoms may occur while the vehicle relearns its adaptive strategy. The vehicle may need to be driven to relearn its strategy.

4. Remove the fuel tank.
5. Clean the area around the fuel pump mounting flange.
6. Using the special tool, remove the fuel tank pump assembly locking retainer ring.

⁂ WARNING

The fuel pump assembly must be removed and handled carefully to avoid damage to the float arm and filter.

7. Remove the fuel pump assembly.

8. Remove and discard the fuel pump mounting gasket.

To install:

9. Clean the fuel pump mounting flange and the fuel tank mounting surface.

10. Install a new fuel pump mounting gasket.

11. Install the fuel pump and sender assembly with the float toward the rear of the tank. Align the arrows molded into the tank and flange.

12. Install the locking ring while compressing the pump assembly into the tank.

13. Using the special tool, tighten the fuel pump assembly locking ring retainer ring until it locks in place.

14. Install the fuel tank.

FUEL TANK

REMOVAL & INSTALLATION

1. Relieve the fuel pressure.

2. Disconnect the fill pipe main hose from the main fill pipe.

3. Drain the fuel tank.

4. If equipped, remove the 4 nuts (two each side) and remove the skid plate.

5. Support the fuel tank.

6. Remove the bolts and pivot the fuel tank support straps away from the fuel tank.

7. Disconnect the fuel pump connector.

8. Lower the fuel tank slightly.

9. Disconnect the fuel pump electrical connector.

10. Disconnect the fuel supply tube, vapor tube and the fill tube vent hose from the fuel pump.

11. Disconnect the vapor tube fitting at the rear of the fuel tank.

12. Lower the fuel tank.

13. To install, reverse the removal procedure.

14. Tighten the skid plate to 22 ft. lbs. (30 Nm).

15. Tighten the fuel tank support straps to 30 ft. lbs. (40 Nm).

FUEL RAIL & INJECTORS

REMOVAL & INSTALLATION

2.3L Engine

See Figures 130 through 132.

1. Before servicing the vehicle, refer to the precautions in the beginning of this section.

2. Properly relieve the fuel system pressure.

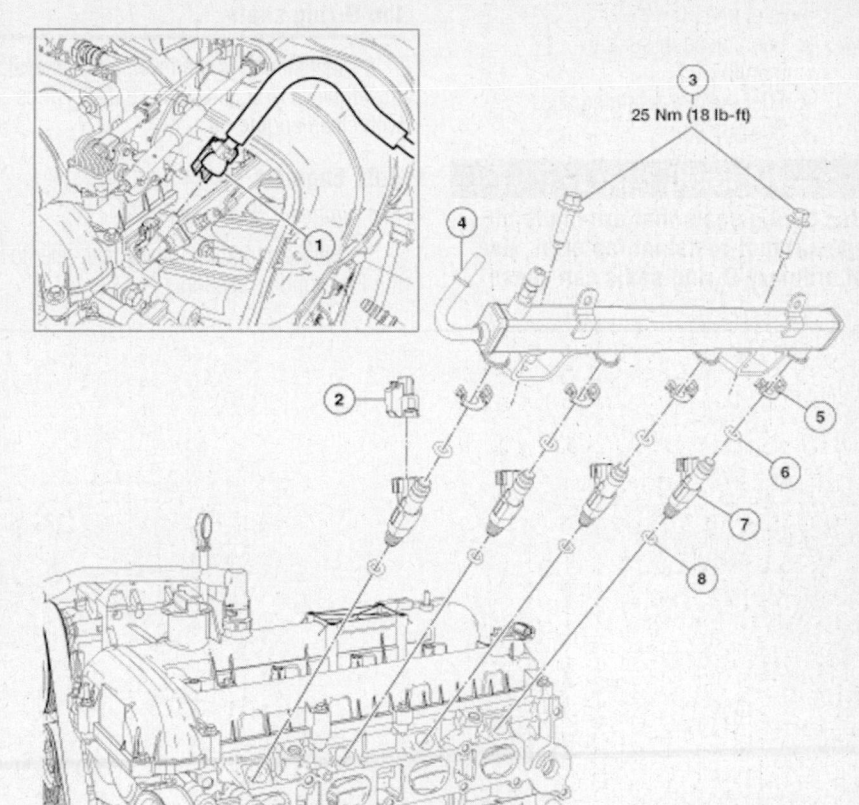

1. Fuel supply tube spring lock coupling
2. Fuel injector electrical connectors
3. Fuel rail bolts (2 required)
4. Fuel rail
5. Fuel injector retaining clip
6. Fuel injector upper O-ring seals (4 required)
7. Fuel injector
8. Fuel injector lower O-ring seals (4 required)

22140_BPUP_G0173

Fig. 130 Fuel injection supply manifold and fuel injector removal and installation exploded component view—2.3L engine

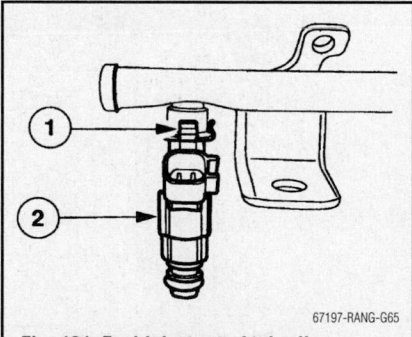

67197-RANG-G65

Fig. 131 Fuel injector-to-fuel rail installation. 1 retaining clip; 2 injector—2.3L engine

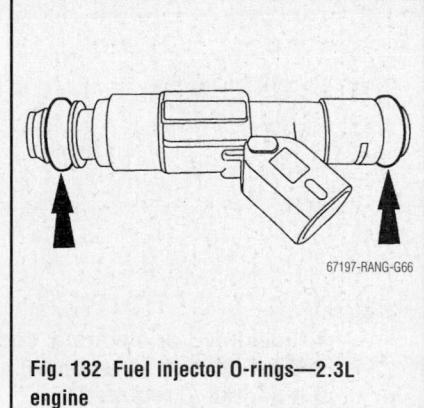

67197-RANG-G66

Fig. 132 Fuel injector O-rings—2.3L engine

3. Remove or disconnect the following:

4. Remove the intake manifold.

5. Disconnect the fuel tube spring lock coupling.

6. Disconnect the fuel injector electrical connectors.

7. Remove or disconnect the following:

- Negative battery cable
- Upper intake manifold
- Fuel injector connectors
- Fuel injector harness from the fuel injector supply manifold

- Fuel line spring lock
- Fuel line
- Fuel injection supply manifold
- Fuel injector retaining clip
- Fuel injector

✳✷ WARNING

Use O-ring seals that are made of special fuel-resistant material. Use of ordinary O-ring seals can cause the fuel system to leak. Do not reuse the O-ring seals.

8. Installation is the reverse of removal. Install new O-rings. Lubricate the O-rings with clean engine oil.

4.0L Engine

See Figures 133 through 135.

1. Before servicing the vehicle, refer to the Precautions Section.

2. Disconnect the negative battery cable.

➡On some vehicles, when the battery cable is disconnected and reconnected, some abnormal drive symptoms may occur while the vehicle relearns its adaptive strategy. The vehicle may need to be driven to relearn its strategy.

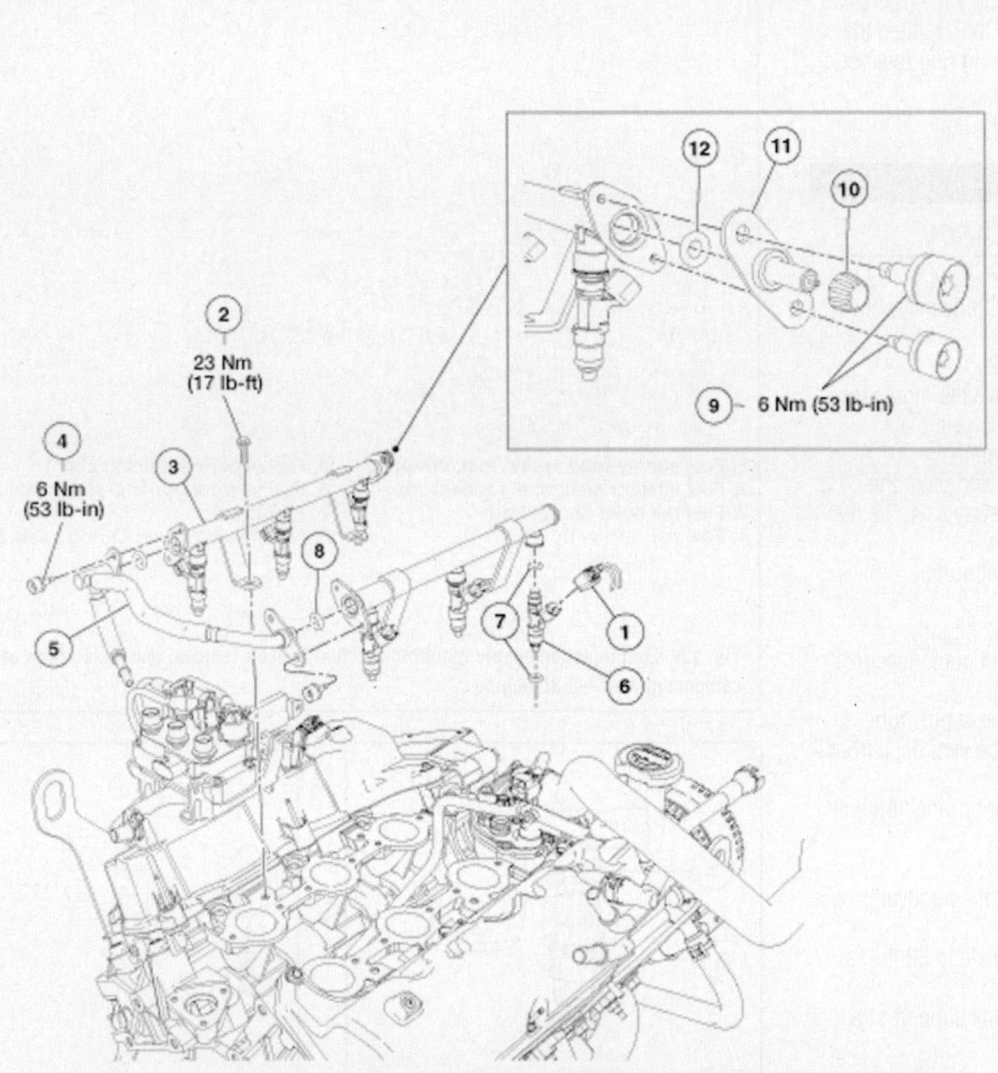

1. Fuel injector electrical connectors
2. Fuel rail bolt (4 required)
3. Fuel rail (2 required)
4. Fuel supply tube-to-valve cover bolt
5. Fuel rail supply tube
6. Fuel injectors (6 required)
7. Fuel injector O-ring seals (12 required)
8. Fuel rail supply tube-to-fuel rail O-ring seals (2 required)
9. Fuel pressure relief valve bolts (2 required)
10. Fuel pressure relief valve cap
11. Fuel pressure relief valve
12. Fuel pressure relief valve O-ring seal

22140_BPUP_G0174

Fig. 133 Fuel injection supply manifold and fuel injector removal and installation exploded component view—4.0L engine

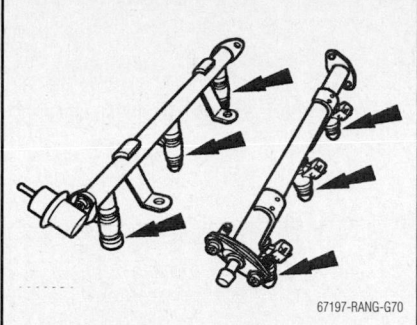

Fig. 134 Fuel rail and injectors—4.0L engine

67197-RANG-G70

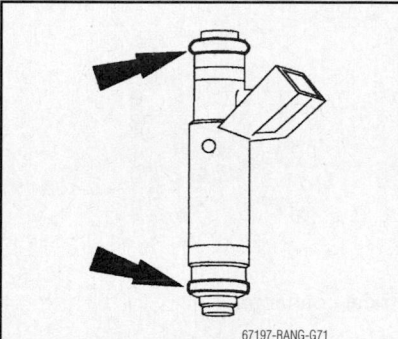

Fig. 135 Fuel injector O-rings—4.0L engine

67197-RANG-G71

3. Properly relieve the fuel system pressure.

4. Remove or disconnect the following:

- Upper intake manifold
- Engine control sensor wiring from the fuel injectors
- Fuel lines
- Fuel injection supply manifold and injectors as an assembly
- Vacuum line
- Fuel injectors from the supply manifold
- Inspect the O-rings and replace them as needed

To install:

5. Install or connect the following:

- Fuel injectors
- Vacuum line
- Fuel injection supply manifold. Torque the bolts to 89 inch lbs. (10 Nm).
- Fuel line
- Engine control sensor wiring to the fuel injectors
- Upper intake manifold
- Negative battery cable

6. Start the vehicle, check for leaks and repair if necessary.

IDLE SPEED

ADJUSTMENT

No adjustments can be made The Electronic Control Module (ECM) controls idle speed dependent on input.

THROTTLE BODY

REMOVAL & INSTALLATION

2.3L Engine

See Figure 136.

✳✳ CAUTION

Do not smoke or carry lighted tobacco or open flame of any type when working on or near any fuel-related components. Highly flammable mixtures are always present and may be ignited. Failure to follow these instructions may result in personal injury.

✳✳ WARNING

The throttle body bore and plate area have a special coating and should not be cleaned.

1. Before servicing the vehicle, refer to the Precautions Section.

2. Remove or disconnect the following:

- Negative battery cable.
- Air cleaner outlet pipe
- If equipped, transmission vent tube and bracket bolt and set aside
- Accelerator cable and, if equipped, the cruise control actuator cable
- Throttle position sensor electrical connector
- Screws and throttle body

➡**The throttle body sealing surfaces are soft materials. Use care if cleaning is necessary.**

3. To install, reverse the removal procedure.

4. Torque throttle body screws to 89 inch lbs. (10 Nm).

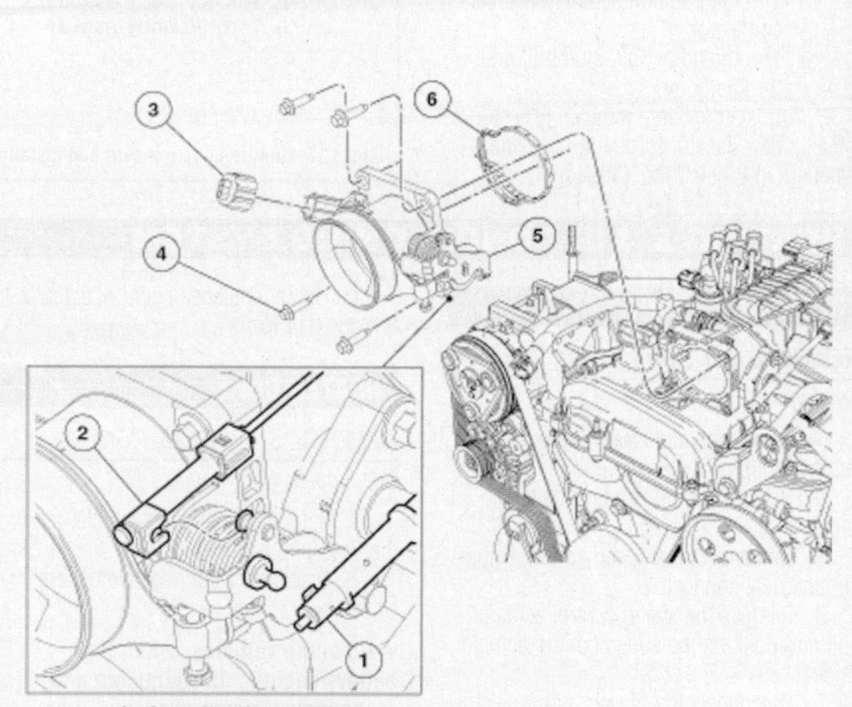

1. Accelerator cable
2. Cruise control cable
3. Throttle position (TP) sensor electrical connector
4. Throttle body bolts (4 required)
5. Throttle body
6. Throttle body gasket

22140_BPUP_G0175

Fig. 136 Exploded view of the throttle body—2.3L engine

4.0L Engine

See Figure 137.

⁂ CAUTION

Throttle body bore and plate area have a special coating and should not be cleaned.

1. Before servicing the vehicle, refer to the Precautions Section.
2. Remove or disconnect the following:
   - Negative battery cable.
   - Air cleaner outlet pipe
   - Accelerator cable and, if equipped, the cruise control actuator cable
   - Throttle position sensor electrical connector
   - Throttle body bolt, stud bolt and throttle body
3. To install, reverse removal procedure. Install a new gasket. Torque throttle body fasteners to 89 inch lbs. (10 Nm).

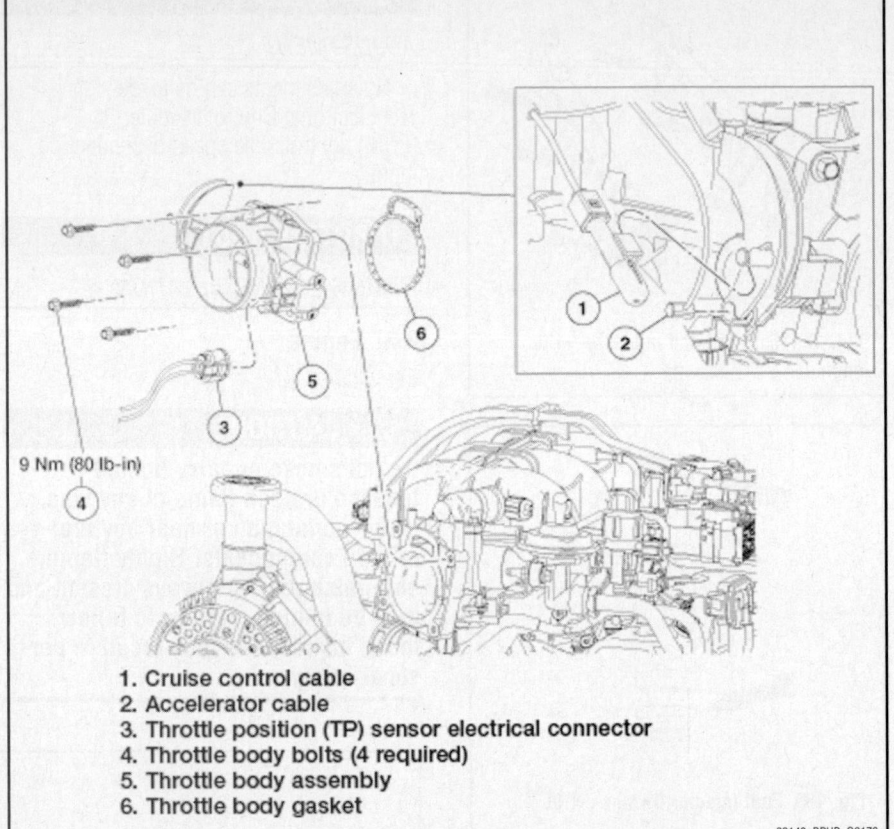

9 Nm (80 lb-in)

1. Cruise control cable
2. Accelerator cable
3. Throttle position (TP) sensor electrical connector
4. Throttle body bolts (4 required)
5. Throttle body assembly
6. Throttle body gasket

22140_BPUP_G0176

Fig. 137 Throttle body removal and installation exploded component view—4.0L engine

HEATING & AIR CONDITIONING SYSTEM

BLOWER MOTOR

REMOVAL & INSTALLATION

See Figure 138.

1. Before servicing the vehicle, refer to the Precautions Section.
2. Disconnect the negative battery cable.
3. Disconnect the speed control actuator electrical connector.
4. Remove the speed control actuator bolt and position the speed control actuator aside.
5. Disconnect the blower motor electrical connector.
6. Remove the blower motor vent tube.
7. Remove the 4 blower motor screws and the blower motor.
8. Remove the blower motor wheel clip.
9. Remove the blower motor wheel.
10. To install, reverse the removal procedure.

11. Tighten the speed control actuator to 8 ft. lbs. (11 Nm).

HEATER CORE

REMOVAL & INSTALLATION

See Figures 139 through 147.

1. Before servicing the vehicle, refer to the Precautions Section.
2. Disconnect the negative battery cable.

➡On some vehicles, when the battery cable is disconnected and reconnected, some abnormal drive symptoms may occur while the vehicle relearns its adaptive strategy. The vehicle may need to be driven to relearn its strategy.

⁂ CAUTION

After disconnecting the negative battery cable, wait for 1 minute for the SRS module to deplete its energy.

3. Depower the SRS system.
4. Discharge the air conditioning system.
5. Drain the cooling system.
6. Remove the suction accumulator.
7. If equipped with the 2.3L engine, remove the A/C compressor and the engine oil indicator and tube.
8. If equipped with the 4.0L engine, position the coolant reservoir and windshield washer reservoir aside.
9. Detach the speed control servo, on vehicles equipped with cruise control.
10. Disconnect the blower motor and blower motor resistor electrical connectors.
11. On 2.3L engine, remove the heater hoses from the heater core.
12. Disconnect the vacuum line from the water control valve.

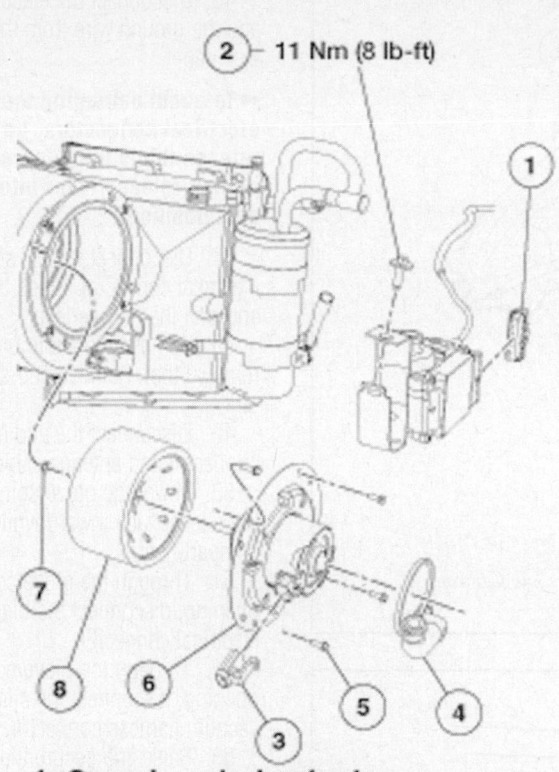

② ─ 11 Nm (8 lb-ft)

1. Speed control actuator electrical connector
2. Speed control actuator bolt
3. Blower motor electrical connector
4. Blower motor vent tube
5. Blower motor screws (4 required)
6. Blower motor
7. Blower motor wheel clip
8. Blower motor wheel

22140_BPUP_G0177

Fig. 138 Cabin blower motor removal and installation exploded component view

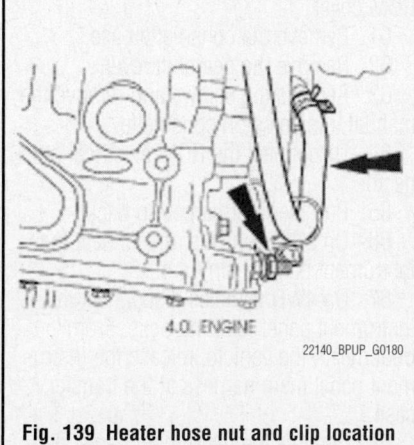

4.0L ENGINE

22140_BPUP_G0180

Fig. 139 Heater hose nut and clip location

13. Remove the heater hose from the rear coolant tube.
14. For all engines, remove the heater hoses from between the heater core and the water control valve.
15. On 2.3L engine, disconnect the heater hose from the water outlet tube and remove the heater hose assembly.
16. For all engines, remove the heater hose from between the water control valve and the water pump.
17. On 4.0L engine, remove the nut from the clip.
18. Remove the heater hose from between the water control valve and the water outlet.
19. Detach the pin-type retainer and position aside the windshield washer hose.
20. Disconnect and detach the heater control valve vacuum hose.
21. Disconnect the vacuum supply hose near the evaporator core housing.

22. Disconnect the condenser to evaporator line spring lock coupling from the evaporator core inlet. Discard the O-ring seals.

➡ **This step is performed at the lower passenger side dash panel, inside the passenger compartment.**

23. Disconnect the vacuum hose connector and remove the nut.
24. If equipped with the 4.0L engine, remove the vehicle splash guard. Remove the splash shield. On the right splash shield, disconnect the vacuum lines from the vacuum storage reservoir.

➡ **Remove the electrical connector locators from the splash shield prior to removal.**

25. Remove the nuts and the evaporator core housing.
26. Lock the steering column.
27. Remove the front seats.
28. Remove the screws and position the parking brake release handle aside.
29. Remove the left and right lower cowl kick panels.
30. Position the parking brake assembly aside.
31. Remove the screws and position the hood release handle aside.
32. Remove the instrument panel steering column cover.
33. Remove the instrument panel steering column opening cover reinforcement.
34. Disconnect the Brake Pedal Position (BPP) switch electrical connector from the steering column shaft.
35. If equipped, disconnect the Clutch Pedal Position (CPP) switch electrical connector.
36. If equipped, disconnect the shift cable from the steering column.
37. Remove the steering column pinch bolt and disconnect the steering column intermediate shaft.

➡ **To avoid damage to the clockspring, do not allow the steering shaft to rotate while the intermediate shaft is disconnected.**

38. Remove the left and right side garnish moldings.
39. Remove the bolt covers and remove the bolts.
40. Remove the assist handle.
41. Remove the windshield side garnish molding.
42. Remove the door moldings.
43. On regular cab vehicles
44. Remove the screws.
45. Remove the scuff plate.

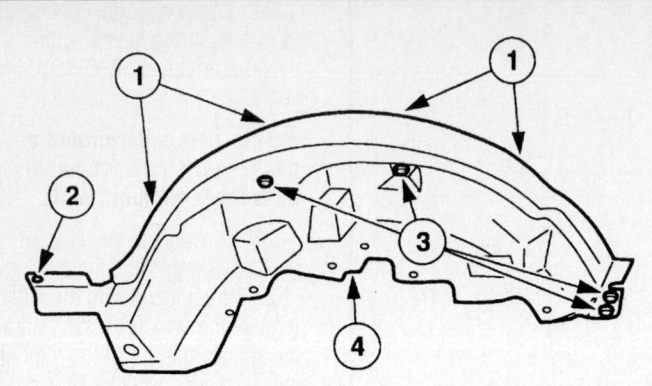

1   Remove the screws.

2   Remove the pushpin.

3   Remove the bolts.

4   Remove the splash shield.

09482_MAZB_G0003

Fig. 140 Splash shield removal points

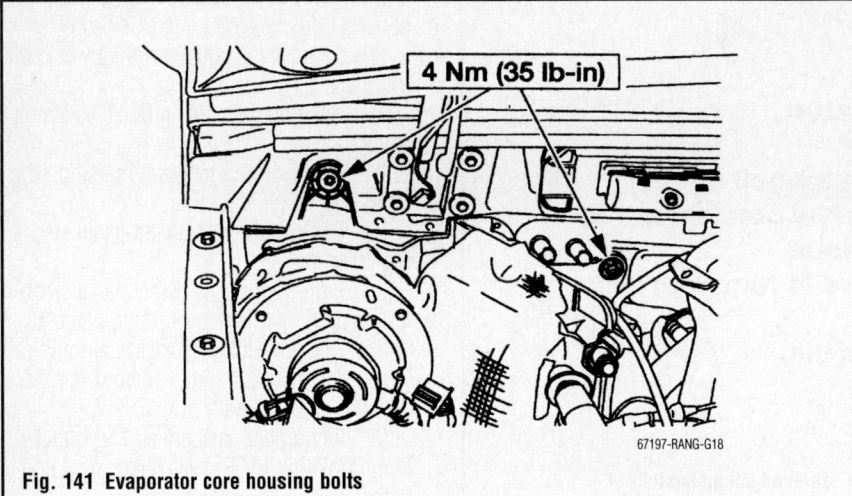

4 Nm (35 lb-in)

67197-RANG-G18

Fig. 141 Evaporator core housing bolts

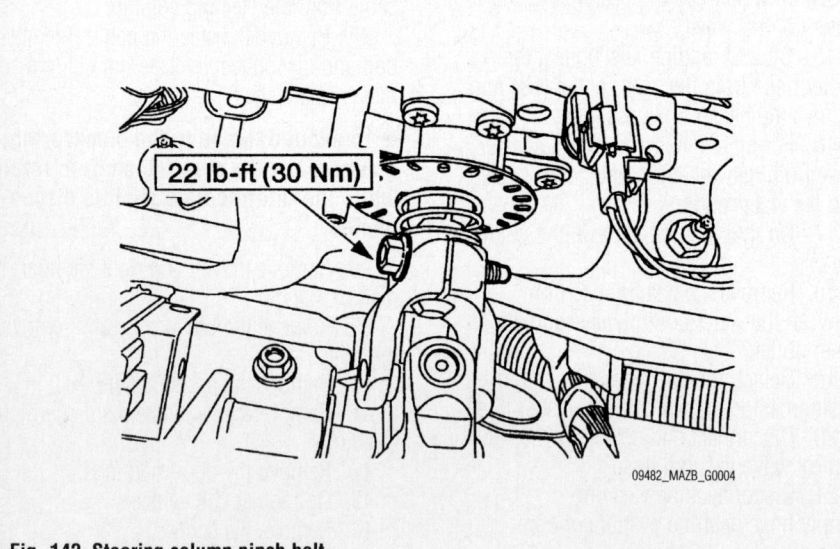

22 lb-ft (30 Nm)

09482_MAZB_G0004

Fig. 142 Steering column pinch bolt

46. Disconnect the electrical connectors and the ground wire from the RH side lower kick panel.

➡**To avoid damaging the bulkhead electrical connectors, be sure the release tab is fully depressed before pulling release lever into the disconnect position.**

47. Disconnect the LH side bulkhead electrical connector. Press the release tab and pull the release lever.

48. Remove the audio unit. Insert the removal tool. Remove and support the audio unit.

49. Disconnect the audio unit electrical connector and antenna cable.

50. Lower the glove compartment. Press the release tabs inward while lowering the compartment.

51. Through the glove compartment opening, disconnect the blend door actuator electrical connector.

52. Through the glove compartment opening, disconnect the climate control vacuum harness connector.

53. Raise and secure the glove compartment. Press the release tabs inward while raising the glove compartment.

54. Remove the instrument panel defroster opening grille.

55. Remove the instrument panel cowl top bolts.

56. If equipped, remove the floor console.

57. If not equipped with the high-series floor console, remove the cup holders.

58. Release the clips and remove the Restraints Control Module (RCM) cover.

59. If not equipped with high-series floor console, remove the consolette mat.

60. Remove the screws and remove the RCM cover.

61. Remove the consolette base.

62. Remove the gearshift lever.

63. Remove the screws and remove the manual transmission consolette.

64. Disconnect the RCM electrical connector.

65. Pull the floor carpeting back.

66. On 2WD vehicles, disconnect the instrument panel main harness.

67. On 4WD vehicles disconnect the instrument panel main harness. From underneath the vehicle, release the instrument panel main harness at the transfer case.

68. Remove the instrument panel side finish panel.

69. Disconnect the door harness electrical connector.

70. Remove the RH side instrument

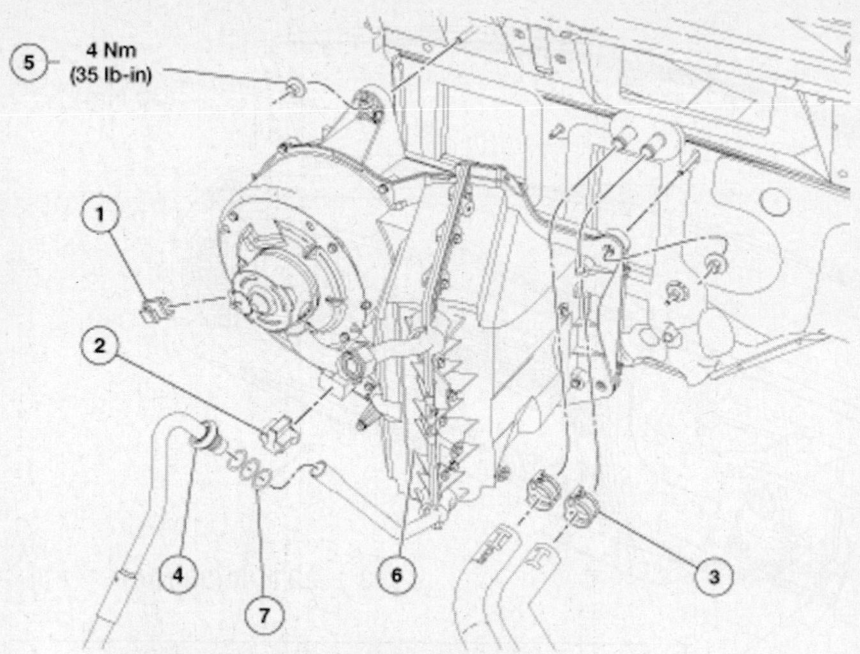

1. **Blower motor electrical connector**
2. **Blower motor resistor electrical connector**
3. **Heater hose clamps (2 required)**
4. **Evaporator inlet line fitting**
5. **Evaporator core housing nuts (2 required)**
6. **Evaporator core housing**
7. **O-ring seals (3 required)**

22140_BPUP_G0179

Fig. 143 Evaporator core housing

panel bolt. If necessary, transfer the components to the new instrument panel.

71. Remove the LH instrument panel cowl side bolts.

72. Remove the instrument panel.

73. Remove the evaporator core housing.

74. Remove the Powertrain Control Module (PCM).

75. Remove the PCM heat sink.

76. Remove the four nuts from the engine side of the dash panel. Position the plenum chamber on the vehicle floor.

77. Remove the heater core cover.

78. Remove the heater core.

To install:

79. During installation, be sure to install a new oval foam seal around the heater core inlet and outlet tubes.

80. Torque the steering column pinch bolt to 21 ft. lbs.

➡**To avoid damage to the clockspring, do not allow the steering shaft to rotate while the intermediate shaft is disconnected.**

81. Lubricate the refrigerant system with the correct amount of clean PAG oil or equivalent. Install new O-ring seals lubricated in

clean mineral oil. Lubricate the coolant hoses with plain water only, if needed.

82. Evacuate, leak test, and charge the refrigerant system.

83. Install the heater core.

84. Install the heater core cover.

85. Install the four nuts from the engine side of the dash panel. Position the plenum chamber on the vehicle.

86. Install the Powertrain Control Module (PCM) heat sink.

87. Install the PCM.

88. Install the evaporator core housing.

89. Install the instrument panel.

90. Install gearshift lever.

91. Install consolette base.

92. Install screws and install the Restraints Control Module (RCM) cover.

93. Install, if equipped with high-series floor console, the consolette mat.

94. Install the clips and the RCM cover.

95. Install, if not equipped with the high-series floor console, the cup holders.

96. Install the floor console.

97. Install the instrument panel cowl top bolts.

98. Install the instrument panel defroster opening grille.

99. Install the glove compartment. Press the release tabs inward while raising the glove compartment.

100. Through the glove compartment opening, connect the climate control vacuum harness connector.

101. Through the glove compartment opening, connect the blend door actuator electrical connector.

102. Install the glove compartment. Press the release tabs inward while lowering the compartment.

103. Connect the audio unit electrical connector and antenna cable.

104. Install the audio unit. Insert the tool. and support the audio unit.

105. Connect the LH side bulkhead electrical connector.

106. Install the audio unit.

➡**To avoid damaging the bulkhead electrical connectors, be sure the release tab is fully depressed.**

107. Connect the electrical connectors and the ground wire from the RH side lower kick panel.

108. Install the scuff plate.

109. Install the screws.

110. On regular cab, vehicles install the door moldings.

111. Install the windshield side garnish molding.

112. Install the assist handle.

113. Install the bolt covers and the bolts.

114. Install the left and right side garnish moldings.

➡**To avoid damage to the clockspring, do not allow the steering shaft to rotate while the intermediate shaft is disconnected.**

115. Install the steering column pinch bolt and disconnect the steering column intermediate shaft.

116. If equipped, connect the shift cable to the steering column.

117. If equipped, connect the Clutch Pedal Position (CPP) switch electrical connector.

118. Connect the Brake Pedal Position (BPP) switch electrical connector from the steering column shaft.

119. Install the instrument panel steering column opening cover reinforcement.

120. Install the screws and position the hood release handle.

121. Install the left and right lower cowl kick panels.

122. Install the screws and position the parking brake release handle.

123. Install the front seats.

124. Lock the steering column.

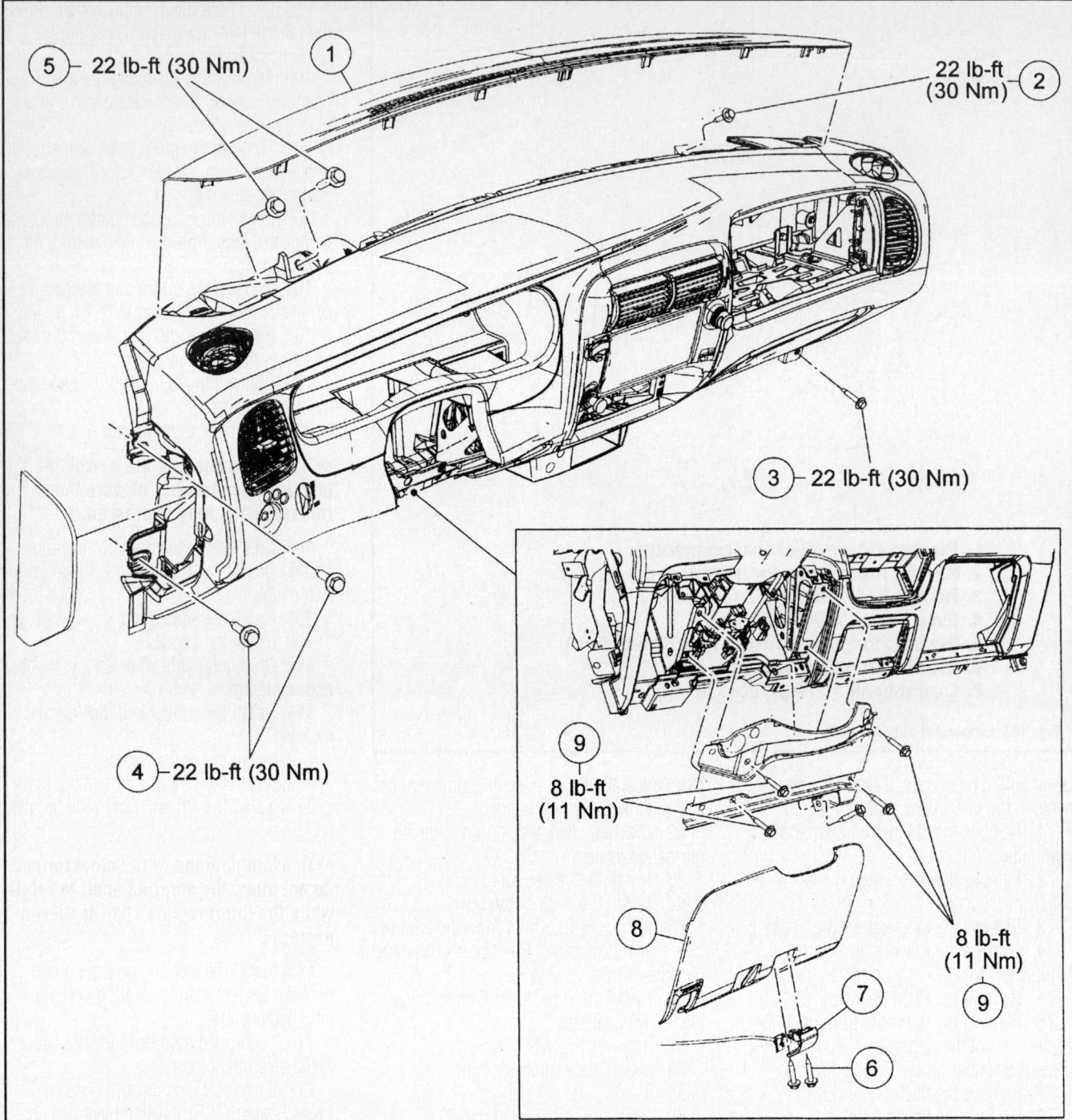

5 — 22 lb-ft (30 Nm)

22 lb-ft (30 Nm) — 2

3 — 22 lb-ft (30 Nm)

4 — 22 lb-ft (30 Nm)

9

8 lb-ft (11 Nm)

8 lb-ft (11 Nm)

8

7

6

09482_MAZB_G0005

1	Instrument panel defroster opening grille	6	Hood release handle bolt (2 required)
2	Instrument panel cowl top bolt	7	Hood release handle
3	Instrument panel bolt	8	Steering column opening panel cover
4	Instrument panel bolts	9	Opening cover reinforcement panel bolts
5	Instrument panel cowl top bolts		

Fig. 144 Instrument panel and related components

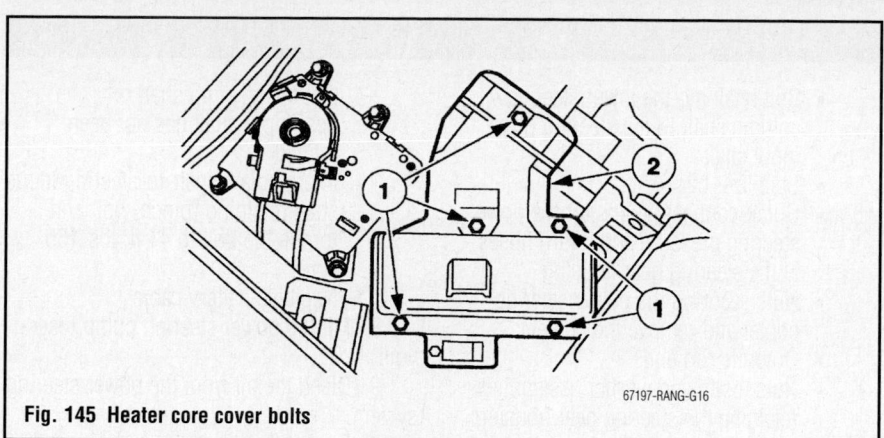

Fig. 145 Heater core cover bolts

67197-RANG-G16

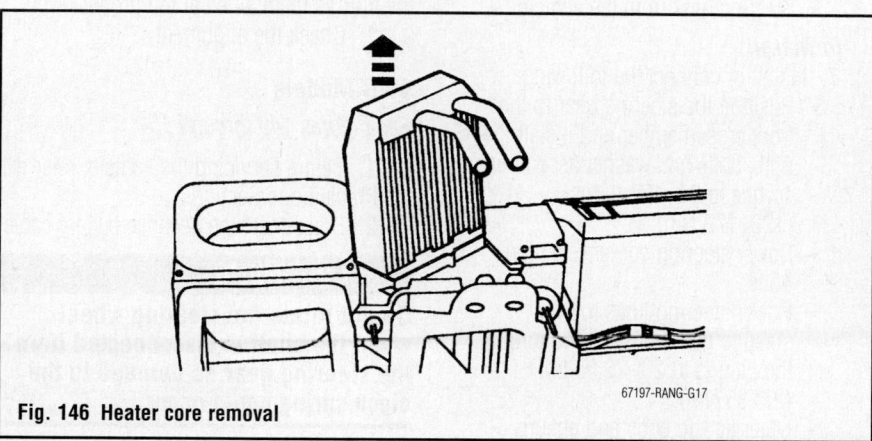

Fig. 146 Heater core removal

67197-RANG-G17

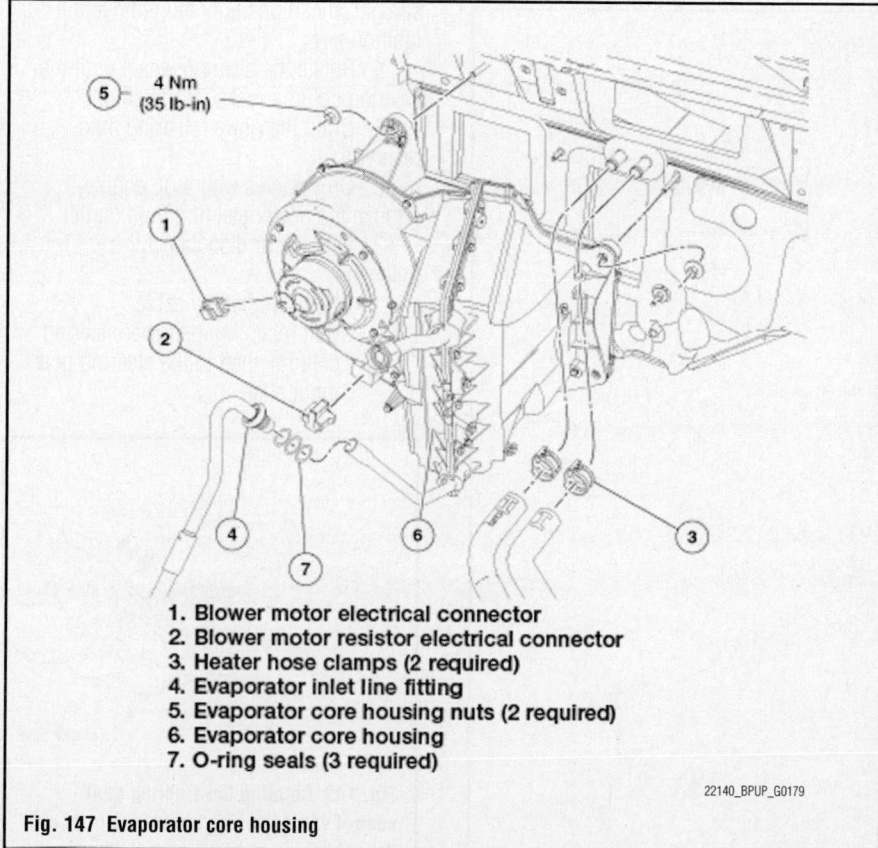

1. Blower motor electrical connector
2. Blower motor resistor electrical connector
3. Heater hose clamps (2 required)
4. Evaporator inlet line fitting
5. Evaporator core housing nuts (2 required)
6. Evaporator core housing
7. O-ring seals (3 required)

22140_BPUP_G0179

Fig. 147 Evaporator core housing

125. Install the nuts and the evaporator core housing.

➡**Remove the electrical connector locators from the splash shield prior to removal.**

126. If equipped with the 4.0L engine, install the vehicle splash guard. Install the splash shield. On the right splash shield, connect the vacuum lines from the vacuum storage reservoir.

127. Connect the vacuum hose connector and remove the nut.

➡**This step is performed at the lower passenger side dash panel, inside the passenger compartment.**

128. Connect the vacuum hose connector and remove the nut.

129. Connect the vacuum supply hose near the evaporator core housing.

130. Connect and detach the heater control valve vacuum hose.

131. Install the pin-type retainer and position the windshield washer hose into place.

132. Install the heater hose from between the water control valve and the water outlet.

133. On 4.0L engine, install the nut from the clip.

134. Install the heater hose from between the water control valve and the water pump.

135. On 2.3L engine, connect the heater hose from the water outlet tube and remove the heater hose assembly.

136. Install the heater hoses from between the heater core and the water control valve.

137. Install the heater hose from the rear coolant tube.

138. On 2.3L engine, install the heater hoses from the heater core.

139. Connect the blower motor and blower motor resistor electrical connectors.

140. Install the speed control servo, on vehicles equipped with cruise control.

141. If equipped with the 3 or 4.0L engines, position the coolant reservoir and windshield washer reservoir.

142. If equipped with the 2.3L engine, install the A/C compressor and the engine oil indicator and tube.

143. Install the suction accumulator.

144. Fill the cooling system and bleed excess air from system.

145. Charge the air conditioning system

146. Power the Supplemental Restraint System (SRS).

147. Connect the battery cable.

STEERING

POWER RACK & PINION STEERING GEAR

REMOVAL & INSTALLATION

2WD Models

See Figure 148.

1. Before servicing the vehicle, refer to the Precautions Section.
2. Disconnect the negative battery cable.
3. Place the steering wheel in the straight-ahead position and remove the ignition key.
4. Rotate the steering wheel until the steering column locks into position.

> ✳ **WARNING**
>
> **Do not rotate the steering wheel when the shaft is disconnected from the steering gear as damage to the clock spring could occur.**

5. Drain the power steering fluid reservoir.
6. Remove or disconnect the following:
   - Negative battery cable

- Bolt retaining the lower steering column shaft to the steering gear input shaft
- Stabilizer bar
- Quick-connect fittings for the power steering pressure and return hoses at the steering gear housing
- Nuts securing the power steering cooler and remove the cooler
- Outer tie rod ends
- Nuts, bolts and washer assemblies retaining the steering gear housing to the front crossmember
- Steering gear from the vehicle

To install:

7. Install or connect the following:
   - Position the steering gear to the front crossmember and install the nuts, bolts and washer assemblies. Torque to 94–127 ft. lbs. (128–172 Nm).
   - Power steering cooler retaining bolts
   - Power steering lines to the steering gear housing and torque the fittings to 20–26 ft. lbs. (27–35 Nm).
   - Outer tie rod ends and ensure

that the steering shaft or gear input shaft has not been rotated
- Intermediate shaft-to-steering input shaft retaining (pinch) bolt and torque the bolt to 41 ft. lbs. (55 Nm).
- Negative battery cable

8. Fill the power steering pump reservoir.
9. Bleed the air from the power steering system.
10. Ensure that there are no leaks and the fluid is maintained at the proper level.
11. Check the alignment.

4WD Models

See Figures 149 through 151.

1. Before servicing the vehicle, refer to the Precautions Section.
2. Disconnect the negative battery cable.

> ✳✳ **WARNING**
>
> **Do not rotate the steering wheel when the shaft is disconnected from the steering gear as damage to the clock spring could occur.**

3. Place the steering wheel in the straight-ahead position and remove the ignition key.
4. Rotate the steering wheel until the steering column locks into position.
5. Drain the power steering fluid reservoir.
6. On vehicles with 4.0L engine, remove the power steering fluid cooler.
7. Remove or disconnect the following:
   - Negative battery cable
   - Bolt retaining the lower steering column shaft to the steering gear input shaft

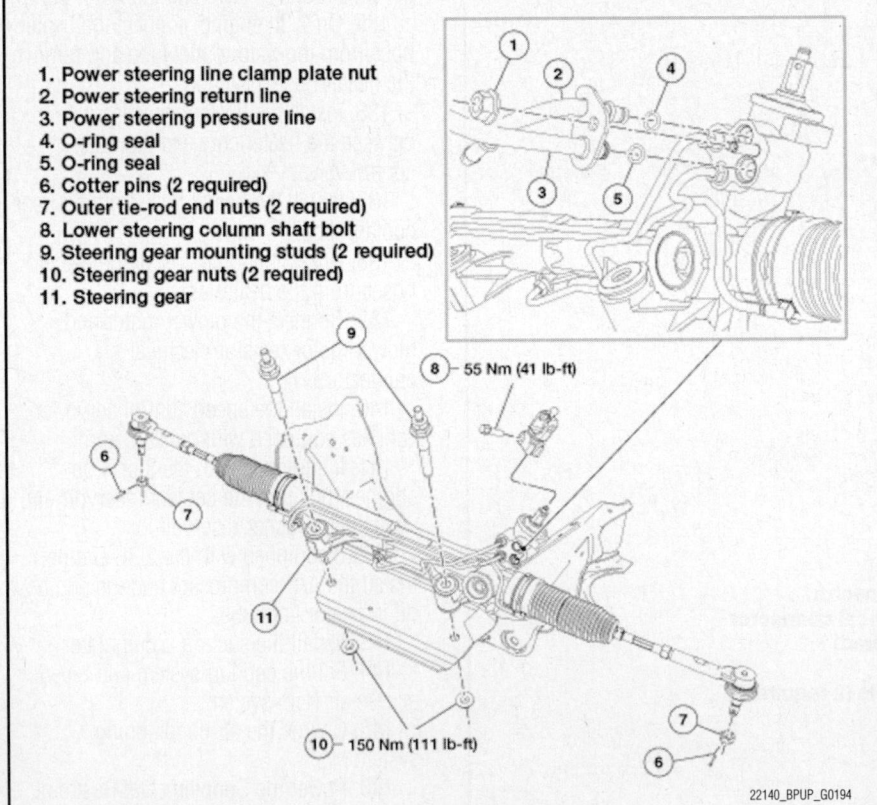

1. Power steering line clamp plate nut
2. Power steering return line
3. Power steering pressure line
4. O-ring seal
5. O-ring seal
6. Cotter pins (2 required)
7. Outer tie-rod end nuts (2 required)
8. Lower steering column shaft bolt
9. Steering gear mounting studs (2 required)
10. Steering gear nuts (2 required)
11. Steering gear

8 — 55 Nm (41 lb-ft)

10 — 150 Nm (111 lb-ft)

22140_BPUP_G0194

Fig. 148 Exploded view of the steering gear—2WD models

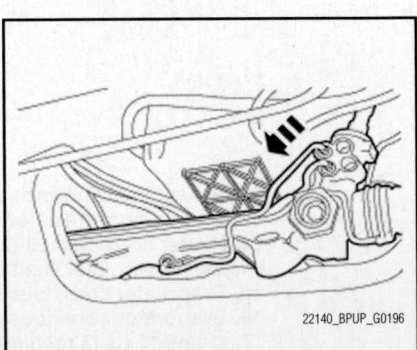

22140_BPUP_G0196

Fig. 149 Rotating the steering gear control valve housing toward the front of the vehicle.

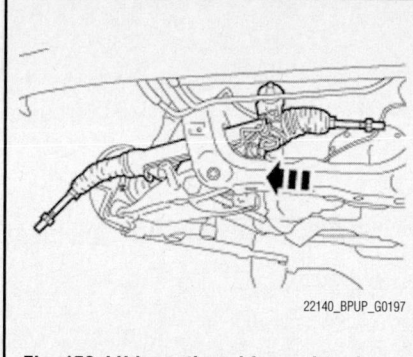

Fig. 150 LH inner tie rod forward to clear the frame crossmember

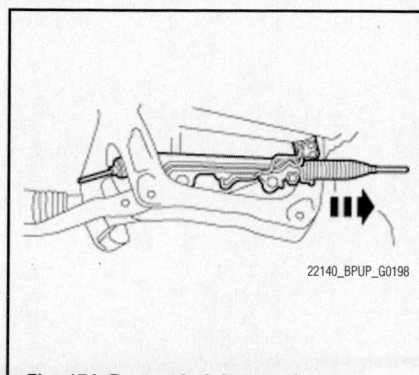

Fig. 151 Removal of the steering gear— 4WD

• Stabilizer bar

➡**Do not allow the steering column shaft to rotate while it is disconnected from the steering gear or damage to the clockspring can result. If there is evidence that the steering column shaft has rotated, the clockspring must be removed and reentered.**

8. Remove the lower steering column shaft to steering gear bolt and disconnect the shaft from the gear. Discard the bolt.

9. Remove the steering line clamp plate nut and disconnect the pressure and return lines. Discard the O-ring seals.

10. Remove or disconnect the following:
• Quick-connect fittings for the power steering pressure and return hoses at the steering gear housing
• Nuts securing the power steering cooler and remove the cooler
• Outer tie rod ends
• Nuts, bolts and washer assemblies retaining the steering gear housing to the front crossmember

11. Rotate the steering gear control valve housing toward the front of the vehicle.

12. Turn the steering gear input shaft to the right until the stop is reached.

13. Move the steering gear as far to the RH side of the vehicle as possible.

14. Remove the steering gear.

To install:

➡**Handle the steering gear with caution to avoid damage to fluid transfer tubes and to avoid dimples in the steering gear bellows boot.**

➡**Make sure the steering gear input shaft is turned to the left until the stop is reached.**

➡**Turn the steering gear input shaft to the right until the stop is reached. Note the number of turns required.**

15. Make sure the steering gear control valve housing is turned toward the front of the vehicle. Install the steering gear into the RH opening of the crossmember.

16. Move the LH inner tie rod into the opening in the crossmember and move the steering gear into position.

17. Move the LH inner tie rod into the opening in the crossmember and move the steering gear into position.

18. To place the steering gear in the straight-ahead position, turn the steering gear input shaft to the left by 1/2 the number of turns previously recorded.

19. Rotate the steering gear control valve housing toward the rear of the vehicle.

20. Install the 2 steering gear stud bolts.

21. Install or connect the following:
• Position the steering gear to the front crossmember and install the nuts, bolts and washer assemblies. Torque to 94–110 ft. lbs. (150–172 Nm).
• Power steering cooler retaining bolts
• Power steering lines to the steering gear housing and torque the fittings to 20–26 ft. lbs. (27–35 Nm)
• Outer tie rod ends and ensure that the steering shaft or gear input shaft has not been rotated
• Intermediate shaft-to-steering input shaft retaining (pinch) bolt and torque the bolt to 41 ft. lbs. (55 Nm)

22. Position the outer tie rods and install the nuts and new cotter pins.

23. Check that the brake disc shields are not bent and are not in contact with the outer tie-rod end boots.

24. Tighten the tie rod nuts to 59 ft. lbs. (80 Nm).

25. On vehicles with 4.0L engine, install the power steering fluid cooler.

26. Connect the negative battery cable.

27. Fill the power steering pump reservoir.

28. Bleed the air from the power steering system.

29. Ensure that there are no leaks and the fluid is maintained at the proper level.

30. Check the alignment.

POWER STEERING PUMP

REMOVAL & INSTALLATION

2.3L Engine

See Figure 152.

1. Before servicing the vehicle, refer to the Precautions Section.

2. Disconnect the negative battery cable.

❈❈ WARNING

Do not allow power steering fluid to contact the accessory drive belt or the belt may be damaged. Using a suitable suction device, remove the power steering fluid from the reservoir.

3. Remove the power steering pump pulley.

4. Compress the clamp and disconnect the supply hose.

5. Disconnect the pressure line-to-pump fitting.

6. Remove and discard the Teflon seal.

7. Remove the 4 power steering pump bolts and the power steering pump.

❈❈ WARNING

A new Teflon seal must be installed any time the pressure line fitting is disconnected from the power steering pump or a fluid leak may occur.

8. To install, reverse the removal procedure.

9. Tighten the pressure line-to-pump fitting to 48 ft. lbs. (65 Nm).

10. Tighten the 4 power steering pump bolts and the power steering pump to 17 ft. lbs. (23 Nm).

❈❈ WARNING

A new Teflon seal must be installed any time the pressure line fitting is disconnected from the power steering pump or a fluid leak may occur.

11. Fill the power steering system.

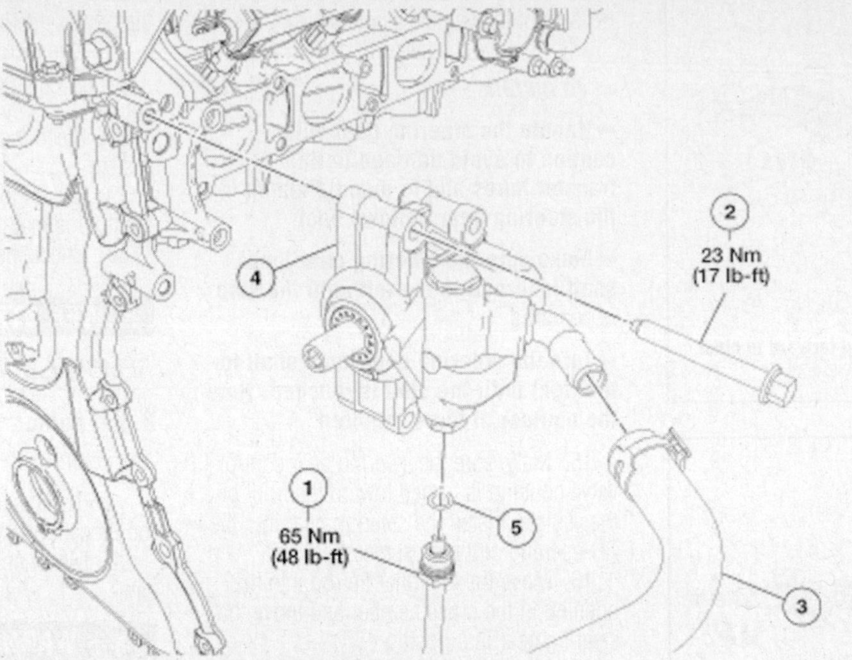

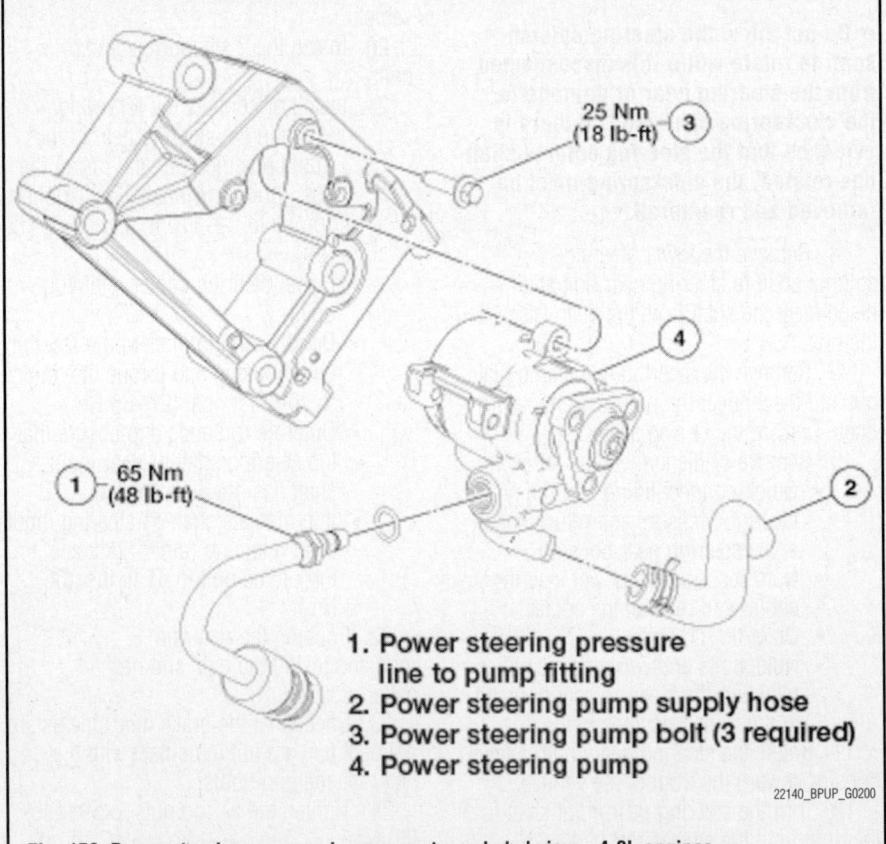

1. Pressure line-to-pump fitting
2. Power steering pump bolts (4 required)
3. Power steering supply hose
4. Power steering pump
5. Teflon® seal

22140_BPUP_G0199

Fig. 152 Power steering pump and components exploded view—2.3L engine

4.0L Engine

See Figure 153.

> ❋❋ **WARNING**
>
> **Do not allow power steering fluid to contact the accessory drive belt or the belt may be damaged.**

1. Using a suitable suction device, remove the power steering fluid from the reservoir.

2. Rotate the accessory drive belt tensioner counterclockwise and remove the accessory drive belt.

3. Remove the engine cooling fan.

4. Remove the power steering pump pulley.

5. Compress the clamp and disconnect the power steering pump supply hose from the power steering fluid reservoir.

6. Disconnect the power steering pressure line to pump fitting.

7. Compress the clamp and disconnect the power steering pump supply hose from the power steering pump.

8. Remove the 3 bolts and the power steering pump.

1. Power steering pressure line to pump fitting
2. Power steering pump supply hose
3. Power steering pump bolt (3 required)
4. Power steering pump

22140_BPUP_G0200

Fig. 153 Power steering pump and components exploded view—4.0L engines

A new Teflon O-ring seal must be installed any time the power steering pressure line is disconnected from the power steering pump or a fluid leak may occur.

9. To install, reverse removal procedure

10. Install a new Teflon O-ring seal on the power steering pressure line

11. Tighten the pressure line-to-pump fitting to 48 ft. lbs. (65 Nm).

12. Tighten the 3 power steering pump bolts and the power steering pump to18 ft. lbs. (25 Nm).

A new Teflon seal must be installed any time the pressure line fitting is disconnected from the power steering pump or a fluid leak may occur.

13. Fill the power steering system.

SUSPENSION

FRONT SUSPENSION

COIL SPRING

REMOVAL & INSTALLATION

See Figures 154 and 155.

Do not apply heat or flame to the shock absorber or strut tube. The shock absorber and strut tube are gas pressurized and could explode if heated. Failure to follow this instruction may result in serious personal injury.

Keep all body parts clear of shock absorbers or strut rods. Shock absorbers or struts can extend unassisted. Failure to follow this instruction may result in serious personal injury.

Suspension fasteners are critical parts because they affect performance of vital components and systems and their failure may result in major service expense. New parts must be installed with the same part number or an equivalent part if replacement is necessary. Do not use a replacement part of lesser quality or substitute design. Torque values must be used as specified during reassembly to make sure of correct retention of these parts.

1. Mark the lower arm to the frame for reassembly.
2. Before servicing the vehicle, refer to the precautions in the beginning of this section.
3. Remove or disconnect the following:
   - Wheel and tire assembly
   - Shock absorber
   - Front stabilizer bar link nut
4. Use a coil spring compressor to compress the coil spring.
5. Using a suitable jack, support the lower control arm.
6. Remove the cotter pin and castellated nut.
7. Separate the lower ball joint from the front wheel spindle.
8. Position the front wheel spindle out of the way and remove the coil spring.

To install:

➡The end of the coil spring must cover the first hole and should not be visible in the second hole.

9. Install the coil spring in the lower arm.

Always install the cotter pin into the lower ball joint castellated nut from outboard to inboard. Failure to do so will result in damage to the wheel and tire assembly.

10. Install the lower ball joint.
11. Tighten the lower ball joint to 98 ft. lbs. (133 Nm).

12. Install a new cotter pin.
13. Install the front stabilizer bar link nut.
14. Install a new stabilizer bar link nut and grommet.
15. Tighten the new stabilizer bar link nut to 26 ft. lbs. (35 Nm).
16. Remove the coil spring compressor.
17. Install the front shock absorber and the two lower nuts.
18. Tighten the front shock absorber lower nuts to 18 ft. lbs. (25 Nm).
19. Install the upper shock absorber bushing and nut/washer assembly.
20. Tighten the front shock absorber upper nuts to 35 ft. lbs. (35 Nm).
21. Position a suitable jack under the lower arm and raise the suspension until the index marks on the lower arm and the frame are aligned (curb height).
22. Tighten the lower arm to 148 ft. lbs. (200 Nm).
23. Remove the jack from under the lower control arm.
24. Install the wheel and tire assembly.

LOWER BALL JOINT

REMOVAL & INSTALLATION

See Figures 156 and 157.

1. Before servicing the vehicle, refer to the Precautions Section.
2. Raise and support the vehicle safely.
3. Remove the tire and wheel assembly.
4. On 2WD vehicles, remove the front spindle. On 4WD vehicles, remove the front knuckle.
5. Remove and discard the lower ball joint snap ring.
6. Using a ball joint removal too, remove the ball joint from its mounting.

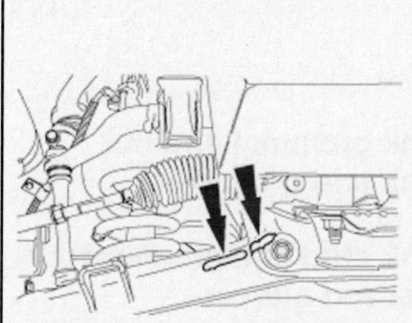

22140_BPUP_G0205

Fig. 154 Index marking on the lower arm to the frame with the vehicle in a static ground position (curb height).

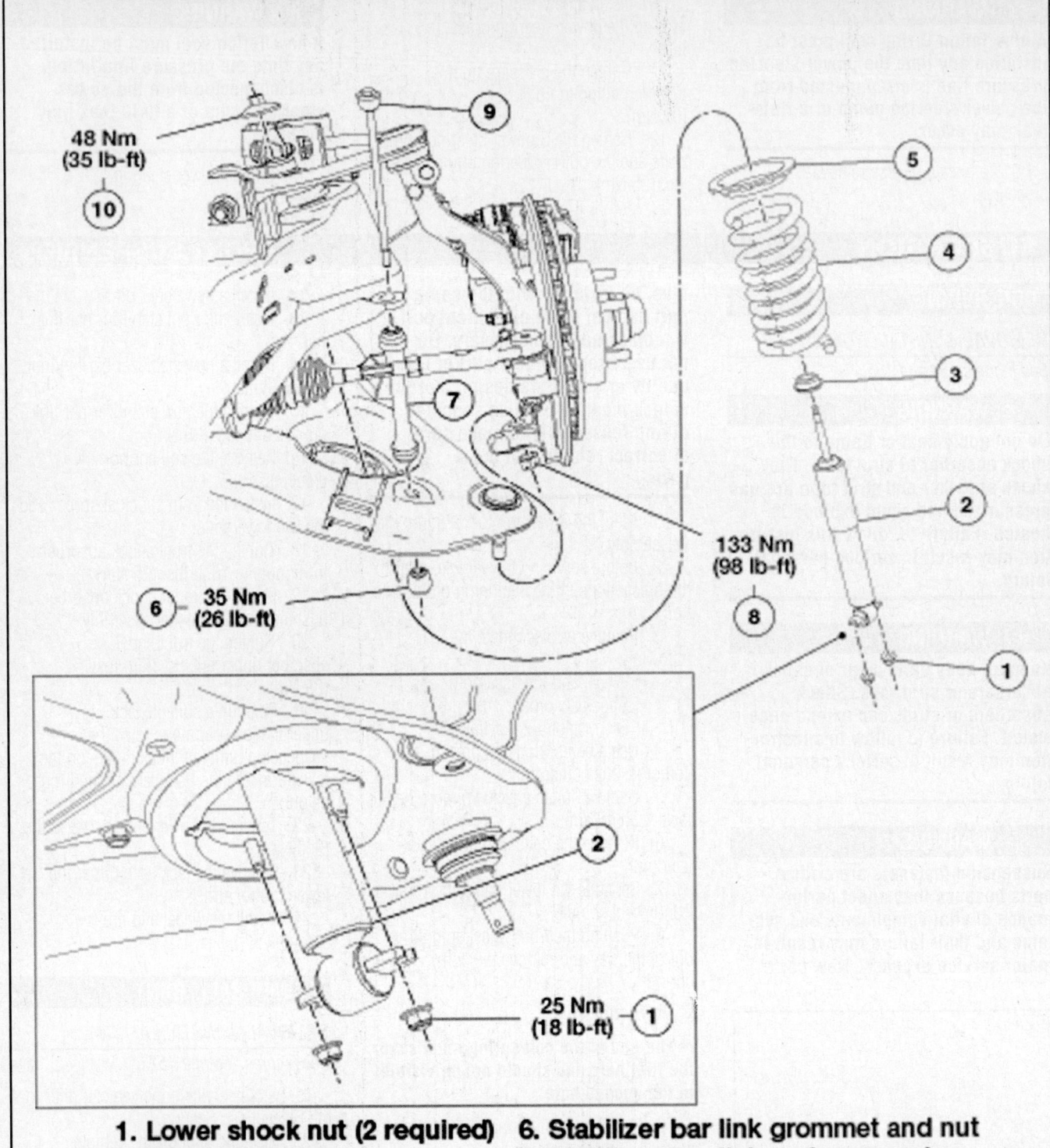

48 Nm
(35 lb-ft)

10

9

5

4

3

2

7

133 Nm
(98 lb-ft)

8

1

6 · 35 Nm
(26 lb-ft)

2

25 Nm
(18 lb-ft) 1

1. Lower shock nut (2 required)
2. Shock absorber
3. Shock absorber insulator
4. Coil spring
5. Coil spring insulator
6. Stabilizer bar link grommet and nut
7. Stabilizer bar link and grommet
8. Lower ball joint nut
9. Stabilizer bar link stud and grommet
10. Shock absorber rod nut

22140_BPUP_G0204

Fig. 155 Front coil spring and components exploded view

To install:

✳✳ WARNING

Do not damage the ball joint boot when installing the special tool.

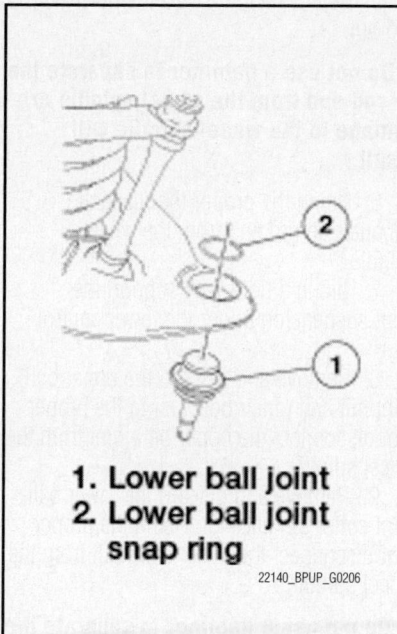

1. **Lower ball joint**
2. **Lower ball joint snap ring**

22140_BPUP_G0206

Fig. 156 Lower ball joint—2WD models

➡**Clean and inspect the control arm ball joint bore for damage before installing a new ball joint.**

➡**Make sure the new ball joint snap ring is fully seated.**

7. Continue the installation in the reverse order of the removal procedure. Always use new bolts and nuts, as required.

8. Check and adjust the front end, as required.

LOWER CONTROL ARM

REMOVAL & INSTALLATION

2WD Models

1. Before servicing the vehicle, refer to the Precautions Section.

2. Raise and support the vehicle safely.

3. Remove the tire and wheel assembly.

4. Remove the coil spring.

5. Remove the lower control arm retaining bolts. Remove the lower control arm from the vehicle.

To install:

6. Installation is the reverse of the removal procedure.

7. Be sure to use new bolts and nuts, as required.

8. Torque the left side lower control arm rearward bolt to 148 ft. lbs. (200 Nm).

9. Torque the right side lower control arm rearward bolt to 129 ft. lbs. (174 Nm).

10. Torque the lower control arm forward bolts to 148 ft. lbs. (200 Nm).

11. Check and adjust the front end alignment, as required.

4WD Models

1. Before servicing the vehicle, refer to the Precautions Section.

2. Raise and support the vehicle safely.

3. Remove the tire and wheel assembly.

4. Remove the torsion bar.

5. Remove and discard the stabilizer bar link nut and grommet.

6. Remove the stabilizer link stud and grommet and the stabilizer link assembly.

7. Remove and discard the two lower shock absorber retaining nuts.

8. Remove and discard the lower ball joint cotter pin and nut.

➡**Do not use a hammer to separate the ball joint from the knuckle or damage to the knuckle will result.**

9. Using the proper tool, separate the ball joint from the knuckle.

10. Remove the lower control arm retaining bolts. Remove the lower control arm from the vehicle.

To install:

11. Installation is the reverse of the removal procedure.

12. Be sure to use new bolts and nuts, as required.

✳✳ WARNING

Always install the cotter pin into the lower ball joint castellated nut from outboard to inboard. Failure to do so will result in damage to the wheel and tire assembly.

13. Torque the lower control arm bolts to 148 ft. lbs. (200 Nm).

14. Check and adjust the front end alignment, as required.

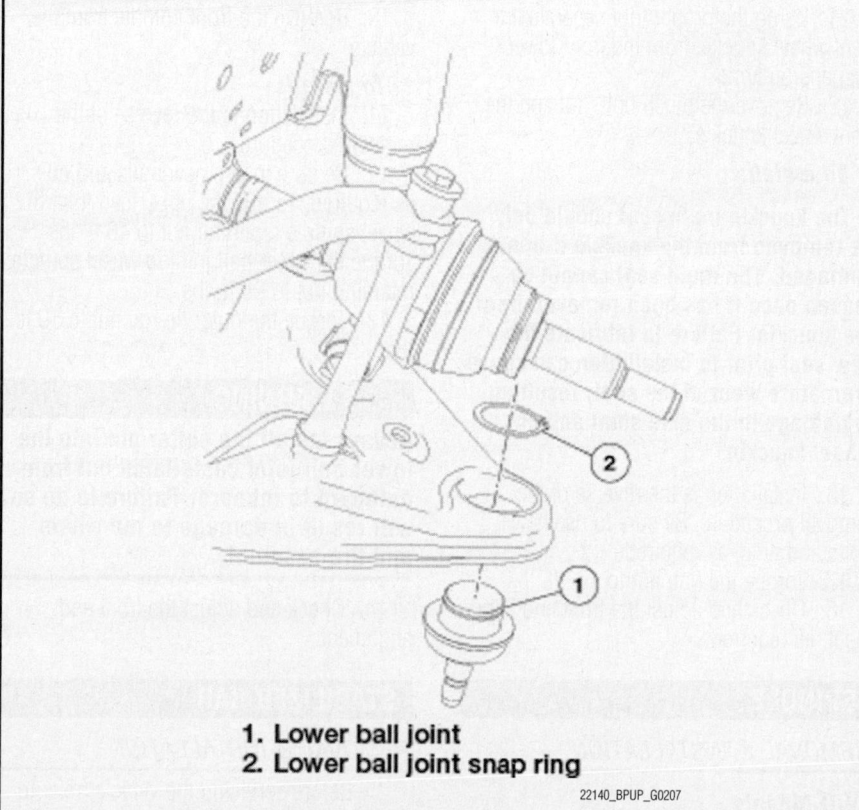

1. **Lower ball joint**
2. **Lower ball joint snap ring**

22140_BPUP_G0207

Fig. 157 Lower ball joint—4WD models

SHOCK ABSORBERS

REMOVAL & INSTALLATION

➡ **Low pressure gas shocks are charged with Nitrogen gas. Do not attempt to open, puncture or apply heat to them. Prior to installing a new shock absorber, hold it upright and extend it fully. Invert it and fully compress and extend it at least 3 times. This will bleed trapped air.**

1. Before servicing the vehicle, refer to the precautions in the beginning of this section.
2. Remove or disconnect the following:
   • Negative battery cable
   • Upper shock-to-frame attaching nut, washer and insulator assembly
   • Lower shock-to-control arm attaching nuts
   • Slightly compress the shock absorber by hand and remove it from the vehicle

To install:

3. Install or connect the following:
   • Position the lower washer and insulator on the shock absorber rod and position the shock absorber to the upper frame bracket mount
   • Position the upper insulator and washer on the shock absorber rod and install the attaching nut loosely.
   • Position the lower shock absorber mounting studs into the control arm and install the attaching nuts loosely.
   • Torque the lower shock attaching nuts to 15–21 ft. lbs. (21–29 Nm), and the upper shock attaching bolts to 30–40 ft. lbs. (40–55 Nm).
   • Negative battery cable

STEERING KNUCKLE

REMOVAL & INSTALLATION

2WD Models

1. Before servicing the vehicle, refer to the Precautions Section.
2. Raise and support the vehicle safely.
3. Remove the tire and wheel assembly.
4. Remove the front disc brake caliper and wire it to the side out of the way.

➡ **Do not allow the caliper to hang by the brake line. Damage may result.**

5. Remove the rotor.
6. Remove the hub nut and washer.

➡ **Do not use a hammer to separate the outer CV joint from the hub. Damage to the outboard CV threads and to internal components may result.**

7. Use special tool 205-D070, or equivalent and separate the outer CV joint from the hub.
8. Remove the brake dust shield. Remove the bolt and detach the wheel speed sensor from the wheel hub.

➡ **Do not overextend the CV joint and boots when removing the hub and bearing assembly.**

9. Remove the three retaining bolts. Remove the wheel and hub.
10. Remove the torsion bar.

➡ **Secure the front axle shaft to prevent it from overextending. Failure to do so can cause damage to the front axle shaft. Suspend the front axle shaft with wire.**

11. Remove the tie rod end cotter pin and nut.
12. Using the proper tool separate the tie rod end from the front wheel knuckle.
13. Remove the lower ball joint cotter pin and nut.
14. Using the proper tool separate the front wheel knuckle from the front lower suspension arm.
15. Remove the pinch bolt, nut and the front wheel knuckle.

To install:

➡ **The knuckle main seal should only be removed from the knuckle if it is damaged. The main seal cannot be reused once it has been removed from the knuckle. Failure to lubricate the new seal prior to installation can cause premature wear of the seal, resulting in damage to the axle shaft and or wheel knuckle.**

16. Installation is the reverse of the removal procedure. Be sure to use new bolts and nuts, as required.
17. Torque the hub nut to 162 ft. lbs.
18. Check and adjust the front end alignment, as required.

SPINDLE

REMOVAL & INSTALLATION

2WD Models

1. Before servicing the vehicle, refer to the Precautions Section.

2. Raise and support the vehicle safely.
3. Remove the tire and wheel assembly.
4. Remove the brake disc shield. Remove and discard the wheel speed sensor harness bracket bolt.
5. Remove and discard the outer tie rod end nut.

➡ **Do not use a hammer to separate the tie rod end from the wheel spindle or damage to the wheel spindle will result.**

6. Using the proper tool separate the outer tie rod end from the wheel spindle.
7. Using a floor jack support the front suspension under the lower control arm.
8. Remove and discard the upper ball joint nut and pinch bolt. Using the proper tool disconnect the upper ball joint from the wheel spindle.
9. Remove and discard the lower ball joint cotter pin and nut. Using the proper tool disconnect the lower ball joint from the wheel spindle.

➡ **Do not use a hammer to separate the ball joint from the wheel spindle or damage to the wheel spindle will result.**

10. Remove the front spindle from the vehicle.

To install:

11. Installation is the reverse of the removal procedure.
12. Be sure to use new bolts and nuts, as required. Torque the upper ball joint to wheel spindle retaining nut to 46 ft. lbs. Torque the lower ball joint to wheel spindle retaining nut to 98 ft. lbs.
13. Torque the outer tie rod nut to 59 ft. lbs.

※※ WARNING

Always install the cotter pin into the lower ball joint castellated nut from outboard to inboard. Failure to do so will result in damage to the wheel and tire assembly.

14. Check and adjust the front end alignment.

STABILIZER BAR

REMOVAL & INSTALLATION

1. Before servicing the vehicle, refer to the Precautions Section.
2. Raise and support the vehicle safely.

3. Remove the wheel and tire assembly.

4. Remove the front stabilizer bar link nuts from the front suspension lower arms. Discard the nuts.

5. Remove the front stabilizer bar link studs and the front stabilizer bar links.

6. Remove the four bolts and two brackets. Discard the bolts.

7. Remove the front stabilizer bar.

8. Remove the stabilizer bar insulator.

To install:

9. Installation is the reverse of removal. Be sure to use new bolts and nuts. Observe the following torques:
- End link nuts: 18 ft. lbs. (25 Nm)
- Bracket bolts: 30 ft. lbs. (40 Nm)

➡**In the event the self-tapping bolts cannot be installed in the frame, there is a kit available with flag nuts.**

TORSION BAR

REMOVAL & INSTALLATION

❊❊ WARNING

The electrical power to the air suspension system must be shut off prior to hoisting, jacking or towing an air suspension vehicle. This can be accomplished by turning off the air suspension switch located in the rear jack storage area. Failure to do so can result in unexpected inflation or deflation of the air springs or shocks, which can result in shifting of the vehicle during these operations.

1. Before servicing the vehicle, refer to the precautions in the beginning of this section.

2. Remove or disconnect the following:

3. Remove the torsion bar cover plate

➡**Before relieving the torsion bar tension, measure and record the measurement of the torsion bar adjustment bolt. This measurement will be used as the preset depth for the new torsion bar adjustment bolt during installation.**

4. Relieve the torsion bar tension.

5. position the torsion bar tool and adapters.

6. Tighten the torsion bar tool until the torsion bar adjuster lifts off the adjustment bolt.

❊❊ WARNING

The torsion bar adjustment bolt is coated with dry adhesive; and must

be replaced if it is backed off or removed. Failure to do so can cause the adjustment bolt to loosen during operation and cause a loss of vehicle alignment.

7. Remove the torsion bar adjustment bolt and nut.

8. Loosen the torsion bar tool until the tension is removed from the torsion bar.

9. Mark the torsion bar and the adjuster for proper installation.

10. Remove the torsion bar insulator.

11. Grasp the torsion bar, and pull it free from the front suspension lower arm.

To install:

12. Position the torsion bar and the torsion bar adjuster.

13. Align the marks on the torsion bar and the torsion bar adjuster, then install the torsion bar adjuster.

14. Position the torsion bar insulator.

15. Install the torsion bar tool and the adapters.

16. Tighten the torsion bar tool until the new adjustment bolt and nut can be installed.

17. Turn the adjustment bolt until the preliminary adjustment measurement (recorded length of the old adjustment bolt) is reached.

18. Install the torsion bar cover plate. Torque the bolts to 46 ft. lbs. (63 Nm).

19. If equipped with air suspension, reactivate the system by turning on the air suspension switch.

20. Lower the vehicle.

21. Adjust the ride height.

22. Check the alignment.

UPPER BALL JOINT

REMOVAL & INSTALLATION

The ball joints are integral with the control arm. If the ball joint is defective, the entire control arm must be replaced.

UPPER CONTROL ARM

REMOVAL & INSTALLATION

1. Before servicing the vehicle, refer to the Precautions Section.

2. Raise and support the vehicle safely.

3. Remove the tire and wheel assembly.

4. Using a suitable jack, support the lower control arm assembly.

➡**To avoid possible damage to the wheel spindle (2WD) or knuckle (4WD) secure the assembly to keep it from tilting before removing the pinch bolt and nut.**

5. Remove and discard the upper ball joint nut and pinch bolt. Separate the ball joint from the wheel spindle (2WD) or the knuckle (4WD).

6. Remove and discard the two upper control arm nuts.

7. Remove the upper control arm from the vehicle.

To install:

8. Installation is the reverse of the removal procedure. Be sure to use new nuts and bolts, as required.

9. The upper control arm retaining nuts must be fully tightened at curb height. Torque them to 98 ft. lbs.

10. Make sure that the shims are installed in their original locations.

11. Torque the upper ball joint nut and pinch bolt to 46 ft. lbs.

12. Check and adjust the front end alignment, as required.

WHEEL BEARINGS

REMOVAL & INSTALLATION

2WD Models

1. Before servicing the vehicle, refer to the precautions in the beginning of this section.

2. Remove or disconnect the following:
- Disc brake caliper anchor plate
- Hub grease cap
- Cotter pin
- Nut retainer
- Spindle nut
- Wheel outer bearing retainer washer
- Outer front wheel bearing
- Brake disc and hub
- Hub grease seal
- Inner wheel bearing

To install:

3. Thoroughly clean and inspect the front wheel bearings and the brake disc and hub.

4. Lubricate the front wheel bearings.

5. Install the inner front wheel bearing.

6. Install a new wheel hub grease seal.

7. Position the brake disc and hub.

8. Assemble all parts and adjust the bearings.

9. Tighten lug nuts to 100 ft. lbs. (135 Nm).

✳✳ CAUTION

Before driving the vehicle, pump the brake pedal several times to restore normal brake pedal travel.

4WD Models

1. Before servicing the vehicle, refer to the precautions in the beginning of this section.

2. Remove or disconnect the following:
- Negative battery cable
- Wheel assembly
- Disc brake caliper anchor plate
- Hub grease cap
- Cotter pin
- Nut retainer
- Spindle nut
- Wheel bearing
- Pull the hub and rotor assembly from spindle.
- Grease seal—discard seal

To install:

3. Use a bearing packer tool and properly repack the wheel bearings with the proper grade and type of grease. If a bearing packer is not available, work as much of the grease as possible between the rollers and cages. Also, grease the cone surfaces.

4. Install or connect the following:
- Inner bearing cone and roller assembly in the inner cup. A light film of grease should be included between the lips of the new grease seal.
- Grease seal by driving in place with Hub Seal Replacer tool T83T-1175-B and Driver Handle T80T-4000-W
- Hub and rotor assembly onto the spindle. Keep the hub centered on the spindle to prevent damage to the spindle and the retainer
- Rotor onto the spindle
- Install retainer, new cotter pin and grease cap.
- Axle shaft spacer
- Install caliper.
- Wheel assembly
- Negative battery cable

5. Tighten lug nuts to 100 ft. lbs. (135 Nm).

✳✳ CAUTION

Before driving the vehicle, pump the brake pedal several times to restore normal brake pedal travel.

ADJUSTMENT

2WD Models

See Figure 158.

1. Before servicing the vehicle, refer to the precautions in the beginning of this section.

2. Remove the grease cap from the hub and wipe the excess grease from the end of the spindle. Remove the cotter pin and retainer. Discard the cotter pin.

3. Loosen the adjusting nut 3 turns.

✳✳ WARNING

Obtain running clearance between the disc brake rotor surface and shoe linings by rocking the entire wheel assembly in and out several times in order to push the caliper and brake pads away from the rotor. An alternate method to obtain proper running clearance is to tap lightly on the caliper housing. Be sure not to tap on any other area that may damage the disc brake rotor or the brake lining surfaces. Do not pry on the phenolic caliper piston. The running clearance must be maintained throughout the adjustment procedure. If proper clearance cannot be maintained, the caliper must be removed from its mounting.

4. While rotating the wheel assembly, tighten the adjusting nut to 17–25 ft. lbs. (23–34 Nm) in order to seat the bearings. Loosen the adjusting nut a half turn. Retighten the adjusting nut 18–20 inch lbs. (2.0–2.2 Nm).

5. Place the retainer on the adjusting nut. The castellations on the retainer must be in alignment with the cotter pin holes in the spindle. Once this is accomplished, install a new cotter pin and bend the ends to insure its being locked in place.

6. Check for proper wheel rotation. If correct, install the grease cap.

7. Lower the vehicle and tighten the lug nuts to 100 ft. lbs., (136 Nm) if the wheel was removed. Before driving the vehicle, pump the brake pedal several times to restore normal brake pedal travel.

✳✳ CAUTION

If the wheel was removed, retighten the wheel lug nuts to specification after about 500 miles (804km) of driving. Failure to do this could result in the wheel coming off while the vehicle is in motion causing loss of vehicle control or collision.

4WD Models

1. If a packer is not available, work as much grease as possible between the rollers and cages. Grease the cone surfaces. Pack the bearing cone and roller assemblies with appropriate grease.

2. Place the bearing cone and roller assembly in the inner cup. A light film of grease should be included between the lips of the new grease seal. Additional grease may be added to the new grease seal if required.

3. When installing seal, be sure it is properly seated.

4. Install the seal.

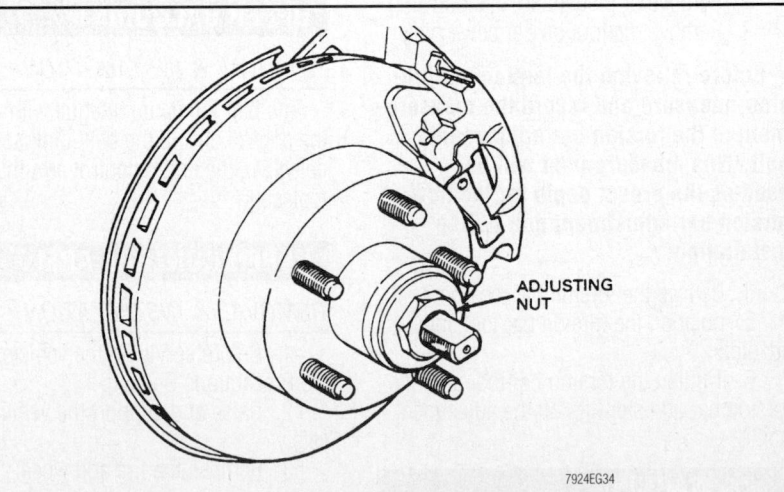

7924EG34

Fig. 158 Loosen the adjusting nut 3 turns, then rock the entire wheel assembly in-and-out to spread the brake pads before attempting to adjust the bearing—2WD vehicles

ADJUSTING NUT

SUSPENSION

REAR SUSPENSION

COIL SPRING

REMOVAL & INSTALLATION

See Figures 156, 159 and 160.

✳✳ WARNING

Suspension fasteners are critical parts because they affect performance of vital components and systems and their failure may result in major service expense. New parts must be installed with the same part numbers or equivalent parts if replacement is necessary. Do not use replacement parts of lesser quality or substitute design. Torque values must be used as specified during reassembly to make sure of correct retention of these parts.

1. With the vehicle in neutral, position it on a hoist.
2. Using a suitable jack, support the rear axle.
3. Remove and discard the lower shock absorber bolt and nut.

➡**When installing new U-bolts, the distance between the U-bolts should be 93 mm (3.66 in) ± 3 mm (0.12 in). This is important for providing the correct U-bolt clamp load and retention of parts. Remove and discard the 4 rear U-bolt nuts.**

4. Remove the U-bolts and the spring plate.
5. Carefully lower the rear axle and remove the rear spring spacer, if equipped.
6. Remove and discard the spring-to-frame bracket bolt and flag nut.
7. To install, tighten to 148 ft. lbs. (200 Nm).
8. Remove and discard the spring shackle-to-spring bolt and nut and remove the spring.
9. If necessary, remove the spring shackle.
10. Remove and discard the spring shackle-to-frame bracket bolt and flag nut.
11. Remove the shackle

To install:

12. To install, reverse the removal procedure and note the following.
13. Torque the rear shackle to 85 ft. lbs. (115 Nm).
14. Torque the spring-to-frame bracket bolt and flag nut rear shackle to 76 ft. lbs. (103 Nm).
15. To install, tighten Torsion axle U-bolt nuts in 4 stages.
    a. Stage 1: Tighten in a cross pattern to 18 ft. lbs. (25 Nm).
    b. Stage 2: Tighten in a cross pattern to 37 ft. lbs. (50 Nm).
    c. Stage 3: Tighten in a cross pattern to 55 ft. lbs. (75 Nm).
    d. Stage 4: Tighten in a cross pattern to 76 ft. lbs. (103 Nm).
16. To install, tighten standard axle U-bolt nuts in 4 stages:
    a. Stage 1: Tighten in a cross pattern to 22 ft. lbs. (30 Nm).
    b. Stage 2: Tighten in a cross pattern to 44 ft. lbs. (60 Nm).
    c. Stage 3: Tighten in a cross pattern to 66 ft. lbs. (90 Nm).
    d. Stage 4: Tighten in a cross pattern to 85 ft. lbs. (115 Nm).

LEAF SPRING

REMOVAL & INSTALLATION

1. Before servicing the vehicle, refer to the precautions in the beginning of this section.
2. Remove or disconnect the following:
   - Negative battery cable
   - Rear wheels
   - U-bolts from the rear spring plate
   - Hardware from the spring to bracket at the front of the rear spring
   - Upper and lower shackle bolts at the rear of the spring
   - Spring and shackle from the bracket

To install:

3. Install or connect the following:
   - Spring and shackle to the bracket
   - Upper and lower shackle bolts at the rear of the spring. Torque the nuts to 87 ft. lbs. (118 Nm).
   - U-bolts to the spring plate. Torque the nuts 87 ft. lbs. (113 Nm).
   - Rear wheels
   - Negative battery cable

SHOCK ABSORBER

REMOVAL & INSTALLATION

➡**Low pressure gas shocks are charged with nitrogen gas. Do not attempt to open, puncture or apply heat to them. Prior to installing a new shock absorber, hold it upright and extend it fully. Invert it and fully compress and extend it at least 3 times. This will bleed trapped air.**

1. Before servicing the vehicle, refer to the Precautions Section.
2. Raise and support the vehicle safely. Remove the tire and wheel assembly.
3. Remove or disconnect the following:
   - Upper shock-to-frame attaching nut
   - Lower shock nut
   - Slightly compress the shock absorber by hand and remove it from the vehicle

To install:

4. Install or connect the following:
   - Shock absorber upper end and nut
   - Shock absorber lower end and nut
   - Torque the upper shock attaching nuts to 46 ft. lbs. (63 Nm)
   - Torque the lower shock attaching nuts to 59 ft. lbs. (80 Nm)

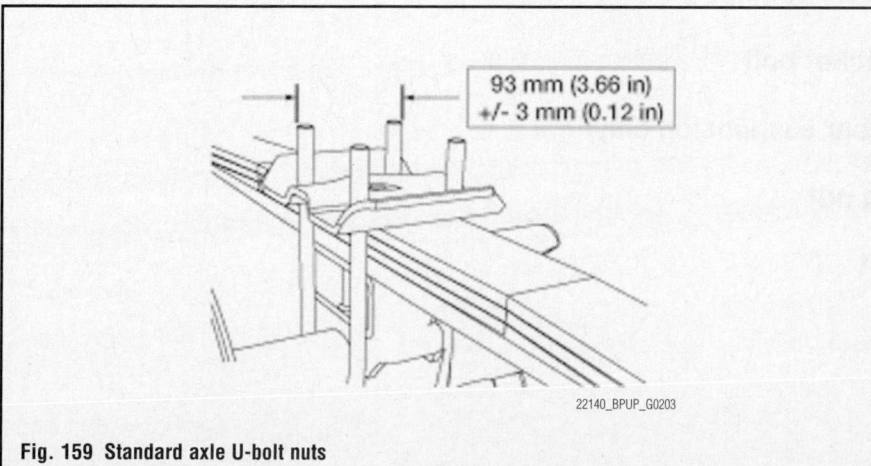

93 mm (3.66 in)
+/- 3 mm (0.12 in)

22140_BPUP_G0203

Fig. 159 Standard axle U-bolt nuts

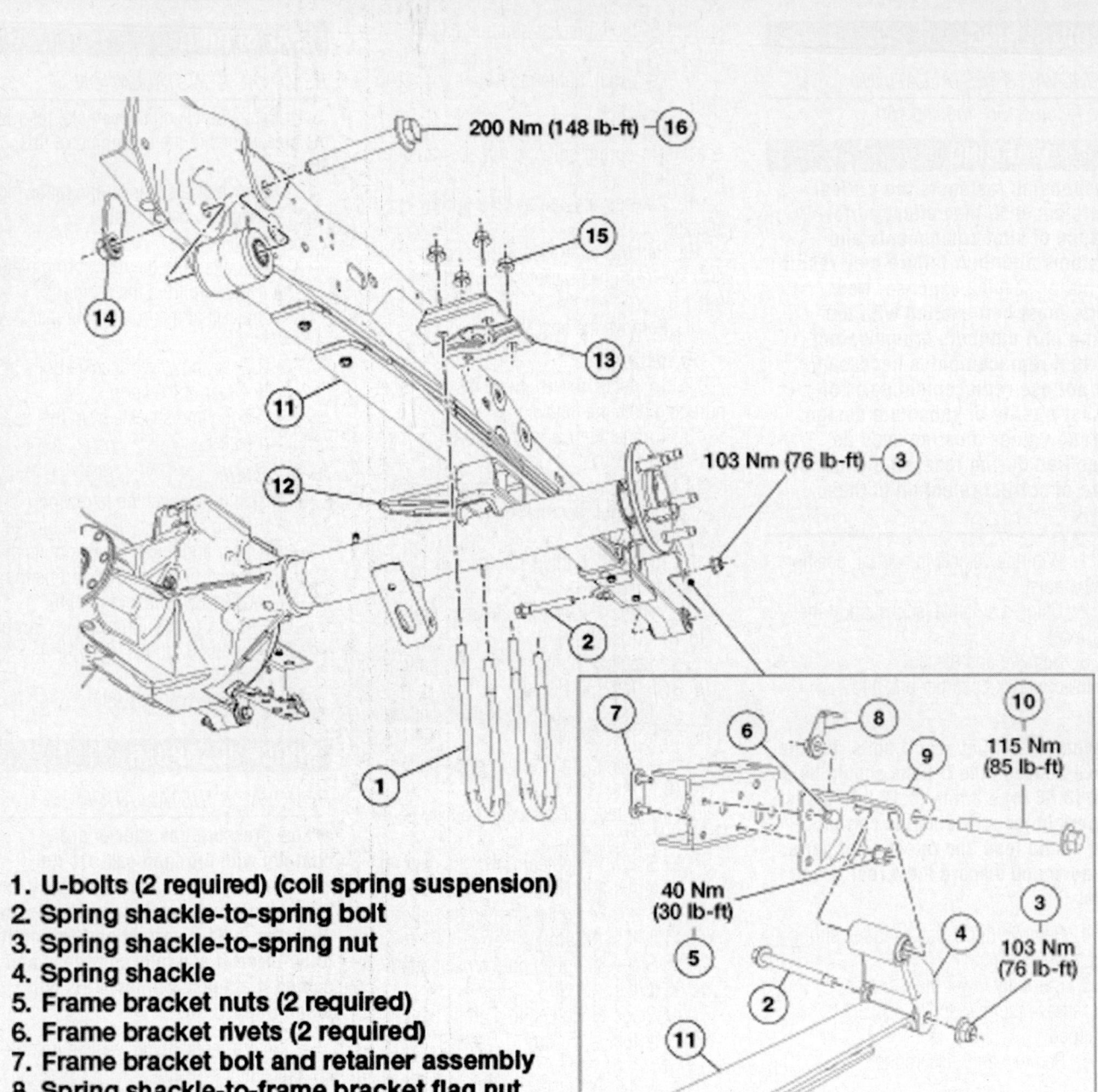

1. U-bolts (2 required) (coil spring suspension)
2. Spring shackle-to-spring bolt
3. Spring shackle-to-spring nut
4. Spring shackle
5. Frame bracket nuts (2 required)
6. Frame bracket rivets (2 required)
7. Frame bracket bolt and retainer assembly
8. Spring shackle-to-frame bracket flag nut
9. Frame bracket
10. Spring shackle-to-frame bracket bolt
11. Leaf spring
12. Rear spring spacer (torsion bar suspension only)
13. Spring plate
14. Spring-to-frame bracket flag nut
15. U-bolt nuts (4 required)
16. Spring-to-frame bracket bolt

22140_BPUP_G0202

Fig. 160 Rear suspension springs exploded view

STABILIZER BAR

REMOVAL & INSTALLATION

See Figure 161.

Suspension fasteners are critical parts because they affect performance of vital components and systems and their failure can result in major service expense. A new part with the same part number or an equivalent part must be installed, if installation is necessary. Do not use a part of lesser quality or substitute design. Torque values must be used as specified during reassembly to make sure of correct retention of these parts.

1. Before servicing the vehicle, refer to the Precautions Section.
2. With the vehicle in **NEUTRAL**, position it on a hoist.
3. Remove the wheel and tire assembly.

4. Remove the rear stabilizer bar link nuts and bolt. Discard the bolts and nuts.
5. Remove the rear stabilizer bar link.
6. Remove the rear stabilizer bar mounting bracket bolts and remove the brackets. Discard the bolts and nuts.
7. Remove the 4 stabilizer bar bracket bolts and then remove the stabilizer bar, bushings and brackets.

To install:
8. Installation is the reverse of the removal procedure.
9. Observe the following torques:
   - Link-to-stabilizer bar bolts and nuts:52 ft. lbs. (70 Nm)
   - Bar link-to-frame bolts and nuts: 52 ft. lbs. (70 Nm)
   - 4 stabilizer bar bracket bolts: 30 ft. lbs. (40 Nm)

WHEEL BEARINGS

REMOVAL & INSTALLATION

1. Before servicing the vehicle, refer to the Precautions Section.
2. Remove the axle shaft.

➡**If only a new seal needs to be installed, use care to avoid damaging the seal bore.**

3. Using a suitable seal remover, remove the axle shaft oil seal. Discard the oil seal.
4. Inspect the rear wheel bearing and axle shaft for wear or damage.
5. If necessary, using the impact slide hammer remove the rear wheel bearing.
   To install:
6. Lubricate the new rear wheel bearing with lubricant.

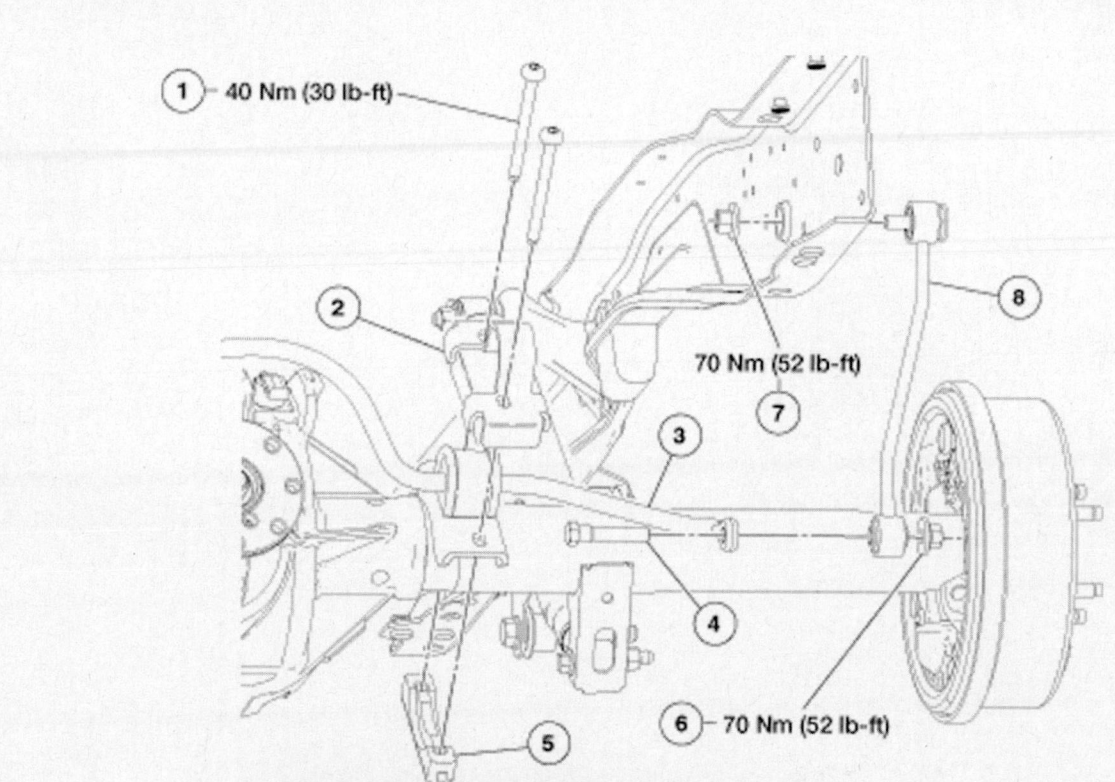

1. 40 Nm (30 lb-ft)

70 Nm (52 lb-ft)

70 Nm (52 lb-ft)

1. **Stabilizer bar bracket bolts (4 required)**
2. **Upper stabilizer bar bracket**
3. **Stabilizer bar assembly**
4. **Stabilizer bar link-to-stabilizer bar bolts (2 required)**
5. **Lower stabilizer bar brackets (2 required)**
6. **Stabilizer bar link-to-stabilizer bar nuts (2 required)**
7. **Stabilizer bar link-to-frame nuts (2 required)**
8. **Stabilizer bar links (2 required)**

22140_BPUP_G0208

Fig. 161 Stabilizer bar and link

7. Use driver handle and axle tube bearing replacer to install the rear wheel bearing.

8. Lubricate the lip of the new wheel bearing oil seal with appropriate grease

9. Use driver handle axle tube bearing replacer to install the inner wheel bearing oil seal.

10. Install the axle shaft.

ADJUSTMENT

There is no adjustment on in the rear bearings.

REPACKING

The rear wheel bearings are not serviceable and are lubricated by differential fluid.

SPECIFICATIONS AND MAINTENANCE CHARTS

ENGINE AND VEHICLE IDENTIFICATION

		Engine					Model Year	
Code ①	Liters (cc)	Cu. In.	Cyl.	Fuel Sys.	Engine Type	Eng. Mfg.	Code ②	Year
L3	2.3 (2260)	137.9	4	EDI	DOHC	Mazda	9	2009
L5	2.5 (2488)	151.8	4	EPI	DOHC	Mazda	A	2010

EDI: Electronic Direct Injection

EPI: Electronic Port Injection

DOHC: Double Over Head Cam

① 8th digit of the Vehicle Identification Number (VIN)

② 10th digit of the Vehicle Identification Number (VIN)

37671_MCX7_C0001

GENERAL ENGINE SPECIFICATIONS

Year	Model	Engine Displacement Liters (VIN)	Net Horsepower @ rpm	Net Torque @ rpm (ft. lbs.)	Bore x Stroke (in.)	Com-pression Ratio	Oil Pressure @ rpm
2009	CX-7	2.3 (L3)	244@5000	258@2500	344x3.70	9.5:1	43-79@3000
2010	CX-7	2.3 (L3)	244@5000	258@2500	344x3.70	9.5:1	43-79@3000
		2.5 (L5)	161@6000	161@3500	350x3.94	9.7:1	57-94@3000

37671_MCX7_C0002

ENGINE TUNE-UP SPECIFICATIONS

Year	Engine Displacement Liters (VIN)	Spark Plug Gap (in.)	Ignition Timing (deg.) MT	Ignition Timing (deg.) AT	Fuel Pump (psi)	Idle Speed (rpm) MT	Idle Speed (rpm) AT	Valve Clearance In.①	Valve Clearance Ex.①
2009	2.3 (L3)	28-31	N/A	10B	60-71	N/A	650-750	0.008-0.011	0.010-0.012
2010	2.3 (L3)	24-27	N/A	10B	60-71	N/A	650-750	0.008-0.011	0.010-0.012
	2.5 (L5)	40-53	N/A	12B	57	N/A	650-750	0.008-0.011	0.010-0.012

① Engine cold

The label figures must be used if they differ from those in this chart.

B: Before top dead center

N/A: Not Applicable

NOTE: The Vehicle Emission Control Information label often reflects specification changes made during production.

37671_MCX7_C0003

CAPACITIES

Year	Model	Engine Displacement Liters (VIN)	Engine Oil with Filter (qts.)	Transmission (pts.)	Transaxle AWD (pts.)	Drive Axle Front (pts.)	Drive Axle Rear (pts.)	Fuel Tank (gal.)	Cooling System (qts.)
2009	CX-7	2.3 (L3)	6.0	14.8	1.10	②	2.12	18.2	9.5
2010	CX-7	2.3 (L3)	6.0	14.8 ①	1.10	②	2.12	18.2	9.5
		2.5 (L5)	5.3	14.8 ①	1.10	②	2.12	16.4	9.5

① 5 Seed transmission 6.4 pts.

② Included in transaxle

NOTE: All capacities are approximate. Add fluid gradually and check to be sure a proper fluid level is obtained.

37671_MCX7_C0004

FLUID SPECIFICATIONS

Year	Model	Engine Displacement Liters (cc)	Engine ID/VIN	Engine Oil	Auto. Trans.	Drive Axle	Transfer Case	Power Steering Fluid	Brake Master Cylinder	Engine Coolant
2009	CX-7	2.3 (2260)	L3	5W-30	ATF M-V	80W-90	80W-90	Dexron®II	DOT 3	FL 22
2010	CX-7	2.3 (2260)	L3	5W-20	ATF M-V	80W-90	80W-90	Dexron®II	DOT 3	FL 22
		2.5 (2488)	L5	5W-20	ATF M-V	80W-90	80W-90	Dexron®II	DOT 3	FL 22

DOT: Department Of Transpotation

37671_MCX7_C0009

VALVE SPECIFICATIONS

Year	Engine Displacement Liters (VIN)	Seat Angle (deg.)	Spring Installed Height (in.)	Seat Contact With (in.)	Stem-to-Guide Clearance (in.) Intake	Stem-to-Guide Clearance (in.) Exhaust	Stem Diameter (in.) Intake	Stem Diameter (in.) Exhaust
2009	2.3 (L3)	45	1.492	0.0511	0.0009-0.0027	0.0009-0.0027	0.2351-0.2356	0.2349-0.2354
2010	2.3 (L3)	45	1.492	0.0511	0.0009-0.0027	0.0009-0.0027	0.2351-0.2356	0.2349-0.2354
	2.5 (L5)	45	1.129	0.039-0.072	0.0010-0.0027	0.0012-0.0029	0.2154	0.2152

37671_MCX7_C0005

CAMSHAFT AND BEARING SPECIFICATIONS CHART

All measurements are given in inches.

Year	Engine Displ. Liters	Engine ID/VIN	Journal Dia.	Brg. Oil Clearance	Shaft End-play	Runout	Journal Bore	Lobe Height Intake	Lobe Height Exhaust
2009	2.3	L3	0.9827-0.9834	0.002-0.003-	0.0034-0.0094	0.0012	N/A	1.667-1.671-	1.618-1.6220
2010	2.3	L3	1.0614-1.0620	0.002-0.003-	0.0035-0.0094	0.0012	N/A	1.667-1.671-	1.618-1.6220
	2.5	L5	0.9827-0.9834	0.002-0.003-	0.004-0.0090	0.0012	N/A	1.666-1.6710	1.616-1.6210

N/A: Not Available

37671_MCX7_C0006

CRANKSHAFT AND CONNECTING ROD SPECIFICATIONS

All measurements are given in inches.

Year	Engine Displacement Liters (VIN)	Crankshaft Main Brg. Journal Dia.	Crankshaft Main Brg. Oil Clearance	Crankshaft Shaft End-play	Crankshaft Thrust on No.	Connecting Rod Journal Diameter	Connecting Rod Oil Clearance	Connecting Rod Side Clearance
2009	2.3 (L3)	2.0465-2.0472	0.007-0.015	0.0087-0.0177	N/A	N/A	0.0011-0.0020	0.006-0.014
2010	2.3 (L3)	2.0465-2.0472	0.007-0.015	0.0087-0.0177	N/A	N/A	0.0011-0.0020	0.006-0.014
	2.5 (L5)	2.0465-2.0472	0.007-0.018	0.0087-0.0177	N/A	N/A	0.0011-0.002	0.006-0.014

N/A: Not Available

37671_MCX7_C0007

PISTON AND RING SPECIFICATIONS

All measurements are given in inches.

Year	Engine Displacement Liters (VIN)	Piston Clearance	Ring Gap Top Compression	Ring Gap Bottom Compression	Ring Gap Oil Control	Ring Side Clearance Top Compression	Ring Side Clearance Bottom Compression	Ring Side Clearance Oil Control
2009	2.3 (L3)	0.0010-0.0017	0.0056-0.0094	0.0134-0.0173	0.0060-0.0157	0.002-0.003	0.0012-0.0027	0.0008-0.0023
2010	2.3 (L3)	0.0010-0.0017	0.0056-0.0094	0.0134-0.0173	0.0060-0.0157	0.002-0.003	0.0012-0.0027	0.0008-0.0023
	2.5 (L5)	0.0010-0.0017	0.0056-0.0094	0.0134-0.0173	0.0060-0.0157	0.002-0.003	0.0012-0.0027	0.0008-0.0023

37671_MCX7_C0008

TORQUE SPECIFICATIONS
All readings in ft. lbs.

Year	Engine Displacement Liters (VIN)	Cylinder Head Bolts	Main Bearing Bolts	Rod Bearing Bolts	Crankshaft Damper Bolts	Flywheel Bolts	Manifold Intake	Manifold Exhaust	Spark Plugs	Oil Pan Drain Plug
2009	2.3 (L3)	①	N/A	N/A	②	80-85	13-16	32-47	8-10	23-30
2010	2.3 (L3)	①	N/A	N/A	②	80-85	13-16	32-47	8-10	23-30
	2.5 (L5)	①	N/A	N/A	③	80-85	12-14	32-47	8-10	23-30

NA: Not Available

① Step 1: 27-97 inch lbs.
　Step 2: Tighten 116-150 inch. lbs
　Step 3: Tighten 32-34 ft. lbs.
　Step 4: Tighten 92 degrees
　Step 5: Tighten an additional 92 degrees

② 71-77 ft. lbs. plus 87-90 degrees

③ 71-77 ft. lbs. plus 87-93 degrees

37671_MCX7_C0010

WHEEL ALIGNMENT

Year	Model		Caster Range (+/-Deg.)	Caster Preferred Setting (Deg.)	Camber Range (+/-Deg.)	Camber Preferred Setting (Deg.)	Toe-in (in.)
2009	CX-7	F	1.00	+2.58	1.00	-0 20 +/- 1.00	0 +/- 0.12
		R	N/A	N/A	1.00	①	②
2010	CX-7	F	1.00	+2.58	1.00	-0 20 +/- 1.00	0 +/- 0.12
		R	N/A	N/A	1.00	①	②

NA: Not Applicable

① Vehicles with 17 inch rims: 0.046
　Vehicles with 18 inch rims: 0.049
　Vehicles with 19 inch rims: 0.051

② Fuel gauge indication
　Empty: 0.51
　1/4: -0.54
　1/2: -0.57
　3/4: -1.00
　Full: -1.03

37671_MCX7_C0011

TIRE, WHEEL AND BALL JOINT SPECIFICATIONS

Year	Model	OEM Tires Standard	OEM Tires Optional	Tire Pressures (psi) Front	Tire Pressures (psi) Rear	Wheel Size	Ball Joint Inspection	Lug Nut (ft. lbs.)
2009	CX-7	P215/70R17 100H	N/A	32	32	17 × 7J	①	66-87
		P235/60R18 102H	N/A	32	32	18 × 71/2J	①	66-87
		235/60R18 103H	N/A	32	32	18 × 71/2J	①	66-87
		P235/55R19 101H	N/A	32	32	19 × 71/2J	①	66-87
2010	CX-7	P215/70R17 100H	N/A	32	32	17 × 7J	①	66-87
		P235/60R18 102H	N/A	32	32	18 × 71/2J	①	66-87
		235/60R18 103H	N/A	32	32	18 × 71/2J	①	66-87
		P235/55R19 101H	N/A	32	32	19 × 71/2J	①	66-87

OEM: Original Equipment Manufacturer

PSI: Pounds Per Square Inch

① Lower arm ball rotation torque: 13-17 inch lbs.

NA: Not Applicable

37671_MCX7_C0012

BRAKE SPECIFICATIONS
All measurements in inches unless noted

Year	Model		Brake Disc Original Thickness	Brake Disc Minimum Thickness	Brake Disc Maximum Runout	Brake Drum Original Inside Diameter	Brake Drum Max. Wear Limit	Brake Drum Maximum Machine Diameter	Minimum Lining Thickness	Brake Caliper Bracket Bolts (ft. lbs.)	Brake Caliper Mounting Bolts (ft. lbs.)
2009	CX-7	F	N/A	1.030	0.002	N/A	N/A	N/A	0.080	80-94	62-69
		R	N/A	0.630	0.002	N/A	N/A	N/A	0.080	58-75	16-23
2010	CX-7	F	N/A	1.030	0.002	N/A	N/A	N/A	0.080	80-94	62-69
		R	N/A	0.630	0.002	N/A	N/A	N/A	0.080	58-75	16-23

NA: Not Applicable

F: Front

R: Rear

37671_MCX7_C0013

SCHEDULED MAINTENANCE INTERVALS
Mazda—CX-7

TO BE SERVICED	TYPE OF SERVICE	7.5	15	22.5	30	37.5	45	52.5	60	67.5	75	82.5	90	97.5
Engine oil & filter	R	✓	✓	✓	✓	✓	✓	✓	✓	✓	✓	✓	✓	✓
Air cleaner element	R					✓					✓			
Engine coolant	R								✓					
Spark plugs	R										✓			
Bolts & nuts on chassis & body	S/I				✓				✓				✓	
Cabin air filter	S/I				✓				✓				✓	
Brake lines, hoses & connections	S/I				✓				✓				✓	
Cooling system	S/I				✓				✓				✓	
Disc brakes	S/I		✓		✓				✓				✓	
Drive belts	S/I								✓		✓			
Drive shaft dust boots	S/I				✓				✓				✓	
Exhaust system heat shield	S/I						✓		✓				✓	
Front & rear ball joints	S/I				✓				✓				✓	
Fuel lines & hoses	S/I				✓				✓				✓	
Steering operation & linkages	S/I				✓				✓				✓	
Valve clearance	S/I										✓			
Hose & tube for emission	S/I								✓					

R: Replace   S/I: Service or Inspect

FREQUENT OPERATION MAINTENANCE (SEVERE SERVICE)

If a vehicle is operated under any of the following conditions it is considered severe service

- Extremely dusty areas.

- 50% or more of the vehicle operation is in 32°C (90°F) or higher temperatures, or constant operation in temperatures below 0°C (32°F).

- Prolonged idling (vehicle operation in stop and go traffic).

- Frequent short running periods (engine does not warm to normal operating temperatures).

- Police, taxi, delivery usage or trailer towing usage.

Oil & oil filter: change every 3000 miles.

Air cleaner element: service or inspect every 15,000 miles

Automatic transaxle fluid: service or inspect every 15,000 miles.

Bolts & nuts on chassis & body: tighten every 15,000 miles.

Disc brakes: service or inspect every 15,000 miles.

37671_MCX7_C0014

BRAKES   INFORMATION AND PRECAUTIONS

ANTI-LOCK SYSTEMS

• Certain components within the ABS system are not intended to be serviced or repaired individually.

• Do not use rubber hoses or other parts not specifically specified for and ABS system. When using repair kits, replace all parts included in the kit. Partial or incorrect repair may lead to functional problems and require the replacement of components.

• Lubricate rubber parts with clean, fresh brake fluid to ease assembly. Do not use shop air to clean parts; damage to rubber components may result.

• Use only DOT 3 brake fluid from an unopened container.

• If any hydraulic component or line is removed or replaced, it may be necessary to bleed the entire system.

• A clean repair area is essential. Always clean the reservoir and cap thoroughly before removing the cap. The slightest amount of dirt in the fluid may plug an orifice and impair the system function. Perform repairs after components have been thoroughly cleaned; use only denatured alcohol to clean components. Do not allow ABS components to come into contact with any substance containing mineral oil; this includes used shop rags.

• The Anti-Lock control unit is a microprocessor similar to other computer units in the vehicle. Ensure that the ignition switch is **OFF** before removing or installing controller harnesses. Avoid static electricity discharge at or near the controller.

• If any arc welding is to be done on the vehicle, the control unit should be unplugged before welding operations begin.

DISC AND DRUM SYSTEMS

> **✲✲ CAUTION**
> Dust and dirt accumulating on brake parts during normal use may contain asbestos fibers from production or aftermarket brake linings. Breathing excessive concentrations of asbestos fibers can cause serious bodily harm. Exercise care when servicing brake parts. Do not sand or grind brake lining unless equipment used is designed to contain the dust residue. Do not clean brake parts with compressed air or by dry brushing. Cleaning should be done by dampening the brake components with a fine mist of water, then wiping the brake components clean with a dampened cloth. Dispose of cloth and all residue containing asbestos fibers in an impermeable container with the appropriate label. Follow practices prescribed by the Occupational Safety and Health Administration (OSHA) and the Environmental Protection Agency (EPA) for the handling, processing, and disposing of dust or debris that may contain asbestos fibers.

BRAKES   BLEEDING THE BRAKE SYSTEM

BLEEDING PROCEDURE

BLEEDING PROCEDURE
See Figure 1.

> **✲✲ CAUTION**
> Brake fluid will damage painted surfaces. Be careful not to spill any on painted surfaces. If it is spilled, wipe it off immediately.

➡ Keep the fluid level in the reserve tank at 3/4 full or more during the air bleeding. Begin air bleeding with the brake caliper that is furthest from the master cylinder. Use DOT-3 brake fluid.

1. Before servicing the vehicle, refer to the Precautions Section.

2. Remove the bleeder cap on the brake caliper, and attach a vinyl tube to the bleeder screw.

3. Place the other end of the vinyl tube in a clear container and fill the container with fluid during air bleeding.

4. Working with two people, one should pump the brake pedal several times and depress and hold the pedal down.

5. While the brake pedal is depressed, the other should loosen the bleeder screw using the SST, drain out any fluid containing air bubbles, and tighten the bleeder screw.

6. Repeat until no air bubbles are seen.

7. Perform air bleeding as described in the above procedures for all brake calipers.

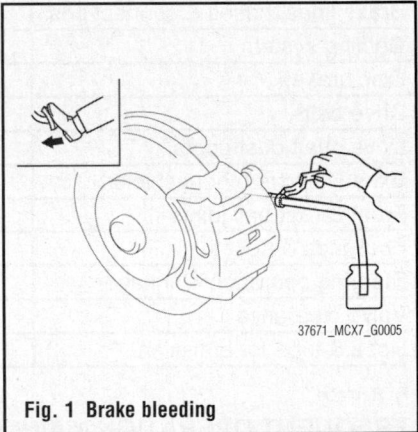

37671_MCX7_G0005

Fig. 1 Brake bleeding

BRAKES   **ANTI-LOCK BRAKE SYSTEM (ABS)**

SPEED SENSORS

REMOVAL & INSTALLATION

Front

See Figure 2.

1. Before servicing the vehicle, refer to the Precautions Section.
2. If removing the left front sensor, remove the battery, battery tray, air cleaner and hose.
3. Disconnect the wiring harness.
4. Unbolt and remove the wheel speed sensor.
5. Installation is the reverse of the removal procedure.

Rear

See Figure 3.

1. Before servicing the vehicle, refer to the Precautions Section.
2. Remove the rear seat.
3. Remove the rear scuff plate inner.
4. Remove the sub trunk box.
5. Remove the trunk end trim.

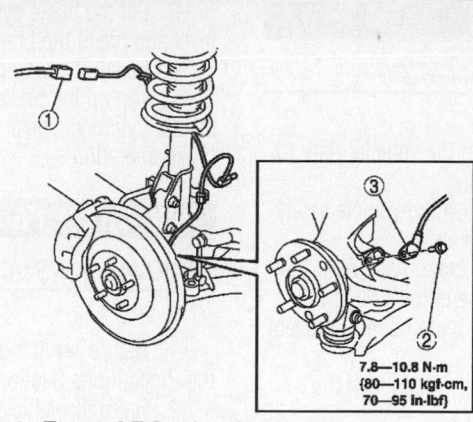

1. **Front ABS wheel-speed sensor connector**
2. **Bolt**
3. **Front ABS wheel-speed sensor**

7.8—10.8 N·m
{80—110 kgf-cm,
70—95 in-lbf}

22140_MCX7_G0089

Fig. 2 Front wheel speed sensor

6. Remove the trunk side trim.
7. Remove the wiring harness connector.

8. Unbolt and remove the rear wheel-speed sensor
9. Install in the reverse order of removal.

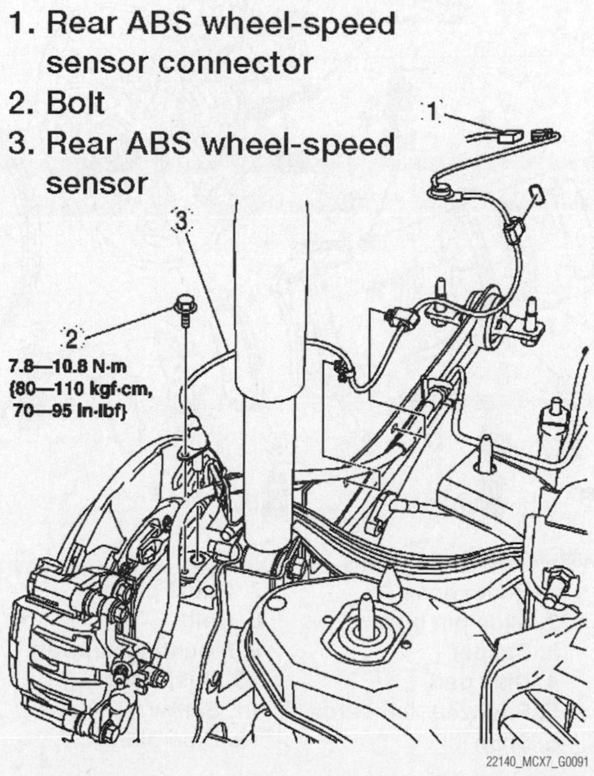

1. **Rear ABS wheel-speed sensor connector**
2. **Bolt**
3. **Rear ABS wheel-speed sensor**

7.8—10.8 N·m
{80—110 kgf-cm,
70—95 in-lbf}

22140_MCX7_G0091

Fig. 3 Rear wheel speed sensor—AWD

BRAKES　　　　　　　　　　　　　　　　　　**FRONT DISC BRAKES**

BRAKE CALIPER

REMOVAL & INSTALLATION

See Figure 4.

1. Before servicing the vehicle, refer to the Precautions Section.
2. Raise and support the vehicle safely.
3. Remove the wheels.
4. Disconnect the brake hose from the caliper by removing the union bolt and 2 gaskets. Plug the end of the hose to prevent loss of fluid.
5. Remove the bolts that attach the caliper to its mounting.
6. Lift the bottom of the caliper up and remove the caliper assembly.

To install:

7. Grease the caliper slides and bolts with lithium grease or equivalent. Install the caliper and secure with the bolts.
8. Connect the brake hose to the caliper, using 2 new washers. Torque the union bolt to 16–21 ft. lbs. (22–29 Nm).

9. Fill the brake system to the proper level and bleed the brake system.
10. Install the tire and wheel assembly.
11. Top off the brake fluid level in the master cylinder. Check for leaks and proper brake operation.

DISC BRAKE PADS

REMOVAL & INSTALLATION

See Figure 5.

1. Before servicing the vehicle, refer to the Precautions Section.
2. Raise the vehicle and support it safely.
3. Remove the wheel and tire assembly.
4. When servicing the front pads, loosen the brake caliper upper side mounting bolt. Loosen and remove the lower side mounting bolt. Lift the caliper and suspend it so the hose is not stretched.
5. If equipped, remove the anti-squeal spring.
6. Remove the brake pads.

To install:

7. Siphon a small amount of brake fluid from the reservoir. Press in the brake caliper piston with the proper tool.
8. Before installing the new pads, check the disc thickness and disc run-out.
9. Install the pad support plates.
10. Install the anti-squeal shims to each pad.

➡**Apply disc brake grease to both sides of the inner anti-squeal shims.**

11. Install the disc pads so the wear indicator plate is facing downward.
12. If removed, install the anti-squeal springs.
13. Carefully install the brake caliper so the boot is not wedged.
14. Install the wheel and tire assembly.
15. Check and adjust the fluid level. Apply the brake pedal several times.
16. Road test the vehicle for proper operation.

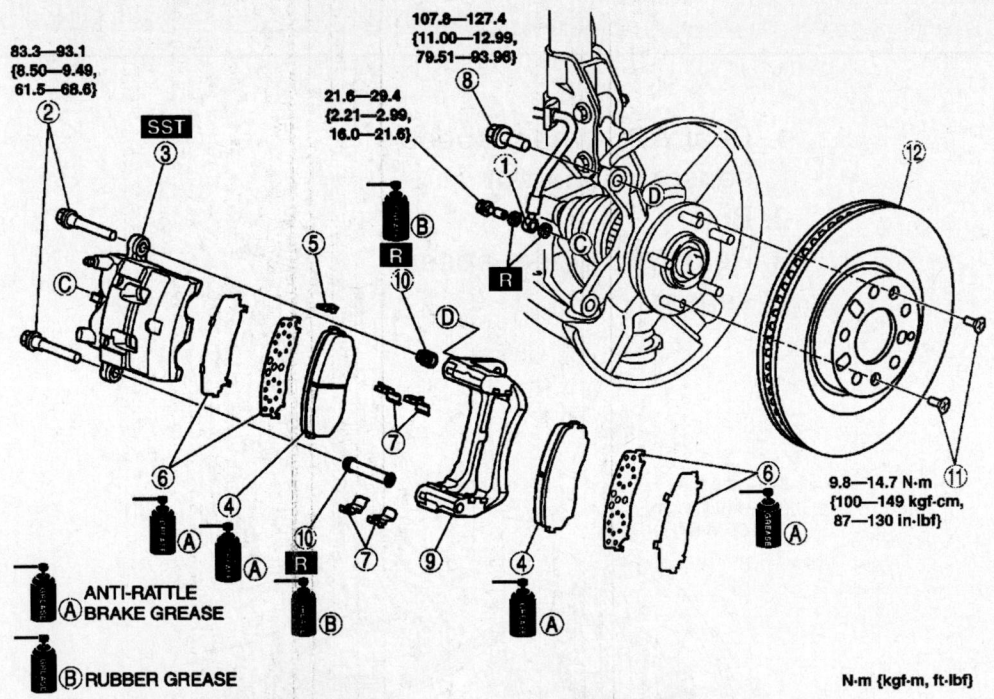

1. Brake hose
2. Slide pin bolt
3. Caliper
4. Disc pad
5. Pad wear indicator
6. Shim
7. Guide plate
8. Bolt
9. Mounting support
10. Dust boot
11. Screw
12. Disc plate

ANTI-RATLLE
Ⓐ BRAKE GREASE

Ⓑ RUBBER GREASE

N·m {kgf-m, ft-lbf}

22140_MCX7_G0105

Fig. 4 Front brake—exploded view

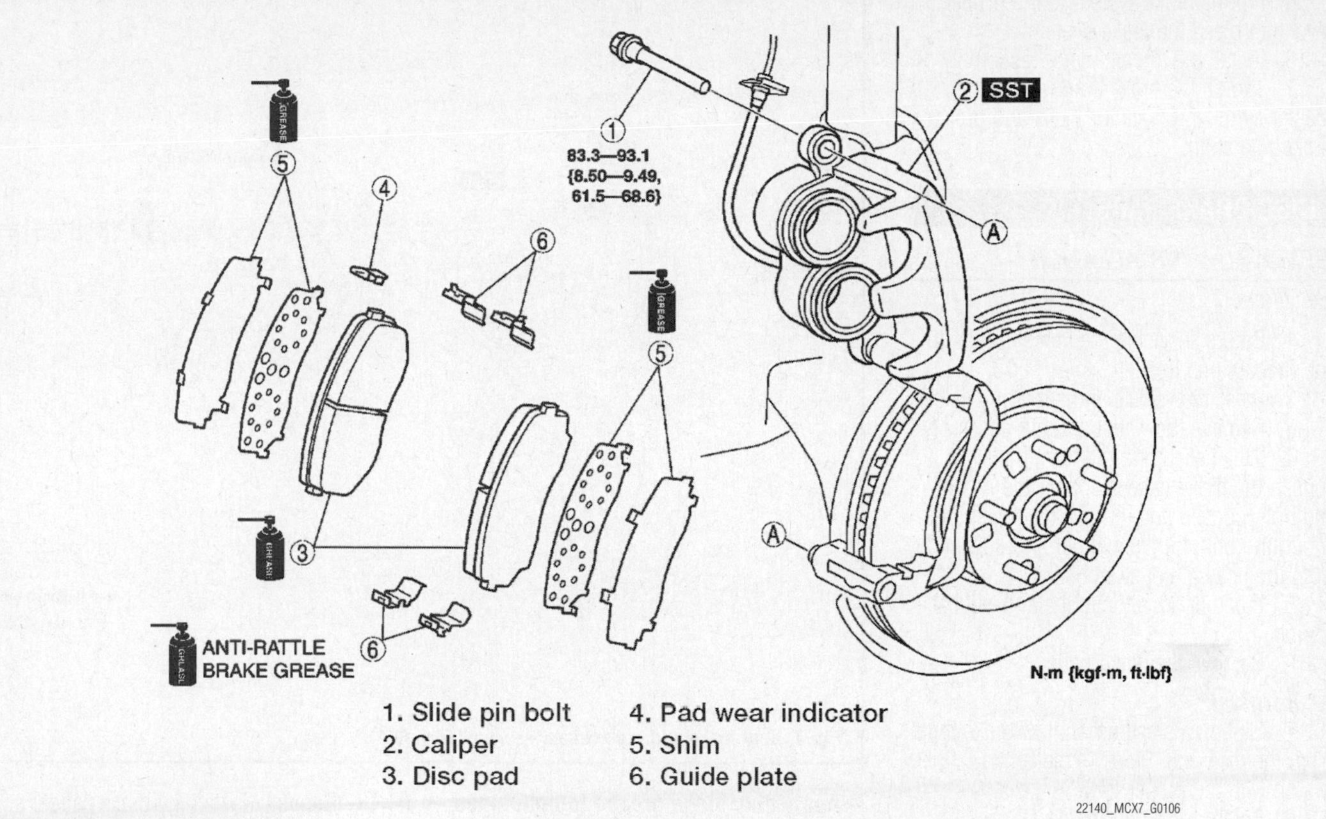

1. Slide pin bolt   4. Pad wear indicator
2. Caliper   5. Shim
3. Disc pad   6. Guide plate

83.3—93.1
{8.50—9.49,
61.5—68.6}

N·m {kgf·m, ft·lbf}

ANTI-RATTLE
BRAKE GREASE

22140_MCX7_G0106

Fig. 5 Brake pad replacement—exploded view

BRAKES   REAR DISC BRAKES

BRAKE CALIPER

REMOVAL & INSTALLATION

See Figure 8.

1. Before servicing the vehicle, refer to the Precautions Section.

2. Raise and support the vehicle safely.

3. Remove the wheels.

4. Disconnect the brake hose from the caliper by removing the union bolt and 2 gaskets. Plug the end of the hose to prevent loss of fluid.

5. Remove the bolts that attach the caliper to its mounting.

6. Lift the bottom of the caliper up and remove the caliper assembly.

To install:

7. Grease the caliper slides and bolts with lithium grease or equivalent. Install the caliper and secure with the bolts.

8. Connect the brake hose to the caliper, using 2 new washers. Torque the union bolt to 16–21 ft. lbs. (22–29 Nm).

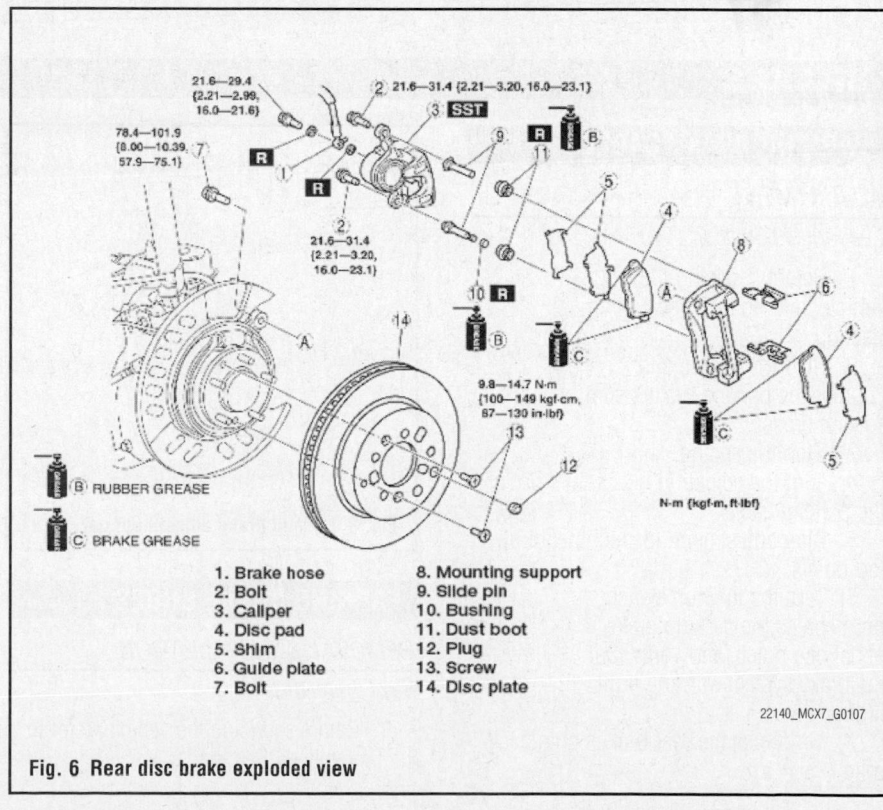

21.6—29.4
{2.21—2.99,
16.0—21.6}

78.4—101.9
{8.00—10.39,
57.9—75.1}

21.6—31.4 {2.21—3.20, 16.0—23.1}

21.6—31.4
{2.21—3.20,
16.0—23.1}

9.8—14.7 N·m
{100—149 kgf·cm,
87—130 in·lbf}

B RUBBER GREASE
C BRAKE GREASE

N·m {kgf·m, ft·lbf}

1. Brake hose   8. Mounting support
2. Bolt   9. Slide pin
3. Caliper   10. Bushing
4. Disc pad   11. Dust boot
5. Shim   12. Plug
6. Guide plate   13. Screw
7. Bolt   14. Disc plate

22140_MCX7_G0107

Fig. 6 Rear disc brake exploded view

9. Fill the brake system to the proper level and bleed the brake system.

10. Install the tire and wheel assembly.

11. Top off the brake fluid level in the master cylinder. Check for leaks and proper brake operation.

DISC BRAKE PADS

REMOVAL & INSTALLATION

See Figure 7.

1. Before servicing the vehicle, refer to the Precautions Section.

2. Raise the vehicle and support it safely.

3. Remove the wheel and tire assembly.

4. When servicing the front pads, loosen the brake caliper upper side mounting bolt. Loosen and remove the lower side mounting bolt. Lift the caliper and suspend it so the hose is not stretched.

5. If equipped, remove the anti-squeal spring.

6. Remove the brake pads.

To install:

7. Siphon a small amount of brake fluid from the reservoir. Press in the brake caliper piston with the proper tool.

8. Before installing the new pads, check the disc thickness and disc run-out.

9. Install the pad support plates.

10. Install the anti-squeal shims to each pad.

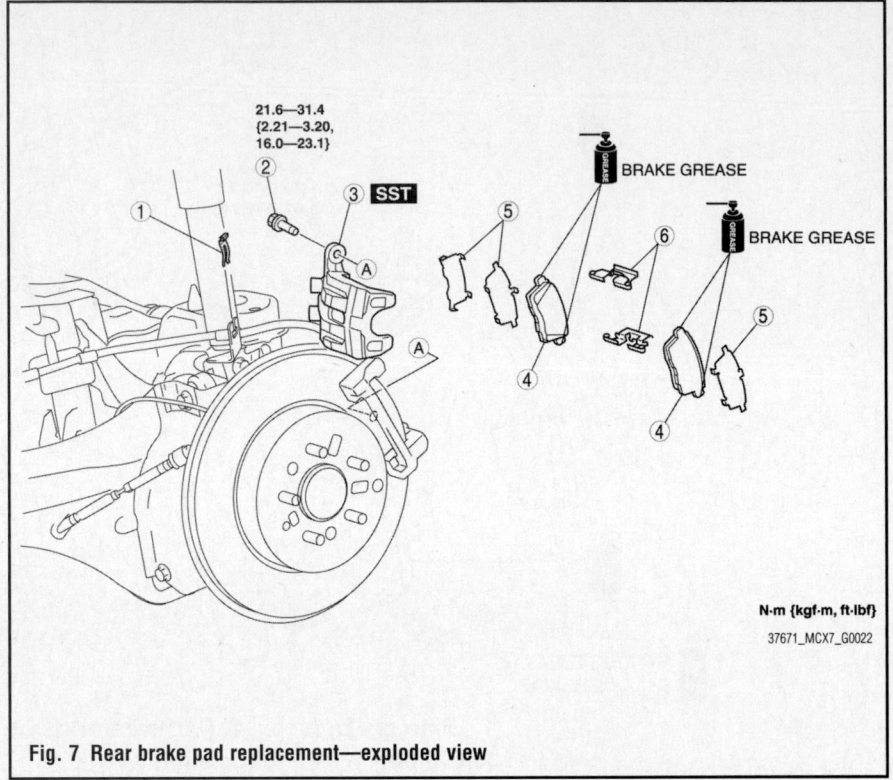

N·m {kgf·m, ft·lbf}

37671_MCX7_G0022

Fig. 7 Rear brake pad replacement—exploded view

→**Apply disc brake grease to both sides of the inner anti-squeal shims.**

11. Install the disc pads so the wear indicator plate is facing downward.

12. If removed, install the anti-squeal springs.

13. Carefully install the brake caliper so the boot is not wedged.

14. Install the wheel and tire assembly.

15. Check and adjust the fluid level. Apply the brake pedal several times.

16. Road test the vehicle for proper operation.

BRAKES

PARKING BRAKE CABLES

ADJUSTMENT

See Figures 8 and 9.

1. Before servicing the vehicle, refer to the Precautions Section.

2. Start the engine and depress the parking brake pedal several times.

3. Stop the engine.

4. Turn the adjusting nut at the front of the parking cable.

5. After adjustment, inspect the following points:

6. Turn the ignition switch on, depress the parking brake pedal one notch, and verify that the brake system warning light illuminates.

7. Verify that the rear brakes do not drag.

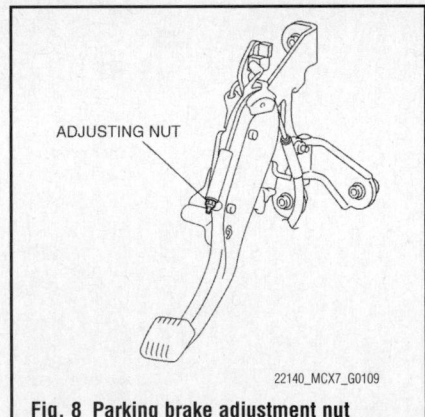

ADJUSTING NUT

22140_MCX7_G0109

Fig. 8 Parking brake adjustment nut

PARKING BRAKE SHOES

REMOVAL & INSTALLATION

See Figure 10.

1. Before servicing the vehicle, refer to the Precautions Section.

PARKING BRAKE

2. Raise the vehicle and support it safely.

3. Remove the wheel and tire assembly.

4. Remove the caliper and support bracket.

5. Remove the rotor retaining screws and remove the rotor.

6. When removing the parking brake shoe, firmly press the adjuster bolt and tappet by hand and slowly remove the parking brake shoe to prevent the adjuster bolt, tappet, operation lever and other parts from flying off.

7. Install in reverse of the removal procedure.

ADJUSTMENT

See Figures 11 and 12.

1. Remove the end cable installation bolts.

2. Disconnect the end cable from the operation lever.

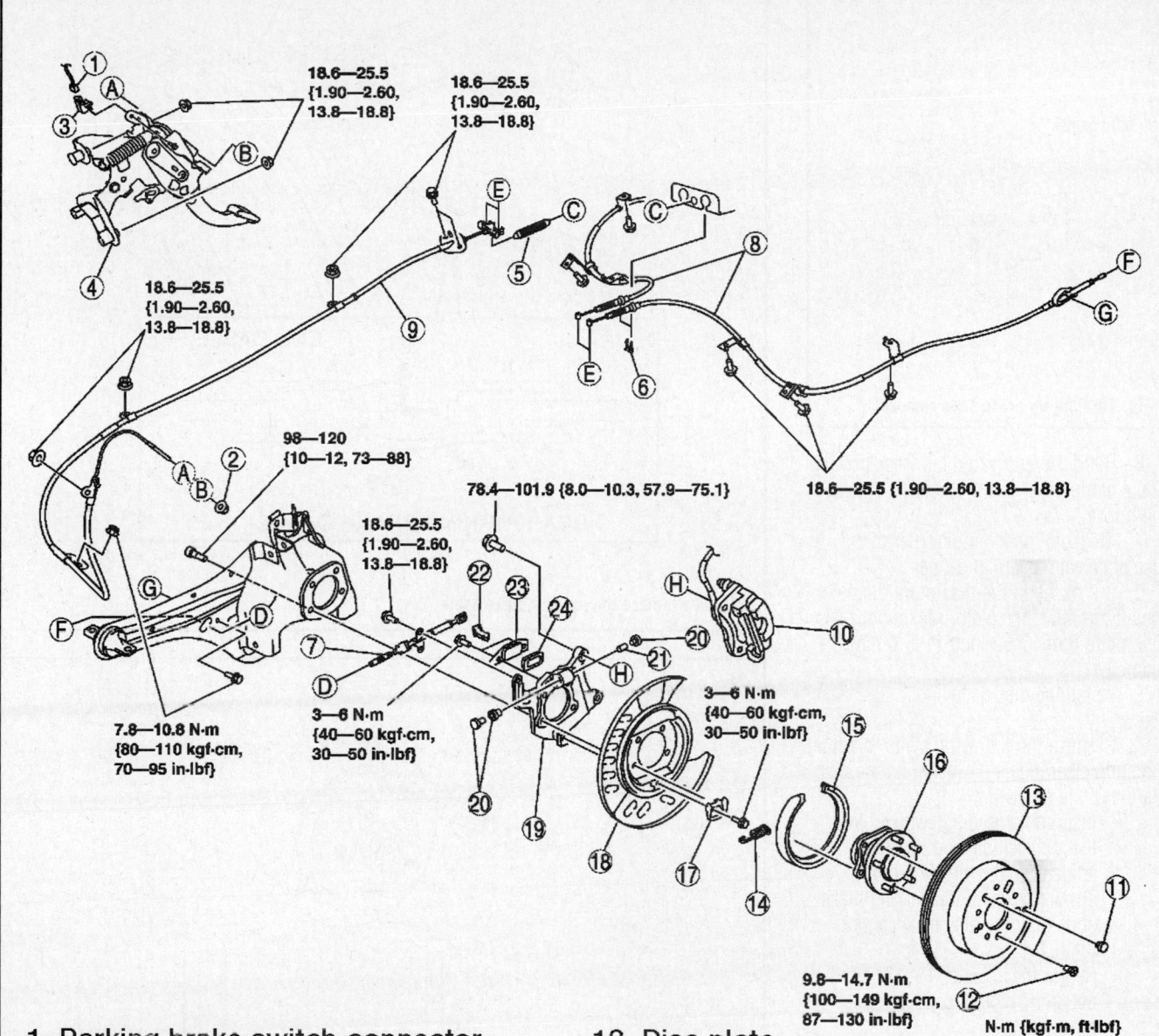

18.6—25.5
{1.90—2.60,
13.8—18.8}

18.6—25.5
{1.90—2.60,
13.8—18.8}

18.6—25.5
{1.90—2.60,
13.8—18.8}

98—120
{10—12, 73—88}

78.4—101.9 {8.0—10.3, 57.9—75.1}

18.6—25.5 {1.90—2.60, 13.8—18.8}

18.6—25.5
{1.90—2.60,
13.8—18.8}

7.8—10.8 N·m
{80—110 kgf·cm,
70—95 in·lbf}

3—6 N·m
{40—60 kgf·cm,
30—50 in·lbf}

3—6 N·m
{40—60 kgf·cm,
30—50 in·lbf}

9.8—14.7 N·m
{100—149 kgf·cm,
87—130 in·lbf}

N·m {kgf·m, ft·lbf}

1. Parking brake switch connector
2. Adjusting nut
3. Parking brake switch
4. Parking brake pedal
5. Spring
6. Clip
7. End cable
8. Rear parking brake cable
9. Front Parking brake cable, equalizer
10. Brake caliper component
11. Plug
12. Screw
13. Disc plate
14. Spring
15. Parking brake shoe
16. Rear wheel hub component
17. Shoe stopper
18. Backing plate
19. Parking brake plate
20. Adjuster bolt and nut, tappet
21. Pin
22. Operation lever
23. Plate
24. Dust boot

22140_MCX7_G0108

Fig. 9 Parking brake system exploded view

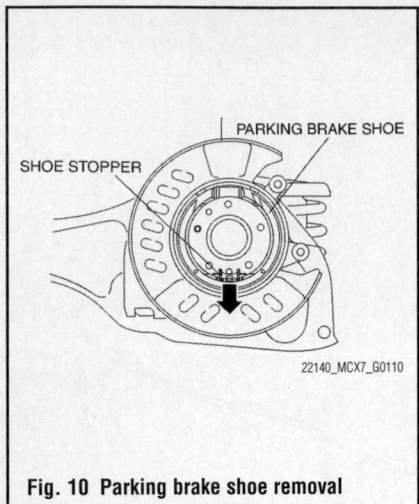

Fig. 10 Parking brake shoe removal

3. Bend the end cable tab (rear parking brake cable side) in the direction shown in the figure.

4. Measure the inner diameter of the disc plate with a vernier caliper.

5. If it exceeds the maximum diameter, install the new disc plate. Maximum rear disc plate inner 7.52 inch {191.0 mm} diameter.

6. Install the disc plate and screw.

7. Perform the following steps to adjust the shoe clearance after installing the disc plate and the screws.

- Insert a flathead screwdriver into the service hole and turn the adjuster in the direction of the arrow to expand the parking brake shoe until the disc plate cannot rotate.
- Return the adjuster 13–17 notches in the direction of the arrow.
- Rotate the disc plate and make sure it does not drag.

➡ **Shoe clearance can be adjusted to 0.15 mm {0.006 in} by returning the adjuster about 15 notches.**

8. Install the end cable to the operation lever.

9. Install the end cable installation bolts.

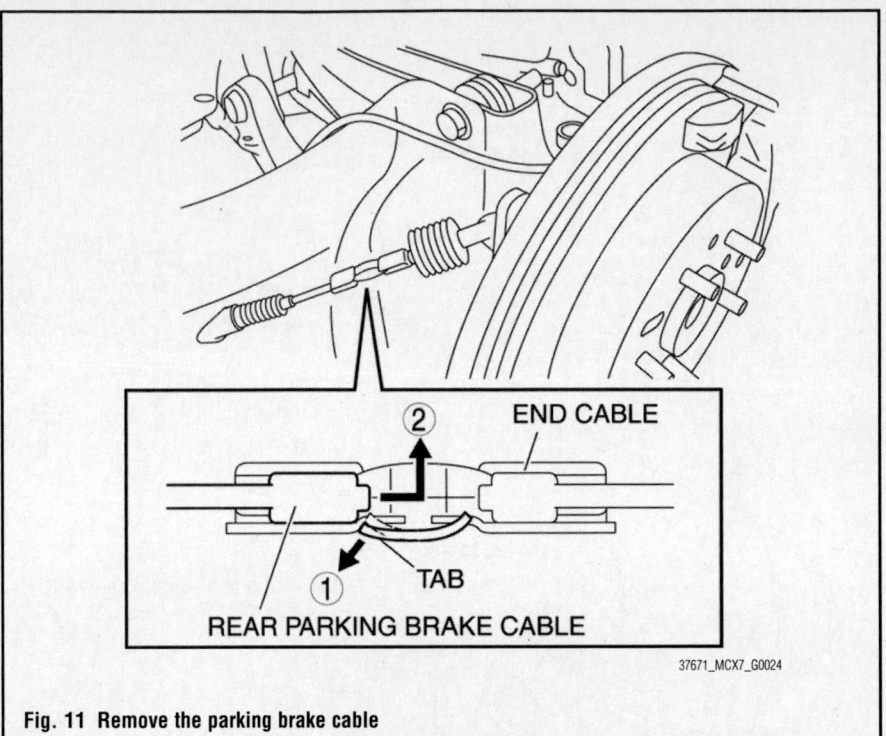

Fig. 11 Remove the parking brake cable

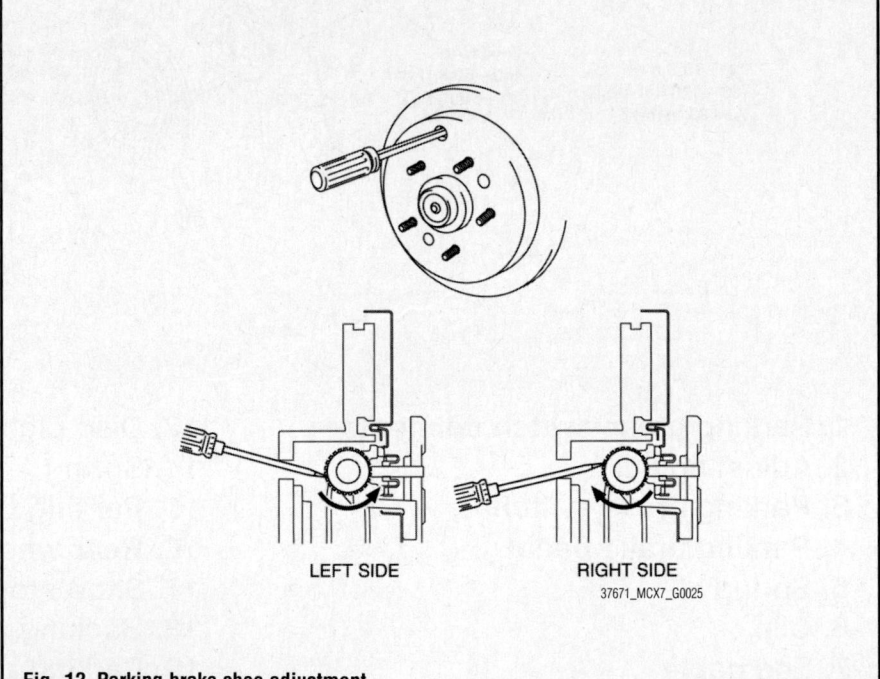

Fig. 12 Parking brake shoe adjustment

CHASSIS ELECTRICAL   AIR BAG (SUPPLEMENTAL RESTRAINT SYSTEM)

GENERAL INFORMATION

> #### ✳✳ CAUTION
>
> **These vehicles are equipped with an air bag system. The system must be disarmed before performing service on, or around, system components, the steering column, instrument panel components, wiring and sensors. Failure to follow the safety precautions and the disarming procedure could result in accidental air bag deployment, possible injury and unnecessary system repairs.**

SERVICE PRECAUTIONS

> #### ✳✳ CAUTION
>
> **Disconnect and isolate the battery negative cable before beginning any airbag system component diagnosis, testing, removal, or installation procedures. Wait at least 90 seconds after the ignition switch is turned off and the negative (-) terminal cable is disconnected from the battery before starting the operation. The SRS is equipped with a backup power source, so if work is started within 90 seconds after disconnecting the negative (-) terminal cable from the battery, the SRS may be deployed. Failure to disable the airbag system may result in accidental airbag deployment, personal injury, or death.**

DISARMING THE SYSTEM

1. Disconnect the negative battery cable and wait for 1 minute or more.

ARMING THE SYSTEM

1. Connect the negative battery cable. Check for proper operation of the air bag warning light.
2. Perform the steering angle sensor reference point setting.

CLOCKSPRING CENTERING

See Figures 13 through 15.

➡ The adjustment procedure is also specified on the caution label of the clock spring.

1. Set the front tires straight-ahead.

> #### ✳✳ WARNING
>
> **The clock spring will break if over-wound. Do not forcibly turn the clock spring.**

2. Turn the clock spring clockwise until it stops.
3. From the stopped position, turn the clock spring counterclockwise 2 3/4 turns
4. Align the marks.

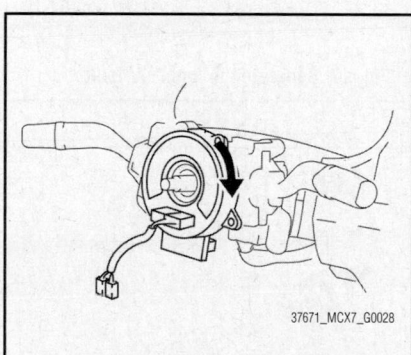

Fig. 13 Turn the clock spring clockwise until it stops

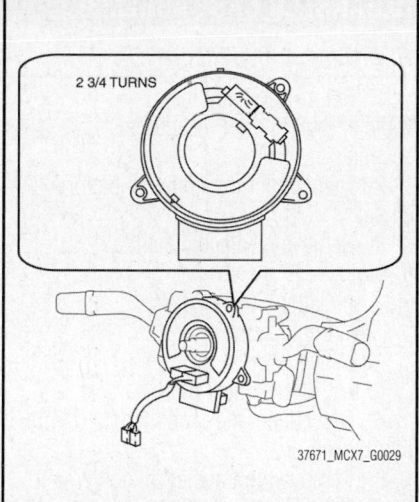

Fig. 14 Turn the clock spring counterclockwise 2 3/4 turns

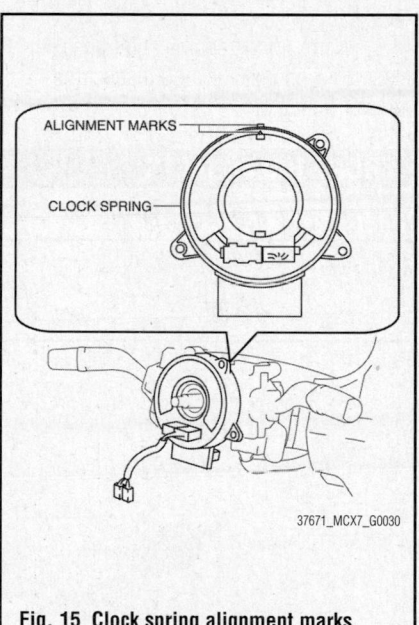

Fig. 15 Clock spring alignment marks

DRIVE TRAIN

TRANSFER CASE ASSEMBLY

REMOVAL & INSTALLATION

1. Before servicing the vehicle, refer to the Precautions section.

2. Disconnect the negative battery cable.

3. Remove or disconnect the following:
   - Battery, battery tray
   - Engine cover
   - Intercooler
   - Front exhaust pipe

4. Disconnect the propeller shaft from the transfer side.

5. Remove the transfer oil cooler with the hose still connected.

6. Remove in the order indicated in the table.

7. Install in the reverse order of removal.

8. Warm up the engine and transaxle, inspect for oil leakage, and inspect the transfer operation.

FRONT HALFSHAFTS

REMOVAL & INSTALLATION

See Figures 16 through 19.

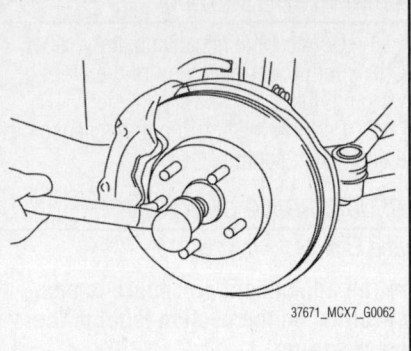

Fig. 16 Separate the outer stub shaft from the wheel hub

37671_MCX7_G0062

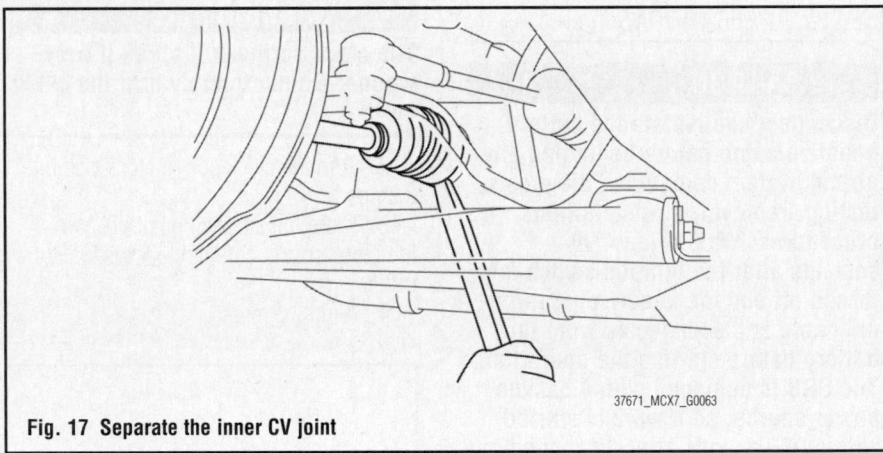

Fig. 17 Separate the inner CV joint

37671_MCX7_G0063

✳✳ WARNING

Performing the following procedures without first removing the ABS wheel-speed sensor may possibly cause an open circuit in the harness if it is pulled by mistake. Before performing the following procedures, remove the ABS wheel-speed sensor (axle side) and fix it to an appropriate place where the sensor will not be pulled by mistake while the vehicle is being serviced.

1. Before servicing the vehicle, refer to the Precautions Section.

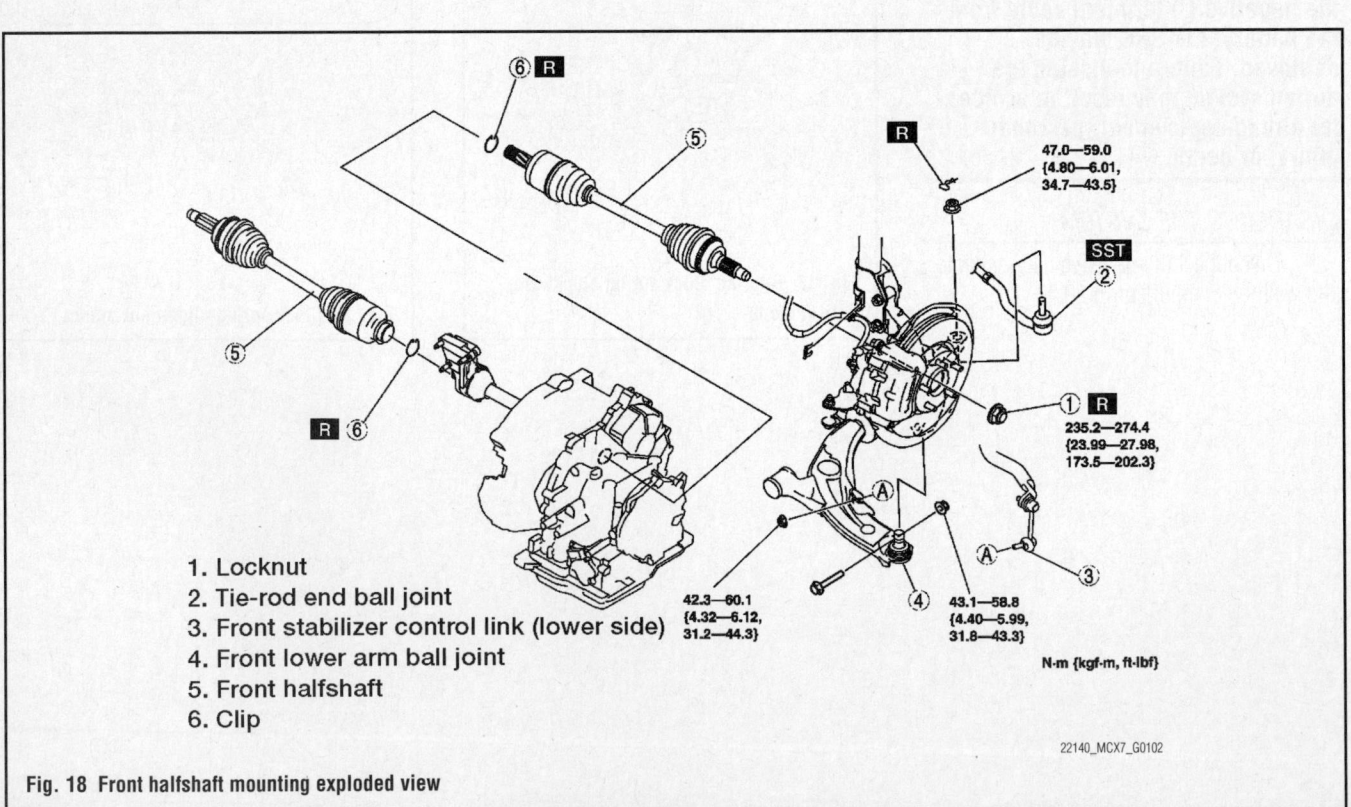

47.0—59.0
{4.80—6.01,
34.7—43.5}

235.2—274.4
{23.99—27.98,
173.5—202.3}

42.3—60.1
{4.32—6.12,
31.2—44.3}

43.1—58.8
{4.40—5.99,
31.8—43.3}

N·m {kgf·m, ft-lbf}

1. Locknut
2. Tie-rod end ball joint
3. Front stabilizer control link (lower side)
4. Front lower arm ball joint
5. Front halfshaft
6. Clip

22140_MCX7_G0102

Fig. 18 Front halfshaft mounting exploded view

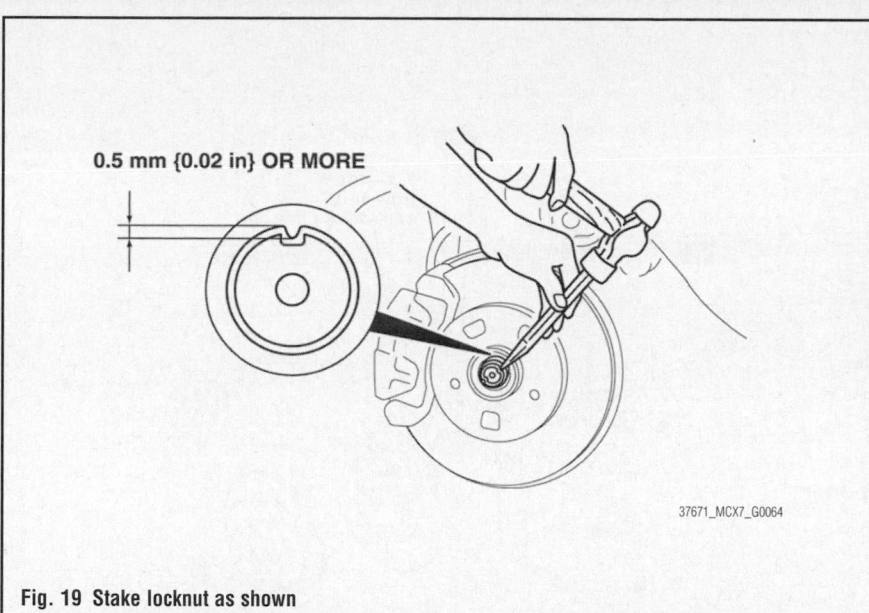

0.5 mm {0.02 in} OR MORE

37671_MCX7_G0064

Fig. 19 Stake locknut as shown

2. Remove the side cover and under-cover.

3. Drain the ATF.

4. Remove the front ABS wheel-speed sensor.

5. Remove the hub locknut.

6. Remove the tie rod end.

7. Remove the lower ball joint.

8. Separate the outer stub shaft from the wheel hub using a plastic-faced hammer.

9. Separate the inner CV joint from the transaxle using a suitable prytool.

10. Install in the reverse order of removal.

11. Refill the ATF.

12. Install a new locknut and stake it as shown.

CV-BOOTS INSPECTION

See Figure 20.

1. Inspect the dust boot on the drive shaft for cracks, damage, leaking grease, and looseness in the boot band.

2. Inspect the drive shaft for bends, cracks, and wear in the joint or splines

3. Repair or replace the drive shaft or boot/band as necessary.

REAR HALFSHAFTS

REMOVAL & INSTALLATION

See Figures 21 and 22.

1. Before servicing the vehicle, refer to the Precautions Section.

2. Drain the rear differential oil into a container.

3. Remove the trailing link, knuckle component.

4. Remove the rear halfshaft.

5. Disengage the rear drive shaft using the proper tools.

➡**Be careful not to damage the rear differential oil seal.**

6. Install in the reverse order of removal.

7. After installation, add 2 pts API service GL-5 SAE 80W-90 rear differential oil.

8. Inspect the rear wheel alignment and adjust it if necessary.

CV-BOOTS INSPECTION

See Figure 20.

1. Inspect the dust boot on the drive shaft for cracks, damage,

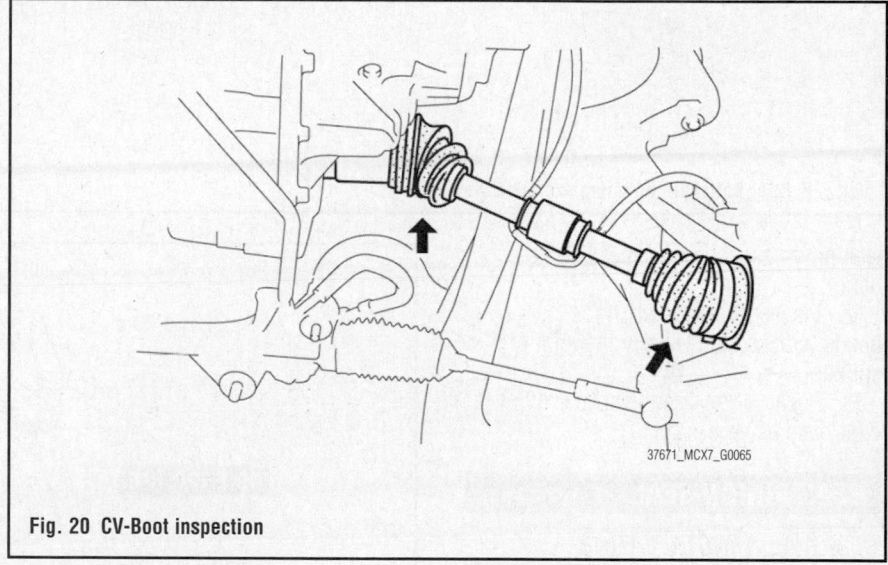

37671_MCX7_G0065

Fig. 20 CV-Boot inspection

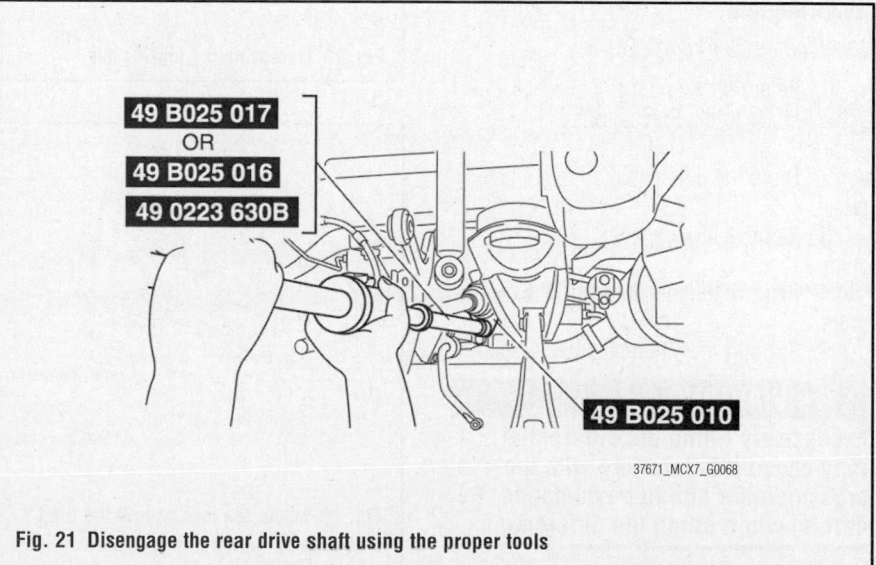

49 B025 017
OR
49 B025 016
49 0223 630B

49 B025 010

37671_MCX7_G0068

Fig. 21 Disengage the rear drive shaft using the proper tools

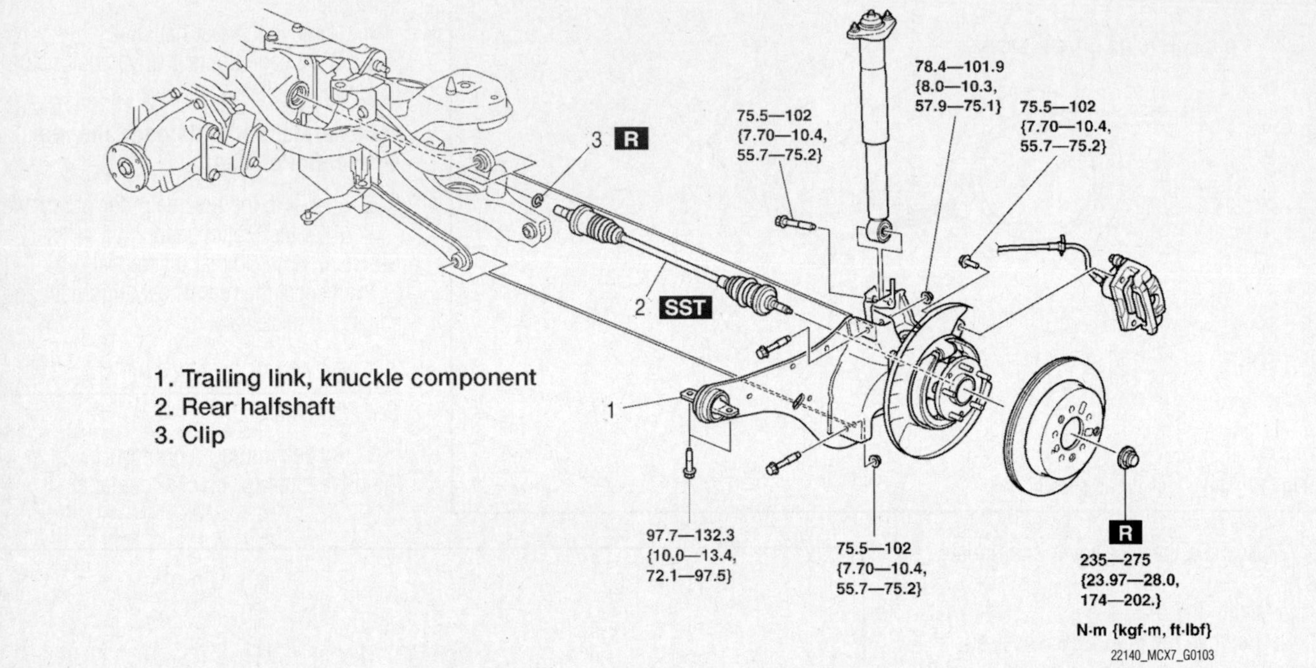

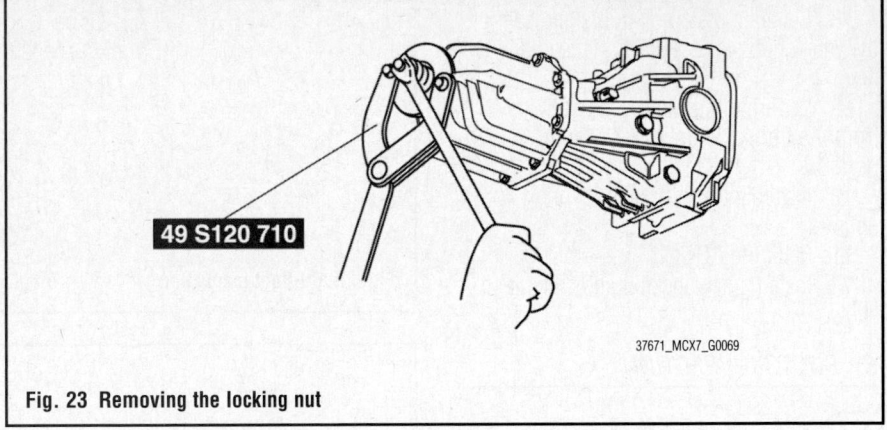

1. Trailing link, knuckle component
2. Rear halfshaft
3. Clip

78.4—101.9
{8.0—10.3,
57.9—75.1}

75.5—102
{7.70—10.4,
55.7—75.2}

75.5—102
{7.70—10.4,
55.7—75.2}

97.7—132.3
{10.0—13.4,
72.1—97.5}

75.5—102
{7.70—10.4,
55.7—75.2}

235—275
{23.97—28.0,
174—202.}

N·m {kgf·m, ft·lbf}

22140_MCX7_G0103

Fig. 22 Rear halfshaft mounting exploded view

leaking grease, and looseness in the boot band.

2. Inspect the drive shaft for bends, cracks, and wear in the joint or splines

3. Repair or replace the drive shaft or boot/band as necessary.

REAR PINION SEAL

REMOVAL & INSTALLATION

AWD Models

See Figures 23 through 26.

1. Before servicing the vehicle, refer to the Precautions Section.

2. Drain the differential oil into a container.

3. Remove the propeller shaft.

4. Remove the locknut using the proper tools.

5. Support the differential using a jack.

✻ WARNING

Excessively tilting the differential may cause it to interfere with the crossmember and to be damaged. Be careful when tilting the differential.

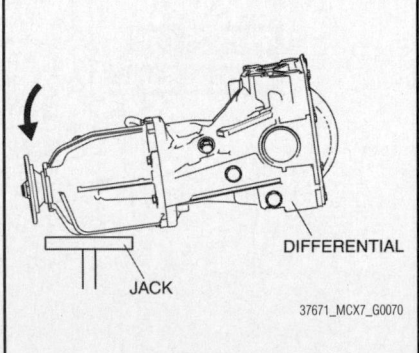

Fig. 23 Removing the locking nut

49 S120 710

37671_MCX7_G0069

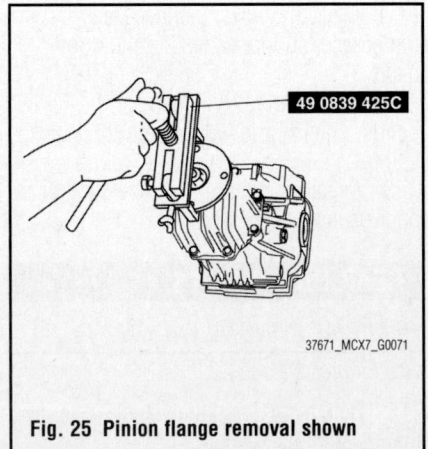

DIFFERENTIAL

JACK

37671_MCX7_G0070

Fig. 24 Lower the jack and tilt the front side of the differential downward

49 0839 425C

37671_MCX7_G0071

Fig. 25 Pinion flange removal shown

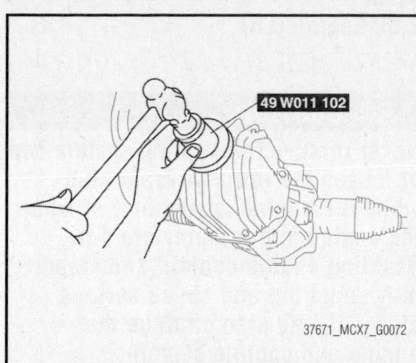

Fig. 26 Carefully installing the new pinion seal

37671_MCX7_G0072

6. After removing the front side of the differential mount, gradually lower the jack and tilt the front side of the differential downward.

7. Pull the pinion flange off using a puller.

8. Remove the oil seal from the differential carrier using a screwdriver or similar tool.

To install:

9. Apply differential oil to the new oil seal lip.

10. Install the oil seal at a straight angle.

11. Install the new oil seal to the differential carrier using a seal driver.

12. Install the companion flange to the drive pinion.

13. Jack up the differential and install the front side of the differential mount. Tighten the mount to 56–75 ft. lbs. (76–102 Nm)

14. Install the washer and a new locknut.

15. Tighten the locknut to 124–166 ft. lbs. (167–226 Nm)

16. Install the propeller shaft.

17. Add the specified differential oil, 2 pts API service GL-5 SAE 80W-90.

ENGINE COOLING

ENGINE FAN

REMOVAL & INSTALLATION

2.3L Engine (L3)

See Figure 27.

1. Before servicing the vehicle, refer to the Precautions section.

2. Disconnect the negative battery cable.

3. Remove the undercover.

4. Drain the engine coolant.

5. Remove the charge air cooler duct, air cleaner and fresh air duct component.

6. Remove the coolant reserve tank.

7. Remove the dipstick pipe.

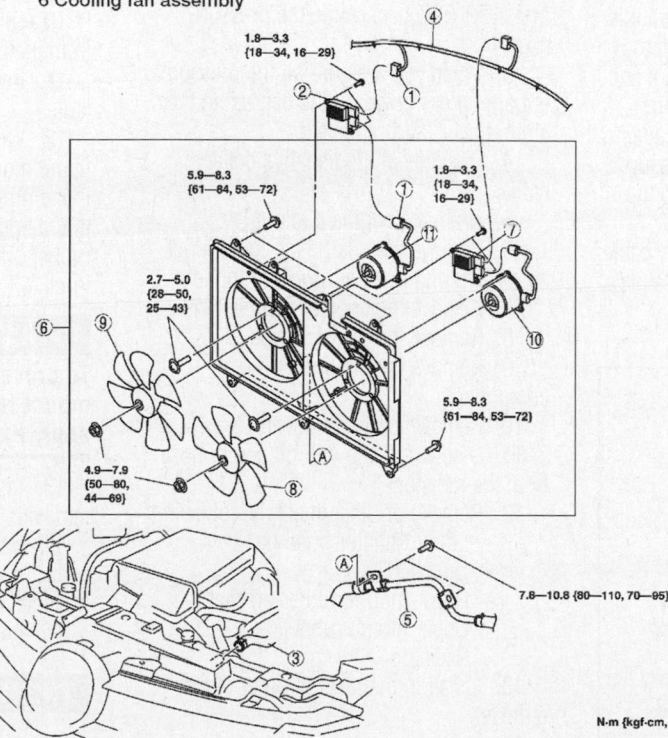

1 Fan control module connectors
2 Fan control module No.2
3 Upper radiator hose
4 Wiring harness
5 ATF oil cooler pipe
6 Cooling fan assembly
7 Fan control module No.1
8 Cooling fan No.1
9 Cooling fan No.2
10 Cooling fan motor No.1
11 Cooling fan motor No.2

N·m {kgf-cm, in-lbf}

22140_MCX7_G0115

Fig. 27 Cooling fans exploded view

8. Remove or disconnect the following:
- Fan control module connectors
- Fan control module No.2
- Upper radiator hose
- Wiring harness
- ATF oil cooler pipe
- Cooling fan assembly
- Fan control module No.1
- Cooling fan No.1
- Cooling fan No.2
- Cooling fan motor No.1
- Cooling fan motor No.2

9. Install in the reverse order of removal.
10. Refill the engine coolant.
11. Inspect for engine coolant leakage.

2.5L Engine (L5)

See Figures 27 and 28.

> ※※ **CAUTION**
>
> Never remove the cooling system cap or loosen the radiator drain plug while the engine is running, or when the engine and radiator are hot. Scalding engine coolant and steam may shoot out and cause serious injury. It may also damage the engine and cooling system.

1. Before servicing the vehicle, refer to the Precautions section.
2. Turn off the engine and wait until it is cool. Even then, be very careful when removing the cap. Wrap a thick cloth around it and slowly turn it counterclockwise to the first stop. Step back while the pressure escapes.
3. When you are sure all the pressure is gone, press down on the cap using the cloth, turn it, and remove it.
4. Disconnect the negative battery cable.
5. Lift off and remove the plug hole plate.
6. Drain the engine coolant.

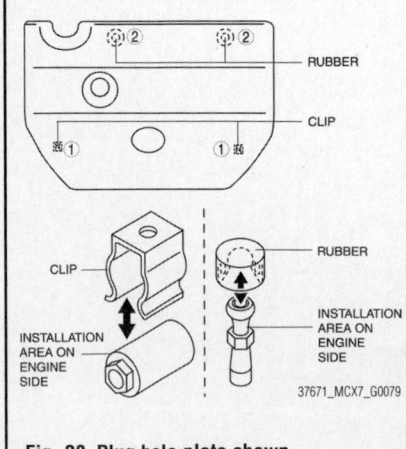

Fig. 28 Plug hole plate shown

RUBBER
CLIP
CLIP
RUBBER
INSTALLATION AREA ON ENGINE SIDE
INSTALLATION AREA ON ENGINE SIDE
37671_MCX7_G0079

7. Remove the air hose, air cleaner and fresh air duct component.
8. Remove the coolant reserve tank.
9. Remove in the order indicated in the table.
10. Install in the reverse order of removal.
11. Refill the engine coolant.
12. Inspect for engine coolant leakage.

RADIATOR

REMOVAL & INSTALLATION

2.3L Engine (L3)

See Figures 29 and 30.

> ※※ **CAUTION**
>
> Never remove the cooling system cap or loosen the radiator drain plug while the engine is running, or when the engine and radiator are hot. Scalding engine coolant and steam may shoot out and cause serious injury. It may also damage the engine and cooling system.

1. Before servicing the vehicle, refer to the Precautions section.
2. Turn off the engine and wait until it is cool. Even then, be very careful when removing the cap. Wrap a thick cloth around it and slowly turn it counterclockwise to the first stop. Step back while the pressure escapes.
3. When you are sure all the pressure is gone, press down on the cap using the cloth, turn it, and remove it.
4. Disconnect the negative battery cable.
5. Drain the engine coolant.
6. Remove the charge air cooler duct, air cleaner and fresh air duct component.
7. Remove the coolant reserve tank.
8. Remove the dipstick pipe.
9. Remove the cooling fan component.
10. Remove the undercover.
11. Disconnect the ATF oil cooler hose from the radiator.
12. Remove or disconnect the following:
- Lower radiator hose
- Duct clip
- Upper mount rubber bracket
- Upper mount rubber
- Radiator

13. Install in the reverse order of the removal.
14. Refill the engine coolant.
15. Inspect for engine coolant leakage.
16. Inspect the ATF level.

2.5L Engine (L5)

See Figures 29, 31 and 32.

> ※※ **CAUTION**
>
> Never remove the cooling system cap or loosen the radiator drain plug while the engine is running, or when the engine and radiator are hot. Scalding engine coolant and steam may shoot out and cause serious injury. It may also damage the engine and cooling system.

1. Before servicing the vehicle, refer to the Precautions section.
2. Turn off the engine and wait until it is cool. Even then, be very careful when removing the cap. Wrap a thick cloth around it and slowly turn it counterclockwise to the first stop. Step back while the pressure escapes.
3. When you are sure all the pressure is gone, press down on the cap using the cloth, turn it, and remove it.
4. Disconnect the negative battery cable.
5. Remove the plug hole plate.
6. Drain the engine coolant.
7. Remove the air hose, air cleaner and fresh air duct component.
8. Remove the coolant reserve tank.
9. Remove the cooling fan component. (Refer to Engine Fan, Removal & Installation)
10. Disconnect the ATF oil cooler hose from the radiator.
11. Remove in the order indicated in the table.
12. While pressing down the upper mount rubber bracket tab lightly in the direction of the arrow shown in the figure, press the upper mount rubber bracket toward the vehicle front.

> ※※ **WARNING**
>
> To prevent damage to the upper mount rubber bracket tab, do not apply excessive force.

13. Install in the reverse order of the removal.
14. Refill the engine coolant.
15. Inspect for engine coolant leakage.
16. Inspect the ATF level.

THERMOSTAT

REMOVAL & INSTALLATION

2.3L Engine (L3)

See Figure 33.

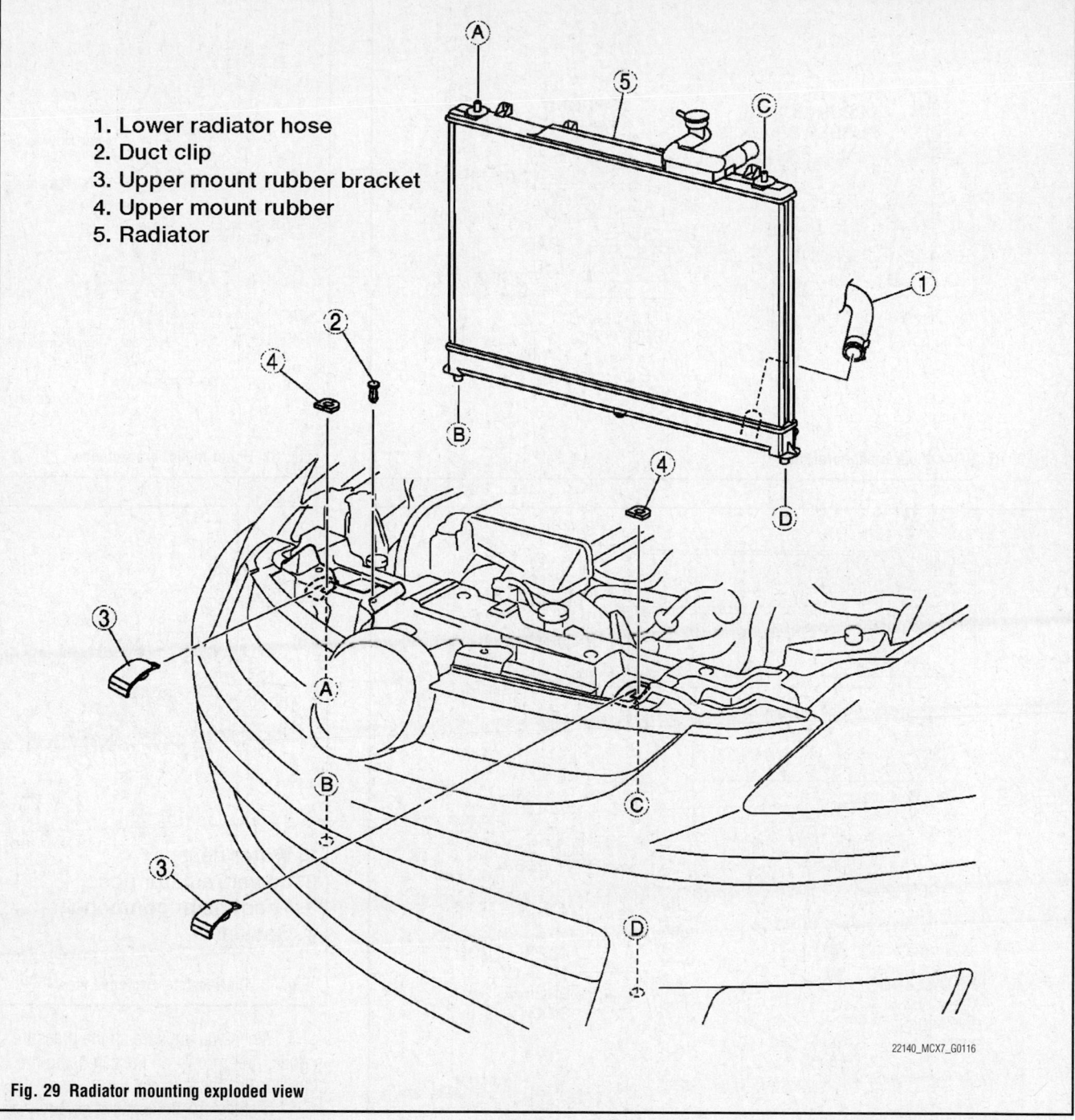

1. Lower radiator hose
2. Duct clip
3. Upper mount rubber bracket
4. Upper mount rubber
5. Radiator

Fig. 29 Radiator mounting exploded view

22140_MCX7_G0116

☀☀ CAUTION

Never remove the cooling system cap or loosen the radiator drain plug while the engine is running, or when the engine and radiator are hot. Scalding engine coolant and steam may shoot out and cause serious injury. It may also damage the engine and cooling system.

1. Before servicing the vehicle, refer to the Precautions section.

2. Turn off the engine and wait until it is cool. Even then, be very careful when removing the cap. Wrap a thick cloth around it and slowly turn it counterclockwise to the first stop. Step back while the pressure escapes.

3. When you are sure all the pressure is gone, press down on the cap using the cloth, turn it, and remove it.

4. Disconnect the negative battery cable.

5. Drain the engine coolant.

6. Remove the front splash shield (RH).

7. Remove the charge air cooler duct.

8. Remove the drive belt.

9. Remove the P/S oil pump with hose and pipe still connected. Position the P/S oil pump out of the way.

10. Remove or disconnect the following:
- Water hose
- Lower radiator hose
- Thermostat component
- Gasket

11. Install in the reverse order of removal.

12. Refill the engine coolant.

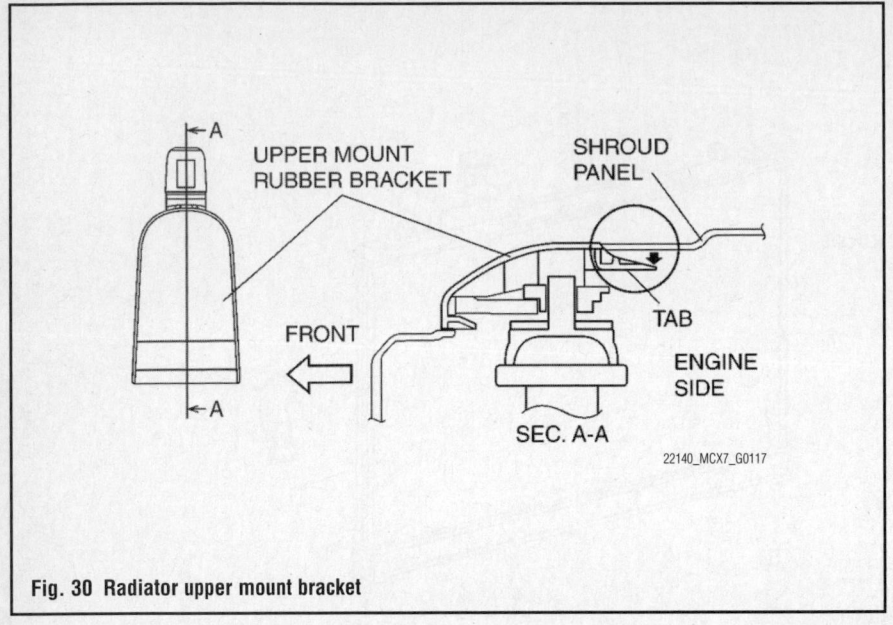

Fig. 30 Radiator upper mount bracket

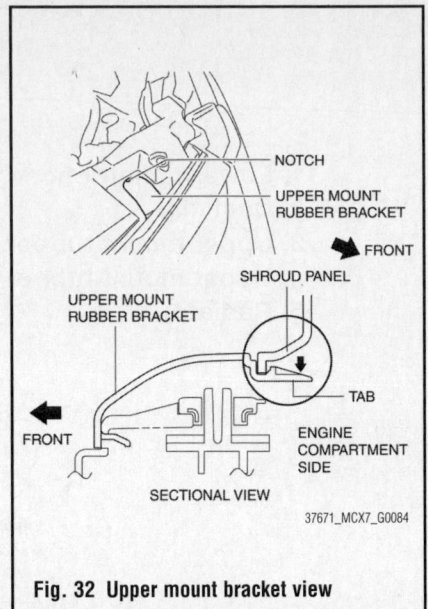

Fig. 32 Upper mount bracket view

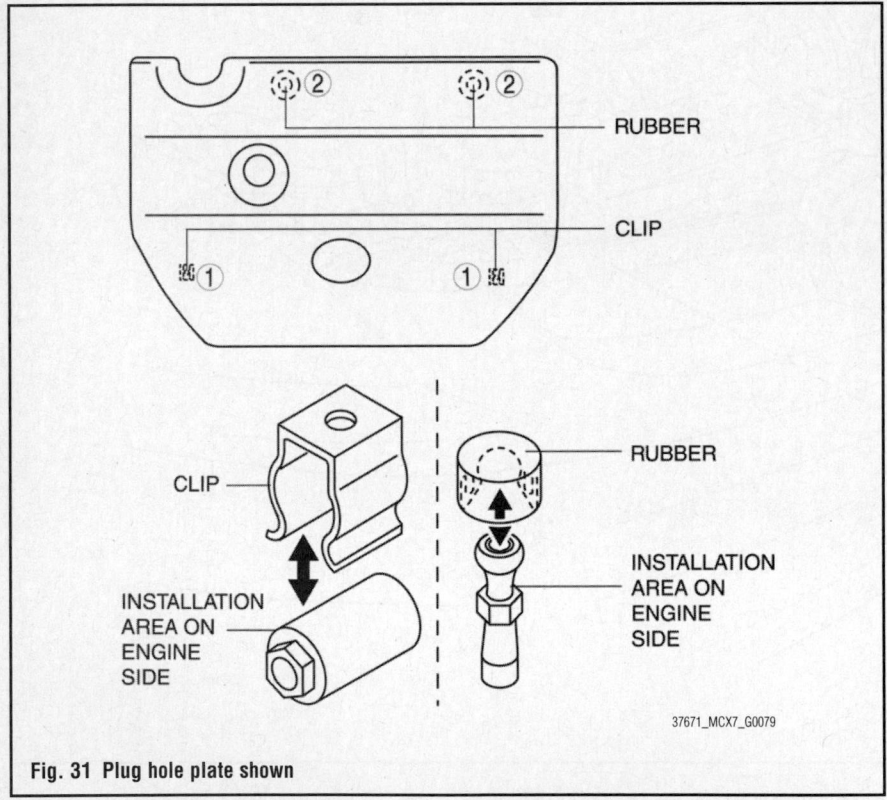

Fig. 31 Plug hole plate shown

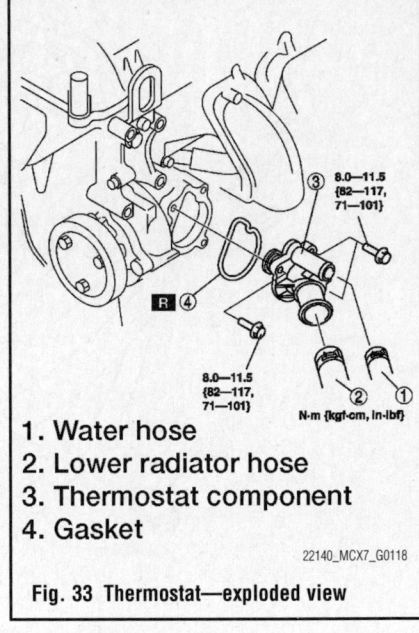

1. Water hose
2. Lower radiator hose
3. Thermostat component
4. Gasket

Fig. 33 Thermostat—exploded view

13. Inspect for engine coolant leakage.

2.5L Engine (L5)

See Figure 33.

✳✳ CAUTION

Never remove the cooling system cap or loosen the radiator drain plug while the engine Is running, or when the engine and radiator are hot. Scalding engine coolant and steam may shoot out and cause serious injury. It may also damage the engine and cooling system.

1. Before servicing the vehicle, refer to the Precautions section.

2. Turn off the engine and wait until it is cool. Even then, be very careful when removing the cap. Wrap a thick cloth around it and slowly turn it counterclockwise to the first stop. Step back while the pressure escapes.

3. When you are sure all the pressure is gone, press down on the cap using the cloth, turn it, and remove it.

4. Disconnect the negative battery cable.

5. Remove the plug hole plate.

6. Drain the engine coolant.

7. Remove the drive belt.

8. Remove the P/S oil pump with the hose and pipe still connected. Position the P/S oil pump out of the way.

9. Remove in the order indicated in the table.

10. Install in the reverse order of removal.

11. Refill the engine coolant.

12. Inspect for engine coolant leakage.

WATER PUMP

REMOVAL & INSTALLATION

2.3L Engine (L3)

See Figure 34.

❋❋ CAUTION

Never remove the cooling system cap or loosen the radiator drain plug while the engine is running, or when the engine and radiator are hot. Scalding engine coolant and steam may shoot out and cause serious injury. It may also damage the engine and cooling system.

1. Before servicing the vehicle, refer to the Precautions section.
2. Disconnect the negative battery cable.
3. Drain the engine coolant.
4. Remove the front splash shield (RH).
5. Remove the charge air cooler duct.
6. Loosen the water pump pulley bolts before removing the drive belt.
7. Remove the drive belt.
8. Remove the water pump pulley.
9. Remove the water pump and O-ring seal.

➡**Use a new O-ring seal for assembly. Lubricate the seal with engine coolant.**

10. Install in the reverse order of removal.

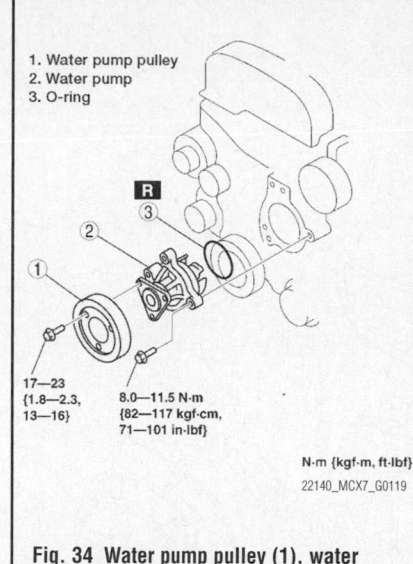

1. Water pump pulley
2. Water pump
3. O-ring

17—23
{1.8—2.3, 13—16}

8.0—11.5 N·m
{82—117 kgf·cm, 71—101 in·lbf}

N·m {kgf·m, ft·lbf}

22140_MCX7_G0119

Fig. 34 Water pump pulley (1), water pump (2) and O-ring (3)

11. Refill the engine coolant.
12. Inspect for engine coolant leakage.

2.5L Engine (L5)

See Figure 39.

❋❋ CAUTION

Never remove the cooling system cap or loosen the radiator drain plug while the engine is running, or when the engine and radiator are hot. Scalding engine coolant and steam may shoot out and

cause serious injury. It may also damage the engine and cooling system.

1. Before servicing the vehicle, refer to the Precautions section.
2. Turn off the engine and wait until it is cool. Even then, be very careful when removing the cap. Wrap a thick cloth around it and slowly turn it counterclockwise to the first stop. Step back while the pressure escapes.
3. When you are sure all the pressure is gone, press down on the cap using the cloth, turn it, and remove it.
4. Disconnect the negative battery cable.
5. Remove the plug hole plate.
6. Drain the engine coolant.
7. Loosen the water pump pulley bolts before removing the drive belt.
8. Remove the drive belt.
9. Remove the P/S oil pump with the hose and pipe still connected. Position the P/S oil pump out of the way.
10. Remove the water pump pulley.
11. Remove the water pump and O-ring seal.

➡**Use a new O-ring seal for assembly. Lubricate the seal with engine coolant.**

12. Install in the reverse order of removal.
13. Refill the engine coolant.
14. Inspect for engine coolant leakage.

ENGINE ELECTRICAL

ALTERNATOR

REMOVAL & INSTALLATION

2.3L Engine (L3)

See Figure 35.

❋❋ CAUTION

Remove and install all parts when the engine is cold, otherwise they can cause severe burns or serious injury.

❋❋ WARNING

When the battery cables are connected, touching the vehicle body with alternator terminal B will generate sparks. This can cause personal injury, fire, and damage to the electrical

components. Always disconnect the negative battery cable before performing the following operation.

❋❋ WARNING

The alternator can be damaged by the heat from the exhaust manifold. Make sure the alternator duct is installed securely.

1. Before servicing the vehicle, refer to the Precautions section.
2. Disconnect the negative battery cable.
3. Remove the charge air cooler duct and charge air cooler cover.
4. Remove the front splash shield (RH).

CHARGING SYSTEM

5. Remove the drive belt.
6. Remove the alternator duct.
7. Remove the wiring harness bracket.
8. Disconnect the terminal B cable.
9. Disconnect the alternator connector.
10. Remove the alternator.
11. Install in the reverse order of removal.

2.5L Engine (L5)

See Figures 35 through 37.

❋❋ CAUTION

Remove and install all parts when the engine is cold, otherwise they can cause severe burns or serious injury.

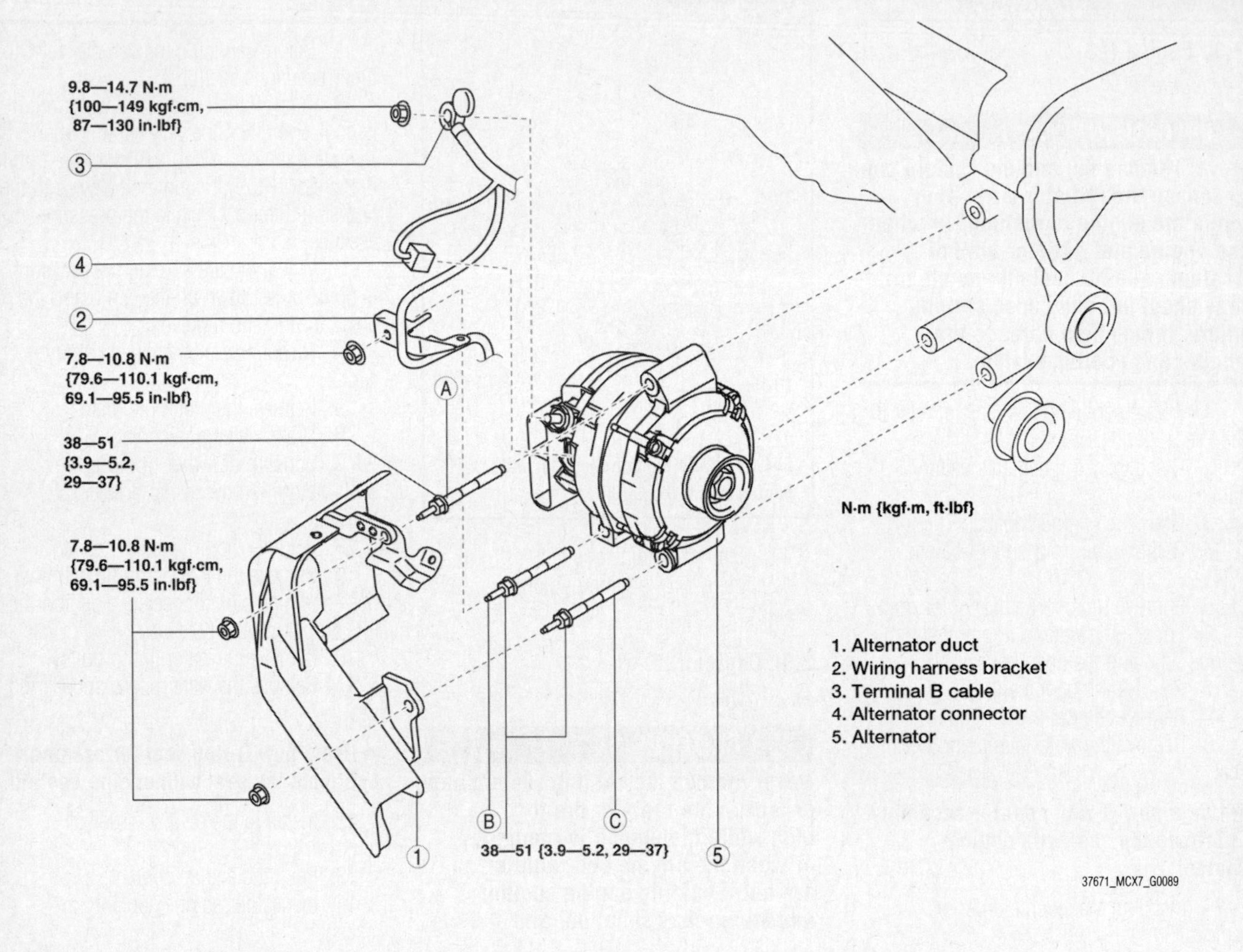

9.8—14.7 N·m
{100—149 kgf·cm,
87—130 in·lbf}

③

④

②

7.8—10.8 N·m
{79.6—110.1 kgf·cm,
69.1—95.5 in·lbf}

38—51
{3.9—5.2,
29—37}

7.8—10.8 N·m
{79.6—110.1 kgf·cm,
69.1—95.5 in·lbf}

A

B   C

38—51 {3.9—5.2, 29—37}

N·m {kgf·m, ft·lbf}

1. Alternator duct
2. Wiring harness bracket
3. Terminal B cable
4. Alternator connector
5. Alternator

37671_MCX7_G0089

Fig. 35 Alternator—exploded view

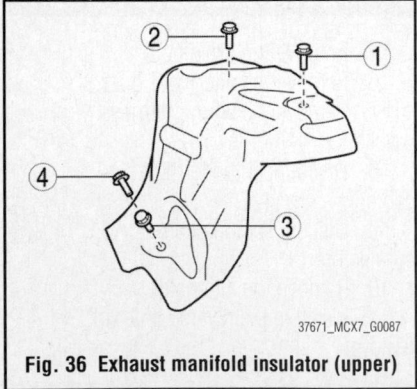

37671_MCX7_G0087

Fig. 36 Exhaust manifold insulator (upper)

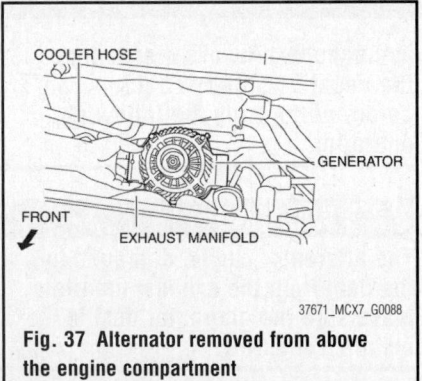

COOLER HOSE

GENERATOR

FRONT

EXHAUST MANIFOLD

37671_MCX7_G0088

Fig. 37 Alternator removed from above the engine compartment

the negative battery cable before performing the following operation.

generator duct is installed securely.

1. Before servicing the vehicle, refer to the Precautions section.
2. Disconnect the negative battery cable.
3. Remove the drive belt.
4. Remove the undercover.
5. Remove the exhaust manifold insulator. (Tighten in the sequence shown)
6. Remove the alternator duct.
7. Remove the wiring harness bracket.
8. Disconnect the terminal B cable.
9. Disconnect the alternator connector.
10. Remove the alternator from above the engine compartment.
11. Install in the reverse order of removal.
12. Connect the negative battery cable.

FIRING ORDER

See Figure 38.

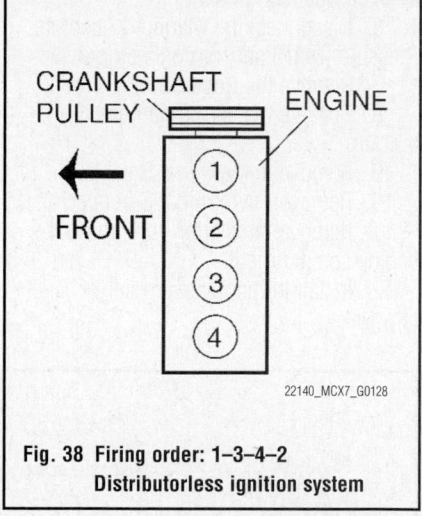

CRANKSHAFT PULLEY   ENGINE

FRONT

① ② ③ ④

22140_MCX7_G0128

**Fig. 38 Firing order: 1–3–4–2
Distributorless ignition system**

IGNITION COIL

REMOVAL & INSTALLATION

See Figure 39.

1. Before servicing the vehicle, refer to the Precautions section.
2. Disconnect the negative battery cable.
3. Remove the charge air cooler.
4. Disconnect the ignition coil connectors.

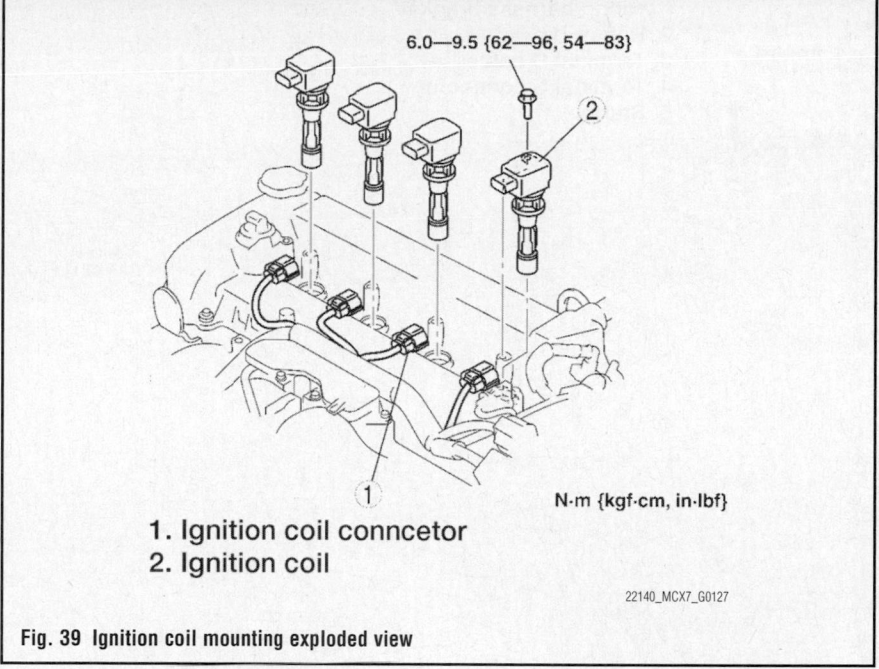

6.0—9.5 {62—96, 54—83}

2

1

**1. Ignition coil conncetor
2. Ignition coil**

N·m {kgf·cm, in·lbf}

22140_MCX7_G0127

Fig. 39 Ignition coil mounting exploded view

5. Remove the ignition coils.
6. Install in the reverse order of removal.

SPARK PLUGS

REMOVAL & INSTALLATION

1. Before servicing the vehicle, refer to the Precautions section.

2. Disconnect the negative battery cable.
3. Remove the charge air cooler.
4. Remove the ignition coils.
5. Remove the spark plugs using a plug-wrench.
6. Install in the reverse order of removal. Tighten the spark plugs to 96–120 inch lbs. (10–14 Nm).

STARTER

REMOVAL & INSTALLATION

2.3L Engine (L3)

See Figure 40.

☀ WARNING

Remove and install all parts when the engine is cold, otherwise they can cause severe burns or serious injury.

☀ WARNING

When the battery cables are connected, touching the vehicle body with starter terminal B will generate sparks. This can cause personal injury, fire, and damage to the electrical components. Always

disconnect the negative battery cable before performing the following operation.

1. Before servicing the vehicle, refer to the Precautions section.
2. Remove the battery and battery tray.
3. Remove the charge air cooler duct, air cleaner and fresh air duct component, and air hose.
4. Remove the undercover.
5. Remove the wiring harness and bracket.
6. Remove the heater pipe with the water hoses still connected. Position the heater pipe so that it is out of the way.
7. Remove the terminal B cable and the terminal S connector.
8. Remove the starter.

9. Install in the reverse order of removal.

2.5L Engine (L5)

See Figures 40 and 41.

☀ CAUTION

Remove and install all parts when the engine is cold, otherwise they can cause severe burns or serious injury.

1. When the battery cables are connected, touching the vehicle body with starter terminal B will generate sparks. This can cause personal injury, fire, and damage to the electrical components. Always disconnect the negative battery cable before performing the following operation.

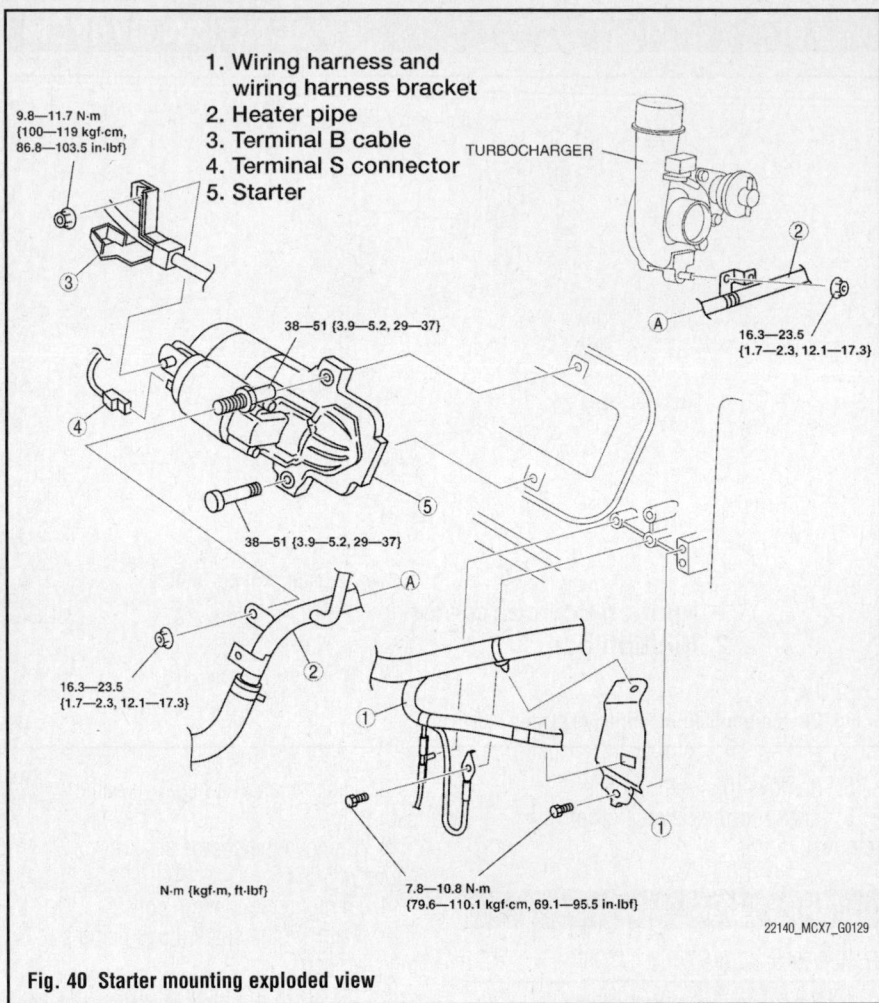

1. Wiring harness and
   wiring harness bracket
2. Heater pipe
3. Terminal B cable
4. Terminal S connector
5. Starter

9.8—11.7 N·m
{100—119 kgf-cm,
86.8—103.5 in-lbf}

TURBOCHARGER

38—51 {3.9—5.2, 29—37}

38—51 {3.9—5.2, 29—37}

16.3—23.5
{1.7—2.3, 12.1—17.3}

16.3—23.5
{1.7—2.3, 12.1—17.3}

N·m {kgf·m, ft-lbf}

7.8—10.8 N·m
{79.6—110.1 kgf-cm, 69.1—95.5 in-lbf}

22140_MCX7_G0129

Fig. 40 Starter mounting exploded view

2. Before servicing the vehicle, refer to the Precautions section.

3. Disconnect the negative battery cable.

4. Remove the plug hole plate.

5. Remove the air hose, air cleaner and fresh air duct component.

6. Disconnect the Manifold Absolute Pressure (MAP) sensor connector.

7. Remove the undercover.

8. Remove the wiring harness bracket.

9. Remove the terminal B cable.

10. Remove the terminal S connector.

11. Remove the starter from above the engine compartment.

12. Install in the reverse order of removal.

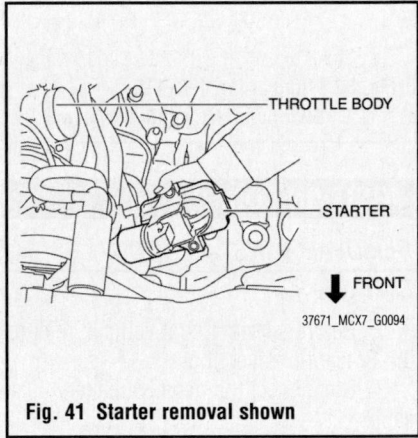

THROTTLE BODY

STARTER

FRONT

37671_MCX7_G0094

Fig. 41 Starter removal shown

ENGINE MECHANICAL

➡Disconnecting the negative battery cable may interfere with the functions of the on board computer systems and may require the computer to undergo a relearning process, once the negative battery cable is reconnected.

ACCESSORY DRIVE BELTS

ACCESSORY BELT ROUTING

See Figure 42.

Refer to the accompanying illustration for belt routing.

INSPECTION

See Figure 43.

Inspect the drive belt for signs of glazing or cracking. A glazed belt will be perfectly smooth from slippage, while a good belt will have a slight texture of fabric visible. Cracks will usually start at the inner edge of the belt and run outward. All worn or damaged drive belts should be replaced immediately.

➡Drive belt deflection/tension inspection is not necessary because of the use of the drive belt auto tensioner.

1. Verify that the drive belt auto tensioner indicator mark does not exceed the limit.

2. If it exceeds the limit, replace the drive belt.

3. Before servicing the vehicle, refer to the Precautions Section.

4. Remove the camshaft from the engine.

5. Check the camshaft bearing journals for damage and binding.

6. If the journals are binding, check the cylinder head for damage.

7. Check the cylinder head for clogged oil holes.

8. Check the camshaft surface for abnormal wear and damage. Replace the camshaft, as required.

9. Measure the camshaft lobe surface

and replace the camshaft if not within specification.

10. Measure the camshaft journal diameter and replace the camshaft if not within specification.

11. Measure the camshaft run out and replace the camshaft if not within specification.

ADJUSTMENT

A drive belt auto tensioner with an embedded coil spring has been adopted to automatically maintain the optimal drive belt tension.

REMOVAL & INSTALLATION

See Figure 44.

1. Before servicing the vehicle, refer to the Precautions section.

2. Remove the splash shield (RH).

3. Rotate the drive belt auto tensioner in the direction shown in the figure and remove the drive belt.

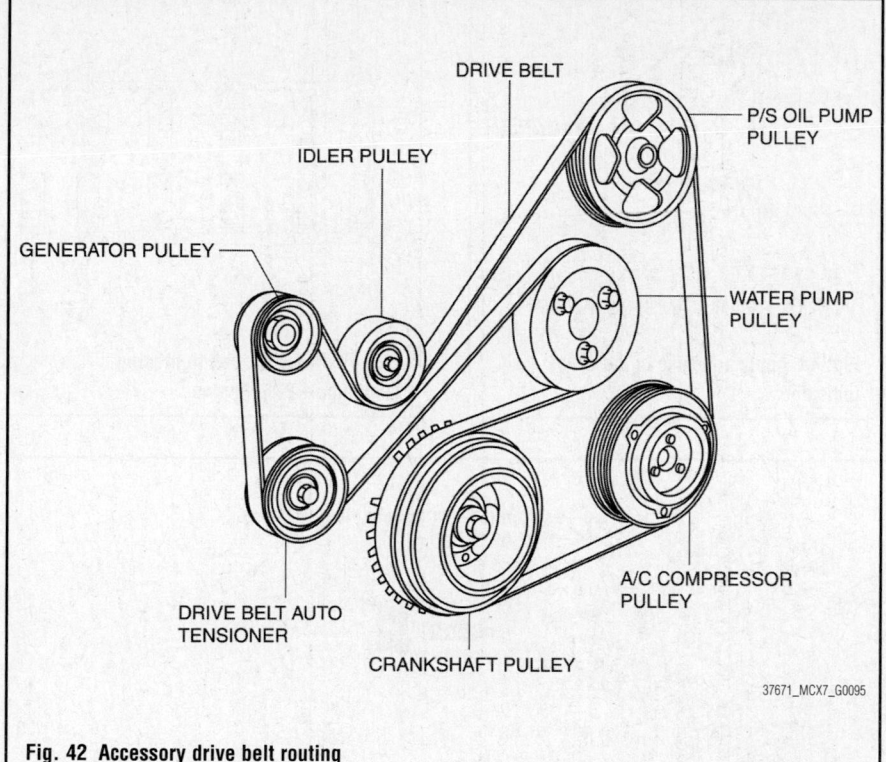

Fig. 42 Accessory drive belt routing

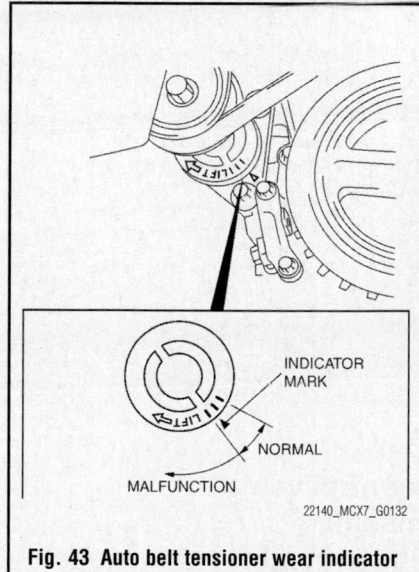

Fig. 43 Auto belt tensioner wear indicator

To install:

4. Install the drive belt.
5. Verify that the drive belt auto tensioner indicator mark does not exceed the limit.
6. Install the splash shield (RH).

CAMSHAFT AND TAPPETS

REMOVAL & INSTALLATION

2.3L Engine (L3)

See Figures 45 through 50.

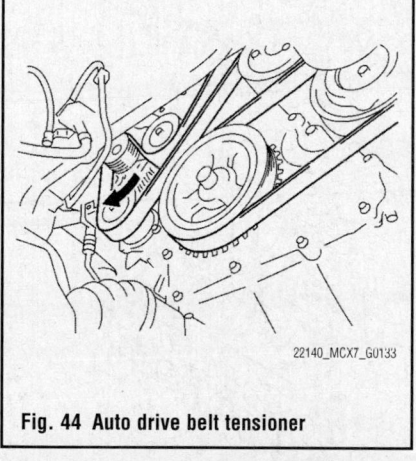

Fig. 44 Auto drive belt tensioner

✳✳ **WARNING**

Fuel vapor is hazardous. It can very easily ignite, causing death, serious injury, or damage. Always keep sparks and flames away from fuel.

✳✳ **WARNING**

Fuel line spills and leakage from the pressurized fuel system are dangerous. Fuel can easily ignite and cause serious injury or death and damage. Fuel can also irritate skin and eyes. To prevent this, always perform the Fuel Line Safety Procedure.

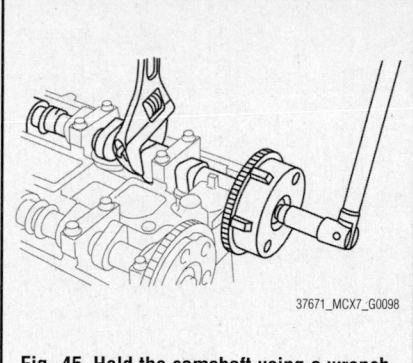

Fig. 45 Hold the camshaft using a wrench on the cast hexagon

1. Before servicing the vehicle, refer to the Precautions section.
2. Remove the timing chain. Refer to Timing Chain Removal & Installation.
3. Before removing the camshaft, hold the camshaft using a wrench on the cast hexagon and loosen the camshaft sprocket installation bolt.

➡**Before removing the camshafts inspect camshaft end play.**

4. Loosen the camshaft cap bolts in two or three passes in the order shown and remove the end caps.

➡**Note the position of each removed end cap for installation.**

5. Remove the camshafts.

To install:

6. Apply a small amount of gear oil to the camshaft journal of the cylinder head before you install the caps.

➡**Install the camshaft with number 1 cylinder cam aligned at TDC of the compression stroke.**

7. Carefully apply adhesive to the area indicated in the graphic, so that it does not leak into the sliding part.
8. Install camshaft caps finger tight, then in two or three passes tighten the caps to:

- First pass: 4.3–79.5 inch lbs. (5–9 Nm) in the order shown.
- Second pass: 10–13 ft. lbs (14–17 Nm) in the order shown.

9. Tighten the camshaft sprocket installation bolts to 51–55 ft. lbs. (69–75 Nm).
10. Install the timing chain. Refer to Timing Chain Removal & Installation.
11. Refill the engine with coolant.
12. Bleed the air from the cooling system.
13. Inspect the compression pressure.

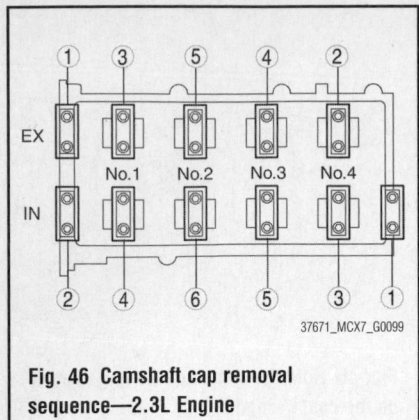

Fig. 46 Camshaft cap removal sequence—2.3L Engine

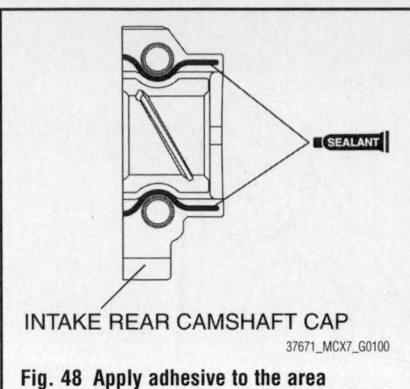

INTAKE REAR CAMSHAFT CAP

37671_MCX7_G0100

Fig. 48 Apply adhesive to the area indicated

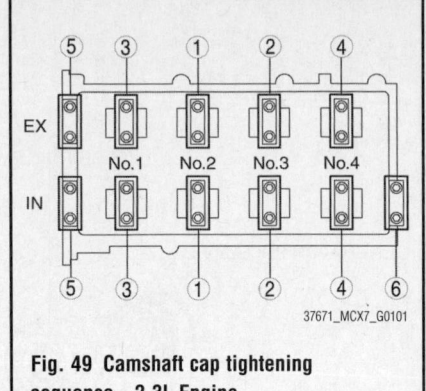

37671_MCX7_G0101

Fig. 49 Camshaft cap tightening sequence—2.3L Engine

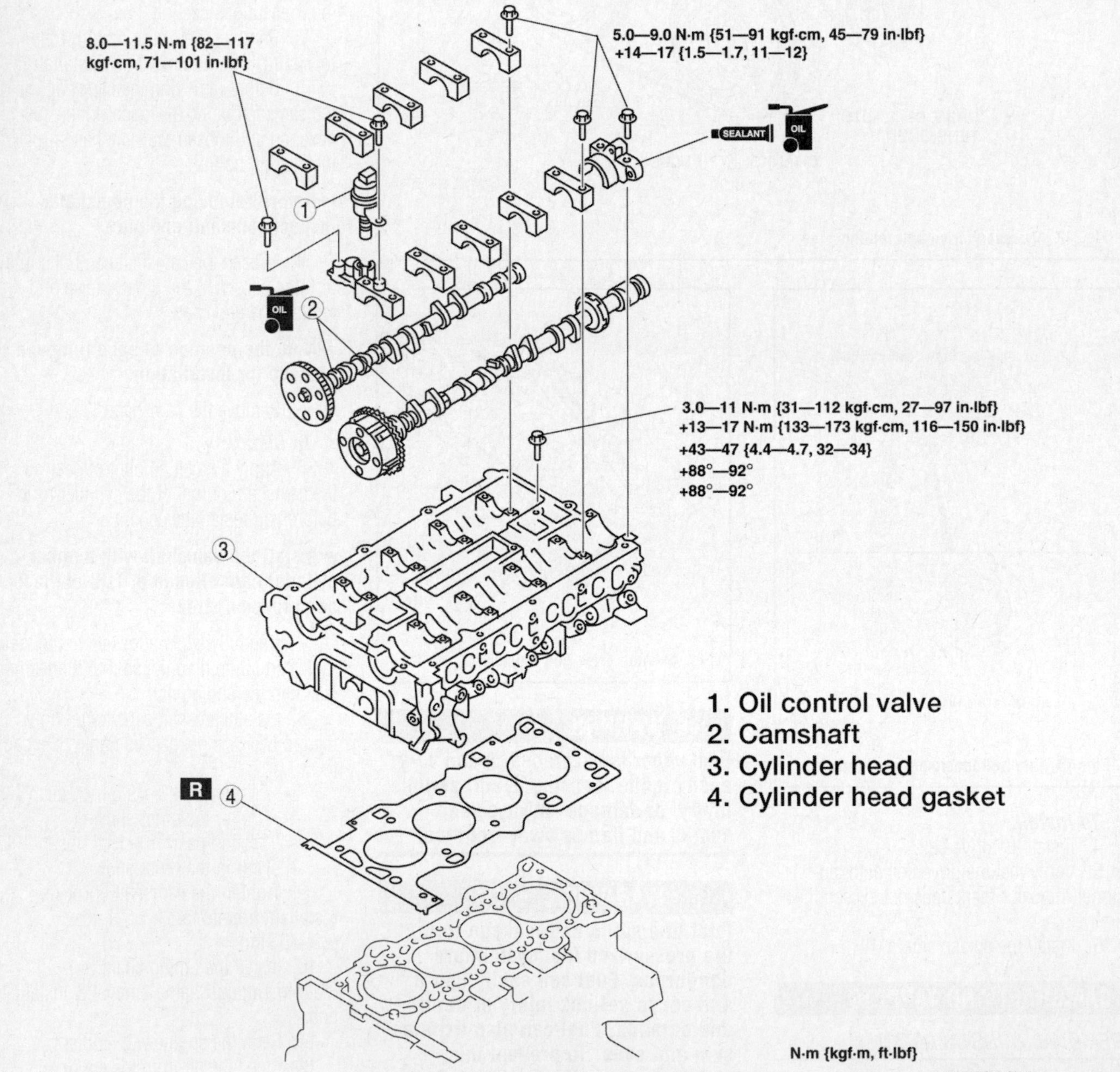

8.0—11.5 N·m {82—117 kgf·cm, 71—101 in·lbf}

5.0—9.0 N·m {51—91 kgf·cm, 45—79 in·lbf} +14—17 {1.5—1.7, 11—12}

SEALANT   OIL

OIL

3.0—11 N·m {31—112 kgf·cm, 27—97 in·lbf}
+13—17 N·m {133—173 kgf·cm, 116—150 in·lbf}
+43—47 {4.4—4.7, 32—34}
+88°—92°
+88°—92°

1. Oil control valve
2. Camshaft
3. Cylinder head
4. Cylinder head gasket

R

N·m {kgf·m, ft·lbf}

22140_MCX7_G0146

Fig. 47 Cylinder head exploded view—2.3L Engine

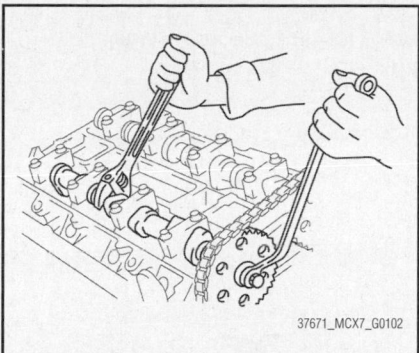

Fig. 50 Tightening the camshaft sprocket retaining bolt

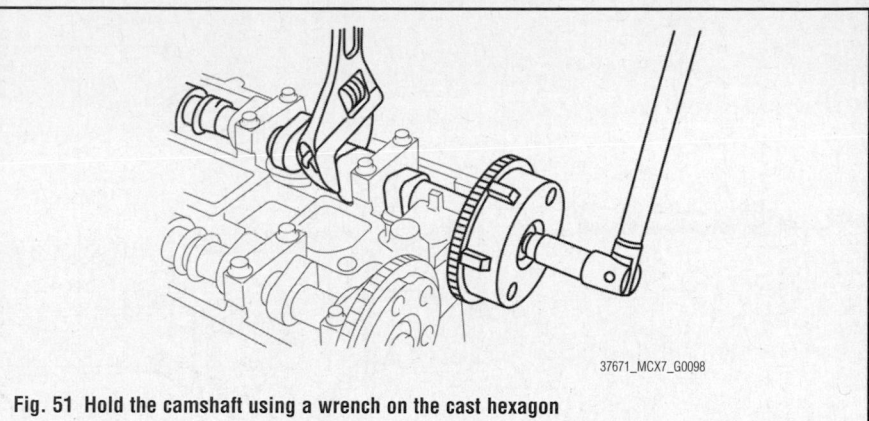

Fig. 51 Hold the camshaft using a wrench on the cast hexagon

2.5L Engine (L5)

See Figures 51 through 55.

> ✳✳ **WARNING**
>
> Fuel vapor is hazardous. It can very easily ignite, causing death, serious injury, or damage. Always keep sparks and flames away from fuel.

> ✳✳ **WARNING**
>
> Fuel line spills and leakage from the pressurized fuel system are dangerous. Fuel can easily ignite and cause serious injury or death and damage. Fuel can also irritate skin and eyes. To prevent this, always perform the Fuel Line Safety Procedure.

1. Before servicing the vehicle, refer to the Precautions section.
2. Remove the timing chain. Refer to Timing Chain Removal & Installation.
3. Before removing the camshaft, hold the camshaft using a wrench on the cast hexagon and loosen the camshaft sprocket installation bolt.

➡ **Before removing the camshafts inspect camshaft end play.**

4. Loosen the camshaft cap bolts in two or three passes in the order shown and remove the end caps.

➡ **Note the position of each removed end cap for installation.**

5. Remove the camshafts.

To install:

➡ **Install the camshaft with number 1 cylinder cam aligned at TDC of the compression stroke.**

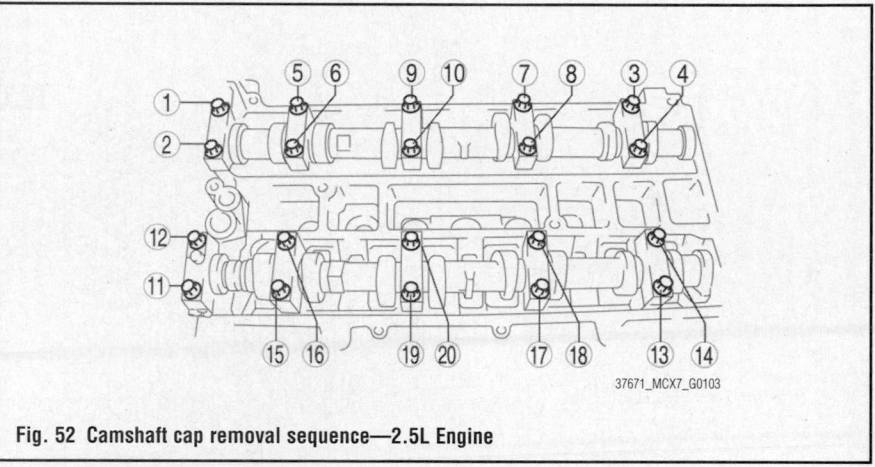

Fig. 52 Camshaft cap removal sequence—2.5L Engine

6. Apply a small amount of gear oil to the camshaft journal of the cylinder head before you install the caps.
7. Install camshaft caps finger tight, then in two or three passes tighten the caps to:

- First pass: 4.3–79.5 inch lbs. (5–9 Nm) in the order shown.
- Second pass: 10–13 ft. lbs (14–17 Nm) in the order shown.

8. Tighten the camshaft sprocket installation bolts to 51–55 ft. lbs. (69–75 Nm).
9. Install the timing chain. Refer to Timing Chain Removal & Installation.
10. Refill the engine with coolant.
11. Bleed the air from the cooling system.
12. Inspect the compression pressure.

CRANKSHAFT FRONT SEAL

REMOVAL & INSTALLATION

See Figures 56 and 57.

1. Before servicing the vehicle, refer to the Precautions section.

2. Remove the crankshaft pulley.
3. Remove the oil seal lip using a razor.
4. Remove the oil seal using a suitable prytool.

To install:

5. Apply clean engine oil to a new oil seal.
6. Insert the oil seal into the engine front cover.
7. Tap in the oil seal using SST 49 H010 401 as shown.
8. Replace the crankshaft pulley.

CYLINDER HEAD

REMOVAL & INSTALLATION

2.3L Engine (L3)

See Figures 58 and 59.

> ✳✳ **WARNING**
>
> Fuel vapor is hazardous. It can very easily ignite, causing death, serious injury, or damage. Always keep sparks and flames away from fuel.

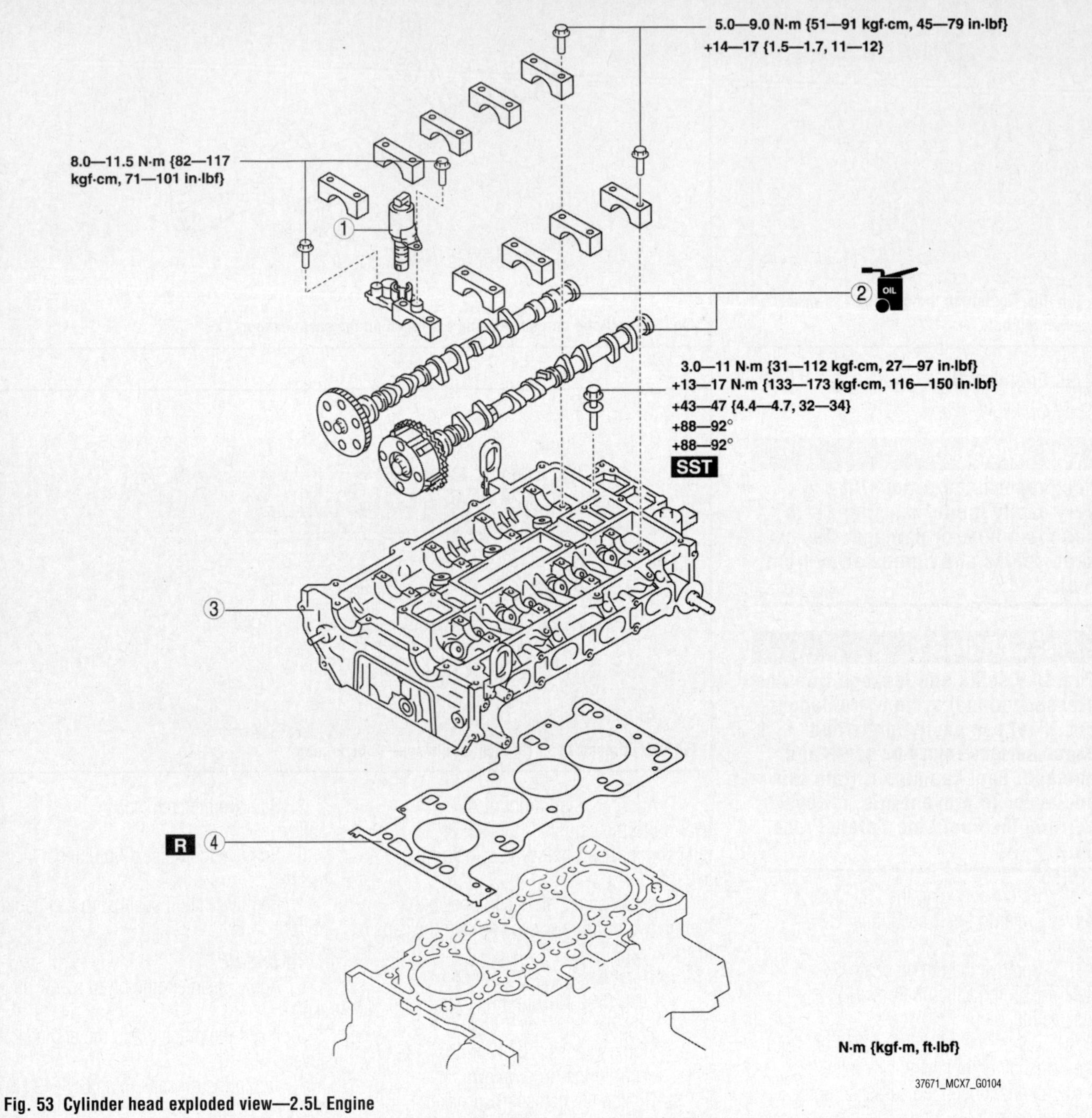

5.0—9.0 N·m {51—91 kgf·cm, 45—79 in·lbf}
+14—17 {1.5—1.7, 11—12}
8.0—11.5 N·m {82—117 kgf·cm, 71—101 in·lbf}
1
2
OIL
3.0—11 N·m {31—112 kgf·cm, 27—97 in·lbf}
+13—17 N·m {133—173 kgf·cm, 116—150 in·lbf}
+43—47 {4.4—4.7, 32—34}
+88—92°
+88—92°
SST
3
R
4
N·m {kgf·m, ft·lbf}
37671_MCX7_G0104

Fig. 53 Cylinder head exploded view—2.5L Engine

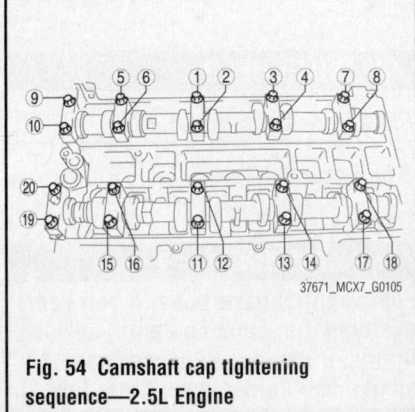

5 6
1 2
3 4
7 8
9
10
20
19
15 16
11 12
13 14
17 18
37671_MCX7_G0105

Fig. 54 Camshaft cap tightening sequence—2.5L Engine

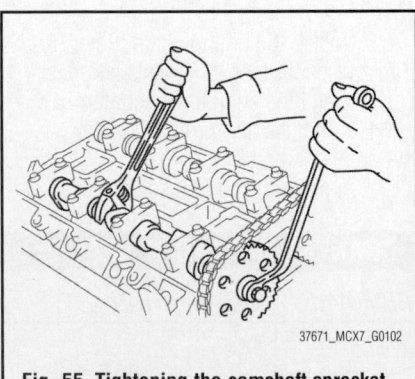

37671_MCX7_G0102

Fig. 55 Tightening the camshaft sprocket retaining bolt

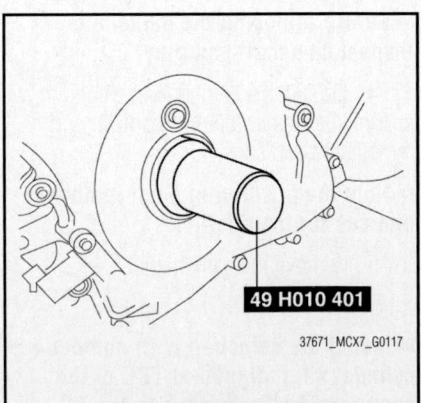

49 H010 401
37671_MCX7_G0117

Fig. 56 SST 49 H010 401 seal driver

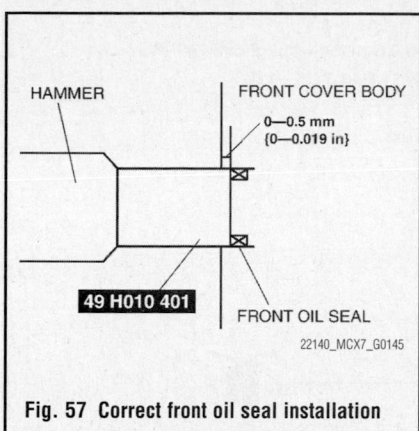

Fig. 57 Correct front oil seal installation

1. Before servicing the vehicle, refer to the Precautions section.
2. Drain the engine coolant
3. Remove the generator. Place the generator out of the way with the wiring harnesses connected.
4. Remove the exhaust manifold.
5. Remove the intake manifold.
6. Disconnect the heater hose and radiator hose.
7. Remove the timing chain. Refer to Timing Chain, Removal & Installation.
8. Remove the Camshafts. Refer to Camshaft, Removal & Installation.
9. Disconnect the wiring harness.
10. To firmly support the engine, first set the engine jack and attachment to the oil pan.
11. Remove the oil control valve
12. Remove the camshafts.
13. Remove the bolts and the cylinder head.

To install:

14. Measure the cylinder head bolts. Replace any bolts that exceed 5.740 inches (146 mm).
15. Install the cylinder head with a new gasket. Tighten the bolts in sequence as follows:
    a. Step 1: 27–97 inch lbs. (3–11 Nm)
    b. Step 2: 116–150 inch lbs. (13–17 Nm)
    c. Step 3: 32–34 ft. lbs. (43–47 Nm)
    d. Step 4: Plus 88°–92°
    e. Step 5: Plus 88°–92°
16. Install in the reverse order of removal.

17. Bleed the air from the cooling system.
18. Inspect the compression pressure.

2.5L Engine (L5)

See Figures 58 and 60.

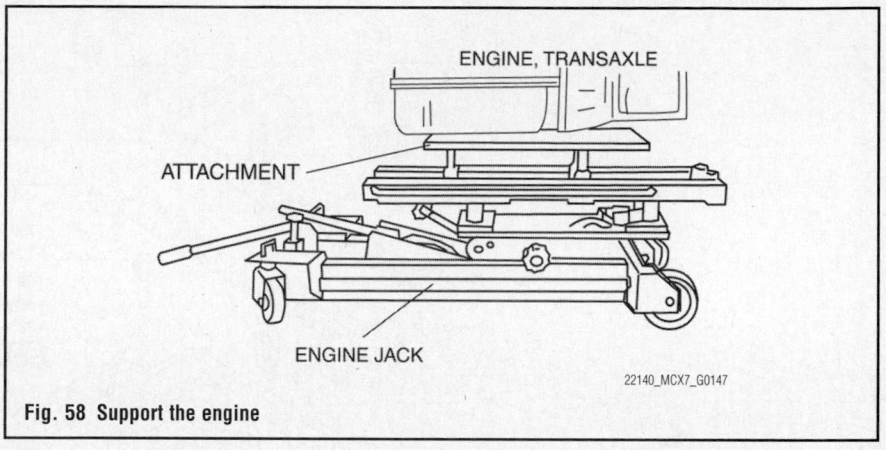

Fig. 58 Support the engine

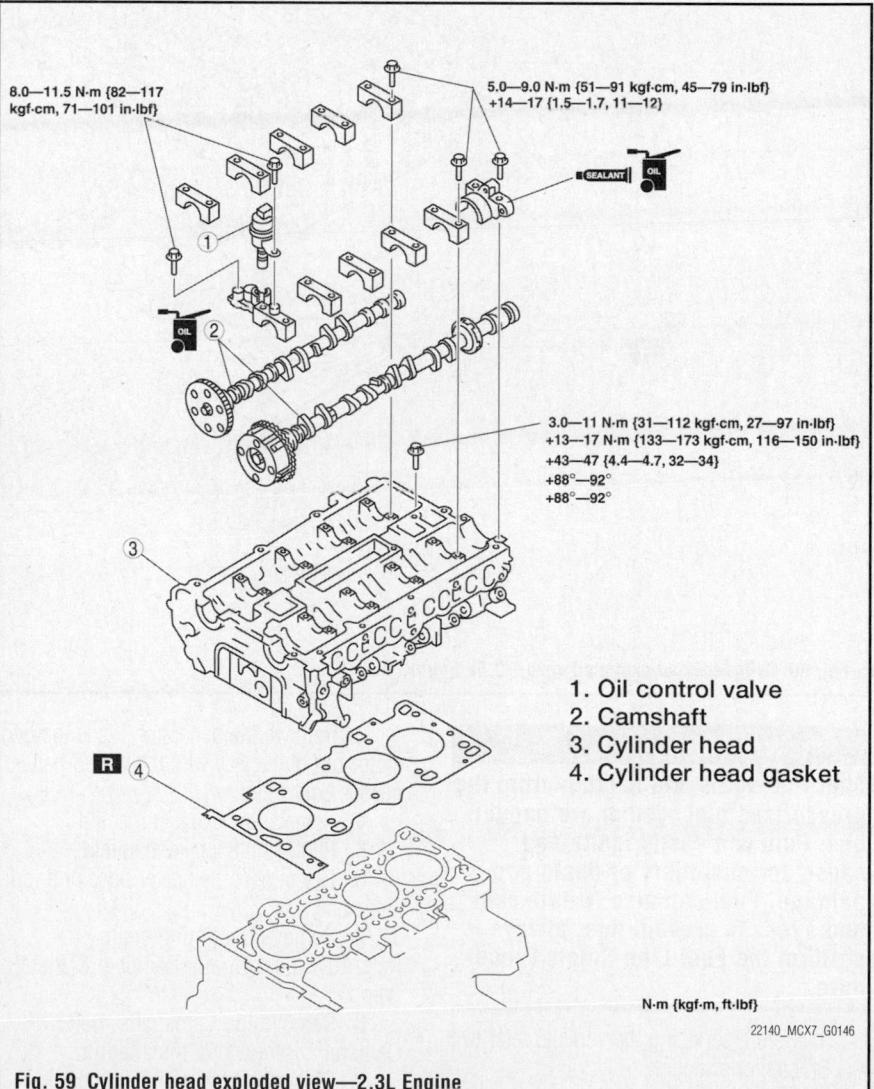

1. Oil control valve
2. Camshaft
3. Cylinder head
4. Cylinder head gasket

N·m {kgf·m, ft·lbf}

Fig. 59 Cylinder head exploded view—2.3L Engine

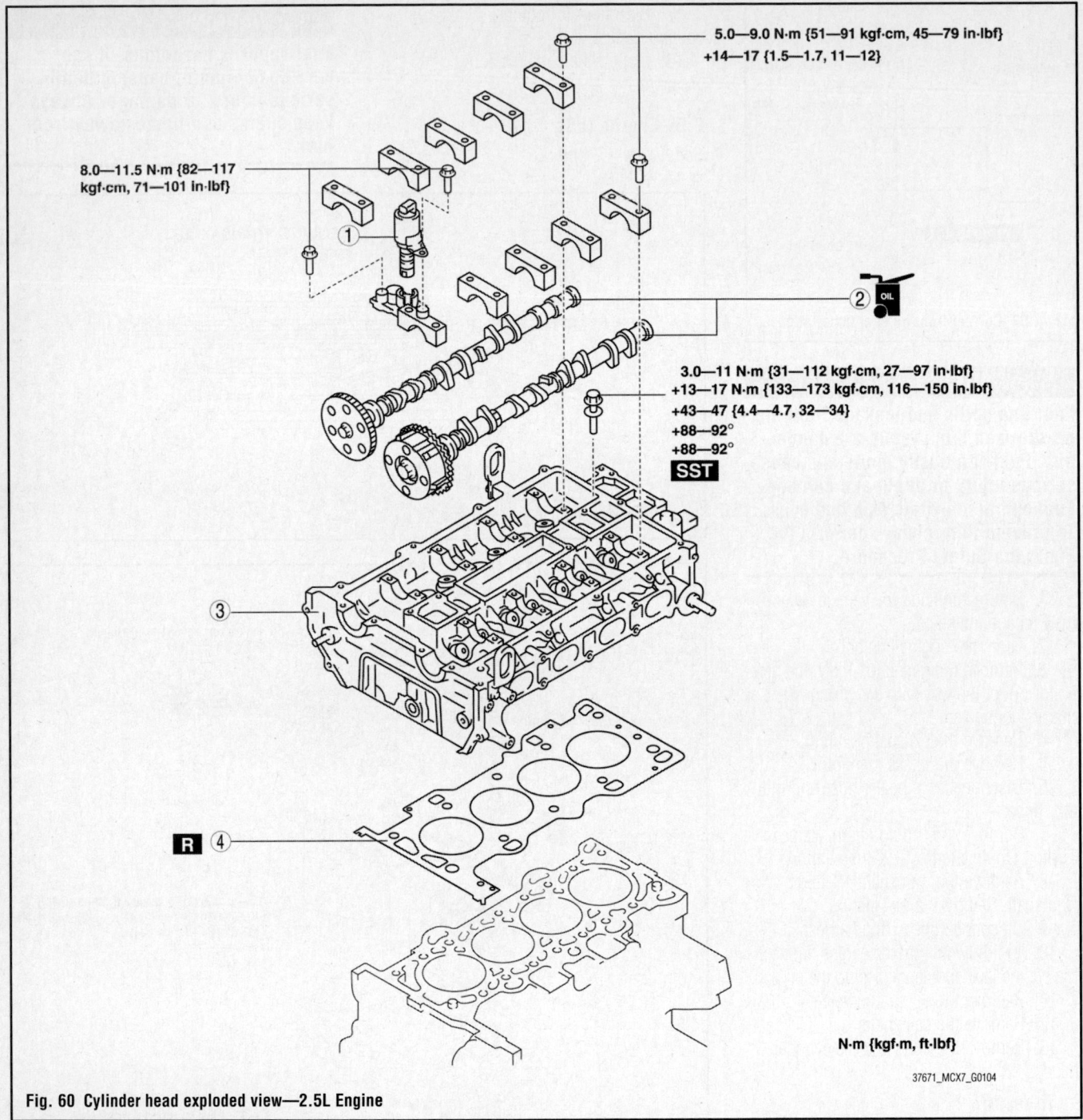

8.0—11.5 N·m {82—117 kgf·cm, 71—101 in·lbf}

5.0—9.0 N·m {51—91 kgf·cm, 45—79 in·lbf}
+14—17 {1.5—1.7, 11—12}

3.0—11 N·m {31—112 kgf·cm, 27—97 in·lbf}
+13—17 N·m {133—173 kgf·cm, 116—150 in·lbf}
+43—47 {4.4—4.7, 32—34}
+88—92°
+88—92°
SST

N·m {kgf·m, ft·lbf}

37671_MCX7_G0104

Fig. 60 Cylinder head exploded view—2.5L Engine

❊❊ WARNING

Fuel line spills and leakage from the pressurized fuel system are dangerous. Fuel can easily ignite and cause serious injury or death and damage. Fuel can also irritate skin and eyes. To prevent this, always perform the Fuel Line Safety Procedure.

1. Before servicing the vehicle, refer to the Precautions section.
2. Drain the engine coolant.
3. Remove the generator. Place the generator out of the way with the wiring harnesses connected.
4. Remove the exhaust manifold.
5. Remove the intake manifold.
6. Disconnect the heater hose and radiator hose.
7. Remove the timing chain. Refer to Timing Chain, Removal & Installation.
8. Remove the Camshafts. Refer to Camshaft, Removal & Installation.
9. Disconnect the wiring harness.
10. To firmly support the engine, first set the engine jack and attachment to the oil pan.
11. Remove the oil control valve
12. Remove the camshafts.
13. Remove the bolts and the cylinder head.

To install:

14. Measure the cylinder head bolts. Replace any bolts that exceed 5.740 inches (146 mm).
15. Install the cylinder head with a new gasket. Tighten the bolts in sequence as follows:

a. Step 1: 27–97 inch lbs. (3–11 Nm)
b. Step 2: 116–150 inch lbs. (13–17 Nm)
c. Step 3: 32–34 ft. lbs. (43–47 Nm)
d. Step 4: Plus 88°–92°
e. Step 5: Plus 88°–92°
16. Install in the reverse order of removal.
17. Bleed the air from the cooling system.

18. Inspect the compression pressure.

EXHAUST MANIFOLD

REMOVAL & INSTALLATION

2.3L Engine (L3)
See Figure 61.

1. Before servicing the vehicle, refer to the Precautions section.

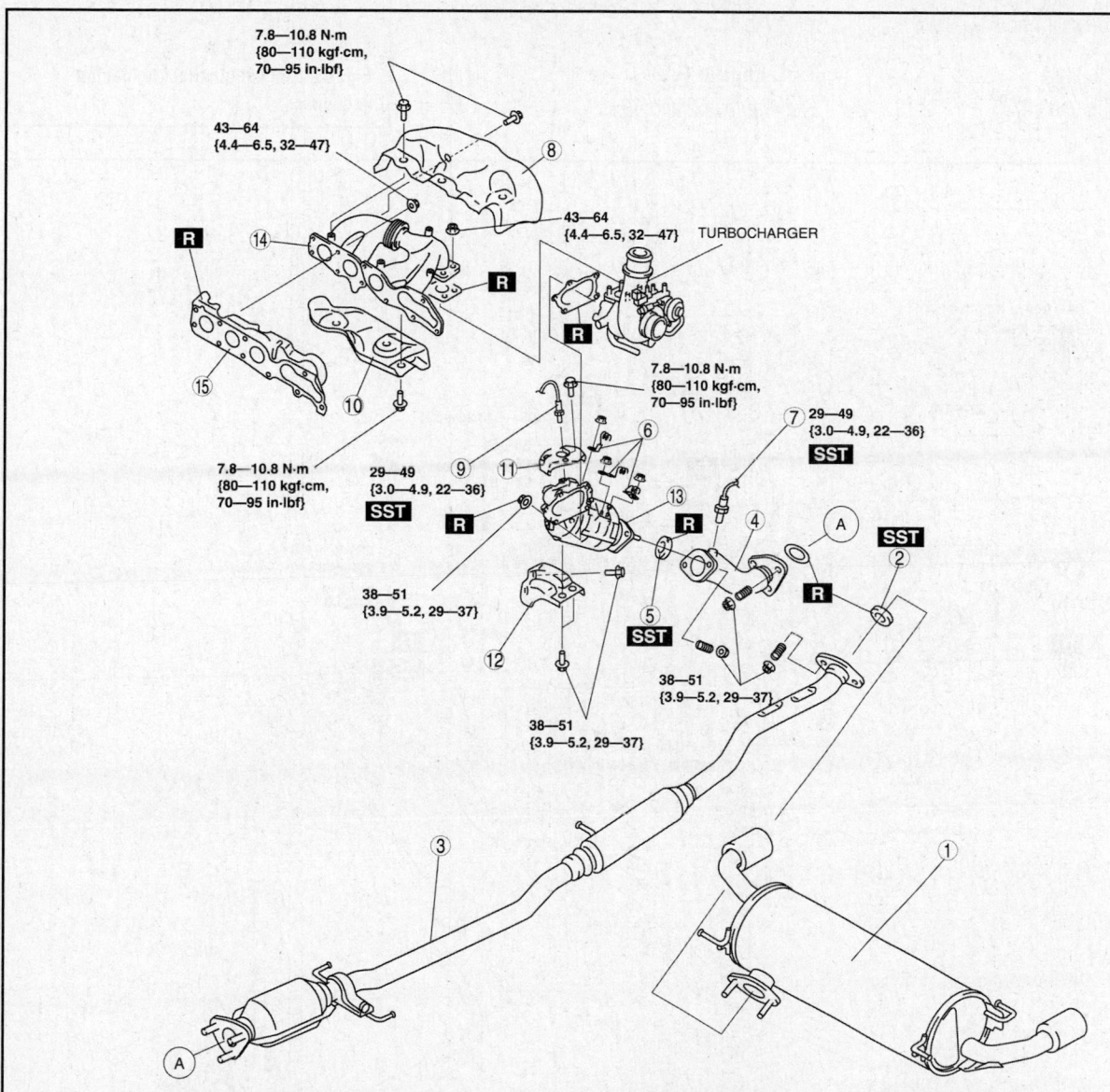

1. Main silencer
2. Seal ring (middle pipe side)
3. Middle pipe
4. Front pipe
5. Seal ring (WU-TWC side)
6. Clip
7. Rear HO2S
8. Exhaust manifold insulator (Upper)
9. Front HO2S
10. Exhaust manifold insulator (Lower)
11. WU-TWC insulator
12. WU-TWC bracket
13. WU-TWC
14. Exhaust manifold
15. Exhaust manifold gasket

N·m {kgf·m, ft·lbf}

Fig. 61 Exhaust system exploded view

22140_MCX7_G0151

2. Disconnect the negative battery cable.

3. Remove the charge air cooler duct.

4. Remove the charge air cooler cover.

5. Remove the charge air cooler.

6. Remove the main silencer

7. Remove the seal ring (middle pipe side)

8. Remove the middle pipe

9. Remove the front pipe

10. Remove the seal ring (WU-TWC side)

11. Remove the clip

12. Remove the rear HO2S

13. Remove the exhaust manifold insulator (Upper)

14. Remove the front HO2S

15. Remove the exhaust manifold insulator (Lower)

16. Remove the WU-TWC insulator

17. Remove the WU-TWC bracket

18. Remove the WU-TWC

19. Remove the exhaust manifold

20. Remove the exhaust manifold gasket

21. Install in the reverse order of removal.

2.5L Engine (L5)

See Figures 62 and 63.

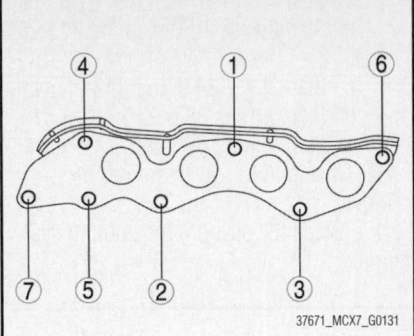

37671_MCX7_G0131

Fig. 63 Exhaust manifold tightening sequence

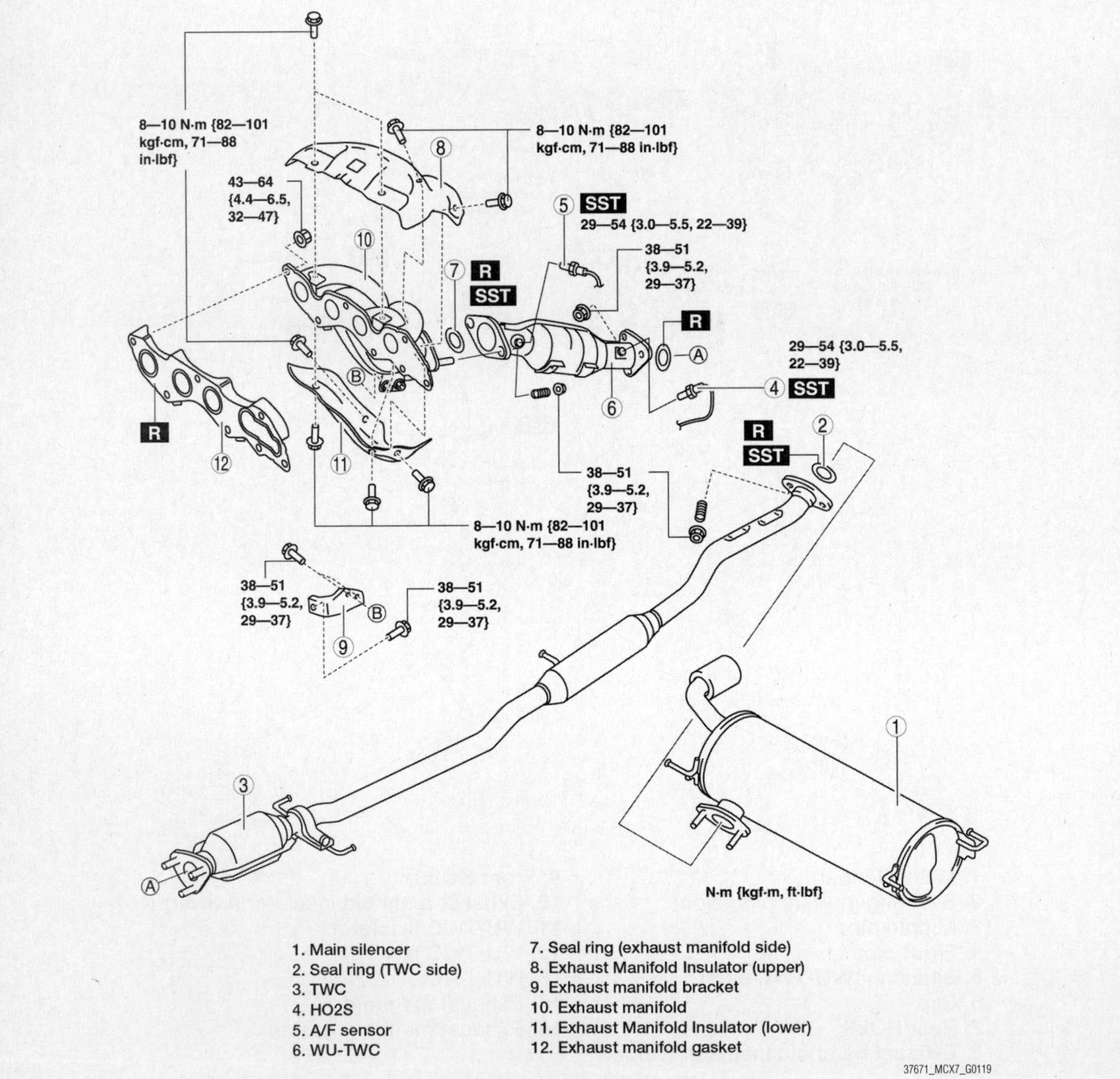

8—10 N·m {82—101 kgf·cm, 71—88 in·lbf}

8—10 N·m {82—101 kgf·cm, 71—88 in·lbf}

43—64 {4.4—6.5, 32—47}

⑤ SST 29—54 {3.0—5.5, 22—39}

38—51 {3.9—5.2, 29—37}

29—54 {3.0—5.5, 22—39}

38—51 {3.9—5.2, 29—37}

8—10 N·m {82—101 kgf·cm, 71—88 in·lbf}

38—51 {3.9—5.2, 29—37}

38—51 {3.9—5.2, 29—37}

N·m {kgf·m, ft·lbf}

1. Main silencer
2. Seal ring (TWC side)
3. TWC
4. HO2S
5. A/F sensor
6. WU-TWC
7. Seal ring (exhaust manifold side)
8. Exhaust Manifold Insulator (upper)
9. Exhaust manifold bracket
10. Exhaust manifold
11. Exhaust Manifold Insulator (lower)
12. Exhaust manifold gasket

37671_MCX7_G0119

Fig. 62 Exhaust system—exploded view 2.5L Engine

✳✳ CAUTION

A hot engine and exhaust system can cause severe burns. Turn off the engine and wait until they are cool before removing the exhaust system.

1. Before servicing the vehicle, refer to the Precautions section.
2. Disconnect the negative battery cable.
3. Remove the plug hole plate.
4. Remove the undercover.
5. Remove the exhaust system insulator.
6. Unbolt the WU-TWC from the exhaust manifold.
7. Unbolt an remove the exhaust manifold.
8. Install in the reverse order of removal.
9. Tighten the exhaust manifold new installation nuts to 32–47 ft. lbs. (43–64 Nm).

➡**Always use new gaskets, and clean off any old gasket material.**

10. Tighten the WU-TWC to the exhaust manifold to 29–37 ft. lbs.(38–51 Nm).
11. Tighten the exhaust manifold insulator to 71–88 inch lbs. 8–10 Nm).
12. Start the engine and check for leaks.

INTAKE MANIFOLD

REMOVAL & INSTALLATION

2.3L Engine (L3)
See Figures 64 and 65.

✳✳ WARNING

A hot engine and intake air system can cause severe burns. Turn off the engine and wait until they are cool before removing the intake air system.

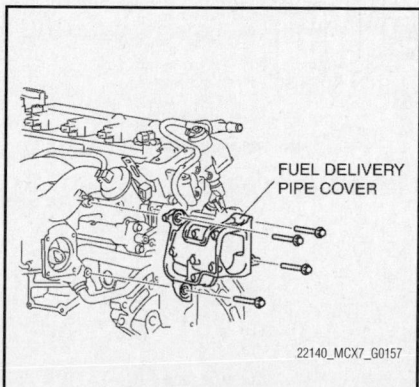

FUEL DELIVERY
PIPE COVER

22140_MCX7_G0157

Fig. 64 Fuel delivery pipe cover

✳✳ WARNING

Fuel vapor is hazardous. It can easily ignite, causing serious injury and damage. Always keep sparks and flames away from fuel.

✳✳ WARNING

Fuel line spills and leakage are dangerous. Fuel can ignite and cause serious injuries or death and damage. Fuel can also irritate skin and eyes.

1. Before servicing the vehicle, refer to the Precautions section.
2. Disconnect the negative battery cable.
3. Drain the cooling system.
4. Remove the charge air cooler duct.
5. Remove the battery and battery tray.
6. Remove the undercover.
7. Remove the Mass Airflow sensor (MAF).
8. Remove the fresh air duct.
9. Remove the air cleaner cover and case.
10. Remove the resonance chamber.
11. Remove the charge air cooler.
12. Remove the air bypass valve.
13. Remove the air hose and duct.
14. Remove the throttle body.
15. Remove the vacuum chamber.
16. Remove the variable swirl solenoid valve.
17. Remove the fuel delivery pipe cover.
18. Disconnect the quick connector connected to the intake manifold.
19. Remove the EGR pipe.
20. Disconnect the OCV connector.
21. Disconnect the PSP switch connector.
22. Remove the oil level gauge pipe.
23. Remove the splash shield (RF).
24. Remove the drive belt.
25. Set the power steering oil pump out of the way.
26. Disconnect the fuel pressure sensor connector.
27. Disconnect the vacuum hose connected between the intake manifold and the master cylinder from the intake manifold.
28. Disconnect the MAP sensor connector.
29. Remove the intake manifold installation bolts.
30. Disconnect the evaporative hose connected between the intake manifold and the PCV valve from the intake manifold.

31. Remove the intake manifold.
32. Install in the reverse order of removal. Tighten the intake manifold bolts to 13–16 ft. lbs. (17–23 Nm).

2.5L Engine (L5)
See Figures 66 through 68.

✳✳ WARNING

A hot engine and intake air system can cause severe burns. Turn off the engine and wait until they are cool before removing the intake air system.

✳✳ WARNING

Fuel vapor is hazardous. It can easily ignite, causing serious injury and damage. Always keep sparks and flames away from fuel.

✳✳ WARNING

Fuel line spills and leakage are dangerous. Fuel can ignite and cause serious injuries or death and damage. Fuel can also irritate skin and eyes.

1. Before servicing the vehicle, refer to the Precautions section.
2. Disconnect the negative battery cable.
3. Drain the cooling system.
4. Remove the plug hole plate.
5. Disconnect the wiring harness.
6. Remove or disconnect the following:
   - MAF/IAT sensor
   - Ventilation hose
   - Fresh-air duct
   - Air cleaner cover
   - Air cleaner element
   - Air cleaner case
   - Remove the mudguard (LF).
   - Resonance chamber
   - Air hose
7. Remove or disconnect the following:
   - Water hose
   - Throttle body
   - Vacuum hose
   - Variable intake air solenoid valve
   - Vacuum hose
   - Variable tumble solenoid valve
   - Vacuum hose
   - Quick release connector
   - Vacuum cap
   - Intake manifold
   - Coolant reserve tank.
   - Fan control module No.2.
   - Engine undercover
   - All clips for securing wiring harnesses from the intake manifold

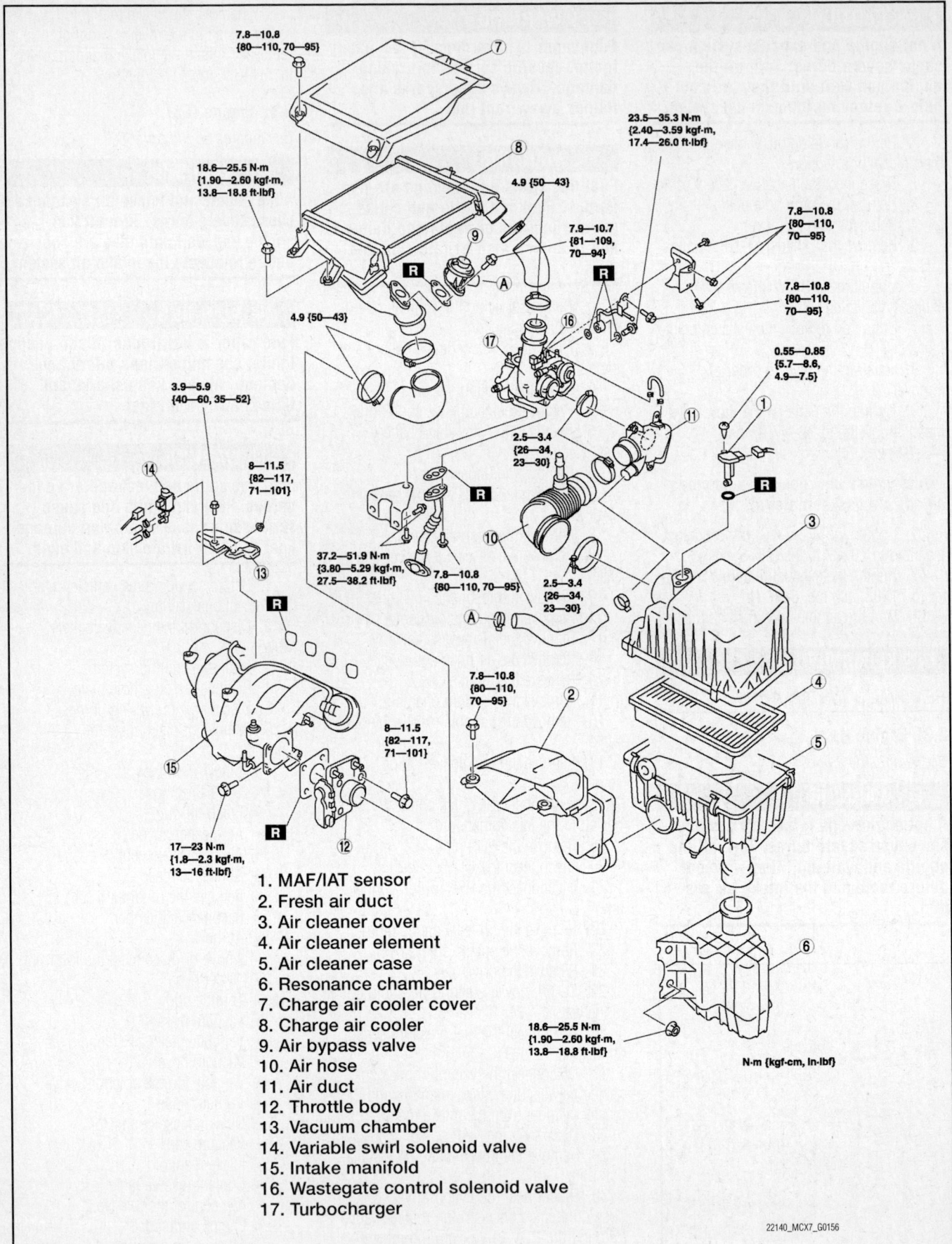

7.8—10.8
{80—110, 70—95}

18.6—25.5 N·m
{1.90—2.60 kgf·m,
13.8—18.8 ft·lbf}

23.5—35.3 N·m
{2.40—3.59 kgf·m,
17.4—26.0 ft·lbf}

4.9 {50—43}

7.9—10.7
{81—109,
70—94}

7.8—10.8
{80—110,
70—95}

7.8—10.8
{80—110,
70—95}

0.55—0.85
{5.7—8.6,
4.9—7.5}

4.9 {50—43}

3.9—5.9
{40—60, 35—52}

2.5—3.4
{26—34,
23—30}

8—11.5
{82—117,
71—101}

37.2—51.9 N·m
{3.80—5.29 kgf·m,
27.5—38.2 ft·lbf}

7.8—10.8
{80—110, 70—95}

2.5—3.4
{26—34,
23—30}

7.8—10.8
{80—110,
70—95}

8—11.5
{82—117,
71—101}

17—23 N·m
{1.8—2.3 kgf·m,
13—16 ft·lbf}

18.6—25.5 N·m
{1.90—2.60 kgf·m,
13.8—18.8 ft·lbf}

N·m {kgf·cm, in·lbf}

1. MAF/IAT sensor
2. Fresh air duct
3. Air cleaner cover
4. Air cleaner element
5. Air cleaner case
6. Resonance chamber
7. Charge air cooler cover
8. Charge air cooler
9. Air bypass valve
10. Air hose
11. Air duct
12. Throttle body
13. Vacuum chamber
14. Variable swirl solenoid valve
15. Intake manifold
16. Wastegate control solenoid valve
17. Turbocharger

22140_MCX7_G0156

Fig. 65 Intake manifold and related parts 2.3L engine—exploded view

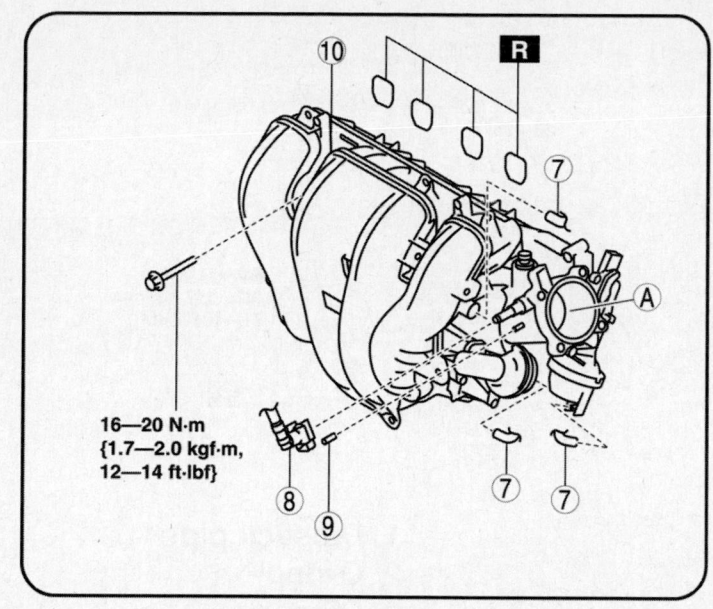
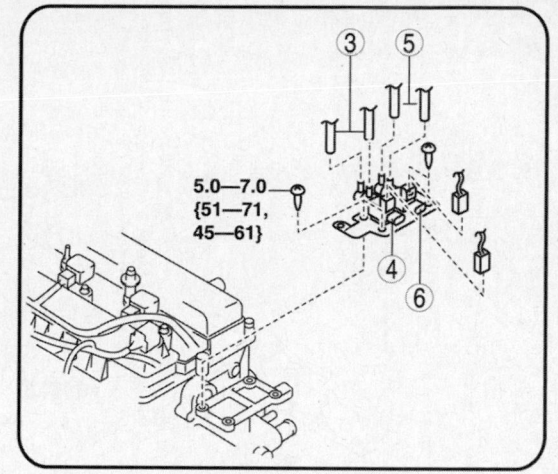

16—20 N·m
{1.7—2.0 kgf·m,
12—14 ft·lbf}

5.0—7.0
{51—71,
45—61}

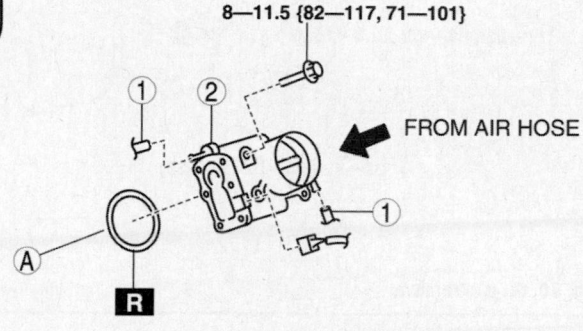

8—11.5 {82—117, 71—101}

FROM AIR HOSE

N·m {kgf·cm, in·lbf}

37671_MCX7_G0134

Fig. 66 Intake manifold and related parts 2.5L engine—exploded view

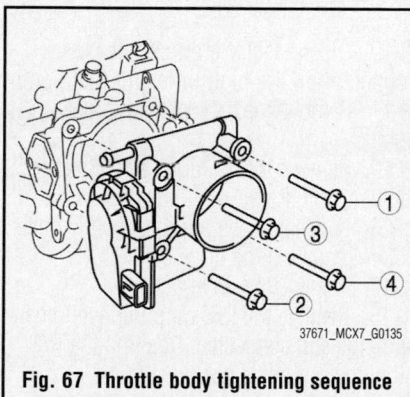

37671_MCX7_G0135

Fig. 67 Throttle body tightening sequence

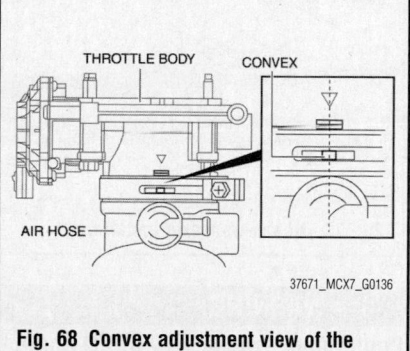

THROTTLE BODY   CONVEX

AIR HOSE

37671_MCX7_G0136

Fig. 68 Convex adjustment view of the throttle body and the air hose

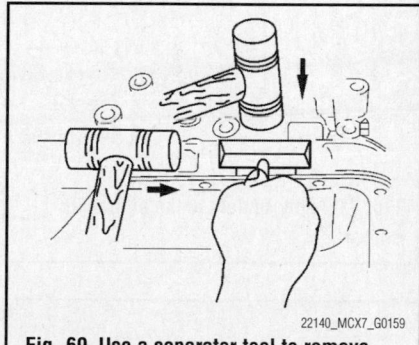

22140_MCX7_G0159

Fig. 69 Use a separator tool to remove the oil pan

8. Remove the intake manifold.

9. Install in the reverse order of removal and note the following:
- Tighten the throttle body to 71–101 inch lbs. (8–11.5 Nm).
- Adjust the convex of the throttle body and the air hose.

10. Bleed the air from the cooling system.

OIL PAN

REMOVAL & INSTALLATION

See Figures 69 through 73.

❊❊ WARNING

Hot engines and engine oil can cause severe burns. Turn off the engine and wait until it and the engine oil have cooled.

❊❊ WARNING

A vehicle that is lifted but not securely supported on safety stands is dangerous. It can slip or fall, causing death or serious injury. Never work around or under a lifted vehicle if it is not securely supported on safety stands.

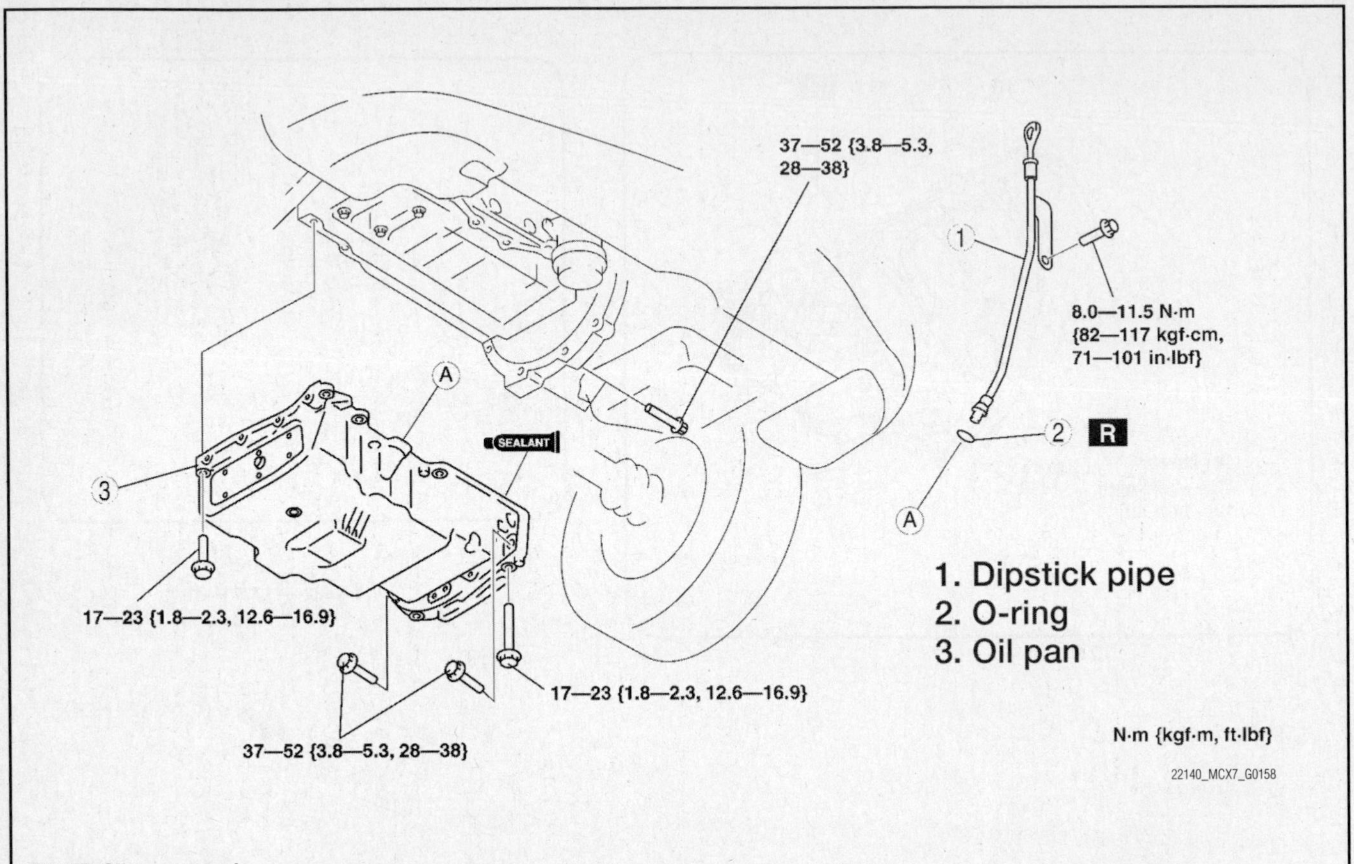

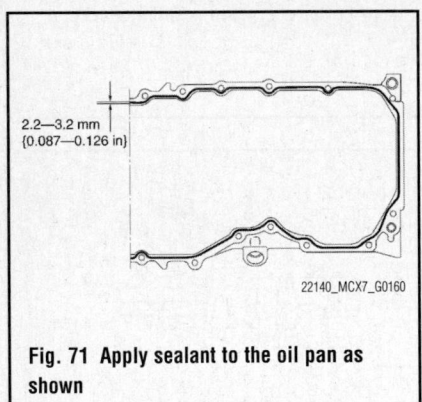

Fig. 70 Oil pan removal

37—52 {3.8—5.3, 28—38}

8.0—11.5 N·m {82—117 kgf·cm, 71—101 in·lbf}

17—23 {1.8—2.3, 12.6—16.9}

17—23 {1.8—2.3, 12.6—16.9}

37—52 {3.8—5.3, 28—38}

1. Dipstick pipe
2. O-ring
3. Oil pan

N·m {kgf·m, ft·lbf}

22140_MCX7_G0158

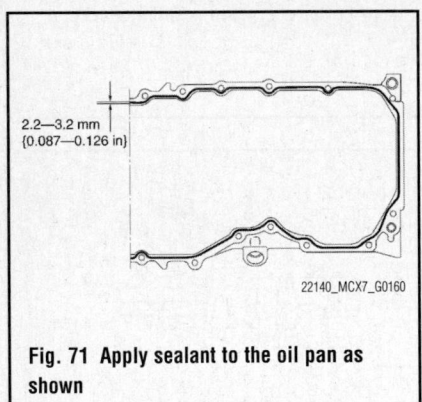

2.2—3.2 mm {0.087—0.126 in}

22140_MCX7_G0160

Fig. 71 Apply sealant to the oil pan as shown

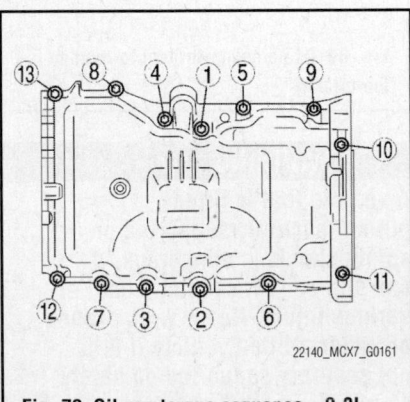

22140_MCX7_G0161

Fig. 72 Oil pan torque sequence—2.3L engine shown—2.5L similar

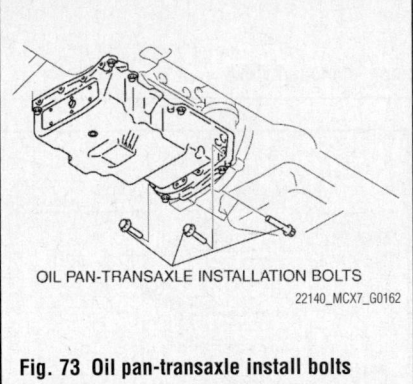

OIL PAN-TRANSAXLE INSTALLATION BOLTS

22140_MCX7_G0162

Fig. 73 Oil pan-transaxle install bolts

> ## ❊❊ WARNING
>
> **Continuous exposure to USED engine oil has caused skin cancer in laboratory mice. Protect your skin by washing with soap and water immediately after working with engine oil.**

1. Before servicing the vehicle, refer to the Precautions section.
2. Relieve the fuel system pressure.
3. Remove the battery and battery tray.
4. Remove the plug hole plate. 2.5L (L5 engines only).
5. Remove the undercover.
6. Remove the splash shield (RH).
7. Drain the engine oil.
8. Drain the engine coolant.
9. Remove the charge air cooler, air cleaner and fresh air duct component, and air hose.
10. Disconnect the quick release connector on the high pressure fuel pump.
11. Remove the high pressure fuel pump.
12. Remove the ignition coils. Refer to Ignition Coil, Removal & Installation.
13. Loosen the water pump pulley bolts before removing the drive belt.
14. Remove the drive belt.
15. Remove the P/S oil pump with hose and pipe still connected. Position the P/S oil pump out of the way.
16. Remove the crankshaft position (CKP) sensor.
17. Remove the engine front cover. Refer to Timing Chain, Removal and Installation.
18. Remove the oil dipstick tube. 2.3L (L3 engines only).
19. Remove the oil pan.

To install:

20. Apply silicone sealant to the oil pan along the inside of the bolt holes as shown in the figure.

21. Install the oil pan. Tighten the bolts in sequence to 13–16 ft. lbs. (17–23 Nm).

22. Install the oil pan-transaxle bolts and tighten to 28–38 ft. lbs. (37–52 Nm).

23. Install in the reverse order of removal.

24. Refill with the specified type and amount of the engine oil.

25. Refill the engine coolant.

26. Start the engine and confirm that there is no oil leakage. If there is oil leakage, repair or replace the applicable part.

27. Refill with the specified type and amount of the engine oil.

28. Refill the engine coolant and check for leaks.

29. Inspect for the ignition timing and idle speed.

OIL PUMP

REMOVAL & INSTALLATION
See Figures 74 through 76.

> **⁂⁂ WARNING**
> **Hot engines and engine oil can cause severe burns. Turn off the engine and wait until it and the engine oil have cooled.**

> **⁂⁂ WARNING**
> **A vehicle that is lifted but not securely supported on safety stands is dangerous. It can slip or fall, causing death or serious injury. Never work around or under a lifted vehicle if it is not securely supported on safety stands.**

> **⁂ WARNING**
> **Continuous exposure to USED engine oil has caused skin cancer in labora-**

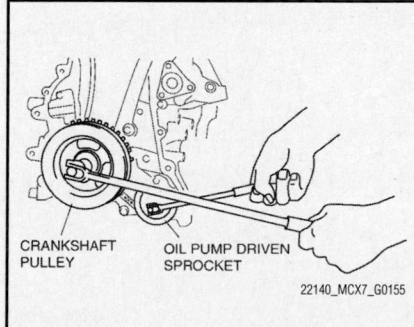

CRANKSHAFT PULLEY · OIL PUMP DRIVEN SPROCKET

22140_MCX7_G0155

Fig. 74 Oil pump driven sprocket removal and installation

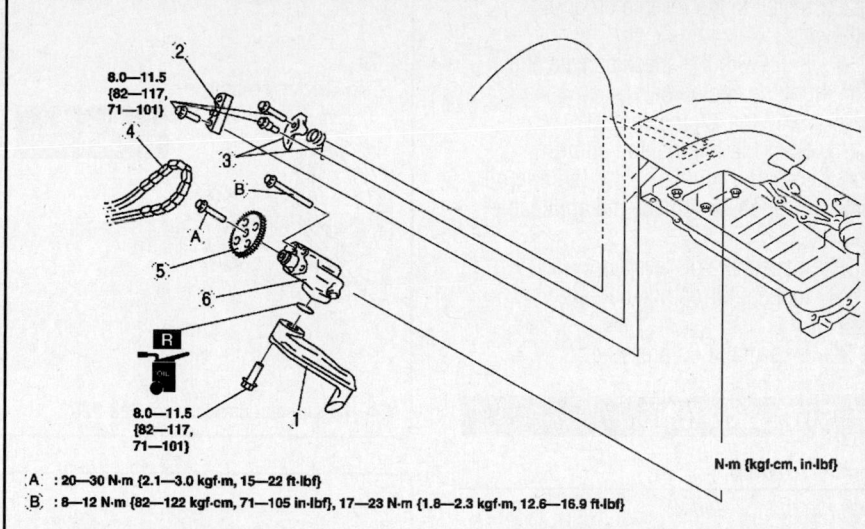

A : 20—30 N·m {2.1—3.0 kgf·m, 15—22 ft·lbf}
B : 8—12 N·m {82—122 kgf·cm, 71—105 in·lbf}, 17—23 N·m {1.8—2.3 kgf·m, 12.6—16.9 ft·lbf}

1. **Oil strainer**
2. **Oil pump chain guide**
3. **Oil pump chain tensioner**
4. **Oil pump chain**
5. **Oil pump sprocket**
6. **Oil pump**

22140_MCX7_G0163

Fig. 75 Oil pump exploded view

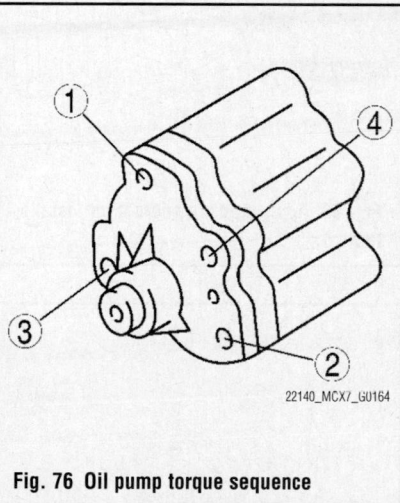

22140_MCX7_G0164

Fig. 76 Oil pump torque sequence

tory mice. Protect your skin by washing with soap and water immediately after working with engine oil.

1. Before servicing the vehicle, refer to the Precautions section.
2. Remove the battery and battery tray.
3. Remove the undercover.
4. Remove the splash shield (RH).
5. Drain the engine oil.
6. Drain the engine coolant.
7. Remove the charge air cooler, air cleaner and fresh air duct component, and air hose.
8. Disconnect the quick release connector on the high pressure fuel pump.

9. Remove the high pressure fuel pump.
10. Remove the ignition coils.
11. Loosen the water pump pulley bolts before removing the drive belt.
12. Remove the drive belt.
13. Remove the P/S oil pump with hose and pipe still connected. Position the P/S oil pump out of the way.
14. Remove the crankshaft position (CKP) sensor.
15. Remove the engine front cover.
16. Remove the oil pan.
17. Temporarily install the crankshaft pulley and crankshaft pulley lock bolt to the crankshaft, and lock the oil pump against rotation as shown in figure.
18. Remove the oil pump.

To install:
19. Install the oil pump. Tighten the oil pump bolts as follows:
   a. Step 1: 71–105 inch lbs. (8–12 Nm)
   b. Step 2: 13–17 ft. lbs. (17–23 Nm)
20. Temporarily install the crankshaft pulley and crankshaft pulley lock bolt to the crankshaft, and lock the oil pump against rotation as shown in figure.
21. Install the oil pump driven sprocket and tighten the bolt to 15–22 ft. lbs. (20–30 Nm).
22. Remove the crankshaft pulley and crankshaft pulley lock bolt.

23. Install in the reverse order of removal.

24. Refill with the specified type and amount of the engine oil.

25. Refill the engine coolant.

26. Start the engine and confirm that there is no oil leakage. If there is oil leakage, repair or replace the applicable part.

27. Inspect the oil level.

28. Inspect for engine coolant leakage.

29. Inspect the oil pressure.

PISTON AND RING

POSITIONING

See Figures 77 and 78.

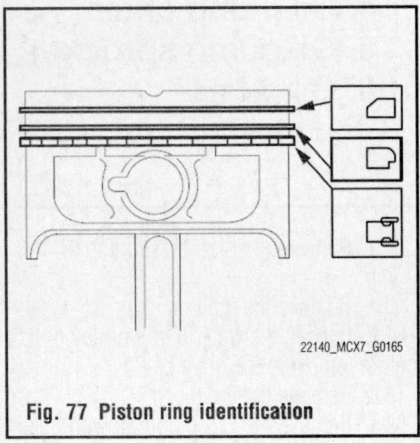

Fig. 77 Piston ring identification

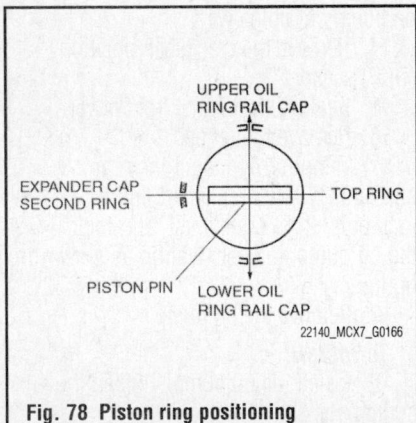

Fig. 78 Piston ring positioning

REAR MAIN SEAL

REMOVAL & INSTALLATION

See Figures 79 through 82.

1. Before servicing the vehicle, refer to the Precautions section.

2. Remove the transaxle and flywheel.

3. Unbolt and remove the oil seal.

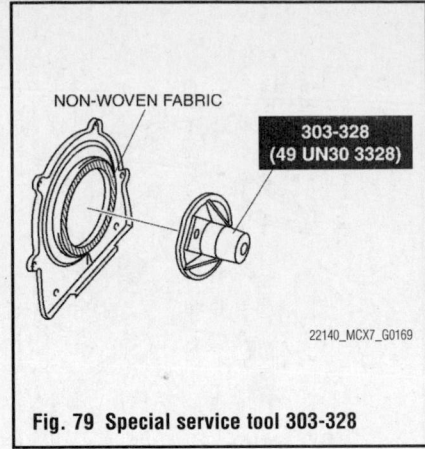

Fig. 79 Special service tool 303-328

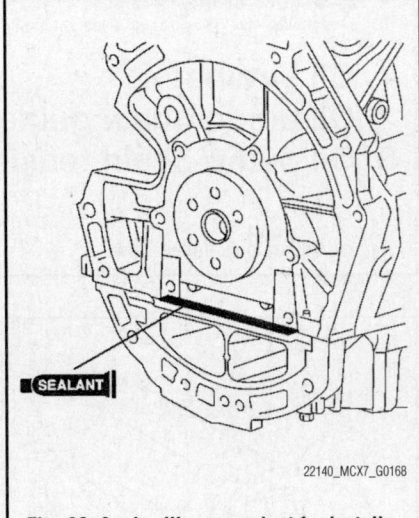

Fig. 80 Apply silicone sealant for installation

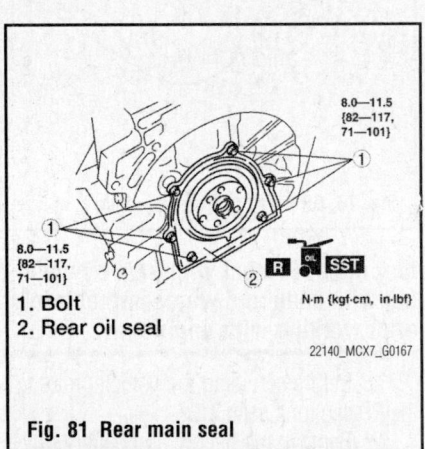

1. Bolt
2. Rear oil seal

Fig. 81 Rear main seal

To install:

4. Apply silicone sealant as shown in the illustration.

5. Install SST 303-328 to the non-woven fabric side of the rear main seal.

6. From the back side of the rear oil seal, verify that there is no damage or separation in the lip area of the rear oil seal.

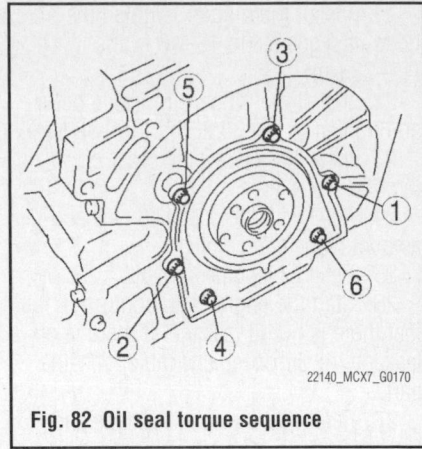

Fig. 82 Oil seal torque sequence

7. Install the rear main seal. Tighten the bolts to 71–101 inch lbs. (8–11.5 Nm).

8. Remove SST 303-328.

9. Install the flywheel and transaxle.

10. Road test and check for leaks.

TIMING CHAIN, SPROCKETS, FRONT COVER & SEAL

REMOVAL & INSTALLATION

2.3L Engine (L3)

See Figures 83 through 96.

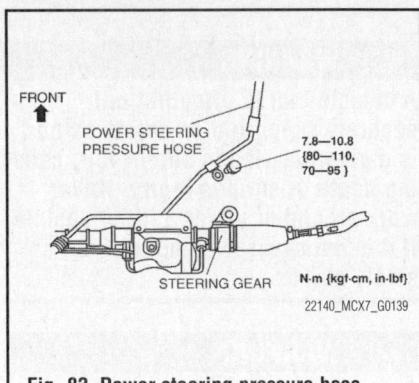

Fig. 83 Power steering pressure hose bracket

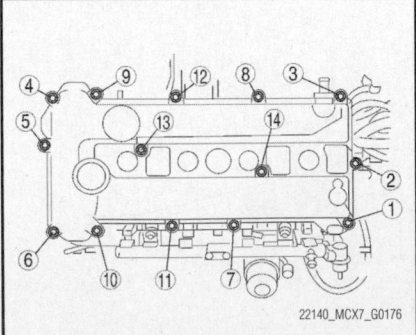

Fig. 84 Valve cover bolt removal sequence

1. Before servicing the vehicle, refer to the Precautions section.

2. Disconnect the negative battery cable.

3. Remove the undercover.

4. Remove the splash shield.

5. Remove the high pressure fuel pump.

6. Remove the ignition coils.

7. Disconnect the wiring harness.

8. Remove the ventilation hose.

9. Remove the drive belt.

10. Remove the Crankshaft Position (CKP) sensor.

11. Remove the P/S oil pump with the hoses and pipes still connected.

➡**Position and secure the P/S oil pump out of the way with a rope or wire.**

12. Remove the bracket shown in the figure, and place the power steering pressure hose outside of the vehicle.

13. Remove the valve cover.

14. Rotate the crankshaft in the direction of the engine rotation and remove the cylinder block lower blind plug when the No. 1 cylinder is at the point prior to Top Dead Center (TDC) of compression, then install SST 303-507.

15. Rotate the crankshaft in the direction of the engine rotation so that the No.1 piston is at TDC of the compression stroke. (Until the crank weight contacts SST and stops.)

16. Install SSTs 205-07202 and 205-126 to the crankshaft pulley and lock the crankshaft against rotation.

17. Remove the crankshaft pulley lock bolt and the crankshaft pulley.

18. Remove the water pump pulley.

19. Remove the front drive belt idler pulley.

20. Install engine support fixture 49 C017 5A0 or equivalent.

21. Remove the No. 3 engine mount.

22. Remove the engine front cover.

23. Remove the front oil seal.

24. Re-set the chain tensioner as follows:

- Press the timing chain tensioner ratchet to the left using a thin flathead screwdriver (precision screwdriver) to unlock the plunger.
- Slowly press the plunger back in the direction shown in the figure while pressing the ratchet.
- Release the ratchet with the plunger still pressed down.
- Press-in the plunger until the ratchet position is as indicated in the figure, and then insert a wire to lock the plunger.

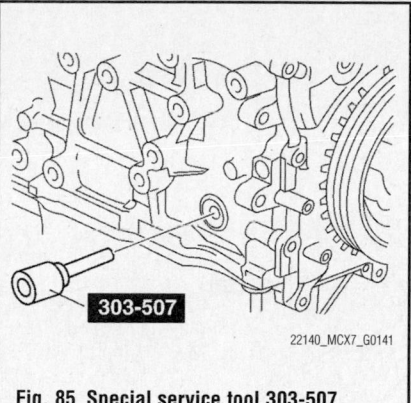

Fig. 85 Special service tool 303-507

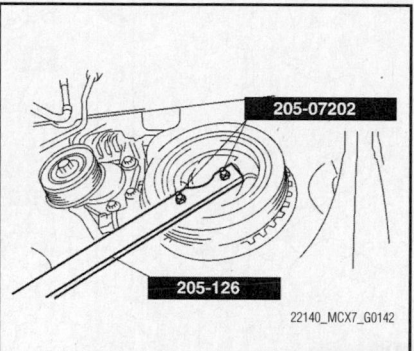

Fig. 86 Special service tools 205-07202 and 205-126

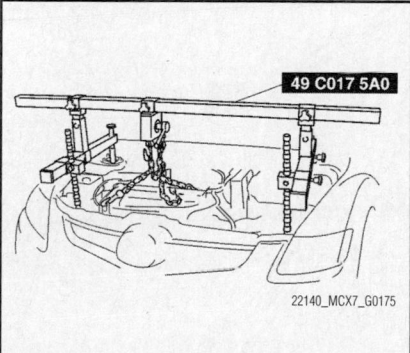

Fig. 87 Engine support fixture 49 C017 5A0

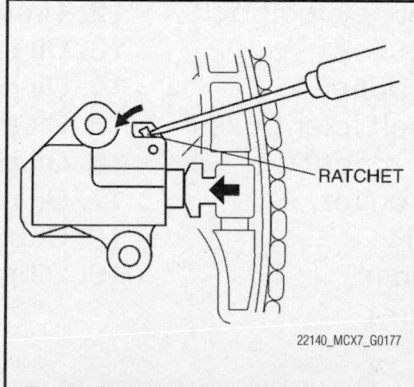

Fig. 88 Chain tensioner removal—1 of 2

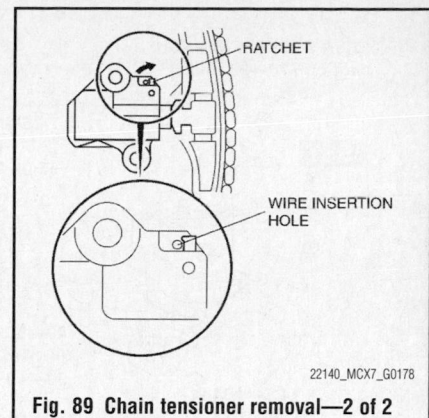

Fig. 89 Chain tensioner removal—2 of 2

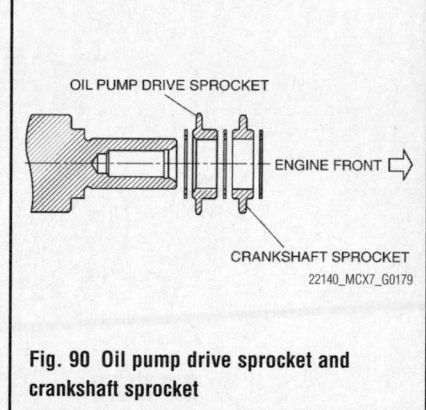

Fig. 90 Oil pump drive sprocket and crankshaft sprocket

25. Remove the tensioner arm.

26. Remove the chain guide.

27. Remove the timing chain.

28. Remove the oil jet.

29. Remove the oil pump chain tensioner.

30. Remove the oil pump chain guide.

31. Remove the oil pump driven sprocket.

32. Remove the oil pump chain.

33. Remove the crankshaft sprocket.

34. Remove the oil pump drive sprocket and the oil pump chain.

To install:

35. Replace the oil pump drive sprocket and the oil pump chain.

36. Replace the crankshaft sprocket.

37. Replace the oil pump chain.

38. Replace the oil pump driven sprocket.

39. Replace the oil pump chain guide.

40. Replace the oil pump chain tensioner.

41. Replace the oil jet.

42. Install the chain guide and the tensioner arm.

43. Install SST 303-1061 to the camshaft as shown in the figure.

44. Verify that cylinder No.1 is at TDC of the compression stroke. (Position crank weight contacts SST 303-507.)

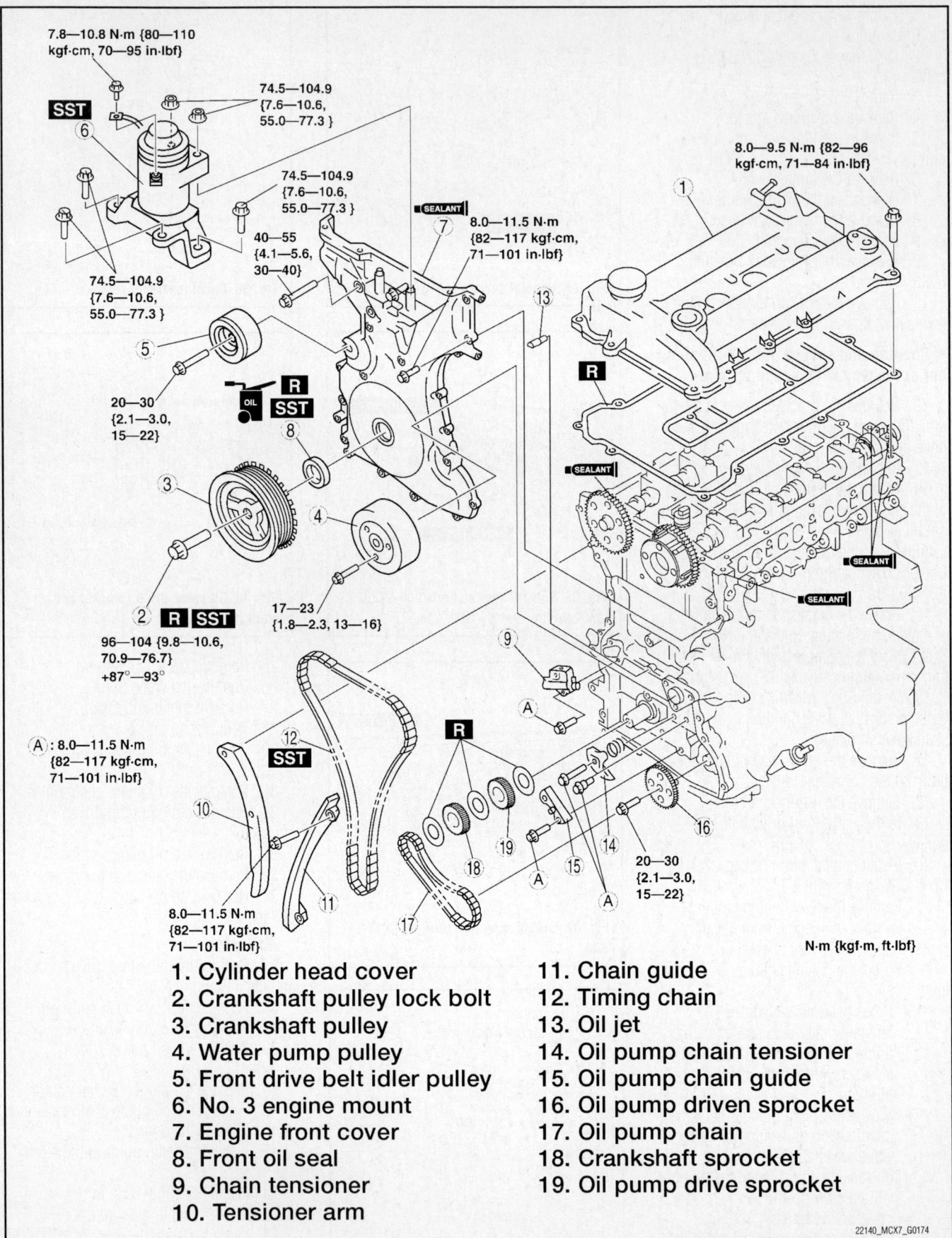

7.8—10.8 N·m {80—110 kgf·cm, 70—95 in·lbf}

SST

74.5—104.9 {7.6—10.6, 55.0—77.3 }

74.5—104.9 {7.6—10.6, 55.0—77.3 }

40—55 {4.1—5.6, 30—40}

74.5—104.9 {7.6—10.6, 55.0—77.3}

SEALANT

8.0—11.5 N·m {82—117 kgf·cm, 71—101 in·lbf}

8.0—9.5 N·m {82—96 kgf·cm, 71—84 in·lbf}

R

SEALANT

20—30 {2.1—3.0, 15—22}

OIL **R** **SST**

SEALANT

SEALANT

17—23 {1.8—2.3, 13—16}

96—104 {9.8—10.6, 70.9—76.7} +87°—93°

(A): 8.0—11.5 N·m {82—117 kgf·cm, 71—101 in·lbf}

SST

20—30 {2.1—3.0, 15—22}

8.0—11.5 N·m {82—117 kgf·cm, 71—101 in·lbf}

R

N·m {kgf·m, ft·lbf}

1. Cylinder head cover
2. Crankshaft pulley lock bolt
3. Crankshaft pulley
4. Water pump pulley
5. Front drive belt idler pulley
6. No. 3 engine mount
7. Engine front cover
8. Front oil seal
9. Chain tensioner
10. Tensioner arm
11. Chain guide
12. Timing chain
13. Oil jet
14. Oil pump chain tensioner
15. Oil pump chain guide
16. Oil pump driven sprocket
17. Oil pump chain
18. Crankshaft sprocket
19. Oil pump drive sprocket

22140_MCX7_G0174

Fig. 91 Timing chain exploded view—2.3L engine

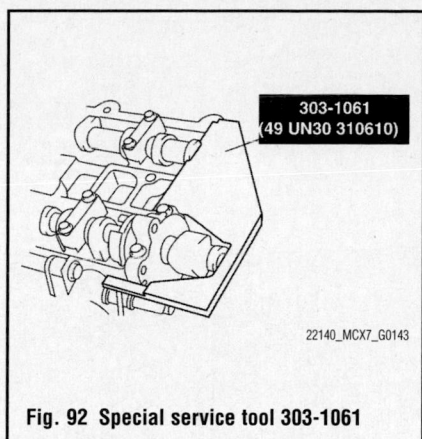

Fig. 92 Special service tool 303-1061

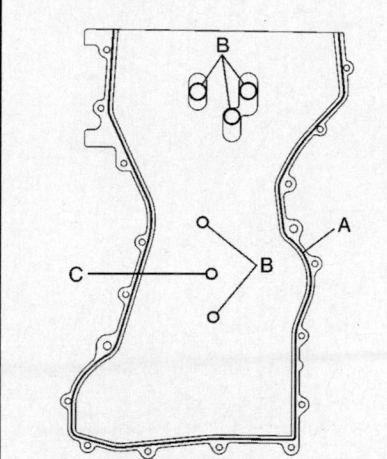

Thickness
A: 2.2—3.2 mm
B: 1.5—2.5 mm
C: No sealant

22140_MCX7_G0180

Fig. 93 Applying sealant to the front cover

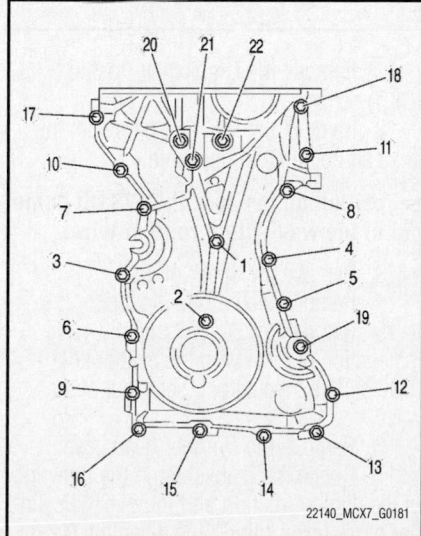

Fig. 94 Front cover torque sequence

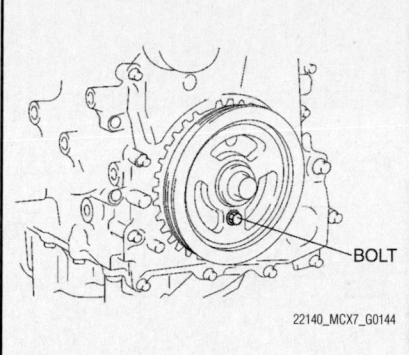

Fig. 95 Position the crankshaft pulley

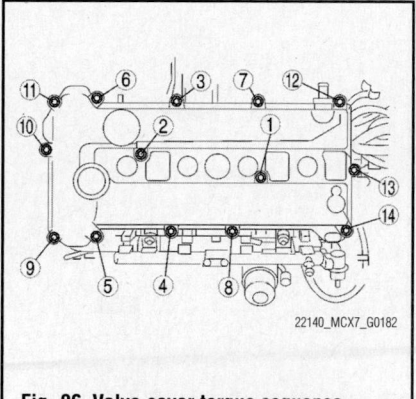

Fig. 96 Valve cover torque sequence

45. Install the timing chain.

46. Remove the wire or paper clip from the chain tensioner piston and apply tension to the timing chain.

➡**Install the engine front cover within 10 min of applying the silicone sealant.**

47. Apply silicone sealant to the engine front cover as shown. Sealant is not needed in area C.

48. Install the front cover. Tighten the bolts in sequence as follows:
- Bolts 1–18: 71–101 inch lbs. (8–11.5 Nm)
- Bolts 19–22: 30–40 ft. lbs. (40–55 Nm)

49. Replace the front oil seal

50. Replace the No. 3 engine mount

51. Replace the front drive belt idler pulley.

52. Replace the water pump pulley.

53. Replace the crankshaft pulley and the crankshaft pulley lock bolt.

54. Install the valve cover. Tighten the bolts in sequence to 71–84 inch lbs. (8–9.5 Nm)

55. Install in the reverse order of removal.

56. Start the engine, and inspect and adjust the following:

- Bleed the air from the cooling system.
- Run-out and contact on pulley and belt
- Leakage of engine oil, engine coolant or fuel.
- Ignition timing, and idle speed.
- Engine accessories operation

57. Perform a road test and verify that there is no abnormal vibration or noise.

2.5L Engine (L5)

See Figures 97 through 108.

1. Disconnect the negative battery cable.

2. Remove the plug hole plate.

3. Remove the ignition coils.

4. Disconnect the ventilation hose from the cylinder head cover.

5. Disconnect the Camshaft Position (CMP) sensor connector.

6. Disconnect the Oil Control Valve (OCV) connector.

7. Remove the front wheel and tire (RH).

8. Set the front mudguard (RH) out of the way.

9. Remove the splash shield (RH).

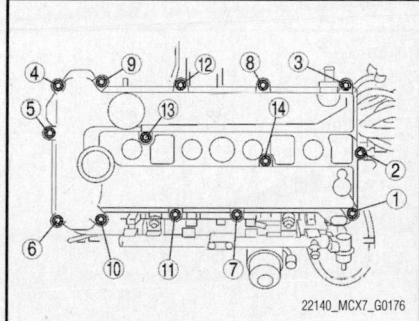

Fig. 97 Valve cover bolt removal sequence—2.3L engine shown 2.5L engine similar

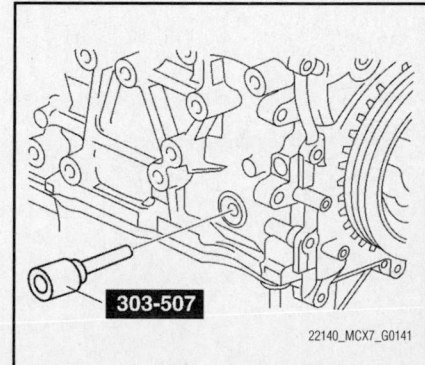

Fig. 98 Special service tool 303-507

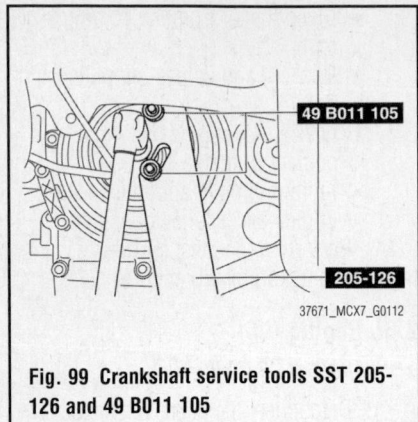

Fig. 99 Crankshaft service tools SST 205-126 and 49 B011 105

`49 B011 105`
`205-126`
37671_MCX7_G0112

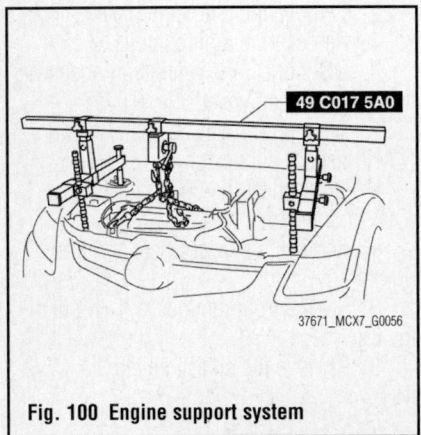

Fig. 100 Engine support system

`49 C017 5A0`
37671_MCX7_G0056

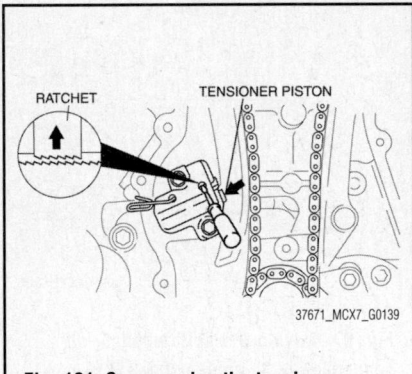

RATCHET   TENSIONER PISTON

Fig. 101 Compressing the tensioner before removal

37671_MCX7_G0139

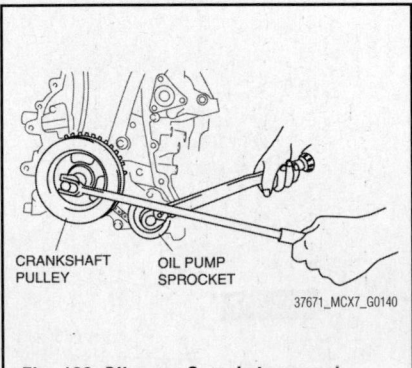

CRANKSHAFT PULLEY   OIL PUMP SPROCKET

Fig. 102 Oil pump Sprocket removal shown

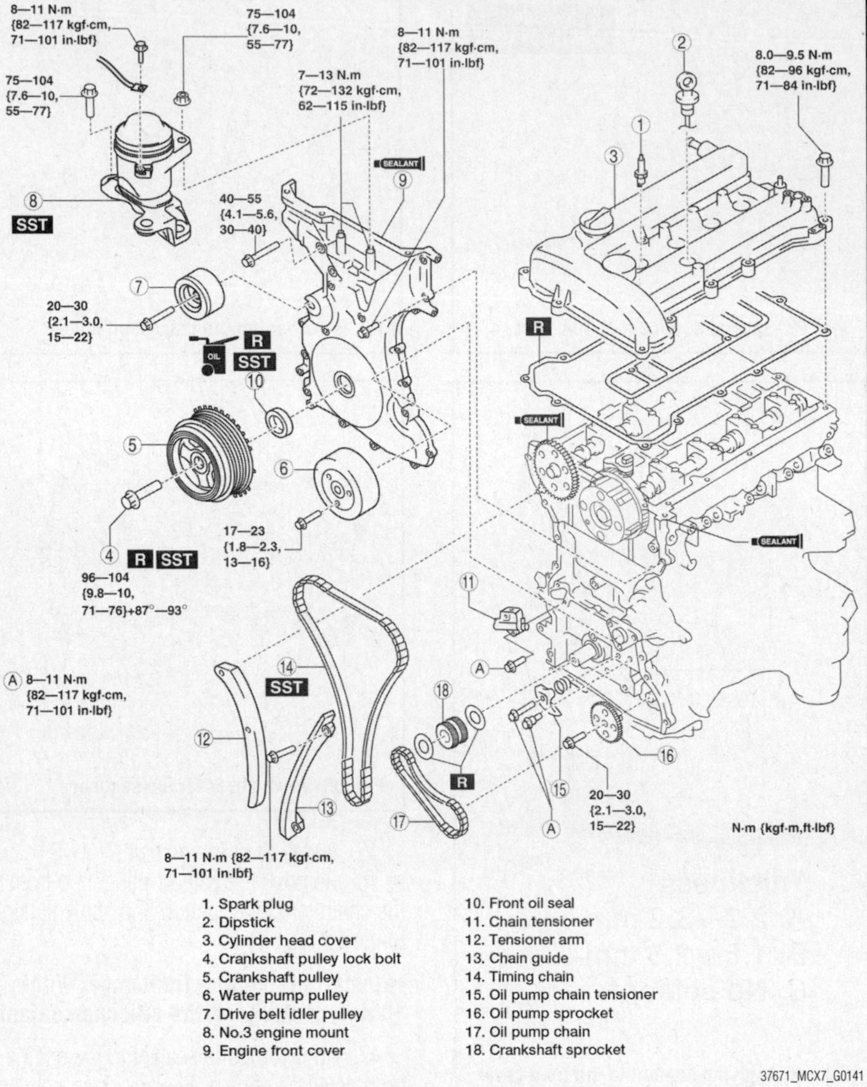

37671_MCX7_G0140

8—11 N·m {82—117 kgf·cm, 71—101 in·lbf}

75—104 {7.6—10, 55—77}

75—104 {7.6—10, 55—77}

7—13 N·m {72—132 kgf·cm, 62—115 in·lbf}

8—11 N·m {82—117 kgf·cm, 71—101 in·lbf}

8.0—9.5 N·m {82—96 kgf·cm, 71—84 in·lbf}

40—55 {4.1—5.6, 30—40}

20—30 {2.1—3.0, 15—22}

17—23 {1.8—2.3, 13—16}

96—104 {9.8—10, 71—76}+87°—93°

Ⓐ 8—11 N·m {82—117 kgf·cm, 71—101 in·lbf}

20—30 {2.1—3.0, 15—22}

N·m {kgf·m,ft·lbf}

8—11 N·m {82—117 kgf·cm, 71—101 in·lbf}

1. Spark plug
2. Dipstick
3. Cylinder head cover
4. Crankshaft pulley lock bolt
5. Crankshaft pulley
6. Water pump pulley
7. Drive belt idler pulley
8. No.3 engine mount
9. Engine front cover
10. Front oil seal
11. Chain tensioner
12. Tensioner arm
13. Chain guide
14. Timing chain
15. Oil pump chain tensioner
16. Oil pump sprocket
17. Oil pump chain
18. Crankshaft sprocket

37671_MCX7_G0141

Fig. 103 Timing chain exploded view—2.5L engine

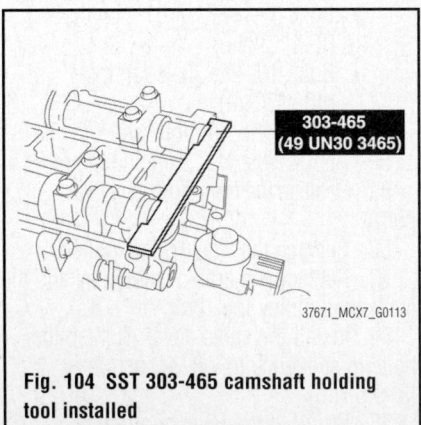

`303-465 (49 UN30 3465)`
37671_MCX7_G0113

Fig. 104 SST 303-465 camshaft holding tool installed

10. Loosen the water pump pulley bolts and remove the drive belt.

11. Remove the Crankshaft Position (CKP) sensor.

12. Remove the P/S oil pump with the hose and pipe still connected.

➡**Position and secure the P/S oil pump out of the way with a rope or wire.**

13. Remove the spark plugs.

14. Remove the engine dipstick.

15. Remove the purge solenoid valve installation bracket with the evaporative hose still connected and set it out of the way

16. Remove the cylinder head cover.

17. Rotate the crankshaft in the direction of the engine rotation and remove the cylinder block lower blind plug when the No. 1 cylinder is at the point prior to Top Dead

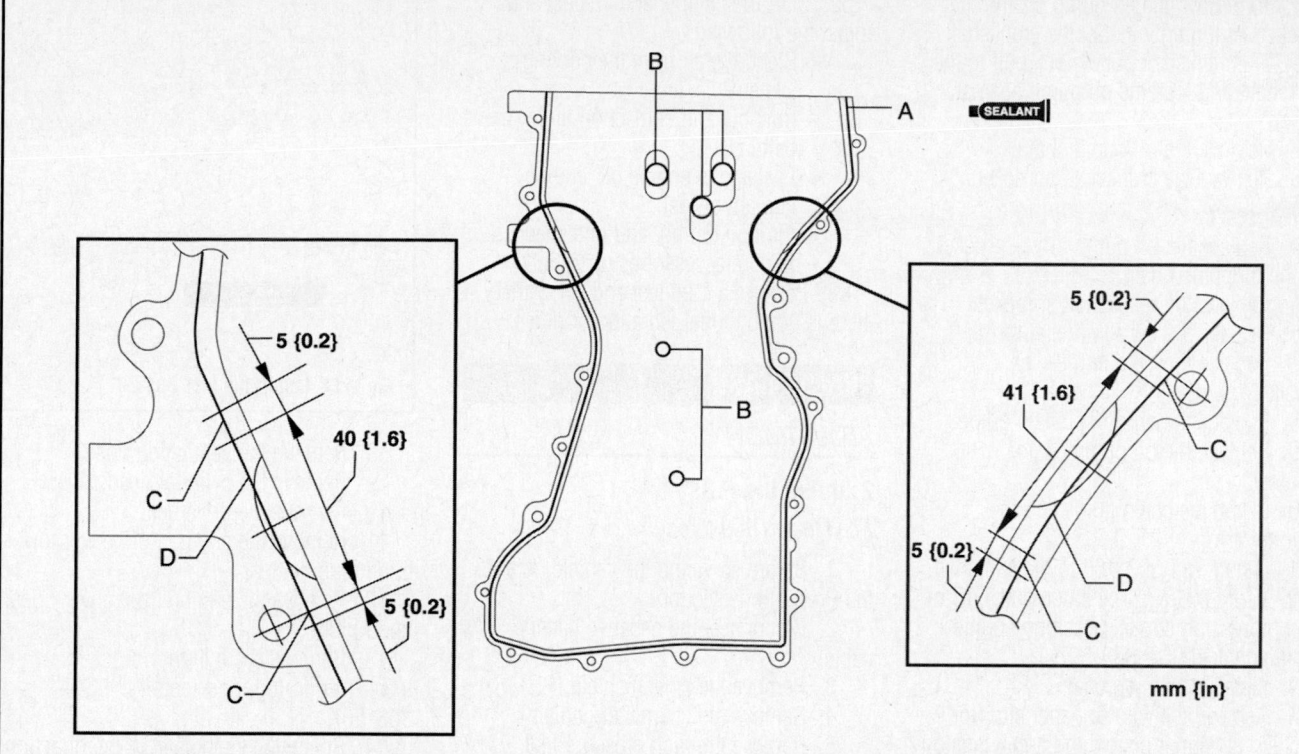

Sealant thickness A: 0.087-0.12 inch (2.2-3.2 mm)
Sealant thickness B: 0.06-0.098 inch (1.5-2.5 mm)

Sealant thickness C: 0.09-0.16 inch (2.2-4.3 mm)
Sealant thickness D: 0.13-0.16 inch (3.3-4.3 mm)

37671_MCX7_G0142

Fig. 105 Engine front cover sealant locations and thickness

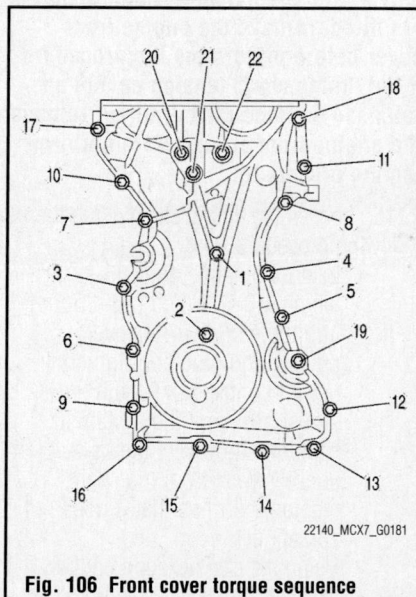

22140_MCX7_G0181

Fig. 106 Front cover torque sequence

Center (TDC) of compression, then install the SST.

18. Remove the crankshaft pulley lock bolt. Hold the crankshaft pulley by using the SSTs.

19. Remove the crankshaft pulley.

20. Remove the water pump pulley.

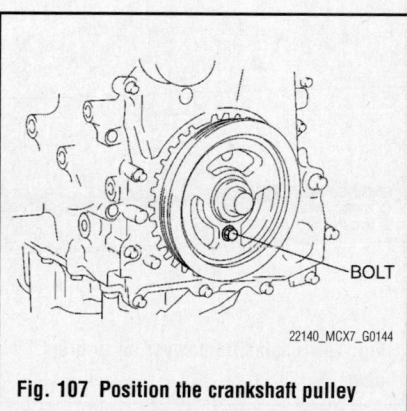

22140_MCX7_G0144

Fig. 107 Position the crankshaft pulley

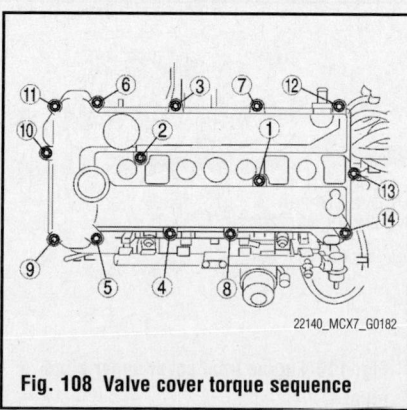

22140_MCX7_G0182

Fig. 108 Valve cover torque sequence

21. Remove the drive belt idler pulley.

22. Support the engine using a engine support system.

23. To remove the No.3 Engine mount remove the following:
  • Windshield wiper arm and blade
  • The cowl grille
  • Air cleaner and air cleaner bracket

24. Remove the No.3 engine mount.

25. Remove the engine front cover.

26. Remove the front oil seal.

27. Remove the chain tensioner as follows:
  • Using a thin screwdriver, hold the chain tensioner ratchet lock mechanism away from the ratchet stem.
  • Slowly compress the tensioner piston
  • Hold the tensioner piston using a 0.059 inch (1.5 mm) wire or paper clip.
  • Remove the mounting bolt and tensioner

28. Remove the tensioner arm and chain guide.

29. Remove the timing chain.

30. Remove the oil pump chain tensioner.

31. To remove the oil pump chain sprocket, temporarily install the crankshaft pulley and crankshaft pulley lock bolt to the crankshaft, and lock the oil pump against rotation.

32. Remove the oil pump chain.

33. Remove the crankshaft sprocket.

To install:

34. Replace the oil pump drive sprocket and the oil pump chain.

35. Replace the crankshaft sprocket.

36. Replace the oil pump chain.

37. Replace the oil pump driven sprocket.

38. Replace the oil pump chain guide.

39. Replace the oil pump chain tensioner.

40. Install the chain guide and the tensioner arm.

41. Install SST 303-465 to the camshaft.

42. Verify that No.1 cylinder is at TDC of the compression stroke. (Position counterweight contacts SST. 303-507)

43. Install the timing chain.

44. Remove the wire or paper clip from the chain tensioner piston and apply tension to the timing chain.

➡**Install the engine front cover within 10 min of applying the silicone sealant.**

45. Apply silicone sealant to the engine front cover. Verify that there is no oil or dust on the seal.

➡**Install the engine front cover before the applied silicone sealant starts to harden.**

46. Install the front cover. Tighten the bolts in sequence as follows:
   - Bolts 1–18: 71–101 inch lbs. (8–11.5 Nm)
   - Bolts 19–22: 30–40 ft. lbs. (40–55 Nm)

47. Verify that No.1 cylinder is at TDC of the compression stroke. (Position counterweight contacts SST.)

48. To position the crankshaft pulley, temporarily tighten it and, using a suitable bolt (M6 X 1.0), fix the crankshaft pulley to the engine front cover.

49. Install SSTs SST 205-126 and 49 B011 105 to the crankshaft pulley and lock the crankshaft against rotation, and tighten the crankshaft pulley lock bolt using the following two steps:
   - Step 1: 71–77 ft. lbs. (96–104 Nm)
   - Step 2: Plus 87°–93°

50. Install the valve cover. Tighten the bolts in sequence to 71–84 inch lbs. (8–9.5 Nm)

51. Install in the reverse order of removal.

52. Start the engine, and inspect and adjust the following:
   - Bleed the air from the cooling system.
   - Run-out and contact on pulley and belt
   - Leakage of engine oil, engine coolant or fuel.
   - Ignition timing, and idle speed.
   - Engine accessories operation

53. Perform a road test and verify that there is no abnormal vibration or noise.

VALVE LASH

ADJUSTMENT

2.3L Engine (L3)

See Figures 109 through 123.

1. Before servicing the vehicle, refer to the Precautions section.

2. Disconnect the negative battery cable.

3. Remove the splash shield (RH).

4. Remove the charge air cooler.

5. Remove the high pressure fuel pump.

6. Remove the ignition coils.

7. Disconnect the wiring harness.

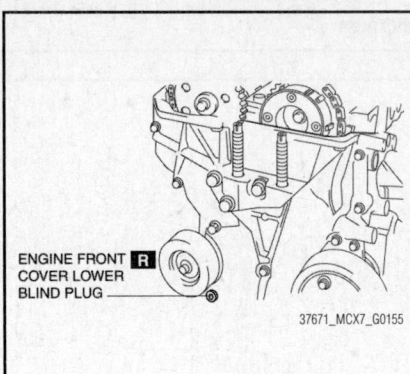

Fig. 109 Engine front cover lower blind plug

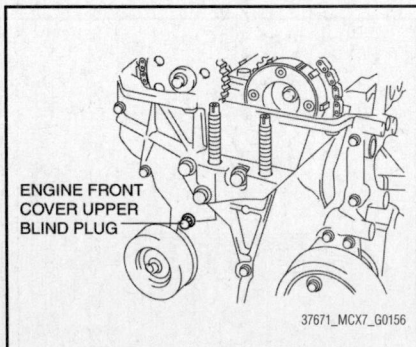

Fig. 110 Engine front cover upper blind plug

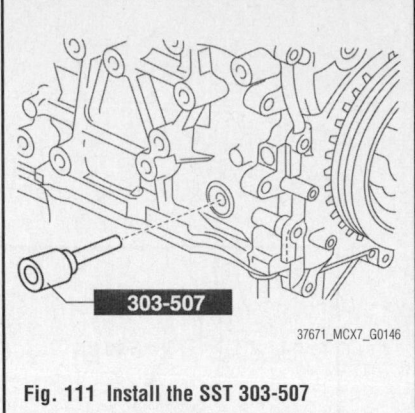

Fig. 111 Install the SST 303-507

8. Remove the ventilation hose.

9. Remove the cylinder head cover.

10. Remove the drive belt.

11. Remove the engine front cover lower blind plug.

12. Remove the engine front cover upper blind plug.

13. Remove the cylinder block lower blind plug, and install the SST 303-507.

14. Rotate the crankshaft in the direction of the engine rotation so that the No.1 piston is at TDC of the compression stroke. (Until the counterweight contacts SST and stops.)

➡**Verify the ratchet position using a mirror before servicing. If the ratchet is not tilted, remove the engine front cover before performing the procedure. If the timing chain tension cannot be released with the ratchet tilted, remove the engine front cover before performing the procedure.**

15. Loosen the timing chain using the following procedure:
   - Insert a suitable bolt M6 X 1.0, length 1.0–1.3 inch (25–35 mm) into the engine front cover upper blind plug and tighten it until it contacts the chain tensioner arm, and then rotate it back one turn. (Set the bolt slightly away from the chain tensioner arm so that it does not contact it.)
   - Using the cast hexagon on the exhaust camshaft, apply force counterclockwise to facilitate unlocking the chain tensioner ratchet.

➡**The chain tensioner rack is compressed using the chain tension generated by applying force to the exhaust camshaft in the direction of the engine rotation.**

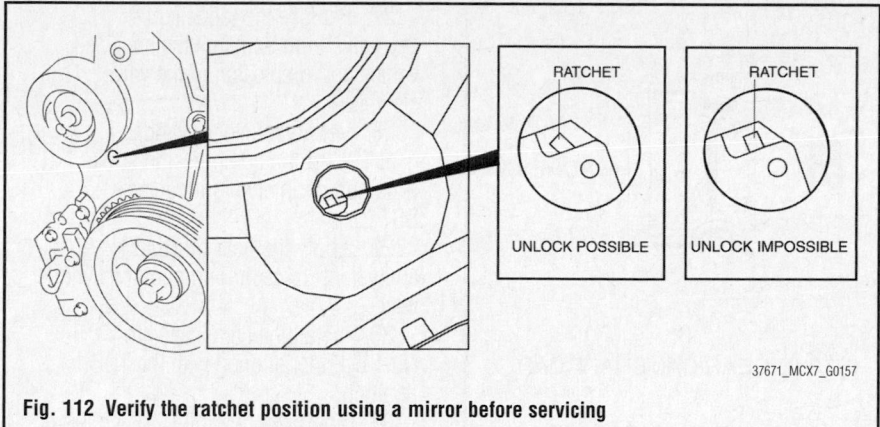

Fig. 112 Verify the ratchet position using a mirror before servicing

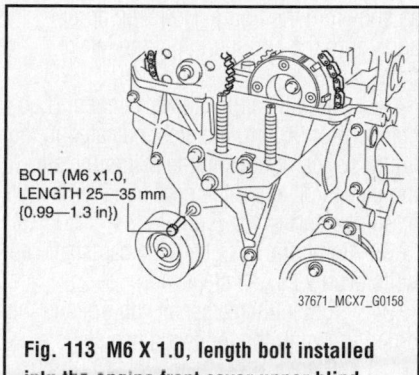

Fig. 113 M6 X 1.0, length bolt installed into the engine front cover upper blind plug

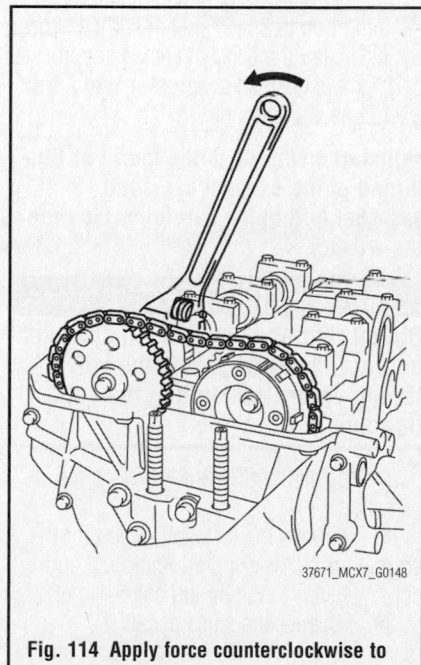

Fig. 114 Apply force counterclockwise to facilitate unlocking the chain tensioner ratchet

➡The ratchet has not been unlocked if the bolt cannot be pressed in approx. 0.2 inch (5 mm).

• Screw in the bolt to approx. 0.2 inch

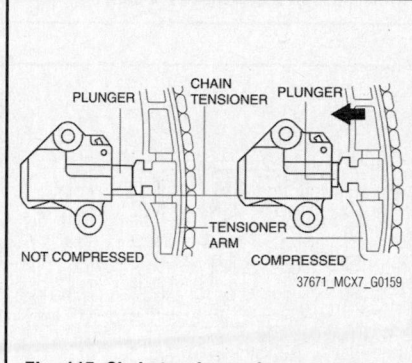

Fig. 115 Chain tensioner shown not compressed and compressed

(5 mm) and secure the tensioner arm with the rack compressed.

➡If the tensioner arm cannot be secured, return the bolt to its original position and repeat the procedure.

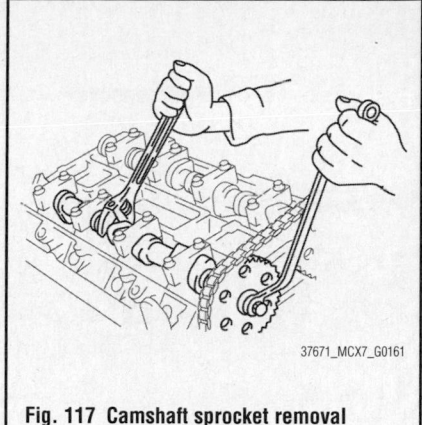

Fig. 117 Camshaft sprocket removal

16. Fix the exhaust camshaft using a wrench on the cast hexagon, and loosen the camshaft sprocket bolt.

17. Remove the exhaust camshaft sprocket bolt, exhaust camshaft sprocket, and washer as a single unit. Perform the work carefully so that the washer does not drop out.

18. Remove the oil control valve (OCV).

✳✳ WARNING

The camshaft caps are to be kept ordered for correct reassembly in their original positions. Do not mix the caps.

19. Loosen the camshaft cap bolts in two or three passes in the order shown in the figure and remove the camshaft cap.

20. Remove the camshafts for the intake and exhaust sides.

21. Remove the tappet.

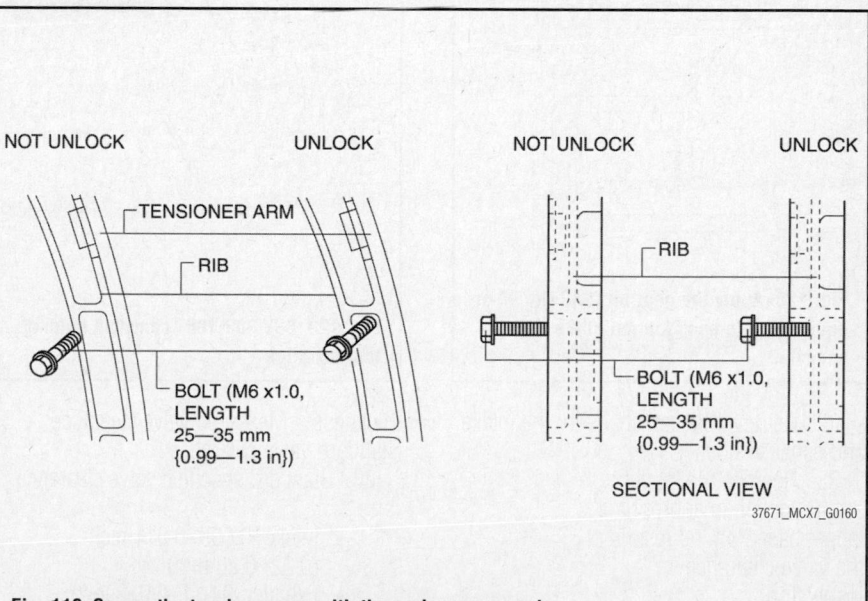

Fig. 116 Secure the tensioner arm with the rack compressed

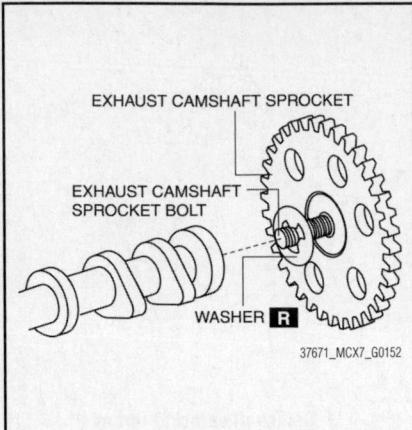

Fig. 118 Exhaust camshaft sprocket, bolt and washer removed as a unit

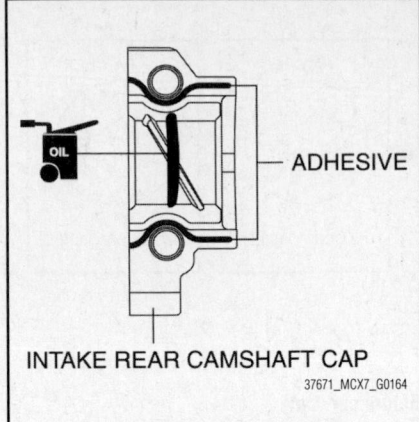

Fig. 121 Apply adhesive as shown, to thickness of 0.039 inch (1.0 mm)

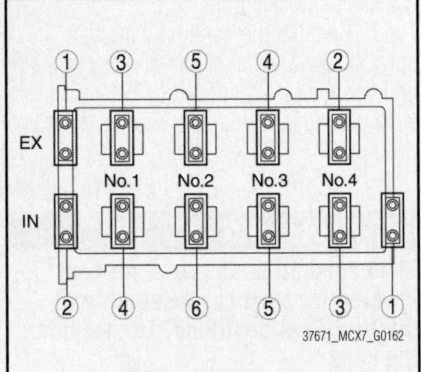

Fig. 119 Camshaft cap removal sequence—2.3L engine

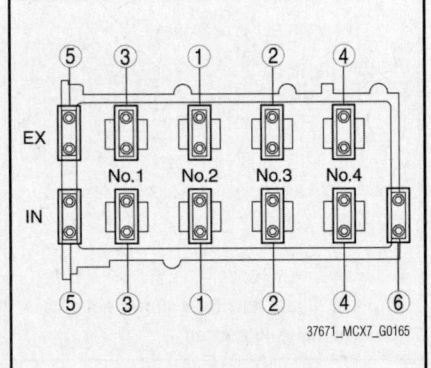

Fig. 122 Camshaft cap tightening sequence—2.3L engine

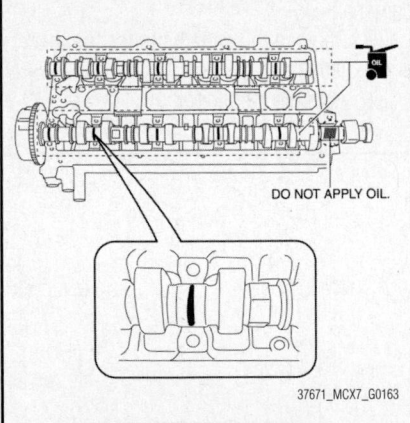

Fig. 120 Apply the gear oil (SAE No. 90 or equivalent) to each journal of the camshaft

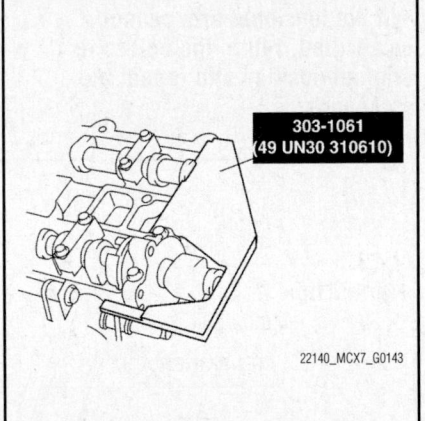

Fig. 123 SST 303-1061 camshaft holding tool installed

22. Remove the camshafts for the intake and exhaust sides.

23. Remove the tappet.

24. Install an appropriate tappet based on the results of the valve clearance inspection.

25. Selected tappet = Removed tappet thickness + Measured valve clearance - Standard valve clearance

26. Standard specified valve clearance is:

- Intake: 0.009–0.011 inch (0.22–0.28 mm)
- Exhaust: 0.011–0.012 inch (0.27–0.33 mm).

27. Replace the tappets of any valve found outside the specified values and premeasured the valve clearance.

28. Verify that No.1 cylinder is at TDC of the compression stroke. (Position counterweight contacts SST.)

29. Apply the gear oil (SAE No.90 or equivalent) to each journal of the cylinder head.

30. Install the camshaft with No.1 cylinder aligned with the TDC position.

31. Apply the gear oil (SAE No. 90 or equivalent) to each journal of the camshaft as shown in the figure. However, do not apply it to the end journal of the intake camshaft.

32. Carefully apply adhesive agent (Loctite® 518 or 962) to the area indicated in the figure so that it does not leak into the sliding part then, apply the gear oil (SAE No. 90 or equivalent) to the journal.

33. Temporarily tighten the camshaft cap bolts evenly in 2–3 steps.

34. Tighten the camshaft cap bolts in the order shown in the following two steps;

- Step 1: 45–79 inch lbs. (5.0–9.0 Nm)
- Step 2: 11–12 ft. lbs. (14–17 Nm)

35. Install the OCV and tighten the retaining bolt to 71–97 inch lbs. (8–11 Nm).

36. Install the exhaust camshaft sprocket bolt, exhaust camshaft sprocket, and a new washer as a single unit.

➡Install a washer to the fourth or fifth thread of the exhaust camshaft sprocket bolt being careful not to drop the washer.

☼ WARNING

Do not tighten the camshaft sprocket bolt at this stage. Verify the valve timing before performing the bolt tightening.

37. Install the SST 303-1061 on the camshaft.

38. Remove the installation bolt for the engine front cover upper blind plug, and apply tension to the timing chain.

39. Fix the exhaust camshaft using a wrench on the cast hexagon, and tighten the sprocket bolt to 51–55 ft. lbs. (69–75 Nm).

40. Remove the SST 303-1061 from the camshaft.

41. Remove the installation bolt for the engine front cover upper blind plug, and apply tension to the timing chain.

42. Rotate the crankshaft clockwise and verify that the No.1 cylinder is at TDC of the compression stroke. (The position counter weight contacts the SST.)

43. Fix the exhaust camshaft using a wrench on the cast hexagon, and tighten the sprocket bolt to 51–55 ft. lbs. (69–75 Nm).

44. Remove the SST 303-1061 from the camshaft.

45. Remove the SST 303-507 installed in the cylinder block lower blind plug hole.

46. Rotate the crankshaft clockwise two turns and inspect the valve timing.

➡ **If not aligned, loosen the camshaft sprocket bolt and repeat the procedure.**

47. Apply the silicone sealant and install the engine front cover upper blind plug. Tighten to 71–101 inch lbs. (8–11 Nm).

48. Install a new engine front cover lower blind plug and tighten to 89–123 inch lbs. (10–14 Nm).

49. Install the cylinder block lower blind plug, and tighten to 14–16 ft. lbs. (18–22 Nm).

50. Install a new engine front cover lower blind plug and tighten to 89–123 inch lbs. (10–14 Nm).

51. Install the drive belt.

52. Install the cylinder head cover.

53. Install the ventilation hose.

54. Connect the wiring harness.

55. Install the ignition coils.

56. Install the high pressure fuel pump.

57. Install the charge air cooler.

58. Install the splash shield (RH).

59. Connect the negative battery cable.

60. Start the engine, and inspect and adjust the following:

- Bleed the air from the cooling system.
- Leakage of engine oil, engine coolant or fuel.
- Ignition timing, and idle speed.

61. Perform a road test and verify that there is no abnormal vibration or noise.

2.5L Engine (L5)

See Figures 124 through 135.

1. Before servicing the vehicle, refer to the Precautions section

2. Disconnect the negative battery cable.

3. Remove the plug hole plate.

4. Disconnect the wiring harness.

5. Remove the ignition coils.

6. Remove the spark plugs.

7. Disconnect the ventilation hose from the cylinder head cover.

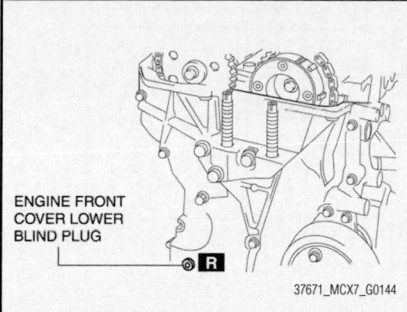

Fig. 124 Engine front cover lower blind plug

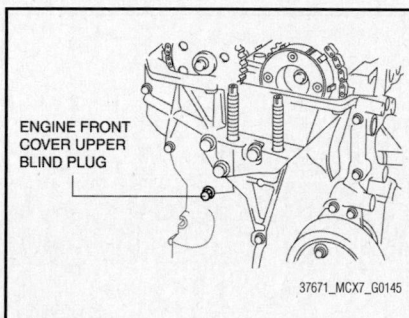

Fig. 125 Engine front cover upper blind plug

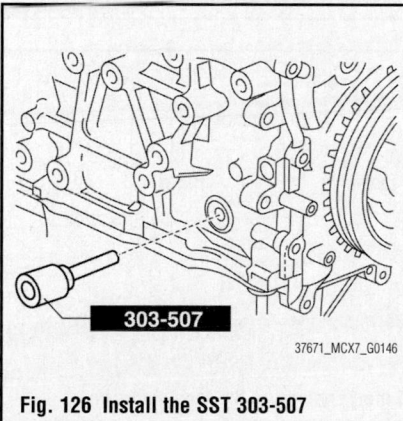

Fig. 126 Install the SST 303-507

8. Remove the dipstick.

9. Remove the cylinder head cover.

10. Remove the front wheel and tire (RH).

11. Remove the undercover.

12. Set the front mudguard (RH) out of the way.

13. Remove the splash shield (RH).

14. Remove the drive belt.

15. Remove the engine front cover lower blind plug.

16. Remove the engine front cover upper blind plug.

17. Remove the cylinder block lower blind plug, and install the SST 303-507.

18. Loosen the timing chain using the following procedure:

- Insert a suitable bolt M6 X 1.0, length 1.0–1.3 inch (25–35 mm) into the

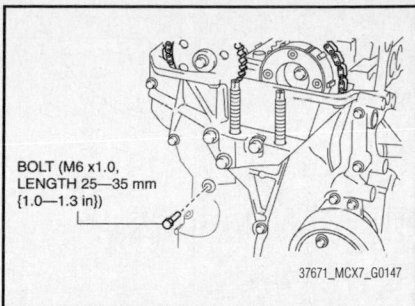

Fig. 127 M6 X 1.0, length bolt installed into the engine front cover upper blind plug

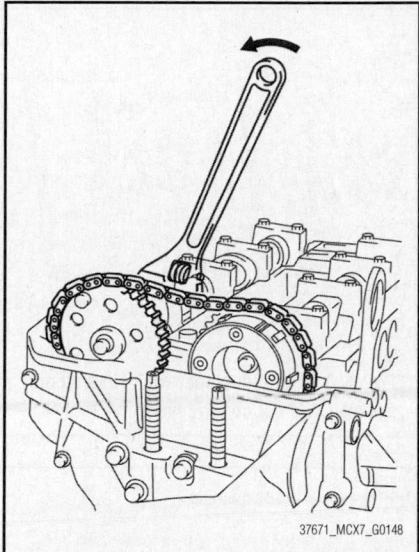

Fig. 128 Apply force counterclockwise to facilitate unlocking the chain tensioner ratchet

engine front cover upper blind plug and tighten it until it contacts the chain tensioner arm, and then rotate it back one turn. (Set the bolt slightly away from the chain tensioner arm so that it does not contact it.)

- Using the cast hexagon on the exhaust camshaft, apply force counterclockwise to facilitate unlocking the chain tensioner ratchet.

- Using a Hex bit socket 0.098 inch (2.5 mm) or T15 Torx® bit socket, unlock the chain tensioner ratchet so that it can be lifted up.

- Using the cast hexagon on the exhaust camshaft, apply force in the direction of the engine rotation to increase tension on the chain

➡ **The chain tensioner rack is compressed using the chain tension generated by applying force to the exhaust camshaft in the direction of the engine rotation.**

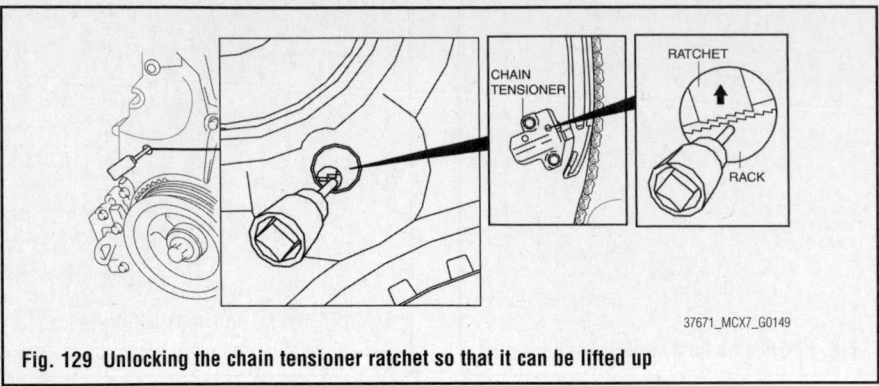

Fig. 129 Unlocking the chain tensioner ratchet so that it can be lifted up

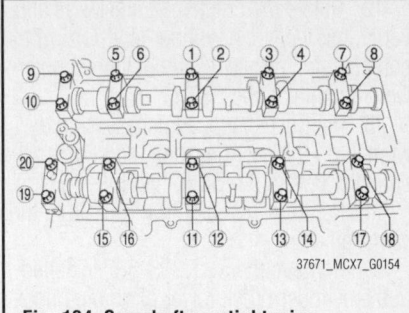

Fig. 134 Camshaft cap tightening sequence—2.5L engine

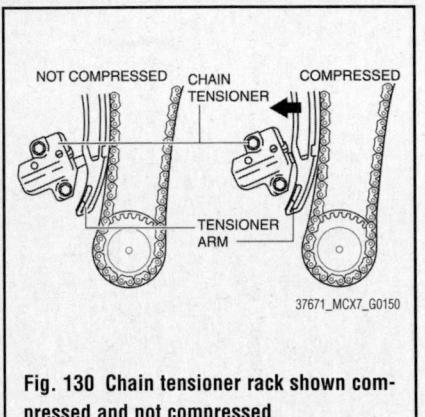

Fig. 130 Chain tensioner rack shown compressed and not compressed

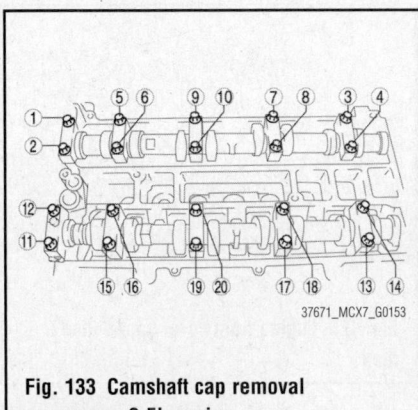

Fig. 133 Camshaft cap removal sequence—2.5L engine

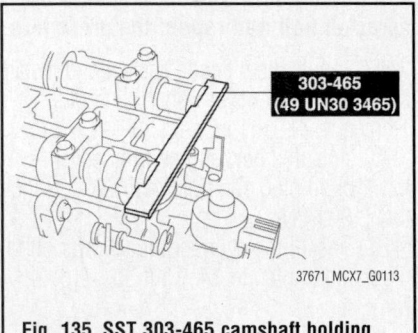

Fig. 135 SST 303-465 camshaft holding tool installed

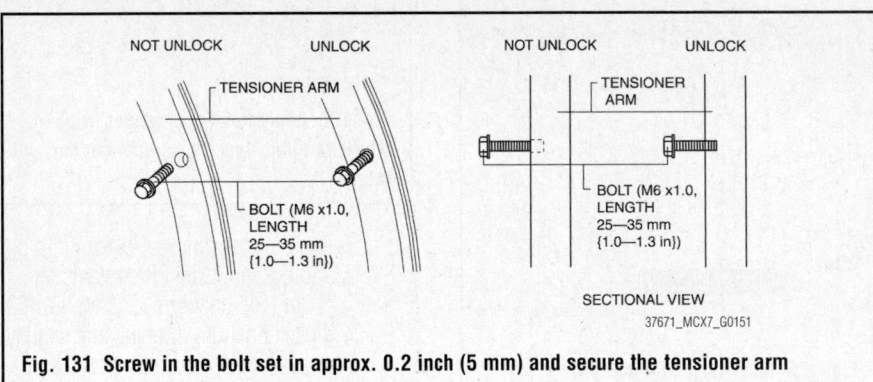

Fig. 131 Screw in the bolt set in approx. 0.2 inch (5 mm) and secure the tensioner arm

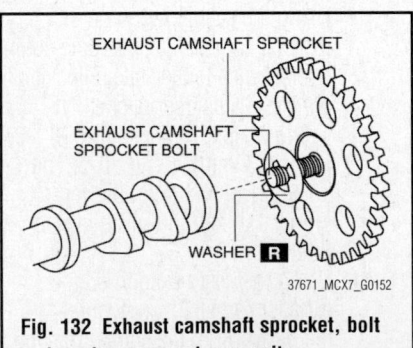

Fig. 132 Exhaust camshaft sprocket, bolt and washer removed as a unit

• Screw in the bolt set in approx. 0.2 inch (5 mm) and secure the tensioner arm with the rack compressed. The racket has not been unlocked if the bolt can-

not be pressed in approx 0.2 inch (5 mm).

➡**If the tensioner arm cannot be secured, return the bolt to its original position and repeat the procedure.**

19. Fix the exhaust camshaft using a wrench on the cast hexagon, and loosen the camshaft sprocket bolt.

20. Remove the exhaust camshaft sprocket bolt, exhaust camshaft sprocket, and washer as a single unit. Perform the work carefully so that the washer does not drop out.

21. Loosen the camshaft cap bolts in two or three steps in the order shown in the figure and remove the camshaft cap.

➡**The camshaft caps are to be kept ordered for correct reassembly in**

their original positions. Do not mix the caps.

22. Remove the camshafts for the intake and exhaust sides.

23. Remove the tappet.

24. Install an appropriate tappet based on the results of the valve clearance inspection.

25. Selected tappet = Removed tappet thickness + Measured valve clearance - Standard valve clearance

26. Standard specified valve clearance is:
   • Intake: 0.009–0.011 inch (0.22–0.28 mm)
   • Exhaust: 0.011–0.012 inch (0.27–0.33 mm).

27. Replace the tappets of any valve found outside the specified values and pre-measured the valve clearance.

28. Verify that No.1 cylinder is at TDC of the compression stroke. (Position counter-weight contacts SST.)

29. Apply the gear oil (SAE No.90 or equivalent) to each journal of the cylinder head.

30. Install the camshaft with No.1 cylinder aligned with the TDC position.

31. Apply the gear oil (SAE No.90 or equivalent) to each camshaft journal of the cylinder head.

32. Temporarily tighten the camshaft cap bolts evenly in 2–3 steps.

33. Tighten the camshaft cap bolts in the order shown in the following two steps;

- Step 1: 45–79 inch lbs. (5.0–9.0 Nm)
- Step 2: 11–12 ft. lbs. (14–17 Nm)

34. Install the OCV and tighten the retaining bolt to 71–97 inch lbs. (8–11 Nm).

35. Install the exhaust camshaft sprocket bolt, exhaust camshaft sprocket, and a new washer as a single unit.

➡**Install a washer to the fourth or fifth thread of the exhaust camshaft sprocket bolt being careful not to drop the washer.**

✵✵ WARNING

Do not tighten the camshaft sprocket bolt at this stage. Verify the valve timing before performing the bolt tightening.

36. Install the SST 303-465 on the camshaft.

37. Remove the installation bolt for the engine front cover upper blind plug, and apply tension to the timing chain.

38. Fix the exhaust camshaft using a wrench on the cast hexagon, and tighten the sprocket bolt to 51–55 ft. lbs. (69–75 Nm).

39. Remove the SST 303-465 from the camshaft.

40. Remove the SST 303-507 installed in the cylinder block lower blind plug hole.

41. Rotate the crankshaft clockwise two turns and inspect the valve timing.

➡**If not aligned, loosen the camshaft sprocket bolt and repeat the procedure.**

42. Install the cylinder block lower blind plug, and tighten to 14–16 ft. lbs. (18–22 Nm).

43. Apply the silicone sealant and install the engine front cover upper blind plug. Tighten to 71–101 inch lbs. (8–11 Nm).

44. Install a new engine front cover lower blind plug and tighten to 89–123 inch lbs. (10–14 Nm).

45. Install the drive belt.

46. Install the splash shield (RH).

47. Install the front mudguard (RH).

48. Install the undercover.

49. Install the front wheel and tire (RH).

50. Install the cylinder head cover.

51. Install the dipstick.

52. Connect the ventilation hose.

53. Install the spark plugs.

54. Install the ignition coils.

55. Connect the wiring harness.

56. Install the plug hole plate.

57. Connect the negative battery cable.

58. Start the engine, and inspect the following:

- Leakage of engine oil, engine coolant or fuel.
- Ignition timing, and idle speed.

59. Perform a road test and verify that there is no abnormal vibration or noise.

ENGINE PERFORMANCE & EMISSION CONTROLS

CAMSHAFT POSITION (CMP) SENSOR

LOCATION

2.3L (L3) Engine

See Figure 136.

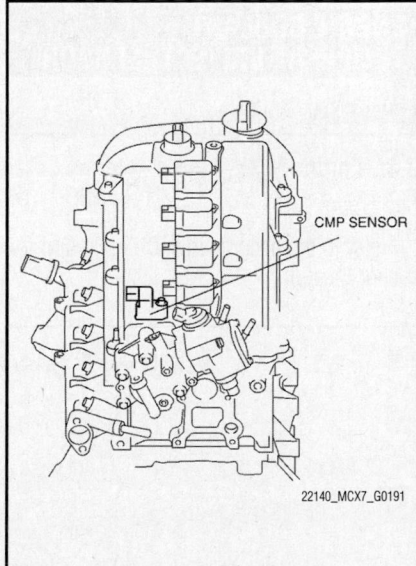

22140_MCX7_G0191

Fig. 136 Camshaft position sensor—2.3L Engine

2.5L (L5) Engine

See Figure 137.

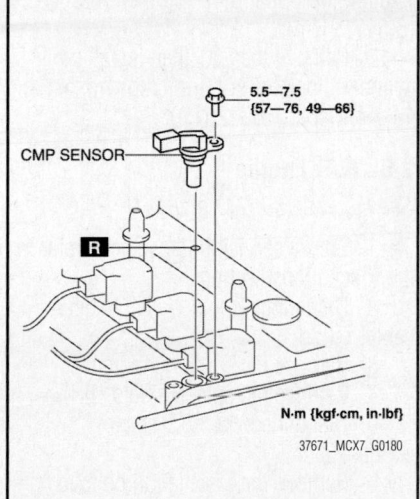

37671_MCX7_G0180

Fig. 137 Camshaft position sensor—2.5L Engine

REMOVAL & INSTALLATION

2.3L (L3) Engine

✵✵ WARNING

When replacing the CMP sensor, make sure there is no foreign material on it such as metal shavings. If it is installed with foreign material, the sensor output signal will malfunction resulting from fluctuation in magnetic flux and cause a deterioration in engine control.

1. Before servicing the vehicle, refer to the Precautions section.

2. Disconnect the negative battery cable.

3. Remove the charge air cooler duct.

4. Remove the charge air cooler cover.

5. Disconnect the CMP sensor connector.

6. Remove the CMP sensor installation bolt.

7. Remove the CMP sensor from the cylinder head cover.

8. Install in the reverse order of removal. Tighten the CMP bolt to 49–66 inch lbs. (5.5–7.5 Nm).

2.5L (L5) Engine

✵✵ WARNING

When replacing the CMP sensor, make sure there is no foreign material on it such as metal shavings. If it is installed with foreign material, the sensor output signal will malfunction resulting from fluctuation in magnetic flux and cause a deterioration in engine control.

1. Before servicing the vehicle, refer to the Precautions section.

2. Disconnect the negative battery cable.

3. Remove the plug hole plate.
4. Disconnect the CMP sensor connector.
5. Remove the CMP sensor.
6. Install in the reverse order of removal. Tighten the CMP bolt to 49–66 inch lbs. (5.5–7.5 Nm).

CRANKSHAFT POSITION (CKP) SENSOR

LOCATION

See Figure 138.

Refer to the accompanying illustration for sensor location.

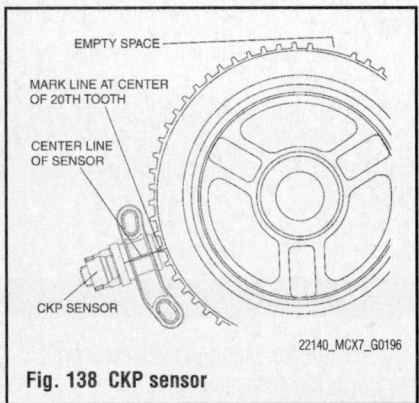

Fig. 138 CKP sensor

REMOVAL & INSTALLATION

2.3L (L3) Engine

See Figure 139.

1. Before servicing the vehicle, refer to the Precautions section.
2. Disconnect the negative battery cable.
3. Remove the charge air cooler duct.
4. Remove the undercover.
5. Remove the splash shield (RH).
6. Disconnect the CKP sensor connector.
7. Remove the installation bolts to remove the CKP sensor.

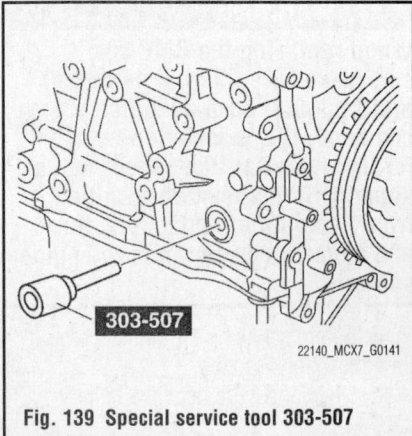

Fig. 139 Special service tool 303-507

To install:

8. Rotate the crankshaft in the direction of the engine rotation and remove the cylinder block lower blind plug when the No. 1 cylinder is at the point prior to top dead center (TDC) of compression, then install SST 303-507.

9. Rotate the crankshaft in the direction of the engine rotation so that the No.1 piston is at TDC of the compression stroke. (Until the crank weight contacts SST and stops.)

10. Using a straight edge, draw a straight line directly in the center of the twentieth tooth of the crankshaft pulley pulse wheel (counting counterclockwise from the empty space).

✳✳ CAUTION

If the line is not accurately drawn, ignition timing, fuel injection and other engine control systems will be adversely affected. Draw the straight line carefully using a straight edge.

11. Align the center line of the crankshaft position sensor and the line drawn on the tooth, then install the sensor. Tighten the bolts to 49–66 inch lbs. (5.5–7.5 Nm).

12. Remove the SST then install the cylinder block lower blind plug. Tighten to 14–16 ft. lbs. (18–22 Nm).

2.5L (L5) Engine

See Figures 139, 140 and 141.

1. Before servicing the vehicle, refer to the Precautions section.
2. Disconnect the negative battery cable.
3. Remove the engine undercover.
4. Remove the splash shield (RH).
5. Disconnect the CKP sensor connector.
6. Remove the installation bolts to remove the CKP sensor.

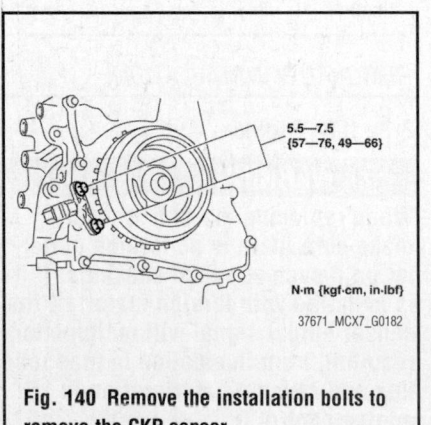

Fig. 140 Remove the installation bolts to remove the CKP sensor

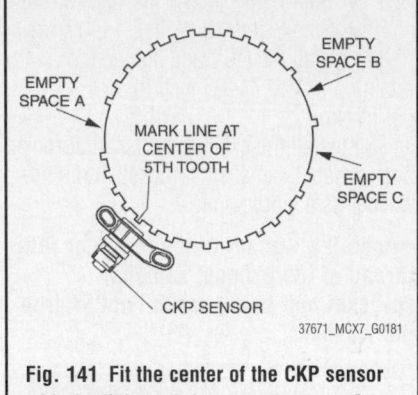

Fig. 141 Fit the center of the CKP sensor with the fifth tooth from empty space A

To install:

7. Rotate the crankshaft in the direction of the engine rotation and remove the cylinder block lower blind plug when the No. 1 cylinder is at the point prior to top dead center (TDC) of compression, then install SST 303-507.

8. Fit the center of the CKP sensor with the fifth tooth (counting counterclockwise from the empty space A as shown in the figure) of the pulse wheel

9. Install the CKP sensor fitting bolts. Tighten the bolts to 49–66 inch lbs. (5.5–7.5 Nm).

10. Remove the SST 303-507, then install the cylinder block lower blind plug. Tighten to 14–16 ft. lbs. (18–22 Nm).

ENGINE COOLANT TEMPERATURE (ECT) SENSOR

LOCATION

2.3L Engine (L3)

See Figure 142.

Refer to the accompanying illustration for sensor location.

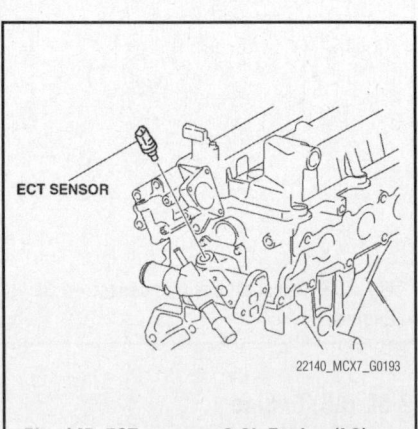

Fig. 142 ECT sensor—2.3L Engine (L3)

2.5L Engine (L5)

See Figure 143.

Refer to the accompanying illustration for sensor location.

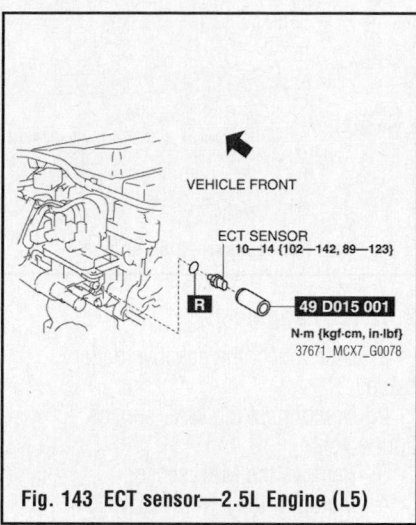

Fig. 143 ECT sensor—2.5L Engine (L5)

REMOVAL & INSTALLATION

2.3L (L3) Engine

> ※※ **CAUTION**
>
> **A hot engine can cause severe burns. Turn off the engine and wait until it is cool before removing the ECT sensor.**

1. Before servicing the vehicle, refer to the Precautions section.
2. Disconnect the negative battery cable.
3. Remove the charge air cooler duct.
4. Remove the battery and battery tray.
5. Remove the air cleaner and air hose.
6. Drain the engine coolant.
7. Disconnect the wiring harness.
8. Remove the ECT sensor.
9. Install in the reverse order of removal. Tighten to 89–123 inch lbs. (10–14 Nm).
10. Refill the engine coolant.
11. Connect the negative battery cable.

2.5L (L5) Engine

> ※※ **CAUTION**
>
> **When the engine is hot, it can badly burn. Turn off the engine and wait until it is cool before removing the ECT sensor.**

1. Before servicing the vehicle, refer to the Precautions section.
2. Disconnect the negative battery cable.
3. Drain the engine coolant.
4. Disconnect the ECT sensor connector.
5. Remove the ECT sensor.
6. Install in the reverse order of removal. Tighten the sensor to 89–123 inch lbs. (10–14 Nm).
7. Refill the engine coolant.
8. Connect the negative battery cable.

HEATED OXYGEN (HO2S) SENSOR

LOCATION

The Heated Oxygen (HO2S) Sensor is the second sensor mounted in the exhaust system.

REMOVAL & INSTALLATION

2.3L (L3) Engine

See Figures 144 and 145.

> ※※ **CAUTION**
>
> **A hot engine and exhaust system can cause severe burns. Turn off the engine and wait until**

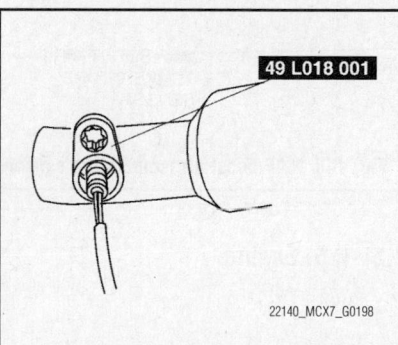

Fig. 144 Special service tool 49 L018 001

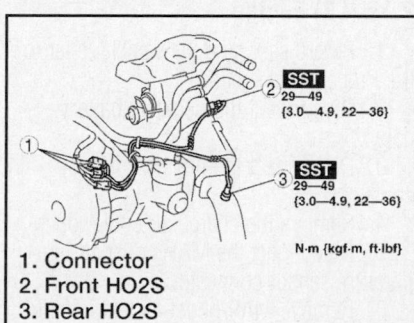

1. Connector
2. Front HO2S
3. Rear HO2S

Fig. 145 Heated Oxygen (HO2S) Sensor— 2.3L engine

they are cool before removing the exhaust system.

1. Before servicing the vehicle, refer to the Precautions section.
2. Disconnect the negative battery cable.
3. Remove the charge air cooler duct.
4. Disconnect the wiring harness connector.
5. Using SST 49 L018 001 or equivalent, remove the HO2S.
6. Install in the reverse order of removal.

2.5L (L5) Engine

See Figure 146.

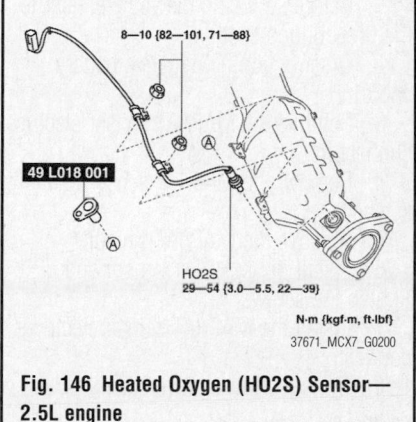

Fig. 146 Heated Oxygen (HO2S) Sensor— 2.5L engine

> ※※ **CAUTION**
>
> **A hot engine and exhaust system can cause severe burns. Turn off the engine and wait until they are cool before removing the exhaust system.**

1. Disconnect the negative battery cable.
2. Remove the undercover.
3. Disconnect the HO2S connector.
4. Remove the HO2S using the SST 49 L018 001.
5. Install in the reverse order of removal.

INTAKE AIR TEMPERATURE (IAT) SENSOR

LOCATION

See Figure 147.

The Mass Air Flow (MAF)/Intake Air Temperature (IAT) sensor is mounted to the air cleaner cover. The Intake Air Temperature (IAT) Sensor is built into the Mass Air Flow (MAF) sensor.

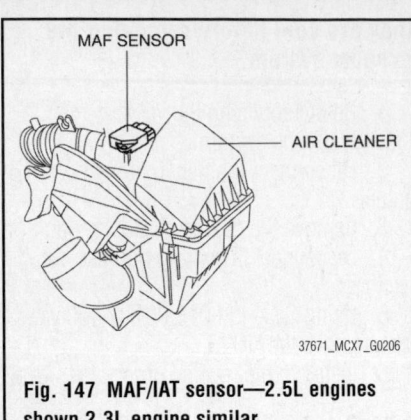

Fig. 147 MAF/IAT sensor—2.5L engines shown 2.3L engine similar

REMOVAL & INSTALLATION

1. Before servicing the vehicle, refer to the Precautions section.
2. Disconnect the negative battery cable.
3. Remove the MAF/IAT sensor electrical connector.
4. Remove the MAF/IAT sensor retaining screws.
5. Remove the MAF/IAT sensor.
6. Install the MAF/IAT sensor and tighten the screws.
7. Install the MAF/IAT sensor electrical connector.
8. Connect the negative battery cable.

KNOCK SENSOR (KS)

REMOVAL & INSTALLATION

1. Disconnect the negative battery cable.
2. Remove the plug hole plate. (2.5L engine only)
3. Remove the intake manifold. Refer to Intake Manifold Removal & Installation, in the Engine Mechanical section.
4. Disconnect the KS connector.
5. Remove the KS.
6. Install in the reverse of the removal procedure and tighten the KS sensor mounting bolt to 12–17 ft. lbs. (16–24 Nm)

MASS AIR FLOW (MAF) SENSOR

LOCATION

See Figure 147.

The Mass Air Flow (MAF)/Intake Air Temperature (IAT) sensor is mounted to the air cleaner cover. The Intake Air Temperature (IAT) Sensor is built into the Mass Air Flow (MAF) sensor.

REMOVAL & INSTALLATION

1. Before servicing the vehicle, refer to the Precautions section.
2. Disconnect the negative battery cable.
3. Remove the MAF/IAT sensor electrical connector.
4. Remove the MAF/IAT sensor retaining screws.
5. Remove the MAF/IAT sensor.
6. Install the MAF/IAT sensor and tighten the screws.
7. Install the MAF/IAT sensor electrical connector.
8. Connect the negative battery cable.

MANIFOLD ABSOLUTE PRESSURE (MAP) SENSOR

LOCATION

2.3L (L3) Engine
See Figure 148.

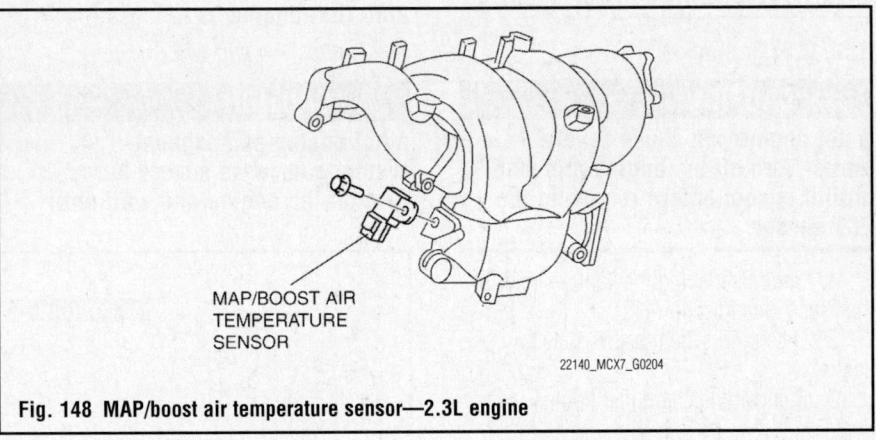

Fig. 148 MAP/boost air temperature sensor—2.3L engine

2.5L (L5) Engine
See Figure 149.

REMOVAL & INSTALLATION

2.3L (L3) Engine

1. Before servicing the vehicle, refer to the Precautions section.
2. Disconnect the negative battery cable.
3. Disconnect the negative battery cable.
4. Remove the charge air cooler duct.
5. Disconnect the MAP/boost air temperature sensor connector.
6. Remove MAP/boost air temperature sensor from the intake manifold.
7. Install in the reverse order of removal and tighten the MAP/boost air temperature bolt to 45–61 ft. lbs. (5–7 Nm).

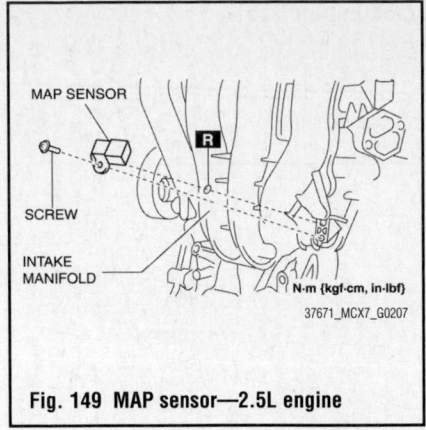

Fig. 149 MAP sensor—2.5L engine

2.5L (L5) Engine

1. Disconnect the negative battery cable.
2. Disconnect the MAP sensor connector.
3. Remove the MAP sensor.
4. Install in the reverse order of removal. Tighten the MAP sensor retaining bolt to 45–61 ft. lbs. (5–7 Nm).

POWERTRAIN CONTROL MODULE (PCM)

LOCATION

The Powertrain Control Module (PCM) is located in the engine compartment, just to the right of the battery.

REMOVAL & INSTALLATION

See Figures 150 through 152.

➡**If the PCM is replaced, the replacement must be programmed with the vehicle parameters..**

1. Disconnect the negative battery cable.
2. Remove the charge air cooler duct. (2.3L engines only)

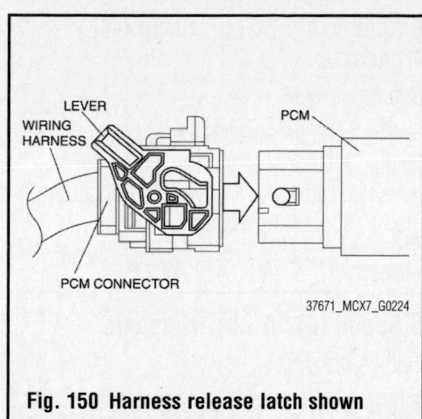

Fig. 150 Harness release latch shown

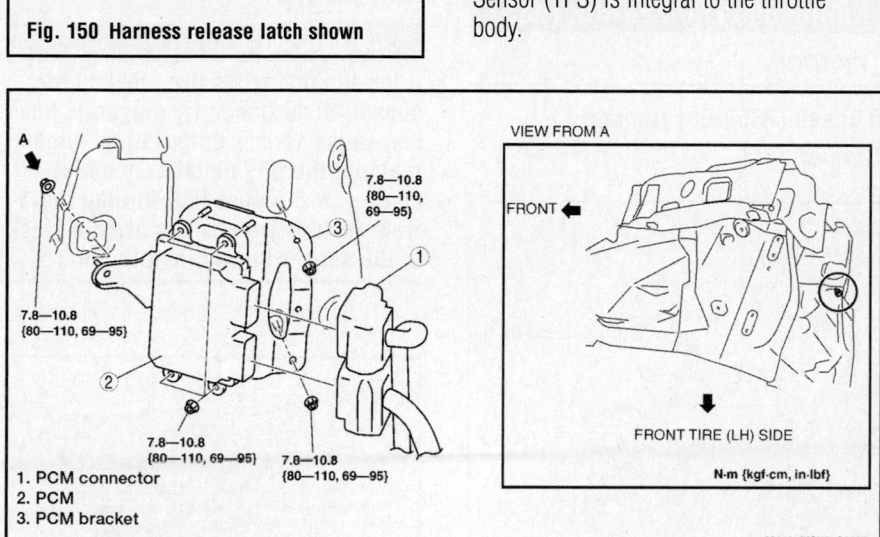

1. PCM connector
2. PCM
3. PCM bracket

Fig. 151 PCM mounting

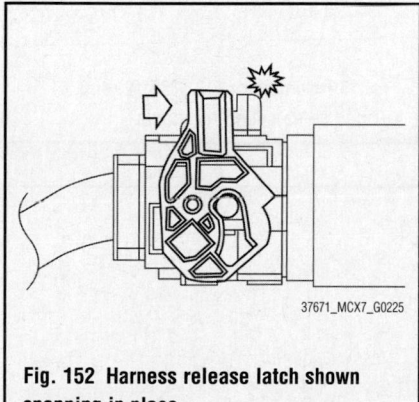

Fig. 152 Harness release latch shown snapping in place

3. Remove the battery and battery tray.
4. Remove the (LH) mudguard if needed.
5. Disconnect the PCM connector. Verify that the PCM connector lever is tilted towards the wiring harness side for removal.
6. Remove the PCM.
7. Install in the reverse order of removal.

➡ **Push the lever until a click is heard.**

8. When replacing the PCM on the vehicles, perform the following:

- IMMOBILIZER SYSTEM-RELATED PARTS PROGRAMMING (ADVANCED KEYLESS ENTRY AND START SYSTEM)

THROTTLE POSITION SENSOR (TPS)

LOCATION

See Figure 153.

The Throttle Position Sensor (TPS) is integral to the throttle body.

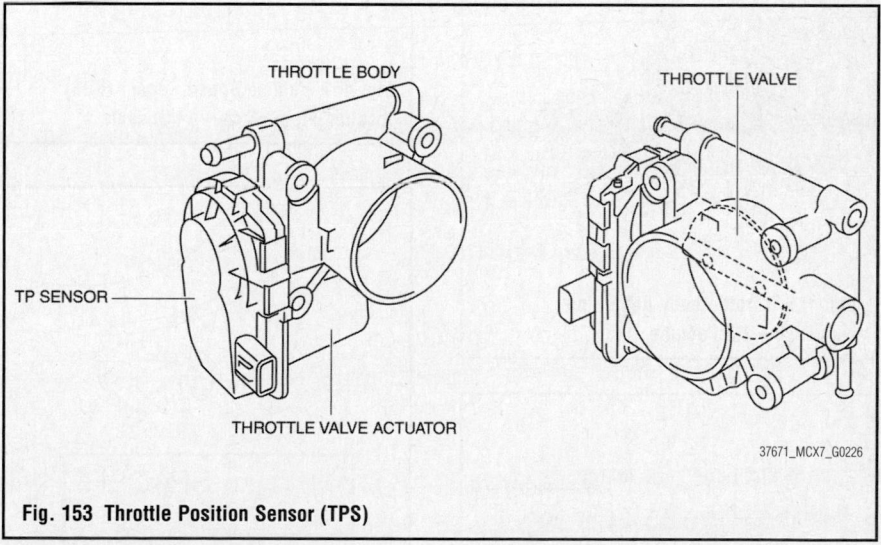

Fig. 153 Throttle Position Sensor (TPS)

REMOVAL & INSTALLATION

2.3L (L3) Engine

See Figure 154.

❊❊ CAUTION

A hot engine and intake air system can cause severe burns. Turn off the

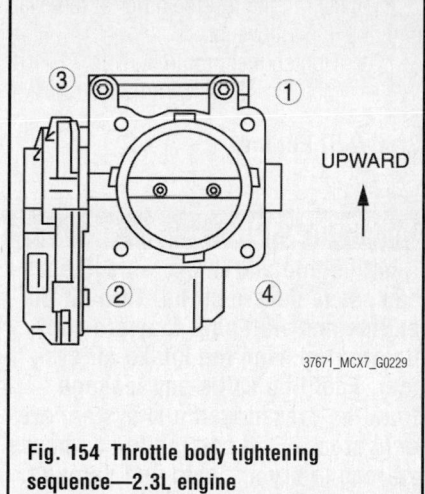

UPWARD

Fig. 154 Throttle body tightening sequence—2.3L engine

engine and wait until they are cool before removing the intake air system. Fuel line spills and leakage from the pressurized fuel system are dangerous. Fuel can ignite and cause serious injury or death and damage. Fuel can also irritate skin and eyes.

1. Before servicing the vehicle, refer to the Precautions section.
2. Disconnect the negative battery cable.
3. Remove the charge air cooler and duct.
4. Disconnect the throttle body connector.
5. Drain the engine coolant before removing the water hose.
6. Remove the throttle body retaining bolts.
7. Carefully remove the throttle body.

8. Install in the reverse order of removal and note the following:
- Tighten the throttle body to 71–101 inch lbs. (8–11.5 Nm).

2.5L (L5) Engine

See Figures 155 and 156.

※※ CAUTION

A hot engine and intake air system can cause severe burns. Turn off the engine and wait until they are cool before removing the intake air system. Fuel line spills and leakage from the pressurized fuel system are dangerous. Fuel can ignite and cause serious injury or death and damage. Fuel can also irritate skin and eyes.

1. Before servicing the vehicle, refer to the Precautions section.
2. Disconnect the negative battery cable.
3. Drain the cooling system.
4. Remove the plug hole plate.
5. Air cleaner cover and MAF sensor connector.
6. Remove the intake fresh air hose.
7. Disconnect the wiring harness for the throttle body.
8. Remove the water hoses.

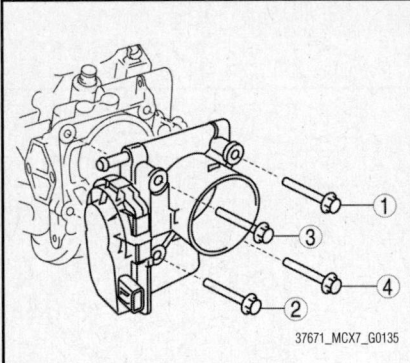

Fig. 155 Throttle body tightening sequence—2.5L engine

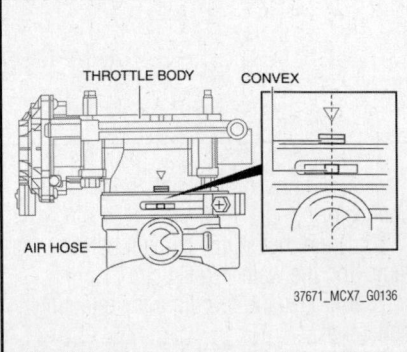

Fig. 156 Convex adjustment view of the throttle body and the air hose

9. Remove the throttle body retaining bolts.
10. Carefully remove the throttle body.
11. Install in the reverse order of removal and note the following:
- Tighten the throttle body to 71–101 inch lbs. (8–11.5 Nm).
- Adjust the convex of the throttle body and the air hose.
- Refill and bleed the cooling system.

VEHICLE SPEED SENSOR (VSS)

LOCATION

5 Speed (FS5A-EL) Transaxle

See Figure 157.

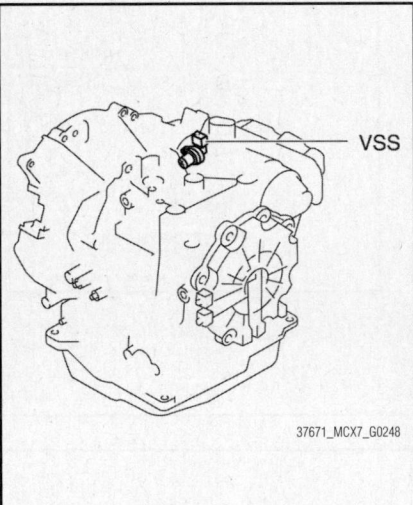

Fig. 157 Vehicle Speed Sensor (VSS) location view—5 speed transaxle

6 Speed (AW6A-EL, AW6AX-EL) Transaxle

See Figure 158.

The Vehicle Speed Sensor (VSS) is located deep inside the transaxle, the transaxle must be removed and disassembled to access the sensor.

REMOVAL & INSTALLATION

5 Speed (FS5A-EL) Transaxle

See Figure 159.

※※ WARNING

If foreign materials are stuck to the sensor, disturbance by magnetic flux can cause sensor output to be abnormal and thereby negatively affect control. Make sure that foreign materials such as iron filings are not stuck to the sensor during installation.

1. Disconnect the negative battery cable.

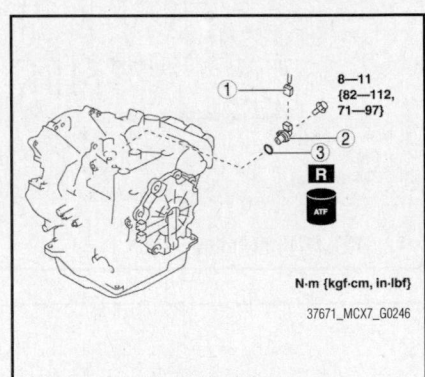

Fig. 159 Connector (1), VSS (2) and O-ring (3)—5 speed transaxle

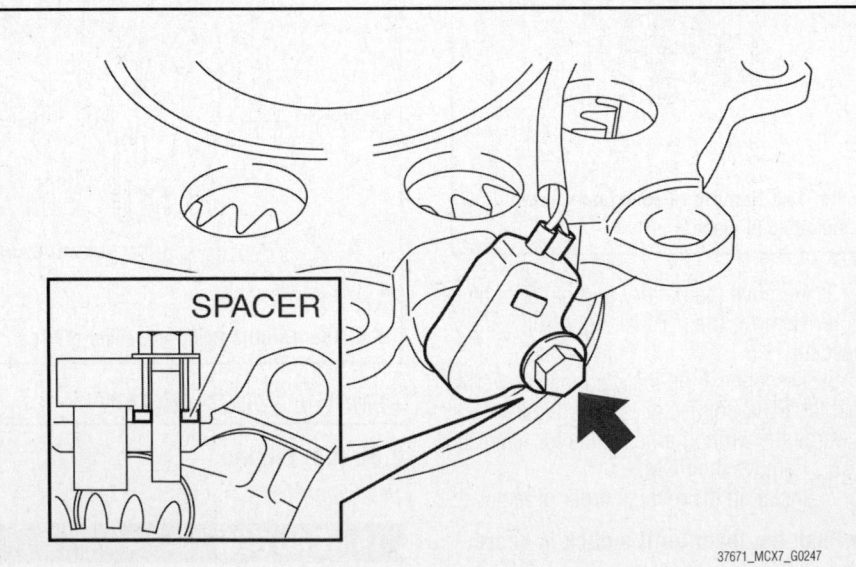

Fig. 158 Vehicle Speed Sensor (VSS) view—6 speed transaxle

2. Remove the undercover.
3. Remove the Vehicle Speed Sensor (VSS) connector.
4. Remove the VSS retaining bolt.
5. Remove the VSS and O-ring.
6. Install in the reverse order of removal.

6 Speed (AW6A-EL, AW6AX-EL) Transaxle

1. Disconnect the negative battery cable.
2. Remove the transaxle.

3. Disassemble the transaxle
4. Remove the Vehicle Speed Sensor (VSS)
5. Reassemble the transaxle and install it.

FUEL   GASOLINE FUEL INJECTION SYSTEM

FUEL SYSTEM SERVICE PRECAUTIONS

Safety is the most important factor when performing not only fuel system maintenance but any type of maintenance. Failure to conduct maintenance and repairs in a safe manner may result in serious personal injury or death. Maintenance and testing of the vehicle's fuel system components can be accomplished safely and effectively by adhering to the following rules and guidelines.

• To avoid the possibility of fire and personal injury, always disconnect the negative battery cable unless the repair or test procedure requires that battery voltage be applied.

• Always relieve the fuel system pressure prior to disconnecting any fuel system component (injector, fuel rail, pressure regulator, etc.), fitting or fuel line connection. Exercise extreme caution whenever relieving fuel system pressure to avoid exposing skin, face and eyes to fuel spray. Please be advised that fuel under pressure may penetrate the skin or any part of the body that it contacts.

• Always place a shop towel or cloth around the fitting or connection prior to loosening to absorb any excess fuel due to spillage. Ensure that all fuel spillage (should it occur) is quickly removed from engine surfaces. Ensure that all fuel soaked cloths or towels are deposited into a suitable waste container.

• Always keep a dry chemical (Class B) fire extinguisher near the work area.

• Do not allow fuel spray or fuel vapors to come into contact with a spark or open flame.

• Always use a back-up wrench when loosening and tightening fuel line connection fittings. This will prevent unnecessary stress and torsion to fuel line piping.

• Always replace worn fuel fitting O-rings with new Do not substitute fuel hose or equivalent where fuel pipe is installed.

Before servicing the vehicle, make sure to also refer to the precautions in the beginning of this section as well.

RELIEVING FUEL SYSTEM PRESSURE

See Figure 160.

1. Disconnect the negative battery cable.
2. Remove the fuel-filler cap and release the pressure in the fuel tank.
3. Remove the fuel pump relay.
4. Connect the negative battery cable.
5. Start the engine.
6. After the engine stalls, crank the engine several times.
7. Turn the ignition switch to the LOCK position.
8. Disconnect the negative battery cable.
9. Install the fuel pump relay.

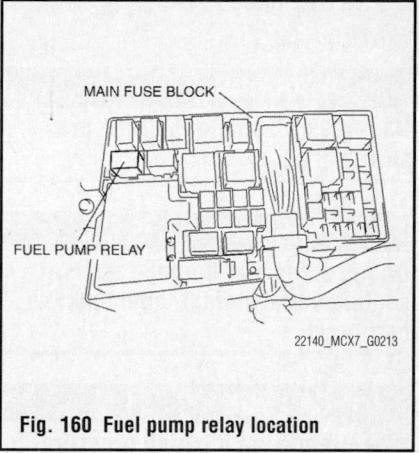

Fig. 160 Fuel pump relay location

FUEL FILTER

REMOVAL & INSTALLATION
See Figure 161.

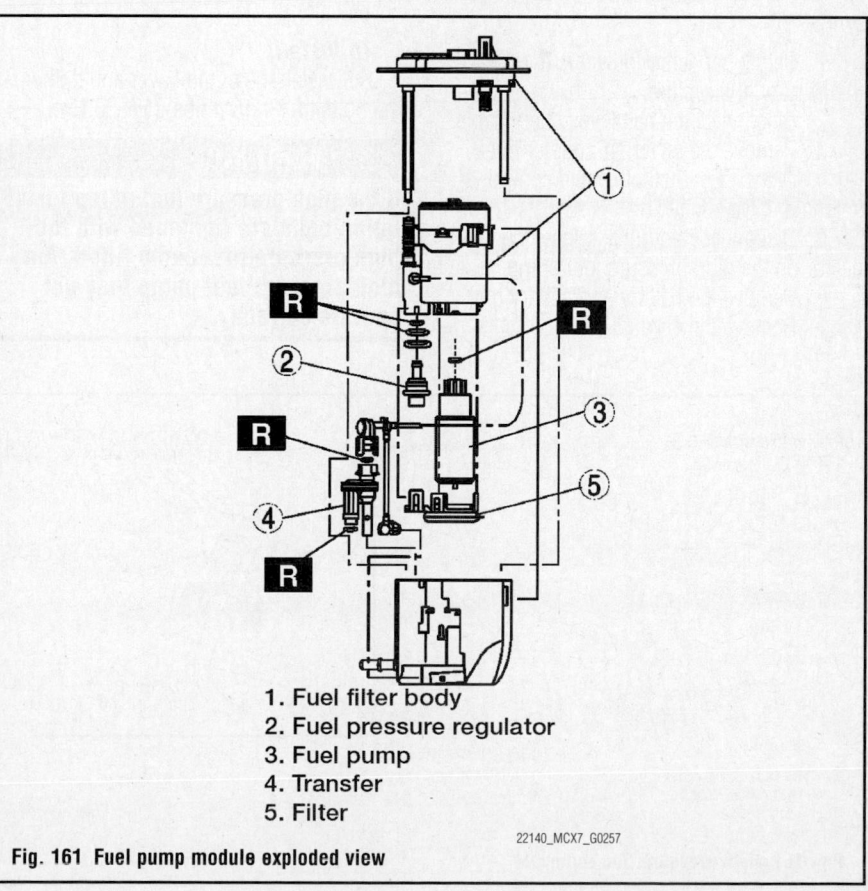

1. Fuel filter body
2. Fuel pressure regulator
3. Fuel pump
4. Transfer
5. Filter

Fig. 161 Fuel pump module exploded view

The fuel filter on this model is located in the fuel tank, integral to the fuel pump. Refer to Fuel Pump Removal & Installation in this section.

HIGH PRESSURE FUEL PUMP

REMOVAL & INSTALLATION

2.3L (L3) Engine
See Figure 162.

✳✳ WARNING
Do not disassemble the high pressure fuel pump.

✳✳ WARNING
Do not scratch or damage the fuel sealing surface of the high and low fuel ports.

✳✳ CAUTION
When removing the high pressure fuel pipe, secure the joint (pump side) so that it does not rotate, and loosen the screw (pipe side).

➡If the high pressure fuel pump is removed, replace the O-ring with a new one.

1. Before servicing the vehicle, refer to the Precautions section.
2. Disconnect the negative battery cable.
3. Remove the charge air cooler duct.
4. Disconnect the spill valve control solenoid valve connector.
5. Disconnect the quick release connector on the high pressure fuel pump.
6. Remove the battery and battery tray.
7. Remove the air duct.

✳✳ WARNING
If the high pressure fuel pump joint nut is loosened, fuel leakage may occur resulting in death or serious injury, or damage to the equipment or the vehicle. Fuel can also irritate the skin and eyes.

✳✳ WARNING
When removing the high pressure line pipe, always tighten the high pressure line pipe installation nut while fixing the high pressure fuel pump joint nut with a wrench. If the high pressure fuel pump joint nut has rotated, replace the high pressure fuel pump with a new one.

8. Fix the joint nut with a wrench on the high pressure fuel pump side as shown in the figure.
9. Loosen the high pressure line pipe installation nut.
10. Drain engine coolant.
11. Loosen the water outlet case installation bolts securing the high pressure line pipe.
12. Remove the high pressure fuel pump.
13. Remove the high pressure fuel pump cover.

To install:
14. Install the pump cover and tighten the bolts to 13–16 ft. lbs. (17–23 Nm).

✳✳ WARNING
If the high pressure fuel pump installation bolts are tightened with the high pressure fuel pump tilted, the high pressure fuel pump may not operate correctly.

15. Install the high pressure pump. Tighten the high pressure fuel pump installation bolts in a few passes to 71–101 inch lbs. (8.5–11.5 Nm).
16. Assemble the high pressure line pipe.
17. Fix the joint nut with a wrench on the high pressure fuel pump side as shown in the figure.
18. Tighten the high pressure line pipe installation nut to 18–26 ft. lbs. (23.5–35.5 Nm).
19. Tighten the water outlet case installation bolts to 71–101 inch lbs. (8.5–11.5 Nm).
20. Install the quick release connector.
21. Verify that the high pressure fuel pump is assembled securely.
22. Drive the vehicle starting from a standstill and brake suddenly five to six times at a low speed.
23. Stop the vehicle and verify from outside the vehicle that there is no fuel leakage around the high pressure fuel pump.

FUEL PUMP

REMOVAL & INSTALLATION
See Figures 163 though 165.

✳✳ CAUTION
Because the fuel tank is constructed such that the fuel level is higher than the installation surface of the fuel pump, fuel leakage could occur. If the fuel gauge indicates a fuel level of half or more, perform the following Steps 1-6 to drain approx. 10 L (11 US gal, 8.8 imp gal) of fuel.

➡Disconnecting/connecting the quick release connector without cleaning it may possibly cause damage to the fuel

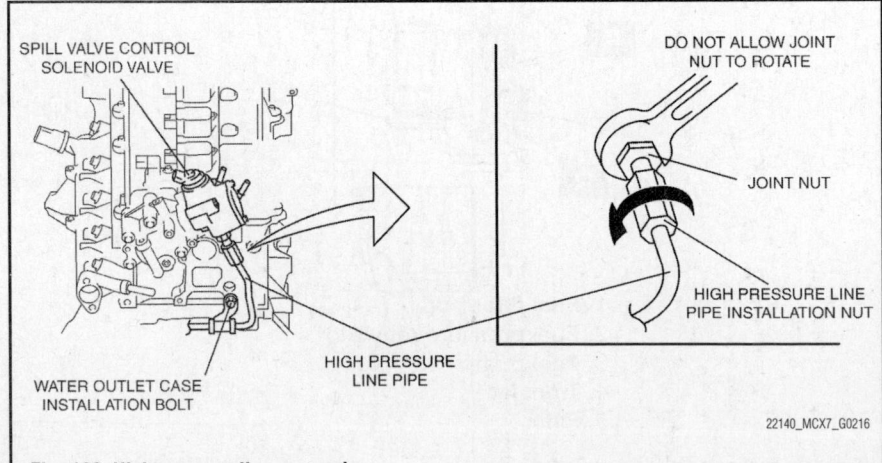

Fig. 162 High pressure line removal

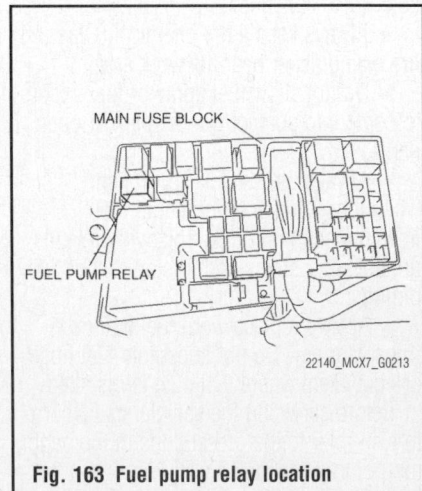

Fig. 163 Fuel pump relay location

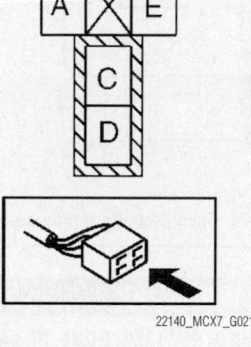

MAIN FUSE BLOCK (FUEL PUMP RELAY)

22140_MCX7_G0212

Fig. 164 Fuel pump relay terminal identification

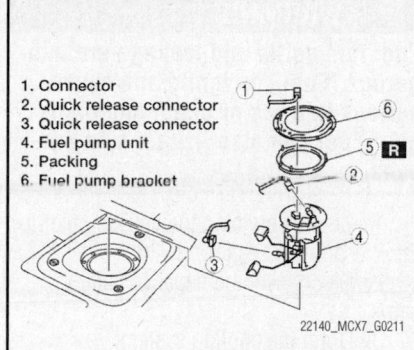

1. Connector
2. Quick release connector
3. Quick release connector
4. Fuel pump unit
5. Packing
6. Fuel pump bracket

22140_MCX7_G0211

Fig. 165 Fuel pump module exploded view

pipe and quick release connector. Always clean the quick release connector joint area before disconnecting/connecting using a cloth or soft brush, and make sure that it is free of foreign material.

1. Before servicing the vehicle, refer to the Precautions section.
2. Remove the rear seat cushion.
3. Remove the service hole cover.
4. Disconnect the quick release connector connected to the fuel pump unit.
5. Connect a long hose to the disconnected quick release connector and drain the fuel into a container used for collecting gasoline.
6. Remove the fuel pump relay.
7. Using a jumper wire, short fuel pump relay terminals C and D in the main fuse block.
8. Turn the ignition key **ON** and operate the fuel pump.

✳✳ CAUTION

The fuel pump could be damaged if it is operated (fuel pump idling) while there is no fuel in the fuel tank. Verify the amount of fuel being discharged from the hose and stop operation of the fuel pump when essentially no fuel is being discharged.

9. When essentially no fuel is being discharged, stop the fuel pump.
10. Turn the ignition switch to LOCK position to stop the fuel pump.
11. Disconnect the jumper wire from the check connector.
12. Disconnect the negative battery cable.
13. Remove the fuel pump bracket installation screws.
14. Remove the connector.
15. Disconnect the quick release connector and remove the fuel pump module.
16. Install in the reverse order of removal.

FUEL RAIL AND INJECTORS

REMOVAL & INSTALLATION

2.3L (L3) Engine

See Figure 166.

1. Before servicing the vehicle, refer to the Precautions section.
2. Disconnect the negative battery cable.
3. Remove the intake manifold.
4. Remove or disconnect the following:
   • High pressure line pipe
   • Fuel delivery pipe
   • Fuel injector bracket
   • Fuel injector
   • Relief valve
5. Install in the reverse order of removal.

2.5L (L5) Engine

See Figures 166 through 169.

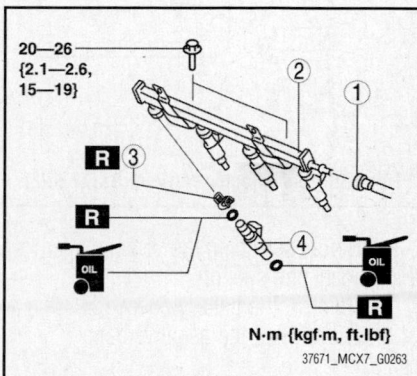

N·m {kgf·m, ft·lbf}

37671_MCX7_G0263

Fig. 167 Fuel rail and injector exploded view—2.5L engine

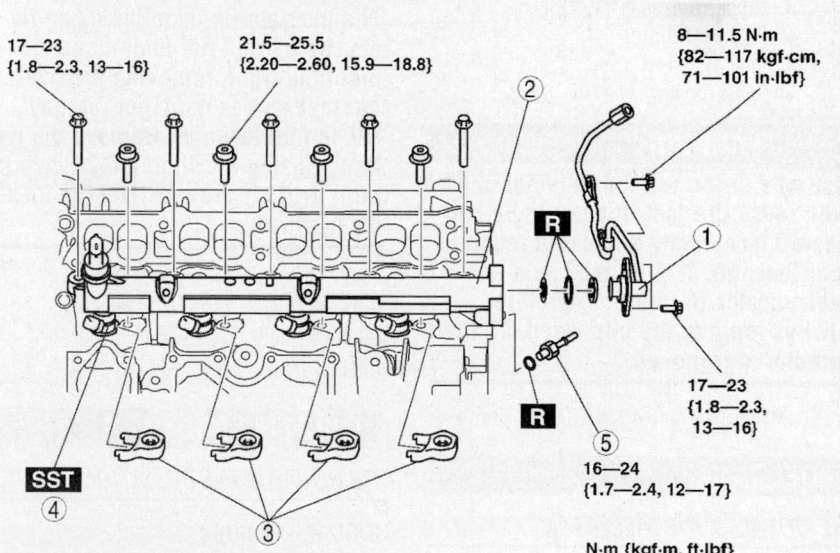

1. High pressure line pipe
2. Fuel delivery pipe
3. Fuel injector bracket
4. Fuel injector
5. Relief valve

22140_MCX7_G0214

Fig. 166 Fuel rail and injector exploded view—2.3L engine

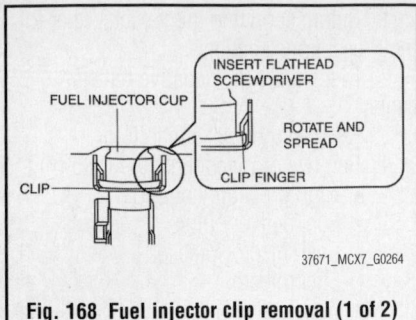

Fig. 168 Fuel injector clip removal (1 of 2)

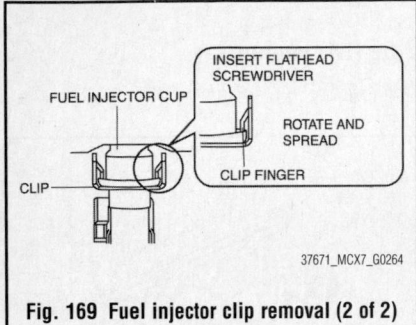

Fig. 169 Fuel injector clip removal (2 of 2)

1. Before servicing the vehicle, refer to the Precautions section.
2. Remove the battery cover.
3. Disconnect the negative battery cable.
4. Remove the plug hole plate.
5. Disconnect the fuel injector connector.
6. Remove or disconnect the following:
   - Quick release connector
   - Fuel distributor
   - Injector clip
   - Fuel injector

✱✱ WARNING

Use of a deformed fuel injector clip will cause the fuel injector to be connected incorrectly and could result in fuel leakage. It will also cause the fuel injector to rotate. Therefore, always replace the clip when the fuel injector is removed.

7. Install in the reverse order of removal.

FUEL TANK

REMOVAL & INSTALLATION
See Figure 170.

1. Before servicing the vehicle, refer to the Precautions section.
2. Level the vehicle.
3. Remove the rear seat cushion.
4. Remove the service hole cover.
5. Drain the fuel tank. Refer to Fuel Tank Draining, in this section.

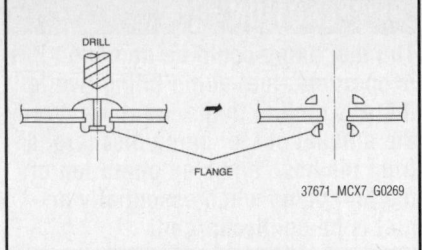

Fig. 170 Rivet flange removal, using a drill and 0.20 inch (5 mm) drill bit

6. Remove the fuel pump unit.
7. Remove the fuel gauge sender unit.
8. Remove the exhaust system..
9. Remove the rear undercover (LR, RR).
10. Remove the rear parking brake cable installation bolts, then move the rear parking brake cable out of the way.
11. Disconnect the joint hose connected between the fuel tank and the fuel-filler pipe from the fuel-filler pipe.
12. Disconnect the breather hose connected between the fuel tank and the fuel-filler pipe from the fuel-filler pipe.
13. Disconnect the evaporative hose connected between the fuel tank and the charcoal canister from the charcoal canister.
14. Disconnect the fuel tank pressure sensor connector.

✱✱ WARNING

The insulator is installed using rivets. Be careful not to damage the fuel tank when removing the rivet. If the fuel tank is damaged, it may cause fuel leakage. Remove the rivet flange using a drill and 0.20 inch (5 mm) drill bit. Bolt or rivet the insulator in place.

15. Install in the reverse order of removal. Tighten the fuel tank strap retaining bolts to 32–45 ft. lbs. (43–61 Nm).

THROTTLE BODY

REMOVAL & INSTALLATION

2.3L (L3) Engine
See Figure 171.

✱✱ WARNING

A hot engine and intake air system can cause severe burns. Turn off the engine and wait until they are cool before removing the intake air system.

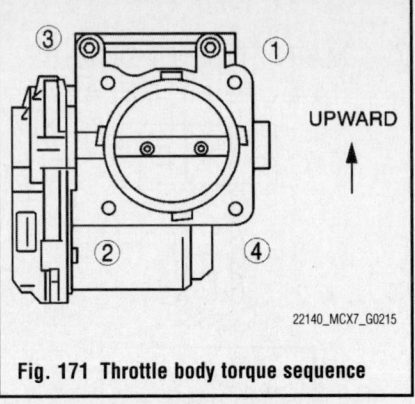

Fig. 171 Throttle body torque sequence

✱✱ WARNING

Fuel vapor is hazardous. It can easily ignite, causing serious injury and damage. Always keep sparks and flames away from fuel.

✱✱ WARNING

Fuel line spills and leakage are dangerous. Fuel can ignite and cause serious injuries or death and damage. Fuel can also irritate skin and eyes.

1. Before servicing the vehicle, refer to the Precautions section.
2. Disconnect the negative battery cable.
3. Drain the cooling system.
4. Remove the charge air cooler duct.
5. Remove the battery and battery tray.
6. Remove the undercover.
7. Remove the Mass Airflow sensor (MAF).
8. Remove the fresh air duct.
9. Remove the air cleaner cover and case.
10. Remove the resonance chamber.
11. Remove the charge air cooler.
12. Remove the air bypass valve.
13. Remove the air hose and duct.
14. Remove the throttle body.
15. Install in the reverse order of removal. Tighten the throttle body bolts in sequence to 71–101 inch lbs. (8–11 Nm).

2.5L (L5) Engine
See Figures 172 and 173.

✱✱ CAUTION

A hot engine and intake air system can cause severe burns. Turn off the engine and wait until they are cool before removing the intake air system. Fuel line spills and leakage

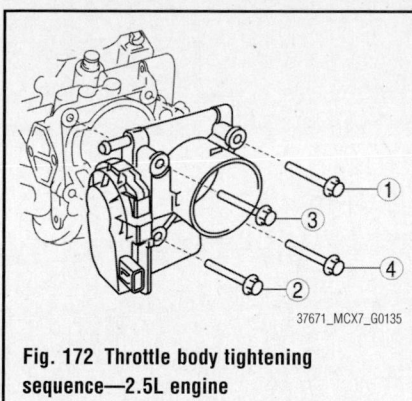

Fig. 172 Throttle body tightening sequence—2.5L engine

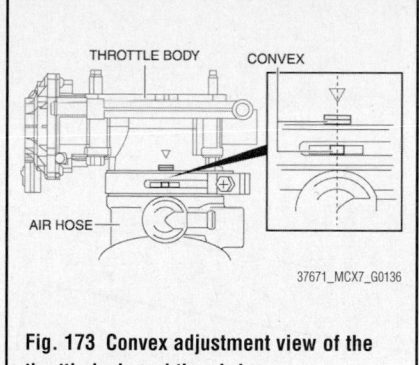

Fig. 173 Convex adjustment view of the throttle body and the air hose

from the pressurized fuel system are **dangerous. Fuel can ignite and cause serious injury or death and damage. Fuel can also irritate skin and eyes.**

1. Before servicing the vehicle, refer to the Precautions section.
2. Disconnect the negative battery cable.
3. Drain the cooling system.

4. Remove the plug hole plate.
5. Air cleaner cover and MAF sensor connector.
6. Remove the intake fresh air hose.
7. Disconnect the wiring harness for the throttle body.
8. Remove the water hoses.
9. Remove the throttle body retaining bolts.
10. Carefully remove the throttle body.
11. Install in the reverse order of removal and note the following:
   - Tighten the throttle body to 71–101 inch lbs. (8–11.5 Nm).
   - Adjust the convex of the throttle body and the air hose.

HEATING & AIR CONDITIONING SYSTEM

PRECAUTIONS

Using a flammable refrigerant, such as OZ-12, in this vehicle is dangerous. In an accident, the refrigerant may catch fire, resulting in serious injury or death. When servicing this vehicle, use only R-134a.

Checking for system leakage on a vehicle that has been serviced with flammable refrigerant, such as OZ-12, is dangerous. Conventional leak detectors use an electronically generated arc which can ignite the refrigerant, causing serious injury or death. If a flammable refrigerant may have been used to service the system, or if you suspect a flammable refrigerant has been used, contact the local fire marshal or EPA office for information on handling the refrigerant.

Avoid breathing air conditioning refrigerant or lubricant vapor. Exposure may irritate eyes, nose and throat. Also, due to environmental concerns, use service equipment certified to meet the requirements of SAE J2210 (R-134a recycling equipment) when draining R-134a from the air conditioning system. If accidental system discharge occurs, ventilate work area before resuming service.

Do not pressure test or leak test R-134a service equipment and/or vehicle air conditioning system with compressed air. Some mixtures of air and R-134a have been shown to be combustible at elevated pressures. These mixtures, if ignited, may cause injury or property damage. Additional health

and safety information may be obtained from refrigerant manufacturers.

Do not allow the refrigerant to leak near fire or any kind of heat. A poisonous gas may be generated if the refrigerant gas contacts fire or heat such as from cigarettes and heaters. When carrying out any operation that can cause refrigerant leakage, extinguish or remove the above-mentioned heat sources and maintain adequate ventilation.

Handling liquid refrigerant is dangerous. A drop of it on the skin can result in localized frostbite. When handling the refrigerant, wear gloves and safety goggles. If refrigerant splashes into the eyes, immediately wash them with clean water and consult a doctor.

The refrigerant container is highly pressurized. If it is subjected to high heat, it could explode, scattering metal fragments and liquid refrigerant that can seriously injure you. Store the refrigerant at temperatures below 40°C (104°F).

BLOWER MOTOR

REMOVAL & INSTALLATION

See Figure 174.

1. Before servicing the vehicle, refer to the Precautions section.
2. Disconnect the negative battery cable.
3. Remove the dashboard undercover (passenger's side).
4. Remove the blower motor connector and the blower motor.
5. Install in the reverse order of removal.

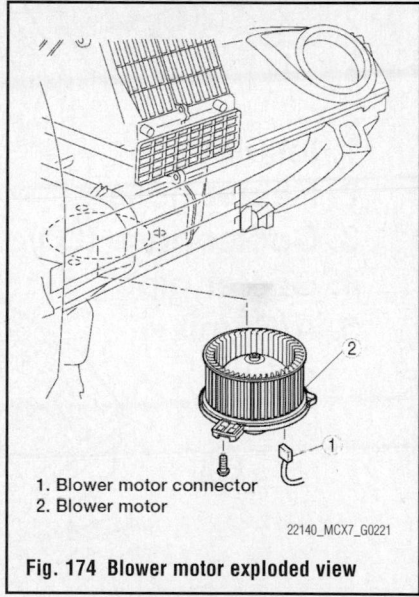

1. Blower motor connector
2. Blower motor

Fig. 174 Blower motor exploded view

HEATER CORE

REMOVAL & INSTALLATION

See Figures 175 through 177.

1. Remove heater unit assembly. Refer to the Heater Unit in this section.
2. Remove or disconnect the following:
   - Insulator
   - Heater hose
   - Cooler hose (LO)
   - Cooler pipe
   - A/C unit
3. To disassemble the A/C unit, remove or disconnect the following:

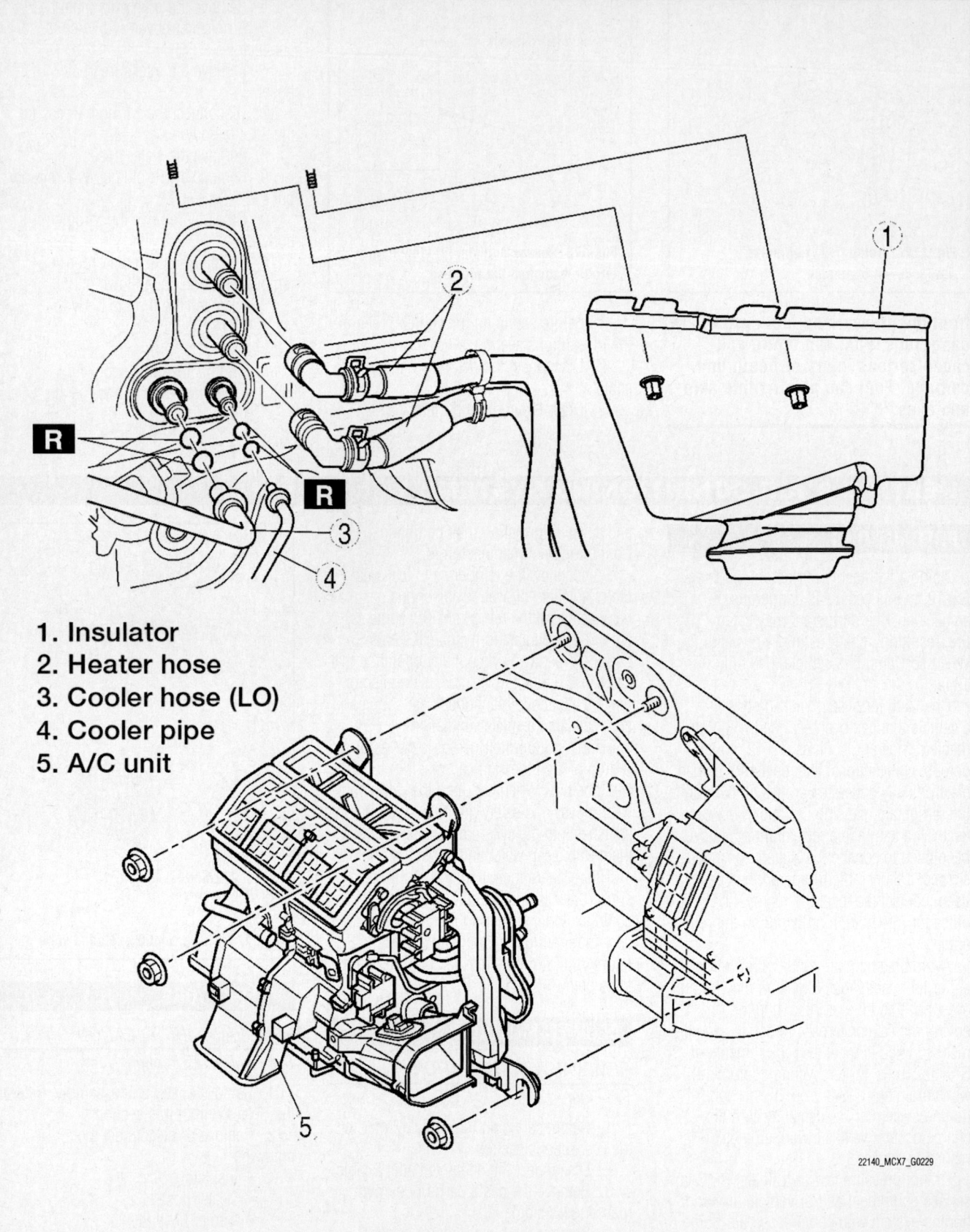

1. Insulator
2. Heater hose
3. Cooler hose (LO)
4. Cooler pipe
5. A/C unit

22140_MCX7_G0229

Fig. 175 A/C unit removal

FULL-AUTO AIR CONDITIONER

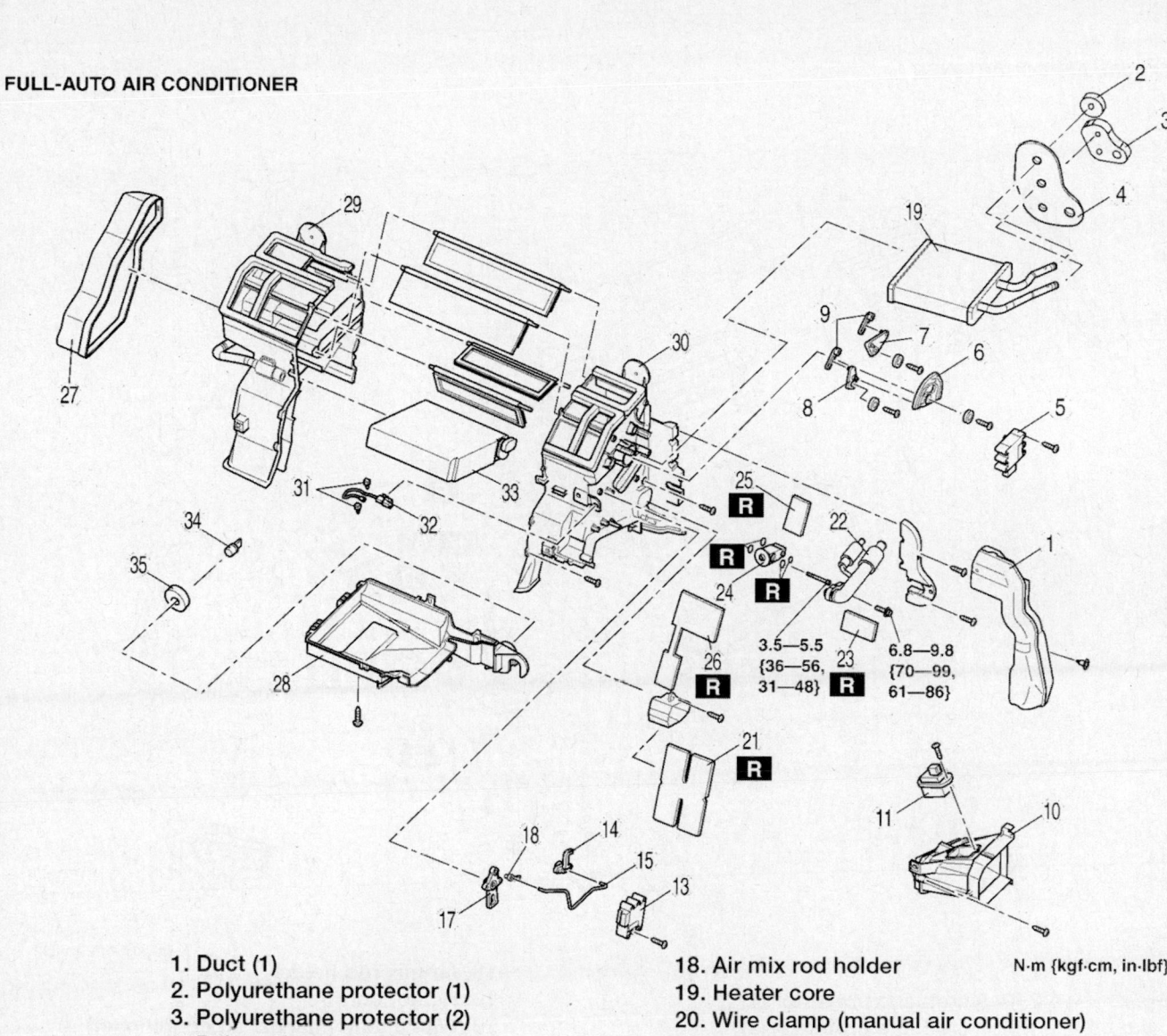

N·m {kgf·cm, in·lbf}

3.5—5.5
{36—56,
31—48}

6.8—9.8
{70—99,
61—86}

1. Duct (1)
2. Polyurethane protector (1)
3. Polyurethane protector (2)
4. Polyurethane protector (3)
5. Airflow mode actuator
6. Airflow mode main link
7. Airflow mode sub link (1)
8. Airflow mode sub link (2)
9. Airflow mode crank
10. Duct (2)
11. Power MOS FET (full-auto air conditioner)
12. Resistor (manual air conditioner)
13. Air mix actuator (full-auto air conditioner)
14. Air mix crank (1)
15. Air mix rod
16. Air mix link (manual air conditioner)
17. Air mix crank (2)
18. Air mix rod holder
19. Heater core
20. Wire clamp (manual air conditioner)
21. Adhesive polyurethane (1)
22. Evaporator pipe
23. Adhesive polyurethane (3)
24. Expansion valve
25. Adhesive polyurethane (4)
26. Adhesive polyurethane (2)
27. Duct (3)
28. A/C case (3)
29. A/C case (1)
30. A/C case (2)
31. Sensor clamp
32. Evaporator temperature sensor
33. Evaporator
34. Drain hose
35. Polyurethane protector (4)

22140_MCX7_G0230

Fig. 176 Full-auto A/C unit exploded view

MANUAL AIR CONDITIONER

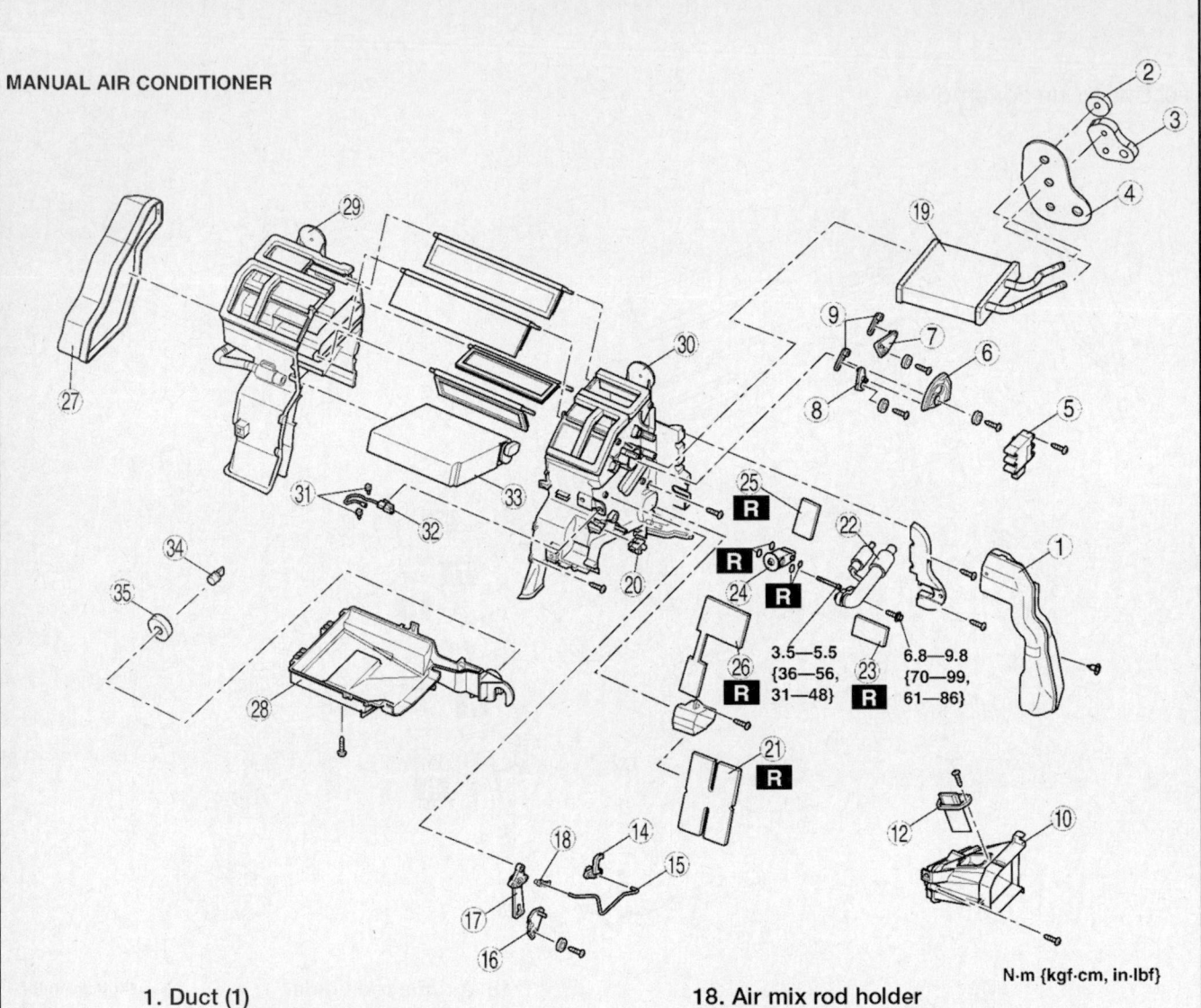

3.5—5.5 {36—56, 31—48}

6.8—9.8 {70—99, 61—86}

N·m {kgf·cm, in·lbf}

1. Duct (1)
2. Polyurethane protector (1)
3. Polyurethane protector (2)
4. Polyurethane protector (3)
5. Airflow mode actuator
6. Airflow mode main link
7. Airflow mode sub link (1)
8. Airflow mode sub link (2)
9. Airflow mode crank
10. Duct (2)
11. Power MOS FET (full-auto air conditioner)
12. Resistor (manual air conditioner)
13. Air mix actuator (full-auto air conditioner)
14. Air mix crank (1)
15. Air mix rod
16. Air mix link (manual air conditioner)
17. Air mix crank (2)
18. Air mix rod holder
19. Heater core
20. Wire clamp (manual air conditioner)
21. Adhesive polyurethane (1)
22. Evaporator pipe
23. Adhesive polyurethane (3)
24. Expansion valve
25. Adhesive polyurethane (4)
26. Adhesive polyurethane (2)
27. Duct (3)
28. A/C case (3)
29. A/C case (1)
30. A/C case (2)
31. Sensor clamp
32. Evaporator temperature sensor
33. Evaporator
34. Drain hose
35. Polyurethane protector (4)

22140_MCX7_G0231

Fig. 177 Manual A/C unit exploded view

- Duct
- Polyurethane protector
- Polyurethane protector
- Polyurethane protector
- Airflow mode actuator
- Airflow mode main link
- Airflow mode sub link
- Airflow mode sub link

- Airflow mode crank
- Duct
- Power MOS FET (full-auto air conditioner)
- Resistor (manual air conditioner)
- Air mix actuator (full-auto air conditioner)
- Air mix crank

- Air mix rod
- Air mix link (manual air conditioner)
- Air mix crank
- Air mix rod holder
- Heater core

4. Install in the reverse order of removal.
5. Fill the cooling system.
6. Vacuum and recharge the A/C system.

STEERING

POWER RACK & PINION STEERING GEAR

REMOVAL & INSTALLATION

See Figures 178 through 181.

✳✳ CAUTION

Performing the following procedures without first removing the ABS wheel-speed sensor may possibly cause an open circuit in the wiring harness if it is pulled by mistake. Before performing the following procedures, disconnect the ABS wheel-speed sensor connector (axle side) and fix the wiring harness to an appropriate place where it will not be pulled by mistake while servicing the vehicle.

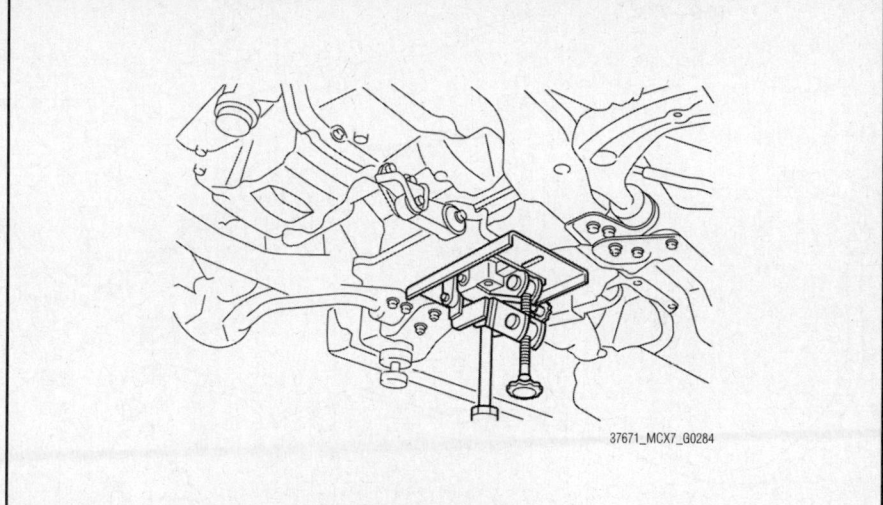

37671_MCX7_G0284

Fig. 179 Support the crossmember component with a jack

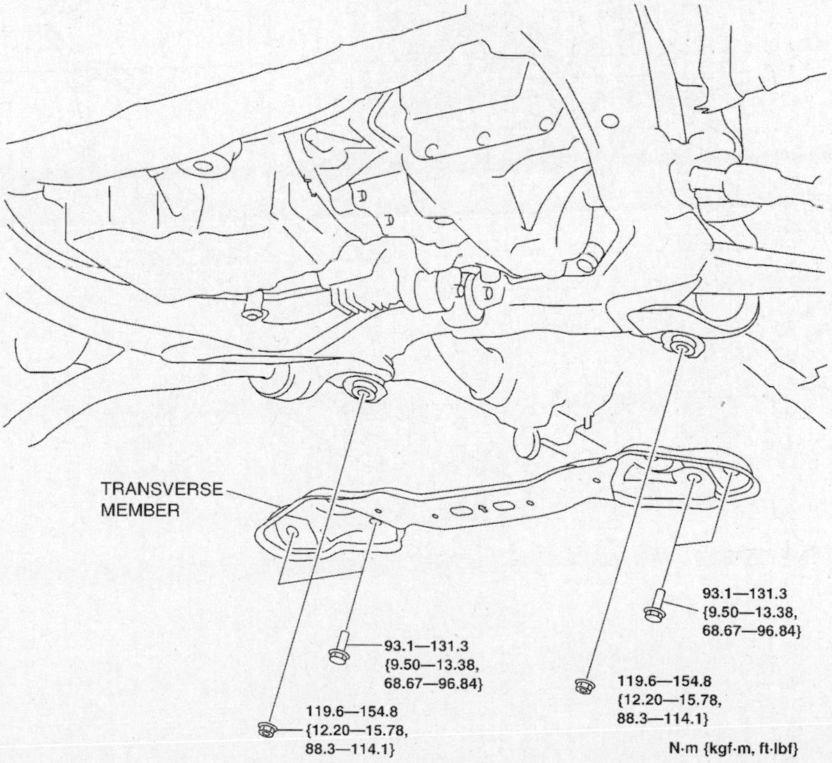

TRANSVERSE MEMBER

93.1—131.3
{9.50—13.38,
68.67—96.84}

119.6—154.8
{12.20—15.78,
88.3—114.1}

93.1—131.3
{9.50—13.38,
68.67—96.84}

119.6—154.8
{12.20—15.78,
88.3—114.1}

N·m {kgf·m, ft·lbf}

Fig. 178 Transverse member removal

22140_MCX7_G0233

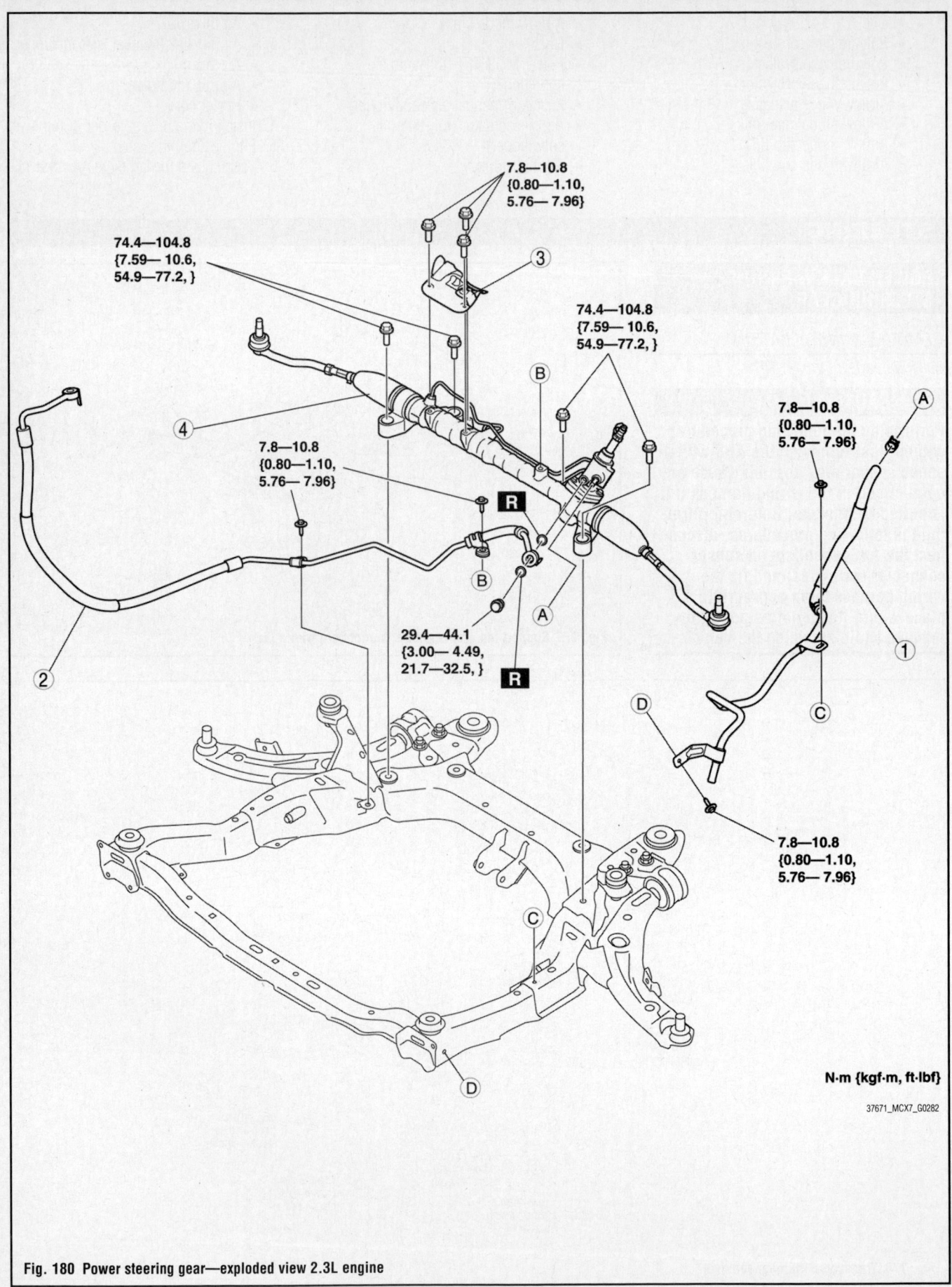

7.8—10.8
{0.80—1.10,
5.76—7.96}

74.4—104.8
{7.59— 10.6,
54.9—77.2, }

3

74.4—104.8
{7.59— 10.6,
54.9—77.2, }

B

7.8—10.8
{0.80—1.10,
5.76—7.96}

A

4

7.8—10.8
{0.80—1.10,
5.76—7.96}

R

B

2

A

R

29.4—44.1
{3.00— 4.49,
21.7—32.5, }

1

C

D

7.8—10.8
{0.80—1.10,
5.76— 7.96}

C

D

N·m {kgf·m, ft·lbf}

37671_MCX7_G0282

Fig. 180 Power steering gear—exploded view 2.3L engine

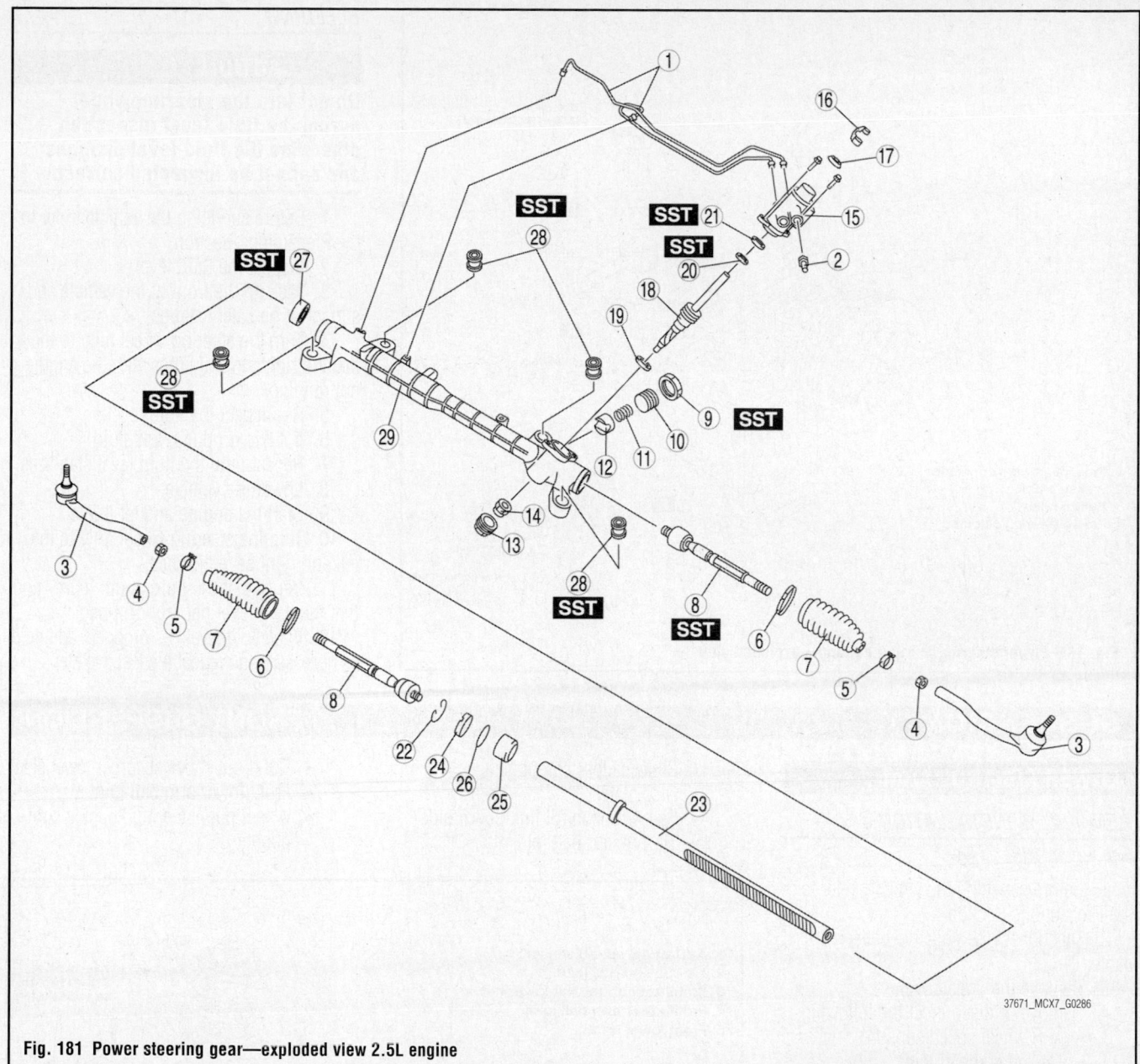

Fig. 181 Power steering gear—exploded view 2.5L engine

37671_MCX7_G0286

1. Disable the air bag system.

2. Disconnect the negative battery cable.

3. Before servicing the vehicle, refer to the Precautions Section.

4. Remove the front wheels.

5. Remove the side cover and under-cover.

6. Drain the power steering fluid.

7. Remove the transverse member.

8. Remove the intermediate shaft installation bolt, and disconnect the steering shaft.

➡**Support the crossmember component with a jack.**

9. Remove the front crossmember, lower arm, front stabilizer, and steering gear and linkage as a single unit.

10. Remove the front stabilizer from the crossmember component.

11. Remove insulator and the power steering gear.

12. Install in the reverse order of removal, and note the tightening specifications in the graphics

13. Refill and bleed the power steering system.

14. After installation, inspect the front wheel alignment and adjust it if necessary.

POWER STEERING PUMP

REMOVAL & INSTALLATION

See Figure 182.

1. Before servicing the vehicle, refer to the Precautions Section.

2. Remove the accessory drive belt.

3. Remove the pressure switch connector.

4. Remove the high pressure line.

5. Disconnect the return hose.

6. Unbolt and remove the power steering pump.

7. Install in reverse of removal. Tighten the power steering pump bolts to 14–18 ft. lbs. (29.4–44.1 Nm).

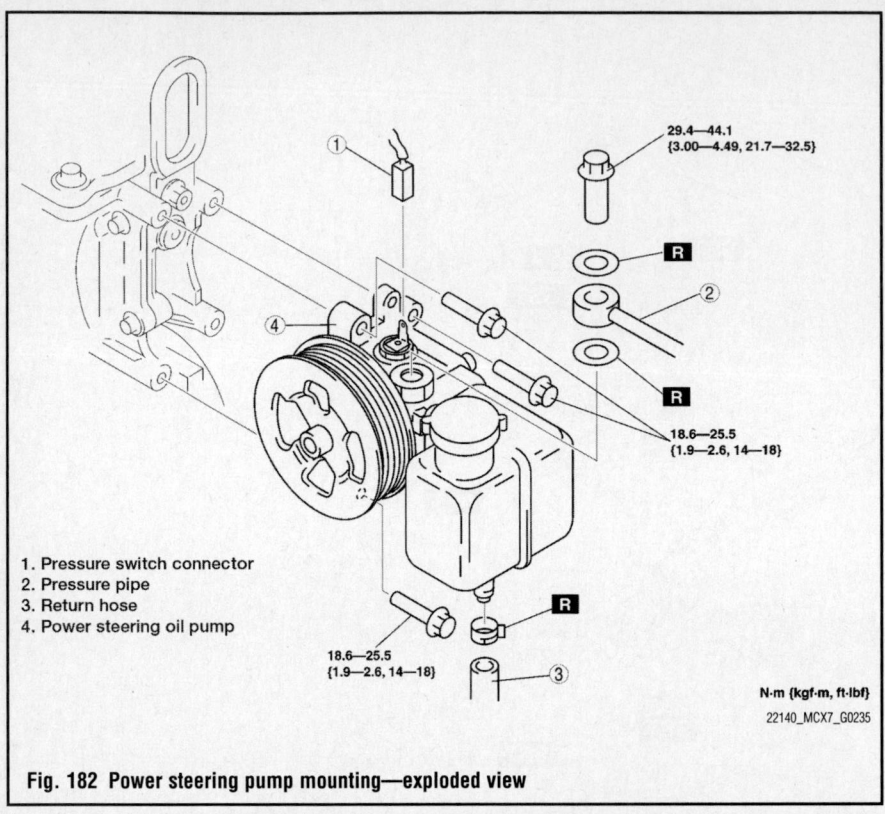

1. Pressure switch connector
2. Pressure pipe
3. Return hose
4. Power steering oil pump

29.4—44.1
{3.00—4.49, 21.7—32.5}

18.6—25.5
{1.9—2.6, 14—18}

18.6—25.5
{1.9—2.6, 14—18}

N·m {kgf·m, ft·lbf}

22140_MCX7_G0235

Fig. 182 Power steering pump mounting—exploded view

BLEEDING

❊❊ CAUTION

Do not turn the steering wheel during the fluid level inspection, otherwise the fluid level changes and cannot be inspected correctly.

1. Before servicing the vehicle, refer to the Precautions Section.
2. Inspect the fluid level.
3. Jack up the front of the vehicle and support it on safety stands.
4. Turn the steering wheel fully to the left and right several times with the engine not running.
5. Re-inspect the fluid level.
6. If it has dropped, add fluid.
7. Repeat until the fluid level stabilizes.
8. Lower the vehicle.
9. Start the engine and let it idle.
10. Turn the steering wheel fully to the left and right several times.
11. Verify that the fluid is not foamy and that the fluid level has not dropped.
12. If the fluid level has dropped, add fluid as necessary and repeat the last two steps.

SUSPENSION

LOWER CONTROL ARM

REMOVAL & INSTALLATION

See Figure 183.

1. Before servicing the vehicle, refer to the Precautions Section.
2. Disconnect the negative battery cable.
3. Remove the undercover.
4. Remove or disconnect the following:
   - ABS wheel speed sensor
   - Tie-rod end ball joint
   - Stabilizer control link lower nut
   - Front lower arm ball joint
   - Front lower arm
5. Install in reverse of removal.

STEERING KNUCKLE

REMOVAL & INSTALLATION

See Figure 184.

1. Before servicing the vehicle, refer to the Precautions Section.
2. Disconnect the negative battery cable.
3. Remove the front ABS wheel-speed sensor.
4. Remove or disconnect the following:
   - Brake hose clip
   - Locknut

- Brake caliper component
- Disc plate
- Stabilizer control link (lower side)
- Tie-rod end ball joint

FRONT SUSPENSION

- Nut (front shock absorber lower side)
- Front lower arm ball joint
- Wheel hub, steering knuckle component

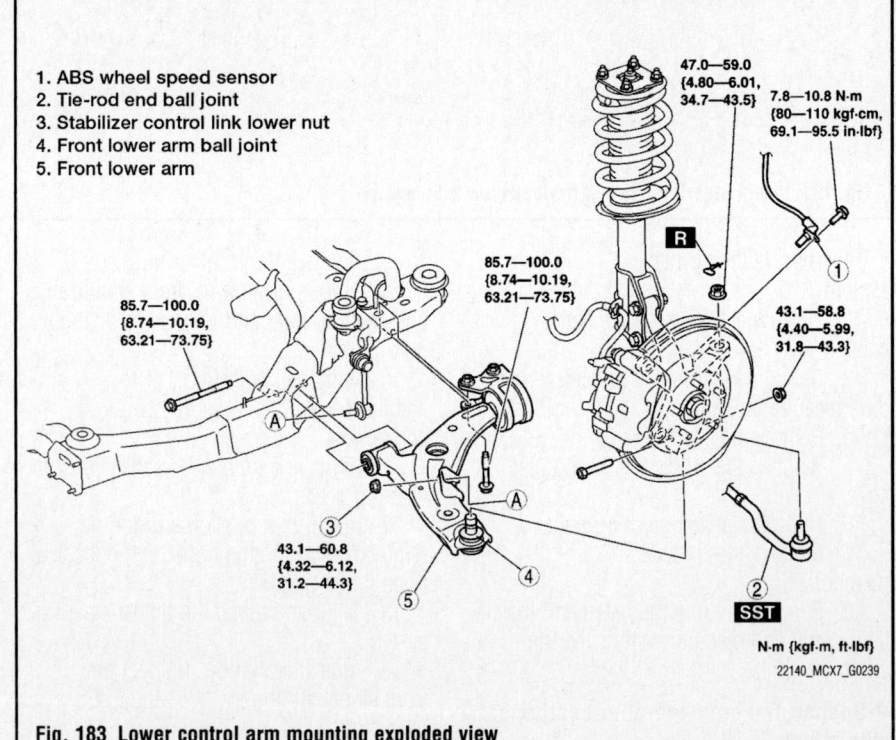

1. ABS wheel speed sensor
2. Tie-rod end ball joint
3. Stabilizer control link lower nut
4. Front lower arm ball joint
5. Front lower arm

47.0—59.0
{4.80—6.01, 34.7—43.5}

7.8—10.8 N·m
{80—110 kgf-cm, 69.1—95.5 in-lbf}

85.7—100.0
{8.74—10.19, 63.21—73.75}

85.7—100.0
{8.74—10.19, 63.21—73.75}

43.1—58.8
{4.40—5.99, 31.8—43.3}

43.1—60.8
{4.32—6.12, 31.2—44.3}

SST

N·m {kgf·m, ft-lbf}

22140_MCX7_G0239

Fig. 183 Lower control arm mounting exploded view

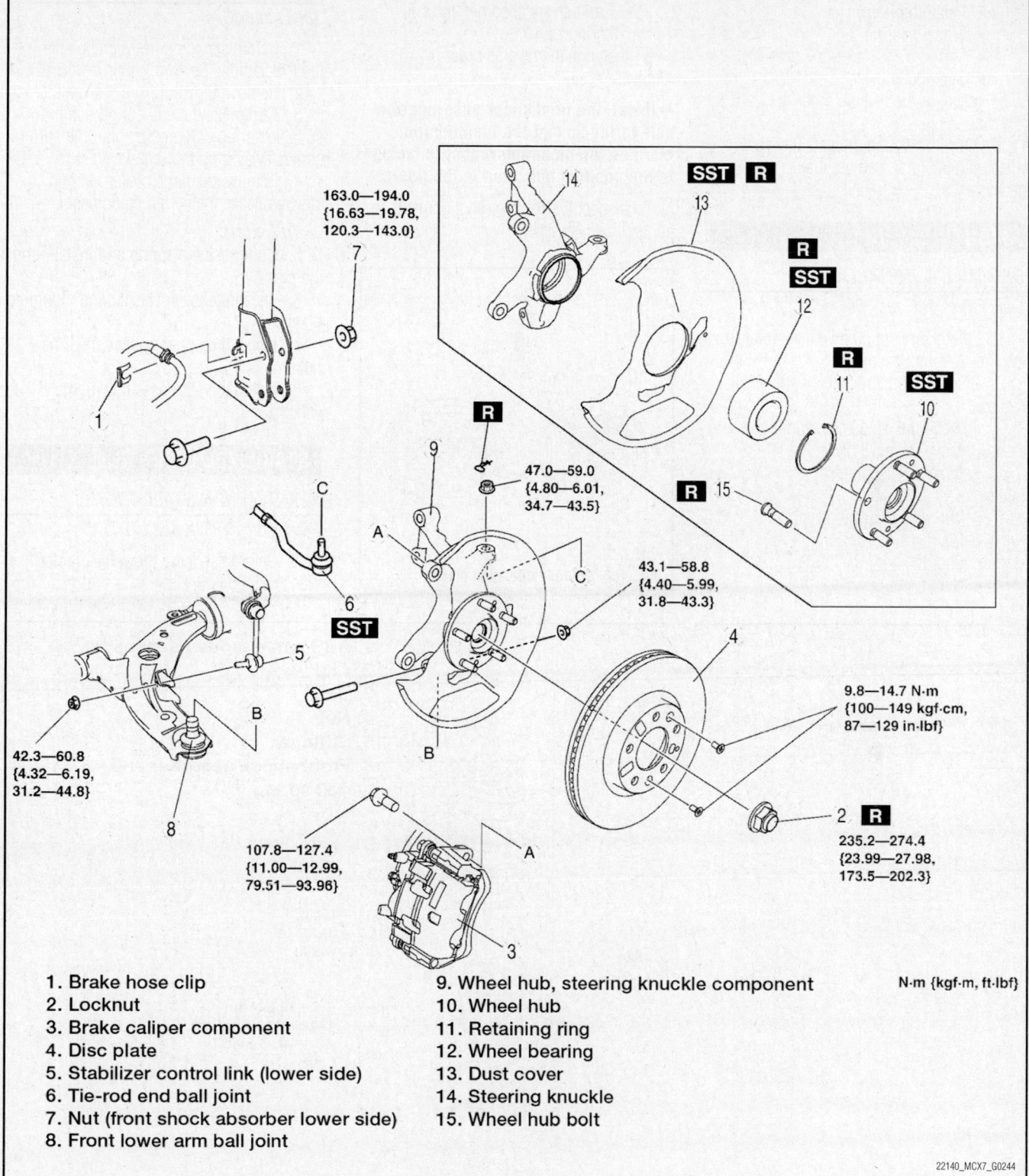

163.0—194.0
{16.63—19.78,
120.3—143.0}

47.0—59.0
{4.80—6.01,
34.7—43.5}

43.1—58.8
{4.40—5.99,
31.8—43.3}

9.8—14.7 N·m
{100—149 kgf·cm,
87—129 in·lbf}

42.3—60.8
{4.32—6.19,
31.2—44.8}

107.8—127.4
{11.00—12.99,
79.51—93.96}

235.2—274.4
{23.99—27.98,
173.5—202.3}

N·m {kgf·m, ft·lbf}

1. Brake hose clip
2. Locknut
3. Brake caliper component
4. Disc plate
5. Stabilizer control link (lower side)
6. Tie-rod end ball joint
7. Nut (front shock absorber lower side)
8. Front lower arm ball joint
9. Wheel hub, steering knuckle component
10. Wheel hub
11. Retaining ring
12. Wheel bearing
13. Dust cover
14. Steering knuckle
15. Wheel hub bolt

22140_MCX7_G0244

Fig. 184 Steering knuckle, hub, and wheel bearing exploded view

- Wheel hub
- Retaining ring
- Wheel bearing
- Dust cover
- Steering knuckle

5. Install in the reverse order of removal.
6. After installation, inspect the front wheel alignment and adjust it if necessary.

STRUT

REMOVAL & INSTALLATION

See Figures 185 and 186.

1. Before servicing the vehicle, refer to the Precautions Section.
2. Disconnect the negative battery cable.
3. Remove the windshield wiper arm.
4. Remove the cowl grille.
5. Remove or disconnect the following:
   - ABS wheel-speed sensor
   - Clip
   - Cap
   - Nut
   - Stiffener
   - Front shock absorber and coil spring
6. Install in the reverse order of removal.

➡Install the front shock absorber and coil spring so that the identification mark on the mounting rubber is facing to the position indicated in the figure.

7. Inspect for front wheel alignment, and adjust it as necessary.

FRONT
LH
RH
IDENTIFICATION MARK
22140_MCX7_G0241

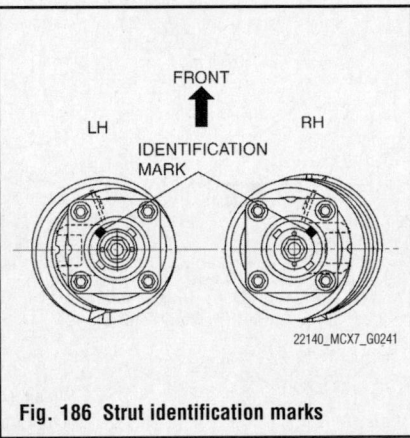

Fig. 186 Strut identification marks

OVERHAUL

See Figure 187.

1. Before servicing the vehicle, refer to the Precautions Section.
2. Remove the strut from the vehicle.
3. Compress the coil spring using a suitable spring compressor until the spring comes away from the seat.
4. Remove the large center nut and slowly release the spring compressor.

To install:

5. Compress the spring and install it on the strut.
6. Install the lower washer and mounting bracket.
7. Install the upper washer and a new nut.
8. Install the strut assembly in the vehicle.

STABILIZER BAR

REMOVAL & INSTALLATION

See Figures 188 through 190.

1. Before servicing the vehicle, refer to the Precautions Section.

1. ABS wheel-speed sensor
2. Clip
3. Cap
4. Nut
5. Stiffener
6. Front shock absorber and coil spring
46.1—62.7 {4.71—6.39, 34.1—46.2}
163.0—194.0 {16.7—19.7, 120.3—143.0}
7.8—10.8 N·m {79.6—110.1 kgf·cm, 69.1—95.5 in·lbf}
N·m {kgf·m, ft·lbf}
22140_MCX7_G0240

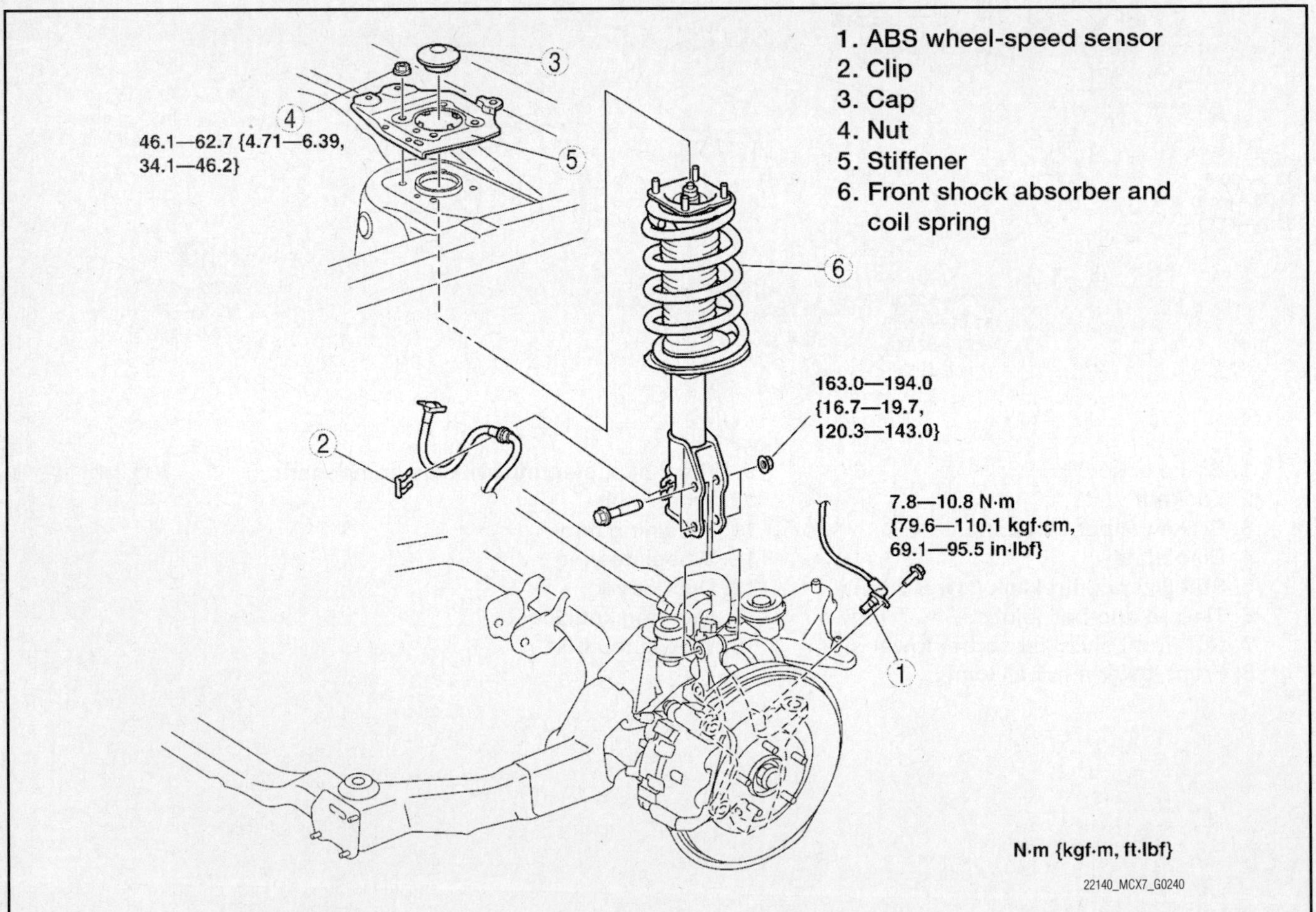

Fig. 185 Strut mounting exploded view

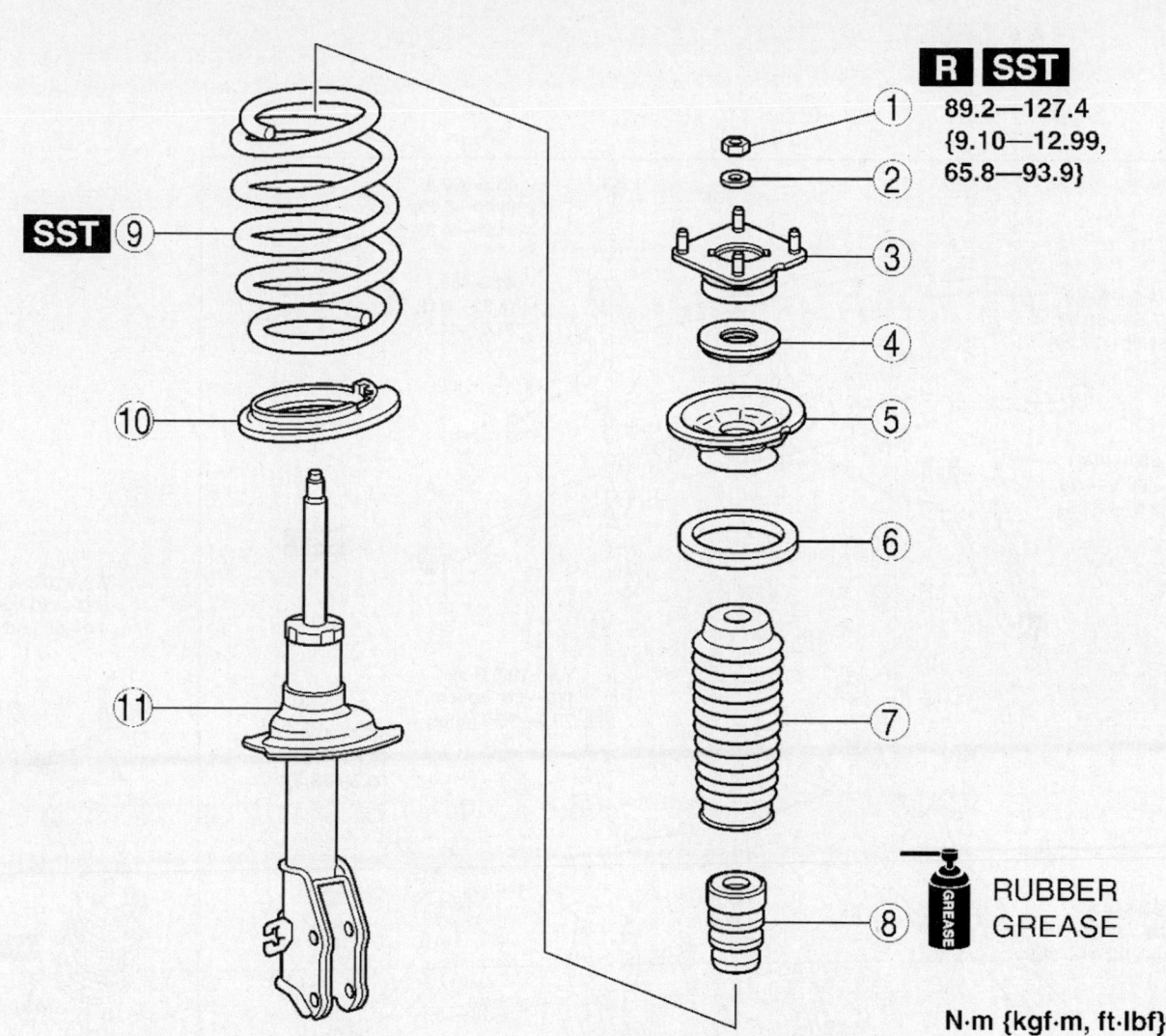

R **SST**
89.2—127.4
{9.10—12.99,
65.8—93.9}

**RUBBER
GREASE**

N·m {kgf·m, ft·lbf}

1. Piston rod nut
2. Washer
3. Mounting rubber
4. Bearing
5. Upper spring seat
6. Upper spring seat rubber
7. Dust boot
8. Bound stopper
9. Coil spring
10. Lower spring seat
11. Front shock absorber

22140_MCX7_G0238

Fig. 187 Shock absorber and spring assembly—exploded view

2. Disconnect the negative battery cable.

3. Remove the side cover and under-cover.

4. Drain the power steering fluid.

5. Remove the transverse member.

6. Remove the front crossmember component.

7. Remove or disconnect the following:
 • Front stabilizer component
 • Front stabilizer control link
 • Stabilizer bracket
 • Stabilizer bushing
 • Front stabilizer

8. Install in the reverse order of removal.

→**Install the stabilizer bar so that the identification mark is on the left side of the vehicle.**

WHEEL HUB & BEARING

REMOVAL & INSTALLATION
See Figures 191 and 192.

1. Before servicing the vehicle, refer to the Precautions Section.

2. Remove the front ABS wheel-speed sensor.

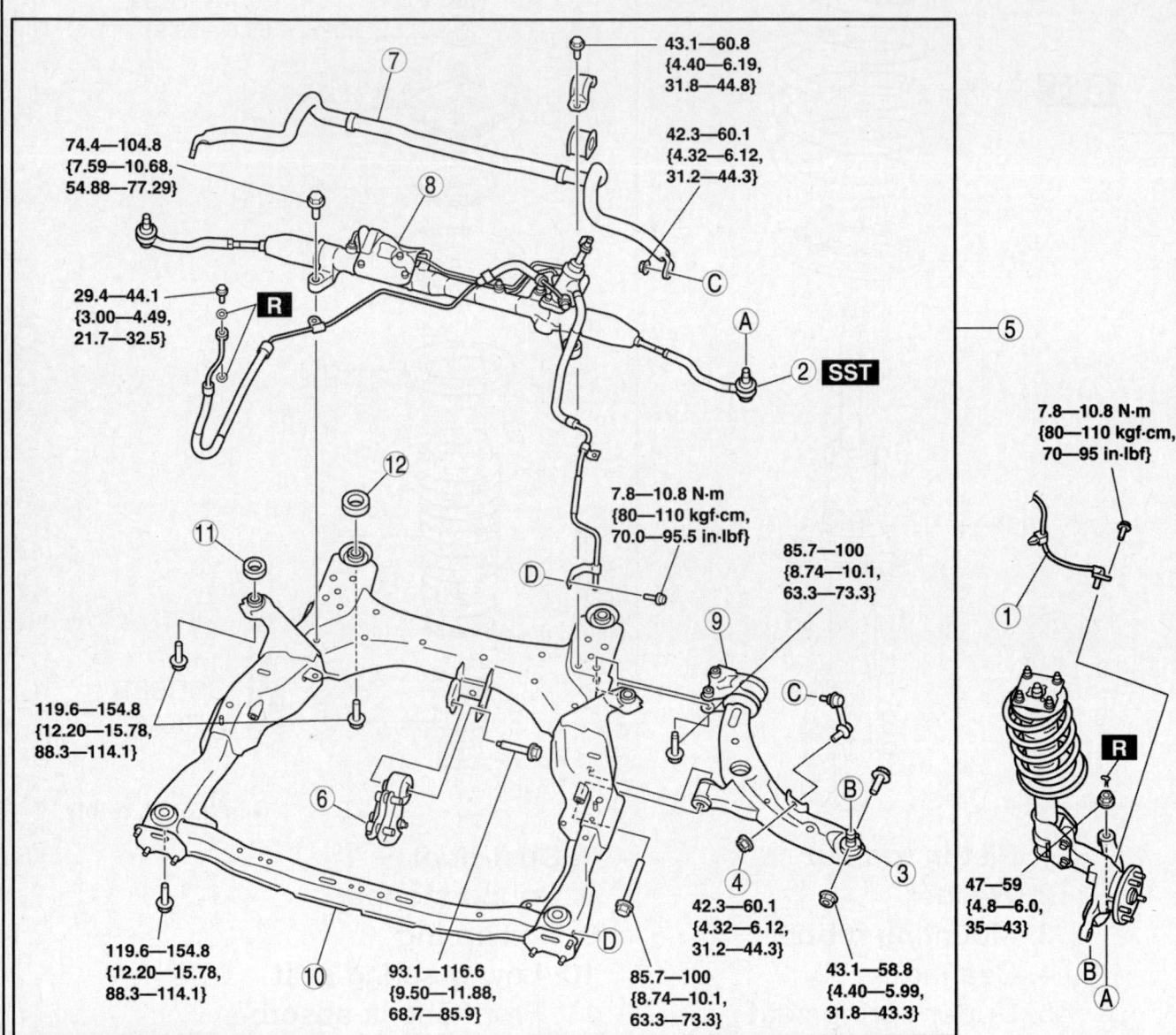

43.1—60.8
{4.40—6.19,
31.8—44.8}

42.3—60.1
{4.32—6.12,
31.2—44.3}

74.4—104.8
{7.59—10.68,
54.88—77.29}

29.4—44.1
{3.00—4.49,
21.7—32.5}

2 SST

7.8—10.8 N·m
{80—110 kgf·cm,
70—95 in·lbf}

7.8—10.8 N·m
{80—110 kgf·cm,
70.0—95.5 in·lbf}

85.7—100
{8.74—10.1,
63.3—73.3}

119.6—154.8
{12.20—15.78,
88.3—114.1}

119.6—154.8
{12.20—15.78,
88.3—114.1}

93.1—116.6
{9.50—11.88,
68.7—85.9}

85.7—100
{8.74—10.1,
63.3—73.3}

42.3—60.1
{4.32—6.12,
31.2—44.3}

43.1—58.8
{4.40—5.99,
31.8—43.3}

47—59
{4.8—6.0,
35—43}

N·m {kgf·m, ft·lbf}

1. ABS wheel speed sensor
2. Tie-rod end ball joint
3. Front lower arm ball joint
4. Stabilizer control link lower side nut
5. Steering gear and linkage, front stabilizer,
   front lower arm and front crossmember component
6. No. 1 engine mount
7. Front stabilizer
8. Steering gear and linkage
9. Front lower arm
10. Front crossmember
11. Front crossmember mounting rubber (front)
12. Front crossmember mounting rubber (rear)

22140_MCX7_G0246

Fig. 188 Front crossmember exploded view

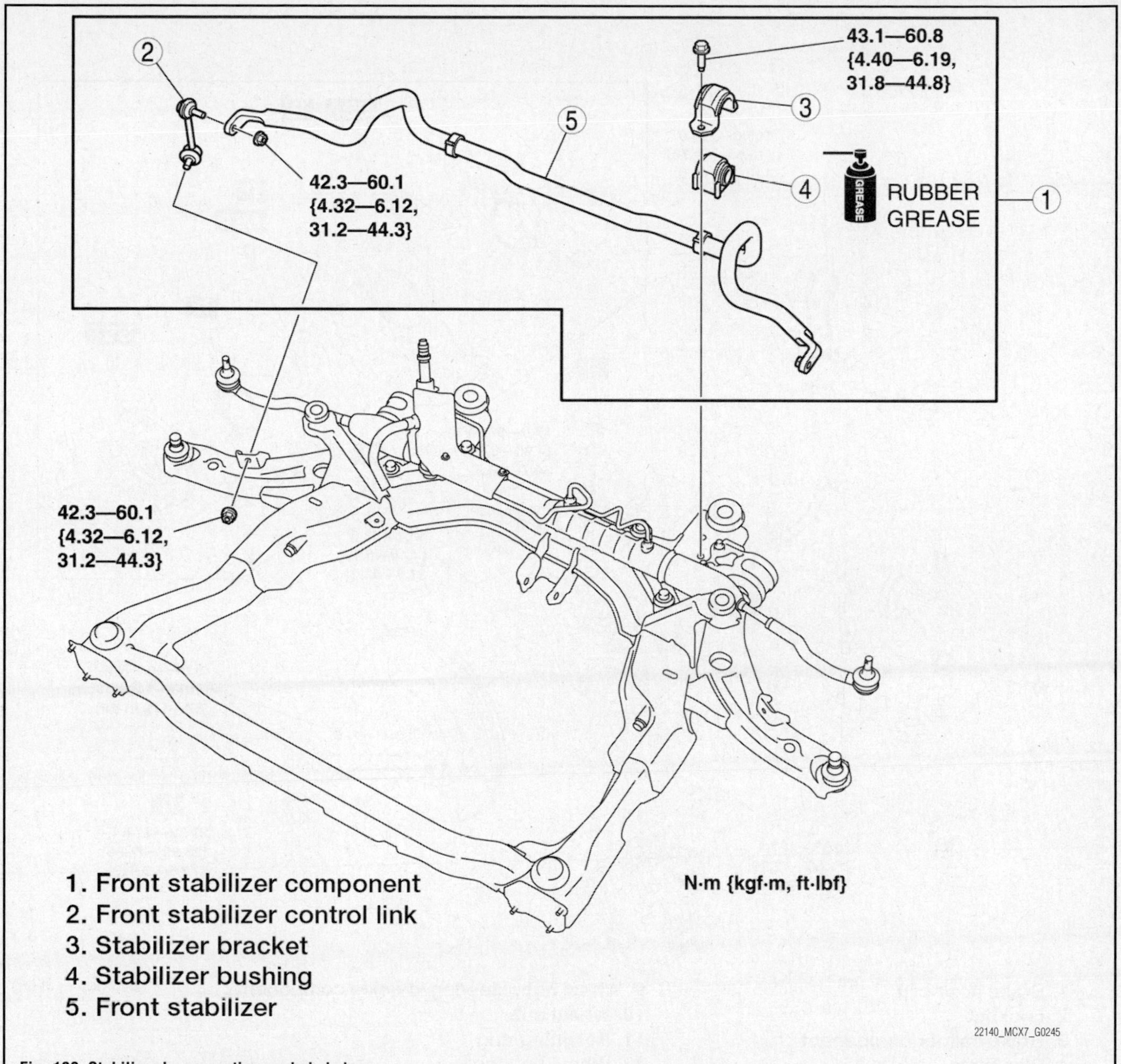

43.1—60.8
{4.40—6.19,
31.8—44.8}

42.3—60.1
{4.32—6.12,
31.2—44.3}

RUBBER
GREASE

42.3—60.1
{4.32—6.12,
31.2—44.3}

1. Front stabilizer component
2. Front stabilizer control link
3. Stabilizer bracket
4. Stabilizer bushing
5. Front stabilizer

N·m {kgf·m, ft·lbf}

22140_MCX7_G0245

Fig. 189 Stabilizer bar mounting exploded view

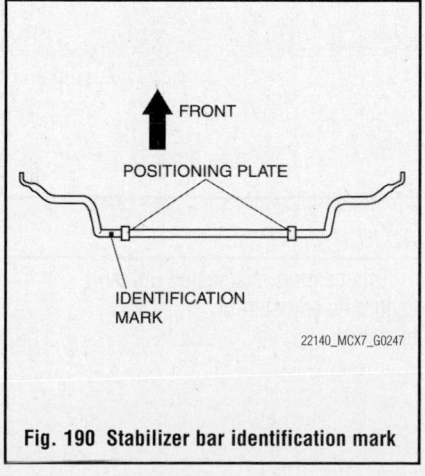

FRONT

POSITIONING PLATE

IDENTIFICATION
MARK

22140_MCX7_G0247

Fig. 190 Stabilizer bar identification mark

3. Remove or disconnect the following:
- Brake hose clip
- Locknut
- Brake caliper component
- Disc plate
- Stabilizer control link (lower side)
- Tie-rod end ball joint
- Nut (front shock absorber lower side)
- Front lower arm ball joint
- Wheel hub, steering knuckle component
- Wheel hub
- Retaining ring
- Wheel bearing
- Dust cover
- Steering knuckle

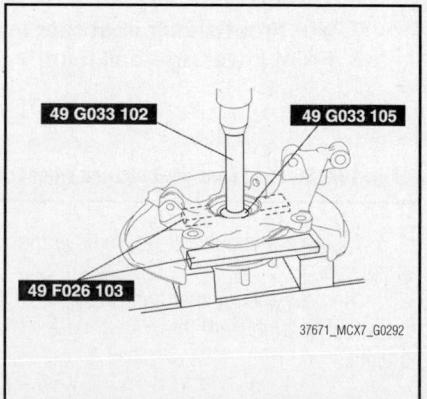

49 G033 102   49 G033 105

49 F026 103

37671_MCX7_G0292

Fig. 191 Using a hydraulic press, remove the wheel hub

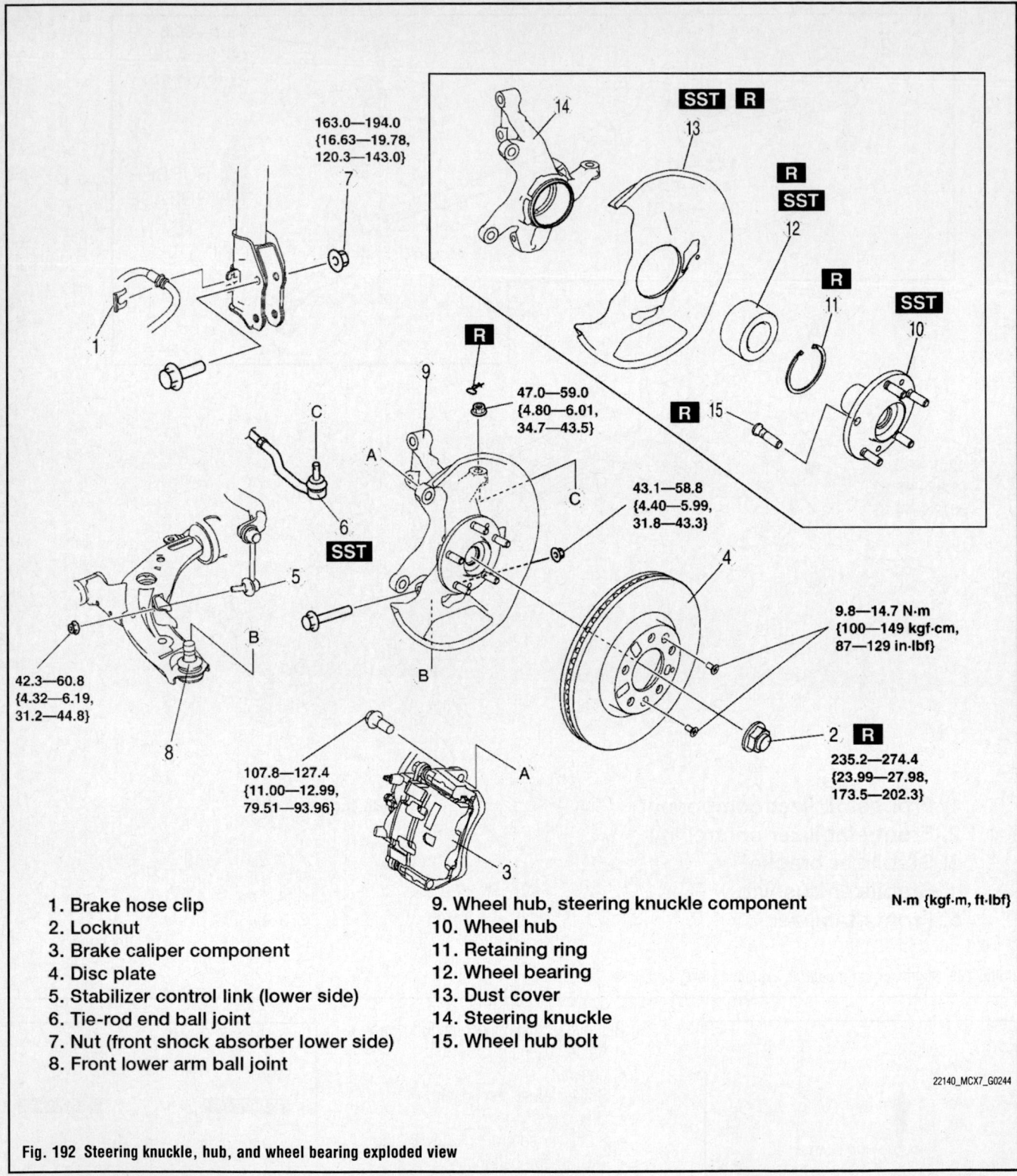

163.0—194.0
{16.63—19.78,
120.3—143.0}

47.0—59.0
{4.80—6.01,
34.7—43.5}

43.1—58.8
{4.40—5.99,
31.8—43.3}

42.3—60.8
{4.32—6.19,
31.2—44.8}

107.8—127.4
{11.00—12.99,
79.51—93.96}

9.8—14.7 N·m
{100—149 kgf·cm,
87—129 in·lbf}

235.2—274.4
{23.99—27.98,
173.5—202.3}

N·m {kgf·m, ft·lbf}

1. Brake hose clip
2. Locknut
3. Brake caliper component
4. Disc plate
5. Stabilizer control link (lower side)
6. Tie-rod end ball joint
7. Nut (front shock absorber lower side)
8. Front lower arm ball joint
9. Wheel hub, steering knuckle component
10. Wheel hub
11. Retaining ring
12. Wheel bearing
13. Dust cover
14. Steering knuckle
15. Wheel hub bolt

22140_MCX7_G0244

Fig. 192 Steering knuckle, hub, and wheel bearing exploded view

4. Using a hydraulic press, remove the wheel hub.

5. Remove the retainer and press the wheel bearing from the knuckle.

6. Install in the reverse order of removal.

7. After installation, inspect the front wheel alignment and adjust it if necessary.

ADJUSTMENT

This bearing is a sealed unit and cannot be adjusted or repacked.

SUSPENSION   **REAR SUSPENSION**

COIL SPRING

REMOVAL & INSTALLATION

See Figures 193 and 194.

1. Before servicing the vehicle, refer to the Precautions Section.
2. Remove the rear wheel.
3. Remove the middle exhaust pipe.
4. Remove the rear stabilizer.
5. Raise the rear trailing link to the unloaded condition with a jack.
6. Loosen the inner and outer bolts of the lateral link.
7. Loosen the outer bolt of the upper arm.
8. Support the lower arm with a jack.
9. Loosen the inner bolt of the lower arm.
10. Remove the outer bolt of the lower arm.
11. Jack down slowly and remove the coil spring.

❋❋ CAUTION

Removing the coil spring is dangerous. The coil spring could fly off, the cause serious injury or death.

To install:

12. Install the spring.
13. Position the jack under the lower arm and jack up slowly.

❋❋ WARNING

Installing the coil spring is dangerous. The coil spring could fly off, and cause serious injury or death.

14. Install the lower arm (outer side) bolt.
15. Raise the rear trailing link to the unloaded condition with a jack.

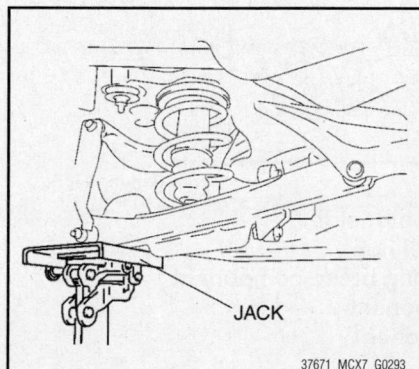

37671_MCX7_G0293

Fig. 193 Supporting the lower arm with a jack

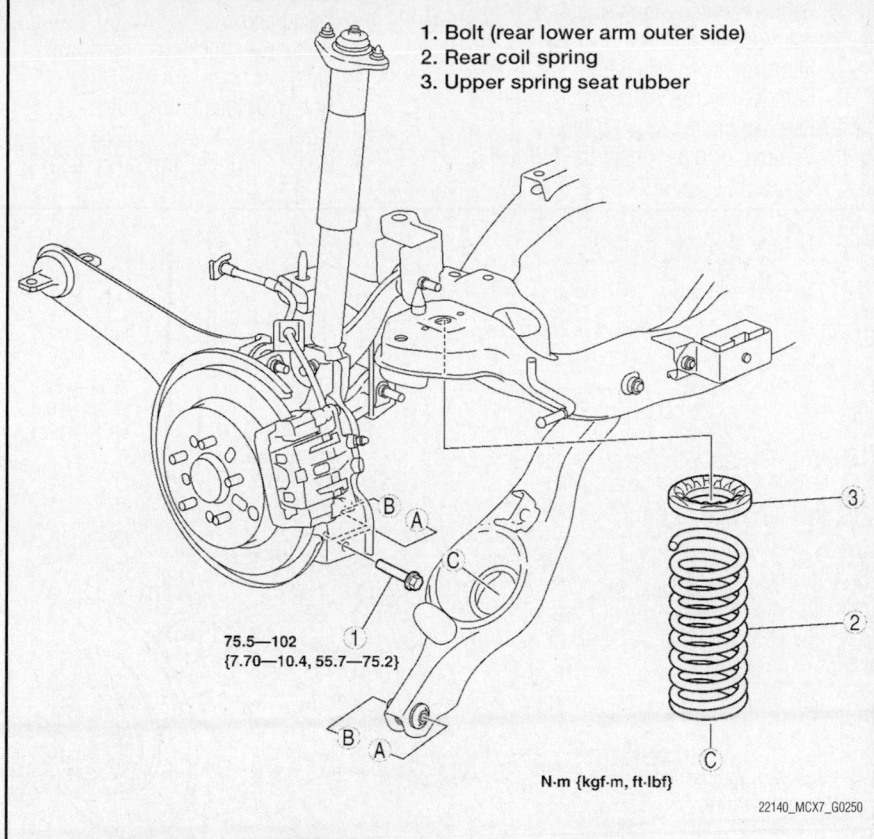

1. Bolt (rear lower arm outer side)
2. Rear coil spring
3. Upper spring seat rubber

75.5—102
{7.70—10.4, 55.7—75.2}

N·m {kgf·m, ft·lbf}

22140_MCX7_G0250

Fig. 194 Coil spring exploded view

16. Tighten the outer bolt of the upper arm to 72.3–75.2 ft. lbs. (75.5–102 Nm).
17. Tighten the inner and outer bolts of the lateral link to 72.3–75.2 ft. lbs. (75.5–102 Nm).
18. Install the rear stabilizer.
19. Install the middle exhaust pipe.
20. Install the rear wheel.
21. Check the rear wheel alignment and adjust if necessary.

CONTROL ARMS/LINKS

REMOVAL & INSTALLATION

Lateral link

See Figure 195.

1. Before servicing the vehicle, refer to the Precautions Section.
2. Remove the rear wheel.
3. Raise the trailing link to the unloaded condition with the jack.

➡ **Jacking up the rear suspension to the no-occupant position will lighten the force on the bushing and make it easier to perform the procedure.**

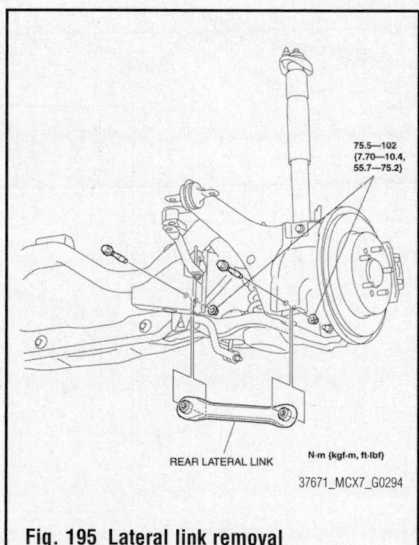

75.5—102
{7.70—10.4, 55.7—75.2}

REAR LATERAL LINK

N·m {kgf·m, ft·lbf}

37671_MCX7_G0294

Fig. 195 Lateral link removal

4. Remove the rear lateral link.

To install:

5. Install the lateral link. Tighten the bolts to 55.7–75.2 ft. lbs. (75.5–102 Nm).
6. Install the rear wheel.
7. Check the rear wheel alignment and adjust if necessary.

Trailing link 2WD Models

See Figure 196.

1. Before servicing the vehicle, refer to the Precautions Section.
2. Remove the middle pipe.
3. Remove the rear stabilizer.
4. Remove the rear coil spring.
5. Remove or disconnect the following:

- ABS wheel-speed sensor
- Parking brake cable
- Caliper component
- Disc plate
- Bolt (rear shock absorber lower side)
- Bolt (rear lateral link outer side)
- Bolt (rear upper arm outer side)

- Trailing link, rear wheel hub component and parking brake component
- Rear wheel hub component
- Parking brake component
- Trailing link

6. Install in the reverse order of removal.
7. Check the rear wheel alignment and adjust if necessary.

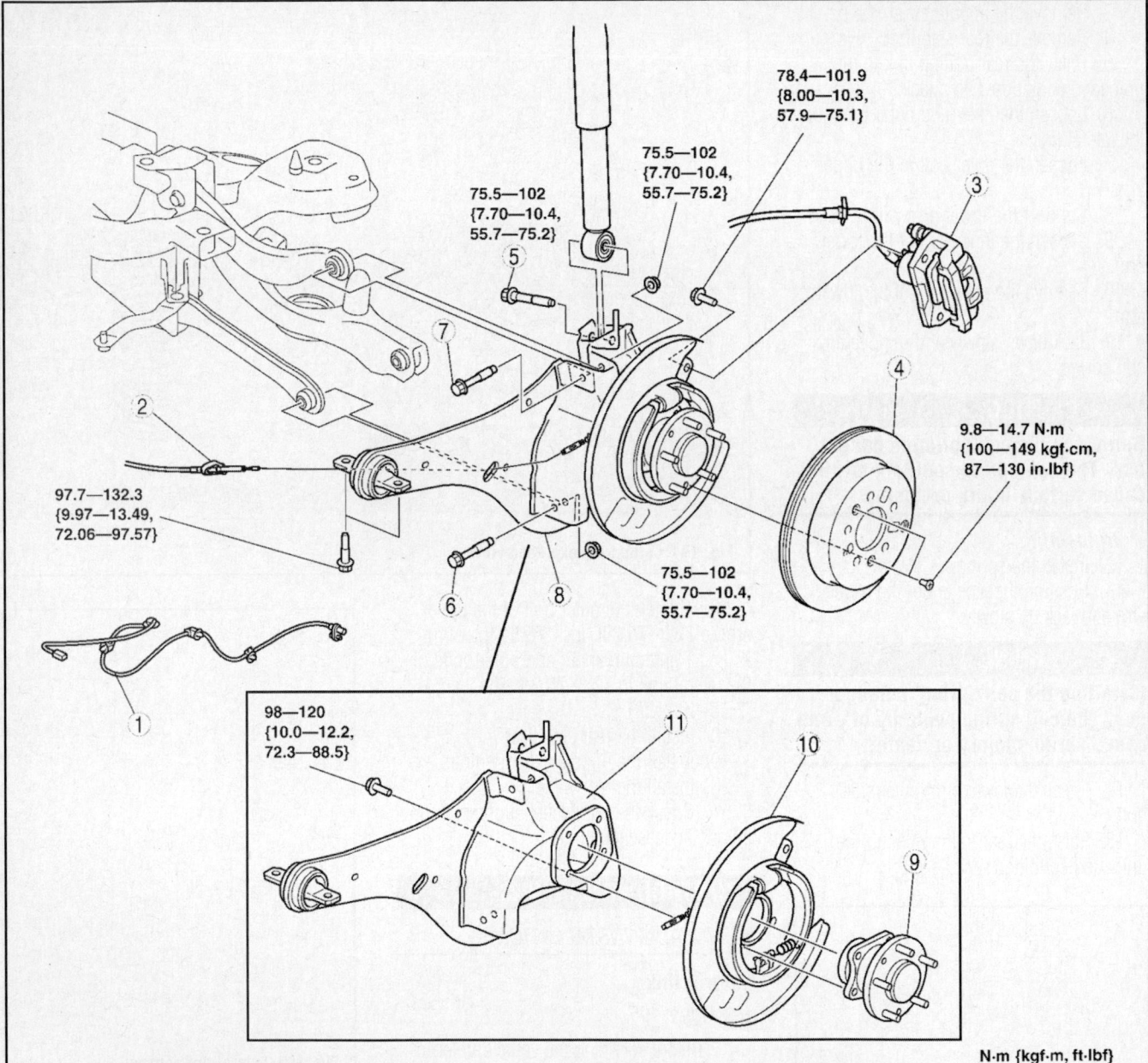

1. ABS wheel-speed sensor
2. Parking brake cable
3. Caliper component
4. Disc plate
5. Bolt (rear shock absorber lower side)
6. Bolt (rear lateral link outer side)
7. Bolt (rear upper arm outer side)
8. Trailing link, rear wheel hub component and parking brake component
9. Rear wheel hub component
10. Parking brake component
11. Trailing link

N·m {kgf·m, ft·lbf}

22140_MCX7_G0253

Fig. 196 Exploded view of the trailing link, and rear hub and related components—2WD

Trailing link AWD Models

See Figure 197.

1. Before servicing the vehicle, refer to the Precautions Section.
2. Remove the middle pipe.
3. Remove the rear stabilizer.
4. Remove the rear coil spring.

5. Remove or disconnect the following:
   - Rear ABS wheel-speed sensor
   - Parking brake cable
   - Locknut
   - Brake caliper component
   - Disc plate

- Bolt (rear shock absorber lower side)
- Bolt (rear lateral link outer side)
- Bolt (rear upper arm outer side)
- Trailing link, rear wheel hub component and parking brake component

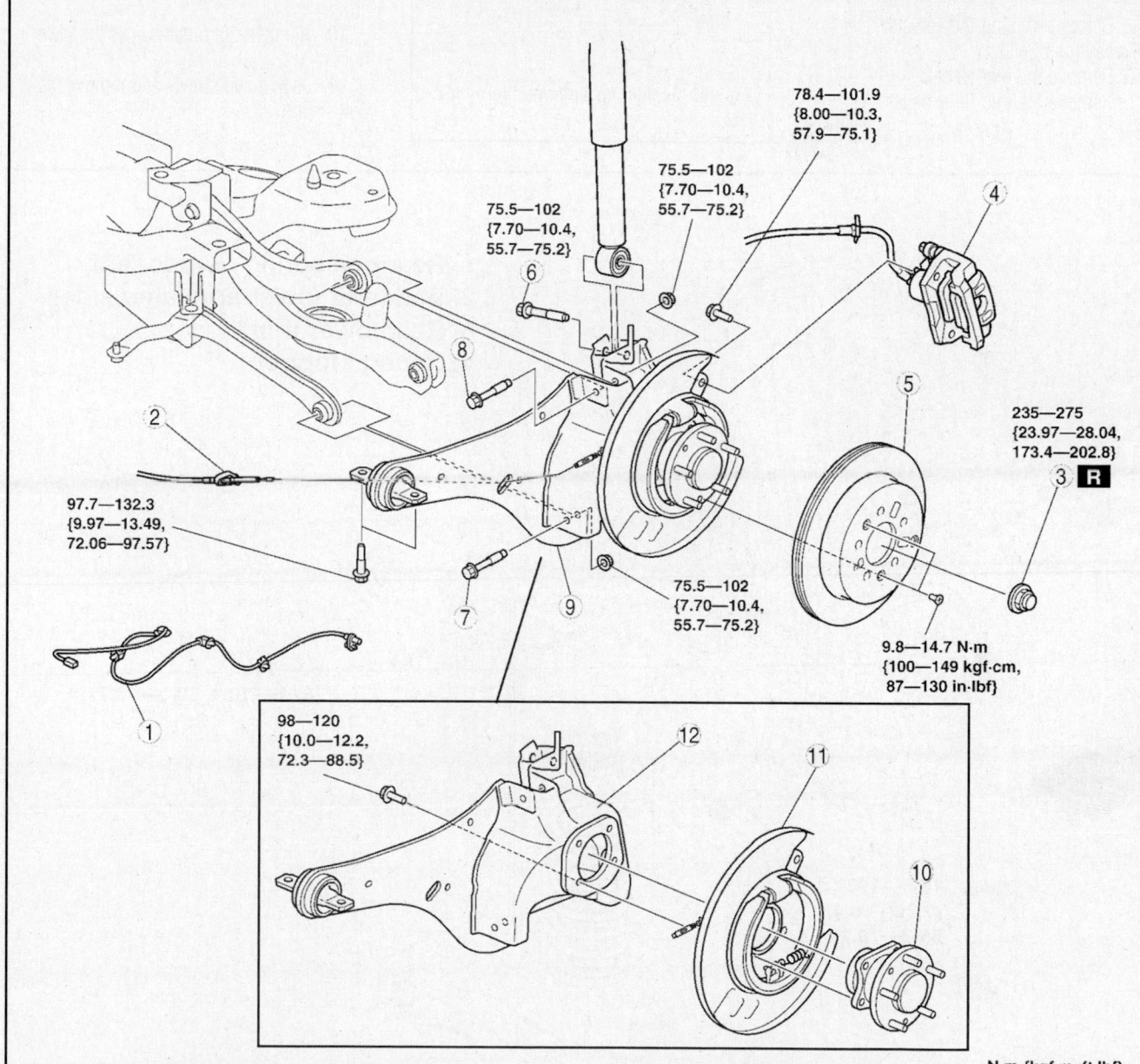

1. Rear ABS wheel-speed sensor
2. Parking brake cable
3. Locknut
4. Brake caliper component
5. Disc plate
6. Bolt (rear shock absorber lower side)
7. Bolt (rear lateral link outer side)
8. Bolt (rear upper arm outer side)
9. Trailing link, rear wheel hub component and parking brake component
10. Rear wheel hub component
11. Parking brake component
12. Trailing link

22140_MCX7_G0254

Fig. 197 Exploded view of the trailing link, and rear hub and related components—AWD

- Rear wheel hub component
- Parking brake component
- Trailing link

6. Install in the reverse order of removal.
7. Check the rear wheel alignment and adjust if necessary.

Lower Control Arm

See Figures 198 and 199.

1. Before servicing the vehicle, refer to the Precautions Section.
2. Remove the rear wheel.
3. Remove the middle exhaust pipe.

JACK
37671_MCX7_G0293

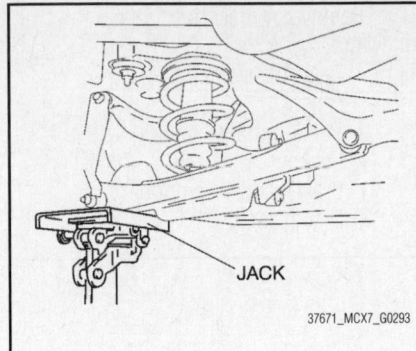

Fig. 198 Supporting the lower arm with a jack

4. Remove the rear stabilizer.
5. Raise the rear trailing link to the unloaded condition with a jack.
6. Loosen the inner and outer bolts of the lateral link.
7. Loosen the outer bolt of the upper arm.
8. Support the lower arm with a jack.
9. Loosen the inner bolt of the lower arm.
10. Remove the outer bolt of the lower arm.
11. Jack down slowly and remove the coil spring.

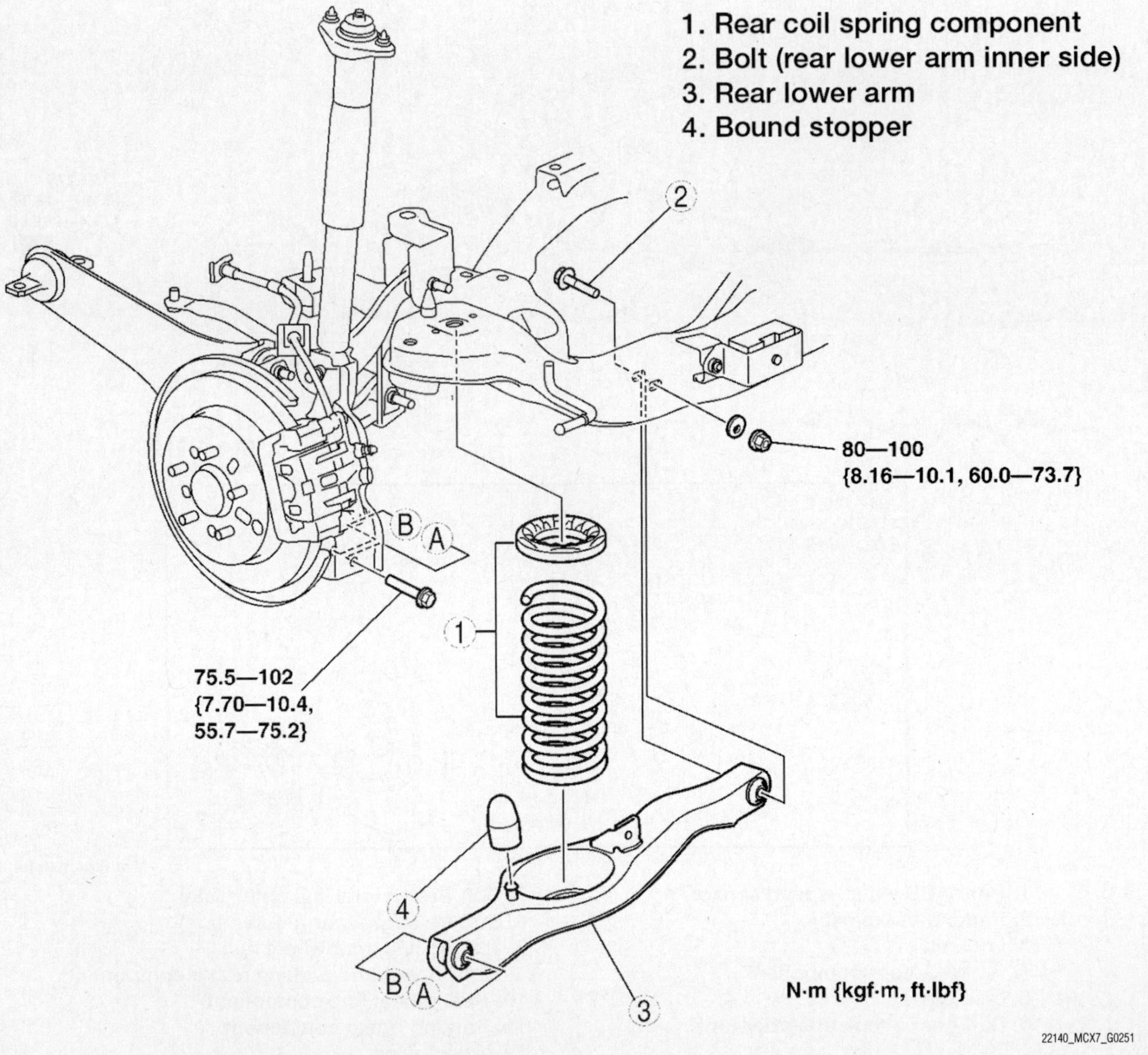

1. Rear coil spring component
2. Bolt (rear lower arm inner side)
3. Rear lower arm
4. Bound stopper
80—100
{8.16—10.1, 60.0—73.7}
75.5—102
{7.70—10.4,
55.7—75.2}
B
A
N·m {kgf·m, ft·lbf}
22140_MCX7_G0251

Fig. 199 Lower arm exploded view

**Removing the coil spring is danger-
ous. The coil spring could fly off, the
cause serious injury or death.**

12. Remove the bolt and remove the
lower arm.

To install:

13. Install the lower arm
14. Install the spring.
15. Position the jack under the lower
arm and jack up slowly.

**Installing the coil spring is danger-
ous. The coil spring could fly off, and
cause serious injury or death.**

16. Install the lower arm (outer side) bolt.
17. Raise the rear trailing link to the
unloaded condition with a jack.
18. Tighten the outer bolt of the upper
arm to 72.3–75.2 ft. lbs. (75.5–102 Nm).
19. Tighten the inner and outer bolts of
the lateral link to 72.3–75.2 ft. lbs.
(75.5–102 Nm).
20. Install the rear stabilizer.
21. Install the middle exhaust pipe.
22. Install the rear wheel.
23. Check the rear wheel alignment and
adjust if necessary.

SHOCK ABSORBER

REMOVAL & INSTALLATION

See Figures 198, 200 and 201.

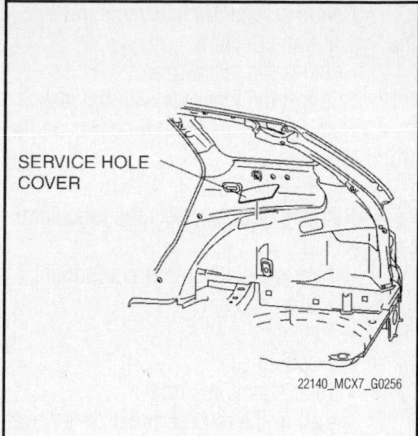

SERVICE HOLE
COVER

22140_MCX7_G0256

**Fig. 200 Trunk side trim service hole
cover**

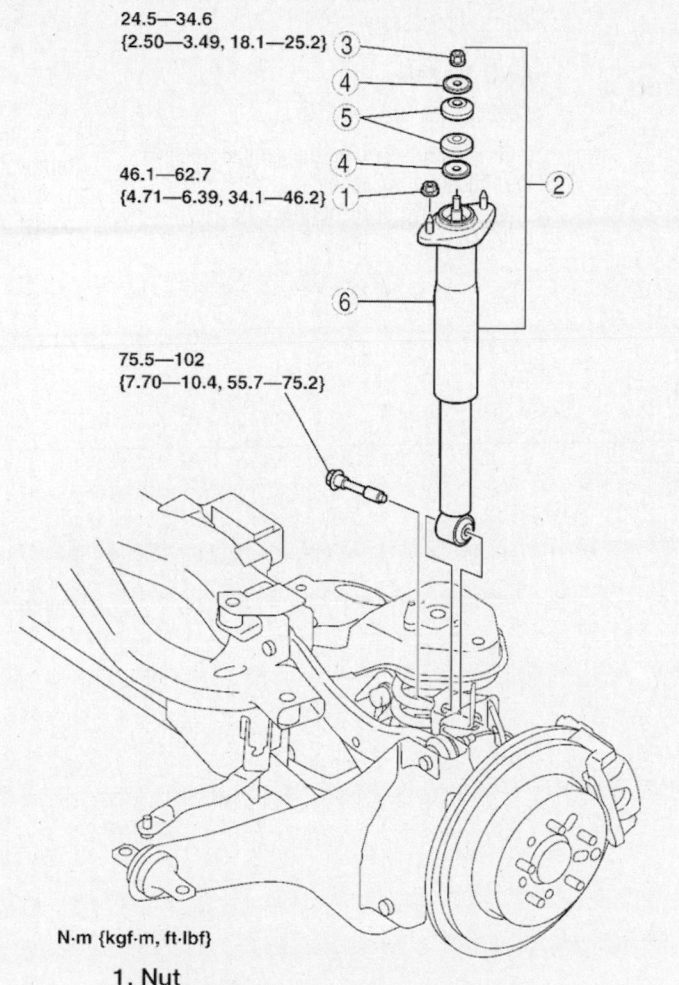

24.5—34.6
{2.50—3.49, 18.1—25.2}

46.1—62.7
{4.71—6.39, 34.1—46.2}

75.5—102
{7.70—10.4, 55.7—75.2}

N·m {kgf·m, ft·lbf}

1. Nut
2. Rear shock absorber component
3. Piston rod nut
4. Retainer
5. Bushing
6. Rear shock absorber

22140_MCX7_G0255

Fig. 201 Shock absorber exploded view

1. Before servicing the vehicle, refer to the Precautions Section.
2. Remove the rear wheel.
3. Support the rear axle with the jack.
4. Remove the service hole cover on the trunk side trim.
5. Remove the nuts.
6. Remove or disconnect the following:
   - Nut
   - Rear shock absorber component
   - Piston rod nut
   - Retainer
   - Bushing
   - Rear shock absorber
7. Install in the reverse order of removal.

WHEEL HUB & BEARING

REMOVAL & INSTALLATION

2WD Models

See Figure 196.

1. Before servicing the vehicle, refer to the Precautions Section.
2. Remove or disconnect the following:
   - ABS wheel-speed sensor
   - Caliper component

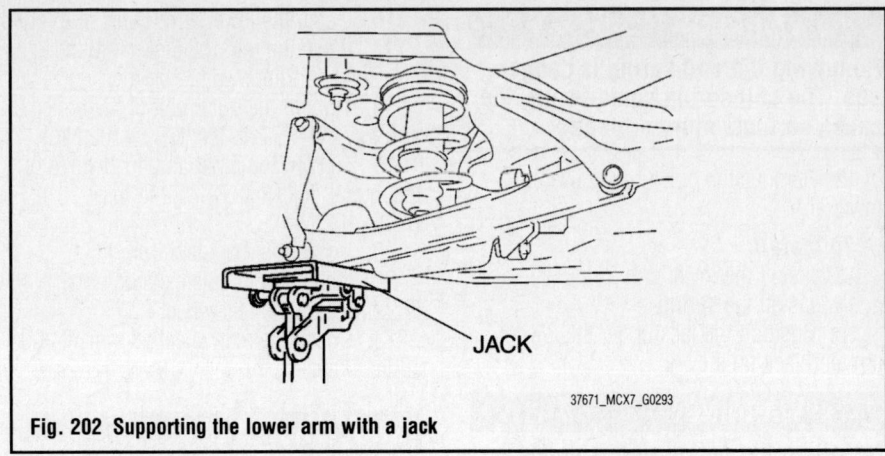

Fig. 202 Supporting the lower arm with a jack

37671_MCX7_G0293

   - Disc plate
   - Rear wheel hub component
3. Install in the reverse order of removal.
4. Check the rear wheel alignment and adjust if necessary.

AWD Models

See Figures 197 and 202.

1. Before servicing the vehicle, refer to the Precautions Section.
2. Remove or disconnect the following:
   - Rear ABS wheel-speed sensor
   - Rear stabilizer
   - Rear coil spring
   - Rear trailing link
   - Locknut
   - Brake caliper component
   - Disc plate
   - Rear wheel hub component
3. Install in the reverse order of removal.
4. Check the rear wheel alignment and adjust if necessary.

MAZDA

CX-9

3

SPECIFICATIONS AND MAINTENANCE CHARTS

ENGINE AND VEHICLE IDENTIFICATION CHART

			Engine Code				Model Year	
Code	Liters (cc)	Cu. In.	Cyl.	Fuel Sys.	Engine Type	Eng. Mfg.	Code ①	Year
A ②	3.7 (3721)	227	6	SMFI	DOHC	Ford Motor Co.	9	2009
V ③	3.7 (3721)	227	6	SMFI	DOHC	Ford Motor Co.	A	2010

DOHC: Double Overhead Cam

SMFI: Sequential Multi-port Fuel Injection

① 10th position of VIN

② Federal/Canada

③ California

37671_MCX9_C0001

GENERAL ENGINE SPECIFICATIONS

Year	Engine Displacement Liters (VIN)	Net Horsepower @ rpm	Net Torque @ rpm (ft. lbs.)	Bore x Stroke (in.)	Compression Ratio	Oil Pressure @ rpm
2009	3.7 (A)	273@6250	270@4250	3.76x3.41	10.0:1	NA
	3.7 (V)	273@6250	270@4250	3.76x3.41	10.0:1	NA
2010	3.7 (A)	273@6250	270@4250	3.76x3.41	10.0:1	NA
	3.7 (V)	273@6250	270@4250	3.76x3.41	10.0:1	NA

NA - Not Available

37671_MCX9_C0002

ENGINE TUNE-UP SPECIFICATIONS

Year	Engine Displacement Liters (VIN)	Spark Plug Gap (in.)	Ignition Timing (deg.) MT	Ignition Timing (deg.) AT	Fuel Pump (psi)	Idle Speed (rpm) MT	Idle Speed (rpm) AT	Valve Clearance (in.) Intake	Valve Clearance (in.) Exhaust
2009	3.7 (A)	0.051-0.057	N/A	13 BTDC	48-70	N/A	570-670	0.006-0.010	0.012-0.016
	3.7 (V)	0.051-0.057	N/A	13 BTDC	48-70	N/A	570-670	0.006-0.010	0.012-0.016
2010	3.7 (A)	0.051-0.057	N/A	13 BTDC	48-70	N/A	570-670	0.006-0.010	0.012-0.016
	3.7 (V)	0.051-0.057	N/A	13 BTDC	48-70	N/A	570-670	0.006-0.010	0.012-0.016

BTDC - Before Top Dead Center

N/A - Not Applicable

37671_MCX9_C0003

CAPACITIES

Year	Model	Engine Displacement Liters (VIN)	Engine Oil with Filter (qts.)	Transmission (pts.) 5-Spd	Transmission (pts.) Auto.	Transfer Case (pts.)	Drive Axle Front (pts.)	Drive Axle Rear (pts.)	Fuel Tank (gal.)	Cooling System (qts.)
2009	CX9	3.7 (A)	5.5	N/A	14.8	1.12	N/A	2.2	20.1	①
	CX9	3.7 (V)	5.5	N/A	14.8	1.12	N/A	2.2	20.1	①
2010	CX9	3.7 (A)	5.5	N/A	14.8	1.12	N/A	2.2	20.1	①
	CX9	3.7 (V)	5.5	N/A	14.8	1.12	N/A	2.2	20.1	①

NOTE: All capacities are approximate. Add fluid gradually and check to be sure a proper fluid level is obtained.

① Single fan: 12.3 qt., Dual fan: 12.9 qt.

N/A - Not Applicable

37671_MCX9_C0004

FLUID SPECIFICATIONS

Year	Engine Displacement Liters (VIN)	Engine Oil	Auto. Trans.	Transfer Case	Drive Axle	Power Steering Fluid	Brake Master Cylinder	Cooling System
2009	3.7 (A)	①	JWS3309	75W-140	80W-90	Mercon® ATF Fluid	DOT 3	②
	3.7 (V)	①	JWS3309	75W-140	80W-90	Mercon® ATF Fluid	DOT 3	②
2010	3.7 (A)	①	JWS3309	75W-140	80W-90	Mercon® ATF Fluid	DOT 3	②
	3.7 (V)	①	JWS3309	75W-140	80W-90	Mercon® ATF Fluid	DOT 3	②

DOT: Department Of Transpotation

① 5W-20 Premium Synthetic Blend Motor Oil (US) or 5W-20 Super Premium Motor Oil (Canada)

② Ethylene-glycol-based coolant, If the "FL22" mark is shown use Mazda Genuine FL22 engine coolant

37671_MCX9_C0005

VALVE SPECIFICATIONS

Year	Engine Displacement Liters (VIN)	Seat Angle (deg.)	Face Angle (deg.)	Spring Test Pressure (lbs. @ in.)	Spring Installed Height (in.)	Stem-to-Guide Clearance (in.) Intake	Stem-to-Guide Clearance (in.) Exhaust	Stem Diameter (in.) Intake	Stem Diameter (in.) Exhaust
2009	3.7 (A)	89.0-91.0	90.50-91.50	115 @ 1.08	1.4500	0.0008-0.0027	0.0013-0.0320	0.2157-0.2164	0.2151-0.2159
	3.7 (V)	89.0-91.0	90.50-91.50	115 @ 1.08	1.4500	0.0008-0.0027	0.0013-0.0320	0.2157-0.2164	0.2151-0.2159
2010	3.7 (A)	89.0-91.0	90.50-91.50	115 @ 1.08	1.4500	0.0008-0.0027	0.0013-0.0320	0.2157-0.2164	0.2151-0.2159
	3.7 (V)	89.0-91.0	90.50-91.50	115 @ 1.08	1.4500	0.0008-0.0027	0.0013-0.0320	0.2157-0.2164	0.2151-0.2159

37671_MCX9_C0006

CAMSHAFT SPECIFICATIONS

All measurements are given in inches

Year	Engine Displacement Liters (VIN)	Journal Diameter	Bearing Oil Clearance	Shaft End-play	Runout	Lobe Height	
						Intake	Exhaust
2009	3.7 (A)	①	②	0.0012-0.0066	0.0015	0.3800	0.3800
	3.7 (V)	①	②	0.0012-0.0066	0.0015	0.3800	0.3800
2010	3.7 (A)	①	②	0.0012-0.0066	0.0015	0.3800	0.3800
	3.7 (V)	①	②	0.0012-0.0066	0.0015	0.3800	0.3800

① 1st journal: 1.2202-1.2209 in.
   Intermediate journals: 1.021-1.022 in.

② 1st journal: 0.0027 MAX
   Intermediate journals: 0.0029 MAX

37671_MCX9_C0007

CRANKSHAFT AND CONNECTING ROD SPECIFICATIONS

All measurements are given in inches

Year	Engine Displacement Liters (VIN)	Crankshaft				Connecting Rod		
		Main Brg. Journal Dia.	Main Brg. Oil Clearance	Shaft End-play	Thrust on No.	Journal Diameter	Oil Clearance	Side Clearance
2009	3.7 (A)	2.657	N/A	0.0039-0.0114	N/A	2.204-2.205	N/A	0.0068-0.0167
	3.7 (V)	2.657	N/A	0.0039-0.0114	N/A	2.204-2.205	N/A	0.0068-0.0167
2010	3.7 (A)	2.657	N/A	0.0039-0.0114	N/A	2.204-2.205	N/A	0.0068-0.0167
	3.7 (V)	2.657	N/A	0.0039-0.0114	N/A	2.204-2.205	N/A	0.0068-0.0167

N/A - Not Available

37671_MCX9_C0008

PISTON AND RING SPECIFICATIONS
All measurements are given in inches

Year	Engine Displacement Liters (VIN)	Piston Clearance	Ring Gap			Ring Side Clearance		
			Top Compression	Bottom Compression	Oil Control	Top Compression	Bottom Compression	Oil Control
2009	3.7 (A)	0.0003-0.0017	0.0059-0.0098	0.0118-0.0216	0.0059-0.0177	NA	NA	NA
	3.7 (V)	0.0003-0.0017	0.0059-0.0098	0.0118-0.0216	0.0059-0.0177	NA	NA	NA
2010	3.7 (A)	0.0003-0.0017	0.0059-0.0098	0.0118-0.0216	0.0059-0.0177	NA	NA	NA
	3.7 (V)	0.0003-0.0017	0.0059-0.0098	0.0118-0.0216	0.0059-0.0177	NA	NA	NA

NA: Not Available

37671_MCX9_C0009

TORQUE SPECIFICATIONS
All readings in ft. lbs.

Year	Engine Displacement Liters (VIN)	Cylinder Head Bolts	Main Bearing Bolts	Rod Bearing Bolts	Crankshaft Damper Bolts	Flywheel Bolts	Manifold		Spark Plugs	Oil Pan Drain Plug
							Intake	Exhaust		
2009	3.7 (A)	①	N/A	N/A	②	59	③	19	11	20
	3.7 (V)	①	N/A	N/A	②	59	③	19	11	20
2010	3.7 (A)	①	N/A	N/A	②	59	③	19	11	20
	3.7 (V)	①	N/A	N/A	②	59	③	19	11	20

① Step 1: Tigthen bolts 1-16 to 15 ft. lbs.

　Step 2: Tigthen bolts 1-16 to 26 ft. lbs.

　Step 3: Tigthen bolts 1-16 to +90 degrees

　Step 4: Tigthen bolts 1-16 to +90 degrees

　Step 4: Tigthen bolts 1-16 to +90 degrees

　Step 5: Tigthen bolts 17-18 to 76-101 inch lbs.

② Step 1: 89 ft. lbs.

　Step 2: Loosen one full turn

　Step 3: 37 ft. lbs.

　Step 4: +90 degrees

③ Upper intake manifold: 89 inch lbs.

　Lower intake manfold: 89 inch lbs.

N/A: Not Available

37671_MCX9_C0010

WHEEL ALIGNMENT

Year	Model		Caster Range (+/-Deg.)	Caster Preferred Setting (Deg.)	Camber Range (+/-Deg.)	Camber Preferred Setting (Deg.)	Toe-in (in.)
2009	CX9	F	±1°	3°04'	±1°	−0°21'	0.08±0.16
		R	±1°	—	±1°	-0°33'	0.08±0.16
2010	CX9	F	±1°	3°04'	±1°	−0°21'	0.08±0.16
		R	±1°	—	±1°	-0°33'	0.08±0.16

NOTE: Perform wheel alignment with the fuel tank full.

37671_MCX9_C0011

TIRE, WHEEL AND BALL JOINT SPECIFICATIONS

Year	Model	OEM Tires Standard	OEM Tires Optional	Tire Pressures (psi) Front	Tire Pressures (psi) Rear	Wheel Size	Ball Joint Inspection	Lug Nut (ft. lbs.)
2009	CX9	P245/60R18	P245/50R20	①	①	②	NS	94
2010	CX9	P245/60R18	P245/50R20	①	①	②	NS	94

OEM: Original Equipment Manufacturer

PSI: Pounds Per Square Inch

NS: Not specified by manufacturer

① See the safety certification label on the driver side door jamb for tire pressures.

② P245/60R18: 18x7.5

   P245/50R20: 20x7.5

37671_MCX9_C0012

BRAKE SPECIFICATIONS
All measurements in inches unless noted

Year	Model		Brake Disc Original Thickness	Brake Disc Minimum Thickness	Brake Disc Maximum Runout	Brake Drum Diameter Original Inside Diameter	Max. Wear Limit	Brake Drum Diameter Maximum Machine Diameter	Minimum Lining Thickness Front	Minimum Lining Thickness Rear	Brake Caliper Bracket Bolts (ft. lbs.)	Brake Caliper Mounting Bolts (ft. lbs.)
2009	CX9	F	1.102	1.030	0.002	N/A	N/A	N/A	0.080	—	98	65
		R	0.708	0.630	0.002	N/A	N/A	N/A	—	0.08	66	19
2010	CX9	F	1.102	1.030	0.002	N/A	N/A	N/A	0.080	—	98	65
		R	0.708	0.630	0.002	N/A	N/A	N/A	—	0.08	66	19

F: Front

R: Rear

N/A: Not Available

37671_MCX9_C0013

SCHEDULED MAINTENANCE INTERVALS

MAZDA CX9

TO BE SERVICED	TYPE OF SERVICE	VEHICLE MILEAGE INTERVAL (x1000)											
		10	20	30	40	50	60	70	80	90	100	110	120
Accessory drive belts	I & A			✓			✓			✓			✓
Air cleaner element	R			✓			✓			✓			✓
Air conditioning filter	R			✓			✓			✓			✓
Brake fluid	R											✓	
Brake hoses & lines (including ABS)	I		✓		✓		✓		✓		✓		
Cooling system hoses & connections	I		✓		✓		✓		✓		✓		
Engine coolant	R												✓
Engine oil	R	✓	✓	✓	✓	✓	✓	✓	✓	✓	✓	✓	✓
Engine oil and coolant levels	I	Inspect at each fuel stop											
Engine oil filter	R		✓		✓		✓		✓		✓		
Exhaust system	I		✓		✓		✓		✓		✓		
Fluid levels and condition	I		✓		✓		✓		✓		✓		
Front and rear brakes	I		✓		✓		✓		✓		✓		
Fuel lines & connection	I		✓		✓		✓		✓		✓		
Halfshaft boots	I		✓		✓		✓		✓		✓		
Idle speed	I & A											✓	
Parking brake system	I & A		✓		✓		✓		✓		✓		
Rear differential fluid	R									✓			
Rotate and inspect tires	I	✓	✓	✓	✓	✓	✓	✓	✓	✓	✓	✓	✓
Spark plugs	R											✓	
Suspension components	I		✓		✓		✓		✓		✓		
Tie rod ends, steering gear box & boots	I		✓		✓		✓		✓		✓		
Transmission fluid	R												✓
Valve clearance	I											✓	

R: Replace   I: Inspect   A: Adjust

FREQUENT OPERATION MAINTENANCE (SEVERE SERVICE)

If a vehicle is operated under any of the following conditions it is considered severe service:

- Towing a trailer or using a camper or car-top carrier.
- Repeated short trips of less than 5 miles in temperatures below freezing, or trips of less than 10 miles in any temperature.
- Extensive idling or low-speed driving for long distances as in heavy commercial use, such as delivery, taxi or police cars.
- Operating on rough, muddy or salt-covered roads.
- Operating on unpaved or dusty roads.
- Driving in extremely hot (over 90°) conditions.

Air cleaner element: replace every 15,000 miles

Engine oil and filter: replace every 3750 miles or 6 months, whichever occurs first.

Timing belt: replace every 60,000 miles if the vehicle is regularly driven in temperatures above 110°F or below -20°F.

Transmission fluid: replace every 30,000 miles.

Rear differential fluid: replace every 60,000 miles.

Front and rear brakes: inspect every 7500 miles or 6 months, whichever occurs first.

Locks and hinges: lubricate every 15,000 miles.

Tie rods, steering gear box, boots: inspect every 7500 miles or 6 months, whichever occurs first.

Suspension components: inspect every 7500 miles or 6 months, whichever occurs first.

Halfshaft boots: inspect every 7500 miles or 6 months, whichever occurs first.

BRAKES — INFORMATION AND PRECAUTIONS

ANTI-LOCK SYSTEMS

- Certain components within the ABS system are not intended to be serviced or repaired individually.
- Do not use rubber hoses or other parts not specifically specified for and ABS system. When using repair kits, replace all parts included in the kit. Partial or incorrect repair may lead to functional problems and require the replacement of components.
- Lubricate rubber parts with clean, fresh brake fluid to ease assembly. Do not use shop air to clean parts; damage to rubber components may result.
- Use only DOT 3 brake fluid from an unopened container.
- If any hydraulic component or line is removed or replaced, it may be necessary to bleed the entire system.
- A clean repair area is essential. Always clean the reservoir and cap thoroughly before removing the cap. The slightest amount of dirt in the fluid may plug an orifice and impair the system function. Perform repairs after components have been thoroughly cleaned; use only denatured alcohol to clean components. Do not allow ABS components to come into contact with any substance containing mineral oil; this includes used shop rags.

- The Anti-Lock control unit is a microprocessor similar to other computer units in the vehicle. Ensure that the ignition switch is **OFF** before removing or installing controller harnesses. Avoid static electricity discharge at or near the controller.
- If any arc welding is to be done on the vehicle, the control unit should be unplugged before welding operations begin.

DISC AND DRUM SYSTEMS

> **✱✱ CAUTION**
>
> Dust and dirt accumulating on brake parts during normal use may contain asbestos fibers from production or aftermarket brake linings. Breathing excessive concentrations of asbestos fibers can cause serious bodily harm. Exercise care when servicing brake parts. Do not sand or grind brake lining unless equipment used is designed to contain the dust residue. Do not clean brake parts with compressed air or by dry brushing. Cleaning should be done by dampening the brake components with a fine mist of water, then wiping the brake components clean with a dampened cloth. Dispose of cloth and all residue containing asbestos fibers in an impermeable container with the appropriate label. Follow practices prescribed by the Occupational Safety and Health Administration (OSHA) and the Environmental Protection Agency (EPA) for the handling, processing, and disposing of dust or debris that may contain asbestos fibers.

BRAKES — BLEEDING THE BRAKE SYSTEM

BLEEDING PROCEDURE

BLEEDING PROCEDURE

1. Before servicing the vehicle, refer to the Precautions Section.

> **✱✱ WARNING**
>
> Keep the fluid level in the reserve tank at 3/4 full or more during the air bleeding.

➡ **Begin air bleeding with the brake caliper that is furthest from the master cylinder.**

2. Remove the bleeder cap on the brake caliper, and attach a vinyl tube to the bleeder screw.

3. Place the other end of the vinyl tube in a clear container and fill the container with fluid during air bleeding.

4. Working with two people, one should pump the brake pedal several times and depress and hold the pedal down.

5. While the brake pedal is depressed, the other should loosen the bleeder screw using a commercially available flare nut wrench, drain out any fluid containing air bubbles, and tighten the bleeder screw to 53–86 inch lbs. 6–10 Nm) for the front caliper and 61–86 inch lbs. 7–10 Nm) for the rear.

6. Repeat the steps no air bubbles are seen.

7. Perform air bleeding as described for all brake calipers.

8. After air bleeding, check for proper brake operation, any fluid leakage and correct fluid level.

BLEEDING THE ABS SYSTEM

1. Before servicing the vehicle, refer to the Precautions Section.
2. Turn the ignition switch **OFF**
3. Connect a scan tool to the diagnostic link connector.
4. After the vehicle is identified, select "ABS Service Bleed".
5. Perform the air bleeding according to the directions on the screen.

BRAKES   **ANTI-LOCK BRAKE SYSTEM (ABS)**

SPEED SENSORS

REMOVAL & INSTALLATION

Front

See Figure 1.

1. Before servicing the vehicle, refer to the Precautions Section.

2. When removing the right front ABS wheel-speed sensor, perform the following:.

   a. Remove the coolant reserve tank installation bolts and move the coolant reserve tank.

   b. Remove the power steering reserve tank installation bolt and nut, and move the power steering reserve tank.

3. When removing the left front ABS wheel-speed sensor, perform the following procedure:

   a. Remove the air cleaner case.

4. Disconnect the front ABS wheel-speed sensor connector.

5. Remove the attaching bolt.

6. Remove the front ABS wheel-speed sensor.

To install:

7. Install the front ABS wheel-speed sensor.

8. Install the attaching bolt and tighten to 70–95 inch lbs. (8–11 Nm).

9. Disconnect the front ABS wheel-speed sensor connector.

   a. Install the air cleaner case.

10. When removing the left front ABS wheel-speed sensor, perform the following procedure:

   a. Install the power steering reserve tank and tighten bolts to 14–19 ft. lbs. (19–26 Nm).

   b. Install the coolant reserve tank and tighten bolts to 62–104 inch lbs. (7–12 Nm).

11. When removing the right front ABS wheel-speed sensor, perform the following:.

Rear

See Figures 2 through 6.

1. Before servicing the vehicle, refer to the Precautions Section.

2. Remove the trunk box.

3. Remove the seat side box.

4. Remove the third-row seat.

5. Remove the rear scuff plate inner.

6. Remove the third-row seat belt lower anchor installation bolt.

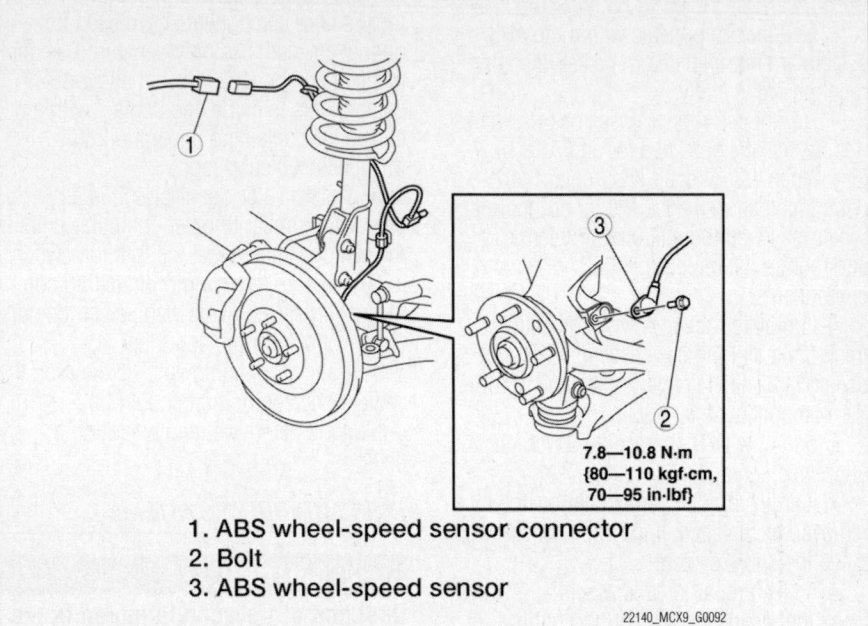

1. ABS wheel-speed sensor connector
2. Bolt
3. ABS wheel-speed sensor

7.8—10.8 N·m
{80—110 kgf·cm,
70—95 in·lbf}

22140_MCX9_G0092

Fig. 1 Front wheel speed sensor assembly

22140_MCX9_G0094

Fig. 2 Removing the trunk box

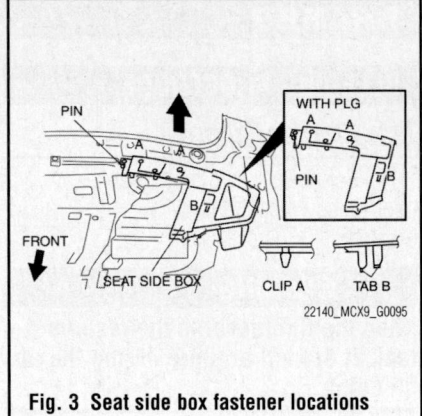

22140_MCX9_G0095

Fig. 3 Seat side box fastener locations

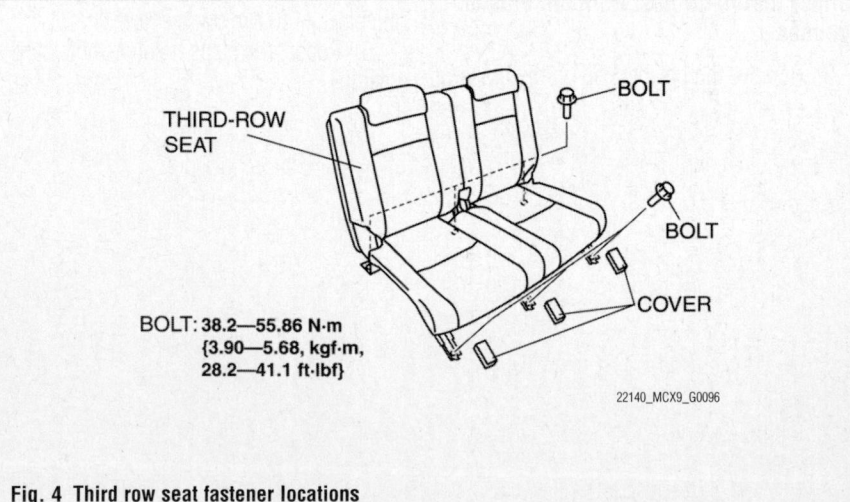

BOLT: 38.2—55.86 N·m
{3.90—5.68, kgf·m,
28.2—41.1 ft·lbf}

22140_MCX9_G0096

Fig. 4 Third row seat fastener locations

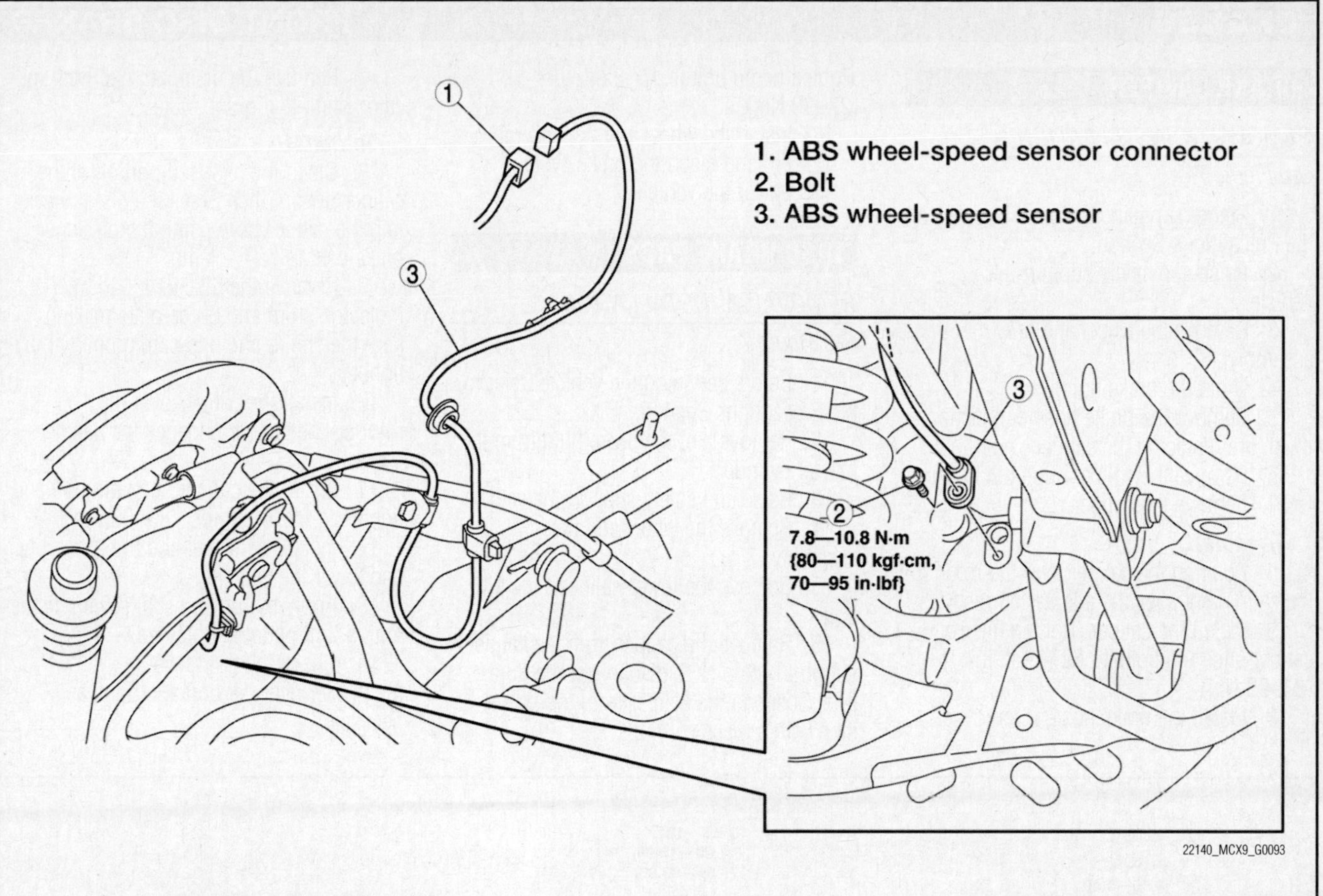

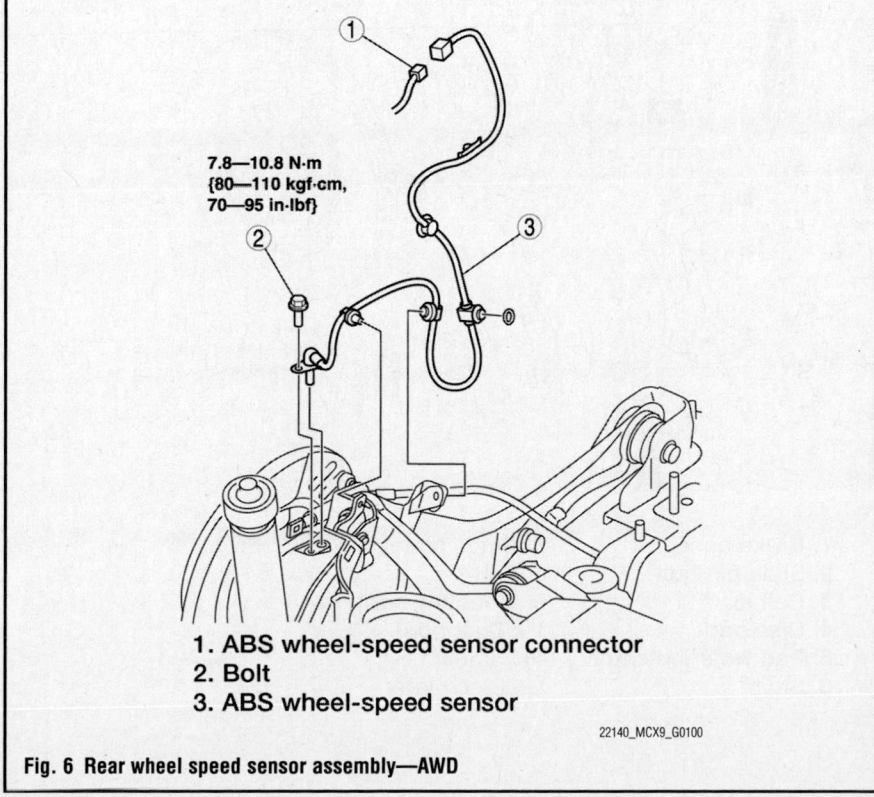

1. ABS wheel-speed sensor connector
2. Bolt
3. ABS wheel-speed sensor

7.8—10.8 N·m
{80—110 kgf·cm,
70—95 in·lbf}

Fig. 5 Rear wheel speed sensor assembly—2WD

22140_MCX9_G0093

7.8—10.8 N·m
{80—110 kgf·cm,
70—95 in·lbf}

1. ABS wheel-speed sensor connector
2. Bolt
3. ABS wheel-speed sensor

22140_MCX9_G0100

Fig. 6 Rear wheel speed sensor assembly—AWD

7. Remove the trunk side trim.

8. Disconnect the rear ABS wheel-speed sensor connector.

9. Remove the attaching bolt.

10. Remove the rear ABS wheel-speed sensor.

To install:

11. Install the rear ABS wheel-speed sensor.

12. Install the attaching bolt and tighten to 70–95 inch lbs. (8–11 Nm).

13. Connect the rear ABS wheel-speed sensor connector.

14. Install the trunk side trim.

15. Install the third-row seat belt lower anchor bolt. Tighten to 28–58 ft. lbs. (38–78 Nm).

16. Install the rear scuff plate inner.

17. Install the third-row seat.

18. Install the seat side box.

19. Install the trunk box.

BRAKE CALIPER

REMOVAL & INSTALLATION

See Figure 7.

1. Before servicing the vehicle, refer to the Precautions Section.
2. Raise and safely support the vehicle.
3. Remove the wheel and tire assembly.
4. Drain the brake fluid.
5. Remove the brake hose connection, if you are replacing the caliper.
6. Remove the caliper slide pin bolts.
7. Remove the caliper.

To install:

8. Position the caliper over the rotor so the caliper engages the adapter correctly.
9. Install the caliper. Tighten the front caliper slide pin bolts to 62–69 ft. lbs. (83–93 Nm).
10. Install the brake hose, if removed.

Tighten banjo bolt to 16–22 ft. lbs. (22–30 Nm).

11. Install the wheel and tire assembly.
12. Fill and bleed the brake system.
13. Lower the vehicle.

DISC BRAKE PADS

REMOVAL & INSTALLATION

See Figure 7.

1. Before servicing the vehicle, refer to the Precautions Section.
2. Remove ½ of the brake fluid from the master cylinder.
3. Raise and safely support the vehicle.
4. Remove the wheel and tire assembly.
5. Remove the lower caliper slide pin bolt.
6. Remove the caliper from the caliper support but do NOT disconnect the fluid line. Suspend the caliper from the suspension with a piece of wire.

7. Remove the disc pad, pad wear indicator, shim and guide plate.

To install:

8. Clean the exposed portion of the caliper piston, then press the piston back into the caliper bore using the old inner brake pad and a C-clamp.
9. Remove the disc pad, pad wear indicator, shim and guide plate making sure the shims and clips are properly positioned.
10. Install the caliper over the rotor so the caliper engages the adapter correctly.
11. Install the caliper slide pin bolt and tighten to 62–69 ft. lbs. (83–93 Nm).
12. Install the wheel and tire assembly and lower the vehicle.
13. Apply the brake pedal several times until a firm pedal is obtained.
14. Check the fluid level in the master cylinder and add fluid, as necessary.

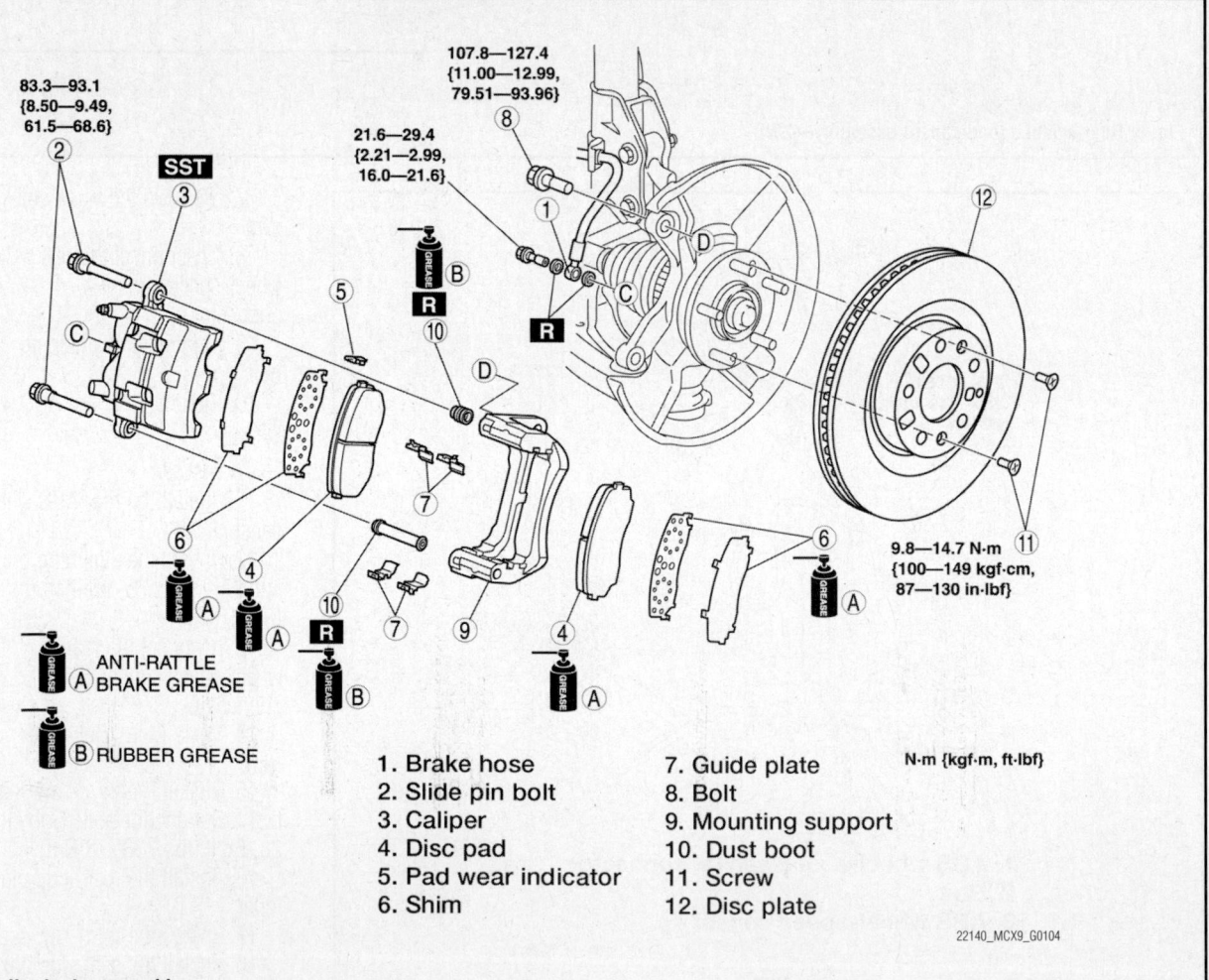

83.3—93.1
{8.50—9.49,
61.5—68.6}

107.8—127.4
{11.00—12.99,
79.51—93.96}

21.6—29.4
{2.21—2.99,
16.0—21.6}

9.8—14.7 N·m
{100—149 kgf·cm,
87—130 in·lbf}

A ANTI-RATTLE
BRAKE GREASE

B RUBBER GREASE

1. Brake hose
2. Slide pin bolt
3. Caliper
4. Disc pad
5. Pad wear indicator
6. Shim
7. Guide plate
8. Bolt
9. Mounting support
10. Dust boot
11. Screw
12. Disc plate

N·m {kgf·m, ft·lbf}

22140_MCX9_G0104

Fig. 7 Front disc brake assembly

BRAKE CALIPER

REMOVAL & INSTALLATION

See Figure 8.

1. Before servicing the vehicle, refer to the Precautions Section.
2. Raise and safely support the vehicle.
3. Remove the wheel and tire assembly.
4. Drain the brake fluid.
5. Remove the brake hose connection, if you are replacing the caliper.
6. Remove the caliper slide pin bolts.
7. Remove the caliper.

To install:

8. Position the caliper over the rotor so the caliper engages the adapter correctly.
9. Install the caliper. Tighten the front caliper slide pin bolts to 16–23 ft. lbs. (22–31 Nm).
10. Install the brake hose, if removed. Tighten banjo bolt to 16–22 ft. lbs. (22–30 Nm).
11. Install the wheel and tire assembly.
12. Fill and bleed the brake system.
13. Lower the vehicle.

DISC BRAKE PADS

REMOVAL & INSTALLATION

See Figure 8.

1. Before servicing the vehicle, refer to the Precautions Section.
2. Remove ½ of the brake fluid from the master cylinder.
3. Raise and safely support the vehicle.
4. Remove the wheel and tire assembly.
5. Remove the lower caliper slide pin bolt.
6. Remove the caliper from the caliper support but do NOT disconnect the fluid

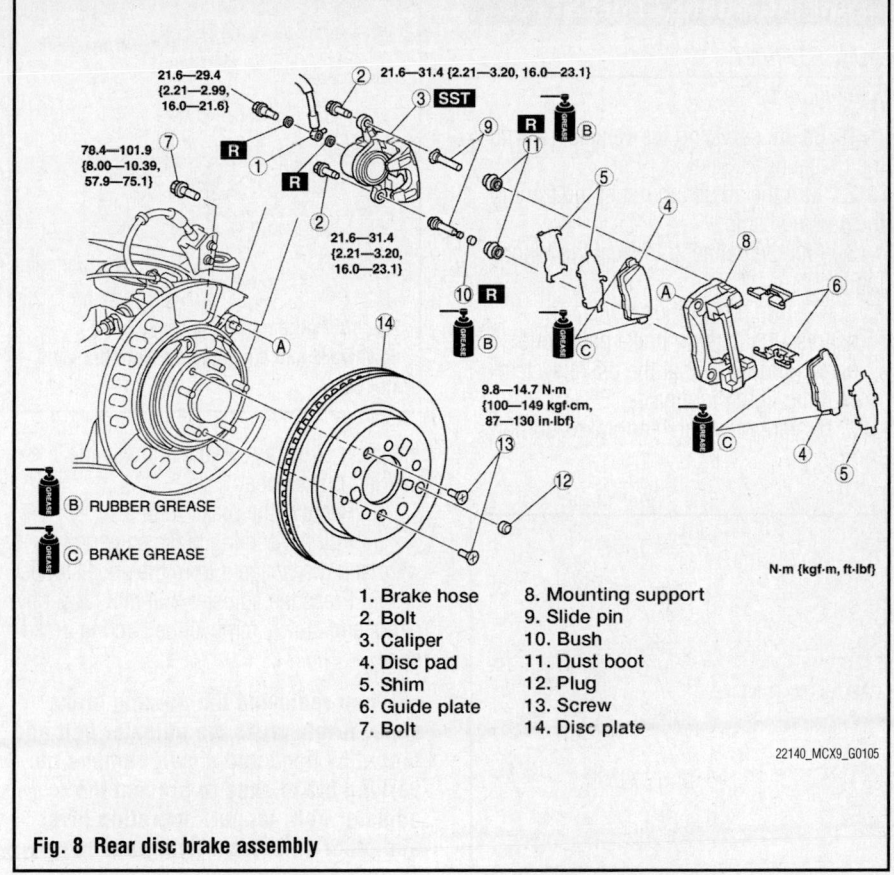

1. Brake hose
2. Bolt
3. Caliper
4. Disc pad
5. Shim
6. Guide plate
7. Bolt
8. Mounting support
9. Slide pin
10. Bush
11. Dust boot
12. Plug
13. Screw
14. Disc plate

B. RUBBER GREASE
C. BRAKE GREASE

N·m {kgf·m, ft-lbf}

22140_MCX9_G0105

Fig. 8 Rear disc brake assembly

line. Suspend the caliper from the suspension with a piece of wire.

7. Remove the disc pad, shim and guide plate.

To install:

8. Clean the exposed portion of the caliper piston, then press the piston back into the caliper bore using the old inner brake pad and a C-clamp.
9. Remove the disc pad, shim and guide plate making sure the shims and clips are properly positioned.

10. Install the caliper over the rotor so the caliper engages the adapter correctly.
11. Install the caliper slide pin bolt and tighten to 16–23 ft. lbs. (22–31 Nm).
12. Install the wheel and tire assembly and lower the vehicle.
13. Apply the brake pedal several times until a firm pedal is obtained.
14. Check the fluid level in the master cylinder and add fluid, as necessary.

PARKING BRAKE CABLES

ADJUSTMENT

See Figure 9.

1. Before servicing the vehicle, refer to the Precautions Section.

2. Turn the adjusting nut at the front of the parking cable.

3. After adjustment, inspect the following points:

   a. Turn the ignition switch on, depress the parking brake pedal one notch, and verify that the brake system warning light illuminates.

   b. Verify that the rear brakes do not drag.

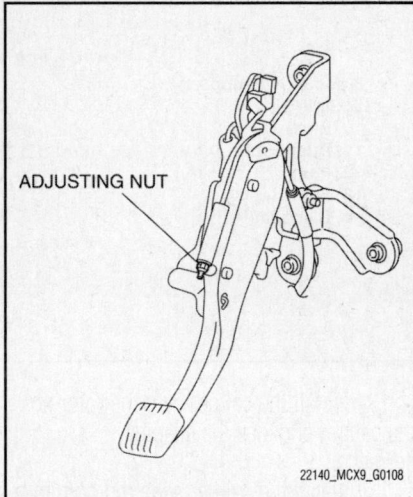

ADJUSTING NUT

22140_MCX9_G0108

Fig. 9 The adjusting nut is located at the front of the parking brake cable

PARKING BRAKE SHOES

REMOVAL & INSTALLATION

See Figures 10 through 12.

1. Before servicing the vehicle, refer to the Precautions Section.

2. Raise and safely support the vehicle securely on jackstands.

3. Remove the wheels.

4. Insert a flathead screwdriver into the service hole and turn the adjuster in the

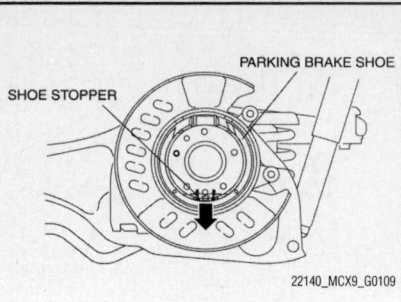

SHOE STOPPER   PARKING BRAKE SHOE

22140_MCX9_G0109

Fig. 10 Pull the parking brake shoe downward and disengage it from the shoe stopper

direction of the arrow to compress the parking brake shoe.

5. Remove the brake rotor.

6. Pull the parking brake shoe downward and disengage it from the shoe stopper.

7. Press the adjuster bolt and tappet by hand, and slowly remove the parking brake shoe.

➥**When removing the parking brake shoe, firmly press the adjuster bolt and tappet by hand and slowly remove the parking brake shoe to prevent the adjuster bolt, tappet, operation lever and other parts from flying off.**

To install:

8. Install the operation lever, pin, adjuster bolt and nut, and tappet so that the

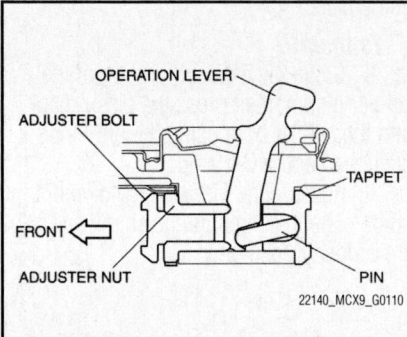

OPERATION LEVER
ADJUSTER BOLT
TAPPET
FRONT
ADJUSTER NUT   PIN

22140_MCX9_G0110

Fig. 11 Install the operation lever, pin, adjuster bolt and nut, and tappet so that the adjuster nut is facing toward the vehicle front

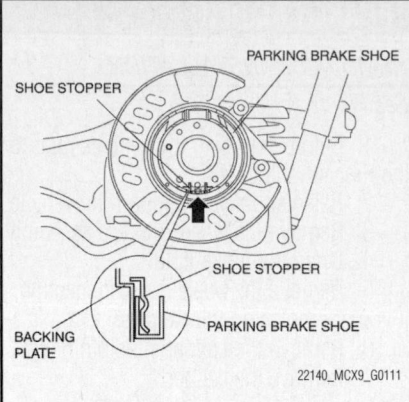

SHOE STOPPER   PARKING BRAKE SHOE
SHOE STOPPER
BACKING   PARKING BRAKE SHOE
PLATE

22140_MCX9_G0111

Fig. 12 After installing the opening of the parking brake shoe to the adjuster bolt and tappet, push the brake shoe upward and attach it to the shoe stopper

adjuster nut is facing toward the vehicle front.

9. Completely tighten the adjuster bolt and nut.

10. Move the operation lever by hand and verify that it operates properly.

➥**If proper operation cannot be verified, reinstall.**

11. Install the brake shoes.

   a. Measure the parking brake lining thickness with a Vernier caliper. If it is less than 0.04 in. (1.0 mm), install a new parking brake shoe.

   b. Apply grease to the contact surface of the parking brake shoe and shoe stopper.

   c. After installing the opening of the parking brake shoe to the adjuster bolt and tappet, push the brake shoe upward and attach it to the shoe stopper.

12. Install the brake rotor.

   a. Measure the inner diameter of the brake rotor with a Vernier caliper.

   b. If it exceeds 7.52 in (191.0 mm) in diameter, install a new brake rotor.

   c. Install the brake rotor and screw. Tighten the screws to 87–130 inch lbs. (10–15

13. Adjust the parking brakes.

14. Install the wheels and lower the vehicle.

CHASSIS ELECTRICAL   **AIR BAG (SUPPLEMENTAL RESTRAINT SYSTEM)**

GENERAL INFORMATION

✳✳ CAUTION

These vehicles are equipped with an air bag system. The system must be disarmed before performing service on, or around, system components, the steering column, instrument panel components, wiring and sensors. Failure to follow the safety precautions and the disarming procedure could result in accidental air bag deployment, possible injury and unnecessary system repairs.

SERVICE PRECAUTIONS

✳✳ CAUTION

Disconnect and isolate the battery negative cable before beginning any airbag system component diagnosis, testing, removal, or installation procedures. Wait at least 90 seconds after the ignition switch is turned off and the negative (-) terminal cable is disconnected from the battery before starting the operation. The SRS is equipped with a backup power source,

so if work is started within 90 seconds after disconnecting the negative (-) terminal cable from the battery, the SRS may be deployed. Failure to disable the airbag system may result in accidental airbag deployment, personal injury, or death.

DISARMING THE SYSTEM

To avoid personal injury when working on vehicles equipped with an air bag, the negative battery cable must be disconnected and at least 1 minute must elapse before working on the system. Failure to do so may result in deployment of the air bag. You should also wrap or isolate the negative battery cable with electrical or other non-conductive tape.

ARMING THE SYSTEM

To arm the system after service is completed, connect the negative battery cable.

CLOCKSPRING CENTERING

See Figure 13.

1. Before servicing the vehicle, refer to the Precautions Section.

2. Set the front tires straight-ahead.

✳✳ WARNING

The clock spring will break if overwound. Do not forcibly turn the clock spring.

3. Turn the clock spring clockwise until it stops.
4. From the stopped position, turn the clock spring counterclockwise 2 3/4 turns.
5. Align the marks.

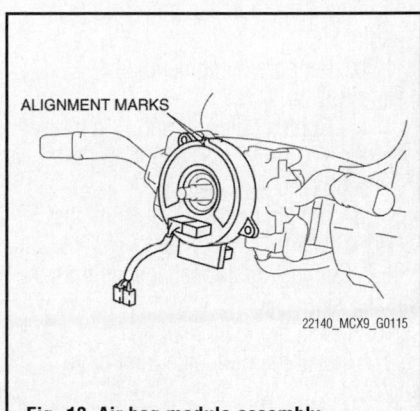

Fig. 13 Air bag module assembly

DRIVE TRAIN

TRANSFER CASE ASSEMBLY

REMOVAL & INSTALLATION

See Figures 14 through 16.

1. Before servicing the vehicle, refer to the Precautions Section.
2. Disconnect the negative battery cable.
3. Remove the middle pipe.
4. Remove the front pipe.
5. Disconnect the stabilizer control link.

6. Disconnect the tie-rod end ball joint.
7. Disconnect the front lower arm ball joint.
8. Remove the right side front drive shaft.
9. Remove the joint shaft.
10. Disconnect the propeller shaft from the transfer side.
11. Remove the transfer case bracket.
12. Remove the transfer case.

  a. Disconnect the front stabilizer control link, and rotate the stabilizer to the front to secure a space from the side.
  b. Remove the No.1 engine mount bolt and nut.
  c. Remove the transfer installation bolts.
  d. Push the engine forward so that the front side of the transfer faces downward, then remove the transfer.

To install:
13. Install the transfer case.
  a. Push the engine forward so that the front side of the transfer case faces downward, then install the transfer case.
  b. Install the transfer case installation bolts. 66–74 ft lbs. (90–100 Nm). Tightening torque
  c. Install the No.1 engine mount bolt and nut. 63–86 ft. lbs. (85–117 Nm).
  d. Rotate the front stabilizer to the rear to set it in its original position.
14. Install the transfer case bracket.

Fig. 14 Disconnect the front stabilizer control link, and rotate the stabilizer to the front to secure a space from the side

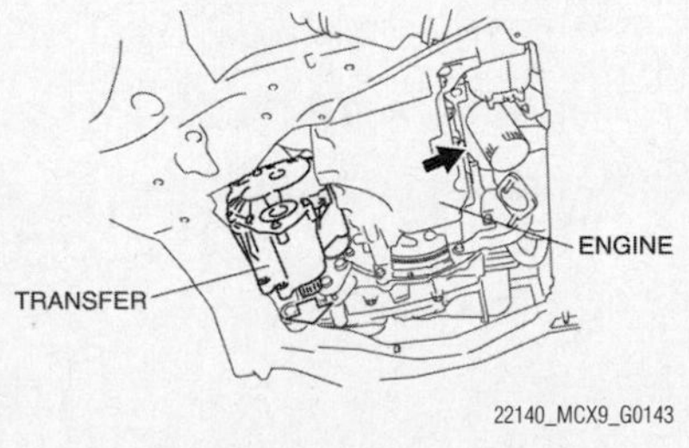

Fig. 15 Push the engine forward so that the front side of the transfer faces downward, then remove the transfer case

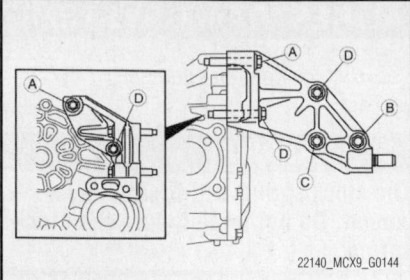

Fig. 16 Tighten bolt A, B and C in the order of C–A–B to 31–45 ft. lbs. (42–62 Nm). Then tighten bolt D to 31–45 ft. lbs. (42–62 Nm)

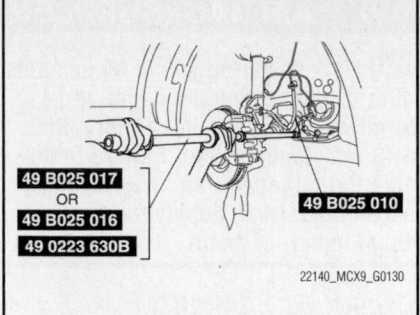

Fig. 17 Separate the right side driveshaft from the joint shaft by using the special tools

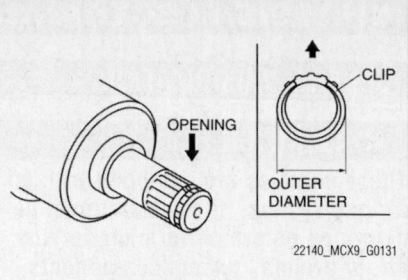

Fig. 19 Install a new clip onto the drive-shaft with the opening facing upward. Ensure that the diameter of the clip does not exceed 1.34 in. (34 mm)

a. Install the transfer bracket to the transfer case, then temporarily tighten bolt C.

b. Temporarily tighten bolt A and B.

c. Tighten bolt A, B and C in the order of C–A–B to 31–45 ft. lbs. (42–62 Nm).

d. Tighten bolt D to 31–45 ft. lbs. (42–62 Nm).

15. Connect the propeller shaft from the transfer side.

16. Install the joint shaft.

17. Install the right side front drive shaft.

18. Connect the front lower arm ball joint.

19. Connect the tie-rod end ball joint.

20. Connect the stabilizer control link.

21. Install the front pipe.

22. Install the middle pipe.

23. Connect the negative battery cable.

24. Lower the vehicle.

25. Warm up the engine and transaxle, inspect for oil leakage, and inspect the transfer operation.

FRONT HALFSHAFTS

REMOVAL & INSTALLATION

See Figures 17 through 19.

❄ WARNING

Performing the following procedures without first removing the ABS wheel speed sensor may possibly cause an open circuit in the harness if it is pulled by mistake. Before performing the following procedures, remove the ABS wheel speed sensor (axle side) and fix it to an appropriate place where the sensor will not be pulled by mistake while the vehicle is being serviced.

1. Before servicing the vehicle, refer to the Precautions Section.

2. Raise and safely support the vehicle securely on jackstands.

3. Remove the side cover.

4. Remove the front ABS wheel-speed sensor.

5. Remove the locknut.

a. Lock the hub by applying the brakes.

b. Knock the crimped portion of the locknut outward using a small chisel and a hammer.

c. Remove the locknut.

6. Disconnect the tie-rod end ball joint.

7. Disconnect the front stabilizer control link (lower side).

8. Disconnect the front lower arm ball joint.

9. Remove the front halfshaft.

a. Install a spare nut onto the drive-shaft so that the nut is flush with the end of the driveshaft.

b. Tap the nut with a copper hammer to loosen the driveshaft from the front wheel hub.

c. Separate the driveshaft from the wheel hub.

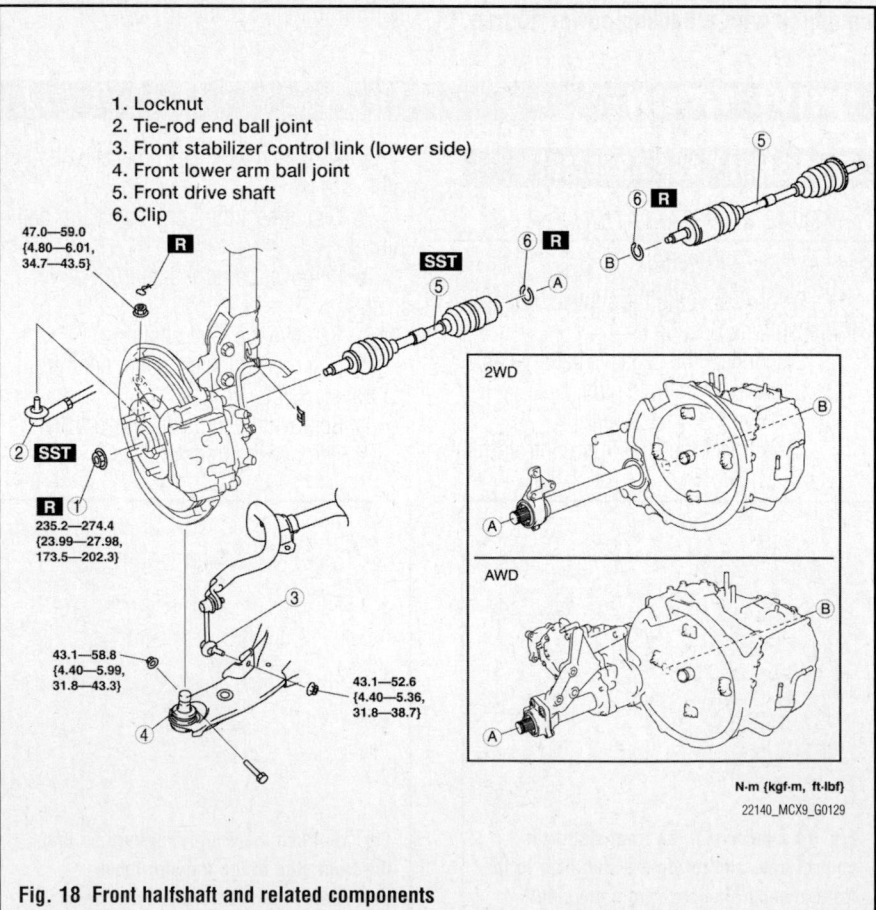

1. Locknut
2. Tie-rod end ball joint
3. Front stabilizer control link (lower side)
4. Front lower arm ball joint
5. Front drive shaft
6. Clip

47.0—59.0
{4.80—6.01,
34.7—43.5}

235.2—274.4
{23.99—27.98,
173.5—202.3}

43.1—58.8
{4.40—5.99,
31.8—43.3}

43.1—52.6
{4.40—5.36,
31.8—38.7}

N·m {kgf·m, ft-lbf}

Fig. 18 Front halfshaft and related components

✳✳ WARNING

The sharp edges of the driveshaft can slice or puncture the oil seal. Be careful when removing the driveshaft from the transaxle.

  d. Separate the left side driveshaft from the transaxle by prying with a bar inserted between the outer ring and the transaxle, as shown.

  e. Separate the right side driveshaft from the joint shaft by using the special tools.

10. Remove the clip.

To install:

11. Install the clip.

  a. Install a new clip onto the driveshaft with the opening facing upward. Ensure that the diameter of the clip does not exceed the specification on installation.

  b. After installation, measure the outer diameter. If it exceeds 1.34 in. (34 mm), repeat steps using a new clip.

12. Install the left side halfshaft as follows:

  a. Insert the driveshaft into the wheel hub.

  b. Apply transaxle oil to the oil seal lip.

  c. Push the driveshaft into the transaxle.

  d. After installation, pull the transaxle side outer ring forward to confirm that the driveshaft is securely held by the clip.

13. Install the right side halfshaft as follows:

  a. Insert the driveshaft into the wheel hub.

  b. Insert the driveshaft into the joint shaft.

  c. After installation, pull the transaxle side outer ring forward to confirm that the driveshaft is securely held by the clip.

14. Connect the front lower arm ball joint.

15. Connect the front stabilizer control link (lower side).

16. Connect the tie-rod end ball joint.

17. Install a new locknut and stake it into place.

18. Install the front ABS wheel-speed sensor.

19. Install the side cover.

20. Lower the vehicle.

CV-BOOT INSPECTION

See Figure 20.

1. Raise and safely support the vehicle securely on jackstands.

2. Turn the steering wheel to full right lock.

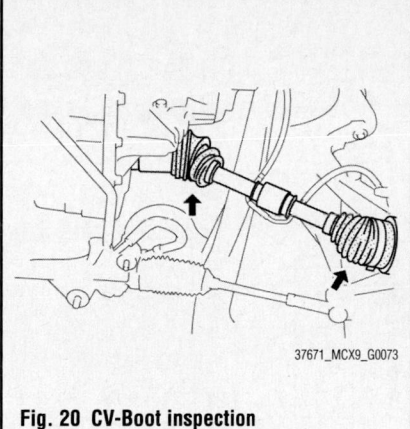

Fig. 20 CV-Boot inspection

3. Inspecting from behind the halfshaft, examine the driver's side CV boot while slowly turning the wheel.

4. Inspect the valleys between the folds. If cracks are apparent, replace the boot.

5. Also inspect the circumference where the boot covers the edge of the hub, as cracks can appear there too. If large cracks (large enough to fit a finger nail in) are apparent, replace the boot.

6. Repeat for the inner boot.

7. Turn the steering wheel to full left lock and examine the passenger side boots in the same manner.

DIFFERENTIAL ASSEMBLY

REMOVAL & INSTALLATION

See Figures 21 through 27.

1. Before servicing the vehicle, refer to the Precautions Section.

2. Raise and safely support the vehicle securely on jackstands.

3. Drain the rear differential oil into a container.

4. Remove the under guard (LH).

5. Remove the pre-silencer and insulator.

6. Remove the propeller shaft.

7. Remove the locknut.

  a. Lock the hub by applying the brakes.

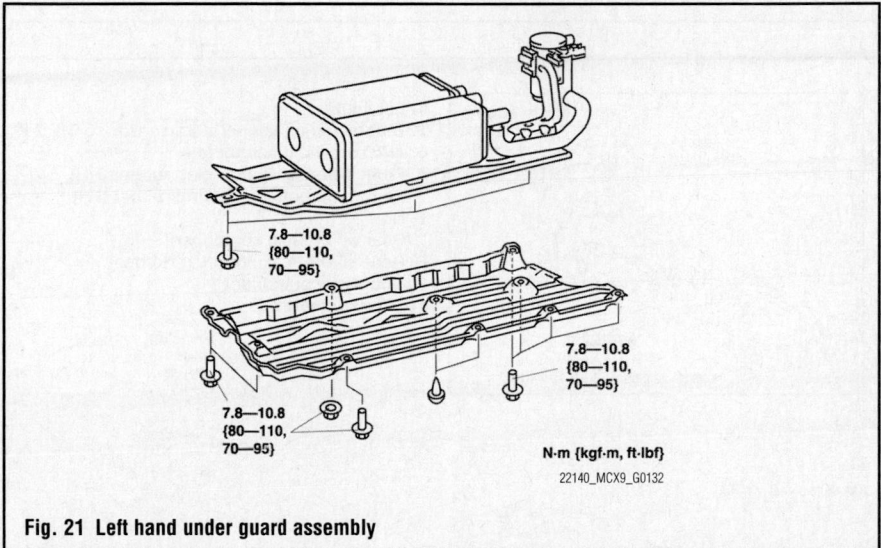

7.8—10.8 {80—110, 70—95}

7.8—10.8 {80—110, 70—95}

7.8—10.8 {80—110, 70—95}

N·m {kgf·m, ft·lbf}

22140_MCX9_G0132

Fig. 21 Left hand under guard assembly

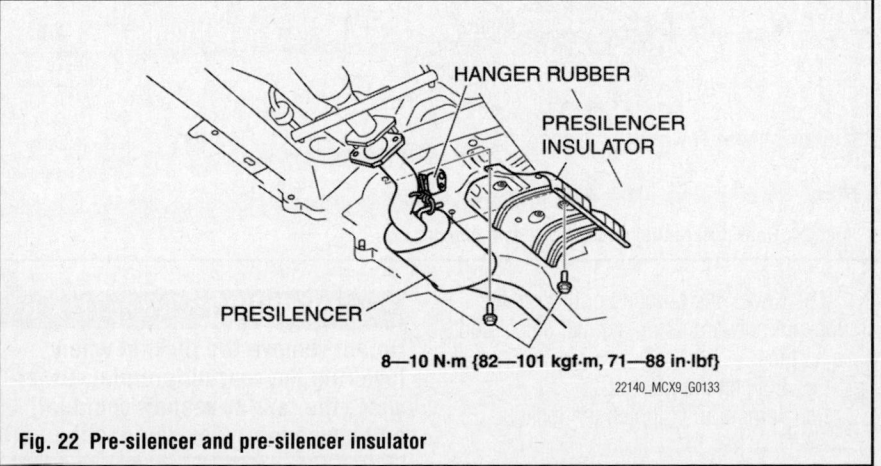

HANGER RUBBER

PRESILENCER INSULATOR

PRESILENCER

8—10 N·m {82—101 kgf·m, 71—88 in·lbf}

22140_MCX9_G0133

Fig. 22 Pre-silencer and pre-silencer insulator

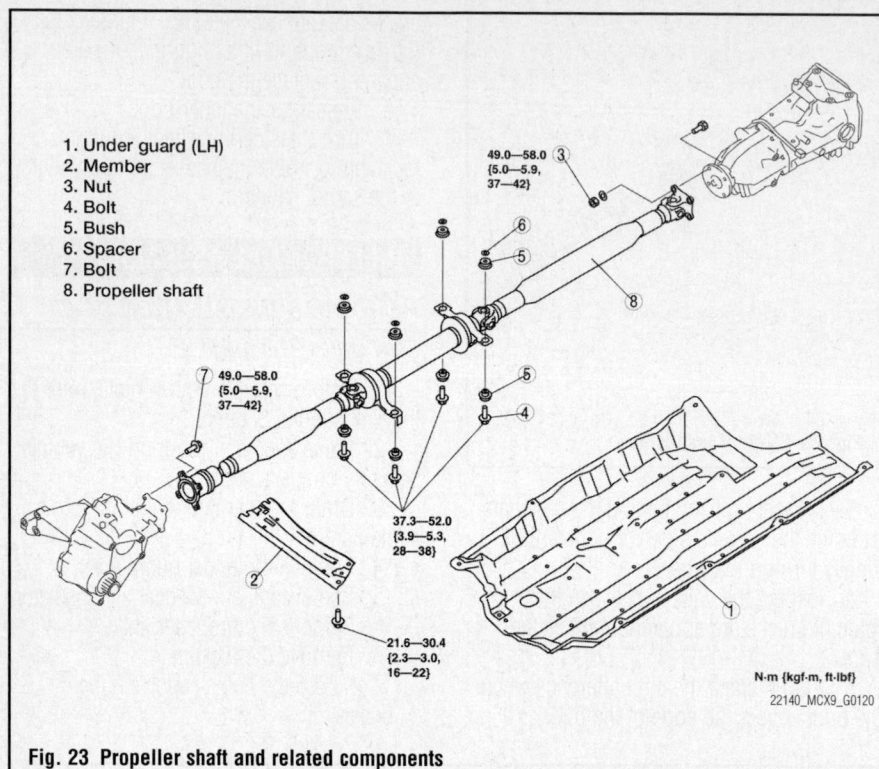

1. Under guard (LH)
2. Member
3. Nut
4. Bolt
5. Bush
6. Spacer
7. Bolt
8. Propeller shaft

49.0—58.0 {5.0—5.9, 37—42}

37.3—52.0 {3.9—5.3, 28—38}

21.6—30.4 {2.3—3.0, 16—22}

N·m {kgf·m, ft·lbf}

22140_MCX9_G0120

Fig. 23 Propeller shaft and related components

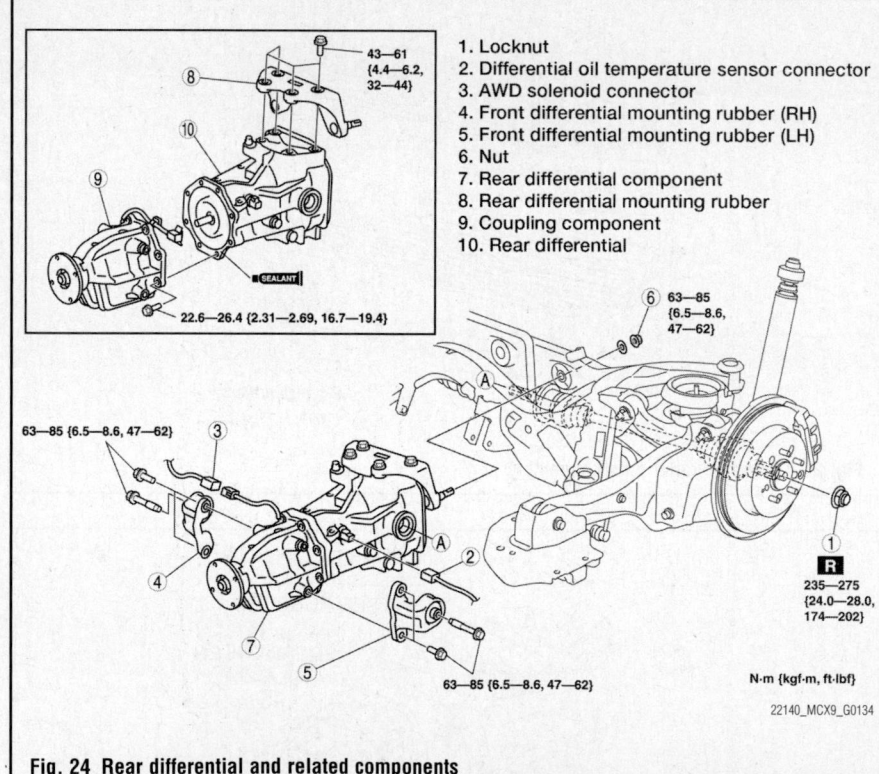

1. Locknut
2. Differential oil temperature sensor connector
3. AWD solenoid connector
4. Front differential mounting rubber (RH)
5. Front differential mounting rubber (LH)
6. Nut
7. Rear differential component
8. Rear differential mounting rubber
9. Coupling component
10. Rear differential

43—61 {4.4—6.2, 32—44}

22.6—26.4 {2.31—2.69, 16.7—19.4}

63—85 {6.5—8.6, 47—62}

63—85 {6.5—8.6, 47—62}

63—85 {6.5—8.6, 47—62}

R
235—275 {24.0—28.0, 174—202}

N·m {kgf·m, ft·lbf}

22140_MCX9_G0134

Fig. 24 Rear differential and related components

b. Knock the crimped portion of the locknut outward using a small chisel and a hammer.

c. Loosen the locknut.

d. Temporarily tighten the locknut.

❋❋ WARNING

Do not remove the locknut when lowering the rear differential. Otherwise, the rear driveshaft could fall and cause injury, or damage the part.

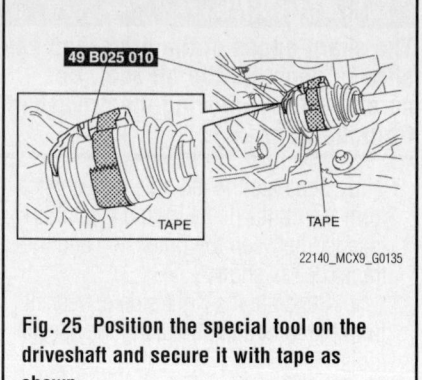

49 B025 010

TAPE

TAPE

22140_MCX9_G0135

Fig. 25 Position the special tool on the driveshaft and secure it with tape as shown

8. Disconnect the differential oil temperature sensor connector.

9. Disconnect the AWD solenoid connector.

10. Remove the right hand front differential mounting rubber.

11. Remove the left hand front differential mounting rubber.

12. Remove the nut.

13. Remove the rear differential.

a. Detach clips as shown.

b. Position the special tool on the driveshaft and secure the special tool with tape as shown.

❋❋ WARNING

Always secure the special tool using tape because the special tool may fall off during the operation and cause injury.

c. Disconnect the rear driveshaft from the rear differential side using the special tool and two bars.

d. Support the rear differential using a jack.

❋❋ WARNING

When lowering the rear differential, be careful not to damage the rear

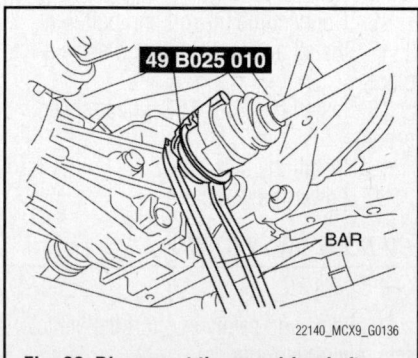

49 B025 010

BAR

22140_MCX9_G0136

Fig. 26 Disconnect the rear driveshaft from the rear differential side using the special tool and two pry bars

Fig. 27 Support the rear differential using a jack

differential oil seal, rear driveshaft, and surrounding parts.

e. Remove the rear driveshaft from the rear differential while lowering the rear differential gradually.

f. Suspend the driveshaft using a cable.

14. Remove the rear differential mounting rubber

To install:

15. Install the rear differential mounting rubber.

16. Install the rear differential.

❋❋ WARNING

When raising the rear differential, be careful not to damage the rear differential oil seal, rear driveshaft, and surrounding parts.

a. Install the rear driveshaft to the rear differential while raising the rear differential gradually.

b. Jack up the rear driveshaft until it is horizontal and engage the rear driveshaft (differential side) and rear differential by tapping the wheel side.

17. Install the nut.

18. Install the left hand front differential mounting rubber.

19. Install the right hand front differential mounting rubber.

20. Disconnect the AWD solenoid connector.

21. Disconnect the differential oil temperature sensor connector.

fall and cause injury, or damage the part.

❋❋ WARNING

Do not Install the locknut when lowering the rear differential. Otherwise, the rear driveshaft could

22. Install the locknut.

a. Lock the hub by applying the brakes.

b. Tighten the locknut.

c. Knock the crimped portion of the locknut inward using a small chisel and a hammer.

23. Install the propeller shaft.

24. Install the pre-silencer and insulator.

25. Install the under guard (LH).

26. Add the specified rear differential oil.

27. Inspect the rear wheel alignment and adjust it if necessary.

28. Lower the vehicle.

REAR HALFSHAFTS

REMOVAL & INSTALLATION
See Figures 28 through 30.

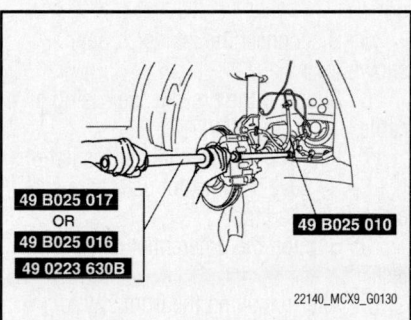

Fig. 28 Separate the right side driveshaft from the joint shaft by using the special tools

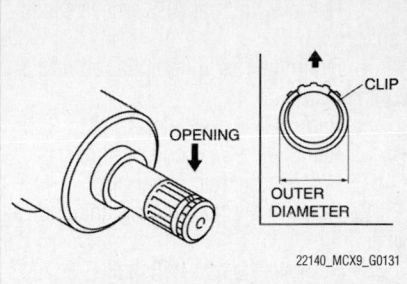

Fig. 30 Install a new clip onto the driveshaft with the opening facing upward. Ensure that the diameter of the clip does not exceed 1.16 in. (29.5 mm)

❋❋ WARNING

Performing the following procedures without first removing the ABS wheel speed sensor may possibly cause an open circuit in the harness if it is pulled by mistake. Before performing the following procedures, remove the ABS wheel speed sensor (wheel side) and fix it to an appropriate place where the sensor will not be pulled by mistake while the vehicle is being serviced.

1. Before servicing the vehicle, refer to the Precautions Section.

2. Raise and safely support the vehicle securely on jackstands.

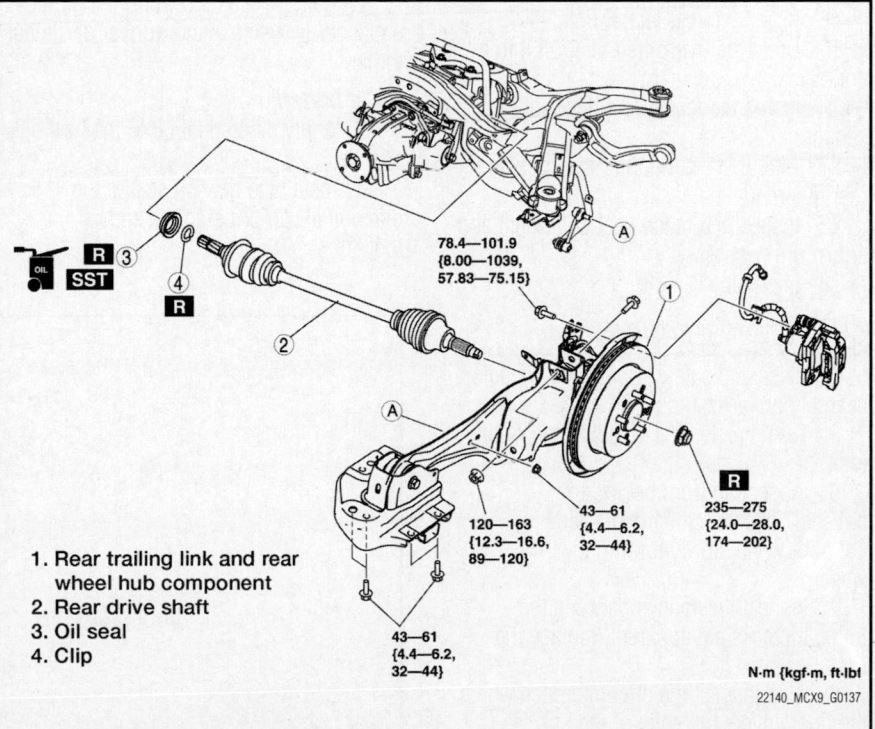

1. Rear trailing link and rear wheel hub component
2. Rear drive shaft
3. Oil seal
4. Clip

Fig. 29 Rear halfshaft and related components

3. Remove the rear ABS wheel-speed sensor.

4. Drain the rear differential oil into a container.

5. Remove the rear lateral link.

6. Remove the rear coil spring.

7. Remove the rear lower arm.

8. Remove the rear trailing link and rear wheel hub.

9. Remove the rear halfshaft.

a. Disengage the rear drive shaft using the special tools.

✼✼ WARNING

Be careful not to damage the rear differential oil seal.

10. Remove the oil seal.

11. Remove the clip.

To install:

12. Install the clip.

a. Install a new clip onto the joint shaft with the opening facing upward. Ensure that the diameter of the clip does not exceed 1.16 in. (29.5 mm) on installation.

b. After installation, measure the outer diameter. If it exceeds the specification, repeat steps using a new clip.

13. Install the oil seal.

14. Install the rear halfshaft.

15. Install the rear trailing link and rear wheel hub.

16. Install the rear lower arm.

17. Install the rear coil spring.

18. Install the rear lateral link.

19. Drain the rear differential oil into a container.

20. Install the rear ABS wheel-speed sensor.

21. After installation, add the specified rear differential oil.

22. Inspect the rear wheel alignment and adjust it if necessary.

23. Lower the vehicle.

CV-BOOTS INSPECTION

1. Raise and safely support the vehicle securely on jackstands.

2. Turn the steering wheel to full right lock.

3. Inspecting from behind the halfshaft, examine the driver's side CV boot while slowly turning the wheel.

4. Inspect the valleys between the folds. If cracks are apparent, replace the boot.

5. Also inspect the circumference where the boot covers the edge of the hub, as cracks can appear there too. If large cracks

(large enough to fit a finger nail in) are apparent, replace the boot.

6. Repeat for the inner boot.

7. Turn the steering wheel to full left lock and examine the passenger side boots in the same manner.

REAR PINION SEAL

REMOVAL & INSTALLATION

See Figures 31 through 34.

1. Before servicing the vehicle, refer to the Precautions Section.

2. Raise and safely support the vehicle securely on jackstands.

3. Drain the differential oil into a container.

4. Remove the pre-silencer installation nuts.

5. Disconnect the hanger rubber as shown.

6. Suspend the pre-silencer using a cable.

7. Remove the pre-silencer insulator.

8. Remove the locknut using the special tool and a pipe.

9. Support the differential using a jack.

10. After removing the front side of the differential mount, gradually lower the jack and tilt the front side of the differential downward.

11. Pull the companion flange off using the special tool.

12. Remove the oil seal from the differential carrier using a screwdriver or similar tool.

To install:

13. Apply differential oil to the new oil seal lip.

14. Install the new oil seal to the differential carrier using the special tool.

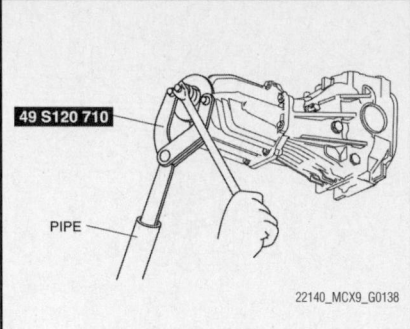

Fig. 32 Remove the locknut using the special tool and a pipe

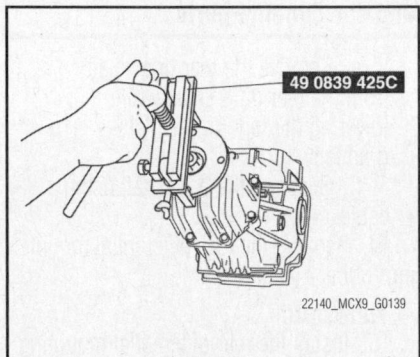

Fig. 33 Pull the companion flange off using the special tool

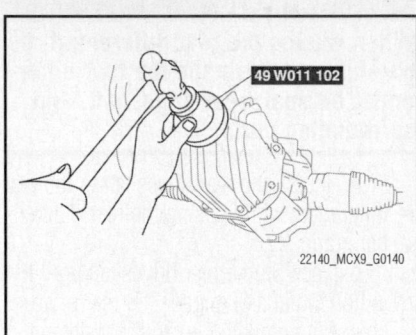

Fig. 34 Install the new oil seal to the differential carrier using the special tool

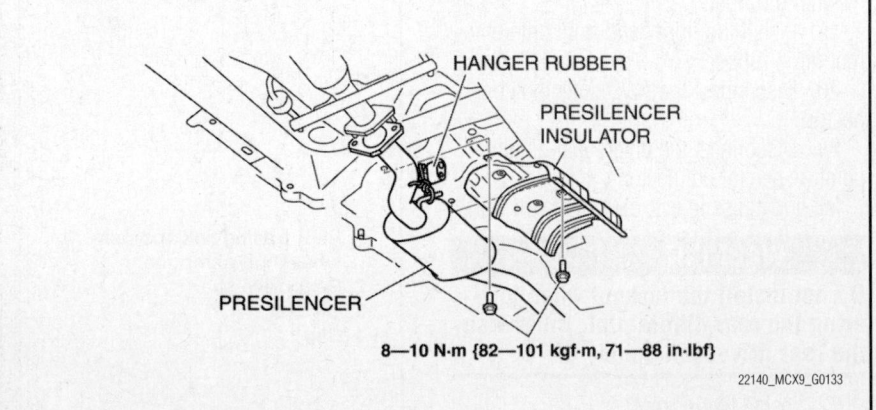

8—10 N·m {82—101 kgf·m, 71—88 in·lbf}

Fig. 31 Pre-silencer and pre-silencer insulator

15. Install the companion flange to the drive pinion.

16. Jack up the differential and install the front side of the differential mount. Tighten to 47–62 ft. lbs. (63–85 Nm).

17. Install the washer and a new locknut to 124–166 ft. lbs. (167–226 Nm).

18. Install the propeller shaft.

19. Add the specified differential oil through the filler plug.

20. Lower the vehicle.

PROPELLER SHAFT

REMOVAL & INSTALLATION

See Figures 31, 35 and 36.

1. Before servicing the vehicle, refer to the Precautions Section.

2. Raise and safely support the vehicle securely on jackstands.

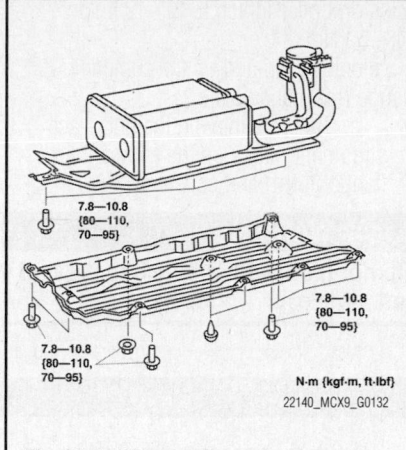

7.8—10.8 {80—110, 70—95}

7.8—10.8 {80—110, 70—95}

7.8—10.8 {80—110, 70—95}

N·m {kgf·m, ft·lbf}

22140_MCX9_G0132

Fig. 35 Left hand under guard assembly

3. Remove the pre-silencer installation nuts.

4. Disconnect the hanger rubber as shown.

5. Suspend the pre-silencer using a cable.

6. Remove the pre-silencer insulator.

7. Remove the under guard (LH).

8. Remove the crossmember.

9. Remove the propeller shaft-to-differential nut.

❈❈ WARNING

Do not mark with a punch to prevent imbalance.

a. Before removing the nut, place alignment marks on the companion flange (front) and constant velocity joint, and on the companion flange (rear) and yoke.

10. Remove the propeller shaft-to-transmission bolt.

11. Remove the bushing.

12. Remove the spacer.

13. Remove the propeller shaft.

1. Under guard (LH)
2. Member
3. Nut
4. Bolt
5. Bush
6. Spacer
7. Bolt
8. Propeller shaft

49.0—58.0 {5.0—5.9, 37—42}

49.0—58.0 {5.0—5.9, 37—42}

37.3—52.0 {3.9—5.3, 28—38}

21.6—30.4 {2.3—3.0, 16—22}

N·m {kgf·m, ft·lbf}

22140_MCX9_G0120

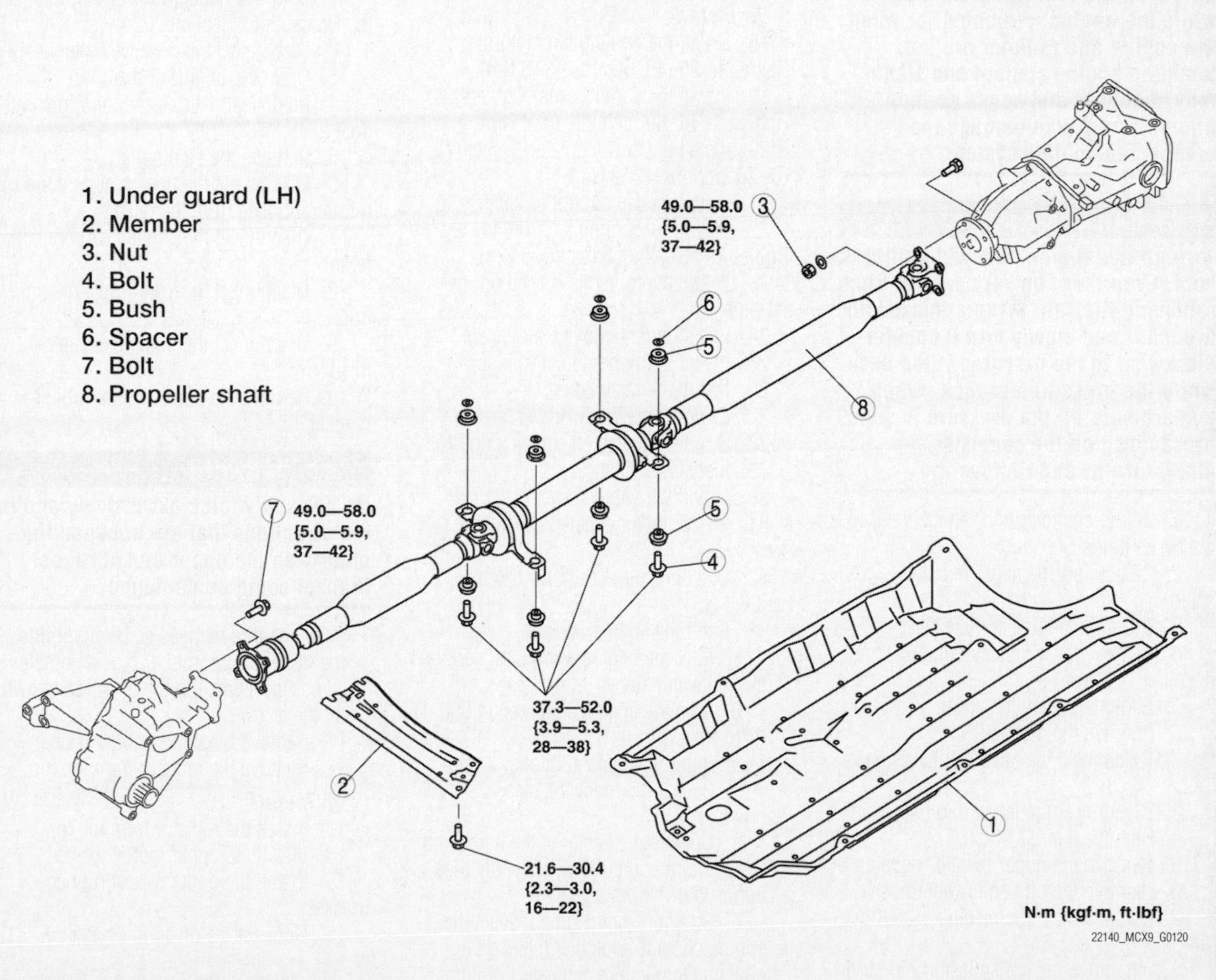

Fig. 36 Propeller shaft and related components

To install:

14. Install the propeller shaft.

a. Align the alignment marks and install the propeller shaft.

b. When installing a new propeller shaft, align the rear differential companion flange mark with the tag on the propeller shaft so that they are at the nearest position, and assemble.

15. Install the spacer.

16. Install the bushing.

17. Install the propeller shaft-to-trans-

mission bolt. Tighten to 37–42 ft. lbs. (49–59 Nm).

velocity joint, and on the companion flange (rear) and yoke.

a. Before removing the nut, place alignment marks on the companion flange (front) and constant

✸✸ WARNING

Do not mark with a punch to prevent imbalance.

18. Install the propeller shaft-to-differential nut. Tighten to 37–42 ft. lbs. (49–59 Nm).

19. Install the crossmember.

20. Install the under guard (LH).

21. Install the pre-silencer insulator.

22. Connect the hanger rubber.

23. Install the pre-silencer installation nuts.

24. Lower the vehicle.

25. Perform a road test and verify that there is no abnormal vibration or noise.

ENGINE COOLING

ENGINE FAN

REMOVAL & INSTALLATION

See Figure 37.

✸✸ CAUTION

Never remove the cooling system cap or loosen the radiator drain plug while the engine is running, or when the engine and radiator are hot. Scalding engine coolant and steam may shoot out and cause serious injury. It may also damage the engine and cooling system.

✸ CAUTION

Turn off the engine and wait until it is cool. Even then, be very careful when removing the cap. Wrap a thick cloth around it and slowly turn it counterclockwise to the first stop. Step back while the pressure escapes. When you are sure all the pressure is gone, press down on the cap using the cloth, turn it, and remove it.

1. Before servicing the vehicle, refer to the Precautions Section.

2. Disconnect the negative battery cable.

3. Drain the engine coolant.

4. Remove the air cleaner and fresh air duct.

5. Remove the engine cover.

6. Remove the dipstick.

7. Disconnect the fan electrical connector.

8. Disconnect the fan wiring harness.

9. Remove the relay box.

10. Remove the upper radiator hose.

11. Remove the oil cooler water hose.

12. Remove the transmission oil refrigerant pipe

a. Remove the transmission oil refrigerant pipe with the hoses still connected.

Position the transmission oil refrigerant pipe so that it is out of the way.

13. Remove the cooling fan from the top.

14. Remove the fan control module(s).

15. Remove the heat insulator.

16. Remove the cooling fans.

17. Remove the cooling fan motors.

To install:

18. Install the cooling fan motors. Tighten to 25–43 inch lbs. (3–5 Nm).

19. Install the cooling fans. Tighten to Tighten to 44–69 inch lbs. (5–8 Nm).

20. Install the heat insulator. Tighten to 16–28 inch lbs. (2–3 Nm).

21. Install the fan control module(s).

22. Install the cooling fan from the top. Tighten to 53–72 inch lbs. (6–8 Nm).

23. Install the transmission oil refrigerant pipe.

24. Install the oil cooler water hose.

25. Install the upper radiator hose.

26. Install the relay box.

27. Connect the fan wiring harness.

28. Connect the fan electrical connector.

29. Install the dipstick.

30. Install the engine cover.

31. Install the air cleaner and fresh air duct.

32. Disconnect the negative battery cable.

33. Refill the engine coolant.

a. After the engine warms up, perform the following steps.

b. Run the engine at approx. 4,000 RPM for 1 min.

c. Run the engine at idle for 1 min.

d. Repeat the following steps 2 times.

e. Operate the heater at the maximum temperature and airflow, and verify that hot air blows from vent.

f. Stop the engine, and inspect the engine coolant level after the engine coolant temperature decreases. If it is low, repeat steps.

34. Start the engine and inspect for engine coolant leakage.

RADIATOR

REMOVAL & INSTALLATION

See Figures 38 and 39.

1. Before servicing the vehicle, refer to the Precautions Section.

2. Disconnect the negative battery cable.

3. Drain the engine coolant.

4. Remove the air cleaner and fresh air duct.

5. Remove the cooling fan.

6. Disconnect the transmission fluid oil refrigerant hose from the radiator.

7. Disconnect the coolant reserve tank hose.

8. Disconnect the lower radiator hose.

9. Disconnect the oil cooler water hose.

10. Remove the upper mount rubber bracket.

✸✸ WARNING

Do not apply force excessively and in the directions that are not specified, otherwise the upper mount rubber bracket could be damaged.

a. Insert a flathead screwdriver into the upper mount rubber bracket notch.

b. Apply force in the direction shown by the arrow.

11. Remove the upper mount rubber

12. Remove the radiator from the top.

To install:

13. Install the radiator from the top.

14. Install the upper mount rubber.

15. Install the upper mount rubber bracket.

16. Connect the oil cooler water hose.

17. Connect the lower radiator hose.

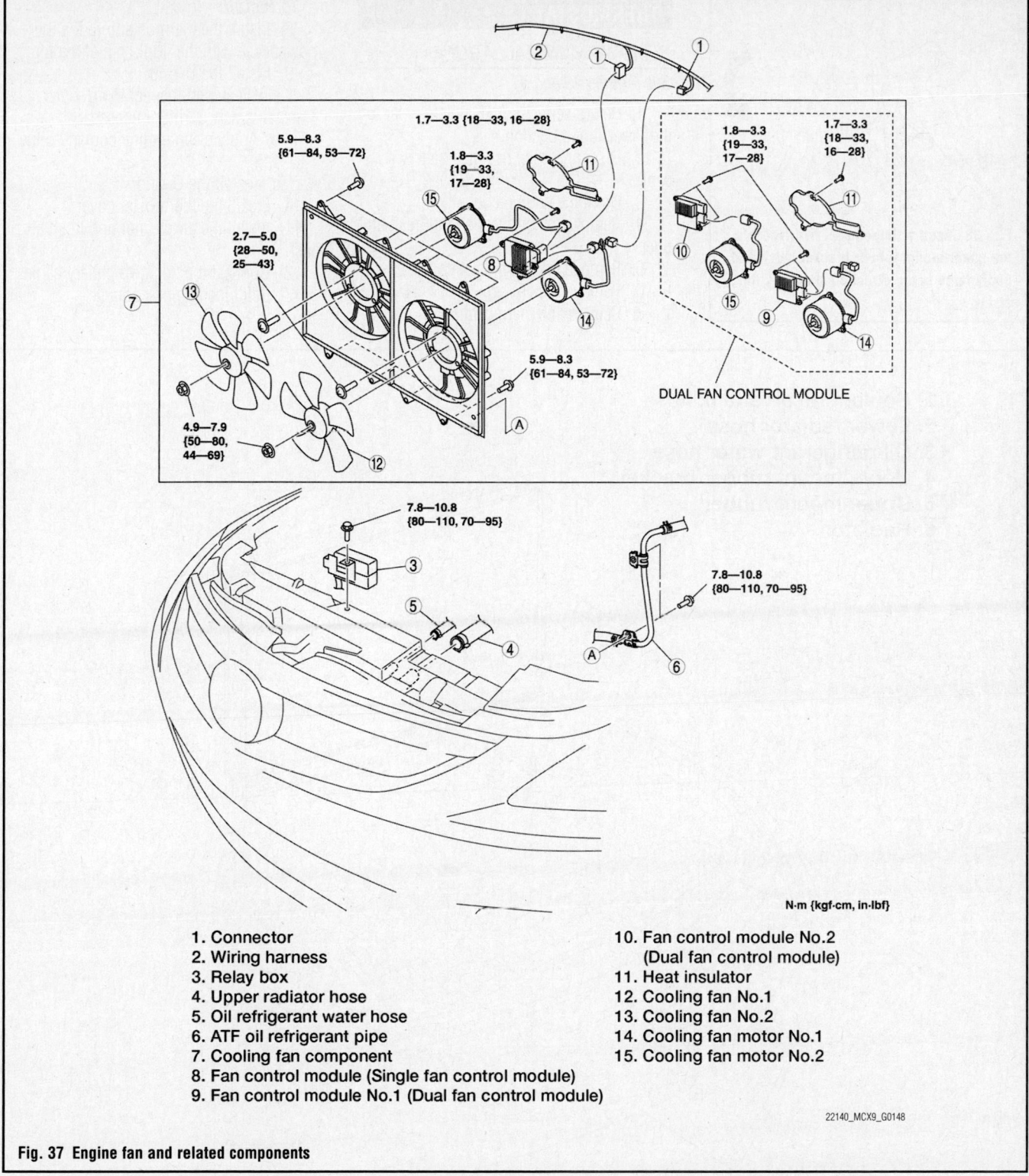

N·m {kgf·cm, in·lbf}

1. Connector
2. Wiring harness
3. Relay box
4. Upper radiator hose
5. Oil refrigerant water hose
6. ATF oil refrigerant pipe
7. Cooling fan component
8. Fan control module (Single fan control module)
9. Fan control module No.1 (Dual fan control module)
10. Fan control module No.2
    (Dual fan control module)
11. Heat insulator
12. Cooling fan No.1
13. Cooling fan No.2
14. Cooling fan motor No.1
15. Cooling fan motor No.2

22140_MCX9_G0148

Fig. 37 Engine fan and related components

18. Connect the coolant reserve tank hose.

19. Connect the transmission fluid oil refrigerant hose from the radiator.

20. Install the cooling fan.

21. Install the air cleaner and fresh air duct.

22. Connect the negative battery cable.

23. Refill the engine coolant.
   a. After the engine warms up, perform the following steps.
   b. Run the engine at approx. 4,000 RPM for 1 min.
   c. Run the engine at idle for 1 min.
   d. Repeat the following steps 2 times.
   e. Operate the heater at the maximum

temperature and airflow, and verify that hot air blows from vent.
   f. Stop the engine, and inspect the engine coolant level after the engine coolant temperature decreases. If it is low, repeat steps.

24. Inspect for engine coolant leakage.

25. Inspect the transmission fluid level.

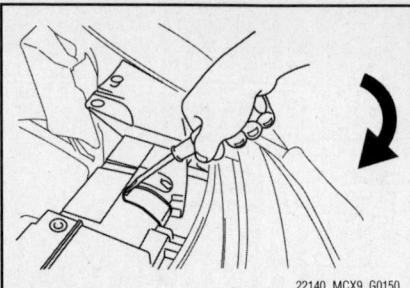

Fig. 38 Insert a flathead screwdriver into the upper mount rubber bracket notch and apply force in the direction shown by the arrow

22140_MCX9_G0150

THERMOSTAT

REMOVAL & INSTALLATION

See Figures 40 and 41.

1. Before servicing the vehicle, refer to the Precautions Section.
2. Disconnect the negative battery cable.
3. Drain the engine coolant.
4. Remove the air cleaner and fresh air duct.
5. Remove the thermostat cover.
6. Remove the O-ring.
7. Remove the thermostat.

To install:

8. Install the thermostat into the thermostat case with the jiggle pin at the top.
9. Install the O-ring.
   a. Clean and inspect the O-ring. Install a new O-ring if necessary.
   b. Apply clean engine coolant to the O-ring.
   c. Install the O-ring.
10. Install the thermostat cover.
11. Install the air cleaner and fresh air duct.
12. Connect the negative battery cable.
13. Refill the engine coolant.

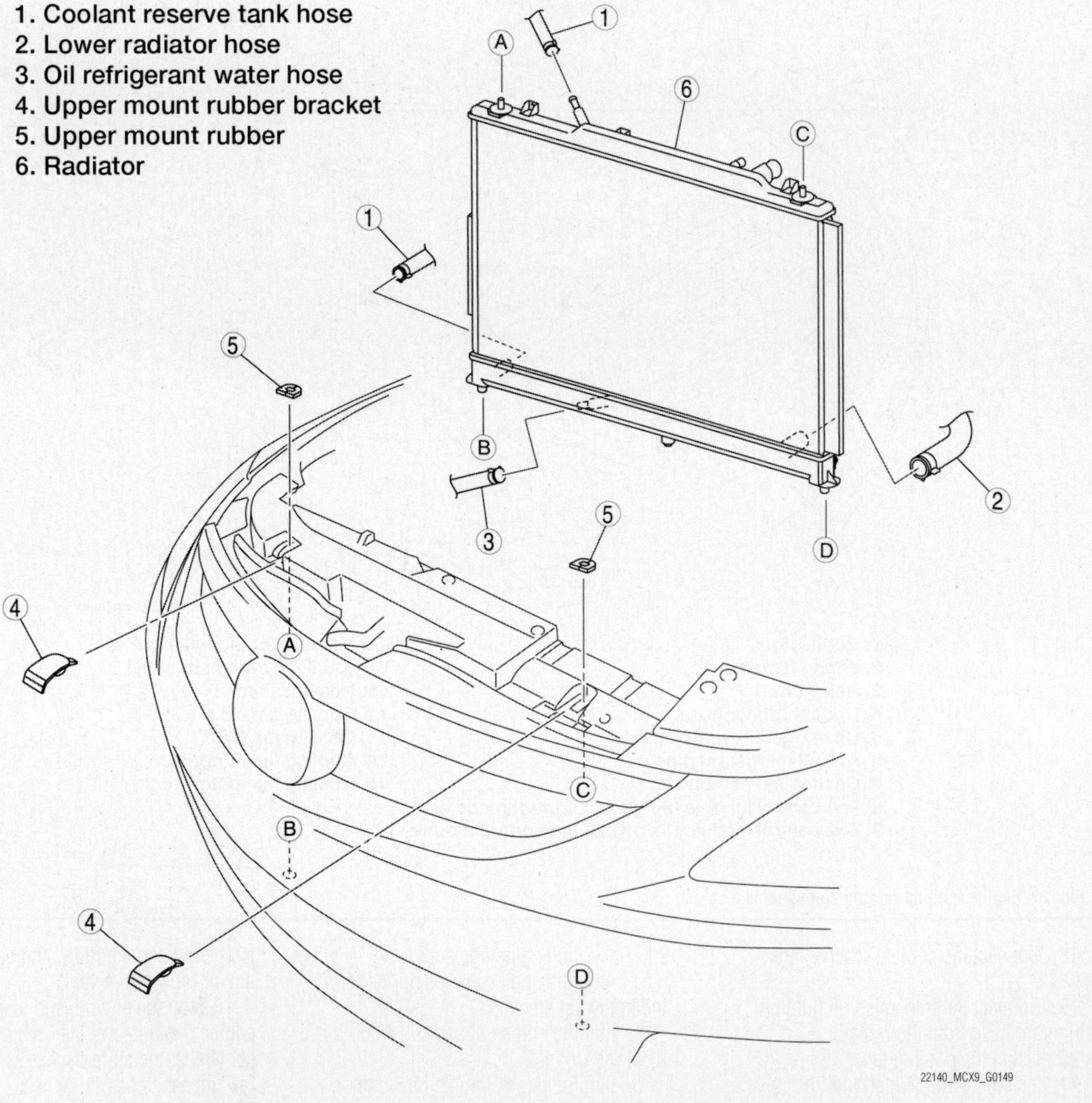

---

1. Coolant reserve tank hose
2. Lower radiator hose
3. Oil refrigerant water hose
4. Upper mount rubber bracket
5. Upper mount rubber
6. Radiator

22140_MCX9_G0149

Fig. 39 Radiator and related components

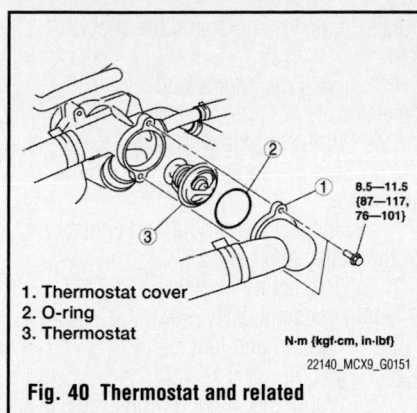

1. Thermostat cover
2. O-ring
3. Thermostat

N·m {kgf·cm, in·lbf}

8.5—11.5 {87—117, 76—101}

22140_MCX9_G0151

Fig. 40 Thermostat and related components

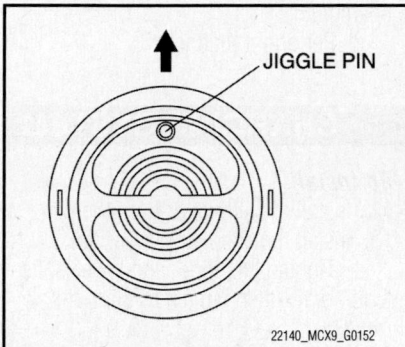

JIGGLE PIN

22140_MCX9_G0152

Fig. 41 Install the thermostat into the thermostat case with the jiggle pin at the top

   a. After the engine warms up, perform the following steps.
   b. Run the engine at approx. 4,000 RPM for 1 min.
   c. Run the engine at idle for 1 min.
   d. Repeat the following steps 2 times.
   e. Operate the heater at the maximum temperature and airflow, and verify that hot air blows from vent.
   f. Stop the engine, and inspect the engine coolant level after the engine coolant temperature decreases. If it is low, repeat steps.
14. Inspect for engine coolant leakage.

WATER PUMP

REMOVAL & INSTALLATION

See Figures 42 through 45.

   Mazda's official procedure for water pump removal and installation requires the engine and transaxle to be removed from the vehicle. It may be possible to perform this procedure with the engine and transaxle in the vehicle.
   1. Before servicing the vehicle, refer to the Precautions Section.
   2. Drain the engine oil.
   3. Remove the engine and transaxle.

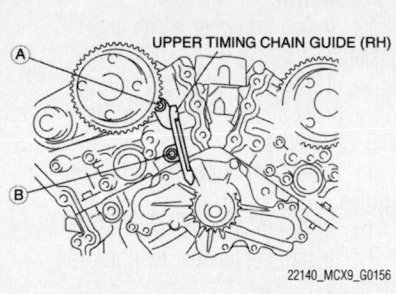

UPPER TIMING CHAIN GUIDE (RH)

22140_MCX9_G0156

Fig. 42 Loosen the upper timing chain guide (RH) bolt A. Remove the upper timing chain guide (RH) bolt B. Move the upper timing chain guide (RH) up and to the right, then retighten the bolt A

   4. Secure the engine and transaxle using a hoist and an engine stand.
   5. Remove the dynamic chamber and throttle body as a single unit.
   6. Remove the ignition coils.
   7. Remove the dipstick.
   8. Remove the power steering oil pump drive belt.
   9. Remove the power steering oil pump.
   10. Remove the timing chain.
   11. Remove the upper timing chain guides.
     a. Loosen the upper timing chain guide (RH) bolt A.
     b. Remove the upper timing chain guide (RH) bolt B.

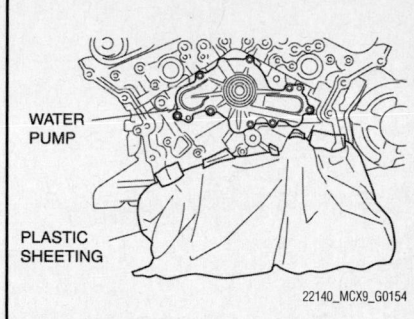

WATER PUMP

PLASTIC SHEETING

22140_MCX9_G0154

Fig. 43 To prevent engine coolant from penetrating the oil pan, line the cylinder block with plastic sheeting as shown. before removing the water pump

     c. Move the upper timing chain guide (RH) up and to the right.
     d. Retighten the bolt A.
   12. Remove the water pump and gasket.

✳✳ WARNING

When the water pump is removed, the engine coolant may flow through the cylinder block and penetrate the oil pan. To prevent the engine coolant from accumulating in the oil pan, remove the oil pan drain plug before removing the water pump.

   a. Remove the oil pan drain plug.

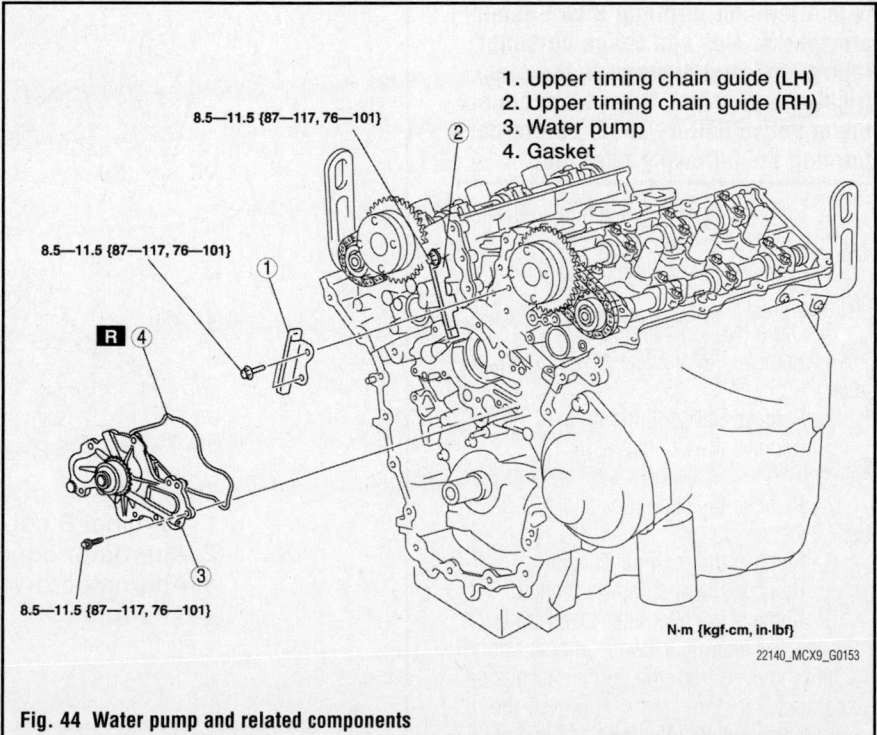

1. Upper timing chain guide (LH)
2. Upper timing chain guide (RH)
3. Water pump
4. Gasket

8.5—11.5 {87—117, 76—101}

8.5—11.5 {87—117, 76—101}

8.5—11.5 {87—117, 76—101}

N·m {kgf·cm, in·lbf}

22140_MCX9_G0153

Fig. 44 Water pump and related components

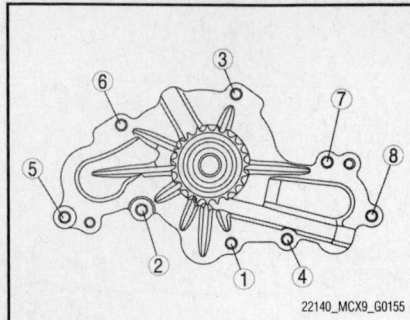

Fig. 45 Tighten the water pump bolts in the order shown

b. To prevent engine coolant from penetrating the oil pan, line the cylinder block with plastic sheeting as shown. before removing the water pump.

ENGINE ELECTRICAL

ALTERNATOR

REMOVAL & INSTALLATION
See Figure 46.

✳✳ WARNING

Remove and install all parts when the engine is cold, otherwise they can cause severe burns or serious injury.

✳✳ WARNING

When the battery cables are connected, touching the vehicle body with alternator terminal B will generate sparks. This can cause personal injury, fire, and damage to the electrical components. Always disconnect the negative battery cable before performing the following operation.

1. Before servicing the vehicle, refer to the Precautions Section.
2. Disconnect the negative battery cable.
3. Drain the engine coolant.
4. Remove the air cleaner and fresh air duct.
5. Remove the cooling fan.
6. Remove the front splash shield (RH).
7. Remove the alternator and A/C drive belt.
8. Remove the terminal B cable.
9. Remove the alternator connector.
10. Remove the alternator lower bolt.
   a. The alternator lower bolt cannot be fully removed from the engine because it contacts the body frame. However, the alternator can be removed/installed with-

To install:
13. Install the water pump and gasket.
14. Tighten the water pump bolts in the order shown. 76–101 ft. lbs. (8.5–11.5 Nm).
15. Install the upper timing chain guides.
16. Install the timing chain.
17. Install the power steering oil pump.
18. Install the power steering oil pump drive belt.
19. Install the dipstick.
20. Install the ignition coils.
21. Install the dynamic chamber and throttle body as a single unit.
22. Secure the engine and

out fully removing the lower bolt because there is a notch at the lower bolt installation part of the alternator.
   b. Fully loosen the alternator lower bolt.
11. Remove the alternator from the top.

transaxle using a hoist and the special tool.
23. Install the engine and transaxle.
24. Fill the engine with oil.
25. Start the engine and perform the following:
   a. Inspect the runout and contact on the pulley and belt.
   b. Inspect for engine oil, engine coolant, ATF, power steering fluid and fuel leakage.
   c. Verify the ignition timing, idle speed and idle mixture.
   d. Inspect engine accessories operation.
   e. Perform a road test.

CHARGING SYSTEM

To install:
12. Install the alternator from the top.
13. Install the alternator upper bolt.
   a. Tighten the upper and lower bolts temporarily, then tighten to 35 ft. lbs. (48 Nm).
14. Install the alternator connector.

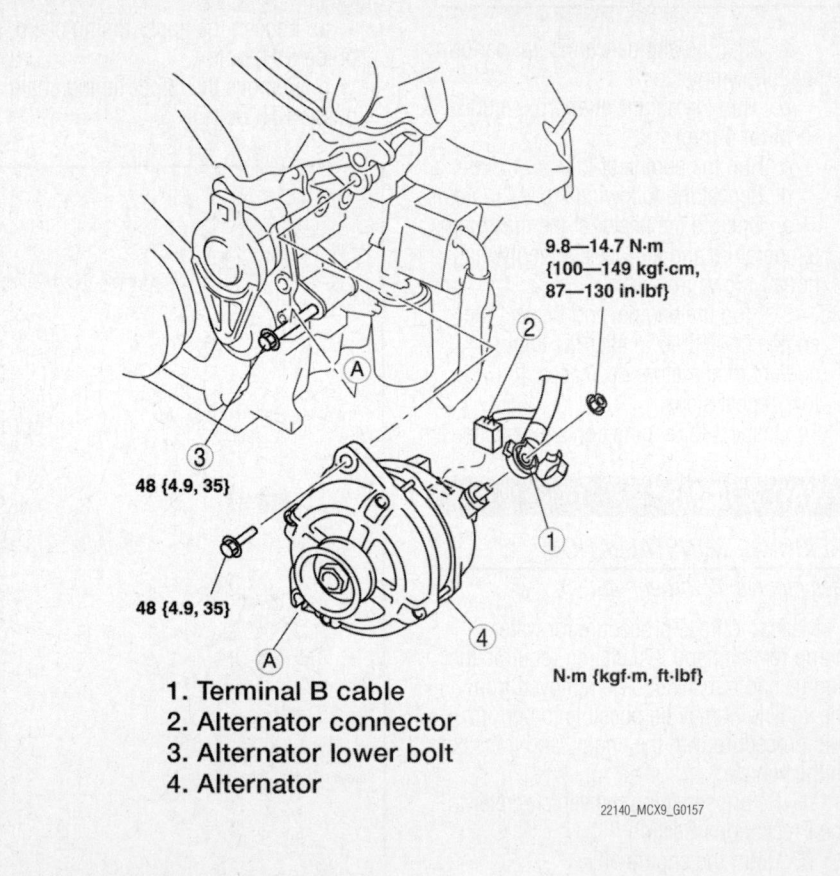

1. Terminal B cable
2. Alternator connector
3. Alternator lower bolt
4. Alternator

9.8—14.7 N·m {100—149 kgf·cm, 87—130 in·lbf}

48 {4.9, 35}

N·m {kgf·m, ft·lbf}

Fig. 46 Alternator and related components

15. Install the terminal B cable.

16. Install the alternator and A/C drive belt.

17. Install the front splash shield (RH).

18. Install the cooling fan.

19. Install the air cleaner and fresh air duct.

20. Connect the negative battery cable.

21. Refill the engine coolant.

　a.　After the engine warms up, perform the following steps.

　b.　Run the engine at approx. 4,000 RPM for 1 min.

　c.　Run the engine at idle for 1 min.

　d.　Repeat the following steps 2 times.

　e.　Operate the heater at the maximum temperature and airflow, and verify that hot air blows from vent.

　f.　Stop the engine, and inspect the engine coolant level after the engine coolant temperature decreases. If it is low, repeat steps.

22. Inspect for engine coolant leakage.

ENGINE ELECTRICAL

FIRING ORDER

See Figure 47.

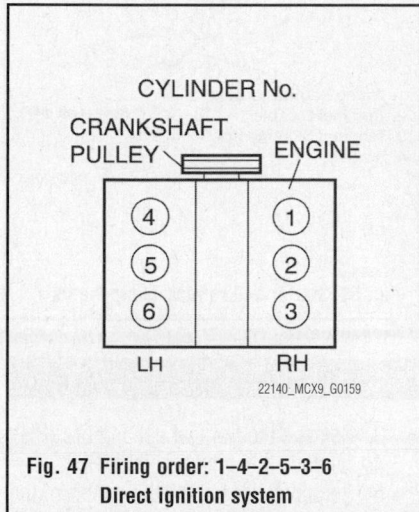

Fig. 47 Firing order: 1–4–2–5–3–6 Direct ignition system

IGNITION COIL

REMOVAL & INSTALLATION

See Figures 47 and 48.

1. Before servicing the vehicle, refer to the Precautions Section.

2. Disconnect the negative battery cable.

3. Remove the engine cover.

4. Remove the intake manifold dynamic chamber.

5. Disconnect the coil connector.

6. Remove the ignition coil.

To install:

7. Apply small amount of Motorcraft XG-3A dielectric grease to the inside of the ignition coils, then install the ignition coils to the spark plugs.

8. Tighten coils to 58–75 inch lbs. (7–9 Nm).

9. Install the intake manifold dynamic chamber.

10. Install the engine cover.

11. Connect the negative battery cable.

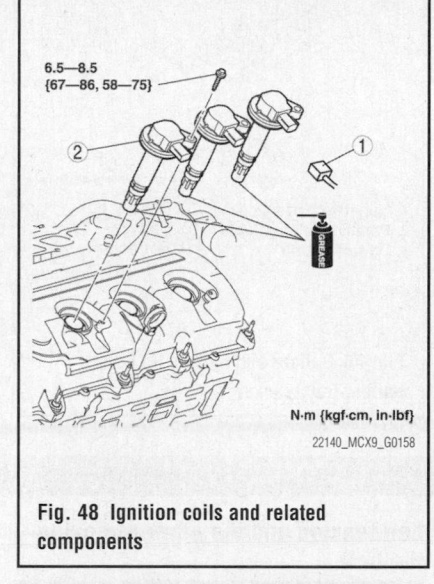

Fig. 48 Ignition coils and related components

SPARK PLUGS

REMOVAL & INSTALLATION

See Figure 49.

IGNITION SYSTEM

1. Before servicing the vehicle, refer to the Precautions Section.

2. Disconnect the negative battery cable.

3. Remove the engine cover.

4. Remove the intake manifold dynamic chamber.

5. Remove the ignition coils.

6. Remove the spark plugs using a plug-wrench.

To install:

7. Install the spark plugs and tighten to 79–177 inch lbs. (9–20 Nm).

8. Apply small amount of Motorcraft XG-3A dielectric grease to the inside of the ignition coils, then install the ignition coils to the spark plugs.

9. Tighten coils to 58–75 inch lbs. (7–9 Nm).

10. Install the intake manifold dynamic chamber.

11. Install the engine cover.

12. Connect the negative battery cable.

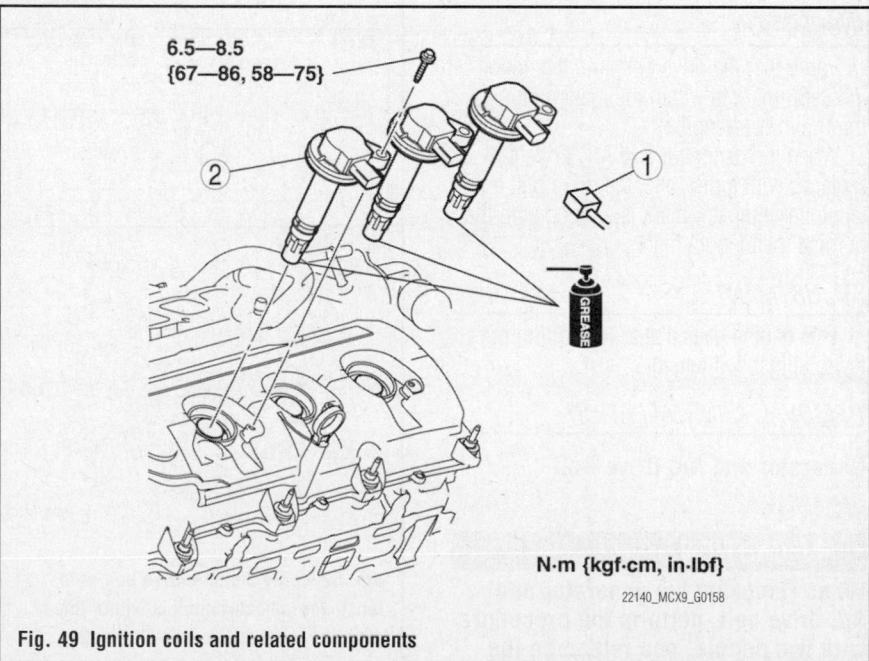

Fig. 49 Ignition coils and related components

STARTER

REMOVAL & INSTALLATION

See Figures 50 and 51.

1. Before servicing the vehicle, refer to the Precautions Section.
2. Remove the battery and battery tray.
3. Position the selector cable out of the way.
4. Remove the wiring harness bracket.
5. Disconnect the terminal B cable.
6. Disconnect the terminal S connector.
7. Remove the starter.

To install:

8. Install the starter. Tighten bolts to 20 ft. lbs. (27 Nm).
9. Connect the terminal S connector.
10. Connect the terminal B cable. Tighten to 87–103 inch lbs. (10–12 Nm).
11. Install the wiring harness bracket. Tighten to 69–96 inch lbs. (8–11 Nm).
12. Position the selector cable out of the way.
13. Install the battery and battery tray.

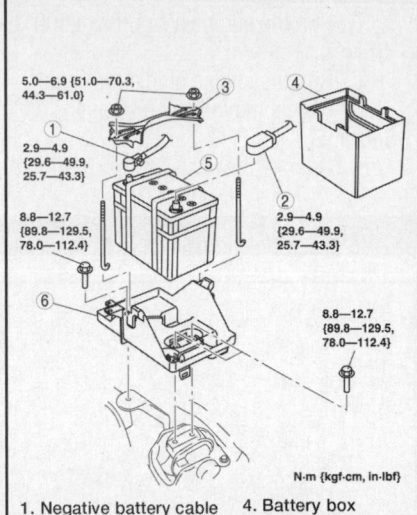

1. Negative battery cable
2. Positive battery cable
3. Battery clamp
4. Battery box
5. Battery
6. Battery tray

N·m {kgf-cm, in-lbf}

22140_MCX9_G0086

Fig. 50 Battery and battery tray and battery tray bracket

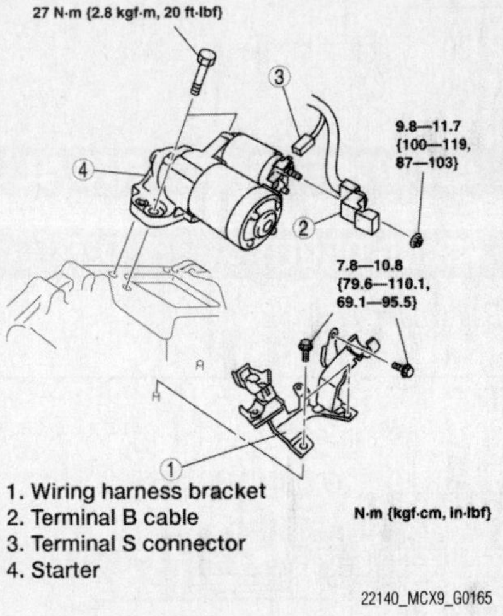

27 N·m {2.8 kgf·m, 20 ft-lbf}

9.8–11.7 {100–119, 87–103}

7.8–10.8 {79.6–110.1, 69.1–95.5}

1. Wiring harness bracket
2. Terminal B cable
3. Terminal S connector
4. Starter

N·m {kgf-cm, in-lbf}

22140_MCX9_G0165

Fig. 51 Starter and related components

ENGINE MECHANICAL

➡ **Disconnecting the negative battery cable may interfere with the functions of the on board computer systems and may require the computer to undergo a relearning process, once the negative battery cable is reconnected.**

ACCESSORY DRIVE BELTS

INSPECTION

Verify that the drive belt auto tensioner indicator mark is within the maximum and minimum belt length.

When the generator and A/C drive belt is replaced with a new one, the drive belt auto tensioner indicator mark is aligned with the nominal belt length mark.

ADJUSTMENT

This engine uses a auto tensioning pulley to adjust belt tension.

REMOVAL & INSTALLATION

Generator and A/C Drive Belt

See Figure 52.

❊ **CAUTION**

When removing the generator and A/C drive belt, perform the procedure with two people, one releasing the belt tension and the other removing the belt.

1. Before servicing the vehicle, refer to the Precautions Section.
2. Remove the engine cover.

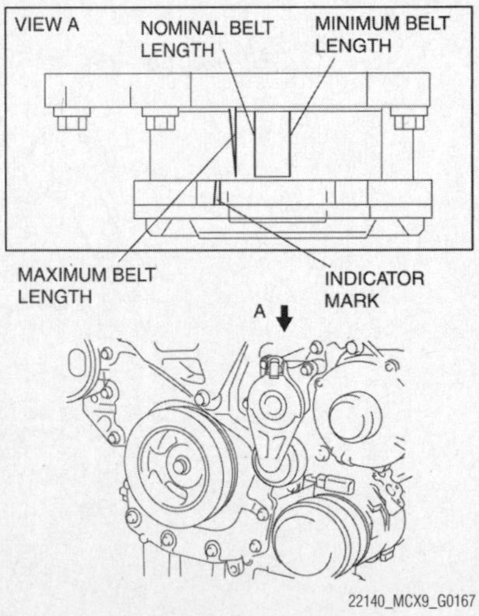

VIEW A   NOMINAL BELT LENGTH   MINIMUM BELT LENGTH

MAXIMUM BELT LENGTH   INDICATOR MARK

A

22140_MCX9_G0167

Fig. 52 Verify that the drive belt auto tensioner indicator mark is within the maximum and minimum belt length

3. Remove the front wheel and tire (RH).
4. Remove the splash shield (RH).
5. Set a breaker bar on the center of the tensioner pulley.
6. Using the breaker bar, turn the center of the tensioner pulley clockwise to release the tension on the generator and A/C drive belt.
7. Remove the generator and A/C drive belt.

To install:

8. Install the generator and A/C drive belt.

➡ **When the generator and A/C drive belt is replaced with a new one, the drive belt auto tensioner indicator mark is aligned with the nominal belt length mark.**

9. Verify that the drive belt auto tensioner indicator mark is within the maximum and minimum belt length.
10. If the drive belt auto tensioner indicator mark is not within the limit, replace the generator and A/C drive belt.
11. Install the splash shield (RH).
12. Install the front wheel and tire (RH).
13. Install the engine cover.

Power Steering Oil Pump Drive Belt

See Figures 53 through 55.

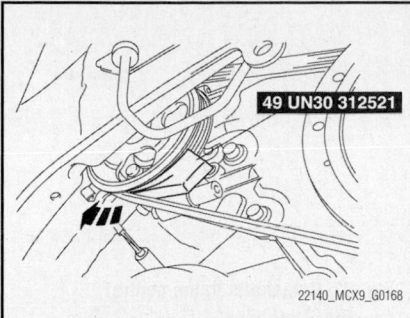

Fig. 53 Install the special tool between the power steering oil pump drive belt and the power steering oil pump pulley

1. Before servicing the vehicle, refer to the Precautions Section.

2. Remove the generator and A/C drive belt.

3. Remove the power steering oil pump drive belt.

　a. Install the special tool between the power steering oil pump drive belt and the power steering oil pump pulley.

　b. Hold the special tool by hand until it is lodges between the power steering oil pump pulley and power steering oil pump drive belt.

　c. Turn the crankshaft clockwise to remove the power steering oil pump drive belt.

　d. If there is power steering oil pump drive belt damage and cracks, replace the power steering oil pump drive belt.

4. Install the power steering oil pump drive belt.

　a. Install the special tool according to the following steps.

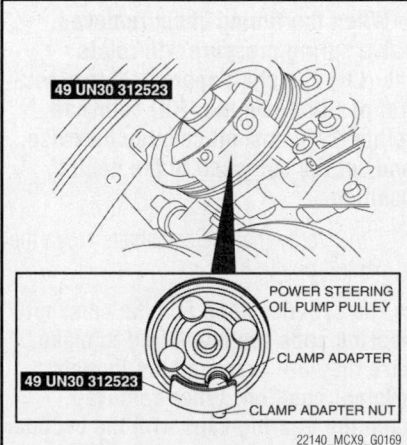

Fig. 54 Position the special tool on the power steering oil pump pulley with the clamp adapter in one of the holes of the pulley

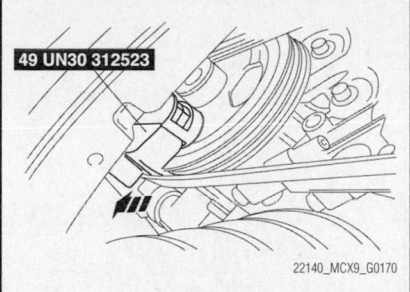

Fig. 55 Position the power steering oil pump drive belt around the special tool and the power steering oil pump pulley

　b. Loosen the clamp adapter nut of the special tool.

　c. There is no need to remove the clamp adapter nut completely when positioning the special tool on the power steering oil pump pulley.

　d. Position the special tool on the power steering oil pump pulley with the clamp adapter in one of the holes of the pulley.

　e. Tighten the clamp adapter nut by hand.

　f. Hand tightening of the clamp adapter nut is sufficient.

　g. Position the power steering oil pump drive belt around the special tool and the power steering oil pump pulley.

　h. Hold the special tool and power steering oil pump drive belt by hand until the power steering oil pump drive belt is properly seated.

　i. Turn the crankshaft clockwise to install the power steering oil pump drive belt.

　j. Verify that the power steering oil pump drive belt is firmly attached to the pulleys.

5. Install the generator and A/C drive belt.

6. Verify that the drive belt auto tensioner indicator mark does not exceeds the maximum belt length.

7. If it exceeds the maximum belt length, replace the generator and A/C drive belt.

CAMSHAFT AND VALVE LIFTERS

REMOVAL & INSTALLATION

See Figures 56 through 73.

Mazda's official procedure for camshaft removal and installation requires the engine and transaxle to be removed from the vehicle. It may be possible to perform this pro-cedure with the engine and transaxle in the vehicle.

1. Before servicing the vehicle, refer to the Precautions Section.

2. Drain the engine oil.

3. Remove the engine and transaxle.

4. Secure the engine and transaxle using a hoist and an engine stand.

5. Remove the intake manifold dynamic chamber and throttle body as a single unit.

6. Remove the fuel injector and fuel distributor together as a single unit.

7. Remove the thermostat and thermo-stat housing together as a single unit.

8. Remove the Intake manifold.

9. Remove both catalytic converters together with both exhaust manifolds as a single unit.

10. Remove the ignition coils.

11. Remove the dipstick.

12. Remove the power steering oil pump drive belt.

13. Remove the power steering oil pump.

14. Remove the alternator.

➡**When removing the timing chain and marking the timing marks on the chain, mark the camshaft timing chain as well.**

15. Remove the timing chain.

16. Remove the camshaft caps.

17. Remove the camshaft timing chain tensioners.

18. Remove the camshaft timing chains.

19. Remove the camshafts.

❊❊ **WARNING**

Do not rotate the crankshaft counter-clockwise. The timing chains may bind, causing engine damage.

　a. Turn the crankshaft clockwise so that the crankshaft keyway is in the 11 o'clock position. This will position the No.1 cylinder at TDC.

Fig. 56 Turn the crankshaft clockwise so that the crankshaft keyway is in the 11 o'clock position. This will position the No.1 cylinder at TDC

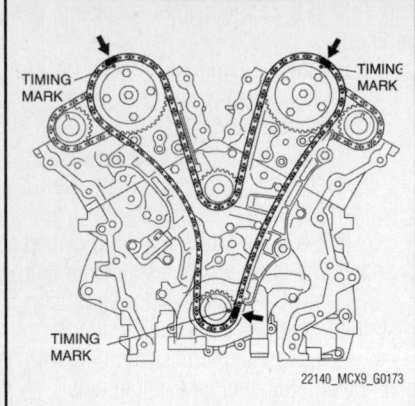

Fig. 57 Mark the timing chain at the position of each timing sprocket timing mark

→Verify that there are timing marks in three locations (Yellow 1, Black 2) on the timing chain. If any timing marks are missing, mark the timing chain. When marking the crankshaft sprocket side timing chain, change the mark color. When the timing chain is replaced with a new one, mark the new timing chain at the same positions as the removed timing chain.

b. Mark the timing chain at the position of each timing sprocket timing mark.

→Verify that there are timing marks in two locations on the camshaft timing chain. If any timing marks are missing, mark the camshaft timing chain. If replacing with a new camshaft timing chain, place alignment marks in the

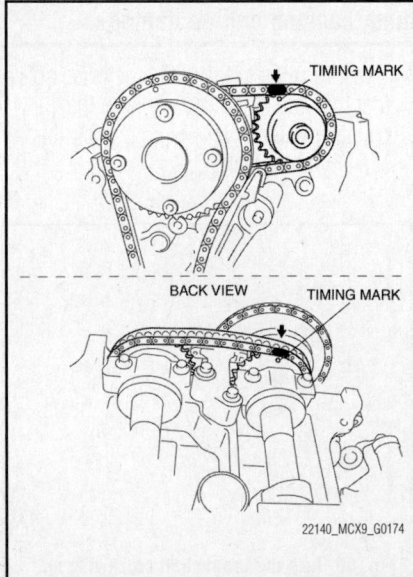

Fig. 58 Mark the camshaft timing chain at the positions where it is aligned with each of the camshaft sprocket on both banks

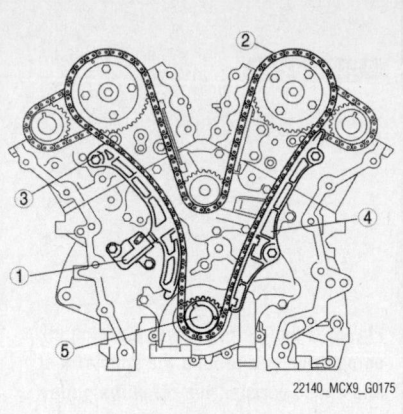

Fig. 59 Remove the timing chain in the following order: (1) Chain tensioner, (2) Timing chain, (3) Tensioner arm, (4) Chain guide and (5) Crankshaft sprocket

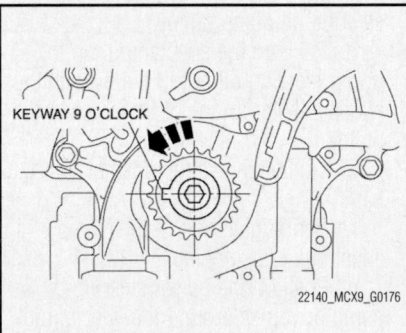

Fig. 60 Rotate the crankshaft counterclockwise until the keyway is in the 9 o'clock position

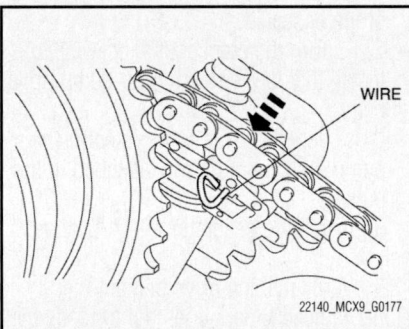

Fig. 61 Insert a paper clip into the camshaft timing chain tensioner to hold the tensioner piston

same positions as those prior to the replacement.

c. Mark the camshaft timing chain at the positions where it is aligned with each of the camshaft sprocket on both banks.

d. Remove the timing chain in the following order.

• Chain tensioner

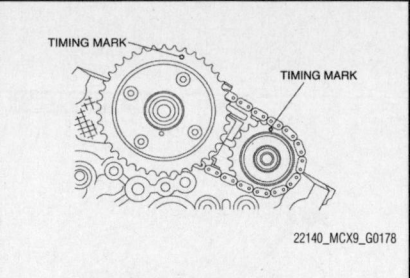

Fig. 62 Camshafts in the neutral position—left hand

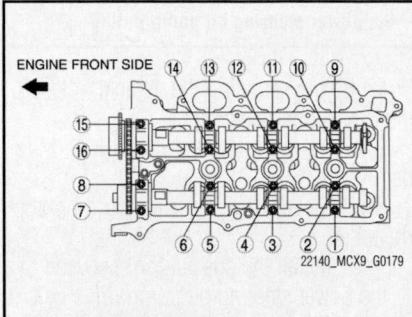

Fig. 63 Camshaft cap loosening sequence—left hand

• Timing chain
• Tensioner arm
• Chain guide
• Crankshaft sprocket

e. Rotate the crankshaft counterclockwise until the keyway is in the 9 o'clock position.

f. Slowly compress the camshaft timing chain tensioner (LH) piston by hand.

g. Insert an approx. 1.0 mm {0.039 in} thin wire or paper clip into the camshaft timing chain tensioner shown. to hold the tensioner piston.

→When the timing chain removed, valve spring pressure will rotate the (LH) camshaft approx. 3° to a neutral position. On the (RH) camshaft, rotate the camshaft counterclockwise, and set the camshaft to the neutral position.

h. Verify that the camshafts are in the neutral position.

→The cylinder head and the camshaft bearing caps are numbered to make sure they are reassembled in their original position. When removed, keep the bearing caps with the cylinder head they were removed from. Do not mix the caps.

i. Loosen the camshaft cap bolts in several passes in the order shown. and remove the camshaft cap.

j. Remove the camshaft, camshaft sprocket, camshaft timing chain, and camshaft timing chain tensioner as a single unit.

To install:

20. Assemble the camshaft timing chain tensioners on both sides. Tighten to 76–101 inch lbs. (8.5–11.5 Nm).

21. Install the right hand camshaft.

➡**Install the camshaft, camshaft sprocket and camshaft timing chain of the both bank as a single unit.**

a. Install the camshafts.

b. Install the camshaft timing chain by aligning the colored links on the camshaft timing chain with the marks on the camshaft sprockets.

c. Position the camshaft component onto the cylinder head in the neutral position as shown.

d. Install the camshaft caps and temporarily tighten the camshaft cap bolts

evenly in the order shown in several passes. Tighten to 76–101 inch lbs. (8.5–11.5 Nm).

e. Remove the retaining wire inserted into the camshaft timing chain tensioner.

22. Using the right hand camshaft sprocket installation bolt, rotate the right hand camshaft clockwise so the timing marks align as shown.

23. Install the left hand camshaft in the same manner as the right hand camshaft.

24. Rotate the crankshaft clockwise so that the crankshaft keyway is in the 11 o'clock position. This will position the No.1 cylinder at TDC.

25. Install the timing chain.

26. Install the alternator.

27. Install the power steering oil pump.

28. Install the power steering oil pump drive belt.

29. Install the dipstick.

30. Install the ignition coils.

31. Install both catalytic

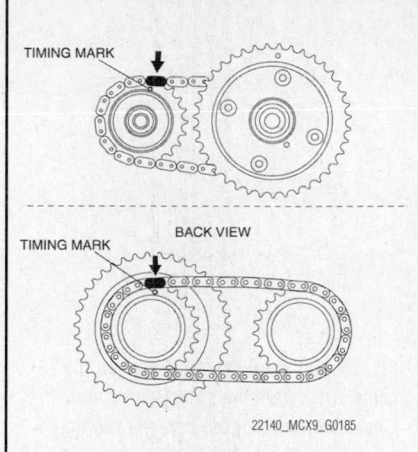

Fig. 65 Install the camshaft timing chain by aligning the colored links on the camshaft timing chain with the marks on the camshaft sprockets—right hand

converters together with both exhaust manifolds as a single unit.

32. Install the Intake manifold.

33. Install the thermostat and thermostat housing together as a single unit.

34. Install the fuel injector and fuel distributor together as a single unit.

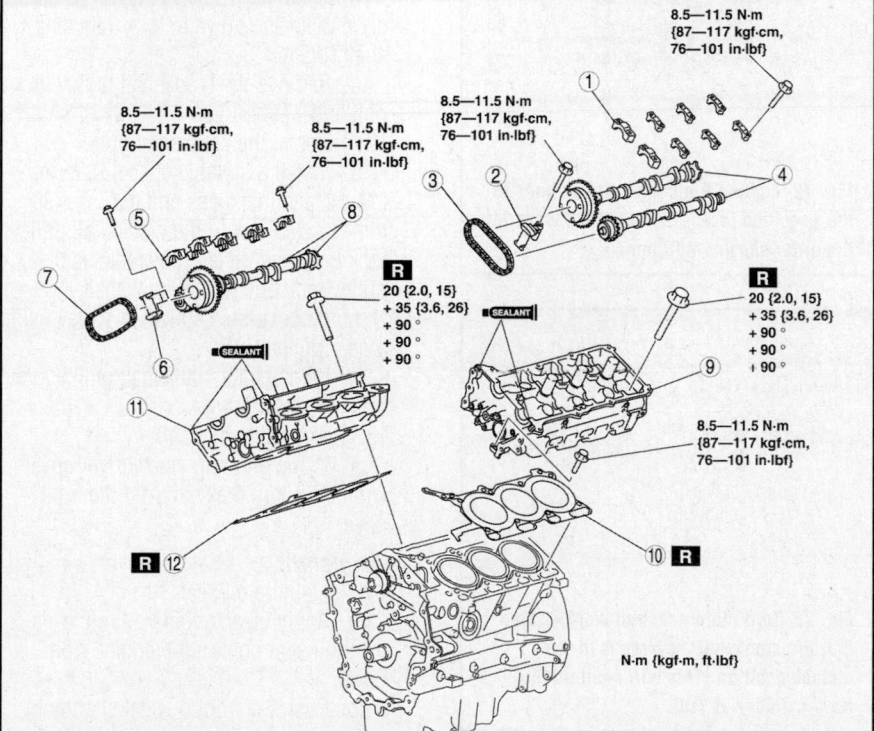

1. Camshaft cap (LH)
2. Camshaft timing chain tensioner (LH)
3. Camshaft timing chain (LH)
4. Camshaft component (LH)
5. Camshaft cap (RH)
6. Camshaft timing chain tensioner (RH)
7. Camshaft timing chain (RH)
8. Camshaft component (RH)
9. Cylinder head (LH)
10. Cylinder head gasket (LH)
11. Cylinder head (RH)
12. Cylinder head gasket (RH)

Fig. 64 Exploded view of the cylinder head assembly

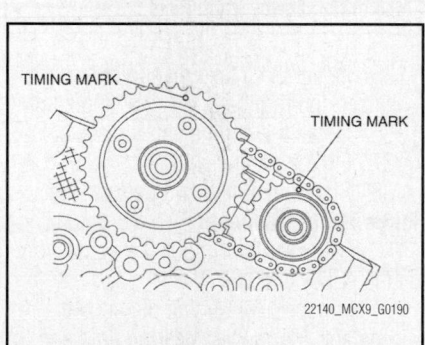

Fig. 66 Position the camshaft component onto the cylinder head in the neutral position as shown—left hand

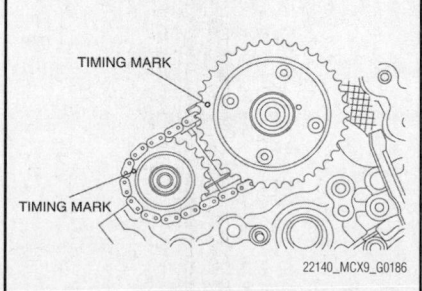

Fig. 67 Position the camshaft component onto the cylinder head in the neutral position as shown—right hand

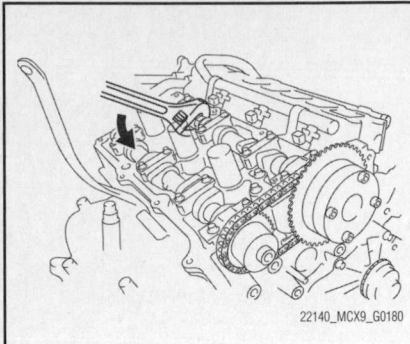

Fig. 68 Put a wrench as shown on the camshaft, rotate the camshaft counter-clockwise, and set the camshaft to the neutral position—right hand

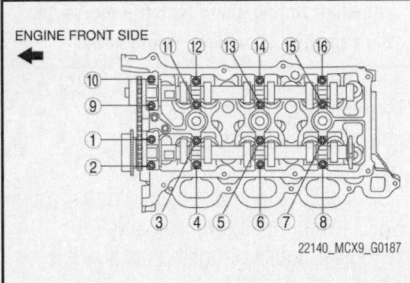

Fig. 69 Tighten the camshaft cap bolts to the specified torque, in several passes, in the order shown—right hand

35. Install the intake manifold dynamic chamber and throttle body as a single unit.
36. Install the engine and transaxle.
37. Fill the engine with oil.
38. Start the engine and perform the following inspections:
   a. Inspect the runout and contact on the pulley and belt.
   b. Inspect for engine oil, engine coolant, ATF, power steering fluid and fuel leakage.
   c. Verify the ignition timing and idle speed.

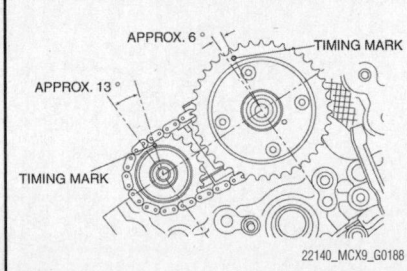

Fig. 70 Using the right hand camshaft sprocket installation bolt, rotate the right hand camshaft clockwise so the timing marks align as shown

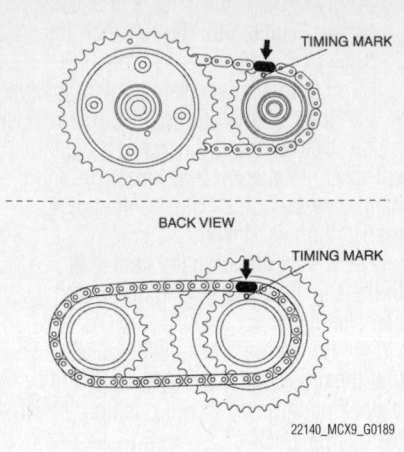

Fig. 71 Install the camshaft timing chain by aligning the colored links on the camshaft timing chain with the marks on the camshaft sprockets—left hand

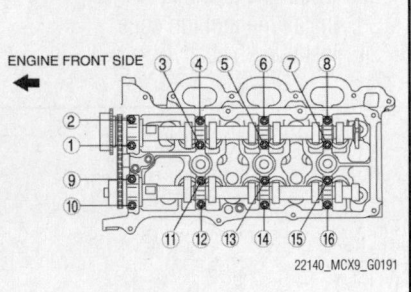

Fig. 72 Tighten the camshaft cap bolts to the specified torque, in several passes, in the order shown—left hand

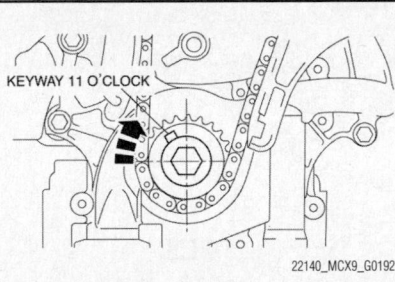

Fig. 73 Turn the crankshaft clockwise so that the crankshaft keyway is in the 11 o'clock position. This will position the No.1 cylinder at TDC.

   d. Inspect engine accessories for proper operation.
39. Perform a road test.

CRANKSHAFT FRONT SEAL

REMOVAL & INSTALLATION

See Figures 74 through 80.

Mazda's official procedure for the crankshaft damper removal and installation requires the engine and transaxle to be removed from the vehicle. It may be possible to perform this procedure with the engine and transaxle in the vehicle.

1. Before servicing the vehicle, refer to the Precautions Section.
2. Drain the engine oil.
3. Remove the engine and transaxle.
4. Secure the engine and transaxle using a hoist and an engine stand.
5. Remove the intake manifold dynamic chamber and throttle body as a single unit.
6. Remove the ignition coils.
7. Remove the dipstick.
8. Remove the power steering oil pump drive belt.
9. Remove the power steering oil pump.
10. Remove the drive belt auto tensioner.
11. Remove the crankshaft pulley lock bolt.
   a. Remove the starter.
   b. Set a flathead screwdriver to the drive plate as shown to lock the crankshaft rotation.
   c. Remove the crankshaft pulley lock bolt and washer.
12. Remove the crankshaft pulley.
   a. Install a suitable spacer 0.55 in. (14 mm) in thickness and 1.18 in. (30 mm) in diameter to the crankshaft pulley lock bolt, and install the crankshaft pulley lock bolt to the crankshaft.
   b. Remove the crankshaft pulley using a gear puller.
   c. Remove the crankshaft pulley lock bolt and spacer.
13. Remove the oil seal.
   a. Using a flat tip screwdriver cover with a rag, pry the seal from the front cover

To install:
14. Install the oil seal.
   a. Apply clean engine oil to the front oil seal bore in the engine front cover.
   b. Push the front oil seal slightly in by hand.
   c. Tap the front oil seal in evenly using the special tool and a hammer.
15. Install the crankshaft pulley.
16. Install the crankshaft pulley lock bolt.
   a. Set a flathead screwdriver to the drive plate in the position indicated. to lock the crankshaft rotation.

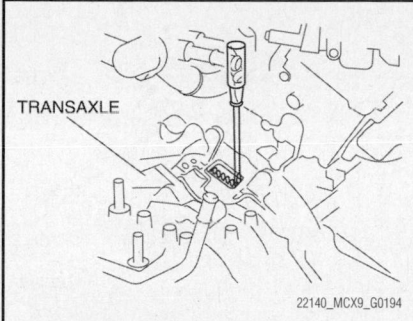

Fig. 74 Set a flathead screwdriver to the drive plate as shown to lock the crankshaft rotation

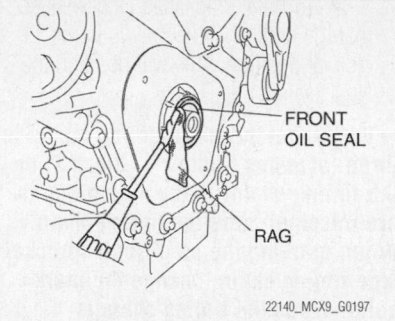

Fig. 77 Using a flat tip screwdriver cover with a rag, pry the seal from the front cover

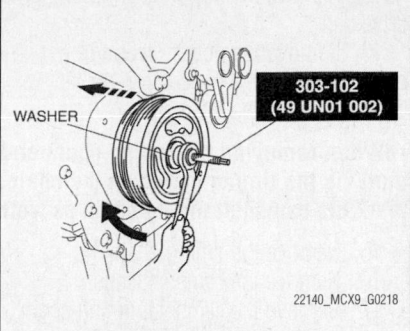

Fig. 80 Tighten the special tool nut and install the crankshaft pulley

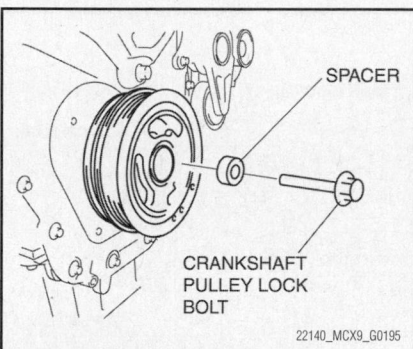

Fig. 75 Install a suitable spacer 0.55 in. (14 mm) in thickness and 1.18 in. (30 mm) in diameter to the crankshaft pulley lock bolt, and install the crankshaft pulley lock bolt to the crankshaft

Fig. 78 Tap the front oil seal in evenly using the special tool and a hammer

CYLINDER HEAD

REMOVAL & INSTALLATION

See Figures 81 through 100.

Mazda's official procedure for cylinder head removal and installation requires the engine and transaxle to be removed from the vehicle. It may be possible to perform this procedure with the engine and transaxle in the vehicle.

1. Before servicing the vehicle, refer to the Precautions Section.
2. Drain the engine oil.
3. Remove the engine and transaxle.
4. Secure the engine and transaxle using a hoist and an engine stand.
5. Remove the intake manifold dynamic chamber and throttle body as a single unit.
6. Remove the fuel injector and fuel distributor together as a single unit.
7. Remove the thermostat and thermostat housing together as a single unit.
8. Remove the Intake manifold.
9. Remove both catalytic converters together with both exhaust manifolds as a single unit.
10. Remove the ignition coils.
11. Remove the dipstick.

Fig. 76 Remove the crankshaft pulley using a gear puller

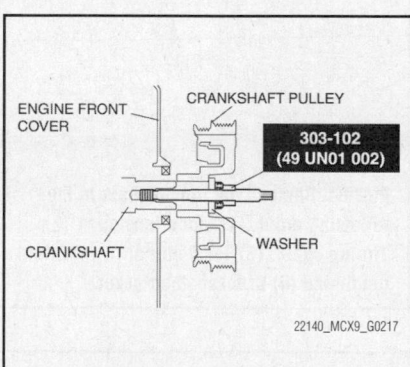

Fig. 79 Install the crankshaft pulley, washer, and special tool to the crankshaft

b. Tighten the new crankshaft pulley lock bolt in four steps.
- Step 1: Tighten to 103 ft. lbs. (140 Nm)
- Step 2: Loosen 360° (one full turn)
- Step 3: Tighten to 35–39 ft. lbs (47–53 Nm)
- Step 4: Tighten an additional 85–95° of rotation
17. Install the power steering oil pump.
18. Install the power steering oil pump drive belt.

19. Install the dipstick.
20. Install the ignition coils.
21. Install the intake manifold dynamic chamber and throttle body as a single unit.
22. Install the engine and transaxle.
23. Start the engine and perform the following inspections:
    a. Inspect the runout and contact on the pulley and belt.
    b. Inspect engine accessories for proper operation.

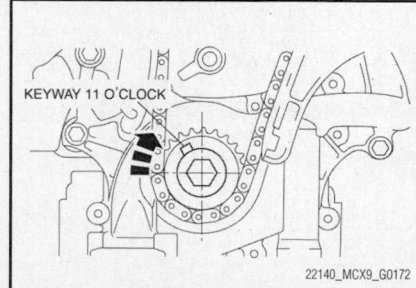

Fig. 81 Turn the crankshaft clockwise so that the crankshaft keyway is in the 11 o'clock position. This will position the No.1 cylinder at TDC

12. Remove the power steering oil pump drive belt.

13. Remove the power steering oil pump.

14. Remove the alternator.

➡ When removing the timing chain and marking the timing marks on the chain, mark the camshaft timing chain as well.

15. Remove the timing chain.

16. Remove the camshaft caps.

17. Remove the camshaft timing chain tensioners.

18. Remove the camshaft timing chains.

19. Remove the camshaft components.

❋ WARNING

Do not rotate the crankshaft counter-clockwise. The timing chains may bind, causing engine damage.

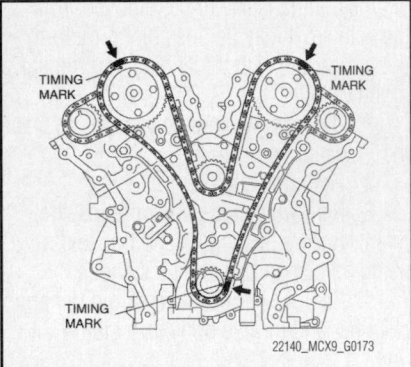

Fig. 82 Mark the timing chain at the position of each timing sprocket timing mark

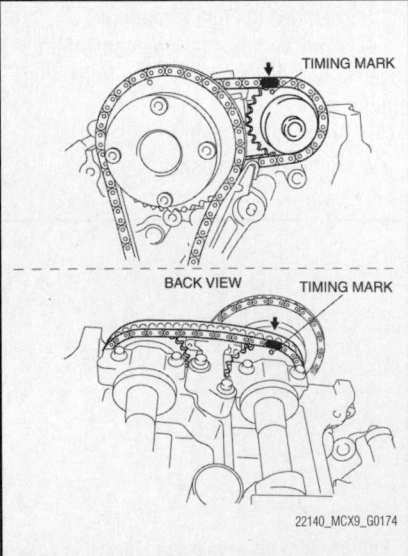

Fig. 83 Mark the camshaft timing chain at the positions where it is aligned with each of the camshaft sprocket on both banks

a. Turn the crankshaft clockwise so that the crankshaft keyway is in the 11 o'clock position. This will position the No.1 cylinder at TDC.

➡ Verify that there are timing marks in three locations (Yellow 1, Black 2) on the timing chain. If any timing marks are missing, mark the timing chain. When marking the crankshaft sprocket side timing chain, change the mark color. When the timing chain is replaced with a new one, mark the new timing chain at the same positions as the removed timing chain.

b. Mark the timing chain at the position of each timing sprocket timing mark.

➡ Verify that there are timing marks in two locations on the camshaft timing chain. If any timing marks are missing, mark the camshaft timing chain. If replacing with a new camshaft timing chain, place alignment marks in the

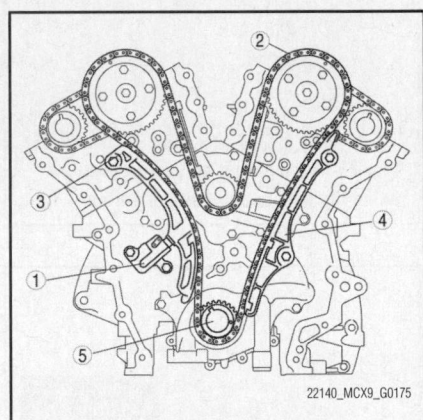

Fig. 84 Remove the timing chain in the following order: (1) Chain tensioner, (2) Timing chain, (3) Tensioner arm, (4) Chain guide and (5) Crankshaft sprocket

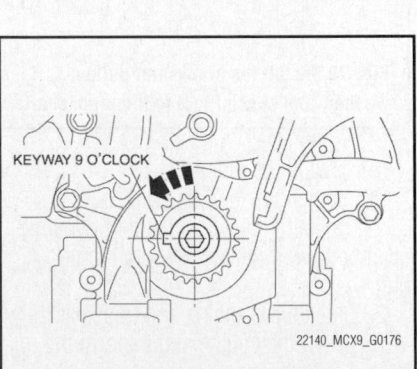

Fig. 85 Rotate the crankshaft counter-clockwise until the keyway is in the 9 o'clock position

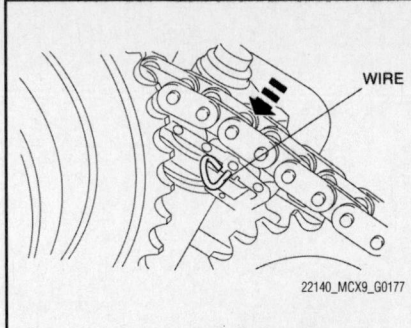

Fig. 86 Insert a paper clip into the camshaft timing chain tensioner to hold the tensioner piston

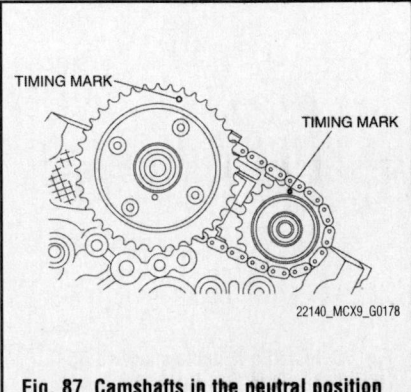

Fig. 87 Camshafts in the neutral position

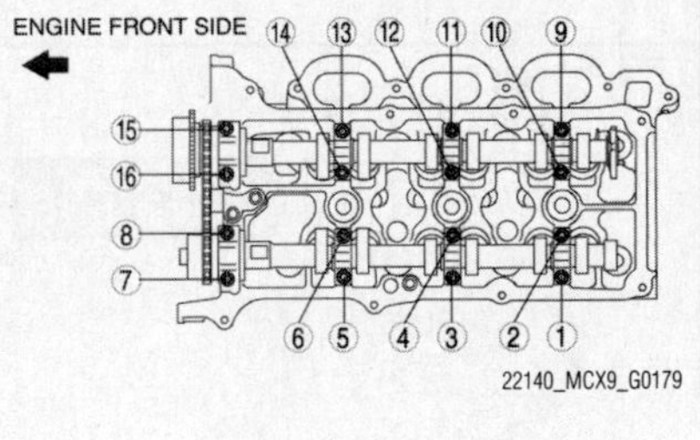

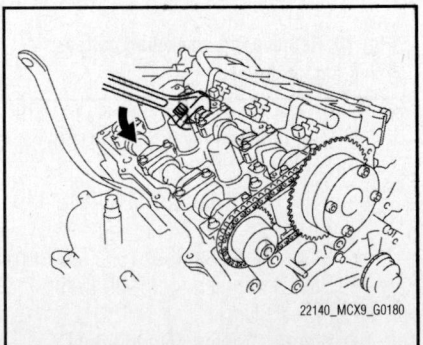

Fig. 88 Camshaft cap loosening sequence

Fig. 89 Put a wrench as shown on the camshaft, rotate the camshaft counter-clockwise, and set the camshaft to the neutral position—right hand

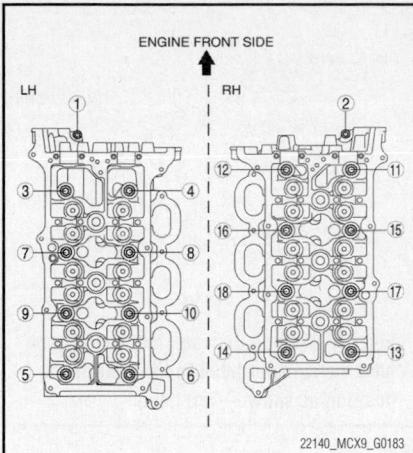

ENGINE FRONT SIDE

Fig. 90 Loosen the cylinder head bolts in several passes in the order shown

same positions as those prior to the replacement.

c. Mark the camshaft timing chain at the positions where it is aligned with each of the camshaft sprocket on both banks.

d. Remove the timing chain in the following order:
- Chain tensioner
- Timing chain
- Tensioner arm
- Chain guide
- Crankshaft sprocket

e. Rotate the crankshaft counterclockwise until the keyway is in the 9 o'clock position.

f. Slowly compress the camshaft timing chain tensioner (LH) piston by hand.

g. Insert an approx. 1.0 mm {0.039 in} thin wire or paper clip into the camshaft timing chain tensioner shown. to hold the tensioner piston.

➡ **When the timing chain removed, valve spring pressure will rotate the (LH) camshaft approx. 3° to a neutral position. On the (RH) camshaft, rotate the camshaft counterclockwise, and set the camshaft to the neutral position.**

h. Verify that the camshafts are in the neutral position.

➡ **The cylinder head and the camshaft bearing caps are numbered to make sure they are reassembled in their original position. When removed, keep the bearing caps with the cylinder head they were removed from. Do not mix the caps.**

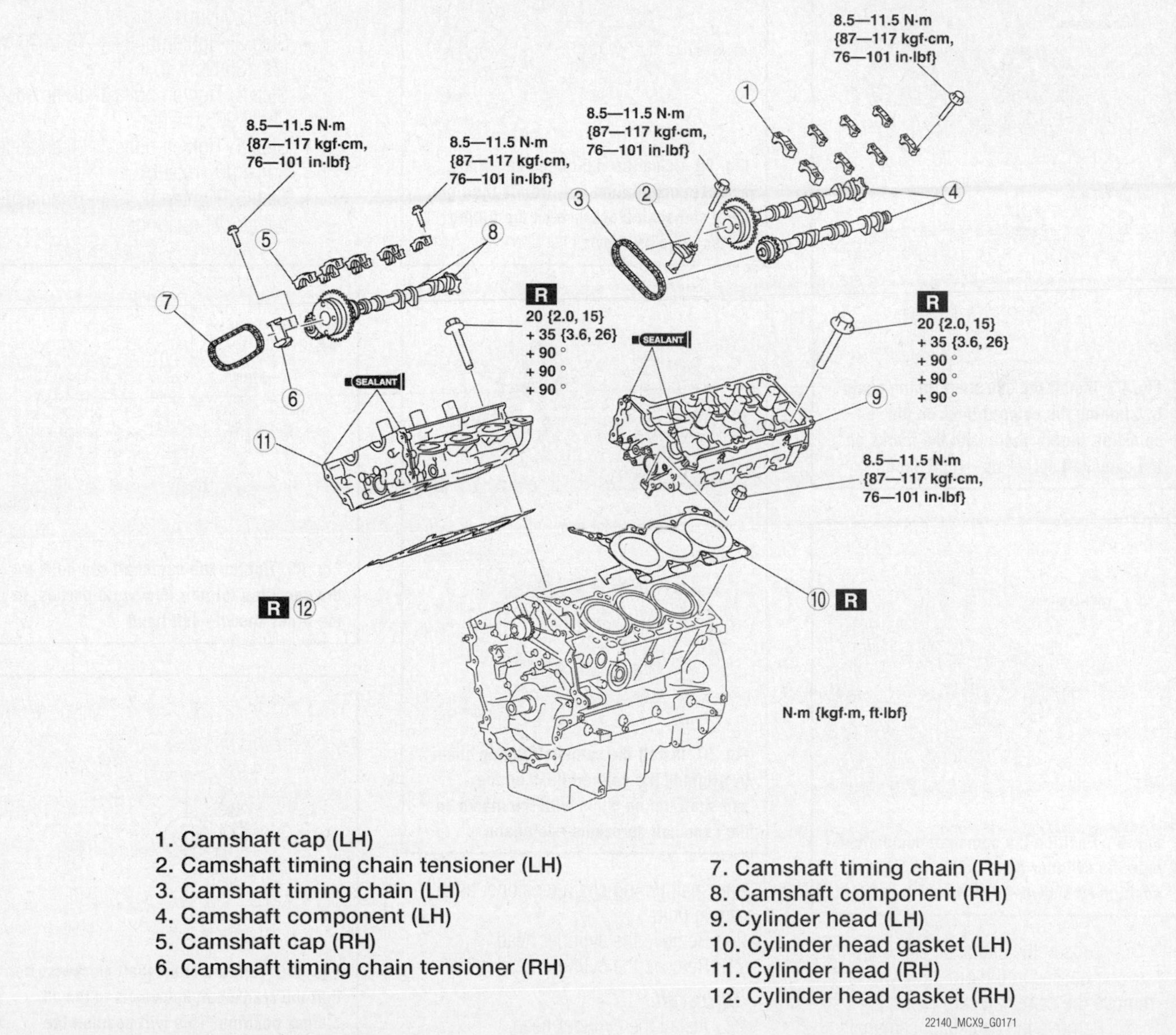

8.5—11.5 N·m
{87—117 kgf·cm,
76—101 in·lbf}

8.5—11.5 N·m
{87—117 kgf·cm,
76—101 in·lbf}

8.5—11.5 N·m
{87—117 kgf·cm,
76—101 in·lbf}

8.5—11.5 N·m
{87—117 kgf·cm,
76—101 in·lbf}

20 {2.0, 15}
+ 35 {3.6, 26}
+ 90 °
+ 90 °
+ 90 °

20 {2.0, 15}
+ 35 {3.6, 26}
+ 90 °
+ 90 °
+ 90 °

8.5—11.5 N·m
{87—117 kgf·cm,
76—101 in·lbf}

N·m {kgf·m, ft·lbf}

1. Camshaft cap (LH)
2. Camshaft timing chain tensioner (LH)
3. Camshaft timing chain (LH)
4. Camshaft component (LH)
5. Camshaft cap (RH)
6. Camshaft timing chain tensioner (RH)
7. Camshaft timing chain (RH)
8. Camshaft component (RH)
9. Cylinder head (LH)
10. Cylinder head gasket (LH)
11. Cylinder head (RH)
12. Cylinder head gasket (RH)

Fig. 91 Exploded view of the cylinder head assembly

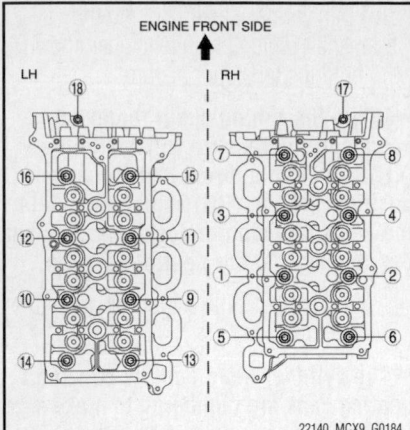

Fig. 92 Tighten the cylinder head bolts to the specified torque, in several passes, in the order shown

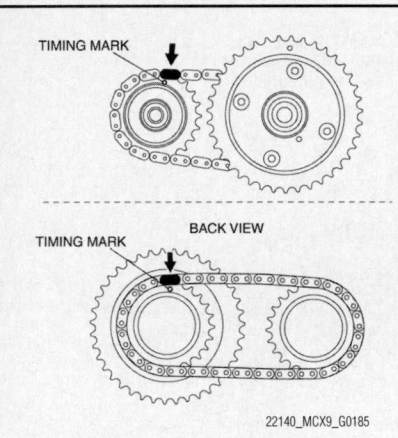

Fig. 93 Install the camshaft timing chain by aligning the colored links on the camshaft timing chain with the marks on the camshaft sprockets—right hand

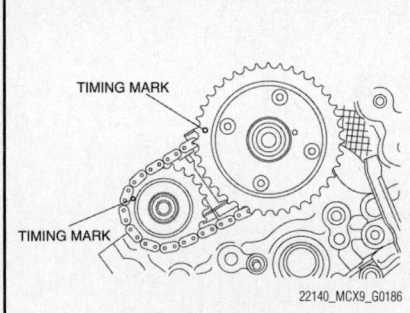

Fig. 94 Position the camshaft component onto the cylinder head in the neutral position as shown—right hand

    i. Loosen the camshaft cap bolts in several passes in the order shown. and remove the camshaft cap.
    j. Remove the camshaft, camshaft sprocket, camshaft timing chain, and

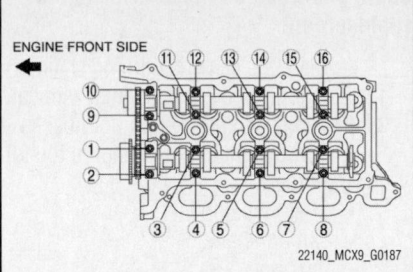

Fig. 95 Tighten the camshaft cap bolts to the specified torque, in several passes, in the order shown—right hand

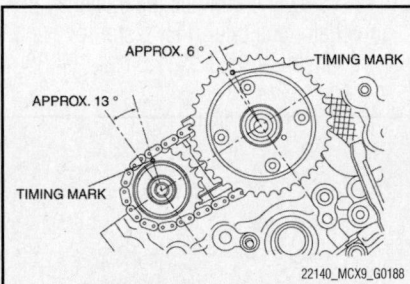

Fig. 96 Using the right hand camshaft sprocket installation bolt, rotate the right hand camshaft clockwise so the timing marks align as shown

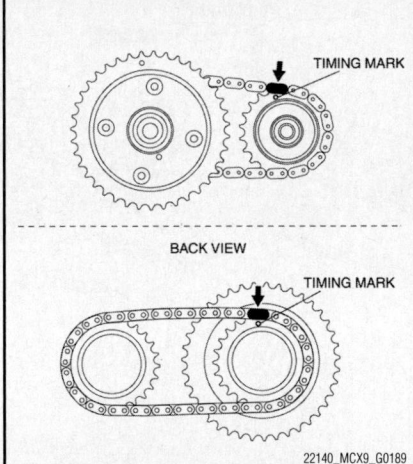

Fig. 97 Install the camshaft timing chain by aligning the colored links on the camshaft timing chain with the marks on the camshaft sprockets—left hand

camshaft timing chain tensioner as a single unit.
20. Remove the cylinder head.
21. Remove the cylinder head gasket.

To install:
22. Install the cylinder head gasket.

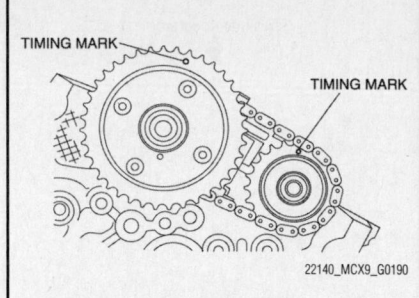

Fig. 98 Position the camshaft component onto the cylinder head in the neutral position as shown—left hand

23. Install the cylinder head.
24. Install the cylinder head bolts and tighten to the following torque in the order shown:

- Step 1: Tighten bolts 1–16 to 15 ft. lbs. (20 Nm).
- Step 2: Tighten bolts 1–16 to 26 ft. lbs. (35 Nm).
- Step 3: Tighten bolts 1–16 an additional 90° rotation.
- Step 4: Tighten bolts 1–16 an additional 90° rotation.
- Step 5: Tighten bolts 1–16 an additional 90° rotation.

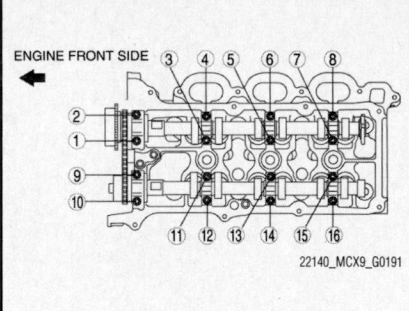

Fig. 99 Tighten the camshaft cap bolts to the specified torque, in several passes, in the order shown—left hand

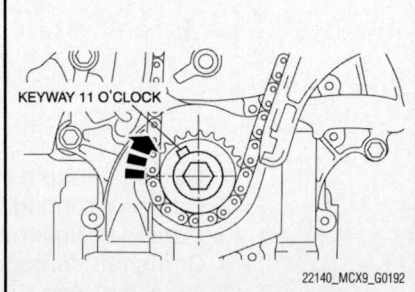

Fig. 100 Turn the crankshaft clockwise so that the crankshaft keyway is in the 11 o'clock position. This will position the No.1 cylinder at TDC.

- Step 6: Tighten bolts 17–18 to 76–101 inch lbs. (9–12 Nm).

25. Assemble the camshaft timing chain tensioners on both sides. Tighten to 76–101 inch lbs. (8.5–11.5 Nm).

26. Install the right hand camshaft.

➡**Install the camshaft, camshaft sprocket and camshaft timing chain of the both bank as a single unit.**

    a. Install the camshafts.

    b. Install the camshaft timing chain by aligning the colored links on the camshaft timing chain with the marks on the camshaft sprockets.

    c. Position the camshaft component onto the cylinder head in the neutral position as shown.

    d. Install the camshaft caps and temporarily tighten the camshaft cap bolts evenly in the order shown in several passes. Tighten to 76–101 inch lbs. (8.5–11.5 Nm).

    e. Remove the retaining wire inserted into the camshaft timing chain tensioner.

27. Using the right hand camshaft sprocket installation bolt, rotate the right hand camshaft clockwise so the timing marks align as shown.

28. Install the left hand camshaft in the same manner as the right hand camshaft.

29. Rotate the crankshaft clockwise so that the crankshaft keyway is in the 11 o'clock position. This will position the No.1 cylinder at TDC.

30. Install the timing chain.

31. Install the alternator.

32. Install the power steering oil pump.

33. Install the power steering oil pump drive belt.

34. Install the dipstick.

35. Install the ignition coils.

36. Install both catalytic converters together with both exhaust manifolds as a single unit.

37. Install the intake manifold.

38. Install the thermostat and thermostat housing together as a single unit.

39. Install the fuel injector and fuel distributor together as a single unit.

40. Install the intake manifold dynamic chamber and throttle body as a single unit.

41. Install the engine and transaxle.

42. Fill the engine with oil.

43. Start the engine and perform the following inspections:

    a. Inspect the runout and contact on the pulley and belt.

    b. Inspect for engine oil, engine coolant, ATF, power steering fluid and fuel leakage.

    c. Verify the ignition timing and idle speed.

    d. Inspect engine accessories for proper operation.

44. Perform a road test.

EXHAUST MANIFOLD

REMOVAL & INSTALLATION

Right Side

See Figure 101.

1. Before servicing the vehicle, refer to the Precautions Section.

2. Disconnect the negative battery cable.

3. Remove the engine cover.

4. On 2WD vehicles:

    a. Remove the steering linkage joint shaft.

    b. Disconnect the warm up-three way catalyst from the exhaust manifold.

5. On AWD vehicles:

    a. Remove the steering linkage joint shaft and insulator.

    b. Remove the No.1 engine mount (transaxle side).

    c. Push up the engine.

    d. Disconnect the warm up-three way catalyst from the exhaust manifold.

6. Remove the exhaust manifold insulators.

7. Disconnect the oxygen sensor connector.

8. Remove the exhaust manifold.

To install:

9. Install the exhaust manifold using a new gasket. Tighten the attaching nuts to 17–20 ft. lbs. (22–28 Nm).

10. Connect the oxygen sensor connector.

11. Install the exhaust manifold insulators. Tighten bolts to 16–21 ft. lbs. (21–29 Nm).

12. On AWD vehicles:

    a. Connect the warm up-three way catalyst from the exhaust manifold.

    b. Push up the engine.

    c. Install the No.1 engine mount (transaxle side).

    d. Install the steering linkage joint shaft and insulator.

13. On 2WD vehicles:

    a. Connect the warm up-three way catalyst from the exhaust manifold.

    b. Install the steering linkage joint shaft.

14. Install the engine cover.

15. Connect the negative battery cable.

Left Side

See Figure 111.

1. Before servicing the vehicle, refer to the Precautions Section.

2. Disconnect the negative battery cable.

3. Remove the engine cover.

4. Disconnect the warm up-three way catalyst from the exhaust manifold.

5. Set the fan control module out of the way.

6. Remove the exhaust manifold insulator.

7. Disconnect the oxygen sensor connector.

8. Remove the exhaust manifold and gasket.

To install:

9. Install the exhaust manifold using a new gasket. Tighten the attaching nuts to 17–20 ft. lbs. (22–28 Nm).

10. Connect the oxygen sensor connector.

11. Install the exhaust manifold insulator. Tighten bolts to 16–21 ft. lbs. (21–29 Nm).

12. Replace the fan control module.

13. Connect the warm up-three way catalyst to the exhaust manifold using an new O-ring. Tighten nuts to 26–33 ft. lbs. (34–46 Nm).

14. Install the engine cover.

15. Connect the negative battery cable.

INTAKE MANIFOLD

REMOVAL & INSTALLATION

See Figures 102 through 111.

1. Before servicing the vehicle, refer to the Precautions Section.

2. Properly relieve the fuel system pressure.

3. Remove the battery.

4. Remove the engine cover.

5. Remove the air cleaner case.

6. Remove the fresh-air duct.

7. Remove the air cleaner element.

8. Disconnect the MAF/IAT sensor connector.

9. Remove the air cleaner cover.

10. Disconnect the vacuum hose to the master back.

11. Disconnect the ventilation hose.

    a. Release the retainer shown in the figure.

    b. Disconnect the quick release connector.

    c. Cover the disconnected quick

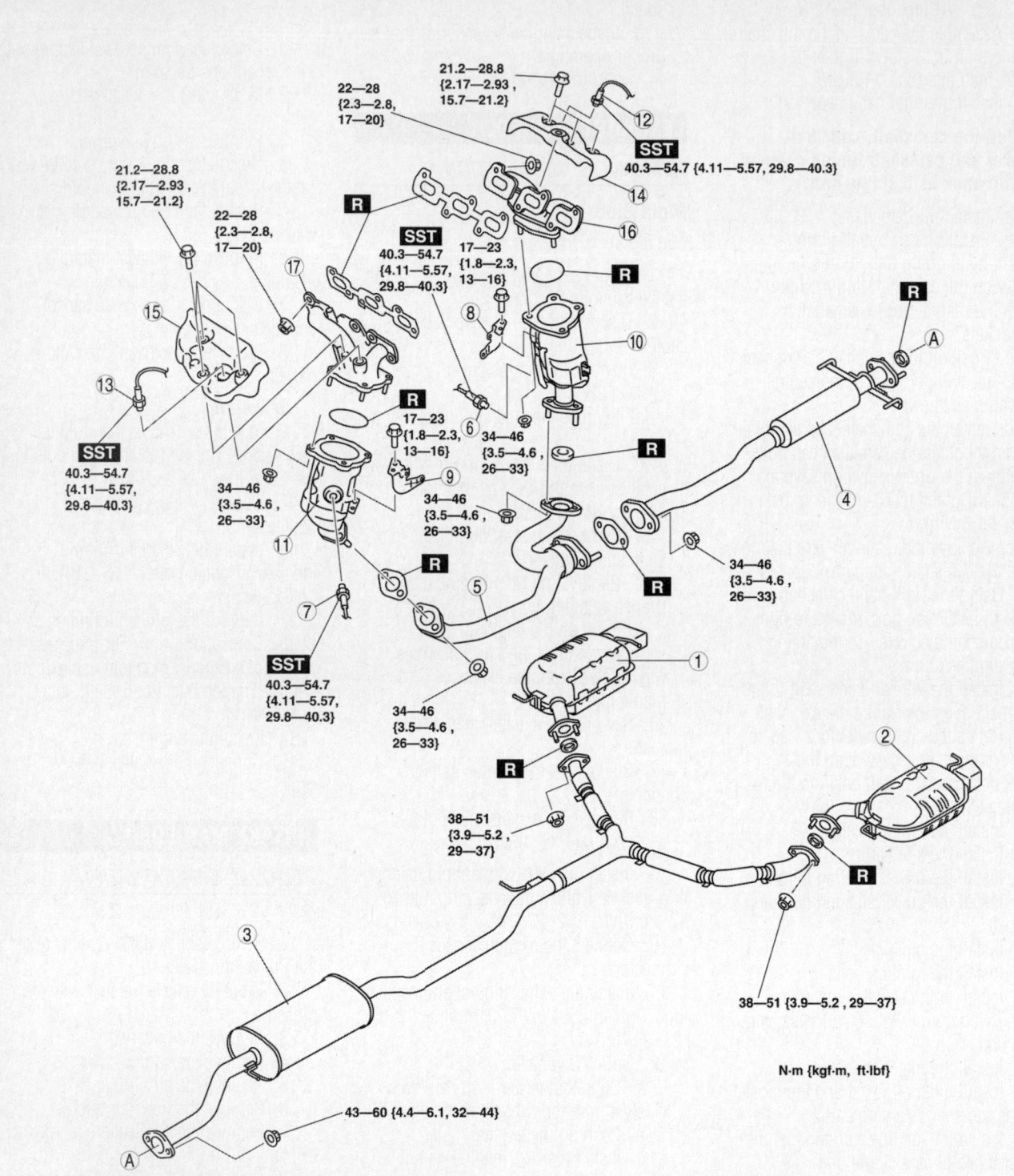

21.2—28.8
{2.17—2.93 ,
15.7—21.2}

22—28
{2.3—2.8,
17—20}

SST
40.3—54.7 {4.11—5.57, 29.8—40.3}

R

17—23
{1.8—2.3,
13—16}

R

21.2—28.8
{2.17—2.93 ,
15.7—21.2}

22—28
{2.3—2.8,
17—20}

R

SST
40.3—54.7
{4.11—5.57,
29.8—40.3}

R

34—46
{3.5—4.6 ,
26—33}

R

SST
40.3—54.7
{4.11—5.57,
29.8—40.3}

34—46
{3.5—4.6 ,
26—33}

17—23
{1.8—2.3,
13—16}

34—46
{3.5—4.6 ,
26—33}

R

34—46
{3.5—4.6 ,
26—33}

R

R

SST
40.3—54.7
{4.11—5.57,
29.8—40.3}

34—46
{3.5—4.6 ,
26—33}

R

38—51
{3.9—5.2 ,
29—37}

R

38—51 {3.9—5.2 , 29—37}

N·m {kgf·m, ft·lbf}

43—60 {4.4—6.1, 32—44}

1. Main silencer (RH)
2. Main silencer (LH)
3. Presilencer
4. Middle pipe
5. Front pipe
6. Rear HO2S (RH)
7. Rear HO2S (LH)

8. Warm Up-Three Way Catalyst
   (RH) bracket (L type)
9. Warm Up-Three Way Catalyst
   (LH) bracket (L type)
10. Warm Up-Three Way Catalyst (RH)
11. Warm Up-Three Way Catalyst (LH)
12. Front HO2S (RH)

13. Front HO2S (LH)
14. Exhaust manifold insulator (RH)
15. Exhaust manifold insulator (LH)
16. Exhaust manifold (RH)
17. Exhaust manifold (LH)

22140_MCX9_G0230

Fig. 101 Exploded view of the exhaust system

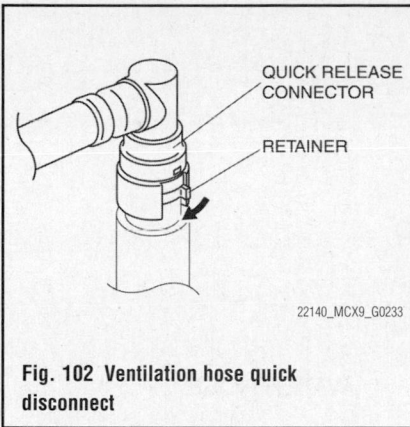

Fig. 102 Ventilation hose quick disconnect

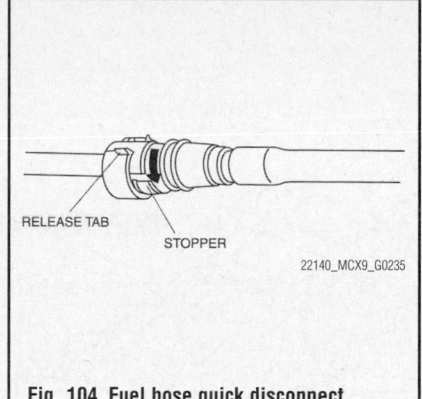

Fig. 104 Fuel hose quick disconnect

release connector and fuel pipe with vinyl sheeting or a similar material to prevent it from scratches or dirt.

12. Remove the resonance chamber.

13. Disconnect the TP sensor connector.

14. Remove the throttle body.

15. Disconnect the vacuum hose to the purge solenoid valve.

   a. Move the retainer upward using a small flathead screwdriver or a similar tool.

   b. Pull out the fuel hose straight from the fuel pipe and disconnect it.

   c. Cover the disconnected quick release connector and fuel pipe with vinyl sheeting or a similar material to prevent it from scratches or dirt.

16. Disconnect the vacuum hose to the master back.

17. Remove the dynamic chamber.

18. Disconnect the fuel hose.

   a. Rotate the release tab on the quick release connector to the stopper position.

   b. Pull out the fuel hose straight from the fuel pipe and disconnect it.

   c. Cover the disconnected quick release connector and fuel pipe with vinyl

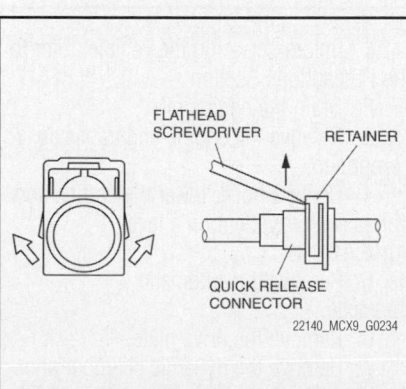

Fig. 103 Vacuum hose to purge solenoid valve quick disconnect

sheeting or a similar material to prevent it from scratches or dirt.

19. Remove the fuel distributor

20. Remove the intake manifold.

   a. Drain the engine coolant.

   b. Disconnect the upper radiator hose.

   c. Disconnect the lower radiator hose.

   d. Disconnect the two water hose from the thermostat component.

   e. Disconnect the thermostat component side of the water inlet pipe which is under the intake manifold.

   f. Remove the intake manifold and thermostat component as a single unit.

   g. Remove the thermostat component from the intake manifold.

To install:

21. Install the intake manifold.

   a. Tighten the intake manifold installation bolts in the order shown.

   b. Tighten the bolts to 76–101 inch lbs. (9–12 Nm).

22. Install the fuel distributor

23. Connect the fuel hose.

   a. If the quick release connector O-ring is damaged or has slipped, replace the fuel hose.

   b. A checker tab is integrated with the quick release connector for new fuel hoses and evaporative hoses. Remove the checker tab from the quick release connector after the connector is completely engaged with the fuel pipe.

   c. Remove the retainer remaining on the charcoal canister pipe.

   d. Install a new retainer to the quick release connector.

   e. Reconnect the hose straight to the pipe until a click is heard.

   f. Lightly pull and push the quick release connector a few times by hand, and then verify that it is connected securely.

24. Install the dynamic chamber.

   a. Tighten the dynamic chamber installation bolts in the order shown.

   b. Tighten the bolts to 76–101 inch lbs. (9–12 Nm).

25. Connect the vacuum hose to the master back.

26. Connect the vacuum hose to the purge solenoid valve.

   a. If the quick release connector O-ring is damaged or has slipped, replace the fuel hose.

   b. Inspect the fuel hose and fuel pipe sealing surface for damage and deformation. If there is any malfunction, replace it with a new one.

   c. Install the quick release connector.

   d. Insert the fuel pipe straight to the end of the quick release connector.

   e. Push down the retainer using a finger.

   f. If the retainer cannot be pushed down, push the fuel pipe further to the quick release connector.

   g. Lightly pull and push the quick release connector a few times by hand, and then verify that it is connected securely.

27. Install the throttle body.

28. Connect the TP sensor connector.

29. Install the resonance chamber.

30. Connect the ventilation hose.

   a. Inspect the fuel hose and fuel pipe sealing surface for damage and deformation.

   If there is any malfunction, replace it with a new one.

   b. Connect the quick release connector.

31. Connect the vacuum hose to the master back.

32. Install the air cleaner cover.

33. Connect the MAF/IAT sensor connector.

34. Install the air cleaner element.

35. Install the fresh-air duct.

36. Install the air cleaner case.

37. Install the engine cover.

38. Install the battery but leave the negative battery cable disconnected.

39. Perform a fuel leakage test as follows:

   a. Remove the fuel pump relay.

✳✳ WARNING

Short the specified terminals because shorting the wrong terminal of the main fuse block may cause malfunctions.

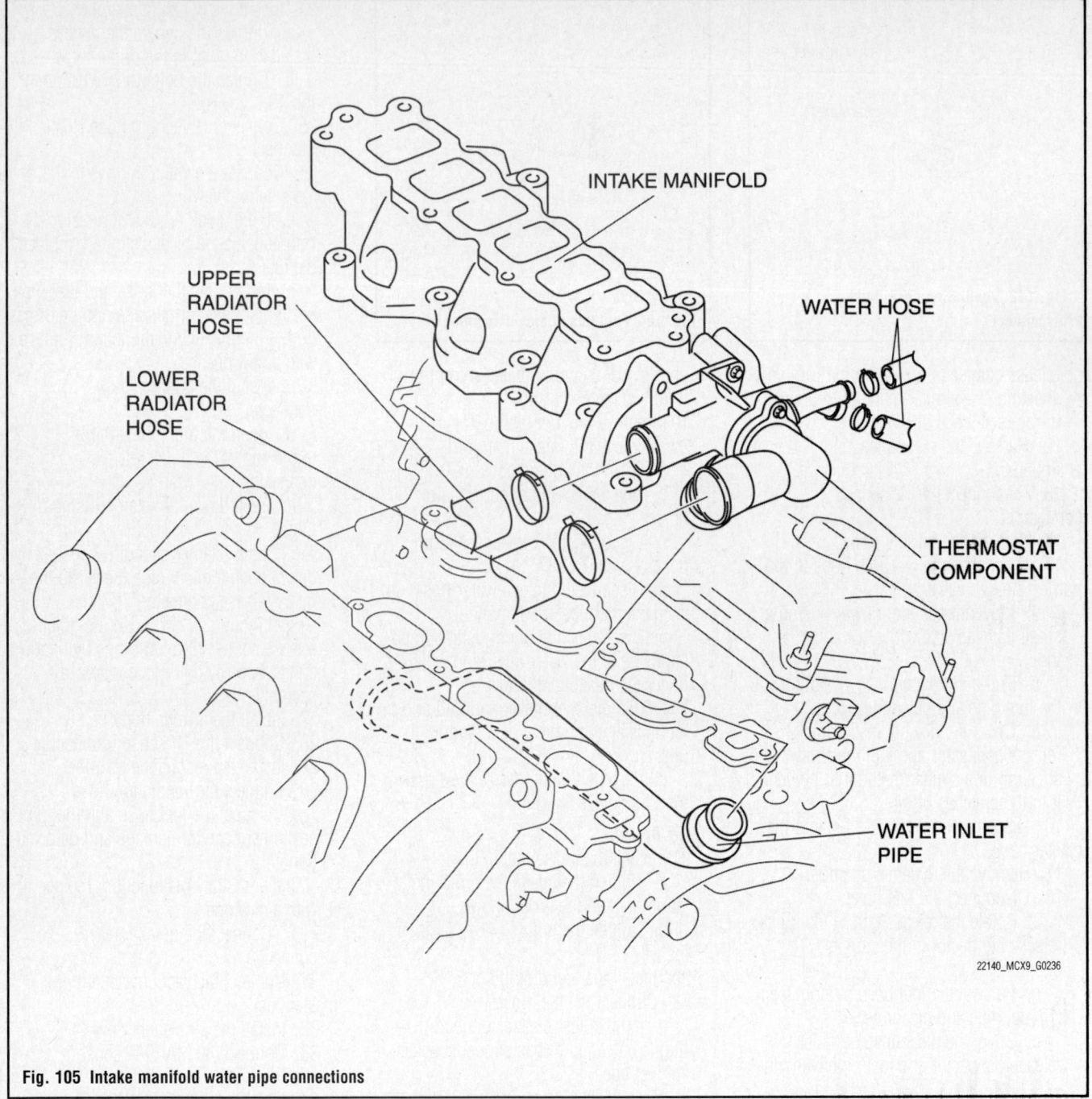

INTAKE MANIFOLD

WATER HOSE

UPPER
RADIATOR
HOSE

LOWER
RADIATOR
HOSE

THERMOSTAT
COMPONENT

WATER INLET
PIPE

22140_MCX9_G0236

Fig. 105 Intake manifold water pipe connections

b. Using a jumper wire, short fuel pump relay terminals C and D in the main fuse block.

c. Connect the negative battery cable and operate the fuel pump.

d. Verify that there is no fuel leakage from the pressurized parts.

e. If there is leakage, replace the fuel hoses and clips.

f. If there is damage on the seal on the fuel pipe side, replace the fuel pipe.

g. There should be no leakage after 5 minutes.

h. After making any repairs, assemble the system and repeat the leakage test.

OIL PAN

REMOVAL & INSTALLATION

See Figures 112 through 117.

Mazda's official procedure for the oil pan removal and installation requires the engine and transaxle to be removed from the vehicle. It may be possible to perform this procedure with the engine and transaxle in the vehicle.

1. Before servicing the vehicle, refer to the Precautions Section.

2. Drain the engine oil.

3. Remove the engine and transaxle component.

4. Using a hoist, lower the engine and transaxle component on a level surface.

5. Remove the automatic transaxle.

6. Remove the drive plate.

7. Remove the dynamic chamber and throttle body as a single unit.

8. Remove the ignition coils.

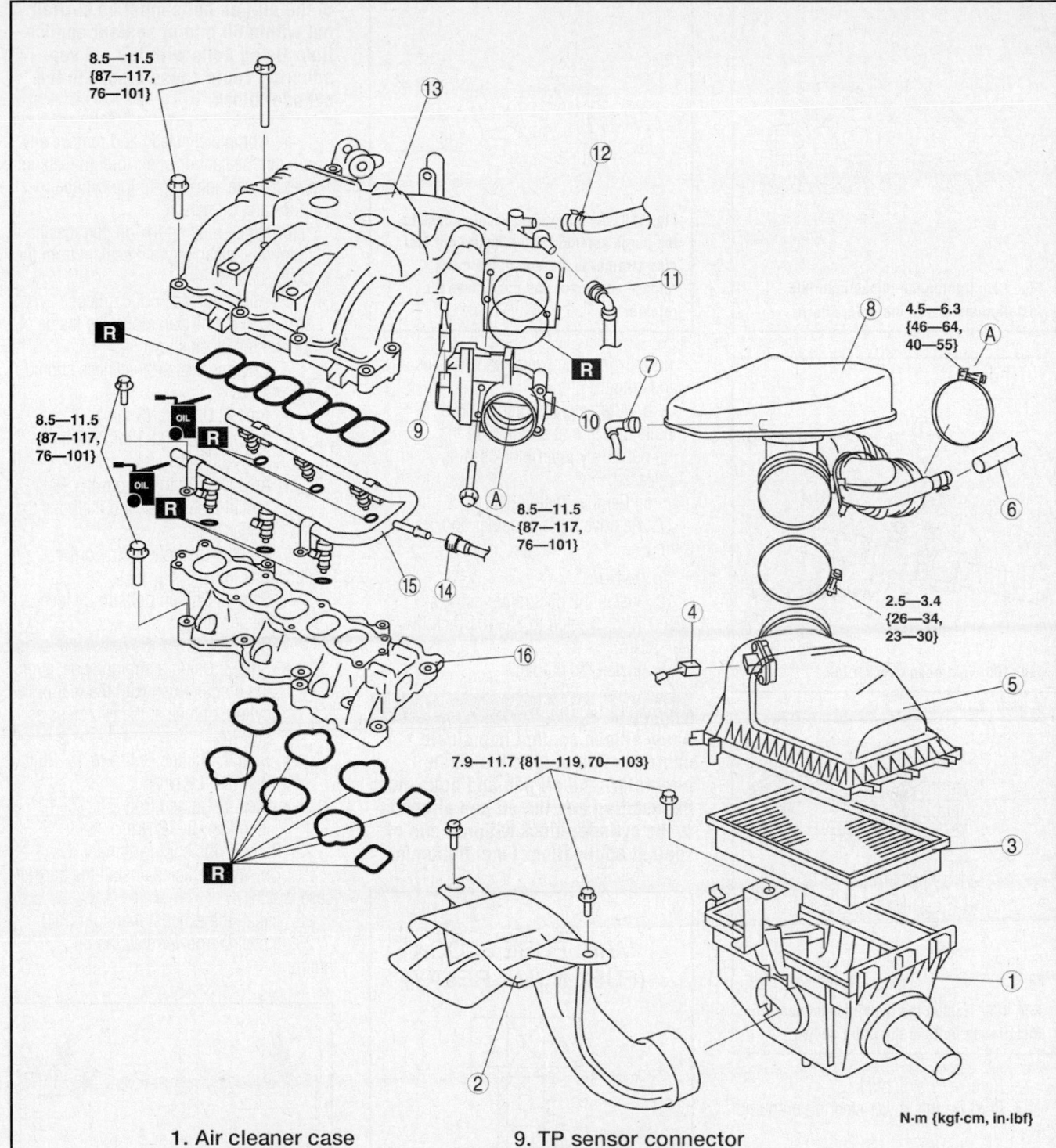

8.5—11.5
{87—117,
76—101}

4.5—6.3
{46—64,
40—55}

8.5—11.5
{87—117,
76—101}

8.5—11.5
{87—117,
76—101}

2.5—3.4
{26—34,
23—30}

7.9—11.7 {81—119, 70—103}

N·m {kgf·cm, in·lbf}

1. Air cleaner case
2. Fresh-air duct
3. Air cleaner element
4. MAF/IAT sensor connector
5. Air cleaner cover
6. Vacuum hose (to master back)
7. Ventilation hose
8. Resonance chamber

9. TP sensor connector
10. Throttle body
11. Vacuum hose (to purge solenoid valve)
12. Vacuum hose (to master back)
13. Dynamic chamber
14. Fuel hose
15. Fuel distributor
16. Intake manifold

22140_MCX9_G0232

Fig. 106 Exploded view of the air intake assembly

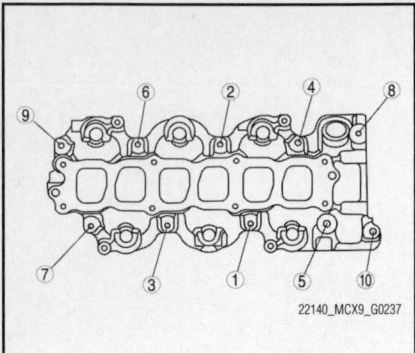

Fig. 107 Tighten the intake manifold installation bolts in the order shown

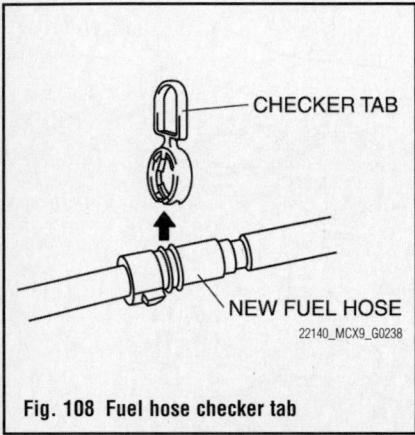

CHECKER TAB

NEW FUEL HOSE

Fig. 108 Fuel hose checker tab

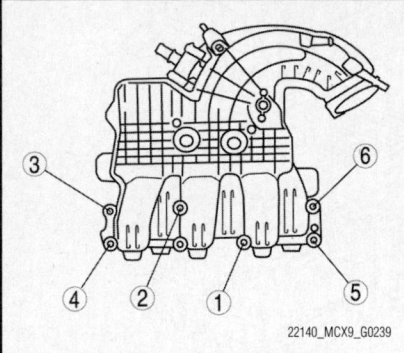

Fig. 109 Tighten the dynamic chamber installation bolts in the order shown

9. Remove the dipstick.
10. Remove the power steering oil pump drive belt.
11. Remove the power steering oil pump.
12. Remove the exhaust manifold (RH).
13. On AWD vehicles, remove the transfer case bracket.
14. Install the engine onto an engine stand.
15. Remove the engine front cover.
16. Remove the oil pan.
   a. Install two of the oil pan bolts tem-

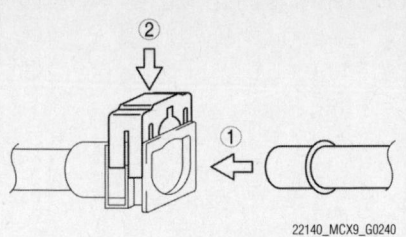

Fig. 110 To connect the vacuum hose to the purge solenoid valve, insert the fuel pipe straight to the end of the quick release connector and push down the retainer

porarily into the two threaded holes in the oil pan.
   b. Alternately tighten the two bolts one turn at a time until the oil pan-to-cylinder block seal is released.
   c. Remove the oil pan.
17. Remove the oil strainer and O-ring.

To install:

18. Install the oil strainer using a new O-ring. Tighten to 76–101 inch lbs. (9–12 Nm).
19. Install the oil pan.

❋❋ WARNING

Apply silicon sealant in a single, unbroken line around the whole perimeter. The oil pan and bolts must be installed and the oil pan aligned to the cylinder block within 5 min of sealant application. Final tightening

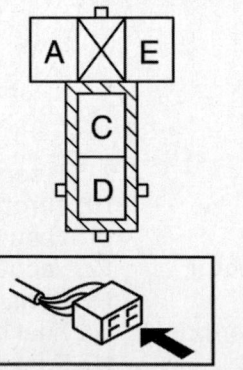

MAIN FUSE BLOCK (FUEL PUMP RELAY)

Fig. 111 Using a jumper wire, short fuel pump relay terminals C and D in the main fuse block to check for fuel leaks

of the oil pan bolts must be carried out within 60 min of sealant application. Using bolts with the old seal adhering could cause cracks in the cylinder block.

   a. Completely clean and remove any oil, dirt, sealant or other foreign material that may be adhering to the cylinder block and oil pan.
   b. When reusing the oil pan installation bolts, clean any old sealant from the bolts.
   c. Apply Loctite® 5900 silicone sealant to the oil pan along the inside of the bolt holes as shown.
   d. Silicone sealant thickness should be as follows:
   • Area A: 0.12 in. (3 mm)
   • Area B: 0.20–0.23 in. 0.12 in (5.0–6.0 mm)
   • Area C: 0.39 in (10 mm)
   e. Install the oil pan and the bolts to the cylinder block.
   f. Tighten the bolts in the order shown as follows:
   • Step 1: Tighten bolts to 27 inch lbs. (3 Nm).
   • Step 2: Loosen the bolts 180°.
   • Step 3: Using a straightedge, align the oil pan flush with the rear of the cylinder block at the two areas as shown.
   • Step 4: Tighten Bolt A to 15–20 ft. lbs. (20–28 Nm).
   • Step 4: Tighten Bolt B to 76–101 inch lbs. (9–12 Nm)
20. Remove the engine stand.
21. On AWD vehicles, install the transfer case bracket.
22. Install the exhaust manifold (RH).
23. Install the power steering oil pump.

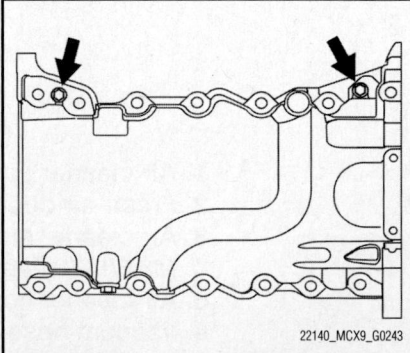

Fig. 112 Install two of the oil pan bolts temporarily into the two threaded holes in the oil pan. Alternately tighten the two bolts one turn at a time until the oil pan-to-cylinder block seal is released

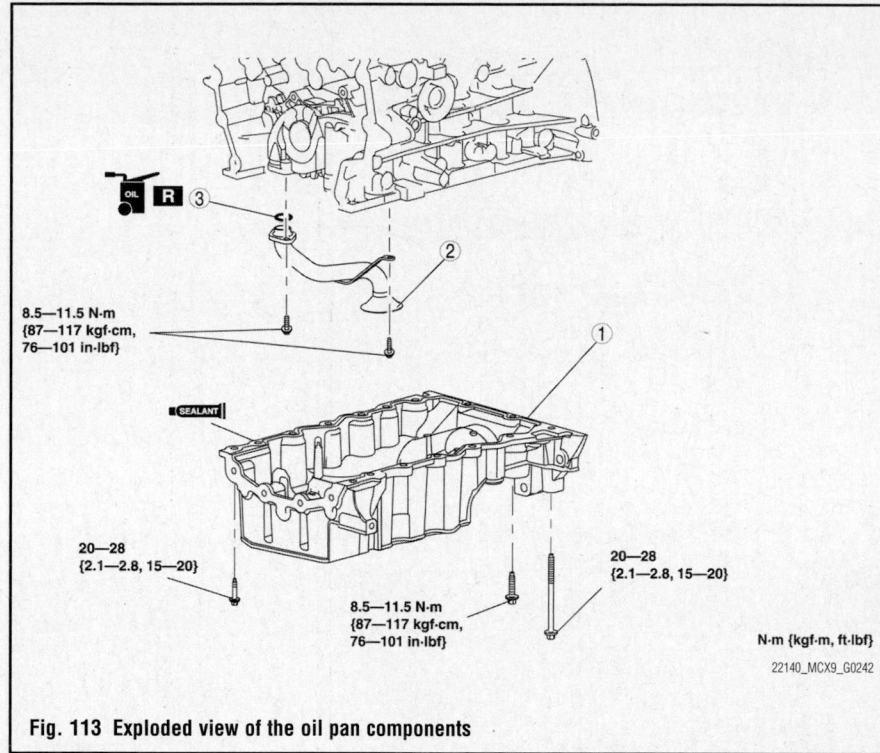

8.5—11.5 N·m
{87—117 kgf·cm,
76—101 in·lbf}

20—28
{2.1—2.8, 15—20}

8.5—11.5 N·m
{87—117 kgf·cm,
76—101 in·lbf}

20—28
{2.1—2.8, 15—20}

N·m {kgf·m, ft·lbf}

22140_MCX9_G0242

Fig. 113 Exploded view of the oil pan components

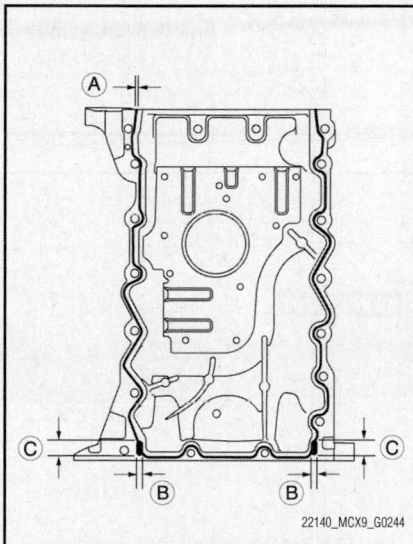

22140_MCX9_G0244

Fig. 114 Apply Loctite® 5900 silicone sealant to the oil pan along the inside of the bolt holes as shown

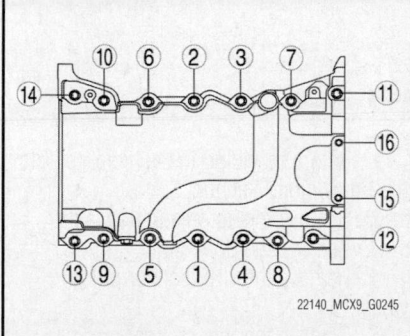

22140_MCX9_G0245

Fig. 115 Tighten the oil pan bolts in the order shown

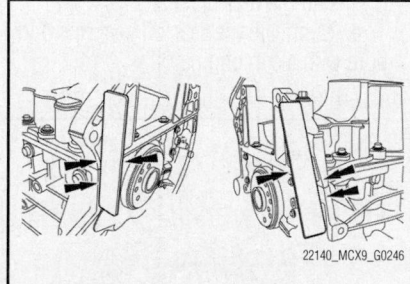

22140_MCX9_G0246

Fig. 116 Using a straightedge, align the oil pan flush with the rear of the cylinder block at the two areas as shown

24. Install the power steering oil pump drive belt.
25. Install the dipstick.
26. Install the ignition coils.
27. Install the dynamic chamber and throttle body as a single unit.
28. Install the drive plate.
29. Install the automatic transaxle.
30. Using a hoist, lower the engine and transaxle component on a level surface.

31. Install the engine and transaxle component.
32. Fill the engine with oil.

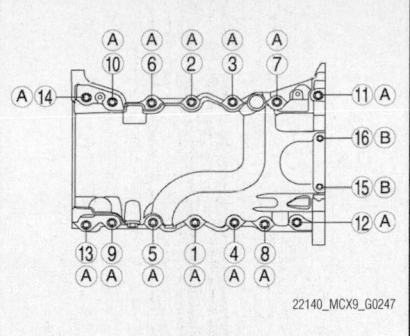

22140_MCX9_G0247

Fig. 117 Final tighten the oil pan bolts in the order shown and to the proper torque

33. Install the engine front cover.
34. Start the engine and perform the following inspections:
  a. Inspect the runout and contact on the pulley and belt.
  b. Inspect for engine oil, engine coolant, ATF, power steering fluid and fuel leakage.
  c. Verify the ignition timing, idle speed.
  d. Inspect engine accessories for proper operation.
35. Perform a road test.

OIL PUMP

REMOVAL & INSTALLATION

See Figure 118.

Mazda's official procedure for the oil pump removal and installation requires the engine and transaxle to be removed from the vehicle. It may be possible to perform this procedure with the engine and transaxle in the vehicle.
1. Before servicing the vehicle, refer to the Precautions Section.
2. Drain the engine oil.
3. Remove the engine and transaxle component.
4. Secure the engine and transaxle using an engine stand.
5. Remove the dynamic chamber and throttle body as a single unit.
6. Remove the ignition coils.
7. Remove the dipstick.
8. Remove the power steering oil pump drive belt.
9. Remove the power steering oil pump.
10. Remove the timing chain.
11. Remove the oil pump and O-ring.

To install:
12. Install the oil pump and O-ring. Tighten to 76–101 inch lbs. (9–12 Nm)
13. Install the timing chain.
14. Install the power steering oil pump.

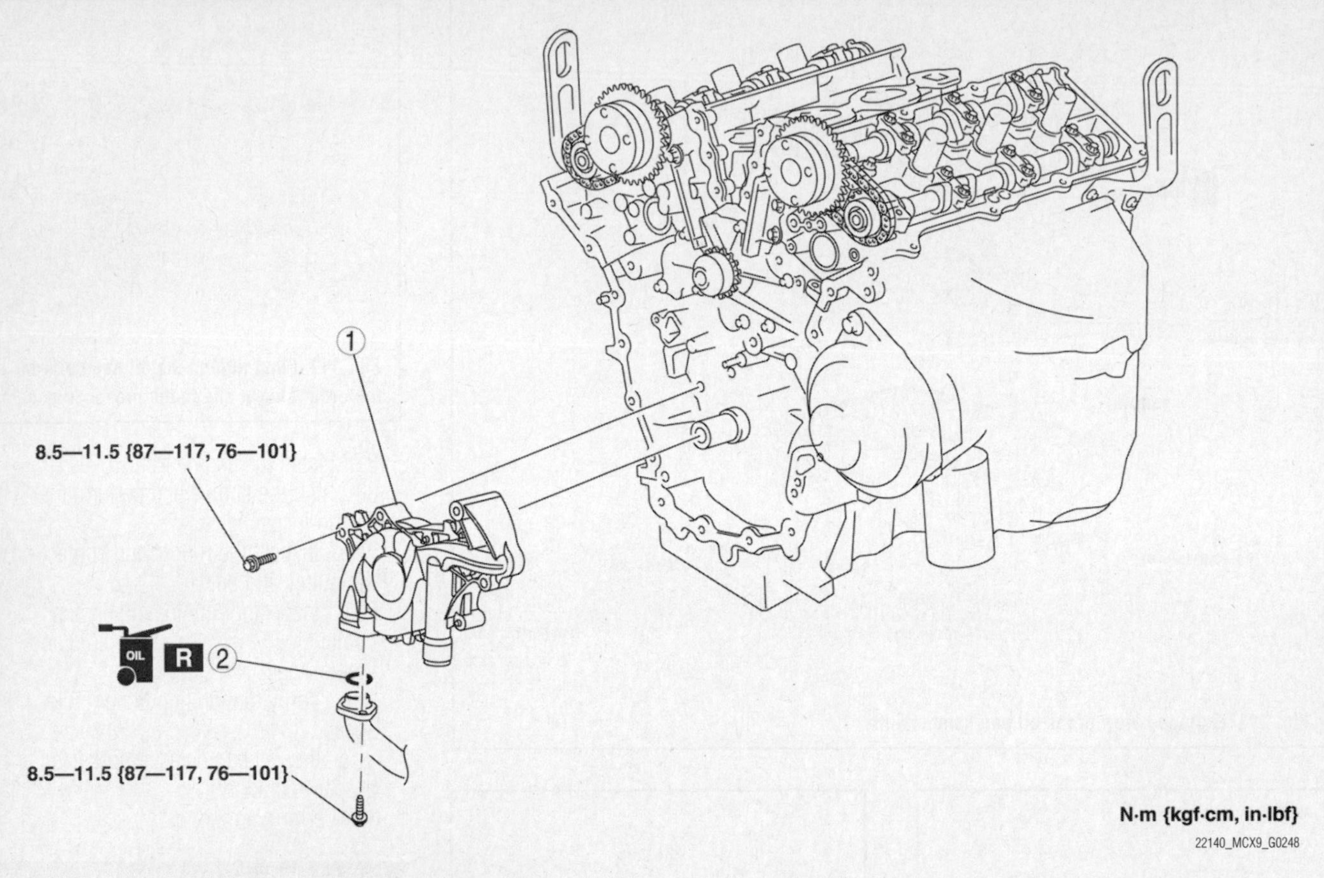

8.5—11.5 {87—117, 76—101}

8.5—11.5 {87—117, 76—101}

N·m {kgf·cm, in·lbf}

22140_MCX9_G0248

Fig. 118 Exploded view of the oil pump (1) components

15. Install the power steering oil pump drive belt.

16. Install the dipstick.

17. Install the ignition coils.

18. Install the dynamic chamber and throttle body as a single unit.

19. Secure the engine and transaxle using an engine stand.

20. Install the engine and transaxle component.

21. Fill the engine with oil.

22. Start the engine and perform the following inspections:

　a. Inspect the runout and contact on the pulley and belt.

　b. Inspect for engine oil, engine coolant, ATF, power steering fluid and fuel leakage.

　c. Verify the ignition timing, idle speed.

　d. Inspect engine accessories for proper operation.

　e. Inspect the oil pressure.

23. Perform a road test.

REAR MAIN SEAL

REMOVAL & INSTALLATION

See Figures 119 and 120.

1. Before servicing the vehicle, refer to the Precautions Section.

2. Remove the automatic transaxle.

3. Remove the drive plate.

4. Remove the CKP sensor pulse wheel.

5. Remove the rear oil seal.

　a. Using a flathead screwdriver, remove the rear oil seal.

To install:

6. Install the rear oil seal.

　a. Lubricate the rear oil seal lips and bore with clean engine oil.

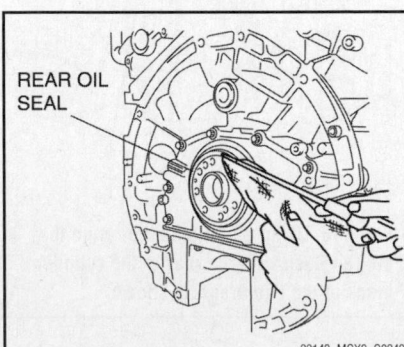

REAR OIL SEAL

22140_MCX9_G0249

Fig. 119 Using a flathead screwdriver, remove the rear oil seal

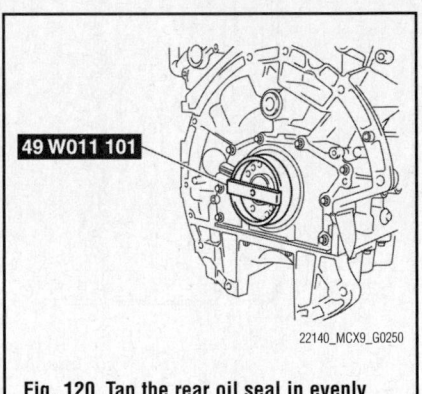

49 W011 101

22140_MCX9_G0250

Fig. 120 Tap the rear oil seal in evenly using the special tool and a hammer

　b. Push the rear oil seal slightly in by hand.

　c. Tap the rear oil seal in evenly using the special tool and a hammer.

7. Install the CKP sensor pulse wheel.

8. Install the drive plate.

9. Install the automatic transaxle.

TIMING CHAIN FRONT COVER & SEAL

REMOVAL & INSTALLATION

See Figures 121 through 138.

Mazda's official procedure for the timing chain cover removal and installation requires the engine and transaxle to be removed from the vehicle. It may be possible to perform this procedure with the engine and transaxle in the vehicle.

1. Before servicing the vehicle, refer to the Precautions Section.
2. Drain the engine oil.
3. Remove the engine and transaxle.
4. Secure the engine and transaxle using a hoist and an engine stand.
5. Remove the intake manifold dynamic chamber and throttle body as a single unit.
6. Remove the ignition coils.
7. Remove the dipstick.
8. Remove the power steering oil pump drive belt.
9. Remove the power steering oil pump.
10. Remove the drive belt auto tensioner.
11. Remove the crankshaft pulley lock bolt.

a. Remove the starter.
b. Set a flathead screwdriver to the drive plate as shown to lock the crankshaft rotation.
c. Remove the crankshaft pulley lock bolt and washer.
12. Remove the crankshaft pulley.
a. Install a suitable spacer 0.55 in. (14 mm) in thickness and 1.18 in. (30 mm) in diameter to the crankshaft pulley lock bolt, and install the crankshaft pulley lock bolt to the crankshaft.
b. Remove the crankshaft pulley using a gear puller.
c. Remove the crankshaft pulley lock bolt and spacer.
13. Remove the front oil seal.
a. Remove the front oil seal using a flathead screwdriver as shown
14. Remove the cylinder head covers.
a. Remove the cylinder head cover bolts in the order shown.
15. Remove the cylinder head cover oil seals.

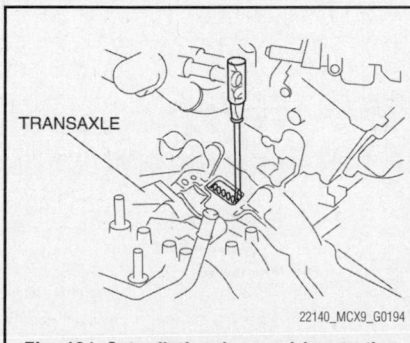
Fig. 121 Set a flathead screwdriver to the drive plate as shown to lock the crankshaft rotation

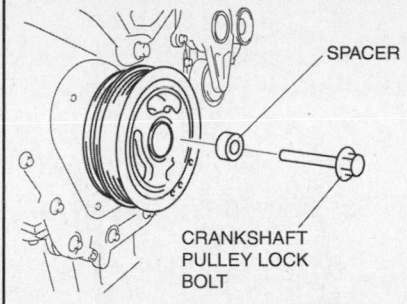
Fig. 122 Install a suitable spacer 0.55 in. (14 mm) in thickness and 1.18 in. (30 mm) in diameter to the crankshaft pulley lock bolt, and install the crankshaft pulley lock bolt to the crankshaft

a. Inspect the OCV attachment hole seals and spark plug tube attachment hole seals. Remove the any damaged seals.
b. The OCV attachment hole seals removal is shown, the spark plug tube attachment hole seals removal procedure is the same.
c. To remove, tap the OCV attachment hole seals and spark plug tube attachment hole seals using the special tool and hammer.
16. Remove the no.3 engine mount bracket and engine front cover.
a. Loosen the engine front cover and No.3 engine mount bracket installation bolts in the order shown.
b. Install 4 of the engine front cover bolts (finger tightened) into the 4 threaded holes in the engine front cover.
c. Tighten the bolts one turn at a time in a criss-cross pattern until the engine front cover-to-cylinder block seal is released.
d. Remove the engine front cover.

To install:
17. Install the engine front cover and No.3 engine mount bracket.

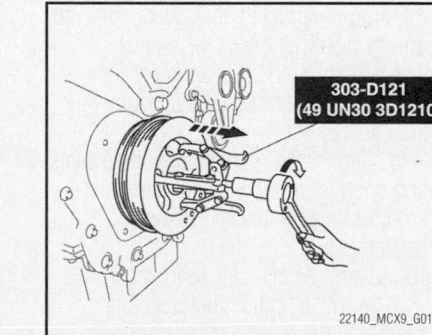
Fig. 123 Remove the crankshaft pulley using a gear puller

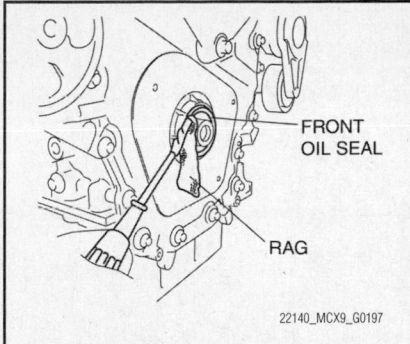
Fig. 124 Using a flat tip screwdriver cover with a rag, pry the seal from the front cover

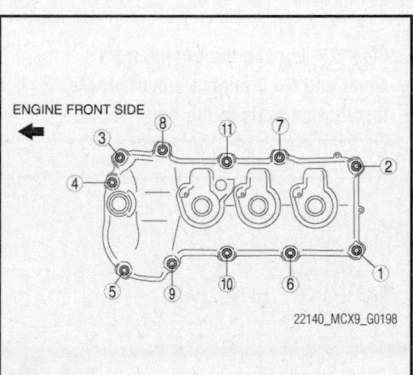
Fig. 125 Remove the cylinder head cover bolts in the order shown—right hand

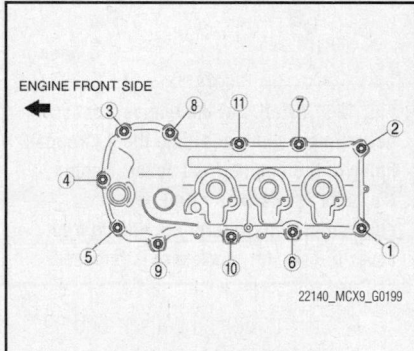
Fig. 126 Remove the cylinder head cover bolts in the order shown—left hand

a. Apply the Loctite® 5900 silicon sealant to the engine front cover as shown.
b. Silicon sealant thickness should be as follows: A—0.197–0.236 in. (5.0–6.0 mm) and B— 0.099–0.137 in. (2.5–3.5 mm).
c. Install the engine front cover and No.3 engine mount bracket installation bolts within 10 min of applying the silicone sealant.
d. Tighten the engine front cover and No.3 engine mount bracket installation bolts in 4 steps in the order shown.

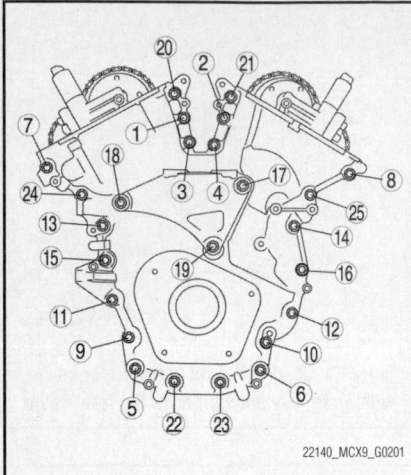

Fig. 127 Loosen the engine front cover and No.3 engine mount bracket installation bolts in the order shown

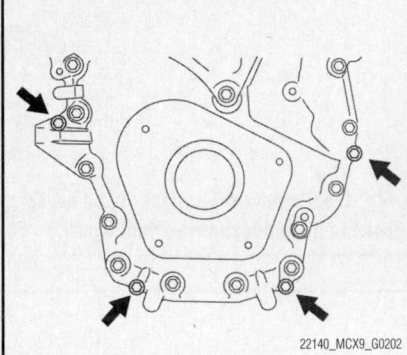

Fig. 128 Install 4 of the engine front cover bolts (finger tightened) into the 4 threaded holes in the engine front cover. Tighten the bolts one turn at a time in a criss-cross pattern until the engine front cover-to-cylinder block seal is released

- Step 1: Tighten bolts 1–6 to 89 inch lbs. (10 Nm).
- Step 2: Tighten bolts 7–9 to 11 ft. lbs. (15 Nm).
- Step 3: Tighten bolts 1–6 to 15–20 ft. lbs. (20–28 Nm).
- Step 4: Tighten bolts 7–9 to 52–59 ft. lbs. (70–80 Nm).

e. Install the engine front cover installation bolts within 60 min of applying the silicone sealant.

f. Tighten the engine front cover installation bolts in the order in 2 steps in the order shown.

- Step 1: Tighten to 89 inch lbs. (10 Nm).
- Step 2: Tighten to 15–20 ft. lbs. (20–28 Nm).

18. Install the cylinder head cover oil seal.

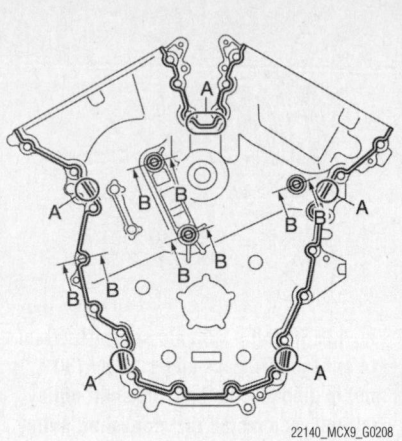

Fig. 129 Apply the Loctite® 5900 silicon sealant to the engine front cover as shown. Silicon sealant thickness should be as follows: A—0.197–0.236 in. (5.0–6.0 mm) and B— 0.099–0.137 in. (2.5–3.5 mm)

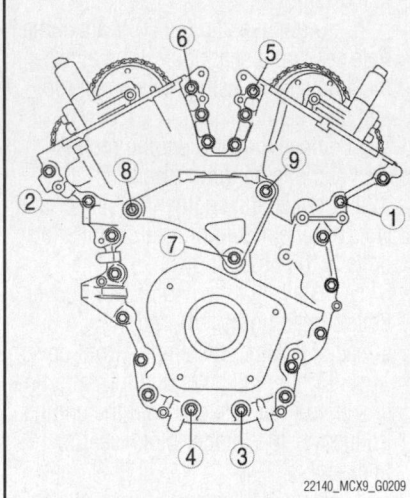

Fig. 130 Tighten the engine front cover and No.3 engine mount bracket installation bolts in the order shown

a. Installation of new seals is only required if damaged seals were removed during disassembly of the engine.

b. Push the OCV attachment hole seals and spark plug tube attachment hole seals slightly in by hand.

c. Tap the OCV attachment hole seals and spark plug tube attachment hole seals using the special tool and hammer.

19. Install the cylinder head cover.

a. Apply Loctite® 5900 silicone sealant to the mating faces as shown.

b. Install the cylinder head cover with a new gasket.

c. Install the cylinder head cover

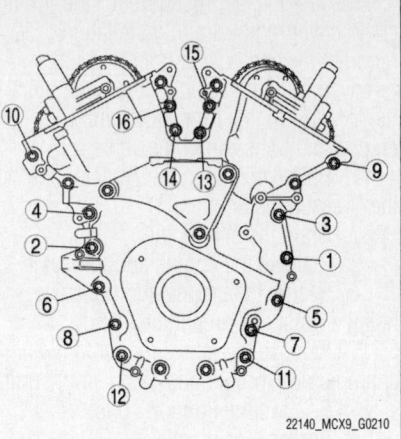

Fig. 131 Tighten the engine front cover installation bolts in the order in the order shown

installation bolt and studs within 5 min of applying the silicone sealant.

d. Tighten the bolts in the order shown to 76–101 inch lbs. (9–12 Nm).

20. Install the front oil seal

a. Apply clean engine oil to the front oil seal bore in the engine front cover.

b. Push the front oil seal slightly in by hand.

c. Tap the front oil seal in evenly using the special tool and a hammer.

21. Install the crankshaft pulley

22. Install the crankshaft pulley lock bolt.

a. Set a flathead screwdriver to the drive plate in the position indicated. to lock the crankshaft rotation.

b. Tighten the new crankshaft pulley lock bolt in four steps.

- Step 1: Tighten to 103 ft. lbs. (140 Nm)
- Step 2: Loosen 360° (one full turn)
- Step 3: Tighten to 35–39 ft. lbs (47–53 Nm)
- Step 4: Tighten an additional 85–95° of rotation

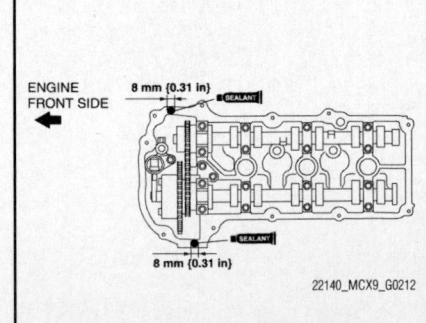

Fig. 132 Apply Loctite® 5900 silicone sealant to the mating faces as shown— right side

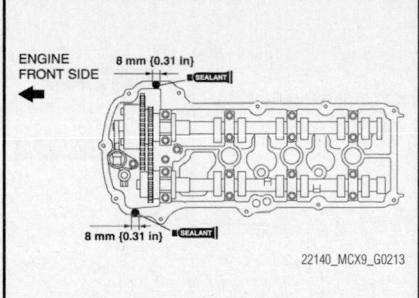

Fig. 133 Apply Loctite 5900 silicone sealant to the mating faces as shown—left side

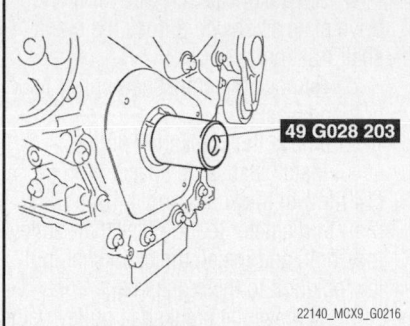

Fig. 136 Tap the front oil seal in evenly using the special tool and a hammer

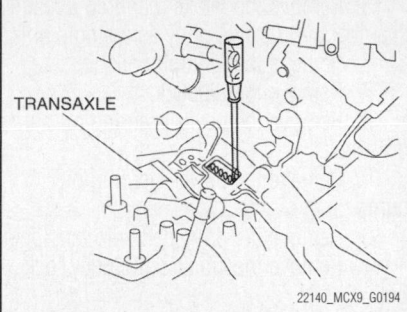

Fig. 139 Set a flathead screwdriver to the drive plate as shown to lock the crankshaft rotation

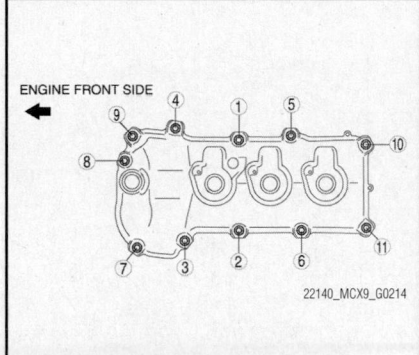

Fig. 134 Tighten the cylinder head cover bolts in the order shown—right side

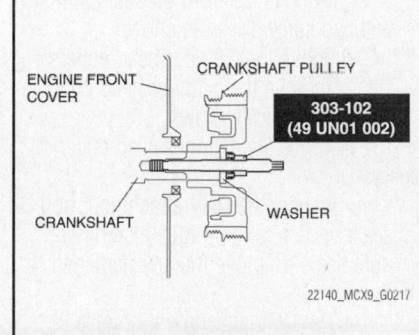

Fig. 137 Install the crankshaft pulley, washer, and special tool to the crankshaft

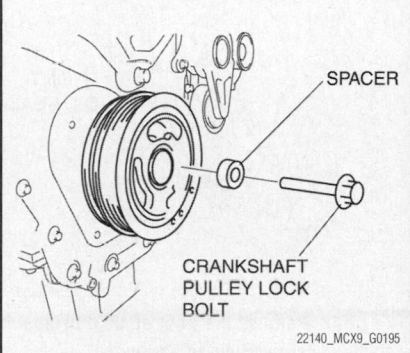

Fig. 140 Install a suitable spacer 0.55 in. (14 mm) in thickness and 1.18 in. (30 mm) in diameter to the crankshaft pulley lock bolt, and install the crankshaft pulley lock bolt to the crankshaft

23. Install the power steering oil pump.
24. Install the power steering oil pump drive belt.
25. Install the dipstick.
26. Install the ignition coils.
27. Install the intake manifold dynamic chamber and throttle body as a single unit.
28. Install the engine and transaxle.
29. Fill the engine with oil.
30. Start the engine and perform the following inspections:
   a. Inspect the runout and contact on the pulley and belt.
   b. Inspect for engine oil, engine

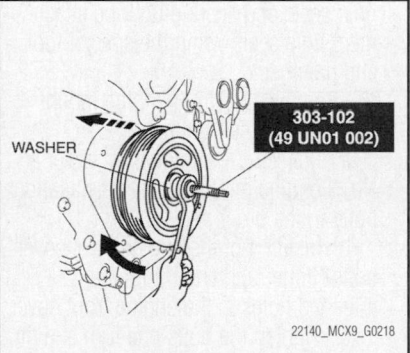

Fig. 138 Tighten the special tool nut and install the crankshaft pulley

coolant, ATF, power steering fluid and fuel leakage.
   c. Verify the ignition timing and idle speed.
   d. Inspect engine accessories for proper operation.
31. Perform a road test.

TIMING CHAIN & SPROCKETS

REMOVAL & INSTALLATION

See Figures 139 through 166.

Mazda's official procedure for timing chain

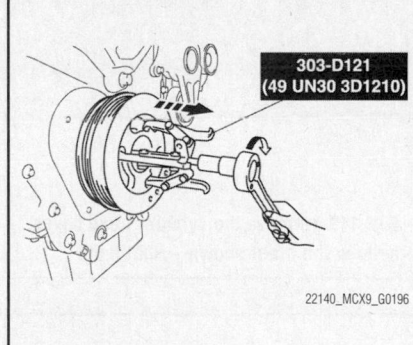

Fig. 141 Remove the crankshaft pulley using a gear puller

removal and installation requires the engine and transaxle to be removed from the vehicle. It may be possible to perform this procedure with the engine and transaxle in the vehicle.
1. Before servicing the vehicle, refer to the Precautions Section.
2. Drain the engine oil.
3. Remove the engine and transaxle.
4. Secure the engine and transaxle using a hoist and an engine stand.

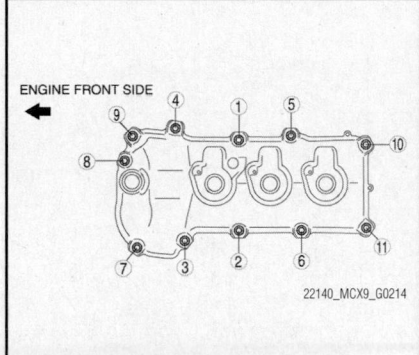

Fig. 135 Tighten the cylinder head cover bolts in the order shown—left side

5. Remove the intake manifold dynamic chamber and throttle body as a single unit.

6. Remove the ignition coils.

7. Remove the dipstick.

8. Remove the power steering oil pump drive belt.

9. Remove the power steering oil pump.

10. Remove the drive belt auto tensioner.

11. Remove the crankshaft pulley lock bolt.

a. Remove the starter.

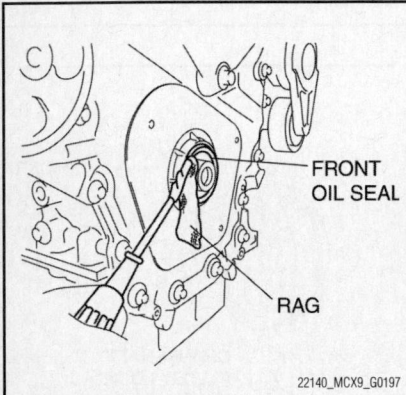

Fig. 142 Remove the front oil seal using a flathead screwdriver as shown

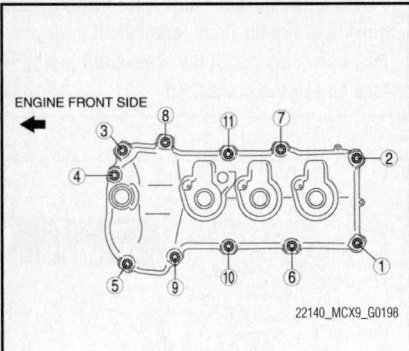

Fig. 143 Remove the cylinder head cover bolts in the order shown—right hand

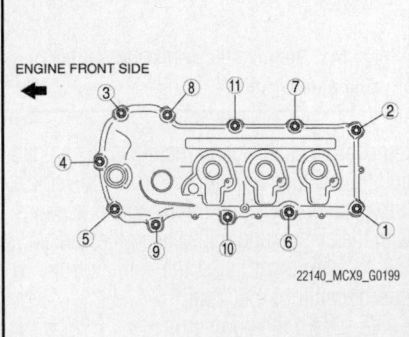

Fig. 144 Remove the cylinder head cover bolts in the order shown—left hand

b. Set a flathead screwdriver to the drive plate as shown to lock the crankshaft rotation.

c. Remove the crankshaft pulley lock bolt and washer.

12. Remove the crankshaft pulley.

a. Install a suitable spacer 0.55 in. (14 mm) in thickness and 1.18 in. (30 mm) in diameter to the crankshaft pulley lock bolt, and install the crankshaft pulley lock bolt to the crankshaft.

b. Remove the crankshaft pulley using a gear puller.

c. Remove the crankshaft pulley lock bolt and spacer.

13. Remove the front oil seal.

a. Remove the front oil seal using a flathead screwdriver as shown

14. Remove the cylinder head covers.

a. Remove the cylinder head cover bolts in the order shown.

15. Remove the cylinder head cover oil seals.

a. Inspect the OCV attachment hole seals and spark plug tube attachment hole seals. Remove the any damaged seals.

b. The OCV attachment hole seals removal is shown, the spark plug tube attachment hole seals removal procedure is the same.

c. To remove, tap the OCV attachment hole seals and spark plug tube attachment hole seals using the special tool and hammer.

16. Remove the No. 3 engine mount bracket and engine front cover.

a. Loosen the engine front cover and No. 3 engine mount bracket installation bolts in the order shown.

b. Install 4 of the engine front cover bolts (finger tightened) into the 4 threaded holes in the engine front cover.

c. Tighten the bolts one turn at a time in a criss-cross pattern until the engine front cover-to-cylinder block seal is released.

d. Remove the engine front cover.

17. Remove the OCV component.

a. Loosen the OCV component installation bolts in the order shown

18. Remove the chain tensioner.

19. Remove the timing chain.

➡When removing the timing chain and marking the timing marks on the chain, mark the camshaft timing chain as well.

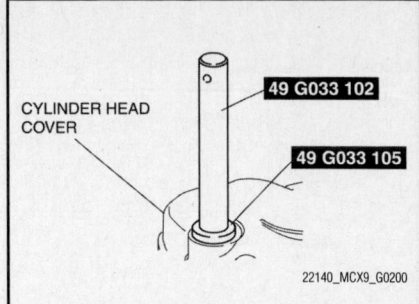

Fig. 145 Tap the OCV attachment hole seals and spark plug tube attachment hole seals using the special tool and hammer

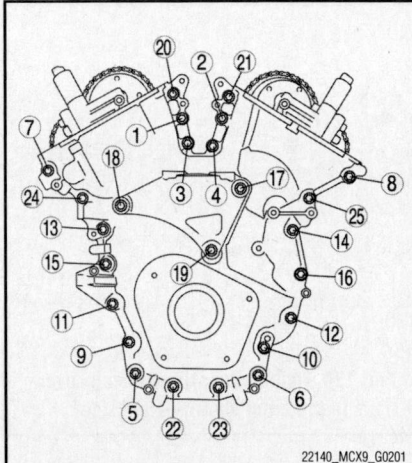

Fig. 146 Loosen the engine front cover and No.3 engine mount bracket installation bolts in the order shown

Do not rotate the crankshaft counterclockwise. The timing chains may bind, causing engine damage.

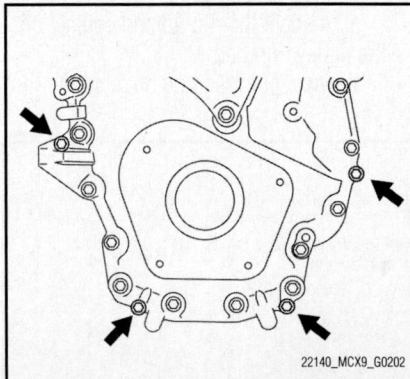

Fig. 147 Install 4 of the engine front cover bolts (finger tightened) into the 4 threaded holes in the engine front cover. Tighten the bolts one turn at a time in a criss-cross pattern until the engine front cover-to-cylinder block seal is released

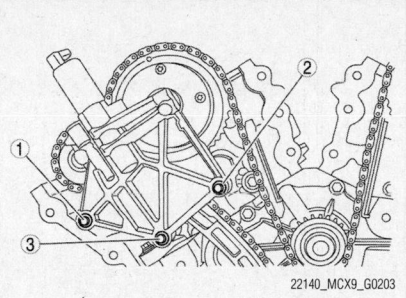

Fig. 148 Loosen the OCV component installation bolts in the order shown—right hand

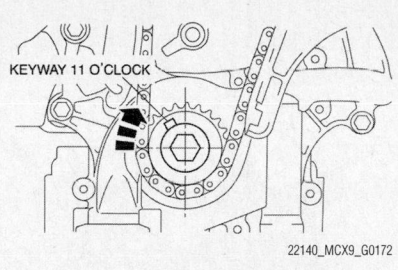

KEYWAY 11 O'CLOCK

Fig. 150 Turn the crankshaft clockwise so that the crankshaft keyway is in the 11 o'clock position. This will position the No.1 cylinder at TDC

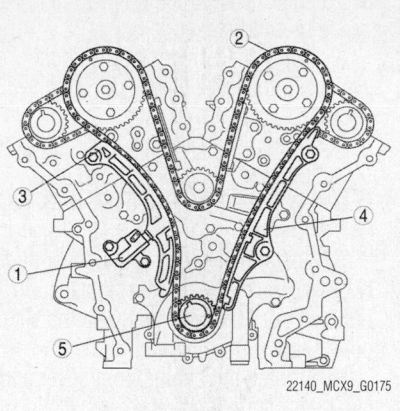

Fig. 152 Remove the timing chain in the following order: (1) Chain tensioner, (2) Timing chain, (3) Tensioner arm, (4) Chain guide and (5) Crankshaft sprocket

a. Turn the crankshaft clockwise so that the crankshaft keyway is in the 11 o'clock position. This will position the No.1 cylinder at TDC.

➡Verify that there are timing marks in three locations (Yellow 1, Black 2) on the timing chain. If any timing marks are missing, mark the timing chain. When marking the crankshaft sprocket side timing chain, change the mark color. When the timing chain is replaced with a new one, mark the new timing chain at the same positions as the removed timing chain.

b. Mark the timing chain at the position of each timing sprocket timing mark.

➡Verify that there are timing marks in two locations on the camshaft timing chain. If any timing marks are missing, mark the camshaft timing chain. If replacing with a new camshaft timing chain, place alignment marks in the same positions as those prior to the replacement.

c. Mark the camshaft timing chain at the positions where it is aligned with each of the camshaft sprocket on both banks.

d. Remove the timing chain in the following order:
- Chain tensioner
- Timing chain
- Tensioner arm
- Chain guide
- Crankshaft sprocket

To install:
20. Install the timing chain
a. Verify that the crankshaft keyway is at 11 o'clock position.

➡Of the three marked locations on the timing chain, align the mark that has a different color to the crankshaft sprocket side timing mark.

b. Place alignment marks on the timing chain corresponding to each of the alignment marks on the timing sprocket.

c. Push down the link plate of the timing chain tensioner and release the plunger lock.

➡The plunger should retract with minimal force. If binding occurs, remove the chain tensioner from the vise and reset it in the vise.

d. Secure the chain tensioner using a vice attached with a soft base, and slowly press the plunger back shown. while pressing down the link plate.

e. Release the pressure slightly from the plunger, and move the plunger back and forth 0.08–0.11 in (2–3 mm).

f. Insert a paper clip where the link plate hole and the tensioner body hole overlap to fix the link plate and lock the plunger.

g. Install the timing chain in the following order:
- Crankshaft sprocket
- Chain guide
- Tensioner arm
- Timing chain
- Chain tensioner

h. Tighten components to 76–101 inch lbs. (9–12 Nm).
i. Remove the retaining wire.
21. Install the chain tensioner
22. Install the OCV components.
a. Tighten the OCV component installation bolts in the order shown

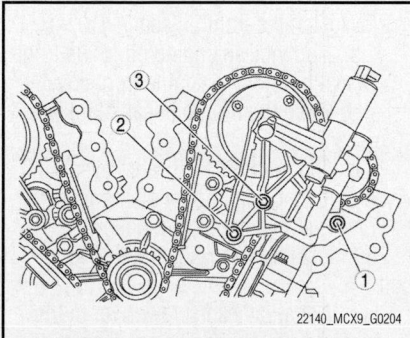

Fig. 149 Loosen the OCV component installation bolts in the order shown—left hand

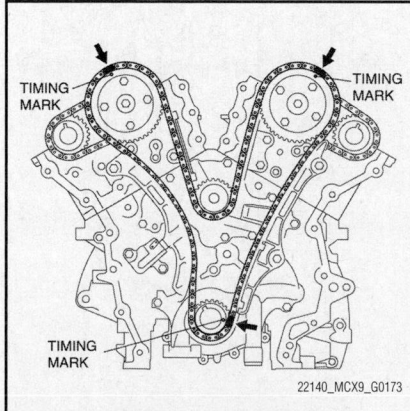

TIMING MARK

TIMING MARK

TIMING MARK

Fig. 151 Mark the timing chain at the position of each timing sprocket timing mark

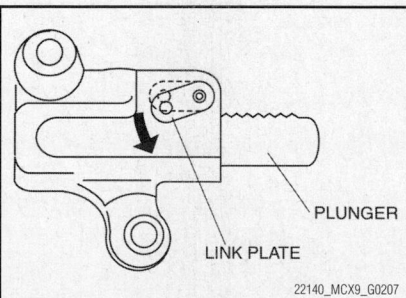

PLUNGER

LINK PLATE

Fig. 153 Insert a paper clip where the link plate hole and the tensioner body hole overlap to fix the link plate and lock the plunger

23. Install the engine front cover and No. 3 engine mount bracket.

a. Apply the Loctite 5900 silicon sealant to the engine front cover as shown.

b. Silicon sealant thickness should be as follows: A—0.197–0.236 in. (5.0–6.0 mm) and B— 0.099–0.137 in. (2.5–3.5 mm).

c. Install the engine front cover and No.3 engine mount bracket installation bolts within 10 min of applying the silicone sealant.

d. Tighten the engine front cover and No.3 engine mount bracket installation bolts in 4 steps in the order shown.

- Step 1: Tighten bolts 1–6 to 89 inch lbs. (10 Nm).
- Step 2: Tighten bolts 7–9 to 11 ft. lbs. (15 Nm).
- Step 3: Tighten bolts 1–6 to 15–20 ft. lbs. (20–28 Nm).
- Step 4: Tighten bolts 7–9 to 52–59 ft. lbs. (70–80 Nm).

e. Install the engine front cover installation bolts within 60 min of applying the silicone sealant.

f. Tighten the engine front cover installation bolts in the order in 2 steps in the order shown.

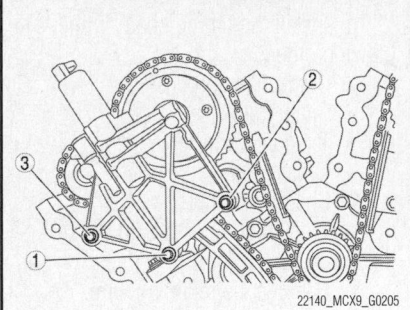

Fig. 154 Tighten the OCV component installation bolts in the order shown— right hand

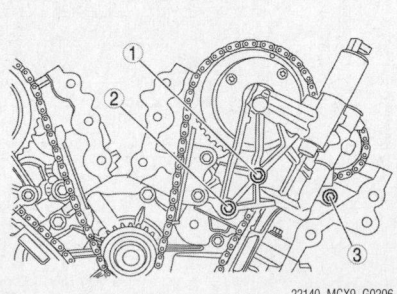

Fig. 155 Tighten the OCV component installation bolts in the order shown—left hand

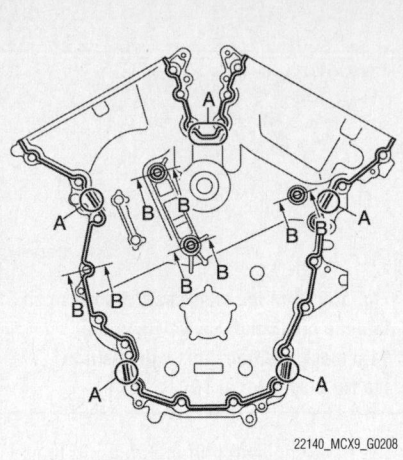

Fig. 156 Apply the Loctite 5900 silicon sealant to the engine front cover as shown. Silicon sealant thickness should be as follows: A—0.197–0.236 in. (5.0–6.0 mm) and B— 0.099–0.137 in. (2.5–3.5 mm)

- Step 1: Tighten to 89 inch lbs. (10 Nm).
- Step 2: Tighten to 15–20 ft. lbs. (20–28 Nm).

24. Install the cylinder head cover oil seal.

a. Installation of new seals is only required if damaged seals were removed during disassembly of the engine.

b. Push the OCV attachment hole seals and spark plug tube attachment hole seals slightly in by hand.

c. Tap the OCV attachment hole seals and spark plug tube attachment hole seals using the special tool and hammer.

25. Install the cylinder head cover.

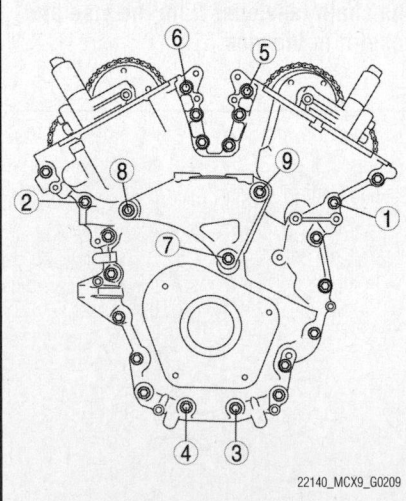

Fig. 157 Tighten the engine front cover and No.3 engine mount bracket installation bolts in the order shown

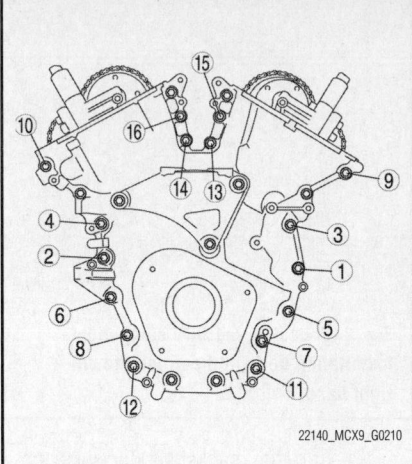

Fig. 158 Tighten the engine front cover installation bolts in the order in the order shown

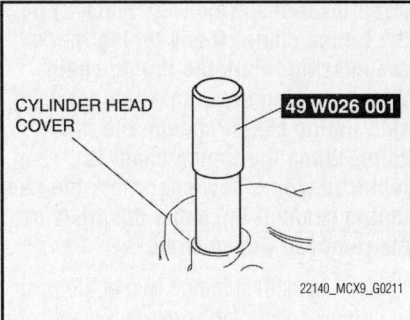

Fig. 159 Tap the OCV attachment hole seals and spark plug tube attachment hole seals using the special tool and hammer

a. Apply Loctite 5900 silicone sealant to the mating faces as shown.

b. Install the cylinder head cover with a new gasket.

c. Install the cylinder head cover installation bolt and studs within 5 min of applying the silicone sealant.

d. Tighten the bolts in the order shown to 76–101 inch lbs. (9–12 Nm).

26. Install the front oil seal

a. Apply clean engine oil to the front oil seal bore in the engine front cover.

b. Push the front oil seal slightly in by hand.

c. Tap the front oil seal in evenly using the special tool and a hammer.

27. Install the crankshaft pulley

28. Install the crankshaft pulley lock bolt.

a. Set a flathead screwdriver to the drive plate in the position indicated. to lock the crankshaft rotation.

b. Tighten the new crankshaft pulley lock bolt in four steps.

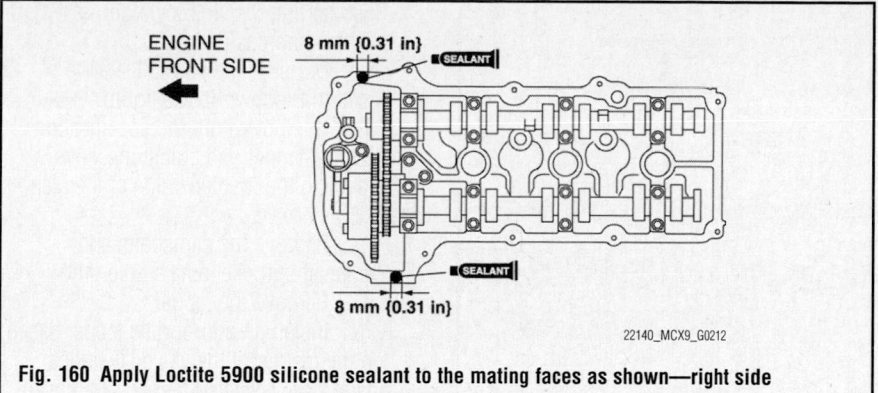

Fig. 160 Apply Loctite 5900 silicone sealant to the mating faces as shown—right side

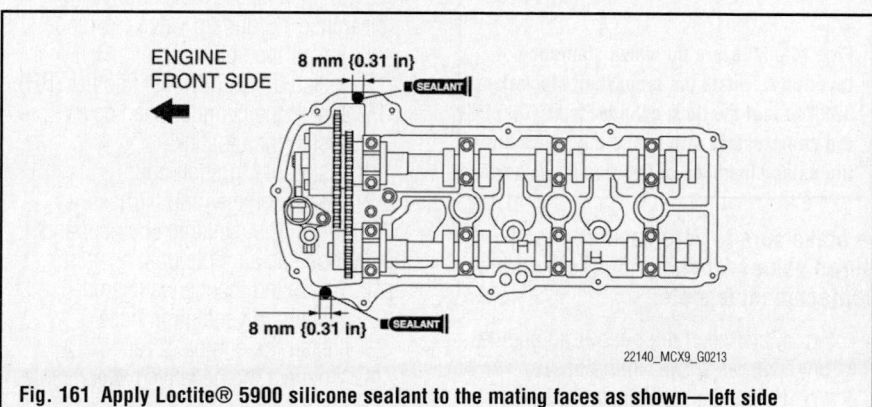

Fig. 161 Apply Loctite® 5900 silicone sealant to the mating faces as shown—left side

- Step 1: Tighten to 103 ft. lbs. (140 Nm)
- Step 2: Loosen 360° (one full turn).
- Step 3: Tighten to 35–39 ft. lbs (47–53 Nm).
- Step 4: Tighten an additional 85–95° of rotation.

29. Install the power steering oil pump.

30. Install the power steering oil pump drive belt.

31. Install the dipstick.

32. Install the ignition coils.

33. Install the intake manifold dynamic chamber and throttle body as a single unit.

34. Install the engine and transaxle.

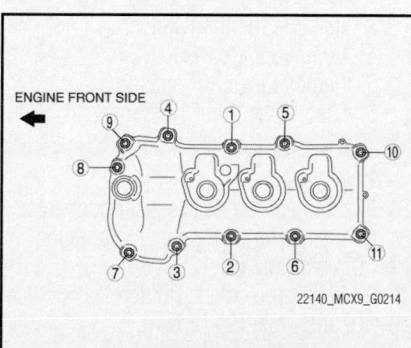

Fig. 162 Tighten the cylinder head cover bolts in the order shown—right side

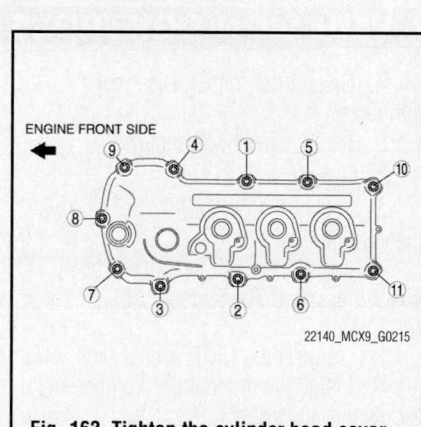

Fig. 163 Tighten the cylinder head cover bolts in the order shown—left side

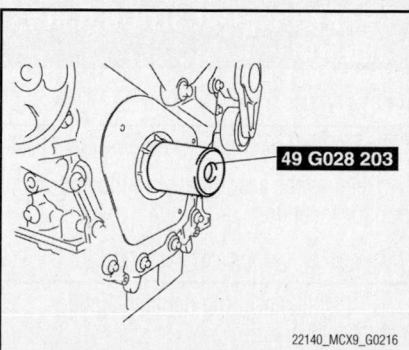

Fig. 164 Tap the front oil seal in evenly using the special tool and a hammer

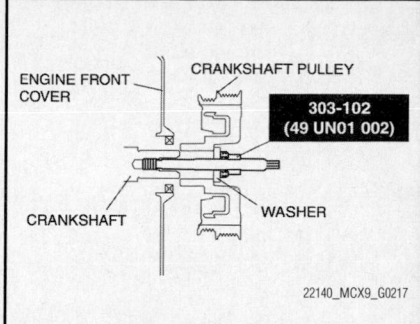

Fig. 165 Install the crankshaft pulley, washer, and special tool to the crankshaft

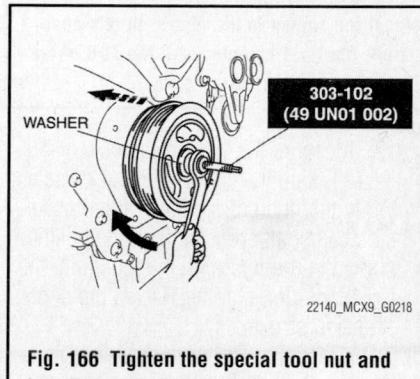

Fig. 166 Tighten the special tool nut and install the crankshaft pulley

35. Fill the engine with oil.

36. Start the engine and perform the following inspections:

  a. Inspect the runout and contact on the pulley and belt.

  b. Inspect for engine oil, engine coolant, ATF, power steering fluid and fuel leakage.

  c. Verify the ignition timing and idle speed.

  d. Inspect engine accessories for proper operation.

37. Perform a road test.

VALVE LASH

ADJUSTMENT

See Figures 167 and 168.

1. Disconnect the negative battery cable.

2. Remove the engine cover.

3. Remove the ventilation hose.

4. Remove the resonance chamber.

5. Remove the dynamic chamber and throttle body as a single unit.

6. Disconnect the wiring harness.

7. Remove the ignition coils.

8. Remove the dipstick.

9. Remove the cylinder head cover.

10. Remove the front wheel and tire (RH).

11. Remove the splash shield (RH).

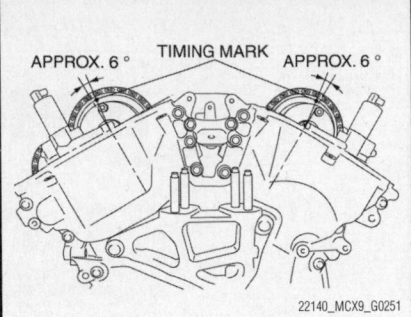

Fig. 167 Rotate the crankshaft clockwise and verify that the timing mark of the intake side camshaft sprocket is in the position shown in the figure. In this position, the No.1 cylinder is at the TDC of the compression stroke

12. Measure the valve clearance.

    a. Rotate the crankshaft clockwise and verify that the timing mark of the intake side camshaft sprocket is in the position shown in the figure. In this position, the No.1 cylinder is at the TDC of the compression stroke.

    b. Measure the valve clearance of location A shown in the figure.

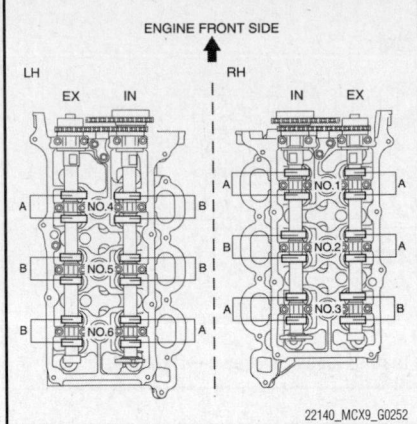

Fig. 168 Measure the valve clearance of location A, rotate the crankshaft clockwise 360° so that the No.5 cylinder is at TDC of the compression stroke and then measure the valve clearance of location B as shown

➡**Make sure to note down the measured values for choosing the suitable replacement tappets.**

    c. Mark the crank pulley and engine front cover as shown in the figure.

    d. Rotate the crankshaft clockwise 360° so that the No.5 cylinder is at TDC of the compression stroke.

    e. Measure the valve clearance of location B shown in the figure.

13. If it is not within the specification, replace the tappet and adjust the valve clearance to the median value of the standard.

    a. Remove the camshafts and camshaft sprocket as a single unit.

    b. Remove the tappet.

    c. Install an appropriate tappet based on the results of the valve clearance inspection. Selected tappet = Removed tappet thickness + Measured valve clearance - Standard valve clearance

14. Install the splash shield (RH).

15. Install the front wheel and tire (RH).

16. Install the cylinder head cover.

17. Install the dipstick.

18. Install the ignition coils.

19. Connect the wiring harness.

20. Install the dynamic chamber and throttle body as a single unit.

21. Install the resonance chamber.

22. Install the ventilation hose.

23. Install the engine cover.

24. Connect the negative battery cable.

ENGINE PERFORMANCE & EMISSION CONTROLS

CAMSHAFT POSITION (CMP) SENSOR

LOCATION

See Figure 169.

Refer to the accompanying illustration for sensor location.

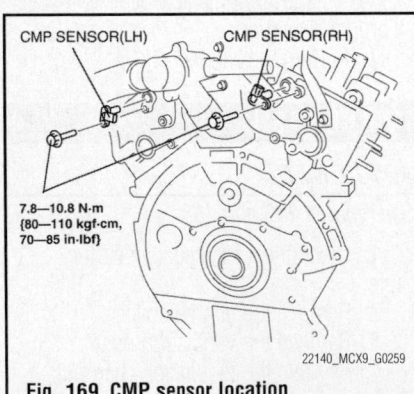

Fig. 169 CMP sensor location

REMOVAL & INSTALLATION

1. Disconnect the negative battery cable.
2. Remove the battery.
3. Remove the resonance chamber and air cleaner assembly.

4. Disconnect the CMP sensor connector
5. Remove the CMP sensor.

To install:

6. Lubricate the CMP O-ring seal with clean engine oil.
7. Install the CMP sensor. Tighten attaching bolt to 70–85 inch lbs. (8–11 Nm).
8. Connect the CMP sensor connector
9. Install the resonance chamber and air cleaner assembly.
10. Install the battery.
11. Connect the negative battery cable.

CRANKSHAFT POSITION (CKP) SENSOR

LOCATION

See Figure 170.

Refer to the accompanying illustration for sensor location.

REMOVAL & INSTALLATION

1. Disconnect the negative battery cable.
2. Remove the left side catalytic converter.

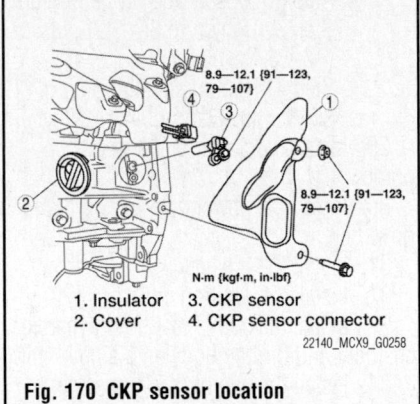

1. Insulator   3. CKP sensor
2. Cover   4. CKP sensor connector

Fig. 170 CKP sensor location

3. Remove the insulator.
4. Remove the cover.
5. Remove the CKP sensor.
6. Disconnect the CKP sensor connector.

To install:

7. Connect the CKP sensor connector.
8. Install the CKP sensor. Tighten to 79–107 inch lbs. (9–12 Nm).
9. Install the cover. Tighten fasteners to 79–107 inch lbs. (9–12 Nm).
10. Install the insulator.
11. Install the left side catalytic converter.

12. Connect the negative battery cable.

CYLINDER HEAD TEMPERATURE (CHT) SENSOR

LOCATION

See Figure 171.

Refer to the accompanying illustration for sensor location.

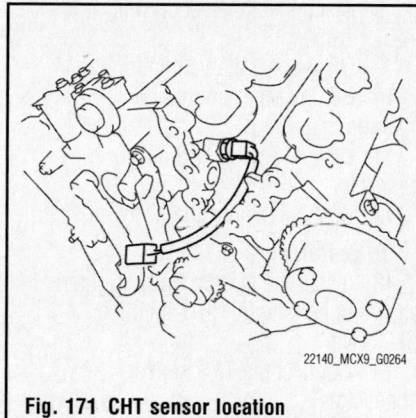

Fig. 171 CHT sensor location

REMOVAL & INSTALLATION

✳✳ WARNING

Do not reuse the CHT sensor. If the sensor is removed, install a new sensor.

1. Disconnect the negative battery cable.
2. Remove the intake manifold.
3. Disconnect the CHT sensor connector.
4. Remove the CHT sensor.

To install:

5. Install the CHT sensor. Tighten to 79–93 inch lbs. (9–11 Nm).
6. Connect the CHT sensor connector.
7. Install the intake manifold.
8. Connect the negative battery cable.

HEATED OXYGEN (HO2S) SENSOR

LOCATION

See Figures 172 and 173.

Refer to the accompanying illustrations for sensor location.

REMOVAL & INSTALLATION

1. Disconnect the negative battery cable.

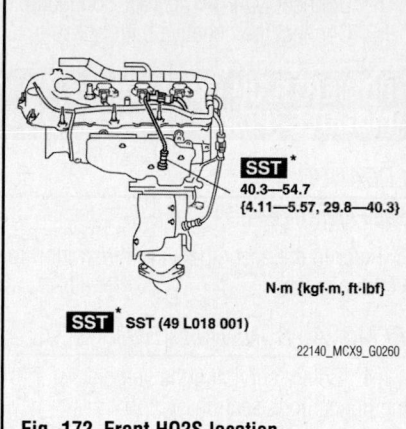

40.3—54.7
{4.11—5.57, 29.8—40.3}

N·m {kgf·m, ft·lbf}

SST SST (49 L018 001)

22140_MCX9_G0260

Fig. 172 Front HO2S location

2. On the left side, remove the engine cover.
3. On the right side, remove the cowl panel.
4. Disconnect the HO2S connector.

➡**If necessary, lubricate the HO2S with Penetrating and Lock Lubricant loosen to aid in removal.**

5. Remove the HO2S using the special tool.

To install:

6. Apply a light coat of anti-seize lubricant to the threads of the HO2S.
7. Install the HO2S using the special tool. Tighten to 30–40 ft. lbs. (40–55 Nm).
8. Connect the HO2S connector.
9. On the right side, Install the cowl panel.
10. On the left side, Install the engine cover.
11. Connect the negative battery cable.

INTAKE AIR TEMPERATURE (IAT) SENSOR

LOCATION

See Figure 174.

Refer to the accompanying illustration for sensor location.

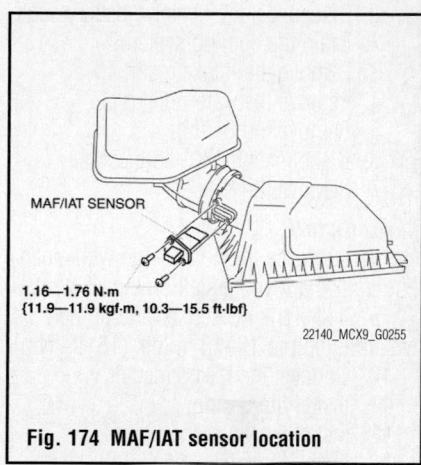

MAF/IAT SENSOR

1.16—1.76 N·m
{11.9—11.9 kgf·m, 10.3—15.5 ft·lbf}

22140_MCX9_G0255

Fig. 174 MAF/IAT sensor location

REMOVAL & INSTALLATION

1. Disconnect the negative battery cable.
2. Disconnect MAF/IAT sensor connector.
3. Remove the MAF/IAT sensor.

To install:

4. Install the MAF/IAT sensor and tighten attaching screws to 10–16 inch lbs. (1–2 Nm).
5. Connect MAF/IAT sensor connector.
6. Connect the negative battery cable.

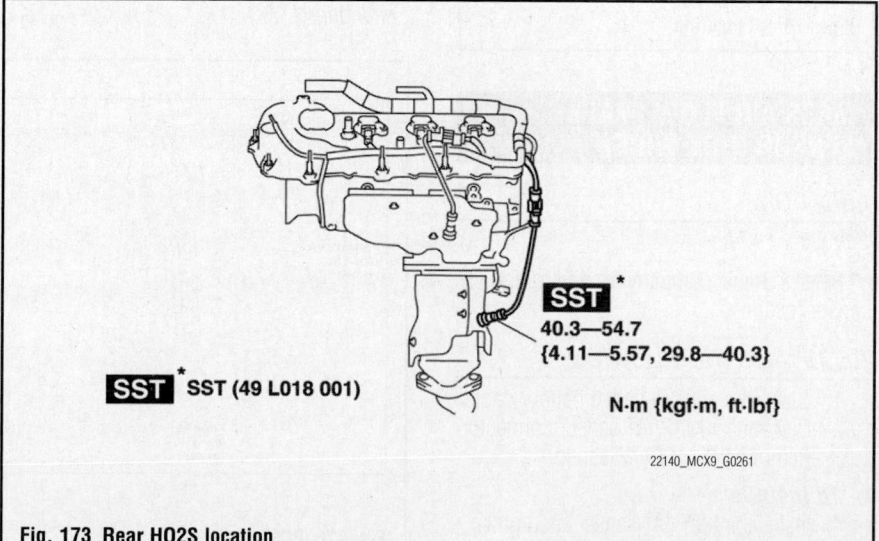

SST SST (49 L018 001)

SST 40.3—54.7
{4.11—5.57, 29.8—40.3}

N·m {kgf·m, ft·lbf}

22140_MCX9_G0261

Fig. 173 Rear HO2S location

KNOCK SENSOR (KS)

LOCATION

See Figure 175.

Refer to the accompanying illustration for sensor location.

REMOVAL & INSTALLATION

1. Disconnect the negative battery cable.
2. Drain the cooling system.
3. Remove the intake manifold.
4. Remove the water inlet pipe.
5. Remove the O-ring.
6. Disconnect the KS connector.
7. Remove the KS sensor

To install:

8. Lubricate the new O-ring with clean engine coolant and install on the KS sensor.
9. Install the KS sensor and tighten attaching bolt to 12–18 ft. lbs. (16–24 Nm).
10. Connect the KS connector.
11. Install the O-ring.
12. Install the water inlet pipe.
13. Install the intake manifold.
14. Drain the cooling system.
15. Connect the negative battery cable.

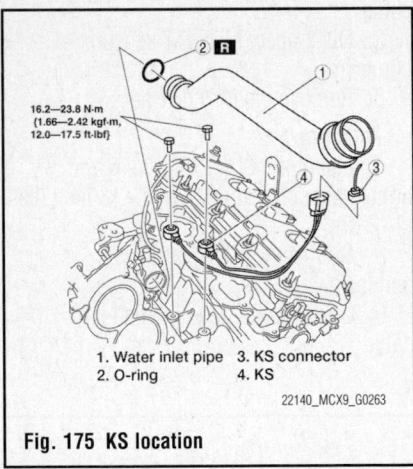

```
16.2—23.8 N·m
{1.66—2.42 kgf-m,
12.0—17.5 ft-lbf}
```

1. Water inlet pipe   3. KS connector
2. O-ring   4. KS

22140_MCX9_G0263

Fig. 175 KS location

MASS AIR FLOW (MAF) SENSOR (HOT WIRE)

LOCATION

See Figure 174.

Refer to the accompanying illustration for sensor location.

REMOVAL & INSTALLATION

1. Disconnect the negative battery cable.
2. Disconnect MAF/IAT sensor connector.
3. Remove the MAF/IAT sensor.

To install:

4. Install the MAF/IAT sensor and tighten attaching screws to 10–16 inch lbs. (1–2 Nm).

5. Connect MAF/IAT sensor connector.
6. Connect the negative battery cable.

POWERTRAIN CONTROL MODULE (PCM)

LOCATION

See Figure 176.

Refer to the accompanying illustration for PCM location.

REMOVAL & INSTALLATION

1. Before servicing the vehicle, refer to the precautions section.
2. Using a scan tool, perform the PCM configuration.
3. Disconnect the negative battery cable.
4. Remove the battery and battery tray.
5. Disconnect the PCM connectors.
6. Remove PCM and bracket as an assembly.
7. Remove PCM from bracket.

To install:

➡ **When replacing the PCM on the vehicles, perform the PCM parameter reset.**

8. Install PCM from bracket. Tighten to 70–95 inch lbs. (8–11 Nm).
9. Install PCM and bracket as an assembly. Tighten to 70–95 inch lbs. (8–11 Nm).
10. Connect the PCM connectors.
11. Install the battery and battery tray.
12. Connect the negative battery cable.
13. Using a scan tool, perform the PCM configuration.

THROTTLE POSITION SENSOR (TPS)

LOCATION

See Figure 177.

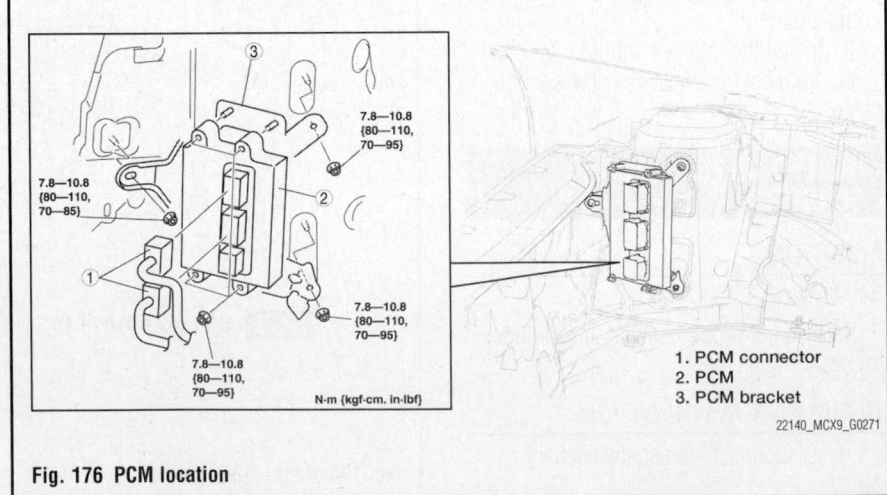

```
7.8—10.8
{80—110,
70—95}

7.8—10.8
{80—110,
70—85}

7.8—10.8
{80—110,
70—95}

7.8—10.8
{80—110,
70—95}

N·m {kgf-cm. in-lbf}
```

1. PCM connector
2. PCM
3. PCM bracket

22140_MCX9_G0271

Fig. 176 PCM location

REMOVAL & INSTALLATION

1. Before servicing the vehicle, refer to the precautions section.
2. Disconnect the negative battery cable.
3. Remove the air cleaner case.
4. Remove the fresh-air duct.
5. Remove the air cleaner element.
6. Disconnect the MAF/IAT sensor connector.
7. Remove the air cleaner cover.
8. Disconnect the vacuum hose to master back.
9. Disconnect the ventilation hose.
10. Remove the resonance chamber.
11. Disconnect the TPS sensor connector.
12. Remove the throttle body.

To install:

13. Install the throttle body. Tighten attaching bolts to 76–101 inch lbs. (9–12 Nm).
14. Connect the TPS sensor connector.
15. Install the resonance chamber.
16. Connect the ventilation hose.
17. Connect the vacuum hose to master back.
18. Install the air cleaner cover.
19. Connect the MAF/IAT sensor connector.
20. Install the air cleaner element.
21. Install the fresh-air duct.
22. Install the air cleaner case.
23. Connect the negative battery cable.

VEHICLE SPEED SENSOR (VSS)

LOCATION

See Figure 178.

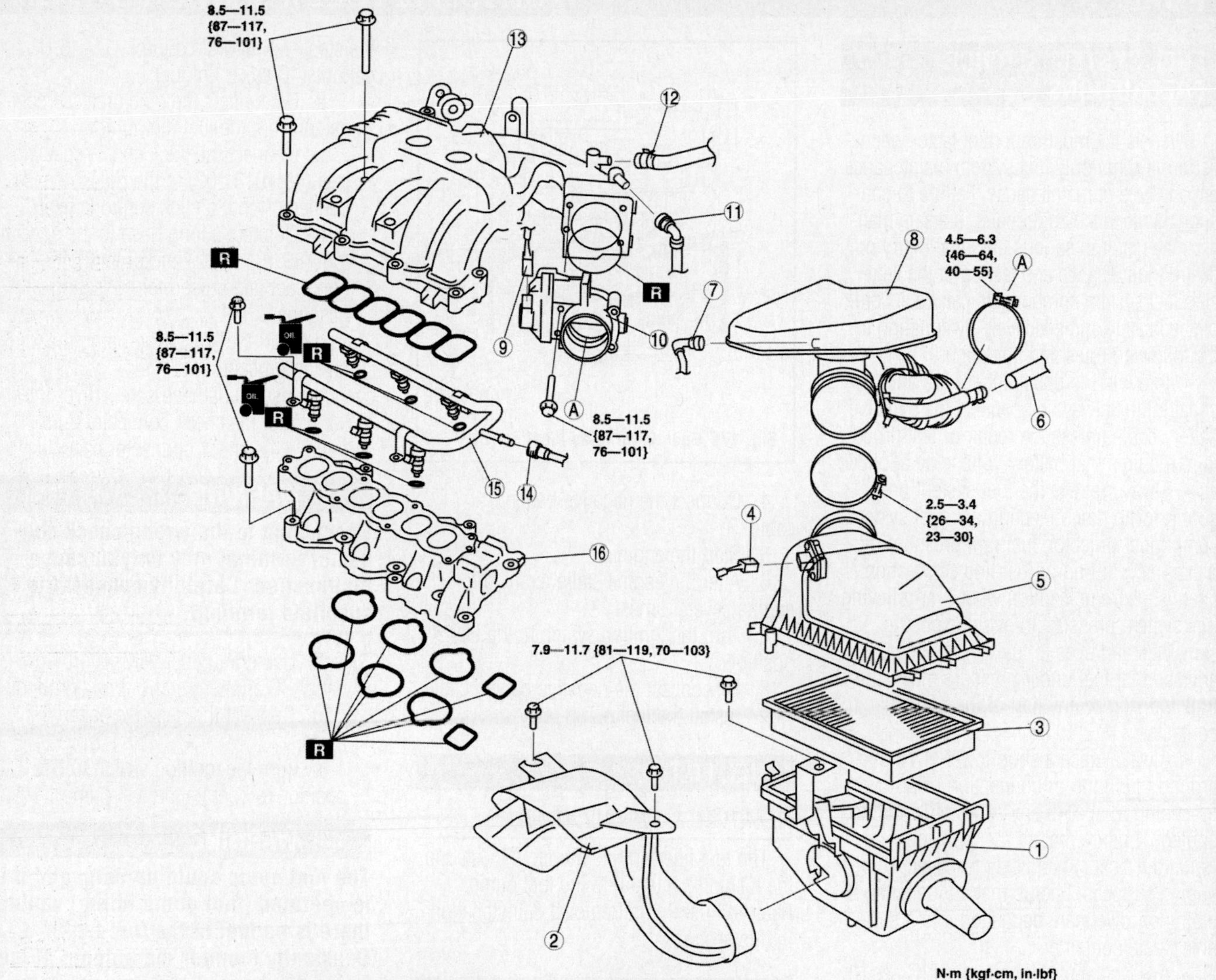

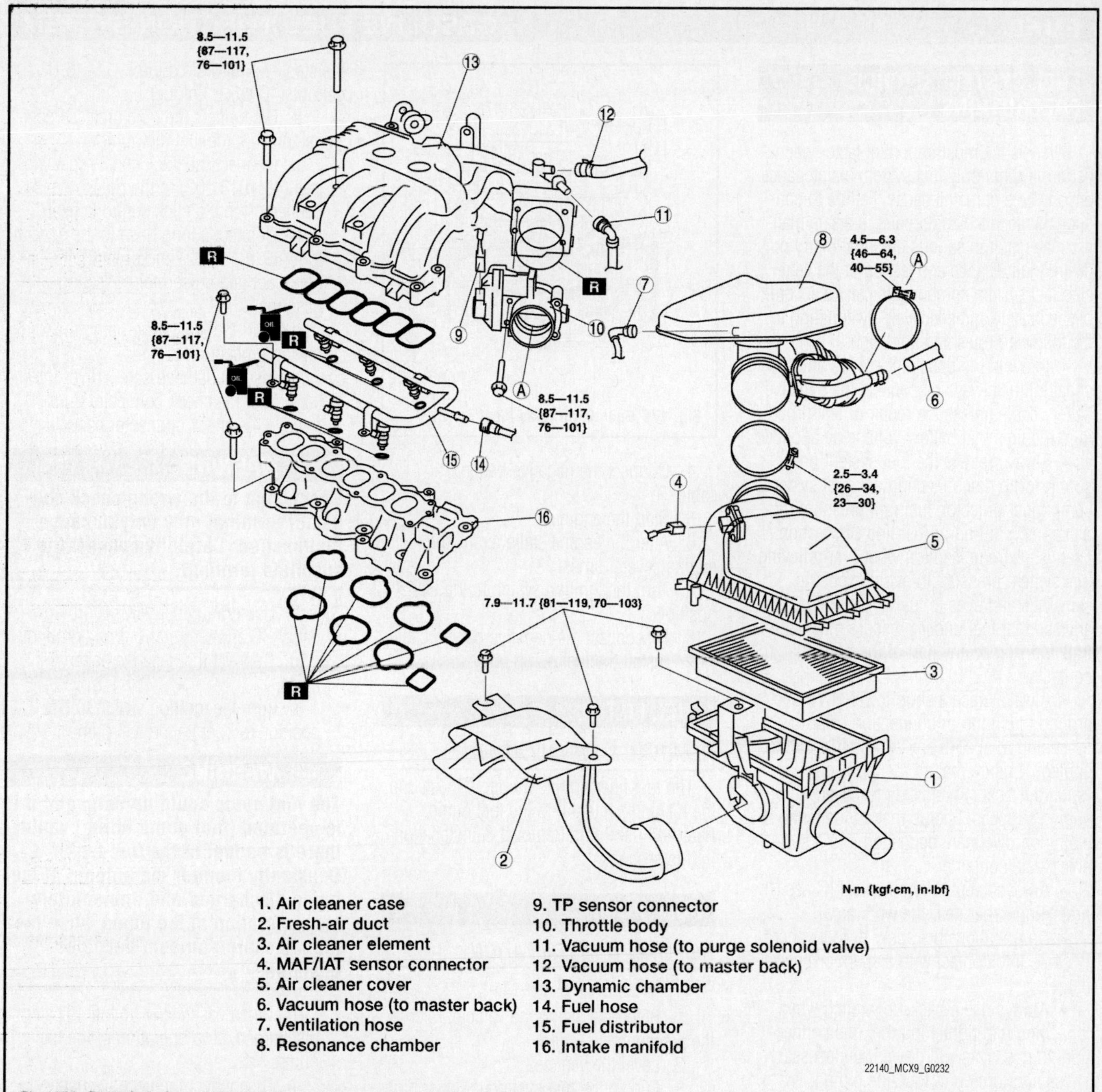

8.5—11.5
{87—117,
76—101}

8.5—11.5
{87—117,
76—101}

R

R

R

R

R

4.5—6.3
{46—64,
40—55}

A

6

2.5—3.4
{26—34,
23—30}

8.5—11.5
{87—117,
76—101}

A

7.9—11.7 {81—119, 70—103}

N·m {kgf·cm, in-lbf}

1. Air cleaner case
2. Fresh-air duct
3. Air cleaner element
4. MAF/IAT sensor connector
5. Air cleaner cover
6. Vacuum hose (to master back)
7. Ventilation hose
8. Resonance chamber
9. TP sensor connector
10. Throttle body
11. Vacuum hose (to purge solenoid valve)
12. Vacuum hose (to master back)
13. Dynamic chamber
14. Fuel hose
15. Fuel distributor
16. Intake manifold

22140_MCX9_G0232

Fig. 177 Exploded view of the throttle position sensor and related components

The Vehicle Speed Sensor (VSS) is located deep inside the transaxle, the transaxle must be removed and disassembled to access the sensor.

REMOVAL & INSTALLATION

1. Disconnect the negative battery cable.
2. Remove the transaxle.
3. Disassemble the transaxle
4. Remove the Vehicle Speed Sensor (VSS)
5. Reassemble the transaxle and install it.

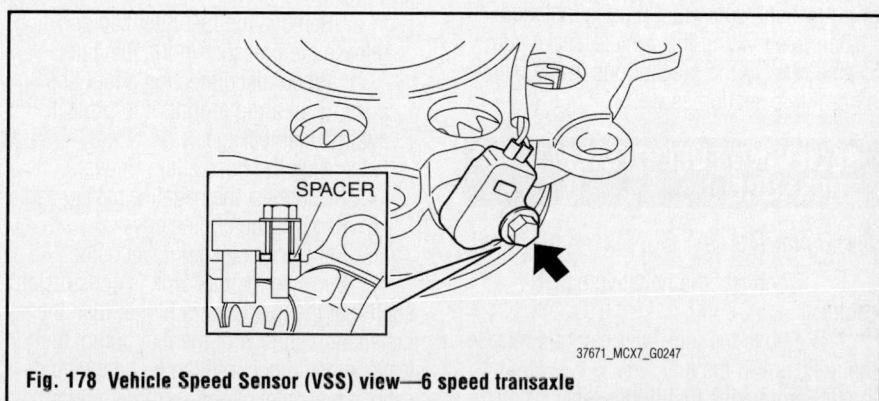

SPACER

37671_MCX7_G0247

Fig. 178 Vehicle Speed Sensor (VSS) view—6 speed transaxle

FUEL   **GASOLINE FUEL INJECTION SYSTEM**

FUEL SYSTEM SERVICE PRECAUTIONS

Safety is the most important factor when performing not only fuel system maintenance but any type of maintenance. Failure to conduct maintenance and repairs in a safe manner may result in serious personal injury or death. Maintenance and testing of the vehicle's fuel system components can be accomplished safely and effectively by adhering to the following rules and guidelines.

- To avoid the possibility of fire and personal injury, always disconnect the negative battery cable unless the repair or test procedure requires that battery voltage be applied.
- Always relieve the fuel system pressure prior to disconnecting any fuel system component (injector, fuel rail, pressure regulator, etc.), fitting or fuel line connection. Exercise extreme caution whenever relieving fuel system pressure to avoid exposing skin, face and eyes to fuel spray. Please be advised that fuel under pressure may penetrate the skin or any part of the body that it contacts.
- Always place a shop towel or cloth around the fitting or connection prior to loosening to absorb any excess fuel due to spillage. Ensure that all fuel spillage (should it occur) is quickly removed from engine surfaces. Ensure that all fuel soaked cloths or towels are deposited into a suitable waste container.
- Always keep a dry chemical (Class B) fire extinguisher near the work area.
- Do not allow fuel spray or fuel vapors to come into contact with a spark or open flame.
- Always use a back-up wrench when loosening and tightening fuel line connection fittings. This will prevent unnecessary stress and torsion to fuel line piping.
- Always replace worn fuel fitting O-rings with new Do not substitute fuel hose or equivalent where fuel pipe is installed.

Before servicing the vehicle, make sure to also refer to the precautions in the beginning of this section as well.

RELIEVING FUEL SYSTEM PRESSURE

See Figure 179.

1. Disconnect the negative battery cable.
2. Remove the fuel-filler cap and release the pressure in the fuel tank.
3. Remove the fuel pump relay.

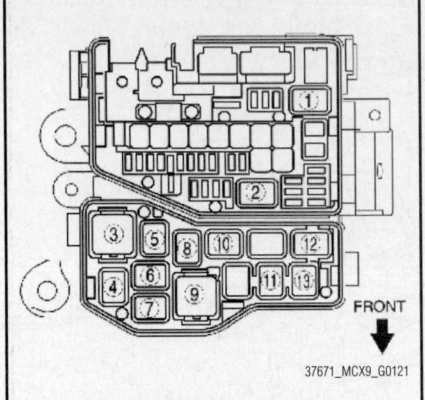

Fig. 179 Fuel pump relay location (2)

37671_MCX9_G0121

4. Connect the negative battery cable.
5. Start the engine.
6. After the engine stalls, crank the engine several times.
7. Turn the ignition switch to the LOCK position.
8. Disconnect the negative battery cable.
9. Install the fuel pump relay.

FUEL FILTER

REMOVAL & INSTALLATION

The fuel filter on this model is located in the fuel tank, integral to the fuel pump. Refer to Fuel Pump Removal & Installation in this section.

FUEL PUMP

REMOVAL & INSTALLATION

See Figures 180 and 181.

1. Before servicing the vehicle, refer to the precautions section.
2. Level the vehicle.
3. Fuel in the fuel system is under high pressure also when the engine is not running.
   a. Remove the fuel filler cap and release the pressure in the fuel tank.
   b. When disconnecting a fuel line hose, wrap a rag around it to protect against fuel leakage.
   c. Plug the hose after removal.
4. Disconnect the negative battery cable.
5. Remove the engine cover.
6. Set the air cleaner cover aside.
7. Because the fuel tank is constructed such that the fuel level is higher than the installation surface of the fuel pump, fuel leakage could occur. If the fuel gauge indicates a fuel level of half or more, perform

the steps 1-6 to drain approx. 11–15 US gallons (10–15 L) of fuel.
   a. Disconnect the quick release connector from the fuel distributor.
   b. Connect the thick end of special tool (49 T013 102) to the quick release connector until a click sound is heard.
   c. Connect a long hose to the special tool (49 T013 102) and drain the fuel into a container used for collecting gasoline.
   d. Start the fuel pump using the following procedure.
   e. Insert a flathead screwdriver into tab part of the check connector and remove the check connector cap.

❊❊ WARNING

Connecting to the wrong check connector terminal may only to cause malfunction. Carefully connect the specified terminal.

   f. Using a jumper wire, short terminals A–C and terminal E–body ground.
   g. Connect the negative battery cable.
   h. Turn the ignition switch to **ON** position to operate the fuel pump.

❊❊ WARNING

The fuel pump could be damaged if it is operated (fuel pump idling) while there is no fuel in the fuel tank. Constantly monitor the amount of fuel being discharged and immediately stop operation of the pump when the fuel discharge amount becomes unstable.

   i. When essentially no fuel is being discharged, stop operation of the fuel pump.
   j. Disconnect the negative battery cable.
8. Remove the second-row seat.
9. Remove the edge cover.
10. Remove the long slider cover.
11. Remove the rear heat duct No.4.
12. Remove the screw.
13. Remove the service hole cover.
14. Disconnect the fuel level sender connector
15. Remove the screw and the fuel pump unit.

To install:

16. Install the fuel pump unit. Tighten the attaching screw to 10–19 inch lbs. (1–2 Nm).

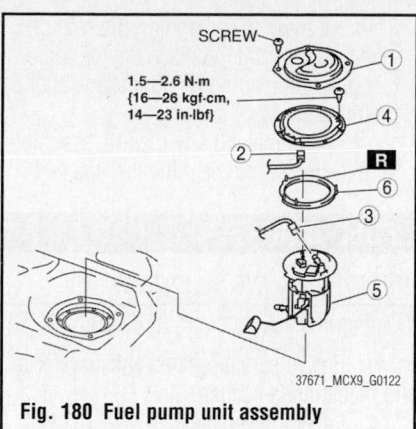

Fig. 180 Fuel pump unit assembly

17. Connect the fuel level sender connector.
18. Install the service hole cover.
19. Install the screw.
20. Install the rear heat duct No.4.
21. Install the long slider cover.
22. Install the edge cover.
23. Install the second-row seat.
24. Set the air cleaner cover aside.
25. Install the engine cover.
26. Connect the fuel line hose.
27. Connect the negative battery cable.
28. Perform a fuel leakage test as follows:
  a. Remove the fuel pump relay.

✳✳ WARNING

Short the specified terminals because shorting the wrong terminal of the main fuse block may cause malfunctions.

  b. Using a jumper wire, short fuel pump relay terminals C and D in the main fuse block.

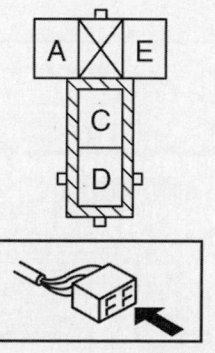

MAIN FUSE BLOCK (FUEL PUMP RELAY)

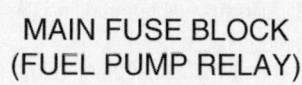

Fig. 181 Using a jumper wire, short fuel pump relay terminals C and D

c. Connect the negative battery cable and operate the fuel pump.
d. Verify that there is no fuel leakage from the pressurized parts.
e. If there is leakage, replace the fuel hoses and clips.
f. If there is damage on the seal on the fuel pipe side, replace the fuel pipe.
g. There should be no leakage after 5 minutes.
h. After making any repairs, assemble the system and repeat the leakage test.

FUEL RAIL AND INJECTOR

REMOVAL & INSTALLATION
See Figures 182 through 185.

1. Before servicing the vehicle, refer to the precautions section.
2. Fuel in the fuel system is under high pressure also when the engine is not running.
  a. Remove the fuel filler cap and release the pressure in the fuel tank.
  b. When disconnecting a fuel line hose, wrap a rag around it to protect against fuel leakage.
  c. Plug the hose after removal.
3. Disconnect the negative battery cable.
4. Remove the engine cover.)
5. Remove the intake manifold dynamic chamber.
6. Disconnect the quick release connector.
  a. Move the retainer upward using a small flathead screwdriver or a similar tool.
  b. Pull out the fuel hose straight from the fuel pipe and disconnect it.
  c. Cover the disconnected quick release connector and fuel pipe

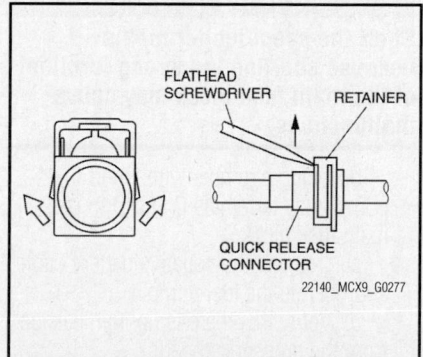

Fig. 182 Disconnecting the quick release connector

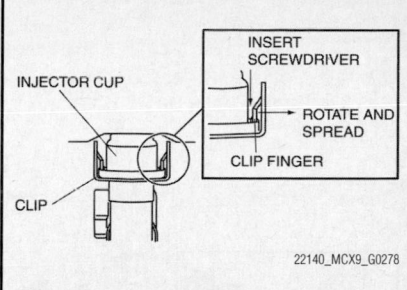

Fig. 183 When rotating the screwdriver to spread the clip fingers, deform them sufficiently to make sure they release from the notched edge of the injector cup

with vinyl sheeting or a similar material to prevent it from scratches or dirt.
7. Remove the wiring harness retainer.
8. Disconnect the fuel injector connectors.
9. Remove the fuel distributor.
10. Remove the clip.
11. Remove the fuel injector.

✳✳ WARNING

Use of a deformed injector retaining clip will cause the injector to not engage correctly. Always use a new clip when reattaching the injector, otherwise it may cause the injector to rotate.

  a. Insert a screwdriver between the injector cup and the clip fingers.
  b. When rotating the screwdriver to spread the clip fingers, deform them sufficiently to make sure they release from the notched edge of the injector cup.
  c. Rotate the screwdriver to spread the clip fingers and remove them from the injector cup.

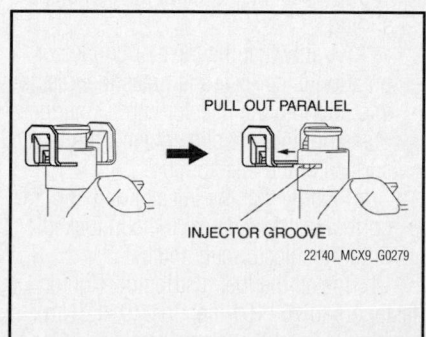

Fig. 184 Pull the clip parallel to the injector groove, and remove it from the injector

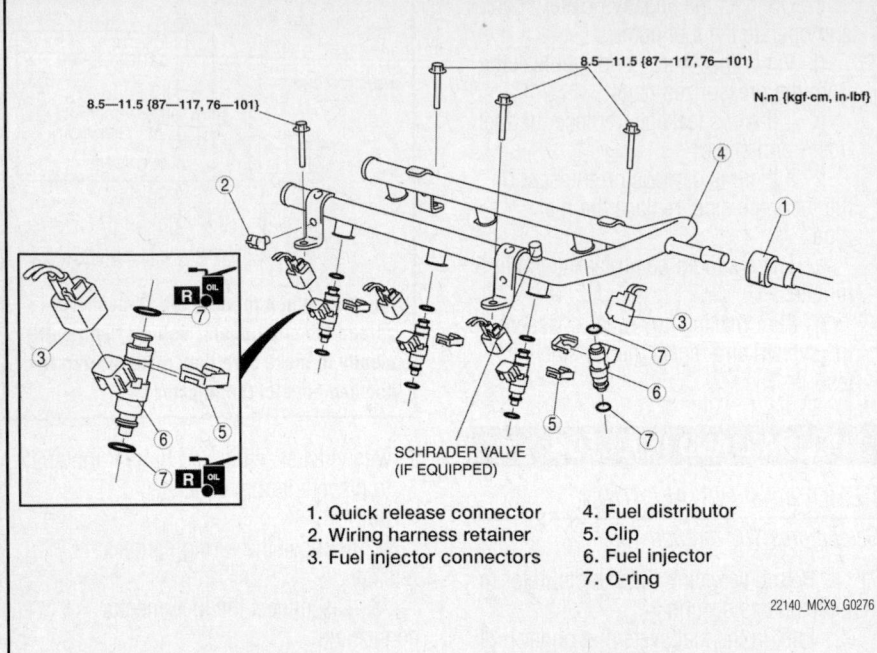

8.5—11.5 {87—117, 76—101}

8.5—11.5 {87—117, 76—101}

N·m {kgf·cm, in-lbf}

SCHRADER VALVE
(IF EQUIPPED)

1. Quick release connector
2. Wiring harness retainer
3. Fuel injector connectors
4. Fuel distributor
5. Clip
6. Fuel injector
7. O-ring

22140_MCX9_G0276

Fig. 185 Fuel rail and injector assembly

d. Pull the injector and clip out of the rail.

e. Remove the clip from the injector according to the following:

f. Grasp the clip with a pair of pliers.

g. Pull the clip parallel to the injector groove, and remove it from the injector.

h. Discard the clip.

12. Remove the O-ring.

To install:

13. Install the fuel injector.

a. Install the new O-ring.

b. Lightly lubricate the injector groove and O-ring.

c. Pre-attach a new clip in the injector groove.

d. When the clip is attached correctly, the central area of the injector and the clip finger positions are aligned.

e. Hold the injector firmly, push the clip into the injector until the clip stops sliding.

f. Verify that the injector connector position is correct, and press the injector and clip into the injector cup. Continue pressing until the clip contacts the lower surface of the injector cup.

g. Verify that the injector and clip are correctly attached with the clip locked onto the injector cup notch.

14. Install the fuel distributor. Tighten fasteners to 76–101 inch lbs. (9–12 Nm).

15. Connect the fuel injector connectors.

16. Install the wiring harness retainer.

17. Connect the quick release connector.

a. Insert the fuel pipe straight to the end of the quick release connector.

b. Push down the retainer using a finger.

c. If the retainer cannot be pushed down, push the fuel pipe further to the quick release connector.

d. Lightly pull and push the quick release connector a few times by hand, and then verify that it is connected securely.

18. Install the intake manifold dynamic chamber.

19. Install the engine cover.

20. Connect the negative battery cable.

21. Perform a fuel leakage test as follows:

a. Remove the fuel pump relay.

❊❊ WARNING

Short the specified terminals because shorting the wrong terminal of the main fuse block may cause malfunctions.

b. Using a jumper wire, short fuel pump relay terminals C and D in the main fuse block.

c. Connect the negative battery cable and operate the fuel pump.

d. Verify that there is no fuel leakage from the pressurized parts.

e. If there is leakage, replace the fuel hoses and clips.

f. If there is damage on the seal on the fuel pipe side, replace the fuel pipe.

g. There should be no leakage after 5 minutes.

h. After making any repairs, assemble the system and repeat the leakage test.

FUEL TANK

REMOVAL & INSTALLATION

See Figures 186 through 190, 00.

1. Before servicing the vehicle, refer to the precautions section.

2. Level the vehicle.

3. Fuel in the fuel system is under high pressure also when the engine is not running.

a. Remove the fuel filler cap and release the pressure in the fuel tank.

b. When disconnecting a fuel line hose, wrap a rag around it to protect against fuel leakage.

c. Plug the hose after removal.

4. Disconnect the negative battery cable.

5. Remove the engine cover.

6. Set the air cleaner cover aside.

7. Because the fuel tank is constructed such that the fuel level is higher than the installation surface of the fuel pump, fuel leakage could occur. If the fuel gauge indicates a fuel level of half or more, perform the steps 1-6 to drain approx. 11–15 US gallons (10–15 L) of fuel.

a. Disconnect the quick release connector from the fuel distributor.

b. Connect the thick end of special tool (49 T013 102) to the quick release connector until a click sound is heard.

c. Connect a long hose to the special tool (49 T013 102) and drain the fuel into a container used for collecting gasoline.

d. Start the fuel pump using the following procedure.

e. Remove the fuel pump relay.

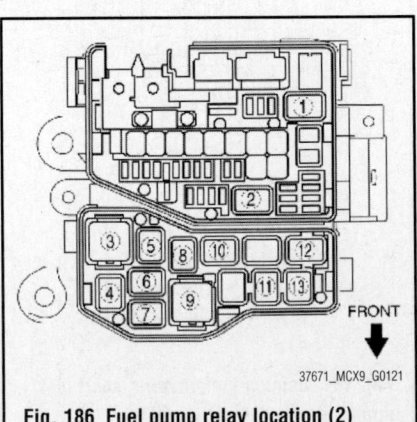

37671_MCX9_G0121

Fig. 186 Fuel pump relay location (2)

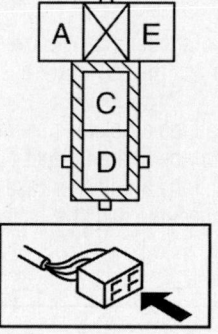

Fig. 187 Using a jumper wire, short fuel pump relay terminals C and D

WARNING

Connecting to the wrong check connector terminal may only to cause malfunction. Carefully connect the specified terminal.

f. Using a jumper wire, short fuel pump relay terminals C and D in the main fuse block.

g. Connect the negative battery cable.

h. Turn the ignition switch to **ON** position to operate the fuel pump.

WARNING

The fuel pump could be damaged if it is operated (fuel pump idling) while there is no fuel in the fuel tank. Constantly monitor the amount of fuel being discharged and immediately stop operation of the pump when the fuel discharge amount becomes unstable.

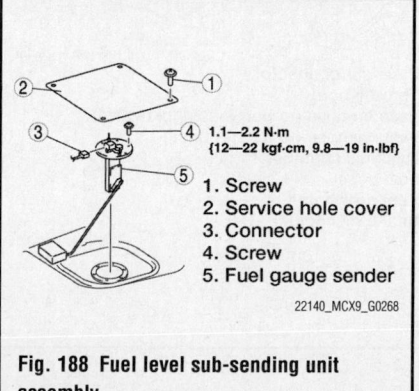

1. Screw
2. Service hole cover
3. Connector
4. Screw
5. Fuel gauge sender

Fig. 188 Fuel level sub-sending unit assembly

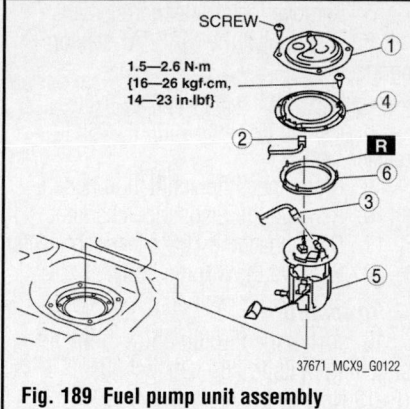

Fig. 189 Fuel pump unit assembly

i. When essentially no fuel is being discharged, stop operation of the fuel pump.

j. Disconnect the negative battery cable.

8. On AWD vehicles, remove the propeller shaft.

9. Remove the second-row seat.

10. Remove the edge cover.

11. Remove the long slider cover.

12. Remove the rear heat duct No.4.

13. Remove the screw.

14. Remove the service hole cover.

15. Disconnect the fuel level sender connector

16. Remove the screw and the fuel pump unit.

17. Remove the joint hose.

18. Disconnect the evaporative hose.

19. Remove the fuel tank.

20. Remove the fuel tank bracket.

21. Remove the fuel tank insulator.

22. Remove the fuel-filler cap.

23. Remove the fuel filler pipe protector.

   a. Remove the rear tire (LH).

   b. Remove the splash shield (LH).

   c. Remove the fuel-filler pipe protector

24. Disconnect the fuel filler pipe.

To install:

25. Connect the fuel filler pipe.

26. Install the fuel filler pipe protector.

   a. Install the splash shield (LH).

   b. Install the rear tire (LH).

27. Install the fuel-filler cap.

28. Install the fuel tank insulator. Tighten fasteners to 18–22 ft. lbs. (22–30 Nm).

29. Install the fuel tank bracket. Tighten fasteners to 18–22 ft. lbs. (22–30 Nm).

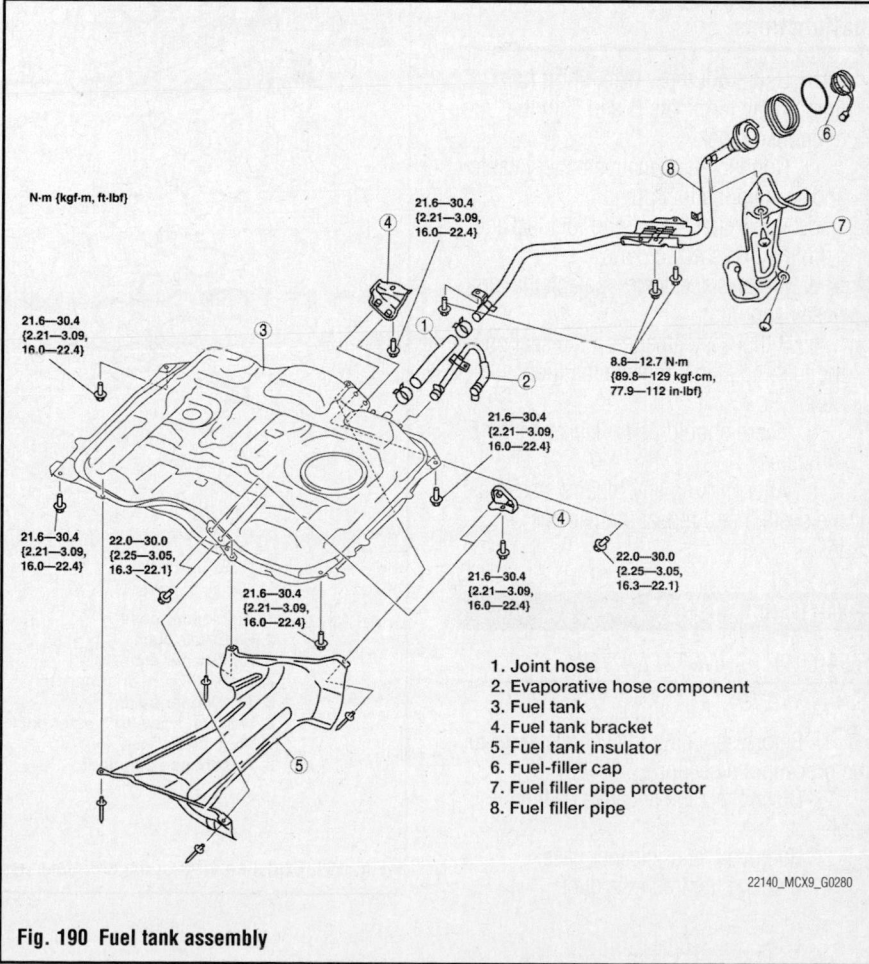

1. Joint hose
2. Evaporative hose component
3. Fuel tank
4. Fuel tank bracket
5. Fuel tank insulator
6. Fuel-filler cap
7. Fuel filler pipe protector
8. Fuel filler pipe

Fig. 190 Fuel tank assembly

30. Install the fuel tank. Tighten fasteners to 18–22 ft. lbs. (22–30 Nm).

31. Connect the evaporative hose.

32. Install the joint hose.

33. Install the fuel pump unit. Tighten the attaching screw to 10–19 inch lbs. (1–2 Nm).

34. Connect the fuel level sender connector.

35. Install the service hole cover.

36. Install the screw.

37. Install the rear heat duct No.4.

38. Install the long slider cover.

39. Install the edge cover.

40. Install the second-row seat.

41. Set the air cleaner cover aside.

42. Install the engine cover.

43. Connect the fuel line hose.

44. On AWD vehicles, install the propeller shaft.

45. Connect the negative battery cable.

46. Perform a fuel leakage test as follows:

   a. Remove the fuel pump relay.

❋❋ WARNING

Short the specified terminals because shorting the wrong terminal of the main fuse block may cause malfunctions.

   b. Using a jumper wire, short fuel pump relay terminals C and D in the main fuse block.

   c. Connect the negative battery cable and operate the fuel pump.

   d. Verify that there is no fuel leakage from the pressurized parts.

   e. If there is leakage, replace the fuel hoses and clips.

   f. If there is damage on the seal on the fuel pipe side, replace the fuel pipe.

   g. There should be no leakage after 5 minutes.

   h. After making any repairs, assemble the system and repeat the leakage test.

THROTTLE BODY

REMOVAL & INSTALLATION

See Figure 191.

1. Before servicing the vehicle, refer to the precautions section.

2. Disconnect the negative battery cable.

3. Remove the air cleaner case.

4. Remove the fresh-air duct.

5. Remove the air cleaner element.

6. Disconnect the MAF/IAT sensor connector.

7. Remove the air cleaner cover.

8. Disconnect the vacuum hose to master back.

9. Disconnect the ventilation hose.

10. Remove the resonance chamber.

11. Disconnect the TPS sensor connector.

12. Remove the throttle body.

To install:

13. Install the throttle body. Tighten attaching bolts to 76–101 inch lbs. (9–12 Nm).

14. Connect the TPS sensor connector.

15. Install the resonance chamber.

16. Connect the ventilation hose.

17. Connect the vacuum hose to master back.

18. Install the air cleaner cover.

19. Connect the MAF/IAT sensor connector.

20. Install the air cleaner element.

21. Install the fresh-air duct.

22. Install the air cleaner case.

23. Connect the negative battery cable.

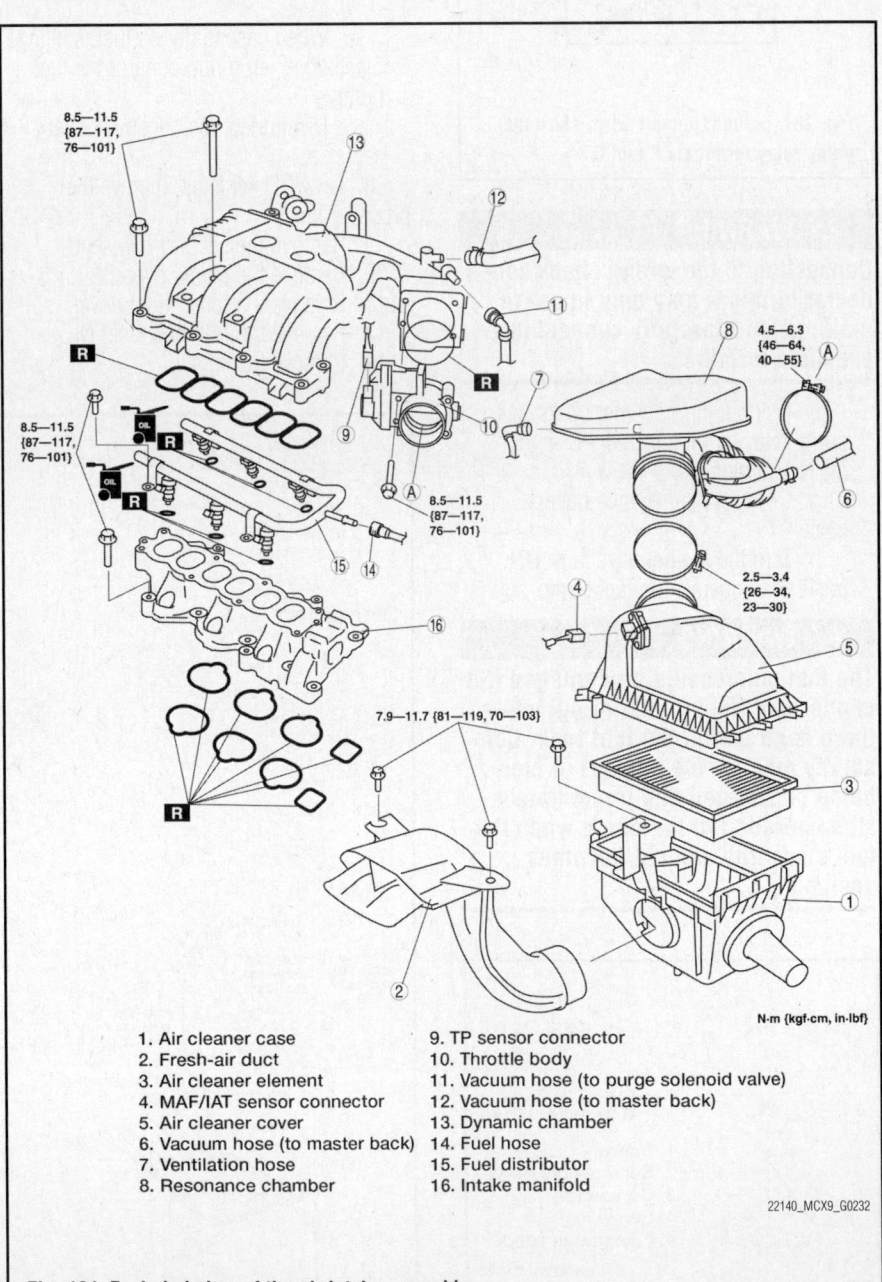

1. Air cleaner case
2. Fresh-air duct
3. Air cleaner element
4. MAF/IAT sensor connector
5. Air cleaner cover
6. Vacuum hose (to master back)
7. Ventilation hose
8. Resonance chamber
9. TP sensor connector
10. Throttle body
11. Vacuum hose (to purge solenoid valve)
12. Vacuum hose (to master back)
13. Dynamic chamber
14. Fuel hose
15. Fuel distributor
16. Intake manifold

N·m {kgf·cm, in-lbf}

22140_MCX9_G0232

Fig. 191 Exploded view of the air intake assembly

HEATING & AIR CONDITIONING SYSTEM

BLOWER MOTOR

REMOVAL & INSTALLATION

See Figure 192.

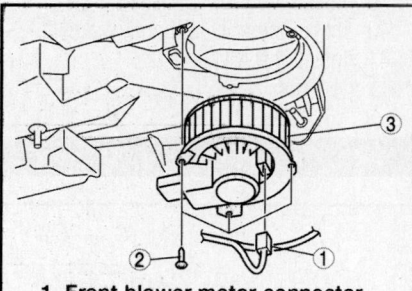

1. Front blower motor connector
2. Screw
3. Front blower motor

22140_MCX9_G0292

Fig. 192 Blower motor assembly

1. Before servicing the vehicle, refer to the precautions section.
2. Disconnect the negative battery cable.
3. Remove the instrument panel under cover. (passenger side)
4. Disconnect the front blower motor connector.
5. Remove the screw.
6. Remove the front blower motor.

To install:

7. Install the screw.
8. Connect the front blower motor connector.
9. Install the instrument panel under cover (Passenger Side).
10. Connect the negative battery cable.
11. Install the front blower motor.

HEATER CORE

REMOVAL & INSTALLATION

See Figure 193.

1. Before servicing the vehicle, refer to the precautions section.
2. Remove the A/C unit.
3. Remove the polyurethane gasket.
4. Remove the blower case.
5. Remove the blower case cover.
6. Remove the air filter.
7. Remove the air intake actuator.
8. Remove the air intake link set.
9. Remove the blower case.
10. Remove the air intake door.
11. Remove the harness.
12. Remove the front power.
13. Remove the front blower motor.

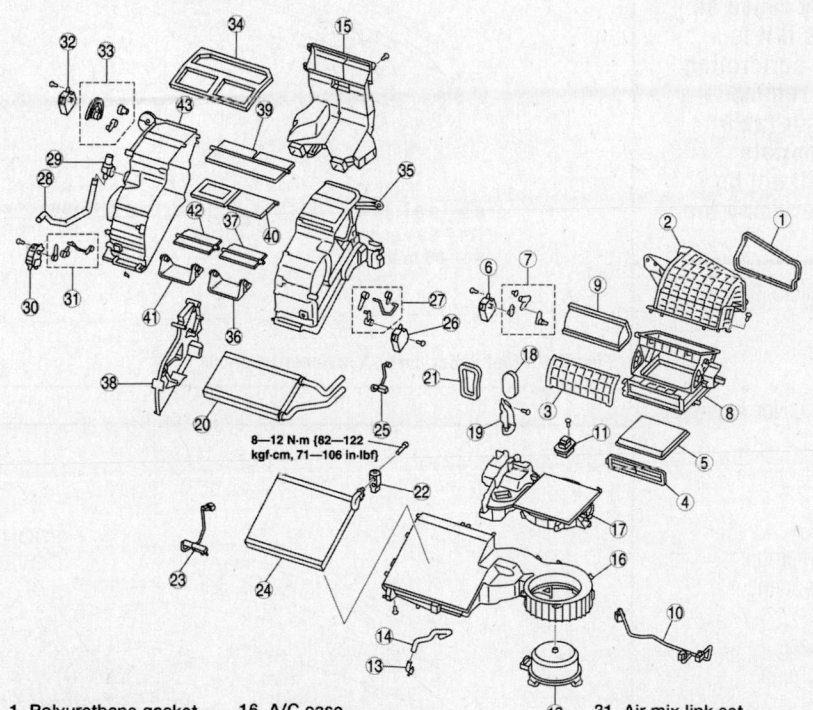

8–12 N·m {82—122 kgf·cm, 71—106 in·lbf}

1. Polyurethane gasket	16. A/C case	31. Air mix link set
2. Blower case	17. Blower case	32. Front airflow mode actuator
3. Blower case cover	18. Polyurethane foam	33. Airflow mode link set
4. Blower case cover	19. A/C case bracket	34. Polyurethane foam
5. Air filter	20. Front heater core	35. A/C case
6. Air intake actuator	21. Polyurethane foam	36. Passenger-side air mix door
7. Air intake link set	22. Front expansion valve	37. Passenger-side air mix door
8. Blower case	23. Front evaporator sensor	38. A/C case
9. Air intake door	24. Front evaporator	39. Front airflow mode door
10. Harness	25. Harness	40. Front airflow mode door
11. Front power	26. Passenger-side front air mix actuator	41. Driver-side air mix door
12. Front blower motor	27. Air mix link set	42. Driver-side air mix door
13. Grommet	28. Air hose	43. A/C case
14. Drain hose	29. Air hose duct	
15. Duct	30. Driver-side front air mix actuator	

22140_MCX9_G0288

Fig. 193 Exploded view of the A/C unit

14. Remove the grommet.
15. Remove the drain hose.
16. Remove the duct.
17. Remove the A/C case.
18. Remove the blower case.
19. Remove the polyurethane foam.
20. Remove the A/C case bracket.
21. Remove the front heater core.

To install:

22. Install the front heater core.
23. Install the A/C case bracket.
24. Install the polyurethane foam.
25. Install the blower case.
26. Install the A/C case.
27. Install the duct.
28. Install the drain hose.
29. Install the grommet.
30. Install the front blower motor.
31. Install the front power.

32. Install the harness.
33. Install the air intake door.
34. Install the blower case.
35. Install the air intake link set.
36. Install the air intake actuator.
37. Install the air filter.
38. Install the blower case cover.
39. Install the blower case.
40. Install the polyurethane gasket.
41. Install the A/C unit.

STEERING

POWER RACK & PINION STEERING GEAR

REMOVAL & INSTALLATION

See Figures 194 through 199.

✳✳ WARNING

Performing the following procedures without first removing the ABS wheel speed sensor may possibly cause an open circuit in the harness if it is pulled by mistake. Before performing the following procedures, remove the ABS wheel speed sensor (axle side) and fix it to an appropriate place where the sensor will not be pulled by mistake while servicing the vehicle.

1. Before servicing the vehicle, refer to the precautions section.
2. Drain the power steering fluid.
3. Remove the front under cover A and front under cover B.
4. Remove the middle pipe.
5. Remove the transverse member.
6. Remove the front crossmember, steering gear and linkage component.
7. Remove the insulator.
8. Remove the front stabilizer.
9. Remove the steering gear and linkage.

To install:

10. Install the steering gear and linkage.
   a. Tighten the mounting bracket bolts to 72–100 ft. lbs. (97–136 Nm) in the order shown.
11. Install the front stabilizer.
12. Install the insulator.
13. Install the front crossmember, steering gear and linkage component.
14. Install the transverse member.
15. Install the middle pipe.

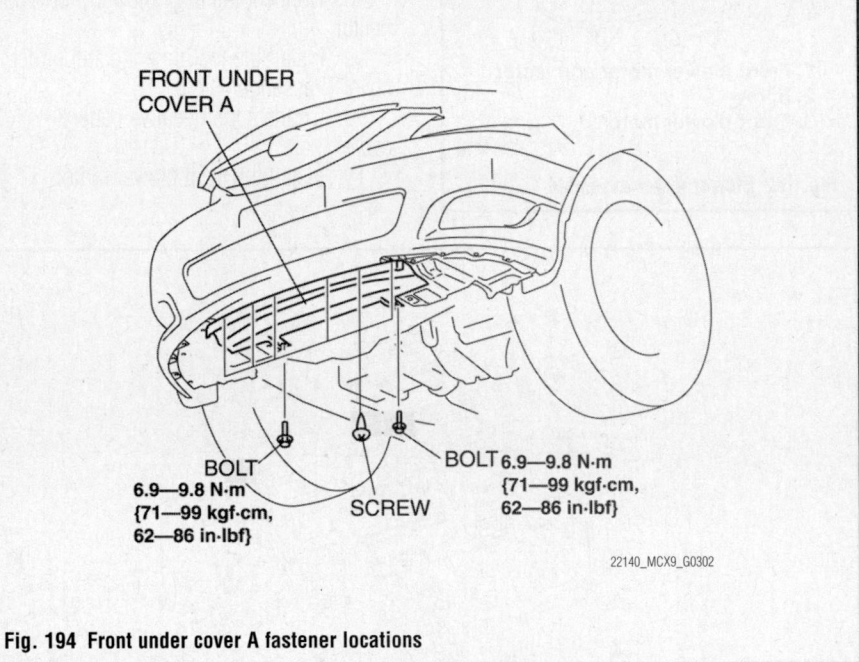

Fig. 194 Front under cover A fastener locations

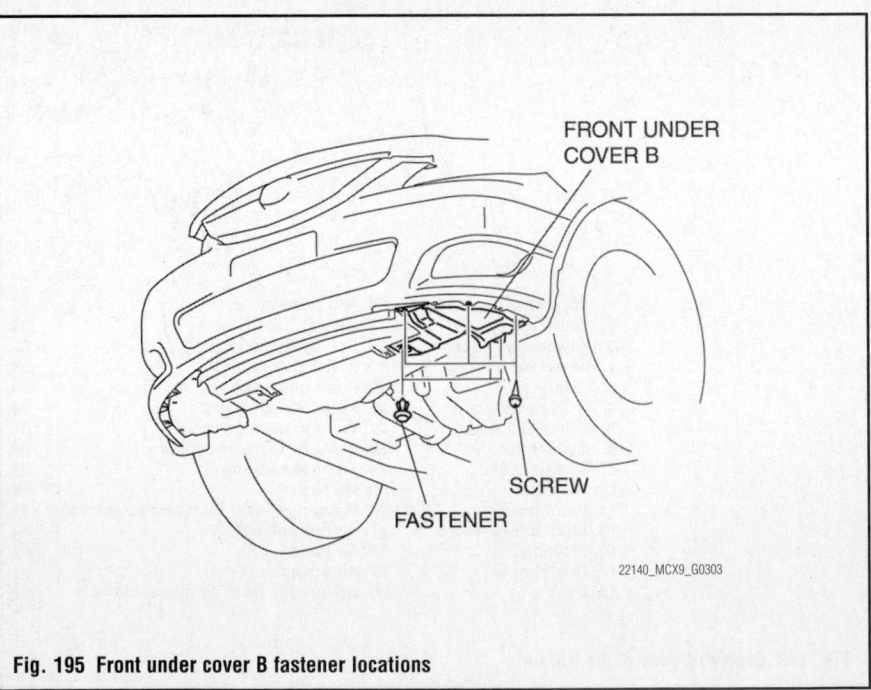

Fig. 195 Front under cover B fastener locations

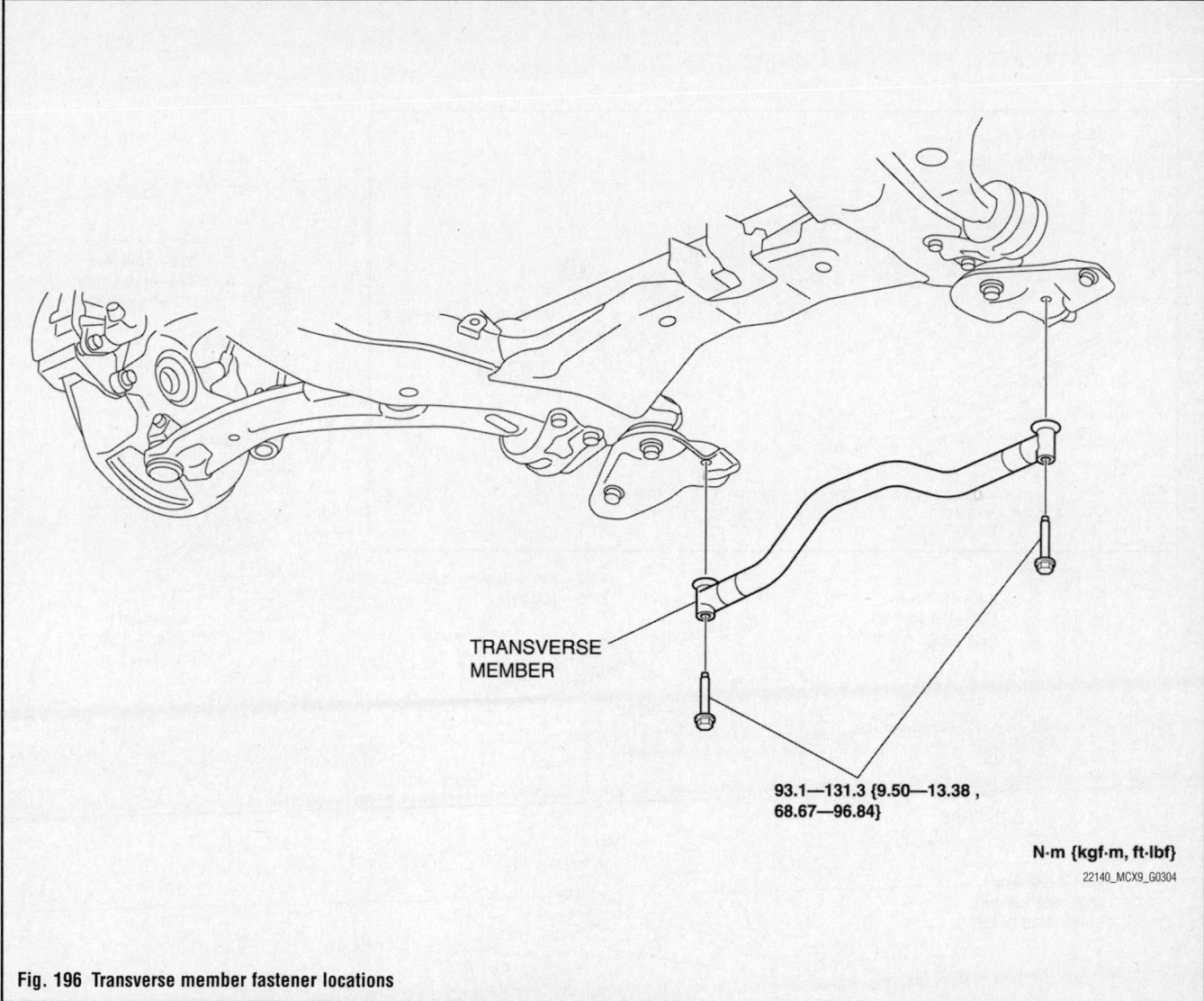

TRANSVERSE
MEMBER

93.1—131.3 {9.50—13.38 ,
68.67—96.84}

N·m {kgf·m, ft·lbf}

22140_MCX9_G0304

Fig. 196 Transverse member fastener locations

16. Install the front under cover A and front under cover B.

17. Fill and bleed the power steering system.

18. Check and adjust the total toe-in.

POWER STEERING PUMP

REMOVAL & INSTALLATION

See Figure 200.

1. Before servicing the vehicle, refer to the precautions section.

2. Drain the power steering fluid.

3. Remove the windshield wiper arm and blade.

4. Remove the cowl grill.

5. Remove the windshield wiper motor.

6. Remove the cowl panel.

7. Remove the right side splashing shield.

8. Remove the drive belt.

9. Disconnect the pressure switch connector.

10. Disconnect the pressure pipe.

11. Remove the bolts.

12. Disconnect the suction hose at the pump. Plug the hose.

13. Remove the power steering pump and suction hose.

To install:

14. Install the power steering pump and suction hose.

15. Connect the suction hose at the pump. Plug the hose.

16. Install the bolts.

17. Connect the pressure pipe.

18. Connect the pressure switch connector.

19. Install the drive belt.

20. Install the right side splashing shield.

21. Install the cowl panel.

22. Install the windshield wiper motor.

23. Install the cowl grill.

24. Install the windshield wiper arm and blade.

25. Fill and bleed the power steering system.

BLEEDING

1. Before servicing the vehicle, refer to the precautions section.

✷✷ WARNING

Do not turn the steering wheel during the fluid level inspection, otherwise the fluid level changes and cannot be inspected correctly.

2. Inspect the fluid level.

3. Raise and safely support the vehicle securely on jackstands.

4. Turn the steering wheel fully to the

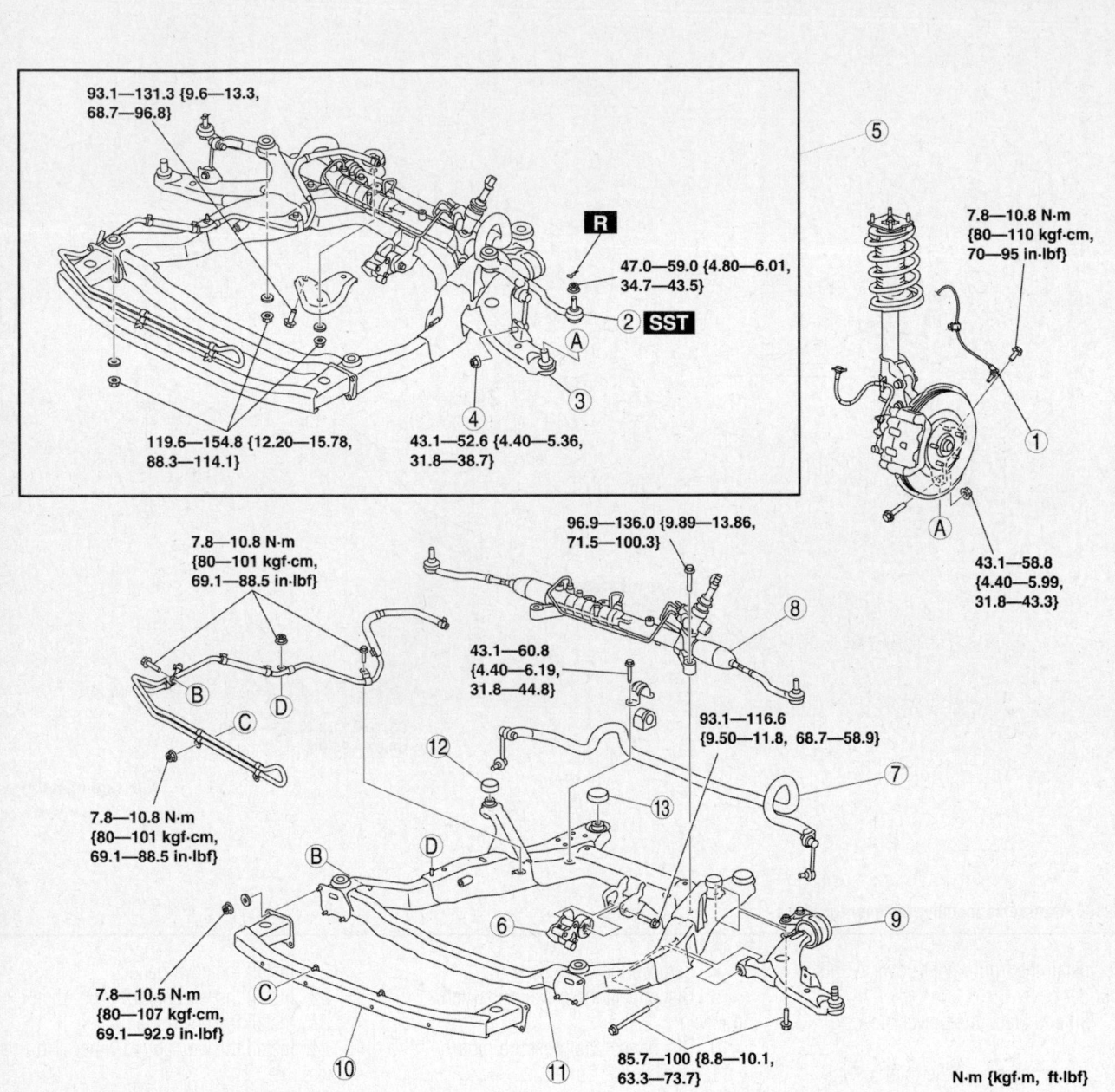

93.1—131.3 {9.6—13.3, 68.7—96.8}

R

47.0—59.0 {4.80—6.01, 34.7—43.5}

2 SST

A

3

4

119.6—154.8 {12.20—15.78, 88.3—114.1}

43.1—52.6 {4.40—5.36, 31.8—38.7}

5

7.8—10.8 N·m {80—110 kgf·cm, 70—95 in·lbf}

1

A

43.1—58.8 {4.40—5.99, 31.8—43.3}

96.9—136.0 {9.89—13.86, 71.5—100.3}

7.8—10.8 N·m {80—101 kgf·cm, 69.1—88.5 in·lbf}

B

C

D

7.8—10.8 N·m {80—101 kgf·cm, 69.1—88.5 in·lbf}

43.1—60.8 {4.40—6.19, 31.8—44.8}

8

12

93.1—116.6 {9.50—11.8, 68.7—58.9}

7

13

B

D

6

9

7.8—10.5 N·m {80—107 kgf·cm, 69.1—92.9 in·lbf}

C

10

11

85.7—100 {8.8—10.1, 63.3—73.7}

N·m {kgf·m, ft·lbf}

1. ABS wheel-speed sensor
2. Tie-rod end ball joint
3. Front lower arm ball joint
4. Stabilizer control link lower side nut
5. Steering gear and linkage, front stabilizer, front lower arm and front crossmember component
6. No.1 engine mount
7. Front stabilizer
8. Steering gear and linkage
9. Front lower arm
10. Front crossmember extension
11. Front crossmember
12. Front crossmember mounting rubber (front)
13. Front crossmember mounting rubber (rear)

22140_MCX9_G0305

Fig. 197 Exploded view of the front crossmember assembly

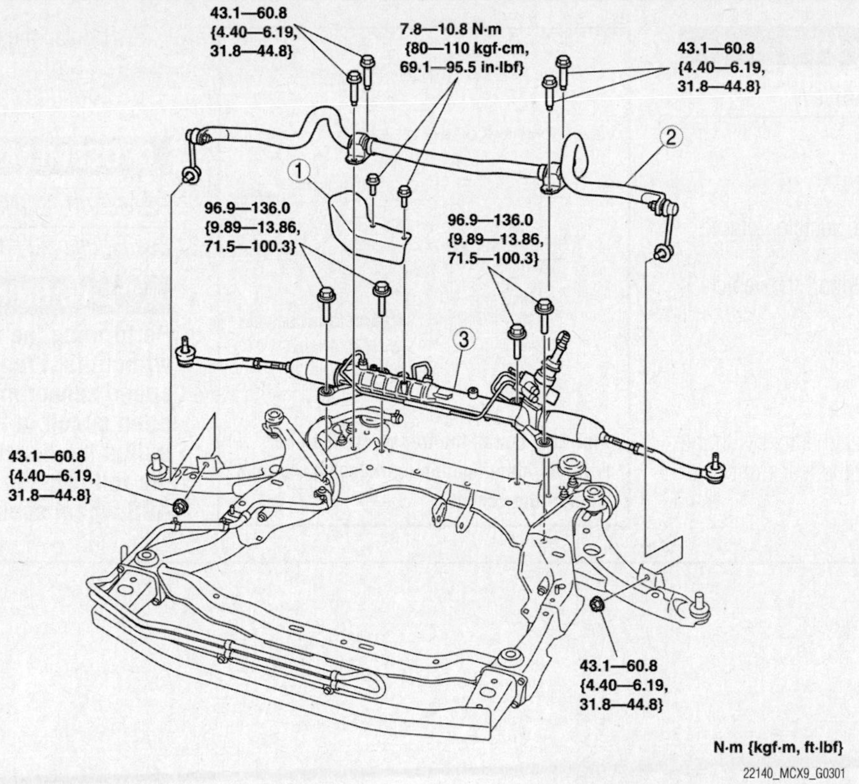

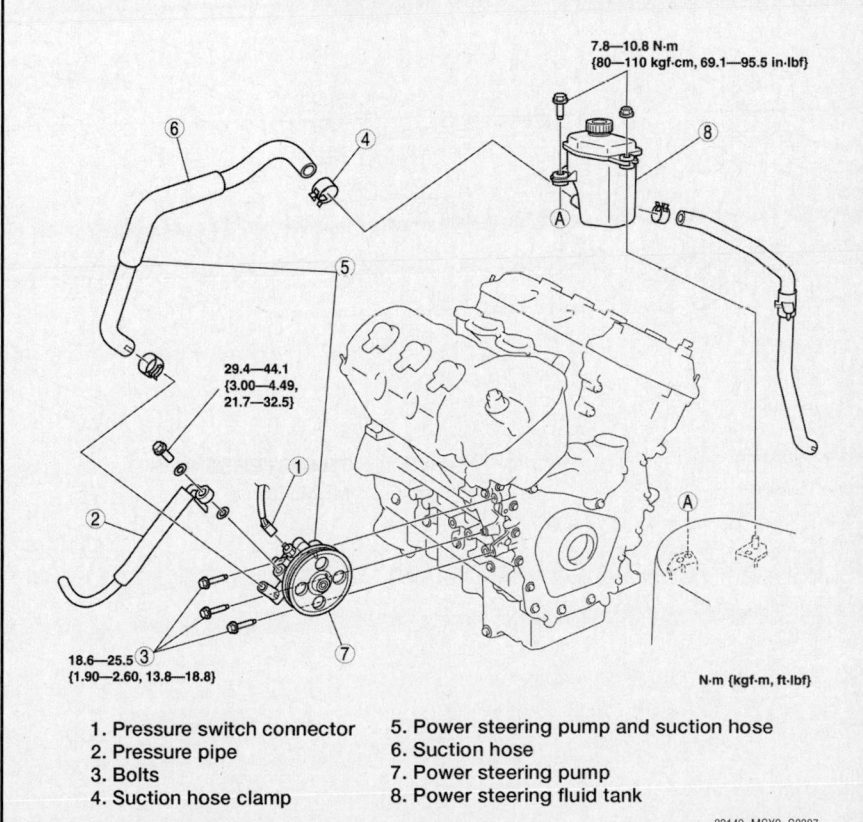

Fig. 198 Power steering gear fastener locations

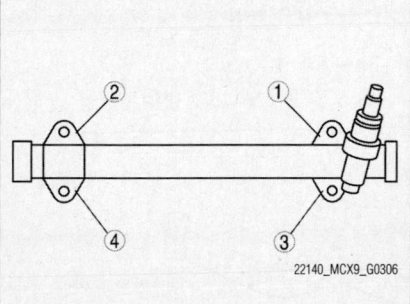

Fig. 199 Tighten the mounting bracket bolts to the specified torque in the order shown

left and right several times with the engine not running.

5. Re-inspect the fluid level. If it has dropped, add fluid.

6. Repeat the previous steps until the fluid level stabilizes.

7. Lower the vehicle.

8. Start the engine and let it idle.

9. Turn the steering wheel fully to the left and right several times.

10. Verify that the fluid is not foamy and that the fluid level has not dropped.

11. If the fluid level has dropped, add fluid as necessary and repeat bleeding.

1. Pressure switch connector
2. Pressure pipe
3. Bolts
4. Suction hose clamp
5. Power steering pump and suction hose
6. Suction hose
7. Power steering pump
8. Power steering fluid tank

Fig. 200 Power steering pump fastener locations

CONTROL LINKS

REMOVAL & INSTALLATION

Transverse Member

See Figures 201 and 202.

1. Before servicing the vehicle, refer to the precautions section.
2. Raise and safely support the vehicle securely on jackstands.
3. Remove the transverse member.

To install:

4. Install the transverse member with the identification mark facing the rear right of the vehicle.

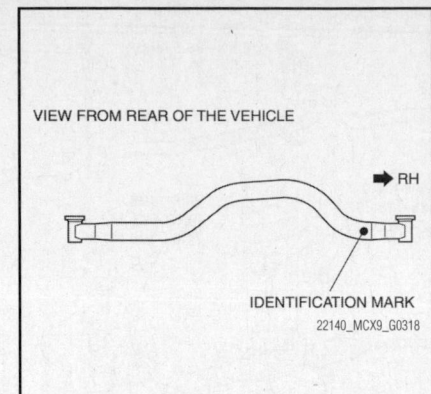

VIEW FROM REAR OF THE VEHICLE

→ RH

IDENTIFICATION MARK

22140_MCX9_G0318

Fig. 202 Install the transverse member with the identification mark facing the rear right of the vehicle

5. Tighten the bolts to 69–97 ft. lbs. (93–131 Nm).
6. Lower the vehicle.

CROSSMEMBER (FRONT)

REMOVAL & INSTALLATION

See Figures 203 through 207.

✳✳ WARNING

Performing the following procedures without first removing the ABS wheel speed sensor may possibly cause an open circuit in the harness if it is pulled by mistake. Before performing the following procedures, remove the ABS wheel speed sensor (axle side)

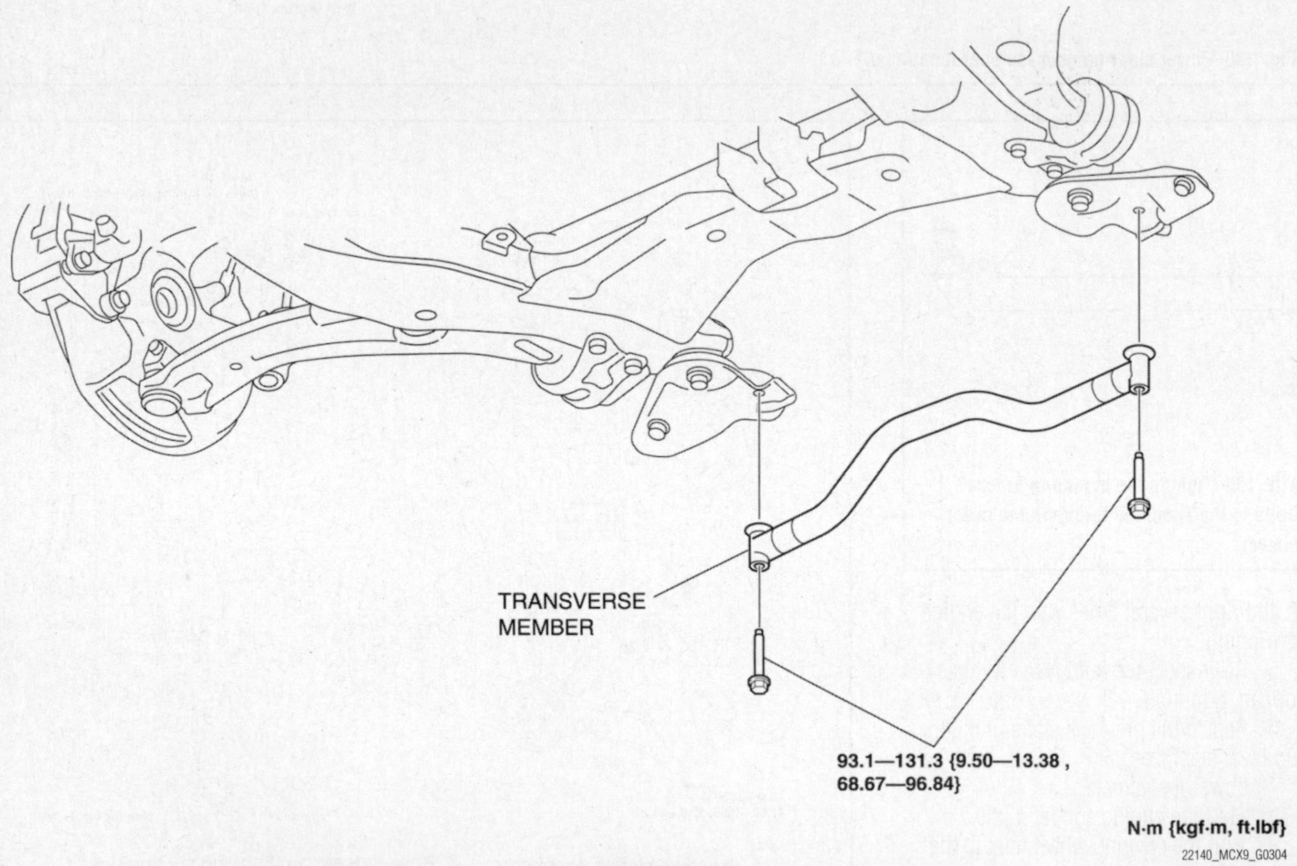

TRANSVERSE MEMBER

93.1—131.3 {9.50—13.38 , 68.67—96.84}

N·m {kgf·m, ft·lbf}

22140_MCX9_G0304

Fig. 201 Transverse link assembly

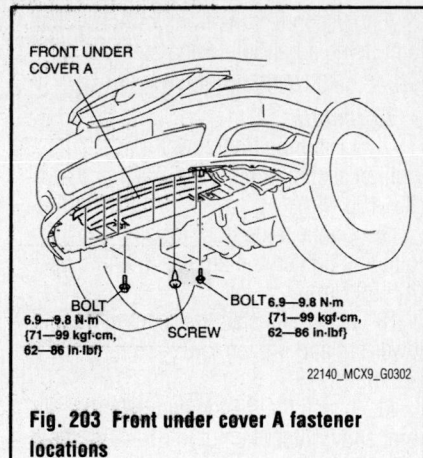

FRONT UNDER COVER A

BOLT
6.9—9.8 N·m
{71—99 kgf·cm,
62—86 in·lbf}

SCREW

BOLT 6.9—9.8 N·m
{71—99 kgf·cm,
62—86 in·lbf}

22140_MCX9_G0302

Fig. 203 Front under cover A fastener locations

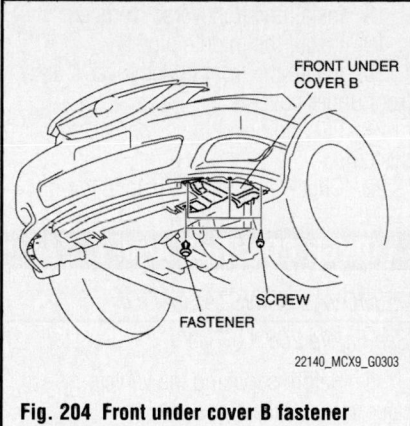

FRONT UNDER COVER B

SCREW

FASTENER

22140_MCX9_G0303

Fig. 204 Front under cover B fastener locations

and fix it to an appropriate place where the sensor will not be pulled by mistake while servicing the vehicle.

1. Before servicing the vehicle, refer to the precautions section.
2. Drain the power steering fluid.
3. Remove the front under cover A and front under cover B.
4. Remove the middle pipe.
5. Remove the transverse member.
6. Remove the intermediate shaft installation bolt, and disconnect the steering shaft.
7. Remove the ABS wheel-speed sensor.
8. Disconnect the tie-rod end ball joint using a ball joint separator.

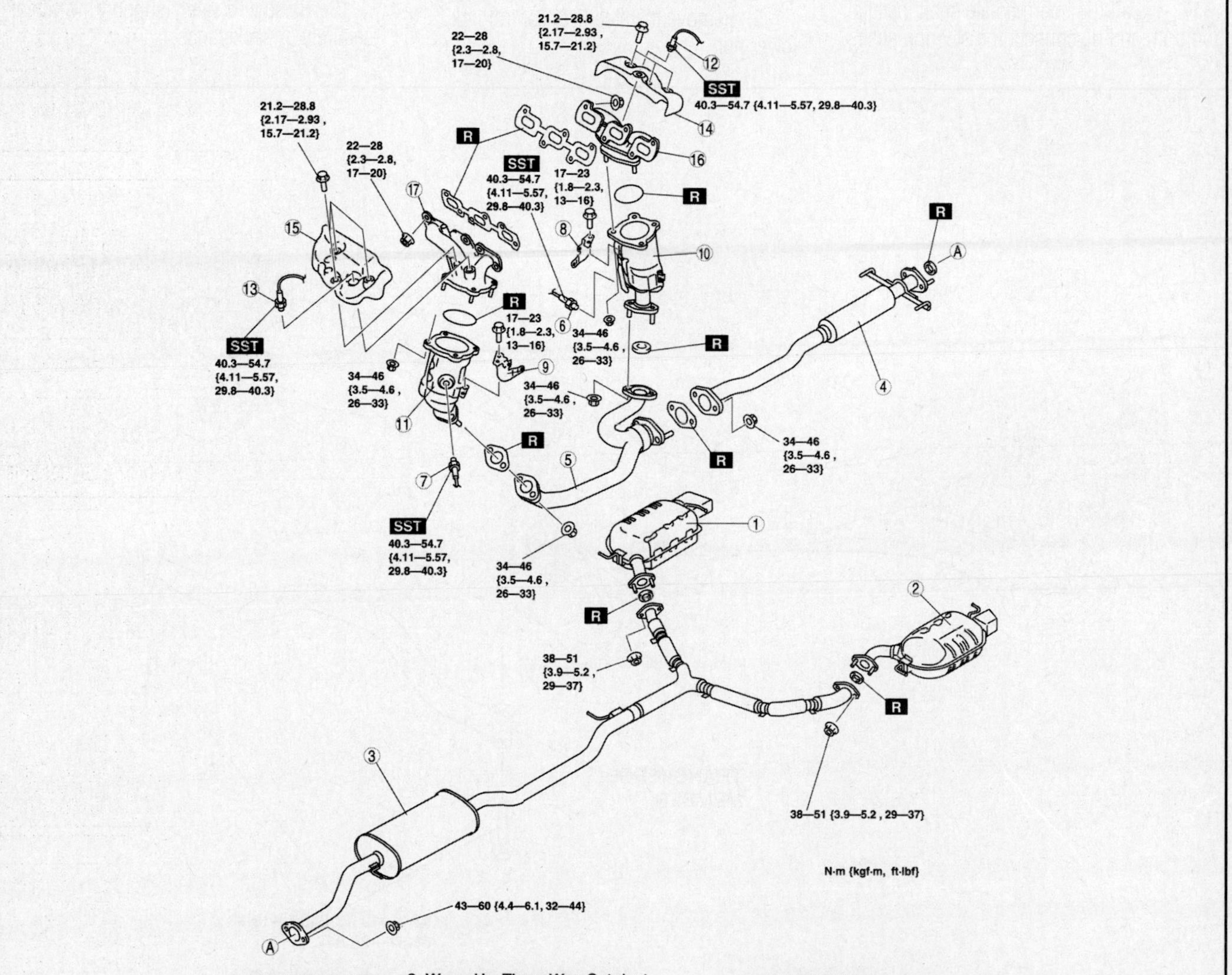

21.2—28.8
{2.17—2.93,
15.7—21.2}

22—28
{2.3—2.8,
17—20}

40.3—54.7 {4.11—5.57, 29.8—40.3}

21.2—28.8
{2.17—2.93,
15.7—21.2}

22—28
{2.3—2.8,
17—20}

SST
40.3—54.7
{4.11—5.57,
29.8—40.3}

17—23
{1.8—2.3,
13—16}

SST
40.3—54.7
{4.11—5.57,
29.8—40.3}

17—23
{1.8—2.3,
13—16}

34—46
{3.5—4.6,
26—33}

34—46
{3.5—4.6,
26—33}

34—46
{3.5—4.6,
26—33}

34—46
{3.5—4.6,
26—33}

SST
40.3—54.7
{4.11—5.57,
29.8—40.3}

34—46
{3.5—4.6,
26—33}

38—51
{3.9—5.2,
29—37}

38—51 {3.9—5.2, 29—37}

N·m {kgf·m, ft·lbf}

43—60 {4.4—6.1, 32—44}

1. Main silencer (RH)
2. Main silencer (LH)
3. Presilencer
4. Middle pipe
5. Front pipe
6. Rear HO2S (RH)
7. Rear HO2S (LH)
8. Warm Up-Three Way Catalyst (RH) bracket (L type)
9. Warm Up-Three Way Catalyst (LH) bracket (L type)
10. Warm Up-Three Way Catalyst (RH)
11. Warm Up-Three Way Catalyst (LH)
12. Front HO2S (RH)
13. Front HO2S (LH)
14. Exhaust manifold insulator (RH)
15. Exhaust manifold insulator (LH)
16. Exhaust manifold (RH)
17. Exhaust manifold (LH)

22140_MCX9_G0230

Fig. 205 Exploded view of the exhaust system

9. Disconnect the front lower ball joint using a ball joint separator.

10. Disconnect the stabilizer lower side control link.

11. Remove the steering gear and linkage, front stabilizer, front lower arm and front crossmember as an assembly.

To install:

12. Install the steering gear and linkage, front stabilizer, front lower arm and front crossmember as an assembly.

13. Connect the stabilizer lower side control link.

14. Connect the front lower ball joint using a ball joint separator.

15. Connect the tie-rod end ball joint using a ball joint separator.

16. Install the ABS wheel-speed sensor.

17. Install the intermediate shaft installation bolt, and disconnect the steering shaft.

18. Install the transverse member.

19. Install the middle pipe.

20. Install the front under cover A and front under cover B.

21. Fill and bleed the power steering system.

22. Check and adjust the total toe-in.

LOWER CONTROL ARM

REMOVAL & INSTALLATION

See Figure 208.

1. Before servicing the vehicle, refer to the precautions section.

2. Raise and safely support the vehicle securely on jackstands.

3. Remove the ABS wheel-speed sensor.

4. Remove the tie-rod end ball joint.

5. Remove the stabilizer control link lower nut.

6. Disconnect the front lower arm ball joint using a ball joint separator.

7. Remove the front lower arm.

To install:

8. Install the front lower arm. Tighten attaching bolts to 63–73 ft. lbs. (86–100 Nm).

9. Connect the front lower arm ball joint and tighten the nut to 35–44 ft. lbs. (47–59 Nm).

10. Install the stabilizer control link lower nut and tighten to 32–39 ft. lbs (43–53 Nm).

11. Install the tie-rod end ball joint and tighten the nut to 35–44 ft. lbs. (47–59 Nm).

12. Install the ABS wheel-speed sensor.

13. Raise and safely support the vehicle securely on jackstands.

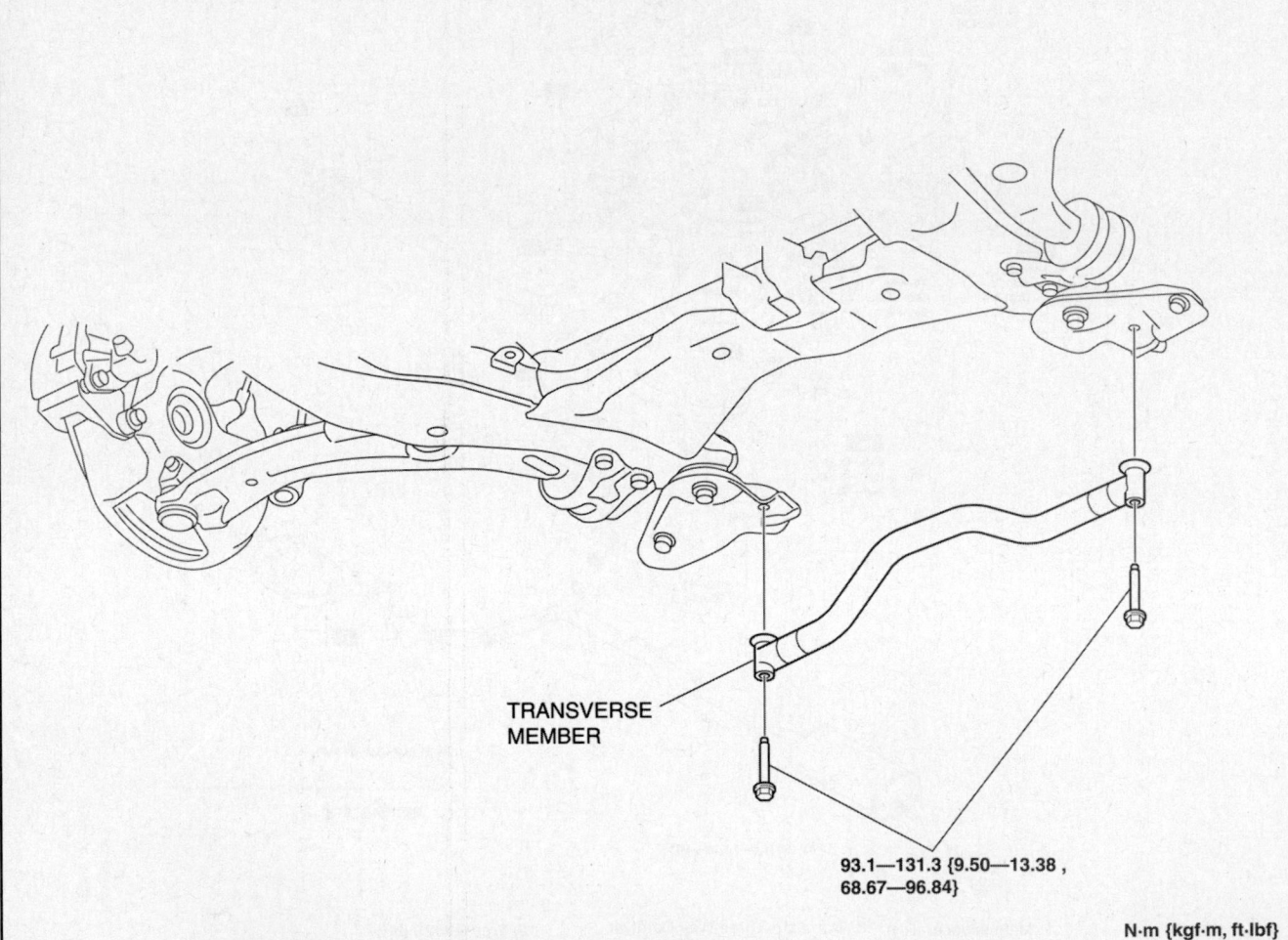

TRANSVERSE MEMBER

93.1—131.3 {9.50—13.38 , 68.67—96.84}

N·m {kgf·m, ft·lbf}

22140_MCX9_G0304

Fig. 206 Transverse member fastener locations

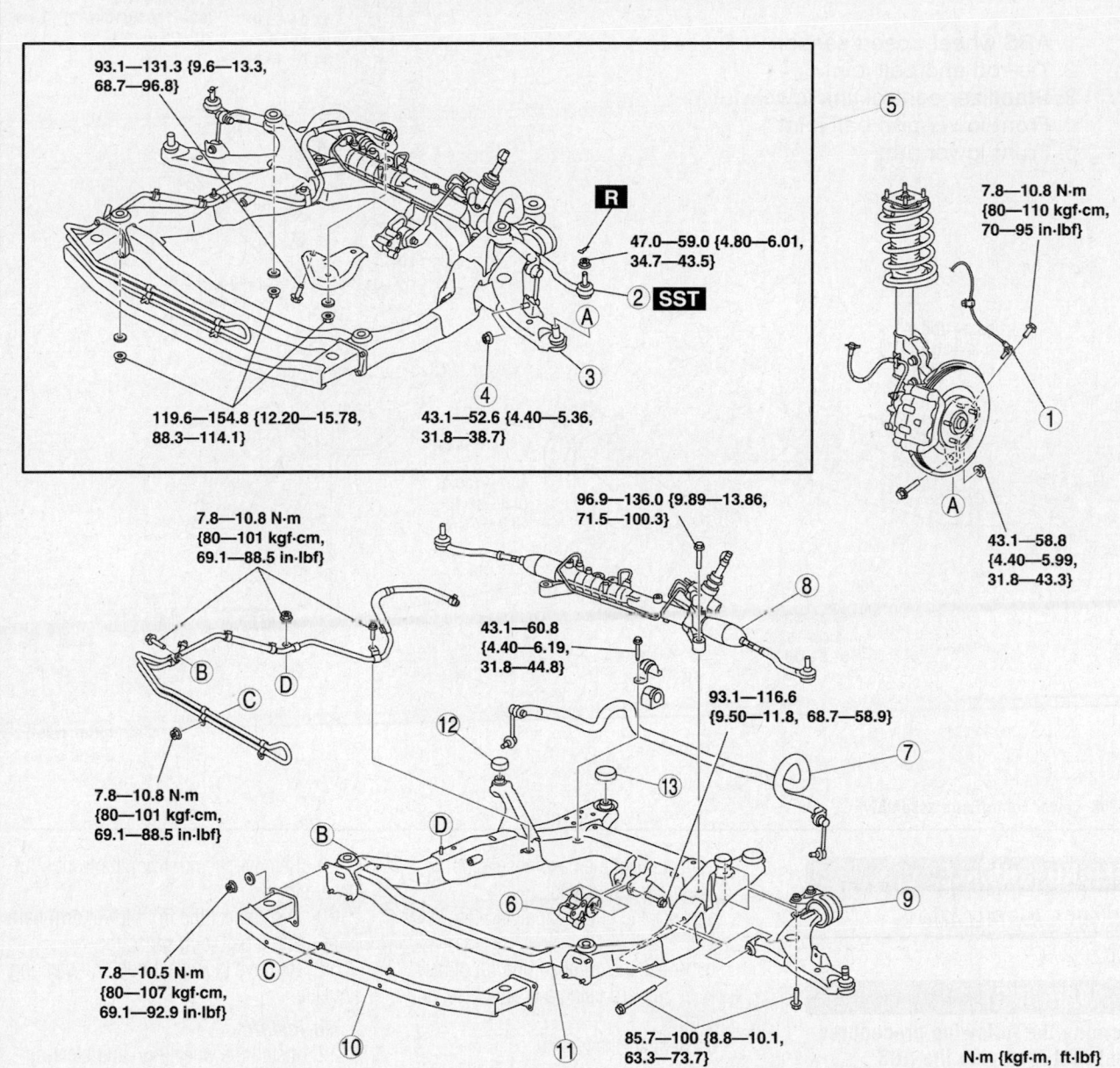

Fig. 207 Exploded view of the front crossmember assembly

1. ABS wheel-speed sensor
2. Tie-rod end ball joint
3. Front lower arm ball joint
4. Stabilizer control link lower side nut
5. Steering gear and linkage, front stabilizer,
   front lower arm and front crossmember component
6. No.1 engine mount
7. Front stabilizer

8. Steering gear and linkage
9. Front lower arm
10. Front crossmember extension
11. Front crossmember
12. Front crossmember mounting rubber (front)
13. Front crossmember mounting rubber (rear)

22140_MCX9_G0305

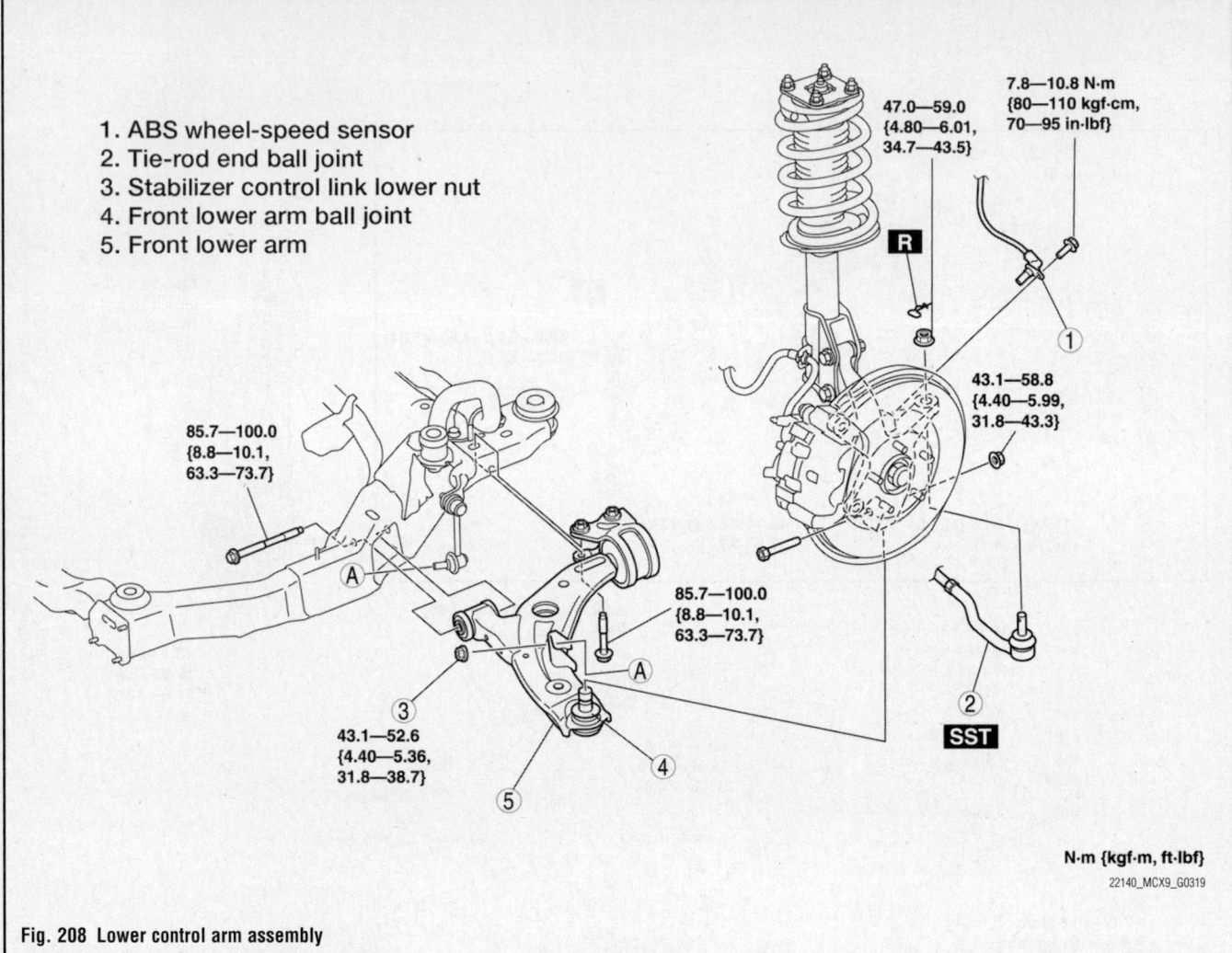

1. ABS wheel-speed sensor
2. Tie-rod end ball joint
3. Stabilizer control link lower nut
4. Front lower arm ball joint
5. Front lower arm

47.0—59.0
{4.80—6.01,
34.7—43.5}

7.8—10.8 N·m
{80—110 kgf·cm,
70—95 in·lbf}

R

85.7—100.0
{8.8—10.1,
63.3—73.7}

43.1—58.8
{4.40—5.99,
31.8—43.3}

85.7—100.0
{8.8—10.1,
63.3—73.7}

43.1—52.6
{4.40—5.36,
31.8—38.7}

SST

N·m {kgf·m, ft·lbf}

22140_MCX9_G0319

Fig. 208 Lower control arm assembly

STEERING KNUCKLE

REMOVAL & INSTALLATION

See Figure 209.

✳✳ WARNING

Performing the following procedures without first removing the ABS wheel-speed sensor may possibly cause an open circuit in the wiring harness if it is pulled by mistake. Before performing the following procedures, disconnect the ABS wheel-speed sensor connector (axle side) and fix the wiring harness to an appropriate place where it will not be pulled by mistake while servicing the vehicle.

1. Before servicing the vehicle, refer to the precautions section.
2. Raise and safely support the vehicle securely on jackstands.
3. Remove the front ABS wheel-speed sensor.

4. Disconnect the bake hose clip.
5. Remove the locknut.
   a. Lock the hub by applying the brakes.
   b. Knock the crimped portion of the locknut outward using a small chisel and a hammer.
   c. Remove the locknut.
   d. Install a spare nut onto the drive shaft so that the nut is flush with the end of the drive shaft.
   e. Tap the nut with a copper hammer to loosen the drive shaft from the front wheel hub.
   f. Separate the drive shaft from the wheel hub.
6. Remove the brake caliper and rotor.
   a. Remove the brake caliper component from the steering knuckle and suspend it out of the way using a cable.
7. Disconnect the stabilizer control link (lower side).
8. Disconnect the tie-rod end ball joint using a ball joint separator.

9. Remove the nut front the front strut lower side
10. Disconnect the front lower arm ball joint using a ball joint separator.
11. Remove the wheel hub and steering knuckle.

To install:

12. Install the wheel hub and steering knuckle.
13. Connect the front lower arm ball joint using a ball joint separator.
14. Install the nut front the front strut lower side
15. Connect the tie-rod end ball joint using a ball joint separator.
16. Connect the stabilizer control link (lower side).
17. Install the brake caliper and rotor.
18. Install a new locknut and stake it into place.
19. connect the bake hose clip.
20. Install the front ABS wheel-speed sensor.
21. Lower the vehicle.
22. Check and adjust the wheel alignment.

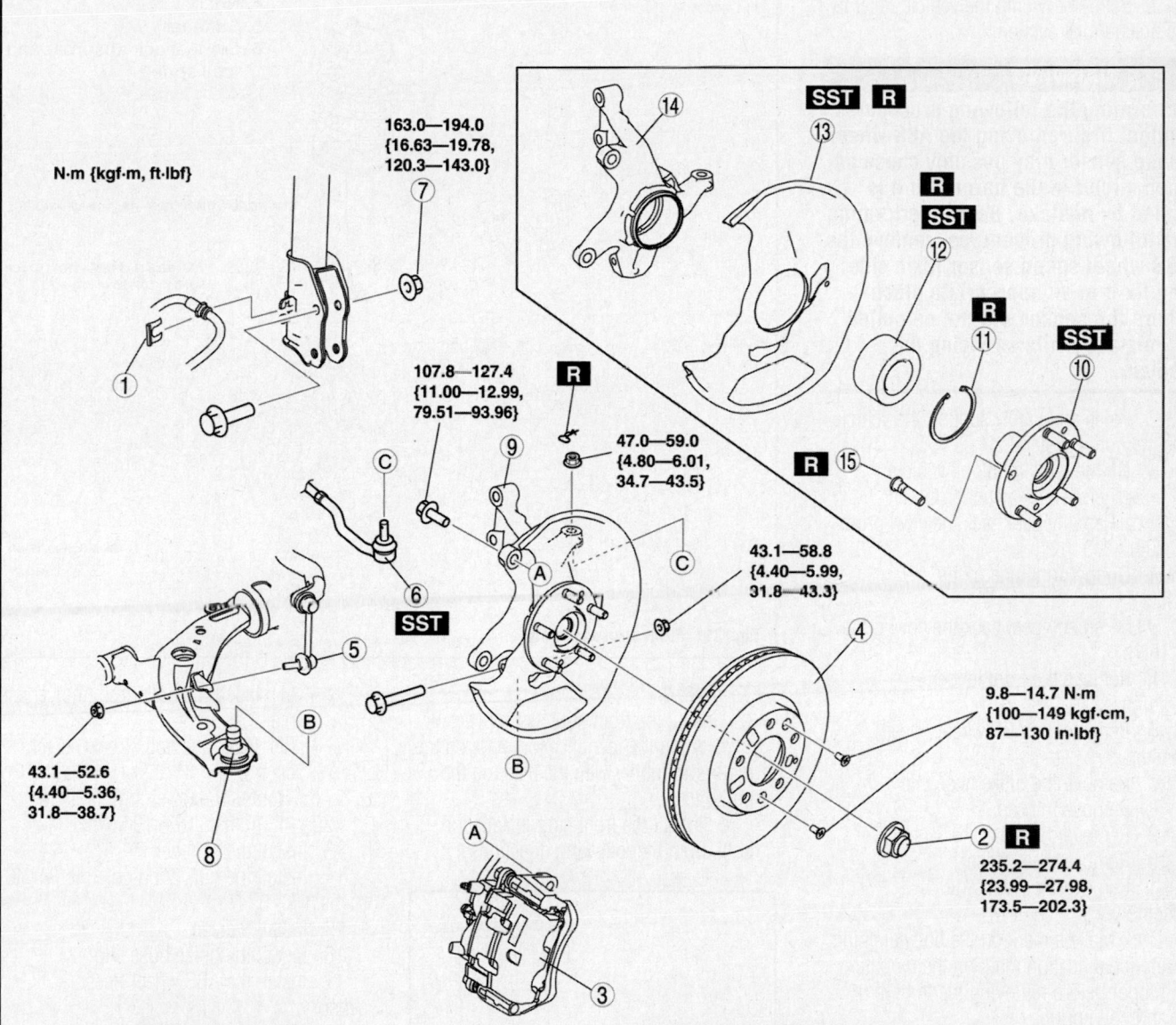

N·m {kgf·m, ft·lbf}

163.0—194.0
{16.63—19.78,
120.3—143.0}

107.8—127.4
{11.00—12.99,
79.51—93.96}

47.0—59.0
{4.80—6.01,
34.7—43.5}

43.1—58.8
{4.40—5.99,
31.8—43.3}

43.1—52.6
{4.40—5.36,
31.8—38.7}

9.8—14.7 N·m
{100—149 kgf·cm,
87—130 in·lbf}

235.2—274.4
{23.99—27.98,
173.5—202.3}

1. Brake hose clip
2. Locknut
3. Brake caliper component
4. Disc plate
5. Stabilizer control link (lower side)
6. Tie-rod end ball joint
7. Nut (front shock absorber lower side)
8. Front lower arm ball joint
9. Wheel hub, steering knuckle component
10. Wheel hub
11. Retaining ring
12. Wheel bearing
13. Dust cover
14. Steering knuckle
15. Wheel hub bolt

22140_MCX9_G0321

Fig. 209 Exploded view of the steering knuckle assembly

STRUT

REMOVAL & INSTALLATION

See Figures 210 through 212.

1. Before servicing the vehicle, refer to the precautions section.

✳✳ WARNING

Performing the following procedures without first removing the ABS wheel speed sensor may possibly cause an open circuit in the harness if it is pulled by mistake. Before performing the following procedures, remove the ABS wheel speed sensor (axle side) and fix it to an appropriate place where the sensor will not be pulled by mistake while servicing the vehicle.

2. Raise and safely support the vehicle securely on jackstands.

3. On the left side of the vehicle:

   a. Remove the wiper arm (LH).

   b. Partially peel back the cowl grille (LH).

4. On the right side of the vehicle:

   a. Partially peel back the cowl grille (RH).

5. Remove the front fender molding.

6. Remove the ABS wheel-speed sensor.

7. Remove the brake hose clip

8. Remove the cap.

9. Remove the nut.

10. Remove the stiffener.

11. Remove the front strut assembly.

   a. Make a mark on the body aligning the identification mark on the mounting rubber before removing the front strut and coil spring.

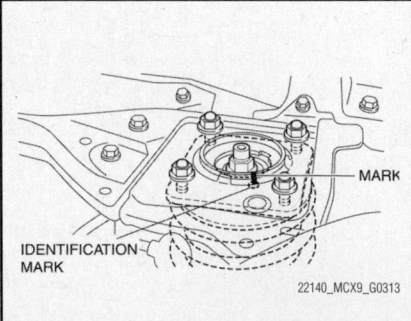

Fig. 210 Make a mark on the body aligning the identification mark on the mounting rubber before removing the front strut and coil spring

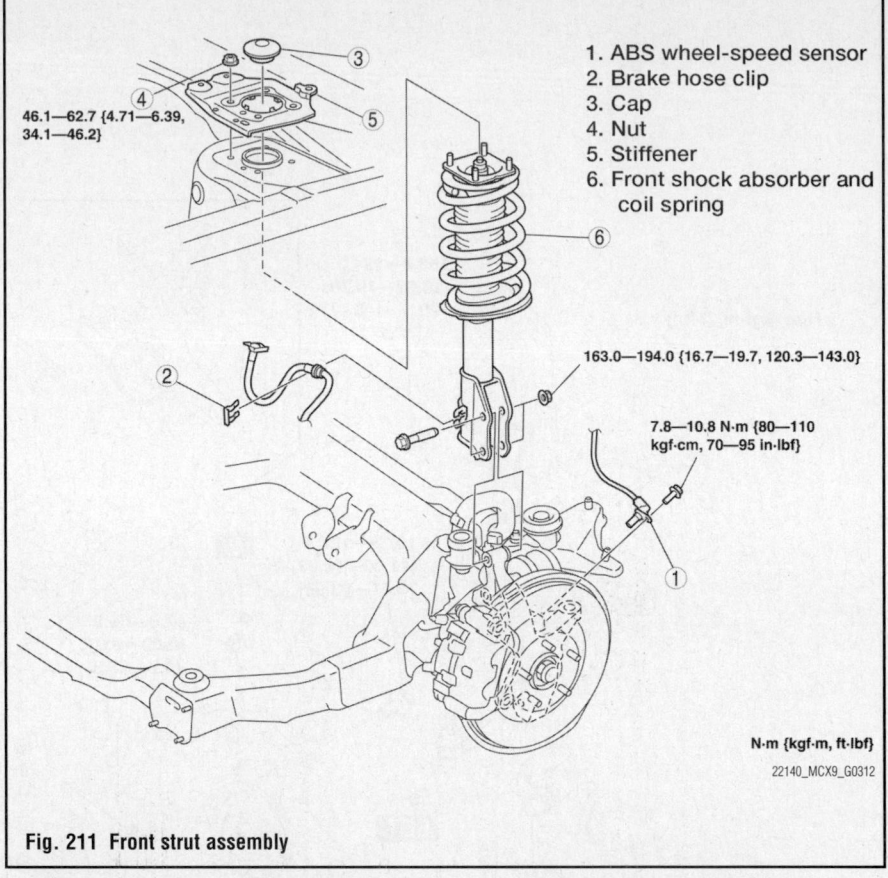

1. ABS wheel-speed sensor
2. Brake hose clip
3. Cap
4. Nut
5. Stiffener
6. Front shock absorber and coil spring

46.1—62.7 {4.71—6.39, 34.1—46.2}

163.0—194.0 {16.7—19.7, 120.3—143.0}

7.8—10.8 N·m {80—110 kgf·cm, 70—95 in·lbf}

N·m {kgf·m, ft·lbf}

22140_MCX9_G0312

Fig. 211 Front strut assembly

To install:

12. Install the front strut assembly.

   a. Align the identification mark on the mounting rubber with the mark on the body.

   b. Install the front strut installation bolts from the following directions:

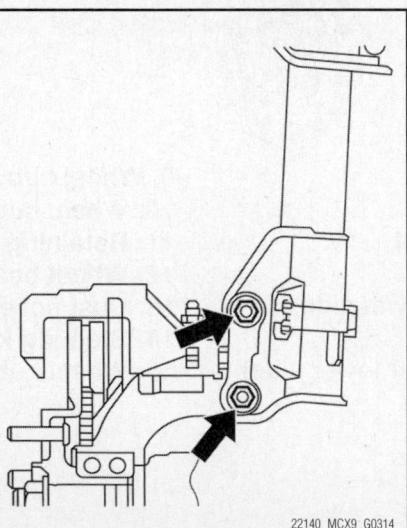

22140_MCX9_G0314

Fig. 212 Install the front strut installation bolts from the following directions: LH - Insert the bolts from the vehicle rear and RH - Insert the bolts from the vehicle front

- LH: Insert the bolts from the vehicle rear
- RH: Insert the bolts from the vehicle front

   c. Tighten the lower strut bolts to 120–143 ft. lbs. (163–194 Nm).

13. Install the stiffener.

14. Install the nuts and tighten to 34–46 ft. lbs. (46–62 Nm).

15. Install the cap.

16. Install the brake hose clip

17. Install the ABS wheel-speed sensor.

18. Install the front fender molding.

19. On the right side of the vehicle:

   a. Position the cowl grille (RH).

20. On the left side of the vehicle:

   a. Position the cowl grille (LH).

   b. Install the wiper arm (LH).

21. Lower the vehicle.

22. Inspect for front wheel alignment, and adjust it as necessary.

OVERHAUL

See Figures 213 through 215.

✳✳ WARNING

Removing/installing the piston rod nut is dangerous. The shock absorber and spring could fly off under

tremendous pressure and cause serious injury or death. Secure the shock absorber in the Special tools before removing the piston rod nut.

1. Before servicing the vehicle, refer to the precautions section.

2. Remove the front shock absorber and coil spring.

3. Protect the coil spring using a piece of cloth, then install a spring compressor.

4. Compress the coil spring, and

remove the piston rod nut, washer and mounting rubber.

5. Remove the bearing, upper spring seat, rubber, dust boot, bound stopper, coil spring and lower spring seat.

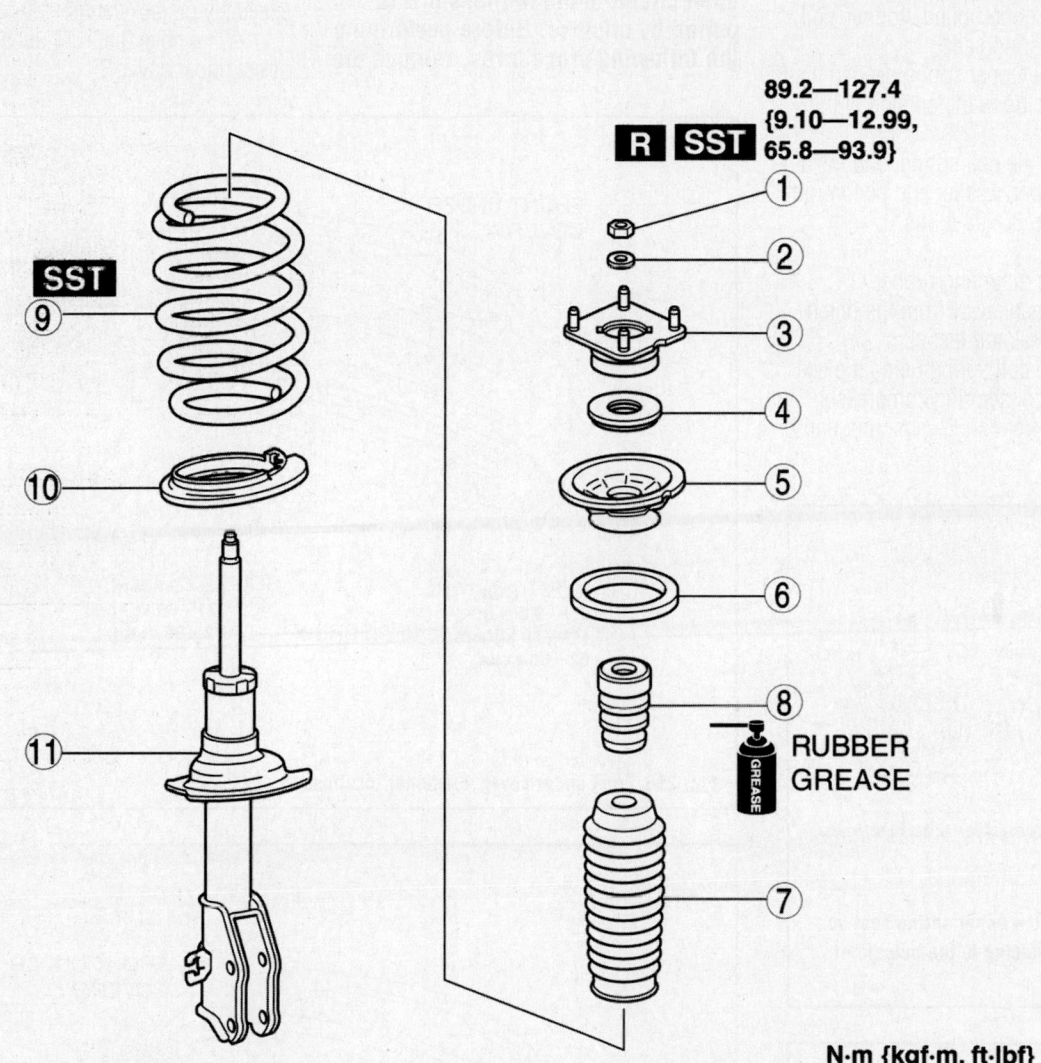

89.2—127.4
{9.10—12.99,
65.8—93.9}

R **SST**

N·m {kgf·m, ft·lbf}

1. Piston rod nut
2. Washer
3. Mounting rubber
4. Bearing
5. Upper spring seat
6. Upper spring seat rubber
7. Dust boot
8. Bound stopper
9. Coil spring
10. Lower spring seat
11. Front shock absorber

22140_MCX9_G0315

Fig. 213 Exploded view of the front strut assembly

To assemble:

6. Inspect the strut for damage and oil leakage.

   a. Compress and extend the strut piston at least three times. Verify that the operational force does not change and that there is no unusual noise.

   b. If not as specified, replace the front strut.

7. Install the bearing, upper spring seat, rubber, dust boot, bound stopper, coil spring and lower spring seat.

   a. Install the upper spring seat so that the notch is facing to the outside of the vehicle.

8. Compress the coil spring, and install the mounting rubber, washer and piston rod nut. Tighten nut to 66–94 ft. lbs. (89–127Nm).

   a. Slide the mounting rubber identification mark away from the notch of the spring seat and install.

9. Protect the coil spring using a piece of cloth, then install a spring compressor.

10. Install the front shock absorber and coil spring.

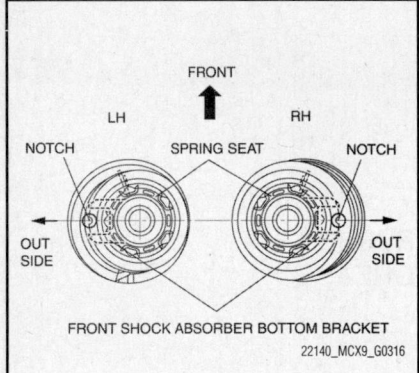

Fig. 214 Install the upper spring seat so that the notch is facing to the outside of the vehicle

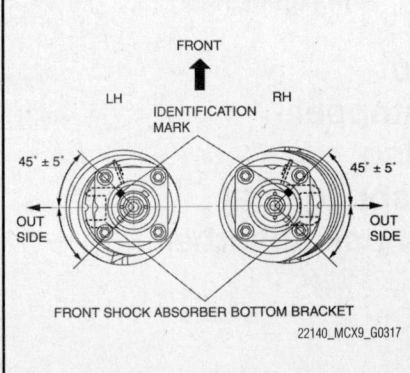

Fig. 215 Slide the mounting rubber identification mark away from the notch of the spring seat and install

STABILIZER BAR

REMOVAL & INSTALLATION
See Figures 216 through 222.

✳✳ WARNING

Performing the following procedures without first removing the ABS wheel speed sensor may possibly cause an open circuit in the harness if it is pulled by mistake. Before performing the following procedures, remove the ABS wheel speed sensor (axle side) and fix it to an appropriate place where the sensor will not be pulled by mistake while servicing the vehicle.

1. Before servicing the vehicle, refer to the precautions section.

2. Drain the power steering fluid.

3. Raise and safely support the vehicle securely on jackstands.

4. Remove the front under cover A and front under cover B.

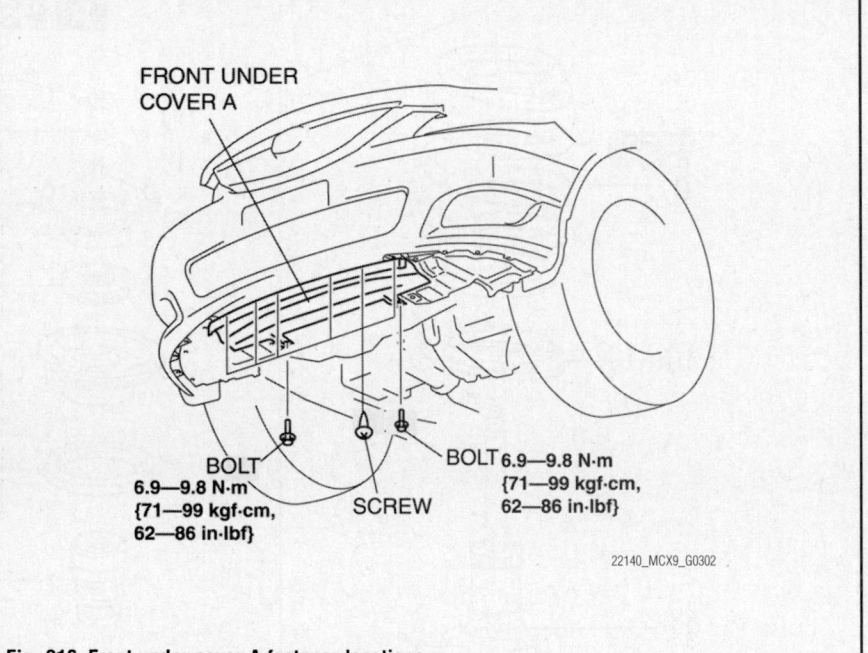

Fig. 216 Front under cover A fastener locations

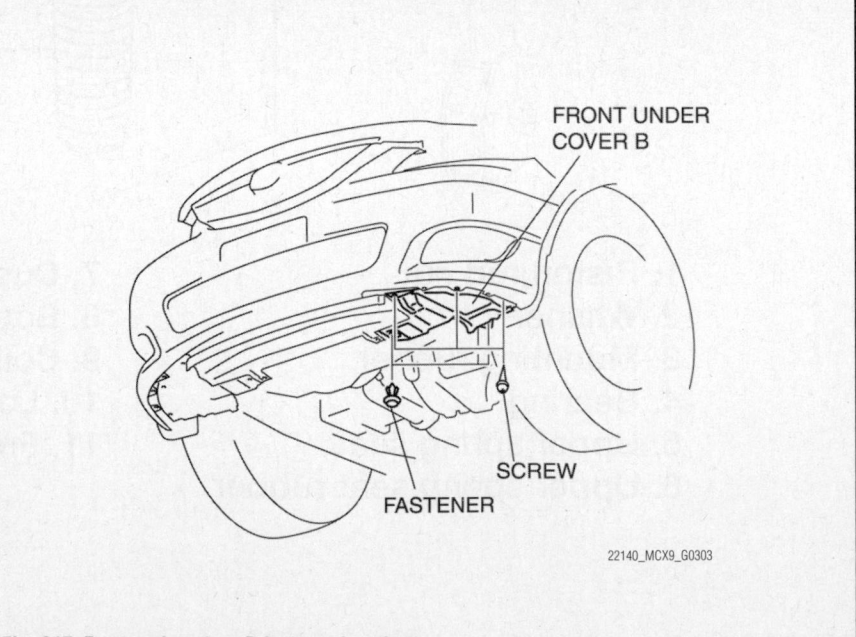

Fig. 217 Front under cover B fastener locations

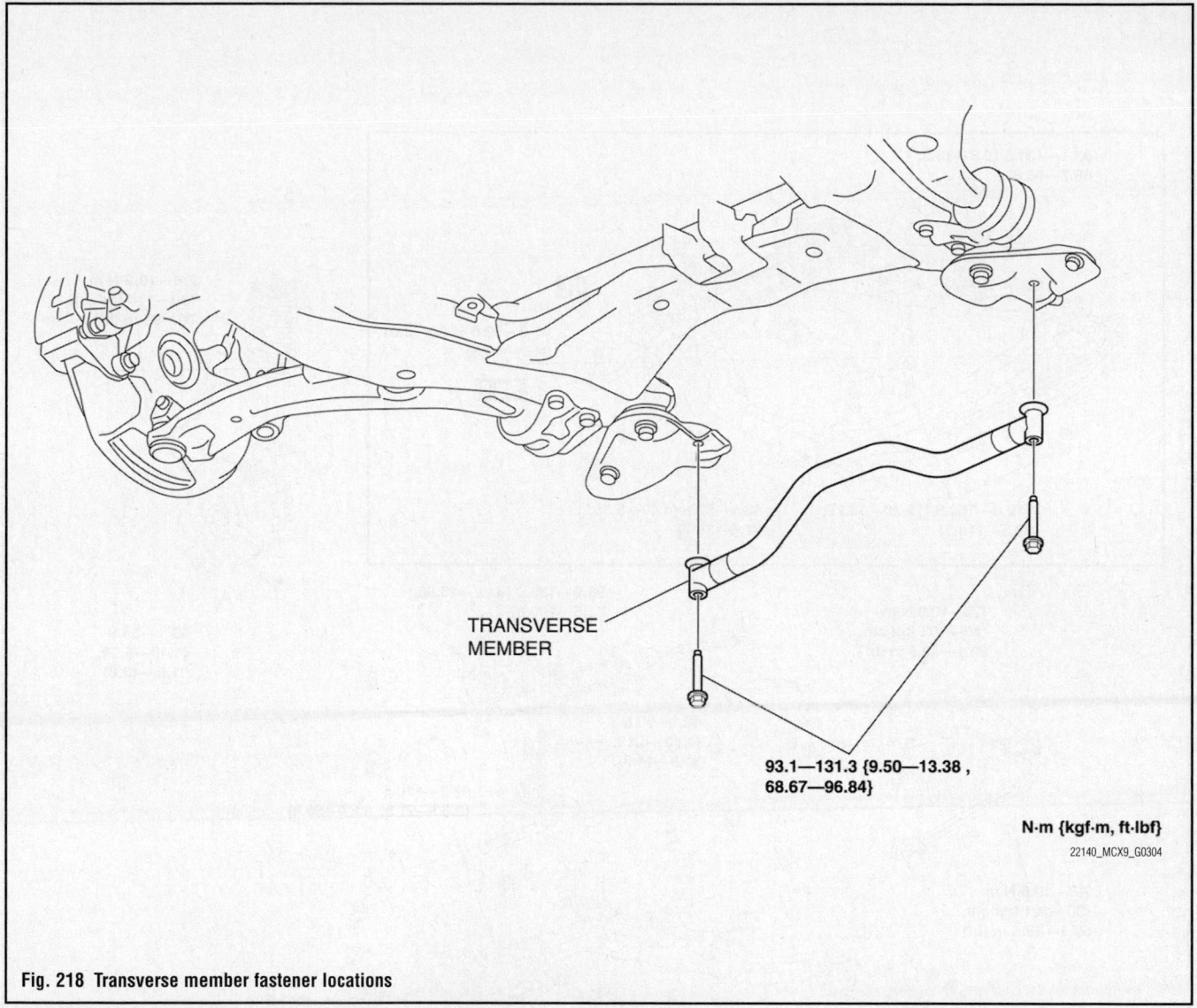

TRANSVERSE
MEMBER

93.1—131.3 {9.50—13.38 ,
68.67—96.84}

N·m {kgf·m, ft·lbf}

22140_MCX9_G0304

Fig. 218 Transverse member fastener locations

5. Remove the middle pipe.

6. Remove the transverse member.

7. Remove the intermediate shaft installation bolt, and disconnect the steering shaft.

8. Remove the ABS wheel-speed sensor.

9. Disconnect the tie-rod end ball joint using a ball joint separator.

10. Disconnect the front lower ball joint using a ball joint separator.

11. Disconnect the stabilizer lower side control link.

12. Remove the steering gear and linkage, front stabilizer, front lower arm and front crossmember as an assembly.

13. Disconnect the front stabilizer control link.

14. Remove the stabilizer bracket and bushing.

15. Remove the stabilizer.

To install:

16. Install the stabilizer.

a. Install the stabilizer bar so that the identification mark is on the left side of the vehicle.

17. Install the stabilizer bracket and bushing.

a. Align the outer side of the positioning plate with the stabilizer bushing.

b. Install the stabilizer bracket so that the arrow is pointed to the front.

18. Connect the front stabilizer control link.

19. Install the steering gear and linkage, front stabilizer, front lower arm and front crossmember as an assembly.

20. Connect the stabilizer lower side control link.

21. Connect the front lower ball joint using a ball joint separator.

22. Connect the tie-rod end ball joint using a ball joint separator.

23. Install the ABS wheel-speed sensor.

24. Install the intermediate shaft installation bolt, and disconnect the steering shaft.

25. Install the transverse member.

26. Install the middle pipe.

27. Install the front under cover A and front under cover B.

28. Lower the vehicle.

29. Fill and bleed the power steering system.

30. Check and adjust the total toe-in.

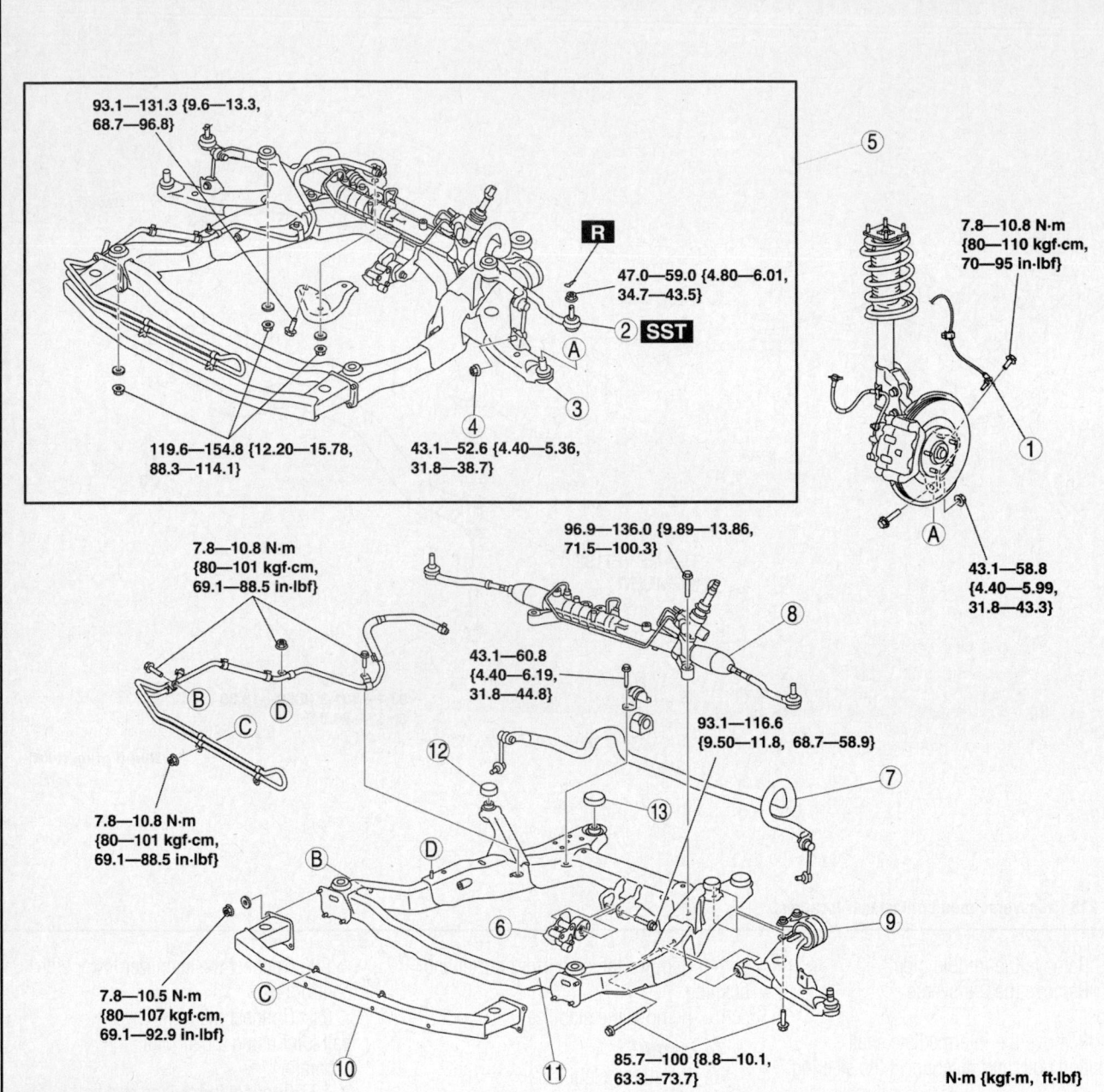

93.1—131.3 {9.6—13.3, 68.7—96.8}

7.8—10.8 N·m {80—110 kgf·cm, 70—95 in·lbf}

R

47.0—59.0 {4.80—6.01, 34.7—43.5}

② **SST**

Ⓐ

③

④

119.6—154.8 {12.20—15.78, 88.3—114.1}

43.1—52.6 {4.40—5.36, 31.8—38.7}

⑤

Ⓐ

43.1—58.8 {4.40—5.99, 31.8—43.3}

①

96.9—136.0 {9.89—13.86, 71.5—100.3}

7.8—10.8 N·m {80—101 kgf·cm, 69.1—88.5 in·lbf}

Ⓑ Ⓒ Ⓓ

43.1—60.8 {4.40—6.19, 31.8—44.8}

⑧

7.8—10.8 N·m {80—101 kgf·cm, 69.1—88.5 in·lbf}

Ⓑ Ⓓ

⑫

93.1—116.6 {9.50—11.8, 68.7—58.9}

⑬

⑦

7.8—10.5 N·m {80—107 kgf·cm, 69.1—92.9 in·lbf}

Ⓒ

⑥

⑨

⑩

⑪

85.7—100 {8.8—10.1, 63.3—73.7}

N·m {kgf·m, ft·lbf}

1. ABS wheel-speed sensor
2. Tie-rod end ball joint
3. Front lower arm ball joint
4. Stabilizer control link lower side nut
5. Steering gear and linkage, front stabilizer, front lower arm and front crossmember component
6. No.1 engine mount
7. Front stabilizer
8. Steering gear and linkage
9. Front lower arm
10. Front crossmember extension
11. Front crossmember
12. Front crossmember mounting rubber (front)
13. Front crossmember mounting rubber (rear)

22140_MCX9_G0305

Fig. 219 Exploded view of the front crossmember assembly

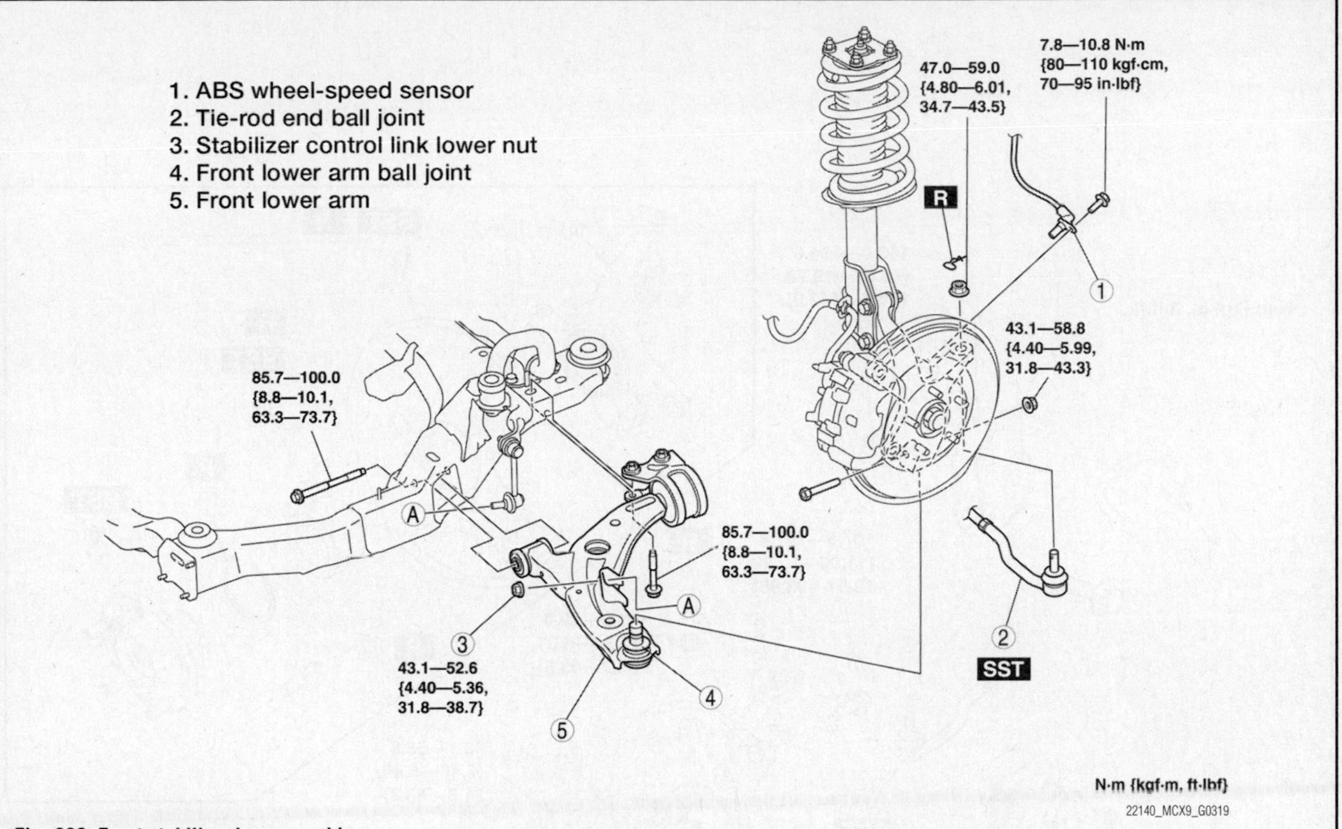

1. ABS wheel-speed sensor
2. Tie-rod end ball joint
3. Stabilizer control link lower nut
4. Front lower arm ball joint
5. Front lower arm

7.8—10.8 N·m
{80—110 kgf·cm,
70—95 in·lbf}

47.0—59.0
{4.80—6.01,
34.7—43.5}

85.7—100.0
{8.8—10.1,
63.3—73.7}

43.1—58.8
{4.40—5.99,
31.8—43.3}

85.7—100.0
{8.8—10.1,
63.3—73.7}

43.1—52.6
{4.40—5.36,
31.8—38.7}

N·m {kgf·m, ft·lbf}

22140_MCX9_G0319

Fig. 220 Front stabilizer bar assembly

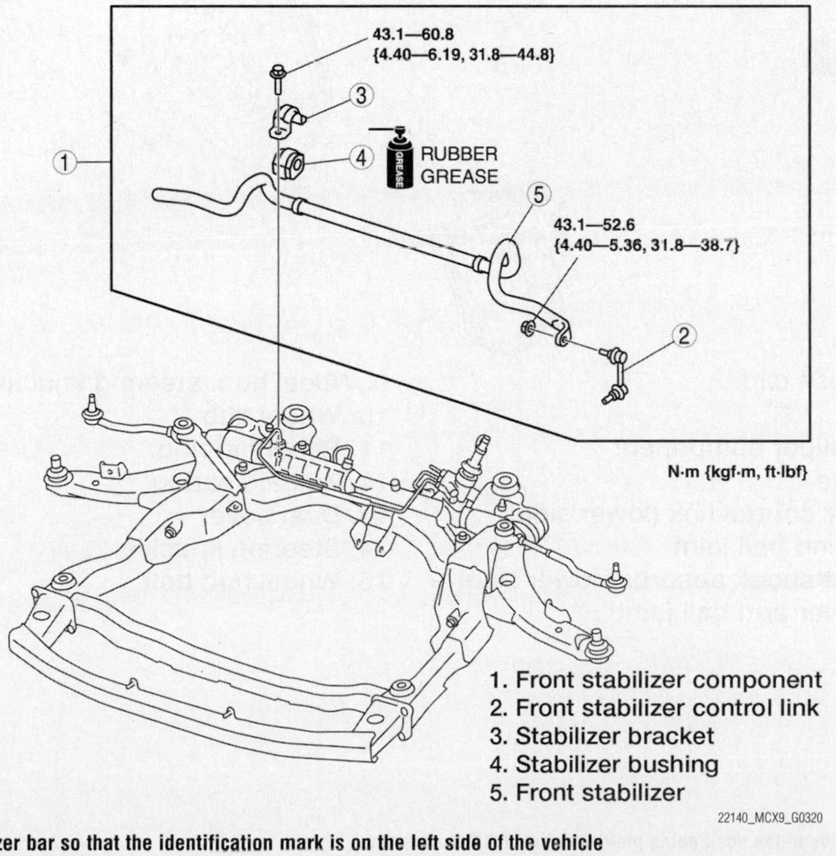

43.1—60.8
{4.40—6.19, 31.8—44.8}

RUBBER
GREASE

43.1—52.6
{4.40—5.36, 31.8—38.7}

N·m {kgf·m, ft·lbf}

1. Front stabilizer component
2. Front stabilizer control link
3. Stabilizer bracket
4. Stabilizer bushing
5. Front stabilizer

22140_MCX9_G0320

Fig. 221 Install the stabilizer bar so that the identification mark is on the left side of the vehicle

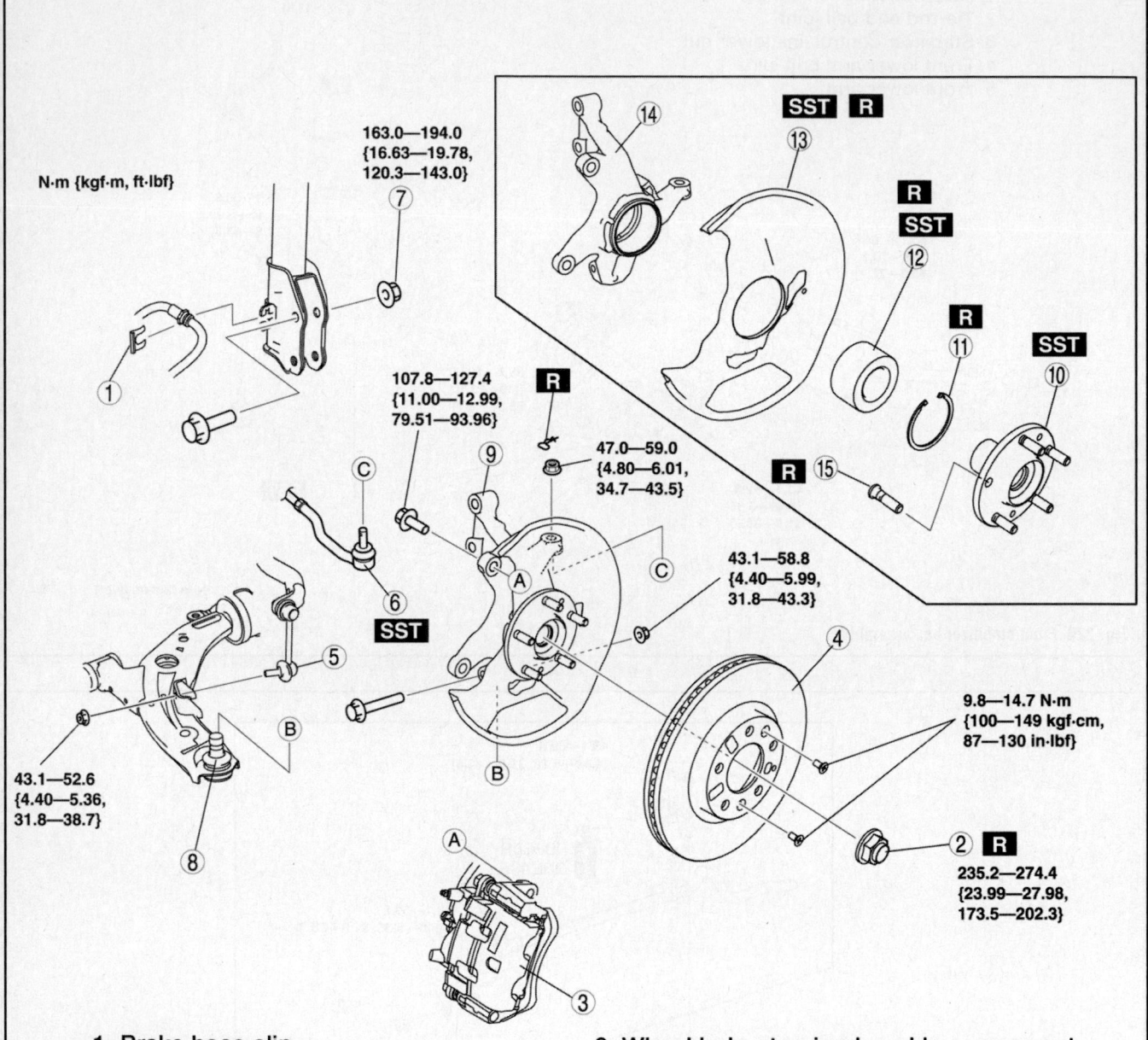

1. Brake hose clip
2. Locknut
3. Brake caliper component
4. Disc plate
5. Stabilizer control link (lower side)
6. Tie-rod end ball joint
7. Nut (front shock absorber lower side)
8. Front lower arm ball joint
9. Wheel hub, steering knuckle component
10. Wheel hub
11. Retaining ring
12. Wheel bearing
13. Dust cover
14. Steering knuckle
15. Wheel hub bolt

22140_MCX9_G0321

Fig. 222 Align the outer side of the positioning plate with the stabilizer bushing

WHEEL BEARINGS

REMOVAL & INSTALLATION

See Figures 223 through 228.

1. Before servicing the vehicle, refer to the precautions section.

✳✳ WARNING

Performing the following procedures without first removing the ABS wheel-speed sensor may possibly cause an open circuit in the wiring harness if it is pulled by mistake. Before performing the following procedures, disconnect the ABS wheel-speed sensor connector (axle side) and fix the wiring harness to an appropriate place where it will not be pulled by mistake while servicing the vehicle.

2. Raise and safely support the vehicle securely on jackstands.
3. Remove the front ABS wheel-speed sensor.
4. Disconnect the bake hose clip.
5. Remove the locknut.
   a. Lock the hub by applying the brakes.
   b. Knock the crimped portion of the locknut outward using a small chisel and a hammer.
   c. Remove the locknut.
   d. Install a spare nut onto the drive shaft so that the nut is flush with the end of the drive shaft.
   e. Tap the nut with a copper hammer to loosen the drive shaft from the front wheel hub.
   f. Separate the drive shaft from the wheel hub.
6. Remove the brake caliper and rotor.
   a. Remove the brake caliper component from the steering knuckle and suspend it out of the way using a cable.
7. Disconnect the stabilizer control link (lower side).
8. Disconnect the tie-rod end ball joint using a ball joint separator.
9. Remove the nut front the front strut lower side
10. Disconnect the front lower arm ball joint using a ball joint separator.
11. Remove the wheel hub and steering knuckle.
12. Remove the wheel hub.
   a. Remove the wheel hub using the special tool.
   b. If the bearing inner race remains

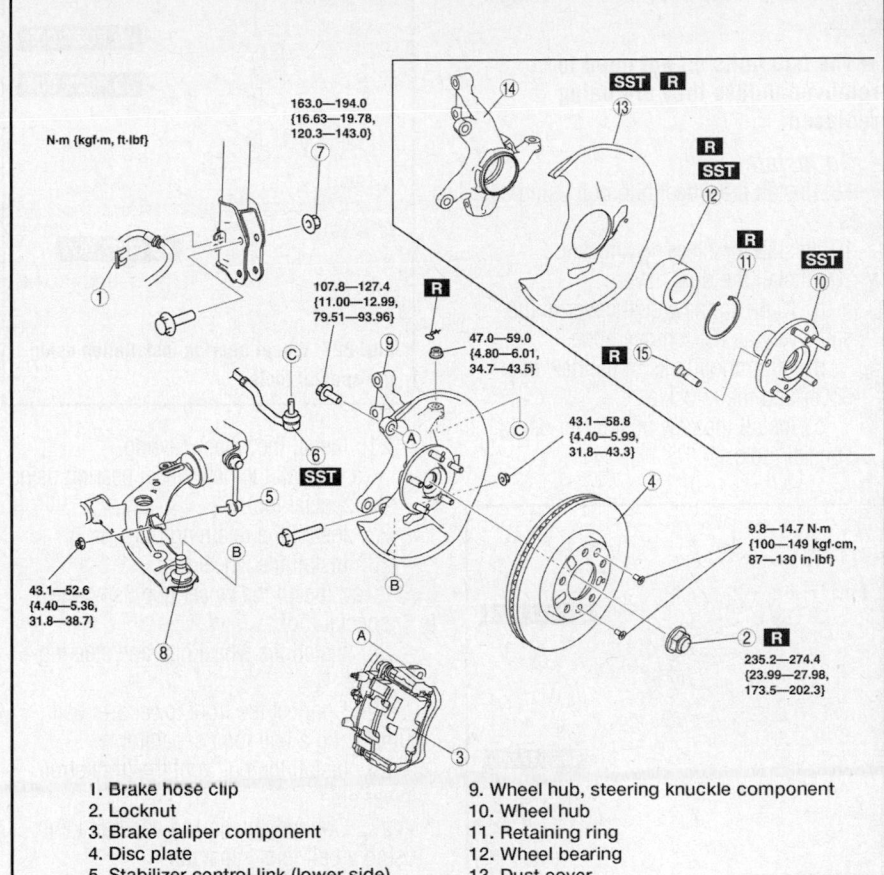

N·m {kgf-m, ft-lbf}

163.0—194.0 {16.63—19.78, 120.3—143.0}

107.8—127.4 {11.00—12.99, 79.51—93.96}

47.0—59.0 {4.80—6.01, 34.7—43.5}

43.1—58.8 {4.40—5.99, 31.8—43.3}

9.8—14.7 N·m {100—149 kgf-cm, 87—130 in-lbf}

43.1—52.6 {4.40—5.36, 31.8—38.7}

235.2—274.4 {23.99—27.98, 173.5—202.3}

1. Brake hose clip
2. Locknut
3. Brake caliper component
4. Disc plate
5. Stabilizer control link (lower side)
6. Tie-rod end ball joint
7. Nut (front shock absorber lower side)
8. Front lower arm ball joint
9. Wheel hub, steering knuckle component
10. Wheel hub
11. Retaining ring
12. Wheel bearing
13. Dust cover
14. Steering knuckle
15. Wheel hub bolt

22140_MCX9_G0321

Fig. 223 Exploded view of the steering knuckle assembly

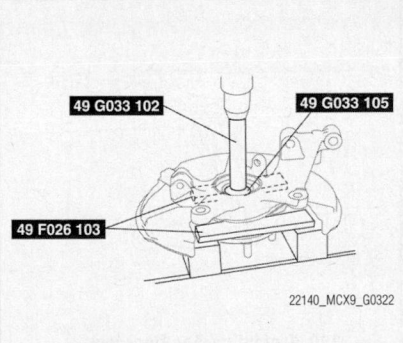

49 G033 102   49 G033 105

49 F026 103

22140_MCX9_G0322

Fig. 224 Wheel hub removal using the special tools

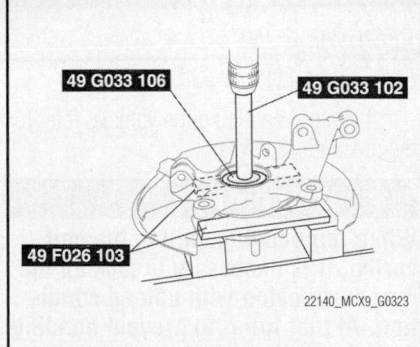

49 G033 106   49 G033 102

49 F026 103

22140_MCX9_G0323

Fig. 225 Wheel bearing removal using the special tools

on the front wheel hub component, grind a section of the bearing inner race until approximately 0.02 in (0.5 mm) remains. Then remove it using a chisel.

13. Remove the retaining ring.
14. Remove the wheel bearing
   a. Remove the wheel bearing using the special tool.
15. Remove the dust cover.

➡**The dust cover does not need to be removed unless it is being replaced.**

   a. Mark the dust cover and steering knuckle for proper installation.
   b. Remove the dust cover using a chisel.
16. Remove the steering knuckle.

17. Remove the wheel hub bolt using a press.

➡ **The hub bolts do not need to be removed unless they are being replaced.**

To install:

18. Install the wheel hub bolt using a press.
19. Install the steering knuckle.
20. Install the dust cover.
   a. Mark the new dust cover in the same way as the removed one.
   b. Align the marks of the new dust cover and the knuckle.
   c. Install the new dust cover using the special tools.

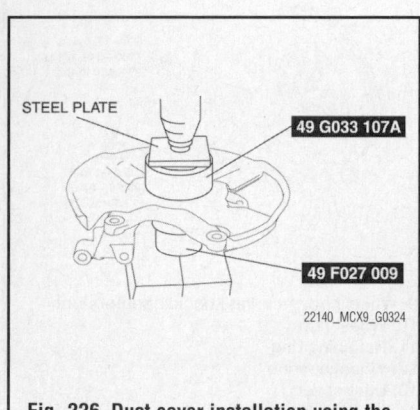

Fig. 226 Dust cover installation using the special tools

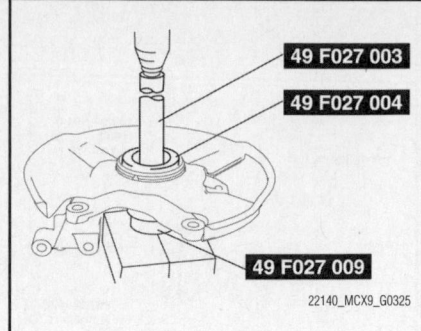

Fig. 227 Wheel bearing installation using the special tools

21. Install the wheel bearing.
   a. Install the new wheel bearing using the special tool.
22. Install the retaining ring.
23. Install the wheel hub.
   a. Install the wheel hub using the special tool.
24. Install the wheel hub and steering knuckle.
25. Connect the front lower arm ball joint using a ball joint separator.
26. Install the nut front the front strut lower side
27. Connect the tie-rod end ball joint using a ball joint separator.
28. Connect the stabilizer control link (lower side).
29. Install the brake caliper and rotor.

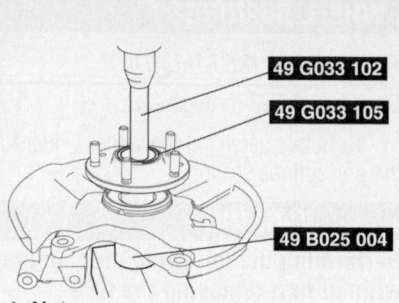

1. Nut
2. Retainer
3. Upper rubber bushing
4. Bolt and nut (shock absorber lower side)
5. Lower rubber bushing
6. Retainer
7. Rear shock absorber

Fig. 228 Wheel hub installation using the special tools

30. Install a new locknut and stake it into place.
31. connect the bake hose clip.
32. Install the front ABS wheel-speed sensor.
33. Lower the vehicle.
34. Check and adjust the wheel alignment.

ADJUSTMENT

The bearings are sealed units and not adjustable.

SUSPENSION

COIL SPRING

REMOVAL & INSTALLATION

See Figures 229 and 230.

1. Before servicing the vehicle, refer to the precautions section.

❋❋ WARNING

When removing/installing the coil spring, it is necessary to jack up the rear suspension with unload condition. At that time, to prevent bending the front of the vehicle, support the jack up point for the front crossmember using a jack.

2. Raise and safely support the vehicle securely on jackstands.
3. Remove the rear ABS wheel-speed sensor.
4. Remove the rear stabilizer control link upper side nut.
5. Remove the rear shock absorber lower side bolt.

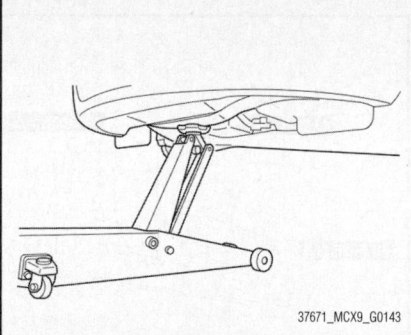

Fig. 229 Typical garage floor jack

   a. Raise the rear trailing link to the unloaded condition with a jack.
   b. Remove the rear shock absorber lower side bolt.
6. Remove the rear coil spring

❋❋ CAUTION

Removing the coil spring is dangerous. The coil spring could fly off, and cause serious injury or death.

REAR SUSPENSION

   a. Support the rear lower arm inner side using a garage jack.
   b. Loosen the rear lower arm outer side bolt.
   c. Lower the rear lower arm inner side slowly.
   d. Remove the rear coil spring.
7. Remove the rear upper and lower spring seats, and the bump stop.

To install:

8. Install the rear upper and lower spring seats, and the bump stop.
9. Install the rear coil spring
   a. Install the rear coil spring so that the lower end of the rear coil spring is seated on the step of the rear lower spring seat.
   b. Position the jack under the lower arm inner side and compress the rear coil spring slowly.
   c. Support the rear lower arm outer side and raise the rear suspension to the

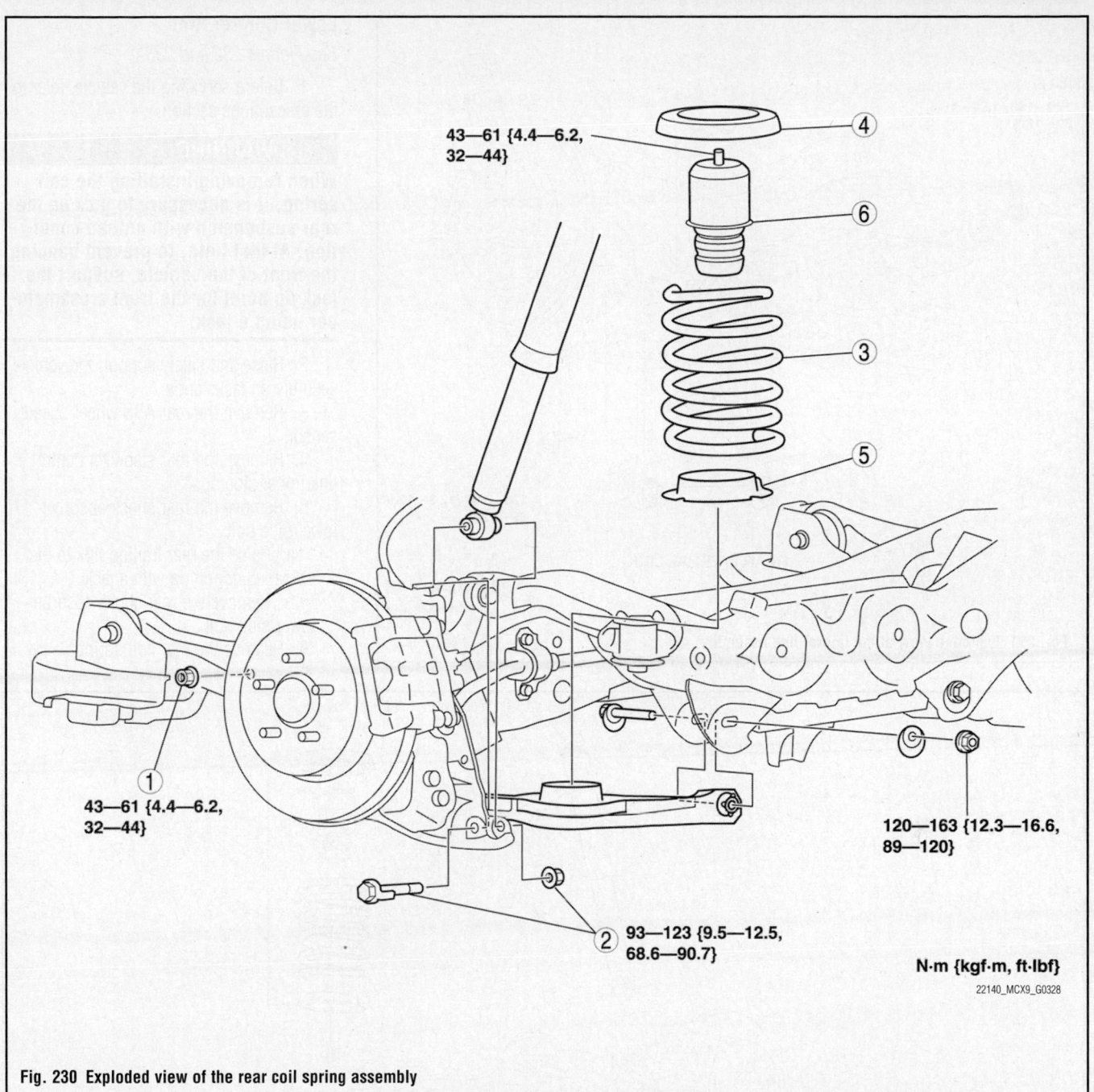

43—61 {4.4—6.2, 32—44}

43—61 {4.4—6.2, 32—44}

120—163 {12.3—16.6, 89—120}

93—123 {9.5—12.5, 68.6—90.7}

N·m {kgf·m, ft·lbf}

22140_MCX9_G0328

Fig. 230 Exploded view of the rear coil spring assembly

unloaded condition with a separate garage jack.

  d. Install the lower arm inner side bolt and tighten to 89–120 ft. lbs. (120–163 Nm).

  e. Install and tighten the rear lower arm outer side bolt to 89–120 ft. lbs. (120–163 Nm).

  10. Install the rear shock absorber lower side bolt. Tighten to 69–91 ft. lbs. (93–123 Nm).

  11. Install the rear stabilizer control link upper side nut.

  12. Install the rear ABS wheel-speed sensor.

  13. Lower the vehicle.

  14. Inspect the rear wheel alignment.

CONTROL ARMS/LINKS

REMOVAL & INSTALLATION

Lateral Link

See Figure 231.

  1. Before servicing the vehicle, refer to the precautions section.

  2. Raise and safely support the vehicle securely on jackstands.

  3. Mark the adjusting cam bolt and cam plate of the rear lateral link.

  4. Jack up the rear suspension with unloaded condition.

  5. Remove the rear lateral link.

  To install:

  6. Install the rear lateral link by aligning it with the mark placed during removal.

  7. Tighten the attaching bolts to 89–120 ft. lbs. (120–163 Nm).

  8. Lower the vehicle.

  9. Inspect for rear wheel alignment, and adjust it as neccssary.

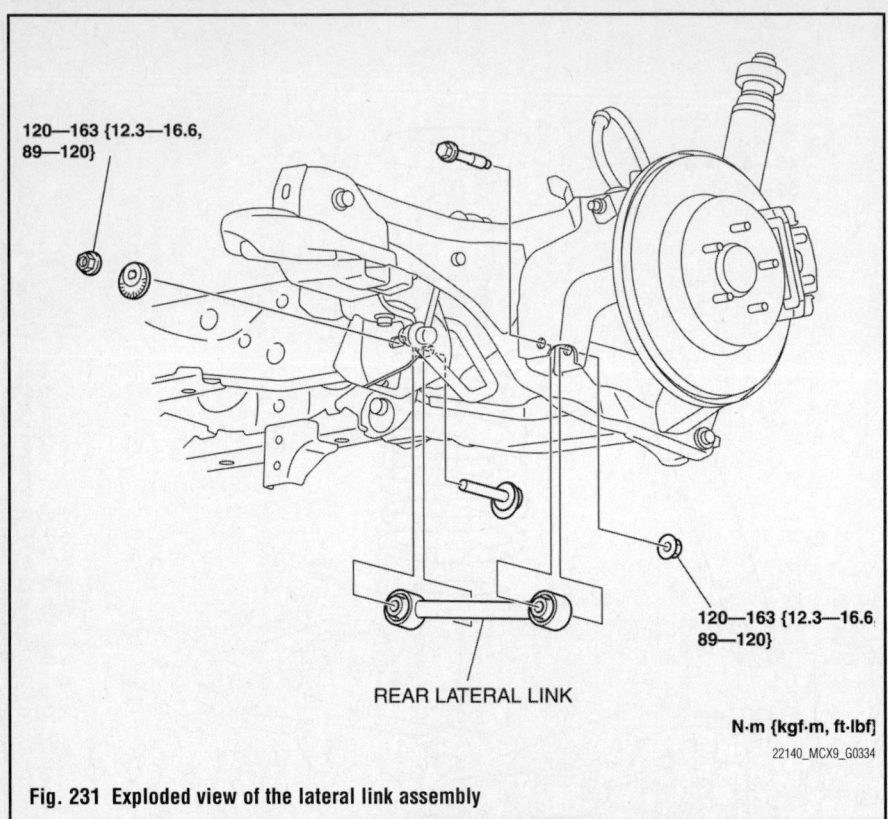

120—163 {12.3—16.6, 89—120}

REAR LATERAL LINK

120—163 {12.3—16.6, 89—120}

N·m {kgf·m, ft·lbf}

22140_MCX9_G0334

Fig. 231 Exploded view of the lateral link assembly

Lower Control Arm

See Figures 232 and 233.

1. Before servicing the vehicle, refer to the precautions section.

❊❊ WARNING

When removing/installing the coil spring, it is necessary to jack up the rear suspension with unload condition. At that time, to prevent bending the front of the vehicle, support the jack up point for the front crossmember using a jack.

2. Raise and safely support the vehicle securely on jackstands.

3. Remove the rear ABS wheel-speed sensor.

4. Remove the rear stabilizer control link upper side nut.

5. Remove the rear shock absorber lower side bolt.

   a. Raise the rear trailing link to the unloaded condition with a jack.

   b. Remove the rear shock absorber lower side bolt.

6. Remove the rear coil spring.

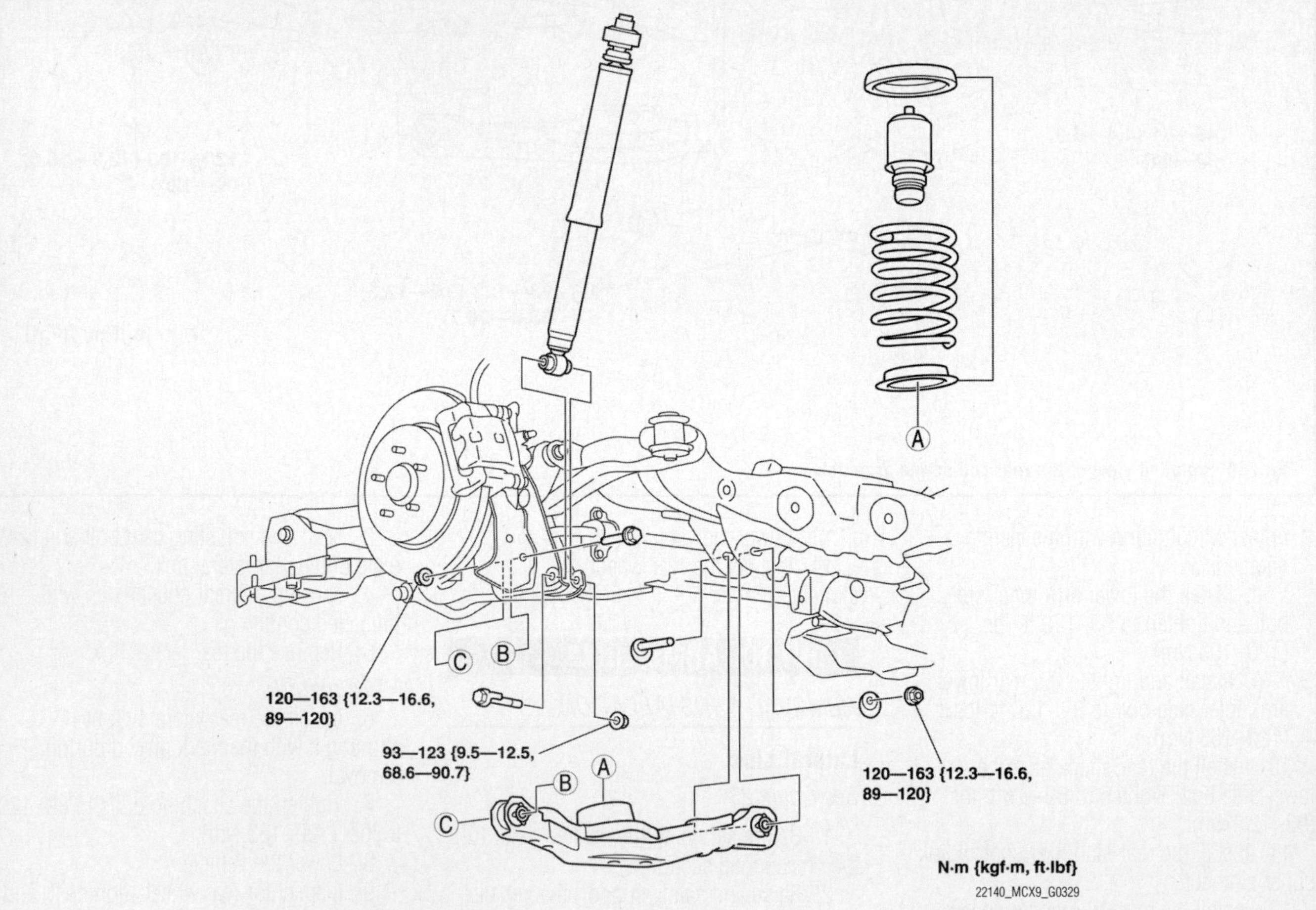

A

120—163 {12.3—16.6, 89—120}

93—123 {9.5—12.5, 68.6—90.7}

B

A

C

120—163 {12.3—16.6, 89—120}

N·m {kgf·m, ft·lbf}

22140_MCX9_G0329

Fig. 232 Exploded view of the lower control arm assembly

7. Remove the lower control arm attaching bolts.

8. Remove the lower control arm.

To install:

9. Install the lower control arm.

10. Install the lower control arm attaching bolts and tighten to 89–120 ft. lbs. (120–163 Nm).

11. Install the rear coil spring.

a. Install the rear coil spring so that the lower end of the rear coil spring is seated on the step of the rear lower spring seat.

b. Position the jack under the lower arm inner side and compress the rear coil spring slowly.

c. Support the rear lower arm outer side and raise the rear suspension to the unloaded condition with a separate garage jack.

d. Install the lower arm inner side bolt and tighten to 89–120 ft. lbs. (120–163 Nm).

e. Install and tighten the rear lower

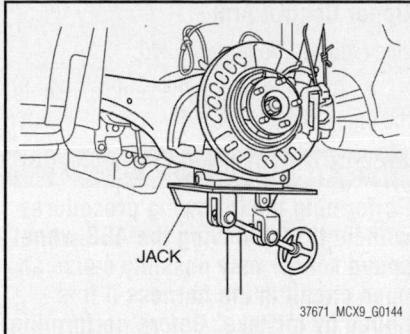

Fig. 233 Supporting the lower arm with a jack

arm outer side bolt to 89–120 ft. lbs. (120–163 Nm).

12. Install the rear shock absorber lower side bolt. Tighten to 69–91 ft. lbs. (93–123 Nm).

13. Install the rear stabilizer control link upper side nut.

14. Install the rear ABS wheel-speed sensor.

15. Lower the vehicle.

16. Inspect the rear wheel alignment.

Trailing Link

See Figures 234 and 235.

1. Before servicing the vehicle, refer to the precautions section.

❋❋ WARNING

When removing/installing the coil spring, it is necessary to jack up the rear suspension with unload condition. At that time, to prevent bending the front of the vehicle, support the jack up point for the front crossmember using a jack.

2. Raise and safely support the vehicle securely on jackstands.

3. Remove the rear ABS wheel-speed sensor.

4. Remove the rear lateral link.

5. Remove the rear coil spring.

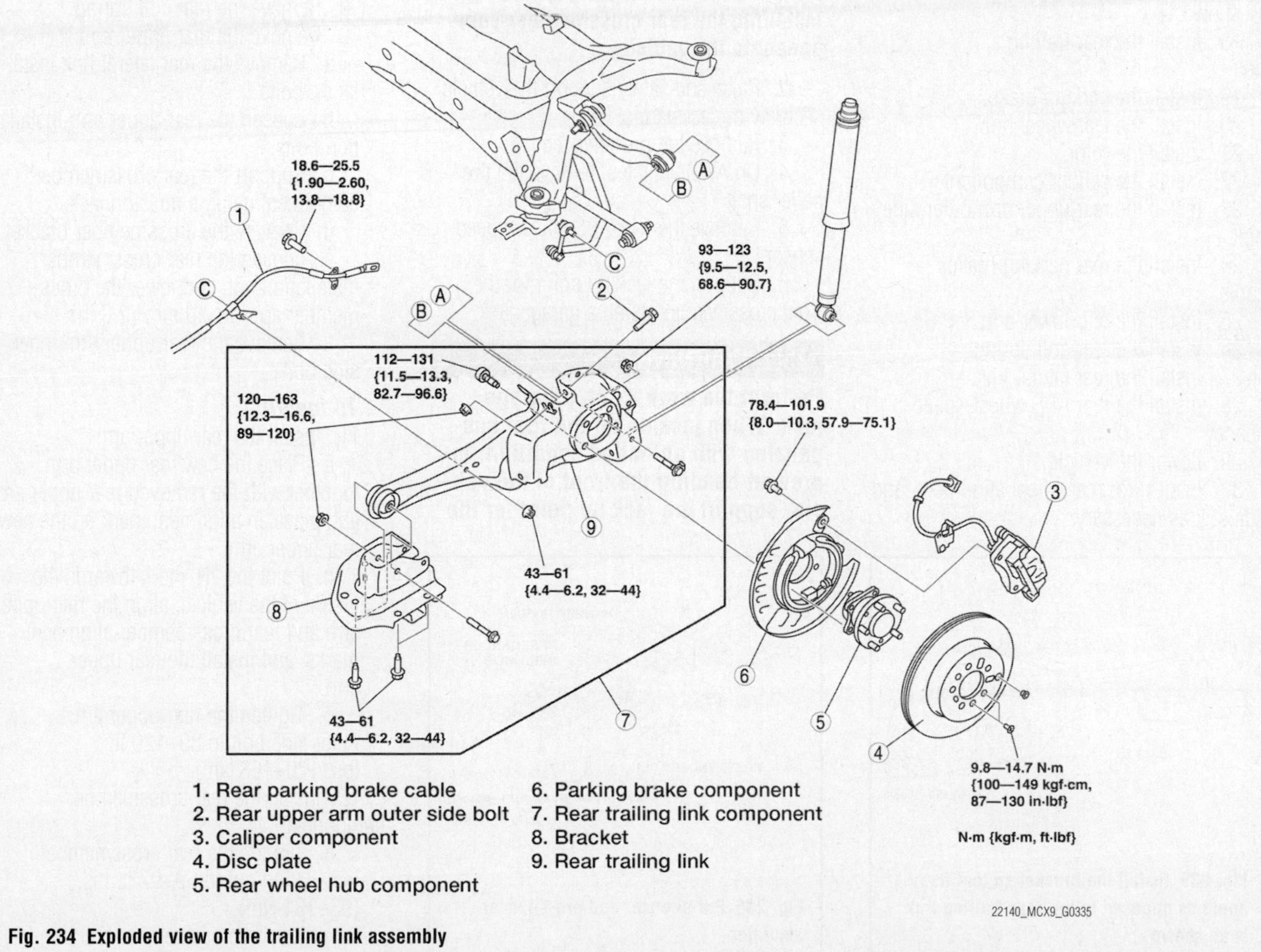

1. Rear parking brake cable
2. Rear upper arm outer side bolt
3. Caliper component
4. Disc plate
5. Rear wheel hub component
6. Parking brake component
7. Rear trailing link component
8. Bracket
9. Rear trailing link

Fig. 234 Exploded view of the trailing link assembly

6. Remove the rear lower arm.
7. Remove the rear parking brake cable.
8. Remove the rear upper arm outer side bolt.
9. Remove the caliper component

   a. Remove the caliper component and suspend it with a cable in a location out of the way.
10. Remove the rotor.
11. Remove the tear wheel hub.
12. Remove the parking brake.
13. Remove the rear trailing link.

   a. Support the rear trailing link using a jack.

   b. Remove the rear trailing link.
14. Remove the bracket.
15. Remove the rear trailing link.

To install:

16. Install the rear trailing link.
17. Install the bracket.

   a. Install the bracket so that its angle as opposed to the rear trailing link is as shown.
18. Install the rear trailing link.
19. Install the parking brake.
20. Install the tear wheel hub.
21. Install the rotor.
22. Install the caliper component
23. Install the rear upper arm outer side bolt.
24. Install the rear parking brake cable.
25. Install the rear lower arm.
26. Install the rear coil spring.
27. Install the rear lateral link.
28. Install the rear ABS wheel-speed sensor.
29. Lower the vehicle.
30. Inspect for rear wheel alignment, and adjust it as necessary.

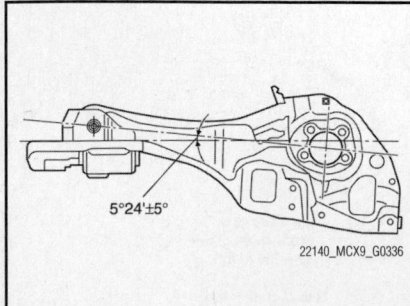

Fig. 235 Install the bracket so that its angle as opposed to the rear trailing link is as shown

Upper Control Arm

See Figures 236 through 240.

   1. Before servicing the vehicle, refer to the precautions section.

> ❋❋ **WARNING**
>
> **Performing the following procedures without first removing the ABS wheel speed sensor may possibly cause an open circuit in the harness if it is pulled by mistake. Before performing the following procedures, remove the ABS wheel speed sensor (axle side) and fix it to an appropriate place where the sensor will not be pulled by mistake while servicing the vehicle.**

➡**The rear upper arm inner bolt cannot be tightened completely with the rear upper arm installed to the vehicle due to the narrow operation space. Therefore, mark the position before removing the rear upper arm, align the rear crossmember and rear upper arm markings when installing the rear upper arm, and tighten the rear upper arm inner bolt completely before installing the rear crossmember component to the vehicle.**

   2. Raise and safely support the vehicle securely on jackstands.
   3. Remove the pre-silencer.
   4. On AWD vehicles, remove the propeller shaft.
   5. Remove the rear ABS wheel-speed sensor.
   6. Support the jack up point for the front crossmember using a garage jack.

> ❋❋ **WARNING**
>
> **Perform the work using a garage jack. When jacking up the rear suspension with unloaded condition, to prevent bending the front of the vehicle, support the jack up point for the**

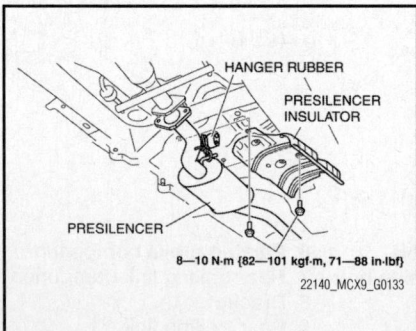

Fig. 236 Pre-silencer and pre-silencer insulator

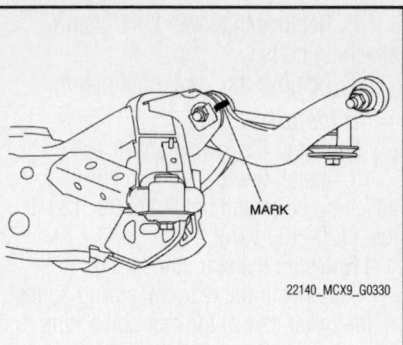

Fig. 237 Mark the rear upper arm and rear crossmember shown

front crossmember using a garage jack.

   7. Jack up the rear suspension with unloaded condition.

➡**Jacking up the rear suspension with unloaded condition will lighten the force on the bushing and make it easier to perform the procedure.**

   8. Mark the rear upper arm and rear crossmember shown in the figure.
   9. Remove the rear coil spring.
   10. Remove the rear upper arm.

   a. Remove the rear lateral link installation bolts.

   b. Loosen the rear upper arm installation bolts.

   c. Support the rear crossmember component using a mission jack.

   d. Remove the crossmember bracket.

   e. Loosen the rear crossmember installation nut, and lower the crossmember approx. 50 mm {2.0 in}.

   f. Remove the rear upper arm inner side bolt.

To install:

11. Install the rear upper arm.

   a. Place the new rear upper arm together with the removed rear upper arm and place an alignment mark on the new rear upper arm.

   b. Point the 'IN' mark towards the inside of the vehicle, align the rear upper arm and rear crossmember alignment marks, and install the rear upper arm.

   c. Tighten the rear upper arm inner side bolt to 89–120 ft. lbs.(120–163 Nm).
12. Install the rear crossmember component.

   a. Tighten the rear crossmember installation nut to 69–97 ft. lbs. (93–131 Nm).

   b.Tighten the crossmember bracket

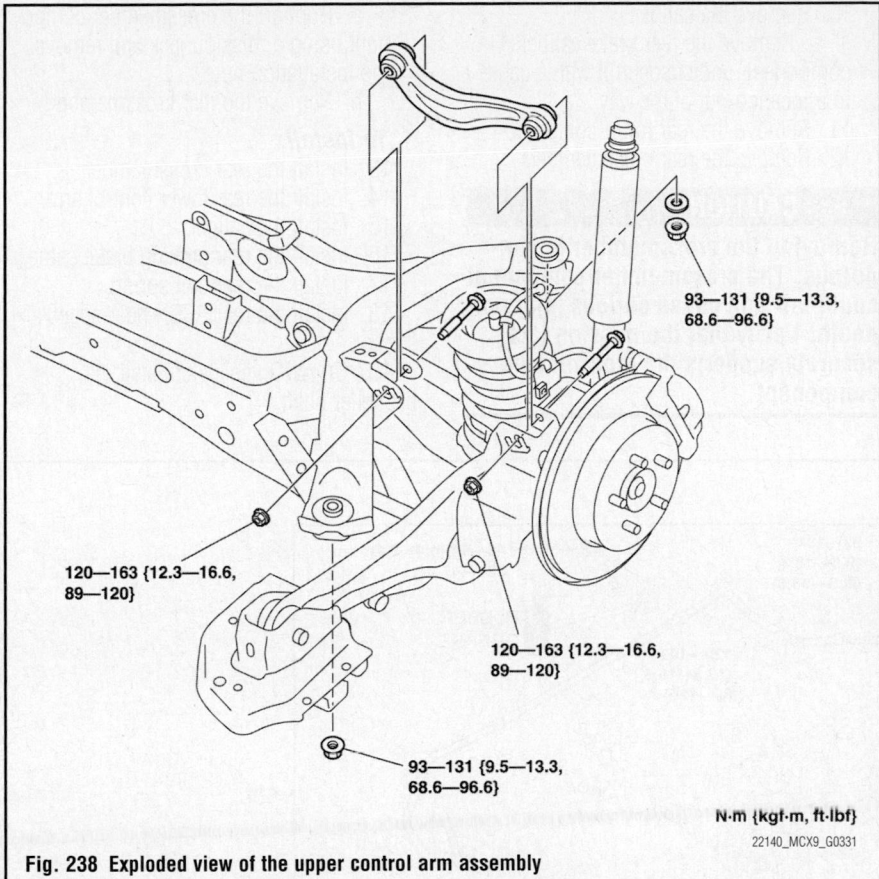

Fig. 238 Exploded view of the upper control arm assembly

93—131 {9.5—13.3, 68.6—96.6}

120—163 {12.3—16.6, 89—120}

120—163 {12.3—16.6, 89—120}

93—131 {9.5—13.3, 68.6—96.6}

N·m {kgf·m, ft·lbf}

22140_MCX9_G0331

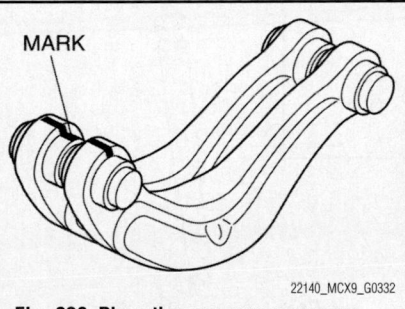

MARK

22140_MCX9_G0332

Fig. 239 Place the new rear upper arm together with the removed rear upper arm and place an alignment mark on the new rear upper arm

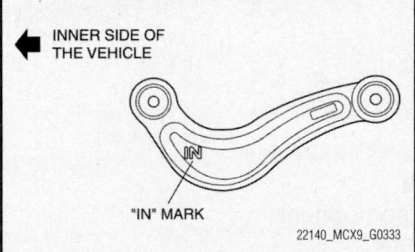

INNER SIDE OF THE VEHICLE

"IN" MARK

22140_MCX9_G0333

Fig. 240 Point the 'IN' mark towards the inside of the vehicle, align the rear upper arm and rear crossmember alignment marks, and install the rear upper arm

installation bolt to 14–19 ft. lbs. (19–26 Nm).

c. Tighten the crossmember bracket installation nut to 14–19 ft. lbs. (19–26 Nm).

13. Install the rear coil spring.

14. Install the rear ABS wheel-speed sensor.

15. On AWD vehicles, install the propeller shaft.

16. Install the pre-silencer.

17. Lower the vehicle.

18. Inspect for rear wheel alignment, and adjust it as necessary.

CROSSMEMBER (REAR)

REMOVAL & INSTALLATION

See Figures 241 through 243.

1. Before servicing the vehicle, refer to the precautions section.

✸✸ WARNING

Performing the following procedures without first removing the ABS wheel speed sensor may possibly cause an open circuit in the harness if it is pulled by mistake. Before performing the following procedures, remove the ABS wheel speed sensor (axle side) and fix it to an appropriate place

where the sensor will not be pulled by mistake while servicing the vehicle.

➡The rear upper arm inner bolt cannot be tightened completely with the rear upper arm installed to the vehicle due to the narrow operation space. Therefore, mark the position before removing the rear upper arm, align the rear crossmember and rear upper arm markings when installing the rear upper arm, and tighten the rear upper arm inner bolt completely before installing the rear crossmember component to the vehicle.

2. Support the jack up point for the front crossmember using a garage jack.

✸✸ WARNING

Perform the work using a garage jack. When jacking up the rear suspension with unloaded condition, to prevent bending the front of the vehicle, support the jack up point for the front crossmember using a garage jack.

3. Jack up the rear suspension with unloaded condition.

➡Jacking up the rear suspension with unloaded condition will lighten the force on the bushing and make it easier to perform the procedure.

4. Mark the rear upper arm and rear crossmember shown.

5. Remove the pre-silencer.

6. On AWD vehicles, remove the propeller shaft.

7. Remove the rear ABS wheel-speed sensor.

8. Remove the rear coil spring.

9. Remove the rear parking brake cable.

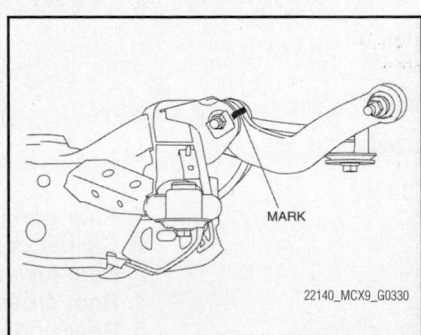

MARK

22140_MCX9_G0330

Fig. 241 Mark the rear upper arm and rear crossmember shown

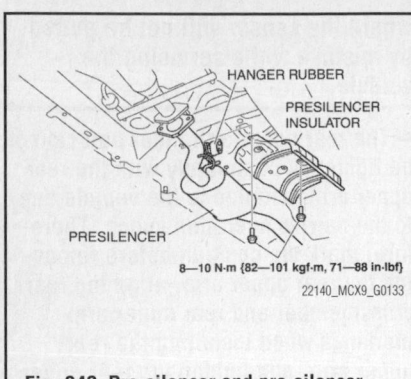

HANGER RUBBER

PRESILENCER
INSULATOR

PRESILENCER

8—10 N·m {82—101 kgf·m, 71—88 in·lbf}

22140_MCX9_G0133

Fig. 242 Pre-silencer and pre-silencer insulator

10. Remove the caliper.
   a. Remove the rear brake caliper component, and suspend it with a cable in a location out of the way.
11. Remove the rear lower control arm.
12. Remove the rear crossmember.

※※ CAUTION

Removing the crossmember is dangerous. The crossmember component could fall and cause serious injury or death. Verify that the mission jack securely supports the crossmember component.

   a. Support the crossmember component using a mission jack and remove the installation nuts.
   b. Remove the rear crossmember.

To install:

13. Install the rear crossmember.
14. Install the rear lower control arm.
15. Install the caliper.
16. Install the rear parking brake cable.
17. Install the rear coil spring.
18. Install the rear ABS wheel-speed sensor.
19. On AWD vehicles, install the propeller shaft.

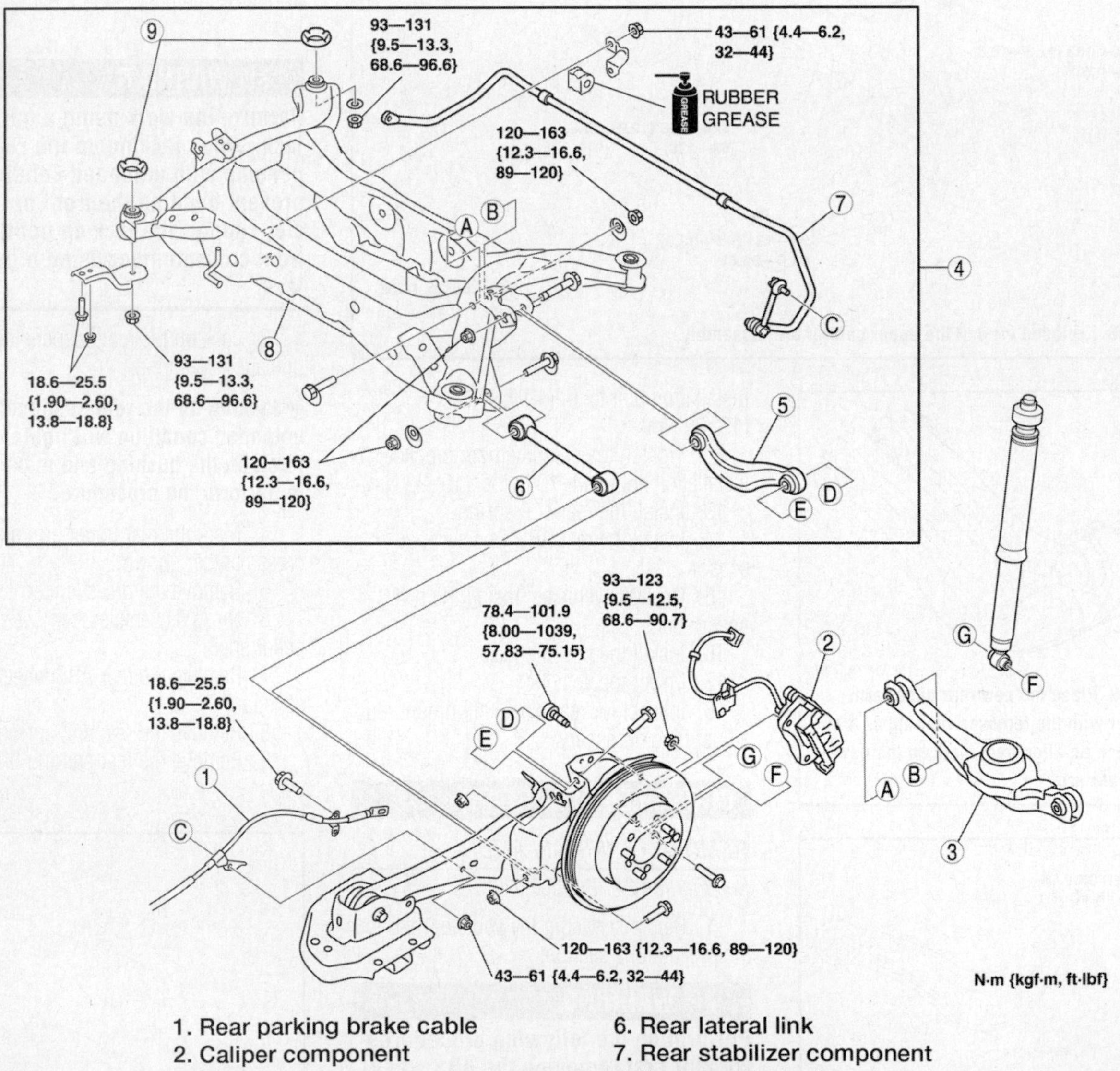

9

93—131
{9.5—13.3,
68.6—96.6}

43—61 {4.4—6.2,
32—44}

RUBBER
GREASE

120—163
{12.3—16.6,
89—120}

7

A B

4

C

18.6—25.5
{1.90—2.60,
13.8—18.8}

93—131
{9.5—13.3,
68.6—96.6}

8

5

120—163
{12.3—16.6,
89—120}

6

E

D

93—123
{9.5—12.5,
68.6—90.7}

78.4—101.9
{8.00—1039,
57.83—75.15}

2

G

18.6—25.5
{1.90—2.60,
13.8—18.8}

E D

1

F

G

C

B

A

3

120—163 {12.3—16.6, 89—120}

43—61 {4.4—6.2, 32—44}

N·m {kgf·m, ft·lbf}

1. Rear parking brake cable
2. Caliper component
3. Rear lower arm
4. Rear crossmember component
5. Rear upper arm
6. Rear lateral link
7. Rear stabilizer component
8. Rear crossmember
9. Stopper rubber

22140_MCX9_G0337

Fig. 243 Exploded view of the rear crossmember assembly

20. Install the pre-silencer.
21. Inspect for rear wheel alignment, and adjust it as necessary.

SHOCK ABSORBER

REMOVAL & INSTALLATION

See Figures 244 through 249, 000.

1. Before servicing the vehicle, refer to the precautions section.
2. Remove the trunk side trim.
3. Remove the rear wheel.
4. Support the rear lower arm using a jack.

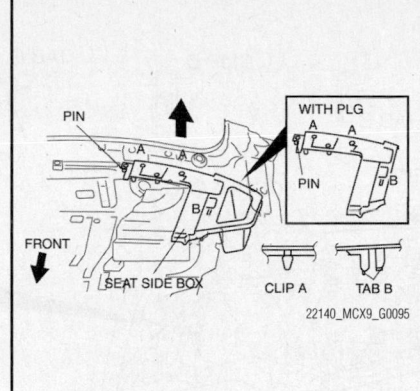

Fig. 245 Seat side box fastener locations

5. Remove the shock top nut, retainer, upper rubber bushing
6. Remove the shock lower side band nut, rubber bushing and retainer.
7. Remove the rear shock absorber.

To install:

8. Install the rear shock absorber.
9. Install the shock lower side band nut, rubber bushing and retainer. Tighten to 69–91 ft. lbs. (93–123 Nm).
10. Install the shock top nut, retainer, upper rubber bushing. Tighten to 16–21 ft. lbs. (21–29 Nm).
11. Support the rear lower arm using a jack.

Fig. 244 Removing the trunk box

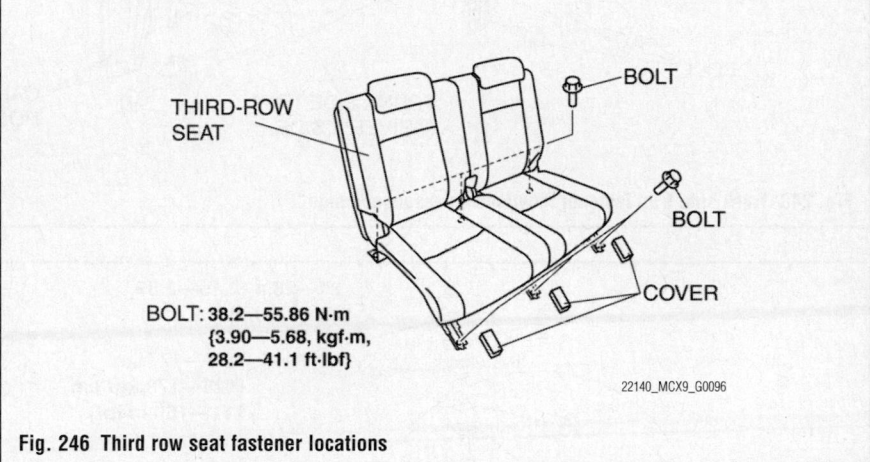

Fig. 246 Third row seat fastener locations

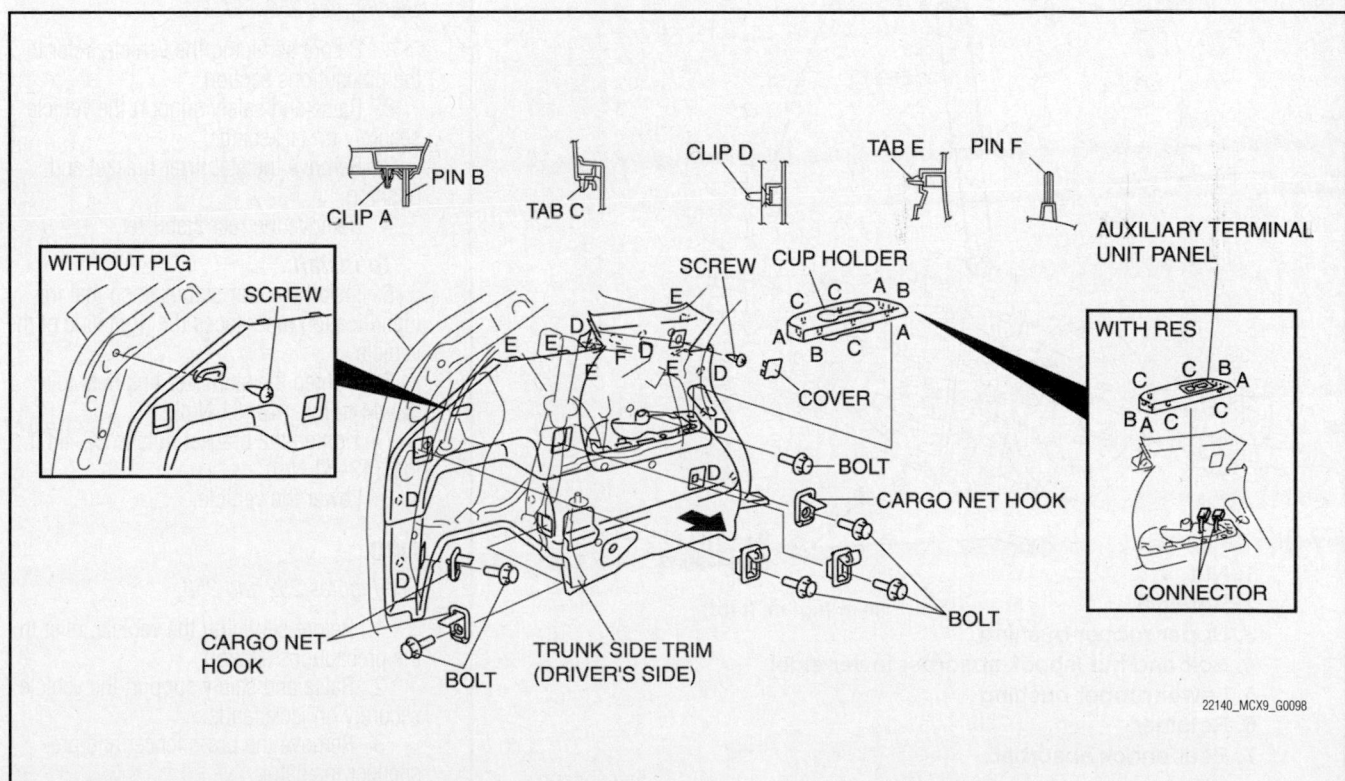

Fig. 247 Trunk side trim fastener locations—driver's side

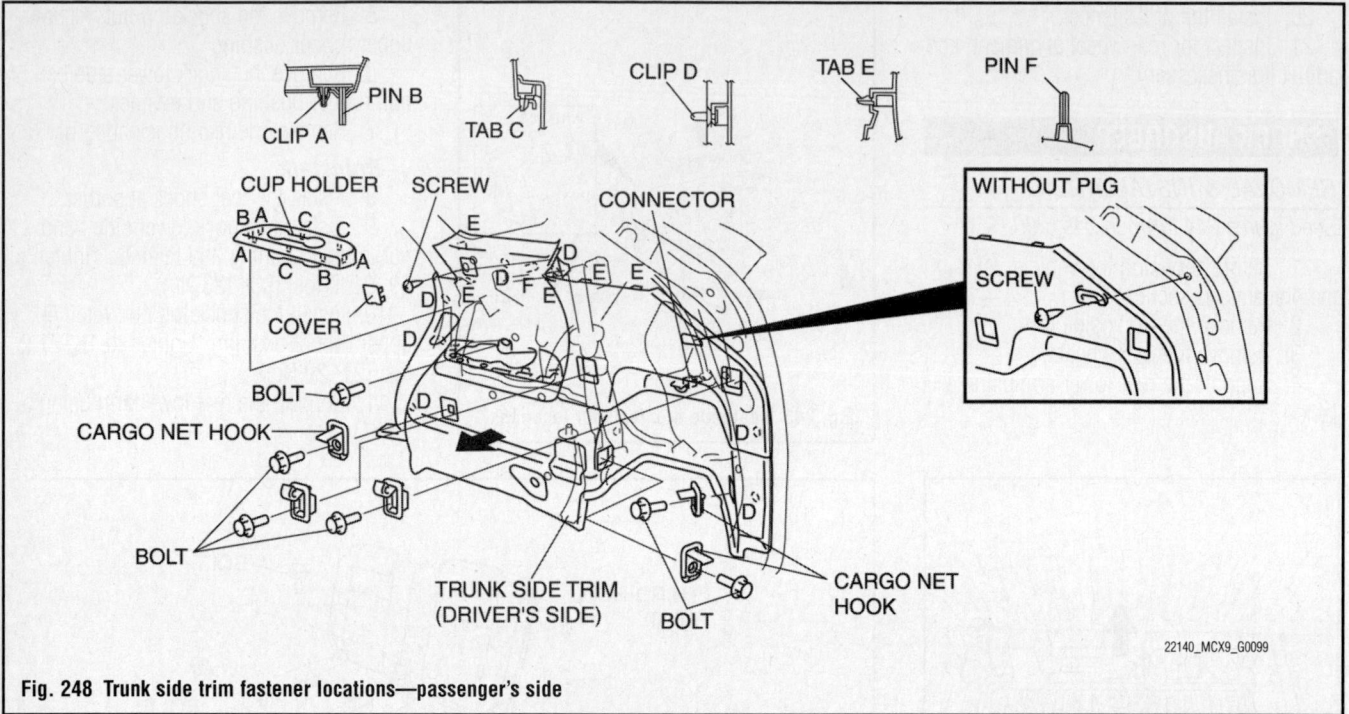

PIN B
CLIP A
TAB C
CLIP D
TAB E
PIN F

CUP HOLDER
SCREW
CONNECTOR
WITHOUT PLG

B A C C
A C B
E
SCREW

COVER
D D E F E E E
D

BOLT

CARGO NET HOOK

BOLT

TRUNK SIDE TRIM
(DRIVER'S SIDE)
BOLT

CARGO NET
HOOK

22140_MCX9_G0099

Fig. 248 Trunk side trim fastener locations—passenger's side

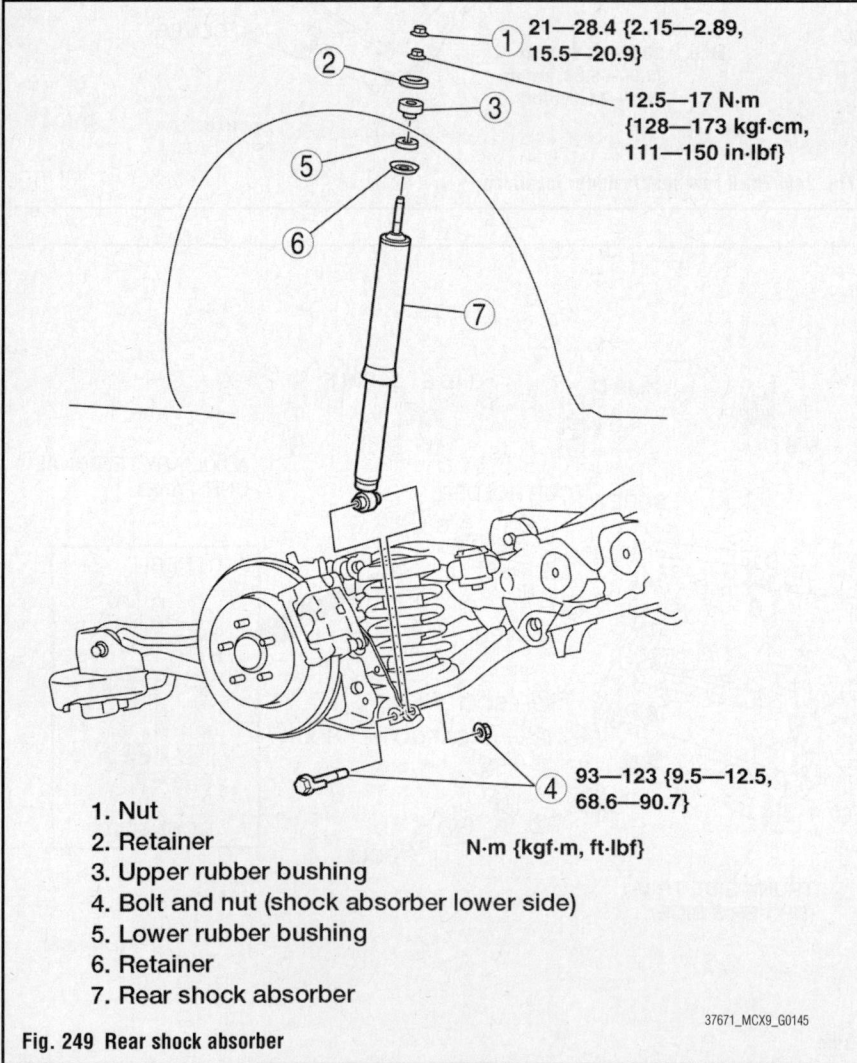

① 21—28.4 {2.15—2.89, 15.5—20.9}

② 12.5—17 N·m {128—173 kgf·cm, 111—150 in·lbf}

③

⑤

⑥

⑦

④ 93—123 {9.5—12.5, 68.6—90.7}

N·m {kgf·m, ft·lbf}

1. Nut
2. Retainer
3. Upper rubber bushing
4. Bolt and nut (shock absorber lower side)
5. Lower rubber bushing
6. Retainer
7. Rear shock absorber

37671_MCX9_G0145

Fig. 249 Rear shock absorber

12. Install the rear wheel.
13. Install the trunk side trim.

STABILIZER

REMOVAL & INSTALLATION

2WD

See Figures 250 and 251.

1. Before servicing the vehicle, refer to the precautions section.
2. Raise and safely support the vehicle securely on jackstands.
3. Remove the stabilizer bracket and bushing.
4. Remove the rear stabilizer.

To install:

5. Install the rear stabilizer so that the identification mark faces the right side of the vehicle.
6. Tighten the stabilizer link nuts to 32–44 ft. lbs. (43–61 Nm).
7. Tighten the bracket nuts to 32–44 ft. lbs. (43–61 Nm).
8. Lower the vehicle.

AWD

See Figures 252 and 253.

1. Before servicing the vehicle, refer to the precautions section.
2. Raise and safely support the vehicle securely on jackstands.
3. Remove the pre-silencer and pre-silencer insulator.
4. Remove the propeller shaft.

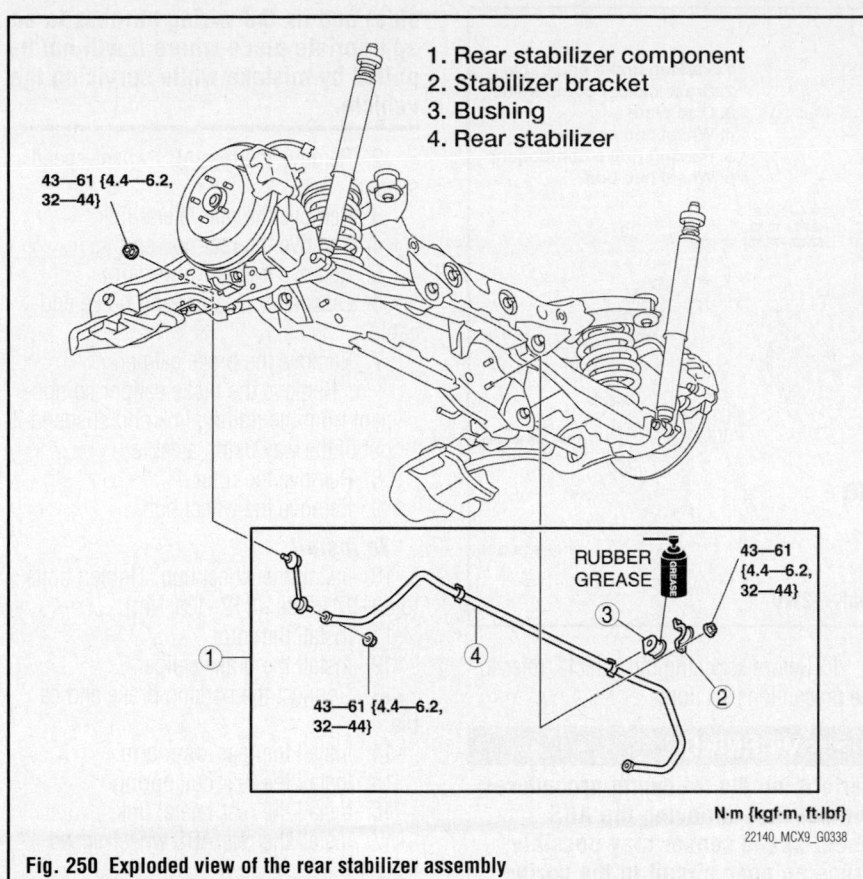

1. Rear stabilizer component
2. Stabilizer bracket
3. Bushing
4. Rear stabilizer

43—61 {4.4—6.2, 32—44}

43—61 {4.4—6.2, 32—44}

43—61 {4.4—6.2, 32—44}

RUBBER GREASE

N·m {kgf·m, ft·lbf}

22140_MCX9_G0338

Fig. 250 Exploded view of the rear stabilizer assembly

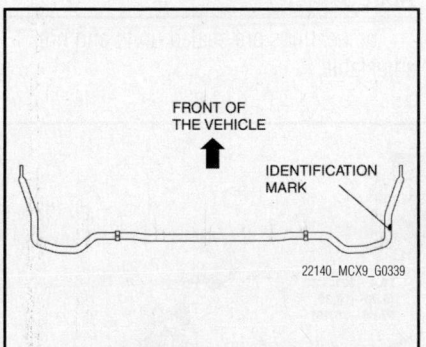

FRONT OF THE VEHICLE

IDENTIFICATION MARK

22140_MCX9_G0339

Fig. 251 Install the rear stabilizer so that the identification mark faces the right side of the vehicle

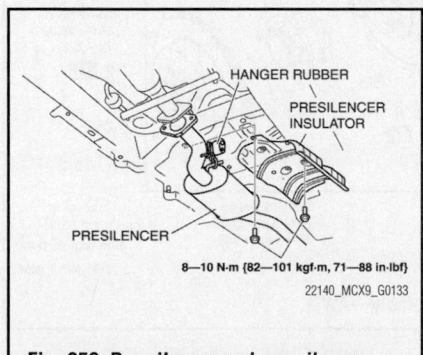

HANGER RUBBER

PRESILENCER INSULATOR

PRESILENCER

8—10 N·m {82—101 kgf·m, 71—88 in·lbf}

22140_MCX9_G0133

Fig. 252 Pre-silencer and pre-silencer insulator

5. Remove the rear ABS wheel-speed sensor.

6. Remove the rear coil spring.

7. Remove the stabilizer bracket and bushing.

8. Remove the rear stabilizer.

To install:

9. Install the rear coil spring.

10. Install the rear ABS wheel-speed sensor.

11. Install the propeller shaft.

12. Install the pre-silencer and pre-silencer insulator.

13. Install the rear stabilizer so that the identification mark faces the right side of the vehicle.

14. Tighten the stabilizer link nuts to 32–44 ft. lbs. (43–61 Nm).

15. Tighten the bracket nuts to 32–44 ft. lbs. (43–61 Nm).

16. Lower the vehicle.

WHEEL HUB & BEARING

REMOVAL & INSTALLATION

2WD

See Figure 254.

1. Before servicing the vehicle, refer to the precautions section.

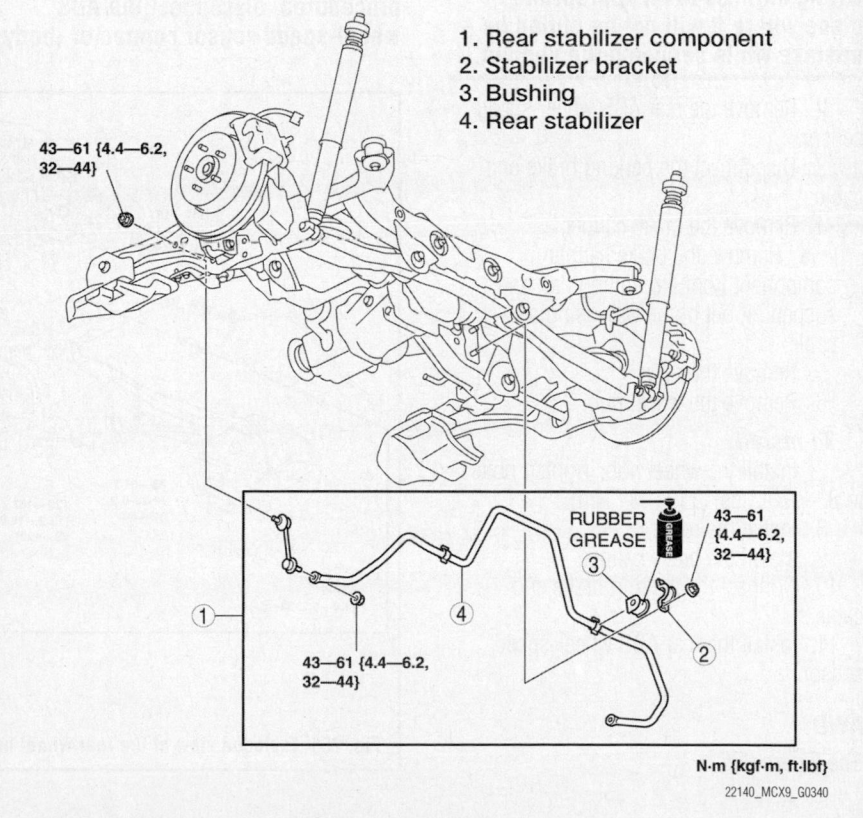

1. Rear stabilizer component
2. Stabilizer bracket
3. Bushing
4. Rear stabilizer

43—61 {4.4—6.2, 32—44}

43—61 {4.4—6.2, 32—44}

43—61 {4.4—6.2, 32—44}

RUBBER GREASE

N·m {kgf·m, ft·lbf}

22140_MCX9_G0340

Fig. 253 Exploded view of the rear stabilizer assembly

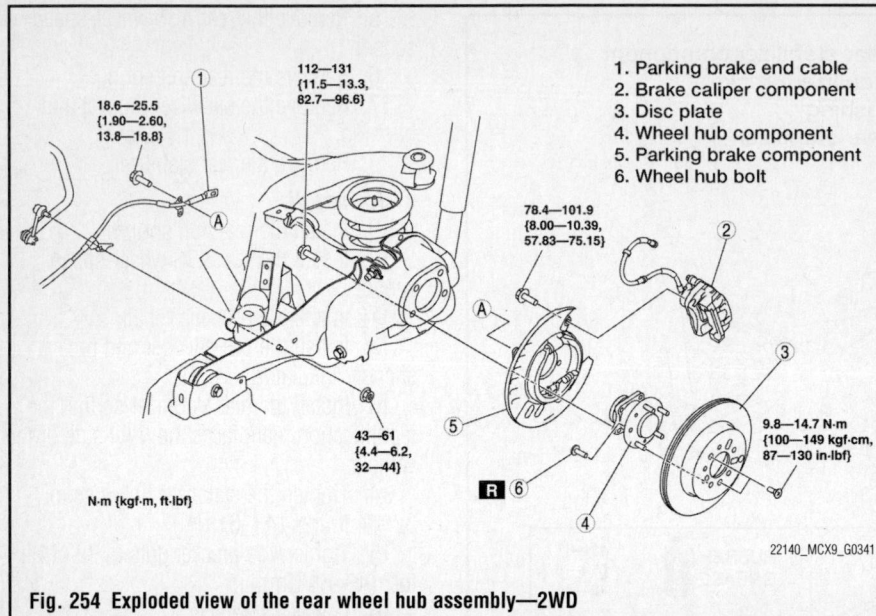

1. Parking brake end cable
2. Brake caliper component
3. Disc plate
4. Wheel hub component
5. Parking brake component
6. Wheel hub bolt

N·m {kgf·m, ft·lbf}

Fig. 254 Exploded view of the rear wheel hub assembly—2WD

⁕⁕ WARNING

Performing the following procedures without first removing the ABS wheel-speed sensor may possibly cause an open circuit in the wiring harness if it is pulled by mistake. Before performing the following procedures, disconnect the ABS wheel-speed sensor connector (body side) and fix the wiring harness to an appropriate place where it will not be pulled by mistake while servicing the vehicle.

2. Remove the rear ABS wheel-speed sensor.
3. Disconnect the parking brake end cable.
4. Remove the brake caliper.
   a. Remove the brake caliper component from the trailing link and suspend it out of the way using a cable.
5. Remove the rotor.
6. Remove the wheel hub.

To install:

7. Install the wheel hub. Tighten bolts to 83–97 ft. lbs. (112–131 Nm).
8. Install the rotor.
9. Install the brake caliper.
10. Connect the parking brake end cable.
11. Install the rear ABS wheel-speed sensor.

AWD

See Figure 255.

1. Before servicing the vehicle, refer to the precautions section.

⁕⁕ WARNING

Performing the following procedures without first removing the ABS wheel-speed sensor may possibly cause an open circuit in the wiring harness if it is pulled by mistake. Before performing the following procedures, disconnect the ABS wheel-speed sensor connector (body

side) and fix the wiring harness to an appropriate place where it will not be pulled by mistake while servicing the vehicle.

2. Remove the rear ABS wheel-speed sensor.
3. Remove the rear lateral link.
4. Remove the rear coil spring.
5. Remove the rear lower arm.
6. Disconnect the parking brake end cable.
7. Remove the brake caliper.
   a. Remove the brake caliper component from the trailing link and suspend it out of the way using a cable.
8. Remove the rotor.
9. Remove the wheel hub.

To install:

10. Install the wheel hub. Tighten bolts to 83–97 ft. lbs. (112–131 Nm).
11. Install the rotor.
12. Install the brake caliper.
13. Connect the parking brake end cable.
14. Install the rear lower arm.
15. Install the rear coil spring.
16. Install the rear lateral link.
17. Install the rear ABS wheel-speed sensor.

ADJUSTMENT

The bearings are sealed units and not adjustable.

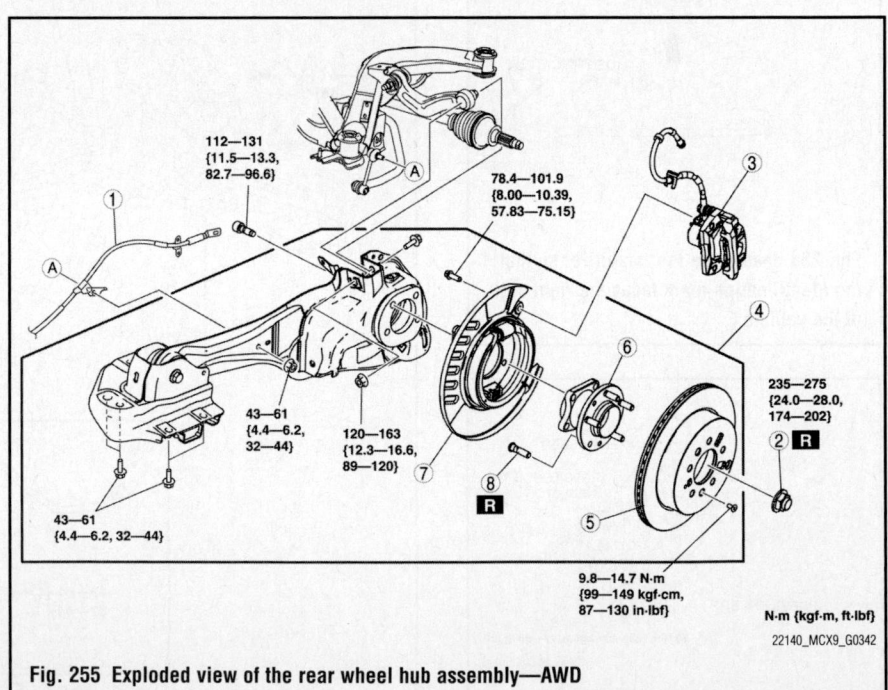

Fig. 255 Exploded view of the rear wheel hub assembly—AWD

MAZDA

Mazda3 • Mazdaspeed3

SPECIFICATIONS AND MAINTENANCE CHARTS

ENGINE AND VEHICLE IDENTIFICATION

Code	Liters (cc)	Cu. In.	Cyl.	Fuel Sys.	Engine Type	Eng. Mfg.	Code ②	Year
①	2.0 (1999)	121.9	4	MPFI	DOHC	Mazda	9	2009
①	2.3 (2261)	137.9	4	MPFI	DOHC	Mazda	A	2010
①	2.5 (2488)	151.8	4	MPFI	DOHC	Mazda		

MPFI: Multi-Point Fuel Injection

DOHC: Dual Over Head Cam

① 2009/2010 2.0L vin: F (fed/Canada) or G (California), code: LF

   2009 2.3L vin: 3 (fed/Canada) or 4 (California), code L3

   2009 2.3L vin: M (California), code L3 w/TC (turbocharger)

   2010 2.3L vin: 3 (fed/Canada) or 4 (California), code L3 w/TC (turbocharger)

   2010 2.5L vin: 5 (fed/Canada) or 6 (California), code L5

② 10th digit of the Vehicle Identification Number (VIN)

37671_MAZ3_C0001

GENERAL ENGINE SPECIFICATIONS

Year	Model	Engine Displacement Liters (VIN) ①	Net Horsepower @ rpm	Net Torque @ rpm (ft. lbs.)	Bore x Stroke (in.)	Compression Ratio	Oil Pressure @ rpm
2009	Mazda3	2.0	②	③	3.44x3.27	10.0:1	34.0-75.5@3000
	Mazda3	2.3	160@6500	150@4500	3.44x3.70	9.7:1	34.0-75.5@3000
	Mazdaspeed3	2.3	263@5500	280@3000	3.44x3.70	9.5:1	43.1-79.9@3000
2010	Mazda3	2.0	②	③	3.44x3.27	10.0:1	34.0-75.5@3000
	Mazda3	2.5	④	⑤	3.44x3.70	9.7:1	57.3-94.1@3000
	Mazdaspeed3	2.3	263@5500	280@3000	3.44x3.70	9.5:1	43.1-79.9@3000

① 2009/2010 2.0L vin: F (fed/Canada) or G (California), code: LF

   2009 2.3L vin: 3 (fed/Canada) or 4 (California), code L3

   2009 2.3L vin: M (California), code L3 w/TC (turbocharger)

   2010 2.3L vin: 3 (fed/Canada) or 4 (California), code L3 w/TC (turbocharger)

   2010 2.5L vin: 5 (fed/Canada) or 6 (California), code L5

② Partial Zero Emission Vehicle (PZEV): 144@6500

   Except PZEV states: 148@6500

③ Partial Zero Emission Vehicle (PZEV): 132@4500

   Except PZEV states: 135@4500

④ Partial Zero Emission Vehicle (PZEV): 165@6000

   Except PZEV states: 167@6000

⑤ Partial Zero Emission Vehicle (PZEV): 167@4000

   Except PZEV states: 168@4000

37671_MAZ3_C0002

ENGINE TUNE-UP SPECIFICATIONS

Year	Engine Displacement Liters (VIN) ①	Spark Plug Gap (in.)	Ignition Timing (deg.)		Fuel Pump (psi)	Idle Speed (rpm)		Valve Clearance	
			MT	AT		MT	AT	In.	Ex.
2009	2.0	0.049-0.053	8B	8B	51-59	600-700	650-750	0.008-0.0011	0.010-0.012
	2.3	0.050-0.053	8B	8B	51-59	700-800	700-800	0.008-0.0011	0.010-0.012
	2.3	0.049-0.053	8B	8B	55-65	650-750	650-750	0.008-0.0011	0.010-0.012
2010	2.0	0.049-0.053	8B	8B	50.8-62.3	600-700	650-750	0.009-0.0011	0.011-0.012
	2.5	0.050-0.053	12B	12B	50.8-62.3	②	③	0.009-0.0011	0.011-0.012
	2.3	0.049-0.053	10B	10B	50.8-62.3	650-750	650-750	0.009-0.0011	0.011-0.012

NOTE: The Vehicle Emission Control Information label often reflects specification changes made during production. The label figures must be used if they differ from those in this chart.

B: Before top dead center

HYD: Hydraulic

① 2009/2010 2.0L vin: F (fed/Canada) or G (California), code: LF

2009 2.3L vin: 3 (fed/Canada) or 4 (California), code L3

2009 2.3L vin: M (California), code L3 w/TC (turbocharger)

2010 2.3L vin: 3 (fed/Canada) or 4 (California), code L3 w/TC (turbocharger)

2010 2.5L vin: 5 (fed/Canada) or 6 (California), code L5

② 700-800 with AC on

③ 650-770 with AC on

37671_MAZ3_C0003

CAPACITIES

Year	Model	Engine Displacement Liters (VIN) ①	Engine Oil with Filter (qts.)	Transmission (pts.)		Drive Axle		Fuel Tank (gal.)	Cooling System (qts.)
				Man	Auto.	Front (pts.)	Rear (pts.)		
2009	Mazda3	2.0	4.5	6.06	15.2	②	—	14.5	7.9
	Mazda3	2.3	4.5	6.06	15.2	②	—	14.5	7.9
	Mazdaspeed3	2.3	6.0	5.38	—	—	2.3	15.9	8.5
2010	Mazda3	2.0	4.5	③	④	N/A	N/A	14.5	7.9
	Mazda3	2.5	5.3	③	④	N/A	N/A	14.5	7.9
	Mazdaspeed3	2.3	6.0	③	④	N/A	N/A	15.9	7.9

NOTE: All capacities are approximate. Add fluid gradually and check to be sure a proper fluid level is obtained.

N/A: Not Available

① 2009/2010 2.0L vin: F (fed/Canada) or G (California), code: LF

2009 2.3L vin: 3 (fed/Canada) or 4 (California), code L3

2009 2.3L vin: M (California), code L3 w/TC (turbocharger)

2010 2.3L vin: 3 (fed/Canada) or 4 (California), code L3 w/TC (turbocharger)

2010 2.5L vin: 5 (fed/Canada) or 6 (California), code L5

② Included in transaxle

③ G35M-R: 5.86, G66M-R: 6.02, A26M-R: 5.2-5.4

④ FS5A-EL 6.4. Overhaul 10.6

37671_MAZ3_C0005

FLUID SPECIFICATIONS

Year	Model	Engine Displacement Liters	Engine ID/VIN	Engine Oil	Auto. Trans.	Manual Trans.	Power Drive Axle	Brake Master Cylinder	Engine Coolant ③ ④
2009	Mazda3	2.0	①	5W-20	②	75W-90	N/A	DOT 3	Ethylene-glycol
	Mazda3	2.3	①	5W-20	②	75W-90	N/A	DOT 3	Ethylene-glycol
	Mazdaspeed3	2.3	①	5W-20	②	75W-90	N/A	DOT 3	Ethylene-glycol
2010	Mazda3	2.0	①	5W-20	ATF M-V	75W-90	N/A	DOT 3	Ethylene-glycol
	Mazda3	2.5	①	5W-20	ATF M-V	75W-90	N/A	DOT 3	Ethylene-glycol
	Mazdaspeed3	2.3	①	5W-20	ATF M-V	75W-90	N/A	DOT 3	Ethylene-glycol

DOT: Department Of Transpotation

N/A: Not Available

① 2009/2010 2.0L vin: F (fed/Canada) or G (California), code: LF

  2009 2.3L vin: 3 (fed/Canada) or 4 (California), code L3

  2009 2.3L vin: M (California), code L3 w/TC (turbocharger)

  2010 2.3L vin: 3 (fed/Canada) or 4 (California), code L3 w/TC (turbocharger)

  2010 2.5L vin: 5 (fed/Canada) or 6 (California), code L5

② The Asian Warner 6-speed transaxle uses JWS3309 a special fluid produced by Exxon/Mobil

③ Do not use coolants containing Alcohol, Methanol, Borate or Silicate. These coolants could damage the cooling system.

④ If the "FL22" mark is shown on or near the cooling system cap, use FL22 type engine coolant.

37671_MAZ3_C0004

VALVE SPECIFICATIONS

Year	Engine Displacement Liters (VIN) ①	Seat Angle (deg.)	Face Angle (deg.)	Maximum out of Square (in.)	Spring Free Length (in.)	Stem-to-Guide Clearance (in.) Intake	Stem-to-Guide Clearance (in.) Exhaust	Stem Diameter (in.) Intake	Stem Diameter (in.) Exhaust
2009	2.0	N/A	N/A	N/A	1.129	0.0009-0.0027	0.0012-0.0029	0.2154-0.2159	0.2152-0.2157
	2.3	N/A	N/A	N/A	1.129	0.0009-0.0027	0.0012-0.0029	0.2154-0.2159	0.2152-0.2157
	2.3	45	45	N/A	②	0.0009-0.0027	0.0012-0.0029	0.2351-0.2356	0.2349-0.2354
2010	2.0	45	45	N/A	③	0.0009-0.0027	0.0012-0.0029	0.2154-0.2159	0.2152-0.2157
	2.5	45	45	N/A	③	0.0009-0.0027	0.0012-0.0029	0.2154-0.2159	0.2152-0.2157
	2.3	45	45	N/A	②	0.0009-0.0027	0.0012-0.0029	0.2351-0.2356	0.2349-0.2354

N/A: Not Available

① 2009/2010 2.0L vin: F (fed/Canada) or G (California), code: LF

  2009 2.3L vin: 3 (fed/Canada) or 4 (California), code L3

  2009 2.3L vin: M (California), code L3 w/TC (turbocharger)

  2010 2.3L vin: 3 (fed/Canada) or 4 (California), code L3 w/TC (turbocharger)

  2010 2.5L vin: 5 (fed/Canada) or 6 (California), code L5

② 1.492 inch at 38.55 lbs. (installed height)

③ 1.129 inch at 87.7 lbs. (installed height)

37671_MAZ3_C0006

CAMSHAFT AND BEARING SPECIFICATIONS CHART

All measurements are given in inches.

Year	Engine Displ. Liters	Engine ID/VIN	Journal Dia.	Brg. Oil Clearance	Shaft End-play	Runout	Journal Bore	Lobe Height Intake	Exhaust
2009	2.0	①	0.9827-0.9834	0.0014-0.0031	0.0035-0.0094	N/A	0.9827-0.9834	②	③
	2.3	①	0.9827-0.9834	0.0014-0.0031	0.0035-0.0094	N/A	0.9827-0.9834	②	③
	2.3	①	0.9827-0.9834	N/A	0.0035-0.0094	N/A	N/A	1.671	1.667
2010	2.0	①	0.9827-0.9834	0.0014-0.0031	0.0035-0.0094	N/A	0.9827-0.9834	②	③
	2.5	①	0.9827-0.9834	0.0020-0.0030-	0.0040-0.0090	0.0012	N/A	1.666-1.671	1.616-1.621
	2.3	①	0.9827-0.9834	N/A	0.0035-0.0094	N/A	N/A	1.671	1.667

N/A: Not Available

① 2009/2010 2.0L vin: F (fed/Canada) or G (California), code: LF

2009 2.3L vin: 3 (fed/Canada) or 4 (California), code L3

2009 2.3L vin: M (California), code L3 w/TC (turbocharger)

2010 2.3L vin: 3 (fed/Canada) or 4 (California), code L3 w/TC (turbocharger)

2010 2.5L vin: 5 (fed/Canada) or 6 (California), code L5

② 1.671 with VVT. 1.621 w/o VVT

③ 1.621 with VVT. 1.618 w/o VVT

37671_MAZ3_C0014

CRANKSHAFT AND CONNECTING ROD SPECIFICATIONS

All measurements are given in inches.

Year	Engine Displacement Liters (VIN) ①	Crankshaft Main Brg. Journal Dia.	Main Brg. Oil Clearance	Shaft End-play	Thrust on No.	Connecting Rod Journal Diameter	Oil Clearance	Side Clearance
2009	2.0	2.0464-2.0472	0.0007-0.0013	0.0087-0.0177	N/A	N/A	0.0011-0.0020	0.0056-0.0141
	2.3	2.0464-2.0472	0.0007-0.0013	0.0087-0.0177	N/A	N/A	0.0011-0.0020	0.0056-0.0141
	2.3	2.0465-2.0472	0.0007-0.0015	0.0087-0.0177	N/A	N/A	0.0011-0.0020	0.0056-0.0141
2010	2.0	2.0465-2.0472	0.0007-0.0018	0.0090-0.0170	N/A	N/A	0.0011-0.0020	0.0060-0.0140
	2.5	2.0465-2.0472	0.0007-0.0018	0.0090-0.0170	N/A	N/A	0.0011-0.0020	0.0060-0.0140
	2.3	2.0465-2.0472	0.0007-0.0015	0.0087-0.0177	N/A	N/A	0.0011-0.0020	0.0056-0.0141

N/A: Not Avilable

① 2009/2010 2.0L vin: F (fed/Canada) or G (California), code: LF

2009 2.3L vin: 3 (fed/Canada) or 4 (California), code L3

2009 2.3L vin: M (California), code L3 w/TC (turbocharger)

2010 2.3L vin: 3 (fed/Canada) or 4 (California), code L3 w/TC (turbocharger)

2010 2.5L vin: 5 (fed/Canada) or 6 (California), code L5

37671_MAZ3_C0007

PISTON AND RING SPECIFICATIONS

All measurements are given in inches.

Year	Engine Displacement Liters (VIN) ①	Piston Clearance	Ring Gap			Ring Side Clearance		
			Top Compression	Bottom Compression	Oil Control	Top Compression	Bottom Compression	Oil Control
2009	2.0	0.0010-0.0017	0.0063-0.0122	0.0130-0.0189	0.0079-0.0275	0.0012-0.0031	0.0012-0.0027	0.0012-0.0027
	2.3	0.0010-0.0017	0.0063-0.0122	0.0130-0.0189	0.0079-0.0275	0.0012-0.0031	0.0012-0.0027	0.0012-0.0027
	2.3	N/A	0.0056-0.0094	0.0134-0.0173	0.0060-0.0157	0.0020-0.0031	0.0012-0.0027	0.0008-0.0023
2010	2.0	N/A	0.0063-0.0100	0.0130-0.0180	0.0060-0.0150	0.0020-0.0030	0.0012-0.0027	0.0030-0.0050
	2.5	N/A	0.0063-0.0100	0.0130-0.0180	0.0060-0.0150	0.0020-0.0030	0.0012-0.0027	0.0030-0.0050
	2.3	N/A	0.0056-0.0094	0.0134-0.0173	0.0060-0.0157	0.0020-0.0031	0.0012-0.0027	0.0008-0.0023

N/A: Not Available

① 2009/2010 2.0L vin: F (fed/Canada) or G (California), code: LF

2009 2.3L vin: 3 (fed/Canada) or 4 (California), code L3

2009 2.3L vin: M (California), code L3 w/TC (turbocharger)

2010 2.3L vin: 3 (fed/Canada) or 4 (California), code L3 w/TC (turbocharger)

2010 2.5L vin: 5 (fed/Canada) or 6 (California), code L5

37671_MAZ3_C0008

TORQUE SPECIFICATIONS

All readings in ft. lbs.

Year	Engine Displacement Liters (VIN) ①	Cylinder Head Bolts	Main Bearing Bolts	Rod Bearing Bolts	Crankshaft Damper Bolts	Flywheel Bolts	Manifold		Spark Plugs	Oil Pan Drain Plug
							Intake	Exhaust		
2009	2.0	②	NA	NA	③	④	⑤	32-47	8-10	22-30
	2.3	②	NA	NA	③	④	⑤	32-47	8-10	19-22
	2.3	⑥	NA	NA	⑦	⑧	⑨	14-18	⑩	19-22
2010	2.0	②	NA	NA	③	④	⑤	32-47	8-10	22-30
	2.5	⑪	NA	NA	⑫	④	13-16	32-47	8-10	23-30
	2.3	⑥	NA	NA	⑦	⑧	⑨	14-18	⑩	19-22

NA: Not Available

① 2009/2010 2.0L vin: F (fed/Canada) or G (California), code: LF

2009 2.3L vin: 3 (fed/Canada) or 4 (California), code L3

2009 2.3L vin: M (California), code L3 w/TC (turbocharger)

2010 2.3L vin: 3 (fed/Canada) or 4 (California), code L3 w/TC (turbo)

2010 2.5L vin: 5 (fed/Canada) or 6 (California), code L5

② Step 1: 97 inch lbs.

Step 2: Tighten12.5 ft. lbs

Step 3: Tighten 34.6 ft. lbs.

Step 4: Tighten 88 degrees

Step 5: Tighten 88 degrees

③ 71-77 ft. lbs. plus 87-93 degrees

④ Manual transmission: 80-85 ft. lbs.

Automatic transmission: 71-76 ft. lbs.

⑤ Mazda 3 with the 2.0L (LF) and 2.3L (L3) engines: 142-177 ft. lbs.

⑥ Step 1: 17-19 ft. lbs.

Step 2: Tighten 85-95 degees

Step 3: Repeat step 2

⑦ Step 1: 88 ft. lbs.

Step 2: loosen one full turn

Step 3: Tighten 35-39 ft. lbs.

Step 4: Tighten 85-95 degrees

⑧ G35M-R A26MX-R manual or FN4A-EL automatic transmission: 80-85 ft. lbs.

A65M-R manual or JA5A-EL automatic transmission: 54-64 ft. lbs.

⑨ Lower intake: 72-106 inch lbs.

Upper intake plenum: 72-106 inch lbs.

⑩ 79-177 inch. lbs.

⑪ Step 1: 27-97 inch lbs.

Step 2: Tighten 116-150 inch. lbs

Step 3: Tighten 32-34 ft. lbs.

Step 4: Tighten 92 degrees

Step 5: Tighten an additional 92 degrees

⑫ 71-77 ft. lbs. plus 87-90 degrees

37671_MAZ3_C0009

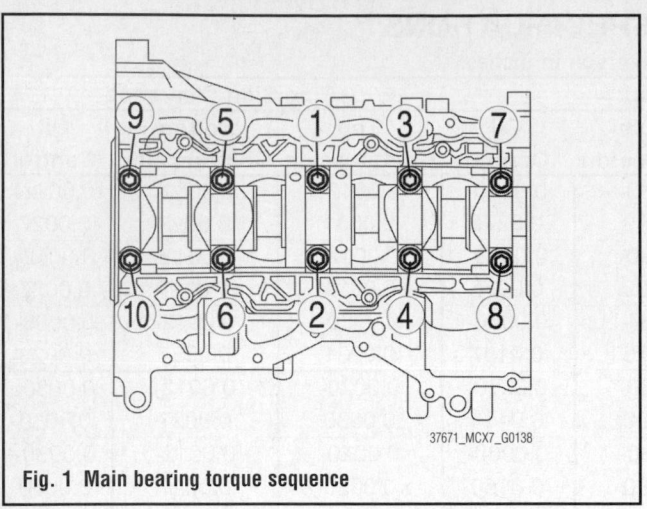

37671_MCX7_G0138

Fig. 1 Main bearing torque sequence

WHEEL ALIGNMENT

Year	Model		Caster Range (+/-Deg.)	Caster Preferred Setting (Deg.)	Camber Range (+/-Deg.)	Camber Preferred Setting (Deg.)	Toe-in (in.)
2009	Mazda3/Mazdaspeed3	F	①	①	②	②	③
		R	N/A	N/A	④	④	③
2010	Mazda3/Mazdaspeed3	F	①	①	②	②	③
		R	N/A	N/A	④	④	③

NA: Not Applicable

① 2.0L, 2.3L and 2.5L engine: fuel tank empty 2 degrees 59'. Fuel tank full 3 degrees 07'. Difference between R & L must not exceed 1 degree 30'.

  2.3L engine with turbocharger: fuel tank empty 3 degrees 06'. Fuel tank full 3 degrees 12'. Difference between R & L must not exceed 1 degree 30'.

② 2.0L, 2.3L and 2.5L engine: fuel tank empty -0 degrees 37'. Fuel tank full -0 degrees 38'. Difference between R & L must not exceed 1 degree 30'.

  2.3L engine with turbocharger: fuel tank empty -0 degrees 53'. Fuel tank full -0 degrees 54'. Difference between R & L must not exceed 1 degree 30'.

③ Tire: 0.08 +/- 0.2

④ 2.0L, 2.3L and 2.5L engine: fuel tank empty 1 degrees 23'. Fuel tank full 1 degrees 30'. Difference between R & L must not exceed 1 degree 30'.

  2.3L engine with turbocharger: fuel tank empty -1 degrees 35'. Fuel tank full -1 degrees 41'. Difference between R & L must not exceed 1 degree 30'.

37671_MAZ3_C0010

TIRE, WHEEL AND BALL JOINT SPECIFICATIONS

| Year | Model | OEM Tires | | Tire Pressures (psi) | | Wheel Size | Ball Joint Inspection | Lug Nut (ft. lbs.) |
		Standard	Optional	Front	Rear			
2009	Mazda 3	①	①	②	②	③	N/A	65-87
	Mazdaspeed3	P2225/40R18 88Y	N/A	35	34	④	N/A	65-87
2010	Mazda 3	①	①	32	32	②	N/A	65-87
	Mazdaspeed3	P205/50R16	N/A	35	34	⑥	N/A	65-87

OEM: Original Equipment Manufacturer

PSI: Pounds Per Square Inch

N/A: Not Available

① P205/55R16 89H

  P205/50R17 88V

② P205/55R16 89H tire: 35. P205/50R17 88V tire: 32

③ 16x6.5J or 17x7.0J

④ 18x7.5J

37671_MAZ3_C0011

BRAKE SPECIFICATIONS
All measurements in inches unless noted

| Year | Model | | Brake Disc | | | Brake Drum | | | Minimum Lining Thickness | Brake Caliper | |
			Original Thickness	Minimum Thickness	Maximum Runout	Original Inside Diameter	Max. Wear Limit	Maximum Machine Diameter		Bracket Bolts (ft. lbs.)	Mounting Bolts (ft. lbs.)
2009	Mazda 3	F	N/A	0.910	0.002	—	—	—	0.079	61-74	19-22
		R	N/A	0.350	0.002	—	—	—	0.079	45-56	19-22
	Mazdaspeed 3	F	N/A	0.910	0.002	—	—	—	0.079	61-74	19-22
		R	N/A	0.350	0.002	—	—	—	0.079	45-56	19-22
2010	Mazda 3	F	N/A	0.910	0.002	—	—	—	0.079	61-74	19-22
		R	N/A	0.350	0.002	—	—	—	0.079	45-56	19-22
	Mazdaspeed 3	F	N/A	0.910	0.002	—	—	—	0.079	61-74	19-22
		R	N/A	0.350	0.002	—	—	—	0.079	45-56	19-22

N/A: Not Avilable

F: Front

R: Rear

37671_MAZ3_C0012

SCHEDULED MAINTENANCE INTERVALS
Mazda 3 and Mazdaspeed3

TO BE SERVICED	TYPE OF SERVICE	VEHICLE MILEAGE INTERVAL (x1000)												
		7.5	15	22.5	30	37.5	45	52.5	60	67.5	75	82.5	90	97.5
Engine oil & filter	R	✓	✓	✓	✓	✓	✓	✓	✓	✓	✓	✓	✓	✓
Air cleaner element	R					✓					✓			
Engine coolant (FL22) ① ② ③	R	120,000 or 10 years, 60,000 thereafter												
Engine coolant ① ② ③	R								✓					
Spark plugs (type 1)	R										✓			
Spark plugs (type 2)	R	every 100,000 miles												
Bolts & nuts on chassis & body	S/I				✓				✓				✓	
Brake lines, hoses & connections	S/I				✓				✓				✓	
Cooling system	S/I				✓				✓				✓	
Cabin filter	R	every 25,000 miles												
Disc brakes	S/I				✓				✓				✓	
Drive belts ④	S/I				✓				✓				✓	
Drive shaft dust boots	S/I				✓				✓				✓	
Exhaust system heat shield	S/I				✓				✓				✓	
Front & rear suspension ball joints	S/I				✓				✓				✓	
Fuel lines & hoses	S/I				✓				✓				✓	
Steering operation & linkages	S/I				✓				✓				✓	
Tire rotation	R	✓	✓	✓	✓	✓	✓	✓	✓	✓	✓	✓	✓	✓
Valve clearance ⑤	I								✓					
Hose & tube for emission	S/I								✓					

R: Replace   S/I: Service or Inspect

① Be sure to use the proper grade and type coolant

② Replace initially at 60,000 miles or 4 years and every 24 months thereafter

③ Models with FL22 long-life coolant can be identified by the FL22 marking on the cooling system cap label. It is recommended that FL22 coolant continue to be used for models originally filled with FL22 coolant from the factory.

④ 2.0L engine and 2.3L engine: inspect every 37, 500 miles. 2.5L engine: inspect every 30,000 miles

⑤ 2.0L, 2.3L and 2.5L engine: Audible inspect every 75, 000 miles and if noisy adjust.

FREQUENT OPERATION MAINTENANCE (SEVERE SERVICE)

If a vehicle is operated under any of the following conditions it is considered severe service

- Extremely dusty areas.

- 50% or more of the vehicle operation is in 32°C (90°F) or higher temperatures, or constant operation in temperatures below 0°C (32°F).

- Prolonged idling (vehicle operation in stop and go traffic).

- Frequent short running periods (engine does not warm to normal operating temperatures).

- Police, taxi, delivery usage or trailer towing usage.

Oil & oil filter: change every 5000 miles.

Oil & oil filter (Puerto Rico): change every 3000 miles.

Air cleaner element: service or inspect every 15,000 miles

Automatic transaxle fluid: service or inspect every 15,000 miles.

Bolts & nuts on chassis & body: tighten every 15,000 miles.

Disc brakes: service or inspect every 15,000 miles.

37671_MAZ3_C0013

BRAKES — INFORMATION AND PRECAUTIONS

ANTI-LOCK SYSTEMS

- Certain components within the ABS system are not intended to be serviced or repaired individually.
- Do not use rubber hoses or other parts not specifically specified for and ABS system. When using repair kits, replace all parts included in the kit. Partial or incorrect repair may lead to functional problems and require the replacement of components.
- Lubricate rubber parts with clean, fresh brake fluid to ease assembly. Do not use shop air to clean parts; damage to rubber components may result.
- Use only DOT 3 brake fluid from an unopened container.
- If any hydraulic component or line is removed or replaced, it may be necessary to bleed the entire system.
- A clean repair area is essential. Always clean the reservoir and cap thoroughly before removing the cap. The slightest amount of dirt in the fluid may plug an orifice and impair the system function. Perform repairs after components have been thoroughly cleaned; use only denatured alcohol to clean components. Do not allow ABS components to come into contact with any substance containing mineral oil; this includes used shop rags.
- The Anti-Lock control unit is a microprocessor similar to other computer units in the vehicle. Ensure that the ignition switch is **OFF** before removing or installing controller harnesses. Avoid static electricity discharge at or near the controller.
- If any arc welding is to be done on the vehicle, the control unit should be unplugged before welding operations begin.

DISC AND DRUM SYSTEMS

> **✳✳ CAUTION**
>
> **Dust and dirt accumulating on brake parts during normal use may contain asbestos fibers from production or** aftermarket brake linings. Breathing excessive concentrations of asbestos fibers can cause serious bodily harm. Exercise care when servicing brake parts. Do not sand or grind brake lining unless equipment used is designed to contain the dust residue. Do not clean brake parts with compressed air or by dry brushing. Cleaning should be done by dampening the brake components with a fine mist of water, then wiping the brake components clean with a dampened cloth. Dispose of cloth and all residue containing asbestos fibers in an impermeable container with the appropriate label. Follow practices prescribed by the Occupational Safety and Health Administration (OSHA) and the Environmental Protection Agency (EPA) for the handling, processing, and disposing of dust or debris that may contain asbestos fibers.

BRAKES — BLEEDING THE BRAKE SYSTEM

BLEEDING PROCEDURE

BLEEDING PROCEDURE

➡ **Do not reuse the drained fluid. Use only clean DOT 3 Brake Fluid from an unopened container.**

> **✳✳ WARNING**
>
> **Make sure no dirt or other foreign matter is allowed to contaminate the brake fluid.**

> **✳✳ WARNING**
>
> **Do not spill brake fluid on the vehicle, it may damage the paint; if brake fluid does contact the paint, wash it off immediately with water.**

1. The reservoir on the master cylinder must be at the MAX level mark at the start of the bleeding procedure and checked after bleeding each brake caliper. Add fluid as required.
2. Make sure the brake fluid level in the reservoir is at the MAX level line.
3. Slide a piece of clear plastic hose over the first bleed screw, and submerge the other end in a container of new brake fluid.
4. Have someone slowly pump the brake pedal several times, then apply steady pressure.
5. Bleed the hydraulic brake system in the following sequence:
   a. Right rear bleeder valve
   b. Left rear bleeder valve
   c. Right front bleeder valve
   d. Left front bleeder valve
6. Repeat the procedure for each wheel in the until air bubbles no longer appear in the fluid.
7. Refill the master cylinder reservoir to the MAX level line.

BLEEDING THE ABS SYSTEM

See Figure 2.

➡ **Do not reuse the drained fluid. Use only clean DOT 3 Brake Fluid from an unopened container.**

> **✳✳ WARNING**
>
> **Make sure no dirt or other foreign matter is allowed to contaminate the brake fluid.**

> **✳✳ WARNING**
>
> **Do not spill brake fluid on the vehicle, it may damage the paint; if brake fluid does contact the paint, wash it off immediately with water.**

1. The reservoir on the master cylinder must be at the MAX level mark at the start of the bleeding procedure and checked after

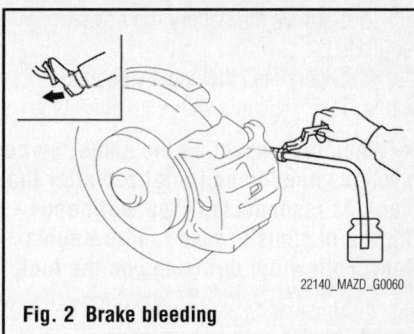

Fig. 2 Brake bleeding

22140_MAZD_G0060

bleeding each brake caliper. Add fluid as required.
2. Make sure the brake fluid level in the reservoir is at the MAX level line.
3. Slide a piece of clear plastic hose over the first bleed screw, and submerge the other end in a container of new brake fluid.
4. Have someone slowly pump the brake pedal several times, then apply steady pressure.
5. Bleed the hydraulic brake system in the following sequence:
   a. Right rear bleeder valve
   b. Left rear bleeder valve
   c. Right front bleeder valve
   d. Left front bleeder valve
6. Repeat the procedure for each wheel in the until air bubbles no longer appear in the fluid.
7. Refill the master cylinder reservoir to the MAX level line.

WHEEL SPEED SENSORS

REMOVAL & INSTALLATION

Front

2009 Vehicles

See Figure 3.

1. Before servicing the vehicle, refer to the Precautions Section.

➡ If working near and/or around the SRS system and components, be sure to disable the SRS system. Tape the negative battery cable with insulating tape. Always disconnect the negative battery cable first.

> ❈ **CAUTION**
>
> **To avoid personal injury when working on vehicles equipped with an air bag, the negative battery cable must be disconnected and at least one minute must elapse before working on the system. Failure to do so may result in deployment of the air bag.**

2. Remove the battery top cover, if equipped.
3. Disconnect the negative battery cable. Tape the cable with insulating tape.

➡ When disconnecting the cable, some systems need to be initialized after the cable is reconnected. You will need Mazda diagnostic scan tool or equivalent. Follow the directions on the tool.

4. Raise and safely support the vehicle, as required.
5. Remove the tire and wheel assembly, as required.
6. Remove the front wheel.
7. Remove the mud guard.
8. Disconnect the wheel speed sensor electrical connector.
9. Remove the fasteners attaching the wheel speed sensor harness.
10. Remove the wheel speed sensor retaining bolts and sensor.

To install:

➡ Be sure to use new fasteners, as required.

11. Installation is the reverse of the removal procedure.
12. Tighten the mounting bolt to 53 inch lbs. (6 Nm)
13. Use the Mazda diagnostic scan tool, or equivalent and reprogram the required systems.

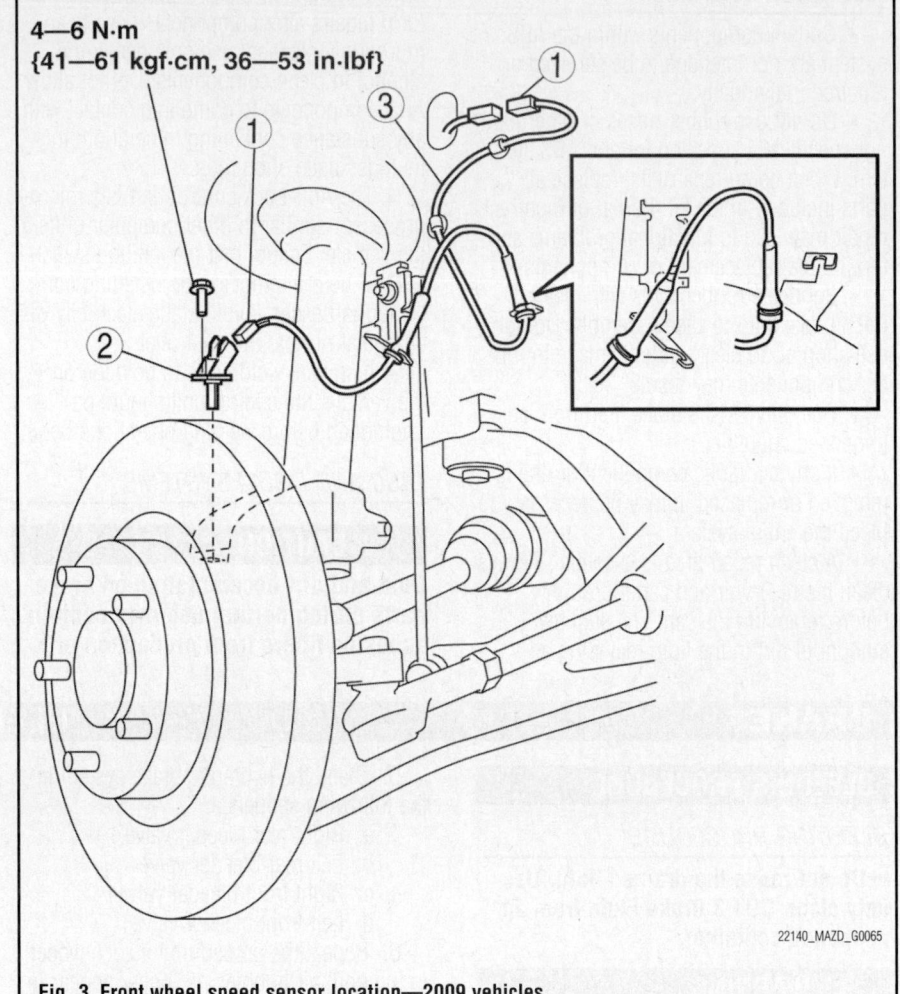

4—6 N·m
{41—61 kgf·cm, 36—53 in·lbf}

Fig. 3 Front wheel speed sensor location—2009 vehicles

22140_MAZD_G0065

2010 Vehicles

See Figure 4.

1. Before servicing the vehicle, refer to the Precautions Section.

➡ If working near and/or around the SRS system and components, be sure to disable the SRS system. Tape the negative battery cable with insulating tape. Always disconnect the negative battery cable first.

> ❈❈ **CAUTION**
>
> **To avoid personal injury when working on vehicles equipped with an air bag, the negative battery cable must be disconnected and at least one minute must elapse before working on the system. Failure to do so may result in deployment of the air bag.**

2. Remove the battery top cover, if equipped.

3. Disconnect the negative battery cable. Tape the cable with insulating tape.

➡ When disconnecting the cable, some systems need to be initialized after the cable is reconnected. You will need Mazda diagnostic scan tool or equivalent. Follow the directions on the tool.

4. Raise and safely support the vehicle, as required.
5. Remove the tire and wheel assembly, as required.
6. Disconnect the electrical connector.
7. Remove the sensor from its mounting.
8. Remove the connector.
9. Remove the wiring harness, as required.

To install:

➡ Be sure to use new fasteners, as required.

10. Installation is the reverse of the removal procedure.

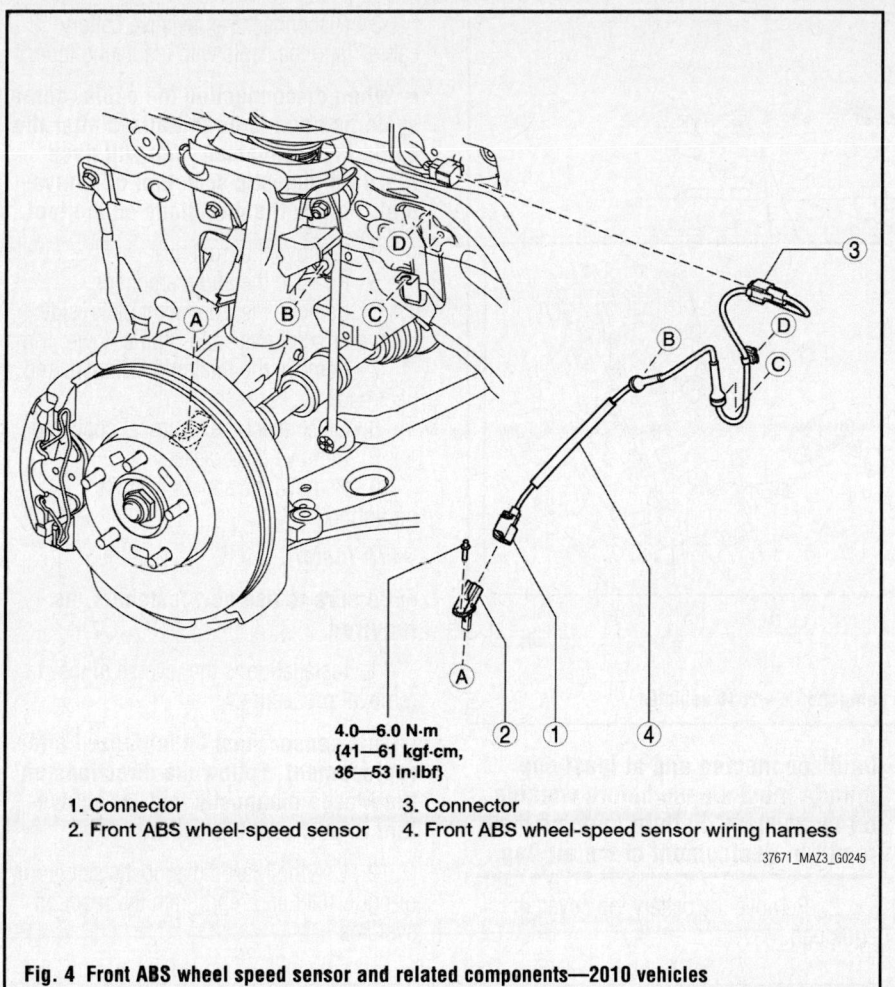

4.0—6.0 N·m
{41—61 kgf-cm,
36—53 in·lbf}

1. Connector
2. Front ABS wheel-speed sensor
3. Connector
4. Front ABS wheel-speed sensor wiring harness

37671_MAZ3_G0245

Fig. 4 Front ABS wheel speed sensor and related components—2010 vehicles

3. Disconnect the negative battery cable. Tape the cable with insulating tape.

➡ **When disconnecting the cable, some systems need to be initialized after the cable is reconnected. You will need Mazda diagnostic scan tool or equivalent. Follow the directions on the tool.**

4. Raise and support the vehicle safely, as required.

5. Remove the tire and wheel assembly, as required.

6. Remove the rear undercover.

7. Disconnect the wheel speed sensor electrical connector.

8. Remove the rear ABS wheel-speed sensor.

9. Press the tab of the ABS hole cover to separate the ABS hole cover from the body and remove the hole cover. Unplug the connector and remove the sensor wiring harness.

To install:

➡ **Be sure to use new fasteners, as required.**

10. Installation is the reverse of the removal procedure.

11. Installation is the reverse of removal, tighten the sensor fasteners to 36–53 inch lbs. (4–6 Nm).

12. Pass the rear wheel speed sensor wiring harness outside the rear parking brake cable.

11. Use the Mazda diagnostic scan tool, or equivalent and reprogram the required systems.

Rear

See Figures 5 and 6.

1. Before servicing the vehicle, refer to the Precautions Section.

➡ **If working near and/or around the SRS system and components, be sure to disable the SRS system. Tape the negative battery cable with insulating tape. Always disconnect the negative battery cable first.**

✳✳ CAUTION

To avoid personal injury when working on vehicles equipped with an air bag, the negative battery cable must be disconnected and at least one minute must elapse before working on the system. Failure to do so may result in deployment of the air bag.

2. Remove the battery top cover, if equipped.

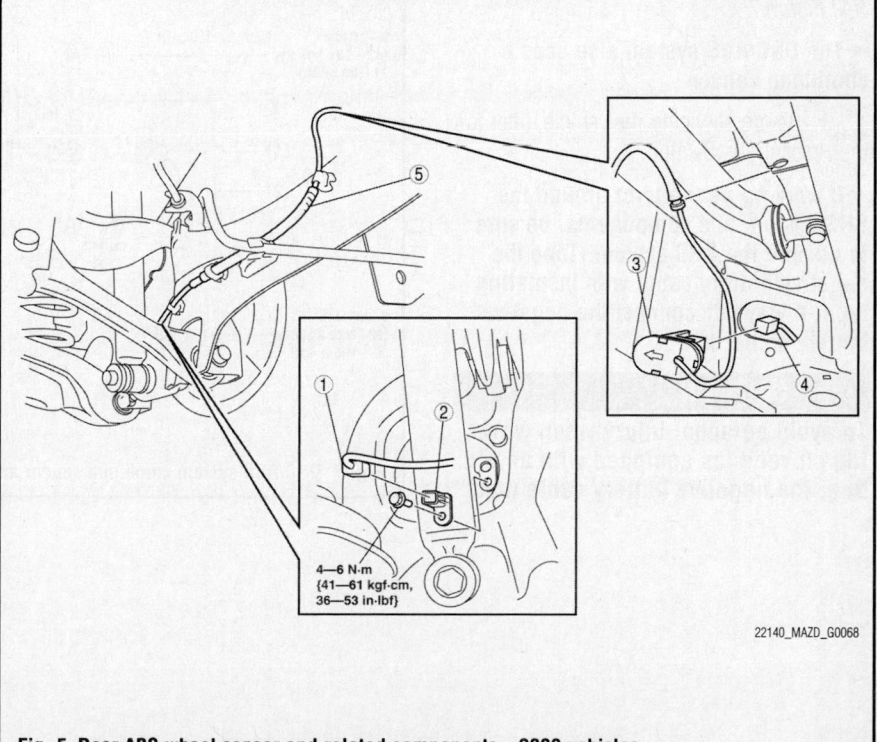

4—6 N·m
{41—61 kgf-cm,
36—53 in·lbf}

22140_MAZD_G0068

Fig. 5 Rear ABS wheel sensor and related components—2009 vehicles

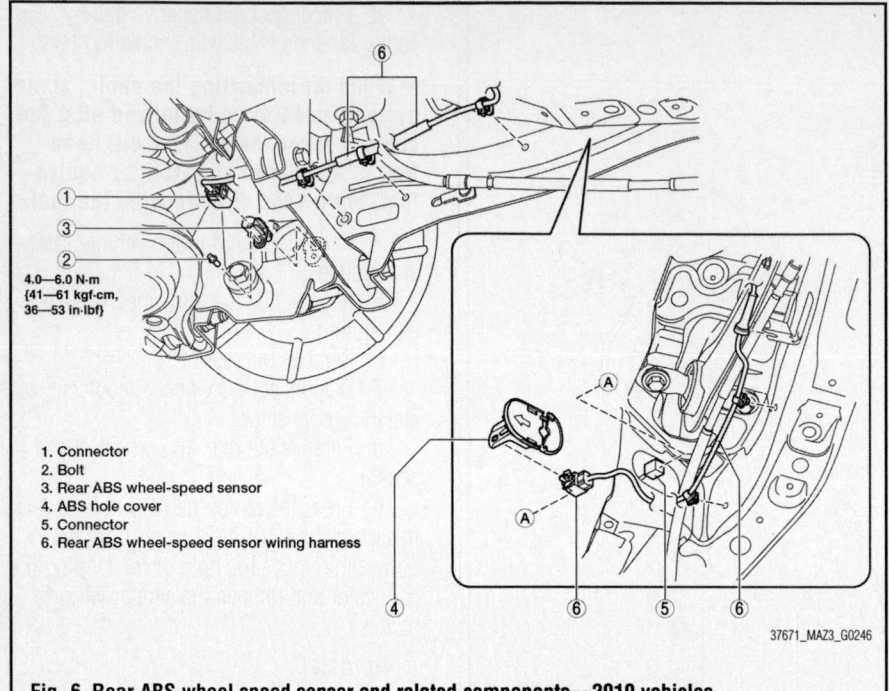

1. Connector
2. Bolt
3. Rear ABS wheel-speed sensor
4. ABS hole cover
5. Connector
6. Rear ABS wheel-speed sensor wiring harness

4.0—6.0 N·m
{41—61 kgf-cm,
36—53 in-lbf}

37671_MAZ3_G0246

Fig. 6 Rear ABS wheel speed sensor and related components—2010 vehicles

13. Install the ABS hole cover into the body so that the arrow on it is facing toward the outer side of the vehicle.

14. Use the Mazda diagnostic scan tool, or equivalent and reprogram the required systems.

Combined

2010 Vehicles

See Figure 7.

➡The DSC/ABS system also uses a combined sensor.

1. Before servicing the vehicle, refer to the Precautions Section.

➡If working near and/or around the SRS system and components, be sure to disable the SRS system. Tape the negative battery cable with insulating tape. Always disconnect the negative battery cable first.

❊❊ CAUTION

To avoid personal injury when working on vehicles equipped with an air bag, the negative battery cable must be disconnected and at least one minute must elapse before working on the system. Failure to do so may result in deployment of the air bag.

2. Remove the battery top cover, if equipped.

3. Disconnect the negative battery cable. Tape the cable with insulating tape.

➡When disconnecting the cable, some systems need to be initialized after the cable is reconnected. You will need Mazda diagnostic scan tool or equivalent. Follow the directions on the tool.

4. Remove the front seat.
5. Remove the audio amplifier.
6. Remove the front scuff plate, side trim, rear scuff plate and B pillar lower trim.
7. Remove the retaining fasteners and peel back the carpeting.
8. Disconnect the electrical connector.
9. Remove the bracket.
10. Remove the sensor from its mounting.

To install:

➡Be sure to use new fasteners, as required.

11. Installation is the reverse of the removal procedure.

➡This sensor must be initialized after replacement. Follow the directions on the Mazda diagnostic tool, or equivalent to perform this operation.

12. Use the Mazda diagnostic scan tool, or equivalent and reprogram the required systems.

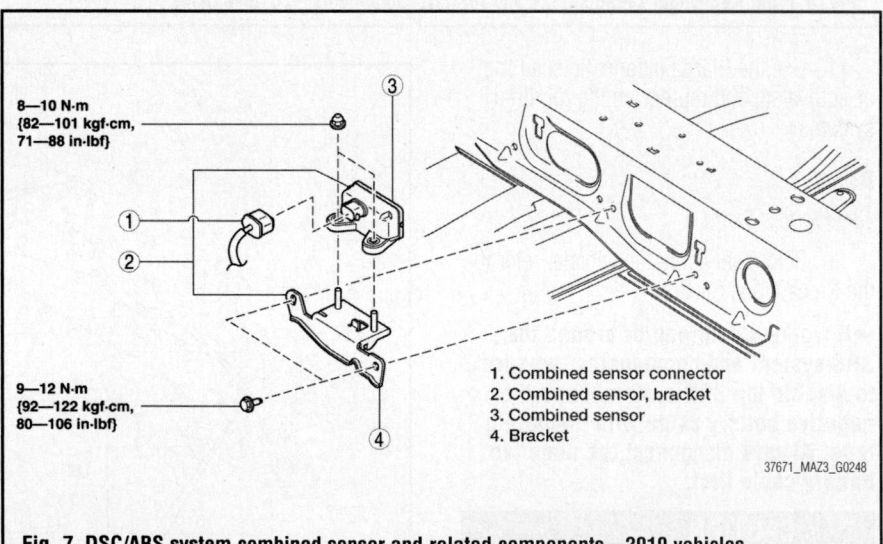

8—10 N·m
{82—101 kgf-cm,
71—88 in-lbf}

9—12 N·m
{92—122 kgf-cm,
80—106 in-lbf}

1. Combined sensor connector
2. Combined sensor, bracket
3. Combined sensor
4. Bracket

37671_MAZ3_G0248

Fig. 7 DSC/ABS system combined sensor and related components—2010 vehicles

BRAKES   **FRONT DISC BRAKES**

BRAKE CALIPER

REMOVAL & INSTALLATION

See Figures 8 through 10.

1. Before servicing the vehicle, refer to the Precautions Section.

➡ If working near and/or around the SRS system and components, be sure to disable the SRS system. Tape the negative battery cable with insulating tape. Always disconnect the negative battery cable first.

✳✳ CAUTION

To avoid personal injury when working on vehicles equipped with an air bag, the negative battery cable must be disconnected and at least one minute must elapse before working on the system. Failure to do so may result in deployment of the air bag.

2. Remove the battery top cover, if equipped.

3. Disconnect the negative battery cable. Tape the cable with insulating tape.

➡ When disconnecting the cable, some systems need to be initialized after the cable is reconnected. You will need Mazda diagnostic scan tool or equivalent. Follow the directions on the tool.

4. Remove or disconnect the following:
- Wheels
- Brake hose
- Retaining clip
- Cap from the caliper bolts
- Caliper mounting bolts and the caliper

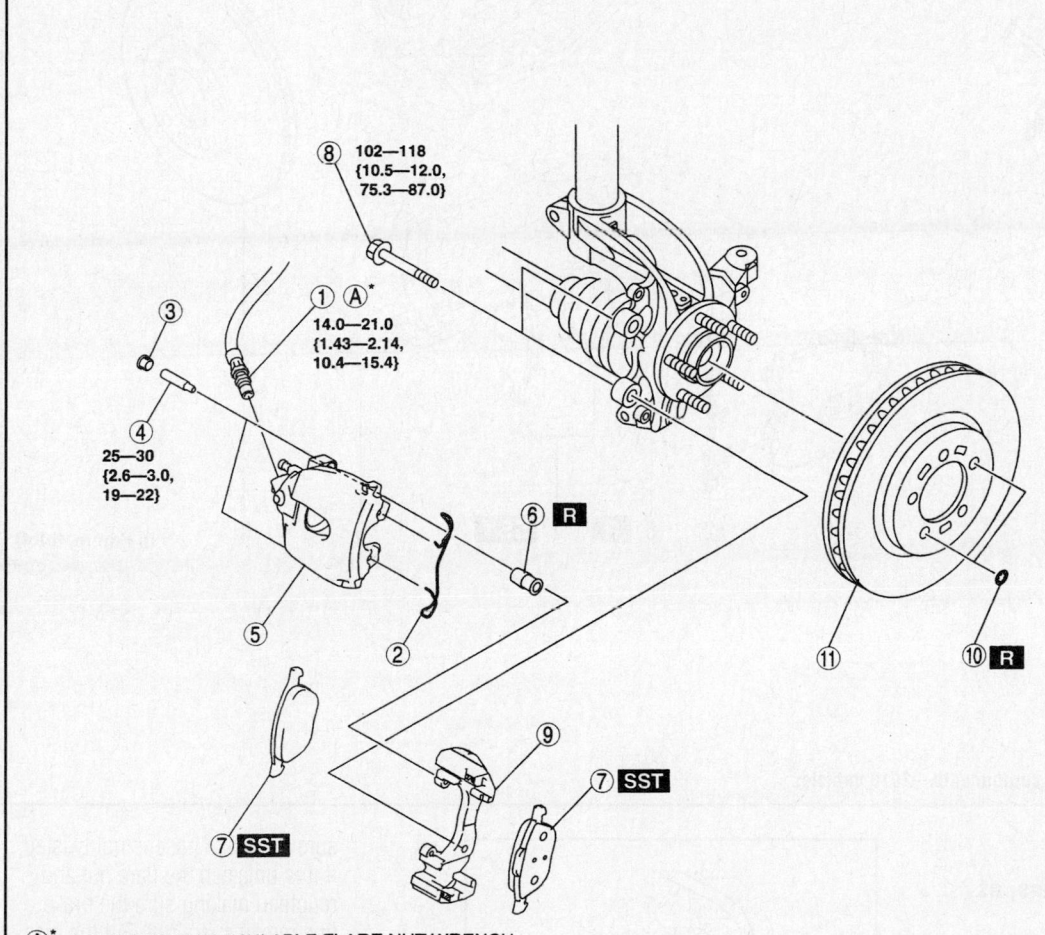

8 102—118 {10.5—12.0, 75.3—87.0}

1 Ⓐ*
14.0—21.0 {1.43—2.14, 10.4—15.4}

25—30 {2.6—3.0, 19—22}

1	Brake hose
2	Retaining clip
3	Cap
4	Bolt
5	Caliper
6	Boot
7	Disc pad
8	Bolt
9	Mounting support
10	Washer
11	Disc plate

Ⓐ* COMMERCIALLY AVAILABLE FLARE NUT WRENCH
(FLARE NUT ACROSS FLAT 13 mm {0.51 in})

N·m {kgf·m, ft·lbf}

67162-MAZC-G104

Fig. 8 Front caliper and related components—2009 vehicles

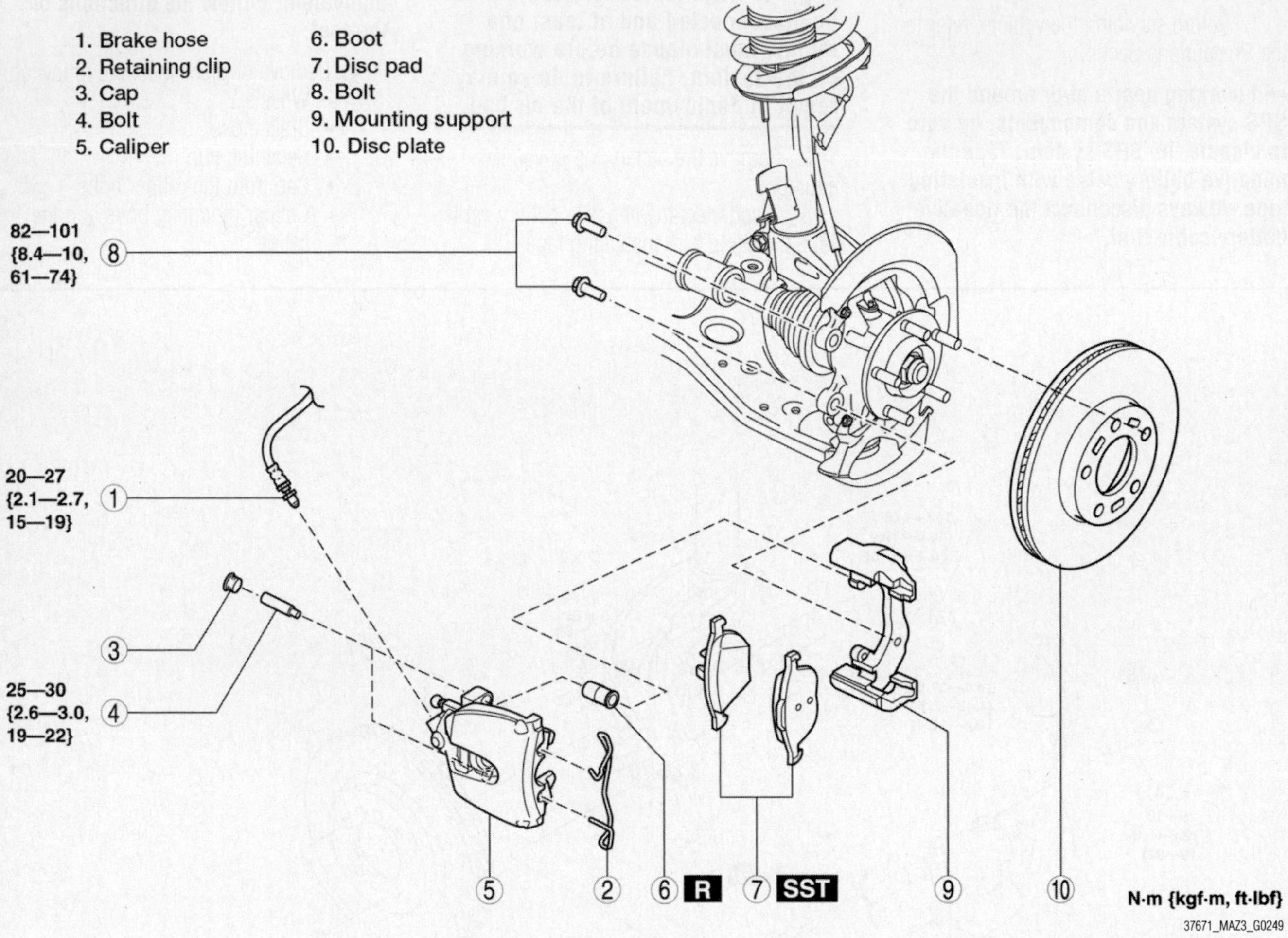

1. Brake hose
2. Retaining clip
3. Cap
4. Bolt
5. Caliper
6. Boot
7. Disc pad
8. Bolt
9. Mounting support
10. Disc plate

82—101
{8.4—10,
61—74} ⑧

20—27
{2.1—2.7, ①
15—19}

25—30
{2.6—3.0, ④
19—22}

③

⑤   ② ⑥ **R** ⑦ **SST**   ⑨   ⑩

N·m {kgf·m, ft·lbf}

37671_MAZ3_G0249

Fig. 9 Front caliper and related components—2010 vehicles

To install:

➡**Be sure to use new fasteners, as required.**

5. Install or connect the following:
 • Caliper. Torque the caliper mounting bolts to specification. Install the bolt caps.
 • Brake hose to the caliper. Tighten the flare nut while holding the hose at location A shown in the accompanying illustration with an open end wrench. Tighten the nut to 10–15 ft. lbs. (14–21 Nm) and make

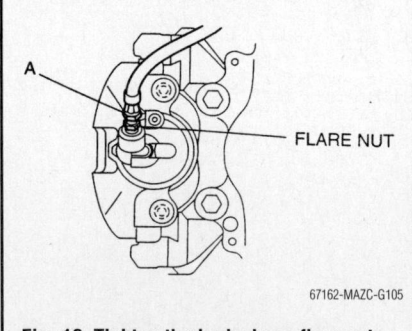

A

FLARE NUT

67162-MAZC-G105

Fig. 10 Tighten the brake hose flare nut while holding the hose at point A

sure the brake hose is not twisted, if it is unfasten the flare nut and retighten making sure the brake line remains straight. Fill the master cylinder with clean brake fluid and bleed the hydraulic system.
 • Retaining clip
 • Wheels
6. Pump the brake pedal several times to seat the pads.
7. Use the Mazda diagnostic scan tool, or equivalent and reprogram the required systems.

DISC BRAKE PADS

REMOVAL & INSTALLATION

See Figures 11 through 14.

1. Before servicing the vehicle, refer to the Precautions Section.

➡ **If working near and/or around the SRS system and components, be sure to disable the SRS system. Tape the negative battery cable with insulating tape. Always disconnect the negative battery cable first.**

❋❋ CAUTION

To avoid personal injury when working on vehicles equipped with an air bag, the negative battery cable must be disconnected and at least one minute must elapse before working on the system. Failure to do so may result in deployment of the air bag.

2. Remove the battery top cover, if equipped.
3. Disconnect the negative battery cable. Tape the cable with insulating tape.

➡ **When disconnecting the cable, some systems need to be initialized after the cable is reconnected. You will need Mazda diagnostic scan tool or equivalent. Follow the directions on the tool.**

4. Remove or disconnect the following:
   • Wheels
   • Pad retaining clip
   • Cap from the bottom caliper bolt
   • Lower caliper mounting bolt and swing the caliper up
   • Brake pads

To install:

➡ **Be sure to use new fasteners, as required.**

5. Press the caliper piston back into the cylinder using tool 49 0221 600C.
6. Install or connect the following:
   • Outer pad to the mounting support and inner pad to the caliper
   • Caliper. Torque the caliper lower mounting bolt to specification. Install the bolt cap.
   • Retaining clip
   • Wheels
7. Pump the brake pedal several times to seat the pads.
8. Use the Mazda diagnostic scan tool, or equivalent and reprogram the required systems.

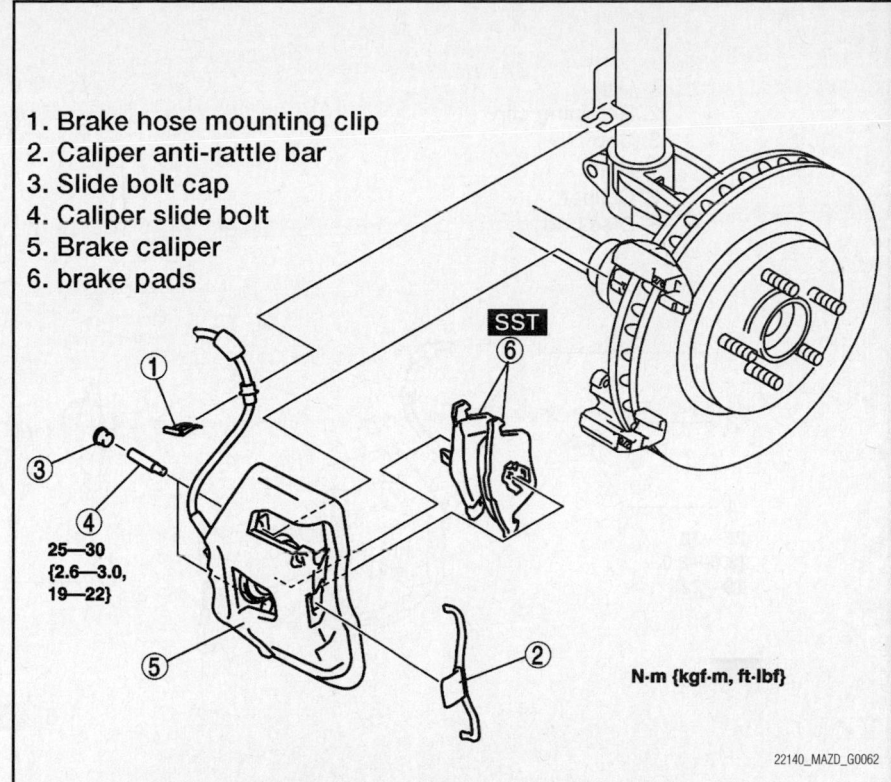

1. Brake hose mounting clip
2. Caliper anti-rattle bar
3. Slide bolt cap
4. Caliper slide bolt
5. Brake caliper
6. brake pads

SST

25—30
{2.6—3.0, 19—22}

N·m {kgf·m, ft·lbf}

22140_MAZD_G0062

Fig. 11 Front brake pads and related components (2.3L engine with turbocharger)—2009 vehicles

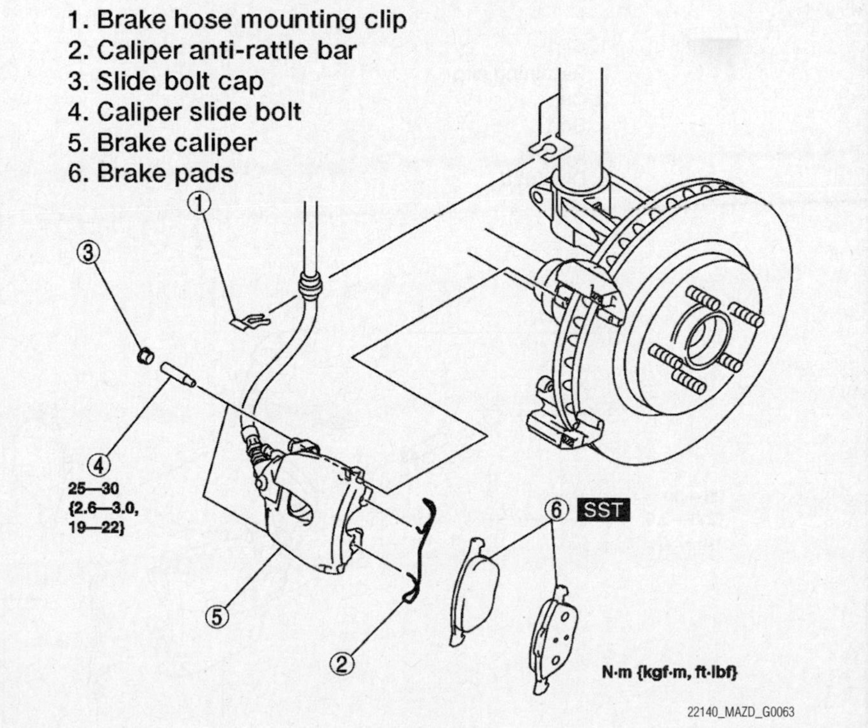

1. Brake hose mounting clip
2. Caliper anti-rattle bar
3. Slide bolt cap
4. Caliper slide bolt
5. Brake caliper
6. Brake pads

25—30
{2.6—3.0, 19—22}

SST

N·m {kgf·m, ft·lbf}

22140_MAZD_G0063

Fig. 12 Front brake pads and related components (2.0L and 2.3L engines)—2009 vehicles

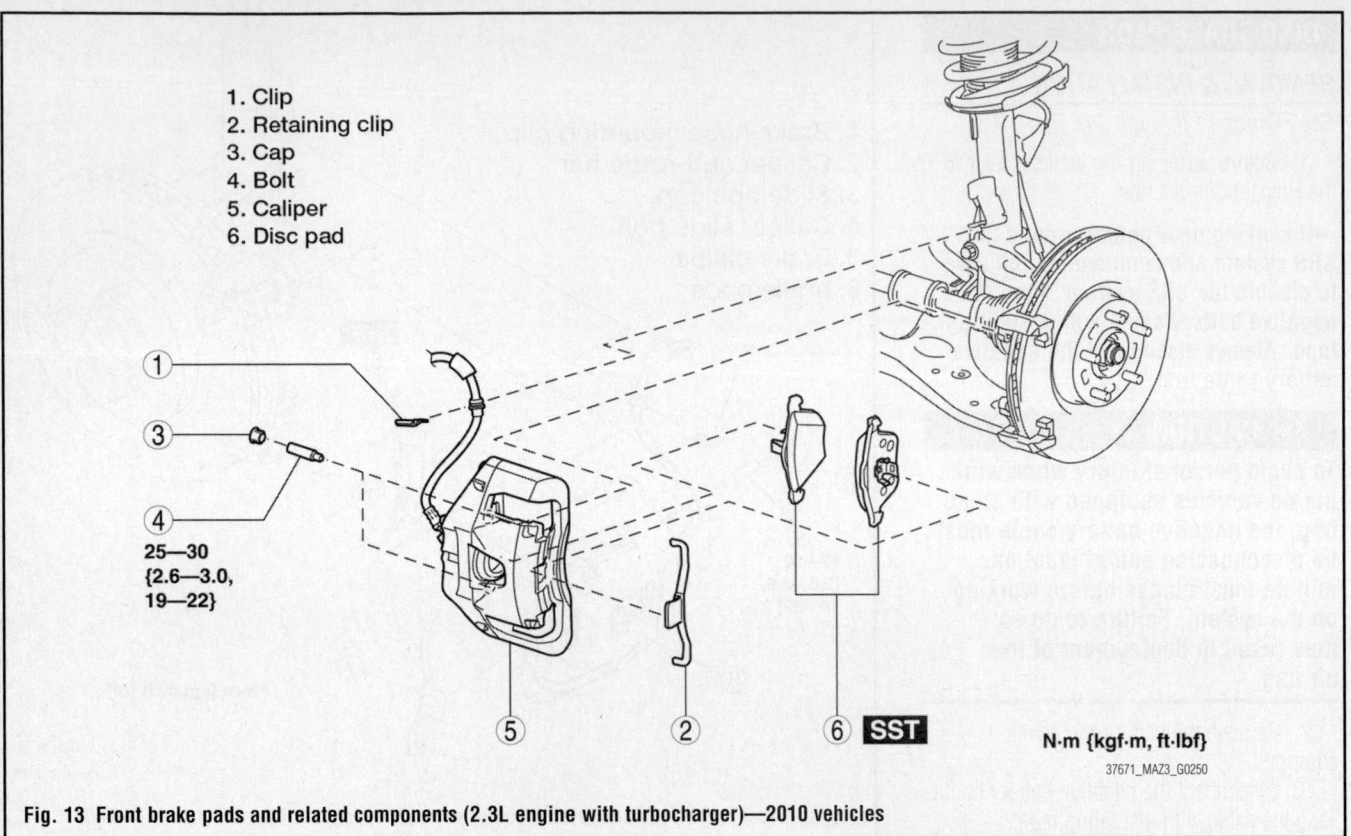

1. Clip
2. Retaining clip
3. Cap
4. Bolt
5. Caliper
6. Disc pad
1
3
4
25—30
{2.6—3.0,
19—22}
5
2
6 SST
N·m {kgf·m, ft·lbf}
37671_MAZ3_G0250

Fig. 13 Front brake pads and related components (2.3L engine with turbocharger)—2010 vehicles

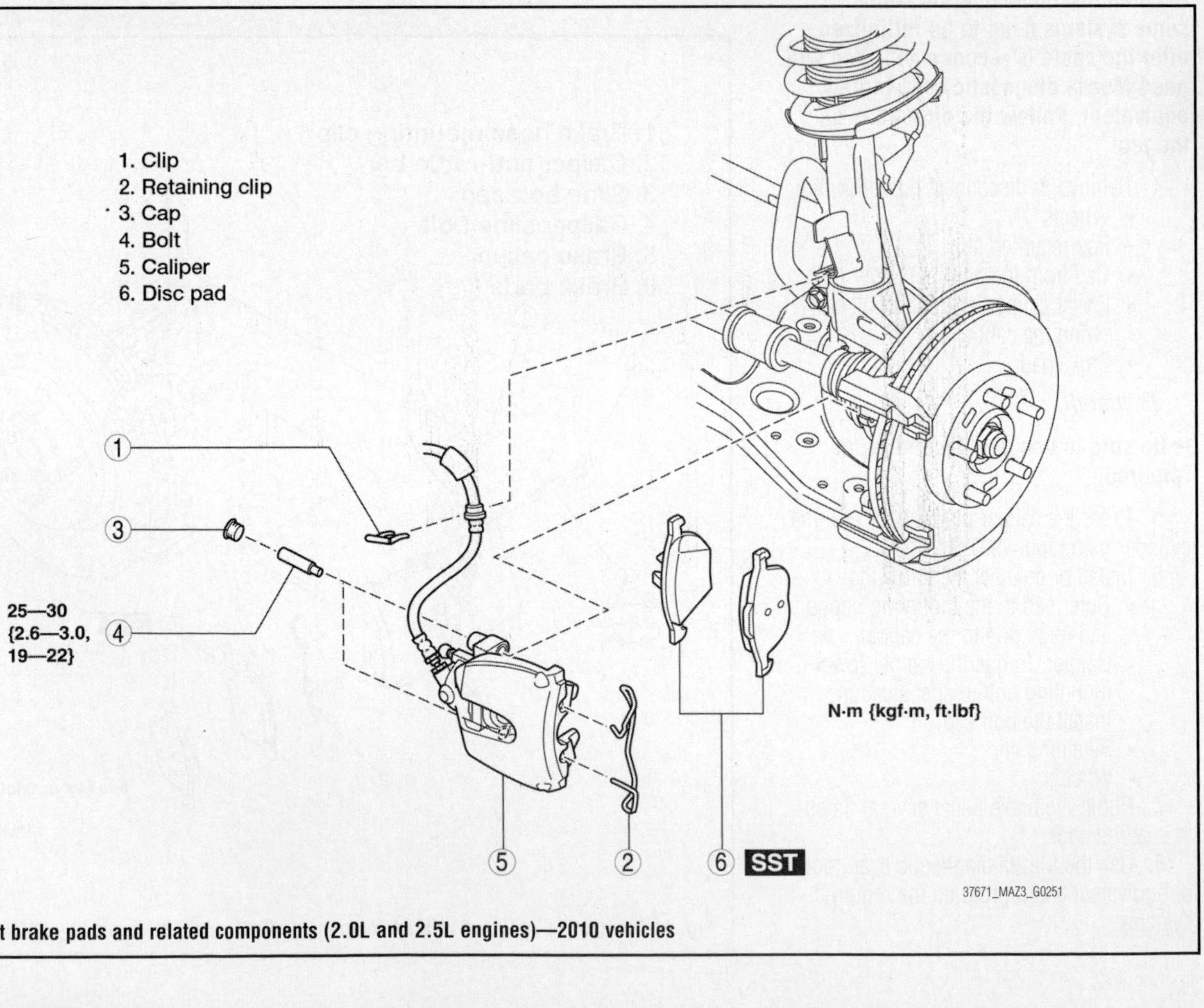

1. Clip
2. Retaining clip
3. Cap
4. Bolt
5. Caliper
6. Disc pad
1
3
25—30
{2.6—3.0,
19—22}
4
N·m {kgf·m, ft·lbf}
5
2
6 SST
37671_MAZ3_G0251

Fig. 14 Front brake pads and related components (2.0L and 2.5L engines)—2010 vehicles

BRAKES   **REAR DISC BRAKES**

BRAKE CALIPER

REMOVAL & INSTALLATION

See Figure 15.

1. Before servicing the vehicle, refer to the Precautions Section.

➡ **If working near and/or around the SRS system and components, be sure** to disable the SRS system. Tape the negative battery cable with insulating tape. Always disconnect the negative battery cable first.

❊ **CAUTION**

To avoid personal injury when working on vehicles equipped with an air bag, the negative battery cable must be disconnected and at least one minute must elapse before working on the system. Failure to do so may result in deployment of the air bag.

2. Remove the battery top cover, if equipped.

3. Disconnect the negative battery cable. Tape the cable with insulating tape.

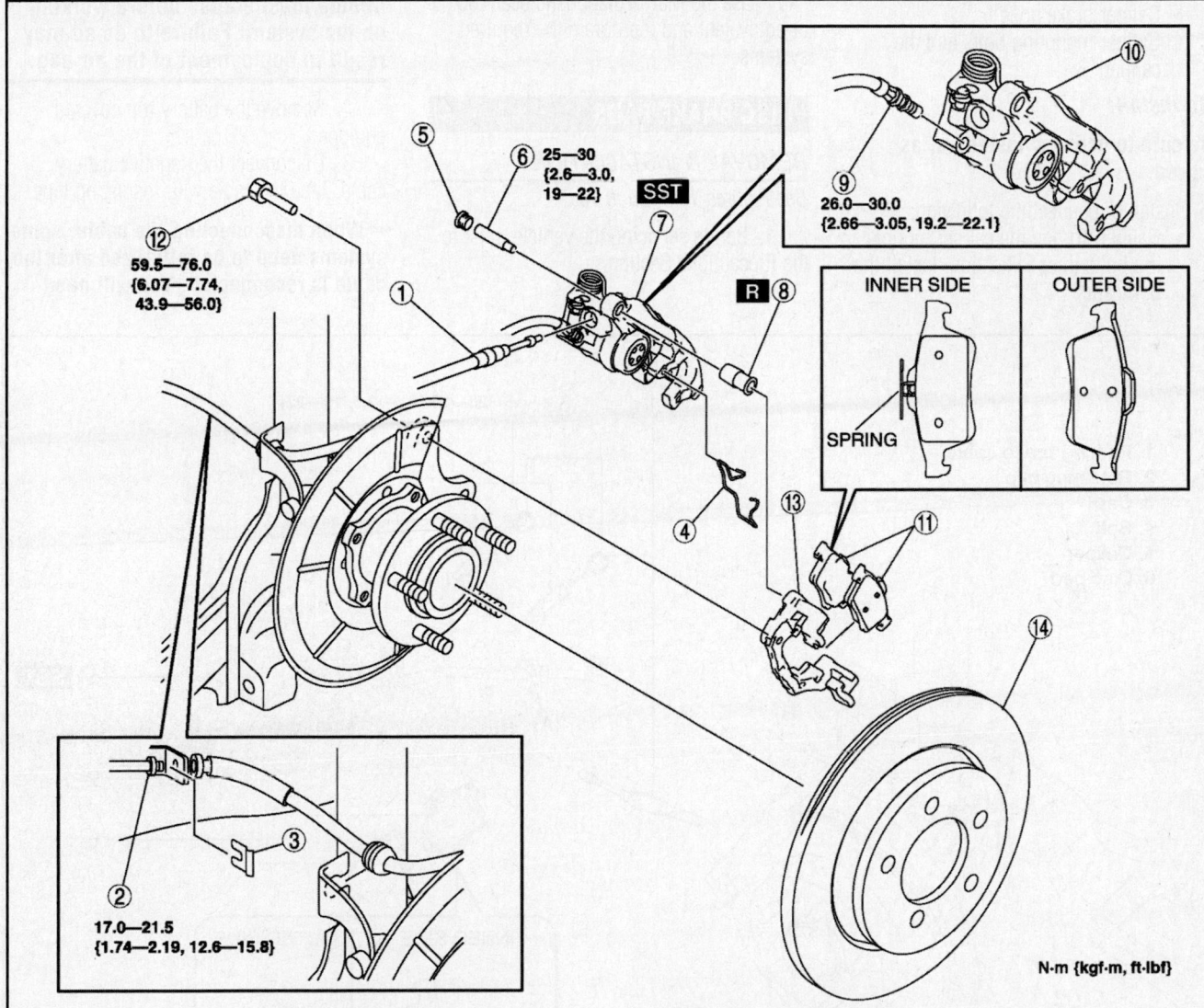

1. Emergency brake
2. Brake line
3. Brake hose clip
4. Anti rattle bar
5. Caliper slide bolt cap
6. Caliper slide bolt
7. Brake caliper
8. Caliper slide pin bushing
9. Brake hose
10. Brake caliper
11. Brake pads
12. Caliper bracket mounting bolt
13. Caliper bracket
14. Brake disc

22140_MAZD_G0072

Fig. 15 Rear caliper and related components

➡**When disconnecting the cable, some systems need to be initialized after the cable is reconnected. You will need Mazda diagnostic scan tool or equivalent. Follow the directions on the tool.**

4. Remove or disconnect the following:
   - Wheels
   - Parking brake cable
   - Brake pipe from the hose and the clip
   - Pad retaining clip
   - Caps from the caliper bolts
   - Caliper brake hose
   - Caliper mounting bolts and the caliper

To install:

➡**Be sure to use new fasteners, as required.**

5. Install or connect the following:
   - Caliper. Torque the caliper mounting bolts to specification. Install the bolt caps.

- Brake hose and pipe and tighten to 14–16 ft. lbs. (19–23 Nm). Reinstall a new clip and fill the master cylinder with clean brake fluid and bleed the hydraulic system.
- Pad retaining clip and parking brake cable
- Wheels

6. Pump the brake pedal several times to seat the pads. Inspect the parking brake lever stroke and brake drag and adjust as necessary.

7. Use the Mazda diagnostic scan tool, or equivalent and reprogram the required systems.

DISC BRAKE PADS

REMOVAL & INSTALLATION

See Figures 16 through 18.

1. Before servicing the vehicle, refer to the Precautions Section.

➡**If working near and/or around the SRS system and components, be sure to disable the SRS system. Tape the negative battery cable with insulating tape. Always disconnect the negative battery cable first.**

✳✳ CAUTION

To avoid personal injury when working on vehicles equipped with an air bag, the negative battery cable must be disconnected and at least one minute must elapse before working on the system. Failure to do so may result in deployment of the air bag.

2. Remove the battery top cover, if equipped.

3. Disconnect the negative battery cable. Tape the cable with insulating tape.

➡**When disconnecting the cable, some systems need to be initialized after the cable is reconnected. You will need**

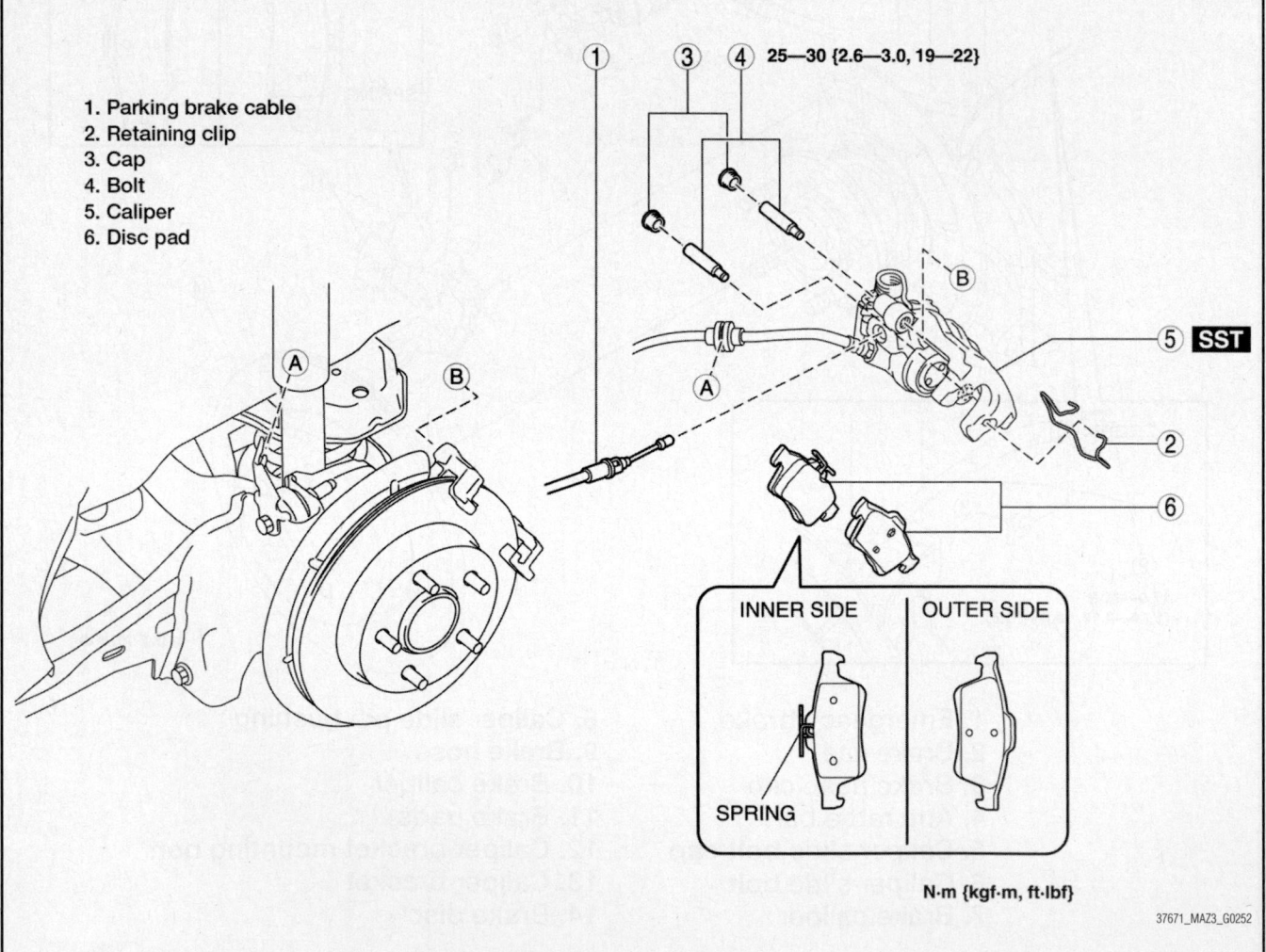

1. Parking brake cable
2. Retaining clip
3. Cap
4. Bolt
5. Caliper
6. Disc pad

25—30 {2.6—3.0, 19—22}

INNER SIDE   OUTER SIDE

SPRING

N·m {kgf·m, ft·lbf}

37671_MAZ3_G0252

Fig. 16 Rear brake pads and related components—2010 vehicles

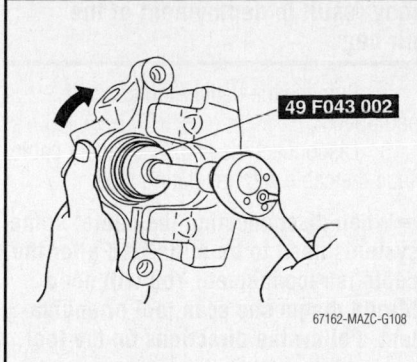

Fig. 17 Turn the rear caliper piston clock-wise slowly until the piston is fully seated in its bore—2009 vehicles

Mazda diagnostic scan tool or equivalent. Follow the directions on the tool.

4. Remove or disconnect the following:
- Wheels
- Parking brake cable
- Pad retaining clip
- Caps from the caliper bolts
- Caliper mounting bolts and the caliper
- Outer brake pad from the mount support and pull the inner pad from the caliper

To install:

➡**Be sure to use new fasteners, as required.**

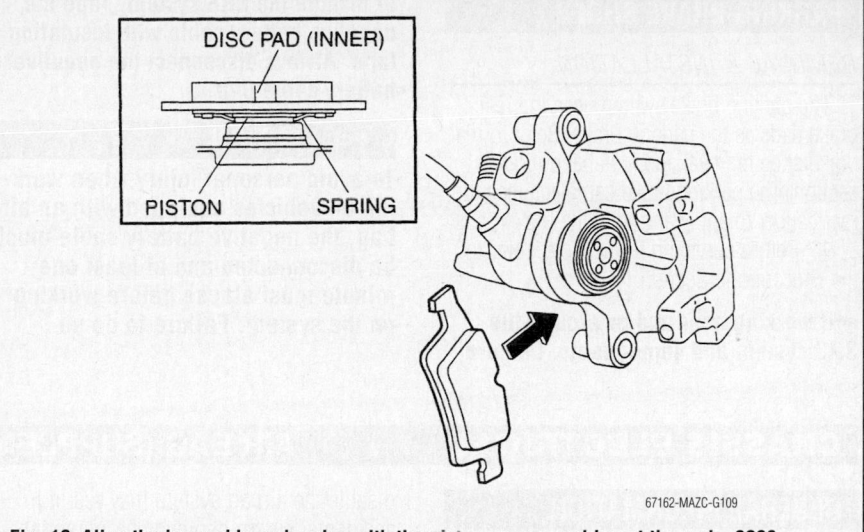

Fig. 18 Align the inner side pad spring with the piston groove and insert the pad—2009 vehicles

5. Install the out pad to the mounting support and clean the piston area.

6. Using tool 49 F043 002, turn the piston clockwise slowly until the piston is fully seated in its bore.

7. Align the inner side pad spring with the piston groove and insert the pad. Refer to the illustration for spring location and inner pad installation arrow.

8. Install or connect the following:
- Caliper. Torque the caliper mounting bolts to 19–22 ft. lbs.

(25–30 Nm). Install the bolt caps.
- Pad retaining clip and parking brake cable
- Wheels

9. Pump the brake pedal several times to seat the pads. Inspect the parking brake lever stroke and brake drag and adjust as necessary.

10. Use the Mazda diagnostic scan tool, or equivalent and reprogram the required systems.

BRAKES   **PARKING BRAKE**

PARKING BRAKE CABLES

ADJUSTMENT

See Figures 19 and 20.

1. Before servicing the vehicle, refer to the Precautions Section.

➡**If working near and/or around the SRS system and components, be sure to disable the SRS system. Tape the negative battery cable with insulating tape. Always disconnect the negative battery cable first.**

✳ CAUTION

To avoid personal injury when working on vehicles equipped with an air bag, the negative battery cable must be disconnected and at least one minute must elapse before working on the system. Failure to do so may result in deployment of the air bag.

2. Remove the battery top cover, if equipped.

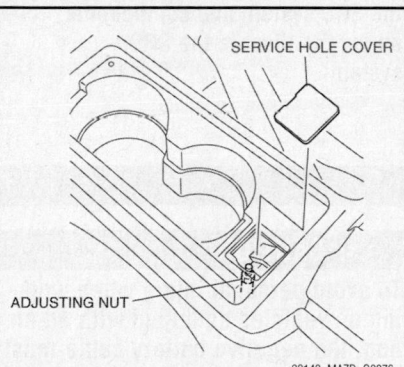

Fig. 19 Remove the service hole cover and adjust the parking brake—2009 vehicles

3. Disconnect the negative battery cable. Tape the cable with insulating tape.

➡**When disconnecting the cable, some systems need to be initialized after the cable is reconnected. You will need Mazda diagnostic scan tool or equivalent. Follow the directions on the tool.**

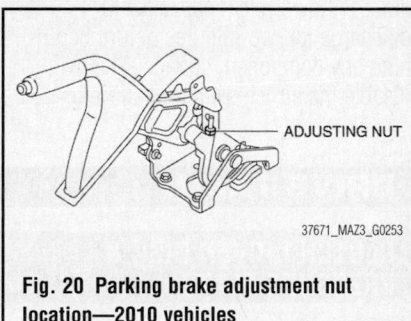

Fig. 20 Parking brake adjustment nut location—2010 vehicles

4. Pump the brake pedal a few times.

5. Remove the service hole cover of the rear console.

6. Turn the adjusting nut and adjust the parking brake lever.

7. After adjustment, pull the parking brake lever one notch and verify that the parking brake warning light illuminates.

8. Verify that the rear brakes do not drag.

9. Use the Mazda diagnostic scan tool, or equivalent and reprogram the required systems.

PARKING BRAKE SHOES

REMOVAL & INSTALLATION

The parking brake system uses the rear brake pads as the parking brake. Refer to rear brake shoe removal and installation in this section. The parking brake cable actuates the rear caliper to apply and hold the brake pads.

1. Before servicing the vehicle, refer to the Precautions Section.

➡If working near and/or around the SRS system and components, be sure

CHASSIS ELECTRICAL

GENERAL INFORMATION

Mazda vehicles are equipped with an air bag system. The system must be disarmed before performing service on, or around, system components, the steering column, instrument panel components, wiring and sensors. Failure to follow the safety precautions and the disarming procedure could result in accidental air bag deployment, possible injury and unnecessary system repairs.

PRECAUTIONS

Disconnect and isolate the battery negative cable before beginning any airbag system component diagnosis, testing, removal, or installation procedures. Allow system capacitor to discharge for two minutes before beginning any component service. This will disable the airbag system. Failure to

DRIVE TRAIN

CLUTCH DRIVEN DISC & PRESSURE PLATE

REMOVAL & INSTALLATION

2009 Vehicles

See Figures 21 through 23.

1. Before servicing the vehicle, refer to the Precautions Section.

➡If working near and/or around the SRS system and components, be sure to disable the SRS system. Tape the negative battery cable with insulating tape. Always disconnect the negative battery cable first.

to disable the SRS system. Tape the negative battery cable with insulating tape. Always disconnect the negative battery cable first.

✳✳ CAUTION

To avoid personal injury when working on vehicles equipped with an air bag, the negative battery cable must be disconnected and at least one minute must elapse before working on the system. Failure to do so

disable the airbag system may result in accidental airbag deployment, personal injury, or death.

DISARMING THE SYSTEM

✳✳ CAUTION

To avoid personal injury when working on vehicles equipped with an air bag, the negative battery cable must be disconnected and at least one minute must elapse before working on the system. Failure to do so may result in deployment of the air bag.

1. Before servicing the vehicle, refer to the Precautions Section.

➡If working near and/or around the SRS system and components, be sure to disable the SRS system.

✳✳ CAUTION

To avoid personal injury when working on vehicles equipped with an air bag, the negative battery cable must be disconnected and at least one minute must elapse before working on the system. Failure to do so may result in deployment of the air bag.

2. Remove the battery top cover, if equipped.
3. Disconnect the negative battery cable. Tape the cable with insulating tape.

➡When disconnecting the cable, some systems need to be initialized after the cable is reconnected. You will need

may result in deployment of the air bag.

2. Remove the battery top cover, if equipped.
3. Disconnect the negative battery cable. Tape the cable with insulating tape.

➡When disconnecting the cable, some systems need to be initialized after the cable is reconnected. You will need Mazda diagnostic scan tool or equivalent. Follow the directions on the tool.

AIR BAG (SUPPLEMENTAL RESTRAINT SYSTEM)

2. Disconnect the negative battery cable.

➡Always disconnect the negative battery cable first.

3. Tape the negative battery cable with insulating tape.
4. If equipped, deactivate the audio anti-theft system.
5. Wait at least one minute before starting any service work so the system can depower itself.

ARMING THE SYSTEM

1. Before servicing the vehicle, refer to the precautions section.
2. Connect the negative battery cable, turn the ignition switch **ON** and verify the air bag warning light cones on for 6 seconds. If the light does not illuminate there are problems with the system.
3. If equipped, activate the audio anti-theft system.

Mazda diagnostic scan tool or equivalent. Follow the directions on the tool.

4. Remove or disconnect the following:
   • Negative battery cable
   • Clutch release cylinder
   • Transaxle
   • Rubber boot
   • Clutch release collar
   • Clutch release fork
   • Pressure plate loosening the bolts one turn each in a criss—cross pattern
   • Clutch disc

5. Inspect the pilot bearing. If it is worn or damaged and does not turn easily by hand, remove it using a puller/slide hammer.

6. Check the flywheel surface for scor-

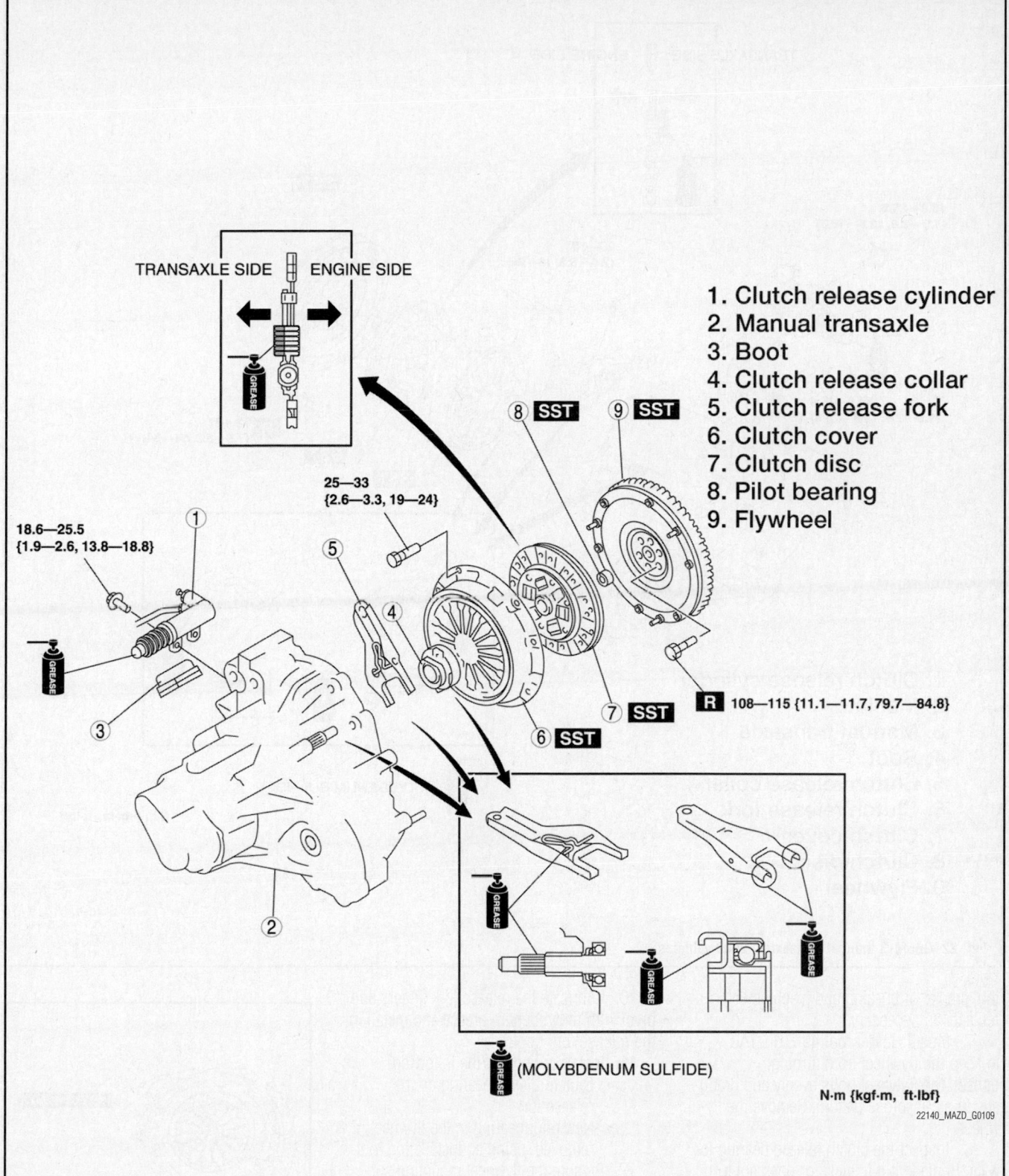

TRANSAXLE SIDE   ENGINE SIDE

1. Clutch release cylinder
2. Manual transaxle
3. Boot
4. Clutch release collar
5. Clutch release fork
6. Clutch cover
7. Clutch disc
8. Pilot bearing
9. Flywheel

8 SST   9 SST

18.6—25.5
{1.9—2.6, 13.8—18.8}

25—33
{2.6—3.3, 19—24}

7 SST

6 SST

R 108—115 {11.1—11.7, 79.7—84.8}

(MOLYBDENUM SULFIDE)

N·m {kgf·m, ft·lbf}

22140_MAZD_G0109

Fig. 21 G35M-R manual transaxle—2009 vehicles

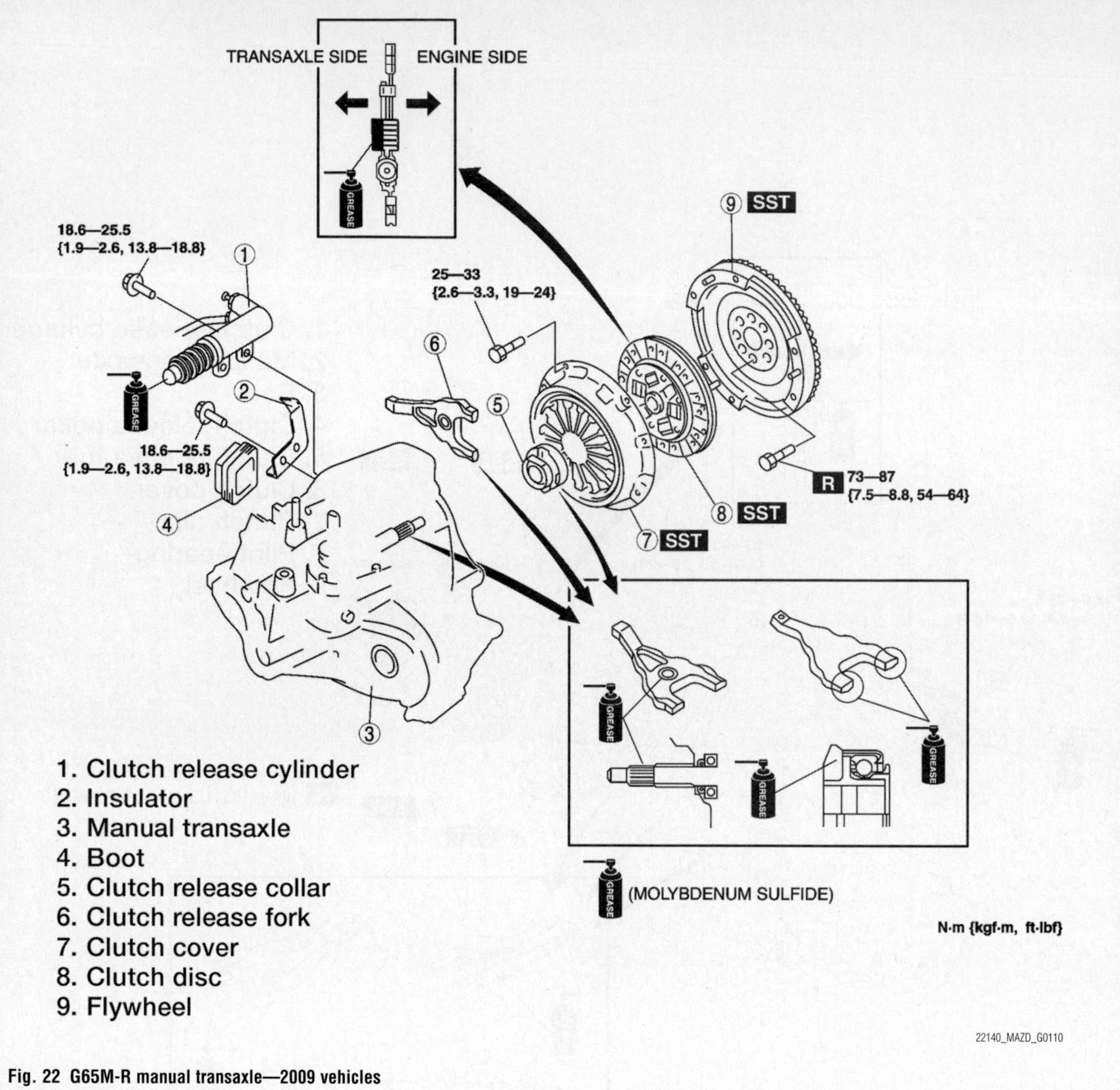

TRANSAXLE SIDE | ENGINE SIDE

GREASE

18.6—25.5
{1.9—2.6, 13.8—18.8}

25—33
{2.6—3.3, 19—24}

9 SST

GREASE

18.6—25.5
{1.9—2.6, 13.8—18.8}

R 73—87
{7.5—8.8, 54—64}

8 SST

7 SST

GREASE

GREASE

GREASE

GREASE

(MOLYBDENUM SULFIDE)

N·m {kgf·m, ft·lbf}

1. Clutch release cylinder
2. Insulator
3. Manual transaxle
4. Boot
5. Clutch release collar
6. Clutch release fork
7. Clutch cover
8. Clutch disc
9. Flywheel

22140_MAZD_G0110

Fig. 22 G65M-R manual transaxle—2009 vehicles

ing, cracks or burning and machine or replace, as necessary.

7. Install Holder tool 49 E011 1A0 to keep the flywheel from turning. Loosen the flywheel bolts evenly and gradually in a crisscross pattern. Remove the flywheel.

8. Inspect the clutch release bearing for wear. Replace it if it sticks or does not turn easily.

9. Inspect the release fork for wear or damage and replace as necessary.

To install:

➡**Be sure to use new fasteners, as required.**

10. Lubricate the release fork fingers and pivot with molybdenum grease and install in the release fork boot.

11. Install or connect the following:
   • Clutch release bearing on the release fork
   • New pilot bearing in the flywheel, if removed, using an installation tool

12. Be sure the flywheel mounting surface and the crankshaft or eccentric shaft mounting surfaces are clean. Remove any old sealant from the flywheel bolt hole threads and the flywheel bolts.
   • Flywheel
   • Sealant to the flywheel bolt threads and install them hand tight

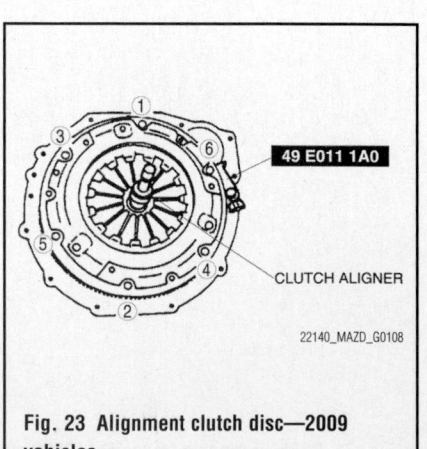

49 E011 1A0

CLUTCH ALIGNER

22140_MAZD_G0108

Fig. 23 Alignment clutch disc—2009 vehicles

- Flywheel holding tool. Tighten the bolts, in a crisscross pattern. Tighten to 79–85 ft. lbs. (108–116 Nm).

13. Install or connect the following:
  - Small amount of molybdenum grease to the clutch disc splines
  - Clutch disc on the flywheel with the spring side toward the transaxle
  - An alignment tool in the pilot bearing to position the clutch disc

- Clutch pressure plate by aligning the dowel holes with the flywheel dowels
- Pressure plate. Gradually, torque the bolts, in a crisscross pattern, to 19 ft. lbs. (26 Nm).

14. Remove the alignment tool.
  - Clutch release fork
  - Clutch release collar
  - Rubber boot

- Transaxle
- Clutch release cylinder
- Negative battery cable

15. Use the Mazda diagnostic scan tool, or equivalent and reprogram the required systems.

2010 Vehicles

See Figures 24 through 27.

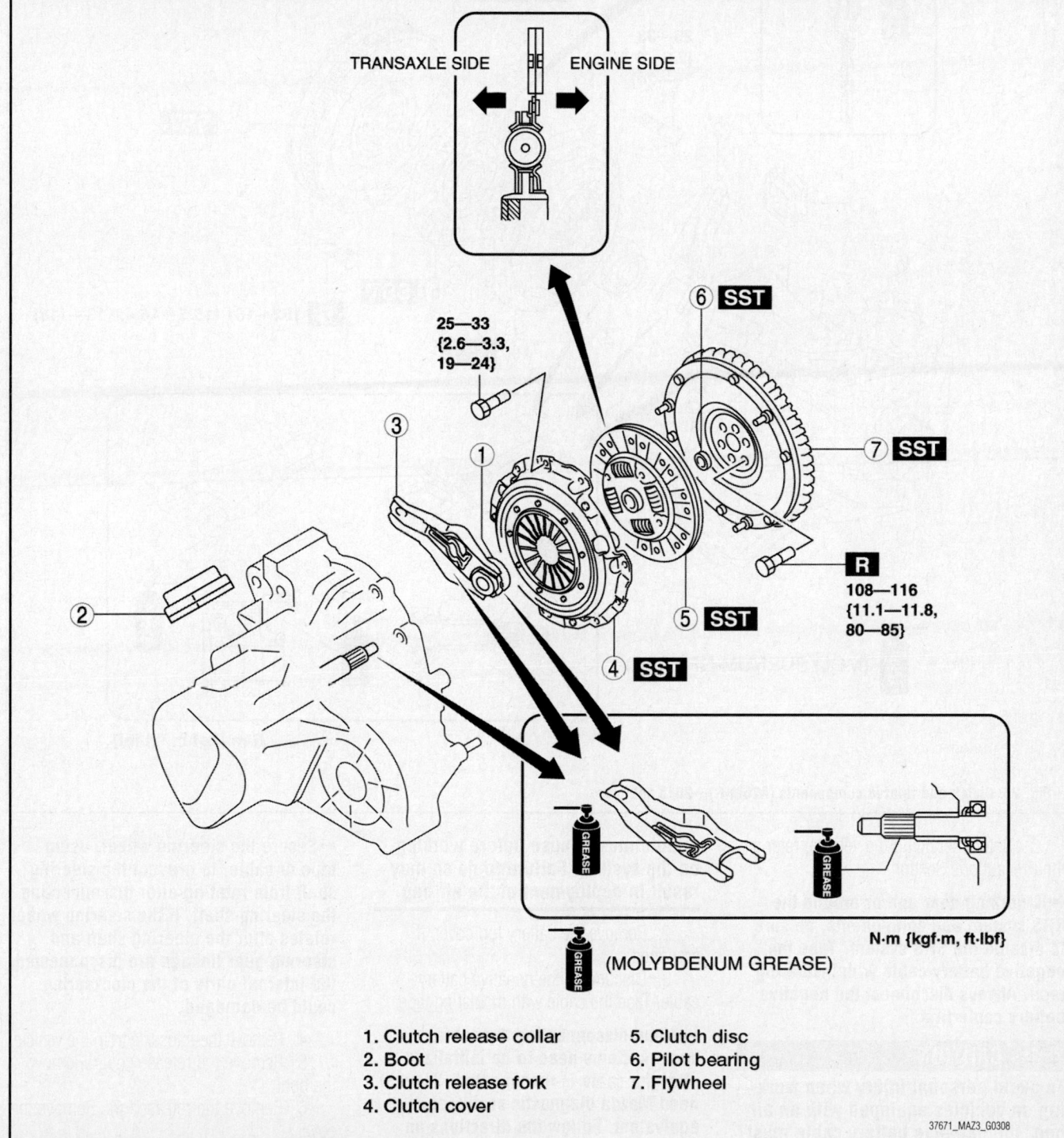

1. Clutch release collar
2. Boot
3. Clutch release fork
4. Clutch cover
5. Clutch disc
6. Pilot bearing
7. Flywheel

Fig. 24 Clutch and related components (G35M-R and G66M-R)—2010 vehicles

37671_MAZ3_G0308

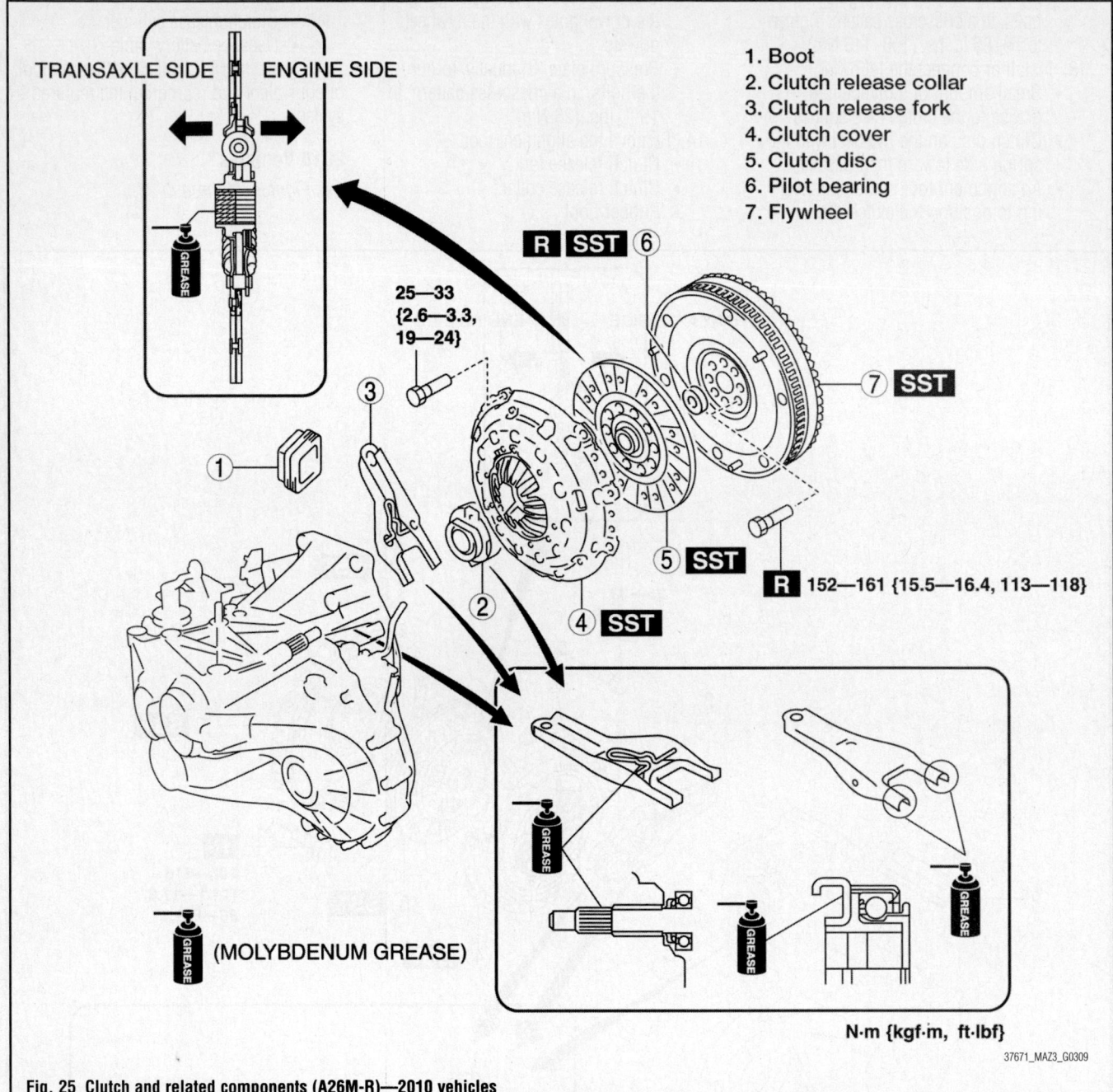

TRANSAXLE SIDE **ENGINE SIDE**

GREASE

25—33
{2.6—3.3,
19—24}

1. Boot
2. Clutch release collar
3. Clutch release fork
4. Clutch cover
5. Clutch disc
6. Pilot bearing
7. Flywheel

R **SST** 6

7 **SST**

5 **SST**

4 **SST**

R 152—161 {15.5—16.4, 113—118}

(MOLYBDENUM GREASE)

GREASE

N·m {kgf·m, ft·lbf}

37671_MAZ3_G0309

Fig. 25 Clutch and related components (A26M-R)—2010 vehicles

1. Before servicing the vehicle, refer to the Precautions Section.

➡If working near and/or around the SRS system and components, be sure to disable the SRS system. Tape the negative battery cable with insulating tape. Always disconnect the negative battery cable first.

✳✳ CAUTION

To avoid personal injury when working on vehicles equipped with an air bag, the negative battery cable must be disconnected and at least one minute must elapse before working on the system. Failure to do so may result in deployment of the air bag.

2. Remove the battery top cover, if equipped.
3. Disconnect the negative battery cable. Tape the cable with insulating tape.

➡When disconnecting the cable, some systems need to be initialized after the cable is reconnected. You will need Mazda diagnostic scan tool or equivalent. Follow the directions on the tool.

➡Secure the steering wheel, using tape or cable, to prevent the steering shaft from rotating after disconnecting the steering shaft. If the steering wheel rotates after the steering shaft and steering gear linkage are disconnected the internal parts of the clockspring could be damaged.

4. Remove the transaxle from the vehicle.
5. Remove the release collar. Remove the boot
6. Remove the release fork. Remove the cover.
7. Remove the clutch disc.

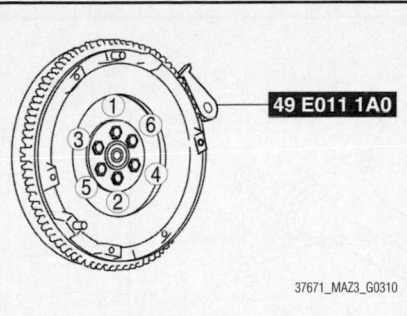

49 E011 1A0

37671_MAZ3_G0310

Fig. 26 Clutch flywheel tightening sequence (G35M-R and G66M-R)—2010 vehicles

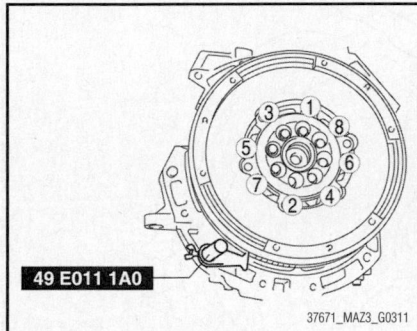

49 E011 1A0

37671_MAZ3_G0311

Fig. 27 Clutch flywheel tightening sequence (A26M-R)—2010 vehicles

8. Remove the pilot bearing
9. Remove the flywheel. Remove the bolts in a crisscross pattern.

To install:

➡**Be sure to use new fasteners, as required.**

10. Installation is the reverse of the removal procedure.
11. Tighten the flywheel retaining bolts to specification and in the proper sequence.
12. Use the Mazda diagnostic scan tool, or equivalent and reprogram the required systems.

ADJUSTMENTS

The hydraulic clutch system does not require any adjustment.

HYDRAULIC SYSTEM BLEEDING

BLEEDING PROCEDURE

See Figure 28.

1. Before servicing the vehicle, refer to the Precautions Section.

➡**If working near and/or around the SRS system and components, be sure**
to disable the SRS system. Tape the negative battery cable with insulating tape. Always disconnect the negative battery cable first.

✳✳ CAUTION

To avoid personal injury when working on vehicles equipped with an air bag, the negative battery cable must be disconnected and at least one minute must elapse before working on the system. Failure to do so may result in deployment of the air bag.

2. Remove the battery top cover, if equipped.
3. Disconnect the negative battery cable. Tape the cable with insulating tape.

➡**When disconnecting the cable, some systems need to be initialized after the cable is reconnected. You will need Mazda diagnostic scan tool or equivalent. Follow the directions on the tool.**

4. On 2010 vehicles, remove the No. 2 undercover.
5. Remove the rubber cap from the bleeder screw on the release cylinder.
6. Place a bleeder tube over the end of the bleeder screw.
7. Submerge the other end of the tube in a jar half filled with clutch fluid.
8. Slowly pump the clutch pedal fully and allow it to return slowly, several times.
9. While pressing the clutch pedal to the floor, loosen the bleeder screw until the fluid starts to run out. Then, close the bleeder screw. Keep repeating this Step, while watching the fluid in the jar. As soon as the air bubbles disappear, close the bleeder screw.
10. During the bleeding procedure the reservoir must be kept at least ¾ full.

11. Fluid specification is SAE J1703 OR FMVSS116 DOT-3.

FRONT HALFSHAFT

REMOVAL & INSTALLATION

2009 Vehicles
See Figure 29.

1. Before servicing the vehicle, refer to the Precautions Section.

➡**If working near and/or around the SRS system and components, be sure to disable the SRS system. Tape the negative battery cable with insulating tape. Always disconnect the negative battery cable first.**

✳✳ CAUTION

To avoid personal injury when working on vehicles equipped with an air bag, the negative battery cable must be disconnected and at least one minute must elapse before working on the system. Failure to do so may result in deployment of the air bag.

2. Remove the battery top cover, if equipped.
3. Disconnect the negative battery cable. Tape the cable with insulating tape.

➡**When disconnecting the cable, some systems need to be initialized after the cable is reconnected. You will need Mazda diagnostic scan tool or equivalent. Follow the directions on the tool.**

4. Drain the transaxle oil.
5. Remove or disconnect the following:
   - Wheels
   - ABS sensor
   - Stabilizer control link upper nut
   - Front lower arm ball joint
   - Joint shaft
   - Clip

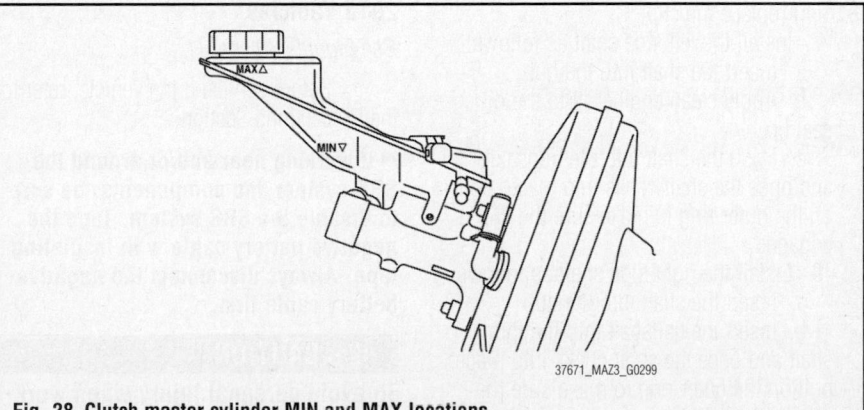

37671_MAZ3_G0299

Fig. 28 Clutch master cylinder MIN and MAX locations

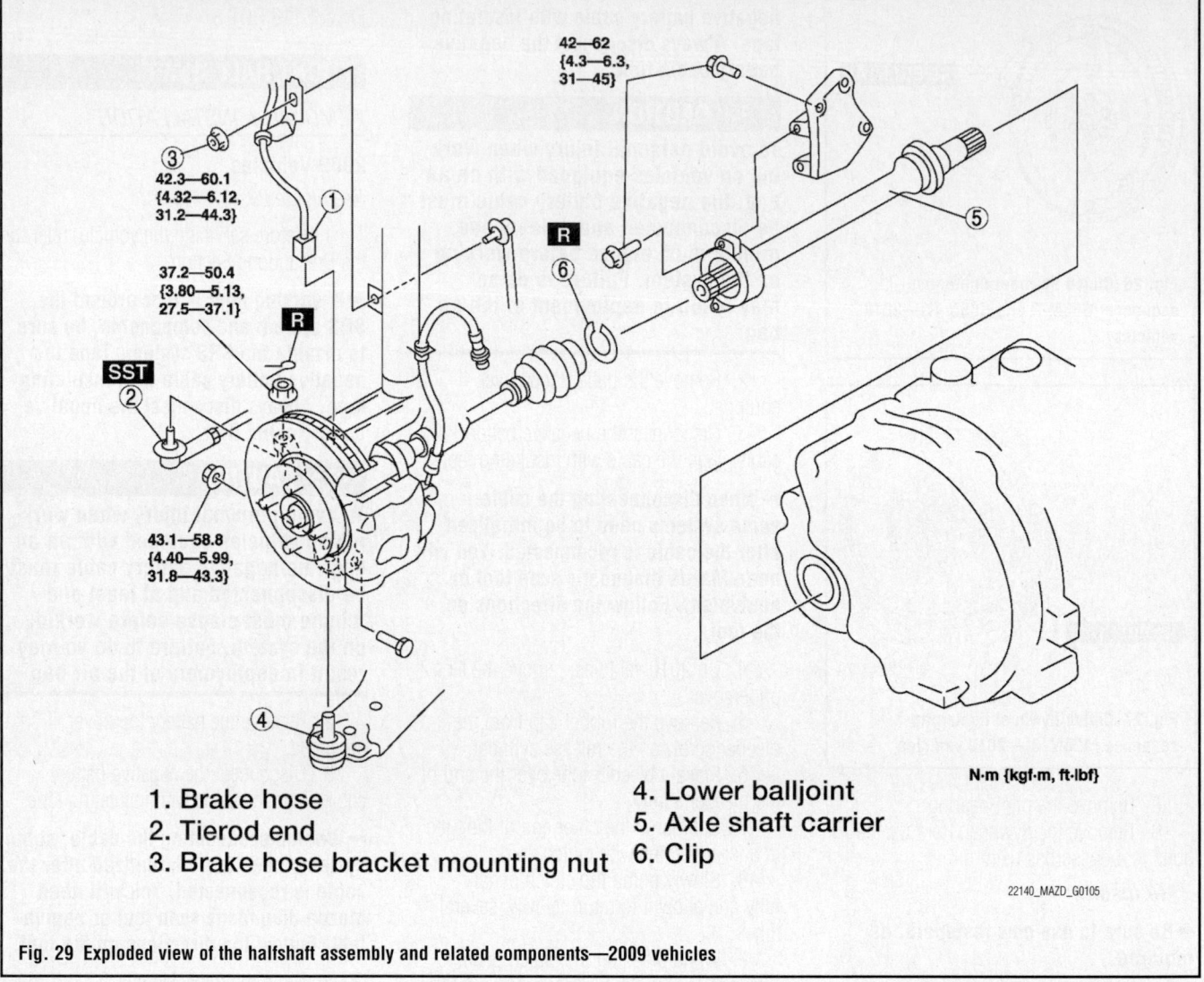

1. Brake hose
2. Tierod end
3. Brake hose bracket mounting nut
4. Lower balljoint
5. Axle shaft carrier
6. Clip

N·m {kgf·m, ft·lbf}

22140_MAZD_G0105

Fig. 29 Exploded view of the halfshaft assembly and related components—2009 vehicles

To install:

➡Be sure to use new fasteners, as required.

6. Install a new circlip on the end of the shaft, if removed, with the end gap facing upward. Measure the outer diameter of the clip, if it exceeds 1.19–1.23 inch (30–31mm) replace the clip.

7. Install the left side shaft as follows:
   a. Insert the shaft into the hub.
   b. Apply clean engine oil to the oil seal lip.
   c. Push the shaft into the transaxle and once the shaft clicks into place, pull on the outer ring to make sure the clip is engaged.

8. Install the right side shaft as follows:
   a. Insert the shaft into the hub.
   b. Insert the halfshaft into the joint shaft and once the shaft clicks into place, pull on the outer ring to make sure the clip is engaged.

9. Install the remaining components in the reverse order of removal. Refer to the exploded view of the halfshaft assembly and related components for torque specifications.

10. Use the Mazda diagnostic scan tool, or equivalent and reprogram the required systems.

2010 Vehicles

See Figures 30 and 31.

1. Before servicing the vehicle, refer to the Precautions Section.

➡If working near and/or around the SRS system and components, be sure to disable the SRS system. Tape the negative battery cable with insulating tape. Always disconnect the negative battery cable first.

❋❋ CAUTION

To avoid personal injury when working on vehicles equipped with an air bag, the negative battery cable must be disconnected and at least one minute must elapse before working on the system. Failure to do so may result in deployment of the air bag.

2. Remove the battery top cover, if equipped.

3. Disconnect the negative battery cable. Tape the cable with insulating tape.

➡When disconnecting the cable, some systems need to be initialized after the cable is reconnected. You will need Mazda diagnostic scan tool or equivalent. Follow the directions on the tool.

➡Performing the following procedure without removing the ABS wheel sensor may cause an open circuit in the harness if it is pulled by mistake. Before performing the procedure remove the ABS wheel speed sensor (axle side) and position it to an appropriate place

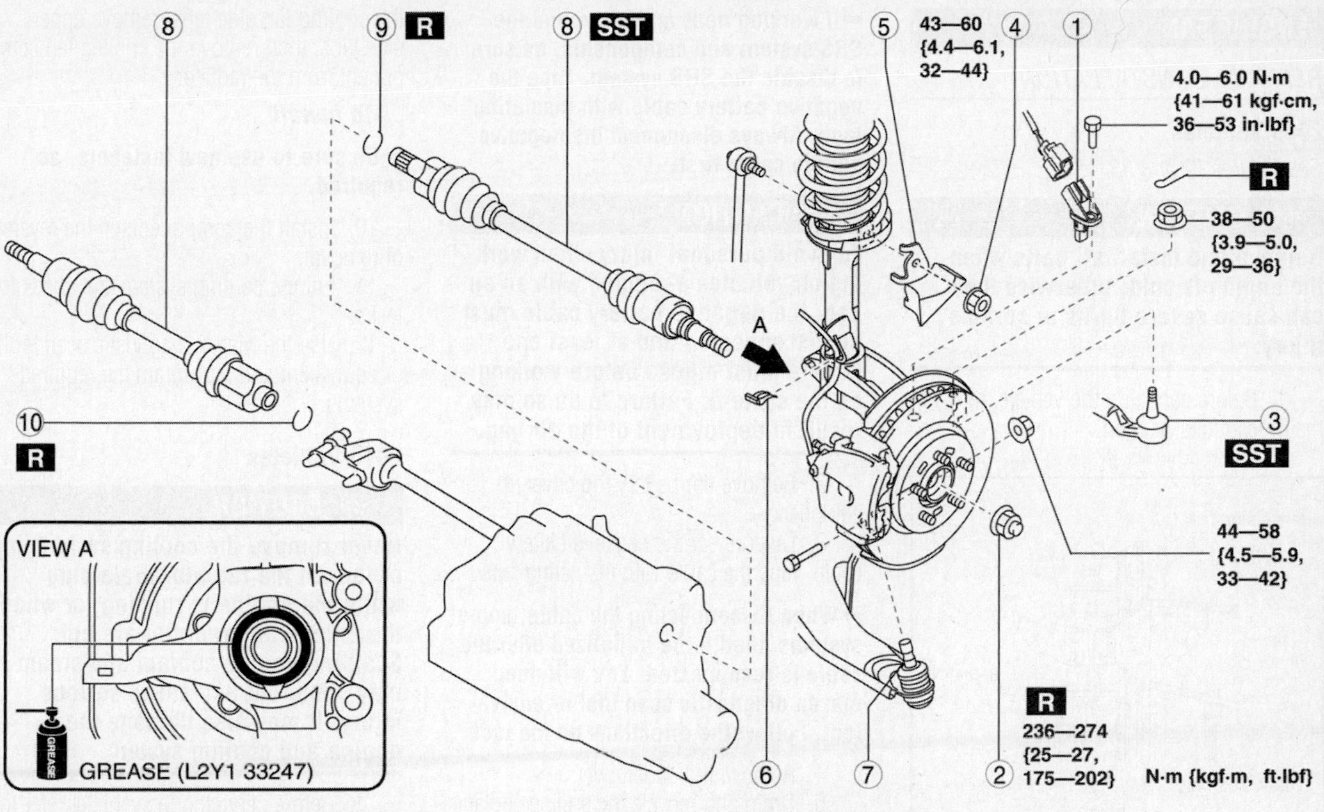

1. ABS wheel-speed sensor
2. Locknut
3. Tie-rod end ball joint
4. Stabilizer control link upper nut
5. Dynamic damper
6. Brake hose clip
7. Front lower arm ball joint
8. Halfshaft
9. Halfshaft clip
10. Joint shaft clip

43—60 {4.4—6.1, 32—44}

4.0—6.0 N·m {41—61 kgf·cm, 36—53 in·lbf}

38—50 {3.9—5.0, 29—36}

44—58 {4.5—5.9, 33—42}

236—274 {25—27, 175—202} N·m {kgf·m, ft·lbf}

VIEW A

GREASE (L2Y1 33247)

37671_MAZ3_G0324

Fig. 30 Front halfshafts and related components—2010 vehicles

where the sensor will not be pulled by mistake while servicing the vehicle.

➡The right side halfshaft on 2.3L engines with turbocharger, cannot be disassembled. Therefore replace the front halfshaft as a single unit if there is any malfunction of the boots, boot bands, or CV joints.

4. Raise and support the vehicle safely.
5. Remove the tire and wheel assemblies.
6. Remove the No. 2 undercover.
7. Drain the transaxle fluid.
8. When working on the left side disconnect the auto sensing sensor link, if equipped.
9. Remove the ABS sensor.
10. Remove the locknut.
11. Disconnect the tie rod end ball joint.
12. Disconnect the stabilizer control link upper nut.

13. Disconnect the dynamic damper.
14. Disconnect the brake hose clip.
15. Disconnect the lower arm ball joint.
16. Using the proper tools, remove the halfshaft and clip. Remove the joint shaft clip.

To install:

➡Be sure to use new fasteners, as required.

17. Installation is the reverse of the removal procedure.
18. Install a new shaft clip to the clip groove at the end of the shaft with the clip opening facing upward and the clip within to specification. Specification is 1.23 inch for 2.0L and 2.5L engines and 1.31 inch for 2.3L engines with turbocharger.

➡The front lower control arm bolt insertion direction bolt can be either

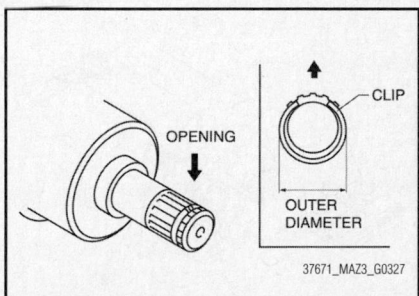

OPENING

CLIP

OUTER DIAMETER

37671_MAZ3_G0327

Fig. 31 Front halfshaft clip installation—2010 vehicles

front or rear of the vehicle. However, keep the insertion direction the same on both left and right sides of the vehicle.

19. Use the Mazda diagnostic scan tool, or equivalent and reprogram the required systems.

ENGINE COOLING

ENGINE FAN

REMOVAL & INSTALLATION

2009 Vehicles

See Figures 32 and 33.

> ※ **CAUTION**
>
> **Remove and install all parts when the engine is cold, otherwise they can cause severe burns or serious injury.**

1. Before servicing the vehicle, refer to the Precautions Section.

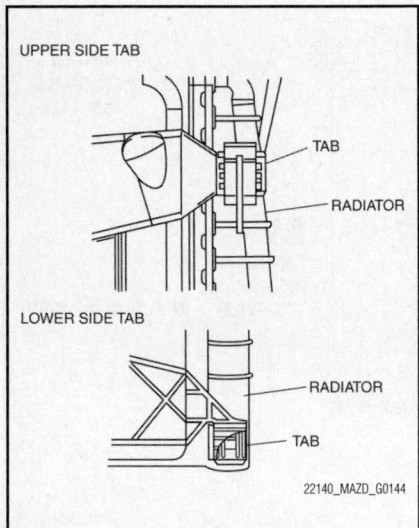

UPPER SIDE TAB

TAB

RADIATOR

LOWER SIDE TAB

RADIATOR

TAB

22140_MAZD_G0144

Fig. 32 Upper and lower side tabs shown—2009 vehicles

➡️**If working near and/or around the SRS system and components, be sure to disable the SRS system. Tape the negative battery cable with insulating tape. Always disconnect the negative battery cable first.**

> ※ **CAUTION**
>
> **To avoid personal injury when working on vehicles equipped with an air bag, the negative battery cable must be disconnected and at least one minute must elapse before working on the system. Failure to do so may result in deployment of the air bag.**

2. Remove the battery top cover, if equipped.
3. Disconnect the negative battery cable. Tape the cable with insulating tape.

➡️**When disconnecting the cable, some systems need to be initialized after the cable is reconnected. You will need Mazda diagnostic scan tool or equivalent. Follow the directions on the tool.**

4. Remove the undercover.
5. Drain and recycle the engine coolant.
6. Disconnect the coolant reservoir hose from the radiator.
7. Position the front wiring harnesses out of the way.
8. Remove the fan control module connector.
9. Unlock the cooling fan component lower side tabs a at two points by pressing

the cooling fan side tabs, remove upper side tabs, then remove the cooling fan component from the radiator

> ***To install:***

➡️**Be sure to use new fasteners, as required.**

10. Install the components in the reverse of removal.
11. Fill the cooling system and check for leaks.
12. Use the Mazda diagnostic scan tool, or equivalent and reprogram the required systems.

2010 Vehicles

> ※ **CAUTION**
>
> **Never remove the cooling system cap or loosen the radiator drain plug while the engine is running, or when the engine and radiator are hot. Scalding engine coolant and steam may shoot out and cause serious injury. It may also damage the engine and cooling system.**

1. Before servicing the vehicle, refer to the Precautions Section.

➡️**If working near and/or around the SRS system and components, be sure to disable the SRS system. Tape the negative battery cable with insulating tape. Always disconnect the negative battery cable first.**

> ※ **CAUTION**
>
> **To avoid personal injury when working on vehicles equipped with an air bag, the negative battery cable must be disconnected and at least one minute must elapse before working on the system. Failure to do so may result in deployment of the air bag.**

2. Disconnect the negative battery cable. Tape the cable with insulating tape.

➡️**When disconnecting the cable, some systems need to be initialized after the cable is reconnected. You will need the Mazda diagnostic scan tool or equivalent. Follow the directions on the tool.**

3. Turn off the engine and wait until it is cool. Even then, be very careful when removing the cap. Wrap a thick cloth around it and slowly turn it counterclock-

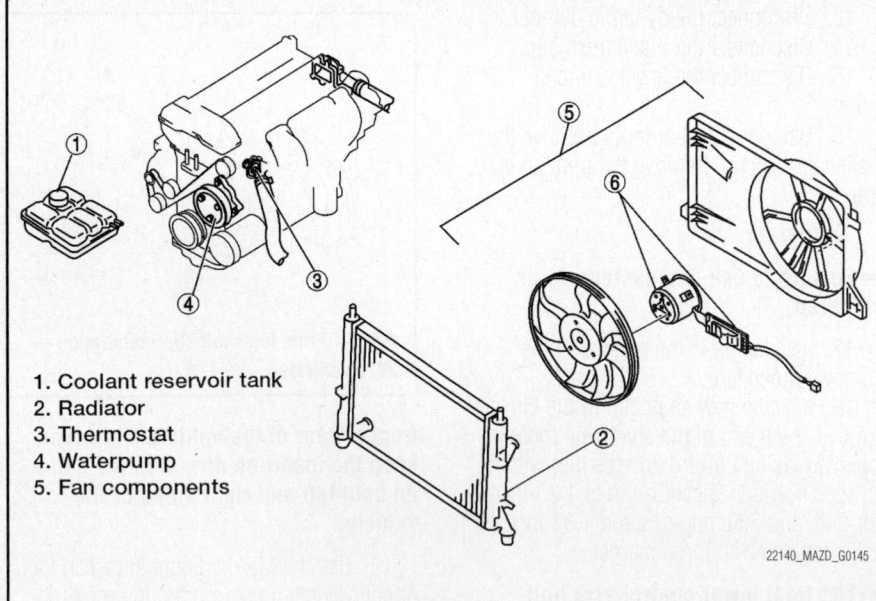

1. Coolant reservoir tank
2. Radiator
3. Thermostat
4. Waterpump
5. Fan components

22140_MAZD_G0145

Fig. 33 Cooling system index view—2009 vehicles

wise to the first stop. Step back while the pressure escapes.

4. When you are sure all the pressure is gone, press down on the cap using the cloth, turn it, and remove it.

5. Drain the coolant. Be sure to properly dispose of used coolant.

6. Remove the aerodynamic cover No. 2.

7. Remove the air cleaner assembly.

8. Remove the oil dipstick, if equipped with 2.3L engine with turbocharger.

9. Remove the connector.

10. Remove the coolant reserve tank hose.

11. Disconnect the electrical connectors.

12. Remove the air cleaner component bracket.

13. Remove the cooling fan component from its mounting.

To install:

➡**Be sure to use new fasteners, as required.**

14. Installation is the reverse of the removal procedure.

15. Fill the cooling system with the proper grade and type coolant.

16. Start the engine and check for leaks.

17. Use the Mazda diagnostic scan tool, or equivalent and reprogram the required systems.

RADIATOR

REMOVAL & INSTALLATION

2009 Vehicles

See Figure 34.

✳✳ CAUTION

Remove and install all parts when the engine is cold, otherwise they can cause severe burns or serious injury.

1. Before servicing the vehicle, refer to the Precautions Section.

➡**If working near and/or around the SRS system and components, be sure to disable the SRS system. Tape the negative battery cable with insulating tape. Always disconnect the negative battery cable first.**

✳ CAUTION

To avoid personal injury when working on vehicles equipped with an air bag, the negative battery cable must be disconnected and at least one minute must elapse before working

on the system. Failure to do so may result in deployment of the air bag.

2. Remove the battery top cover, if equipped.

3. Disconnect the negative battery cable. Tape the cable with insulating tape.

➡**When disconnecting the cable, some systems need to be initialized after the cable is reconnected. You will need Mazda diagnostic scan tool or equivalent. Follow the directions on the tool.**

4. Remove the undercover.

5. Drain and recycle the engine coolant.

6. Disconnect the coolant reservoir hose from the radiator.

7. Position the front wiring harnesses out of the way.

8. Remove the fan control module connector.

9. Unlock the cooling fan component lower side tabs a at two points by pressing the cooling fan side tabs, remove upper side tabs, then remove the cooling fan component from the radiator

10. Disconnect the radiator hoses.

11. Remove the radiator mount.

12. Remove the radiator as follows:

a. Remove the condenser from the radiator with the pipes still connected, by pressing the radiator side tabs to unlock tabs b on the condenser.

b. Remove rubber mount A from the mount installation hole.

c. Tilt the radiator to the engine side.

d. Remove rubber mount B from the mount installation hole.

e. Remove the radiator from the bottom of the vehicle.

To install:

➡**Be sure to use new fasteners, as required.**

13. Install the radiator, mount and hoses.

14. Install the condenser to the radiator by aligning lower side tabs with the radiator side tabs, install upper tabs b, then install lower side tabs.

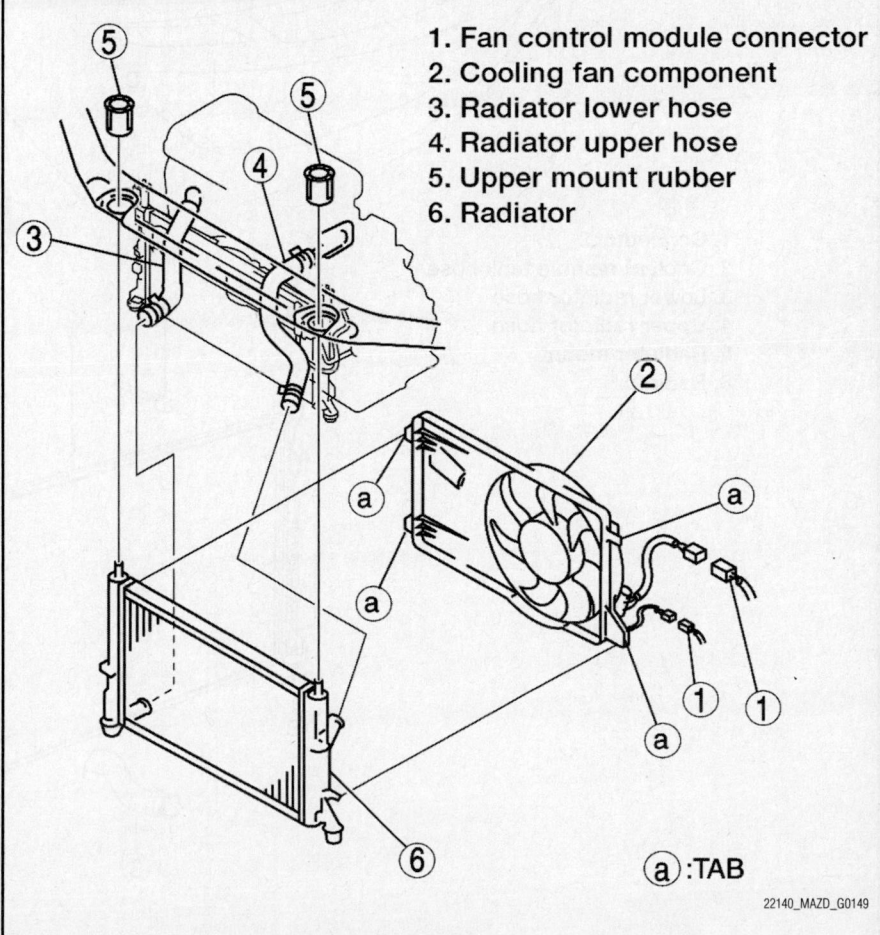

1. Fan control module connector
2. Cooling fan component
3. Radiator lower hose
4. Radiator upper hose
5. Upper mount rubber
6. Radiator

(a) :TAB

22140_MAZD_G0149

Fig. 34 Radiator and related components—2009 vehicles

15. Insert the cooling fan component tabs to the radiator to install the cooling fan component.

16. Install the remaining components in the reverse of removal.

17. Fill the cooling system and check for leaks.

18. Use the Mazda diagnostic scan tool, or equivalent and reprogram the required systems.

2010 Vehicles

See Figures 35 and 36.

※※ **CAUTION**

Never remove the cooling system cap or loosen the radiator drain plug while the engine is running, or when the engine and radiator are hot. Scalding engine coolant and steam

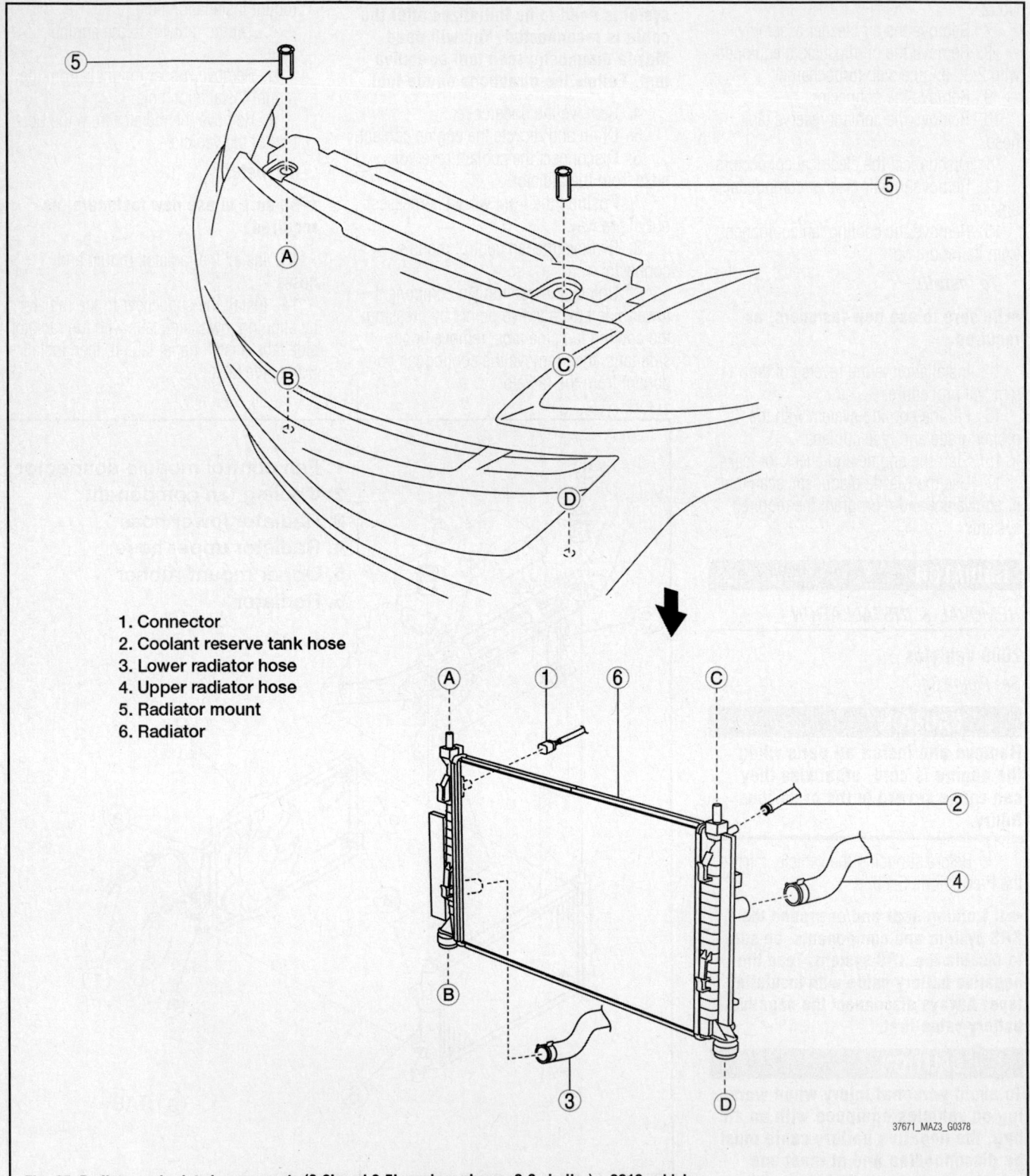

1. Connector
2. Coolant reserve tank hose
3. Lower radiator hose
4. Upper radiator hose
5. Radiator mount
6. Radiator

37671_MAZ3_G0378

Fig. 35 Radiator and related components (2.0L and 2.5L engines shown, 2.3 similar)—2010 vehicles

may shoot out and cause serious injury. It may also damage the engine and cooling system.

1. Before servicing the vehicle, refer to the Precautions Section.

➡If working near and/or around the SRS system and components, be sure to disable the SRS system. Tape the negative battery cable with insulating tape. Always disconnect the negative battery cable first.

✳✳ CAUTION

To avoid personal injury when working on vehicles equipped with an air bag, the negative battery cable must be disconnected and at least one minute must elapse before working on the system. Failure to do so may result in deployment of the air bag.

2. Disconnect the negative battery cable. Tape the cable with insulating tape.

➡When disconnecting the cable, some systems need to be initialized after the cable is reconnected. You will need the Mazda diagnostic scan tool or equivalent. Follow the directions on the tool.

3. Turn off the engine and wait until it is cool. Even then, be very careful when removing the cap. Wrap a thick cloth around it and slowly turn it counterclockwise to the first stop. Step back while the pressure escapes.

4. When you are sure all the pressure is gone, press down on the cap using the cloth, turn it, and remove it.

5. Drain the coolant. Be sure to properly dispose of used coolant.

6. Remove the aerodynamic cover No. 2.

7. Remove the plug hole plate, 2.0L and 2.5L engines.

8. Remove the air cleaner assembly.

9. Drain the engine coolant. Be sure to properly dispose of used coolant.

10. Disconnect the connector.

11. Remove the reservoir tank.

12. Remove the radiator hoses.

13. Remove the radiator mount.

14. Remove the radiator from its mounting.

To install:

➡Be sure to use new fasteners, as required.

15. Installation is the reverse of the removal procedure.

16. Fill the cooling system with the proper grade and type coolant.

17. Start the engine and check for leaks.

18. Use the Mazda diagnostic scan tool, or equivalent and reprogram the required systems.

THERMOSTAT

REMOVAL & INSTALLATION

2009 Vehicles

See Figures 37 and 38.

✳✳ CAUTION

Remove and install all parts when the engine is cold, otherwise they

can cause severe burns or serious injury.

1. Before servicing the vehicle, refer to the Precautions Section.

➡If working near and/or around the SRS system and components, be sure to disable the SRS system. Tape the negative battery cable with insulating tape. Always disconnect the negative battery cable first.

✳✳ CAUTION

To avoid personal injury when working on vehicles equipped with an air bag, the negative battery cable must be disconnected and at least one minute must elapse before working on the system. Failure to do so may result in deployment of the air bag.

2. Remove the battery top cover, if equipped.

3. Disconnect the negative battery cable. Tape the cable with insulating tape.

➡When disconnecting the cable, some systems need to be initialized after the cable is reconnected. You will need Mazda diagnostic scan tool or equivalent. Follow the directions on the tool.

4. Remove the undercover and splash shield as a single unit.

5. Drain and recycle the engine coolant.

6. Position the coolant reserve tank out of the way.

7. Remove the plug hole plate by lifting off and removing the plug hole plate from the areas shown in the accompanying illustration.

8. Position the drive belt out of the way.

9. Remove the drive belt tensioner.

10. Disconnect the water hose and lower radiator hose from the thermostat.

11. Remove the thermostat retainers, thermostat housing, thermostat and gasket.

To install:

➡Be sure to use new fasteners, as required.

12. Installation is the reverse of removal. Use a new thermostat housing gasket and tighten the thermostat housing retainers to 71–101 inch lbs. (8–11.5 Nm).

13. Fill the cooling system and check for leaks.

14. Use the Mazda diagnostic scan tool, or equivalent and reprogram the required systems.

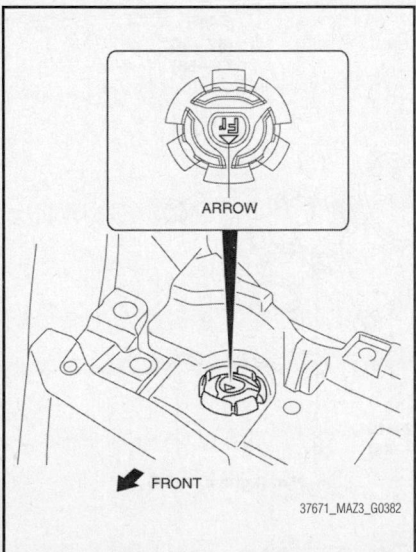

Fig. 36 Radiator mount installation—2010 vehicles

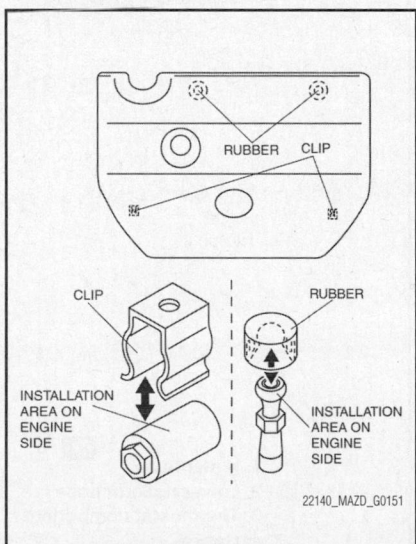

Fig. 37 Plug hole plate locations—2009 vehicles

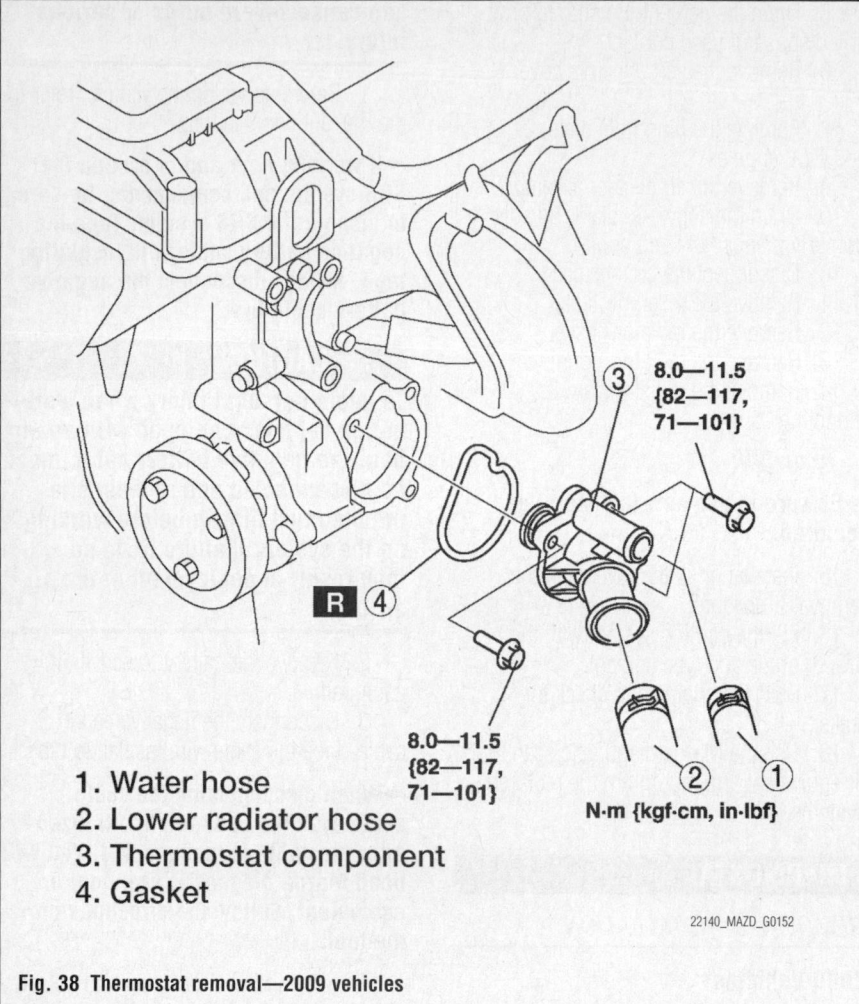

1. Water hose
2. Lower radiator hose
3. Thermostat component
4. Gasket

8.0—11.5
{82—117,
71—101}

8.0—11.5
{82—117,
71—101}

N·m {kgf·cm, in·lbf}

22140_MAZD_G0152

Fig. 38 Thermostat removal—2009 vehicles

bag, the negative battery cable must be disconnected and at least one minute must elapse before working on the system. Failure to do so may result in deployment of the air bag.

2. Disconnect the negative battery cable. Tape the cable with insulating tape.

➡When disconnecting the cable, some systems need to be initialized after the cable is reconnected. You will need the Mazda diagnostic scan tool or equivalent. Follow the directions on the tool.

3. Turn off the engine and wait until it is cool. Even then, be very careful when removing the cap. Wrap a thick cloth around it and slowly turn it counterclockwise to the first stop. Step back while the pressure escapes.

4. When you are sure all the pressure is gone, press down on the cap using the cloth, turn it, and remove it.

5. Drain the coolant. Be sure to properly dispose of used coolant.

6. Remove the plug hole plate.

7. Remove the drive belt. Remove the auto tensioner, 2.0L and 2.5L engines. Remove the idler pulley.

8. Remove the retaining bolts.

2010 Vehicles

See Figure 39.

✴ **CAUTION**

Never remove the cooling system cap or loosen the radiator drain plug while the engine is running, or when the engine and radiator are hot. Scalding engine coolant and steam may shoot out and cause serious injury. It may also damage the engine and cooling system.

1. Before servicing the vehicle, refer to the Precautions Section.

➡If working near and/or around the SRS system and components, be sure to disable the SRS system. Tape the negative battery cable with insulating tape. Always disconnect the negative battery cable first.

✴ **CAUTION**

To avoid personal injury when working on vehicles equipped with an air

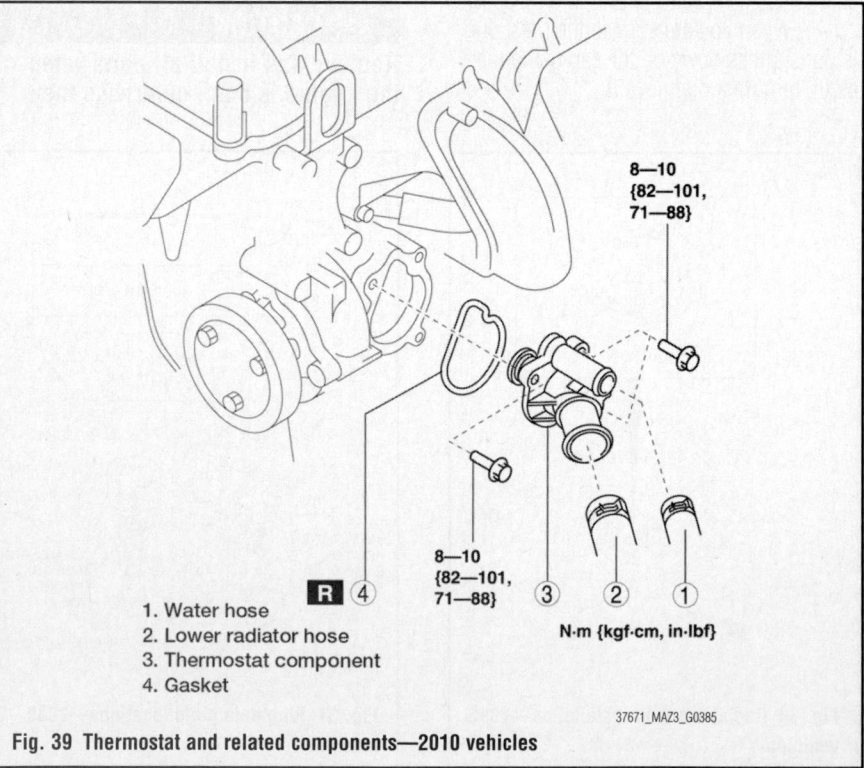

1. Water hose
2. Lower radiator hose
3. Thermostat component
4. Gasket

8—10
{82—101,
71—88}

8—10
{82—101,
71—88}

N·m {kgf·cm, in·lbf}

37671_MAZ3_G0385

Fig. 39 Thermostat and related components—2010 vehicles

9. Remove the component from its mounting.

10. Discard the gasket.

To install:

➡**Be sure to use new fasteners, as required.**

11. Installation is the reverse of the removal procedure.

12. Be sure to use a gasket.

13. Fill the cooling system with the proper grade and type coolant.

14. Start the engine and check for leaks.

15. Use the Mazda diagnostic scan tool, or equivalent and reprogram the required systems.

WATER PUMP

REMOVAL & INSTALLATION

2009 Vehicles

See Figure 40.

1. Before servicing the vehicle, refer to the Precautions Section.

➡**If working near and/or around the SRS system and components, be sure to disable the SRS system. Tape the negative battery cable with insulating tape. Always disconnect the negative battery cable first.**

❊❊ CAUTION

To avoid personal injury when working on vehicles equipped with an air bag, the negative battery cable must be disconnected and at least one minute must elapse before working on the system. Failure to do so may result in deployment of the air bag.

---

2. Remove the battery top cover, if equipped.

3. Disconnect the negative battery cable. Tape the cable with insulating tape.

➡**When disconnecting the cable, some systems need to be initialized after the cable is reconnected. You will need Mazda diagnostic scan tool or equivalent. Follow the directions on the tool.**

4. Remove the undercover and splash shield as an assembly.

5. Drain the cooling system.

6. Position the coolant reservoir tank aside with the hose still attached.

7. Remove the plug hole plate.

8. Loosen the water pump pulley bolt and position the drive belt aside.

9. Remove the water pump pulley.

10. Remove the water pump bolts, the pump and the O-ring.

To install:

➡**Be sure to use new fasteners, as required.**

11. Clean the water pump mating surfaces.

12. Install a new O-ring and the water pump. Tighten the bolts to 71-101 inch lbs. (8-11 Nm).

13. Install the water pump pulley and tighten the tighten the bolts to 12-17 ft. lbs. (17-23 Nm).

14. Install the drive belt.

15. Install the plug hole plate

16. Install the coolant reservoir tank.

17. Install the splash shield and undercover.

18. Connect the battery cable and install the cover.

19. Fill the cooling system.

20. Start the engine, check for leaks and repair if necessary.

21. Use the Mazda diagnostic scan tool, or equivalent and reprogram the required systems.

2010 Vehicles

See Figure 41.

❊❊ CAUTION

Never remove the cooling system cap or loosen the radiator drain plug while the engine is running, or when the engine and radiator are hot. Scalding engine coolant and steam may shoot out and cause serious injury. It may also damage the engine and cooling system.

---

1. Before servicing the vehicle, refer to the Precautions Section.

➡**If working near and/or around the SRS system and components, be sure to disable the SRS system. Tape the negative battery cable with insulating tape. Always disconnect the negative battery cable first.**

❊❊ CAUTION

To avoid personal injury when working on vehicles equipped with an air bag, the negative battery cable must be disconnected and at least one minute must elapse before working on the system. Failure to do so may result in deployment of the air bag.

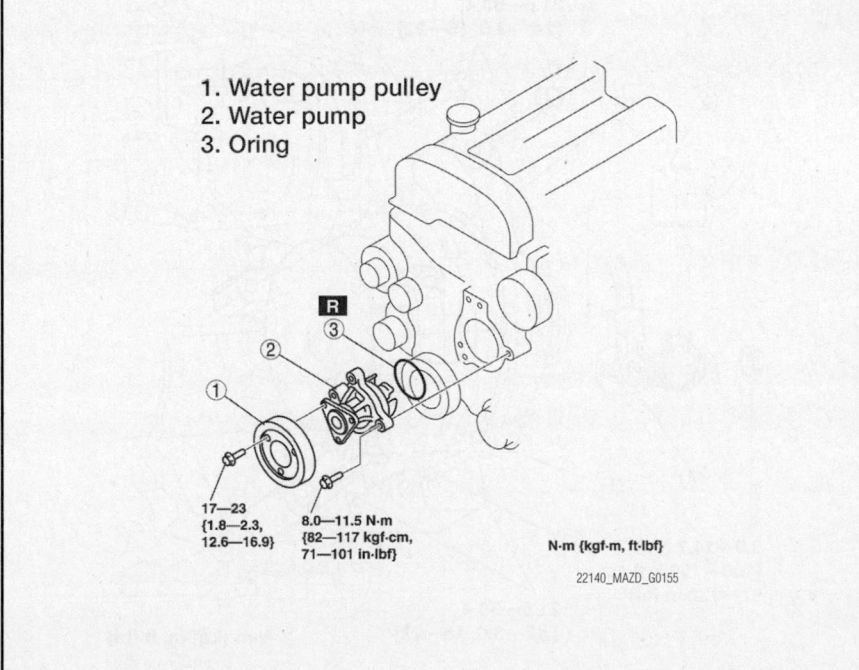

1. Water pump pulley
2. Water pump
3. Oring

17—23
{1.8—2.3,
12.6—16.9}

8.0—11.5 N·m
{82—117 kgf-cm,
71—101 in-lbf}

N·m {kgf·m, ft-lbf}

22140_MAZD_G0155

Fig. 40 Water pump mounting and related components—2009 vehicles

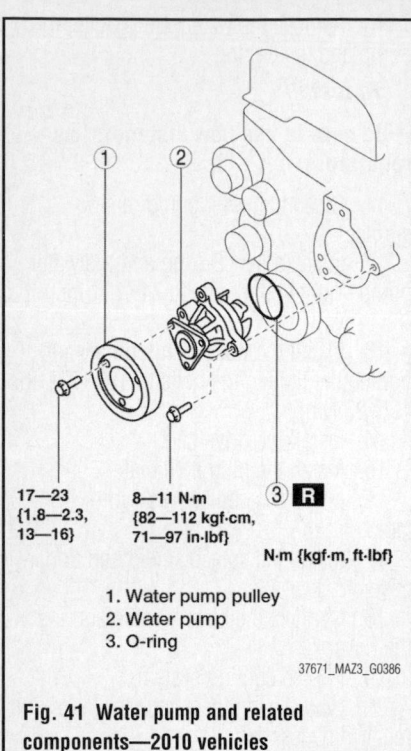

17—23
{1.8—2.3,
13—16}

8—11 N·m
{82—112 kgf·cm,
71—97 in·lbf}

③ **R**

N·m {kgf·m, ft·lbf}

1. Water pump pulley
2. Water pump
3. O-ring

37671_MAZ3_G0386

Fig. 41 Water pump and related components—2010 vehicles

2. Disconnect the negative battery cable. Tape the cable with insulating tape.

➡**When disconnecting the cable, some systems need to be initialized after the cable is reconnected. You will need the Mazda diagnostic scan tool or equivalent. Follow the directions on the tool.**

3. Turn off the engine and wait until it is cool. Even then, be very careful when removing the cap. Wrap a thick cloth around it and slowly turn it counterclockwise to the first stop. Step back while the pressure escapes.

4. When you are sure all the pressure is gone, press down on the cap using the cloth, turn it, and remove it.

5. Remove the plug hole plate.

6. Remove the No. 2 undercover and splash shield as a single unit.

7. Remove the alternator belt with the A/C belt still installed. Position it out of the way.

8. Remove the drive belt. Remove the auto tensioner, 2.0L and 2.5L engines. Remove the idler pulley.

9. Drain the engine coolant.

10. Remove the water pump pulley.

11. Remove the pump retaining bolts.

12. Remove the pump from its mounting.

13. Discard the O-ring.

To install:

➡**Be sure to use new fasteners, as required.**

14. Installation is the reverse of the removal procedure.

15. Be sure to use a new O-ring. Coat the O-ring with clean engine coolant.

16. Fill the cooling system with the proper grade and type coolant.

17. Start the engine and check for leaks.

18. Use the Mazda diagnostic scan tool, or equivalent and reprogram the required systems.

ENGINE ELECTRICAL

■ ALTERNATOR

REMOVAL & INSTALLATION

2009 Vehicles

2.0L & 2.3L Engines

See Figure 42.

1. Before servicing the vehicle, refer to the Precautions Section.

➡**If working near and/or around the SRS system and components, be sure to disable the SRS system. Tape the negative battery cable with insulating tape. Always disconnect the negative battery cable first.**

❖ CAUTION

To avoid personal injury when working on vehicles equipped with an air bag, the negative battery cable must be disconnected and at least one minute must elapse before working on the system. Failure to do so may result in deployment of the air bag.

2. Remove the battery top cover, if equipped.

3. Disconnect the negative battery cable. Tape the cable with insulating tape.

➡**When disconnecting the cable, some systems need to be initialized after the cable is reconnected. You will need Mazda diagnostic scan tool or**

CHARGING SYSTEM

equivalent. Follow the directions on the tool.

4. Remove the undercover and splash shield as an assembly.

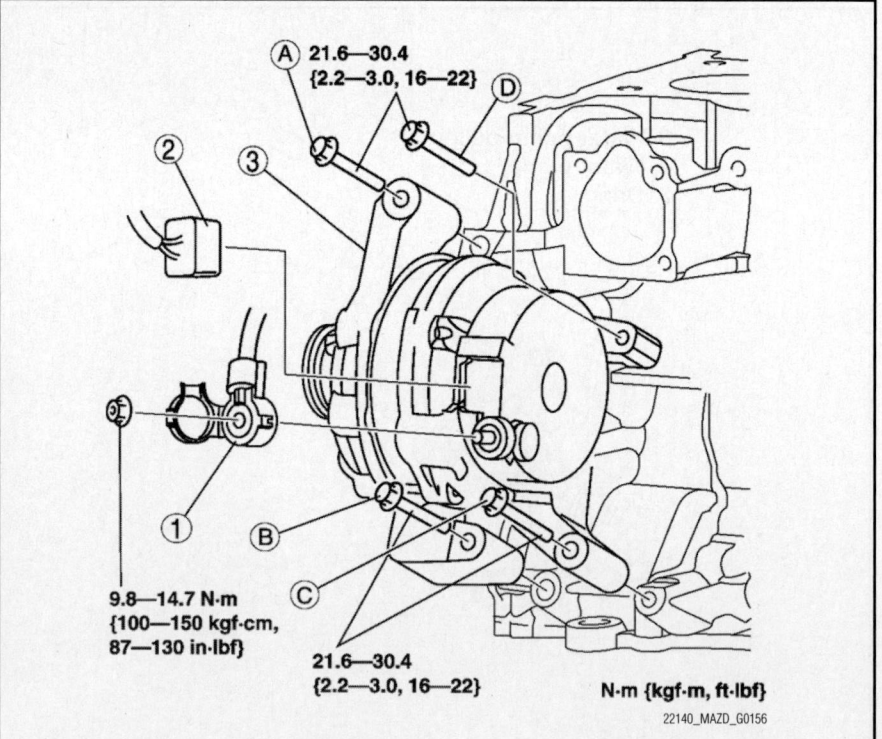

Ⓐ 21.6—30.4
{2.2—3.0, 16—22}

9.8—14.7 N·m
{100—150 kgf·cm,
87—130 in·lbf}

21.6—30.4
{2.2—3.0, 16—22}

N·m {kgf·m, ft·lbf}

22140_MAZD_G0156

Fig. 42 Alternator and mounting bolts A, B, C, and D locating points—2009 vehicles

5. Remove the plug hole plate.

6. Remove the drive belt.

7. Disconnect the alternator electrical connections.

8. Position the coolant overflow tank to one side to facilitate alternator removal.

9. Remove the bolts A, B, C, and D in order and the alternator. Refer to the illustration for bolt location.

To install:

10. Before servicing the vehicle, refer to the precautions section.

11. Place the alternator in position matching the fixing and engine side holes and hand tighten the bolts A, B, C and D in alphabetical order. Refer to the illustration for bolt locations.

12. Attach the alternator electrical connections. Tighten bolts A, B, C and D in alphabetical order to 16–22 ft. lbs. (21–30 Nm). Refer to the illustration for bolt locations.

13. Install the drive belt.

14. Install the undercover and splash shield.

15. Install the plug hole plate.

16. Connect the negative battery cable and install the battery cover.

17. Use the Mazda diagnostic scan tool, or equivalent and reprogram the required systems.

2.3L Engine With Turbocharger

See Figure 43.

1. Before servicing the vehicle, refer to the Precautions Section.

➡**If working near and/or around the SRS system and components, be sure to disable the SRS system. Tape the negative battery cable with insulating tape. Always disconnect the negative battery cable first.**

2. Remove the battery top cover, if equipped.

3. Disconnect the negative battery cable. Tape the cable with insulating tape.

➡**When disconnecting the cable, some systems need to be initialized after the cable is reconnected. You will need Mazda diagnostic scan tool or equivalent. Follow the directions on the tool.**

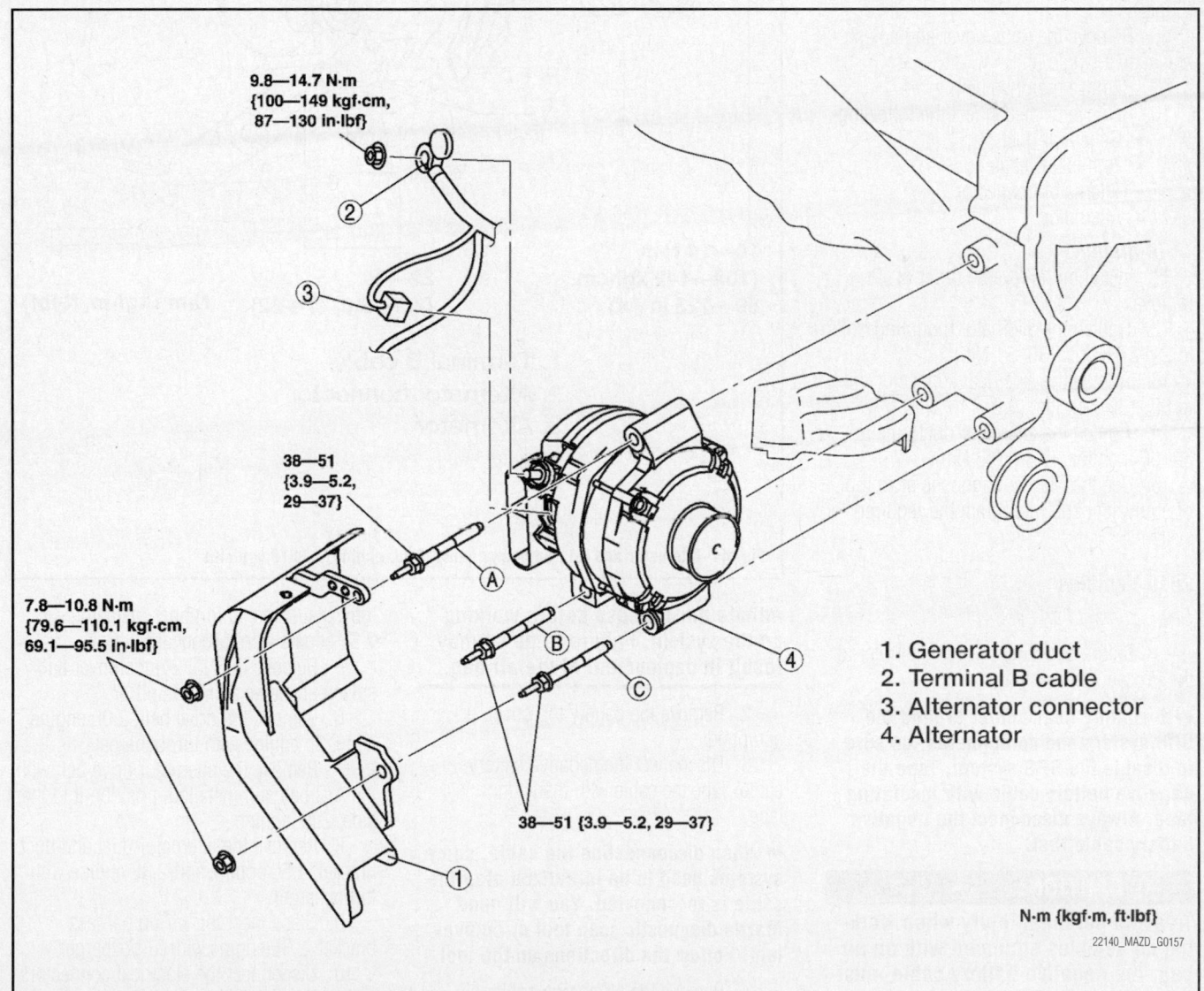

9.8—14.7 N·m
{100—149 kgf·cm,
87—130 in·lbf}

38—51
{3.9—5.2,
29—37}

7.8—10.8 N·m
{79.6—110.1 kgf·cm,
69.1—95.5 in·lbf}

38—51 {3.9—5.2, 29—37}

1. Generator duct
2. Terminal B cable
3. Alternator connector
4. Alternator

N·m {kgf·m, ft·lbf}

22140_MAZD_G0157

Fig. 43 Alternator and related components (2.3L engine with turbocharger)—2009 vehicles

When the battery cables are connected, touching the vehicle body with generator terminal B will generate sparks. This can cause personal injury, fire, and damage to the electrical components. Always disconnect the negative battery cable before performing the following operation.

➡The generator can be damaged by the heat from the exhaust manifold. Make sure the generator duct is installed securely.

4. Remove the battery cover.
5. Disconnect the negative battery cable.
6. Remove the charge air cooler and charge air cooler bracket.
7. Remove the heat insulator (body side) and exhaust manifold insulator (upper).
8. Remove the undercover and splash shield as a single unit.
9. Remove the drive belt.
10. Remove or disconnect the following:
- Generator duct
- Terminal B cable
- Alternator connector
- Alternator

To install:

11. Install in the reverse order of removal.
12. Tighten the alternator mounting bolts to 29–37 ft. lbs. (38–51 Nm).
13. Tighten the electrical connector nut to 87–130 inch lbs. (9.8–14.7 Nm).
14. Tighten the alternator duct nuts to 60–96 inch lbs. (7.8–10.8 Nm).
15. Use the Mazda diagnostic scan tool, or equivalent and reprogram the required systems.

2010 Vehicles

See Figures 44 and 45.

1. Before servicing the vehicle, refer to the Precautions Section.

➡If working near and/or around the SRS system and components, be sure to disable the SRS system. Tape the negative battery cable with insulating tape. Always disconnect the negative battery cable first.

To avoid personal injury when working on vehicles equipped with an air bag, the negative battery cable must be disconnected and at least one minute must elapse before working on the system. Failure to do so may result in deployment of the air bag.

2. Remove the battery top cover, if equipped.
3. Disconnect the negative battery cable. Tape the cable with insulating tape.

➡When disconnecting the cable, some systems need to be initialized after the cable is reconnected. You will need Mazda diagnostic scan tool or equivalent. Follow the directions on the tool.

4. Remove the air charge cooler air charge bracket, insulator (body side) and the exhaust manifold upper insulator, on 2.3L engine with turbocharger.
5. Remove the No. 2 undercover and splash shield as a single unit.
6. Remove the drive belt, 2.0L engine and 2.3L engine with turbocharger.
7. Remove the alternator drive belt with the A/C belt still installed, position it to the side, 2.5L engine.
8. Remove the alternator duct and duct bracket, 2.5L engine and 2.3L engine with turbocharger.
9. Disconnect the wiring harness bracket, 2.3L engine with turbocharger.
10. Disconnect the electrical connectors.
11. Remove the retaining bolts.

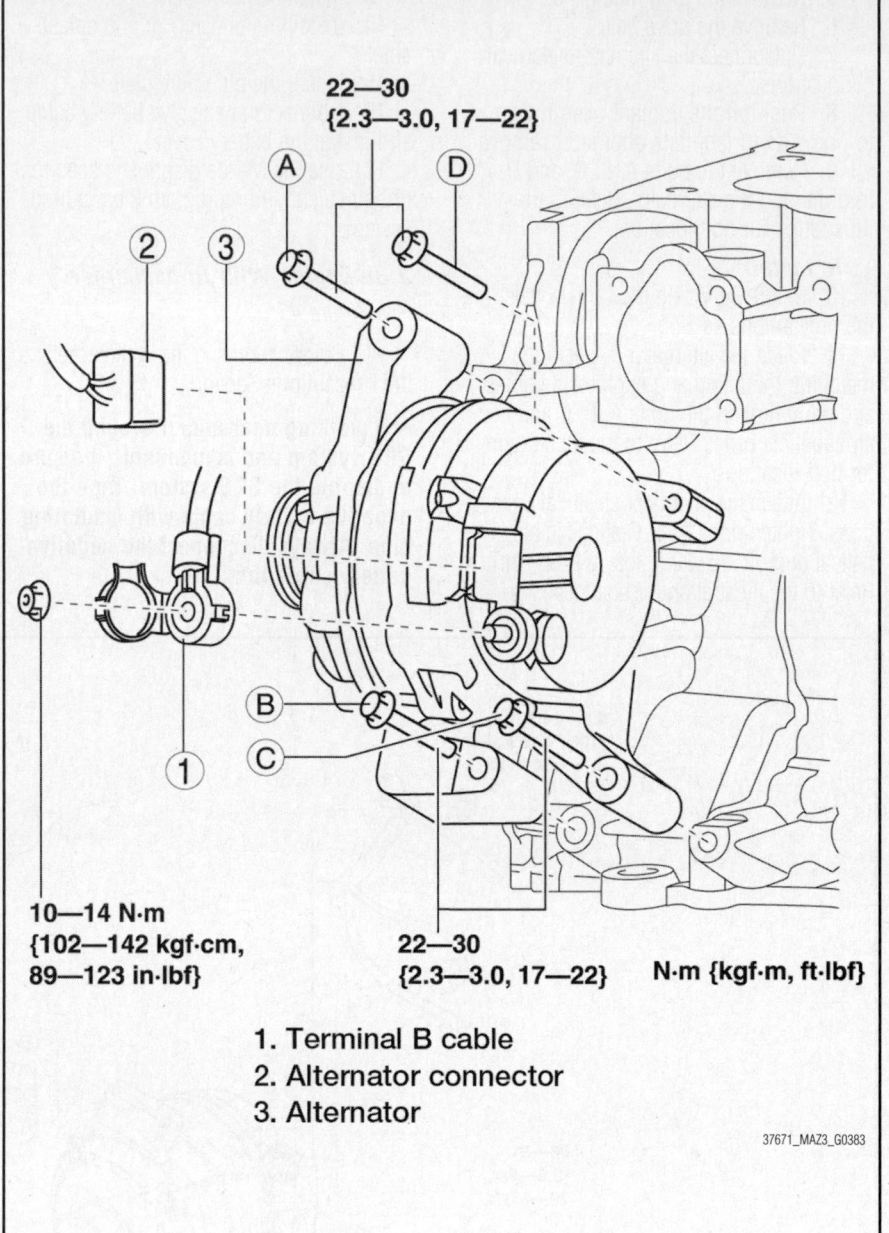

22—30
{2.3—3.0, 17—22}

10—14 N·m
{102—142 kgf·cm,
89—123 in·lbf}

22—30
{2.3—3.0, 17—22}

N·m {kgf·m, ft·lbf}

1. Terminal B cable
2. Alternator connector
3. Alternator

37671_MAZ3_G0383

Fig. 44 Alternator and related components (2.0L engine)—2010 vehicles

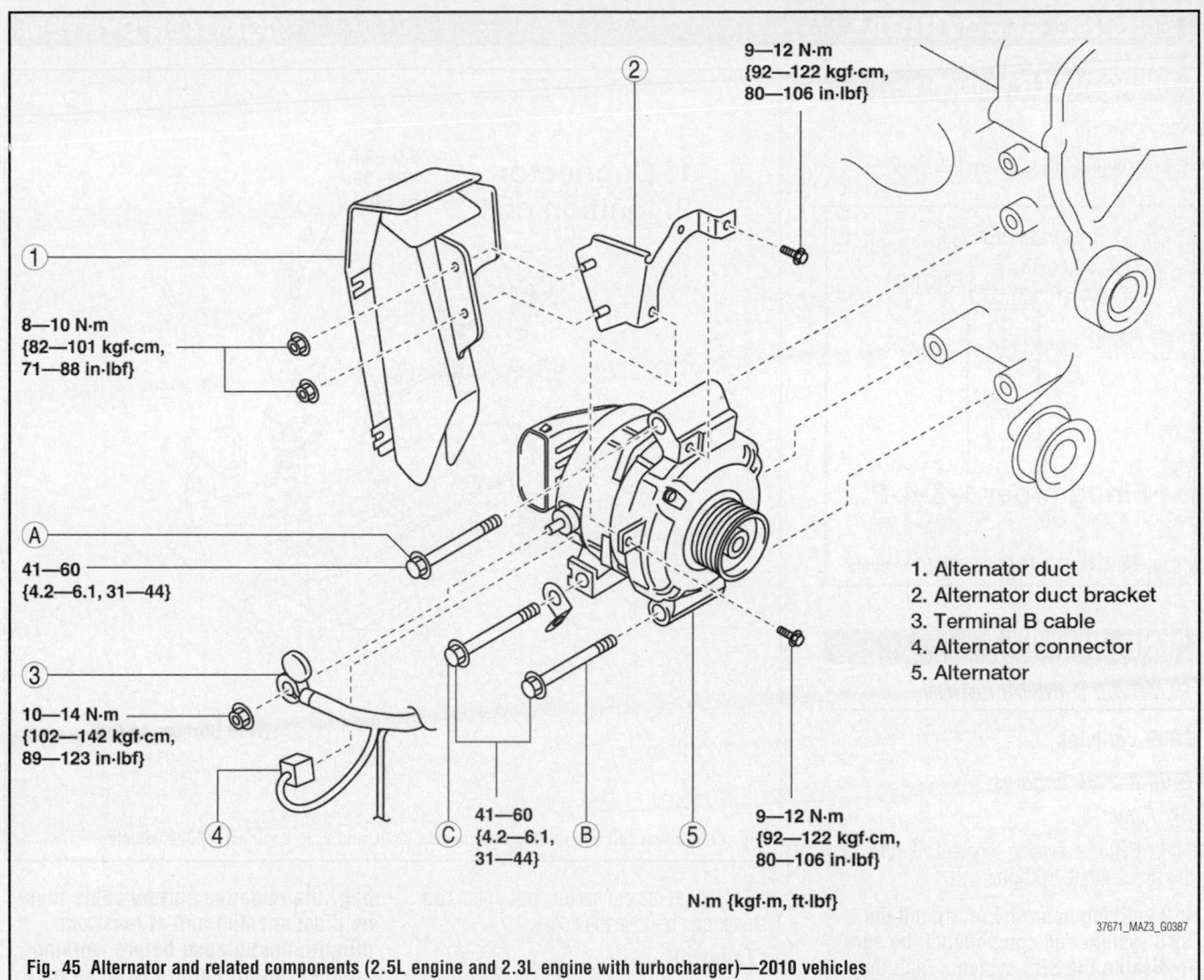

9—12 N·m
{92—122 kgf·cm,
80—106 in·lbf}

8—10 N·m
{82—101 kgf·cm,
71—88 in·lbf}

41—60
{4.2—6.1, 31—44}

10—14 N·m
{102—142 kgf·cm,
89—123 in·lbf}

41—60
{4.2—6.1,
31—44}

9—12 N·m
{92—122 kgf·cm,
80—106 in·lbf}

N·m {kgf·m, ft·lbf}

1. Alternator duct
2. Alternator duct bracket
3. Terminal B cable
4. Alternator connector
5. Alternator

37671_MAZ3_G0387

Fig. 45 Alternator and related components (2.5L engine and 2.3L engine with turbocharger)—2010 vehicles

12. Remove the component from its mounting.

To install:

➡**Be sure to use new fasteners, as required.**

13. Installation is the reverse of the removal procedure.

14. On 2.0L engine, align the alternator retaining hole with the engine side hole, then temporarily tighten the retaining bolts in the order shown in the illustration. Final tighten in the order shown in the illustration.

15. On the 2.5L engine and 2.3L engine with turbocharger temporarily

tighten bolt A. Tighten bolts B, C and D to specification. Final tighten bolt A to specification. See illustration for tightening sequence.

16. Use the Mazda diagnostic scan tool, or equivalent and reprogram the required systems.

FIRING ORDER

See Figure 46.

2.3L engine firing order 1–3–4–2

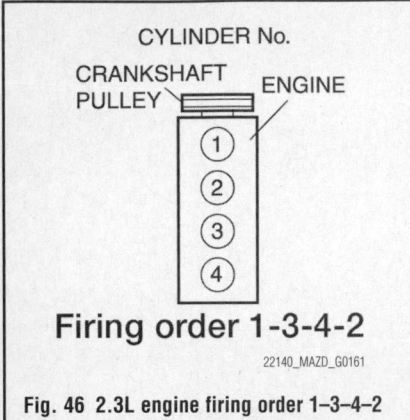

Fig. 46 2.3L engine firing order 1–3–4–2

IGNITION COIL

REMOVAL & INSTALLATION

2009 Vehicles

2.0L & 2.3L Engines

See Figure 47.

1. Before servicing the vehicle, refer to the Precautions Section.

➡ **If working near and/or around the SRS system and components, be sure to disable the SRS system. Tape the negative battery cable with insulating tape. Always disconnect the negative battery cable first.**

> ☀️ **CAUTION**
>
> **To avoid personal injury when working on vehicles equipped with an air bag, the negative battery cable must be disconnected and at least one minute must elapse before working on the system. Failure to do so may result in deployment of the air bag.**

2. Remove the battery top cover, if equipped.
3. Disconnect the negative battery cable. Tape the cable with insulating tape.

➡ **When disconnecting the cable, some systems need to be initialized after the cable is reconnected. You will need Mazda diagnostic scan tool or equivalent. Follow the directions on the tool.**

4. Remove the plug hole plate by lifting off and removing the plug hole plate.

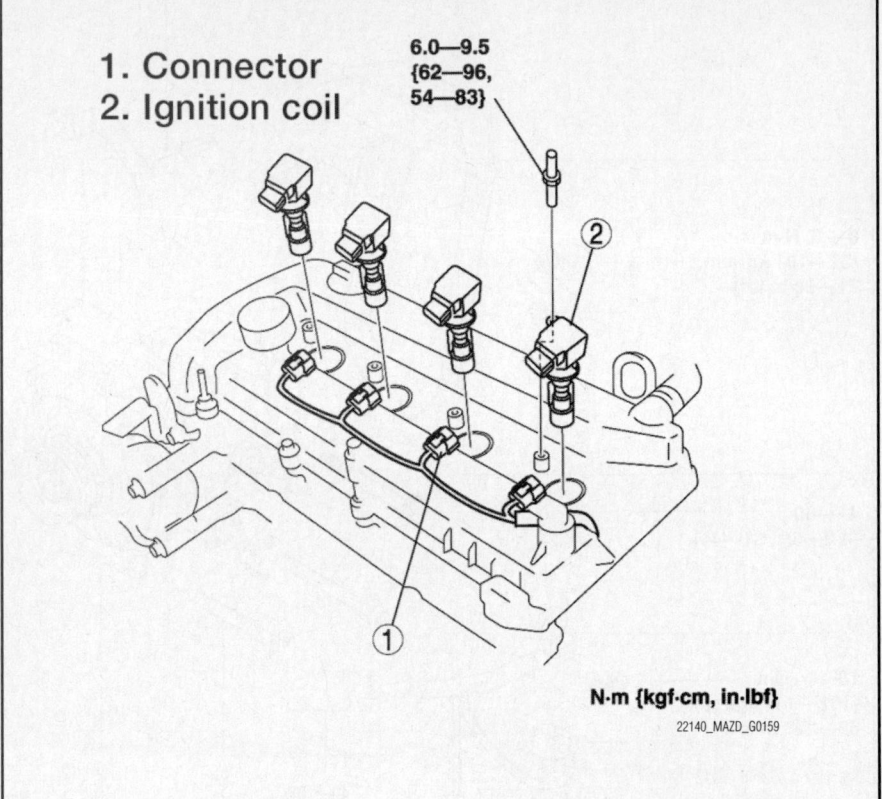

1. Connector
2. Ignition coil

6.0—9.5
{62—96,
54—83}

N·m {kgf·cm, in·lbf}

Fig. 47 Ignition coil and related components (2.0L and 2.3L engines)—2009 vehicles

5. Unplug the connector, loosen the coil fastener and remove the coil.

To install:

➡ **Be sure to use new fasteners, as required.**

6. Installation is the reverse of the removal procedure.
7. Tighten the fastener to 54 inch lbs. (83 Nm).
8. Use the Mazda diagnostic scan tool, or equivalent and reprogram the required systems.

2.3L Engine With Turbocharger

1. Before servicing the vehicle, refer to the Precautions Section.

➡ **If working near and/or around the SRS system and components, be sure to disable the SRS system. Tape the negative battery cable with insulating tape. Always disconnect the negative battery cable first.**

> ☀️ **CAUTION**
>
> **To avoid personal injury when working on vehicles equipped with an air bag, the negative battery cable must be disconnected and at least one minute must elapse before working on the system. Failure to do so may result in deployment of the air bag.**

2. Remove the battery top cover, if equipped.
3. Disconnect the negative battery cable. Tape the cable with insulating tape.

➡ **When disconnecting the cable, some systems need to be initialized after the cable is reconnected. You will need Mazda diagnostic scan tool or equivalent. Follow the directions on the tool.**

4. Remove the charge air cooler.
5. Unplug the connector, loosen the coil fastener and remove the coil.

To install:

➡ **Be sure to use new fasteners, as required.**

6. Installation is the reverse of the removal procedure.
7. Tighten the fastener to 54 inch lbs. (8 Nm).
8. Use the Mazda diagnostic scan tool,

or equivalent and reprogram the required systems.

2010 Vehicles

See Figures 48 and 49.

1. Before servicing the vehicle, refer to the Precautions Section.

➡ **If working near and/or around the SRS system and components, be sure to disable the SRS system. Tape the negative battery cable with insulating tape. Always disconnect the negative battery cable first.**

✳✳ CAUTION

To avoid personal injury when working on vehicles equipped with an air bag, the negative battery cable must be disconnected and at least one minute must elapse before working on the system. Failure to do so may result in deployment of the air bag.

2. Remove the battery top cover, if equipped.
3. Disconnect the negative battery cable. Tape the cable with insulating tape.

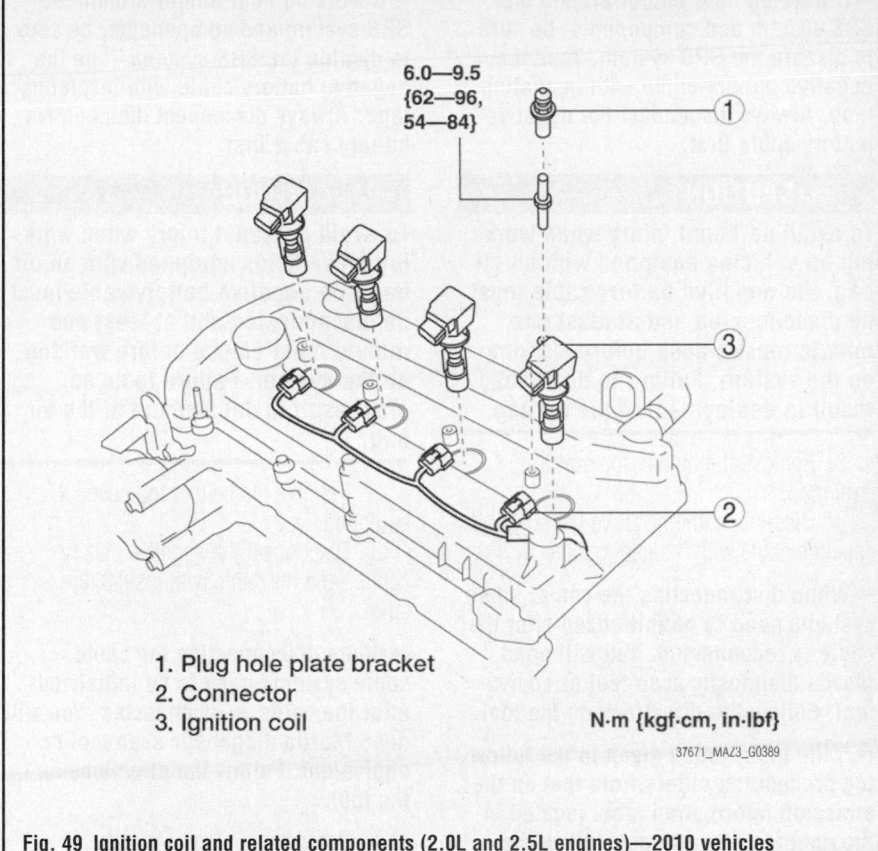

1. Plug hole plate bracket
2. Connector
3. Ignition coil

N·m {kgf·cm, in·lbf}

37671_MAZ3_G0389

Fig. 49 Ignition coil and related components (2.0L and 2.5L engines)—2010 vehicles

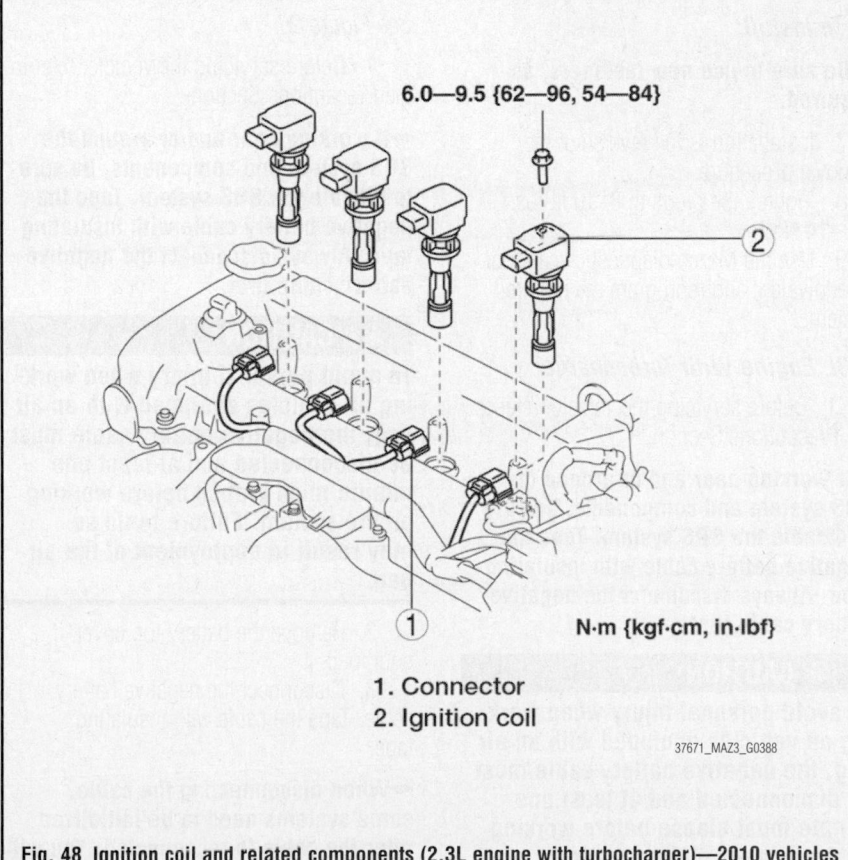

1. Connector
2. Ignition coil

N·m {kgf·cm, in·lbf}

37671_MAZ3_G0388

Fig. 48 Ignition coil and related components (2.3L engine with turbocharger)—2010 vehicles

➡ **When disconnecting the cable, some systems need to be initialized after the cable is reconnected. You will need Mazda diagnostic scan tool or equivalent. Follow the directions on the tool.**

4. Remove the air charge cooler, 2.3L engine with turbocharger.
5. Remove the plug hole plate, 2.0L and 2.5L engines
6. Disconnect the electrical connectors.
7. Remove the coils from their mountings.

To install:

➡ **Be sure to use new fasteners, as required.**

8. Installation is the reverse of the removal procedure.
9. Use the Mazda diagnostic scan tool, or equivalent and reprogram the required systems.

IGNITION TIMING

ADJUSTMENT

1. Before servicing the vehicle, refer to the Precautions Section.

➡If working near and/or around the SRS system and components, be sure to disable the SRS system. Tape the negative battery cable with insulating tape. Always disconnect the negative battery cable first.

✳✳ CAUTION

To avoid personal injury when working on vehicles equipped with an air bag, the negative battery cable must be disconnected and at least one minute must elapse before working on the system. Failure to do so may result in deployment of the air bag.

2. Remove the battery top cover, if equipped.

3. Disconnect the negative battery cable. Tape the cable with insulating tape.

➡When disconnecting the cable, some systems need to be initialized after the cable is reconnected. You will need Mazda diagnostic scan tool or equivalent. Follow the directions on the tool.

➡If the information given in the following procedures differs from that on the emission information label located in the engine compartment, follow the directions given on the label. The label often reflects production changes made during the model year.

The timing is controlled by the computer. Ignition timing adjustment is not possible or necessary.

4. If the timing is still not within specification. the following components may be defective:

- Camshaft position (CMP) sensor
- Crankshaft Position (CKP) sensor
- Throttle Position (TP) sensor
- Engine Coolant Temperature (ECT) sensor
- Neutral switch if equipped with a manual transaxle
- Clutch switch if equipped with a manual transaxle
- Transaxle range switch if equipped with an automatic transaxle

5. If the above components are normal, replace the Powertrain Control Module (PCM).

SPARK PLUGS

REMOVAL & INSTALLATION

2009 Vehicles

2.0L & 2.3L Engines

1. Before servicing the vehicle, refer to the Precautions Section.

➡If working near and/or around the SRS system and components, be sure to disable the SRS system. Tape the negative battery cable with insulating tape. Always disconnect the negative battery cable first.

✳✳ CAUTION

To avoid personal injury when working on vehicles equipped with an air bag, the negative battery cable must be disconnected and at least one minute must elapse before working on the system. Failure to do so may result in deployment of the air bag.

2. Remove the battery top cover, if equipped.

3. Disconnect the negative battery cable. Tape the cable with insulating tape.

➡When disconnecting the cable, some systems need to be initialized after the cable is reconnected. You will need Mazda diagnostic scan tool or equivalent. Follow the directions on the tool.

4. Remove the plug hole plate.
5. Remove the ignition coils.
6. Remove the spark plugs.

To install:

➡Be sure to use new fasteners, as required.

7. Installation is the reverse of removal procedure.

8. Tighten the plugs to 8–10 ft. lbs. (10–14 Nm).

9. Use the Mazda diagnostic scan tool, or equivalent and reprogram the required systems.

2.3L Engine With Turbocharger

1. Before servicing the vehicle, refer to the Precautions Section.

➡If working near and/or around the SRS system and components, be sure to disable the SRS system. Tape the negative battery cable with insulating tape. Always disconnect the negative battery cable first.

✳✳ CAUTION

To avoid personal injury when working on vehicles equipped with an air bag, the negative battery cable must be disconnected and at least one minute must elapse before working on the system. Failure to do so

may result in deployment of the air bag.

2. Remove the battery top cover, if equipped.

3. Disconnect the negative battery cable. Tape the cable with insulating tape.

➡When disconnecting the cable, some systems need to be initialized after the cable is reconnected. You will need Mazda diagnostic scan tool or equivalent. Follow the directions on the tool.

4. Remove the charge air cooler.
5. Remove the plug hole plate.
6. Remove the ignition coils.
7. Remove the spark plugs.

To install:

➡Be sure to use new fasteners, as required.

8. Installation is the reverse of the removal procedure.

9. Tighten the plugs to 8–10 ft. lbs. (10–14 Nm).

10. Use the Mazda diagnostic scan tool, or equivalent and reprogram the required systems.

2010 Vehicles

See Figure 25.

1. Before servicing the vehicle, refer to the Precautions Section.

➡If working near and/or around the SRS system and components, be sure to disable the SRS system. Tape the negative battery cable with insulating tape. Always disconnect the negative battery cable first.

✳✳ CAUTION

To avoid personal injury when working on vehicles equipped with an air bag, the negative battery cable must be disconnected and at least one minute must elapse before working on the system. Failure to do so may result in deployment of the air bag.

2. Remove the battery top cover, if equipped.

3. Disconnect the negative battery cable. Tape the cable with insulating tape.

➡When disconnecting the cable, some systems need to be initialized after the cable is reconnected. You will need Mazda diagnostic scan tool or equivalent. Follow the directions on the tool.

4. Remove the air charge cooler and ignition coils, 2.3L engine with turbocharger.

5. Remove the plug hole plate, 2.0L and 2.5L engines

6. Disconnect the electrical connectors.

7. Remove the coils from their mountings.

8. Remove the spark plugs.

To install:

➡**Be sure to use new fasteners, as required.**

9. Installation is the reverse of the removal procedure.

10. Use the Mazda diagnostic scan tool, or equivalent and reprogram the required systems.

ENGINE ELECTRICAL

STARTER

REMOVAL & INSTALLATION

2009 Vehicles

See Figure 50.

1. Before servicing the vehicle, refer to the Precautions Section.

➡**If working near and/or around the SRS system and components, be sure to disable the SRS system. Tape the negative battery cable with insulating tape. Always disconnect the negative battery cable first.**

❄ CAUTION

To avoid personal injury when working on vehicles equipped with an air bag, the negative battery cable must be disconnected and at least one minute must elapse before working on the system. Failure to do so may result in deployment of the air bag.

2. Remove the battery top cover, if equipped.

3. Disconnect the negative battery cable. Tape the cable with insulating tape.

➡**When disconnecting the cable, some systems need to be initialized after the cable is reconnected. You will need Mazda diagnostic scan tool or equivalent. Follow the directions on the tool.**

4. Remove the undercover.

STARTING SYSTEM

5. If equipped with a manual transaxle, remove the clutch release cylinder with the line still attached.

6. Remove the starter electrical connections and wiring harness bracket.

7. Remove the starter bolts and the starter.

To install:

➡**Be sure to use new fasteners, as required.**

8. Installation is the reverse of the removal procedure.

9. Tighten the wiring harness bracket retainers to 70–104 inch lbs. (8–11 Nm) and the starter bolts to 29–37 ft. lbs. (38–51 Nm).

10. Use the Mazda diagnostic scan tool, or equivalent and reprogram the required systems.

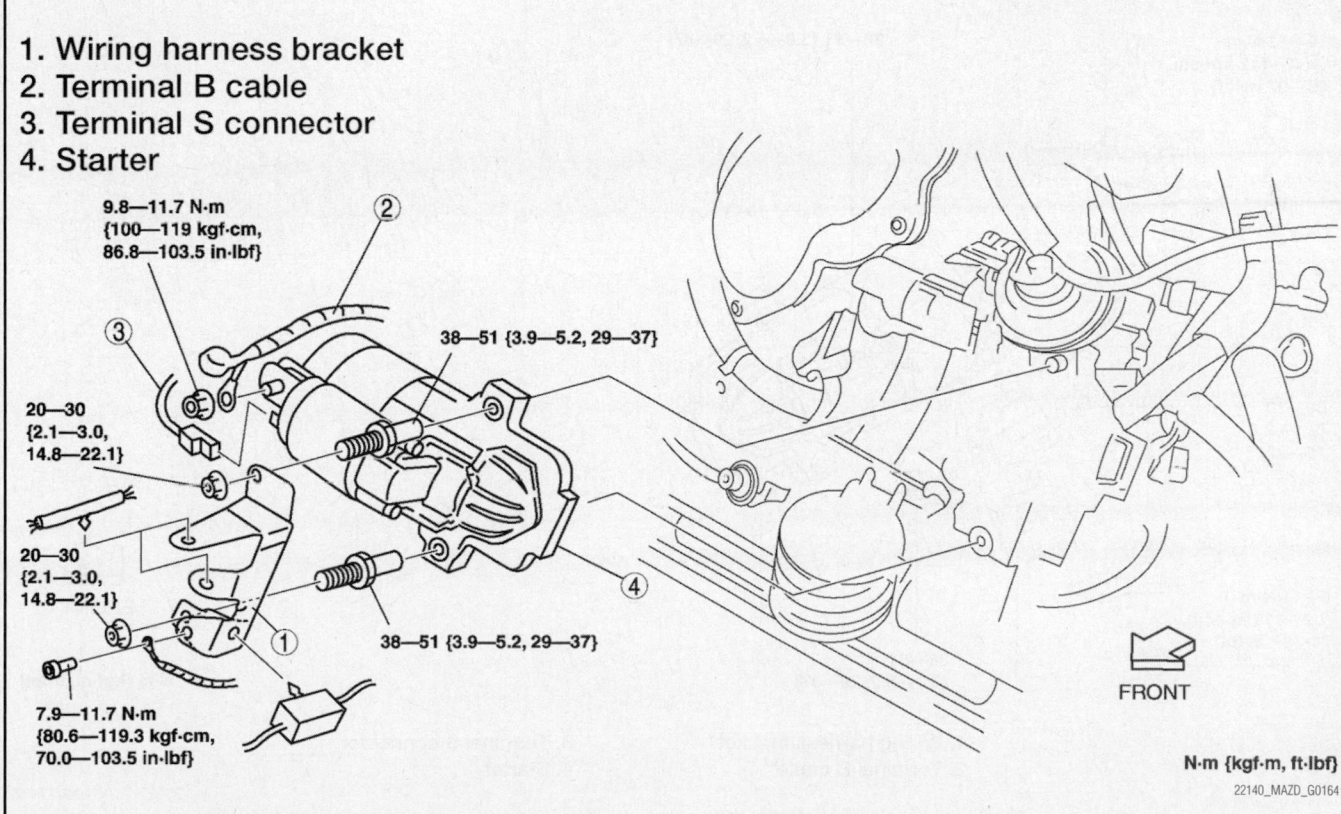

1. Wiring harness bracket
2. Terminal B cable
3. Terminal S connector
4. Starter

9.8—11.7 N·m {100—119 kgf·cm, 86.8—103.5 in·lbf}

38—51 {3.9—5.2, 29—37}

20—30 {2.1—3.0, 14.8—22.1}

20—30 {2.1—3.0, 14.8—22.1}

38—51 {3.9—5.2, 29—37}

7.9—11.7 N·m {80.6—119.3 kgf·cm, 70.0—103.5 in·lbf}

FRONT

N·m {kgf·m, ft·lbf}

22140_MAZD_G0164

Fig. 50 Starter and related components—2009 vehicles

2010 Vehicles

See Figures 51 and 52.

1. Before servicing the vehicle, refer to the Precautions Section.

➡ If working near and/or around the SRS system and components, be sure to disable the SRS system. Tape the negative battery cable with insulating tape. Always disconnect the negative battery cable first.

✷✷ CAUTION

To avoid personal injury when working on vehicles equipped with an air bag, the negative battery cable must be disconnected and at least one minute must elapse before working on the system. Failure to do so may result in deployment of the air bag.

2. Remove the battery top cover, if equipped.

3. Disconnect the negative battery cable. Tape the cable with insulating tape.

➡ When disconnecting the cable, some systems need to be initialized after the cable is reconnected. You will need

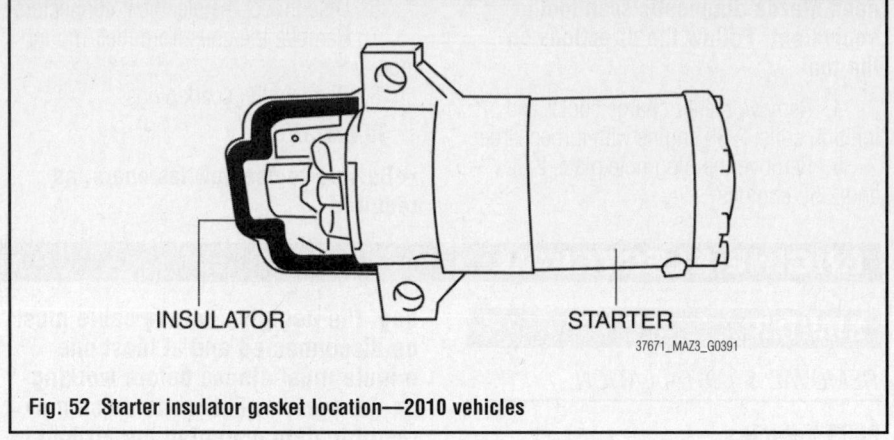

Fig. 52 Starter insulator gasket location—2010 vehicles

INSULATOR   STARTER

37671_MAZ3_G0391

Mazda diagnostic scan tool or equivalent. Follow the directions on the tool.

4. Remove the battery, battery tray, PCM, air cleaner, air charge cooler cover, and duct on 2.3L engine with turbocharger.

5. Remove the No. 2 undercover.

6. Remove the wiring harness bracket.

7. Remove the heater pipe, on 2.3L engine with turbocharger. Do not disconnect the hoses, position them to the side.

8. Disconnect the electrical connectors.

9. Remove the starter retaining bolts.

10. Remove the component from its mounting.

To install:

➡ **Be sure to use new fasteners, as required.**

11. Installation is the reverse of the removal procedure.

12. Be sure to install a new starter insulator (gasket), as required.

13. Use the Mazda diagnostic scan tool, or equivalent and reprogram the required systems.

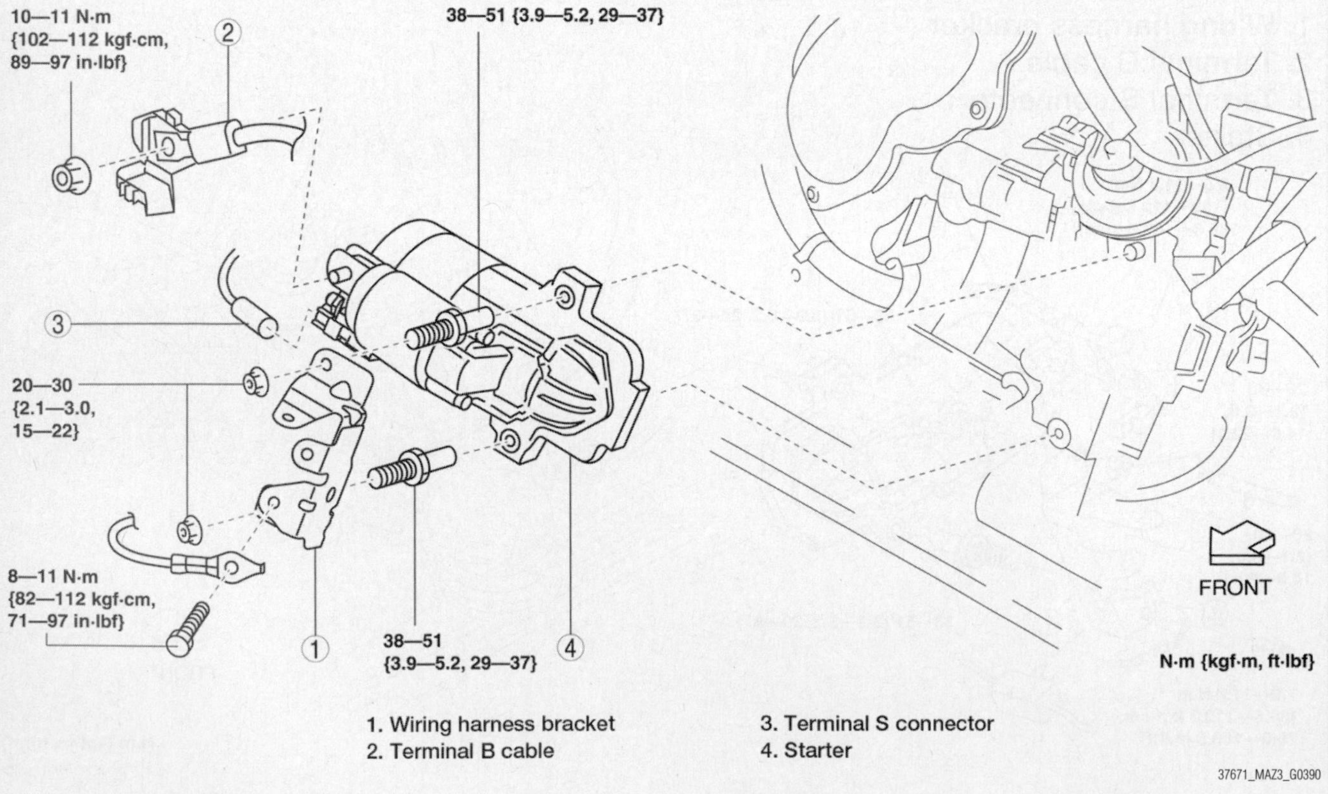

10—11 N·m
{102—112 kgf·cm,
89—97 in·lbf}

38—51 {3.9—5.2, 29—37}

20—30
{2.1—3.0,
15—22}

8—11 N·m
{82—112 kgf·cm,
71—97 in·lbf}

38—51
{3.9—5.2, 29—37}

FRONT

N·m {kgf·m, ft·lbf}

1. Wiring harness bracket
2. Terminal B cable
3. Terminal S connector
4. Starter

37671_MAZ3_G0390

Fig. 51 Starter and related components (2.0L and 2.5L engines shown, 2.3L engine similar)—2010 vehicles

ENGINE MECHANICAL

ACCESSORY DRIVE BELTS

ACCESSORY BELT ROUTING

See Figures 53 and 54.

Refer to the accompanying illustrations.

INSPECTION

See Figure 55.

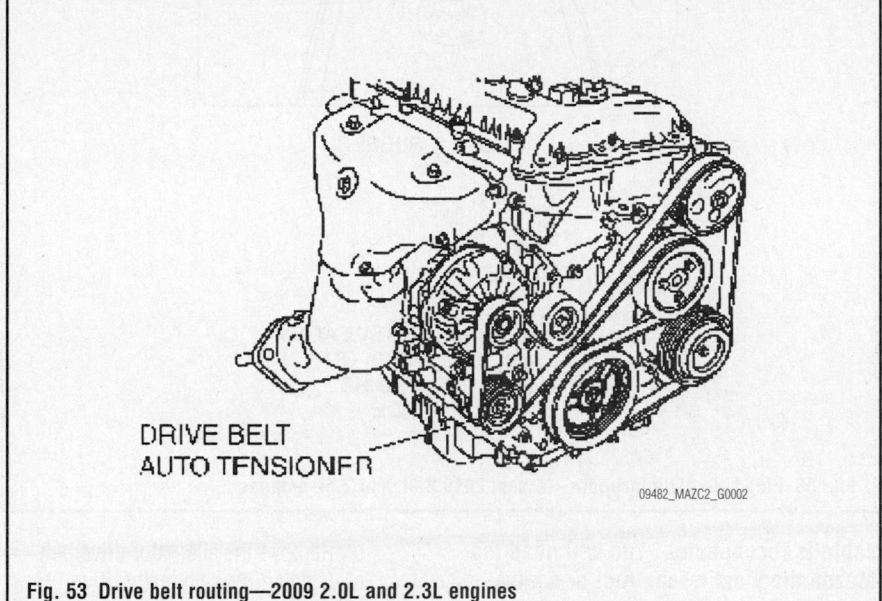

Fig. 53 Drive belt routing—2009 2.0L and 2.3L engines

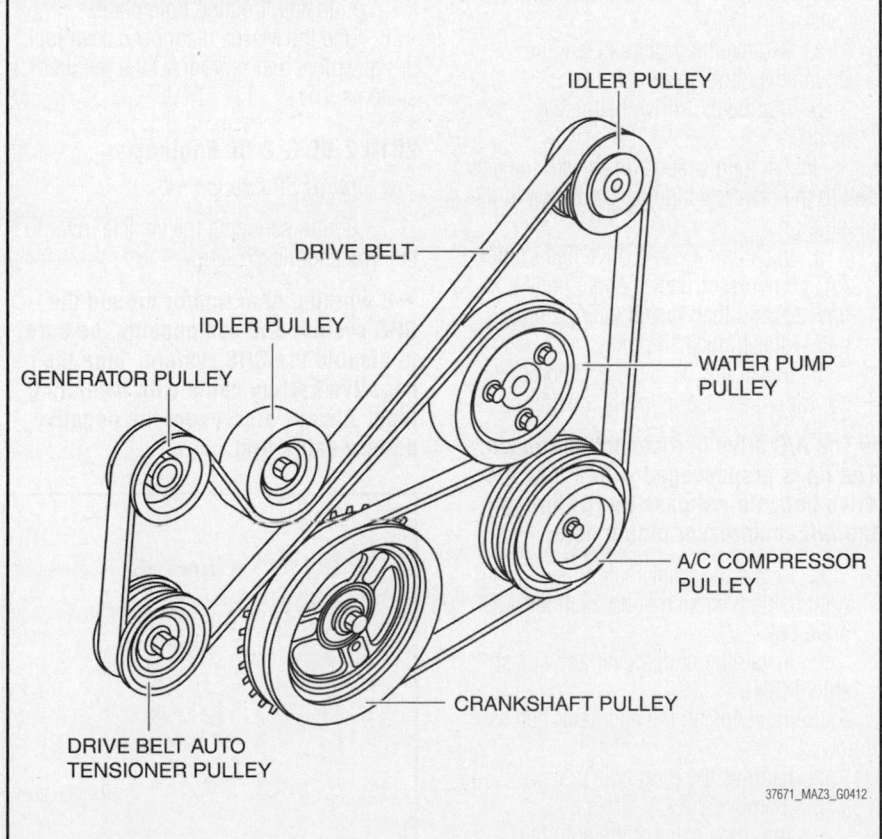

Fig. 54 Drive belt routing—2009 2.3L engines with turbocharger, and all 2010 engines

Fig. 55 Drive belt inspection—2010 2.0L and 2.5L engines

Inspect the drive belt for signs of glazing or cracking. A glazed belt will be perfectly smooth from slippage, while a good belt will have a slight texture of fabric visible. Cracks will usually start at the inner edge of the belt and run outward. All worn or damaged drive belts should be replaced immediately

➡**Drive belt deflection/tension inspection is not necessary because of the use of the drive belt auto tensioner.**

1. Verify that the drive belt auto tensioner indicator mark docs not exceed the limit.
2. If it exceeds the limit, replace the drive belt.

ADJUSTMENT

The belt adjustment is maintained by an auto tensioner. Inspect the tensioner as outlined below.

1. Rcmove the drive bell.
2. Verify that the alternator drive belt auto tensioner moves smoothly in a clockwise and counterclockwise direction.
3. If it does not move smoothly, replace the drive belt auto tensioner.
4. Turn the drive belt auto tensioner pulley by hand and verify that it rotates smoothly.
5. If it does not move smoothly, replace the drive belt auto tensioner.

REMOVAL & INSTALLATION

Except 2010 2.0L & 2.5L Engines

See Figures 56 through 58.

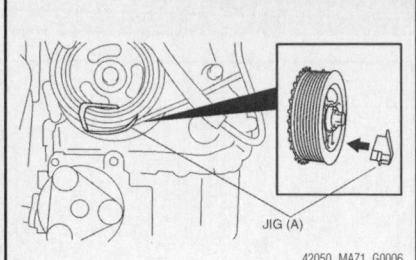

Fig. 56 Install jig (A) which comes with the new A/C belt to the crankshaft pulley as shown—except 2010 2.0L and 2.5L engines

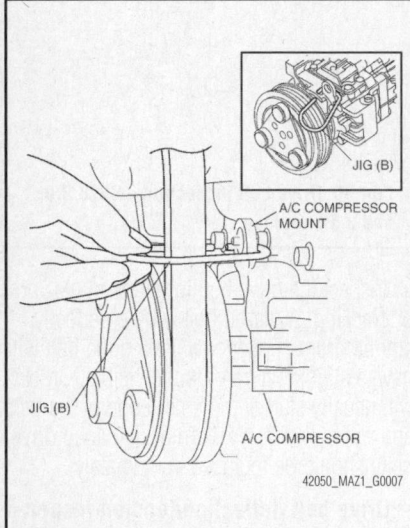

Fig. 57 Install jig (B) to the A/C compressor mount as shown in the illustration—except 2010 2.0L and 2.5L engines

1. Before servicing the vehicle, refer to the Precautions Section.

➡If working near and/or around the SRS system and components, be sure to disable the SRS system. Tape the negative battery cable with insulating tape. Always disconnect the negative battery cable first.

✷✷ CAUTION

To avoid personal injury when working on vehicles equipped with an air bag, the negative battery cable must be disconnected and at least one minute must elapse before working on the system. Failure to do so may result in deployment of the air bag.

2. Disconnect the negative battery cable. Tape the cable with insulating tape.

➡When disconnecting the cable, some systems need to be initialized after the

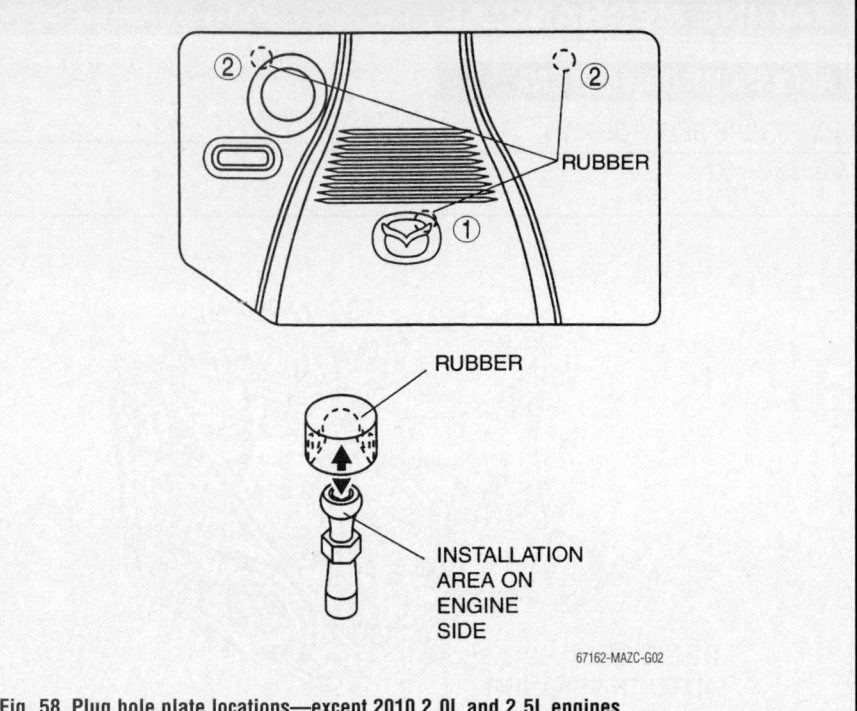

Fig. 58 Plug hole plate locations—except 2010 2.0L and 2.5L engines

cable is reconnected. You will need the Mazda diagnostic scan tool or equivalent. Follow the directions on the tool.

3. Remove/install the A/C belt as follows:
   a. Remove the right side engine undercover and splash shield.
   b. Cut the A/C drive belt using scissors.

4. Install a jig which comes with the new belt to the crankshaft pulley as shown in the illustration.
   a. Install a new A/C drive belt to the A/C compressor pulley, move jig (A) upward, and then install the A/C drive belt to the crankshaft pulley.
   b. Install jig (B) to the A/C compressor mount as shown in the illustration.

➡The A/C drive belt cannot be reused. The jig is prepackaged with a new A/C drive belt. Do not pass jig (B) through the A/C compressor mount hole.

   c. Rotate the crankshaft pulley to the right using a wrench and install the A/C drive belt.
   d. Install the undercover and splash shield (RH).

5. Remove/Install the alternator belt as follows:
   a. Remove the plug hole plate.
   b. Remove the A/C drive belt.
   c. Turn the center of the auto tensioner pulley counterclockwise to release tension to the drive belt tension.

   d. Remove the alternator drive belt.
   e. Install a new alternator drive belt.
   f. Install the A/C drive belt.
   g. Install the plug hole plate.

6. Use the Mazda diagnostic scan tool, or equivalent and reprogram the required systems.

2010 2.0L & 2.5L Engines

See Figures 59 through 64.

1. Before servicing the vehicle, refer to the Precautions Section.

➡If working near and/or around the SRS system and components, be sure to disable the SRS system. Tape the negative battery cable with insulating tape. Always disconnect the negative battery cable first.

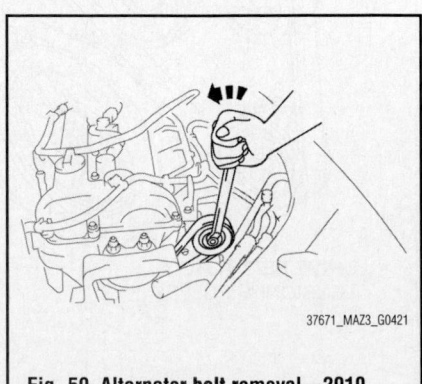

Fig. 59 Alternator belt removal—2010 2.0L engine

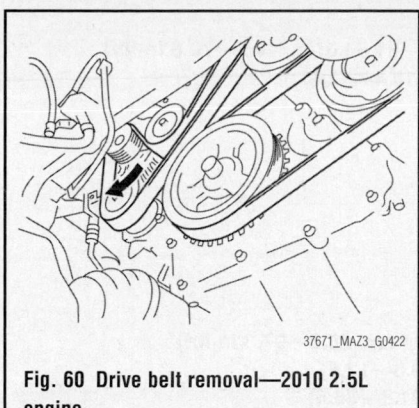

Fig. 60 Drive belt removal—2010 2.5L engine

✳✳ CAUTION

To avoid personal injury when working on vehicles equipped with an air bag, the negative battery cable must be disconnected and at least one minute must elapse before working on the system. Failure to do so may result in deployment of the air bag.

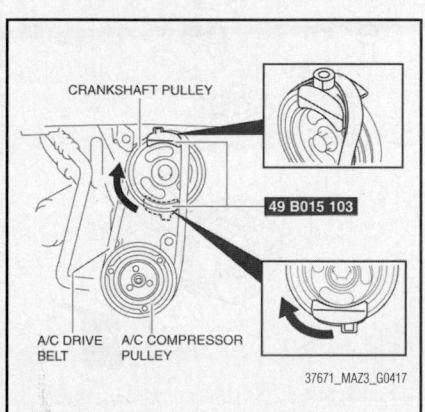

Fig. 61 A/C drive belt installation 1 of 4—2010 2.0L engine

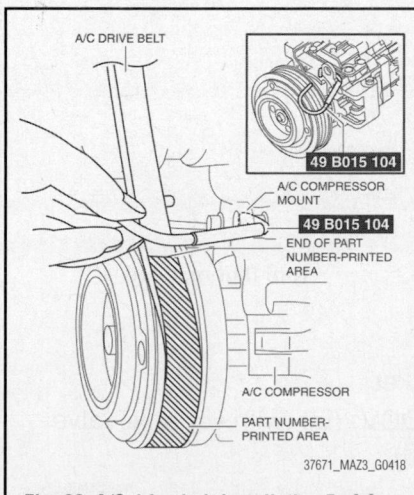

Fig. 62 A/C drive belt installation 2 of 4—2010 2.0L engine

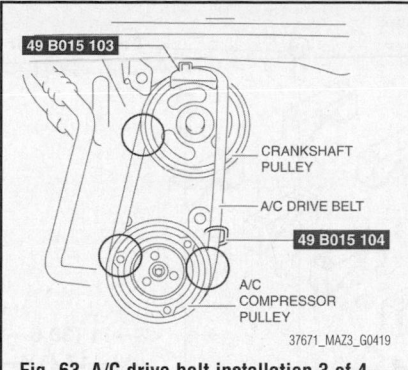

Fig. 63 A/C drive belt installation 3 of 4—2010 2.0L engine

2. Disconnect the negative battery cable. Tape the cable with insulating tape.

➡**When disconnecting the cable, some systems need to be initialized after the cable is reconnected. You will need the Mazda diagnostic scan tool or equivalent. Follow the directions on the tool.**

3. Remove the aerodynamic undercover No. 2 .

4. On the 2.0L engine, remove the A/C drive belt, remove the nut at the power steering pipe component and position the power steering component out of the way. Position special tool SST: 49B015-102, or equivalent and remove the belt.

5. On the 2.0L engine, to remove the alternator belt, turn the center of the auto tensioner pulley counterclockwise to release the tension. Remove the belt.

6. On the 2.5L engine, to remove the drive belt rotate the tensioner and remove the belt.

To install:

➡**Be sure to use new fasteners, as required.**

7. Installation is the reverse of the removal procedure.

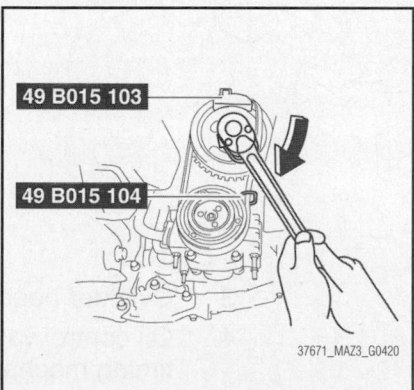

Fig. 64 A/C drive belt installation 4 of 4—2010 2.0L engine

➡On the 2.0L engine, during A/C belt installation, if the SST touches the part numbered printed area on the back of the drive belt, the SST may not be able to slide smoothly and the belt will be pilled inward excessively. This will make it impossible to correctly install the belt to the groove of the compressor pulley. To prevent this adjust the belt to the position shown in the illustrations. After setting the SST, verify that the three parts of the belt are hung on each pulley. Rotate the crankshaft pulley clockwise and install the belt. Remove the SST: 49B015-102, or equivalent immediately after it reaches the lower side to prevent it from falling down and being damaged. Rotate the crankshaft one more time to verify correct belt installation. Do not turn the pulley counterclockwise.

8. As required, verify that the drive belt auto tensioner indicator mark does not exceed the limit.

9. Use the Mazda diagnostic scan tool, or equivalent and reprogram the required systems.

CAMSHAFT AND VALVE LIFTERS

REMOVAL & INSTALLATION

Except 2010 2.0L & 2.5L Engines
See Figures 65 through 68.

1. Before servicing the vehicle, refer to the Precautions Section.

➡**If working near and/or around the SRS system and components, be sure to disable the SRS system. Tape the negative battery cable with insulating tape. Always disconnect the negative battery cable first.**

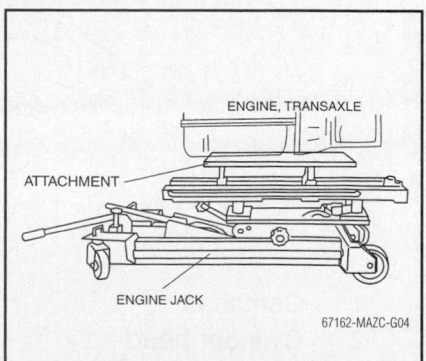

Fig. 65 Support the engine using an engine jack and attachment as shown—except 2010 2.0L and 2.5L engines

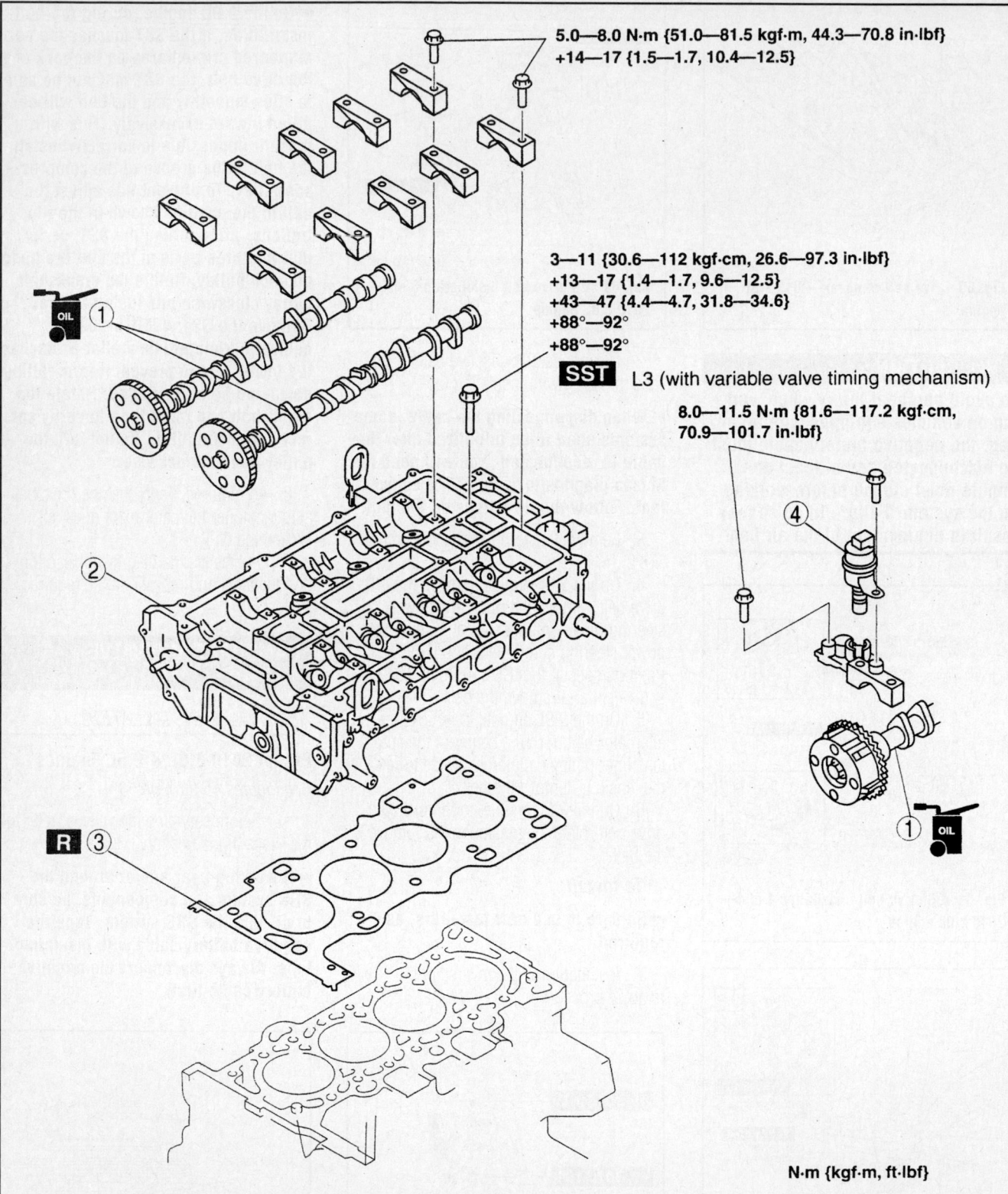

5.0—8.0 N·m {51.0—81.5 kgf·m, 44.3—70.8 in·lbf}
+14—17 {1.5—1.7, 10.4—12.5}

3—11 {30.6—112 kgf·cm, 26.6—97.3 in·lbf}
+13—17 {1.4—1.7, 9.6—12.5}
+43—47 {4.4—4.7, 31.8—34.6}
+88°—92°
+88°—92°

SST L3 (with variable valve timing mechanism)

8.0—11.5 N·m {81.6—117.2 kgf·cm, 70.9—101.7 in·lbf}

N·m {kgf·m, ft·lbf}

1	Camshaft	3	Cylinder head gasket
2	Cylinder head	4	Oil control valve (OCV) (L3 (with variable valve timing mechanism))

67162-MAZC-G34

Fig. 66 Exploded view of the camshaft and cylinder head components—except 2010 2.0L and 2.5L engines

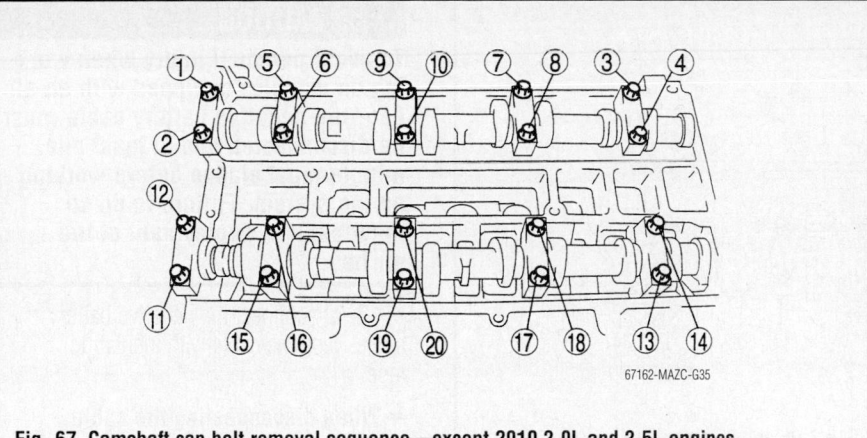

Fig. 67 Camshaft cap bolt removal sequence—except 2010 2.0L and 2.5L engines

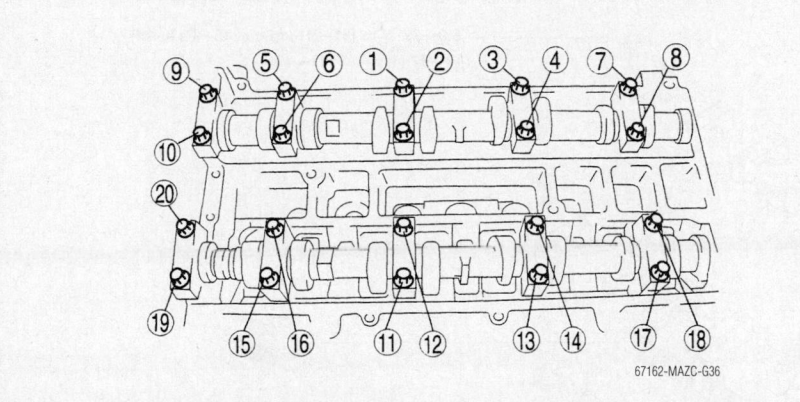

Fig. 68 Camshaft cap bolt tightening sequence—except 2010 2.0L and 2.5L engines

✳✳ CAUTION

To avoid personal injury when working on vehicles equipped with an air bag, the negative battery cable must be disconnected and at least one minute must elapse before working on the system. Failure to do so may result in deployment of the air bag.

2. Remove the battery top cover, if equipped.

3. Disconnect the negative battery cable. Tape the cable with insulating tape.

➡**When disconnecting the cable, some systems need to be initialized after the cable is reconnected. You will need Mazda diagnostic scan tool or equivalent. Follow the directions on the tool.**

4. Remove the timing chain.

5. Remove the intake manifold.

6. Disconnect the following components:
- Warm Up–Three Way Converter (WU–TWC)
- Upper radiator hose
- Water and heater hose
- Wiring harness

7. Support the engine using an engine jack and attachment as shown in the illustration.

8. Remove the camshafts.

➡**The cylinder head and camshaft caps are numbered and must be reassembled in their original locations. When removing a when removed, keep the caps with the cylinder head they were removed from, do not switch the caps.**

9. Loosen the camshaft cap bolts using 2–3 passes in the sequence illustrated.

10. Once all the bolts have been removed, remove the caps, keeping them in the original positions and remove the camshafts.

To install:

11. Set the cam position of the number 1 cylinder at Top Dead center (TDC) and install the camshaft.

12. Temporarily install the camshaft caps in their original positions.

13. Tighten the camshaft cap bolt in two steps:

a. Step 1:44–71 inch lbs. (5–8 Nm).

b. Step 2:10.4–12.5 ft. lbs. (14–17 Nm).

14. Remove the engine jack and attachment.

15. Connect the following components:
- Wiring harness
- Water and heater hose
- Upper radiator hose
- WU–TWC converter

16. Install the intake manifold.

17. Install the timing chain.

18. Connect the negative battery cable.

19. Fill and bleed the cooling system.

20. Change the oil and filter.

21. Run the engine and check for proper operation.

22. Use the Mazda diagnostic scan tool, or equivalent and reprogram the required systems.

2010 2.0L & 2.5L Engines

See Figures 69 through 73.

✳✳ WARNING

Fuel vapor is hazardous. It can very easily ignite, causing death, serious injury, or damage. Always keep sparks and flames away from fuel.

✳✳ WARNING

Fuel line spills and leakage from the pressurized fuel system are dangerous. Fuel can easily ignite and cause serious injury or death and damage. Fuel can also irritate skin and eyes.

1. Before servicing the vehicle, refer to the Precautions Section.

➡**If working near and/or around the SRS system and components, be sure to disable the SRS system. Tape the negative battery cable with insulating tape. Always disconnect the negative battery cable first.**

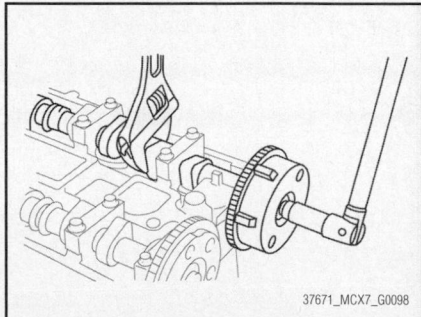

Fig. 69 Hold the camshaft using a wrench on the cast hexagon—except 2010 2.0L and 2.5L engines

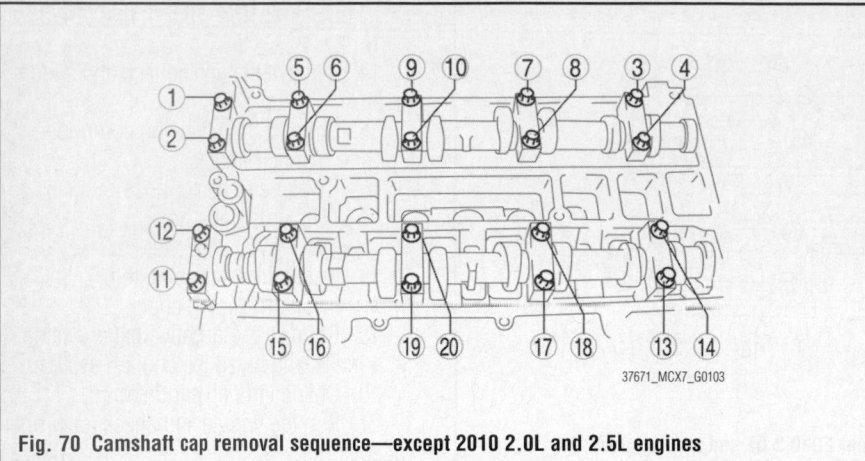

Fig. 70 Camshaft cap removal sequence—except 2010 2.0L and 2.5L engines

37671_MCX7_G0103

2. Disconnect the negative battery cable. Tape the cable with insulating tape.

➡ When disconnecting the cable, some systems need to be initialized

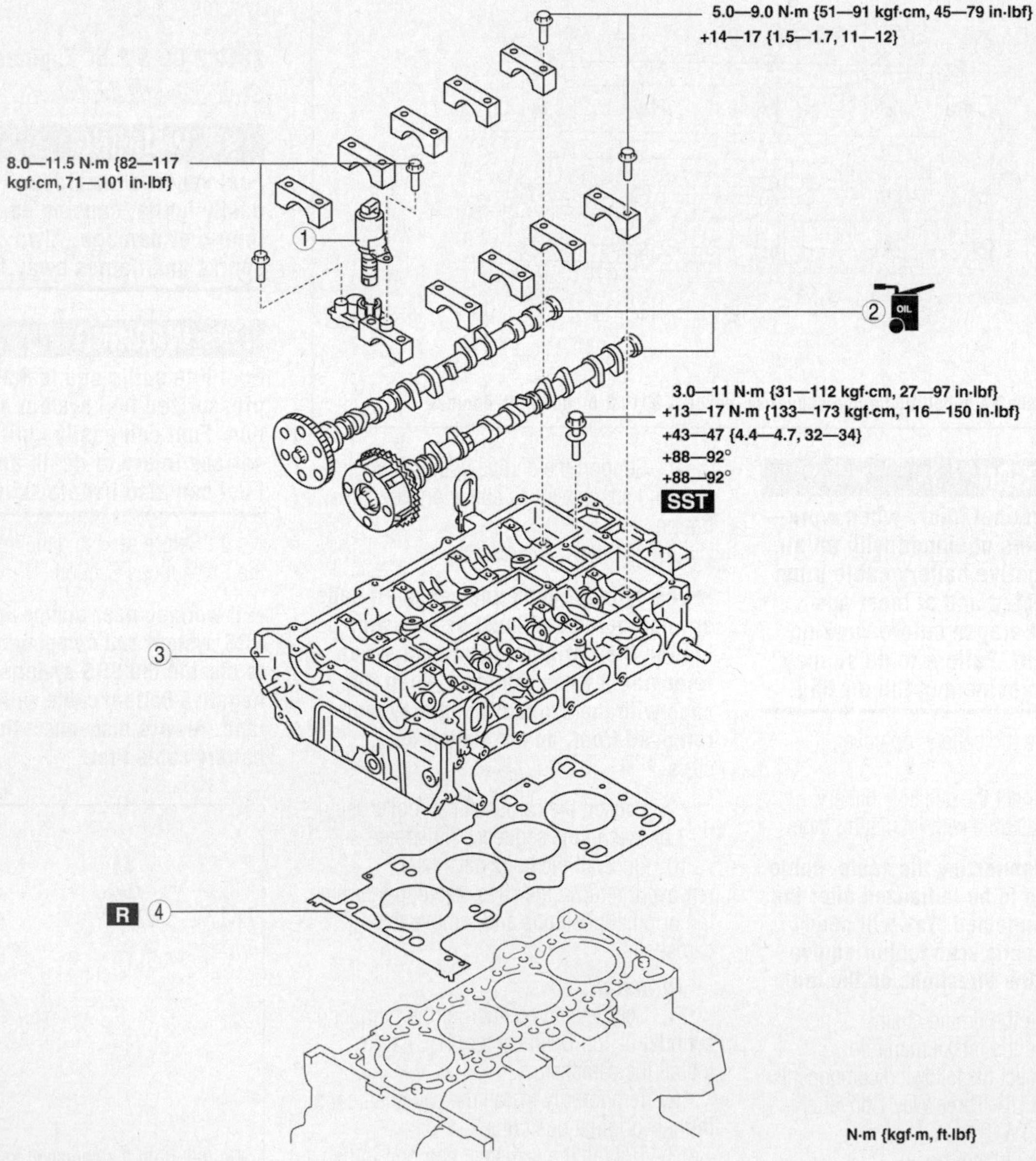

5.0—9.0 N·m {51—91 kgf·cm, 45—79 in·lbf}
+14—17 {1.5—1.7, 11—12}

8.0—11.5 N·m {82—117 kgf·cm, 71—101 in·lbf}

3.0—11 N·m {31—112 kgf·cm, 27—97 in·lbf}
+13—17 N·m {133—173 kgf·cm, 116—150 in·lbf}
+43—47 {4.4—4.7, 32—34}
+88—92°
+88—92°

SST

N·m {kgf·m, ft·lbf}

37671_MCX7_G0104

Fig. 71 Cylinder head exploded view—except 2010 2.0L and 2.5L engines

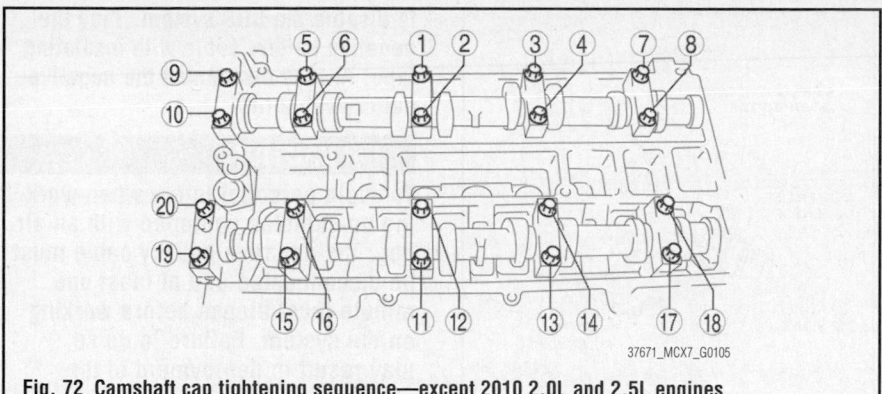

Fig. 72 Camshaft cap tightening sequence—except 2010 2.0L and 2.5L engines

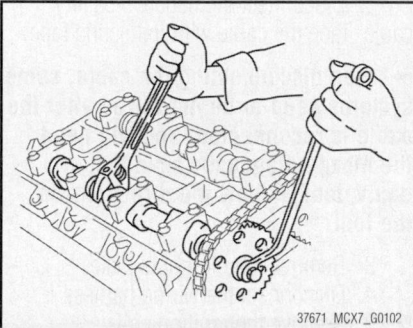

Fig. 73 Tightening the camshaft sprocket retaining bolt—except 2010 2.0L and 2.5L engines

after the cable is reconnected. You will need the Mazda diagnostic scan tool or equivalent. Follow the directions on the tool.

3. Remove the timing chain.

4. Before removing the camshaft, hold the camshaft using a wrench on the cast hexagon and loosen the camshaft sprocket installation bolt.

➡**Before removing the camshafts inspect camshaft end play.**

5. Loosen the camshaft cap bolts in two or three passes in the order shown and remove the end caps.

➡**Note the position of each removed end cap for installation.**

6. Remove the camshafts.

To install:

➡**Be sure to use new fasteners, as required.**

➡**Install the camshaft with number 1 cylinder cam aligned at TDC of the compression stroke.**

7. Apply a small amount of gear oil to the camshaft journal of the cylinder head before you install the caps.

8. Install camshaft caps finger tight,

then in two or three passes tighten the caps to:

- First pass: 4.3–79.5 inch lbs. (5–9 Nm) in the order shown.
- Second pass: 10–13 ft. lbs (14–17 Nm) in the order shown.

9. Tighten the camshaft sprocket installation bolts to 51–55 ft. lbs. (69–75 Nm).

10. Install the timing chain.

11. Refill the engine with coolant.

12. Bleed the air from the cooling system.

13. Inspect the compression pressure.

14. Use the Mazda diagnostic scan tool, or equivalent and reprogram the required systems.

CRANKSHAFT DAMPER

REMOVAL & INSTALLATION

Except 2010 2.0L & 2.5L Engines
See Figures 74 through 78.

1. Before servicing the vehicle, refer to the Precautions Section.

➡**If working near and/or around the SRS system and components, be sure to disable the SRS system. Tape the negative battery cable with insulating tape. Always disconnect the negative battery cable first.**

✳✳ CAUTION

To avoid personal injury when working on vehicles equipped with an air bag, the negative battery cable must be disconnected and at least one minute must elapse before working on the system. Failure to do so may result in deployment of the air bag.

2. Disconnect the negative battery cable. Tape the cable with insulating tape.

➡**When disconnecting the cable, some**

systems need to be initialized after the cable is reconnected. You will need the Mazda diagnostic scan tool or equivalent. Follow the directions on the tool.

3. Remove the plug hole plate and bracket.

4. Disconnect the wiring harness.

5. Remove the ignition coils, spark plugs, valve cover and accessory drive belt.

6. Remove the front wheels.

7. Remove the undercover and splash shield, if equipped.

8. Remove the CKP sensor.

9. Remove the crankshaft pulley bolt and pulley.

10. Remove the cylinder block lower blind plug.

11. Install tool 49 JE01 061.

12. Turn the crankshaft clockwise until the crankshaft is in the number 1 cylinder Top Dead Center position (the balance weight should be attached to tool 49 JE01 061).

13. Hold the crankshaft pulley using the tools illustrated and remove the bolt.

14. Remove the crankshaft pulley.

To install:

15. Install the crankshaft pulley.

16. Install tool 49 UN30 3465 as illustrated on the camshaft.

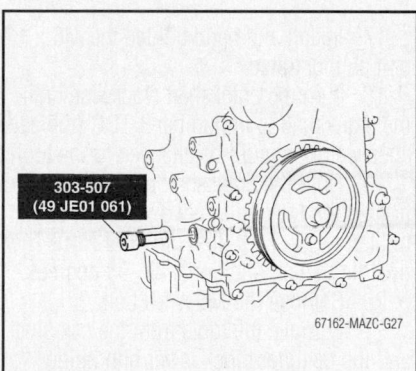

Fig. 74 Install tool 49 JE01 061—except 2010 2.0L and 2.5L engines

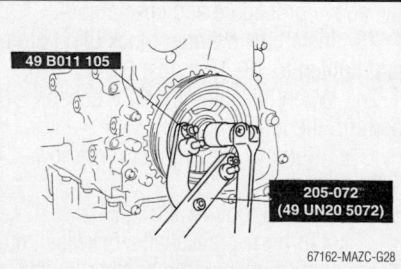

Fig. 75 Install tool 49 JE01 061 Hold the crankshaft pulley using the tools illustrated and remove the bolt—except 2010 2.0L and 2.5L engines

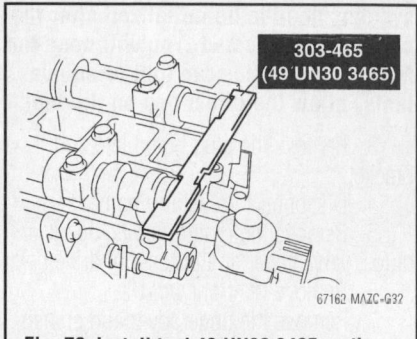

Fig. 76 Install tool 49 UN30 3465 on the camshaft—except 2010 2.0L and 2.5L engines

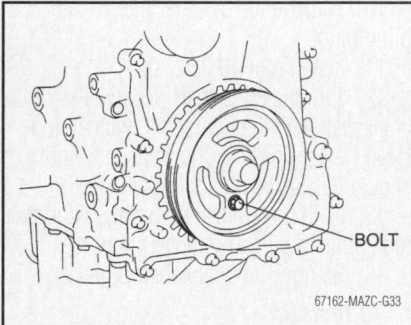

Fig. 77 Install and hand tighten the M6 x 1.0 bolt—except 2010 2.0L and 2.5L engines

17. Install and hand tighten the M6 x 1.0 bolt as illustrated.

18. Turn the crankshaft clockwise until the crankshaft is in number 1 TDC (the balance weight should be attached to the tool).

19. Hold the crankshaft pulley and tighten the lock bolt in two steps. First tighten to 71–77 ft. lbs. (96–104 Nm). Then final tighten an additional 87–93 degrees.

20. Remove the M6 x 1.0 bolt.

21. Remove the tools from the camshaft and the cylinder block lower blind plug.

22. Rotate the crankshaft clockwise 2 turns until you reach TDC. If not aligned properly, loosen the crankshaft pulley lock bolt and reinstall the lock bolt again using the above procedure and special tools.

23. Install the cylinder block blind plug and tighten to 13–16 ft. lbs. (18–22 Nm).

24. When installing the CKP sensor perform the following:

   a. Remove the cylinder block lower blind plug.

   b. Install tool 49 JE01 061.

   c. Turn the crankshaft clockwise until the crankshaft is in the number 1 cylinder Top Dead Center position (the balance weight should be attached to tool 49 JE01 061).

   d. Using a ruler, mark the center line

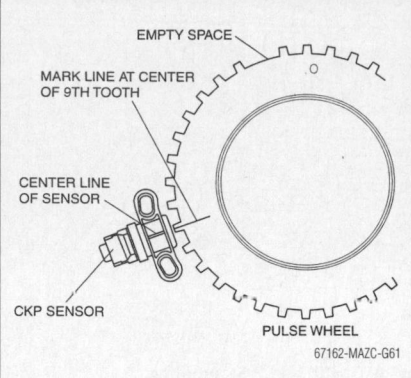

Fig. 78 Using a ruler, mark the center line on the pulse wheel teeth on the crank pulley which is located at the 9th tooth counting counterclockwise from the empty space—except 2010 2.0L and 2.5L engines

on the pulse wheel teeth on the crank pulley which is located at the 9th tooth counting counterclockwise from the empty space. Refer to the illustration for more detail.

✳✳ CAUTION

If you do not mark the center line correctly this will cause improper engine control for the ignition and fuel system, so be sure to mark the line carefully.

   e. Install the CKP sensor making sure the mark on the sensor is aligned with the mark on the pulse wheel. Tighten the sensor bolt to 49–66 inch lbs. (5.5–7.5 Nm) and attach the electrical connector.

   f. Remove the tool from the cylinder block lower blind plug.

   g. Rotate the crankshaft clockwise 2 turns until you reach TDC. If not aligned properly, loosen the crankshaft pulley lock bolt and reinstall the lock bolt.

   h. Install the cylinder block blind plug and tighten to 13–16 ft. lbs. (18–22 Nm).

25. Install the remaining components in the reverse order of removal.

26. Start the vehicle and check for leaks, repair if necessary.

27. Use the Mazda diagnostic scan tool, or equivalent and reprogram the required systems.

2010 2.0L & 2.5L Engines

See Figures 79 through 84.

1. Before servicing the vehicle, refer to the Precautions Section.

➡ **If working near and/or around the SRS system and components, be sure**

to disable the SRS system. Tape the negative battery cable with insulating tape. Always disconnect the negative battery cable first.

✳✳ CAUTION

To avoid personal injury when working on vehicles equipped with an air bag, the negative battery cable must be disconnected and at least one minute must elapse before working on the system. Failure to do so may result in deployment of the air bag.

2. Disconnect the negative battery cable. Tape the cable with insulating tape.

➡ **When disconnecting the cable, some systems need to be initialized after the cable is reconnected. You will need the Mazda diagnostic scan tool or equivalent. Follow the directions on the tool.**

3. Remove the plug hole plate.

4. Disconnect the wiring harness.

5. Remove the ignition coils.

6. Remove the spark plugs.

7. Disconnect the ventilation hose from the cylinder head cover.

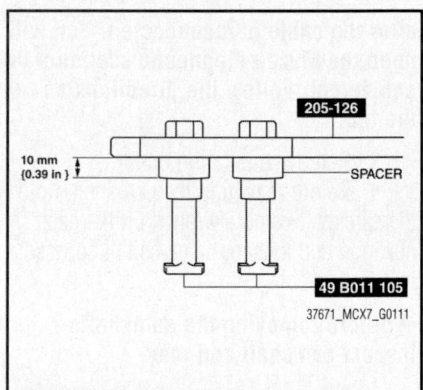

Fig. 79 SST 205-126 and spacer shown installed—2010 2.0L and 2.5L engines

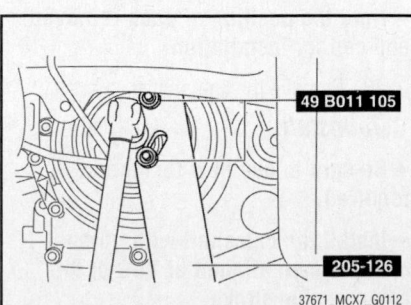

Fig. 80 Crankshaft service tools SST 205-126 and 49 B011 105—2010 2.0L and 2.5L engines

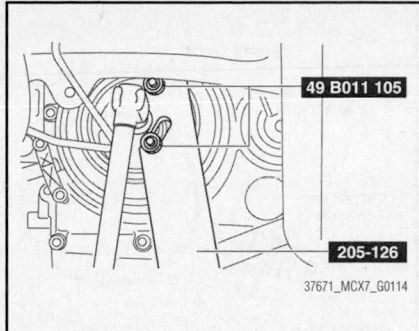

Fig. 81 Hold the crankshaft pulley using the SST 205-126—2010 2.0L and 2.5L engines

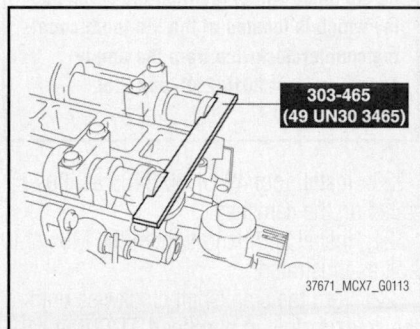

Fig. 82 SST 303-465 camshaft holding tool installed—2010 2.0L and 2.5L engines

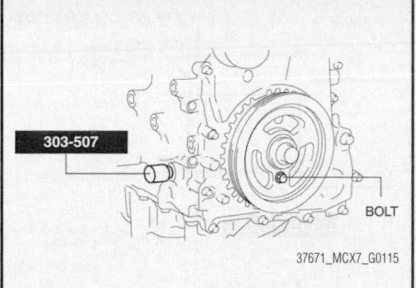

Fig. 83 Fix the crankshaft pulley to the engine front cover—2010 2.0L and 2.5L engines

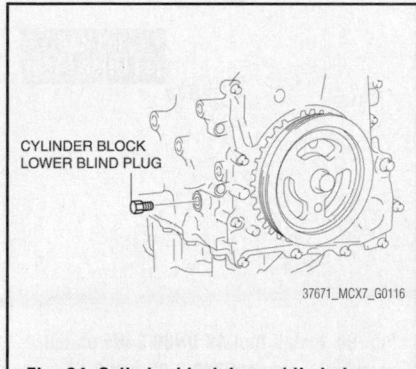

Fig. 84 Cylinder block lower blind plug—2010 2.0L and 2.5L engines

8. Remove the dipstick.

9. Remove the cylinder head cover.

10. Set the front mudguard (RH) out of the way.

11. Remove the splash shield (RH).

12. Remove the drive belt.

13. Set the brake pipe out of the way.

14. Remove the cylinder block lower blind plug.

15. Install the SST 303-507.

16. Turn the crankshaft clockwise until the crankshaft is in the No.1 cylinder top dead center (TDC) position (until the balance weight is attached to the SST).

17. Install the suitable spacer 0.39 inch (10 mm) thickness to the SST 205-126, 49 B011 105 as shown in the figure.

➡️**If the spacer is not installed, the crankshaft pulley cannot be secured because the length for the SST (49 B011 105) is not sufficient.**

18. Hold the crankshaft pulley using the SST 205-126.

19. Remove the crankshaft lock bolt and pulley.

To install:

➡️**Be sure to use new fasteners, as required.**

20. Install SST 303-465 to the camshaft as shown in the figure.

➡️**Verify that No.1 cylinder is at TDC of the compression stroke. (Position counterweight contacts SST 303-507.)**

21. To position the crankshaft pulley, temporarily tighten it and, using a suitable bolt (M6 x 1.0), fix the crankshaft pulley to the engine front cover.

22. Install SSTs 205-126 and 49 B011 105 to the crankshaft pulley and lock the crankshaft against rotation, and tighten the crankshaft pulley lock bolt using the following two steps:

a. Step 1: 71–76 ft. lbs. (96–104 Nm)

b. Step 2: Plus 87°–93°

23. Remove the crankshaft pulley installation bolt (M6 X 1.0).

24. Remove the SST from the camshaft.

25. Remove the SST installed in the cylinder block lower blind plug hole.

26. Verify that the teeth of the crankshaft pulley pulse wheel and the detection area of the Crankshaft Position (CKP) sensor are installed to the correct positions.

27. Rotate the crankshaft clockwise 2 turns until the TDC position. If not aligned, loosen the crankshaft pulley lock bolt and repeat.

28. Install the cylinder block lower blind plug and tighten it to 14–16 ft. lbs. (18–22 Nm).

29. Use the Mazda diagnostic scan tool, or equivalent and reprogram the required systems.

CRANKSHAFT FRONT SEAL

REMOVAL & INSTALLATION

Except 2010 2.0L & 2.5L Engines

See Figures 85 through 92.

1. Before servicing the vehicle, refer to the Precautions Section.

➡️**If working near and/or around the SRS system and components, be sure to disable the SRS system. Tape the negative battery cable with insulating tape. Always disconnect the negative battery cable first.**

✳️ CAUTION

To avoid personal injury when working on vehicles equipped with an air bag, the negative battery cable must be disconnected and at least one minute must elapse before working on the system. Failure to do so may result in deployment of the air bag.

2. Disconnect the negative battery cable. Tape the cable with insulating tape.

➡️**When disconnecting the cable, some systems need to be initialized after the cable is reconnected. You will need the Mazda diagnostic scan tool or equivalent. Follow the directions on the tool.**

3. Remove the plug hole plate and bracket.

4. Disconnect the wiring harness.

5. Remove the ignition coils, spark plugs, valve cover and accessory drive belt.

6. Remove the front wheels.

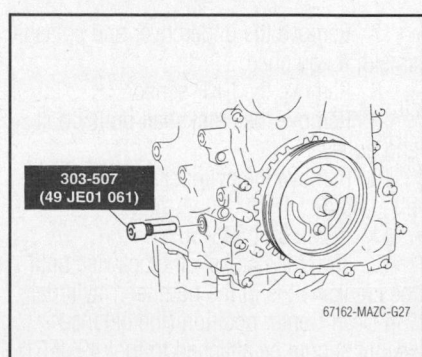

Fig. 85 Install tool 49 JE01 061—except 2010 2.0L and 2.5L engines

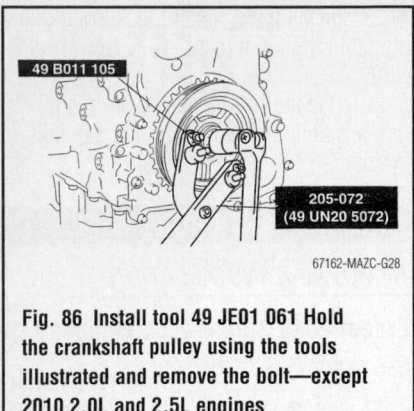

Fig. 86 Install tool 49 JE01 061 Hold the crankshaft pulley using the tools illustrated and remove the bolt—except 2010 2.0L and 2.5L engines

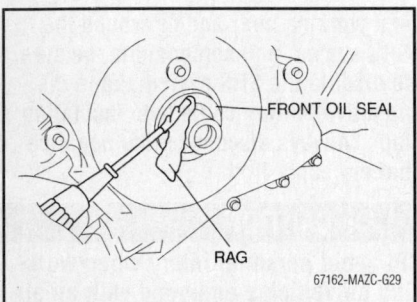

Fig. 87 Cut the seal lip using a suitable knife and using a suitable pry tool wrapped in a rag, remove the seal—except 2010 2.0L and 2.5L engines

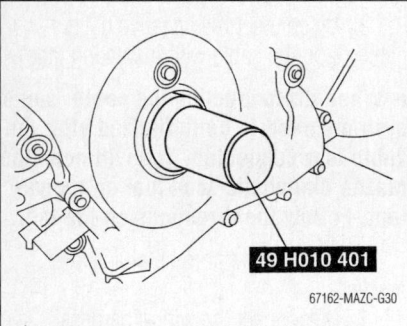

Fig. 88 Use seal installer tool 49 H010 401 and a hammer to install the seal—except 2010 2.0L and 2.5L engines

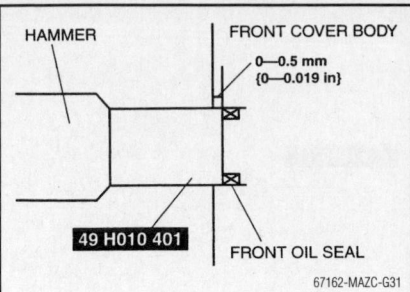

Fig. 89 The seal should be recessed 0–0.019 inch (0–0.5mm) as shown when properly installed—except 2010 2.0L and 2.5L engines

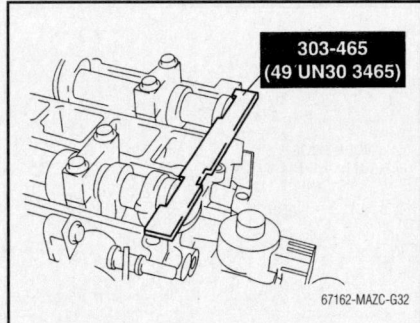

Fig. 90 Install tool 49 UN30 3465 on the camshaft—except 2010 2.0L and 2.5L engines

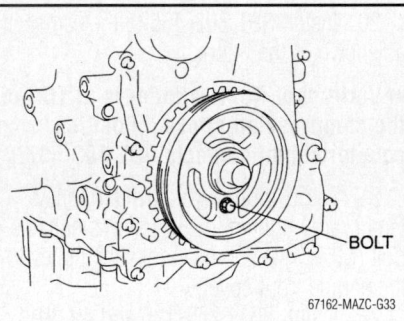

Fig. 91 Install and hand tighten the M6 x 1.0 bolt—except 2010 2.0L and 2.5L engines

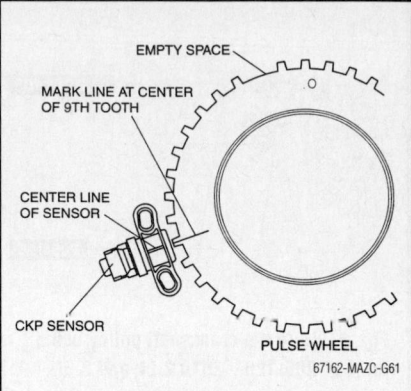

Fig. 92 Using a ruler, mark the center line on the pulse wheel teeth on the crank pulley which is located at the 9th tooth counting counterclockwise from the empty space—except 2010 2.0L and 2.5L engines

7. Remove the undercover and splash shield, if equipped.

8. Remove the CKP sensor.

9. Remove the crankshaft pulley bolt and pulley.

10. Remove the cylinder block lower blind plug.

11. Install tool 303-507.

12. Turn the crankshaft clockwise until the crankshaft is in the number 1 cylinder Top Dead Center position (the balance weight should be attached to tool 49 JE01 061).

13. Hold the crankshaft pulley using the tools illustrated and remove the bolt.

14. Remove the crankshaft pulley.

15. Cut the seal lip using a suitable knife and using a suitable pry tool wrapped in a rag, remove the seal.

To install:

16. Coat the new seal with clean engine oil.

17. Push the seal in by hand to get it started.

18. Use seal installer tool 49 H010 401 and a hammer to install the seal so that it is recessed 0–0.019 inch (0–0.5mm) as shown in the accompanying illustration.

19. Install the crankshaft pulley.

20. Install tool 49 UN30 3465 as illustrated on the camshaft.

21. Install and hand tighten the M6 x 1.0 bolt as illustrated.

22. Turn the crankshaft clockwise until the crankshaft is in number 1 TDC (the balance weight should be attached to the tool).

23. Hold the crankshaft pulley and tighten the lock bolt in two steps. First tighten to 71–77 ft. lbs. (96–104 Nm). Then final tighten an additional 87–93 degrees.

24. Remove the M6 x 1.0 bolt.

25. Remove the tools from the camshaft and the cylinder block lower blind plug.

26. Rotate the crankshaft clockwise 2 turns until you reach TDC. If not aligned properly, loosen the crankshaft pulley lock bolt and reinstall the lock bolt again using the above procedure and special tools.

27. Install the cylinder block blind plug and tighten to 13–16 ft. lbs. (18–22 Nm).

28. When installing the CKP sensor perform the following:

a. Remove the cylinder block lower blind plug.

b. Install tool 49 JE01 061.

c. Turn the crankshaft clockwise until the crankshaft is in the number 1 cylinder Top Dead Center position (the balance weight should be attached to tool 49 JE01 061).

d. Using a ruler, mark the center line on the pulse wheel teeth on the crank pulley which is located at the 9th tooth counting counterclockwise from the empty space. Refer to the illustration for more detail.

✳✳ CAUTION

If you do not mark the center line correctly this will cause improper engine

control for the ignition and fuel system, so be sure to mark the line carefully.

  e. Install the CKP sensor making sure the mark on the sensor is aligned with the mark on the pulse wheel. Tighten the sensor bolt to 49–66 inch lbs. (5.5–7.5 Nm)and attach the electrical connector.

  f. Remove the tool from the cylinder block lower blind plug.

  g. Rotate the crankshaft clockwise 2 turns until you reach TDC. If not aligned properly, loosen the crankshaft pulley lock bolt and reinstall the lock bolt.

  h. Install the cylinder block blind plug and tighten to 13–16 ft. lbs. (18–22 Nm).

29. Install the remaining components in the reverse order of removal.

30. Start the vehicle and check for leaks, repair if necessary.

31. Use the Mazda diagnostic scan tool, or equivalent and reprogram the required systems.

2010 2.0L & 2.5L Engines

See Figures 93 and 94.

1. Before servicing the vehicle, refer to the Precautions Section.

➥If working near and/or around the SRS system and components, be sure to disable the SRS system. Tape the negative battery cable with insulating tape. Always disconnect the negative battery cable first.

2. Disconnect the negative battery cable. Tape the cable with insulating tape.

➥When disconnecting the cable, some systems need to be initialized after the cable is reconnected. You will need the Mazda diagnostic scan tool or equivalent. Follow the directions on the tool.

3. Remove the crankshaft pulley.
4. Remove the oil seal lip using a razor.
5. Remove the oil seal using a suitable prytool.

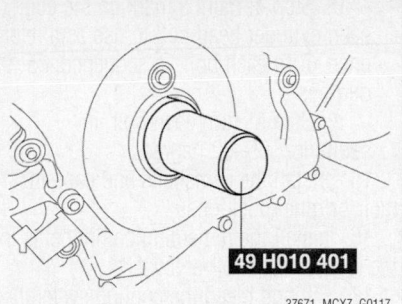

Fig. 93 SST 49 H010 401 seal driver—2010 2.0L and 2.5L engines

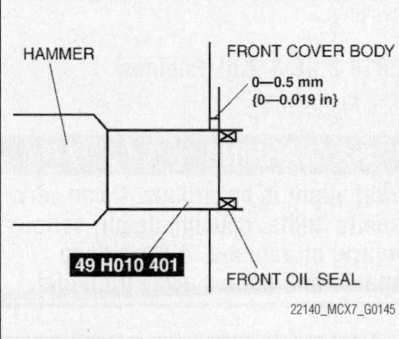

Fig. 94 Correct front oil seal installation—2010 2.0L and 2.5L engines

To install:

➥Be sure to use new fasteners, as required.

6. Apply clean engine oil to a new oil seal.

7. Insert the oil seal into the engine front cover.

8. Tap in the oil seal using SST 49 H010 401 as shown.

9. Replace the crankshaft pulley.

10. Use the Mazda diagnostic scan tool, or equivalent and reprogram the required systems.

CYLINDER HEAD

REMOVAL & INSTALLATION

Except 2010 2.0L & 2.5L Engines

See Figures 95 through 98.

1. Before servicing the vehicle, refer to the Precautions Section.

➥If working near and/or around the SRS system and components, be sure to disable the SRS system. Tape the negative battery cable with insulating tape. Always disconnect the negative battery cable first.

2. Disconnect the negative battery cable. Tape the cable with insulating tape.

➥When disconnecting the cable, some systems need to be initialized after the cable is reconnected. You will need the Mazda diagnostic scan tool or equivalent. Follow the directions on the tool.

3. Remove the timing chain.
4. Remove the intake manifold.
5. Disconnect the following components:
  • Warm Up–Three Way Converter (WU–TWC)
  • Upper radiator hose
  • Water and heater hose
  • Wiring harness

6. Support the engine using an engine jack and attachment as shown in the illustration.

7. Remove the camshafts.

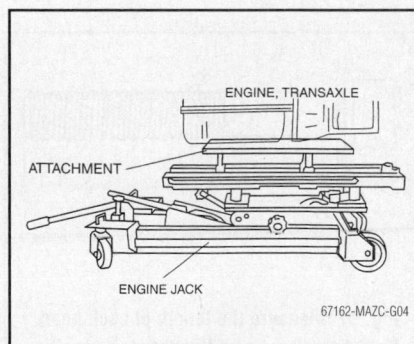

Fig. 95 Support the engine using an engine jack and attachment as shown— except 2010 2.0L and 2.5L engines

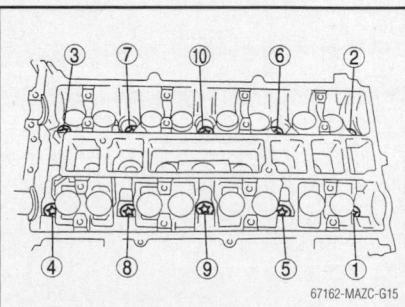

Fig. 96 Cylinder head bolt removal sequence—except 2010 2.0L and 2.5L engines

➡The cylinder head and camshaft caps are numbered and must be reassembled in their original locations. When removing a when removed, keep the caps with the cylinder head they were removed from, do not switch the caps.

8. Loosen the cylinder head bolts using 2–3 passes in the sequence illustrated.

9. Once all the bolts have been removed, remove the cylinder head and gasket.

To install:

10. If reusing old bolts, which is not recommended; measure the length of each head bolt at the locations illustrated. If any bolt exceeds 5.717–5.740 inch (145.2–145.8mm) replace the bolt.

11. Install a new cylinder head gasket.

12. Install the cylinder head.

13. Apply clean engine oil to the bolt threads and seating faces.

14. Install new cylinder head bolts and torque in 2–3 steps, in sequence as follows:

   a. Step 1: Tighten the bolts in sequence to 27–97 inch lbs. (3–11 Nm).

   b. Step 2: Tighten the bolts in sequence to 9.5–12.5 ft. lbs. (13–17 Nm).

   c. Step 3: Tighten the bolts in sequence to 32–34.5 ft. lbs. (43–47 Nm).

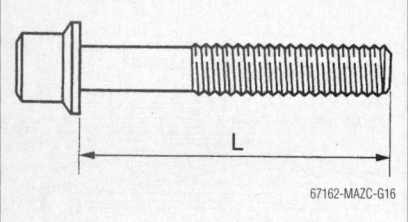

Fig. 97 Measure the length of each head bolt at the locations illustrated, if reusing the old bolts. If any bolt exceeds 5.717–5.740 inch (145.2–145.8mm) replace the bolt—except 2010 2.0L and 2.5L engines

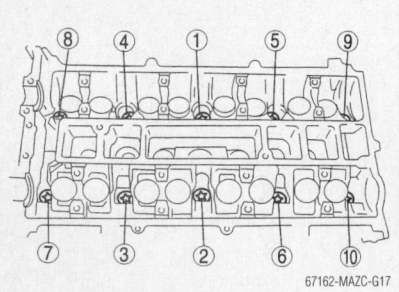

Fig. 98 Cylinder head bolt tightening sequence—except 2010 2.0L and 2.5L engines

   d. Step 4: Paint a mark on the edge of each cylinder head bolt to use as a reference. Turn each bolt, in sequence, 85–92 degrees.

   e. Step 5: Turn each bolt, in sequence, 85–92 degrees.

15. Install the camshafts and caps in their original positions.

16. Install the remaining components in the reverse order of removal.

17. Fill and bleed the cooling system.

18. Change the oil and filter.

19. Run the engine and check for proper operation.

20. Use the Mazda diagnostic scan tool, or equivalent and reprogram the required systems.

2010 2.0L & 2.5L Engines

See Figure 99.

> ❈❈ **WARNING**
>
> **Fuel vapor is hazardous. It can very easily ignite, causing death, serious injury, or damage. Always keep sparks and flames away from fuel.**

> ❈❈ **WARNING**
>
> **Fuel line spills and leakage from the pressurized fuel system are dangerous. Fuel can easily ignite and cause serious injury or death and damage. Fuel can also irritate skin and eyes.**

1. Before servicing the vehicle, refer to the Precautions Section.

➡If working near and/or around the SRS system and components, be sure to disable the SRS system. Tape the negative battery cable with insulating tape. Always disconnect the negative battery cable first.

> ❈❈ **CAUTION**
>
> **To avoid personal injury when working on vehicles equipped with an air bag, the negative battery cable must be disconnected and at least one minute must elapse before working on the system. Failure to do so may result in deployment of the air bag.**

2. Disconnect the negative battery cable. Tape the cable with insulating tape.

➡When disconnecting the cable, some systems need to be initialized after the cable is reconnected. You will need the Mazda diagnostic scan tool or

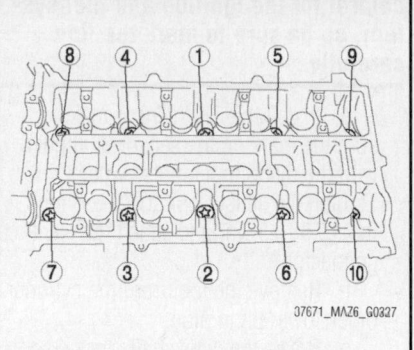

Fig. 99 Cylinder head bolt tightening sequence—2010 2.0L and 2.5L engines

equivalent. Follow the directions on the tool.

3. Drain the engine coolant.

4. Remove the alternator. Place the alternator out of the way with the wiring harnesses connected.

5. Remove the exhaust manifold.

6. Remove the intake manifold.

7. Disconnect the heater hose and radiator hose.

8. Remove the timing chain.

9. Remove the camshafts.

10. Disconnect the wiring harness.

11. To firmly support the engine, first set the engine jack and attachment to the oil pan.

12. Remove the oil control valve.

13. Remove the camshafts.

14. Remove the bolts and the cylinder head.

To install:

➡Be sure to use new fasteners, as required.

15. Measure the cylinder head bolts. Replace any bolts that exceed 5.740 inches (146 mm).

16. Install the cylinder head with a new gasket. Tighten the bolts in sequence as follows:

   a. Step 1: 27–97 inch lbs. (3–11 Nm)

   b. Step 2: 116–150 inch lbs. (13–17 Nm)

   c. Step 3: 32–34 ft. lbs. (43–47 Nm)

   d. Step 4: Plus 88–92°

   e. Step 5: Plus 88–92°

17. Continue the installation in the reverse of the removal procedure.

18. Bleed the air from the cooling system.

19. Inspect the compression pressure.

20. Use the Mazda diagnostic scan tool, or equivalent and reprogram the required systems.

EXHAUST MANIFOLD

REMOVAL & INSTALLATION

Except 2010 2.0L & 2.5L Engines

See Figures 100 and 101.

1. Before servicing the vehicle, refer to the Precautions Section.

➡ **If working near and/or around the SRS system and components, be sure to disable the SRS system. Tape the negative battery cable with insulating tape. Always disconnect the negative battery cable first.**

> ✳✳ **CAUTION**
>
> **To avoid personal injury when working on vehicles equipped with an air bag, the negative battery cable must be disconnected and at least one minute must elapse before working on the system. Failure to do so may result in deployment of the air bag.**

2. Disconnect the negative battery cable. Tape the cable with insulating tape.

➡ **When disconnecting the cable, some systems need to be initialized after the cable is reconnected. You will need the Mazda diagnostic scan tool or equivalent. Follow the directions on the tool.**

3. Remove the plug hole plate.
4. Remove the rear and front tunnel members.
5. If necessary, remove the main silencer as follows:
    a. Disconnect the ABS sensor wiring harness connector.
    b. Disengage the brake pipe mount from the bracket and remove the lower shock absorber bolts.
    c. Loosen the rear crossmember bolts

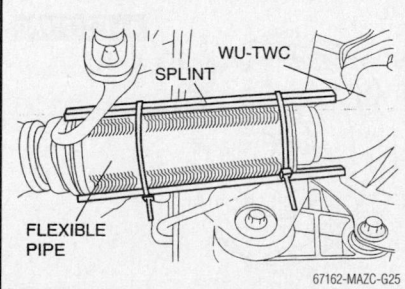

Fig. 100 Support the flexible pipe with a support wrap or splint—except 2010 2.0L and 2.5L engines

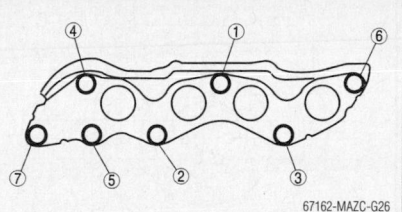

Fig. 101 Tighten the manifold retainers in this sequence—except 2010 2.0L and 2.5L engines

and lower the crossmember approximately 2.8 inches (70mm) and remove the main silencer.

6. Unplug both oxygen sensor (O_2S) connectors.
7. Remove the exhaust manifold bracket, heat shield and clip.
8. Remove the front wheels.
9. Disconnect the steering shaft from the steering gear and linkage side.
10. Support the engine and remove the No. 1 engine mount.
11. Loosen the exhaust manifold bolts.
12. Remove the front stabilizer and stabilizer control link bolts.
13. Loosen the front crossmember bolts and lower the crossmember approximately 3.94 inches (100mm).
14. Support the flexible pipe with a support wrap or splint as illustrated.
15. Remove the exhaust manifold and gasket.

To install:

16. Installation a new exhaust manifold gasket and the manifold . Tighten the manifold retainers in the sequence illustrated to 32–47 ft. lbs. (43–64 Nm).
17. Install the remaining components in the reverse order of removal, using the exploded view of the exhaust system for component location and torque specifications.
18. Use the Mazda diagnostic scan tool, or equivalent and reprogram the required systems.

2010 2.0L & 2.5L Engines

See Figure 102.

> ✳✳ **CAUTION**
>
> **A hot engine and exhaust system can cause severe burns. Turn off the engine and wait until they are cool before removing the exhaust system.**

1. Before servicing the vehicle, refer to the Precautions Section.

➡ **If working near and/or around the SRS system and components, be sure to disable the SRS system. Tape the negative battery cable with insulating tape. Always disconnect the negative battery cable first.**

> ✳✳ **CAUTION**
>
> **To avoid personal injury when working on vehicles equipped with an air bag, the negative battery cable must be disconnected and at least one minute must elapse before working on the system. Failure to do so may result in deployment of the air bag.**

2. Disconnect the negative battery cable. Tape the cable with insulating tape.

➡ **When disconnecting the cable, some systems need to be initialized after the cable is reconnected. You will need the Mazda diagnostic scan tool or equivalent. Follow the directions on the tool.**

3. Remove the plug hole plate.
4. Remove the undercover, as required.
5. Remove the exhaust system insulator.
6. Unbolt the WU-TWC from the exhaust manifold.
7. Unbolt an remove the exhaust manifold.

To install:

➡ **Be sure to use new fasteners, as required.**

8. Installation is the reverse of the removal procedure.
9. Tighten the exhaust manifold new installation nuts to 32–47 ft. lbs. (43–64 Nm).

➡ **Always use new gaskets, and clean off any old gasket material.**

10. Tighten the WU-TWC to the exhaust manifold to 29–37 ft. lbs.(38–51 Nm).

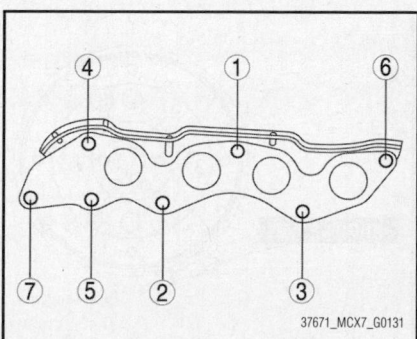

Fig. 102 Exhaust manifold tightening sequence—2010 2.0L and 2.5L engines

11. Tighten the exhaust manifold insulator to 71–88 inch lbs. 8–10 Nm).

12. Start the engine and check for leaks.

13. Use the Mazda diagnostic scan tool, or equivalent and reprogram the required systems.

FLEXPLATE

REMOVAL & INSTALLATION

See Figures 103 through 105.

1. Before servicing the vehicle, refer to the Precautions Section.

➡If working near and/or around the SRS system and components, be sure to disable the SRS system. Tape the negative battery cable with insulating tape. Always disconnect the negative battery cable first.

✳✳ CAUTION

To avoid personal injury when working on vehicles equipped with an air bag, the negative battery cable must be disconnected and at least one minute must elapse before working on the system. Failure to do so may result in deployment of the air bag.

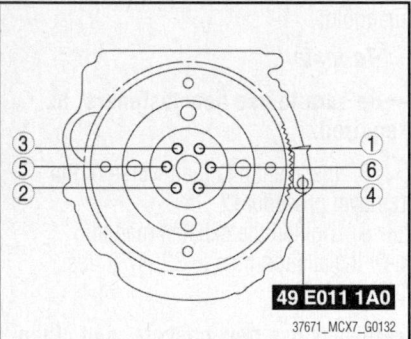

Fig. 103 Special service tool 49 E011 1A0 in place for bolt removal–2010 2.0L and 2.5L engines

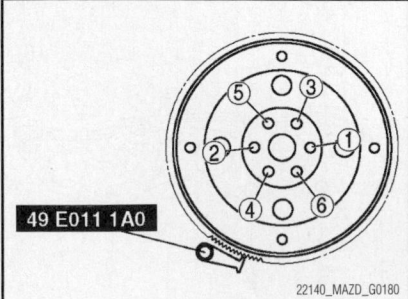

Fig. 104 Flexplate bolt tightening sequence–except 2010 2.0L and 2.5L engines

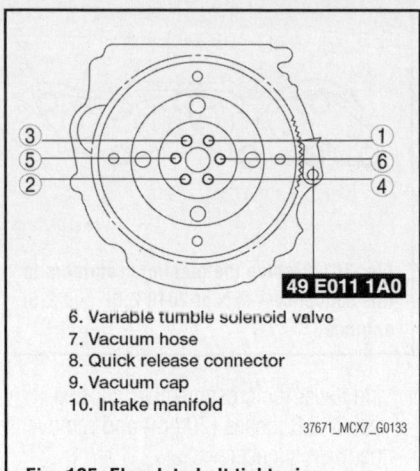

6. Variable tumble solenoid valve
7. Vacuum hose
8. Quick release connector
9. Vacuum cap
10. Intake manifold

Fig. 105 Flexplate bolt tightening sequence–2010 2.0L and 2.5L engines

2. Disconnect the negative battery cable. Tape the cable with insulating tape.

➡When disconnecting the cable, some systems need to be initialized after the cable is reconnected. You will need the Mazda diagnostic scan tool or equivalent. Follow the directions on the tool.

3. Remove the transaxle.

4. Install SST 49 E011 1A0 or equivalent as shown to hold the flexplate from turning, 2010 2.0L and 2.5L engines.

5. Remove the bolts and remove the flexplate.

To install:

➡Be sure to use new fasteners, as required.

6. Remove the sealant from the bolt holes in the crankshaft and from the drive plate mounting bolts.

7. Install the drive plate.

8. Install the backing plate.

9. Set the SST 49 E011 1A0 or equivalent against the flexplate as shown.

10. Tighten the flexplate bolts in sequence and in two or three passes to specification.

11. Install the transaxle.

12. Use the Mazda diagnostic scan tool, or equivalent and reprogram the required systems.

FLYWHEEL

REMOVAL & INSTALLATION

See Figures 106 and 107.

1. Before servicing the vehicle, refer to the Precautions Section.

➡If working near and/or around the SRS system and components, be sure

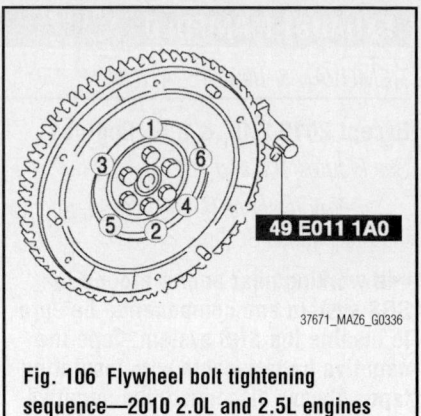

Fig. 106 Flywheel bolt tightening sequence—2010 2.0L and 2.5L engines

to disable the SRS system. Tape the negative battery cable with insulating tape. Always disconnect the negative battery cable first.

✳✳ CAUTION

To avoid personal injury when working on vehicles equipped with an air bag, the negative battery cable must be disconnected and at least one minute must elapse before working on the system. Failure to do so may result in deployment of the air bag.

2. Disconnect the negative battery cable. Tape the cable with insulating tape.

➡When disconnecting the cable, some systems need to be initialized after the cable is reconnected. You will need the Mazda diagnostic scan tool or equivalent. Follow the directions on the tool.

3. Remove the transaxle.

4. Remove the flywheel retaining bolts.

5. Remove the component from its mounting.

To install:

➡Be sure to use new fasteners, as required.

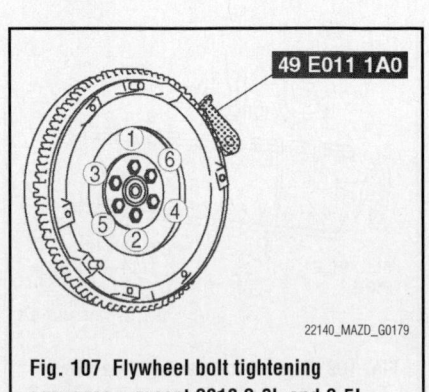

Fig. 107 Flywheel bolt tightening sequence—except 2010 2.0L and 2.5L engines

6. Installation is the reverse of the removal procedure.

7. Tighten the retaining bolts to specification.

8. Use the Mazda diagnostic scan tool, or equivalent and reprogram the required systems.

INTAKE MANIFOLD

REMOVAL & INSTALLATION

2009 2.0L & 2.3L Engines (except Turbocharged Engine)

See Figures 108 through 113.

1. Before servicing the vehicle, refer to the Precautions Section.

➡ If working near and/or around the SRS system and components, be sure to disable the SRS system. Tape the negative battery cable with insulating tape. Always disconnect the negative battery cable first.

✳ CAUTION

To avoid personal injury when working on vehicles equipped with an air bag, the negative battery cable must be disconnected and at least one

minute must elapse before working on the system. Failure to do so may result in deployment of the air bag.

2. Disconnect the negative battery cable. Tape the cable with insulating tape.

➡ When disconnecting the cable, some systems need to be initialized after the cable is reconnected. You will need the Mazda diagnostic scan tool or equivalent. Follow the directions on the tool.

3. Drain the cooling system.
4. Relieve the fuel system pressure.

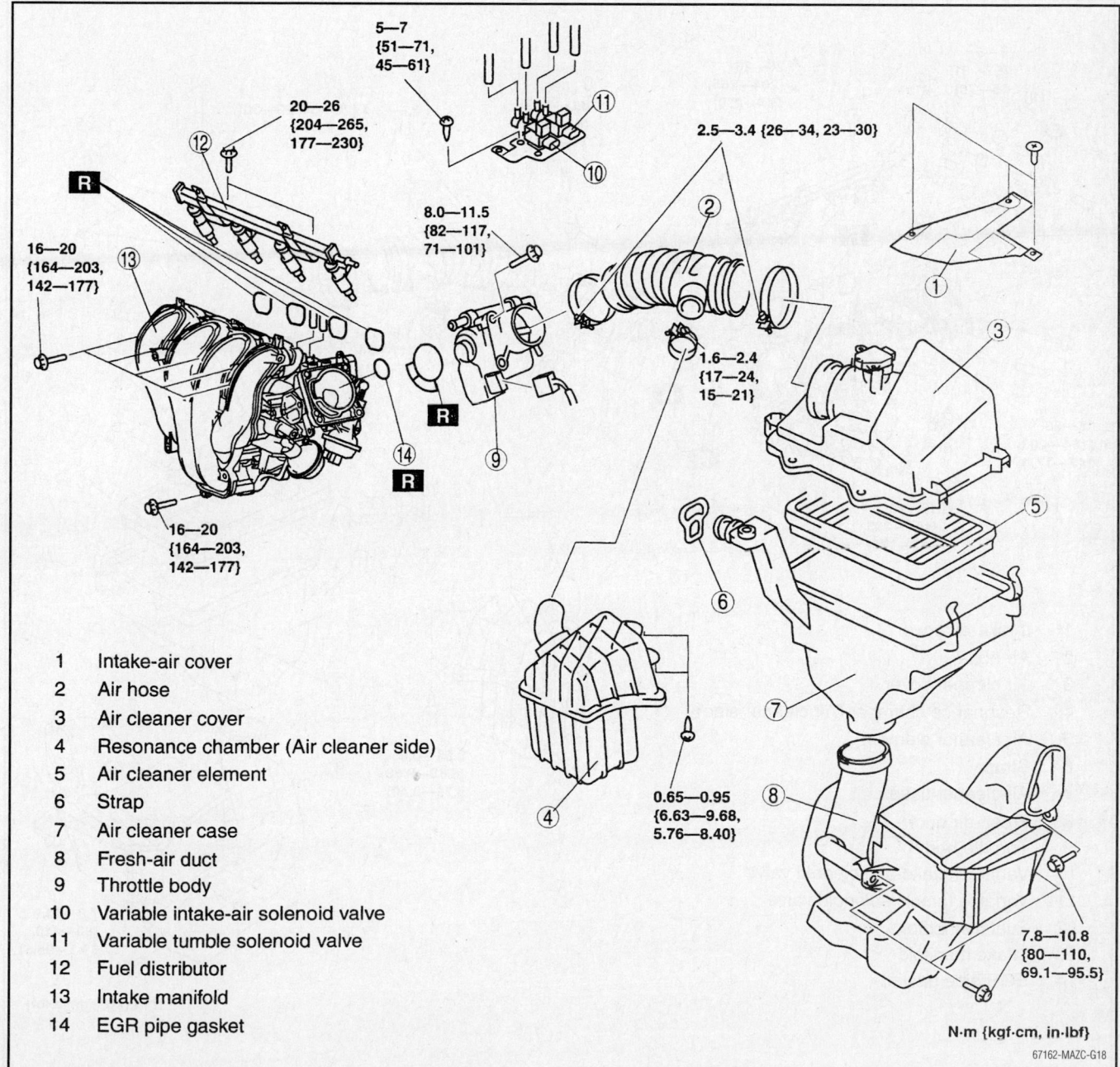

1 Intake-air cover
2 Air hose
3 Air cleaner cover
4 Resonance chamber (Air cleaner side)
5 Air cleaner element
6 Strap
7 Air cleaner case
8 Fresh-air duct
9 Throttle body
10 Variable intake-air solenoid valve
11 Variable tumble solenoid valve
12 Fuel distributor
13 Intake manifold
14 EGR pipe gasket

N·m {kgf·cm, in·lbf}

67162-MAZC-G18

Fig. 108 Exploded view of the intake manifold assembly and related components—2009 2.0L engine

5. Remove the plug hole plate by lifting off and removing the plug hole plate from the areas shown in the accompanying illustration.

6. Remove the battery cover and battery duct.

7. Remove the undercover and disconnect the negative battery cable.

8. Remove the intake air cover, the air hose and the air cleaner element.

9. Remove the resonance chamber and air filter element.

10. Remove the strap and the air cleaner case.

11. Remove the fresh air duct.

12. Remove the throttle body.

13. Remove the variable intake air solenoid valve and the variable tumble solenoid valve.

14. Remove the fuel rail and injectors as an assembly.

15. Disconnect the vacuum hose from the intake manifold.

16. Remove the engine oil dipstick tube.

17. Remove the intake manifold bolts, the manifold and gasket.

18. Discard the gasket.

19. Clean the mating surfaces of any gasket material.

20. If necessary, remove the Exhaust Gas Recirculation (EGR) pipe gasket.

To install:

21. If removed, install the EGR pipe gasket.

22. Install a new gasket and the manifold. Tighten the manifold bolts to 142–177 inch lbs. (16–20 Nm).

23. Install the engine oil dipstick tube.

24. Connect the vacuum hose to the intake manifold.

25. Install the fuel rail and injectors.

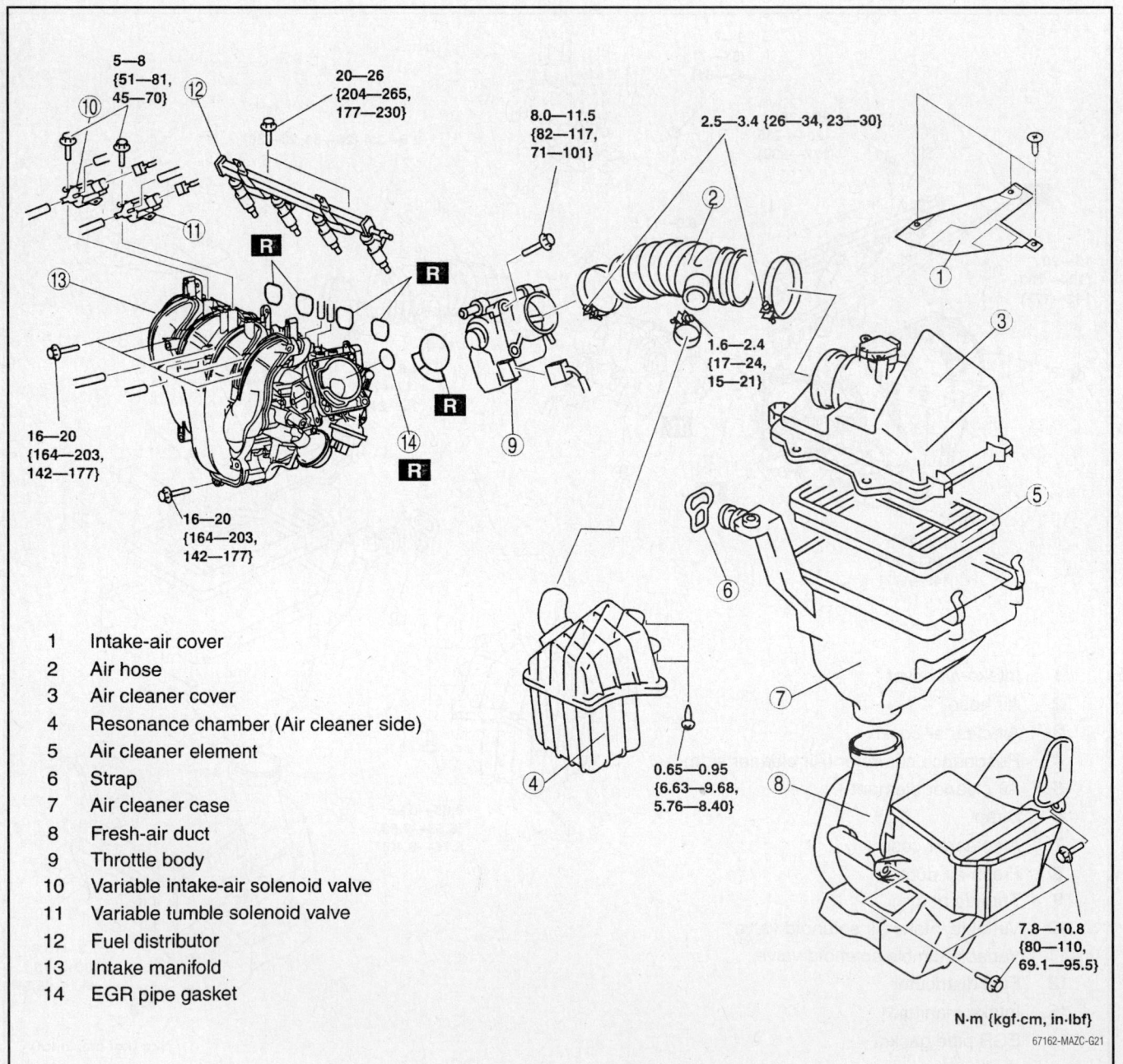

1	Intake-air cover
2	Air hose
3	Air cleaner cover
4	Resonance chamber (Air cleaner side)
5	Air cleaner element
6	Strap
7	Air cleaner case
8	Fresh-air duct
9	Throttle body
10	Variable intake-air solenoid valve
11	Variable tumble solenoid valve
12	Fuel distributor
13	Intake manifold
14	EGR pipe gasket

N·m {kgf·cm, in·lbf}

67162-MAZC-G21

Fig. 109 Exploded view of the intake manifold assembly and related components—2009 2.3L engine

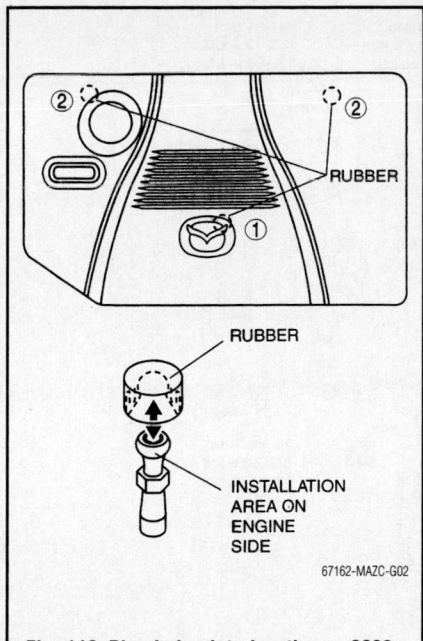

Fig. 110 Plug hole plate locations—2009 2.0L and 2.3L engines

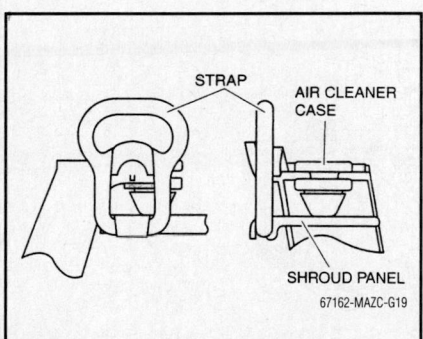

Fig. 111 Use the strap to secure the shroud panel and the air cleaner case—2009 2.0L and 2.3L engines

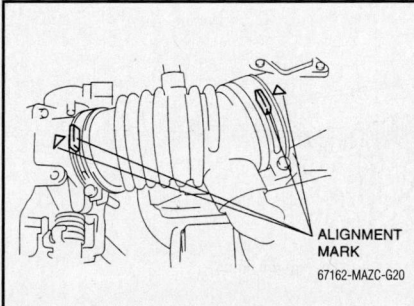

Fig. 112 Make sure to align the alignment marks on the throttle body and the air hose—2009 2.0L and 2.3L engines

26. Install the variable intake air solenoid valve and the variable tumble solenoid valve.

27. Install the throttle body and tighten the bolts to 71–101 inch lbs. (8–11.5 Nm).

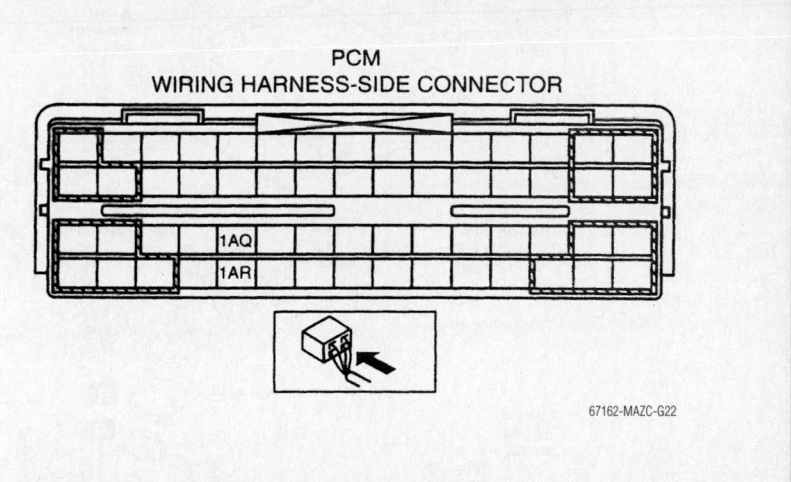

Fig. 113 If equipped with an immobilizer, ground PCM terminal 1AR, if not equipped with an immobilizer, ground PCM terminal 1AQ—2009 2.0L and 2.3L engines

28. Install the fresh air duct.

29. Verify the rubber mounts on the battery support bracket are still in place. Insert the air cleaner case into the rubber mounts, using soapy water if necessary to ease installation.

30. Use the strap to secure the shroud panel and the air cleaner case as shown in the accompanying illustration.

31. Install the air cleaner element and the resonance chamber.

32. Install the air cleaner cover.

33. Install the air hose, make sure to align the alignment marks on the throttle body and the air hose.

34. Install the intake air cover.

35. Install the undercover.

36. Connect the negative battery cable.

37. Install the battery duct and cover.

38. Install the plug hole plate.

39. Fill the cooling system.

40. Start the vehicle and check for leaks as follows:

   a. Using a jumper wire, ground the Powertrain Control Module (PCM) terminals. If equipped with an immobilizer, ground terminal 1AR, if not equipped with an immobilizer, ground terminal 1AQ. Refer to the illustration for terminal location.

   b. Turn the ignition switch the ON position to activate the fuel pump.

   c. Check the hoses, clips and other fuel system components for leaks.

   d. If there are any leaks, replace the fuel hoses and clips. If the is damage to the seal on the fuel pipe side, replace the pipe.

   e. The system must be leak free for five minutes with the terminal grounded.

If any component is replaced because of a system, leak, turn the ignition key OFF; remove the jumper wire from the terminal. reapply the jumper wire, turn the ignition On and check for leaks.

41. Use the Mazda diagnostic scan tool, or equivalent and reprogram the required systems.

2009–10 2.3L Engine with Turbocharger

See Figure 114.

1. Before servicing the vehicle, refer to the Precautions Section.

➡**If working near and/or around the SRS system and components, be sure to disable the SRS system. Tape the negative battery cable with insulating tape. Always disconnect the negative battery cable first.**

✶✶ CAUTION

To avoid personal injury when working on vehicles equipped with an air bag, the negative battery cable must be disconnected and at least one minute must elapse before working on the system. Failure to do so may result in deployment of the air bag.

2. Disconnect the negative battery cable. Tape the cable with insulating tape.

➡**When disconnecting the cable, some systems need to be initialized after the cable is reconnected. You will need the Mazda diagnostic scan tool or equivalent. Follow the directions on the tool.**

3. Remove the charge air cooler cover.

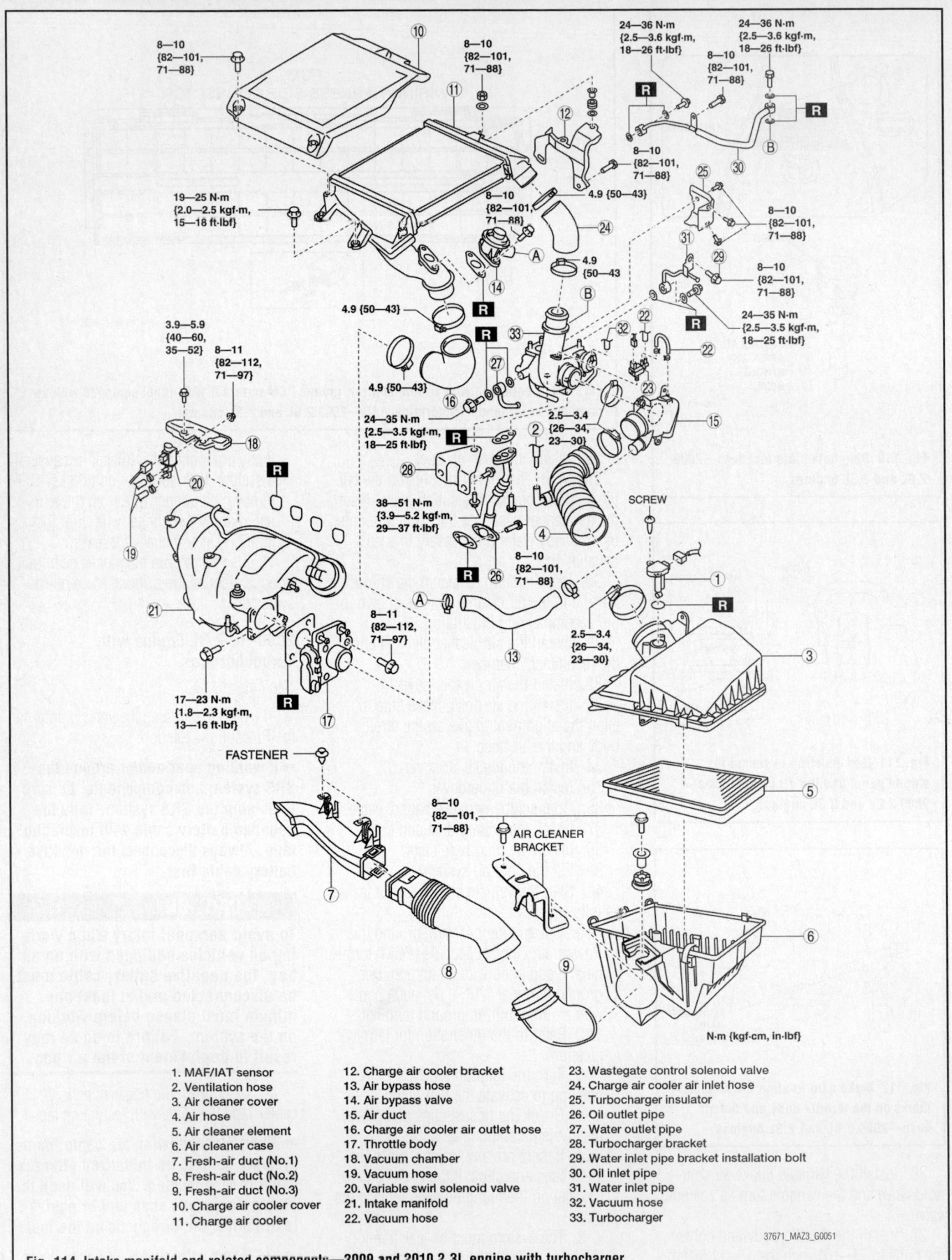

1. MAF/IAT sensor
2. Ventilation hose
3. Air cleaner cover
4. Air hose
5. Air cleaner element
6. Air cleaner case
7. Fresh-air duct (No.1)
8. Fresh-air duct (No.2)
9. Fresh-air duct (No.3)
10. Charge air cooler cover
11. Charge air cooler

12. Charge air cooler bracket
13. Air bypass hose
14. Air bypass valve
15. Air duct
16. Charge air cooler air outlet hose
17. Throttle body
18. Vacuum chamber
19. Vacuum hose
20. Variable swirl solenoid valve
21. Intake manifold
22. Vacuum hose

23. Wastegate control solenoid valve
24. Charge air cooler air inlet hose
25. Turbocharger insulator
26. Oil outlet pipe
27. Water outlet pipe
28. Turbocharger bracket
29. Water inlet pipe bracket installation bolt
30. Oil inlet pipe
31. Water inlet pipe
32. Vacuum hose
33. Turbocharger

37671_MAZ3_G0051

Fig. 114 Intake manifold and related components—2009 and 2010 2.3L engine with turbocharger

4. Remove the battery and tray.

5. Remove the air cleaner.

6. Remove the resonance chamber, remove the left hand front mudguard before removing the resonance chamber.

7. Disconnect the MAF/IAT sensor.

8. Remove the charge air cooler.

9. Remove the air bypass valve.

10. Remove the throttle body.

11. Remove the air hose.

12. Remove the intake manifold.

13. Remove the air bypass valve.

14. Drain and recycle the engine coolant.

15. Remove the throttle body.

16. Disconnect the EGR valve connector.

17. Remove the air hose.

18. Remove the high pressure fuel pump bracket and the EGR pipe bracket.

19. Remove the fuel delivery pipe cover.

20. Remove the oil level gauge pipe.

21. Remove the drive belt.

22. Remove the power steering pump out of the way with the lines still attached.

23. Remove the vacuum hose.

24. Remove the intake manifold.

To install:

25. Installation is the reverse of removal, note the following:

   a. Tighten the intake manifold bolts to 13–16 ft. lbs. (17–23 Nm).

   b. Tighten the fuel delivery pipe cover bolts to 13–16 ft. lbs. (17–23 Nm).

   c. Tighten the charge air cooler bracket to 31–35 ft. lbs. (42–48 Nm).

   d. Tighten the charge air cooler to 14–19 ft. lbs. (42–48 Nm).

✳✳ CAUTION

When installing the charge air cooler cover, be careful not to damage the charge air cooler cover clips.

   e. Tighten the throttle body installation bolts to 71–101 inch lbs. (8–11.5 Nm).

   f. Perform the following when installing the air cleaner case:

   - Verify that the rubber mounts are set in the air cleaner bracket
   - Install the projections on the frame side.
   - Verify that the projections on the frame side are installed securely.
   - Install the projection on the engine side.
   - Verify that the projection on the engine side installed securely.

26. Use the Mazda diagnostic scan tool, or equivalent and reprogram the required systems.

2010 2.0L & 2.5L Engines

See Figures 115 through 117.

✳✳ WARNING

A hot engine and intake air system can cause severe burns. Turn off the engine and wait until they are cool before removing the intake air system.

✳✳ WARNING

Fuel vapor is hazardous. It can easily ignite, causing serious injury and damage. Always keep sparks and flames away from fuel.

✳✳ WARNING

Fuel line spills and leakage are dangerous. Fuel can ignite and cause serious injuries or death and damage. Fuel can also irritate skin and eyes.

1. Before servicing the vehicle, refer to the Precautions Section.

➡If working near and/or around the SRS system and components, be sure to disable the SRS system. Tape the negative battery cable with insulating tape. Always disconnect the negative battery cable first.

✳✳ CAUTION

To avoid personal injury when working on vehicles equipped with an air bag, the negative battery cable must be disconnected and at least one minute must elapse before working on the system. Failure to do so may result in deployment of the air bag.

2. Disconnect the negative battery cable. Tape the cable with insulating tape.

➡When disconnecting the cable, some systems need to be initialized after the cable is reconnected. You will need the Mazda diagnostic scan tool or equivalent. Follow the directions on the tool.

3. Drain the cooling system.

4. Remove the plug hole plate.

5. Disconnect the wiring harness.

6. Remove or disconnect the following:

   - MAF/IAT sensor
   - Ventilation hose
   - Fresh-air duct
   - Air cleaner cover
   - Air cleaner element
   - Air cleaner case
   - Remove the mudguard (LF), as required.
   - Resonance chamber
   - Air hose

7. Remove or disconnect the following:

   - Water hose
   - Throttle body
   - Vacuum hose
   - Variable intake air solenoid valve
   - Vacuum hose
   - Variable tumble solenoid valve
   - Vacuum hose
   - Quick release connector
   - Vacuum cap
   - Intake manifold
   - Coolant reserve tank.
   - Fan control module No.2.
   - Engine under cover
   - All clips for securing wiring harnesses from the intake manifold

8. Remove the intake manifold.

To install:

➡Be sure to use new fasteners, as required.

9. Installation is the reverse of the removal procedure.

   - Tighten the throttle body to 71–101 inch lbs. (8–11.5 Nm).
   - Adjust the convex of the throttle body and the air hose.

10. Bleed the air from the cooling system.

11. Use the Mazda diagnostic scan tool, or equivalent and reprogram the required systems.

OIL PAN

REMOVAL & INSTALLATION

2009 Vehicles

See Figures 118 through 123.

1. Before servicing the vehicle, refer to the Precautions Section.

➡If working near and/or around the SRS system and components, be sure to disable the SRS system. Tape the negative battery cable with insulating tape. Always disconnect the negative battery cable first.

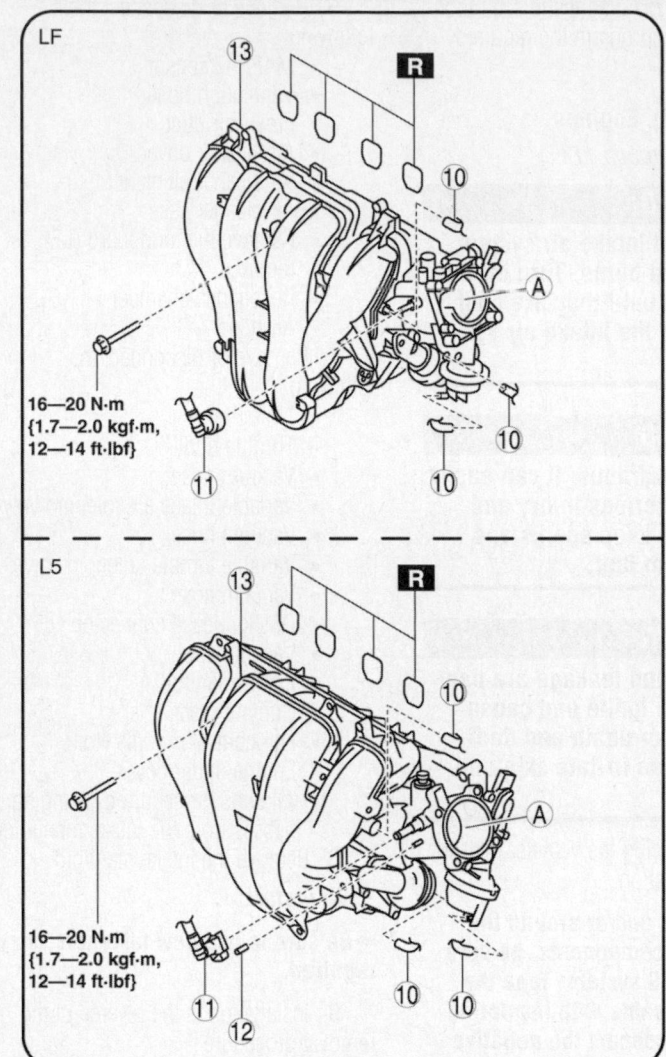

16—20 N·m
{1.7—2.0 kgf·m,
12—14 ft·lbf}

5.0—7.0
{51—71,
45—61}

8—11 {82—112, 71—97}

FROM AIR HOSE

N·m {kgf·cm, in·lbf}

1. Water hose
2. Throttle body connector
3. Throttle body
4. Variable intake air solenoid valve connector
5. Vacuum hose
6. Variable intake air solenoid valve
7. Variable tumble solenoid valve connector
8. Vacuum hose
9. Variable tumble solenoid valve
10. Vacuum hose
11. Quick release connector
12. Vacuum cap (L5)
13. Intake manifold

37671_MAZ3_G0053

Fig. 115 Intake manifold and related components—2010 2.0L and 2.5L engines

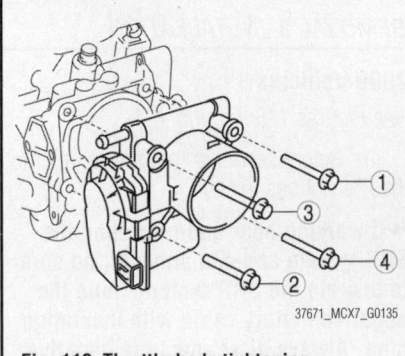

37671_MCX7_G0135

Fig. 116 Throttle body tightening sequence—2010 2.0L and 2.5L engines

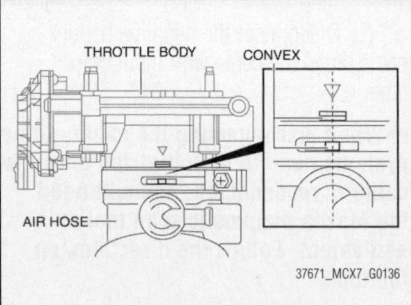

THROTTLE BODY   CONVEX

AIR HOSE

37671_MCX7_G0136

Fig. 117 Convex adjustment view of the throttle body and the air hose—2010 2.0L and 2.5L engines

❊❊ CAUTION

To avoid personal injury when working on vehicles equipped with an air bag, the negative battery cable must be disconnected and at least one minute must elapse before working on the system. Failure to do so may result in deployment of the air bag.

2. Disconnect the negative battery cable. Tape the cable with insulating tape.

➡When disconnecting the cable, some systems need to be initialized after the cable is reconnected. You will need the

Mazda diagnostic scan tool or equivalent. Follow the directions on the tool.

3. Remove the engine undercover and splash shield as an assembly.

4. Remove the right front wheel.

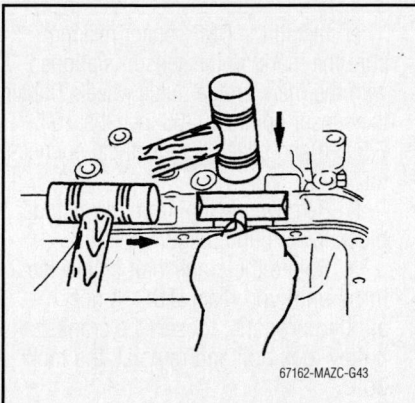

67162-MAZC-G43

Fig. 118 Use a separator tool such as the one illustrated to separate the oil pan from the block—2009 vehicles

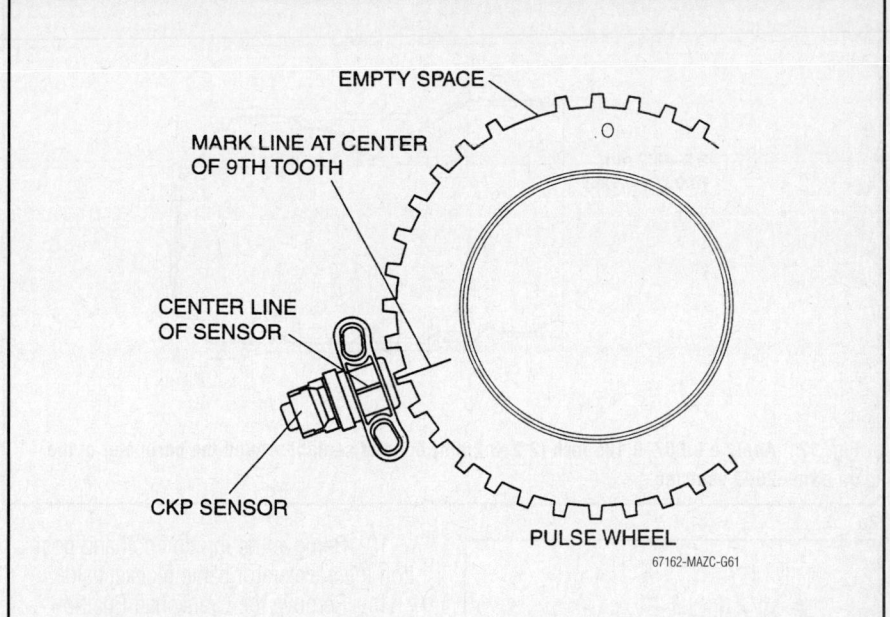

EMPTY SPACE

MARK LINE AT CENTER OF 9TH TOOTH

CENTER LINE OF SENSOR

CKP SENSOR

PULSE WHEEL

67162-MAZC-G61

Fig. 120 Using a ruler, mark the center line on the pulse wheel teeth on the crank pulley which is located at the 9th tooth counting counterclockwise from the empty space—2009 vehicles

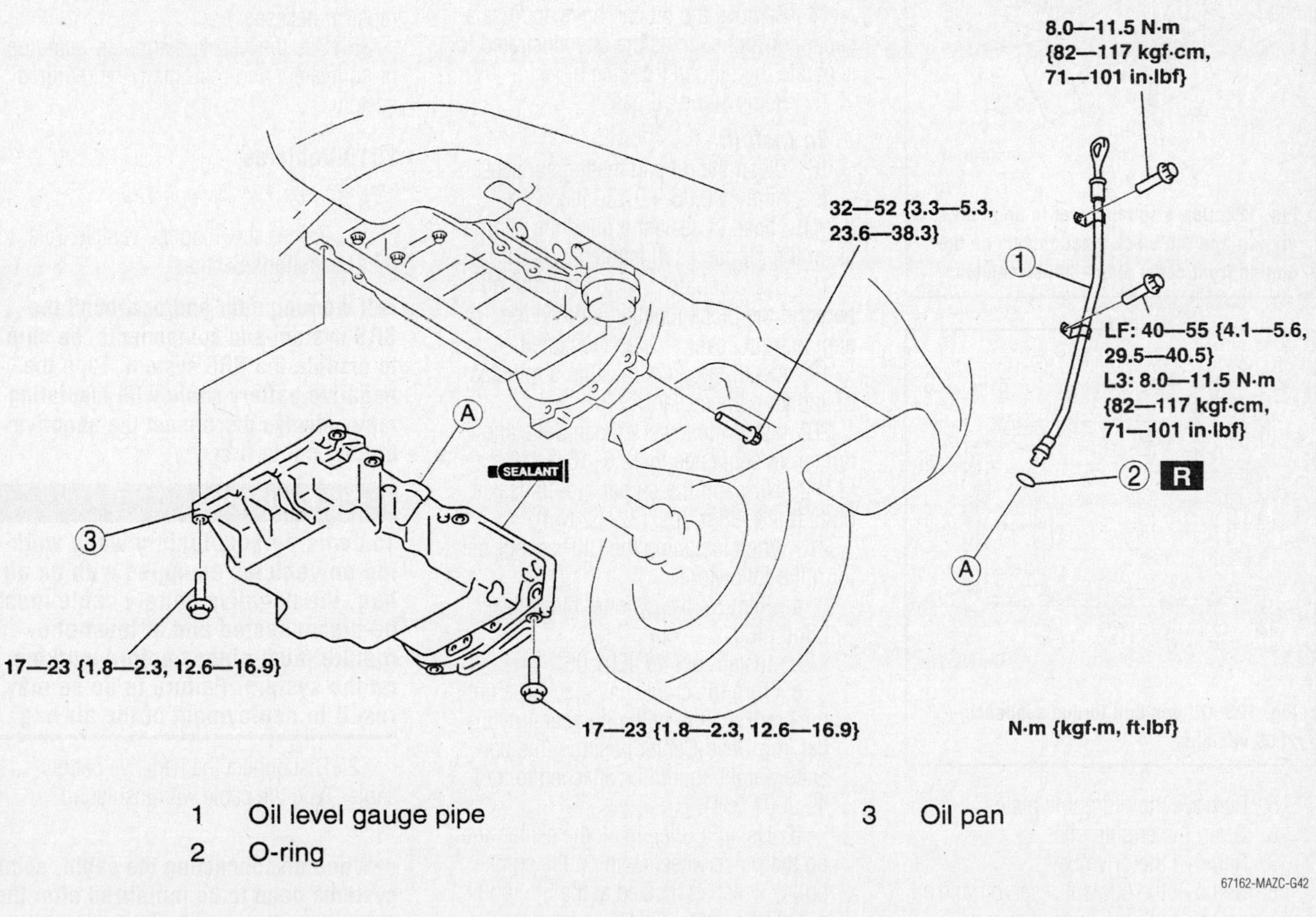

8.0—11.5 N·m {82—117 kgf·cm, 71—101 in·lbf}

32—52 {3.3—5.3, 23.6—38.3}

LF: 40—55 {4.1—5.6, 29.5—40.5}
L3: 8.0—11.5 N·m {82—117 kgf·cm, 71—101 in·lbf}

SEALANT

17—23 {1.8—2.3, 12.6—16.9}

17—23 {1.8—2.3, 12.6—16.9}

N·m {kgf·m, ft·lbf}

1   Oil level gauge pipe
2   O-ring

3   Oil pan

67162-MAZC-G42

Fig. 119 Exploded view of the oil pan and related components—2009 vehicles

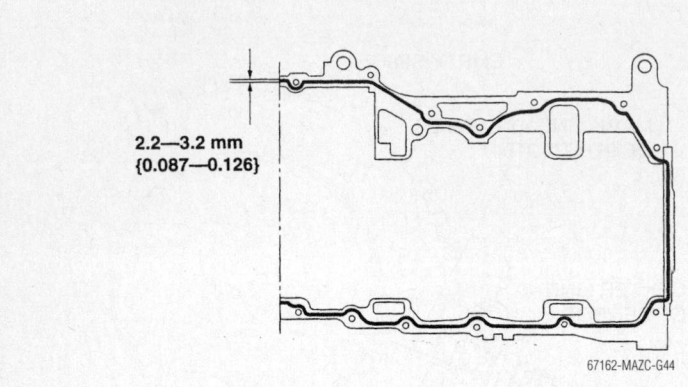

Fig. 121 Apply a 0.087–0.126 inch (2.2–3.2mm) bead of sealant around the perimeter of the oil pan—2009 vehicles

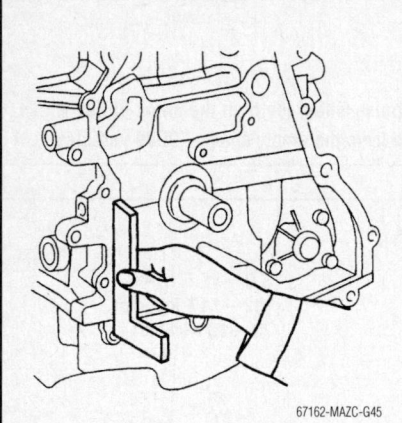

Fig. 122 Use a square ruler to align the oil pan and the block junction side on the engine front cover side—2009 vehicles

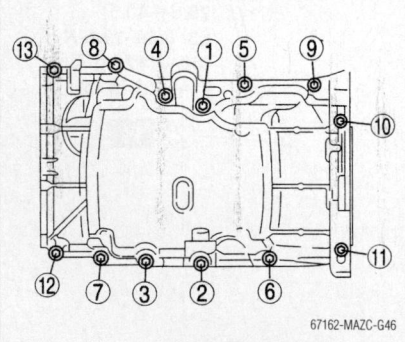

Fig. 123 Oil pan bolt torque sequence—2009 vehicles

5. Remove the plug hole plate.
6. Drain the engine oil.
7. Remove the drive belt.
8. Remove the coolant reservoir and position aside with the lines attached.
9. Remove the A/C compressor and position aside with the lines attached.

10. Remove the ignition coil and position the accelerator cable bracket aside.
11. Remove the Crankshaft Position (CKP) sensor.
12. Remove the engine front cover. Refer to the timing chain removal procedure in this section.
13. Remove the dipstick tube pipe and O–ring.
14. Remove the oil pan bolts and use a separator tool such as the one illustrated to separate the oil pan from the block.
15. Remove the oil pan.

To install:
16. Clean the oil pan mating surfaces.
17. Apply a 0.087–0.126 inch (2.2–3.2mm) bead of sealant around the perimeter of the oil pan as illustrated.
18. Use a square ruler to align the oil pan and the block junction side on the engine front cover side as illustrated.
19. Install the oil pan within 5 minutes of applying the sealant.
20. Install the lower oil pan bolts and tighten in sequence to 12.6–16.9 ft. lbs. (17–23 Nm) and the oil pan–to–transaxle bolts to 23–38 ft. lbs. (32–52 Nm).
21. When installing the CKP sensor perform the following:
   a. Remove the cylinder block lower blind plug.
   b. Install tool 49 JE01 061.
   c. Turn the crankshaft clockwise until the crankshaft is in the number 1 cylinder Top Dead Center position (the balance weight should be attached to tool 49 JE01 061).
   d. Using a ruler, mark the center line on the pulse wheel teeth on the crank pulley which is located at the 9th tooth counting counterclockwise from the empty space. Refer to the illustration for more detail.

✳✳ CAUTION

If you do not mark the center line correctly this will cause improper engine control for the ignition and fuel system, so be sure to mark the line carefully.

   e. Install the CKP sensor making sure the mark on the sensor is aligned with the mark on the pulse wheel. Tighten the sensor bolt to 49–66 inch lbs. (5.5–7.5 Nm)and attach the electrical connector.
   f. Remove the tool from the cylinder block lower blind plug.
   g. Rotate the crankshaft clockwise 2 turns until you reach TDC. If not aligned properly, loosen the crankshaft pulley lock bolt and reinstall the lock bolt.
   h. Install the cylinder block blind plug and tighten to 13–16 ft. lbs. (18–22 Nm).
22. Install the remaining components in the reverse order of removal.
23. Fill the engine with clean oil.
24. Start the vehicle, check for leaks and repair if necessary.
25. Use the Mazda diagnostic scan tool, or equivalent and reprogram the required systems.

2010 Vehicles
See Figures 124 through 129.

1. Before servicing the vehicle, refer to the Precautions Section.

➡**If working near and/or around the SRS system and components, be sure to disable the SRS system. Tape the negative battery cable with insulating tape. Always disconnect the negative battery cable first.**

✳✳ CAUTION

To avoid personal injury when working on vehicles equipped with an air bag, the negative battery cable must be disconnected and at least one minute must elapse before working on the system. Failure to do so may result in deployment of the air bag.

2. Disconnect the negative battery cable. Tape the cable with insulating tape.

➡**When disconnecting the cable, some systems need to be initialized after the cable is reconnected. You will need the Mazda diagnostic scan tool or equivalent. Follow the directions on the tool.**

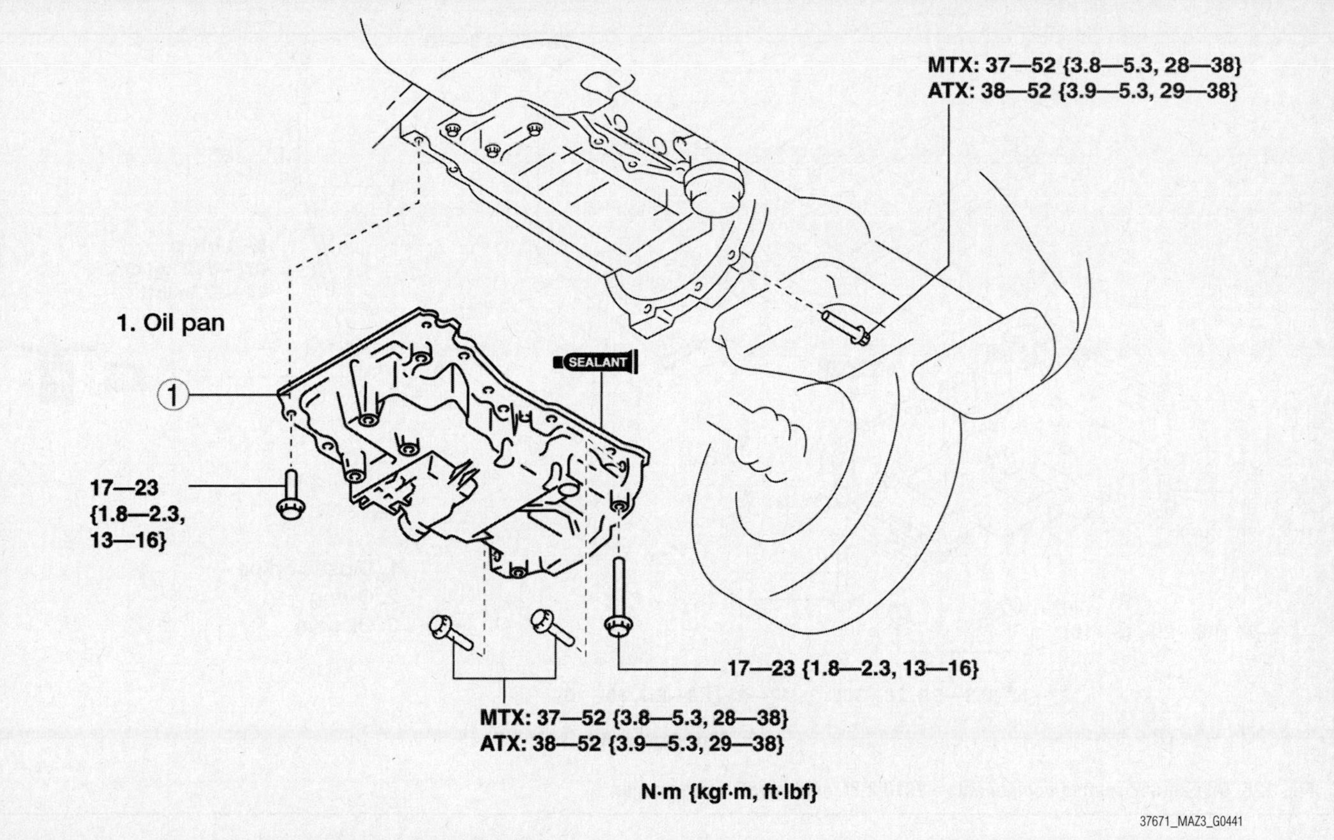

MTX: 37—52 {3.8—5.3, 28—38}
ATX: 38—52 {3.9—5.3, 29—38}

1. Oil pan

SEALANT

17—23
{1.8—2.3,
13—16}

17—23 {1.8—2.3, 13—16}

MTX: 37—52 {3.8—5.3, 28—38}
ATX: 38—52 {3.9—5.3, 29—38}

N·m {kgf·m, ft·lbf}

37671_MAZ3_G0441

Fig. 124 Oil pan and related components—2010 2.0L engine

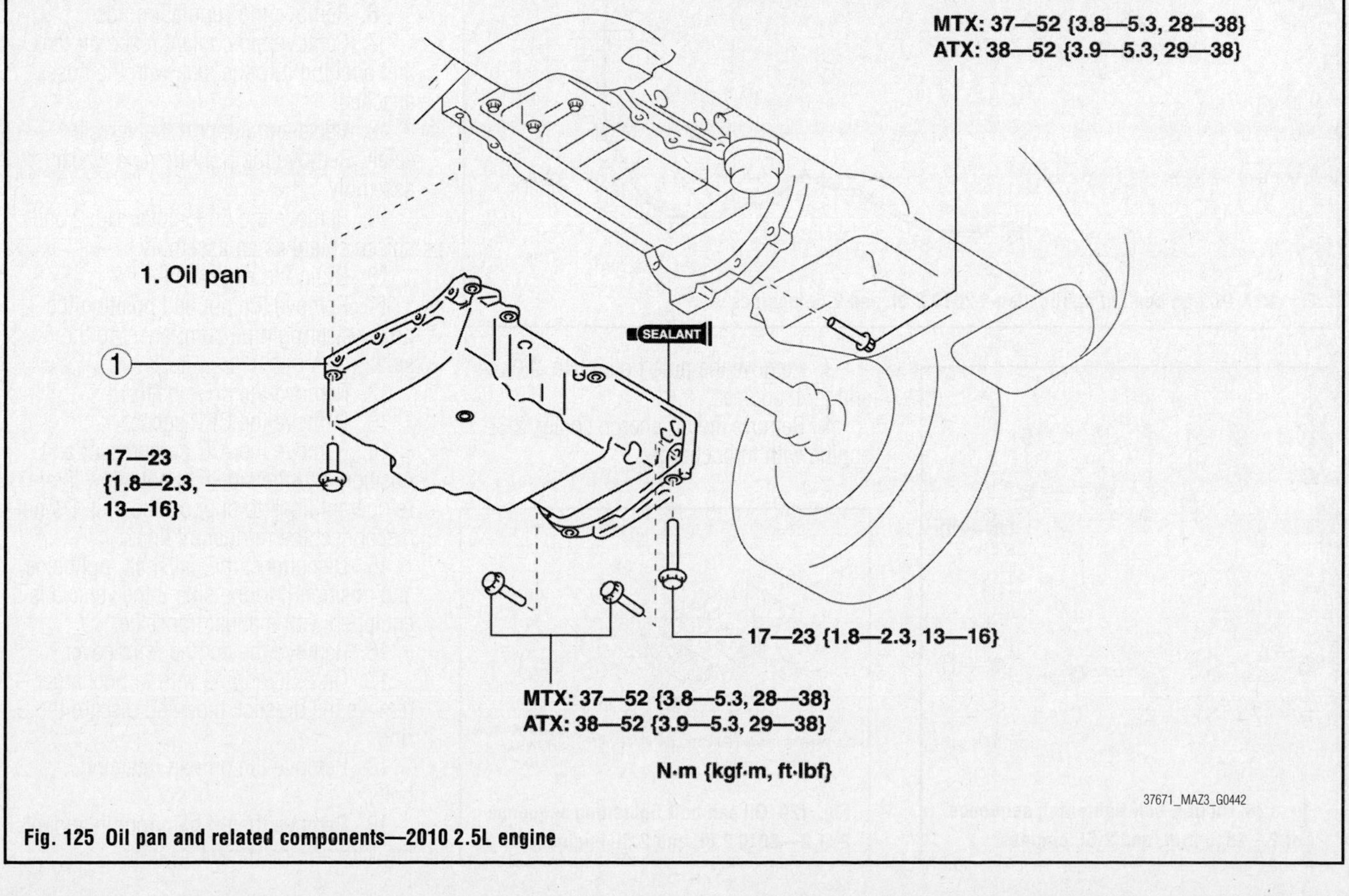

MTX: 37—52 {3.8—5.3, 28—38}
ATX: 38—52 {3.9—5.3, 29—38}

1. Oil pan

SEALANT

17—23
{1.8—2.3,
13—16}

17—23 {1.8—2.3, 13—16}

MTX: 37—52 {3.8—5.3, 28—38}
ATX: 38—52 {3.9—5.3, 29—38}

N·m {kgf·m, ft·lbf}

37671_MAZ3_G0442

Fig. 125 Oil pan and related components—2010 2.5L engine

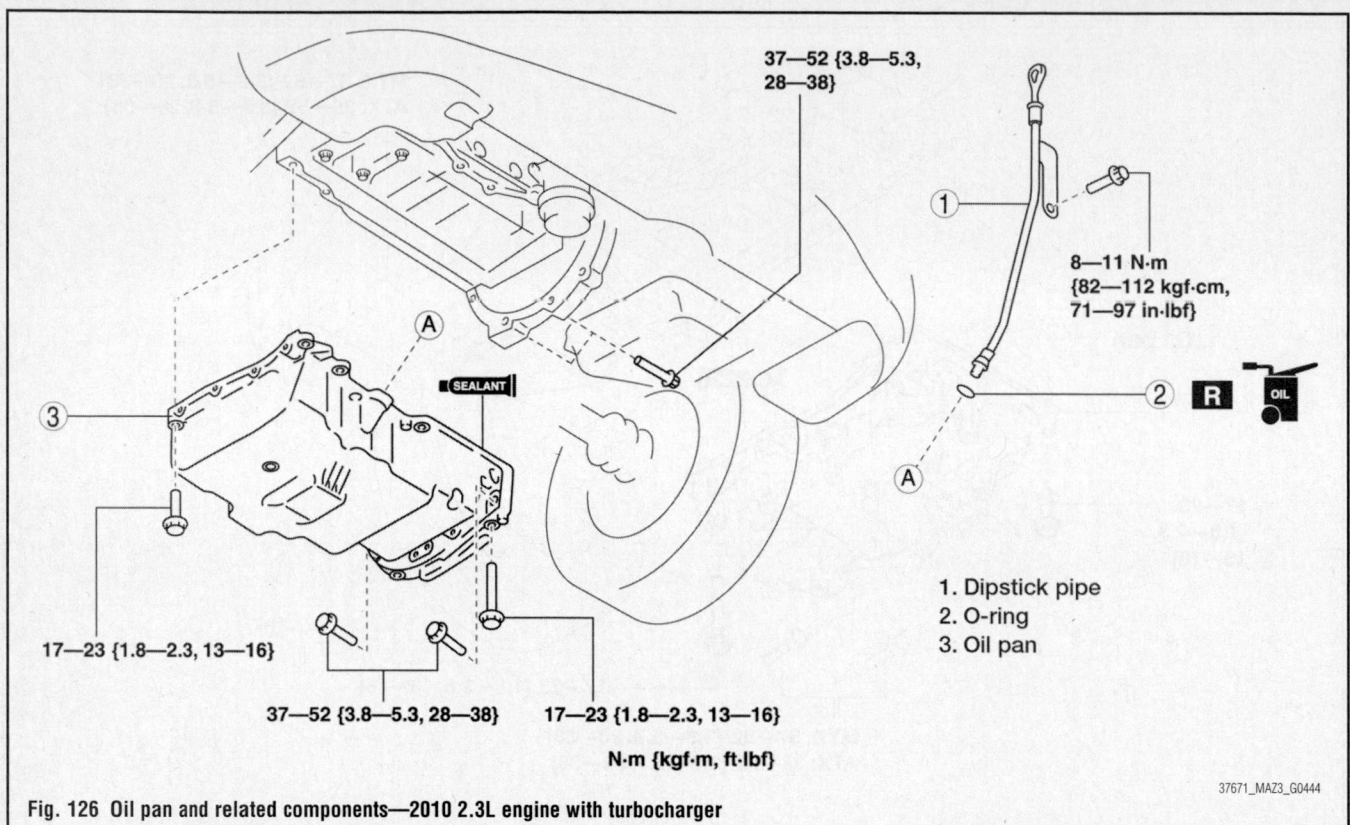

Fig. 126 Oil pan and related components—2010 2.3L engine with turbocharger

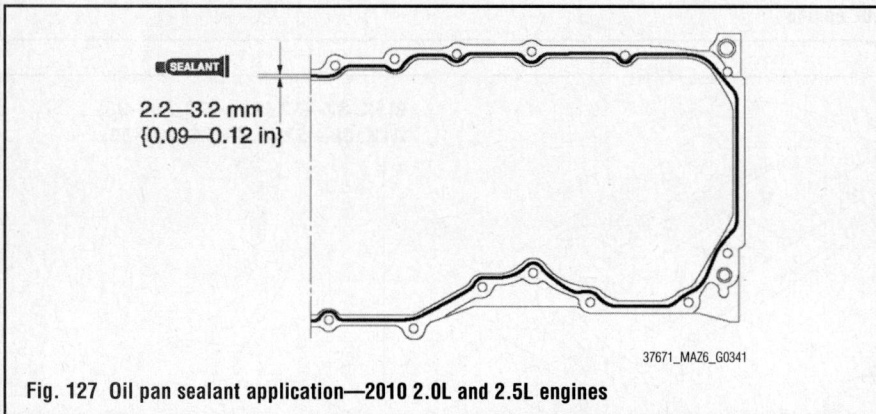

Fig. 127 Oil pan sealant application—2010 2.0L and 2.5L engines

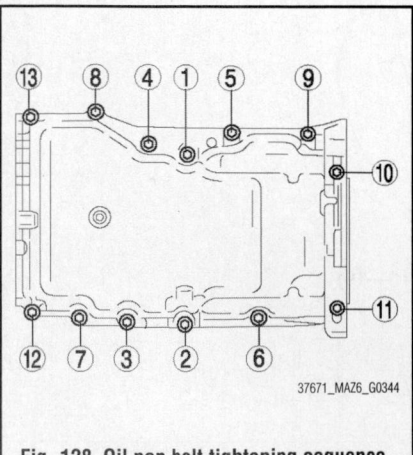

Fig. 128 Oil pan bolt tightening sequence 1 of 2—2010 2.0L and 2.5L engines

3. Remove the plug hole plate, 2.0L and 2.5L engines.

4. Remove the air charge cooler, 2.3L engine with turbocharger.

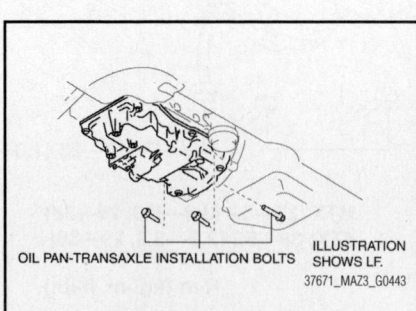

Fig. 129 Oil pan bolt tightening sequence 2 of 2—2010 2.0L and 2.5L engines

5. Disconnect the wiring harness. Remove the ignition coils. Remove the spark plugs.

6. Remove the ventilation hose.

7. Remove the coolant reservoir tank and position it to the side with the hoses attached.

8. Raise and support the vehicle safely. Remove the right tire and wheel assembly.

9. Remove the undercover No. 2 and splash shield as an assembly.

10. Drain the engine oil.

11. Remove the nut and position the power steering pipe component to the side.

12. Remove the drive belt(s).

13. Remove the CKP sensor.

14. Remove the A/C compressor and position it to the side. Do not allow the refrigerant lines to support the unit. Do not disconnect the refrigerant lines.

15. Disconnect the halfshaft, right side, and position it to the side if the vehicle is equipped with manual transaxle.

16. Remove the engine front cover.

17. On 2.3L engine with turbocharger, remove the dipstick tube and discard the O-ring.

18. Remove the oil pan retaining bolts.

19. Remove the oil pan from its mounting. Discard the gasket.

To install:

➡ **Be sure to use new fasteners, as required.**

20. Installation is the reverse of the removal procedure.

➡ **Using old bolts with old sealant adhering could cause cracks in the cylinder block. Be sure the bolts are free of old sealant. If in doubt, replace them.**

➡ **Do not allow gasket residue to enter the oil pan. Using cleaning fluid, completely remove the oil from the joint. Be sure to remove oil and gasket residue completely from the corner section of the front cover and the cylinder block.**

21. Apply the sealant in a single unbroken line.

22. Apply sealant (0.09–0.12 inch) to the oil pan along the inside of the bolt holes, as indicated in the illustration.

23. Install the oil pan to the cylinder block.

24. Tighten the bolts to specification and in the proper sequence. See illustration. Tighten the three oil pan/transaxle bolts to 28–38 ft. lbs.

25. Be sure to fill the engine with clean engine oil. Be sure to use the proper grade and type engine oil.

26. Start the engine and check for leaks. Correct as required.

27. Use the Mazda diagnostic scan tool, or equivalent and reprogram the required systems.

OIL PUMP

REMOVAL & INSTALLATION

2009 2.0L & 2.3L Engines (except Turbocharged Engine)

See Figures 130 through 132.

1. Before servicing the vehicle, refer to the Precautions Section.

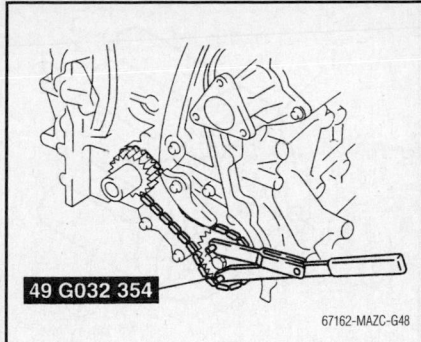

Fig. 130 Remove and install the oil pump sprocket using tool 49 G032 354 to stop the pump rotating—2009 2.0L and 2.3L engines

➡ **If working near and/or around the SRS system and components, be sure to disable the SRS system. Tape the negative battery cable with insulating tape. Always disconnect the negative battery cable first.**

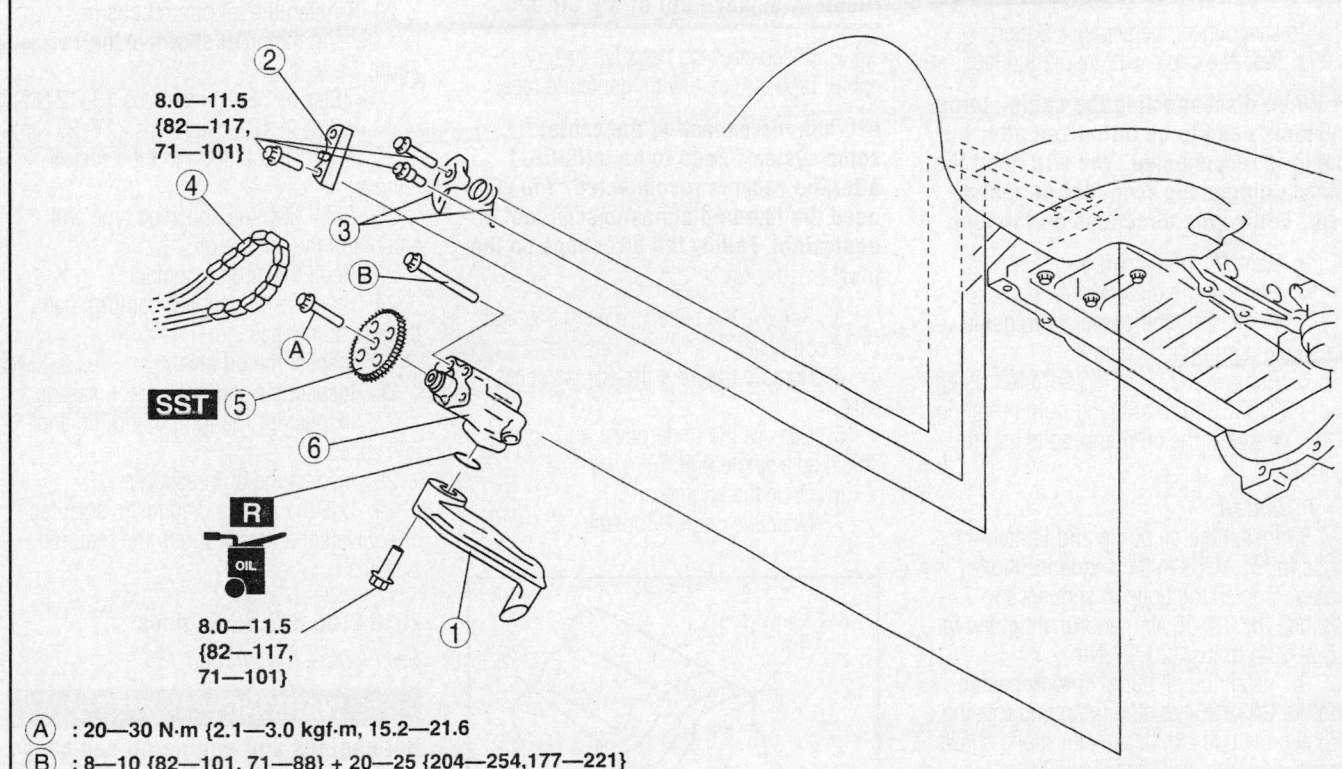

A : 20—30 N·m {2.1—3.0 kgf·m, 15.2—21.6
B : 8—10 {82—101, 71—88} + 20—25 {204—254,177—221}

N·m {kgf·cm, in·lbf}

1	Oil strainer	4	Oil pump chain
2	Oil pump chain guide	5	Oil pump sprocket
3	Oil pump chain tensioner and spring component	6	Oil pump

Fig. 131 Exploded view of the oil pump and related components—2009 2.0L and 2.3L engines

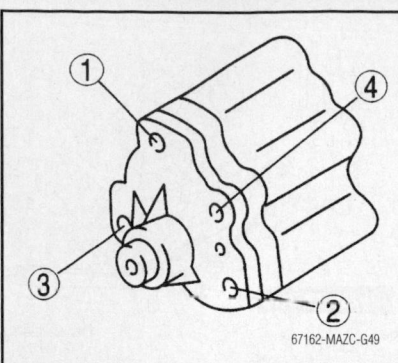

Fig. 132 Oil pump torque sequence—2009 2.0L and 2.3L engines

✱✱ CAUTION

To avoid personal injury when working on vehicles equipped with an air bag, the negative battery cable must be disconnected and at least one minute must elapse before working on the system. Failure to do so may result in deployment of the air bag.

2. Disconnect the negative battery cable. Tape the cable with insulating tape.

➡When disconnecting the cable, some systems need to be initialized after the cable is reconnected. You will need the Mazda diagnostic scan tool or equivalent. Follow the directions on the tool.

3. Remove the oil pan.
4. Remove the oil strainer.
5. Remove the oil pump chain guide, tensioner, spring and chain.
6. Remove the oil pump sprocket using tool 49 G032 354 to stop the pump rotating.
7. Remove the oil pump bolts and the pump.

To install:
8. Install the oil pump and tighten the bolts in two steps in the sequence illustrated. Tighten the bolts in sequence to 71–88 inch lbs. (8–10 Nm) and final tighten to 15.2–18.4 ft. lbs. (20–27 Nm).
9. Install the oil pump sprocket using tool 49 G032 354 to stop the pump rotating. Tighten the bolts to the specifications shown in the oil pump exploded view illustration.
10. Install the oil pump chain, spring, tensioner and guide. Tighten the bolts to the specifications shown in the oil pump exploded view illustration.
11. Install the oil strainer. Tighten the bolts to the specifications shown in the oil pump exploded view illustration.
12. Install the oil pan.
13. Fill the engine with clean oil.

14. Start the vehicle, check for leaks and repair if necessary.
15. Use the Mazda diagnostic scan tool, or equivalent and reprogram the required systems.

2.3L Engine With Turbocharger
See Figure 133.

1. Before servicing the vehicle, refer to the Precautions Section.

➡If working near and/or around the SRS system and components, be sure to disable the SRS system. Tape the negative battery cable with insulating tape. Always disconnect the negative battery cable first.

✱✱ CAUTION

To avoid personal injury when working on vehicles equipped with an air bag, the negative battery cable must be disconnected and at least one minute must elapse before working on the system. Failure to do so may result in deployment of the air bag.

2. Disconnect the negative battery cable. Tape the cable with insulating tape.

➡When disconnecting the cable, some systems need to be initialized after the cable is reconnected. You will need the Mazda diagnostic scan tool or equivalent. Follow the directions on the tool.

3. Remove the battery, battery tray and PCM component.
4. Remove the front tire RH, as necessary.
5. Remove the undercover and splash shield as a single unit.
6. Drain the engine oil.
7. Drain the engine coolant.

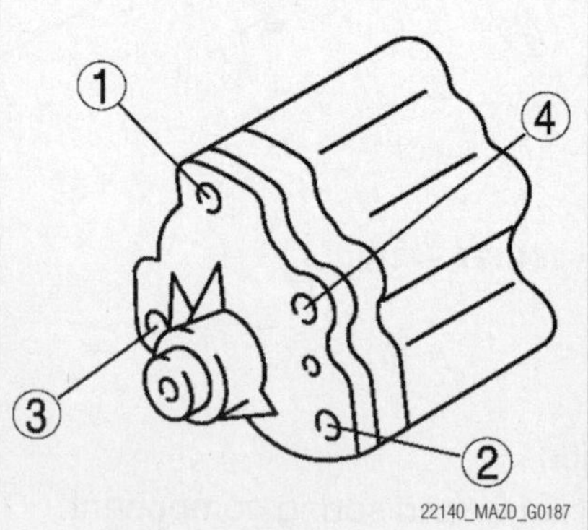

Fig. 133 Oil pump tightening sequence—2009 and 2010 2.3L engine with turbocharger

8. Remove the air cleaner, charge air cooler, and air hose.
9. Disconnect the quick release connector on the high pressure fuel pump.
10. Remove the high pressure fuel pump.
11. Remove the ignition coils.
12. Loosen the water pump pulley bolts before removing the drive belt.
13. Remove the drive belt.
14. Remove the P/S oil pump with hose and pipe still connected. Position the P/S oil pump out of the way.
15. Remove the crankshaft position sensor.
16. Remove the engine front cover.
17. Remove the oil pan.
18. Remove the oil pump as follows:
   • Oil strainer
   • Oil pump chain guide
   • Oil pump chain tensioner
   • Oil pump chain
   • Oil pump sprocket
   • Oil pump

To install:
19. Install the oil pump.
20. Tighten the oil pump bolts in two steps in the order shown in the figure.
   • Step 1: 71–105 inch lbs. (8–12 Nm).
   • Step 2: 13–17 inch lbs. (17–23 Nm).
21. Finish installation in the order of removal.
22. Refill with the specified type and amount of the engine oil.
23. Refill the engine coolant.
24. Start the engine and confirm that there is no oil leakage.
25. Inspect the oil level.
26. Inspect for engine coolant leakage.
27. Inspect for the ignition timing and idle speed.
28. Inspect the oil pressure.
29. Use the Mazda diagnostic scan tool, or equivalent and reprogram the required systems.

2010 2.0L & 2.5L Engines
See Figures 134 through 136.

✱✱ WARNING

Hot engines and engine oil can cause severe burns. Turn off the engine and wait until it and the engine oil have cooled.

✱✱ WARNING

A vehicle that is lifted but not securely supported on safety stands is dangerous. It can slip or fall, causing death or serious injury. Never work around or under a lifted vehicle if it is not securely supported on safety stands.

✳✳ WARNING

Continuous exposure to USED engine oil has caused skin cancer in laboratory mice. Protect your skin by washing with soap and water immediately after working with engine oil.

1. Before servicing the vehicle, refer to the Precautions Section.

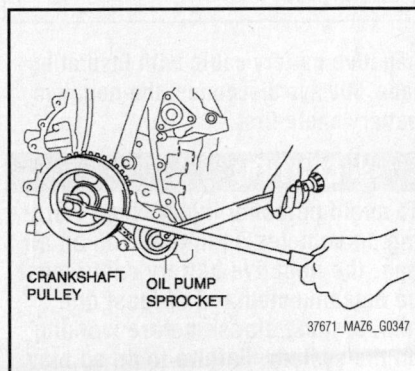

CRANKSHAFT PULLEY   OIL PUMP SPROCKET

37671_MAZ6_G0347

Fig. 134 Oil pump sprocket removal—2010 2.0L and 2.5L engines

➡ If working near and/or around the SRS system and components, be sure to disable the SRS system. Tape the negative battery cable with insulating tape. Always disconnect the negative battery cable first.

✳✳ CAUTION

To avoid personal injury when working on vehicles equipped with an air bag, the negative battery cable must be disconnected and at least one minute must elapse before working on the system. Failure to do so may result in deployment of the air bag.

2. Disconnect the negative battery cable. Tape the cable with insulating tape.

➡ When disconnecting the cable, some systems need to be initialized after the cable is reconnected. You will need the Mazda diagnostic scan tool or equivalent. Follow the directions on the tool.

3. Remove the engine front cover.
4. Remove the oil pan.
5. Remove the oil strainer.
6. Remove the oil pump chain tensioner.
7. Remove the oil pump chain.
8. Remove the oil pump socket.

➡ To remove and/or install the sprocket, temporarily install the crankshaft pulley and lock bolt to the crankshaft and lock the pump against rotation, see figure.

9. Remove the oil pump.

To install:

➡ Be sure to use new fasteners, as required.

10. Installation is the reverse of the removal procedure.
11. Tighten the pump bolts in two stages and in the proper sequence. Tighten bolts to

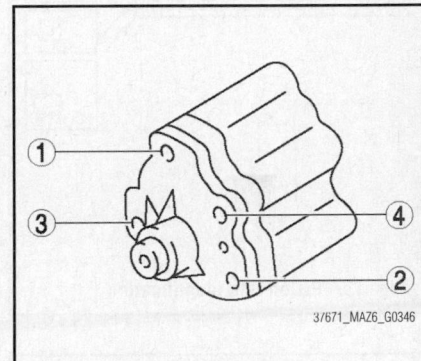

37671_MAZ6_G0346

Fig. 136 Oil pump bolt tightening sequence)—2010 2.0L and 2.5L engines

② 8—11 {82—112, 71—97}

③

8—12 N·m {82—122 kgf·cm, 71—106 in·lbf},
17—23 N·m {1.8—2.3 kgf·m, 13—16 ft·lbf}

20—30 N·m {2.1—3.0 kgf·m, 15—22 ft·lbf}

④
⑤

R

8—11 {82—112, 71—97}

N·m {kgf·cm, in·lbf}

37671_MAZ6_G0345

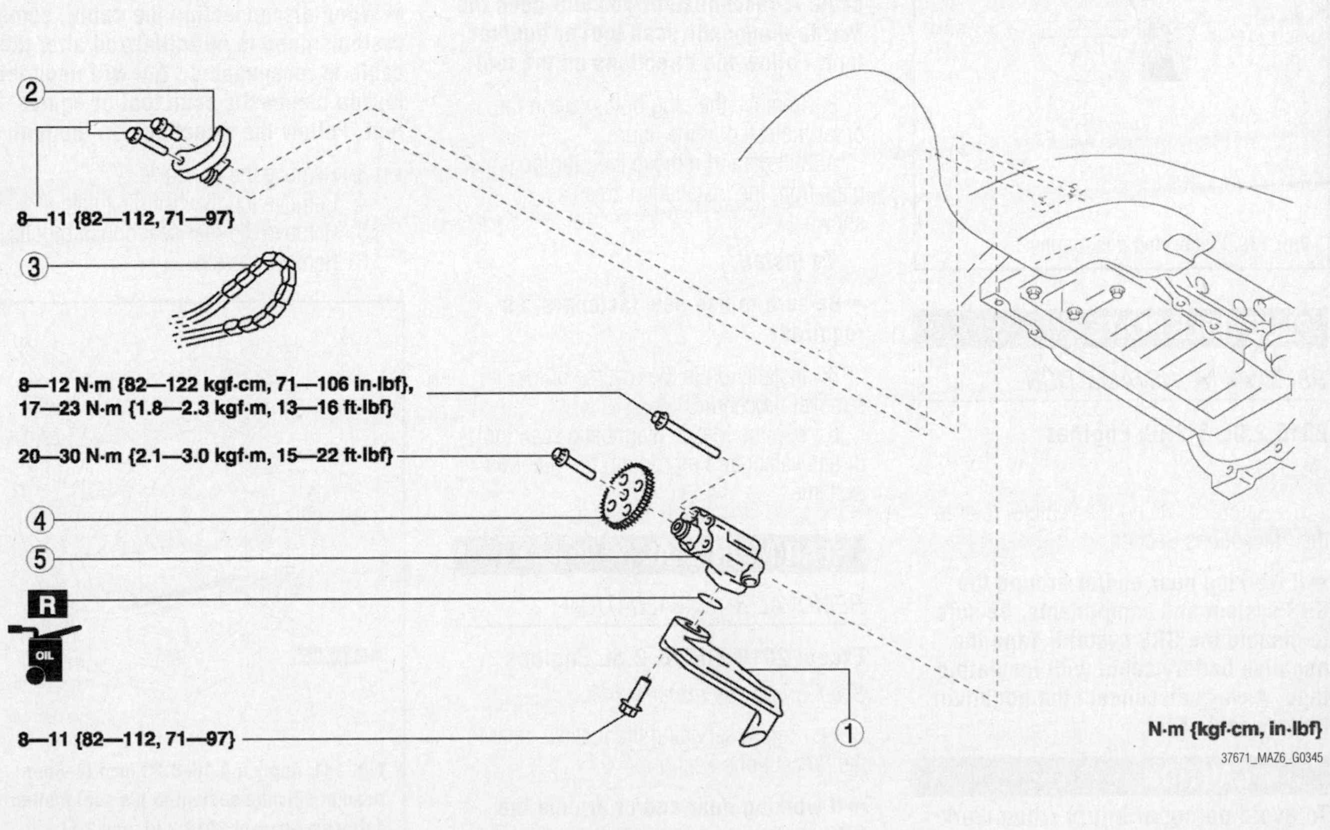

Fig. 135 Oil pump and related components)—2010 2.0L and 2.5L engines

71–106 inch lbs, step one and 13–16 ft. lbs, step two.

12. Use the Mazda diagnostic scan tool, or equivalent and reprogram the required systems.

PISTON AND RING

POSITIONING

See Figures 137 and 138.

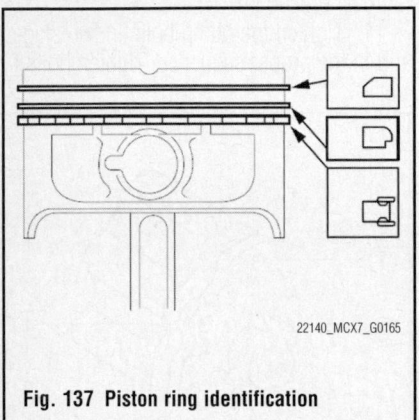

Fig. 137 Piston ring identification

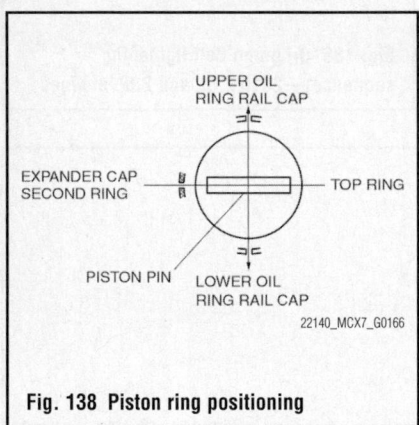

Fig. 138 Piston ring positioning

PLUG HOLE PLATE

REMOVAL & INSTALLATION

2010 2.0L & 2.5L Engines

See Figure 139.

1. Before servicing the vehicle, refer to the Precautions Section.

➡If working near and/or around the SRS system and components, be sure to disable the SRS system. Tape the negative battery cable with insulating tape. Always disconnect the negative battery cable first.

✳✳ CAUTION

To avoid personal injury when working on vehicles equipped with an air

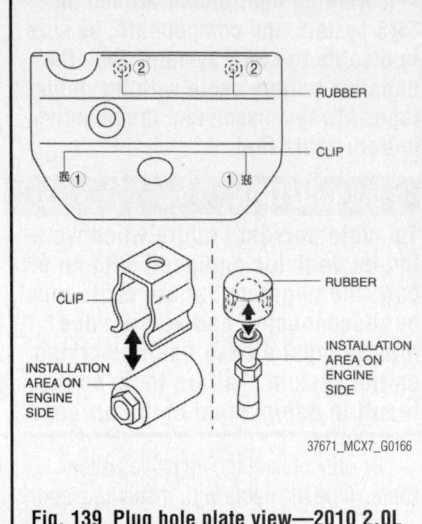

Fig. 139 Plug hole plate view—2010 2.0L and 2.5L engines

bag, the negative battery cable must be disconnected and at least one minute must elapse before working on the system. Failure to do so may result in deployment of the air bag.

2. Disconnect the negative battery cable. Tape the cable with insulating tape.

➡When disconnecting the cable, some systems need to be initialized after the cable is reconnected. You will need the Mazda diagnostic scan tool or equivalent. Follow the directions on the tool.

3. Remove the plug hole plate in the order indicated in the figure.

4. Lift off and remove the plug hole plate from the installation areas as shown.

To install:

➡Be sure to use new fasteners, as required.

5. Installation is the reverse of the removal procedure.

6. Use the Mazda diagnostic scan tool, or equivalent and reprogram the required systems.

REAR MAIN SEAL

REMOVAL & INSTALLATION

Except 2010 2.0L & 2.5L Engines

See Figures 140 through 143.

1. Before servicing the vehicle, refer to the Precautions Section.

➡If working near and/or around the SRS system and components, be sure to disable the SRS system. Tape the

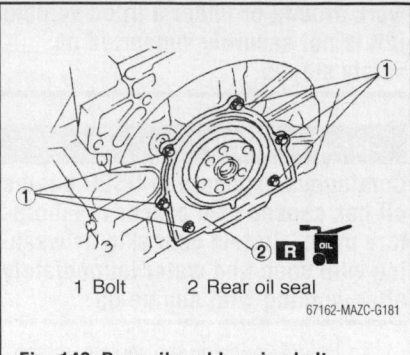

1 Bolt   2 Rear oil seal

Fig. 140 Rear oil seal housing bolt removal sequence—except 2010 2.0L and 2.5L engines

negative battery cable with insulating tape. Always disconnect the negative battery cable first.

✳✳ CAUTION

To avoid personal injury when working on vehicles equipped with an air bag, the negative battery cable must be disconnected and at least one minute must elapse before working on the system. Failure to do so may result in deployment of the air bag.

2. Disconnect the negative battery cable. Tape the cable with insulating tape.

➡When disconnecting the cable, some systems need to be initialized after the cable is reconnected. You will need the Mazda diagnostic scan tool or equivalent. Follow the directions on the tool.

3. Remove the transaxle.
4. Remove the flywheel/flexplate.
5. Remove the rear seal housing bolts.
6. Remove the seal.

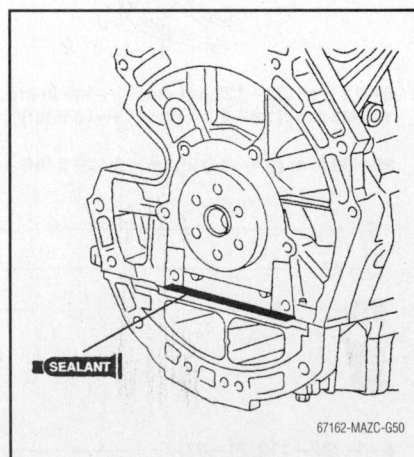

Fig. 141 Apply a 0.16–0.23 inch (4–6mm) bead of silicone sealant to the seal mating surfaces—except 2010 2.0L and 2.5L engines

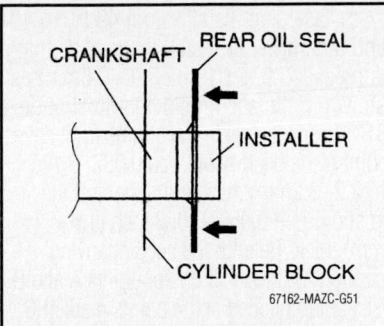

Fig. 142 Using a suitable installer tool, install the seal as shown—except 2010 2.0L and 2.5L engines

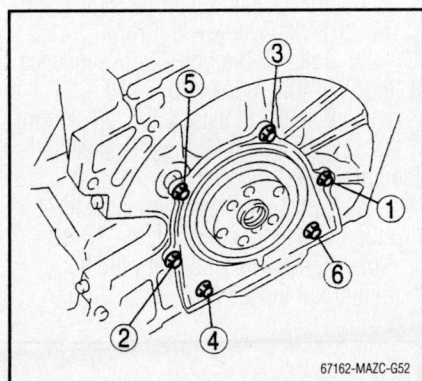

Fig. 143 Rear oil seal housing bolt torque sequence—except 2010 2.0L and 2.5L engines

To install:

7. Apply a 0.16–0.23 inch (4–6mm) bead of silicone sealant to the seal mating surfaces as illustrated. Install the seal within 10 minutes of applying the sealant.

8. Apply a coat of clean engine oil to the new seal lip.

9. Using a suitable installer tool, install the seal as illustrated.

10. Tighten the seal housing bolts in sequence to 71–101 inch lbs. (8–11.5 Nm).

11. Install the flywheel, use tool 49 E011 1A0 to prevent the unit from turning. If reusing the old bolts coat them with a thread locking compound. If install new bolts no locking compound is needed. Bolts to proper specification and in the correct sequence.

12. Install the transaxle.

2010 2.0L & 2.5L Engines

See Figures 144 through 147.

1. Before servicing the vehicle, refer to the Precautions Section.

➡ **If working near and/or around the SRS system and components, be sure to disable the SRS system. Tape the negative battery cable with insulating**

tape. Always disconnect the negative battery cable first.

⁂ CAUTION

To avoid personal injury when working on vehicles equipped with an air bag, the negative battery cable must be disconnected and at least one minute must elapse before working on the system. Failure to do so may result in deployment of the air bag.

2. Disconnect the negative battery cable. Tape the cable with insulating tape.

➡ **When disconnecting the cable, some systems need to be initialized after the cable is reconnected. You will need the Mazda diagnostic scan tool or equivalent. Follow the directions on the tool.**

3. Remove the transaxle and flexplate/flywheel.

4. Unbolt and remove the oil seal.

To install:

➡ **Be sure to use new fasteners, as required.**

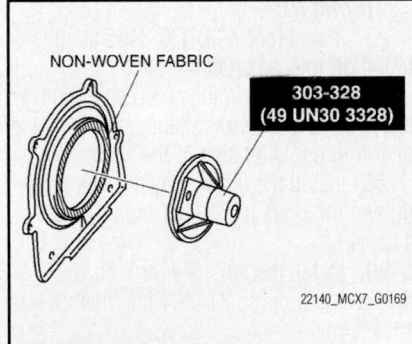

Fig. 144 Special service tool 303-328—2010 2.0L and 2.5L engines

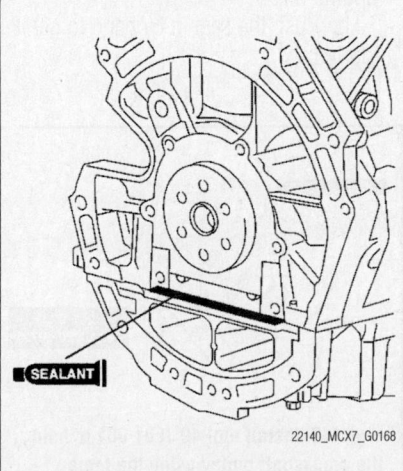

Fig. 145 Apply silicone sealant for installation—2010 2.0L and 2.5L engines

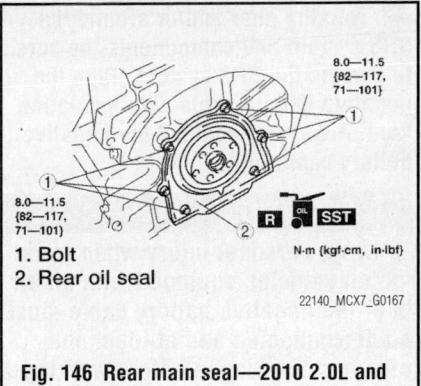

1. Bolt
2. Rear oil seal

Fig. 146 Rear main seal—2010 2.0L and 2.5L engines

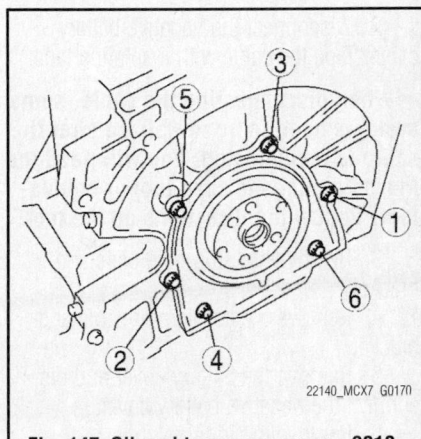

Fig. 147 Oil seal torque sequence—2010 2.0L and 2.5L engines

5. Apply silicone sealant as shown in the illustration.

6. Install SST 303-328 to the non-woven fabric side of the rear main seal.

7. From the back side of the rear oil seal, verify that there is no damage or separation in the lip area of the rear oil seal.

8. Install the rear main seal. Tighten the bolts to 71–101 inch lbs. (8–11.5 Nm).

9. Remove SST 303-328.

10. Install the flexplate/flywheel and transaxle.

11. Road test and check for leaks.

12. Use the Mazda diagnostic scan tool, or equivalent and reprogram the required systems.

TIMING CHAIN & SPROCKETS

REMOVAL & INSTALLATION

2009 2.0L & 2.3L Engines (except Turbocharged Engine)

See Figures 148 through 163.

1. Before servicing the vehicle, refer to the Precautions Section.

➡If working near and/or around the SRS system and components, be sure to disable the SRS system. Tape the negative battery cable with insulating tape. Always disconnect the negative battery cable first.

⁂ CAUTION

To avoid personal injury when working on vehicles equipped with an air bag, the negative battery cable must be disconnected and at least one minute must elapse before working on the system. Failure to do so may result in deployment of the air bag.

2. Disconnect the negative battery cable. Tape the cable with insulating tape.

➡When disconnecting the cable, some systems need to be initialized after the cable is reconnected. You will need the Mazda diagnostic scan tool or equivalent. Follow the directions on the tool.

3. Remove the plug hole plate and bracket.
4. Remove the accelerator cable and bracket.
5. Remove the battery cover and disconnect the negative battery cover.
6. Remove the ignition coils.
7. Remove the right hand front wheel.
8. Remove the engine undercover and splash shields.
9. Remove the Crankshaft Position (CKP) sensor.
10. Remove the accessory drive belt.
11. Remove the A/C compressor and set aside with the lines attached.
12. Remove the coolant reservoir tank and set aside with the lines still attached.
13. Remove the cylinder head cover.
14. Remove the cylinder block lower blind plug.
15. Install tool 49 JE01 061.
16. Turn the crankshaft clockwise until the crankshaft is in the number 1 cylinder Top Dead Center position (the balance weight should be attached to tool 49 JE01 061).
17. Hold the crankshaft pulley using the tools illustrated and remove the bolt.
18. Remove the crankshaft pulley.
19. Remove the water pump pulley.
20. Remove the drive belt tensioner.
21. Remove the No. 3 engine mount and joint bracket as follows:
    a. Install two suitable pieces of wood between the front fender panel and upper apron reinforcement as illustrated. The wood size should be approximately 1.38 inch (35mm) on 4 door models or 2.36 inch (60mm) on 5 door models.
22. Install an engine support device such as 49 E017 5A0.
23. Remove the engine front cover and oil seal.
24. Unlock the chain tensioner using a suitable tool to slowly compress the tensioner piston. Insert a 0.059 inch (1.5mm) wire or a paper clip to hold the piston in its compressed position.
25. Remove the tensioner arm, chain guide and timing chain.

To install:

26. Install tool 49 UN30 3465 as illustrated on the camshaft.
27. Install the timing chain and remove the paper clip or wire retaining the tensioner piston to apply tension to the chain.
28. Install the timing chain guide and tighten the bolts to 71–101 inch lbs. (8–11.5 Nm).
29. Install the tensioner arm and tighten the bolts to 71–101 inch lbs. (8–11.5 Nm).
30. Install a new oil seal in the front cover as follows:
    a. Coat the new seal with clean engine oil.
    b. Push the seal in by hand to get it started.
    c. Use seal installer tool 49 H010 401 and a hammer to install the seal so that it is recessed 0–0.019 inch (0–0.5mm) as shown in the accompanying illustration.
31. Apply sealant to the engine front cover. At point A the bead should be 0.087–0.125 inch (2.2–3.2mm) thick and at point B the bead should be 0.059–0.098 inch (1.5–2.5mm) thick. Refer to the accompanying illustration for the locations of points A and B. No sealant is needed at the points marked C on 2.3L engine with variable valve timing.
32. Install the cover within 10 minutes of apply the sealant. Tighten the cover bolts as follows:
    a. Bolts 1 through 18: In sequence to 71–101 inch lbs. (8–11.5 Nm).
    b. Bolts 19 through 22 : In sequence to 29.7–40.5 ft. lbs. (40–55 Nm).
    c. Bolt 23 : 14.8–22 ft. lbs. (20–30 Nm).
33. Install the No. 3 engine mount and joint bracket as follows:
    a. Tighten the No. 3 mount bracket stud bolts to 62–115 inch lbs. (7–13 Nm). Tighten the stud bolt with the mount nut loosened.

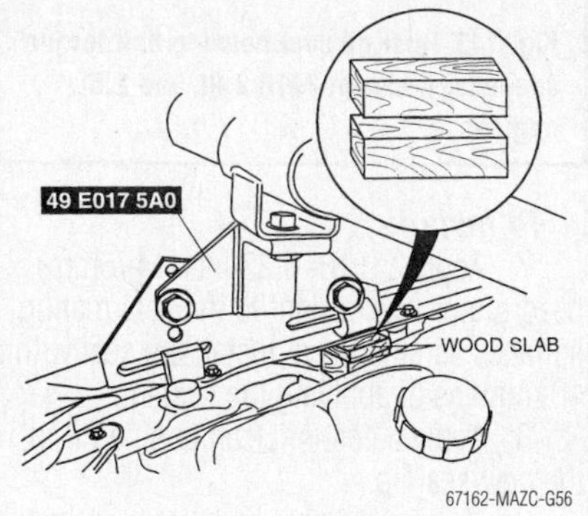

Fig. 150 Install two suitable pieces of wood between the front fender panel and upper apron reinforcement—2009 2.0Land 2.3L engines

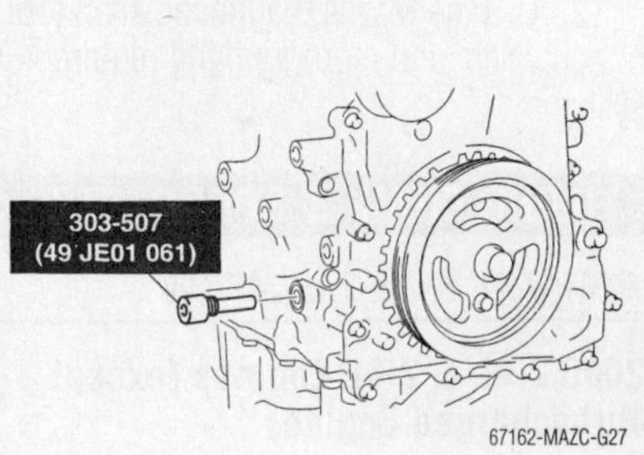

Fig. 148 Install tool 49 JE01 061—2009 2.0Land 2.3L engines

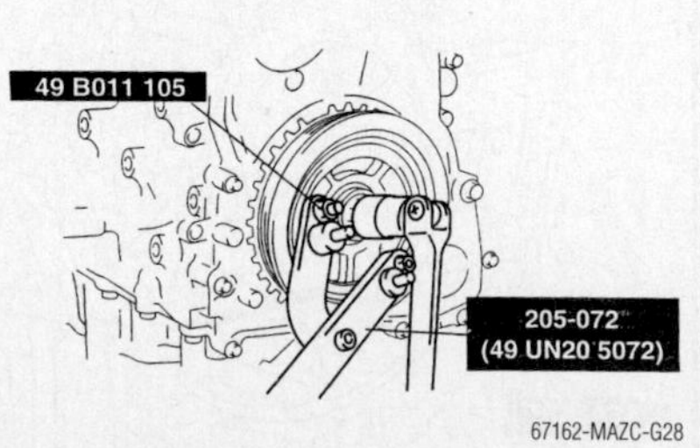

Fig. 149 Install tool 49 JE01 061 to hold the crankshaft pulley using the tools illustrated and remove the bolt—2009 2.0Land 2.3L engines

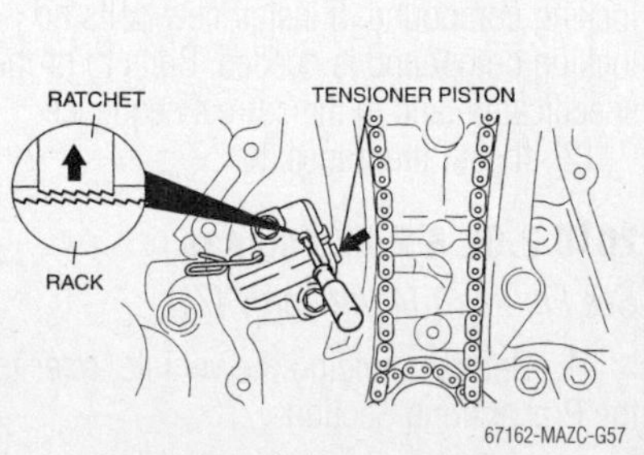

Fig. 151 Compressing and retaining the chain tensioner piston—2009 2.0Land 2.3L engines

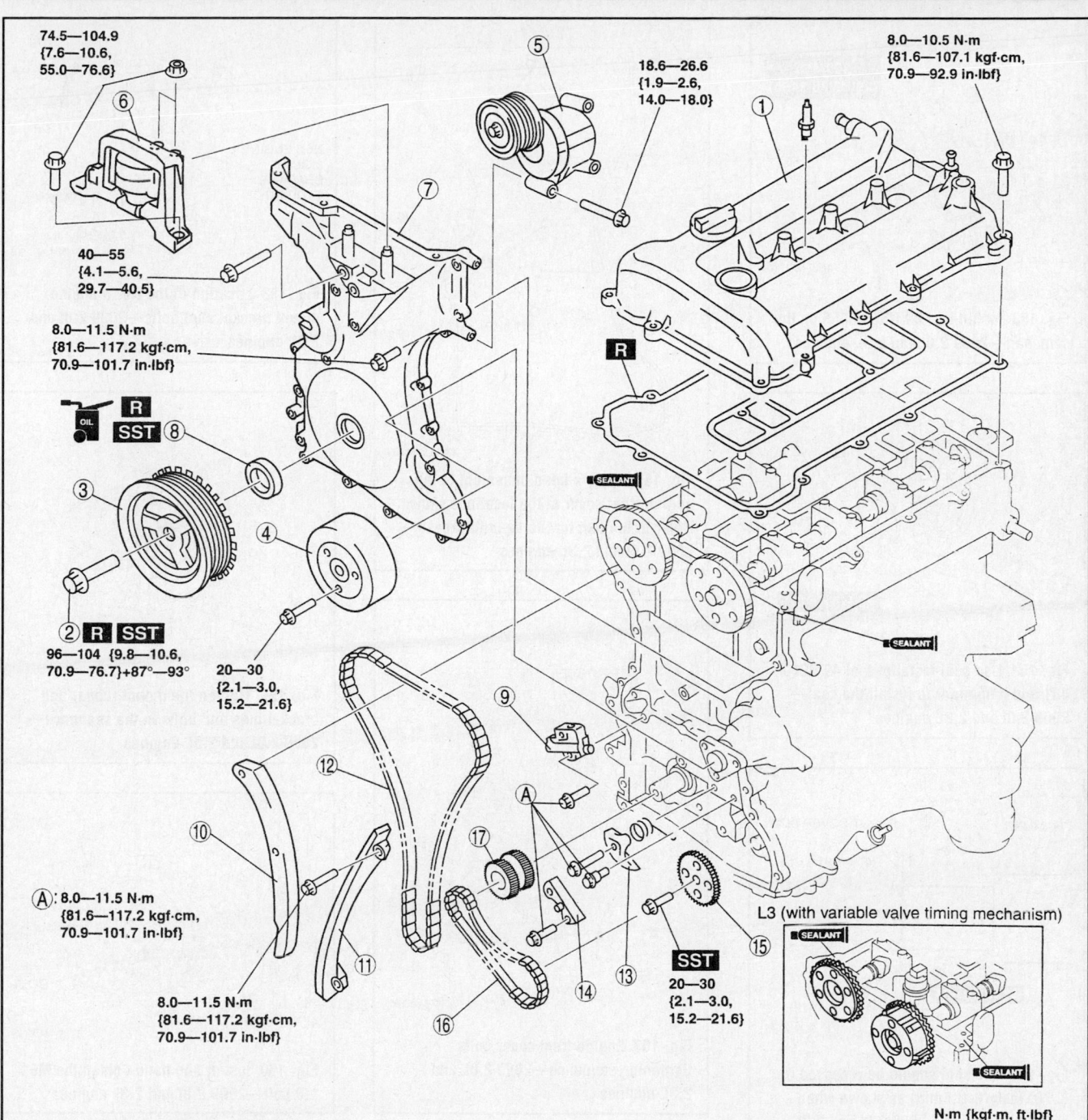

74.5—104.9
{7.6—10.6,
55.0—76.6}

8.0—10.5 N·m
{81.6—107.1 kgf·cm,
70.9—92.9 in·lbf}

18.6—26.6
{1.9—2.6,
14.0—18.0}

40—55
{4.1—5.6,
29.7—40.5}

8.0—11.5 N·m
{81.6—117.2 kgf·cm,
70.9—101.7 in·lbf}

96—104 {9.8—10.6,
70.9—76.7}+87°—93°

20—30
{2.1—3.0,
15.2—21.6}

Ⓐ: 8.0—11.5 N·m
{81.6—117.2 kgf·cm,
70.9—101.7 in·lbf}

8.0—11.5 N·m
{81.6—117.2 kgf·cm,
70.9—101.7 in·lbf}

20—30
{2.1—3.0,
15.2—21.6}

L3 (with variable valve timing mechanism)

N·m {kgf·m, ft·lbf}

1	Cylinder head cover	9	Chain tensioner
2	Crankshaft pulley lock bolt	10	Tensioner arm
3	Crankshaft pulley	11	Chain guide
4	Water pump pulley	12	Timing chain
5	Drive belt auto tensioner	13	Oil pump chain tensioner
6	No.3 engine mount rubber and No.3 engine joint bracket	14	Oil pump chain guide
7	Engine front cover	15	Oil pump sprocket
8	Front oil seal	16	Oil pump chain
		17	Crankshaft sprocket

67162-MAZC-G55

Fig. 152 Exploded view of the timing chain assembly and related components—2009 2.0L and 2.3L engines

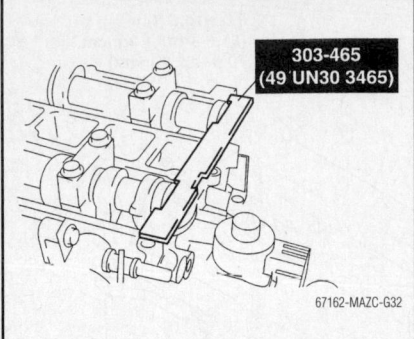

Fig. 153 Install tool 49 UN30 3465 on the camshaft—2009 2.0Land 2.3L engines

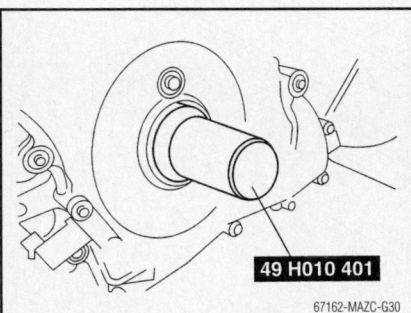

Fig. 154 Use seal installer tool 49 H010 401 and a hammer to install the seal—2009 2.0Land 2.3L engines

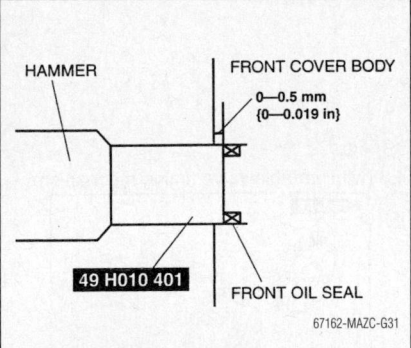

Fig. 155 The seal should be recessed 0–0.019 inch (0–0.5mm) as shown when properly installed—2009 2.0Land 2.3L engines

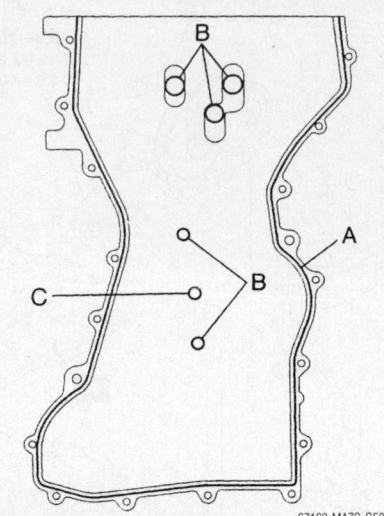

Fig. 156 Apply a bead of sealant to the engine front cover at the locations shown. Refer to the text for the bead thickness—2009 2.0Land 2.3L engines

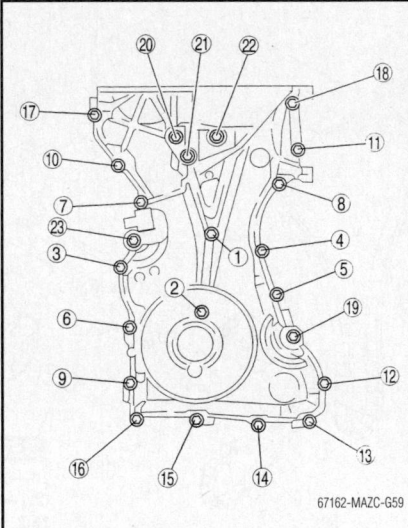

Fig. 157 Engine front cover bolts tightening sequence—2009 2.0L and 2.3L engines

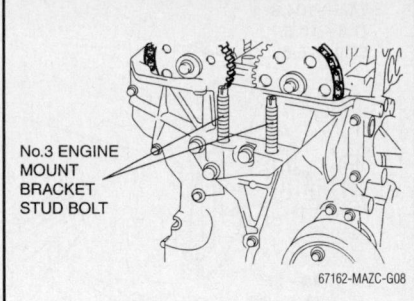

Fig. 158 Location of the No. 3 engine mount bracket stud bolts—2009 2.0Land 2.3L engines

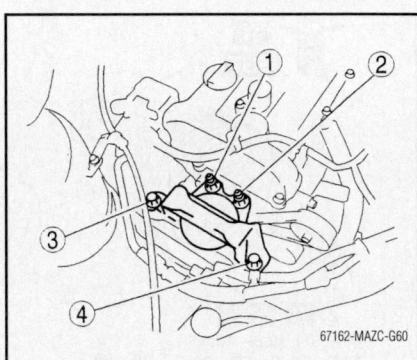

Fig. 159 Tighten the mount rubber and bracket nuts and bolts in the sequence—2009 2.0Land 2.3L engines

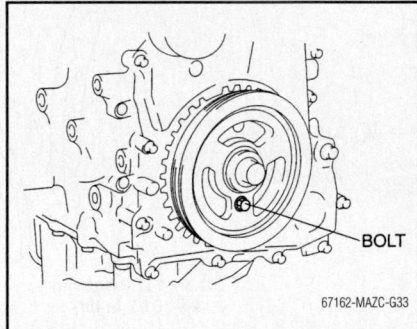

Fig. 160 Install and hand tighten the M6 x 1.0 bolt—2009 2.0Land 2.3L engines

b. Hand tighten the mount rubber and bracket nuts and bolts, then tighten in the sequence illustrated to 55–77 ft. lbs. (74–105 Nm).

34. Install tool 49 UN30 3465 as illustrated on the camshaft.

35. Install and hand tighten the M6 x 1.0 bolt as illustrated.

36. Turn the crankshaft clockwise until the crankshaft is in number 1 TDC (the balance weight should be attached to the tool).

37. Hold the crankshaft pulley and tighten the lock bolt in two steps. First tighten to 71–77 ft. lbs. (96–104 Nm). Then final tighten an additional 87–93 degrees.

38. Remove the M6 x 1.0 bolt.

39. Remove the tools from the camshaft and the cylinder block lower blind plug.

40. Rotate the crankshaft clockwise 2 turns until you reach TDC. If not aligned properly, loosen the crankshaft pulley lock bolt and reinstall the lock bolt again using the above procedure and special tools.

41. Install the cylinder block blind plug and tighten to 13–16 ft. lbs. (18–22 Nm).

42. Apply a 0.16–0.24 inch (4–7mm) bead of silicone sealant to cylinder head at the areas illustrated. Make sure to install the cover within 10 minutes of applying the sealant.

43. Install the cylinder head cover with a new gasket. Torque the bolts in the sequence illustrated to 71–93 inch lbs. (8–11 Nm).

44. Install the remaining components in the reverse order of removal.

45. When installing the CKP sensor perform the following:

a. Remove the cylinder block lower blind plug.

b. Install tool 49 JE01 061.

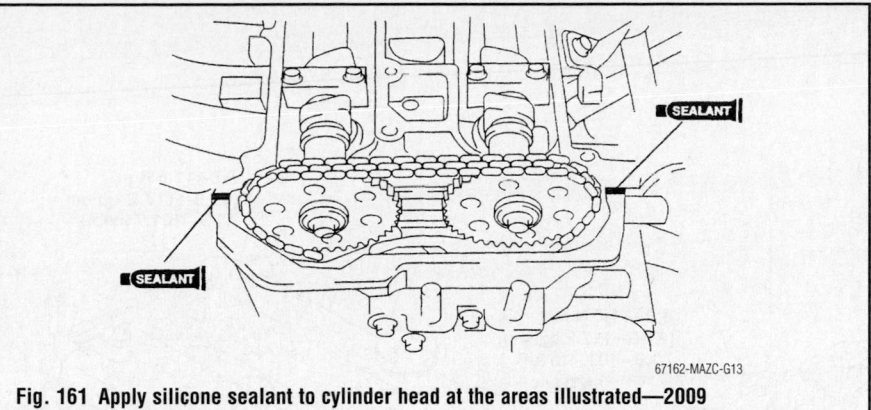

Fig. 161 Apply silicone sealant to cylinder head at the areas illustrated—2009 2.0Land 2.3L engines

c. Turn the crankshaft clockwise until the crankshaft is in the number 1 cylinder Top Dead Center position (the balance weight should be attached to tool 49 JE01 061).

d. Using a ruler, mark the center line on the pulse wheel teeth on the crank pulley which is located at the 9th tooth counting counterclockwise from the empty space. Refer to the illustration for more detail.

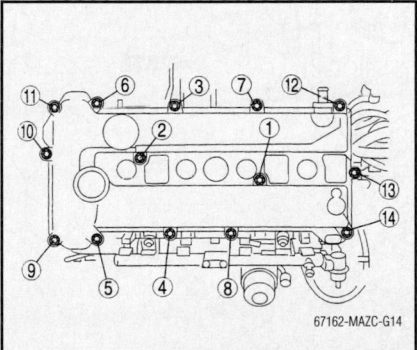

Fig. 162 Tighten the cylinder head cover bolts in the sequence shown—2009 2.0Land 2.3L engines

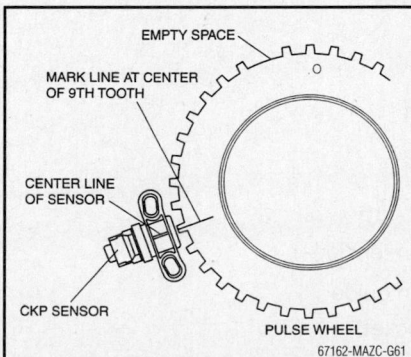

Fig. 163 Using a ruler, mark the center line on the pulse wheel teeth on the crank pulley which is located at the 9th tooth counting counterclockwise from the empty space—2009 2.0Land 2.3L engines

※※ CAUTION

If you do not mark the center line correctly this will cause improper engine control for the ignition and fuel system, so be sure to mark the line carefully.

e. Install the CKP sensor making sure the mark on the sensor is aligned with the mark on the pulse wheel. Tighten the sensor bolt to 49–66 inch lbs. (5.5–7.5 Nm)and attach the electrical connector.

f. Remove the tool from the cylinder block lower blind plug.

g. Rotate the crankshaft clockwise 2 turns until you reach TDC. If not aligned properly, loosen the crankshaft pulley lock bolt and reinstall the lock bolt.

h. Install the cylinder block blind plug and tighten to 13–16 ft. lbs. (18–22 Nm).

46. Change the engine oil.

47. Start the engine.

48. Inspect for the following:
- Pulley and belt for run–out and contact
- Any leaking fluids.
- Ignition timing, idle speed and exhaust emissions
- All remaining components for proper operation.

49. Use the Mazda diagnostic scan tool, or equivalent and reprogram the required systems.

2009–10 2.3L Engine With Turbocharger

See Figures 164 through 175.

1. Before servicing the vehicle, refer to the Precautions Section.

➡**If working near and/or around the SRS system and components, be sure to disable the SRS system. Tape the negative battery cable with insulating tape. Always disconnect the negative battery cable first.**

※※ CAUTION

To avoid personal injury when working on vehicles equipped with an air bag, the negative battery cable must be disconnected and at least one minute must elapse before working on the system. Failure to do so may result in deployment of the air bag.

2. Disconnect the negative battery cable. Tape the cable with insulating tape.

➡**When disconnecting the cable, some systems need to be initialized after the**

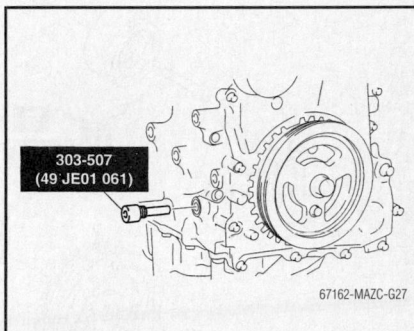

Fig. 164 Install tool 49 JE01 061—2009 and 2010 2.3L engine with turbocharger

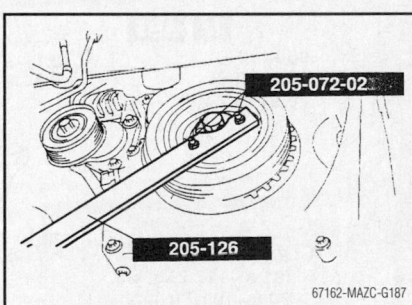

Fig. 165 Install tools 205-072-02 and 205-126 to hold the crankshaft pulley using the tools illustrated and remove the bolt—2009 and 2010 2.3L engine with turbocharger

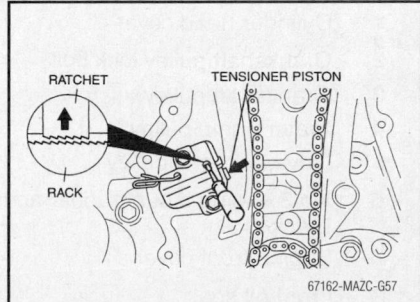

Fig. 166 Compressing and retaining the chain tensioner piston—2009 and 2010 2.3L engine with turbocharger

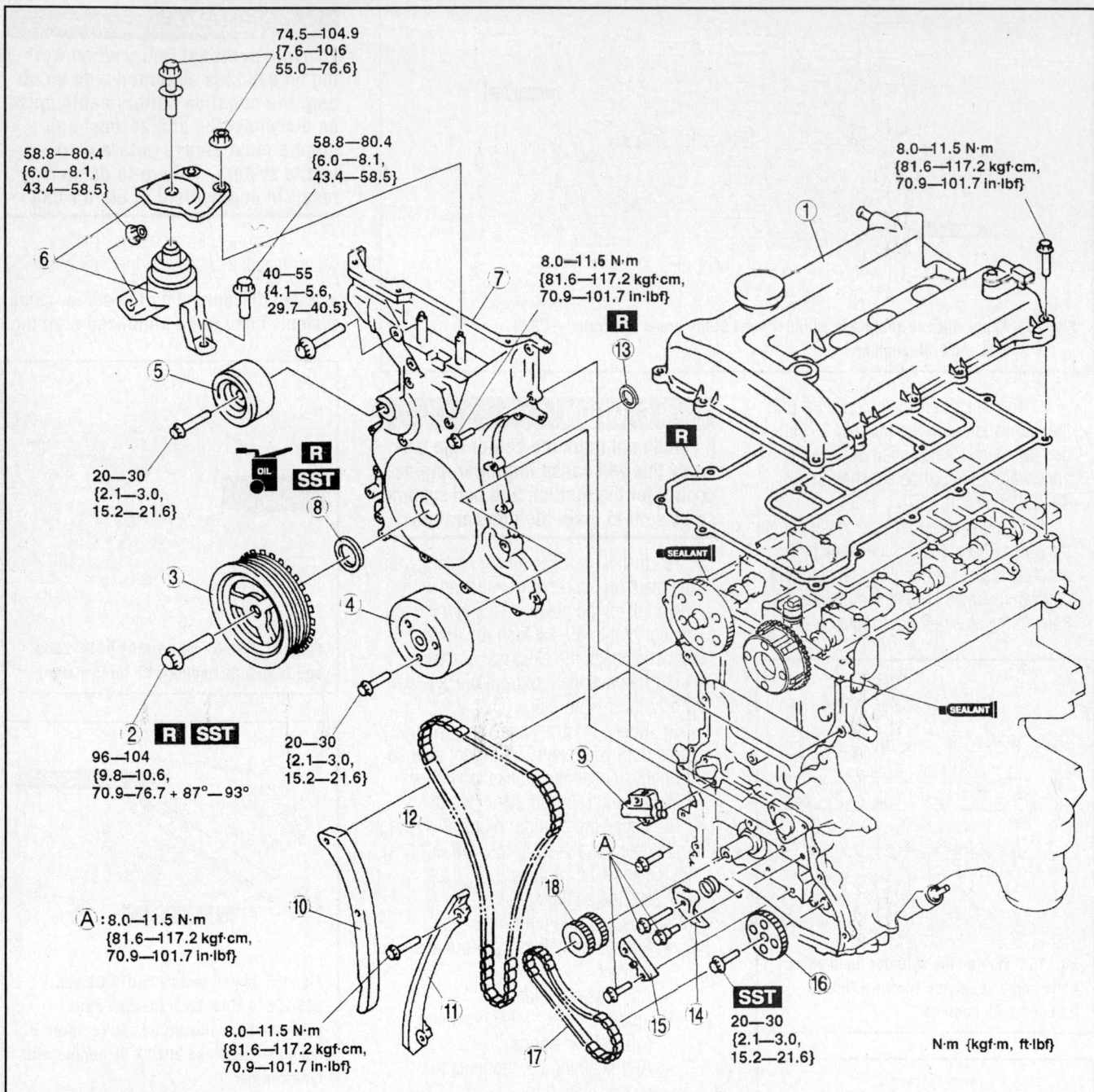

74.5—104.9
{7.6—10.6,
55.0—76.6}

58.8—80.4
{6.0—8.1,
43.4—58.5}

58.8—80.4
{6.0—8.1,
43.4—58.5}

8.0—11.5 N·m
{81.6—117.2 kgf·cm,
70.9—101.7 in·lbf}

40—55
{4.1—5.6,
29.7—40.5}

8.0—11.5 N·m
{81.6—117.2 kgf·cm,
70.9—101.7 in·lbf}

20—30
{2.1—3.0,
15.2—21.6}

96—104
{9.8—10.6,
70.9—76.7 + 87°—93°

20—30
{2.1—3.0,
15.2—21.6}

(A): 8.0—11.5 N·m
{81.6—117.2 kgf·cm,
70.9—101.7 in·lbf}

8.0—11.5 N·m
{81.6—117.2 kgf·cm,
70.9—101.7 in·lbf}

20—30
{2.1—3.0,
15.2—21.6}

N·m {kgf·m, ft·lbf}

1	Cylinder head cover	10	Tensioner arm
2	Crankshaft pulley lock bolt	11	Chain guide
3	Crankshaft pulley	12	Timing chain
4	Water pump pulley	13	Seal
5	Drive belt idler pulley	14	Oil pump chain tensioner
6	No.3 engine mount rubber and No.3 engine joint bracket	15	Oil pump chain guide
7	Engine front cover	16	Oil pump sprocket
8	Front oil seal	17	Oil pump chain
9	Chain tensioner	18	Crankshaft sprocket

67162-MAZC-G186

Fig. 167 Exploded view of the timing chain assembly and related components—2009 and 2010 2.3L engine with turbocharger

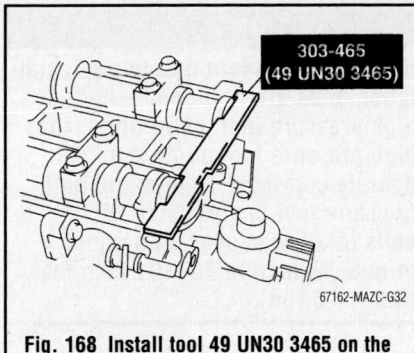

Fig. 168 Install tool 49 UN30 3465 on the camshaft—2009 and 2010 2.3L engine with turbocharger

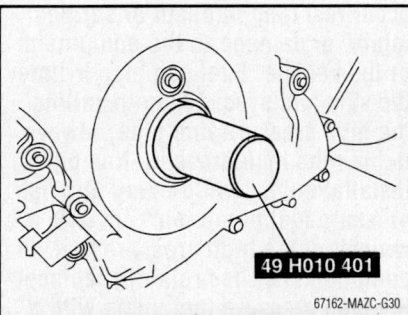

Fig. 169 Use seal installer tool 49 H010 401 and a hammer to install the seal—2009 and 2010 2.3L engine with turbocharger

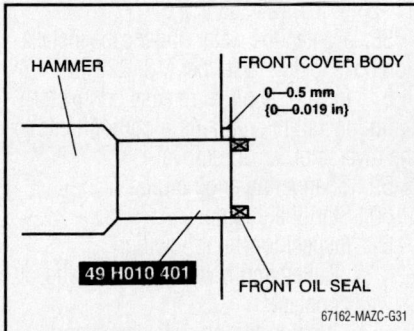

Fig. 170 The seal should be recessed 0–0.019 inch (0–0.5mm) as shown when properly installed—2009 and 2010 2.3L engine with turbocharger

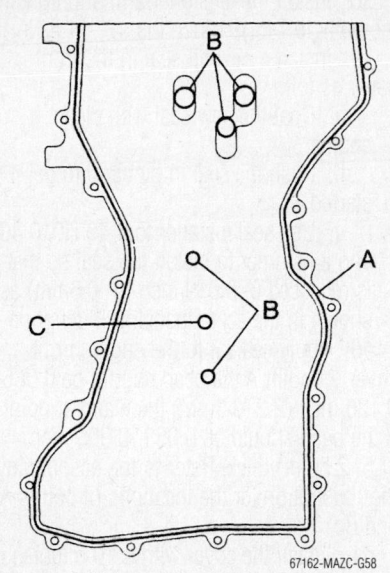

Fig. 171 Apply a bead of sealant to the engine front cover at the locations shown. Refer to the text for the bead thickness—2009 and 2010 2.3L engine with turbocharger

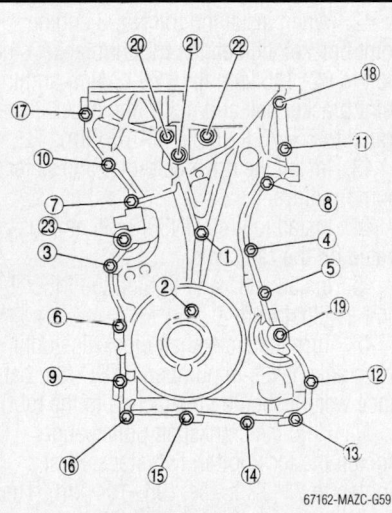

Fig. 172 Engine front cover bolts tightening sequence—2009 and 2010 2.3L engine with turbocharger

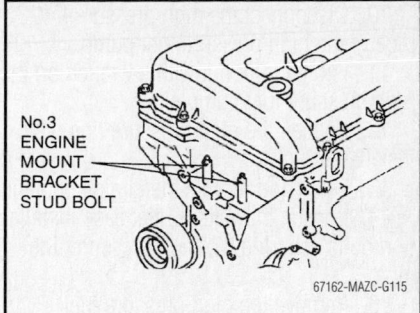

Fig. 173 No. 3 engine mount bracket stud bolt locations—2009 and 2010 2.3L engine with turbocharger

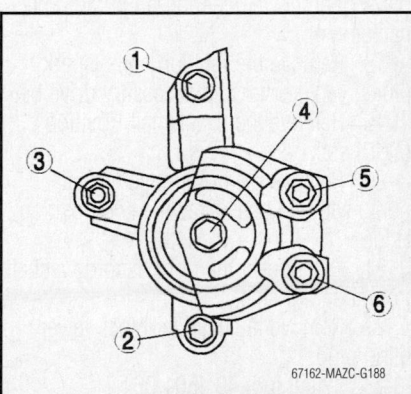

Fig. 174 Tighten the No. 3 joint bracket bolts and nuts in the order shown—2009 and 2010 2.3L engine with turbocharger

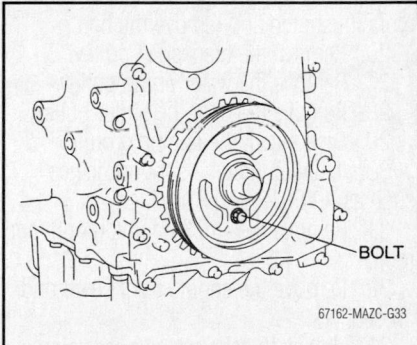

Fig. 175 Install and hand tighten the M6 x 1.0 bolt—2009 and 2010 2.3L engine with turbocharger

cable is reconnected. You will need the Mazda diagnostic scan tool or equivalent. Follow the directions on the tool.

3. Remove the right side front tire.
4. Remove the undercover.
5. Remove the right side splash shield.

➡ If the high pressure fuel pump is removed, replace the O-ring with a new one.

6. Remove the charge air cooler cover.
7. Disconnect the spill valve control solenoid valve connector.
8. Disconnect the quick release connector on the high pressure fuel pump.
9. Remove the battery and battery tray.

❋❋ **WARNING**

If the high pressure fuel pump joint nut is loosened, fuel leakage may occur resulting in death or serious

injury, or damage to the equipment or the vehicle. Fuel can also irritate the skin and eyes. When removing the high pressure line pipe, always tighten the high pressure line pipe installation nut while fixing the high pressure fuel pump joint nut with a wrench. If the high pressure fuel pump joint nut has rotated, replace the high pressure fuel pump with a new one.

10. Disconnect the high pressure line pipe of the high pressure fuel pump.

11. Fix the joint nut with a wrench on the high pressure fuel pump side.

12. Loosen the high pressure line pipe installation nut.

13. Drain and recycle the engine coolant.

14. Loosen the water outlet case installation bolts securing the high pressure line pipe.

15. Remove the high pressure fuel pump.

16. Remove the high pressure fuel pump cover.

17. Disconnect the Camshaft Position (CMP) and power steering oil pump connectors.

18. Remove the ignition coils, spark plugs, valve cover and accessory drive belt.

19. Remove the Crankshaft Position (CKP) sensor.

20. Remove the power steering oil pump with the lines attached and position aside.

21. Disconnect the right hand driveshaft from the joint shaft.

22. Remove the cylinder block lower blind plug.

23. Install tool 49 JE01 061.

24. Turn the crankshaft clockwise until the crankshaft is in the number 1 cylinder Top Dead Center position (the balance weight should be attached to tool 49 JE01 061).

25. Hold the crankshaft pulley using the tools illustrated and remove the bolt.

26. Remove the crankshaft pulley.

27. Remove the water pump pulley.

28. Remove the drive belt idler pulley.

29. Disconnect the engine ground.

30. Install an engine support device such as 49 E017 5A0.

31. Remove the No. 3 engine mount and joint.

32. Remove the engine front cover and oil seal.

33. Unlock the chain tensioner using a suitable tool to slowly compress the tensioner piston. Insert a 0.059 inch (1.5mm) wire or a paper clip to hold the piston in its compressed position.

34. Remove the tensioner arm, chain guide and timing chain.

To install:

35. Install tool 49 UN30 3465 as illustrated on the camshaft.

36. Install the timing chain and remove the paper clip or wire retaining the tensioner piston to apply tension to the chain.

37. Install the timing chain guide and tighten the bolts to 71–101 inch lbs. (8–11.5 Nm).

38. Install the tensioner arm and tighten the bolts to 71–101 inch lbs. (8–11.5 Nm).

39. Install a new oil seal in the front cover as follows:

a. Coat the new seal with clean engine oil.

b. Push the seal in by hand to get it started.

c. Use seal installer tool 49 H010 401 and a hammer to install the seal so that it is recessed 0–0.019 inch (0 0.5mm) as shown in the accompanying illustration.

40. Apply sealant to the engine front cover. At point A the bead should be 0.086–0.126 inch (2.2–3.2mm) thick and at point B the bead should be 0.059–0.098 inch (1.5–2.5mm) thick. Refer to the accompanying illustration for the locations of points A and B.

41. Install the cover within 10 minutes of apply the sealant. Tighten the cover bolts as follows:

a. Bolts 1 through 18: In sequence to 71–101 inch lbs. (8–11.5 Nm).

b. Bolts 19 through 22 : In sequence to 29.7–40.5 ft. lbs. (40–55 Nm).

c. Bolt 23 : 14.8–22 ft. lbs. (20–30 Nm).

42. When installing the No. 3 engine joint bracket, tighten the mount bracket stud bolt to 62–115 inch lbs. (7–13 Nm) and the joint bracket bolt and nut in the order illustrated to 55–76 ft. lbs. (74–105 Nm).

43. Install the drive belt idler and water pump pulleys.

44. Install tool 49 UN30 3465 as illustrated on the camshaft.

45. Install and hand tighten the M6 x 1.0 bolt as illustrated.

46. Turn the crankshaft clockwise until the crankshaft is in number 1 TDC (the balance weight should be attached to the tool).

47. Hold the crankshaft pulley and tighten the lock bolt in two steps. First tighten to 71–77 ft. lbs. (96–104 Nm). Then final tighten an additional 87–93 degrees.

48. Remove the M6 x 1.0 bolt.

49. Remove the tools from the camshaft and the cylinder block lower blind plug.

50. Rotate the crankshaft clockwise 2 turns until you reach TDC. If not aligned properly, loosen the crankshaft pulley lock bolt and reinstall the lock bolt again using the above procedure and special tools.

51. Install the cylinder block blind plug and tighten to 13–16 ft. lbs. (18–22 Nm).

52. Install the cylinder head cover with a new gasket. Torque the bolts in the sequence illustrated to 71–93 inch lbs. (8–11 Nm).

53. Install the high pressure fuel pump cover and tighten to 13–16 ft. lbs. (18–22 Nm).

❊❊ CAUTION

If the high pressure fuel pump installation bolts are tightened with the high pressure fuel pump tilted, the high pressure fuel pump may not operate correctly. Tighten the high pressure fuel pump installation bolts in a few passes with equal torque. Tighten to 76–101 inch lbs. (8.5–11.5 Nm).

❊❊ WARNING

If the high pressure fuel pump joint nut is loosened, fuel leakage may occur resulting in death or serious injury, or damage to the equipment or the vehicle. Fuel can also irritate the skin and eyes. When installing the high pressure line pipe, always tighten the high pressure line pipe installation nut while fixing the high pressure fuel pump joint nut with a wrench. If the high pressure fuel pump joint nut has rotated, replace the high pressure fuel pump with a new one.

54. Assemble the high pressure line pipe.

55. Fix the joint nut with a wrench on the high pressure fuel pump side. Tighten the high pressure line pipe installation nut to 17–26 ft. lbs. (23–35 Nm).

56. Tighten the water outlet case installation bolts to 13–16 ft. lbs. (18–22 Nm).

57. Install the quick release connector.

58. Install the remaining components in the reverse order of removal.

59. Change the engine oil.

60. Start the engine.

61. Inspect for the following:
- Pulley and belt for run–out and contact
- Ignition timing, idle speed and exhaust emissions
- All remaining components for proper operation

62. Verify that the high pressure fuel pump is assembled securely.

63. Drive the vehicle starting from a standstill and brake suddenly five to six times at a low speed.

64. Stop the vehicle and verify from outside the vehicle that there is no fuel leakage around the high pressure fuel pump.

65. Use the Mazda diagnostic scan tool, or equivalent and reprogram the required systems.

2010 2.0L & 2.5L Engines

See Figures 176 through 189.

1. Before servicing the vehicle, refer to the Precautions Section.

➡ **If working near and/or around the SRS system and components, be sure to disable the SRS system. Tape the negative battery cable with insulating tape. Always disconnect the negative battery cable first.**

❋❋ CAUTION

To avoid personal injury when working on vehicles equipped with an air bag, the negative battery cable must be disconnected and at least one minute must elapse before working on the system. Failure to do so may result in deployment of the air bag.

2. Disconnect the negative battery cable. Tape the cable with insulating tape.

➡ **When disconnecting the cable, some systems need to be initialized after the cable is reconnected. You will need the Mazda diagnostic scan tool or equivalent. Follow the directions on the tool.**

3. Remove the plug hole plate.
4. Remove the ignition coils. Remove the spark plugs.

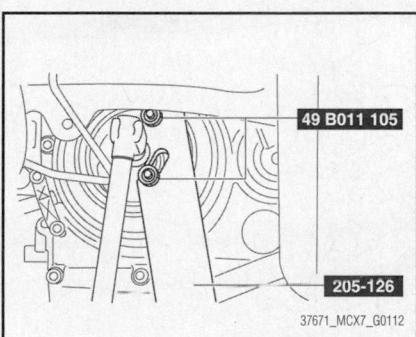

Fig. 176 Crankshaft service tools SST 205-126 and 49 B011 105—2010 2.0L and 2.5L engines

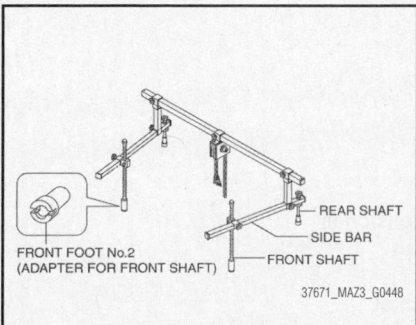

Fig. 177 Engine support system 1 of 5— 2010 2.0L and 2.5L engines

5. Disconnect the ventilation hose from the cylinder head cover.
6. Disconnect the Camshaft Position (CMP) sensor connector.
7. Disconnect the Oil Control Valve (OCV) connector.
8. Remove the front wheel and tire assemblies, as required.
9. Remove the coolant reservoir tank.
10. Remove the aerodynamic No. 2

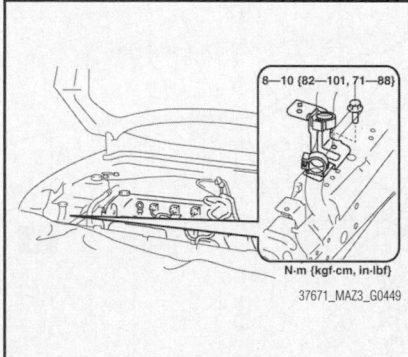

Fig. 178 Engine support system 2 of 5— 2010 2.0L and 2.5L engines

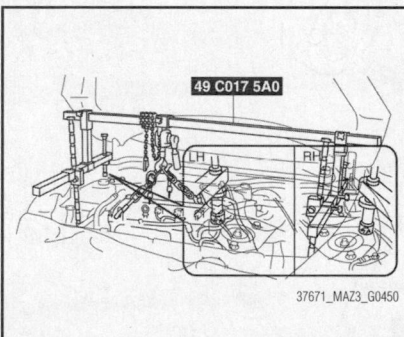

Fig. 179 Engine support system 3 of 5— 2010 2.0L and 2.5L engines

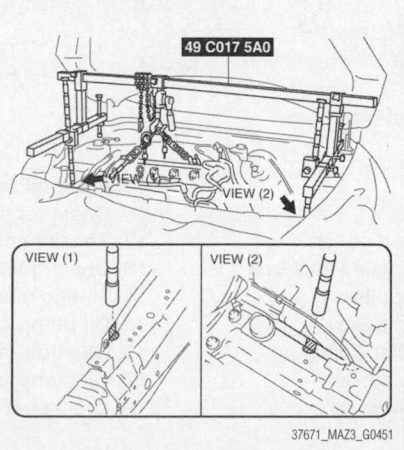

Fig. 180 Engine support system 4 of 5— 2010 2.0L and 2.5L engines

undercover and splash shield as a single unit.

11. Remove the air conditioning compressor and position it to the side. Do not disconnect the refrigerant lines. Do not allow the component to hang by the hoses.
12. Loosen the water pump pulley bolts and remove the drive belt.
13. Remove the Crankshaft Position (CKP) sensor.
14. Disconnect the halfshaft from the joint shaft, right side. Position the component out of the way. Do not allow it to hang unsupported.

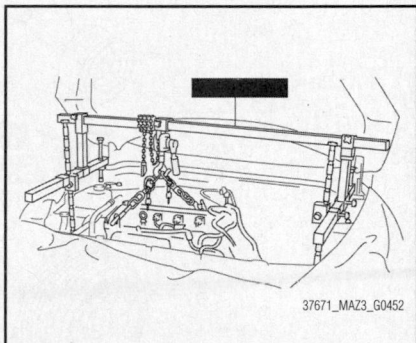

Fig. 181 Engine support system 5 of 5— 2010 2.0L and 2.5L engines

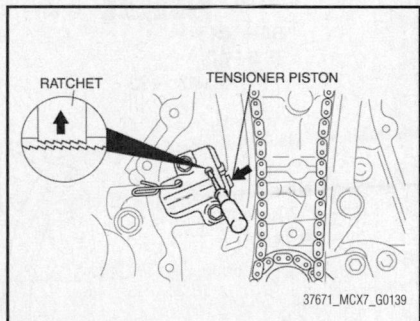

Fig. 182 Compressing the tensioner before removal—2010 2.0L and 2.5L engines

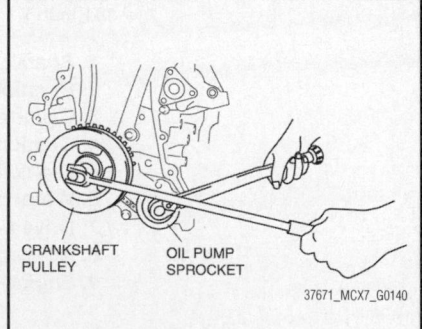

Fig. 183 Oil pump sprocket removal— 2010 2.0L and 2.5L engines

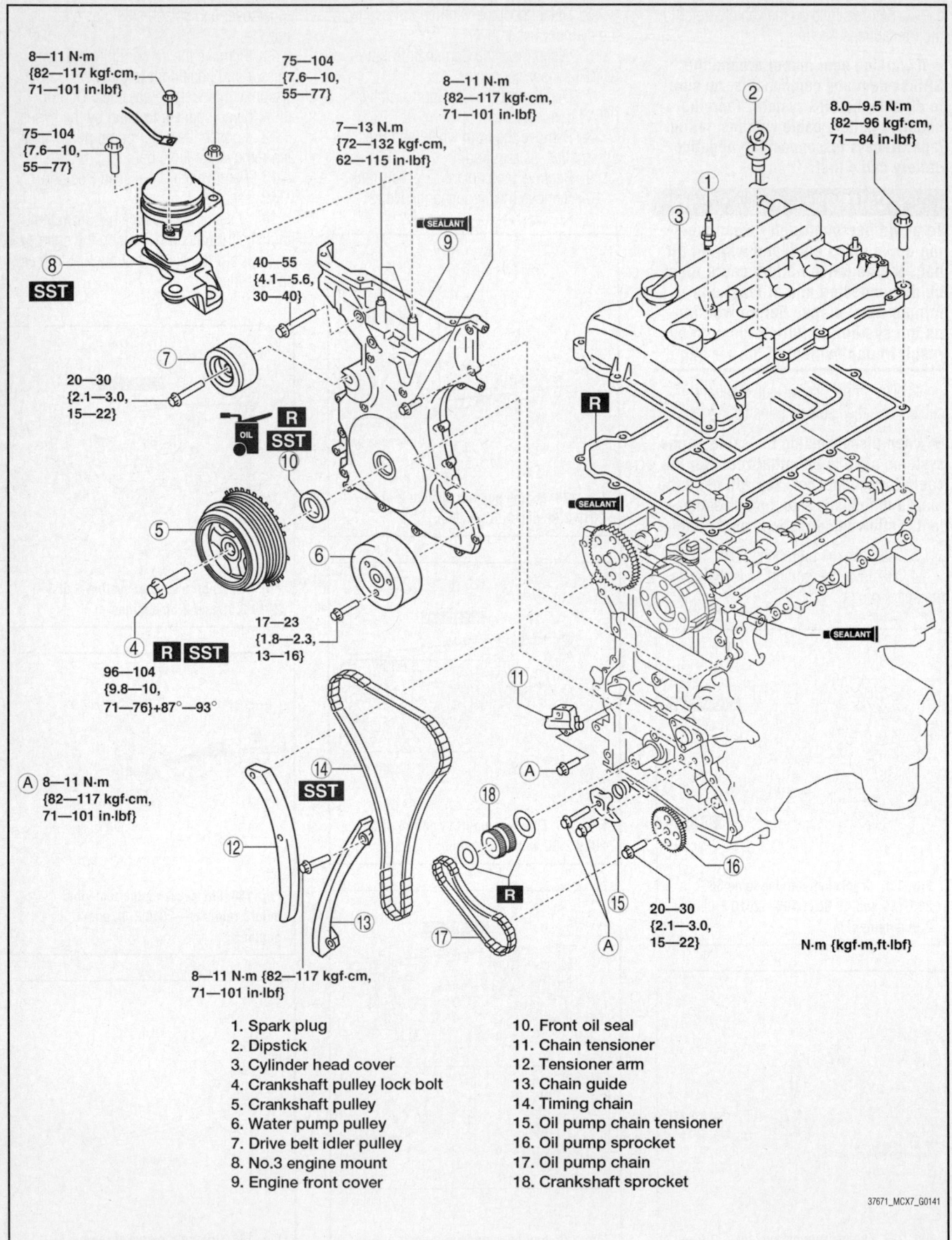

8—11 N·m
{82—117 kgf·cm,
71—101 in·lbf}

75—104
{7.6—10,
55—77}

75—104
{7.6—10,
55—77}

7—13 N·m
{72—132 kgf·cm,
62—115 in·lbf}

8—11 N·m
{82—117 kgf·cm,
71—101 in·lbf}

8.0—9.5 N·m
{82—96 kgf·cm,
71—84 in·lbf}

SEALANT

40—55
{4.1—5.6,
30—40}

SST

20—30
{2.1—3.0,
15—22}

OIL R SST

96—104
{9.8—10,
71—76}+87°—93°

17—23
{1.8—2.3,
13—16}

R SST

R

SEALANT

SEALANT

Ⓐ 8—11 N·m
{82—117 kgf·cm,
71—101 in·lbf}

SST

R

20—30
{2.1—3.0,
15—22}

8—11 N·m {82—117 kgf·cm,
71—101 in·lbf}

N·m {kgf·m,ft·lbf}

1. Spark plug
2. Dipstick
3. Cylinder head cover
4. Crankshaft pulley lock bolt
5. Crankshaft pulley
6. Water pump pulley
7. Drive belt idler pulley
8. No.3 engine mount
9. Engine front cover

10. Front oil seal
11. Chain tensioner
12. Tensioner arm
13. Chain guide
14. Timing chain
15. Oil pump chain tensioner
16. Oil pump sprocket
17. Oil pump chain
18. Crankshaft sprocket

37671_MCX7_G0141

Fig. 184 Timing chain and related components—2010 2.0L and 2.5L engine

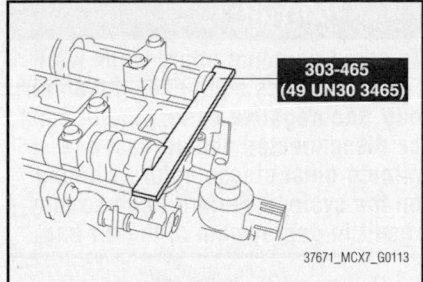

Fig. 185 SST 303-465 camshaft holding tool installed—2010 2.0L and 2.5L engines

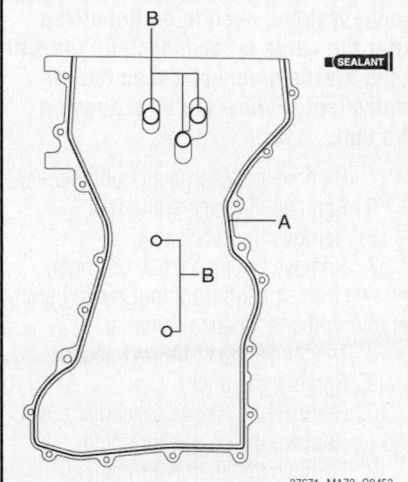

Fig. 186 Engine front cover sealant locations and thickness—2010 2.0L engine

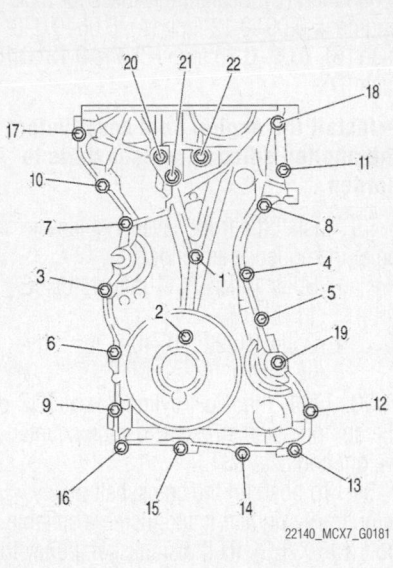

Fig. 188 Front cover torque sequence—2010 2.0L and 2.5L engines

15. Remove the P/S oil pump with the hose and pipe still connected.

➡ **Position and secure the P/S oil pump out of the way with a rope or wire.**

16. Remove the spark plugs.

17. Remove the engine dipstick.

18. Remove the purge solenoid valve installation bracket with the evaporative hose still connected and set it out of the way

19. Remove the cylinder head cover.

20. Rotate the crankshaft in the direction of the engine rotation and remove the cylinder block lower blind plug when the No. 1 cylinder is at the point prior to Top Dead Center (TDC) of compression, then install the SST.

21. Remove the crankshaft pulley lock bolt. Hold the crankshaft pulley by using the SSTs.

22. Remove the crankshaft pulley.

23. Remove the water pump pulley.

24. Remove the drive belt idler pulley.

25. Support the engine using a engine support system.

26. Remove the No. 3 engine mount, remove the following:
- Windshield wiper arm and blade, as required
- The cowl grille, as required
- Air cleaner and air cleaner bracket

27. Remove the No.3 engine mount.

28. Remove the engine front cover.

29. Remove the front oil seal.

30. Remove the chain tensioner as follows:
- Using a thin flat bladed tool, hold the chain tensioner ratchet lock mechanism away from the ratchet stem.
- Slowly compress the tensioner piston
- Hold the tensioner piston using a 0.059 inch (1.5 mm) wire or paper clip.

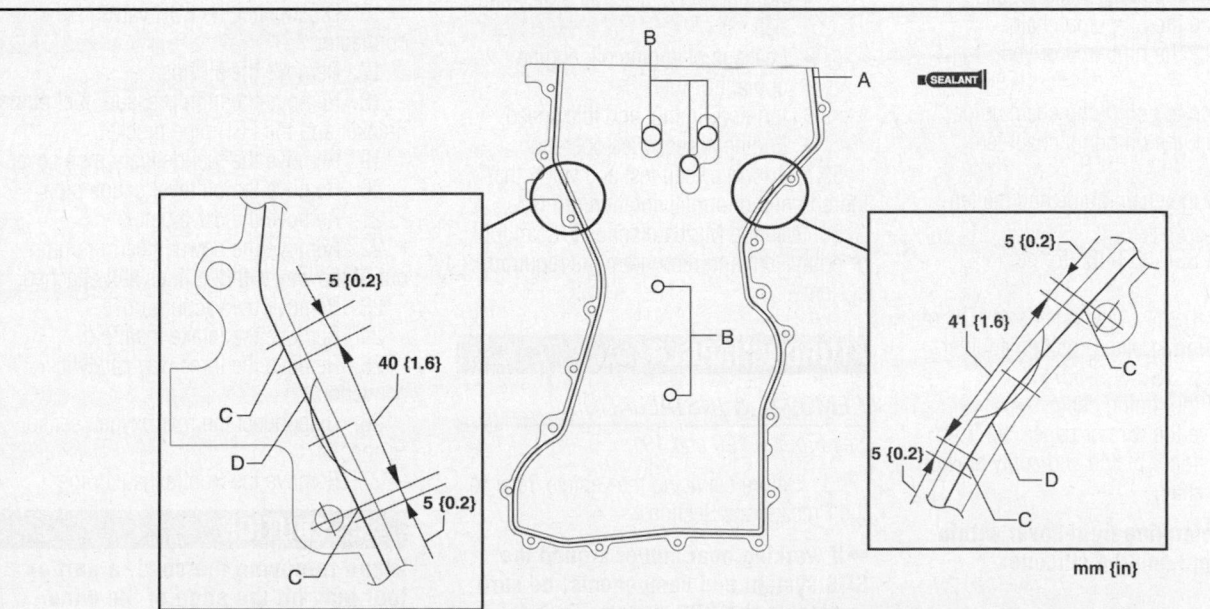

Sealant thickness A: 0.087-0.12 inch (2.2-3.2 mm)
Sealant thickness B: 0.06-0.098 inch (1.5-2.5 mm)
Sealant thickness C: 0.09-0.16 inch (2.2-4.3 mm)
Sealant thickness D: 0.13-0.16 inch (3.3-4.3 mm)

Fig. 187 Engine front cover sealant locations and thickness—2010 2.5L engine

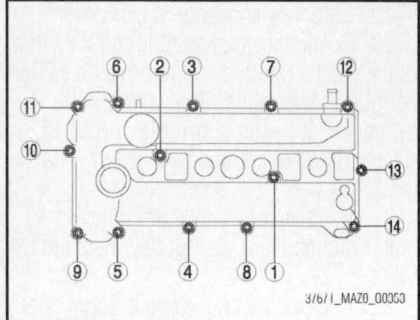

Fig. 189 Valve cover bolt installation sequence—2010 2.0L and 2.5L engines

- Remove the mounting bolt and tensioner

31. Remove the tensioner arm and chain guide.

32. Remove the timing chain.

33. Remove the oil pump chain tensioner.

34. To remove the oil pump chain sprocket, temporarily install the crankshaft pulley and crankshaft pulley lock bolt to the crankshaft, and lock the oil pump against rotation.

35. Remove the oil pump chain.

36. Remove the crankshaft sprocket.

To install:

➡ **Be sure to use new fasteners, as required.**

37. Replace the oil pump drive sprocket and the oil pump chain.

38. Replace the crankshaft sprocket.

39. Replace the oil pump chain.

40. Replace the oil pump driven sprocket.

41. Replace the oil pump chain guide.

42. Replace the oil pump chain tensioner.

43. Install the chain guide and the tensioner arm.

44. Install SST 303-465 to the camshaft.

45. Verify that No.1 cylinder is at TDC of the compression stroke. (Position counterweight contacts SST. 303-507)

46. Install the timing chain.

47. Remove the wire or paper clip from the chain tensioner piston and apply tension to the timing chain.

➡ **Install the engine front cover within 10 min of applying the silicone sealant.**

48. Apply silicone sealant to the engine front cover. Verify that there is no oil or dust on the seal. Sealant thickness for 2.0L engine is 0.087–0.125 inch (A) and 0.059–

0.098 inch (B). Sealant thickness for 2.5L engine is 0.09–0.12 inch (A), 0.06–0.098 inch (B), 0.09–0.16 inch (C) and 0.13–0.16 inch (D).

➡ **Install the engine front cover before the applied silicone sealant starts to harden.**

49. Install the front cover. Tighten the bolts in sequence as follows:
- Bolts 1–18: 71–101 inch lbs. (8–11.5 Nm)
- Bolts 19–22: 30–40 ft. lbs. (40–55 Nm)

50. Verify that No.1 cylinder is at TDC of the compression stroke. (Position counterweight contacts SST.)

51. To position the crankshaft pulley, temporarily tighten it and, using a suitable bolt (M6 X 1.0), fix the crankshaft pulley to the engine front cover.

52. Install SSTs SST 205-126 and 49 B011 105 to the crankshaft pulley and lock the crankshaft against rotation, and tighten the crankshaft pulley lock bolt using the following two steps:
- Step 1: 71–77 ft. lbs. (96–104 Nm)
- Step 2: Plus 87°–93°

53. Install the valve cover. Tighten the bolts in sequence to 71–84 inch lbs. (8–9.5 Nm)

54. Start the engine, and inspect and adjust the following:
- Bleed the air from the cooling system.
- Runout and contact on pulley and belt
- Leakage of engine oil, engine coolant or fuel.
- Ignition timing, and idle speed.
- Engine accessories operation

55. Perform a road test and verify that there is no abnormal vibration or noise.

56. Use the Mazda diagnostic scan tool, or equivalent and reprogram the required systems.

TURBOCHARGER

REMOVAL & INSTALLATION

See Figures 190 and 191.

1. Before servicing the vehicle, refer to the Precautions Section.

➡ **If working near and/or around the SRS system and components, be sure to disable the SRS system. Tape the negative battery cable with insulating tape. Always disconnect the negative battery cable first.**

> ✳✳ **CAUTION**
>
> **To avoid personal injury when working on vehicles equipped with an air bag, the negative battery cable must be disconnected and at least one minute must elapse before working on the system. Failure to do so may result in deployment of the air bag.**

2. Remove the battery top cover, if equipped.

3. Disconnect the negative battery cable. Tape the cable with insulating tape.

➡ **When disconnecting the cable, some systems need to be initialized after the cable is reconnected. You will need Mazda diagnostic scan tool or equivalent. Follow the directions on the tool.**

4. Remove the charge air cooler cover.

5. Remove the battery and tray.

6. Remove the air cleaner.

7. Remove the resonance chamber, remove the left hand front mudguard before removing the resonance chamber.

8. Disconnect the MAF/IAT sensor.

9. Remove the duct.

10. Remove the charge air cooler cover.

11. Remove the charge air cooler.

12. Remove the charge air cooler bracket.

13. Remove the air bypass valve.

14. Drain and recycle the engine coolant.

15. Remove the throttle body.

16. Disconnect the EGR valve connector.

17. Remove the air hose.

18. Remove the high pressure fuel pump bracket and the EGR pipe bracket.

19. Remove the fuel delivery pipe cover.

20. Remove the oil level gauge pipe.

21. Remove the drive belt.

22. Remove the power steering pump out of the way with the lines still attached.

23. Remove the vacuum hose.

24. Remove the intake manifold.

25. Remove the three way catalytic converter.

26. Disconnect the rear oxygen sensor connector.

27. Remove the front exhaust pipe.

> ✳✳ **CAUTION**
>
> **When removing the cowl, a part or tool may hit the edge of the windshield and could damage it. Protect the windshield by covering it with a clean rag to prevent damage to the windshield.**

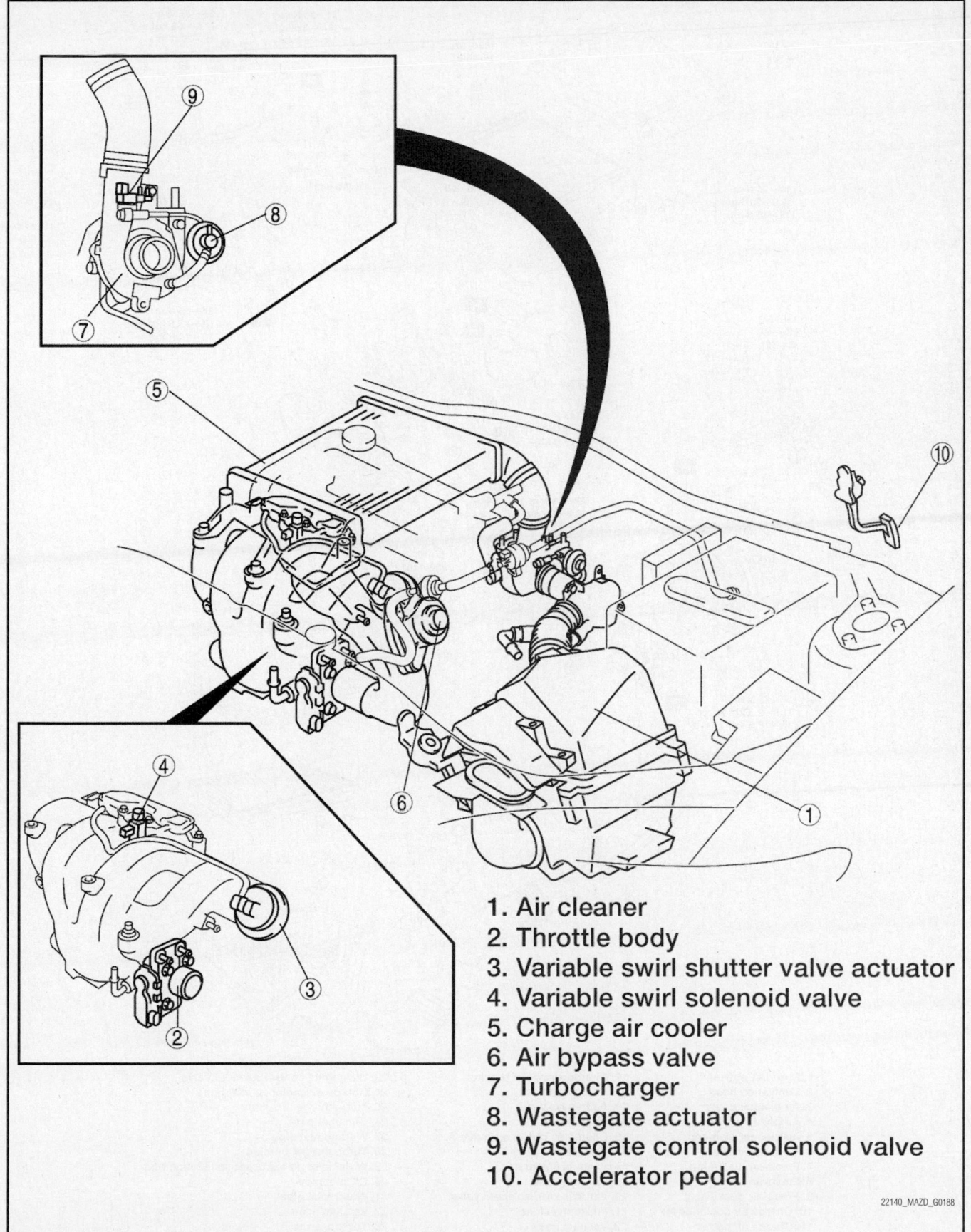

1. Air cleaner
2. Throttle body
3. Variable swirl shutter valve actuator
4. Variable swirl solenoid valve
5. Charge air cooler
6. Air bypass valve
7. Turbocharger
8. Wastegate actuator
9. Wastegate control solenoid valve
10. Accelerator pedal

22140_MAZD_G0188

Fig. 190 Turbocharger and related components (2.3L engine with turbocharger)—2009 vehicles

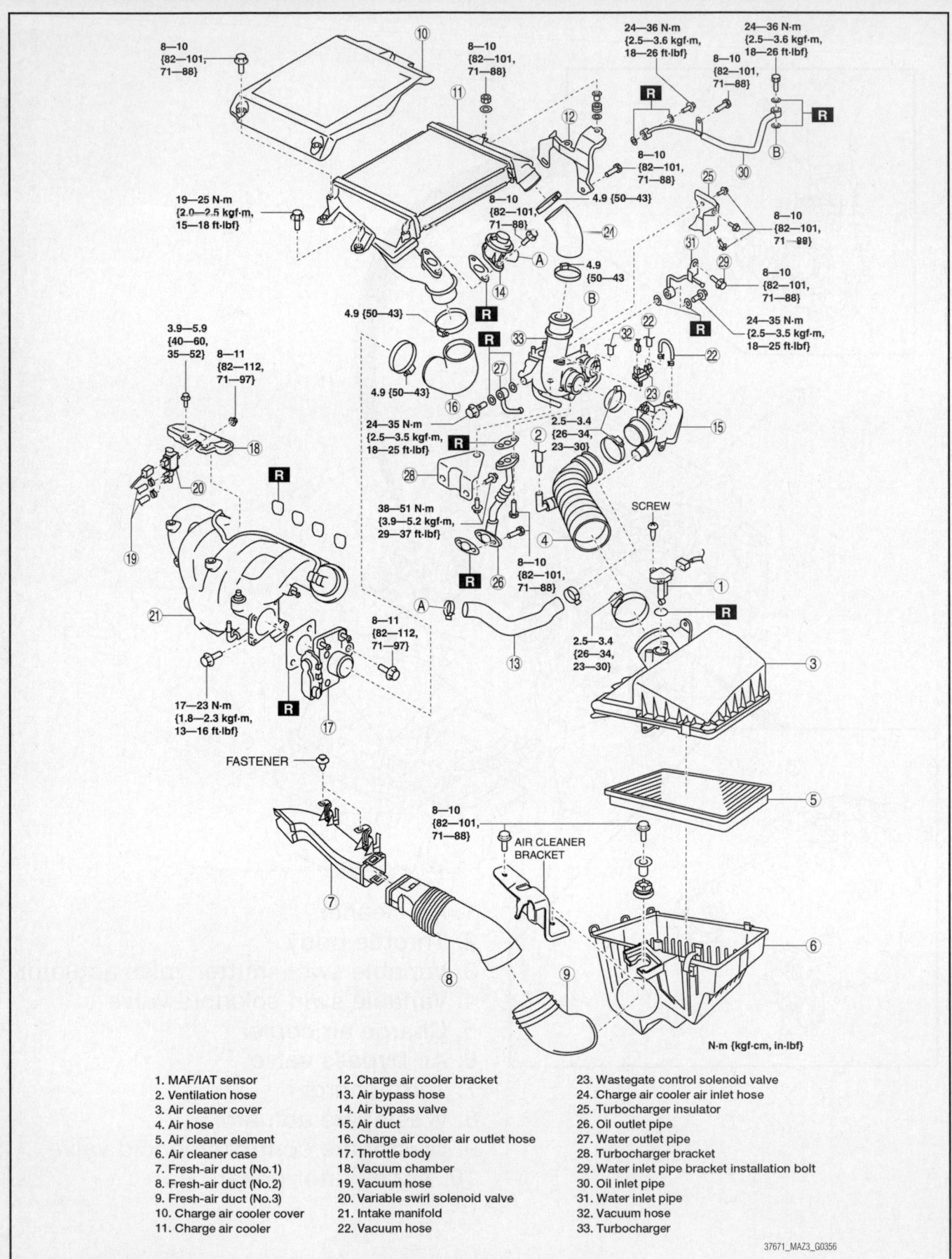

Fig. 191 Turbocharger and related components (2.3L engine with turbocharger)—2010 vehicles

1. MAF/IAT sensor
2. Ventilation hose
3. Air cleaner cover
4. Air hose
5. Air cleaner element
6. Air cleaner case
7. Fresh-air duct (No.1)
8. Fresh-air duct (No.2)
9. Fresh-air duct (No.3)
10. Charge air cooler cover
11. Charge air cooler

12. Charge air cooler bracket
13. Air bypass hose
14. Air bypass valve
15. Air duct
16. Charge air cooler air outlet hose
17. Throttle body
18. Vacuum chamber
19. Vacuum hose
20. Variable swirl solenoid valve
21. Intake manifold
22. Vacuum hose

23. Wastegate control solenoid valve
24. Charge air cooler air inlet hose
25. Turbocharger insulator
26. Oil outlet pipe
27. Water outlet pipe
28. Turbocharger bracket
29. Water inlet pipe bracket installation bolt
30. Oil inlet pipe
31. Water inlet pipe
32. Vacuum hose
33. Turbocharger

37671_MAZ3_G0356

28. Remove the cowl and the insulator under the cowl.

29. Remove the exhaust manifold upper side insulator.

30. Remove the alternator.

31. Remove the exhaust manifold lower side insulator.

32. Disconnect the front oxygen sensor connector.

33. Remove the warm up three-way catalytic converter top side insulator.

34. Remove the brake master back side vacuum hose.

35. Remove the oil pipes.

36. Remove the front oxygen sensor.

37. Remove the warm up three-way catalytic converter.

38. Remove the turbocharger.

To install:

39. Installation is the reverse of removal, note the following:

　a. Tighten the turbocharger bolts to 29–37 ft. lbs. (38–51 Nm).

　b. Tighten the oil pipe installation bolt while the stopper of the oil pipe is faced to the turbocharger.

　c. Tighten the waste gate Control Solenoid Valve bolts to 70–95 inch lbs. (8–11 Nm).

　d. Tighten the intake manifold bolts to 13–16 ft. lbs. (17–23 Nm).

　e. Tighten the fuel delivery pipe cover bolts to 13–16 ft. lbs. (17–23 Nm).

　f. Tighten the charge air cooler bracket to 31–35 ft. lbs. (42–48 Nm).

　g. Tighten the charge air cooler to 14–19 ft. lbs. (42–48 Nm).

✱✱ CAUTION

When installing the charge air cooler cover, be careful not to damage the charge air cooler cover clips.

　h. Tighten the throttle body installation bolts to 71–101 inch lbs. (8–11.5 Nm).

　i. Perform the following when installing the air cleaner case:

- Verify that the rubber mounts are set in the air cleaner bracket
- Install the projections on the frame side.
- Verify that the projections on the frame side are installed securely.
- Install the projection on the engine side.
- Verify that the projection on the engine side installed securely.

40. Use the Mazda diagnostic scan tool, or equivalent and reprogram the required systems.

VALVE LASH

ADJUSTMENT

Except 2010 2.0L & 2.5L Engines

These engines use solid cam followers with a removable adjustment shim. The valve lash clearance is measured with the original shim installed and checked against the specification. If adjustment is necessary, the original shim is removed, and a thicker or thinner shim is installed to obtain the proper clearance. Special tools are required in order to adjust the shim without removing the camshaft.

2010 2.0L & 2.5L Engines

See Figures 192 through 203.

1. Before servicing the vehicle, refer to the Precautions Section.

➡**If working near and/or around the SRS system and components, be sure to disable the SRS system. Tape the negative battery cable with insulating tape. Always disconnect the negative battery cable first.**

✱✱ CAUTION

To avoid personal injury when working on vehicles equipped with an air bag, the negative battery cable must be disconnected and at least one minute must elapse before working on the system. Failure to do so may result in deployment of the air bag.

2. Disconnect the negative battery cable. Tape the cable with insulating tape.

➡**When disconnecting the cable, some systems need to be initialized after the cable is reconnected. You will need the Mazda diagnostic scan tool or equivalent. Follow the directions on the tool.**

3. Remove the plug hole plate.

4. Disconnect the wiring harness.

5. Remove the ignition coils.

6. Remove the spark plugs.

7. Disconnect the ventilation hose from the cylinder head cover.

8. Remove the CMP sensor connector.

9. Remove the cylinder head cover.

10. Set the front mudguard (RH) out of the way.

11. Remove the splash shield (RH).

12. Remove the No. 2 aerodynamic undercover.

13. Remove the drive belt.

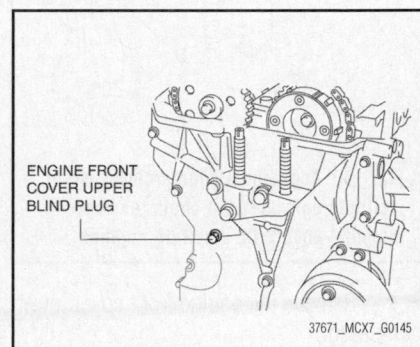

Fig. 193 Engine front cover upper blind plug—2010 2.0L and 2.5L engines

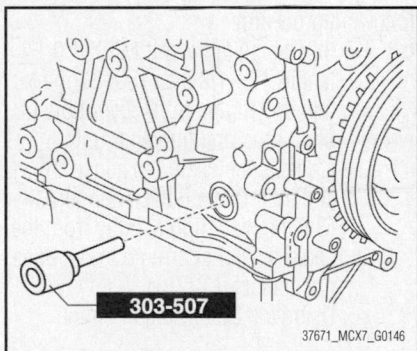

Fig. 194 Install the SST 303-507—2010 2.0L and 2.5L engines

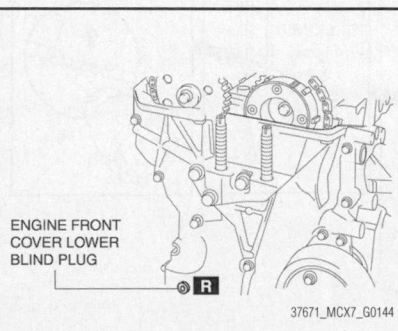

Fig. 192 Engine front cover lower blind plug—2010 2.0L and 2.5L engines

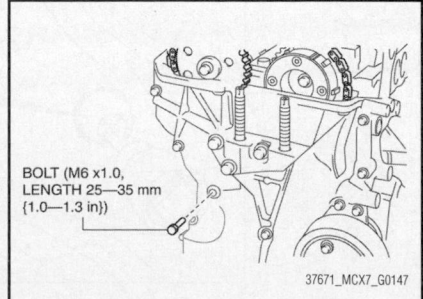

Fig. 195 M6 X 1.0, length bolt installed into the engine front cover upper blind plug—2010 2.0L and 2.5L engines

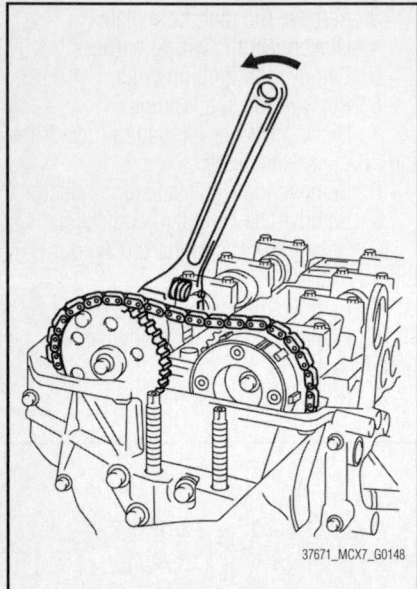

Fig. 196 Apply force counterclockwise to facilitate unlocking the chain tensioner ratchet—2010 2.0L and 2.5L engines

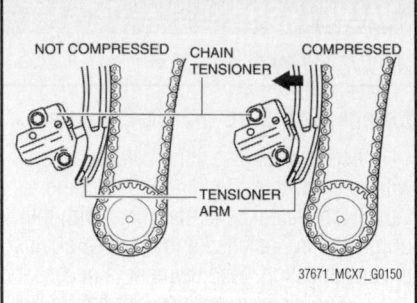

Fig. 198 Chain tensioner rack shown compressed and not compressed—2010 2.0L and 2.5L engines

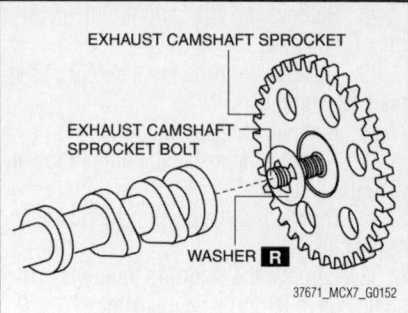

Fig. 200 Exhaust camshaft sprocket, bolt and washer removed as a unit—2010 2.0L and 2.5L engines

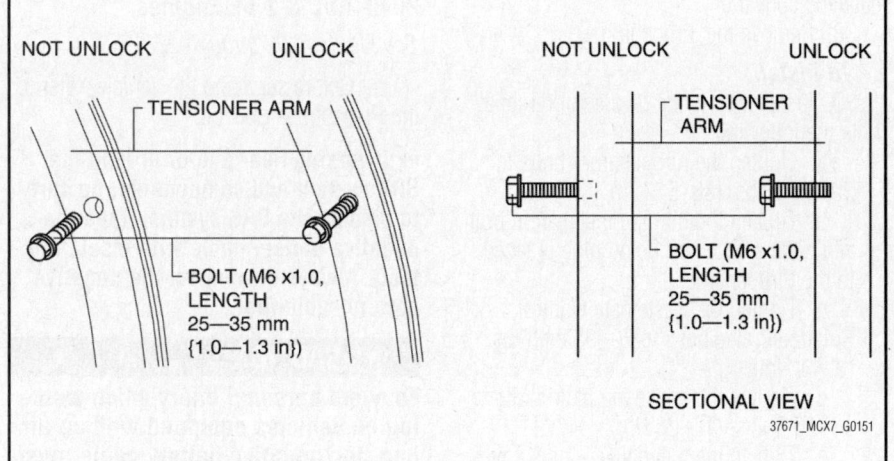

Fig. 199 Screw in the bolt set in approx. 0.2 inch (5 mm) and secure the tensioner arm—2010 2.0L and 2.5L engines

14. Remove the engine front cover lower blind plug.

15. Remove the engine front cover upper blind plug.

16. Remove the cylinder block lower blind plug, and install the SST 303-507.

17. Loosen the timing chain using the following procedure:

- Insert a suitable bolt M6 X 1.0, length 1.0–1.3 inch (25–35 mm) into the engine front cover upper blind plug and tighten it until it contacts the chain tensioner arm, and then rotate it back one turn. (Set the bolt slightly away from the chain tensioner arm so that it does not contact it.)
- Using the cast hexagon on the exhaust camshaft, apply force counterclockwise to facilitate unlocking the chain tensioner ratchet.
- Using a Hex bit socket 0.098 inch (2.5 mm) or T15 Torx® bit socket, unlock the chain tensioner ratchet so that it can be lifted up.
- Using the cast hexagon on the exhaust camshaft, apply force in the direction of the engine rotation to increase tension on the chain

➡The chain tensioner rack is compressed using the chain tension generated by applying force to the exhaust camshaft in the direction of the engine rotation.

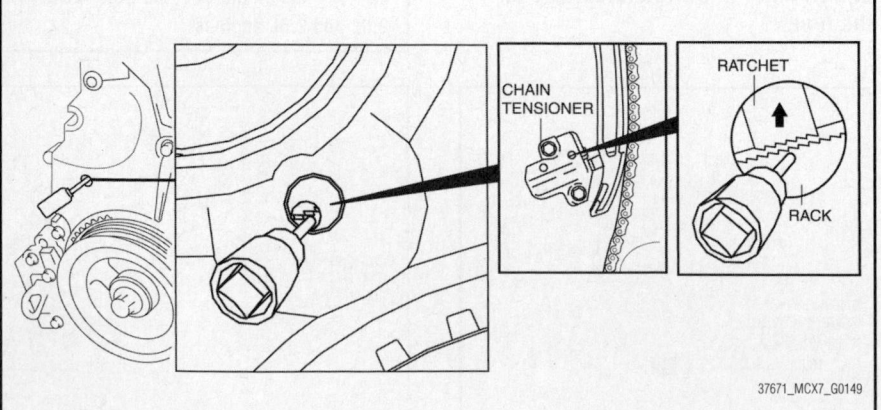

Fig. 197 Unlocking the chain tensioner ratchet so that it can be lifted up—2010 2.0L and 2.5L engines

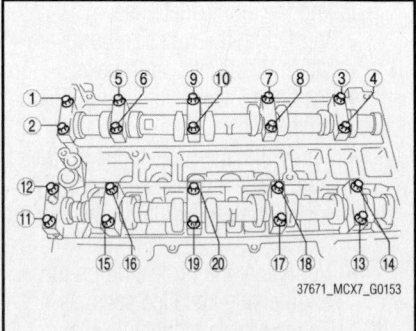

Fig. 201 Camshaft cap removal sequence—2010 2.0L and 2.5L engines

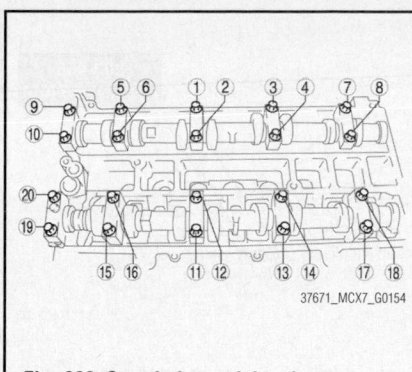

Fig. 202 Camshaft cap tightening sequence—2010 2.0L and 2.5L engines

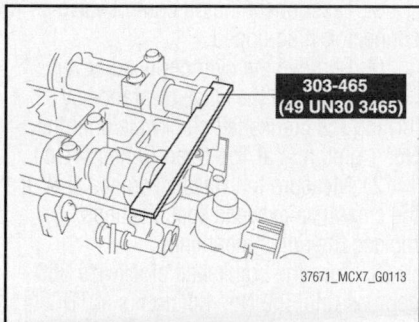

303-465
(49 UN30 3465)

Fig. 203 SST 303-465 camshaft holding tool installed—2010 2.0L and 2.5L engines

- Screw in the bolt set in approx. 0.2 inch (5 mm) and secure the tensioner arm with the rack compressed. The racket has not been unlocked if the bolt cannot be pressed in approx 0.2 inch (5 mm).

➡**If the tensioner arm cannot be secured, return the bolt to its original position and repeat the procedure.**

18. Fix the exhaust camshaft using a wrench on the cast hexagon, and loosen the camshaft sprocket bolt.

19. Remove the exhaust camshaft sprocket bolt, exhaust camshaft sprocket, and washer as a single unit. Perform the work carefully so that the washer does not drop out.

20. Loosen the camshaft cap bolts in two or three steps in the order shown in the figure and remove the camshaft cap.

➡**The camshaft caps are to be kept ordered for correct reassembly in their original positions. Do not mix the caps.**

21. Remove the camshafts for the intake and exhaust sides.

22. Remove the tappet.

23. Install an appropriate tappet based on the results of the valve clearance inspection.

24. Selected tappet = Removed tappet thickness + Measured valve clearance - Standard valve clearance

25. Standard specified valve clearance is:
- Intake: 0.009–0.011 inch (0.22–0.28 mm)
- Exhaust: 0.011–0.012 inch (0.27–0.33 mm).

26. Replace the tappets of any valve found outside the specified values and pre-measured the valve clearance.

27. Verify that No.1 cylinder is at TDC of the compression stroke. (Position counter-weight contacts SST.)

28. Apply the gear oil (SAE No.90 or equivalent) to each journal of the cylinder head.

29. Install the camshaft with No.1 cylinder aligned with the TDC position.

30. Apply the gear oil (SAE No.90 or equivalent) to each camshaft journal of the cylinder head.

31. Temporarily tighten the camshaft cap bolts evenly in 2–3 steps.

32. Tighten the camshaft cap bolts in the order shown in the following two steps;
- Step 1: 45–79 inch lbs. (5.0–9.0 Nm)
- Step 2: 11–12 ft. lbs. (14–17 Nm)

33. Install the OCV and tighten the retaining bolt to 71–97 inch lbs. (8–11 Nm).

34. Install the exhaust camshaft sprocket bolt, exhaust camshaft sprocket, and a new washer as a single unit.

➡**Install a washer to the fourth or fifth thread of the exhaust camshaft sprocket bolt being careful not to drop the washer.**

✳✳ WARNING

Do not tighten the camshaft sprocket bolt at this stage. Verify the valve timing before performing the bolt tightening.

35. Install the SST 303-465 on the camshaft.

36. Remove the installation bolt for the engine front cover upper blind plug, and apply tension to the timing chain.

37. Fix the exhaust camshaft using a wrench on the cast hexagon, and tighten the sprocket bolt to 51–55 ft. lbs. (69–75 Nm).

38. Remove the SST 303-465 from the camshaft.

39. Remove the SST 303-507 installed in the cylinder block lower blind plug hole.

40. Rotate the crankshaft clockwise two turns and inspect the valve timing.

➡**If not aligned, loosen the camshaft sprocket bolt and repeat the procedure.**

41. Install the cylinder block lower blind plug, and tighten to 14–16 ft. lbs. (18–22 Nm).

42. Apply the silicone sealant and install the engine front cover upper blind plug. Tighten to 71–101 inch lbs. (8–11 Nm).

43. Install a new engine front cover lower blind plug and tighten to 89–123 inch lbs. (10–14 Nm).

44. Install the drive belt.

45. Install the splash shield (RH).

46. Install the front mudguard (RH).

47. Install the undercover.

48. Install the front wheel and tire (RH).

49. Install the cylinder head cover.

50. Connect the ventilation hose.

51. Install the spark plugs.

52. Install the ignition coils.

53. Connect the wiring harness.

54. Install the plug hole plate.

55. Connect the negative battery cable.

56. Start the engine, and inspect the following:
- Leakage of engine oil, engine coolant or fuel.
- Ignition timing, and idle speed.

57. Perform a road test and verify that there is no abnormal vibration or noise.

58. Use the Mazda diagnostic scan tool, or equivalent and reprogram the required systems.

INSPECTION

2009 Vehicles

2.0L Engine

See Figures 170, 204 through 211.

➡**With the engine cold, standard valve clearance is 0.087–0.0110 inch (0.22–0.28mm) on the intake side and 0.0107–0.0129 inch (0.27–0.33mm) on the exhaust side.**

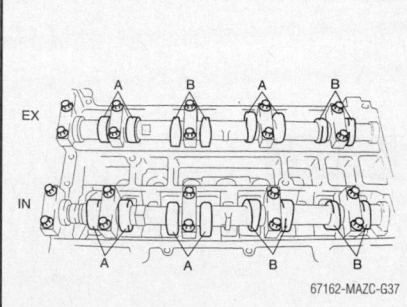

Fig. 204 Valve clearance checking positions—2009 2.0L engine

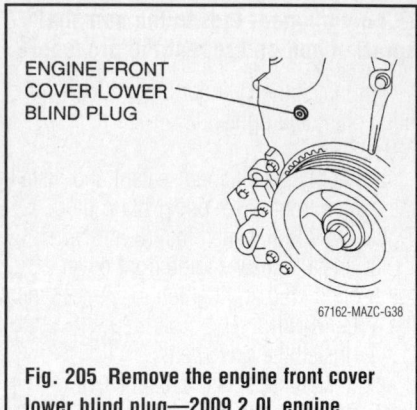

Fig. 205 Remove the engine front cover lower blind plug—2009 2.0L engine

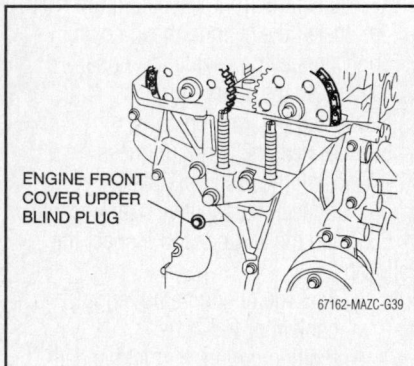

Fig. 206 Remove the engine front cover upper blind plug—2009 2.0L engine

1. Before servicing the vehicle, refer to the Precautions Section.

➡ If working near and/or around the SRS system and components, be sure to disable the SRS system. Tape the negative battery cable with insulating tape. Always disconnect the negative battery cable first.

✳✳ CAUTION

To avoid personal injury when working on vehicles equipped with an air bag, the negative battery cable must be disconnected and at least one

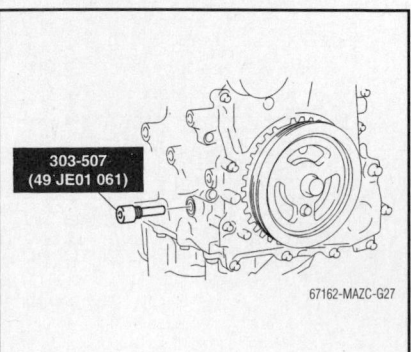

Fig. 207 Install tool 49 JE01 06—2009 2.0L engine

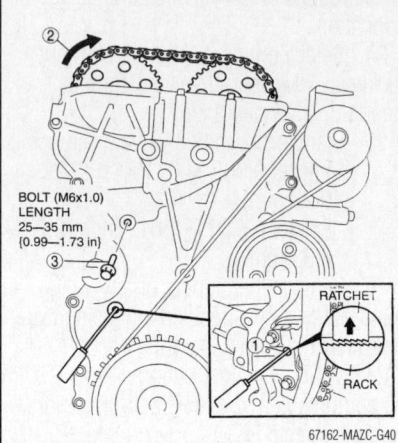

Fig. 208 Loosen the timing chain tension, refer to the text for procedure—2009 2.0L engine

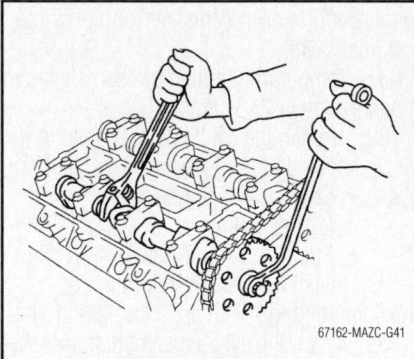

Fig. 209 Hold the exhaust camshaft using a wrench on the cast hexagon portion of the camshaft and remove the exhaust camshaft sprocket—2009 2.0L engine

minute must elapse before working on the system. Failure to do so may result in deployment of the air bag.

2. Remove the battery top cover, if equipped.

3. Disconnect the negative battery cable. Tape the cable with insulating tape.

➡ When disconnecting the cable, some systems need to be initialized after the cable is reconnected. You will need Mazda diagnostic scan tool or equivalent. Follow the directions on the tool.

4. Remove the plug hole plate.
5. Disconnect the wiring harness.
6. Remove the right front wheel, engine undercover and splash shield.
7. Remove the ignition coils.
8. Remove the Positive Crankcase Ventilation (PCV) hose, if equipped.

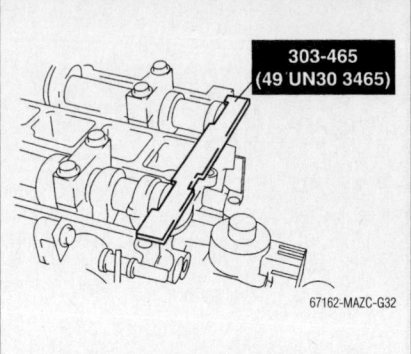

Fig. 210 Install tool 49 UN30 3465 on the camshaft—2009 2.0L engine

9. Disconnect the oil control valve connector, if equipped.

10. Remove the cylinder head cover.

11. Measure the valve clearance by turning the crankshaft clockwise until the No. 1 piston is at Top Dead Center (TDC).

12. Measure the valve clearance at **A**. If the clearance exceeds specifications, replace the adjustment shim.

13. Turn the crankshaft clockwise 360 degrees until the No. 4 piston is at TDC. Measure the valve clearance at **B**. If the clearance exceeds specifications, replace the adjustment shim.

14. Repeat this procedure for all the camshafts.

15. Remove the engine front cover lower and upper blind plugs and the cylinder block lower blind plug.

16. Install tool 49 JE01 061.

17. Turn the crankshaft clockwise until the crankshaft is in the number 1 cylinder Top Dead Center position (the balance weight should be attached to tool 49 JE01 061).

18. Loosen the timing chain as follows:
   a. Unlock the chain tensioner ratchet using a suitable tool.
   b. Turn the exhaust camshaft clockwise using a wrench on the cast hexagon

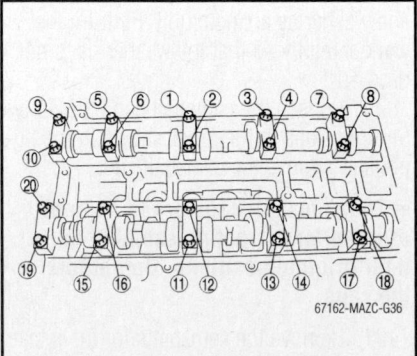

Fig. 211 Camshaft cap bolt tightening sequence—2009 2.0L engine

portion of the camshaft and loosen the timing chain.

c. Place a M6 x 1.0 length bolt: 0.99–1.37 inch (25–35mm) at the engine front cover upper blind plug and secure the timing chain guide at the location where the chain tension is released.

19. Hold the exhaust camshaft using a wrench on the cast hexagon portion of the camshaft and remove the exhaust camshaft sprocket.

➡️**The cylinder head and camshaft caps are numbered and must be reassembled in their original locations. When removing a when removed, keep the caps with the cylinder head they were removed from, do not switch the caps.**

20. Loosen the camshaft cap bolts using 2–3 passes in the sequence illustrated.

21. Once all the bolts have been removed, remove the caps, keeping them in the original positions and remove the camshafts.

22. Remove the tappets and select the proper adjustment shim. The formula for selecting a shim is as follows:

a. Removed shim thickness + measured valve clearance–standard valve clearance 0.0098 inch (0.25mm) for the intake and 0.0118 inch (0.30mm) for the exhaust.

23. Install the adjustment shim.

24. Set the cam position of the number 1 cylinder at Top Dead center (TDC) and install the camshaft.

25. Temporarily install the camshaft caps in their original positions.

26. Tighten the camshaft cap bolt in two steps:

a. Step 1: 44–71 inch lbs. (5–8 Nm).
b. Step 2 : 10.4–12.5 ft. lbs. (14–17 Nm).

27. Install the exhaust camshaft sprocket and hand tighten the bolt.

28. Install tool 49 UN30 3465 as illustrated on the camshaft.

29. Remove the M6 x 1.0 bolt from the engine front cover to tension the timing chain.

30. Turn the crankshaft clockwise until the crankshaft is in the number 1 cylinder Top Dead Center position (the balance weight should be attached to tool 49 JE01 061).

31. Hold the exhaust camshaft using a wrench on the cast hexagon portion of the camshaft and tighten the exhaust camshaft sprocket bolt to 51–55 ft. lbs. (69–75 Nm).

32. Remove the tools from the camshaft and from the block lower blind plug.

33. Rotate the engine two full turns clockwise to TDC. If not aligned, remove the crankshaft pulley lockbolt, remove the tappet and repeat the adjustment shim selection and replacement procedure.

34. Apply sealant to the engine front cover upper blind plug and tighten to 71–101 inch lbs. (8–11 Nm).

35. Install the cylinder block lower blind plug and tighten to 13–16 ft. lbs. (18–22 Nm).

36. Install a new engine front cover blind plug and tighten to 7.4–10 ft. lbs. (10–14 Nm).

37. Install the cylinder head cover.

38. Install the PCV hose, if equipped.

39. Connect the oil control valve connector, if equipped.

40. Install the ignition coils.

41. Connect the wiring harness.

42. Connect the negative battery cable and install the battery cover.

43. Install the plug hole plate.

44. Install the splash shield, undercover and wheel.

45. Use the Mazda diagnostic scan tool, or equivalent and reprogram the required systems.

2.3L Engine

➡️**With the engine cold, standard valve clearance is 0.087–0.0110 inch (0.22–0.28mm) on the intake side and 0.0107–0.0129 inch (0.27–0.33mm) on the exhaust side.**

1. Before servicing the vehicle, refer to the Precautions Section.

➡️**If working near and/or around the SRS system and components, be sure to disable the SRS system. Tape the negative battery cable with insulating tape. Always disconnect the negative battery cable first.**

❄️ CAUTION

To avoid personal injury when working on vehicles equipped with an air bag, the negative battery cable must be disconnected and at least one minute must elapse before working on the system. Failure to do so may result in deployment of the air bag.

2. Remove the battery top cover, if equipped.

3. Disconnect the negative battery cable. Tape the cable with insulating tape.

➡️**When disconnecting the cable, some systems need to be initialized after the cable is reconnected. You will need**

Mazda diagnostic scan tool or equivalent. Follow the directions on the tool.

4. Remove the plug hole plate.

5. Disconnect the wiring harness.

6. Remove the ignition coils.

7. Disconnect the Oil Control Valve (OCV) connector.

8. Remove the right front wheel, undercover and splash shield.

9. Remove the Positive Crankcase Ventilation (PCV) hose.

10. Remove the cylinder head cover.

11. Measure the valve clearance by turning the crankshaft clockwise until the No. 1 piston is at Top Dead Center (TDC).

12. Measure the valve clearance at**A**. If the clearance exceeds specifications, replace the adjustment shim.

13. Turn the crankshaft clockwise 360 degrees until the No. 4 piston is at TDC. Measure the valve clearance at**B**. If the clearance exceeds specifications, replace the adjustment shim.

14. Repeat this procedure for all the camshafts.

15. Remove the engine front cover lower and upper blind plugs and the cylinder block lower blind plug.

16. Install tool 49 JE01 061.

17. Turn the crankshaft clockwise until the crankshaft is in the number 1 cylinder Top Dead Center position (the balance weight should be attached to tool 49 JE01 061).

18. Loosen the timing chain as follows:

a. Unlock the chain tensioner ratchet using a suitable tool.

b. Turn the exhaust camshaft clockwise using a wrench on the cast hexagon portion of the camshaft and loosen the timing chain.

c. Place a M6 x 1.0 length bolt: 0.99–1.37 inch (25–35mm) at the engine front cover upper blind plug and secure the timing chain guide at the location where the chain tension is released.

19. Hold the exhaust camshaft using a wrench on the cast hexagon portion of the camshaft and remove the exhaust camshaft sprocket.

➡️**The cylinder head and camshaft caps are numbered and must be reassembled in their original locations. When removing a when removed, keep the caps with the cylinder head they were removed from, do not switch the caps.**

20. Loosen the camshaft cap bolts using 2–3 passes in the sequence illustrated.

21. Once all the bolts have been removed, remove the caps, keeping them in the original positions and remove the camshafts.

22. Remove the tappets and select the proper adjustment shim. The formula for selecting a shim is as follows:

    a. Removed shim thickness + measured valve clearance - standard valve clearance 0.0098 inch (0.25mm) for the intake and 0.0118 inch (0.30mm) for the exhaust.

23. Install the adjustment shim.

24. Set the cam position of the number 1 cylinder at Top Dead center (TDC) and install the camshaft.

25. Temporarily install the camshaft caps in their original positions.

26. Tighten the camshaft cap bolt in two steps:

    a. Step 1: 44–71 inch lbs. (5–8 Nm).

    b. Step 2 : 10.4–12.5 ft. lbs. (14–17 Nm).

27. Install the exhaust camshaft sprocket and hand tighten the bolt.

28. Install tool 49 UN30 3465 as illustrated on the camshaft.

29. Remove the M6 x 1.0 bolt from the engine front cover to tension the timing chain.

30. Turn the crankshaft clockwise until the crankshaft is in the number 1 cylinder Top Dead Center position (the balance weight should be attached to tool 49 JE01 061).

31. Hold the exhaust camshaft using a wrench on the cast hexagon portion of the camshaft and tighten the exhaust camshaft sprocket bolt to 51–55 ft. lbs. (69–75 Nm).

32. Remove the tools from the camshaft and from the block lower blind plug.

33. Rotate the engine two full turns clockwise to TDC. If not aligned, remove the crankshaft pulley lockbolt, remove the tappet and repeat the adjustment shim selection and replacement procedure.

34. Apply sealant to the engine front cover upper blind plug and tighten to 71–101 inch lbs. (8–11 Nm).

35. Install the cylinder block lower blind plug and tighten to 13–16 ft. lbs. (18–22 Nm).

36. Install a new engine front cover blind plug and tighten to 7.4–10 ft. lbs. (10–14 Nm).

37. Install the cylinder head cover.

38. Install the PCV hose.

39. Connect the OCV connector.

40. Install the ignition coils.

41. Connect the wiring harness.

42. Connect the negative battery cable and install the battery cover.

43. Install the plug hole plate.

44. Install the undercover, splash shield and wheel.

45. Use the Mazda diagnostic scan tool, or equivalent and reprogram the required systems.

2009–10 2.3L Engine With Turbocharger

See Figures 212 through 217.

➡ **With the engine cold, standard valve clearance is 0.087–0.0110 inch (0.22–0.28mm) on the intake side and 0.0107–0.0129 inch (0.27–0.33mm) on the exhaust side.**

1. Before servicing the vehicle, refer to the Precautions Section.

➡ **If working near and/or around the SRS system and components, be sure to disable the SRS system. Tape the negative battery cable with insulating tape. Always disconnect the negative battery cable first.**

✳✳ CAUTION

To avoid personal injury when working on vehicles equipped with an air bag, the negative battery cable must be disconnected and at least one minute must elapse before working on the system. Failure to do so may result in deployment of the air bag.

2. Remove the battery top cover, if equipped.

3. Disconnect the negative battery cable. Tape the cable with insulating tape.

➡ **When disconnecting the cable, some systems need to be initialized after the cable is reconnected. You will need Mazda diagnostic scan tool or equivalent. Follow the directions on the tool.**

4. Disconnect the negative battery cable.

5. Remove the right side front tire.

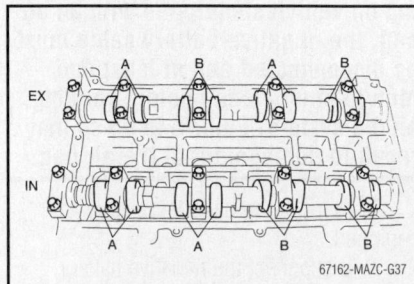

Fig. 212 Valve clearance checking positions—2009 and 2010 2.3L engine with turbocharger

6. Remove the undercover.

7. Remove the right side splash shield.

➡ **If the high pressure fuel pump is removed, replace the O-ring with a new one.**

8. Remove the charge air cooler cover.

9. Disconnect the spill valve control solenoid valve connector.

10. Disconnect the quick release connector on the high pressure fuel pump.

11. Remove the battery and battery tray.

✳✳ WARNING

If the high pressure fuel pump joint nut is loosened, fuel leakage may occur resulting in death or serious injury, or damage to the equipment or the vehicle. Fuel can also irritate the skin and eyes. When removing the high pressure line pipe, always tighten the high pressure line pipe installation nut while fixing the high pressure fuel pump joint nut with a wrench. If the high pressure fuel pump joint nut has

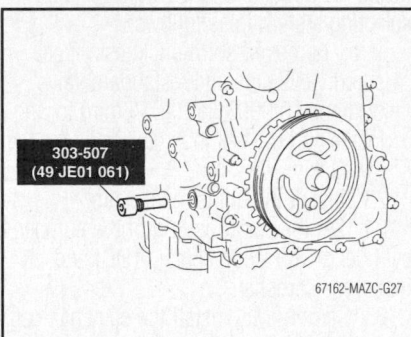

Fig. 213 Install tool 49 JE01 061—2009 and 2010 2.3L engine with turbocharger

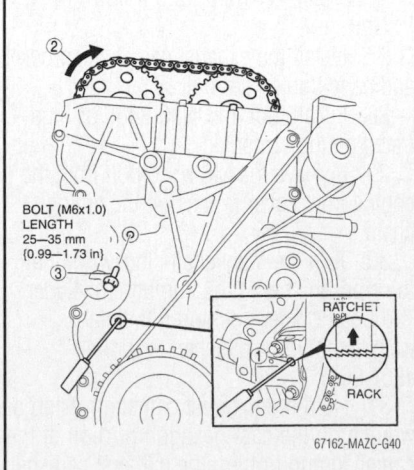

Fig. 214 Loosen the timing chain tension, refer to the text for procedure—2009 and 2010 2.3L engine with turbocharger

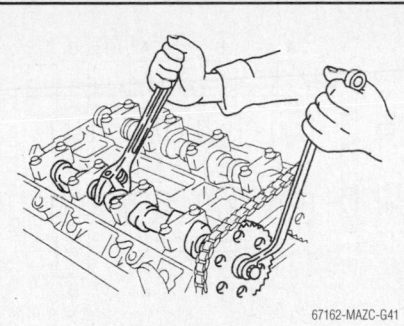

Fig. 215 Hold the exhaust camshaft using a wrench on the cast hexagon portion of the camshaft and remove the exhaust camshaft sprocket—2009 and 2010 2.3L engine with turbocharger

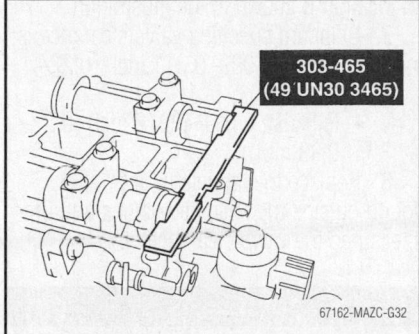

Fig. 216 Install tool 49 UN30 3465 on the camshaft—2009 and 2010 2.3L engine with turbocharger

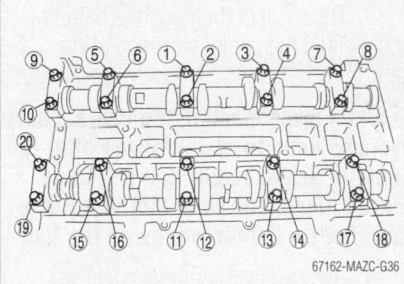

Fig. 217 Camshaft cap bolt tightening sequence—2009 and 2010 2.3L engine with turbocharger

rotated, replace the high pressure fuel pump with a new one.

12. Disconnect the high pressure line pipe of the high pressure fuel pump.

13. Fix the joint nut with a wrench on the high pressure fuel pump side.

14. Loosen the high pressure line pipe installation nut.

15. Drain and recycle the engine coolant.

16. Loosen the water outlet case installation bolts securing the high pressure line pipe.

17. Remove the high pressure fuel pump.

18. Remove the high pressure fuel pump cover.

19. Remove the ignition coils, spark plugs, valve cover and accessory drive belt.

20. Measure the valve clearance by turning the crankshaft clockwise until the No. 1 piston is at Top Dead Center (TDC).

21. Measure the valve clearance at **A**. If the clearance exceeds specifications, replace the adjustment shim.

22. Turn the crankshaft clockwise 360 degrees until the No. 4 piston is at TDC. Measure the valve clearance at **B**. If the clearance exceeds specifications, replace the adjustment shim.

23. Repeat this procedure for all the camshafts.

24. Remove the engine front cover lower and upper blind plugs and the cylinder block lower blind plug.

25. Install tool 49 JE01 061.

26. Turn the crankshaft clockwise until the crankshaft is in the number 1 cylinder Top Dead Center position (the balance weight should be attached to tool 49 JE01 061).

27. Loosen the timing chain as follows:

   a. Unlock the chain tensioner ratchet using a suitable tool.

   b. Turn the exhaust camshaft clockwise using a wrench on the cast hexagon portion of the camshaft and loosen the timing chain.

   c. Place a M6 x 1.0 length bolt: 0.99–1.37 inch (25–35mm) at the engine front cover upper blind plug and secure the timing chain guide at the location where the chain tension is released.

28. Hold the exhaust camshaft using a wrench on the cast hexagon portion of the camshaft and remove the exhaust camshaft sprocket.

➡**The cylinder head and camshaft caps are numbered and must be reassembled in their original locations. When removing a when removed, keep the caps with the cylinder head they were removed from, do not switch the caps.**

29. Loosen the camshaft cap bolts using 2–3 passes in the sequence illustrated.

30. Once all the bolts have been removed, remove the caps, keeping them in the original positions and remove the camshafts.

31. Remove the tappets and select the proper adjustment shim. The formula for selecting a shim is as follows:

   a. Removed shim thickness + measured valve clearance - standard valve clearance 0.0098 inch (0.25mm) for the intake and 0.0118 inch (0.30mm) for the exhaust.

32. Install the adjustment shim.

33. Set the cam position of the number 1 cylinder at Top Dead center (TDC) and install the camshaft.

34. Temporarily install the camshaft caps in their original positions.

35. Tighten the camshaft cap bolt in two steps:

   a. Step 1: 44–71 inch lbs. (5–8 Nm).

   b. Step 2 : 10.4–12.5 ft. lbs. (14–17 Nm).

36. Install the exhaust camshaft sprocket and hand tighten the bolt.

37. Install tool 49 UN30 3465 as illustrated on the camshaft.

38. Remove the M6 x 1.0 bolt from the engine front cover to tension the timing chain.

39. Turn the crankshaft clockwise until the crankshaft is in the number 1 cylinder Top Dead Center position (the balance weight should be attached to tool 49 JE01 061).

40. Hold the exhaust camshaft using a wrench on the cast hexagon portion of the camshaft and tighten the exhaust camshaft sprocket bolt to 51–55 ft. lbs. (69–75 Nm).

41. Remove the tools from the camshaft and from the block lower blind plug.

42. Rotate the engine two full turns clockwise to TDC. If not aligned, remove the crankshaft pulley lockbolt, remove the tappet and repeat the adjustment shim selection and replacement procedure.

43. Apply sealant to the engine front cover upper blind plug and tighten to 71–101 inch lbs. (8–11 Nm).

44. Install the cylinder block lower blind plug and tighten to 13–16 ft. lbs. (18–22 Nm).

45. Install a new engine front cover blind plug and tighten to 7.4–10 ft. lbs. (10–14 Nm).

46. Install the high pressure fuel pump cover and tighten to 13–16 ft. lbs. (18–22 Nm).

❋❋ CAUTION

If the high pressure fuel pump installation bolts are tightened with the high pressure fuel pump tilted, the high pressure fuel pump may not operate correctly. Tighten the high pressure fuel pump installation bolts in a few passes with equal torque. Tighten to 76–101 inch lbs. (8.5–11.5 Nm).

❊❊ WARNING

If the high pressure fuel pump joint nut is loosened, fuel leakage may occur resulting in death or serious injury, or damage to the equipment or the vehicle. Fuel can also irritate the skin and eyes. When installing the high pressure line pipe, always tighten the high pressure line pipe installation nut while fixing the high pressure fuel pump joint nut with a wrench. If the high pressure fuel pump joint nut has rotated, replace the high pressure fuel pump with a new one.

47. Assemble the high pressure line pipe.

48. Fix the joint nut with a wrench on the high pressure fuel pump side. Tighten the high pressure line pipe installation nut to 17–26 ft. lbs. (23–35 Nm).

49. Tighten the water outlet case installation bolts to 13–16 ft. lbs. (18–22 Nm).

50. Install the quick release connector.

51. Install the remaining components in the reverse order of removal.

52. Verify that the high pressure fuel pump is assembled securely.

53. Drive the vehicle starting from a standstill and brake suddenly five to six times at a low speed.

54. Stop the vehicle and verify from outside the vehicle that there is no fuel leakage around the high pressure fuel pump.

55. Use the Mazda diagnostic scan tool, or equivalent and reprogram the required systems.

2010 2.0L & 2.5L Engines

See Figure 218.

➡Check valve clearance with the engine cold.

1. Before servicing the vehicle, refer to the Precautions section.

2. Remove the valve cover.

3. Set the No. 1 cylinder to Top Dead Center (TDC) of the compression stroke.

4. Measure and note the valve clearances of location **A** shown in the illustration.

5. Rotate the crankshaft 1 full turn so

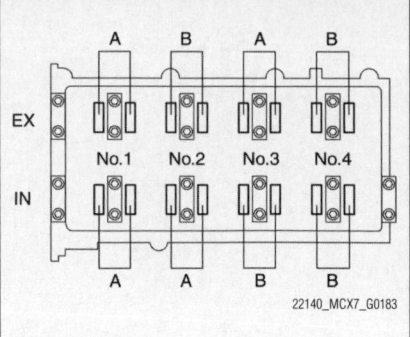

Fig. 218 Measuring valve clearance—2010 2.0L and 2.5L engines

that cylinder No. 4 is at TDC of the compression stroke.

6. Measure and note the valve clearances of location **B** shown in the illustration.

7. Standard specified valve clearance is:
   - Intake: 0.009–0.011 inch (0.22–0.28 mm)
   - Exhaust: 0.011–0.012 inch (0.27–0.33 mm)

8. Replace the tappets of any valve found outside the specified values and pre-measured the valve clearance.

ENGINE PERFORMANCE & EMISSION CONTROLS

CAMSHAFT POSITION (CMP) SENSOR

LOCATION

See Figures 219 through 221.

REMOVAL & INSTALLATION

See Figures 219 through 221.

1. Before servicing the vehicle, refer to the Precautions Section.

➡If working near and/or around the SRS system and components, be sure to disable the SRS system. Tape the negative battery cable with insulating tape. Always disconnect the negative battery cable first.

❊❊ CAUTION

To avoid personal injury when working on vehicles equipped with an air bag, the negative battery cable must be disconnected and at least one minute must elapse before working on the system. Failure to do so

may result in deployment of the air bag.

2. Remove the battery top cover, if equipped.

3. Disconnect the negative battery cable. Tape the cable with insulating tape.

➡When disconnecting the cable, some systems need to be initialized after the cable is reconnected. You will need Mazda diagnostic scan tool or equivalent. Follow the directions on the tool.

4. Remove the plug hole cover, as required.

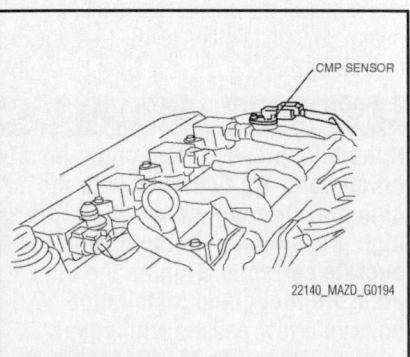

Fig. 219 Camshaft Position (CMP) sensor location—2009 vehicles

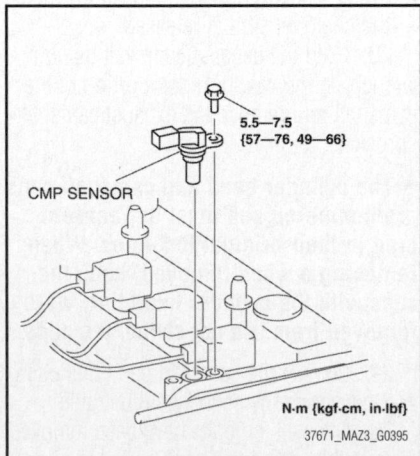

Fig. 220 Camshaft Position (CMP) sensor location (2.0L and 2.5L engines)—2010 vehicles

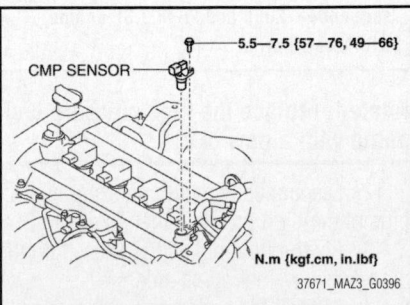

Fig. 221 Camshaft Position (CMP) sensor location (2.3L engine with turbocharger)—2010 vehicles

5. Remove the charge air cooler, if equipped with 2.3L engine with turbocharger.

6. Disconnect the CMP sensor connector.

7. Remove the CMP sensor installation bolt.

8. Remove the CMP sensor.

9. Make sure that the CMP sensor is free of any metallic shavings or particles.

➡If metallic shavings or particles are found on the sensor, clean them off.

To install:

➡Be sure to use new fasteners, as required.

10. Install the CMP sensor in the reverse order of removal.

11. Tightening torque is 49–66 inch lbs. (5.5—7.5 Nm).

12. Use the Mazda diagnostic scan tool, or equivalent and reprogram the required systems.

CRANKSHAFT POSITION (CKP) SENSOR

LOCATION

See Figures 222 through 224.

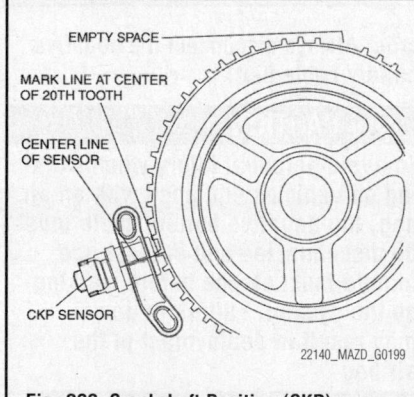

Fig. 222 Crankshaft Position (CKP) sensor location—2009 vehicles

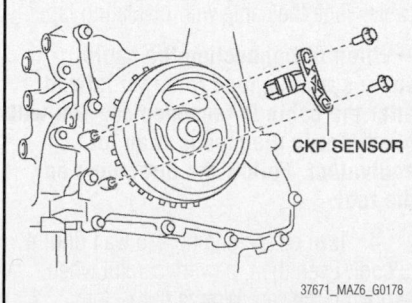

Fig. 223 Crankshaft Position (CKP) sensor location (2.0L and 2.5L engines—2010 vehicles

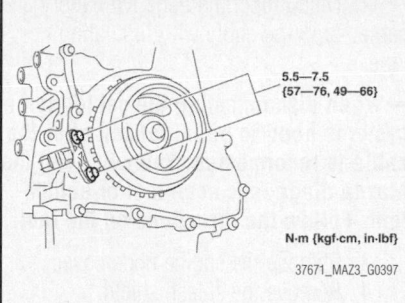

N·m {kgf·cm, in-lbf}

Fig. 224 Crankshaft Position (CKP) sensor location (2.3L engine with turbocharger)—2010 vehicles

OPERATION

The Crankshaft Position (CKP) Sensor is a permanent magnet output coil device that monitors the speed and position of the crankshaft. A reluctor is attached directly to the crankshaft, and is used to generate a constant signal as it passes the CKP sensor magnetic coil. The CKP Sensor utilizes system voltage (12 volts DC), or reference voltage (5 volts DC) to generate a digital output signal to the PCM that is based upon a 0.5 volt AC reference. The alternating magnetic field is used by the sensor output electronics to produce a digital pulse. The CKP sensor returns a digital ON/OFF signal pulse during each revolution of the crankshaft, with a momentary signal interrupt for the top dead center (TDC) position of the #1 cylinder. The Pulse Width Modulation pattern is used by the Powertrain Control Module (PCM) to calculate ignition timing, fuel injector timing, misfire diagnostics, and the tachometer display.

REMOVAL & INSTALLATION

2009 Vehicles
See Figures 222 and 225.

1. Before servicing the vehicle, refer to the Precautions Section.

➡If working near and/or around the SRS system and components, be sure to disable the SRS system. Tape the negative battery cable with insulating tape. Always disconnect the negative battery cable first.

❋❋ CAUTION

To avoid personal injury when working on vehicles equipped with an air bag, the negative battery cable must be disconnected and at least one minute must elapse before working on the system. Failure to do so may result in deployment of the air bag.

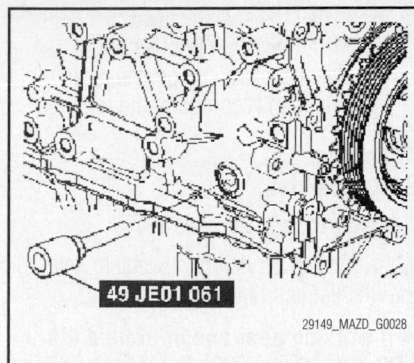

Fig. 225 Crankshaft Position (CKP) sensor TDC locator–2009 vehicles

2. Remove the battery top cover, if equipped.

3. Disconnect the negative battery cable. Tape the cable with insulating tape.

➡When disconnecting the cable, some systems need to be initialized after the cable is reconnected. You will need Mazda diagnostic scan tool or equivalent. Follow the directions on the tool.

4. Remove the right front wheel.

5. Remove the splash shield.

6. Disconnect the CKP sensor connector.

7. Remove the installation bolts to remove the CKP sensor.

To install:

➡Be sure to use new fasteners, as required.

➡Perform the following procedure so that piston No.1 is at the top dead center.

8. Remove the cylinder block lower blind plug and install the special service tool or equivalent.

9. Turn the crankshaft pulley to the clockwise until it stops.

10. Using a straight edge, draw a straight line directly in the center of the ninth tooth of the crankshaft pulley pulse wheel (counting counterclockwise from the empty space).

➡If the line is not accurately drawn, ignition timing, fuel injection and other engine control systems will be adversely affected. Drawn the straight line carefully using a straight edge.

11. Align the centerline of the Crankshaft Position (CKP) sensor and the line drawn, then install the sensor.

12. Install the CKP sensor fitting bolts. Tighten to 4.1–5.5 ft lbs

13. Remove the special service tool or equivalent then install the cylinder block lower blind plug. Tightening torque 15 ft lbs

14. Install the splash shield.
15. Install the right front wheel
16. Use the Mazda diagnostic scan tool, or equivalent and reprogram the required systems.

2010 Vehicles

See Figures 224, 226 and 227.

1. Before servicing the vehicle, refer to the Precautions Section.

➡ **If working near and/or around the SRS system and components, be sure to disable the SRS system. Tape the negative battery cable with insulating tape. Always disconnect the negative battery cable first.**

❊❊ CAUTION

To avoid personal injury when working on vehicles equipped with an air bag, the negative battery cable must be disconnected and at least one minute must elapse before working on the system. Failure to do so may result in deployment of the air bag.

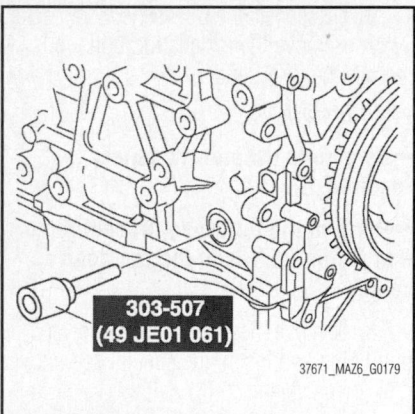

Fig. 226 Special service tool 303-507—2010 vehicles

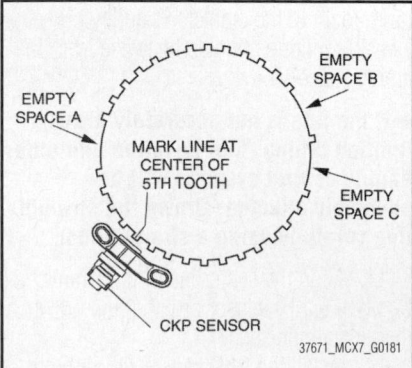

Fig. 227 Fit the center of the CKP sensor with the fifth tooth from empty space A—2010 vehicles

2. Disconnect the negative battery cable. Tape the cable with insulating tape.

➡ **When disconnecting the cable, some systems need to be initialized after the cable is reconnected. You will need the Mazda diagnostic scan tool or equivalent. Follow the directions on the tool.**

3. Remove the engine undercover.
4. Remove the splash shield.
5. Disconnect the CKP sensor connector.
6. Remove the installation bolts to remove the CKP sensor.

To install:

7. Rotate the crankshaft in the direction of the engine rotation and remove the cylinder block lower blind plug when the No. 1 cylinder is at the point prior to top dead center (TDC) of compression, then install SST 303-507.

8. Fit the center of the CKP sensor with the fifth tooth (counting counterclockwise from the empty space A as shown in the figure) of the pulse wheel

9. Install the CKP sensor fitting bolts. Tighten the bolts to 49–66 inch lbs. (5.5–7.5 Nm).

10. Remove the SST 303-507, then install the cylinder block lower blind plug. Tighten to 14–16 ft. lbs. (18–22 Nm).

ENGINE COOLANT TEMPERATURE (ECT) SENSOR

LOCATION

See Figures 228 and 229.

Engine Coolant Temperature (ECT) Sensor location.

REMOVAL & INSTALLATION

See Figures 228 and 229.

❊❊ CAUTION

Never remove the cooling system cap or loosen the radiator drain plug while the engine is running, or when the engine and radiator are hot. Scalding engine coolant and steam may shoot out and cause serious injury. It may also damage the engine and cooling system.

1. Before servicing the vehicle, refer to the Precautions Section.

➡ **If working near and/or around the SRS system and components, be sure to disable the SRS system. Tape the negative battery cable with insulating**

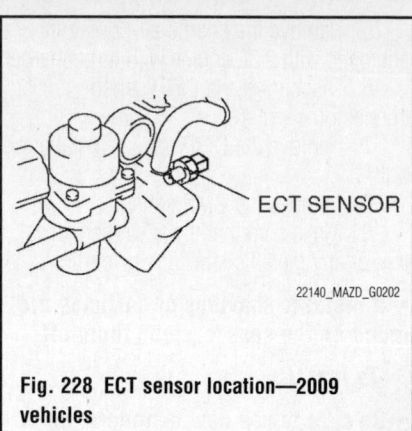

Fig. 228 ECT sensor location—2009 vehicles

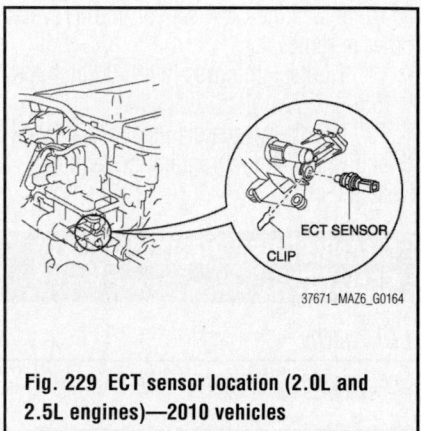

Fig. 229 ECT sensor location (2.0L and 2.5L engines)—2010 vehicles

tape. Always disconnect the negative battery cable first.

❊❊ CAUTION

To avoid personal injury when working on vehicles equipped with an air bag, the negative battery cable must be disconnected and at least one minute must elapse before working on the system. Failure to do so may result in deployment of the air bag.

2. Remove the battery top cover, if equipped.
3. Disconnect the negative battery cable. Tape the cable with insulating tape.

➡ **When disconnecting the cable, some systems need to be initialized after the cable is reconnected. You will need Mazda diagnostic scan tool or equivalent. Follow the directions on the tool.**

4. Turn off the engine and wait until it is cool. Even then, be very careful when removing the cap. Wrap a thick cloth around it and slowly turn it counterclockwise to the first stop. Step back while the pressure escapes.

5. When you are sure all the pressure is gone, press down on the cap using the cloth, turn it, and remove it.

6. Drain the coolant. Be sure to properly dispose of used coolant.

7. Remove the battery and the battery tray.

8. Disconnect the sensor electrical connector.

9. Remove the sensor from its mounting.

10. Discard the O-ring.

To install:

➡ Be sure to use new fasteners, as required.

11. Installation is the reverse of the removal procedure.

12. Fill the cooling system with the proper grade and type coolant.

13. Start the engine and check for leaks.

14. Use the Mazda diagnostic scan tool, or equivalent and reprogram the required systems.

HEATED OXYGEN SENSOR (HO2S)

LOCATION

See Figures 230 and 231.

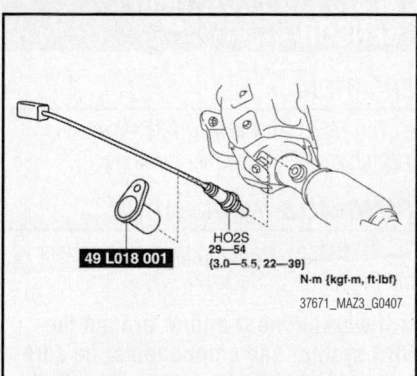

Fig. 230 Heated oxygen location (2.0L and 2.5L engines)—2010 vehicles

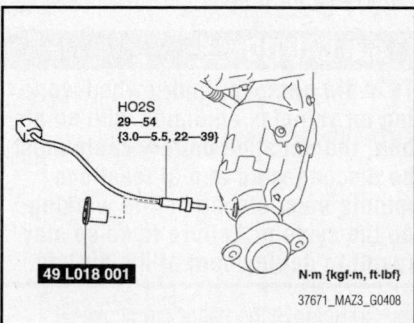

Fig. 231 Heated oxygen sensor location (2.3L engine with turbocharger)—2010 vehicles

REMOVAL & INSTALLATION

See Figures 230 and 231.

> **※※ CAUTION**
>
> **The temperature of the exhaust system is extremely high after the engine has been run. To prevent personal injury, allow the exhaust system to cool completely before removing sensor from the exhaust system.**

1. Before servicing the vehicle, refer to the Precautions Section.

➡ If working near and/or around the SRS system and components, be sure to disable the SRS system. Tape the negative battery cable with insulating tape. Always disconnect the negative battery cable first.

> **※※ CAUTION**
>
> **To avoid personal injury when working on vehicles equipped with an air bag, the negative battery cable must be disconnected and at least one minute must elapse before working on the system. Failure to do so may result in deployment of the air bag.**

2. Disconnect the negative battery cable. Tape the cable with insulating tape.

➡ When disconnecting the cable, some systems need to be initialized after the cable is reconnected. You will need the Mazda diagnostic scan tool or equivalent. Follow the directions on the tool.

3. As required, raise and support the vehicle safely.

4. Remove the plug hole plate, and air cleaner assembly, if equipped with 2.0L, 2.3L or 2.5L engines.

5. For 2009, remove the charge air cooler, 2.3L engine with turbocharger.

6. Disconnect the sensor wiring harness.

7. Remove the sensor with the proper tool.

To install:

➡ Be sure to use new fasteners, as required.

8. Installation is the reverse of the removal procedure.

9. Use the Mazda diagnostic scan tool, or equivalent and reprogram the required systems.

INTAKE AIR TEMPERATURE (IAT) SENSOR

LOCATION

The Intake Air Temperature (IAT) Sensor is integral to the Mass Air Flow (MAF) sensor. It is located in the air cleaner housing.

REMOVAL & INSTALLATION

1. Before servicing the vehicle, refer to the Precautions Section.

➡ If working near and/or around the SRS system and components, be sure to disable the SRS system. Tape the negative battery cable with insulating tape. Always disconnect the negative battery cable first.

> **※※ CAUTION**
>
> **To avoid personal injury when working on vehicles equipped with an air bag, the negative battery cable must be disconnected and at least one minute must elapse before working on the system. Failure to do so may result in deployment of the air bag.**

2. Remove the battery top cover, if equipped.

3. Disconnect the negative battery cable. Tape the cable with insulating tape.

➡ When disconnecting the cable, some systems need to be initialized after the cable is reconnected. You will need Mazda diagnostic scan tool or equivalent. Follow the directions on the tool.

4. Remove the MAF/IAT sensor electrical connector.

5. Remove the MAF/IAT sensor retaining screws.

6. Remove the MAF/IAT sensor.

To install:

➡ Be sure to use new fasteners, as required.

7. Installation is the reverse of the removal procedure.

8. Install the MAF/IAT sensor and tighten the screws.

9. Install the MAF/IAT sensor electrical connector.

10. Use the Mazda diagnostic scan tool, or equivalent and reprogram the required systems.

KNOCK SENSOR (KS)

LOCATION

See Figures 232 and 233.

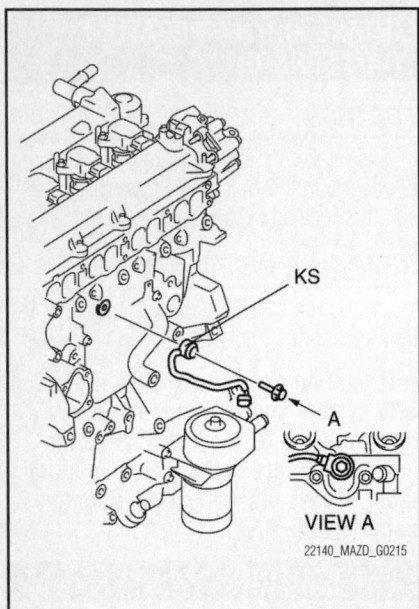

Fig. 232 Knock sensor location—2009 vehicles

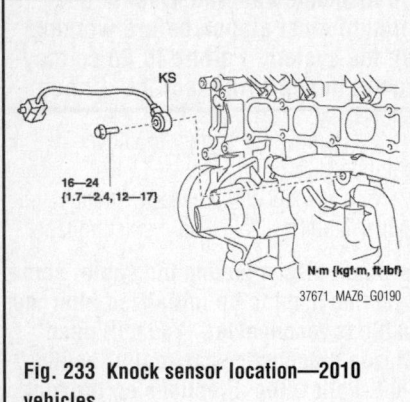

Fig. 233 Knock sensor location—2010 vehicles

The Knock Sensor (KS), is located at the rear of the engine block.

REMOVAL & INSTALLATION

See Figures 232 and 233.

1. Before servicing the vehicle, refer to the Precautions Section.

➡**If working near and/or around the SRS system and components, be sure to disable the SRS system. Tape the negative battery cable with insulating tape. Always disconnect the negative battery cable first.**

✳✳ CAUTION

To avoid personal injury when working on vehicles equipped with an air bag, the negative battery cable must be disconnected and at least one minute must elapse before working on the system. Failure to do so

may result in deployment of the air bag.

2. Remove the battery top cover, if equipped.
3. Disconnect the negative battery cable. Tape the cable with insulating tape.

➡**When disconnecting the cable, some systems need to be initialized after the cable is reconnected. You will need Mazda diagnostic scan tool or equivalent. Follow the directions on the tool.**

4. Remove the intake manifold.
5. Disconnect the sensor connector.
6. Remove the sensor from the engine block.

To install:

➡**Be sure to use new fasteners, as required.**

7. Installation is the reverse of the removal procedure.
8. Tighten the sensor mounting bolt to 12–17 ft. lbs. (16–24 Nm).
9. Use the Mazda diagnostic scan tool, or equivalent and reprogram the required systems.

MALFUNCTION INDICATOR LIGHT (MIL)

RESET PROCEDURE

See Figure 234.

➡**The Malfunction Indicator Light (MIL) has to be reset with the use of a scanner.**

1. Before servicing the vehicle, refer to the Precautions Section.

➡**If working near and/or around the SRS system and components, be sure to disable the SRS system. Tape the negative battery cable with insulating**

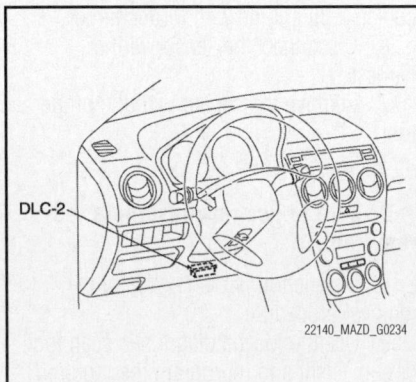

Fig. 234 DLC-2 connector location

tape. Always disconnect the negative battery cable first.

✳✳ CAUTION

To avoid personal injury when working on vehicles equipped with an air bag, the negative battery cable must be disconnected and at least one minute must elapse before working on the system. Failure to do so may result in deployment of the air bag.

2. Disconnect the negative battery cable. Tape the cable with insulating tape.

➡**When disconnecting the cable, some systems need to be initialized after the cable is reconnected. You will need the Mazda diagnostic scan tool or equivalent. Follow the directions on the tool.**

3. Connect the scan tool to the DLC-2.
4. Verify the DTC according to the directions on the scan tool screen.
5. Press the clear button on the DTC screen to clear the DTC.
6. Verify that no DTCs are displayed.
7. Remove the scan tool.

MASS AIR FLOW (MAF) SENSOR

LOCATION

The Mass Air Flow (MAF) sensor is located in the air cleaner housing.

REMOVAL & INSTALLATION

1. Before servicing the vehicle, refer to the Precautions Section.

➡**If working near and/or around the SRS system and components, be sure to disable the SRS system. Tape the negative battery cable with insulating tape. Always disconnect the negative battery cable first.**

✳✳ CAUTION

To avoid personal injury when working on vehicles equipped with an air bag, the negative battery cable must be disconnected and at least one minute must elapse before working on the system. Failure to do so may result in deployment of the air bag.

2. Remove the battery top cover, if equipped.
3. Disconnect the negative battery cable. Tape the cable with insulating tape.

➥When disconnecting the cable, some systems need to be initialized after the cable is reconnected. You will need Mazda diagnostic scan tool or equivalent. Follow the directions on the tool.

4. Remove the MAF/IAT sensor electrical connector.
5. Remove the MAF/IAT sensor retaining screws.
6. Remove the MAF/IAT sensor.

To install:

➥Be sure to use new fasteners, as required.

7. Installation is the reverse of removal procedure.
8. Install the MAF/IAT sensor and tighten the screws.
9. Install the MAF/IAT sensor electrical connector.
10. Use the Mazda diagnostic scan tool, or equivalent and reprogram the required systems.

MANIFOLD ABSOLUTE PRESSURE (MAP) SENSOR

LOCATION

See Figures 235 through 237.

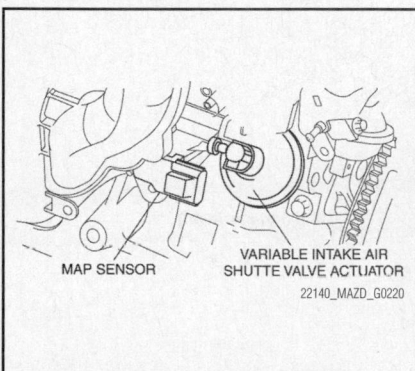

MAP SENSOR   VARIABLE INTAKE AIR
                    SHUTTE VALVE ACTUATOR
22140_MAZD_G0220

Fig. 235 MAP sensor location—2009 vehicles

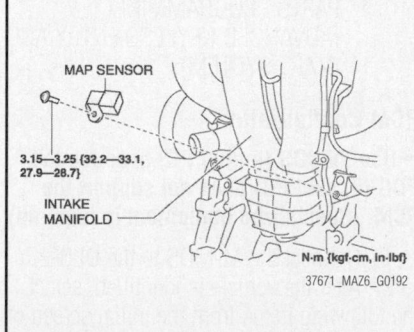

MAP SENSOR

3.15—3.25 {32.2—33.1,
27.9—28.7}

INTAKE
MANIFOLD

N·m {kgf-cm, in-lbf}
37671_MAZ6_G0192

Fig. 236 MAP sensor location (2.0L and 2.5L engines)—2010 vehicles

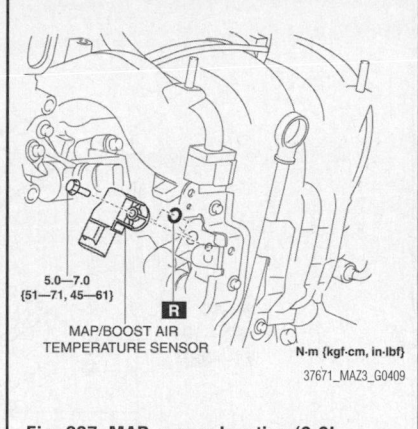

5.0—7.0
{51—71, 45—61}

MAP/BOOST AIR
TEMPERATURE SENSOR   N·m {kgf-cm, in-lbf}

R

37671_MAZ3_G0409

Fig. 237 MAP sensor location (2.3L engine with turbocharger)—2010 vehicles

REMOVAL & INSTALLATION

2009 Vehicles

See Figure 235.

1. Before servicing the vehicle, refer to the Precautions Section.

➥If working near and/or around the SRS system and components, be sure to disable the SRS system. Tape the negative battery cable with insulating tape. Always disconnect the negative battery cable first.

> ❊❊ **CAUTION**
>
> To avoid personal injury when working on vehicles equipped with an air bag, the negative battery cable must be disconnected and at least one minute must elapse before working on the system. Failure to do so may result in deployment of the air bag.

2. Remove the battery top cover, if equipped.
3. Disconnect the negative battery cable. Tape the cable with insulating tape.

➥When disconnecting the cable, some systems need to be initialized after the cable is reconnected. You will need Mazda diagnostic scan tool or equivalent. Follow the directions on the tool.

4. Remove the oil level gauge pipe.
5. Remove the plug hole cover.
6. Disconnect the vacuum hose.
7. Disconnect sensor connector.
8. Remove sensor mounting screw.
9. Remove the sensor from intake manifold.

To install:

➥Be sure to use new fasteners, as required.

10. Installation is the reverse of removal procedure.
11. Tighten to 24—32 inch lbs. (2.7—3.7 Nm).
12. Use the Mazda diagnostic scan tool, or equivalent and reprogram the required systems.

2010 Vehicles

See Figures 236 and 237.

1. Before servicing the vehicle, refer to the Precautions Section.

➥If working near and/or around the SRS system and components, be sure to disable the SRS system. Tape the negative battery cable with insulating tape. Always disconnect the negative battery cable first.

> ❊❊ **CAUTION**
>
> To avoid personal injury when working on vehicles equipped with an air bag, the negative battery cable must be disconnected and at least one minute must elapse before working on the system. Failure to do so may result in deployment of the air bag.

2. Remove the battery top cover, if equipped.
3. Disconnect the negative battery cable. Tape the cable with insulating tape.

➥When disconnecting the cable, some systems need to be initialized after the cable is reconnected. You will need Mazda diagnostic scan tool or equivalent. Follow the directions on the tool.

4. Remove the oil gauge level pipe, 2.3L engine with turbocharger.
5. Disconnect the MAP sensor connector.
6. Remove the MAP sensor.

To install:

➥Be sure to use new fasteners, as required.

7. Installation is the reverse of removal procedure.
8. Use the Mazda diagnostic scan tool, or equivalent and reprogram the required systems.

POWERTRAIN CONTROL MODULE (PCM)

LOCATION

The Powertrain Control Module (PCM) is located in the engine compartment, just to the right of the battery.

REMOVAL & INSTALLATION

See Figures 238 through 240.

1. Before servicing the vehicle, refer to the Precautions Section.

➡ **If working near and/or around the SRS system and components, be sure to disable the SRS system. Tape the negative battery cable with insulating tape. Always disconnect the negative battery cable first.**

❊❊ CAUTION

To avoid personal injury when working on vehicles equipped with an air bag, the negative battery cable must be disconnected and at least one minute must elapse before working on the system. Failure to do so may result in deployment of the air bag.

2. Remove the battery top cover, if equipped.

3. Disconnect the negative battery cable. Tape the cable with insulating tape.

➡ **When disconnecting the cable, some systems need to be initialized after the cable is reconnected. You will need Mazda diagnostic scan tool or equivalent. Follow the directions on the tool.**

4. Remove the battery duct.

5. Remove the battery and battery tray with the PCM.

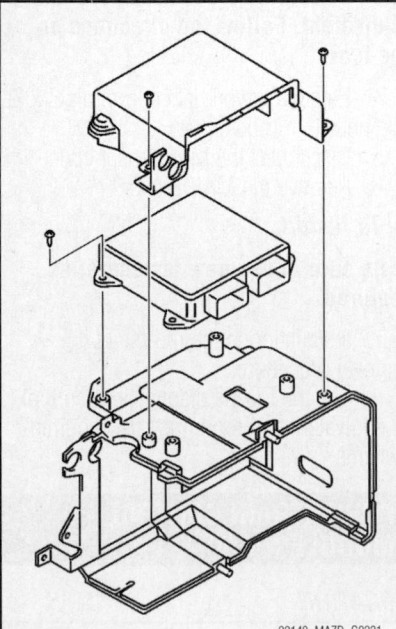

Fig. 238 PCM and related components—2009 vehicles

22140_MAZD_G0221

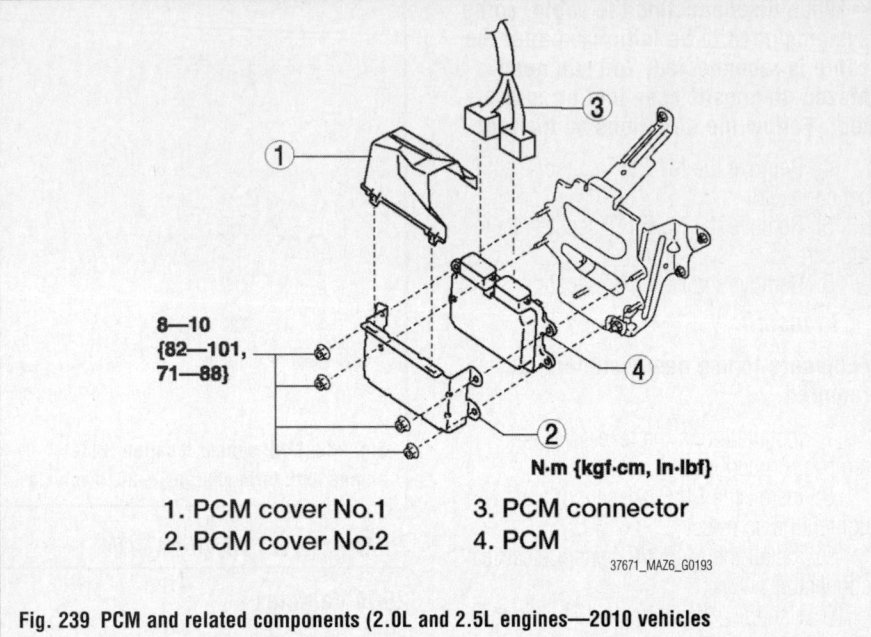

8—10
{82—101,
71—88}

N·m {kgf·cm, in·lbf}

1. PCM cover No.1
2. PCM cover No.2
3. PCM connector
4. PCM

37671_MAZ6_G0193

Fig. 239 PCM and related components (2.0L and 2.5L engines—2010 vehicles

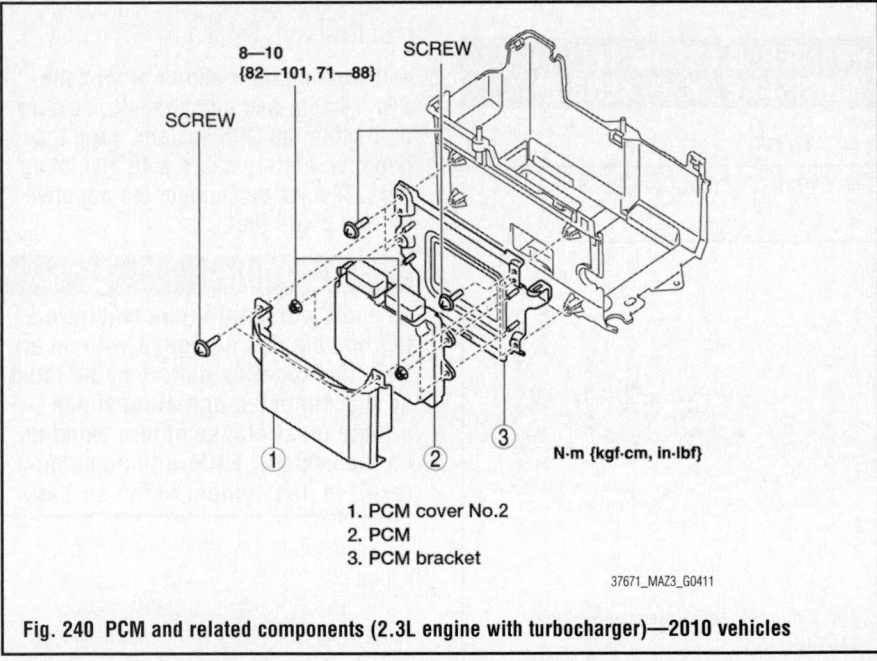

8—10
{82—101, 71—88}

SCREW

SCREW

N·m {kgf·cm, in·lbf}

1. PCM cover No.2
2. PCM
3. PCM bracket

37671_MAZ3_G0411

Fig. 240 PCM and related components (2.3L engine with turbocharger)—2010 vehicles

6. Remove the PCM harness connectors.

7. Remove the PCM from the battery tray.

To install:

➡ **Be sure to use new fasteners, as required.**

8. Installation is the reverse of the removal procedure.

9. Use the Mazda diagnostic scan tool, or equivalent and reprogram the required systems.

10. When replacing the PCM on the vehicles, perform the following:

• IMMOBILIZER SYSTEM-RELATED PARTS PROGRAMMING (ADVANCED KEYLESS ENTRY AND START SYSTEM)

PCM Configuration

➡ **Use the IDS (laptop PC) because the PDS (Pocket PC) does not support the PCM. (Specialized equipment is required)**

1. Connect the M-MDS to the DLC-2.

2. After the vehicle is identified, select the following items from the initial screen of the M-MDS.

3. When using the IDS (laptop PC)

• Select the "Module Programming".

4. Then, select items from the screen menu in the following order:
- Select "Programmable Module Installation".
- Select "PCM".

5. Perform the configuration according to the directions on the screen.

6. Retrieve DTCs by the M-MDS, then verify that there is no DTC present.

7. If a DTC(s) is detected, perform the applicable DTC inspection.

8. Remove the M-MDS to the DLC-2.

IMMOBILIZER SYSTEM-RELATED PARTS PROGRAMMING (ADVANCED KEYLESS ENTRY AND START SYSTEM)

1. Connect the scan tool to the DLC-2.

2. Turn the ignition switch to the ON position using the valid key or the valid advanced key.

➡**Although the security light remains illuminated and DTC 23 is displayed after approximately 1 minute, continue to perform the procedure as indicated.**

3. Verify that the keyless warning light illuminates for approximately 3 seconds, and then turns off.

4. After vehicle identification, select the following from the M-MDS initial screen. Using a PDS (Pocket PC):
- Select the "All Tests and Calibrations"
- Select the "PATS Functions"
- Select "Parameter Reset" from the M-MDS screen menu.
- Perform the security access according to the directions on the M-MDS screen.
- Select the replaced parts (PCM) according to the directions on the M-MDS screen.

- At this time, do not select the other parts from the M-MDS menu.
- After verifying that the PATS function menu is displayed again on the M-MDS screen, select "Finish (this menu)".
- Turn the ignition switch to the LOCK position.
- Turn the ignition switch to the ON position.
- Verify that the security light illuminates for approximately 3 seconds, and then turns off.
- Turn the ignition switch to the LOCK position.

5. Verify that the engine can be started with all the keys.

6. Procedure is completed.

TESTING

See Figure 241.

The PCM terminal voltage can vary with the conditions when measuring and changes due to aged deterioration on the vehicle, causing false diagnosis. Therefore determine comprehensively where the malfunction occurs among the input systems, output systems, and the PCM.

1. With the ignition on check the PCM for voltage at the following connectors:
- 1AY
- 1AW
- 1AZ

2. Check the PCM for ground at the following connectors:
- 1AZ
- 1BB
- 1BD
- 1BH

3. If the voltage or ground is not present repair open or short.

4. If the voltage and ground checks okay, suspect faulty PCM.

THROTTLE POSITION SENSOR (TPS)

REMOVAL & INSTALLATION

The TPS sensor is integral to the throttle body assembly. Refer to the Throttle Body procedure in the Fuel Section.

VEHICLE SPEED SENSOR (VSS)

LOCATION

The Vehicle Speed Sensor (VSS) is located to the right rear of the transaxle.

REMOVAL & INSTALLATION

See Figure 242.

❊❊ WARNING

If foreign materials are stuck to the sensor, disturbance by magnetic flux can cause sensor output to be abnormal and thereby negatively affect control. Make sure that foreign materials such as iron filings are not stuck to the sensor during installation.

1. Before servicing the vehicle, refer to the Precautions Section.

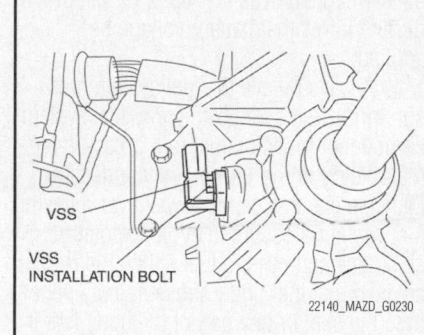

Fig. 242 VSS sensor and related components

PCM HARNESS SIDE CONNECTOR

2BE	2BA	2AW	2AS	2AO	2AK	2AG	2AC	2Y	2U	2Q	2M	2I	2E	2A
2BF	2BB	2AX	2AT	2AP	2AL	2AH	2AD	2Z	2V	2R	2N	2J	2F	2B

2BG	2BC	2AY	2AU	2AQ	2AM	2AI	2AE	2AA	2W	2S	2O	2K	2G	2C
2BH	2BD	2AZ	2AV	2AR	2AN	2AJ	2AF	2AB	2X	2T	2P	2L	2H	2D

1BE	1BA	1AW	1AS	1AO	1AK	1AG	1AC	1Y	1U	1Q	1M	1I	1E	1A
1BF	1BB	1AX	1AT	1AP	1AL	1AH	1AD	1Z	1V	1R	1N	1J	1F	1B

1BG	1BC	1AY	1AU	1AQ	1AM	1AI	1AE	1AA	1W	1S	1O	1K	1G	1C
1BH	1BD	1AZ	1AV	1AR	1AN	1AJ	1AF	1AB	1X	1T	1P	1L	1H	1D

22140_MAZD_G0224

Fig. 241 PCM harness connector

➡If working near and/or around the SRS system and components, be sure to disable the SRS system. Tape the negative battery cable with insulating tape. Always disconnect the negative battery cable first.

※※ CAUTION

To avoid personal injury when working on vehicles equipped with an air bag, the negative battery cable must be disconnected and at least one

minute must elapse before working on the system. Failure to do so may result in deployment of the air bag.

2. Remove the battery top cover, if equipped.
3. Disconnect the negative battery cable. Tape the cable with insulating tape.

➡When disconnecting the cable, some systems need to be initialized after the cable is reconnected. You will need Mazda diagnostic scan tool or equiva-

lent. Follow the directions on the tool.

To install:

4. Apply ATF to a new O-ring and install it on a VSS.
5. Install the VSS and tighten the mounting bolt to 71—97 inch lbs. (8—11 Nm).
6. Connect the VSS connector.
7. Connect the negative battery cable.
8. Use the Mazda diagnostic scan tool, or equivalent and reprogram the required systems.

FUEL   GASOLINE FUEL INJECTION SYSTEM

FUEL SYSTEM SERVICE PRECAUTIONS

Safety is the most important factor when performing not only fuel system maintenance, but any type of maintenance. Failure to conduct maintenance and repairs in a safe manner may result in serious personal injury or death. Work on a vehicle's fuel system components can be accomplished safely and effectively by adhering to the following rules and guidelines.

• To avoid the possibility of fire and personal injury, always disconnect the negative battery cable unless the repair or test procedure requires that battery voltage be applied.

• Always relieve the fuel system pressure prior to disconnecting any fuel system component (injector, fuel rail, pressure regulator, etc.) fitting or fuel line connection. Exercise extreme caution whenever relieving fuel system pressure to avoid exposing skin, face and eyes to fuel spray. Please be advised that fuel under pressure may penetrate the skin or any part of the body that it contacts.

• Always place a shop towel or cloth around the fitting or connection prior to loosening to absorb any excess fuel due to spillage. Ensure that all fuel spillage is quickly removed from engine surfaces. Ensure that all fuel-soaked cloths or towels are deposited into a flame-proof waste container with a lid.

• Always keep a dry chemical (Class B) fire extinguisher near the work area.

• Do not allow fuel spray or fuel vapors to come into contact with a spark or open flame.

• Always use a second wrench when loosening or tightening fuel line connection fittings. This will prevent unnecessary stress and torsion on fuel piping. Always follow the proper torque specifications.

• Always replace worn fuel fitting O-

rings with new ones. Do not substitute fuel hose where rigid pipe is installed.

FUEL SYSTEM PRESSURE

RELIEVING

See Figures 243 through 245.

1. Before servicing the vehicle, refer to the Precautions Section.

➡If working near and/or around the SRS system and components, be sure to disable the SRS system. Tape the negative battery cable with insulating tape. Always disconnect the negative battery cable first.

※※ CAUTION

To avoid personal injury when working on vehicles equipped with an air bag, the negative battery cable must be disconnected and at least one minute must elapse before working on the system. Failure to do so

may result in deployment of the air bag.

2. Disconnect the negative battery cable. Tape the cable with insulating tape.

➡When disconnecting the cable, some systems need to be initialized after the cable is reconnected. You will need the Mazda diagnostic scan tool or equivalent. Follow the directions on the tool.

3. Remove the fuel-filler cap and release the pressure in the fuel tank.
4. Remove the fuel pump relay.
5. Connect the battery cable.
6. Start the engine.
7. After the engine stalls, crank the engine several times.
8. Turn the ignition switch to the LOCK position.
9. Install the fuel pump relay fuse.
10. Use the Mazda diagnostic scan tool,

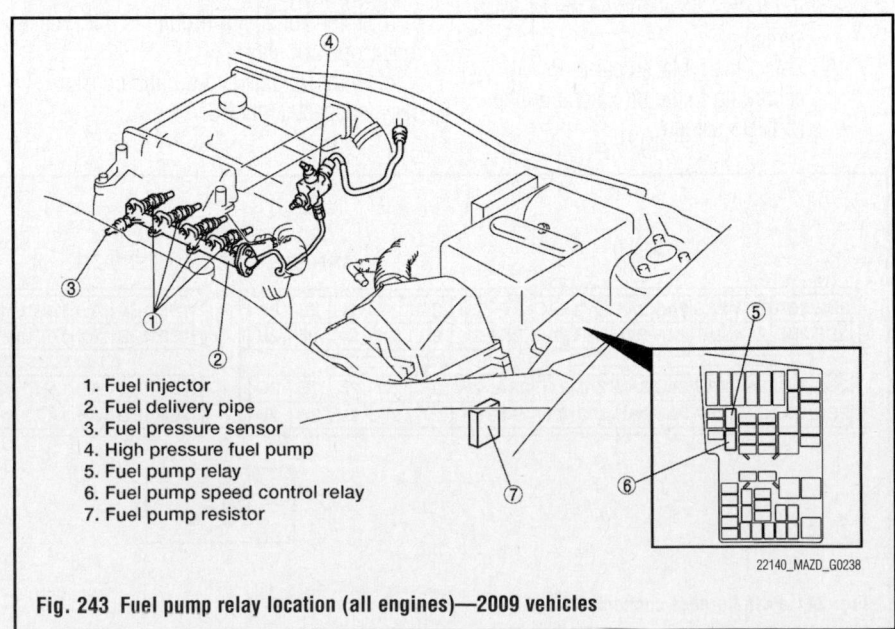

1. Fuel injector
2. Fuel delivery pipe
3. Fuel pressure sensor
4. High pressure fuel pump
5. Fuel pump relay
6. Fuel pump speed control relay
7. Fuel pump resistor

22140_MAZD_G0238

Fig. 243 Fuel pump relay location (all engines)—2009 vehicles

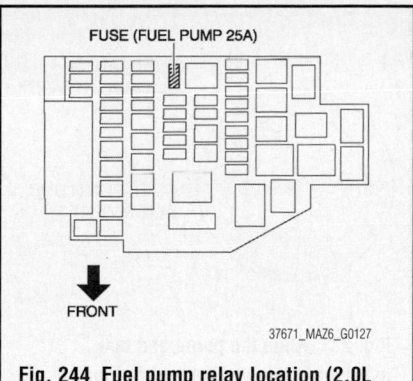

Fig. 244 Fuel pump relay location (2.0L and 2.5L engine)—2010 vehicles

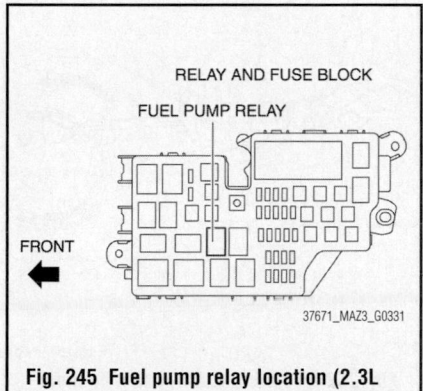

Fig. 245 Fuel pump relay location (2.3L engine with turbocharger)—2010 vehicles

or equivalent and reprogram the required systems.

FUEL PUMP MODULE

REMOVAL & INSTALLATION

2009 Vehicles

See Figures 246 through 253.

1. Before servicing the vehicle, refer to the Precautions Section.

➡ **If working near and/or around the SRS system and components, be sure to disable the SRS system. Tape the negative battery cable with insulating tape. Always disconnect the negative battery cable first.**

✳ CAUTION

To avoid personal injury when working on vehicles equipped with an air bag, the negative battery cable must be disconnected and at least one minute must elapse before working on the system. Failure to do so may result in deployment of the air bag.

2. Remove the battery top cover, if equipped.

3. Disconnect the negative battery cable. Tape the cable with insulating tape.

➡ **When disconnecting the cable, some systems need to be initialized after the cable is reconnected. You will need Mazda diagnostic scan tool or equivalent. Follow the directions on the tool.**

4. Relive the fuel system pressure.
5. Disconnect the quick release connector in the engine compartment.
6. Attach a long hose to the disconnect fuel pipe and drain the fuel into a suitable container as follows:

a. Using a jumper wire, ground the Powertrain Control Module (PCM) terminals. If equipped with an immobilizer, ground terminal 1AR, if not equipped with an immobilizer, ground terminal 1AQ. Refer to the illustration for terminal location.

b. Turn the ignition switch the ON position to activate the fuel pump.

✳ CAUTION

The fuel pump can be damaged if all fuel is removed from the tank, monitor the hose and stop when no fuel is be discharged from the tank.

c. Once no more fuel is being discharged, turn the ignition OFF.
7. Disconnect the negative battery cable.
8. Remove the rear seat cushion and remove the pump service hole cover.
9. Disconnect the fuel pump connector and remove the charcoal canister.
10. Lower the exhaust main silencer so the insulator can be removed and remove the insulator.

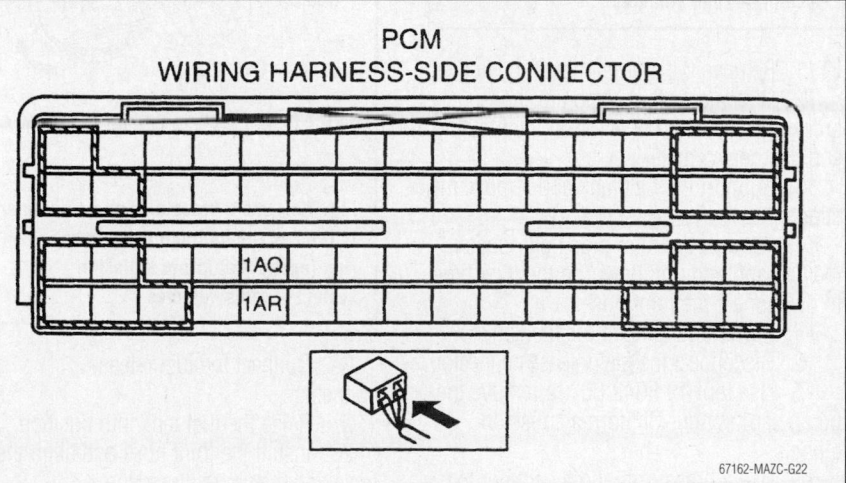

Fig. 246 If equipped with an immobilizer, ground PCM terminal 1AR, if not equipped with an immobilizer, ground PCM terminal 1AQ—2009 vehicles

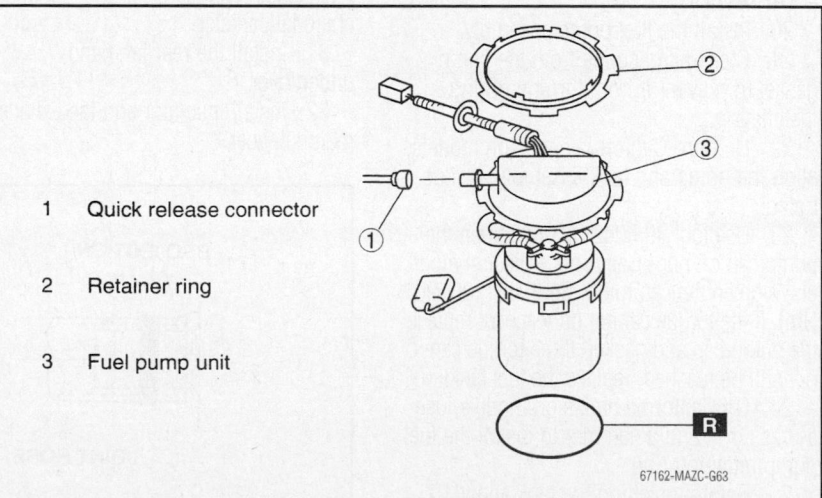

1. Quick release connector
2. Retainer ring
3. Fuel pump unit

Fig. 247 Fuel pump and related components (except California emission system)—2009 vehicles

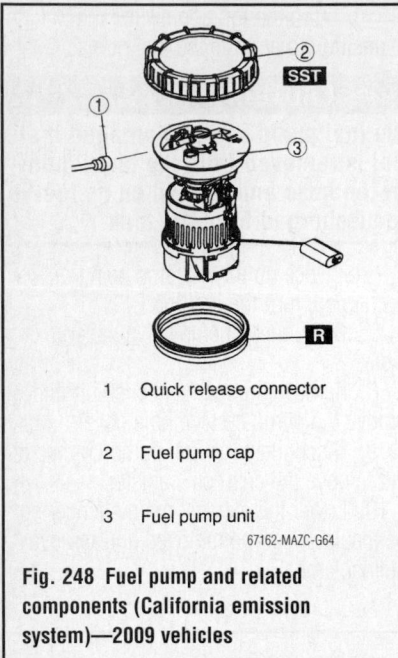

1 Quick release connector

2 Fuel pump cap

3 Fuel pump unit

67162-MAZC-G64

Fig. 248 Fuel pump and related components (California emission system)—2009 vehicles

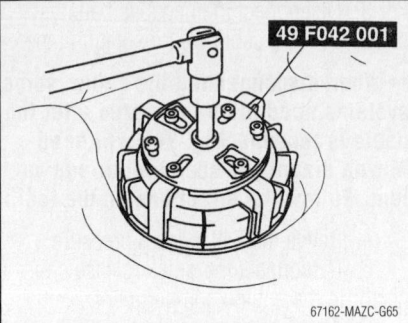

67162-MAZC-G65

Fig. 249 Use tool 49 F042 001 to remove the pump cap (except California emission system)—2009 vehicles

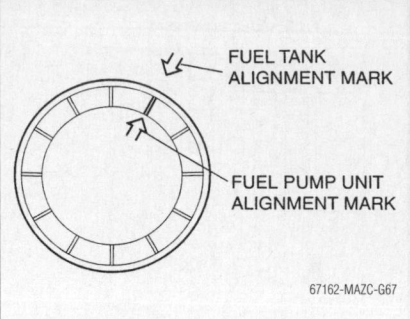

67162-MAZC-G67

Fig. 251 Align the pump and tank assembly as shown (except California emission system)—2009 vehicles

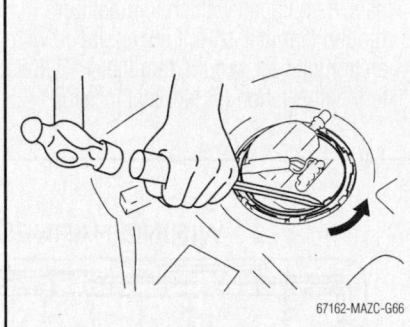

67162-MAZC-G66

Fig. 250 Use a brass drift and a hammer to remove the fuel pump retaining ring (except California emission system)—2009 vehicles

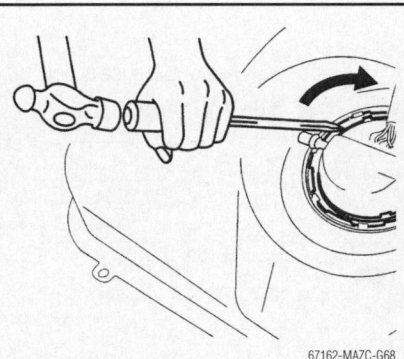

67162-MAZC-G68

Fig. 252 Use a brass drift and a hammer install the fuel pump retaining ring (except California emission system)—2009 vehicles

11. Remove the rear left hand undercover.

12. Disconnect the evaporator hose, quick release fuel connector on the fuel line and charcoal canister.

13. Support the fuel tank and remove the strap.

14. Remove the filler pipe bolt, loosen the tie band and pull down on the filler pipe to disconnect the joint hose.

15. Lower and remove the fuel tank.

16. Disconnect the fuel release connector.

17. Use tool 49 F042 001 to remove the pump cap on non–California emission models.

18. Use a brass drift and a hammer to remove the fuel pump retaining ring on California emission models.

19. Remove the fuel pump assembly.

To install:

20. Install the fuel pump assembly.

21. Clean all gasoline from the pump gasket to prevent it from turning during installation.

22. On non–California emission models, align the pump and tank assembly as illustrated.

23. Use tool 49 F042 001 to tighten the pump cap on non–California emission models. Tighten the cap to 59–66 ft. lbs. (80–90 Nm). If the torque cannot be reached, replace the pump cap and gasket. If the torque cannot still be reached, replace the fuel tank.

24. On California emission models, use a brass drift and a hammer to install the fuel pump retaining ring

25. Before installing the tank apply 1.7 inch Hg (5.9 kPa) of pressure to the tank to check for leakage around the pump.

26. Connect the fuel release connector.

27. Raise the fuel tank into position.

28. Install the joint hose and align the hose and clamp as illustrated.

29. Install the strap. Tighten the bolts to 16–22 ft. lbs. (22–30 Nm).

30. Connect the evaporator hose, quick release fuel connector on the fuel line and charcoal canister.

31. Install the rear left hand undercover.

32. Install insulator and the exhaust main silencer.

33. Install the charcoal canister and connect the fuel pump connector.

34. Install the pump service hole cover and rear seat.

35. Disconnect the negative battery cable.

36. Start the vehicle and check for leaks as follows:

a. Using a jumper wire, ground the Powertrain Control Module (PCM) terminals. If equipped with an immobilizer, ground terminal 1AR, if not equipped with an immobilizer, ground terminal 1AQ. Refer to the illustration for terminal location.

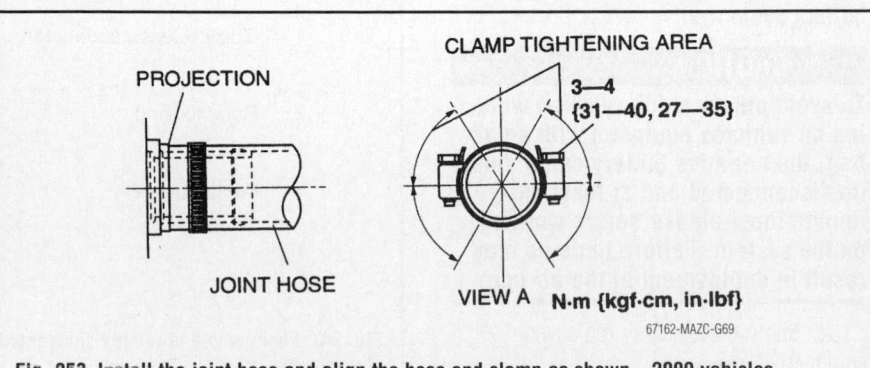

67162-MAZC-G69

Fig. 253 Install the joint hose and align the hose and clamp as shown—2009 vehicles

b. Turn the ignition switch the ON position to activate the fuel pump.

c. Check the hoses, clips and other fuel system components for leaks.

d. If there are any leaks, replace the fuel hoses and clips. If the is damage to the seal on the fuel pipe side, replace the pipe.

e. The system must be leak free for five minutes with the terminal grounded. If any component is replaced because of a system, leak, turn the ignition key OFF; remove the jumper wire from the terminal. reapply the jumper wire, turn the ignition On and check for leaks.

37. Use the Mazda diagnostic scan tool, or equivalent and reprogram the required systems.

2010 Vehicles

See Figure 254.

1. Before servicing the vehicle, refer to the Precautions Section.

➡**If working near and/or around the SRS system and components, be sure to disable the SRS system. Tape the negative battery cable with insulating tape. Always disconnect the negative battery cable first.**

✳✳ CAUTION

To avoid personal injury when working on vehicles equipped with an air bag, the negative battery cable must be disconnected and at least one minute must elapse before working on the system. Failure to do so may result in deployment of the air bag.

2. Remove the battery top cover, if equipped.

3. Disconnect the negative battery cable. Tape the cable with insulating tape.

➡**When disconnecting the cable, some systems need to be initialized after the cable is reconnected. You will need Mazda diagnostic scan tool or equivalent. Follow the directions on the tool.**

4. Properly relieve the fuel system pressure.

5. Remove the rear seat cushion.

6. Remove the service hole cover.

7. Disconnect the electrical connectors.

8. Disconnect the fuel hose.

9. Remove the unit from its mounting. Discard the O-ring.

To install:

➡**Be sure to use new fasteners, as required.**

10. Installation is the reverse of the removal procedure.

11. Be sure to use a new O-ring.

12. Start the engine and check for leaks, correct as required.

13. Use the Mazda diagnostic scan tool, or equivalent and reprogram the required systems.

FUEL RAIL AND INJECTOR

REMOVAL & INSTALLATION

2009 Vehicles

See Figure 255.

✳✳ CAUTION

Fuel injection systems remain under pressure after the engine has been turned OFF. Properly relieve fuel pressure before disconnecting any fuel lines. Failure to do so may result in fire or personal injury. Do not allow fuel spray or fuel vapors to come in contact with a spark or open flame. Keep a dry chemical fire extinguisher nearby. Never store fuel in an open container due to risk of fire or explosion.

1. Before servicing the vehicle, refer to the Precautions Section.

➡**If working near and/or around the SRS system and components, be sure to disable the SRS system. Tape the negative battery cable with insulating tape. Always disconnect the negative battery cable first.**

✳✳ CAUTION

To avoid personal injury when working on vehicles equipped with an air bag, the negative battery cable must be disconnected and at least one minute must elapse before working on the system. Failure to do so may result in deployment of the air bag.

2. Remove the battery top cover, if equipped.

3. Disconnect the negative battery cable. Tape the cable with insulating tape.

➡**When disconnecting the cable, some systems need to be initialized after the cable Is reconnected. You will need Mazda diagnostic scan tool or equivalent. Follow the directions on the tool.**

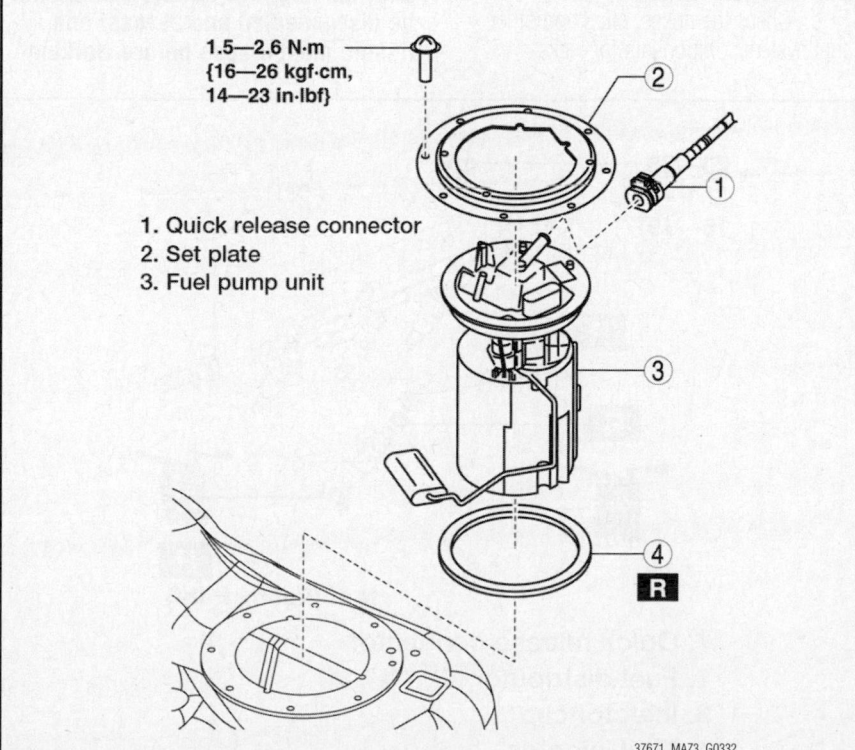

1.5—2.6 N·m
{16—26 kgf·cm,
14—23 in·lbf}

1. Quick release connector
2. Set plate
3. Fuel pump unit

37671_MAZ3_G0332

Fig. 254 Fuel pump and related components (Federal and California emission systems)—2010 vehicles

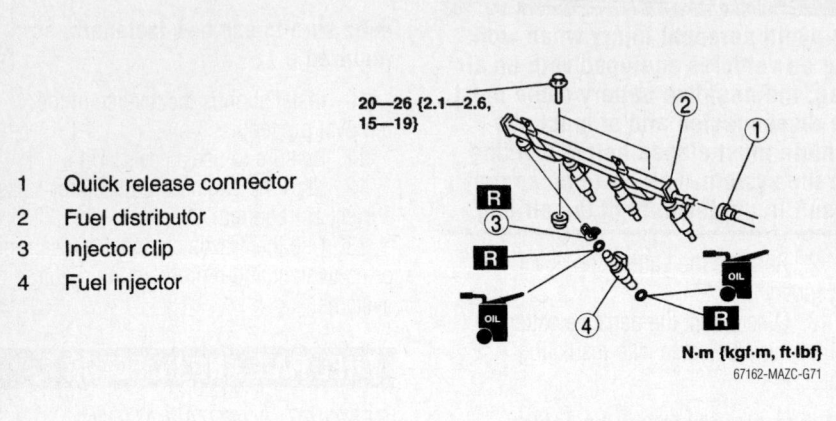

20—26 {2.1—2.6, 15—19}

1 Quick release connector
2 Fuel distributor
3 Injector clip
4 Fuel injector

N·m {kgf·m, ft·lbf}

67162-MAZC-G71

Fig. 255 Exploded view of the fuel rail and injector assembly—2009 vehicles

4. Properly relieve the fuel system pressure.

5. Remove the plug hole plate.

6. Remove the battery cover and disconnect the negative battery cable.

7. Unplug the injector electrical connections.

8. Disconnect the fuel lines from the rail.

9. Remove the fuel rail bolts.

10. Remove the rail and injectors as an assembly.

11. Insert a flat head tool between the injector cup and clip finger as illustrated.

12. Push the clip finger out using the suitable tool. Deform the finger until it is completely removed from the cup notch.

13. Remove the injector from the clip.

14. Use pliers to pull the clip parallel to the injector groove and remove the clip from the injector. Discard the clip.

To install:

➡ Be sure to use new fasteners, as required.

15. Use new O-rings and coat them with clean engine oil.

16. Temporarily attach a new clip to the injector groove.

➡ When the clip is correctly installed, the central area of the injector and clip finger positions are aligned.

17. While firmly holding the injector, push the clip into the injector until the clip stops. Make sure the injector connector is correctly positioned.

18. Press the injector into the cup until the cup contacts the lower surface of the cup. Make sure the injector and clip are installed properly and the clip is hooked into the injector cup notch.

19. Install the injectors and fuel rail as an assembly.

20. Install the fuel rail bolts and tighten to 15–19 ft. lbs. (20–26 Nm).

21. Attach the fuel lines to the rail.

22. Start the vehicle and check for leaks as follows:

a. Using a jumper wire, ground the Powertrain Control Module (PCM) terminals. If equipped with an immobilizer, ground terminal 1AR, if not equipped with an immobilizer, ground terminal 1AQ. Refer to illustration 67162-mazc-g22 for terminal location.

b. Turn the ignition switch the ON position to activate the fuel pump.

c. Check the hoses, clips and other fuel system components for leaks.

d. If there are any leaks, replace the fuel hoses and clips. If the is damage to the seal on the fuel pipe side, replace the pipe.

e. The system must be leak free for five minutes with the terminal grounded. If any component is replaced because of a system, leak, turn the ignition key OFF; remove the jumper wire from the terminal. reapply the jumper wire, turn the ignition On and check for leaks.

23. Install the remaining components.

24. Use the Mazda diagnostic scan tool, or equivalent and reprogram the required systems.

2010 Vehicles

2.5L Engine

See Figure 256.

1. Before servicing the vehicle, refer to the Precautions Section.

➡ If working near and/or around the SRS system and components, be sure to disable the SRS system. Tape the negative battery cable with insulating tape. Always disconnect the negative battery cable first.

✳✳ CAUTION

To avoid personal injury when working on vehicles equipped with an air bag, the negative battery cable must be disconnected and at least one minute must elapse before working

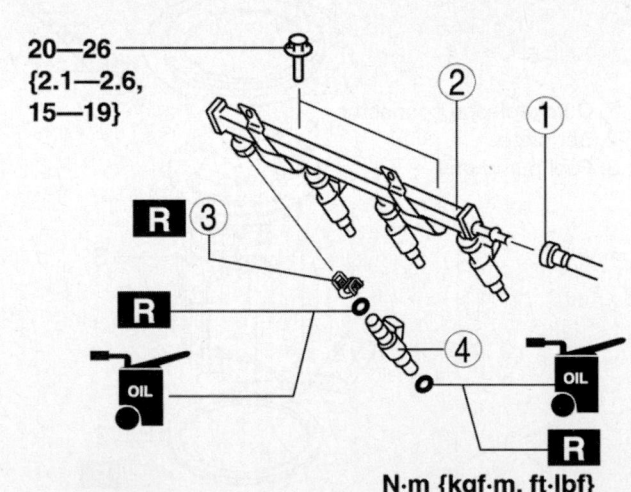

20—26 {2.1—2.6, 15—19}

N·m {kgf·m, ft·lbf}

1. Quick release connector
2. Fuel distributor
3. Injector clip
4. Fuel injector

37671_MAZ3_G0340

Fig. 256 Fuel injectors and related components (2.0L and 2.5L engines)—2010 vehicles

on the system. Failure to do so may result in deployment of the air bag.

2. Disconnect the negative battery cable. Tape the cable with insulating tape.

➡**When disconnecting the cable, some systems need to be initialized after the cable is reconnected. You will need the Mazda diagnostic scan tool or equivalent. Follow the directions on the tool.**

3. Properly relieve the fuel system pressure.

4. Disconnect the negative battery cable.

5. Remove the air cleaner assembly, as required for access.

6. Remove the plug hole plate.

7. Disconnect the electrical connectors.

8. Disconnect the quick release connector.

9. Remove the fuel distributor.

10. Remove the injector. Discard the O-rings. Discard the clips.

To install:

➡**Be sure to use new fasteners, as required.**

11. Installation is the reverse of the removal procedure.

12. Start the engine and check for leaks. Correct as required.

13. Use the Mazda diagnostic scan tool, or equivalent and reprogram the required systems.

2.3L Engine with Turbocharger
See Figures 257 through 260.

1. Before servicing the vehicle, refer to the Precautions Section.

➡**If working near and/or around the SRS system and components, be sure to disable the SRS system. Tape the negative battery cable with insulating tape. Always disconnect the negative battery cable first.**

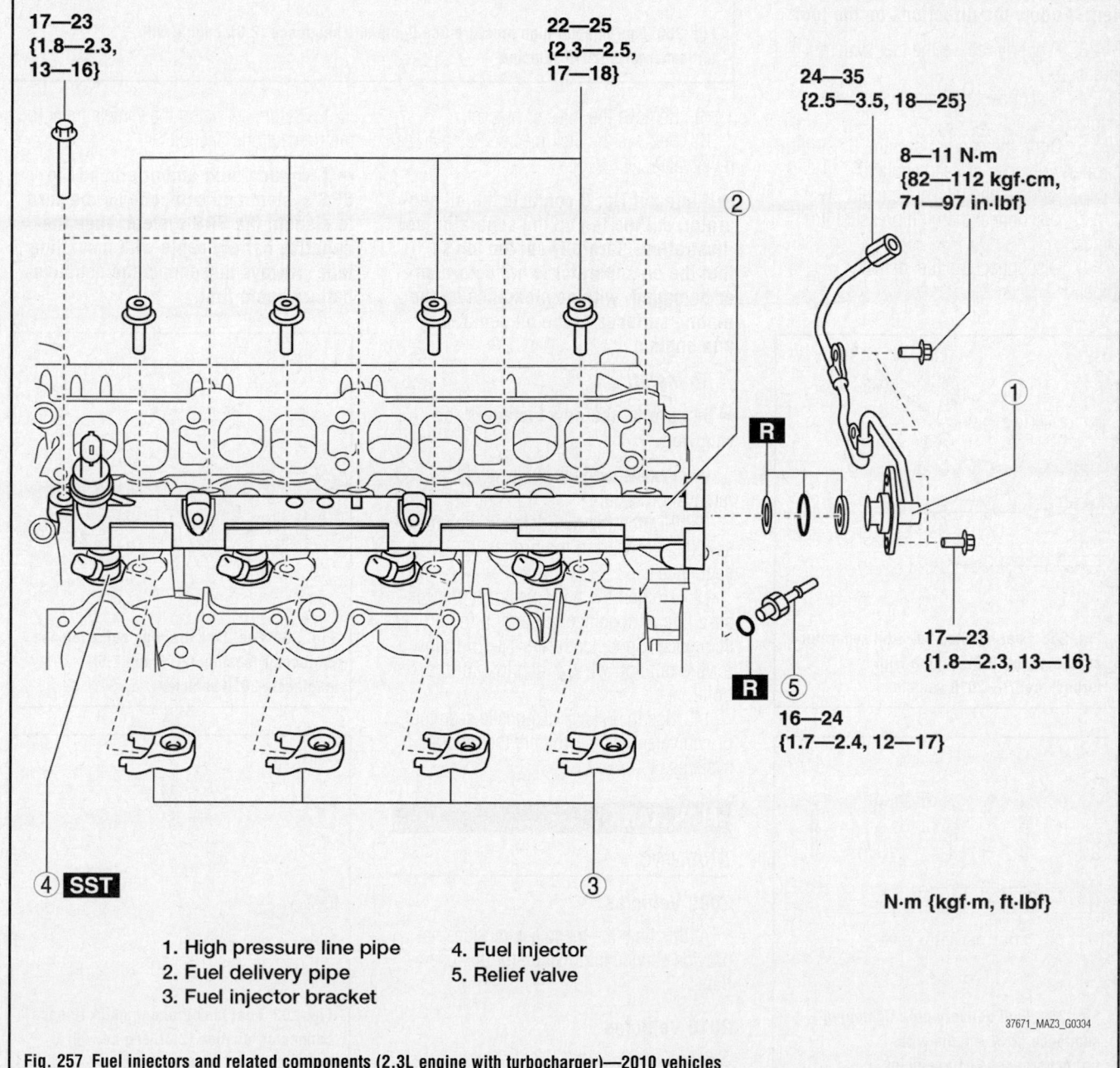

1. High pressure line pipe
2. Fuel delivery pipe
3. Fuel injector bracket
4. Fuel injector
5. Relief valve

N·m {kgf·m, ft·lbf}

37671_MAZ3_G0334

Fig. 257 Fuel injectors and related components (2.3L engine with turbocharger)—2010 vehicles

2. Disconnect the negative battery cable. Tape the cable with insulating tape.

➡When disconnecting the cable, some systems need to be initialized after the cable is reconnected. You will need the Mazda diagnostic scan tool or equivalent. Follow the directions on the tool.

3. Properly relieve the fuel system pressure.
4. Disconnect the negative battery cable.
5. Drain the cooling system. Be sure to properly dispose of used coolant.
6. Remove the intake manifold.
7. Disconnect the high pressure line pipe.
8. Disconnect the fuel delivery pipe.

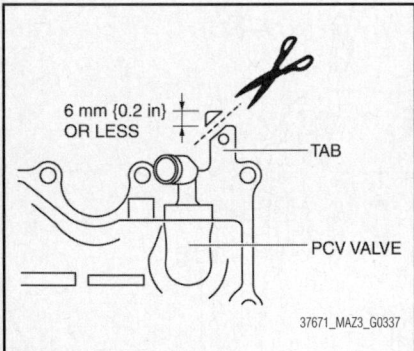

6 mm {0.2 in} OR LESS
TAB
PCV VALVE

37671_MAZ3_G0337

Fig. 258 Fuel injector No. 3 oil separator contact point (2.3L engine with turbocharger)—2010 vehicles

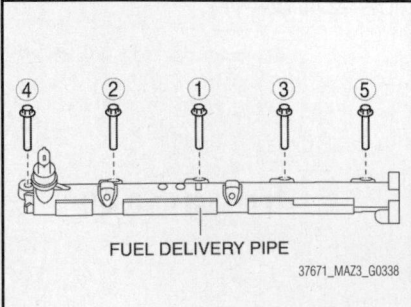

FUEL DELIVERY PIPE

37671_MAZ3_G0338

Fig. 259 Fuel delivery pipe tightening sequence (2.3L engine with turbocharger)—2010 vehicles

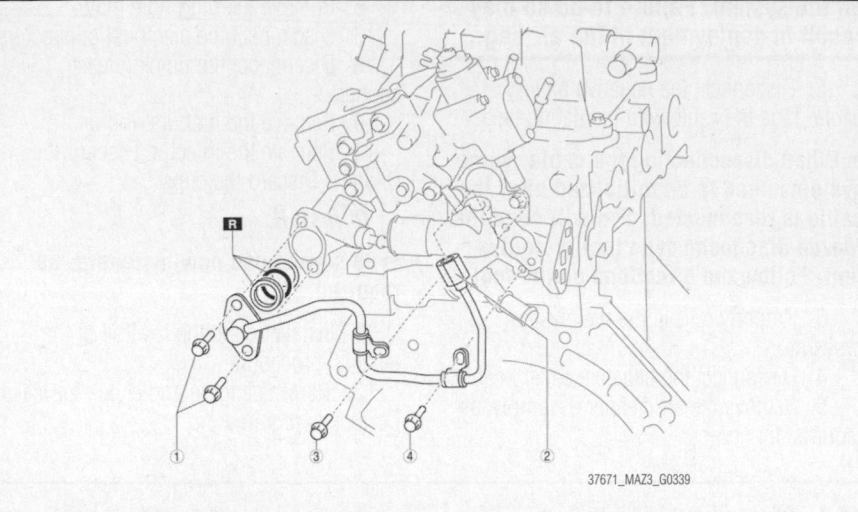

37671_MAZ3_G0339

Fig. 260 Fuel injector high pressure line tightening sequence (2.3L engine with turbocharger)—2010 vehicles

9. Remove the injector bracket.
10. Remove the fuel injector. Remove the relief valve.

➡If injector No. 3 contacts the oil separator, cut the tab on the separator, see illustration. Carefully cut the tab so that the oil separator is not deformed or damaged, with no clearance on the mating surfaces of the oil separator and engine.

To install:

➡Be sure to use new fasteners, as required.

11. Installation is the reverse of the removal procedure.
12. Tighten the fuel delivery pipe to specification and in the proper sequence.
13. Tighten the fuel injector high pressure line to specification and in the proper sequence. Specification is 18–25 ft. lbs for the nut and 71–97 inch lbs. for the bolts.
14. Use the Mazda diagnostic scan tool, or equivalent and reprogram the required systems.

FUEL TANK

DRAINING

2009 Vehicles

At this time the manufacturer does not provide service information for this operation.

2010 Vehicles

See Figures 261 and 262.

1. Before servicing the vehicle, refer to the Precautions Section.

➡If working near and/or around the SRS system and components, be sure to disable the SRS system. Tape the negative battery cable with insulating tape. Always disconnect the negative battery cable first.

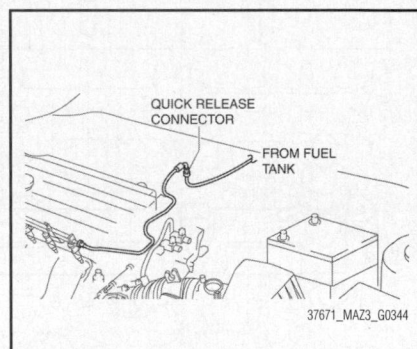

QUICK RELEASE CONNECTOR
FROM FUEL TANK

37671_MAZ3_G0344

Fig. 261 Fuel tank draining quick release connector location (2.0L and 2.5L engines—2010 vehicles

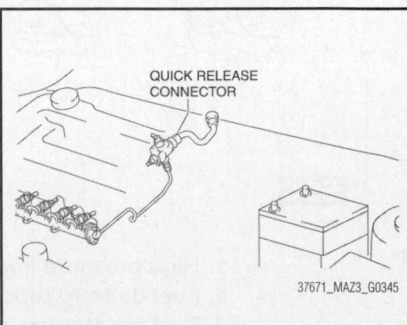

QUICK RELEASE CONNECTOR

37671_MAZ3_G0345

Fig. 262 Fuel tank draining quick release connector location (2.3L engine with turbocharger—2010 vehicles

To avoid personal injury when working on vehicles equipped with an air bag, the negative battery cable must be disconnected and at least one minute must elapse before working on the system. Failure to do so may result in deployment of the air bag.

2. Disconnect the negative battery cable. Tape the cable with insulating tape.

➡**When disconnecting the cable, some systems need to be initialized after the cable is reconnected. You will need the Mazda diagnostic scan tool or equivalent. Follow the directions on the tool.**

3. Level the vehicle.
4. Properly relieve the fuel system pressure.
5. Disconnect the quick release connector.
6. Connect a hose to the disconnected quick release connector to drain the required fuel. Be sure to use an approved container for collecting gasoline.
7. Connect the Mazda diagnostic scan tool, or equivalent, to the DLC-2 connector.
8. Using the simulation function "FP", start the fuel pump to drain the fuel. Be sure to use an approved container for collecting gasoline. Discard the fuel.
9. Using the simulation function "FP", stop the fuel pump to drain the fuel. Disconnect the negative battery cable.

REMOVAL & INSTALLATION

2009 Vehicles

See Figures 263 and 264.

1. Before servicing the vehicle, refer to the Precautions Section.

➡**If working near and/or around the SRS system and components, be sure to disable the SRS system. Tape the negative battery cable with insulating tape. Always disconnect the negative battery cable first.**

To avoid personal injury when working on vehicles equipped with an air bag, the negative battery cable must be disconnected and at least one minute must elapse before working on the system. Failure to do so may result in deployment of the air bag.

2. Disconnect the negative battery cable. Tape the cable with insulating tape.

➡**When disconnecting the cable, some systems need to be initialized after the cable is reconnected. You will need the Mazda diagnostic scan tool or equivalent. Follow the directions on the tool.**

3. Level the vehicle.
4. Properly relieve the fuel system pressure.
5. Drain the fuel tank to an acceptable level.
6. Remove the rear seat cushion. Remove the service hole cover.
7. Disconnect the fuel pump unit connector.
8. Remove the charcoal canister protector.
9. Lower the main silencer so that the insulator can be removed. Remove the insulator.
10. Remove the rear undercover, left side.
11. Remove the EVAP hose. Remove the quick release connectors.
12. As required, remove the filler pipe. If removing this component you will have to remove the tire and wheel assembly and mudguards. Properly support the rear crossmember. Remove the lower shock bolts. Loosen the six crossmember installation nuts about 1.2 inch.

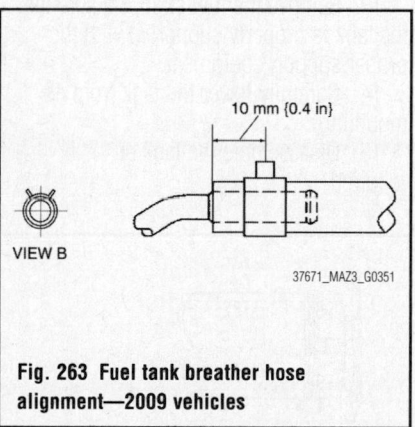

Fig. 263 Fuel tank breather hose alignment—2009 vehicles

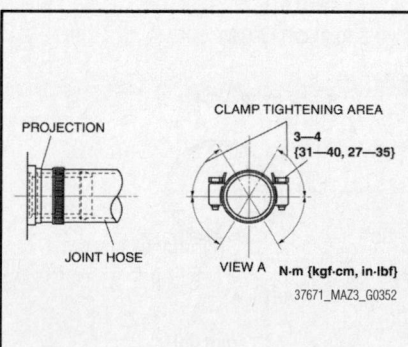

Fig. 264 Fuel tank joint hose alignment— 2009 vehicles

13. Be sure that all components, wires, hoses and lines are disconnected from the tank before removing the retaining straps.
14. Remove the fuel strap. Be sure that the tank is properly supported with the proper support equipment.
15. Carefully lower the tank from its mounting.
16. Disassemble the tank as required.

To install:

➡**Be sure to use new fasteners, as required.**

17. Installation is the reverse of the removal procedure.
18. Be sure that the hoses are properly installed and aligned.
19. Use the Mazda diagnostic scan tool, or equivalent and reprogram the required systems.

2010 Vehicles

See Figures 265 and 266.

1. Before servicing the vehicle, refer to the Precautions Section.

➡**If working near and/or around the SRS system and components, be sure to disable the SRS system. Tape the negative battery cable with insulating tape. Always disconnect the negative battery cable first.**

To avoid personal injury when working on vehicles equipped with an air bag, the negative battery cable must be disconnected and at least one minute must elapse before working on the system. Failure to do so may result in deployment of the air bag.

2. Disconnect the negative battery cable. Tape the cable with insulating tape.

➡**When disconnecting the cable, some systems need to be initialized after the cable is reconnected. You will need the Mazda diagnostic scan tool or equivalent. Follow the directions on the tool.**

3. Level the vehicle.
4. Properly relieve the fuel system pressure.
5. Drain the fuel tank to an acceptable level.
6. Remove the fuel pump.
7. On 2.0L and 2.5L engines, remove the three way catalytic converter.

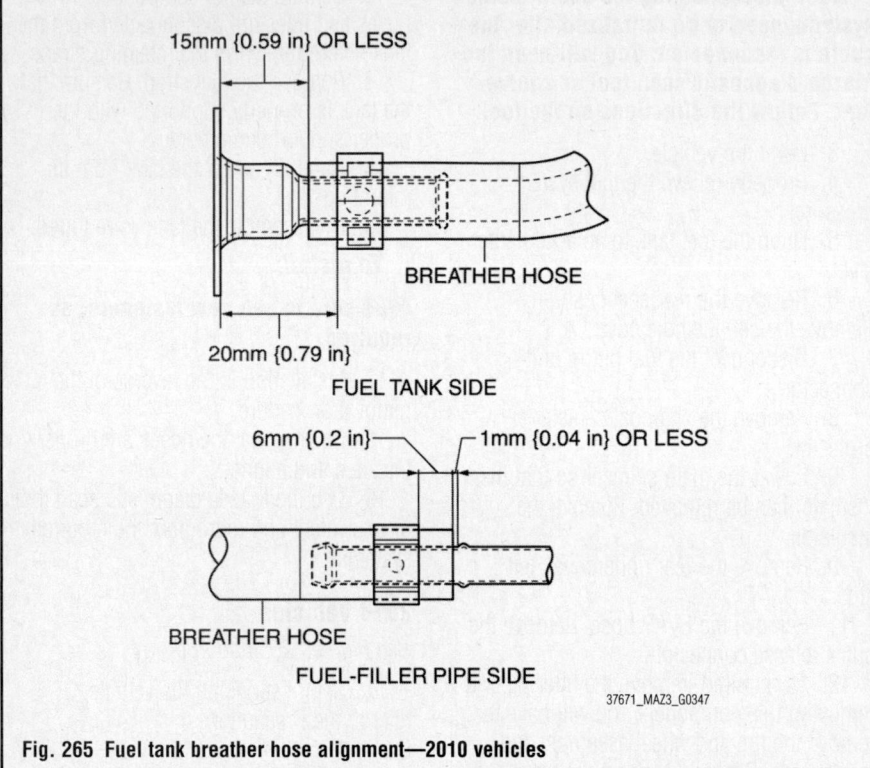

Fig. 265 Fuel tank breather hose alignment—2010 vehicles

8. On 2.3L engine with turbocharger, remove the undercover. Remove the middle pipe.

9. Remove the No. 1 middle insulator.

10. Disconnect the quick release connectors.

11. As required, remove the filler pipe. If removing this component you will have to remove the tire and wheel assembly.

12. Be sure that all components, wires, hoses and lines are disconnected from the tank before removing the retaining straps.

13. Remove the fuel strap. Be sure that the tank is properly supported with the proper support equipment.

14. Carefully lower the tank from its mounting.

15. Disassemble the tank as required.

To install:

➡ Be sure to use new fasteners, as required.

16. Installation is the reverse of the removal procedure.

17. Be sure that the hoses are properly installed and aligned.

18. Use the Mazda diagnostic scan tool, or equivalent and reprogram the required systems.

IDLE SPEED

ADJUSTMENT

The idle speed is controlled by the PCM.

THROTTLE BODY

REMOVAL & INSTALLATION

2009 Vehicles

2.0L & 2.3L Engines

See Figures 267 through 270.

1. Before servicing the vehicle, refer to the Precautions Section.

➡ If working near and/or around the SRS system and components, be sure to disable the SRS system. Tape the negative battery cable with insulating tape. Always disconnect the negative battery cable first.

✳✳ CAUTION

To avoid personal injury when working on vehicles equipped with an air bag, the negative battery cable must

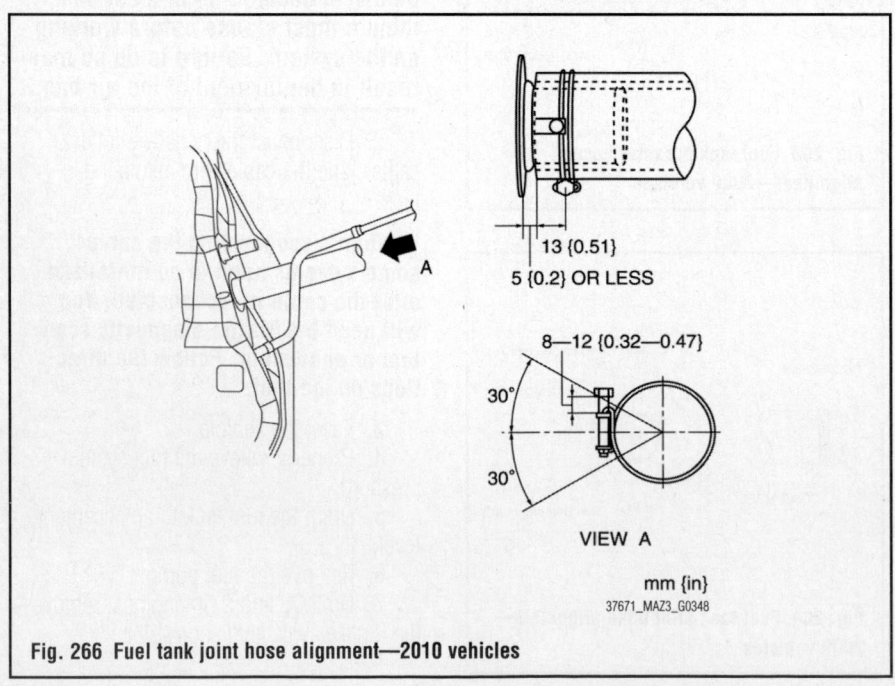

Fig. 266 Fuel tank joint hose alignment—2010 vehicles

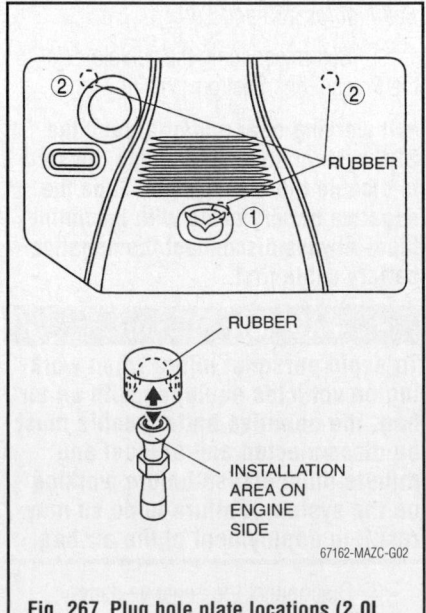

Fig. 267 Plug hole plate locations (2.0L and 2.3L engines—2009 vehicles

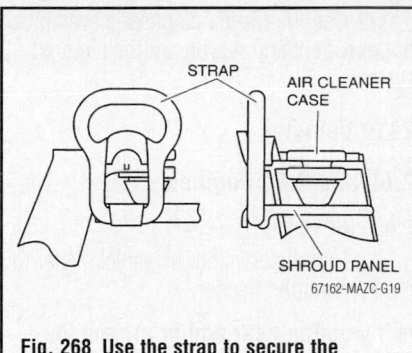

Fig. 268 Use the strap to secure the shroud panel and the air cleaner case (2.0L and 2.3L engines—2009 vehicles

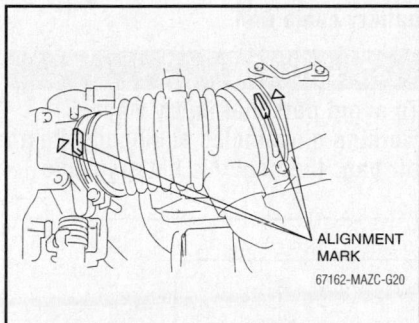

Fig. 269 Make sure to align the alignment marks on the throttle body and the air hose (2.0L and 2.3L engines—2009 vehicles

be disconnected and at least one minute must elapse before working on the system. Failure to do so may result in deployment of the air bag.

2. Remove the battery top cover, if equipped.

3. Disconnect the negative battery cable. Tape the cable with insulating tape.

➡**When disconnecting the cable, some systems need to be initialized after the cable is reconnected. You will need Mazda diagnostic scan tool or equivalent. Follow the directions on the tool.**

4. Drain the cooling system.

5. Relieve the fuel system pressure.

6. Remove the plug hole plate by lifting off and removing the plug hole plate from the areas shown in the accompanying illustration.

7. Remove the battery cover and battery duct.

8. Remove the undercover and disconnect the negative battery cable.

9. Remove the intake air cover, the air hose and the air cleaner element.

10. Remove the resonance chamber and air filter element.

11. Remove the strap and the air cleaner case.

12. Remove the fresh air duct.

13. Remove the throttle body.

To install:

➡**Be sure to use new fasteners, as required.**

14. Install the throttle body and tighten the bolts to 71–101 inch lbs. (8–11.5 Nm).

15. Install the fresh air duct.

16. Verify the rubber mounts on the battery support bracket are still in place. Insert the air cleaner case into the rubber mounts, using soapy water if necessary to ease installation.

17. Use the strap to secure the shroud panel and the air cleaner case as shown in the accompanying illustration.

18. Install the air cleaner element and the resonance chamber.

19. Install the air cleaner cover.

20. Install the air hose, make sure to align the alignment marks on the throttle body and the air hose.

21. Install the intake air cover.

22. Install the undercover.

23. Connect the negative battery cable.

24. Install the battery duct and cover.

25. Install the plug hole plate.

26. Fill the cooling system.

27. Start the vehicle and check for leaks as follows:

a. Using a jumper wire, ground the Powertrain Control Module (PCM) terminals. If equipped with an immobilizer, ground terminal 1AR, if not equipped with an immobilizer, ground terminal 1AQ. Refer to the illustration for terminal location.

b. Turn the ignition switch the ON position to activate the fuel pump.

c. Check the hoses, clips and other fuel system components for leaks.

d. If there are any leaks, replace the fuel hoses and clips. If the is damage to the seal on the fuel pipe side, replace the pipe.

e. The system must be leak free for five minutes with the terminal grounded. If any component is replaced because of a system, leak, turn the ignition key OFF; remove the jumper wire from the terminal. reapply the jumper wire, turn the ignition On and check for leaks.

28. Use the Mazda diagnostic scan tool, or equivalent and reprogram the required systems.

2.3L Engine with Turbocharger

1. Before servicing the vehicle, refer to the Precautions Section.

➡**If working near and/or around the SRS system and components, be sure to disable the SRS system. Tape the negative battery cable with insulating tape. Always disconnect the negative battery cable first.**

❋❋ CAUTION

To avoid personal injury when working on vehicles equipped with an air bag, the negative battery cable must be disconnected and at least one minute must elapse before working on the system. Failure to do so may result in deployment of the air bag.

2. Remove the battery top cover, if equipped.

3. Disconnect the negative battery cable. Tape the cable with insulating tape.

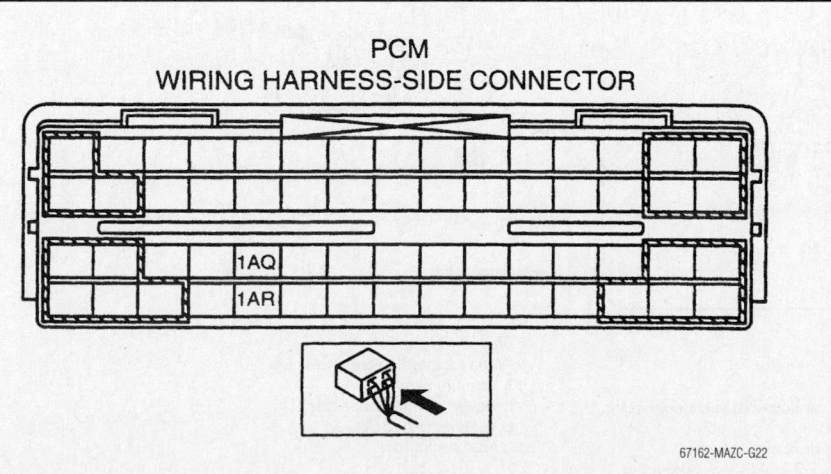

Fig. 270 If equipped with an immobilizer, ground PCM terminal 1AR, if not equipped with an immobilizer, ground PCM terminal 1AQ (2.0L and 2.3L engines)—2009 vehicles

➡ When disconnecting the cable, some systems need to be initialized after the cable is reconnected. You will need Mazda diagnostic scan tool or equivalent. Follow the directions on the tool.

4. Remove the charge air cooler cover.

5. Remove the battery and tray.

6. Remove the air cleaner.

7. Remove the resonance chamber, remove the left hand front mudguard before removing the resonance chamber.

8. Disconnect the MAF/IAT sensor.

9. Remove the duct.

10. Remove the charge air cooler cover.

11. Remove the charge air cooler.

12. Remove the charge air cooler bracket.

13. Remove the air bypass valve.

14. Drain and recycle the engine coolant.

15. Remove the throttle body.

To install:

➡ Be sure to use new fasteners, as required.

16. Installation is the reverse of removal, note the following:

a. Tighten the throttle body installation bolts to 71–101 inch lbs. (8–11.5 Nm).

b. Perform the following when installing the air cleaner case:

- Verify that the rubber mounts are set in the air cleaner bracket
- Install the projections on the frame side.
- Verify that the projections on the frame side are installed securely.
- Install the projection on the engine side.
- Verify that the projection on the engine side installed securely.

17. Use the Mazda diagnostic scan tool, or equivalent and reprogram the required systems.

2010 Vehicles

2.0L and 2.5L Engines

See Figures 271 and 272.

1. Before servicing the vehicle, refer to the Precautions Section.

➡ If working near and/or around the SRS system and components, be sure to disable the SRS system. Tape the negative battery cable with insulating tape. Always disconnect the negative battery cable first.

✳✳ CAUTION

To avoid personal injury when working on vehicles equipped with an air bag, the negative battery cable

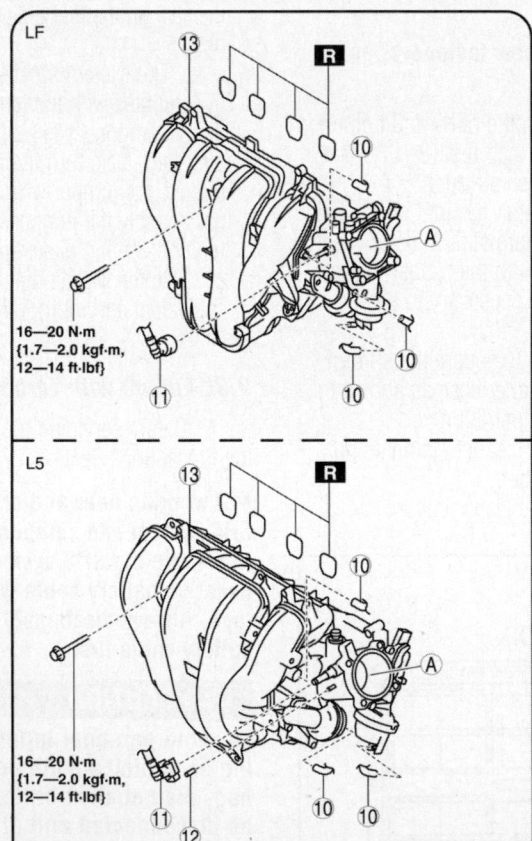

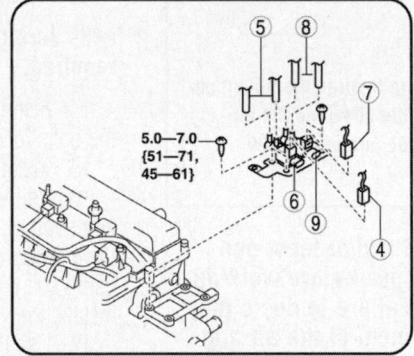

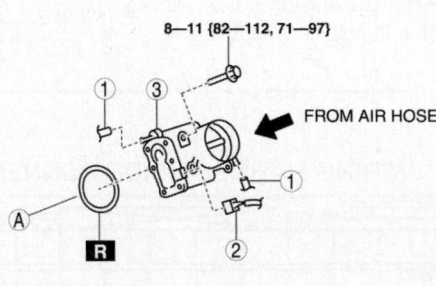

1. Water hose
2. Throttle body connector
3. Throttle body
4. Variable intake air solenoid valve connector
5. Vacuum hose
6. Variable intake air solenoid valve
7. Variable tumble solenoid valve connector
8. Vacuum hose
9. Variable tumble solenoid valve
10. Vacuum hose
11. Quick release connector
12. Vacuum cap (2.5L)
13. Intake manifold

37671_MAZ3_G0353

Fig. 271 Throttle body and related components (2.0L and 2.5L engines)—2010 vehicles

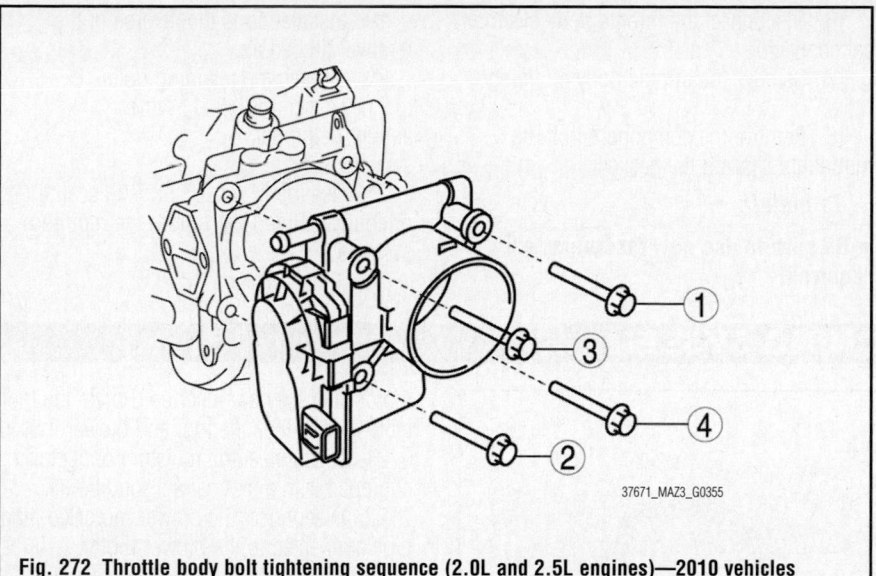

Fig. 272 Throttle body bolt tightening sequence (2.0L and 2.5L engines)—2010 vehicles

must be disconnected and at least one minute must elapse before working on the system. Failure to do so may result in deployment of the air bag.

✳✳ WARNING

A hot engine and intake air system can cause severe burns. Turn off the engine and wait until they are cool before removing the intake air system.

✳✳ WARNING

Fuel vapor is hazardous. It can easily ignite, causing serious injury and damage. Always keep sparks and flames away from fuel.

✳✳ WARNING

Fuel line spills and leakage are dangerous. Fuel can ignite and cause serious injuries or death and damage. Fuel can also irritate skin and eyes.

2. Disconnect the negative battery cable. Tape the cable with insulating tape.

➡When disconnecting the cable, some systems need to be initialized after the cable is reconnected. You will need the Mazda diagnostic scan tool or equivalent. Follow the directions on the tool.

3. Remove the air cleaner assembly.
4. Remove the hole plate cover.

5. Drain the cooling system. Be sure to properly dispose of used coolant.
6. Remove the water hose.
7. Disconnect the required electrical connectors.
8. Remove the throttle body retaining bolts.
9. Remove the component from its mounting. Discard the gasket.

To install:

➡Be sure to use new fasteners, as required.

10. Installation is the reverse of the removal procedure.
11. Be sure to use a new gasket.
12. Tighten the retaining bolts to 71–97 inch lbs. (8–11 Nm) and in the proper sequence.
13. Use the Mazda diagnostic scan tool, or equivalent and reprogram the required systems.

2.3L Engine with Turbocharger
See Figure 273.

1. Before servicing the vehicle, refer to the Precautions Section.

➡If working near and/or around the SRS system and components, be sure to disable the SRS system. Tape the negative battery cable with insulating tape. Always disconnect the negative battery cable first.

✳✳ CAUTION

To avoid personal injury when working on vehicles equipped with an air bag, the negative battery cable must be disconnected and at least one minute must elapse before working on the system. Failure to do so may result in deployment of the air bag.

2. Remove the battery top cover, if equipped.
3. Disconnect the negative battery cable. Tape the cable with insulating tape.

➡When disconnecting the cable, some systems need to be initialized after the cable is reconnected. You will need Mazda diagnostic scan tool or equivalent. Follow the directions on the tool.

4. Remove the battery and battery tray.
5. Drain the cooling system. Be sure to properly dispose of used coolant.
6. Remove the right side splash shield and No. 2 undercover as a single unit.
7. Disconnect the wiring harness.
8. Remove the MAT/IAT sensor.
9. Remove the ventilation hose.
10. Remove the air cleaner cover. Remove the air hose.

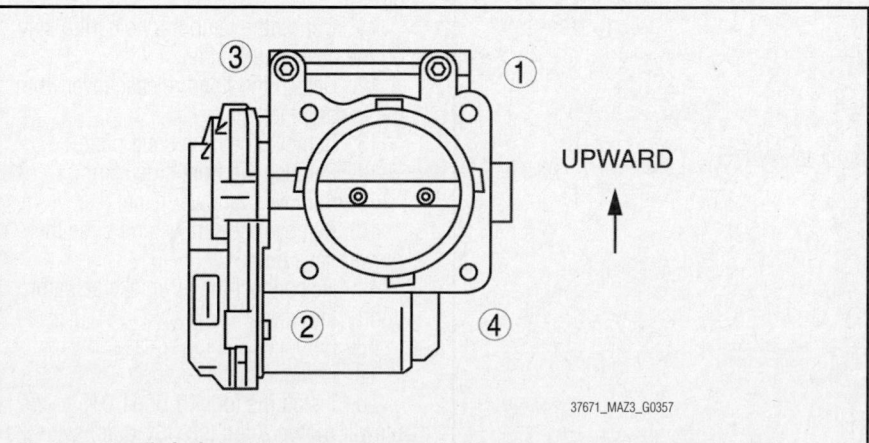

Fig. 273 Throttle body bolt tightening sequence (2.3L engine with turbocharger)—2010 vehicles

11. Remove the air cleaner element. Remove the air cleaner case.

12. Remove the fresh air ducts.

13. Remove the charge air cooler cover. Remove the air cooler. Remove the cooler bracket.

14. Remove the air bypass hose, valve and duct.

15. Remove the air cooler air outlet hose.

16. Disconnect the throttle body electrical connector.

17. Remove the throttle body retaining bolts.

18. Remove the component from its mounting. Discard the gasket.

To install:

➡ Be sure to use new fasteners, as required.

19. Installation is the reverse of the removal procedure.

20. Tighten the retaining bolts to 71–97 inch lbs. (8–11 Nm) and in the proper sequence.

21. Use the Mazda diagnostic scan tool, or equivalent and reprogram the required systems.

HEATING & AIR CONDITIONING SYSTEM

BLOWER MOTOR

REMOVAL & INSTALLATION

2009 Vehicles

See Figures 274 through 277.

1. Before servicing the vehicle, refer to the Precautions Section.

➡ If working near and/or around the SRS system and components, be sure to disable the SRS system. Tape the negative battery cable with insulating tape. Always disconnect the negative battery cable first.

✳✳ CAUTION

To avoid personal injury when working on vehicles equipped with an air bag, the negative battery cable must be disconnected and at least one minute must elapse before working on the system. Failure to do so may result in deployment of the air bag.

2. Remove the battery top cover, if equipped.

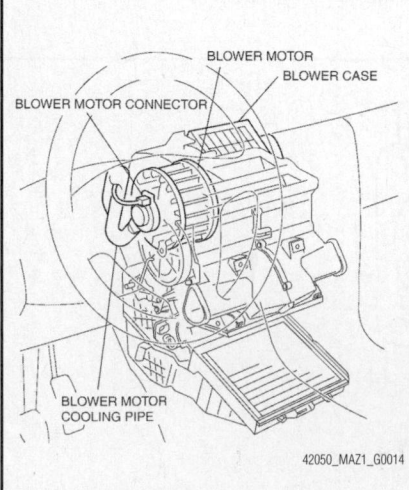

Fig. 274 Exploded view of the blower motor assembly—2009 vehicles

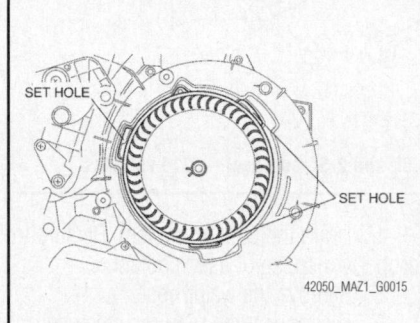

Fig. 275 Install the tool 49 B061 015 to the blower motor—2009 vehicles

3. Disconnect the negative battery cable. Tape the cable with insulating tape.

➡ When disconnecting the cable, some systems need to be initialized after the cable is reconnected. You will need Mazda diagnostic scan tool or equivalent. Follow the directions on the tool.

4. Set the air intake mode to FRESH.

5. Remove or disconnect the following:

6. Right side front side trim.

7. Right side front scuff plate.

8. Decoration panel.

9. Glove compartment.

10. Passenger junction box and bracket.

11. Cut off the temporary steering stay on the passenger's side.

12. Detach the hood release lever from the lower panel.

13. Remove the front scuff plate.

14. Remove the front side trim.

15. Remove the lower panel.

16. Remove the screws and slide the blower case out.

17. Disconnect the air intake actuator connector.

18. Remove the blower case.

19. Remove the air guide.

20. Install the tool 49 B061 015 to the blower motor. Align the SST guide with the sirocco fan clip position and press the tool tabs into the three set holes on the blower

motor until they are inserted. Rotate the tool clockwise to lock the tool and blower motor.

21. Disconnect the resistor connector if equipped with a manual air conditioner.

22. Disconnect the blower motor cooling pipe connected to the blower motor.

23. Remove the airflow mode actuator.

24. Disconnect the blower motor connector.

25. Remove the blower motor cover by pulling the lock on the top of the blower motor cover and rotating the blower motor cover.

✳✳ CAUTION

When the blower motor cover is removed, the blower motor could fall in the A/C unit case causing the sirocco fan to be damaged. Therefore another person must hold the blower motor at the installation position. To prevent damage to the sirocco fan, pull the blower motor out being careful that the blower motor does not interfere with the A/C unit.

26. Remove the blower motor by pulling it out.

To install:

➡ Be sure to use new fasteners, as required.

27. Installation is the reverse of removal, please note the following:

➡ Position the blower motor projection upward and install the blower motor to the A/C unit.

✳✳ CAUTION

To prevent damage to the sirocco fan, install the blower motor being careful that the blower motor does not interfere with the A/C unit. Also, another person must hold the blower motor at the installation position.

a. Install the tool to the blower motor.

b. Install the blower motor with the tool installed, to the A/C unit.

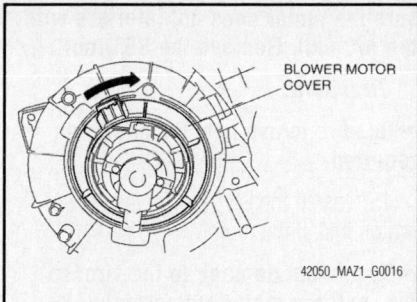

Fig. 276 Rotate the blower motor cover until a click is heard—2009 vehicles

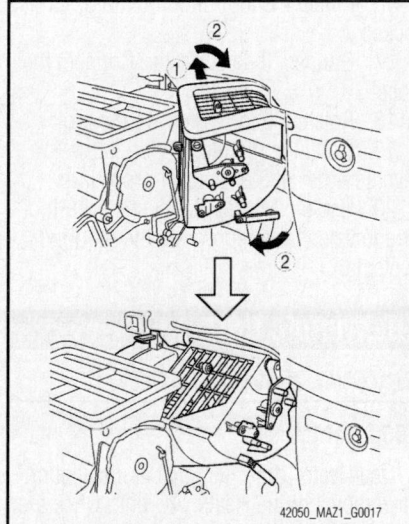

Fig. 277 Install the blower case by inserting and rotating it in the directions of the arrows—2009

c. Install the blower motor cover from the driver's side. Rotate the blower motor cover until a click is heard.

d. Connect the blower motor connector.

e. Install the blower motor cooling pipe.

f. Remove the tool from the blower motor.

28. If not replacing the blower case, replace the adhesive polyurethane on the fresh-air inlet of the blower case.

> ※※ **CAUTION**
>
> **To adhere new polyurethane properly, be sure to remove the adhesive agent and adhesive polyurethane completely.**

➡**If the blower case is removed or installed, the adhesive polyurethane can be damaged. Damaged adhesive polyurethane could cause abnormal noise or other malfunctions, therefore replace it.**

a. Insert the screw into the blower case and install the case to the A/C unit.

b. Install the blower case by inserting and rotating it in the directions of the arrows shown in the illustration

29. Use the Mazda diagnostic scan tool, or equivalent and reprogram the required systems.

2010 Vehicles

See Figures 278 through 283.

1. Before servicing the vehicle, refer to the Precautions Section.

➡**If working near and/or around the SRS system and components, be sure to disable the SRS system. Tape the negative battery cable with insulating tape. Always disconnect the negative battery cable first.**

> ※※ **CAUTION**
>
> **To avoid personal injury when working on vehicles equipped with an air bag, the negative battery cable must be disconnected and at least one minute must elapse before working on the system. Failure to do so may result in deployment of the air bag.**

2. Remove the battery top cover, if equipped.

3. Disconnect the negative battery cable. Tape the cable with insulating tape.

➡**When disconnecting the cable, some systems need to be initialized after the cable is reconnected. You will need Mazda diagnostic scan tool or equivalent. Follow the directions on the tool.**

➡**The blower motor is located on the A/C unit.**

4. Set the air intake mode to FRESH.

5. Remove the scuff plate, front side trim, upper panel, shift knob or selector lever knob, shift panel and side wall.

6. Remove the console.

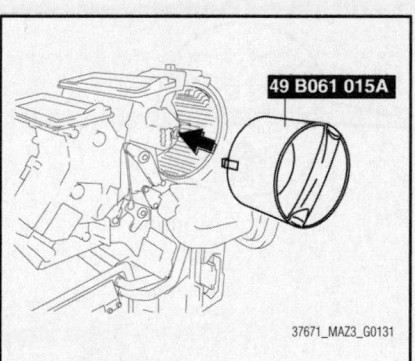

Fig. 278 Blower motor removal 1 of 5—2010 vehicles

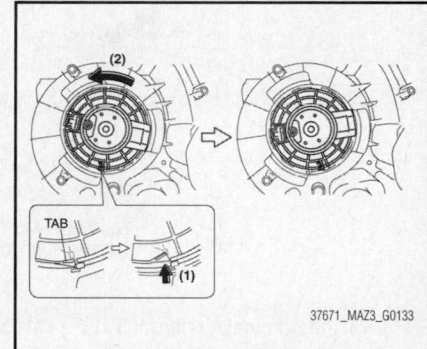

Fig. 280 Blower motor removal 3 of 5—2010 vehicles

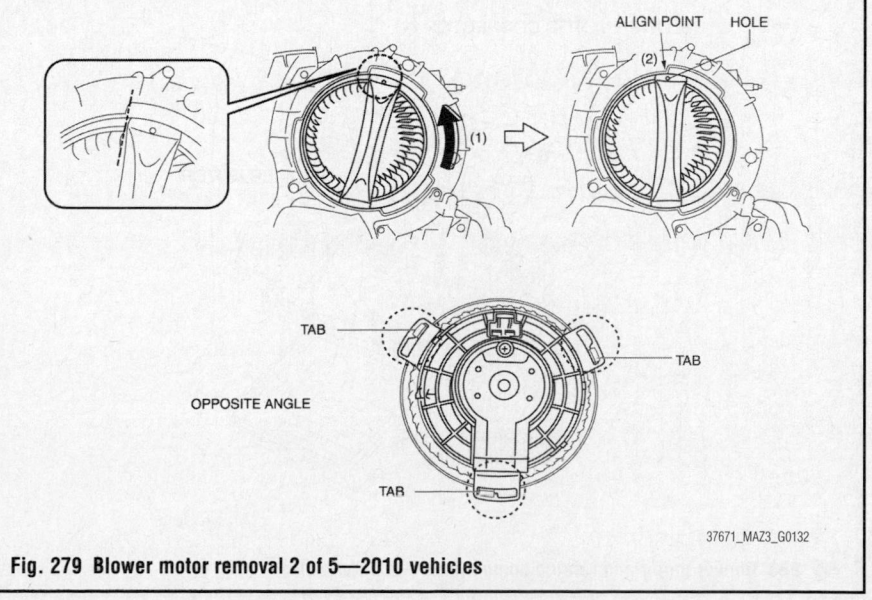

Fig. 279 Blower motor removal 2 of 5—2010 vehicles

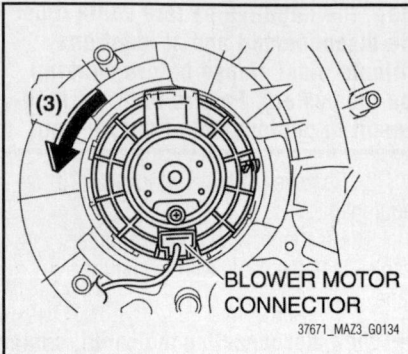

Fig. 281 Blower motor removal 4 of 5—2010 vehicles

7. Remove the dashboard undercover, glove compartment, hood release lever, lower panel and shower duct. Remove the accelerator pedal.

8. Disconnect the air intake actuator electrical connector.

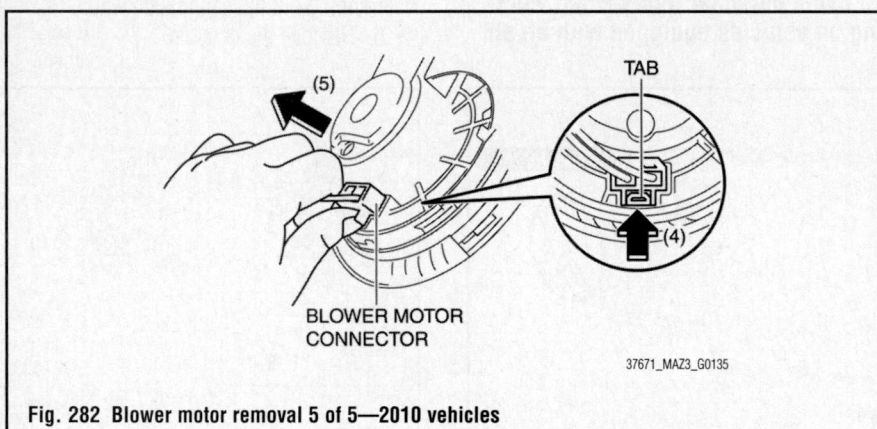

Fig. 282 Blower motor removal 5 of 5—2010 vehicles

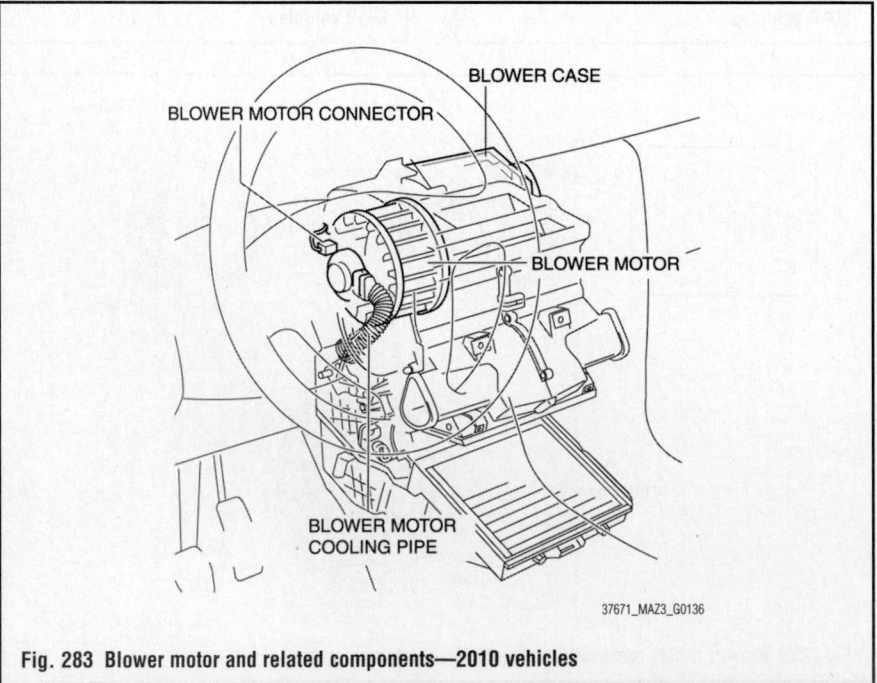

Fig. 283 Blower motor and related components—2010 vehicles

9. Detach the harness clip from the blower case.

10. Remove the retaining screws and slide the blower case. Remove the blower case.

11. Disconnect the blower motor hose.

12. Install tool SST:49 B061 015a to the blower motor. Rotate the tool and align the SST hole with the align point (2). Confirm that the SST tabs into the three set holes on the blower motor, see illustration.

13. Press the tab (1) and rotate the motor (2) and push the motor into the A/C case slightly.

14. Rotate the motor and connector (3). Press tab (4) and disconnect the connector. Remove the blower motor as shown in the illustration.

➡**To prevent damage to the sirocco fan, pull the motor out carefully. Be** sure the motor does not interfere with the A/C unit. Remove the SST tool.

To install:

➡**Be sure to use new fasteners, as required.**

15. Install the SST tool to the blower motor. Install the motor.

➡**To prevent damage to the sirocco fan, pull the motor out carefully. Be sure the motor does not interfere with the A/C unit, an assistant must hold the motor at the installation position.**

16. Rotate the tool until the motor tab is locked.

17. Connect the connector. Connect the hose.

18. Rotate the tool. Remove the tool.

19. Continue the installation in the reverse order of the removal procedure.

20. Use the Mazda diagnostic scan tool, or equivalent and reprogram the required systems.

HEATER/AC UNIT

REMOVAL & INSTALLATION

2009 Vehicles

See Heater Core Removal & Installation for removal of the Heater/AC unit. Once the Heater/AC unit is removed the heater core can be serviced.

2010 Vehicles

See Figures 284 through 291.

1. Before servicing the vehicle, refer to the Precautions Section.

➡**If working near and/or around the SRS system and components, be sure to disable the SRS system. Tape the negative battery cable with insulating tape. Always disconnect the negative battery cable first.**

✳✳ CAUTION

To avoid personal injury when working on vehicles equipped with an air bag, the negative battery cable must be disconnected and at least one minute must elapse before working on the system. Failure to do so may result in deployment of the air bag.

2. Remove the battery top cover, if equipped.

3. Disconnect the negative battery cable. Tape the cable with insulating tape.

➡**When disconnecting the cable, some systems need to be initialized after the**

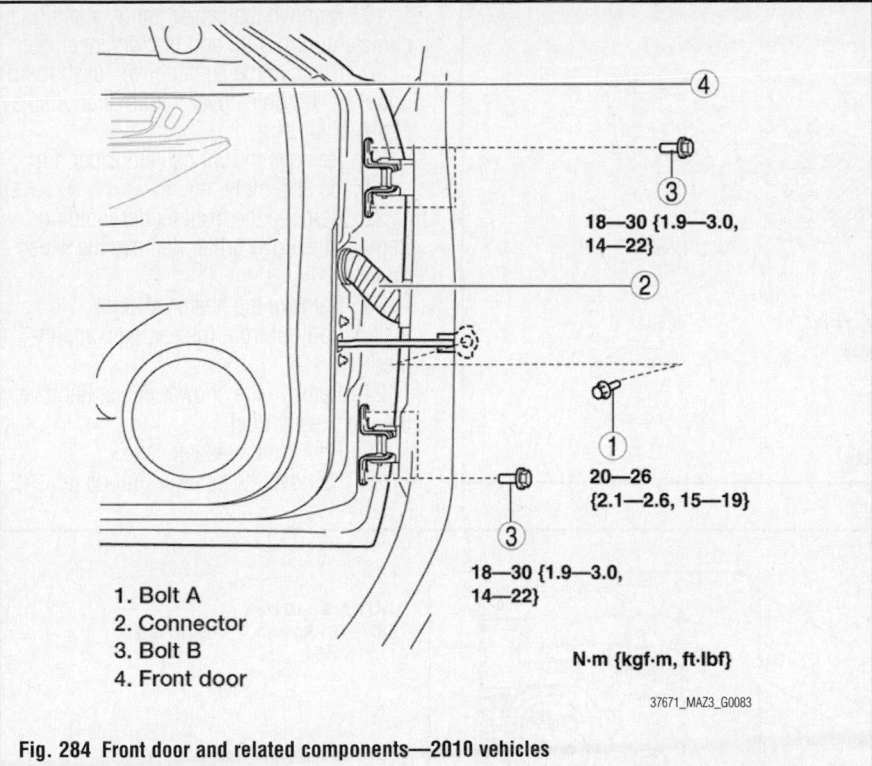

1. Bolt A
2. Connector
3. Bolt B
4. Front door

18—30 {1.9—3.0, 14—22}

20—26 {2.1—2.6, 15—19}

18—30 {1.9—3.0, 14—22}

N·m {kgf·m, ft·lbf}

37671_MAZ3_G0083

Fig. 284 Front door and related components—2010 vehicles

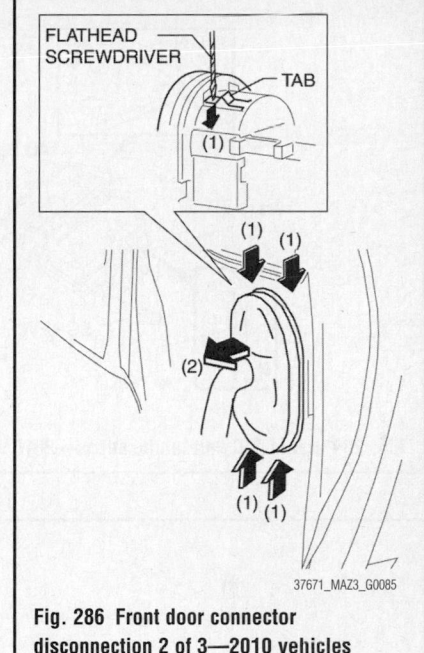

Fig. 286 Front door connector disconnection 2 of 3—2010 vehicles

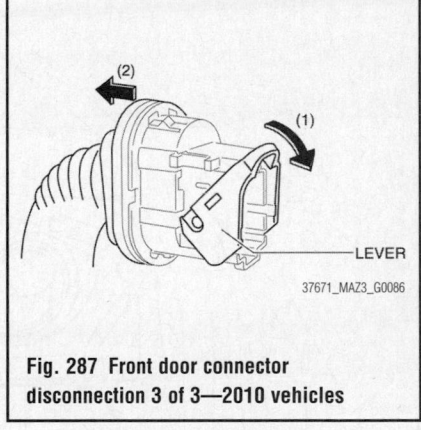

Fig. 287 Front door connector disconnection 3 of 3—2010 vehicles

cable is reconnected. You will need Mazda diagnostic scan tool or equivalent. Follow the directions on the tool.

4. Set the air mix mode to MAX COLD.

5. Drain the engine coolant. Be sure to properly dispose of used coolant.

6. Properly discharge the A/C system.

7. Remove the engine cover and insulator if equipped with 2.3L turbocharged engine.

8. Remove the bolts and nuts holding the refrigerant hoses in place, in the engine compartment.

9. Disconnect and plug the refrigerant lines.

10. Disconnect and plug the heater hoses.

11. Remove the front doors.

➡**Removing the door without supporting it could cause the door to fall and** cause serious injury. Be sure that the door is properly supported before removing it. To remove the door, remove bolt (A), disconnect the electrical connector. Remove bolt (B). Remove the door from its mounting.

12. Remove the scuff plate, side trim, dashboard undercover, glove compartment, upper panel, shift knob or selector lever, shift panel and side wall.

13. Remove the console.

14. Remove the shift lever component or selector lever, hood release lever, lower panel and knee bolster.

15. Remove the driver's side air bag module.

16. Remove the steering wheel.

17. Remove the steering column cover and combination switch.

18. Remove the joint cover and steering shaft.

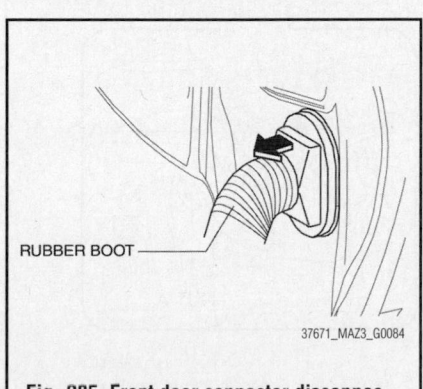

Fig. 285 Front door connector disconnection 1 of 3—2010 vehicles

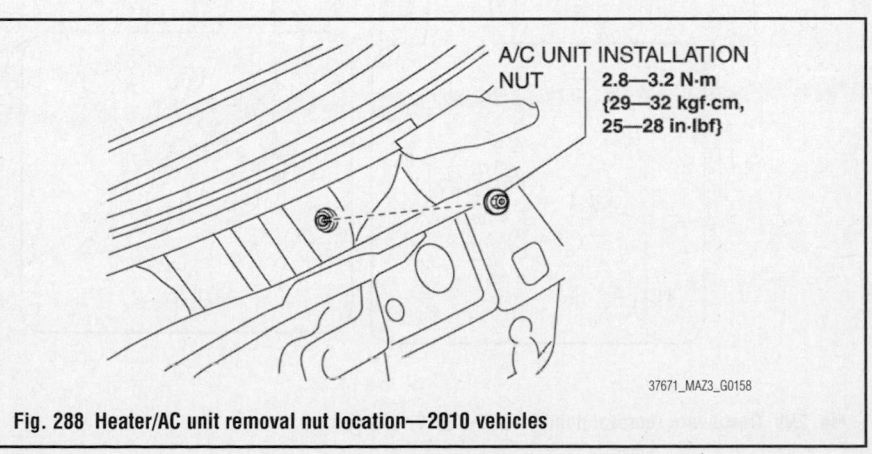

Fig. 288 Heater/AC unit removal nut location—2010 vehicles

TAB B
A/C UNIT
TAB B
TAB A
A/C UNIT
SCREW
HEATER
CASE
A/C UNIT
37671_MAZ3_G0159

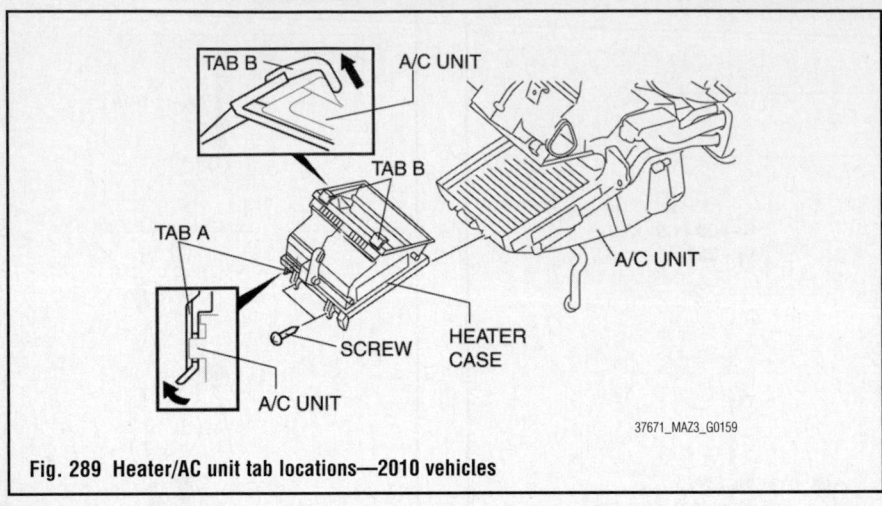

Fig. 289 Heater/AC unit tab locations—2010 vehicles

19. Remove the center panel, audio unit, climate control unit and instrument cluster.
20. Remove the center cover, dashboard upper panel, hole cover, information display and A pillar trim.
21. Remove the windshield wiper arm and blade assembly.
22. Remove the front fender molding. Remove the cowl grille. Remove the wiper motor.
23. Remove the heater/AC unit installation nut from the engine compartment.
24. Remove the shower ducts. Remove the rear heater duct.
25. Remove the heater case.
26. Remove the screws. Pull up tab (A)

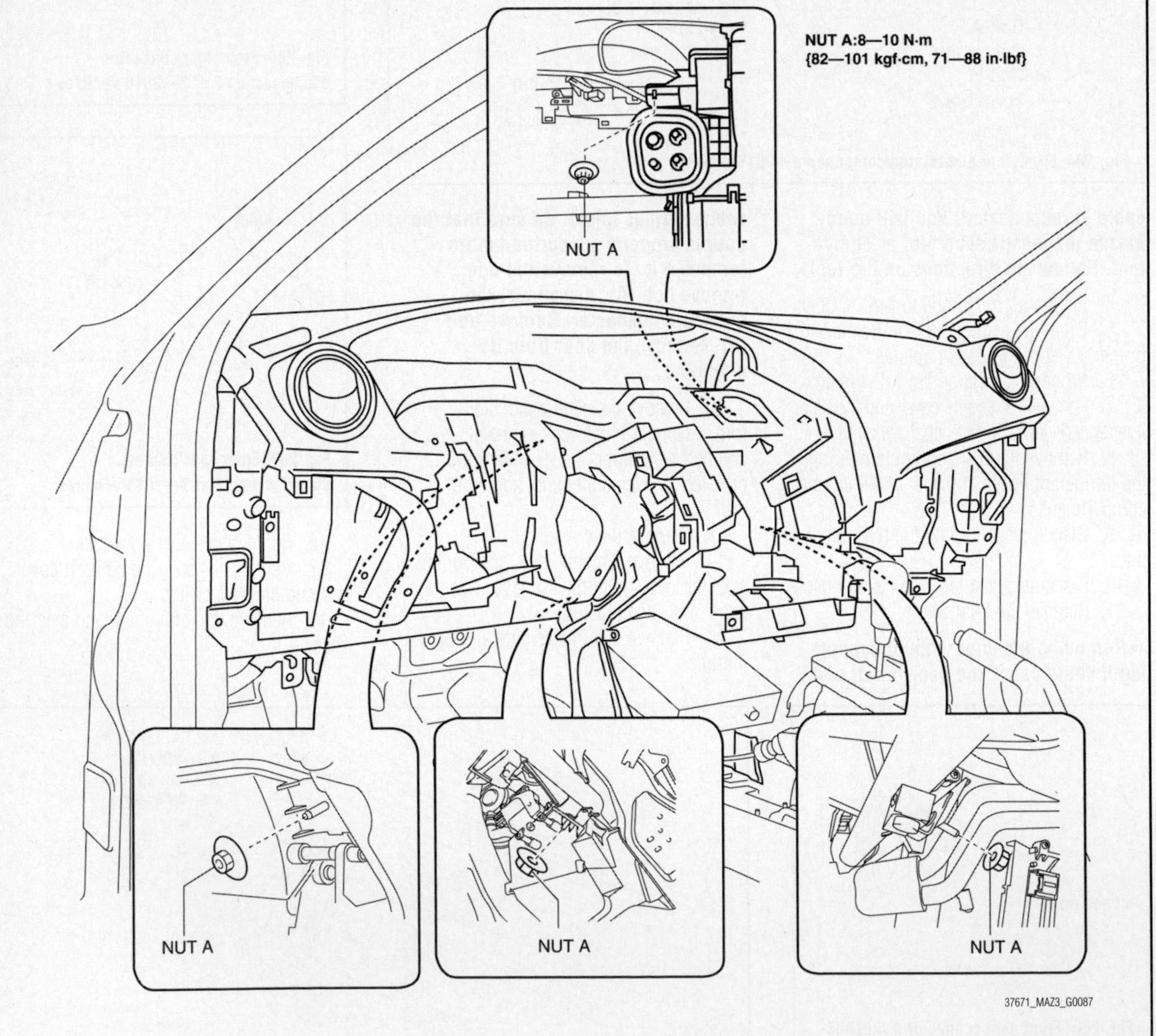

NUT A:8—10 N·m
{82—101 kgf·cm, 71—88 in·lbf}
NUT A
NUT A
NUT A
NUT A
37671_MAZ3_G0087

Fig. 290 Dashboard removal points 1 of 6—2010 vehicles

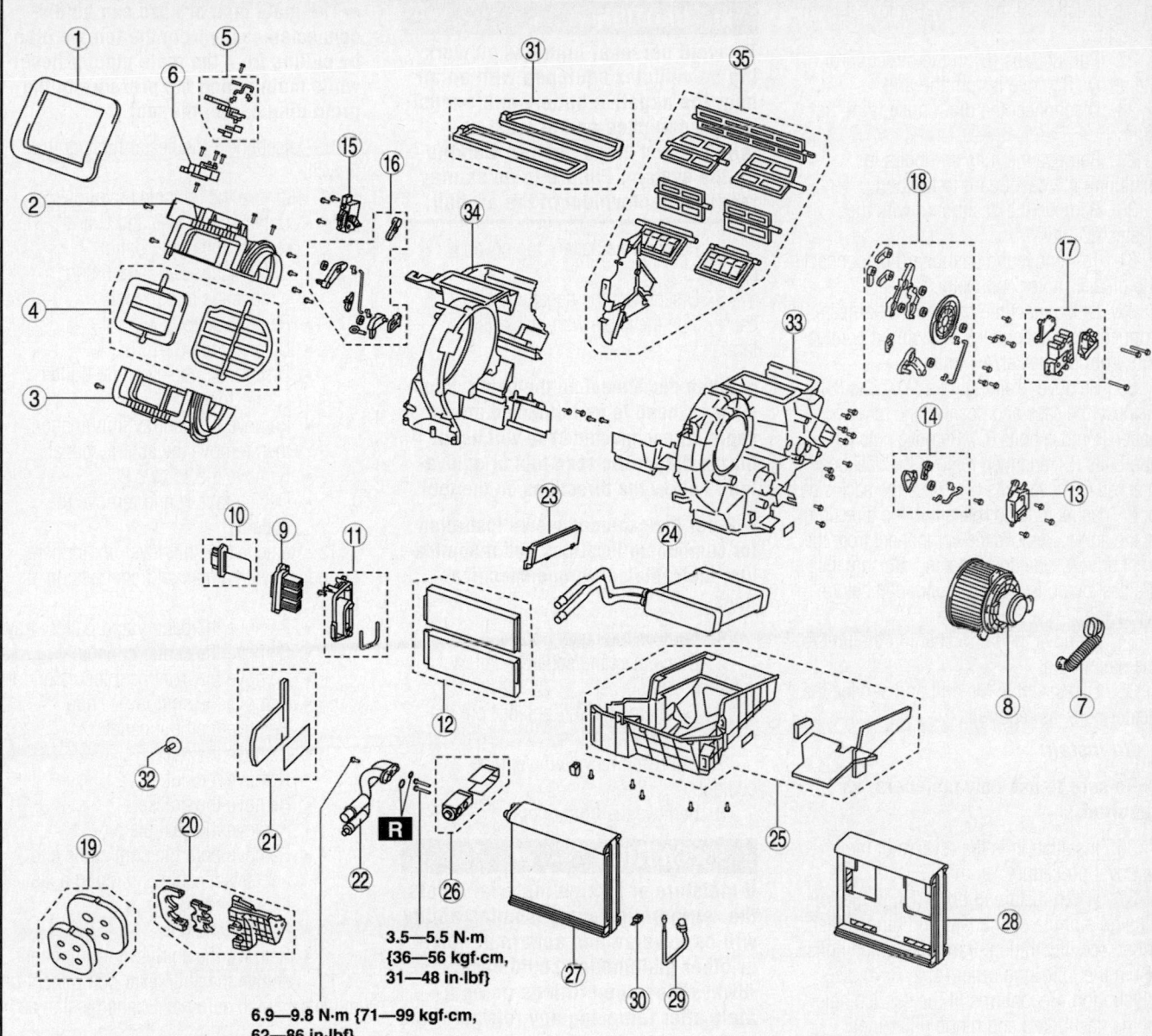

3.5—5.5 N·m
{36—56 kgf·cm,
31—48 in·lbf}

6.9—9.8 N·m {71—99 kgf·cm,
62—86 in·lbf}

1. Adhesive polyurethane
2. Blower case
3. Blower case
4. Air intake door
5. Air intake link set
6. Air intake actuator
7. Blower motor pipe
8. Blower motor
9. Power MOS FET (Full-auto air conditioner)
10. Resistor (Manual air conditioner
11. Plate
12. Air filter
13. Driver-side air mix actuator (Full-auto air conditioner)
14. Driver-side air mix link set (Full-auto air conditioner)
15. Passenger-side air mix actuator (Full-auto air conditioner)
16. Passenger-side air mix link set (Full-auto air conditioner)
17. Airflow mode actuator (Full-auto air conditioner)

18. Airflow mode link set
19. Polyurethane foam
20. Plate cover
21. Adhesive polyurethane
22. Evaporator pipe
23. Cover
24. Heater core
25. A/C case
26. Expansion valve
27. Evaporator
28. A/C case
29. Evaporator temperature sensor
30. Clip
31. Polyurethane foam
32. Bolt
33. A/C case
34. A/C case
35. Door damper

37671_MAZ3_G0162

Fig. 291 Heater/AC unit and related components—2010 vehicles

in the direction of the arrow. Remove it from the unit.

27. Pull up tabs (B) in the direction of the arrow. Remove it from the unit.

28. Disconnect the drain hose from the unit.

29. Remove the nuts and bolts for installing the dashboard to the body.

30. Remove the dashboard with the heater/AC unit.

31. To remove the dashboard disconnect the blower motor connector.

32. Disconnect the dashboard harness connectors. Disconnect all required connectors from the heater/AC unit.

33. Remove retaining nuts (A), bolts (B). Remove the caps and position the rolled paper. Remove bolts (C). Remove nuts (D) and bolts (E), and then remove the dashboard bracket. Remove bolts (F). Raise the back side of the dashboard and rotate it in the direction of the arrow. Remove the component from the front driver's side door opening. Be sure that the dashboard is properly supported before removing it. See illustrations.

34. Remove the nuts retaining the unit to the dashboard.

35. Disassemble the unit and service the heater core, as required.

To install:

➡ **Be sure to use new fasteners, as required.**

36. Installation is the reverse of the removal procedure.

37. When installing bolts (F), fit the service hole of the dashboard and hinge pillar inner, see illustration. Partially tighten bolts (F) in the following order (1, 2, 3, 4), see illustration. Reconfirms fit the service hole of the dashboard and hinge pillar inner, tighten the upper side bolt (F). Tighten all bolts (F). See illustration.

38. Use the Mazda diagnostic scan tool, or equivalent and reprogram the required systems.

HEATER CORE

REMOVAL & INSTALLATION

2009 Vehicles

See Figures 292 through 329.

1. Before servicing the vehicle, refer to the Precautions Section.

➡ **If working near and/or around the SRS system and components, be sure to disable the SRS system. Tape the negative battery cable with insulating tape. Always disconnect the negative battery cable first.**

❋❋ CAUTION

To avoid personal injury when working on vehicles equipped with an air bag, the negative battery cable must be disconnected and at least one minute must elapse before working on the system. Failure to do so may result in deployment of the air bag.

2. Remove the battery top cover, if equipped.

3. Disconnect the negative battery cable. Tape the cable with insulating tape.

➡ **When disconnecting the cable, some systems need to be initialized after the cable is reconnected. You will need Mazda diagnostic scan tool or equivalent. Follow the directions on the tool.**

➡ **Refer the exploded view illustration for component locations and if applicable, their retainer torque specifications.**

4. Remove the battery cover.

5. Disconnect the negative battery cable.

6. Discharge the refrigerant from the system.

7. Drain and recycle the engine coolant.

8. Remove the front doors.

❋❋ CAUTION

If moisture or foreign material enters the refrigeration cycle, cooling ability will be lowered and abnormal noise or other malfunction could occur. Always plug open fittings immediately after removing any refrigeration cycle parts.

9. To disconnect the different types of refrigerant line connectors, perform the following:
- On block joint types. disconnect the block joint type pipes by grasping female side of the block with pliers or similar tool and holding firmly, then remove the connection bolt or nut.
- On spring-lock coupling type , set the service tool shown in the illustration and while looking through the inspection hole of the tool , insert the protruding part of the tool until it makes contact with the cage section. Use the tool to disconnect the male pipe or hose from the female by pulling the male pipe or hose.

➡ **The male pipe or hose can be disconnected easily from the female pipe or hose by pulling from the male pipe or hose while maintaining the pressure of the protruding part of the tool.**

10. Disconnect the heater hose at the firewall.

11. Remove the console as follows:
- Detach clips A, B and C and remove the upper panel.
- Detach clips D, E and F and remove the boot panel.
- Remove screws.
- Detach clips G.
- Disconnect the cigarette lighter connector.
- Remove the ashtray illumination, then remove the ashtray panel.
- Remove the screws B.
- Detach tabs H and remove the console.

12. Remove the shift lever component on models with a manual transmission as follows:
- Remove the battery and battery tray
- Remove the center console.
- Remove the front heat insulator.
- Remove the shift lever knob.
- Remove the boot panel.
- Remove the nuts.
- Remove the nut.
- Remove the bracket.
- Remove the seal plate.
- Remove both the shift cable end and select cable end using a fastener remover.
- Remove the nuts.
- Remove the shift lever component.

13. Remove the shift lever component on models with an automatic transmission as follows:
- Remove the air cleaner component.
- Remove the console.
- Remove the front and center heat insulator.
- Remove the selector lever component connector.
- Remove the clip and selector cable.

14. Remove the selector lever component.

15. Pull the decoration panel outward and detach the clips A and tab B.

16. Remove the decoration panel.

17. Pull the front scuff plate upward, detach clips A, locator pins B and C from the body, and then remove the front scuff plate.

18. Remove the front side trim fastener.

19. Pull the front side trim in the direction of the arrow and detach clip A and locator pin B.

20. Remove the glove box screws.

21. Pull the glove compartment outward and detach clips A and tab B.

22. Remove the glove compartment.

23. Remove the shower ducts.

24. Remove the Passenger Junction Box (PJB) as follows:

   a. Remove the cover.

   b. On connector A, push the release tab in the direction of the arrow, then rotate the lever in the direction of the arrow and remove connector A.

   c. On connector B, rotate the lever in the direction of the arrow and remove connector B.

   d. Turn the screws counterclockwise to remove the PJB.

➡**The screws cannot be removed from the PJB.**

   e. Remove the PJB as shown in the illustration.

25. Remove the car navigation unit.

26. Remove the lower panel as follows:

   a. Detach the hood release lever from the lower panel.

   b. Remove the lever on 5 door models.

   c. Remove the hood latch.

   d. Remove the hood latch switch connector.

   e. Pull the hood release lever. While pushing the tab in the direction of the arrow using a tape-wrapped, small flat bladed tool, detach it from the lower panel.

   f. Remove the hood release cable.

✳✳ CAUTION

Be careful not to damage the hood release cable when removing the hood release lever with the flat bladed tool.

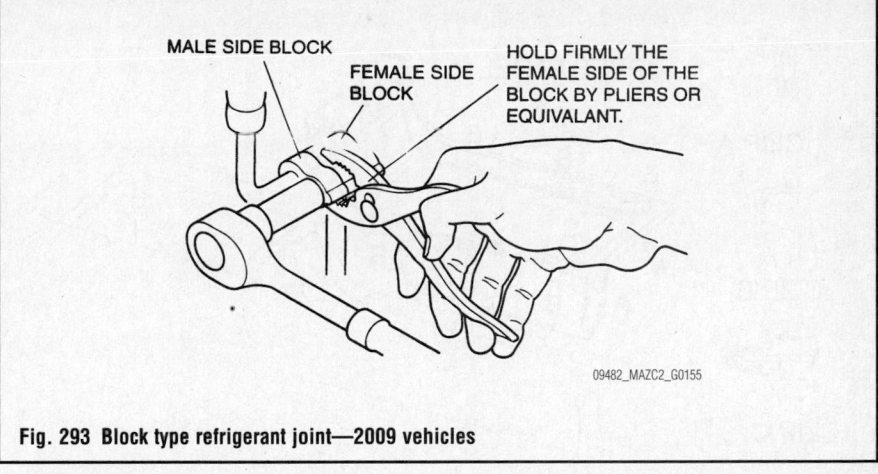

Fig. 293 Block type refrigerant joint—2009 vehicles

   g. Pull the hood release lever outward, then remove it from the lower panel.

   h. Remove the front scuff plate.

   i. Remove the front side trim.

   j. Remove the screw.

   k. Pull the lower panel outward, and detach clips A and tab.

   l. Disconnect the panel light control switch connector and the headlight leveling switch connector.

   m. Remove the lower panel.

27. Remove the column cover as follows:

   a. Detach the fit of the upper column cover from the meter hood rubber.

   b. Remove the upper column cover.

   c. Remove the ignition key illumination.

   d. Remove the screws.

   e. Remove the lower column cover.

28. Remove the steering shaft as follows:

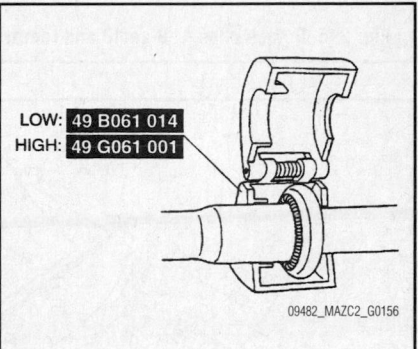

Fig. 294 Spring lock type refrigerant joint—2009 vehicles

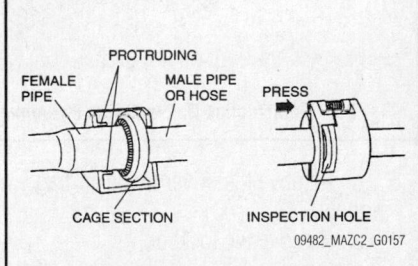

Fig. 295 Looking through the inspection hole of the tool, insert the protruding part of the tool until it makes contact with the cage on spring lock type refrigerant joints—2009 vehicles

✳✳ WARNING

Handling the air bag module improperly can accidentally operate (deploy) the air bag module, which may seriously injure you.

   a. Remove the air bag cover.

   b. Remove the bolt.

   c. Using a flathead tool, pry out the connector stopper plate and unplug the connector.

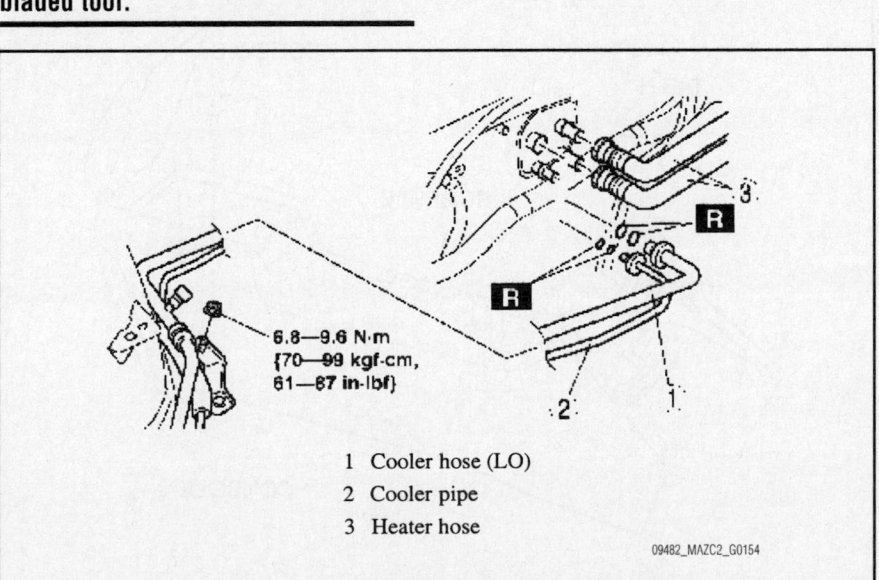

1 Cooler hose (LO)
2 Cooler pipe
3 Heater hose

Fig. 292 Cooler and heater hose connections at the firewall—2009 vehicles

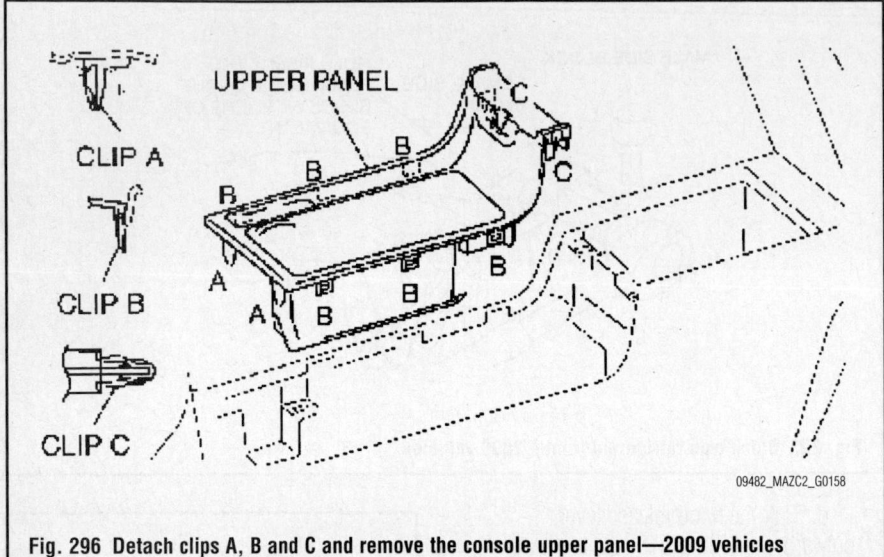

Fig. 296 Detach clips A, B and C and remove the console upper panel—2009 vehicles

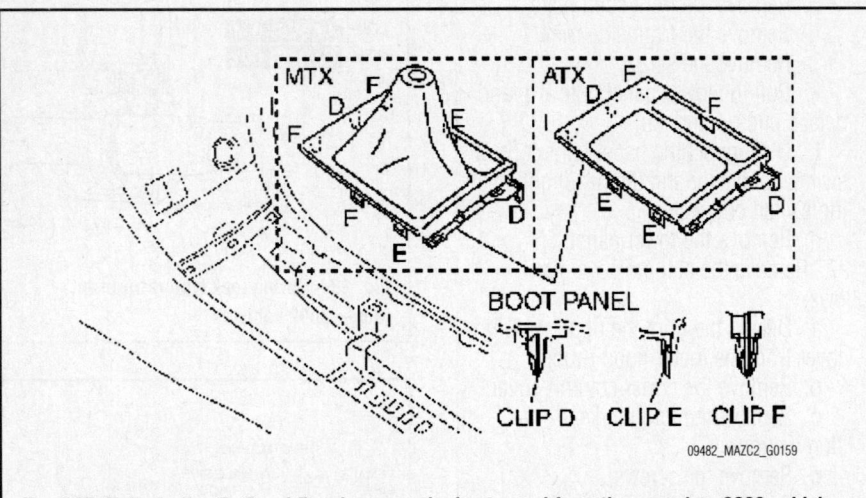

Fig. 297 Detach clips D, E and F and remove the boot panel from the console—2009 vehicles

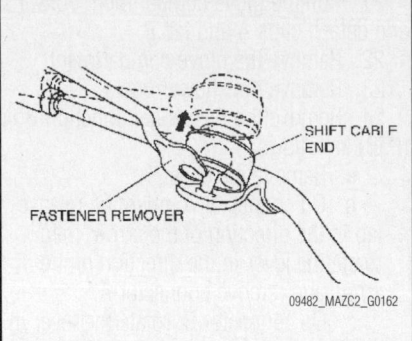

Fig. 299 Remove both the shift cable end and select cable end using a fastener remover—2009 vehicles

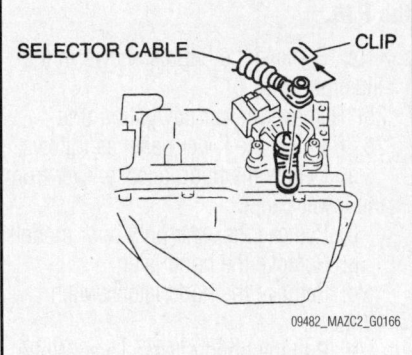

Fig. 300 Remove the clip and selector cable on models with an automatic transmission—2009 vehicles

j. Remove the clock spring connector.

k. Remove the steering angle sensor connector, if equipped with a steering angle sensor.

d. Remove the driver's side air bag module.

e. Remove the lockbolt.

> ✶✶ **WARNING**
>
> Handling the air bag module improperly can accidentally deploy the air bag module, which may seriously injure you.

f. Remove the locknut.

> ✶✶ **CAUTION**
>
> Do not try to remove the steering wheel by hitting the shaft with a hammer. The column will collapse.

g. Set the vehicle in the straight-ahead position.

h. Remove the steering wheel using a suitable puller.

i. Remove the column cover.

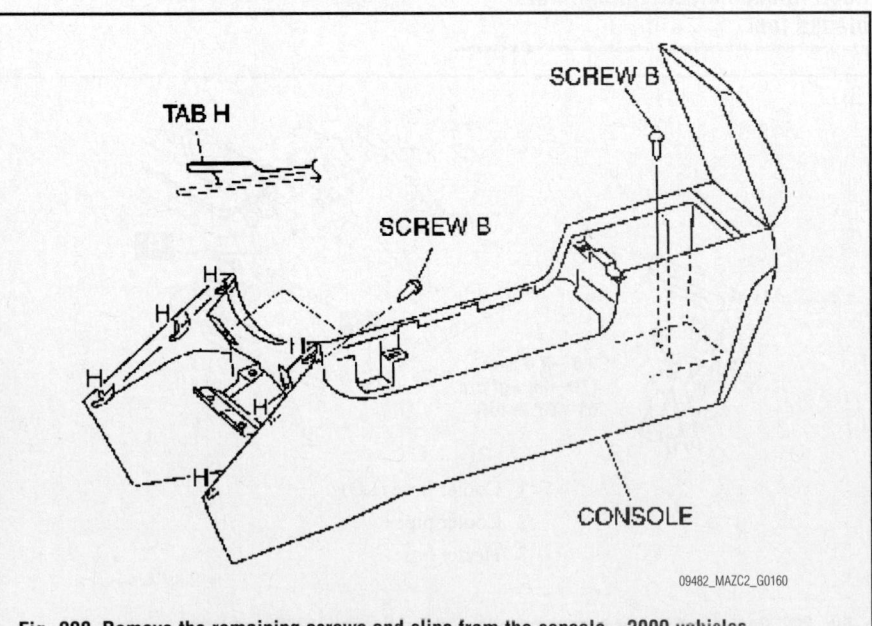

Fig. 298 Remove the remaining screws and clips from the console—2009 vehicles

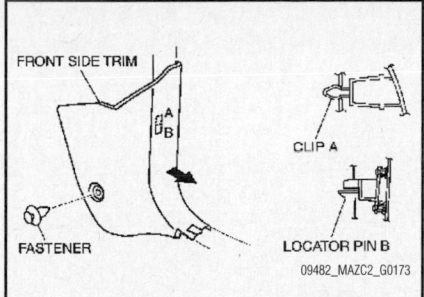

Fig. 301 Exploded view of the front side trim assembly—2009 vehicles

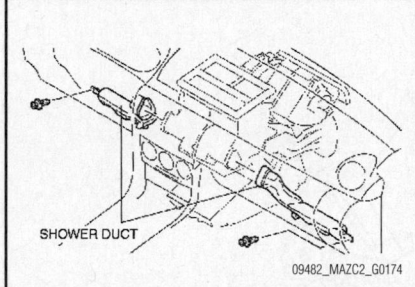

Fig. 302 Exploded view of the shower duct assembly—2009 vehicles

l. Remove the screw.

m. Remove the clock spring.

n. Remove the steering angle sensor by detaching the tabs at the four locations shown in the figure and remove the steering angle sensor.

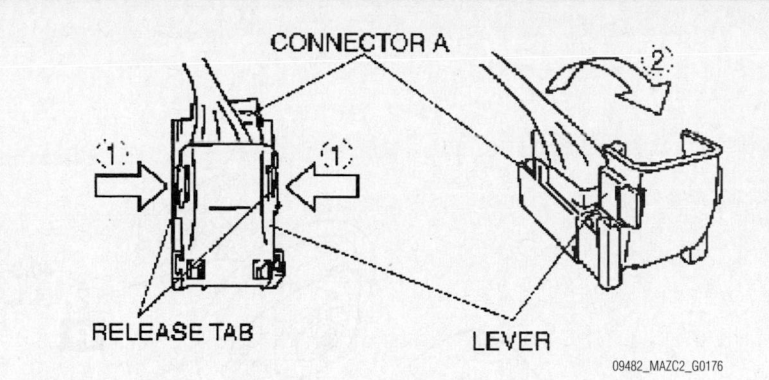

Fig. 304 On connector A, push the release tab in the direction of the arrow, then rotate the lever in the direction of the arrow and remove connector A—2009 vehicles

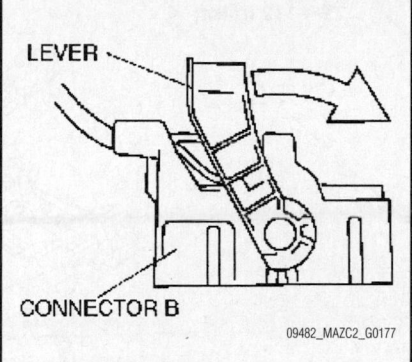

Fig. 305 On connector B, rotate the lever in the direction of the arrow and remove connector B—2009 vehicles

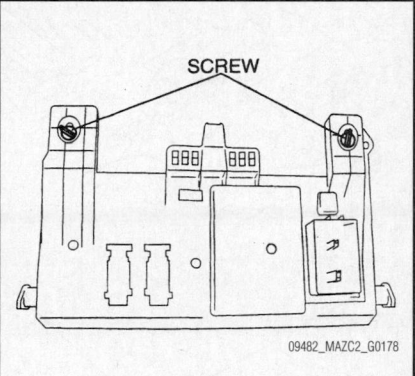

Fig. 306 Turn the screws counterclockwise to remove the PJB—2009 vehicles

o. Remove the combination switch.

p. Remove the dust cover.

q. Remove the steering shaft.

29. Remove the A-pillar trim shaft as follows:

a. Partially peel back the seaming welt.

b. Detach clips A using a fastener remover.

c. Pull the A-pillar trim and detach clip B (1).

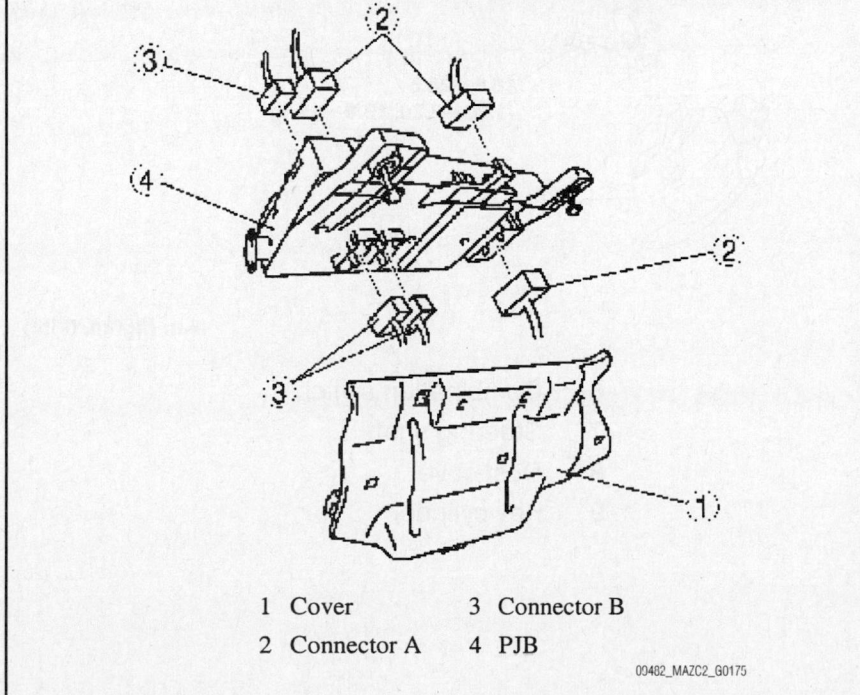

| 1 | Cover | 3 | Connector B |
| 2 | Connector A | 4 | PJB |

Fig. 303 Exploded view of the passenger junction box assembly—2009 vehicles

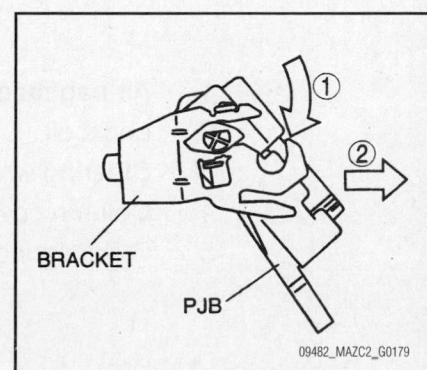

Fig. 307 Remove the PJB as shown—2009 vehicles

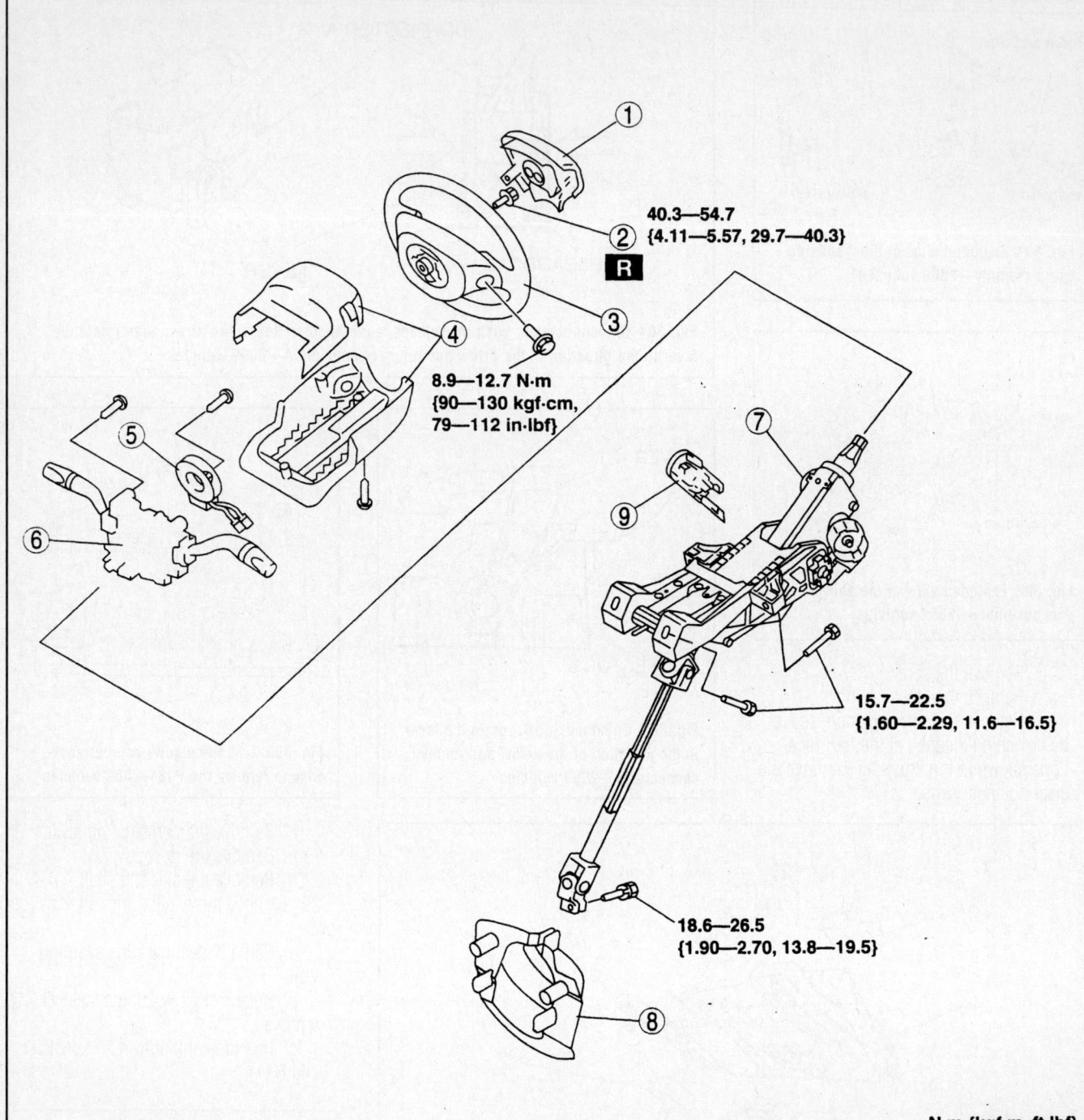

40.3—54.7
{4.11—5.57, 29.7—40.3}

R

8.9—12.7 N·m
{90—130 kgf·cm,
79—112 in·lbf}

15.7—22.5
{1.60—2.29, 11.6—16.5}

18.6—26.5
{1.90—2.70, 13.8—19.5}

N·m {kgf·m, ft·lbf}

1	Air bag module	6	Combination switch
2	Lockbolt	7	Steering shaft
3	Steering wheel	8	Dust cover
4	Column cover	9	Key cylinder
5	Clock spring		

09482_MAZC2_G0182

Fig. 308 Exploded view of the steering shaft—2009 vehicles

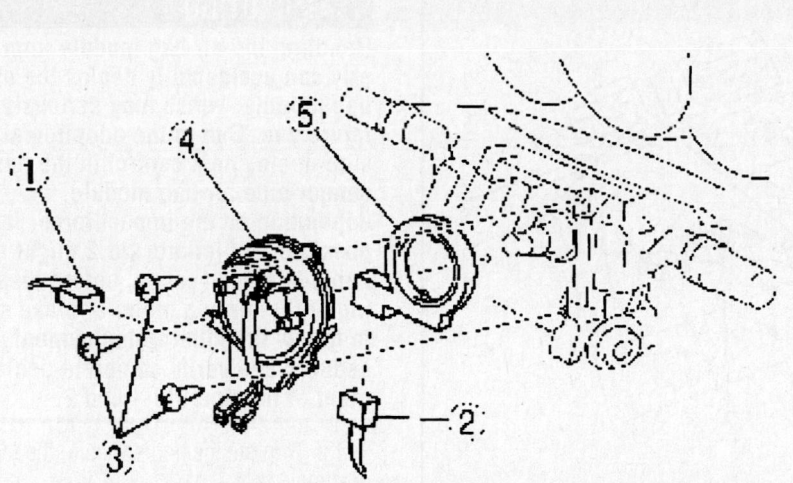

1   Clock spring connector

2   Steering angle sensor connector (With steering angle sensor)

3   Screw

4   Clock spring

5   Steering angle sensor (With steering angle sensor)

09482_MAZC2_G0192

Fig. 309 Exploded view of clock spring assembly—2009 vehicles

d. Pull the A-pillar trim upward and remove clip B from the A-pillar trim (2).

e. Pull clip B out and rotate it 45 degrees.

f. Remove clip B from the grommet by pulling it upward.

30. Remove the center panel module as follows:

a. Remove the screw and connector.

b. Remove the antenna feeder and the center panel module. Pull the center panel module outward, detach clip A from the dashboard, and then remove the center panel module.

31. Remove the wiper arm and blade.

32. Remove the cowl grille.

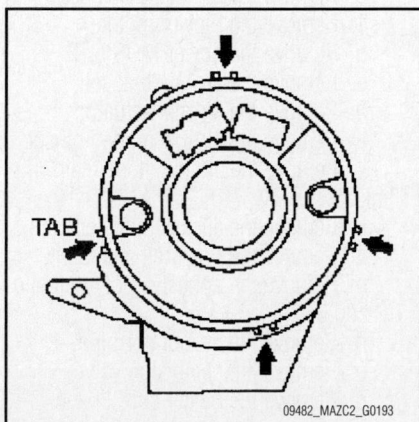

09482_MAZC2_G0193

Fig. 310 Remove the steering angle sensor by detaching the tabs at the four locations shown—2009 vehicles

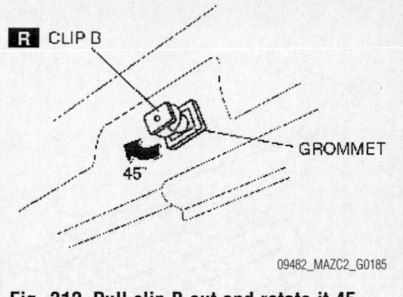

09482_MAZC2_G0185

Fig. 312 Pull clip B out and rotate it 45 degrees to remove the A-pillar trim—2009 vehicles

33. Remove the cowl panel.

34. Remove the wiper motor bolts.

35. Move the wiper motor in the direction of the arrow, remove the securing rubber from the stud pin (for connecting the motor securing rubber), and then remove the windshield wiper motor.

36. Disconnect the windshield wiper motor connector.

37. Remove the A/C unit installation nut from the engine compartment, then remove the A/C unit.

38. Remove the rear heat duct.

39. Disconnect the drain hose connected to the A/C unit.

40. Remove the nuts and bolts for installing the dashboard to the body.

41. Remove the climate control unit.

42. Remove the LCD unit.

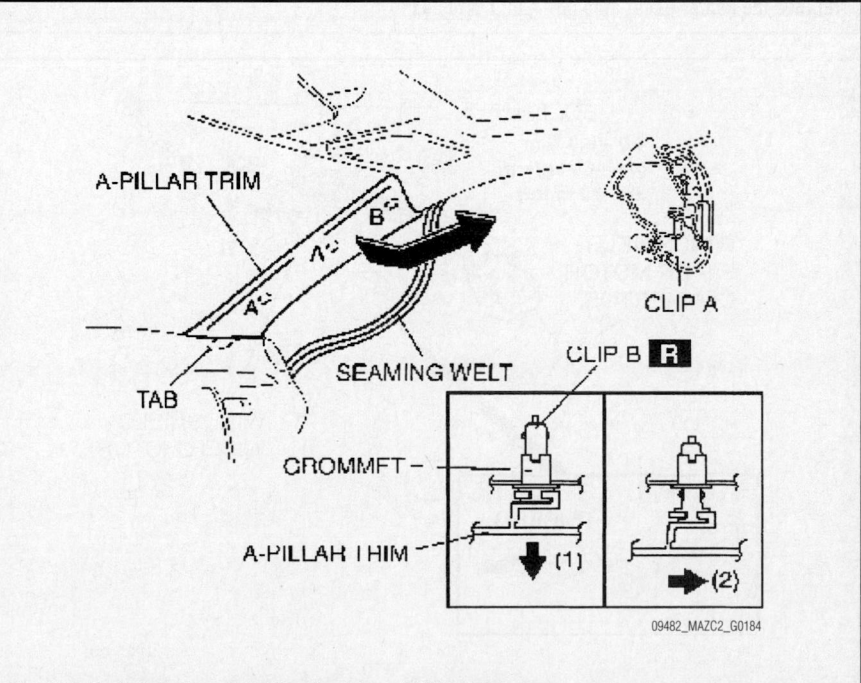

09482_MAZC2_G0184

Fig. 311 Exploded view of the A-pillar trim—2009 vehicles

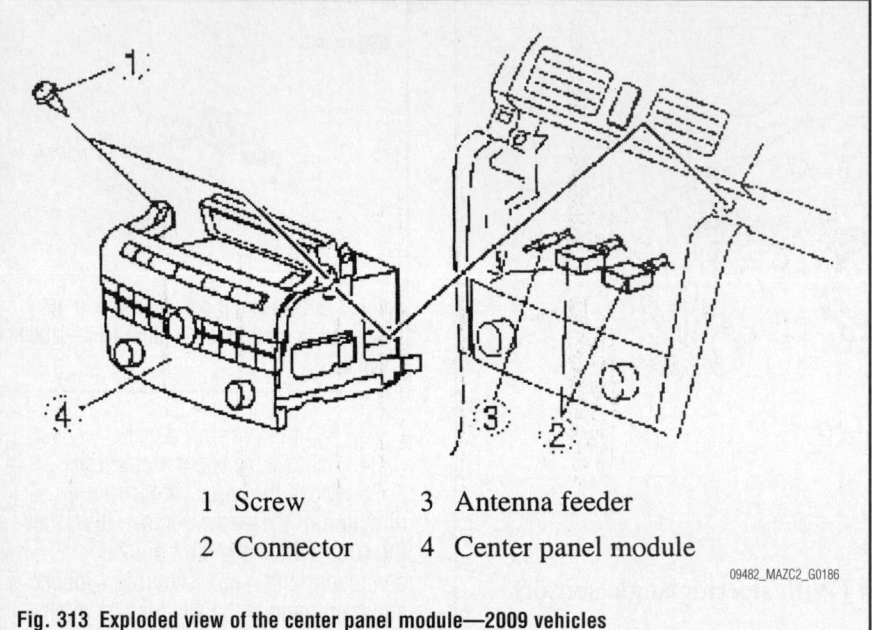

1 Screw
2 Connector
3 Antenna feeder
4 Center panel module

09482_MAZC2_G0186

Fig. 313 Exploded view of the center panel module—2009 vehicles

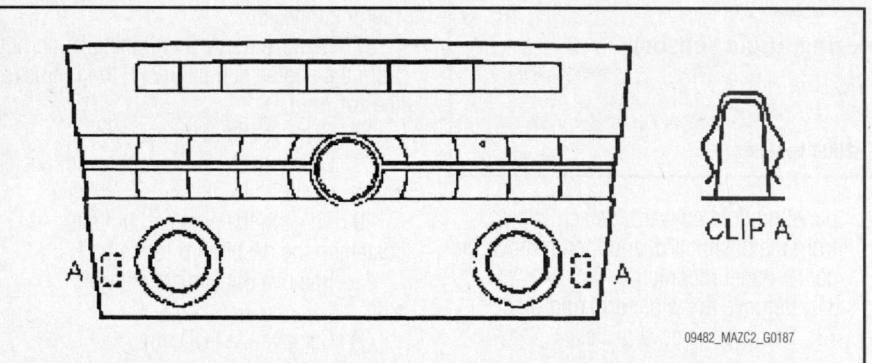

09482_MAZC2_G0187

Fig. 314 Pull the center panel module outward, detach clip A from the dashboard, and then remove the center panel module—2009 vehicles

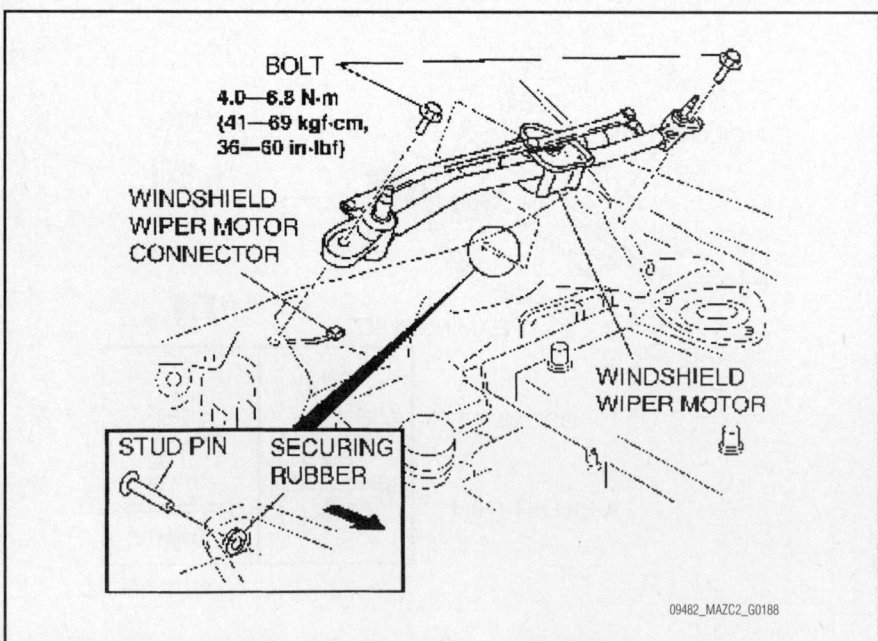

09482_MAZC2_G0188

Fig. 315 Exploded view of the wiper motor mounting—2009 vehicles

✳✳ CAUTION

Handling the air bag module improperly can accidentally deploy the air bag module, which may seriously injure you. Due to the adoption of 2-step deployment control in the passenger-side air bag module, depending on the impact force, it is possible that inflator No.2 might not deploy. In such cases, before disposing of the air bag module, make sure to follow the inflator deployment procedures and verify complete deployment of inflators No.1 and 2.

43. Turn the ignition switch to the LOCK position.

44. Remove the dashboard garnish.

45. Remove the bolt, the passenger side air bag module, the nut and bracket.

46. Remove the heater case and duct.

47. Remove the power MOS FET connector.

48. Remove the evaporator temperature sensor connector.

49. Remove the air intake actuator connector.

50. Remove the air mix actuator connector.

51. Remove the airflow mode actuator connector.

52. Remove the resistor connector on models with manual air conditioner system.

53. Remove the blower motor connector.

54. Remove the upper and lower dashboard brackets.

55. Remove the cap and bolt.

56. Remove the dashboard.

57. Remove the heater core as follows:

  a. Remove the adhesive polyurethane (1).

  b. Remove the blower case 1 and 2.

  c. Remove the air intake actuator.

  d. Remove the air intake link set.

  e. Remove the blower motor.

  f. Remove the power MOS FET.

  g. Remove the air mix link set.

  h. Remove the air mix actuator.

  i. Remove the airflow mode link set.

  j. Remove the airflow mode main link.

  k. Remove the airflow mode actuator.

  l. Remove the polyurethane foam.

  m. Remove the adhesive polyurethane (2 and 3).

  n. Remove the evaporator pipe.

  o. Remove the expansion valve.

  p. Remove the heater core.

To install:

➡Be sure to use new fasteners, as required.

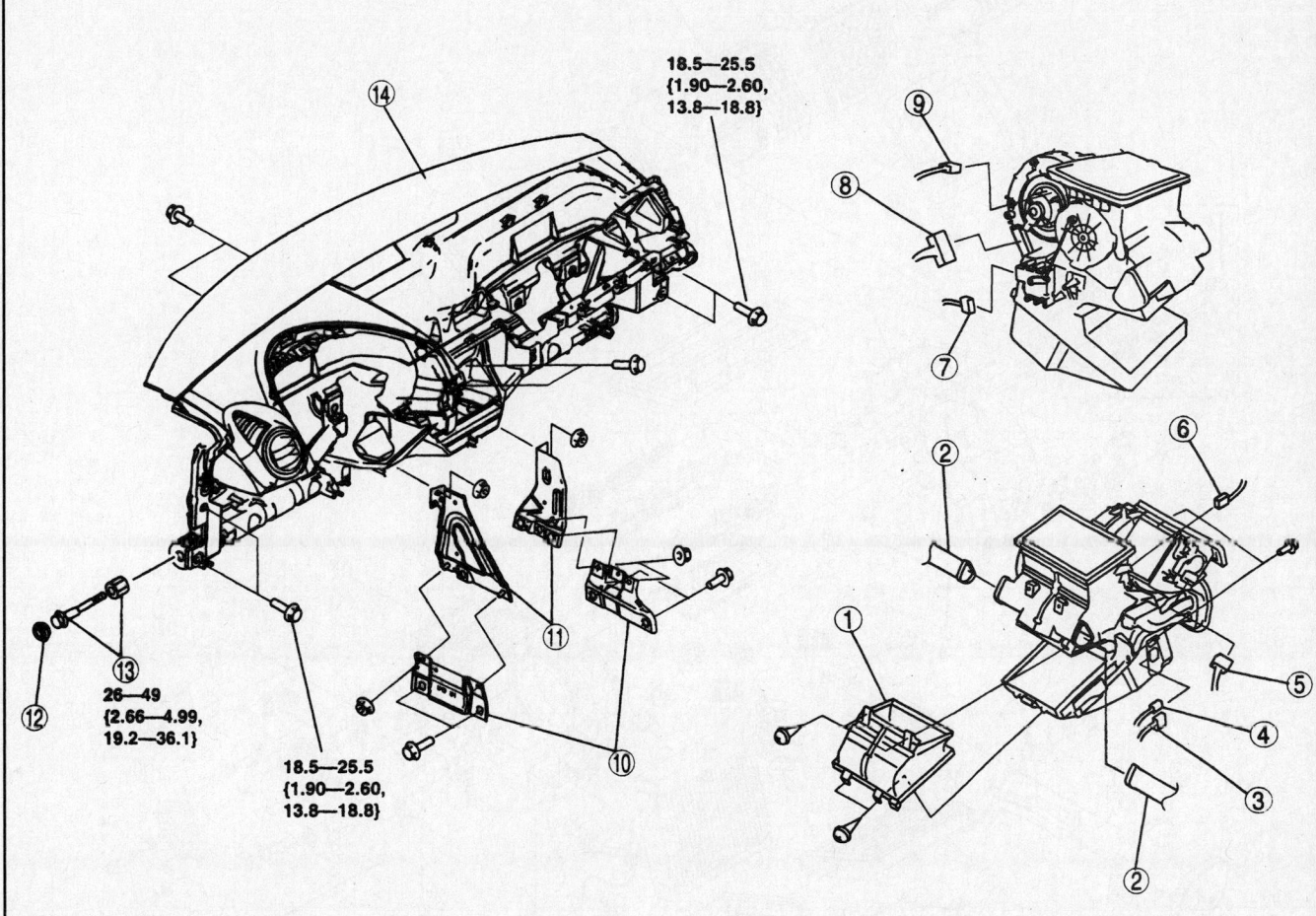

N·m {kgf·m, ft·lbf}

1	Heater case
2	Duct
3	Power MOS FET connector (with full-auto air conditioner system)
4	Evaporator temperature sensor connector
5	Air intake actuator connector
6	Air mix actuator connector (with full-auto air conditioner system)
7	Airflow mode actuator connector (with full-auto air conditioner system)
8	Resistor connector (with manual air conditioner system)
9	Blower motor connector
10	Dashboard bracket (lower)
11	Dashboard bracket (upper)
12	Cap
13	Bolt
14	Dashboard

09482_MAZC2_G0189

Fig. 316 Exploded view of the dashboard mounting—2009 vehicles

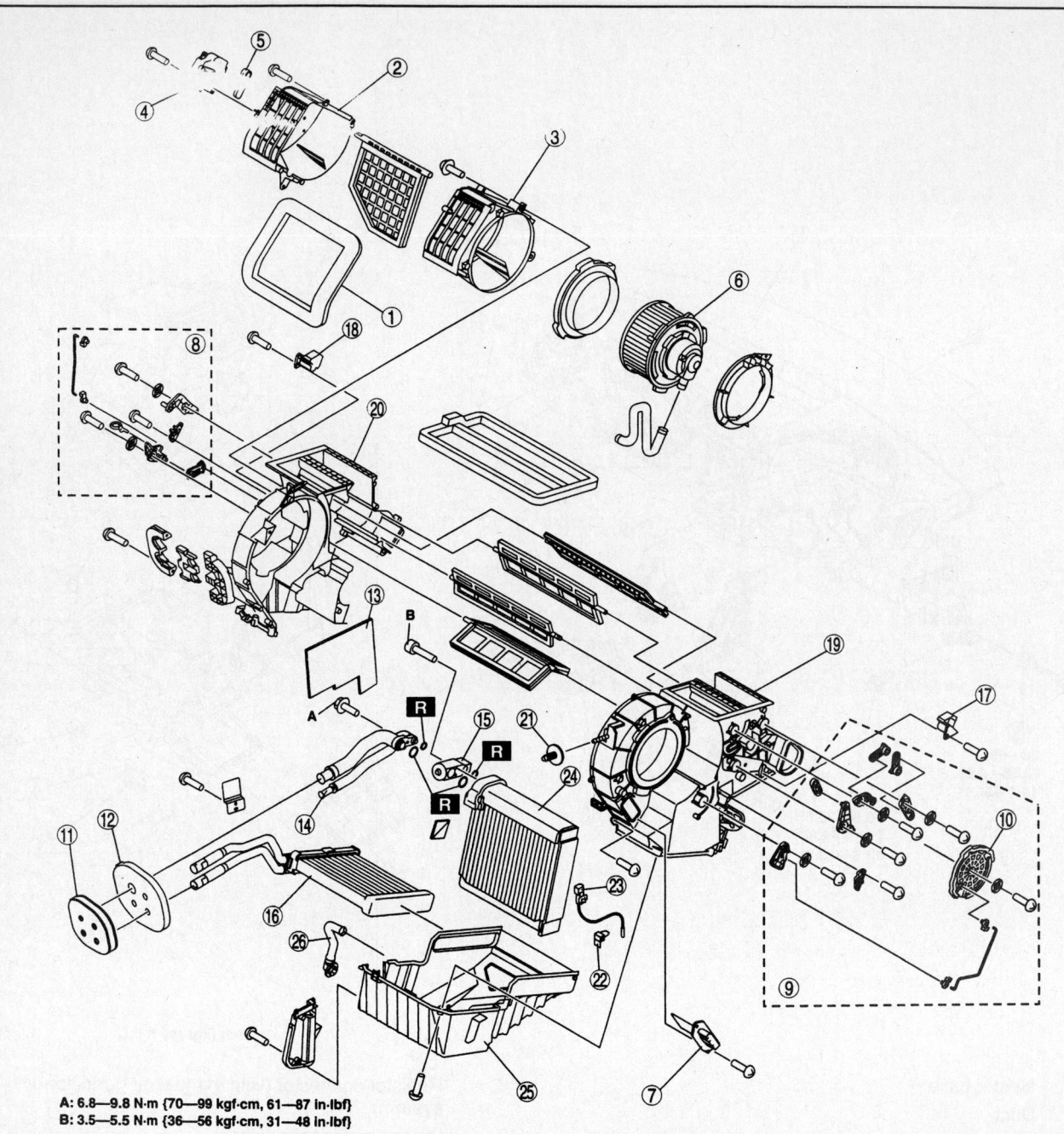

A: 6.8—9.8 N·m {70—99 kgf·cm, 61—87 in·lbf}
B: 3.5—5.5 N·m {36—56 kgf·cm, 31—48 in·lbf}

1	Adhesive polyurethane (1)	14	Evaporator pipe
2	Blower case (1)	15	Expansion valve
3	Blower case (2)	16	Heater core
4	Air intake actuator	17	Bracket (1)
5	Air intake link set	18	Bracket (2)
6	Blower motor	19	A/C case (1)
7	Resistor	20	A/C case (2)
8	Air mix link set	21	Bolt
9	Airflow mode link set	22	Sensor clamp
10	Airflow mode main link	23	Evaporator temperature sensor
11	Polyurethane foam	24	Evaporator
12	Adhesive polyurethane (2)	25	A/C case (3)
13	Adhesive polyurethane (3)	26	Drain hose

09482_MAZC2_G0190

Fig. 317 Exploded view of Type (A) A/C unit assembly—2009 vehicles

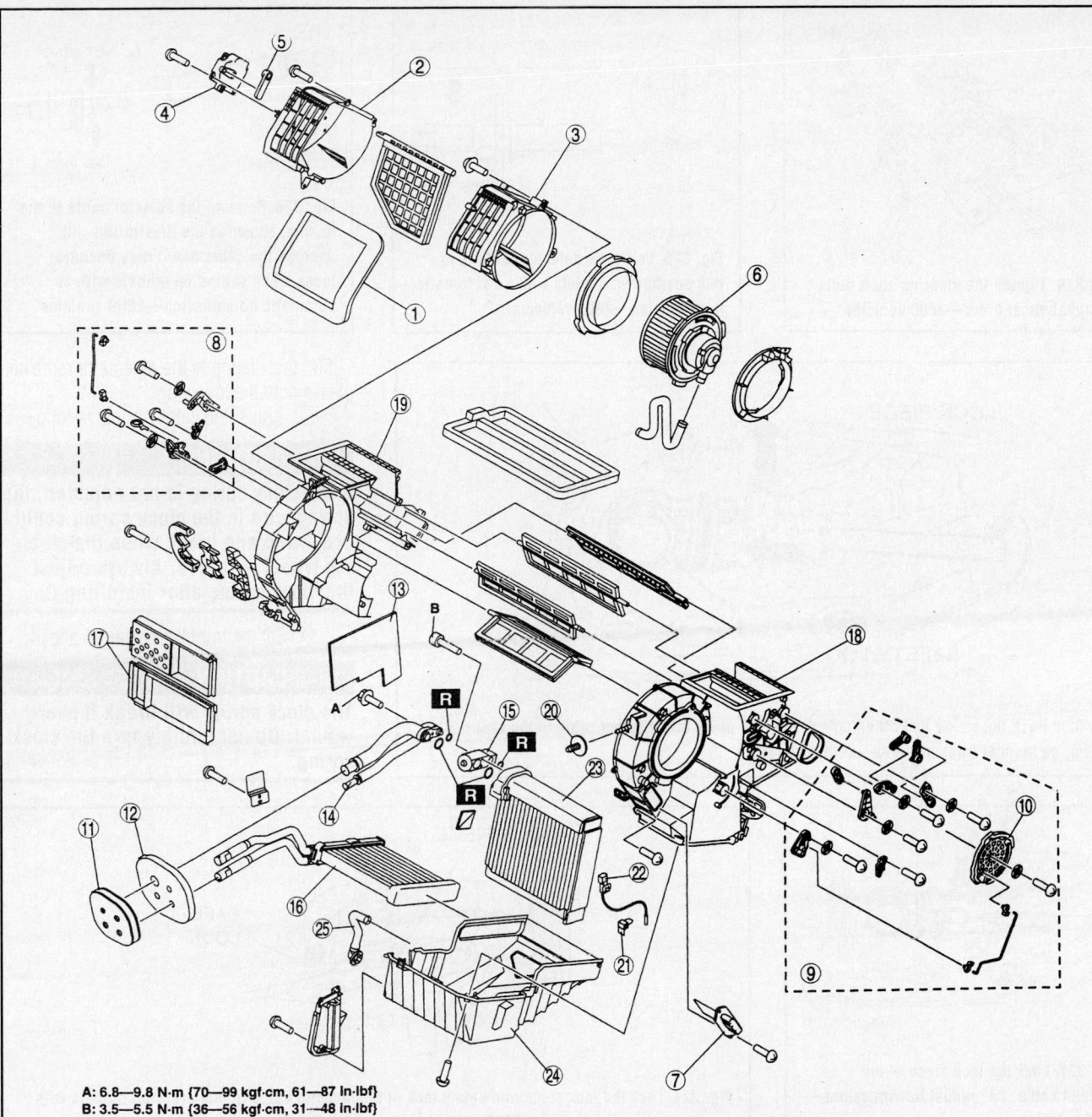

A: 6.8—9.8 N·m {70—99 kgf·cm, 61—87 in·lbf}
B: 3.5—5.5 N·m {36—56 kgf·cm, 31—48 in·lbf}

1	Adhesive polyurethane (1)	14	Evaporator pipe
2	Blower case (1)	15	Expansion valve
3	Blower case (2)	16	Heater core
4	Air intake actuator	17	Air filter cover
5	Air intake link set	18	A/C case (1)
6	Blower motor	19	A/C case (2)
7	Resistor	20	Bolt
8	Air mix link set	21	Sensor clamp
9	Airflow mode link set	22	Evaporator temperature sensor
10	Airflow mode main link	23	Evaporator
11	Polyurethane foam	24	A/C case (3)
12	Adhesive polyurethane (2)	25	Drain hose
13	Adhesive polyurethane (3)		

09482_MAZC2_G0191

Fig. 318 Exploded view of Type (B) A/C unit assembly—2009 vehicles

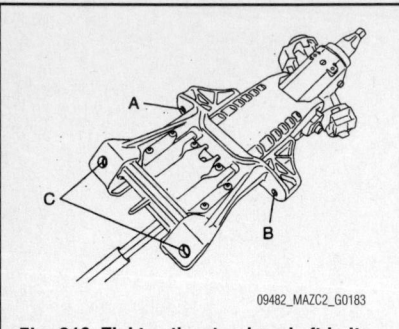

Fig. 319 Tighten the steering shaft bolts in alphabetical order—2009 vehicles

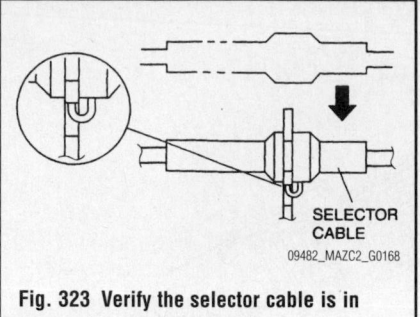

SELECTOR CABLE
09482_MAZC2_G0168

Fig. 323 Verify the selector cable is in this position on models with an automatic transmission—2009 vehicles

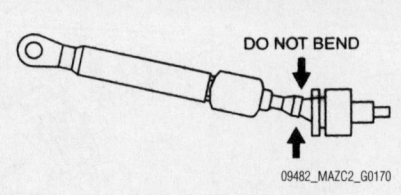

DO NOT BEND
09482_MAZC2_G0170

Fig. 325 Bending the selector cable in the manner shown in the illustration will damage the cable and it may become loose when shifted on models with an automatic transmission—2009 vehicles

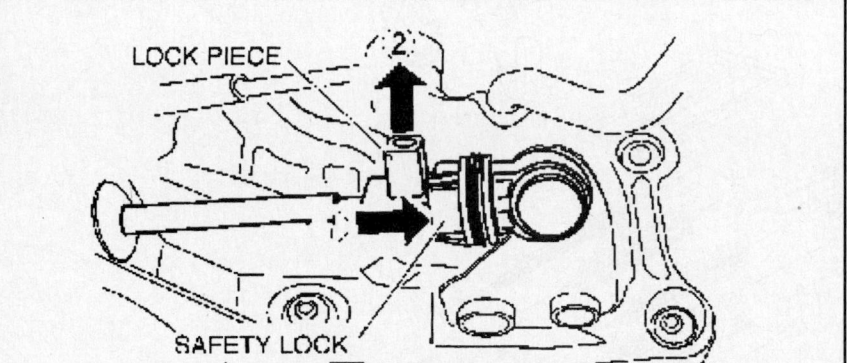

Fig. 320 Push the safety lock, then unlock the lock piece of the select cable in the order shown, on manual transmissions—2009 vehicles

58. Installation is the reverse of removal, please note the following:

  a. Adjust the clock spring as follows:

❊❊ CAUTION

If the clock spring is not adjusted, the spring wire in the clock spring could over-wind and break when the steering wheel is turned. Always adjust the clock spring after installing it.

• Set the front tires straight-ahead

❊❊ CAUTION

The clock spring will break if over-wound. Do not forcibly turn the clock spring.

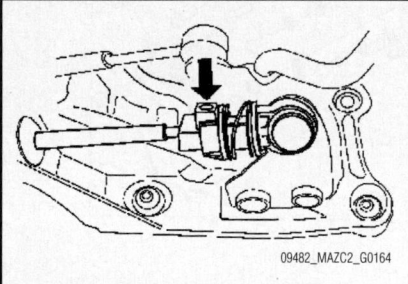

Fig. 321 Lock the lock piece of the selector cable, on manual transmissions—2009 vehicles

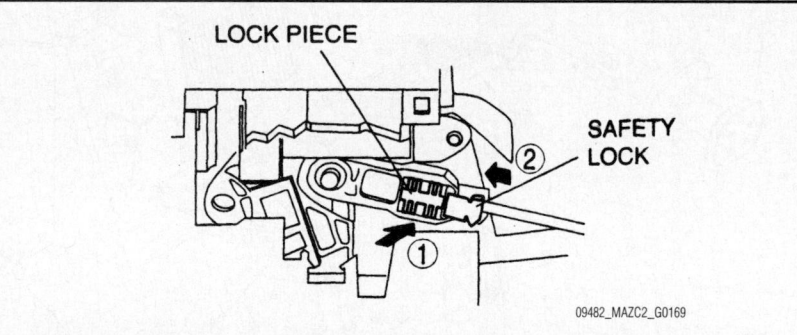

Fig. 324 Lock the lock piece and safety lock of the selector cable in this order on models with an automatic transmission—2009 vehicles

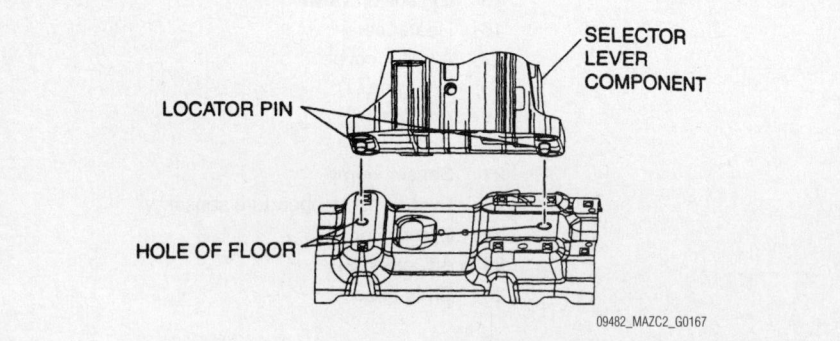

Fig. 322 Insert the location pin of selector lever component to the hole of the floor on models with an automatic transmission—2009 vehicles

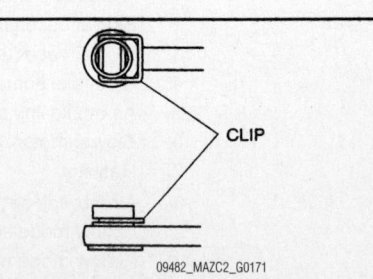

CLIP
09482_MAZC2_G0171

Fig. 326 Install the selector cable to the manual shaft cable with the clip side of the selector cable end facing the front of the vehicle on models with an automatic transmission—2009 vehicles

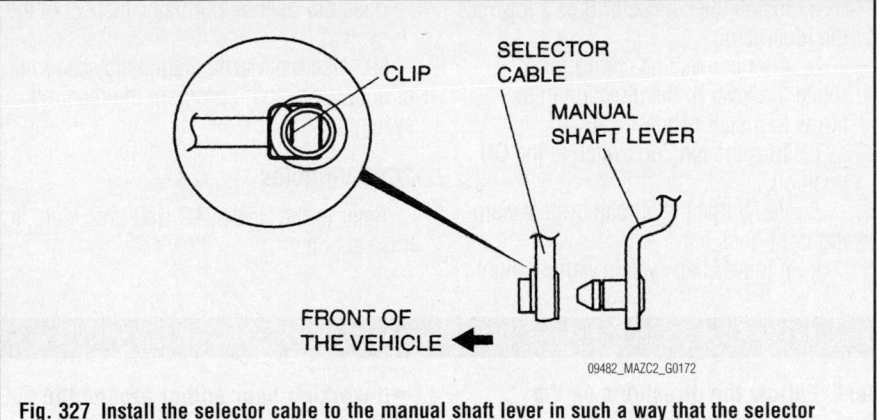

Fig. 327 Install the selector cable to the manual shaft lever in such a way that the selector cable does not bear a load on models with an automatic transmission—2009 vehicles

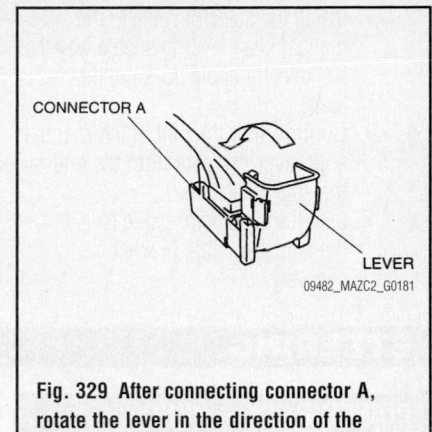

Fig. 329 After connecting connector A, rotate the lever in the direction of the arrow—2009 vehicles

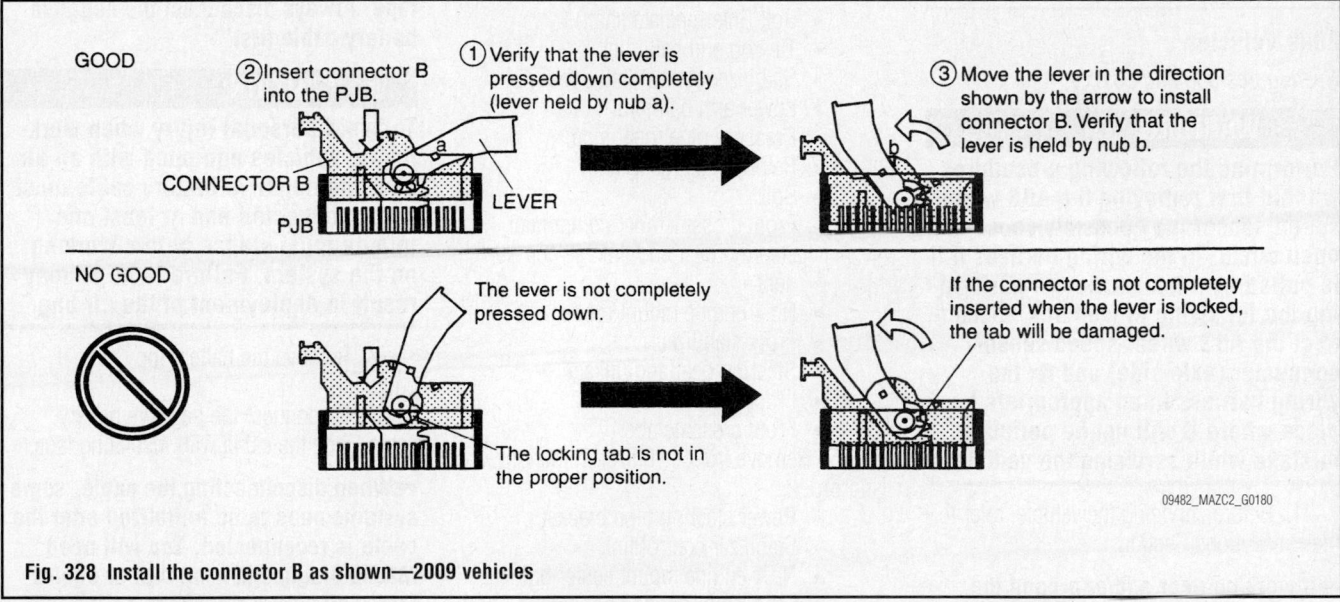

Fig. 328 Install the connector B as shown—2009 vehicles

- Turn the clock spring clockwise until it stops
- From the stopped position, turn the clock spring counterclockwise 2 ¾ turns
- Align the marks
- Verify that the tilt / telescope lever is in the LOCK position

b. Tighten the steering shaft bolts in alphabetical order.

c. Set the wheels in the straight-ahead position and install the steering wheel

✴✴ CAUTION

When installing the center panel module, make sure that the wiring harness and antenna feeder are not caught between the unit and dashboard. If the wiring harness or the antenna feeder is caught between the unit and dashboard, it may cause malfunctions.

d. Select Cable for models with an manual transmission should be installed as follows:
- Make sure that the shift lever (transaxle side) is in neutral
- Push the safety lock, then unlock the lock piece of the select cable in the order shown in the illustration.
- Shift the shift lever to neutral.
- Lock the lock piece of the selector cable in the illustration.
- Shift the shift lever from neutral to other position, and make sure that there are no other components in that area to interfere with the lever.

e. Selector lever for models with an automatic transmission should be installed as follows:
- Insert the location pin of selector lever component to the hole of the floor
- Tighten the selector lever compo-

nent installation bolts to 7.8–10.8 Nm (69.5–95.4 inch lbs.).

f. Selector cable for models with an automatic transmission should be installed as follows:
- Install the selector cable to the selector lever securely.
- Install the selector cable to the bracket securely.

✴✴ CAUTION

Bending the selector cable in the manner shown in the illustration will damage the cable and it may become loose when shifted. When installing the selector cable, hold it straight.

- Install the clip as shown in the illustration.

➡Install the selector cable to the manual shaft cable with the clip side of the selector cable end facing the front of the vehicle.

- Install the selector cable to the manual shaft lever in such a way that the selector cable does not bear a load.
- Confirm that the end of the manual shift lever projects from the end of the selector cable.
- Install the selector cable to the selector cable bracket securely.

g. Install the connector B as shown in the illustration.

h. After connecting connector A, rotate the lever in the direction of the arrow to install connector A.

i. Turn the ignition switch to the ON position.

j. Verify that the air bag system warning light goes out.

k. If the air bag system warning light does not operate normally, inspect of the system.

59. Use the Mazda diagnostic scan tool, or equivalent and reprogram the required systems.

2010 Vehicles

Refer to the Heater/AC Unit procedure in this section.

STEERING

POWER RACK & PINION STEERING GEAR

REMOVAL & INSTALLATION

2009 Vehicles

See Figures 330 and 331.

⁂ WARNING

Performing the following procedures without first removing the ABS wheel-speed sensor may possibly cause an open circuit in the wiring harness if it is pulled by mistake. Before performing the following procedures, disconnect the ABS wheel-speed sensor connector (axle side) and fix the wiring harness to an appropriate place where it will not be pulled by mistake while servicing the vehicle.

1. Before servicing the vehicle, refer to the Precautions Section.

➡If working near and/or around the SRS system and components, be sure to disable the SRS system. Tape the negative battery cable with insulating tape. Always disconnect the negative battery cable first.

⁂ CAUTION

To avoid personal injury when working on vehicles equipped with an air bag, the negative battery cable must be disconnected and at least one minute must elapse before working on the system. Failure to do so may result in deployment of the air bag.

2. Remove the battery top cover, if equipped.
3. Disconnect the negative battery cable. Tape the cable with insulating tape.

➡When disconnecting the cable, some systems need to be initialized after the cable is reconnected. You will need Mazda diagnostic scan tool or equivalent. Follow the directions on the tool.

4. Remove front crossmember in the following order:
   - Bolt (intermediate shaft)
   - Tie-rod end ball joint
   - Stabilizer control link upper nut
   - Lower arm ball joint
   - Pressure pipe (gear side)
   - Return hose (gear side)
   - Bolt
   - Front crossmember component, steering gear and linkage component
   - No.1 engine mount rubber
   - Front stabilizer
   - Steering gear and linkage
   - Front lower arm
   - Front crossmember
5. Remove front stabilizer in the following order:
   - Power steering pipe bracket
   - Stabilizer control link
   - No.1 engine mount center bolt
   - Front stabilizer component
   - Stabilizer bracket
   - Stabilizer bushing
   - Stabilizer plate
6. Remove the insulator.
7. Remove the steering gear and linkage.

To install:

➡Be sure to use new fasteners, as required.

8. Installation is the reverse of the removal procedure.
9. Check and/or adjust the front end alignment.
10. Use the Mazda diagnostic scan tool, or equivalent and reprogram the required systems.

2010 Vehicles

See Figures 332 through 334.

1. Before servicing the vehicle, refer to the Precautions Section.

➡If working near and/or around the SRS system and components, be sure to disable the SRS system. Tape the negative battery cable with insulating tape. Always disconnect the negative battery cable first.

⁂ CAUTION

To avoid personal injury when working on vehicles equipped with an air bag, the negative battery cable must be disconnected and at least one minute must elapse before working on the system. Failure to do so may result in deployment of the air bag.

2. Remove the battery top cover, if equipped.
3. Disconnect the negative battery cable. Tape the cable with insulating tape.

➡When disconnecting the cable, some systems need to be initialized after the cable is reconnected. You will need Mazda diagnostic scan tool or equivalent. Follow the directions on the tool.

➡Performing the following procedure without removing the ABS wheel sensor may cause an open circuit in the harness if it is pulled by mistake. Before performing the procedure remove the ABS wheel speed sensor (axle side) and position it to an appropriate place where the sensor will not be pulled by mistake while servicing the vehicle.

➡Secure the steering wheel, using tape or cable, to prevent the steering shaft from rotating after disconnecting the steering shaft. If the steering wheel rotates after the steering shaft and steering gear linkage are disconnected the internal parts of the clockspring could be damaged.

4. Drain the power steering fluid.Be sure to properly dispose of used fluid.
5. Remove the joint cover.
6. Disconnect the steering shaft from

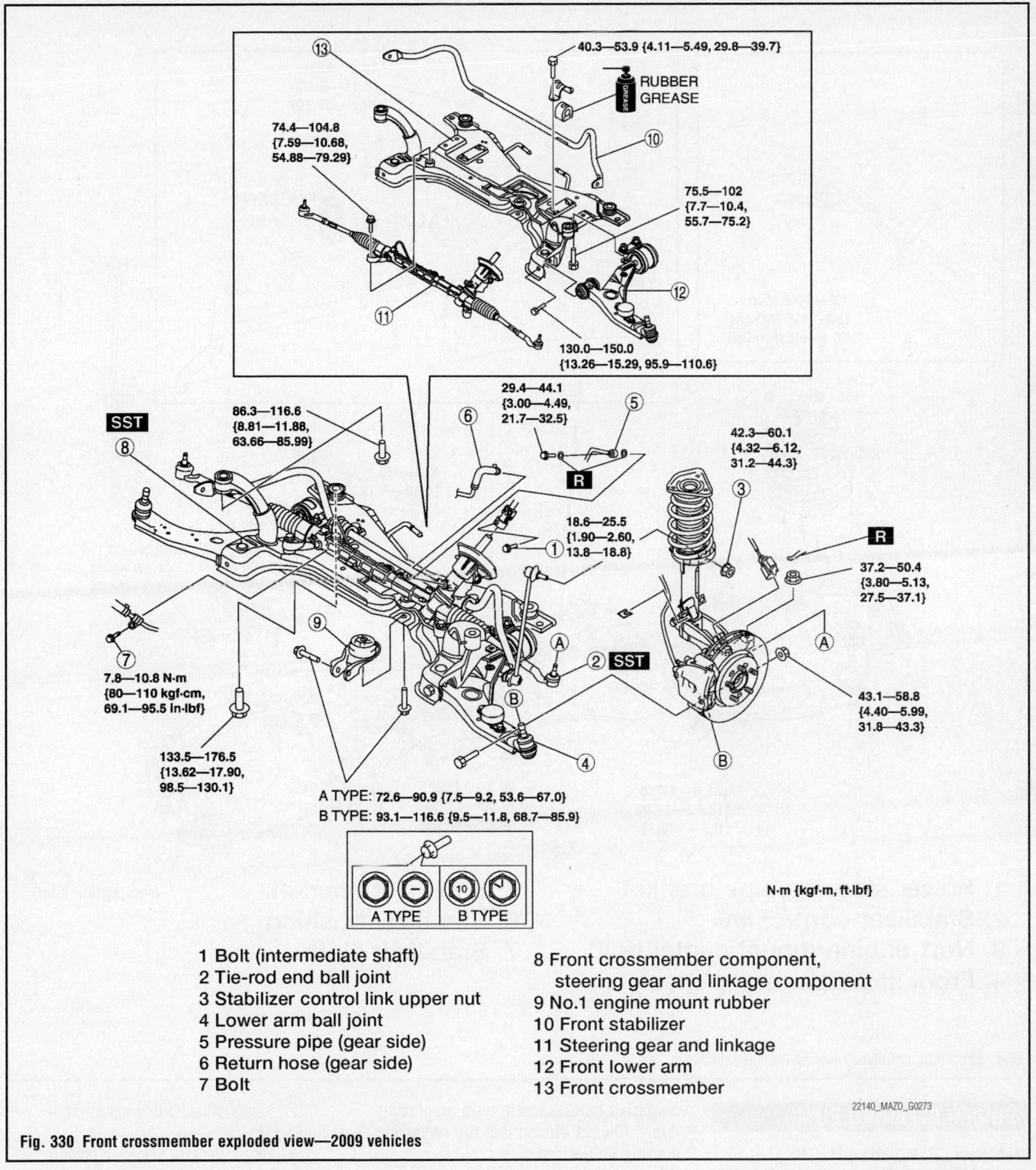

40.3—53.9 {4.11—5.49, 29.8—39.7}

RUBBER GREASE

74.4—104.8
{7.59—10.68,
54.88—79.29}

75.5—102
{7.7—10.4,
55.7—75.2}

130.0—150.0
{13.26—15.29, 95.9—110.6}

86.3—116.6
{8.81—11.88,
63.66—85.99}

SST

29.4—44.1
{3.00—4.49,
21.7—32.5}

42.3—60.1
{4.32—6.12,
31.2—44.3}

R

18.6—25.5
{1.90—2.60,
13.8—18.8}

R

37.2—50.4
{3.80—5.13,
27.5—37.1}

7.8—10.8 N·m
{80—110 kgf-cm,
69.1—95.5 in-lbf}

133.5—176.5
{13.62—17.90,
98.5—130.1}

SST

43.1—58.8
{4.40—5.99,
31.8—43.3}

A TYPE: 72.6—90.9 {7.5—9.2, 53.6—67.0}
B TYPE: 93.1—116.6 {9.5—11.8, 68.7—85.9}

A TYPE   B TYPE

N·m {kgf-m, ft-lbf}

1 Bolt (intermediate shaft)
2 Tie-rod end ball joint
3 Stabilizer control link upper nut
4 Lower arm ball joint
5 Pressure pipe (gear side)
6 Return hose (gear side)
7 Bolt
8 Front crossmember component,
　 steering gear and linkage component
9 No.1 engine mount rubber
10 Front stabilizer
11 Steering gear and linkage
12 Front lower arm
13 Front crossmember

22140_MAZD_G0273

Fig. 330 Front crossmember exploded view—2009 vehicles

the steering gear linkage. Matchmark before disconnecting.

7. Remove the auto leveling sensor, if equipped.

8. Disconnect and plug the fluid lines.

9. Remove the front crossmember.

10. Remove the insulator.

11. Remove the component retaining bolts.

12. Remove the component from its mounting.

To install:

➡ **Be sure to use new fasteners, as required.**

13. Installation is the reverse of the removal procedure.

14. Check and adjust alignment, as required.

15. Be sure to fill the system with the proper grade and type power steering fluid.

16. Bleed the system.

17. Use the Mazda diagnostic scan tool, or equivalent and reprogram the required systems.

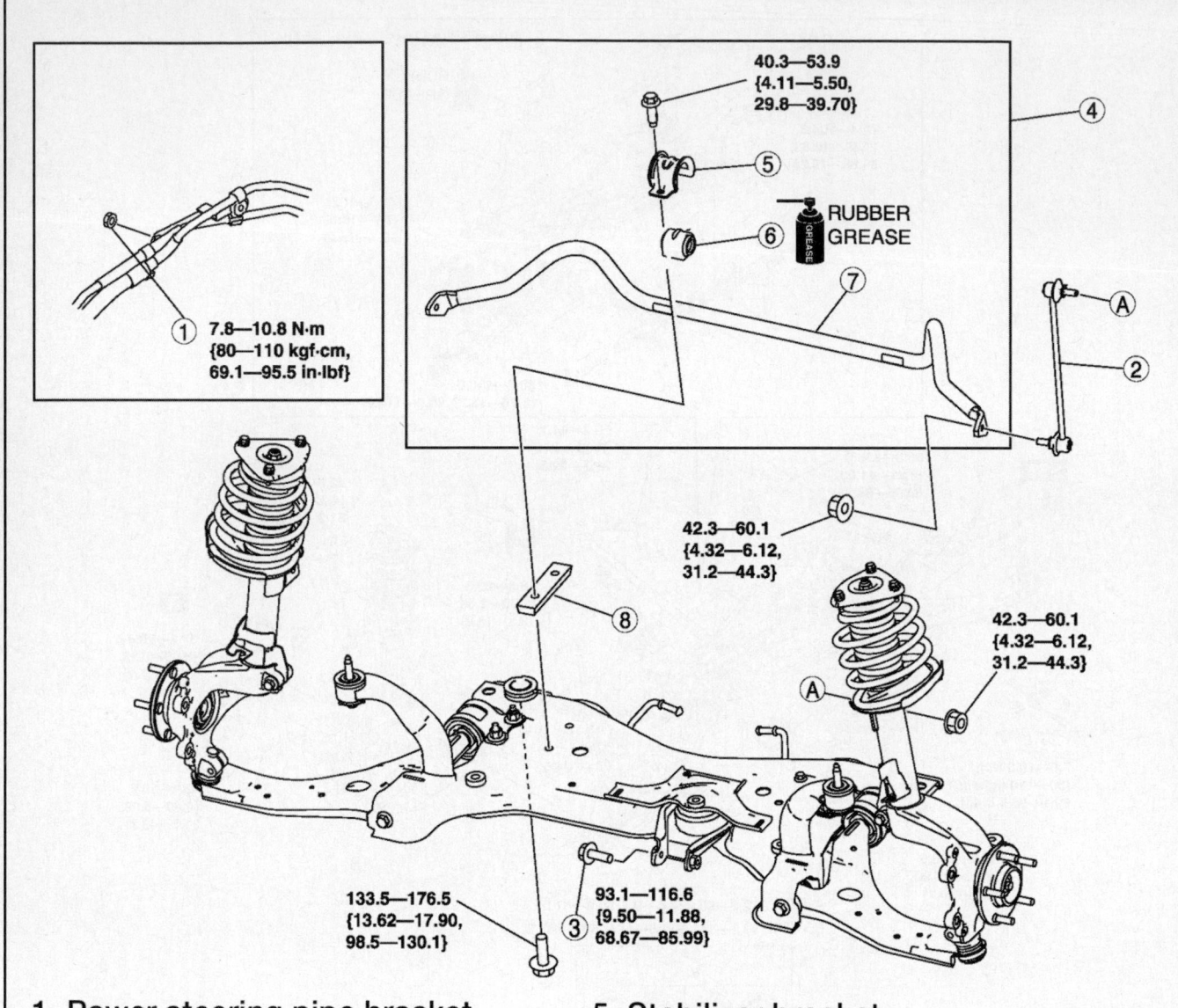

40.3—53.9
{4.11—5.50,
29.8—39.70}

RUBBER GREASE

7.8—10.8 N·m
{80—110 kgf·cm,
69.1—95.5 in·lbf}

42.3—60.1
{4.32—6.12,
31.2—44.3}

42.3—60.1
{4.32—6.12,
31.2—44.3}

133.5—176.5
{13.62—17.90,
98.5—130.1}

93.1—116.6
{9.50—11.88,
68.67—85.99}

N·m {kgf·m, ft·lbf}

1. Power steering pipe bracket
2. Stabilizer control link
3. No.1 engine mount center bolt
4. Front stabilizer component
5. Stabilizer bracket
6. Stabilizer bushing
7. Stabilizer plate

22140_MAZD_G0274

Fig. 331 Front stabilizer removal—2009 vehicles

POWER STEERING PUMP

REMOVAL & INSTALLATION

2009 Vehicles

See Figure 335.

1. Before servicing the vehicle, refer to the Precautions Section.

➡If working near and/or around the SRS system and components, be sure to disable the SRS system. Tape the negative battery cable with insulating tape. Always disconnect the negative battery cable first.

✲✲ CAUTION

To avoid personal injury when working on vehicles equipped with an air bag, the negative battery cable must be disconnected and at least one minute must elapse before working on the system. Failure to do so may result in deployment of the air bag.

2. Remove the battery top cover, if equipped.

3. Disconnect the negative battery cable. Tape the cable with insulating tape.

➡When disconnecting the cable, some systems need to be initialized after the cable is reconnected. You will need Mazda diagnostic scan tool or equivalent. Follow the directions on the tool.

4. Remove the undercover, splash shield and mudguard.

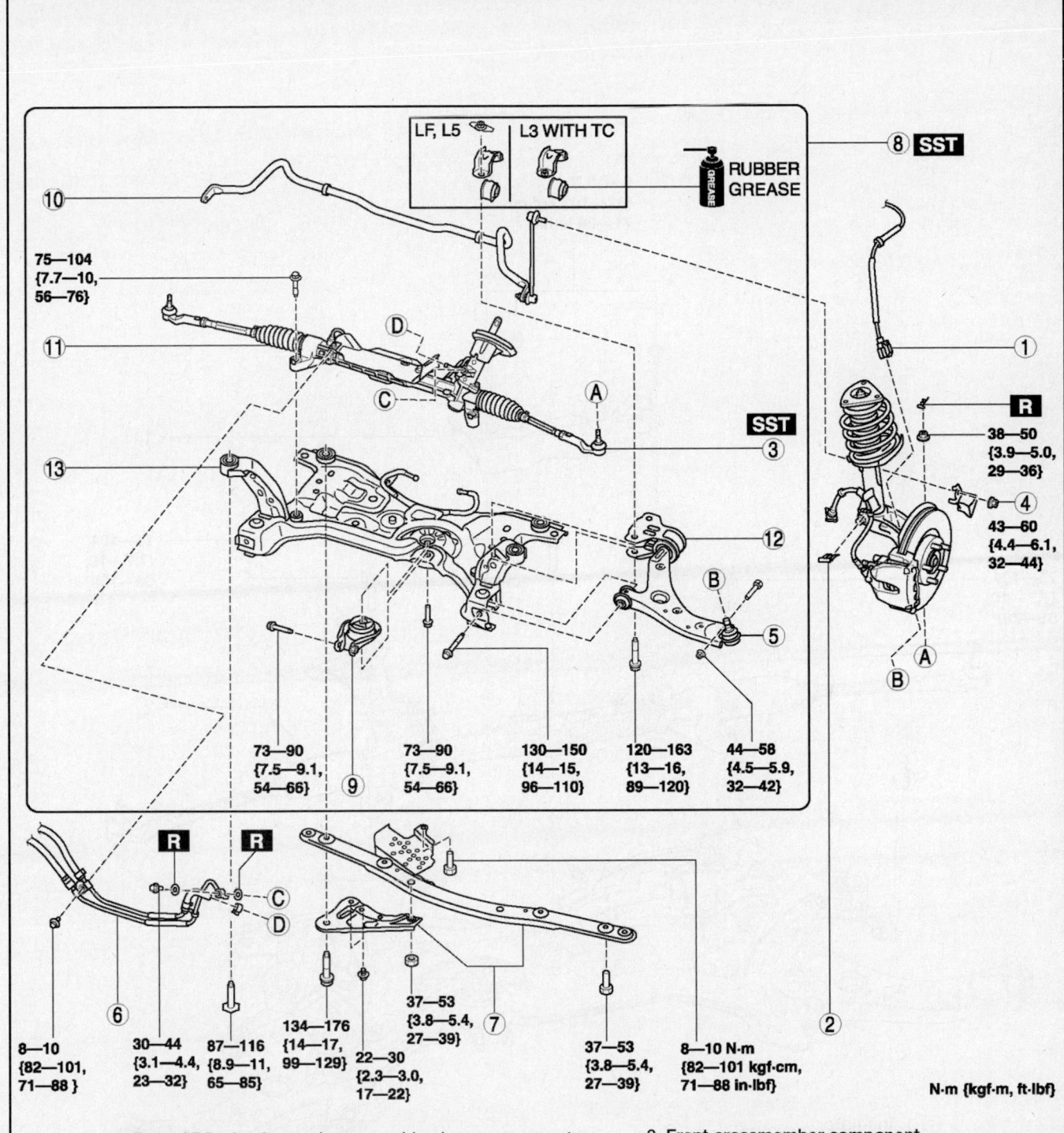

1. Front ABS wheel-speed sensor wiring harness connector
2. Brake hose clip
3. Tie-rod end ball joint
4. Stabilizer control link upper nut
5. Front lower arm ball joint
6. Power steering pipe component
7. Crossmember bracket

8. Front crossmember component
9. No.1 engine mount rubber
10. Front stabilizer
11. Steering gear and linkage
12. Front lower arm
13. Front crossmember

37671_MAZ3_G0165

Fig. 332 Front crossmember and related components—2010 vehicles

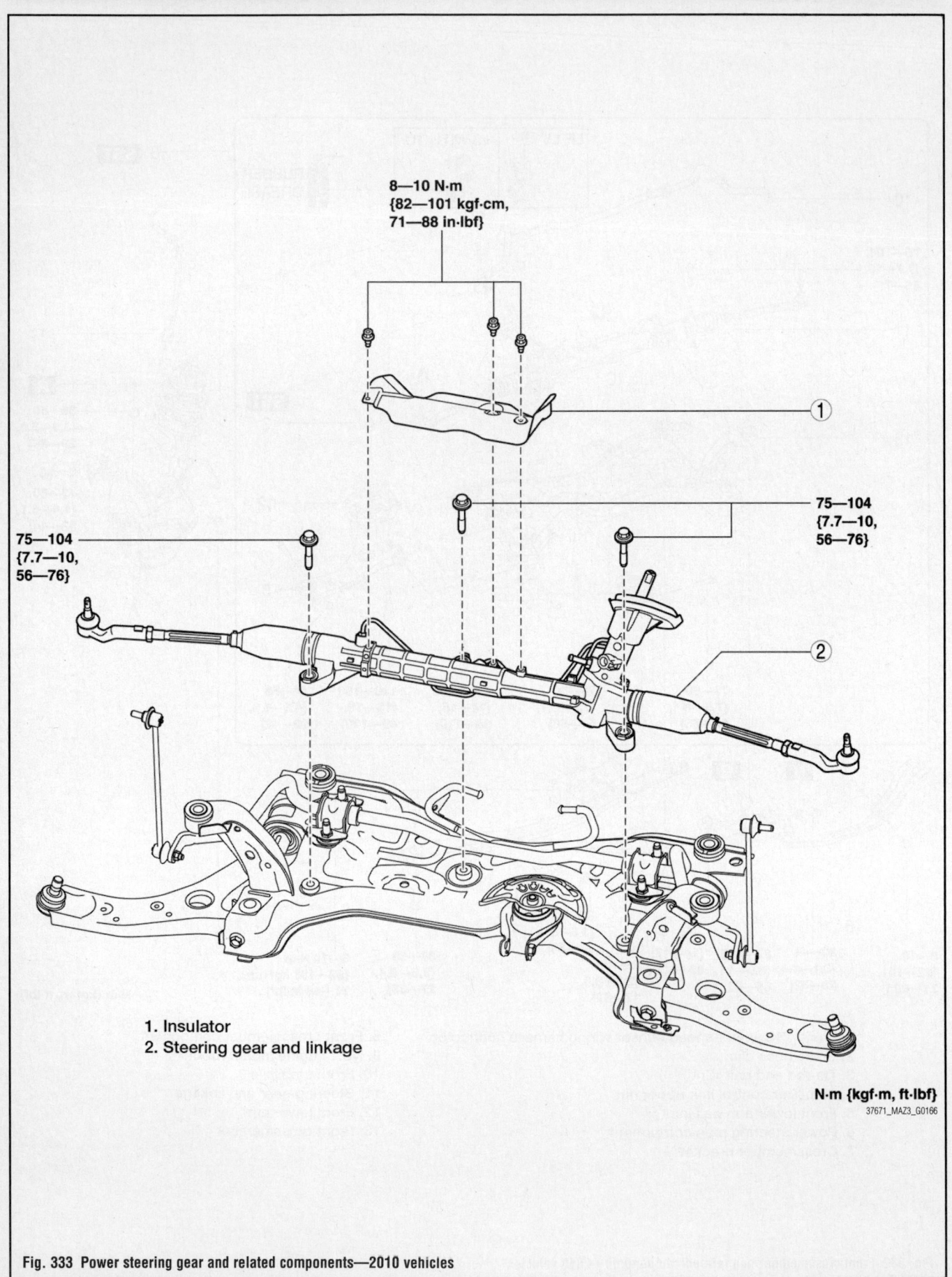

8—10 N·m
{82—101 kgf·cm,
71—88 in·lbf}

75—104
{7.7—10,
56—76}

75—104
{7.7—10,
56—76}

1. Insulator
2. Steering gear and linkage

N·m {kgf·m, ft·lbf}

37671_MAZ3_G0166

Fig. 333 Power steering gear and related components—2010 vehicles

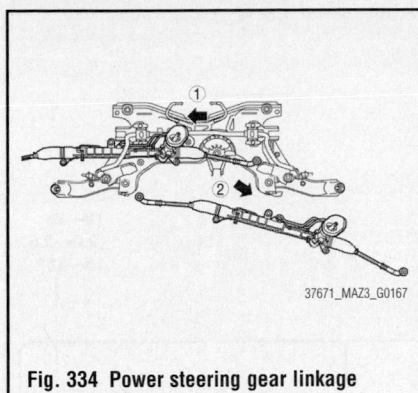

Fig. 334 Power steering gear linkage removal points—2010 vehicles

5. Remove the front bumper.
6. Remove the washer tank.
7. Unplug the electrical connector from the pump.
8. Disconnect the pressure pipe.
9. Remove the power steering fluid reserve tank.
10. Remove the electric power steering oil pump and bracket assembly.

11. Remove the bracket and disconnect to the return hose.
12. Remove the pump.

To install:

➡**Be sure to use new fasteners, as required.**

13. Installation is the reverse of the removal procedure.
14. Be sure to fill the system with the proper grade and type power steering fluid.
15. Bleed the system.
16. Use the Mazda diagnostic scan tool, or equivalent and reprogram the required systems.

2010 Vehicles

See Figure 336.

1. Before servicing the vehicle, refer to the Precautions Section.

➡**If working near and/or around the SRS system and components, be sure** to disable the SRS system. Tape the negative battery cable with insulating tape. Always disconnect the negative battery cable first.

> **✳✳ CAUTION**
>
> **To avoid personal injury when working on vehicles equipped with an air bag, the negative battery cable must be disconnected and at least one minute must elapse before working on the system. Failure to do so may result in deployment of the air bag.**

2. Remove the battery top cover, if equipped.
3. Disconnect the negative battery cable. Tape the cable with insulating tape.

➡**When disconnecting the cable, some systems need to be initialized after the cable is reconnected. You will need Mazda diagnostic scan tool or equivalent. Follow the directions on the tool.**

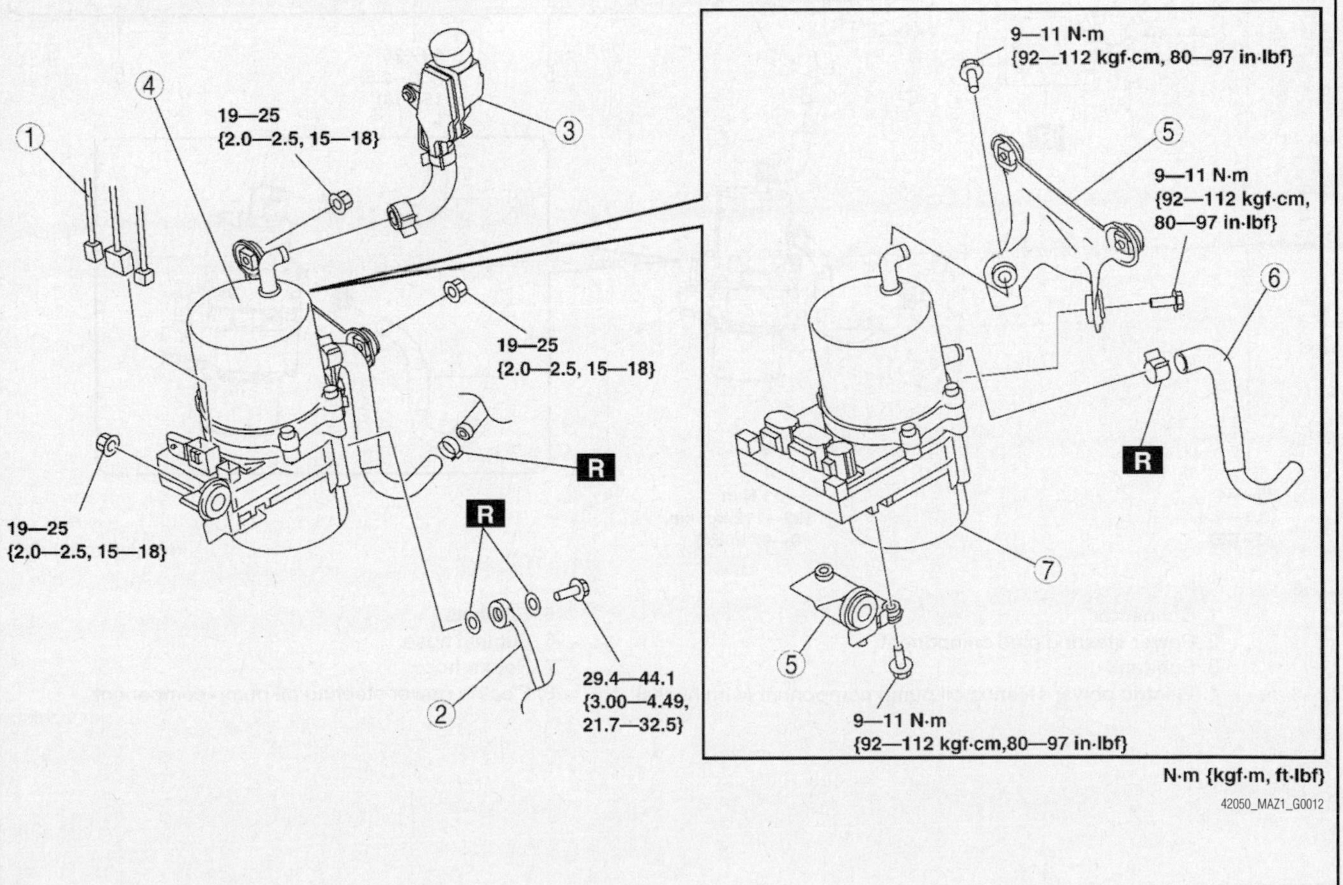

19—25
{2.0—2.5, 15—18}

19—25
{2.0—2.5, 15—18}

19—25
{2.0—2.5, 15—18}

29.4—44.1
{3.00—4.49, 21.7—32.5}

9—11 N·m
{92—112 kgf·cm, 80—97 in·lbf}

9—11 N·m
{92—112 kgf·cm, 80—97 in·lbf}

9—11 N·m
{92—112 kgf·cm, 80—97 in·lbf}

N·m {kgf·m, ft·lbf}

Fig. 335 Exploded view of the power steering pump assembly–2009 vehicles

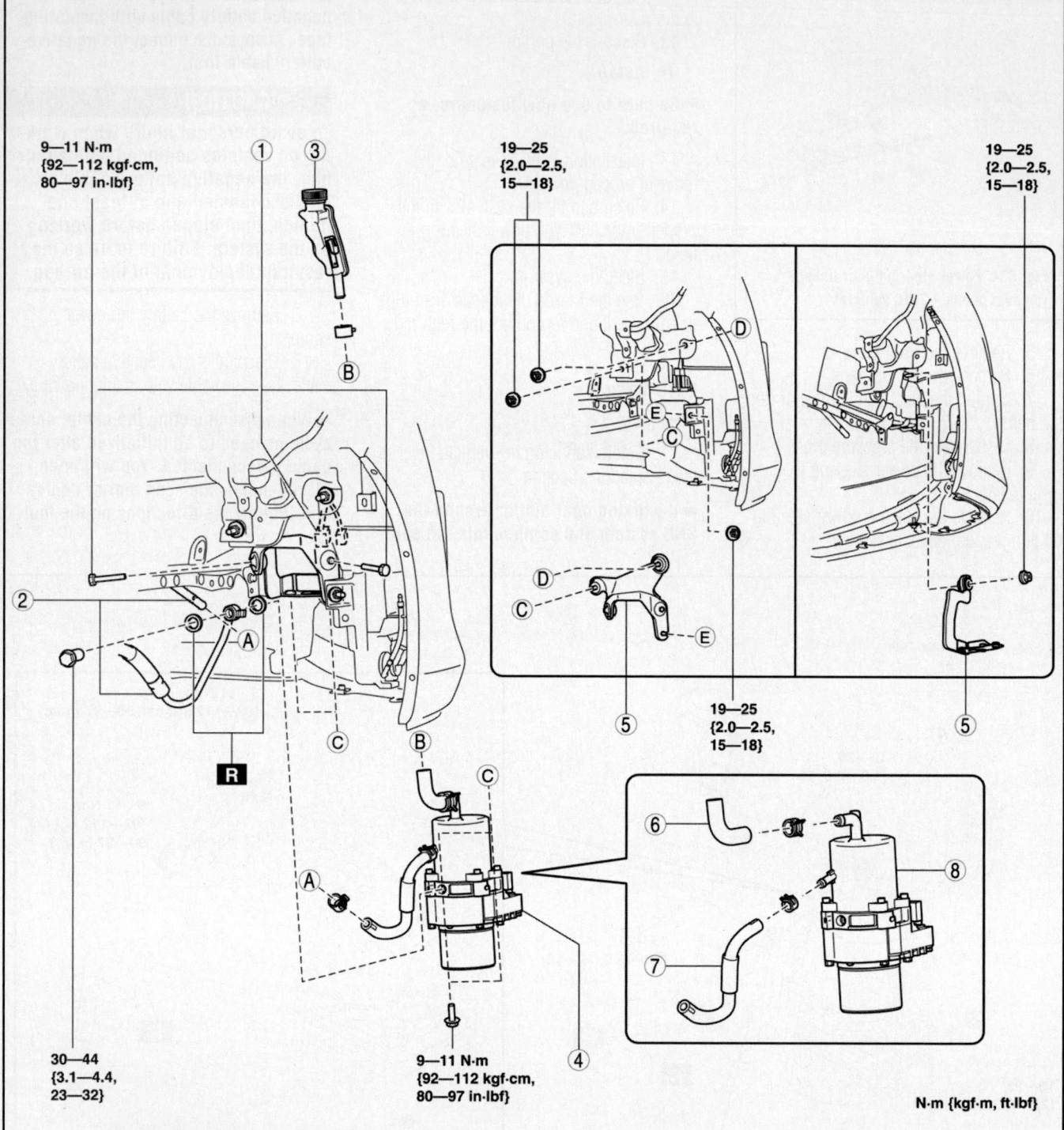

9—11 N·m
{92—112 kgf·cm,
80—97 in·lbf}

19—25
{2.0—2.5,
15—18}

19—25
{2.0—2.5,
15—18}

19—25
{2.0—2.5,
15—18}

30—44
{3.1—4.4,
23—32}

9—11 N·m
{92—112 kgf·cm,
80—97 in·lbf}

N·m {kgf·m, ft·lbf}

1. Connector
2. Power steering pipe component
3. Sub tank
4. Electric power steering oil pump component (with hoses)

5. Brackets
6. Suction hose
7. Return hose
8. Electric power steering oil pump component

37671_MAZ3_G0172

Fig. 336 Power steering pump and related components—2010 vehicles

➡Be careful not to drop the oil pump as the internal parts of the EHPAS CM could be damaged. Replace the component if it is subjected to impact.

4. Remove the front mudguard.
5. Remove the aerodynamic No. 2 undercover.
6. Remove the splash shield, right side.
7. Drain the power steering fluid. Be sure to properly dispose of used fluid. Be sure to plug the fluid lines.
8. Disconnect the connector.
9. Disconnect the steering pipe component. Remove the sub tank.
10. Disconnect the electric power steering oil pump component, with hoses.
11. Remove the brackets.
12. Remove the hoses.
13. Remove the component from its mounting.

To install:

➡Be sure to use new fasteners, as required.

14. Installation is the reverse of the removal procedure.
15. Be sure to perform the EHPAS CM configuration procedure. Use the Mazda diagnostic scan tool, or equivalent.
16. Be sure to fill the system with the proper grade and type power steering fluid.
17. Bleed the system.
18. Use the Mazda diagnostic scan tool, or equivalent and reprogram the required systems.

BLEEDING

Vehicles Without Strainer

1. Before servicing the vehicle, refer to the Precautions Section.

➡If working near and/or around the SRS system and components, be sure to disable the SRS system. Tape the negative battery cable with insulating tape. Always disconnect the negative battery cable first.

✱✱ CAUTION

To avoid personal injury when working on vehicles equipped with an air bag, the negative battery cable must be disconnected and at least one

minute must elapse before working on the system. Failure to do so may result in deployment of the air bag.

2. Remove the battery top cover, if equipped.
3. Disconnect the negative battery cable. Tape the cable with insulating tape.

➡When disconnecting the cable, some systems need to be initialized after the cable is reconnected. You will need Mazda diagnostic scan tool or equivalent. Follow the directions on the tool.

➡Do not turn the steering wheel during fluid inspection.

4. Be sure that the fluid level is full.
5. Raise and support the vehicle safely.
6. Turn the steering wheel fully to the left and right several times with the engine not running.
7. Inspect the fluid level. Correct as required.
8. Repeat the above until the level stabilizes.
9. Lower the vehicle.
10. Start the engine and let it idle.
11. Verify that the fluid is not foamy and that the level has not dropped.

Vehicles With Strainer

See Figure 337.

1. Before servicing the vehicle, refer to the Precautions Section.

➡If working near and/or around the SRS system and components, be sure to disable the SRS system. Tape the negative battery cable with insulating tape. Always disconnect the negative battery cable first.

✱✱ CAUTION

To avoid personal injury when working on vehicles equipped with an air bag, the negative battery cable must be disconnected and at least one minute must elapse before working on the system. Failure to do so may result in deployment of the air bag.

2. Remove the battery top cover, if equipped.
3. Disconnect the negative battery cable. Tape the cable with insulating tape.

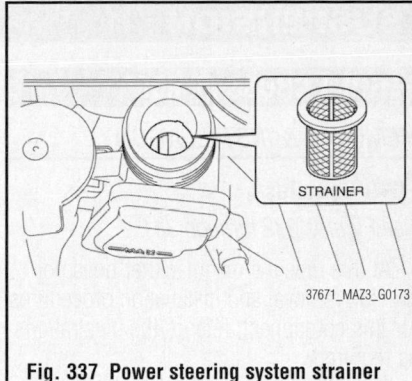

37671_MAZ3_G0173

Fig. 337 Power steering system strainer location

➡When disconnecting the cable, some systems need to be initialized after the cable is reconnected. You will need Mazda diagnostic scan tool or equivalent. Follow the directions on the tool.

➡Verify that the strainer is installed to the sub tank, because the procedure differs depending on whether the vehicle has a strainer. If the strainer is removed, impurities may enter the system. Always bleed the system with the strainer installed.

➡Do not maintain the steering wheel fully turned for more than five seconds as fluid temperature could rise and cause damage to the pump.

➡You will need an assistant to complete this procedure.

4. Working with two people, one person protects the area around the sub tank using a cloth and adds fluid to the sub tank. Do not spread the cloth near the drive belt, as it (cloth) could get caught when starting the engine.
5. Start the engine and allow it to idle.
6. Turn the steering wheel to the full left detent, than to the full right detent several times. Be sure to turn the wheel slowly, if steering wheel operation speed is too fast, fluid will splatter from the filler port. Do not exceed the steering wheel operation reference value (90 degrees/s).
7. Inspect the fluid level, be sure it is within specification (between MAX and MIN marking).
8. Correct fluid level as required.

SUSPENSION **FRONT SUSPENSION**

FRONT CROSSMEMBER

REMOVAL & INSTALLATION

2010 Vehicles

See Figures 332 through 341.

At this time the manufacturer does not provide removal and installation procedures for this component, refer to the illustrations as required.

1. Before servicing the vehicle, refer to the Precautions Section.

➡ If working near and/or around the SRS system and components, be sure to disable the SRS system. Tape the negative battery cable with insulating tape. Always disconnect the negative battery cable first.

※ CAUTION

To avoid personal injury when working on vehicles equipped with an air bag, the negative battery cable must be disconnected and at least one minute must elapse before working on the system. Failure to do so may result in deployment of the air bag.

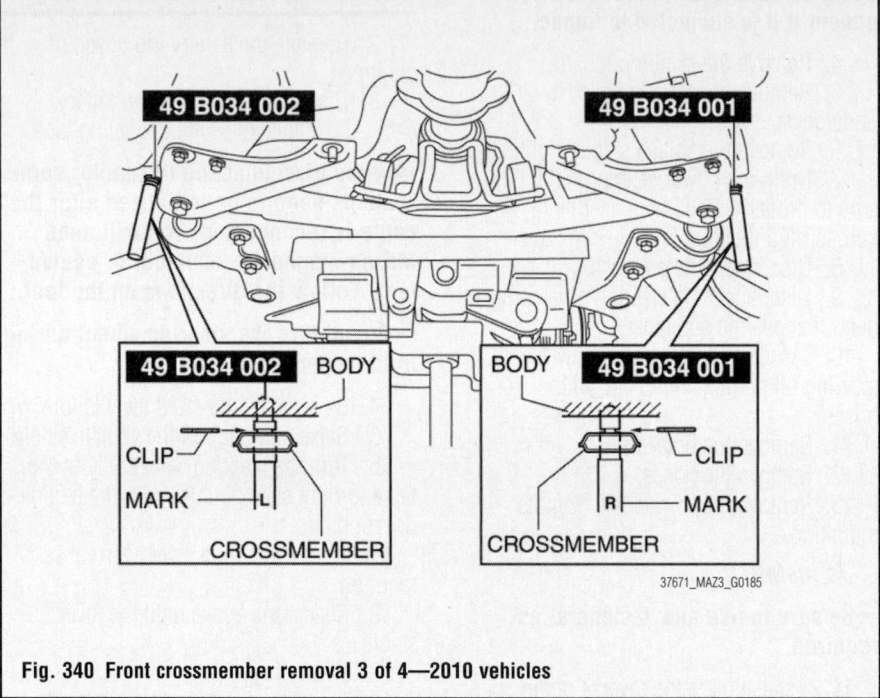

Fig. 340 Front crossmember removal 3 of 4—2010 vehicles

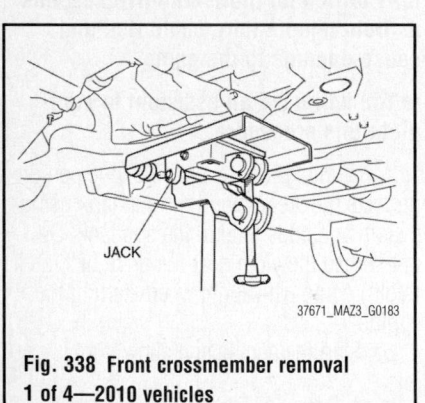

Fig. 338 Front crossmember removal 1 of 4—2010 vehicles

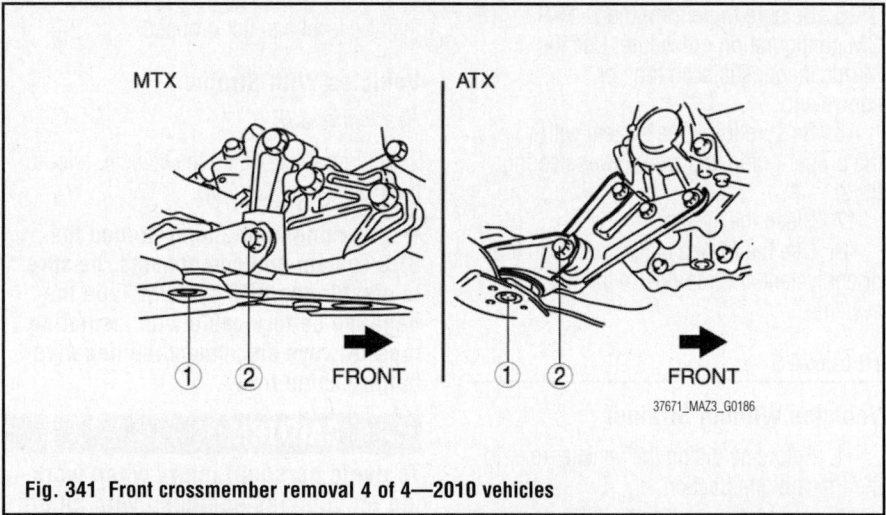

Fig. 341 Front crossmember removal 4 of 4—2010 vehicles

2. Remove the battery top cover, if equipped.

3. Disconnect the negative battery cable. Tape the cable with insulating tape.

➡ When disconnecting the cable, some systems need to be initialized after the cable is reconnected. You will need Mazda diagnostic scan tool or equivalent. Follow the directions on the tool.

➡ Removing the crossmember is dangerous. Be sure that the component is properly supported, before removing it. Special tools will be required to remove this component.

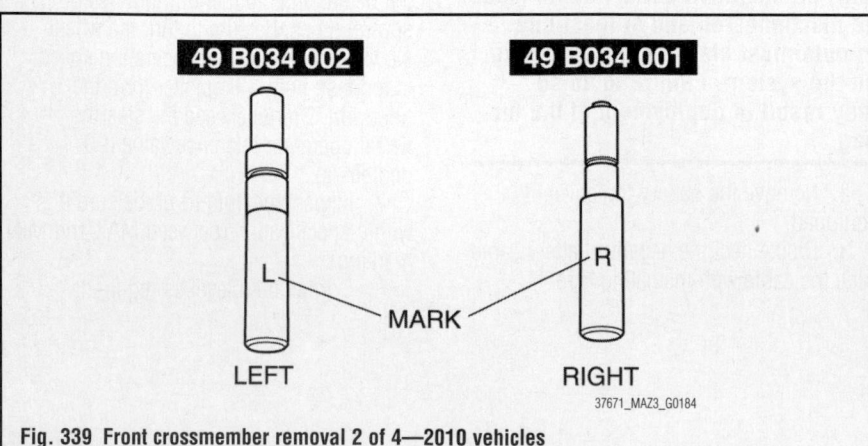

Fig. 339 Front crossmember removal 2 of 4—2010 vehicles

LOWER BALL JOINT

REMOVAL & INSTALLATION

The lower ball joint is an integral part of the lower control and cannot be replaced separately. If the lower ball joint is defective, the entire lower control arm must be replaced. Refer to the lower control arm procedure.

LOWER CONTROL ARM

REMOVAL & INSTALLATION

2009 Vehicles

See Figure 342.

1. Before servicing the vehicle, refer to the Precautions Section.

➡If working near and/or around the SRS system and components, be sure to disable the SRS system. Tape the negative battery cable with insulating tape. Always disconnect the negative battery cable first.

✳✳ CAUTION

To avoid personal injury when working on vehicles equipped with an air bag, the negative battery cable must be disconnected and at least one

minute must elapse before working on the system. Failure to do so may result in deployment of the air bag.

2. Remove the battery top cover, if equipped.
3. Disconnect the negative battery cable. Tape the cable with insulating tape.

➡When disconnecting the cable, some systems need to be initialized after the cable is reconnected. You will need Mazda diagnostic scan tool or equivalent. Follow the directions on the tool.

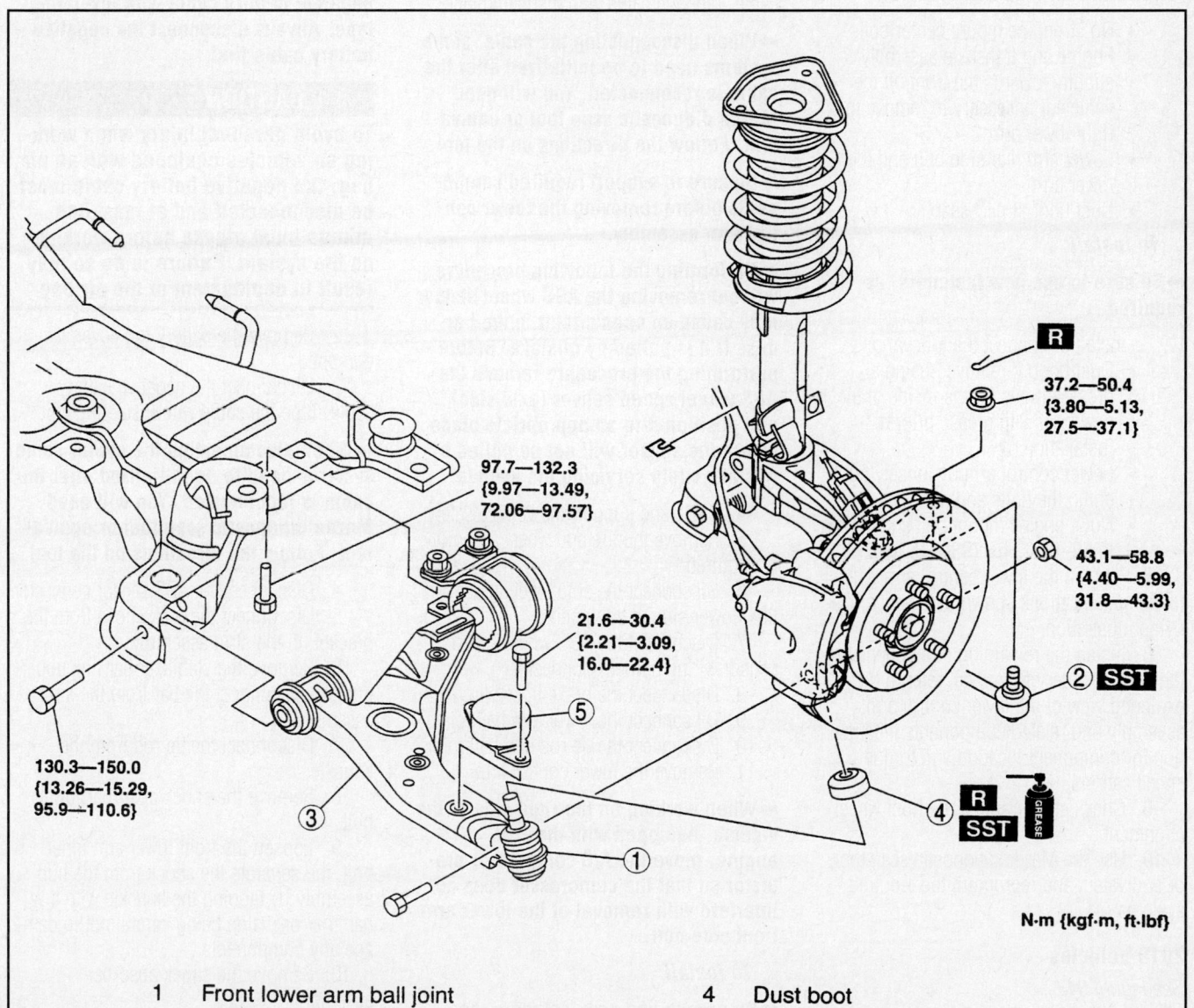

97.7—132.3
{9.97—13.49,
72.06—97.57}

21.6—30.4
{2.21—3.09,
16.0—22.4}

130.3—150.0
{13.26—15.29,
95.9—110.6}

37.2—50.4
{3.80—5.13,
27.5—37.1}

43.1—58.8
{4.40—5.99,
31.8—43.3}

N·m {kgf·m, ft·lbf}

1   Front lower arm ball joint
2   Tie-rod end ball joint
3   Front lower arm

4   Dust boot
5   Dynamic damper

67162-MAZC-G96

Fig. 342 Exploded view of the front lower control arm assembly and related components—2009 vehicles

4. Remove or disconnect the following:
- Wheel
- Cotter pin
- Lower ball joint by loosening it

5. With the nut protecting the ball joint stud, separate the stud from the knuckle. Remove the nut.
- Cotter pin and nut from the tie-rod end
- Tie-rod end from the steering knuckle

➡ If removing the right side lower arm, move the engine and transaxle slightly towards the front side of the vehicle so the engine does not interfere with the lower arm rear side bolt removal.

- No. 1 engine mount center bolt.
- Engine and transaxle assembly slightly towards the front off the vehicle if necessary to remove the right lower arm
- Lower arm rear side bolt and the lower arm.
- Dust boot, if necessary

To install:

➡ Be sure to use new fasteners, as required.

6. Install or connect the following:
- Dust boot, if removed using a press. Always fill the inside of the new boot with grease prior to installation.
- Lower control arm by loosely tightening the bolts and nuts
- No. 1 engine mount bolt and tighten to 68–86 ft. lbs. (93–116 Nm).

7. Tighten the lower control arm bolts to the specifications shown in the accompanying illustration.

8. Install the remaining components in the reverse order of removal, refer to the exploded view of the lower control arm assembly and related components illustration for component location and torque specifications.

9. Check and/or adjust the front wheel alignment.

10. Use the Mazda diagnostic scan tool, or equivalent and reprogram the required systems.

2010 Vehicles

See Figure 343.

1. Before servicing the vehicle, refer to the Precautions Section.

➡ If working near and/or around the SRS system and components, be sure to disable the SRS system. Tape the negative battery cable with insulating tape. Always disconnect the negative battery cable first.

> ✳✳ **CAUTION**
>
> To avoid personal injury when working on vehicles equipped with an air bag, the negative battery cable must be disconnected and at least one minute must elapse before working on the system. Failure to do so may result in deployment of the air bag.

2. Remove the battery top cover, if equipped.

3. Disconnect the negative battery cable. Tape the cable with insulating tape.

➡ When disconnecting the cable, some systems need to be initialized after the cable is reconnected. You will need Mazda diagnostic scan tool or equivalent. Follow the directions on the tool.

➡ Be sure to support required components before removing the lower control arm assembly.

➡ Performing the following procedure without removing the ABS wheel sensor may cause an open circuit in the harness if it is pulled by mistake. Before performing the procedure remove the ABS wheel speed sensor (axle side) and position it to an appropriate place where the sensor will not be pulled by mistake while servicing the vehicle.

4. Raise and support the vehicle safely.

5. Remove the tire and wheel assembly, as required.

6. Disconnect the auto leveling sensor link, lower side, if equipped.

7. Disconnect the ABS wheel speed sensor wiring harness connector.

8. Disconnect the brake hose clip.

9. Disconnect the lower arm ball joint.

10. Disconnect the tie rod end ball joint.

11. Remove the lower control arm.

➡ When working on the right side of the vehicle, equipped with the 2.3L engine, move the A/C compressor protector so that the compressor does not interfere with removal of the lower arm front side bolt.

To install:

➡ Be sure to use new fasteners, as required.

12. Installation is the reverse of the removal procedure.

13. Check and adjust alignment, as required.

14. Use the Mazda diagnostic scan tool, or equivalent and reprogram the required systems.

MACPHERSON STRUT

REMOVAL & INSTALLATION

2009 Vehicles

See Figures 344 and 345.

1. Before servicing the vehicle, refer to the Precautions Section.

➡ If working near and/or around the SRS system and components, be sure to disable the SRS system. Tape the negative battery cable with insulating tape. Always disconnect the negative battery cable first.

> ✳✳ **CAUTION**
>
> To avoid personal injury when working on vehicles equipped with an air bag, the negative battery cable must be disconnected and at least one minute must elapse before working on the system. Failure to do so may result in deployment of the air bag.

2. Remove the battery top cover, if equipped.

3. Disconnect the negative battery cable. Tape the cable with insulating tape.

➡ When disconnecting the cable, some systems need to be initialized after the cable is reconnected. You will need Mazda diagnostic scan tool or equivalent. Follow the directions on the tool.

4. Disconnect the ABS sensor connector.

5. Disconnect the brake hose from the bracket on the strut assembly.

6. Remove the stabilizer bar link upper nut and disconnect the bar from the strut assembly.

7. Disconnect the tie rod from the knuckle.

8. Remove the shock absorber lower bolt.

9. Loosen the front lower arm inner bolt, the separate the shock from the hub assembly by tapping the knuckle with a hammer or mallet being careful not to damage any components.

10. Remove the shock absorber assembly.

To install:

➡ Be sure to use new fasteners, as required.

11. Align the piston rod nut with the center part where the shock absorber is

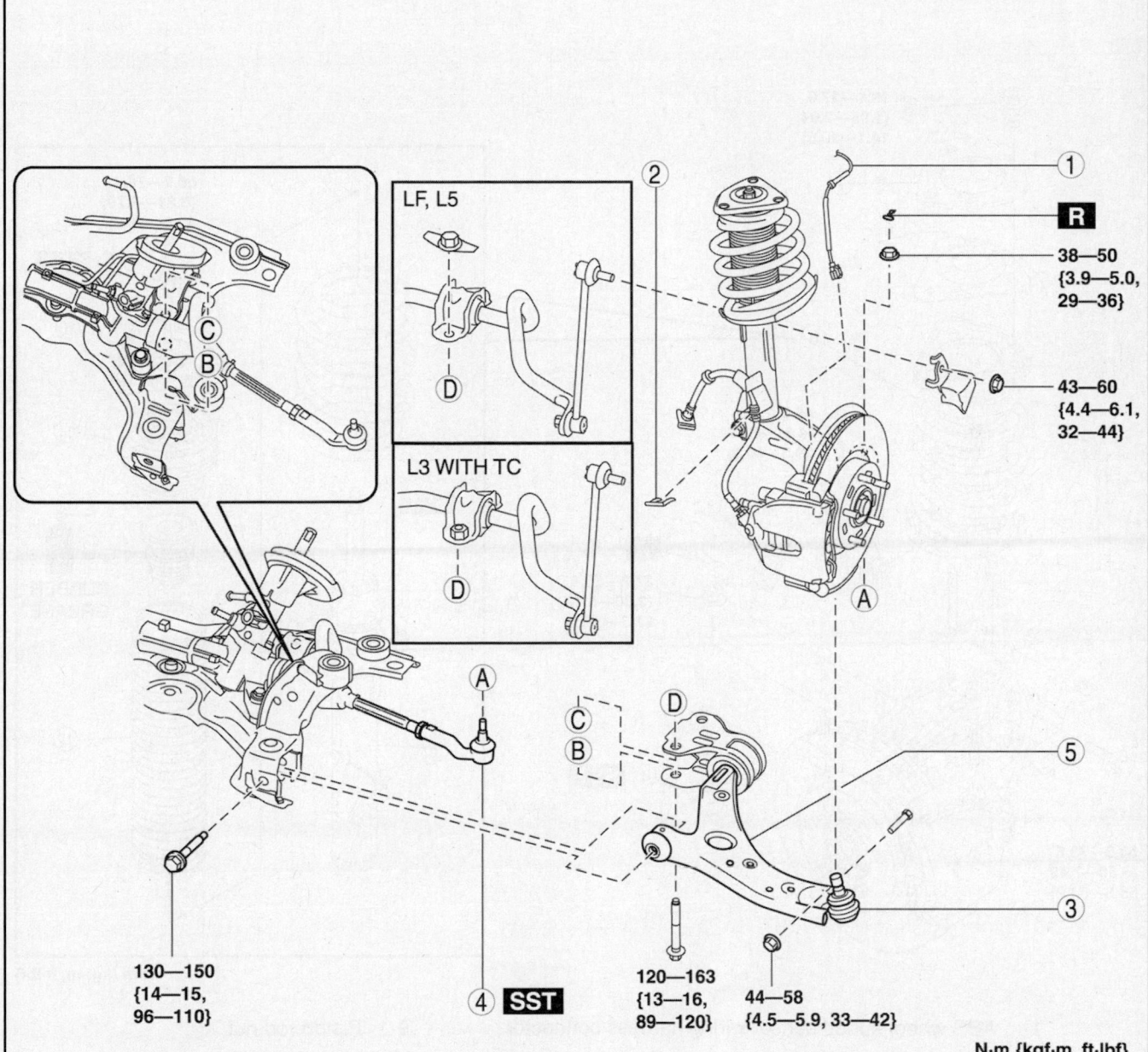

LF, L5

L3 WITH TC

R

38—50
{3.9—5.0,
29—36}

43—60
{4.4—6.1,
32—44}

130—150
{14—15,
96—110}

SST

120—163
{13—16,
89—120}

44—58
{4.5—5.9, 33—42}

N·m {kgf·m, ft·lbf}

1. Front ABS wheel-speed sensor wiring harness connector
2. Brake hose clip
3. Front lower arm ball joint
4. Tie-rod end ball joint
5. Front lower arm

37671_MAZ3_G0178

Fig. 343 Front lower control arm and related components—2010 vehicles

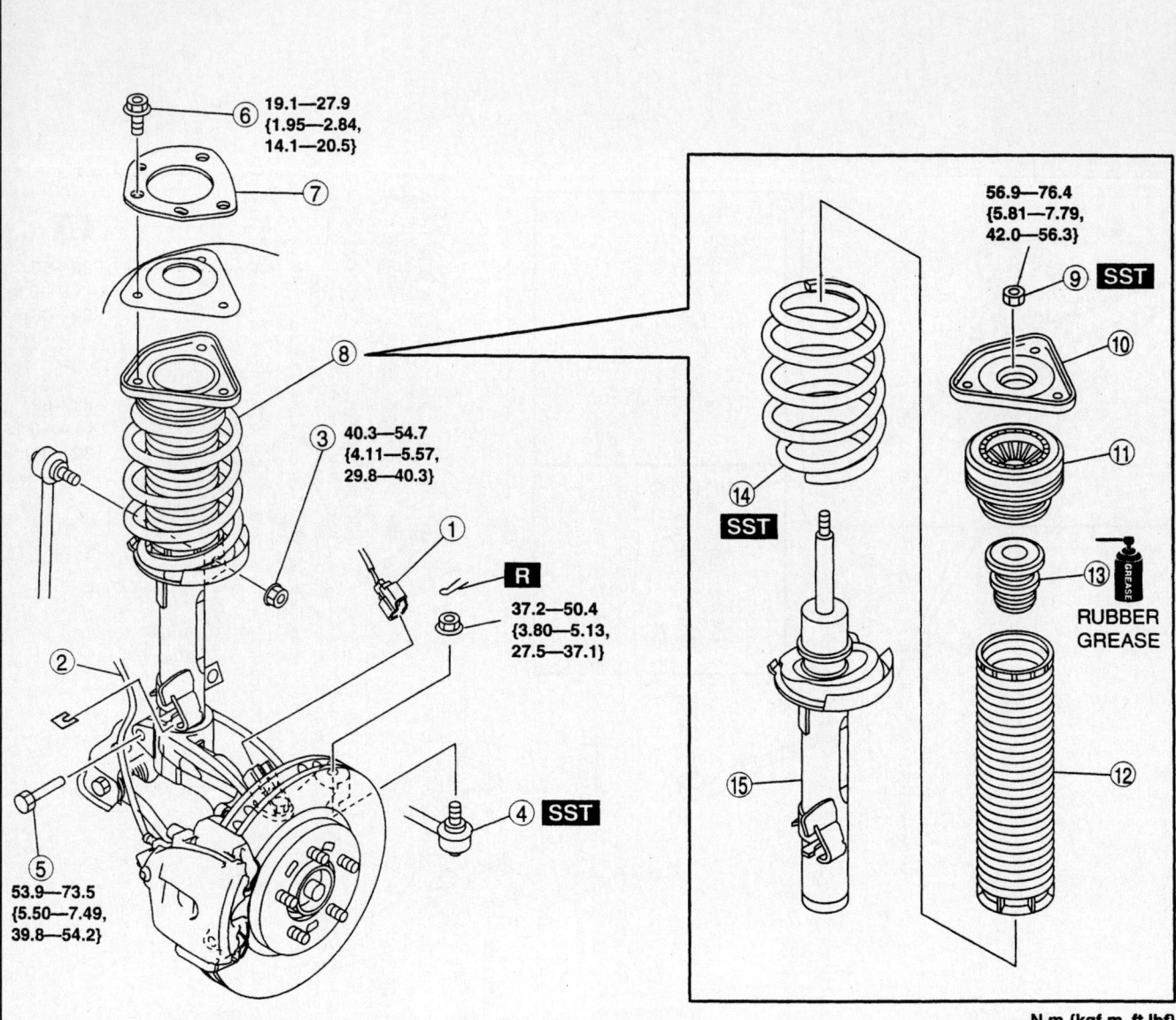

1   ABS wheel-speed sensor wiring harness connector
2   Brake hose
3   Stabilizer control link upper nut
4   Tie-rod end ball joint
5   Shock absorber lower bolt
6   Shock absorber upper bolt
7   Stiffener
8   Shock absorber and coil spring

9   Piston rod nut
10 Mounting rubber
11 Bearing
12 Dust boot
13 Bound stopper
14 Coil spring
15 Front shock absorber

67162-MAZC-G88

Fig. 344 Exploded view of the front strut and spring assembly components—2009 vehicles

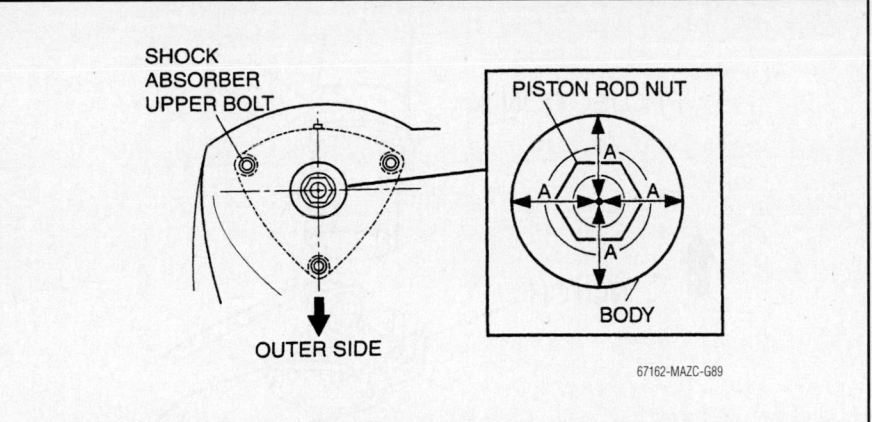

Fig. 345 Align the piston rod nut with the center part where the shock absorber is installed by placing the piston rod nut with lengths (A) all the same—2009 vehicles

installed by placing the piston rod nut with lengths (A) all the same and tighten the shock absorber upper bolts. Refer to the accompanying illustration.

12. Use a jack to raise the lower control arm, attach the shock absorber and tighten the bolts

13. Tighten the upper shock nuts to 14–20 ft. lbs. (19–28 Nm) and the lower shock nut and bolt to 40–54 ft. lbs. (54–74 Nm).

14. Tighten the tie rod end nut to 27–37 ft. lbs. (37–50 Nm), install a new cotter pin. Tighten the nut, if necessary, to align the ball stud hole with the nut castellation.

15. Install the stabilizer bar link and tighten the upper nut to 30–40 ft. lbs. (40–54 Nm).

16. Install the brake hose and ABS sensor connector.

17. Check and/or adjust the front end alignment.

18. Use the Mazda diagnostic scan tool, or equivalent and reprogram the required systems.

SHOCK ABSORBERS

REMOVAL & INSTALLATION

2010 Vehicles

See Figures 346 and 347.

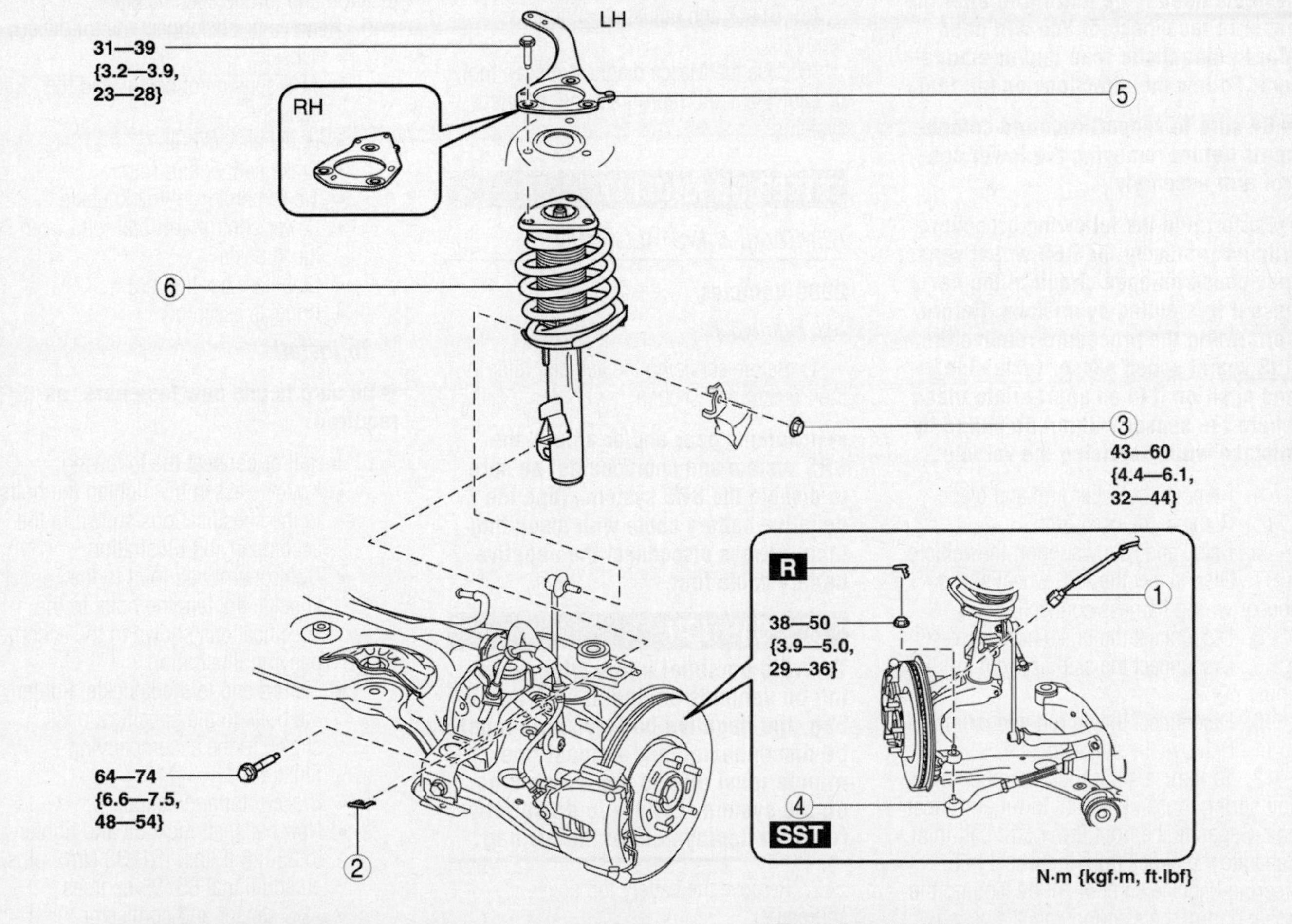

1. ABS wheel-speed sensor wiring harness connector
2. Brake hose clip
3. Stabilizer control link upper nut
4. Tie-rod end ball joint
5. Stiffener
6. Shock absorber and coil spring

Fig. 346 Front shock absorber and related components—2010 vehicles

1. Before servicing the vehicle, refer to the Precautions Section.

➡ **If working near and/or around the SRS system and components, be sure to disable the SRS system. Tape the negative battery cable with insulating tape. Always disconnect the negative battery cable first.**

✳✳ CAUTION

To avoid personal injury when working on vehicles equipped with an air bag, the negative battery cable must be disconnected and at least one minute must elapse before working on the system. Failure to do so may result in deployment of the air bag.

2. Remove the battery top cover, if equipped.
3. Disconnect the negative battery cable. Tape the cable with insulating tape.

➡ **When disconnecting the cable, some systems need to be initialized after the cable is reconnected. You will need Mazda diagnostic scan tool or equivalent. Follow the directions on the tool.**

➡ **Be sure to support required components before removing the lower control arm assembly.**

➡ **Performing the following procedure without removing the ABS wheel sensor may cause an open circuit in the harness if it is pulled by mistake. Before performing the procedure remove the ABS wheel speed sensor (axle side) and position it to an appropriate place where the sensor will not be pulled by mistake while servicing the vehicle.**

4. Remove the wiper arm and blade.
5. Remove the cowl grille.
6. Raise and safely support the vehicle.
7. Disconnect the ABS wheel speed sensor wiring harness connector.
8. Disconnect the brake hose clip.
9. Disconnect the stabilizer control link upper nut.
10. Disconnect the tie rod end ball joint.
11. Remove the stiffener.
12. To remove the shock absorber and coil spring loosen the front lower arm inner bolt. Separate the front lower arm ball joint. Separate the shock from the wheel hub, steering knuckle component by tapping the upper part of the steering knuckle with a hammer.

To install:

➡ **Be sure to use new fasteners, as required.**

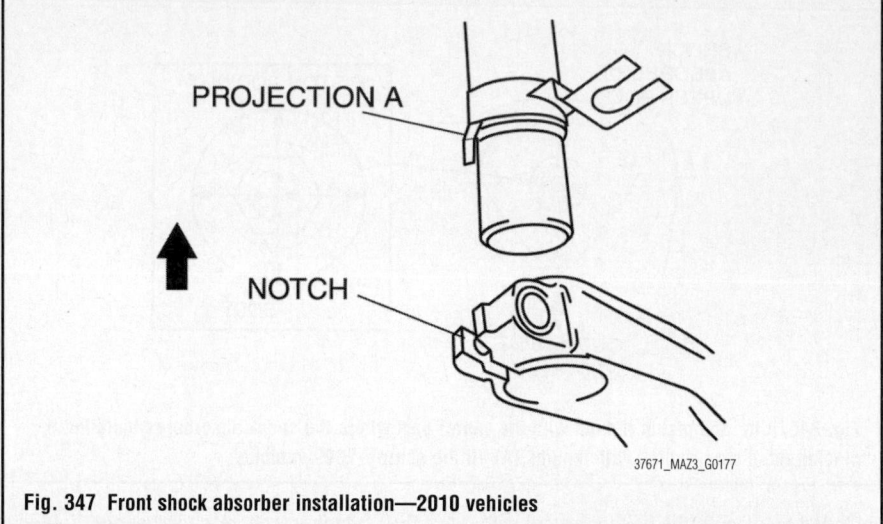

Fig. 347 Front shock absorber installation—2010 vehicles

13. Installation is the reverse of the removal procedure.
14. When installing, raise the front lower arm using a jack and install the shock and coil spring.
15. Check and adjust alignment, as required.
16. Use the Mazda diagnostic scan tool, or equivalent and reprogram the required systems.

STEERING KNUCKLE

REMOVAL & INSTALLATION

2009 Vehicles
See Figure 348.

1. Before servicing the vehicle, refer to the Precautions Section.

➡ **If working near and/or around the SRS system and components, be sure to disable the SRS system. Tape the negative battery cable with insulating tape. Always disconnect the negative battery cable first.**

✳✳ CAUTION

To avoid personal injury when working on vehicles equipped with an air bag, the negative battery cable must be disconnected and at least one minute must elapse before working on the system. Failure to do so may result in deployment of the air bag.

2. Remove the battery top cover, if equipped.
3. Disconnect the negative battery cable. Tape the cable with insulating tape.

➡ **When disconnecting the cable, some systems need to be initialized after the**

cable is reconnected. You will need Mazda diagnostic scan tool or equivalent. Follow the directions on the tool.

4. Refer to the illustration for component location and torque specifications.
5. Remove or disconnect the following:
   - Wheels
   - ABS sensor connector and the sensor
   - Halfshaft lockbolt
   - Brake caliper and rotor
   - Tie rod end from the knuckle
   - Lower control arm ball joint from the knuckle
   - Stabilizer bar link nut
   - Knuckle assembly

To install:

➡ **Be sure to use new fasteners, as required.**

6. Install or connect the following:
   - Knuckle assembly, tighten the bolts to the specifications shown in the accompanying illustration
   - Control arm ball joint to the knuckle tighten, the bolts to the specifications shown in the accompanying illustration
   - Tie rod end to the knuckle, tighten the bolts to the specifications shown in the accompanying illustration
   - Brake rotor and caliper
   - New halfshaft lockbolt and tighten to 23–28 ft. lbs. (31–38 Nm), plus an additional 85–95 degrees
   - ABS sensor and connector
   - Wheels

7. Use the Mazda diagnostic scan tool, or equivalent and reprogram the required systems.

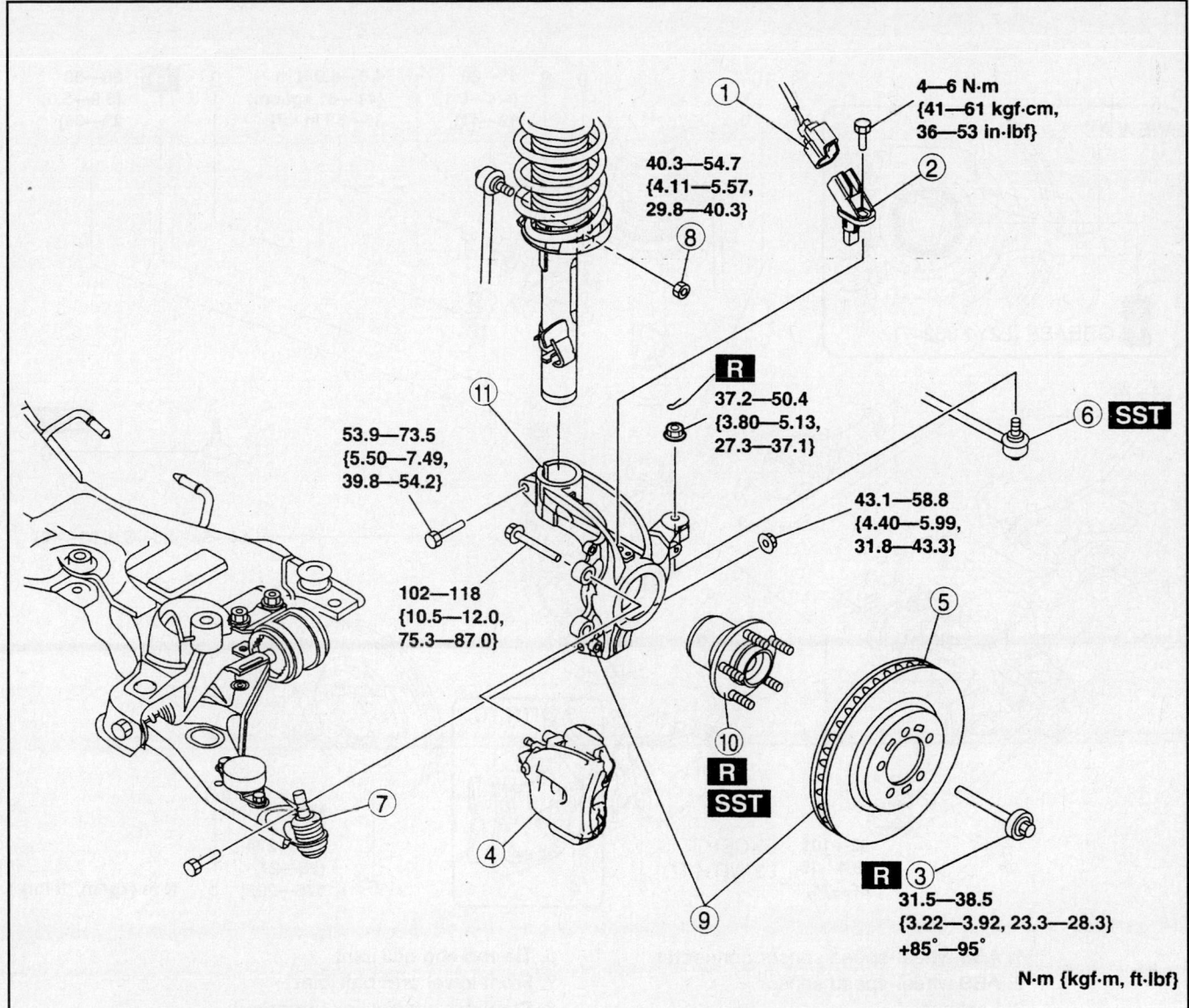

Fig. 348 Exploded view of the front wheel bearing and knuckle assembly—2009 vehicles

1	ABS wheel-speed sensor connector	7	Front lower arm ball joint
2	ABS wheel-speed sensor	8	Stabilizer control link upper nut
3	Lockbolt	9	Wheel hub, steering knuckle component
4	Brake caliper component	10	Wheel hub component
5	Disc plate	11	Steering knuckle
6	Tie-rod end ball joint		

67162-MAZC-G99

2010 Vehicles

See Figure 349.

1. Before servicing the vehicle, refer to the Precautions Section.

➡ **If working near and/or around the SRS system and components, be sure to disable the SRS system. Tape the** negative battery cable with insulating tape. Always disconnect the negative battery cable first.

✲✲ CAUTION

To avoid personal injury when working on vehicles equipped with an air bag, the negative battery cable must be disconnected and at least one minute must elapse before working on the system. Failure to do so may result in deployment of the air bag.

2. Remove the battery top cover, if equipped.

3. Disconnect the negative battery cable. Tape the cable with insulating tape.

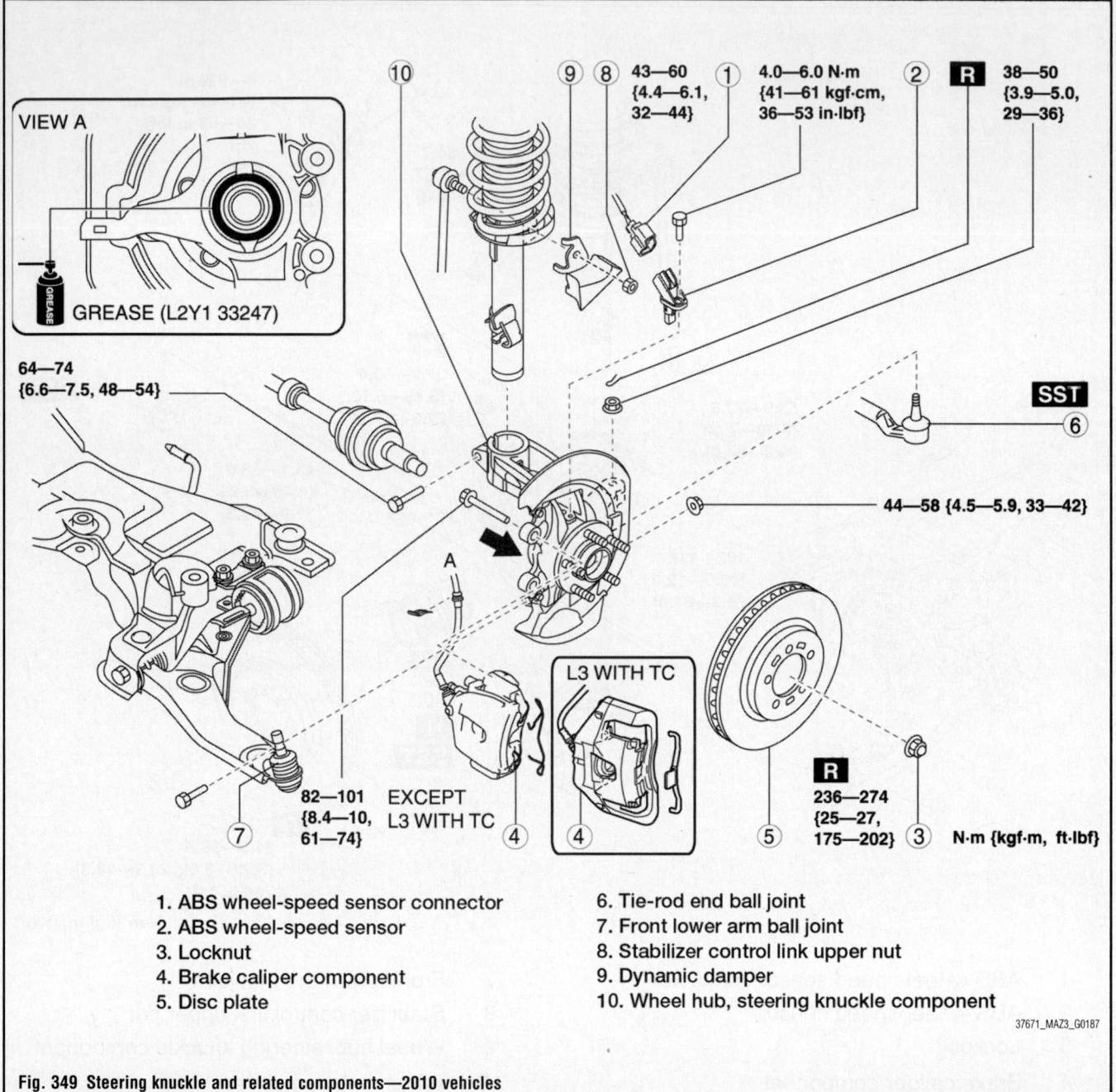

VIEW A

GREASE (L2Y1 33247)

64—74
{6.6—7.5, 48—54}

43—60
{4.4—6.1,
32—44}

4.0—6.0 N·m
{41—61 kgf·cm,
36—53 in·lbf}

38—50
{3.9—5.0,
29—36}

R

SST

44—58 {4.5—5.9, 33—42}

82—101
{8.4—10,
61—74}

EXCEPT
L3 WITH TC

L3 WITH TC

R 236—274
{25—27,
175—202}

N·m {kgf·m, ft·lbf}

1. ABS wheel-speed sensor connector
2. ABS wheel-speed sensor
3. Locknut
4. Brake caliper component
5. Disc plate
6. Tie-rod end ball joint
7. Front lower arm ball joint
8. Stabilizer control link upper nut
9. Dynamic damper
10. Wheel hub, steering knuckle component

37671_MAZ3_G0187

Fig. 349 Steering knuckle and related components—2010 vehicles

➡When disconnecting the cable, some systems need to be initialized after the cable is reconnected. You will need Mazda diagnostic scan tool or equivalent. Follow the directions on the tool.

➡Performing the following procedure without removing the ABS wheel sensor may cause an open circuit in the harness if it is pulled by mistake. Before performing the procedure remove the ABS wheel speed sensor (axle side) and position it to an appropriate place where the sensor will not be pulled by mistake while servicing the vehicle.

4. Raise and support the vehicle safely.
5. Remove the tire and wheel assembly.
6. When servicing the left component, disconnect the auto leveling sensor link, if equipped.
7. Disconnect the ABS wheel speed sensor electrical connector. Remove the sensor.
8. Remove the locknut.
9. Remove the brake caliper. Do not allow the caliper to hang by the hose.
10. Remove the rotor.
11. Disconnect the tie rod end ball joint.
12. Disconnect the lower arm ball joint.
13. Disconnect the stabilizer control link upper nut.

14. Remove the dynamic damper.
15. Remove the wheel hub and steering gear component from the vehicle.

To install:

➡Be sure to use new fasteners, as required.

16. Installation is the reverse of the removal procedure.
17. Check and adjust alignment, as required.
18. Use the Mazda diagnostic scan tool, or equivalent and reprogram the required systems.

STABILIZER BAR

REMOVAL & INSTALLATION

2009 Vehicles

See Figures 350 and 351.

1. Before servicing the vehicle, refer to the Precautions Section.

➡If working near and/or around the SRS system and components, be sure to disable the SRS system. Tape the negative battery cable with insulating tape. Always disconnect the negative battery cable first.

✳✳ CAUTION

To avoid personal injury when working on vehicles equipped with an air bag, the negative battery cable must be disconnected and at least one minute must elapse before working on the system. Failure to do so may result in deployment of the air bag.

2. Remove the battery top cover, if equipped.
3. Disconnect the negative battery cable. Tape the cable with insulating tape.

➡When disconnecting the cable, some systems need to be initialized after the cable is reconnected. You will need Mazda diagnostic scan tool or equivalent. Follow the directions on the tool.

4. Raise and support the vehicle safely.
5. Remove the tire and wheel assembly, as required.
6. Detach the steering shaft from the steering gear and linkage.
7. Remove the power steering pipe bracket.
8. Remove the stabilizer control link.
9. Support the front crossmember using a jack and remove the front crossmember bracket.
10. Remove the No.1 engine mount center bolt.
11. Detach the silencer hangers on the middle pipe from the front crossmember.
12. Lower the front crossmember slowly approx. 90 mm {3.5 in} and remove the front stabilizer .component.
13. Secure the stabilizer bracket flange using a vise.
14. Remove the front stabilizer.

   To install:

➡**Be sure to use new fasteners, as required.**

15. Apply grease to the stabilizer bushing.

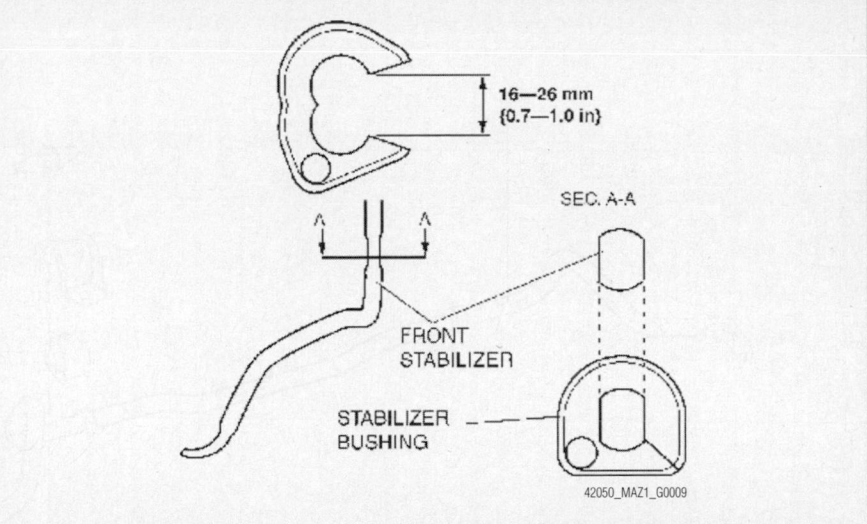

Fig. 350 Widen the stabilizer bushing opening 0.7–1 inch (16–26 mm) and install the bushing to the front stabilizer —2009 vehicles

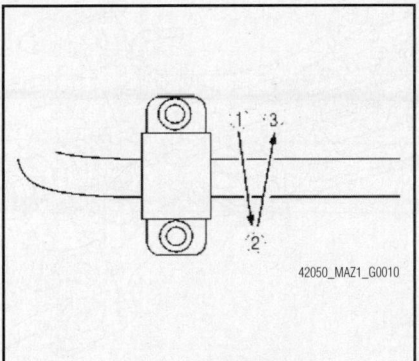

Fig. 351 Tighten the bolts in order indicated in the illustration—2009 vehicles

16. Install the stabilizer bracket using a vise
17. Widen the stabilizer bushing opening 0.7–1 inch (16–26 mm) and install the bushing to the front stabilizer as shown in the illustration.
18. Install the remaining components in the reverse of removal
19. Tighten the bolts in order indicated in the illustration.
20. Use the Mazda diagnostic scan tool, or equivalent and reprogram the required systems.

2010 Vehicles

See Figures 352 through 355.

1. Before servicing the vehicle, refer to the Precautions Section.

➡If working near and/or around the SRS system and components, be sure to disable the SRS system. Tape the negative battery cable with insulating tape. Always disconnect the negative battery cable first.

✳✳ CAUTION

To avoid personal injury when working on vehicles equipped with an air bag, the negative battery cable must be disconnected and at least one minute must elapse before working on the system. Failure to do so may result in deployment of the air bag.

2. Remove the battery top cover, if equipped.
3. Disconnect the negative battery cable. Tape the cable with insulating tape.

➡When disconnecting the cable, some systems need to be initialized after the cable is reconnected. You will need Mazda diagnostic scan tool or equivalent. Follow the directions on the tool.

➡Secure the steering wheel, using tape or cable, to prevent the steering shaft from rotating after disconnecting the steering shaft. If the steering wheel rotates after the steering shaft and steering gear linkage are disconnected the internal parts of the clockspring could be damaged.

4. Raise and support the vehicle safely.
5. Remove the tire and wheel assembly, as required.
6. Remove the joint cover.
7. Matchmark and disconnect the steering shaft from the steering gear and linkage.
8. Remove the front auto leveling sensor, if equipped.
9. Remove the front crossmember component.

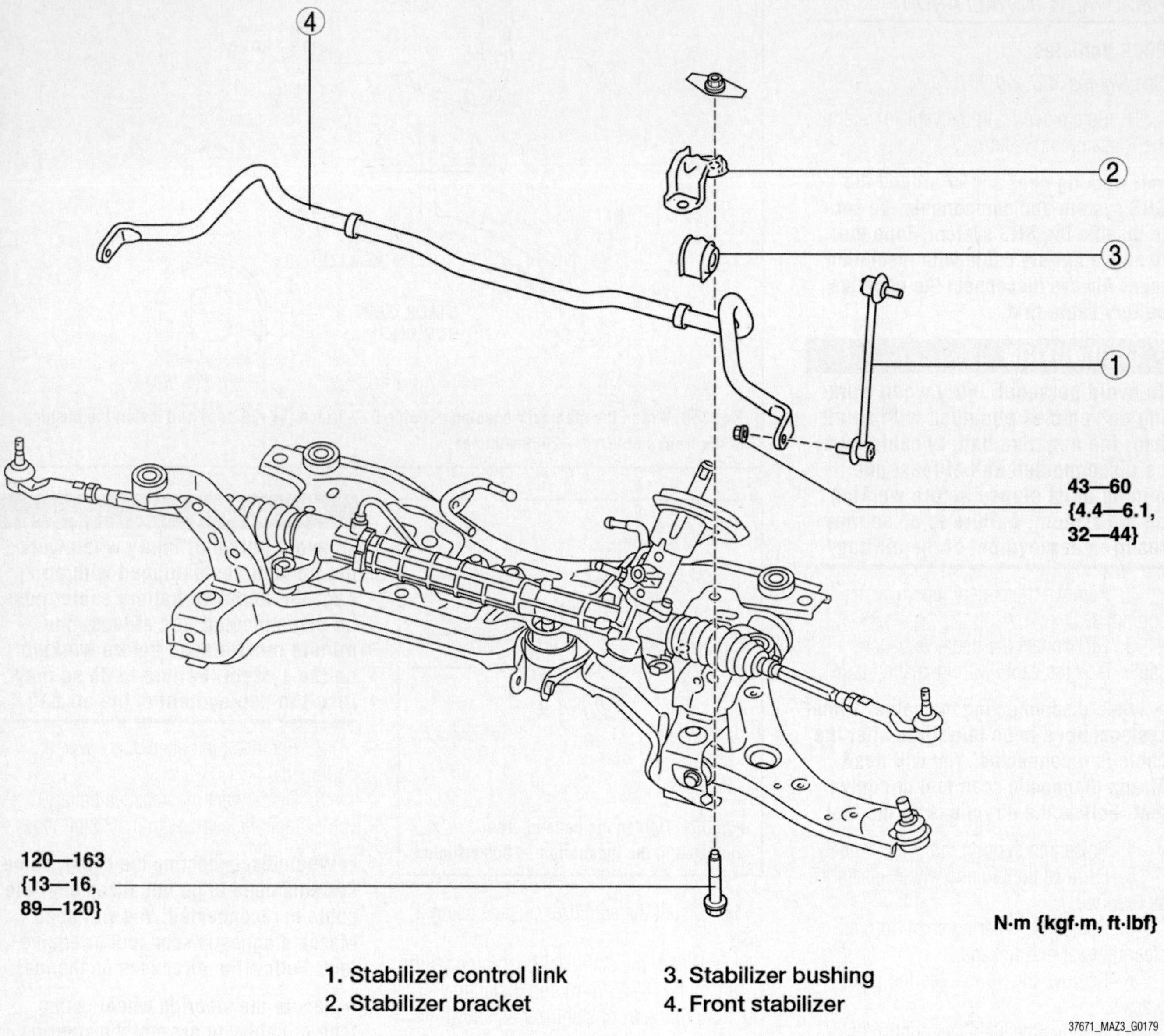

43—60
{4.4—6.1,
32—44}

120—163
{13—16,
89—120}

N·m {kgf·m, ft·lbf}

1. Stabilizer control link
2. Stabilizer bracket
3. Stabilizer bushing
4. Front stabilizer

37671_MAZ3_G0179

Fig. 352 Front stabilizer bar arm and related components—2010 vehicles

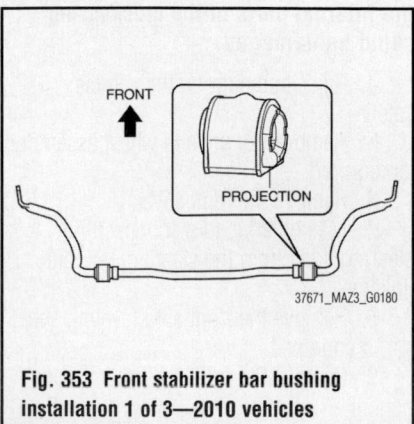

FRONT

PROJECTION

37671_MAZ3_G0180

Fig. 353 Front stabilizer bar bushing
installation 1 of 3—2010 vehicles

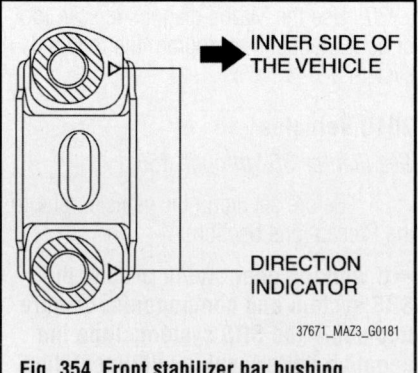

INNER SIDE OF
THE VEHICLE

DIRECTION
INDICATOR

37671_MAZ3_G0181

Fig. 354 Front stabilizer bar bushing
installation 1 of 3—2010 vehicles

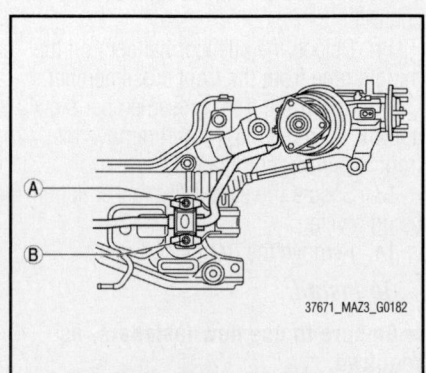

A

B

37671_MAZ3_G0182

Fig. 355 Front stabilizer bar bushing
installation 1 of 3—2010 vehicles

10. Remove the stabilizer control link.

11. Remove the stabilizer bracket.

12. Remove the stabilizer bushing.

13. Remove the stabilizer from its mounting.

To install:

➡**Be sure to use new fasteners, as required.**

14. Installation is the reverse of the removal procedure.

15. Tighten the stabilizer bushing bolts in sequence, see illustration

16. Be sure to check and adjust alignment, as required.

17. Use the Mazda diagnostic scan tool, or equivalent and reprogram the required systems.

WHEEL HUB & BEARING (SEALED UNIT)

REMOVAL & INSTALLATION

2009 Vehicles

See Figures 356 through 358.

1. Before servicing the vehicle, refer to the Precautions Section.

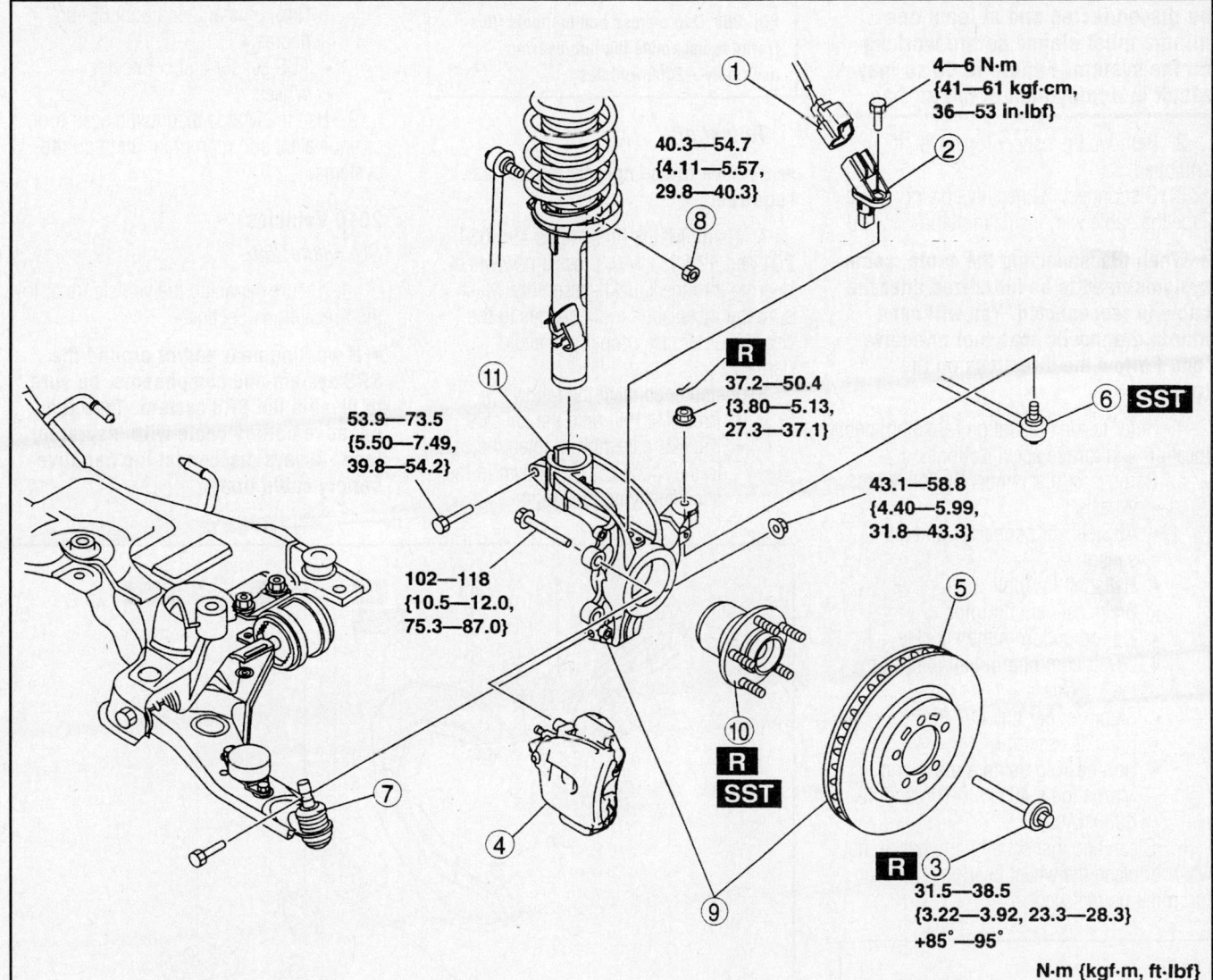

1	ABS wheel-speed sensor connector	7	Front lower arm ball joint
2	ABS wheel-speed sensor	8	Stabilizer control link upper nut
3	Lockbolt	9	Wheel hub, steering knuckle component
4	Brake caliper component	10	Wheel hub component
5	Disc plate	11	Steering knuckle
6	Tie-rod end ball joint		

67162-MAZC-G99

Fig. 356 Exploded view of the front wheel bearing and knuckle assembly—2009 vehicles

➡️**If working near and/or around the SRS system and components, be sure to disable the SRS system. Tape the negative battery cable with insulating tape. Always disconnect the negative battery cable first.**

✳️ CAUTION

To avoid personal injury when working on vehicles equipped with an air bag, the negative battery cable must be disconnected and at least one minute must elapse before working on the system. Failure to do so may result in deployment of the air bag.

2. Remove the battery top cover, if equipped.

3. Disconnect the negative battery cable. Tape the cable with insulating tape.

➡️**When disconnecting the cable, some systems need to be initialized after the cable is reconnected. You will need Mazda diagnostic scan tool or equivalent. Follow the directions on the tool.**

4. Refer to the illustration for component location and torque specifications.

5. Remove or disconnect the following:
- Wheels
- ABS sensor connector and the sensor
- Halfshaft lockbolt
- Brake caliper and rotor
- Tie rod end from the knuckle
- Lower control arm ball joint from the knuckle
- Stabilizer bar link nut
- Knuckle assembly
- Hub bearing using a press and Mazda tools 49 G030 795 and 49 B033 1A0

6. Clean and inspect all parts but do not wash or clean the wheel bearing. The bearing must be replaced.

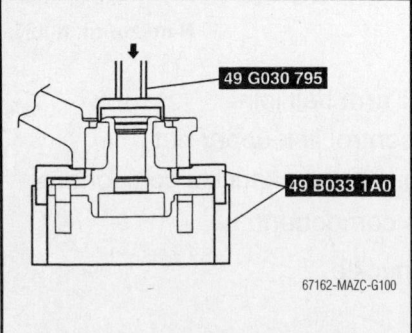

Fig. 357 Use a press and the tools illustrated to disassemble the hub/bearing assembly—2009 vehicles

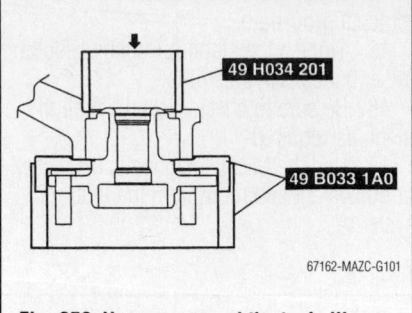

Fig. 358 Use a press and the tools illustrated to assemble the hub/bearing assembly—2009 vehicles

To install:

➡️**Be sure to use new fasteners, as required.**

7. Using Mazda Press tools 49 H034 201 and 49 B033 1A0, press a new wheel bearing into the knuckle assembly. Make sure the installation tool engages to the bearing outer race properly to avoid damage.

8. Install or connect the following:
- Press the hub onto the knuckle
- Knuckle assembly, tighten the bolts to the specifications shown in the accompanying illustration

- Control arm ball joint to the knuckle tighten, the bolts to the specifications shown in the accompanying illustration
- Tie rod end to the knuckle, tighten the bolts to the specifications shown in the accompanying illustration
- Brake rotor and caliper
- New halfshaft lockbolt and tighten to 23–28 ft. lbs. (31–38 Nm), plus an additional 85–95 degrees
- ABS sensor and connector
- Wheels

9. Use the Mazda diagnostic scan tool, or equivalent and reprogram the required systems.

2010 Vehicles

See Figure 359.

1. Before servicing the vehicle, refer to the Precautions Section.

➡️**If working near and/or around the SRS system and components, be sure to disable the SRS system. Tape the negative battery cable with insulating tape. Always disconnect the negative battery cable first.**

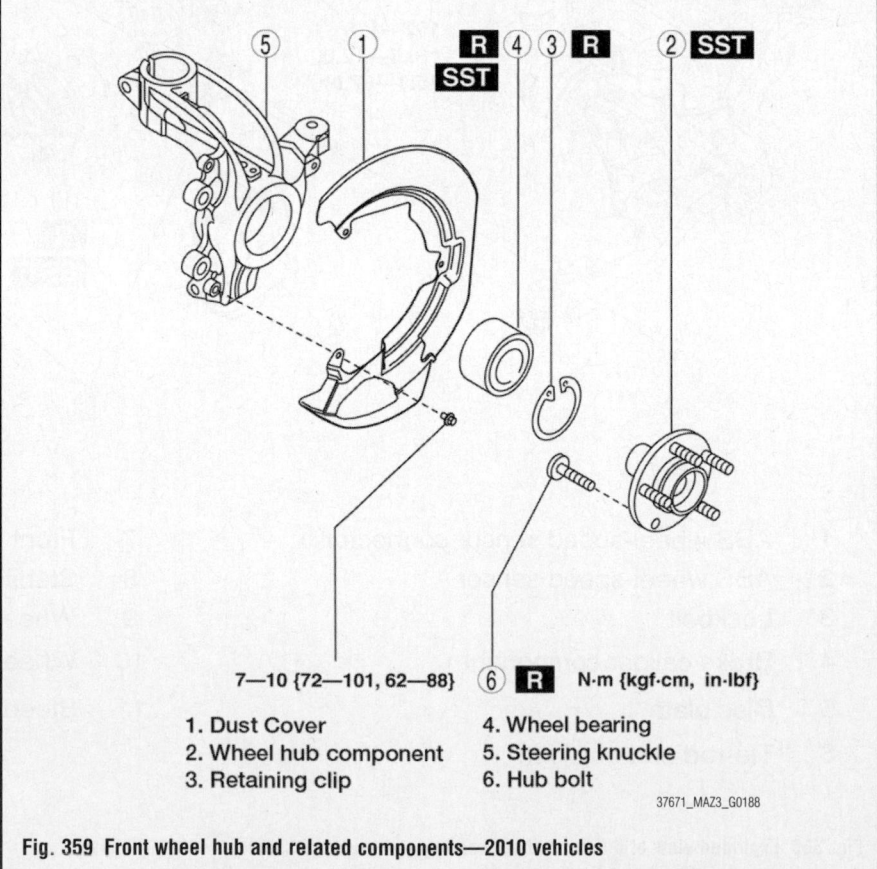

7—10 {72—101, 62—88}   ⑥ **R**   N·m {kgf·cm, in·lbf}

1. Dust Cover
2. Wheel hub component
3. Retaining clip
4. Wheel bearing
5. Steering knuckle
6. Hub bolt

Fig. 359 Front wheel hub and related components—2010 vehicles

To avoid personal injury when working on vehicles equipped with an air bag, the negative battery cable must be disconnected and at least one minute must elapse before working on the system. Failure to do so may result in deployment of the air bag.

2. Remove the battery top cover, if equipped.

3. Disconnect the negative battery cable. Tape the cable with insulating tape.

➡**When disconnecting the cable, some systems need to be initialized after the cable is reconnected. You will need Mazda diagnostic scan tool or equivalent. Follow the directions on the tool.**

➡**Performing the following procedure without removing the ABS wheel sensor may cause an open circuit in the harness if it is pulled by mistake. Before performing the procedure remove the ABS wheel speed sensor (axle side) and position it to an appropriate place where the sensor will not be pulled by mistake while servicing the vehicle.**

4. Raise and support the vehicle safely.

5. Remove the tire and wheel assembly.

6. When servicing the left component, disconnect the auto leveling sensor link, if equipped.

7. Disconnect the ABS wheel speed sensor electrical connector. Remove the sensor.

8. Remove the locknut.

9. Remove the brake caliper. Do not allow the caliper to hang by the hose.

10. Remove the rotor.

11. Disconnect the tie rod end ball joint.

12. Disconnect the lower arm ball joint.

13. Disconnect the stabilizer control link upper nut.

14. Remove the dynamic damper.

15. Remove the wheel hub and steering gear component from the vehicle.

16. As required, service the hub and bearing assembly.

To install:

➡**Be sure to use new fasteners, as required.**

17. Installation is the reverse of the removal procedure.

18. Check and adjust alignment, as required.

19. Use the Mazda diagnostic scan tool, or equivalent and reprogram the required systems.

ADJUSTMENT

The front wheel bearings are not adjustable. If the bearings become loose or make noise, they must be replaced.

SUSPENSION   REAR SUSPENSION

COIL SPRING

REMOVAL & INSTALLATION

2009 Vehicles

See Figures 360 through 362.

1. Before servicing the vehicle, refer to the Precautions Section.

➡**If working near and/or around the SRS system and components, be sure to disable the SRS system. Tape the negative battery cable with insulating tape. Always disconnect the negative battery cable first.**

To avoid personal injury when working on vehicles equipped with an air bag, the negative battery cable must be disconnected and at least one minute must elapse before working on the system. Failure to do so may result in deployment of the air bag.

2. Remove the battery top cover, if equipped.

3. Disconnect the negative battery cable. Tape the cable with insulating tape.

➡**When disconnecting the cable, some systems need to be initialized after the cable is reconnected. You will need Mazda diagnostic scan tool or equivalent. Follow the directions on the tool.**

4. Remove the rear stabilizer bar.

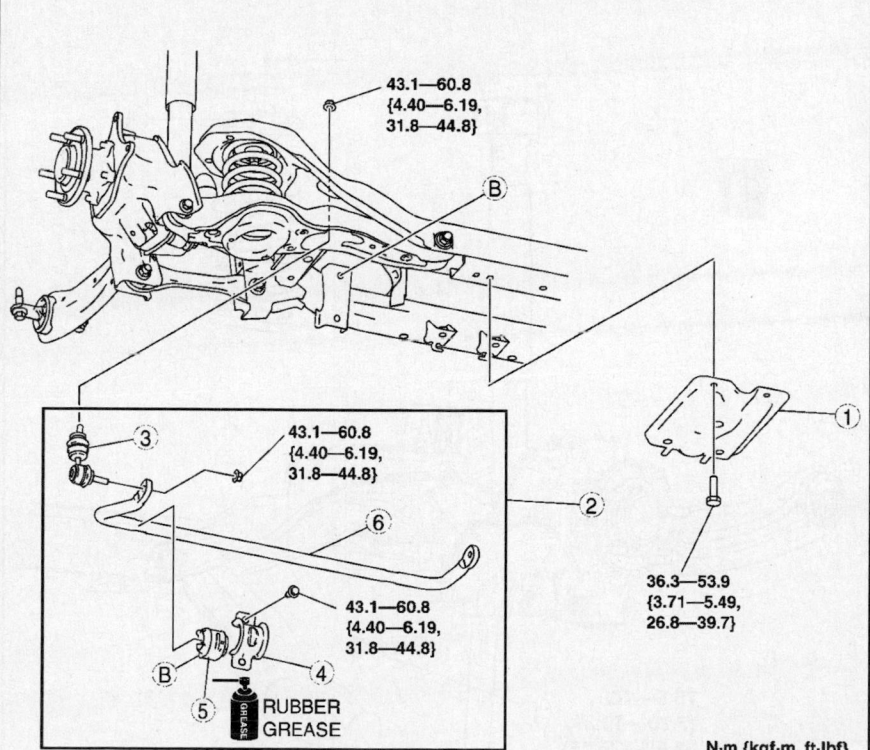

N·m {kgf·m, ft·lbf}

1. Rear crossmember bracket
2. Rear stabilizer component
3. Stabilizer control link
4. Stabilizer bracket
5. Stabilizer bushing
6. Rear stabilizer

22140_MAZD_G0287

Fig. 360 Stabilizer bar and related components—2009 vehicles

5. Support the rear lower arm using a jack.

6. Loosen the rear lower arm inner bolt.

7. Rear lower arm outer bolt.

8. Position the jack under the rear lower arm and lower slowly.

9. Remove the rear coil spring.

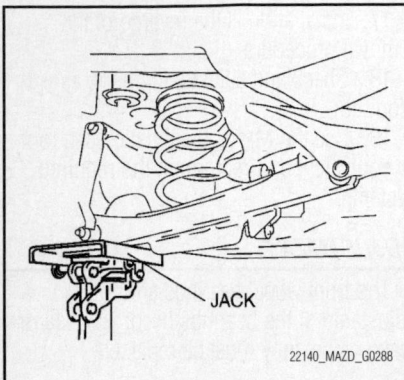

Fig. 361 Support the rear lower arm using a jack —2009 vehicles

To install:

➡ Be sure to use new fasteners, as required.

➡ Before installation of the coil spring check the spring seat rubber. Replace as needed.

✳✳ CAUTION

Installing the coil spring is dangerous. The coil spring could fly off and cause serious injury or death, and damage the vehicle.

10. Align the upper end of the rear coil spring with the step of the upper spring seat rubber.

11. Align the lower end of the rear coil spring with the step of the lower spring seat rubber.

12. Install the lower arm outer bolt and tighten to 56—75 ft. lbs. (75—102 Nm).

13. Tighten the inner bolt to 56—75 ft. lbs. (75—102 Nm).

14. Inspect rear wheel alignment.

15. Use the Mazda diagnostic scan tool, or equivalent and reprogram the required systems.

2010 Vehicles

See Figure 363.

1. Before servicing the vehicle, refer to the Precautions Section.

➡ If working near and/or around the SRS system and components, be sure to disable the SRS system. Tape the negative battery cable with insulating tape. Always disconnect the negative battery cable first.

✳✳ CAUTION

To avoid personal injury when working on vehicles equipped with an air bag, the negative battery cable must be disconnected and at least one minute must elapse before working on the system. Failure to do so may result in deployment of the air bag.

2. Remove the battery top cover, if equipped.

3. Disconnect the negative battery cable. Tape the cable with insulating tape.

➡ When disconnecting the cable, some systems need to be initialized after the cable is reconnected. You will need Mazda diagnostic scan tool or equivalent. Follow the directions on the tool.

4. Raise and safely support the vehicle.

5. Remove the tire and wheel assembly, as required.

6. Disconnect the auto leveling sensor, lower side, if equipped.

7. Disconnect the stabilizer control link upper nut.

8. Disconnect the lower arm outer bolt. Be sure to properly support the lower arm using a suitable jack.

9. Remove the spring from its mounting. Be sure to properly support the lower arm using a suitable jack.

➡ Removing and installing the coil spring is dangerous. The spring could fly off and cause serious injury or death and damage to the vehicle. Be sure to properly support the lower arm before spring removal, using the proper service jack.

10. Remove the upper spring seat.

To install:

➡ Be sure to use new fasteners, as required.

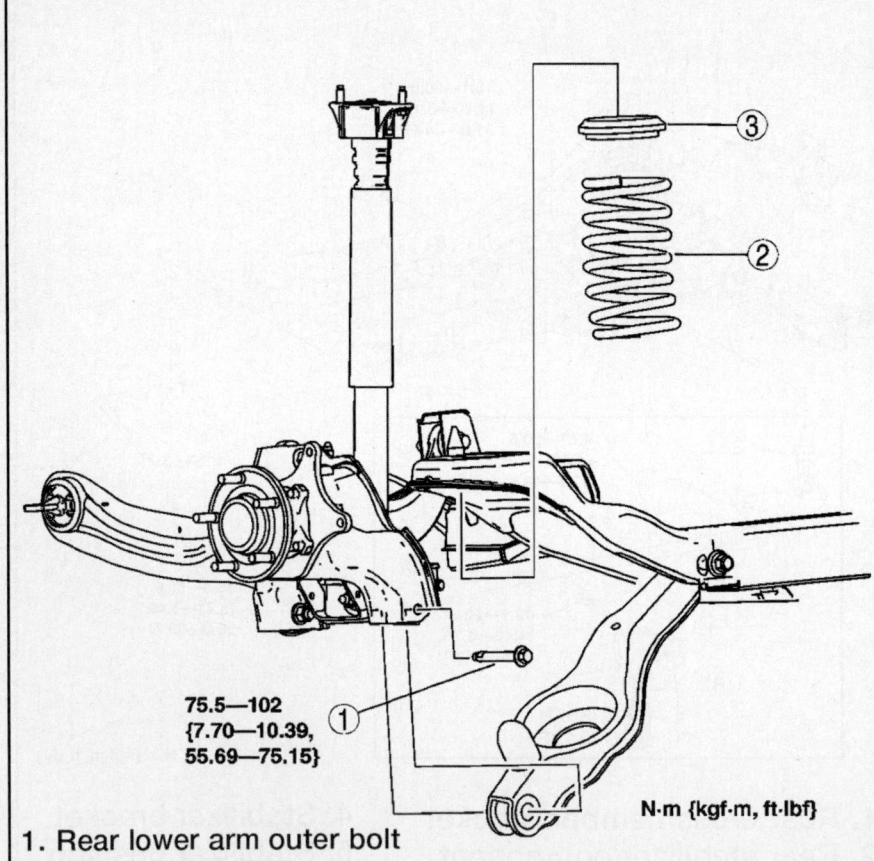

75.5—102
{7.70—10.39,
55.69—75.15}

N·m {kgf·m, ft·lbf}

1. Rear lower arm outer bolt
2. Rear coil spring
3. Upper spring seat rubber

Fig. 362 Rear coil spring exploded view—2009 vehicles

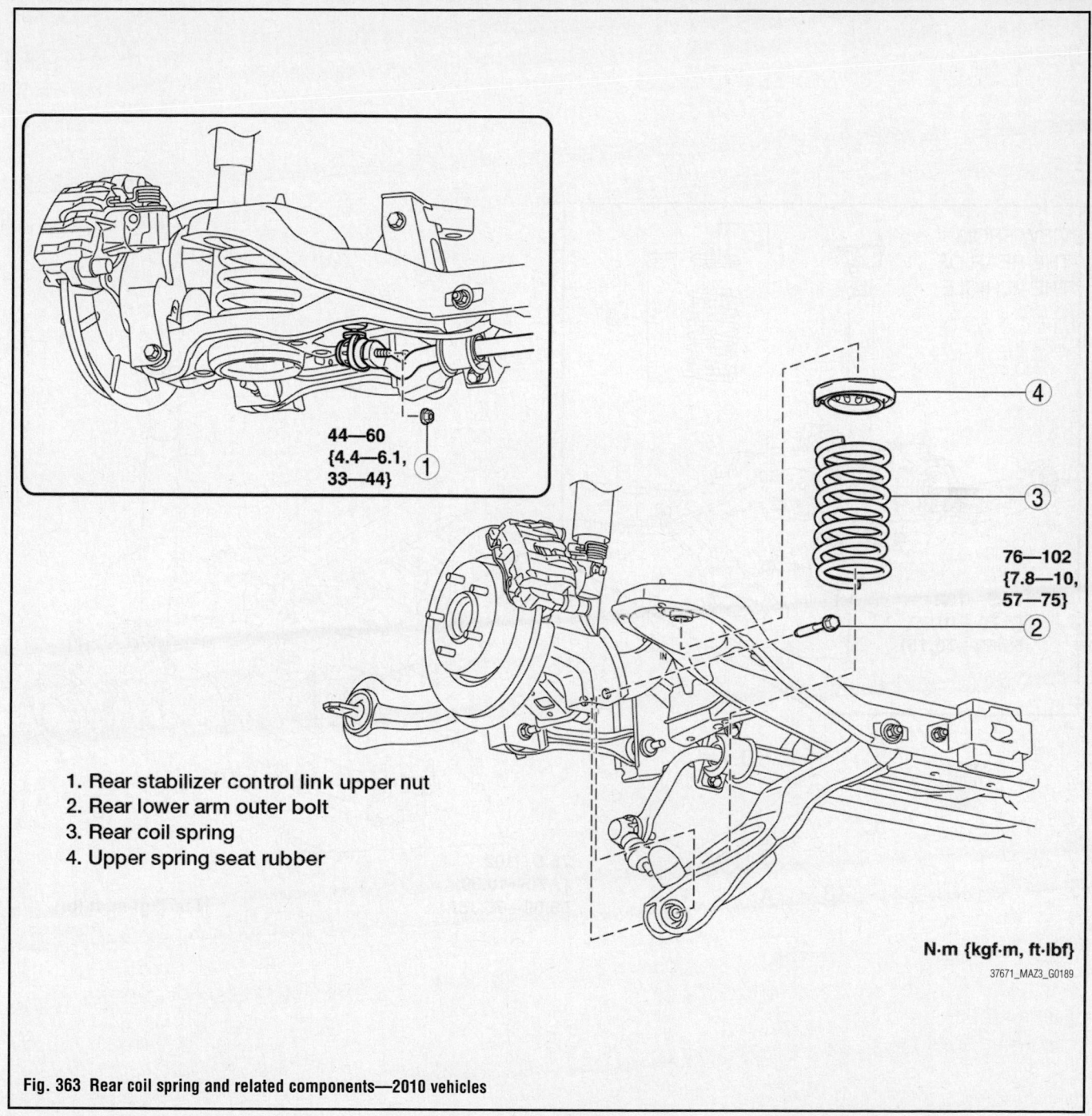

1. Rear stabilizer control link upper nut
2. Rear lower arm outer bolt
3. Rear coil spring
4. Upper spring seat rubber

44—60
{4.4—6.1,
33—44}

76—102
{7.8—10,
57—75}

N·m {kgf·m, ft·lbf}

37671_MAZ3_G0189

Fig. 363 Rear coil spring and related components—2010 vehicles

11. Installation is the reverse of the removal procedure.

12. Check and adjust alignment, as required.

13. Use the Mazda diagnostic scan tool, or equivalent and reprogram the required systems.

CONTROL ARMS/LINKS

REMOVAL & INSTALLATION

Lateral Link

See Figures 364 through 366.

1. Before servicing the vehicle, refer to the Precautions Section.

➡**If working near and/or around the SRS system and components, be sure to disable the SRS system. Tape the negative battery cable with insulating tape. Always disconnect the negative battery cable first.**

❄ CAUTION

To avoid personal injury when working on vehicles equipped with an air bag, the negative battery cable must

be disconnected and at least one minute must elapse before working on the system. Failure to do so may result in deployment of the air bag.

2. Remove the battery top cover, if equipped.

3. Disconnect the negative battery cable. Tape the cable with insulating tape.

➡**When disconnecting the cable, some systems need to be initialized after the cable is reconnected. You will need Mazda diagnostic scan tool or equivalent. Follow the directions on the tool.**

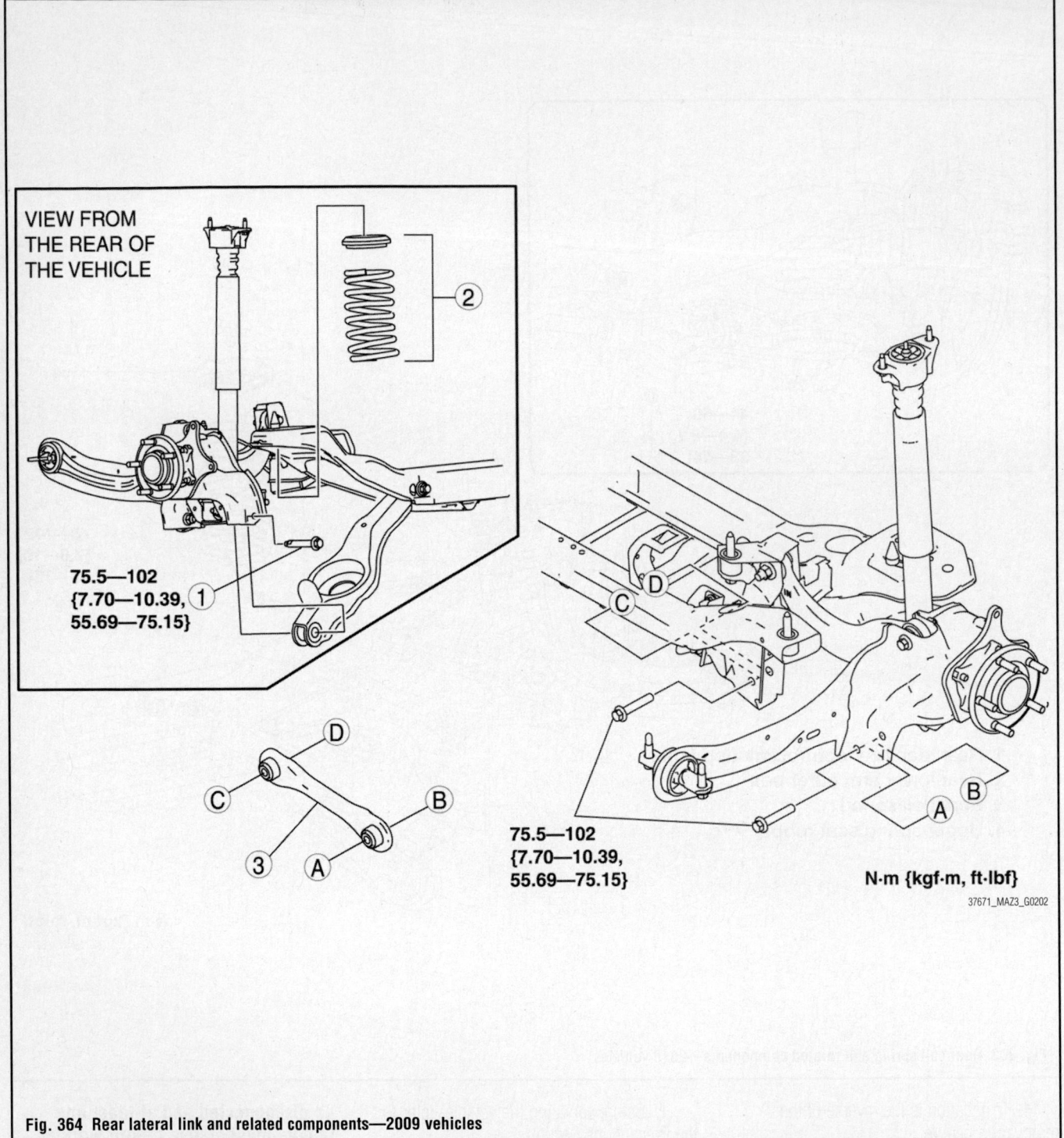

VIEW FROM
THE REAR OF
THE VEHICLE

75.5—102
{7.70—10.39,
55.69—75.15}

N·m {kgf·m, ft·lbf}

37671_MAZ3_G0202

Fig. 364 Rear lateral link and related components—2009 vehicles

➡Performing the following procedure without removing the ABS wheel sensor may cause an open circuit in the harness if it is pulled by mistake. Before performing the procedure remove the ABS wheel speed sensor (axle side) and position it to an appropriate place where the sensor will not be pulled by mistake while servicing the vehicle.

4. Raise and support the vehicle safely.

5. Remove the tire and wheel assembly, as required.

6. On 2009 vehicles, remove the stabilizer bar.

7. Remove the stabilizer control link upper nut.

8. Support the trailing link using a suitable jack.

9. Remove the component from its mounting.

To install:

➡Be sure to use new fasteners, as required.

10. Installation is the reverse of the removal procedure.

11. Install the link so that the rib is facing toward the front of the vehicle.

12. Check and adjust alignment, as required.

13. Use the Mazda diagnostic scan tool,

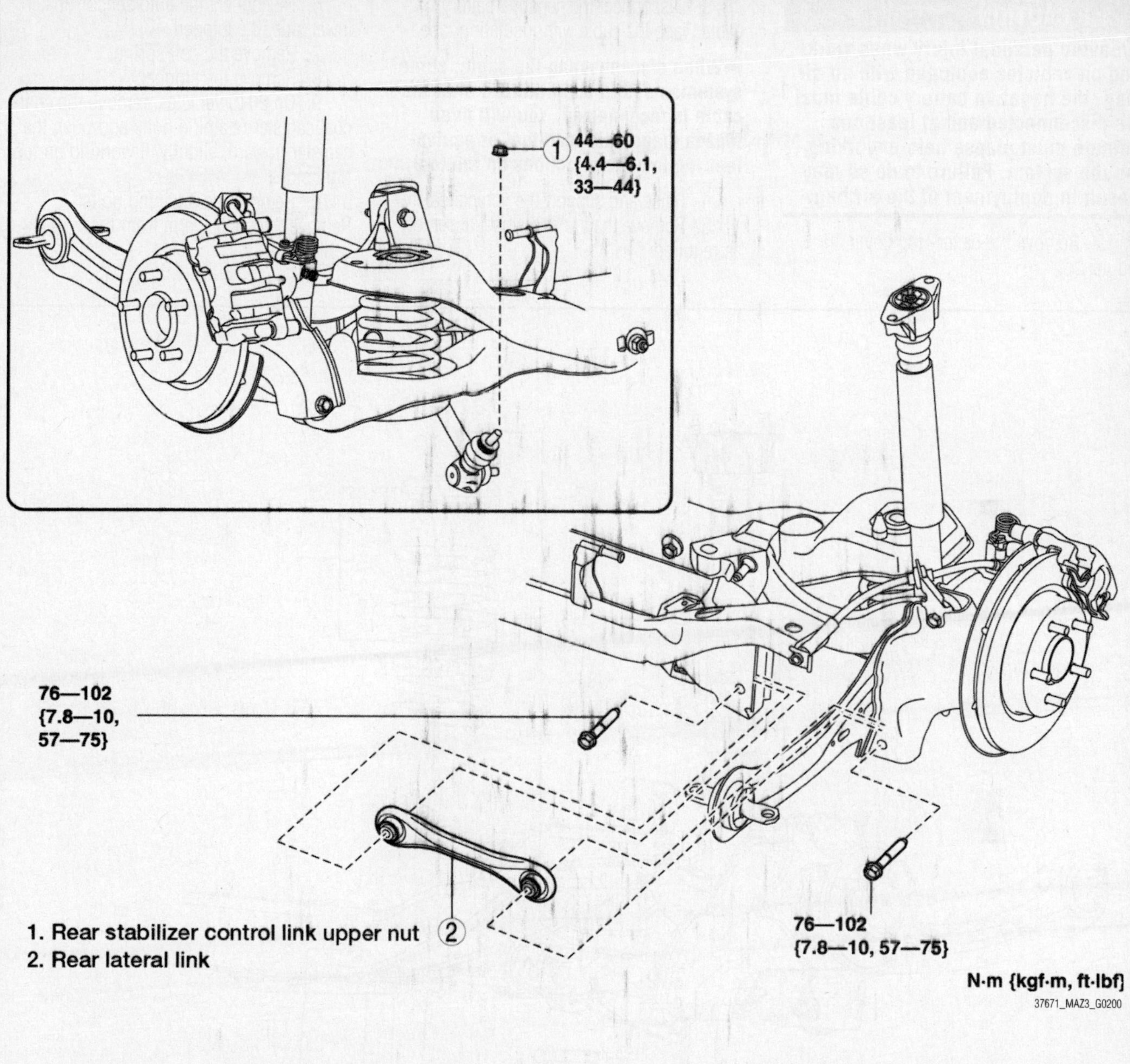

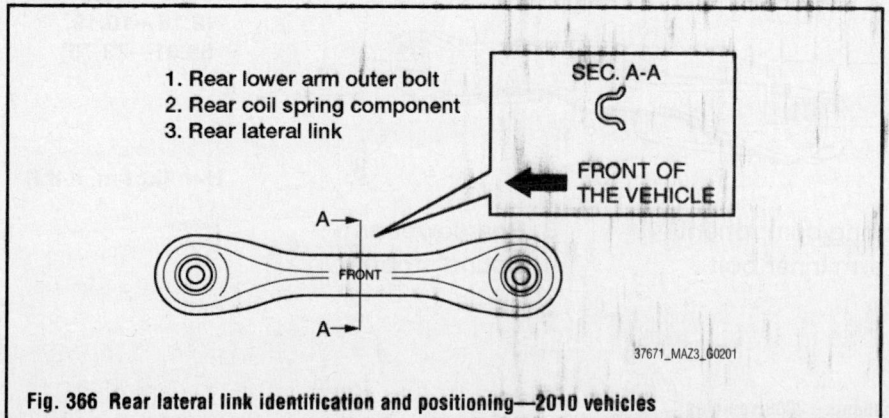

44—60
{4.4—6.1,
33—44}

76—102
{7.8—10,
57—75}

76—102
{7.8—10, 57—75}

1. Rear stabilizer control link upper nut ②
2. Rear lateral link

N·m {kgf·m, ft·lbf}

37671_MAZ3_G0200

Fig. 365 Rear lateral link and related components—2010 vehicles

1. Rear lower arm outer bolt
2. Rear coil spring component
3. Rear lateral link

SEC. A-A

FRONT OF
THE VEHICLE

A→

FRONT

A→

37671_MAZ3_G0201

Fig. 366 Rear lateral link identification and positioning—2010 vehicles

or equivalent and reprogram the required systems.

Lower Control Arm

See Figures 367 and 368.

1. Before servicing the vehicle, refer to the Precautions Section.

➡**If working near and/or around the SRS system and components, be sure to disable the SRS system. Tape the negative battery cable with insulating tape. Always disconnect the negative battery cable first.**

2. Remove the battery top cover, if equipped.

3. Disconnect the negative battery cable. Tape the cable with insulating tape.

➡**When disconnecting the cable, some systems need to be initialized after the cable is reconnected. You will need Mazda diagnostic scan tool or equivalent. Follow the directions on the tool.**

4. Raise and support the vehicle safely.

5. Remove the tire and wheel assembly, as required.

6. Disconnect the auto sensor link, lower side, if equipped.

7. Remove the coil spring.

8. Remove the stopper.

9. On 2010 vehicles, remove the charcoal canister retaining bolts and push the canister upward, slightly, if working on the right side.

10. Remove the retaining bolts. Remove the control arm from the vehicle.

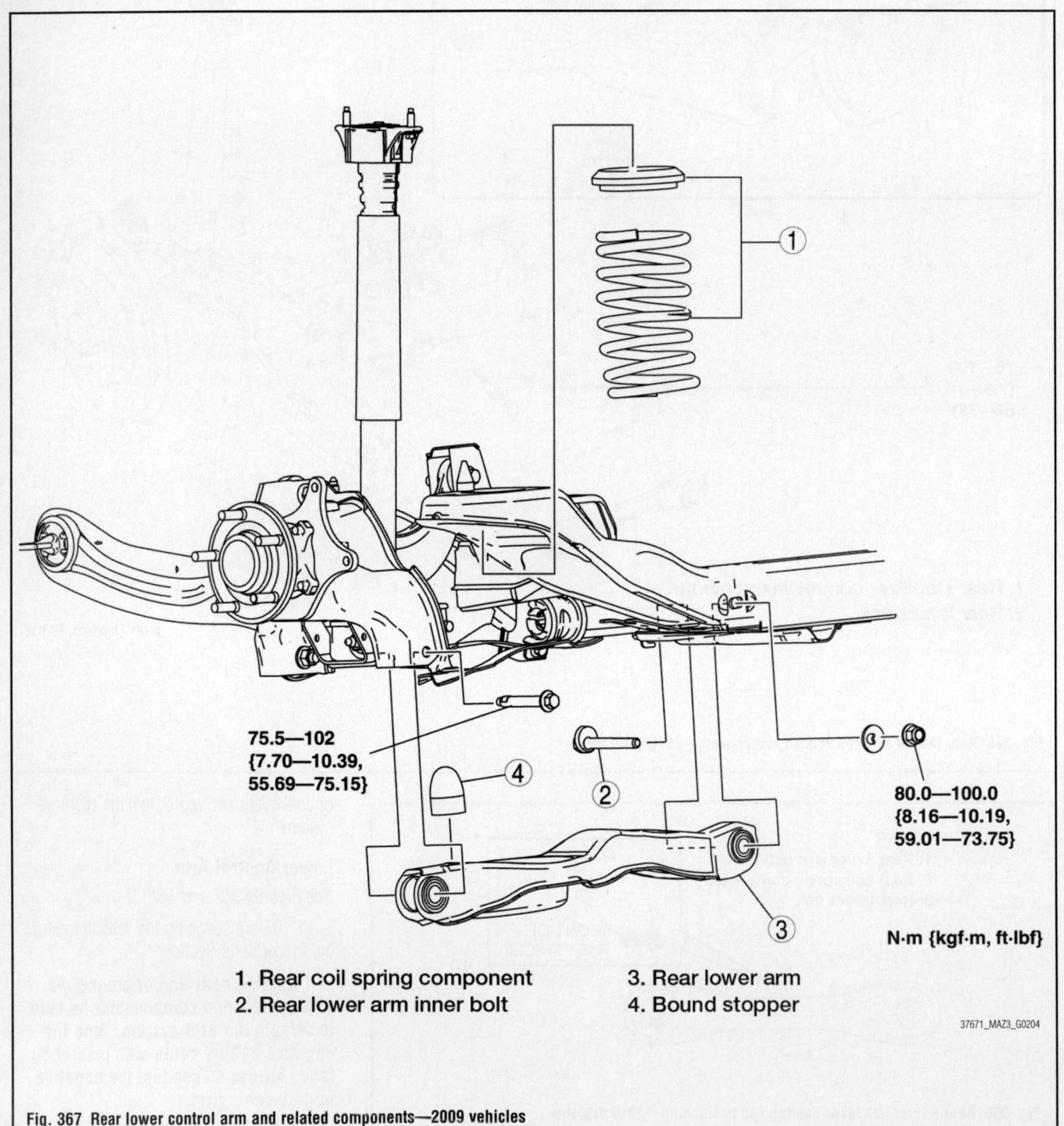

75.5—102
{7.70—10.39,
55.69—75.15}

80.0—100.0
{8.16—10.19,
59.01—73.75}

N·m {kgf·m, ft·lbf}

1. Rear coil spring component
2. Rear lower arm inner bolt
3. Rear lower arm
4. Bound stopper

37671_MAZ3_G0204

Fig. 367 Rear lower control arm and related components—2009 vehicles

➡Be sure to support the component using a suitable jack prior to removal.

To install:

➡Be sure to use new fasteners, as required.

11. Installation is the reverse of the removal procedure.

12. Check and adjust alignment, as required.

13. Use the Mazda diagnostic scan tool, or equivalent and reprogram the required systems.

Trailing Link

2009 Vehicles

See Figure 369.

1. Before servicing the vehicle, refer to the Precautions Section.

➡If working near and/or around the SRS system and components, be sure to disable the SRS system. Tape the negative battery cable with insulating tape. Always disconnect the negative battery cable first.

✳✳ CAUTION

To avoid personal injury when working on vehicles equipped with an air bag, the negative battery cable must be disconnected and at least one minute must elapse before working on the system. Failure to do so may result in deployment of the air bag.

---

2. Remove the battery top cover, if equipped.

3. Disconnect the negative battery cable. Tape the cable with insulating tape.

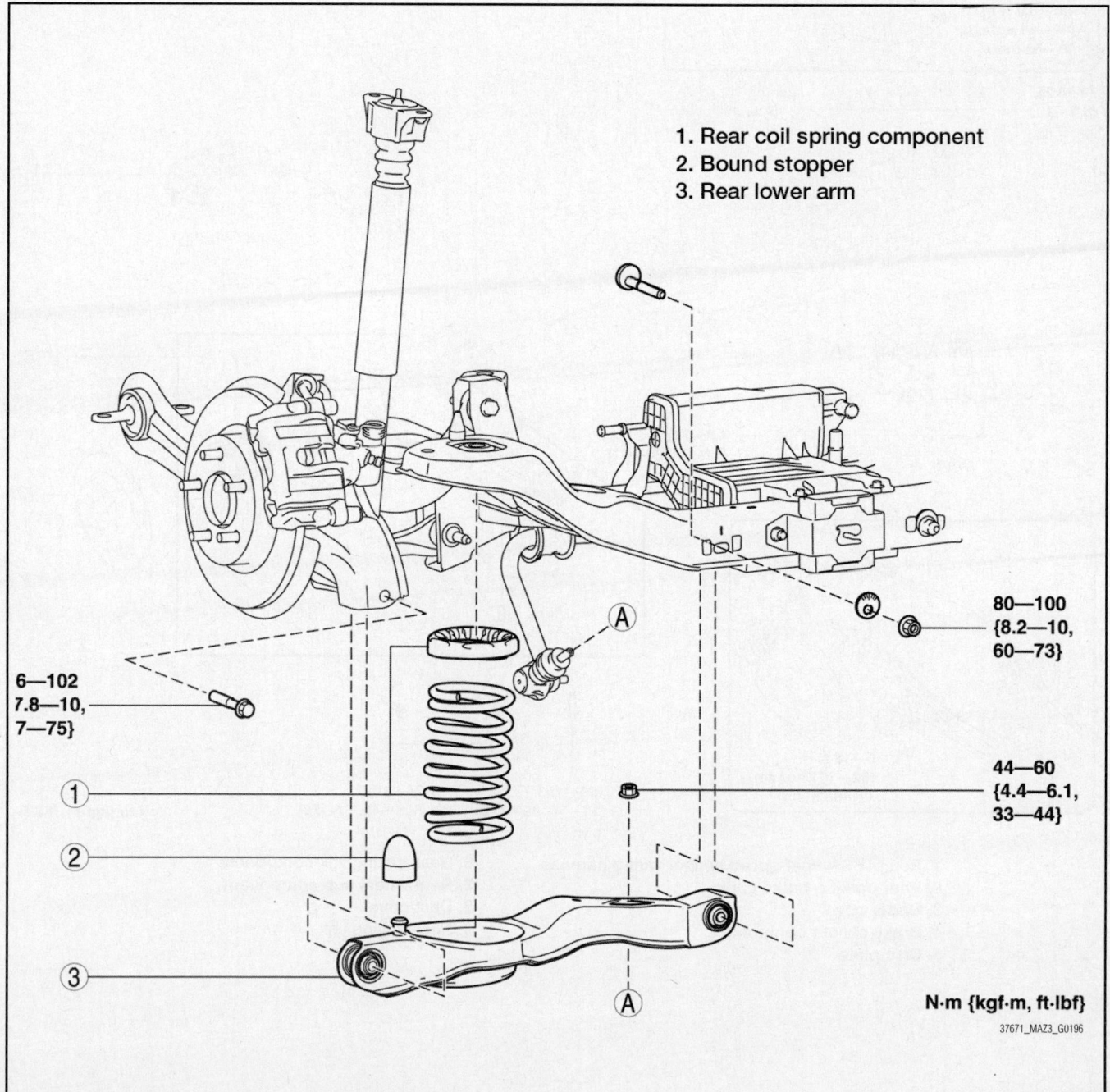

1. Rear coil spring component
2. Bound stopper
3. Rear lower arm

6—102
7.8—10,
7—75}

80—100
{8.2—10,
60—73}

44—60
{4.4—6.1,
33—44}

N·m {kgf·m, ft·lbf}

37671_MAZ3_G0196

Fig. 368 Rear lower control arm and related components—2010 vehicles

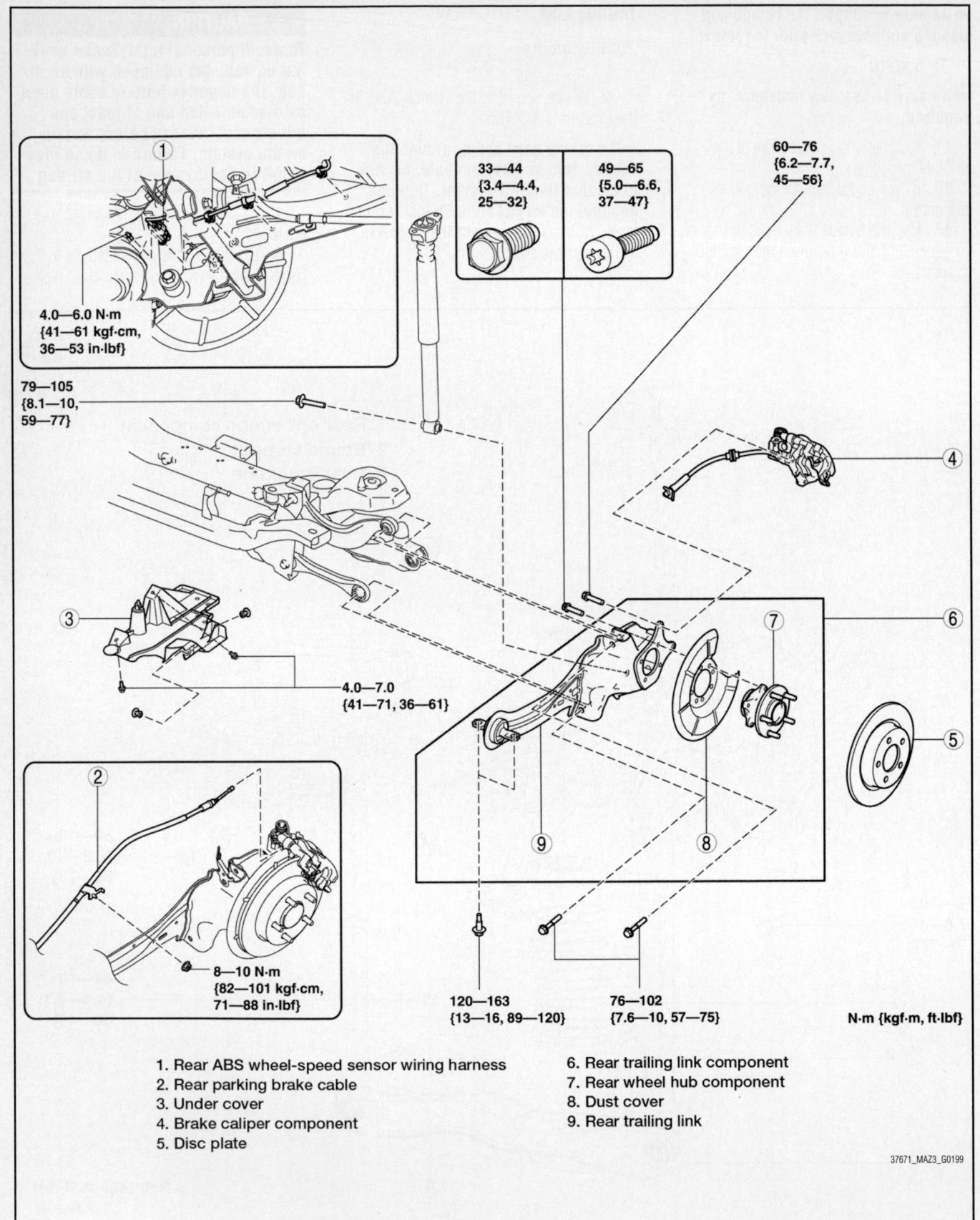

Fig. 369 Rear trailing link and related components—2010 vehicles

1. Rear ABS wheel-speed sensor wiring harness
2. Rear parking brake cable
3. Under cover
4. Brake caliper component
5. Disc plate
6. Rear trailing link component
7. Rear wheel hub component
8. Dust cover
9. Rear trailing link

37671_MAZ3_G0199

➡When disconnecting the cable, some systems need to be initialized after the cable is reconnected. You will need Mazda diagnostic scan tool or equivalent. Follow the directions on the tool.

➡Performing the following procedure without removing the ABS wheel sensor may cause an open circuit in the harness if it is pulled by mistake. Before performing the procedure remove the ABS wheel speed sensor (axle side) and position it to an appropriate place where the sensor will not be pulled by mistake while servicing the vehicle.

4. Raise and support the vehicle safely.
5. Remove the tire and wheel assembly, as required.
6. Disconnect the ABS sensor wiring harness connector.
7. Disconnect the rear parking brake cable.
8. Remove the brake caliper. Do not allow the caliper to hang by the hose.
9. Remove the rotor.
10. Remove the rear hub.
11. Remove the stud cover.
12. Remove the rear spring.
13. Remove the rear lateral link outer bolt.

➡Removing this component is dangerous. This component could fall and cause serious injury or death. Verify that the jack securely supports the trailing link.

14. Remove the upper arm outer bolt.
15. Remove the shock lower bolt.
16. Remove the rear trailing link.
17. Remove the bushing.

➡Removing the trailing arm is dangerous. This component could fall and cause serious injury or death. Verify that the jack securely supports the trailing link.

18. Support the trailing link using a suitable jack. Remove the front side bolts.
19. Remove the shock lower side bolt.
20. Remove the lateral link outer side bolt.
21. Remove the upper arm outer side bolt.
22. Remove the component from its mounting.

To install:

➡Be sure to use new fasteners, as required.

23. Installation is the reverse of the removal procedure.
24. Check and adjust alignment, as required.

25. Use the Mazda diagnostic scan tool, or equivalent and reprogram the required systems.

2010 Vehicles

See Figure 369.

1. Before servicing the vehicle, refer to the Precautions Section.

➡If working near and/or around the SRS system and components, be sure to disable the SRS system. Tape the negative battery cable with insulating tape. Always disconnect the negative battery cable first.

✳✳ CAUTION

To avoid personal injury when working on vehicles equipped with an air bag, the negative battery cable must be disconnected and at least one minute must elapse before working on the system. Failure to do so may result in deployment of the air bag.

2. Remove the battery top cover, if equipped.
3. Disconnect the negative battery cable. Tape the cable with insulating tape.

➡When disconnecting the cable, some systems need to be initialized after the cable is reconnected. You will need Mazda diagnostic scan tool or equivalent. Follow the directions on the tool.

➡Performing the following procedure without removing the ABS wheel sensor may cause an open circuit in the harness if it is pulled by mistake. Before performing the procedure remove the ABS wheel speed sensor (axle side) and position it to an appropriate place where the sensor will not be pulled by mistake while servicing the vehicle.

4. Raise and support the vehicle safely.
5. Remove the tire and wheel assembly, as required.
6. Disconnect the auto sensor link, lower side, if equipped.
7. Remove the spring.
8. Disconnect the ABS sensor wiring harness connector.
9. Disconnect the rear parking brake cable.
10. Remove the undercover.
11. Remove the brake caliper. Do not allow the caliper to hang by the hose.
12. Remove the rotor.

➡Removing the trailing arm is dangerous. This component could fall and

cause serious injury or death. Verify that the jack securely supports the trailing link.

13. Support the trailing link using a suitable jack. Remove the front side bolts.
14. Remove the shock lower side bolt.
15. Remove the lateral link outer side bolt.
16. Remove the upper arm outer side bolt.
17. Remove the component from its mounting.

To install:

➡Be sure to use new fasteners, as required.

18. Installation is the reverse of the removal procedure.
19. Check and adjust alignment, as required.
20. Use the Mazda diagnostic scan tool, or equivalent and reprogram the required systems.

Upper Control Arm

2009 Vehicles

See Figures 370 and 371.

1. Before servicing the vehicle, refer to the Precautions Section.

➡If working near and/or around the SRS system and components, be sure to disable the SRS system. Tape the negative battery cable with insulating tape. Always disconnect the negative battery cable first.

✳✳ CAUTION

To avoid personal injury when working on vehicles equipped with an air bag, the negative battery cable must be disconnected and at least one minute must elapse before working on the system. Failure to do so may result in deployment of the air bag.

2. Remove the battery top cover, if equipped.
3. Disconnect the negative battery cable. Tape the cable with insulating tape.

➡When disconnecting the cable, some systems need to be initialized after the cable is reconnected. You will need Mazda diagnostic scan tool or equivalent. Follow the directions on the tool.

4. Raise and support the vehicle safely.
5. Remove the tire and wheel assembly, as required.

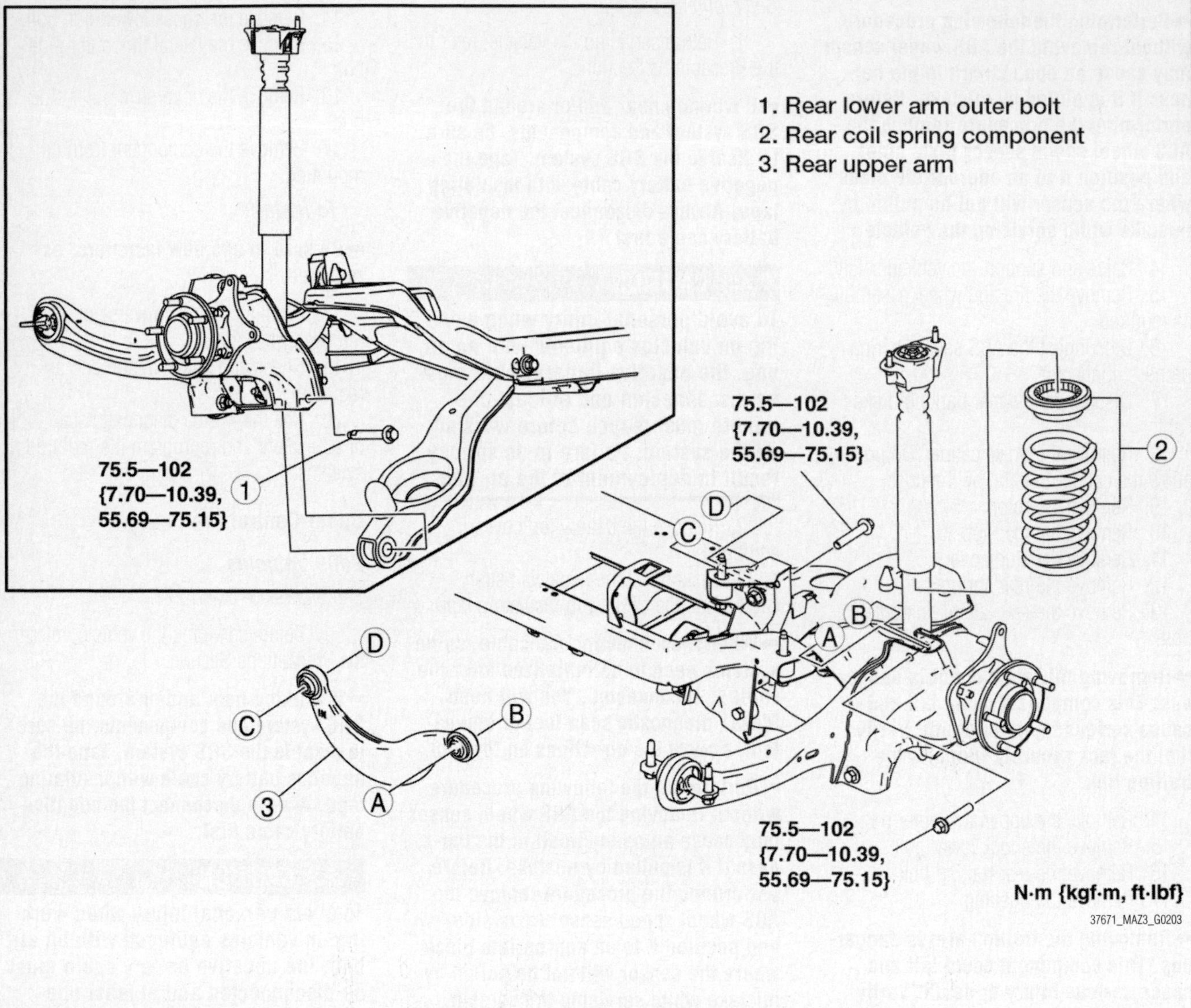

1. Rear lower arm outer bolt
2. Rear coil spring component
3. Rear upper arm

75.5—102
{7.70—10.39,
55.69—75.15}

75.5—102
{7.70—10.39,
55.69—75.15}

75.5—102
{7.70—10.39,
55.69—75.15}

N·m {kgf·m, ft·lbf}

37671_MAZ3_G0203

Fig. 370 Rear upper control arm and related components—2009 vehicles

6. Remove the EVAP system leak detection pump.

7. Remove the rear lower arm outer bolt. Loosen the inner bolt. Remove the outer bolt.

➡**Be sure to support the lower arm using a suitable jack.**

8. Remove the rear spring.

9. Remove the upper arm from its mounting.

10. Disconnect the stabilizer control link upper side nut.

11. Remove the retaining bolts. Remove the control arm from the vehicle.

➡**Be sure the vehicle is in the unloaded condition and properly support the trailing link using a suitable jack.**

To install:

➡**Be sure to use new fasteners, as required.**

12. Installation is the reverse of the removal procedure.

13. Check and adjust alignment, as required.

14. Use the Mazda diagnostic scan tool, or equivalent and reprogram the required systems.

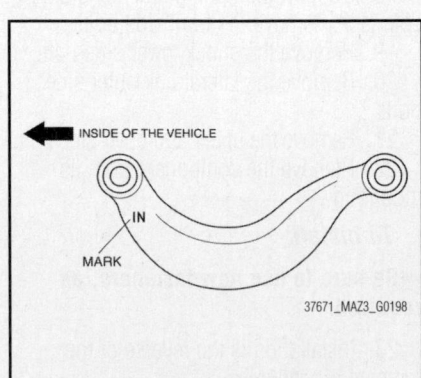

INSIDE OF THE VEHICLE

IN

MARK

37671_MAZ3_G0198

Fig. 371 Rear upper control arm identification and positioning—2009 vehicles

2010 Vehicles

See Figures 372 and 373.

1. Before servicing the vehicle, refer to the Precautions Section.

➡ If working near and/or around the SRS system and components, be sure to disable the SRS system. Tape the negative battery cable with insulating tape. Always disconnect the negative battery cable first.

⚹⚹ CAUTION

To avoid personal injury when working on vehicles equipped with an air bag, the negative battery cable must be disconnected and at least one minute must elapse before working on the system. Failure to do so may result in deployment of the air bag.

2. Remove the battery top cover, if equipped.

3. Disconnect the negative battery cable. Tape the cable with insulating tape.

➡ When disconnecting the cable, some systems need to be initialized after the cable is reconnected. You will need Mazda diagnostic scan tool or equivalent. Follow the directions on the tool.

4. Raise and support the vehicle safely.
5. Remove the tire and wheel assembly, as required.

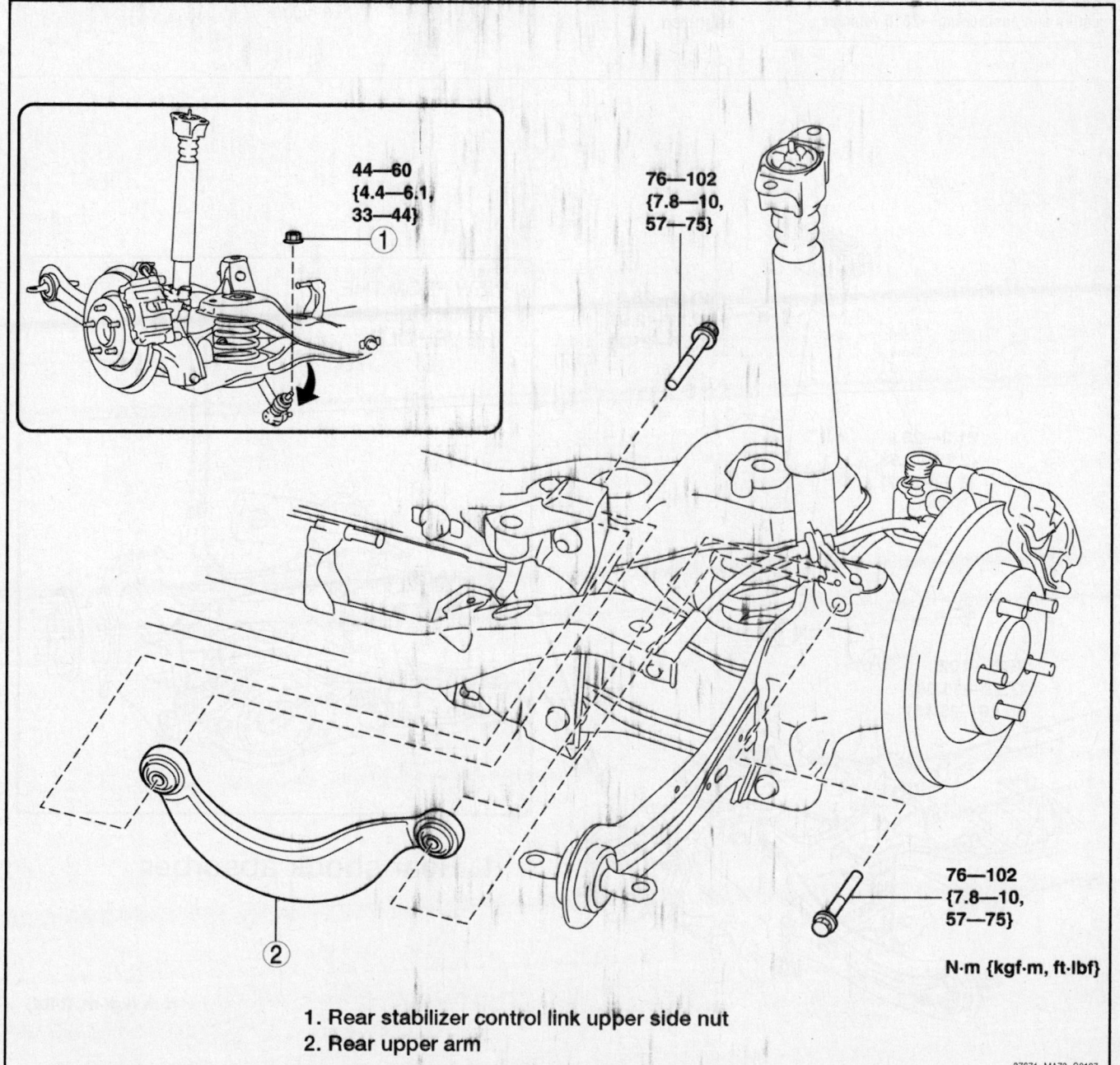

44—60
{4.4—6.1,
33—44}
①

76—102
{7.8—10,
57—75}

76—102
{7.8—10,
57—75}

N·m {kgf·m, ft·lbf}

1. Rear stabilizer control link upper side nut
2. Rear upper arm

37671_MAZ3_G0197

Fig. 372 Rear upper control arm and related components—2010 vehicles

INSIDE OF THE VEHICLE
IN
MARK
37671_MAZ3_G0198

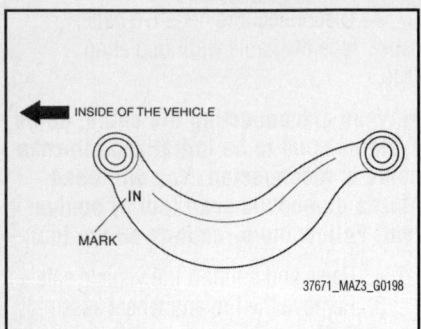

Fig. 373 Rear upper control arm identification and positioning—2010 vehicles

6. Disconnect the charcoal canister vent solenoid valve from the rear crossmember, if working on the right side.
7. Disconnect the stabilizer control link upper side nut.
8. Remove the retaining bolts. Remove the control arm from the vehicle.

➡**Be sure the vehicle is in the unloaded condition and properly support the trailing link using a suitable jack.**

To install:

➡**Be sure to use new fasteners, as required.**

9. Installation is the reverse of the removal procedure.
10. Check and adjust alignment, as required.
11. Use the Mazda diagnostic scan tool, or equivalent and reprogram the required systems.

SHOCK ABSORBER

REMOVAL & INSTALLATION

2009 Vehicles

See Figure 374.

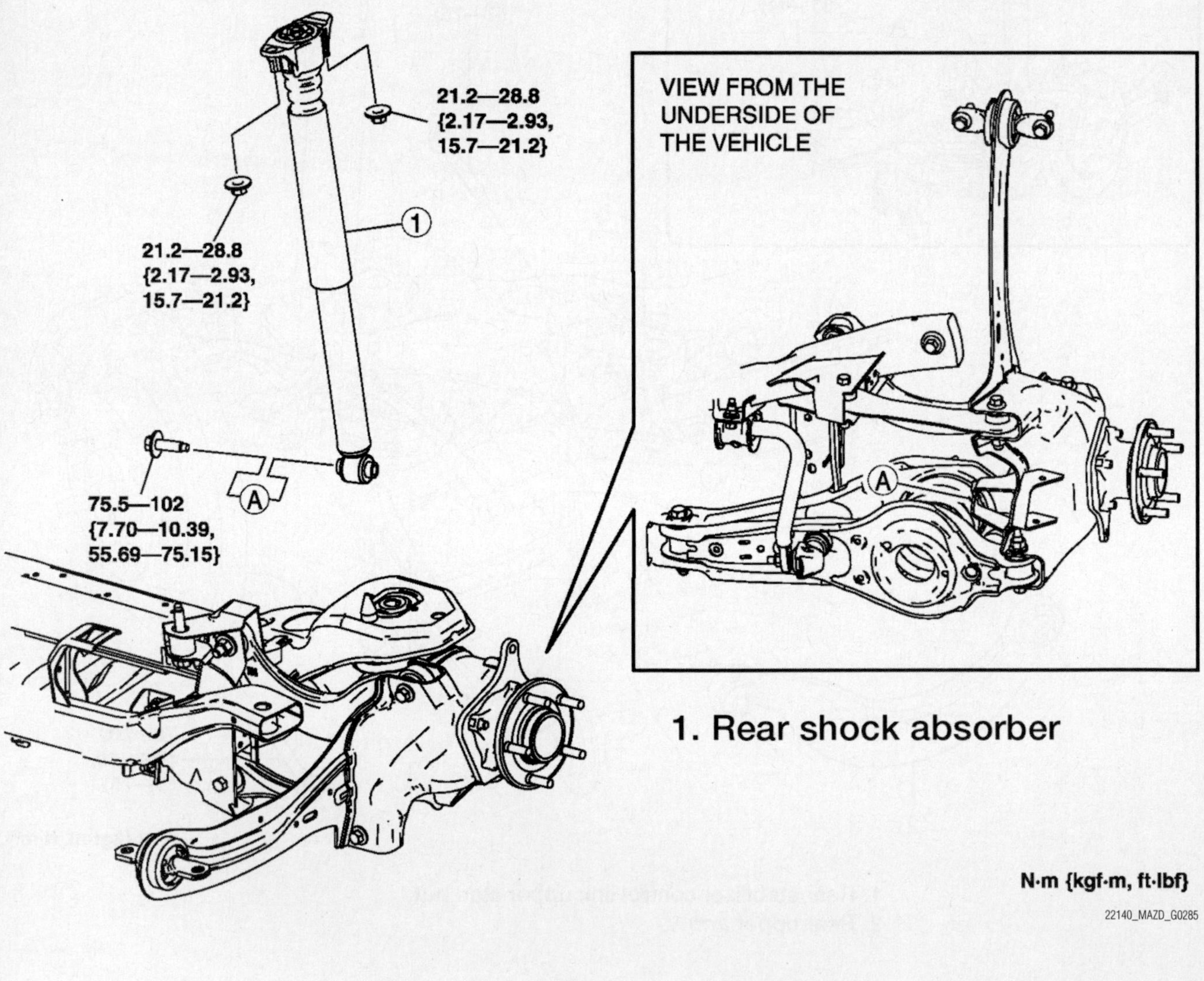

21.2—28.8
{2.17—2.93,
15.7—21.2}
21.2—28.8
{2.17—2.93,
15.7—21.2}
1
75.5—102
{7.70—10.39,
55.69—75.15}
A
VIEW FROM THE
UNDERSIDE OF
THE VEHICLE
A
1. Rear shock absorber
N·m {kgf·m, ft·lbf}
22140_MAZD_G0285

Fig. 374 Rear shock absorber and related components—2009 vehicles

1. Before servicing the vehicle, refer to the Precautions Section.

➡**If working near and/or around the SRS system and components, be sure to disable the SRS system. Tape the negative battery cable with insulating tape. Always disconnect the negative battery cable first.**

✴✴ CAUTION

To avoid personal injury when working on vehicles equipped with an air bag, the negative battery cable must be disconnected and at least one minute must elapse before working on the system. Failure to do so may result in deployment of the air bag.

2. Remove the battery top cover, if equipped.

3. Disconnect the negative battery cable. Tape the cable with insulating tape.

➡**When disconnecting the cable, some systems need to be initialized after the cable is reconnected. You will need Mazda diagnostic scan tool or equivalent. Follow the directions on the tool.**

4. Support the rear axle assembly with a jack.

5. Remove or disconnect the following:
  • Rear wheel(s)
  • Top shock absorber nuts
  • Bottom nut and bolt
  • Shock absorber

To install:

➡**Be sure to use new fasteners, as required.**

6. Install or connect the following:
  • Shock absorber
  • Bottom nut and bolt and tighten to 56—75 ft. lbs. (76—102Nm).
  • Upper mounting nuts and tighten to 16—21 ft. lbs. (21–29 Nm)
  • Rear wheel(s)

7. Use the Mazda diagnostic scan tool, or equivalent and reprogram the required systems.

2010 Vehicles
See Figure 375.

1. Before servicing the vehicle, refer to the Precautions Section.

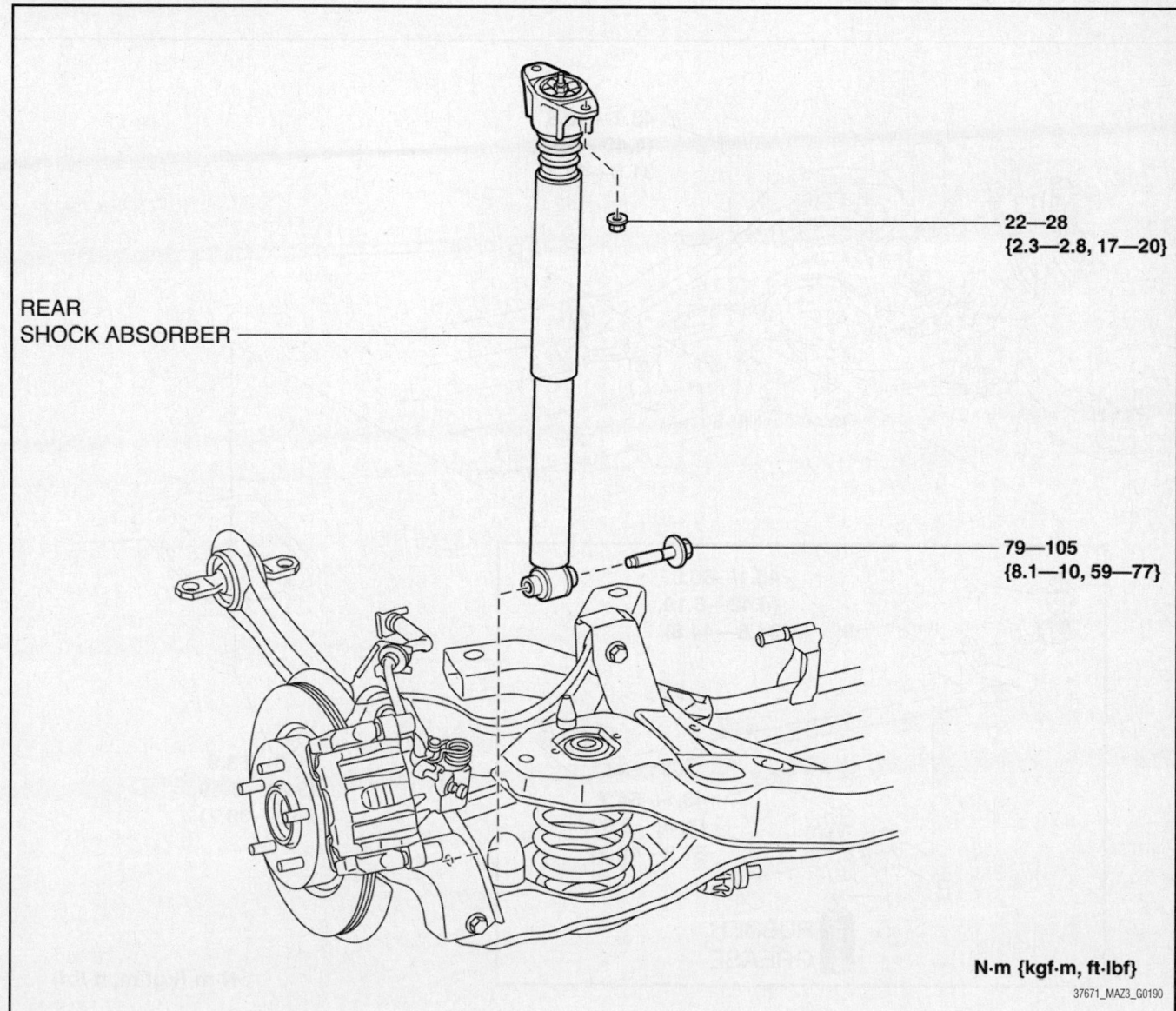

REAR SHOCK ABSORBER

22—28 {2.3—2.8, 17—20}

79—105 {8.1—10, 59—77}

N·m {kgf·m, ft·lbf}

37671_MAZ3_G0190

Fig. 375 Rear shock absorber and related components—2010 vehicles

➡If working near and/or around the SRS system and components, be sure to disable the SRS system. Tape the negative battery cable with insulating tape. Always disconnect the negative battery cable first.

❊❊ CAUTION

To avoid personal injury when working on vehicles equipped with an air bag, the negative battery cable must be disconnected and at least one minute must elapse before working on the system. Failure to do so may result in deployment of the air bag.

2. Remove the battery top cover, if equipped.

3. Disconnect the negative battery cable. Tape the cable with insulating tape.

➡When disconnecting the cable, some systems need to be initialized after the cable is reconnected. You will need Mazda diagnostic scan tool or equivalent. Follow the directions on the tool.

4. Raise and safely support the vehicle.
5. Remove the tire and wheel assembly, as required.
6. Properly support the rear axle using a suitable jack.
7. Remove the shock retaining bolts.
8. Remove the shock from the vehicle.

To install:

➡Be sure to use new fasteners, as required.

9. Installation is the reverse of the removal procedure.

10. Use the Mazda diagnostic scan tool, or equivalent and reprogram the required systems.

TESTING

Road test the vehicle to check for excessive bouncing rolling, or unusual noises.

1. Raise the vehicle safely on a lift and check for the following:
- Fluid leaking from strut or shock
- Worn bushings
- Broken spring
- loose or missing bolts
2. Replace or tighten as needed.

STABILIZER BAR

REMOVAL & INSTALLATION
See Figures 376 through 380.

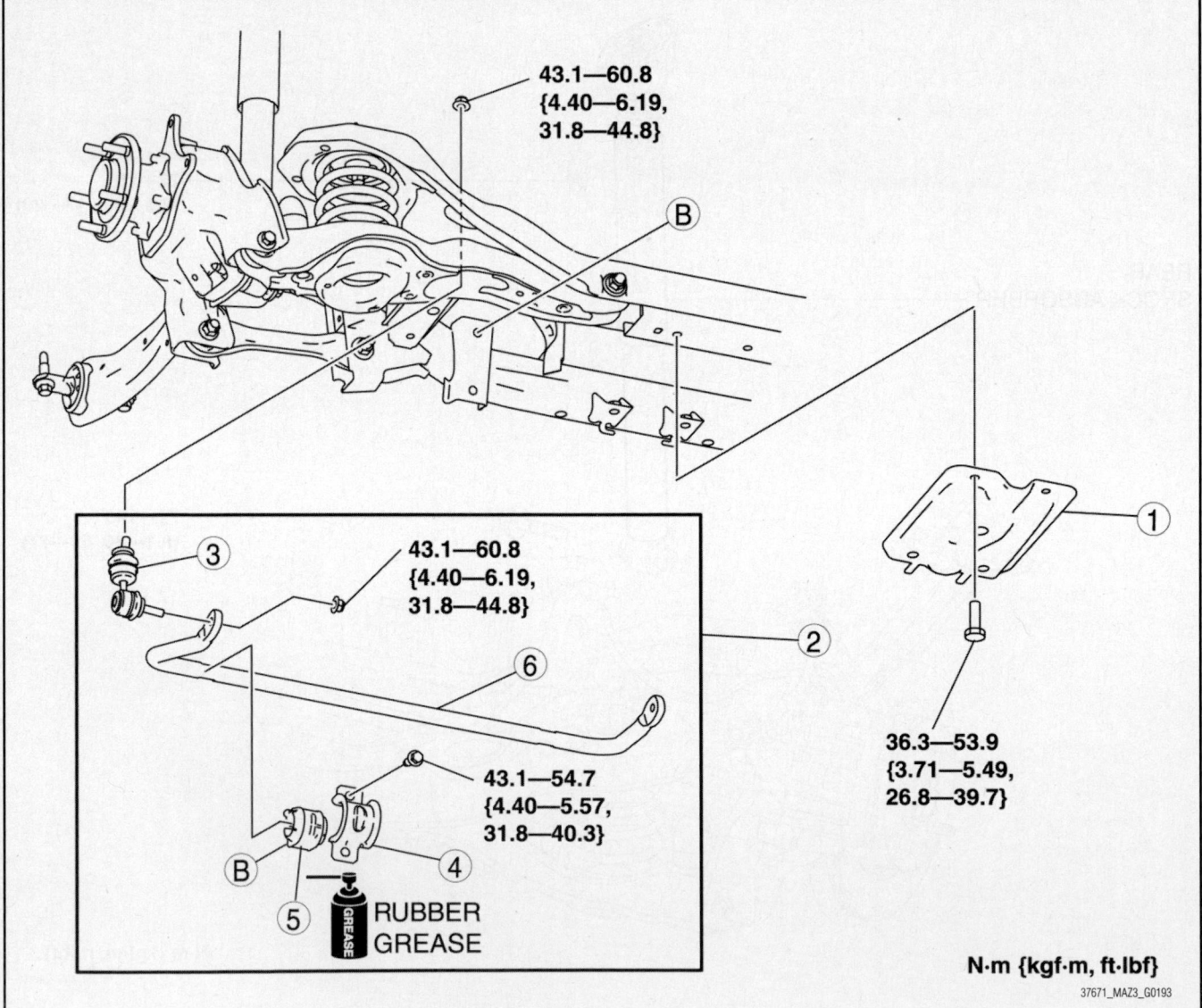

43.1—60.8
{4.40—6.19,
31.8—44.8}

43.1—60.8
{4.40—6.19,
31.8—44.8}

43.1—54.7
{4.40—5.57,
31.8—40.3}

36.3—53.9
{3.71—5.49,
26.8—39.7}

RUBBER GREASE

N·m {kgf·m, ft·lbf}

37671_MAZ3_G0193

Fig. 376 Rear stabilizer bar and related components—2009 vehicles

1. Before servicing the vehicle, refer to the Precautions Section.

➡If working near and/or around the SRS system and components, be sure to disable the SRS system. Tape the negative battery cable with insulating tape. Always disconnect the negative battery cable first.

✳✳ CAUTION

To avoid personal injury when working on vehicles equipped with an air bag, the negative battery cable must be disconnected and at least one minute must elapse before working on the system. Failure to do so may result in deployment of the air bag.

2. Remove the battery top cover, if equipped.

3. Disconnect the negative battery cable. Tape the cable with insulating tape.

➡When disconnecting the cable, some systems need to be initialized after the cable is reconnected. You will need Mazda diagnostic scan tool or equivalent. Follow the directions on the tool.

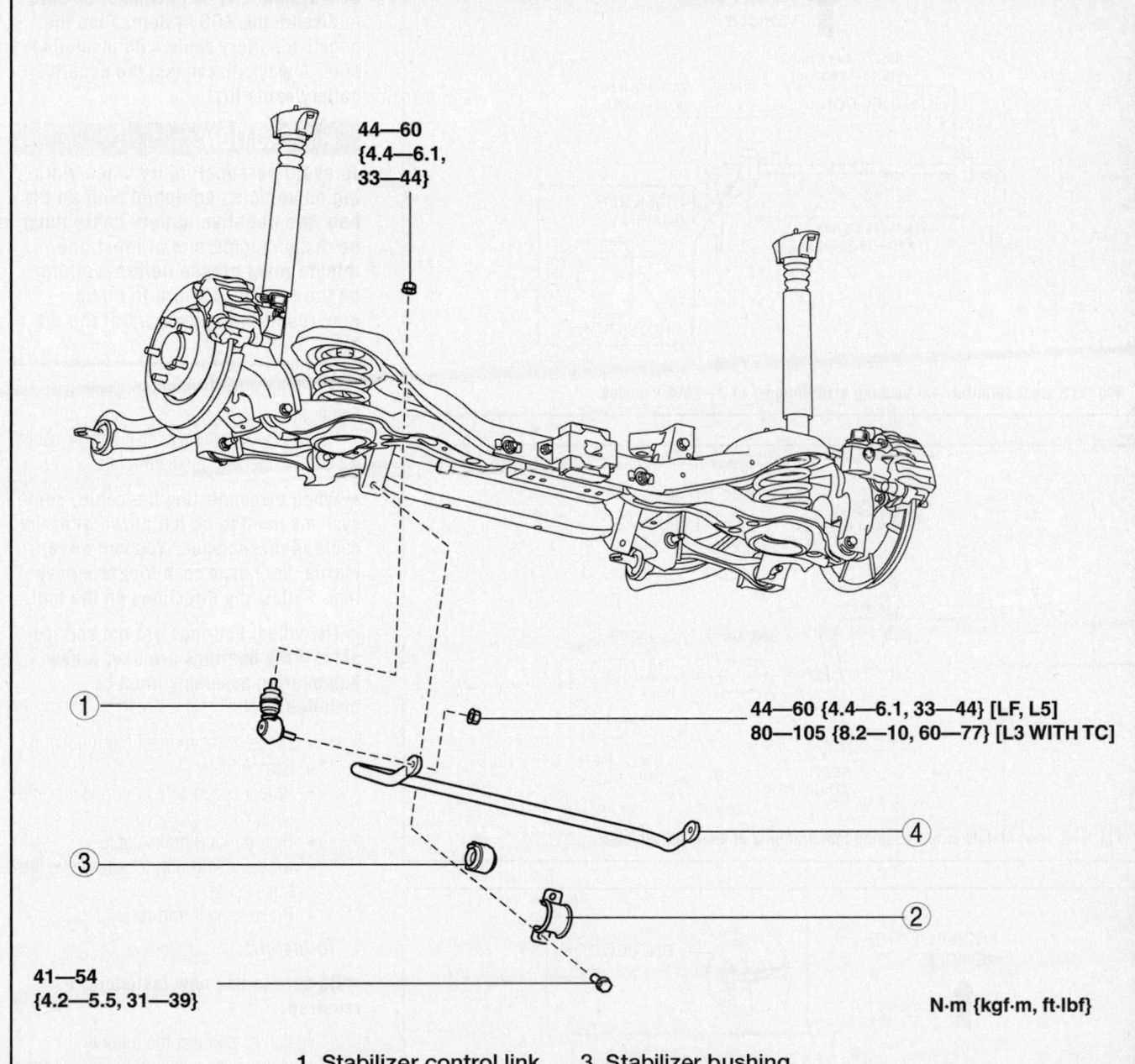

44—60
{4.4—6.1,
33—44}

44—60 {4.4—6.1, 33—44} [LF, L5]
80—105 {8.2—10, 60—77} [L3 WITH TC]

41—54
{4.2—5.5, 31—39}

N·m {kgf·m, ft·lbf}

1. Stabilizer control link
2. Stabilizer bracket
3. Stabilizer bushing
4. Rear stabilizer

37671_MAZ3_G0191

Fig. 377 Rear stabilizer bar and related components—2010 vehicles

4. Raise and safely support the vehicle.

5. Remove the tire and wheel assembly, as required.

6. Remove the stabilizer control link.

7. Remove the stabilizer bracket.

8. Remove the stabilizer bushing.

9. Remove the stabilizer bar from the vehicle.

To install:

➡ **Be sure to use new fasteners, as required.**

10. Installation is the reverse of the removal procedure.

11. Use the Mazda diagnostic scan tool, or equivalent and reprogram the required systems.

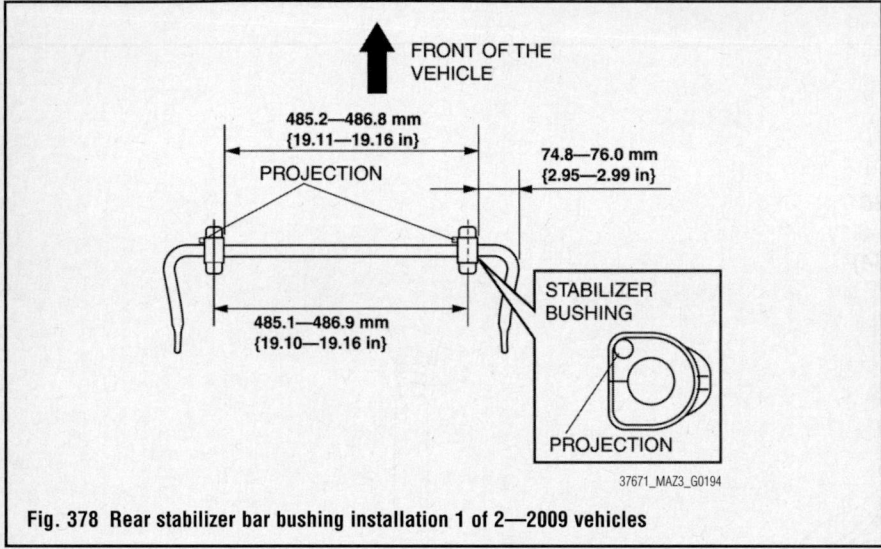

Fig. 378 Rear stabilizer bar bushing installation 1 of 2—2009 vehicles

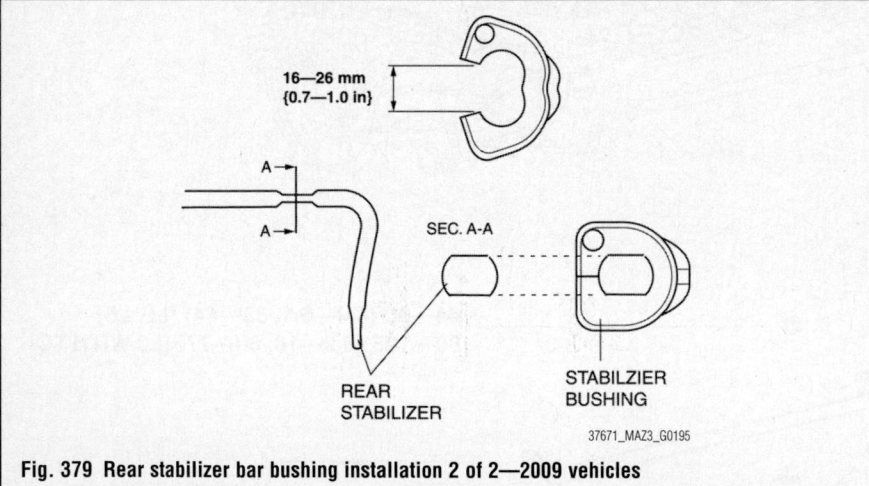

Fig. 379 Rear stabilizer bar bushing installation 2 of 2—2009 vehicles

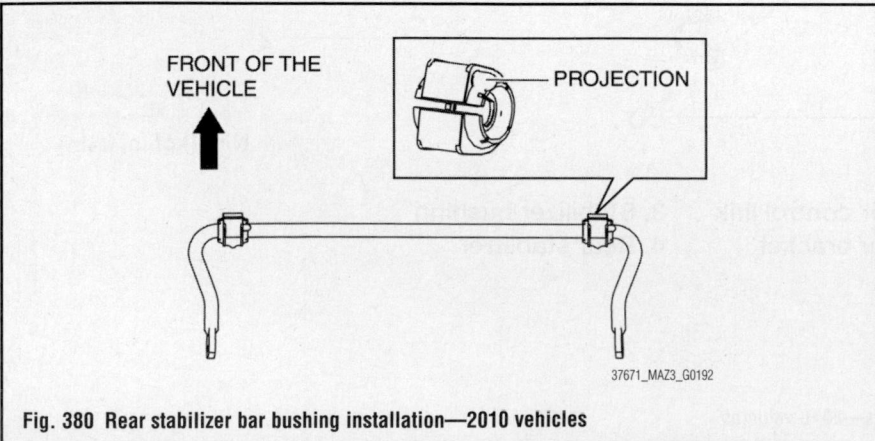

Fig. 380 Rear stabilizer bar bushing installation—2010 vehicles

WHEEL HUB & BEARING

REMOVAL & INSTALLATION

2009 Vehicles

See Figures 381 and 382.

1. Before servicing the vehicle, refer to the Precautions Section.

➡ **If working near and/or around the SRS system and components, be sure to disable the SRS system. Tape the negative battery cable with insulating tape. Always disconnect the negative battery cable first.**

❊❊ CAUTION

To avoid personal injury when working on vehicles equipped with an air bag, the negative battery cable must be disconnected and at least one minute must elapse before working on the system. Failure to do so may result in deployment of the air bag.

2. Remove the battery top cover, if equipped.

3. Disconnect the negative battery cable. Tape the cable with insulating tape.

➡ **When disconnecting the cable, some systems need to be initialized after the cable is reconnected. You will need Mazda diagnostic scan tool or equivalent. Follow the directions on the tool.**

➡ **The wheel bearings are not serviceable. If the bearings are bad, a new hub/bearing assembly must be installed.**

4. Remove or disconnect the following:
   - Rear wheels
   - Wheel speed sensor connector and sensor
   - Rear parking brake cable
   - Brake hose grommet and move the hose aside
   - Brake caliper and rotor

To install:

➡ **Be sure to use new fasteners, as required.**

5. Install or connect the following:
   - Hub/bearing assembly. Torque the new nut to 36—48 ft. lbs. (49—65 Nm).
   - Brake assembly
   - Rear parking brake cable. Pass the cable inside the rear wheel speed sensor wiring harness as illustrated.

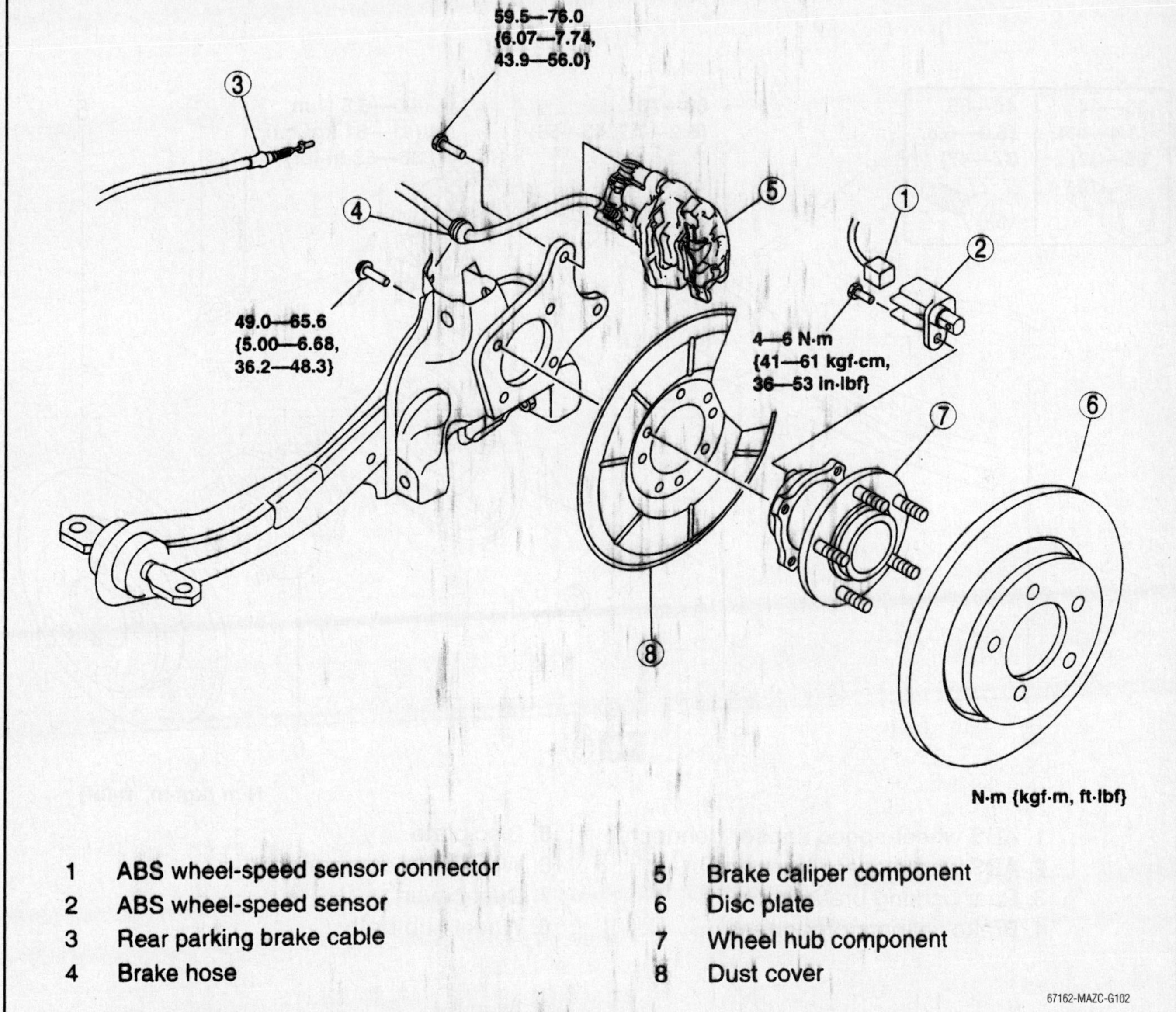

59.5—76.0
{6.07—7.74,
43.9—56.0}

49.0—65.6
{5.00—6.68,
36.2—48.3}

4—6 N·m
{41—61 kgf·cm,
36—53 in·lbf}

N·m {kgf·m, ft·lbf}

1 ABS wheel-speed sensor connector
2 ABS wheel-speed sensor
3 Rear parking brake cable
4 Brake hose
5 Brake caliper component
6 Disc plate
7 Wheel hub component
8 Dust cover

67162-MAZC-G102

Fig. 381 Exploded view of the rear wheel bearing assembly and related components—2009 vehicles

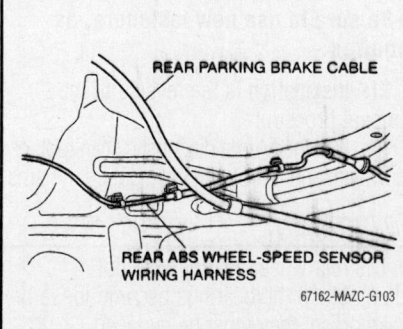

REAR PARKING BRAKE CABLE

REAR ABS WHEEL-SPEED SENSOR
WIRING HARNESS

67162-MAZC-G103

Fig. 382 Pass the cable inside the rear wheel speed sensor wiring harness when installing—2009 vehicles

• Wheel speed sensor and connector
• Rear wheel

6. Use the Mazda diagnostic scan tool, or equivalent and reprogram the required systems.

2010 Vehicles

See Figure 383.

1. Before servicing the vehicle, refer to the Precautions Section.

➡️If working near and/or around the SRS system and components, be sure to disable the SRS system. Tape the negative battery cable with insulating tape. Always disconnect the negative battery cable first.

※ CAUTION

To avoid personal injury when working on vehicles equipped with an air bag, the negative battery cable must be disconnected and at least one minute must elapse before working on the system. Failure to do so may result in deployment of the air bag.

2. Remove the battery top cover, if equipped.

3. Disconnect the negative battery cable. Tape the cable with insulating tape.

➡️When disconnecting the cable, some systems need to be initialized after the

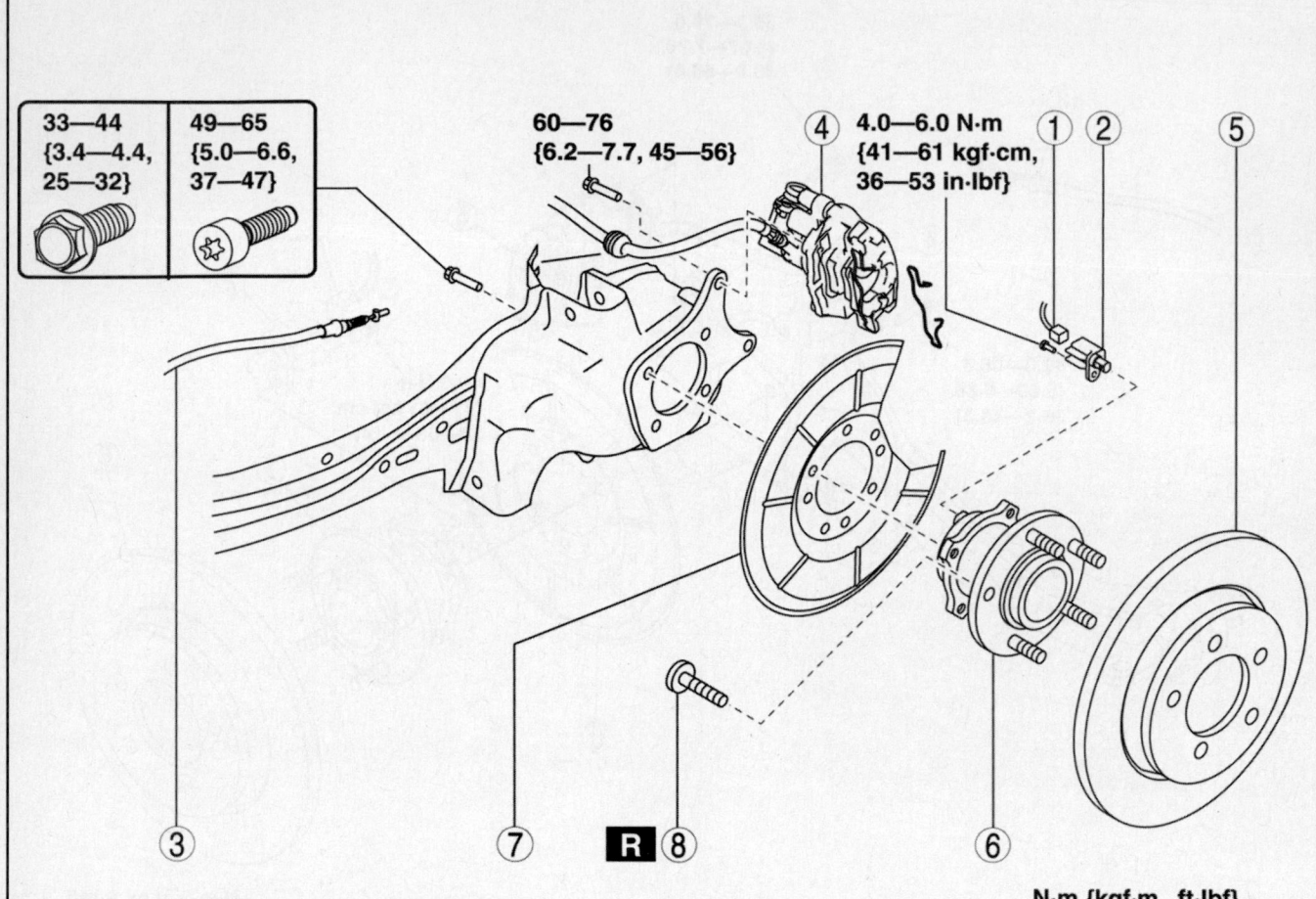

33—44
{3.4—4.4,
25—32}

49—65
{5.0—6.6,
37—47}

60—76
{6.2—7.7, 45—56}

④ 4.0—6.0 N·m
{41—61 kgf·cm,
36—53 in·lbf}

① ② ⑤

③   ⑦   **R** ⑧   ⑥

N·m {kgf·m, ft·lbf}

1. ABS wheel-speed sensor connector
2. ABS wheel-speed sensor
3. Rear parking brake cable
4. Brake caliper component
5. Disc plate
6. Wheel hub component
7. Dust cover
8. Wheel hub bolt

37671_MAZ3_G0206

Fig. 383 Rear hub and related components—2010 vehicles

cable is reconnected. You will need Mazda diagnostic scan tool or equivalent. Follow the directions on the tool.

➡Performing the following procedure without removing the ABS wheel sensor may cause an open circuit in the harness if it is pulled by mistake. Before performing the procedure remove the ABS wheel speed sensor (axle side) and position it to an appropriate place where the sensor will not be pulled by mistake while servicing the vehicle.

4. Raise and support the vehicle safely.
5. Remove the tire and wheel assembly, as required.
6. Disconnect the ABS sensor electrical connector. Remove the sensor.
7. Disconnect the rear parking brake cable.
8. Remove the caliper. Do not allow the caliper to hang by the brake hose.
9. Remove the rotor.
10. Remove the wheel hub assembly.

To install:

➡Be sure to use new fasteners, as required.

11. Installation is the reverse of the removal procedure.
12. Use the Mazda diagnostic scan tool, or equivalent and reprogram the required systems.

ADJUSTMENT

The rear wheel bearings are not adjustable. If the bearings become loose or make noise, they must be replaced.

MAZDA

Mazda 5

SPECIFICATIONS AND MAINTENANCE CHARTS

ENGINE AND VEHICLE IDENTIFICATION

	Engine						Model Year	
Code ①	Liters (cc)	Cu. In.	Cyl.	Fuel Sys.	Engine Type	Eng. Mfg.	Code ②	Year
L3	2.3 (2261)	137.9	4	MPFI	DOHC	Mazda	6	2006
							7	2007
MPFI: Multi-Point Fuel Injection							8	2008
DOHC: Double Over Head Cam							9	2009
① Located above the starter							A	2010
② 10th digit of the Vehicle Identification Number (VIN)								

37671_MAZ5_C0001

GENERAL ENGINE SPECIFICATIONS

Year	Model	Engine Displacement Liters (VIN)	Net Horsepower @ rpm	Net Torque @ rpm (ft. lbs.)	Bore x Stroke (in.)	Com-pression Ratio	Oil Pressure @ rpm
2006	Mazda 5	2.3 (L3)	157@6500	148@4500	3.44x3.27	10.0:1	49.0-85.8@3000
2007	Mazda 5	2.3 (L3)	157@6500	148@4500	3.44x3.27	10.0:1	49.0-85.8@3000
2008	Mazda 5	2.3 (L3)	157@6500	148@4500	3.44x3.27	10.0:1	49.0-85.8@3000
2009	Mazda 5	2.3 (L3)	157@6500	148@4500	3.44x3.27	10.0:1	49.0-85.8@3000
2010	Mazda 5	2.3 (L3)	157@6500	148@4500	3.44x3.27	10.0:1	49.0-85.8@3000

NA: Not Available
EFI: Electronic Fuel Injection

37671_MAZ5_C0002

ENGINE TUNE-UP SPECIFICATIONS

Year	Engine Displacement Liters (VIN)	Spark Plug Gap (in.)	Ignition Timing (deg.) MT	Ignition Timing (deg.) AT	Fuel Pump (psi)	Idle Speed (rpm) MT	Idle Speed (rpm) AT	Valve Clearance In.	Valve Clearance Ex.
2006	2.3 (L3)	0.049-0.053	8B	8B	55-65	600-700	650-750	0.008-0.0011	0.010-0.012
2007	2.3 (L3)	0.049-0.053	8B	8B	55-65	600-700	650-750	0.008-0.0011	0.010-0.012
2008	2.3 (L3)	0.049-0.053	8B	8B	55-65	600-700	650-750	0.008-0.0011	0.010-0.012
2009	2.3 (L3)	0.049-0.053	8B	8B	55-65	600-700	650-750	0.008-0.0011	0.010-0.012
2010	2.3 (L3)	0.049-0.053	8B	8B	55-65	600-700	650-750	0.008-0.0011	0.010-0.012

NOTE: The Vehicle Emission Control Information label often reflects specification changes made during production. The label figures must be used if they differ from those in this chart.

NA: Not Available
B: Before top dead center

37671_MAZ5_C0003

CAPACITIES

Year	Model	Engine Displacement Liters (VIN)	Engine Oil with Filter (qts.)	Transmission (pts.) Man	Transmission (pts.) Auto.	Drive Axle Front (pts.)	Drive Axle Rear (pts.)	Fuel Tank (gal.)	Cooling System (qts.)
2006	Mazda 5	2.3 (L3)	4.5	6.06	15.2	①	—	15.9	7.4
2007	Mazda 5	2.3 (L3)	4.5	6.06	15.2	①	—	15.9	7.4
2008	Mazda 5	2.3 (L3)	4.5	5.86	17.2	①	—	15.9	7.9
2009	Mazda 5	2.3 (L3)	4.5	5.86	17.2	①	—	15.9	7.9
2010	Mazda 5	2.3 (L3)	4.5	5.86	17.2	①	—	15.9	7.9

NOTE: All capacities are approximate. Add fluid gradually and check to be sure a proper fluid level is obtained.

① Included in transaxle

37671_MAZ5_C0005

FLUID SPECIFICATIONS

Year	Model	Engine Displacement Liters	Engine ID/VIN	Engine Oil	Auto. Trans.	Manual Trans.	Power Drive Axle	Brake Master Cylinder	Engine Coolant ② ③
2006	Mazda 5	2.3 (L3)	C	5W-20	①	75W-90	75W-90	DOT 3	Ethylene-glycol
2007	Mazda 5	2.3 (L3)	L	5W-20	①	75W-90	75W-90	DOT 3	Ethylene-glycol
2008	Mazda 5	2.3 (L3)	C	5W-20	①	75W-90	75W-90	DOT 3	Ethylene-glycol
2009	Mazda 5	2.3 (L3)	L	5W-20	①	75W-90	75W-90	DOT 3	Ethylene-glycol
2010	Mazda 5	2.3 (L3)	F	5W-20	①	75W-90	75W-90	DOT 3	Ethylene-glycol

DOT: Department Of Transpotation

37671_MAZ5_C0004

VALVE SPECIFICATIONS

Year	Engine Displacement Liters (VIN)	Seat Angle (deg.)	Face Angle (deg.)	Maximum out of Square (in.)	Spring Free Length (in.)	Stem-to-Guide Clearance (in.) Intake	Stem-to-Guide Clearance (in.) Exhaust	Stem Diameter (in.) Intake	Stem Diameter (in.) Exhaust
2006	2.3 (L3)	NA	NA	NA	NA	NA	NA	NA	NA
2007	2.3 (L3)	NA	NA	NA	NA	NA	NA	NA	NA
2008	2.3 (L3)	NA	NA	NA	NA	NA	NA	NA	NA
2009	2.3 (L3)	NA	NA	NA	NA	NA	NA	NA	NA
2010	2.3 (L3)	NA	NA	NA	NA	NA	NA	NA	NA

NA: Not Available

37671_MAZ5_C0006

CAMSHAFT AND BEARING SPECIFICATIONS

All measurements are given in inches.

Year	Engine Displacement Liters	Engine VIN	Journal Diameter	Brg. Oil Clearance	Shaft End-play	Runout	Journal Bore	Lobe Lift	
								Intake	Exhaust
2006	2.3	L3	NA	NA	NA	NA	NA	NA	NA
2007	2.3	L3	NA	NA	NA	NA	NA	NA	NA
2008	2.3	L3	NA	NA	NA	NA	NA	NA	NA
2009	2.3	L3	NA	NA	NA	NA	NA	NA	NA
2010	2.3	L3	NA	NA	NA	NA	NA	NA	NA

37671_MAZ5_C0014

CRANKSHAFT AND CONNECTING ROD SPECIFICATIONS

All measurements are given in inches.

Year	Engine Displacement Liters (VIN)	Crankshaft				Connecting Rod		
		Main Brg. Journal Dia.	Main Brg. Oil Clearance	Shaft End-play	Thrust on No.	Journal Diameter	Oil Clearance	Side Clearance
2006	2.3 (L3)	NA	NA	NA	NA	NA	NA	NA
2007	2.3 (L3)	NA	NA	NA	NA	NA	NA	NA
2008	2.3 (L3)	NA	NA	NA	NA	NA	NA	NA
2009	2.3 (L3)	NA	NA	NA	NA	NA	NA	NA
2010	2.3 (L3)	NA	NA	NA	NA	NA	NA	NA

NA: Not Avilable

37671_MAZ5_C0007

PISTON AND RING SPECIFICATIONS

All measurements are given in inches.

Year	Engine Displacement Liters (VIN)	Piston Clearance	Ring Gap			Ring Side Clearance		
			Top Compression	Bottom Compression	Oil Control	Top Compression	Bottom Compression	Oil Control
2006	2.3 (L3)	NA	NA	NA	NA	NA	NA	NA
2007	2.3 (L3)	NA	NA	NA	NA	NA	NA	NA
2008	2.3 (L3)	NA	NA	NA	NA	NA	NA	NA
2009	2.3 (L3)	NA	NA	NA	NA	NA	NA	NA
2010	2.3 (L3)	NA	NA	NA	NA	NA	NA	NA

NA: Not Available

37671_MAZ5_C0008

TORQUE SPECIFICATIONS

All readings in ft. lbs.

Year	Engine Displacement Liters (VIN)	Cylinder Head Bolts	Main Bearing Bolts	Rod Bearing Bolts	Crankshaft Damper Bolts	Flywheel Bolts	Manifold		Spark Plugs	Oil Pan Drain Plug
							Intake	Exhaust		
2006	2.3 (L3)	①	NA	NA	②	③	④	32-47	8-10	19-22
2007	2.3 (L3)	①	NA	NA	②	③	④	32-47	8-10	19-22
2008	2.3 (L3)	①	NA	NA	②	③	④	32-47	8-10	19-22
2009	2.3 (L3)	①	NA	NA	②	③	④	32-47	8-10	19-22
2010	2.3 (L3)	①	NA	NA	②	③	④	32-47	8-10	19-22

NA: Not Available

① Step 1: 97 inch lbs.

　Step 2: Tighten 12.5 ft. lbs

　Step 3: Tighten 34.6 ft. lbs.

　Step 4: Tighten 92 degrees

　Step 5: Tighten 92 degrees

② 71-77 ft. lbs. plus 87-93 degrees

③ Manual transmission: 80-85 ft. lbs.

　Automatic transmission: 71-76 ft. lbs.

④ Mazda 3 with the 2.0L (LF) and 2.3L (L3) engines: 142-177 ft. lbs.

　Mazda 6 with the 2.3L (L3) engine: 71-101 inch lbs.

37671_MAZ5_C0009

WHEEL ALIGNMENT

Year	Model		Caster Range (+/-Deg.)	Caster Preferred Setting (Deg.)	Camber Range (+/-Deg.)	Camber Preferred Setting (Deg.)	Toe-in (in.)
2006	Mazda 5	F	1.00	+2.58	1.00	-0.41	0 +/- 0 22'
		R	—	—	1.00	-1.15	0 +/- 0 27'
2007	Mazda 5	F	1.00	+2.58	1.00	-0.41	0 +/- 0 22'
		R	—	—	1.00	-1.15	0 +/- 0 27'
2008	Mazda 5	F	1.00	+2.58	1.00	-0.41	0 +/- 0 22'
		R	—	—	1.00	-1.15	0 +/- 0 27'
2009	Mazda 5	F	1.00	+2.58	1.00	-0.41	0 +/- 0 22'
		R	—	—	1.00	-1.15	0 +/- 0 27'
2010	Mazda 5	F	1.00	+2.58	1.00	-0.41	0 +/- 0 22'
		R	—	—	1.00	-1.15	0 +/- 0 27'

37671_MAZ5_C0010

TIRE, WHEEL AND BALL JOINT SPECIFICATIONS

Year	Model	OEM Tires Standard	OEM Tires Optional	Tire Pressures (psi) Front	Tire Pressures (psi) Rear	Wheel Size	Ball Joint Inspection	Lug Nut
2006	Mazda 5	P205/55R16	205/50R17	34	34	①	⑭①	65-87
2007	Mazda 5	P205/55R16	205/50R17	34	34	①	⑭①	65-87
2008	Mazda 5	P205/55R16	205/50R17	34	34	①	⑭①	65-87
2009	Mazda 5	P205/55R16	205/50R17	34	34	①	⑭①	65-87
2010	Mazda 5	P205/55R16	205/50R17	34	34	①	⑭①	65-87

OEM: Original Equipment Manufacturer

PSI: Pounds Per Square Inch

STD: Standard

OPT: Optional

NA: Not Available

① Aluminum Alloy: 16 x 6 1/2J

   Aluminum alloy: 17 x 6 1/2J

37671_MAZ5_C0011

BRAKE SPECIFICATIONS

All measurements in inches unless noted

Year	Model		Brake Disc Original Thickness	Brake Disc Minimum Thickness	Brake Disc Maximum Runout	Brake Drum Original Inside Diameter	Brake Drum Max. Wear Limit	Brake Drum Maximum Machine Diameter	Minimum Lining Thickness	Brake Caliper Bracket Bolts (ft. lbs.)	Brake Caliper Mounting Bolts (ft. lbs.)
2006	Mazda 5	F	NA	0.910	0.002	—	—	—	0.079	51-75	19-22
		R	NA	0.350	0.002	—	—	—	0.079	44-56	19-22
2007	Mazda 5	F	NA	0.910	0.002	—	—	—	0.079	51-75	19-22
		R	NA	0.350	0.002	—	—	—	0.079	44-56	19-22
2008	Mazda 5	F	NA	0.910	0.002	—	—	—	0.079	51-75	19-22
		R	NA	0.350	0.002	—	—	—	0.079	44-56	19-22
2009	Mazda 5	F	NA	0.910	0.002	—	—	—	0.079	51-75	19-22
		R	NA	0.350	0.002	—	—	—	0.079	44-56	19-22
2010	Mazda 5	F	NA	0.910	0.002	—	—	—	0.079	51-75	19-22
		R	NA	0.350	0.002	—	—	—	0.079	44-56	19-22

NA: Not Avilable

F: Front

R: Rear

37671_MAZ5_C0012

SCHEDULED MAINTENANCE INTERVALS
2006-2010 Mazda 5

TO BE SERVICED	TYPE OF SERVICE	VEHICLE MILEAGE INTERVAL (x1000)												
		7.5	15	22.5	30	37.5	45	52.5	60	67.5	75	82.5	90	97.5
Engine oil & filter	R	✓	✓	✓	✓	✓	✓	✓	✓	✓	✓	✓	✓	✓
Air cleaner element	R					✓					✓			
Engine coolant ①	R													
Spark plugs	R										✓			
Bolts & nuts on chassis & body	S/I				✓				✓				✓	
Brake lines, hoses & connections	S/I				✓				✓				✓	
Cooling system	S/I				✓				✓				✓	
Cabin filter	R			✓										
Disc brakes	S/I				✓				✓				✓	
Drive belts	S/I				✓				✓				✓	
Drive shaft dust boots	S/I				✓				✓				✓	
Exhaust system heat shield	S/I				✓				✓				✓	
Front & rear suspension ball joints	S/I				✓				✓				✓	
Fuel lines & hoses	S/I				✓				✓				✓	
Steering operation & linkages	S/I				✓				✓				✓	
Fuel filter	R								✓					
Valve clearance	I								✓					
Hose & tube for emission	S/I								✓					

R: Replace   S/I: Service or Inspect

① Models with FL22 long-life coolant can be identified by the FL22 marking on the cooling system cap label. It is recommended that FL22 coolant continue to be used for models originally filled with FL22 coolant from the factory.

FREQUENT OPERATION MAINTENANCE (SEVERE SERVICE)

If a vehicle is operated under any of the following conditions it is considered severe service

- Extremely dusty areas.

- 50% or more of the vehicle operation is in 32°C (90°F) or higher temperatures, or constant operation in temperatures below 0°C (32°F).

- Prolonged idling (vehicle operation in stop and go traffic).

- Frequent short running periods (engine does not warm to normal operating temperatures).

- Police, taxi, delivery usage or trailer towing usage.

Oil & oil filter: change every 5000 miles.

Oil & oil filter (Puerto Rico): change every 3000 miles.

Air cleaner element: service or inspect every 15,000 miles

Automatic transaxle fluid: service or inspect every 15,000 miles.

Bolts & nuts on chassis & body: tighten every 15,000 miles.

Disc brakes: service or inspect every 15,000 miles.

37671_MAZ5_C0013

BRAKES   INFORMATION AND PRECAUTIONS

ANTI-LOCK SYSTEMS

• Certain components within the ABS system are not intended to be serviced or repaired individually.

• Do not use rubber hoses or other parts not specifically specified for and ABS system. When using repair kits, replace all parts included in the kit. Partial or incorrect repair may lead to functional problems and require the replacement of components.

• Lubricate rubber parts with clean, fresh brake fluid to ease assembly. Do not use shop air to clean parts; damage to rubber components may result.

• Use only DOT 3 brake fluid from an unopened container.

• If any hydraulic component or line is removed or replaced, it may be necessary to bleed the entire system.

• A clean repair area is essential. Always clean the reservoir and cap thoroughly before removing the cap. The slightest amount of dirt in the fluid may plug an ori-

fice and impair the system function. Perform repairs after components have been thoroughly cleaned; use only denatured alcohol to clean components. Do not allow ABS components to come into contact with any substance containing mineral oil; this includes used shop rags.

• The Anti-Lock control unit is a microprocessor similar to other computer units in the vehicle. Ensure that the ignition switch is **OFF** before removing or installing controller harnesses. Avoid static electricity discharge at or near the controller.

• If any arc welding is to be done on the vehicle, the control unit should be unplugged before welding operations begin.

DISC AND DRUM SYSTEMS

✳✳ CAUTION

Dust and dirt accumulating on brake parts during normal use may contain asbestos fibers from production or

aftermarket brake linings. Breathing excessive concentrations of asbestos fibers can cause serious bodily harm. Exercise care when servicing brake parts. Do not sand or grind brake lining unless equipment used is designed to contain the dust residue. Do not clean brake parts with compressed air or by dry brushing. Cleaning should be done by dampening the brake components with a fine mist of water, then wiping the brake components clean with a dampened cloth. Dispose of cloth and all residue containing asbestos fibers in an impermeable container with the appropriate label. Follow practices prescribed by the Occupational Safety and Health Administration (OSHA) and the Environmental Protection Agency (EPA) for the handling, processing, and disposing of dust or debris that may contain asbestos fibers.

BRAKES   BLEEDING THE BRAKE SYSTEM

BLEEDING PROCEDURE

BLEEDING PROCEDURE

See Figure 1.

✳✳ CAUTION

Note the following:

• When replacing the ABS HU/CM, the configuration procedure must be done before removing the ABS HU/CM. If the configuration is not completed before removing the ABS HU/CM, ABS will not work properly after installation of the ABS HU/CM.

• Do not separate the ABS HU and ABS CM unless replacing them, otherwise the ABS HU/CM may not function properly. When replacing them with new ones, always perform procedures according to the instructions included with the new parts.

• The internal parts of the ABS HU/CM could be damaged if dropped. Be careful not to drop the ABS HU/CM. Replace the ABS HU/CM if it is subjected to an impact.

1. Remove the battery and battery tray.

2. Remove the reserve hose (M/T vehicles).

3. Remove the reserve tank hose.

4. Remove in the order indicated in the table.

To install:

5. Install in the reverse order of removal.

6. After installation, add brake fluid, bleed the air, and inspect for fluid leakage.

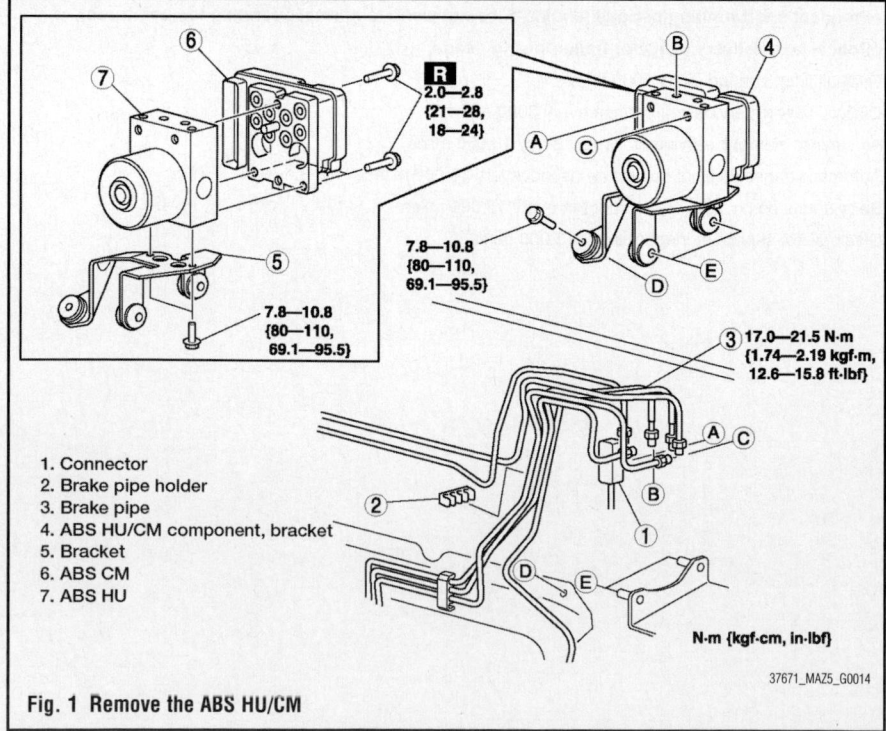

1. Connector
2. Brake pipe holder
3. Brake pipe
4. ABS HU/CM component, bracket
5. Bracket
6. ABS CM
7. ABS HU

N·m {kgf·cm, in·lbf}

Fig. 1 Remove the ABS HU/CM

37671_MAZ5_G0014

7. Configure the ABS HU/CM (only when replacing it).

BLEEDING THE ABS SYSTEM

✴ CAUTION

Brake fluid will damage painted surfaces. Be careful not to spill any on painted surfaces. If it is spilled, wipe it off immediately.

- Keep the fluid level in the reserve tank at ¾ full or more during the air bleeding.

- Begin air bleeding with the brake caliper that is furthest from the master cylinder.

1. Remove the bleeder cap on the brake caliper, and attach a vinyl tube to the bleeder screw.

2. Place the other end of the vinyl tube in a clear container and fill the container with fluid during air bleeding.

3. Working with two people, one should pump the brake pedal several times and depress and hold the pedal down.

4. While the brake pedal is depressed, the other should loosen the bleeder screw using a commercially available flare nut wrench, drain out any fluid containing air bubbles, and tighten the bleeder screw.

5. Repeat Steps 3 and 4 until no air bubbles are seen.

6. Perform air bleeding as described in the above procedures for all brake calipers.

7. After air bleeding, inspect the following:
- Brake operation
- Fluid leakage
- Fluid level

| **BRAKES** | **ANTI-LOCK BRAKE SYSTEM (ABS)** |

SPEED SENSORS

REMOVAL & INSTALLATION

Front

See Figure 2.

1. Remove the mudguard.
2. Remove in the order indicated.
   a. Disconnect the connector.
   b. Remove the Front ABS wheel-speed sensor

c. Remove the Front ABS wheel-speed sensor wiring harness

3. Install in the reverse order of removal.

Rear

See Figure 3.

1. Remove the rear under cover.
2. Remove in the order indicated:

a. Remove the connector.
b. Remove the rear wheel speed sensor.
c. Remove the ABS hole cover.
d. Disconnect the connector.
e. Remove the rear ABS wheel speed sensor wiring harness.

3. Install in the reverse order of removal.

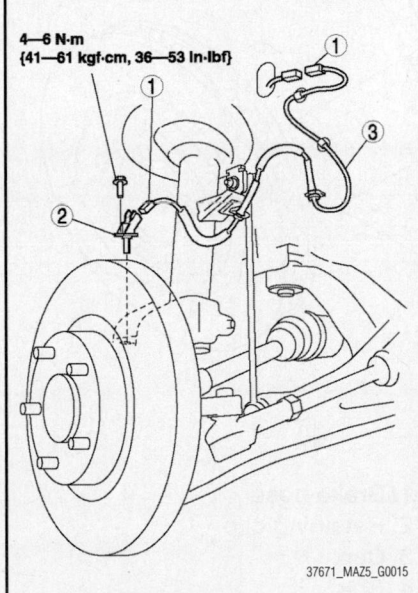

4—6 N·m
{41—61 kgf·cm, 36—53 in·lbf}

37671_MAZ5_G0015

Fig. 2 Remove the connector (1), the front ABS Wheel speed sensor (2) and the front ABS wheel speed sensor wiring harness (3)

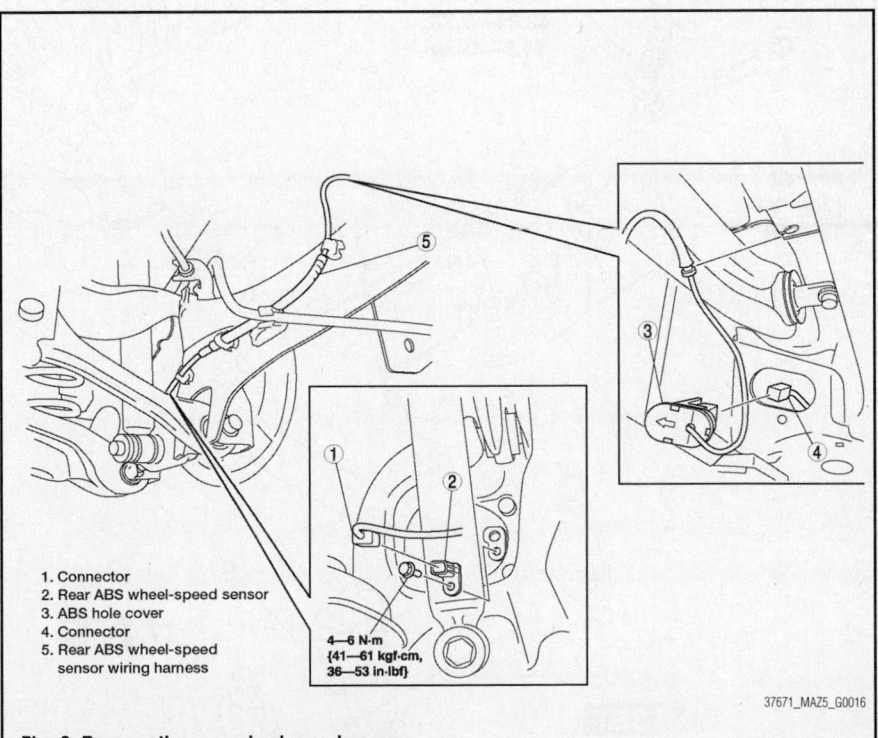

1. Connector
2. Rear ABS wheel-speed sensor
3. ABS hole cover
4. Connector
5. Rear ABS wheel-speed sensor wiring harness

4—6 N·m
{41—61 kgf·cm, 36—53 in·lbf}

37671_MAZ5_G0016

Fig. 3 Remove the rear wheel speed sensor

BRAKE CALIPER

REMOVAL & INSTALLATION

See Figure 4.

1. Remove the front caliper by removing components in the following sequence:
   - Brake hose
   - Retaining clip
   - Cap
   - Bolt
   - Caliper
   - Boot
   - Disc pad
   - Bolt
   - Mounting support
   - Disc plate

To install:

2. Install in the reverse order of removal.

3. After installation, add brake fluid, bleed the air, and inspect for fluid leakage.

4. After installation, pump the brake pedal a few times and verify that the brakes do not drag.

DISC BRAKE PADS

REMOVAL & INSTALLATION

See Figure 5.

1. Remove the front brake disc pads by removing components in the following sequence:
   - Clip
   - Retaining clip
   - Cap
   - Bolt
   - Caliper
   - Disc pad

To install:

2. Install in the reverse order of removal.

3. After installation, pump the brake pedal a few times and verify that the brakes do not drag.

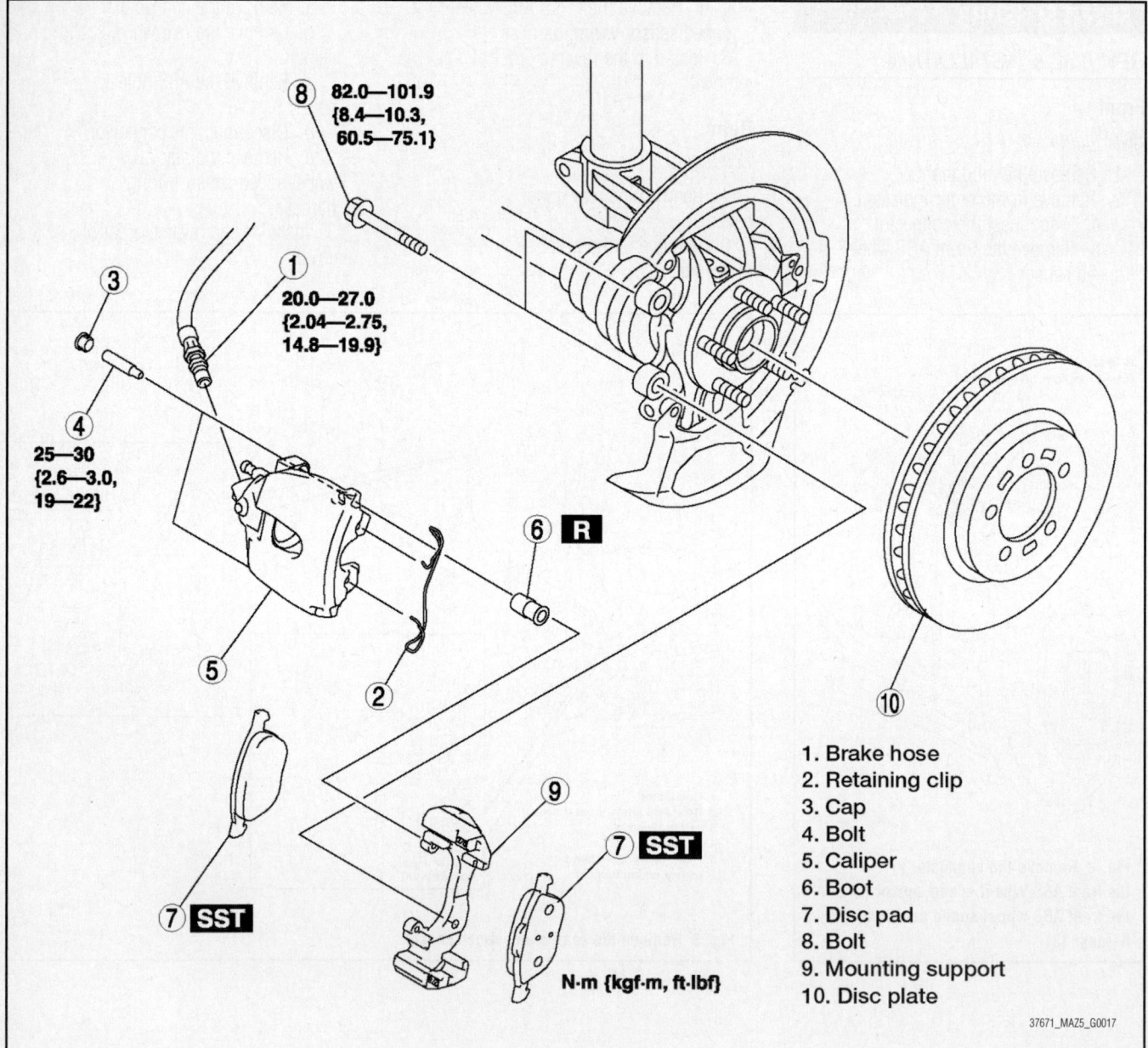

⑧ 82.0—101.9 {8.4—10.3, 60.5—75.1}

① 20.0—27.0 {2.04—2.75, 14.8—19.9}

④ 25—30 {2.6—3.0, 19—22}

⑥ **R**

⑦ **SST**

⑨

⑦ **SST**

N·m {kgf·m, ft·lbf}

1. Brake hose
2. Retaining clip
3. Cap
4. Bolt
5. Caliper
6. Boot
7. Disc pad
8. Bolt
9. Mounting support
10. Disc plate

37671_MAZ5_G0017

Fig. 4 Exploded view of front brake assembly

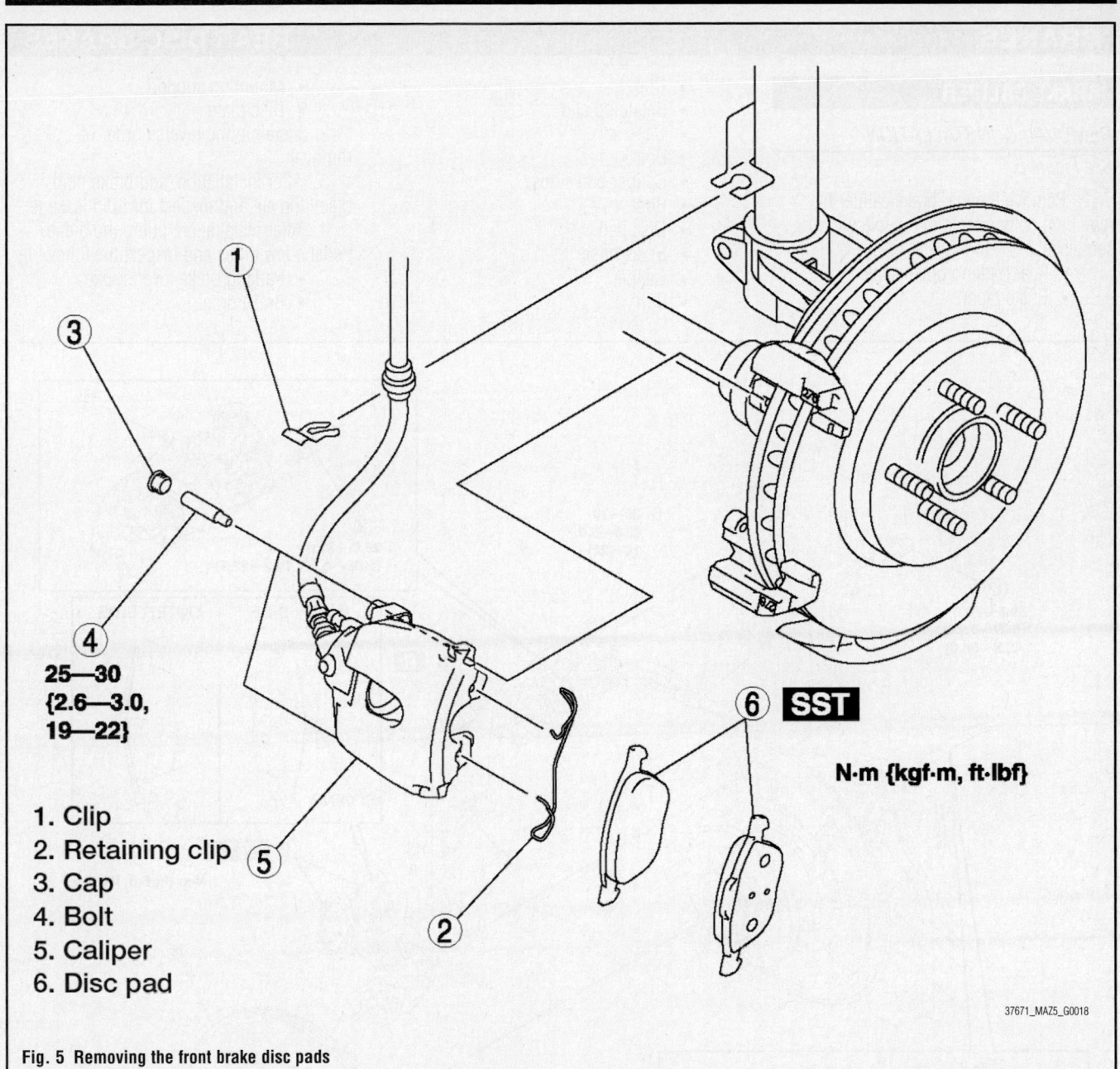

25—30
{2.6—3.0,
19—22}

1. Clip
2. Retaining clip
3. Cap
4. Bolt
5. Caliper
6. Disc pad

SST

N·m {kgf·m, ft·lbf}

37671_MAZ5_G0018

Fig. 5 Removing the front brake disc pads

BRAKE CALIPER

REMOVAL & INSTALLATION

See Figure 6.

1. Remove the rear brake caliper by removing components in the following sequence:
- Rear parking brake cable
- Brake pipe
- Clip
- Retaining clip
- Cap
- Bolt
- Caliper brake hose
- Boot
- Disc pad
- Brake hose
- Caliper
- Bolt
- Mounting support
- Disc plate

2. Install in the reverse order of removal.

3. After installation, add brake fluid, bleed the air, and inspect for fluid leakage.

4. After installation, pump the brake pedal a few times and inspect the following:
- Parking brake lever stroke
- Brake drag

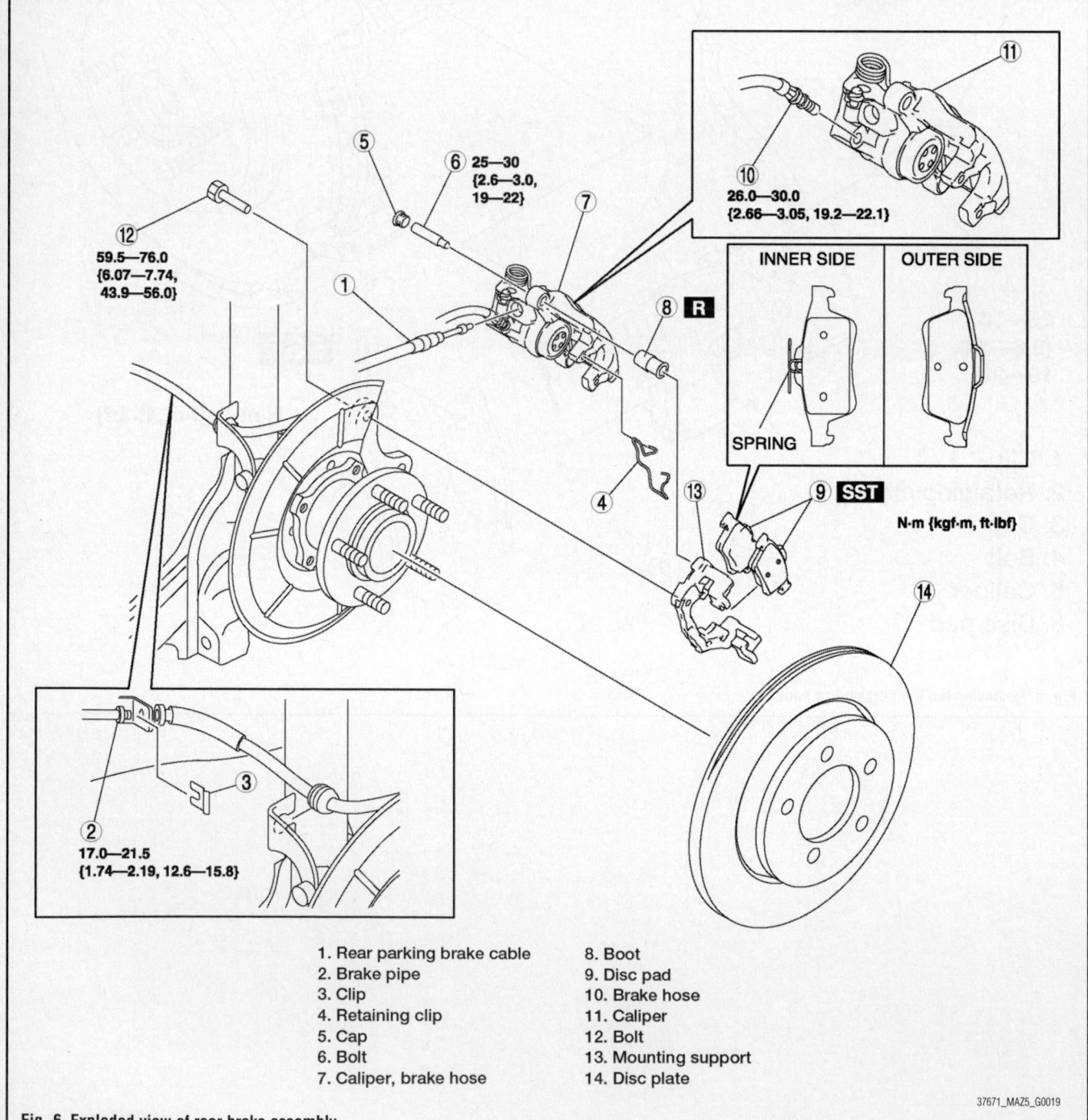

1. Rear parking brake cable
2. Brake pipe
3. Clip
4. Retaining clip
5. Cap
6. Bolt
7. Caliper, brake hose
8. Boot
9. Disc pad
10. Brake hose
11. Caliper
12. Bolt
13. Mounting support
14. Disc plate

37671_MAZ5_G0019

Fig. 6 Exploded view of rear brake assembly

DISC BRAKE PADS

REMOVAL & INSTALLATION

See Figure 7.

1. Remove the rear brake disc pads by removing

components in the following sequence:
- Rear parking brake cable
- Retaining clip
- Cap
- Bolt
- Caliper
- Disc pad

To install:

2. Install in the reverse order of removal.

3. After installation, pump the brake pedal a few times and verify that the brakes do not drag.

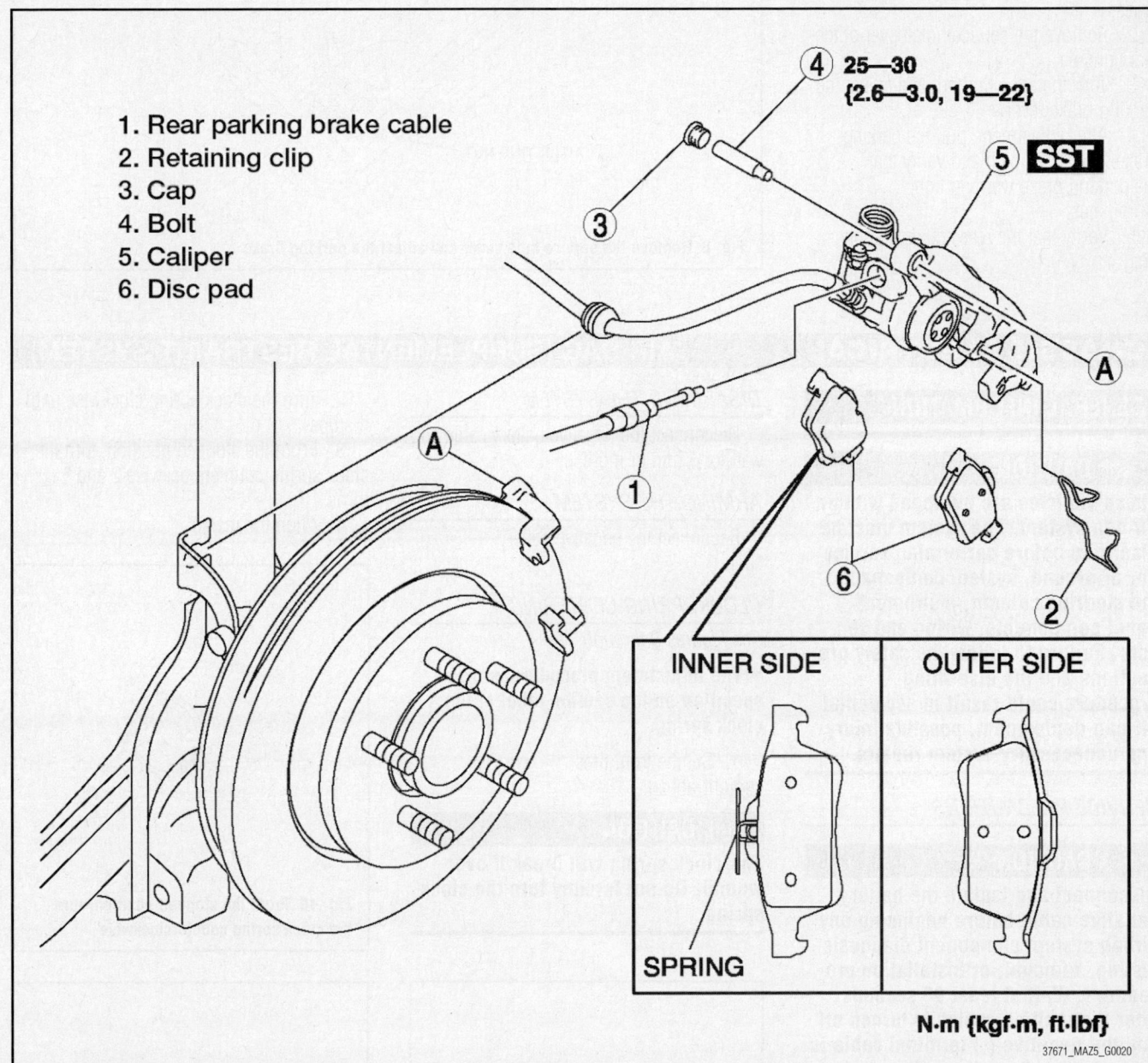

1. Rear parking brake cable
2. Retaining clip
3. Cap
4. Bolt
5. Caliper
6. Disc pad

④ 25—30
{2.6—3.0, 19—22}

⑤ **SST**

INNER SIDE

OUTER SIDE

SPRING

N·m {kgf·m, ft·lbf}

37671_MAZ5_G0020

Fig. 7 Removing the rear brake disc pads

BRAKES
PARKING BRAKE

PARKING BRAKE CABLES

ADJUSTMENT

See Figure 8.

1. Pump the brake pedal a few times.
2. Remove the service hole cover of the rear console.
3. Turn the adjusting nut and adjust the parking brake lever.
4. After adjustment, pull the parking brake lever one notch and verify that the parking brake warning light illuminates.
5. Verify that the rear brakes do not drag.

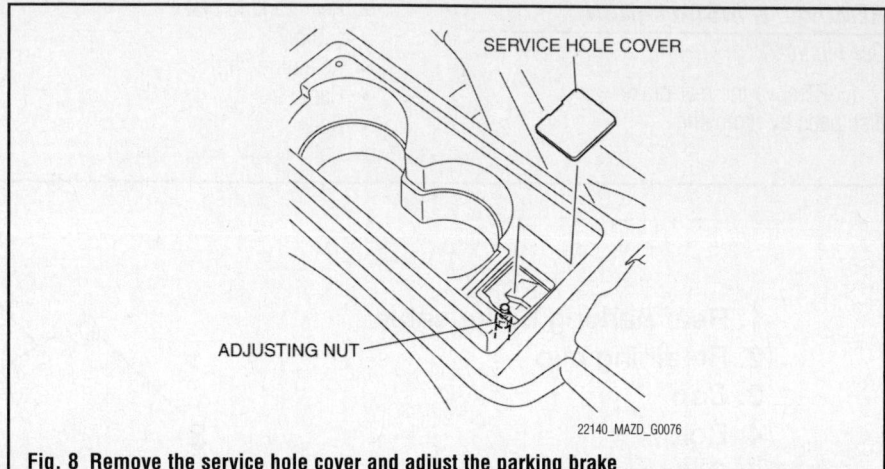

Fig. 8 Remove the service hole cover and adjust the parking brake

CHASSIS ELECTRICAL
AIR BAG (SUPPLEMENTAL RESTRAINT SYSTEM)

GENERAL INFORMATION

> ✳✳ **CAUTION**
>
> **These vehicles are equipped with an air bag system. The system must be disarmed before performing service on, or around, system components, the steering column, instrument panel components, wiring and sensors. Failure to follow the safety precautions and the disarming procedure could result in accidental air bag deployment, possible injury and unnecessary system repairs.**

SERVICE PRECAUTIONS

> ✳✳ **CAUTION**
>
> **Disconnect and isolate the battery negative cable before beginning any airbag system component diagnosis, testing, removal, or installation procedures. Wait at least 90 seconds after the ignition switch is turned off and the negative (-) terminal cable is disconnected from the battery before starting the operation. The SRS is equipped with a backup power source, so if work is started within 90 seconds after disconnecting the negative (-) terminal cable from the battery, the SRS may be deployed. Failure to disable the airbag system may result in accidental airbag deployment, personal injury, or death.**

DISARMING THE SYSTEM

Disconnect the negative battery cable and wait for 1 min or more.

ARMING THE SYSTEM

Reconnect the negative battery cable.

CLOCKSPRING CENTERING

See Figures 9 through 11.

➡ The adjustment procedure is also specified on the caution label of the clock spring.

1. Set the front tires straight-ahead.

> ✳✳ **CAUTION**
>
> **The clock spring will break if overwound. Do not forcibly turn the clock spring.**

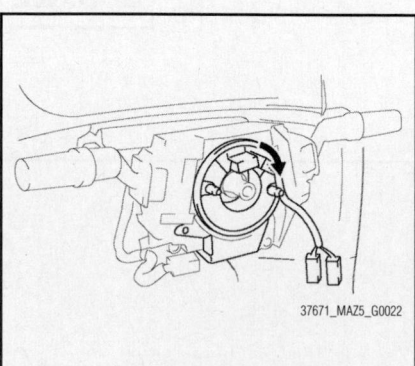

Fig. 9 Turn the clock spring clockwise until it stops

2. Turn the clock spring clockwise until it stops.
3. From the stopped position, turn the clock spring counterclockwise 2 and ¾ turns.
4. Align the marks.

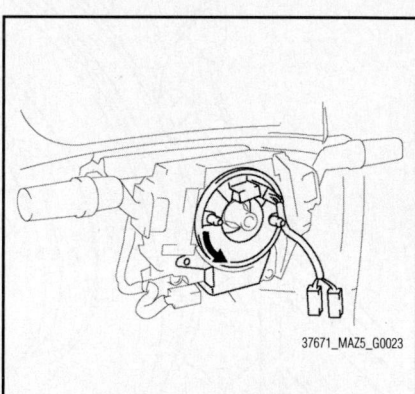

Fig. 10 From the stopped position, turn the clock spring counterclockwise

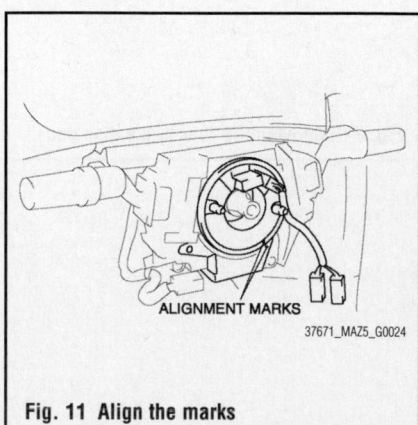

Fig. 11 Align the marks

DRIVE TRAIN

CLUTCH

REMOVAL & INSTALLATION

See Figure 12.

1. Remove the clutch in the order indicated:

- Clutch release cylinder
- Manual transaxle
- Clutch release collar
- Boot
- Clutch release fork
- Clutch cover
- Clutch disc
- Pilot bearing

2. Install in the reverse order of removal.

HALFSHAFTS

REMOVAL & INSTALLATION

See Figures 13 through 15.

※ CAUTION

Performing the following procedures without first removing the ABS wheel-speed sensor may possibly cause an open circuit in the wiring harness if it is pulled by mistake. Before perform-ing the following procedures, disconnect the ABS wheel-speed sensor connector (axle side) and attach the wiring harness to an appropriate place where it will not be pulled by mistake while servicing the vehicle.

➡ **When replacing the drive shaft (double offset joint type), replace it as a single unit because of it cannot be disassembled.**

1. Drain the transaxle oil.
2. Disconnect the ABS wheel-speed sensor connector.

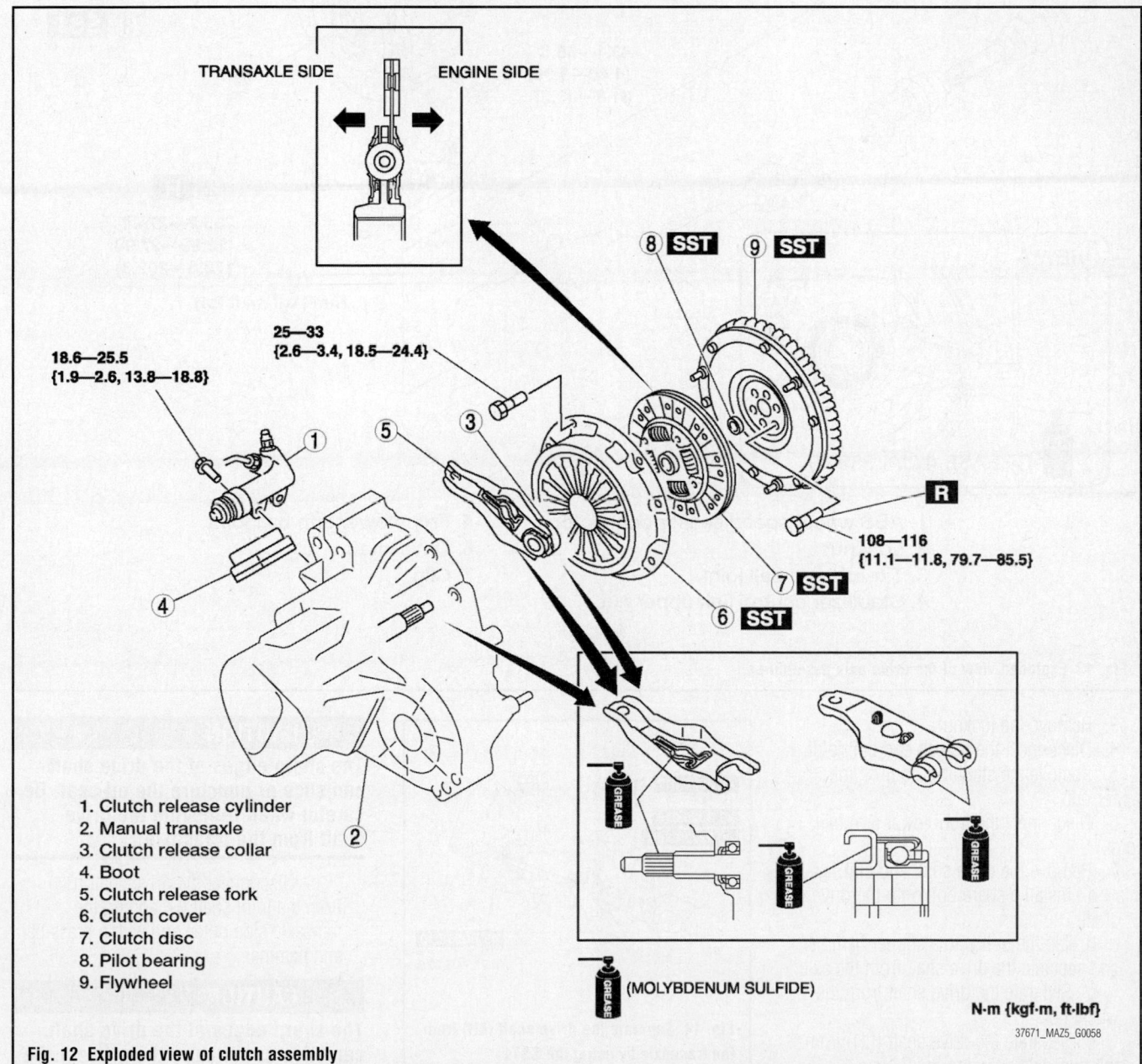

1. Clutch release cylinder
2. Manual transaxle
3. Clutch release collar
4. Boot
5. Clutch release fork
6. Clutch cover
7. Clutch disc
8. Pilot bearing
9. Flywheel

18.6—25.5 {1.9—2.6, 13.8—18.8}

25—33 {2.6—3.4, 18.5—24.4}

108—116 {11.1—11.8, 79.7—85.5}

(MOLYBDENUM SULFIDE)

N·m {kgf·m, ft-lbf}

Fig. 12 Exploded view of clutch assembly

37671_MAZ5_G0058

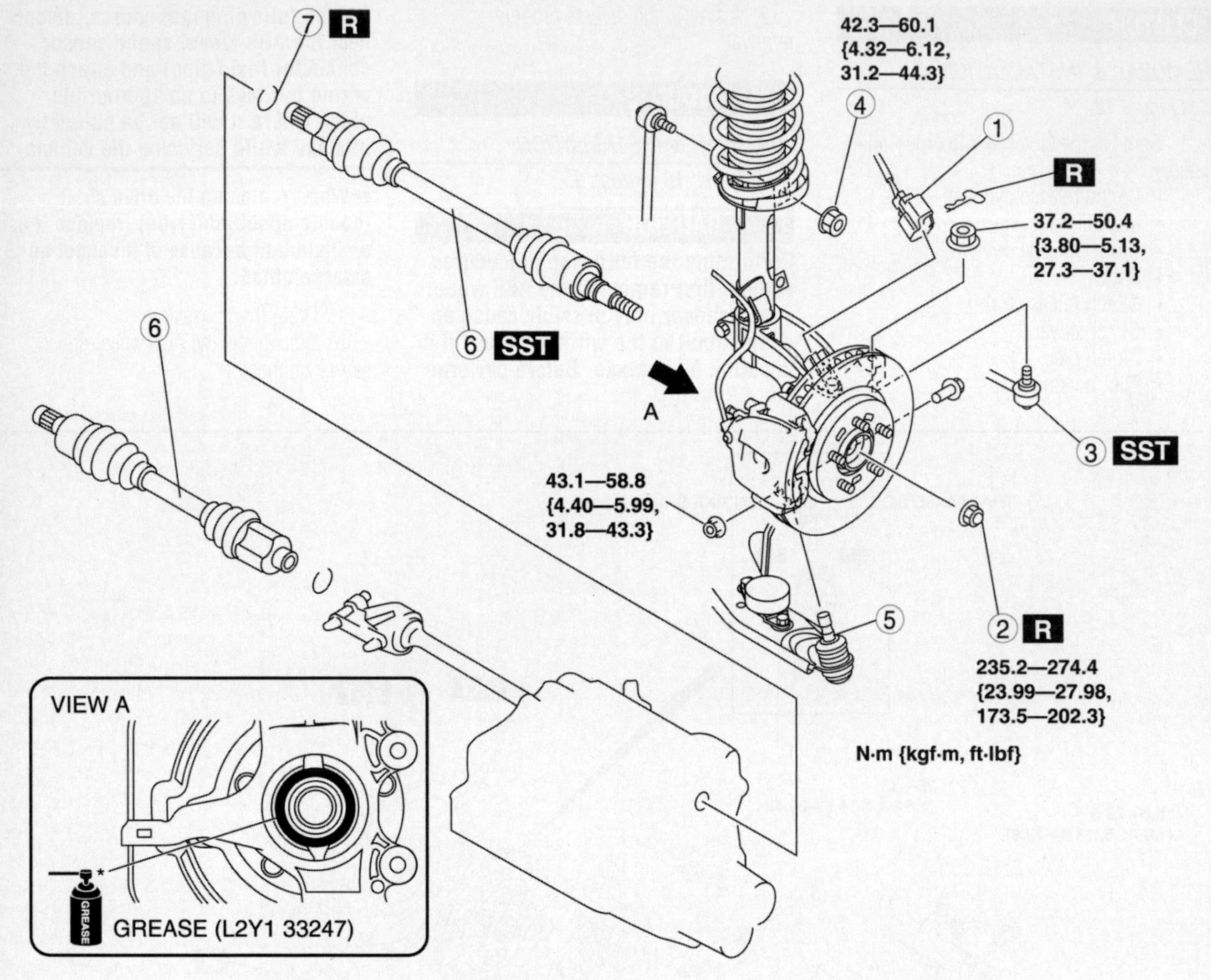

42.3—60.1
{4.32—6.12,
31.2—44.3}

R

37.2—50.4
{3.80—5.13,
27.3—37.1}

43.1—58.8
{4.40—5.99,
31.8—43.3}

SST

R

235.2—274.4
{23.99—27.98,
173.5—202.3}

N·m {kgf·m, ft·lbf}

VIEW A

GREASE (L2Y1 33247)

1. ABS wheel-speed sensor connector
2. Locknut
3. Tie-rod end ball joint
4. Stabilizer control link upper nut
5. Front lower arm ball joint
6. Drive shaft
7. Clip

37671_MAZ5_G0064

Fig. 13 Exploded view of the drive axle assemblies

3. Remove the locknut.
4. Disconnect the tie-rod end ball joint.
5. Remove the stabilizer control link upper nut.
6. Disconnect the front lower arm ball joint.
7. Remove the drive shaft and clip.
   a. Install a spare bolt onto the drive shaft.
   b. Tap the bolt with a copper hammer and separate the drive shaft from the axle.
   c. Separate the drive shaft from the wheel hub.
   d. Separate the drive shaft (LH) from the transaxle by using the SSTs.

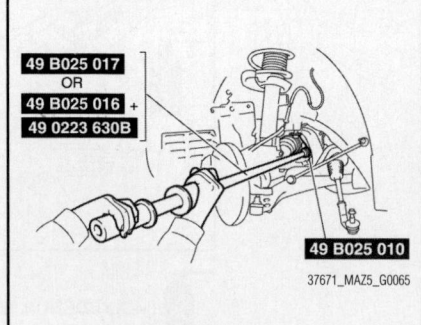

49 B025 017
OR
49 B025 016 +
49 0223 630B

49 B025 010

37671_MAZ5_G0065

Fig. 14 Separate the drive shaft (LH) from the transaxle by using the SSTs

✳✳ CAUTION

The sharp edges of the drive shaft can slice or puncture the oil seal. Be careful when removing the drive shaft from the transaxle.

   e. Disconnect the drive shaft (RH) from the joint shaft by tapping the transaxle side outer ring with a brass bar and hammer.

✳✳ CAUTION

The sharp edges of the drive shaft can slice or puncture the oil seal. Be

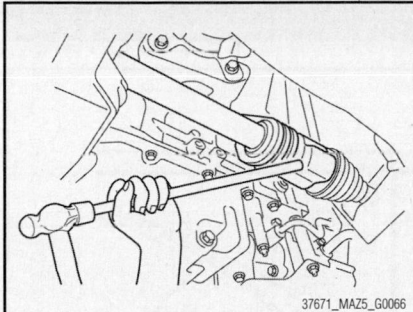

37671_MAZ5_G0066

Fig. 15 Disconnect the drive shaft (RH) from the joint shaft by tapping the transaxle side outer ring with a brass bar and hammer

careful when removing the drive shaft from the transaxle.

To install:

8. Install in the reverse order of removal.

9. Install a new drive shaft clip to the clip groove at the end of the drive shaft with the clip opening facing upward.

Left Side

See Figure 16.

✳✳ CAUTION

The sharp edges of the drive shaft can slice or puncture the oil seal. Be careful when installing the drive shaft from the transaxle.

1. Apply grease (L2Y1 33247) to the wheel bearing inner race and drive shaft contact surface.

2. Insert the drive shaft into the wheel hub.

3. Apply transaxle oil to the oil seal lip.

4. Install the drive shaft to the transaxle.

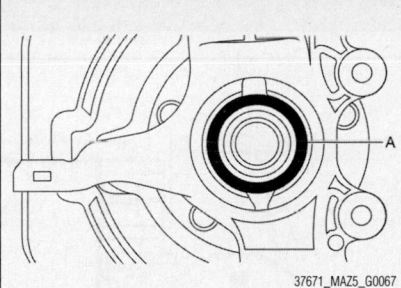

37671_MAZ5_G0067

Fig. 16 Apply grease (L2Y1 33247) to the wheel bearing inner race and drive shaft contact surface (A)

5. After installation, pull the transaxle side outer ring forward to confirm that the drive shaft is securely held by the clip.

Right Side

1. Install a new clip onto the joint shaft.

2. Apply grease (L2Y1 33247) to the wheel bearing inner race and drive shaft contact surface.

3. Insert the drive shaft to the wheel hub.

4. Insert the drive shaft to the joint shaft.

5. After installation, pull the transaxle side outer ring forward to confirm that the drive shaft is securely held by the clip.

PINION SEAL

REMOVAL & INSTALLATION

Manual Transaxle

1. On level ground, jack up the vehicle and support it evenly on safety stands.

2. Drain the oil from the transaxle.

3. Remove the front wheels and splash shields.

4. Separate the drive shaft and joint shaft from the transaxle.

5. Remove the oil seals using a screwdriver.

To install:

6. Using the SST 49G030795 and a hammer, tap each new oil seal in evenly until the SST contacts the transaxle case.

7. Coat the lip of each oil seal with transaxle oil.

8. Insert the drive shaft and joint shaft to the transaxle.

9. Install the wheels and splash shields.

10. Add the specified amount and type of oil.

Automatic Transaxle

1. Remove the battery cover.

2. Disconnect the negative battery cable.

3. Remove the undercover.

4. Drain the ATF.

✳✳ CAUTION

The oil seal is easily damaged by the sharp edges of the drive shaft splines. Do not let the splines contact the oil seal.

5. Remove the drive shaft and joint shaft.

6. Remove the oil seal using the screwdriver.

To install:

7. Using the SST 49G030795 and a hammer, tap a new oil seal in evenly until the SST contacts the transaxle case.

8. Coat the lip of the oil seal with ATF.

9. Install the drive shaft and joint shaft.

10. Add ATF to the specified level.

11. Perform the mechanical system test.

ENGINE COOLING

ENGINE FAN

REMOVAL & INSTALLATION

See Figures 17 through 19.

1. Remove the battery cover and battery duct.
2. Disconnect the negative battery cable.
3. Remove the undercover.
4. Remove in the order indicated in the table.
5. When removing the fan, note the following:

   a. Release the left and right tabs on the upper side of the radiator by pressing them in the direction shown in the figure.

   b. Lift the cooling fan component to remove it from the left and right insertion areas on the lower side of the radiator.

   c. Remove the cooling fan component from below.
6. Installation is the reverse of the removal procedure.

RADIATOR

REMOVAL & INSTALLATION

> ✳✳ **WARNING**
>
> **Never remove the cooling system cap or loosen the radiator drain plug while the engine is running, or when the engine and radiator are hot. Scalding engine coolant and steam may shoot out and cause serious injury. It may also damage the engine and cooling system.**

> ✳✳ **CAUTION**
>
> **Turn off the engine and wait until it is cool. Even then, be very careful when removing the cap. Wrap a thick cloth around it and slowly turn it counterclockwise to the first stop. Step back while the pressure escapes.**

1. When all the pressure is gone, press down on the cap using the cloth, turn it, and remove it.
2. Remove the battery cover and battery duct.
3. Disconnect the negative battery cable.
4. Remove the undercover.

5. Remove the cooling fan component.
6. Drain the engine coolant.
7. Remove the air cleaner.
8. Remove the front bumper.
9. Remove the upper and lower radiator hoses.
10. Remove the upper mount rubber bracket and rubber.

11. Remove the coolant filler neck and remove the radiator.

To install:
12. Install in the reverse order of removal.
13. Refill the engine coolant.
14. Inspect for engine coolant leakage.

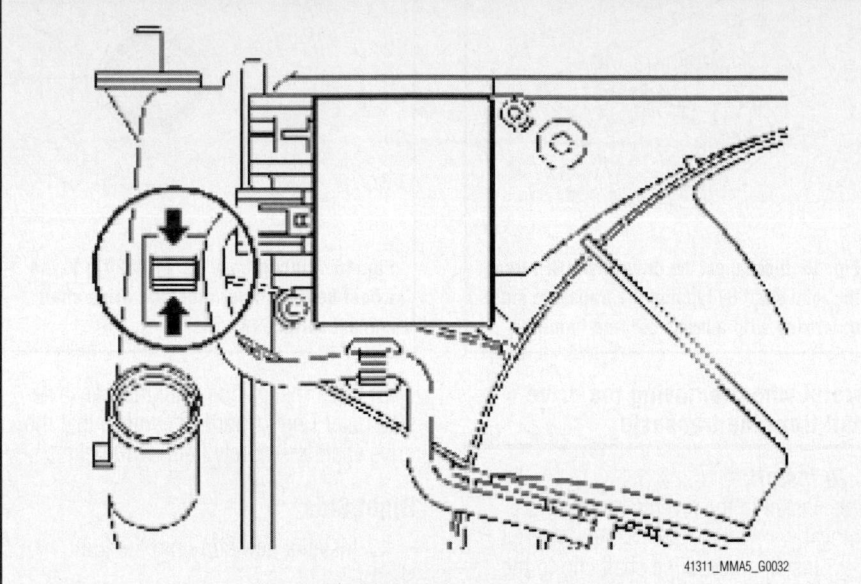

41311_MMA5_G0032

Fig. 17 Release the left and right tabs on the upper side of the radiator by pressing them in the direction shown.

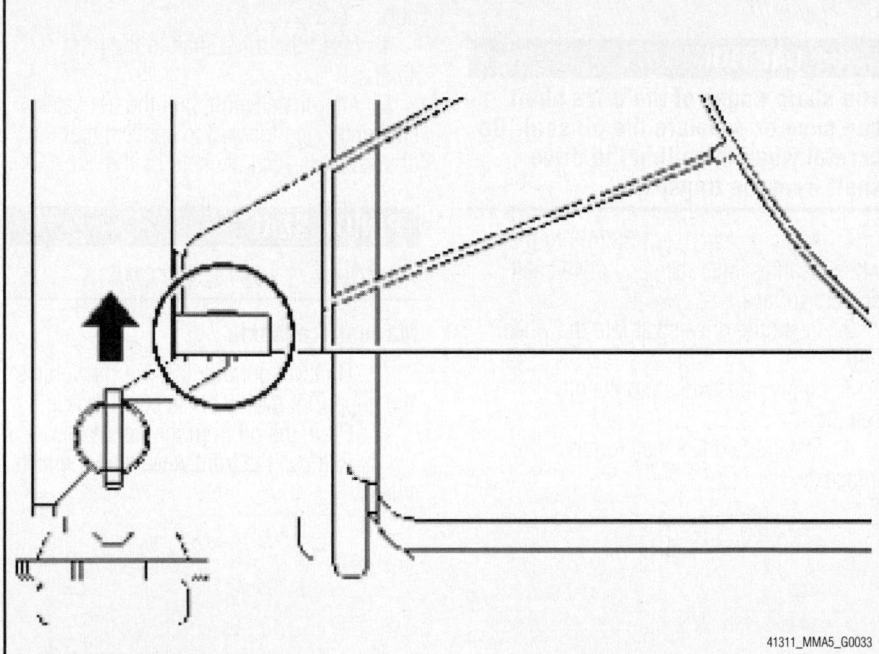

41311_MMA5_G0033

Fig. 18 Lift the cooling fan component to remove it from the left and right insertion areas on the lower side of the radiator.

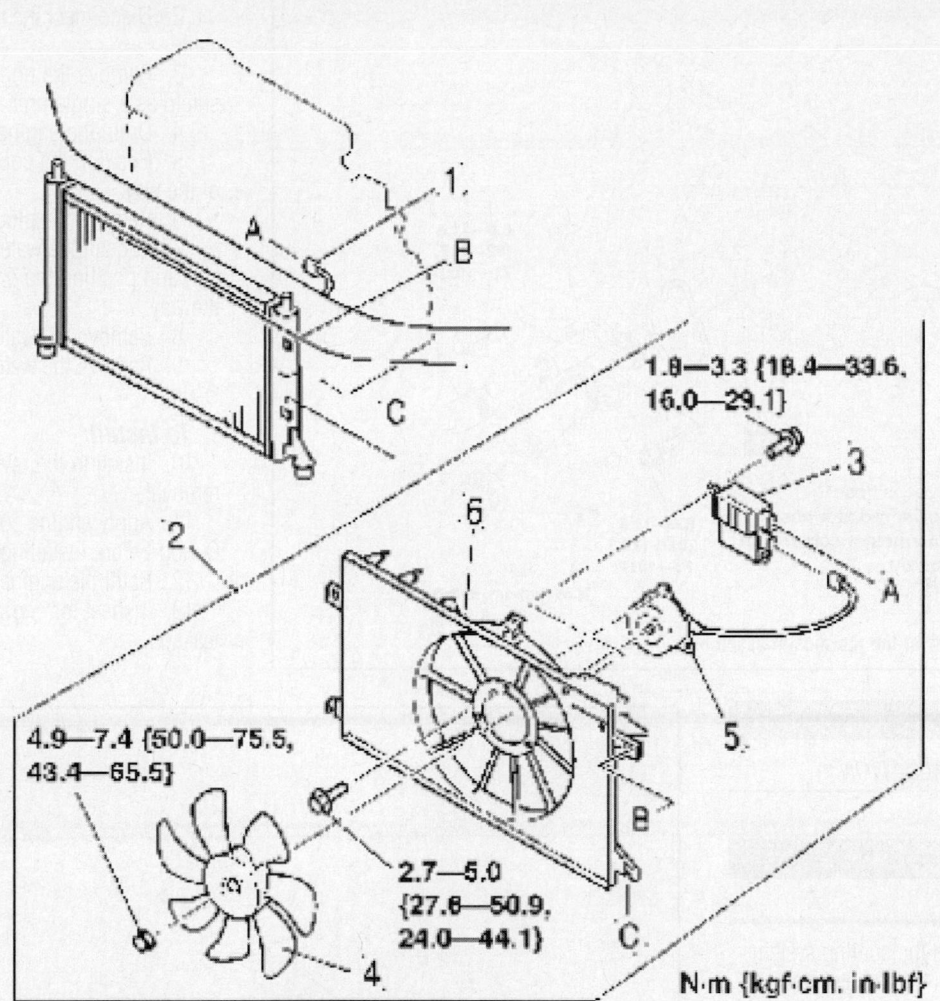

1. Fan control module connector
2. Cooling fan component
3. Fan control module
4. Cooling fan
5. Cooling fan motor
6. Radiator cowling

41311_MMA5_G0034

Fig. 19 Remove the cooling fan and system components, following the numerical sequence shown.

THERMOSTAT

REMOVAL & INSTALLATION

See Figure 20.

✳✳ WARNING

Note the following:

- Never remove the cooling system cap or loosen the radiator drain plug while the engine is running, or when the engine and radiator are hot. Scalding engine coolant and steam may shoot out and cause serious injury. It may also damage the engine and cooling system.

- Turn off the engine and wait until it is cool. Even then, be very careful when removing the cap. Wrap a thick cloth around it and slowly turn it counterclockwise to the first stop. Step back while the pressure escapes.
- When positive all the pressure is gone, press down on the cap using the cloth, turn it, and remove it.

1. Remove the battery cover.
2. Disconnect the negative battery cable.
3. Remove the undercover and splash shield as a single unit.
4. Drain the engine coolant.

5. Position the coolant reserve tank out of the way.
6. Remove the plug hole plate.
7. Position the drive belt out of the way.
8. Remove the drive belt tensioner.
9. Remove the bypass hose.
10. Remove the lower radiator hose.
11. Remove the thermostat component.
12. Remove the gasket.

To install:

13. Install in the reverse order of removal.
14. Refill the engine coolant.
15. Inspect for engine coolant leakage.

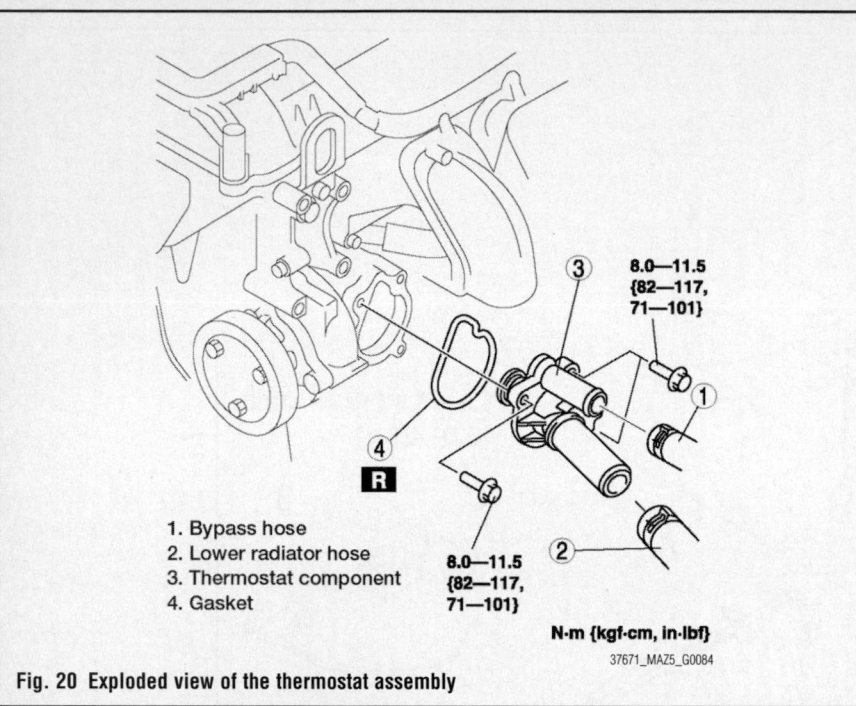

1. Bypass hose
2. Lower radiator hose
3. Thermostat component
4. Gasket

8.0—11.5 {82—117, 71—101}

8.0—11.5 {82—117, 71—101}

N·m {kgf·cm, in·lbf}

37671_MAZ5_G0084

Fig. 20 Exploded view of the thermostat assembly

1. Remove the battery cover.
2. Disconnect the negative battery cable.
3. Remove the undercover and splash shield as a single unit.
4. Drain the engine coolant.
5. Position the coolant reserve tank out of the way.
6. Remove the plug hole plate.
7. Loosen the water pump pulley bolt and position the drive belt out of the way.
8. Remove the water pump pulley.
9. Remove the water pump and O-ring.

To install:

10. Install in the reverse order of removal.
11. Apply engine coolant to a new O-ring before installing.
12. Refill the engine coolant.
13. Inspect for engine coolant leakage.

WATER PUMP

REMOVAL & INSTALLATION

See Figure 21.

✳✳ WARNING

Note the following:

- Never remove the cooling system cap or loosen the radiator drain plug while the engine is running, or when the engine and radiator are hot. Scalding engine coolant and steam may shoot out and cause serious injury. It may also damage the engine and cooling system.
- Turn off the engine and wait until it is cool. Even then, be very careful when removing the cap. Wrap a thick cloth around it and slowly turn it counterclockwise to the first stop. Step back while the pressure escapes.
- When you are sure all the pressure is gone, press down on the cap using the cloth, turn it, and remove it.

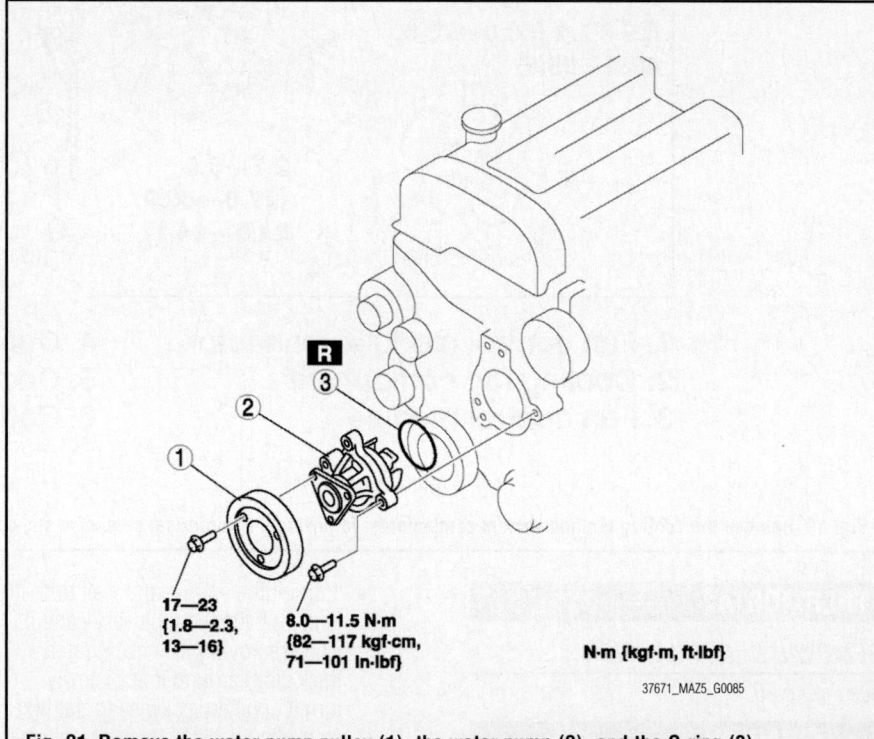

17—23 {1.8—2.3, 13—16}

8.0—11.5 N·m {82—117 kgf·cm, 71—101 in·lbf}

N·m {kgf·m, ft·lbf}

37671_MAZ5_G0085

Fig. 21 Remove the water pump pulley (1), the water pump (2), and the O-ring (3)

ENGINE ELECTRICAL | CHARGING SYSTEM

ALTERNATOR

REMOVAL & INSTALLATION
See Figure 22.

✳✳ WARNING
Note the following:

- Remove and install all parts when the engine is cold, otherwise they can cause severe burns or serious injury.
- When the battery cables are connected, touching the vehicle body with alternator terminal B will generate sparks. This can cause personal injury, fire, and damage to the electrical components. Always disconnect the battery negative cable before performing the following operation.

1. Remove the battery cover.
2. Disconnect the negative battery cable.
3. Remove the undercover and splash shield as a single unit.
4. Remove the plug hole plate.
5. Position the coolant reserve tank out of the way.
6. Position the drive belt out of the way.

7. Remove the B terminal cable, the alternator connector, and the alternator.

➡ **The letters A, B, C, and D are designators for the order of installation**

To install:
8. Install in the reverse order of removal.

9. Match the alternator mounting hole and engine side hole, then temporarily tighten the alternator installation bolts in the order A, B, C, and D.
10. Tighten the alternator installation bolts in the order A, B, C, and D.

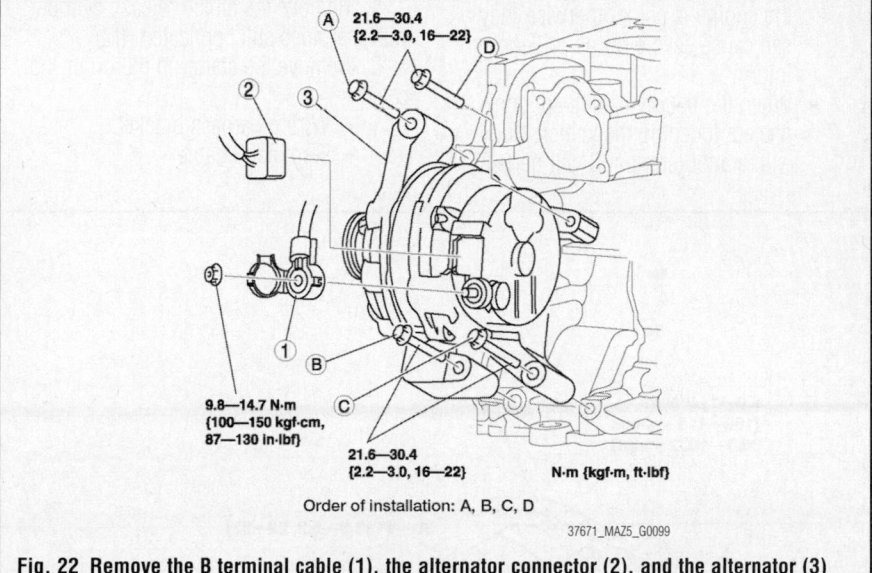

Fig. 22 Remove the B terminal cable (1), the alternator connector (2), and the alternator (3)

ENGINE ELECTRICAL | IGNITION SYSTEM

FIRING ORDER

See Figure 23.

The firing order for the 2.3L engine is 1–3–4–2.

CYLINDER No.

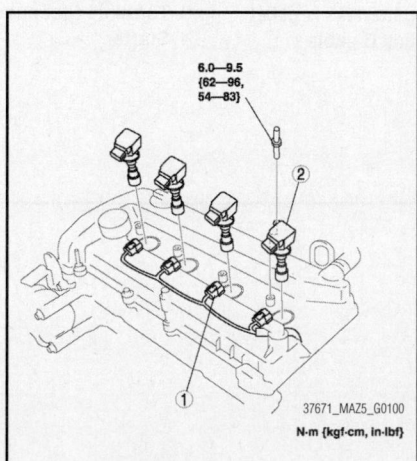

Fig. 23 2.3L engine cylinder locations

IGNITION COIL PACK

REMOVAL & INSTALLATION
See Figure 24.

1. Remove the battery cover.
2. Disconnect the negative battery cable.
3. Remove the plug hole plate.

Fig. 24 Remove the connectors (1) and the ignition coils (2)

N·m {kgf·cm, in·lbf}

4. Remove the connectors and the ignition coils.
5. Install in the reverse order of removal.

IGNITION TIMING

ADJUSTMENT

Ignition timing is controlled by the PCM. No adjustment is necessary or possible.

SPARK PLUGS

REMOVAL & INSTALLATION

✳✳ CAUTION
If a spark plug that is not as specified is installed, engine performance will be deteriorated. Install only the specified spark plug when replacing.

1. Remove the battery cover.
2. Disconnect the negative battery cable.
3. Remove the plug hole plate.
4. Remove the ignition coils.
5. Remove the spark plugs using a plug-wrench.
6. Install in the reverse order of removal.

ENGINE ELECTRICAL   **STARTING SYSTEM**

STARTER

REMOVAL & INSTALLATION

See Figure 25.

⚙ **WARNING**

Note the following:

- Remove and install all parts when the engine is cold, otherwise they can cause severe burns or serious injury.
- When the battery cables are connected, touching the vehicle body with starter terminal B will generate sparks. This can cause personal injury, fire, and damage to the electrical components. Always disconnect the negative battery cable before performing the following operation.

1. Remove the battery cover.
2. Disconnect the negative battery cable.
3. Remove the undercover.
4. Remove the clutch release cylinder with the pipes still connected. (M/T)
5. Remove the starter in the order indicated:

- Wiring harness bracket
- Terminal B cable
- Terminal S connector
- Starter

To install:

6. Install in the reverse order of removal.

➡**If there is peeling on or damage to the insulator, attach a new insulator using the following procedure:**

7. Peel off the insulator from the starter completely using a scraper.
8. Degrease the insulator attachment area.
9. Attach a new insulator to the starter.

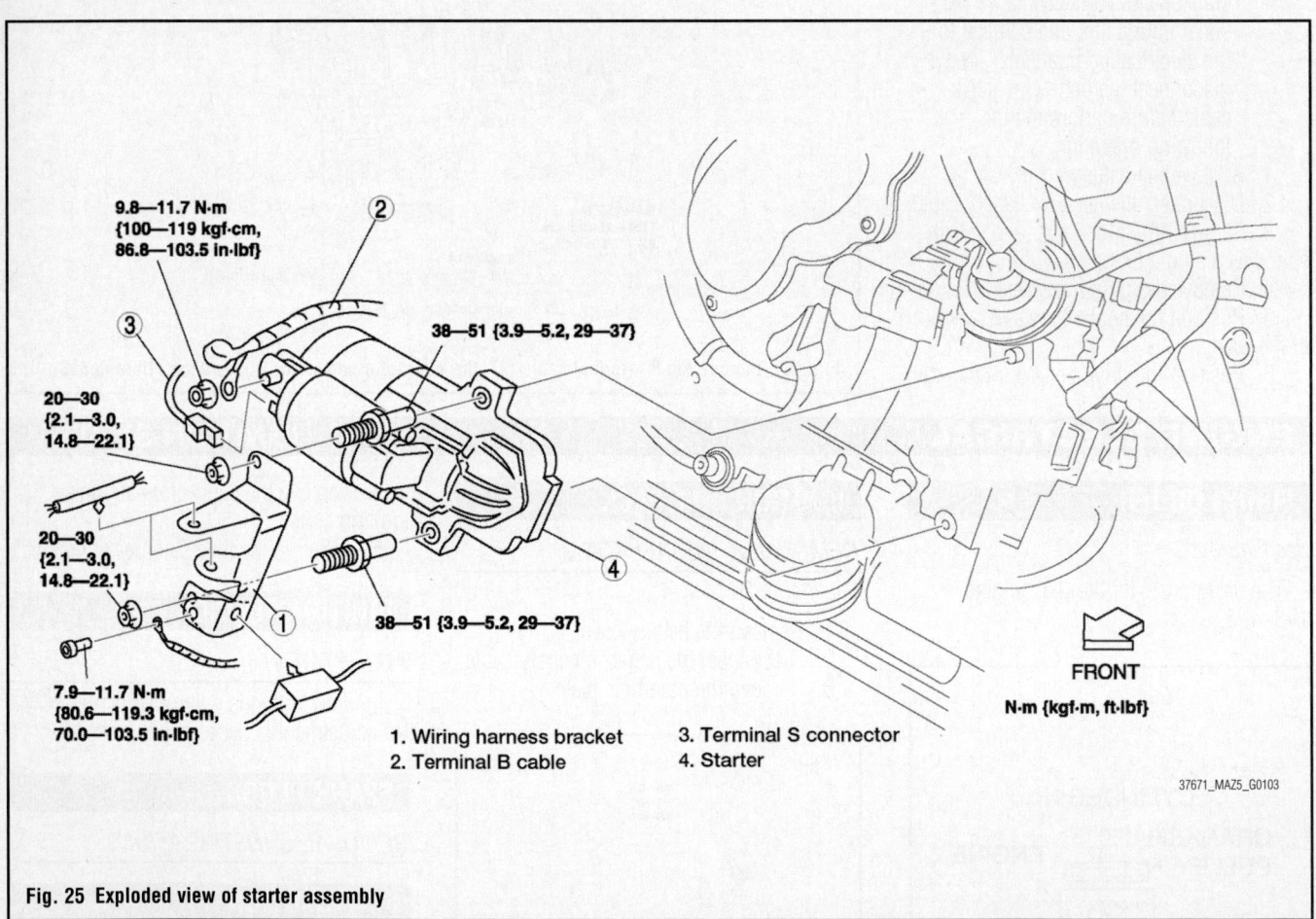

Fig. 25 Exploded view of starter assembly

1. Wiring harness bracket
2. Terminal B cable
3. Terminal S connector
4. Starter

ENGINE MECHANICAL

➡Disconnecting the negative battery cable may interfere with the functions of the on board computer systems and may require the computer to undergo a relearning process, once the negative battery cable is reconnected.

ACCESSORY DRIVE BELTS

INSPECTION

Alternator Drive Belt

See Figure 26.

➡Drive belt deflection/tension inspection is not necessary because of the use of the alternator drive belt auto tensioner.

1. Verify that the drive belt auto tensioner indicator mark does not exceed the limit.
2. If it exceeds the limit, replace the drive belt.

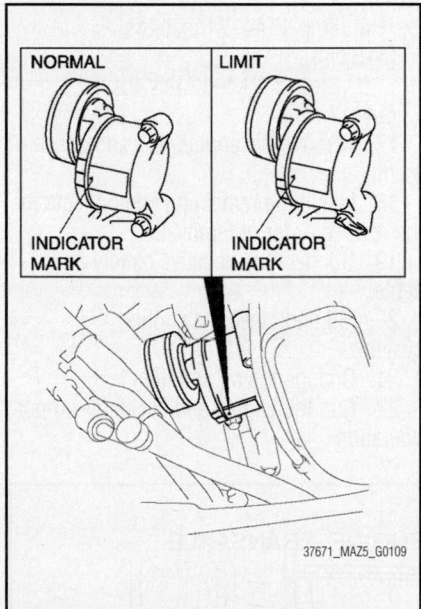

Fig. 26 Verify that the drive belt auto tensioner indicator mark does not exceed the limit

A/C Drive Belt

➡Drive belt deflection/tension inspection is not necessary because of the use of the maintenance-free type A/C drive belt.

➡Replace the drive belt if it is found to be damaged during visual Inspection, or if there is a malfunction or noise in the A/C compressor.

ADJUSTMENT

Belt tension is applied by automatic tensioners. No adjustment is necessary.

REMOVAL & INSTALLATION

A/C Drive Belt Removal

See Figure 27.

1. Remove the engine under cover and splash shield (RH).
2. Set the SST 49B015102 as shown.

➡Hold the SST by hand until it fits between the A/C drive belt and the crankshaft pulley.

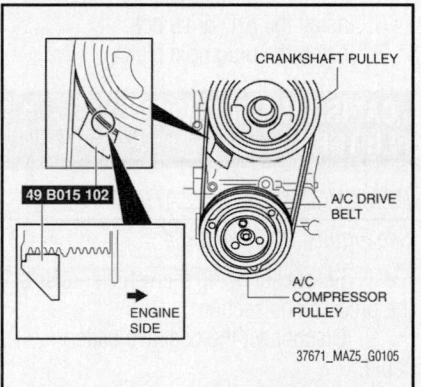

Fig. 27 Set the SST 49B015102 as shown

3. Rotate the crankshaft pulley clockwise using a wrench and remove the A/C drive belt.

A/C Drive Belt Installation

See Figures 28 through 30.

1. Set the SST 49B015103 to the crankshaft pulley as shown.

➡Hang the A/C drive belt on to the A/C compressor pulley, then move the SST upward along the crankshaft pulley.

2. Set the SST 49105104 to the A/C compressor mount as shown.

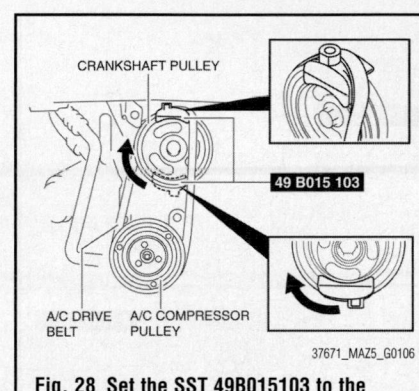

Fig. 28 Set the SST 49B015103 to the crankshaft pulley as shown

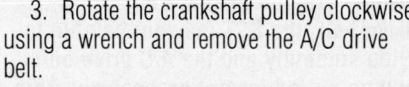

Fig. 29 Set the SST 49105104 to the A/C compressor mount as shown

➡During the A/C drive belt installation, if the SST touches the part number-printed area on the back of the A/C drive belt, the SST may not be able to slide smoothly and the A/C drive belt will be pulled inward excessively. This will make it impossible to correctly install the A/C drive belt to the groove of the A/C compressor pulley. To prevent this, adjust the A/C drive belt to the position shown, then start the procedure.

3. After setting the SST, verify that the three parts of A/C drive belt shown are hung on each pulley.

4. Rotate the crankshaft pulley clockwise using a wrench and install the A/C drive belt.

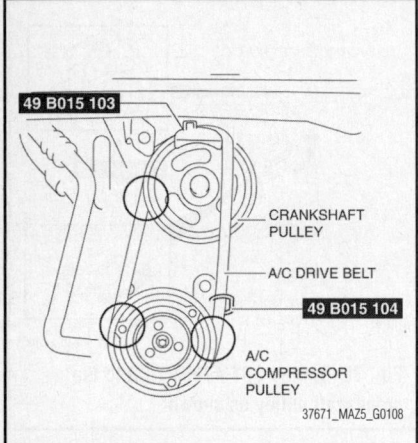

Fig. 30 After setting the SST, verify that the three parts of A/C drive belt shown are hung on each pulley

✳✳ CAUTION

Remove the SST (49 B015 103) immediately after it reaches the lower side to prevent it from falling down and being damaged.

5. Remove the SST (49 B015 104)
6. Rotate the crankshaft pulley clockwise one time or more to verify that the A/C drive belt is installed correctly.

✳✳ CAUTION

Note the following:

- Do not turn the crankshaft pulley counter clockwise, otherwise the crankshaft pulley bolt could loosen and the timing could change causing engine damage.

- If the A/C belt is not seated properly, push/pull the belt with a suitable tool in the direction where it would be seated properly and then rotate the crank pulley clockwise until the belt is correctly seated. If this does not correct the problem, remove the belt and install it again.

7. Install the undercover and splash shield (RH).

Alternator Drive Belt

1. Remove the plug hole plate.
2. Remove the A/C drive belt.
3. Turn the center of the auto tensioner pulley counterclockwise to release tension to the drive belt tension.
4. Remove the alternator drive belt.
5. Install a new alternator drive belt.
6. Install the A/C drive belt.
7. Install the plug hole plate.

CAMSHAFT AND VALVE LIFTERS

REMOVAL & INSTALLATION

See Figures 31 through 34.

1. Before servicing the vehicle, refer to the precautions section.
2. Disconnect the negative battery cable.
3. Remove the timing chain.
4. Remove the ignition coil and spark plug wires.
5. Disconnect the alternator and move it aside.
6. Remove the front exhaust pipe.
7. Remove the intake manifold.
8. Remove the upper radiator hose, bypass and heater hoses.

9. Support the engine using an engine jack and attachment as shown in the illustration.
10. Remove the oil control valve.
11. Remove the camshafts.

➡The cylinder head and camshaft caps are numbered and must be reassembled in their original locations. When removing a when removed, keep the caps with the cylinder head they were removed from, do not switch the caps.

12. Loosen the camshaft cap bolts using 2–3 passes in the sequence illustrated.
13. Once all the bolts have been removed, remove the caps, keeping them in the original positions and remove the camshafts.

To install:

14. Set the cam position of the number 1 cylinder at Top Dead center (TDC) and install the camshaft.
15. Temporarily install the camshaft caps in their original positions.
16. Tighten the camshaft cap bolt in two steps:
   a. Step 1: 44–71 inch lbs. (5–8 Nm).
   b. Step 2: 10.4–12.5 ft. lbs. (14–17 Nm).
17. Remove the engine jack and attachment.
18. Install the remaining components in the reverse order of removal.
19. Connect the negative battery cable.
20. Fill and bleed the cooling system.
21. Change the oil and filter.
22. Run the engine and check for proper operation.

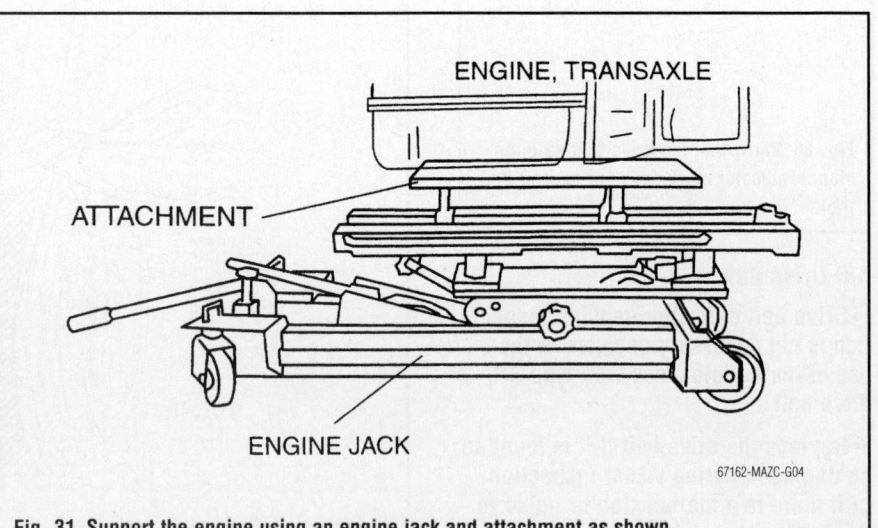

Fig. 31 Support the engine using an engine jack and attachment as shown

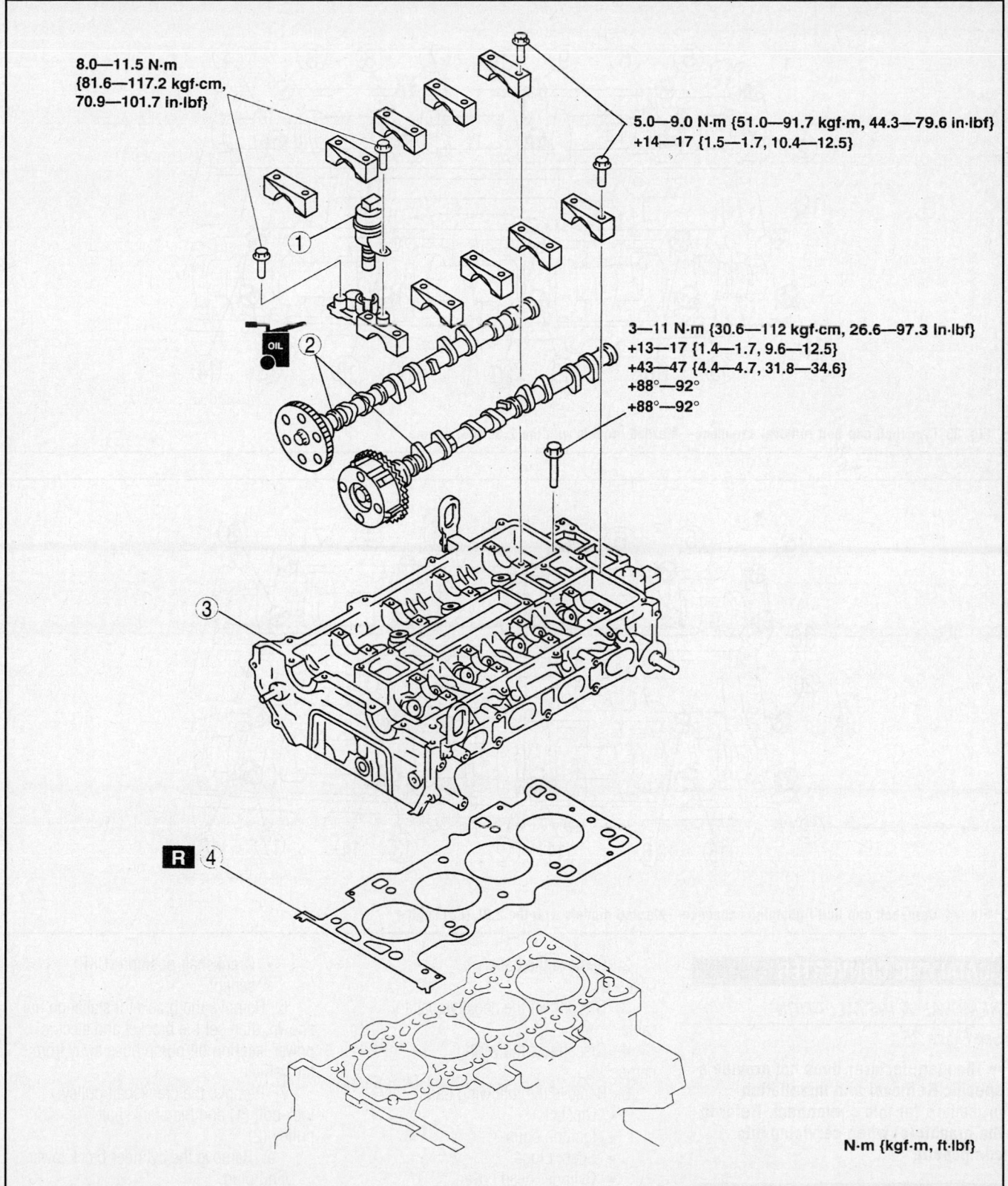

8.0—11.5 N·m
{81.6—117.2 kgf·cm,
70.9—101.7 in·lbf}

5.0—9.0 N·m {51.0—91.7 kgf·m, 44.3—79.6 in·lbf}
+14—17 {1.5—1.7, 10.4—12.5}

OIL

3—11 N·m {30.6—112 kgf·cm, 26.6—97.3 in·lbf}
+13—17 {1.4—1.7, 9.6—12.5}
+43—47 {4.4—4.7, 31.8—34.6}
+88°—92°
+88°—92°

N·m {kgf·m, ft·lbf}

1	Oil control valve (OCV)	3	Cylinder head
2	Camshaft	4	Cylinder head gasket

67162-MAZC-G168

Fig. 32 Exploded view of the camshaft and cylinder head components—Mazda6 models with the 2.3L (L3) engine

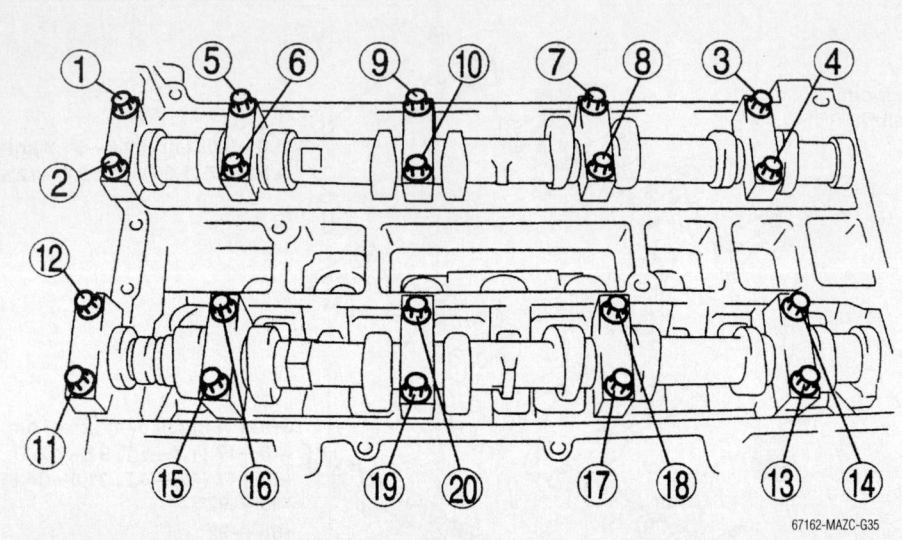

Fig. 33 Camshaft cap bolt removal sequence—Mazda6 models with the 2.3L (L3) engine

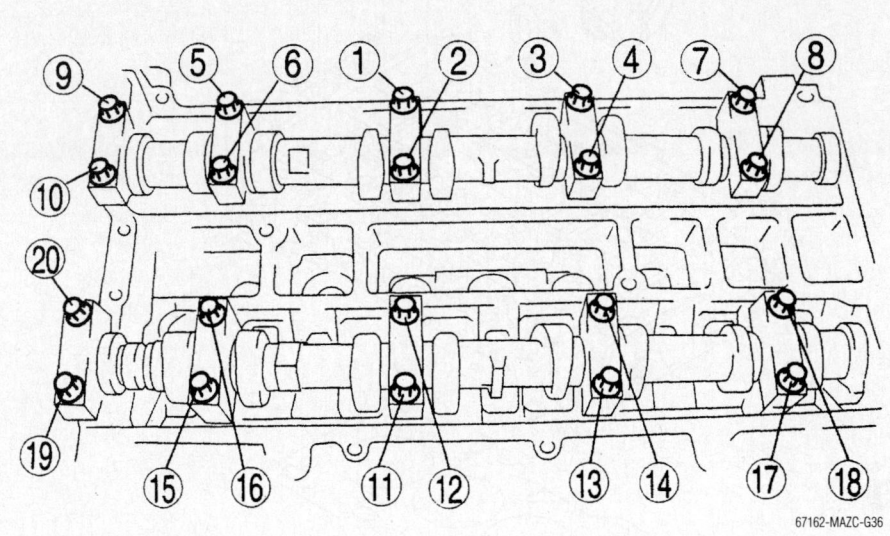

Fig. 34 Camshaft cap bolt tightening sequence—Mazda6 models with the 2.3L (L3) engine

CATALYTIC CONVERTER

REMOVAL & INSTALLATION

See Figure 35.

➡ **The manufacturer does not provide a specific Removal and Installation procedure for this component. Refer to the graphic(s) when servicing this component.**

CRANKSHAFT FRONT SEAL

REMOVAL & INSTALLATION

See Figures 36 through 40.

1. Remove the plug hole plate.

2. Remove the battery cover.

3. Disconnect the negative battery cable.

4. Disconnect the wiring harness.

5. Remove the following parts.:
- Dipstick
- Ignition coils
- Spark plugs
- Cylinder head cover
- Drive belt
- Under cover and splash shield (RH)
- Front wheel and tire (RH)
- Disconnect the drive shaft (RH) from joint shaft, set the drive shaft (RH) out of the way.

- Crankshaft position (CKP) sensor

6. Remove the bracket installation nut shown, then set the bracket and electric power steering oil pump hose away from the vehicle.

7. Remove the crankshaft pulley lock bolt (1) and the crankshaft pulley (2).

   a. Remove the cylinder block lower blind plug.

   b. Install the SST 303-507.

   c. Turn the crankshaft clockwise until the crankshaft is in the No.1 cylinder TDC position (until the balance weight is attached to the SST).

   d. Hold the crankshaft pulley using the SSTs.

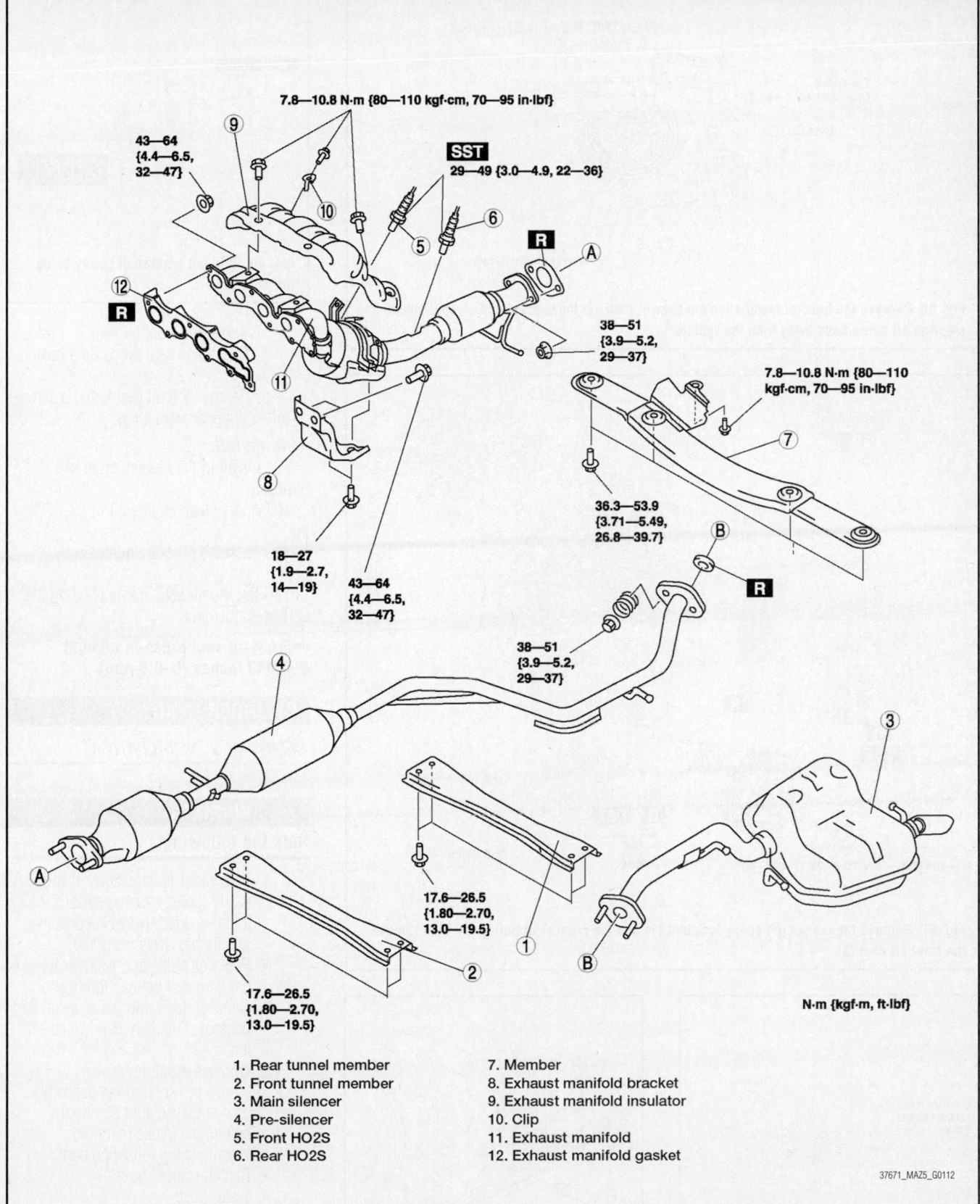

7.8—10.8 N·m {80—110 kgf·cm, 70—95 in·lbf}

43—64
{4.4—6.5,
32—47}

SST

29—49 {3.0—4.9, 22—36}

R

R

38—51
{3.9—5.2,
29—37}

7.8—10.8 N·m {80—110
kgf·cm, 70—95 in·lbf}

18—27
{1.9—2.7,
14—19}

43—64
{4.4—6.5,
32—47}

36.3—53.9
{3.71—5.49,
26.8—39.7}

R

38—51
{3.9—5.2,
29—37}

17.6—26.5
{1.80—2.70,
13.0—19.5}

17.6—26.5
{1.80—2.70,
13.0—19.5}

N·m {kgf·m, ft·lbf}

1. Rear tunnel member
2. Front tunnel member
3. Main silencer
4. Pre-silencer
5. Front HO2S
6. Rear HO2S
7. Member
8. Exhaust manifold bracket
9. Exhaust manifold insulator
10. Clip
11. Exhaust manifold
12. Exhaust manifold gasket

37671_MAZ5_G0112

Fig. 35 Exploded view of exhaust system

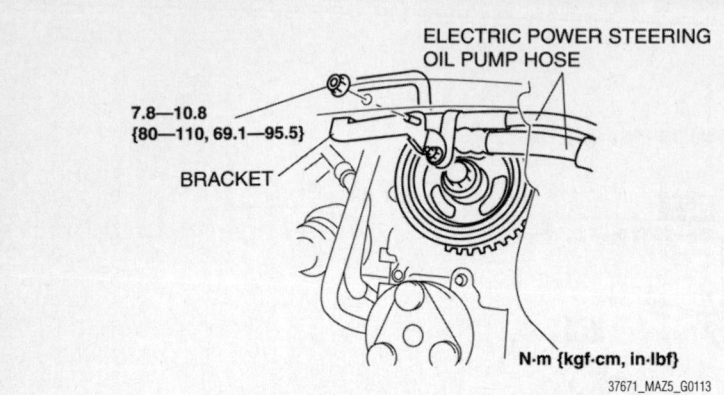

Fig. 36 Remove the bracket installation nut shown, then set the bracket and electric power steering oil pump hose away from the vehicle

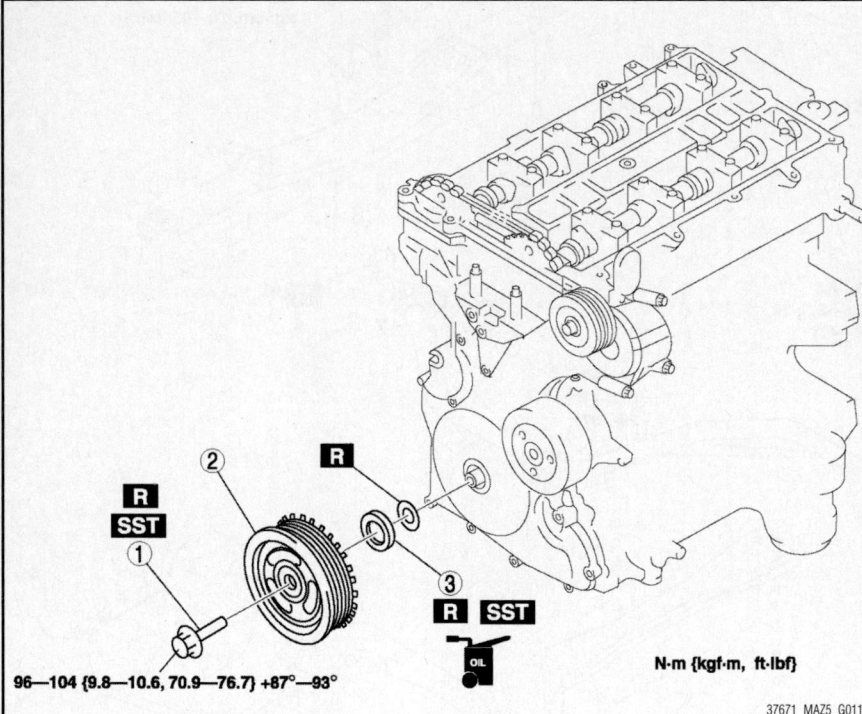

Fig. 37 Remove the crankshaft pulley lock bolt (1) and the crankshaft pulley (2), then remove the front oil seal (3)

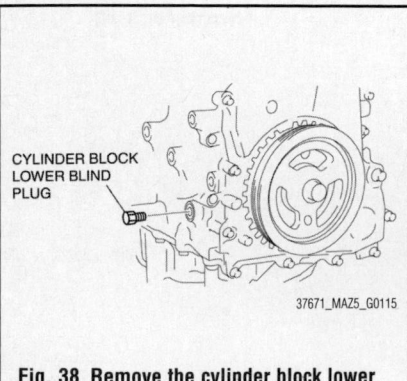

Fig. 38 Remove the cylinder block lower blind plug

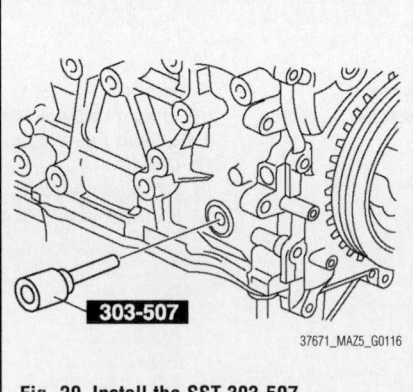

Fig. 39 Install the SST 303-507

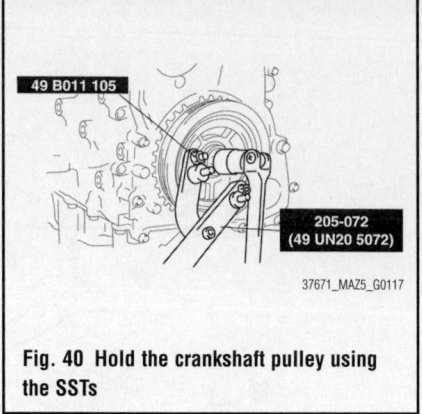

Fig. 40 Hold the crankshaft pulley using the SSTs

8. Remove the front oil seal.
   a. Cut the oil seal lip using a razor knife.
   b. Remove the oil seal using a screwdriver wrapped with a rag.

To install:

9. Install in the reverse order of removal.

10. Apply clean engine oil to the oil seal lip.

11. Push the oil seal slightly in by hand.

12. Tap the oil seal in evenly using the SST and a hammer.

➡️**Front oil seal press-in amount 0–0.019 inches (0–0.5 mm)**

CYLINDER HEAD

REMOVAL & INSTALLATION

See Figures 41 through 43.

❄❄ WARNING

Note the following:

- Fuel vapor is hazardous. It can very easily ignite, causing serious injury and damage. Always keep sparks and flames away from fuel.
- Fuel line spills and leakage are dangerous. Fuel can ignite and cause serious injuries or death and damage. Fuel can also irritate skin and eyes.

1. Remove the timing chain.
2. Remove the cooling fan assembly.
3. Remove the exhaust manifold.
4. Remove the intake manifold.
5. Disconnect the following parts:
   - Radiator upper hose
   - Water hose
   - Heater hose
   - Wiring harness
6. To firmly support the engine, first set the engine jack to the oil pan.

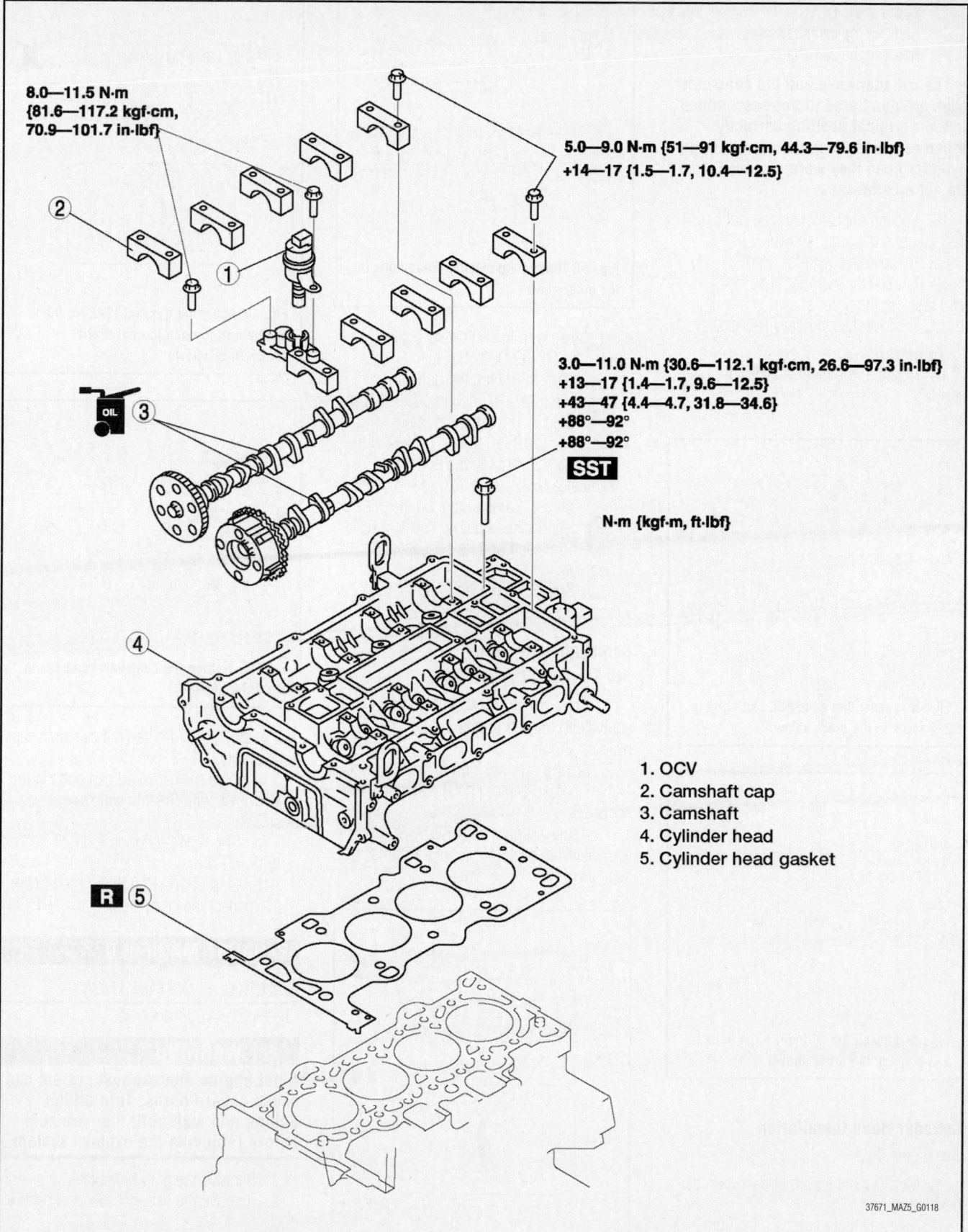

8.0—11.5 N·m
{81.6—117.2 kgf·cm,
70.9—101.7 in·lbf}

5.0—9.0 N·m {51—91 kgf·cm, 44.3—79.6 in·lbf}
+14—17 {1.5—1.7, 10.4—12.5}

3.0—11.0 N·m {30.6—112.1 kgf·cm, 26.6—97.3 in·lbf}
+13—17 {1.4—1.7, 9.6—12.5}
+43—47 {4.4—4.7, 31.8—34.6}
+88°—92°
+88°—92°

SST

N·m {kgf·m, ft·lbf}

1. OCV
2. Camshaft cap
3. Camshaft
4. Cylinder head
5. Cylinder head gasket

R 5

37671_MAZ5_G0118

Fig. 41 Exploded view of cylinder head assembly

7. Remove the OCV.
8. Remove the camshaft caps.
9. Remove the camshafts.

➡ The cylinder head and the camshaft caps are numbered to be reassembled in their original position correctly. When removed, keep the caps with the cylinder head they were removed from. Do not mix the caps.

10. Loosen the camshaft cap bolts in 2–3 steps in the order shown.
11. Remove the cylinder head.
12. Loosen the cylinder head bolts in 2–3 steps in the order shown.
13. Remove the cylinder head gasket

To install:

14. Install in the reverse order of removal.

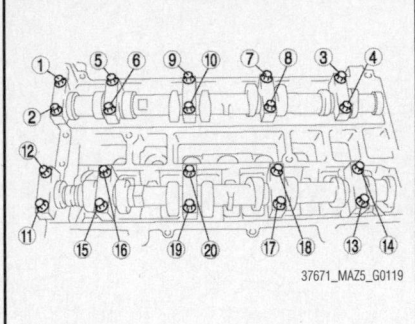

Fig. 42 Loosen the camshaft cap bolts in 2–3 steps in the order shown

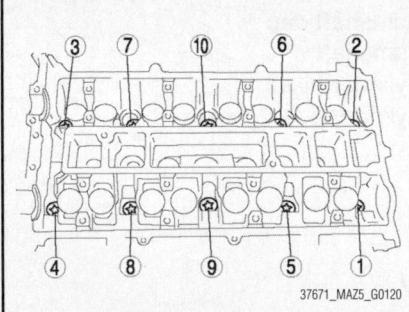

Fig. 43 Loosen the cylinder head bolts in 2–3 steps in the order shown

Cylinder Head Installation

See Figure 44.

1. Measure the length of each cylinder head bolt.
   Replace any that exceeds maximum length. Cylinder Head Bolt Maximum: 5.768 inches (146.5 mm)
2. Tighten the cylinder head bolts in the

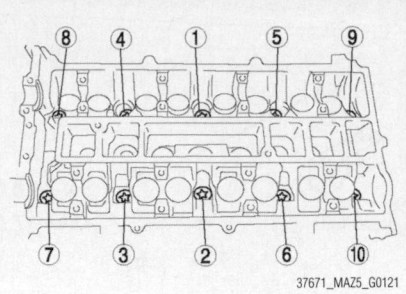

Fig. 44 Tighten the cylinder head bolts in the order shown

order shown with the following 5 steps using the SST (49 D032 316).

   a. Step 1: Tighten the bolts in sequence to 27–97 inch lbs. (3–11 Nm).

   b. Step 2: Tighten the bolts in sequence to 10–13 ft. lbs. (13–17 Nm).

   c. Step 3: Tighten the bolts in sequence to 32–35 ft. lbs. (43–47 Nm).

   d. Step 4: Paint a mark on the edge of each cylinder head bolt to use as a reference. Turn each bolt, in sequence, 88–92 degrees.

   e. Step 5: Turn each bolt, in sequence, 88–92 degrees.

Camshaft Installation

See Figures 45 through 47.

1. Apply the gear oil (SAE No.90 or equivalent) to each journal of the cylinder head as shown.
2. Set the cam position of No.1 cylinder at the top dead center (TDC) and install the camshaft.
3. Apply the gear oil (SAE No.90 or equivalent) to each journal of the camshaft as shown.

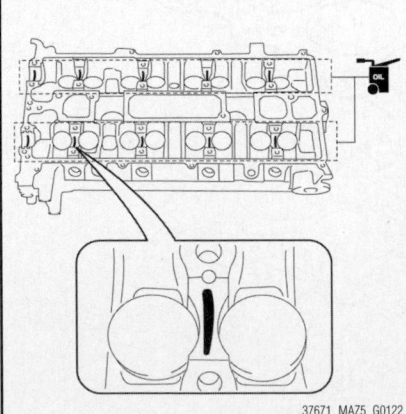

Fig. 45 Apply the gear oil (SAE No.90 or equivalent) to each journal of the cylinder head as shown

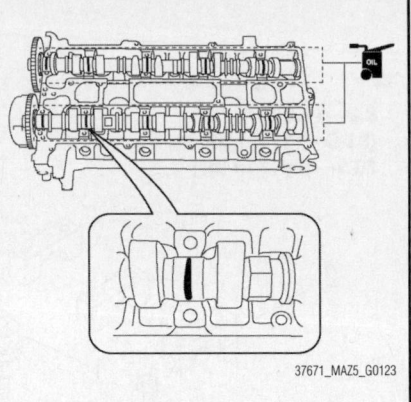

Fig. 46 Apply the gear oil (SAE No.90 or equivalent) to each journal of the camshaft as shown

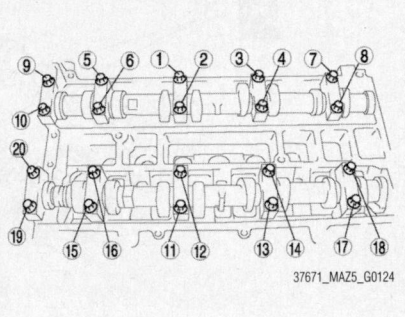

Fig. 47 Tighten the camshaft cap bolts in the order shown

4. Temporarily tighten the camshaft cap bolts evenly in 2–3 steps.
5. Tighten the camshaft cap bolts in the order shown with the following two steps.

   a. Step 1: 44–80 inch lbs. (5.0–9.0 Nm)

   b. Step 2: 10–13 ft. lbs. (14–17 Nm)

6. Inspect the compression.

EXHAUST MANIFOLD

REMOVAL & INSTALLATION

See Figures 48 through 50.

✳✳ WARNING

A hot engine and exhaust system can cause severe burns. Turn off the engine and wait until they are cool before removing the exhaust system.

1. Remove the plug hole plate.
2. Remove the battery cover and battery duct.
3. Disconnect the negative battery cable.
4. Remove the undercover.

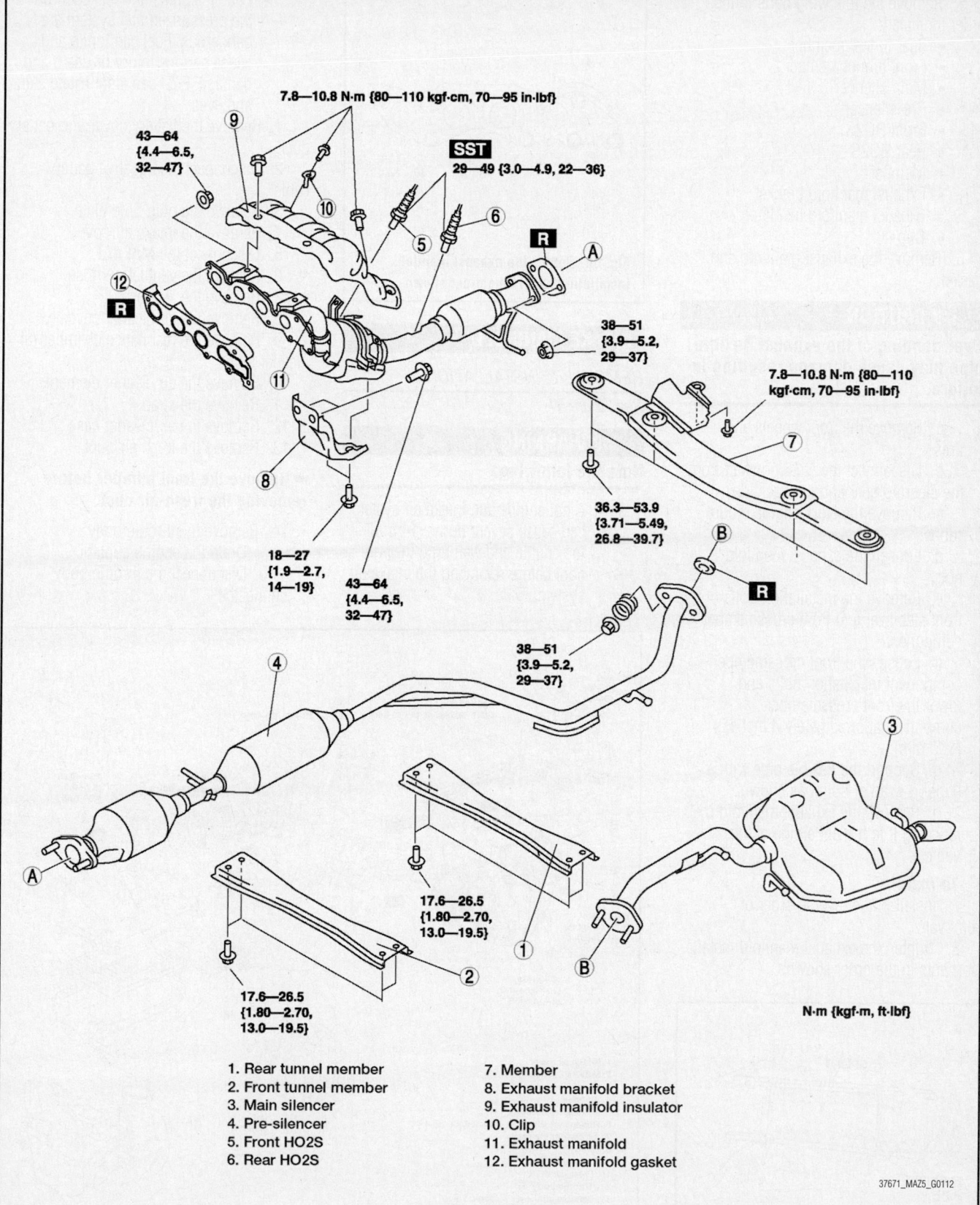

7.8—10.8 N·m {80—110 kgf·cm, 70—95 in·lbf}

43—64 {4.4—6.5, 32—47}

SST
29—49 {3.0—4.9, 22—36}

38—51 {3.9—5.2, 29—37}

7.8—10.8 N·m {80—110 kgf·cm, 70—95 in·lbf}

36.3—53.9 {3.71—5.49, 26.8—39.7}

18—27 {1.9—2.7, 14—19}

43—64 {4.4—6.5, 32—47}

38—51 {3.9—5.2, 29—37}

17.6—26.5 {1.80—2.70, 13.0—19.5}

17.6—26.5 {1.80—2.70, 13.0—19.5}

N·m {kgf·m, ft·lbf}

1. Rear tunnel member
2. Front tunnel member
3. Main silencer
4. Pre-silencer
5. Front HO2S
6. Rear HO2S
7. Member
8. Exhaust manifold bracket
9. Exhaust manifold insulator
10. Clip
11. Exhaust manifold
12. Exhaust manifold gasket

37671_MAZ5_G0112

Fig. 48 Exploded view of exhaust system

5. Remove the following parts in the order indicated:
- Rear tunnel member
- Front tunnel member
- Main silencer
- Pre-silencer
- Front HO2S
- Rear HO2S
- Member
- Exhaust manifold bracket
- Exhaust manifold insulator
- Clip

6. Remove the exhaust manifold and gasket.

✳✳ CAUTION

Over bending of the exhaust flexible pipe may cause damage resulting in failure.

a. Remove the front wheels and tires.

b. Disconnect the steering shaft from the steering gear and linkage side.

c. Remove the No.1 engine mount rubber.

d. Loosen the exhaust manifold nuts.

e. Remove the installation bolts of the front stabilizer and front crossmember component.

f. Loosen the front crossmember component installation bolts and lower the front crossmember component approximately 4 inches (100 mm).

g. Support the flexible pipe with a support wrap or splint as shown.

h. Remove the exhaust manifold by lowering it to the underside of the vehicle.

To install:

7. Install in the reverse order of removal.

8. Tighten the exhaust manifold installation nuts in the order shown.

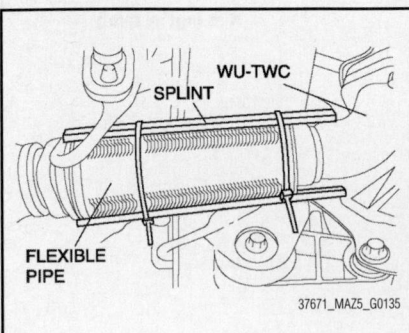

Fig. 49 Support the flexible pipe with a support wrap or splint as shown

WU-TWC
SPLINT
FLEXIBLE PIPE
37671_MAZ5_G0135

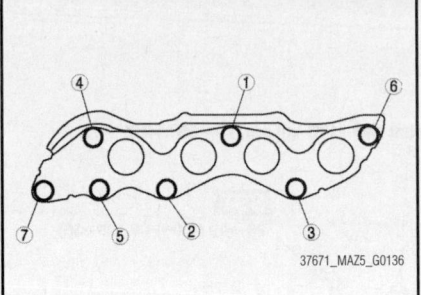

Fig. 50 Tighten the exhaust manifold installation nuts in the order shown

37671_MAZ5_G0136

INTAKE MANIFOLD

REMOVAL & INSTALLATION
See Figure 51.

✳✳ WARNING

Note the following:

- A hot engine and intake air system can cause severe burns. Turn off the engine and wait until they are cool before removing the intake air system.

- Fuel line spills and leakage from the pressurized fuel system are dangerous. Fuel can ignite and cause serious injury or death and damage. Fuel can also irritate skin and eyes.

1. Remove the battery cover and battery duct.

2. Disconnect the negative battery cable.

3. Remove the plug hole plate.

4. Remove the intake air cover.

5. Disconnect the MAF/IAT.

6. Remove the ventilation hose.

7. Remove the air hose.

8. Remove the air cleaner cover.

9. Remove the resonance chamber (air cleaner side).

10. Remove the air cleaner element.

11. Remove the strap.

12. Remove the air cleaner case.

13. Remove the fresh air duct.

➡ **Remove the front bumper before removing the fresh-air duct.**

14. Remove the throttle body.

a. Drain the engine coolant.

b. Disconnect the throttle body connector.

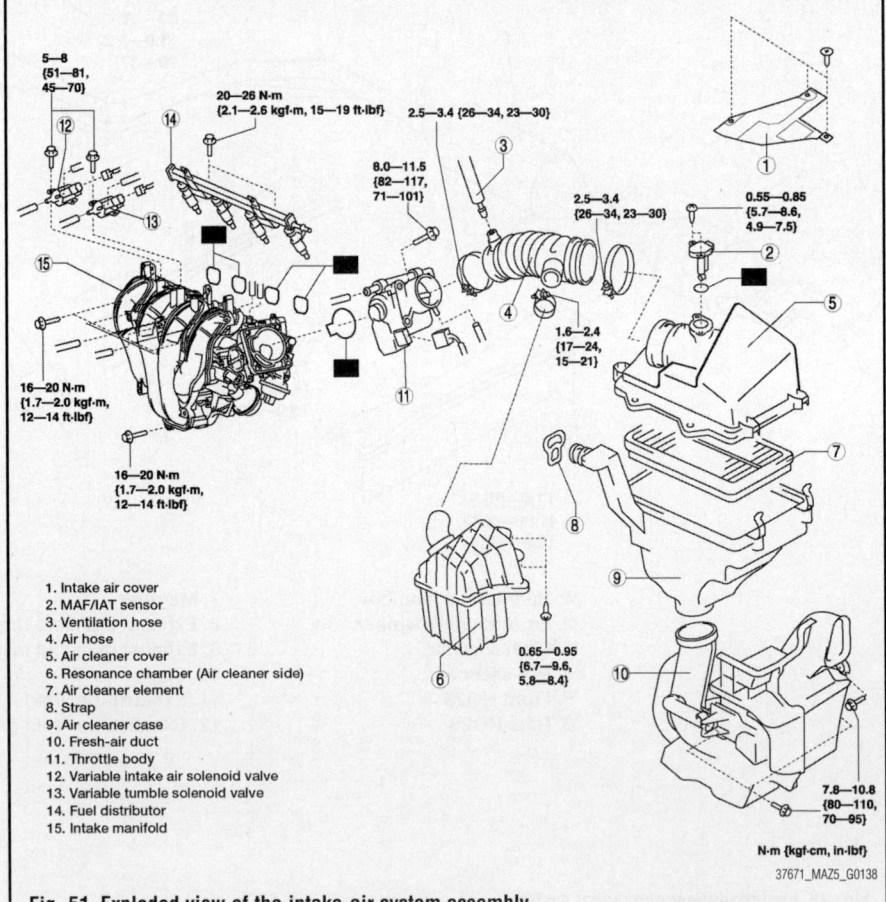

5—8
{51—81, 45—70}

20—26 N·m
{2.1—2.6 kgf·m, 15—19 ft·lbf}

2.5—3.4 {26—34, 23—30}

8.0—11.5
{82—117, 71—101}

2.5—3.4
{26—34, 23—30}

0.55—0.85
{5.7—8.6, 4.9—7.5}

16—20 N·m
{1.7—2.0 kgf·m, 12—14 ft·lbf}

1.6—2.4
{17—24, 15—21}

16—20 N·m
{1.7—2.0 kgf·m, 12—14 ft·lbf}

0.65—0.95
{6.7—9.6, 5.8—8.4}

7.8—10.8
{80—110, 70—95}

1. Intake air cover
2. MAF/IAT sensor
3. Ventilation hose
4. Air hose
5. Air cleaner cover
6. Resonance chamber (Air cleaner side)
7. Air cleaner element
8. Strap
9. Air cleaner case
10. Fresh-air duct
11. Throttle body
12. Variable intake air solenoid valve
13. Variable tumble solenoid valve
14. Fuel distributor
15. Intake manifold

N·m {kgf·cm, in·lbf}

37671_MAZ5_G0138

Fig. 51 Exploded view of the intake-air system assembly

c. Disconnect the water hose connecting the throttle body.

d. Remove the throttle body.

15. Remove the variable intake air solenoid valve.

16. Remove the variable tumble solenoid valve.

17. Remove the fuel distributor.

18. Remove the intake manifold.

a. Remove the battery, battery tray and PCM component.

b. Remove the undercover.

c. Disconnect the vacuum hose connecting the intake manifold.

To install:

19. Install in the reverse order of removal.

Air Cleaner Case Installation

➡**Before installing the air cleaner case, verify that the rubber mounts on the battery support bracket have not fallen off. When inserting the air cleaner case into the rubber mounts, applying soapy water aids the operation.**

1. Verify that two rubber mounts are installed on the battery support bracket.

2. Install the air cleaner case into the rubber mounts.

Strap Installation

See Figure 52.

1. Using the strap, secure the shroud panel and the air cleaner case as shown.

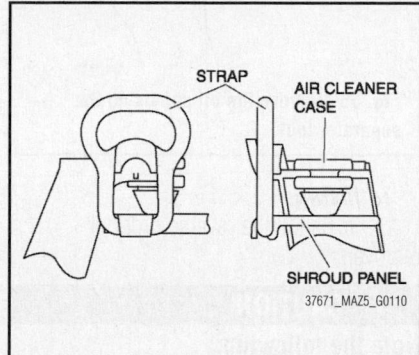

Fig. 52 Using the strap, secure the shroud panel and the air cleaner case as shown

Air Hose Installation

See Figure 53.

1. Align the alignment marks on the throttle body and the air hose.

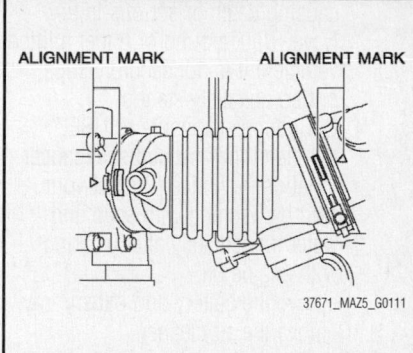

Fig. 53 Align the alignment marks on the throttle body and the air hose

OIL PAN

REMOVAL & INSTALLATION

See Figures 54 through 59.

✳✳ WARNING

Note the following:

- Hot engine and engine oil can cause severe burns. Turn off the engine and wait until it and the engine oil have cooled.
- A vehicle that is lifted but not securely supported on safety stands is dangerous. It can slip or fall,

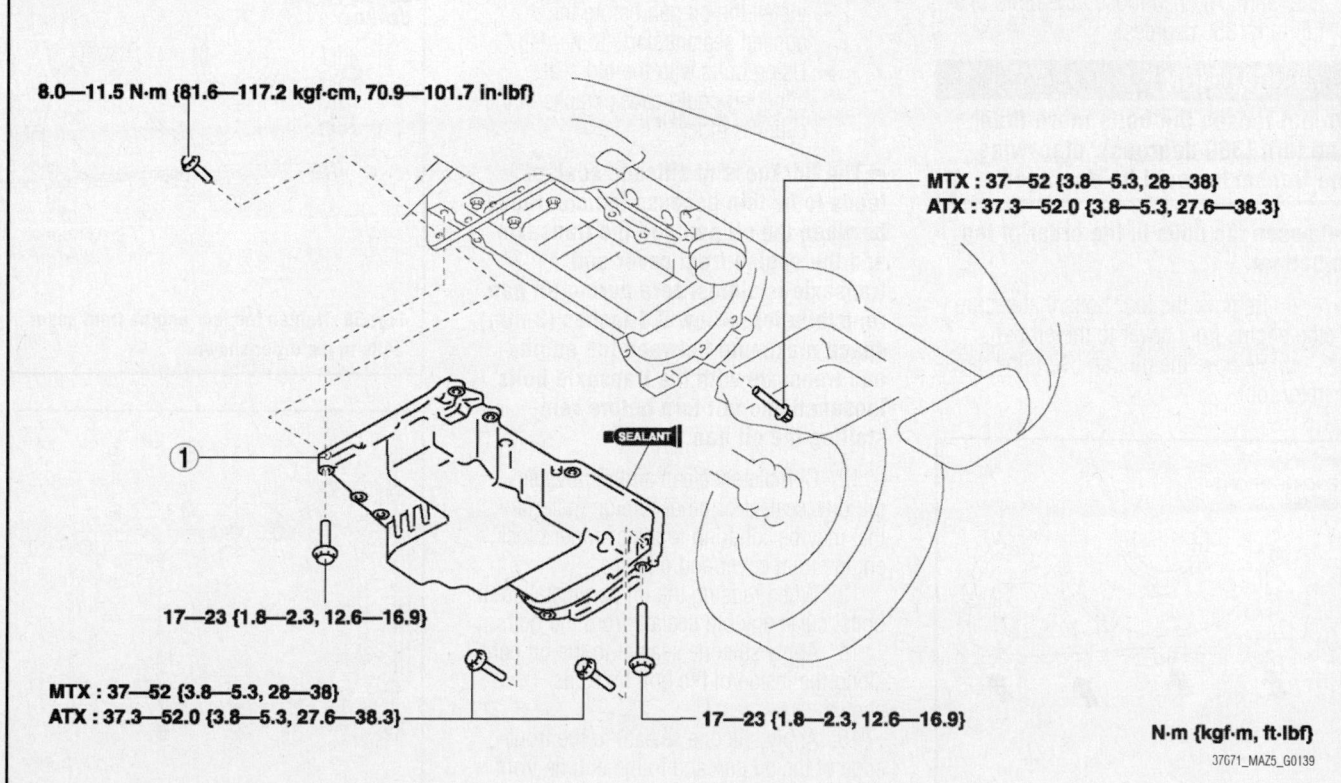

8.0—11.5 N·m {81.6—117.2 kgf·cm, 70.9—101.7 in·lbf}

MTX : 37—52 {3.8—5.3, 28—38}
ATX : 37.3—52.0 {3.8—5.3, 27.6—38.3}

17—23 {1.8—2.3, 12.6—16.9}

MTX : 37—52 {3.8—5.3, 28—38}
ATX : 37.3—52.0 {3.8—5.3, 27.6—38.3}

17—23 {1.8—2.3, 12.6—16.9}

N·m {kgf·m, ft·lbf}

Fig. 54 Remove the oil pan (1)

causing death or serious injury. Never work around or under a lifted vehicle if it is not securely supported on safety stands.

- Continuous exposure to USED engine oil has caused skin cancer in laboratory mice. Protect your skin by washing with soap and water immediately after working with engine oil.

1. Remove the battery and battery tray.
2. Remove the air cleaner.
3. Remove the undercover and splash shield as a single unit.
4. Remove the front tire (RH).
5. Drain the engine oil.
6. Remove the exhaust manifold insulator.
7. Remove the A/C drive belt.
8. Remove the A/C compressor with the pipes still connected.
9. Remove the ATF oil cooler with the water hose and ATF oil cooler hose still connected. (A/T)
10. Remove the oil pan.

   a. Set the wiring harness and harness bracket out of the way to access the transaxle upper fitting bolts.

   b. Before loosening the bolts securing the engine and transaxle, put a mark on each transaxle bolt and the transaxle (to determine when they have been turned one full turn), then loosen the bolts one full turn (360 degrees).

✳ CAUTION

Do not loosen the bolts more than one turn (360 degrees), otherwise the transaxle could be damaged.

➡Loosen the bolts in the order of top to bottom.

   c. Remove the four bolts that secure the engine front cover to the oil pan.

   d. Remove the oil pan using the separator tool.

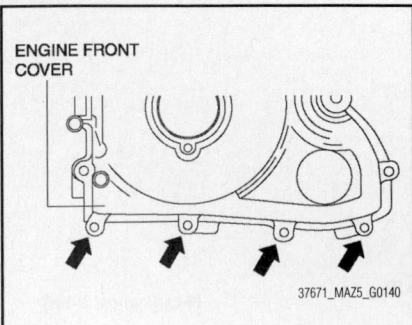

Fig. 55 Remove the four bolts that secure the engine front cover to the oil pan

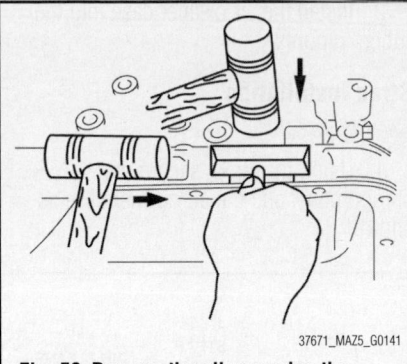

Fig. 56 Remove the oil pan using the separator tool

To install:

11. Install in the reverse order of removal.

✳ CAUTION

Note the following:

- Do not allow the gasket residue to enter the oil pan. Using cleaning fluid, completely remove the oil from the joint. Especially be sure to completely remove oil and gasket residue from the corner section of the engine front cover and the cylinder block.
- Apply the silicon sealant in a single, unbroken line.
- Install the oil pan before the applied sealant starts to harden.
- Using bolts with the old seal adhering could cause cracks in the cylinder block.

➡The thickness of silicone sealant tends to be thin because the clearance between the oil pan and the transaxle and the engine front cover and transaxle is almost zero during oil pan reinstallation. Allow 0.1 inches (3 mm) space maximum between the engine and transaxle with the transaxle bolts loosened one full turn before reinstalling the oil pan.

12. Completely clean and remove any oil, dirt, sealant or other foreign material that may be adhering to the cylinder block, engine front cover and oil pan.
13. When reusing the oil pan installation bolts, clean any old sealant from the bolts.
14. Apply silicone sealant to the oil pan along the inside of the bolt holes as shown.
15. Apply silicone sealant to the front edge of the oil pan and to the engine front cover where the two pieces attach together.
16. Apply an ample amount to the

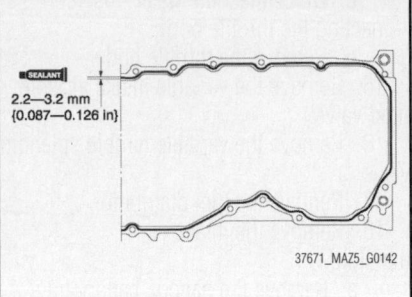

Fig. 57 Apply silicone sealant to the oil pan along the inside of the bolt holes as shown

engine front cover where the oil pan is fitted, and also apply an extra amount of silicone sealant to the corners of the engine front cover and the cylinder block.

17. Install the oil pan to the cylinder block.
18. Tighten the four engine front cover bolts in the order shown.
19. Tighten the oil pan bolts in the order shown. Tightening torque: 13–17 ft. lbs. (17–23 Nm).
20. Tighten all the bolts securing the engine and transaxle. Tightening torque: 28–38 ft. lbs. (37–52 Nm)

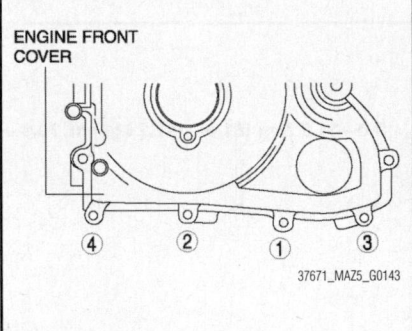

Fig. 58 Tighten the four engine front cover bolts in the order shown

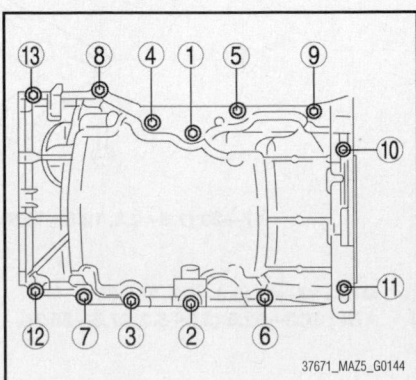

Fig. 59 Tighten the oil pan bolts in the order shown

➡**Tighten the bolts from the bottom of the transaxle to the top of the transaxle.**

21. Refill the specified type and amount of the engine oil.

22. Start the engine and confirm that there is no oil leakage.

✳✳ CAUTION

If there is oil leakage, repair or replace the applicable part.

23. Inspect the oil level.

OIL PUMP

REMOVAL & INSTALLATION

See Figures 60 through 62.

✳✳ WARNING

Note the following:

- Hot engine and engine oil can cause severe burns. Turn off the engine and wait until it and the engine oil have cooled.

- A vehicle that is lifted but not securely supported on safety stands is dangerous. It can slip or fall, causing death or serious injury. Never work around or under a lifted vehicle if it is not securely supported on safety stands.

- Continuous exposure to USED engine oil has caused skin cancer in laboratory mice. Protect your skin by washing with soap and water immediately after working with engine oil.

1. Remove the battery cover.

2. Disconnect the negative battery cable.

3. Remove the undercover and splash shield as a single unit.

4. Remove the front tire (RH).

5. Drain the engine oil.

6. Remove the plug hole plate.

7. Remove the drive belt.

8. Position the coolant reserve tank out of the way.

9. Remove the A/C compressor with the pipes still connected.

10. Remove the ignition coils.

11. Remove the Crankshaft Position (CKP) sensor.

12. Remove the engine front cover.

13. Remove the oil pan.

14. Remove the oil strainer.

15. Remove the oil pump chain tensioner.

16. Remove the oil pump chain.

17. Remove the oil pump sprocket

   a. Temporarily install the crankshaft pulley and crankshaft pulley lock bolt to the crankshaft, and lock the oil pump against rotation as shown.

   b. Remove the oil pump sprocket, and then remove the crankshaft pulley and crankshaft pulley lock bolt.

18. Remove the oil pump.

To install:

19. Install in the reverse order of removal.

20. Tighten the oil pump bolts in two steps in the order indicated.

   a. Step 1: 71–105 inch lbs. (8–12 Nm)

   b. Step 2: 13–17 ft. lbs. (17–23 Nm).

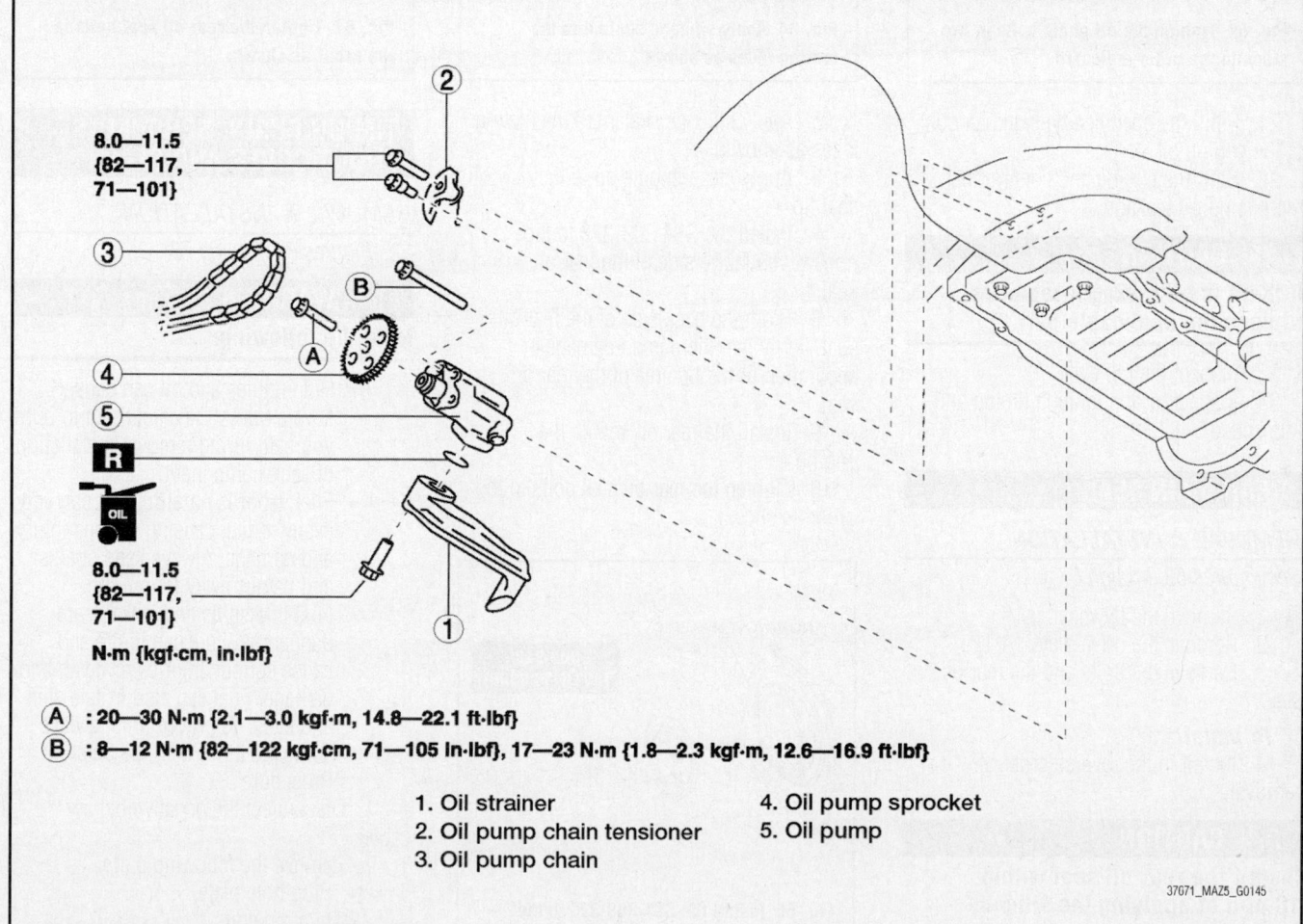

8.0—11.5
{82—117, 71—101}

8.0—11.5
{82—117, 71—101}

N·m {kgf·cm, in·lbf}

Ⓐ : 20—30 N·m {2.1—3.0 kgf·m, 14.8—22.1 ft·lbf}

Ⓑ : 8—12 N·m {82—122 kgf·cm, 71—105 in·lbf}, 17—23 N·m {1.8—2.3 kgf·m, 12.6—16.9 ft·lbf}

1. Oil strainer
2. Oil pump chain tensioner
3. Oil pump chain
4. Oil pump sprocket
5. Oil pump

37671_MAZ5_G0145

Fig. 60 Exploded view of oil pump assembly

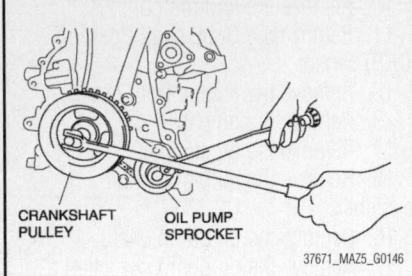

Fig. 61 Temporarily install the crankshaft pulley and crankshaft pulley lock bolt to the crankshaft, and lock the oil pump against rotation

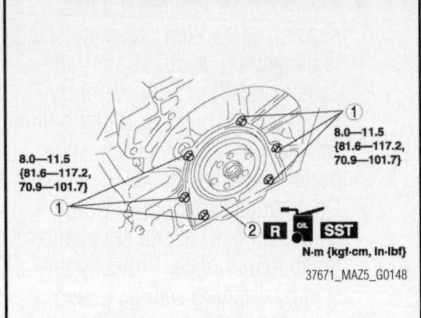

Fig. 63 Remove the bolts (1) and the rear oil seal (2)

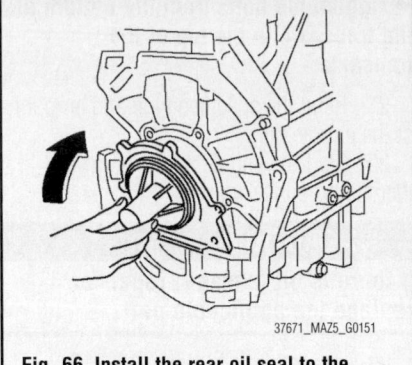

Fig. 66 Install the rear oil seal to the engine

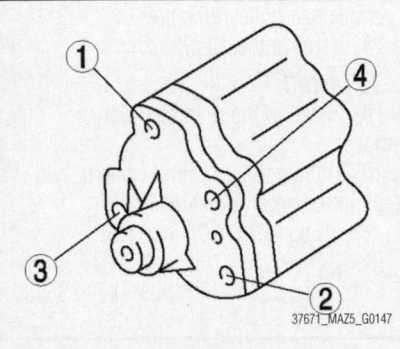

Fig. 62 Tighten the oil pump bolts in two steps in the order indicated

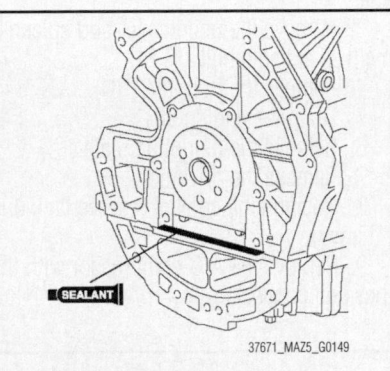

Fig. 64 Apply silicone sealant to the mating faces as shown

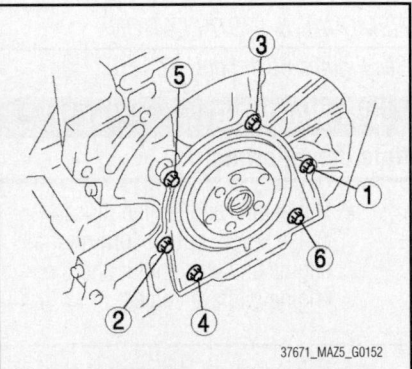

Fig. 67 Tighten the rear oil seal bolts in the order as shown

21. Refill the specified type and amount of the engine oil.

22. Start the engine and confirm that there is no oil leakage.

> ⁂ **CAUTION**
>
> **If there is oil leakage, repair or replace the applicable part.**

23. Inspect the oil level.

24. Inspect for the ignition timing and idle speed.

REAR MAIN SEAL

REMOVAL & INSTALLATION

See Figures 63 through 67.

1. Remove the flywheel. (M/T)
2. Remove the drive plate. (A/T)
3. Remove the bolts and the rear oil seal.

To install:

4. Install in the reverse order of removal.

> ⁂ **CAUTION**
>
> **Install the rear oil seal within 10 min of applying the silicone sealant.**

5. Apply silicone sealant to the mating faces as shown.

6. Apply clean engine oil to the new oil seal lip.

7. Install the SST 303-328 to the non-woven fabric side of the rear oil seal.

8. From the back side of the rear oil seal, verify that there is no damage or separation in the lip area of the rear oil seal.

9. Install the rear oil seal to the engine.

10. Tighten the rear oil seal bolts in the order as shown.

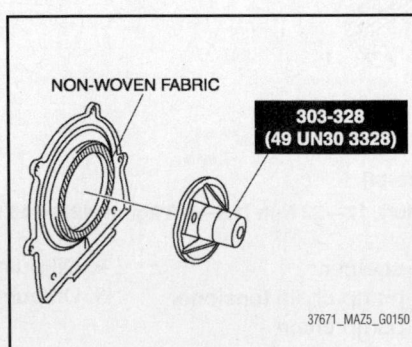

Fig. 65 Install the SST 303-328 to the non-woven fabric side of the rear oil seal

TIMING CHAIN, SPROCKETS, FRONT COVER AND SEAL

REMOVAL & INSTALLATION

See Figures 68 through 78.

> ⁂⁂ **WARNING**
>
> **Note the following:**

- Hot engines and oil can cause severe burns. Be careful not to burn yourself during removal/installation of each component.
- Fuel vapor is hazardous. It can very easily ignite, causing serious injury and damage. Always keep sparks and flames away from fuel.
- Fuel line spills and leakage are dangerous. Fuel can ignite and cause serious injuries or death and damage. Fuel can also irritate skin and eyes. To prevent this, always complete the "Fuel Line Safety Procedure".

1. Disconnect the negative battery cable.

2. Remove the following parts:
- Plug hole plate
- Ignition coils
- Drive belt

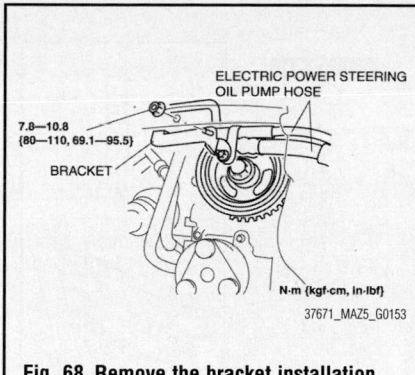

Fig. 68 Remove the bracket installation nut shown

- Under cover and splash shields (RH)
- Front wheel and tire (RH)
- Disconnect the drive shaft (RH) from joint shaft, set the drive shaft (RH) out of the way.
- Crankshaft Position (CKP) sensor
- A/C compressor with the oil hose still connected and position the A/C compressor so that it is out of the way
- Coolant reserve tank with the hose still connected and position the coolant reserve tank so that it is out of the way

3. Remove the bracket installation nut shown, then set the bracket and electric power steering oil pump hose away from the vehicle.

4. Remove the following items:
- Dipstick
- Spark plug
- Cylinder head cover

5. Remove the crankshaft pulley lock bolt.

a. Remove the cylinder block lower blind plug.

b. Install the SST 303-507.

c. Turn the crankshaft clockwise the crankshaft is in the No.1 cylinder TDC

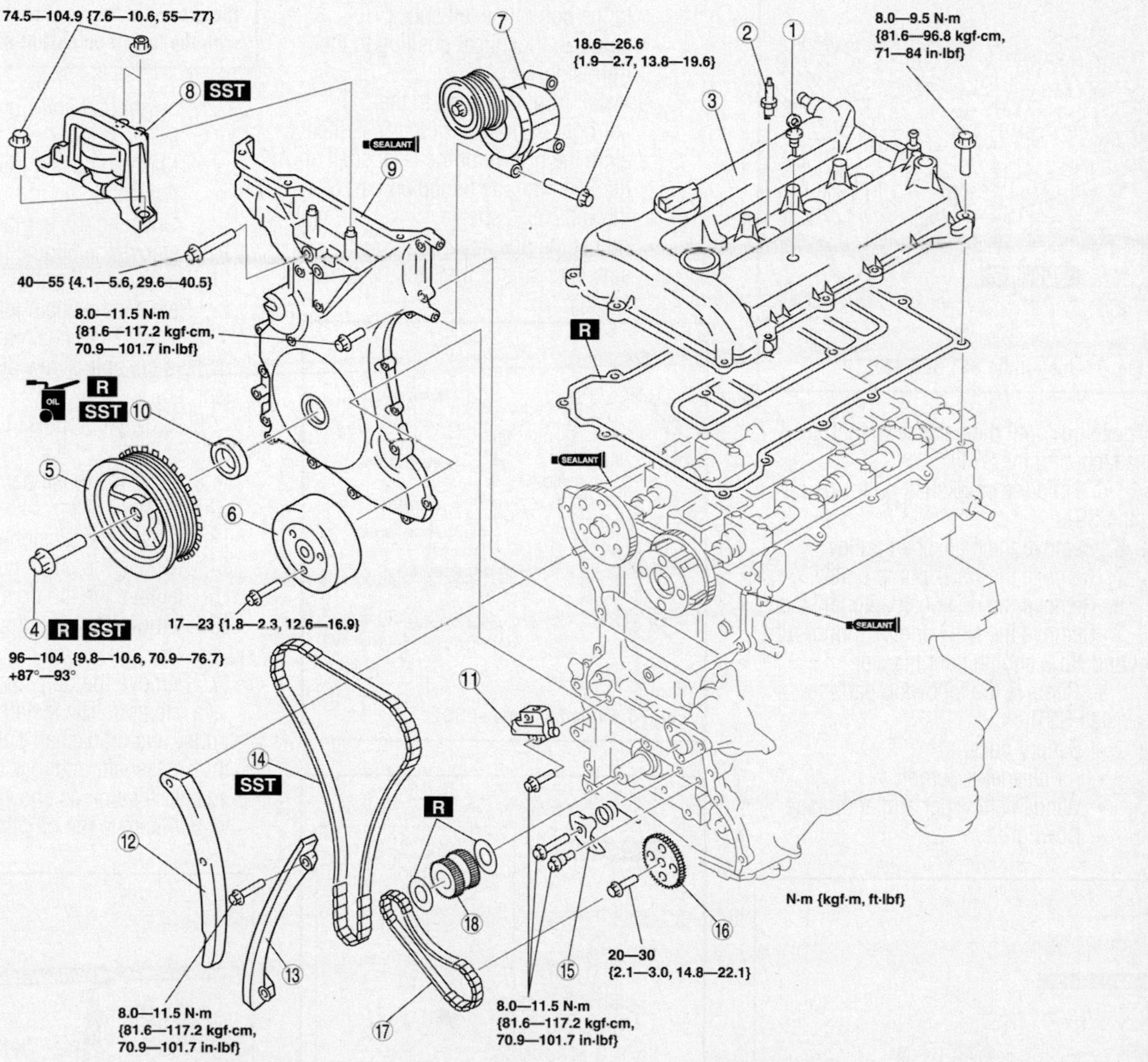

1. Dipstick
2. Spark plug
3. Cylinder head cover
4. Crankshaft pulley lock bolt
5. Crankshaft pulley
6. Water pump pulley
7. Drive belt auto tensioner
8. No.3 engine mount rubber and No.3 engine joint bracket
9. Engine front cover
10. Front oil seal
11. Chain tensioner
12. Tensioner arm
13. Chain guide
14. Timing chain
15. Oil pump chain tensioner
16. Oil pump sprocket
17. Oil pump chain
18. Crankshaft sprocket

Fig. 69 Exploded view of timing chain assembly

37671_MAZ5_G0154

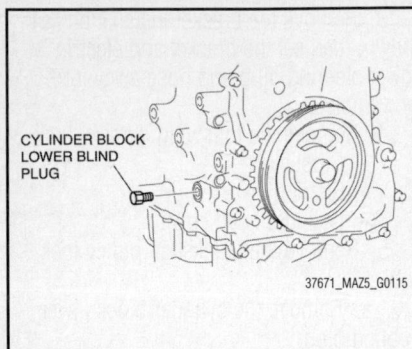

CYLINDER BLOCK
LOWER BLIND
PLUG

37671_MAZ5_G0115

Fig. 70 Remove the cylinder block lower blind plug

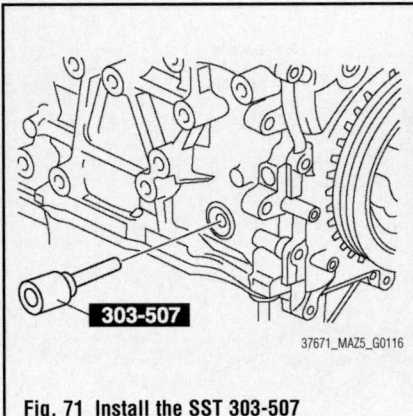

303-507

37671_MAZ5_G0116

Fig. 71 Install the SST 303-507

position (until the balance weight is attached to the SST).

   d. Hold the crankshaft pulley using the SSTs.

  6. Remove the crankshaft pulley.

  7. Remove the water pump pulley.

  8. Remove the drive belt auto tensioner.

  9. Remove the No.3 engine mount rubber and No.3 engine joint bracket.

   a. Remove the following parts to install SST:
- Battery duct
- Air cleaner assembly
- Windshield wiper arm and blade
- Cowl grille

49 B011 105

205-072
(49 UN20 5072)

37671_MAZ5_G0117

Fig. 72 Hold the crankshaft pulley using the SSTs

- Cowl panel
- No.1 reserve tank with the hose still connected and position the No.1 reserve tank aside so that it is out of the way
- Reserve tank bracket

  b. Install the SST using the following procedure:

✳✳ CAUTION

Refer to the SST instruction manual for the basic handing procedure.

- As shown, set the rear shafts of the SST to the left and right shock absorber bolts.
- Install the left rear shaft of the SST to the bolt of the left shock absorber. (identical position to the right side)
- Install front foot No.2 to the left/right front shaft of the SST, then align the groove of the front shaft of the SST with the folded up part of the vehicle as shown.
- Adjust the positions of the SST side bars so that they are the same

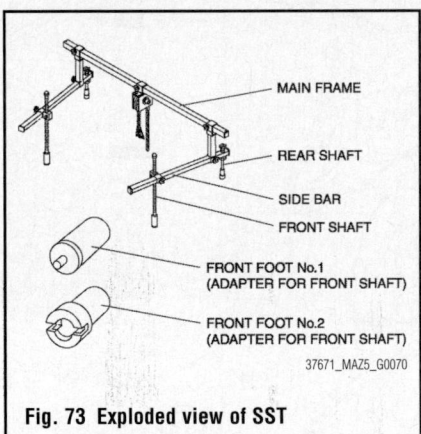

MAIN FRAME

REAR SHAFT

SIDE BAR

FRONT SHAFT

FRONT FOOT No.1
(ADAPTER FOR FRONT SHAFT)

FRONT FOOT No.2
(ADAPTER FOR FRONT SHAFT)

37671_MAZ5_G0070

Fig. 73 Exploded view of SST

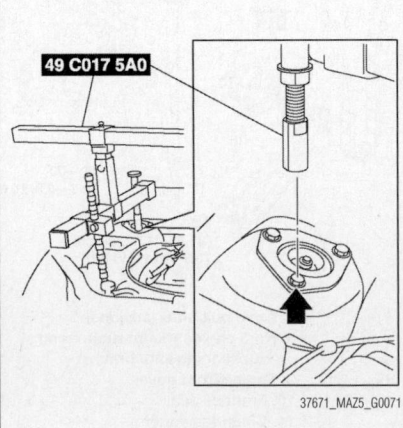

49 C017 5A0

37671_MAZ5_G0071

Fig. 74 Set the rear shafts of the SST to the left and right shock absorber bolts

---

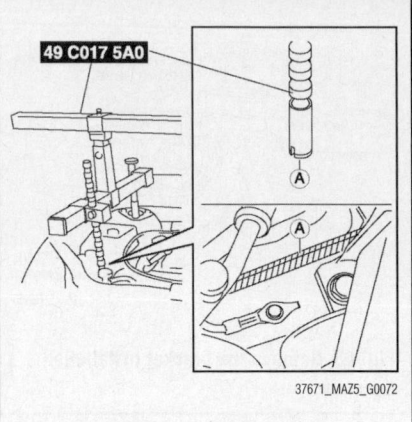

49 C017 5A0

A

A

37671_MAZ5_G0072

Fig. 75 Install front foot No.2 to the left/right front shaft of the SST, then align the groove of the front shaft of the SST with the folded up part of the vehicle

height (left and right) and horizontal.

- Make sure each joint is securely tightened.

  c. Suspend the engine using the SST.

  10. Remove the engine front cover.

  11. Remove the front oil seal.

  12. Remove the chain tensioner

   a. Unlock the chain tensioner ratchet using a suitable screw driver or equivalent tool.

   b. Slowly compress the tensioner piston.

   c. Hold the tensioner piston using a wire or paper clip.

  13. Remove the tensioner arm.

  14. Remove the chain guide.

  15. Remove the timing chain.

  16. Remove the oil pump chain tensioner.

  17. Remove the oil pump sprocket

   a. Temporarily install the crankshaft pulley and crankshaft pulley lock bolt to the crankshaft, and lock the oil pump against rotation as shown.

   b. Remove the oil pump sprocket, and

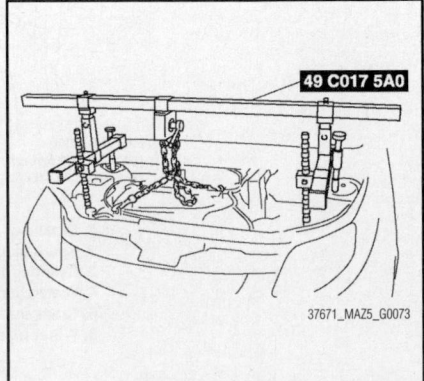

49 C017 5A0

37671_MAZ5_G0073

Fig. 76 Suspend the engine using the SST

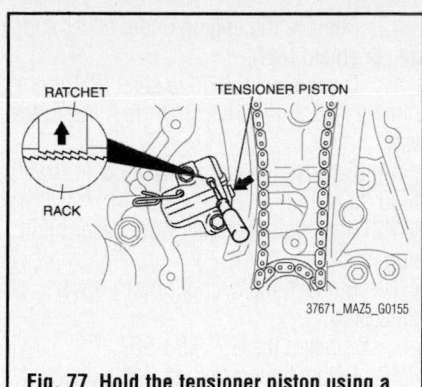

Fig. 77 Hold the tensioner piston using a wire or paper clip

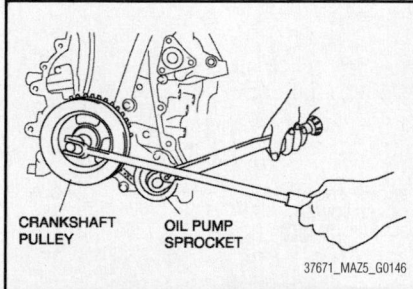

Fig. 78 Temporarily install the crankshaft pulley and crankshaft pulley lock bolt to the crankshaft, and lock the oil pump against rotation

then remove the crankshaft pulley and crankshaft pulley lock bolt. Tightening torque for installation: 15–22 ft. lbs. (20–30 Nm)

18. Remove the oil pump chain.
19. Remove the crankshaft sprocket.

To install:

20. Install in the reverse order of removal.

Timing Chain Installation

See Figure 79.

1. Install the SST to the camshaft as shown.

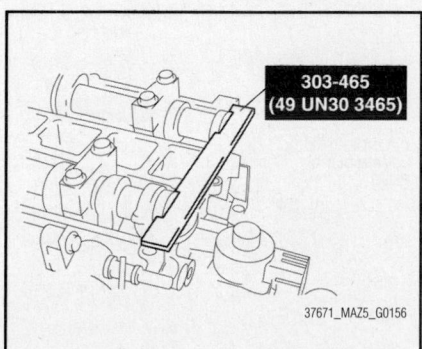

Fig. 79 Install the SST to the camshaft as shown

2. Install the timing chain.
3. Remove the retaining wire or paper clip from the auto tensioner to apply tension to the timing chain.

Engine Front Cover Installation

See Figures 80 and 81.

> ※※ **CAUTION**
>
> **Install the engine front cover before the applied silicone sealant starts to harden.**

1. Apply silicone sealant to the engine front cover as shown.

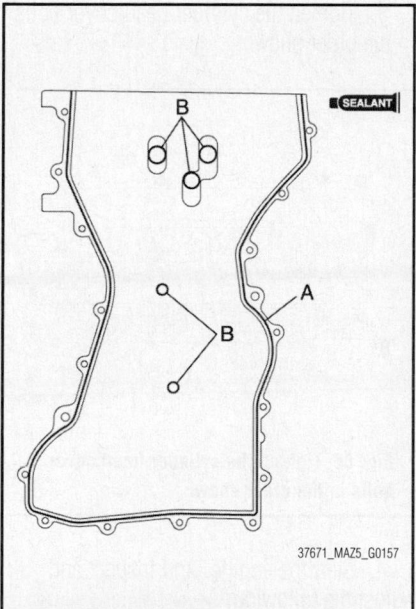

Fig. 80 Apply silicone sealant to the engine front cover as shown

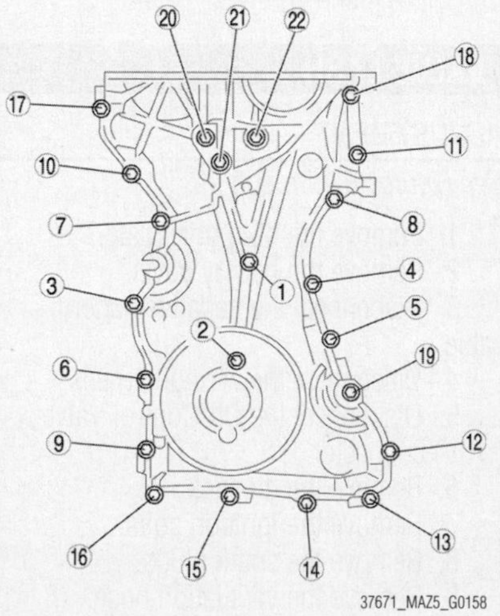

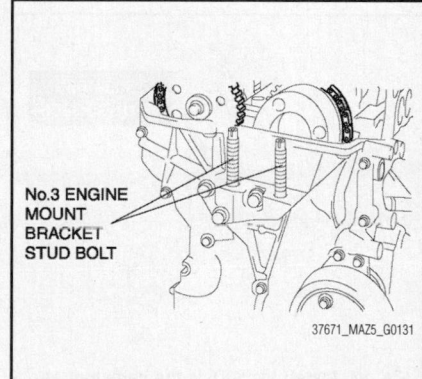

Fig. 81 Install the engine front cover bolts in the order shown

2. Install the engine front cover bolts in the order shown.
   a. Bolt Nos. 1–18: 71–102 inch lbs. (8.0–11.5 Nm)
   b. Bolt Nos. 19–22: 30–41 ft. lbs. (40–55 Nm)

Front Oil Seal Installation

1. Apply clean engine oil to the oil seal.
2. Push the oil seal slightly in by hand.
3. Compress the oil seal using the SST and a hammer.

No. 3 Engine Mount Installation

See Figures 82 and 83.

➡ **If the No.3 engine mount bracket and the engine are removed, retighten the No.3 engine mount stud bolts.**

1. Tighten the No. 3 engine mount stud bolts.
2. Install the No .3 engine mount rubber hand-tighten.
3. Tighten the No. 3 engine mount bracket bolts and nuts in the order as shown. Tightening torque: 55–77 ft. lbs. (75–105 Nm).

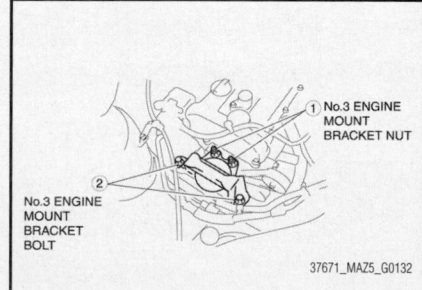

Fig. 82 Tighten the No. 3 engine mount bracket stud bolts

Fig. 83 Tighten the No. 3 engine mount bracket bolts and nuts in the order as shown

Crankshaft Pulley Lock Bolt Installation

See Figures 84 and 85.

1. Install the SST to the camshaft.
2. Verify that No.1 cylinder is at TDC of the compression stroke. (Position counter-weight contacts SST.)
3. To position the crankshaft pulley, temporarily tighten it and, using a suitable bolt, secure the crankshaft pulley to the engine front cover.
4. Hold the crankshaft pulley using the SST.
5. Tighten the crankshaft pulley lock bolt in the order shown with the following two steps using the SST (49 D032 316).
    a. Step 1: Tighten to 71–77 ft. lbs. (96–104 Nm).
    b. Step 2: Tighten an additional 87–93 degrees.
6. Remove the bolt installed to the crankshaft pulley.
7. Remove the SST from the camshaft.
8. Remove the SST installed in the cylinder block lower blind plug hole.
9. Rotate the crankshaft clockwise two turns and inspect the valve timing. If not aligned, loosen the crankshaft pulley lock bolt and repeat from Step 1.

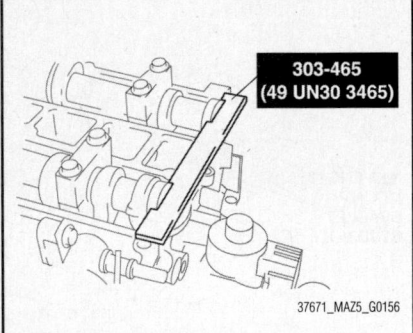

Fig. 84 Install the SST to the camshaft as shown

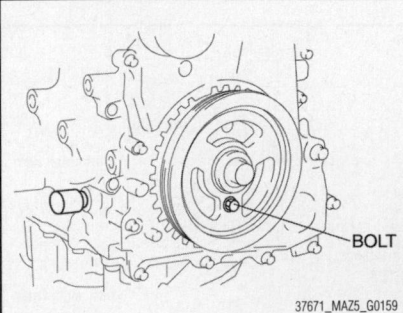

Fig. 85 To position the crankshaft pulley, temporarily tighten it and, using a suitable bolt, secure the crankshaft pulley to the engine front cover

10. Install the cylinder block lower blind plug. Tightening torque: 13–16 ft. lbs. (18–22 Nm)

Cylinder Head Cover Installation

See Figure 86.

※※ CAUTION

Install the cylinder head cover within 10 min of applying the silicone sealant.

1. Apply silicone sealant to the mating faces.
2. Install the cylinder head cover with a new gasket.
3. Tighten the cylinder head cover bolts in the order shown.

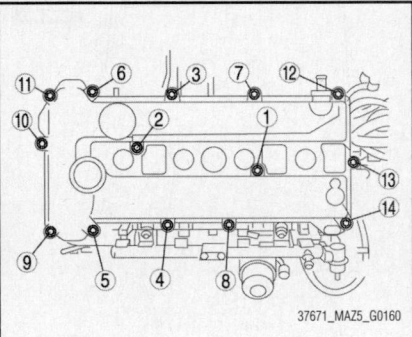

Fig. 86 Tighten the cylinder head cover bolts in the order shown

4. Start the engine. And inspect and adjust the following:
    - Leakage of engine oil.
    - Runout and contact of pulley and belt.
    - Verify the ignition timing, idle speed and idle mixture.

VALVE LASH

ADJUSTMENT

See Figures 87 through 99.

1. Remove the plug hole plate.
2. Remove the battery cover.
3. Disconnect the negative battery cable.
4. Disconnect the wiring harness.
5. Disconnect the Oil Control Valve (OCV) connector.
6. Remove the dipstick.
7. Remove the Ignition coils.
8. Remove the spark plugs.
9. Remove the ventilation hose.
10. Remove the cylinder head cover.
11. Remove the front wheel and tire (RH).

12. Remove the engine under cover and splash shield (RH).
13. Disconnect the drive shaft (RH) from joint shaft, set the drive shaft (RH) out of the way.
14. Remove the engine front cover lower blind plug.
15. Remove the engine front cover blind plug.
16. Remove the cylinder block lower blind plug.
    a. Install the SST 303-507.
17. Rotate the crankshaft in the direction of the engine rotation so that the No.1

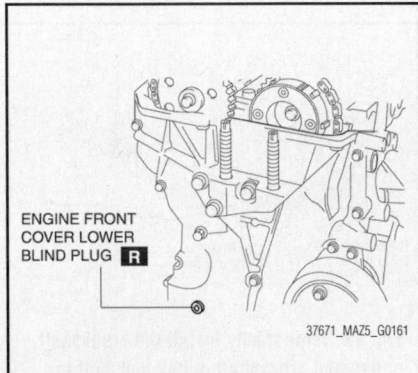

Fig. 87 Remove the engine front cover lower blind plug

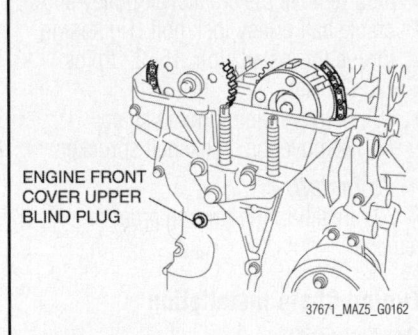

Fig. 88 Remove the engine front cover blind plug

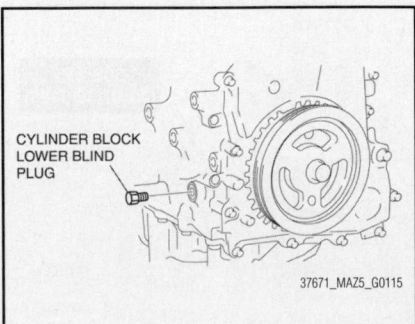

Fig. 89 Remove the cylinder block lower blind plug

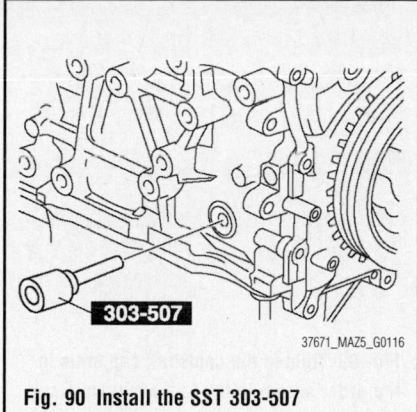

Fig. 90 Install the SST 303-507

piston is at TDC of the compression stroke. (Until the counterweight contacts SST and stops.)

18. Loosen the timing chain using the following procedure.

    a. Insert a suitable bolt into the engine front cover upper blind plug and tighten it until it contacts the chain tensioner arm, and then rotate it back one turn. (Set the bolt slightly away from the chain tensioner arm so that it does not contact it.)

    b. Using the cast hexagon on the exhaust camshaft, apply force counterclockwise to facilitate unlocking the chain tensioner ratchet.

    c. Using a Hex bit socket or T15 Torx bit socket, unlock the chain tensioner ratchet so that it can be lifted up.

    d. Using the cast hexagon on the exhaust camshaft, apply force in the direction of the engine rotation to increase tension on the chain.

➡**The chain tensioner rack is compressed using the chain tension generated by applying force to the exhaust camshaft in the direction of the engine rotation.**

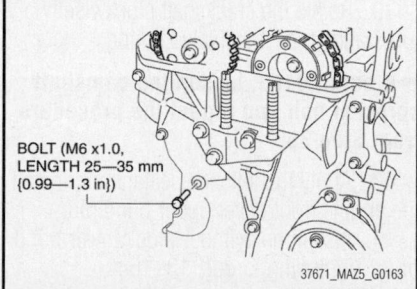

Fig. 91 Insert a suitable bolt into the engine front cover upper blind plug and tighten it until it contacts the chain tensioner arm, and then rotate it back one turn

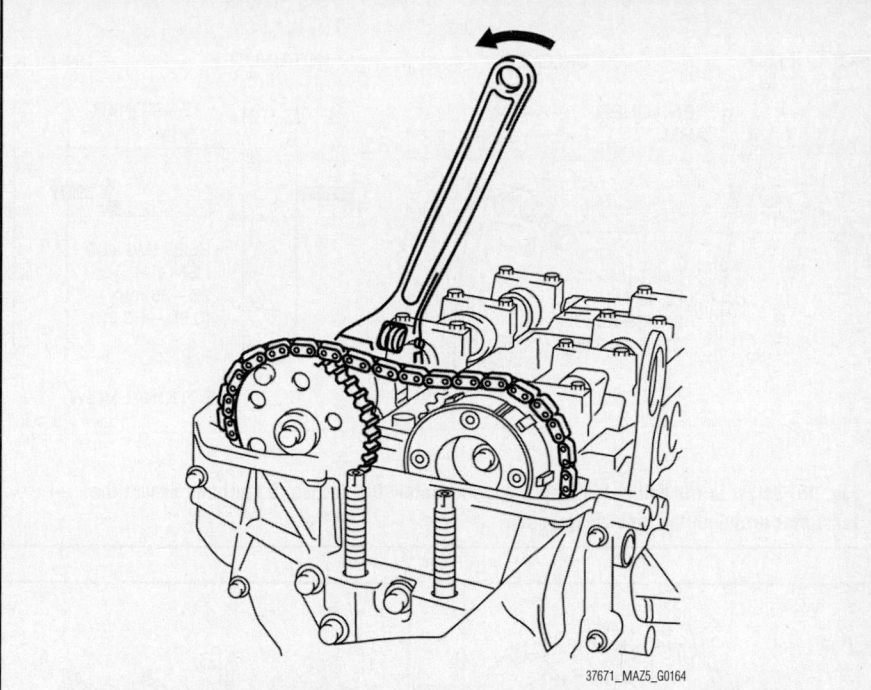

Fig. 92 Using the cast hexagon on the exhaust camshaft, apply force counterclockwise to facilitate unlocking the chain tensioner ratchet

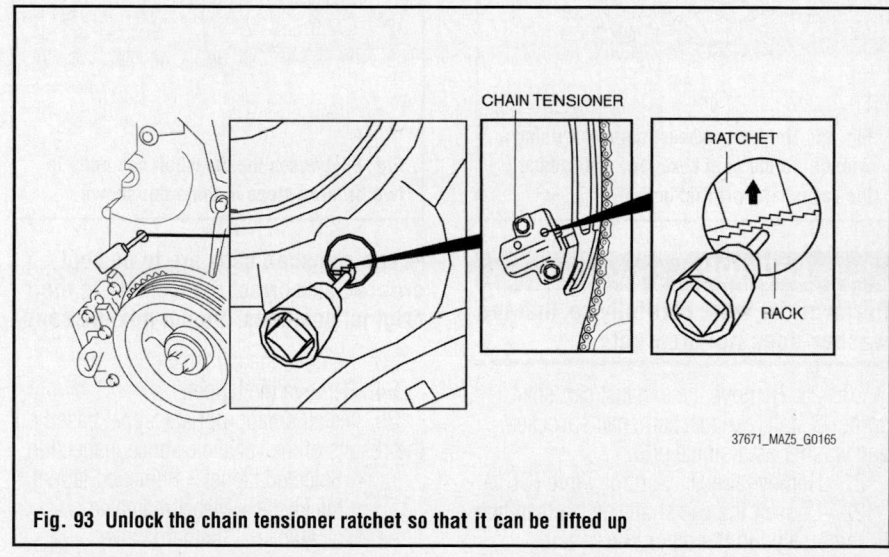

Fig. 93 Unlock the chain tensioner ratchet so that it can be lifted up

➡**The ratchet has not been unlocked if the bolt cannot be pressed in approximately 0.2 inches (5 mm).**

    e. Screw in the bolt set in Step 1 approximately 0.2 inches (5 mm) and secure the tensioner arm with the rack compressed.

➡**If the tensioner arm cannot be secured, return the bolt to its original position and repeat the procedure from Step 3.**

19. Hold the exhaust camshaft using a wrench on the cast hexagon, and loosen the camshaft sprocket bolt.

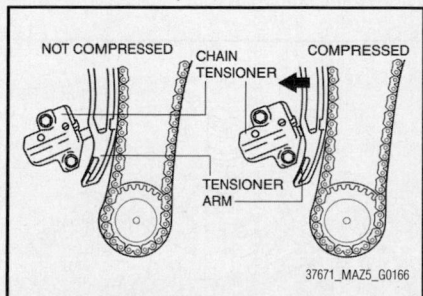

Fig. 94 The chain tensioner rack is compressed using the chain tension generated by applying force to the exhaust camshaft

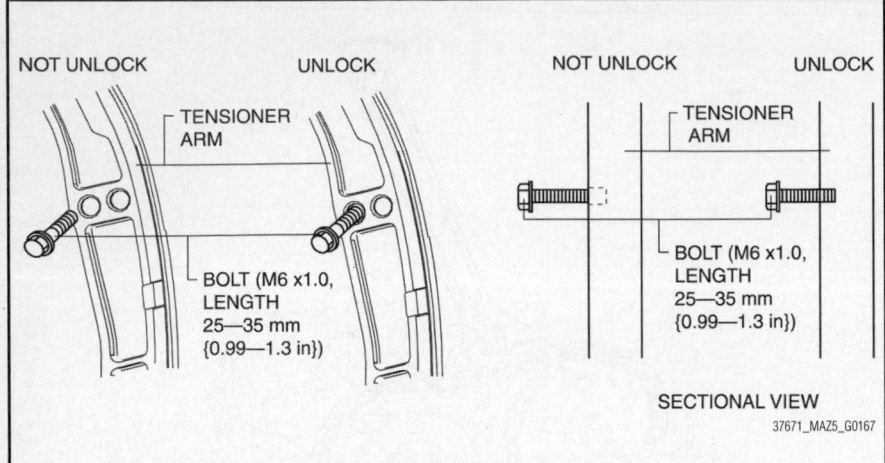

Fig. 95 Screw in the bolt set in Step 1 approximately 0.2 inches (5 mm) and secure the tensioner arm with the rack compressed

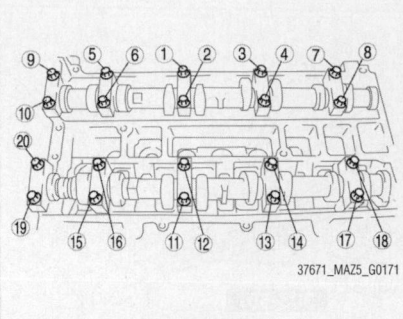

Fig. 99 Tighten the camshaft cap bolts in the order shown in the following two steps

30. Temporarily tighten the camshaft cap bolts evenly in 2-3 steps.
31. Tighten the camshaft cap bolts in the order shown in the following two steps.
   a. Step 1: 44–80 inch lbs.
(5.0–9.0 Nm)
   b. Step 2: 10–13 ft. lbs. (14–17 Nm)
32. Install the OCV.

✳✳ CAUTION

Install a washer to the fourth or fifth thread of the exhaust camshaft sprocket bolt being careful not to drop the washer. Do not tighten the camshaft sprocket bolt at this stage. Verify the valve timing before performing the bolt tightening.

33. Install the exhaust camshaft sprocket bolt, exhaust camshaft sprocket, and a new washer as a single unit.
34. 35. Install the SST on the camshafts.
35. Remove the installation bolt for the engine front cover upper blind plug, and apply tension to the timing chain.
36. Hold the exhaust camshaft using a wrench on the cast hexagon, and tighten the sprocket bolt. Tightening torque: 51–55 ft. lbs. (69–75 Nm)
37. Remove the SST from the camshafts.
38. Remove the SST installed in the cylinder block lower blind plug hole.
39. Rotate the crankshaft clockwise two turns and inspect the valve timing.

➡ **If not aligned, loosen the camshaft sprocket bolt and repeat the procedure from Step 35.**

40. Apply the silicone sealant and install the engine front cover upper blind plug.
41. Install the cylinder block lower blind plug. Tightening torque: 13–16 ft. lbs.(18–22 Nm).
42. Install a new engine front cover lower blind plug.
43. Connect the drive shaft (RH) to joint shaft.

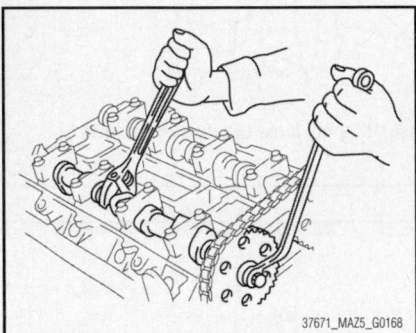

Fig. 96 Hold the exhaust camshaft using a wrench on the cast hexagon, and loosen the camshaft sprocket bolt

✳✳ CAUTION

Perform the work carefully so that the washer does not drop out.

20. 21. Remove the exhaust camshaft sprocket bolt, exhaust camshaft sprocket, and washer as a single unit.
21. Remove the Oil Control Valve (OCV).
22. Loosen the camshaft cap bolts in two or three steps in the order shown and remove the camshaft cap.

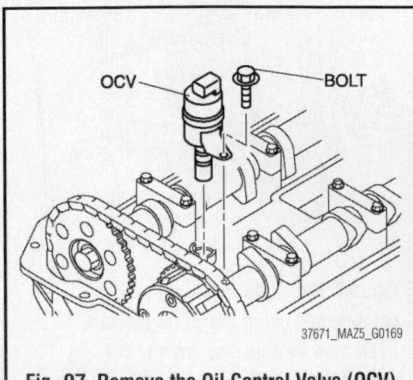

Fig. 97 Remove the Oil Control Valve (OCV)

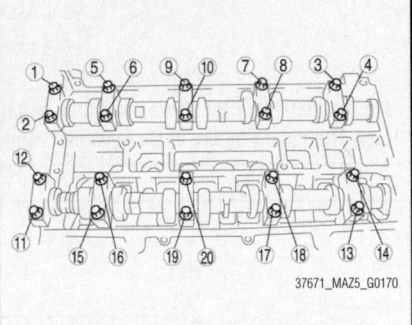

Fig. 98 Loosen the camshaft cap bolts in two or three steps in the order shown

➡ **The camshaft caps are to be kept ordered for correct reassembly in their original positions. Do not mix the caps.**

23. Remove the camshafts.
24. Remove the tappet.
25. Install an appropriate tappet based on the results of the valve clearance inspection.
- Selected tappet = Removed tappet thickness + Measured valve clearance - Standard valve clearance
- Valve clearance [Engine cold]
- Intake: 0.009–0.011 inches (0.22–0.28 mm)
- Exhaust: 0.011–0.012 inches (0.27–0.33 mm)
26. Verify that No.1 cylinder is at TDC of the compression stroke. (Position counterweight contacts SST.)
27. Apply the gear oil (SAE No.90 or equivalent) to each journal of the cylinder head.
28. Install the camshaft with No.1 cylinder aligned with the TDC position.
29. Apply the gear oil (SAE No.90 or equivalent) to each journal of the camshaft.

44. Install the engine under cover and splash shield (RH).
45. Install the front wheel and tire (RH).
46. Install the cylinder head cover.

47. Install the ventilation hose.
48. Install the spark plugs.
49. Install the ignition coils.
50. Install the dipstick.
51. Connect the OCV connector.

52. Connect the wiring harness.
53. Connect the negative battery cable.
54. Install the battery cover.
55. Install the plug hole plate.

ENGINE PERFORMANCE & EMISSION CONTROLS

COMPONENT LOCATIONS

See Figures 100 and 101.

CAMSHAFT POSITION (CMP) SENSOR

LOCATION

See Figure 102.

Refer to the accompanying illustration for sensor location.

REMOVAL & INSTALLATION

1. Remove the battery cover.
2. Disconnect the negative battery cable.
3. Remove the plug hole plate.

❄❄ CAUTION

When replacing the CMP sensor, make sure there is no foreign material on it such as metal shavings. If it is installed with foreign material, the sensor output signal will malfunction resulting from fluctuation in magnetic flux and cause a deterioration in engine control.

4. Disconnect the CMP sensor connector.
5. Remove the CMP sensor installation bolt.
6. Remove the CMP sensor from the cylinder head cover.
7. Install in the reverse order of removal.

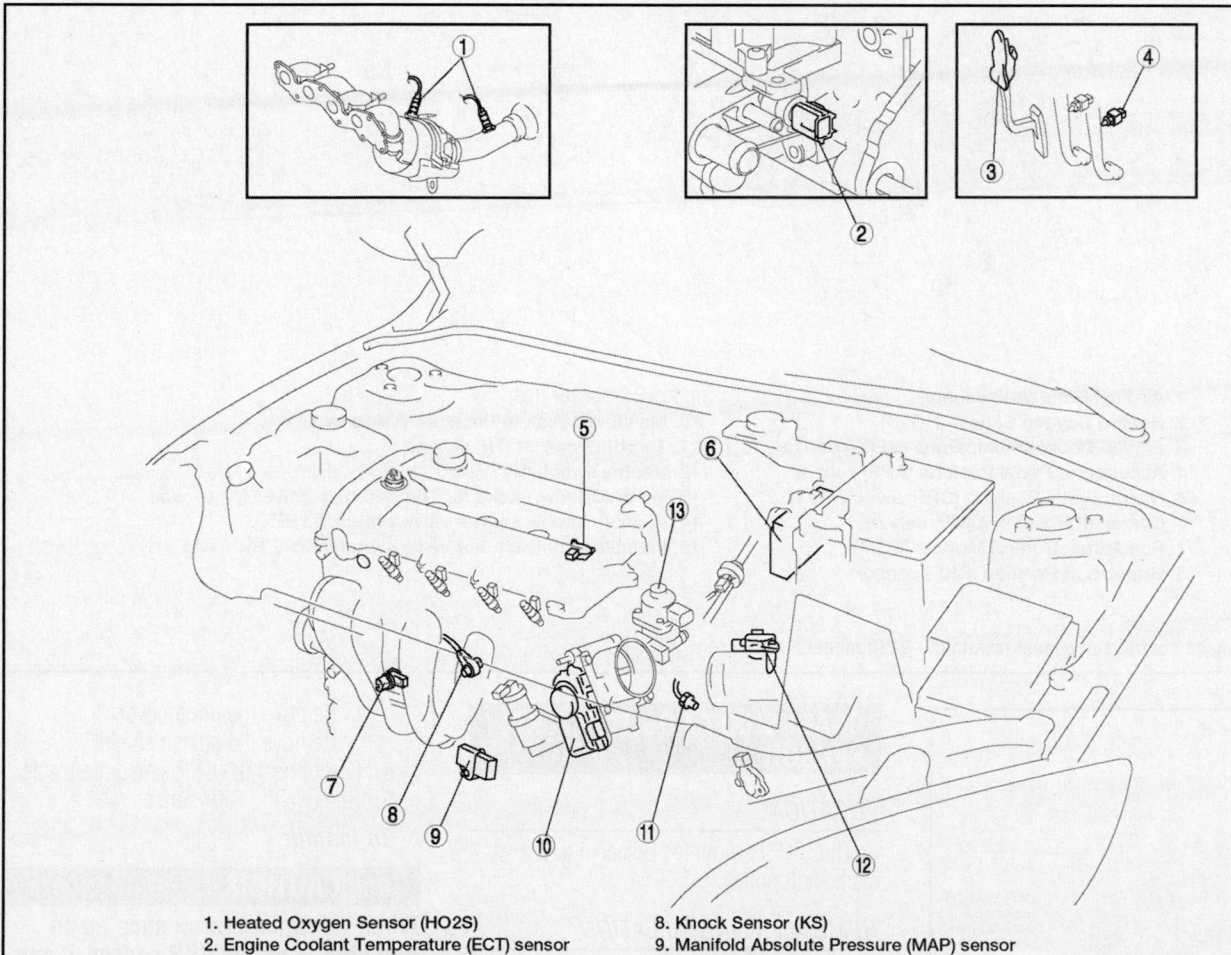

1. Heated Oxygen Sensor (HO2S)
2. Engine Coolant Temperature (ECT) sensor
3. Accelerator Pedal Position (APP) sensor
4. Clutch Pedal Position (CPP) switch (M/T)
5. Camshaft Position (CMP) sensor
6. Powertrain Control Module (PCM)
7. Crankshaft Position (CKP) sensor
8. Knock Sensor (KS)
9. Manifold Absolute Pressure (MAP) sensor
10. Throttle Position (TP) sensor
11. Neutral switch (M/T)
12. Mass Air Flow/Intake Air Temperature (MAF/IAT) sensor
13. Exhaust Gas Recirculation (EGR) valve

37671_MAZ5_G0173

Fig. 100 Engine control component locations—2006–09 models

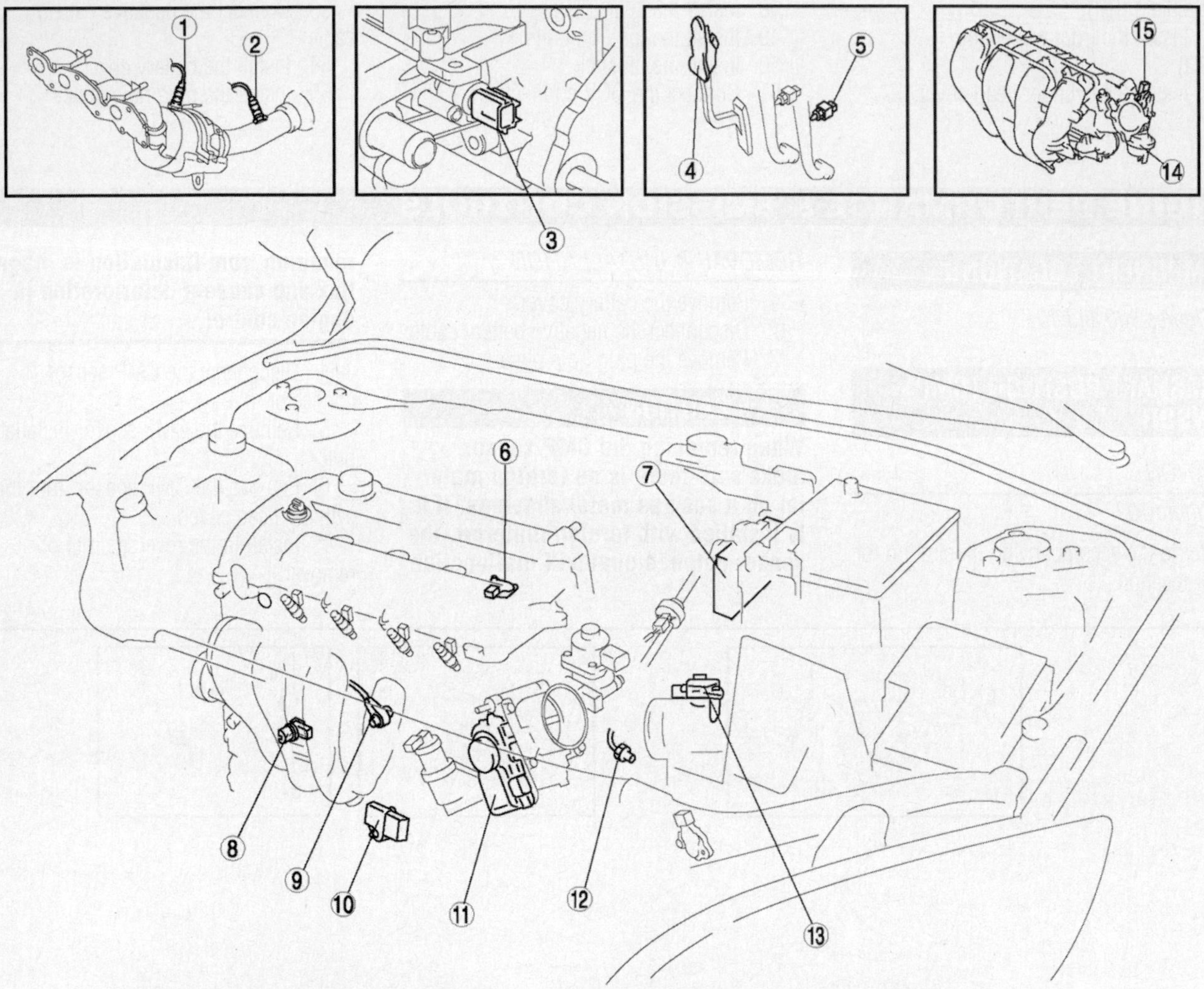

Fig. 101 Engine control component locations—2010 models

1. Air Fuel Ratio (A/F) sensor
2. Heated Oxygen Sensor (HO2S)
3. Engine Coolant Temperature (ECT) sensor
4. Accelerator Pedal Position (APP) sensor
5. Clutch Pedal Position (CPP) switch (M/T)
6. Camshaft Position (CMP) sensor
7. Powertrain Control Module (PCM)
8. Crankshaft Position (CKP) sensor
9. Knock Sensor (KS)
10. Manifold Absolute Pressure (MAP) sensor
11. Throttle Position (TP) sensor
12. Neutral switch (M/T)
13. Mass Air Flow/Intake Air Temperature (MAF/IAT) sensor
14. Variable tumble shutter valve switch (TYPE A)
15. Variable tumble shutter valve switch (TYPE B)

37671_MAZ5_G0174

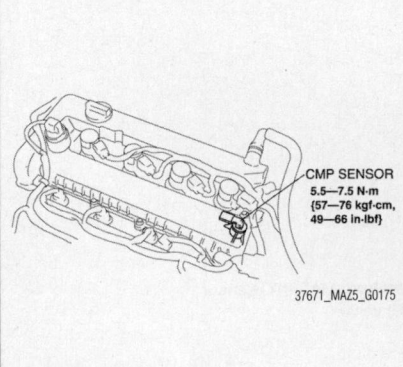

CMP SENSOR
5.5—7.5 N·m
{57—76 kgf-cm,
49—66 in-lbf}

37671_MAZ5_G0175

Fig. 102 CMP sensor location

CRANKSHAFT POSITION (CKP) SENSOR

LOCATION

The CKP is mounted adjacent to the crankshaft pulley.

REMOVAL & INSTALLATION

See Figures 103 through 106.

1. Remove the battery cover.
2. Disconnect the negative battery cable.
3. Perform the following procedure for easier access.

   a. Remove the undercover.
   b. Remove the splash shield.
4. Disconnect the CKP sensor connector.
5. Remove the CKP sensor.

To install:

✴✴ CAUTION

When foreign material such as an iron chip is on the CKP sensor, it can cause abnormal output from the sensor because of flux turbulence and adversely affect the engine control. Be sure there is no foreign material on the CKP sensor when replacing.

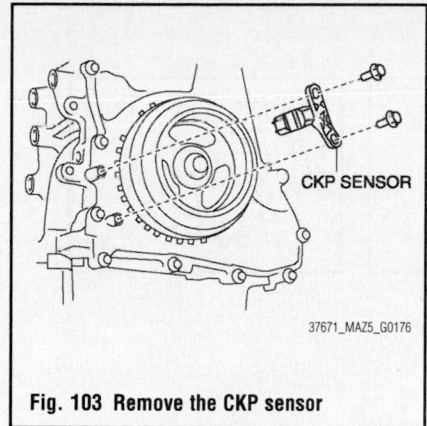

Fig. 103 Remove the CKP sensor

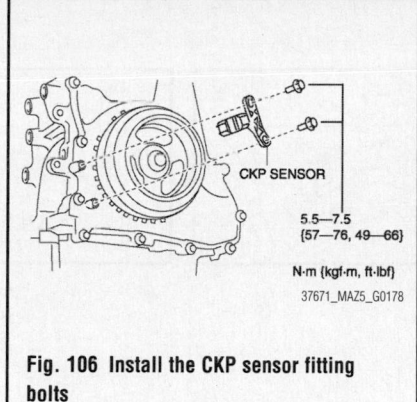

Fig. 106 Install the CKP sensor fitting bolts

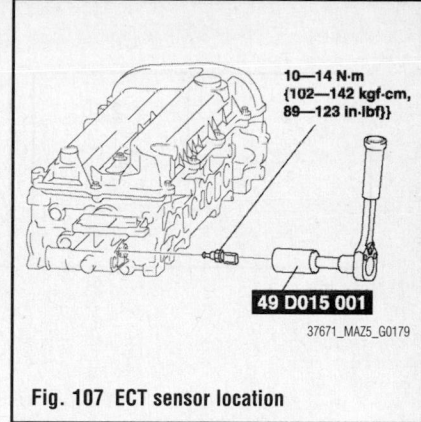

Fig. 107 ECT sensor location

6. Perform the following procedure so that cylinder No.1 is at TDC.

　a. Disconnect the drive shaft (RH) from joint shaft, set the drive shaft (RH) out of the way.

　b. Remove the cylinder block lower blind plug and install the SST.

　c. Rotate the crankshaft pulley clockwise until the crank weight contacts the SST so that cylinder No.1 is at TDC.

7. Fit the center of the CKP sensor with the fifth tooth (counting counterclockwise from the empty space A as shown) of the pulse wheel.

8. Install the CKP sensor fitting bolts.

9. Remove the SST then install the cylinder block lower blind plug.

ENGINE COOLANT TEMPERATURE (ECT) SENSOR

LOCATION

See Figure 107.

Refer to the accompanying illustration for sensor location.

REMOVAL & INSTALLATION

1. Remove the battery cover.
2. Disconnect the negative battery cable.

3. Remove the plug hole plate.
4. Drain the engine coolant.
5. Disconnect the ECT sensor connector.
6. Remove the ECT sensor using the SST 49D015001.
7. Install in the reverse order of removal.

HEATED OXYGEN (HO2S) SENSOR

LOCATION

See Figures 108 and 109.

Refer to the accompanying illustrations for sensor location.

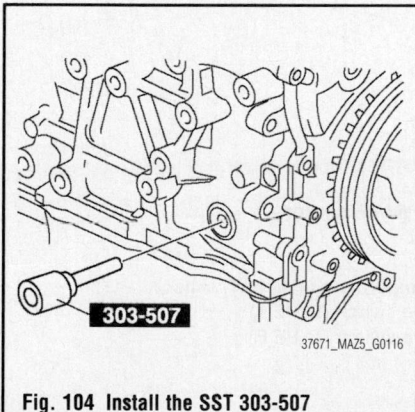

Fig. 104 Install the SST 303-507

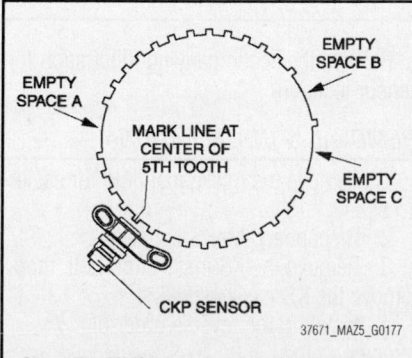

Fig. 105 Fit the center of the CKP sensor with the fifth tooth of the pulse wheel

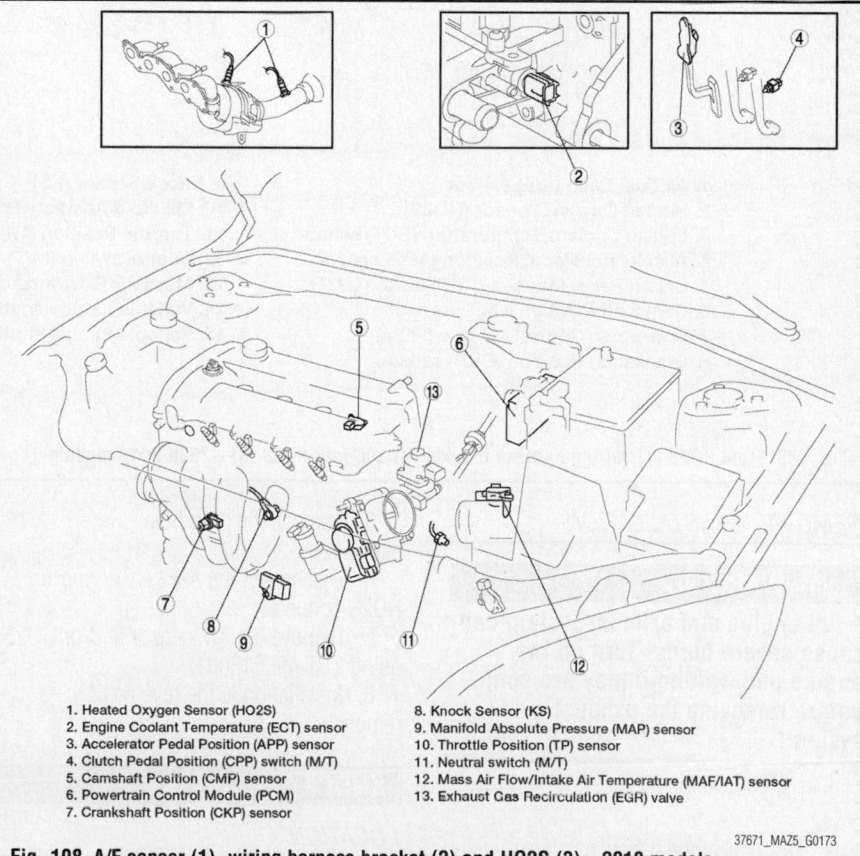

1. Heated Oxygen Sensor (HO2S)
2. Engine Coolant Temperature (ECT) sensor
3. Accelerator Pedal Position (APP) sensor
4. Clutch Pedal Position (CPP) switch (M/T)
5. Camshaft Position (CMP) sensor
6. Powertrain Control Module (PCM)
7. Crankshaft Position (CKP) sensor
8. Knock Sensor (KS)
9. Manifold Absolute Pressure (MAP) sensor
10. Throttle Position (TP) sensor
11. Neutral switch (M/T)
12. Mass Air Flow/Intake Air Temperature (MAF/IAT) sensor
13. Exhaust Gas Recirculation (EGR) valve

Fig. 108 A/F sensor (1), wiring harness bracket (2) and HO2S (3)—2010 models

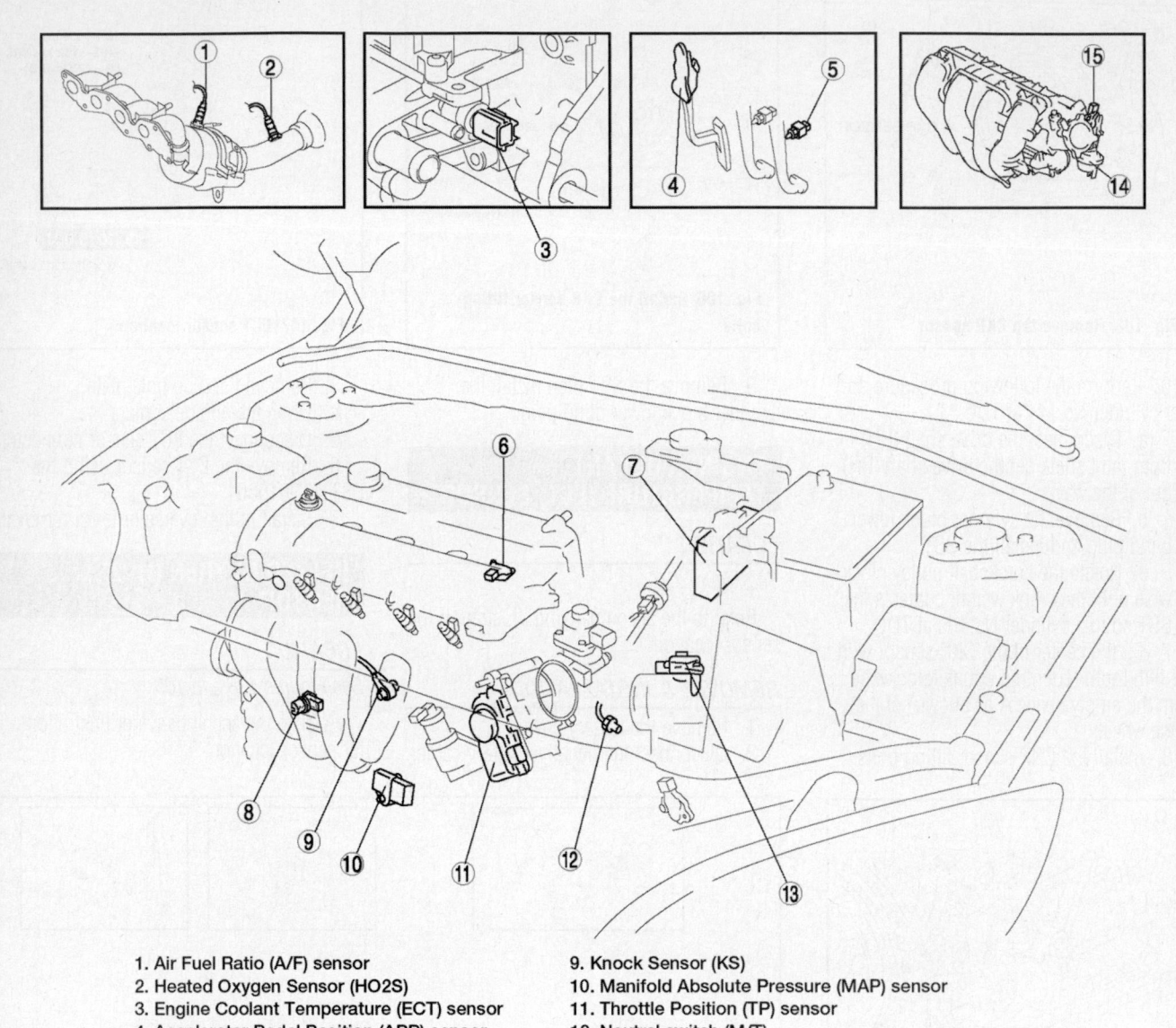

1. Air Fuel Ratio (A/F) sensor
2. Heated Oxygen Sensor (HO2S)
3. Engine Coolant Temperature (ECT) sensor
4. Accelerator Pedal Position (APP) sensor
5. Clutch Pedal Position (CPP) switch (M/T)
6. Camshaft Position (CMP) sensor
7. Powertrain Control Module (PCM)
8. Crankshaft Position (CKP) sensor
9. Knock Sensor (KS)
10. Manifold Absolute Pressure (MAP) sensor
11. Throttle Position (TP) sensor
12. Neutral switch (M/T)
13. Mass Air Flow/Intake Air Temperature (MAF/IAT) sensor
14. Variable tumble shutter valve switch (TYPE A)
15. Variable tumble shutter valve switch (TYPE B)

37671_MAZ5_G0174

Fig. 109 Front HO2S (1), wiring harness bracket (2) and rear HO2S (3)—2006-2010 models

REMOVAL & INSTALLATION

❊❊ WARNING

A hot engine and exhaust system can cause severe burns. Turn off the engine and wait until they are cool before removing the exhaust system.

1. Remove the battery cover.
2. Disconnect the negative battery cable.

3. Remove the plug hole plate.
4. Disconnect the A/F sensor and/or HO2S connector.
5. Remove the A/F sensor and/or HO2S using SST 49L018001.
6. Installation is the reverse of removal.

KNOCK SENSOR (KS)

LOCATION

See Figure 110.

Refer to the accompanying illustration for sensor location.

REMOVAL & INSTALLATION

1. Remove the intake manifold for easier access.
2. Disconnect the KS connector.
3. Remove the KS installation bolt, then remove the KS.
4. Install in the reverse order of removal.
5. Tightening torque for KS installation bolt 12–17 ft. lbs. (16–24 Nm)

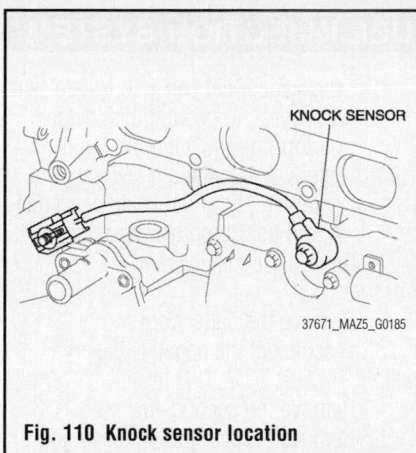

Fig. 110 Knock sensor location

MANIFOLD ABSOLUTE PRESSURE (MAP) SENSOR

LOCATION

See Figure 111.

Refer to the accompanying illustration for sensor location.

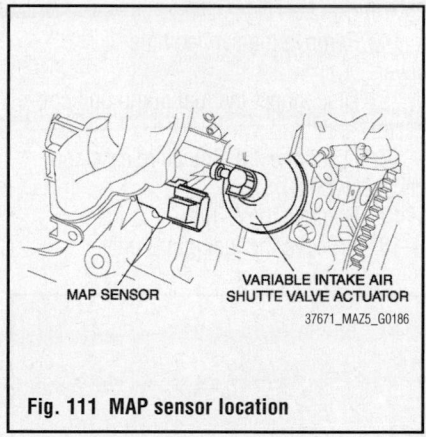

Fig. 111 MAP sensor location

REMOVAL & INSTALLATION

1. Remove the battery cover.
2. Disconnect the negative battery cable.
3. Remove the plug hole plate.
4. Remove the vacuum hose.
5. Disconnect the MAP sensor connector.
6. Remove the MAP sensor installation screw.
7. Remove MAP sensor from the intake manifold.
8. Install in the reverse order of removal.

POWERTRAIN CONTROL MODULE (PCM)

LOCATION

The PCM is attached to the battery tray.

REMOVAL & INSTALLATION

See Figure 112.

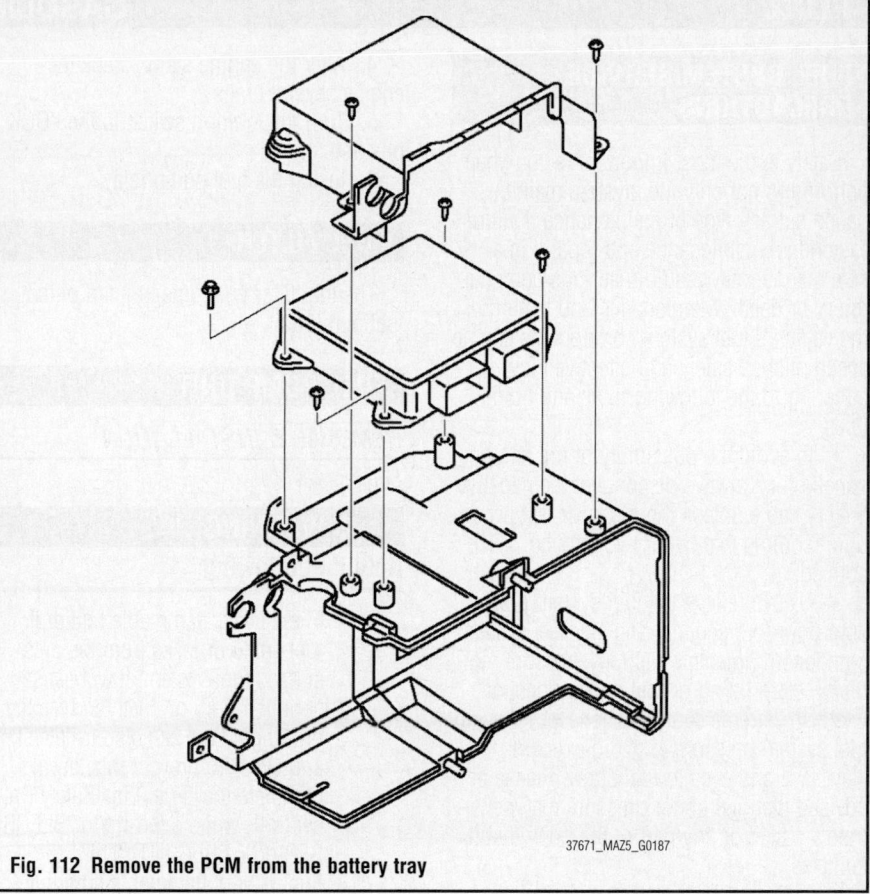

Fig. 112 Remove the PCM from the battery tray

1. When replacing the PCM, perform the PCM configuration.
2. Remove the battery cover, battery duct, battery and battery tray with PCM.
3. Disconnect the PCM connector.
4. Remove the PCM from the battery tray.
5. When replacing the PCM on the vehicles, perform the PCM parameter reset.
6. Install in the reverse order of removal.

VEHICLE SPEED SENSOR (VSS)

REMOVAL & INSTALLATION

See Figure 113.

❊❊ CAUTION
Note the following:

- Water or foreign objects entering the connector can cause a poor connection or corrosion. Be sure not to drop water or foreign objects on the connector when disconnecting it.
- If foreign materials are stuck to the VSS, disturbance by magnetic flux can cause sensor output to be abnormal and thereby negatively affect control. Make sure that foreign materials such as iron filings are not stuck to the VSS during installation.

1. Remove the battery cover.
2. Disconnect the negative battery cable.
3. Disconnect the VSS connector.
4. Remove the VSS.
5. Apply ATF to a new O-ring and install it on VSS.
6. Install the VSS.
7. Connect the VSS connector.
8. Connect the negative battery cable.
9. Install the battery cover.

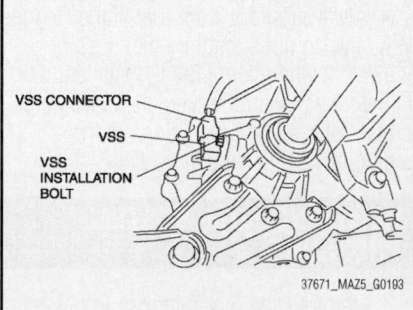

Fig. 113 Disconnect the VSS connector and remove the VSS

FUEL SYSTEM SERVICE PRECAUTIONS

Safety is the most important factor when performing not only fuel system maintenance but any type of maintenance. Failure to conduct maintenance and repairs in a safe manner may result in serious personal injury or death. Maintenance and testing of the vehicle's fuel system components can be accomplished safely and effectively by adhering to the following rules and guidelines.

• To avoid the possibility of fire and personal injury, always disconnect the negative battery cable unless the repair or test procedure requires that battery voltage be applied.

• Always relieve the fuel system pressure prior to disconnecting any fuel system component (injector, fuel rail, pressure regulator, etc.), fitting or fuel line connection. Exercise extreme caution whenever relieving fuel system pressure to avoid exposing skin, face and eyes to fuel spray. Please be advised that fuel under pressure may penetrate the skin or any part of the body that it contacts.

• Always place a shop towel or cloth around the fitting or connection prior to loosening to absorb any excess fuel due to spillage. Ensure that all fuel spillage (should it occur) is quickly removed from engine surfaces. Ensure that all fuel soaked cloths or towels are deposited into a suitable waste container.

• Always keep a dry chemical (Class B) fire extinguisher near the work area.

• Do not allow fuel spray or fuel vapors to come into contact with a spark or open flame.

• Always use a back-up wrench when loosening and tightening fuel line connection fittings. This will prevent unnecessary stress and torsion to fuel line piping.

• Always replace worn fuel fitting O-rings with new Do not substitute fuel hose or equivalent where fuel pipe is installed.

Before servicing the vehicle, make sure to also refer to the precautions in the beginning of this section as well.

RELIEVING FUEL SYSTEM PRESSURE

1. Remove the fuel-filler cap to release the pressure inside the fuel tank.
2. Remove the fuel pump relay.
3. Start the engine.

4. After the engine stalls, crank the engine several times.
5. Turn the ignition switch to the LOCK position.
6. Install the fuel pump relay.

FUEL FILTER

The fuel filter is an integral part of the fuel pump unit.

FUEL PUMP UNIT

REMOVAL & INSTALLATION
See Figure 114.

✳✳ WARNING
Note the following:

• Fuel is very flammable liquid. If fuel spills or leaks from the pressurized fuel system, it will cause serious injury or death and facility breakage. Fuel can also irritate skin and eyes. To prevent this, always complete the "Fuel Line Safety Procedure", while referring to "BEFORE SERVICE PRECAUTIONS".

• Fuel is very flammable liquid. If fuel spills or leaks from the pressurized fuel system, it will cause serious injury or death and facility

breakage. Fuel can also irritate skin and eyes. To prevent this, before performing the fuel pump unit removal/installation, always complete the "Fuel Leak Inspection After Fuel Pump Unit Installation".

1. Relieve the fuel system pressure.
2. Remove the battery cover.
3. Disconnect the negative battery cable.
4. Remove the second-row seat.
5. Remove the rear package tray lid.
6. Remove the sub-trunk
7. Remove the third-row seat.
8. Remove the rear scuff plate.
9. Remove the trunk end trim.
10. Remove the third-row seat belt lower anchor installation bolt.
11. Remove the cargo compartment light.
12. Remove the trunk side trim.
13. Partially peel back the floor covering.
14. Remove the service hole cover.
15. Disconnect the fuel pump unit connector.
16. Disconnect the quick release connector (Type B).
17. Remove the O-ring.
18. Remove the fuel pump unit

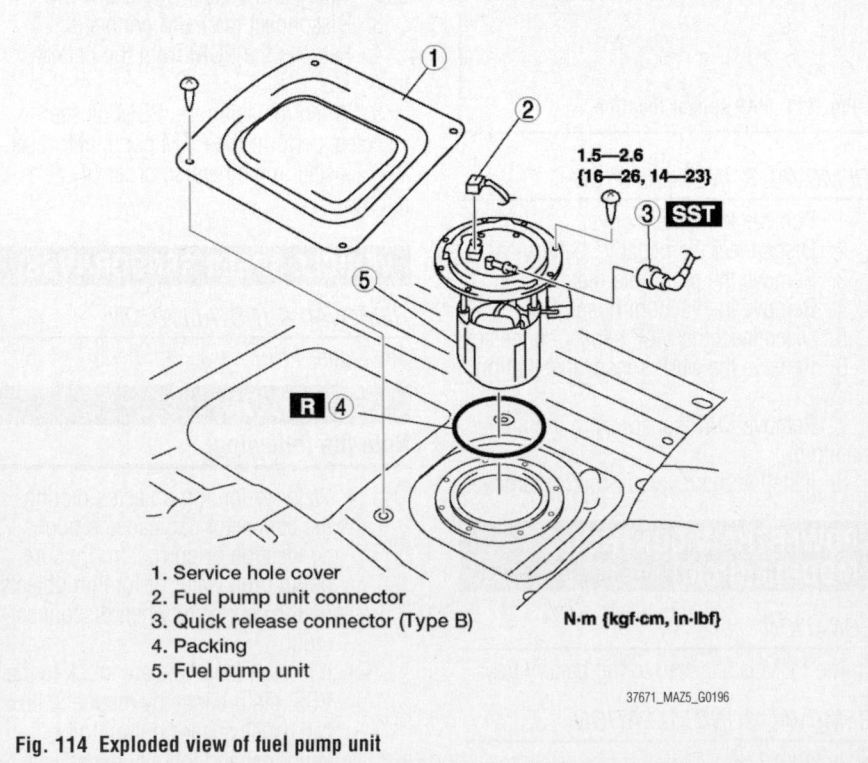

1.5—2.6
{16—26, 14—23}

N·m {kgf-cm, in·lbf}

1. Service hole cover
2. Fuel pump unit connector
3. Quick release connector (Type B)
4. Packing
5. Fuel pump unit

Fig. 114 Exploded view of fuel pump unit

To install:

19. Install in the reverse order of removal.

20. Inspect all related parts by performing "AFTER SERVICE PRECAUTIONS".

FUEL RAIL AND INJECTOR

REMOVAL & INSTALLATION

See Figures 115 through 117.

1. Follow "BEFORE SERVICE PRECAUTIONS" before performing any work operations to prevent fuel from spilling from the fuel system.
2. Remove the plug hole plate.
3. Remove the battery cover.
4. Disconnect the negative battery cable.
5. Disconnect the fuel injector connector.
6. Disconnect the quick release connector.
7. Remove the fuel rail (distributor).
8. Remove the injector clip.

➡ **The clip will not be reused.**

  a. Hold the clip using pliers.
  b. Pull the clip parallel to the injector groove and remove it from the injector.
9. Remove the fuel injector.

✳✳ CAUTION

Use of a deformed injector clip will cause the fuel injector to be connected incorrectly and could result in fuel leakage. It will also cause the injector to rotate. Therefore, always replace the clip when the injector is removed.

  a. Insert a flathead screwdriver between the injector cup and clip finger.

➡ **When pushing the clip finger outward, deform the finger until it is**

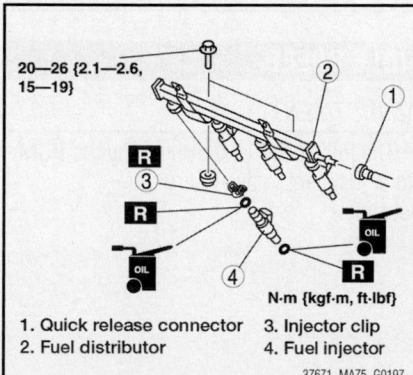

1. Quick release connector
2. Fuel distributor
3. Injector clip
4. Fuel injector

37671_MAZ5_G0197

Fig. 115 Exploded view of fuel rail and injector assembly

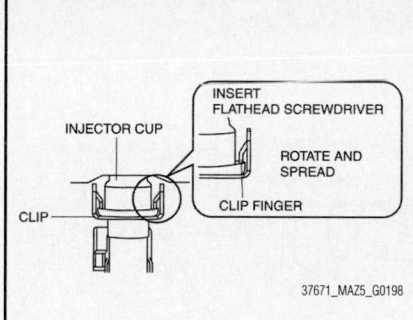

37671_MAZ5_G0198

Fig. 116 Insert a flathead screwdriver between the injector cup and clip finger

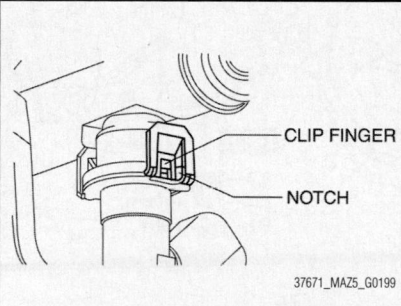

37671_MAZ5_G0199

Fig. 117 When pushing the clip finger outward, deform the finger until it is removed completely from the cup notch

removed completely from the cup notch.

  b. Push the clip finger outward using a flathead screwdriver.
  c. Remove the injector with the clip.

To install:

10. Install in the reverse order of removal.

Fuel Injector Installation

1. Apply a small amount of clean oil to the injector groove and the O-ring.
2. Temporarily attach a new clip to the injector groove.

➡ **When the clip is attached correctly, the central area of the injector and the clip finger positions are aligned.**

3. Hold the injector firmly and push the clip into the injector until the clip stops sliding.
4. Verify that the injector connector position is correct.
5. Press the injector into the injector cup. Continue pressing until the clip contacts the lower surface of the injector cup.
6. Verify that the injector and clip are correctly installed with the clip locked onto the injector cup notch.

7. Inspect all related parts by performing "AFTER SERVICE PRECAUTIONS".

FUEL TANK

REMOVAL & INSTALLATION

See Figure 118.

1. Relieve the fuel system pressure.
2. Remove the battery cover.
3. Disconnect the negative battery cable.
4. Remove or disconnect the following:
   - Second–row seat.
   - Rear package tray lid.
   - Sub–trunk.
   - Third–row seat.
   - Rear scuff plate.
   - Trunk end trim.
   - Third–row seat belt lower anchor installation bolt.
   - Cargo compartment light.
   - Trunk side trim.
   - Floor covering (partially peel back)
   - Service hole cover.
   - Fuel pump unit connector.
   - Quick release connector.
5. Park the vehicle on a level surface.

✳✳ WARNING

A person charged with static electricity could cause a fire or explosion, resulting in death or serious injury. Before draining fuel, make sure to discharge static electricity by touching a vehicle.

6. Drain the fuel.
7. Remove the flange nuts between the catalytic converter and pre–silencer, and between the pre–silencer and main silencer, to allow the insulator to be removed.
8. Remove the rear under cover.
9. Remove the insulator.
10. Remove in the order indicated in the figure.
11. When removing the fuel filler pipe, proceed as follows:
  a. Remove the rear tire (RH).
  b. Remove the fuel–filler pipe.
12. When removing the protector, proceed as follows:

✳✳ CAUTION

Be careful not to damage the fuel tank when removing the rivet. If the fuel tank is damaged, it may cause fuel leakage.

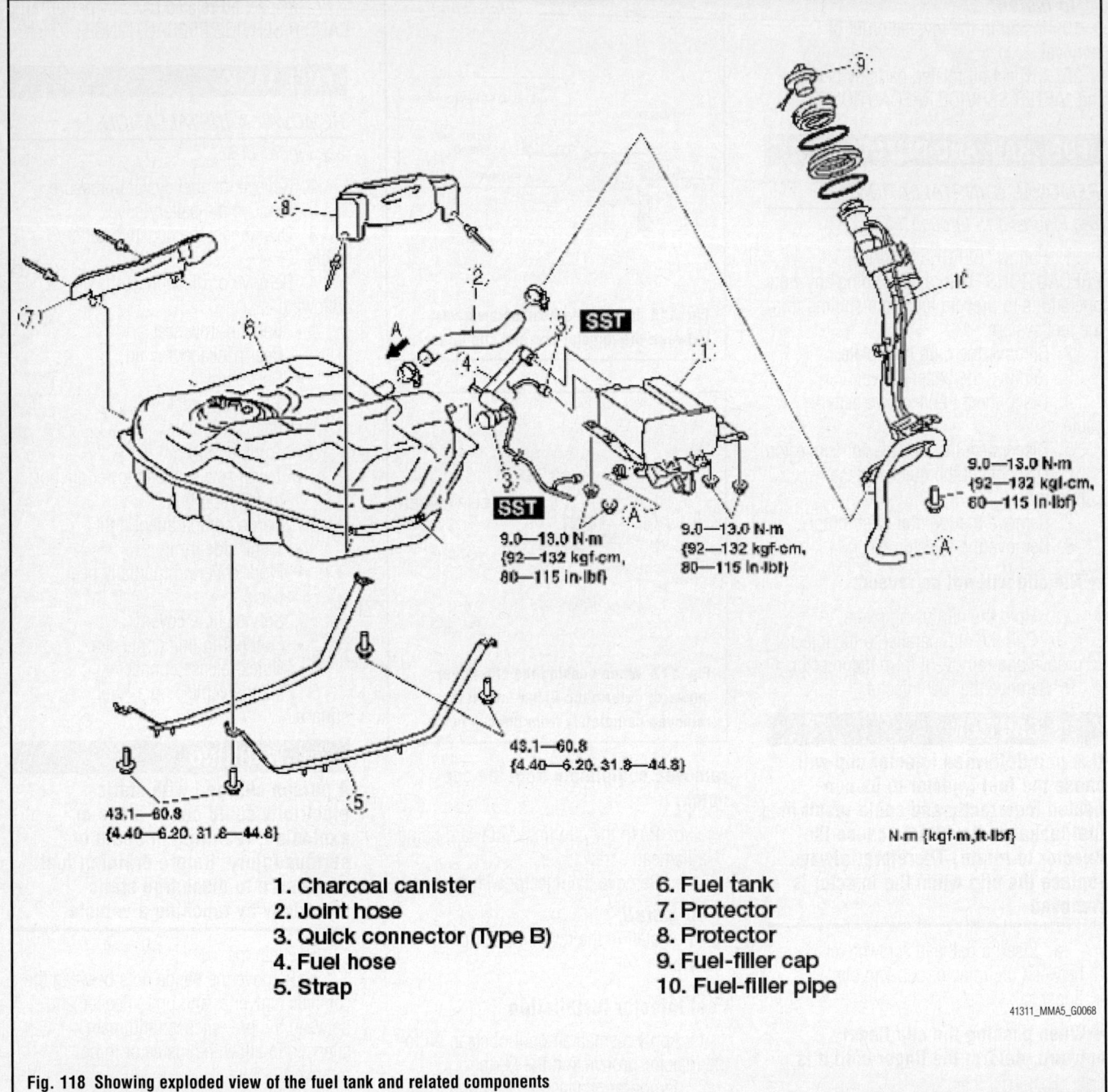

1. Charcoal canister
2. Joint hose
3. Quick connector (Type B)
4. Fuel hose
5. Strap
6. Fuel tank
7. Protector
8. Protector
9. Fuel-filler cap
10. Fuel-filler pipe

41311_MMA5_G0068

Fig. 118 Showing exploded view of the fuel tank and related components

a. The protector is installed using rivets.

b. Push out the mandrel using a hammer and punch.

c. Remove the flange using a drill.

To install:

13. Install in the reverse order of removal, noting the following.

14. When installing the joint hose and clamp, ensure proper installation position.

IDLE SPEED

ADJUSTMENT

The idle speed is controlled by the PCM. No adjustment is necessary or possible.

HEATING & AIR CONDITIONING SYSTEM

BLOWER MOTOR

REMOVAL & INSTALLATION

See Figures 119 through 122.

➡**The front blower motor is located on the A/C unit.**

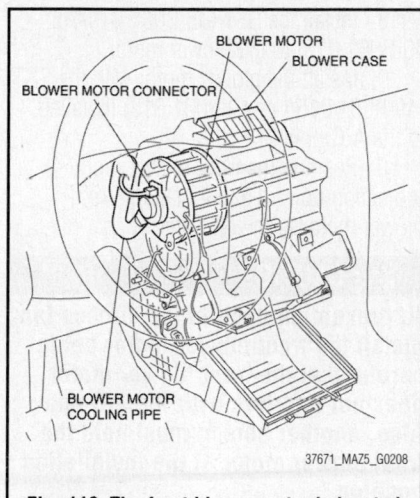

Fig. 119 The front blower motor is located on the A/C unit

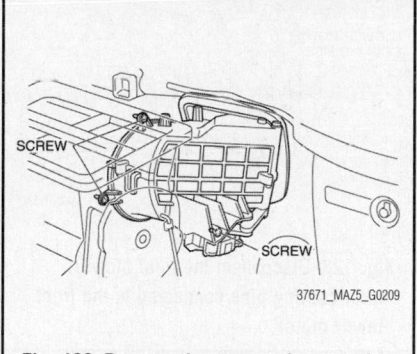

Fig. 120 Remove the screws shown and slide the front blower case

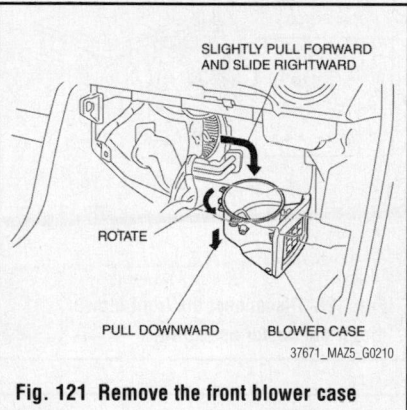

Fig. 121 Remove the front blower case shown

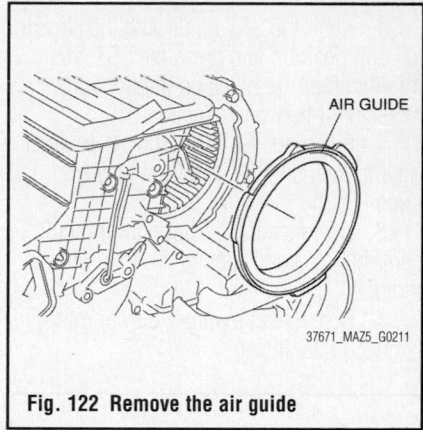

Fig. 122 Remove the air guide

1. Set the air intake mode to FRESH.
2. Disconnect the negative battery cable.
3. Remove the following parts:
   - Side wall
   - Selector lever component (A/T)
   - Shift lever component (M/T)
   - Front console
   - Front scuff plate inner (passenger's side)
   - Front side trim (passenger's side)
   - Side panel. (passenger's side)
   - Glove compartment
   - Hood release lever
   - Hands-Free telephone unit
   - Lower panel
   - Airflow mode actuator
   - Accelerator pedal
   - Brake pedal
4. Remove the BCM wiring harness grommets.
5. Remove the screws shown and slide the front blower case.

✳✳ CAUTION

Slide the front blower case while pressing the dashboard insulator, otherwise the front blower case could be damaged.

6. Disconnect the air intake actuator connector.

7. Remove the front blower case shown.
8. Remove the air guide.
9. Install the SST (49 B061 015A or 49 B061 015) to the blower motor.

SST (49 B061 015A) Installation

See Figure 123.

1. Align the SST hole with the align point and rotate the SST and then confirm that the SST tabs into the three set holes on the blower motor they are inserted as shown.

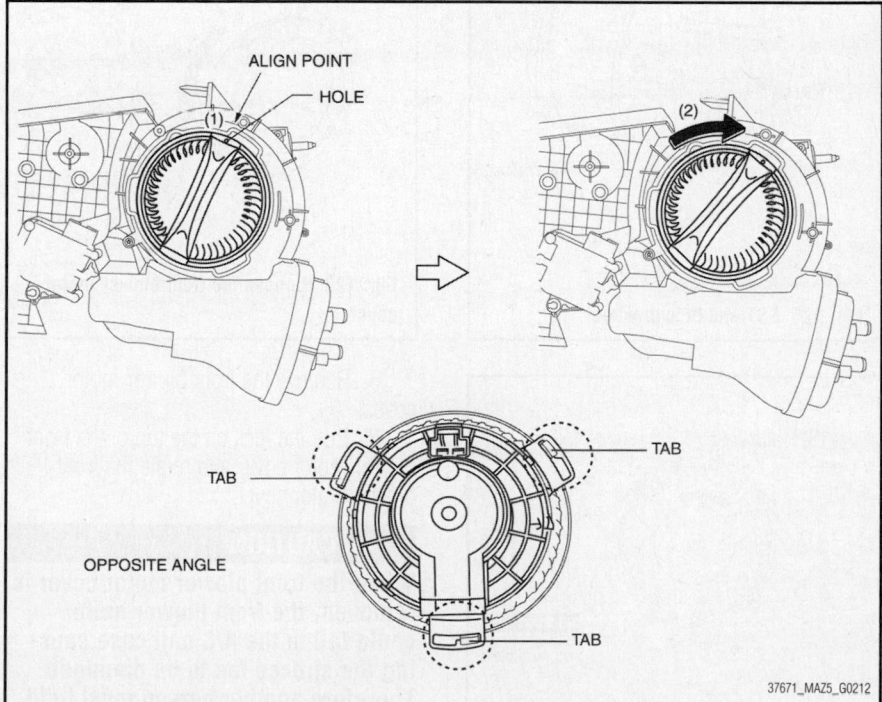

Fig. 123 Align the SST hole with the align point (1) and rotate the SST (2) and then confirm that the SST tabs into the three set holes on the blower motor they are inserted as shown

SST (49 B061 015) Installation

See Figures 124 through 133.

1. Align the SST guide with the sirocco fan clip position and press the SST tabs into the three set holes on the front blower motor until they are inserted.

2. Rotate the SST (49 B061 015) clockwise to lock the SST and front blower motor.

3. Disconnect the front blower motor cooling pipe connected to the front blower motor.

4. Disconnect the front blower motor connector as shown.

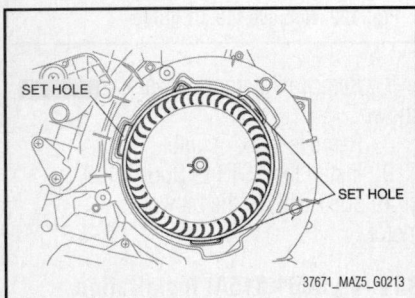

Fig. 124 Align the SST guide with the sirocco fan clip position and press the SST tabs into the three set holes on the front blower motor until they are inserted

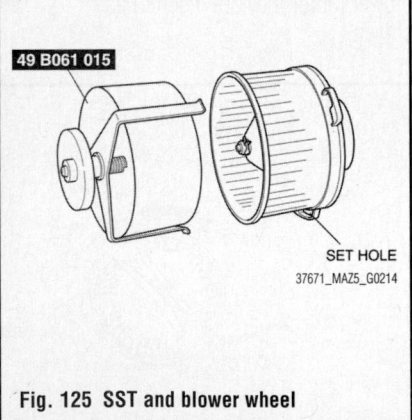

Fig. 125 SST and blower wheel

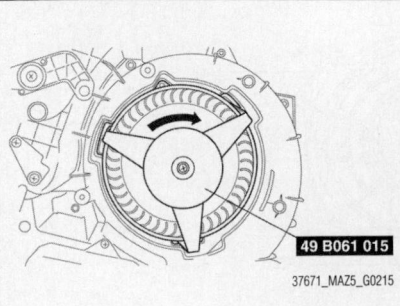

Fig. 126 Rotate the SST (49 B061 015) clockwise to lock the SST and front blower motor

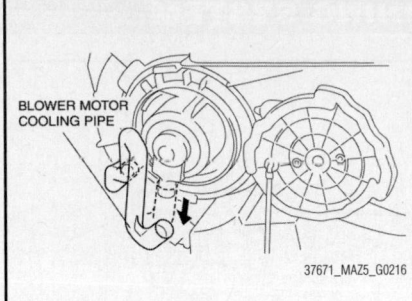

Fig. 127 Disconnect the front blower motor cooling pipe connected to the front blower motor

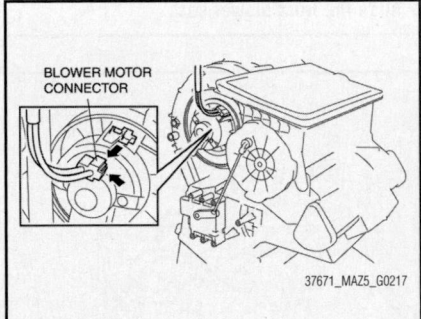

Fig. 128 Disconnect the front blower motor connector as shown

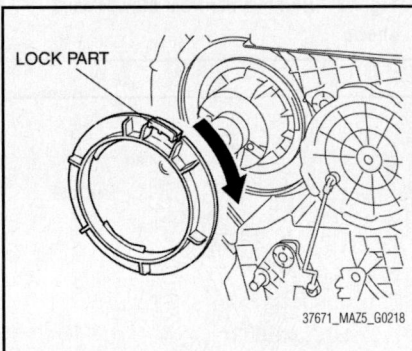

Fig. 129 Remove the front blower motor cover

5. Remove the front blower motor cover.

6. Pull the lock on the top of the front blower motor cover and rotate the front blower motor cover.

❊❊ CAUTION

When the front blower motor cover is removed, the front blower motor could fall in the A/C unit case causing the sirocco fan to be damaged. Therefore another person must hold the front blower motor at the installation position.

7. Remove the front blower motor by pulling it out.

❊❊ CAUTION

To prevent damage to the sirocco fan, pull the front blower motor out being careful that the front blower motor does not interfere with the A/C unit.

To install:

8. Install the SST (49 B061 015A or 49 B061 015) to the blower motor.

9. Install the blower motor with the SST (49 B061 015A or 49 B061 015) installed, to the A/C unit.

10. Position the front blower motor projection upward and install the front blower motor to the A/C unit.

❊❊ CAUTION

To prevent damage to the sirocco fan, install the front blower motor being careful that the front blower motor does not interfere with the A/C unit. Also, another person must hold the front blower motor at the installation position.

11. Install the front blower motor cover from the driver's side.

12. To install, rotate the front blower motor cover until a click is heard.

13. Connect the front blower motor connector.

14. Install the front blower motor cooling pipe.

15. Rotate the SST (49 B061 015A or 49 B061 015) counterclockwise.

16. Remove the SST (49 B061 015A or 49 B061 015) from the blower motor.

17. Install the air guide.

❊❊ CAUTION

Install the front blower case while pressing the dashboard insulator, otherwise the front blower case could be damaged.

18. Temporarily install the front blower case.

19. Connect the air intake actuator connector.

20. Install the front blower case.

   a. If not replacing the front blower case, replace the adhesive polyurethane on the fresh-air inlet of the front blower case.

❊❊ CAUTION

To adhere new polyurethane properly, be sure to remove the adhesive agent and adhesive polyurethane completely.

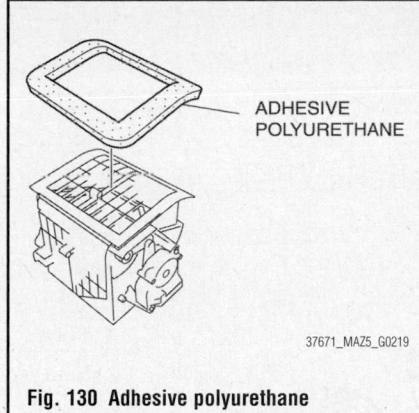

Fig. 130 Adhesive polyurethane

➡ **If the front blower case is removed or installed, the adhesive polyurethane can be damaged. Damaged adhesive polyurethane could cause abnormal noise or other malfunctions, therefore replace it.**

    b. Insert the screw into the front blower case.

    c. Insert the front blower case in the direction shown.

    d. Insert and rotate it in the directions of the arrows shown.

21. Install the following parts:

- Brake pedal
- Accelerator pedal
- Airflow mode actuator
- Lower panel
- Hood release lever
- BCM harness grommet
- HF/TEL unit
- Glove compartment
- Side panel. (passenger's side)
- Front side trim (passenger's side)
- Front scuff plate inner (passenger's side)
- Front console

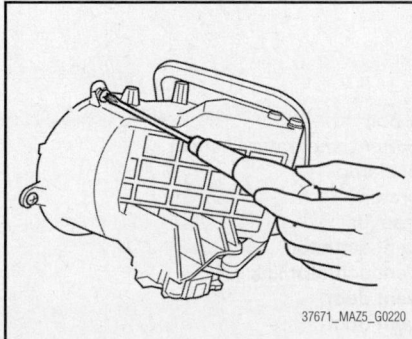

Fig. 131 Insert the screw into the front blower case

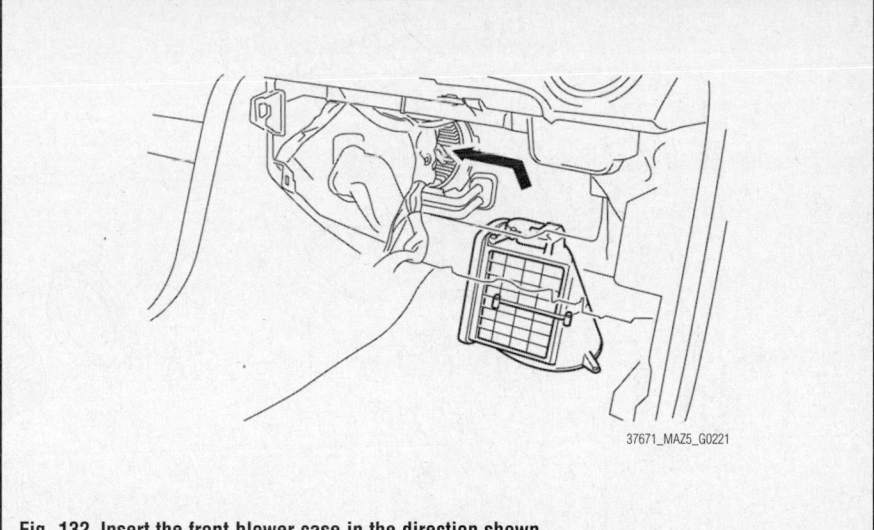

Fig. 132 Insert the front blower case in the direction shown

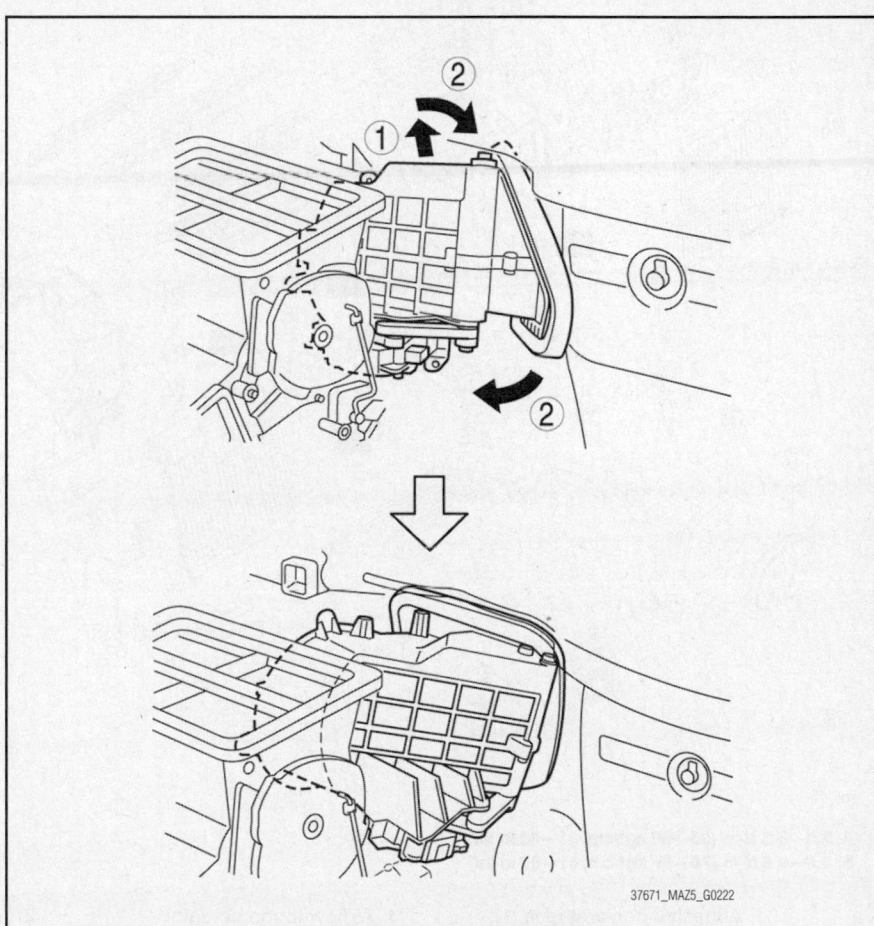

Fig. 133 Insert and rotate it in the directions of the arrows shown

- Shift lever component (M/T)
- Selector lever component (A/T)
- Side wall

HEATER CORE

REMOVAL & INSTALLATION

See Figures 134 and 135.

➡ **The manufacturer does not provide a specific Removal and Installation procedure for this component. Refer to the graphic(s) when servicing this component. Disassemble the unit in numerical order.**

37671_MAZ5_G0226

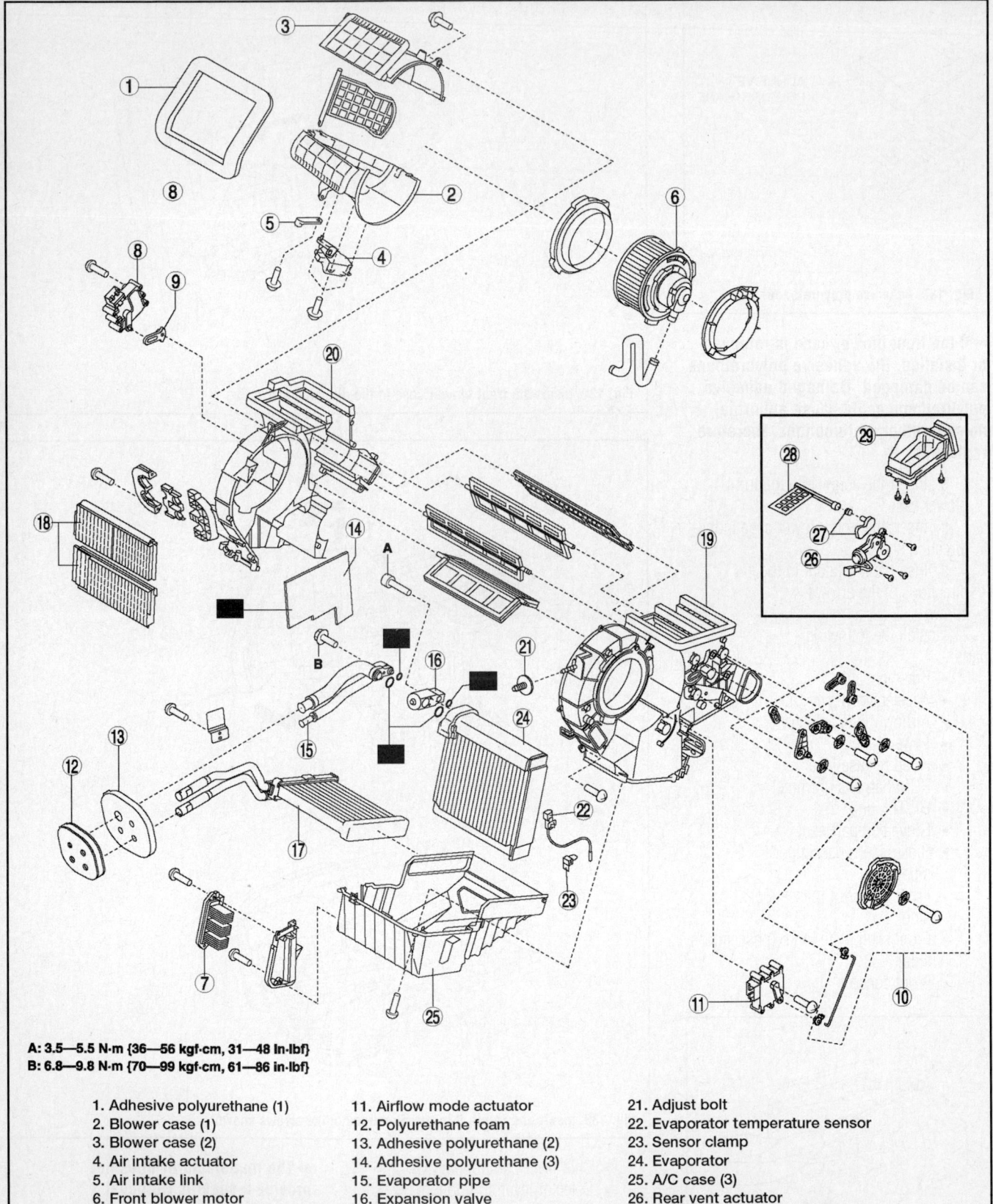

A: 3.5—5.5 N·m {36—56 kgf·cm, 31—48 in·lbf}
B: 6.8—9.8 N·m {70—99 kgf·cm, 61—86 in·lbf}

1. Adhesive polyurethane (1)	11. Airflow mode actuator	21. Adjust bolt
2. Blower case (1)	12. Polyurethane foam	22. Evaporator temperature sensor
3. Blower case (2)	13. Adhesive polyurethane (2)	23. Sensor clamp
4. Air intake actuator	14. Adhesive polyurethane (3)	24. Evaporator
5. Air intake link	15. Evaporator pipe	25. A/C case (3)
6. Front blower motor	16. Expansion valve	26. Rear vent actuator
7. Front power MOS FET	17. Heater core	27. Rear vent actuator link set
8. Air mix actuator	18. Air filter	28. Rear vent door
9. Air mix link	19. A/C case (1)	29. Rear vent duct
10. Airflow mode link set	20. A/C case (2)	

Fig. 134 Exploded view of HVAC components (Full-Auto Air Conditioner)

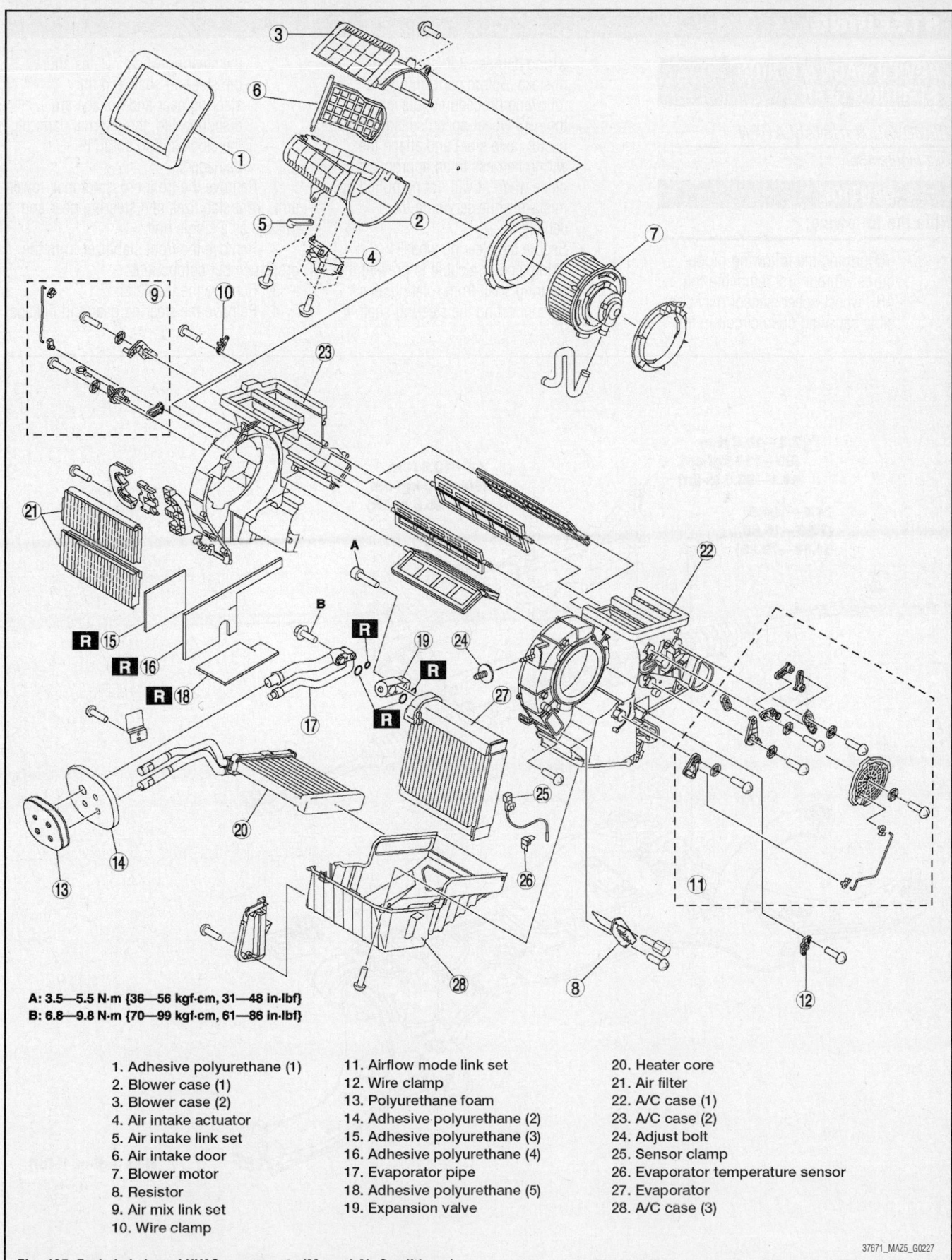

A: 3.5—5.5 N·m {36—56 kgf·cm, 31—48 in·lbf}
B: 6.8—9.8 N·m {70—99 kgf·cm, 61—86 in·lbf}

1. Adhesive polyurethane (1)
2. Blower case (1)
3. Blower case (2)
4. Air intake actuator
5. Air intake link set
6. Air intake door
7. Blower motor
8. Resistor
9. Air mix link set
10. Wire clamp

11. Airflow mode link set
12. Wire clamp
13. Polyurethane foam
14. Adhesive polyurethane (2)
15. Adhesive polyurethane (3)
16. Adhesive polyurethane (4)
17. Evaporator pipe
18. Adhesive polyurethane (5)
19. Expansion valve

20. Heater core
21. Air filter
22. A/C case (1)
23. A/C case (2)
24. Adjust bolt
25. Sensor clamp
26. Evaporator temperature sensor
27. Evaporator
28. A/C case (3)

37671_MAZ5_G0227

Fig. 135 Exploded view of HVAC components (Manual Air Conditioner)

STEERING

POWER RACK & PINION STEERING GEAR

REMOVAL & INSTALLATION

See Figure 136.

✳✳ CAUTION

Note the following:

- Performing the following procedures without first removing the ABS wheel-speed sensor may possibly cause an open circuit in the wiring harness if it is pulled by mistake. Before performing the following procedures, disconnect the ABS wheel-speed sensor connector (axle side) and attach the wiring harness to an appropriate place where it will not be pulled by mistake while servicing the vehicle.
- Secure the steering wheel using tape or a cable to prevent the steering shaft from rotating after disconnecting the steering shaft. If the steering wheel rotates after the steering shaft and the steering gear and linkage are disconnected, the internal parts of the clock spring could be damaged.

1. Remove the front crossmember, lower arm, front stabilizer, and steering gear and linkage as a single unit.
2. Remove the front stabilizer from the crossmember component.
3. Remove the insulator.
4. Remove the steering gear and linkage.

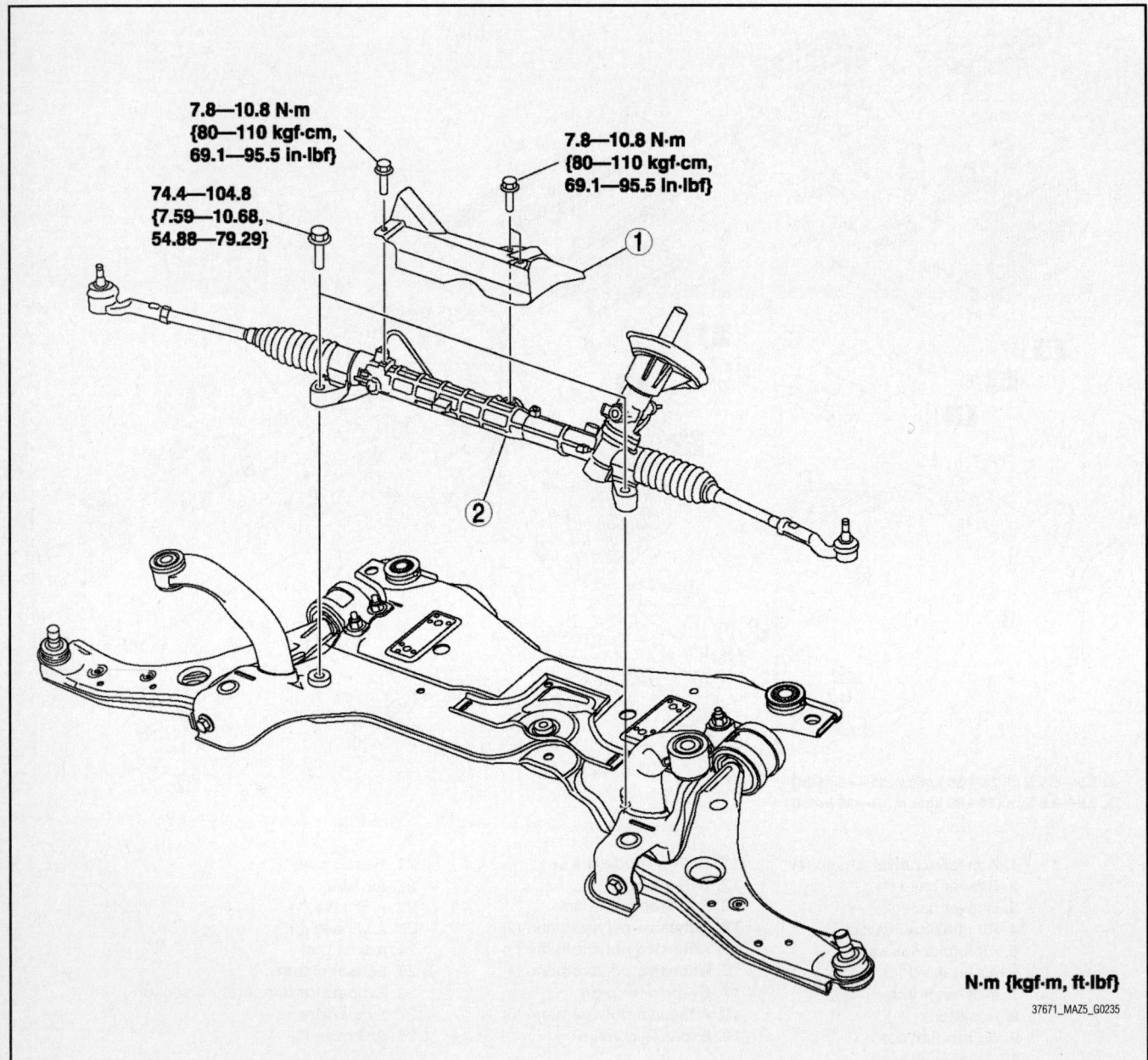

7.8—10.8 N·m
{80—110 kgf·cm,
69.1—95.5 in·lbf}

7.8—10.8 N·m
{80—110 kgf·cm,
69.1—95.5 in·lbf}

74.4—104.8
{7.59—10.68,
54.88—79.29}

N·m {kgf·m, ft·lbf}

37671_MAZ5_G0235

Fig. 136 Exploded view of steering gear and linkage (2) showing insulator (1)

5. Install in the reverse order of removal.

6. After installation, inspect the front wheel alignment and adjust it if necessary.

POWER STEERING PUMP

REMOVAL & INSTALLATION

See Figure 137.

❄❄ CAUTION

Be careful not to drop the electric power steering oil pump component as the internal parts of the EHPAS CM could be damaged. Replace the electric power steering oil pump

component if it is subjected to an impact.

1. Remove the undercover, splash shield (RH) and mudguard (RH).

2. Remove the washer tank. (Vehicles with the washer fluid-level sensor)

3. Remove the connector.

   a. Before disconnecting the connector, clean the connector with a brush or cloth to remove foreign matter.

   b. Disconnect the connector.

   c. After removing the connector, install the cap included in the new pump

kit package being careful that fluid or foreign matter does not penetrate the connector area of the control module.

➡ **If a new pump kit is not available, adhere protective tape to the connector.**

4. Disconnect the pressure pipe.

5. Remove the power steering fluid reserve tank.

6. Remove the electric power steering oil pump and bracket component.

7. Remove the brackets.

8. Disconnect the return hose.

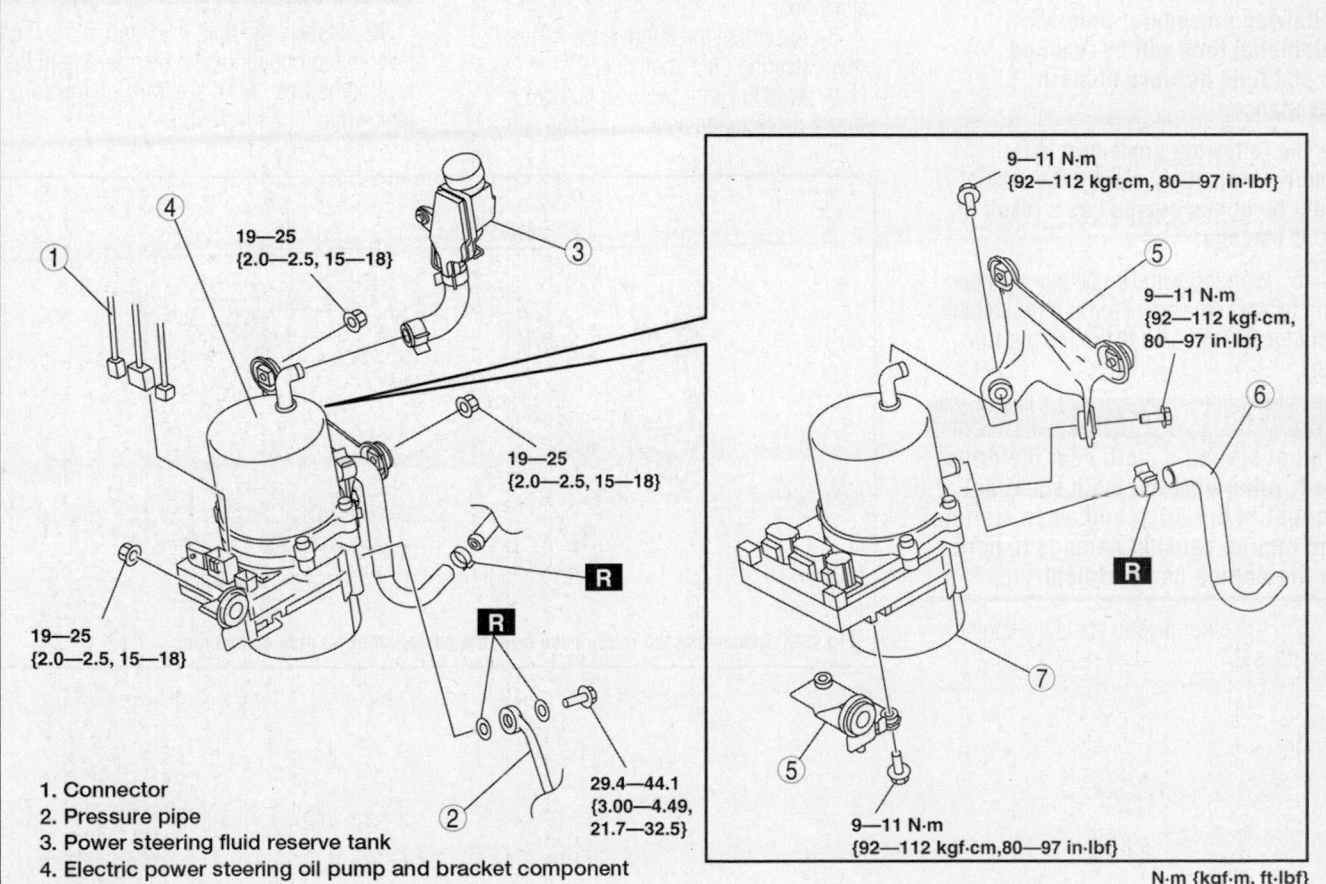

1. Connector
2. Pressure pipe
3. Power steering fluid reserve tank
4. Electric power steering oil pump and bracket component
5. Bracket
6. Return hose
7. Electric power steering oil pump component

37671_MAZ5_G0236

Fig. 137 Exploded view of the electric power steering oil pump assembly

9. Remove the electric power steering oil pump component.

10. Install in the reverse order of removal.

BLEEDING

See Figure 138.

> **✳✳ CAUTION**
>
> **Note the following:**

- If the strainer is removed, impurities may penetrate the power steering system and damage it. To prevent this, always bleed air with the strainer installed.
- Do not maintain the steering wheel fully turned for 5 seconds or more. The oil temperature could rise and damage the oil pump.

➡**Add fluid and bleed air using the following procedure. Otherwise, additional time will be required to add fluid because of mesh resistance.**

➡**The following procedure is for adding (adjusting) fluid if the lack of fluid level has dropped as a result of fluid leakage.**

1. Working with two people, one person protects the area around the sub tank using a cloth and adds fluid to the sub tank.

> **✳✳ CAUTION**
>
> **Do not spread a cloth near the drive belt, otherwise the cloth could get caught in the drive belt when starting the engine causing damage to parts in the engine compartment.**

2. The other person starts the engine and idles it.

3. Turn the steering wheel fully to the left and right slowly several times.

If the steering wheel operation speed is too fast, the fluid may spatter from the filler port. Do not exceed the steering wheel operation speed reference value of 90 degrees/second.

4. Inspect the fluid level and, if the fluid level has dropped, repeat Step 1–3 until the level is between MAX and MIN on the sub tank while idling the engine.

➡**The following procedure is for adding fluid to the tank (full amount) after replacing the electric power steering oil pump component or steering gear and linkage.**

5. Remove the undercover.

6. Disconnect the return hose from the power steering pipe component.

7. Add fluid until fluid escapes from the return hose.

8. Assemble the return hose to the power steering pipe component.

9. Working with two people, one person protects the area around the sub tank using a cloth and adds fluid to the sub tank.

> **✳✳ CAUTION**
>
> **Do not spread a cloth near the drive belt, otherwise the cloth could get caught in the drive belt when starting the engine causing damage to parts in the engine compartment.**

10. The other person starts the engine and idles it.

11. Turn the steering wheel fully to the left and right slowly several times.

> **✳✳ CAUTION**
>
> **If the steering wheel operation speed is too fast, the fluid may spatter from the filler port. Do not exceed the steering wheel operation speed reference value of 90 degrees/second.**

12. Inspect the fluid level and, if the fluid level has dropped, repeat Step 5–7 until the level is between MAX and MIN on the sub tank while idling the engine.

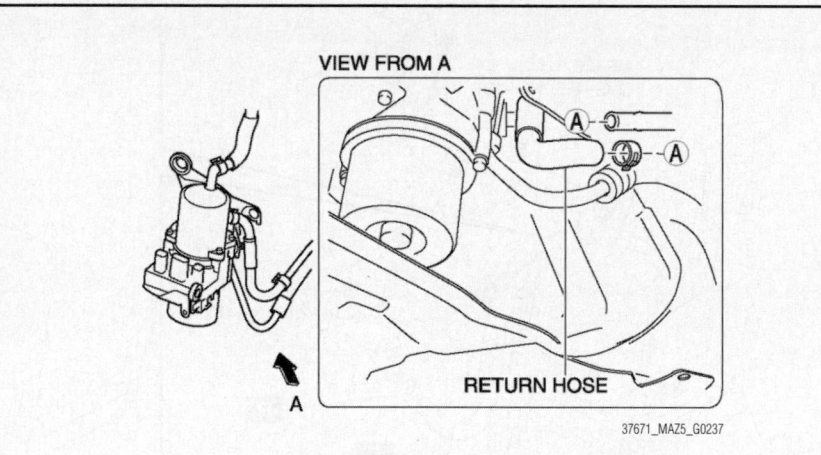

Fig. 138 Disconnect the return hose from the power steering pipe component

LOWER BALL JOINT AND CONTROL ARM

REMOVAL & INSTALLATION

See Figures 139 and 140.

1. Remove the stabilizer control link lower nut.
2. Remove the front lower arm ball joint.
3. Remove the tie-rod end ball joint.
4. Remove the front lower arm.

➡**Move the engine and transaxle slightly towards the front side of the** vehicle so that the engine does not interfere with the removal of the lower arm front side bolt.

    a. Remove the No.1 engine mount center bolt.
    b. Move the engine and transaxle slightly towards the front side of the vehicle using a jack.
    c. Remove the front lower arm rear side bolt.
    d. Remove the front lower arm.
5. Remove the dynamic damper.
6. Install in the reverse order of removal.

✳✳ CAUTION

Install the front lower arm according to the following procedures for optimal installation. Tighten the lower arm installation bolt properly when the vehicle is lifted.

➡**Move the engine and transaxle slightly towards the front side of the vehicle install of the lower arm front side bolt.**

7. Temporarily install the front lower arm.
8. Tighten the No.1 engine mount rubber installation bolt.
    a. A Type: 77–98 ft. lbs. (104–133 Nm)
    b. B Type: 54–67 ft. lbs. (73–91 Nm)

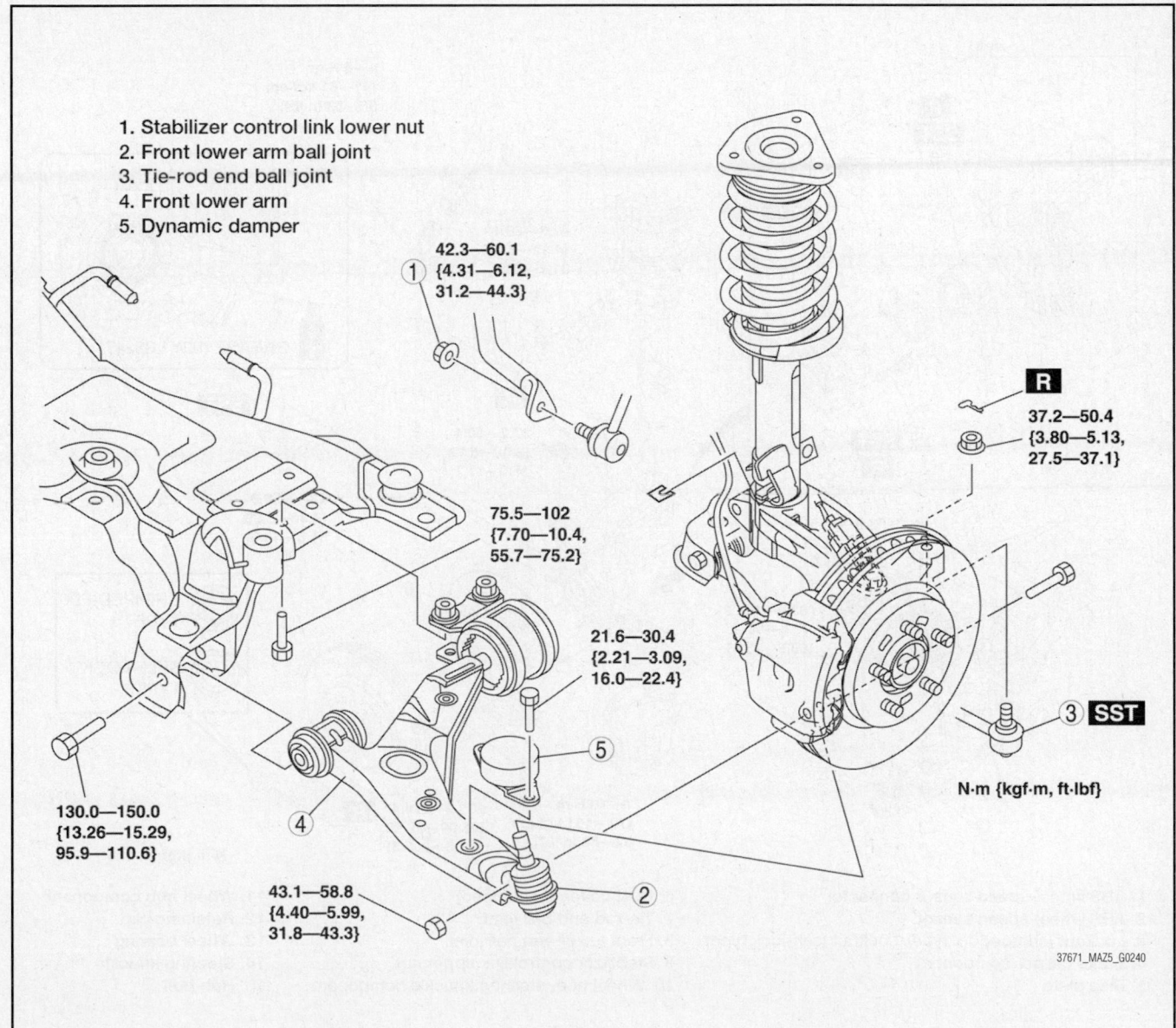

1. Stabilizer control link lower nut
2. Front lower arm ball joint
3. Tie-rod end ball joint
4. Front lower arm
5. Dynamic damper

42.3—60.1
{4.31—6.12,
31.2—44.3}

75.5—102
{7.70—10.4,
55.7—75.2}

21.6—30.4
{2.21—3.09,
16.0—22.4}

130.0—150.0
{13.26—15.29,
95.9—110.6}

43.1—58.8
{4.40—5.99,
31.8—43.3}

37.2—50.4
{3.80—5.13,
27.5—37.1}

N·m {kgf·m, ft·lbf}

37671_MAZ5_G0240

Fig. 139 Exploded view of lower control arm

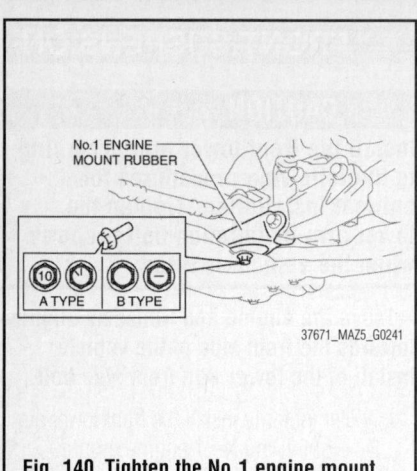

No.1 ENGINE
MOUNT RUBBER

A TYPE   B TYPE

37671_MAZ5_G0241

**Fig. 140 Tighten the No.1 engine mount
rubber installation bolt**

9. Tighten the front lower arm rear side
bolt.
10. Tighten the front lower arm front side
bolt.
11. Tighten the nut (front lower arm ball
joint).
12. Inspect the wheel alignment and
adjust it if necessary.

STEERING KNUCKLE

REMOVAL & INSTALLATION
See Figures 141 through 151.

✳✳ CAUTION

**Performing the following procedures
without first removing the ABS**
wheel-speed sensor may possibly
cause an open circuit in the wiring
harness if it is pulled by mistake.
Before performing the following
procedures, disconnect the ABS
wheel-speed sensor connector (axle
side) and attach the wiring harness
to an appropriate place where it will
not be pulled by mistake while
servicing the vehicle.

➡The locknuts include two types; a
self-lock type and a crimped-fix type.
Install a self-lock type locknut to the
drive shaft locknut installation area
without groove, or a crimped-fix type
to the one with a groove.

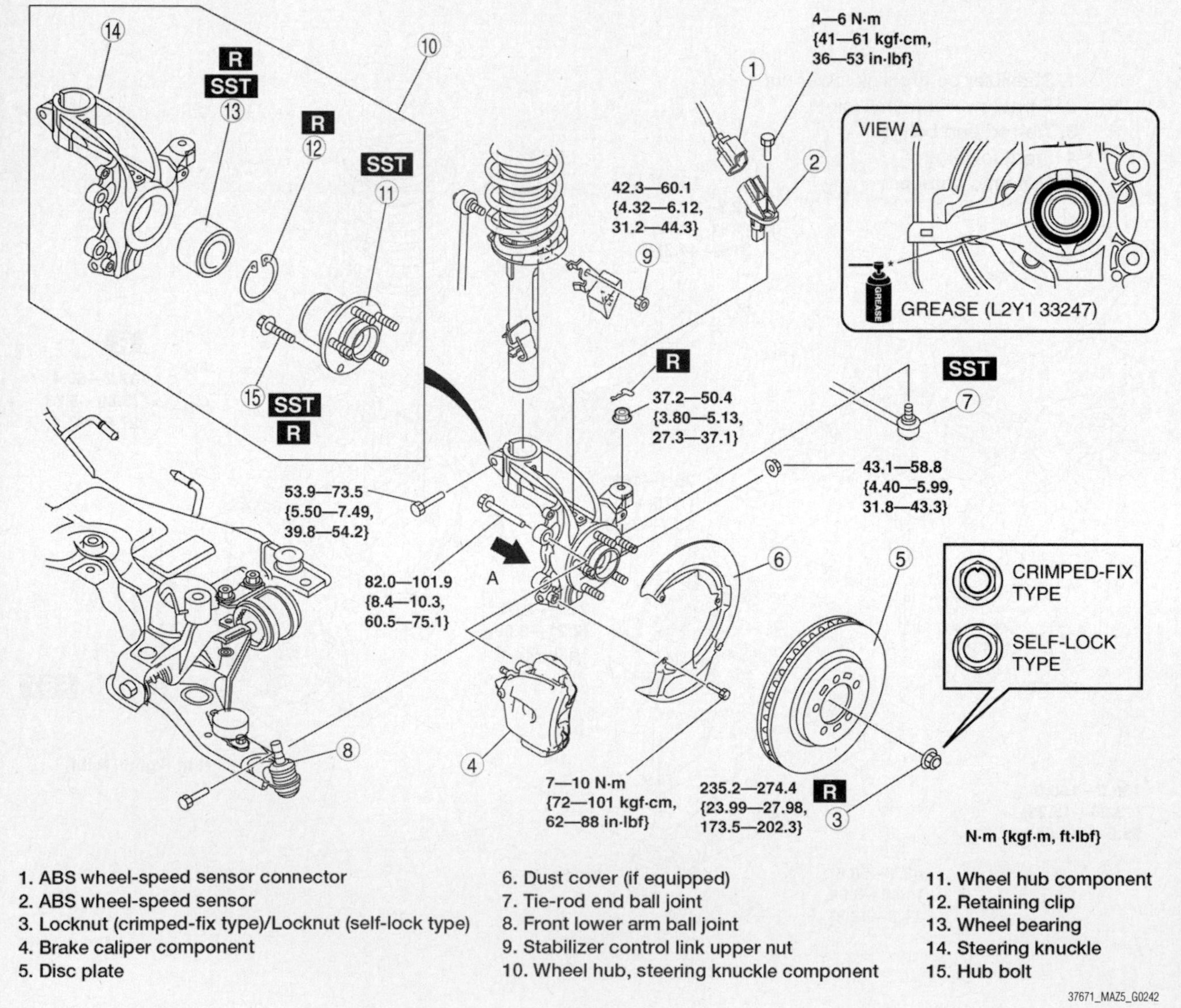

4—6 N·m
{41—61 kgf·cm,
36—53 in·lbf}

VIEW A

GREASE (L2Y1 33247)

42.3—60.1
{4.32—6.12,
31.2—44.3}

37.2—50.4
{3.80—5.13,
27.3—37.1}

43.1—58.8
{4.40—5.99,
31.8—43.3}

53.9—73.5
{5.50—7.49,
39.8—54.2}

82.0—101.9
{8.4—10.3,
60.5—75.1}

CRIMPED-FIX
TYPE

SELF-LOCK
TYPE

7—10 N·m
{72—101 kgf·cm,
62—88 in·lbf}

235.2—274.4
{23.99—27.98,
173.5—202.3}

N·m {kgf·m, ft·lbf}

1. ABS wheel-speed sensor connector
2. ABS wheel-speed sensor
3. Locknut (crimped-fix type)/Locknut (self-lock type)
4. Brake caliper component
5. Disc plate
6. Dust cover (if equipped)
7. Tie-rod end ball joint
8. Front lower arm ball joint
9. Stabilizer control link upper nut
10. Wheel hub, steering knuckle component
11. Wheel hub component
12. Retaining clip
13. Wheel bearing
14. Steering knuckle
15. Hub bolt

37671_MAZ5_G0242

Fig. 141 Exploded view of wheel hub and steering knuckle

1. Disconnect the ABS wheel-speed sensor connector.
2. Remove the ABS wheel-speed sensor.
3. Remove the Locknut (crimped-fix type)/Locknut (self-lock type).

➡ **The locknuts include two types; a self-lock type and a crimped-fix type. Verify which type of locknut is used on the vehicle being serviced. If it is a crimped-fix type locknut, remove it using the following procedure.**

  a. Knock the crimped portion of the locknut outward using a small chisel and a hummer.
  b. Lock the hub by applying the brakes.
  c. Remove the locknut.
4. Remove the brake caliper component from the steering knuckle and suspend it out of the way using a cable.
5. Remove the disc plate.
6. Remove the dust cover (if equipped).
7. Remove the tie-rod end ball joint.
8. Remove the front lower arm ball joint.
9. Remove the stabilizer control link upper nut.
10. Remove the wheel hub, steering knuckle component.

➡ **Separate the shock absorber from the wheel hub, steering knuckle component by tapping the upper part of the steering knuckle with a hammer.**

11. Remove the wheel hub component.
  a. Remove the wheel hub component using the SSTs.
  b. If the bearing inner race remains on the front wheel hub component, perform the following procedure:
  c. Without using SST: Grind a section of the bearing inner race until approximately 0.02 inches (0.5 mm) remains.
  d. Remove the bearing inner race using a chisel.
  e. With using SST: Remove the part as shown.

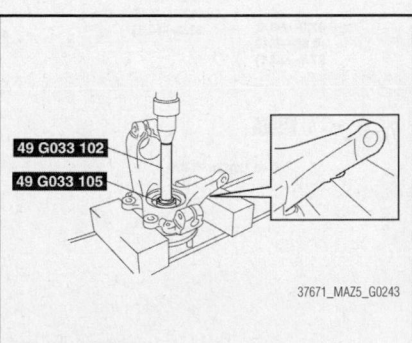

Fig. 142 Remove the wheel hub component using the SSTs

  f. Position the SSTs and a spare bolt as shown.

✳✳ CAUTION

When tightening section A, tighten section B securely because the engagement of the SST tab is shallow and it can come off easily.

  g. Tighten SST section A until the space between the bearing inner race and

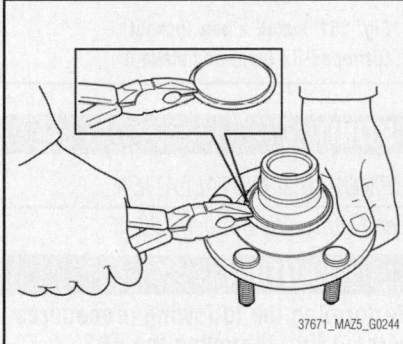

Fig. 143 With using SST: Remove the part as shown

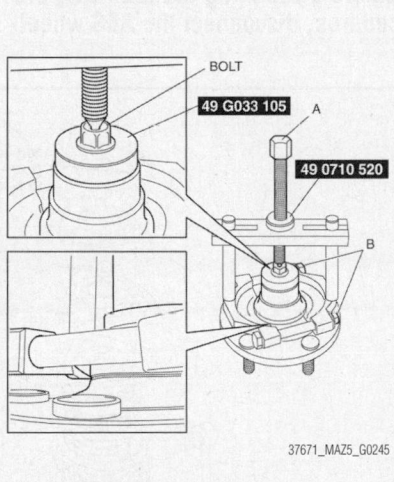

Fig. 144 Position the SSTs and a spare bolt as shown

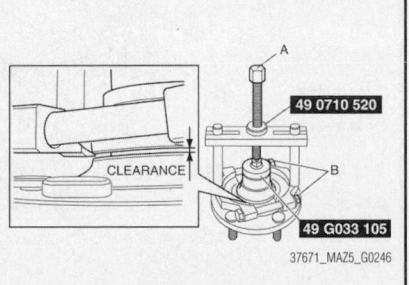

Fig. 145 Tighten SST section A until the space between the bearing inner race and wheel hub component

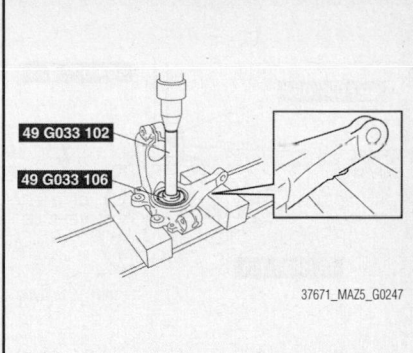

Fig. 146 Remove the wheel bearing using the SSTs

wheel hub component is 0.04–0.07 inches (1–2 mm).
  h. Temporarily loosen SST section A and position the SST again.
  i. Tighten SST section A and remove the bearing inner race.
12. Remove the retaining clip.
13. Remove the wheel bearing using the SSTs.
14. Remove the steering knuckle.
15. Remove the hub bolt(s) using a press.

➡ **The hub bolts do not need to be removed unless they are being replaced.**

  To install:
16. Install in the reverse order of removal.
17. Install the new hub bolt(s) using a press.

✳✳ CAUTION

Install the wheel bearing with the ABS encoder (tea color) facing inside of the vehicle.

18. Install the new wheel bearing using the SSTs.

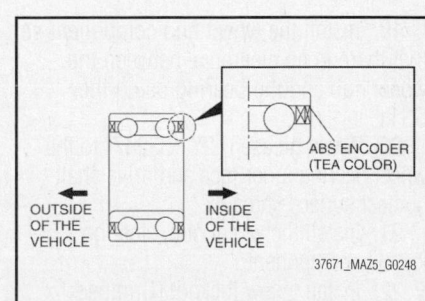

Fig. 147 Install the wheel bearing with the ABS encoder (tea color) facing inside of the vehicle

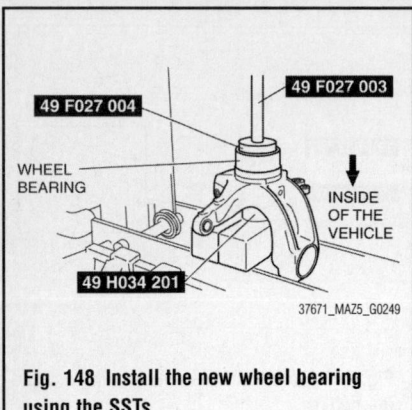

Fig. 148 Install the new wheel bearing using the SSTs

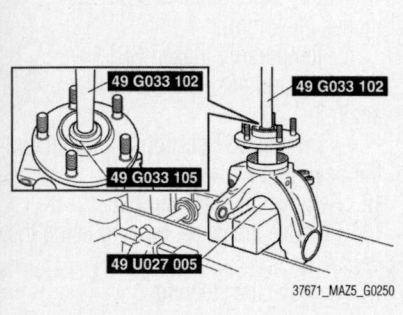

Fig. 149 Install the wheel hub component so that there is no clearance between the wheel hub and the bearing using the SSTs

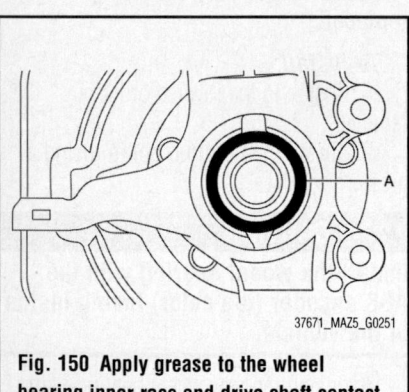

Fig. 150 Apply grease to the wheel bearing inner race and drive shaft contact surface (Area A)

19. Install the wheel hub component so that there is no clearance between the wheel hub and the bearing using the SSTs.

20. Apply grease (L2Y1 33247) to the wheel bearing inner race and drive shaft contact surface (Area A).

21. Install the wheel hub, steering knuckle component.

22. Install a new locknut (Crimped-fix Type) and stake it.

23. After installation, inspect the front wheel alignment and adjust it if necessary.

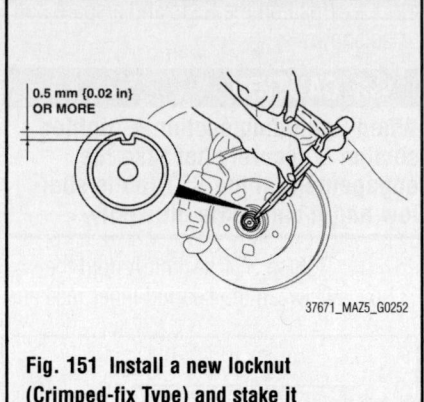

Fig. 151 Install a new locknut (Crimped-fix Type) and stake it

STRUT

REMOVAL & INSTALLATION
See Figures 152 through 156.

> ⁂ **CAUTION**
>
> Performing the following procedures without first removing the ABS wheel-speed sensor may possibly cause an open circuit in the wiring harness if it is pulled by mistake. Before performing the following procedures, disconnect the ABS wheel-

speed sensor wiring harness connector (axle side) and attach the wiring harness to an appropriate place where it will not be pulled by mistake while servicing the vehicle.

1. Disconnect the ABS wheel-speed sensor wiring harness connector.
2. Remove the brake hose.
3. Remove the stabilizer control link upper nut.
4. Disconnect the ABS wheel-speed sensor wiring harness.
5. Remove the dynamic damper.
6. Remove the tie-rod end ball joint.
7. Remove the lower arm ball joint.
   a. Loosen the installation bolt on the inner side of the front lower arm.
   b. Separate the lower arm ball joint.
8. Remove the shock absorber lower bolt.
9. Remove the shock absorber upper bolts.
10. Remove the stiffener.
11. Remove the shock absorber and coil spring.

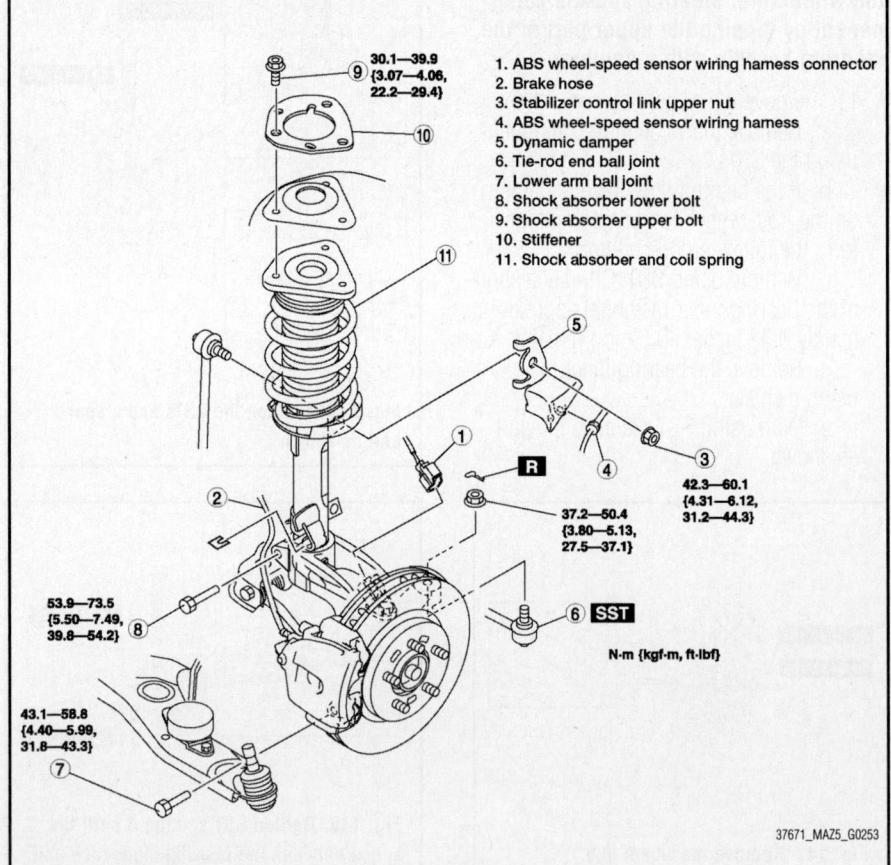

1. ABS wheel-speed sensor wiring harness connector
2. Brake hose
3. Stabilizer control link upper nut
4. ABS wheel-speed sensor wiring harness
5. Dynamic damper
6. Tie-rod end ball joint
7. Lower arm ball joint
8. Shock absorber lower bolt
9. Shock absorber upper bolt
10. Stiffener
11. Shock absorber and coil spring

N·m {kgf·m, ft-lbf}

Fig. 152 Exploded view of shock absorber and coil spring assembly

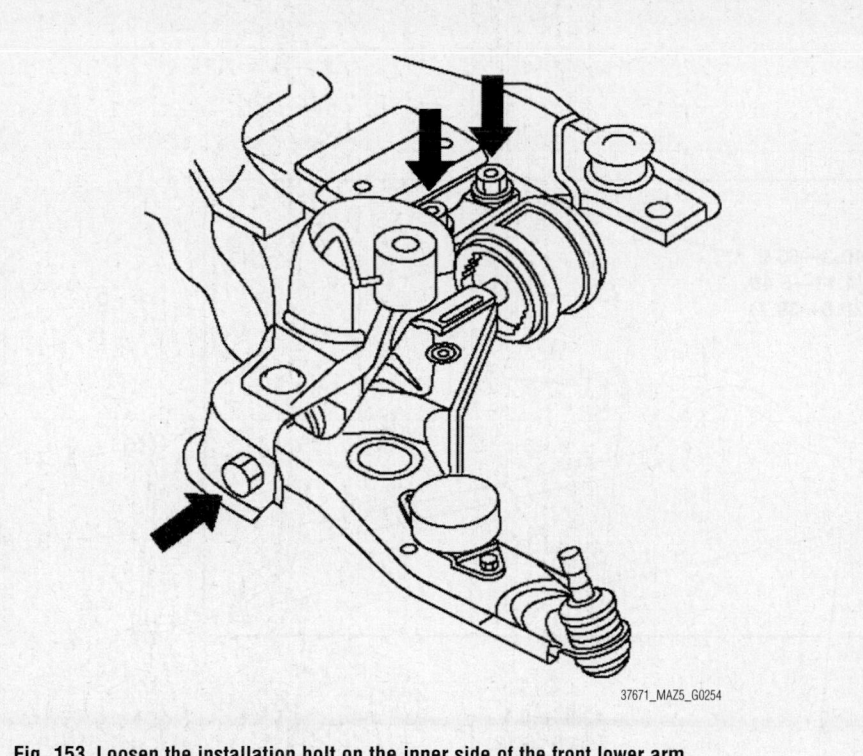

Fig. 153 Loosen the installation bolt on the inner side of the front lower arm

➡Separate the shock absorber from the wheel hub, steering knuckle component by tapping the upper part of the steering knuckle with a hammer.

✴✴ CAUTION

When removing the steering knuckle by lightly tapping, the steering knuckle could fall and cause injury, or damage to the part. When performing the removal, support the bottom of the steering knuckle with a jack.

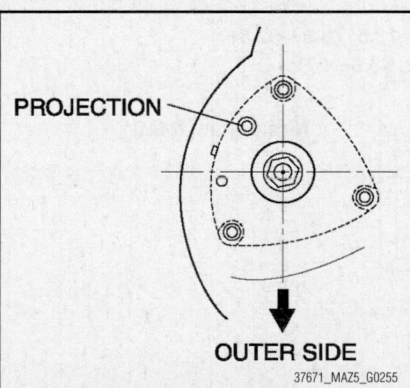

Fig. 154 Install the shock absorber and coil spring so that the positioning projection on the mounting rubber is aligned with the positioning hole on the body

To install:

12. Install in the reverse order of removal.

13. Install the shock absorber and coil spring so that the positioning projection on the mounting rubber is aligned with the positioning hole on the body.

14. Align the knuckle notch with projection A on the lower side of the shock absorber.

15. Raise the front lower arm using a jack and install the shock absorber and coil spring.

16. Install the stiffener so that the mark (RH or LH) faces upward.

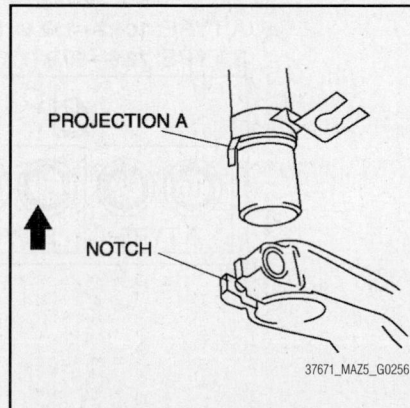

Fig. 155 Align the knuckle notch with projection A on the lower side of the shock absorber

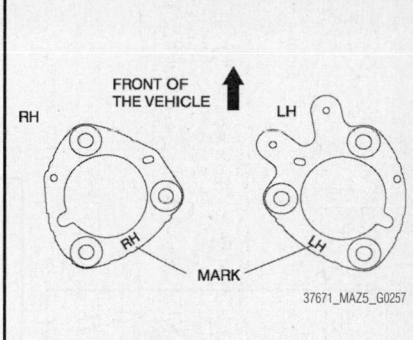

Fig. 156 Install the stiffener so that the mark (RH or LH) faces upward

17. Install the lower arm ball joint to the steering knuckle.

18. Tighten the installation bolts on the inner side of the front lower arm.

   a. Front side: 96–111 ft. lbs. (130–150 Nm)

   b. Rear side: 72–98 ft. lbs. (98–132 Nm)

19. Inspect the front wheel alignment and adjust it if necessary.

OVERHAUL

1. Remove the front shock absorber.

2. Inspect for damage and oil leakage.

3. Compress and extend the shock absorber piston rod at least three times at a steady rate. From the fourth compression stroke, verify that the operational force does not change and that there is no unusual noise.

4. If there is any malfunction, replace the shock absorber.

STABILIZER BAR

REMOVAL & INSTALLATION

See Figures 157 through 160.

1. Detach the steering shaft.

2. Remove the power steering pipe nut.

3. Remove the dynamic damper.

4. Remove the stabilizer control link.

5. Remove the No.1 engine mount center bolt.

   a. Support the front crossmember using a jack.

   b. Remove the No. 1 engine mount center bolt.

6. Remove the front stabilizer component.

   a. Detach the silencer hangers on the middle pipe from the front crossmember.

   b. Lower the front crossmember slowly approximately 3.5 inches (90 mm) and remove the front stabilizer component.

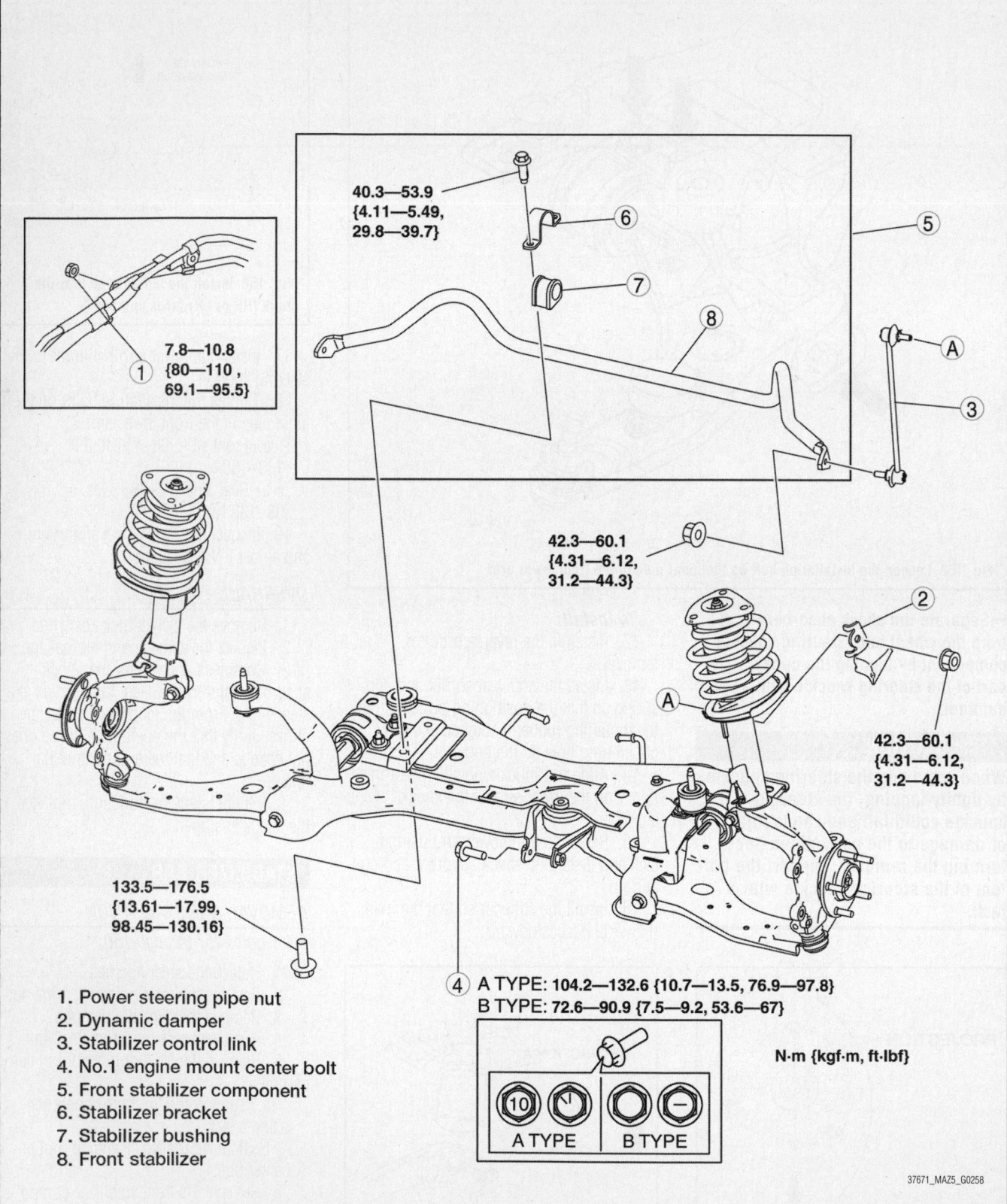

40.3—53.9
{4.11—5.49,
29.8—39.7}

7.8—10.8
{80—110 ,
69.1—95.5}

42.3—60.1
{4.31—6.12,
31.2—44.3}

42.3—60.1
{4.31—6.12,
31.2—44.3}

133.5—176.5
{13.61—17.99,
98.45—130.16}

A TYPE: 104.2—132.6 {10.7—13.5, 76.9—97.8}
B TYPE: 72.6—90.9 {7.5—9.2, 53.6—67}

N·m {kgf·m, ft·lbf}

A TYPE   B TYPE

1. Power steering pipe nut
2. Dynamic damper
3. Stabilizer control link
4. No.1 engine mount center bolt
5. Front stabilizer component
6. Stabilizer bracket
7. Stabilizer bushing
8. Front stabilizer

37671_MAZ5_G0258

Fig. 157 Exploded view of front stabilizer assembly

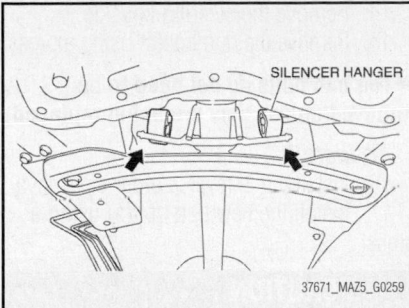

Fig. 158 Detach the silencer hangers on the middle pipe from the front crossmember

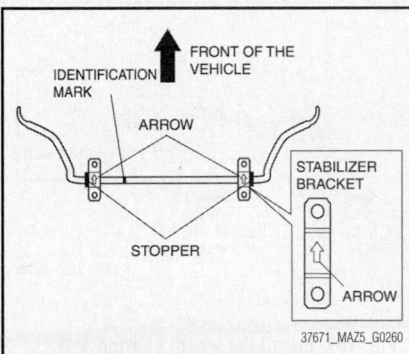

Fig. 159 Verify the installation direction of the stabilizer bracket (the arrow should point to the front)

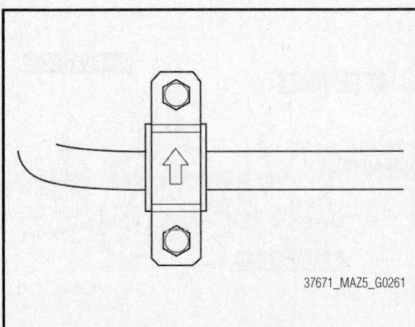

Fig. 160 With the arrow on the stabilizer bracket facing the vehicle front, tighten the bolts

7. Remove the stabilizer brackets.
8. Remove the stabilizer bushings.
9. Remove the front stabilizer.

To install:

10. Install in the reverse order of removal.

11. Verify the installation direction of the stabilizer bracket (the arrow should point to the front) by placing the stabilizer bar so that the identification mark faces the left side of the vehicle.

12. Install the stabilizer bushings and brackets so that they contact the stabilizer stoppers.

13. With the arrow on the stabilizer bracket facing the vehicle front, tighten the bolts.

14. Inspect the wheel alignment and adjust it if necessary.

WHEEL HUB & BEARING

REMOVAL & INSTALLATION

See Figures 161 through 171.

✳✳ CAUTION

Performing the following procedures without first removing the ABS wheel-speed sensor may possibly cause an open circuit in the wiring harness if it is pulled by mistake. Before performing the following procedures, disconnect the ABS wheel-speed sensor connector (axle side) and attach the wiring harness to an appropriate place where it will not be pulled by mistake while servicing the vehicle.

➡**The locknuts include two types; a self-lock type and a crimped-fix type. Install a self-lock type locknut to the drive shaft locknut installation area without groove, or a crimped-fix type to the one with a groove.**

1. Disconnect the ABS wheel-speed sensor connector.
2. Remove the ABS wheel-speed sensor.
3. Remove the Locknut (crimped-fix type)/Locknut (self-lock type).

➡**The locknuts include two types; a self-lock type and a crimped-fix type. Verify which type of locknut is used on the vehicle being serviced. If it is a crimped-fix type locknut, remove it using the following procedure.**

   a. Knock the crimped portion of the locknut outward using a small chisel and a hummer.
   b. Lock the hub by applying the brakes.
   c. Remove the locknut.

4. Remove the brake caliper component from the steering knuckle and suspend it out of the way using a cable.
5. Remove the disc plate.
6. Remove the dust cover (if equipped).
7. Remove the tie-rod end ball joint.
8. Remove the front lower arm ball joint.
9. Remove the stabilizer control link upper nut.

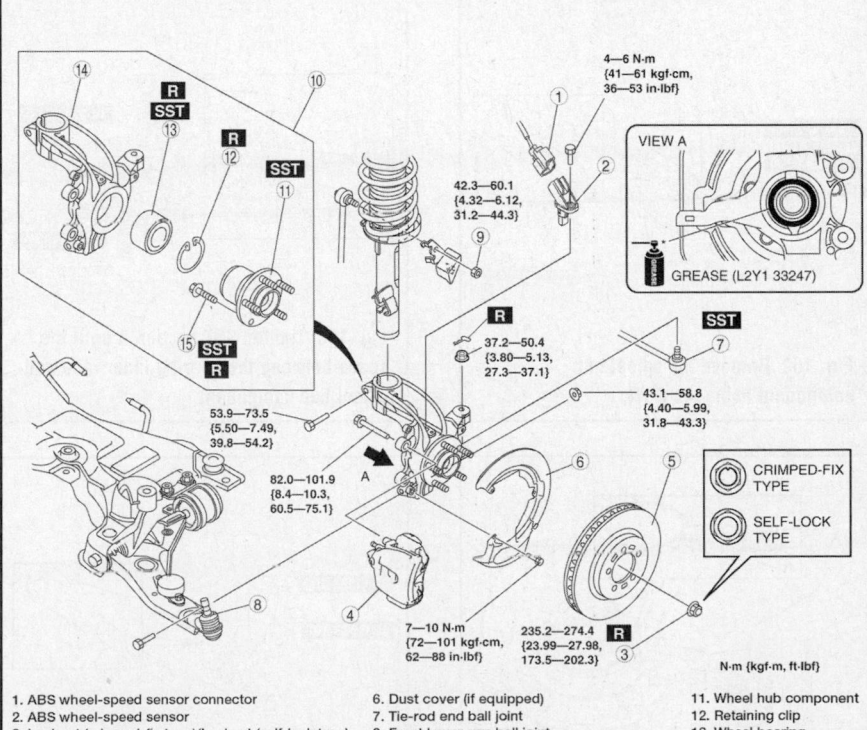

1. ABS wheel-speed sensor connector
2. ABS wheel-speed sensor
3. Locknut (crimped-fix type)/Locknut (self-lock type)
4. Brake caliper component
5. Disc plate
6. Dust cover (if equipped)
7. Tie-rod end ball joint
8. Front lower arm ball joint
9. Stabilizer control link upper nut
10. Wheel hub, steering knuckle component
11. Wheel hub component
12. Retaining clip
13. Wheel bearing
14. Steering knuckle
15. Hub bolt

Fig. 161 Exploded view of wheel hub and steering knuckle

10. Remove the wheel hub, steering knuckle component.

➡**Separate the shock absorber from the wheel hub, steering knuckle component by tapping the upper part of the steering knuckle with a hammer.**

11. Remove the wheel hub component.

a. Remove the wheel hub component using the SSTs.

b. If the bearing inner race remains on the front wheel hub component, perform the following procedure:

c. Without using SST: Grind a section of the bearing inner race until approximately 0.02 inches (0.5 mm) remains.

d. Remove the bearing inner race using a chisel.

e. With using SST: Remove the part as shown.

f. Position the SSTs and a spare bolt as shown.

✳✳ CAUTION

When tightening section A, tighten section B securely because the engagement of the SST tab is shallow and it can come off easily.

g. Tighten SST section A until the space between the bearing inner race and wheel hub component is 0.04–0.07 inches (1–2 mm).

h. Temporarily loosen SST section A and position the SST again.

i. Tighten SST section A and remove the bearing inner race.

12. Remove the retaining clip.

13. Remove the wheel bearing using the SSTs.

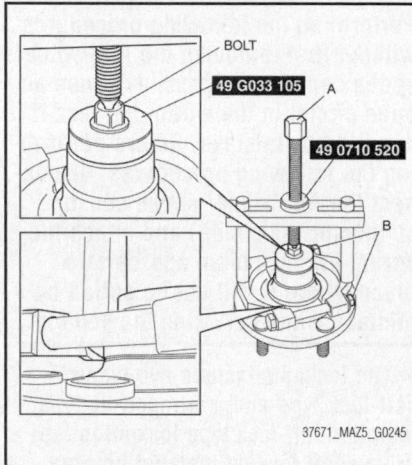

Fig. 164 Position the SSTs and a spare bolt as shown

14. Remove the steering knuckle.

15. Remove the hub bolt(s) using a press.

➡**The hub bolts do not need to be removed unless they are being replaced.**

To install:

16. Install in the reverse order of removal.

17. Install the new hub bolt(s) using a press.

✳✳ CAUTION

Install the wheel bearing with the ABS encoder (tea color) facing inside of the vehicle.

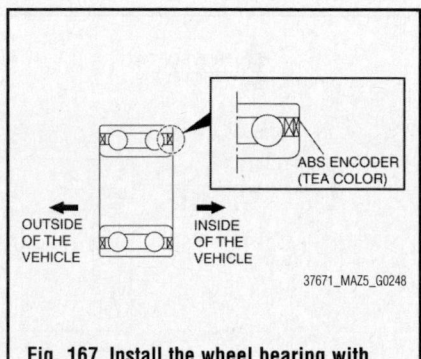

Fig. 167 Install the wheel bearing with the ABS encoder (tea color) facing inside of the vehicle

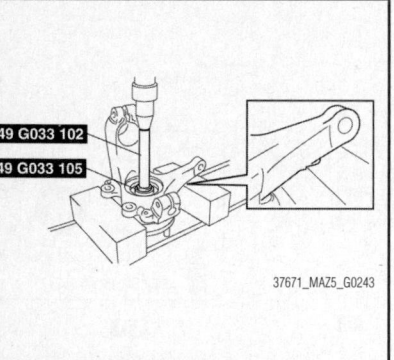

Fig. 162 Remove the wheel hub component using the SSTs

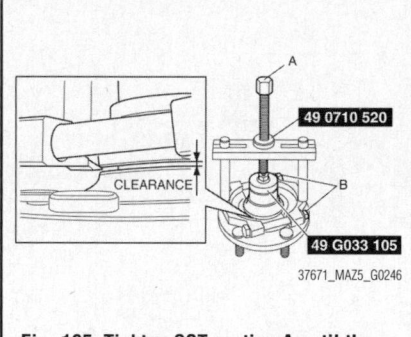

Fig. 165 Tighten SST section A until the space between the bearing inner race and wheel hub component

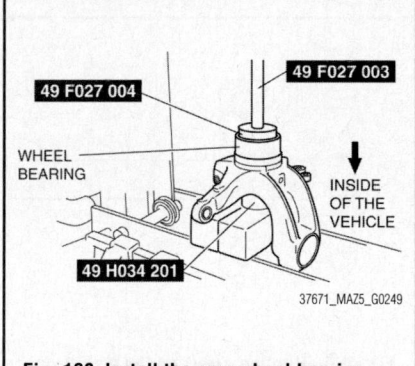

Fig. 168 Install the new wheel bearing using the SSTs

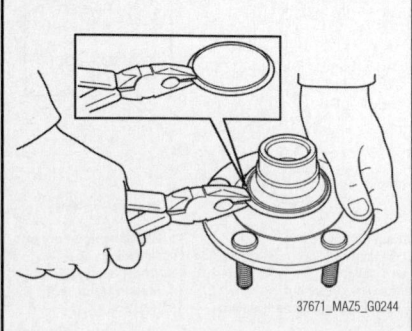

Fig. 163 With using SST: Remove the part as shown

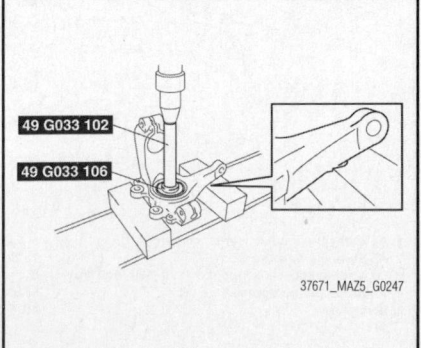

Fig. 166 Remove the wheel bearing using the SSTs

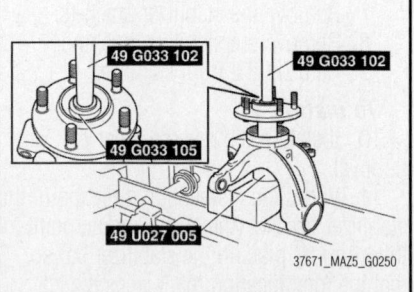

Fig. 169 Install the wheel hub component so that there is no clearance between the wheel hub and the bearing using the SSTs

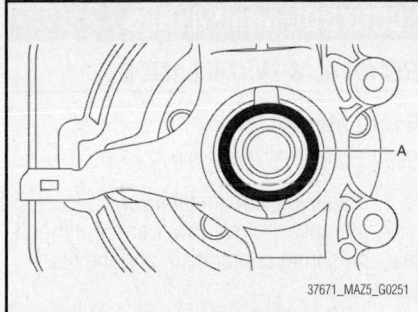

Fig. 170 Apply grease to the wheel bearing inner race and drive shaft contact surface (Area A)

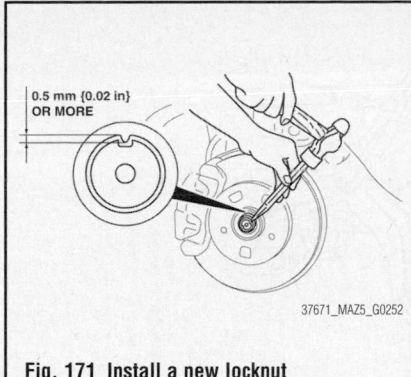

Fig. 171 Install a new locknut (Crimped-fix Type) and stake it

18. Install the new wheel bearing using the SSTs.

19. Install the wheel hub component so that there is no clearance between the wheel hub and the bearing using the SSTs.

20. Apply grease (L2Y1 33247) to the wheel bearing inner race and drive shaft contact surface (Area A).

21. Install the wheel hub, steering knuckle component.

22. Install a new locknut (Crimped-fix Type) and stake it.

23. After installation, inspect the front wheel alignment and adjust it if necessary.

SUSPENSION

REAR SUSPENSION

COIL SPRING

REMOVAL & INSTALLATION

See Figures 172 through 174.

✳✳ WARNING

Installing/removing the coil spring is dangerous. The coil spring could fly off and cause serious injury or death, and damage the vehicle.

1. Remove the rear stabilizer.

2. Remove the rear lower arm outer bolt, rear coil spring, and upper spring seat rubber.

   a. Support the rear lower arm using a jack.

   b. Loosen the rear lower arm inner bolt.

   c. Remove the rear lower arm outer bolt.

To install:

3. Install in the reverse order of removal.

4. Position the jack under the rear lower arm and jack up slowly.

✳✳ WARNING

Installing the coil spring is dangerous. The coil spring could fly off and cause serious injury or death, and damage the vehicle.

5. Align the upper end of the rear coil spring with the step of the upper spring seat rubber.

6. Align the lower end of the rear coil spring with the step of the lower spring seat rubber.

7. Install the lower arm outer bolt.

8. Inspect the wheel alignment and adjust it if necessary.

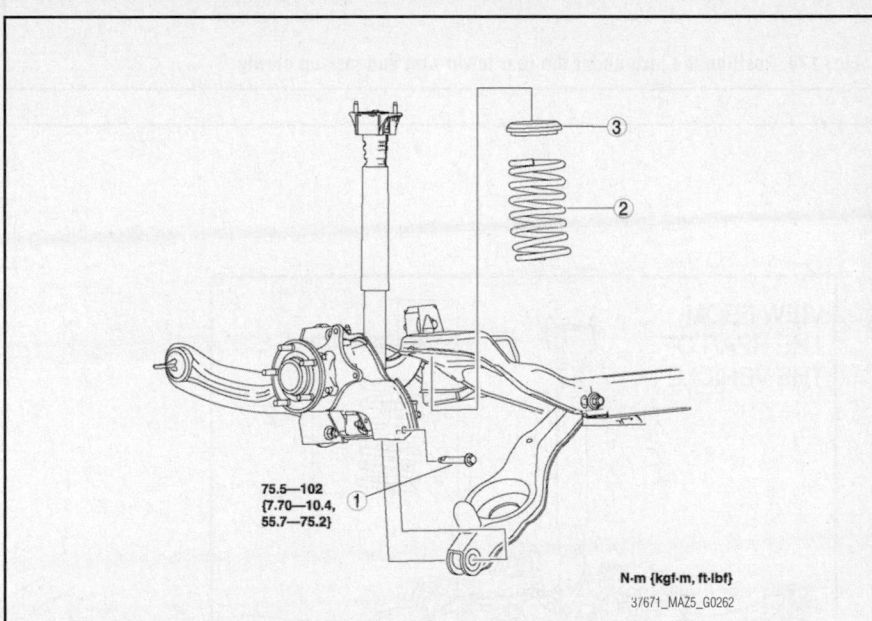

75.5—102
{7.70—10.4,
55.7—75.2}

N·m {kgf·m, ft-lbf}

Fig. 172 Remove the rear lower arm outer bolt (1), rear coil spring (2), and upper spring seat rubber (3)

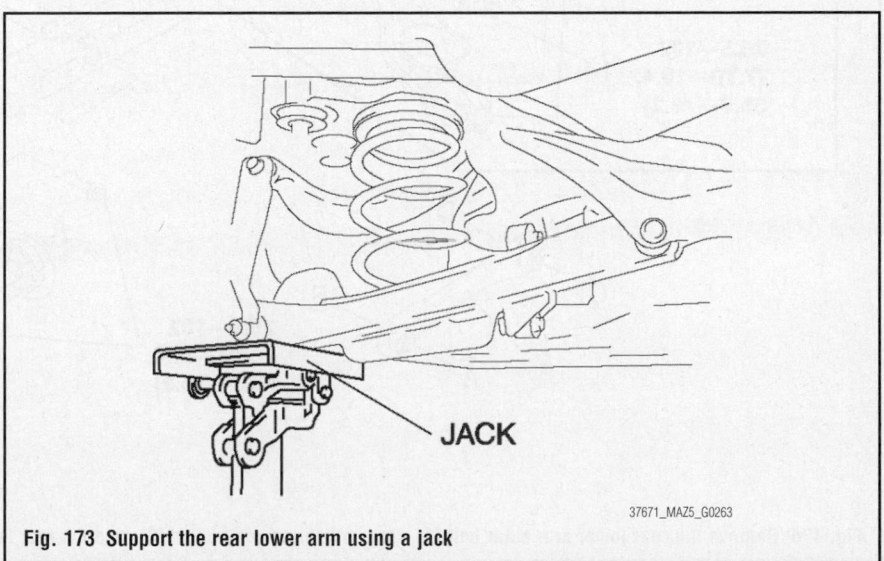

JACK

Fig. 173 Support the rear lower arm using a jack

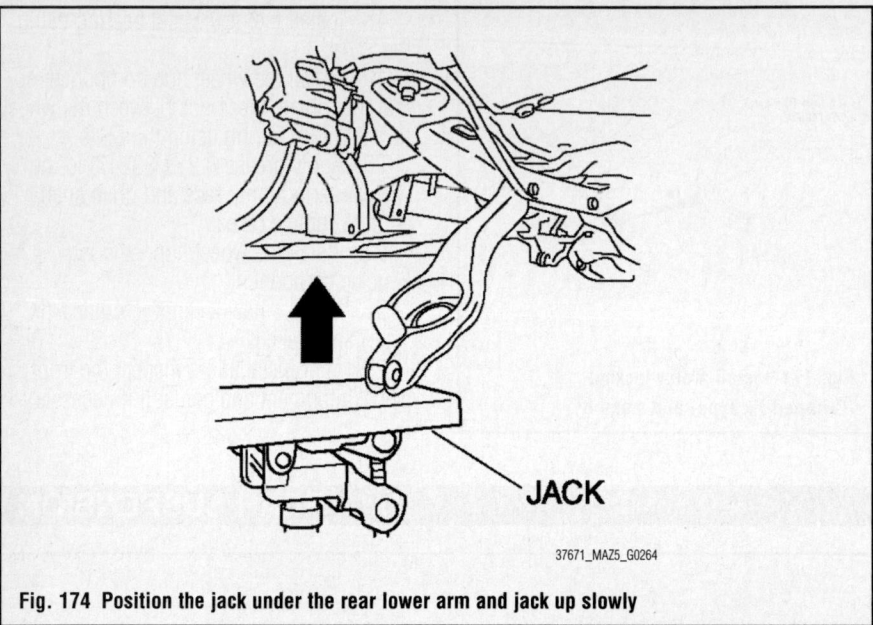

Fig. 174 Position the jack under the rear lower arm and jack up slowly

37671_MAZ5_G0264

CONTROL ARMS/LINKS

REMOVAL & INSTALLATION

Rear Lateral Link

See Figures 175 through 177.

1. Remove the rear stabilizer.
2. Remove the rear lower arm outer bolt, rear coil spring component, and the rear lateral link.

   a. Support the rear lower arm using a jack.

   b. Loosen the rear lower arm inner bolt.

   c. Remove the rear lower arm outer bolt.

To install:

3. Install in the reverse order of removal.
4. Install the rear lateral link with "FRONT" facing the vehicle front and the

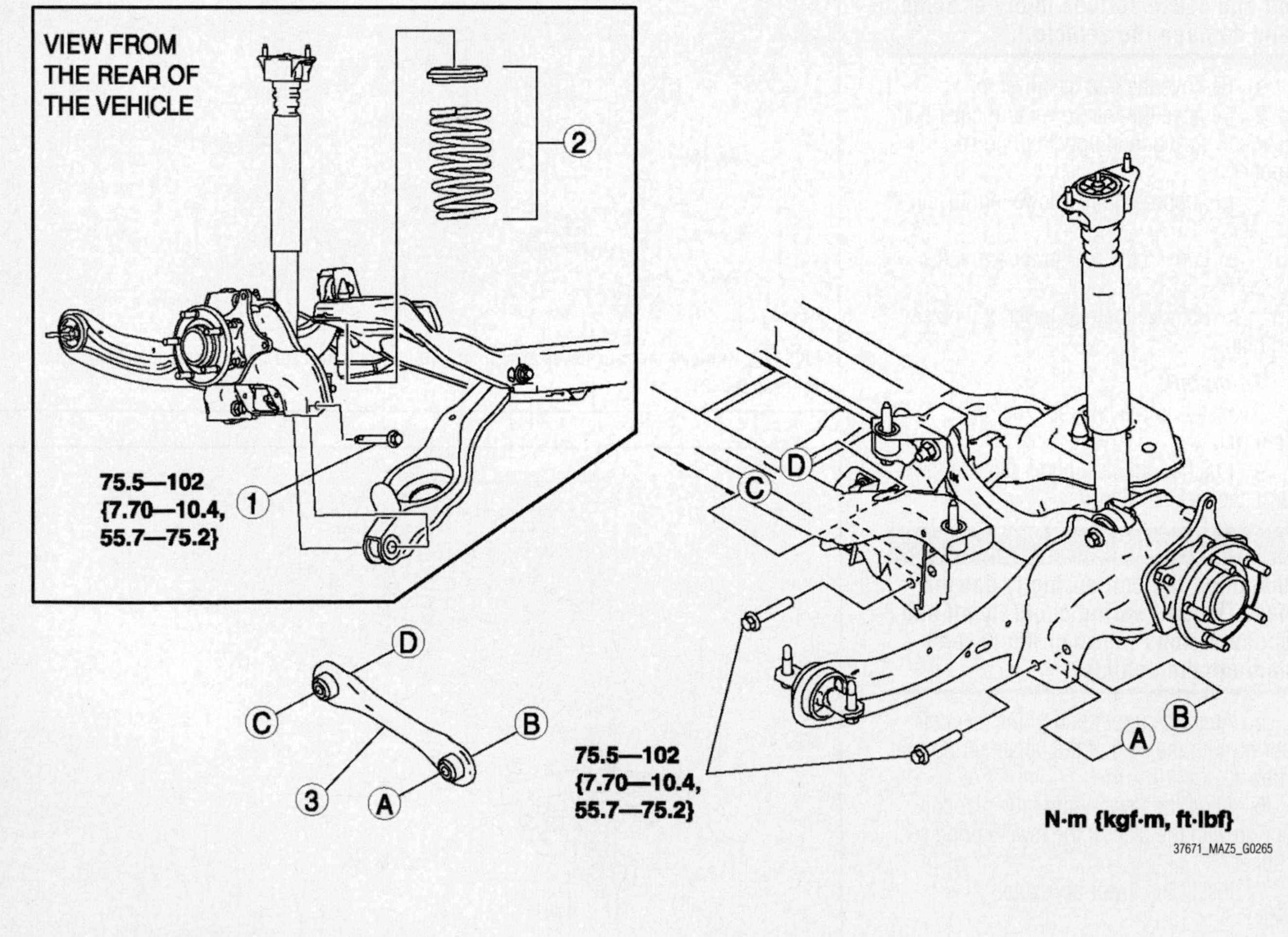

VIEW FROM THE REAR OF THE VEHICLE

75.5—102 {7.70—10.4, 55.7—75.2}

75.5—102 {7.70—10.4, 55.7—75.2}

N·m {kgf·m, ft·lbf}

37671_MAZ5_G0265

Fig. 175 Remove the rear lower arm outer bolt (1), rear coil spring component (2), and the rear lateral link (3)

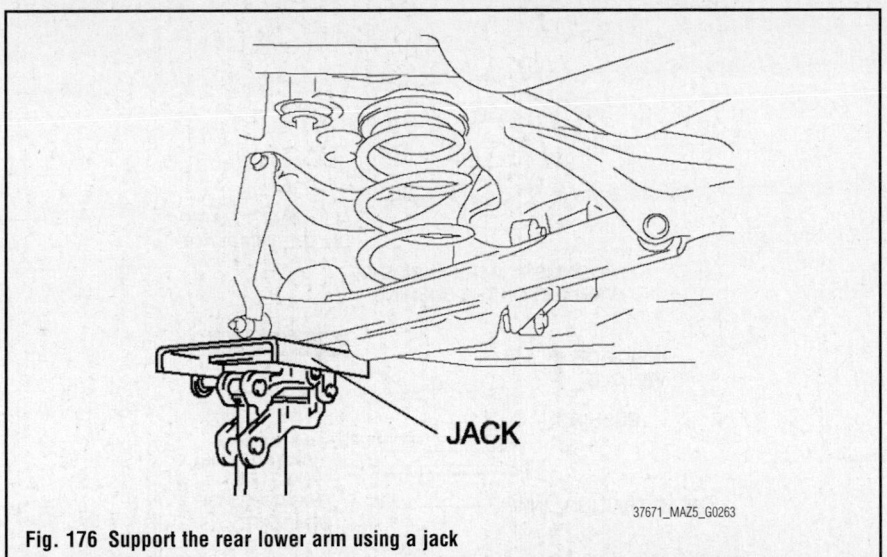

Fig. 176 Support the rear lower arm using a jack

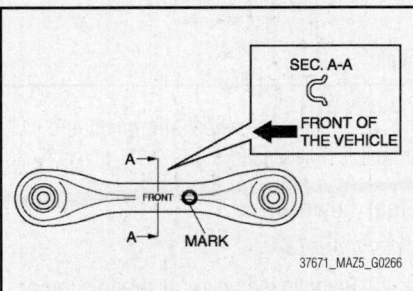

Fig. 177 Install the rear lateral link with "FRONT" facing the vehicle front and the painted mark facing the inside of the vehicle

painted mark facing the inside of the vehicle.

5. Inspect the wheel alignment and adjust it if necessary.

Rear Trailing Link

See Figures 178 through 183.

✳✳ CAUTION

Performing the following procedures without first removing the ABS wheel-speed sensor may possibly cause an open circuit in the wiring harness if it is pulled by mistake. Before performing the following procedures, disconnect the ABS wheel-speed sensor wiring harness connector (axle side) and attach the wiring harness to an appropriate place where it will not be pulled by mistake while servicing the vehicle.

1. Disconnect the ABS wheel-speed sensor wiring harness connector.
2. Disconnect the parking brake cable.
3. Remove the brake caliper compo-

nent. Suspend the caliper component using a cable and move aside.
4. Remove the brake pipe.
5. Remove the disc plate.
6. Remove the rear hub component.
7. Remove the dust cover.
8. Remove the rear coil spring component.
9. Remove the rear lateral link outer bolt.

a. Support the trailing link using a jack.

✳✳ CAUTION

Verify that the trailing link is securely supported by a jack. If the trailing link falls off, it can cause serious injury or death, and damage to the vehicle.

b. Remove the rear lateral link outer bolt.
10. Remove the rear upper arm outer bolt.
11. Remove the rear shock absorber lower bolt.

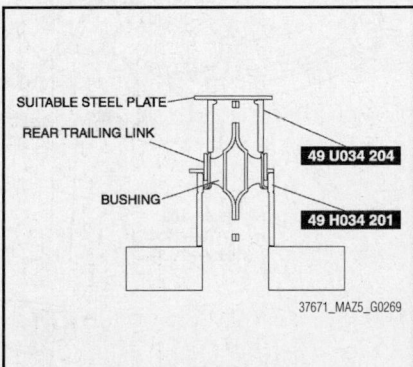

Fig. 178 Remove the bushing using the SSTs

12. Remove the rear trailing link.
13. Remove the bushing using the SSTs.

To install:

14. Install in the reverse order of removal.
15. Support the trailing link using a jack.
16. Tighten the trailing link front side bolts.
17. Install the bushing to the rear trailing arm link.

a. Verify that there is no dirt on the bushing of the rear trailing link, and clean it if it has become soiled.

b. Place alignment marks (outside/inside of vehicle) on the center line of the bushing as shown.

➥The alignment mark (outside of vehicle) is used for positioning before press-fitting. The alignment mark (inside of vehicle) is used for positioning verification after press-fitting and as an indication that press-fitting is completed.

c. Set the template included in the packaging for the bushing to the rear trailing link using tape or a similar material as shown.

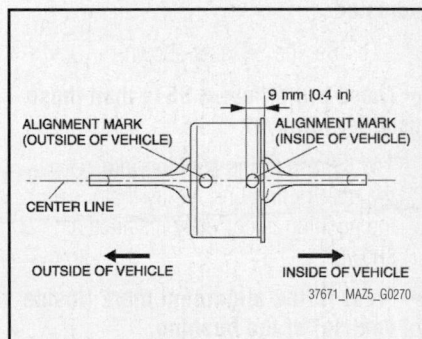

Fig. 179 Alignment marks used for positioning before press-fitting

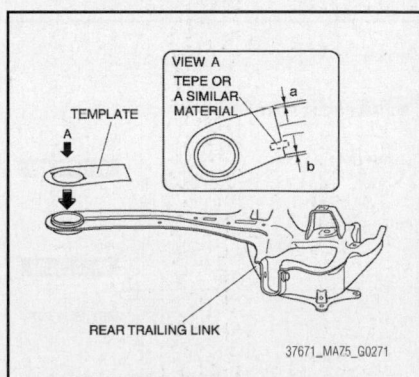

Fig. 180 Set the template for the bushing to the rear trailing link using tape

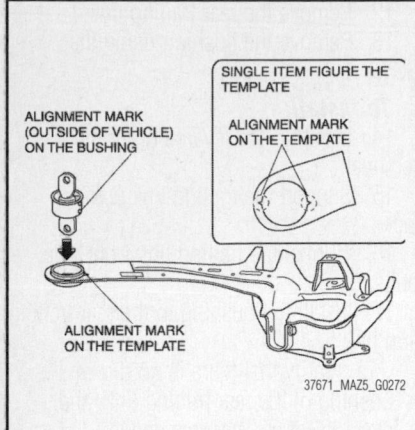

Fig. 181 Align the alignment mark on the template with the alignment mark (outside of vehicle) on the bushing as shown

➡ **If the rear trailing link shape does not correspond to the template shape, adjust the length of a and b shown so that they are equal in size.**

d. Align the alignment mark on the template with the alignment mark (outside of vehicle) on the bushing as shown.

➡ **There are 2 alignment marks on the template, and either can be adjusted.**

e. Set the SSTs as shown.

➡ **These are different SSTs than those used for removal.**

f. Press fit the bushing into the rear trailing link. Verify that the bushing is correctly installed as shown.

➡ **Press fit the alignment mark (inside of vehicle) of the bushing.**

g. If the bushing is not correctly installed, remove and re-install it.

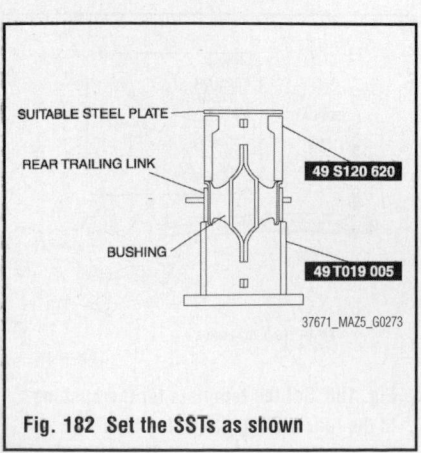

Fig. 182 Set the SSTs as shown

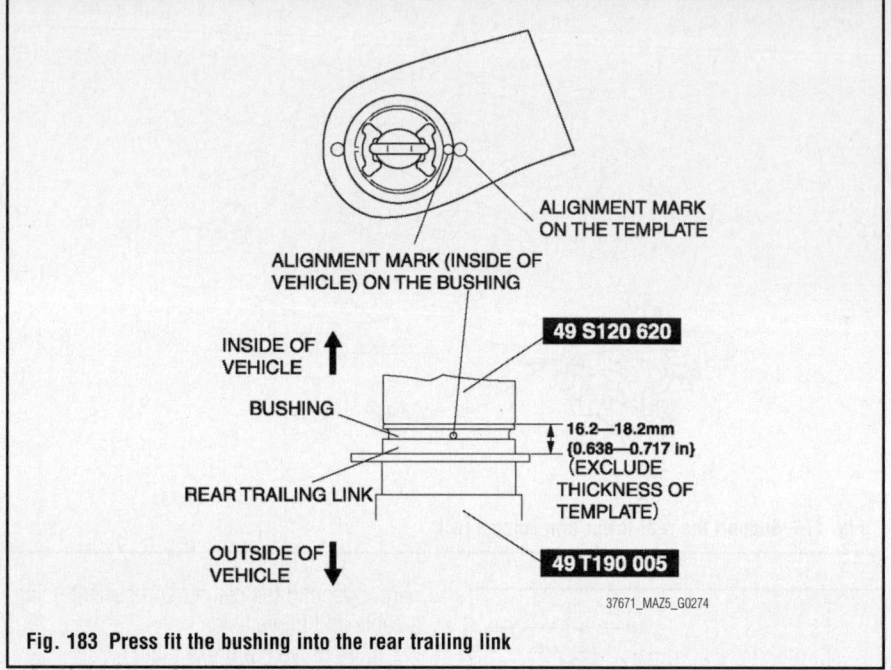

Fig. 183 Press fit the bushing into the rear trailing link

✳✳ CAUTION

If the bushing is re-press-fit, its holding strength decreases and its proper functioning cannot be guaranteed. Therefore, do not press-fit the bushing more than 3 times.

h. Remove the template from the rear trailing link.

18. Inspect the wheel alignment and adjust it if necessary.

Rear Lower Arm

See Figure 184.

1. Remove the rear coil spring component.
2. Remove the rear lower arm inner bolt.
3. Remove the rear lower arm.

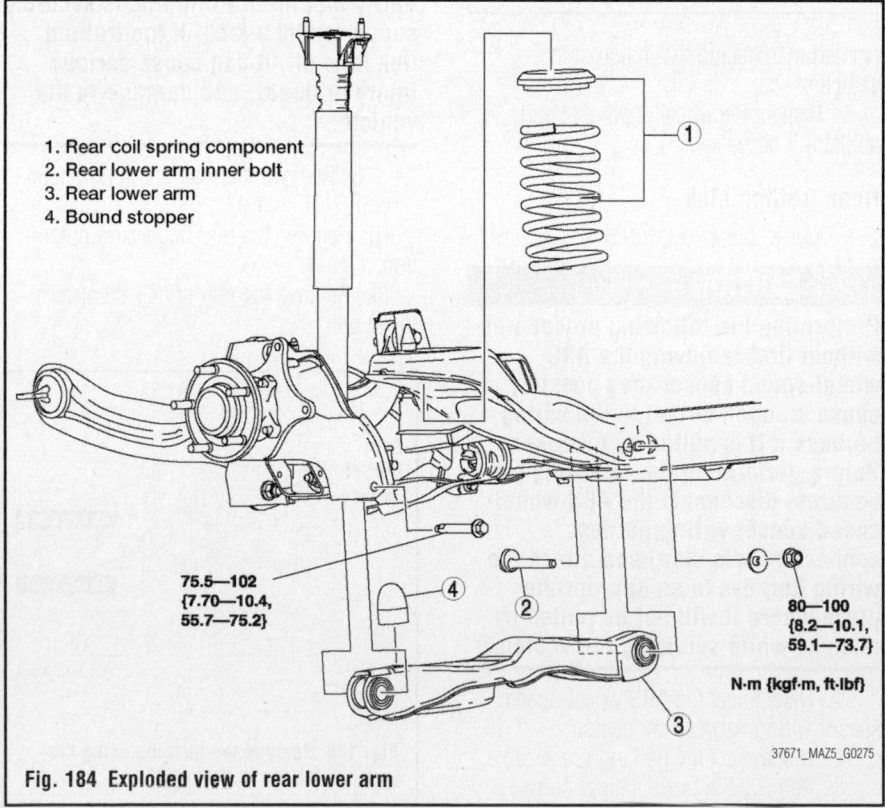

1. Rear coil spring component
2. Rear lower arm inner bolt
3. Rear lower arm
4. Bound stopper

75.5—102
{7.70—10.4,
55.7—75.2}

80—100
{8.2—10.1,
59.1—73.7}

N·m {kgf·m, ft·lbf}

Fig. 184 Exploded view of rear lower arm

4. Remove the bound stopper

5. Install in the reverse order of removal.

6. Inspect the wheel alignment and adjust it if necessary.

Rear Upper Arm

See Figures 185 and 186.

1. Remove the rear lower arm outer bolt.

a. Support the rear lower arm using a jack.

b. Loosen the rear lower arm inner bolt.

c. Remove the rear lower arm outer bolt.

2. Remove the rear coil spring component.

3. Remove the rear upper arm.

To install:

4. Install in the reverse order of removal.

5. Install the rear upper arm so that IN mark is facing toward the inside of the vehicle.

6. Inspect the wheel alignment and adjust it if necessary.

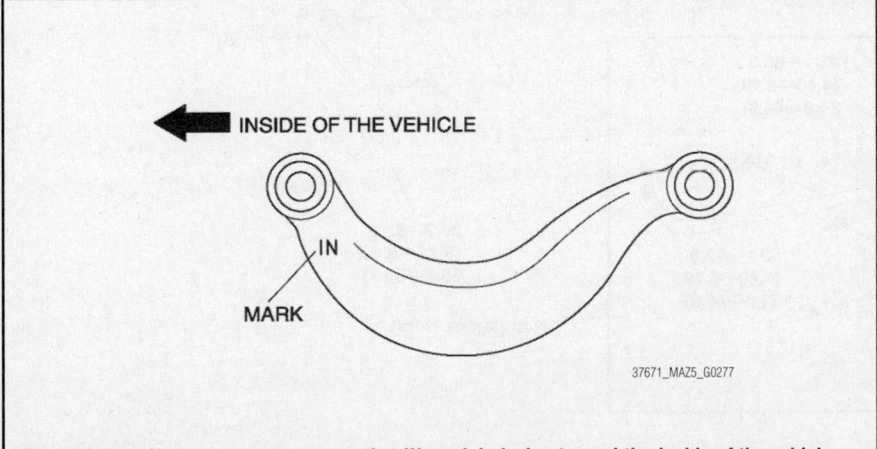

75.5—102
{7.70—10.4,
55.7—75.2}

75.5—102
{7.70—10.4,
55.7—75.2}

75.5—102
{7.70—10.4,
55.7—75.2}

N·m {kgf·m, ft·lbf}

37671_MAZ5_G0276

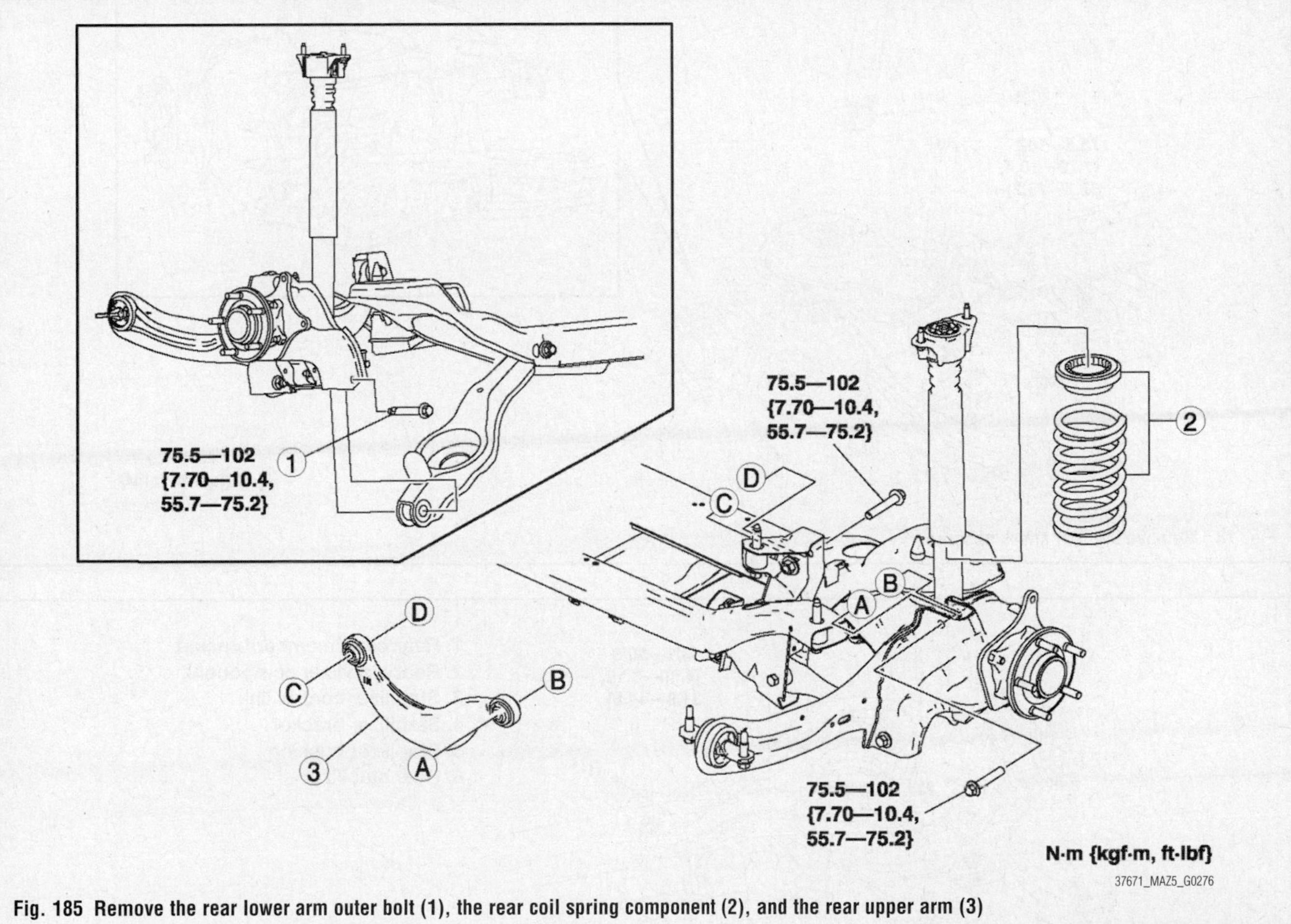

Fig. 185 Remove the rear lower arm outer bolt (1), the rear coil spring component (2), and the rear upper arm (3)

← INSIDE OF THE VEHICLE

IN

MARK

37671_MAZ5_G0277

Fig. 186 Install the rear upper arm so that IN mark is facing toward the inside of the vehicle

SHOCK ABSORBER

REMOVAL & INSTALLATION

See Figures 185 and 187.

1. Support the rear axle using a jack.

2. Remove the rear shock absorber.

3. Install the rear shock absorber.

STABILIZER BAR

REMOVAL & INSTALLATION

See Figures 188 through 190.

1. Remove the rear crossmember bracket.

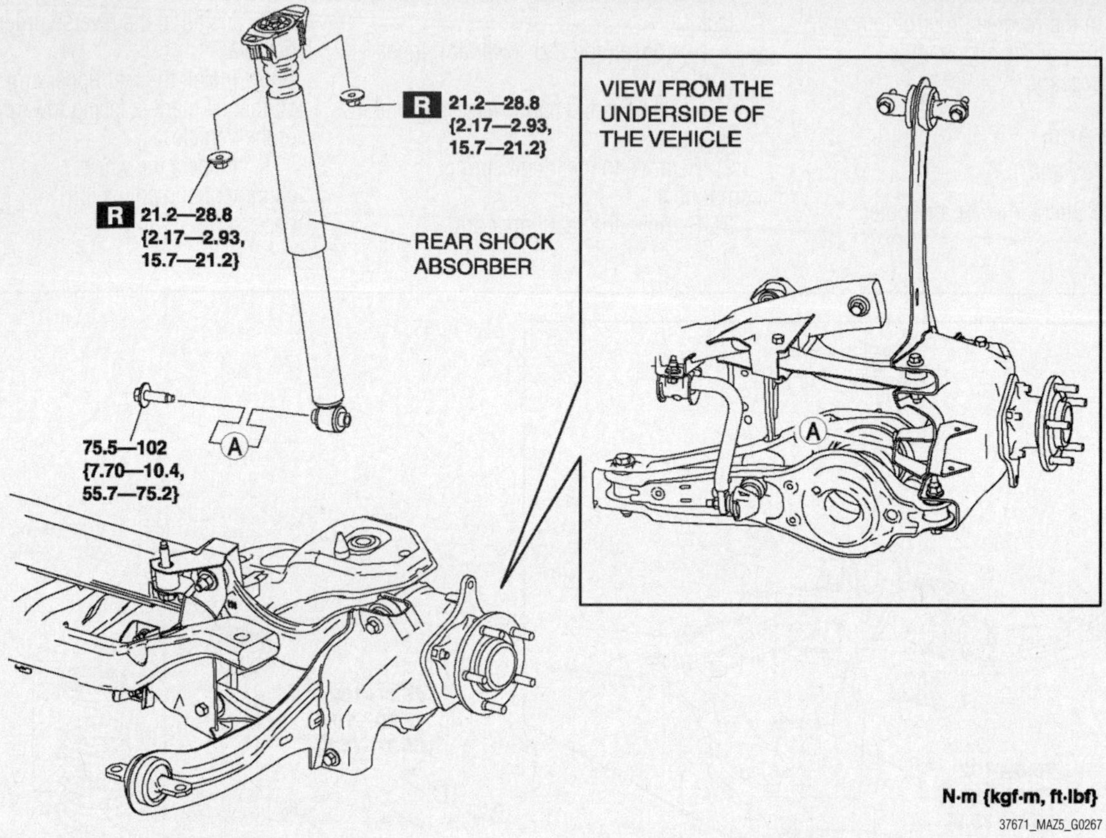

R 21.2—28.8
{2.17—2.93,
15.7—21.2}

R 21.2—28.8
{2.17—2.93,
15.7—21.2}

VIEW FROM THE
UNDERSIDE OF
THE VEHICLE

REAR SHOCK
ABSORBER

75.5—102
{7.70—10.4,
55.7—75.2}

N·m {kgf·m, ft·lbf}

37671_MAZ5_G0267

Fig. 187 Remove the rear shock absorber

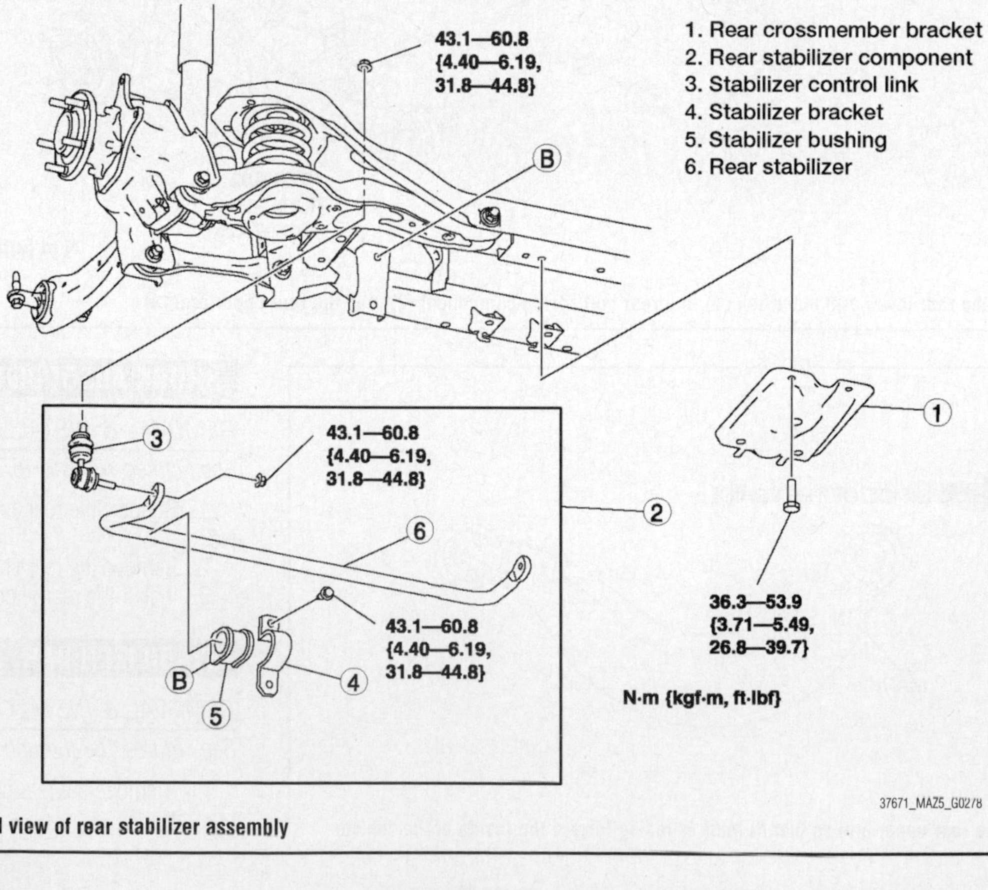

43.1—60.8
{4.40—6.19,
31.8—44.8}

1. Rear crossmember bracket
2. Rear stabilizer component
3. Stabilizer control link
4. Stabilizer bracket
5. Stabilizer bushing
6. Rear stabilizer

43.1—60.8
{4.40—6.19,
31.8—44.8}

43.1—60.8
{4.40—6.19,
31.8—44.8}

36.3—53.9
{3.71—5.49,
26.8—39.7}

N·m {kgf·m, ft·lbf}

37671_MAZ5_G0278

Fig. 188 Exploded view of rear stabilizer assembly

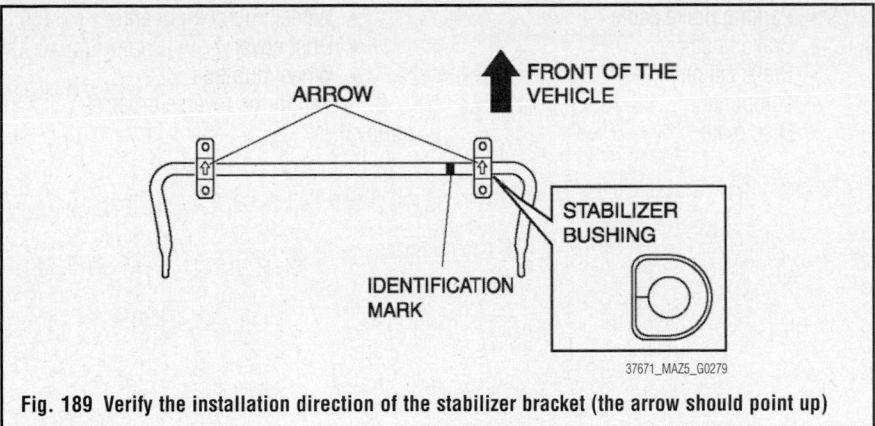

Fig. 189 Verify the installation direction of the stabilizer bracket (the arrow should point up)

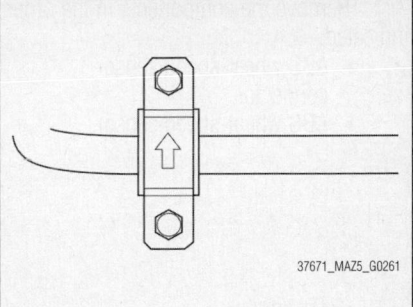

Fig. 190 With the arrow on the stabilizer bracket facing the vehicle front, tighten the bolts

2. Remove the rear stabilizer component.

3. Remove the stabilizer control link.

4. Remove the stabilizer bracket.

5. Remove the stabilizer bushing.

6. Remove the rear stabilizer.

To install:

7. Install in the reverse order of removal.

8. Verify the installation direction of the stabilizer bracket (the arrow should point up) by placing the stabilizer bar so that the identification mark faces the right side of the vehicle.

9. Install the stabilizer bush bracket at the position shown.

10. With the arrow on the stabilizer bracket facing the vehicle front, tighten the bolts.

WHEEL HUB & BEARING

REMOVAL & INSTALLATION

See Figure 191.

✲✲ CAUTION

Performing the following procedures without first removing the ABS wheel-speed sensor may possibly cause an open circuit in the wiring harness if it is pulled by mistake. Before performing the following procedures, disconnect the ABS wheel-speed sensor connector (body side) and attach the wiring harness to an appropriate place where it will not be pulled by mistake while servicing the vehicle.

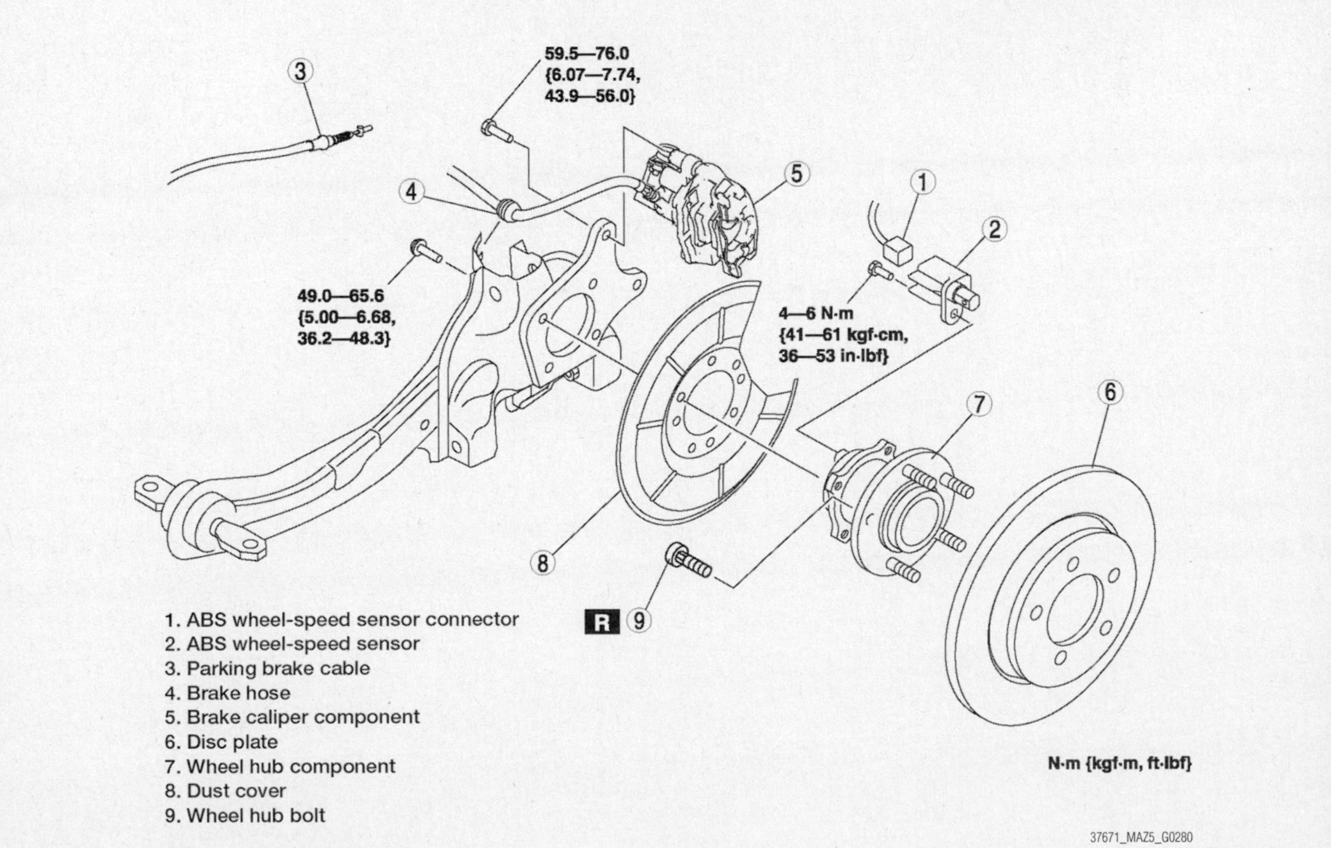

59.5—76.0
{6.07—7.74,
43.9—56.0}

49.0—65.6
{5.00—6.68,
36.2—48.3}

4—6 N·m
{41—61 kgf·cm,
36—53 in·lbf}

1. ABS wheel-speed sensor connector
2. ABS wheel-speed sensor
3. Parking brake cable
4. Brake hose
5. Brake caliper component
6. Disc plate
7. Wheel hub component
8. Dust cover
9. Wheel hub bolt

R 9

N·m {kgf·m, ft·lbf}

Fig. 191 Exploded view of rear hub assembly

1. Remove the components in the order indicated:
   - ABS wheel-speed sensor connector
   - ABS wheel-speed sensor

   - Parking brake cable
   - Brake hose
   - Brake caliper component
   - Disc plate

   - Wheel hub component
   - Dust cover
   - Wheel hub bolt

2. Install in the reverse order of removal.

MAZDA

Mazda6

SPECIFICATIONS AND MAINTENANCE CHARTS

ENGINE AND VEHICLE IDENTIFICATION

		Engine					Model Year	
Code ① ②	Liters (cc)	Cu. In.	Cyl.	Fuel Sys.	Engine Type	Eng. Mfg.	Code ③	Year
A	2.5 (2488)	151.8	4	EPI	DOHC	Mazda	9	2009
H	2.5 (2488)	151.8	4	EPI	DOHC	Mazda	A	2010
B	3.7 (3721)	227.0	6	SMPI	DOHC	Mazda		

EPI: Electronic Port Injection

DOHC: Double Over Head Cam

① 8th digit of the Vehicle Identification Number (VIN)

② Vin A engine made in mexico. Vin H engine made in Hiroshima

③ 10th digit of the Vehicle Identification Number (VIN)

37671_MAZ6_C0001

GENERAL ENGINE SPECIFICATIONS

Year	Model	Engine Displacement Liters (VIN)	Net Horsepower @ rpm	Net Torque @ rpm (ft. lbs.)	Bore x Stroke (in.)	Com-pression Ratio	Oil Pressure @ rpm
2009	Mazda6	2.5 (A, H)	①	②	3.50x3.94	9.7:1	57-94@3000
		3.7L (B)	272@6250	269@4250	3.76x3.41	10.3:1	45-90@2000
2010	Mazda6	2.5 (A, H)	①	②	3.50x3.94	9.7:1	57-94@3000
		3.7L (B)	272@6250	269@4250	3.76x3.41	10.3:1	45-90@2000

① 170@6000. 168@6000 for PZEV

② 167@4000. 166@4000 for PZEV.

37671_MAZ6_C0002

ENGINE TUNE-UP SPECIFICATIONS

Year	Engine Displacement Liters (VIN)	Spark Plug Gap (in.)	Ignition Timing (deg.) MT	Ignition Timing (deg.) AT	Fuel Pump (psi)	Idle Speed (rpm) MT	Idle Speed (rpm) AT	Valve Clearance In.①	Valve Clearance Ex.①
2009	2.5 (A, H)	49-57	12B	12B	54.4-65.2	600-750	650-750	0.009-0.011	0.011-0.012
	3.7L (B)	51-57	15B	15B	62.4-73.9	570-670	570-670	N/A	N/A
2010	2.5 (A, H)	49-57	12B	12B	54.4-65.2	600-750	650-750	0.009-0.011	0.011-0.012
	3.7L (B)	51-57	15B	15B	62.4-73.9	570-670	570-670	N/A	N/A

The label figures must be used if they differ from those in this chart.

B: Before top dead center

N/A: Not Applicable

NOTE: The Vehicle Emission Control Information label often reflects specification changes made during production.

① Engine cold

37671_MAZ6_C0003

CAPACITIES

Year	Model	Engine Displacement Liters (VIN)	Engine Oil with Filter (qts.)	Transmission (pts.)	Transaxle AWD (pts.)	Drive Axle Front (pts.)	Drive Axle Rear (pts.)	Fuel Tank (gal.)	Cooling System (qts.)
2009	Mazda6	2.5 (A, H)	5.3	①	N/A	N/A	N/A	18.5	9.9
		3.7L (B)	5.5	①	N/A	N/A	N/A	18.5	10.0
2010	Mazda6	2.5 (A, H)	5.3	①	N/A	N/A	N/A	18.5	9.9
		3.7L (B)	5.5	①	N/A	N/A	N/A	18.5	10.0

NOTE: All capacities are approximate. Add fluid gradually and check to be sure a proper fluid level is obtained.

N/A: Not Applicable

① Manual: 6.2. Automatic: 6.4 for the FS5A-EL, 14.8 for the AW6A-EL.

37671_MAZ6_C0004

FLUID SPECIFICATIONS

Year	Model	Engine Displacement Liters	Engine ID/VIN	Engine Oil	Auto. Trans.	Drive Axle	Power Steering Fluid	Brake Master Cylinder	Engine Coolant
2009	Mazda6	2.5	A, H	5W-20	①	80W-90	Dexron®II	DOT 3	FL 22
		3.7	B	5W-20	①	80W-90	Dexron®II	DOT 3	FL 22
2010	Mazda6	2.5	A, H	5W-20	①	80W-90	Dexron®II	DOT 3	FL 22
		3.7	B	5W-20	①	80W-90	Dexron®II	DOT 3	FL 22

DOT: Department Of Transpotation

N/A: Not Avilable

① Manual: SAE 75W-90 viscosity, API service GL-4 or GL-5 grade. Automatic: ATF M-V for the FS5A-EL, JWS3309 for the AW6A-EL.

37671_MAZ6_C0009

VALVE SPECIFICATIONS

Year	Engine Displacement Liters (VIN)	Seat Angle (deg.)	Spring Installed Height (in.)	Seat Contact With (in.)	Stem-to-Guide Clearance (in.) Intake	Stem-to-Guide Clearance (in.) Exhaust	Stem Diameter (in.) Intake	Stem Diameter (in.) Exhaust
2009	2.5 (A, H)	45	1.129	0.039-0.072	0.0010-0.0027	0.0012-0.0029	0.2154	0.2152
	3.7L (B)	45	NA	①	0.0008-0.0027	0.0018-0.0037	0.2352-0.2360	0.2343-0.2350
2010	2.5 (A, H)	45	1.129	0.039-0.072	0.0010-0.0027	0.0012-0.0029	0.2154	0.2152
	3.7L (B)	45	NA	①	0.0008-0.0027	0.0018-0.0037	0.2352-0.2360	0.2343-0.2350

NA: Not Applicable

① Intake: 0.043-0.055. Exhaust: 0.056-0.066

37671_MAZ6_C0005

CAMSHAFT AND BEARING SPECIFICATIONS CHART

All measurements are given in inches.

Year	Engine Displ. Liters	Engine ID/VIN	Journal Dia.	Brg. Oil Clearance	Shaft End-play	Runout	Journal Bore	Lobe Height Intake	Lobe Height Exhaust
2009	2.5	A, H	0.9827-0.9834	0.002-0.003-	0.004-0.0090	0.0012	N/A	1.6660-1.6710	1.6160-1.6210
	3.7	B	1.0640-1.0615	0.0010-0.0029	0.0010-0.0064	N/A	N/A	0.1890	0.1890
2010	2.5	A, H	0.9827-0.9834	0.002-0.003-	0.004-0.0090	0.0012	N/A	1.6660-1.6710	1.6160-1.6210
	3.7	B	1.0640-1.0615	0.0010-0.0029	0.0010-0.0064	N/A	N/A	0.1890	0.1890

N/A: Not Available

37671_MAZ6_C0006

CRANKSHAFT AND CONNECTING ROD SPECIFICATIONS

All measurements are given in inches.

Year	Engine Displacement Liters (VIN)	Crankshaft Main Brg. Journal Dia.	Crankshaft Main Brg. Oil Clearance	Crankshaft Shaft End-play	Crankshaft Thrust on No.	Connecting Rod Journal Diameter	Connecting Rod Oil Clearance	Connecting Rod Side Clearance
2009	2.5 (A, H)	2.0465-2.0472	0.007-0.018	0.0087-0.0177	N/A	N/A	0.0011-0.002	0.006-0.014
	3.7L (B)	2.4791-2.4799	0.0010-0.0017	0.00434-0.00913	N/A	N/A	0.0012-0.0025	0.004-0.011
2010	2.5 (A, H)	2.0465-2.0472	0.007-0.018	0.0087-0.0177	N/A	N/A	0.0011-0.002	0.006-0.014
	3.7L (B)	2.4791-2.4799	0.0010-0.0017	0.00434-0.00913	N/A	N/A	0.0012-0.0025	0.004-0.011

N/A: Not Avilable

37671_MAZ6_C0007

PISTON AND RING SPECIFICATIONS

All measurements are given in inches.

Year	Engine Displacement Liters (VIN)	Piston Clearance	Ring Gap Top Compression	Ring Gap Bottom Compression	Ring Gap Oil Control	Ring Side Clearance Top Compression	Ring Side Clearance Bottom Compression	Ring Side Clearance Oil Control
2009	2.5 (A, H)	0.0010-0.0017	0.0056-0.0094	0.0134-0.0173	0.0060-0.0157	0.002-0.003	0.0012-0.0027	0.0008-0.0023
	3.7L (B)	0.0004-0.0008	0.0040-0.0090	0.0011-0.0160	0.0060-0.0250	N/A	N/A	N/A
2010	2.5 (A, H)	0.0010-0.0017	0.0056-0.0094	0.0134-0.0173	0.0060-0.0157	0.002-0.003	0.0012-0.0027	0.0008-0.0023
	3.7L (B)	0.0004-0.0008	0.0040-0.0090	0.0011-0.0160	0.0060-0.0250	N/A	N/A	N/A

N/A: Not Avilable

37671_MAZ6_C0008

TORQUE SPECIFICATIONS
All readings in ft. lbs.

Year	Engine Displacement Liters (VIN)	Cylinder Head Bolts	Main Bearing Bolts	Rod Bearing Bolts	Crankshaft Damper Bolts	Flywheel Bolts	Manifold		Spark Plugs	Oil Pan Drain Plug
							Intake	Exhaust		
2009	2.5 (A, H)	①	N/A	N/A	②	80-85	13-16	32 47	8-10	23-30
	3.7L (B)	③	④	⑤	⑥	80-85	⑦	17-20	⑧	17-22
2010	2.5 (A, H)	①	N/A	N/A	②	80-85	13-16	32-47	8-10	23-30
	3.7L (B)	③	④	⑤	⑥	80-85	⑦	17-20	⑧	17-22

NA: Not Available

① Step 1: 27-97 inch lbs.
   Step 2: Tighten 116-150 inch. lbs
   Step 3: Tighten 32-34 ft. lbs.
   Step 4: Tighten 92 degrees
   Step 5: Tighten an additional 92 degrees

② 71-77 ft. lbs. plus 87-90 degrees

③ Step 1: Tighten 23-28
   Step 2: Tighten 85-95 degrees
   Step 3: Loosen 360 degrees
   Step 4: Tighten 23-28
   Step 5: Tighten 85-95 degrees
   Step 6: Tighten 85-95 degrees

④ Lower bolts 1-8: Tighten 16-20
   Lower bolts 9-16: Tighten 27-31
   Lower bolts 1-16: Tighten 85-95 degrees
   Lower bolts 17-22: Tighten 15-22
   Verify crankshaft rotates freely

⑤ Step 1: Tighten 15-18
   Step 2: Tighten 29-33
   Step 3: Tighten 90-120 degrees

⑥ Step 1: Tighten 88.5
   Step 2: Loosen 360 degrees
   Step 3: Tighten 35-39
   Step 4: Tighten 85-95 degrees

⑦ 80-97 inch lbs.

⑧ 80-177 inch lbs.

37671_MAZ6_C0010

WHEEL ALIGNMENT

Year	Model		Caster		Camber		Toe-in (in.)
			Range (+/-Deg.)	Preferred Setting (Deg.)	Range (+/-Deg.)	Preferred Setting (Deg.)	
2009	Mazda6	F	①	①	②	②	③
		R	N/A	N/A	④	④	③
2010	Mazda6	F	①	①	②	②	③
		R	N/A	N/A	④	④	③

NA: Not Applicable

① 2.5L engine: fuel tank empty 3 degrees 37'. Fuel tank full 3 degrees 42'. Difference between R & L must not exceed 1 degree 30'.
   3.7L engine: fuel tank empty 3 degrees 34'. Fuel tank full 3 degrees 40'. Difference between R & L must not exceed 1 degree 30'.

② 2.5L engine: fuel tank empty 0 degrees 0.9'. Fuel tank full 0 degrees 15'. Difference between R & L must not exceed 1 degree 30'.
   3.7L engine: fuel tank empty 0 degrees 10'. Fuel tank full 0 degrees 16'. Difference between R & L must not exceed 1 degree 30'.

③ Tire: 0.08 +/- 0.16. Rim: 0.04 +/- 0.12.

④ Fuel tank empty 0 degrees 48'. Fuel tank full 0 degrees 57'. Difference between R & L must not exceed 1 degree 30'.

37671_MAZ6_C0011

TIRE, WHEEL AND BALL JOINT SPECIFICATIONS

Year	Model	OEM Tires Standard	OEM Tires Optional	Tire Pressures (psi) Front	Tire Pressures (psi) Rear	Wheel Size	Ball Joint Inspection	Lug Nut (ft. lbs.)
2009	Mazda6	P205/65R16 94H	N/A	32	32	16 × 6.5J	N/A	65-87
		P215/55R17 93V	N/A	32	32	17 × 7.0J	N/A	6587
		P235/45R18 94W	N/A	32	32	18 × 8.0J	N/A	65-87
2010	Mazda6	P205/65R16 94H	N/A	32	32	16 × 6.5J	N/A	65-87
		P215/55R17 93V	N/A	32	32	17 × 7.0J	N/A	6587
		P235/45R18 94W	N/A	32	32	18 × 8.0J	N/A	65-87

OEM: Original Equipment Manufacturer

PSI: Pounds Per Square Inch

NA: Not Applicable

37671_MAZ6_C0012

BRAKE SPECIFICATIONS
All measurements in inches unless noted

Year	Model		Brake Disc Original Thickness	Brake Disc Minimum Thickness	Brake Disc Maximum Runout	Brake Drum Original Inside Diameter	Brake Drum Max. Wear Limit	Brake Drum Maximum Machine Diameter	Minimum Lining Thickness	Brake Caliper Bracket Bolts (ft. lbs.)	Brake Caliper Mounting Bolts (ft. lbs.)
2009	Mazda6	F	N/A	0.910	0.002	N/A	N/A	N/A	0.080	N/A	N/A
		R	N/A	0.300	0.002	N/A	N/A	N/A	0.079	N/A	N/A
2010	Mazda6	F	N/A	0.910	0.002	N/A	N/A	N/A	0.080	N/A	N/A
		R	N/A	0.300	0.002	N/A	N/A	N/A	0.079	N/A	N/A

NA: Not Applicable

F: Front

R: Rear

37671_MAZ6_C0013

SCHEDULED MAINTENANCE INTERVALS
Mazda6

TO BE SERVICED	TYPE OF SERVICE	VEHICLE MILEAGE INTERVAL (x1000)												
		7.5	15	22.5	30	37.5	45	52.5	60	67.5	75	82.5	90	97.5
Engine oil & filter	R	✓	✓	✓	✓	✓	✓	✓	✓	✓	✓	✓	✓	✓
Air cleaner element	R					✓					✓			
Engine coolant- F22 type	R	**120,000 or 10 years, 60,000 thereafter**												
Engine coolant- except F22 type	R								✓					
Spark plugs- 2.5L	R								✓					
Spark plugs- 3.7L	R	**every 100,000 miles**												
Bolts & nuts on chassis & body	S/I				✓				✓				✓	
Cabin air filter	R	**every 25,000 miles**												
Brake lines, hoses & connections	S/I				✓				✓				✓	
Cooling system	S/I				✓				✓				✓	
Disc brakes	S/I		✓		✓				✓				✓	
Drive belts- 2.5L	S/I				✓				✓		✓			
Drive belts- 3.7L	R	**every 150,000 miles**												
Drive shaft dust boots	S/I				✓				✓				✓	
Exhaust system heat shield	S/I						✓		✓				✓	
Front & rear ball joints	S/I				✓				✓				✓	
Fuel lines & hoses	S/I				✓				✓				✓	
Steering operation & linkages	S/I				✓				✓				✓	
Tire rotation	R	✓	✓	✓	✓	✓	✓	✓	✓	✓	✓	✓	✓	✓
Valve clearance- 2.5L	S/I										✓			
Hose & tube for emission	S/I								✓					

R: Replace   S/I: Service or Inspect

FREQUENT OPERATION MAINTENANCE (SEVERE SERVICE)

If a vehicle is operated under any of the following conditions it is considered severe service

- Extremely dusty areas.

- 50% or more of the vehicle operation is in 32°C (90°F) or higher temperatures, or constant operation in temperatures below 0°C (32°F).

- Prolonged idling (vehicle operation in stop and go traffic).

- Frequent short running periods (engine does not warm to normal operating temperatures).

- Police, taxi, delivery usage or trailer towing usage.

Oil & oil filter: change every 3000 miles.

Air cleaner element: service or inspect every 15,000 miles.

Automatic transaxle fluid: service or inspect every 15,000 miles.

Bolts & nuts on chassis & body: tighten every 15,000 miles.

Disc brakes: service or inspect every 15,000 miles.

37671_MAZ6_C0014

BRAKES — INFORMATION AND PRECAUTIONS

ANTI-LOCK SYSTEMS

- Certain components within the ABS system are not intended to be serviced or repaired individually.
- Do not use rubber hoses or other parts not specifically specified for and ABS system. When using repair kits, replace all parts included in the kit. Partial or incorrect repair may lead to functional problems and require the replacement of components.
- Lubricate rubber parts with clean, fresh brake fluid to ease assembly. Do not use shop air to clean parts; damage to rubber components may result.
- Use only DOT 3 brake fluid from an unopened container.
- If any hydraulic component or line is removed or replaced, it may be necessary to bleed the entire system.
- A clean repair area is essential. Always clean the reservoir and cap thoroughly before removing the cap. The slightest amount of dirt in the fluid may plug an orifice and impair the system function.

Perform repairs after components have been thoroughly cleaned; use only denatured alcohol to clean components. Do not allow ABS components to come into contact with any substance containing mineral oil; this includes used shop rags.

- The Anti-Lock control unit is a microprocessor similar to other computer units in the vehicle. Ensure that the ignition switch is **OFF** before removing or installing controller harnesses. Avoid static electricity discharge at or near the controller.
- If any arc welding is to be done on the vehicle, the control unit should be unplugged before welding operations begin.

DISC AND DRUM SYSTEMS

> ❋❋ **CAUTION**
>
> **Dust and dirt accumulating on brake parts during normal use may contain asbestos fibers from production or aftermarket brake linings. Breathing excessive concentrations of asbestos fibers can cause serious bodily harm. Exercise care when servicing brake parts. Do not sand or grind brake lining unless equipment used is designed to contain the dust residue. Do not clean brake parts with compressed air or by dry brushing. Cleaning should be done by dampening the brake components with a fine mist of water, then wiping the brake components clean with a dampened cloth. Dispose of cloth and all residue containing asbestos fibers in an impermeable container with the appropriate label. Follow practices prescribed by the Occupational Safety and Health Administration (OSHA) and the Environmental Protection Agency (EPA) for the handling, processing, and disposing of dust or debris that may contain asbestos fibers.**

BRAKES — BLEEDING THE BRAKE SYSTEM

BLEEDING PROCEDURE

BLEEDING PROCEDURE

> ❋❋ **CAUTION**
>
> **Brake fluid will damage painted surfaces. Be careful not to spill any on painted surfaces. If it is spilled, wipe it off immediately.**

➡**Keep the fluid level in the reserve tank at 3/4 full or more during the air bleeding. Begin air bleeding with the brake caliper that is furthest from the master cylinder. Use DOT-3 brake fluid.**

1. Before servicing the vehicle, refer to the Precautions Section.
2. Remove the bleeder cap on the brake caliper, and attach a vinyl tube to the bleeder screw.
3. Place the other end of the vinyl tube in a clear container and fill the container with fluid during air bleeding.
4. Working with two people, one should pump the brake pedal several times and depress and hold the pedal down.
5. While the brake pedal is depressed, the other should loosen the bleeder screw using the SST, drain out any fluid containing air bubbles, and tighten the bleeder screw.
6. Repeat until no air bubbles are seen.
7. Perform air bleeding as described in the above procedures for all brake calipers.

BLEEDING THE ABS SYSTEM

See Figure 1.

> ❋❋ **CAUTION**
>
> **Brake fluid will damage painted surfaces. Be careful not to spill any on painted surfaces. If it is spilled, wipe it off immediately.**

➡**Keep the fluid level in the reserve tank at 3/4 full or more during the air bleeding. Begin air bleeding with the brake caliper that is furthest from the master cylinder. Use DOT-3 brake fluid.**

1. Before servicing the vehicle, refer to the Precautions Section.
2. Remove the bleeder cap on the brake caliper, and attach a vinyl tube to the bleeder screw.
3. Place the other end of the vinyl tube in a clear container and fill the container with fluid during air bleeding.

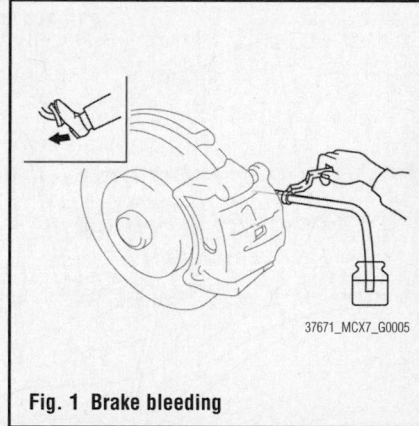

37671_MCX7_G0005

Fig. 1 Brake bleeding

4. Working with two people, one should pump the brake pedal several times and depress and hold the pedal down.
5. While the brake pedal is depressed, the other should loosen the bleeder screw using the SST, drain out any fluid containing air bubbles, and tighten the bleeder screw.
6. Repeat until no air bubbles are seen.
7. Perform air bleeding as described in the above procedures for all brake calipers.

SPEED SENSORS

REMOVAL & INSTALLATION

Front

See Figure 2.

1. Before servicing the vehicle, refer to the Precautions Section.

➡If working near and/or around the SRS system and components, be sure to disable the SRS system. Tape the negative battery cable with insulating tape. Always disconnect the negative battery cable first.

✳ CAUTION

To avoid personal injury when working on vehicles equipped with an air bag, the negative battery cable must be disconnected and at least one minute must elapse before working on the system. Failure to do so may result in deployment of the air bag.

2. Disconnect the negative battery cable. Tape the cable with insulating tape.

➡When disconnecting the cable, some systems need to be initialized after the cable is reconnected. You will need the Mazda diagnostic scan tool or equivalent. Follow the directions on the tool.

3. Raise and safely support the vehicle.
4. Remove the tire and wheel assembly.
5. Disconnect the electrical connector.
6. Remove the senor from its mounting.

To install:

➡Be sure to use new fasteners, as required.

7. Installation is the reverse of the removal procedure.
8. Use the Mazda diagnostic scan tool, or equivalent and reprogram the required systems.

Rear

See Figure 3.

1. Before servicing the vehicle, refer to the Precautions Section.

➡If working near and/or around the SRS system and components, be sure to disable the SRS system. Tape the negative battery cable with insulating tape. Always disconnect the negative battery cable first.

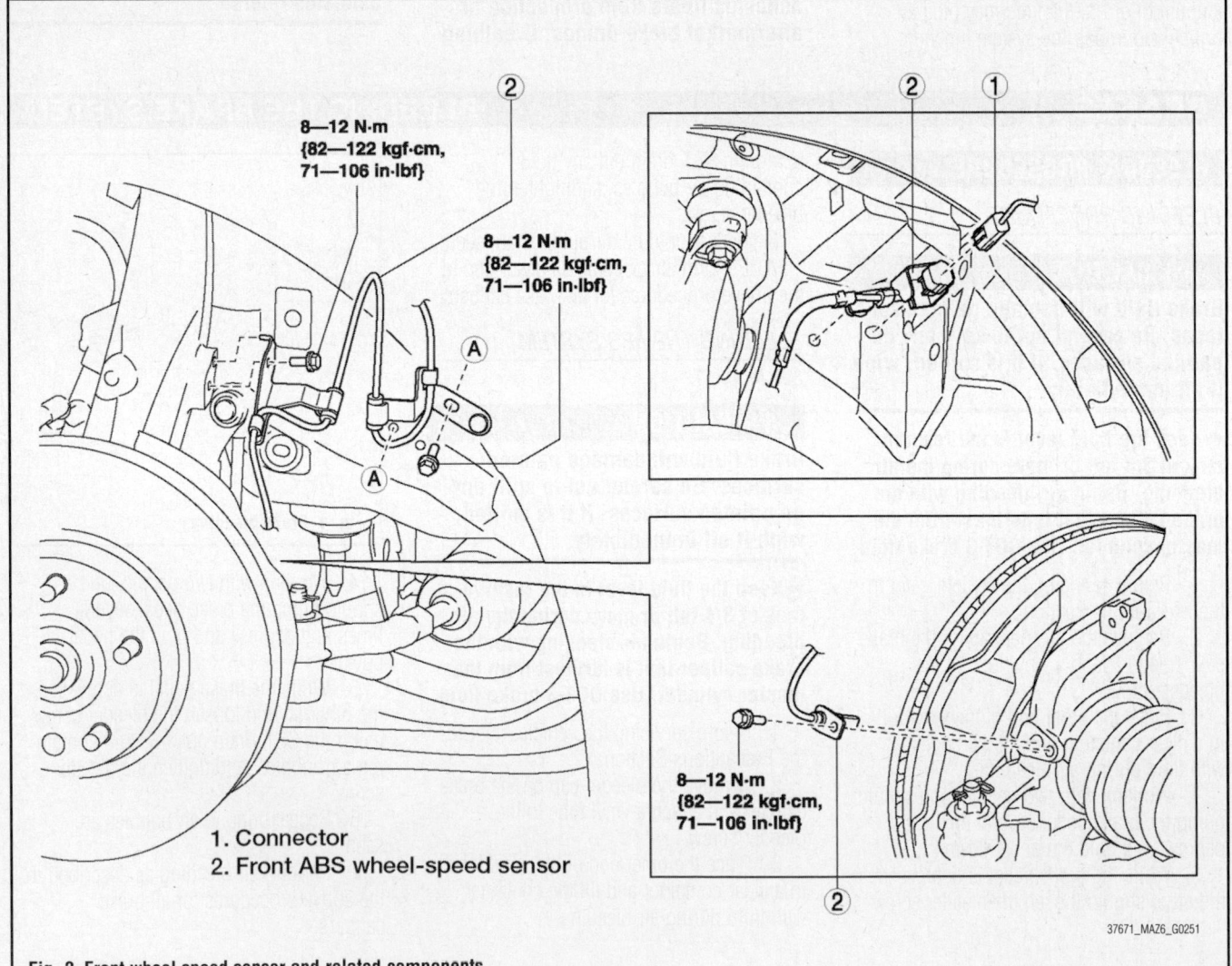

8—12 N·m
{82—122 kgf·cm,
71—106 in·lbf}

8—12 N·m
{82—122 kgf·cm,
71—106 in·lbf}

8—12 N·m
{82—122 kgf·cm,
71—106 in·lbf}

1. Connector
2. Front ABS wheel-speed sensor

Fig. 2 Front wheel speed sensor and related components

37671_MAZ6_G0251

✳ CAUTION

To avoid personal injury when working on vehicles equipped with an air bag, the negative battery cable must be disconnected and at least one minute must elapse before working on the system. Failure to do so may result in deployment of the air bag.

2. Disconnect the negative battery cable. Tape the cable with insulating tape.

➡ When disconnecting the cable, some systems need to be initialized after the cable is reconnected. You will need the Mazda diagnostic scan tool or equivalent. Follow the directions on the tool.

3. Remove the rear seat cushion. Remove the rear seat side back.

4. Remove the trunk end trim. Remove the trunk side trim.

5. Disconnect the sensor electrical connector.

6. Press the sensor tabs and press out the sensor toward the outside of the vehicle.

7. Raise and safely support the vehicle.

8. Remove the tire and wheel assembly.

9. Remove the retaining bolt and nut.

10. Remove the sensor from its mounting.

To install:

➡ Be sure to use new fasteners, as required.

11. Installation is the reverse of the removal procedure.

12. Use the Mazda diagnostic scan tool, or equivalent and reprogram the required systems.

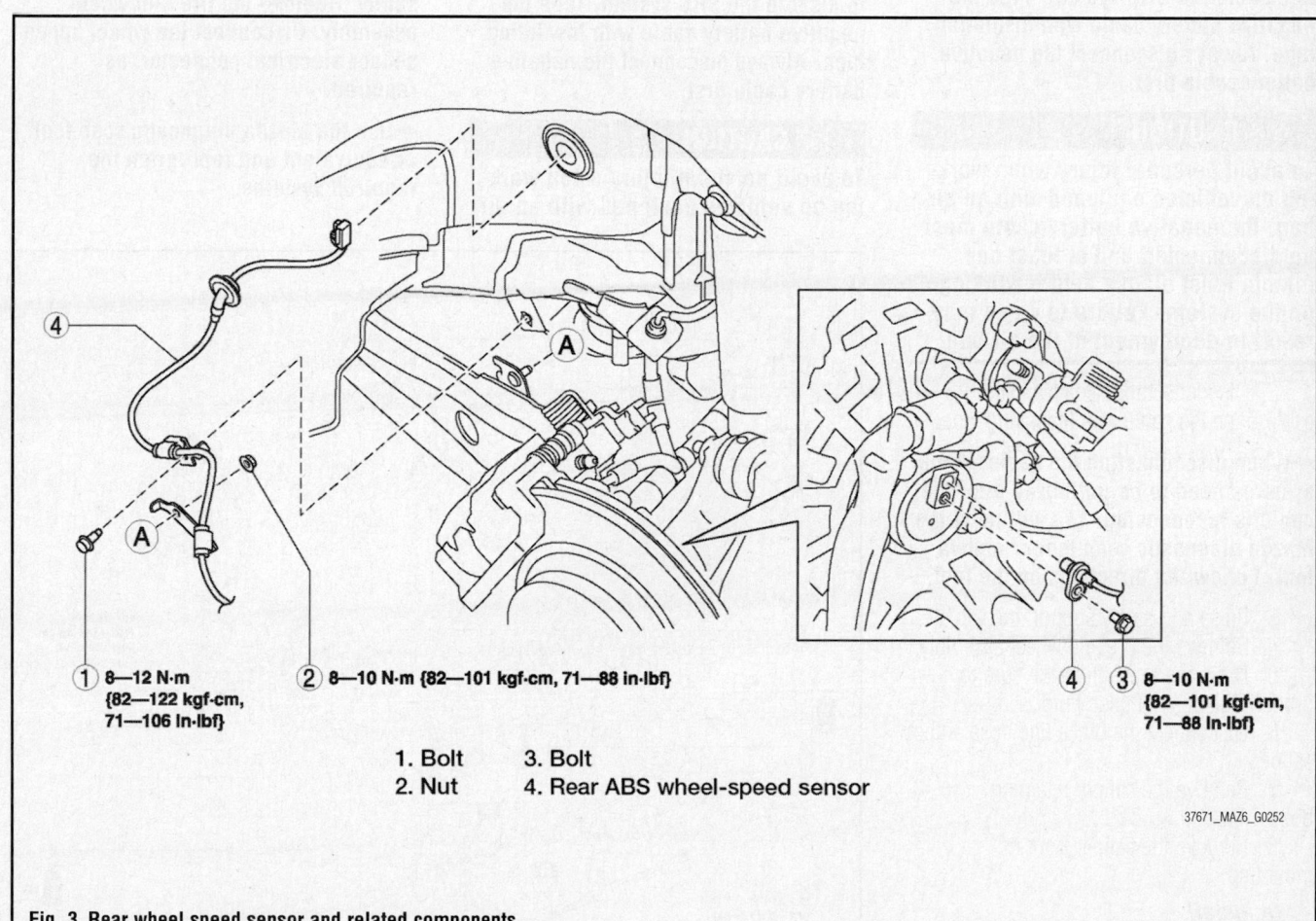

① 8—12 N·m
{82—122 kgf·cm,
71—106 in·lbf}

② 8—10 N·m {82—101 kgf·cm, 71—88 in·lbf}

③ 8—10 N·m
{82—101 kgf·cm,
71—88 in·lbf}

1. Bolt   3. Bolt
2. Nut   4. Rear ABS wheel-speed sensor

37671_MAZ6_G0252

Fig. 3 Rear wheel speed sensor and related components

BRAKE CALIPER

REMOVAL & INSTALLATION

At this time the manufacturer does not provide removal and installation procedures for this component. The following procedure is a guideline and may differ from the vehicle you are servicing.

1. Before servicing the vehicle, refer to the Precautions Section.

➡️ If working near and/or around the SRS system and components, be sure to disable the SRS system. Tape the negative battery cable with insulating tape. Always disconnect the negative battery cable first.

❊❊ CAUTION

To avoid personal injury when working on vehicles equipped with an air bag, the negative battery cable must be disconnected and at least one minute must elapse before working on the system. Failure to do so may result in deployment of the air bag.

2. Disconnect the negative battery cable. Tape the cable with insulating tape.

➡️ When disconnecting the cable, some systems need to be initialized after the cable is reconnected. You will need the Mazda diagnostic scan tool or equivalent. Follow the directions on the tool.

3. Raise and safely support the vehicle.
4. Remove the tire and wheel assembly.
5. Drain the brake fluid. Be sure to properly dispose of used fluid.
6. Disconnect the brake line hose at the caliper.
7. Remove the caliper retaining bolts.
8. Remove the caliper from its mounting.

To install:

➡️ Be sure to use new fasteners, as required.

9. Installation is the reverse of the removal procedure.
10. Be sure to tighten the caliper retaining bolts to specification.
11. Be sure to fill the brake system with the proper grade and type fluid.
12. Bleed the brakes. Correct fluid level, as required.
13. Use the Mazda diagnostic scan tool, or equivalent and reprogram the required systems.

DISC BRAKE PADS

REMOVAL & INSTALLATION

See Figures 4 and 5.

At this time the manufacturer does not provide removal and installation procedures for this component, refer to the illustration as required.

1. Before servicing the vehicle, refer to the Precautions Section.

➡️ If working near and/or around the SRS system and components, be sure to disable the SRS system. Tape the negative battery cable with insulating tape. Always disconnect the negative battery cable first.

❊❊ CAUTION

To avoid personal injury when working on vehicles equipped with an air bag, the negative battery cable must be disconnected and at least one minute must elapse before working on the system. Failure to do so may result in deployment of the air bag.

➡️ When disconnecting the cable, some systems need to be initialized after the cable is reconnected. You will need the Mazda diagnostic scan tool or equivalent. Follow the directions on the tool.

➡️ Raise and support the vehicle safely. Remove the tire and wheel assembly. Disconnect the wheel speed sensor electrical connector, as required.

➡️ Use the Mazda diagnostic scan tool, or equivalent and reprogram the required systems.

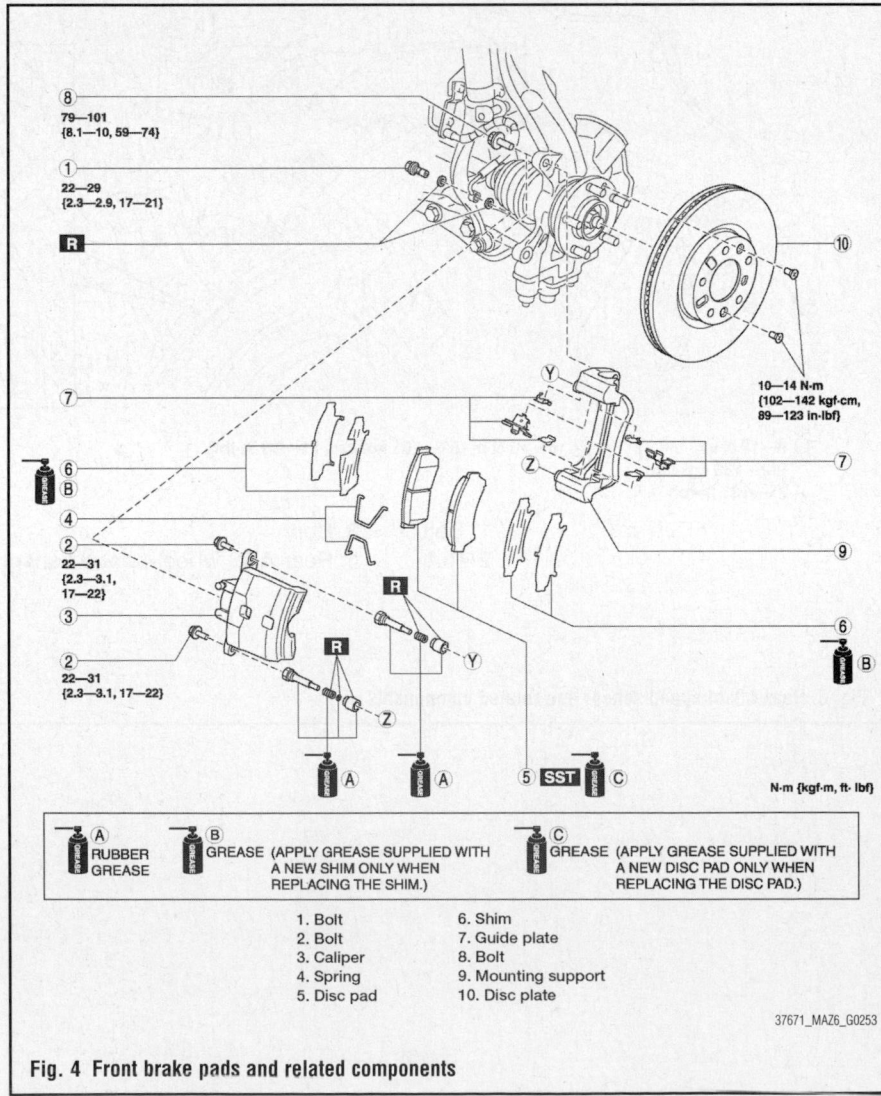

1. Bolt
2. Bolt
3. Caliper
4. Spring
5. Disc pad
6. Shim
7. Guide plate
8. Bolt
9. Mounting support
10. Disc plate

Ⓐ RUBBER GREASE

Ⓑ GREASE (APPLY GREASE SUPPLIED WITH A NEW SHIM ONLY WHEN REPLACING THE SHIM.)

Ⓒ GREASE (APPLY GREASE SUPPLIED WITH A NEW DISC PAD ONLY WHEN REPLACING THE DISC PAD.)

N·m {kgf·m, ft·lbf}

37671_MAZ6_G0253

Fig. 4 Front brake pads and related components

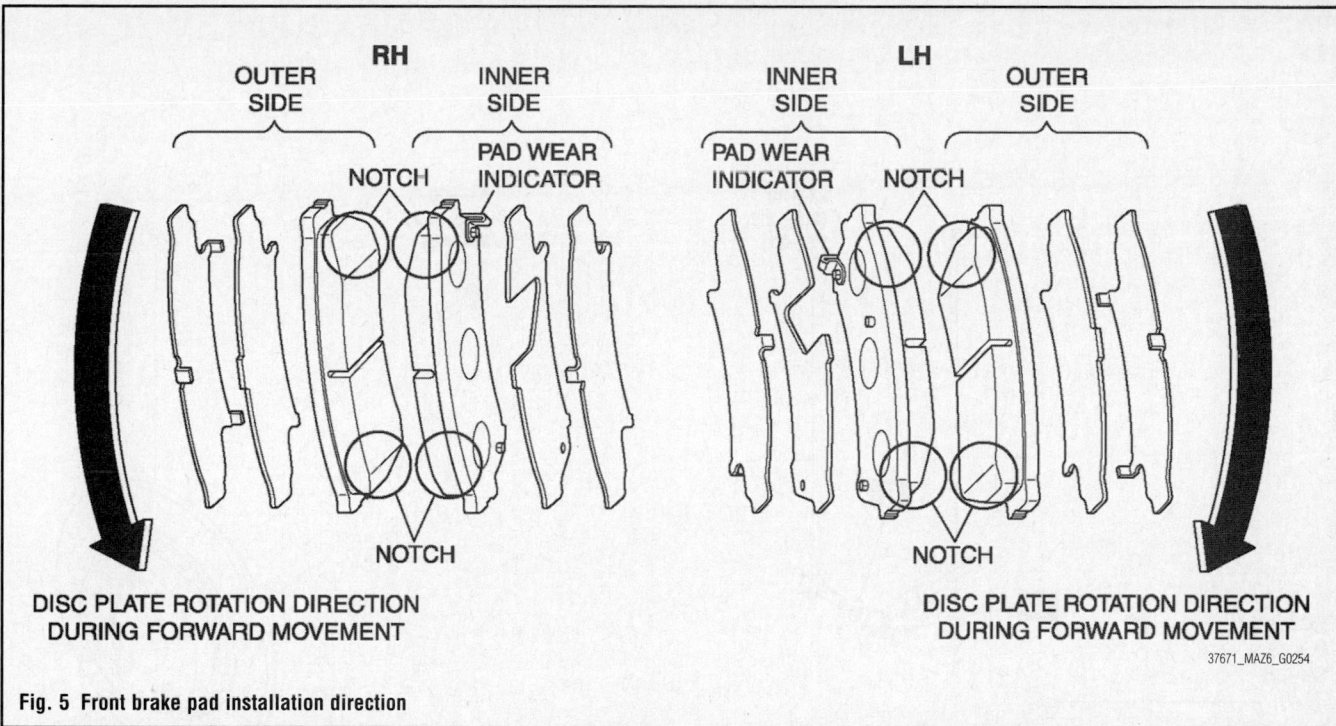

RH
OUTER SIDE — INNER SIDE

LH
INNER SIDE — OUTER SIDE

NOTCH — PAD WEAR INDICATOR

PAD WEAR INDICATOR — NOTCH

NOTCH

NOTCH

DISC PLATE ROTATION DIRECTION DURING FORWARD MOVEMENT

DISC PLATE ROTATION DIRECTION DURING FORWARD MOVEMENT

37671_MAZ6_G0254

Fig. 5 Front brake pad installation direction

BRAKES

BRAKE CALIPER

REMOVAL & INSTALLATION

At this time the manufacturer does not provide removal and installation procedures for this component. The following procedure is a guideline and may differ from the vehicle you are servicing.

1. Before servicing the vehicle, refer to the Precautions Section.

➡ **If working near and/or around the SRS system and components, be sure to disable the SRS system. Tape the negative battery cable with insulating tape. Always disconnect the negative battery cable first.**

❋ CAUTION

To avoid personal injury when working on vehicles equipped with an air bag, the negative battery cable must be disconnected and at least one minute must elapse before working on the system. Failure to do so may result in deployment of the air bag.

2. Disconnect the negative battery cable. Tape the cable with insulating tape.

➡ **When disconnecting the cable, some systems need to be initialized after the cable is reconnected. You will need the Mazda diagnostic scan tool or**

equivalent. **Follow the directions on the tool.**

3. Raise and safely support the vehicle.
4. Remove the tire and wheel assembly.
5. Drain the brake fluid. Be sure to properly dispose of used fluid.
6. Disconnect the brake line hose at the caliper.
7. Remove the caliper retaining bolts.
8. Remove the caliper from its mounting.

To install:

➡ **Be sure to use new fasteners, as required.**

9. Installation is the reverse of the removal procedure.
10. Be sure to tighten the caliper retaining bolts to specification.
11. Be sure to fill the brake system with the proper grade and type fluid.
12. Bleed the brakes. Correct fluid level, as required.
13. Use the Mazda diagnostic scan tool, or equivalent and reprogram the required systems.

DISC BRAKE PADS

REMOVAL & INSTALLATION

See Figure 6.

At this time the manufacturer does not provide removal and installation procedures

for this component, refer to the illustration as required.

1. Before servicing the vehicle, refer to the Precautions Section.

➡ **If working near and/or around the SRS system and components, be sure to disable the SRS system. Tape the negative battery cable with insulating tape. Always disconnect the negative battery cable first.**

❋ CAUTION

To avoid personal injury when working on vehicles equipped with an air bag, the negative battery cable must be disconnected and at least one minute must elapse before working on the system. Failure to do so may result in deployment of the air bag.

➡ **When disconnecting the cable, some systems need to be initialized after the cable is reconnected. You will need the Mazda diagnostic scan tool or equivalent. Follow the directions on the tool.**

➡ **Raise and support the vehicle safely. Remove the tire and wheel assembly. Disconnect the wheel speed sensor electrical connector, as required.**

➡ **Use the Mazda diagnostic scan tool, or equivalent and reprogram the required systems.**

REAR DISC BRAKES

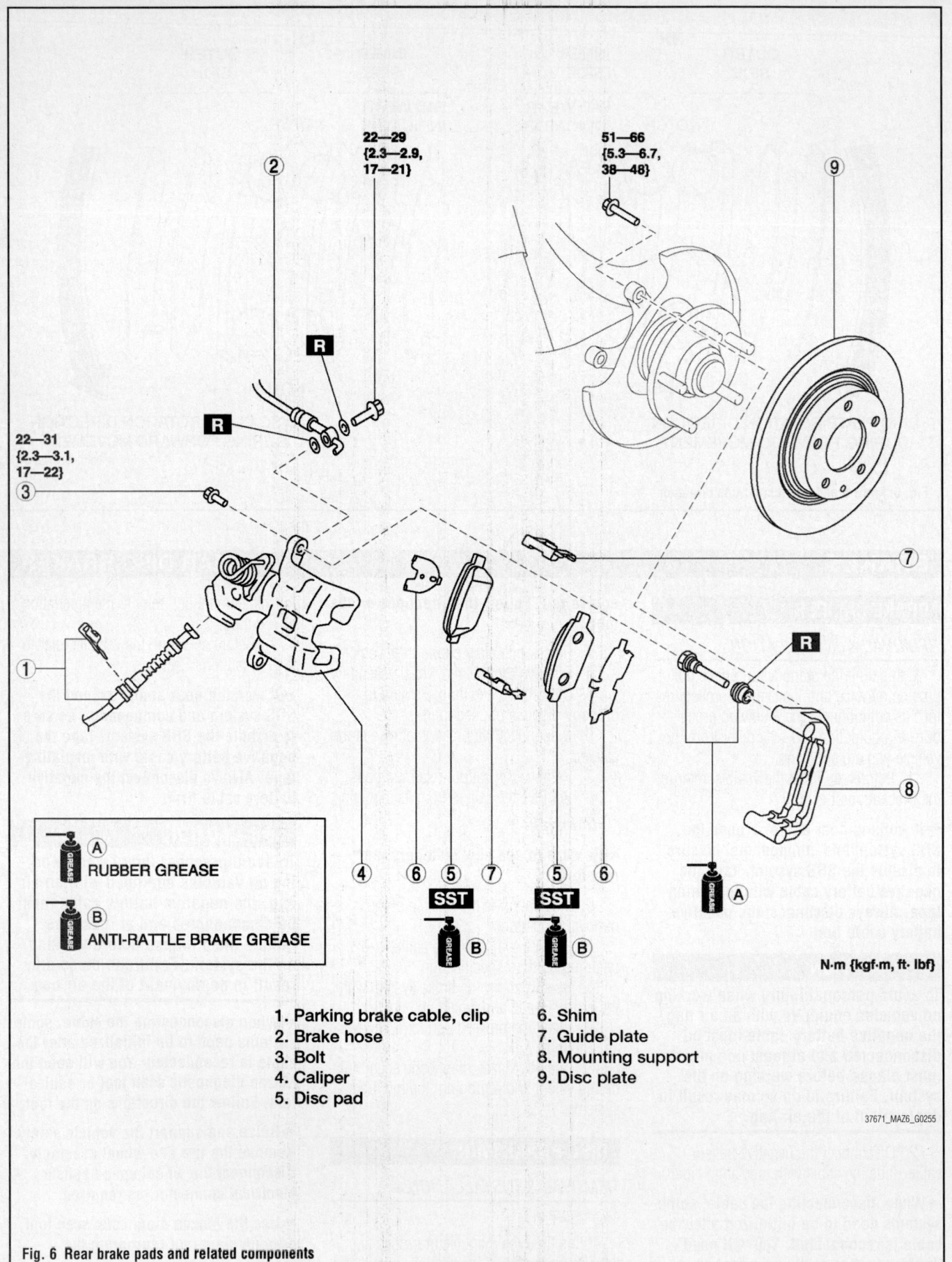

22—29
{2.3—2.9,
17—21}

51—66
{5.3—6.7,
38—48}

22—31
{2.3—3.1,
17—22}

A RUBBER GREASE

B ANTI-RATTLE BRAKE GREASE

N·m {kgf·m, ft· lbf}

1. Parking brake cable, clip
2. Brake hose
3. Bolt
4. Caliper
5. Disc pad
6. Shim
7. Guide plate
8. Mounting support
9. Disc plate

37671_MAZ6_G0255

Fig. 6 Rear brake pads and related components

BRAKES   PARKING BRAKE

PARKING BRAKE CABLES

ADJUSTMENT

See Figure 241.

1. Before servicing the vehicle, refer to the Precautions Section.

➡️**If working near and/or around the SRS system and components, be sure to disable the SRS system. Tape the negative battery cable with insulating tape. Always disconnect the negative battery cable first.**

> ✳️ **CAUTION**
>
> **To avoid personal injury when working on vehicles equipped with an air**

bag, the negative battery cable must be disconnected and at least one minute must elapse before working on the system. Failure to do so may result in deployment of the air bag.

2. Disconnect the negative battery cable. Tape the cable with insulating tape.

➡️**When disconnecting the cable, some systems need to be initialized after the cable is reconnected. You will need the Mazda diagnostic scan tool or equivalent. Follow the directions on the tool.**

3. Pull the parking brake lever two or three times.

4. Inspect the brake stroke by slowly pulling at a point 2.0 inches from the end of the parking brake lever with a force of about twenty-two lbs. and counting the number of clicks.

5. If not within specification, adjust the lever.

6. Specification should be 2–4 notches.

To install:

➡️**Be sure to use new fasteners, as required.**

7. Installation is the reverse of the removal procedure.

8. Use the Mazda diagnostic scan tool, or equivalent and reprogram the required systems.

CHASSIS ELECTRICAL   AIR BAG (SUPPLEMENTAL RESTRAINT SYSTEM)

GENERAL INFORMATION

> ✳️ **CAUTION**
>
> **These vehicles are equipped with an air bag system. The system must be disarmed before performing service on, or around, system components, the steering column, instrument panel components, wiring and sensors. Failure to follow the safety precautions and the disarming procedure could result in accidental air bag deployment, possible injury and unnecessary system repairs.**

SERVICE PRECAUTIONS

> ✳️ **CAUTION**
>
> **Disconnect and isolate the battery negative cable before beginning any airbag system component diagnosis, testing, removal, or installation procedures. Wait at least 90 seconds after the ignition switch is turned off and the negative (-) terminal cable is disconnected from the battery before starting the operation. The SRS is equipped with a backup power source, so if work is started within 90 seconds after disconnecting the negative (-) terminal cable from the battery, the SRS may be deployed. Failure to disable the airbag system may result in accidental airbag deployment, personal injury, or death.**

DISARMING THE SYSTEM

> ✳️✳️ **CAUTION**
>
> **Handling an air bag module improperly can accidentally deploy the module, which may seriously injure you. Due to the adoption of 2-step deployment control in the driver-side air bag module, depending on the impact force, it is possible that inflator No. 2 might not deploy. In such cases, before disposing of the air bag module, make sure to follow the inflator deployment procedures and verify complete deployment of inflator's No. 1 and No. 2.**

1. Disconnect the negative battery cable and wait for 1 minute or more.

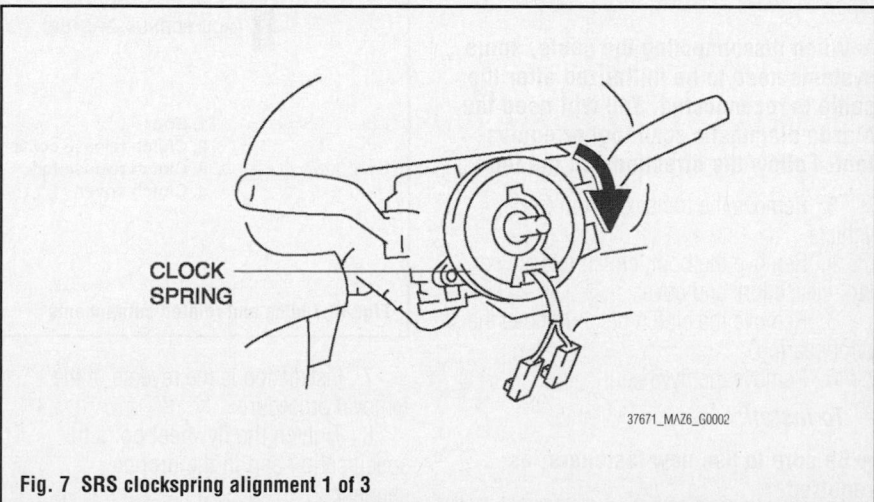

CLOCK
SPRING

37671_MAZ6_G0002

Fig. 7 SRS clockspring alignment 1 of 3

ARMING THE SYSTEM

1. Connect the negative battery cable. Check for proper operation of the air bag warning light.

2. Verify that the air bag system warning light illuminates for approximately 6 seconds and goes out.

3. If the air bag system warning light does not operate normally, perform inspection of the system.

CLOCKSPRING CENTERING

See Figures 7 through 9.

➡️**The adjustment procedure is also specified on the caution label of the clock spring.**

1. Set the front tires straight-ahead.

The clock spring will break if over-wound. Do not forcibly turn the clock spring.

2. Turn the clock spring clockwise until it stops.

3. From the stopped position, turn the clock spring counterclockwise 2 3/4 turns

4. Align the marks.

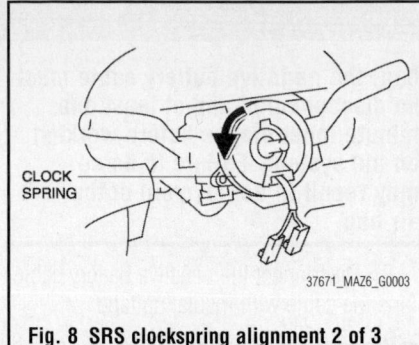

Fig. 8 SRS clockspring alignment 2 of 3

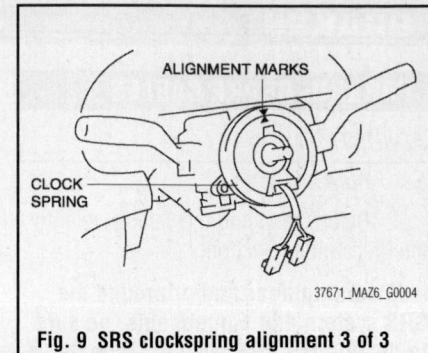

Fig. 9 SRS clockspring alignment 3 of 3

DRIVE TRAIN

DRIVEN DISC & PRESSURE PLATE

REMOVAL & INSTALLATION

See Figures 10 through 12.

1. Before servicing the vehicle, refer to the Precautions Section.

➡ If working near and/or around the SRS system and components, be sure to disable the SRS system. Tape the negative battery cable with insulating tape. Always disconnect the negative battery cable first.

To avoid personal injury when working on vehicles equipped with an air bag, the negative battery cable must be disconnected and at least one minute must elapse before working on the system. Failure to do so may result in deployment of the air bag.

2. Disconnect the negative battery cable. Tape the cable with insulating tape.

➡ When disconnecting the cable, some systems need to be initialized after the cable is reconnected. You will need the Mazda diagnostic scan tool or equivalent. Follow the directions on the tool.

3. Remove the transaxle from the vehicle.

4. Remove the boot, clutch release collar, release fork and cover.

5. Remove the clutch disc. Remove the pilot bearing.

6. Remove the flywheel.

To install:

➡ Be sure to use new fasteners, as required.

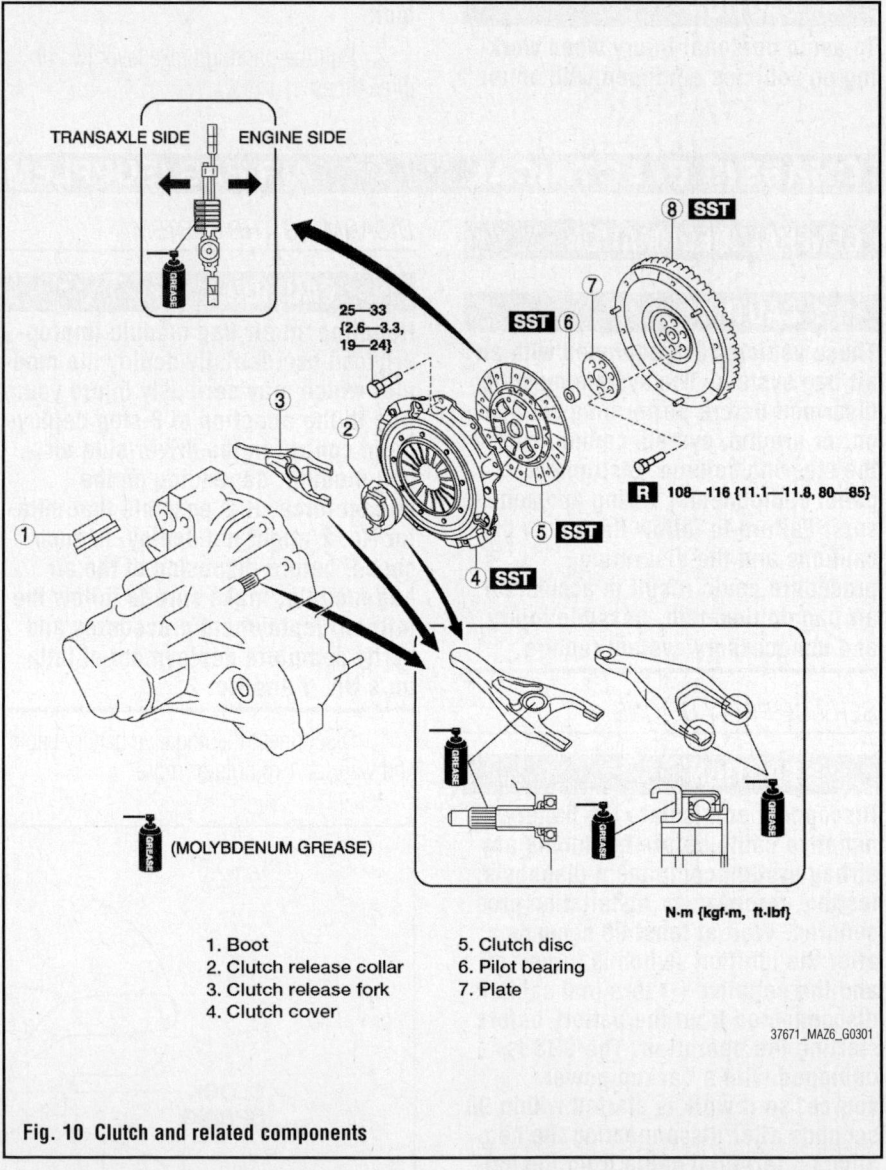

1. Boot
2. Clutch release collar
3. Clutch release fork
4. Clutch cover
5. Clutch disc
6. Pilot bearing
7. Plate

N·m {kgf·m, ft·lbf}

Fig. 10 Clutch and related components

7. Installation is the reverse of the removal procedure.

8. Tighten the flywheel bolts to specification and in the proper sequence.

9. Be sure to properly install the pilot bushing.

10. Use the Mazda diagnostic scan tool, or equivalent and reprogram the required systems.

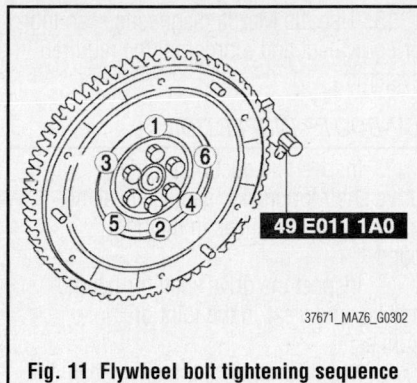

Fig. 11 Flywheel bolt tightening sequence

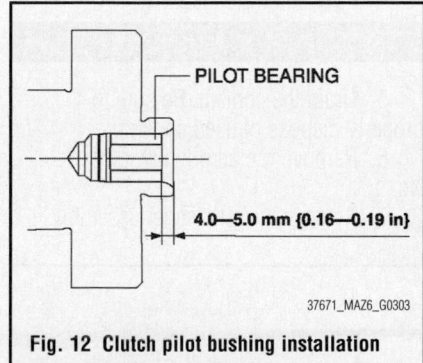

PILOT BEARING

4.0—5.0 mm {0.16—0.19 in}

Fig. 12 Clutch pilot bushing installation

FRONT HALFSHAFTS

REMOVAL & INSTALLATION
See Figures 00, 13 through 16.

✻✻ WARNING

Performing the following procedures without first removing the ABS wheel-speed sensor may possibly cause an open circuit in the harness if it is pulled by mistake. Before performing the following procedures, remove the ABS wheel-speed sensor (axle side) and fix it to an appropriate place where the sensor will not be pulled by mistake while the vehicle is being serviced.

1. Before servicing the vehicle, refer to the Precautions Section.

➡If working near and/or around the SRS system and components, be sure to disable the SRS system. Tape the negative battery cable with insulating tape. Always disconnect the negative battery cable first.

✻✻ CAUTION

To avoid personal injury when working on vehicles equipped with an air bag, the negative battery cable must be disconnected and at least one minute must elapse before working

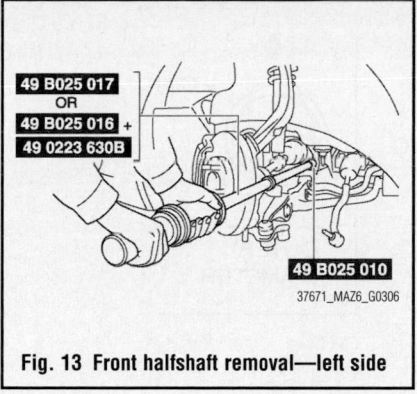

Fig. 13 Front halfshaft removal—left side

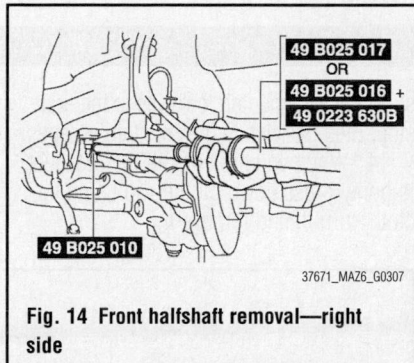

Fig. 14 Front halfshaft removal—right side

on the system. Failure to do so may result in deployment of the air bag.

2. Disconnect the negative battery cable. Tape the cable with insulating tape.

➡When disconnecting the cable, some systems need to be initialized after the cable is reconnected. You will need the Mazda diagnostic scan tool or equivalent. Follow the directions on the tool.

3. Set the mudguard out of the way.
4. Remove the splash shield.
5. Drain the transaxle oil. Be sure to properly dispose of used oil.
6. Remove the locknut. Remove the bolt (brake hose).
7. Remove the tie rod end ball joint.
8. Remove the front stabilizer control link lower side nut.
9. Remove the lower arm nut.
10. Remove the damper fork.
11. Remove the halfshaft using the removal tools. See illustration
12. Remove the halfshaft clip. Remove the joint shaft clip.

To install:

➡**Be sure to use new fasteners, as required.**

13. Installation is the reverse of the removal procedure.
14. Be sure to properly install the halfshaft clip to the clip groove. Specification is 1.23 inch or less for 2.5L engine and 1.34 inch or less for 3.7L engine.

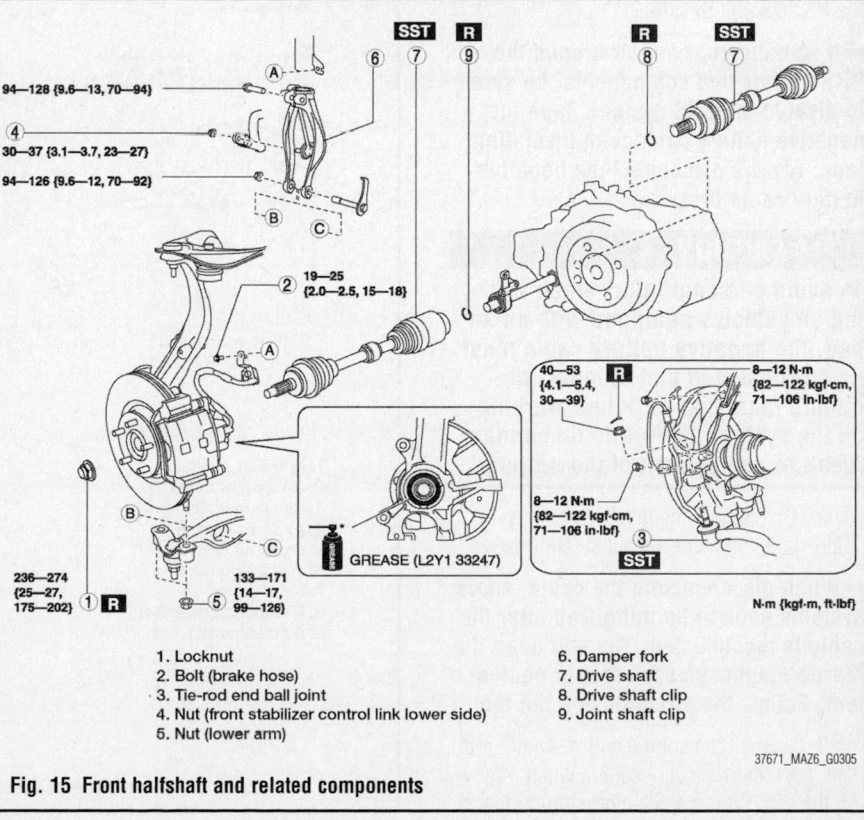

1. Locknut
2. Bolt (brake hose)
3. Tie-rod end ball joint
4. Nut (front stabilizer control link lower side)
5. Nut (lower arm)
6. Damper fork
7. Drive shaft
8. Drive shaft clip
9. Joint shaft clip

Fig. 15 Front halfshaft and related components

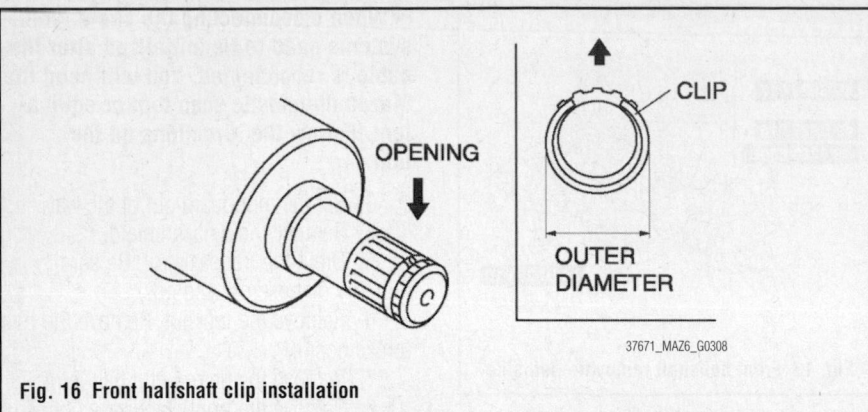

Fig. 16 Front halfshaft clip installation

15. Use the Mazda diagnostic scan tool, or equivalent and reprogram the required systems.

CV-BOOTS INSPECTION

1. Inspect the dust boot on the drive shaft for cracks, damage, leaking grease, and looseness in the boot band.

2. Inspect the drive shaft for bends, cracks, and wear in the joint or splines

3. Repair or replace the drive shaft or boot/band as necessary.

ENGINE COOLING

ENGINE FAN

REMOVAL & INSTALLATION

See Figures 17 through 19.

> **❉ CAUTION**
>
> **Never remove the cooling system cap or loosen the radiator drain plug while the engine is running, or when the engine and radiator are hot. Scalding engine coolant and steam may shoot out and cause serious injury. It may also damage the engine and cooling system.**

1. Before servicing the vehicle, refer to the Precautions Section.

➡ **If working near and/or around the SRS system and components, be sure to disable the SRS system. Tape the negative battery cable with insulating tape. Always disconnect the negative battery cable first.**

> **❉ CAUTION**
>
> **To avoid personal injury when working on vehicles equipped with an air bag, the negative battery cable must be disconnected and at least one minute must elapse before working on the system. Failure to do so may result in deployment of the air bag.**

2. Disconnect the negative battery cable. Tape the cable with insulating tape.

➡ **When disconnecting the cable, some systems need to be initialized after the cable is reconnected. You will need the Mazda diagnostic scan tool or equivalent. Follow the directions on the tool.**

3. Turn off the engine and wait until it is cool. Even then, be very careful when removing the cap. Wrap a thick cloth around it and slowly turn it counterclockwise to the first stop. Step back while the pressure escapes.

4. When you are sure all the pressure is gone, press down on the cap using the cloth, turn it, and remove it.

5. Drain the coolant. Be sure to properly dispose of used coolant.

6. Remove the aerodynamic cover No. 1.

7. Remove the air cleaner assembly.

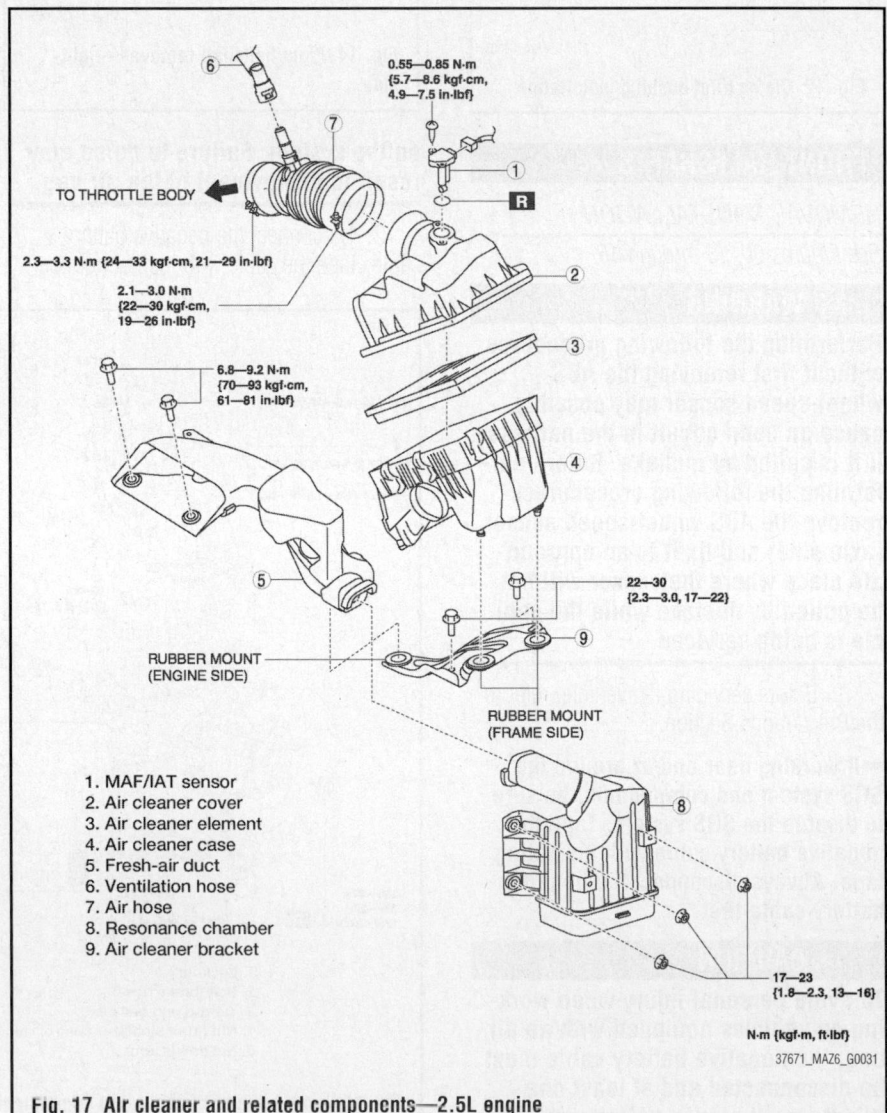

1. MAF/IAT sensor
2. Air cleaner cover
3. Air cleaner element
4. Air cleaner case
5. Fresh-air duct
6. Ventilation hose
7. Air hose
8. Resonance chamber
9. Air cleaner bracket

Fig. 17 Air cleaner and related components—2.5L engine

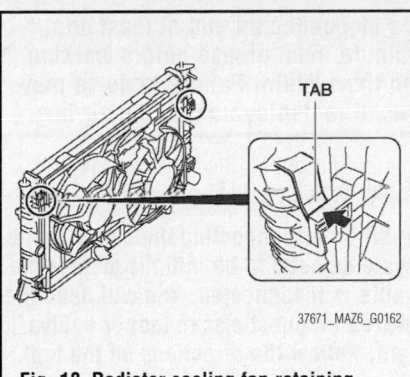

Fig. 18 Radiator cooling fan retaining tabs 1 of 2

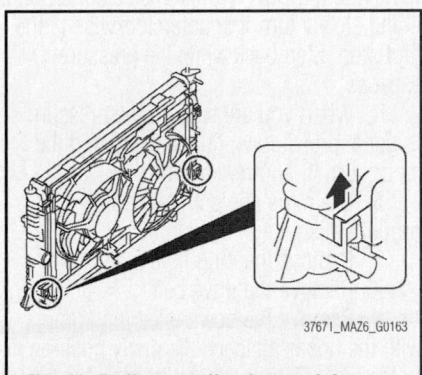

Fig. 19 Radiator cooling fan retaining tabs 2 of 2

8. Remove the engine cover, 3.7L engine.

9. Remove the coolant reserve tank hose.

10. Disconnect the electrical connectors.

11. Remove the wiring harness.

12. Remove the upper radiator hose. Remove the lower radiator hose.

13. Remove the ATF oil cooler pipes from the clips installed to the underside of the radiator, if equipped.

14. Remove the cooling fan component from its mounting.

To install:

➡**Be sure to use new fasteners, as required.**

15. Installation is the reverse of the removal procedure.

16. Fill the cooling system with the proper grade and type coolant.

17. Start the engine and check for leaks.

18. Use the Mazda diagnostic scan tool, or equivalent and reprogram the required systems.

RADIATOR

REMOVAL & INSTALLATION

See Figures 17, and 20 through 23.

❋❋ **CAUTION**

Never remove the cooling system cap or loosen the radiator drain plug while the engine is running, or when the engine and radiator are hot. Scalding engine coolant and steam may shoot out and cause serious injury. It may also damage the engine and cooling system.

1. Before servicing the vehicle, refer to the Precautions Section.

➡**If working near and/or around the SRS system and components, be sure to disable the SRS system. Tape the negative battery cable with insulating tape. Always disconnect the negative battery cable first.**

❋❋ **CAUTION**

To avoid personal injury when working on vehicles equipped with an air bag, the negative battery cable must be disconnected and at least one minute must elapse before working on the system. Failure to

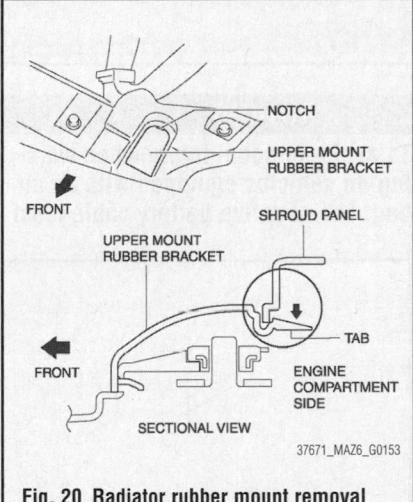

Fig. 20 Radiator rubber mount removal

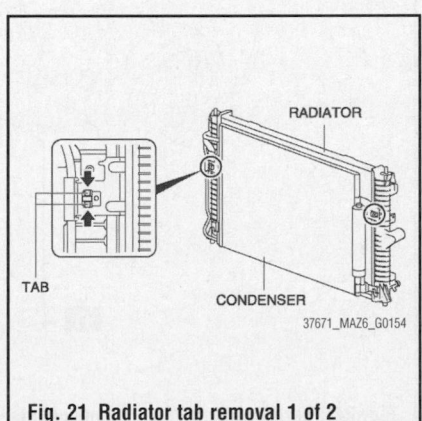

Fig. 21 Radiator tab removal 1 of 2

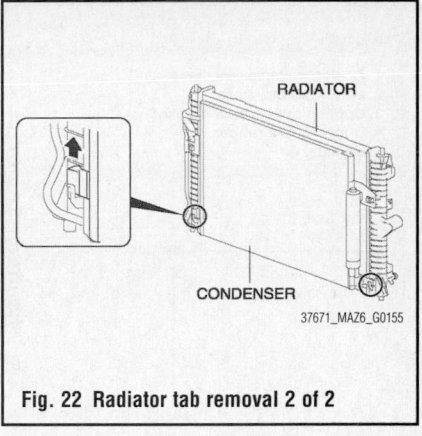

Fig. 22 Radiator tab removal 2 of 2

do so may result in deployment of the air bag.

2. Disconnect the negative battery cable. Tape the cable with insulating tape.

➡**When disconnecting the cable, some systems need to be initialized after the cable is reconnected. You will need the Mazda diagnostic scan tool or equivalent. Follow the directions on the tool.**

3. Turn off the engine and wait until it is cool. Even then, be very careful when removing the cap. Wrap a thick cloth around it and slowly turn it counterclockwise to the first stop. Step back while the pressure escapes.

4. When you are sure all the pressure is gone, press down on the cap using the cloth, turn it, and remove it.

5. Drain the coolant. Be sure to properly dispose of used coolant.

6. Remove the aerodynamic cover No. 1.

7. Remove the air cleaner assembly.

8. Remove the engine cover, 3.7L engine.

9. Remove the cooling fan component.

10. Disconnect the ATF oil cooler pipe and hose at the radiator, if equipped.

11. Remove the front bumper.

12. Remove the coolant reserve tank hose.

13. Remove the radiator hoses.

14. Remove the upper mounts.

15. Remove the radiator from its mounting.

To install:

➡**Be sure to use new fasteners, as required.**

16. Installation is the reverse of the removal procedure.

17. Fill the cooling system with the proper grade and type coolant.

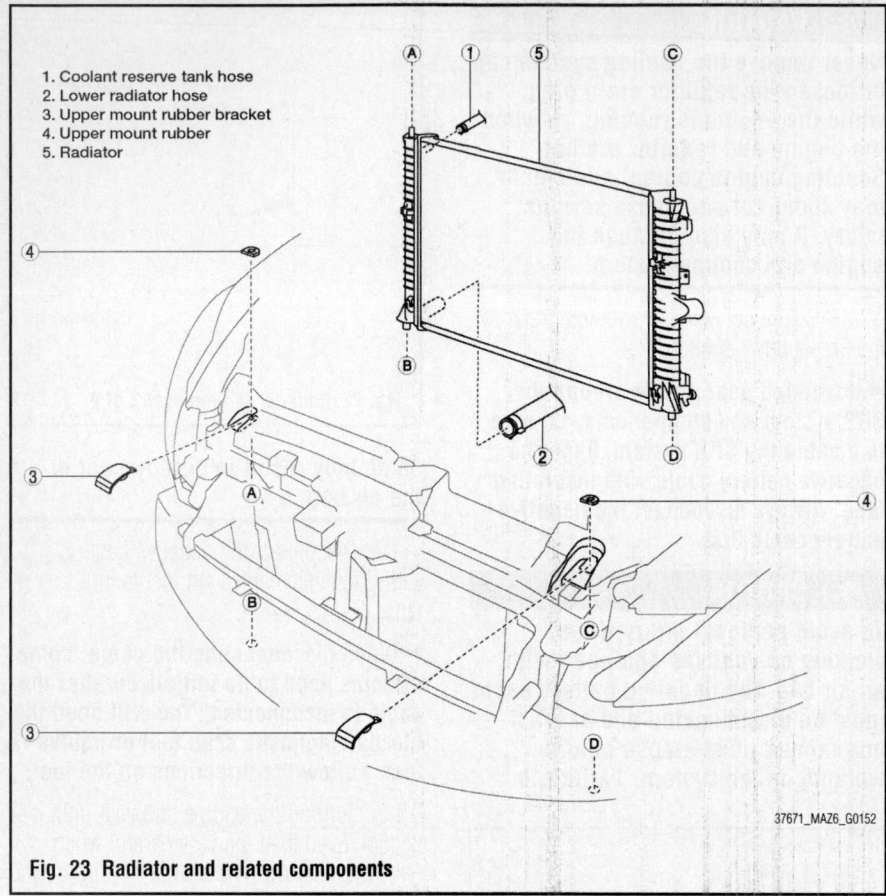

1. Coolant reserve tank hose
2. Lower radiator hose
3. Upper mount rubber bracket
4. Upper mount rubber
5. Radiator

37671_MAZ6_G0152

Fig. 23 Radiator and related components

18. Start the engine and check for leaks.
19. Use the Mazda diagnostic scan tool, or equivalent and reprogram the required systems.

THERMOSTAT

REMOVAL & INSTALLATION

2.5L Engine

See Figure 24.

> ※ **CAUTION**
>
> **Never remove the cooling system cap or loosen the radiator drain plug while the engine is running, or when the engine and radiator are hot. Scalding engine coolant and steam may shoot out and cause serious injury. It may also damage the engine and cooling system.**

1. Before servicing the vehicle, refer to the Precautions Section.

➡ **If working near and/or around the SRS system and components, be sure to disable the SRS system. Tape the negative battery cable with insulating tape. Always disconnect the negative battery cable first.**

> ※ **CAUTION**
>
> **To avoid personal injury when working on vehicles equipped with an air bag, the negative battery cable must**

be disconnected and at least one minute must elapse before working on the system. Failure to do so may result in deployment of the air bag.

2. Disconnect the negative battery cable. Tape the cable with insulating tape.

➡ **When disconnecting the cable, some systems need to be initialized after the cable is reconnected. You will need the Mazda diagnostic scan tool or equivalent. Follow the directions on the tool.**

3. Turn off the engine and wait until it is cool. Even then, be very careful when removing the cap. Wrap a thick cloth around it and slowly turn it counterclockwise to the first stop. Step back while the pressure escapes.
4. When you are sure all the pressure is gone, press down on the cap using the cloth, turn it, and remove it.
5. Drain the coolant. Be sure to properly dispose of used coolant.
6. Remove the plug hole plate.
7. Remove the drive belt.
8. Remove the power steering pump, with the hoses attached. Properly position it to the side. Do not allow it to hang by the hoses.
9. Disconnect the hoses from the housing.
10. Remove the retaining bolts.
11. Remove the component from its mounting.
12. Discard the gasket.

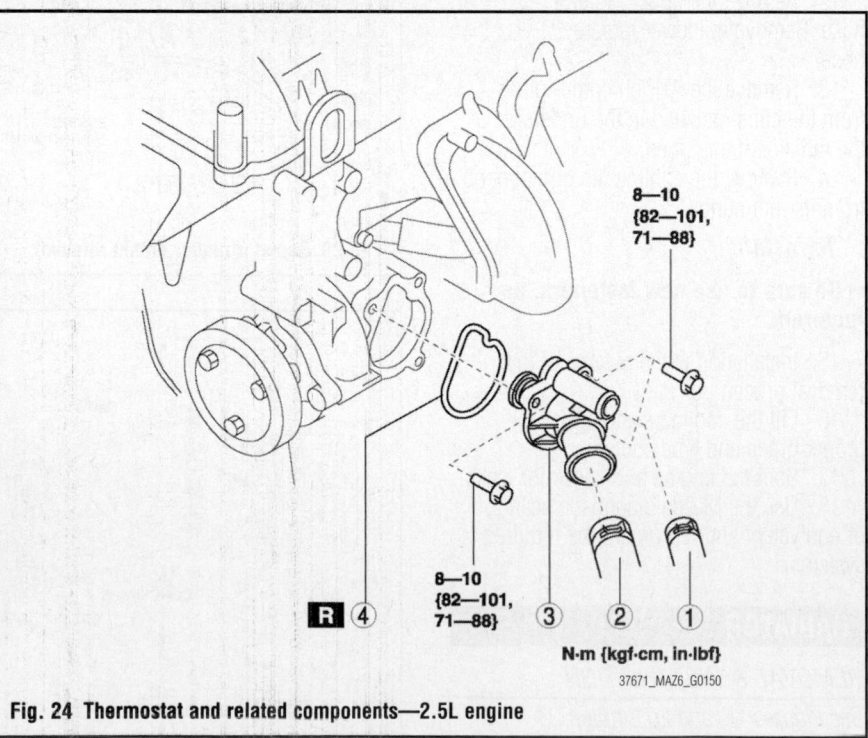

8—10
{82—101, 71—88}

8—10
{82—101, 71—88}

R 4   3   2   1

N·m {kgf·cm, in·lbf}

37671_MAZ6_G0150

Fig. 24 Thermostat and related components—2.5L engine

To install:

➡**Be sure to use new fasteners, as required.**

13. Installation is the reverse of the removal procedure.

14. Be sure to use a gasket.

15. Fill the cooling system with the proper grade and type coolant.

16. Start the engine and check for leaks.

17. Use the Mazda diagnostic scan tool, or equivalent and reprogram the required systems.

3.7L Engine

See Figure 25.

> ⁕ **CAUTION**
>
> **Never remove the cooling system cap or loosen the radiator drain plug while the engine is running, or when the engine and radiator are hot. Scalding engine coolant and steam may shoot out and cause serious injury. It may also damage the engine and cooling system.**

1. Before servicing the vehicle, refer to the Precautions Section.

➡**If working near and/or around the SRS system and components, be sure to disable the SRS system. Tape the negative battery cable with insulating tape. Always disconnect the negative battery cable first.**

> ⁕ **CAUTION**
>
> **To avoid personal injury when working on vehicles equipped with an air bag, the negative battery cable must be disconnected and at least one minute must elapse before working on the system. Failure to do so may result in deployment of the air bag.**

2. Disconnect the negative battery cable. Tape the cable with insulating tape.

➡**When disconnecting the cable, some systems need to be initialized after the cable is reconnected. You will need the Mazda diagnostic scan tool or equivalent. Follow the directions on the tool.**

3. Turn off the engine and wait until it is cool. Even then, be very careful when removing the cap. Wrap a thick cloth around it and slowly turn it counterclockwise to the first stop. Step back while the pressure escapes.

4. When you are sure all the pressure is gone, press down on the cap using the cloth, turn it, and remove it.

5. Disconnect the positive battery cable. Remove the battery. Remove the battery tray.

6. Drain the coolant. Be sure to properly dispose of used coolant.

7. Remove the retaining bolts.

8. Remove the thermostat housing from its mounting.

9. Discard the O-ring.

To install:

➡**Be sure to use new fasteners, as required.**

10. Installation is the reverse of the removal procedure.

11. Be sure to use a new O-ring. Coat the O-ring with clean engine coolant.

12. Be sure to install the thermostat with the jiggle pin facing upward.

13. Fill the cooling system with the proper grade and type coolant.

14. Start the engine and check for leaks.

15. Use the Mazda diagnostic scan tool, or equivalent and reprogram the required systems.

WATER PUMP

REMOVAL & INSTALLATION

2.5L Engine

See Figure 26.

> ⁕⁕ **CAUTION**
>
> **Never remove the cooling system cap or loosen the radiator drain plug while the engine is running, or when the engine and radiator are hot. Scalding engine coolant and steam may shoot out and cause serious injury. It may also damage the engine and cooling system.**

1. Before servicing the vehicle, refer to the Precautions Section.

➡**If working near and/or around the SRS system and components, be sure to disable the SRS system. Tape the negative battery cable with insulating tape. Always disconnect the negative battery cable first.**

> ⁕⁕ **CAUTION**
>
> **To avoid personal injury when working on vehicles equipped with an air bag, the negative battery cable must be disconnected and at least one minute must elapse before working on the system. Failure to do so may result in deployment of the air bag.**

2. Disconnect the negative battery cable. Tape the cable with insulating tape.

➡**When disconnecting the cable, some systems need to be initialized after the cable is reconnected. You will need the Mazda diagnostic scan tool or equivalent. Follow the directions on the tool.**

3. Turn off the engine and wait until it is cool. Even then, be very careful when removing the cap. Wrap a thick cloth around it and slowly turn it counterclockwise to the first stop. Step back while the pressure escapes.

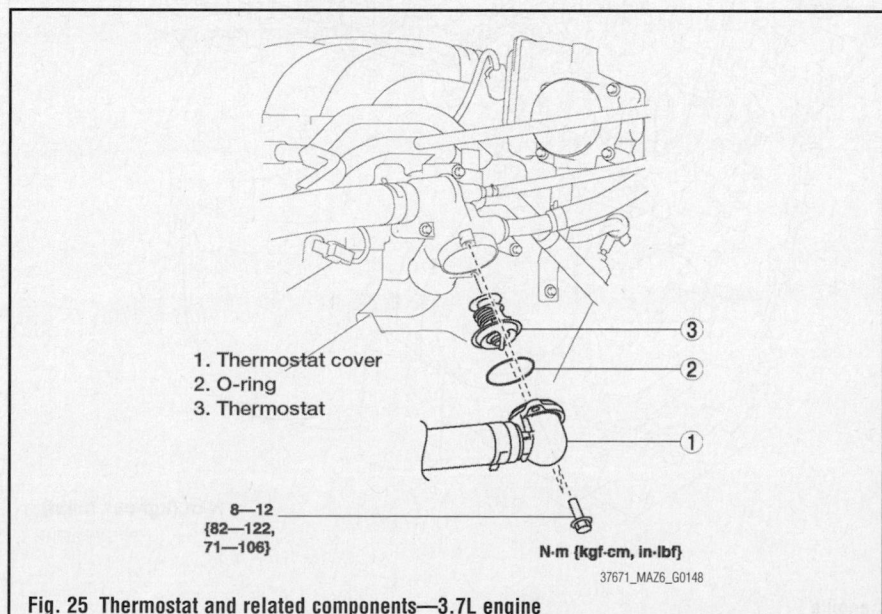

1. Thermostat cover
2. O-ring
3. Thermostat

8—12
{82—122,
71—106}

N·m {kgf-cm, in·lbf}

37671_MAZ6_G0148

Fig. 25 Thermostat and related components—3.7L engine

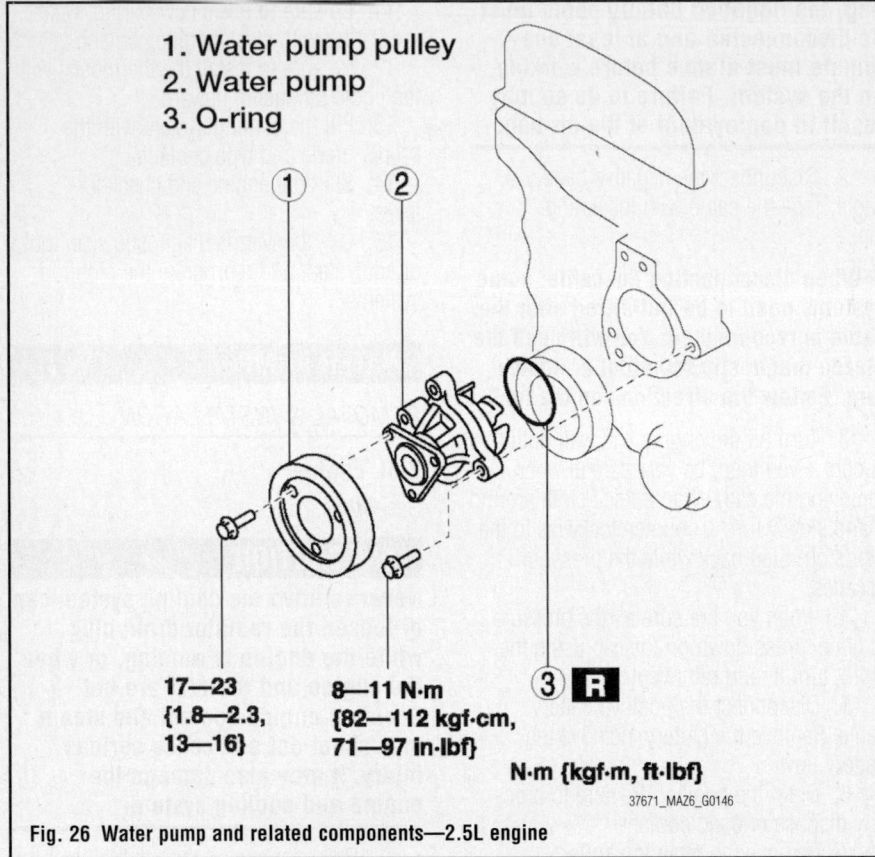

1. Water pump pulley
2. Water pump
3. O-ring

17—23
{1.8—2.3,
13—16}

8—11 N·m
{82—112 kgf·cm,
71—97 in·lbf}

③ **R**

N·m {kgf·m, ft·lbf}

37671_MAZ2_G0146

Fig. 26 Water pump and related components—2.5L engine

4 When you are sure all the pressure is gone, press down on the cap using the cloth, turn it, and remove it.

5. Remove the plug hole plate.

6. Drain the engine coolant.

7. Loosen the water pump pulley bolts before removing the drive belt.

8. Remove the drive belt.

9. Remove the water pump pulley.

10. Remove the pump retaining bolts.

11. Remove the pump from its mounting.

12. Discard the O-ring.

To install:

➡**Be sure to use new fasteners, as required.**

13. Installation is the reverse of the removal procedure.

14. Be sure to use a new O-ring. Coat the O-ring with clean engine coolant.

15. Fill the cooling system with the proper grade and type coolant.

16. Start the engine and check for leaks.

17. Use the Mazda diagnostic scan tool, or equivalent and reprogram the required systems.

3.7L Engine

See Figure 27.

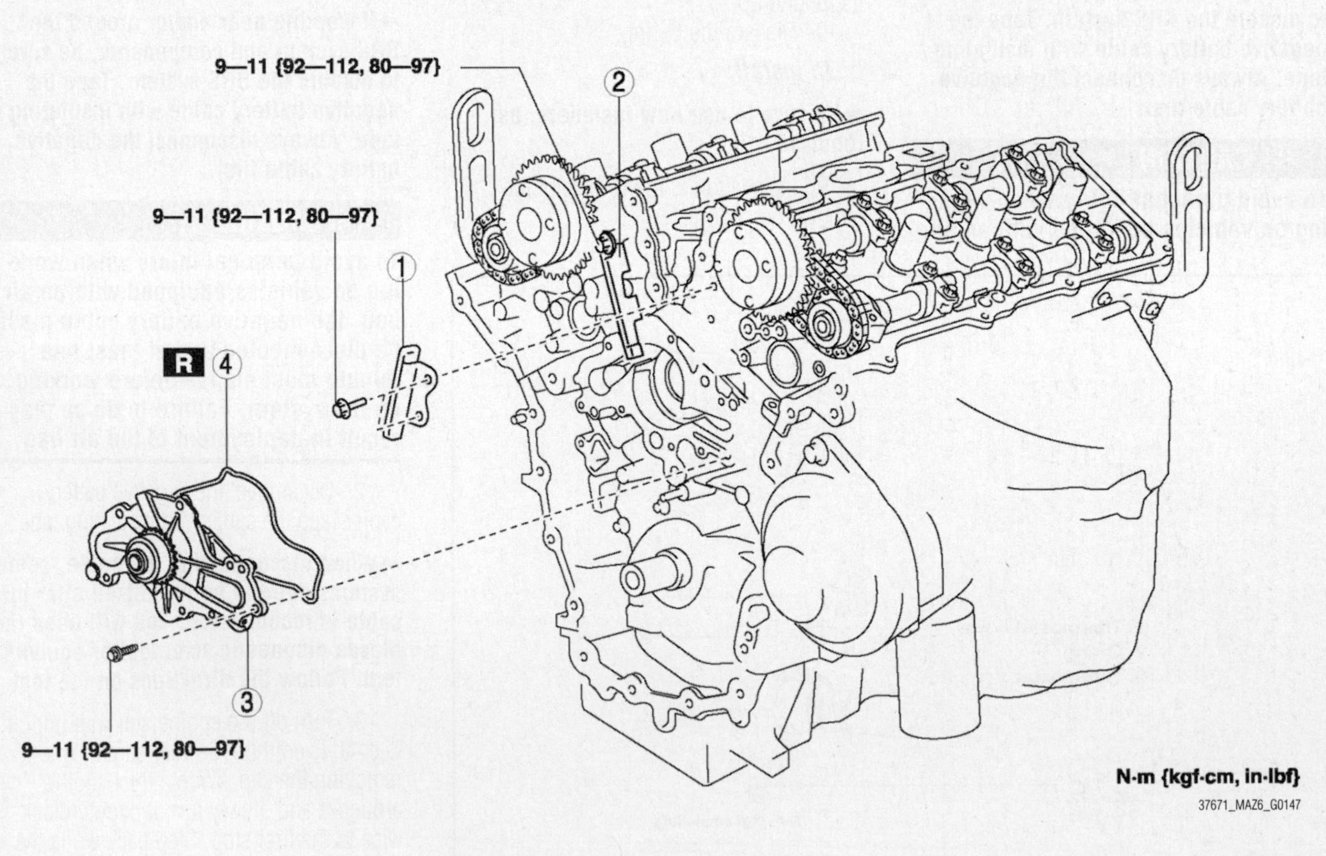

9—11 {92—112, 80—97}

9—11 {92—112, 80—97}

R ④

9—11 {92—112, 80—97}

N·m {kgf·cm, in·lbf}

37671_MAZ6_G0147

Fig. 27 Water pump and related components—3.7L engine

➡Mazda's official procedure for removal and installation of this component requires the engine/transaxle assembly to be removed from the vehicle. Remove the engine/transaxle assembly and position it in a suitable holding fixture.

❄❄ **CAUTION**

Never remove the cooling system cap or loosen the radiator drain plug while the engine is running, or when the engine and radiator are hot. Scalding engine coolant and steam may shoot out and cause serious injury. It may also damage the engine and cooling system.

1. Before servicing the vehicle, refer to the Precautions Section.

➡If working near and/or around the SRS system and components, be sure to disable the SRS system. Tape the negative battery cable with insulating tape. Always disconnect the negative battery cable first.

❄ **CAUTION**

To avoid personal injury when working on vehicles equipped with an air bag, the negative battery cable must be disconnected and at least one minute must elapse before working on the system. Failure to do so may result in deployment of the air bag.

2. Disconnect the negative battery cable. Tape the cable with insulating tape.

➡When disconnecting the cable, some systems need to be initialized after the cable is reconnected. You will need the Mazda diagnostic scan tool or equivalent. Follow the directions on the tool.

3. Turn off the engine and wait until it is cool. Even then, be very careful when removing the cap. Wrap a thick cloth around it and slowly turn it counterclockwise to the first stop. Step back while the pressure escapes.
4. When you are sure all the pressure is gone, press down on the cap using the cloth, turn it, and remove
5. Remove the engine/transaxle assembly and position it in a suitable holding fixture.
6. Remove the dynamic chamber and throttle body as a single unit.

7. Remove the ignition coils.
8. Remove the oil dipstick.
9. Remove the power steering pump drive belt.
10. Remove the power steering pump.
11. Remove the timing chain component.
12. Remove the upper timing chain guides.
13. Remove the water pump retaining bolts.
14. Remove the pump from its mounting.
15. Discard the gasket.

To install:

➡Be sure to use new fasteners, as required.

16. Installation is the reverse of the removal procedure.
17. Be sure to use a new gasket.
18. Fill the cooling system with the proper grade and type coolant.
19. Start the engine and check for leaks.
20. Use the Mazda diagnostic scan tool, or equivalent and reprogram the required systems.

ENGINE ELECTRICAL

ALTERNATOR

REMOVAL & INSTALLATION
See Figures 28 and 29.

1. Before servicing the vehicle, refer to the Precautions Section.

➡If working near and/or around the SRS system and components, be sure to disable the SRS system. Tape the negative battery cable with insulating tape. Always disconnect the negative battery cable first.

❄ **CAUTION**

To avoid personal injury when working on vehicles equipped with an air bag, the negative battery cable must be disconnected and at least one minute must elapse before working on the system. Failure to do so may result in deployment of the air bag.

2. Disconnect the negative battery cable. Tape the cable with insulating tape.

CHARGING SYSTEM

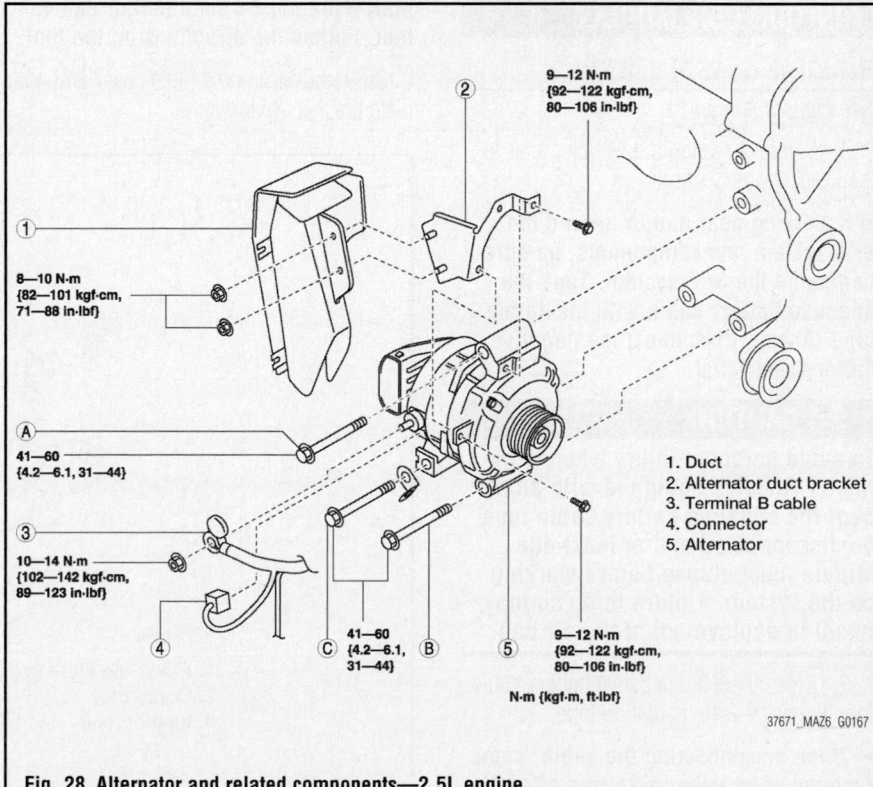

1. Duct
2. Alternator duct bracket
3. Terminal B cable
4. Connector
5. Alternator

N·m {kgf·m, ft·lbf}

37671_MAZ6 G0167

Fig. 28 Alternator and related components—2.5L engine

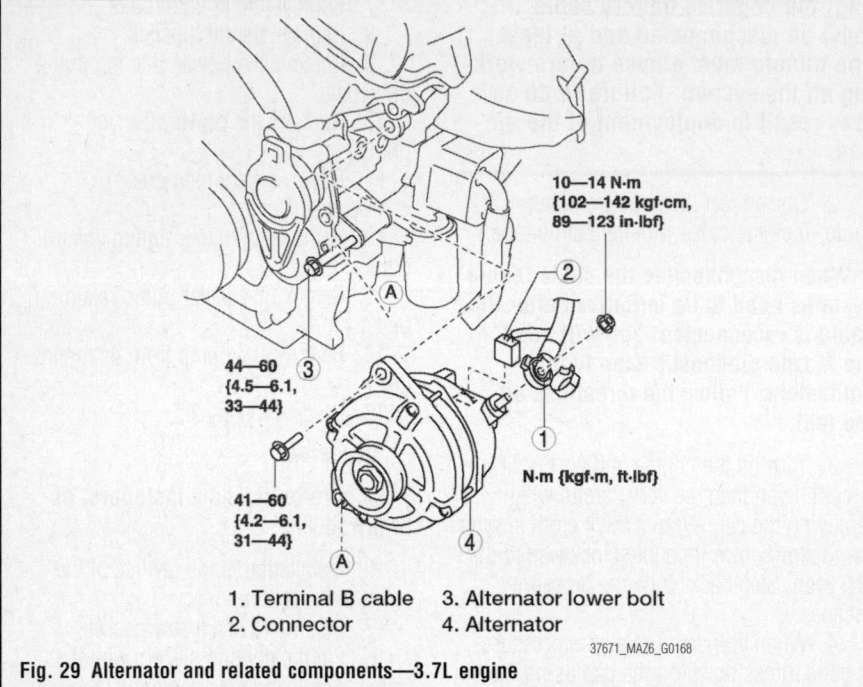

1. Terminal B cable   3. Alternator lower bolt
2. Connector   4. Alternator

37671_MAZ6_G0168

Fig. 29 Alternator and related components—3.7L engine

➡ **When disconnecting the cable, some systems need to be initialized after the cable is reconnected. You will need the Mazda diagnostic scan tool or equivalent. Follow the directions on the tool.**

3. Remove the plug hole plate, 2.5L engine.

4. Remove the aerodynamic cover No. 2., 2.5L engine

5. Remove the air cleaner assembly,

remove the engine cover and remove the cooling fan assembly, 3.7L engine.

6. Remove the drive belt, 2.5L engine.

7. Remove the alternator and air conditioning drive belts, 3.7L engine. Position the compressor to the side. Do not allow it to hang by the refrigerant hoses.

8. Remove the alternator duct. Remove the bracket.

9. Disconnect the terminal B cable.

10. Disconnect the electrical connector.

11. Remove the component retaining bolts.

12. Remove the component from its mounting.

To install:

➡ **Be sure to use new fasteners, as required.**

13. Installation is the reverse of the removal procedure.

14. Tighten retaining bolts to specification.

15. Use the Mazda diagnostic scan tool, or equivalent and reprogram the required systems.

ENGINE ELECTRICAL

IGNITION COIL MODULE

REMOVAL & INSTALLATION

See Figures 30 and 31.

1. Before servicing the vehicle, refer to the Precautions Section.

➡ **If working near and/or around the SRS system and components, be sure to disable the SRS system. Tape the negative battery cable with insulating tape. Always disconnect the negative battery cable first.**

❄ CAUTION

To avoid personal injury when working on vehicles equipped with an air bag, the negative battery cable must be disconnected and at least one minute must elapse before working on the system. Failure to do so may result in deployment of the air bag.

2. Disconnect the negative battery cable. Tape the cable with insulating tape.

➡ **When disconnecting the cable, some systems need to be initialized after the cable is reconnected. You will need the**

Mazda diagnostic scan tool or equivalent. Follow the directions on the tool.

3. Remove the hole plate cover and hole plate bracket, 2.5L engine

IGNITION SYSTEM

4. Remove the engine cover and air cleaner assembly with dynamic chamber, 3.7L engine.

5. Disconnect the electrical connectors.

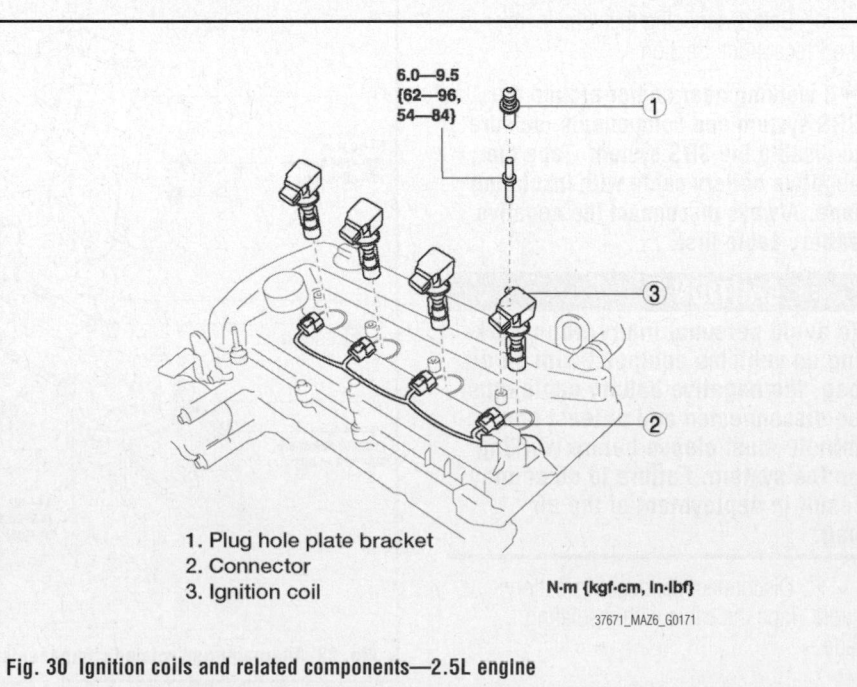

1. Plug hole plate bracket
2. Connector
3. Ignition coil

N·m {kgf·cm, in·lbf}

37671_MAZ6_G0171

Fig. 30 Ignition coils and related components—2.5L engine

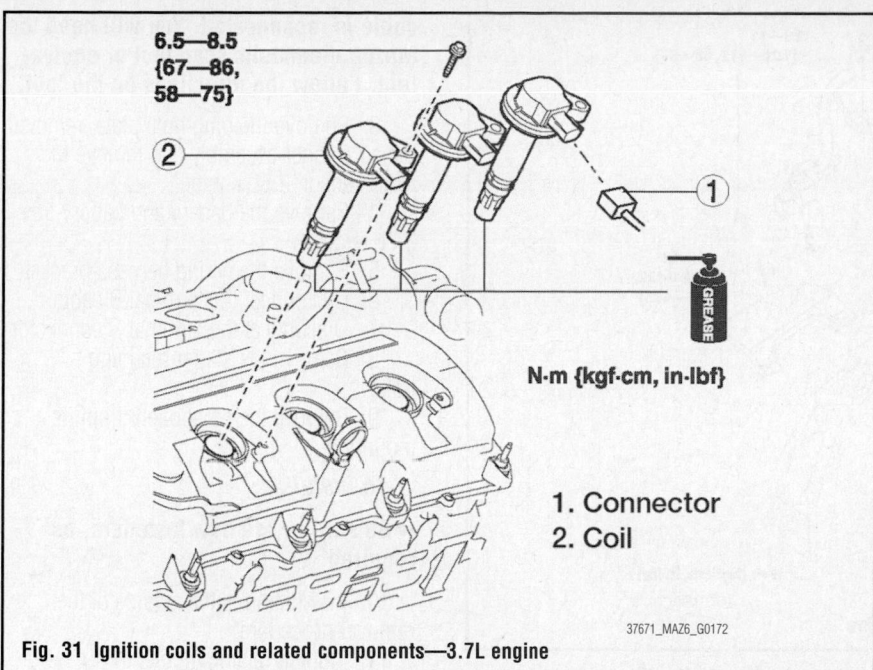

Fig. 31 Ignition coils and related components—3.7L engine

N·m {kgf·cm, in·lbf}

1. Connector
2. Coil

6. Remove the component from its mounting.

To install:

➡**Be sure to use new fasteners, as required.**

7. Installation is the reverse of the removal procedure.

8. On the 3.7L engine, apply a small amount of dielectric grease to the inside of the ignition coils, then install the coils to the spark plugs.

9. Use the Mazda diagnostic scan tool, or equivalent and reprogram the required systems.

SPARK PLUGS

REMOVAL & INSTALLATION

1. Before servicing the vehicle, refer to the Precautions Section.

➡**If working near and/or around the SRS system and components, be sure to disable the SRS system. Tape the negative battery cable with insulating tape. Always disconnect the negative battery cable first.**

❊❊ CAUTION

To avoid personal injury when working on vehicles equipped with an air

bag, the negative battery cable must be disconnected and at least one minute must elapse before working on the system. Failure to do so may result in deployment of the air bag.

2. Disconnect the negative battery cable. Tape the cable with insulating tape.

➡**When disconnecting the cable, some systems need to be initialized after the cable is reconnected. You will need the Mazda diagnostic scan tool or equivalent. Follow the directions on the tool.**

3. Remove the hole plate cover and hole plate bracket, 2.5L engine

4. Remove the engine cover and air cleaner assembly with dynamic chamber, 3.7L engine.

5. Disconnect the electrical connectors.

6. Remove the ignition coils.

7. Using the proper spark plug removal tool, remove the spark plugs.

To install:

➡**Be sure to use new fasteners, as required.**

8. Installation is the reverse of the removal procedure.

9. Tighten the plug to specification.

10. On the 3.7L engine, apply a small amount of dielectric grease to the inside of the ignition coils, then install the coils to the spark plugs.

11. Use the Mazda diagnostic scan tool, or equivalent and reprogram the required systems.

ENGINE ELECTRICAL

STARTING SYSTEM

STARTER

REMOVAL & INSTALLATION

See Figures 32 and 33.

1. Before servicing the vehicle, refer to the Precautions Section.

➡**If working near and/or around the SRS system and components, be sure to disable the SRS system. Tape the negative battery cable with insulating tape. Always disconnect the negative battery cable first.**

❊❊ CAUTION

To avoid personal injury when working on vehicles equipped with an air bag, the negative battery cable must

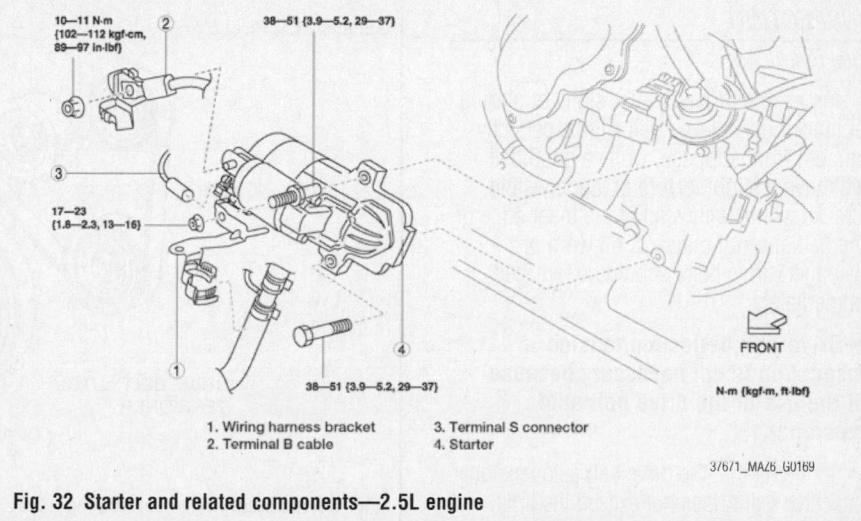

1. Wiring harness bracket
2. Terminal B cable
3. Terminal S connector
4. Starter

N·m {kgf·m, ft·lbf}

Fig. 32 Starter and related components—2.5L engine

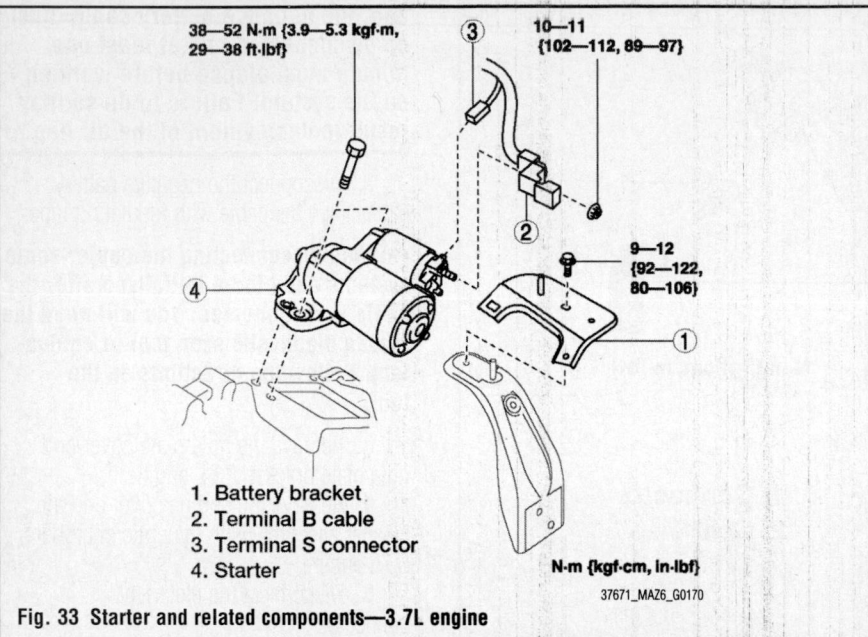

1. Battery bracket
2. Terminal B cable
3. Terminal S connector
4. Starter

38—52 N·m {3.9—5.3 kgf·m, 29—38 ft·lbf}

10—11 {102—112, 89—97}

9—12 {92—122, 80—106}

N·m {kgf·cm, in·lbf}

37671_MAZ6_G0170

Fig. 33 Starter and related components—3.7L engine

be disconnected and at least one minute must elapse before working on the system. Failure to do so may result in deployment of the air bag.

2. Disconnect the negative battery cable. Tape the cable with insulating tape.

➡**When disconnecting the cable, some systems need to be initialized after the** cable is reconnected. You will need the Mazda diagnostic scan tool or equivalent. Follow the directions on the tool.

3. Remove the plug hole plate, remove the air cleaner assembly and remove the MAP sensor, 2.5L engine.

4. Remove the battery and battery tray, 3.7L engine.

5. Remove the wiring harness bracket.

6. Disconnect the terminal B cable.

7. Disconnect the terminal S connector.

8. Remove the starter retaining bolts.

9. Remove the component from its mounting.

To install:

➡**Be sure to use new fasteners, as required.**

10. Installation is the reverse of the removal procedure.

11. Tighten retaining bolts to specification.

12. Use the Mazda diagnostic scan tool, or equivalent and reprogram the required systems.

ENGINE MECHANICAL

➡**Disconnecting the negative battery cable may interfere with the functions of the on board computer systems and may require the computer to undergo a relearning process, once the negative battery cable is reconnected.**

ACCESSORY DRIVE BELTS

ACCESSORY BELT ROUTING

See Figure 34.

Refer to the accompanying illustration.

INSPECTION

See Figure 35.

Inspect the drive belt for signs of glazing or cracking. A glazed belt will be perfectly smooth from slippage, while a good belt will have a slight texture of fabric visible. Cracks will usually start at the inner edge of the belt and run outward. All worn or damaged drive belts should be replaced immediately.

➡**Drive belt deflection/tension inspection is not necessary because of the use of the drive belt auto tensioner.**

1. Verify that the drive belt auto tensioner indicator mark does not exceed the limit.

2. If it exceeds the limit, replace the drive belt.

Verify that the drive belt auto tensioner indicator mark is within the maximum and minimum belt length.

When the generator and A/C drive belt is replaced with a new one, the drive belt auto tensioner indicator mark is aligned with the nominal belt length mark.

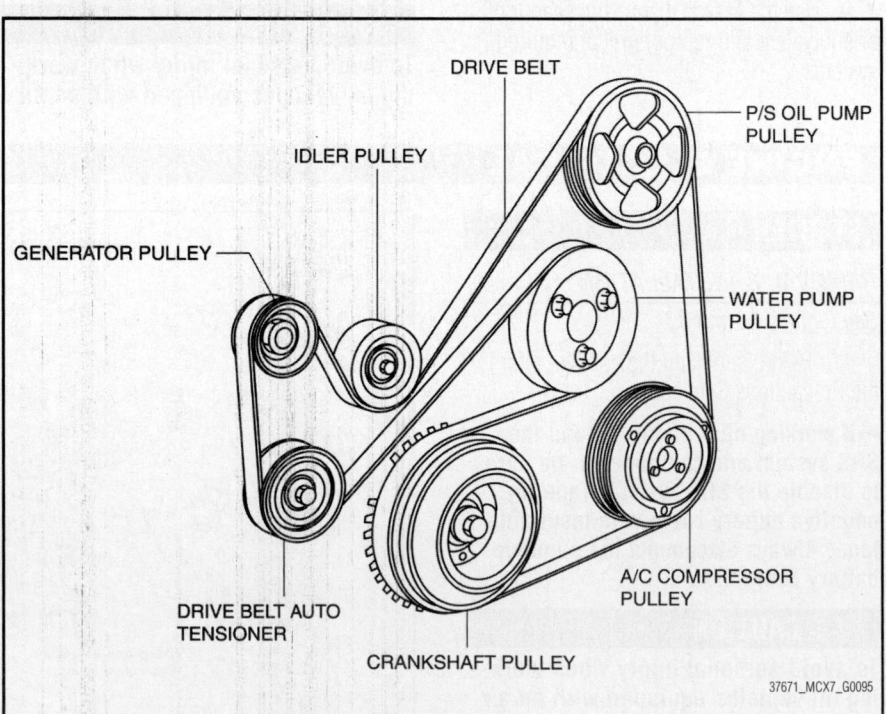

37671_MCX7_G0095

Fig. 34 Accessory drive belt routing—2.5L engine

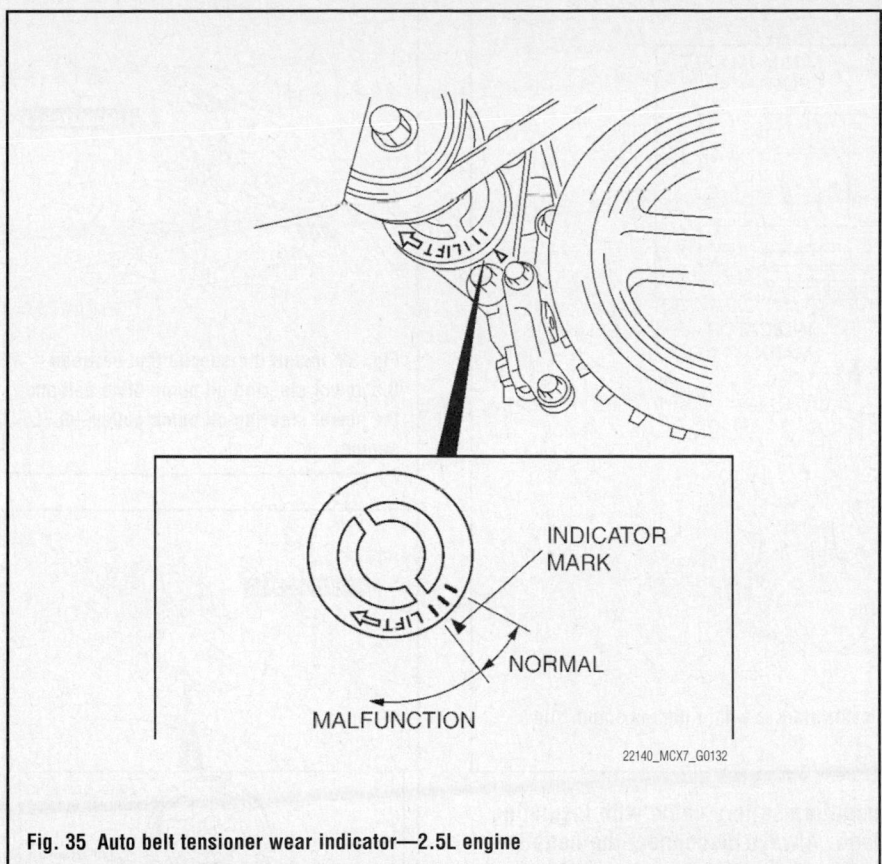

Fig. 35 Auto belt tensioner wear indicator—2.5L engine

ADJUSTMENT

On the 2.5L engine, a drive belt auto tensioner with an embedded coil spring has been adopted to automatically maintain the optimal drive belt tension.

On the 3.7L engine an auto tensioning pulley to adjust belt tension.

REMOVAL & INSTALLATION

2.5L Engine

See Figure 36.

1. Before servicing the vehicle, refer to the Precautions Section.

➡ **If working near and/or around the SRS system and components, be sure to disable the SRS system. Tape the negative battery cable with insulating tape. Always disconnect the negative battery cable first.**

✳✳ CAUTION

To avoid personal injury when working on vehicles equipped with an air bag, the negative battery cable must be disconnected and at least one minute must elapse before working on the system. Failure to do so may result in deployment of the air bag.

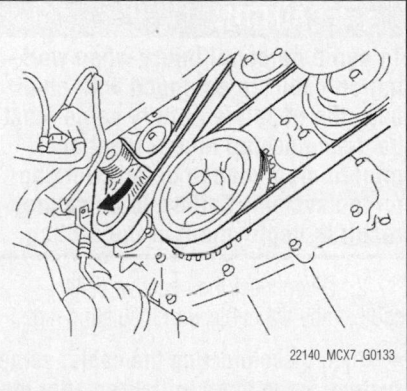

Fig. 36 Auto drive belt tensioner—2.5L engine

2. Disconnect the negative battery cable. Tape the cable with insulating tape.

➡ **When disconnecting the cable, some systems need to be initialized after the cable is reconnected. You will need the Mazda diagnostic scan tool or equivalent. Follow the directions on the tool.**

3. Set the mudguard out of the way (RH).
4. Remove the splash shield (RH).
5. Rotate the drive belt auto tensioner in the direction shown in the figure and remove the drive belt.

To install:

➡ **Be sure to use new fasteners, as required.**

6. Install the drive belt.
7. Verify that the drive belt auto tensioner indicator mark does not exceed the limit.
8. Install the splash shield (RH).
9. Use the Mazda diagnostic scan tool, or equivalent and reprogram the required systems.

3.7L Engine

Alternator And A/C Belt

See Figure 37.

✳✳ CAUTION

When removing the generator and A/C drive belt, perform the procedure with two people, one releasing the belt tension and the other removing the belt.

1. Before servicing the vehicle, refer to the Precautions Section.

➡ **If working near and/or around the SRS system and components, be sure to disable the SRS system. Tape the negative battery cable with insulating tape. Always disconnect the negative battery cable first.**

✳✳ CAUTION

To avoid personal injury when working on vehicles equipped with an air bag, the negative battery cable must be disconnected and at least one minute must elapse before working on the system. Failure to do so may result in deployment of the air bag.

2. Disconnect the negative battery cable. Tape the cable with insulating tape.

➡ **When disconnecting the cable, some systems need to be initialized after the cable is reconnected. You will need the Mazda diagnostic scan tool or equivalent. Follow the directions on the tool.**

3. Remove the engine cover.
4. Move the mudguard out of the way (RH).
5. Remove the splash shield (RH).
6. Disconnect the cooler hose.
7. Set a breaker bar on the center of the tensioner pulley.
8. Using the breaker bar, turn the center of the tensioner pulley clockwise to release the tension on the generator and A/C drive belt.

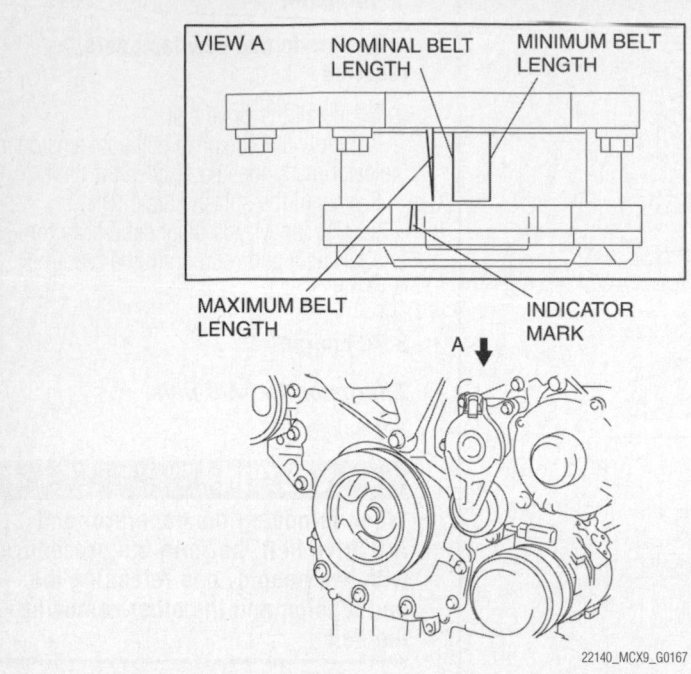

Fig. 37 Verify that the drive belt auto tensioner indicator mark is within the maximum and minimum belt length—3.7L engine

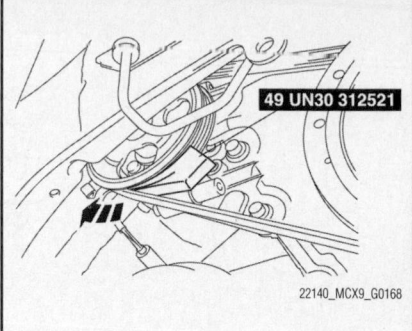

Fig. 38 Install the special tool between the power steering oil pump drive belt and the power steering oil pump pulley—3.7L engine

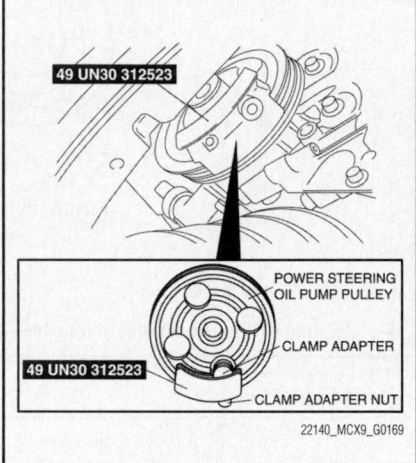

Fig. 39 Position the special tool on the power steering oil pump pulley with the clamp adapter in one of the holes of the pulley—3.7L engine

9. Remove the alternator and A/C drive belt.

To install:

➡ Be sure to use new fasteners, as required.

10. Install the alternator and A/C drive belt.

➡ When the alternator and A/C drive belt is replaced with a new one, the drive belt auto tensioner indicator mark is aligned with the nominal belt length mark.

11. Verify that the drive belt auto tensioner indicator mark is within the maximum and minimum belt length.

12. If the drive belt auto tensioner indicator mark is not within the limit, replace the alternator and A/C drive belt.

13. Install the splash shield (RH).

14. Install the engine cover.

15. Use the Mazda diagnostic scan tool, or equivalent and reprogram the required systems.

Power Steering Pump Belt

See Figures 38 through 40.

1. Before servicing the vehicle, refer to the Precautions Section.

➡ If working near and/or around the SRS system and components, be sure to disable the SRS system. Tape the negative battery cable with insulating tape. Always disconnect the negative battery cable first.

❊❊ CAUTION

To avoid personal injury when working on vehicles equipped with an air bag, the negative battery cable must be disconnected and at least one minute must elapse before working on the system. Failure to do so may result in deployment of the air bag.

2. Disconnect the negative battery cable. Tape the cable with insulating tape.

➡ When disconnecting the cable, some systems need to be initialized after the cable is reconnected. You will need the Mazda diagnostic scan tool or equivalent. Follow the directions on the tool.

3. Remove the alternator and A/C drive belt.

4. Remove the power steering oil pump drive belt.

5. Install the special tool between the power steering oil pump drive belt and the power steering oil pump pulley.

6. Hold the special tool by hand until it is lodges between the power steering oil pump pulley and power steering oil pump drive belt.

7. Turn the crankshaft clockwise to remove the power steering oil pump drive belt.

8. If there is power steering oil pump drive belt damage and cracks, replace the power steering oil pump drive belt.

9. Install the power steering oil pump drive belt.

10. Install the special tool according to the following steps.

11. Loosen the clamp adapter nut of the special tool.

12. There is no need to remove the clamp adapter nut completely when positioning the special tool on the power steering oil pump pulley.

13. Position the special tool on the power steering oil pump pulley with the clamp adapter in one of the holes of the pulley.

14. Tighten the clamp adapter nut by hand.

15. Hand tightening of the clamp adapter nut is sufficient.

16. Position the power steering oil pump drive belt around the special tool and the power steering oil pump pulley.

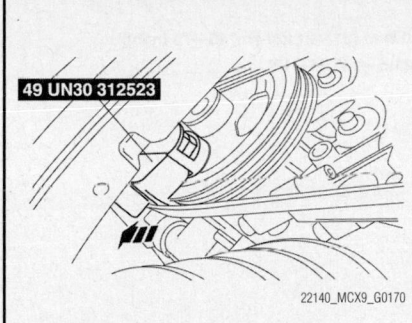

Fig. 40 Position the power steering oil pump drive belt around the special tool and the power steering oil pump pulley—3.7L engine

17. Hold the special tool and power steering oil pump drive belt by hand until the power steering oil pump drive belt is properly seated.

18. Turn the crankshaft clockwise to install the power steering oil pump drive belt.

19. Verify that the power steering oil pump drive belt is firmly attached to the pulleys.

20. Install the alternator and A/C drive belt.

21. Verify that the drive belt auto tensioner indicator mark does not exceeds the maximum belt length.

22. If it exceeds the maximum belt length, replace the generator and A/C drive belt.

23. Use the Mazda diagnostic scan tool, or equivalent and reprogram the required systems.

CAMSHAFT AND TAPPETS

REMOVAL & INSTALLATION

2.5L Engine

See Figures 41 through 45.

❊❊ WARNING

Fuel vapor is hazardous. It can very easily ignite, causing death, serious injury, or damage. Always keep sparks and flames away from fuel.

❊❊ WARNING

Fuel line spills and leakage from the pressurized fuel system are dangerous. Fuel can easily ignite and cause serious injury or death and damage. Fuel can also irritate skin and eyes.

1. Before servicing the vehicle, refer to the Precautions Section.

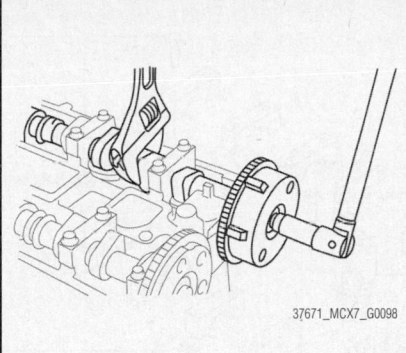

Fig. 41 Hold the camshaft using a wrench on the cast hexagon—2.5L engine

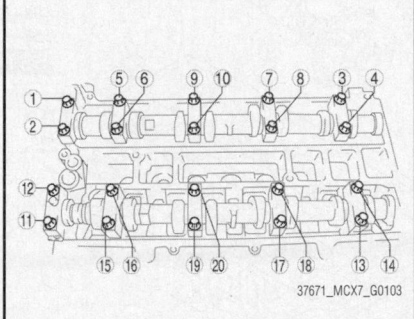

Fig. 42 Camshaft cap removal sequence—2.5L engine

➡If working near and/or around the SRS system and components, be sure to disable the SRS system. Tape the negative battery cable with insulating tape. Always disconnect the negative battery cable first.

❊❊ CAUTION

To avoid personal injury when working on vehicles equipped with an air bag, the negative battery cable must be disconnected and at least one minute must elapse before working on the system. Failure to do so may result in deployment of the air bag.

2. Disconnect the negative battery cable. Tape the cable with insulating tape.

➡When disconnecting the cable, some systems need to be initialized after the cable is reconnected. You will need the Mazda diagnostic scan tool or equivalent. Follow the directions on the tool.

3. Remove the timing chain.

4. Before removing the camshaft, hold the camshaft using a wrench on the cast

hexagon and loosen the camshaft sprocket installation bolt.

➡**Before removing the camshafts inspect camshaft end play.**

5. Loosen the camshaft cap bolts in two or three passes in the order shown and remove the end caps.

➡**Note the position of each removed end cap for installation.**

6. Remove the camshafts.

To install:

➡**Be sure to use new fasteners, as required.**

➡**Install the camshaft with number 1 cylinder cam aligned at TDC of the compression stroke.**

7. Apply a small amount of gear oil to the camshaft journal of the cylinder head before you install the caps.

8. Install camshaft caps finger tight, then in two or three passes tighten the caps to:

- First pass: 4.3–79.5 inch lbs. (5–9 Nm) in the order shown.
- Second pass: 10–13 ft. lbs (14–17 Nm) in the order shown.

9. Tighten the camshaft sprocket installation bolts to 51–55 ft. lbs. (69–75 Nm).

10. Install the timing chain.

11. Refill the engine with coolant.

12. Bleed the air from the cooling system.

13. Inspect the compression pressure.

14. Use the Mazda diagnostic scan tool, or equivalent and reprogram the required systems.

3.7L Engine

See Figures 46 through 62.

➡**Mazda's official procedure for removal and installation of this component requires the engine/transaxle assembly to be removed from the vehicle. Remove the engine/transaxle assembly and position it in a suitable holding fixture.**

1. Before servicing the vehicle, refer to the Precautions Section.

➡**If working near and/or around the SRS system and components, be sure to disable the SRS system. Tape the negative battery cable with insulating tape. Always disconnect the negative battery cable first.**

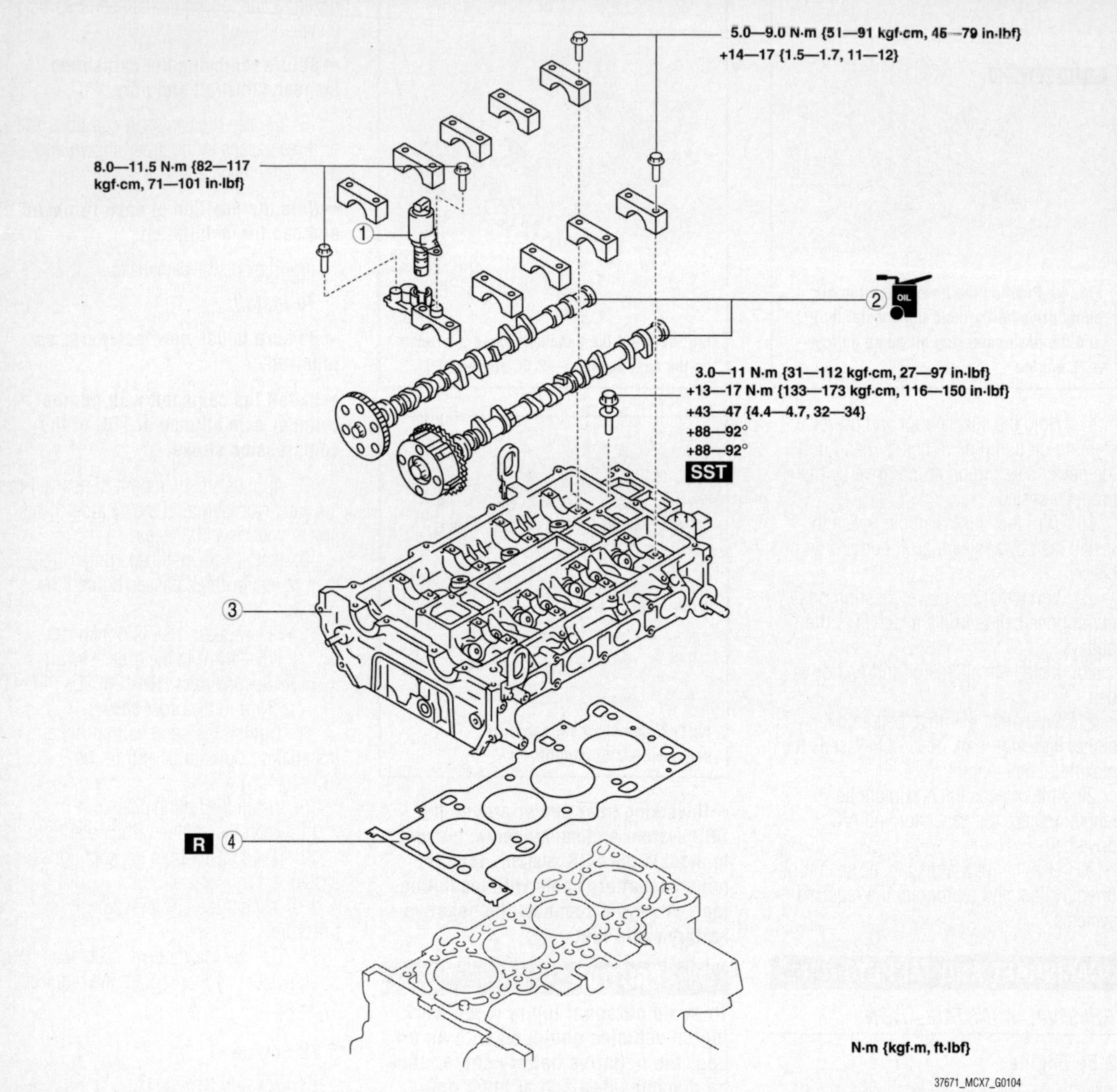

5.0—9.0 N·m {51—91 kgf·cm, 45—79 in·lbf}
+14—17 {1.5—1.7, 11—12}
8.0—11.5 N·m {82—117
kgf·cm, 71—101 in·lbf}
1
2
OIL
3.0—11 N·m {31—112 kgf·cm, 27—97 in·lbf}
+13—17 N·m {133—173 kgf·cm, 116—150 in·lbf}
+43—47 {4.4—4.7, 32—34}
+88—92°
+88—92°
SST
3
R 4
N·m {kgf·m, ft·lbf}
37671_MCX7_G0104

Fig. 43 Cylinder head exploded view—2.5L engine

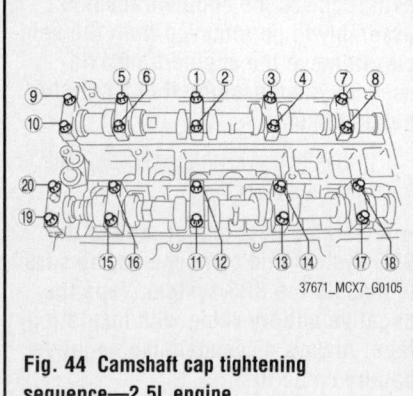

5 6 1 2 3 4 7 8
9
10
20
19
15 16 11 12 13 14 17 18
37671_MCX7_G0105

Fig. 44 Camshaft cap tightening sequence—2.5L engine

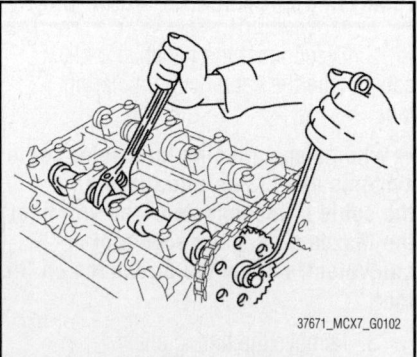

37671_MCX7_G0102

Fig. 45 Tightening the camshaft sprocket retaining bolt—2.5L engine

**** CAUTION**

To avoid personal injury when working on vehicles equipped with an air bag, the negative battery cable must be disconnected and at least one minute must elapse before working on the system. Failure to do so may result in deployment of the air bag.

2. Disconnect the negative battery cable. Tape the cable with insulating tape.

➡When disconnecting the cable, some systems need to be initialized after the cable is reconnected. You will need the

Mazda diagnostic scan tool or equivalent. Follow the directions on the tool.

3. Drain the engine oil. Be sure to properly dispose of used oil.

4. Drain the cooling system. Be sure to properly dispose of used coolant.

5. Drain the transaxle. Be sure to dispose of used fluid.

6. Properly relieve the fuel system pressure.

7. Remove the engine/transaxle assembly. Position the unit in a suitable holding fixture.

8. Remove the intake manifold dynamic chamber and throttle body as a single unit.

9. Remove the fuel injector and fuel distributor together as a single unit.

10. Remove the thermostat and thermostat housing together as a single unit.

11. Remove the intake manifold.

12. Remove both catalytic converters together with both exhaust manifolds as a single unit.

13. Remove the ignition coils.

14. Remove the dipstick.

15. Remove the power steering oil pump drive belt.

16. Remove the power steering oil pump.

17. Remove the alternator.

➡**When removing the timing chain and marking the timing marks on the chain, mark the camshaft timing chain as well.**

18. Remove the timing chain.

19. Remove the camshaft caps.

20. Remove the camshaft timing chain tensioners.

21. Remove the camshaft timing chains.

22. Remove the camshafts.

❋❋ WARNING

Do not rotate the crankshaft counterclockwise. The timing chains may bind, causing engine damage.

a. Turn the crankshaft clockwise so that the crankshaft keyway is in the 11 o'clock position. This will position the No.1 cylinder at TDC.

➡**Verify that there are timing marks in three locations (Yellow 1, Black 2) on the timing chain. If any timing marks are missing, mark the timing chain. When marking the crankshaft sprocket side timing chain, change the mark color. When the timing chain is replaced with a new one, mark the new timing chain at the same positions as the removed timing chain.**

b. Mark the timing chain at the position of each timing sprocket timing mark.

➡**Verify that there are timing marks in two locations on the camshaft timing chain. If any timing marks are missing, mark the camshaft timing chain. If replacing with a new camshaft timing chain, place alignment marks in the same positions as those prior to the replacement.**

c. Mark the camshaft timing chain at the positions where it is aligned with each of the camshaft sprocket on both banks.

d. Remove the timing chain in the following order.
- Chain tensioner
- Timing chain
- Tensioner arm
- Chain guide
- Crankshaft sprocket

e. Rotate the crankshaft counterclockwise until the keyway is in the 9 o'clock position.

f. Slowly compress the camshaft timing chain tensioner (LH) piston by hand.

g. Insert an approx. 1.0 mm {0.039 in} thin wire or paper clip into the camshaft timing chain tensioner shown. to hold the tensioner piston.

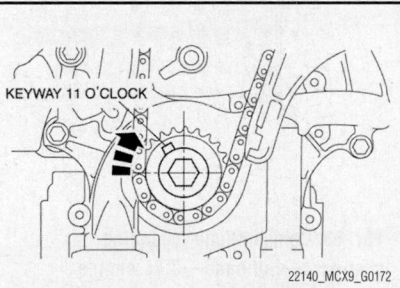

Fig. 46 Turn the crankshaft clockwise so that the crankshaft keyway is in the 11 o'clock position. This will position the No.1 cylinder at TDC—3.7L engine

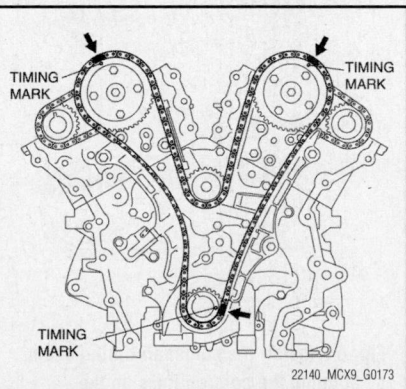

Fig. 47 Mark the timing chain at the position of each timing sprocket timing mark—3.7L engine

➡**When the timing chain removed, valve spring pressure will rotate the (LH) camshaft approx. 3° to a neutral position. On the (RH) camshaft, rotate the camshaft counterclockwise, and set the camshaft to the neutral position.**

h. Verify that the camshafts are in the neutral position.

➡**The cylinder head and the camshaft bearing caps are numbered to make sure they are reassembled in their original position. When removed, keep the bearing caps with the cylinder head they were removed from. Do not mix the caps.**

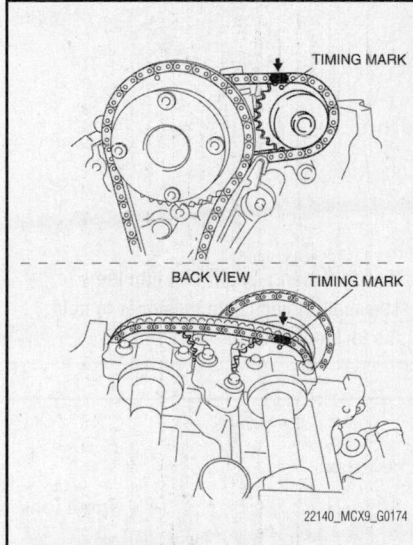

Fig. 48 Mark the camshaft timing chain at the positions where it is aligned with each of the camshaft sprocket on both banks—3.7L engine

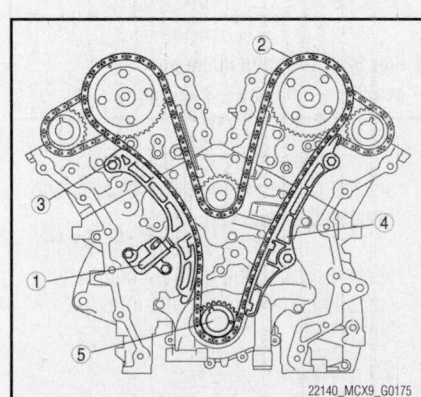

Fig. 49 Remove the timing chain in the following order: (1) Chain tensioner, (2) Timing chain, (3) Tensioner arm, (4) Chain guide and (5) Crankshaft sprocket—3.7L engine

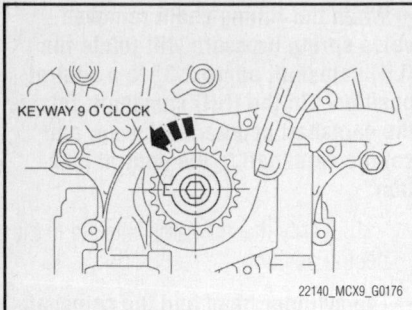

Fig. 50 Rotate the crankshaft counterclockwise until the keyway is in the 9 o'clock position—3.7L engine

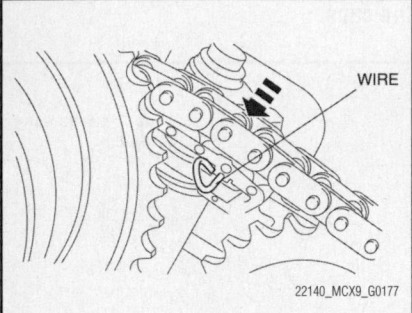

Fig. 51 Insert a paper clip into the camshaft timing chain tensioner to hold the tensioner piston—3.7L engine

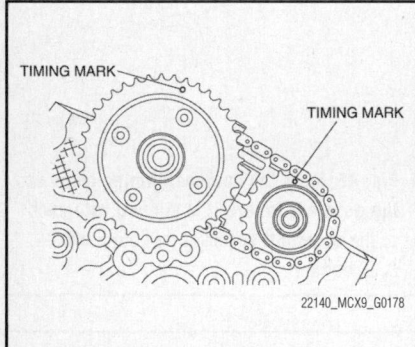

Fig. 52 Camshafts in the neutral position—left hand—3.7L engine

i. Loosen the camshaft cap bolts in several passes in the order shown. and remove the camshaft cap.

j. Remove the camshaft, camshaft sprocket, camshaft timing chain, and camshaft timing chain tensioner as a single unit.

To install:

➡**Be sure to use new fasteners, as required.**

23. Assemble the camshaft timing chain tensioners on both sides. Tighten to 76–101 inch lbs. (8.5–11.5 Nm).

24. Install the right hand camshaft.

➡**Install the camshaft, camshaft sprocket and camshaft timing chain of the both bank as a single unit.**

   a. Install the camshafts.

   b. Install the camshaft timing chain by aligning the colored links on the camshaft timing chain with the marks on the camshaft sprockets.

   c. Position the camshaft component onto the cylinder head in the neutral position as shown.

   d. Install the camshaft caps and temporarily tighten the camshaft cap bolts evenly in the order shown in several passes. Tighten to 76–101 inch lbs. (8.5–11.5 Nm).

   e. Remove the retaining wire inserted into the camshaft timing chain tensioner.

25. Using the right hand camshaft sprocket installation bolt, rotate the right

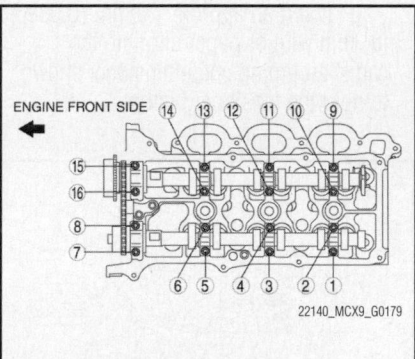

Fig. 53 Camshaft cap loosening sequence—left hand—3.7L engine

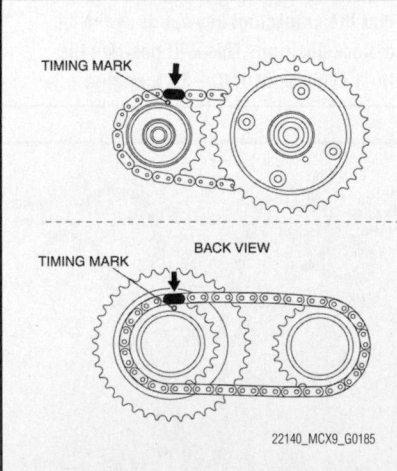

Fig. 54 Install the camshaft timing chain by aligning the colored links on the camshaft timing chain with the marks on the camshaft sprockets—right hand—3.7L engine

hand camshaft clockwise so the timing marks align as shown.

26. Install the left hand camshaft in the same manner as the right hand camshaft.

27. Rotate the crankshaft clockwise so that the crankshaft keyway is in the 11 o'clock position. This will position the No.1 cylinder at TDC.

28. Install the timing chain.

29. Install the alternator.

30. Install the power steering oil pump.

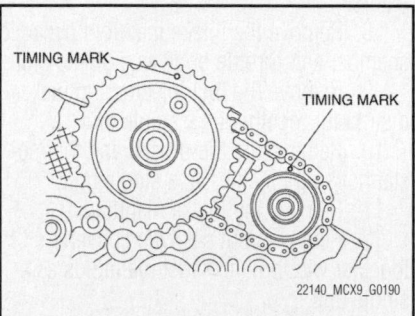

Fig. 55 Position the camshaft component onto the cylinder head in the neutral position as shown—left hand—3.7L engine

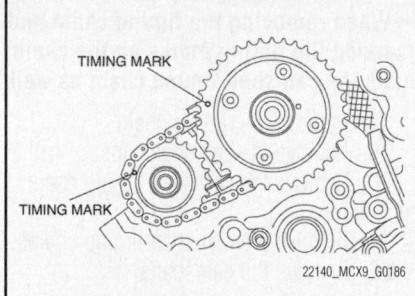

Fig. 56 Position the camshaft component onto the cylinder head in the neutral position as shown—right hand—3.7L engine

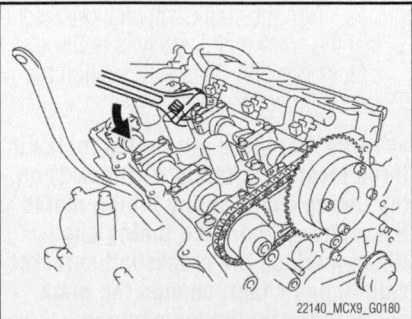

Fig. 57 Put a wrench as shown on the camshaft, rotate the camshaft counterclockwise, and set the camshaft to the neutral position—right hand—3.7L engine

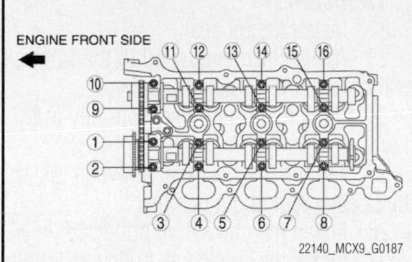

Fig. 58 Tighten the camshaft cap bolts to the specified torque, in several passes, in the order shown—right hand—3.7L engine

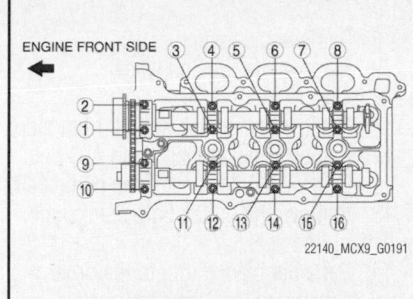

Fig. 61 Tighten the camshaft cap bolts to the specified torque, in several passes, in the order shown—left hand—3.7L engine

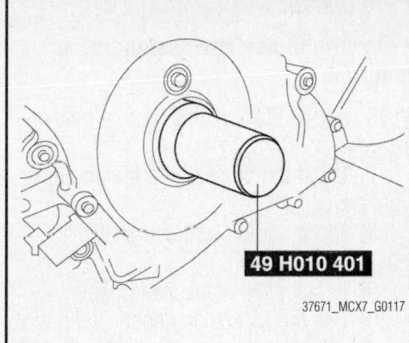

Fig. 63 SST 49 H010 401 seal driver—2.5L engine

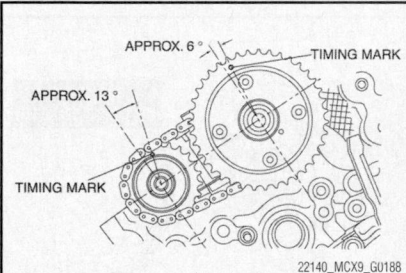

Fig. 59 Using the right hand camshaft sprocket installation bolt, rotate the right hand camshaft clockwise so the timing marks align as shown—3.7L engine

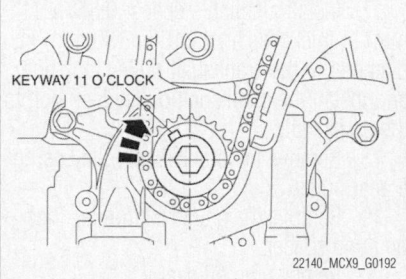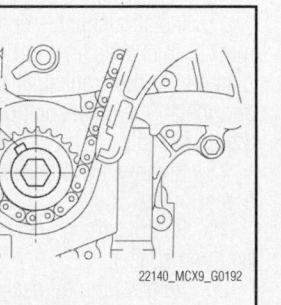

Fig. 62 Turn the crankshaft clockwise so that the crankshaft keyway is in the 11 o'clock position. This will position the No.1 cylinder at TDC. —3.7L engine

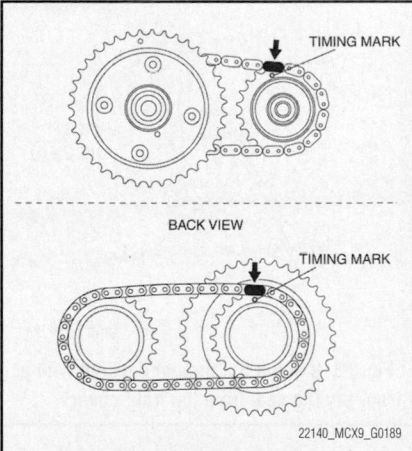

Fig. 60 Install the camshaft timing chain by aligning the colored links on the camshaft timing chain with the marks on the camshaft sprockets—left hand—3.7L engine

31. Install the power steering oil pump drive belt.
32. Install the dipstick.
33. Install the ignition coils.
34. Install both catalytic converters together with both exhaust manifolds as a single unit.
35. Install the Intake manifold.

36. Install the thermostat and thermostat housing together as a single unit.
37. Install the fuel injector and fuel distributor together as a single unit.
38. Install the intake manifold dynamic chamber and throttle body as a single unit.
39. Install the engine and transaxle.
40. Fill the engine with oil.
41. Start the engine and perform the following inspections:
   a. Inspect the runout and contact on the pulley and belt.
   b. Inspect for engine oil, engine coolant, ATF, power steering fluid and fuel leakage.
   c. Verify the ignition timing and idle speed.
   d. Inspect engine accessories for proper operation.
42. Perform a road test.
43. Use the Mazda diagnostic scan tool, or equivalent and reprogram the required systems.

CRANKSHAFT FRONT SEAL

REMOVAL & INSTALLATION

2.5L Engine
See Figures 63 and 64.

1. Before servicing the vehicle, refer to the Precautions Section.

➡ If working near and/or around the SRS system and components, be sure to disable the SRS system. Tape the negative battery cable with insulating tape. Always disconnect the negative battery cable first.

✳✳ CAUTION

To avoid personal injury when working on vehicles equipped with an air bag, the negative battery cable must be disconnected and at least one minute must elapse before working on the system. Failure to do so may result in deployment of the air bag.

2. Disconnect the negative battery cable. Tape the cable with insulating tape.

➡ When disconnecting the cable, some systems need to be initialized after the cable is reconnected. You will need the Mazda diagnostic scan tool or equivalent. Follow the directions on the tool.

3. Remove the crankshaft pulley.
4. Remove the oil seal lip using a razor.
5. Remove the oil seal using a suitable prytool.

To install:

➡ **Be sure to use new fasteners, as required.**

6. Apply clean engine oil to a new oil seal.

7. Insert the oil seal into the engine front cover.

8. Tap in the oil seal using SST 49 H010 401 as shown.

9. Replace the crankshaft pulley.

10. Use the Mazda diagnostic scan tool, or equivalent and reprogram the required systems.

3.7L Engine

See Figures 65 through 71.

➡ **Mazda's official procedure for removal and installation of this component requires the engine/transaxle assembly to be removed from the vehicle. Remove the engine/transaxle assembly and position it in a suitable holding fixture.**

1. Before servicing the vehicle, refer to the Precautions Section.

➡ **If working near and/or around the SRS system and components, be sure to disable the SRS system. Tape the negative battery cable with insulating tape. Always disconnect the negative battery cable first.**

※※ CAUTION

To avoid personal injury when working on vehicles equipped with an air bag, the negative battery cable must be disconnected and at least one minute must elapse before working on the system. Failure to do so may result in deployment of the air bag.

2. Disconnect the negative battery cable. Tape the cable with insulating tape.

➡ **When disconnecting the cable, some systems need to be initialized after the cable is reconnected. You will need the Mazda diagnostic scan tool or equivalent. Follow the directions on the tool.**

3. Drain the engine oil. Be sure to properly dispose of used oil.

4. Drain the cooling system. Be sure to properly dispose of used coolant.

5. Drain the transaxle. Be sure to dispose of used fluid.

6. Properly relieve the fuel system pressure.

7. Remove the engine/transaxle assem-

bly. Position the unit in a suitable holding fixture.

8. Remove the engine cover.

9. Remove the tire and wheel assemblies.

10. Remove the splash shield, right side.

11. Remove the alternator and A/C belt.

12. Remove the power steering pump belt.

13. Remove the pulley bolt, crankshaft pulley and oil seal.

14. Set a flat bladed tool to the drive plate as shown to lock the crankshaft rotation.

15. Remove the crankshaft pulley lock bolt and washer.

16. Remove the crankshaft pulley.

17. Install a suitable spacer 0.55 in. (14 mm) in thickness and 1.18 in. (30 mm) in diameter to the crankshaft pulley lock bolt, and install the crankshaft pulley lock bolt to the crankshaft.

18. Remove the crankshaft pulley using a gear puller.

19. Remove the crankshaft pulley lock bolt and spacer.

20. Remove the oil seal.

21. Using a flat tip screwdriver cover with a rag, pry the seal from the front cover

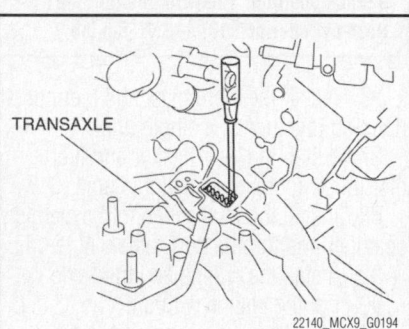

Fig. 65 Set a flathead screwdriver to the drive plate as shown to lock the crankshaft rotation

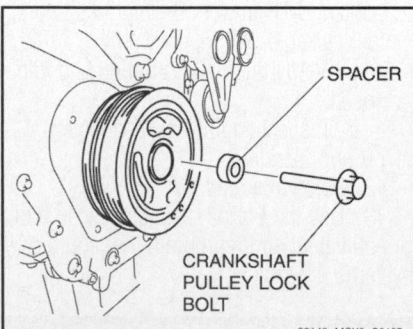

Fig. 66 Install a suitable spacer 0.55 in. (14 mm) in thickness and 1.18 in. (30 mm) in diameter to the crankshaft pulley lock bolt, and install the crankshaft pulley lock bolt to the crankshaft

To install:

22. Install the oil seal.

23. Apply clean engine oil to the front oil seal bore in the engine front cover.

24. Push the front oil seal slightly in by hand.

25. Tap the front oil seal in evenly using the special tool and a hammer.

26. Install the crankshaft pulley.

27. Install the crankshaft pulley lock bolt.

28. Set a flat tip tool to the drive plate in the position indicated. to lock the crankshaft rotation.

29. Tighten the new crankshaft pulley lock bolt in four steps.

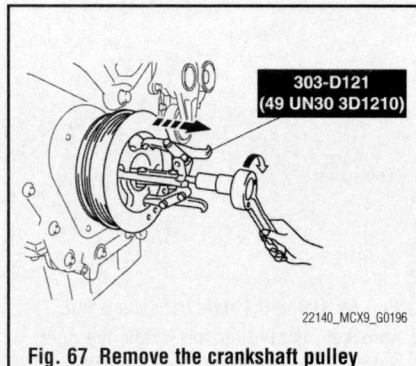

Fig. 67 Remove the crankshaft pulley using a gear puller

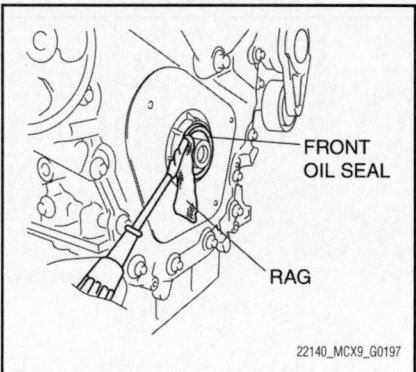

Fig. 68 Using a flat tip tool covered with a rag, pry the seal from the front cover

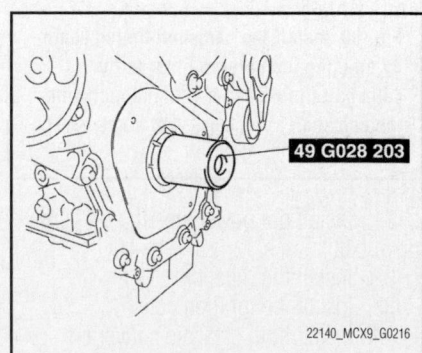

Fig. 69 Tap the front oil seal in evenly using the special tool and a hammer

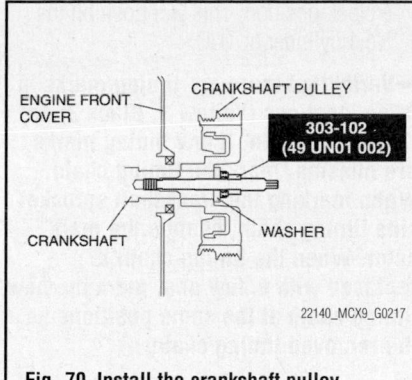

Fig. 70 Install the crankshaft pulley, washer, and special tool to the crankshaft

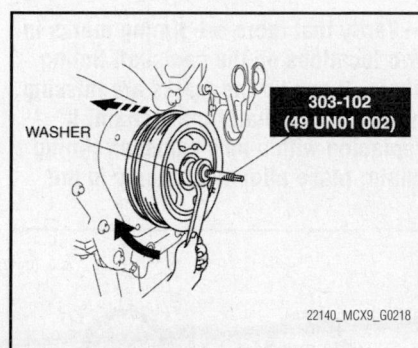

Fig. 71 Tighten the special tool nut and install the crankshaft pulley

- Step 1: Tighten to 103 ft. lbs. (140 Nm)
- Step 2: Loosen 360° (one full turn)
- Step 3: Tighten to 35–39 ft. lbs (47–53 Nm)
- Step 4: Tighten an additional 85–95° of rotation

30. Install the power steering oil pump.

31. Install the power steering oil pump drive belt.

32. Install the dipstick.

33. Install the ignition coils.

34. Install the intake manifold dynamic chamber and throttle body as a single unit.

35. Install the engine and transaxle.

36. Start the engine and perform the following inspections:

    a. Inspect the runout and contact on the pulley and belt.

    b. Inspect engine accessories for proper operation.

37. Use the Mazda diagnostic scan tool, or equivalent and reprogram the required systems.

CYLINDER HEAD

REMOVAL & INSTALLATION

2.5L Engine

See Figures 43 and 72.

> ✳✳ **WARNING**
>
> **Fuel vapor is hazardous. It can very easily ignite, causing death, serious injury, or damage. Always keep sparks and flames away from fuel.**

> ✳✳ **WARNING**
>
> **Fuel line spills and leakage from the pressurized fuel system are dangerous. Fuel can easily ignite and cause serious injury or death and damage. Fuel can also irritate skin and eyes.**

1. Before servicing the vehicle, refer to the Precautions Section.

➡If working near and/or around the SRS system and components, be sure to disable the SRS system. Tape the negative battery cable with insulating tape. Always disconnect the negative battery cable first.

> ✳✳ **CAUTION**
>
> **To avoid personal injury when working on vehicles equipped with an air bag, the negative battery cable must be disconnected and at least one minute must elapse before working on the system. Failure to do so may result in deployment of the air bag.**

2. Disconnect the negative battery cable. Tape the cable with insulating tape.

➡When disconnecting the cable, some systems need to be initialized after the cable is reconnected. You will need the Mazda diagnostic scan tool or equivalent. Follow the directions on the tool.

3. Drain the engine coolant.

4. Remove the generator. Place the

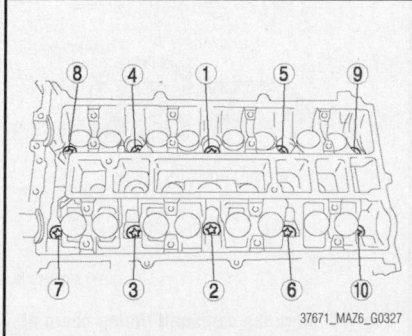

Fig. 72 Cylinder head bolt tightening sequence—2.5L engine

---

generator out of the way with the wiring harnesses connected.

5. Remove the exhaust manifold.

6. Remove the intake manifold.

7. Disconnect the heater hose and radiator hose.

8. Remove the timing chain.

9. Remove the camshafts.

10. Disconnect the wiring harness.

11. To firmly support the engine, first set the engine jack and attachment to the oil pan.

12. Remove the oil control valve

13. Remove the camshafts.

14. Remove the bolts and the cylinder head.

To install:

➡Be sure to use new fasteners, as required.

15. Measure the cylinder head bolts. Replace any bolts that exceed 5.740 inches (146 mm).

16. Install the cylinder head with a new gasket. Tighten the bolts in sequence as follows:

    a. Step 1: 27–97 inch lbs. (3–11 Nm)

    b. Step 2: 116–150 inch lbs. (13–17 Nm)

    c. Step 3: 32–34 ft. lbs. (43–47 Nm)

    d. Step 4: Plus 88°–92°

    e. Step 5: Plus 88°–92°

17. Continue the installation in the reverse of the removal procedure.

18. Bleed the air from the cooling system.

19. Inspect the compression pressure.

20. Use the Mazda diagnostic scan tool, or equivalent and reprogram the required systems.

3.7L Engine

See Figures 73 through 92.

➡Mazda's official procedure for removal and installation of this component requires the engine/transaxle assembly to be removed from the vehicle. Remove the engine/transaxle assembly and position it in a suitable holding fixture.

1. Before servicing the vehicle, refer to the Precautions Section.

➡If working near and/or around the SRS system and components, be sure to disable the SRS system. Tape the negative battery cable with insulating tape. Always disconnect the negative battery cable first.

✳✳ CAUTION

To avoid personal injury when working on vehicles equipped with an air bag, the negative battery cable must be disconnected and at least one minute must elapse before working on the system. Failure to do so may result in deployment of the air bag.

2. Disconnect the negative battery cable. Tape the cable with insulating tape.

➡When disconnecting the cable, some systems need to be initialized after the cable is reconnected. You will need the Mazda diagnostic scan tool or equivalent. Follow the directions on the tool.

3. Drain the engine oil. Be sure to properly dispose of used oil.

4. Drain the cooling system. Be sure to properly dispose of used coolant.

5. Drain the transaxle. Be sure to dispose of used fluid.

6. Properly relieve the fuel system pressure.

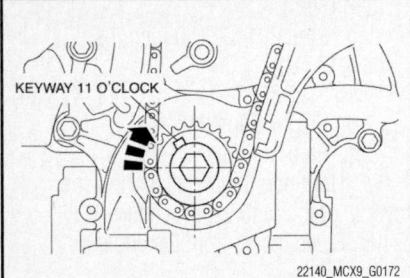

Fig. 73 Turn the crankshaft clockwise so that the crankshaft keyway is in the 11 o'clock position. This will position the No.1 cylinder at TDC—3.7L engine

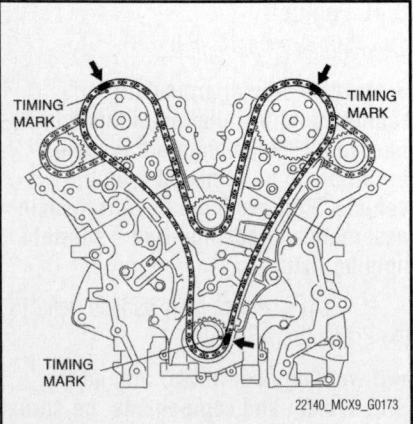

Fig. 74 Mark the timing chain at the position of each timing sprocket timing mark—3.7L engine

7. Remove the engine/transaxle assembly. Position the unit in a suitable holding fixture.

8. Remove the intake manifold dynamic chamber and throttle body as a single unit.

9. Remove the fuel injector and fuel distributor together as a single unit.

10. Remove the thermostat and thermostat housing together as a single unit.

11. Remove the intake manifold.

12. Remove both catalytic converters together with both exhaust manifolds as a single unit.

13. Remove the ignition coils.

14. Remove the dipstick.

15. Remove the power steering oil pump drive belt.

16. Remove the power steering oil pump.

17. Remove the alternator.

➡When removing the timing chain and marking the timing marks on the chain, mark the camshaft timing chain as well.

18. Remove the timing chain.

19. Remove the camshaft caps.

20. Remove the camshaft timing chain tensioners.

21. Remove the camshaft timing chains.

22. Remove the camshaft components.

✳✳ WARNING

Do not rotate the crankshaft counterclockwise. The timing chains may bind, causing engine damage.

a. Turn the crankshaft clockwise so that the crankshaft keyway is in the 11

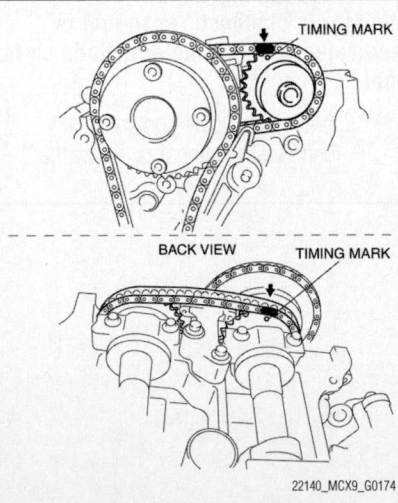

Fig. 75 Mark the camshaft timing chain at the positions where it is aligned with each of the camshaft sprocket on both banks—3.7L engine

o'clock position. This will position the No.1 cylinder at TDC.

➡Verify that there are timing marks in three locations (Yellow 1, Black 2) on the timing chain. If any timing marks are missing, mark the timing chain. When marking the crankshaft sprocket side timing chain, change the mark color. When the timing chain is replaced with a new one, mark the new timing chain at the same positions as the removed timing chain.

b. Mark the timing chain at the position of each timing sprocket timing mark.

➡Verify that there are timing marks in two locations on the camshaft timing chain. If any timing marks are missing, mark the camshaft timing chain. If replacing with a new camshaft timing chain, place alignment marks in the

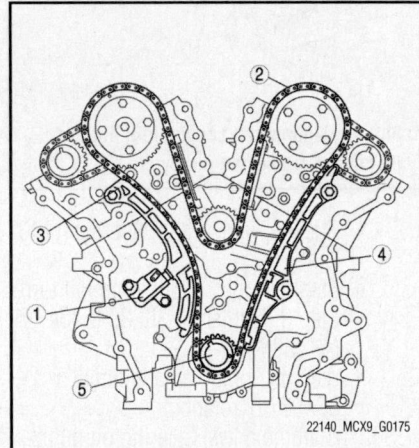

Fig. 76 Remove the timing chain in the following order: (1) Chain tensioner, (2) Timing chain, (3) Tensioner arm, (4) Chain guide and (5) Crankshaft sprocket—3.7L engine

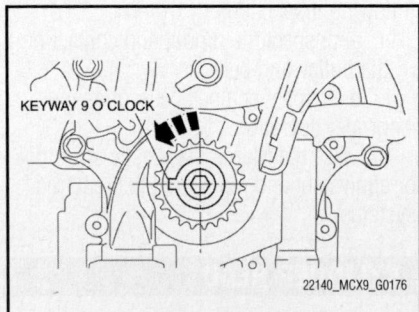

Fig. 77 Rotate the crankshaft counterclockwise until the keyway is in the 9 o'clock position—3.7L engine

same positions as those prior to the replacement.

   c. Mark the camshaft timing chain at the positions where it is aligned with each of the camshaft sprocket on both banks.

   d. Remove the timing chain in the following order:
- Chain tensioner
- Timing chain
- Tensioner arm
- Chain guide
- Crankshaft sprocket

   e. Rotate the crankshaft counterclockwise until the keyway is in the 9 o'clock position.

   f. Slowly compress the camshaft timing chain tensioner (LH) piston by hand.

   g. Insert an approx. 1.0 mm {0.039 in} thin wire or paper clip into the camshaft timing chain tensioner shown. to hold the tensioner piston.

➡ **When the timing chain removed, valve spring pressure will rotate the (LH) camshaft approx. 3° to a neutral position. On the (RH) camshaft, rotate the camshaft counterclockwise, and set the camshaft to the neutral position.**

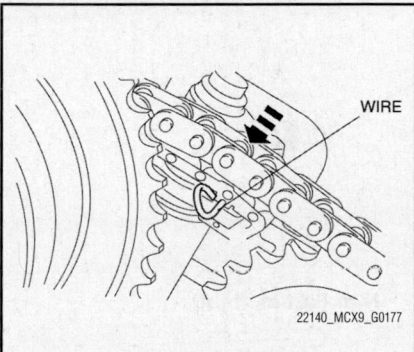

Fig. 78 Insert a paper clip into the camshaft timing chain tensioner to hold the tensioner piston—3.7L engine

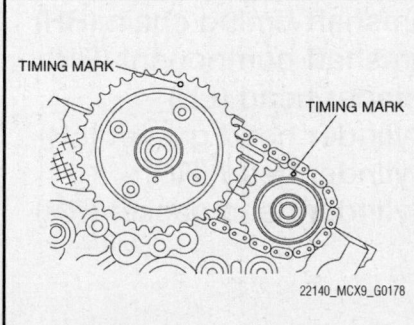

Fig. 79 Camshafts in the neutral position—3.7L engine

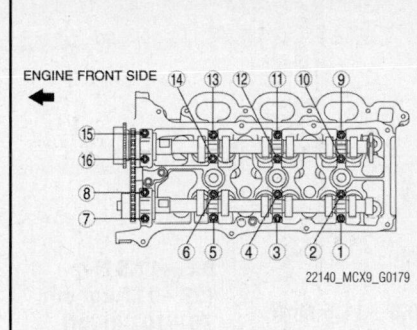

Fig. 80 Camshaft cap loosening sequence—3.7L engine

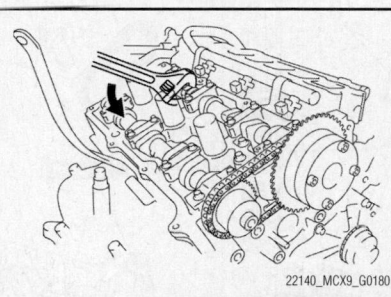

Fig. 81 Put a wrench as shown on the camshaft, rotate the camshaft counterclockwise, and set the camshaft to the neutral position—right hand—3.7L engine

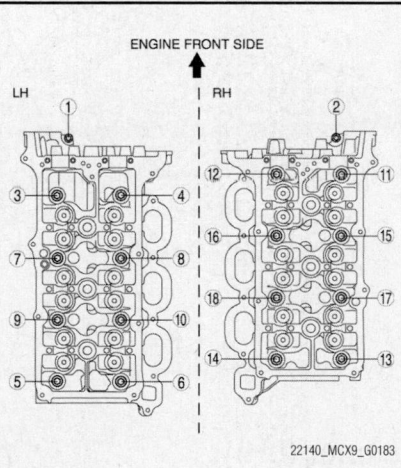

Fig. 82 Loosen the cylinder head bolts in several passes in the order shown—3.7L engine

   h. Verify that the camshafts are in the neutral position.

➡ **The cylinder head and the camshaft bearing caps are numbered to make sure they are reassembled in their original position. When removed, keep the bearing caps with the cylinder head**

they were removed from. Do not mix the caps.

   i. Loosen the camshaft cap bolts in several passes in the order shown. and remove the camshaft cap.

   j. Remove the camshaft, camshaft sprocket, camshaft timing chain, and camshaft timing chain tensioner as a single unit.

23. Remove the cylinder head.
24. Remove the cylinder head gasket.

To install:

25. Install the cylinder head gasket.
26. Install the cylinder head.
27. Install the cylinder head bolts and tighten to the following torque in the order shown:
- Step 1: Tighten bolts 1–16 to 15 ft. lbs. (20 Nm).
- Step 2: Tighten bolts 1–16 to 26 ft. lbs. (35 Nm).
- Step 3: Tighten bolts 1–16 an additional 90° rotation.
- Step 4: Tighten bolts 1–16 an additional 90° rotation.
- Step 5: Tighten bolts 1–16 an additional 90° rotation.
- Step 6: Tighten bolts 17–18 to 76–101 inch lbs. (9–12 Nm).

28. Assemble the camshaft timing chain tensioners on both sides. Tighten to 76–101 inch lbs. (8.5–11.5 Nm).
29. Install the right hand camshaft.

➡ **Install the camshaft, camshaft sprocket and camshaft timing chain of the both bank as a single unit.**

   a. Install the camshafts.

   b. Install the camshaft timing chain by aligning the colored links on the camshaft timing chain with the marks on the camshaft sprockets.

   c. Position the camshaft component onto the cylinder head in the neutral position as shown.

   d. Install the camshaft caps and temporarily tighten the camshaft cap bolts evenly in the order shown in several passes. Tighten to 76–101 inch lbs. (8.5–11.5 Nm).

   e. Remove the retaining wire inserted into the camshaft timing chain tensioner.

30. Using the right hand camshaft sprocket installation bolt, rotate the right hand camshaft clockwise so the timing marks align as shown.
31. Install the left hand camshaft in the same manner as the right hand camshaft.
32. Rotate the crankshaft clockwise so that the crankshaft keyway is in the 11

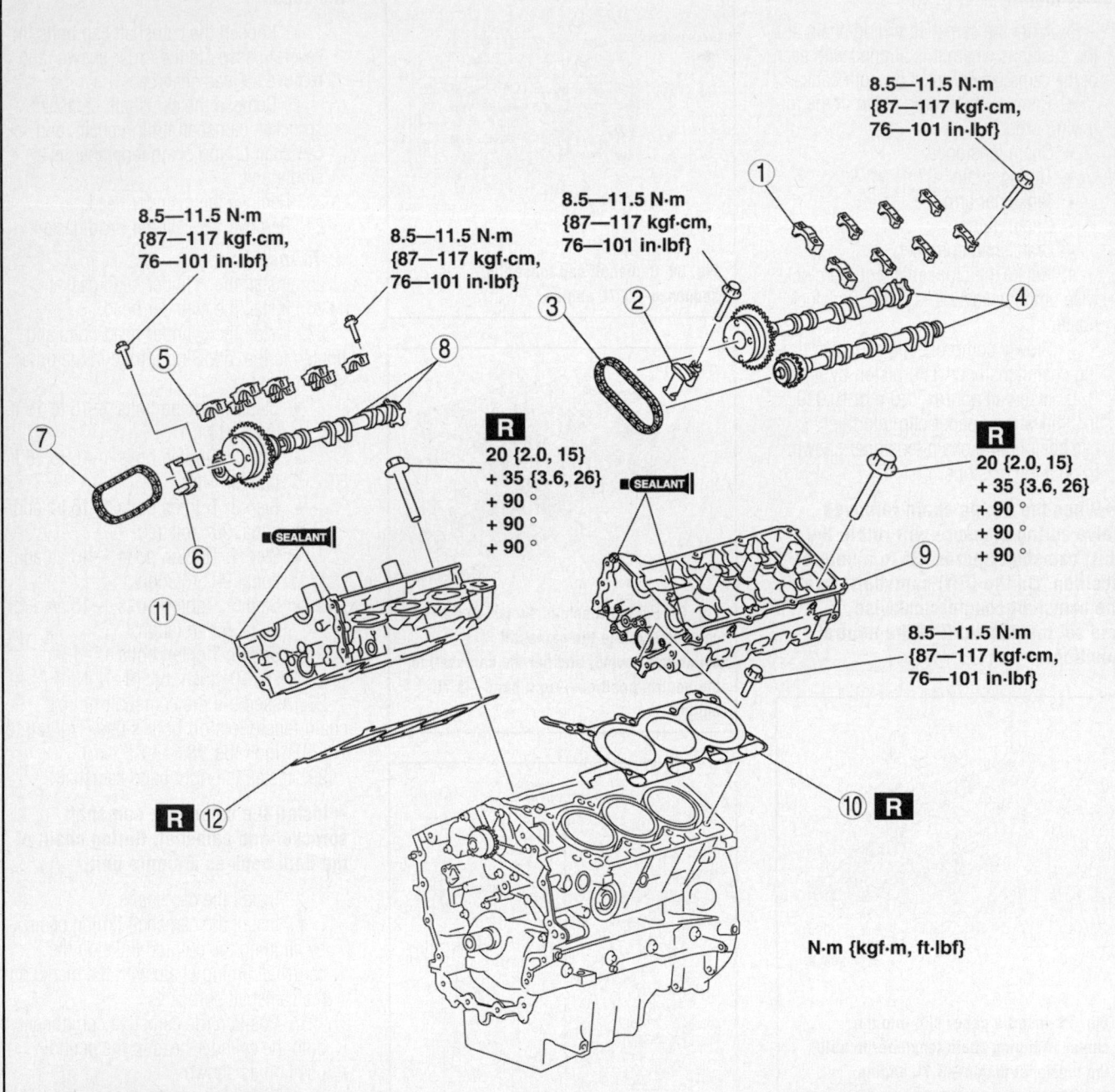

8.5—11.5 N·m
{87—117 kgf·cm,
76—101 in·lbf}

8.5—11.5 N·m
{87—117 kgf·cm,
76—101 in·lbf}

8.5—11.5 N·m
{87—117 kgf·cm,
76—101 in·lbf}

8.5—11.5 N·m
{87—117 kgf·cm,
76—101 in·lbf}

R
20 {2.0, 15}
+ 35 {3.6, 26}
+ 90 °
+ 90 °
+ 90 °

R
20 {2.0, 15}
+ 35 {3.6, 26}
+ 90 °
+ 90 °
+ 90 °

SEALANT

SEALANT

8.5—11.5 N·m
{87—117 kgf·cm,
76—101 in·lbf}

R

N·m {kgf·m, ft·lbf}

22140_MCX9_G0171

1. Camshaft cap (LH)
2. Camshaft timing chain tensioner (LH)
3. Camshaft timing chain (LH)
4. Camshaft component (LH)
5. Camshaft cap (RH)
6. Camshaft timing chain tensioner (RH)

7. Camshaft timing chain (RH)
8. Camshaft component (RH)
9. Cylinder head (LH)
10. Cylinder head gasket (LH)
11. Cylinder head (RH)
12. Cylinder head gasket (RH)

Fig. 83 Exploded view of the cylinder head assembly—3.7L engine

o'clock position. This will position the No.1 cylinder at TDC.

33. Install the timing chain.
34. Install the alternator.
35. Install the power steering oil pump.
36. Install the power steering oil pump drive belt.
37. Install the dipstick.
38. Install the ignition coils.
39. Install both catalytic converters together with both exhaust manifolds as a single unit.

40. Install the intake manifold.
41. Install the thermostat and thermostat housing together as a single unit.
42. Install the fuel injector and fuel distributor together as a single unit.
43. Install the intake manifold dynamic chamber and throttle body as a single unit.
44. Install the engine and transaxle.
45. Fill the engine with oil.
46. Start the engine and perform the following inspections:
   a. Inspect the runout and contact on the pulley and belt.

b. Inspect for engine oil, engine coolant, ATF, power steering fluid and fuel leakage.

c. Verify the ignition timing and idle speed.

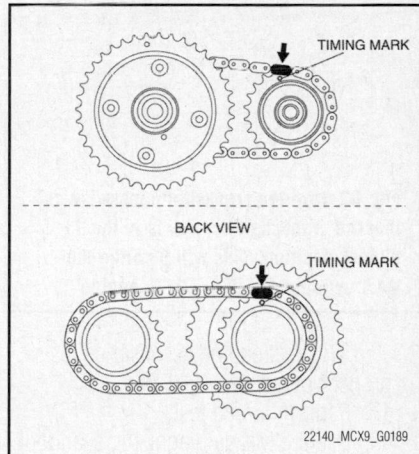

Fig. 89 Install the camshaft timing chain by aligning the colored links on the camshaft timing chain with the marks on the camshaft sprockets—left hand—3.7L engine

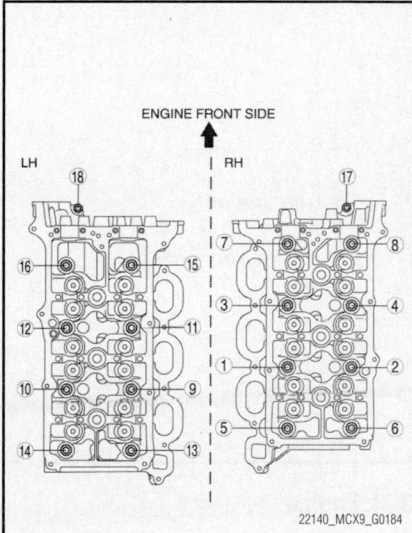

Fig. 84 Tighten the cylinder head bolts to the specified torque, in several passes, in the order shown—3.7L engine

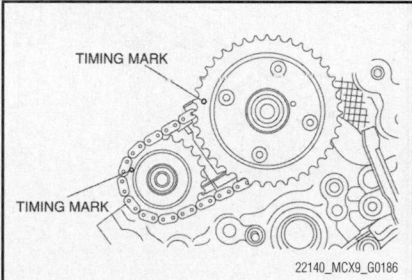

Fig. 86 Position the camshaft component onto the cylinder head in the neutral position as shown—right hand—3.7L engine

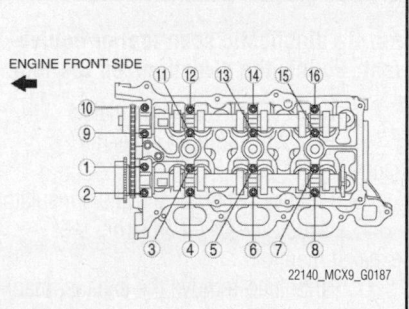

Fig. 87 Tighten the camshaft cap bolts to the specified torque, in several passes, in the order shown—right hand—3.7L engine

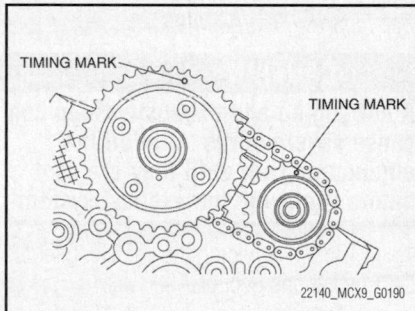

Fig. 90 Position the camshaft component onto the cylinder head in the neutral position as shown—left hand—3.7L engine

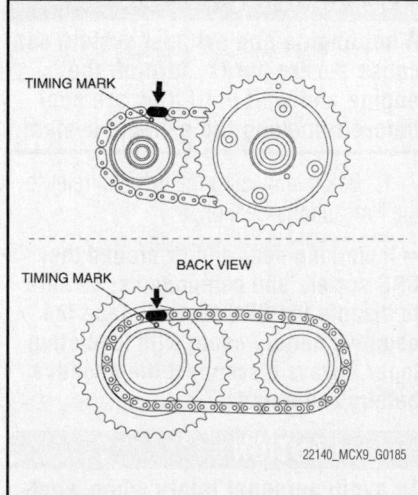

Fig. 85 Install the camshaft timing chain by aligning the colored links on the camshaft timing chain with the marks on the camshaft sprockets—right hand—3.7L engine

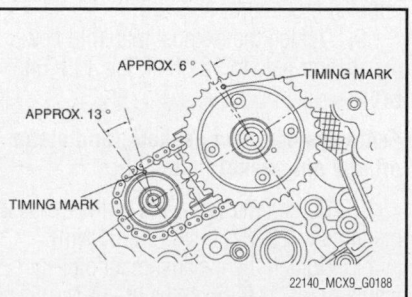

Fig. 88 Using the right hand camshaft sprocket installation bolt, rotate the right hand camshaft clockwise so the timing marks align as shown—3.7L engine

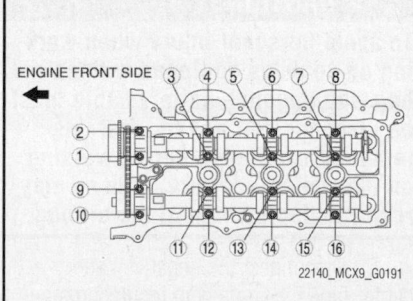

Fig. 91 Tighten the camshaft cap bolts to the specified torque, in several passes, in the order shown—left hand—3.7L engine

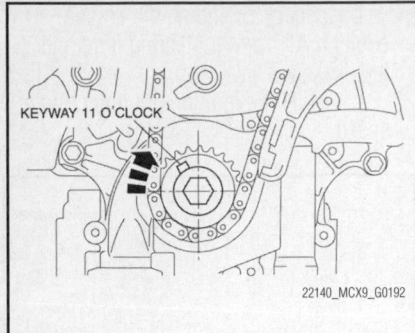

Fig. 92 Turn the crankshaft clockwise so that the crankshaft keyway is in the 11 o'clock position. This will position the No.1 cylinder at TDC. —3.7L engine

KEYWAY 11 O'CLOCK

d. Inspect engine accessories for proper operation.

47. Perform a road test.

48. Use the Mazda diagnostic scan tool, or equivalent and reprogram the required systems.

EXHAUST MANIFOLD

REMOVAL & INSTALLATION

2.5L Engine

See Figures 93 through 95.

> **✳✳ CAUTION**
>
> **A hot engine and exhaust system can cause severe burns. Turn off the engine and wait until they are cool before removing the exhaust system.**

1. Before servicing the vehicle, refer to the Precautions Section.

➡**If working near and/or around the SRS system and components, be sure to disable the SRS system. Tape the negative battery cable with insulating tape. Always disconnect the negative battery cable first.**

> **✳✳ CAUTION**
>
> **To avoid personal injury when working on vehicles equipped with an air bag, the negative battery cable must be disconnected and at least one minute must elapse before working on the system. Failure to do so may result in deployment of the air bag.**

2. Disconnect the negative battery cable. Tape the cable with insulating tape.

➡**When disconnecting the cable, some systems need to be initialized after the cable is reconnected. You will need the**

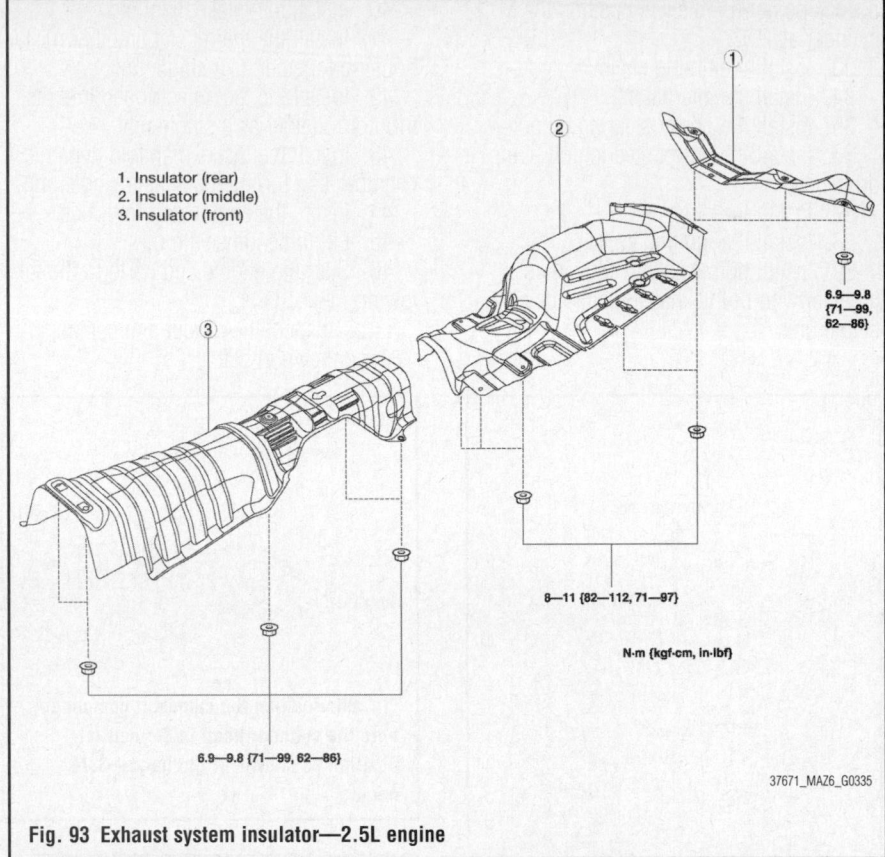

1. Insulator (rear)
2. Insulator (middle)
3. Insulator (front)

6.9—9.8 {71—99, 62—86}

8—11 {82—112, 71—97}

N-m {kgf-cm, in-lbf}

6.9—9.8 {71—99, 62—86}

37671_MAZ6_G0335

Fig. 93 Exhaust system insulator—2.5L engine

Mazda diagnostic scan tool or equivalent. Follow the directions on the tool.

3. Remove the plug hole plate.

4. Remove the under cover, as required.

5. Remove the exhaust system insulator.

6. Unbolt the WU-TWC from the exhaust manifold.

7. Unbolt an remove the exhaust manifold.

To install:

➡**Be sure to use new fasteners, as required.**

8. Installation is the reverse of the removal procedure.

9. Tighten the exhaust manifold new installation nuts to 32–47 ft. lbs. (43–64 Nm).

➡**Always use new gaskets, and clean off any old gasket material.**

10. Tighten the WU-TWC to the exhaust manifold to 29–37 ft. lbs.(38–51 Nm).

11. Tighten the exhaust manifold insulator to 71–88 inch lbs. 8–10 Nm).

12. Start the engine and check for leaks.

13. Use the Mazda diagnostic scan tool, or equivalent and reprogram the required systems.

3.7L Engine

See Figure 96.

At this time the manufacturer does not provide removal and installation procedures for this component, refer to the illustration as required.

> **✳✳ CAUTION**
>
> **A hot engine and exhaust system can cause severe burns. Turn off the engine and wait until they are cool before removing the exhaust system.**

1. Before servicing the vehicle, refer to the Precautions Section.

➡**If working near and/or around the SRS system and components, be sure to disable the SRS system. Tape the negative battery cable with insulating tape. Always disconnect the negative battery cable first.**

> **✳✳ CAUTION**
>
> **To avoid personal injury when working on vehicles equipped with an air bag, the negative battery cable must be disconnected and at least one minute must elapse before working on the system. Failure to do so may result in deployment of the air bag.**

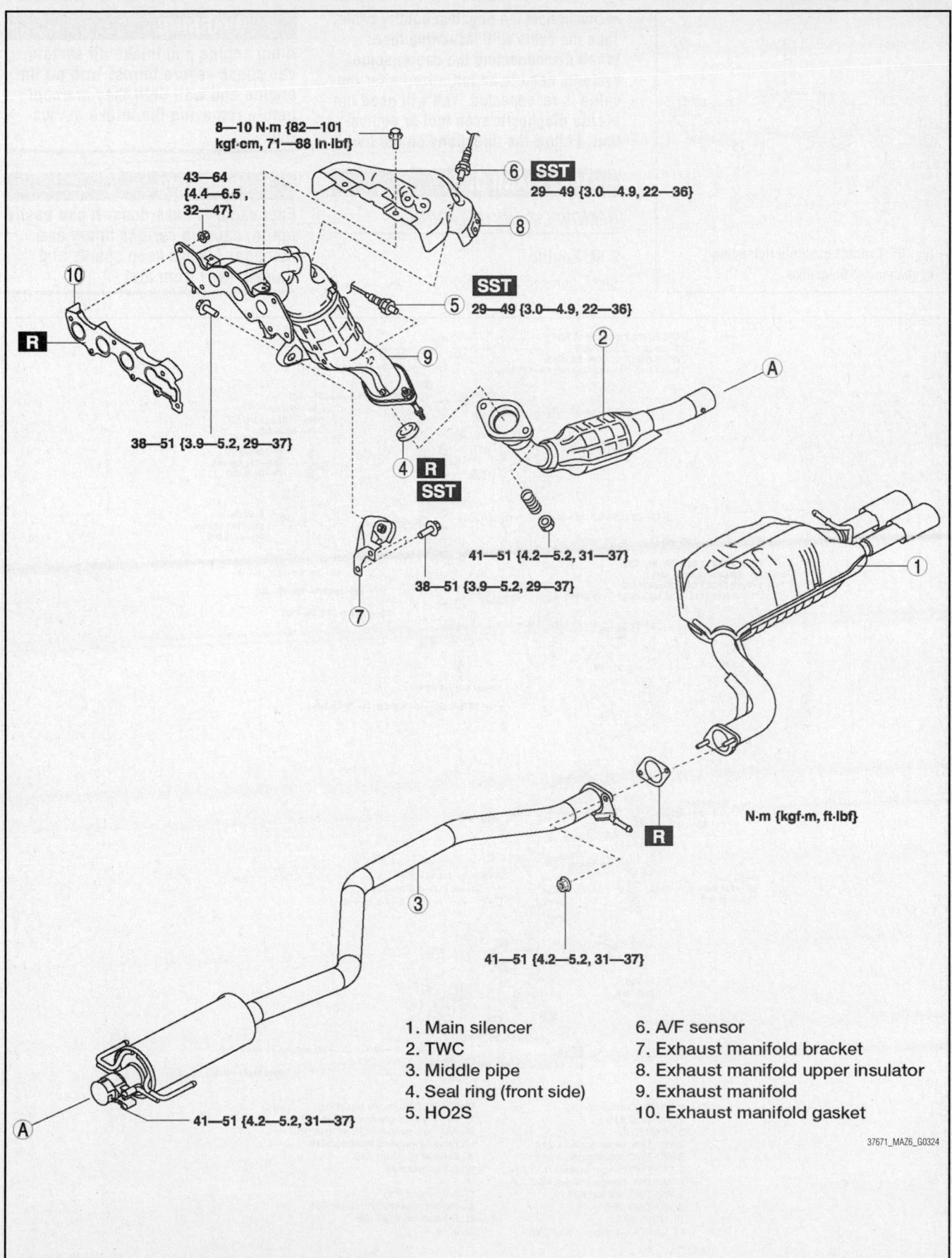

8—10 N·m {82—101
kgf·cm, 71—88 In·lbf}

43—64
{4.4—6.5 ,
32—47}

6　**SST**
29—49 {3.0—4.9, 22—36}

8

5　**SST**
29—49 {3.0—4.9, 22—36}

10　**R**

2

A

9

38—51 {3.9—5.2, 29—37}

4　**R**
SST

41—51 {4.2—5.2, 31—37}

38—51 {3.9—5.2, 29—37}

7

1

N·m {kgf·m, ft·lbf}

3

R

41—51 {4.2—5.2, 31—37}

A

41—51 {4.2—5.2, 31—37}

1. Main silencer
2. TWC
3. Middle pipe
4. Seal ring (front side)
5. HO2S

6. A/F sensor
7. Exhaust manifold bracket
8. Exhaust manifold upper insulator
9. Exhaust manifold
10. Exhaust manifold gasket

37671_MAZ6_G0324

Fig. 94 Exhaust manifold and related components—2.5L engine

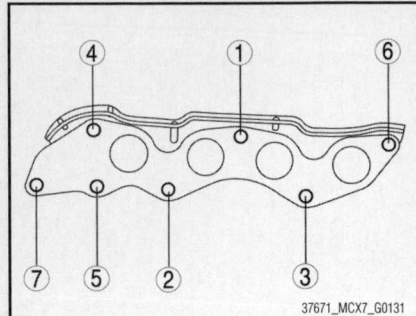

Fig. 95 Exhaust manifold tightening sequence—2.5L engine

➡Disconnect the negative battery cable. Tape the cable with insulating tape. When disconnecting the cable, some systems need to be initialized after the cable is reconnected. You will need the Mazda diagnostic scan tool or equivalent. Follow the directions on the tool.

INTAKE MANIFOLD

REMOVAL & INSTALLATION

2.5L Engine

See Figures 97 through 99.

TYPE A (BOLT COLOR: BLACK):
22—28 {2.3—2.8, 17—20}
TYPE B (BOLT COLOR: SILVER):
11—13 N·m {113—132 kgf·cm, 98—115 in·lbf}

41—54 {4.2—5.5, 31—37}

22—28 {2.3—2.8, 17—20}

41—54 {4.2—5.5, 31—37}

8—10 N·m {82—101 kgf·cm, 71—88 in·lbf}

8—10 N·m {82—101 kgf·cm, 71—88 in·lbf}

17—23 {1.8—2.3, 13—16}

34—46 {3.5—4.6, 26—33}

8—10 N·m {82—101 kgf·cm, 71—88 in·lbf}

TYPE A (BOLT COLOR: BLACK):
22—28 {2.3—2.8, 17—20}
TYPE B (BOLT COLOR: SILVER):
11—13 N·m {113—132 kgf·cm, 98—115 in·lbf}

17—23 {1.8—2.3, 13—16}

22—28 {2.3—2.8, 17—20}

34—46 {3.5—4.6, 26—33}

SST 41—54 {4.2—5.5, 31—39}

FROM FRONT PIPE

8—10 N·m {82—101 kgf·cm, 71—88 in·lbf}

8—10 N·m {82—101 kgf·cm, 71—88 in·lbf}

22—28 {2.3—2.8, 17—20}

8—10 N·m {82—101 kgf·cm, 71—88 in·lbf}

22—28 {2.3—2.8, 17—20}

34—46 {3.5—4.6, 26—33}

17—23 {1.8—2.3, 13—16}

17—23 {1.8—2.3, 13—16}

FROM FRONT PIPE

SST 41—54 {4.2—5.5, 31—39}

8—10 N·m {82—101 kgf·cm, 71—88 in·lbf}

N·m {kgf·m, ft·lbf}

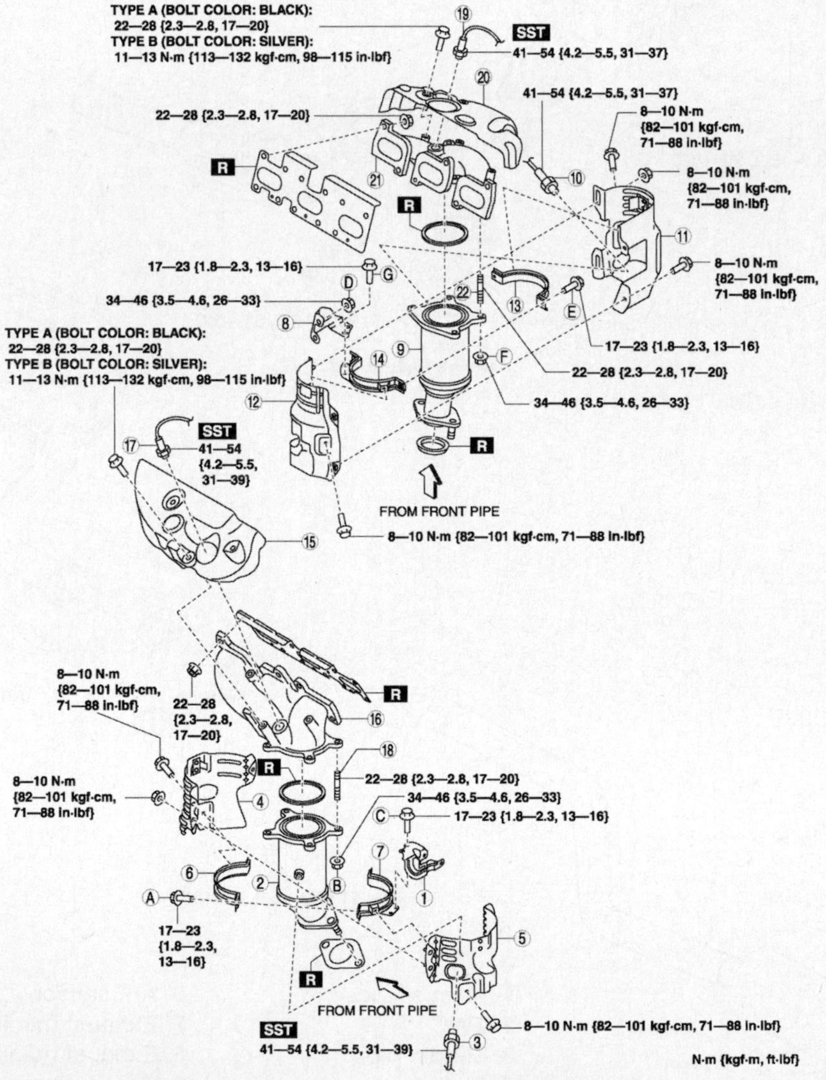

1. WU-TWC bracket (LH)
2. WU-TWC (LH)
3. HO2S (LH)
4. WU-TWC protector No.1 (LH)
5. WU-TWC protector No.2 (LH)
6. WU-TWC hanger bracket No.1 (LH)
7. WU-TWC hanger bracket No.2 (LH)
8. WU-TWC bracket (RH)
9. WU-TWC (RH)
10. HO2S (RH)
11. WU-TWC protector No.1 (RH)
12. WU-TWC protector No.2 (RH)
13. WU-TWC hanger bracket No.1 (RH)
14. WU-TWC hanger bracket No.2 (RH)
15. Exhaust manifold insulator (LH)
16. Exhaust manifold (LH)
17. A/F sensor (LH)
18. Stud
19. A/F sensor (RH)
20. Exhaust manifold insulator (RH)
21. Exhaust manifold (RH)
22. Stud

Fig. 96 Exhaust manifold and related components—3.7L engine

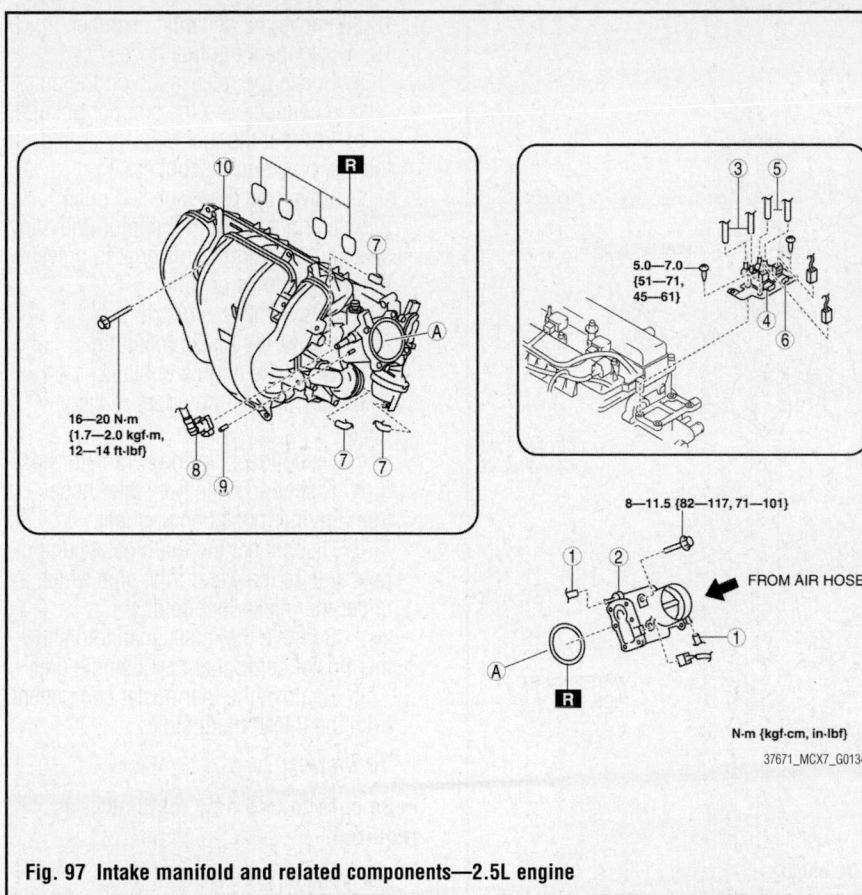

Fig. 97 Intake manifold and related components—2.5L engine

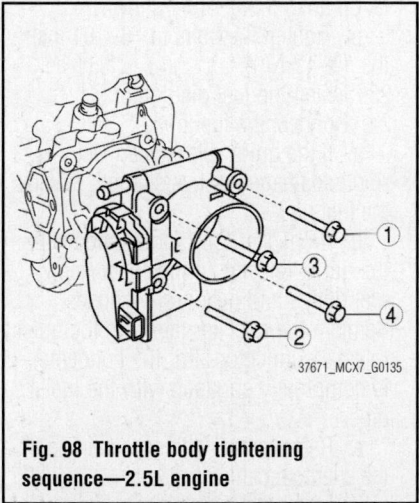

Fig. 98 Throttle body tightening sequence—2.5L engine

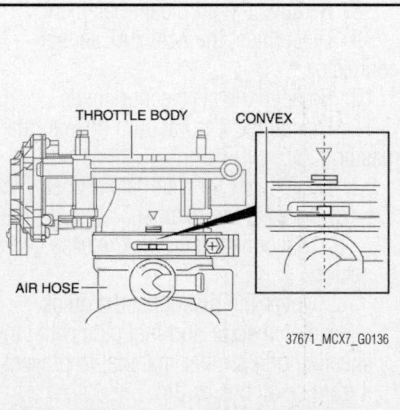

Fig. 99 Convex adjustment view of the throttle body and the air hose—2.5L engine

※※ WARNING

Fuel line spills and leakage are dangerous. Fuel can ignite and cause serious injuries or death and damage. Fuel can also irritate skin and eyes.

1. Before servicing the vehicle, refer to the Precautions Section.

➡**If working near and/or around the SRS system and components, be sure to disable the SRS system. Tape the**

negative battery cable with insulating tape. Always disconnect the negative battery cable first.

※※ CAUTION

To avoid personal injury when working on vehicles equipped with an air bag, the negative battery cable must be disconnected and at least one minute must elapse before working on the system. Failure to do so may result in deployment of the air bag.

2. Disconnect the negative battery cable. Tape the cable with insulating tape.

➡**When disconnecting the cable, some systems need to be initialized after the cable is reconnected. You will need the Mazda diagnostic scan tool or equivalent. Follow the directions on the tool.**

3. Drain the cooling system.
4. Remove the plug hole plate.
5. Disconnect the wiring harness.
6. Remove or disconnect the following:
   - MAF/IAT sensor
   - Ventilation hose
   - Fresh-air duct
   - Air cleaner cover
   - Air cleaner element
   - Air cleaner case
   - Remove the mudguard (LF), as required.
   - Resonance chamber
   - Air hose
7. Remove or disconnect the following:
   - Water hose
   - Throttle body
   - Vacuum hose
   - Variable intake air solenoid valve
   - Vacuum hose
   - Variable tumble solenoid valve
   - Vacuum hose
   - Quick release connector
   - Vacuum cap
   - Intake manifold
   - Coolant reserve tank.
   - Fan control module No.2.
   - Engine under cover
   - All clips for securing wiring harnesses from the intake manifold
8. Remove the intake manifold.

To install:

➡**Be sure to use new fasteners, as required.**

9. Installation is the reverse of the removal procedure.
   - Tighten the throttle body to 71–101 inch lbs. (8–11.5 Nm).
   - Adjust the convex of the throttle body and the air hose.
10. Bleed the air from the cooling system.
11. Use the Mazda diagnostic scan tool, or equivalent and reprogram the required systems.

3.7L Engine

See Figures 100 through 103.

1. Before servicing the vehicle, refer to the Precautions Section.

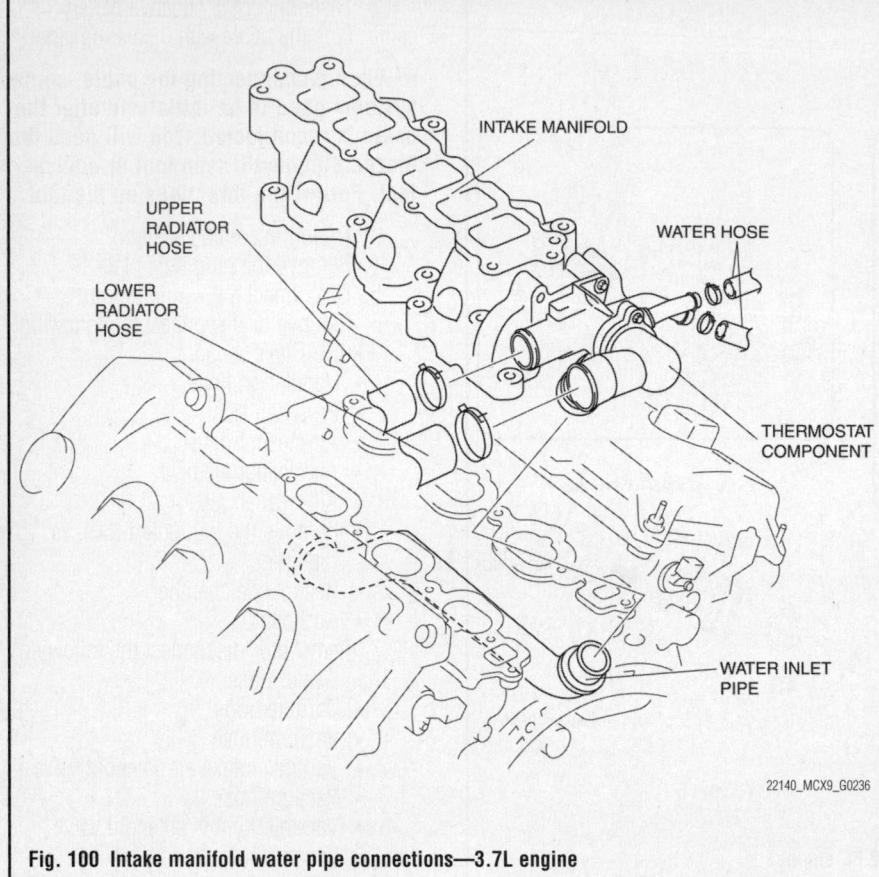

UPPER RADIATOR HOSE

LOWER RADIATOR HOSE

INTAKE MANIFOLD

WATER HOSE

THERMOSTAT COMPONENT

WATER INLET PIPE

22140_MCX9_G0236

Fig. 100 Intake manifold water pipe connections—3.7L engine

➡️If working near and/or around the SRS system and components, be sure to disable the SRS system. Tape the negative battery cable with insulating tape. Always disconnect the negative battery cable first.

※※ CAUTION

To avoid personal injury when working on vehicles equipped with an air bag, the negative battery cable must be disconnected and at least one minute must elapse before working on the system. Failure to do so may result in deployment of the air bag.

2. Disconnect the negative battery cable. Tape the cable with insulating tape.

➡️When disconnecting the cable, some systems need to be initialized after the cable is reconnected. You will need the Mazda diagnostic scan tool or equivalent. Follow the directions on the tool.

3. Properly relieve the fuel system pressure.
4. Remove the battery.
5. Remove the engine cover.
6. Remove the air cleaner case.
7. Remove the fresh-air duct.

8. Remove the air cleaner element.
9. Disconnect the MAF/IAT sensor connector.
10. Remove the air cleaner cover.
11. Disconnect the vacuum hose to the master back.
12. Disconnect the ventilation hose.
   a. Release the retainer.
   b. Disconnect the quick release connector.
   c. Cover the disconnected quick release connector and fuel pipe with vinyl sheeting or a similar material to prevent it from scratches or dirt.
13. Remove the resonance chamber.
14. Disconnect the TP sensor connector.
15. Remove the throttle body.
16. Disconnect the vacuum hose to the purge solenoid valve.
   a. Move the retainer upward using a small flathead screwdriver or a similar tool.
   b. Pull out the fuel hose straight from the fuel pipe and disconnect it.
   c. Cover the disconnected quick release connector and fuel pipe with vinyl sheeting or a similar material to prevent it from scratches or dirt.
17. Disconnect the vacuum hose to the master back.

18. Remove the dynamic chamber.
19. Disconnect the fuel hose.
   a. Rotate the release tab on the quick release connector to the stopper position.
   b. Pull out the fuel hose straight from the fuel pipe and disconnect it.
   c. Cover the disconnected quick release connector and fuel pipe with vinyl sheeting or a similar material to prevent it from scratches or dirt.
20. Remove the fuel distributor
21. Remove the intake manifold.
   a. Drain the engine coolant.
   b. Disconnect the upper radiator hose.
   c. Disconnect the lower radiator hose.
   d. Disconnect the two water hose from the thermostat component.
   e. Disconnect the thermostat component side of the water inlet pipe which is under the intake manifold.
   f. Remove the intake manifold and thermostat component as a single unit.
   g. Remove the thermostat component from the intake manifold.

To install:

➡️**Be sure to use new fasteners, as required.**

22. Install the intake manifold.
   a. Tighten the intake manifold installation bolts in the order shown.
   b. Tighten the bolts to 76–101 inch lbs. (9–12 Nm).
23. Install the fuel distributor
24. Connect the fuel hose.
   a. If the quick release connector O-ring is damaged or has slipped, replace the fuel hose.
   b. A checker tab is integrated with the quick release connector for new fuel hoses and evaporative hoses. Remove the checker tab from the quick release connector after the connector is completely engaged with the fuel pipe.
   c. Remove the retainer remaining on the charcoal canister pipe.
   d. Install a new retainer to the quick release connector.
   e. Reconnect the hose straight to the pipe until a click is heard.
   f. Lightly pull and push the quick release connector a few times by hand, and then verify that it is connected securely.
25. Install the dynamic chamber.
   a. Tighten the dynamic chamber installation bolts in the order shown.
   b. Tighten the bolts to 76–101 inch lbs. (9–12 Nm).

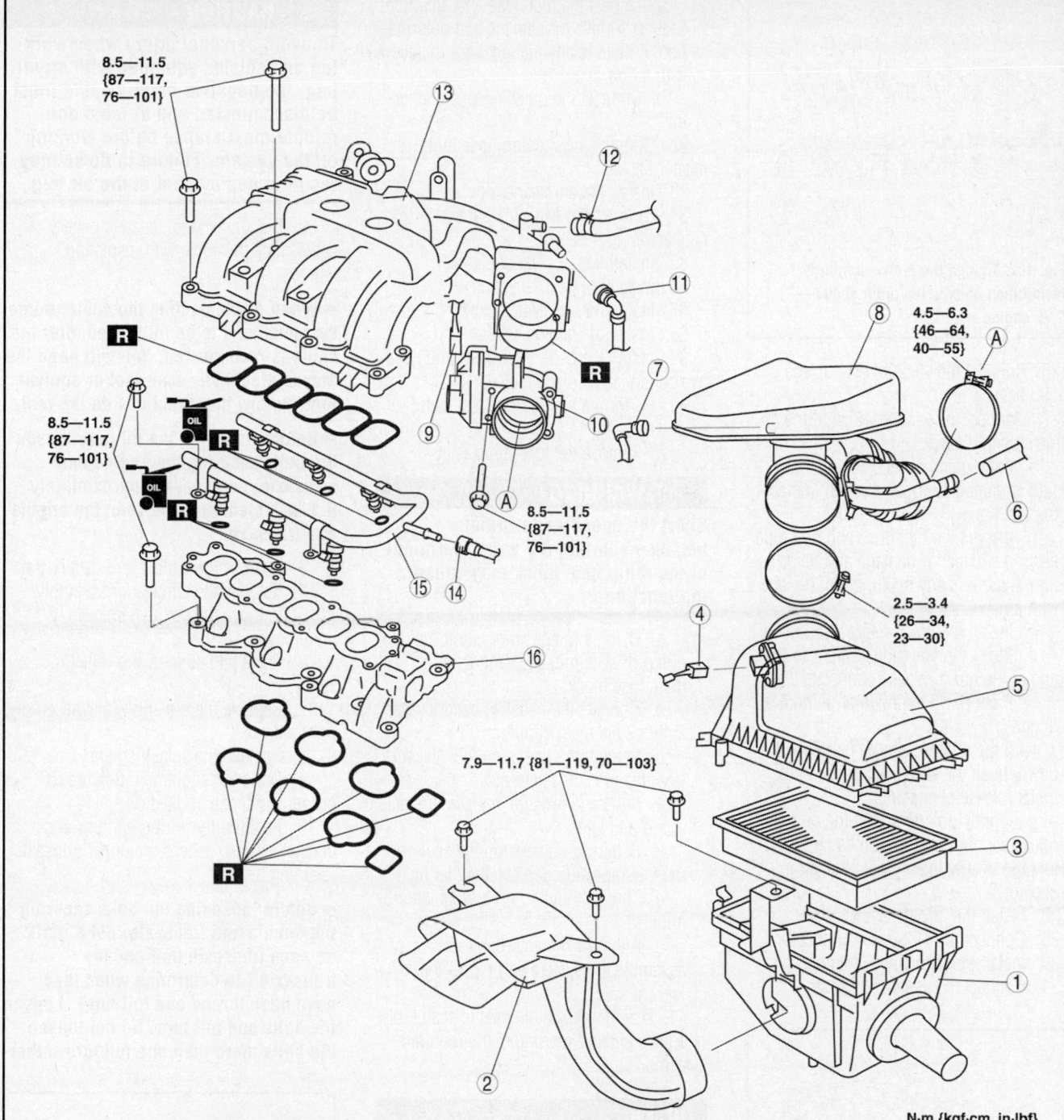

N·m {kgf·cm, in·lbf}

1. Air cleaner case
2. Fresh-air duct
3. Air cleaner element
4. MAF/IAT sensor connector
5. Air cleaner cover
6. Vacuum hose (to master back)
7. Ventilation hose
8. Resonance chamber
9. TP sensor connector
10. Throttle body
11. Vacuum hose (to purge solenoid valve)
12. Vacuum hose (to master back)
13. Dynamic chamber
14. Fuel hose
15. Fuel distributor
16. Intake manifold

22140_MCX9_G0232

Fig. 101 Intake manifold and related components—3.7L engine

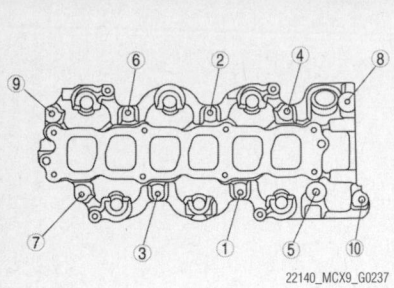

Fig. 102 Tighten the intake manifold installation bolts in the order shown— 3.7L engine

26. Connect the vacuum hose to the master back.

27. Connect the vacuum hose to the purge solenoid valve.

  a. If the quick release connector O-ring is damaged or has slipped, replace the fuel hose.

  b. Inspect the fuel hose and fuel pipe sealing surface for damage and deformation. If there is any malfunction, replace it with a new one.

  c. Install the quick release connector.

  d. Insert the fuel pipe straight to the end of the quick release connector.

  e. Push down the retainer using a finger.

  f. If the retainer cannot be pushed down, push the fuel pipe further to the quick release connector.

  g. Lightly pull and push the quick release connector a few times by hand, and then verify that it is connected securely.

28. Install the throttle body.

29. Connect the TP sensor connector.

30. Install the resonance chamber.

31. Connect the ventilation hose.

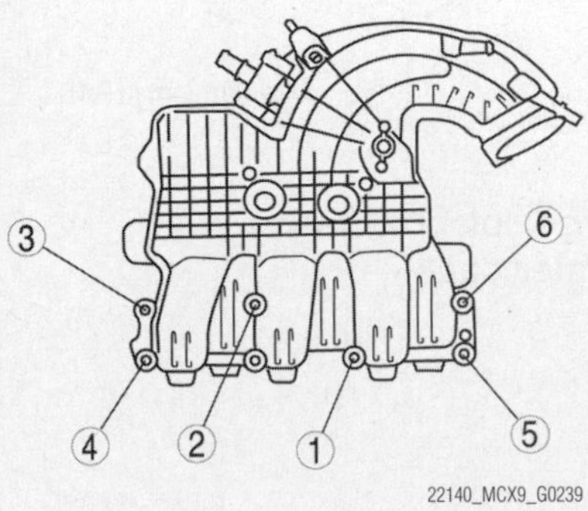

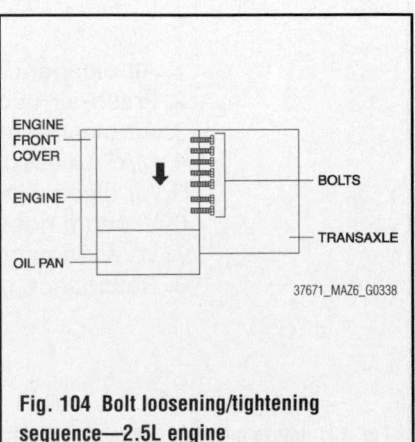

Fig. 103 Tighten the dynamic chamber installation bolts in the order shown— 3.7L engine

  a. Inspect the fuel hose and fuel pipe sealing surface for damage and deformation. If there is any malfunction, replace it with a new one.

  b. Connect the quick release connector.

32. Connect the vacuum hose to the master back.

33. Install the air cleaner cover.

34. Connect the MAF/IAT sensor connector.

35. Install the air cleaner element.

36. Install the fresh-air duct.

37. Install the air cleaner case.

38. Install the engine cover.

39. Install the battery but leave the negative battery cable disconnected.

40. Perform a fuel leakage test as follows:

  a. Remove the fuel pump relay.

✳✳ WARNING

Short the specified terminals because shorting the wrong terminal of the main fuse block may cause malfunctions.

  b. Using a jumper wire, short fuel pump relay terminals C and D in the main fuse block.

  c. Connect the negative battery cable and operate the fuel pump.

  d. Verify that there is no fuel leakage from the pressurized parts.

  e. If there is leakage, replace the fuel hoses and clips.

  f. If there is damage on the seal on the fuel pipe side, replace the fuel pipe.

  g. There should be no leakage after 5 minutes.

  h. After making any repairs, assemble the system and repeat the leakage test.

41. Use the Mazda diagnostic scan tool, or equivalent and reprogram the required systems.

OIL PAN

REMOVAL & INSTALLATION

2.5L Engine

See Figures 104 through 111.

1. Before servicing the vehicle, refer to the Precautions Section.

➡If working near and/or around the SRS system and components, be sure to disable the SRS system. Tape the negative battery cable with insulating tape. Always disconnect the negative battery cable first.

✳✳ CAUTION

To avoid personal injury when working on vehicles equipped with an air bag, the negative battery cable must be disconnected and at least one minute must elapse before working on the system. Failure to do so may result in deployment of the air bag.

2. Disconnect the negative battery cable. Tape the cable with insulating tape.

➡When disconnecting the cable, some systems need to be initialized after the cable is reconnected. You will need the Mazda diagnostic scan tool or equivalent. Follow the directions on the tool.

➡Before removing the oil pan, loosen the bolts securing the engine and transaxle, and secure approximately 0.1 inch clearance between the engine and transaxle.

3. Remove the battery and battery tray.

4. Remove the air cleaner assembly.

5. Remove the aerodynamic No. 2 undercover.

6. Raise and support the vehicle safely.

7. Remove the tire and wheel assembly, right side.

8. Remove the splash shield.

9. Drain the engine oil. Be sure to properly dispose of used oil.

10. Position the wiring harness and bracket aside to access the upper transaxle retaining bolts.

➡Before loosening the bolts securing the engine and transaxle, put a mark on each transaxle bolt and the transaxle (to determine when they have been turned one full turn). Loosen the bolts one full turn. Do not loosen the bolts more than one full turn, other-

Fig. 104 Bolt loosening/tightening sequence—2.5L engine

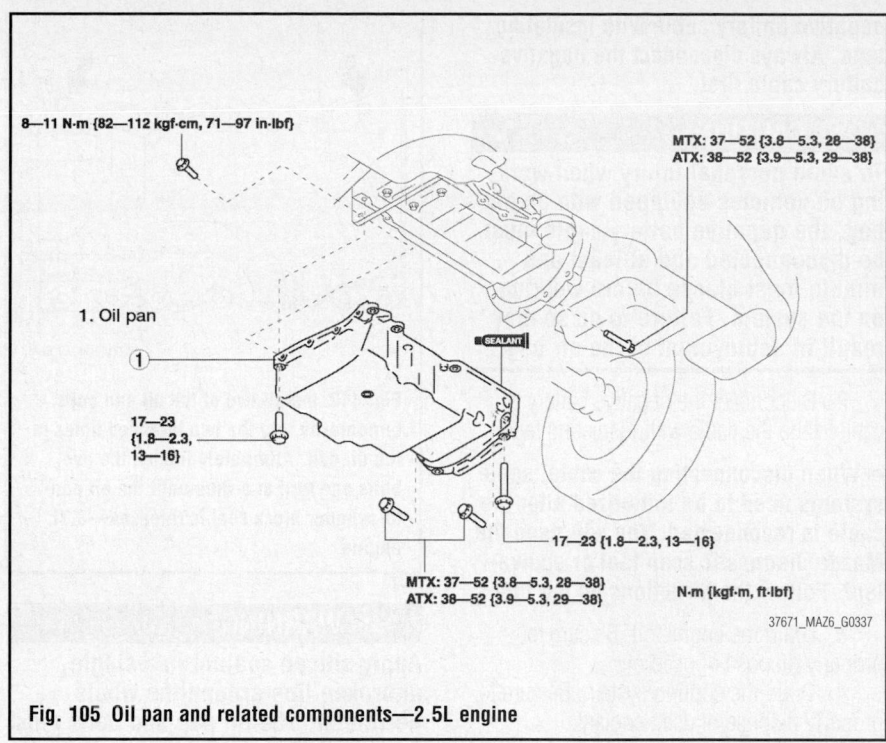

1. Oil pan

8—11 N·m {82—112 kgf·cm, 71—97 in·lbf}

17—23 {1.8—2.3, 13—16}

17—23 {1.8—2.3, 13—16}

MTX: 37—52 {3.8—5.3, 28—38}
ATX: 38—52 {3.9—5.3, 29—38}

MTX: 37—52 {3.8—5.3, 28—38}
ATX: 38—52 {3.9—5.3, 29—38}

N·m {kgf·m, ft·lbf}

37671_MAZ6_G0337

Fig. 105 Oil pan and related components—2.5L engine

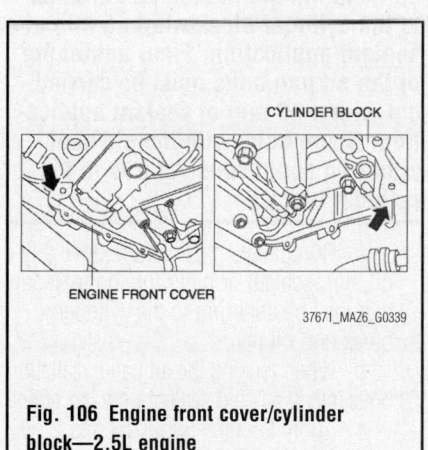

CYLINDER BLOCK

ENGINE FRONT COVER

37671_MAZ6_G0339

Fig. 106 Engine front cover/cylinder block—2.5L engine

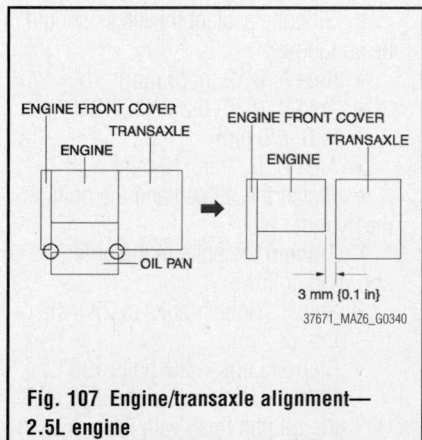

ENGINE FRONT COVER
TRANSAXLE
ENGINE

ENGINE FRONT COVER
TRANSAXLE
ENGINE

OIL PAN

3 mm {0.1 in}

37671_MAZ6_G0340

Fig. 107 Engine/transaxle alignment—2.5L engine

wise the transaxle could be damaged. Loosen the bolts from top to bottom, see illustration.

11. Remove the four bolts that secure the engine front cover to the oil pan.

12. Remove the No. 1 engine mount bracket.

13. Remove the oil pan.

14. Discard the gasket.

To install:

➡ **Be sure to use new fasteners, as required.**

15. Installation is the reverse of the removal procedure.

➡ **Using old bolts with old sealant adhering could cause cracks in the cylinder block. Be sure the bolts are free of old sealant. If in doubt, replace them.**

➡ **Do not allow gasket residue to enter**

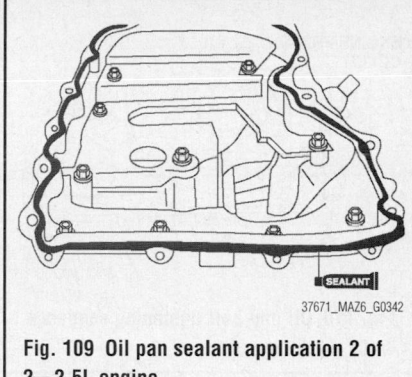

37671_MAZ6_G0342

Fig. 109 Oil pan sealant application 2 of 2—2.5L engine

the oil pan. Using cleaning fluid, completely remove the oil from the joint. Be sure to remove oil and gasket residue completely from the corner section of the front cover and the cylinder block.

16. Apply the sealant in a single unbroken line.

17. The thickness of the sealant tends to be thin because the clearance between the oil pan and the transaxle and the engine front cover and transaxle is almost zero during oil pan installation. Allow 0.1 inch space maximum between the engine and transaxle with the transaxle bolts loosened one full turn before reinstalling the oil pan. See illustration.

18. Apply sealant (0.09–0.12 inch) to the oil pan along the inside of the bolt holes, as indicated in the illustration.

19. Apply sealant to the front edge of the oil pan and to the front cover where the two pieces attach together.

20. Apply an ample amount of sealant to the front cover where the pan is fitted, and also apply an extra amount (a ball of about 0.2 inch in diameter) to the corners of the engine front cover and the cylinder block.

21. Install the oil pan to the cylinder block.

22. Tighten the bolts to specification and

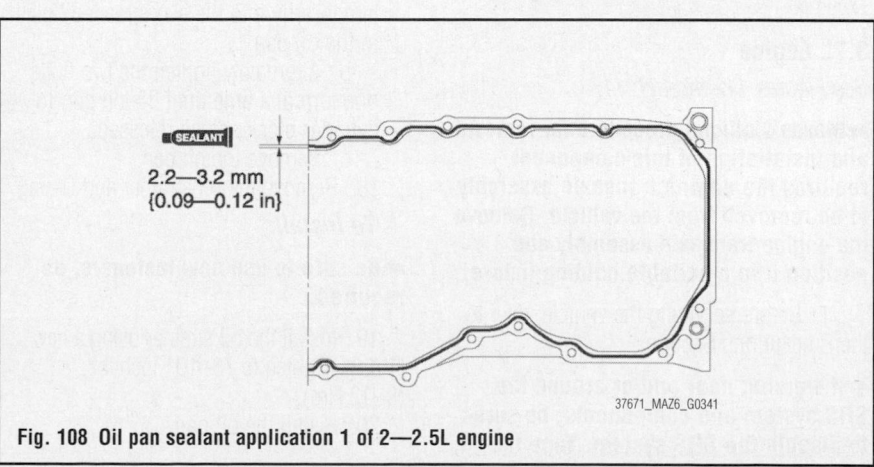

2.2—3.2 mm {0.09—0.12 in}

37671_MAZ6_G0341

Fig. 108 Oil pan sealant application 1 of 2—2.5L engine

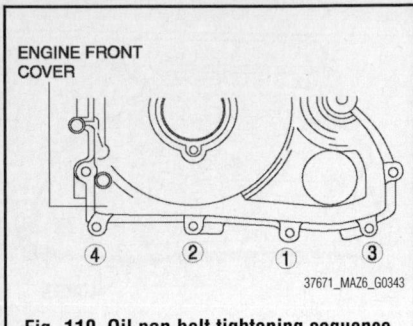

Fig. 110 Oil pan bolt tightening sequence 1 of 2—2.5L engine

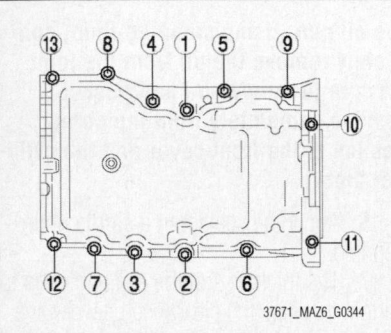

Fig. 111 Oil pan bolt tightening sequence 2 of 2—2.5L engine

in the proper sequence. See illustration. Front cover bolts, there are four, to 71–97 inch lbs. Oil pan bolts to 13–16 ft. lbs.

23. Tighten the bolts securing the engine and transaxle. Tighten the bolts from the bottom of the transaxle to the top of the transaxle. Tighten the bolts to 28–38 ft. lbs. for manual transaxle and 29–38 ft. lbs. for automatic transaxle.

24. Be sure to fill the engine with clean engine oil. Be sure to use the proper grade and type engine oil.

25. Start the engine and check for leaks. Correct as required.

26. Use the Mazda diagnostic scan tool, or equivalent and reprogram the required systems.

3.7L Engine

See Figures 112 through 117.

➡Mazda's official procedure for removal and installation of this component requires the engine/transaxle assembly to be removed from the vehicle. Remove the engine/transaxle assembly and position it in a suitable holding fixture.

1. Before servicing the vehicle, refer to the Precautions Section.

➡If working near and/or around the SRS system and components, be sure to disable the SRS system. Tape the negative battery cable with insulating tape. Always disconnect the negative battery cable first.

2. Disconnect the negative battery cable. Tape the cable with insulating tape.

➡When disconnecting the cable, some systems need to be initialized after the cable is reconnected. You will need the Mazda diagnostic scan tool or equivalent. Follow the directions on the tool.

3. Drain the engine oil. Be sure to properly dispose of used oil.

4. Drain the cooling system. Be sure to properly dispose of used coolant.

5. Drain the transaxle. Be sure to dispose of used fluid.

6. Properly relieve the fuel system pressure.

7. Remove the engine/transaxle assembly. Position the unit in a suitable holding fixture.

8. Remove the automatic transaxle.

9. Remove the drive plate.

10. Remove the dynamic chamber and throttle body as a single unit.

11. Remove the ignition coils.

12. Remove the dipstick.

13. Remove the power steering oil pump drive belt.

14. Remove the power steering oil pump.

15. Remove the exhaust manifold (RH).

16. Remove the engine front cover.

17. Remove the oil pan.
   a. Install two of the oil pan bolts temporarily into the two threaded holes in the oil pan.
   b. Alternately tighten the two bolts one turn at a time until the oil pan-to-cylinder block seal is released.
   c. Remove the oil pan.

18. Remove the oil strainer and O-ring.

To install:

➡Be sure to use new fasteners, as required.

19. Install the oil strainer using a new O-ring. Tighten to 76–101 inch lbs. (9–12 Nm).

20. Install the oil pan.

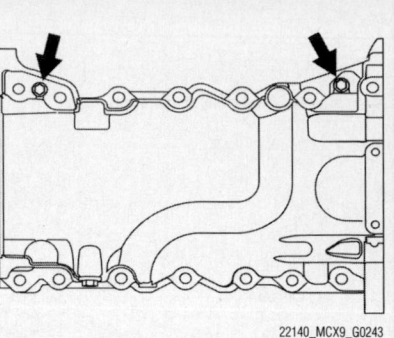

Fig. 112 Install two of the oil pan bolts temporarily into the two threaded holes in the oil pan. Alternately tighten the two bolts one turn at a time until the oil pan-to-cylinder block seal is released—3.7L engine

   a. Completely clean and remove any oil, dirt, sealant or other foreign material that may be adhering to the cylinder block and oil pan.
   b. When reusing the oil pan installation bolts, clean any old sealant from the bolts.
   c. Apply Loctite® 5900 silicone sealant to the oil pan along the inside of the bolt holes as shown.
   d. Silicone sealant thickness should be as follows:
   • Area A: 0.12 in. (3 mm)
   • Area B: 0.20–0.23 in. 0.12 in (5.0–6.0 mm)
   • Area C: 0.39 in (10 mm)
   e. Install the oil pan and the bolts to the cylinder block.
   f. Tighten the bolts in the order shown as follows:
   • Step 1: Tighten bolts to 27 inch lbs. (3 Nm).
   • Step 2: Loosen the bolts 180°.
   • Step 3: Using a straightedge, align the oil pan flush with the rear of the cylinder block at the two areas as shown.
   • Step 4: Tighten Bolt A to 15–20 ft. lbs. (20–28 Nm)

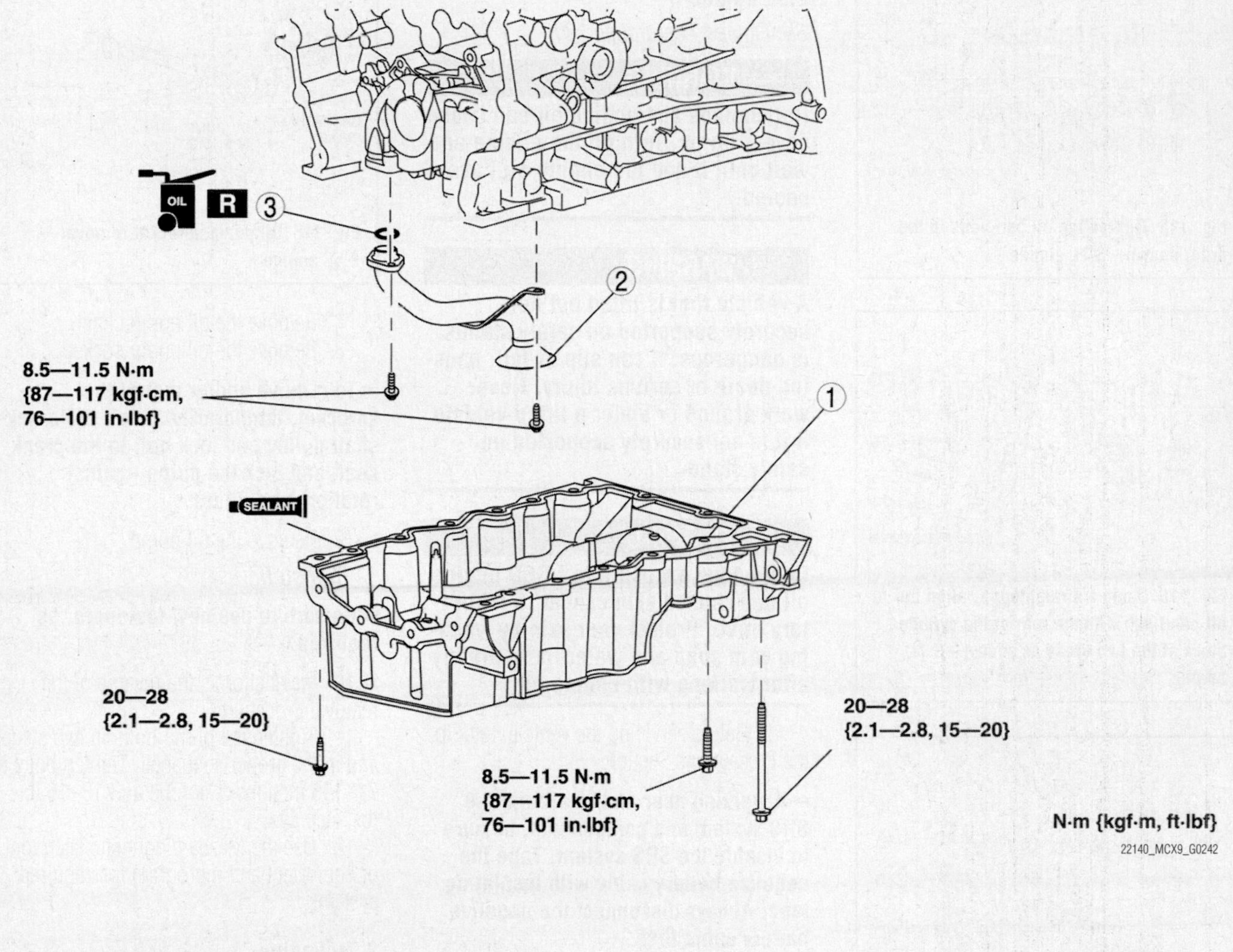

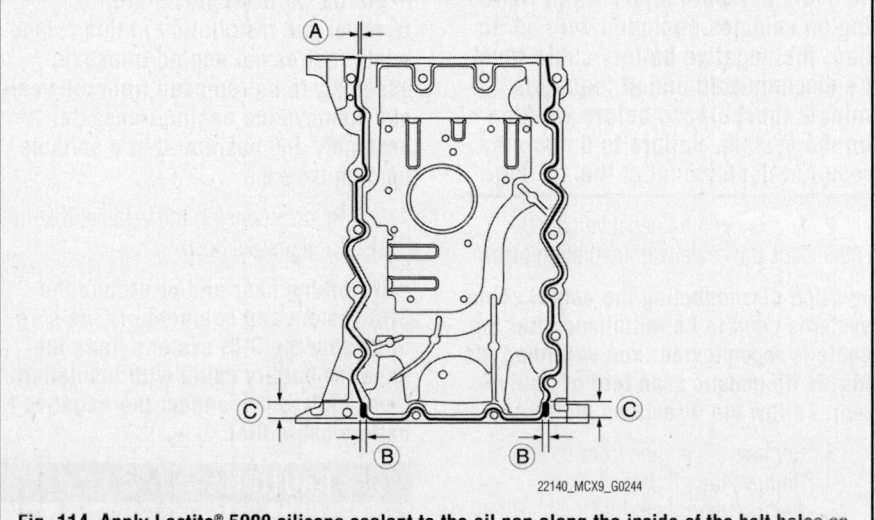

8.5—11.5 N·m
{87—117 kgf·cm,
76—101 in·lbf}

SEALANT

20—28
{2.1—2.8, 15—20}

20—28
{2.1—2.8, 15—20}

8.5—11.5 N·m
{87—117 kgf·cm,
76—101 in·lbf}

N·m {kgf·m, ft·lbf}

22140_MCX9_G0242

Fig. 113 Exploded view of the oil pan components—3.7L engine

A

C   C

B   B

22140_MCX9_G0244

Fig. 114 Apply Loctite® 5900 silicone sealant to the oil pan along the inside of the bolt holes as shown—3.7L engine

- Step 4: Tighten Bolt B to 76–101 inch lbs. (9–12 Nm)
21. Remove the engine stand.
22. Install the exhaust manifold (RH).
23. Install the power steering oil pump.
24. Install the power steering oil pump drive belt.
25. Install the dipstick.
26. Install the ignition coils.
27. Install the dynamic chamber and throttle body as a single unit.
28. Install the drive plate.
29. Install the automatic transaxle.
30. Using a hoist, lower the engine and transaxle component on a level surface.
31. Install the engine and transaxle component.
32. Fill the engine with oil.
33. Install the engine front cover.

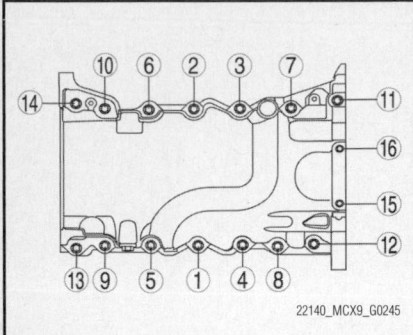

Fig. 115 Tighten the oil pan bolts in the order shown—3.7L engine

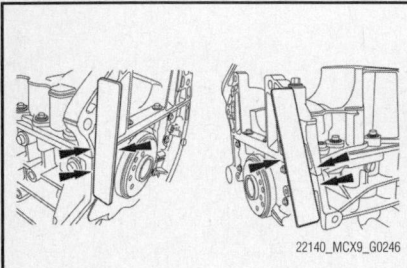

Fig. 116 Using a straightedge, align the oil pan flush with the rear of the cylinder block at the two areas as shown—3.7L engine

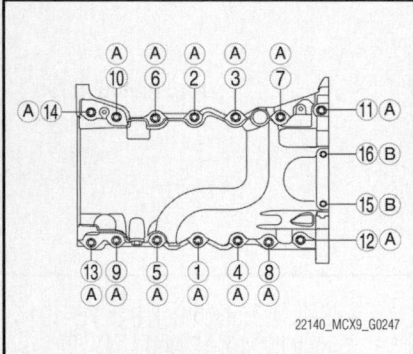

Fig. 117 Final tighten the oil pan bolts in the order shown and to the proper torque

34. Start the engine and perform the following inspections:

   a. Inspect the runout and contact on the pulley and belt.

   b. Inspect for engine oil, engine coolant, ATF, power steering fluid and fuel leakage.

   c. Verify the ignition timing, idle speed.

   d. Inspect engine accessories for proper operation.

35. Perform a road test.

36. Use the Mazda diagnostic scan tool, or equivalent and reprogram the required systems.

OIL PUMP

REMOVAL & INSTALLATION

2.5L Engine

See Figures 118 through 120.

> **✳✳ WARNING**
>
> **Hot engines and engine oil can cause severe burns. Turn off the engine and wait until it and the engine oil have cooled.**

> **✳✳ WARNING**
>
> **A vehicle that is lifted but not securely supported on safety stands is dangerous. It can slip or fall, causing death or serious injury. Never work around or under a lifted vehicle if it is not securely supported on safety stands.**

> **✳✳ WARNING**
>
> **Continuous exposure to USED engine oil has caused skin cancer in laboratory mice. Protect your skin by washing with soap and water immediately after working with engine oil.**

1. Before servicing the vehicle, refer to the Precautions Section.

➡️If working near and/or around the SRS system and components, be sure to disable the SRS system. Tape the negative battery cable with insulating tape. Always disconnect the negative battery cable first.

> **✳✳ CAUTION**
>
> **To avoid personal injury when working on vehicles equipped with an air bag, the negative battery cable must be disconnected and at least one minute must elapse before working on the system. Failure to do so may result in deployment of the air bag.**

2. Disconnect the negative battery cable. Tape the cable with insulating tape.

➡️When disconnecting the cable, some systems need to be initialized after the cable is reconnected. You will need the Mazda diagnostic scan tool or equivalent. Follow the directions on the tool.

3. Remove the engine front cover.

4. Remove the oil pan.

5. Remove the oil strainer.

6. Remove the oil pump chain tensioner.

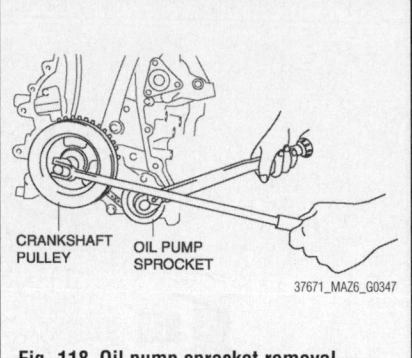

Fig. 118 Oil pump sprocket removal— 2.5L engine

7. Remove the oil pump chain.

8. Remove the oil pump socket.

➡️To remove and/or install the sprocket, temporarily install the crankshaft pulley and lock bolt to the crankshaft and lock the pump against rotation, see figure.

9. Remove the oil pump.

To install:

➡️Be sure to use new fasteners, as required.

10. Installation is the reverse of the removal procedure.

11. Tighten the pump bolts in two stages and in the proper sequence. Tighten bolts to 71–106 inch lbs, step one and 13–16 ft. lbs, step two.

12. Use the Mazda diagnostic scan tool, or equivalent and reprogram the required systems.

3.7L Engine

See Figure 121.

➡️Mazda's official procedure for removal and installation of this component requires the engine/transaxle assembly to be removed from the vehicle. Remove the engine/transaxle assembly and position it in a suitable holding fixture.

1. Before servicing the vehicle, refer to the Precautions Section.

➡️If working near and/or around the SRS system and components, be sure to disable the SRS system. Tape the negative battery cable with insulating tape. Always disconnect the negative battery cable first.

> **✳✳ CAUTION**
>
> **To avoid personal injury when working on vehicles equipped with an air**

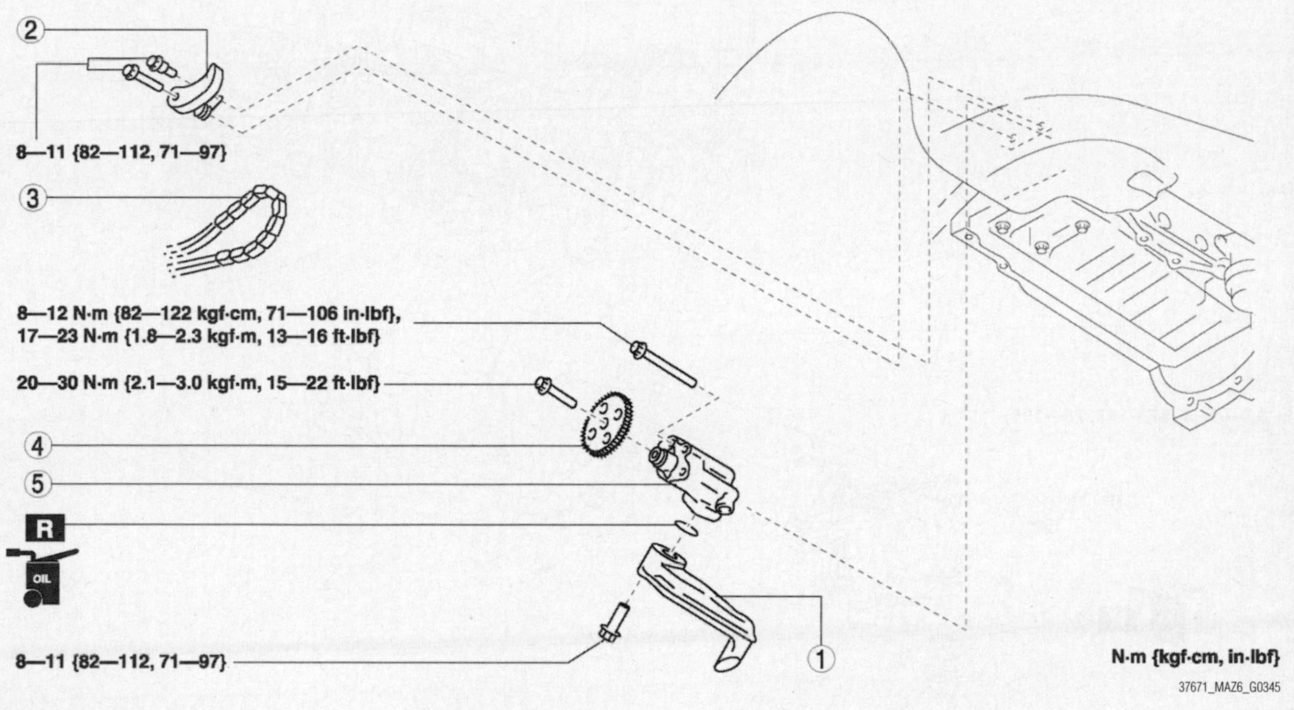

8—11 {82—112, 71—97}

8—12 N·m {82—122 kgf·cm, 71—106 in·lbf},
17—23 N·m {1.8—2.3 kgf·m, 13—16 ft·lbf}

20—30 N·m {2.1—3.0 kgf·m, 15—22 ft·lbf}

8—11 {82—112, 71—97}

N·m {kgf·cm, in·lbf}

37671_MAZ6_G0345

Fig. 119 Oil pump and related components—2.5L engine

bag, the negative battery cable must be disconnected and at least one minute must elapse before working on the system. Failure to do so may result in deployment of the air bag.

2. Disconnect the negative battery cable. Tape the cable with insulating tape.

➡**When disconnecting the cable, some systems need to be initialized after the cable is reconnected. You will need the Mazda diagnostic scan tool or**

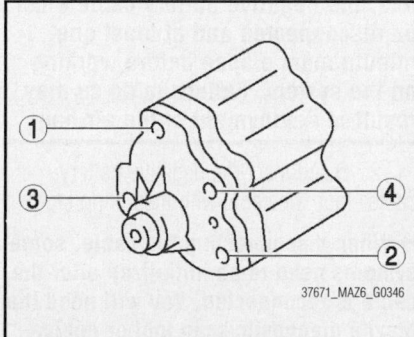

Fig. 120 Oil pump bolt tightening sequence—2.5L engine

equivalent. Follow the directions on the tool.

3. Drain the engine oil. Be sure to properly dispose of used oil.

4. Drain the cooling system. Be sure to properly dispose of used coolant.

5. Drain the transaxle. Be sure to dispose of used fluid.

6. Properly relieve the fuel system pressure.

7. Remove the engine/transaxle assembly. Position the unit in a suitable holding fixture.

8. Remove the dynamic chamber and throttle body as a single unit.

9. Remove the ignition coils.

10. Remove the dipstick.

11. Remove the power steering oil pump drive belt.

12. Remove the power steering oil pump.

13. Remove the timing chain.

14. Remove the oil pump and O-ring.

To install:

➡**Be sure to use new fasteners, as required.**

15. Install the oil pump and O-ring. Tighten to 76–101 inch lbs. (9–12 Nm)

16. Install the timing chain.

17. Install the power steering oil pump.

18. Install the power steering oil pump drive belt.

19. Install the dipstick.

20. Install the ignition coils.

21. Install the dynamic chamber and throttle body as a single unit.

22. Secure the engine and transaxle using an engine stand.

23. Install the engine and transaxle component.

24. Fill the engine with oil.

25. Start the engine and perform the following inspections:

   a. Inspect the runout and contact on the pulley and belt.

   b. Inspect for engine oil, engine coolant, ATF, power steering fluid and fuel leakage.

   c. Verify the ignition timing, idle speed.

   d. Inspect engine accessories for proper operation.

   e. Inspect the oil pressure.

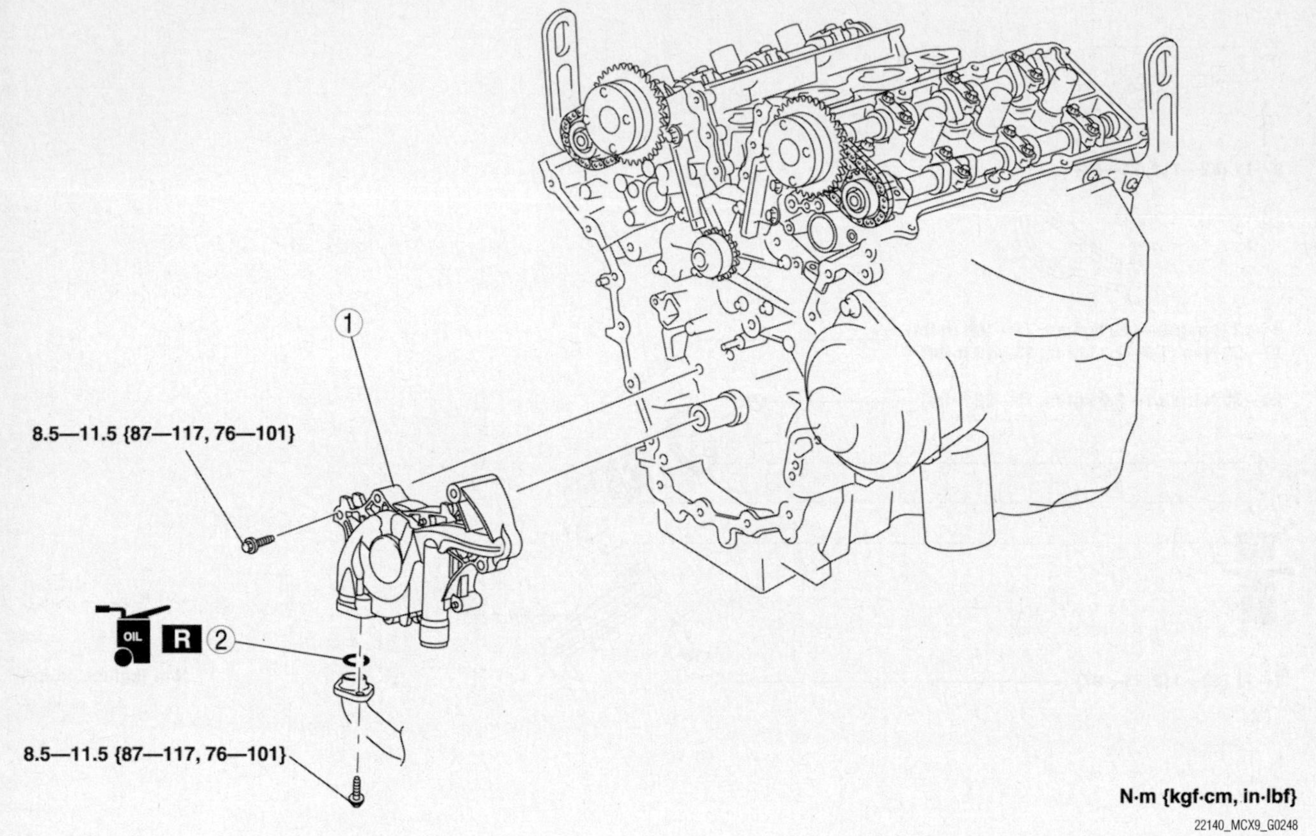

8.5—11.5 {87—117, 76—101}

8.5—11.5 {87—117, 76—101}

N·m {kgf·cm, in·lbf}

22140_MCX9_G0248

Fig. 121 Exploded view of the oil pump components—3.7L engine

26. Perform a road test.
27. Use the Mazda diagnostic scan tool, or equivalent and reprogram the required systems.

PISTON AND RING

POSITIONING

See Figures 122 and 123.

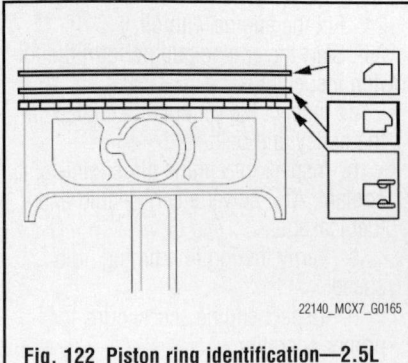

22140_MCX7_G0165

Fig. 122 Piston ring identification—2.5L engine

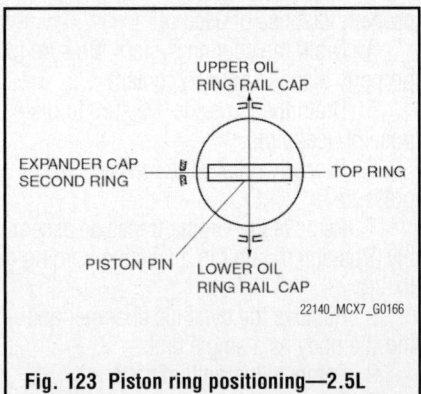

UPPER OIL
RING RAIL CAP

EXPANDER CAP
SECOND RING

TOP RING

PISTON PIN

LOWER OIL
RING RAIL CAP

22140_MCX7_G0166

Fig. 123 Piston ring positioning—2.5L engine

REAR MAIN SEAL

REMOVAL & INSTALLATION

2.5L Engine

See Figures 124 through 127.

1. Before servicing the vehicle, refer to the Precautions Section.

➡ If working near and/or around the SRS system and components, be sure to disable the SRS system. Tape the negative battery cable with insulating tape. Always disconnect the negative battery cable first.

❊❊ CAUTION

To avoid personal injury when working on vehicles equipped with an air bag, the negative battery cable must be disconnected and at least one minute must elapse before working on the system. Failure to do so may result in deployment of the air bag.

2. Disconnect the negative battery cable. Tape the cable with insulating tape.

➡ When disconnecting the cable, some systems need to be initialized after the cable is reconnected. You will need the Mazda diagnostic scan tool or equivalent. Follow the directions on the tool.

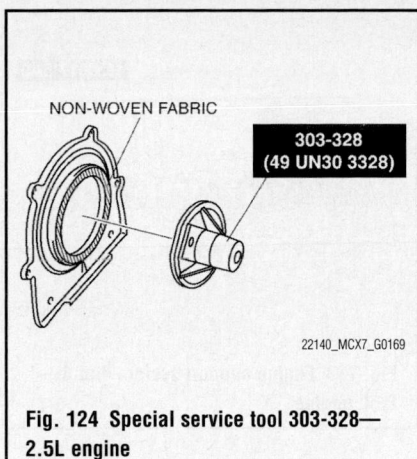

**Fig. 124 Special service tool 303-328—
2.5L engine**

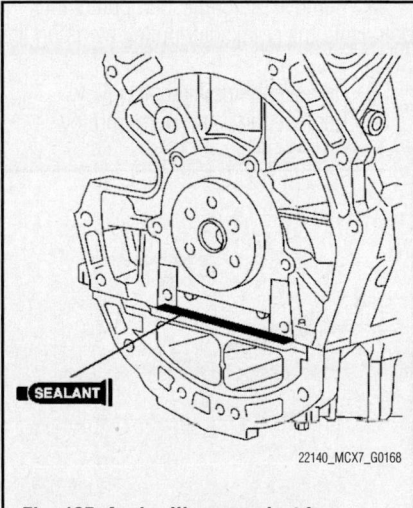

**Fig. 125 Apply silicone sealant for
installation—2.5L engine**

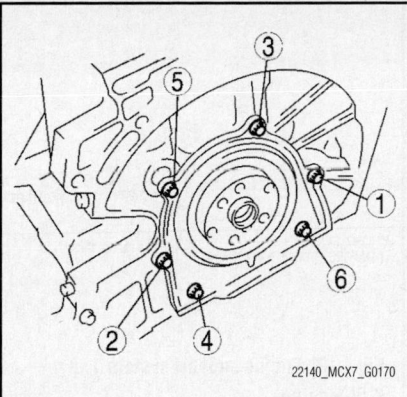

**Fig. 127 Oil seal torque sequence—2.5L
engine**

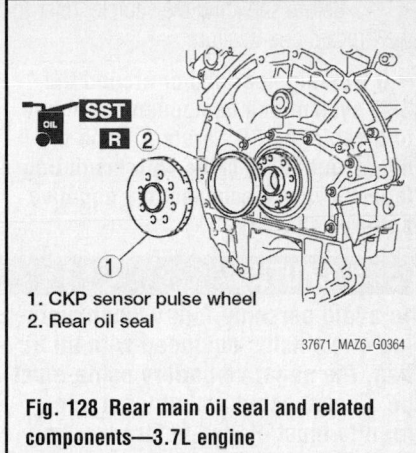

1. CKP sensor pulse wheel
2. Rear oil seal

**Fig. 128 Rear main oil seal and related
components—3.7L engine**

3. Remove the transaxle and flexplate/
flywheel.

4. Unbolt and remove the oil seal.

To install:

➡**Be sure to use new fasteners, as
required.**

5. Apply silicone sealant as shown in
the illustration.

6. Install SST 303-328 to the non-
woven fabric side of the rear main seal.

7. From the back side of the rear oil
seal, verify that there is no damage or separation in the lip area of the rear oil seal.

8. Install the rear main seal. Tighten
the bolts to 71–101 inch lbs. (8–11.5
Nm).

9. Remove SST 303-328.

10. Install the flexplate/flywheel and
transaxle.

11. Road test and check for leaks.

12. Use the Mazda diagnostic scan tool,
or equivalent and reprogram the required
systems.

3.7L Engine

See Figure 128.

1. Before servicing the vehicle, refer to
the Precautions Section.

➡**If working near and/or around the
SRS system and components, be sure
to disable the SRS system. Tape the
negative battery cable with insulating**

tape. Always disconnect the negative
battery cable first.

> **✳✳ CAUTION**
>
> **To avoid personal injury when
> working on vehicles equipped with an
> air bag, the negative battery cable
> must be disconnected and at least
> one minute must elapse before
> working on the system. Failure to do
> so may result in deployment of the
> air bag.**

2. Disconnect the negative battery cable.
Tape the cable with insulating tape.

➡**When disconnecting the cable,
some systems need to be initialized
after the cable is reconnected. You will
need the Mazda diagnostic scan tool or
equivalent. Follow the directions on the
tool.**

3. Remove the transaxle assembly.

4. Remove the drive plate.

5. Remove the CKP sensor pulse
wheel.

6. Remove the rear main oil seal.

To install:

➡**Be sure to use new fasteners, as
required.**

7. Installation is the reverse of the
removal procedure.

8. Use the Mazda diagnostic scan tool,
or equivalent and reprogram the required
systems.

> **TIMING CHAIN, SPROCKETS,
> FRONT COVER & SEAL**

REMOVAL & INSTALLATION

2.5L Engine

See Figures 129 through 139.

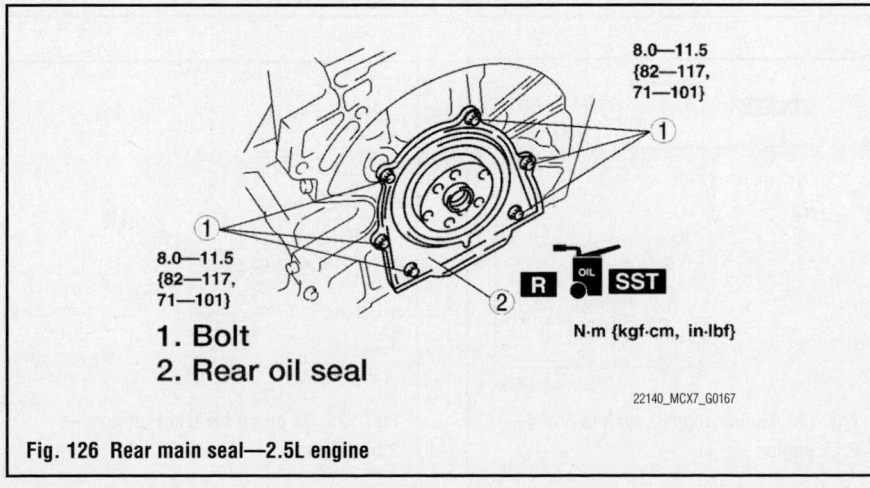

1. **Bolt**
2. **Rear oil seal**

N·m {kgf·cm, in·lbf}

Fig. 126 Rear main seal—2.5L engine

1. Before servicing the vehicle, refer to the Precautions Section.

➡If working near and/or around the SRS system and components, be sure to disable the SRS system. Tape the negative battery cable with insulating tape. Always disconnect the negative battery cable first.

✳✳ CAUTION

To avoid personal injury when working on vehicles equipped with an air bag, the negative battery cable must be disconnected and at least one minute must elapse before working on the system. Failure to do so may result in deployment of the air bag.

2. Disconnect the negative battery cable. Tape the cable with insulating tape.

➡When disconnecting the cable, some systems need to be initialized after the cable is reconnected. You will need the Mazda diagnostic scan tool or equivalent. Follow the directions on the tool.

3. Remove the plug hole plate.
4. Remove the ignition coils.
5. Disconnect the ventilation hose from the cylinder head cover.
6. Disconnect the Camshaft Position (CMP) sensor connector.
7. Disconnect the Oil Control Valve (OCV) connector.
8. Remove the front wheel and tire assemblies, as required.
9. Remove the coolant reservoir tank.
10. Remove the aerodynamic No. 2 cover.
11. Set the front mudguard out of the way.
12. Remove the splash shield.
13. Loosen the water pump pulley bolts and remove the drive belt.

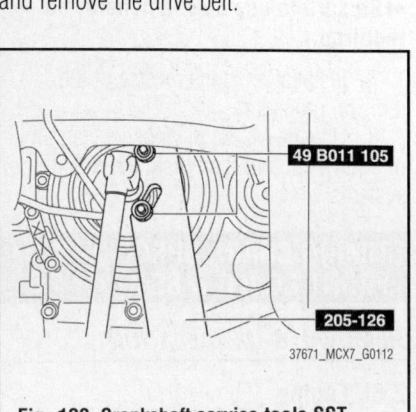

Fig. 129 Crankshaft service tools SST 205-126 and 49 B011 105—2.5L engine

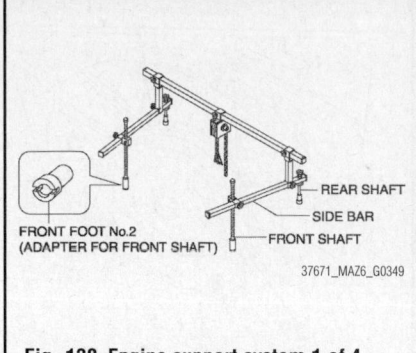

Fig. 130 Engine support system 1 of 4— 2.5L engine

14. Remove the Crankshaft Position (CKP) sensor.
15. Disconnect the halfshaft from the joint shaft, right side. Position the component out of the way. Do not allow it to hang unsupported.
16. Remove the P/S oil pump with the hose and pipe still connected.

➡Position and secure the P/S oil pump out of the way with a rope or wire.

17. Remove the spark plugs.
18. Remove the engine dipstick.
19. Remove the purge solenoid valve

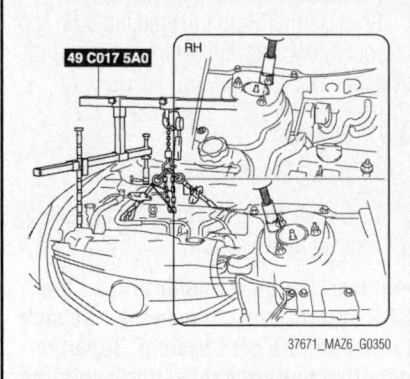

Fig. 131 Engine support system 2 of 4— 2.5L engine

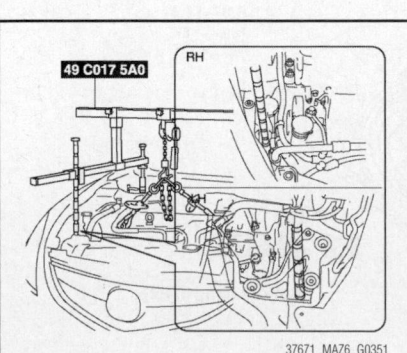

Fig. 132 Engine support system 3 of 4— 2.5L engine

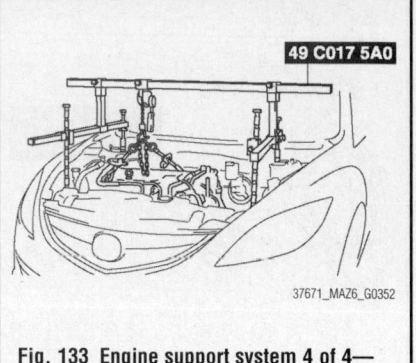

Fig. 133 Engine support system 4 of 4— 2.5L engine

installation bracket with the evaporative hose still connected and set it out of the way

20. Remove the cylinder head cover.
21. Rotate the crankshaft in the direction of the engine rotation and remove the cylinder block lower blind plug when the No. 1 cylinder is at the point prior to Top Dead Center (TDC) of compression, then install the SST.
22. Remove the crankshaft pulley lock bolt. Hold the crankshaft pulley by using the SSTs.
23. Remove the crankshaft pulley.
24. Remove the water pump pulley.

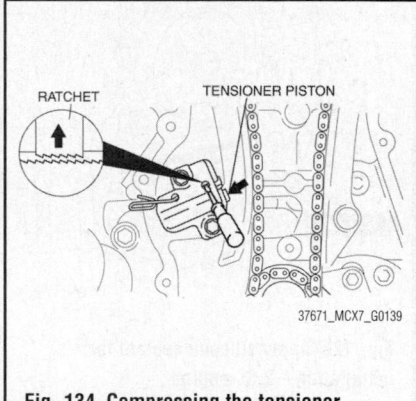

Fig. 134 Compressing the tensioner before removal—2.5L engine

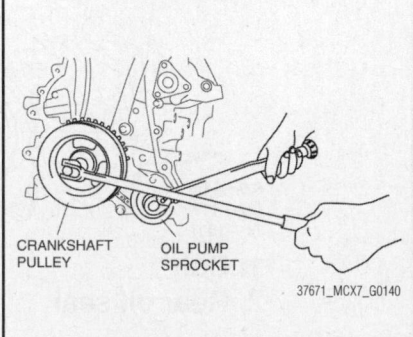

Fig. 135 Oil pump sprocket removal— 2.5L engine

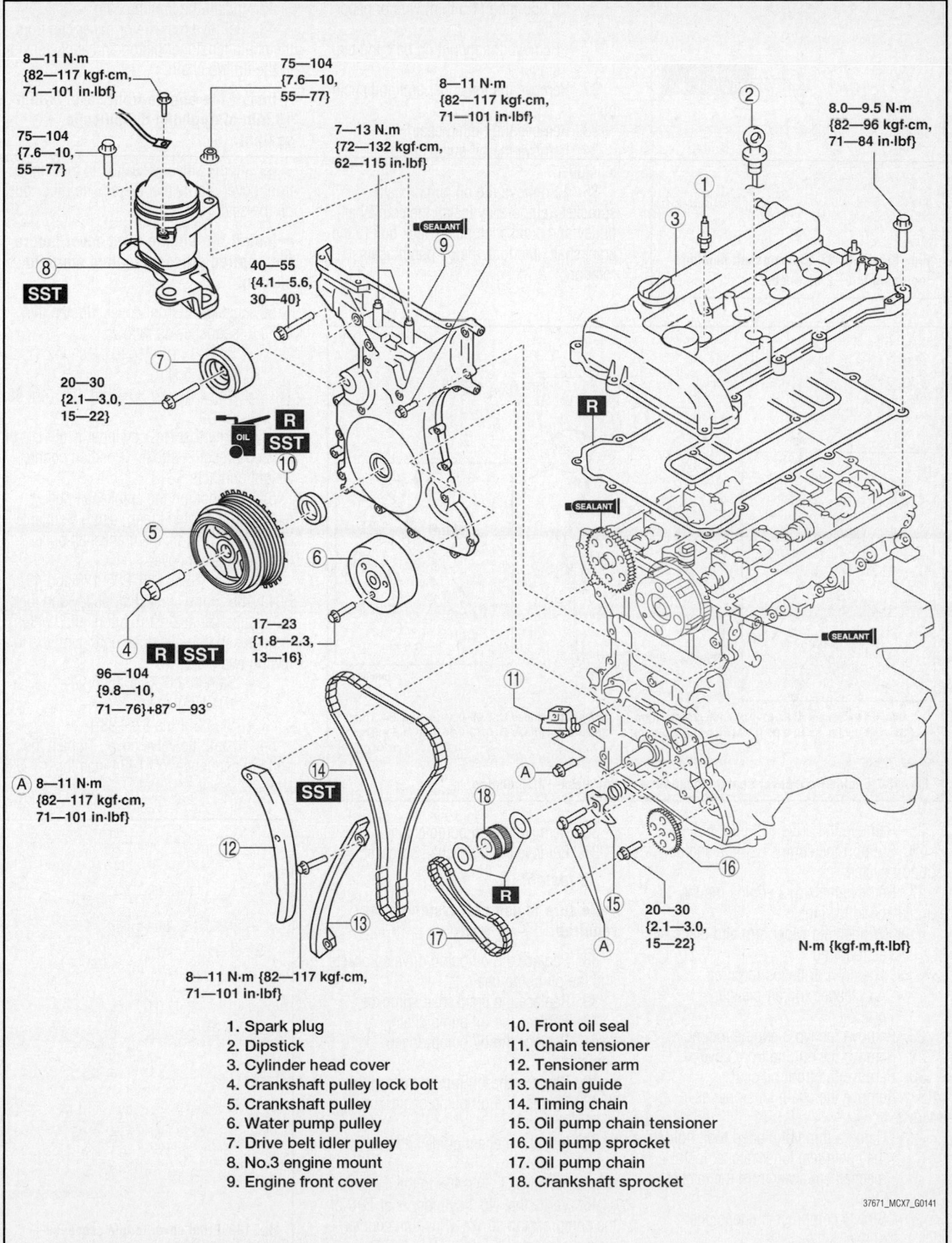

8—11 N·m
{82—117 kgf·cm,
71—101 in·lbf}

75—104
{7.6—10,
55—77}

75—104
{7.6—10,
55—77}

7—13 N·m
{72—132 kgf·cm,
62—115 in·lbf}

8—11 N·m
{82—117 kgf·cm,
71—101 in·lbf}

8.0—9.5 N·m
{82—96 kgf·cm,
71—84 in·lbf}

SST

40—55
{4.1—5.6,
30—40}

SEALANT

R

SEALANT

20—30
{2.1—3.0,
15—22}

OIL **R**
SST

SEALANT

17—23
{1.8—2.3,
13—16}

R **SST**

96—104
{9.8—10,
71—76}+87°—93°

Ⓐ 8—11 N·m
{82—117 kgf·cm,
71—101 in·lbf}

SST

R

20—30
{2.1—3.0,
15—22}

8—11 N·m {82—117 kgf·cm,
71—101 in·lbf}

N·m {kgf·m,ft·lbf}

1. Spark plug
2. Dipstick
3. Cylinder head cover
4. Crankshaft pulley lock bolt
5. Crankshaft pulley
6. Water pump pulley
7. Drive belt idler pulley
8. No.3 engine mount
9. Engine front cover
10. Front oil seal
11. Chain tensioner
12. Tensioner arm
13. Chain guide
14. Timing chain
15. Oil pump chain tensioner
16. Oil pump sprocket
17. Oil pump chain
18. Crankshaft sprocket

37671_MCX7_G0141

Fig. 136 Timing chain and related components—2.5L engine

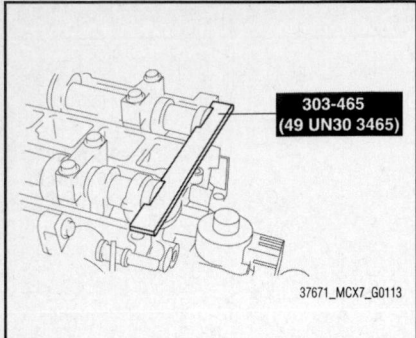

Fig. 137 SST 303-465 camshaft holding tool installed—2.5L engine

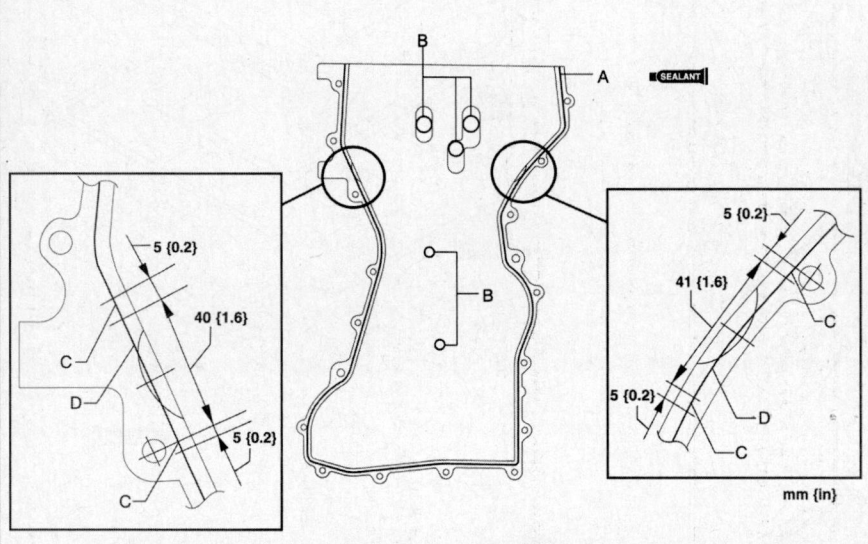

Sealant thickness A: 0.087-0.12 inch (2.2-3.2 mm)
Sealant thickness B: 0.06-0.098 inch (1.5-2.5 mm)

Sealant thickness C: 0.09-0.16 inch (2.2-4.3 mm)
Sealant thickness D: 0.13-0.16 inch (3.3-4.3 mm)

37671_MCX7_G0142

Fig. 138 Engine front cover sealant locations and thickness—2.5L engine

0.059 inch (1.5 mm) wire or paper clip.

- Remove the mounting bolt and tensioner

32. Remove the tensioner arm and chain guide.

33. Remove the timing chain.

34. Remove the oil pump chain tensioner.

35. To remove the oil pump chain sprocket, temporarily install the crankshaft pulley and crankshaft pulley lock bolt to the crankshaft, and lock the oil pump against rotation.

25. Remove the drive belt idler pulley.

26. Support the engine using a engine support system.

27. Remove the No. 3 engine mount, remove the following:

- Windshield wiper arm and blade, as required
- The cowl grille, as required
- Air cleaner and air cleaner bracket

28. Remove the No.3 engine mount.

29. Remove the engine front cover.

30. Remove the front oil seal.

31. Remove the chain tensioner as follows:

- Using a thin flat bladed tool, hold the chain tensioner ratchet lock mechanism away from the ratchet stem.
- Slowly compress the tensioner piston
- Hold the tensioner piston using a

36. Remove the oil pump chain.

37. Remove the crankshaft sprocket.

To install:

➡**Be sure to use new fasteners, as required.**

38. Replace the oil pump drive sprocket and the oil pump chain.

39. Replace the crankshaft sprocket.

40. Replace the oil pump chain.

41. Replace the oil pump driven sprocket.

42. Replace the oil pump chain guide.

43. Replace the oil pump chain tensioner.

44. Install the chain guide and the tensioner arm.

45. Install SST 303-465 to the camshaft.

46. Verify that No.1 cylinder is at TDC of the compression stroke. (Position counterweight contacts SST. 303-507)

47. Install the timing chain.

48. Remove the wire or paper clip from the chain tensioner piston and apply tension to the timing chain.

➡**Install the engine front cover within 10 min of applying the silicone sealant.**

49. Apply silicone sealant to the engine front cover. Verify that there is no oil or dust on the seal.

➡**Install the engine front cover before the applied silicone sealant starts to harden.**

50. Install the front cover. Tighten the bolts in sequence as follows:

- Bolts 1–18: 71–101 inch lbs. (8–11.5 Nm)
- Bolts 19–22: 30–40 ft. lbs. (40–55 Nm)

51. Verify that No.1 cylinder is at TDC of the compression stroke. (Position counterweight contacts SST.)

52. To position the crankshaft pulley, temporarily tighten it and, using a suitable bolt (M6 X 1.0), fix the crankshaft pulley to the engine front cover.

53. Install SSTs SST 205-126 and 49 B011 105 to the crankshaft pulley and lock the crankshaft against rotation, and tighten the crankshaft pulley lock bolt using the following two steps:

- Step 1: 71–77 ft. lbs. (96–104 Nm)
- Step 2: Plus 87°–93°

54. Install the valve cover. Tighten the bolts in sequence to 71–84 inch lbs. (8–9.5 Nm)

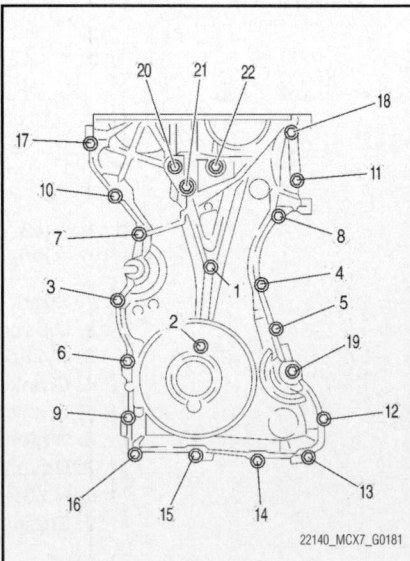

Fig. 139 Front cover torque sequence—2.5L engine

55. Start the engine, and inspect and adjust the following:

- Bleed the air from the cooling system.
- Runout and contact on pulley and belt
- Leakage of engine oil, engine coolant or fuel.
- Ignition timing, and idle speed.
- Engine accessories operation

56. Perform a road test and verify that there is no abnormal vibration or noise.

57. Use the Mazda diagnostic scan tool, or equivalent and reprogram the required systems.

TIMING CHAIN FRONT COVER & SEAL

REMOVAL & INSTALLATION

3.7L Engine

See Figures 140 through 157.

➡ **Mazda's official procedure for removal and installation of this component requires the engine/transaxle assembly to be removed from the vehicle. Remove the engine/transaxle assembly and position it in a suitable holding fixture.**

1. Before servicing the vehicle, refer to the Precautions Section.

➡ **If working near and/or around the SRS system and components, be sure to disable the SRS system. Tape the negative battery cable with insulating tape. Always disconnect the negative battery cable first.**

❋❋ CAUTION

To avoid personal injury when working on vehicles equipped with an air bag, the negative battery cable must be disconnected and at least one minute must elapse before working on the system. Failure to do so may result in deployment of the air bag.

2. Disconnect the negative battery cable. Tape the cable with insulating tape.

➡ **When disconnecting the cable, some systems need to be initialized after the cable is reconnected. You will need the Mazda diagnostic scan tool or equivalent. Follow the directions on the tool.**

3. Drain the engine oil. Be sure to properly dispose of used oil.

4. Drain the cooling system. Be sure to properly dispose of used coolant.

5. Drain the transaxle. Be sure to dispose of used fluid.

6. Properly relieve the fuel system pressure.

7. Remove the engine/transaxle assembly. Position the unit in a suitable holding fixture.

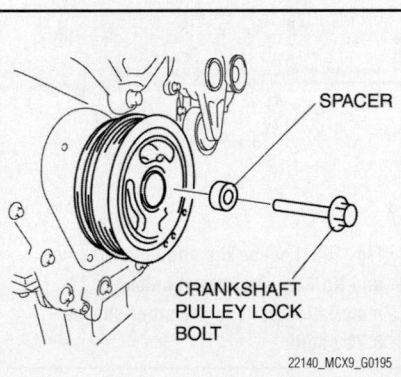

Fig. 141 Install a suitable spacer 0.55 in. (14 mm) in thickness and 1.18 in. (30 mm) in diameter to the crankshaft pulley lock bolt, and install the crankshaft pulley lock bolt to the crankshaft

8. Remove the intake manifold dynamic chamber and throttle body as a single unit.

9. Remove the ignition coils.

10. Remove the dipstick.

11. Remove the power steering oil pump drive belt.

12. Remove the power steering oil pump.

13. Remove the drive belt auto tensioner.

14. Remove the crankshaft pulley lock bolt.

  a. Remove the starter.

  b. Set a flat blade tool to the drive plate as shown to lock the crankshaft rotation.

  c. Remove the crankshaft pulley lock bolt and washer.

15. Remove the crankshaft pulley.

  a. Install a suitable spacer 0.55 in. (14 mm) in thickness and 1.18 in. (30 mm) in diameter to the crankshaft pulley lock bolt, and install the crankshaft pulley lock bolt to the crankshaft.

  b. Remove the crankshaft pulley using a gear puller.

  c. Remove the crankshaft pulley lock bolt and spacer.

16. Remove the front oil seal.

  a. Remove the front oil seal using a flathead screwdriver as shown

17. Remove the cylinder head covers.

  a. Remove the cylinder head cover bolts in the order shown.

18. Remove the cylinder head cover oil seals.

  a. Inspect the OCV attachment hole seals and spark plug tube attachment hole seals. Remove the any damaged seals.

  b. The OCV attachment hole seals removal is shown, the spark plug tube attachment hole seals removal procedure is the same.

  c. To remove, tap the OCV attachment hole seals and spark plug tube attachment hole seals using the special tool and hammer.

19. Remove the no.3 engine mount bracket and engine front cover.

  a. Loosen the engine front cover and No.3 engine mount bracket installation bolts in the order shown.

  b. Install 4 of the engine front cover bolts (finger tightened) into the 4 threaded holes in the engine front cover.

  c. Tighten the bolts one turn at a time in a criss-cross pattern until the engine front cover-to-cylinder block seal is released.

  d. Remove the engine front cover.

To install:

➡ **Be sure to use new fasteners, as required.**

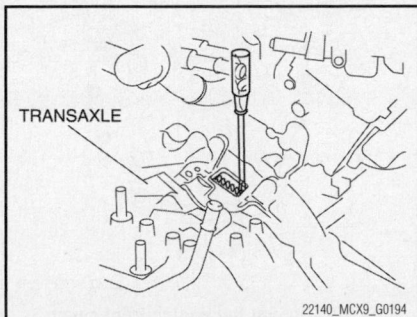

Fig. 140 Set a flat blade tool to the drive plate as shown to lock the crankshaft rotation—3.7L engine

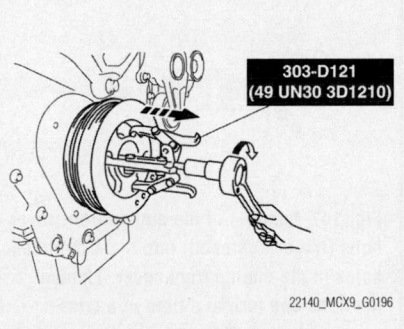

Fig. 142 Remove the crankshaft pulley using a gear puller—3.7L engine

Fig. 143 Using a flat tip tool cover with a rag, pry the seal from the front cover—3.7L engine

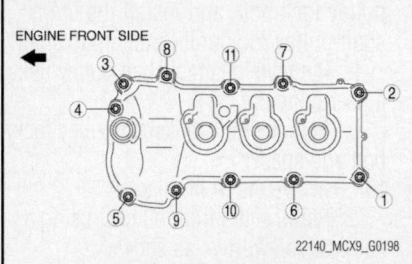

Fig. 144 Remove the cylinder head cover bolts in the order shown—right hand—3.7L engine

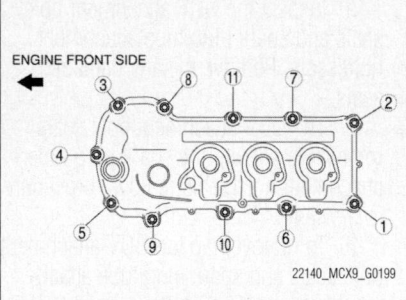

Fig. 145 Remove the cylinder head cover bolts in the order shown—left hand—3.7L engine

20. Install the engine front cover and No.3 engine mount bracket.

a. Apply the Loctite® 5900 silicon sealant to the engine front cover as shown.

b. Silicon sealant thickness should be as follows: A—0.197–0.236 in. (5.0–6.0 mm) and B— 0.099–0.137 in. (2.5–3.5 mm).

c. Install the engine front cover and No.3 engine mount bracket installation bolts within 10 min of applying the silicone sealant.

d. Tighten the engine front cover and No.3 engine mount bracket installation bolts in 4 steps in the order shown.

- Step 1: Tighten bolts 1–6 to 89 inch lbs. (10 Nm).
- Step 2: Tighten bolts 7–9 to 11 ft. lbs. (15 Nm).
- Step 3: Tighten bolts 1–6 to 15–20 ft. lbs. (20–28 Nm).
- Step 4: Tighten bolts 7–9 to 52–59 ft. lbs. (70–80 Nm).

e. Install the engine front cover installation bolts within 60 min of applying the silicone sealant.

f. Tighten the engine front cover installation bolts in the order in 2 steps in the order shown.

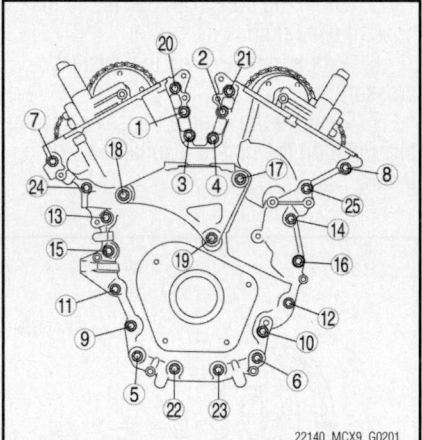

Fig. 146 Loosen the engine front cover and No.3 engine mount bracket installation bolts in the order shown—3.7L engine

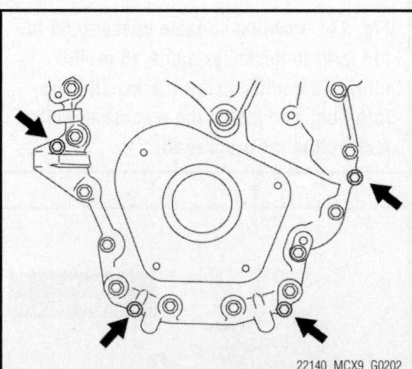

Fig. 147 Install 4 of the engine front cover bolts (finger tightened) into the 4 threaded holes in the engine front cover. Tighten the bolts one turn at a time in a criss-cross pattern until the engine front cover-to-cylinder block seal is released—3.7L engine

- Step 1: Tighten to 89 inch lbs. (10 Nm).
- Step 2: Tighten to 15–20 ft. lbs. (20–28 Nm).

21. Install the cylinder head cover oil seal.

a. Installation of new seals is only required if damaged seals were removed during disassembly of the engine.

b. Push the OCV attachment hole seals and spark plug tube attachment hole seals slightly in by hand.

c. Tap the OCV attachment hole seals and spark plug tube attachment hole seals using the special tool and hammer.

22. Install the cylinder head cover.

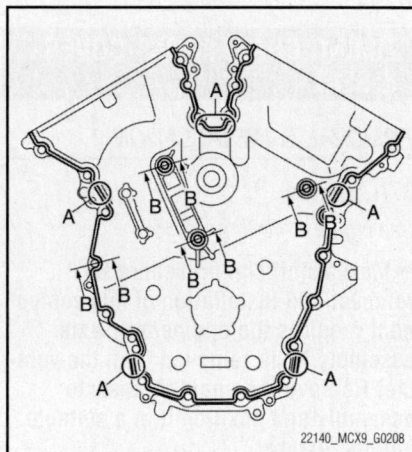

Fig. 148 Apply the Loctite® 5900 silicon sealant to the engine front cover as shown. Silicon sealant thickness should be as follows: A—0.197–0.236 in. (5.0–6.0 mm) and B— 0.099–0.137 in. (2.5–3.5 mm)—3.7L engine

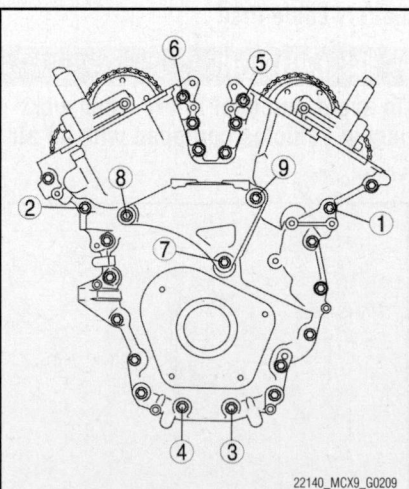

Fig. 149 Tighten the engine front cover and No.3 engine mount bracket installation bolts in the order shown—3.7L engine

a. Apply Loctite® 5900 silicone sealant to the mating faces as shown.

b. Install the cylinder head cover with a new gasket.

c. Install the cylinder head cover installation bolt and studs within 5 min of applying the silicone sealant.

d. Tighten the bolts in the order shown to 76–101 inch lbs. (9–12 Nm).

23. Install the front oil seal

a. Apply clean engine oil to the front oil seal bore in the engine front cover.

b. Push the front oil seal slightly in by hand.

c. Tap the front oil seal in evenly using the special tool and a hammer.

24. Install the crankshaft pulley

25. Install the crankshaft pulley lock bolt.

a. Set a flat tip tool to the drive plate in the position indicated. to lock the crankshaft rotation.

b. Tighten the new crankshaft pulley lock bolt in four steps.

- Step 1: Tighten to 103 ft. lbs. (140 Nm)
- Step 2: Loosen 360° (one full turn)
- Step 3: Tighten to 35–39 ft. lbs (47–53 Nm)
- Step 4: Tighten an additional 85–95° of rotation

26. Install the power steering oil pump.

27. Install the power steering oil pump drive belt.

28. Install the dipstick.

29. Install the ignition coils.

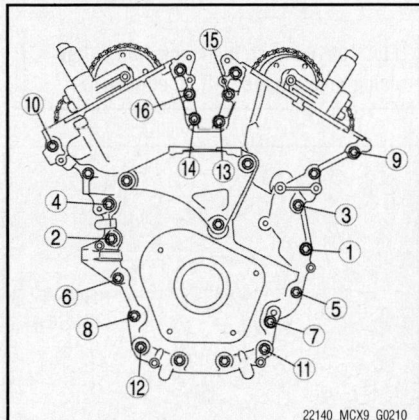
Fig. 150 Tighten the engine front cover installation bolts in the order in the order shown—3.7L engine

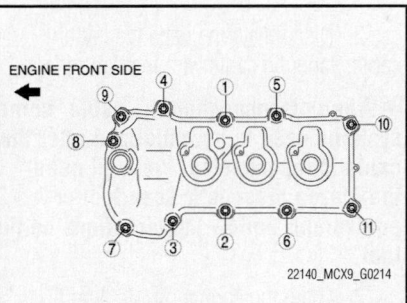
Fig. 153 Tighten the cylinder head cover bolts in the order shown—right side—3.7L engine

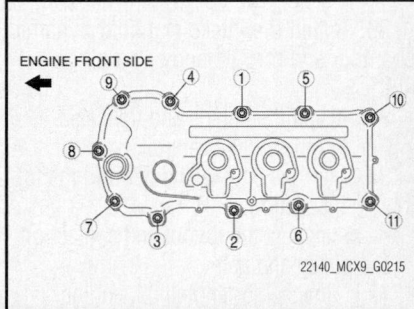
Fig. 154 Tighten the cylinder head cover bolts in the order shown—left side—3.7L engine

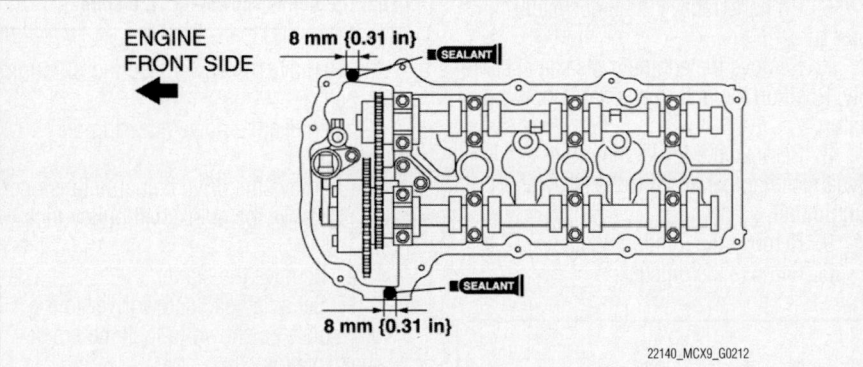
Fig. 151 Apply Loctite® 5900 silicone sealant to the mating faces as shown—right side—3.7L engine

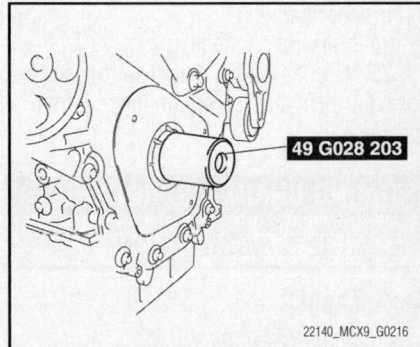
Fig. 155 Tap the front oil seal in evenly using the special tool and a hammer—3.7L engine

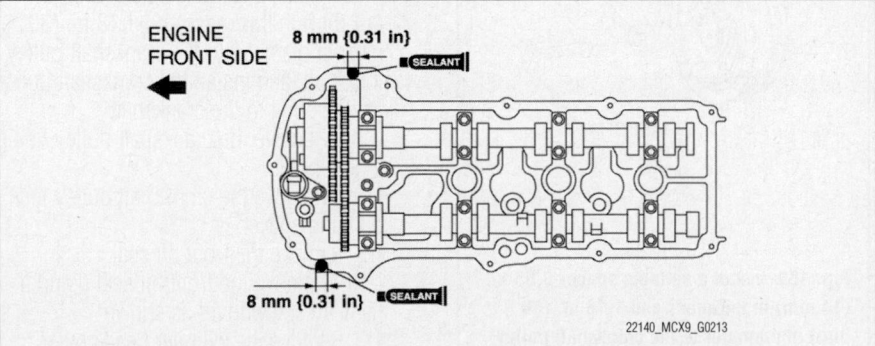
Fig. 152 Apply Loctite 5900 silicone sealant to the mating faces as shown—left side—3.7L engine

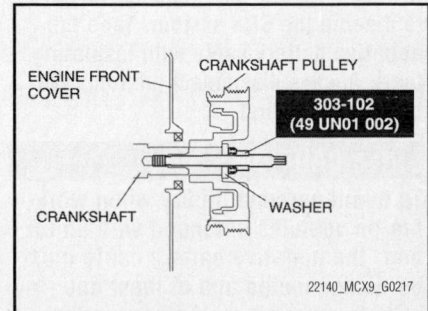
Fig. 156 Install the crankshaft pulley, washer, and special tool to the crankshaft—3.7L engine

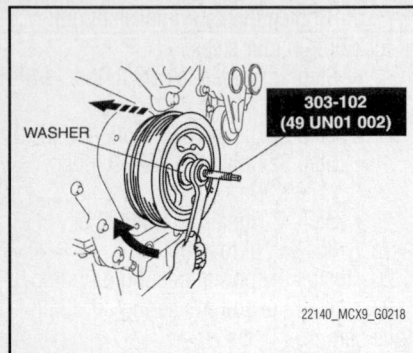

Fig. 157 Tighten the special tool nut and install the crankshaft pulley—3.7L engine

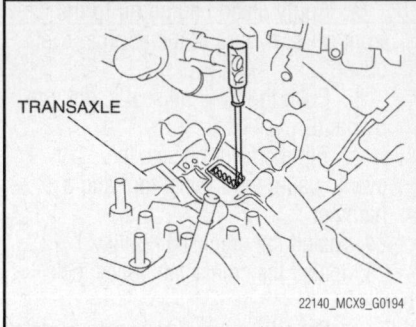

Fig. 158 Set a flat tip tool to the drive plate as shown to lock the crankshaft rotation—3.7L engine

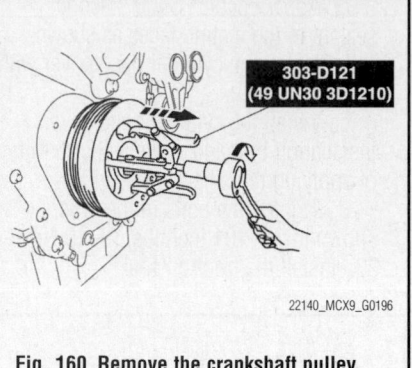

Fig. 160 Remove the crankshaft pulley using a gear puller—3.7L engine

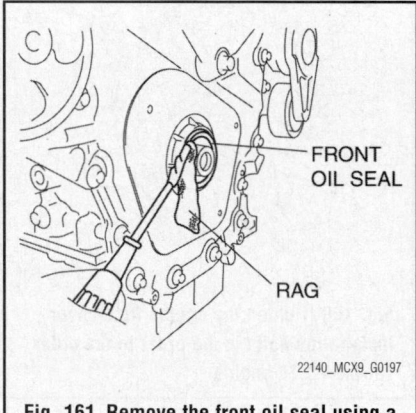

Fig. 161 Remove the front oil seal using a flat tip tool as shown—3.7L engine

30. Install the intake manifold dynamic chamber and throttle body as a single unit.

31. Install the engine and transaxle.

32. Fill the engine with oil.

33. Start the engine and perform the following inspections:

a. Inspect the runout and contact on the pulley and belt.

b. Inspect for engine oil, engine coolant, ATF, power steering fluid and fuel leakage.

c. Verify the ignition timing and idle speed.

d. Inspect engine accessories for proper operation.

34. Perform a road test.

35. Use the Mazda diagnostic scan tool, or equivalent and reprogram the required systems.

TIMING CHAIN & SPROCKETS

REMOVAL & INSTALLATION

3.7L Engine

See Figures 158 through 185.

1. Before servicing the vehicle, refer to the Precautions Section.

➡If working near and/or around the SRS system and components, be sure to disable the SRS system. Tape the negative battery cable with insulating tape. Always disconnect the negative battery cable first.

✳✳ CAUTION

To avoid personal injury when working on vehicles equipped with an air bag, the negative battery cable must be disconnected and at least one minute must elapse before working on the system. Failure to do so may result in deployment of the air bag.

2. Disconnect the negative battery cable. Tape the cable with insulating tape.

➡When disconnecting the cable, some systems need to be initialized after the cable is reconnected. You will need the Mazda diagnostic scan tool or equivalent. Follow the directions on the tool.

3. Drain the engine oil. Be sure to properly dispose of used oil.

4. Drain the cooling system. Be sure to properly dispose of used coolant.

5. Drain the transaxle. Be sure to dispose of used fluid.

6. Properly relieve the fuel system pressure.

7. Remove the engine/transaxle assembly. Position the unit in a suitable holding fixture.

8. Remove the intake manifold dynamic chamber and throttle body as a single unit.

9. Remove the ignition coils.

10. Remove the dipstick.

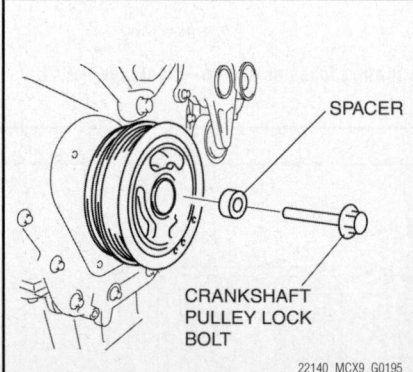

Fig. 159 Install a suitable spacer 0.55 in. (14 mm) in thickness and 1.18 in. (30 mm) in diameter to the crankshaft pulley lock bolt, and install the crankshaft pulley lock bolt to the crankshaft—3.7L engine

11. Remove the power steering oil pump drive belt.

12. Remove the power steering oil pump.

13. Remove the drive belt auto tensioner.

14. Remove the crankshaft pulley lock bolt.

a. Remove the starter.

b. Set a flathead screwdriver to the drive plate as shown to lock the crankshaft rotation.

c. Remove the crankshaft pulley lock bolt and washer.

15. Remove the crankshaft pulley.

a. Install a suitable spacer 0.55 in. (14 mm) in thickness and 1.18 in. (30 mm) in diameter to the crankshaft pulley lock bolt, and install the crankshaft pulley lock bolt to the crankshaft.

b. Remove the crankshaft pulley using a gear puller.

c. Remove the crankshaft pulley lock bolt and spacer.

16. Remove the front oil seal.

a. Remove the front oil seal using a flathead screwdriver as shown

17. Remove the cylinder head covers.

a. Remove the cylinder head cover bolts in the order shown.

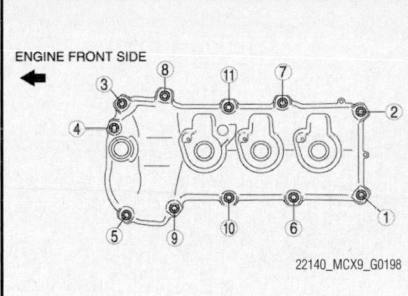

Fig. 162 Remove the cylinder head cover bolts in the order shown—right hand—3.7L engine

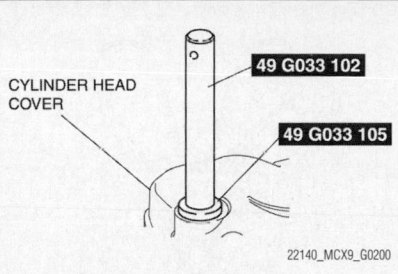

Fig. 164 Tap the OCV attachment hole seals and spark plug tube attachment hole seals using the special tool and hammer—3.7L engine

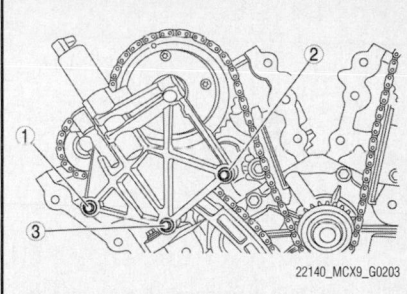

Fig. 167 Loosen the OCV component installation bolts in the order shown—right hand—3.7L engine

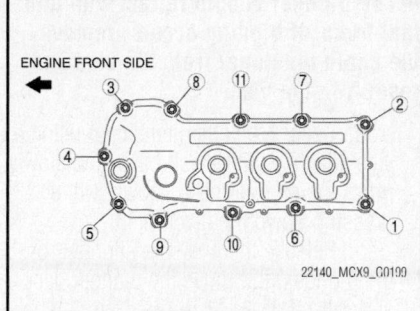

Fig. 163 Remove the cylinder head cover bolts in the order shown—left hand—3.7L engine

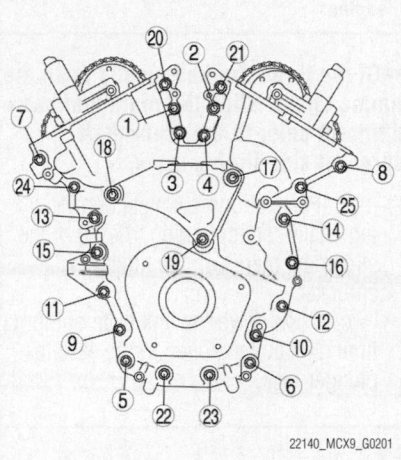

Fig. 165 Loosen the engine front cover and No.3 engine mount bracket installation bolts in the order shown—3.7L engine

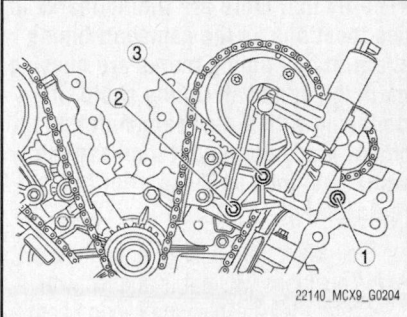

Fig. 168 Loosen the OCV component installation bolts in the order shown—left hand—3.7L engine

18. Remove the cylinder head cover oil seals.

   a. Inspect the OCV attachment hole seals and spark plug tube attachment hole seals. Remove the any damaged seals.

   b. The OCV attachment hole seals removal is shown, the spark plug tube attachment hole seals removal procedure is the same.

   c. To remove, tap the OCV attachment hole seals and spark plug tube attachment hole seals using the special tool and hammer.

19. Remove the No. 3 engine mount bracket and engine front cover.

   a. Loosen the engine front cover and No. 3 engine mount bracket installation bolts in the order shown.

   b. Install 4 of the engine front cover bolts (finger tightened) into the 4 threaded holes in the engine front cover.

   c. Tighten the bolts one turn at a time in a criss-cross pattern until the engine front cover-to-cylinder block seal is released.

   d. Remove the engine front cover.

20. Remove the OCV component.

   a. Loosen the OCV component installation bolts in the order shown

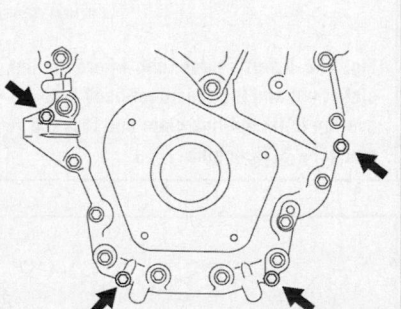

Fig. 166 Install 4 of the engine front cover bolts (finger tightened) into the 4 threaded holes in the engine front cover. Tighten the bolts one turn at a time in a criss-cross pattern until the engine front cover-to-cylinder block seal is released—3.7L engine

21. Remove the chain tensioner.
22. Remove the timing chain.

➥When removing the timing chain and marking the timing marks on the chain, mark the camshaft timing chain as well.

✳✳ WARNING

Do not rotate the crankshaft counterclockwise. The timing chains may bind, causing engine damage.

   a. Turn the crankshaft clockwise so that the crankshaft keyway is in the 11 o'clock position. This will position the No.1 cylinder at TDC.

➥Verify that there are timing marks in three locations (Yellow 1, Black 2) on the timing chain. If any timing marks are missing, mark the timing chain. When marking the crankshaft sprocket side timing chain, change the mark color. When the timing chain is replaced with a new one, mark the new timing chain at the same positions as the removed timing chain.

   b. Mark the timing chain at the position of each timing sprocket timing mark.

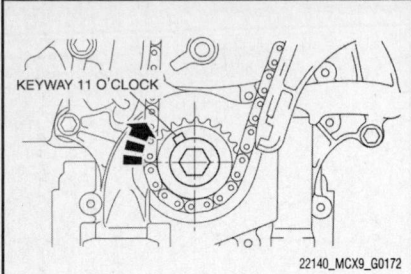

Fig. 169 Turn the crankshaft clockwise so that the crankshaft keyway is in the 11 o'clock position. This will position the No.1 cylinder at TDC—3.7L engine

→ Verify that there are timing marks in two locations on the camshaft timing chain. If any timing marks are missing, mark the camshaft timing chain. If replacing with a new camshaft timing chain, place alignment marks in the same positions as those prior to the replacement.

c. Mark the camshaft timing chain at the positions where it is aligned with each of the camshaft sprocket on both banks.

d. Remove the timing chain in the following order:
- Chain tensioner
- Timing chain
- Tensioner arm
- Chain guide
- Crankshaft sprocket

To install:

→ Be sure to use new fasteners, as required.

23. Install the timing chain
a. Verify that the crankshaft keyway is at 11 o'clock position.

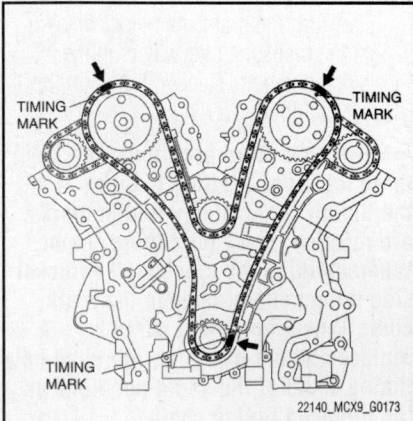

Fig. 170 Mark the timing chain at the position of each timing sprocket timing mark—3.7L engine

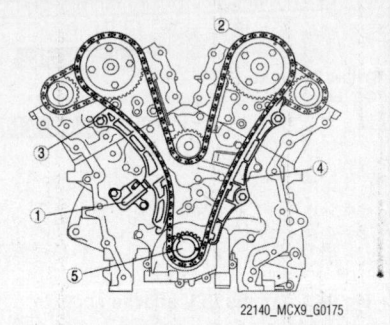

Fig. 171 Remove the timing chain in the following order: (1) Chain tensioner, (2) Timing chain, (3) Tensioner arm, (4) Chain guide and (5) Crankshaft sprocket—3.7L engine

→ Of the three marked locations on the timing chain, align the mark that has a different color to the crankshaft sprocket side timing mark.

b. Place alignment marks on the timing chain corresponding to each of the alignment marks on the timing sprocket.

c. Push down the link plate of the timing chain tensioner and release the plunger lock.

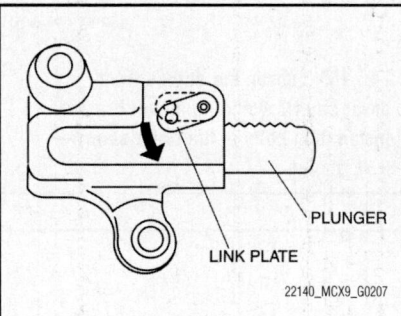

Fig. 172 Insert a paper clip where the link plate hole and the tensioner body hole overlap to fix the link plate and lock the plunger—3.7L engine

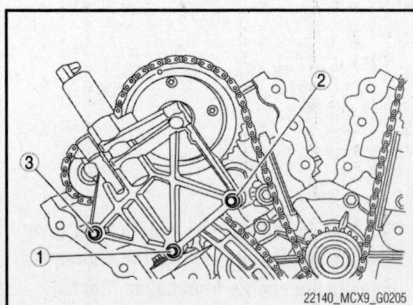

Fig. 173 Tighten the OCV component installation bolts in the order shown—right hand—3.7L engine

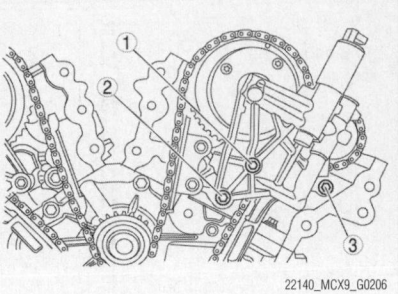

Fig. 174 Tighten the OCV component installation bolts in the order shown—left hand—3.7L engine

→ The plunger should retract with minimal force. If binding occurs, remove the chain tensioner from the vise and reset it in the vise.

d. Secure the chain tensioner using a vice attached with a soft base, and slowly press the plunger back shown. while pressing down the link plate.

e. Release the pressure slightly from the plunger, and move the plunger back and forth 0.08–0.11 in (2–3 mm).

f. Insert a paper clip where the link plate hole and the tensioner body hole overlap to fix the link plate and lock the plunger.

g. Install the timing chain in the following order:
- Crankshaft sprocket
- Chain guide
- Tensioner arm
- Timing chain
- Chain tensioner

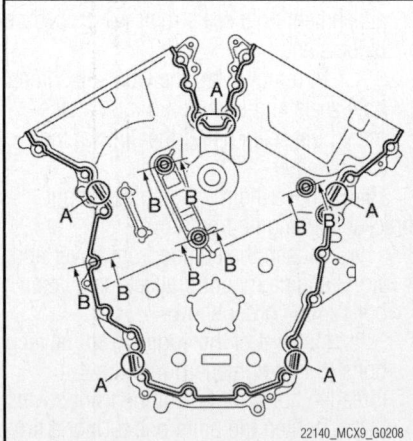

Fig. 175 Apply the Loctite 5900 silicon sealant to the engine front cover as shown. Silicon sealant thickness should be as follows: A—0.197–0.236 in. (5.0–6.0 mm) and B— 0.099–0.137 in. (2.5–3.5 mm)—3.7L engine

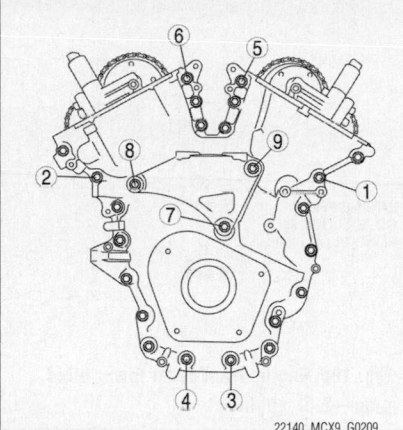

Fig. 176 Tighten the engine front cover and No.3 engine mount bracket installation bolts in the order shown—3.7L engine

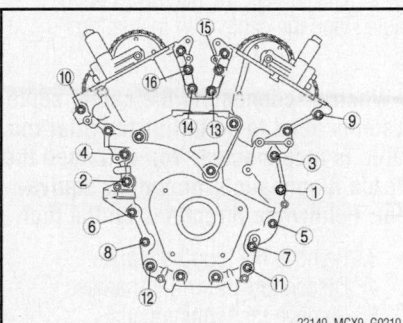

Fig. 177 Tighten the engine front cover installation bolts in the order in the order shown—3.7L engine

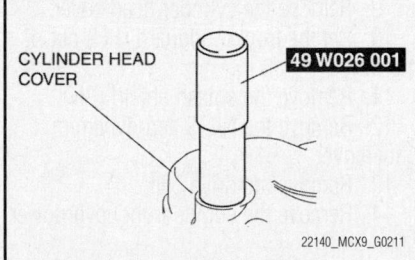

Fig. 178 Tap the OCV attachment hole seals and spark plug tube attachment hole seals using the special tool and hammer—3.7L engine

h. Tighten components to 76–101 inch lbs. (9–12 Nm).

i. Remove the retaining wire.

24. Install the chain tensioner

25. Install the OCV components.

a. Tighten the OCV component installation bolts in the order shown

26. Install the engine front cover and No. 3 engine mount bracket.

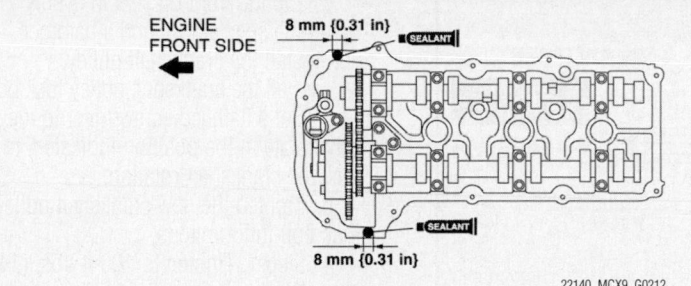

Fig. 179 Apply Loctite 5900 silicone sealant to the mating faces as shown—right side—3.7L engine

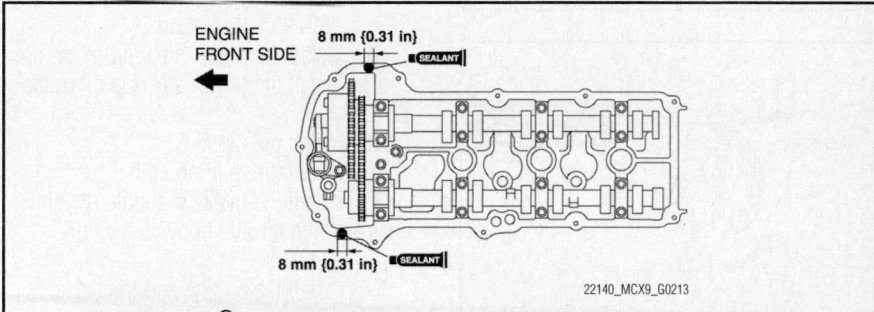

Fig. 180 Apply Loctite® 5900 silicone sealant to the mating faces as shown—left side—3.7L engine

a. Apply the Loctite 5900 silicon sealant to the engine front cover as shown.

b. Silicon sealant thickness should be as follows: A—0.197–0.236 in. (5.0–6.0 mm) and B— 0.099–0.137 in. (2.5–3.5 mm).

c. Install the engine front cover and No.3 engine mount bracket installation bolts within 10 min of applying the silicone sealant.

d. Tighten the engine front cover and No.3 engine mount bracket installation bolts in 4 steps in the order shown.

- Step 1: Tighten bolts 1–6 to 89 inch lbs. (10 Nm).
- Step 2: Tighten bolts 7–9 to 11 ft. lbs. (15 Nm).
- Step 3: Tighten bolts 1–6 to 15–20 ft. lbs. (20–28 Nm).

- Step 4: Tighten bolts 7–9 to 52–59 ft. lbs. (70–80 Nm).

e. Install the engine front cover installation bolts within 60 min of applying the silicone sealant.

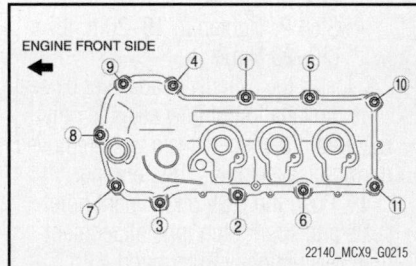

Fig. 182 Tighten the cylinder head cover bolts in the order shown—left side—3.7L engine

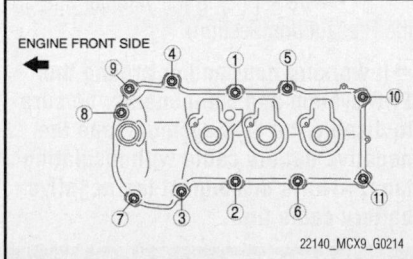

Fig. 181 Tighten the cylinder head cover bolts in the order shown—right side—3.7L engine

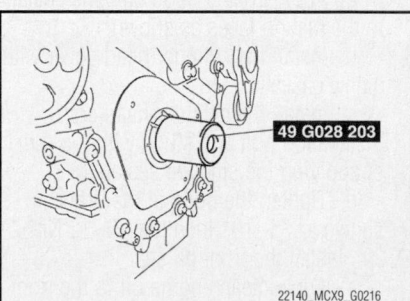

Fig. 183 Tap the front oil seal in evenly using the special tool and a hammer

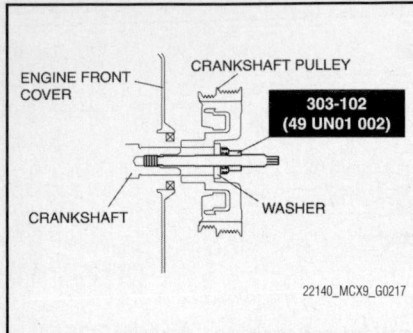

Fig. 184 Install the crankshaft pulley, washer, and special tool to the crankshaft—3.7L engine

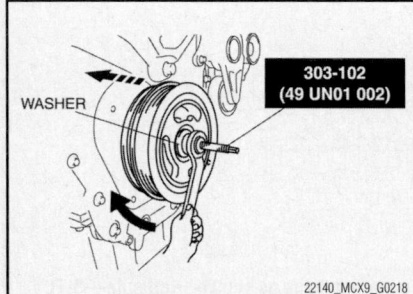

Fig. 185 Tighten the special tool nut and install the crankshaft pulley—3.7L engine

f. Tighten the engine front cover installation bolts in the order in 2 steps in the order shown.

- Step 1: Tighten to 89 inch lbs. (10 Nm).
- Step 2: Tighten to 15–20 ft. lbs. (20–28 Nm).

27. Install the cylinder head cover oil seal.

a. Installation of new seals is only required if damaged seals were removed during disassembly of the engine.

b. Push the OCV attachment hole seals and spark plug tube attachment hole seals slightly in by hand.

c. Tap the OCV attachment hole seals and spark plug tube attachment hole seals using the special tool and hammer.

28. Install the cylinder head cover.

a. Apply Loctite 5900 silicone sealant to the mating faces as shown.

b. Install the cylinder head cover with a new gasket.

c. Install the cylinder head cover installation bolt and studs within 5 min of applying the silicone sealant.

d. Tighten the bolts in the order shown to 76–101 inch lbs. (9–12 Nm).

29. Install the front oil seal

a. Apply clean engine oil to the front oil seal bore in the engine front cover.

b. Push the front oil seal slightly in by hand.

c. Tap the front oil seal in evenly using the special tool and a hammer.

30. Install the crankshaft pulley

31. Install the crankshaft pulley lock bolt.

a. Set a flathead screwdriver to the drive plate in the position indicated. to lock the crankshaft rotation.

b. Tighten the new crankshaft pulley lock bolt in four steps.

- Step 1: Tighten to 103 ft. lbs. (140 Nm)
- Step 2: Loosen 360° (one full turn).
- Step 3: Tighten to 35–39 ft. lbs (47–53 Nm).
- Step 4: Tighten an additional 85–95° of rotation.

32. Install the power steering oil pump.

33. Install the power steering oil pump drive belt.

34. Install the dipstick.

35. Install the ignition coils.

36. Install the intake manifold dynamic chamber and throttle body as a single unit.

37. Install the engine and transaxle.

38. Fill the engine with oil.

39. Start the engine and perform the following inspections:

a. Inspect the runout and contact on the pulley and belt.

b. Inspect for engine oil, engine coolant, ATF, power steering fluid and fuel leakage.

c. Verify the ignition timing and idle speed.

d. Inspect engine accessories for proper operation.

40. Perform a road test.

41. Use the Mazda diagnostic scan tool, or equivalent and reprogram the required systems.

VALVE LASH

ADJUSTMENT

2.5L Engine

See Figures 186 through 197.

1. Before servicing the vehicle, refer to the Precautions Section.

➡**If working near and/or around the SRS system and components, be sure to disable the SRS system. Tape the negative battery cable with insulating tape. Always disconnect the negative battery cable first.**

✳✳ CAUTION

To avoid personal injury when working on vehicles equipped with an air bag, the negative battery cable must

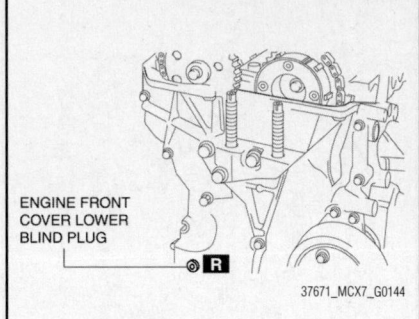

Fig. 186 Engine front cover lower blind plug—2.5L engine

be disconnected and at least one minute must elapse before working on the system. Failure to do so may result in deployment of the air bag.

2. Disconnect the negative battery cable. Tape the cable with insulating tape.

➡**When disconnecting the cable, some systems need to be initialized after the cable is reconnected. You will need the Mazda diagnostic scan tool or equivalent. Follow the directions on the tool.**

3. Remove the plug hole plate.

4. Disconnect the wiring harness.

5. Remove the ignition coils.

6. Remove the spark plugs.

7. Disconnect the ventilation hose from the cylinder head cover.

8. Remove the CMP sensor connector.

9. Remove the cylinder head cover.

10. Set the front mudguard (RH) out of the way.

11. Remove the splash shield (RH).

12. Remove the No. 2 aerodynamic undercover.

13. Remove the drive belt.

14. Remove the engine front cover lower blind plug.

15. Remove the engine front cover upper blind plug.

16. Remove the cylinder block lower blind plug, and install the SST 303-507.

17. Loosen the timing chain using the following procedure:

- Insert a suitable bolt M6 X 1.0, length 1.0–1.3 inch (25–35 mm) into the engine front cover upper blind plug and tighten it until it contacts the chain tensioner arm, and then rotate it back one turn. (Set the bolt slightly away from the chain tensioner arm so that it does not contact it.)

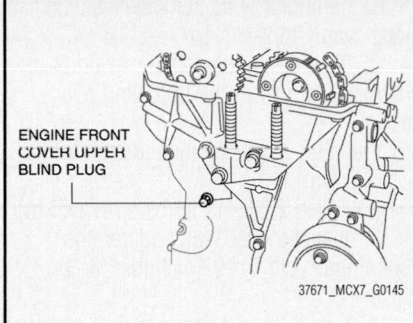

Fig. 187 Engine front cover upper blind plug—2.5L engine

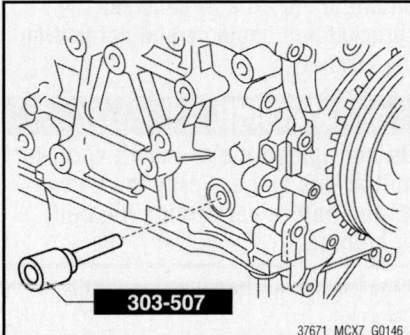

Fig. 188 Install the SST 303-507—2.5L engine

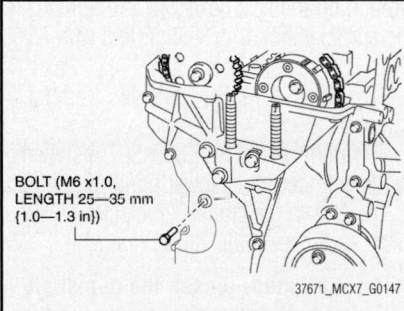

Fig. 189 M6 X 1.0, length bolt installed into the engine front cover upper blind plug—2.5L engine

- Using the cast hexagon on the exhaust camshaft, apply force counterclockwise to facilitate unlocking the chain tensioner ratchet.
- Using a Hex bit socket 0.098 inch (2.5 mm) or T15 Torx® bit socket, unlock the chain tensioner ratchet so that it can be lifted up.
- Using the cast hexagon on the exhaust camshaft, apply force in the direction of the engine rotation to increase tension on the chain

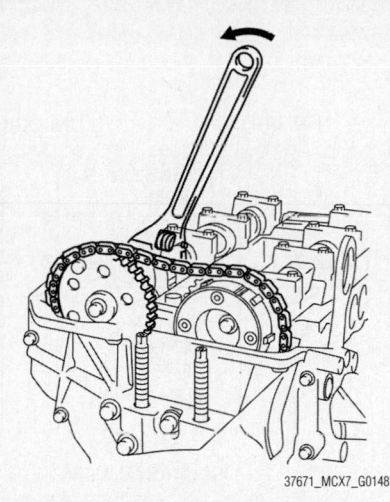

Fig. 190 Apply force counterclockwise to facilitate unlocking the chain tensioner ratchet—2.5L engine

➡**The chain tensioner rack is compressed using the chain tension generated by applying force to the exhaust camshaft in the direction of the engine rotation.**

- Screw in the bolt set in approx. 0.2 inch (5 mm) and secure the tensioner arm with the rack compressed. The racket has not been unlocked if the bolt cannot be pressed in approx 0.2 inch (5 mm).

➡**If the tensioner arm cannot be secured, return the bolt to its original position and repeat the procedure.**

18. Fix the exhaust camshaft using a wrench on the cast hexagon, and loosen the camshaft sprocket bolt.

19. Remove the exhaust camshaft sprocket bolt, exhaust camshaft sprocket,

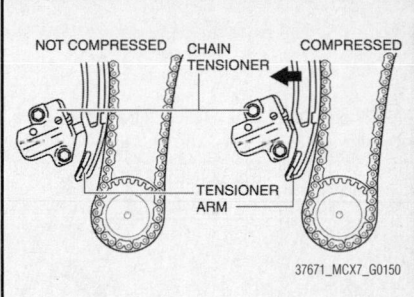

Fig. 192 Chain tensioner rack shown compressed and not compressed—2.5L engine

and washer as a single unit. Perform the work carefully so that the washer does not drop out.

20. Loosen the camshaft cap bolts in two or three steps in the order shown in the figure and remove the camshaft cap.

➡**The camshaft caps are to be kept ordered for correct reassembly in their original positions. Do not mix the caps.**

21. Remove the camshafts for the intake and exhaust sides.

22. Remove the tappet.

23. Install an appropriate tappet based on the results of the valve clearance inspection.

24. Selected tappet = Removed tappet thickness + Measured valve clearance - Standard valve clearance

25. Standard specified valve clearance is:
- Intake: 0.009–0.011 inch (0.22–0.28 mm)
- Exhaust: 0.011–0.012 inch (0.27–0.33 mm).

26. Replace the tappets of any valve found outside the specified values and pre-measured the valve clearance.

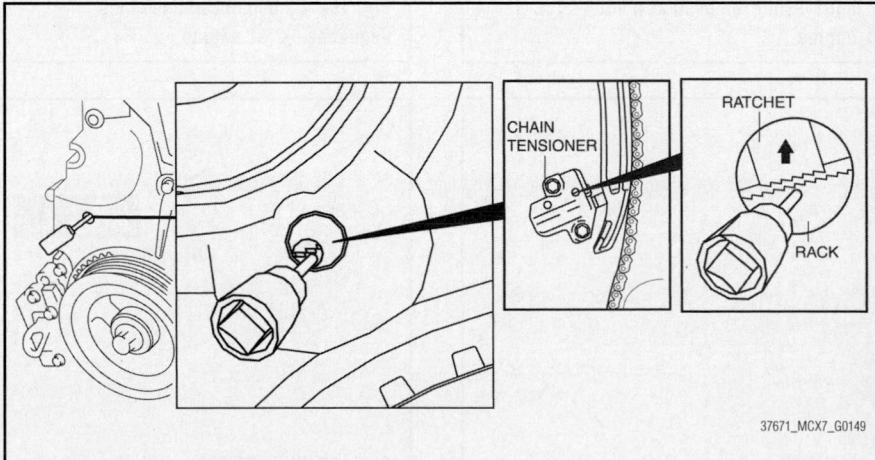

Fig. 191 Unlocking the chain tensioner ratchet so that it can be lifted up—2.5L engine

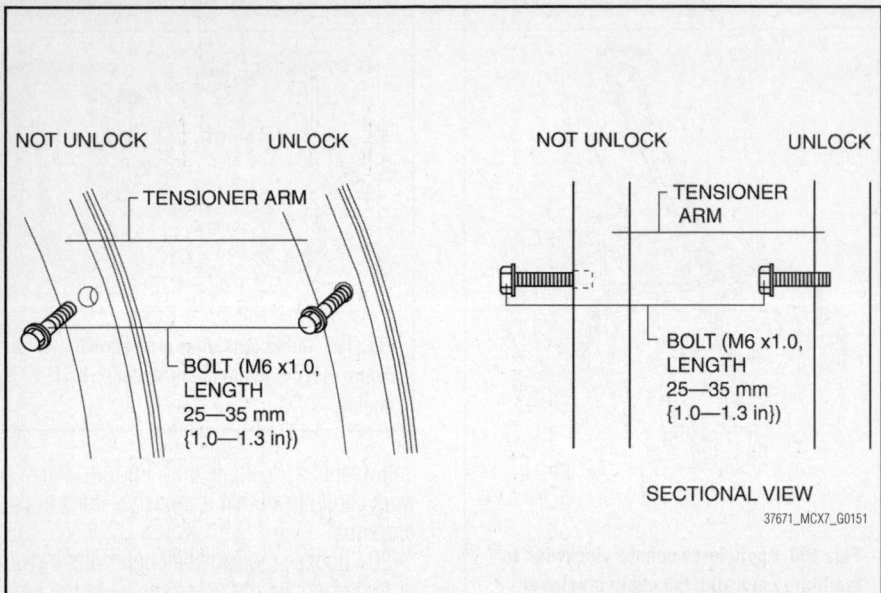

Fig. 193 Screw in the bolt set in approx. 0.2 inch (5 mm) and secure the tensioner arm—2.5L engine

27. Verify that No.1 cylinder is at TDC of the compression stroke. (Position counterweight contacts SST.)

28. Apply the gear oil (SAE No.90 or equivalent) to each journal of the cylinder head.

29. Install the camshaft with No.1 cylinder aligned with the TDC position.

30. Apply the gear oil (SAE No.90 or equivalent) to each camshaft journal of the cylinder head.

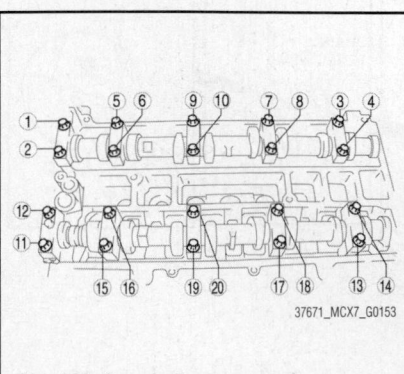

Fig. 194 Exhaust camshaft sprocket, bolt and washer removed as a unit—2.5L engine

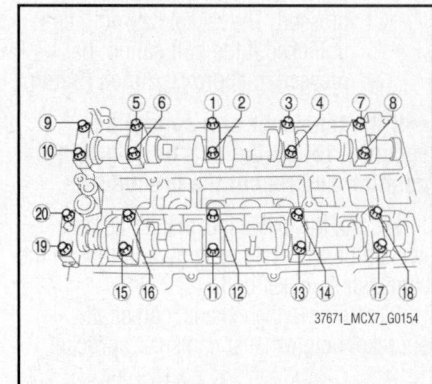

Fig. 196 Camshaft cap tightening sequence—2.5L engine

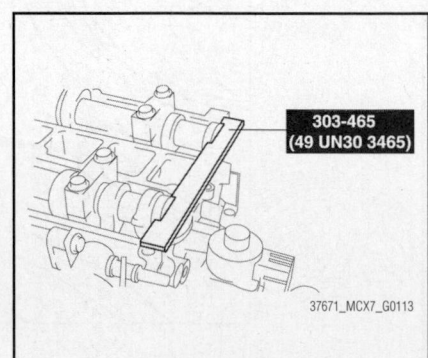

Fig. 195 Camshaft cap removal sequence—2.5L engine

Fig. 197 SST 303-465 camshaft holding tool installed—2.5L engine

31. Temporarily tighten the camshaft cap bolts evenly in 2–3 steps.

32. Tighten the camshaft cap bolts in the order shown in the following two steps;

- Step 1: 45–79 inch lbs. (5.0–9.0 Nm)
- Step 2: 11–12 ft. lbs. (14–17 Nm)

33. Install the OCV and tighten the retaining bolt to 71–97 inch lbs. (8–11 Nm).

34. Install the exhaust camshaft sprocket bolt, exhaust camshaft sprocket, and a new washer as a single unit.

➡**Install a washer to the fourth or fifth thread of the exhaust camshaft sprocket bolt being careful not to drop the washer.**

✳✳ WARNING

Do not tighten the camshaft sprocket bolt at this stage. Verify the valve timing before performing the bolt tightening.

35. Install the SST 303-465 on the camshaft.

36. Remove the installation bolt for the engine front cover upper blind plug, and apply tension to the timing chain.

37. Fix the exhaust camshaft using a wrench on the cast hexagon, and tighten the sprocket bolt to 51–55 ft. lbs. (69–75 Nm).

38. Remove the SST 303-465 from the camshaft.

39. Remove the SST 303-507 installed in the cylinder block lower blind plug hole.

40. Rotate the crankshaft clockwise two turns and inspect the valve timing.

➡**If not aligned, loosen the camshaft sprocket bolt and repeat the procedure.**

41. Install the cylinder block lower blind plug, and tighten to 14–16 ft. lbs. (18–22 Nm).

42. Apply the silicone sealant and install the engine front cover upper blind plug. Tighten to 71–101 inch lbs. (8–11 Nm).

43. Install a new engine front cover lower blind plug and tighten to 89–123 inch lbs. (10–14 Nm).

44. Install the drive belt.
45. Install the splash shield (RH).
46. Install the front mudguard (RH).
47. Install the under cover.
48. Install the front wheel and tire (RH).
49. Install the cylinder head cover.
50. Connect the ventilation hose.
51. Install the spark plugs.

52. Install the ignition coils.
53. Connect the wiring harness.
54. Install the plug hole plate.
55. Connect the negative battery cable.
56. Start the engine, and inspect the following:
- Leakage of engine oil, engine coolant or fuel.
- Ignition timing, and idle speed.

57. Perform a road test and verify that there is no abnormal vibration or noise.
58. Use the Mazda diagnostic scan tool, or equivalent and reprogram the required systems.

3.7L Engine

See Figures 198 and 199.

1. Before servicing the vehicle, refer to the Precautions Section.

➡**If working near and/or around the SRS system and components, be sure to disable the SRS system. Tape the negative battery cable with insulating tape. Always disconnect the negative battery cable first.**

✳✳ CAUTION

To avoid personal injury when working on vehicles equipped with an air bag, the negative battery cable must be disconnected and at least one minute must elapse before working on the system. Failure to do so may result in deployment of the air bag.

2. Disconnect the negative battery cable. Tape the cable with insulating tape.

➡**When disconnecting the cable, some systems need to be initialized after the cable is reconnected. You will need the Mazda diagnostic scan tool or equivalent. Follow the directions on the tool.**

3. Remove the engine cover.
4. Remove the ventilation hose.
5. Remove the resonance chamber.
6. Remove the dynamic chamber and throttle body as a single unit.

7. Disconnect the wiring harness.
8. Remove the ignition coils.
9. Remove the dipstick.
10. Remove the cylinder head cover.
11. Remove the front wheel and tire (RH).
12. Remove the splash shield (RH).
13. Measure the valve clearance.

a. Rotate the crankshaft clockwise and verify that the timing mark of the intake side camshaft sprocket is in the position shown in the figure. In this position, the No.1 cylinder is at the TDC of the compression stroke.

b. Measure the valve clearance of location A shown in the figure.

➡**Make sure to note down the measured values for choosing the suitable replacement tappets.**

c. Mark the crank pulley and engine front cover as shown in the figure.

d. Rotate the crankshaft clockwise 360° so that the No.5 cylinder is at TDC of the compression stroke.

e. Measure the valve clearance of location B shown in the figure.

14. If it is not within the specification,

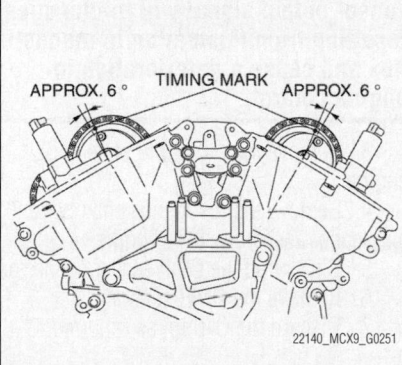

Fig. 198 Rotate the crankshaft clockwise and verify that the timing mark of the intake side camshaft sprocket is in the position shown in the figure. In this position, the No.1 cylinder is at the TDC of the compression stroke—3.7L engine

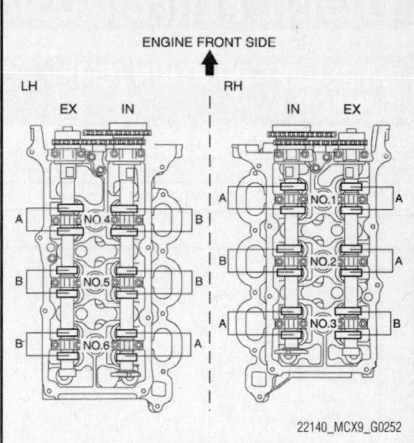

Fig. 199 Measure the valve clearance of location A, rotate the crankshaft clockwise 360° so that the No.5 cylinder is at TDC of the compression stroke and then measure the valve clearance of location B as shown—3.7L engine

replace the tappet and adjust the valve clearance to the median value of the standard.

a. Remove the camshafts and camshaft sprocket as a single unit.

b. Remove the tappet.

c. Install an appropriate tappet based on the results of the valve clearance inspection.

Selected tappet = Removed tappet thickness + Measured valve clearance - Standard valve clearance

15. Install the splash shield (RH).
16. Install the front wheel and tire (RH).
17. Install the cylinder head cover.
18. Install the dipstick.
19. Install the ignition coils.
20. Connect the wiring harness.
21. Install the dynamic chamber and throttle body as a single unit.
22. Install the resonance chamber.
23. Install the ventilation hose.
24. Install the engine cover.
25. Connect the negative battery cable.

ENGINE PERFORMANCE & EMISSION CONTROLS

CAMSHAFT POSITION (CMP) SENSOR

LOCATION

See Figures 200 and 201.

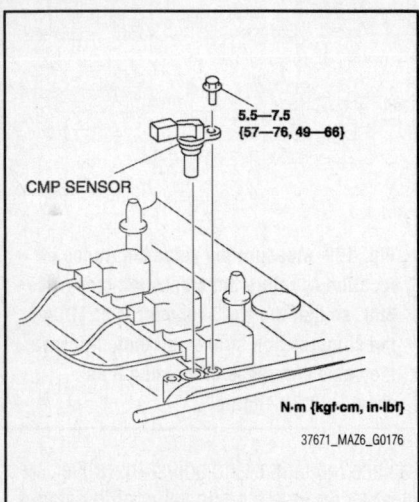

Fig. 200 Camshaft position sensor location—2.5L engine

REMOVAL & INSTALLATION

See Figure 229.

1. Before servicing the vehicle, refer to the Precautions Section.

→ **If working near and/or around the SRS system and components, be sure to disable the SRS system. Tape the negative battery cable with insulating tape. Always disconnect the negative battery cable first.**

✳✳ CAUTION

To avoid personal injury when working on vehicles equipped with an air bag, the negative battery cable must be disconnected and at least one minute must elapse before working on the system. Failure to do so may result in deployment of the air bag.

2. Disconnect the negative battery cable. Tape the cable with insulating tape.

→ **When disconnecting the cable, some systems need to be initialized after the cable is reconnected. You will need the Mazda diagnostic scan tool or equivalent. Follow the directions on the tool.**

✳✳ WARNING

When replacing the CMP sensor, make sure there is no foreign material on it such as metal shavings. If it is installed with foreign material, the sensor output signal will malfunction resulting from fluctuation in magnetic flux and cause a deterioration in engine control.

3. Remove the plug hole plate, 2.5L engine
4. Remove the resonance chamber and air cleaner assembly, 3.7L engine.
5. Disconnect the CMP sensor connector.
6. Remove the CMP sensor.
7. Discard the O-ring, as required.

To install:

→ **Be sure to use new fasteners, as required.**

8. Installation is the reverse of the removal procedure.
9. Lubricate the new O-ring with clean engine oil.
10. Use the Mazda diagnostic scan tool, or equivalent and reprogram the required systems.

CRANKSHAFT POSITION (CKP) SENSOR

LOCATION

See Figures 202 and 203.

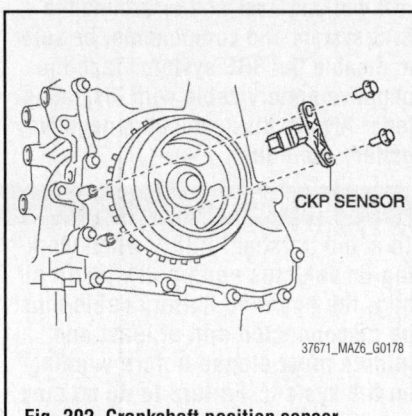

Fig. 202 Crankshaft position sensor location—2.5L engine

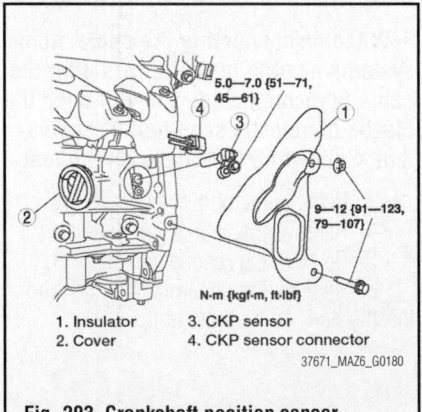

1. Insulator
2. Cover
3. CKP sensor
4. CKP sensor connector

Fig. 203 Crankshaft position sensor location—3.7L engine

REMOVAL & INSTALLATION

2.5L Engine

See Figures 204 and 205.

1. Before servicing the vehicle, refer to the Precautions Section.

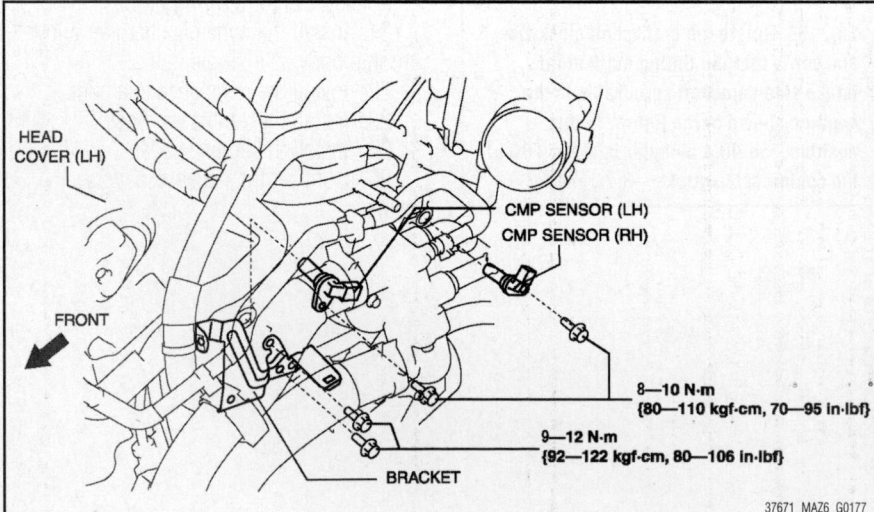

Fig. 201 Camshaft position sensor location—3.7L engine

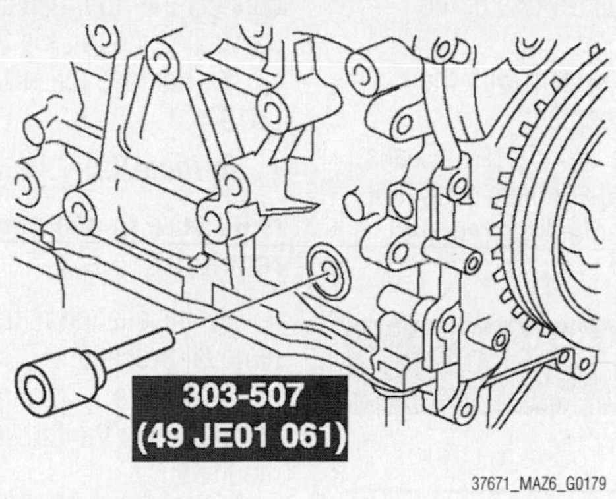

Fig. 204 Special service tool 303-507

➡**If working near and/or around the SRS system and components, be sure to disable the SRS system. Tape the negative battery cable with insulating tape. Always disconnect the negative battery cable first.**

✲✲ CAUTION

To avoid personal injury when working on vehicles equipped with an air bag, the negative battery cable must be disconnected and at least one minute must elapse before working on the system. Failure to do so may result in deployment of the air bag.

2. Disconnect the negative battery cable. Tape the cable with insulating tape.

➡**When disconnecting the cable, some systems need to be initialized after the cable is reconnected. You will need the Mazda diagnostic scan tool or equivalent. Follow the directions on the tool.**

3. Remove the engine undercover.
4. Remove the splash shield.
5. Disconnect the CKP sensor connector.
6. Remove the installation bolts to remove the CKP sensor.

To install:

7. Rotate the crankshaft in the direction of the engine rotation and remove the cylinder block lower blind plug when the No. 1 cylinder is at the point prior to top dead center (TDC) of compression, then install SST 303-507.
8. Fit the center of the CKP sensor with the fifth tooth (counting counterclockwise from the empty space A as shown in the figure) of the pulse wheel
9. Install the CKP sensor fitting bolts. Tighten the bolts to 49–66 inch lbs. (5.5–7.5 Nm).
10. Remove the SST 303-507, then install the cylinder block lower blind plug. Tighten to 14–16 ft. lbs. (18–22 Nm).

3.7L Engine

See Figure 206.

1. Before servicing the vehicle, refer to the Precautions Section.

➡**If working near and/or around the SRS system and components, be sure to disable the SRS system. Tape the negative battery cable with insulating tape. Always disconnect the negative battery cable first.**

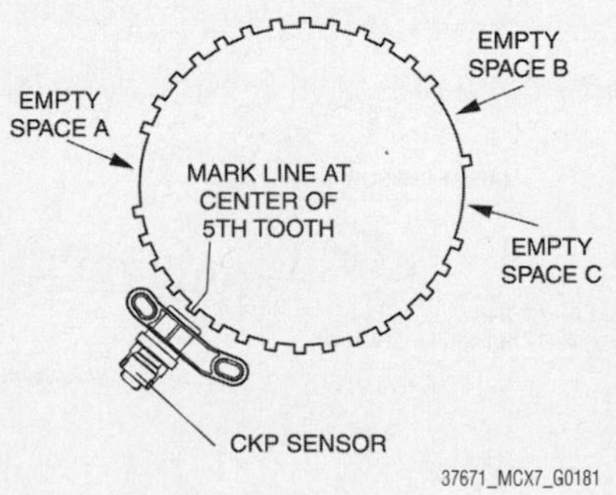

Fig. 205 Fit the center of the CKP sensor with the fifth tooth from empty space A

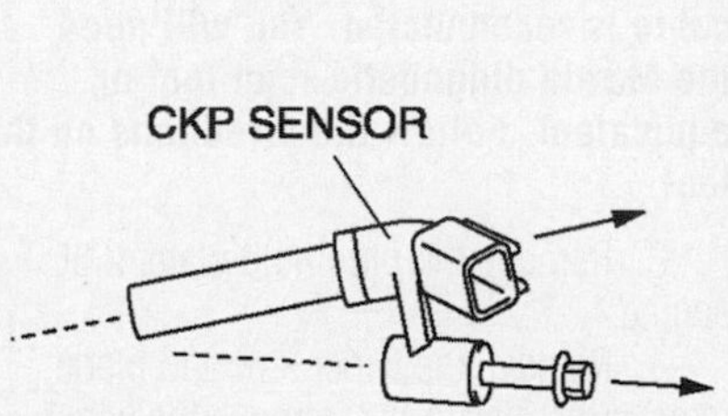

Fig. 206 Crankshaft position sensor removal—3.7L engine

✲✲ CAUTION

To avoid personal injury when working on vehicles equipped with an air bag, the negative battery cable must be disconnected and at least one minute must elapse before working on the system. Failure to do so may result in deployment of the air bag.

2. Disconnect the negative battery cable. Tape the cable with insulating tape.

➡**When disconnecting the cable, some systems need to be initialized after the cable is reconnected. You will need the Mazda diagnostic scan tool or equivalent. Follow the directions on the tool.**

➡**Disconnect the connector after the sensor is removed to prevent the sensor from falling off. The bolt and the sensor are integrated. When removing the sensor be careful not to drop the tool. The bolt and sensor must be pulled in different directions. See illustration.**

3. Remove the WU-TWC catalytic converter for easier access when servicing the left side.
4. Remove the insulator. Remove the cover.
5. Remove the sensor retaining bolt. Remove the sensor from its mounting.
6. Disconnect the electrical connector.

To install:

➡**Be sure to use new fasteners, as required.**

7. Installation is the reverse of the removal procedure.
8. Use the Mazda diagnostic scan tool, or equivalent and reprogram the required systems.

ENGINE COOLANT TEMPERATURE (ECT) SENSOR

LOCATION

See Figure 207.

REMOVAL & INSTALLATION

✲✲ CAUTION

Never remove the cooling system cap or loosen the radiator drain plug while the engine is running, or when the engine and radiator are hot. Scalding engine coolant and steam may shoot out and cause serious injury. It may also damage the engine and cooling system.

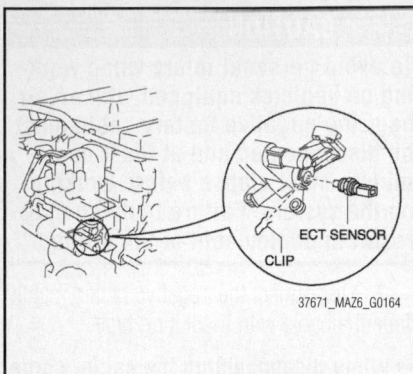

Fig. 207 Engine coolant temperature sensor location—2.5L engine

1. Before servicing the vehicle, refer to the Precautions Section.

➡If working near and/or around the SRS system and components, be sure to disable the SRS system. Tape the negative battery cable with insulating tape. Always disconnect the negative battery cable first.

※※ CAUTION

To avoid personal injury when working on vehicles equipped with an air bag, the negative battery cable must be disconnected and at least one minute must elapse before working on the system. Failure to do so may result in deployment of the air bag.

2. Disconnect the negative battery cable. Tape the cable with insulating tape.

➡When disconnecting the cable, some systems need to be initialized after the cable is reconnected. You will need the Mazda diagnostic scan tool or equivalent. Follow the directions on the tool.

3. Turn off the engine and wait until it is cool. Even then, be very careful when removing the cap. Wrap a thick cloth around it and slowly turn it counterclockwise to the first stop. Step back while the pressure escapes.

4. When you are sure all the pressure is gone, press down on the cap using the cloth, turn it, and remove it.

5. Drain the coolant. Be sure to properly dispose of used coolant.

6. Remove the battery and the battery tray.

7. Disconnect the sensor electrical connector.

8. Remove the sensor from its mounting.

9. Discard the O-ring.

To install:

➡Be sure to use new fasteners, as required.

10. Installation is the reverse of the removal procedure.

11. Fill the cooling system with the proper grade and type coolant.

12. Start the engine and check for leaks.

13. Use the Mazda diagnostic scan tool, or equivalent and reprogram the required systems.

HEATED OXYGEN SENSOR (HO2S)

LOCATION

This sensor is the second sensor mounted in the exhaust system. The 2.5L engine uses one and the 3.7L engine uses two.

REMOVAL & INSTALLATION

※※ CAUTION

A hot engine and exhaust system can cause severe burns. Turn off the engine and wait until they are cool before removing the exhaust system.

1. Before servicing the vehicle, refer to the Precautions Section.

➡If working near and/or around the SRS system and components, be sure to disable the SRS system. Tape the negative battery cable with insulating tape. Always disconnect the negative battery cable first.

※※ CAUTION

To avoid personal injury when working on vehicles equipped with an air bag, the negative battery cable must be disconnected and at least one minute must elapse before working on the system. Failure to do so may result in deployment of the air bag.

2. Disconnect the negative battery cable. Tape the cable with insulating tape.

➡When disconnecting the cable, some systems need to be initialized after the cable is reconnected. You will need the Mazda diagnostic scan tool or equivalent. Follow the directions on the tool.

3. Remove the plug hole plate, 2.5L engine.

4. Remove the wiper arm and blade, remove the cowl grille, remove the wiper motor and cowl panel on the 3.7L engine to allow for easier access to the sensor.

5. Disconnect the HO2S connector.

6. Raise and support the vehicle safely, remove the tire and wheel assembly, dis-

connect the right halfshaft from the transaxle side, on the 3.7L engine.

7. Remove the HO2S using the SST 49 L018 001.

To install:

➡Be sure to use new fasteners, as required.

8. Installation is the reverse of the removal procedure.

9. On the 3.7L engine apply anti-seize compound to the threads of the sensor, as required.

10. Use the Mazda diagnostic scan tool, or equivalent and reprogram the required systems.

INTAKE AIR TEMPERATURE (IAT) SENSOR

LOCATION

See Figures 208 and 209.

The Mass Air Flow (MAF)/Intake Air Temperature (IAT) sensor is mounted to the air cleaner cover. The Intake Air Temperature (IAT) Sensor is built into the Mass Air Flow (MAF) sensor.

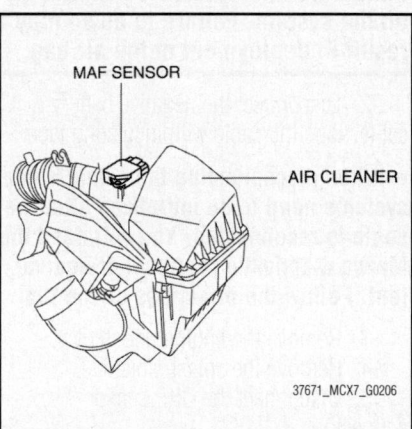

Fig. 208 MAF/IAT sensor location—2.5L engine

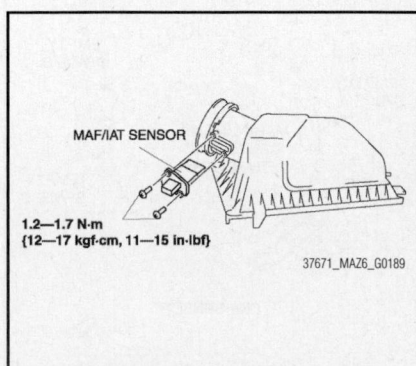

Fig. 209 MAF/IAT sensor location—3.7L engine

REMOVAL & INSTALLATION

1. Before servicing the vehicle, refer to the Precautions Section.

➡**If working near and/or around the SRS system and components, be sure to disable the SRS system. Tape the negative battery cable with insulating tape. Always disconnect the negative battery cable first.**

❋❋ **CAUTION**

To avoid personal injury when working on vehicles equipped with an air bag, the negative battery cable must be disconnected and at least one minute must elapse before working on the system. Failure to do so may result in deployment of the air bag.

2. Disconnect the negative battery cable. Tape the cable with insulating tape.

➡**When disconnecting the cable, some systems need to be initialized after the cable is reconnected. You will need the Mazda diagnostic scan tool or equivalent. Follow the directions on the tool.**

3. Remove the MAF/IAT sensor electrical connector.
4. Remove the MAF/IAT sensor retaining screws.
5. Remove the MAF/IAT sensor.

To install:

➡**Be sure to use new fasteners, as required.**

6. Installation is the reverse of the removal procedure.
7. Install the MAF/IAT sensor and tighten the screws.
8. Install the MAF/IAT sensor electrical connector.
9. Use the Mazda diagnostic scan tool, or equivalent and reprogram the required systems.

KNOCK SENSOR (KS)

LOCATION

See Figures 210 and 211.

REMOVAL & INSTALLATION

2.5L Engine

1. Before servicing the vehicle, refer to the Precautions Section.

➡**If working near and/or around the SRS system and components, be sure to disable the SRS system. Tape the**

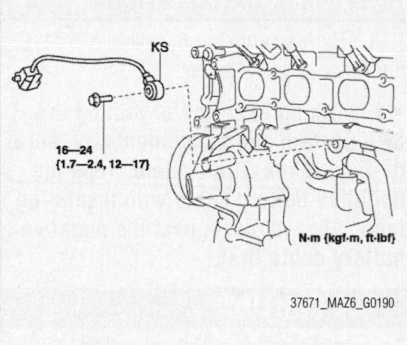

Fig. 210 Knock sensor location—2.5L engine

negative battery cable with insulating tape. Always disconnect the negative battery cable first.

❋❋ **CAUTION**

To avoid personal injury when working on vehicles equipped with an air bag, the negative battery cable must be disconnected and at least one minute must elapse before working on the system. Failure to do so may result in deployment of the air bag.

2. Disconnect the negative battery cable. Tape the cable with insulating tape.

➡**When disconnecting the cable, some systems need to be initialized after the cable is reconnected. You will need the Mazda diagnostic scan tool or equivalent. Follow the directions on the tool.**

3. Remove the plug hole plate.
4. Remove the intake manifold.
5. Disconnect the KS connector.
6. Remove the KS.

To install:

➡**Be sure to use new fasteners, as required.**

7. Installation is the reverse of the removal procedure.
8. Tighten the sensor mounting bolt to 12–17 ft. lbs. (16–24 Nm).
9. Use the Mazda diagnostic scan tool, or equivalent and reprogram the required systems.

3.7L Engine

1. Before servicing the vehicle, refer to the Precautions Section.

➡**If working near and/or around the SRS system and components, be sure to disable the SRS system. Tape the negative battery cable with insulating tape. Always disconnect the negative battery cable first.**

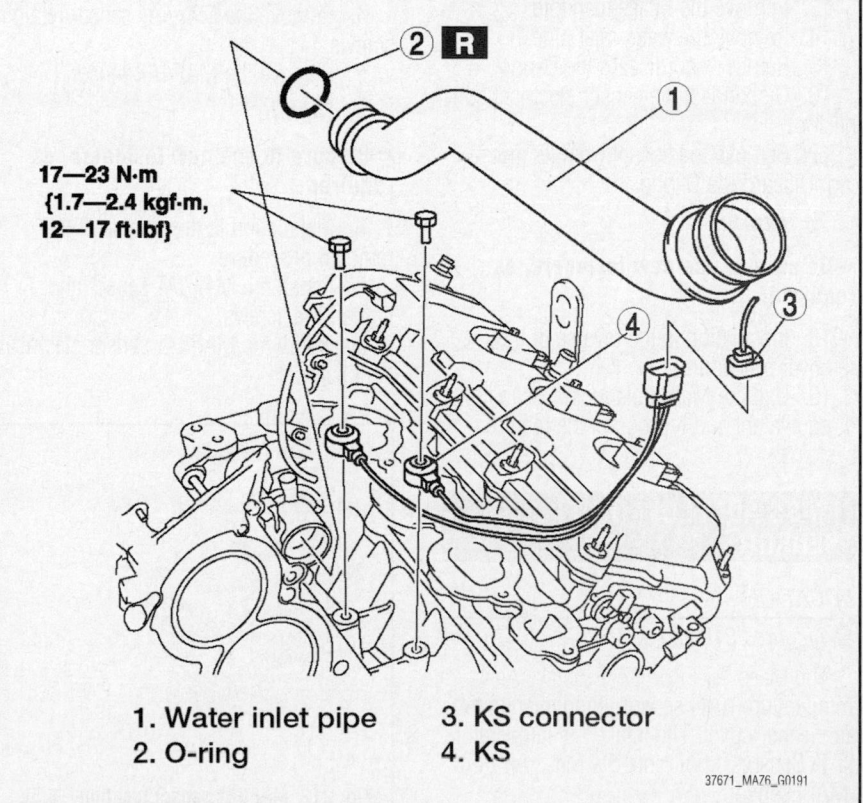

1. Water inlet pipe
2. O-ring
3. KS connector
4. KS

Fig. 211 Knock sensor location—3.7L engine

✳✳ CAUTION

To avoid personal injury when working on vehicles equipped with an air bag, the negative battery cable must be disconnected and at least one minute must elapse before working on the system. Failure to do so may result in deployment of the air bag.

2. Disconnect the negative battery cable. Tape the cable with insulating tape.

➡ **When disconnecting the cable, some systems need to be initialized after the cable is reconnected. You will need the Mazda diagnostic scan tool or equivalent. Follow the directions on the tool.**

3. Properly discharge the fuel system.
4. Drain the cooling system, as required. Be sure to properly dispose of used coolant.
5. Remove the engine cover.
6. Remove the ventilation hose.
7. Remove the throttle position sensor connector.
8. Remove the vacuum hose to the purge solenoid valve.
9. Remove the vacuum hose.
10. Remove the dynamic chamber.
11. Remove the fuel hose.
12. Remove the fuel distributor.
13. Remove the intake manifold.
14. Remove the water inlet pipe.
15. Remove and discard the O-ring.
16. Disconnect the sensor electrical connector.
17. Remove the sensor from its mounting. Discard the O-ring.

To install:

➡ **Be sure to use new fasteners, as required.**

18. Installation is the reverse of the removal procedure.
19. Use the Mazda diagnostic scan tool, or equivalent and reprogram the required systems.

MASS AIR FLOW (MAF) SENSOR

LOCATION

See Figures 212 and 213.

The Mass Air Flow (MAF)/Intake Air Temperature (IAT) sensor is mounted to the air cleaner cover. The Intake Air Temperature (IAT) Sensor is built into the Mass Air Flow (MAF) sensor.

REMOVAL & INSTALLATION

1. Before servicing the vehicle, refer to the Precautions Section.

➡ **If working near and/or around the SRS system and components, be sure to disable the SRS system. Tape the negative battery cable with insulating tape. Always disconnect the negative battery cable first.**

✳✳ CAUTION

To avoid personal injury when working on vehicles equipped with an air bag, the negative battery cable must be disconnected and at least one minute must elapse before working on the system. Failure to do so may result in deployment of the air bag.

2. Disconnect the negative battery cable. Tape the cable with insulating tape.

➡ **When disconnecting the cable, some systems need to be initialized after the cable is reconnected. You will need the Mazda diagnostic scan tool or equivalent. Follow the directions on the tool.**

3. Remove the MAF/IAT sensor electrical connector.
4. Remove the MAF/IAT sensor retaining screws.
5. Remove the MAF/IAT sensor.

To install:

➡ **Be sure to use new fasteners, as required.**

6. Installation is the reverse of the removal procedure.
7. Install the MAF/IAT sensor and tighten the screws.
8. Install the MAF/IAT sensor electrical connector.

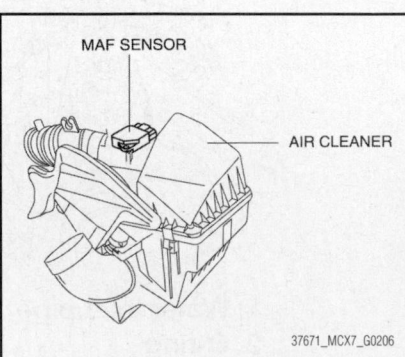

Fig. 212 MAF/IAT sensor location—2.5L engine

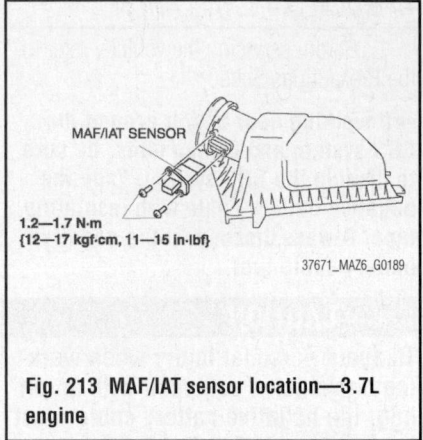

Fig. 213 MAF/IAT sensor location—3.7L engine

9. Use the Mazda diagnostic scan tool, or equivalent and reprogram the required systems.

MANIFOLD ABSOLUTE PRESSURE (MAP) SENSOR

LOCATION

See Figure 214.

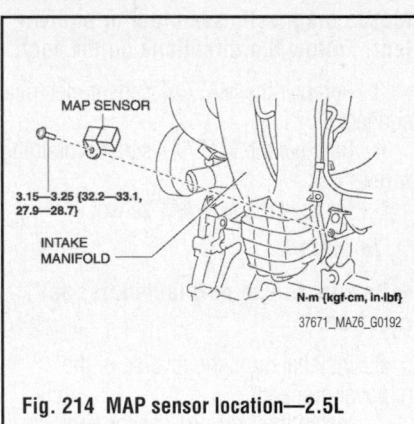

Fig. 214 MAP sensor location—2.5L engine

REMOVAL & INSTALLATION

1. Before servicing the vehicle, refer to the Precautions Section.

➡ **If working near and/or around the SRS system and components, be sure to disable the SRS system. Tape the negative battery cable with insulating tape. Always disconnect the negative battery cable first.**

✳✳ CAUTION

To avoid personal injury when working on vehicles equipped with an air bag, the negative battery cable must be disconnected and at least one minute must elapse before working on the system. Failure to do so may result in deployment of the air bag.

2. Disconnect the negative battery cable. Tape the cable with insulating tape.

➡When disconnecting the cable, some systems need to be initialized after the cable is reconnected. You will need the Mazda diagnostic scan tool or equivalent. Follow the directions on the tool.

3. Disconnect the MAP sensor connector.
4. Remove the MAP sensor.

To install:

➡Be sure to use new fasteners, as required.

5. Installation is the reverse of the removal procedure.
6. Use the Mazda diagnostic scan tool, or equivalent and reprogram the required systems.

POWERTRAIN CONTROL MODULE (PCM)

LOCATION

The Powertrain Control Module (PCM) is located in the engine compartment, just to the right of the battery.

REMOVAL & INSTALLATION

See Figures 215 and 216.

1. Before servicing the vehicle, refer to the Precautions Section.

➡If working near and/or around the SRS system and components, be sure to disable the SRS system. Tape the negative battery cable with insulating tape. Always disconnect the negative battery cable first.

❋❋ CAUTION

To avoid personal injury when working on vehicles equipped with an air bag, the negative battery cable must be disconnected and at least one minute must elapse before working on the system. Failure to do so may result in deployment of the air bag.

2. Disconnect the negative battery cable. Tape the cable with insulating tape.

➡When disconnecting the cable, some systems need to be initialized after the cable is reconnected. You will need the Mazda diagnostic scan tool or equivalent. Follow the directions on the tool.

➡If the PCM is replaced, the replacement must be programmed with the vehicle parameters.

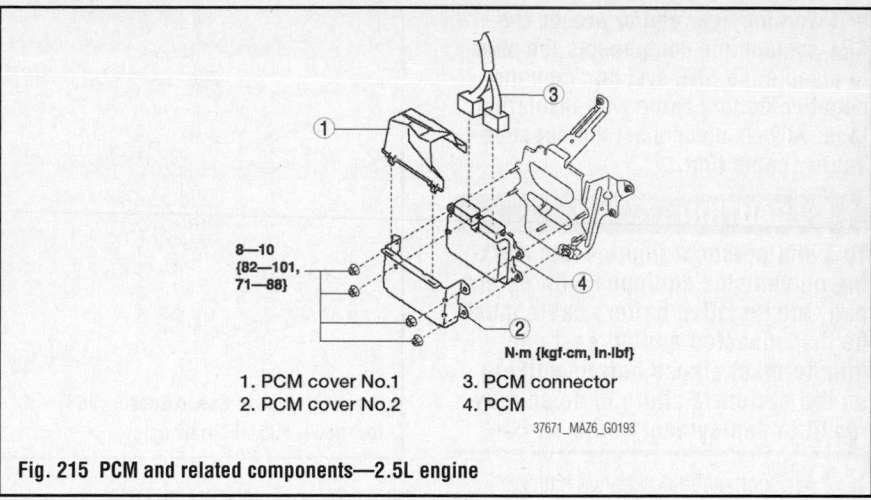

1. PCM cover No.1
2. PCM cover No.2
3. PCM connector
4. PCM

37671_MAZ6_G0193

Fig. 215 PCM and related components—2.5L engine

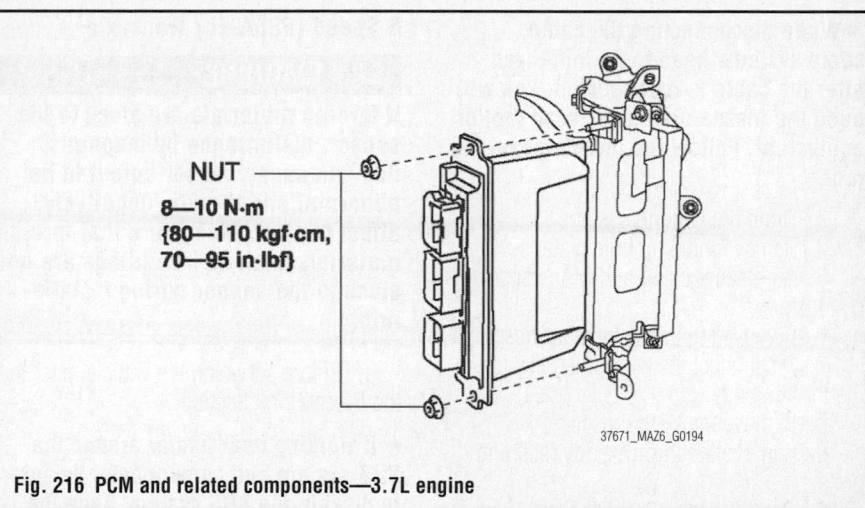

NUT
8—10 N·m
{80—110 kgf·cm,
70—95 in·lbf}

37671_MAZ6_G0194

Fig. 216 PCM and related components—3.7L engine

3. Remove the battery and battery tray.
4. On 3.7L engines, remove all connectors and clips that may interfere with removal of the PCM, loosen the bracket nuts and remove the PCM from the bracket.
5. On 2.5L engine disconnect the electrical connectors.
6. Remove the PCM.

To install:

➡Be sure to use new fasteners, as required.

7. Installation is the reverse of the removal procedure.
8. Use the Mazda diagnostic scan tool, or equivalent and reprogram the required systems.
9. When replacing the PCM on the vehicles, perform the following:
 • IMMOBILIZER SYSTEM-RELATED PARTS PROGRAMMING (ADVANCED KEYLESS ENTRY AND START SYSTEM)

THROTTLE POSITION SENSOR (TPS)

LOCATION

On the 2.5L engine, the Throttle Position Sensor (TPS) is integral to the throttle body.

REMOVAL & INSTALLATION

2.5L Engine

❋❋ CAUTION

A hot engine and intake air system can cause severe burns. Turn off the engine and wait until they are cool before removing the intake air system. Fuel line spills and leakage from the pressurized fuel system are dangerous. Fuel can ignite and cause serious injury or death and damage. Fuel can also irritate skin and eyes.

1. Before servicing the vehicle, refer to the Precautions Section.

➡If working near and/or around the SRS system and components, be sure to disable the SRS system. Tape the negative battery cable with insulating tape. Always disconnect the negative battery cable first.

✳✳ CAUTION

To avoid personal injury when working on vehicles equipped with an air bag, the negative battery cable must be disconnected and at least one minute must elapse before working on the system. Failure to do so may result in deployment of the air bag.

2. Disconnect the negative battery cable. Tape the cable with insulating tape.

➡When disconnecting the cable, some systems need to be initialized after the cable is reconnected. You will need the Mazda diagnostic scan tool or equivalent. Follow the directions on the tool.

3. Drain the cooling system.
4. Remove the plug hole plate.
5. Air cleaner cover and MAF sensor connector.
6. Remove the intake fresh air hose.
7. Disconnect the wiring harness for the throttle body.
8. Remove the water hoses.
9. Remove the throttle body retaining bolts.
10. Carefully remove the throttle body.
11. Install in the reverse order of removal and note the following:
   • Tighten the throttle body to 71–101 inch lbs. (8–11.5 Nm).
   • Refill and bleed the cooling system.
12. Use the Mazda diagnostic scan tool, or equivalent and reprogram the required systems.

VEHICLE SPEED SENSOR (VSS)

LOCATION

5 Speed (FS5A-EL) Transaxle
See Figure 217.

6 Speed (AW6A-EL) Transaxle
See Figure 218.

The Vehicle Speed Sensor (VSS) is located deep inside the transaxle, the transaxle must be removed and disassembled to access the sensor.

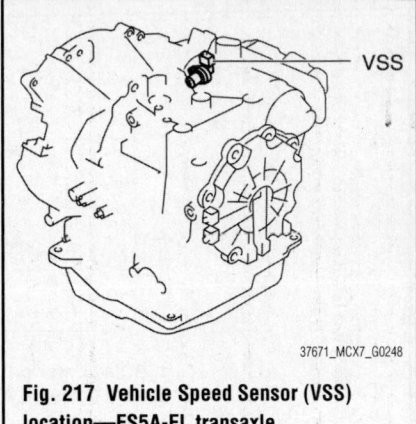

Fig. 217 Vehicle Speed Sensor (VSS) location—FS5A-EL transaxle

REMOVAL & INSTALLATION

5 Speed (FS5A-EL) Transaxle

✳✳ WARNING

If foreign materials are stuck to the sensor, disturbance by magnetic flux can cause sensor output to be abnormal and thereby negatively affect control. Make sure that foreign materials such as iron filings are not stuck to the sensor during installation.

1. Before servicing the vehicle, refer to the Precautions Section.

➡If working near and/or around the SRS system and components, be sure to disable the SRS system. Tape the negative battery cable with insulating tape. Always disconnect the negative battery cable first.

✳✳ CAUTION

To avoid personal injury when working on vehicles equipped with an air bag, the negative battery cable must be disconnected and at least one minute must elapse before working on the system. Failure to do so may result in deployment of the air bag.

2. Disconnect the negative battery cable. Tape the cable with insulating tape.

➡When disconnecting the cable, some systems need to be initialized after the cable is reconnected. You will need the Mazda diagnostic scan tool or equivalent. Follow the directions on the tool.

3. Remove the under cover.
4. Remove the Vehicle Speed Sensor (VSS) connector.
5. Remove the VSS retaining bolt.
6. Remove the VSS and O-ring.

To install:

➡Be sure to use new fasteners, as required.

7. Installation is the reverse of the removal procedure.
8. Use the Mazda diagnostic scan tool, or equivalent and reprogram the required systems.

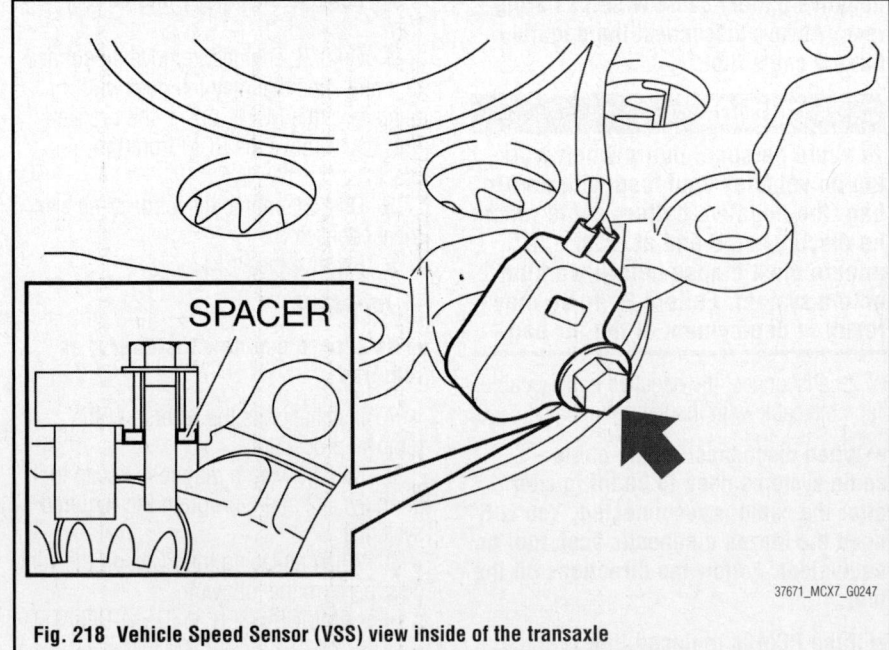

Fig. 218 Vehicle Speed Sensor (VSS) view inside of the transaxle

FUEL SYSTEM SERVICE PRECAUTIONS

Safety is the most important factor when performing not only fuel system maintenance but any type of maintenance. Failure to conduct maintenance and repairs in a safe manner may result in serious personal injury or death. Maintenance and testing of the vehicle's fuel system components can be accomplished safely and effectively by adhering to the following rules and guidelines.

• To avoid the possibility of fire and personal injury, always disconnect the negative battery cable unless the repair or test procedure requires that battery voltage be applied.

• Always relieve the fuel system pressure prior to disconnecting any fuel system component (injector, fuel rail, pressure regulator, etc.), fitting or fuel line connection. Exercise extreme caution whenever relieving fuel system pressure to avoid exposing skin, face and eyes to fuel spray. Please be advised that fuel under pressure may penetrate the skin or any part of the body that it contacts.

• Always place a shop towel or cloth around the fitting or connection prior to loosening to absorb any excess fuel due to spillage. Ensure that all fuel spillage (should it occur) is quickly removed from engine surfaces. Ensure that all fuel soaked cloths or towels are deposited into a suitable waste container.

• Always keep a dry chemical (Class B) fire extinguisher near the work area.

• Do not allow fuel spray or fuel vapors to come into contact with a spark or open flame.

• Always use a back-up wrench when loosening and tightening fuel line connection fittings. This will prevent unnecessary stress and torsion to fuel line piping.

• Always replace worn fuel fitting O-rings with new Do not substitute fuel hose or equivalent where fuel pipe is installed.

Before servicing the vehicle, make sure to also refer to the precautions in the beginning of this section as well.

RELIEVING FUEL SYSTEM PRESSURE

See Figure 219.

1. Before servicing the vehicle, refer to the Precautions Section.

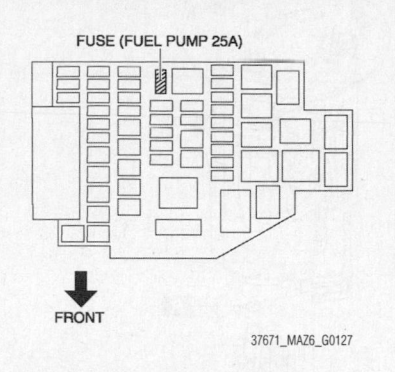

Fig. 219 Fuel pump relay location—2.5L engine

➡**If working near and/or around the SRS system and components, be sure to disable the SRS system. Tape the negative battery cable with insulating tape. Always disconnect the negative battery cable first.**

⁂ **CAUTION**

To avoid personal injury when working on vehicles equipped with an air bag, the negative battery cable must be disconnected and at least one minute must elapse before working on the system. Failure to do so may result in deployment of the air bag.

2. Disconnect the negative battery cable. Tape the cable with insulating tape.

➡**When disconnecting the cable, some systems need to be initialized after the cable is reconnected. You will need the Mazda diagnostic scan tool or equivalent. Follow the directions on the tool.**

3. Remove the fuel-filler cap and release the pressure in the fuel tank.
4. On 3.7L engine, disconnect the fuel pump unit connector.
5. Remove the fuel pump relay, 2.5L engine.
6. Connect the battery cable.
7. Start the engine.
8. After the engine stalls, crank the engine several times.
9. Turn the ignition switch to the LOCK position.
10. Install the fuel pump relay fuse, as required.

FUEL LEVEL SENDING UNIT

LOCATION
See Figure 220.

REMOVAL & INSTALLATION

1. Before servicing the vehicle, refer to the Precautions Section.

➡**If working near and/or around the SRS system and components, be sure to disable the SRS system. Tape the negative battery cable with insulating tape. Always disconnect the negative battery cable first.**

⁂ **CAUTION**

To avoid personal injury when working on vehicles equipped with an air bag, the negative battery cable must be disconnected and at least one minute must elapse before working on the system. Failure to do so may result in deployment of the air bag.

2. Disconnect the negative battery cable. Tape the cable with insulating tape.

➡**When disconnecting the cable, some systems need to be initialized after the cable is reconnected. You will need the Mazda diagnostic scan tool or equivalent. Follow the directions on the tool.**

3. Properly relieve the fuel system pressure.
4. Disconnect the negative battery cable.
5. Remove the rear seat cushion.
6. Remove the service hole cover.
7. Disconnect the electrical connectors.
8. Disconnect the fuel hose.
9. Remove the set plate, using tool SST:310-123, or equivalent.
10. Remove the unit from its mounting. Discard the O-ring.

To install:

➡**Be sure to use new fasteners, as required.**

11. Installation is the reverse of the removal procedure.
12. Be sure to use a new O-ring.
13. Start the engine and check for leaks, correct as required.
14. Use the Mazda diagnostic scan tool, or equivalent and reprogram the required systems.

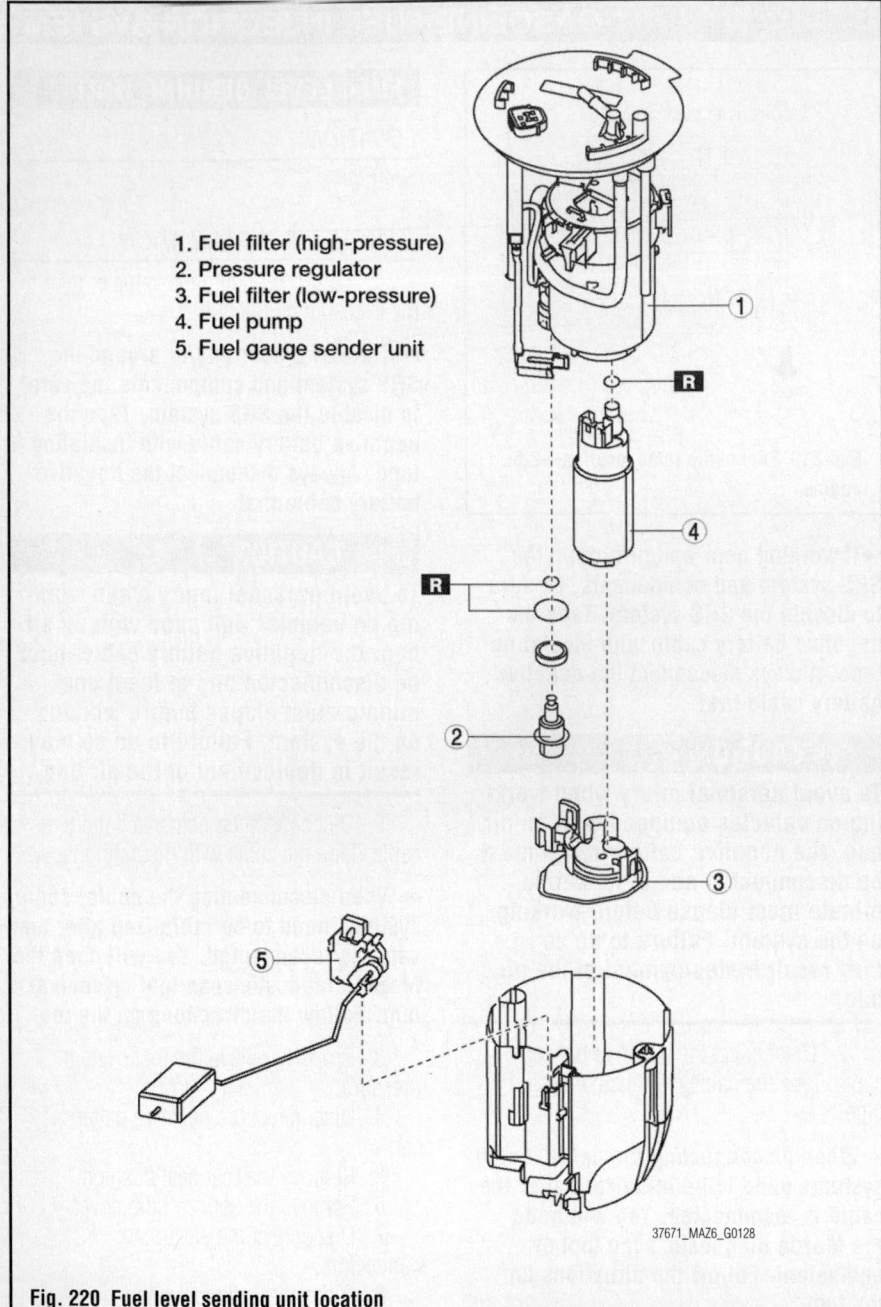

1. Fuel filter (high-pressure)
2. Pressure regulator
3. Fuel filter (low-pressure)
4. Fuel pump
5. Fuel gauge sender unit

37671_MAZ6_G0128

Fig. 220 Fuel level sending unit location

FUEL PUMP MODULE

REMOVAL & INSTALLATION

1. Before servicing the vehicle, refer to the Precautions Section.

➥If working near and/or around the SRS system and components, be sure to disable the SRS system. Tape the negative battery cable with insulating tape. Always disconnect the negative battery cable first.

✴✴ CAUTION

To avoid personal injury when working on vehicles equipped with an air bag, the negative battery cable must be disconnected and at least one minute must elapse before working on the system. Failure to do so may result in deployment of the air bag.

2. Disconnect the negative battery cable. Tape the cable with insulating tape.

➥When disconnecting the cable, some systems need to be initialized after the cable is reconnected. You will need the Mazda diagnostic scan tool or equivalent. Follow the directions on the tool.

3. Properly relieve the fuel system pressure.

4. Disconnect the negative battery cable.
5. Remove the rear seat cushion.
6. Remove the service hole cover.
7. Disconnect the electrical connectors.
8. Disconnect the fuel hose.
9. Remove the set plate, using tool SST:310-123, or equivalent.
10. Remove the unit from its mounting. Discard the O-ring.

To install:

➥Be sure to use new fasteners, as required.

11. Installation is the reverse of the removal procedure.
12. Be sure to use a new O-ring.
13. Start the engine and check for leaks, correct as required.
14. Use the Mazda diagnostic scan tool, or equivalent and reprogram the required systems.

FUEL RAIL AND INJECTOR

REMOVAL & INSTALLATION

See Figures 221 through 226.

1. Before servicing the vehicle, refer to the Precautions Section.

➥If working near and/or around the SRS system and components, be sure to disable the SRS system. Tape the negative battery cable with insulating tape. Always disconnect the negative battery cable first.

✴✴ CAUTION

To avoid personal injury when working on vehicles equipped with an air bag, the negative battery cable must be disconnected and at least one minute must elapse before working on the system. Failure to do so may result in deployment of the air bag.

2. Disconnect the negative battery cable. Tape the cable with insulating tape.

➥When disconnecting the cable, some systems need to be initialized after the cable is reconnected. You will need the Mazda diagnostic scan tool or equivalent. Follow the directions on the tool.

3. Properly relieve the fuel system pressure.
4. Disconnect the negative battery cable.
5. Remove the air cleaner assembly, as required for access.
6. Remove the plug hole plate, if equipped with 2.5L engine.

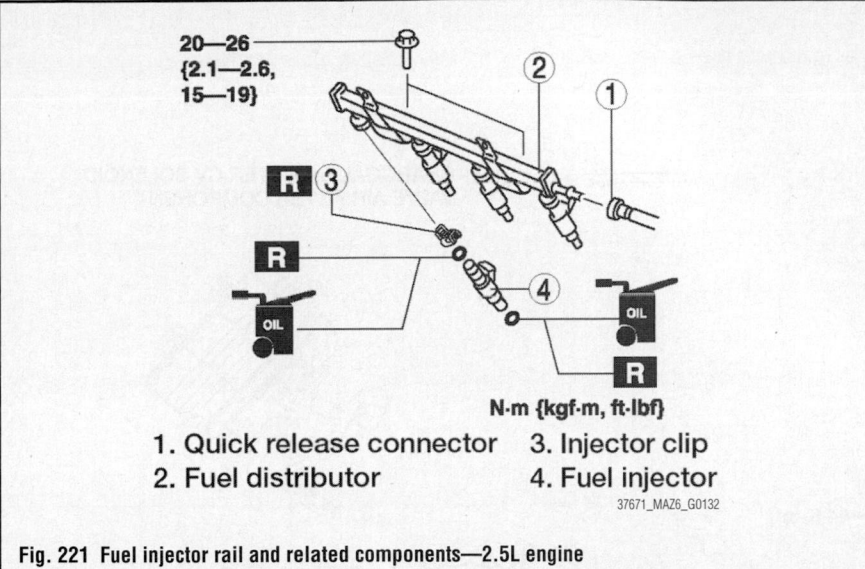

1. Quick release connector
2. Fuel distributor
3. Injector clip
4. Fuel injector

N·m {kgf·m, ft-lbf}

37671_MAZ6_G0132

Fig. 221 Fuel injector rail and related components—2.5L engine

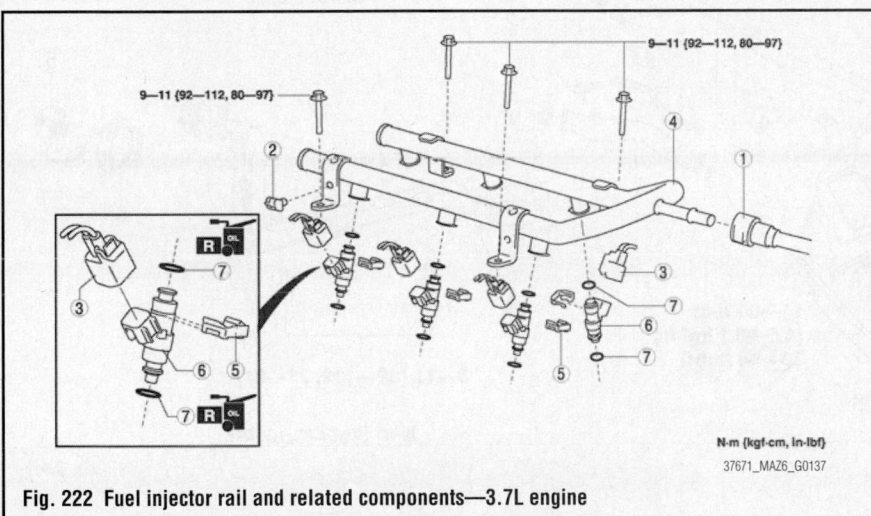

N·m {kgf·cm, in-lbf}

37671_MAZ6_G0137

Fig. 222 Fuel injector rail and related components—3.7L engine

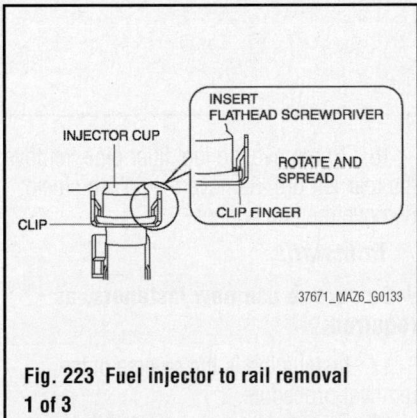

INJECTOR CUP

INSERT FLATHEAD SCREWDRIVER

ROTATE AND SPREAD

CLIP

CLIP FINGER

37671_MAZ6_G0133

Fig. 223 Fuel injector to rail removal 1 of 3

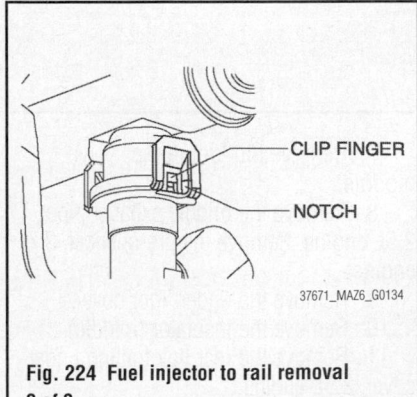

CLIP FINGER

NOTCH

37671_MAZ6_G0134

Fig. 224 Fuel injector to rail removal 2 of 3

7. Remove the engine cover. Remove the dynamic chamber, if equipped with 3.7L engine.
8. Disconnect the electrical connectors.
9. Disconnect and plug the fuel line.
10. Remove the retaining bolts.

11. Remove the assembly.
12. Remove the fuel injectors. Discard the O-rings. Discard the clips.

To install:

➡Be sure to use new fasteners, as required.

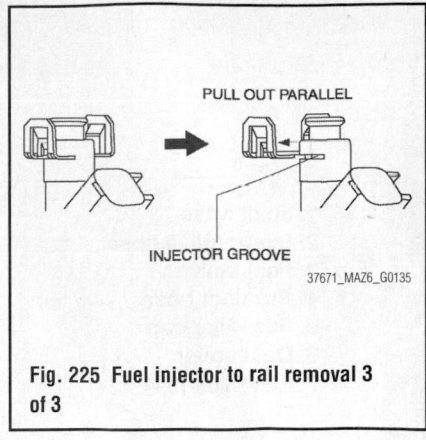

PULL OUT PARALLEL

INJECTOR GROOVE

37671_MAZ6_G0135

Fig. 225 Fuel injector to rail removal 3 of 3

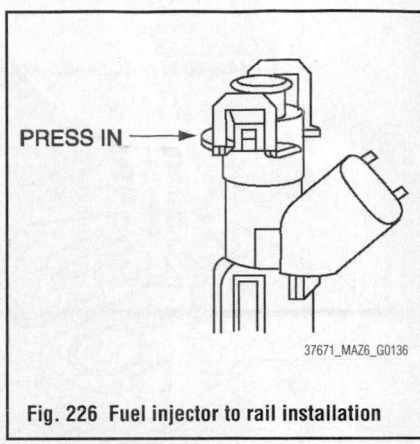

PRESS IN

37671_MAZ6_G0136

Fig. 226 Fuel injector to rail installation

13. Installation is the reverse of the removal procedure.
14. Start the engine and check for leaks. Correct as required.
15. Use the Mazda diagnostic scan tool, or equivalent and reprogram the required systems.

FUEL TANK

REMOVAL & INSTALLATION
See Figures 227 and 228.

1. Before servicing the vehicle, refer to the Precautions Section.

➡If working near and/or around the SRS system and components, be sure to disable the SRS system. Tape the negative battery cable with insulating tape. Always disconnect the negative battery cable first.

✳✳ CAUTION

To avoid personal injury when working on vehicles equipped with an air bag, the negative battery cable must be disconnected and at least one minute must elapse before working on the system. Failure to do so may result in deployment of the air bag.

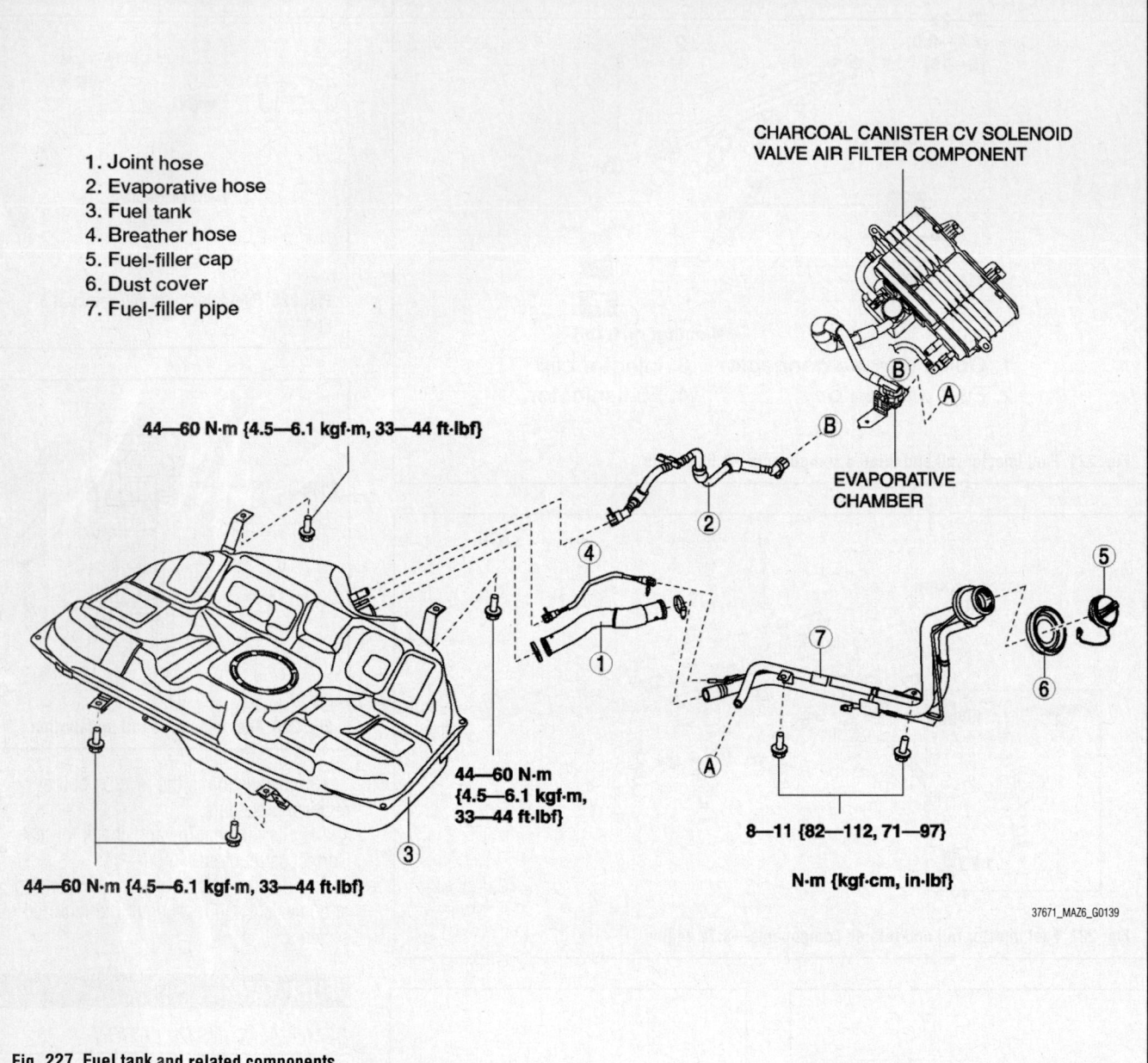

1. Joint hose
2. Evaporative hose
3. Fuel tank
4. Breather hose
5. Fuel-filler cap
6. Dust cover
7. Fuel-filler pipe

CHARCOAL CANISTER CV SOLENOID
VALVE AIR FILTER COMPONENT

EVAPORATIVE
CHAMBER

44—60 N·m {4.5—6.1 kgf·m, 33—44 ft·lbf}

44—60 N·m
{4.5—6.1 kgf·m,
33—44 ft·lbf}

44—60 N·m {4.5—6.1 kgf·m, 33—44 ft·lbf}

8—11 {82—112, 71—97}

N·m {kgf·cm, in·lbf}

37671_MAZ6_G0139

Fig. 227 Fuel tank and related components

2. Disconnect the negative battery cable. Tape the cable with insulating tape.

➡**When disconnecting the cable, some systems need to be initialized after the cable is reconnected. You will need the Mazda diagnostic scan tool or equivalent. Follow the directions on the tool.**

3. Properly relieve the fuel system pressure.

4. Disconnect the negative battery cable.

5. Properly drain the fuel tank to an acceptable level, in order to remove it.

6. Remove the rear seat cushion.

7. Remove the fuel pump module.

8. Remove the middle exhaust pipe, 2.5L engine. Remove the pre-silencer, 3.7L engine.

9. Remove the under floor cover.

10. Remove the insulator (middle).

11. Remove the rear link trailing under-cover, 2.5L engine.

12. Remove the rear insulator, 3.7L engine.

13. Remove the joint hose and evaporator hose. Disconnect any electrical connectors.

14. Remove the fuel tank.

15. Remove the breather hose. Fuel filler cap. Dust cover.

16. To remove the fuel filler pipe, remove the rear left tire. Remove the splash shield. Remove the component.

To install:

➡**Be sure to use new fasteners, as required.**

17. Installation is the reverse of the removal procedure.

18. Be sure to properly install the joint hose and clamp.

19. Start the engine and check for leaks, correct as required.

20. Use the Mazda diagnostic scan tool, or equivalent and reprogram the required systems.

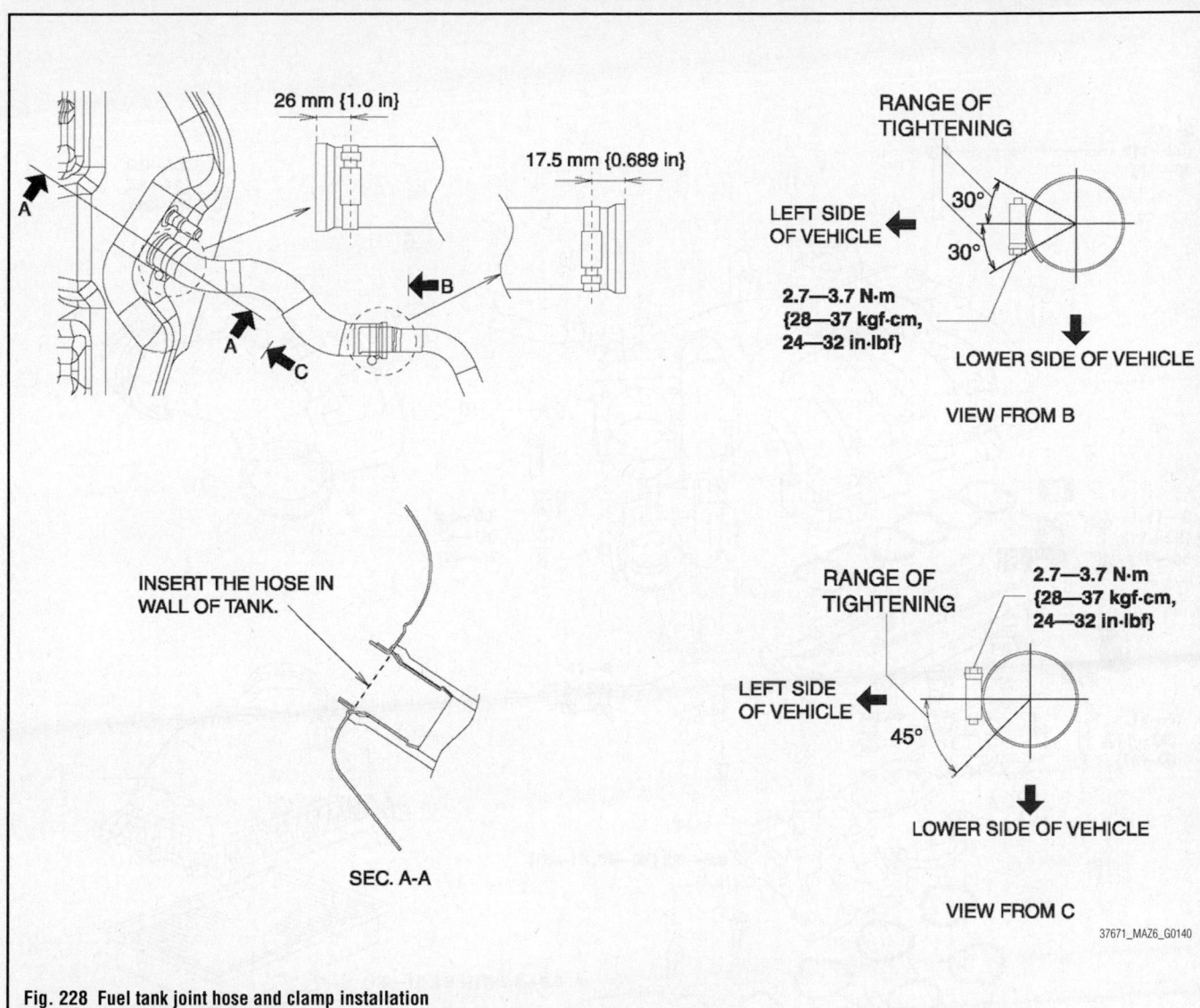

26 mm {1.0 in}

17.5 mm {0.689 in}

A

B

A

C

RANGE OF
TIGHTENING

LEFT SIDE
OF VEHICLE

30°

30°

2.7—3.7 N·m
{28—37 kgf·cm,
24—32 in·lbf}

LOWER SIDE OF VEHICLE

VIEW FROM B

INSERT THE HOSE IN
WALL OF TANK.

SEC. A-A

RANGE OF
TIGHTENING

2.7—3.7 N·m
{28—37 kgf·cm,
24—32 in·lbf}

LEFT SIDE
OF VEHICLE

45°

LOWER SIDE OF VEHICLE

VIEW FROM C

37671_MAZ6_G0140

Fig. 228 Fuel tank joint hose and clamp installation

THROTTLE BODY

REMOVAL & INSTALLATION

2.5L Engine

See Figures 229 through 231.

1. Before servicing the vehicle, refer to the Precautions Section.

➡ If working near and/or around the SRS system and components, be sure to disable the SRS system. Tape the negative battery cable with insulating tape. Always disconnect the negative battery cable first.

☀ CAUTION

To avoid personal injury when working on vehicles equipped with an air bag, the negative battery cable must be disconnected and at least one minute must elapse before working

on the system. Failure to do so may result in deployment of the air bag.

☀ WARNING

A hot engine and intake air system can cause severe burns. Turn off the engine and wait until they are cool before removing the intake air system.

☀ WARNING

Fuel vapor is hazardous. It can easily ignite, causing serious injury and damage. Always keep sparks and flames away from fuel.

☀ WARNING

Fuel line spills and leakage are dangerous. Fuel can ignite and cause

serious injuries or death and damage. Fuel can also irritate skin and eyes.

2. Disconnect the negative battery cable. Tape the cable with insulating tape.

➡ When disconnecting the cable, some systems need to be initialized after the cable is reconnected. You will need the Mazda diagnostic scan tool or equivalent. Follow the directions on the tool.

3. Remove the air cleaner assembly.
4. Drain the cooling system. Be sure to properly dispose of used coolant.
5. Remove the water hose.
6. Disconnect the required electrical connectors.
7. Remove the throttle body retaining bolts.
8. Remove the component from its mounting. Discard the gasket.

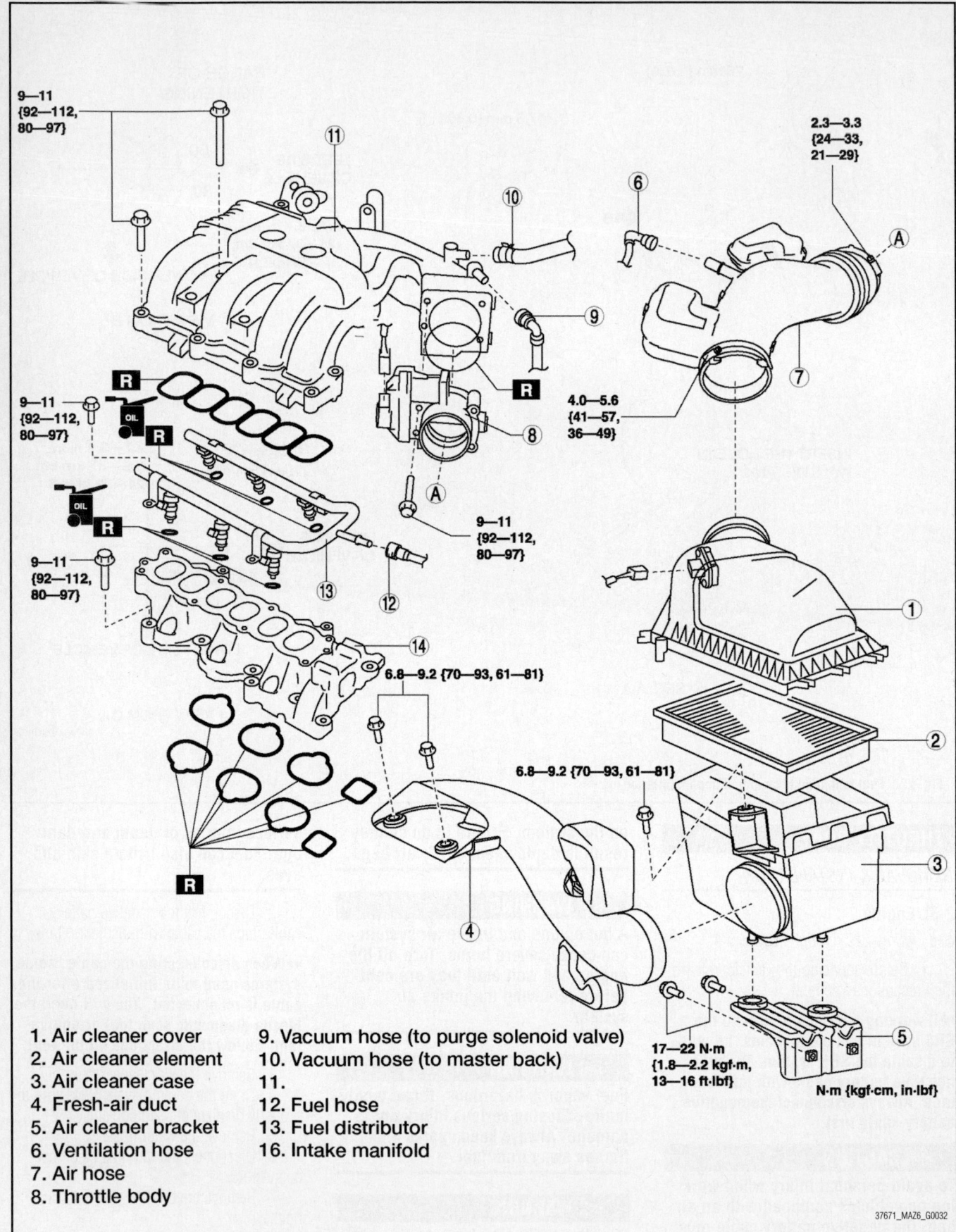

1. Air cleaner cover
2. Air cleaner element
3. Air cleaner case
4. Fresh-air duct
5. Air cleaner bracket
6. Ventilation hose
7. Air hose
8. Throttle body
9. Vacuum hose (to purge solenoid valve)
10. Vacuum hose (to master back)
11.
12. Fuel hose
13. Fuel distributor
16. Intake manifold

37671_MAZ6_G0032

Fig. 229 Air cleaner and related components—3.7L engine

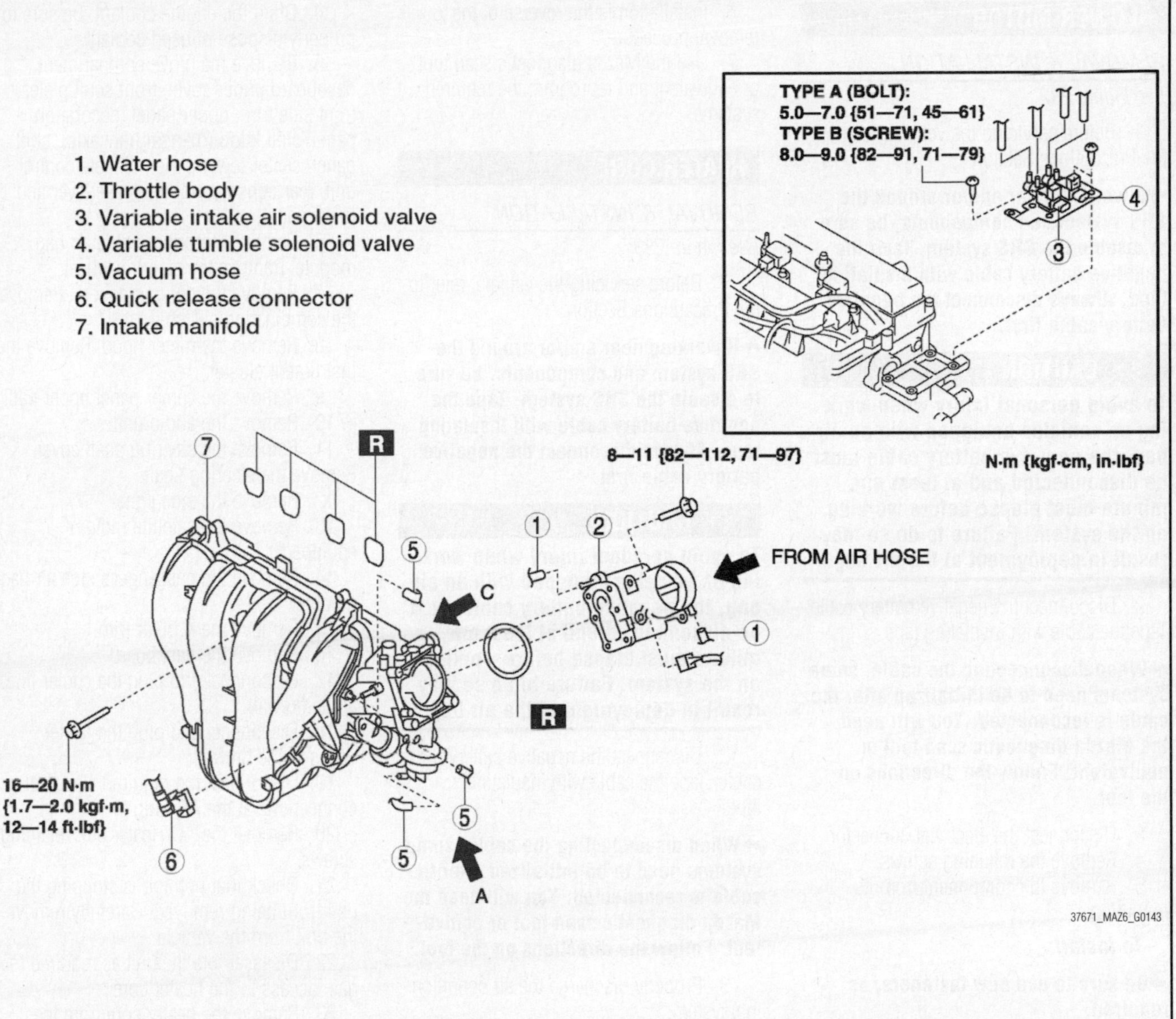

1. Water hose
2. Throttle body
3. Variable intake air solenoid valve
4. Variable tumble solenoid valve
5. Vacuum hose
6. Quick release connector
7. Intake manifold

TYPE A (BOLT):
5.0—7.0 {51—71, 45—61}
TYPE B (SCREW):
8.0—9.0 {82—91, 71—79}

N·m {kgf·cm, in·lbf}

8—11 {82—112, 71—97}

FROM AIR HOSE

16—20 N·m
{1.7—2.0 kgf·m,
12—14 ft·lbf}

37671_MAZ6_G0143

Fig. 230 Throttle body and related components—2.5L engine

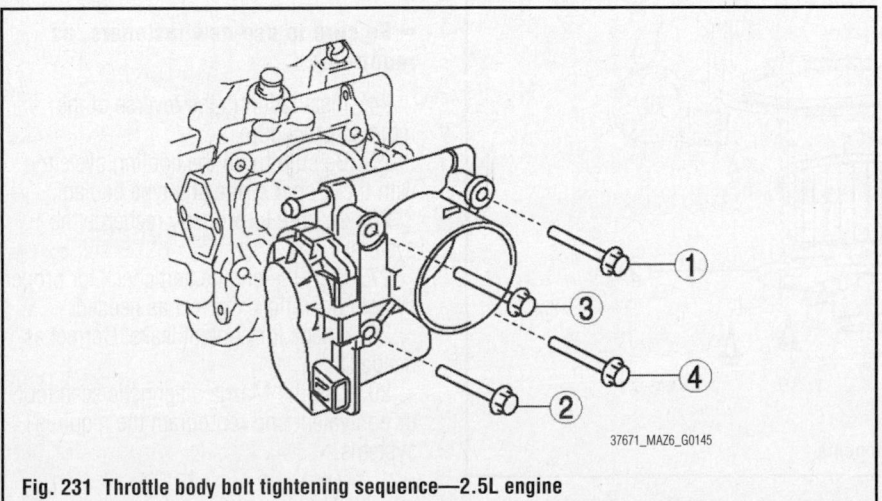

37671_MAZ6_G0145

Fig. 231 Throttle body bolt tightening sequence—2.5L engine

To install:

→ **Be sure to use new fasteners, as required.**

9. Installation is the reverse of the removal procedure.

10. Be sure to use a new gasket.

11. Tighten the retaining bolts to 71–97 inch lbs. (8–11 Nm) and in the proper sequence.

12. Use the Mazda diagnostic scan tool, or equivalent and reprogram the required systems.

3.7L Engine

At this time the manufacturer does not provide removal and installation procedures for this component.

HEATING & AIR CONDITIONING

BLOWER MOTOR

REMOVAL & INSTALLATION

See Figure 232.

1. Before servicing the vehicle, refer to the Precautions Section.

➡**If working near and/or around the SRS system and components, be sure to disable the SRS system. Tape the negative battery cable with insulating tape. Always disconnect the negative battery cable first.**

✳✳ CAUTION

To avoid personal injury when working on vehicles equipped with an air bag, the negative battery cable must be disconnected and at least one minute must elapse before working on the system. Failure to do so may result in deployment of the air bag.

2. Disconnect the negative battery cable. Tape the cable with insulating tape.

➡**When disconnecting the cable, some systems need to be initialized after the cable is reconnected. You will need the Mazda diagnostic scan tool or equivalent. Follow the directions on the tool.**

3. Disconnect the electrical connector.
4. Remove the retaining screws.
5. Remove the component from its mounting.

To install:

➡**Be sure to use new fasteners, as required.**

6. Installation is the reverse of the removal procedure.
7. Use the Mazda diagnostic scan tool, or equivalent and reprogram the required systems.

HEATER CORE

REMOVAL & INSTALLATION

See Figure 233.

1. Before servicing the vehicle, refer to the Precautions Section.

➡**If working near and/or around the SRS system and components, be sure to disable the SRS system. Tape the negative battery cable with insulating tape. Always disconnect the negative battery cable first.**

✳✳ CAUTION

To avoid personal injury when working on vehicles equipped with an air bag, the negative battery cable must be disconnected and at least one minute must elapse before working on the system. Failure to do so may result in deployment of the air bag.

2. Disconnect the negative battery cable. Tape the cable with insulating tape.

➡**When disconnecting the cable, some systems need to be initialized after the cable is reconnected. You will need the Mazda diagnostic scan tool or equivalent. Follow the directions on the tool.**

3. Properly discharge the air conditioning system.

4. Drain the engine coolant. Be sure to properly dispose of used coolant.
5. Remove the glove compartment, dashboard under cover, front scuff plates, front side trim, upper panel, decoration panel, shift knob (manual transaxle), shift panel, center lower panel, climate control unit, rear console, hood release lever and lower panel.
6. Remove the driver's side air bag module. Remove the steering wheel.
7. Remove the column covers. Remove the combination switch.
8. Remove the meter hood. Remove the instrument cluster.
9. Remove the center panel upper pad.
10. Remove the audio unit.
11. Remove the steering shaft cover. Remove the steering shaft.
12. Remove the side panel.
13. Remove the satellite radio, if equipped.
14. Remove the passenger's side air bag module.
15. Remove the A pillar trim.
16. Remove the dashboard.
17. Disconnect and plug the cooler lines at the firewall.
18. Disconnect and plug the heater hoses at the firewall.
19. Disconnect the required electrical connections at the AC unit.
20. Remove the AC/Heater unit retaining screws.
21. Check that nothing is stopping the unit from being removed. Carefully remove the unit from the vehicle.
22. Disassemble the unit as required to gain access to the heater core.
23. Remove the heater core from the case.

To install:

➡**Be sure to use new fasteners, as required.**

24. Installation is the reverse of the removal procedure.
25. Be sure to fill the cooling system with the proper grade and type coolant.
26. Be sure to properly recharge the system.
27. Start the engine and check for proper system operation. Correct as needed.
28. Check for coolant leaks. Correct as needed.
29. Use the Mazda diagnostic scan tool, or equivalent and reprogram the required systems.

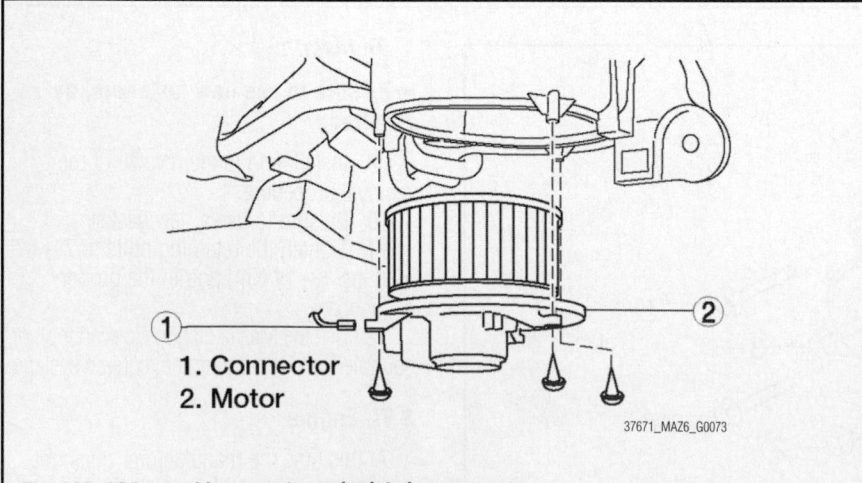

1. Connector
2. Motor

37671_MAZ6_G0073

Fig. 232 AC/heater blower motor and related components

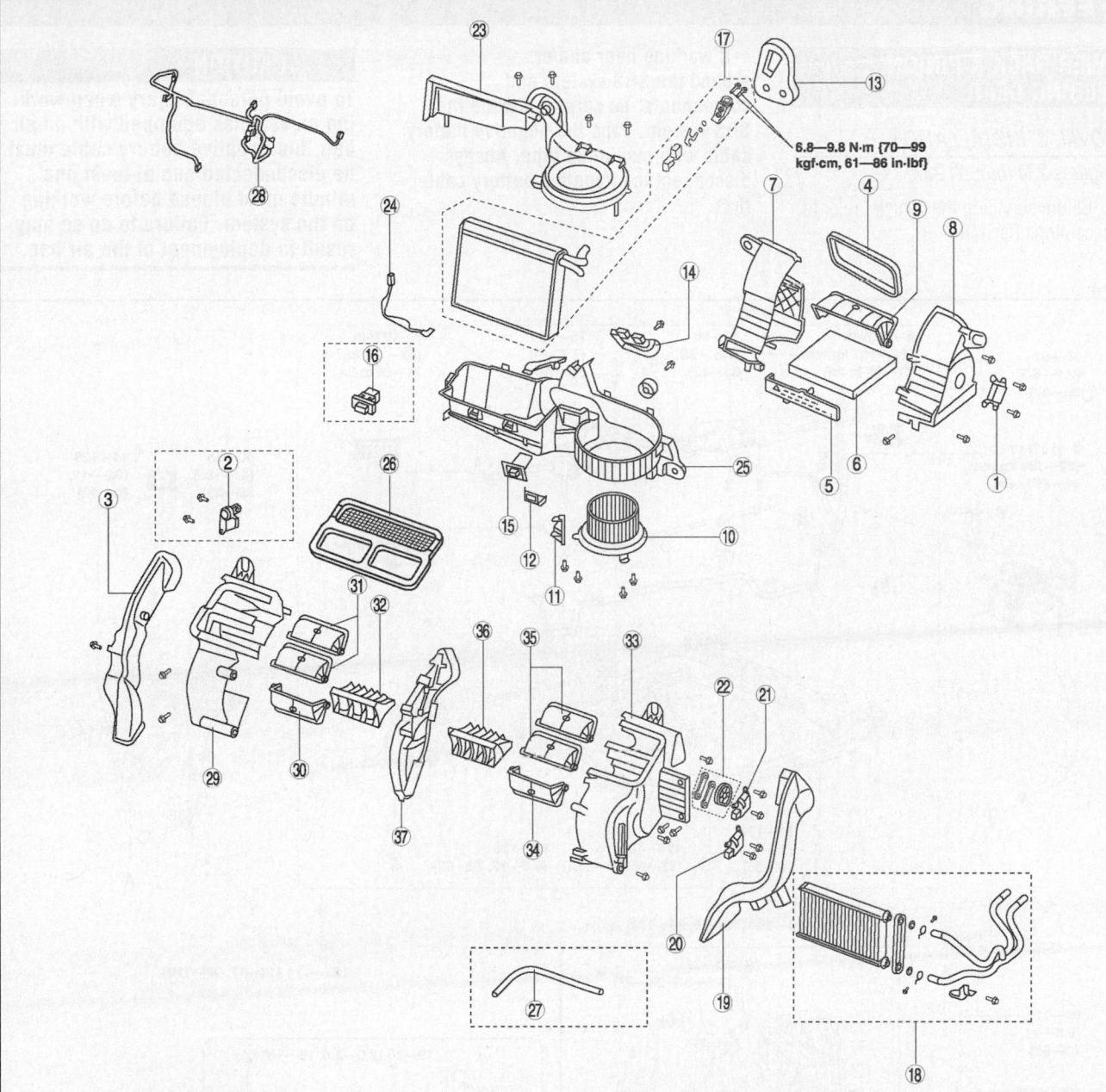

6.8—9.8 N·m {70—99
kgf·cm, 61—86 in·lbf}

1. Air intake actuator
2. Driver's side air mix actuator (Full-auto air conditioner)
3. Side heat duct (LH)
4. Polyurethane form (1)
5. Air filter cover
6. Air filter
7. Blower case (1)
8. Blower case (2)
9. Air intake door
10. Blower motor
11. Bracket
12. Bracket
13. Polyurethane form (2)
14. Plate cover
15. Resistor (Manual air conditioner)
16. Front power MOS FET (Full-auto air conditioner)
17. Evaporator
18. Heater core
19. Side heat duct (RH)

20. Passenger's side air mix actuator (Full-auto air conditioner)
    Air mix actuator (Manual air conditioner)
21. Airflow mode actuator
22. Airflow mode link set
23. A/C case (1)
24. Evaporator temperature sensor
25. A/C case (2)
26. Polyurethane form (3)
27. Air hose (Full-auto air conditioner)
28. Sub harness
29. A/C case (3)
30. Air mix door (LH)
31. Airflow mode door (LH)
32. Baffle (LH)
33. A/C case (4)
34. Air mix door (RH)
35. Airflow mode door (RH)
36. Baffle RH
37. Divider

37671_MAZ6_G0068

Fig. 233 Heater core and related components

STEERING

POWER RACK & PINION STEERING GEAR

REMOVAL & INSTALLATION
See Figures 234 through 237.

1. Before servicing the vehicle, refer to the Precautions Section.

➡If working near and/or around the SRS system and components, be sure to disable the SRS system. Tape the negative battery cable with insulating tape. Always disconnect the negative battery cable first.

✳✳ CAUTION

To avoid personal injury when working on vehicles equipped with an air bag, the negative battery cable must be disconnected and at least one minute must elapse before working on the system. Failure to do so may result in deployment of the air bag.

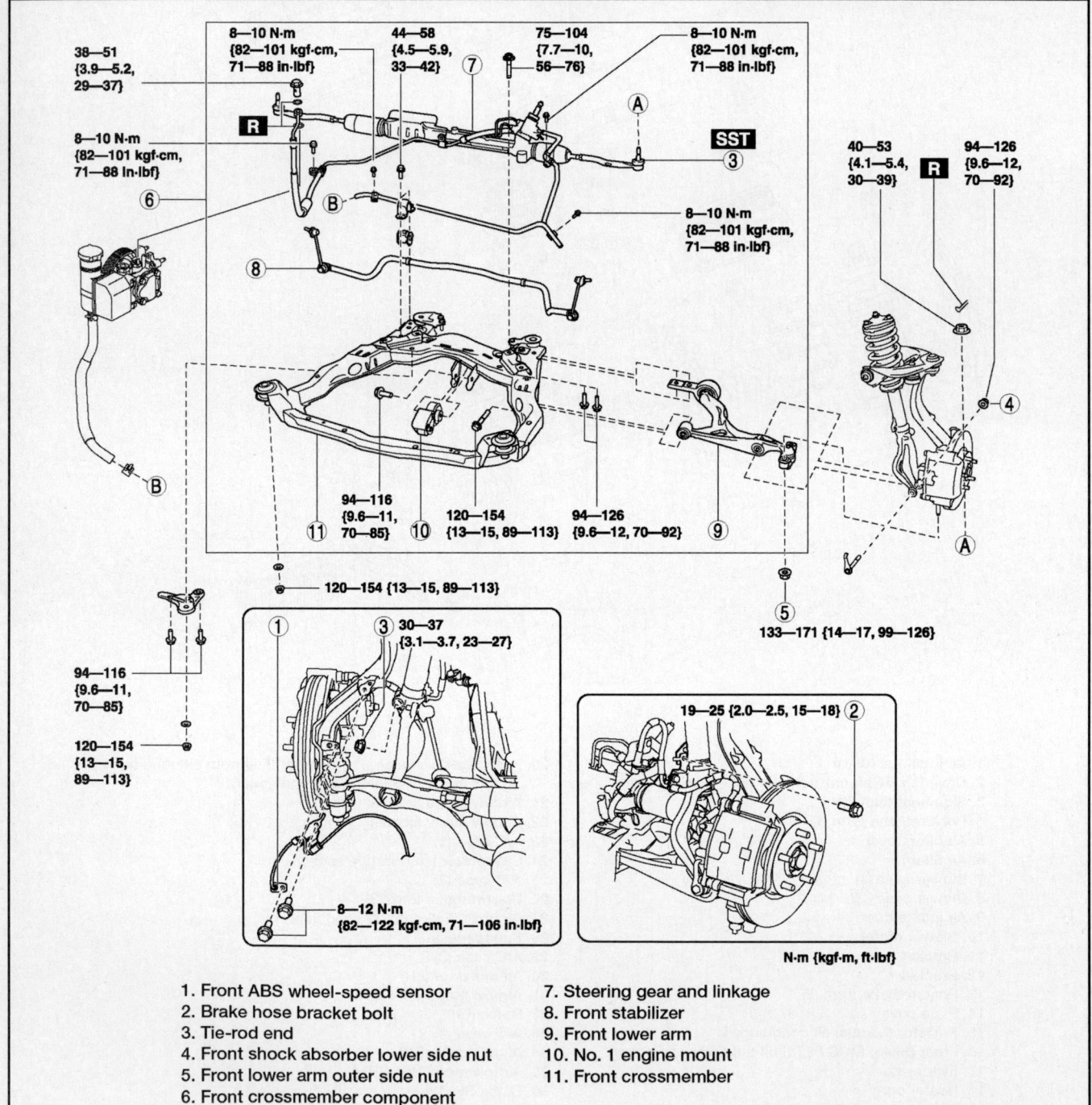

1. Front ABS wheel-speed sensor
2. Brake hose bracket bolt
3. Tie-rod end
4. Front shock absorber lower side nut
5. Front lower arm outer side nut
6. Front crossmember component
7. Steering gear and linkage
8. Front stabilizer
9. Front lower arm
10. No. 1 engine mount
11. Front crossmember

Fig. 234 Front crossmember and related components—2.5L engine

37671_MAZ6_G0084

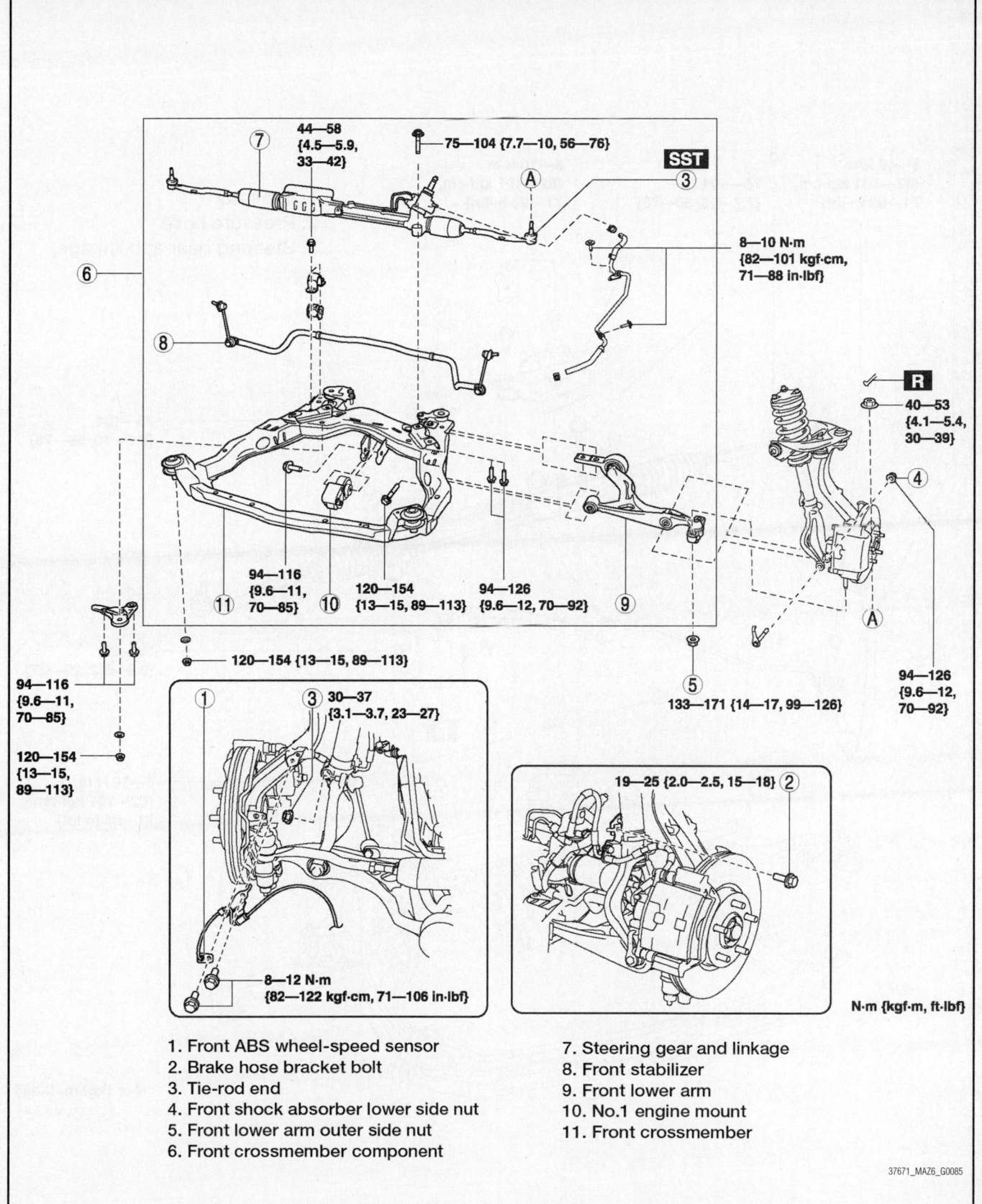

44—58 {4.5—5.9, 33—42}

75—104 {7.7—10, 56—76}

SST

8—10 N·m {82—101 kgf·cm, 71—88 in·lbf}

R

40—53 {4.1—5.4, 30—39}

94—116 {9.6—11, 70—85}

120—154 {13—15, 89—113}

94—126 {9.6—12, 70—92}

94—126 {9.6—12, 70—92}

120—154 {13—15, 89—113}

94—116 {9.6—11, 70—85}

120—154 {13—15, 89—113}

133—171 {14—17, 99—126}

30—37 {3.1—3.7, 23—27}

8—12 N·m {82—122 kgf·cm, 71—106 in·lbf}

19—25 {2.0—2.5, 15—18}

N·m {kgf·m, ft·lbf}

1. Front ABS wheel-speed sensor
2. Brake hose bracket bolt
3. Tie-rod end
4. Front shock absorber lower side nut
5. Front lower arm outer side nut
6. Front crossmember component
7. Steering gear and linkage
8. Front stabilizer
9. Front lower arm
10. No.1 engine mount
11. Front crossmember

37671_MAZ6_G0085

Fig. 235 Front crossmember and related components—3.7L engine

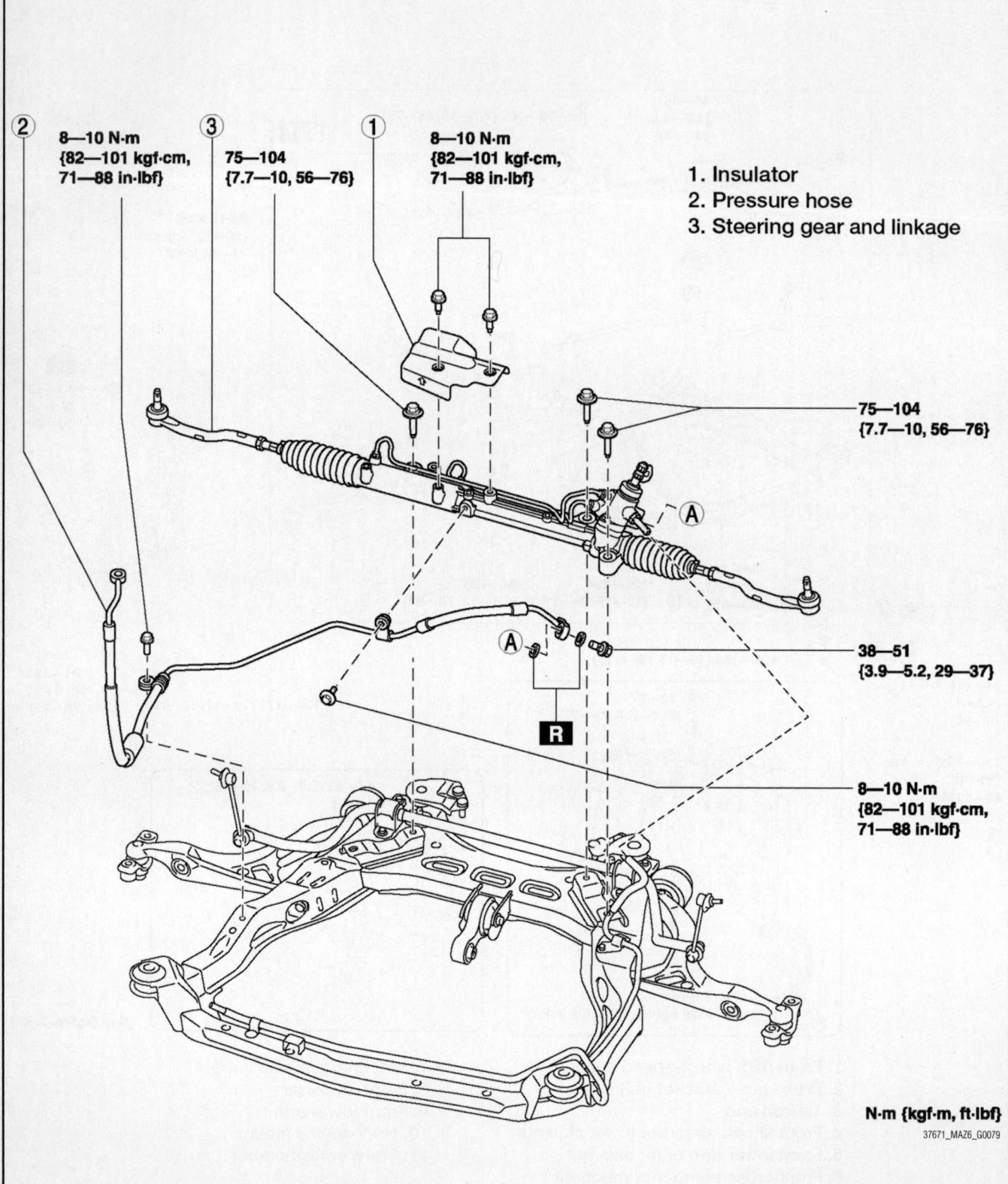

8—10 N·m
{82—101 kgf·cm,
71—88 in·lbf}

75—104
{7.7—10, 56—76}

8—10 N·m
{82—101 kgf·cm,
71—88 in·lbf}

1. Insulator
2. Pressure hose
3. Steering gear and linkage

75—104
{7.7—10, 56—76}

38—51
{3.9—5.2, 29—37}

8—10 N·m
{82—101 kgf·cm,
71—88 in·lbf}

N·m {kgf·m, ft·lbf}

37671_MAZ6_G0079

Fig. 236 Steering gear and related components

2. Disconnect the negative battery cable. Tape the cable with insulating tape.

➡When disconnecting the cable, some systems need to be initialized after the cable is reconnected. You will need the Mazda diagnostic scan tool or equivalent. Follow the directions on the tool.

➡Performing the following procedure without removing the ABS wheel sensor may cause an open circuit in the harness if it is pulled by mistake. Before performing the procedure remove the ABS wheel speed sensor (axle side) and position it to an appropriate place where the sensor will not be pulled by mistake while servicing the vehicle.

➡Secure the steering wheel, using tape or cable, to prevent the steering shaft from rotating after disconnecting the steering shaft. If the steering wheel rotates after the steering shaft and steering gear linkage are disconnected the internal parts of the clockspring could be damaged.

3. Remove the steering shaft cover.
4. Matchmark and disconnect the steering shaft from the steering gear and linkage.
5. Remove the steering gear dust cover.
6. Position the front mudguard out of the way.
7. On 3.7L engine, remove the front pipe.
8. Remove the splash shield.
9. On 2.5L engine, remove the aerodynamic undercover No. 2.
10. Remove the transverse member.
11. Remove the front crossmember component.
12. Remove the gear dust cover.
13. Disconnect and plug the fluid lines.
14. Remove the retaining bolts.
15. Remove the gear from its mounting.

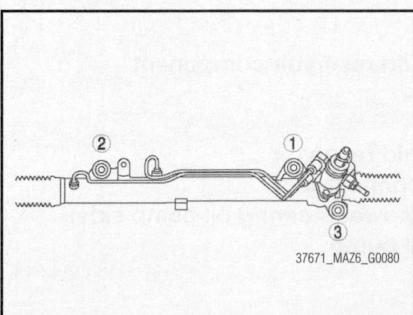

Fig. 237 Steering gear bolt tightening sequence

To install:

➡Be sure to use new fasteners, as required.

16. Installation is the reverse of the removal procedure.
17. Tighten the retaining bolts to 56–76 ft. lbs. (75–104 Nm), and in the proper sequence.
18. Be sure to fill the system with the proper grade and type fluid.
19. Use the Mazda diagnostic scan tool, or equivalent and reprogram the required systems.

POWER STEERING PUMP

REMOVAL & INSTALLATION
See Figures 238 and 239.

1. Before servicing the vehicle, refer to the Precautions Section.

➡If working near and/or around the SRS system and components, be sure to disable the SRS system. Tape the negative battery cable with insulating tape. Always disconnect the negative battery cable first.

✳✳ CAUTION

To avoid personal injury when working on vehicles equipped with an air bag, the negative battery cable must be disconnected and at least one minute must elapse before working on the system. Failure to do so may result in deployment of the air bag.

2. Disconnect the negative battery cable. Tape the cable with insulating tape.

➡When disconnecting the cable, some systems need to be initialized after the cable is reconnected. You will need the Mazda diagnostic scan tool or equivalent. Follow the directions on the tool.

3. On 3.7L engine, remove the windshield wiper motor. Remove the cowl grille.

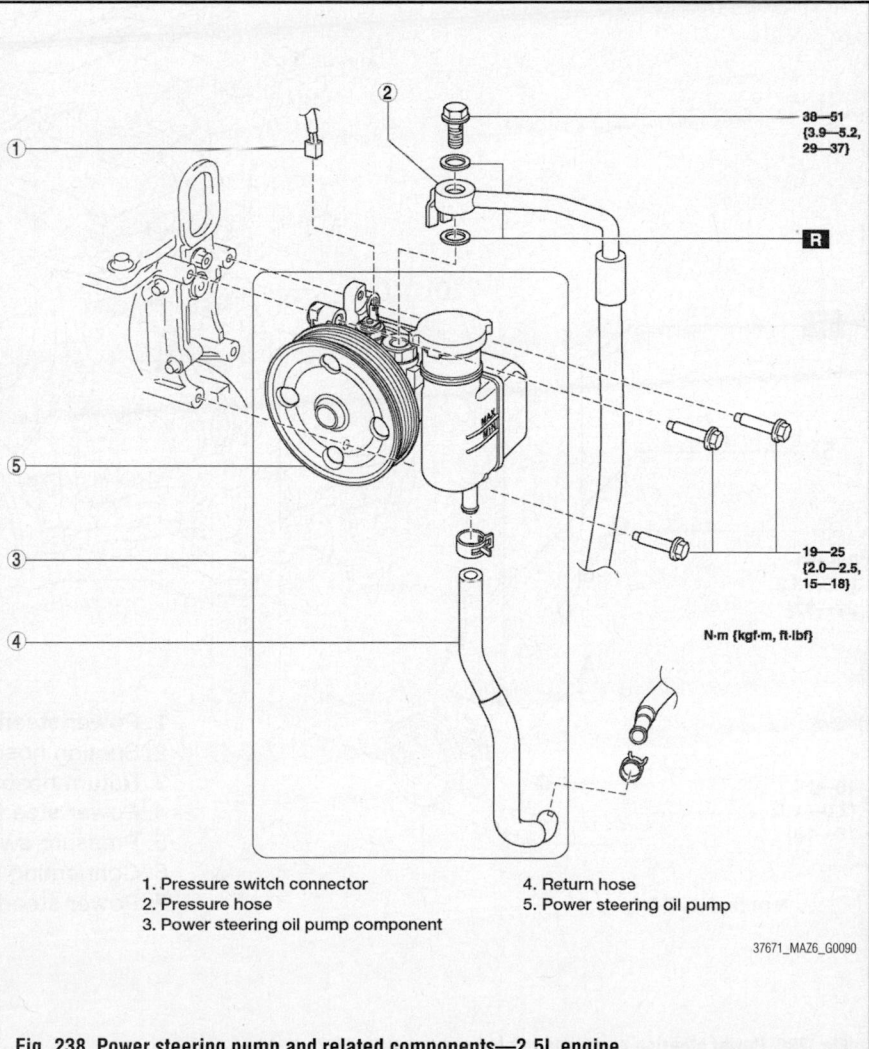

1. Pressure switch connector
2. Pressure hose
3. Power steering oil pump component
4. Return hose
5. Power steering oil pump

Fig. 238 Power steering pump and related components—2.5L engine

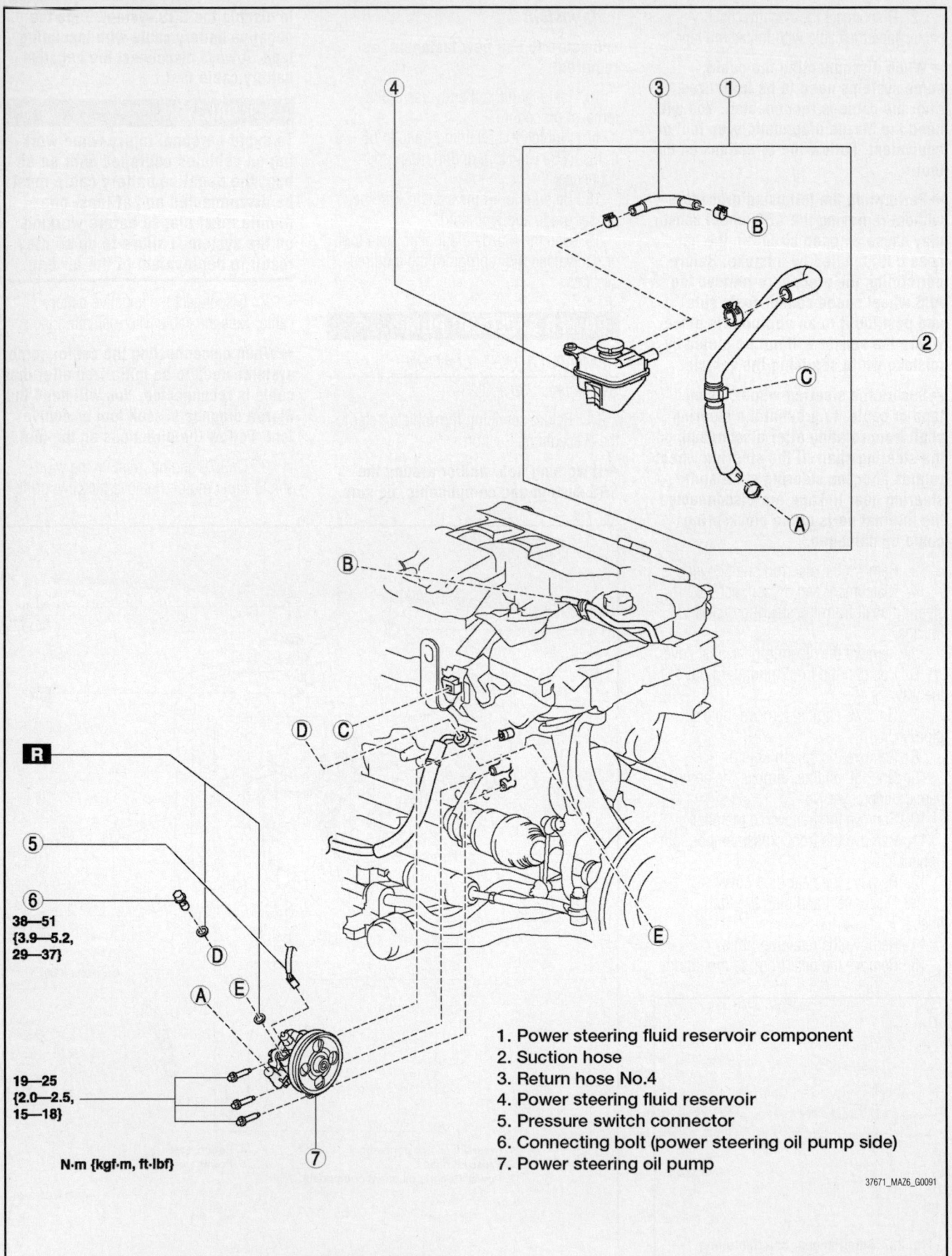

N·m {kgf·m, ft·lbf}

38—51
{3.9—5.2,
29—37}

19—25
{2.0—2.5,
15—18}

1. Power steering fluid reservoir component
2. Suction hose
3. Return hose No.4
4. Power steering fluid reservoir
5. Pressure switch connector
6. Connecting bolt (power steering oil pump side)
7. Power steering oil pump

37671_MAZ6_G0091

Fig. 239 Power steering pump and related components—3.7L engine

4. On 3.7L engine, remove the reservoir from the tank bracket and then remove the tank bracket from the cowl panel. Remove the cowl panel.

5. Position the front mudguard out of the way.

6. Remove the splash shield.

7. Remove the aerodynamic undercover No. 2.

8. Remove the drive belt.

9. Disconnect and plug the fluid lines.

10. Disconnect the electrical connectors.

11. Remove the retaining bolts.

12. Remove the pump from its mounting.

To install:

➡ **Be sure to use new fasteners, as required.**

13. Installation is the reverse of the removal procedure.

14. Be sure to fill the system with the proper grade and type fluid.

15. Use the Mazda diagnostic scan tool, or equivalent and reprogram the required systems.

BLEEDING

1. Before servicing the vehicle, refer to the Precautions Section.

➡ **If working near and/or around the SRS system and components, be sure to disable the SRS system. Tape the negative battery cable with insulating tape. Always disconnect the negative battery cable first.**

✳✳ CAUTION

To avoid personal injury when working on vehicles equipped with an air bag, the negative battery cable must be disconnected and at least one minute must elapse before working on the system. Failure to do so may result in deployment of the air bag.

2. Disconnect the negative battery

cable. Tape the cable with insulating tape.

➡ **When disconnecting the cable, some systems need to be initialized after the cable is reconnected. You will need the Mazda diagnostic scan tool or equivalent. Follow the directions on the tool.**

➡ **Do not turn the steering wheel during fluid inspection.**

3. Be sure that the fluid level is full.

4. Raise and support the vehicle safely.

5. Turn the steering wheel fully to the left and right several times with the engine not running.

6. Inspect the fluid level. Correct as required.

7. Repeat the above until the level stabilizes.

8. Lower the vehicle.

9. Start the engine and let it idle.

10. Verify that the fluid is not foamy and that the level has not dropped.

SUSPENSION

FRONT SUSPENSION

LOWER BALL JOINT

REMOVAL & INSTALLATION

See Figure 240.

At this time the manufacturer does not provide removal and installation procedures for this component. Refer to lower control arm removal and installation for additional information.

LOWER CONTROL ARM

REMOVAL & INSTALLATION

See Figures 241 and 242.

1. Before servicing the vehicle, refer to the Precautions Section.

➡ **If working near and/or around the SRS system and components, be sure to disable the SRS system. Tape the negative battery cable with insulating tape. Always disconnect the negative battery cable first.**

✳✳ CAUTION

To avoid personal injury when working on vehicles equipped with an air bag, the negative battery cable must be disconnected and at least one minute must elapse before working on the system. Failure to do so may result in deployment of the air bag.

2. Disconnect the negative battery cable. Tape the cable with insulating tape.

➡ **When disconnecting the cable, some systems need to be initialized after the cable is reconnected. You will need the Mazda diagnostic scan tool or equivalent. Follow the directions on the tool.**

➡ **Performing the following procedure without removing the ABS wheel sensor may cause an open circuit in the harness if it is pulled by mistake. Before performing the procedure remove the ABS wheel speed sensor (axle side) and position it to an appropriate place where the sensor will not be pulled by mistake while servicing the vehicle.**

✳✳ CAUTION

If the stud bolt of the front knuckle is loosened when removing the lower arm outer nut, replace the stud bolt. If the bolt is reused, it could fall off while driving, resulting in injury or death and damage to the vehicle.

3. Remove the aerodynamic undercover No. 2, if equipped with 2.5L engine.

4. Remove the front exhaust pipe when servicing the right side, if equipped with 3.7L engine.

5. Raise and support the vehicle safely, as required.

6. Properly support the component, using a suitable jack.

7. Remove the ABS speed sensor.

8. Remove the brake hose bracket.

9. Remove the stabilizer control link upper side nut.

10. Remove the lower arm outer side nut.

11. Remove the damper fork.

12. Remove the lower arm inner side bolt (rear).

13. Remove the lower arm inner side bolt (front).

14. Remove the lower arm.

➡ **On 2.5L engine, if working on the left side lower the front crossmember slightly so that the oil pan does not interfere with the removal of the lower arm front side bolt. Remove the radiator hose bracket bolt. Disconnect the steering pipe from the crossmember. Remove the crossmember bracket bolts A and loosen crossmember nut B. (see illustration). Support the crossmember using a suitable jack. Lower the jack and tilt the crossmember towards the front. Remove the lower arm inner side bolt (front).**

➡ **On 3.7L engine, if working on the right side move the engine and transaxle slightly towards the front side of the vehicle so that the engine does not interfere with removal of the lower arm rear side nut. Disconnect the No. 1**

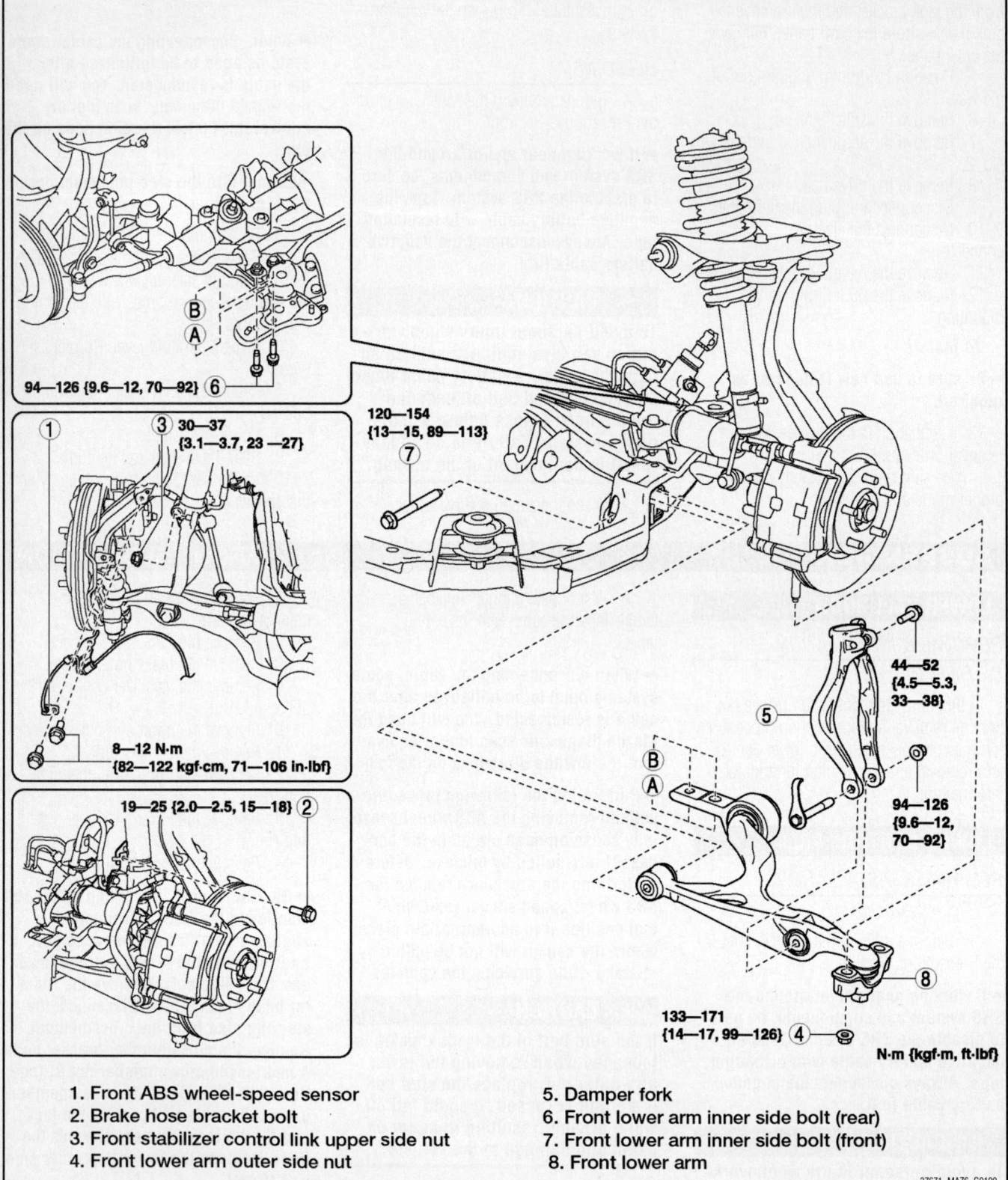

94—126 {9.6—12, 70—92} 6

30—37
{3.1—3.7, 23—27}

120—154
{13—15, 89—113}

8—12 N·m
{82—122 kgf·cm, 71—106 in·lbf}

19—25 {2.0—2.5, 15—18} 2

44—52
{4.5—5.3, 33—38}

94—126
{9.6—12, 70—92}

133—171
{14—17, 99—126} 4

N·m {kgf·m, ft·lbf}

1. Front ABS wheel-speed sensor
2. Brake hose bracket bolt
3. Front stabilizer control link upper side nut
4. Front lower arm outer side nut
5. Damper fork
6. Front lower arm inner side bolt (rear)
7. Front lower arm inner side bolt (front)
8. Front lower arm

37671_MAZ6_G0100

Fig. 240 Front lower control arm and related components

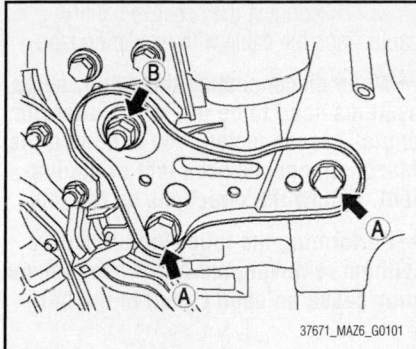

Fig. 241 Front crossmember bolt identification points

37671_MAZ6_G0101

engine mount (engine side). Move the engine/transaxle slightly toward the front side of the vehicle. Remove the lower arm inner side bolt (front).

To install:

➡ Be sure to use new fasteners, as required.

15. Installation is the reverse of the removal procedure.

※ CAUTION

If the stud bolt of the front knuckle is loosened when removing the lower

arm outer nut, replace the stud bolt. If the bolt is reused, it could fall off while driving, resulting in injury or death and damage to the vehicle.

16. Tighten crossmember bolts A to 70–85 ft. lbs. and nut B to 89–113 ft. lbs.

17. Check and adjust wheel alignment, as required.

18. Use the Mazda diagnostic scan tool, or equivalent and reprogram the required systems.

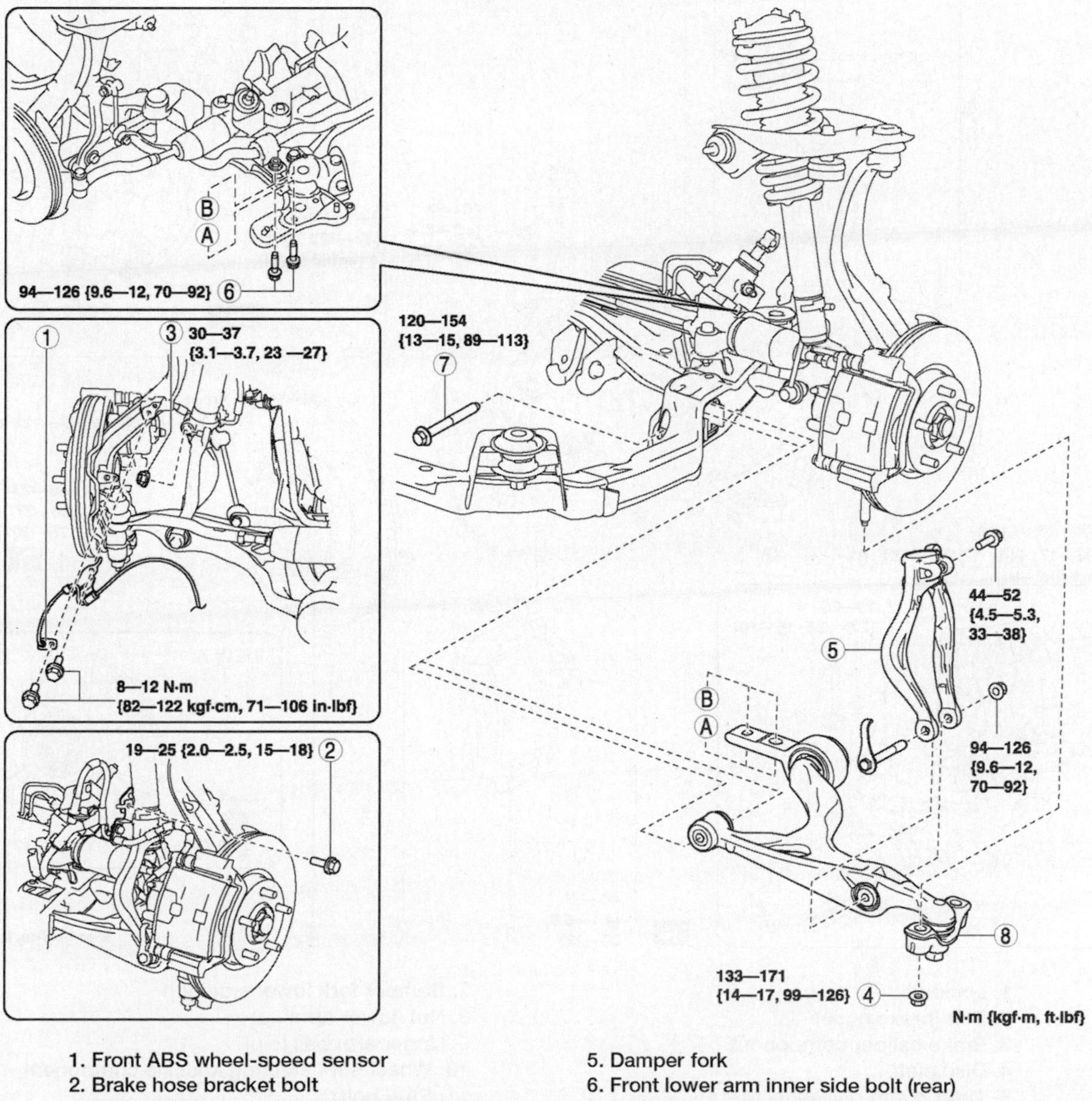

94—126 {9.6—12, 70—92} ⑥

120—154 {13—15, 89—113} ⑦

30—37 {3.1—3.7, 23 —27} ③

①

8—12 N·m {82—122 kgf·cm, 71—106 in·lbf}

19—25 {2.0—2.5, 15—18} ②

44—52 {4.5—5.3, 33—38} ⑤

94—126 {9.6—12, 70—92}

133—171 {14—17, 99—126} ④

⑧

N·m {kgf·m, ft·lbf}

1. Front ABS wheel-speed sensor
2. Brake hose bracket bolt
3. Front stabilizer control link upper side nut
4. Front lower arm outer side nut
5. Damper fork
6. Front lower arm inner side bolt (rear)
7. Front lower arm inner side bolt (front)
8. Front lower arm

37671_MAZ6_G0100

Fig. 242 Front lower control arm and related components

STEERING KNUCKLE

REMOVAL & INSTALLATION

See Figure 243.

1. Before servicing the vehicle, refer to the Precautions Section.

➥If working near and/or around the SRS system and components, be sure to disable the SRS system. Tape the negative battery cable with insulating tape. Always disconnect the negative battery cable first.

✳✳ CAUTION

To avoid personal injury when working on vehicles equipped with an air bag, the negative battery cable must be disconnected and at least one minute must elapse before working on the system. Failure to do so may result in deployment of the air bag.

2. Disconnect the negative battery cable. Tape the cable with insulating tape.

➥When disconnecting the cable, some systems need to be initialized after the cable is reconnected. You will need the Mazda diagnostic scan tool or equivalent. Follow the directions on the tool.

➥Performing the following procedure without removing the ABS wheel sensor may cause an open circuit in the har-

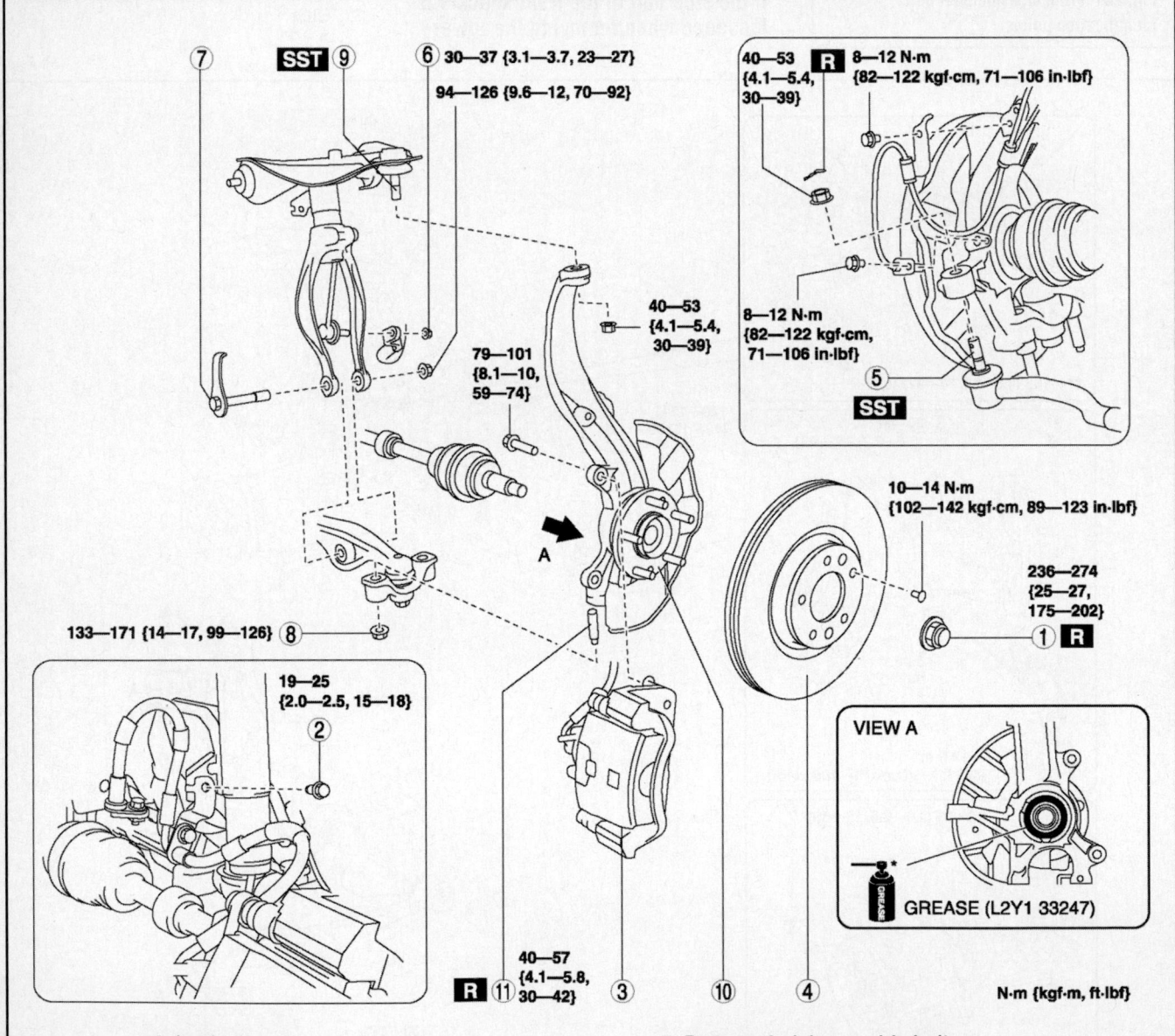

1. Locknut
2. Bolt (brake hose)
3. Brake caliper component
4. Disc plate
5. Tie-rod end ball joint
6. Front stabilizer control link lower side bolt
7. Damper fork lower side bolt
8. Nut (lower arm)
9. Upper arm ball joint
10. Wheel hub, steering knuckle component
11. Stud bolt

37671_MAZ6_G0107

Fig. 243 Front steering knuckle removal points

ness if it is pulled by mistake. Before performing the procedure remove the ABS wheel speed sensor (axle side) and position it to an appropriate place where the sensor will not be pulled by mistake while servicing the vehicle.

✳✳ CAUTION

If the stud bolt of the front knuckle is loosened when removing the lower arm outer nut, replace the stud bolt. If the bolt is reused, it could fall off while driving, resulting in injury or death and damage to the vehicle.

   3. Raise and support the vehicle safely.
   4. Remove the tire and wheel assembly.
   5. Remove the locknut.
   6. Remove the brake hose bolt.
   7. Position the caliper out of the way. Do not allow the caliper to hang by the brake hose.

   8. Remove the rotor.
   9. Remove the tie rod end ball joint.
  10. Remove the stabilizer control link lower side bolt.
  11. Remove the damper fork lower side bolt.
  12. Remove the lower arm nut.

✳✳ CAUTION

If the stud bolt of the front knuckle is loosened when removing the lower arm outer nut, replace the stud bolt. If the bolt is reused, it could fall off while driving, resulting in injury or death and damage to the vehicle.

  13. Remove the upper arm ball joint.
  14. Remove the hub and steering knuckle component.
  15. Remove the stud bolt.

   To install:

➡**Be sure to use new fasteners, as required.**

  16. Installation is the reverse of the removal procedure.

✳✳ CAUTION

If the stud bolt of the front knuckle is loosened when removing the lower arm outer nut, replace the stud bolt. If the bolt is reused, it could fall off while driving, resulting in injury or death and damage to the vehicle.

  17. Check and adjust wheel alignment, as required.
  18. Use the Mazda diagnostic scan tool, or equivalent and reprogram the required systems.

STRUT

REMOVAL & INSTALLATION

See Figures 244 through 247.

   1. Before servicing the vehicle, refer to the Precautions Section.

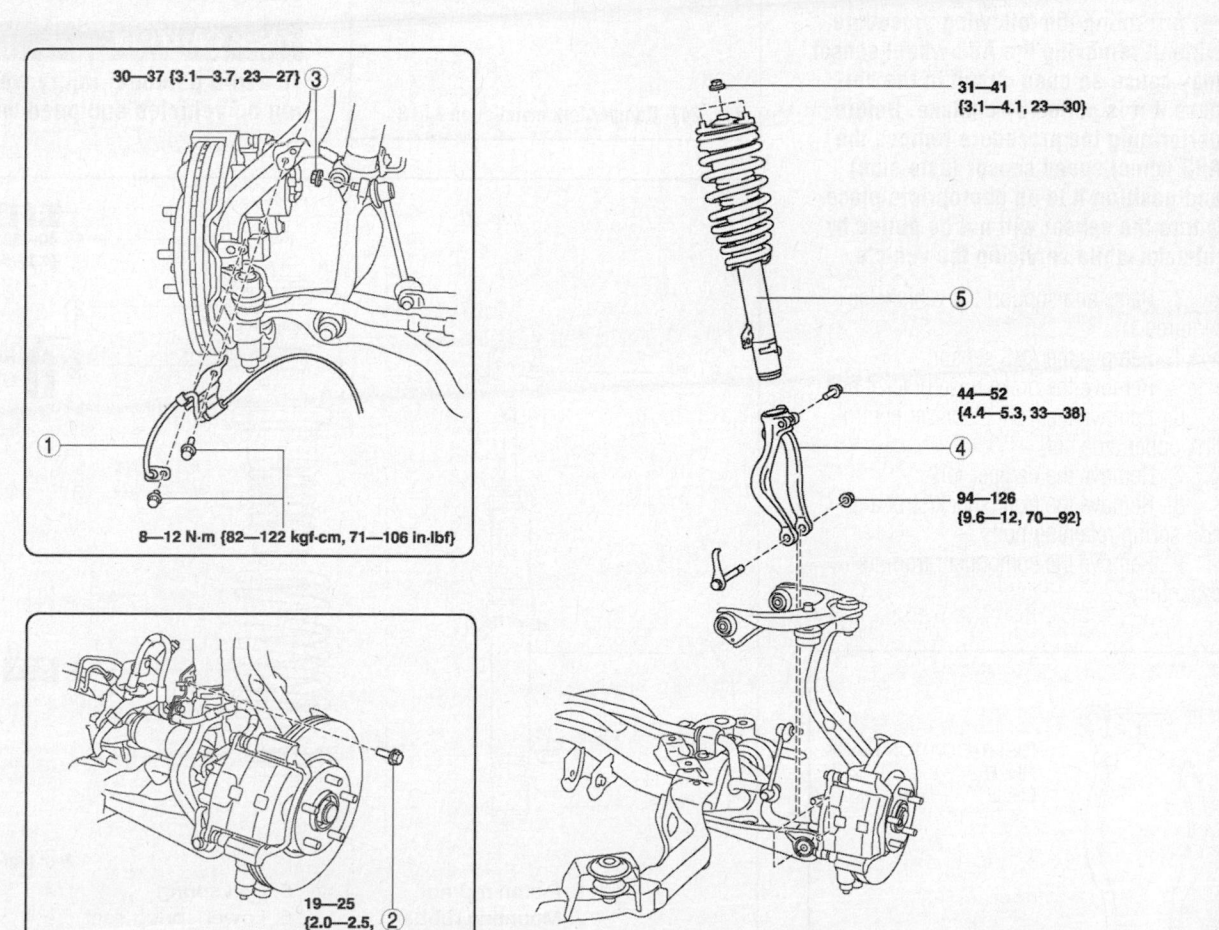

30—37 {3.1—3.7, 23—27}

8—12 N·m {82—122 kgf·cm, 71—106 in·lbf}

19—25 {2.0—2.5, 15—18}

31—41 {3.1—4.1, 23—30}

44—52 {4.4—5.3, 33—38}

94—126 {9.6—12, 70—92}

N·m {kgf·m,ft·lbf}

37671_MAZ6_G0096

Fig. 244 Front shock/strut and related components

➡If working near and/or around the SRS system and components, be sure to disable the SRS system. Tape the negative battery cable with insulating tape. Always disconnect the negative battery cable first.

✸✸ CAUTION

To avoid personal injury when working on vehicles equipped with an air bag, the negative battery cable must be disconnected and at least one minute must elapse before working on the system. Failure to do so may result in deployment of the air bag.

2. Disconnect the negative battery cable. Tape the cable with insulating tape.

➡When disconnecting the cable, some systems need to be initialized after the cable is reconnected. You will need the Mazda diagnostic scan tool or equivalent. Follow the directions on the tool.

➡Performing the following procedure without removing the ABS wheel sensor may cause an open circuit in the harness if it is pulled by mistake. Before performing the procedure remove the ABS wheel speed sensor (axle side) and position it to an appropriate place where the sensor will not be pulled by mistake while servicing the vehicle.

3. Raise and support the vehicle, as required.
4. Remove the ABS sensor.
5. Remove the brake hose bracket bolt.
6. Remove the front stabilizer control link upper side nut.
7. Remove the damper fork.
8. Remove the front shock/strut and coil spring retaining bolts.
9. Remove the component from its mounting.

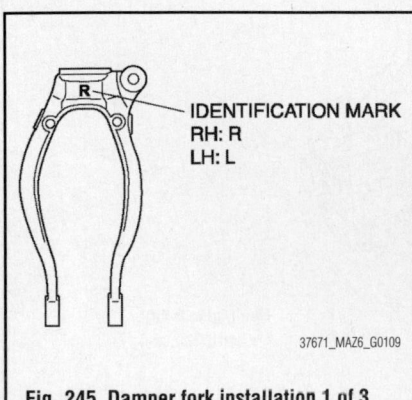

IDENTIFICATION MARK
RH: R
LH: L

37671_MAZ6_G0109

Fig. 245 Damper fork installation 1 of 3

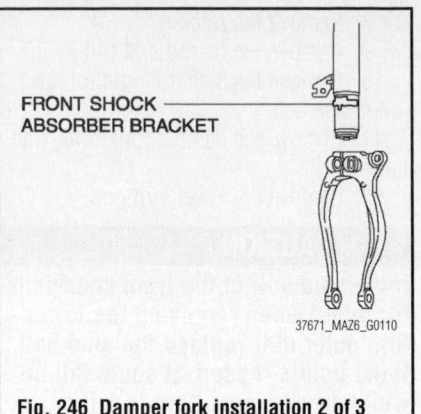

FRONT SHOCK ABSORBER BRACKET

37671_MAZ6_G0110

Fig. 246 Damper fork installation 2 of 3

NOTCH

37671_MAZ6_G0111

Fig. 247 Damper fork installation 3 of 3

To install:

➡Be sure to use new fasteners, as required.

10. Installation is the reverse of the removal procedure.
11. When installing the damper fork be sure to check for proper identification. Set the shock/strut against the fork. Install the bolt aligned with the shock/strut notch. Tighten the bolt to specification.
12. Use the Mazda diagnostic scan tool, or equivalent and reprogram the required systems.

OVERHAUL

See Figures 248 through 250.

1. Before servicing the vehicle, refer to the Precautions Section.

➡If working near and/or around the SRS system and components, be sure to disable the SRS system. Tape the negative battery cable with insulating tape. Always disconnect the negative battery cable first.

✸✸ CAUTION

To avoid personal injury when working on vehicles equipped with an air

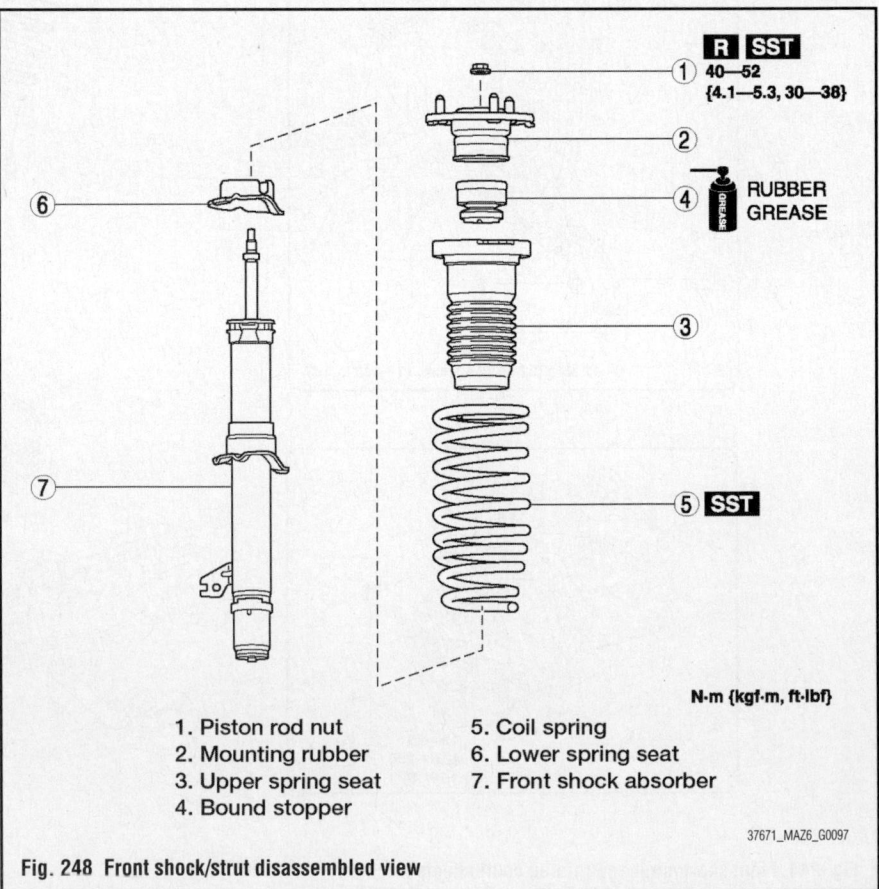

R SST 40—52 {4.1—5.3, 30—38}

RUBBER GREASE

SST

N·m {kgf·m, ft·lbf}

1. Piston rod nut
2. Mounting rubber
3. Upper spring seat
4. Bound stopper
5. Coil spring
6. Lower spring seat
7. Front shock absorber

37671_MAZ6_G0097

Fig. 248 Front shock/strut disassembled view

bag, the negative battery cable must be disconnected and at least one minute must elapse before working on the system. Failure to do so may result in deployment of the air bag.

2. Disconnect the negative battery cable. Tape the cable with insulating tape.

➥When disconnecting the cable, some systems need to be initialized after the cable is reconnected. You will need the Mazda diagnostic scan tool or equivalent. Follow the directions on the tool.

3. Remove the strut from the vehicle.
4. Compress the coil spring using a suitable spring compressor until the spring comes away from the seat.
5. Remove the large center nut and slowly release the spring compressor.

To install:

➥Be sure to use new fasteners, as required.

6. Compress the spring and install it on the strut.
7. Install the lower washer and mounting bracket.

8. Install the upper washer and a new nut.
9. Install the strut assembly in the vehicle.
10. Use the Mazda diagnostic scan tool, or equivalent and reprogram the required systems.

STABILIZER BAR

REMOVAL & INSTALLATION
See Figures 251 through 253.

1. Before servicing the vehicle, refer to the Precautions Section.

➥If working near and/or around the SRS system and components, be sure to disable the SRS system. Tape the negative battery cable with insulating tape. Always disconnect the negative battery cable first.

✳✳ CAUTION

To avoid personal injury when working on vehicles equipped with an air bag, the negative battery cable must be disconnected and at least one minute must elapse before working on the system. Failure to do so may result in deployment of the air bag.

2. Disconnect the negative battery cable. Tape the cable with insulating tape.

➥When disconnecting the cable, some systems need to be initialized after the cable is reconnected. You will need the Mazda diagnostic scan tool or equivalent. Follow the directions on the tool.

➥Secure the steering wheel, using tape or cable, to prevent the steering shaft from rotating after disconnecting the steering shaft. If the steering wheel rotates after the steering shaft and steering gear linkage are disconnected the internal parts of the clockspring could be damaged.

3. Drain the power steering fluid.
4. Remove the steering shaft cover.
5. Matchmark and disconnect the steering shaft from the gear and linkage.
6. Remove the front mudguard.
7. Remove the aerodynamic undercover No.2, if equipped with 2.5L engine.
8. Remove the front pipe, if equipped with 3.7L engine.
9. Remove the transverse member.
10. Remove the lower arm rear bolt.
11. Remove the front crossmember component.
12. Remove the stabilizer control link.

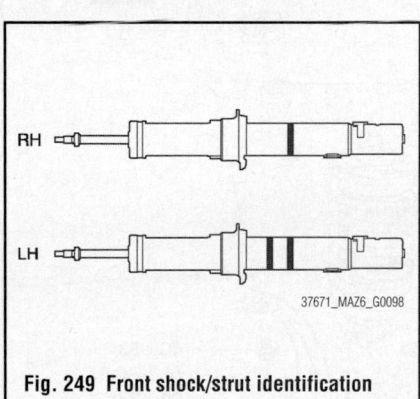

Fig. 249 Front shock/strut identification

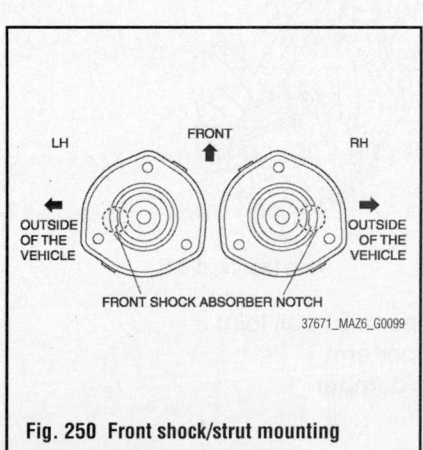

Fig. 250 Front shock/strut mounting rubber installation alignment

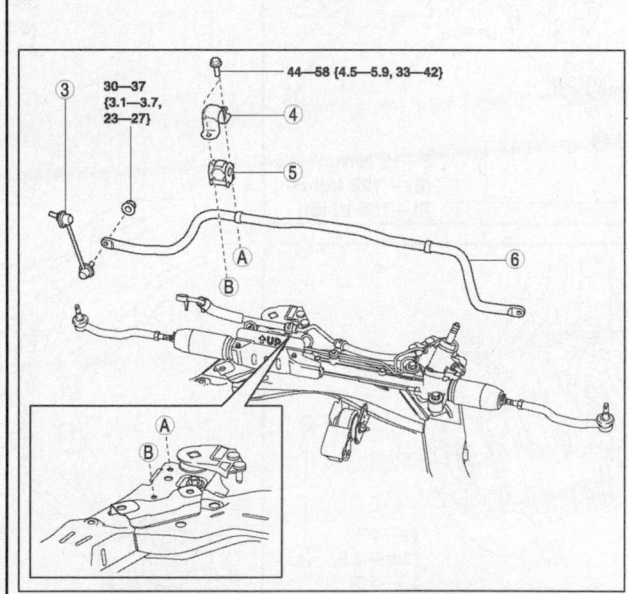

1. Front lower arm (rear) bolt
2. Front crossmember component
3. Stabilizer control link
4. Stabilizer bracket
5. Bushing
6. Front stabilizer

Fig. 251 Front stabilizer bar and related components

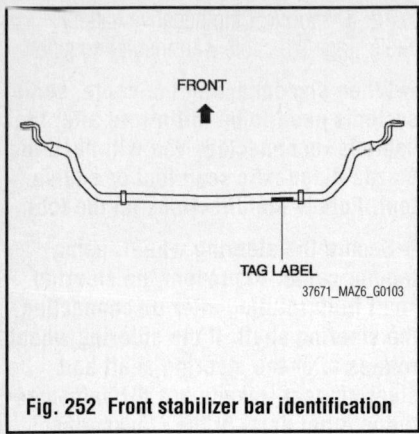

FRONT

TAG LABEL

37671_MAZ6_G0103

Fig. 252 Front stabilizer bar identification

13. Remove the stabilizer bracket.
14. Remove the bushing.
15. Remove the stabilizer from its mounting.

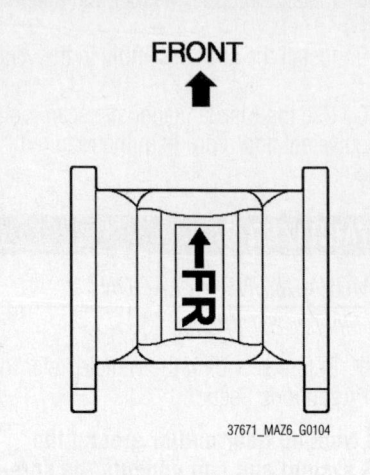

FRONT

FR

37671_MAZ6_G0104

Fig. 253 Front stabilizer bushing identification

To install:

➡ **Be sure to use new fasteners, as required.**

16. Installation is the reverse of the removal procedure.
17. Use the Mazda diagnostic scan tool, or equivalent and reprogram the required systems.

UPPER CONTROL ARM

REMOVAL & INSTALLATION

See Figure 254.

1. Before servicing the vehicle, refer to the Precautions Section.

➡ **If working near and/or around the SRS system and components, be sure**

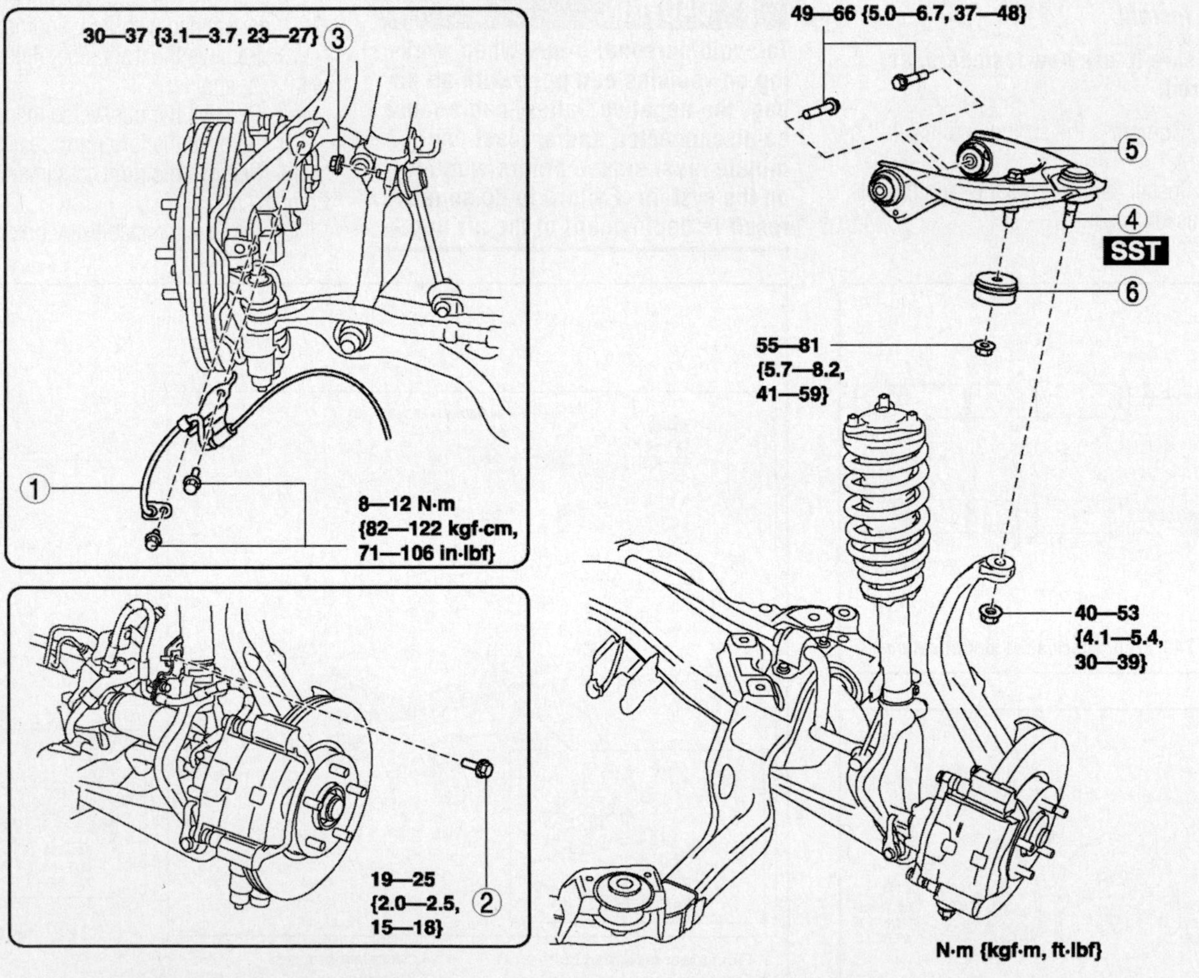

30—37 {3.1—3.7, 23—27}

49—66 {5.0—6.7, 37—48}

SST

55—81
{5.7—8.2,
41—59}

8—12 N·m
{82—122 kgf·cm,
71—106 in·lbf}

40—53
{4.1—5.4,
30—39}

19—25
{2.0—2.5,
15—18}

N·m {kgf·m, ft·lbf}

1. Front ABS wheel-speed sensor
2. Brake hose bracket bolt
3. Front stabilizer control link upper side nut
4. Front upper arm ball joint
5. Front upper arm
6. Dynamic damper

37671_MAZ6_G0105

Fig. 254 Front upper control arm and related components

to disable the SRS system. Tape the negative battery cable with insulating tape. Always disconnect the negative battery cable first.

> ### ✳✳ CAUTION
>
> **To avoid personal injury when working on vehicles equipped with an air bag, the negative battery cable must be disconnected and at least one minute must elapse before working on the system. Failure to do so may result in deployment of the air bag.**

2. Disconnect the negative battery cable. Tape the cable with insulating tape.

➡**When disconnecting the cable, some systems need to be initialized after the cable is reconnected. You will need the Mazda diagnostic scan tool or equivalent. Follow the directions on the tool.**

➡**Performing the following procedure without removing the ABS wheel sensor may cause an open circuit in the harness if it is pulled by mistake. Before performing the procedure remove the ABS wheel speed sensor (axle side) and position it to an appropriate place where the sensor will not be pulled by mistake while servicing the vehicle.**

3. Raise and safely support the vehicle.
4. Properly support the component, using a suitable jack.
5. Remove the ABS sensor.
6. Remove the brake hose bracket.
7. Remove the stabilizer control link upper side nut.
8. Support the knuckle. Using SST tool 49T0283A0, remove the upper ball joint.

➡**Pull the shock/strut outward of the vehicle to assure a space for the front arm bolt removal. Remove the nuts from the upper part of the shock/strut. Pull the unit out of the way toward the outer side of the vehicle.**

9. Remove the upper arm retaining bolts.
10. Remove the component from the vehicle.
11. Remove the dynamic damper.

To install:

➡**Be sure to use new fasteners, as required.**

12. Installation is the reverse of the removal procedure.
13. Check the alignment. Correct as required.
14. Use the Mazda diagnostic scan tool, or equivalent and reprogram the required systems.

SUSPENSION

REAR SUSPENSION

COIL SPRING

REMOVAL & INSTALLATION

See Figures 255 and 256.

1. Before servicing the vehicle, refer to the Precautions Section.

➡**If working near and/or around the SRS system and components, be sure to disable the SRS system. Tape the negative battery cable with insulating tape. Always disconnect the negative battery cable first.**

> ### ✳✳ CAUTION
>
> **To avoid personal injury when working on vehicles equipped with an air bag, the negative battery cable must be disconnected and at least one minute must elapse before working on the system. Failure to do so may result in deployment of the air bag.**

2. Disconnect the negative battery cable. Tape the cable with insulating tape.

➡**When disconnecting the cable, some systems need to be initialized after the cable is reconnected. You will need the Mazda diagnostic scan tool or equivalent. Follow the directions on the tool.**

3. Raise and safely support the vehicle.
4. Remove the tire and wheel assembly.
5. Remove the dynamic damper, if equipped.

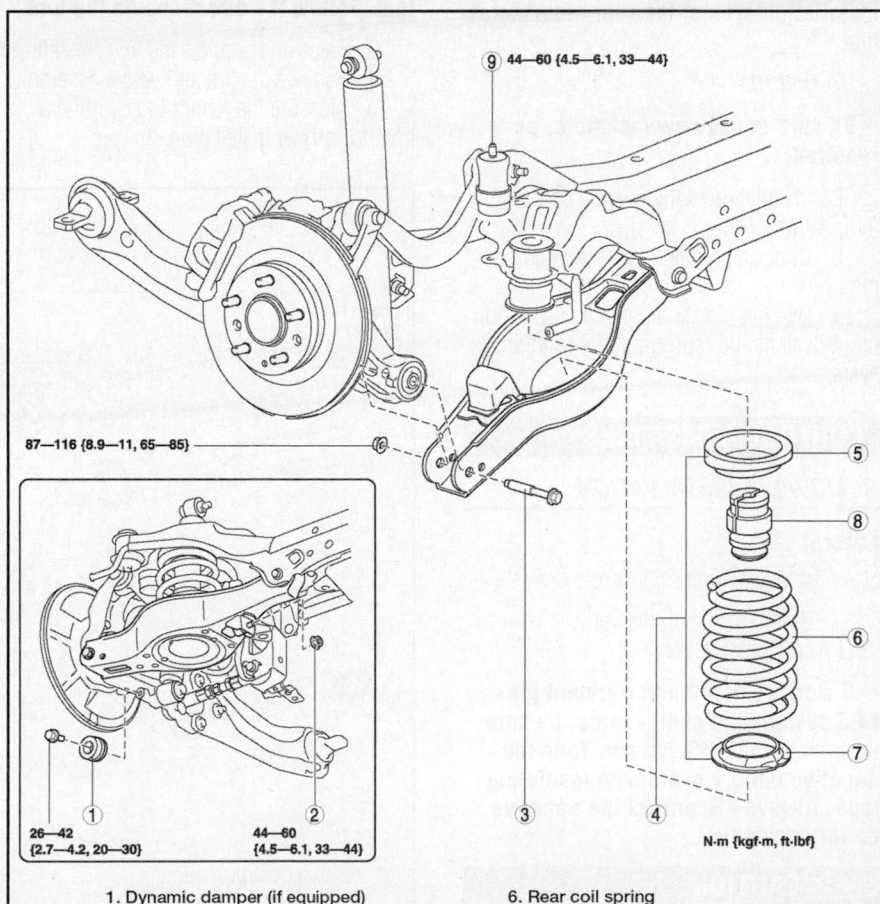

87—116 {8.9—11, 65—85}

⑨ 44—60 {4.5—6.1, 33—44}

26—42 {2.7—4.2, 20—30}

44—60 {4.5—6.1, 33—44}

N·m {kgf·m, ft-lbf}

1. Dynamic damper (if equipped)
2. Rear stabilizer control link lower side nut
3. Rear lower arm outer side bolt
4. Rear coil spring component
5. Upper spring seat rubber
6. Rear coil spring
7. Lower spring seat
8. Bound stopper
9. Bound stopper (body side)

37671_MAZ6_G0113

Fig. 255 Rear coil spring and related components

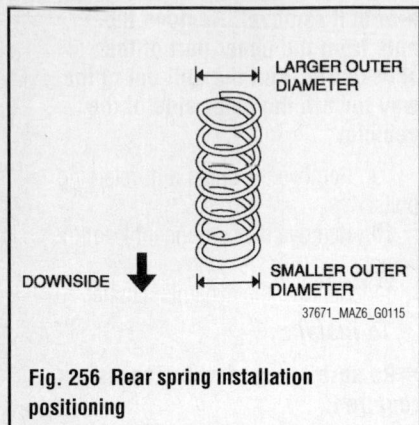

Fig. 256 Rear spring installation positioning

6. Remove the stabilizer control link lower side nut.

7. Remove the lower arm outer side bolt. Be sure to support the arm using a suitable jack before removal. Loosen the inner bolt of the lower arm. Remove the outer bolt.

8. Remove the coil spring component.

➡**When removing the spring apply tape to the edge of the rebound stopper installation area of the rear crossmember.**

To install:

➡**Be sure to use new fasteners, as required.**

9. Installation is the reverse of the removal procedure.

10. Check and adjust the alignment, as required.

11. Use the Mazda diagnostic scan tool, or equivalent and reprogram the required systems.

CONTROL ARMS/LINKS

REMOVAL & INSTALLATION

Lateral Link

See Figures 257 and 258.

1. Before servicing the vehicle, refer to the Precautions Section.

➡**If working near and/or around the SRS system and components, be sure to disable the SRS system. Tape the negative battery cable with insulating tape. Always disconnect the negative battery cable first.**

✳✳ CAUTION

To avoid personal injury when working on vehicles equipped with an air bag, the negative battery cable must be disconnected and at least one minute must elapse before working

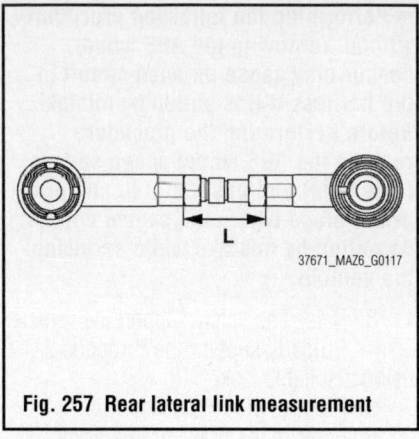

Fig. 257 Rear lateral link measurement

on the system. Failure to do so may result in deployment of the air bag.

2. Disconnect the negative battery cable. Tape the cable with insulating tape.

➡**When disconnecting the cable, some systems need to be initialized after the cable is reconnected. You will need the Mazda diagnostic scan tool or equivalent. Follow the directions on the tool.**

3. Raise and support the vehicle safely.

4. Remove the tire and wheel assembly.

5. Measure the length L, see illustration, for proper installation.

6. Support the trailing link using a suitable jack.

7. Remove the retaining bolts.

8. Remove the component from its mounting.

To install:

➡**Be sure to use new fasteners, as required.**

9. Installation is the reverse of the removal procedure.

10. Be sure to adjust the component as removed.

11. Check and adjust the alignment, as required.

12. Use the Mazda diagnostic scan tool, or equivalent and reprogram the required systems.

Lower Control Arm

See Figure 259.

1. Before servicing the vehicle, refer to the Precautions Section.

➡**If working near and/or around the SRS system and components, be sure to disable the SRS system. Tape the negative battery cable with insulating tape. Always disconnect the negative battery cable first.**

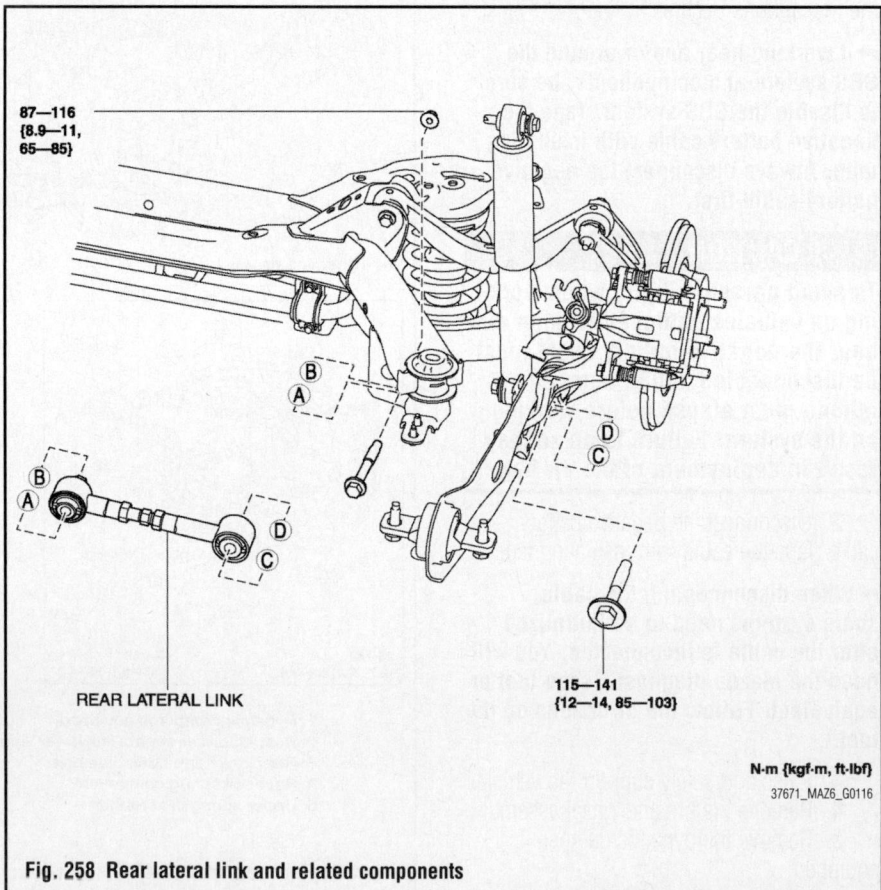

87—116
{8.9—11, 65—85}

115—141
{12—14, 85—103}

REAR LATERAL LINK

N·m {kgf-m, ft-lbf}
37671_MAZ6_G0116

Fig. 258 Rear lateral link and related components

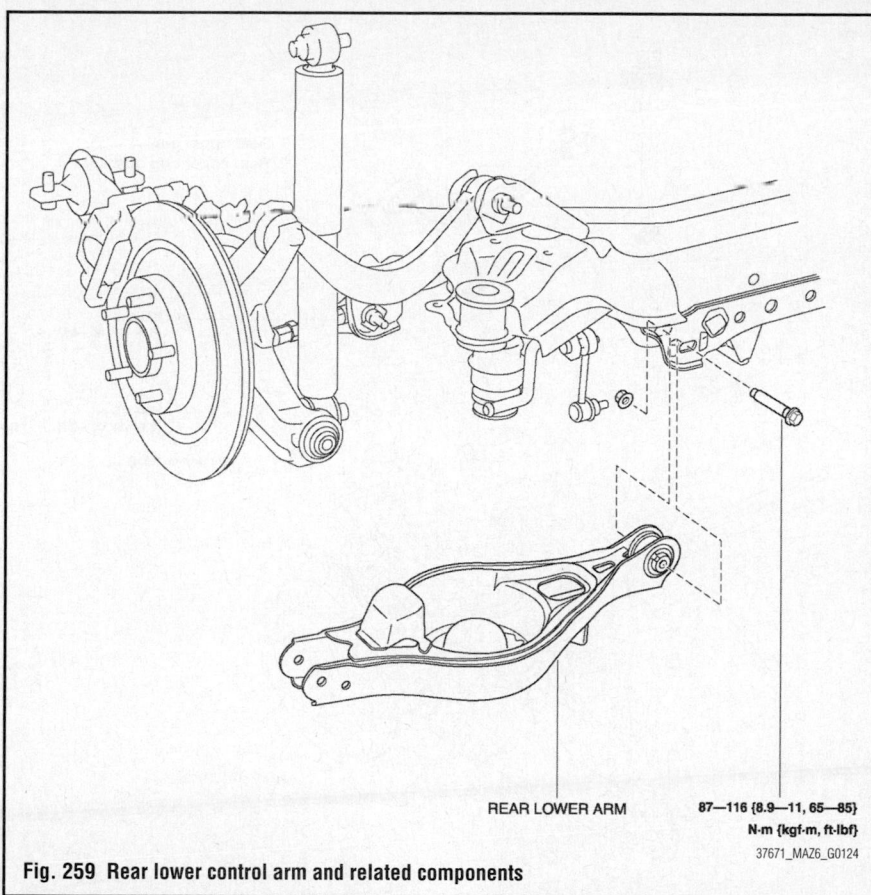

REAR LOWER ARM

87—116 {8.9—11, 65—85}
N·m {kgf·m, ft·lbf}

37671_MAZ6_G0124

Fig. 259 Rear lower control arm and related components

※※ **CAUTION**

To avoid personal injury when working on vehicles equipped with an air bag, the negative battery cable must be disconnected and at least one minute must elapse before working on the system. Failure to do so may result in deployment of the air bag.

2. Disconnect the negative battery cable. Tape the cable with insulating tape.

➡**When disconnecting the cable, some systems need to be initialized after the cable is reconnected. You will need the Mazda diagnostic scan tool or equivalent. Follow the directions on the tool.**

3. Raise and support the vehicle safely.
4. Remove the tire and wheel assembly.
5. Remove the spring.
6. Be sure to properly support the assembly using the proper support jack.
7. Remove the component retaining bolts.
8. Remove the component from the vehicle.

To install:

➡**Be sure to use new fasteners, as required.**

9. Installation is the reverse of the removal procedure.
10. Check and adjust alignment, as required.
11. Use the Mazda diagnostic scan tool, or equivalent and reprogram the required systems.

Trailing Link

See Figure 260.

1. Before servicing the vehicle, refer to the Precautions Section.

➡**If working near and/or around the SRS system and components, be sure to disable the SRS system. Tape the negative battery cable with insulating tape. Always disconnect the negative battery cable first.**

※※ **CAUTION**

To avoid personal injury when working on vehicles equipped with an air bag, the negative battery cable must be disconnected and at least one minute must elapse before working on the system. Failure to do so may result in deployment of the air bag.

2. Disconnect the negative battery cable. Tape the cable with insulating tape.

94—131 {9.6—13, 70—96}

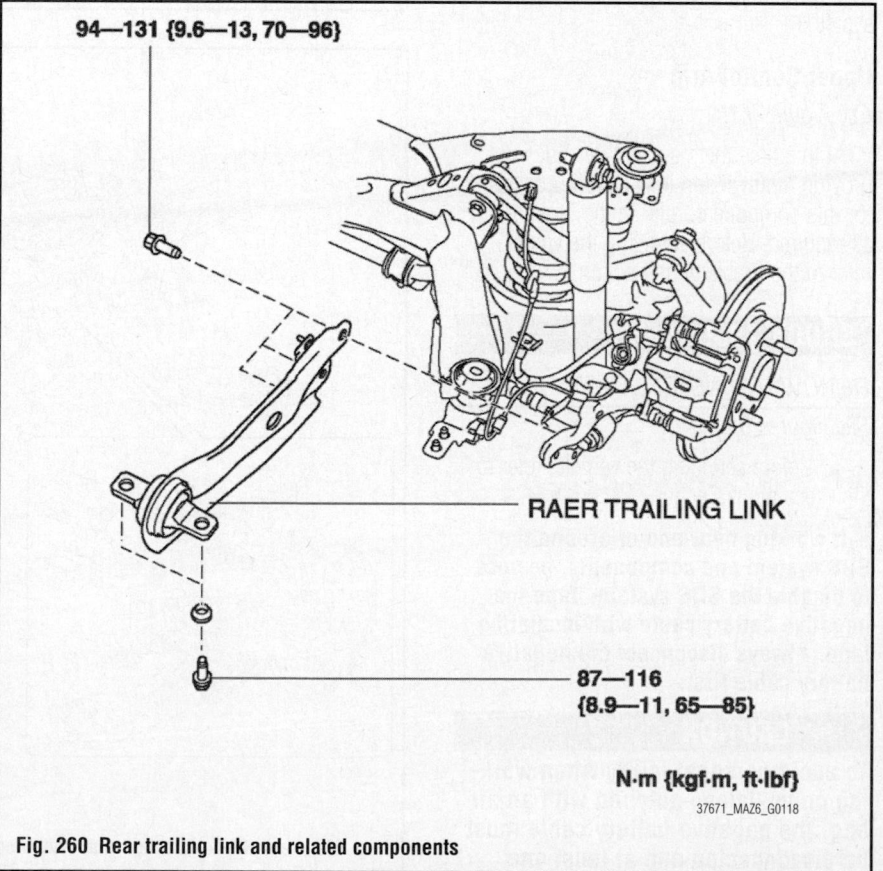

RAER TRAILING LINK

87—116
{8.9—11, 65—85}

N·m {kgf·m, ft·lbf}

37671_MAZ6_G0118

Fig. 260 Rear trailing link and related components

➡When disconnecting the cable, some systems need to be initialized after the cable is reconnected. You will need the Mazda diagnostic scan tool or equivalent. Follow the directions on the tool.

➡Performing the following procedure without removing the ABS wheel sensor may cause an open circuit in the harness if it is pulled by mistake. Before performing the procedure remove the ABS wheel speed sensor and position it to an appropriate place where the sensor will not be pulled by mistake while servicing the vehicle.

3. Raise and support the vehicle safely.

4. Remove the tire and wheel assembly.

5. Remove the under floor undercovers No. 1 and No. 2.

6. Support the trailing link using a suitable jack.

7. Remove the retaining bolts.

8. Remove the component from its mounting.

To install:

➡Be sure to use new fasteners, as required.

9. Installation is the reverse of the removal procedure.

10. Use the Mazda diagnostic scan tool, or equivalent and reprogram the required systems.

Upper Control Arm

See Figure 261.

At this time the manufacturer does not provide removal and installation procedures for this component, refer to the illustration as required. Before servicing the vehicle, refer to the Precautions Section.

SHOCK ABSORBER

REMOVAL & INSTALLATION

See Figure 262.

1. Before servicing the vehicle, refer to the Precautions Section.

➡If working near and/or around the SRS system and components, be sure to disable the SRS system. Tape the negative battery cable with insulating tape. Always disconnect the negative battery cable first.

✷✷ CAUTION

To avoid personal injury when working on vehicles equipped with an air bag, the negative battery cable must be disconnected and at least one

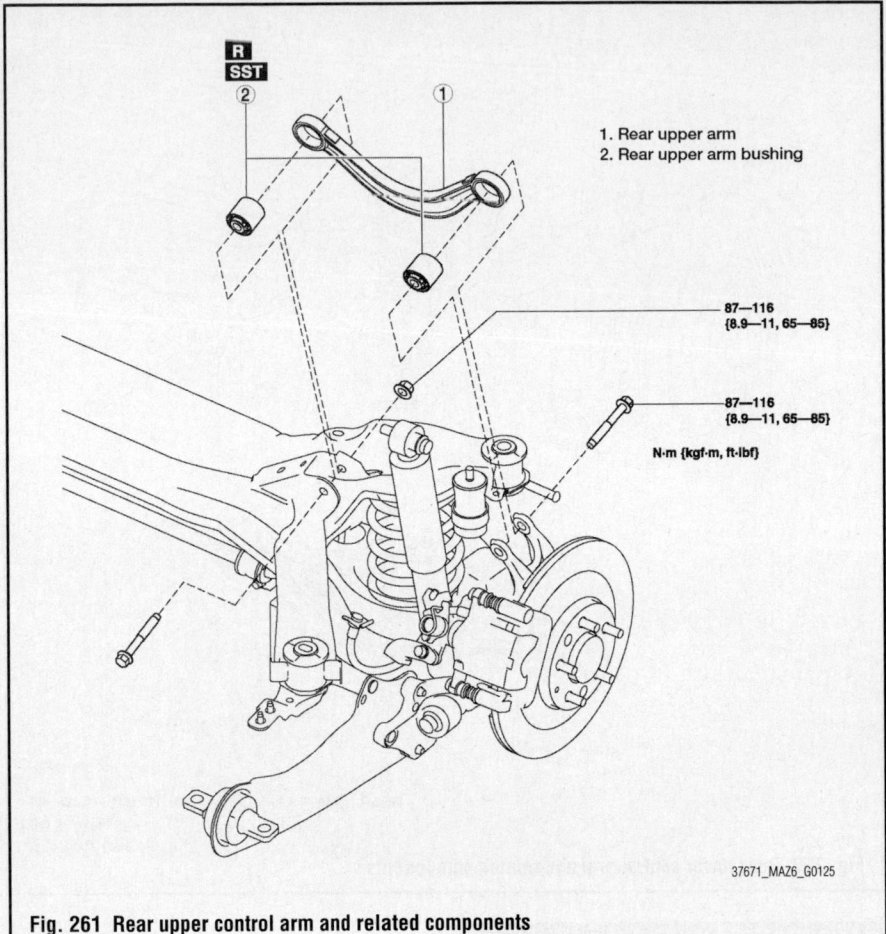

1. Rear upper arm
2. Rear upper arm bushing

87—116
{8.9—11, 65—85}

87—116
{8.9—11, 65—85}

N·m {kgf·m, ft-lbf}

37671_MAZ6_G0125

Fig. 261 Rear upper control arm and related components

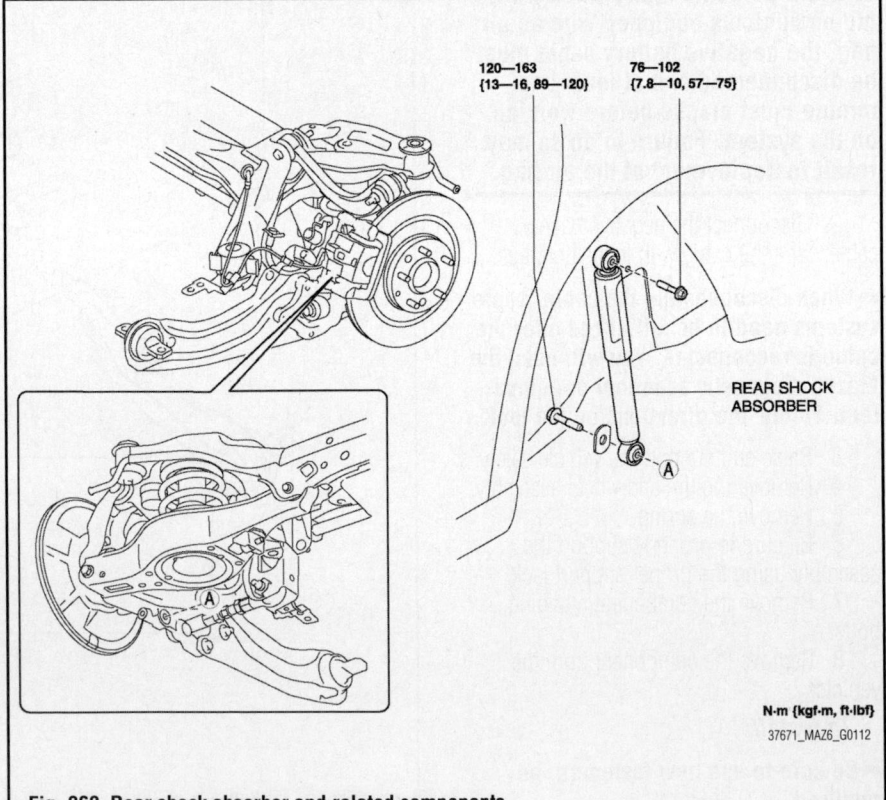

120—163
{13—16, 89—120}

76—102
{7.8—10, 57—75}

REAR SHOCK ABSORBER

N·m {kgf·m, ft-lbf}

37671_MAZ6_G0112

Fig. 262 Rear shock absorber and related components

minute must elapse before working on the system. Failure to do so may result in deployment of the air bag.

2. Disconnect the negative battery cable. Tape the cable with insulating tape.

➡When disconnecting the cable, some systems need to be initialized after the cable is reconnected. You will need the Mazda diagnostic scan tool or equivalent. Follow the directions on the tool.

3. Raise and safely support the vehicle.
4. Remove the tire and wheel assembly.
5. Using a suitable jack, properly support the rear axle.
6. Remove the shock retaining bolts.
7. Remove the component from the vehicle.

To install:

➡Be sure to use new fasteners, as required.

8. Installation is the reverse of the removal procedure.
9. Use the Mazda diagnostic scan tool, or equivalent and reprogram the required systems.

STABILIZER BAR

REMOVAL & INSTALLATION

See Figures 263 through 266.

1. Before servicing the vehicle, refer to the Precautions Section.

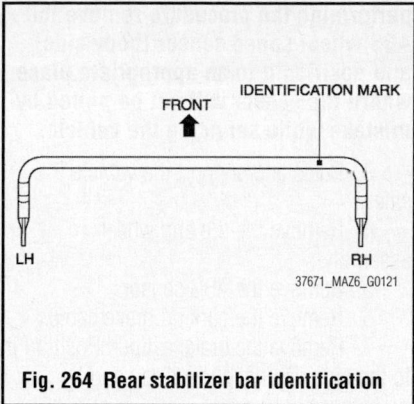

Fig. 264 Rear stabilizer bar identification

➡If working near and/or around the SRS system and components, be sure to disable the SRS system. Tape the negative battery cable with insulating tape. Always disconnect the negative battery cable first.

✷✷ CAUTION

To avoid personal injury when working on vehicles equipped with an air bag, the negative battery cable must be disconnected and at least one minute must elapse before working on the system. Failure to do so may result in deployment of the air bag.

2. Disconnect the negative battery cable. Tape the cable with insulating tape.

➡When disconnecting the cable, some systems need to be initialized after the cable is reconnected. You will need the Mazda diagnostic scan tool or equiva-

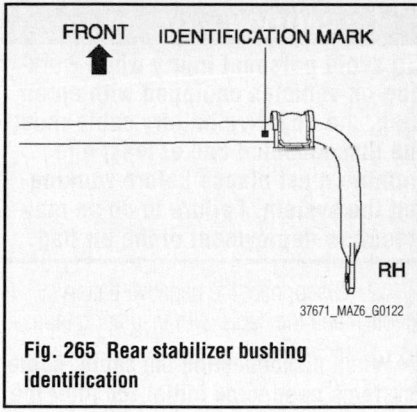

Fig. 265 Rear stabilizer bushing identification

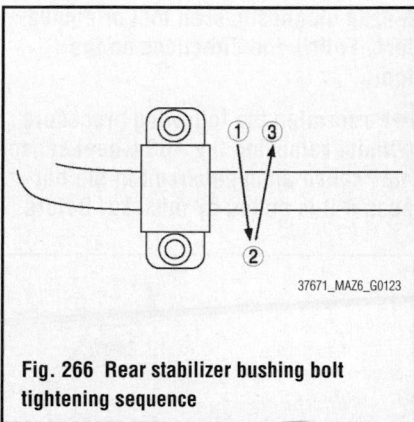

Fig. 266 Rear stabilizer bushing bolt tightening sequence

lent. Follow the directions on the tool.

3. Raise and safely support the vehicle.
4. Remove the tire and wheel assembly.
5. Remove the stabilizer control link.
6. Remove the stabilizer bracket.
7. Remove the bushing.
8. Remove the bar from its mounting.

To install:

➡Be sure to use new fasteners, as required.

9. Installation is the reverse of the removal procedure.
10. Use the Mazda diagnostic scan tool, or equivalent and reprogram the required systems.

WHEEL HUB & BEARING

REMOVAL & INSTALLATION

See Figure 267.

1. Before servicing the vehicle, refer to the Precautions Section.

➡If working near and/or around the SRS system and components, be sure to disable the SRS system. Tape the negative battery cable with insulating tape. Always disconnect the negative battery cable first.

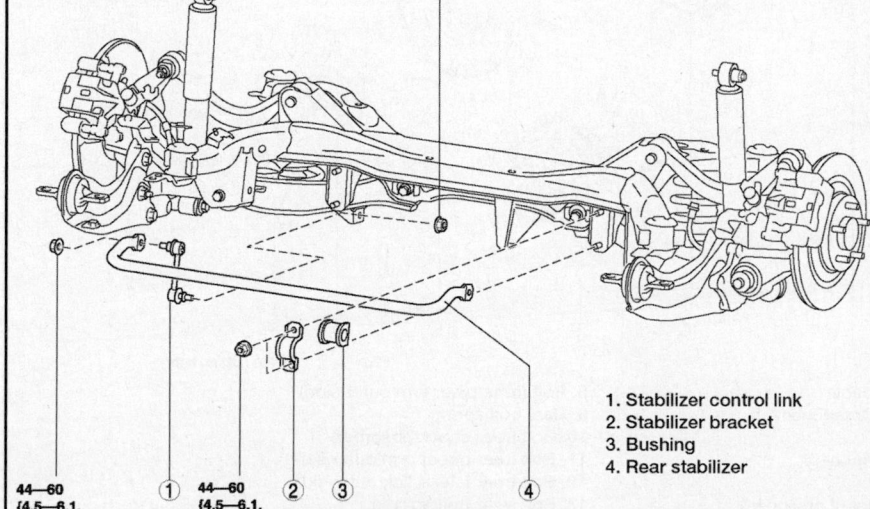

44—60
{4.5—6.1, 33—44}

44—60
{4.5—6.1, 33—44}

44—60
{4.5—6.1, 33—44}

N·m {kgf·m, ft-lbf}

1. Stabilizer control link
2. Stabilizer bracket
3. Bushing
4. Rear stabilizer

Fig. 263 Rear stabilizer bar and related components

✻✻ CAUTION

To avoid personal injury when working on vehicles equipped with an air bag, the negative battery cable must be disconnected and at least one minute must elapse before working on the system. Failure to do so may result in deployment of the air bag.

2. Disconnect the negative battery cable. Tape the cable with insulating tape.

➡**When disconnecting the cable, some systems need to be initialized after the cable is reconnected. You will need the Mazda diagnostic scan tool or equivalent. Follow the directions on the tool.**

➡**Performing the following procedure without removing the ABS wheel sensor may cause an open circuit in the harness if it is pulled by mistake. Before** performing the procedure remove the ABS wheel speed sensor (body side) and position it to an appropriate place where the sensor will not be pulled by mistake while servicing the vehicle.

3. Raise and support the vehicle safely.

4. Remove the tire and wheel assembly.

5. Remove the ABS sensor.

6. Remove the parking brake cable.

7. Remove the brake caliper. Position it to the side. Do not allow it to hang by the brake hose.

8. Remove the rotor.

9. Remove the wheel hub.

10. Remove the dust cover.

11. Remove the dynamic damper, if equipped.

12. Remove the stabilizer control link lower side bolt.

13. Remove the rear lower arm outer side bolt. Be sure to properly support the assembly using a suitable jack before removal.

14. Remove the rear spring.

15. Remove the shock bolt.

16. Remove the upper arm outer side bolt.

17. Remove the lateral link outer side bolt.

18. Remove the trailing link bolt.

19. Remove the hub assembly retaining bolts.

20. Remove the component from the vehicle.

To install:

➡**Be sure to use new fasteners, as required.**

21. Installation is the reverse of the removal procedure.

22. Check and adjust the alignment, as required.

23. Use the Mazda diagnostic scan tool, or equivalent and reprogram the required systems.

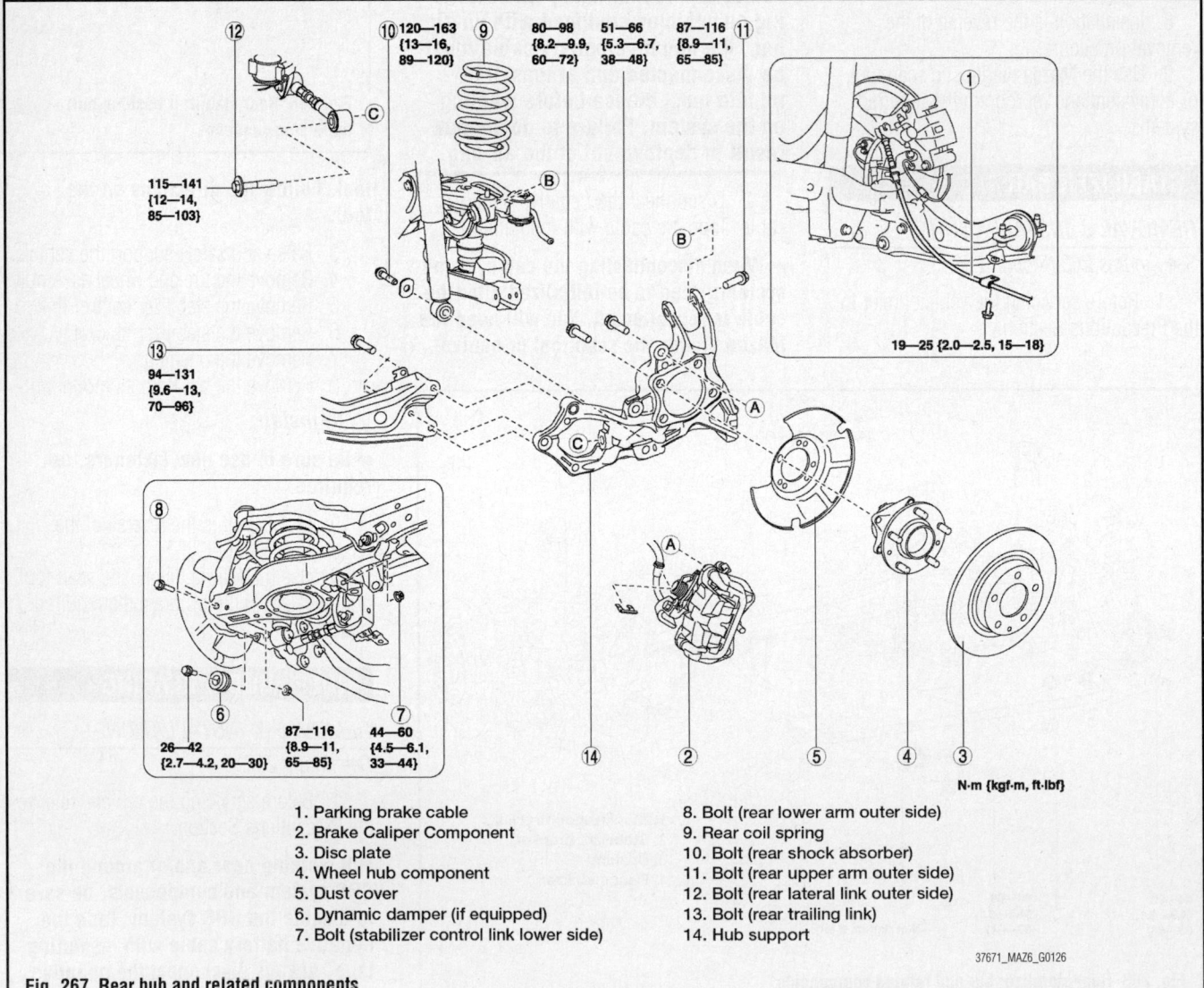

N·m {kgf·m, ft·lbf}

1. Parking brake cable
2. Brake Caliper Component
3. Disc plate
4. Wheel hub component
5. Dust cover
6. Dynamic damper (if equipped)
7. Bolt (stabilizer control link lower side)
8. Bolt (rear lower arm outer side)
9. Rear coil spring
10. Bolt (rear shock absorber)
11. Bolt (rear upper arm outer side)
12. Bolt (rear lateral link outer side)
13. Bolt (rear trailing link)
14. Hub support

37671_MAZ6_G0126

Fig. 267 Rear hub and related components

MAZDA

MX-5 Miata

SPECIFICATIONS AND MAINTENANCE CHARTS

ENGINE AND VEHICLE IDENTIFICATION

Engine							Model Year	
Code ①	Liters (cc)	Cu. In.	Cyl.	Fuel Sys.	Engine Type	Eng. Mfg.	Code ②	Year
LF	2.0 (1999)	121.9	4	MPFI	DOHC	Mazda	9	2009
							A	2010

MPFI: Multi-Point Fuel Injection

DOHC: Double Over Head Cam

① Located above the starter

② 10th digit of the Vehicle Identification Number (VIN)

37671_MIAT_C0001

GENERAL ENGINE SPECIFICATIONS

Year	Model	Engine Displacement Liters (VIN)	Net Horsepower @ rpm	Net Torque @ rpm (ft. lbs.)	Bore x Stroke (in.)	Com-pression Ratio	Oil Pressure @ rpm
2009	MX5-Miata	2.0 (LF)	167@7000	140@5000	3.44x3.27	10.8:1	49-85@3000
2010	MX5-Miata	2.0 (LF)	167@7000	140@5000	3.44x3.27	10.8:1	49-85@3000

37671_MIAT_C0002

ENGINE TUNE-UP SPECIFICATIONS

Year	Engine Displacement Liters (VIN)	Spark Plug Gap (in.)	Ignition Timing (deg.) MT	Ignition Timing (deg.) AT	Fuel Pump (psi)	Idle Speed (rpm) MT	Idle Speed (rpm) AT	Valve Clearance In.	Valve Clearance Ex.
2009	2.0 (LF)	0.050-0.053	8B	8B	51-59	700-800	700-800	0.008-0.0011	0.010-0.012
2010	2.0 (LF)	0.050-0.053	8B	8B	51-59	700-800	700-800	0.008-0.0011	0.010-0.012

NOTE: The Vehicle Emission Control Information label often reflects specification changes made during production.

The label figures must be used if they differ from those in this chart.

NA: Not Available

B: Before top dead center

37671_MIAT_C0003

CAPACITIES

| Year | Model | Engine Displacement Liters (VIN) | Engine Oil with Filter (qts.) | Transmission (pts.) | | Drive Axle | | Fuel Tank (gal.) | Cooling System (qts.) |
				Man	Auto.	Front (pts.)	Rear (pts.)		
2009	MX5-Miata	2.0 (LF)	4.7	4.2	15.6	—	1.7	12.7	7.9
2010	MX5-Miata	2.0 (LF)	4.7	4.2	15.6	—	1.7	12.7	7.9

NOTE: All capacities are approximate. Add fluid gradually and check to be sure a proper fluid level is obtained.

37671_MIAT_C0005

FLUID SPECIFICATIONS

Year	Model	Engine Displacement Liters	Engine ID/VIN	Engine Oil	Auto. Trans.	Manual Trans.	Rear Diff.	Brake Master Cylinder	Engine Coolant
2009	MX5-Miata	2.0 (LF)	F	5W-20	①	75W-90	75W-90	DOT 3	Ethylene-glycol
2010	MX5-Miata	2.0 (LF)	F	5W-20	①	75W-90	75W-90	DOT 3	Ethylene-glycol

DOT: Department Of Transpotation
A Mazda Genuine JWS3309

37671_MIAT_C0004

VALVE SPECIFICATIONS

| Year | Engine Displacement Liters (VIN) | Seat Angle (deg.) | Face Angle (deg.) | Maximum out of Square (in.) | Spring Free Length (in.) | Stem-to-Guide Clearance (in.) | | Stem Diameter (in.) | |
						Intake	Exhaust	Intake	Exhaust
2009	2.0 (LF)	NA	NA	NA	NA	NA	NA	NA	NA
2010	2.0 (LF)	NA	NA	NA	NA	NA	NA	NA	NA

NA: Not Available

37671_MIAT_C0006

CAMSHAFT AND BEARING SPECIFICATIONS

All measurements are given in inches.

Year	Engine Displacement Liters	Engine VIN	Journal Diameter	Brg. Oil Clearance	Shaft End-play	Runout	Journal Bore	Lobe Lift Intake	Lobe Lift Exhaust
2009	2.0 (LF)	LF	NA	NA	NA	NA	NA	NA	NA
2010	2.0 (LF)	LF	NA	NA	NA	NA	NA	NA	NA

NA Not Available

37671_MIAT_C0014

CRANKSHAFT AND CONNECTING ROD SPECIFICATIONS

All measurements are given in inches.

Year	Engine Displacement Liters (VIN)	Crankshaft Main Brg. Journal Dia.	Crankshaft Main Brg. Oil Clearance	Crankshaft Shaft End-play	Crankshaft Thrust on No.	Connecting Rod Journal Diameter	Connecting Rod Oil Clearance	Connecting Rod Side Clearance
2009	2.0 (LF)	NA	NA	NA	NA	NA	NA	NA
2010	2.0 (LF)	NA	NA	NA	NA	NA	NA	NA

NA: Not Avilable

37671_MIAT_C0007

PISTON AND RING SPECIFICATIONS

All measurements are given in inches.

Year	Engine Displacement Liters (VIN)	Piston Clearance	Ring Gap Top Compression	Ring Gap Bottom Compression	Ring Gap Oil Control	Ring Side Clearance Top Compression	Ring Side Clearance Bottom Compression	Ring Side Clearance Oil Control
2009	2.0 (LF)	NA	NA	NA	NA	NA	NA	NA
2010	2.0 (LF)	NA	NA	NA	NA	NA	NA	NA

NA: Not Available

37671_MIAT_C0008

TORQUE SPECIFICATIONS
All readings in ft. lbs.

Year	Engine Displacement Liters (VIN)	Cylinder Head Bolts	Main Bearing Bolts	Rod Bearing Bolts	Crankshaft Damper Bolts	Flywheel Bolts	Manifold Intake	Manifold Exhaust	Spark Plugs	Oil Pan Drain Plug
2009	2.0 (LF)	①	NA	NA	②	③	⑭①	32-47	8-10	23-30
2010	2.0 (LF)	①	NA	NA	②	③	⑭①	32-47	8-10	23-30

NA: Not Available

① Step 1: 44.3 inch lbs.

Step 2: Tighten 10.0 - 12.5 ft. lbs

Step 3: Tighten 32.5 - 33.9 ft. lbs.

Step 4: Tighten 88 - 92 degrees

Step 5: Tighten 88 - 92 degrees

② 71-77 ft. lbs. plus 87-93 degrees

③ Manual transmission: 80-85 ft. lbs.

Automatic transmission: 71-76 ft. lbs.

37671_MIAT_C0009

WHEEL ALIGNMENT

Year	Model		Caster Range (+/-Deg.)	Caster Preferred Setting (Deg.)	Camber Range (+/-Deg.)	Camber Preferred Setting (Deg.)	Toe-in (in.)
2009	MX5-Miata	F	1.00	①	1.00	②	0.063+/-0.15
		R	—	—	1.00	③	0.12+/-0.15
2010	MX5-Miata	F	1.00	①	1.00	②	0.063+/-0.15
		R	—	—	1.00	③	0.12+/-0.15

NOTE: Unloaded vehicle: Fuel tank is full. Engine coolant and engine oil are at specified level. Jack and tools are in designated position.

① Vehicle height: From the end of the rear fender to the center of the wheel (17 inch wheels)

13.8-14.0 inches: 6 Deg. 32'

14.2-14.4 inches: 6 Deg. 18'

14.6-14.8 inches: 6 Deg. 04'

15.0-15.2 inches: 5 Deg. 50'

15.4-15.6 inches: 5 Deg. 36'

② Vehicle height: From the end of the front fender to the center of the wheel (17 inch wheels)

13.8-14.1 inches: -0 Deg. 42'

14.3-14.5 inches: -0 Deg. 26'

14.7-14.9 inches: -0 Deg. 13'

15.0-15.3 inches: -0 Deg. 01'

15.4-15.7 inches: 0 Deg. 08'

③ Vehicle height: From the end of the front fender to the center of the wheel (17 inch wheels)

14.0-14.2 inches: -1 Deg.33'

14.4-14.6 inches: -1 Deg. 18'

14.8-15.0 inches: -1 Deg. 04'

15.2-15.4 inches: -0 Deg. 54'

15.6-15.8 inches: -0 Deg. 45'

37671_MIAT_C0010

TIRE, WHEEL AND BALL JOINT SPECIFICATIONS

| Year | Model | OEM Tires | | Tire Pressures (psi) | | Wheel Size | Ball Joint Inspection | Lug Nut |
		Standard	Optional	Front	Rear			
2009	MX5-Miata	P205/50R16	P205/45R17	29	29	①	⑭①	65-87
2010	MX5-Miata	P205/50R16	P205/45R17	29	29	①	⑭①	65-87

① Standard rim: 16 x 0 1/2 J

Optional rim:17 x 7 J

NA Not Available

37671_MIAT_C0011

BRAKE SPECIFICATIONS

All measurements in inches unless noted

| Year | Model | | Brake Disc | | | Brake Drum | | | Minimum Lining Thickness | Brake Caliper | |
			Original Thickness	Minimum Thickness	Maximum Runout	Original Inside Diameter	Max. Wear Limit	Maximum Machine Diameter		Bracket Bolts (ft. lbs.)	Mounting Bolts (ft. lbs.)
2009	MX5-Miata	F	NA	0.790	0.002	—	—	—	0.079	58-75	15-23
		R	NA	0.310	0.002	—	—	—	0.079	36-50	14-18
2010	MX5-Miata	F	NA	0.790	0.002	—	—	—	0.079	58-75	15-23
		R	NA	0.310	0.002	—	—	—	0.079	36-50	14-18

NA: Not Avilable

F: Front

R: Rear

37671_MIAT_C0012

SCHEDULED MAINTENANCE INTERVALS
2009-10 Mazda MX5-Miata

TO BE SERVICED	TYPE OF SERVICE	VEHICLE MILEAGE INTERVAL (x1000)												
		7.5	15	22.5	30	37.5	45	52.5	60	67.5	75	82.5	90	97.5
Engine oil & filter	R	✓	✓	✓	✓	✓	✓	✓	✓	✓	✓	✓	✓	✓
Air cleaner element	R					✓					✓			
Engine coolant ①	R													
Spark plugs	R										✓			
Bolts & nuts on chassis & body	S/I				✓				✓				✓	
Brake lines, hoses & connections	S/I				✓				✓				✓	
Cooling system	S/I				✓				✓				✓	
Cabin filter	R		✓											
Disc brakes	S/I				✓				✓				✓	
Drive belts B	S/I				✓				✓				✓	
Drive shaft dust boots	S/I				✓				✓				✓	
Exhaust system heat shield	S/I				✓				✓				✓	
Front & rear suspension ball joints	S/I				✓				✓				✓	
Fuel lines & hoses	S/I				✓				✓				✓	
Steering operation & linkages	S/I				✓				✓				✓	
Fuel filter	R								✓					
Valve clearance ③	I								✓					
Hose & tube for emission	S/I								✓					

R: Replace   S/I: Service or Inspect

① Models with FL22 long-life coolant can be identified by the FL22 marking on the cooling system cap label. It is recommended that FL22 coolant continue to be used for vehicles originally filled with FL22 coolant from the factory.

② 2.0L LF engine: Inspect every 37, 500 miles

③ 2.0L LF: Audible inspect every 75, 000 miles and if noisy adjust.

FREQUENT OPERATION MAINTENANCE (SEVERE SERVICE)

If a vehicle is operated under any of the following conditions it is considered severe service

- Extremely dusty areas.

- 50% or more of the vehicle operation is in 32°C (90°F) or higher temperatures, or constant operation in temperatures below 0°C (32°F).

- Prolonged idling (vehicle operation in stop and go traffic).

- Frequent short running periods (engine does not warm to normal operating temperatures).

- Police, taxi, delivery usage or trailer towing usage.

Oil & oil filter: change every 5000 miles.

Oil & oil filter (Puerto Rico): change every 3000 miles.

Air cleaner element: service or inspect every 15,000 miles

Automatic transaxle fluid: service or inspect every 15,000 miles.

Bolts & nuts on chassis & body: tighten every 15,000 miles.

Disc brakes: service or inspect every 15,000 miles.

37671_MIAT_C0013

BRAKES   INFORMATION AND PRECAUTIONS

ANTI-LOCK SYSTEMS

• Certain components within the ABS system are not intended to be serviced or repaired individually.

• Do not use rubber hoses or other parts not specifically specified for and ABS system. When using repair kits, replace all parts included in the kit. Partial or incorrect repair may lead to functional problems and require the replacement of components.

• Lubricate rubber parts with clean, fresh brake fluid to ease assembly. Do not use shop air to clean parts; damage to rubber components may result.

• Use only DOT 3 brake fluid from an unopened container.

• If any hydraulic component or line is removed or replaced, it may be necessary to bleed the entire system.

• A clean repair area is essential. Always clean the reservoir and cap thoroughly before removing the cap. The slightest amount of dirt in the fluid may plug an orifice and impair the system function. Perform repairs after components have been thoroughly cleaned; use only denatured alcohol to clean components. Do not allow ABS components to come into contact with any substance containing mineral oil; this includes used shop rags.

• The Anti-Lock control unit is a microprocessor similar to other computer units in the vehicle. Ensure that the ignition switch is **OFF** before removing or installing controller harnesses. Avoid static electricity discharge at or near the controller.

• If any arc welding is to be done on the vehicle, the control unit should be unplugged before welding operations begin.

DISC AND DRUM SYSTEMS

✳✳ CAUTION

Dust and dirt accumulating on brake parts during normal use may contain asbestos fibers from production or aftermarket brake linings. Breathing excessive concentrations of asbestos fibers can cause serious bodily harm. Exercise care when servicing brake parts. Do not sand or grind brake lining unless equipment used is designed to contain the dust residue. Do not clean brake parts with compressed air or by dry brushing. Cleaning should be done by dampening the brake components with a fine mist of water, then wiping the brake components clean with a dampened cloth. Dispose of cloth and all residue containing asbestos fibers in an impermeable container with the appropriate label. Follow practices prescribed by the Occupational Safety and Health Administration (OSHA) and the Environmental Protection Agency (EPA) for the handling, processing, and disposing of dust or debris that may contain asbestos fibers.

BRAKES   BLEEDING THE BRAKE SYSTEM

BLEEDING PROCEDURE

BLEEDING THE ABS SYSTEM
See Figure 1.

➡ **Do not reuse the drained fluid. Use only clean DOT 3 Brake Fluid from an unopened container.**

✳✳ WARNING
Make sure no dirt or other foreign matter is allowed to contaminate the brake fluid.

✳✳ WARNING
Do not spill brake fluid on the vehicle, it may damage the paint; if brake fluid does contact the paint, wash it off immediately with water.

1. The reservoir on the master cylinder must be at the MAX level mark at the start of the bleeding procedure and checked after bleeding each brake caliper. Add fluid as required.

2. Make sure the brake fluid level in the reservoir is at the MAX level line.

3. Slide a piece of clear plastic hose over the first bleed screw, and submerge the other end in a container of new brake fluid.

4. Have someone slowly pump the brake pedal several times, then apply steady pressure.

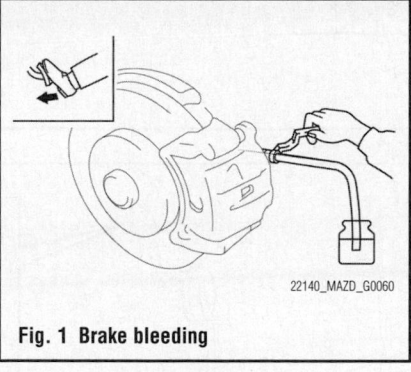
22140_MAZD_G0060

Fig. 1 Brake bleeding

5. Bleed the hydraulic brake system in the following sequence:
   a. Right rear bleeder valve
   b. Left rear bleeder valve
   c. Right front bleeder valve
   d. Left front bleeder valve

6. Repeat the procedure for each wheel in the until air bubbles no longer appear in the fluid.

7. Refill the master cylinder reservoir to the MAX level line.

➡ **Do not reuse the drained fluid. Use only clean DOT 3 Brake Fluid from an unopened container.**

✳✳ WARNING
Make sure no dirt or other foreign matter is allowed to contaminate the brake fluid.

✳✳ WARNING
Do not spill brake fluid on the vehicle, it may damage the paint; if brake fluid does contact the paint, wash it off immediately with water.

8. The reservoir on the master cylinder must be at the MAX level mark at the start of the bleeding procedure and checked after bleeding each brake caliper. Add fluid as required.

9. Make sure the brake fluid level in the reservoir is at the MAX level line.

10. Slide a piece of clear plastic hose over the first bleed screw, and submerge the other end in a container of new brake fluid.

11. Have someone slowly pump the brake pedal several times, then apply steady pressure.

12. Bleed the hydraulic brake system in the following sequence:
   a. Right rear bleeder valve
   b. Left rear bleeder valve
   c. Right front bleeder valve
   d. Left front bleeder valve

13. Repeat the procedure for each wheel in the until air bubbles no longer appear in the fluid.

14. Refill the master cylinder reservoir to the MAX level line.

BRAKES   ANTI-LOCK BRAKE SYSTEM (ABS)

SPEED SENSORS

REMOVAL & INSTALLATION

Front
See Figure 2.

1. Remove or disconnect the following:
   - Wheel
   - Mudguard

- Connector
- Bolt
- Front ABS speed sensor

To install:
2. Install or connect the following:
   - Front ABS speed sensor
   - Bolt
   - Connector
   - Mudguard
   - Wheel

Rear
See Figure 3.

1. Remove the trunk end trim.
2. Remove the partition board.
3. Remove the trunk side trim.
4. Remove the fuel-filler pipe protector.
5. Remove in the order indicated in graphic.
6. Install in the reverse order of removal.

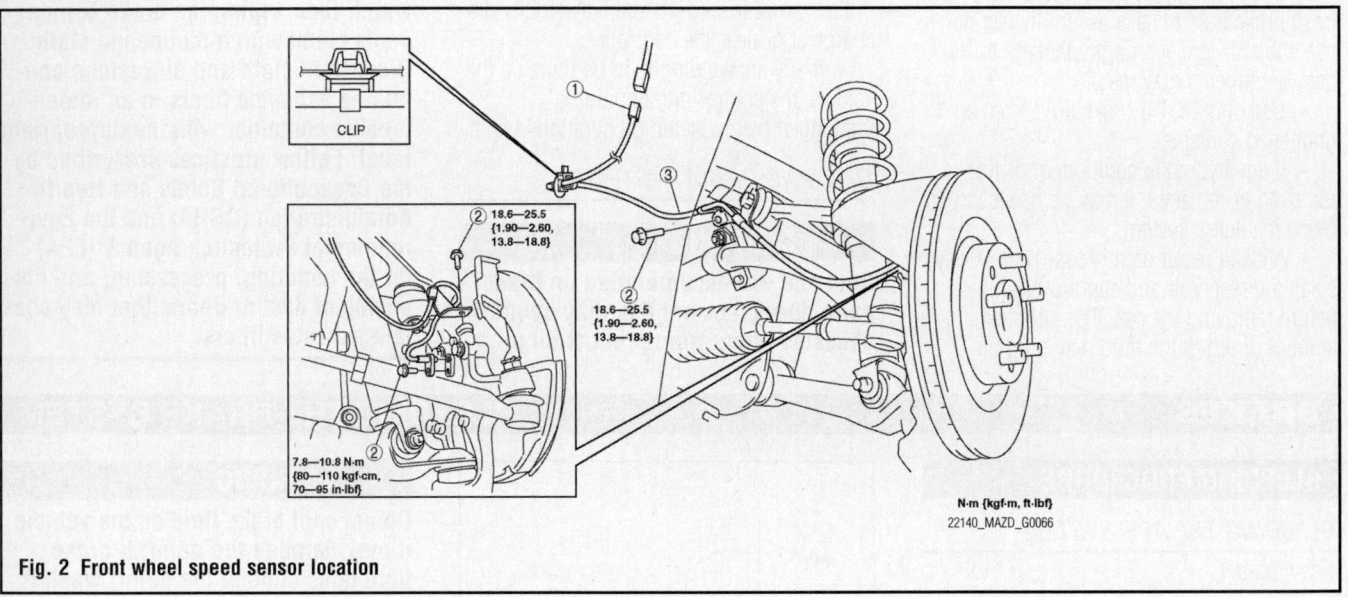

N·m {kgf·m, ft·lbf}
22140_MAZD_G0066

Fig. 2 Front wheel speed sensor location

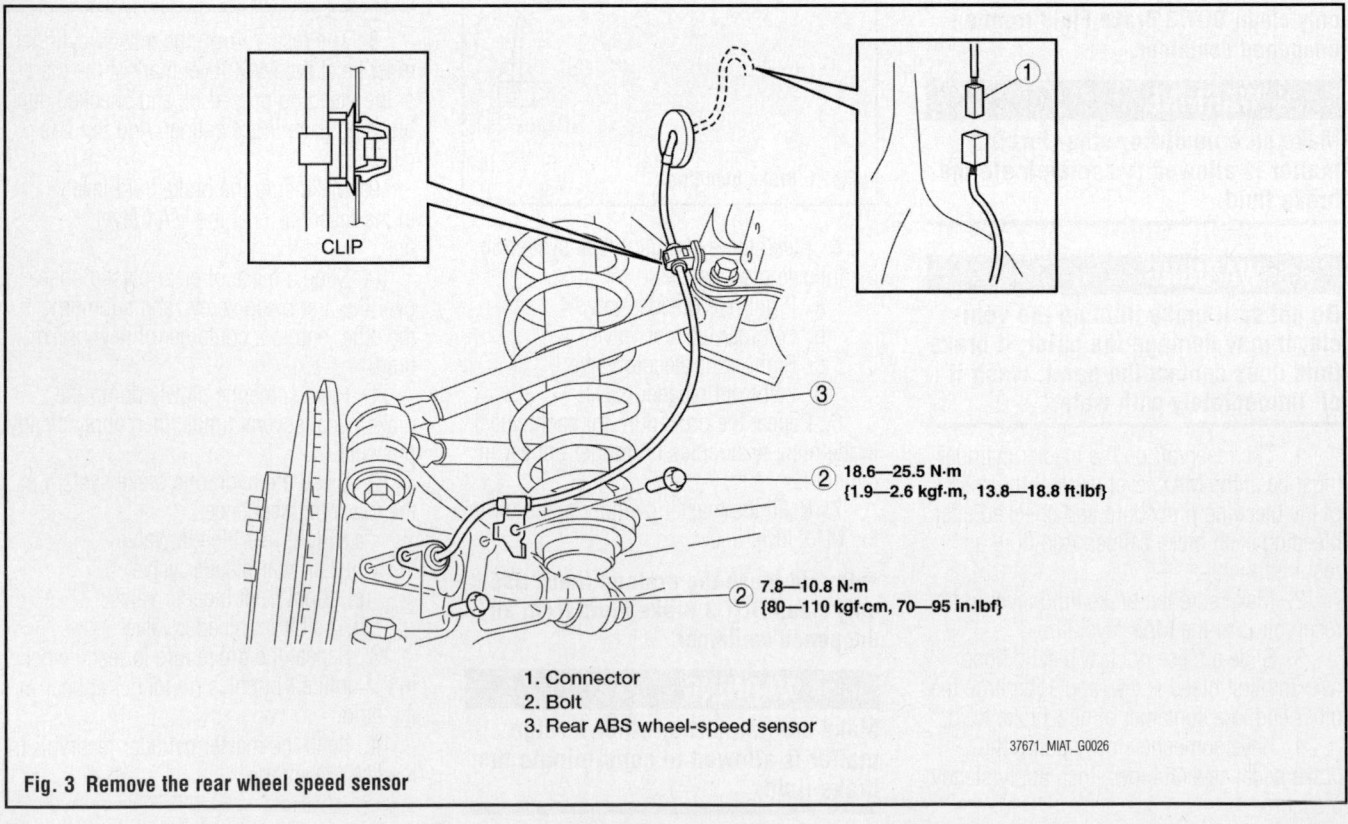

1. Connector
2. Bolt
3. Rear ABS wheel-speed sensor

37671_MIAT_G0026

Fig. 3 Remove the rear wheel speed sensor

BRAKES   **FRONT DISC BRAKES**

BRAKE CALIPER

REMOVAL & INSTALLATION

See Figures 4 and 5.

➡**Refer the exploded view illustration for component locations and if applicable, their retainer torque specifications**

1. Disconnect the brake hose.
2. Remove the caliper bolts and the caliper.

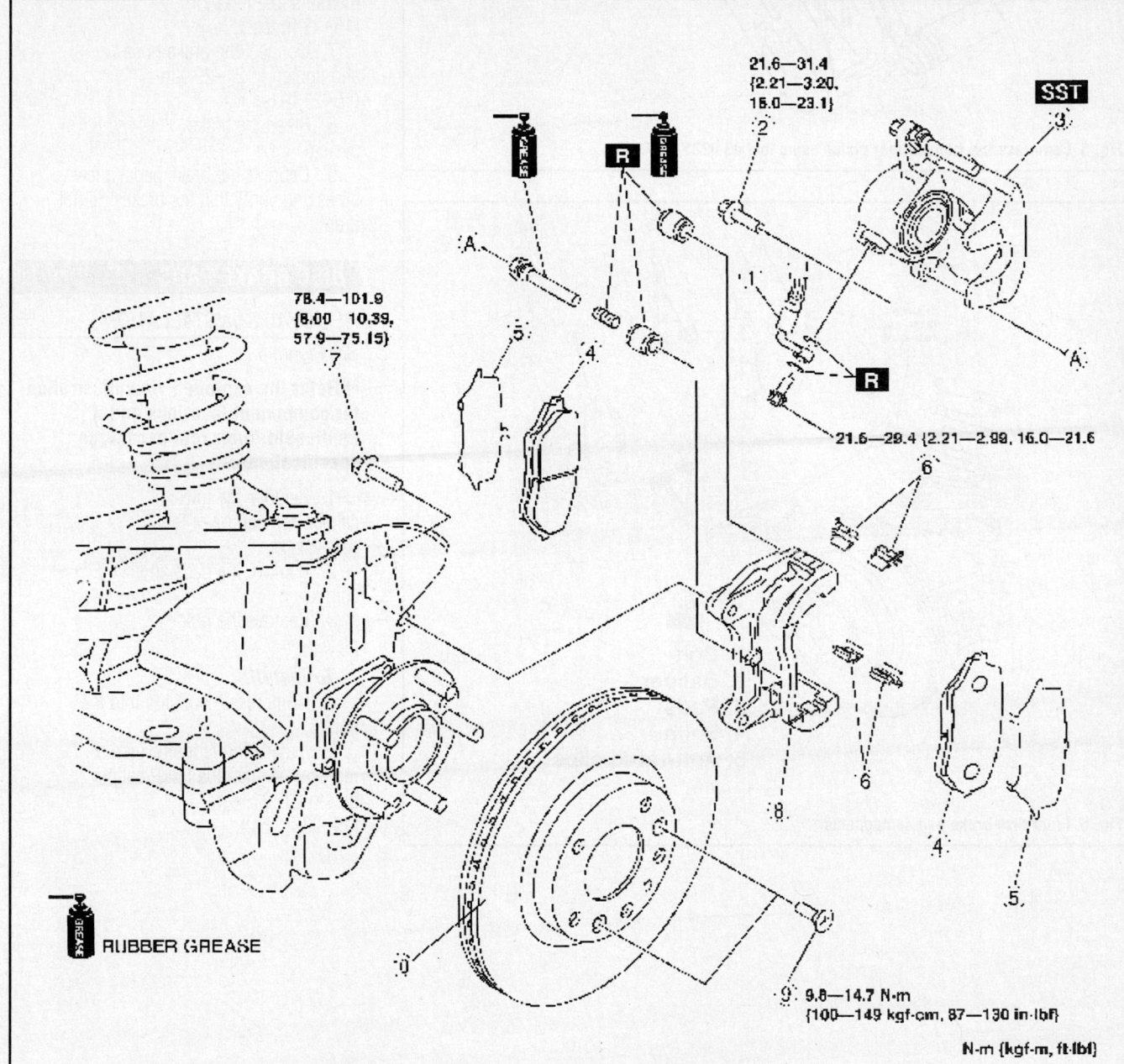

1	Brake hose	6	Guide plate
2	Bolt	7	Bolt
3	Caliper	8	Mounting support
4	Disc pad	9	Screw
5	Shim	10	Disc plate

09482_MAZC2_G0123

Fig. 4 Exploded view of the front disc brake assembly

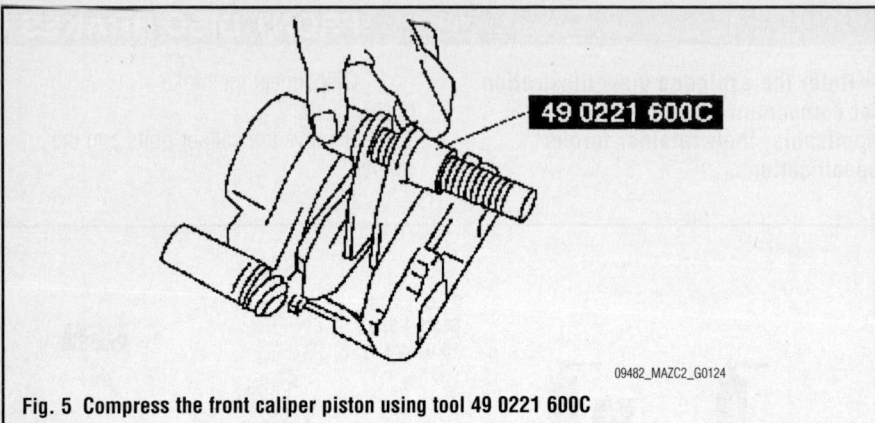

09482_MAZC2_G0124

Fig. 5 Compress the front caliper piston using tool 49 0221 600C

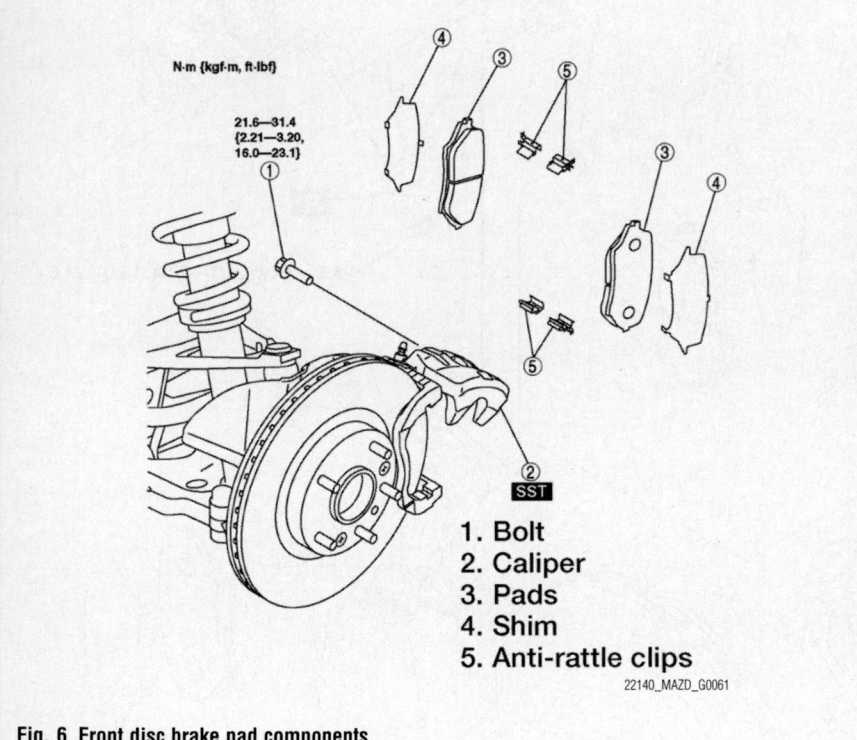

N·m {kgf·m, ft·lbf}

21.6—31.4
{2.21—3.20,
16.0—23.1}

1. Bolt
2. Caliper
3. Pads
4. Shim
5. Anti-rattle clips

22140_MAZD_G0061

Fig. 6 Front disc brake pad components

3. Remove the disc pads and shims.

To install:

4. Clean the exposed area of the piston.
5. Compress the piston using tool 49 0221 600C
6. Install the caliper. Tighten the bolts to 21—31 Nm (15—23 ft. lbs.).
7. Connect the brake hose and tighten to 21—29 Nm (15—21 ft. lbs.).
8. Bleed the brake system.
9. Depress the brake pedal a few times and verify that the brakes do not drag

DISC BRAKE PADS

REMOVAL & INSTALLATION

See Figure 6.

➡Refer the exploded view illustration for component locations and if applicable, their retainer torque specifications.

1. Remove the lower caliper bolt and pivot the caliper up.
2. Remove the pads and shims.
3. Remove the guide plate.

To install:

4. Compress the piston into the bore.
5. Installation is the reverse of removal. Tighten the lower caliper bolt to 21—31 Nm (15—23 ft. lbs.).

BRAKES   **REAR DISC BRAKES**

BRAKE CALIPER

REMOVAL & INSTALLATION

See Figures 7 through 9.

➡**Refer the exploded view illustration for component locations and if applicable, their retainer torque specifications.**

1. Disconnect the parking brake cable.

2. Disconnect the brake hose.
3. Remove the caliper bolts and the caliper.
4. Remove the disc pads and shims.

To install:

5. Clean the exposed area of the piston.

6. Rotate the piston clockwise slowly using the tool shown to push the piston completely until the piston grooves are in the position shown .

7. Install the caliper. Tighten the bolts to 20–25 Nm (14–18 ft. lbs.).

8. Connect the brake hose and tighten to 21–29 Nm (15–21 ft. lbs.).

9. Bleed the brake system.

10. Install the parking brake cable. After installation the parking brake cable, verify that the operating lever returns to the stopper nut with the parking brake lever released.

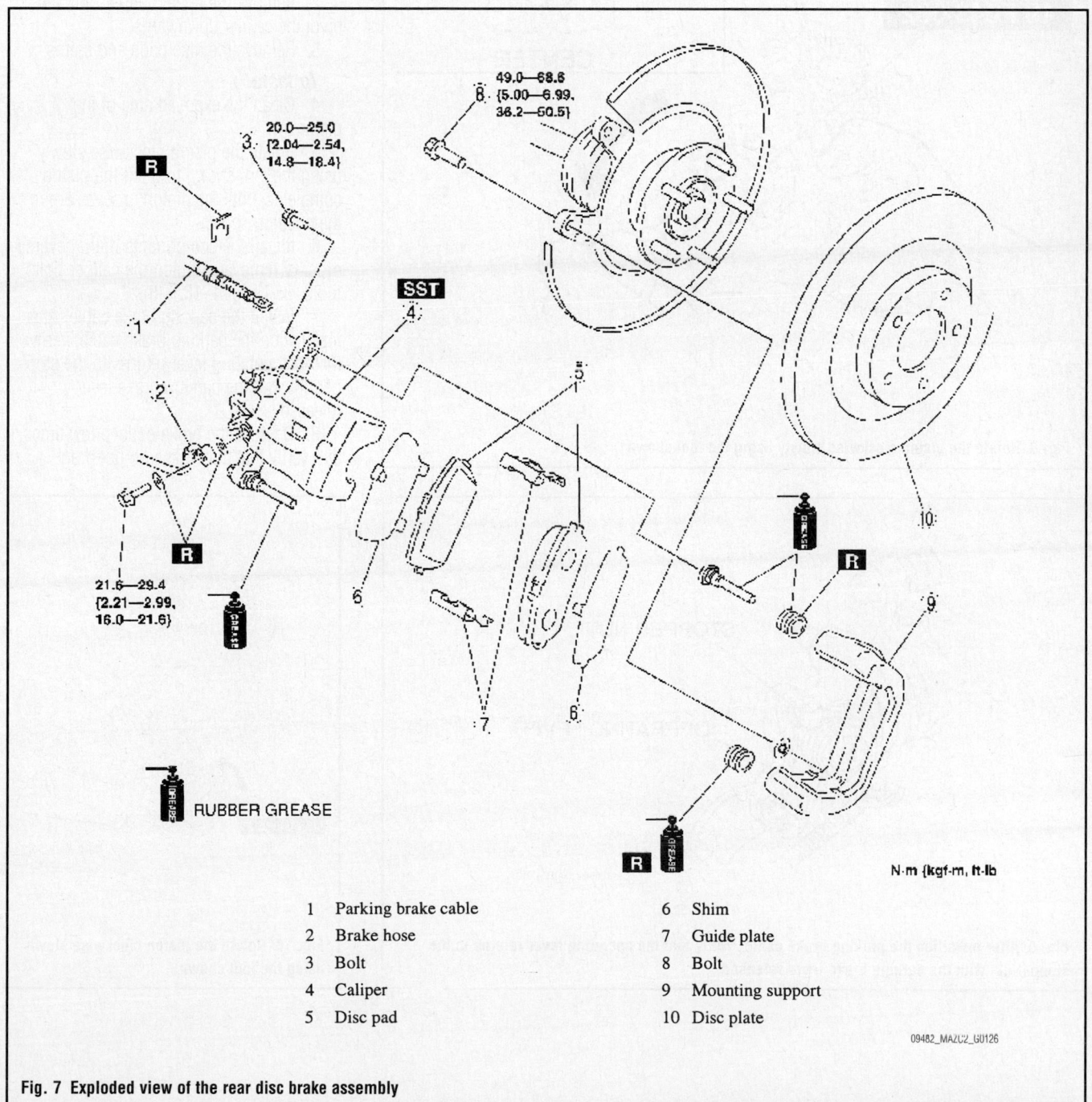

1	Parking brake cable	6	Shim
2	Brake hose	7	Guide plate
3	Bolt	8	Bolt
4	Caliper	9	Mounting support
5	Disc pad	10	Disc plate

09482_MAZC2_G0126

Fig. 7 Exploded view of the rear disc brake assembly

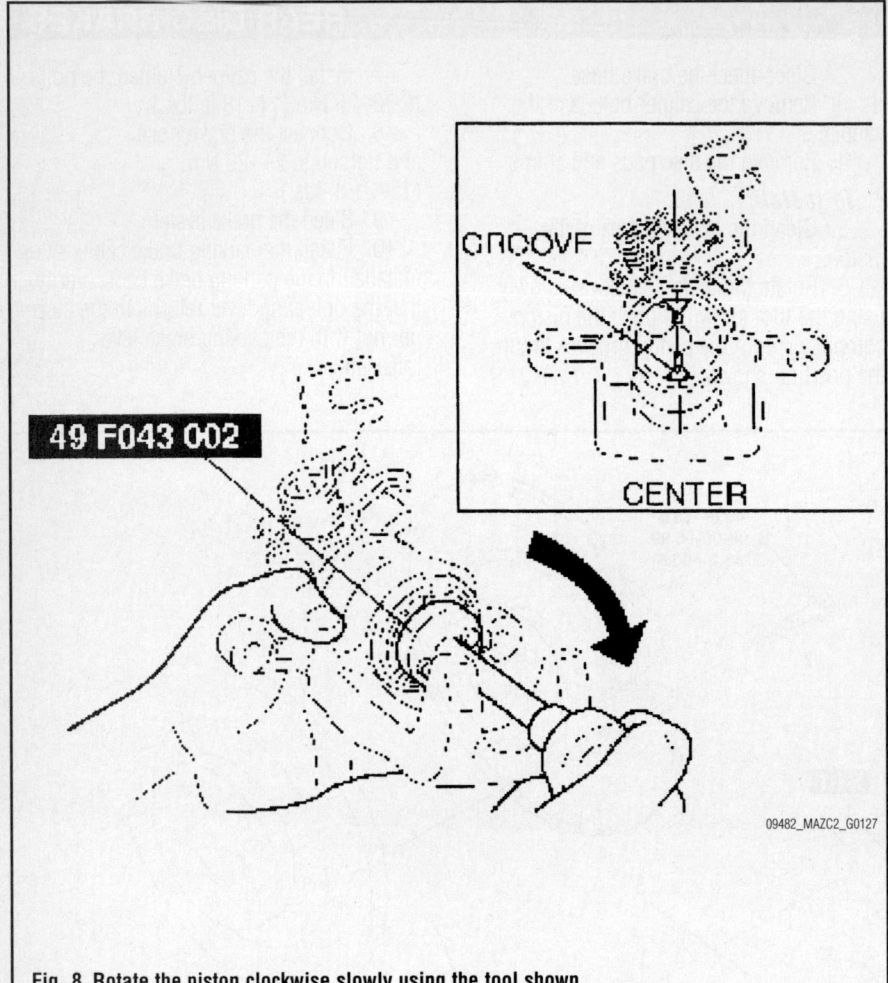

Fig. 8 Rotate the piston clockwise slowly using the tool shown

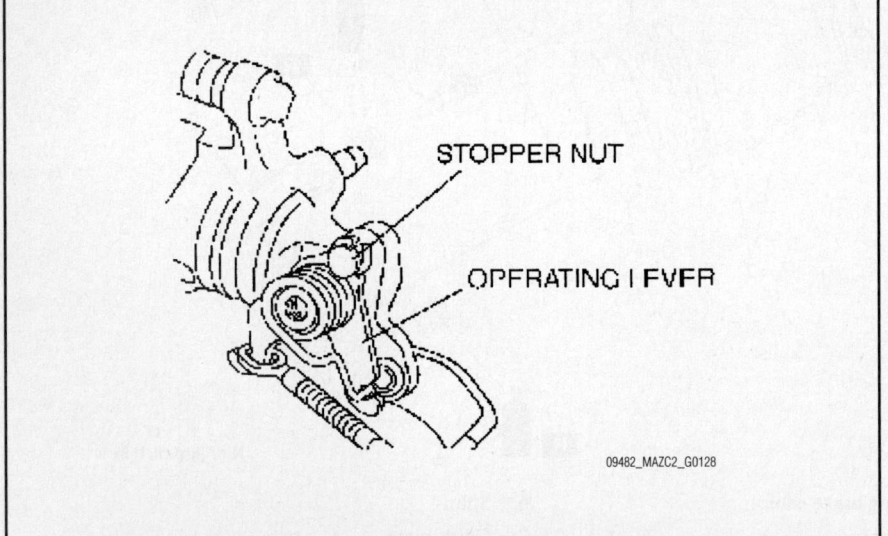

Fig. 9 After installing the parking brake cable, verify that the operating lever returns to the stopper nut with the parking brake lever released

11. Depress the brake pedal a few times and verify that the brakes do not drag.

DISC BRAKE PADS

REMOVAL & INSTALLATION
See Figure 10.

➡**Refer the exploded view illustration for component locations and if applicable, their retainer torque specifications.**

1. Disconnect the parking brake cable.
2. Remove the upper caliper bolt and pivot the caliper downwards.
3. Remove the disc pads and shims.

To install:

4. Clean the exposed area of the piston.
5. Rotate the piston clockwise slowly using the tool shown to push the piston completely until the piston grooves are in the position shown .
6. Install the components in the reverse order of removal. Tighten the caliper bolts to 20–25 Nm (14–18 ft. lbs.).
7. Install the parking brake cable. After installation the parking brake cable, verify that the operating lever returns to the stopper nut with the parking brake lever released.
8. Depress the brake pedal a few times and verify that the brakes do not drag.

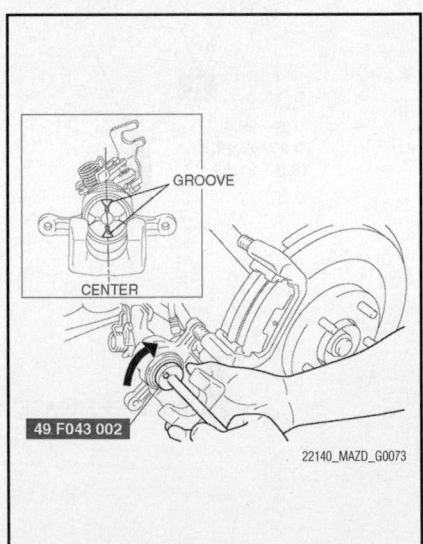

Fig. 10 Rotate the piston clockwise slowly using the tool shown

BRAKES   **PARKING BRAKE**

PARKING BRAKE CABLES

ADJUSTMENT

See Figure 11.

1. Before servicing the vehicle, refer to the precautions section.
2. Depress the brake pedal several times.
3. Remove the parking brake lever boot
4. Turn the adjusting nut and adjust the parking brake lever.
5. After adjustment, pull the parking brake lever one notch and verify that the parking brake warning light illuminates.
6. Verify that the rear brakes do not drag.

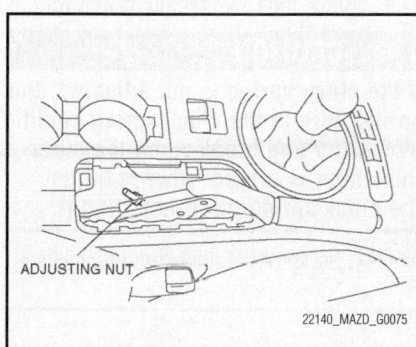

ADJUSTING NUT

22140_MAZD_G0075

Fig. 11 Turn the adjusting nut and adjust the parking brake

PARKING BRAKE SHOES

REMOVAL & INSTALLATION

See Figure 12.

The parking brake system uses the rear brake pads as the parking brake. Refer to rear brake shoe removal and installation in this section. The parking brake cable actuates the rear caliper to apply and hold the brake pads.

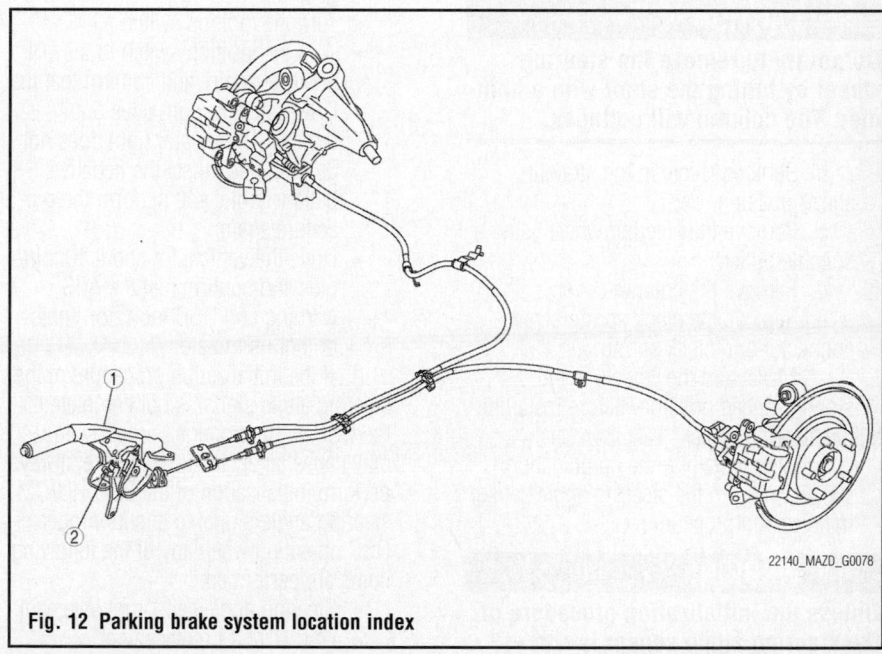

22140_MAZD_G0078

Fig. 12 Parking brake system location index

1. Before servicing the vehicle, refer to the precautions section.
2. Remove or disconnect the following:
   - Wheels
   - Parking brake cable at the lever and caliper

To install:

3. Installation is the reverse of removal.

CHASSIS ELECTRICAL   **AIR BAG (SUPPLEMENTAL RESTRAINT SYSTEM)**

GENERAL INFORMATION

✳✳ CAUTION

These vehicles are equipped with an air bag system. The system must be disarmed before performing service on, or around, system components, the steering column, instrument panel components, wiring and sensors. Failure to follow the safety precautions and the disarming procedure could result in accidental air bag deployment, possible injury and unnecessary system repairs.

SERVICE PRECAUTIONS

✳ CAUTION

Disconnect and isolate the battery negative cable before beginning

any airbag system component diagnosis, testing, removal, or installation procedures. Wait at least 90 seconds after the ignition switch is turned off and the negative (-) terminal cable is disconnected from the battery before starting the operation. The SRS is equipped with a backup power source, so if work is started within 90 seconds after disconnecting the negative (-) terminal cable from the battery, the SRS may be deployed. Failure to disable the airbag system may result in accidental airbag deployment, personal injury, or death.

DISARMING THE SYSTEM

1. Before servicing the vehicle, refer to the precautions section.

2. If equipped, deactivate the audio anti-theft system.
3. Turn the ignition switch to LOCK.
4. Disconnect and isolate the negative battery cable and wait for more than 1 minute to allow the backup power supply to deplete its stored power.

ARMING THE SYSTEM

1. Before servicing the vehicle, refer to the precautions section.
2. Connect the negative battery cable, turn the ignition switch **ON** and verify the air bag warning light cones on for 6 seconds. If the light does not illuminate there are problems with the system.
3. If equipped, activate the audio anti-theft system.

CLOCKSPRING CENTERING

See Figures 13 through 15.

> **⁂ WARNING**
>
> **Handling the air bag module improperly can accidentally deploy the air bag module, which may seriously injure you.**

1. Before servicing the vehicle, refer to the precautions section.
2. Remove the steering wheel as follows:
    a. Remove the locknut.

> **⁂ CAUTION**
>
> **Do not try to remove the steering wheel by hitting the shaft with a hammer. The column will collapse.**

b. Set the vehicle in the straight-ahead position.

c. Remove the steering wheel using a suitable puller.

d. Remove the column cover.

e. Remove the clock spring screw, connector and clock spring.

f. Make sure the wheels in the straight-ahead position, before installing the steering wheel.

g. After installing the combination switch, perform the steering angle sensor initialization procedure

> **⁂ WARNING**
>
> **Unless the initialization procedure of the steering angle sensor is completed, the DSC will not operate, causing an unexpected accident. Therefore, always perform the initialization procedure to ensure DSC operation if the power supply to the steering angle sensor has been cut off due to disconnection of the steering angle sensor connector or negative battery cable, or any other cause.**

➡The initialization value of the steering angle sensor is stored using the battery power supply. Therefore, the battery power supply of the steering angle sensor is cut and the stored initialization value is cleared when any of the following items are performed.

- Negative battery cable disconnection
- Steering angle sensor connector disconnection
- Fuse (ROOM 15A) removal
- Wiring harness disconnection between battery and steering angle sensor connector
- Inspect the wheel alignment, infla-

tion pressure, and the installation condition of the steering wheel
- If there is any malfunction, adjust the applicable part
- Connect the negative battery cable
- Turn the ignition switch to the ON position
- Confirm that the DSC indicator light illuminates and that the DSC OFF light flashes
- Turn the steering wheel to full right lock, then turn it to full left lock
- Confirm that the DSC OFF light goes out
- Turn the ignition switch off
- Turn the ignition switch to the ON position again, and confirm that the DSC indicator light goes out
- If the DSC indicator light does not go out, disconnect the negative battery cable, and perform the procedure again
- Drive the vehicle for about 10 minutes and confirm that the ABS warning and DSC indicator lights do not illuminate

h. If the initialization procedure of the steering angle signal is not completed, the DSC will not operate properly and may cause an accident. Therefore, always perform initialization of the DSC HU/CM steering angle signal to ensure proper DSC operation when any of the following items are performed.

- Steering angle sensor replacement
- DSC HU/CM replacement
- Inspect the wheel alignment and inflation pressure.
- Park the vehicle on level ground
- Turn the ignition switch off
- Connect the WDS or equivalent to the DLC-2
- Access the active command mode, select the following commands, and then follow the indication on the monitor

- Drive the vehicle forward
- After 5 min of driving, verify that the DSC system is normal

i. When installing the center panel unit, make sure that the wiring harness and antenna feeder are not caught between the unit and dashboard. If the wiring harness or the antenna feeder is caught between the unit and dashboard, it may cause malfunctions.

j. Turn the ignition switch to the ON position.

k. Verify that the air bag system warning light illuminates for approximately 6 seconds and goes out.

l. If the air bag system warning light does not operate normally, inspect the system.

To install:

3. Install the clock spring.
4. Adjust the clock spring as follows:

> **⁂ CAUTION**
>
> **If the clock spring is not adjusted, the spring wire in the clock spring could over-wind and break when the steering wheel is turned. Always adjust the clock spring after installing it.**

- Set the front tires straight-ahead

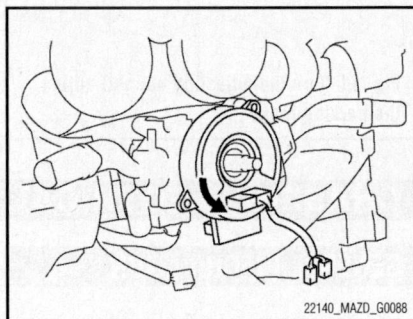

22140_MAZD_G0088

Fig. 14 From the stopped position, turn the clock spring counterclockwise 2 ¾ turns

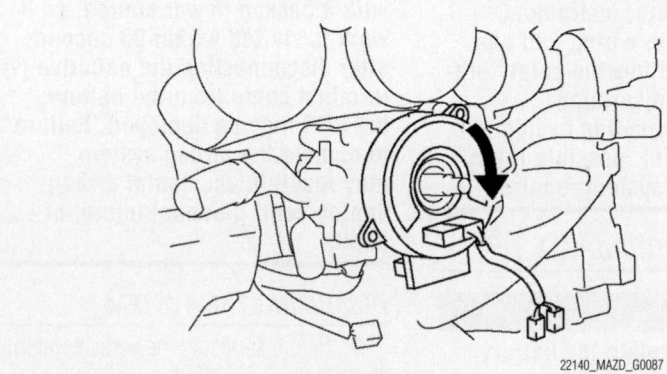

22140_MAZD_G0087

Fig. 13 Turn the clock spring clockwise until it stops

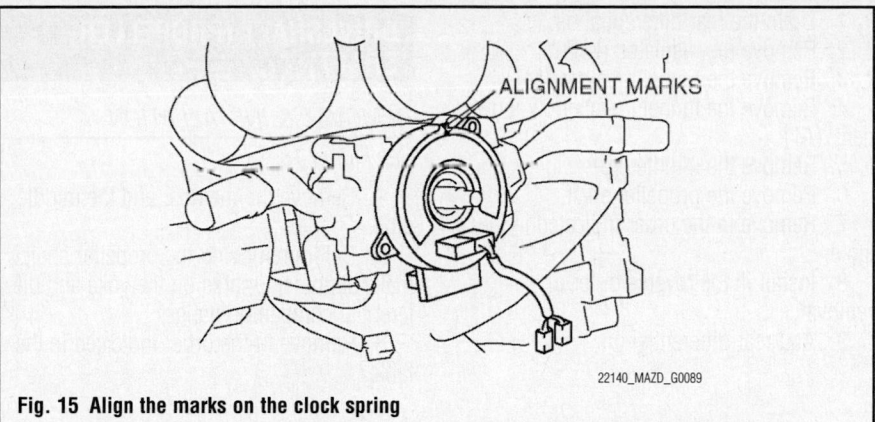

ALIGNMENT MARKS
22140_MAZD_G0089
Fig. 15 Align the marks on the clock spring

**** CAUTION**

The clock spring will break if overwound. Do not forcibly turn the clock spring.

- Turn the clock spring clockwise until it stops
- From the stopped position, turn the clock spring counterclockwise 2 ¾ turns
- Align the marks.

5. Install the remaining components in the reverse of removal.

DRIVE TRAIN

CLUTCH

REMOVAL & INSTALLATION

See Figure 16.

1. Remove in the numerical order shown.
2. Installation is the reverse of removal.

DIFFERENTIAL

REMOVAL & INSTALLATION

See Figure 17.

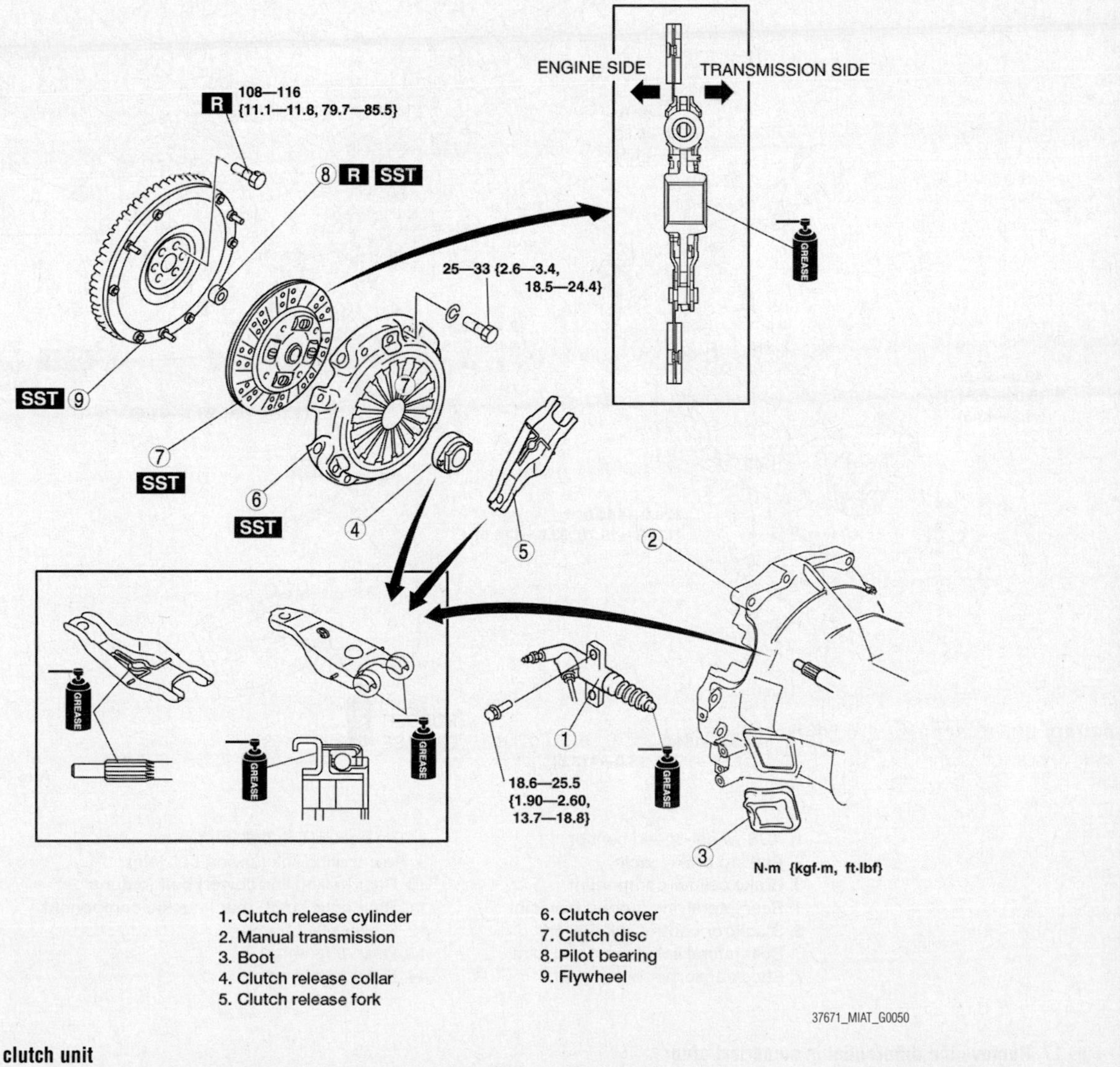

ENGINE SIDE
TRANSMISSION SIDE
R 108—116 {11.1—11.8, 79.7—85.5}
8 R SST
25—33 {2.6—3.4, 18.5—24.4}
GREASE
SST 9
7
SST
7
6
SST
4
5
2
1
18.6—25.5 {1.90—2.60, 13.7—18.8}
3
N·m {kgf·m, ft·lbf}
1. Clutch release cylinder
2. Manual transmission
3. Boot
4. Clutch release collar
5. Clutch release fork
6. Clutch cover
7. Clutch disc
8. Pilot bearing
9. Flywheel
37671_MIAT_G0050
Fig. 16 Remove the clutch unit

1. Drain the rear differential oil.
2. Remove the insulator. (MT)
3. Remove the tunnel member. (MT)
4. Remove the tunnel member component. (AT)
5. Remove the middle pipe.
6. Remove the propeller shaft.
7. Remove in the order indicated in the table.
8. Install in the reverse order of removal.
9. Add rear differential oil.

DRIVESHAFT (PROPELLER SHAFT)

REMOVAL & INSTALLATION

See Figure 18.

1. Remove the member and the middle pipe.
2. Before removing the propeller shaft, make alignment marks on the yoke and differential companion flange.
3. Remove in the order indicated in the table.

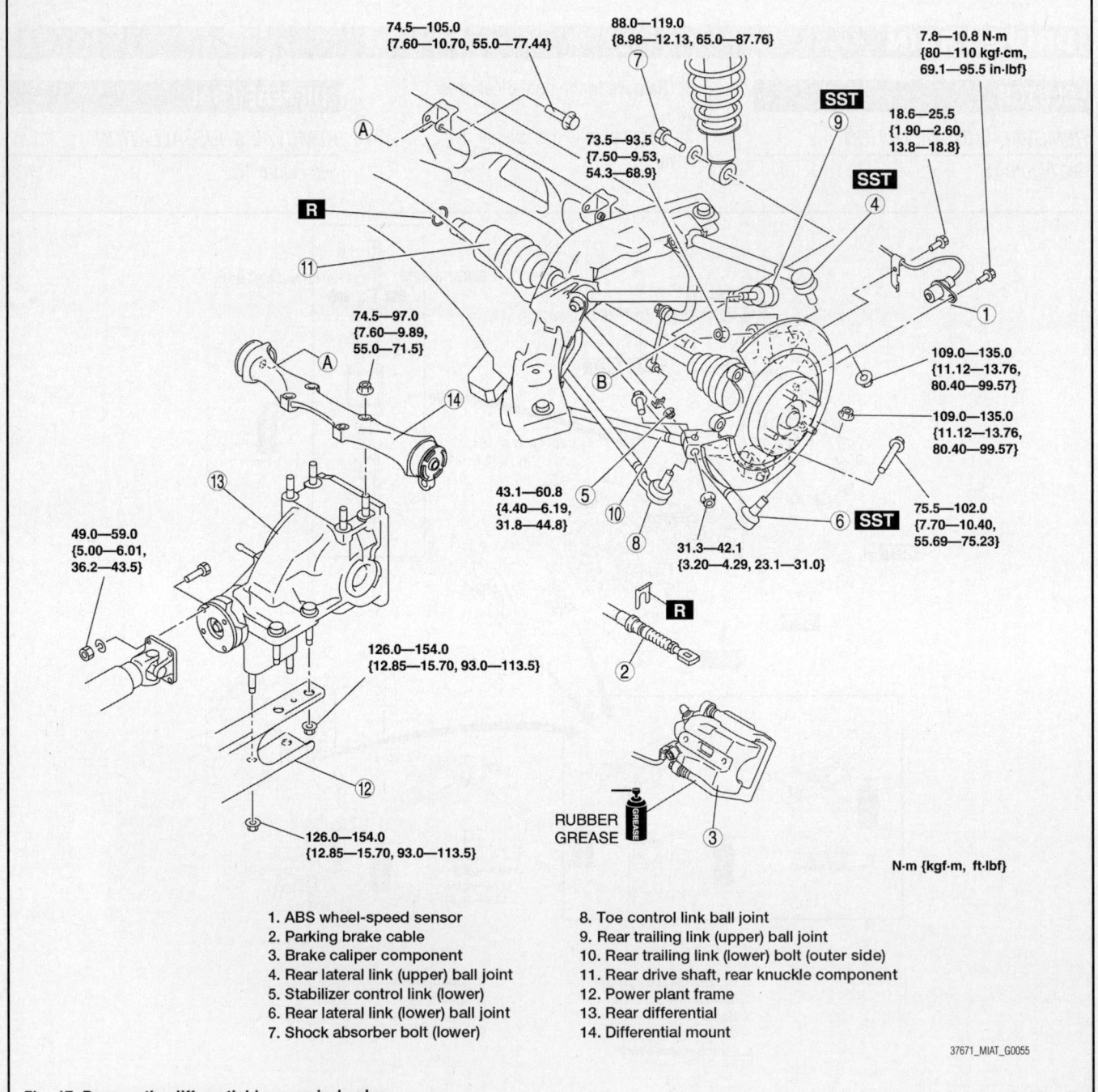

1. ABS wheel-speed sensor
2. Parking brake cable
3. Brake caliper component
4. Rear lateral link (upper) ball joint
5. Stabilizer control link (lower)
6. Rear lateral link (lower) ball joint
7. Shock absorber bolt (lower)
8. Toe control link ball joint
9. Rear trailing link (upper) ball joint
10. Rear trailing link (lower) bolt (outer side)
11. Rear drive shaft, rear knuckle component
12. Power plant frame
13. Rear differential
14. Differential mount

Fig. 17 Remove the differential in numerical order

4. Install in the reverse order of removal.

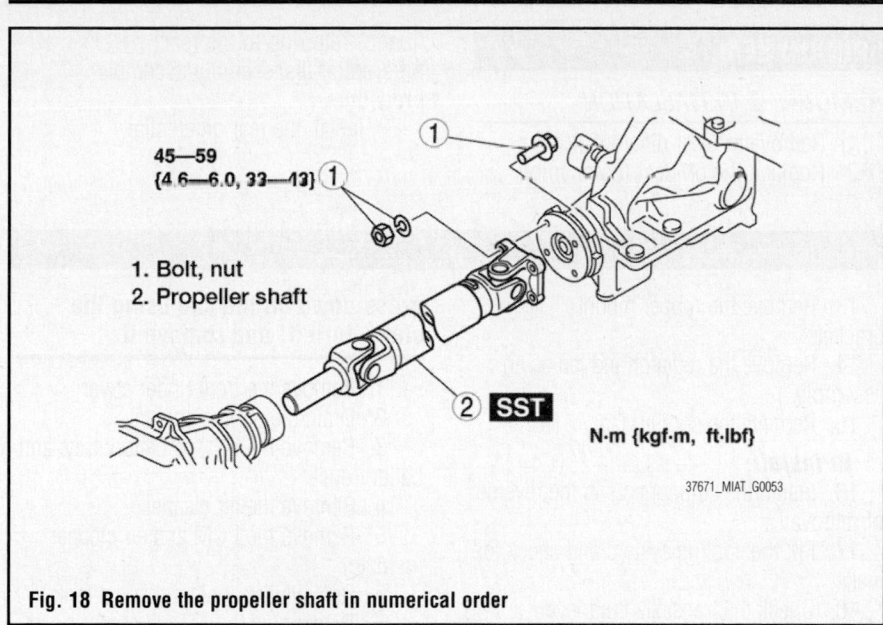

45—59
{4.6—6.0, 33—43}

1. Bolt, nut
2. Propeller shaft

N·m {kgf·m, ft·lbf}

37671_MIAT_G0053

Fig. 18 Remove the propeller shaft in numerical order

HALFSHAFTS

REMOVAL & INSTALLATION

See Figure 19.

✳✳ CAUTION

Performing the following procedures without first removing the ABS wheel-speed sensor may possibly cause an open circuit in the wiring harness if it is pulled by mistake. Before performing the following procedures, remove the ABS wheel-speed sensor (axle side) and fix it to an appropriate place where the sensor will not be pulled by mistake while servicing the vehicle.

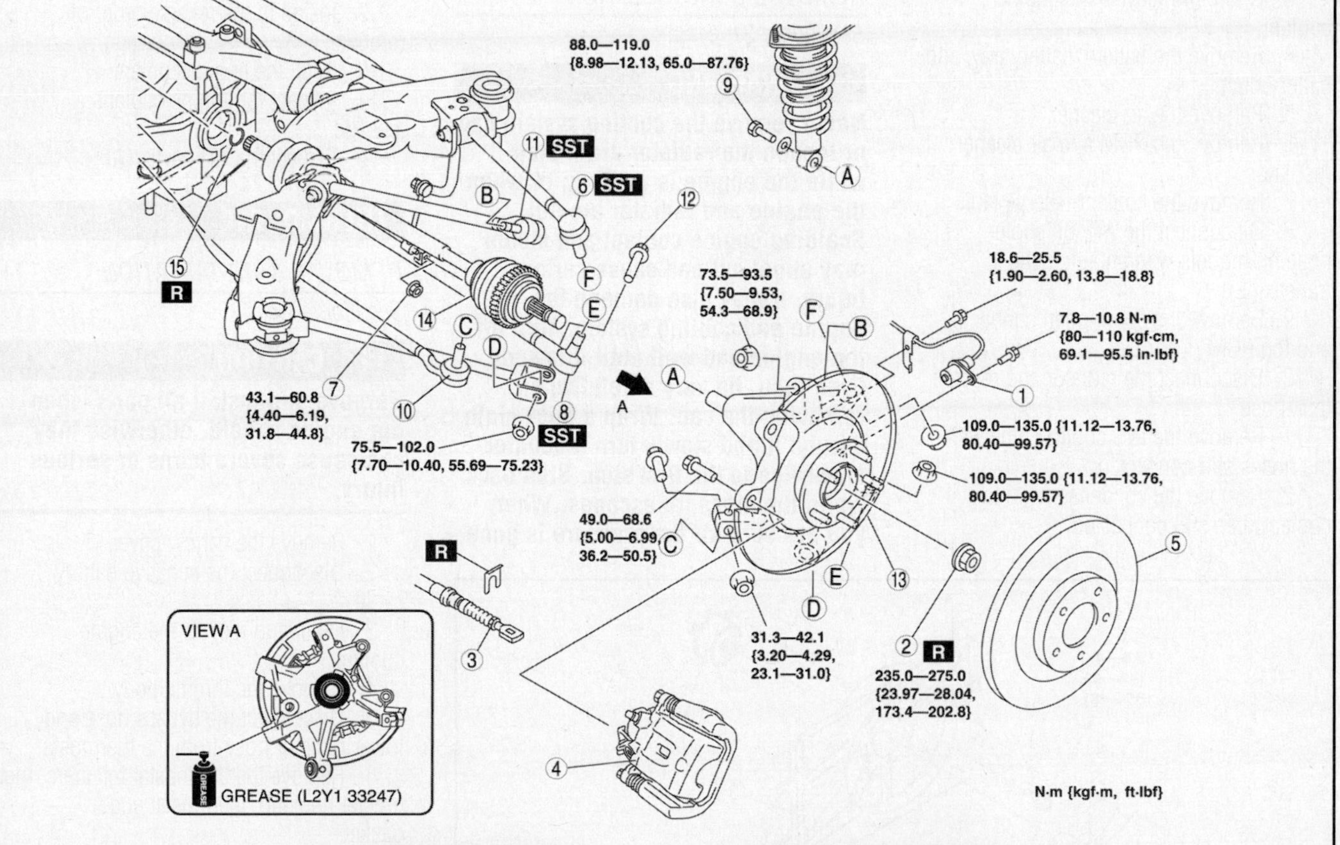

88.0—119.0
{8.98—12.13, 65.0—87.76}

18.6—25.5
{1.90—2.60, 13.8—18.8}

7.8—10.8 N·m
{80—110 kgf·cm,
69.1—95.5 in·lbf}

109.0—135.0 {11.12—13.76,
80.40—99.57}

109.0—135.0 {11.12—13.76,
80.40—99.57}

73.5—93.5
{7.50—9.53,
54.3—68.9}

43.1—60.8
{4.40—6.19,
31.8—44.8}

75.5—102.0
{7.70—10.40, 55.69—75.23}

49.0—68.6
{5.00—6.99,
36.2—50.5}

31.3—42.1
{3.20—4.29,
23.1—31.0}

235.0—275.0
{23.97—28.04,
173.4—202.8}

VIEW A

GREASE (L2Y1 33247)

N·m {kgf·m, ft·lbf}

1. ABS wheel-speed sensor
2. Locknut
3. Parking brake cable
4. Brake caliper component
5. Disc plate
6. Rear lateral link (upper) ball joint
7. Stabilizer control link (lower)
8. Rear lateral link (lower) ball joint
9. Shock absorber bolt (lower)
10. Toe control link ball joint
11. Rear trailing link (upper) ball joint
12. Rear trailing link (lower) bolt (outer side)
13. Rear knuckle component
14. Rear drive shaft
15. Clip

37671_MIAT_G0054

Fig. 19 Remove the rear drive axle in numerical order

1. Drain the rear differential oil.
2. Remove in the order indicated in the table.
3. Install in the reverse order of removal.
4. Add rear differential oil.

PINION SEAL

REMOVAL & INSTALLATION

1. Remove the rear differential.
2. Replace the oil seal (companion flange) referring to the rear differential disassembly/assembly procedure.
3. Install the rear differential.

ENGINE COOLING

ENGINE FAN

REMOVAL & INSTALLATION

See Figure 20.

✳✳ CAUTION

Remove and install all parts when the engine is cold, otherwise they can cause severe burns or serious injury.

1. Disconnect the negative battery cable.
2. Remove the undercover.
3. Drain and recycle the engine coolant.
4. Remove the battery, battery tray, and battery duct.
5. Remove the air cleaner.
6. Remove the PCM and air cleaner insulator.
7. Remove the coolant reserve tank.
8. Disconnect the ATF oil cooler hose on models with an automatic transaxle.
9. Remove the cooling fan connector and the PCM duct.
10. Disconnect the radiator and reserve tank hose.
11. Remove the P/S cooling pipe with the hoses still connected.
12. Remove the condenser with the cooler pipes still connected

13. Remove the rubber mount brackets.
14. Remove the radiator and fan as an assembly.
15. Remove the cooling fan.

To install:

16. Install the components in the reverse of removal.
17. Fill the cooling system and check for leaks.
18. Check the transaxle fluid level.

RADIATOR

REMOVAL & INSTALLATION

See Figure 21.

✳✳ WARNING

Never remove the cooling system cap or loosen the radiator drain plug while the engine is running, or when the engine and radiator are hot. Scalding engine coolant and steam may shoot out and cause serious injury. It may also damage the engine and cooling system. Turn off the engine and wait until it is cool. Even then, be very careful when removing the cap. Wrap a thick cloth around it and slowly turn it counter-clockwise to the first stop. Step back while the pressure escapes. When you are sure all the pressure is gone,

press down on the cap using the cloth, turn it, and remove it.

1. Remove the front under cover.
2. Drain the engine coolant.
3. Remove the battery, battery tray, and battery duct.
4. Remove the air cleaner.
5. Remove the PCM and air cleaner insulator.
6. Remove the coolant reserve tank.
7. Disconnect the ATF oil cooler hose from the radiator. (AT)
8. Remove in the order indicated in the table.
9. Install in the reverse order of removal.
10. Refill the engine coolant.
11. Inspect for engine coolant leakage.
12. Inspect the ATF level. (AT)

THERMOSTAT

REMOVAL & INSTALLATION

See Figure 22.

✳✳ CAUTION

Remove and install all parts when the engine is cold, otherwise they can cause severe burns or serious injury.

1. Remove the battery cover.
2. Disconnect the negative battery cable.
3. Drain and recycle the engine coolant.
4. Remove the throttle body.
5. Disconnect the bypass hose and lower radiator hose from the thermostat.
6. Remove the thermostat retainers, thermostat housing, thermostat and gasket.

To install:

7. Installation is the reverse of removal. Use a new thermostat housing gasket and tighten the thermostat housing retainers to 71–101 inch lbs. (8–11.5 Nm).
8. Fill the cooling system and check for leaks.

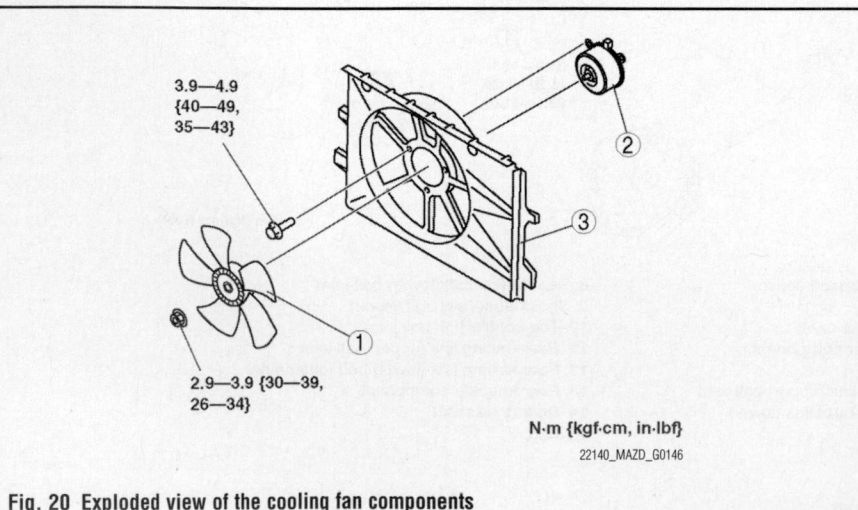

3.9—4.9
{40—49,
35—43}

2.9—3.9 {30—39,
26—34}

N·m {kgf·cm, in·lbf}

22140_MAZD_G0146

Fig. 20 Exploded view of the cooling fan components

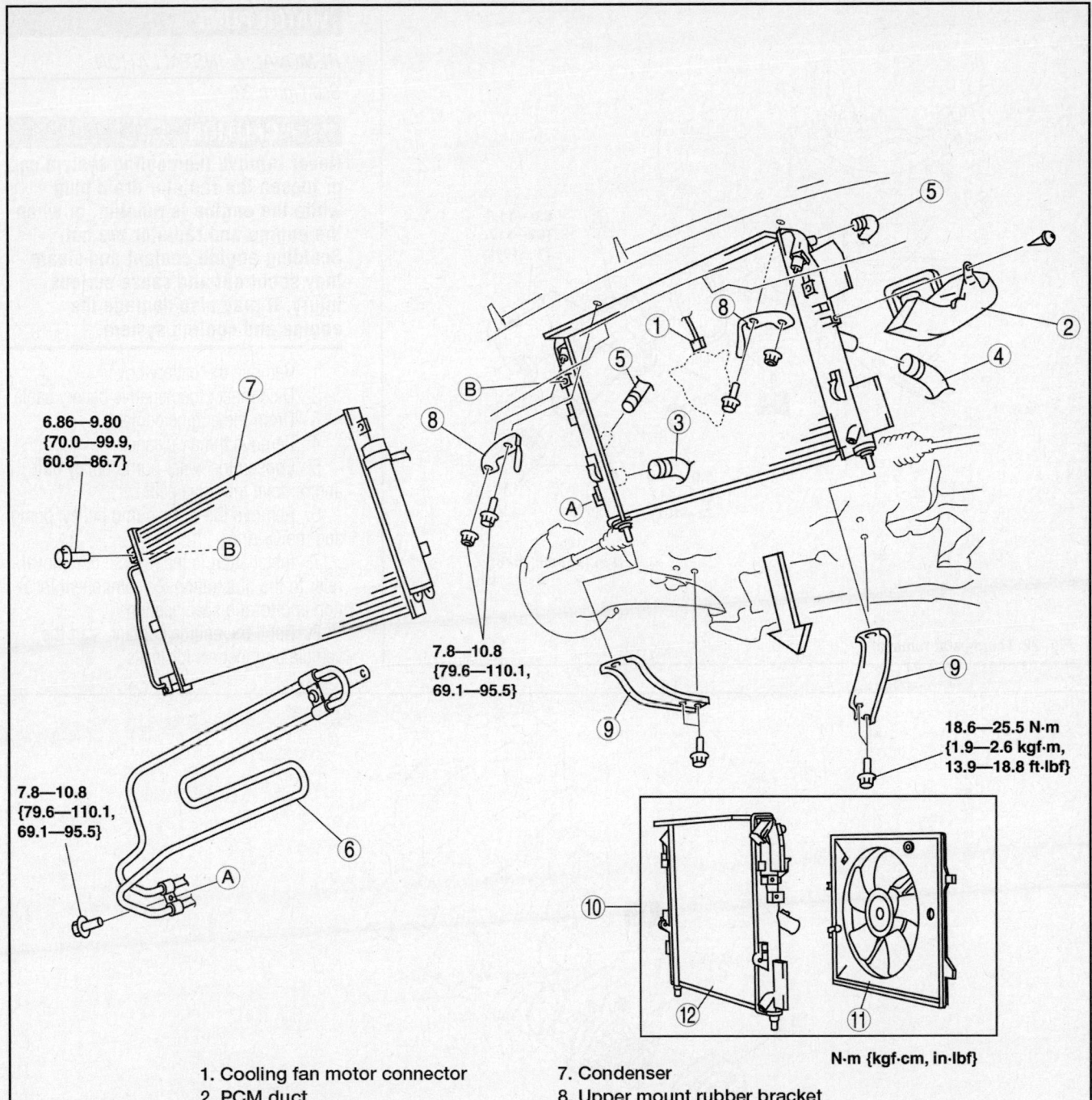

6.86—9.80
{70.0—99.9,
60.8—86.7}

7.8—10.8
{79.6—110.1,
69.1—95.5}

7.8—10.8
{79.6—110.1,
69.1—95.5}

18.6—25.5 N·m
{1.9—2.6 kgf·m,
13.9—18.8 ft·lbf}

N·m {kgf·cm, in·lbf}

1. Cooling fan motor connector
2. PCM duct
3. Radiator lower hose
4. Radiator upper hose
5. Coolant reserve tank hose
6. P/S cooling pipe
7. Condenser
8. Upper mount rubber bracket
9. Lower mount rubber bracket
10. Radiator and cooling fan component
11. Cooling fan component
12. Radiator

37671_MIAT_G0067

Fig. 21 Remove the radiator in numerical order shown

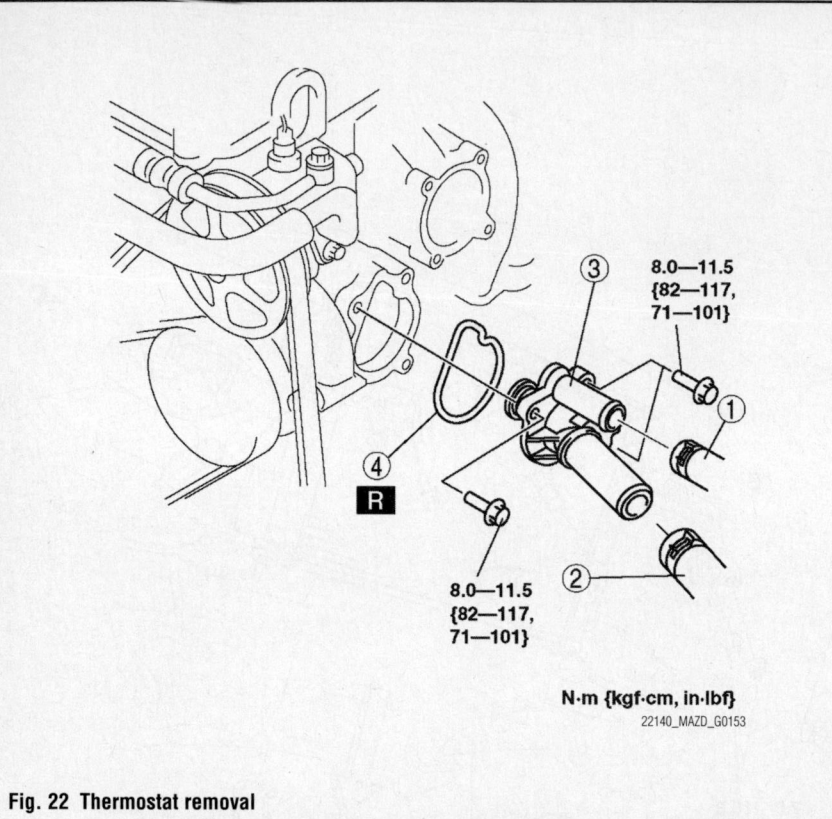

8.0—11.5
{82—117,
71—101}

8.0—11.5
{82—117,
71—101}

N·m {kgf·cm, in·lbf}

22140_MAZD_G0153

Fig. 22 Thermostat removal

WATER PUMP

REMOVAL & INSTALLATION

See Figure 23.

✳✳ CAUTION

Never remove the cooling system cap or loosen the radiator drain plug while the engine is running, or when the engine and radiator are hot. Scalding engine coolant and steam may shoot out and cause serious injury. It may also damage the engine and cooling system.

1. Remove the battery cover.
2. Disconnect the negative battery cable.
3. Drain the engine coolant.
4. Remove the air cleaner.
5. Loosen the water pump pulley bolt and remove the drive belt.
6. Remove the water pump pulley, pump and the O–ring
7. Installation is the reverse of removal, refer to the illustration for component location and torque specifications.
8. Refill the engine coolant, start the vehicle and inspect for leaks.

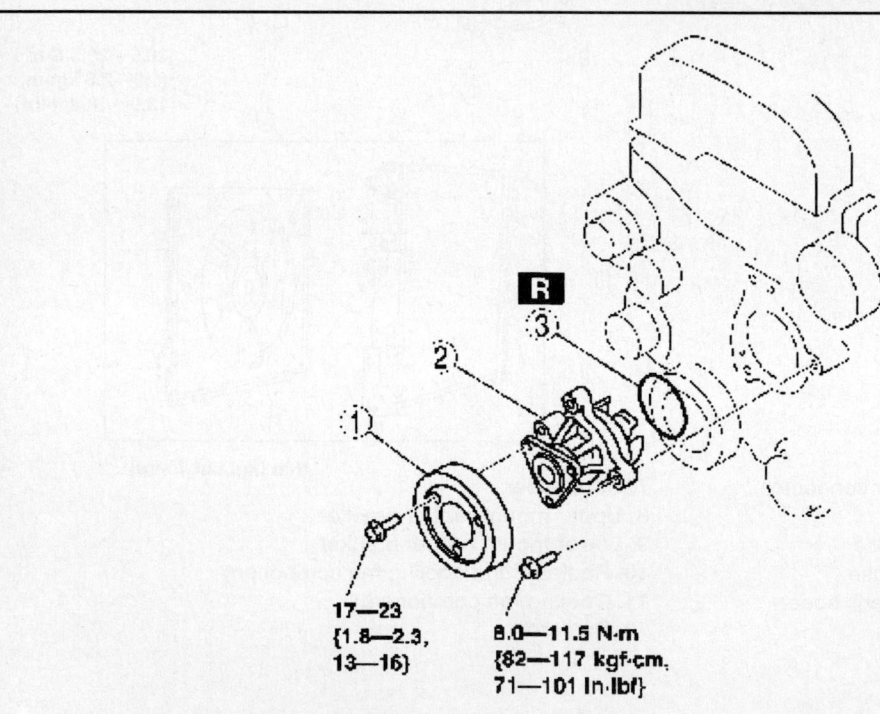

17—23
{1.8—2.3,
13—16}

8.0—11.5 N·m
{82—117 kgf·cm,
71—101 In·lbf}

N·m {kgf·m, ft·lbf}

1 Water pump pulley
2 Water pump
3 O-ring

09482_MAZC2_G0014

Fig. 23 Exploded view of the water pump assembly

ENGINE ELECTRICAL   CHARGING SYSTEM

ALTERNATOR

REMOVAL & INSTALLATION
See Figure 24.

> ✳✳ **CAUTION**
>
> **Remove and install all parts when the engine is cold, otherwise they can cause severe burns or serious injury.**

> ✳✳ **CAUTION**
>
> **The alternator can be damaged by the heat from the exhaust manifold. Make sure the alternator duct is installed securely.**

1. Remove the battery and battery tray.
2. Remove the drive belt.
3. Remove the P/S pressure hose bracket.
4. Remove the alternator bracket.
5. Remove the alternator duct.
6. Disconnect the B terminal cable Alternator connector.
7. Remove the alternator.

To install:

8. Before servicing the vehicle, refer to the precautions section.
9. Installation is the reverse of removal. Please note the following:

    a. Tighten bolt A temporarily.

    b. Tighten bolt B, C to 29–37 ft. lbs. (38–51 Nm).

    c. Tighten bolt A to 29–37 ft. lbs. (38–51 Nm).

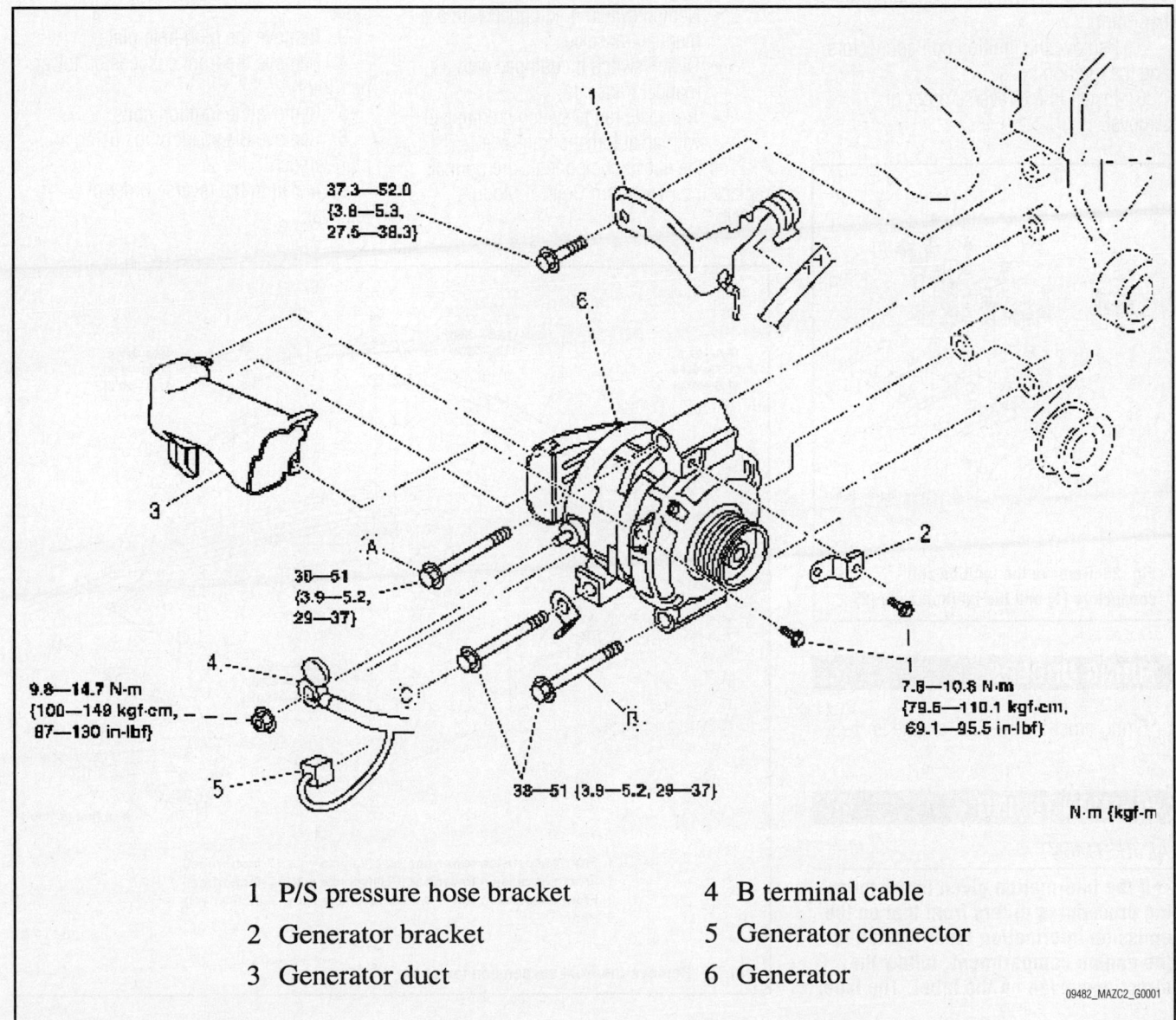

37.3—52.0
{3.8—5.3,
27.5—38.3}

38—51
{3.9—5.2,
29—37}

9.8—14.7 N·m
{100—149 kgf·cm,
87—130 in·lbf}

38—51 {3.9—5.2, 29—37}

7.8—10.8 N·m
{79.6—110.1 kgf·cm,
69.1—95.5 in·lbf}

N·m {kgf·m}

1 P/S pressure hose bracket	4 B terminal cable
2 Generator bracket	5 Generator connector
3 Generator duct	6 Generator

09482_MAZC2_G0001

Fig. 24 Alternator mounting

FIRING ORDER

Firing order for the 2.0L engine is: 1–3–4–2.

IGNITION COIL

REMOVAL & INSTALLATION

See Figure 25.

1. Remove the battery cover.
2. Disconnect the negative battery cable.
3. Remove the plug hole plate.
4. Remove the front suspension tower bar (joint).
5. Remove the ignition coil connectors and the ignition coils.
6. Install in the reverse order of removal.

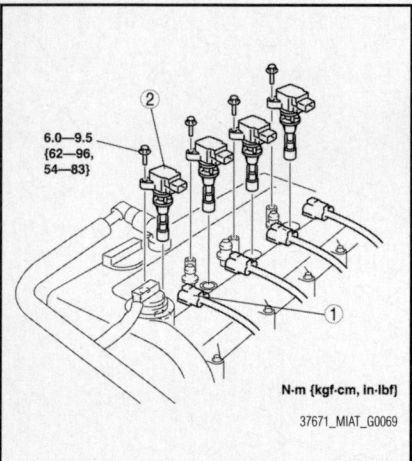

6.0—9.5
{62—96,
54—83}

N·m {kgf·cm, in·lbf}

37671_MIAT_G0069

Fig. 25 Remove the ignition coil connectors (1) and the ignition coils (2)

FIRING ORDERS

Firing order for the 2.0L engine is: 1–3–4–2.

IGNITION TIMING

ADJUSTMENT

➡ **If the information given in the following procedures differs from that on the emission information label located in the engine compartment, follow the directions given on the label. The label**

often reflects production changes made during the model year.

The timing is controlled by the computer. Ignition timing adjustment is not possible or necessary.

1. If the timing is still not within specification. the following components may be defective:

- Camshaft position (CMP) sensor
- Crankshaft Position (CKP) sensor
- Throttle Position (TP) sensor
- Engine Coolant Temperature (ECT) sensor
- Neutral switch if equipped with a manual transaxle
- Clutch switch if equipped with a manual transaxle
- Transaxle range switch if equipped with an automatic transaxle

2. If the above components are normal, replace the Powertrain Control Module (PCM).

SPARK PLUGS

REMOVAL & INSTALLATION

See Figure 26.

❊❊ CAUTION

If a spark plug that is not as specified is installed, engine performance will be deteriorated. Install only the specified spark plug when replacing.

1. Remove the battery cover.
2. Disconnect the negative battery cable.
3. Remove the plug hole plate.
4. Remove the front suspension tower bar (joint).
5. Remove the ignition coils.
6. Remove the spark plugs using a plug-wrench.
7. Install in the reverse order of removal.

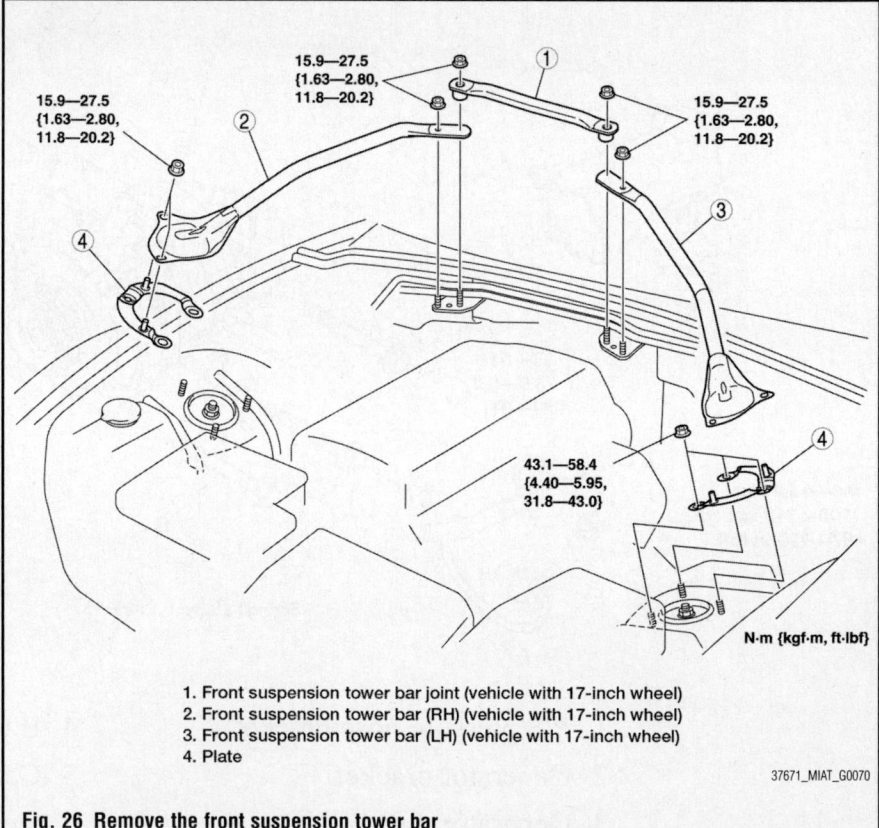

15.9—27.5
{1.63—2.80,
11.8—20.2}

15.9—27.5
{1.63—2.80,
11.8—20.2}

15.9—27.5
{1.63—2.80,
11.8—20.2}

43.1—58.4
{4.40—5.95,
31.8—43.0}

N·m {kgf·m, ft·lbf}

1. Front suspension tower bar joint (vehicle with 17-inch wheel)
2. Front suspension tower bar (RH) (vehicle with 17-inch wheel)
3. Front suspension tower bar (LH) (vehicle with 17-inch wheel)
4. Plate

37671_MIAT_G0070

Fig. 26 Remove the front suspension tower bar

ENGINE ELECTRICAL

STARTER

REMOVAL & INSTALLATION

See Figure 27

> **✻✻ CAUTION**
>
> **Remove and install all parts when the engine is cold, otherwise they can cause severe burns or serious injury.**

> **✻✻ CAUTION**
>
> **When the battery cables are connected, touching the vehicle body with starter terminal B will cause sparks. This can cause personal injury, fire, and damage to the electrical components. Make sure to always disconnect the negative battery cable before removing any electrical component.**

1. Remove the battery cover.
2. Disconnect the negative battery cable.
3. Remove the left hand side cover.
4. Remove the undercover on models equipped with a manual transmission.
5. Remove the oil filter. (Vehicles with oil cooler)
6. On models equipped with a manual transmission, remove the clutch release cylinder with the pipes still connected. Position the clutch release cylinder so that it is out of the way.
7. Remove the wiring from the starter, wiring harness bracket, starter bolts and the starter.

To install:

8. Installation is the reverse of removal.
9. Tighten the starter bolts to 38–51 Nm (29–37 ft. lbs.)

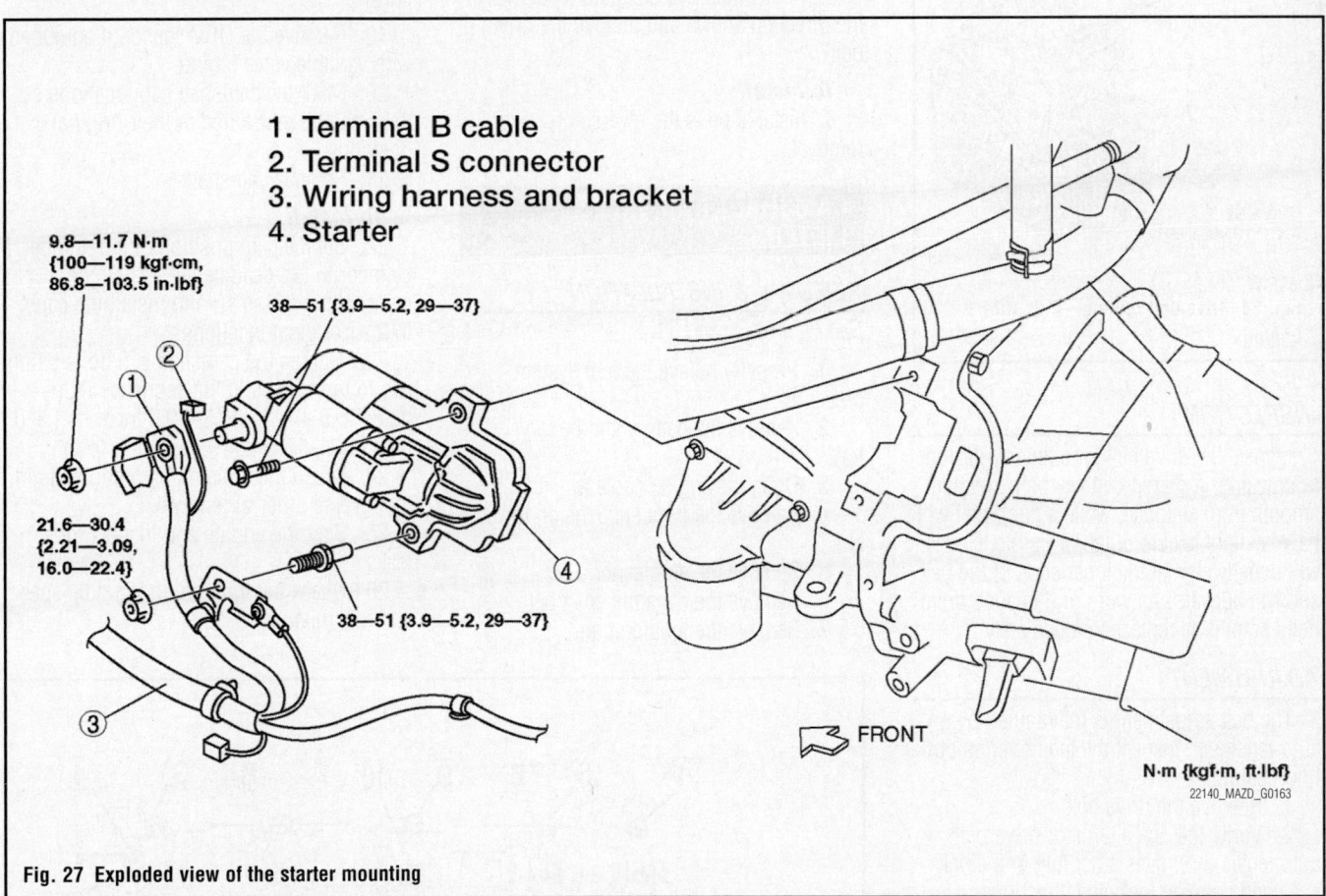

1. Terminal B cable
2. Terminal S connector
3. Wiring harness and bracket
4. Starter

9.8—11.7 N·m {100—119 kgf·cm, 86.8—103.5 in·lbf}

38—51 {3.9—5.2, 29—37}

21.6—30.4 {2.21—3.09, 16.0—22.4}

38—51 {3.9—5.2, 29—37}

⇨ FRONT

N·m {kgf·m, ft·lbf}

22140_MAZD_G0163

Fig. 27 Exploded view of the starter mounting

ENGINE MECHANICAL

→Disconnecting the negative battery cable may interfere with the functions of the on board computer systems and may require the computer to undergo a relearning process, once the negative battery cable is reconnected.

ACCESSORY DRIVE BELTS

ACCESSORY BELT ROUTING

See Figure 28.

Refer to the accompanying illustration for belt routing.

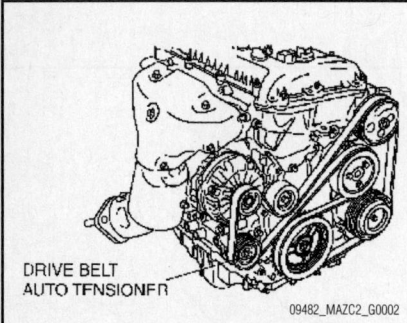

DRIVE BELT
AUTO TENSIONER

09482_MAZC2_G0002

Fig. 28 Drive belt routing—2.0L (LF) Engine

INSPECTION

Inspect the drive belt for signs of glazing or cracking. A glazed belt will be perfectly smooth from slippage, while a good belt will have a slight texture of fabric visible. Cracks will usually start at the inner edge of the belt and run outward. All worn or damaged drive belts should be replaced immediately

ADJUSTMENT

The belt adjustment is maintained by an auto tensioner. Inspect the tensioner as outlined below.

1. Remove the drive belt.
2. Verify that the alternator drive belt auto tensioner moves smoothly in a clockwise and counterclockwise direction.
3. If it does not move smoothly, replace the drive belt auto tensioner.
4. Turn the drive belt auto tensioner pulley by hand and verify that it rotates smoothly.
5. If it does not move smoothly, replace the drive belt auto tensioner.

REMOVAL & INSTALLATION

See Figure 29.

1. Remove the battery and battery tray.

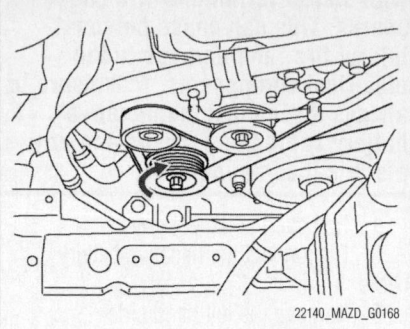

22140_MAZD_G0168

Fig. 29 Rotate the drive belt auto tensioner in the direction shown

2. Rotate the drive belt auto tensioner in the direction shown and remove the drive belt.

To install:
3. Installation is the reverse of removal.

CAMSHAFT AND VALVE LIFTERS

REMOVAL & INSTALLATION

See Figures 30 and 31.

1. Properly relieve the fuel system pressure.
2. Remove the battery and battery tray.
3. Drain the engine coolant.
4. Remove the front suspension tower bar.
5. Remove the air cleaner.
6. Remove the dynamic chamber.
7. Remove the ignition coil.

8. Remove the drive belt.
9. Remove the CKP sensor.
10. Remove the P/S oil pump with the oil hose still connected and position the P/S oil pump so that it is out of the way.
11. Remove the timing chain.
12. Remove the wiper arm.
13. Remove the cowl grille.
14. Remove the side cowl grille.
15. Remove the service hole cover.
16. Disconnect the alternator, but do not remove it from the vehicle.
17. Remove the exhaust manifold.
18. Remove the cylinder head cover.
19. Remove the OCV sensor, if equipped with variable valve timing.
20. Mark the camshaft cap locations so they may be reinstalled in their original positions.
21. Remove camshafts.

To install:
22. Set the cam position of the number 1 cylinder to top dead center, install the camshafts, making sure to install the caps in their original locations.
23. Tighten the camshaft cap bolts using two to three steps in the sequence illustrated to 5–9 Nm (44.3–79.6 inch lbs.) and then to 14–17 Nm (10.4–12.5 ft. lbs.)
24. Install the remaining components in the reverse order of removal
25. Start the engine and inspect for leaks.
26. Check the ignition timing, idle speed and idle mixture.

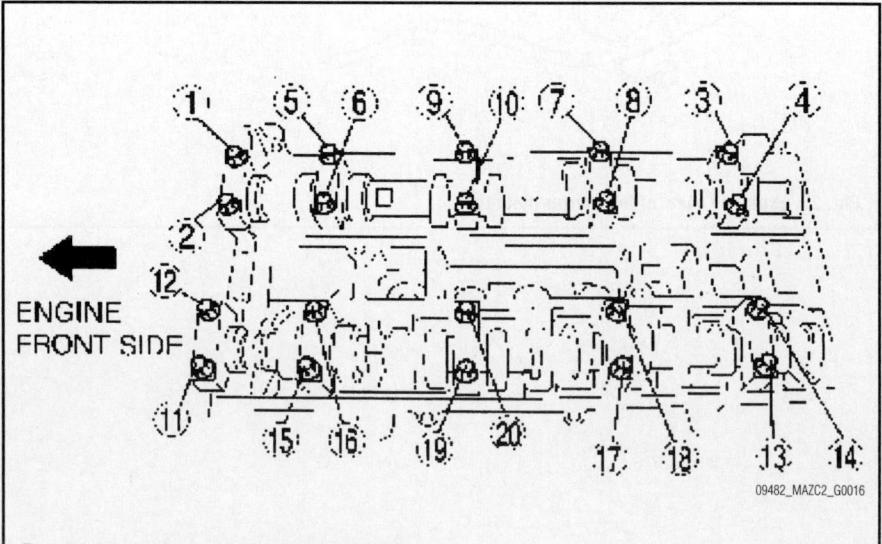

ENGINE FRONT SIDE

09482_MAZC2_G0016

Fig. 30 Camshaft cap bolt loosening sequence

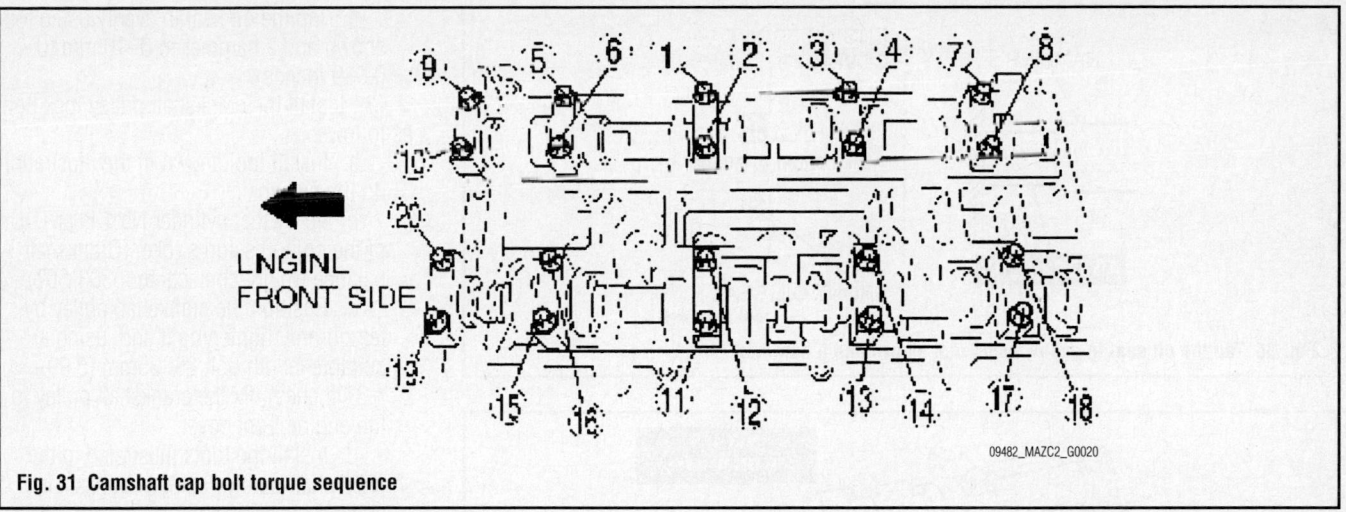

Fig. 31 Camshaft cap bolt torque sequence

CRANKSHAFT FRONT SEAL

REMOVAL & INSTALLATION

See Figures 32 through 38.

1. Remove the battery and battery tray.
2. Remove the air cleaner.
3. Remove the drive belt.
4. Remove the undercover.
5. Remove the front suspension tower bar.
6. Remove the ignition coil.
7. Remove the OCV connector.
8. Remove the cylinder head cover.
9. Remove the CKP sensor.
10. Remove the crankshaft pulley lock bolt as follows:
   a. Remove the cylinder block lower blind plug.
   b. Install tool 303-507.
   c. Turn the crankshaft clockwise until the crankshaft is in the No.1 cylinder TDC position (until the balance weight contacts tool 303-507).
   d. Hold the crankshaft pulley using the tools shown in the illustration.
   e. Remove the crankshaft pulley lock bolt.
11. Remove the front oil seal as follows:
   a. Cut the oil seal lip using a razor knife.
   b. Remove the oil seal using a screwdriver wrapped with a rag.

To install:

12. Install the oil seal as follows:
   a. Apply clean engine oil to a new oil seal.
   b. Push the front oil seal in the engine front cover by hand.

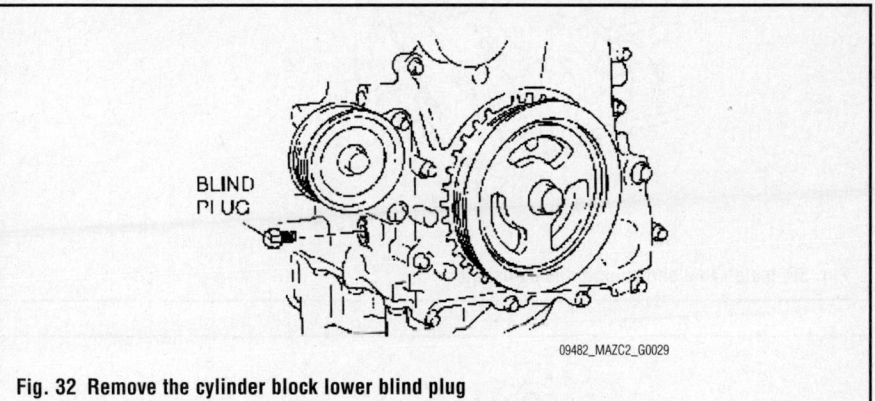

Fig. 32 Remove the cylinder block lower blind plug

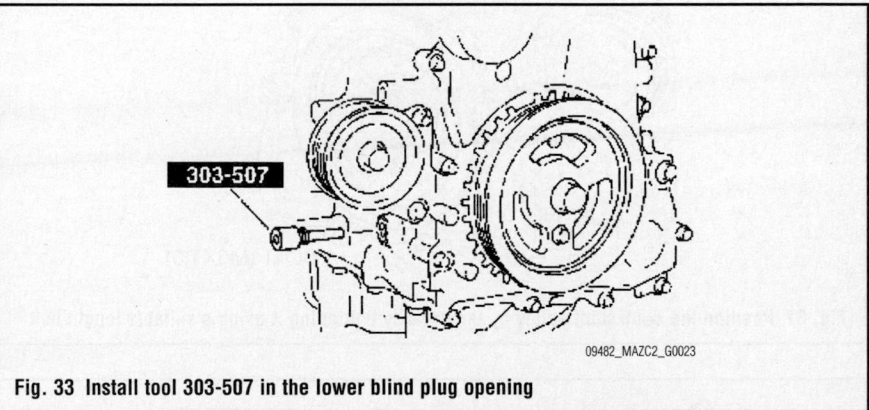

Fig. 33 Install tool 303-507 in the lower blind plug opening

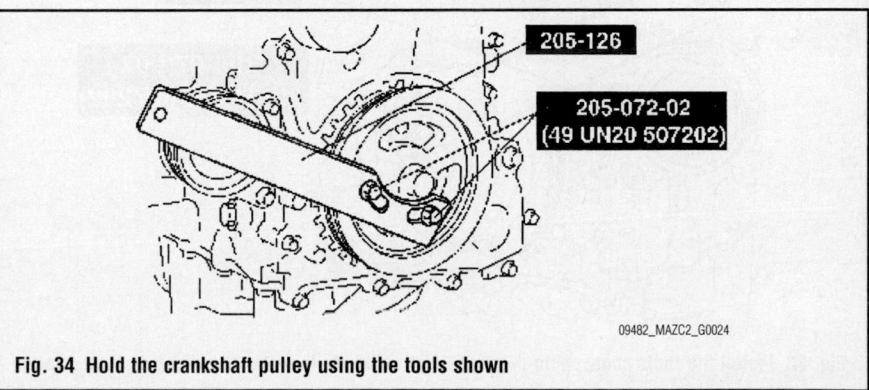

Fig. 34 Hold the crankshaft pulley using the tools shown

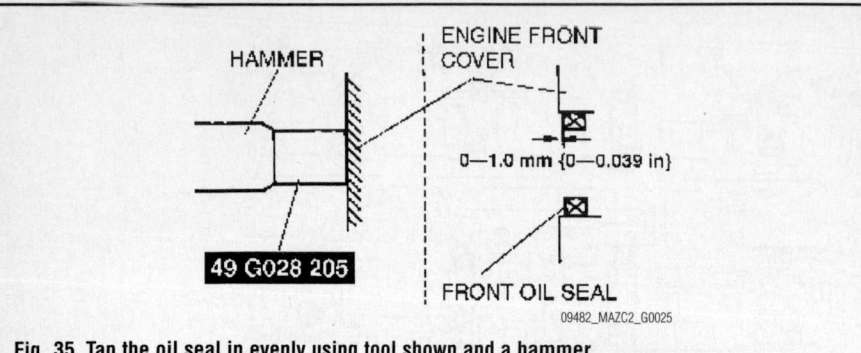

Fig. 35 Tap the oil seal in evenly using tool shown and a hammer

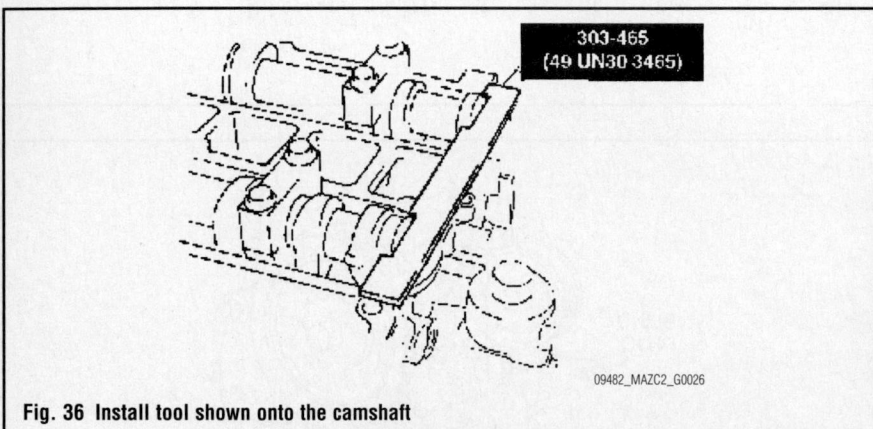

Fig. 36 Install tool shown onto the camshaft

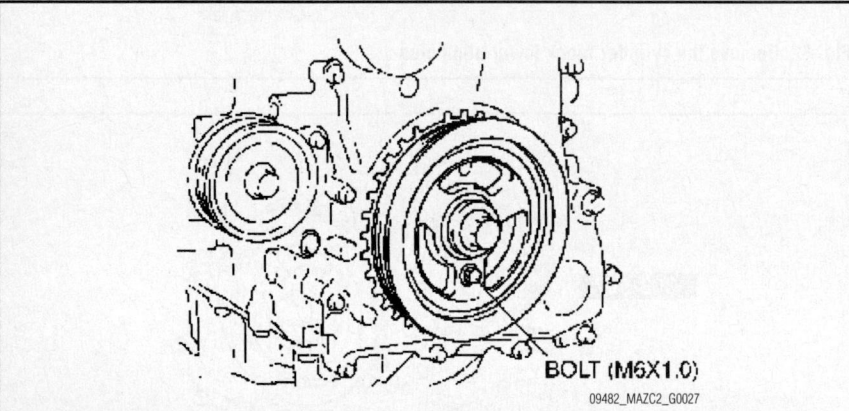

Fig. 37 Position the crankshaft pulley by temporarily tightening it using a suitable length bolt

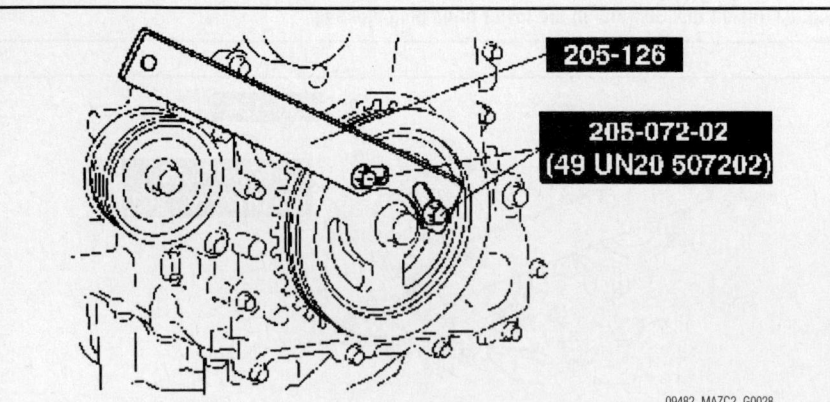

Fig. 38 Install the tools shown onto the crankshaft pulley to lock the crankshaft against rotation

c. Tap the oil seal in evenly using tool shown and a hammer to 0–10 mm (0–0.039 inches).

13. Install the crankshaft pulley lock bolt as follows:

a. Install tool shown in the illustration on the camshaft.

b. Verify that cylinder No.1 is at TDC of the compression stroke. (Crankshaft balance weight contacts tool 303-507)

c. Position the crankshaft pulley by temporarily tightening it and, using a suitable length bolt 25–35mm (0.99–1.37 inches), fix the crankshaft pulley to the engine front cover.

d. Install the tools illustrated to the crankshaft pulley, lock the crankshaft against rotation.

e. Tighten the crankshaft pulley lock bolt to 96–104 Nm (71–77 ft. lbs.), plus 87–93 degrees.

f. Remove the bolt.

g. Remove the tool the camshaft.

h. Remove tool from the cylinder block lower blind plug.

i. Remove the tool from the crankshaft pulley.

j. Rotate the crankshaft clockwise two turns until the TDC position.

k. If not aligned, loosen the crankshaft pulley lock bolt and repeat from the first step.

l. Install the cylinder block lower blind plug and tighten to 18–22 Nm (13–16 ft. lbs.)

14. Install the remaining components in the reverse order of removal.

CYLINDER HEAD

REMOVAL & INSTALLATION

See Figures 39 through 44.

1. Properly relieve the fuel system pressure.

2. Remove the battery and battery tray.

3. Drain the engine coolant.

4. Remove the front suspension tower bar.

5. Remove the air cleaner.

6. Remove the dynamic chamber.

7. Remove the ignition coil.

8. Remove the drive belt.

9. Remove the CKP sensor.

10. Remove the P/S oil pump with the oil hose still connected and position the P/S oil pump so that it is out of the way.

11. Remove the timing chain.

12. Remove the wiper arm.

13. Remove the cowl grille.

14. Remove the side cowl grille.

15. Remove the service hole cover.

16. Disconnect the alternator, but do not remove it from the vehicle.

17. Remove the exhaust manifold.

18. Remove the OCV sensor, if equipped with variable valve timing.

19. Mark the camshaft cap locations so they may be reinstalled in their original positions.

20. Remove camshafts.

21. Mark the cylinder head bolt locations so they may be reinstalled in their original positions.

22. Loosen the cylinder head bolts in the sequence illustrated using two or three passes.

23. Remove the cylinder head.

24. Measure the length of the cylinder head bolts and replace any bolts that exceed the maximum length specification of 146. mm (5.77 inches).

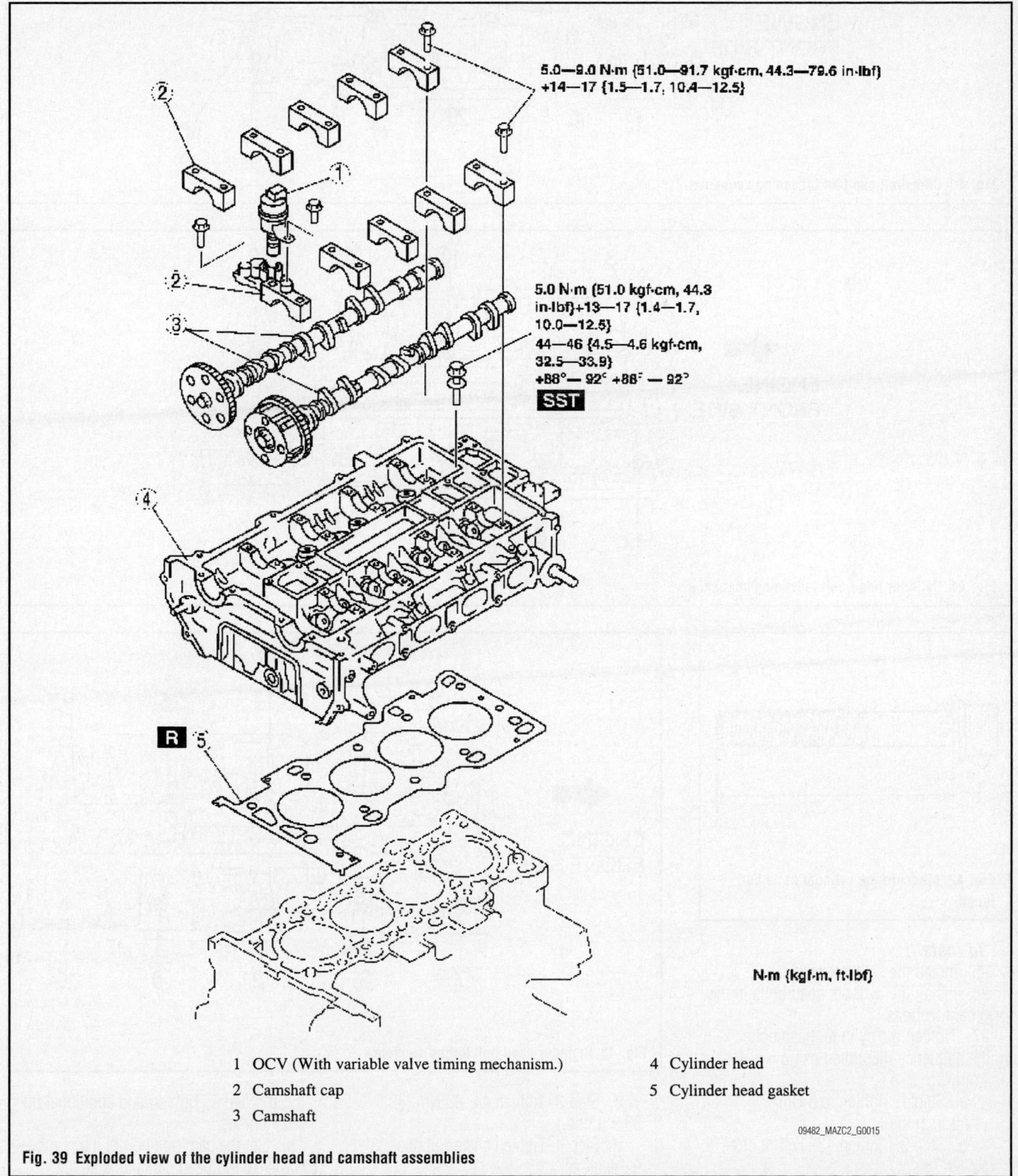

5.0—9.0 N·m {51.0—91.7 kgf·cm, 44.3—79.6 in·lbf}
+14—17 {1.5—1.7, 10.4—12.5}

5.0 N·m {51.0 kgf·cm, 44.3 in·lbf}+13—17 {1.4—1.7, 10.0—12.5}
44—46 {4.5—4.6 kgf·cm, 32.5—33.9}
+88°—92° +88°—92°
SST

R

N·m {kgf·m, ft·lbf}

1 OCV (With variable valve timing mechanism.)	4 Cylinder head
2 Camshaft cap	5 Cylinder head gasket
3 Camshaft	

09482_MAZC2_G0015

Fig. 39 Exploded view of the cylinder head and camshaft assemblies

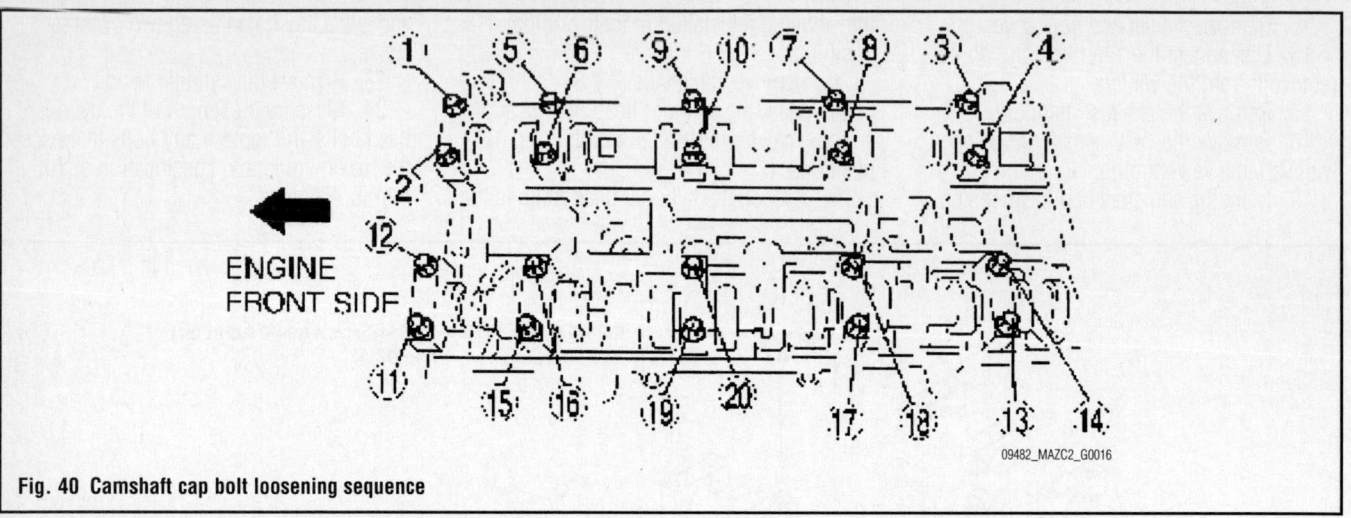

Fig. 40 Camshaft cap bolt loosening sequence

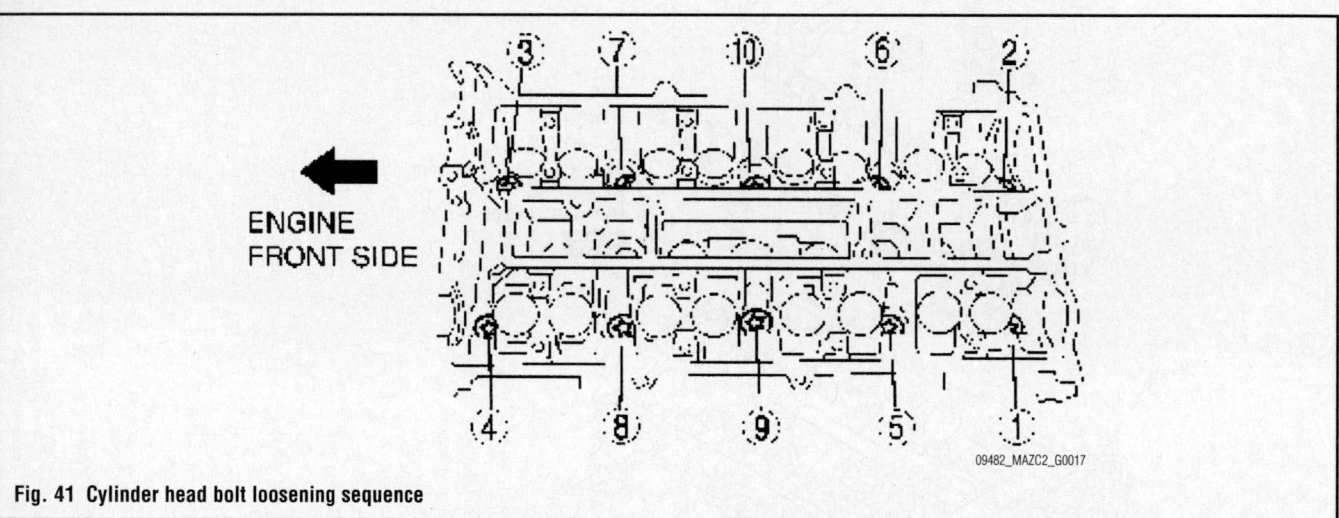

Fig. 41 Cylinder head bolt loosening sequence

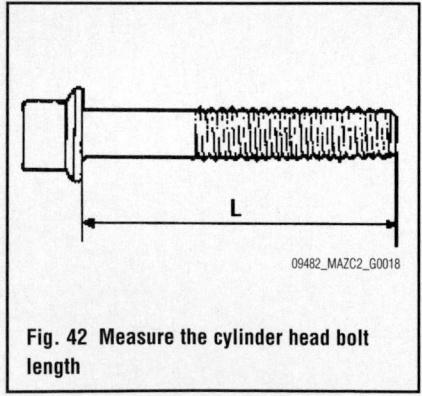

Fig. 42 Measure the cylinder head bolt length

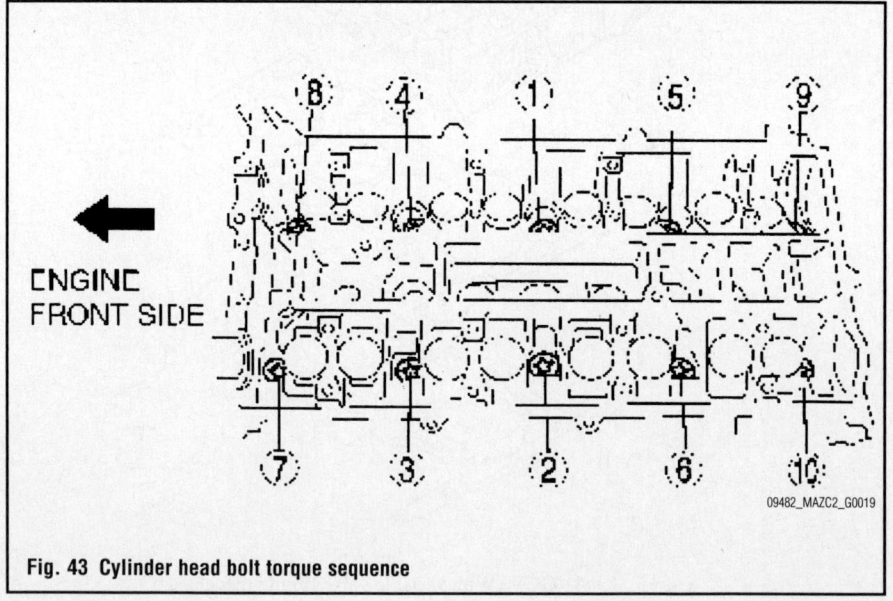

Fig. 43 Cylinder head bolt torque sequence

To install:

25. Install the cylinder heads.

26. Install the cylinder head bolts in their original locations.

27. Tighten the cylinder head bolts in the sequence illustrated using 5 passes as follows:

    a. Step 1: Tighten to 5 Nm (44.3 inch lbs.)

    b. Step 2: Tighten 13–17 Nm (10–12.5 ft. lbs.)

    c. Step 3: Tighten 44–46 Nm (32.5–33.9 ft. lbs.)

    d. Step 4: Tighten and additional 88–92 degrees.

    e. Step 5: Tighten and additional 88–92 degrees.

28. Set the cam position of the number 1 cylinder to top dead center, install the

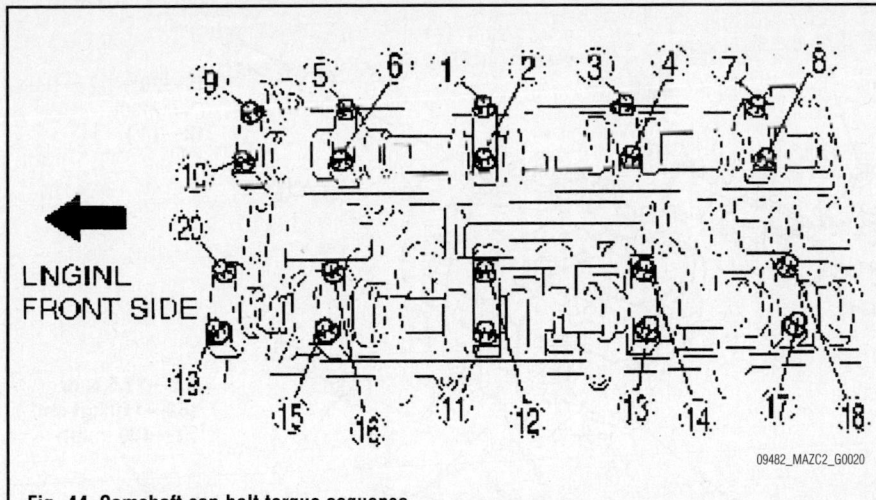

Fig. 44 Camshaft cap bolt torque sequence

camshafts, making sure to install the caps in their original locations.

29. Tighten the camshaft cap bolts using two to three steps in the sequence illustrated to 5–9 Nm (44.3–79.6 inch lbs.) and then to 14–17 Nm (10.4–12.5 ft. lbs.)

30. Install the remaining components in the reverse order of removal

31. Start the engine and inspect for leaks.

32. Check the ignition timing, idle speed and idle mixture.

EXHAUST MANIFOLD

REMOVAL & INSTALLATION

See Figure 45.

1. Disconnect the front and oxygen sensor connectors.

2. Move the water pipe and heater hose slightly out of the way.

3. Disconnect the ventilation hose from the cylinder head

4. Remove the upper exhaust manifold insulator as follows:

   a. Remove the battery and battery tray.

   b. Remove the drive belt.

   c. Remove the undercover.

   d. Remove the suspension tower bar (joint) and (right side).

   e. Remove the alternator.

   f. Remove the upper insulator.

5. Remove the lower exhaust manifold insulator.

6. Remove the exhaust manifold bracket.

7. Remove the exhaust manifold.

To install:

8. Installation is the reverse of removal, please note the following:

   a. Install the manifold and tighten the retainers to 43–64 Nm (32–47 ft. lbs.)

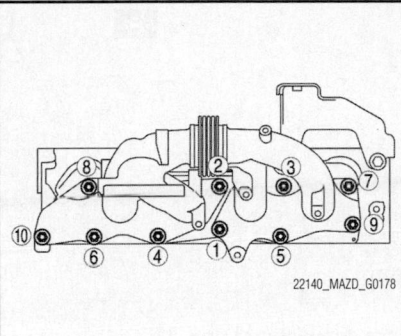

Fig. 45 Exhaust manifold and related components

   b. Tighten the exhaust manifold bracket to 38–52 Nm (29–38 ft. lbs.).

   c. Tighten the upper and lower insulators to 8–11 Nm (70–95 inch lbs.).

   d. If removed, tighten the front and oxygen sensors to 29–39 Nm (22–36 ft. lbs.)

INTAKE MANIFOLD

REMOVAL & INSTALLATION

See Figures 46 and 47.

1. Properly relieve the fuel system pressure.

2. Remove the battery cover.

3. Disconnect the negative battery cable.

4. Remove the air cleaner cover, this involves first removing the MAF/IAT sensor.

5. Remove the air cleaner element.

   a. Remove the air cleaner case.

6. Disconnect the quick connect fitting from the air cleaner hose.

7. Remove the air hose by moving the purge solenoid valve slightly out of the way.

8. Remove the fresh-air duct by first removing the bumper.

9. Drain the cooling system.

10. Remove the throttle body.

11. Disconnect the quick connect fitting from the dynamic chamber.

12. Remove the plug hole plate.

13. Remove the service hole cover.

14. Remove the front suspension tower bar.

15. Remove the wiper arm.

16. Remove the cowl grille.

17. Remove the side cowl grille.

18. Move the cooler pipe No.3 and heater pipe slightly out of the way.

19. Remove the service hole cover.

20. Disconnect the heater hose and move the heater pipe slightly out of the way.

21. Disconnect the heater hose and move the heater pipe slightly out of the way.

22. Remove the harness bracket.

23. Remove the undercover.

24. Disconnect the variable intake air solenoid valve, EGR valve, CMP sensor and PSP switch connectors.

25. Disconnect the ignition coil and fuel injector connectors and move the harness aside.

26. Disconnect the quick release connector from the fuel rail.

27. Remove the fuel rail.

28. Disconnect the water hose from the EGR valve.

29. Disconnect two water hoses from the thermostat.

30. Remove the heater hose and heater pipe from the dynamic chamber.

31. Remove the variable intake air solenoid valve.

32. Remove the dynamic chamber installation bolts.

33. Remove the EGR pipe.

34. Disconnect the connector from the A/C compressor.

35. Disconnect the knock sensor connector.

36. Move the vacuum hose between the purge solenoid valve and the charcoal canister aside.

37. Move the clutch release cylinder aside.

38. Disconnect the evaporative hose with the dynamic chamber raised.

39. Remove the dynamic chamber.

40. Remove the intake manifold.

To install:

41. Install the intake manifold and tighten the bolts to 15–25 Nm (12–18 ft. lbs.).

42. Install the dynamic chamber and tighten the bolts to 16–20 Nm (12–14 ft. lbs.)

43. Installation is the reverse order of

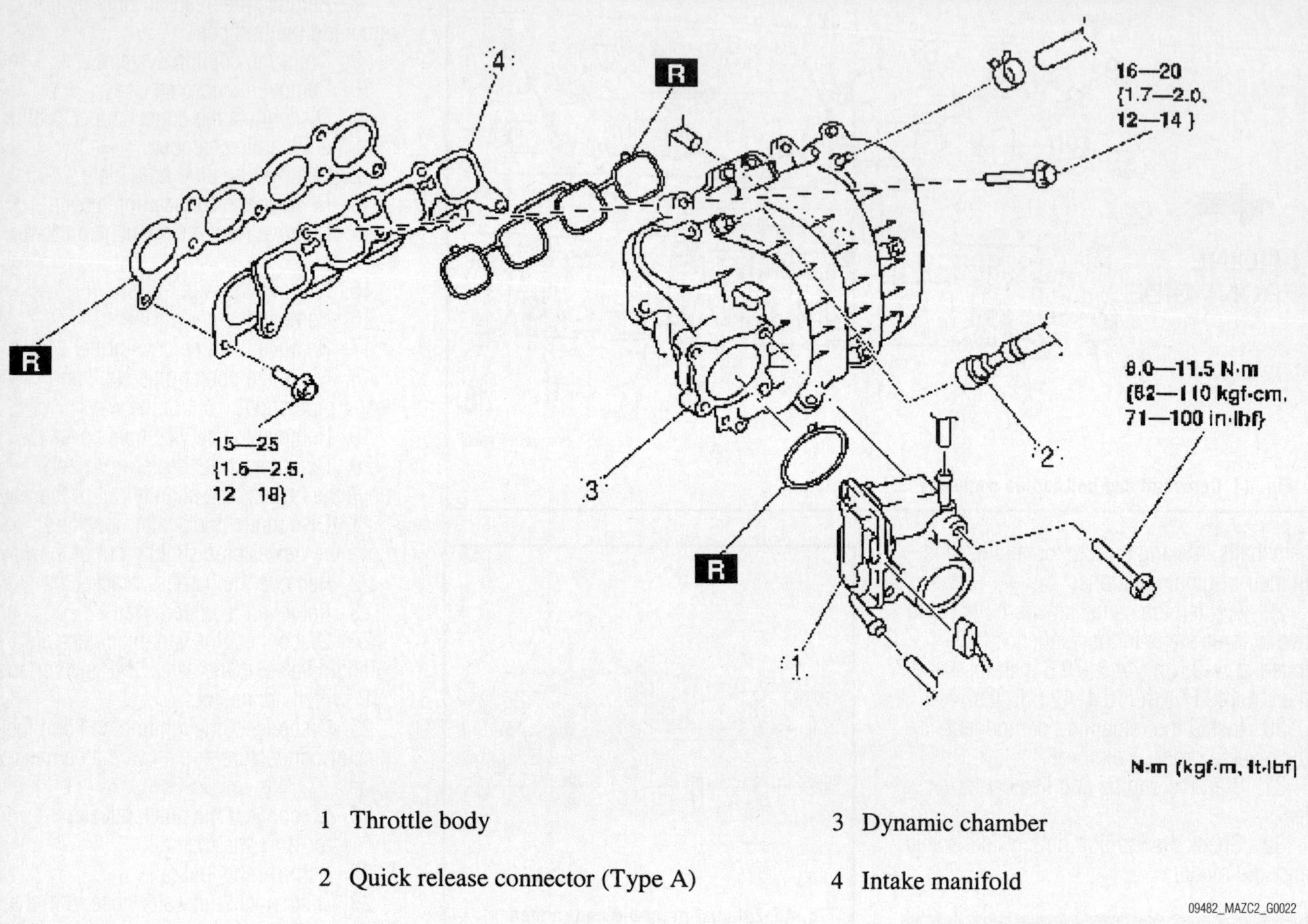

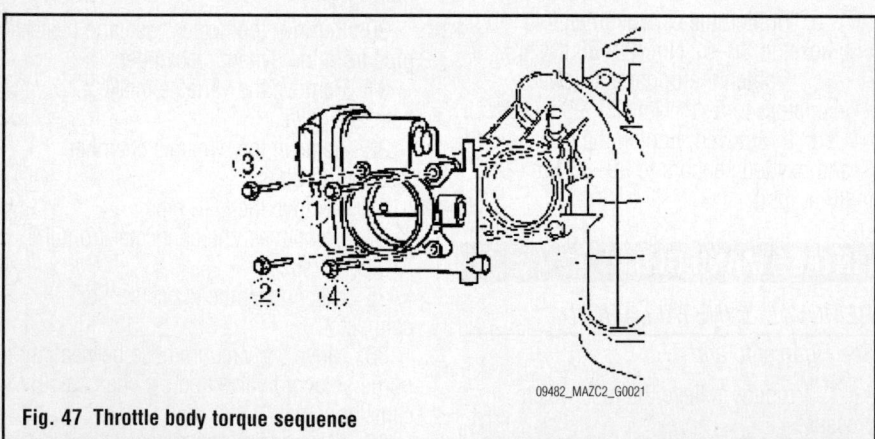

1 Throttle body

2 Quick release connector (Type A)

3 Dynamic chamber

4 Intake manifold

09482_MAZC2_G0022

Fig. 46 Exploded view of the dynamic chamber and intake manifold

Fig. 47 Throttle body torque sequence

09482_MAZC2_G0021

removal, make sure to tighten the throttle body bolts in the sequence shown to 8–11 Nm (71–100 inches lbs).

44. Check and adjust all fluid levels, start the vehicle and check for proper operation.

OIL PAN

REMOVAL & INSTALLATION

See Figures 48 through 55.

✳✳ **CAUTION**

Remove and install all parts when the engine is cold, otherwise they can cause severe burns or serious injury.

1. Remove the battery and battery tray.
2. Remove the air cleaner.
3. Drain the engine oil.
4. Loosen the water pump pulley bolt and remove the drive belt.

5. Remove the front suspension tower bar from the joint, right side and left side.
6. Remove the plug hole plate.
7. Remove the ignition coils.
8. Remove the P/S oil pump with the hose and pipe sill connected. Position the P/S oil pump out of the way.
9. Remove the Crankshaft Position (CKP) sensor.
10. Remove the engine front cover.
11. Remove the transverse member.
12. Remove the member bracket if equipped with a manual transmission.
13. Remove the windshield wiper arm.
14. Remove the cowl grille.
15. Remove the side cowl grille.
16. Remove the engine compartment service hole cover.
17. Remove the front tires.
18. Support the engine using the an engine support tool such as 49 ED17 5AO.
19. Remove the engine mount rubber installation nuts
20. Lift up the engine approximately 25 mm (0.98 inches) to assure clearance for the oil pan removal, then remove the oil pan bolts.

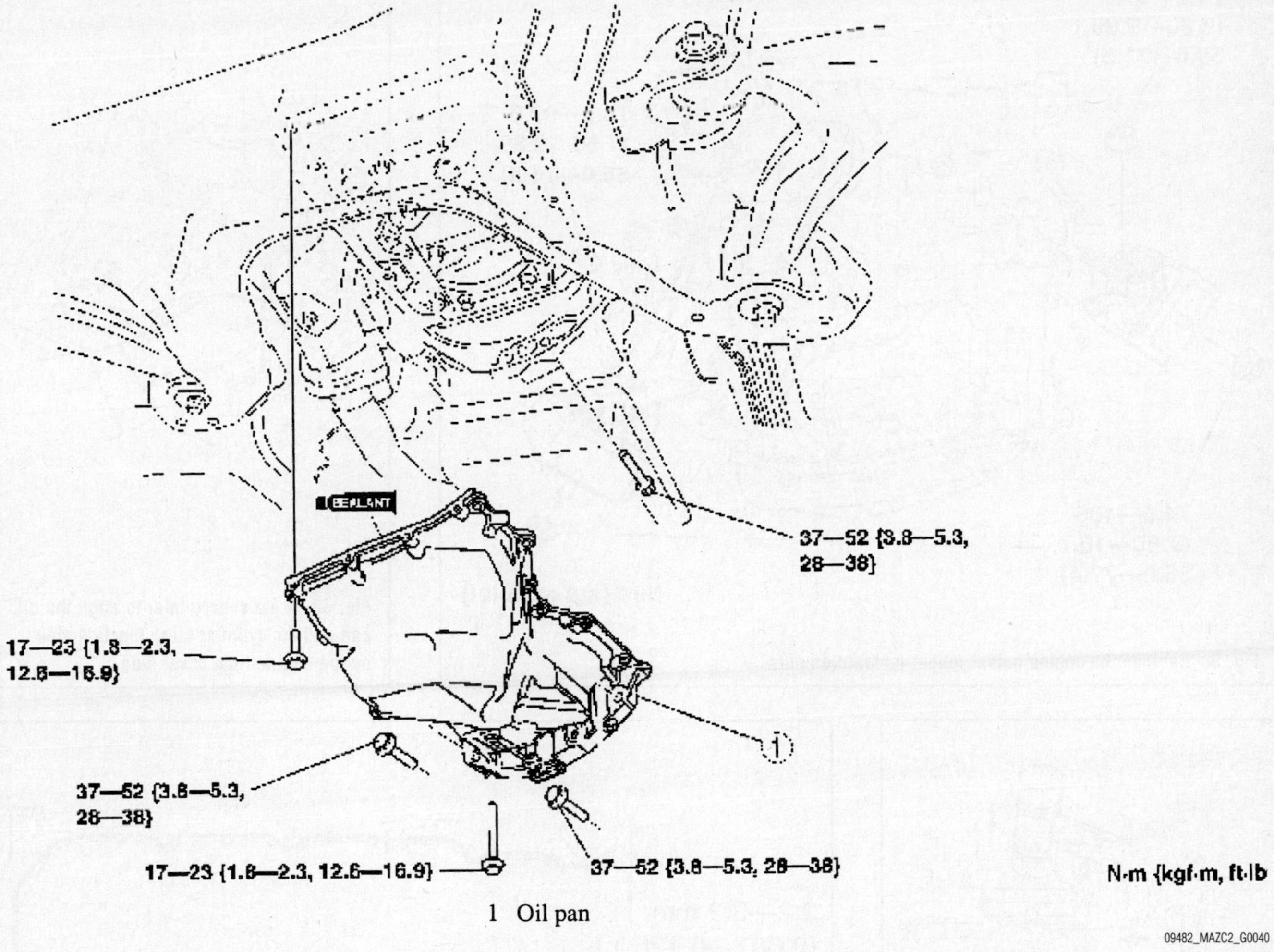

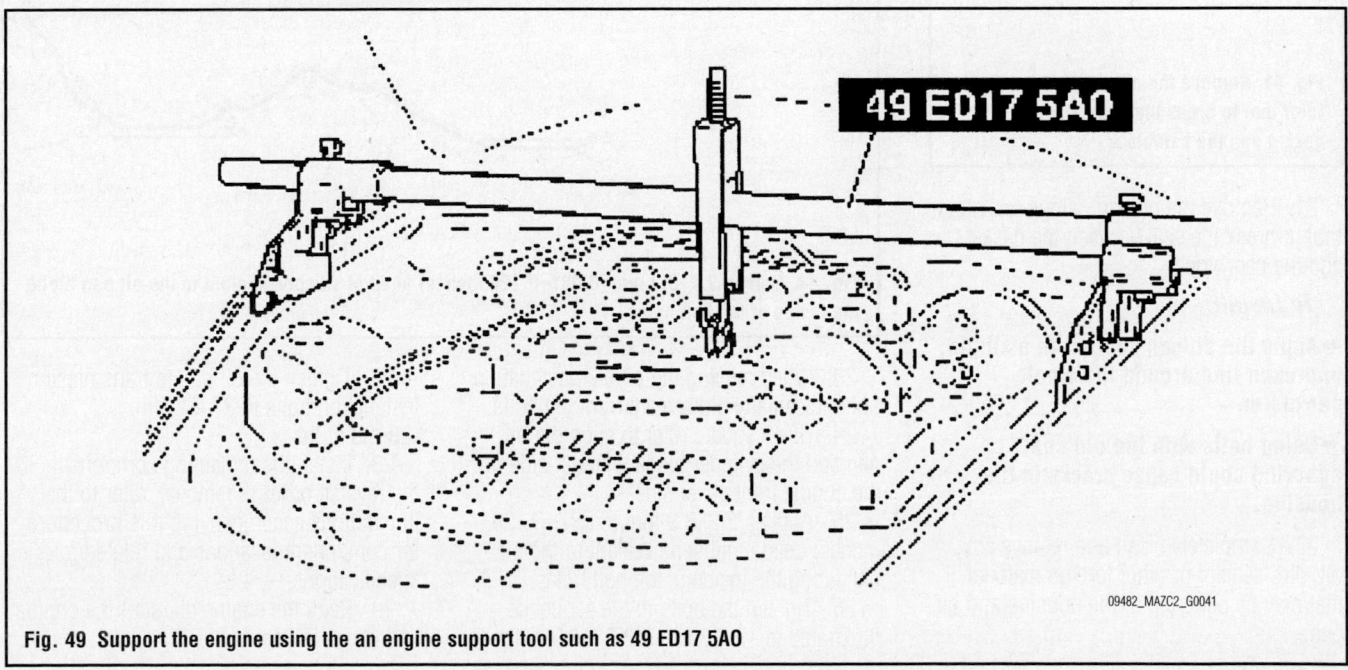

37—52 {3.8—5.3, 28—38}

17—23 {1.8—2.3, 12.8—16.9}

SEALANT

37—52 {3.8—5.3, 28—38}

17—23 {1.8—2.3, 12.8—16.9}

37—52 {3.8—5.3, 28—38}

N·m {kgf·m, ft·lb}

1   Oil pan

09482_MAZC2_G0040

Fig. 48 Exploded view of the oil pan mounting

49 E017 5AO

09482_MAZC2_G0041

Fig. 49 Support the engine using the an engine support tool such as 49 ED17 5AO

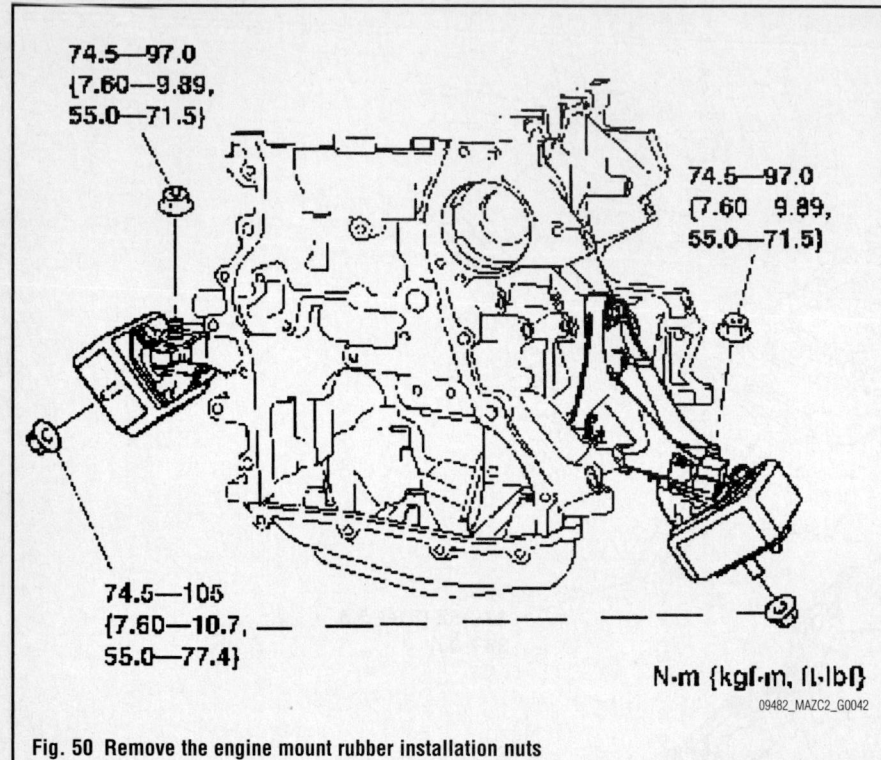

N·m {kgf·m, ft·lbf}

Fig. 50 Remove the engine mount rubber installation nuts

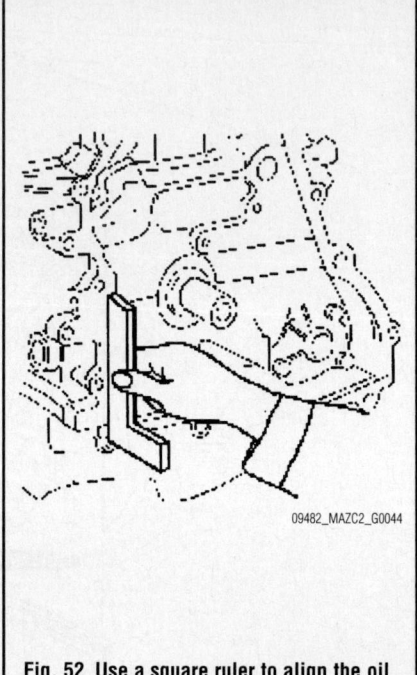

Fig. 52 Use a square ruler to align the oil pan and the cylinder block junction side on the engine front cover side

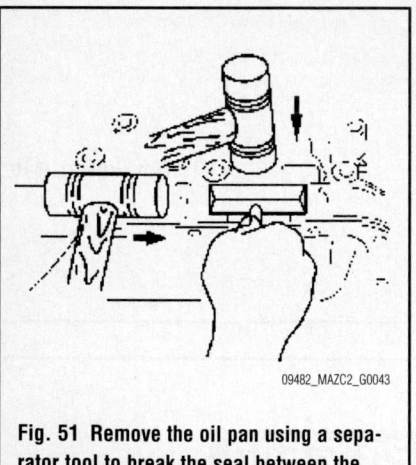

Fig. 51 Remove the oil pan using a separator tool to break the seal between the gasket and the pan/block

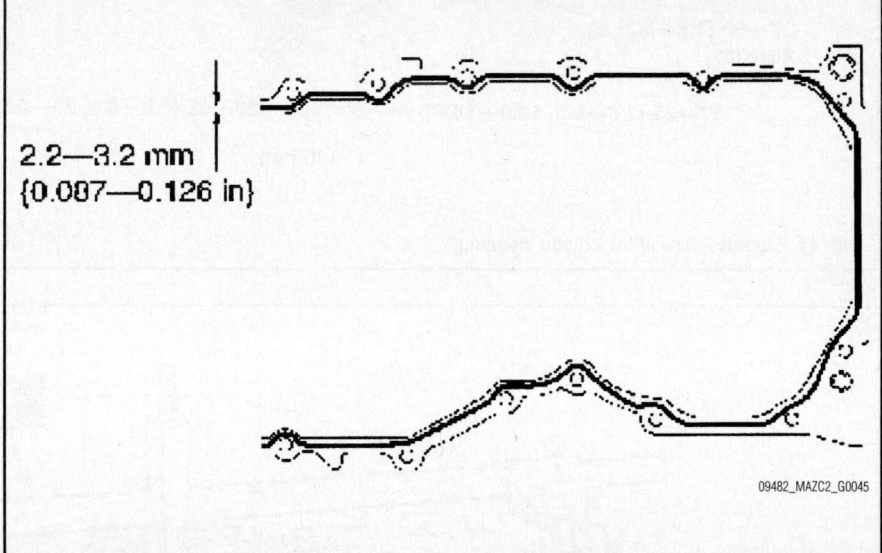

2.2—3.2 mm {0.087—0.126 in}

Fig. 53 Apply a 2.2–3.2 mm (0.087–0.126 inches) bead of silicone sealant to the oil pan along the inside of the bolt holes

21. Remove the oil pan using a separator tool to break the seal between the gasket and the pan/block.

To install:

➡ **Apply the silicon sealant in a single, unbroken line around the whole perimeter.**

➡ **Using bolts with the old seal adhering could cause cracks in the housing.**

22. Completely clean and remove any oil, dirt, sealant or other foreign material that may be adhering to the housing and oil pan.

23. When reusing the oil pan installation bolts, clean any old sealant from the bolts.

24. Use a square ruler to align the oil pan and the cylinder block junction side on the engine front cover side.

25. Apply a 2.2–3.2 mm (0.087–0.126 inches) bead of silicone sealant to the oil pan along the inside of the bolt holes.

26. Tighten the bolts in the sequence illustrated to 17–23 Nm (12.6–16.9 ft. lbs.).

27. Tighten the oil pan to transmission installation bolts to 37–52 Nm (28–38 ft. lbs.).

28. Install the remaining components in the reverse order of removal, refer to the illustrations accompanying this procedure for component locations and related torque specifications.

29. Refill the engine oil, start the engine and check for leaks.

8
4
1
5
9
3
10
11
12
7
3
2
6
09482_MAZC2_G0046

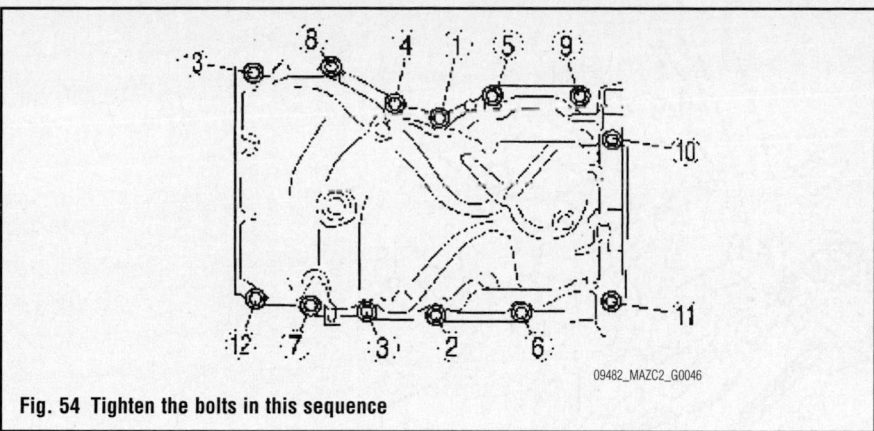

Fig. 54 Tighten the bolts in this sequence

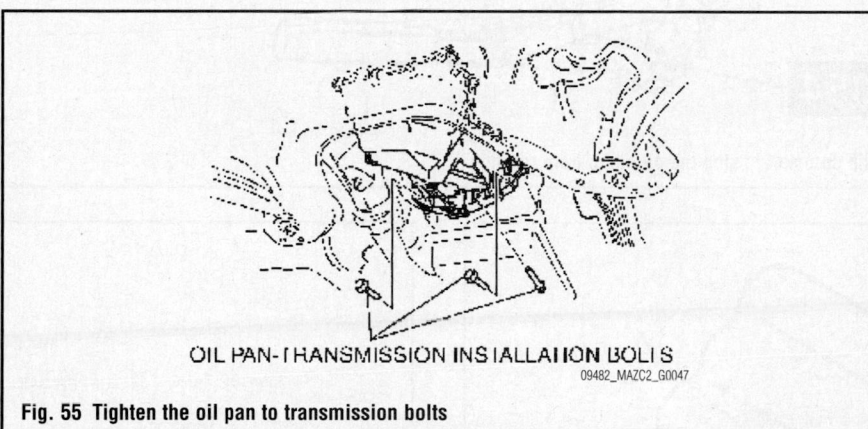

OIL PAN-TRANSMISSION INSTALLATION BOLTS
09482_MAZC2_G0047

Fig. 55 Tighten the oil pan to transmission bolts

OIL PUMP

REMOVAL & INSTALLATION

See Figures 56 through 58.

**** CAUTION**

Remove and install all parts when the engine is cold, otherwise they can cause severe burns or serious injury.

1. Remove the oil pan.
2. Remove the oil strainer.
3. Remove the oil pump chain guide.
4. Remove the oil pump chain tensioner.
5. Remove the oil pump chain.
6. Install the tool 49 G032 354 onto the oil pump sprocket to stop the oil pump from rotating.
7. Remove the oil pump sprocket.
8. Remove the oil pump bolts and pump.

To install:

9. Install the oil pump, tighten the bolts in the sequence illustrated to 8–12 Nm

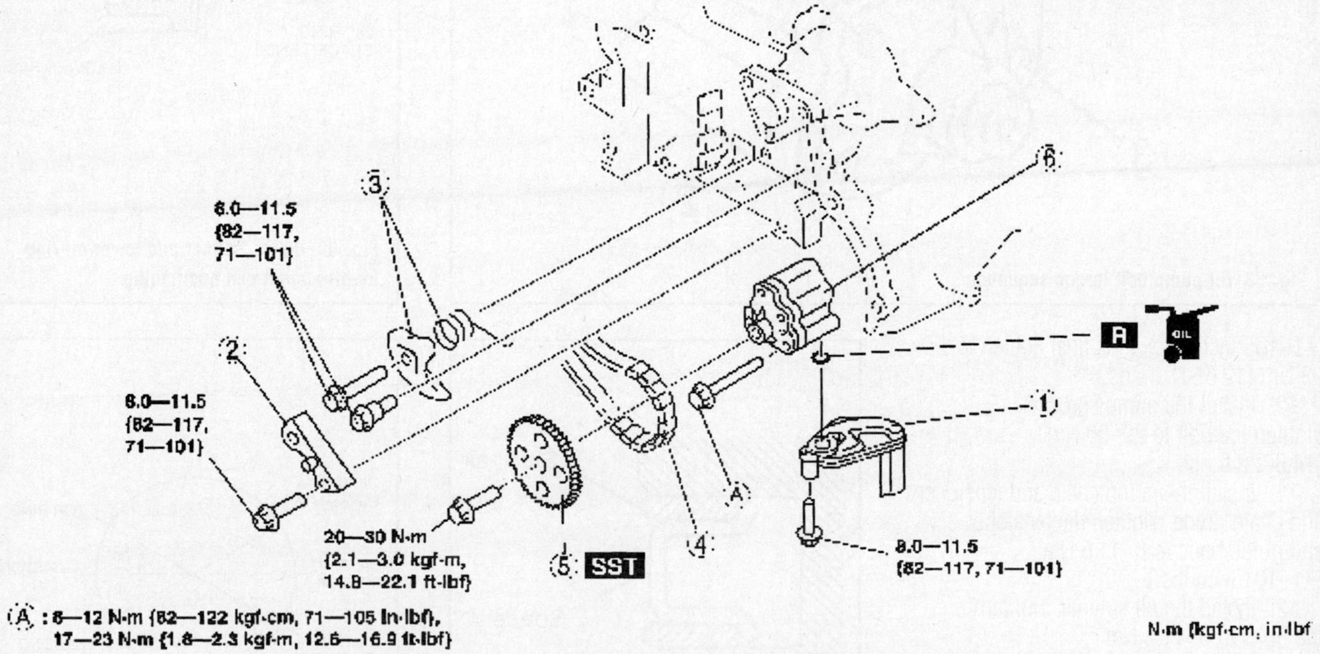

6
3
8.0—11.5 {82—117, 71—101}
2
8.0—11.5 {82—117, 71—101}
R
OIL
1
A
20—30 N·m {2.1—3.0 kgf·m, 14.8—22.1 ft·lbf}
5 SST
4
8.0—11.5 {82—117, 71—101}
A : 8—12 N·m {82—122 kgf·cm, 71—105 in·lbf}, 17—23 N·m {1.8—2.3 kgf·m, 12.6—16.9 ft·lbf}
N·m {kgf·cm, in·lbf}

1 Oil strainer
2 Oil pump chain guide
3 Oil pump chain tensioner
4 Oil pump chain
5 Oil pump sprocket
6 Oil pump

09482_MAZC2_G0048

Fig. 56 Exploded view of the oil pump and related components

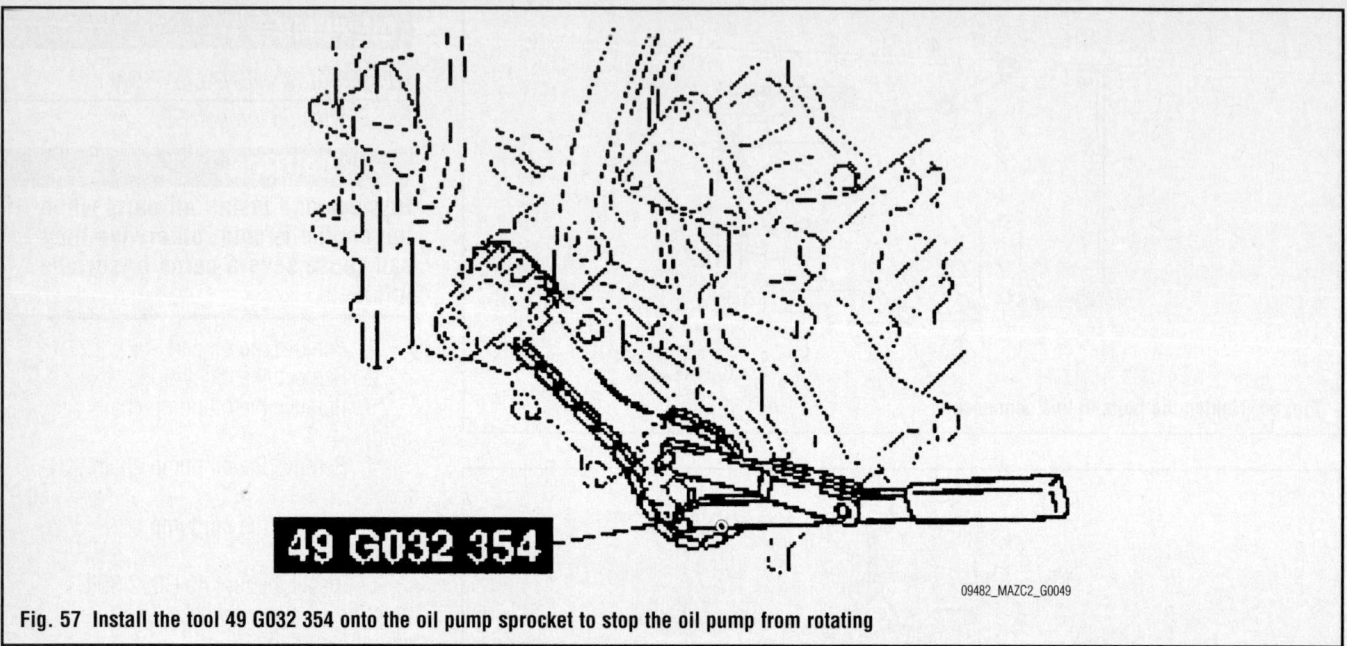

Fig. 57 Install the tool 49 G032 354 onto the oil pump sprocket to stop the oil pump from rotating

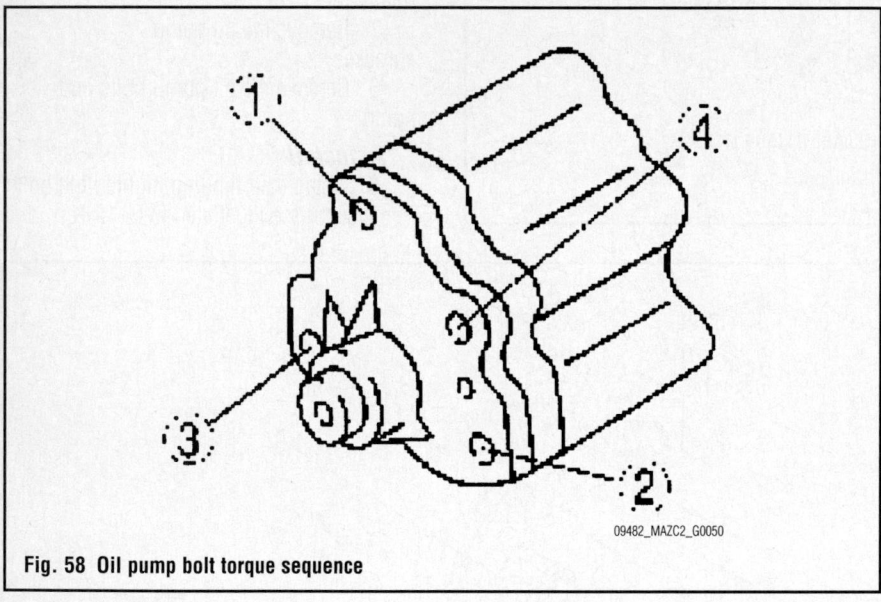

Fig. 58 Oil pump bolt torque sequence

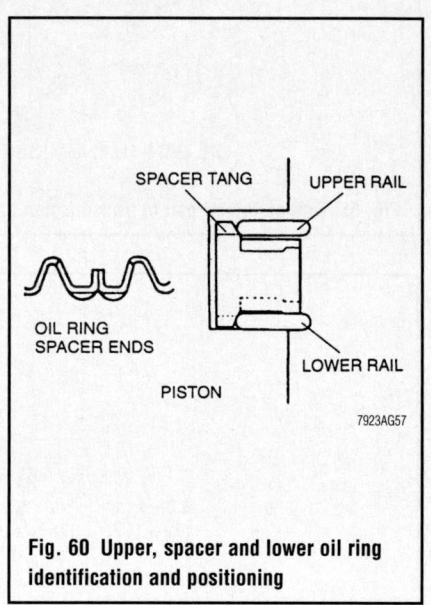

Fig. 60 Upper, spacer and lower oil ring identification and positioning

(71–105 inch lbs.), then final tighten to 17–23 Nm (12.6–17 ft. lbs.).

10. Install the pump sprocket. Tighten the bolt to 20–30 Nm (14.8–22 ft. lbs.).

11. Install the pump chain and tensioner and chain guide. Tighten the tensioner and guide bolts to 8–11.5 Nm (71–101 inch lbs.).

12. Install the oil strainer. Tighten the bolts to 8–11.5 Nm (71–101 inch lbs.).

13. Install the oil pan.

PISTON AND RING

POSITIONING

See Figures 59 through 61.

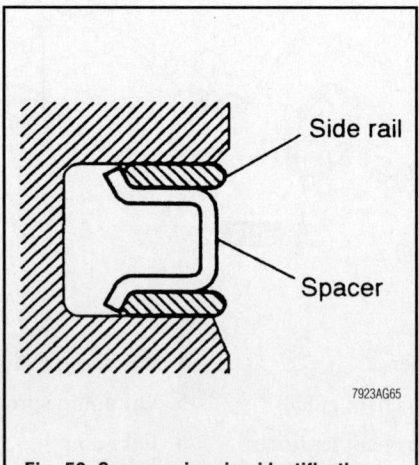

Fig. 59 Compression ring identification and positioning

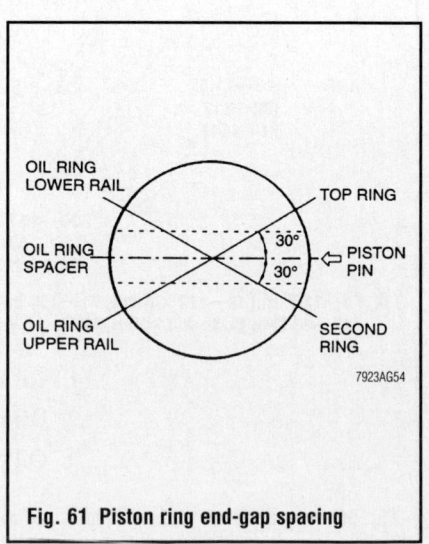

Fig. 61 Piston ring end-gap spacing

REAR MAIN SEAL

REMOVAL & INSTALLATION

See Figures 62 through 65.

1. Remove the flywheel on manual transmissions, or the drive plate on automatic transmissions.

2. Remove the bolt and the rear oil seal.

To install:

3. Apply a 4–6 mm bead of silicone sealant to the mating faces as illustrated.

4. Apply clean engine oil to the new oil seal lip.

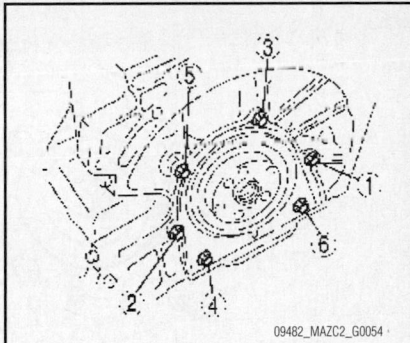

09482_MAZC2_G0054

Fig. 65 Tighten the rear oil seal bolts in this sequence

5. Install the rear oil seal using a tool such as 303-328.

6. Tighten the rear oil seal bolts in the sequence illustrated to 8–11.5 Nm (71–101 inch lbs.).

7. Install the flywheel or drive plate.

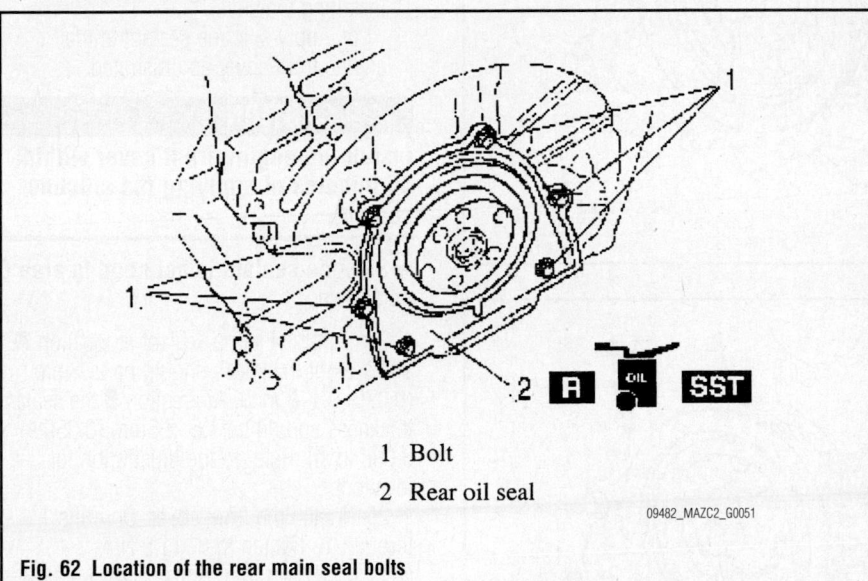

1 Bolt

2 Rear oil seal

09482_MAZC2_G0051

Fig. 62 Location of the rear main seal bolts

TIMING CHAIN FRONT COVER

REMOVAL & INSTALLATION

See Figures 66 through 70.

1. Remove the battery and battery tray.

2. Remove the air cleaner.

3. Disconnect the ventilation hose.

4. Loosen the water pump pulley bolt and removal the drive belt.

5. Remove the front suspension tower bar.

6. Remove the camshaft position sensor.

7. Disconnect the OCV connector.

8. Remove the ignition coils.

9. Remove the drive belt.

10. Remove the undercover.

11. Remove the crankshaft position sensor.

12. Remove the P/S oil pump with the oil hose still connected and position the P/S oil pump so that it is out of the way.

13. Move the cooler pipe No.3 and heater pipe slightly out of the way.

14. Remove the dipstick.

15. Remove the cylinder head cover.

16. Remove the crankshaft pulley lock bolt as follows:

    a. Remove the cylinder block lower blind plug.

    b. Install tool 303-507.

    c. Turn the crankshaft clockwise until the crankshaft is in the No.1 cylinder TDC position (until the balance weight contacts tool 303-507).

    d. Hold the crankshaft pulley using the tools shown in the illustration.

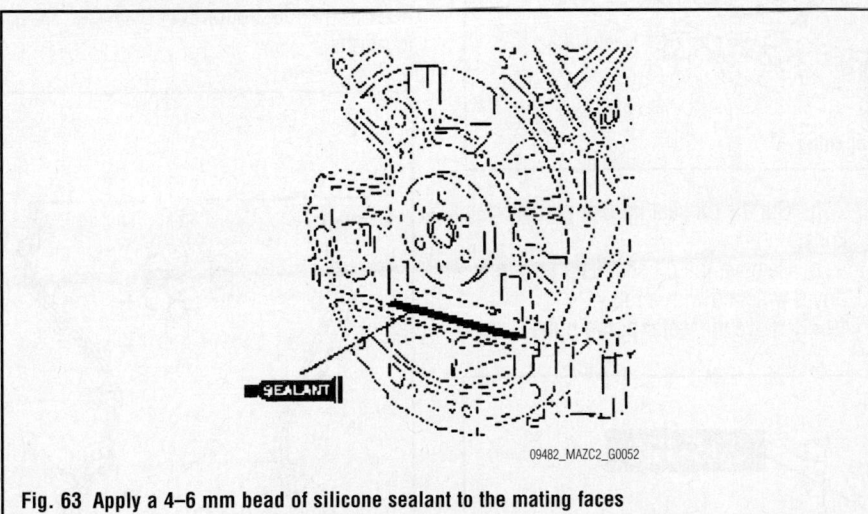

09482_MAZC2_G0052

Fig. 63 Apply a 4–6 mm bead of silicone sealant to the mating faces

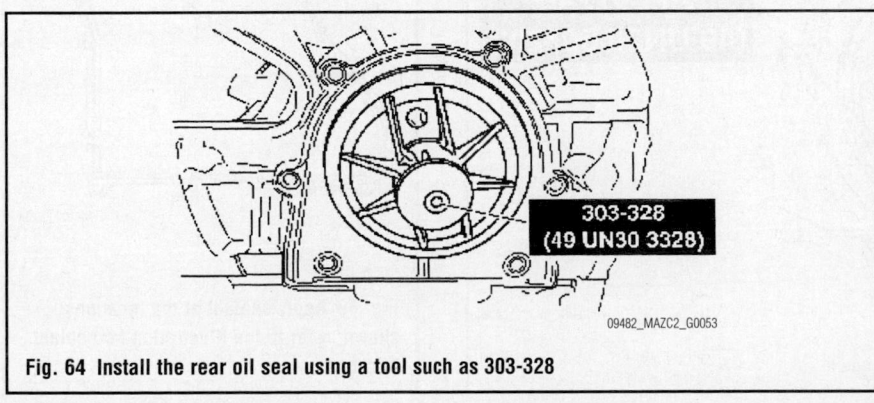

303-328
(49 UN30 3328)

09482_MAZC2_G0053

Fig. 64 Install the rear oil seal using a tool such as 303-328

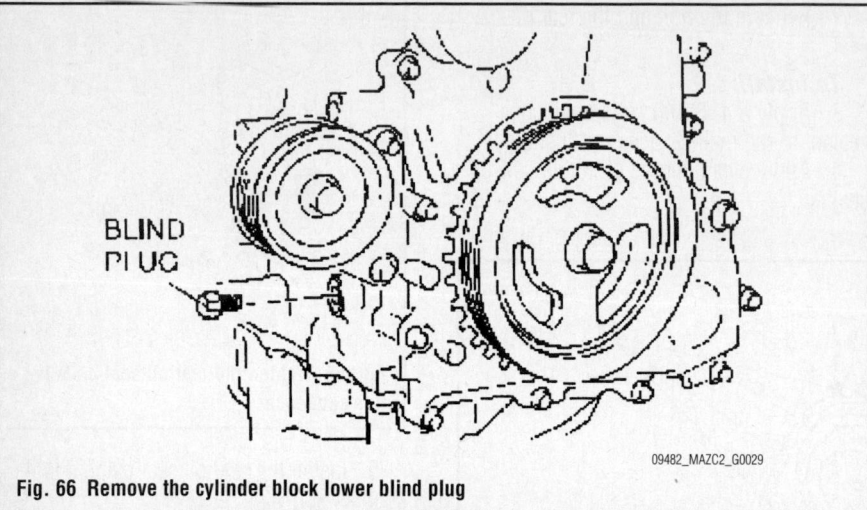

Fig. 66 Remove the cylinder block lower blind plug

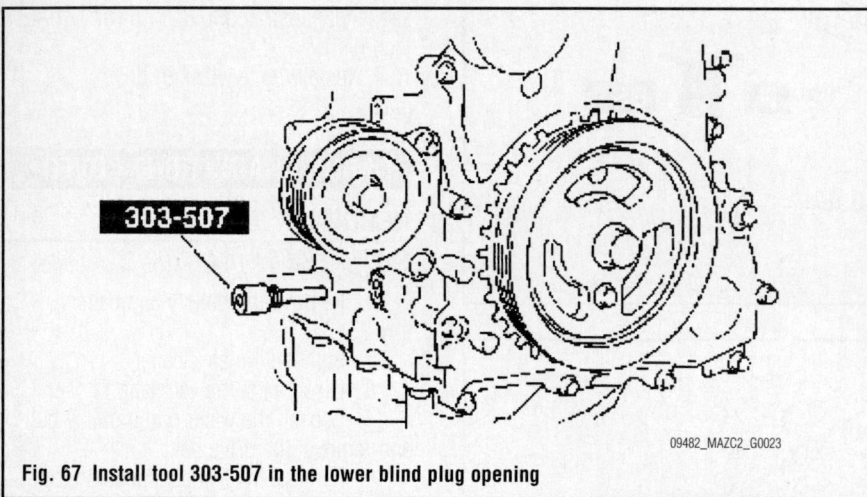

Fig. 67 Install tool 303-507 in the lower blind plug opening

e. Remove the crankshaft pulley lock bolt.

17. Remove the crankshaft pulley.

18. Remove the front oil seal as follows:

a. Cut the oil seal lip using a razor knife.

b. Remove the oil seal using a screwdriver wrapped with a rag.

19. Remove the water pump pulley.

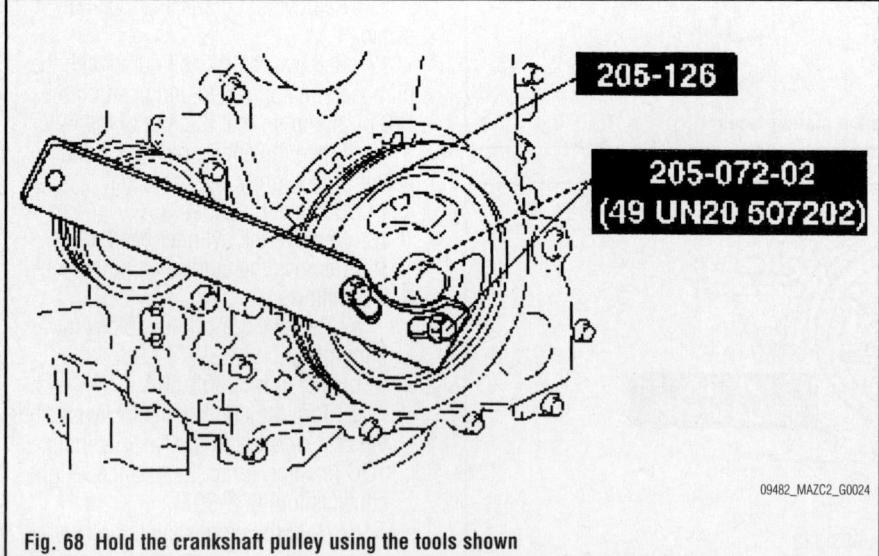

Fig. 68 Hold the crankshaft pulley using the tools shown

20. Remove the drive belt idler pulley.

21. Remove the engine front cover.

22. Remove the oil seal using a screwdriver.

To install:

23. Install the engine front cover as follows:

a. Install the front oil seal before installing the cover.

b. Apply silicone sealant to the engine front cover as illustrated.

❋❋ CAUTION

Install the engine front cover within 10 minutes of applying the silicone sealant.

➡Silicone sealant is not need in area **C** as shown.

24. Apply silicone sealant at location **A**. The sealant thickness should be 2–3mm (0.079–0.118 inch. At location **B** the sealant thickness should be 1.5–2.5mm (0.059–0.098 inch). Refer to the illustration for locations.

25. Install the front cover. On bolts 1 through 18 tighten to 8–11.5 Nm (71–101 inches lbs). Bolts 19 through 22, tighten to 20–55 Nm (30–40 ft. lbs.). Refer to the illustration for locations.

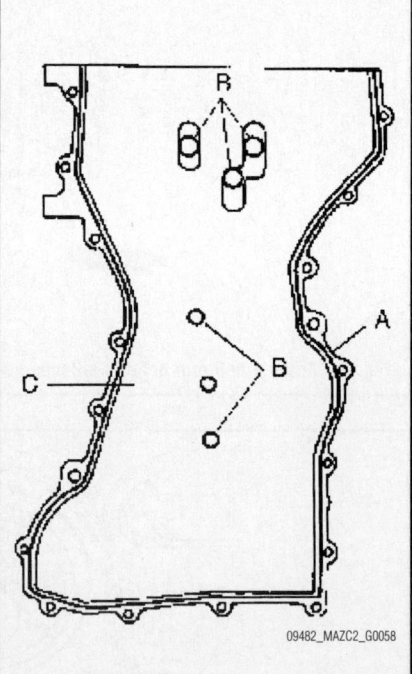

Fig. 69 Apply sealant at the locations shown; refer to the illustration for sealant thickness

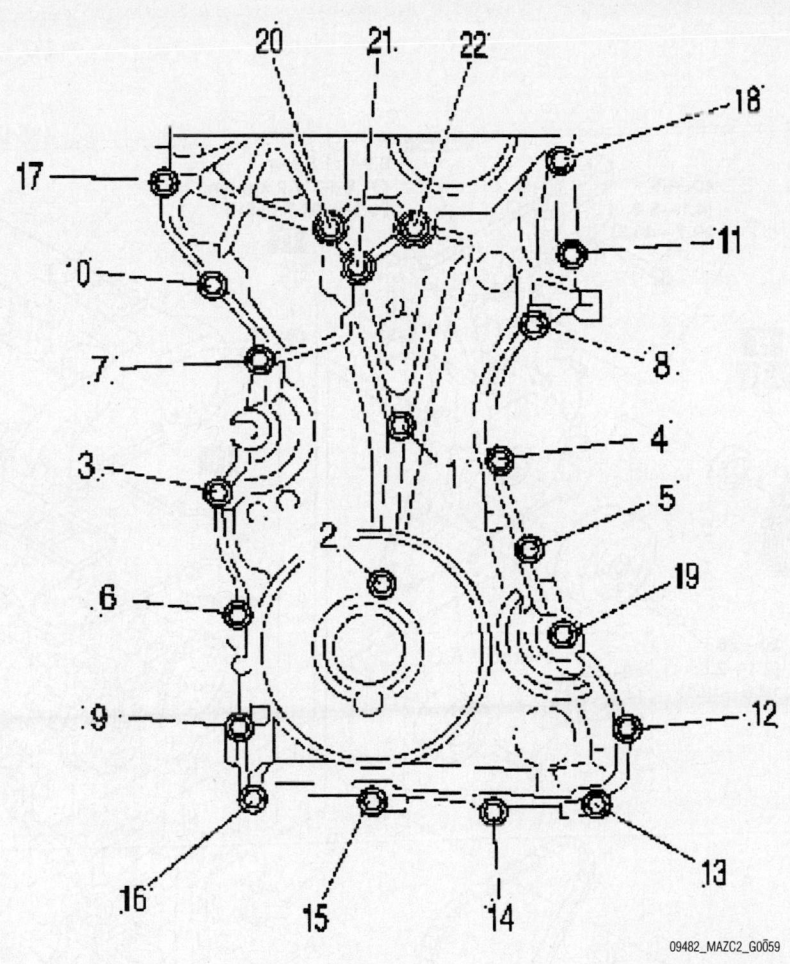

Fig. 70 Front cover bolt locations

TIMING CHAIN & SPROCKETS

REMOVAL & INSTALLATION

See Figures 66 through 68, 71 through 80.

1. Remove the battery and battery tray.
2. Remove the air cleaner.
3. Disconnect the ventilation hose.
4. Loosen the water pump pulley bolt and removal the drive belt.
5. Remove the front suspension tower bar.
6. Remove the camshaft position sensor.
7. Disconnect the OCV connector.
8. Remove the ignition coils.
9. Remove the drive belt.
10. Remove the undercover.
11. Remove the crankshaft position sensor.
12. Remove the P/S oil pump with the oil hose still connected and position the P/S oil pump so that it is out of the way.
13. Move the cooler pipe No.3 and heater pipe slightly out of the way.

14. Remove the dipstick.
15. Remove the cylinder head cover.
16. Remove the crankshaft pulley lock bolt as follows:

a. Remove the cylinder block lower blind plug.
b. Install tool 303-507.
c. Turn the crankshaft clockwise until the crankshaft is in the No.1

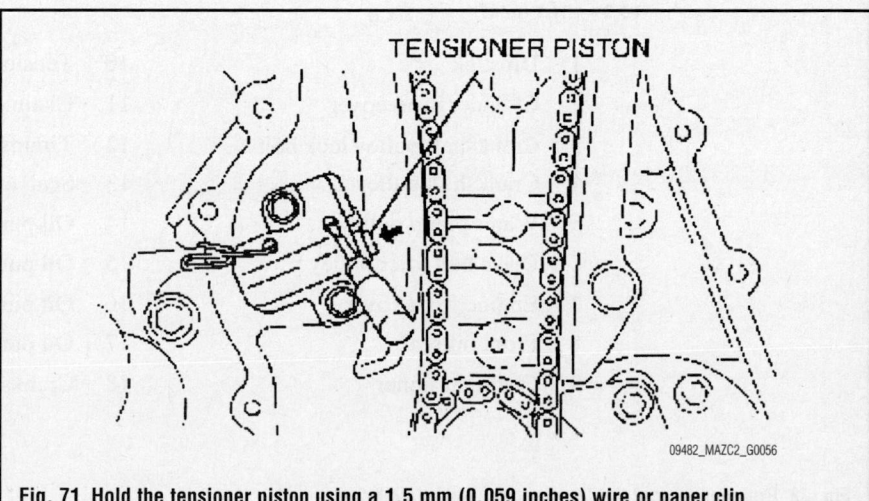

Fig. 71 Hold the tensioner piston using a 1.5 mm (0.059 inches) wire or paper clip

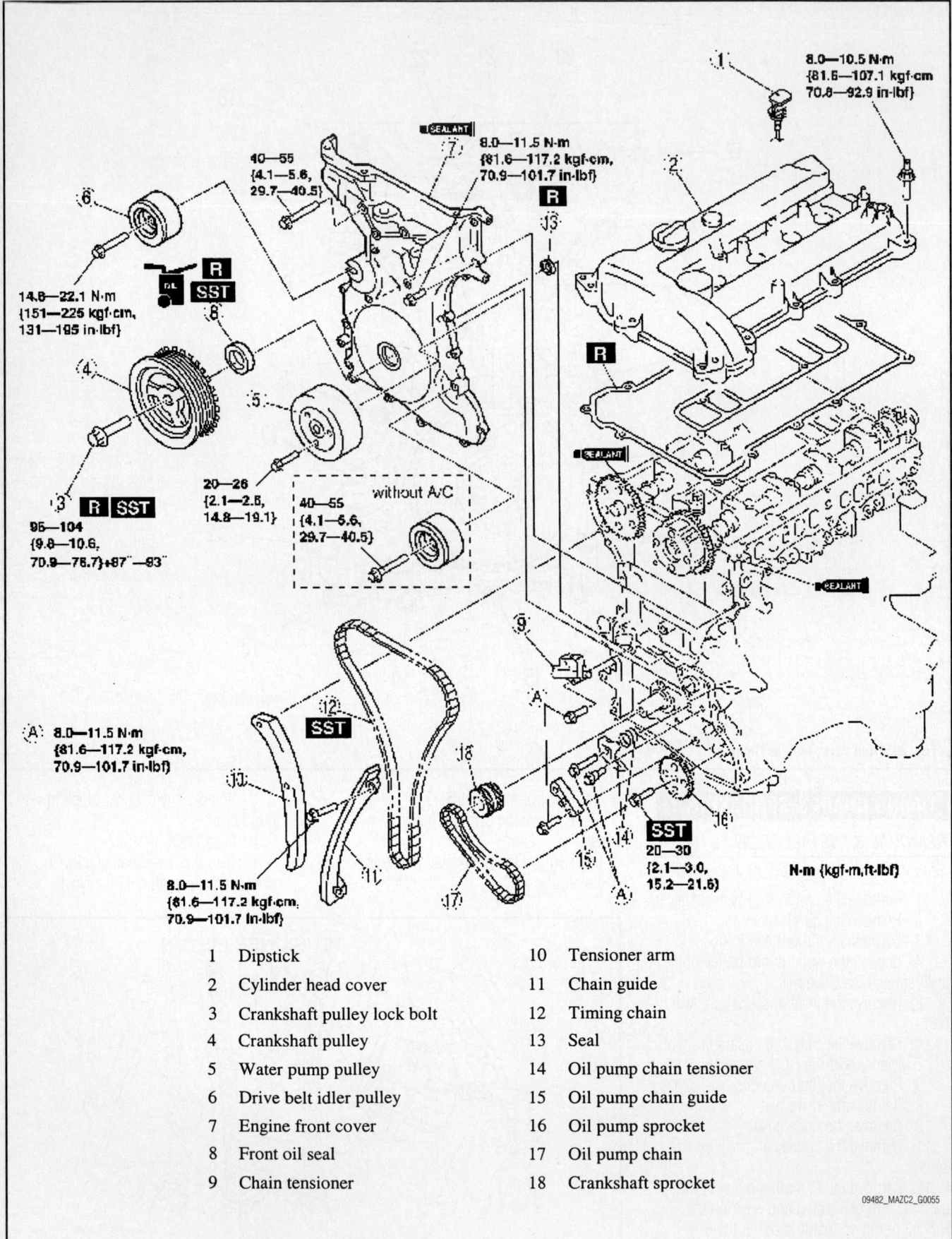

1	Dipstick	10	Tensioner arm
2	Cylinder head cover	11	Chain guide
3	Crankshaft pulley lock bolt	12	Timing chain
4	Crankshaft pulley	13	Seal
5	Water pump pulley	14	Oil pump chain tensioner
6	Drive belt idler pulley	15	Oil pump chain guide
7	Engine front cover	16	Oil pump sprocket
8	Front oil seal	17	Oil pump chain
9	Chain tensioner	18	Crankshaft sprocket

09482_MAZC2_G0055

Fig. 72 Exploded view of the timing chain and related components

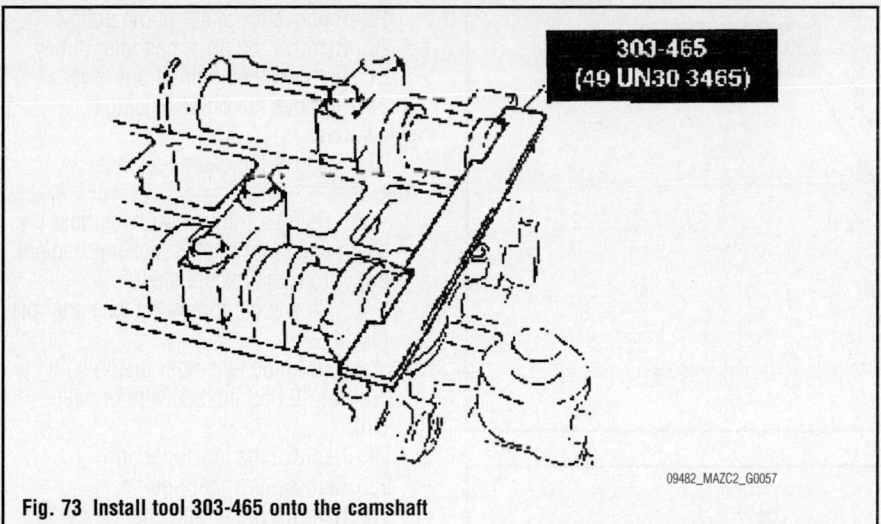

Fig. 73 Install tool 303-465 onto the camshaft

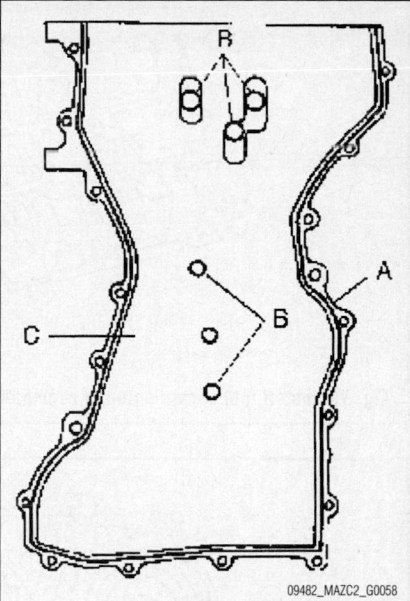

Fig. 74 Apply sealant at the locations shown in the illustration. Refer to the illustration for sealant thickness

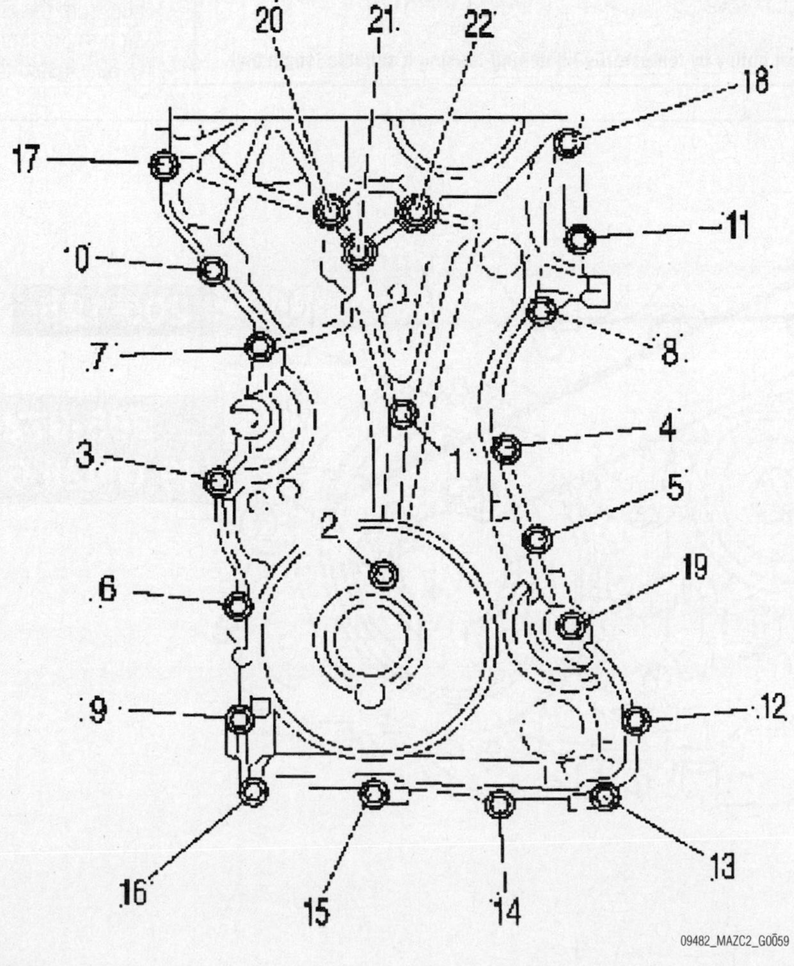

cylinder TDC position (until the balance weight contacts tool 303-507).

d. Hold the crankshaft pulley using the tools shown in the illustration.

e. Remove the crankshaft pulley lock bolt.

17. Remove the crankshaft pulley.
18. Remove the front oil seal as follows:

a. Cut the oil seal lip using a razor knife.

b. Remove the oil seal using a screwdriver wrapped with a rag.

Fig. 75 Front cover bolt locations

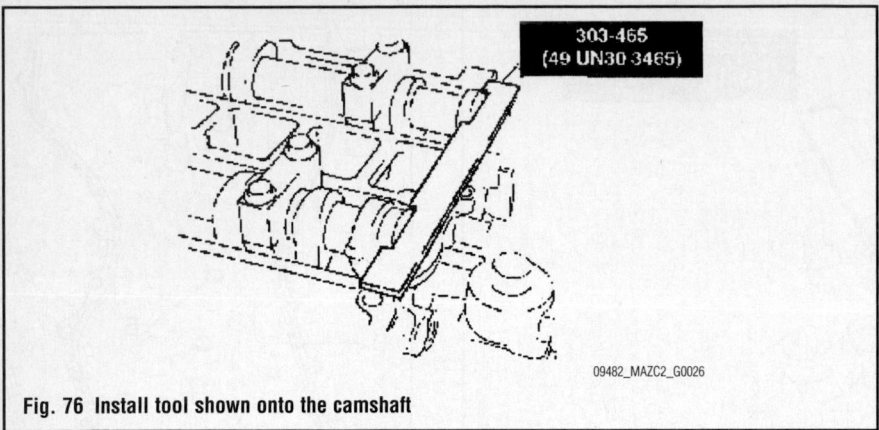

Fig. 76 Install tool shown onto the camshaft

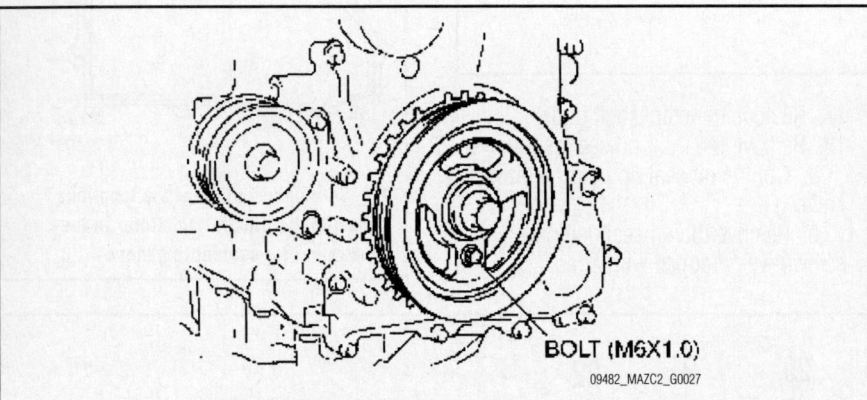

Fig. 77 Position the crankshaft pulley by temporarily tightening it using a suitable length bolt

19. Remove the water pump pulley.
20. Remove the drive belt idler pulley.
21. Remove the engine front cover.
22. Remove the oil seal using a screwdriver.
23. Remove the chain tensioner as follows:

   a. Using a thin screwdriver, hold the chain tensioner ratchet lock mechanism away from the ratchet stem.

   b. Slowly compress the tensioner piston.

   c. Hold the tensioner piston using a 1.5 mm (0.059 inches) wire or paper clip.

24. Remove the tensioner arm.
25. Remove the chain guide.
26. Remove the timing chain

To install:

27. Install tool 303-465 onto the camshaft as illustrated.
28. Install the timing chain.
29. Install the chain guide. Tighten the bolt to 8–11.5 Nm (71–101 inch lbs.).
30. Install the tensioner arm. Tighten the bolt to 8–11.5 Nm (71–101 inch lbs.).
31. Remove the retaining wire or paper clip from the chain tensioner to apply tension to the timing chain.
32. Install the oil seal.

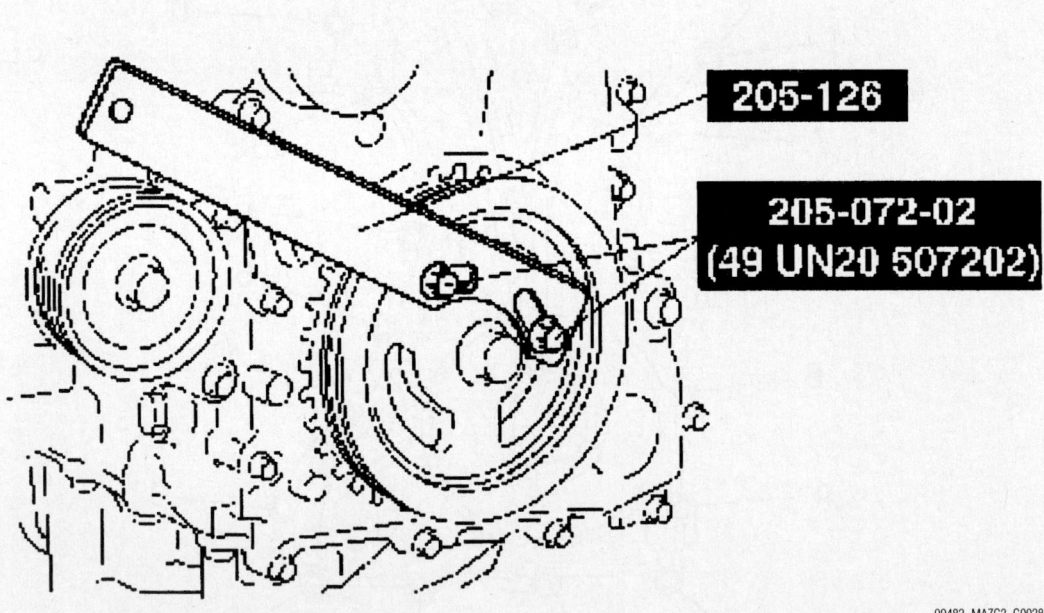

Fig. 78 Install the tools shown onto the crankshaft pulley to lock the crankshaft against rotation

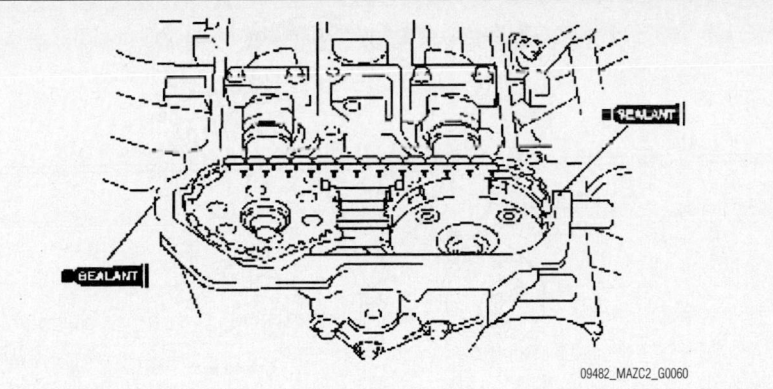

Fig. 79 Prior to install the cylinder head cover, apply a 4–6mm (0.16–0.23 in) bead of sealant at these locations

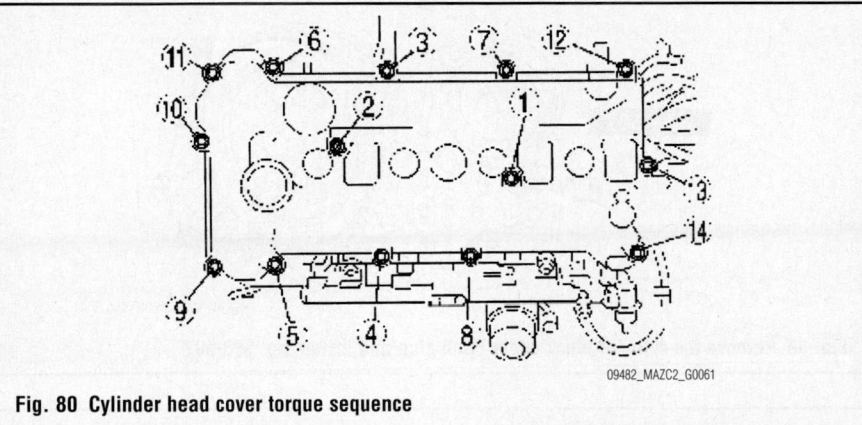

Fig. 80 Cylinder head cover torque sequence

33. Install the engine front cover as follows:

a. Install the front oil seal before installing the cover.

b. Apply silicone sealant to the engine front cover as illustrated.

✳✳ CAUTION

Install the engine front cover within 10 minutes of applying the silicone sealant.

➡**Silicone sealant is not need in area C as shown.**

34. Apply silicone sealant at location **A**. The sealant thickness should be 2–3mm (0.079–0.118 inch. At location **B** the sealant thickness should be 1.5–2.5mm (0.059–0.098 inch). Refer to the illustration for locations.

35. Install the front cover. On bolts 1 through 18 tighten to 8–11.5 Nm (71–101 inch lbs). Bolts 19 through 22, tighten to 20–55 Nm (30–40 ft. lbs.). Refer to the illustration for locations.

36. Install the crankshaft pulley lock bolt a3 follows:

a. Install tool shown in the illustration on the camshaft.

b. Verify that cylinder No.1 is at TDC of the compression stroke. (Crankshaft balance weight contacts tool 303-507)

c. Position the crankshaft pulley by temporarily tightening it and, using a suitable length bolt 25–35mm (0.99–1.37 inches), fix the crankshaft pulley to the engine front cover.

d. Install the tools illustrated to the crankshaft pulley, lock the crankshaft against rotation.

e. Tighten the crankshaft pulley lock bolt to 96–104 Nm (71–77 ft. lbs.), plus 87–93 degrees.

f. Remove the bolt.

g. Remove the tool the camshaft.

h. Remove tool from the cylinder block lower blind plug.

i. Remove the tool from the crankshaft pulley.

j. Rotate the crankshaft clockwise two turns until the TDC position.

k. If not aligned, loosen the crankshaft pulley lock bolt and repeat from the first step.

l. Install the cylinder block lower blind plug and tighten to 18–22 Nm (13–16 ft. lbs.).

37. Install the cylinder head cover bolts and tighten in the sequence illustrated to 8–10.5 Nm (70–93 inch lbs.).

38. Install the remaining components in the reverse order of removal.

VALVE LASH

ADJUSTMENT

See Figures 81 through 88.

These engines use solid cam followers with a removable adjustment shim. The valve lash clearance is measured with the original shim installed and checked against the specification. If adjustment is necessary, the original shim is removed, and a thicker or thinner shim is installed to obtain the proper clearance. Special tools are required in order to adjust the shim without removing the camshaft.

The valve clearance should be measured with the engine cold, the specifications are:

• Intake: 22–28mm (0.0087-0.0110 inches)

• Exhaust: 0.27–0.33 (0.0107-0.0129 inches)

1. Before servicing the vehicle, refer to the precautions section.

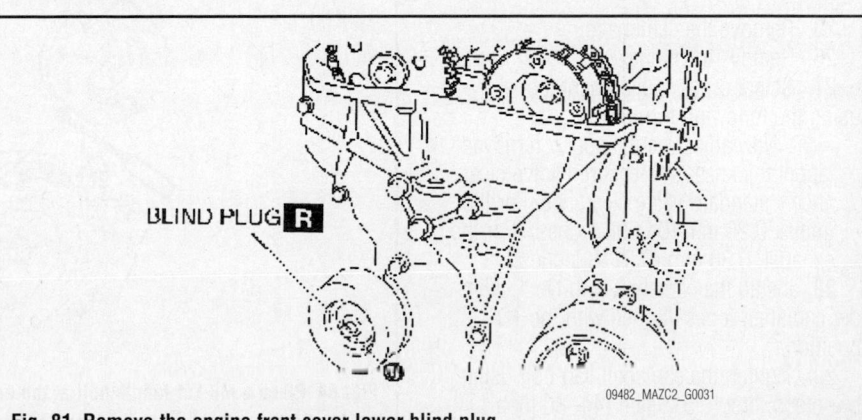

Fig. 81 Remove the engine front cover lower blind plug

2. Remove the battery cover.

3. Disconnect the negative battery cable.

4. Remove the plug hole plate.

5. Disconnect the ventilation hose.

6. Remove the front suspension tower bar.

7. Remove the CMP sensor.

8. Disconnect the OCV connector.

9. Disconnect the P/S pressure switch connector.

10. Remove the ignition coils.

11. Remove the cylinder head cover.

12. Remove the drive belt.

13. Remove the engine front cover lower blind plug.

14. Remove the engine front cover upper blind plug.

15. Remove the cylinder block lower blind plug.

16. Install tool 303-507 as shown.

17. Turn the crankshaft clockwise until the crankshaft is in the No.1 cylinder TDC position.

18. Loosen the timing chain.

19. Using a suitable screwdriver or equivalent tool, unlock the chain tensioner ratchet.

20. Turn the exhaust camshaft clockwise using a suitable wrench on the cast hexagon and loosen the timing chain.

21. Place a M6 X 1 length bolt at the engine front cover upper blind plug, secure the chain guide at the position where the tension is released.

22. Hold the exhaust camshaft using a suitable wrench on the cast hexagon as illustrated.

23. Remove the exhaust camshaft sprocket.

24. Loosen the camshaft cap bolts in several passes in the order shown.

➡**The cylinder head and the camshaft caps are numbered to make sure they are reassembled in their original position. Do not mix the caps.**

25. Remove the camshafts.

26. Remove the tappet.

27. Select proper adjustment tappet using the following formula:

 a. New adjustment tappet = removed tappet thickness + Measured valve clearance - standard valve clearance. For the intake: 0.25 mm (0.0098 inches). For the exhaust: 0.30 mm (0.0118 inches).

28. Install the camshaft with No.1 cylinder camshaft lobes aligned with the TDC position.

29. Tighten the camshaft cap bolt using two steps, first to 5–9 Nm (44–80 inch lbs.), then to 14–17 Nm (10–12.5 ft. lbs.).

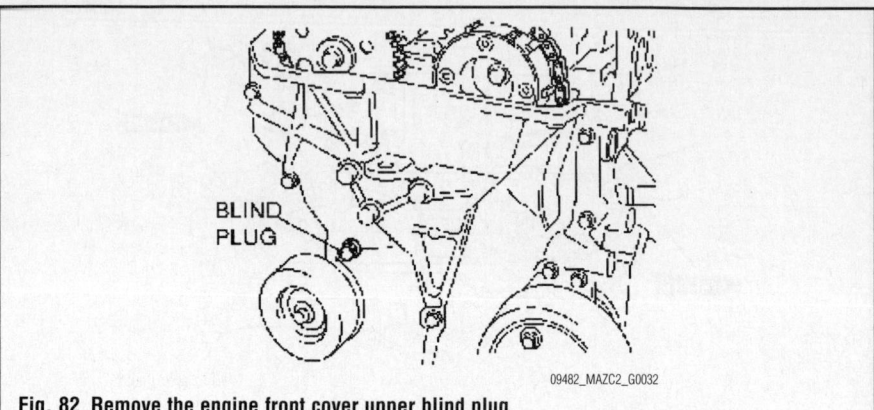

Fig. 82 Remove the engine front cover upper blind plug

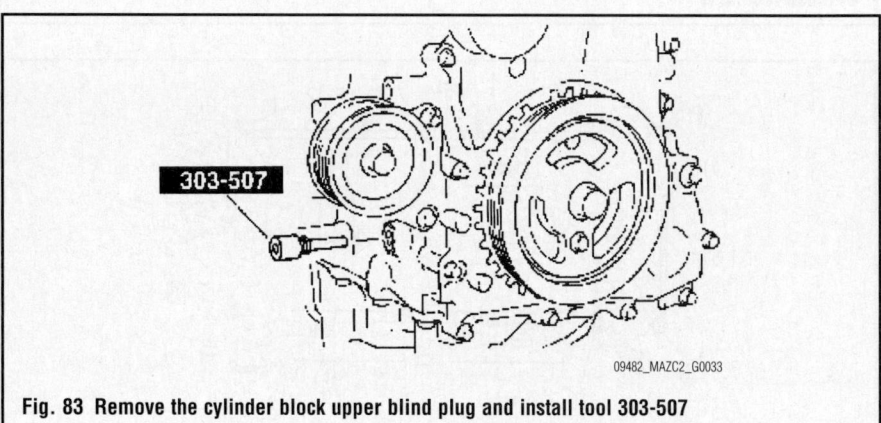

Fig. 83 Remove the cylinder block upper blind plug and install tool 303-507

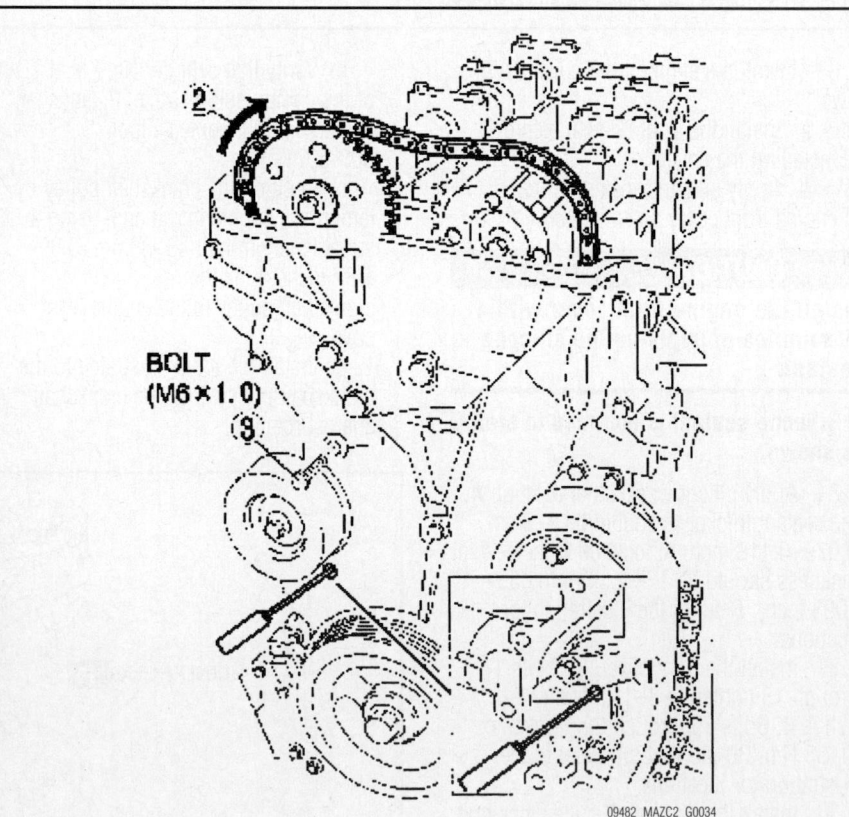

Fig. 84 Place a M6 X 1 length bolt at the engine front cover upper blind plug, secure the chain guide at the position where the tension is released

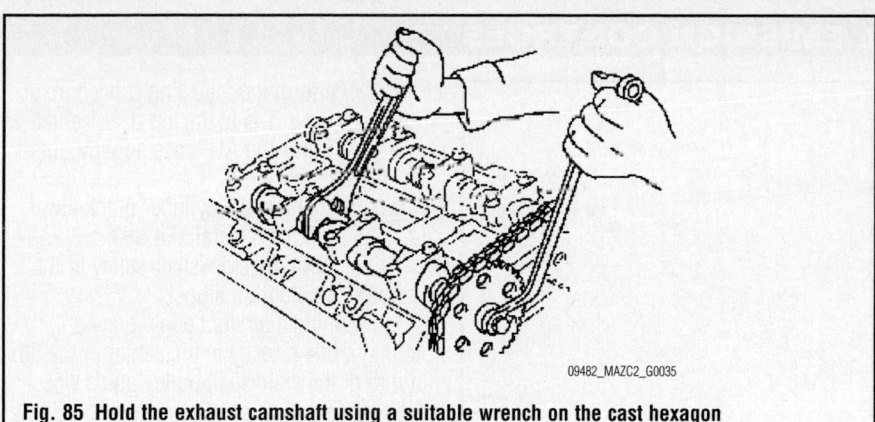

09482_MAZC2_G0035

Fig. 85 Hold the exhaust camshaft using a suitable wrench on the cast hexagon

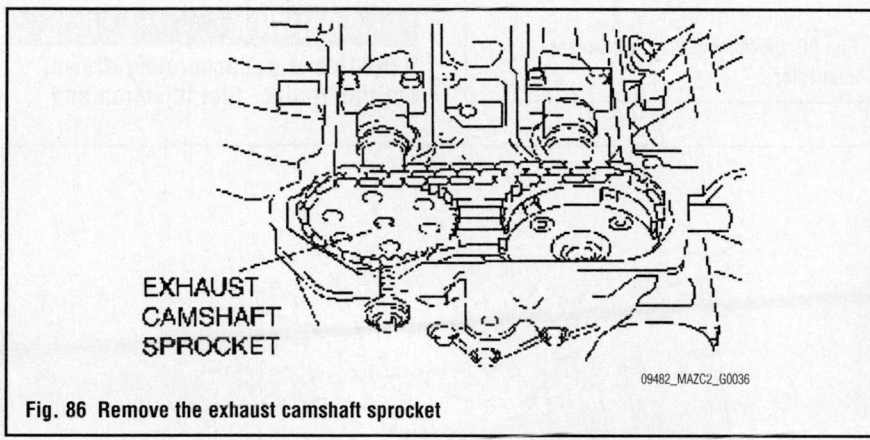

EXHAUST
CAMSHAFT
SPROCKET

09482_MAZC2_G0036

Fig. 86 Remove the exhaust camshaft sprocket

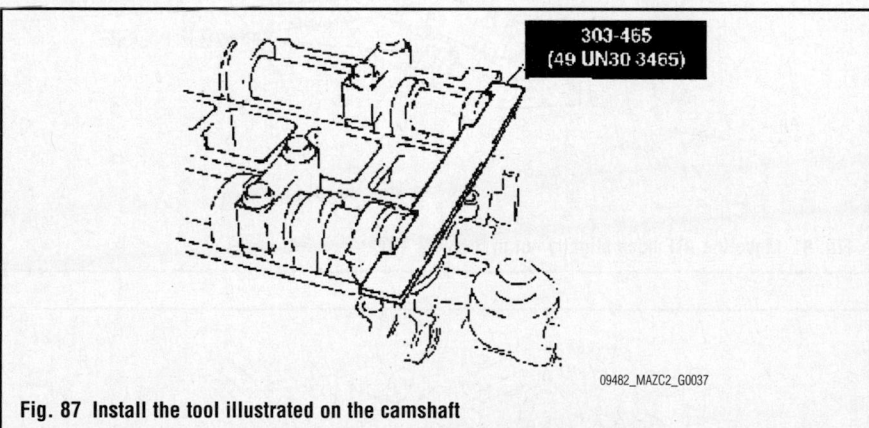

303-465
(49 UN30 3465)

09482_MAZC2_G0037

Fig. 87 Install the tool illustrated on the camshaft

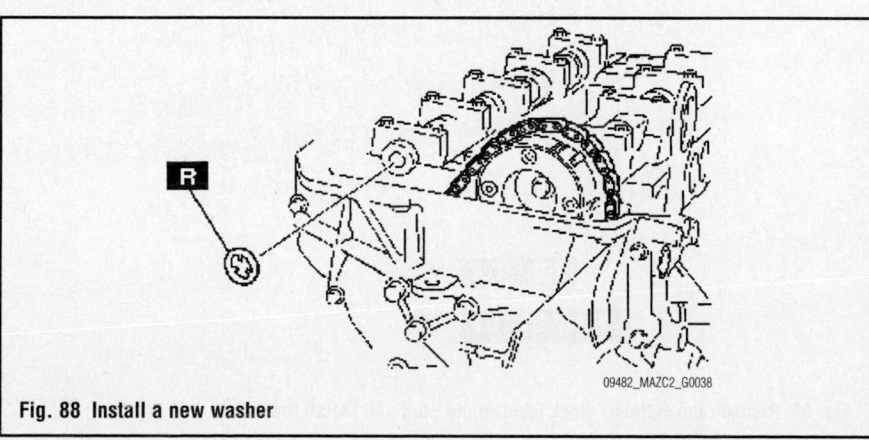

R

09482_MAZC2_G0038

Fig. 88 Install a new washer

30. Install a new washer.

31. Install the exhaust camshaft sprocket.

➡**Do not tighten the bolt for the camshaft sprocket during this step. First confirm the valve timing, then tighten the bolt.**

32. Install the tool on the camshaft as shown.

33. Remove the M6 X 1.0 bolt from the engine front cover to apply tension to the timing chain.

34. Turn the crankshaft clockwise until the crankshaft is in the No.1 cylinder TDC position.

35. Hold the exhaust camshaft using a suitable wrench on the cast hexagon as shown.

36. Tighten the exhaust camshaft sprocket lock bolt to 51–55 ft. lbs. (69–75 Nm).

37. Remove the tool from the camshaft.

38. Remove the tool from the block lower blind plug.

39. Rotate the crankshaft clockwise two turns until the TDC position.

40. If not aligned, loosen the crankshaft pulley lock bolt and repeat the procedure.

41. Apply silicone sealant to the engine front cover upper blind plug.

42. Install the engine front cover upper blind plug and tighten to 71–101 inch lbs. (8–11 Nm.)

43. Install the cylinder block lower blind plug and tighten to 13–16 ft. lbs. (18–23 Nm)

44. Install the new engine front cover lower blind plug and tighten to 89–123 ft. lbs. (10–14 Nm.)

45. Install the drive belt.

46. Measure the valve clearance.

47. Turn the crankshaft clockwise so that the No.1 piston is at TDC of the compression stroke.

48. Measure the valve clearance at point A in the illustration.

49. If the valve clearance is out of the specification, adjust it.

50. Turn the crankshaft 360 degrees clockwise so that the No.4 piston is at TDC of the compression stroke.

51. Measure the valve clearance at B in the illustration.

52. If the valve clearance is out of the specification, adjust it.

53. Install the cylinder head cover.

54. Install the ignition coils.

55. Connect the OCV connector.

56. Install the CMP sensor.

57. Install the front suspension tower bar.

58. Connect the ventilation hose.

59. Install the air cleaner.

60. Install the plug hole plate.

61. Install the battery and battery tray.

ENGINE PERFORMANCE & EMISSION CONTROLS

CAMSHAFT POSITION (CMP) SENSOR

REMOVAL & INSTALLATION

See Figure 90.

1. Remove the battery cover.
2. Disconnect the negative battery cable.

> ※※ **CAUTION**
>
> **When replacing the camshaft position sensor, make sure there are no metal shavings adhering as they can cause the sensor output signal to malfunction from fluctuation in magnetic flux resulting in a deterioration in engine control. Cover a removed CMP sensor in plastic to protect it from foreign material adhering to it.**

3. Disconnect the CMP sensor connector.
4. Remove the CMP sensor installation bolt.
5. Remove the CMP sensor from the cylinder head cover.
6. Install in the reverse order of removal.

CRANKSHAFT POSITION (CKP) SENSOR

REMOVAL & INSTALLATION

See Figures 91 through 93.

1. Remove the battery and battery tray.
2. Remove the air cleaner.
3. Remove the drive belt.
4. Remove the undercover.
5. Disconnect the CKP sensor connector.
6. Remove the installation bolts to remove the CKP sensor.

To install:

> ※※ **CAUTION**
>
> **When foreign material, such as an iron chips, gets on the CKP sensor, it can cause abnormal output from the sensor because of flux turbulence and adversely affect engine control. Be sure there is no foreign material on the CKP sensor when replacing.**

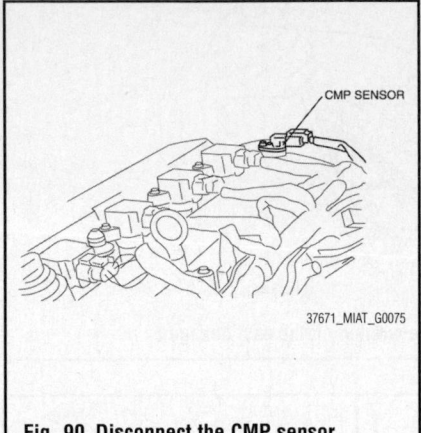

Fig. 90 Disconnect the CMP sensor connector

7. Perform the following procedure so that piston No.1 is at the top dead center.
  a. Move the ATF hose slightly out of the way. (AT)
  b. Remove the cylinder block lower blind plug and install the SST.
  c. Turn the crankshaft pulley to the clockwise until it stops.
8. Using a straight edge, draw a straight line directly in the center of the 5th tooth of the crankshaft pulley pulse wheel (counting counterclockwise from the empty space).

> ※※ **CAUTION**
>
> **If the line is not accurately drawn, ignition timing, fuel injection and**

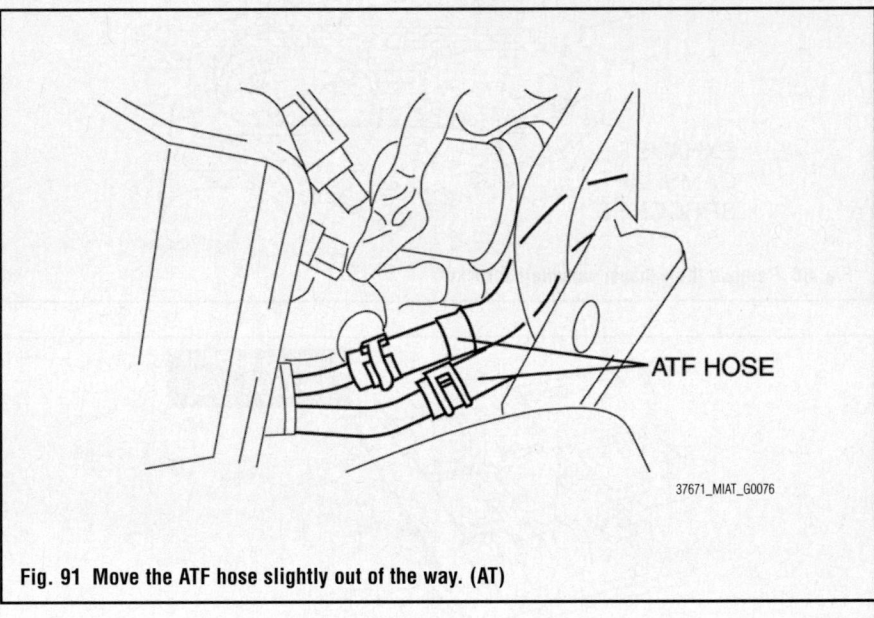

Fig. 91 Move the ATF hose slightly out of the way. (AT)

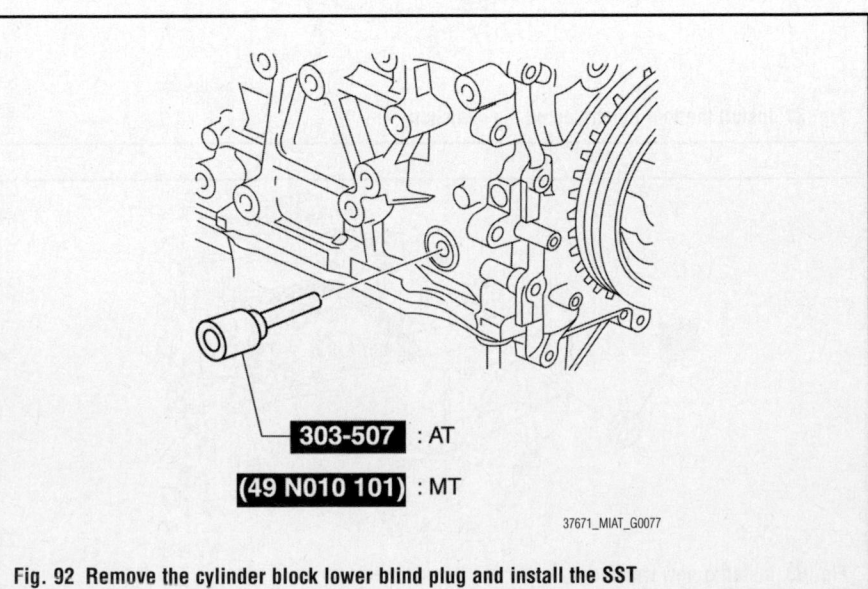

Fig. 92 Remove the cylinder block lower blind plug and install the SST

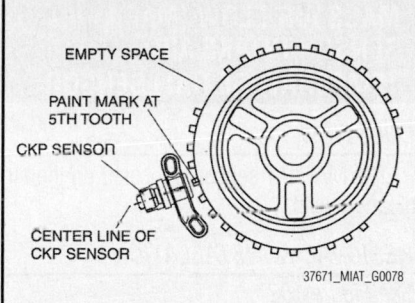

Fig. 93 Draw a straight line directly in the center of the 5th tooth of the crankshaft pulley pulse wheel

other engine control systems will be adversely effected. Draw the straight line carefully using a straight edge.

9. Align the center line of the CKP sensor and the line drawn in Step 2, then install the CKP sensor.

10. Install the CKP sensor fitting bolts.

11. Remove the SST then install the cylinder block lower blind plug.

12. Install in the reverse order of removal.

ENGINE COOLANT TEMPERATURE (ECT) SENSOR

REMOVAL & INSTALLATION

See Figures 94 through 99.

1. Remove the battery cover.
2. Disconnect the negative battery cable.
3. Drain the engine coolant from the radiator.
4. Remove the service hole cover.
   a. Remove the suspension tower bar (joint), (right side) and (left side).
   b. Remove the wiper arm.
   c. Remove the cowl grille.
   d. Remove the side cowl grille.
   e. Move the cooler pipe No.3 and heater pipe slightly out of the way.
   f. Remove the service hole cover.
5. Disconnect the heater hose and move the heater hose slightly out of the way.
6. Disconnect the ECT sensor connector.
7. Remove the ECT sensor.
8. Install in the reverse order of removal.

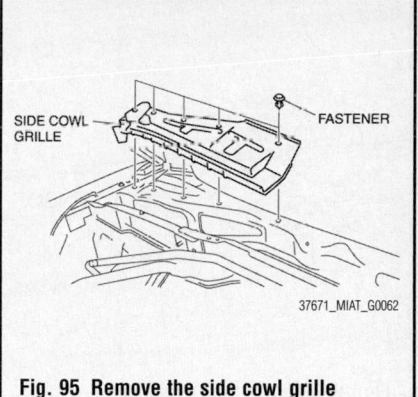

Fig. 95 Remove the side cowl grille

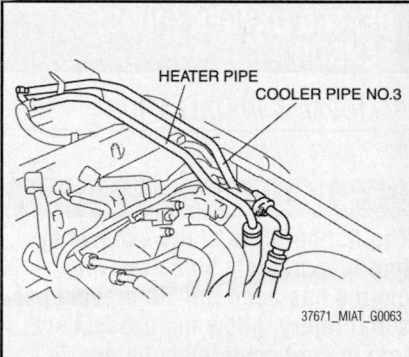

Fig. 96 Move the cooler pipe No.3 and heater pipe slightly out of the way

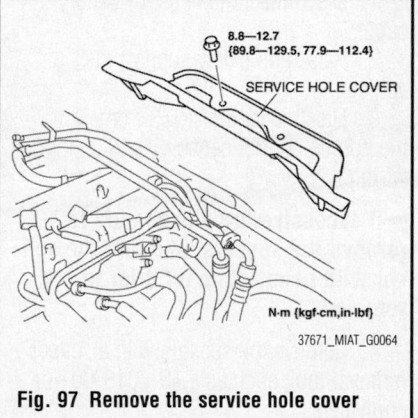

Fig. 97 Remove the service hole cover

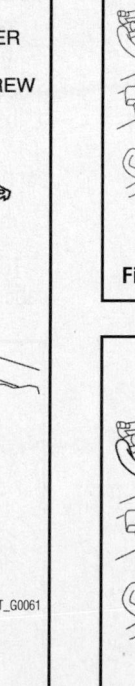

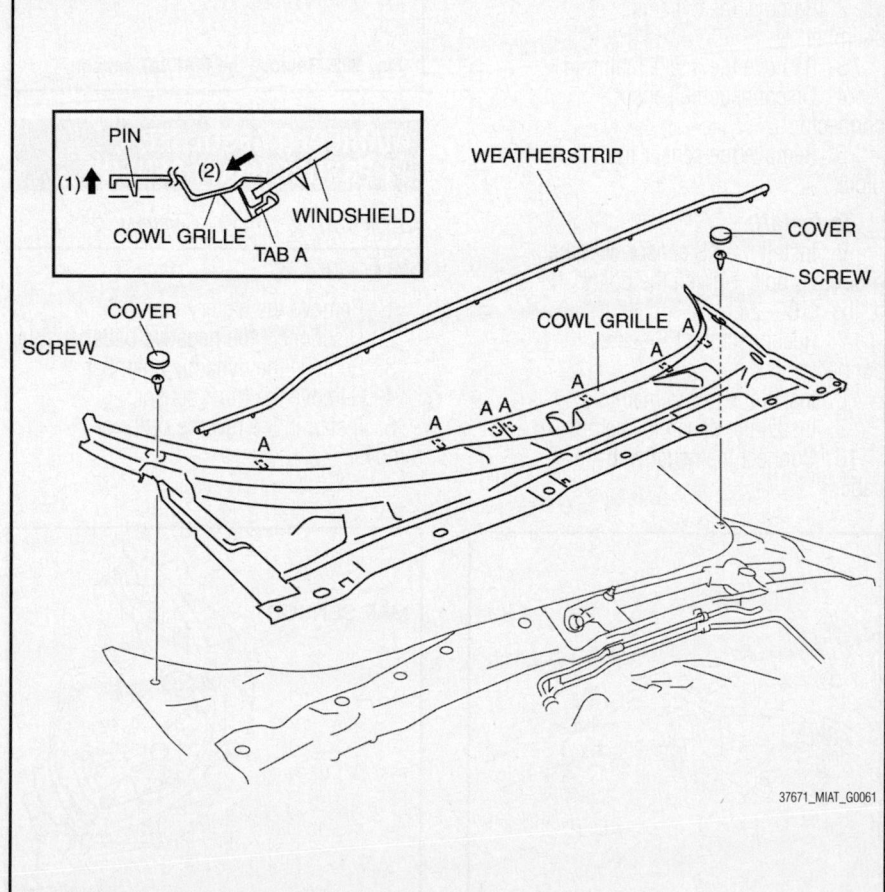

Fig. 94 Remove the cowl grille

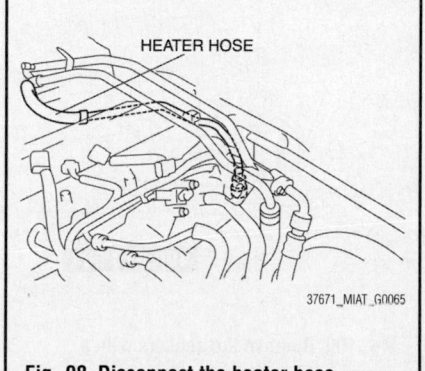

Fig. 98 Disconnect the heater hose

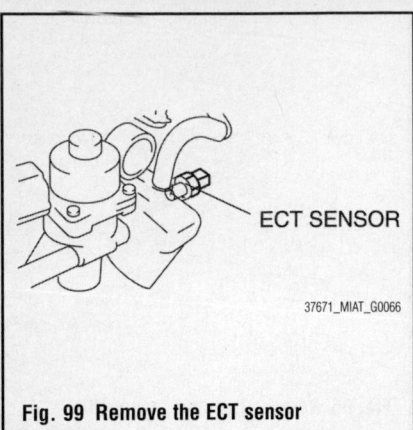

Fig. 99 Remove the ECT sensor

HEATED OXYGEN (HO2S) SENSOR

REMOVAL & INSTALLATION

See Figure 100.

❄ CAUTION

The temperature of the exhaust system is extremely high after the engine has been run. To prevent personal injury, allow the exhaust system to cool completely before removing sensor from the exhaust system.

1. Disconnect the negative battery cable.
2. Raise and safely support the vehicle.
3. Disconnect the HO2S from the engine control sensor wiring.

➡**If excessive force is needed to remove the sensors lubricate the sensor with penetrating oil prior to removal.**

4. Remove the sensors with a sensor removal tool, such as a 49-L018-001 or equivalent.

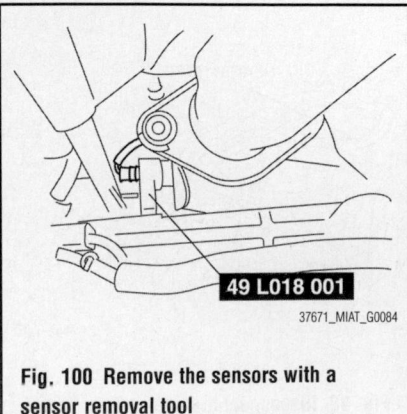

Fig. 100 Remove the sensors with a sensor removal tool

To install:

5. Install the sensor, and then tighten it to 22–36 ft. lbs. (29—49 Nm).
6. Connect the sensor electrical wiring connector to the engine wiring harness.
7. Lower the vehicle.
8. Connect the negative battery cable.

INTAKE AIR TEMPERATURE (IAT) SENSOR

LOCATION

The Intake Air Temperature (IAT) sensor is an integral part of the Mass Air Flow (MAF) sensor/Intake Air Temperature (IAT) sensor assembly which is mounted on the air intake duct. Refer to the Mass Air Flow section for information regarding servicing this component.

KNOCK SENSOR (KS)

REMOVAL & INSTALLATION

See Figure 101.

1. Disconnect the negative battery cable.
2. Remove the dynamic chamber.
3. Remove the intake manifold.
4. Disconnect the sensor connector.
5. Remove the sensor from the engine block.

To install:

6. Install the KS sensor with the mounting bolt. Tighten the bolt to 12—17 ft. lbs. (16—24 Nm).
7. Reconnect the KS sensor connector.
8. Install the intake manifold.
9. Install the dynamic chamber.
10. Connect the negative battery cable.

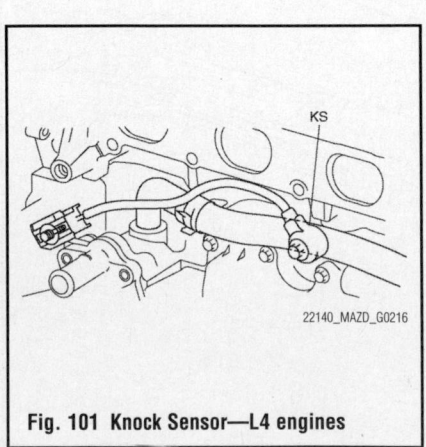

Fig. 101 Knock Sensor—L4 engines

MASS AIR FLOW (MAF) SENSOR/INTAKE AIR TEMPERATURE (IAT) SENSOR

LOCATION

The MAF/IAT sensor is located on the air cleaner cover.

REMOVAL & INSTALLATION

See Figure 102.

1. Remove the battery cover.
2. Disconnect the negative battery cable.
3. Disconnect MAF/IAT sensor connector.
4. Remove the MAF/IAT sensor.
5. Install in the reverse order of removal.

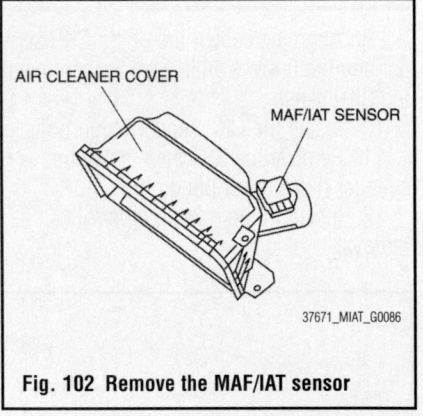

Fig. 102 Remove the MAF/IAT sensor

MANIFOLD ABSOLUTE PRESSURE (MAP) SENSOR

REMOVAL & INSTALLATION

See Figure 103.

1. Remove the battery cover.
2. Disconnect the negative battery cable.
3. Remove the dynamic chamber.
4. Remove the MAP sensor.
5. Install in the reverse order of removal.

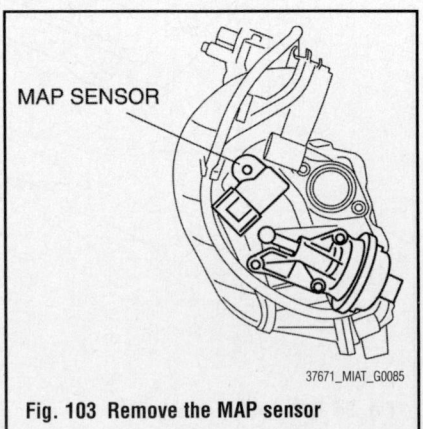

Fig. 103 Remove the MAP sensor

POWERTRAIN CONTROL MODULE (PCM)

LOCATION

The PCM for the MX-5 Miata is located at the front of the engine compartment to the left of the air filter housing.

REMOVAL & INSTALLATION

See Figures 104 through 107.

1. When replacing the PCM, perform the following: PCM configuration
2. Remove the battery cover.
3. Disconnect the negative battery cable.
4. Remove the air cleaner case.
5. Move the water hose from the PCM cover slightly out of the way.

6. Remove in the numerical order indicated.

To install:

7. Install in the reverse order of removal.
8. When replacing the PCM on the vehicles, perform the following: PCM parameter reset

✳✳ CAUTION

If the PCM connector is inserted at an angle and the lever is moved, the connector could be damaged. Verify that the PCM connector is inserted straight.

9. Verify that the PCM connector lever is tilted towards the wiring harness side as shown.

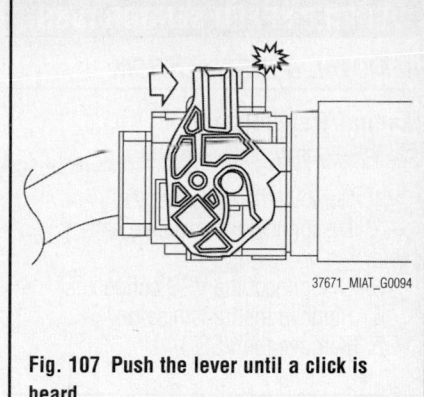

Fig. 107 Push the lever until a click is heard

10. Insert the PCM connector straight until it contacts the PCM and verify that the lever reverts upward naturally.
11. Push the lever until a click is heard.

Reset Procedure

1. Connect the M-MDS to DLC-2.
2. After the vehicle is identified, select the following items from the initialization screen of the M-MDS.
   a. When using the IDS (laptop PC); Select the "Module Programming".
   b. When using the PDS (Pocket PC); Select "Programming".
   c. Select "Module Programming".
3. Then, select items from the screen menu in the following order.
   a. Select "Programmable Module Installation".
   b. Select "PCM".
4. Perform the configuration according to the directions on the screen.
5. Retrieve DTCs by the M-MDS, then verify that there is no DTC present.
6. If a DTC(s) is detected, perform the applicable DTC inspection.

THROTTLE POSITION SENSOR (TPS)

LOCATION

The TPS is an integral part of the throttle body assembly.

REMOVAL & INSTALLATION

1. Disconnect the negative battery cable.
2. Remove the necessary air intake components to access the throttle body.
3. Detach the electrical wire harness plug from the throttle body.
4. Remove the sensor attaching bolts.
5. Remove the throttle body.
6. Installation is the reverse of the removal procedure. Tighten the mounting bolts to 71—101 inch. lbs. (8—11.5 Nm).

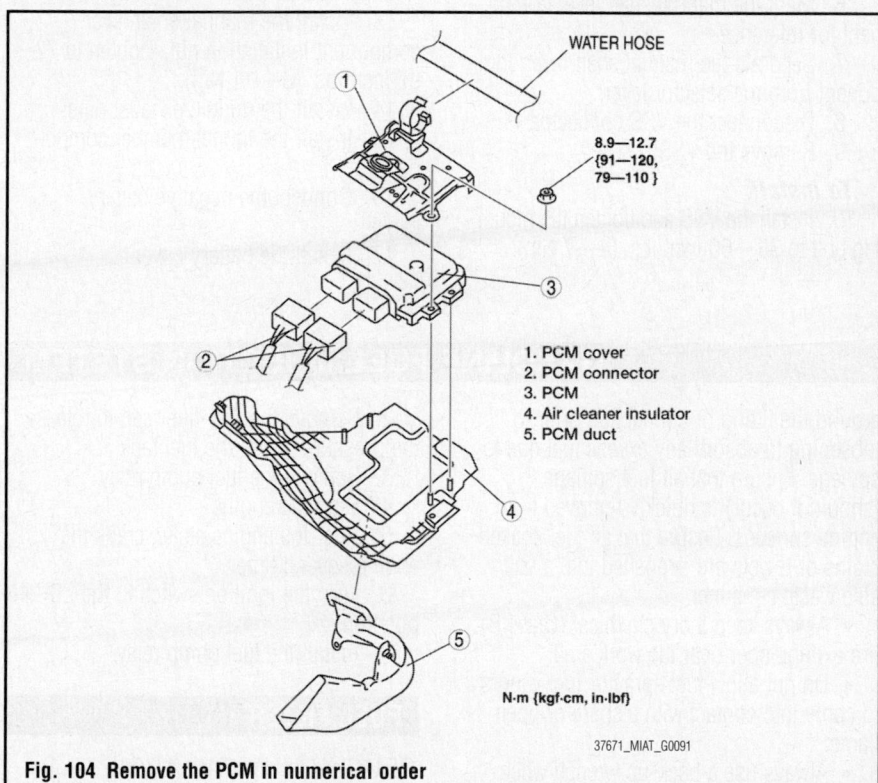

WATER HOSE

8.9—12.7
{91—120,
79—110 }

1. PCM cover
2. PCM connector
3. PCM
4. Air cleaner insulator
5. PCM duct

N·m {kgf·cm, in·lbf}

37671_MIAT_G0091

Fig. 104 Remove the PCM in numerical order

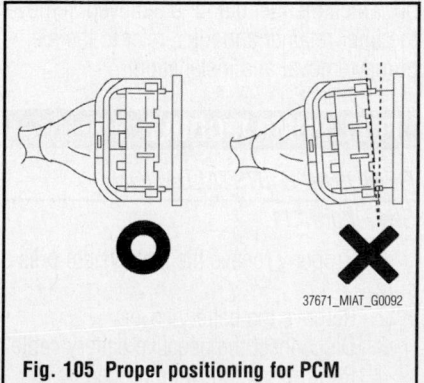

37671_MIAT_G0092

Fig. 105 Proper positioning for PCM connector

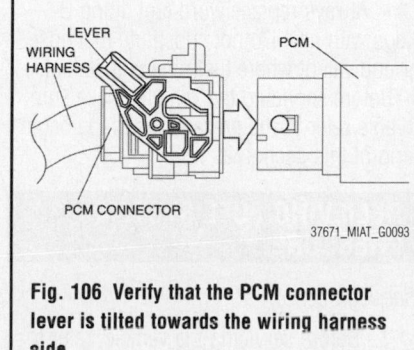

LEVER
WIRING HARNESS
PCM
PCM CONNECTOR

37671_MIAT_G0093

Fig. 106 Verify that the PCM connector lever is tilted towards the wiring harness side

VEHICLE SPEED SENSUR (VSS)

REMOVAL & INSTALLATION

Manual Transmission

See Figure 108.

1. Remove the battery cover.
2. Disconnect the negative battery cable.
3. Disconnect the VSS connector.
4. Remove the transmission.
5. Remove the VSS.

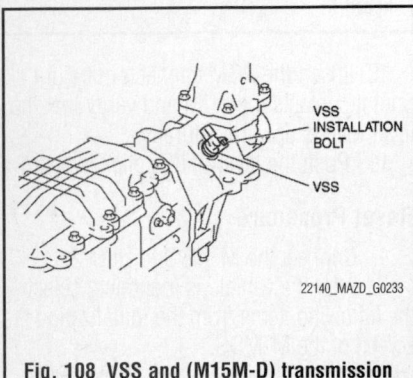

22140_MAZD_G0233

Fig. 108 VSS and (M15M-D) transmission location view

To install:

6. Apply the specified oil to a new O-ring and install it on a new VSS.
7. Install the VSS and tighten the mounting bolt to 79—97 inch lbs. (8—11 Nm).
8. Install the transmission.
9. Connect the VSS connector.
10. Connect the negative battery cable.
11. Install the battery cover.

Automatic Transmission

See Figure 109.

1. Remove the battery cover.
2. Disconnect the negative battery cable.
3. Remove the tunnel member component.
4. Remove the middle exhaust pipe.
5. Remove the insulator.
6. Mark the manual shaft lever component for reference.
7. Separate the manual shaft lever component from the selector lever.
8. Disconnect the VSS connector.
9. Remove the VSS.

To install:

10. Install the VSS and tighten the mounting bolt to 35—60 inch lbs. (4—7 Nm).

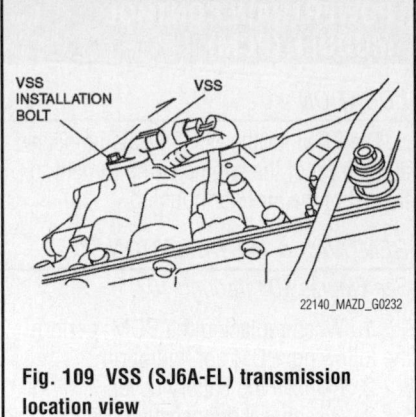

22140_MAZD_G0232

Fig. 109 VSS (SJ6A-EL) transmission location view

11. Connect the VSS connector.
12. Align the mark of the manual shaft lever component.
13. Install the manual shaft lever component installation nut. Tighten to 72—97 inch lbs. (8—11 Nm).
14. Install the middle exhaust pipe.
15. Install the tunnel member component.
16. Connect the negative battery cable.
17. Install the battery cover.

FUEL

GASOLINE FUEL INJECTION SYSTEM

FUEL SYSTEM SERVICE PRECAUTIONS

Safety is the most important factor when performing not only fuel system maintenance but any type of maintenance. Failure to conduct maintenance and repairs in a safe manner may result in serious personal injury or death. Maintenance and testing of the vehicle's fuel system components can be accomplished safely and effectively by adhering to the following rules and guidelines.

• To avoid the possibility of fire and personal injury, always disconnect the negative battery cable unless the repair or test procedure requires that battery voltage be applied.

• Always relieve the fuel system pressure prior to disconnecting any fuel system component (injector, fuel rail, pressure regulator, etc.), fitting or fuel line connection. Exercise extreme caution whenever relieving fuel system pressure to avoid exposing skin, face and eyes to fuel spray. Please be advised that fuel under pressure may penetrate the skin or any part of the body that it contacts.

• Always place a shop towel or cloth around the fitting or connection prior to loosening to absorb any excess fuel due to spillage. Ensure that all fuel spillage (should it occur) is quickly removed from engine surfaces. Ensure that all fuel soaked cloths or towels are deposited into a suitable waste container.

• Always keep a dry chemical (Class B) fire extinguisher near the work area.

• Do not allow fuel spray or fuel vapors to come into contact with a spark or open flame.

• Always use a back-up wrench when loosening and tightening fuel line connection fittings. This will prevent unnecessary stress and torsion to fuel line piping.

• Always replace worn fuel fitting O-rings with new Do not substitute fuel hose or equivalent where fuel pipe is installed.

Before servicing the vehicle, make sure to also refer to the precautions in the beginning of this section as well.

RELIEVING FUEL SYSTEM PRESSURE

See Figure 110.

1. Before servicing the vehicle, refer to the precautions section.

2. Remove the fuel-filler cap to release the pressure inside the fuel tank.
3. Remove the fuel pump relay
4. Start the engine.
5. After the engine stalls, crank the engine several times.
6. Turn the ignition switch to the LOCK position.
7. Install the fuel pump relay.

FUEL FILTER

REMOVAL & INSTALLATION

The fuel filter is part of the pump assembly. Once the fuel pump is removed, remove the filter retainer and filter. refer to the fuel pump removal and installation.

FUEL PUMP MODULE

REMOVAL & INSTALLATION

See Figure 111.

1. Properly relieve the fuel system pressure.
2. Remove the battery cover.
3. Disconnect the negative battery cable.
4. Perform the following procedure to remove the service hole cover.

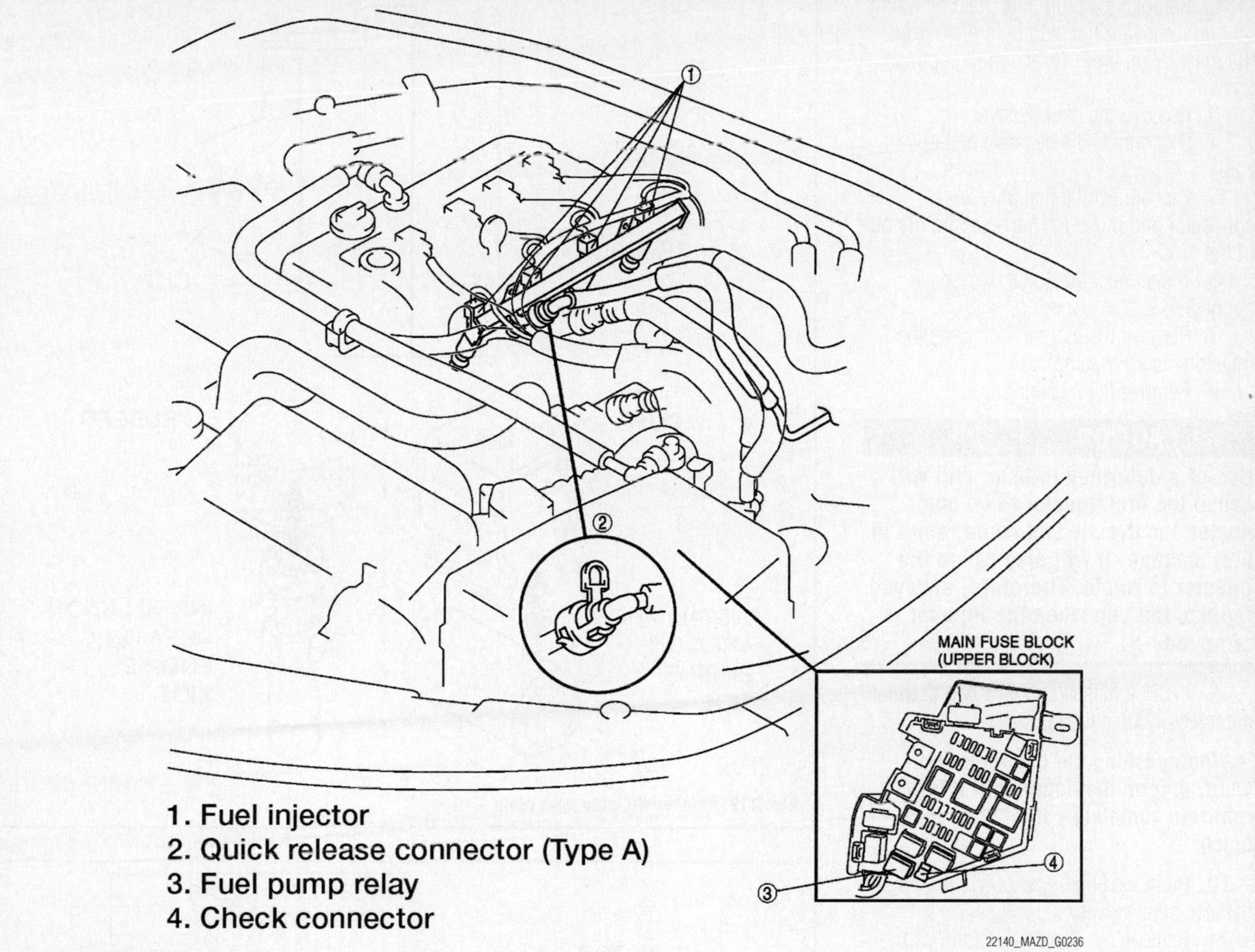

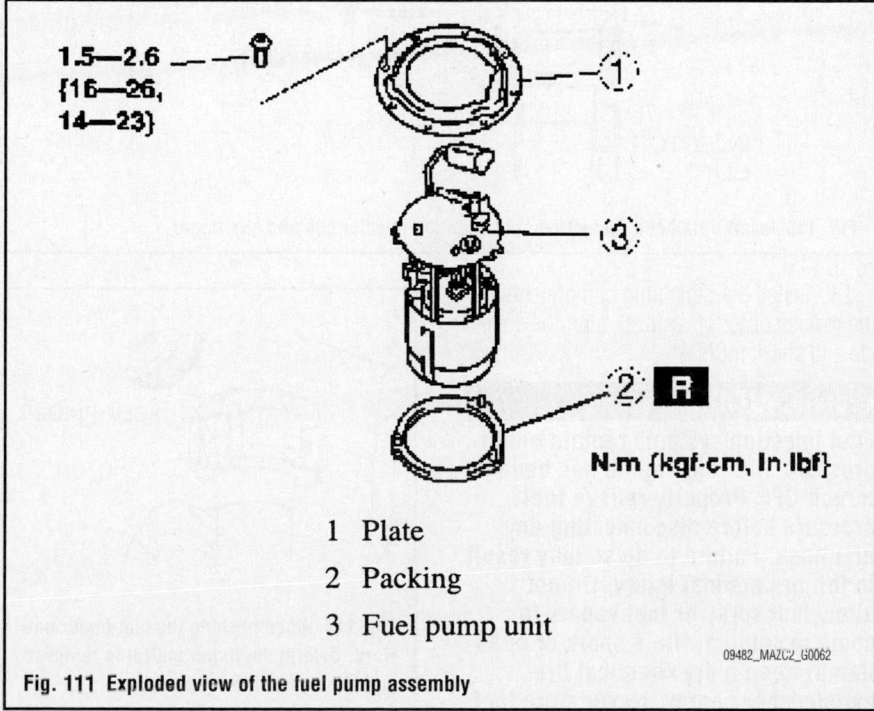

1. Fuel injector
2. Quick release connector (Type A)
3. Fuel pump relay
4. Check connector

MAIN FUSE BLOCK
(UPPER BLOCK)

22140_MAZD_G0236

Fig. 110 Fuel pump relay location

1.5—2.6
{16—26,
14—23}

N·m {kgf·cm, In·lbf}

1   Plate

2   Packing

3   Fuel pump unit

09482_MAZC2_G0062

Fig. 111 Exploded view of the fuel pump assembly

a. Remove the console.
b. Remove the quarter trim.
c. Remove the scuff plate.
d. Remove the tire house trim.
e. Remove the aeroboard.
f. Remove the front seat back bar garnish.
g. Remove the remove the back trim.
h. Remove the service hole cover.

5. Disconnect the quick release connector from the fuel pump unit.

6. Disconnect the fuel pump unit connector.

7. Remove plate, packing and pump assembly.

To install:

8. Installation is the reverse of removal.

FUEL RAIL AND INJECTOR

REMOVAL & INSTALLATION

See Figures 112 through 116.

1. Properly relieve the fuel system pressure.

2. Remove the plug hole plate by lifting off and removing the plug hole plate from the installation areas (rubber and clips) as shown.

3. Remove the battery cover.

4. Disconnect the negative battery cable.

5. Disconnect the fuel injector connector and move the harness slightly out of the way.

6. Disconnect the quick release connector.

7. Remove the fuel rail bolts and the injectors as an assembly.

8. Remove the injector clip.

※ CAUTION

Use of a deformed injector clip will cause the fuel injector to be connected incorrectly and could result in fuel leakage. It will also cause the injector to rotate. Therefore, always replace the clip when the injector is removed.

9. Insert a flathead screwdriver between the injector cup and clip finger.

➡ **When pushing the clip finger outward, deform the finger until it is removed completely from the cup notch.**

10. Push the clip finger outward using a flathead screwdriver.

11. Remove the injector with the clip.

12. Hold the clip using pliers.

13. Pull the clip parallel to the injector groove and remove it from the injector. Discard the clip.

To install:

14. Apply a small amount of clean oil to the injector groove and the O-ring.

15. Temporarily attach a new clip to the injector groove.

➡ **When the clip is attached correctly, the central area of the injector and the clip finger positions are aligned.**

16. Hold the injector firmly and push the clip into the injector until the clip stops sliding.

17. Verify that the injector connector position is correct.

18. Press the injector into the injector cup. Continue pressing until the clip contacts the lower surface of the injector cup.

19. Verify that the injector and clip are correctly installed with the clip locked onto the injector cup notch.

20. Install the fuel rail and tighten the bolts to 20–26 Nm (15–19 ft. lbs.).

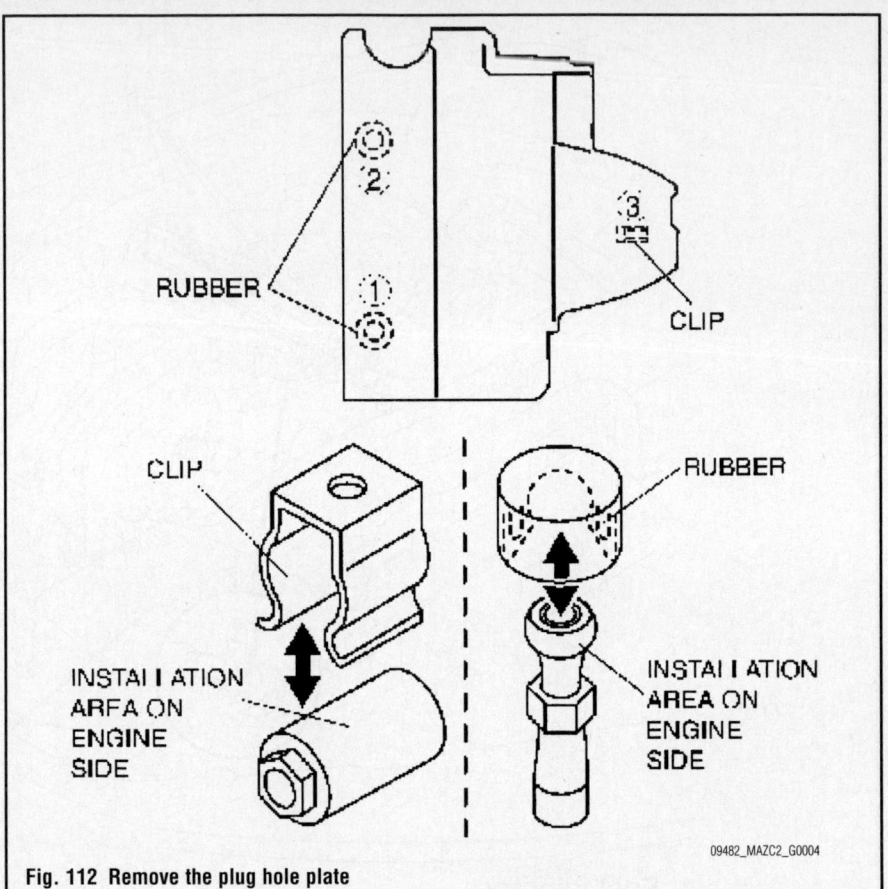

Fig. 112 Remove the plug hole plate

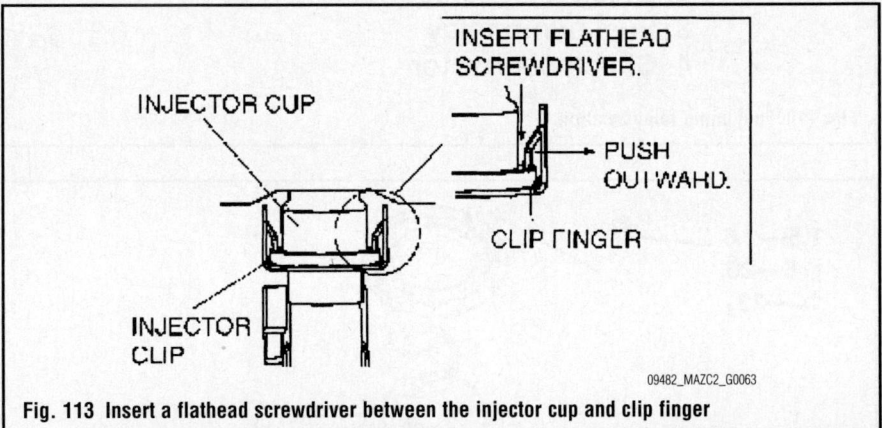

Fig. 113 Insert a flathead screwdriver between the injector cup and clip finger

21. Install the remaining components in the reverse order of removal, start the vehicle and check for leaks.

※ CAUTION

Fuel injection systems remain under pressure after the engine has been turned OFF. Properly relieve fuel pressure before disconnecting any fuel lines. Failure to do so may result in fire or personal injury. Do not allow fuel spray or fuel vapors to come in contact with a spark or open flame. Keep a dry chemical fire extinguisher nearby. Never store fuel

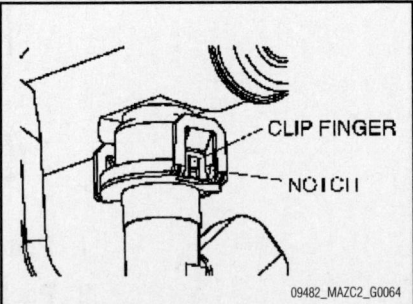

Fig. 114 When pushing the clip finger outward, deform the finger until it is removed completely from the cup notch

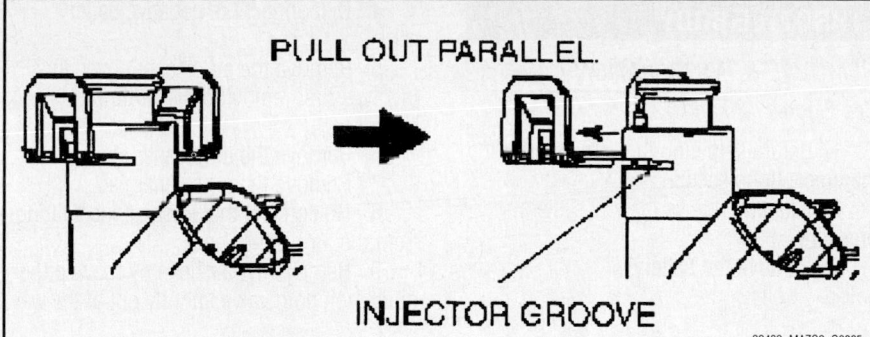

Fig. 115 Pull the clip parallel to the injector groove and remove it from the injector

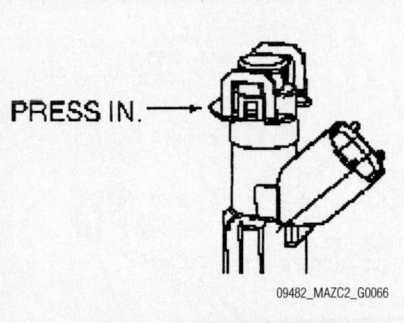

Fig. 116 Hold the injector firmly and push the clip into the injector until the clip stops sliding

in an open container due to risk of fire or explosion.

FUEL TANK

REMOVAL & INSTALLATION

See Figures 117 through 119.

❊❊ WARNING

Heed the following Warnings during removal of the fuel tank:

- Fuel is very flammable liquid. If fuel spills or leaks from the pressurized fuel system, it will cause serious injury or death and facility breakage. Fuel can also irritate skin and eyes. To prevent this, always complete the "Fuel Line Safety Procedure", while referring to "BEFORE SERVICE PRECAUTION".
- Fuel is very flammable liquid. If fuel spills or leaks from the pressurized fuel system, it will cause serious injury or death and facility breakage. Fuel can also irritate skin and eyes. To prevent this, before performing the fuel pump unit removal/installation, always complete the "Fuel Leak

Inspection After Fuel Pump Unit Installation".

- A person charged with static electricity could cause a fire or explosion, resulting in death or serious injury. Before draining fuel, make sure to discharge static electricity by touching the vehicle body.

1. Park the vehicle on a level surface.

2. Follow "BEFORE SERVICE PRECAUTION" before performing any work operations to prevent fuel from spilling from the fuel system.

3. Remove the battery cover.

4. Disconnect the negative battery cable.

5. Remove the following parts:
- Middle pipe
- Propeller shaft
- Power plant frame
- Rear drive shaft
- Rear differential
- Rear crossmember component

6. Perform the following procedure to remove the service hole cover.

a. To remove the back trim, remove the following parts:
- Console
- Quarter trim
- Scuff plate
- Tire house trim
- Aeroboard
- Front seat bar garnish

b. Remove the back trim.

c. Remove the service hole cover.

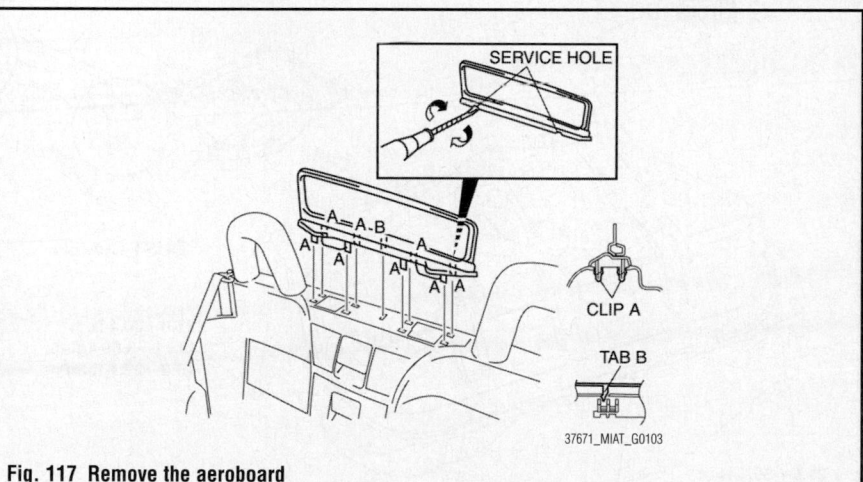

Fig. 117 Remove the aeroboard

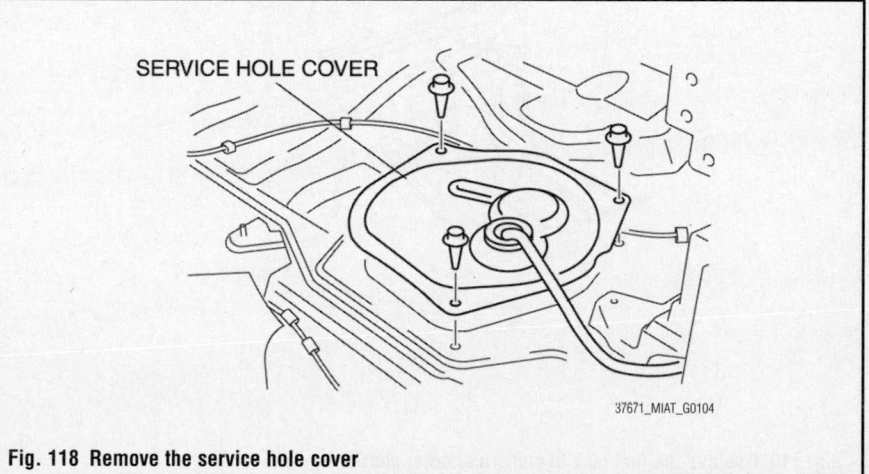

Fig. 118 Remove the service hole cover

7. Disconnect the quick release connector using the SST from the fuel pump unit.

8. Remove the fuel pump unit.

9. Siphon the fuel from the fuel tank.

10. Remove in the order indicated in the table.

11. Install in the reverse order of removal.

12. Complete the "AFTER SERVICE PRECAUTION".

THROTTLE BODY

REMOVAL & INSTALLATION

See Figures 120 and 121.

1. Before servicing the vehicle, refer to the precautions section.

2. Properly relieve the fuel system pressure.

3. Remove the battery cover.

4. Disconnect the negative battery cable.

5. Remove the air cleaner cover, this involves first removing the MAF/IAT sensor.

6. Remove the air cleaner element.

7. Remove the air cleaner case.

8. Disconnect the quick connect fitting from the air cleaner hose.

9. Remove the air hose by moving the purge solenoid valve slightly out of the way.

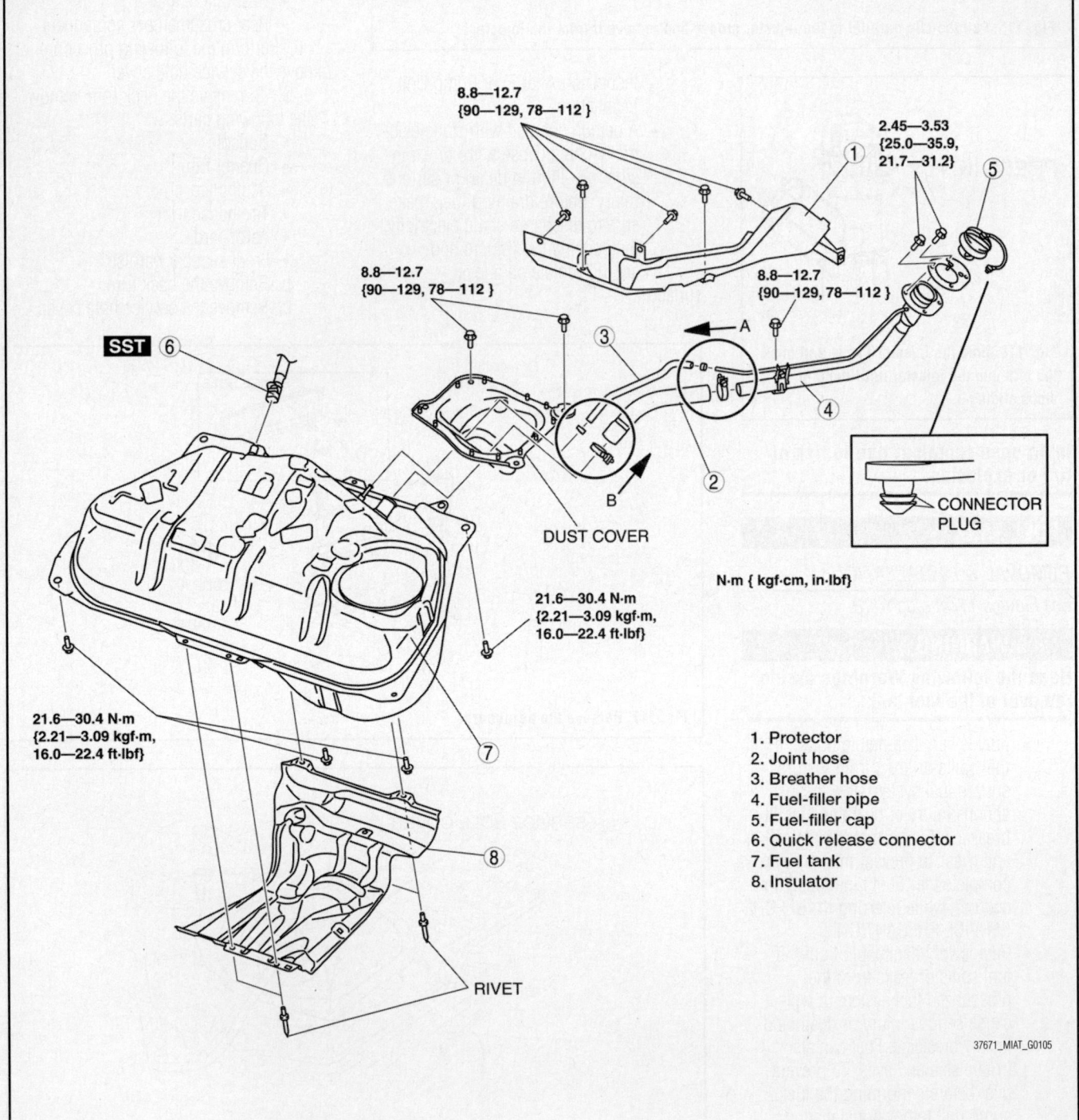

1. Protector
2. Joint hose
3. Breather hose
4. Fuel-filler pipe
5. Fuel-filler cap
6. Quick release connector
7. Fuel tank
8. Insulator

N·m { kgf·cm, in·lbf}

CONNECTOR PLUG

DUST COVER

RIVET

8.8—12.7 {90—129, 78—112 }

2.45—3.53 {25.0—35.9, 21.7—31.2}

21.6—30.4 N·m {2.21—3.09 kgf·m, 16.0—22.4 ft·lbf}

37671_MIAT_G0105

Fig. 119 Remove the fuel tank In numerical order shown

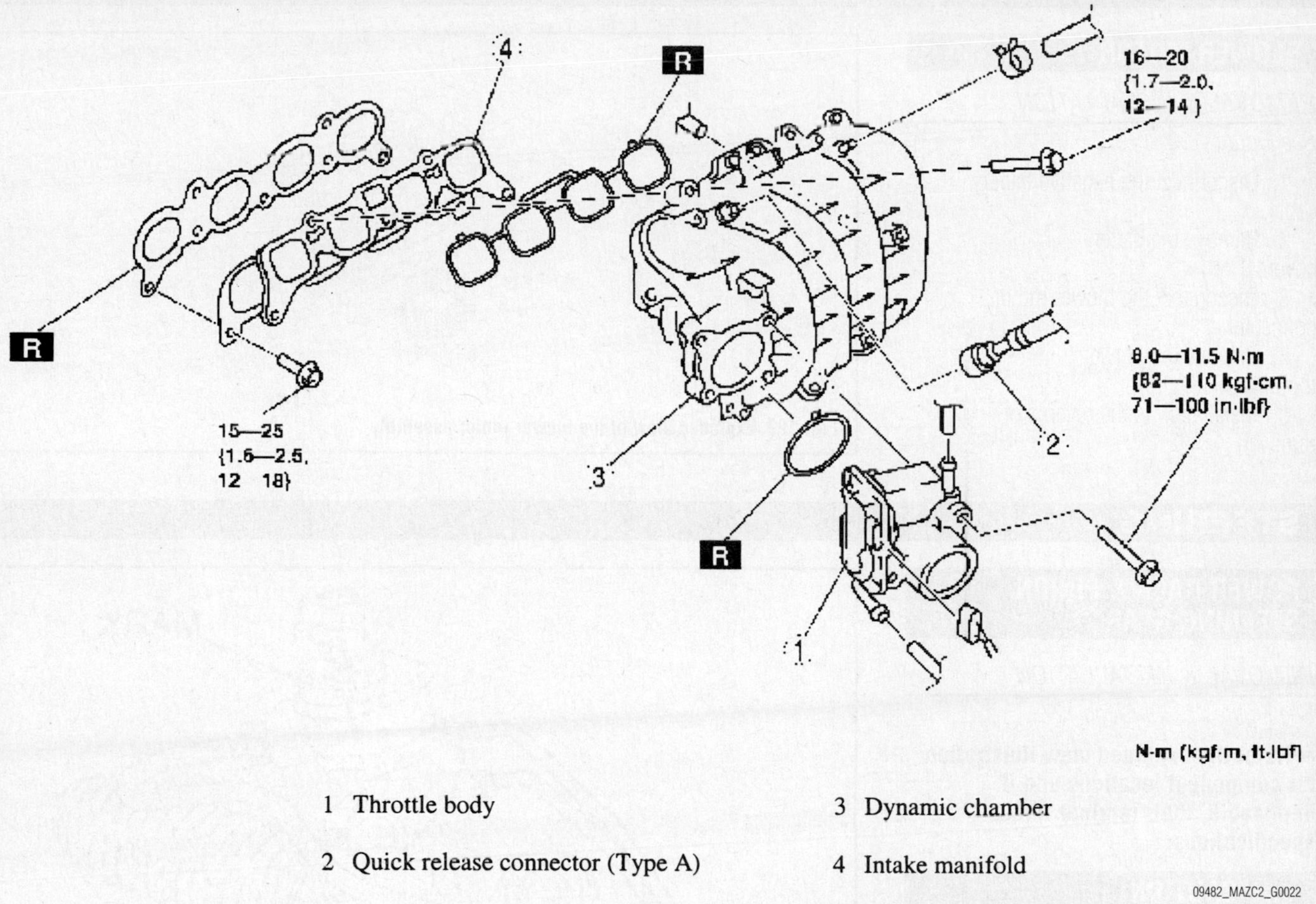

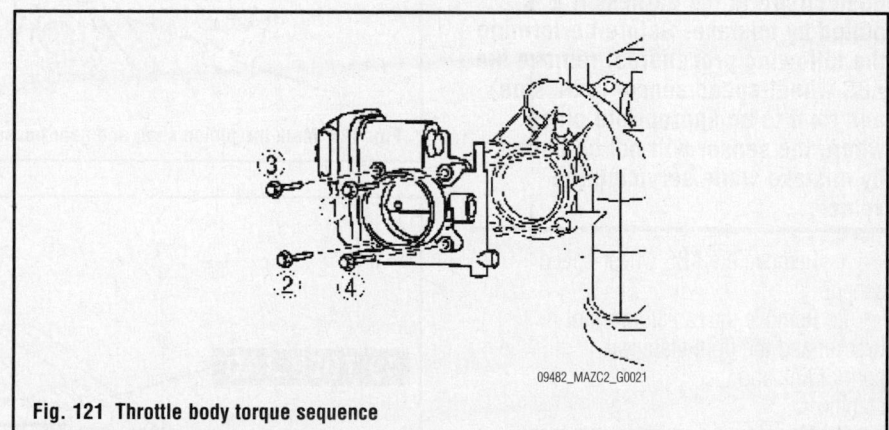

16—20
{1.7—2.0,
12—14 }

8.0—11.5 N·m
{82—110 kgf·cm,
71—100 in·lbf}

15—25
{1.6—2.5,
12—18}

N·m {kgf·m, ft·lbf}

1   Throttle body   3   Dynamic chamber

2   Quick release connector (Type A)   4   Intake manifold

09482_MAZC2_G0022

Fig. 120 Exploded view of the dynamic chamber and intake manifold

10. Remove the fresh-air duct by first removing the bumper.

11. Drain the cooling system.

12. Remove the throttle body.

To install:

13. Installation is the reverse order of removal, make sure to tighten the throttle body bolts in the sequence shown to 71–100 inch lbs. (8–11 Nm).

14. Check and adjust all fluid levels, start the vehicle and check for proper operation.

09482_MAZC2_G0021

Fig. 121 Throttle body torque sequence

HEATING & AIR CONDITIONING COMPONENTS

BLOWER MOTOR

REMOVAL & INSTALLATION

See Figure 122.

1. Disconnect the negative battery cable.
2. Remove the battery cover.
3. Disconnect the blower motor connector.
4. Remove the blower motor.
5. Installation is the reverse of removal.

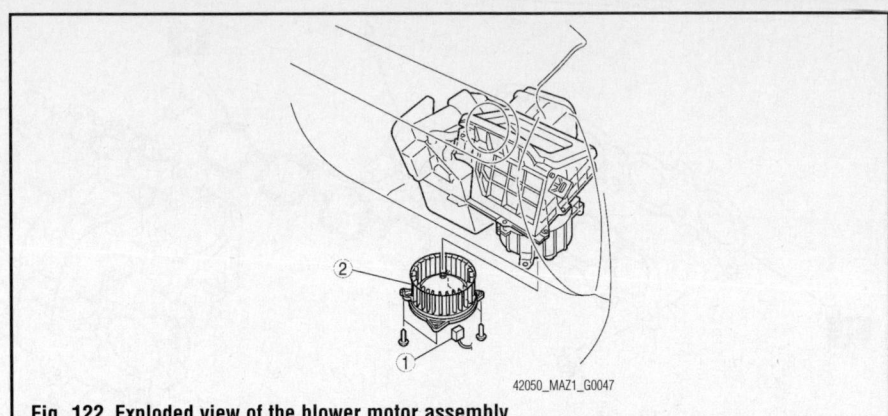

Fig. 122 Exploded view of the blower motor assembly

STEERING

POWER RACK & PINION STEERING GEAR

REMOVAL & INSTALLATION

See Figures 123 through 127.

➡Refer the exploded view illustration for component locations and if applicable, their retainer torque specifications.

✳✳ CAUTION

Performing the following procedures without first removing the ABS wheel-speed sensor may possibly cause an open circuit in the harness if it is pulled by mistake. Before performing the following procedures, remove the ABS wheel-speed sensor (axle side) and fix it to an appropriate place where the sensor will not be pulled by mistake while servicing the vehicle.

1. Remove the ABS wheel-speed sensor.
2. Remove the radiator mount bracket and the front stabilizer control link and the stabilizer.
3. Mark the pinion shaft and gear housing prior to removal.
4. Remove the cotter pin.
5. Remove the tie-rod end and ball joint nuts.
6. Remove the tie-rod nut.
7. Separate the tie-rod end from the steering knuckle using the a suitable puller.
8. Remove the lower mounting rubber bracket.

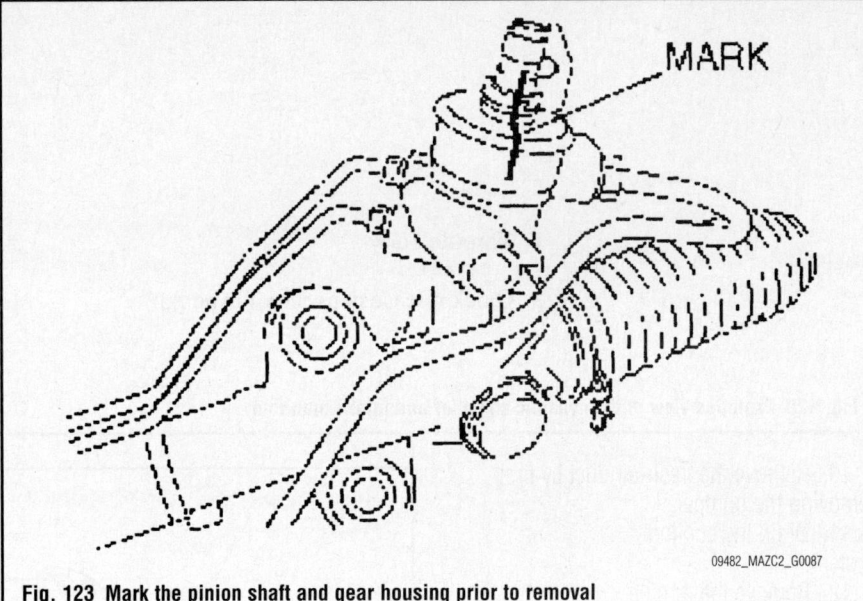

Fig. 123 Mark the pinion shaft and gear housing prior to removal

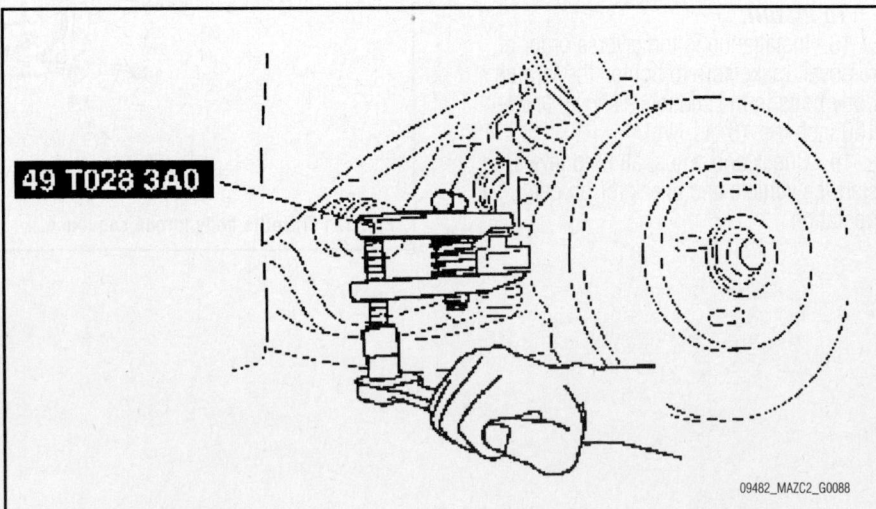

Fig. 124 Separate the tie-rod end from the steering knuckle using the a suitable puller

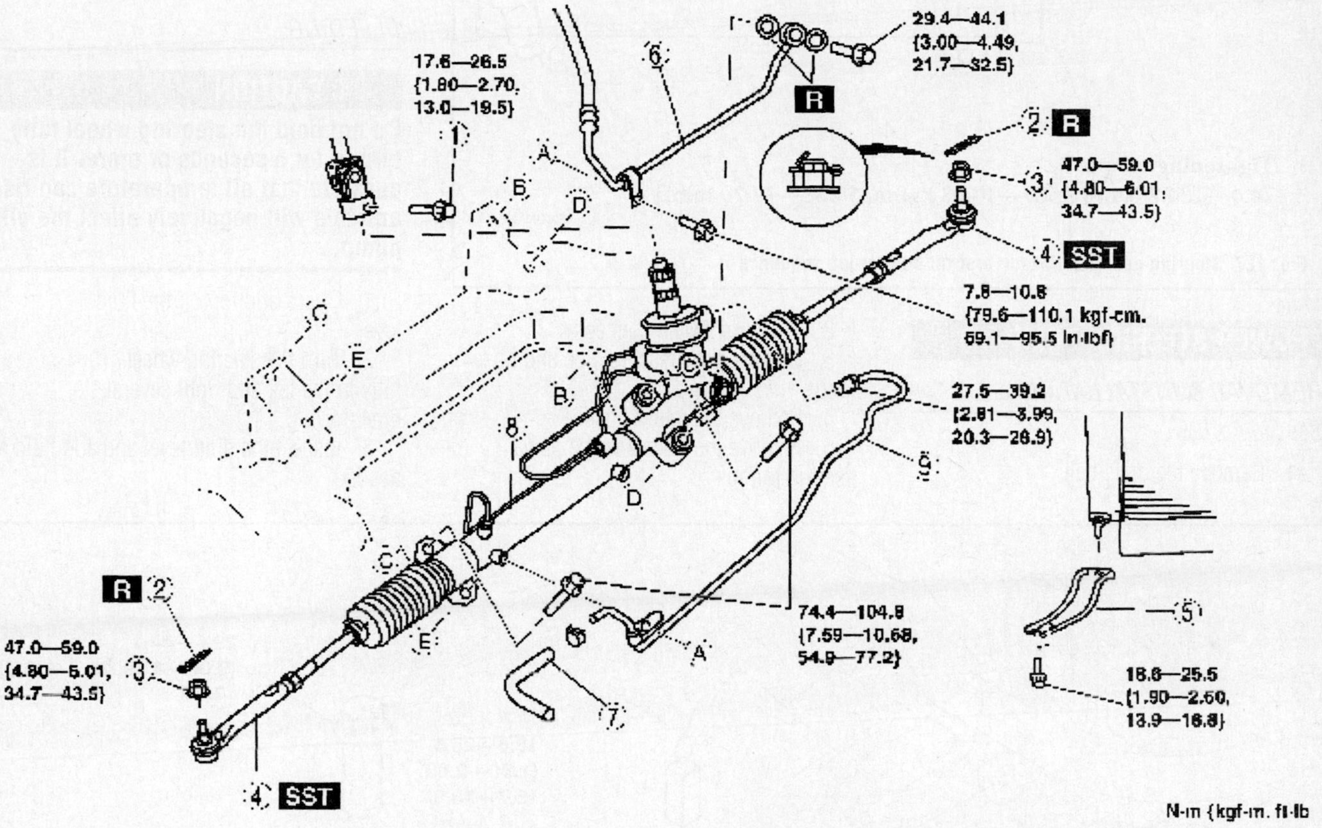

17.6—26.5
{1.80—2.70,
13.0—19.5}

29.4—44.1
{3.00—4.49,
21.7—32.5}

R

47.0—59.0
{4.80—6.01,
34.7—43.5}

SST

7.8—10.8
{79.6—110.1 kgf-cm.
59.1—95.5 in-lbf}

27.5—39.2
{2.81—3.99,
20.3—28.9}

74.4—104.8
{7.59—10.68,
54.9—77.2}

18.6—25.5
{1.90—2.60,
13.9—18.8}

47.0—59.0
{4.80—6.01,
34.7—43.5}

R

SST

N-m {kgf-m. ft-lb}

1 Bolt (intermediate shaft)	6 Pressure pipe
2 Cotter pin	7 Return hose
3 Nuts (tie-rod end ball joint)	8 Steering gear and linkage
4 Tie-rod end ball joint	9 Return pipe
5 Lower mounting rubber bracket	

09482_MAZC2_G0086

Fig. 125 Exploded view of the steering gear and linkage assembly

9. Remove the pressure pipe and return hose

10. Remove the steering gear and linkage retainers.

11. Remove the steering gear and linkage by pulling it from the right side of the vehicle.

To install:

12. Install the gear and linkage assembly and loosely tighten the bolts.

13. Assemble the mounting bracket with the mark on the bracket facing the vehicle rear.

14. Tighten the mounting bracket bolts to the 74.4–104.8 Nm (54.88–77.29 ft. lbs.) in the order shown.

15. Install the remaining components in the reverse order of removal using the accompanying illustration for torque values.

16. When installing the intermediate shaft. Align the marks and install the intermediate shaft and bolt. Tighten to 17–26 Nm (13–19 ft. lbs.).

17. After installation, adjust alignment.

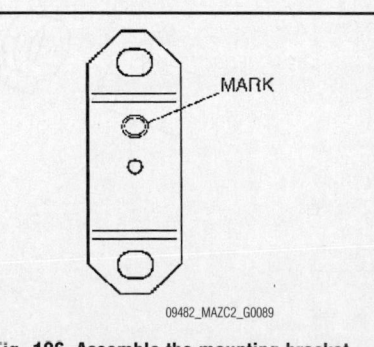

MARK

09482_MAZC2_G0089

Fig. 126 Assemble the mounting bracket with the mark on the bracket facing the vehicle rear

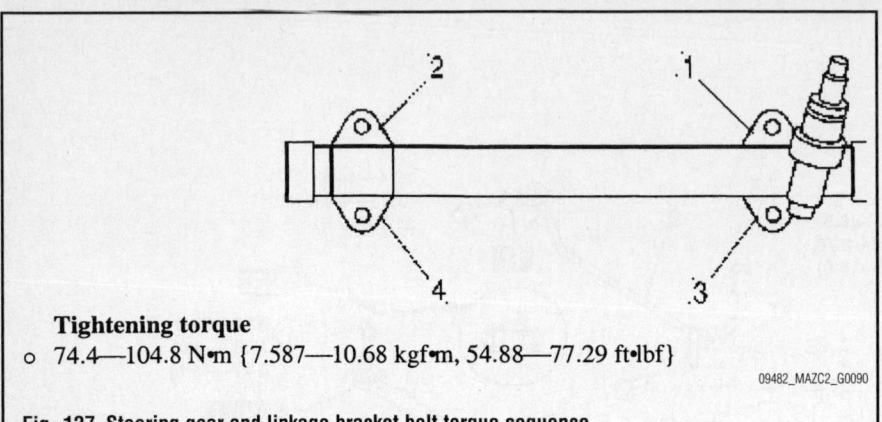

Tightening torque

o 74.4—104.8 N•m {7.587—10.68 kgf•m, 54.88—77.29 ft•lbf}

09482_MAZC2_G0090

Fig. 127 Steering gear and linkage bracket bolt torque sequence

POWER STEERING PUMP

REMOVAL & INSTALLATION

See Figure 128.

1. Remove the drive belt.

2. Pressure switch connector.
3. Remove the pressure pipe and hose band.
4. Remove the return hose.
5. Remove the bolts and the power steering pump.

To install:

6. Installation is the reverse of removal, refer to the illustration for components location and torque values.

BLEEDING

> ✳✳ **CAUTION**
>
> **Do not hold the steering wheel fully turned for 5 seconds or more. It is possible that oil temperature can rise and this will negatively affect the oil pump.**

1. Check and top off the fluid level.
2. Turn the steering wheel fully to the left and right several times.
3. Inspect the fluid level and add fluid as needed.

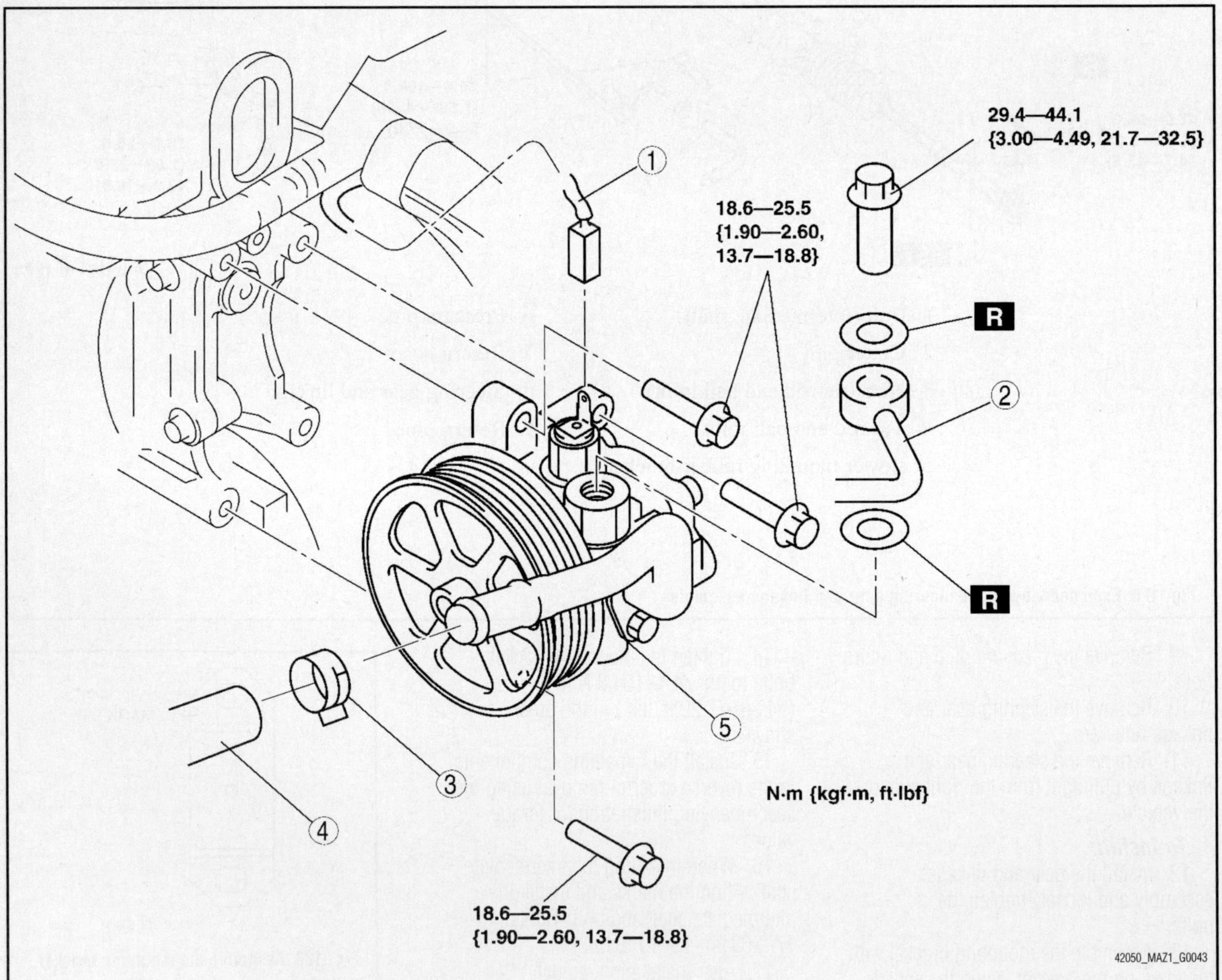

29.4—44.1
{3.00—4.49, 21.7—32.5}

18.6—25.5
{1.90—2.60,
13.7—18.8}

N·m {kgf·m, ft·lbf}

18.6—25.5
{1.90—2.60, 13.7—18.8}

42050_MAZ1_G0043

Fig. 128 Exploded view of the power steering pump assembly

4. Turn the steering wheel fully to the left and right several times until the fluid level stabilizes.

5. Start the engine and let it idle.

6. Turn the steering wheel fully to the left and right several times.

7. Turn the steering wheel fully to the left and right several

times until the fluid is not foamy and the fluid level has not dropped.

8. Check and top off the fluid level.

SUSPENSION

FRONT SUSPENSION

LOWER BALL JOINT

REMOVAL & INSTALLATION

The lower ball joint is an integral part of the lower control and cannot be replaced separately. If the lower ball joint is defective, the entire lower control arm must be replaced. Refer to the lower control arm procedure.

LOWER CONTROL ARM

REMOVAL & INSTALLATION

See Figure 129.

➡**Refer the exploded view illustration for component locations and if**

applicable, their retainer torque specifications.

❋❋ CAUTION

Performing the following procedures without first removing the ABS wheel-speed sensor may possibly cause an open circuit in the wiring harness if it is pulled by mistake. Before performing the following procedures, remove the ABS wheel-speed sensor (axle side) and fix it to an appropriate place where the sensor will not be pulled while servicing the vehicle.

1. Remove the caliper and mounting support from the steering knuckle and

suspend it with a cable in a location out of the way.

2. Separate the front lower arm ball joint from the knuckle.

➡**When removing the front lower arm ball joint, the steering knuckle bushing may also come off. If it comes off, replace the steering knuckle.**

3. Disconnect the tie-rod end.

4. Separate the front upper arm ball joint from the knuckle

5. Remove the front hub and steering knuckle assembly.

6. Remove the stabilizer control link nut on the front lower arm side.

7. Remove the front lower arm.

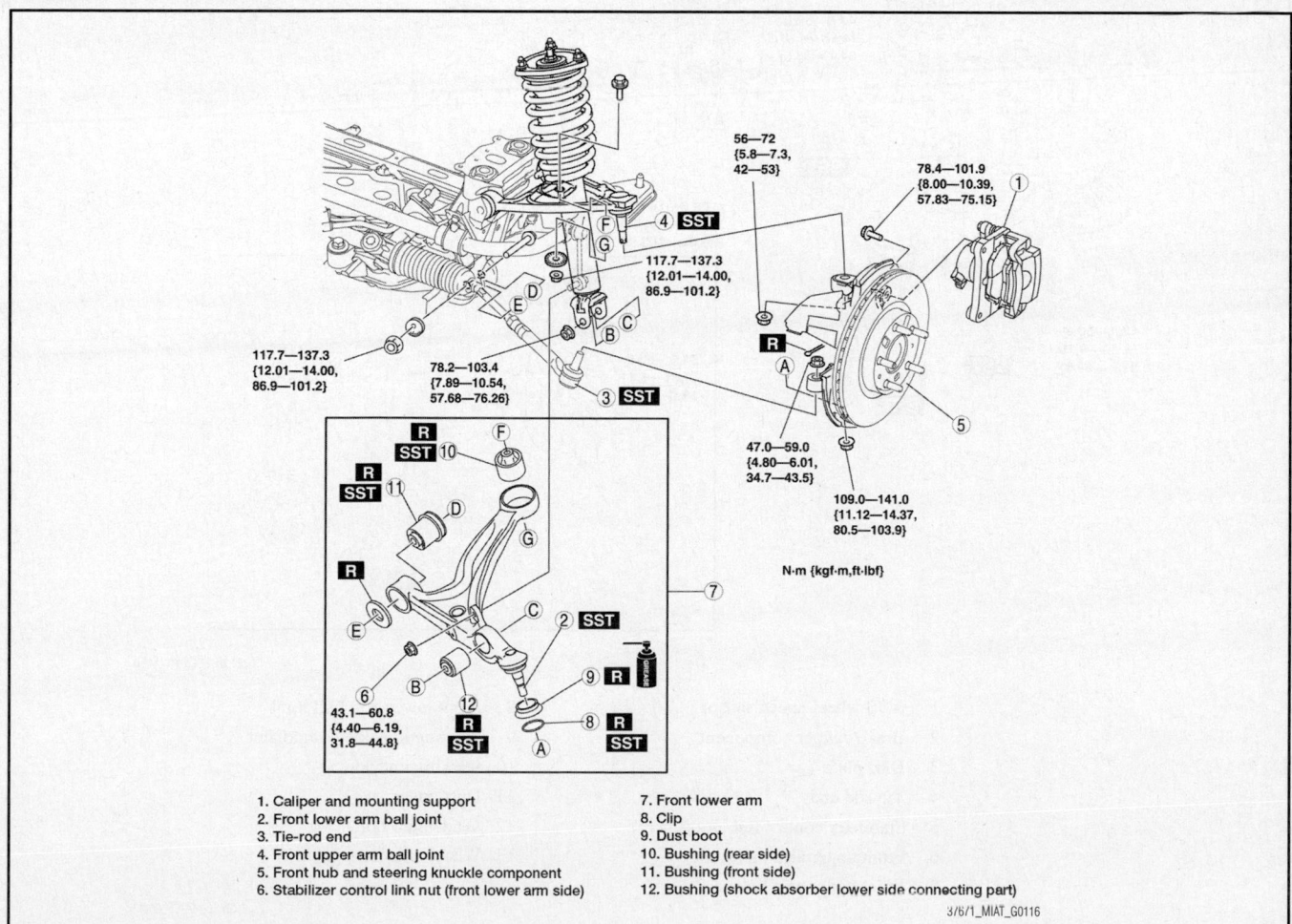

1. Caliper and mounting support
2. Front lower arm ball joint
3. Tie-rod end
4. Front upper arm ball joint
5. Front hub and steering knuckle component
6. Stabilizer control link nut (front lower arm side)
7. Front lower arm
8. Clip
9. Dust boot
10. Bushing (rear side)
11. Bushing (front side)
12. Bushing (shock absorber lower side connecting part)

Fig. 129 Exploded view of the front upper control arm assembly

To install:

8. Install all components in the reverse of removal.

9. Refer to illustration for tightening specifications.

STEERING KNUCKLE

REMOVAL & INSTALLATION

See Figure 130.

➡ **Refer the exploded view illustration for component locations and if applicable, their retainer torque specifications.**

✱✱ CAUTION

Performing the following procedures without first removing the ABS wheel-speed sensor may possibly cause an open circuit in the wiring harness if it is pulled by mistake. Before operations, remove the ABS wheel-speed sensor (axle side) and move the sensor away from the harnesses.

1. Disconnect the ABS wheel-speed sensor.

2. Remove the caliper with the hose still attached and support the assembly with wire so as to not place strain on the hose.

3. Remove the rotor.

4. Disconnect the tie-rod end from the knuckle.

5. Disconnect the stabilizer control link nut (lower).

6. Disconnect the front upper arm ball joint from the knuckle.

7. Remove the front upper arm bolt.

8. Disconnect the front lower arm ball joint from the knuckle.

9. Remove the steering knuckle assembly.

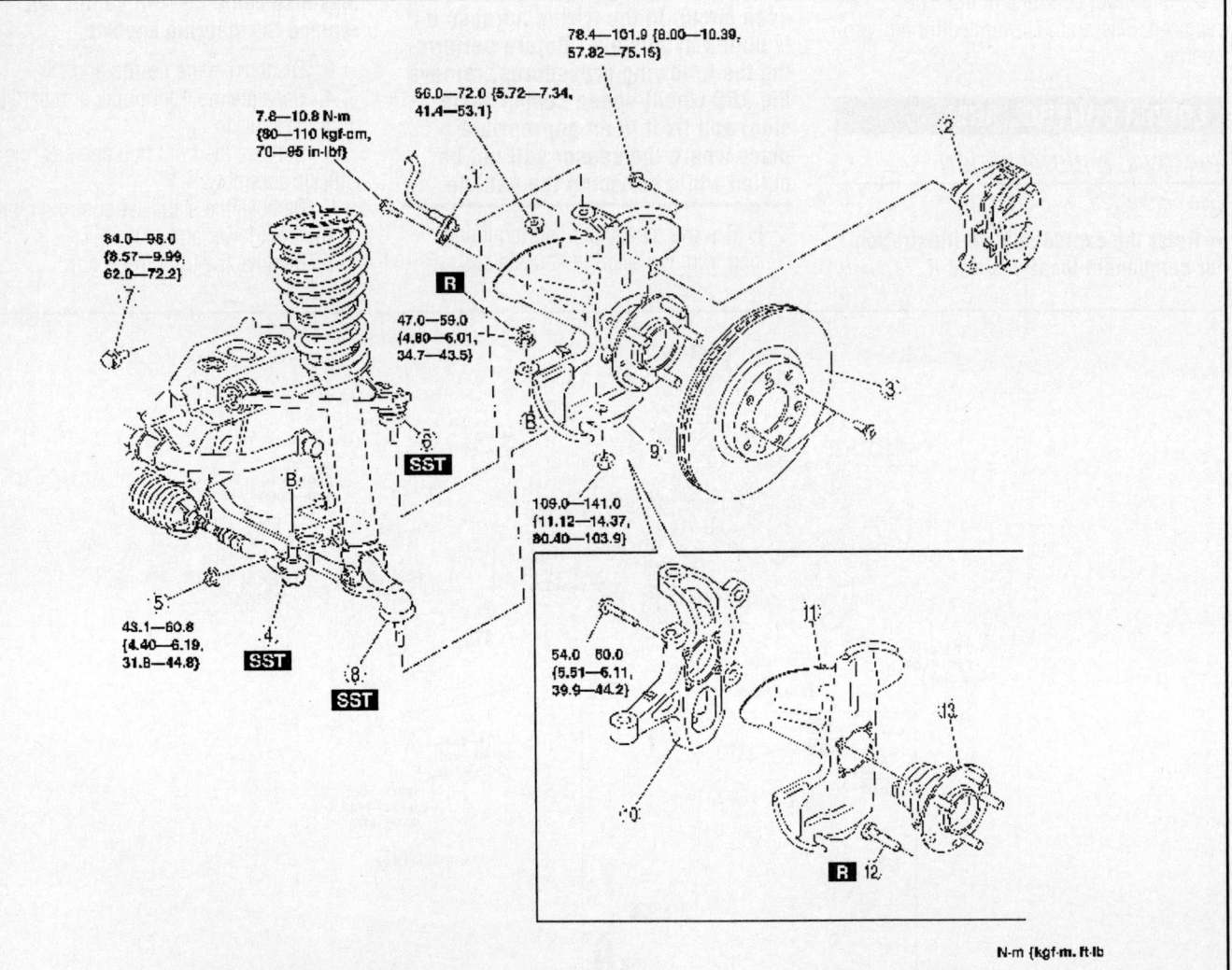

N-m {kgf·m, ft-lb

1	ABS wheel-speed sensor	8	Front lower arm ball joint
2	Brake caliper component	9	Steering knuckle component
3	Disc plate	10	Steering knuckle
4	Tie-rod end	11	Dust cover
5	Stabilizer control link nut (lower)	12	Wheel hub bolt
6	Front upper arm ball joint	13	Wheel hub component
7	Front upper arm bolt		

09482_MAZC2_G0117

Fig. 130 Exploded view of front hub and knuckle assembly

To install:

10. Install the remaining components in the reverse order of removal using the accompanying illustration for torque values.
11. Inspect the wheel alignment.

STRUT

REMOVAL & INSTALLATION

See Figure 131.

➡**Refer the exploded view illustration for component locations and if** applicable, their retainer torque specifications.

❈❈ CAUTION

Performing the following procedures without first removing the ABS wheel-speed sensor may possibly cause an open circuit in the wiring harness if it is pulled by mistake. Before performing the following procedures, remove the ABS wheel-speed sensor (axle side) and fix it to an appropriate place where the sensor will not be pulled while servicing the vehicle.

1. Remove the front suspension tower bar.
2. Remove the brake hose bracket.
3. Remove the front upper arm ball joint from the knuckle.
4. Remove the front shock absorber, coil spring and front upper arm.
5. Remove the front shock absorber and coil spring.

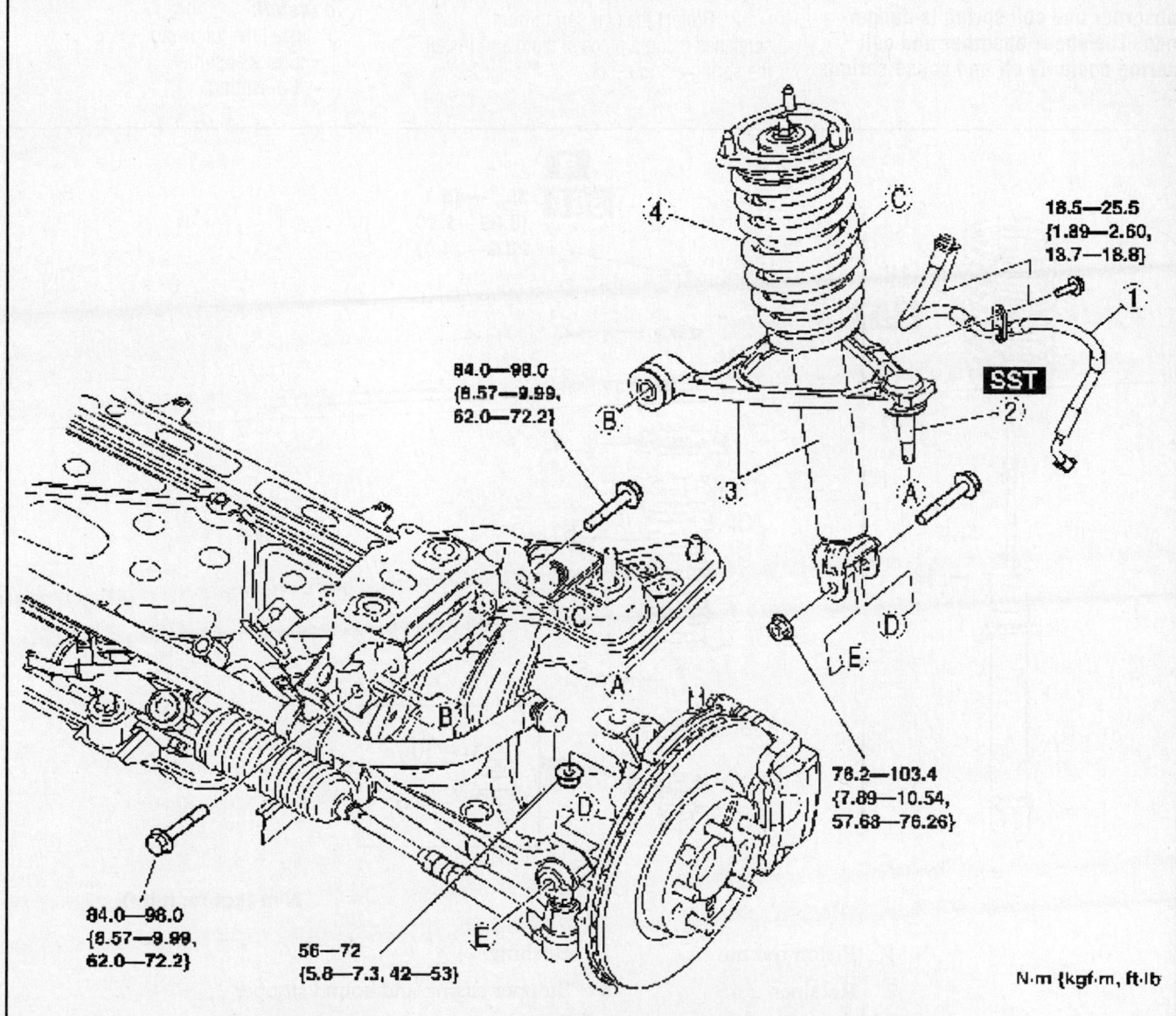

18.5—25.5
{1.89—2.60,
13.7—18.8}

84.0—98.0
{8.57—9.99,
62.0—72.2}

SST

78.2—103.4
{7.89—10.54,
57.63—76.26}

84.0—98.0
{8.57—9.99,
62.0—72.2}

56—72
{5.8—7.3, 42—53}

N·m {kgf·m, ft·lb}

1	Brake hose bracket	3	Front shock absorber, coil spring and front upper arm
2	Front upper arm ball joint	4	Front shock absorber and coil spring

09482_MAZC2_G0091

Fig. 131 Exploded view of the front strut assembly and related components

To install:

6. Install the components in the reverse order of removal using the accompanying illustration for torque values.

OVERHAUL

See Figures 132 through 135.

➡Refer the exploded view illustration for component locations and if applicable, their retainer torque specifications.

✲✲ CAUTION

Removing or installing the shock absorber and coil spring is dangerous. The shock absorber and coil spring could fly off and cause serious injury or death, and damage the vehicle.

1. Remove the front shock absorber and coil spring.

✲✲ WARNING

Before removing the piston rod nut, secure the shock absorber and spring in a suitable spring compressor. Otherwise, the shock absorber and spring could fly off under tremendous pressure and cause serious injury or death, or damage to vehicle parts.

2. Protect the coil spring from scratches using a piece of cloth and install the spring compressor.

3. Compress the coil spring and remove the piston rod nut.

4. Remove the following:
- Retainer
- Bushing
- Upper spring seat
- Dust boot
- Spacer
- Bushing
- Stopper casing and bound stopper
- Bound stopper
- Stopper casing
- Coil spring
- Shock absorber

To install:

5. Install the following:
- Shock absorber
- Coil spring

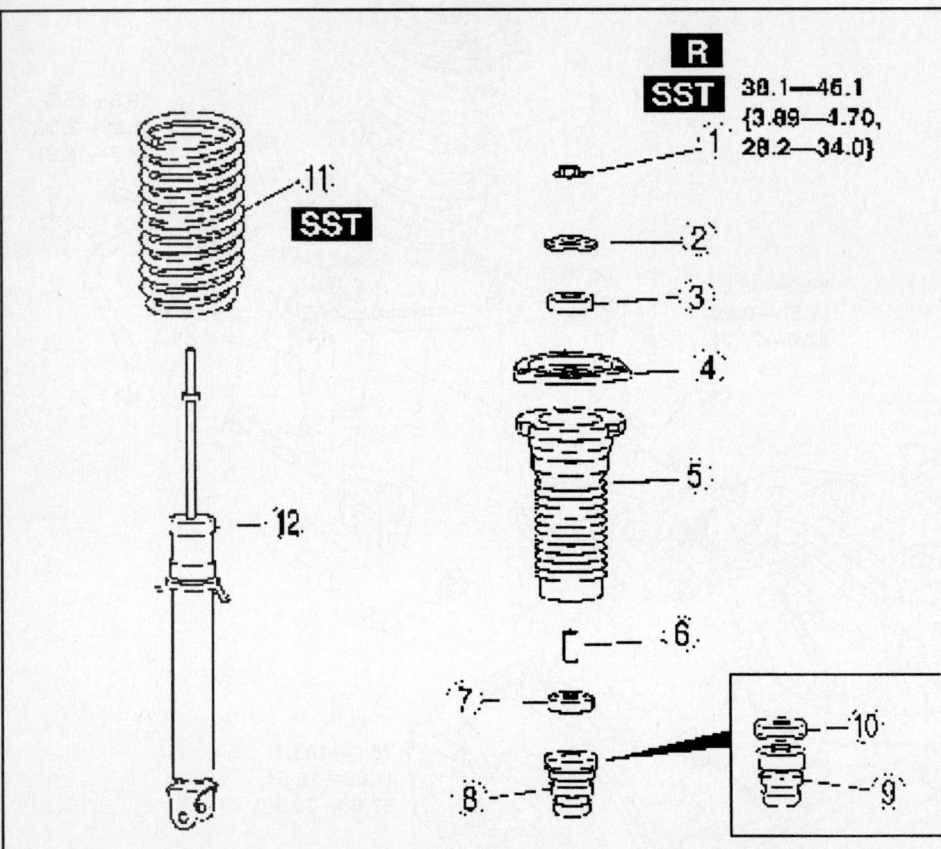

1	Piston rod nut	7	Bushing
2	Retainer	8	Stopper casing and bound stopper
3	Bushing	9	Bound stopper
4	Upper spring seat	10	Stopper casing
5	Dust boot	11	Coil spring
6	Spacer	12	Front shock absorber

09482_MAZC2_G0092

Fig. 132 Exploded view of the front coil spring assembly

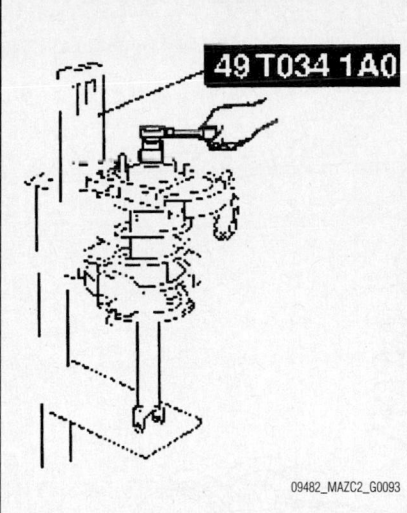

Fig. 133 Compress the front coil spring and remove the piston rod nut

- Stopper casing
- Bound stopper
- Stopper casing and bound stopper

- Bushing
- Spacer
- Dust boot
- Upper spring seat
- Bushing
- Retainer

6. Protect the coil spring from scratches using a piece of cloth and install the spring compressor.

7. Compress the coil spring using the spring compressor.

8. Install the shock absorber so that the lower end of the coil spring is seated on the step of the lower spring seat.

9. When installing the upper spring seat:

a. Align the mark on the upper spring seat with the dust boot projection.

b. Install the upper spring seat so that the upper spring seat stud is at a 27–33 degree angle to the shock absorber installation shaft (lower side).

10. Install the piston rod nut and tighten to 38–46 Nm (28–34 ft. lbs.).

11. Install the strut assembly.

STABILIZER BAR

REMOVAL & INSTALLATION

See Figure 136.

1. Remove the radiator mount bracket.

2. Remove the stabilizer control link.

3. Remove the stabilizer bracket.

4. Remove the stabilizer front stabilizer bushing and stabilizer,

To install:

5. Apply grease to the stabilizer bushing.

6. Align the outer side of the stabilizer slide stopper with the stabilizer bushing.

7. Install the stabilizer bracket.

8. Install the remaining components in the reverse of removal

9. Tighten the bolts in order indicated in the illustration.

UPPER BALL JOINT

REMOVAL & INSTALLATION

The upper ball joint is an integral part of the upper control and cannot be replaced separately. If the upper ball joint is defective, the entire upper control arm must be replaced. Refer to the upper control arm procedure.

UPPER CONTROL ARM

REMOVAL & INSTALLATION

See Figures 137 and 138.

➡**Refer the exploded view illustration for component locations and if applicable, their retainer torque specifications.**

✳✳ CAUTION

Performing the following procedures without first removing the ABS wheel-speed sensor may possibly cause an open circuit in the wiring harness if it is pulled by mistake. Before performing the following procedures, remove the ABS wheel-speed sensor (axle side) and fix it to an appropriate place where the sensor will not be pulled while servicing the vehicle.

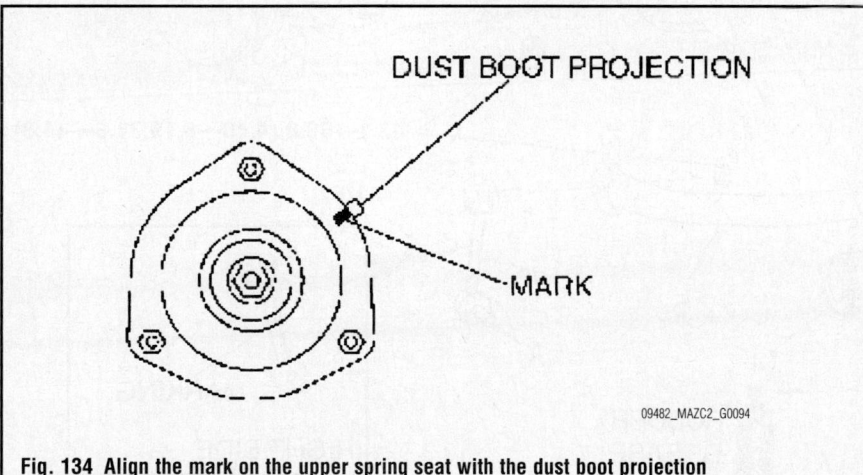

Fig. 134 Align the mark on the upper spring seat with the dust boot projection

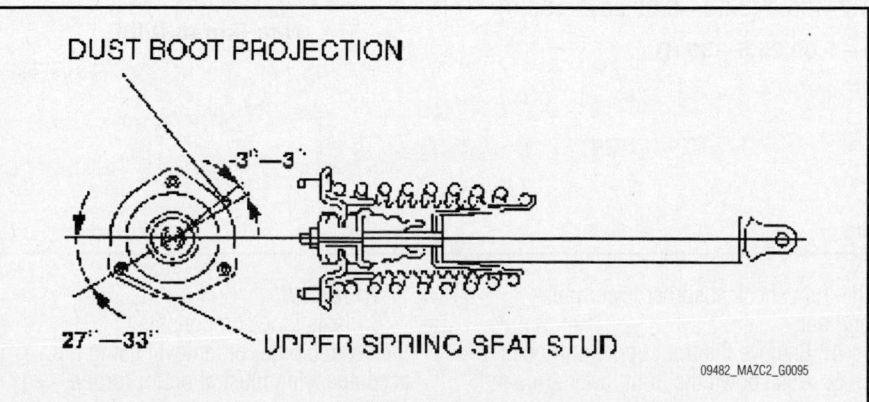

Fig. 135 Install the front strut upper spring seat so that the upper spring seat stud is at a 27–33 degree angle to the shock absorber installation shaft

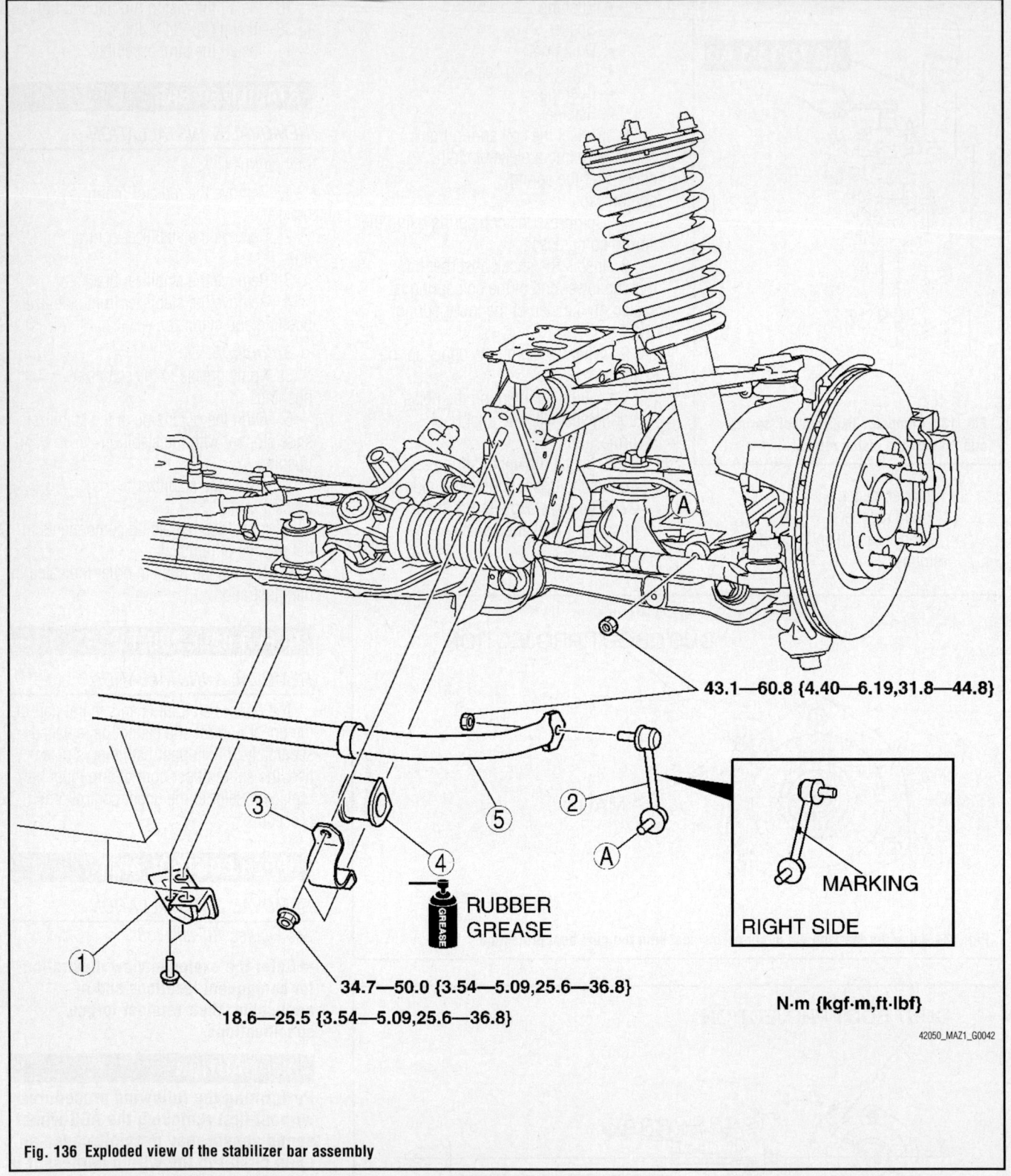

43.1—60.8 {4.40—6.19,31.8—44.8}

RUBBER GREASE

MARKING

RIGHT SIDE

34.7—50.0 {3.54—5.09,25.6—36.8}

18.6—25.5 {3.54—5.09,25.6—36.8}

N·m {kgf·m,ft·lbf}

42050_MAZ1_G0042

Fig. 136 Exploded view of the stabilizer bar assembly

1. Remove the brake hose bracket.

2. Separate the front upper arm ball joint from the knuckle.

3. Remove the front shock absorber, coil spring and front upper arm, by loosening the shock absorber upper nuts, then remove the front shock absorber lower bolt and nut.

4. Remove the front upper arm bolts.

5. Push down the front lower arm, and then remove the front upper arm from the gap between the shock absorber lower end and the front lower arm.

To install:

6. Install the components in the reverse order of removal using the accompanying illustration for torque values.

7. Inspect the front wheel alignment and adjust as necessary.

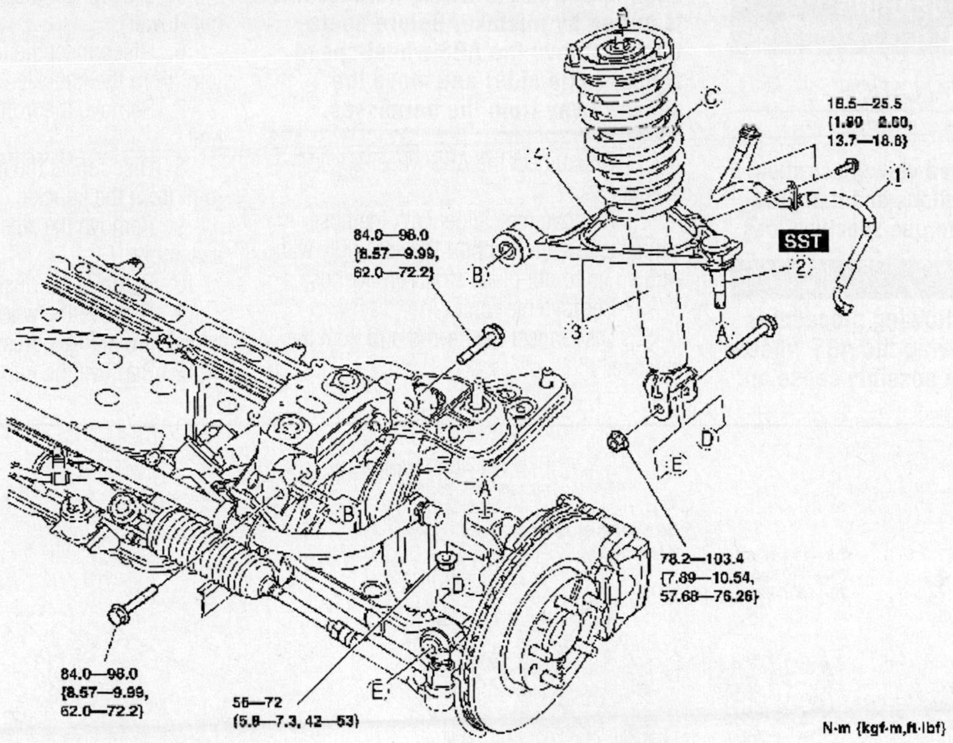

1   Brake hose bracket
2   Front upper arm ball joint
3   Front shock absorber, coil spring and front upper arm
4   Front upper arm

09482 MAZC2_G0101

Fig. 137 Exploded view of the front upper control arm assembly

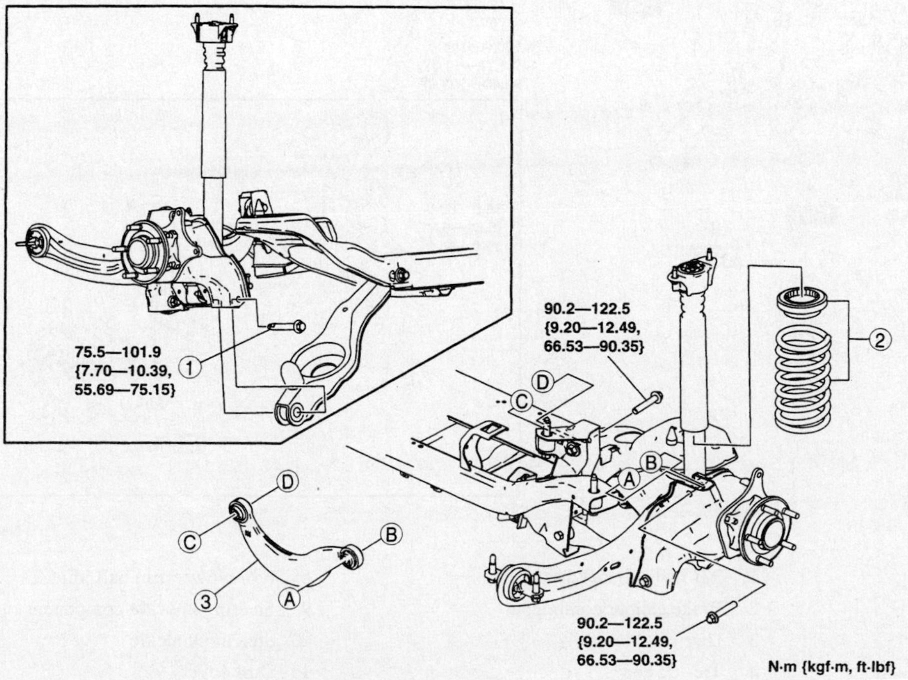

1   Rear lower arm outer bolt

2   Rear coil spring component

3   Rear upper arm

67162-MAZC-G95

Fig. 138 Exploded view of the rear upper control arm assembly and related components

WHEEL HUB & BEARING (SEALED UNIT)

REMOVAL & INSTALLATION
See Figure 139.

➡**Refer the exploded view illustration for component locations and if applicable, their retainer torque specifications.**

✳✳ CAUTION

Performing the following procedures without first removing the ABS wheel-speed sensor may possibly cause an open circuit in the wiring harness if it is pulled by mistake. Before operations, remove the ABS wheel-speed sensor (axle side) and move the sensor away from the harnesses.

1. Disconnect the ABS wheel-speed sensor.
2. Remove the caliper with the hose still attached and support the assembly with wire so as to not place strain on the hose.
3. Remove the rotor.
4. Disconnect the tie-rod end from the knuckle.
5. Disconnect the stabilizer control link nut (lower).
6. Disconnect the front upper arm ball joint from the knuckle.
7. Remove the front upper arm bolt.
8. Disconnect the front lower arm ball joint from the knuckle.
9. Remove the steering knuckle assembly.
10. Remove the dust cover.
11. Remove the wheel hub bolts from the wheel hub using a press.
12. Remove the hub assembly.

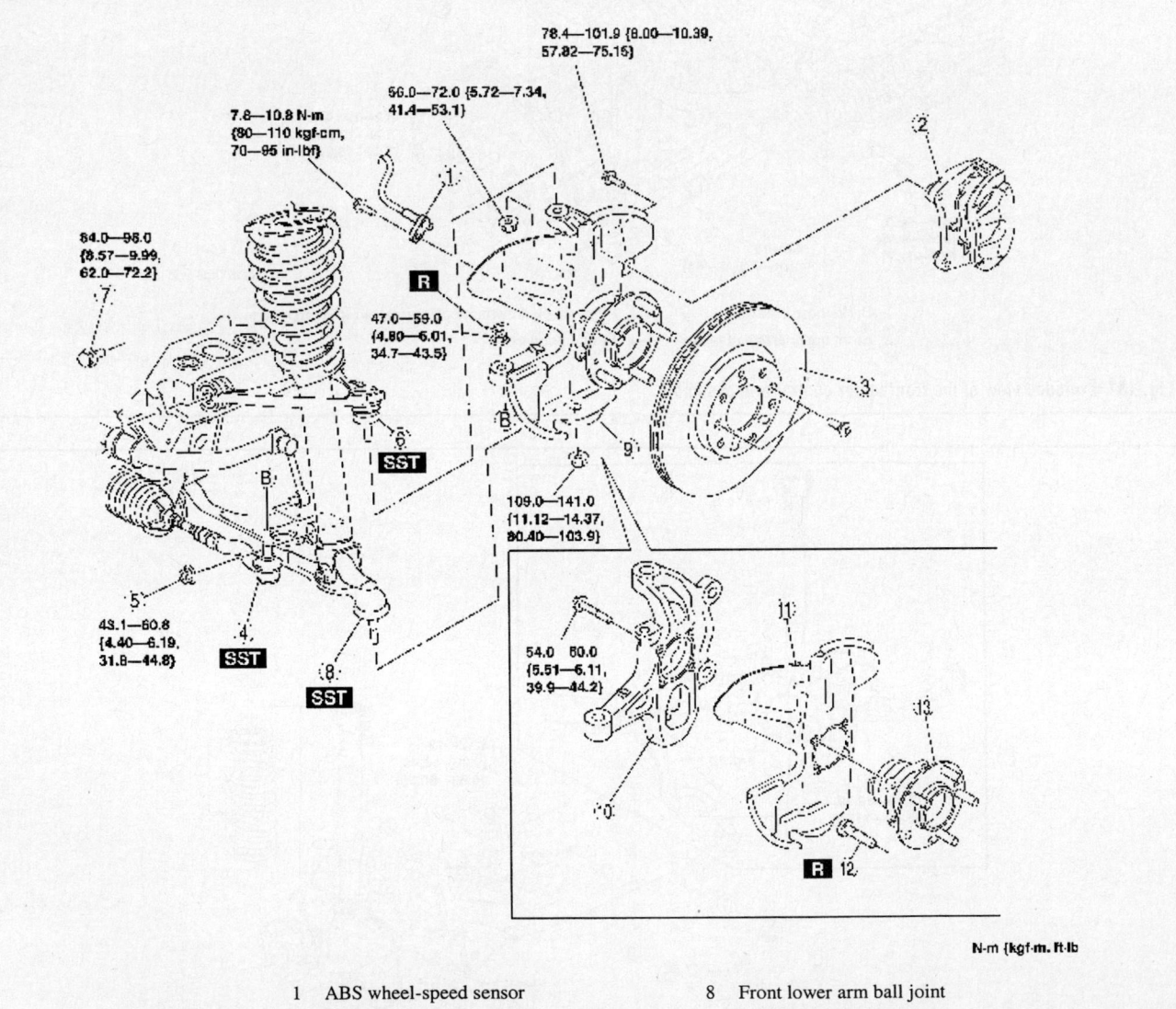

1	ABS wheel-speed sensor	8	Front lower arm ball joint
2	Brake caliper component	9	Steering knuckle component
3	Disc plate	10	Steering knuckle
4	Tie-rod end	11	Dust cover
5	Stabilizer control link nut (lower)	12	Wheel hub bolt
6	Front upper arm ball joint	13	Wheel hub component
7	Front upper arm bolt		

Fig. 139 Exploded view of front hub and knuckle assembly

09482_MAZC2_G0117

To install:

13. Install the hub assembly. tighten the bolts to 54–60 Nm (40–44 ft. lbs.).

14. Press in new wheel hub bolts into the wheel hub using a press.

15. Install the remaining components in the reverse order of removal using the accompanying illustration for torque values.

16. Inspect the wheel alignment.

SUSPENSION

CONTROL ARMS/LINKS

REMOVAL & INSTALLATION

Lower Rear Lateral Link

See Figures 140 and 141.

➡**Refer the exploded view illustration for component locations and if applicable, their retainer torque specifications.**

✳✳ CAUTION

Performing the following procedures without first removing the ABS wheel-speed sensor may possibly cause an open circuit in the wiring harness if it

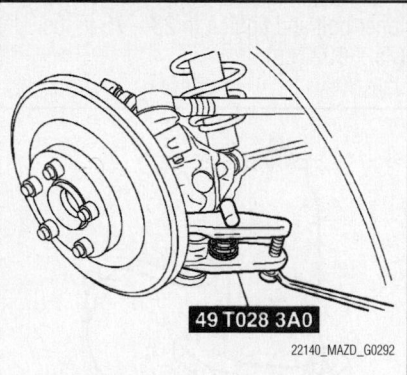

49 T028 3A0

22140_MAZD_G0292

Fig. 140 Using the removal tool remove the lateral link (lower) ball joint

REAR SUSPENSION

is pulled by mistake. Before operations, remove the ABS wheel-speed sensor (axle side) and move the sensor away from the harnesses.

1. remove the lower rear lateral link as follow:

- Stabilizer control link lower nut
- Rear lateral link (lower) ball joint
- Rear lateral link (lower)
- Dust boot

To install:

2. Wipe the grease off the ball joint stud.

3. Fill the inside of the new dust boot with grease.

ADJUSTMENT

The front wheel bearings are not adjustable. If the bearings become loose or make noise, they must be replaced.

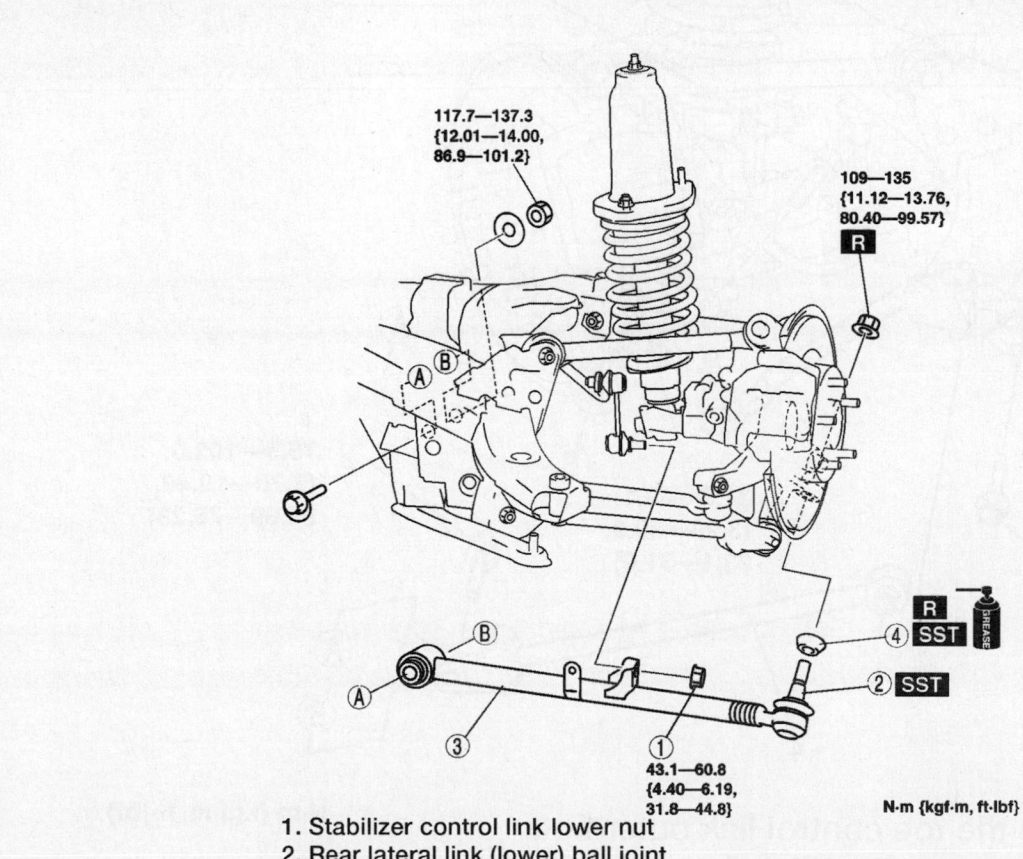

117.7—137.3
{12.01—14.00,
86.9—101.2}

109—135
{11.12—13.76,
80.40—99.57}
R

43.1—60.8
{4.40—6.19,
31.8—44.8}

N·m {kgf·m, ft·lbf}

1. Stabilizer control link lower nut
2. Rear lateral link (lower) ball joint
3. Rear lateral link (lower)
4. Dust boot

22140_MAZD_G0294

Fig. 141 Lateral link illustration

4. Install the dust boot to the ball joint.

5. Wipe off the excess grease.

6. Install the remaining components in the reverse order of removal using the accompanying illustration for torque values.

7. Inspect the rear wheel alignment.

Lower Rear Trailing Link

See Figure 142.

1. Remove the toe control link outer bolt.

2. Remove rear trailing link (lower) outer bolt.

3. Remove rear trailing link (lower) inner bolt.

4. Remove rear trailing link (lower).

To install:

5. Install the rear trailing link (lower).

6. Install the rear trailing link (lower) inner bolt and tighten to 23—75 ft. lbs. (75—102 Nm).

7. Install the rear trailing link (lower) outer bolt and tighten to 56—75 ft. lbs. (75—102 Nm).

8. Install the toe control link outer bolt and tighten to 23—31 ft. lbs. (31—42 Nm).

Upper Rear Lateral Link

See Figures 143 and 144.

➡Refer the exploded view illustration for component locations and if applicable, their retainer torque specifications.

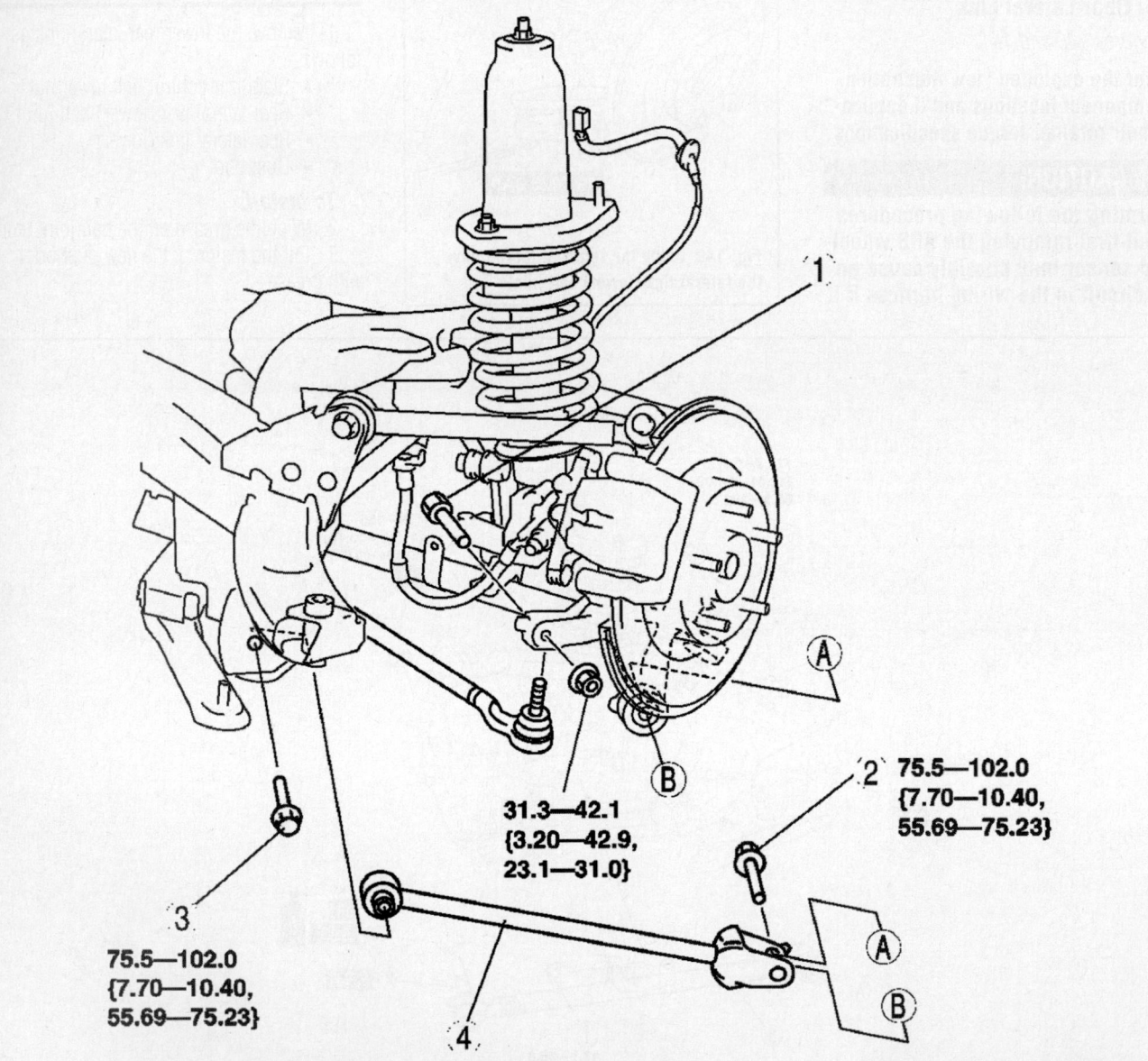

31.3—42.1
{3.20—42.9,
23.1—31.0}

2 75.5—102.0
{7.70—10.40,
55.69—75.23}

75.5—102.0
{7.70—10.40,
55.69—75.23}

N·m {kgf·m, ft·lbf}

1. Remove the toe control link outer bolt.

2. Remove rear trailing link (lower) outer bolt.

3. Remove rear trailing link (lower) inner bolt.

4. Remove rear trailing link (lower).

Fig. 142 Rear lower trailing link removal

22140_MAZD_G0295

⁜⁜ CAUTION

Performing the following procedures without first removing the ABS wheel-speed sensor may possibly cause an open circuit in the wiring harness if it is pulled by mistake. Before operations, remove the ABS wheel-speed sensor (axle side) and move the sensor away from the harnesses.

1. Using the tools illustrated, disconnect the rear lateral link (upper) ball joint.

➡ When removing the rear lateral link (upper) ball joint, the rear knuckle bushing may also come off. If it comes off, replace the rear knuckle.

2. Remove the rear lateral link (upper).
3. Remove the dust boot.

To install:

4. Wipe the grease off the ball joint stud.
5. Fill the inside of the new dust boot with grease.
6. Using the tool shown, install the dust boot to the ball joint.
7. Wipe off the excess grease.
8. Install the remaining components in the reverse order of removal using the

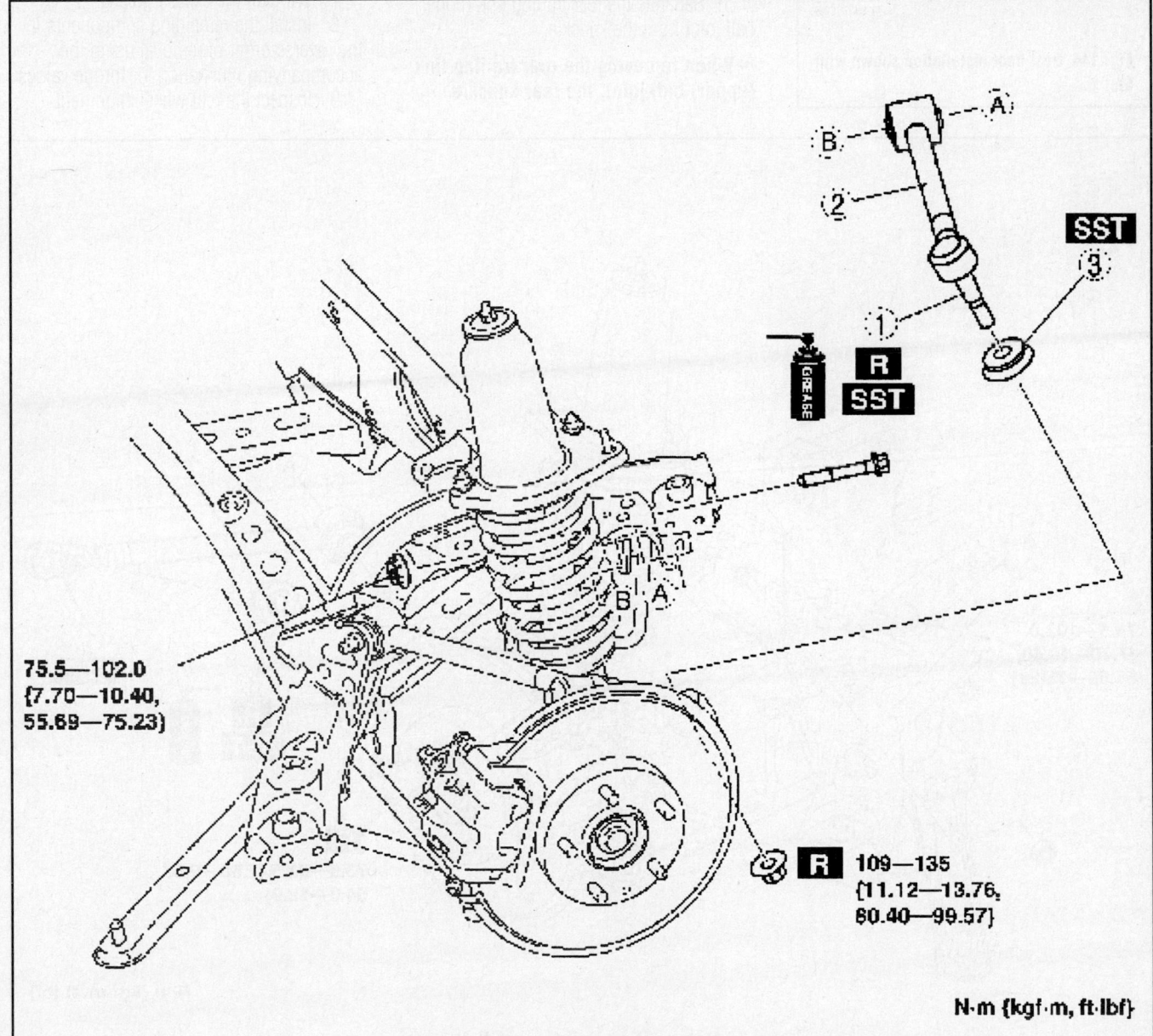

75.5—102.0
{7.70—10.40,
55.69—75.23}

R 109—135
{11.12—13.76,
80.40—99.57}

N·m {kgf·m, ft·lbf}

1 Rear lateral link (upper) ball joint
2 Rear lateral link (upper)
3 Dust boot

09482_MAZC2_G0116

Fig. 143 Exploded view of upper rear lateral link assembly

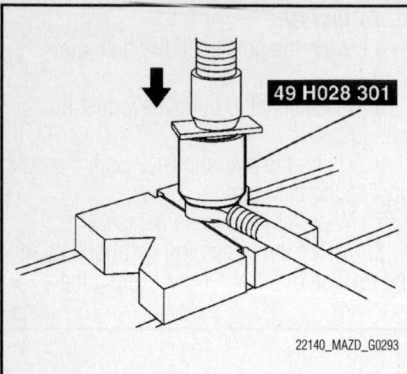

Fig. 144 Dust boot installation shown with tool

accompanying illustration for torque values.

9. Inspect the rear wheel alignment.

Upper Rear Trailing Link

See Figures 145 and 146.

➡ **Refer the exploded view illustration for component locations and if applicable, their retainer torque specifications.**

1. Separate the rear trailing link (upper) ball joint from the knuckle.

➡ **When removing the rear trailing link (upper) ball joint, the rear knuckle bushing may also come off. If it comes off, replace the rear knuckle.**

2. Remove the rear upper trailing link.
3. Remove the dust boot.

To install:

4. Wipe the grease off the ball joint stud.
5. Fill the inside of the new dust boot with grease.
6. Using the tool shown, install the dust boot to the ball joint.
7. Wipe off the excess grease.
8. Install the remaining components in the reverse order of removal using the accompanying illustration for torque values.
9. Inspect the rear wheel alignment.

75.5—102.0
{7.70—10.40,
55.69—75.23}

73.5—93.5 {7.50—9.53,
54.3—68.9}

N·m {kgf·m, ft·lbf}

1　Rear trailing link (upper) ball joint
2　Rear trailing link (upper)
3　Dust boot

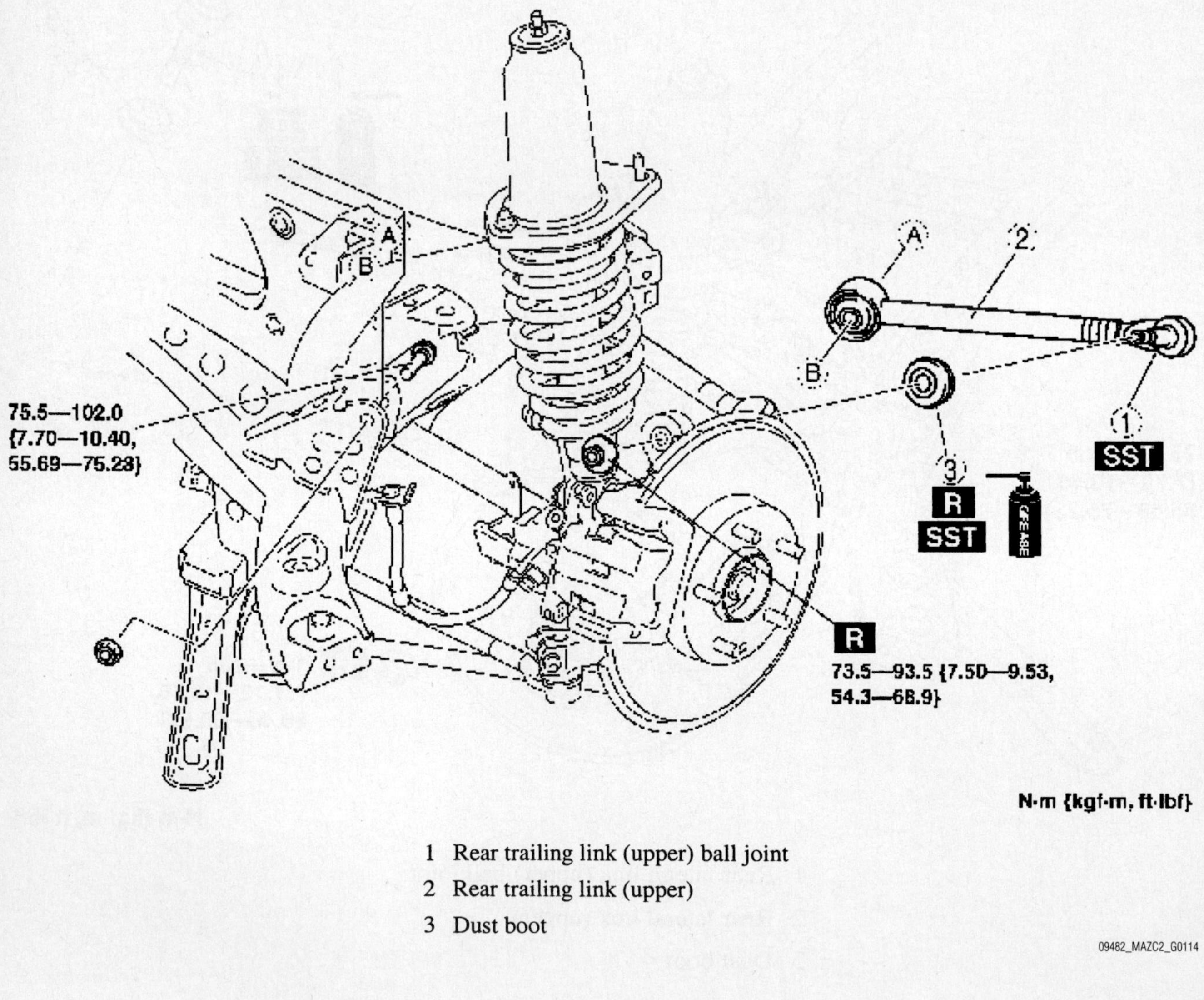

Fig. 145 Exploded view of upper rear trailing link assembly

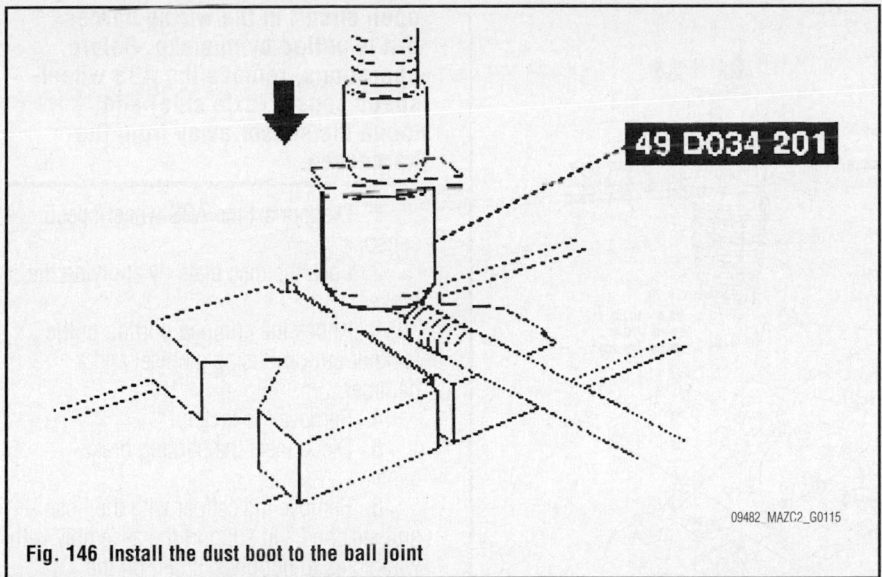

Fig. 146 Install the dust boot to the ball joint

STABILIZER BAR

REMOVAL & INSTALLATION

See Figures 147 through 150.

1. Remove in the numerical order indicated.

To install:

2. Install in the reverse order of removal.
3. Install the rear stabilizer so that the identification mark is on the right side of the vehicle.

4. Install the stabilizer control link in the proper angle as shown.

✳✳ CAUTION

Be sure to install the stabilizer control link in the proper position. If it is not installed properly, the stabilizer control link may interfere with

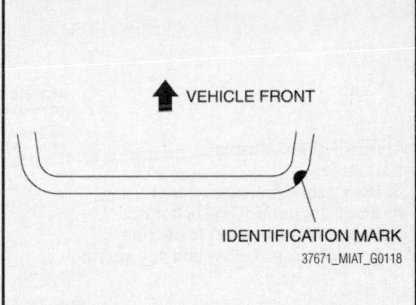

Fig. 148 Install the rear stabilizer so that the identification mark is on the right side of the vehicle

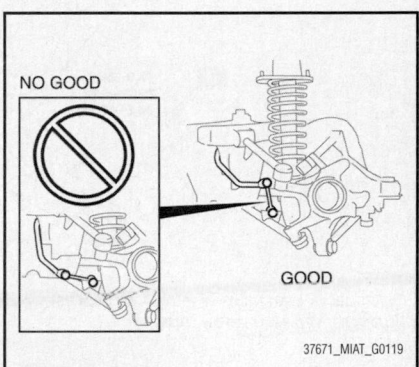

Fig. 149 Install the stabilizer control link in the proper angle

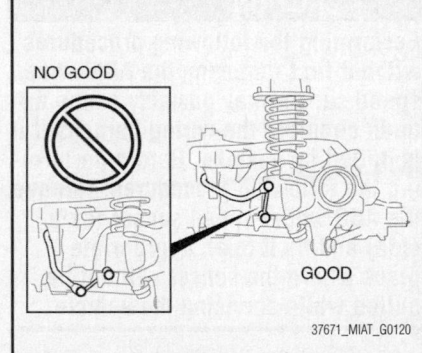

Fig. 150 Verify that the stabilizer control link is installed in the angle shown

peripheral components when driving, causing damage to each other.

5. Place the vehicle on the ground and verify that the stabilizer control link is installed in the angle shown.

STRUT

REMOVAL & INSTALLATION

See Figure 151.

➡Refer the exploded view illustration for component locations and if applicable, their retainer torque specifications.

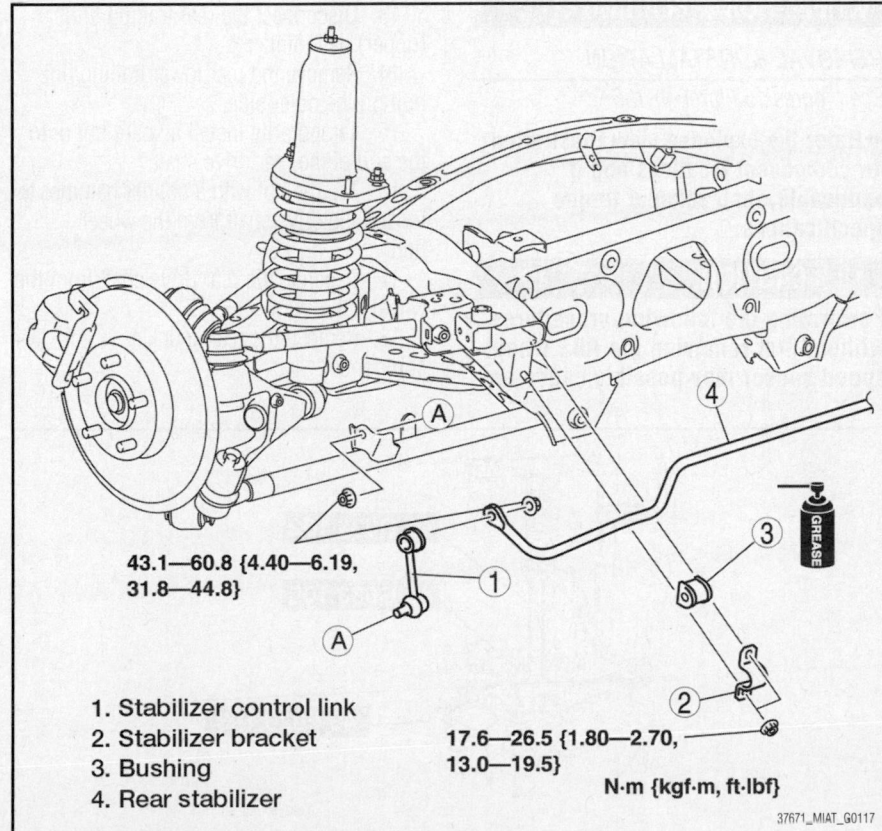

43.1—60.8 {4.40—6.19, 31.8—44.8}

17.6—26.5 {1.80—2.70, 13.0—19.5}

N·m {kgf·m, ft·lbf}

1. Stabilizer control link
2. Stabilizer bracket
3. Bushing
4. Rear stabilizer

Fig. 147 Remove in the numerical order indicated

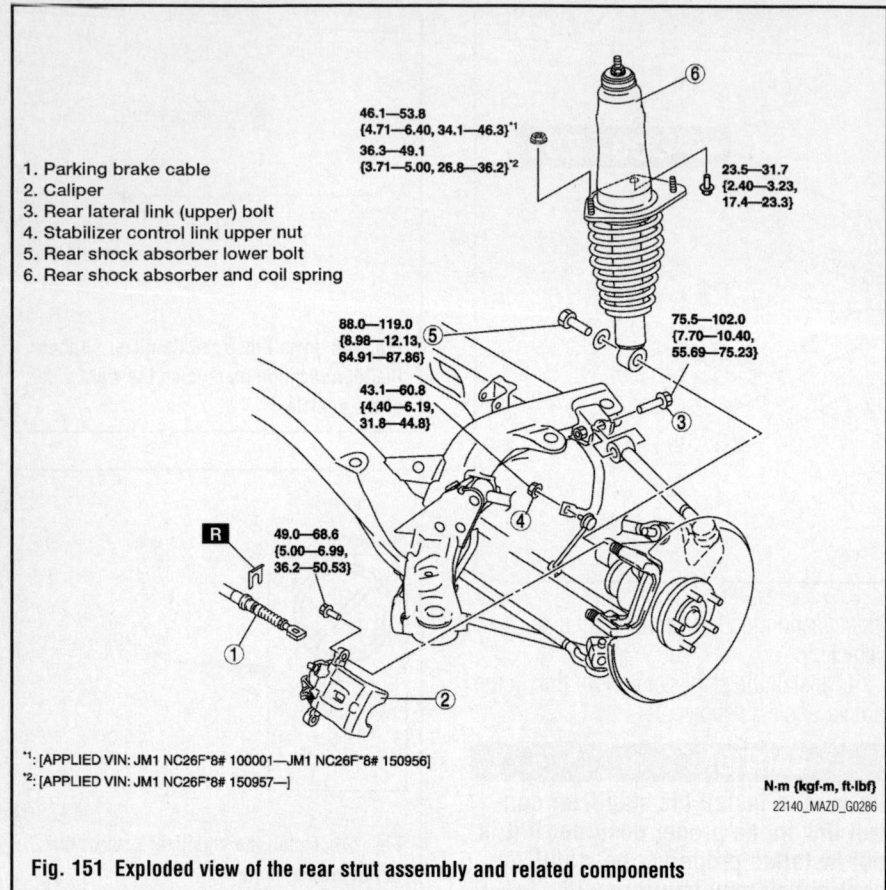

1. Parking brake cable
2. Caliper
3. Rear lateral link (upper) bolt
4. Stabilizer control link upper nut
5. Rear shock absorber lower bolt
6. Rear shock absorber and coil spring

46.1—53.8
{4.71—6.40, 34.1—46.3}*1
36.3—49.1
{3.71—5.00, 26.8—36.2}*2

23.5—31.7
{2.40—3.23, 17.4—23.3}

88.0—119.0
{8.98—12.13, 64.91—87.86}

75.5—102.0
{7.70—10.40, 55.69—75.23}

43.1—60.8
{4.40—6.19, 31.8—44.8}

49.0—68.6
{5.00—6.99, 36.2—50.53}

*1: [APPLIED VIN: JM1 NC26F*8# 100001—JM1 NC26F*8# 150956]
*2: [APPLIED VIN: JM1 NC26F*8# 150957—]

N·m {kgf·m, ft·lbf}
22140_MAZD_G0286

Fig. 151 Exploded view of the rear strut assembly and related components

❊❊ CAUTION
Performing the following procedures without first removing the ABS wheel-speed sensor may possibly cause an open circuit in the wiring harness if it is pulled by mistake. Before performing the following procedures, remove the ABS wheel-speed sensor (axle side) and fix it to an appropriate place where the sensor will not be pulled while servicing the vehicle.

1. If removing the left side, remove the fuel tank protector.
2. Remove the parking brake cable.
3. Remove the caliper without disconnecting the hose, and support the caliper with wire.
4. Remove the rear lateral link upper bolt.
5. Remove the stabilizer control link upper nut.
6. Remove the rear shock absorber lower bolt.
7. Remove the rear shock absorber and coil spring.

To install:
8. Install the components in the reverse order of removal using the accompanying illustration for torque values.

WHEEL HUB & BEARING

REMOVAL & INSTALLATION
See Figures 152 through 156.

➡**Refer the exploded view illustration for component locations and if applicable, their retainer torque specifications.**

❊❊ CAUTION
Performing the following procedures without first removing the ABS wheel-speed sensor may possibly cause an

open circuit in the wiring harness if it is pulled by mistake. Before operations, remove the ABS wheel-speed sensor (axle side) and move the sensor away from the harnesses.

1. Disconnect the ABS wheel-speed sensor.
2. Lock the disc plate by applying the brakes.
3. Knock the crimped portion of the locknut outward using a chisel and a hammer.
4. Remove the locknut.
5. Disconnect the parking brake cable.
6. Remove the caliper with the hose still attached and support the assembly with wire so as to not place strain on the hose.
7. Remove the rotor.
8. Disconnect the rear lateral link (upper) ball joint from the knuckle.
9. Remove the stabilizer control link lower nut.
10. Disconnect the rear lateral link lower ball joint from the knuckle.
11. Remove the lower shock absorber bolt.
12. Disconnect the toe control link ball joint.
13. Disconnect the rear trailing link (upper) ball joint.
14. Remove the rear lower trailing link bolt on the outer side.
15. Temporarily install a spare nut onto the end of the rear drive shaft.
16. Tap the nut with a copper hammer to loosen the drive shaft from the wheel hub.
17. Separate the rear drive shaft from the wheel hub.
18. Remove the rear knuckle component.

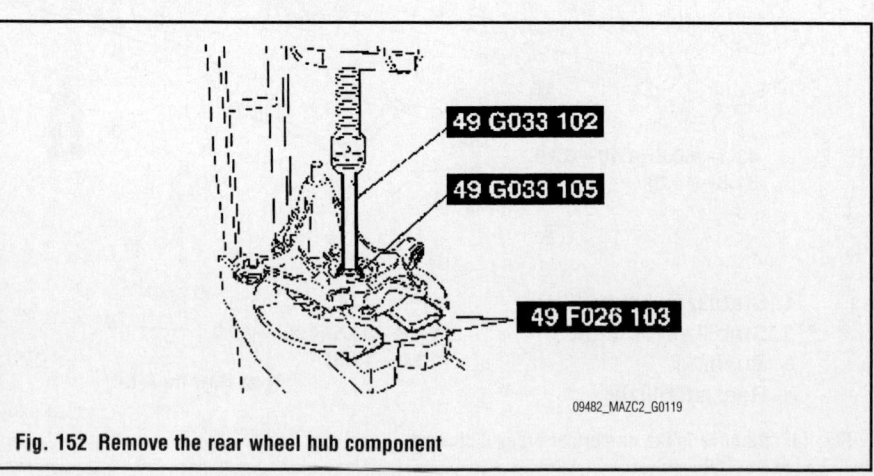

49 G033 102
49 G033 105
49 F026 103

09482_MAZC2_G0119

Fig. 152 Remove the rear wheel hub component

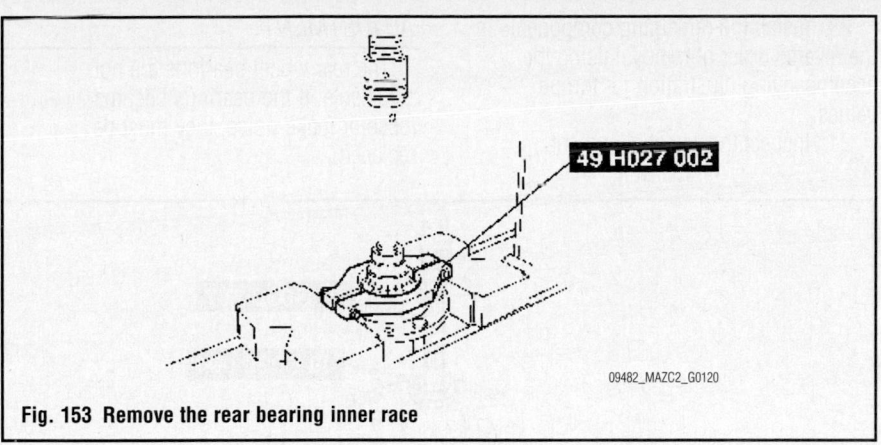

49 H027 002

09482_MAZC2_G0120

Fig. 153 Remove the rear bearing inner race

19. Remove the wheel Hub Component as follows:

  a. Wind the tool illustrated and backing plate contact area with packing tape two times.

  b. Remove the wheel hub component using the tools illustrated.

  c. If the bearing inner race remains on the wheel hub component, use a chisel to secure a sufficient space for installing the service tool between wheel hub components and bearing inner race.

20. Remove the bearing inner race using the tool illustrated.

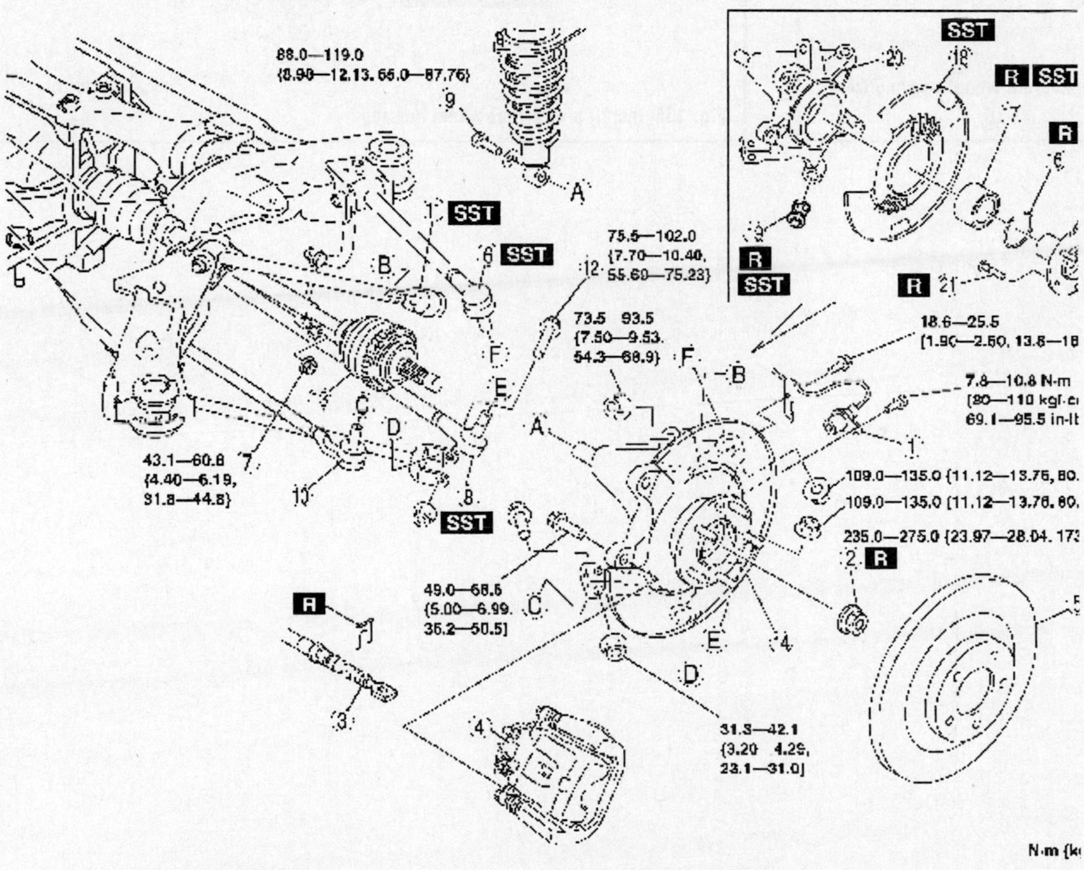

1	ABS wheel-speed sensor	12	Rear trailing link (lower) bolt (outer side)
2	Locknut	13	Rear drive shaft
3	Parking brake cable	14	Rear knuckle component
4	Brake caliper component	15	Wheel hub component
5	Disc plate	16	Retaining ring
6	Rear lateral link (upper) ball joint	17	Wheel bearing
7	Stabilizer control link nut (lower)	18	Dust cover
8	Rear lateral link (lower) ball joint	19	Bushing
9	Shock absorber bolt (lower)	20	Rear knuckle
10	Toe control link ball joint	21	Wheel hub bolt
11	Rear trailing link (upper) ball joint		

09482_MAZC2_G0118

Fig. 154 Exploded view of rear hub and knuckle assembly and related components

To install:

21. Install a new wheel bearing using the tool illustrated.

22. Install the wheel hub component using the tool illustrated.

23. Install the remaining components in the reverse order of removal using the accompanying illustration for torque values.

24. Inspect the wheel alignment.

ADJUSTMENT

The rear wheel bearings are not adjustable. If the bearings become loose or make noise, they must be replaced.

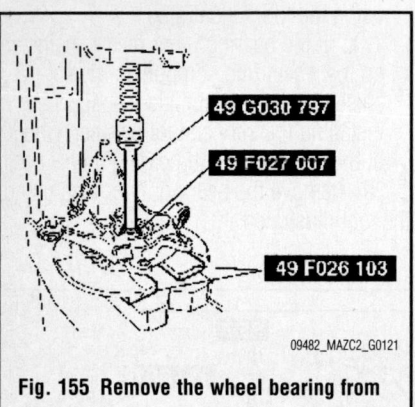

49 G030 797
49 F027 007
49 F026 103

09482_MAZC2_G0121

Fig. 155 Remove the wheel bearing from the knuckle

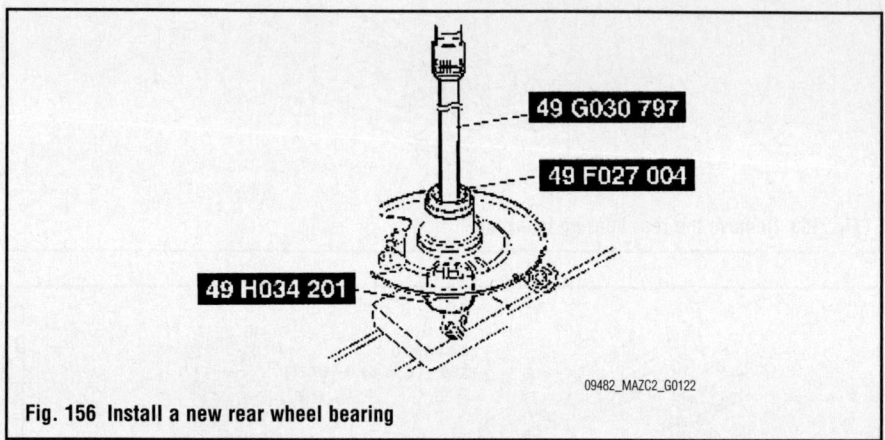

49 G030 797
49 F027 004
49 H034 201

09482_MAZC2_G0122

Fig. 156 Install a new rear wheel bearing

SPECIFICATIONS AND MAINTENANCE CHARTS

VEHICLE AND ENGINE IDENTIFICATION CHART

Engine							Model Year	
Code ①	Liters	Cu. In.	Cyl.	Fuel Sys.	Engine Type	Eng. Mfg.	Code ②	Year
M	1.3	79	③	MPI	Rotary	Mazda	9	2009
2	1.3	79	③	MPI	Rotary	Mazda	A	2010

MPI: Multi point fuel injection

① 8th position of VIN

② 10th position of VIN

③ Twin rotary

37671_MRX8_C0001

GENERAL ENGINE SPECIFICATIONS

Year	Engine Displacement Liters	Engine VIN	Net Horsepower @ rpm	Net Torque @ rpm (ft. lbs.)	Bore x Stroke (in.)	Compression Ratio	Oil Pressure @ rpm
2009	1.3	M	212@7500	159@5500	①	①	51@3000
	1.3	2	232@8500	159@5500	①	①	51@3000
2010	1.3	M	212@7500	159@5500	①	①	51@3000
	1.3	2	232@8500	159@5500	①	①	51@3000

① Rotary chamber not measured.

37671_MRX8_C0002

ENGINE TUNE-UP SPECIFICATIONS

Year	Engine Displacement Liters	Engine VIN	Spark Plugs Gap (in.)	Ignition Timing (deg.) MT	AT	Fuel Pump (psi)	Idle Speed (rpm) MT	AT	Valve Clearance In.	Ex.
2009	1.3	①	0.046-0.049	②	②	③	750-850	760-860	NA	NA
	1.3	①	0.046-0.049	②	②	③	750-850	760-860	NA	NA
2010	1.3	①	0.046-0.049	②	②	③	750-850	760-860	NA	NA
	1.3	①	0.046-0.049	②	②	③	750-850	760-860	NA	NA

NOTE: The Vehicle Emission Control Information label often reflects specification changes changes made during production.

The label figures must be used if they differ from those in this chart.

NA: Not applicable.

① M & 2 Engines

② Controlled by the Powertrain Control Module and cannot be adjusted.

③ Fuel line hold pressure 55-65 psi

37671_MRX8_C0003

CAPACITIES

Year	Model	Engine Displacement Liters	Engine ID/VIN	Engine Oil with Filter (qts.)	Transmission (pts.)		Drive Axle		Fuel Tank (gal.)	Cooling System (qts.)
					6-Spd	Auto.	Front (pts.)	Rear (pts.)		
2009	RX-8	1.3	①	3.7	3.7	18.4	—	2.7	15.9	8.7
2010	RX-8	1.3	①	3.7	3.7	18.4	—	2.7	15.9	8.7

NOTE: All capacities are approximate. Add fluid gradually and check to be sure a proper fluid level is obtained.

① M and 2 Engines

37671_MRX8_C0005

FLUID SPECIFICATIONS

Year	Model	Engine Displacement Liters	Engine Oil	Auto. Trans.	Manual Trans.	Rear Diff.	Brake Master Cylinder	Engine Coolant
2009	RX-8	1.3	5W-20	①	75W-90	75W-90	DOT 3	Ethylene-glycol
2010	RX-8	1.3	5W-20	①	75W-90	75W-90	DOT 3	Ethylene-glycol

DOT: Department Of Transportation

① Mazda Genuine JWS3309

37671_MRX8_C0004

VALVE SPECIFICATIONS

Year	Engine Displacement Liters	Engine VIN	Seat Angle (deg.)	Face Angle (deg.)	Spring Test Pressure (lbs. @ in.)	Spring Installed Height (in.)	Stem-to-Guide Clearance (in.)		Stem Diameter (in.)	
							Intake	Exhaust	Intake	Exhaust
2009	1.3	M & 2	①	①	①	①	①	①	①	①
2010	1.3	M & 2	①	①	①	①	①	①	①	①

①: No valves are utilized in this engine configuration

37671_MRX8_C0008

CAMSHAFT AND BEARING SPECIFICATIONS

All measurements are given in inches.

Year	Engine Displacement Liters	Engine VIN	Journal Diameter	Brg. Oil Clearance	Shaft End-play	Runout	Journal Bore	Lobe Lift	
								Intake	Exhaust
2009	1.3	M & 2	①	①	①	①	①	①	①
2010	1.3	M & 2	①	①	①	①	①	①	①

① This engine does not use a camshaft

37671_MRX8_C0009

CRANKSHAFT AND CONNECTING ROD SPECIFICATIONS

All measurements are given in inches.

Year	Engine Displacement Liters	Engine VIN	Crankshaft				Connecting Rod		
			Main Brg. Journal Dia.	Main Brg. Oil Clearance	Shaft End-play	Thrust on No.	Journal Diameter	Oil Clearance	Side Clearance
2009	1.3	M & 2	①	①	①	①	①	①	①
2010	1.3	M & 2	①	①	①	①	①	①	①

①: No crankshaft or conecting rods are used.

37671_MRX8_C0007

PISTON AND RING SPECIFICATIONS

All measurements are given in inches.

Year	Engine Displ. Liters	Engine VIN	Piston Clearance	Ring Gap			Ring Side Clearance		
				Top Comp.	Bottom Comp.	Oil Control	Top Comp.	Bottom Comp.	Oil Control
2009	1.3	①	②	②	②	②	②	②	②
2010	1.3	①	②	②	②	②	②	②	②

① M and 2 Engines

② No pistons are utilized in this engine configuration

37671_MRX8_C0006

TORQUE SPECIFICATIONS
All readings in ft. lbs.

Year	Engine VIN	Engine Displacement Liters	Cylinder Head Bolts	Main Bearing Bolts	Rod Bearing Bolts	Crankshaft Damper Bolts	Flywheel Locknut	Manifold Intake	Manifold Exhaust	Spark Plugs	Oil Pan Drain Plug
2009	①	1.3	②	②	②	②	30-42	14-19	31-44	9-13	22-29
2010	①	1.3	②	②	②	②	30-42	14-19	31-44	9-13	22-29

① M and 2 Engines

② This engine does not use cylinder heads, main bearings, rod bearings or a crankshaft damper.

37671_MRX8_C0010

WHEEL ALIGNMENT

Year	Model		Caster Range (+/-Deg.)	Caster Preferred Setting (Deg.)	Camber Range (+/-Deg.)	Camber Preferred Setting (Deg.)	Toe-in (Deg.)
2009	RX-8	Front	1.0	① ②	1.0	③ ④	0° 11'±21'
		Rear	—	—	1.0	⑤ ⑥	0° 16'±20'
2010	RX-8	Front	1.0	① ②	1.0	③ ④	0° 11'±21'
		Rear	—	—	1.0	⑤ ⑥	0° 16'±20'

Note Unloaded vehicle: Fuel tank is full. Engine coolant and engine oil are at specified level. Jack and tools are in designated position.

The following specifications are for the front Caster settings for Standard Suspension

① 1: 6°31'±1° - Vehicle height* - 14.2—14.5 inches

  2: 6°18'±1 - Vehicle height* - 14.6—14.9 inches

  3: 6°06'±1° - Vehicle height* - 15.0—15.3 inches

  4: 5°53'±1° - Vehicle height* - 15.4—15.7 inches

  5: 5°40'±1° - Vehicle height* - 15.8—16.1 inches

The following specifications are for the front Caster settings for Sport Suspension

② 1: 6°41'±1° - Vehicle height* - 13.9—14.2 inches

  2: 6°28'±1 - Vehicle height* - 14.3—14.6inches

  3: 6°16'±1° - Vehicle height* - 14.7—15.0 inches

  4: 6°03'±1° - Vehicle height* - 15.1—15.4 inches

  5: 5°50'±1° - Vehicle height* - 15.5—15.8 inches

The following specifications are for the front Camber settings on Standard Suspension

③ 1: -0°33'±1° - Vehicle height* - 14.4—14.8 inches

  2: -0°13'±1 - Vehicle height* - 14.9—15.1 inches

  3: 0°04'±1° - Vehicle height* - 15.2—15.5 inches

  4: -0°20'±1° - Vehicle height* - 15.6—15.9 inches

  5: -0°33'±1° - Vehicle height* -16.0—16.3 inches

The following specifications are for the front Camber settings on Sport Suspension

④ 1: 0°45'±1° - Vehicle height* - 1 4.2—14.5 inches

  2: 0°25'±1 - Vehicle height* - 14.6—14.9 inches

  3: 0°06'±1° - Vehicle height* - 15.0—15.3 inches

  4: -0°11'±1° - Vehicle height* - 15.4—15.7 inches

  5: -0°26'±1° - Vehicle height* - 15.8—16.1 inches

The following specifications are for the rear Camber settings for Standard Suspension

⑤ 1: -1°30'±1° - Vehicle height* - 14.2—14.5 inches □

  2: -1°12'±1 - Vehicle height* - 14.6—14.9 inches

  3: 0°56'±1° - Vehicle height* - 15.0—15.3 inches

  4: -0°43'±1° - Vehicle height* - 15.4—15.7 inches

  5: -0°33'±1° - Vehicle height* - 15.8—16.1 inches

The following specifications are for the rear Camber settings on the Sport Suspension

⑥ 1: -1°44'±1° - Vehicle height* - 14.2—14.5 inches

  2: -1°24'±1 - Vehicle height* - 14.6—14.9 inches

  3: 0°07'±1° - Vehicle height* - 15.0—15.3 inches

  4: -0°52'±1° - Vehicle height* - 15.4—15.7 inches

  5: -0°40'±1° - Vehicle height* - 15.8—16.1 inches

* : From the end of the front fender to the center of the wheel inches

TIRE AND WHEEL SPECIFICATIONS

| Year | Model | OEM Tires | | Tire Pressures (psi) | | Wheel | Lug Nut |
		Standard	Optional	Front	Rear	Size	Torque (ft. lbs.)
2009	RX-8	P225/55R16	P225/45R18	32	32	① ②	65-87
2010	RX-8	P225/55R16	P225/45R18	32	32	① ②	65-87

OEM: Original Equipment Manufacturer

PSI: Pounds Per Square Inch

① 16 X 7 inches standard

② 18X8 optional

37671_MRX8_C0013

BRAKE SPECIFICATIONS
All measurements in inches unless noted

| Year | Model | | Brake Disc | | | Minimum Lining Thickness | | Brake Caliper | |
			Original Thickness	Minimum Thickness	Maximum Runout	Front	Rear	Bracket Bolts (ft. lbs.)	Mounting Bolts (ft. lbs.)
2009	RX-8	F	①	0.870	0.002	0.790	—	58-75	16-23
		R	①	0.630	0.002	—	0.079	36-51	16-23
2010	RX-8	F	①	0.870	0.002	0.790	—	58-75	16-23
		R	①	0.630	0.002	—	0.079	36-51	16-23

① Not available

37671_MRX8_C0012

SCHEDULED MAINTENANCE INTERVALS
MAZDA RX-8 2009- 2010

| TO BE SERVICED | TYPE OF SERVICE | VEHICLE MILEAGE INTERVAL (x1000) | | | | | | | | | | | | | | | |
		7.5	15	22.5	30	37.5	45	52.5	60	67.5	75	82.5	90	97.5	105	113	120
Engine oil & filter	R	✓	✓	✓	✓	✓	✓	✓	✓	✓	✓	✓	✓	✓	✓	✓	✓
Cabin air filter	R			✓			✓					✓			✓			✓
Engine coolant strength hoses & clamps	S/I				✓				✓				✓				✓
Air cleaner filter	R					✓					✓				✓		
Brake fluid	R			✓				✓				✓					✓
Engine coolant ① ②	R								✓								✓
Spark plugs	R					✓					✓				✓		
Drive belts	S/I				✓				✓				✓				✓
Exhaust system & heat shields	S/I				✓				✓				✓				✓
Manual transmission oil	R								✓								✓
Rear differential oil	R								✓								✓
Front & rear brakes	S/I		✓				✓			✓			✓				✓
Fuel filter	R													✓			

R: Replace   S/I: Service or Inspect

① Engine coolant: Use FL22 type coolant in vehicles with the inscription "FL22" on the radiator cap itself or the surrounding area. Use FL22 when replacing the coolant. For F22 change initially at 120,000 miles or 10 ten years and every 60,000 miles or years thereafter.

② All other coolants must initally be changes at 60,000 miles or 4 years or 10 ten years and every 2 years years thereafter.

37671_MRX8_C0014

BRAKES — INFORMATION AND PRECAUTIONS

ANTI-LOCK SYSTEMS

- Certain components within the ABS system are not intended to be serviced or repaired individually.
- Do not use rubber hoses or other parts not specifically specified for and ABS system. When using repair kits, replace all parts included in the kit. Partial or incorrect repair may lead to functional problems and require the replacement of components.
- Lubricate rubber parts with clean, fresh brake fluid to ease assembly. Do not use shop air to clean parts; damage to rubber components may result.
- Use only DOT 3 brake fluid from an unopened container.
- If any hydraulic component or line is removed or replaced, it may be necessary to bleed the entire system.
- A clean repair area is essential. Always clean the reservoir and cap thoroughly before removing the cap. The slightest amount of dirt in the fluid may plug an orifice and impair the system function. Perform repairs after components have been thoroughly cleaned; use only denatured alcohol to clean components. Do not allow ABS components to come into contact with any substance containing mineral oil; this includes used shop rags.
- The Anti-Lock control unit is a microprocessor similar to other computer units in the vehicle. Ensure that the ignition switch is **OFF** before removing or installing controller harnesses. Avoid static electricity discharge at or near the controller.
- If any arc welding is to be done on the vehicle, the control unit should be unplugged before welding operations begin.

DISC AND DRUM SYSTEMS

> ※ **CAUTION**
>
> Dust and dirt accumulating on brake parts during normal use may contain asbestos fibers from production or aftermarket brake linings. Breathing excessive concentrations of asbestos fibers can cause serious bodily harm. Exercise care when servicing brake parts. Do not sand or grind brake lining unless equipment used is designed to contain the dust residue. Do not clean brake parts with compressed air or by dry brushing. Cleaning should be done by dampening the brake components with a fine mist of water, then wiping the brake components clean with a dampened cloth. Dispose of cloth and all residue containing asbestos fibers in an impermeable container with the appropriate label. Follow practices prescribed by the Occupational Safety and Health Administration (OSHA) and the Environmental Protection Agency (EPA) for the handling, processing, and disposing of dust or debris that may contain asbestos fibers.

BRAKES — BLEEDING THE BRAKE SYSTEM

BLEEDING PROCEDURE

BLEEDING PROCEDURE

See Figure 1.

> ※ **WARNING**
>
> Brake fluid will damage painted surfaces. Be careful not to spill any on painted surfaces. If it is spilled, wipe it off immediately.

> ※ **WARNING**
>
> Keep the fluid level in the reserve tank at ¾ full or more during the air bleeding.

1. Begin air bleeding with the master cylinder and then continue with the brake caliper that is furthest away from the master cylinder. Finish by bleeding air from the master cylinder again.

➡ Only use SAE J1703, FMVSS 116 DOT-3 Brake Fluid.

2. Remove the bleeder cap from the brake caliper, and connect a vinyl tube to the bleeder screw.

3. Place the other end of the vinyl tube in a clear container, and fill the container with fluid during air bleeding.

4. Working with two people, one should depress the brake pedal a few times and then depress and hold the pedal down.

5. While the brake pedal is being held down, the other person should loosen the bleeder screw using the SST, and bleed any fluid containing air bubbles. Once completed, tighten the bleeder screw to 62–86 inch lbs. (6.9–9.8 Nm).

6. Repeat Steps 3 and 4 until no air bubbles are seen.

7. Perform the bleeding procedure outlined above for each of the brake calipers.

8. After air bleeding, inspect the following:
   a. Brake operation
   b. Fluid leakage
   c. Fluid level

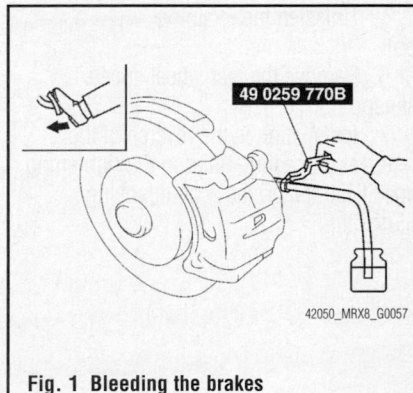

49 0259 770B

42050_MRX8_G0057

Fig. 1 Bleeding the brakes

SPEED SENSORS

REMOVAL & INSTALLATION

Front Wheel

See Figure 2.

➡**If there is any malfunction in the front ABS wheel-speed sensor unit, replace the wheel hub, as outlined in the Suspension Section.**

1. Raise and safely support the vehicle.
2. Remove the wheel and tire assembly.
3. Remove the mudguard.
4. Detach the ABS speed sensor connector.
5. Remove the bolt and the wire harness.
6. Installation is the reverse of the removal procedure. Refer to the tightening specifications on the accompanying illustration.

Rear Wheel

See Figure 3.

1. Raise and safely support the vehicle.
2. Remove the wheel and tire assembly.
3. Remove the trunk side trim.
4. Detach the sensor electrical connector.
5. Unfasten the retaining bolt.
6. Remove the rear wheel speed sensor.
7. Installation is the reverse of the removal procedure. Refer to the tightening specifications on the accompanying illustration.

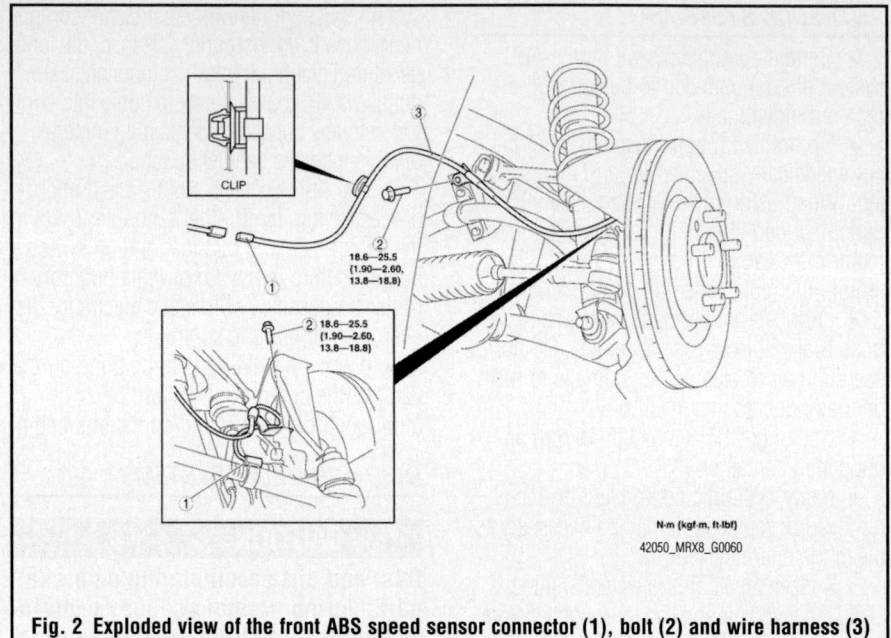

Fig. 2 Exploded view of the front ABS speed sensor connector (1), bolt (2) and wire harness (3)

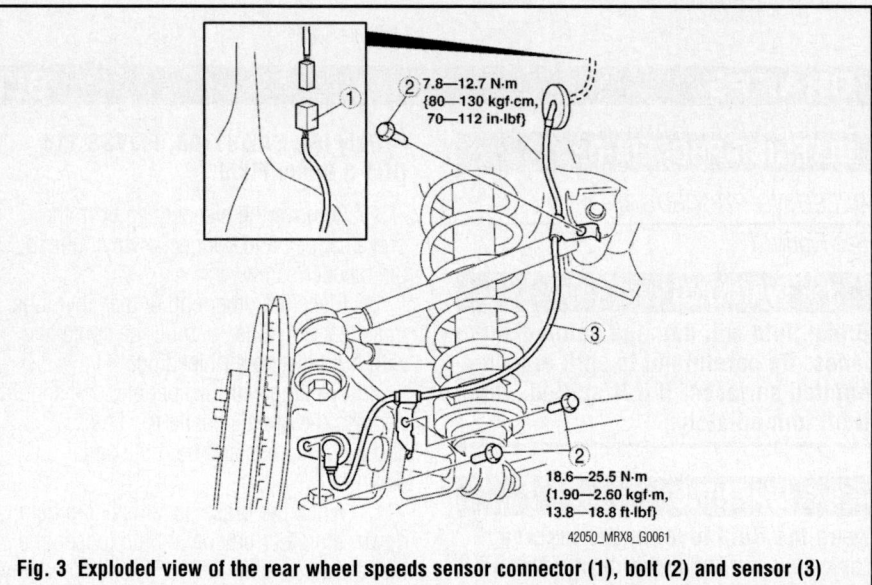

Fig. 3 Exploded view of the rear wheel speeds sensor connector (1), bolt (2) and sensor (3)

BRAKE CALIPER

REMOVAL & INSTALLATION
See Figure 4.

✳✳ CAUTION

Dust and dirt accumulating on brake parts during normal use may contain asbestos fibers from production or aftermarket brake linings. Breathing excessive concentrations of asbestos fibers can cause serious bodily harm. Exercise care when servicing brake parts. Do not sand or grind brake lining unless equipment used is designed to contain the dust residue. Do not clean brake parts with compressed air or by dry brushing. Cleaning should be done by dampening the brake components with a fine mist of water, then wiping the brake components clean with a dampened cloth. Dispose of cloth and all residue containing asbestos fibers in an impermeable container with the appropriate label. Follow practices prescribed by the Occupational Safety and Health Administration (OSHA) and the Environmental Protection Agency (EPA) for the handling, processing, and disposing of dust or debris that may contain asbestos fibers.

1. Before servicing the vehicle, refer to the Precautions Section.
2. Remove or disconnect the following:
   - Wheels
   - Flexible brake hose from the caliper
   - Caliper bolt
   - Caliper

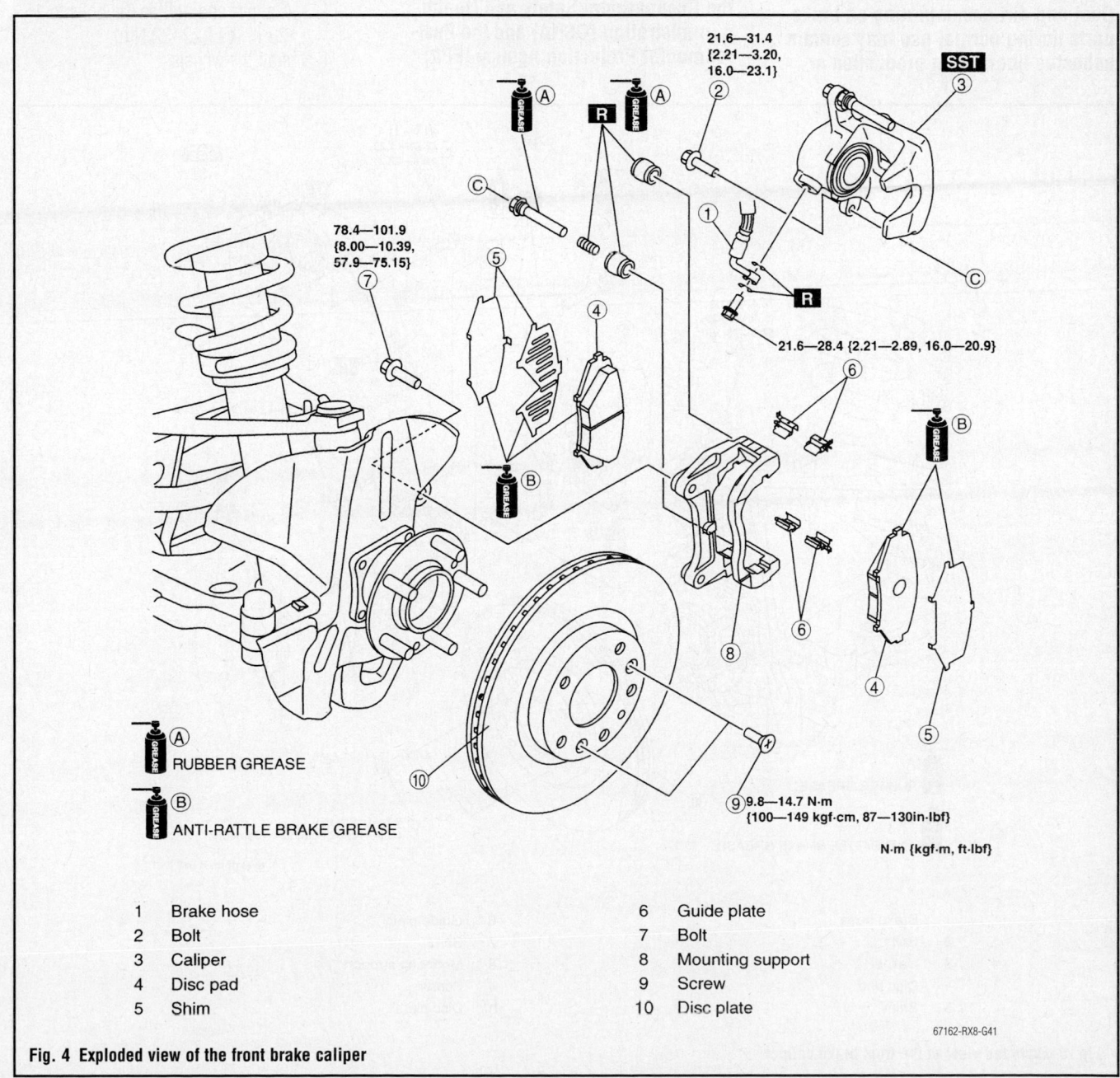

1	Brake hose
2	Bolt
3	Caliper
4	Disc pad
5	Shim
6	Guide plate
7	Bolt
8	Mounting support
9	Screw
10	Disc plate

Ⓐ RUBBER GREASE

Ⓑ ANTI-RATTLE BRAKE GREASE

21.6—31.4 {2.21—3.20, 16.0—23.1}

78.4—101.9 {8.00—10.39, 57.9—75.15}

21.6—28.4 {2.21—2.89, 16.0—20.9}

9.8—14.7 N·m {100—149 kgf·cm, 87—130in·lbf}

N·m {kgf·m, ft·lbf}

67162-RX8-G41

Fig. 4 Exploded view of the front brake caliper

To install:
3. Install or connect the following:
   - Caliper on the brake disc
   - Caliper mounting bolts and tighten the bolts 16–23 ft. lbs. (22–31 Nm)
   - Brake hose to the caliper and tighten the hose nut to 16–21 ft. lbs. (22–29 Nm)
4. Bleed the brake system.
5. Install the wheels.

DISC BRAKE PADS

REMOVAL & INSTALLATION

See Figure 5.

✳✳ CAUTION

Dust and dirt accumulating on brake parts during normal use may contain asbestos fibers from production or aftermarket brake linings. Breathing excessive concentrations of asbestos fibers can cause serious bodily harm. Exercise care when servicing brake parts. Do not sand or grind brake lining unless equipment used is designed to contain the dust residue. Do not clean brake parts with compressed air or by dry brushing. Cleaning should be done by dampening the brake components with a fine mist of water, then wiping the brake components clean with a dampened cloth. Dispose of cloth and all residue containing asbestos fibers in an impermeable container with the appropriate label. Follow practices prescribed by the Occupational Safety and Health Administration (OSHA) and the Environmental Protection Agency (EPA) for the handling, processing, and disposing of dust or debris that may contain asbestos fibers.

1. Before servicing the vehicle, refer to the Precautions Section.
2. Remove or disconnect the following:
   - Wheels
   - Caliper
   - Brake pads
   - Shim
   - Guide plate

To install:
3. Install or connect the following:
   - Guide plate
   - Shim
   - Brake pads
   - Caliper and tighten the bolts to 16–23 ft. lbs. (22–32 Nm).
4. Install the wheels.

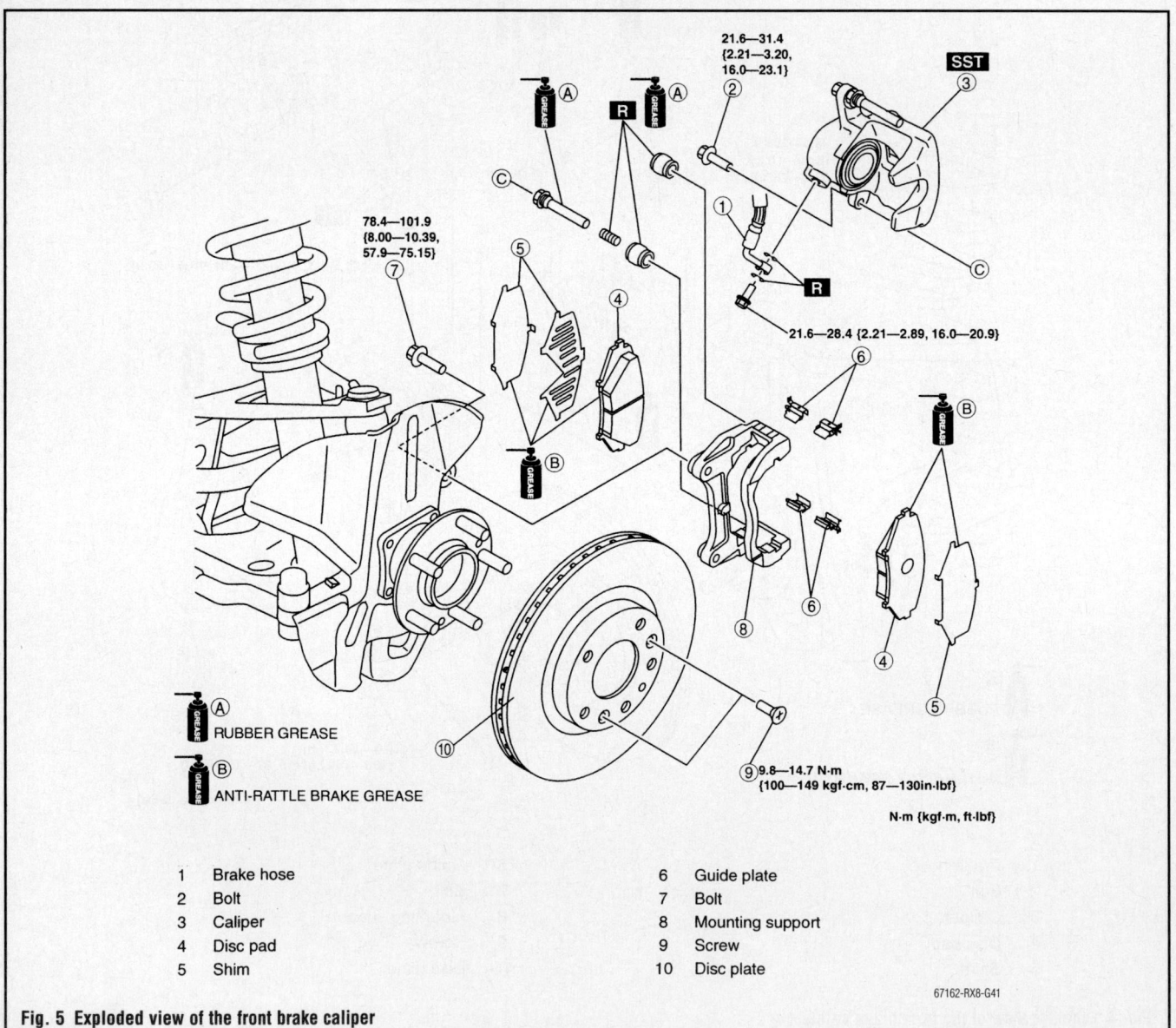

21.6—31.4
{2.21—3.20, 16.0—23.1}

78.4—101.9
{8.00—10.39, 57.9—75.15}

21.6—28.4 {2.21—2.89, 16.0—20.9}

Ⓐ RUBBER GREASE

Ⓑ ANTI-RATTLE BRAKE GREASE

⑨ 9.8—14.7 N·m
{100—149 kgf·cm, 87—130in·lbf}

N·m {kgf·m, ft·lbf}

1	Brake hose	6	Guide plate
2	Bolt	7	Bolt
3	Caliper	8	Mounting support
4	Disc pad	9	Screw
5	Shim	10	Disc plate

67162-RX8-G41

Fig. 5 Exploded view of the front brake caliper

BRAKE CALIPER

REMOVAL & INSTALLATION

See Figure 6.

✳✳ CAUTION

Dust and dirt accumulating on brake parts during normal use may contain asbestos fibers from production or aftermarket brake linings. Breathing excessive concentrations of asbestos fibers can cause serious bodily harm. Exercise care when servicing brake parts. Do not sand or grind brake lining unless equipment used is designed to contain the dust residue. Do not clean brake parts with compressed air or by dry brushing. Cleaning should be done by dampening the brake components with a fine mist of water, then wiping the brake components clean with a dampened cloth. Dispose of cloth and all residue containing asbestos fibers in an impermeable container with the appropriate label. Follow practices prescribed by the Occupational Safety and Health Administration (OSHA) and the Environmental Protection Agency (EPA) for the handling, processing, and disposing of dust or debris that may contain asbestos fibers.

1. Before servicing the vehicle, refer to the Precautions Section.
2. Remove or disconnect the following:
   - Wheels
   - Flexible brake line from the caliper assembly
   - Caliper mounting bolts
   - Caliper

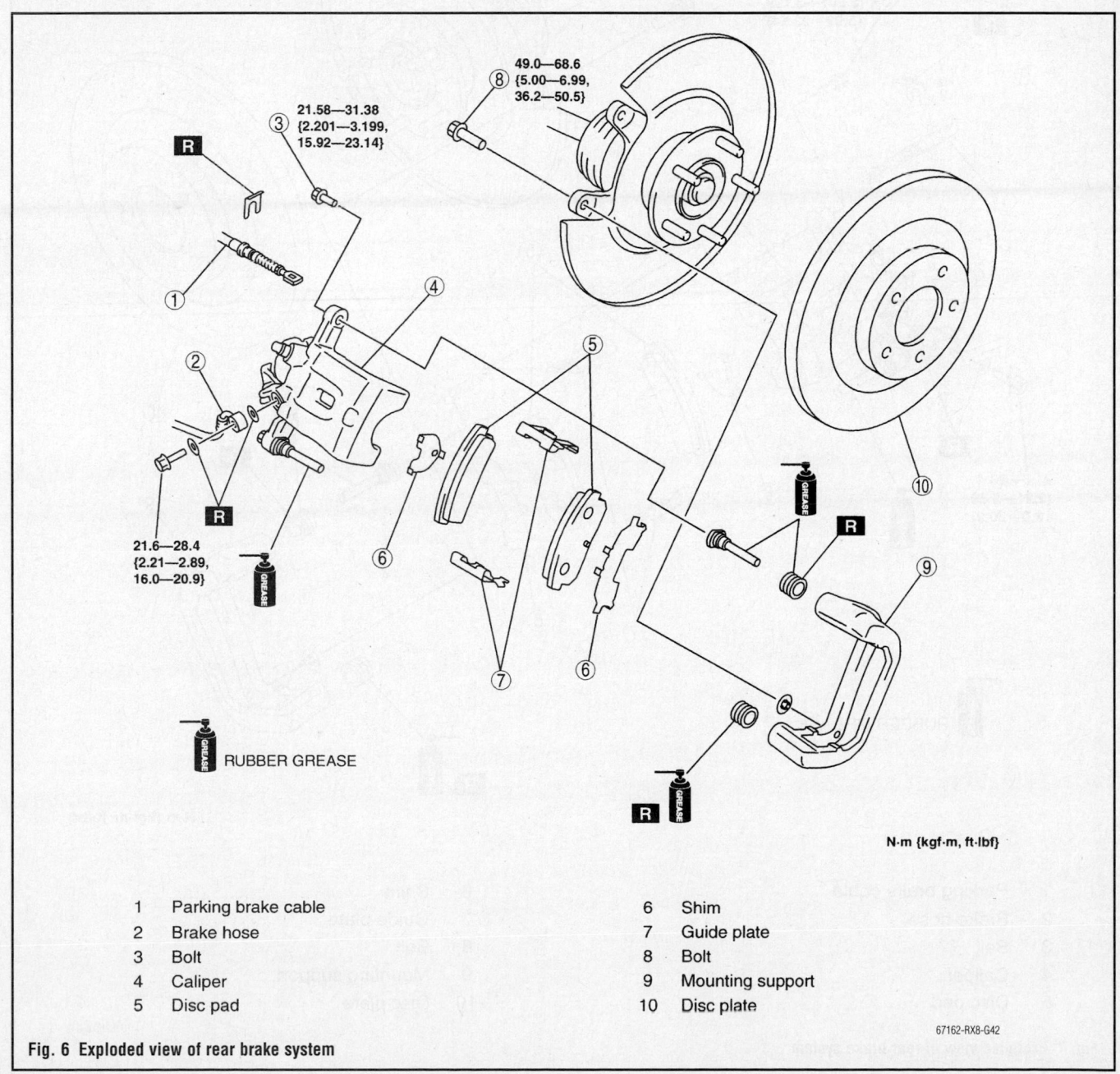

N·m {kgf·m, ft·lbf}

1	Parking brake cable	6	Shim
2	Brake hose	7	Guide plate
3	Bolt	8	Bolt
4	Caliper	9	Mounting support
5	Disc pad	10	Disc plate

RUBBER GREASE

Fig. 6 Exploded view of rear brake system

67162-RX8-G42

To install:

3. Install or connect the following:

- Caliper. Torque the caliper mount bolts to 16–23 ft. lbs. (22–32 Nm).
- Brake hose. Torque the line bolt to 16–22 ft. lbs. (22–30 Nm).
- Parking brake cable

4. Install the wheels and bleed the brake system.

DISC BRAKE PADS

REMOVAL & INSTALLATION

See Figure 7.

1. Before servicing the vehicle, refer to the Precautions Section.

2. Remove or disconnect the following:

- Wheels
- Caliper
- Brake pads

- Shim
- Guide plate

To install:

3. Install or connect the following:

- Guide plate
- Shim
- Brake pads
- Caliper and tighten the bolt to 16–23 ft. lbs. (22–32 Nm).

4. Install the wheels.

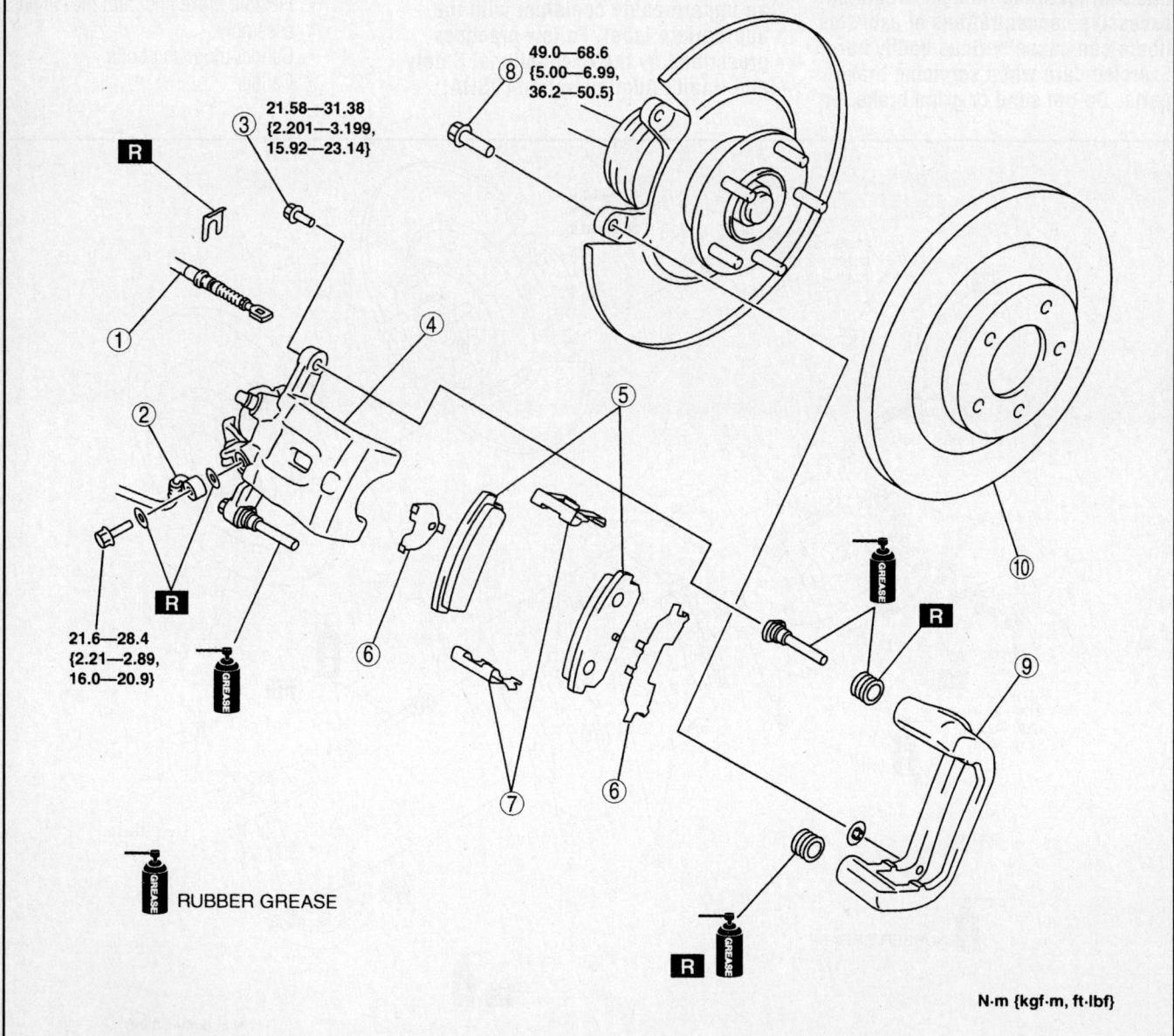

N·m {kgf·m, ft·lbf}

RUBBER GREASE

1	Parking brake cable	6	Shim
2	Brake hose	7	Guide plate
3	Bolt	8	Bolt
4	Caliper	9	Mounting support
5	Disc pad	10	Disc plate

Fig. 7 Exploded view of rear brake system

67162-RX8-G42

BRAKES

PARKING BRAKE

ADJUSTMENTS

PARKING BRAKE LEVER

See Figure 8.

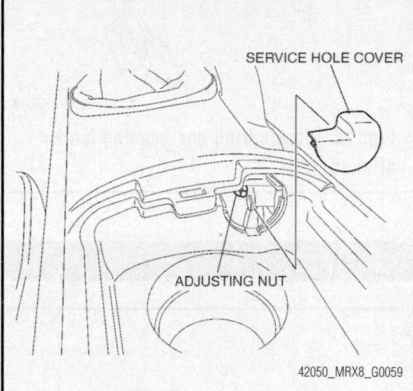

Fig. 8 Remove the service hole cover from the rear console to adjust the parking brake lever

1. Depress the brake pedal several times.
2. Remove the service hole cover from the rear console.
3. Turn the adjusting nut and adjust the parking brake lever.
4. After adjustment, pull the parking brake lever one notch and check that the parking brake warning light illuminates.
5. Make sure that the rear brakes do not drag.

STROKE INSPECTION

See Figure 9.

1. Depress the brake pedal several times.
2. Pull the parking brake lever up 2–3 times.
3. Check the parking brake stroke by slowly pulling at point A (50 mm {1.97 in}) from the end of the parking brake lever with about 22 lbs. (98 N) of force, and counting the number of notches (clicking sound). The proper specification is between 1–3 notches. If not within specifications, adjust the parking brake lever, as outlined in this section.

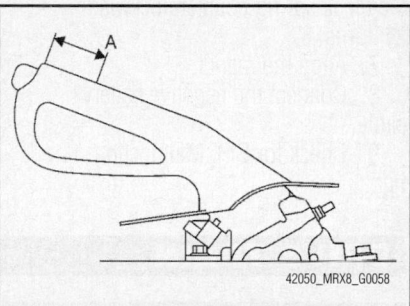

Fig. 9 Check the parking brake stroke by slowly pulling at point A (50 mm {1.97 in}) from the end of the parking brake lever with about 22 lbs. (98 N) of force, and counting the number of notches (clicking sound)

CHASSIS ELECTRICAL

AIR BAG (SUPPLEMENTAL RESTRAINT SYSTEM)

GENERAL INFORMATION

✳✳ CAUTION

These vehicles are equipped with an air bag system. The system must be disarmed before performing service on, or around, system components, the steering column, instrument panel components, wiring and sensors. Failure to follow the safety precautions and the disarming procedure could result in accidental air bag deployment, possible injury and unnecessary system repairs.

SERVICE PRECAUTIONS

✳✳ CAUTION

Disconnect and isolate the battery negative cable before beginning any airbag system component diagnosis, testing, removal, or installation procedures. Wait at least 90 seconds after the ignition switch is turned off and the negative (-) terminal cable is disconnected from the battery before starting the operation. The SRS is equipped with a backup power source, so if work is started within 90 seconds after disconnecting the neg-

ative (-) terminal cable from the battery, the SRS may be deployed. Failure to disable the airbag system may result in accidental airbag deployment, personal injury, or death.

DISARMING THE SYSTEM

1. Before servicing the vehicle, refer to the Precautions Section.
2. If equipped, deactivate the audio anti-theft system.
3. Turn the ignition switch to **LOCK**.
4. Disconnect and isolate the negative battery cable and wait for more than 1 minute to allow the backup power supply to deplete its stored power.

ARMING THE SYSTEM

1. Before servicing the vehicle, refer to the Precautions Section.
2. Connect the negative battery cable, turn the ignition switch **ON** and verify the air bag warning light cones on for 6 seconds. If the light does not illuminate there are problems with the system.
3. If equipped, activate the audio anti-theft system.

CLOCKSPRING CENTERING

See Figures 10 through 12.

1. Before servicing the vehicle, refer to the Precautions Section.
2. Disconnect the negative battery.

➡The adjustment procedure is also specified on the caution label of the clock spring.

3. Set the front tires straight-ahead.

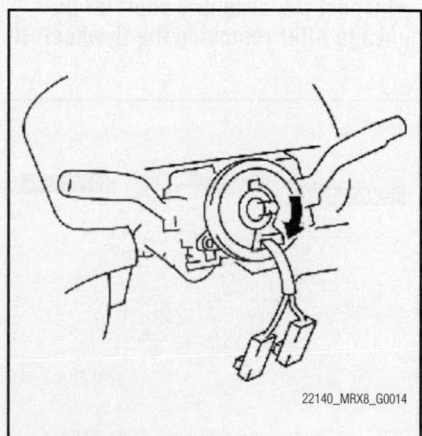

Fig. 10 Clock spring clockwise until it stops— not forcibly

4. The clock spring will break if over-wound. Do not forcibly turn the clock spring.

5. Turn the clock spring clockwise until it stops— not forcibly.

6. From the stopped position, turn the clock spring counterclockwise 2¾turns.

7. Align the marks.

8. Connect the negative battery cable.

9. Check for SIR. Malfunction light.

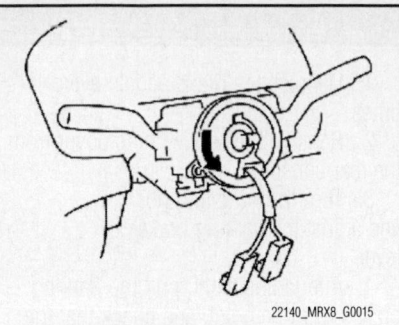

Fig. 11 Clock spring in the stopped position, turn the clock spring counterclockwise 2¾turns.

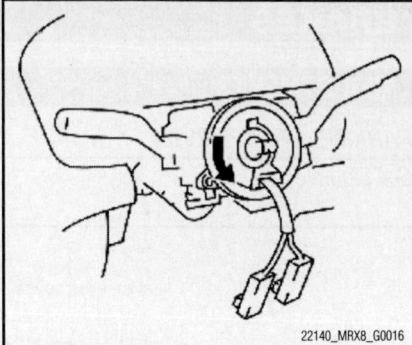

Fig. 12 Clock spring and housing marks aligned

DRIVE TRAIN

CLUTCH

REMOVAL & INSTALLATION

See Figures 13 through 20.

1. Clutch Cover, Clutch Disc
   a. Install the SSTs.
   b. In criss-cross pattern, loosen the bolts one rotation at a time until there is no remaining spring pressure.
   c. Remove the clutch cover and the clutch disc.
2. Oil Seal, Pilot Bearing

➡**Remove the pilot bearing only if there is a malfunction.**

   a. Remove the pilot bearing and the oil seal together using the SST.
3. Flywheel
   a. Install the SST to the flywheel.
   b. Remove the locknut using the SST.

➡**Inspect the eccentric shaft for oil leakage after removing the flywheel. If**

necessary, replace the oil seal. **(See REAR OIL SEAL REPLACEMENT [13B-MSP].)**

   c. Remove the flywheel using the SST.
   d. Remove the key from the eccentric shaft.

To install:
4. Flywheel
   a. Install the key to the eccentric shaft.

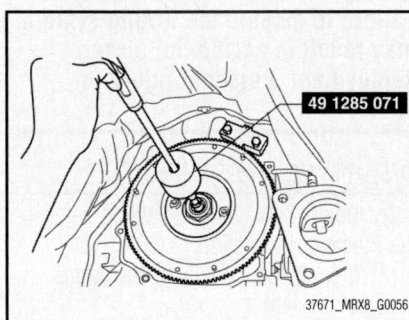

Fig. 14 Removal of the pilot bearing and the oil seal together using the SST 49 1285 071

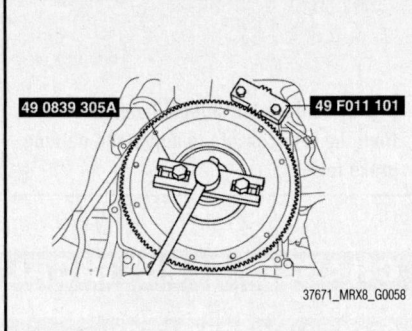

Fig. 16 Remove the flywheel using the SST's 49 0839 305A & 49 F011 101

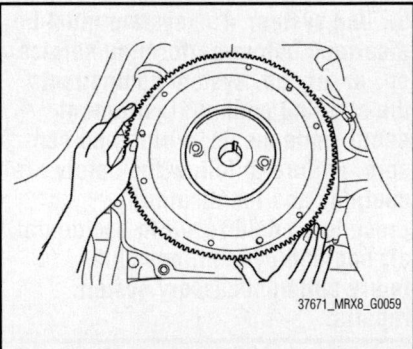

Fig. 17 Align the flywheel key groove with the eccentric shaft key and install

   b. Align the flywheel key groove with the eccentric shaft key and install.
   c. Install the SST to the flywheel.
   d. Tighten the locknut using the SST. Tighten locknut to 289–361 ft. lb. (392–490 Nm).

❋❋ CAUTION

Remove the seal protruding from the threads without it becoming caught in the pilot bearing.

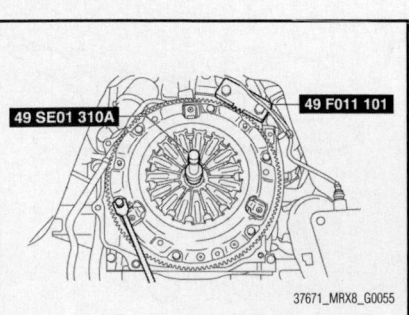

Fig. 13 Install the SST's 49 SE01 310A & 49 F011 101 for to the clutch cover and clutch disc removal

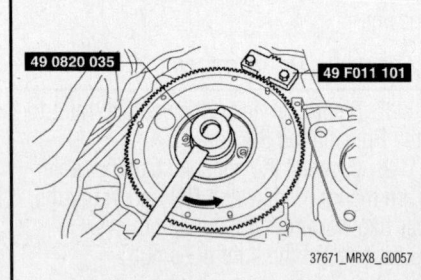

Fig. 15 Removal of the locknut using the SST's 49 0820 035 & 49 F011 101. Tighten locknut using the same SST.

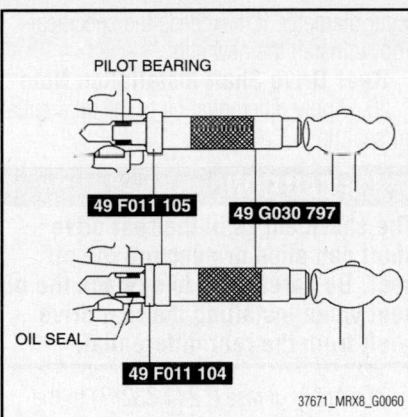

Fig. 18 Install the pilot bearing using the SSTs 49F011 105, 49 G030 797, & 49 F011 F011 104

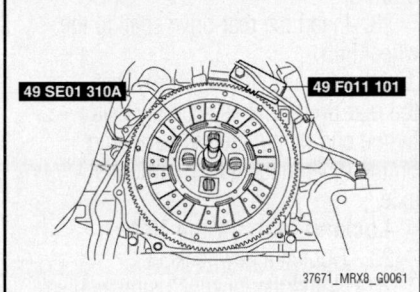

Fig. 19 Secure the clutch disc to the flywheel using the SSTs 49 SE01 310A & 49 F011 101

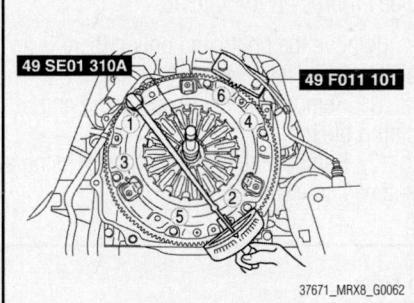

Fig. 20 Tighten the bolts evenly and gradually in the order shown

5. Pilot Bearing, Oil Seal
   a. Install the pilot bearing using the SSTs.

➡**Bearing outer diameter: 0.787 in. (20 mm). Press-in depth: 0.453–0.482 in. (11.5–12.25 mm).**

   b. Install a new oil seal using the SSTs.
6. Clutch Disc
   a. Clean the splines of the clutch disc and the main drive gear with a brush.

   b. Spread a thin layer of clutch grease on the splines.
   c. Secure the clutch disc to the flywheel using the SSTs.
7. Clutch Cover
   a. Align the clutch cover with the flywheel knock pin and install.
   b. Tighten the bolts evenly and gradually in the order shown. Tighten bolts to 16.3–24.4 ft. lb. (22.1–33.1 Nm).

DRIVESHAFT

REMOVAL & INSTALLATION

See Figures 21 through 27.

✳✳ CAUTION

Performing the following procedures without first removing the ABS wheel-speed sensor may possibly cause an open circuit in the wiring harness if it is pulled by mistake. Before performing the following procedures, remove the ABS wheel-speed sensor (axle side) and fix it to an appropriate place where the sensor will not be pulled by mistake while servicing the vehicle.

1. Drain the rear differential oil.
2. Remove the rear auto leveling sensor.
3. Remove in the order indicated in the table.
4. Install in the reverse order of removal.
5. Add rear differential oil.
6. After installation, inspect the rear wheel alignment.
Locknut Removal Note
7. Lock the disc plate by applying the brakes.
8. Knock the crimped portion of the locknut outward using a chisel and a hammer.
9. Remove the locknut.
Brake Caliper Component Removal Note
10. Suspend the brake caliper component using a cable or equivalent.
11. Temporarily tighten the wheel nut to prevent the disc plate from falling off.
Rear Drive Shaft Removal Note
12. Temporarily install a spare nut to the end of the rear drive shaft.
13. Knock the nut with copper hammer lightly and remove the rear drive shaft from the wheel hub.
14. Separate the rear drive shaft from the wheel hub.
15. Insert a tire lever or equivalent between the rear differential and differential side outer ring, and then remove the rear drive shaft.

✳✳ CAUTION

The sharp edges of the drive shaft can slice or puncture the oil seal. Be careful not to damage the oil seal when removing the drive shaft from the differential.

16. Pull the rear drive shaft to the outer side of the vehicle and disconnect it from the rear differential.

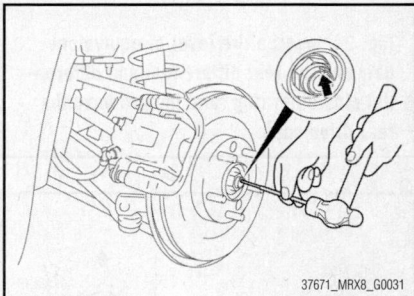

Fig. 21 Knock the crimped portion of the locknut outward using a chisel and a hammer

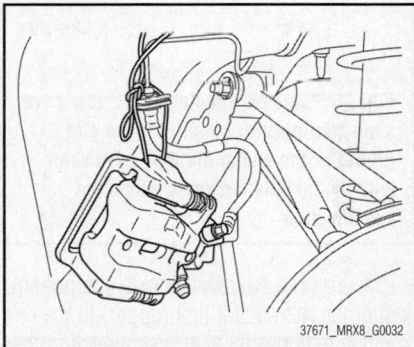

Fig. 22 Suspend the brake caliper component using a cable or equivalent

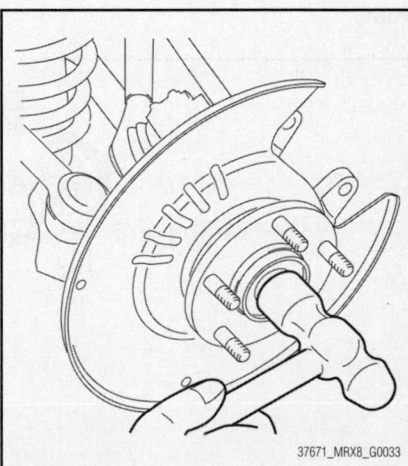

Fig. 23 Knock the nut with copper hammer lightly and remove the rear drive shaft from the wheel hub

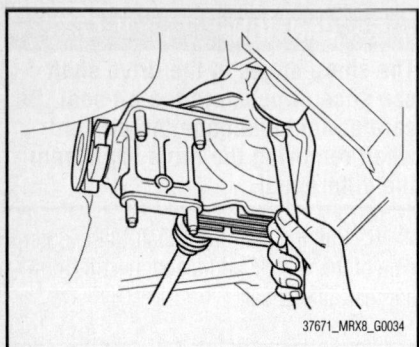

Fig. 24 Insert a tire lever or equivalent between the rear differential and differential side outer ring, and then remove the rear drive shaft

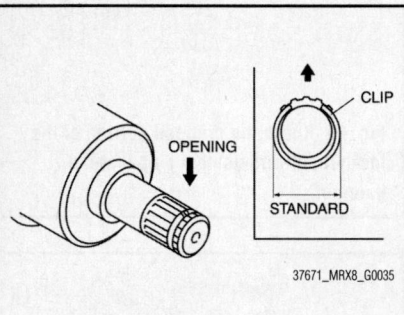

Fig. 25 Point the opening of the new drive shaft clip upward, install it to the clip groove at the end of the rear drive shaft with the installation width within the specification

17. To hold the rear knuckle component, install the rear lateral link (upper) to the rear knuckle temporarily after disconnecting the rear drive shaft.

To install:
Rear Drive Shaft Clip Installation Note

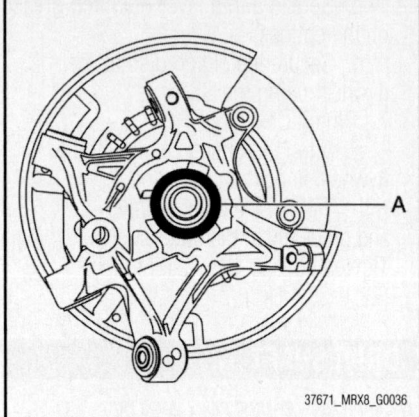

Fig. 26 Apply grease (L2Y1 33247) to the wheel bearing inner race and rear drive shaft contact surface (Area A in figure)

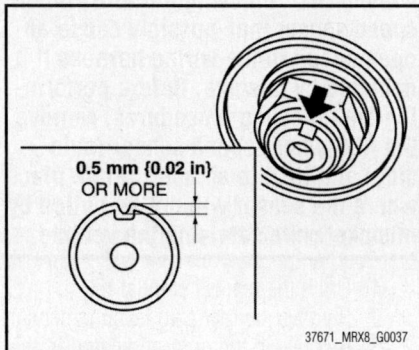

Fig. 27 Tighten a new locknut. Crimp the locknut, using a chisel and hammer.

18. Point the opening of the new drive shaft clip upward, install it to the clip groove at the end of the rear drive shaft with the installation width within the specification.
• Standard 1.3 in. (32 mm).
19. After installing the clip, measure the

outer diameter. If it exceeds the specification, reinstall the new clip.
Rear Drive Shaft Installation Note
20. Apply differential oil to the differential oil seal lip.

✳✳ CAUTION

The sharp edges of the rear drive shaft can slice or puncture the oil seal. Be careful not to damage the oil seal when installing the rear drive shaft from the rear differential.

21. Apply grease (L2Y1 33247) to the wheel bearing inner race and rear drive shaft contact surface (Area A in figure).
22. Insert the rear drive shaft into the rear differential with the clip opening facing upward.
23. Insert the rear drive shaft to the wheel hub.
24. After installation, verify that the rear drive shaft is securely held by the clip by pulling the outer ring on the differential side towards the axle.
Locknut Installation Note
25. Tighten a new locknut.
26. Crimp the locknut, using a chisel and hammer.

FRONT HALFSHAFTS

CV-JOINT OVERHAUL

See Figures 28 through 38.

Remove the boot band only if there is an abnormality.
1. Remove the boot band using end clamp pliers.
2. Remove the crimp of the clip using a flathead screwdriver.

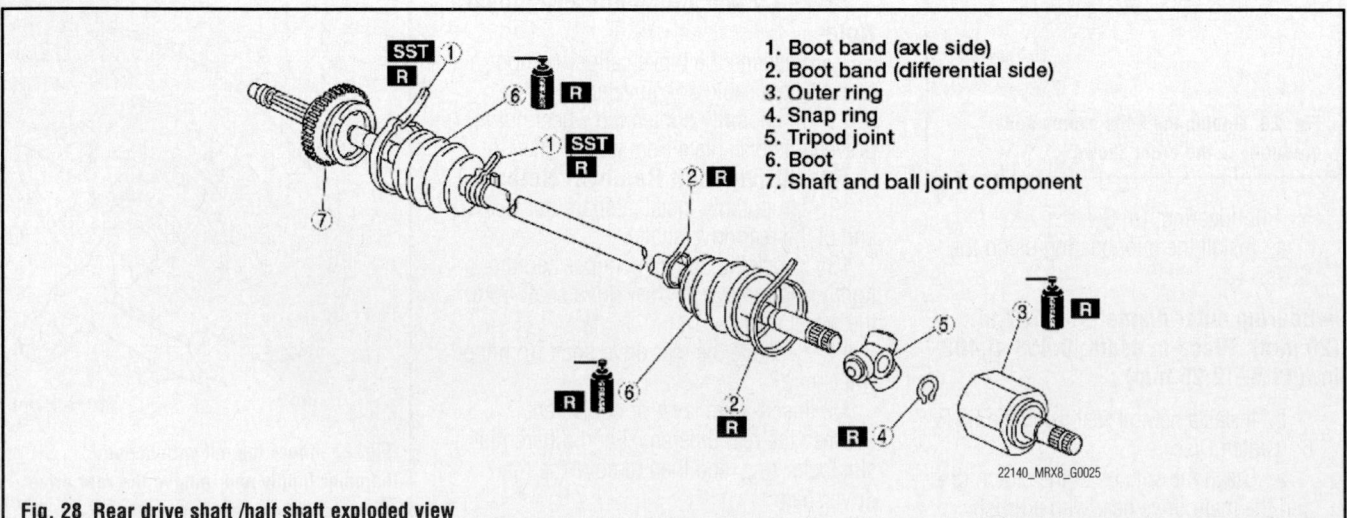

1. Boot band (axle side)
2. Boot band (differential side)
3. Outer ring
4. Snap ring
5. Tripod joint
6. Boot
7. Shaft and ball joint component

Fig. 28 Rear drive shaft /half shaft exploded view

3. Place an alignment mark on the drive shaft and the outer ring.

4. Remove the outer ring.

5. Place an alignment mark on the shaft and tripod joint.

6. Remove the snap ring using a snap ring pliers.

7. Remove the tripod joint from the shaft.

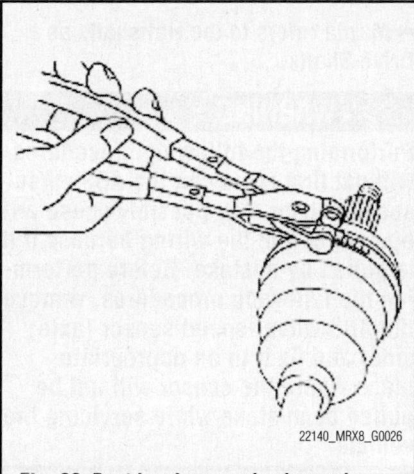

Fig. 29 Boot band removal using end clamp pliers.

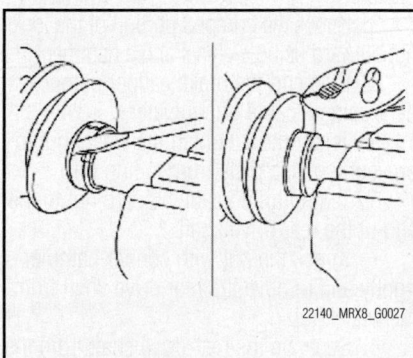

Fig. 30 Band crimp removal using a flathead screwdriver

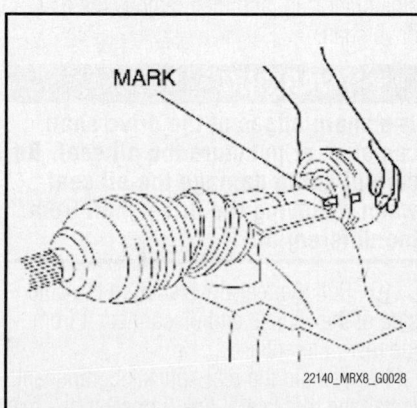

Fig. 31 Alignment mark on the drive shaft and the outer ring

✳ WARNING

To prevent damage to the component, do not use a hammer when removing it.

➡**Remove the axle side boot only if there is an abnormality.**

8. Wrap the shaft spline with vinyl tape.

➡**The boot shapes on the axle side and the differential side are different so do not mis-install them.**

9. Fill the inside of the new dust boot (wheel side) with grease.

➡**Do not touch the grease with your hand. Apply it from the tube to prevent foreign matter from entering the boot.**

10. Grease amount
  • Manual Transmission (Y16M-D): 4.06–4.76 oz (115–135 g)
  • Automatic Transmission (SJ6A-EL): 4.06–4.76 oz (115–135 g)
  • Automatic Transmission (RC4-EL): 3.18–3.88 oz (90–110 g)

11. Install the boot with the drive shaft spline still wrapped with vinyl tape.

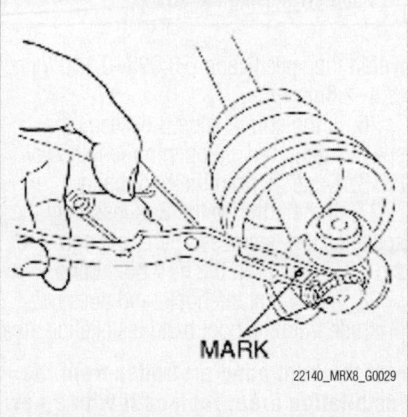

Fig. 32 Remove of the snap ring using a snap ring pliers

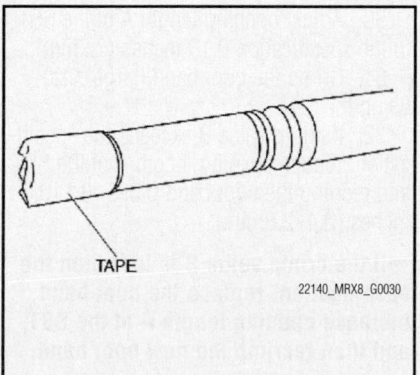

Fig. 33 The shaft spline with vinyl tape

---

12. Remove the vinyl tape.

13. Align the tripod joint with the shaft mark and insert it using a brass bar.

➡**To prevent damage to the component, do not tap the roller part when installing.**

14. Install the new snap ring to the shaft installation slot securely using a snap ring pliers.

15. Fill the outer ring and boot (differential side) with the repair kit grease.

➡**Do not touch the grease with your hand. Apply it from the tube to prevent foreign matter from entering the boot.**

16. Grease amount .
  • Manual Transmission (Y16M-D): 6.18–6.87oz (175–195 g)
  • Automatic Transmission (SJ6A-EL): 6.18–6.87oz (175–195 g)
  • Automatic Transmission (RC4A-EL): 4.77–5.46 oz (135–155 g)

17. Assemble the outer ring.

18. Release any trapped air from the boots by carefully lifting up the small end of each boot with a cloth wrapped screwdriver.

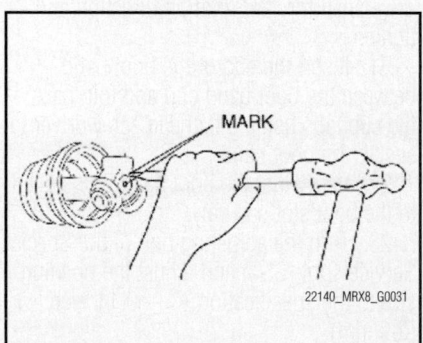

Fig. 34 Alignment of the tripod joint with the shaft mark

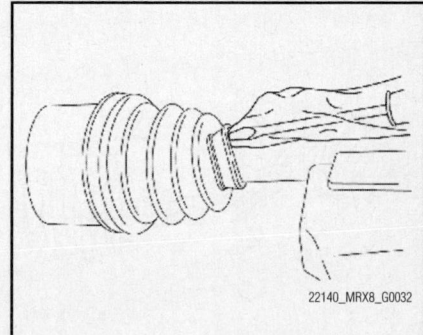

Fig. 35 Releasing trapped air from the boots

➡ **Do not let the grease leak.**

➡ **Do not damage the boot.**

19. Set the drive shaft length to the specification when the inside of the boots is at ambient pressure. Rear drive shaft standard length.

- Manual Transmission (Y16M-D): Left side 31.21–31.59 inches (792.6–802.6mm)
- Manual Transmission (Y16M-D): Right side 32.78–33.17 inches (832.6–842.6mm)
- Automatic Transmission (SJ6A-EL): Left side 31.21–31.59 inches (792.6–802.6mm)
- Automatic Transmission (SJ6A-EL): Right side 32.78–33.17 inches (832.6–842.6mm)
- Automatic Transmission (RC4A-EL): Left side 31.15–31.53 inches (791.1–801.1mm)
- Automatic Transmission (RC4A-EL): Right side 32.71–33.11 inches (831.1–841.1mm)

➡ **After installation, verify that there is no boot damage or grease leakage.**

20. Using pliers, pull the boot band around the boot slot in opposite direction of drive shaft forward rotation direction and tighten.

21. Insert the end of the boot band between the boot band clip and fold back the clip tabs using a flathead screwdriver to secure the boot band.

22. Verify that the boot band is installed to the boot slot securely.

23. Turn the adjusting bolt of the Special Service Tool (SST) and adjust the opening size to the specification A.— 0.11 inch (2.9 mm).

24. Crimp the boot band (small-size) using the SST.

25. Verify that the crimp value B is

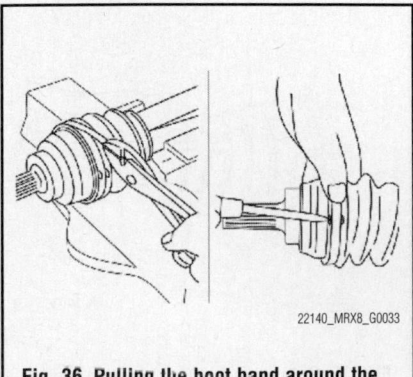

Fig. 36 Pulling the boot band around the CV boot slot

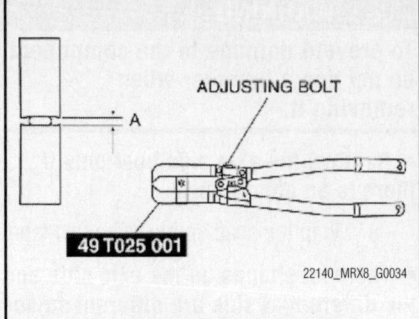

Fig. 37 Adjusting bolt of the Special Service Tool (SST) and adjust the opening size to the specification A

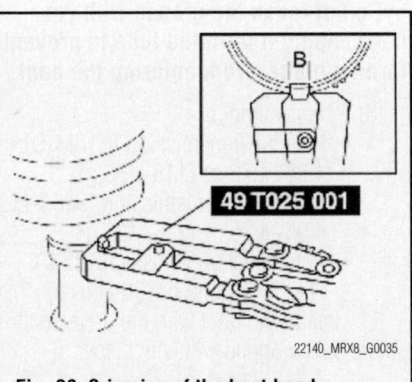

Fig. 38 Crimping of the boot band (small-size) using the SST

within the specification. 0.095–0.110 inches (2.4–2.8 mm)/

26. If the crimp value B exceeds the specification, reduce opening length A of the SST and recrimp the boot band.

27. If the crimp value B is less than the specification, increase opening length A of the SST and crimp the new boot band.

28. Verify that the boot band does not protrude from the boot band installation area.

➡ **If the boot band protrudes from the installation area, replace it with a new band and repeat Step 2–4.**

29. Fill the boot with the repair kit grease.

30. Adjust opening length A of the SST to the specification 0.13 inches (3.2mm)

31. Crimp the boot band (large-size) using the SST.

32. If crimp value B exceeds the specification, reduce opening length A of the SST and recrimp the boot band 0.095–0.110 inches (2.4–2.8 mm)

➡ **If the crimp value B is less than the specification, replace the boot band, increase opening length A of the SST, and then recrimp the new boot band.**

33. Verify that the boot band does not

protrude from the boot band installation area.

34. If the boot band protrudes from the installation area, replace it with a new band and repeat installation.

REAR HALFSHAFTS

REMOVAL & INSTALLATION

See Figures 39 through 44.

➡ **Mazda refers to the Halfshafts as Drive Shafts.**

☀ WARNING

Performing the following procedures without first removing the ABS wheel-speed sensor may possibly cause an open circuit in the wiring harness if it is pulled by mistake. Before performing the following procedures, remove the ABS wheel-speed sensor (axle side) and fix it to an appropriate place where the sensor will not be pulled by mistake while servicing the vehicle.

1. Drain the rear differential oil.
2. Lock the disc plate by applying the brakes.
3. Knock the crimped portion of the lock-nut outward using a chisel and a hammer.
4. Suspend the brake caliper component using a cable or equivalent.
5. Temporarily tighten the wheel nut to prevent the disc plate from falling off.
6. Temporarily install a spare nut to the end of the rear drive shaft.
7. Knock the nut with copper hammer lightly and remove the rear drive shaft from the wheel hub.
8. Separate the rear drive shaft from the wheel hub.
9. Insert a tire lever or equivalent between the rear differential and differential side outer ring, and then remove the rear drive shaft.

☀ WARNING

The sharp edges of the drive shaft can slice or puncture the oil seal. Be careful not to damage the oil seal when removing the drive shaft from the differential.

10. Pull the rear drive shaft to the outer side of the vehicle and disconnect it from the rear differential.

11. To hold the rear knuckle component, install the rear lateral link (upper) to the rear knuckle temporarily after disconnecting the rear drive shaft.

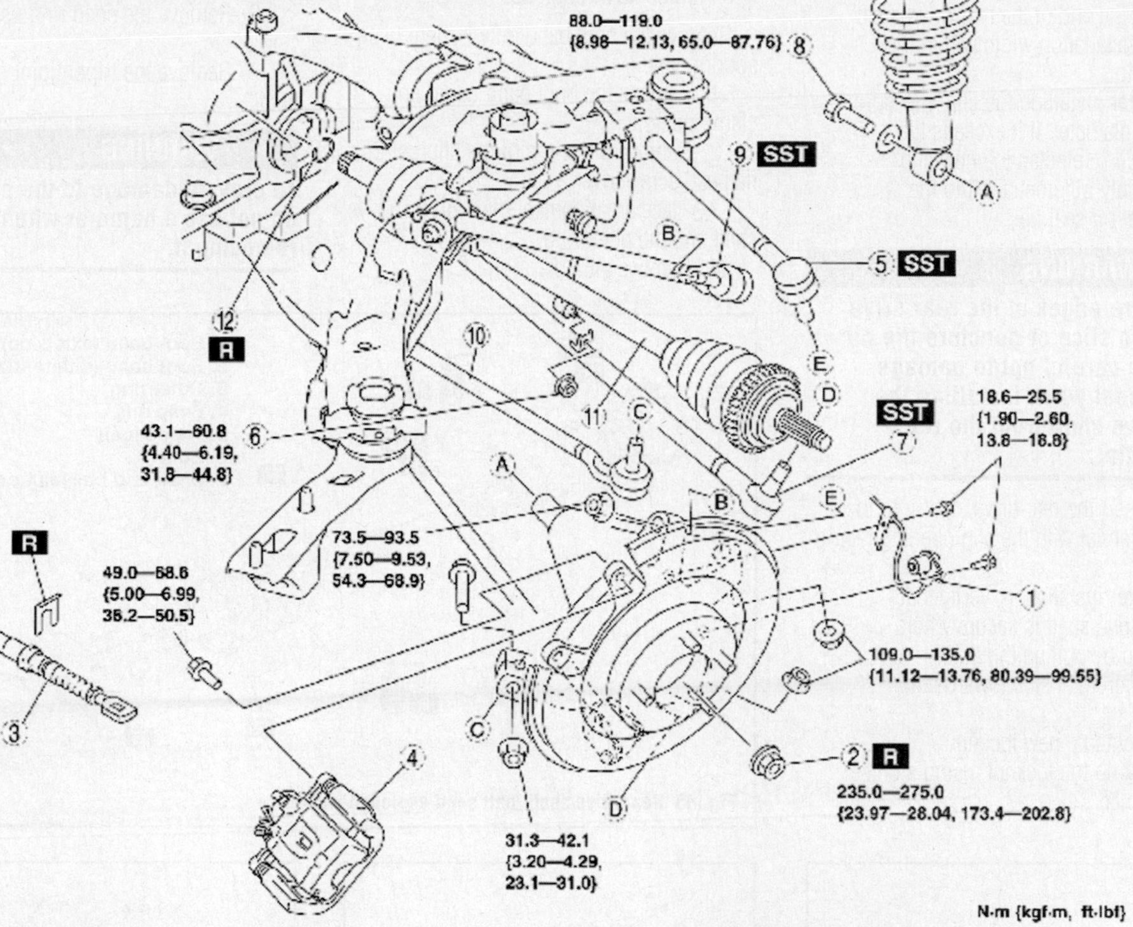

88.0—119.0
{8.98—12.13, 65.0—87.76}

18.6—25.5
{1.90—2.60,
13.8—18.8}

43.1—60.8
{4.40—6.19,
31.8—44.8}

49.0—68.6
{5.00—6.99,
35.2—50.5}

73.5—93.5
{7.50—9.53,
54.3—68.9}

109.0—135.0
{11.12—13.76, 80.39—99.55}

31.3—42.1
{3.20—4.29,
23.1—31.0}

235.0—275.0
{23.97—28.04, 173.4—202.8}

N·m {kgf·m, ft·lbf}

1. ABS wheel-speed sensor
2. Locknut
3. Parking brake cable
4. Brake caliper component
5. Rear lateral link (upper) ball joint.)
6. Stabilizer control link nut (lower)

7. Rear lateral link (lower) ball joint
8. Shock absorber bolt (lower)
9. Rear trailing link (upper) ball joint
10. Toe control link (outer)
11. Rear drive shaft
12. Clip

22140_MRX8_G0019

Fig. 39 Rear halfshaft/drive shaft and components —exploded view

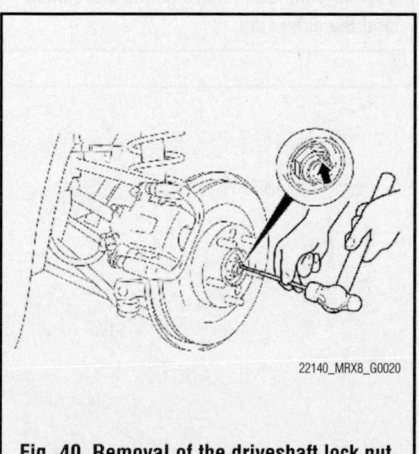

22140_MRX8_G0020

Fig. 40 Removal of the driveshaft lock nut at the rear brake rotor

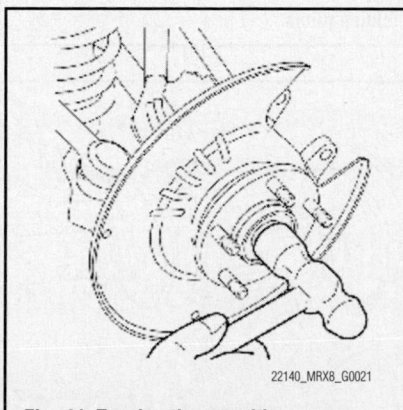

22140_MRX8_G0021

Fig. 41 Tapping the nut with a copper hammer lightly to remove the rear drive shaft from the wheel hub/rotor

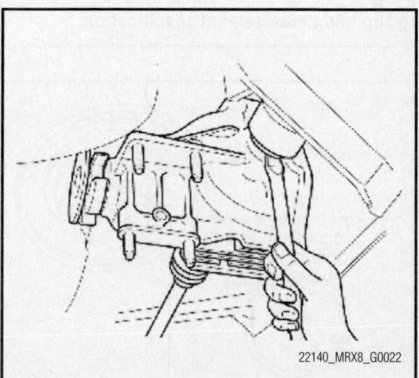

22140_MRX8_G0022

Fig. 42 Prying between the inner driveshaft and the differential side outer ring to remove the rear drive shaft from the differential

12. Point the opening of the new drive shaft clip upward, install it to the clip groove at the end of the rear drive shaft with the installation width within the specification.

13. After installing the clip, measure the outer diameter. If it exceeds the specification, reinstall the new clip.

14. Apply differential oil to the differential oil seal lip.

❋❋ WARNING

The sharp edges of the rear drive shaft can slice or puncture the oil seal. Be careful not to damage the oil seal when installing the rear drive shaft from the rear differential.

15. Insert the rear drive shaft into the rear differential with the clip opening facing upward.

16. After installation, verify that the rear drive shaft is securely held by the clip by pulling the outer ring on the differential side towards the axle

17. Tighten a new locknut.

18. Crimp the locknut, using a chisel and hammer.

CV-BOOTS INSPECTION

See Figures 45 through 55.

Remove the boot band only if there is an abnormality.

1. Remove the boot band using end clamp pliers.

2. Remove the crimp of the clip using a flathead screwdriver.

3. Place an alignment mark on the drive shaft and the outer ring.

4. Remove the outer ring.

5. Place an alignment mark on the shaft and tripod joint.

6. Remove the snap ring using a snap ring pliers.

7. Remove the tripod joint from the shaft.

❋❋ WARNING

To prevent damage to the component, do not use a hammer when removing it.

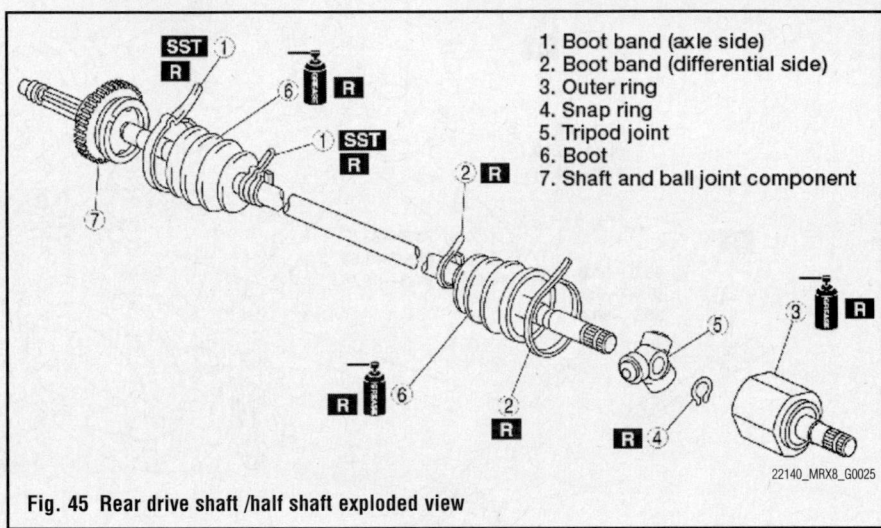

1. Boot band (axle side)
2. Boot band (differential side)
3. Outer ring
4. Snap ring
5. Tripod joint
6. Boot
7. Shaft and ball joint component

22140_MRX8_G0025

Fig. 45 Rear drive shaft /half shaft exploded view

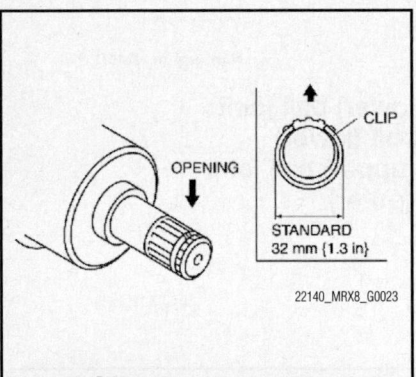

22140_MRX8_G0023

Fig. 43 Orientation of the new drive shaft clip and measurement specification

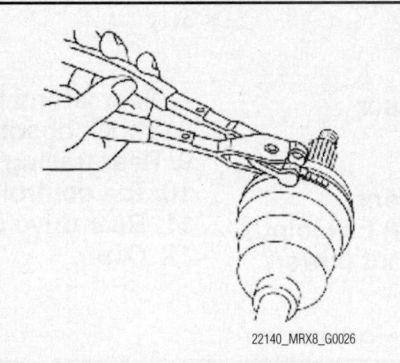

22140_MRX8_G0026

Fig. 46 Boot band removal using end clamp pliers

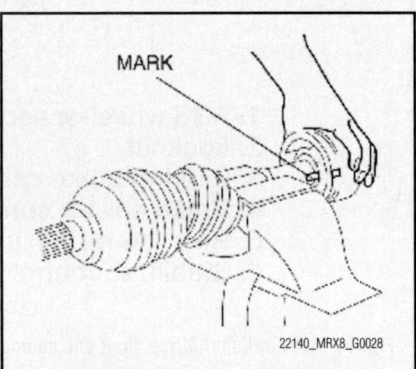

22140_MRX8_G0028

Fig. 48 Alignment mark on the drive shaft and the outer ring

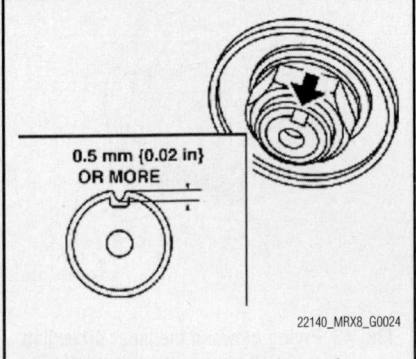

22140_MRX8_G0024

Fig. 44 Axle locknut installation and specification

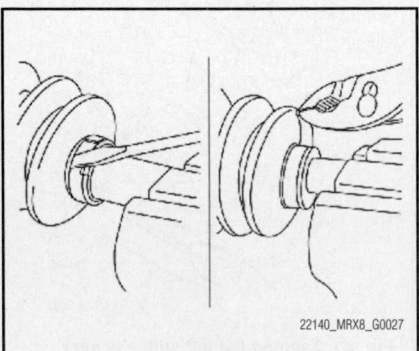

22140_MRX8_G0027

Fig. 47 Band crimp removal using a flathead screwdriver

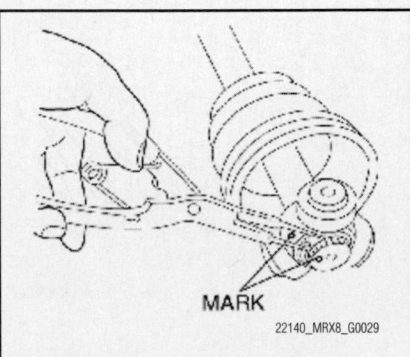

22140_MRX8_G0029

Fig. 49 Remove of the snap ring using a snap ring pliers

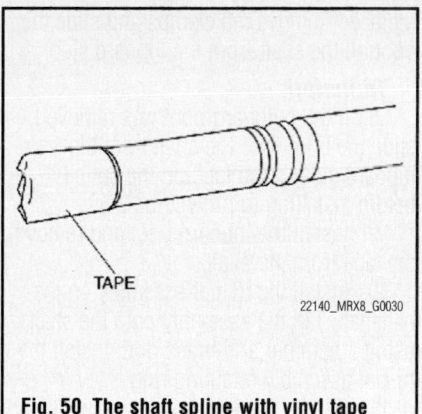

Fig. 50 The shaft spline with vinyl tape

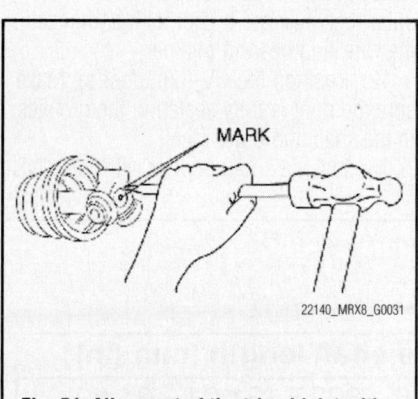

Fig. 51 Alignment of the tripod joint with the shaft mark

➡️**Remove the axle side boot only if there is an abnormality.**

8. Wrap the shaft spline with vinyl tape.

➡️**The boot shapes on the axle side and the differential side are different so do not mis-install them.**

9. Fill the inside of the new dust boot (wheel side) with grease.

➡️**Do not touch the grease with your hand. Apply it from the tube to prevent foreign matter from entering the boot.**

10. Grease amount
   - Manual Transmission (Y16M-D): 4.06–4.76 oz (115–135 g)
   - Automatic Transmission (SJ6A-EL): 4.06–4.76 oz (115–135 g)
   - Automatic Transmission (RC4A-EL): 3.18–3.88 oz (90–110 g)

11. Install the boot with the drive shaft spline still wrapped with vinyl tape.

12. Remove the vinyl tape.

13. Align the tripod joint with the shaft mark and insert it using a brass bar.

➡️**To prevent damage to the component, do not tap the roller part when installing.**

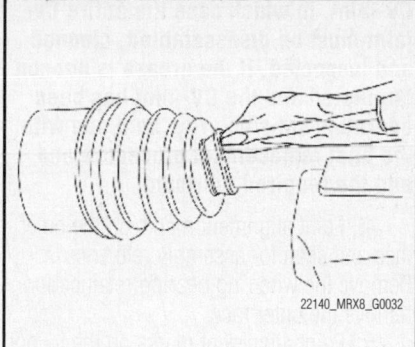

Fig. 52 Releasing trapped air from the boots

14. Install the new snap ring to the shaft installation slot securely using a snap ring pliers.

15. Fill the outer ring and boot (differential side) with the repair kit grease.

➡️**Do not touch the grease with your hand. Apply it from the tube to prevent foreign matter from entering the boot.**

16. Grease amount .
   - Manual Transmission (Y16M-D): 6.18–6.87oz (175–195 g)
   - Automatic Transmission (SJ6A-EL): 6.18–6.87oz (175–195 g)
   - Automatic Transmission (RC4A-EL): 4.77–5.46 oz (135–155 g)

17. Assemble the outer ring.

18. Release any trapped air from the boots by carefully lifting up the small end of each boot with a cloth wrapped screwdriver.

➡️**Do not let the grease leak.**

➡️**Do not damage the boot.**

19. Set the drive shaft length to the specification when the inside of the boots is at ambient pressure. Rear drive shaft standard length.
   - Manual Transmission (Y16M-D): Left side 31.21–31.59 inches (792.6–802.6mm)
   - Manual Transmission (Y16M-D): Right side 32.78–33.17 inches (832.6–842.6mm)
   - Automatic Transmission (SJ6A-EL): Left side 31.21–31.59 inches (792.6–802.6mm)
   - Automatic Transmission (SJ6A-EL): Right side 32.78–33.17 inches (832.6–842.6mm)
   - Automatic Transmission (RC4A-EL): Left side 31.15–31.53 inches (791.1–801.1mm)
   - Automatic Transmission (RC4A-

EL): Right side 32.71–33.11 inches (831.1–841.1mm)

➡️**After installation, verify that there is no boot damage or grease leakage.**

20. Using pliers, pull the boot band around the boot slot in opposite direction of drive shaft forward rotation direction and tighten.

21. Insert the end of the boot band between the boot band clip and fold back the clip tabs using a flathead screwdriver to secure the boot band.

22. Verify that the boot band is installed to the boot slot securely.

23. Turn the adjusting bolt of the Special Service Tool (SST) and adjust the opening size to the specification A.— 0.11 inch (2.9 mm).

24. Crimp the boot band (small-size) using the SST

25. Verify that the crimp value B is within the specification. 0.095–0.110 inches (2.4–2.8 mm)/

26. If the crimp value B exceeds the specification, reduce opening length A of the SST and recrimp the boot band.

27. If the crimp value B is less than the specification, increase opening length A of the SST and crimp the new boot band.

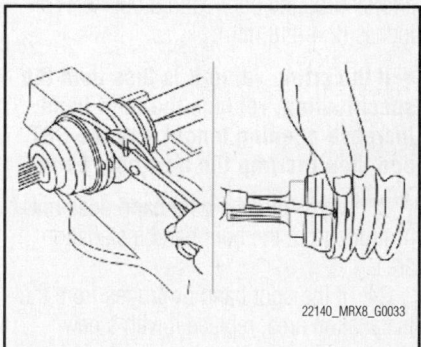

Fig. 53 Pulling the boot band around the CV boot slot

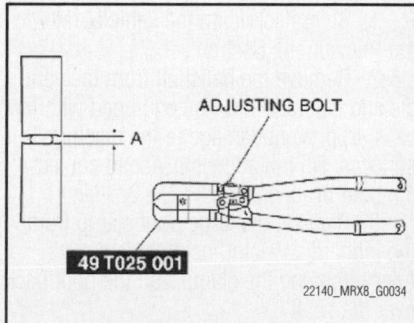

Fig. 54 Adjusting bolt of the Special Service Tool (SST) and adjust the opening size to the specification A

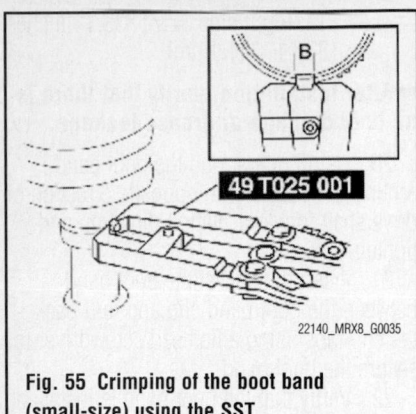

Fig. 55 Crimping of the boot band (small-size) using the SST

22140_MRX8_G0035

28. Verify that the boot band does not protrude from the boot band installation area.

➡**If the boot band protrudes from the installation area, replace it with a new band and repeat Step 2–4.**

29. Fill the boot with the repair kit grease.

30. Adjust opening length A of the SST to the specification 0.13 inches (3.2mm)

31. Crimp the boot band (large-size) using the SST.

32. If crimp value B exceeds the specification, reduce opening length A of the SST and recrimp the boot band 0.095–0.110 inches (2.4–2.8 mm)

➡**If the crimp value B is less than the specification, replace the boot band, increase opening length A of the SST, and then recrimp the new boot band.**

33. Verify that the boot band does not protrude from the boot band installation area.

34. If the boot band protrudes from the installation area, replace it with a new band.

OVERHAUL

See Figures 56 and 57.

1. Before servicing the vehicle, refer to the Precautions Section.

2. Remove the halfshaft from the vehicle and clamp it in a vise equipped with jaw caps, to prevent damage to the machined surfaces. Do not allow the vise to contact the boot or its clamps.

3. Remove the large boot clamp from the inboard CV-joint, using side cutters. After removing the clamp, roll the boot back over the shaft.

➡**Check the grease for contamination by rubbing it between 2 fingers. Any gritty feeling indicates a contaminated**

CV-joint, in which case the entire CV-joint must be disassembled, cleaned and inspected. If the grease is not contaminated and the CV-joint has been operating satisfactorily, continue with the boot replacement procedure and add the required lubricant.

4. Paint alignment marks on the outer race and shaft for assembly reference. Remove the wire ring bearing retainer and remove the outer race.

5. Paint alignment marks on the tri-pot bearing and shaft for assembly reference. Remove the tri-pot bearing snap ring and, using a brass drift and hammer, remove the tri-pot bearing from the shaft.

6. Remove the small clamp and remove the inner boot from the halfshaft. If the boot is to be reused, wrap the shaft splines with tape before removing.

7. If the outer CV-joint boot is to be replaced, remove the clamps and slide the boot off the shaft from the inboard side.

To install:

8. If the outboard boot was removed, slide the boot onto the shaft from the inboard side. Wrap tape on the splines before installing to protect the boot.

9. Install the inboard boot and remove the tape from the shaft.

10. Install the tri-pot assembly on the halfshaft. Tap the assembly onto the shaft using a hammer and brass drift. Install the tri-pot assembly retaining ring.

11. Fill the CV-joint outer race with high temperature CV-joint grease. Install the outer race over the tri-pot joint and install the wire ring bearing retainer.

12. Position the CV-joint boot(s). Make sure the boot is fully seated in the grooves in the shaft and outer race.

13. Insert a small pry bar with rounded

Standard

		Drive shaft length (mm {in})
MT	Left side	792.6—802.6 {31.21—31.59}
	Right side	832.6—842.6 {32.78—33.17}
AT	Left side	791.1—801.1 {31.15—31.53}
	Right side	831.1—841.1 {32.71—33.11}

67162-RX8-G28

Fig. 56 Halfshaft length specifications

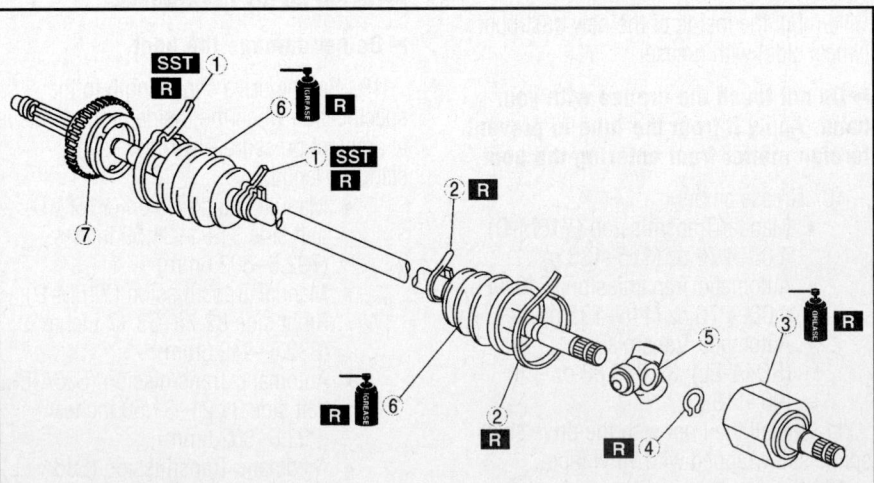

1) BOOT BANDS—AXLE SIDE
2) BOOT BANDS—DIFFERENTIAL SIDE
3) OUTER RING
4) SNAP RING
5) TRIPOD JOINT
6) BOOT
7) SHAFT AND BALL JOINT COMPONENT

09482_RX-8_G0002

Fig. 57 Exploded view of the halfshaft

edges between the boot and the outer bearing race to allow trapped air to escape from the boot. Install new boot clamps.

14. Wrap the clamps around the boots in a clockwise direction,

pull tight with pliers and bend the locking tabs to secure in position.

15. Set the halfshaft length to the specification as shown.

16. Work the CV-joint through its full

range of travel at various angles. The joint should flex, extend and compress smoothly.

17. Install the halfshaft into the vehicle.

ENGINE COOLING

RADIATOR

REMOVAL & INSTALLATION

See Figures 58 and 59.

✳✳ CAUTION

Only perform this procedure if the engine is cold, or personal injury and serious burns can occur.

1. Remove the engine cover.
2. Remove the battery cover.
3. Disconnect the negative battery cable.
4. Drain the engine coolant.
5. Remove the following components.
   a. Splash shield
   b. Under cover

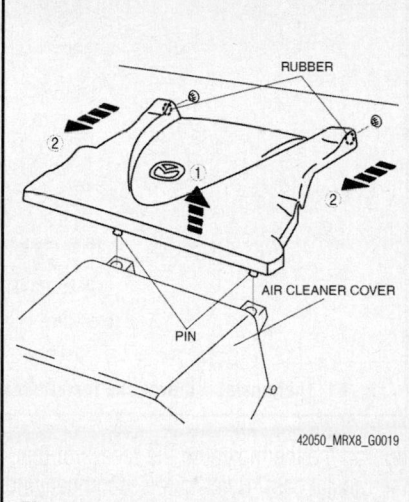

42050_MRX8_G0019

Fig. 58 Remove the engine cover in the order shown

   c. Battery and battery box.
   d. Air cleaner component and air cleaner insulator.
   e. PCM duct
6. Position the coolant reserve tank out of the way.
7. Disconnect the ATF oil cooler hose, if equipped with A/T.
8. Remove the following (NOTE: the numbers correspond with the illustration):
   • Connector.
   • Upper radiator hose
   • Lower radiator hose
   • Condenser mounting bolts
   • Bracket
   • Radiator bracket
   • Radiator
   • Cooling fan assembly

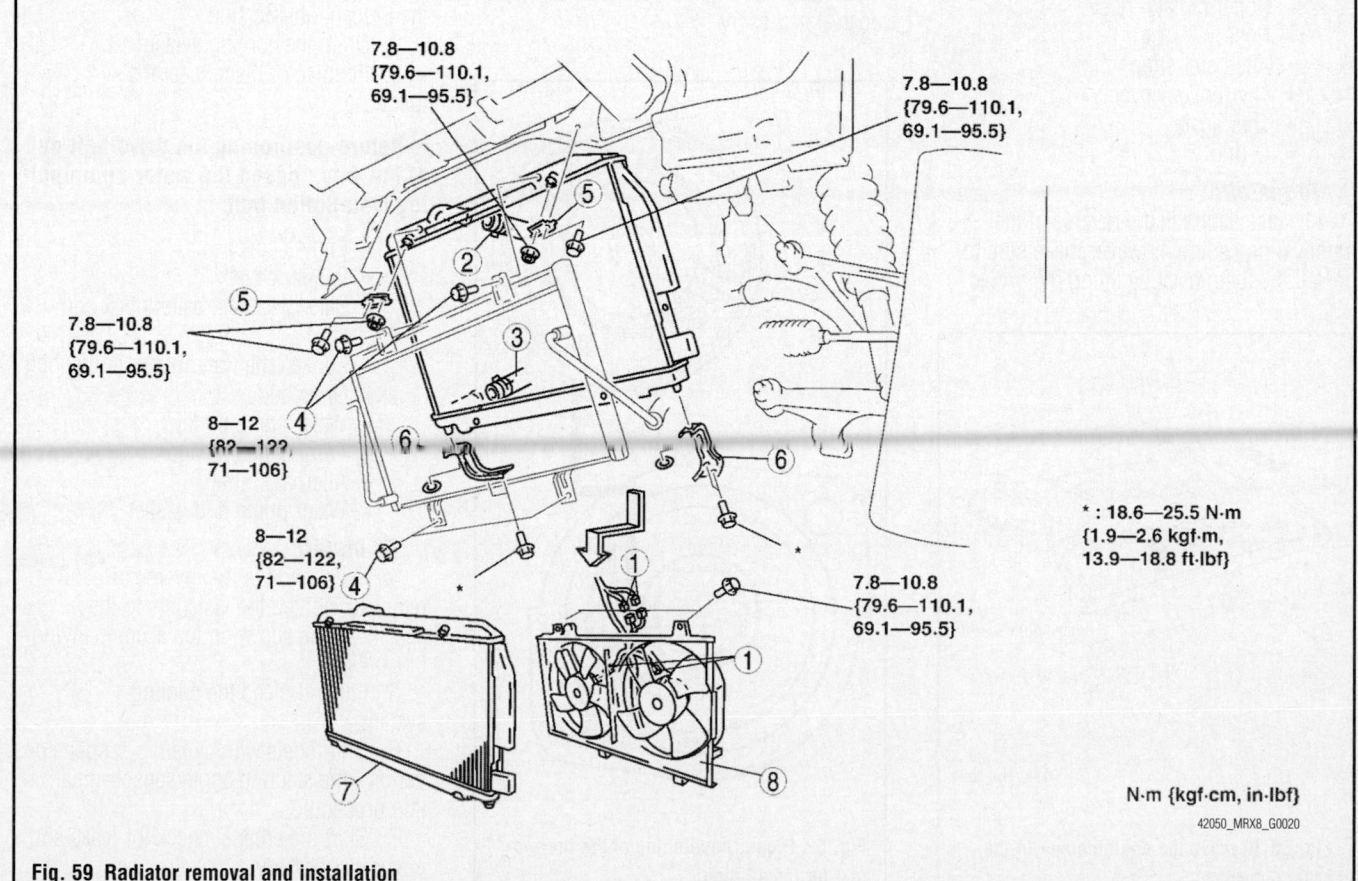

42050_MRX8_G0020

Fig. 59 Radiator removal and installation

To install:

9. Installation is the reverse of the removal procedure.

10. Add engine coolant the proper type and amount of engine coolant.

11. Start the engine, then check for coolant leaks. Check the coolant level again, and add if necessary.

THERMOSTAT

REMOVAL & INSTALLATION

See Figures 59, 60 through 62.

1. Remove the engine cover.
2. Remove the battery cover.
3. Disconnect the negative battery cable.
4. Drain the engine coolant.
5. Remove the battery and battery box.
6. Remove the secondary air control valve.
7. Before positioning the drive belt out of the way, loosen the water pump pulley installation bolt.
8. Position the drive belt out of the way. If necessary, refer to the Accessory Drive Belt procedures in this section.
9. Remove the water pump pulley.
10. Remove the following (NOTE: the numbers correspond with the illustration):
- Upper radiator hose
- Hose
- Alternator strap
- Thermostat cover
- Thermostat
- O-ring

To install:

11. Installation is the reverse of the removal procedure, however make sure to install the thermostat by fitting the projec-

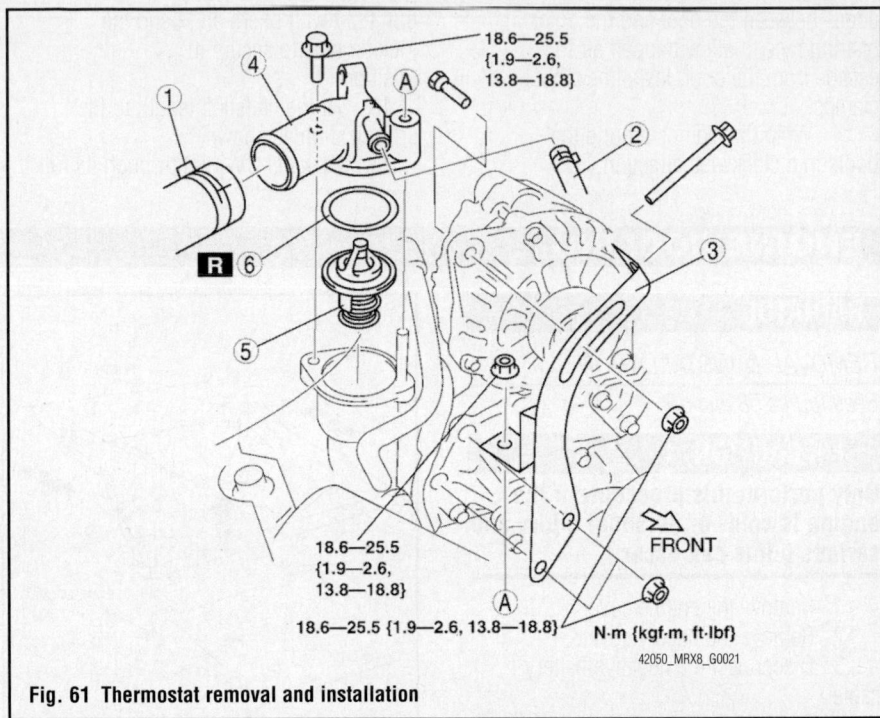

Fig. 61 Thermostat removal and installation

tion on the thermostat to the recess of the thermostat case. Refer to the accompanying illustration for details.

12. Fill and bleed the cooling system.

13. Start the engine, check for leaks and repair if necessary.

WATER PUMP

REMOVAL & INSTALLATION

See Figures 60 and 63.

1. Before servicing the vehicle, refer to the Precautions Section.
2. Drain the cooling system.
3. Remove or disconnect the following:

➡ **Before positioning the drive belt out of the way, loosen the water pump pulley installation bolt.**

- Engine cover
- Battery cover
- Battery cables, battery box and tray
- Loosen the water pump pulley bolt
- Drive belt
- Water pump pulley
- Front engine hangar
- Alternator strap
- Water pump and gasket

To install:

4. Installation is the reverse of removal. Tighten the fasteners to the specifications shown in the accompanying illustration.
5. Fill and bleed the cooling system.
6. On models with dynamic suspension, perform the steering angle sensor initialization procedure.
7. Start the engine, check for leaks and repair if necessary.

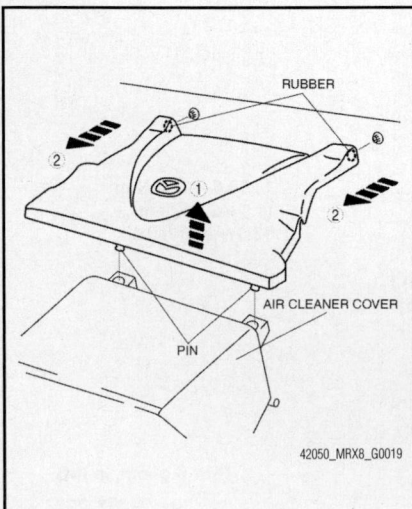

Fig. 60 Remove the engine cover in the order shown

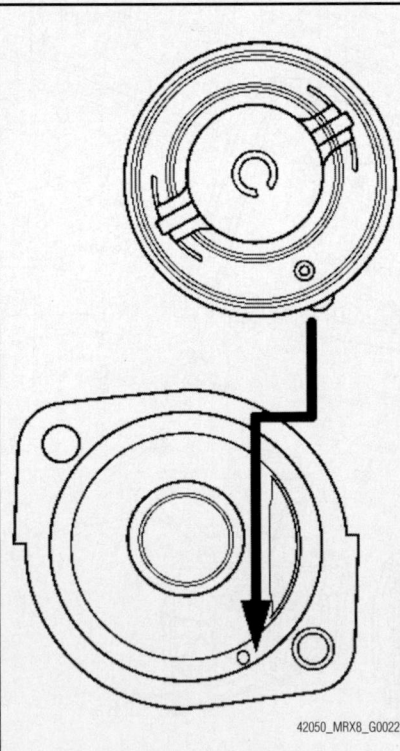

Fig. 62 Proper positioning of the thermostat for installation

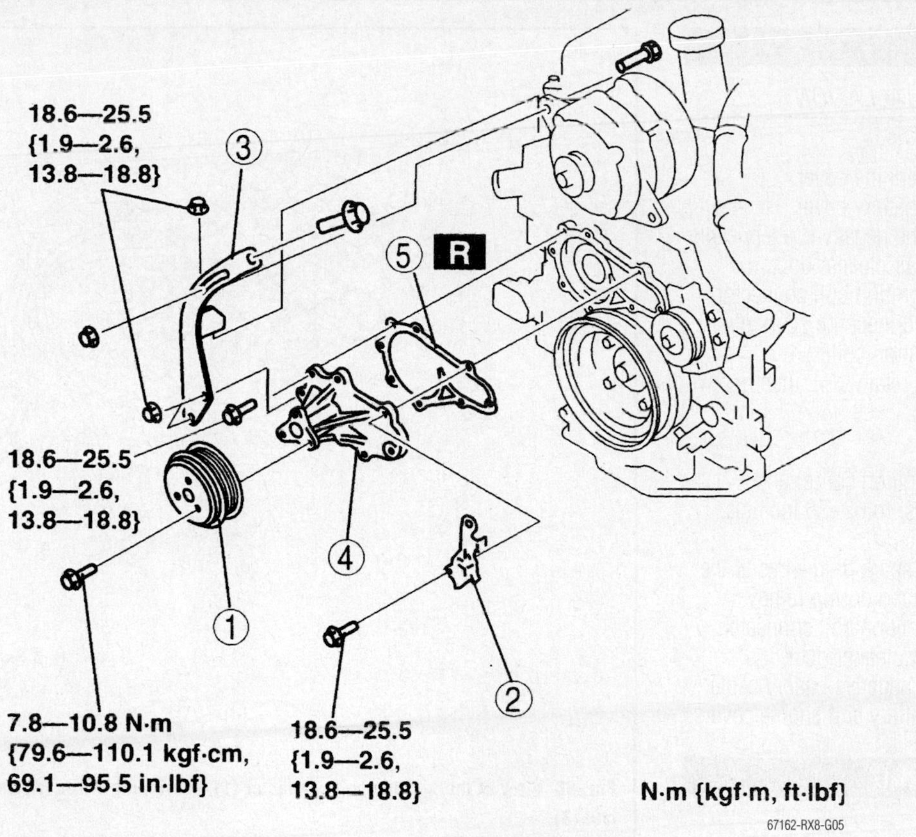

18.6—25.5
{1.9—2.6,
13.8—18.8}

③

⑤ R

18.6—25.5
{1.9—2.6,
13.8—18.8}

④

②

7.8—10.8 N·m
{79.6—110.1 kgf·cm,
69.1—95.5 in·lbf}

18.6—25.5
{1.9—2.6,
13.8—18.8}

①

N·m {kgf·m, ft·lbf}

67162-RX8-G05

Fig. 63 Exploded view of the water pump assembly

ENGINE ELECTRICAL   CHARGING SYSTEM

ALTERNATOR

REMOVAL & INSTALLATION

See Figure 64.

1. Before servicing the vehicle, refer to the Precautions Section.
2. Remove or disconnect the following:

- Negative battery cable
- Engine cover
- Rear engine cross brace
- Intake air duct
- Accessory drive belt
- Electrical connectors from the alternator
- Alternator bolts
- Alternator

To install:

3. Installation is the reverse of the removal procedure, noting the following:

   a. Tighten the left side engine cross brace nut to 40 ft. lbs. (45 Nm) and the right side nut to 16 ft. lbs. (22 Nm)

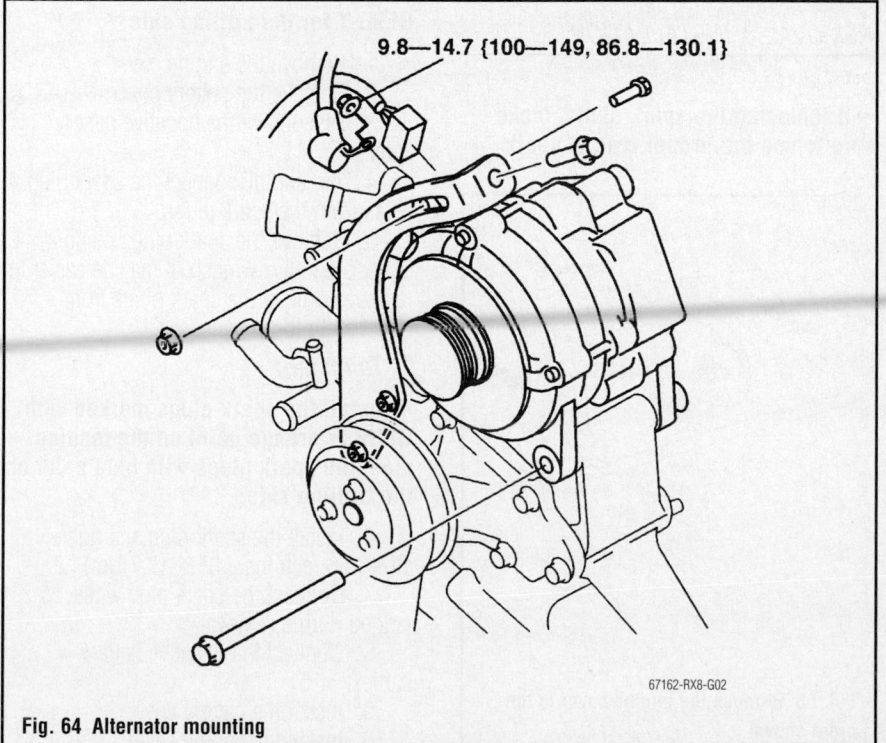

9.8—14.7 {100—149, 86.8—130.1}

67162-RX8-G02

Fig. 64 Alternator mounting

IGNITION COIL

REMOVAL & INSTALLATION

See Figures 65 and 66.

1. Remove the engine cover.
2. Remove the battery cover.
3. Disconnect the negative battery cable.
4. Remove the air cleaner duct.
5. Detach the ignition coil connector.
6. Tag and disconnect the spark plug wire(s) from the ignition coil(s).
7. Unfasten the retainer(s), then remove the ignition coil(s).

To install:

8. Install the ignition coil(s) and tighten the retainer(s) to 62–86 inch lbs. (6.9–9.8 Nm).
9. Connect the spark plug wires to the ignition coils as tagged during removal.
10. Attach the ignition coil connector.
11. Install the air cleaner duct.
12. Connect the negative battery cable.
13. Install the battery and engine covers.

IGNITION TIMING

ADJUSTMENT

The ignition timing cannot be adjusted. To inspect the ignition timing, refer to the procedure in this section.

SPARK PLUGS

REMOVAL & INSTALLATION

See Figures 65, 67 and 68.

➡If replacing the spark plugs, make sure to use the proper type of spark

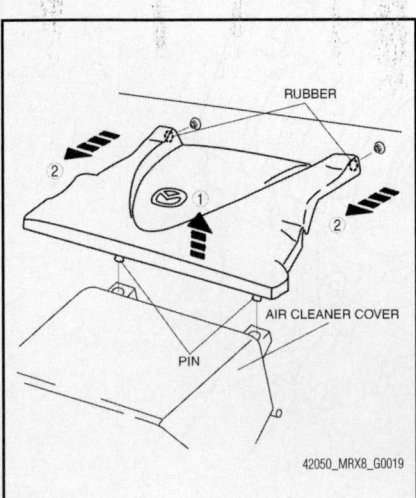

Fig. 65 Remove the engine cover in the order shown

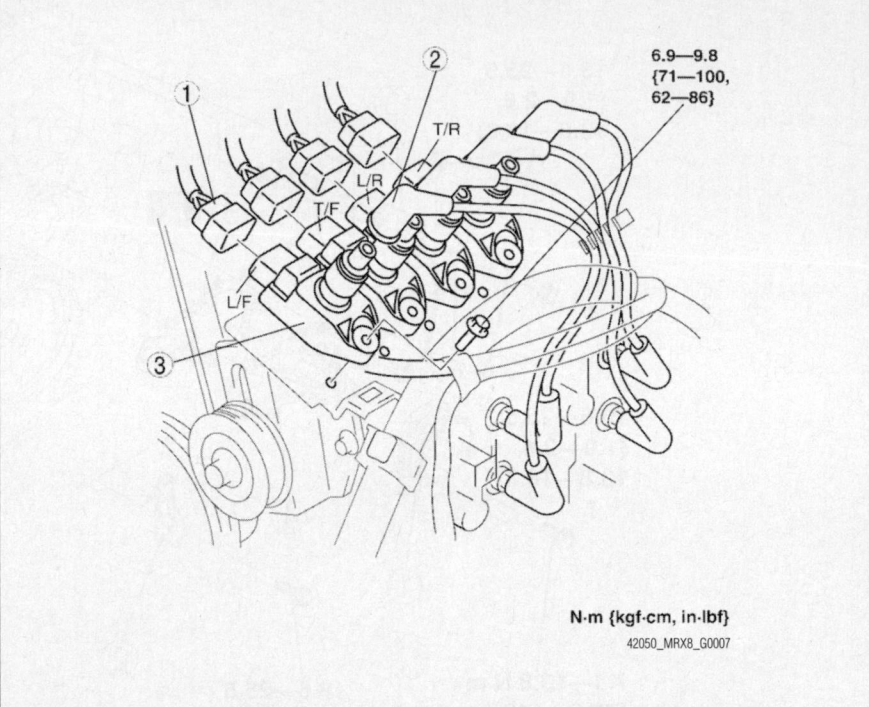

Fig. 66 View of the ignition coil connector (1), spark plug wires (2) and ignition coil (3)

plug. Using the incorrect type may result in improper sealing and reduced engine performance. The proper type of spark plug for the Mazda RX-8 is NGK RE7A-L for the leading side and NGK RE9B-T for the trailing side.

1. Remove the engine cover.
2. Remove the battery cover.
3. Disconnect the negative battery cable.
4. Tag and disconnect the spark plug wires from the spark plugs.
5. Remove the spark plugs using a spark plug plug-wrench. It may be easier to remove some of the spark plugs from underneath the vehicle.

To install:

➡Install the spark plugs marked with white or orange paint on the leading side and spark plugs with blue paint on the trailing side.

6. Install the spark plug and tighten to 114–156 inch lbs. (12.8–17.7 Nm).
7. Connect the spark plug wires, as tagged during removal.
8. Connect the negative battery cable.
9. Install the battery cover.
10. Install the engine cover.

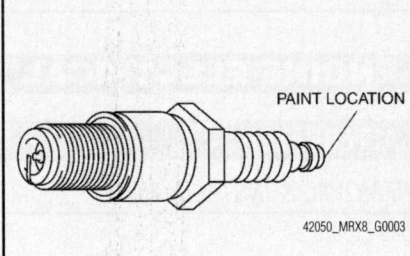

Fig. 67 When you install the spark plugs, make sure the white or orange paint is on the leading side and spark plugs with blue paint on the trailing side

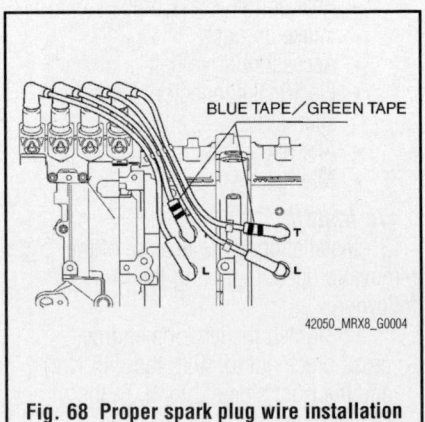

Fig. 68 Proper spark plug wire installation

STARTER

REMOVAL & INSTALLATION

See Figure 69.

1. Remove or disconnect the following:
   - Engine cover
   - Negative battery cable
   - Air cleaner

- Starter electrical connectors
- Starter

To install:

2. Install or connect the following:
   - Starter and loosely tighten the lower starter mounting bolt
   - Starter electrical connectors
   - Starter bolts. Torque the bolts 14–18 ft. lbs. (19–25 Nm) on

automatic transmission model, or 29–37 ft. lbs. (38–51 Nm) on manual transmission model.
   - Air cleaner
   - Negative battery cable
   - Engine cover

3. On models with dynamic suspension, perform the steering angle sensor initialization procedure.

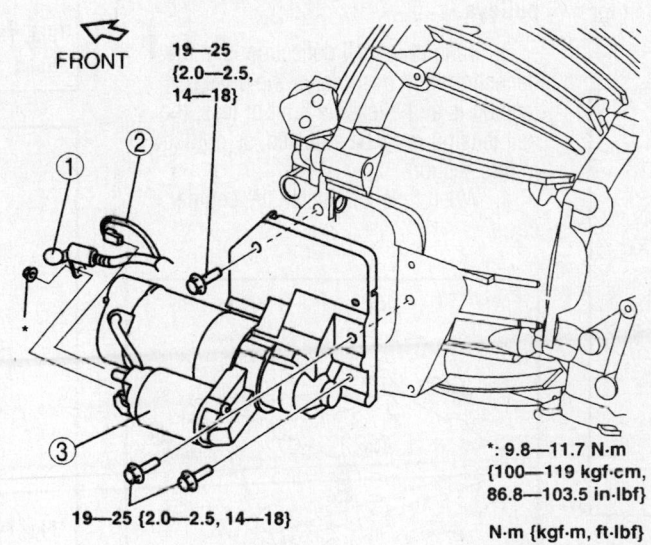

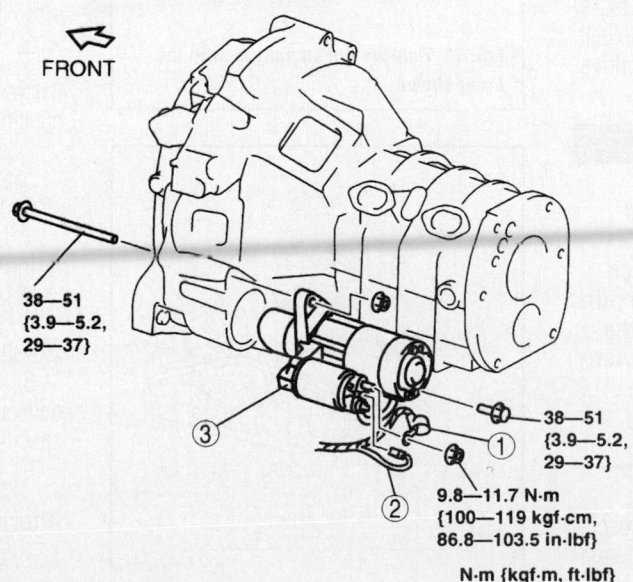

Fig. 69 Starter mounting—A/T top; M/T bottom

ENGINE MECHANICAL

➡ **Disconnecting the negative battery cable may interfere with the functions of the on board computer systems and may require the computer to undergo a relearning process, once the negative battery cable is reconnected.**

ACCESSORY DRIVE BELTS

ACCESSORY BELT ROUTING

See Figure 70.

Refer to the accompanying illustration for belt routing.

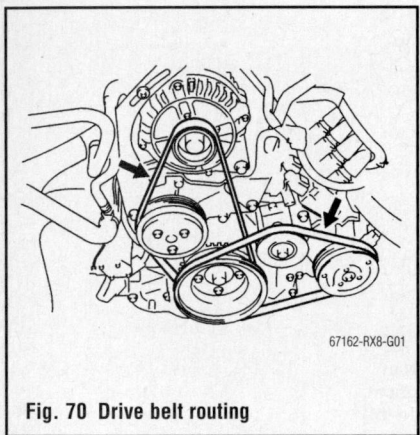

Fig. 70 Drive belt routing

INSPECTION

See Figures 71 through 74.

Inspect the drive belt for signs of glazing or cracking. A glazed belt will be perfectly smooth from slippage, while a good belt will have a slight texture of fabric visible. Cracks will usually start at the inner edge of the belt and run outward. All worn or damaged drive belts should be replaced immediately.

❋ WARNING

The drive belt deflection can be inspected only between specified pulleys. Perform the drive belt deflection/tension inspection when the engine is cold, or at least 30 min after the engine has stopped. If the drive belt exceeds the deflection/tension limit, adjust it to the deflection/tension used when adjusting, as outlined in this section..

After replacing with a new drive belt, assemble with the deflection/tension specified for a new drive belt. Operate the alternator drive belt for 1 min or more and the A/C drive belt for 5 min or more while idling the engine. Then adjust it to the used belt deflection/tension specifications.

1. Remove the engine cover.
2. Apply a pressure of about 22 lbs. (98 Nm) to the back of the drive belt, in the middle of the pulleys shown in the figure and inspect the deflection. Otherwise, inspect the tension using the Special Service Tool (SST), as shown in the accompanying illustration.

➡ **The drive belt tension can be inspected anywhere between the pulleys. The drive belt deflection can be inspected only between specified pulleys.**

3. If the drive belt deflection is at the deflection limit or more, or the drive belt tension is at the tension limit or less, the belt tension must be adjusted, as outlined in this section.

4. When finished, install the engine cover.

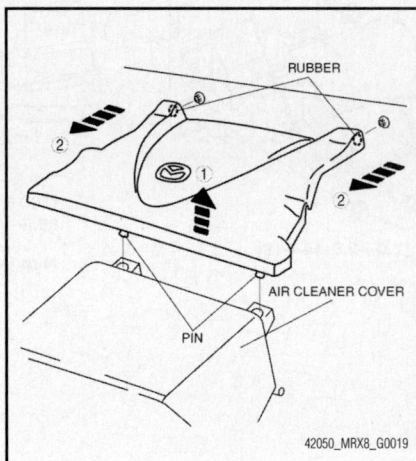

Fig. 71 Remove the engine cover in the order shown

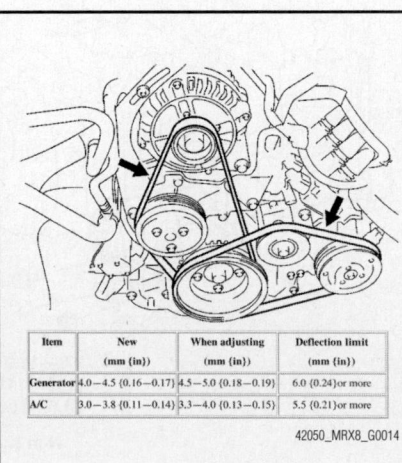

Item	New (mm {in})	When adjusting (mm {in})	Deflection limit (mm {in})
Generator	4.0—4.5 {0.16—0.17}	4.5—5.0 {0.18—0.19}	6.0 {0.24} or more
A/C	3.0—3.8 {0.11—0.14}	3.3—4.0 {0.13—0.15}	5.5 {0.21} or more

42050_MRX8_G0014

Fig. 72 Measuring the belt tension/deflection

Item	New (mm {in})	When adjusting (mm {in})	Deflection limit (mm {in})
Generator	4.0—4.5 {0.16—0.17}	4.5—5.0 {0.18—0.19}	6.0 {0.24} or more
A/C	3.0—3.8 {0.11—0.14}	3.3—4.0 {0.13—0.15}	5.5 {0.21} or more

42050_MRX8_G0015

Fig. 73 Drive belt deflection specifications

Item	New (N {kgf, lbf})	When adjusting (N {kgf, lbf})	Tension limit (N {kgf, lbf})
Generator	620—767 {63.3—78.2, 140—172}	519—666 {53.0—67.9, 117—149}	344 {35.1, 77.3} or less
A/C	559—706 {57.1—71.9, 126—158}	519—617 {53.0—62.9, 117—138}	265 {27.1, 59.6} or less

42050_MRX8_G0016

Fig. 74 Drive belt tension specifications

ADJUSTMENT

A/C Compressor Belt

See Figures 71 and 75.

1. Remove the engine cover.
2. Loosen idle pulley locknut.
3. Adjust the drive belt deflection and tension by turning adjusting bolt B to the specification.

   a. Drive belt deflection (with pressure of 22 lbs. [98 Nm]): 0.13—0.15 inch lbs. (3.3—4.0mm).

   b. Drive belt tension (when using the SST): 117—138 lbs. (519—617 Nm).

4. Tighten idle pulley locknut A to 27—38 ft. lbs. (37—52 Nm).
5. Crank the engine, then measure the deflection and tension again. If not within specifications, repeat Steps 2—5.
6. Install the engine cover

Alternator Belt

See Figures 71 and 76.

1. Remove the engine cover.
2. Loosen generator installation bolt, and locknut.

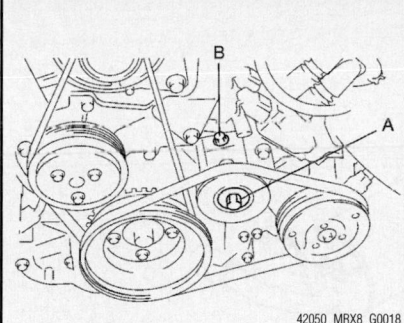

Fig. 75 View of the idle pulley locknut A and adjusting bolt B

3. Adjust the drive belt deflection and tension by turning adjusting bolt C to the specification:

    a. Drive belt deflection (with pressure of 22 lbs. [98 Nm]): 0.18—0.19 inch lbs. (4.5–5.0mm).

    b. Drive belt tension (when using the SST): 117–149 lbs. (519–666 Nm).

4. Tighten generator installation bolt A and locknut B, to the following:

    a. Bolt A: 29–37 ft. lbs. (38–51 Nm).

    b. Locknut B: 15–22 ft. lbs. (20–30 Nm).

5. Crank the engine and measure the deflection and tension again. If not within the specification, repeat Steps 2 & 3.

6. Install the engine cover

REMOVAL & INSTALLATION

A/C Compressor Belt

See Figure 75.

1. Remove the engine cover.
2. Loosen idle pulley locknut.
3. Loosen adjusting bolt B, then remove the A/C compressor drive belt.

To install:

4. Install the drive belt and adjust the drive belt deflection by tightening adjusting bolt B to the specifications given in the Adjustment procedure in this section.

> **❋❋ WARNING**
>
> **When installing a new drive belt, assemble it with the deflection/tension specifications for new drive belt, then start the engine and operate the drive belt for 5 min or longer while idling the engine. Then adjust the belt to the used deflection/tension specifications.**

5. Tighten idle pulley locknut A to 27–38 ft. lbs. (37–52 Nm).

6. Crank the engine and measure the deflection and tension again, as outlined in this section. If not within specifications, go back to Step 2 and repeat the steps.

7. Install the engine cover

Alternator Belt

See Figures 70, 71 and 76.

1. Remove the engine cover.

2. Remove the A/C compressor belt, as outlined later in this section.

3. Loosen generator installation bolt A and locknut B.

4. Loosen adjusting bolt C, then remove the drive belt.

To install:

5. Install the drive belt and adjust the drive belt deflection by tightening adjusting bolt to the specifications given in the Adjustment procedure in this section.

> **❋❋ WARNING**
>
> **When replacing with a new drive belt, assemble with the deflection/tension for a new drive belt, then start the engine and operate the drive belt for 1 minute or longer while idling the engine. Then, adjust the belt deflection/tension to the used belt specifications.**

6. Tighten generator installation bolt A and locknut B, as follows:

    a. A: 29–37 ft. lbs. (38–51 Nm)

    b. B: 15–22 ft. lbs. (20–30 Nm)

7. Crank the engine and measure the deflection and tension again, as outlined in this section. If not within specifications, go back to Step 3 and repeat the steps.

8. Install the A/C compressor belt.

9. Install the engine cover

EXHAUST MANIFOLD

REMOVAL & INSTALLATION

See Figures 77 through 79.

1. Before servicing the vehicle, refer to the Precautions Section.

2. Remove or disconnect the following:

- Negative battery cable
- Front and rear tunnel crossmembers
- Main silencer
- Middle exhaust pipe
- Protector
- Rear oxygen sensor
- Catalytic converter
- Bracket
- Front oxygen sensor
- Air Injection Reactor (AIR pipe)

3. Use a suitable overhead engine lift and support the engine.

4. Remove the right side engine mounting rubber and bracket.

5. Remove the exhaust manifold.

To install:

6. Clean all gasket mating surfaces.

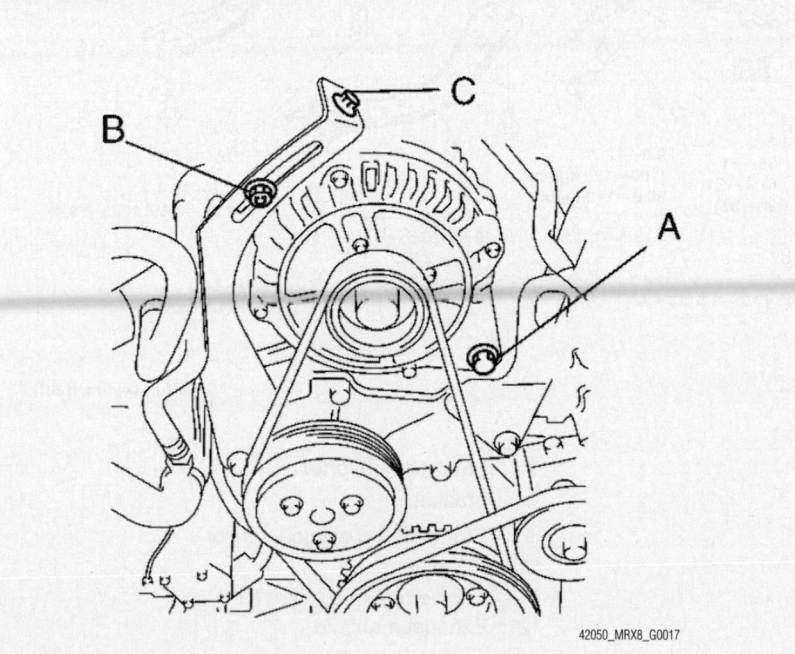

Fig. 76 View of the alternator bolts A, locknut B and adjusting bolt C

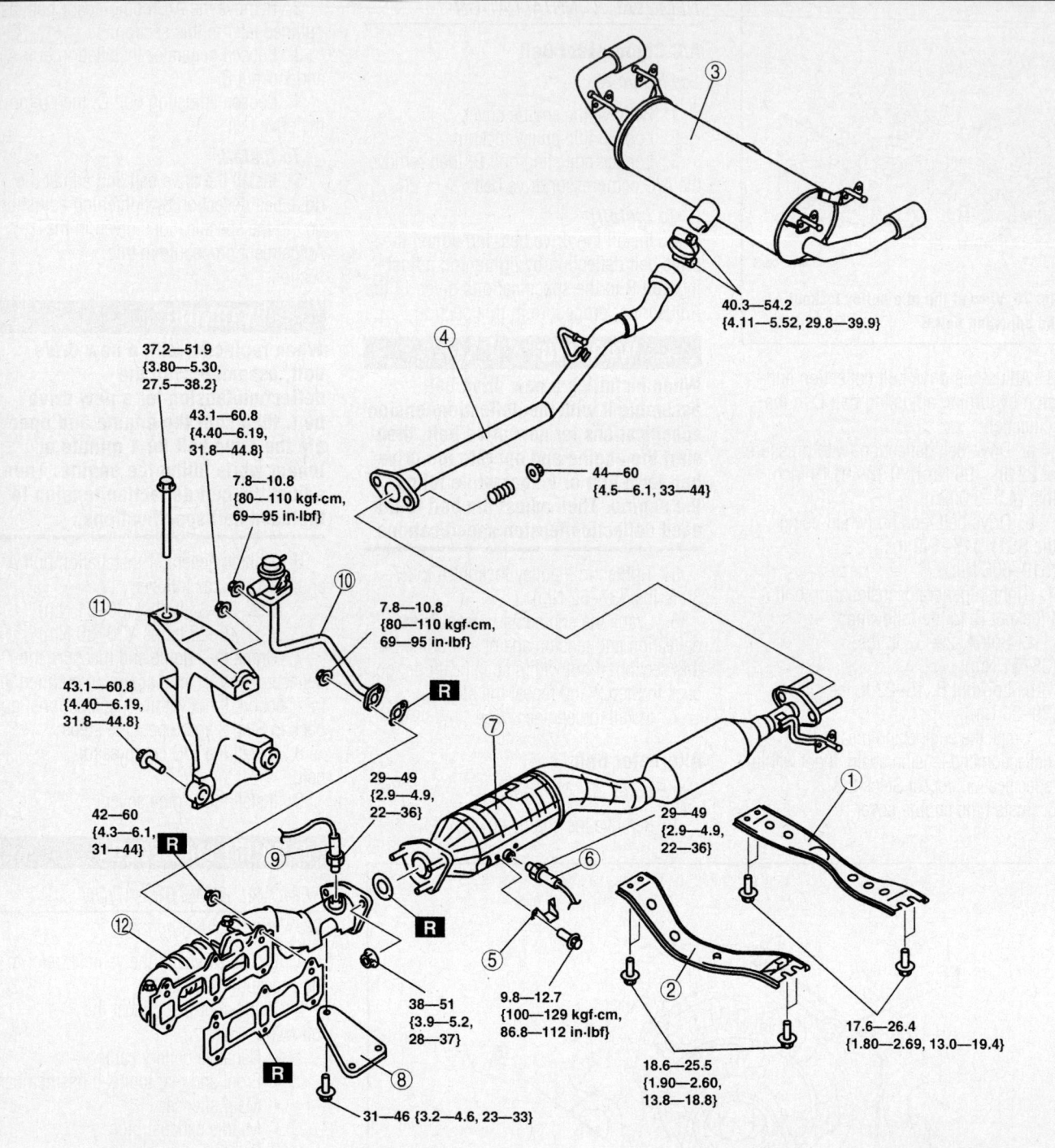

40.3—54.2
{4.11—5.52, 29.8—39.9}

37.2—51.9
{3.80—5.30,
27.5—38.2}

43.1—60.8
{4.40—6.19,
31.8—44.8}

7.8—10.8
{80—110 kgf·cm,
69—95 in·lbf}

44—60
{4.5—6.1, 33—44}

7.8—10.8
{80—110 kgf·cm,
69—95 in·lbf}

R

43.1—60.8
{4.40—6.19,
31.8—44.8}

42—60
{4.3—6.1,
31—44}

R

29—49
{2.9—4.9,
22—36}

29—49
{2.9—4.9,
22—36}

R

38—51
{3.9—5.2,
28—37}

9.8—12.7
{100—129 kgf·cm,
86.8—112 in·lbf}

17.6—26.4
{1.80—2.69, 13.0—19.4}

18.6—25.5
{1.90—2.60,
13.8—18.8}

R

31—46 {3.2—4.6, 23—33}

N·m {kgf·m, ft·lbf}

1	Rear tunnel member	7	Catalytic converter
2	Front tunnel member	8	Bracket
3	Main silencer	9	Front heated oxygen sensor
4	Middle pipe	10	AIR pipe
5	Protector	11	Engine mount bracket (RH)
6	Rear heated oxygen sensor	12	Exhaust manifold

67162-RX8-G12

Fig. 77 Exploded view of the exhaust system

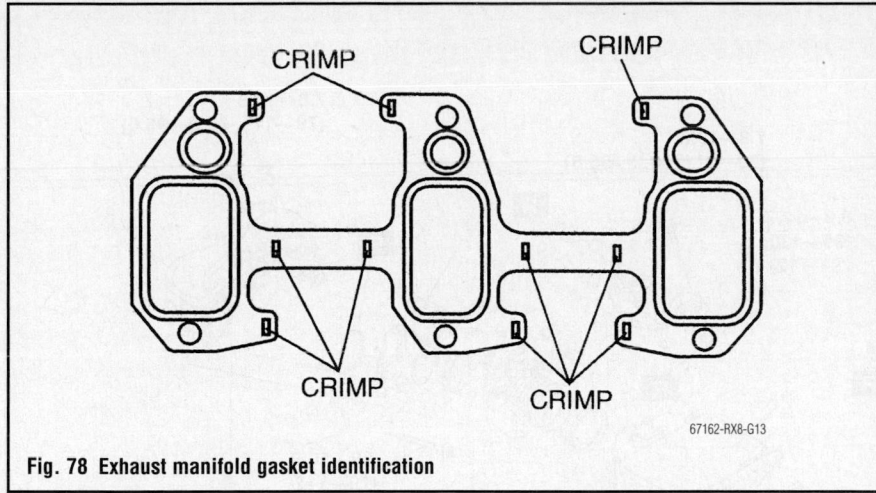

Fig. 78 Exhaust manifold gasket identification

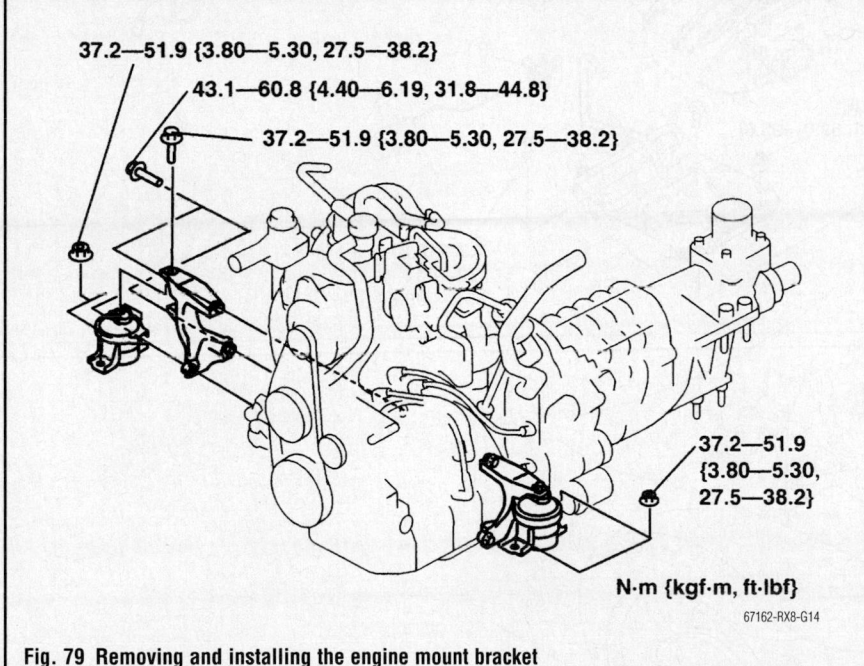

Fig. 79 Removing and installing the engine mount bracket

➡**Use new self-locking nuts. The exhaust manifold gasket has crimps attached to it. Ensure that all the crimps are in place when installing the gasket, or the gasket will leak.**

7. Install or connect the following:
   - Exhaust manifold. torque the nuts to 31–44 ft. lbs. (43–61 Nm).
   - Right side engine mounting bracket. Torque the bolts as shown.
   - Air Injection Reactor (AIR) pipe
   - Bracket
   - Catalytic converter
   - Rear oxygen sensor
   - Protector
   - Middle exhaust pipe
   - Main silencer
   - Front and rear tunnel crossmembers and torque the bolts to 14–19 ft. lbs. (19–26 Nm)
   - Negative battery cable

8. On models with dynamic suspension, perform the steering angle sensor initialization procedure.

INTAKE MANIFOLD

REMOVAL & INSTALLATION
See Figures 80 through 83.

✳✳ **WARNING**

A hot engine and intake-air system can cause severe burns. Turn off the engine and wait until they are cool before removing the intake-air system.

✳✳ **WARNING**

Fuel line spills and leakage from the pressurized fuel system are dangerous. Fuel can ignite and cause serious injury or death and damage. Fuel can also irritate skin and eyes. To prevent this, always complete the "Fuel Line Safety Procedure," while referring to the SERVICE PRECAUTION.

1. Relieve the fuel system pressure.
2. Drain the cooling system.
3. Remove or disconnect the following:
   - Remove the front bumper.
   - Air hose
   - Air cleaner cover
   - Variable Fresh Air Duct (VFAD) solenoid valve on high output engines
   - Vacuum chamber on high output engines
   - Air cleaner case
   - Throttle body
   - Upper extension manifold
   - Lower extension manifold on high output engines
   - Oil filler pipe
   - Air Injection Reactor (AIR) solenoid valve
   - Secondary Shutter Valve (SSV) solenoid
   - Variable Dynamic Effect Intake (VDI) solenoid valve.
   - Air cleaner insulator
   - Auxiliary Port Valve (APV) bracket and motor on high output engines
   - Fresh air intake duct
   - Fuel distributors
   - Intake manifold and discard the gasket

To install:
4. Clean all gasket mating surfaces.
5. Apply clean oil to the Auxiliary Port Valve (APV) valves.
6. Install or connect the following:
   - Intake manifold using a new gasket in the sequence shown. Torque the bolts to 14–19 ft. lbs. (19–26 Nm) re-tighten bolt no. 1.
   - Fuel distributors and torque the bolts to 14–19 ft. lbs. (19–26 Nm)
   - Fresh air intake duct and torque the bolts to 69–96 inch lbs. (8–11 Nm)
   - APV bracket and motor on high output engines
   - Air cleaner insulator and torque the bolts to 69–96 inch lbs. (8–11 Nm)

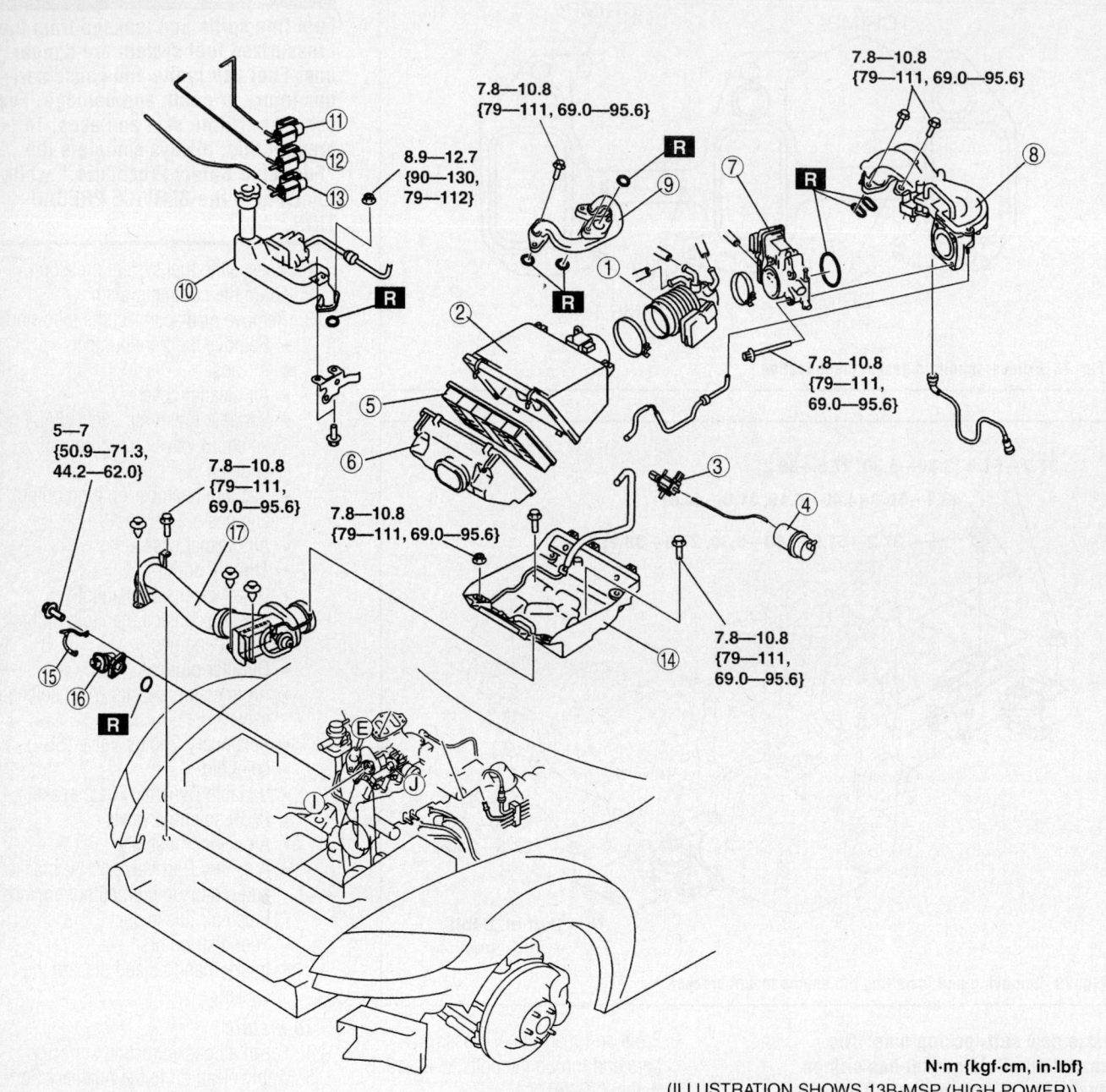

7.8—10.8
{79—111, 69.0—95.6}

7.8—10.8
{79—111, 69.0—95.6}

8.9—12.7
{90—130,
79—112}

7.8—10.8
{79—111,
69.0—95.6}

5—7
{50.9—71.3,
44.2—62.0}

7.8—10.8
{79—111,
69.0—95.6}

7.8—10.8
{79—111, 69.0—95.6}

7.8—10.8
{79—111,
69.0—95.6}

N·m {kgf·cm, in·lbf}
(ILLUSTRATION SHOWS 13B-MSP (HIGH POWER))

1	Air hose	10	Oil filler pipe
2	Air cleaner cover	11	AIR solenoid valve
3	VFAD solenoid valve (13B-MSP (High power))	12	SSV solenoid valve
4	Vacuum chamber (13B-MSP (High power))	13	VDI solenoid valve
5	Air cleaner element	14	Air cleaner insulator
6	Air cleaner case	15	Bracket (13B-MSP (High power))
7	Throttle body	16	APV motor (13B-MSP (High power))
8	Extension manifold (upper)	17	Fresh-air duct
9	Extension manifold (lower) (13B-MSP (High power))		

67162-RX8-G08

Fig. 80 Exploded view of the air intake system

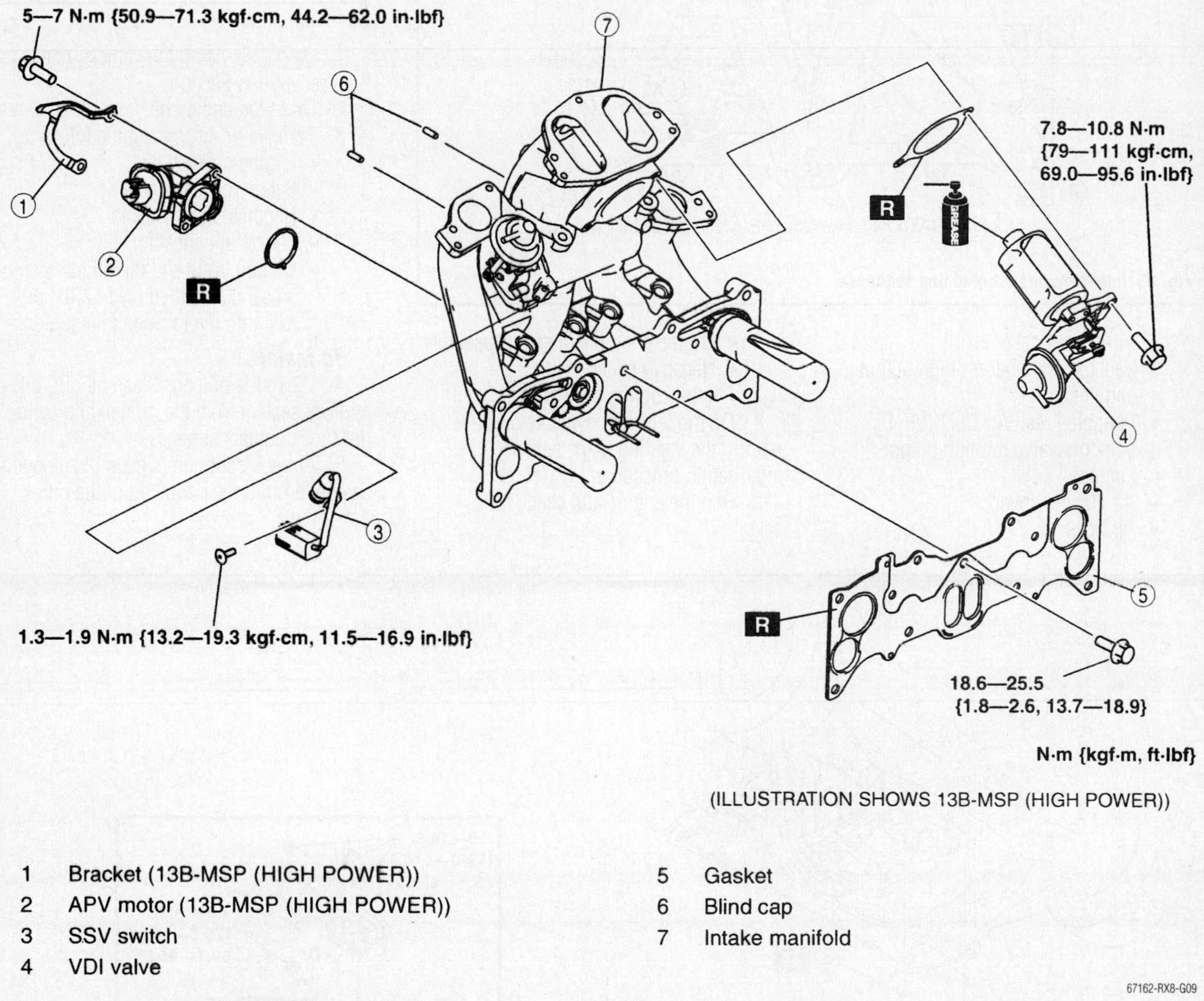

5—7 N·m {50.9—71.3 kgf·cm, 44.2—62.0 in·lbf}

7.8—10.8 N·m
{79—111 kgf·cm,
69.0—95.6 in·lbf}

1.3—1.9 N·m {13.2—19.3 kgf·cm, 11.5—16.9 in·lbf}

18.6—25.5
{1.8—2.6, 13.7—18.9}

N·m {kgf·m, ft·lbf}

(ILLUSTRATION SHOWS 13B-MSP (HIGH POWER))

1	Bracket (13B-MSP (HIGH POWER))	5	Gasket
2	APV motor (13B-MSP (HIGH POWER))	6	Blind cap
3	SSV switch	7	Intake manifold
4	VDI valve		

67162-RX8-G09

Fig. 81 Exploded view of the intake manifold

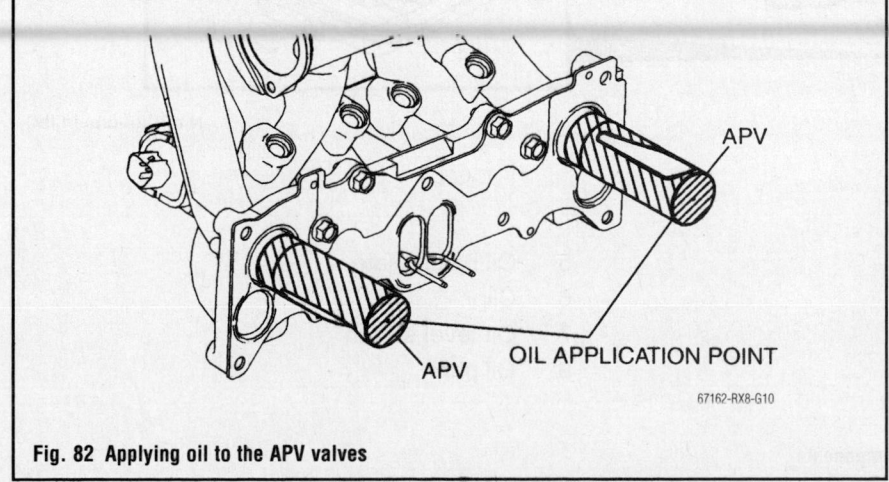

APV

APV   OIL APPLICATION POINT

67162-RX8-G10

Fig. 82 Applying oil to the APV valves

- Variable Dynamic Effect Intake (VDI) solenoid valve.
- Secondary Shutter Valve (SSV) solenoid
- Air Injection Reactor (AIR) solenoid valve
- Oil filler pipe and torque the bolts to 79–112 inch lbs. (9–13 Nm)
- Lower extension manifold on high output engines and torque the bolts to 79–112 inch lbs. (9–13 Nm)
- Upper extension manifold and torque the bolts to 69–96 inch lbs. (8–11 Nm)
- Throttle body and torque the bolts to 69–96 inch lbs. (8–11 Nm)

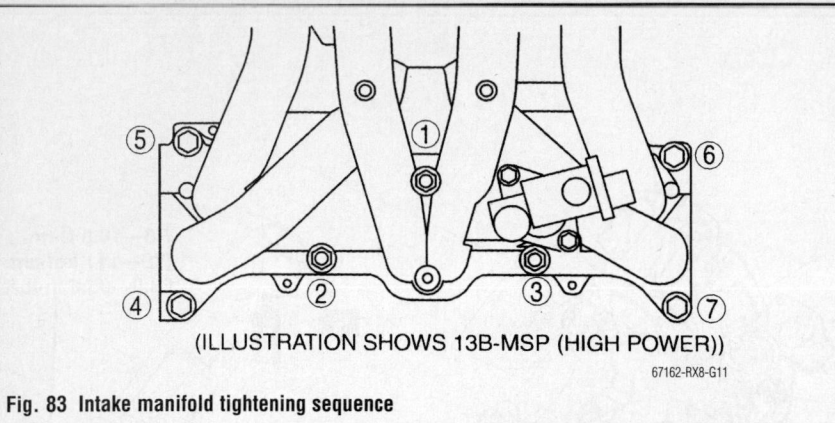

(ILLUSTRATION SHOWS 13B-MSP (HIGH POWER))

67162-RX8-G11

Fig. 83 Intake manifold tightening sequence

- Air cleaner case
- Vacuum chamber on high output engines
- Variable Fresh Air Duct (VFAD) solenoid valve on high output engines
- Air cleaner cover
- Air hose
- Engine and transmission assembly
- Negative battery cable
7. Fill the cooling system.
8. On models with dynamic suspension, perform the steering angle sensor initialization procedure.
9. Run the engine and check for leaks.

OIL PAN

REMOVAL & INSTALLATION

See Figure 84.

1. Before servicing the vehicle, refer to the Precautions Section.
2. Drain the engine oil.
3. Remove or disconnect the following:
   - Engine cover
   - Battery cover
   - Negative battery cable
   - Electrical connector
   - Oil pan bolts and the oil pan using a seal cutter, then insert a flat pry tool into the locations illustrated.

To install:

4. Clean the oil pan. Clean all dirt, oil and old sealant from the oil pan and cylinder block contact surfaces.
5. Apply a continuous bead of silicone sealant around the perimeter of the oil pan.

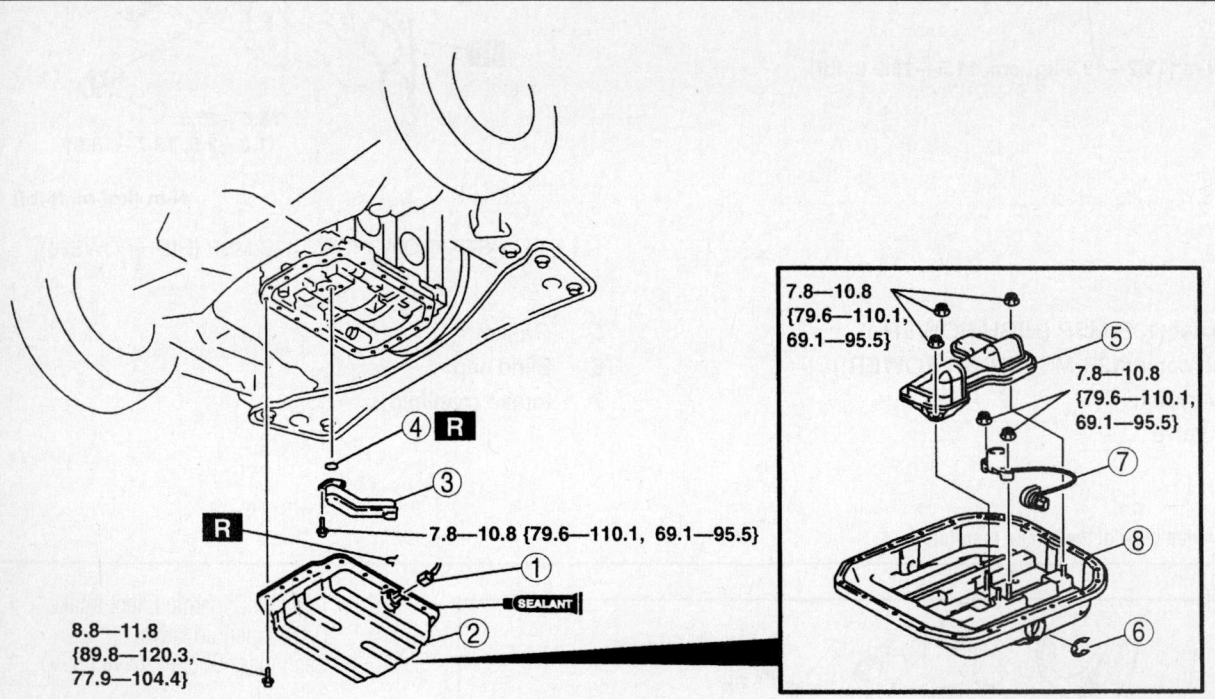

N·m {kgf·cm, in·lbf}

1	Connector	5	Oil baffle plate
2	Oil pan component	6	Clip
3	Oil strainer	7	Oil-level switch
4	O-ring	8	Oil pan

67162-RX8-G16

Fig. 84 Exploded view of the oil pan and related components

6. Install the oil pan and tighten the bolts to 78–104 inch lbs. (9–12 Nm).
- Electrical connector
- Negative battery cable
- Battery cover
- Engine cover

7. Fill the engine with clean oil.

8. Start the vehicle, check for leaks and repair if necessary.

9. On models with dynamic suspension, perform the steering angle sensor initialization procedure.

OIL PUMP

REMOVAL & INSTALLATION

See Figure 85.

1. Before servicing the vehicle, refer to the Precautions Section.

2. Drain the engine oil.

3. Remove or disconnect the following:
- Engine cover
- Battery cables
- Battery cover, battery box and tray
- Upper and lower extension manifolds
- Electrical connectors
- Oil pipe
- Gasket
- Oil pump and o-ring

To install:

4. Clean the oil, dirt and old sealant from all contact surfaces.

5. Install or connect the following:
- New O-rings on the oil pump
- Oil pump. Torque the bolts to 87–122 inch lbs. (10–14 Nm).
- Gasket
- Oil pipe
- Electrical connectors
- Upper and lower extension manifolds
- Battery cover, battery box and tray
- Battery cables
- Engine cover

6. Fill the engine with clean oil.

7. Start the vehicle, check for leaks and repair if necessary.

8. On models with dynamic suspension, perform the steering angle sensor initialization procedure.

REAR MAIN SEAL

REMOVAL & INSTALLATION

See Figure 86.

1. Before servicing the vehicle, refer to the Precautions Section.

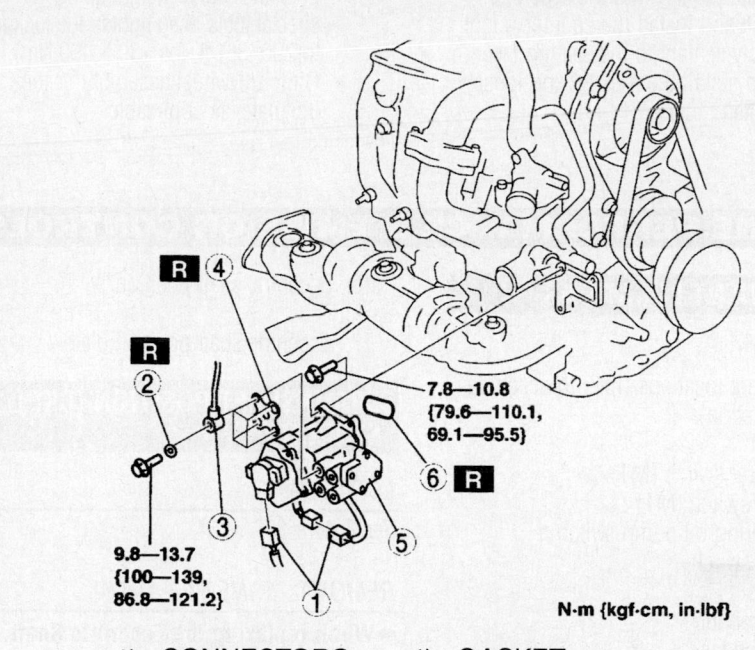

7.8—10.8
{79.6—110.1,
69.1—95.5}

9.8—13.7
{100—139,
86.8—121.2}

N·m {kgf·cm, in·lbf}

1) CONNECTORS   4) GASKET
2) WASHER   5) METERING OIL PUMP
3) OIL PIPE   6) O-RING

09482_RX-8_G0001

Fig. 85 Exploded view of the oil pump mounting

2. Remove or disconnect the following:
- Negative battery cable
- Transmission assembly
- Clutch/flywheel assembly, if equipped with a manual transaxle/transmission
- Flexplate/shim plates, if equipped with an automatic transaxle/transmission
- Using special tools 49 1881-055A and 49-0820-035, remove the counterweight locknut.
- Using special tools 49 1881-055A and 49-0839-305A, remove the counterweight

- Place a rag over the eccentric shaft and using a pry tool, carefully pry the oil seal from the oil seal housing.

3. Clean the gasket mounting surfaces.

To install:

4. Clean the oil seal housing. Coat the lip of the oil seal and the housing with clean engine oil.

5. Install or connect the following:
- New oil seal into the housing by tapping it evenly into place with a hammer and a seal installer until it is flush with the edge of the rear cover

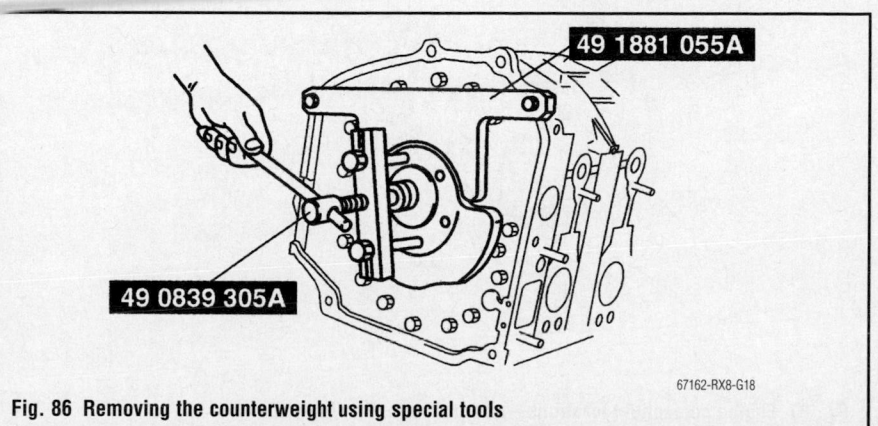

49 1881 055A

49 0839 305A

67162-RX8-G18

Fig. 86 Removing the counterweight using special tools

- Install the key into the eccentric shaft and install the counterweight
- Apply sealant to the seating face then install the locknut and loosely tighten.

- Lock the counterweight using the special tools, then tighten the locknut to 290–361 ft. lbs. (392–490 Nm).
- Clutch/flywheel assembly or the flexplate, as applicable

- Transaxle/transmission
- Negative battery cable

6. On models with dynamic suspension, perform the steering angle sensor initialization procedure.

ENGINE PERFORMANCE & EMISSION CONTROLS

COMPONENT LOCATIONS

See Figure 87.

Component locations (refer to accompanying illustration):
1. PCM
2. Neutral switch (MT)
3. CPP switch (MT)
4. Metering oil pump switch
5. SSV switch
6. ECT
7. IAT sensor
8. TP sensor
9. APP sensor
10. APV position sensor
11. MAF sensor
12. Front HO2S
13. Rear HO2S
14. BARO sensor
15. KS

16. Eccentric shaft position sensor
17. Eccentric shaft position plate

CRANKSHAFT POSITION (CKP) SENSOR

LOCATION
See Figure 88.

REMOVAL & INSTALLATION

➡**When replacing the Eccentric Shaft Position commonly known as the Crankshaft Position Sensor (CKP), make sure there is no foreign material on it such as metal shavings. If it is installed with foreign material, the sensor output signal will malfunction resulting from fluctuation in magnetic**

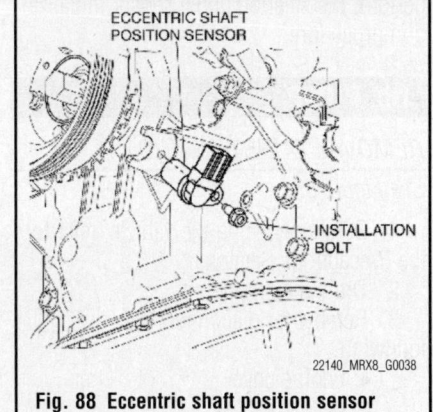

Fig. 88 Eccentric shaft position sensor

flux and cause a deterioration in engine control.

1. Disconnect the eccentric shaft position sensor connector.
2. Remove the eccentric shaft position sensor installation bolt and remove the eccentric shaft position sensor.
3. Install in the reverse order of removal.
4. Tighten the eccentric shaft position sensor bolt to 72 inch lbs. (8.3 Nm).

ENGINE COOLANT TEMPERATURE (ECT) SENSOR

REMOVAL & INSTALLATION
See Figure 89.

1. Remove the battery cover using the following procedure:

➡**When pulling the clips on the rear, do not apply excessive force to a clip using the screwdriver or pair of pliers.**

✳ WARNING

If excessive force is applied to the clips on the rear, they could be damaged. When removing the battery cover, use the following procedure to prevent damaging the clips.

  a. Pull up the clips on the rear and disengage the battery box tabs.
  b. Slightly move the battery cover to the vehicle front side and disengage the battery box tabs on the front.
  c. Remove the battery cover.

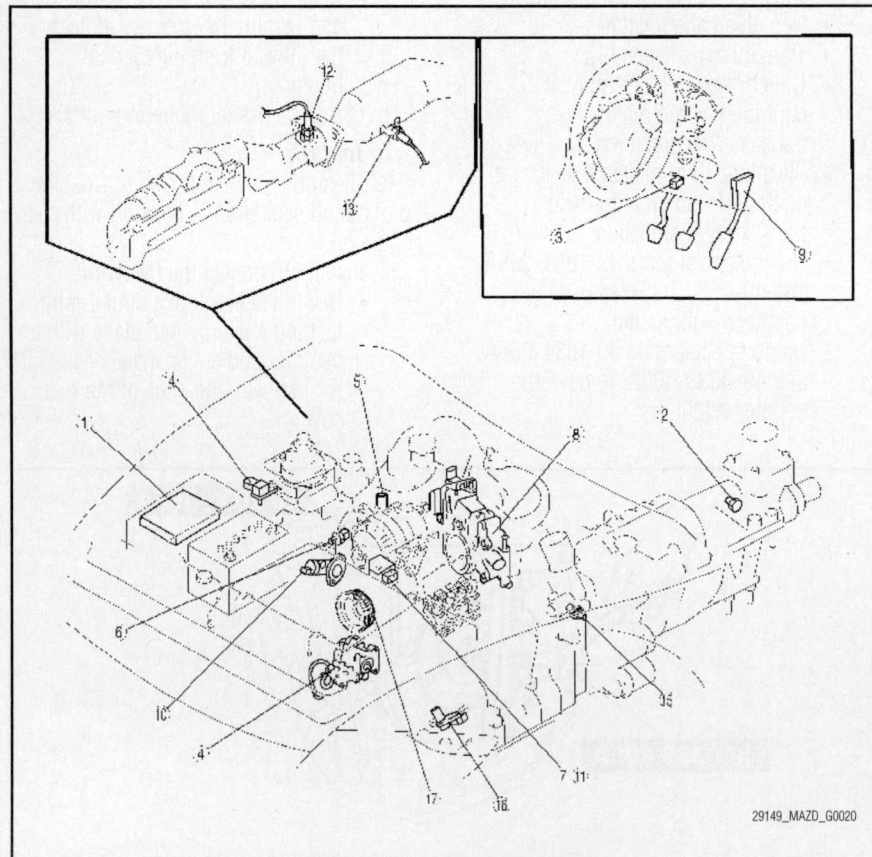

Fig. 87 Engine component locations—refer to the text for keylist

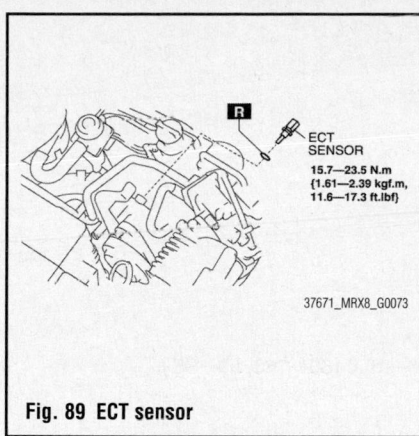

Fig. 89 ECT sensor

2. Disconnect the negative battery cable.

3. Remove the generator.

4. Drain the engine coolant.

5. Disconnect the ECT sensor connector.

6. Remove the ECT sensor.

To install:

7. Install the ECT sensor.

8. Connect the ECT sensor connector.

9. Fill the engine coolant.

10. Install the generator.

11. Connect the negative battery cable.

12. Install the battery cover.

13. Install the battery cover using the following procedure.

 a. Remove the battery cover.

 b. Slightly move the battery cover to the vehicle front side and disengage the battery box tabs on the front.

 c. Pull up the clips on the rear and disengage the battery box tabs.

HEATED OXYGEN SENSOR (HO2S)

REMOVAL & INSTALLATION

✳✳ CAUTION

The temperature of the exhaust system is extremely high after the engine has been run. To prevent personal injury, allow the exhaust system to cool completely before removing sensor from the exhaust system.

1. Disconnect the negative battery cable.

2. Raise and safely support the vehicle on jack stands.

3. Disconnect the HO2S from the engine control sensor wiring.

➡**If excessive force is needed to remove the sensors lubricate the sensor with penetrating oil prior to removal.**

4. Remove the sensors with a sensor removal tool, such as Ford Tool T94P-9472-A or equivalent.

To install:

5. Install the sensor in the mounting boss, and then tighten it to 27–33 ft. lbs.

6. Connect the sensor electrical wiring connector to the engine wiring harness.

7. Lower the vehicle.

8. Connect the negative battery cable.

INTAKE AIR TEMPERATURE (IAT) SENSOR

REMOVAL & INSTALLATION

1. Disconnect the negative battery cable.

2. Remove the IAT sensor from the air cleaner housing or intake air pipe.

To install:

3. Install a new sealing washer to the thermo sensor.

KNOCK SENSOR (KS)

LOCATION

The Knock Sensor (KS) is located on the rear of the engine block.

REMOVAL & INSTALLATION

See Figure 90.

1. Disconnect the negative battery cable.

2. Disconnect the sensor connector.

3. Remove the installation bolt and sensor from the engine block.

To install:

4. Installation is the reverse of the removal procedure.

5. Tightening torque 12–18 ft. lbs. (15–25 Nm).

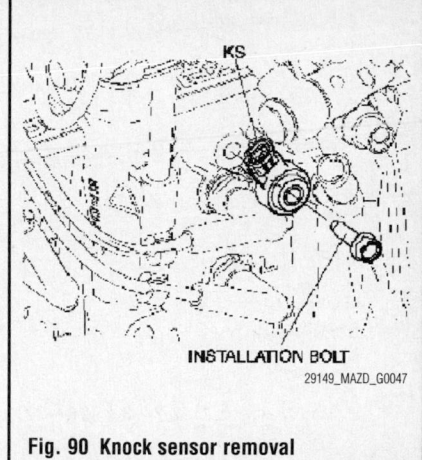

Fig. 90 Knock sensor removal

MASS AIR FLOW (MAF) SENSOR

REMOVAL & INSTALLATION

See Figure 91.

1. Remove the air cleaner outlet tube.

2. Disconnect the MAF sensor electrical connector.

3. Remove the attaching four nuts and the MAF sensor.

4. Installation is the reverse of the removal procedure.

5. Tighten to 6.5 inch lbs. (.70 Nm).

POWERTRAIN CONTROL MODULE (PCM)

REMOVAL & INSTALLATION

See Figure 92.

1. Remove the battery cover.

2. Remove the PCM cover.

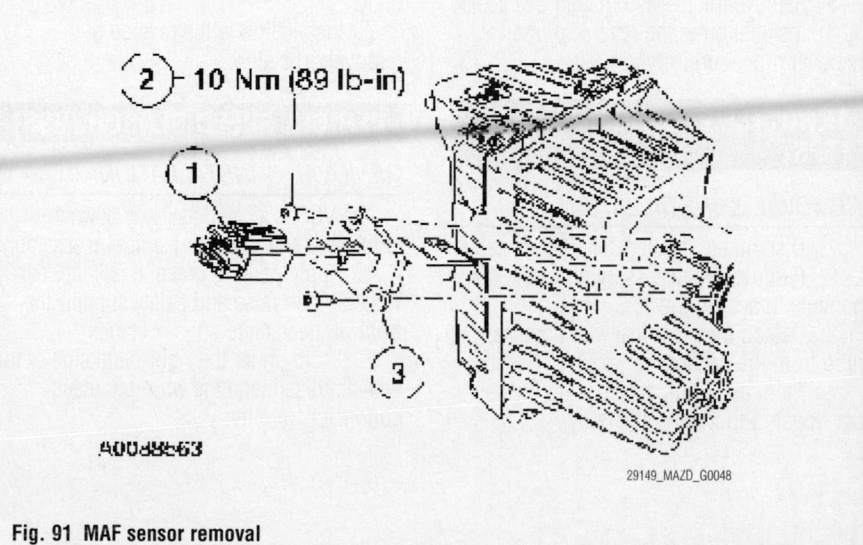

Fig. 91 MAF sensor removal

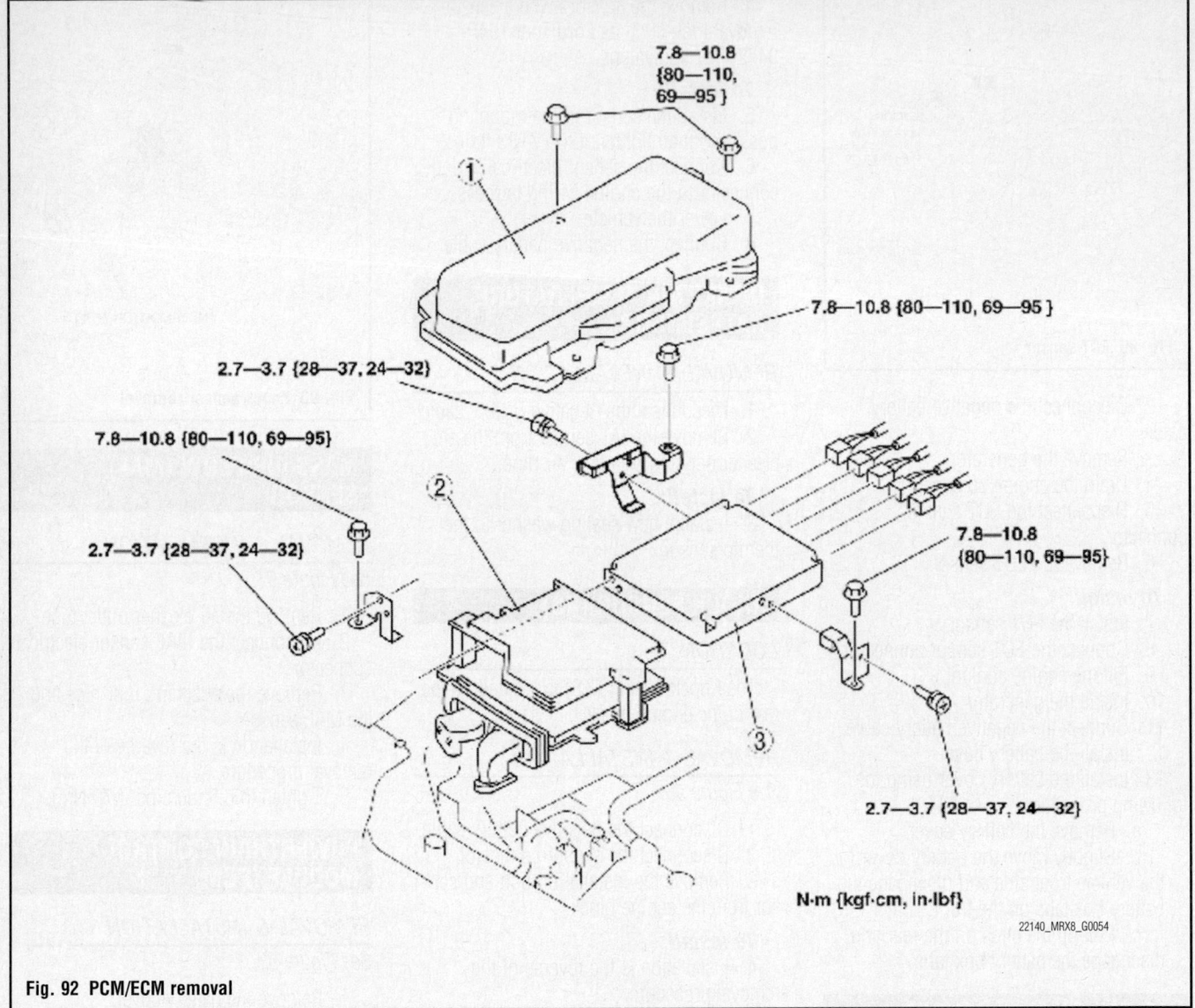

7.8—10.8
{80—110,
69—95}

7.8—10.8 {80—110, 69—95}

2.7—3.7 {28—37, 24—32}

7.8—10.8 {80—110, 69—95}

2.7—3.7 {28—37, 24—32}

7.8—10.8
{80—110, 69—95}

2.7—3.7 {28—37, 24—32}

N·m {kgf·cm, in·lbf}

22140_MRX8_G0054

Fig. 92 PCM/ECM removal

3. Remove the PCM cooler.
4. Remove the PCM harness connectors.
5. Remove the PCM from the PCM cooler.
6. Installation is the reverse of the removal procedure.

THROTTLE POSITION SENSOR (TPS)

REMOVAL & INSTALLATION

1. Disconnect the negative battery cable.
2. Remove the necessary air intake components to access the TPS.
3. Detach the electrical wire harness plug from the sensor.
4. Paint an alignment mark on the sensor housing to the throttle body.

5. Remove the sensor attaching bolts.
6. Remove the sensor from the throttle body.
7. Installation is the reverse of the removal procedure.

VEHICLE SPEED SENSOR (VSS)

REMOVAL & INSTALLATION

The VSS is located half-way down the right-hand side of the transmission assembly.
1. Apply parking brake, block the rear wheels, then raise and safely support the front of the vehicle on jack stands.
2. From under the right-hand side of the vehicle, disengage the wiring harness connector from the VSS.

3. Loosen the VSS hold-down bolt, then pull the VSS out of the transmission housing.

To install:

4. If a new sensor is being installed, transfer the driven gear retainer and gear to the new sensor.
5. Ensure that the O-ring is properly seated in the VSS housing.
6. For ease of assembly, engage the wiring harness connector to the VSS, then insert the VSS into the transmission assembly.
7. Install and tighten the VSS hold-down bolt to 62–88 in lbs.
8. Lower the vehicle and remove the wheel blocks.

FUEL   GASOLINE FUEL INJECTION SYSTEM

FUEL SYSTEM SERVICE PRECAUTIONS

Safety is the most important factor when performing not only fuel system maintenance but any type of maintenance. Failure to conduct maintenance and repairs in a safe manner may result in serious personal injury or death. Maintenance and testing of the vehicle's fuel system components can be accomplished safely and effectively by adhering to the following rules and guidelines.

• To avoid the possibility of fire and personal injury, always disconnect the negative battery cable unless the repair or test procedure requires that battery voltage be applied.

• Always relieve the fuel system pressure prior to disconnecting any fuel system component (injector, fuel rail, pressure regulator, etc.), fitting or fuel line connection. Exercise extreme caution whenever relieving fuel system pressure to avoid exposing skin, face and eyes to fuel spray. Please be advised that fuel under pressure may penetrate the skin or any part of the body that it contacts.

• Always place a shop towel or cloth around the fitting or connection prior to loosening to absorb any excess fuel due to spillage. Ensure that all fuel spillage (should it occur) is quickly removed from engine surfaces. Ensure that all fuel soaked cloths or towels are deposited into a suitable waste container.

• Always keep a dry chemical (Class B) fire extinguisher near the work area.

• Do not allow fuel spray or fuel vapors to come into contact with a spark or open flame.

• Always use a back-up wrench when loosening and tightening fuel line connection fittings. This will prevent unnecessary stress and torsion to fuel line piping.

• Always replace worn fuel fitting O-rings with new Do not substitute fuel hose or equivalent where fuel pipe is installed.

Before servicing the vehicle, make sure to also refer to the precautions in the beginning of this section as well.

RELIEVING FUEL SYSTEM PRESSURE

See Figure 93.

> ⁂ **WARNING**
>
> **Fuel is extremely flammable. Always keep sparks and flame away from fuel. Ignition may cause death or serious injury, or damage to equipment.**

> ⁂ **WARNING**
>
> **Fuel line spills and leakage from the pressurized fuel system are dangerous. Fuel can ignite and cause serious injury or death and damage. Fuel can also irritate skin and eyes. To prevent this, always complete the "Fuel Line Safety Procedure".**

> ⁂ **WARNING**
>
> **If there is foreign material on the connecting area of the quick release connector, it might damage the connector or fuel pipe. To prevent this, when the quick release connector has been disconnected, clean the connecting area before reconnecting it.**

1. Before servicing the vehicle, refer to the Precautions Section.
2. Remove the filler cap.
3. Remove the fuel pump relay from the relay box, located in the main fuse block.
4. Start the engine.
5. After the engine stalls, turn the ignition switch **OFF**.
6. After servicing the vehicle, reinstall the relay.

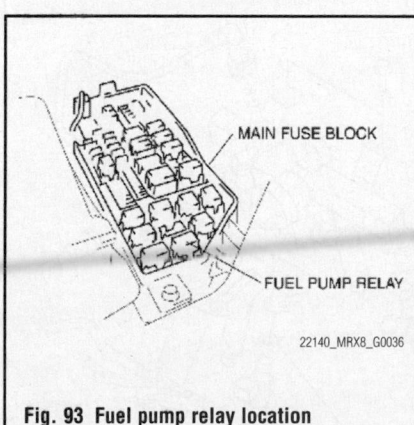

Fig. 93 Fuel pump relay location

MAIN FUSE BLOCK

FUEL PUMP RELAY

22140_MRX8_G0036

FUEL FILTER

REMOVAL & INSTALLATION

The fuel filter is part of the pump assembly, see the removal and installation procedure in this section.

FUEL PUMP

REMOVAL & INSTALLATION

See Figure 94.

1. Before servicing the vehicle, refer to the Precautions Section.
2. Relieve the fuel system pressure.
3. Remove or disconnect the following:
   • Negative battery cable
   • Rear seat cushion
4. Drain the fuel from the tank.
   • Service hole cover
   • Fuel pump electrical connector
   • All fuel hoses from the fuel pump unit
   • Fuel pump ring using tool 49 T042 001
   • Fuel pump and gaskets from the fuel tank

To install:

5. Align the fuel pump alignment mark with the notch in the retainer and install the fuel pump using a new gasket.
6. Align the cap with the retainer and tighten one full turn by hand.
7. Tighten the cap with the special tool to 75 ft. lbs. (102 Nm).
8. Install or connect the following:
   • Fuel hoses to the fuel pump
   • Fuel pump electrical connector
   • Service hole cover
   • Rear seat cushion
   • Negative battery cable
9. Add a minimum of 10 gallons of fuel to the tank and check for leaks.
10. On models with dynamic suspension, perform the steering angle sensor initialization procedure.

FUEL RAIL & INJECTOR

REMOVAL & INSTALLATION

See Figure 95.

> ⁂ **CAUTION**
>
> **Fuel injection systems remain under pressure after the engine has been turned OFF. Properly relieve fuel pressure before disconnecting any fuel lines. Failure to do so may result in fire or personal injury. Do not allow fuel spray or fuel vapors to come in contact with a spark or open flame. Keep a dry chemical fire extinguisher nearby. Never store fuel in an open container due to risk of fire or explosion.**

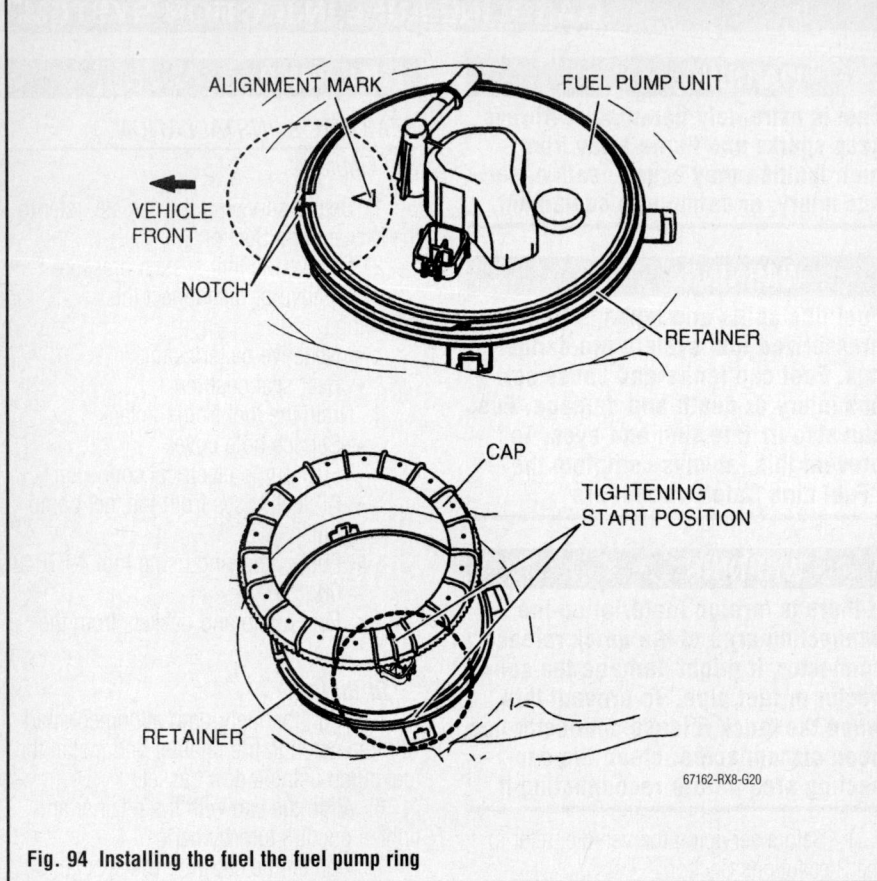

Fig. 94 Installing the fuel the fuel pump ring

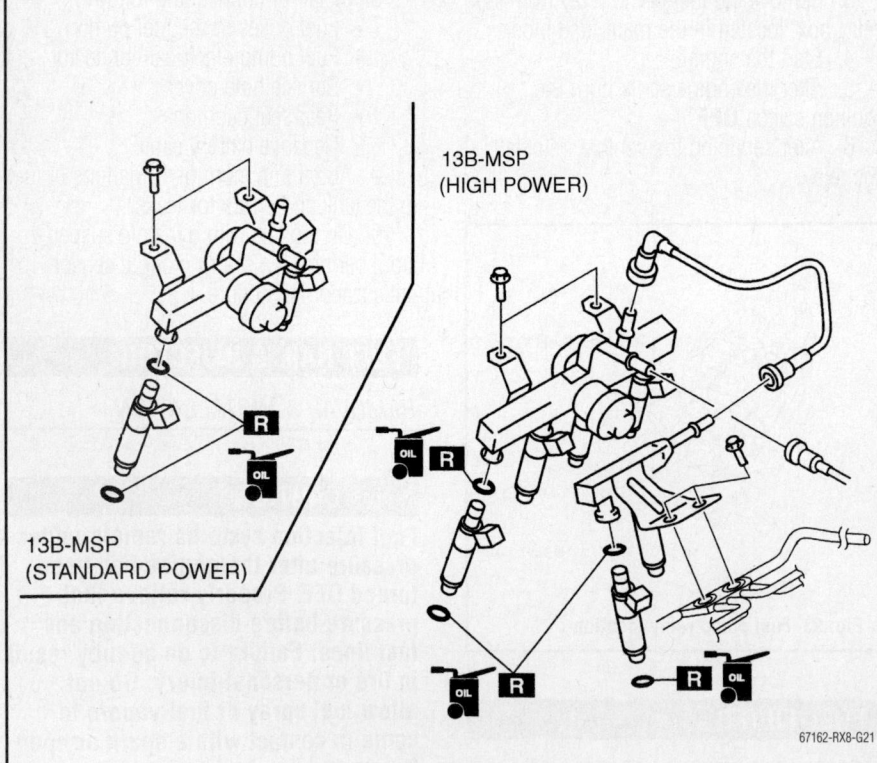

Fig. 95 Exploded view of the fuel rail and Injector assembly

1. Before servicing the vehicle, refer to the Precautions Section.
2. Relieve the fuel system pressure.
3. Remove or disconnect the following:
   - Negative battery cable
   - Upper and lower extension manifolds
   - Variable Dynamic Effect (VDI) actuator and position out of the way
   - Fuel injector wiring harness
   - Fuel lines at the fuel rail
   - Fuel distributor from intake manifold and housing sides
   - Fuel rail with the injectors attached
   - Fuel injectors, grommets and O-rings from the fuel rail
   - O-rings from the fuel injectors

To install:

4. Install or connect the following:
   - New O-rings and grommets lubricated with engine oil on the fuel injectors.
   - Insulators and injectors on the intake manifold
   - Grommets and the fuel rail onto the injectors. Torque the bolts to 14–18 ft. lbs. (19–25 Nm).
   - Fuel lines to the fuel rail
   - Fuel distributors
   - Fuel injector wiring harness
   - VDI actuator
   - Negative battery cable

5. Turn the ignition switch **ON** to pressurize the fuel system.
6. Check for leaks and correct as necessary, before starting the engine.
7. On models with dynamic suspension, perform the steering angle sensor initialization procedure.

FUEL TANK

REMOVAL & INSTALLATION

See Figures 96 through 99.

✳✳ CAUTION

Fuel tank removal, installation and repair containing fuel is dangerous. Explosion or fire may cause death or serious injury. Always properly steam clean a fuel tank before repairing it.

1. Park the vehicle on a level surface.
2. Relieve the fuel system pressure, as outlined in this section.
3. Remove the rear seat.
4. Drain fuel from the fuel tank, as follows:

✳✳ CAUTION

A person charged with static electricity could cause a fire or explosion,

resulting in death or serious injury. Before draining fuel, make sure to discharge static electricity by touching a vehicle.

❋ WARNING

When the fuel gauge is at a level of ¾ full or more, the fuel level is higher than the installation surface of the fuel pump and the fuel suction pipe bracket. Due to this condition, fuel may spill or leak out when performing this procedure. Before performing this procedure, always drain out fuel so that the fuel tank is half full or less (according to the fuel gauge needle).

   a. Disconnect the quick release connector (engine compartment side
   b. Attach a long hose to the disconnected fuel pipe and drain the fuel into a proper receptacle.
   c. Ground check connector terminal F/P to the body using a jumper wire

❋ WARNING

Shorting the wrong terminal of the check connector may cause malfunctions. Make sure to short only the specified terminal.

   d. Turn the ignition switch to the **ON** position and operate the fuel pump for about. 20 min.

❋❋ WARNING

The fuel pump may malfunction if it is operated without any fuel in the fuel tank (fuel pump idling). Constantly monitor the amount of fuel being discharged and immediately stop operation of the pump when essentially no fuel is being discharged.

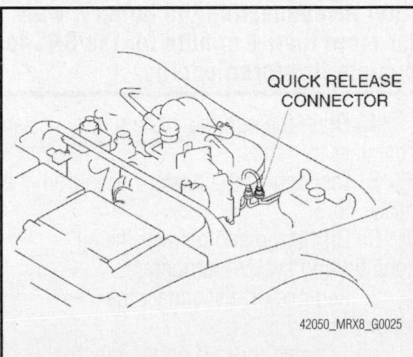

Fig. 96 Detach the quick release connector

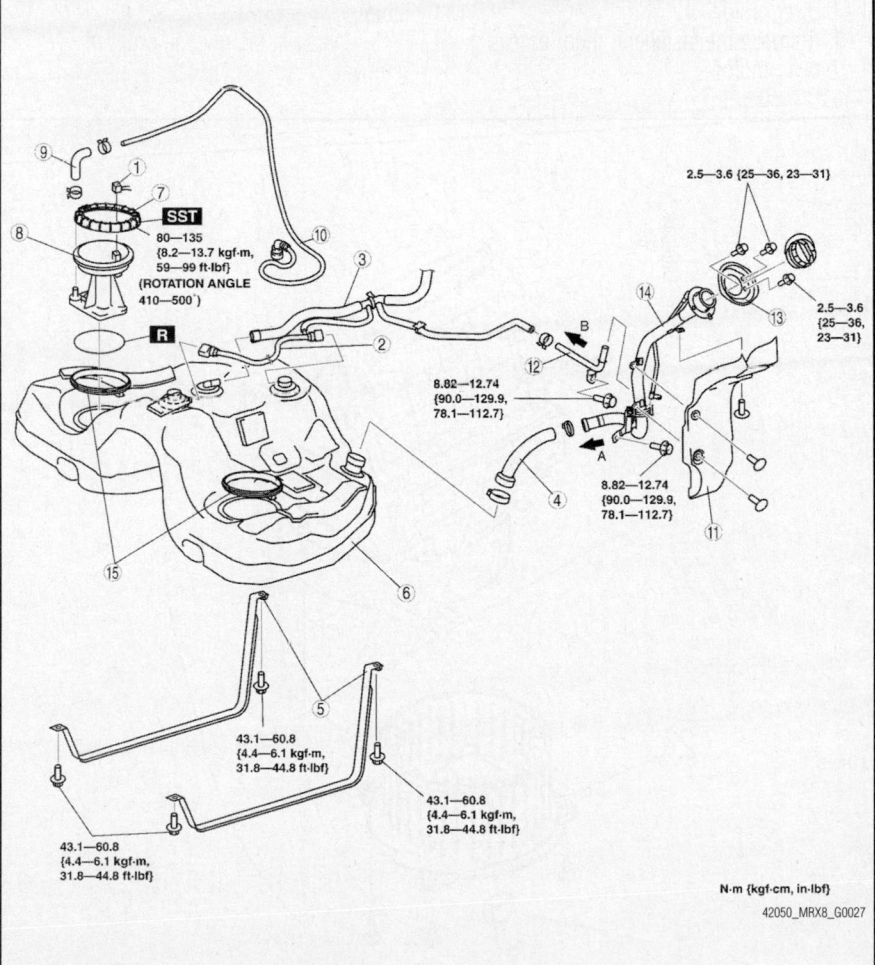

Fig. 97 Ground check connector terminal F/P to the body using a jumper wire

   e. When essentially no fuel is being discharged from the hose, turn the ignition switch to the **LOCK** position.

➡ When operating the fuel pump with a full fuel tank, fuel discharge will become erratic after approx. 10 min but will continue for approx. 10 min more and then essentially no fuel will

be discharged. At this time, the fuel gauge needle will be at the halfway position.

   f. Disconnect the jumper wire.
   g. Disconnect the negative battery cable.
5. Remove the following components:
   a. Fuel pump unit.
   b. Main silencer and middle pipe.
   c. Power plant frame.
   d. Propeller shaft.
   e. Position the parking brake cable out of the way.
6. Remove the fuel tank and components as shown in the accompanying illustration.

To install:

7. Installation is the reverse of the removal procedure.
8. After installation is complete, inspect all parts by performing the following Fuel Line Inspection procedure:

Fig. 98 Exploded view of the fuel tank and related components—Note that the removal order corresponds with the key list number

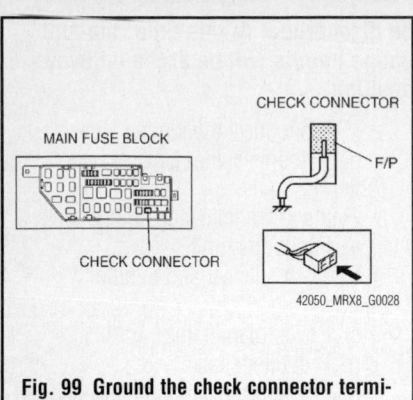

MAIN FUSE BLOCK

CHECK CONNECTOR

CHECK CONNECTOR

F/P

42050_MRX8_G0028

Fig. 99 Ground the check connector terminal F/P to the body using a jumper wire

✳✳ CAUTION

Fuel vapor is hazardous. It can very easily ignite, causing serious injury and damage. Fuel can also irritate skin and eyes. To prevent this, always complete the "Fuel Line Inspection".

✳✳ WARNING

Shorting the wrong terminal of the check connector may cause malfunctions. Make sure to short only the specified terminal.

a. Ground the check connector terminal F/P to the body using a jumper wire.

b. Turn the ignition switch to the **ON** position and operate the fuel pump.

c. Check that that there is no fuel leakage from the pressurized parts.
- If there is leakage, replace the fuel hoses.
- If there is damage to the seal on the fuel pipe side, replace the fuel pipe.
- There should not be any fuel leakage after 5 minutes.

9. After reinstallation, repeat Step a.–c. of the fuel leakage inspection.

HEATING & AIR CONDITIONING COMPONENTS

BLOWER MOTOR

REMOVAL & INSTALLATION

See Figure 100.

1. Disconnect the negative battery cable.
2. Detach the electrical connector from the blower motor.
3. Remove the fasteners, then remove the blower motor.

CABIN AIR FILTER

REMOVAL & INSTALLATION

See Figure 101.

1. Remove the glove compartment.
2. Remove in the order indicated in the table.
3. Install in the reverse order of removal.

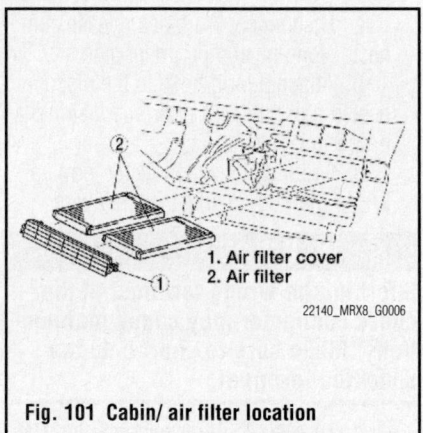

1. Air filter cover
2. Air filter

22140_MRX8_G0006

Fig. 101 Cabin/ air filter location

HEATER CORE

REMOVAL & INSTALLATION

See Figures 102 and 101.

1. Before servicing the vehicle, refer to the Precautions Section.
2. Place the ignition switch in the **LOCK** position.
3. Disconnect the negative battery cable.

✳✳ CAUTION

After disconnecting the battery, wait for more than 1 minute for the SAS to deplete its stored energy.

4. Drain the cooling system into a clean container for reuse.
5. Disconnect the heater hoses from the heater core.
6. Discharge and recover the air conditioning system refrigerant.
7. Remove or disconnect the following:
- Center console upper panel
- Ash tray panel
- Cigar lighter connector

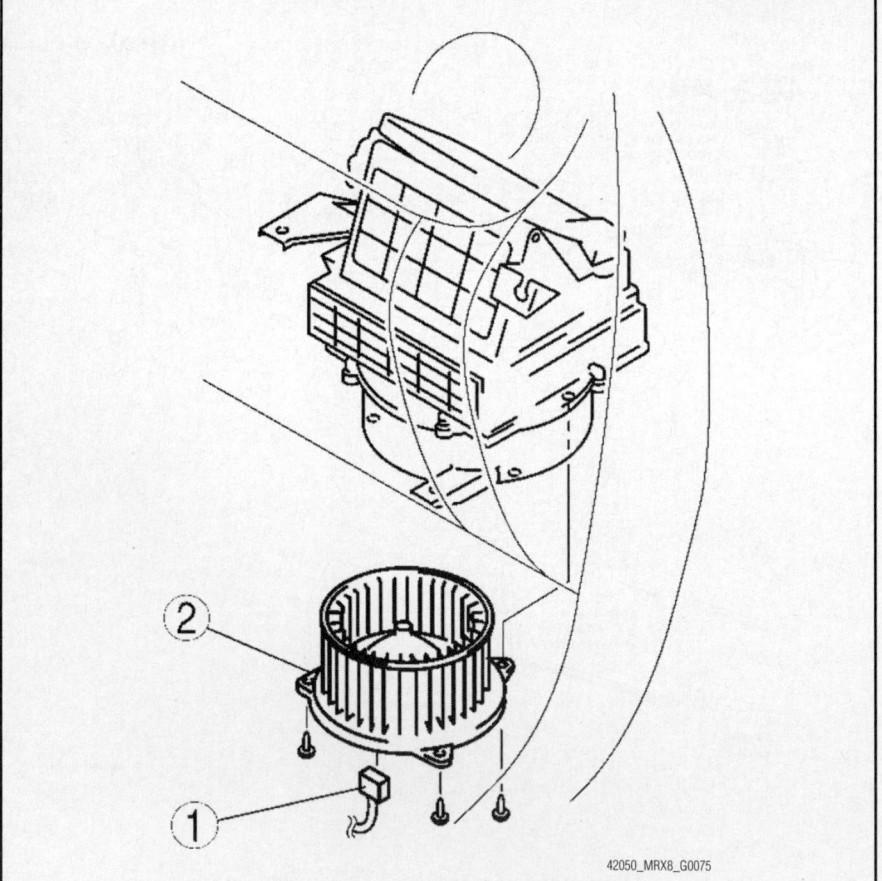

42050_MRX8_G0075

Fig. 100 Exploded view of the blower motor connector (1) and blower motor (2)

- Ash tray light
- Storage compartment
- Front and rear center consoles
- Console under cover
- Glove box
- Lower scuff plate
- Lower door side trim plate
- Lower dashboard side panel
- Lower dashboard front panel
- Steering column upper cover
- Ignition key light
- Steering column lower cover

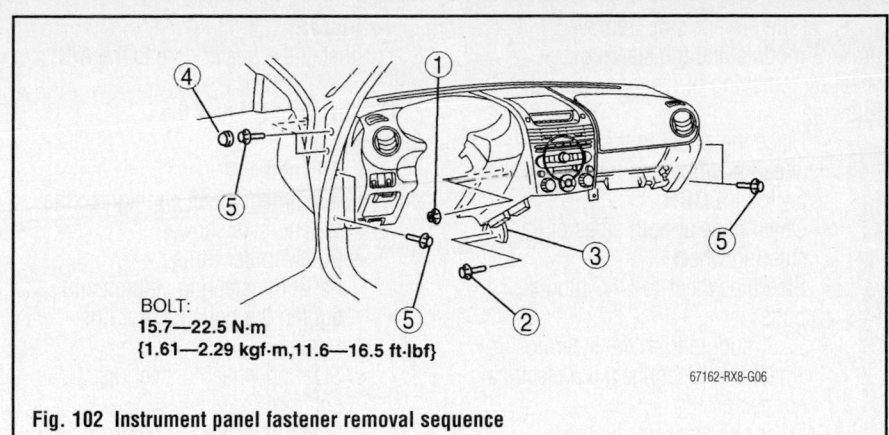

BOLT:
15.7—22.5 N·m
{1.61—2.29 kgf·m, 11.6—16.5 ft·lbf}

67162-RX8-G06

Fig. 102 Instrument panel fastener removal sequence

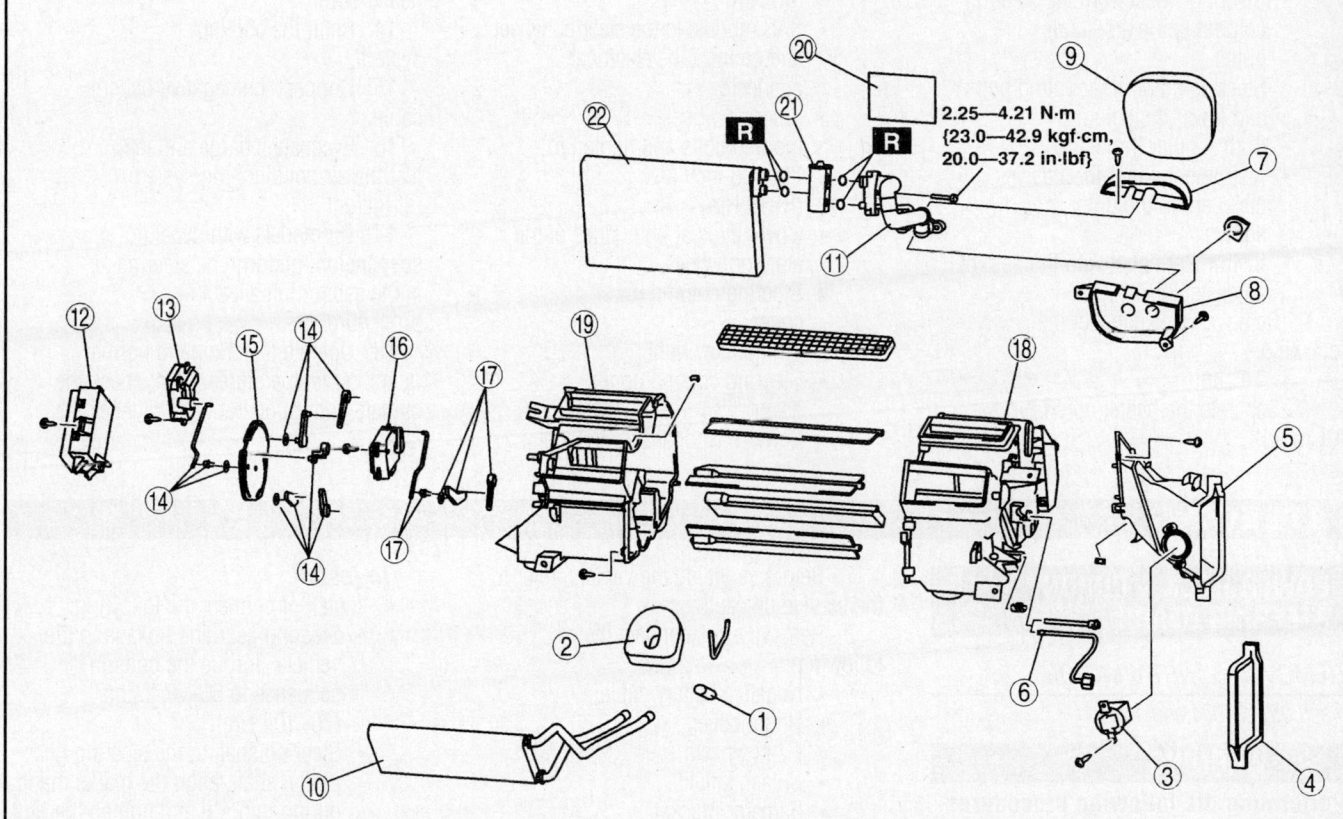

2.25—4.21 N·m
{23.0—42.9 kgf·cm, 20.0—37.2 in·lbf}

1	Drain hose	12	A/C amplifier
2	Polyurethane foam (1)	13	Airflow mode actuator
3	Resistor	14	Airflow mode link set
4	Adhesive polyurethane (1)	15	Airflow mode main link
5	Air duct	16	Air mix actuator
6	Evaporator temperature sensor	17	Air mix link set
7	Polyurethane foam (2)	18	A/C case (1)
8	Bracket (1)	19	A/C case (2)
9	Bracket (2)	20	Adhesive polyurethane (2)
10	Heater core	21	Expansion valve
11	Evaporator pipe	22	Evaporator

67162-RX8-G07

Fig. 103 Exploded view of the A/C unit with heater core

8. At the driver's side, remove the SAS module and the steering wheel by removing or disconnecting the following:

- Place the wheel in the straight-ahead position and turn the ignition switch to **LOCK** .
- Cover clips at both sides of the steering wheel
- Steering wheel-to-SAS module bolts
- SAS module from the steering wheel and disconnect the electrical connector
- Steering wheel-to-column nut
- Steering wheel from the steering column using a suitable puller
- Steering column mounting bolts and lower the column
- Both A pillar trims
- Instrument panel-to-chassis fasteners in the order shown
- Instrument panel with the help of an assistant

9. Remove or disconnect the following:

- A/C unit

10. Separate the heater core from the A/C unit.

To install:

11. Install the heater core to the A/C unit.

12. Install or connect the following:

- A/C unit
- Instrument panel and tighten the fasteners as shown
- Both A pillar trims
- Raise the steering column and tighten the bolts to 14 ft. lbs. (19 Nm)
- Steering wheel to steering column
- Steering wheel-to-column nut and tighten to 33 ft. lbs. (45 Nm)
- SAS module to the steering wheel and connect the electrical connector
- Steering wheel-to-SAS module bolts and tighten to 70–103 inch lbs. (8–12 Nm)
- Cover clips at both sides of the steering wheel
- Steering column lower cover
- Ignition key light
- Steering column upper cover
- Lower dashboard front panel
- Lower dashboard side panel
- Lower door side trim plate
- Lower scuff plate
- Glove box
- Console under cover
- Front and rear center consoles
- Storage compartment
- Ash tray light
- Cigar lighter connector
- Ash tray panel
- Center console upper panel

13. Connect the heater hoses to the heater core.

14. Refill the cooling system.

15. Connect the negative battery cable.

16. Evacuate, charge and leak test the air conditioning system refrigerant.

17. On models with dynamic suspension, perform the steering angle sensor initialization procedure.

18. Operate the engine to normal operating temperatures; then, check the climate control operation and check for leaks.

STEERING

POWER RACK & PINION STEERING GEAR

REMOVAL & INSTALLATION
See Figures 104 and 105.

❋❋ WARNING

Performing the following procedures without first removing the ABS wheel-speed sensor may possibly cause an open circuit in the wiring harness if it is pulled by mistake. Before performing the following procedures, remove the ABS wheel-speed sensor harness (axle side) and fix it to an appropriate place where the harness will not be pulled by mistake while servicing the vehicle.

❋❋ WARNING

After replacing the steering gear and linkage, always set the EPS system to the neutral position to prevent system malfunction

1. Before servicing the vehicle, refer to the service precautions.

2. Remove or disconnect the following:

- Negative battery cable
- Under cover.
- Front wheels
- Splash shield
- Radiator bracket
- Cotter pins and nuts from both steering tie rod ends

3. Press the tie rod out from the knuckle arm.

- Intermediate shaft to steering gear pinion shaft bolt. Mark the shaft-to-gear location.
- Shaft from the steering gear
- Torque sensor connector
- Electronic Power Steering (ESP) motor connector
- Steering gear mounting nuts
- Steering gear and linkage from the vehicle

To install:

4. Install or connect the following:

- Steering gear and linkage to the vehicle. Torque the bolts in sequence to 55–77 ft. lbs. (75–104 Nm).
- Steering shaft to the steering gear pinion shaft, align the marks made during removal and tighten the bolt to 19 ft. lbs. (26 Nm)
- Tie rod ends to the knuckle arm. Torque the nuts to 27–36 ft. lbs. (37–49 Nm).
- Electronic Power Steering (ESP) motor connector
- Torque sensor connector
- New cotter pins
- Radiator bracket
- Splash shield
- Wheels
- Negative battery cable

5. Check and/or adjust the front end alignment.

6. On models with dynamic suspension, perform the steering angle sensor initialization procedure.

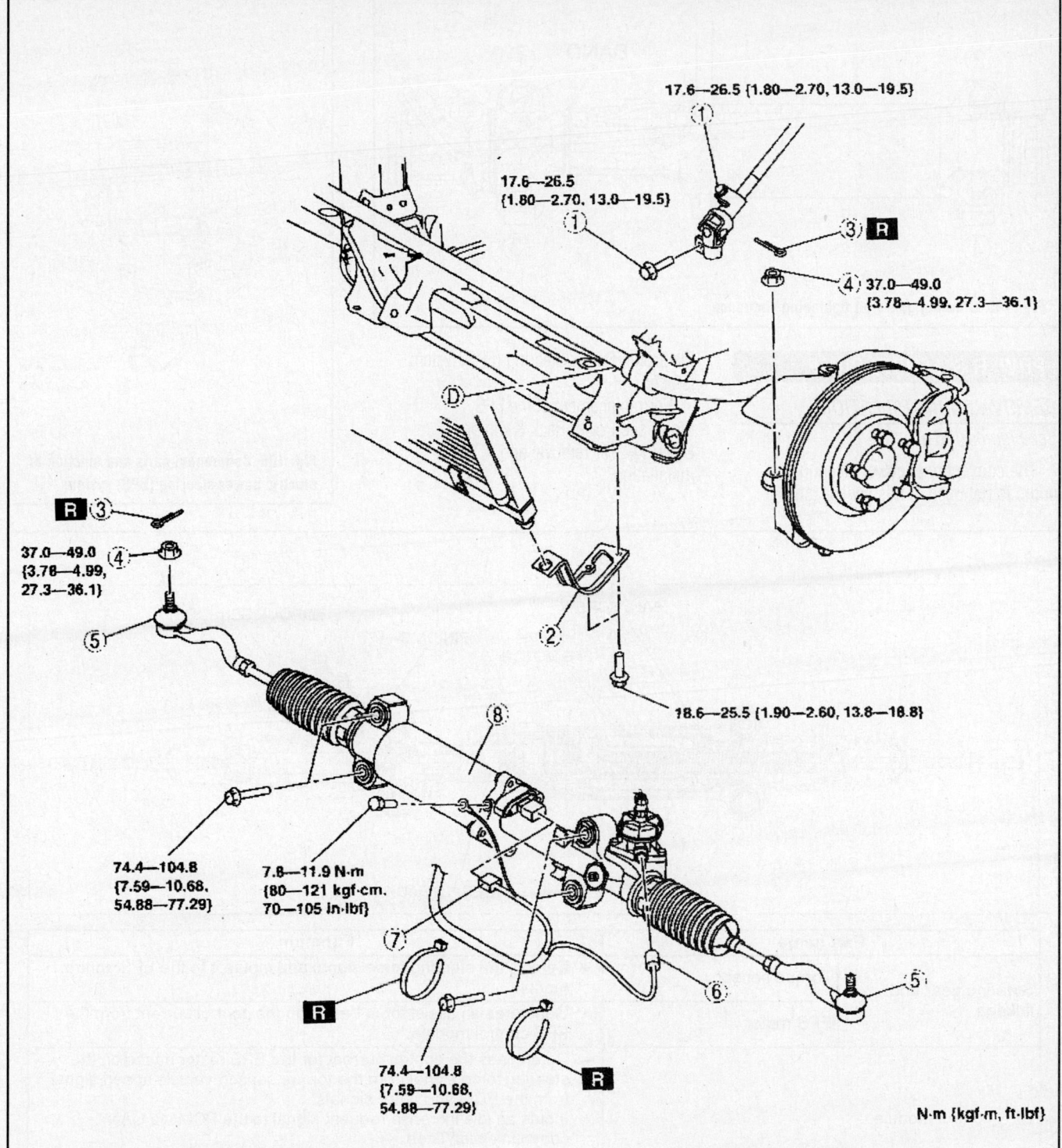

17.6—26.5 {1.80—2.70, 13.0—19.5}

17.6—26.5 {1.80—2.70, 13.0—19.5}

37.0—49.0 {3.78—4.99, 27.3—36.1}

37.0—49.0 {3.78—4.99, 27.3—36.1}

18.6—25.5 {1.90—2.60, 13.8—18.8}

74.4—104.8 {7.59—10.68, 54.88—77.29}

7.8—11.9 N·m {80—121 kgf·cm, 70—105 in·lbf}

74.4—104.8 {7.59—10.68, 54.88—77.29}

N·m {kgf·m, ft·lbf}

1	Bolt (intermediate shaft)	5	Tie-rod end
2	Radiator bracket	6	Torque sensor connector
3	Cotter pin	7	EPS motor connector
4	Locknut (tie-rod end)	8	Steering gear and linkage

67162-RX8-G29

Fig. 104 Exploded view of steering gear mounting

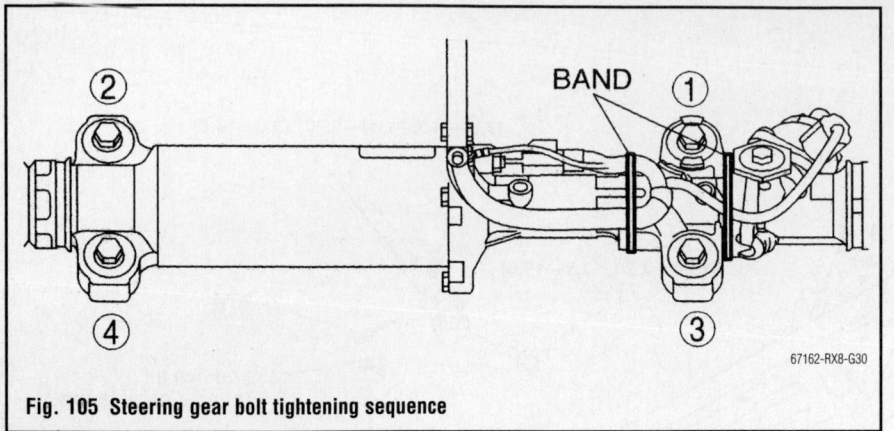

Fig. 105 Steering gear bolt tightening sequence

POWER STEERING PUMP

REMOVAL & INSTALLATION

See Figures 106 and 107.

The conventional power steering pump is not used in this vehicle, an Electronic Power Steering (EPS)system is used.

For replacement of the EPS motor see Power Rack & Pinion Steering Gear removal and installation

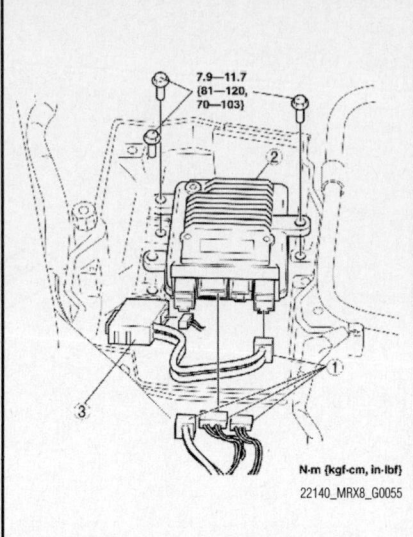

Fig. 106 Component parts and function of electric power steering (EPS) system

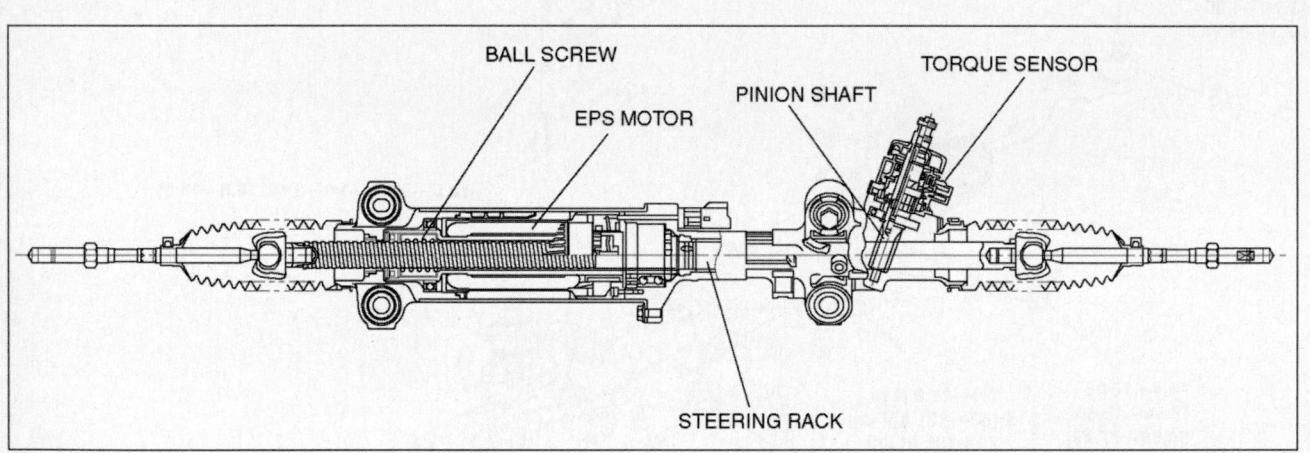

Part name		Function
Steering gear and linkage	Torque sensor	• Detects the steering force signal and inputs it to the EPS control module.
	EPS motor	• Generates an assist force based on the control current from the EPS control module.
EPS control module		• Determines the control current for the EPS motor based on the steering force signal from the torque sensor, vehicle speed signal from the PCM and other signals. • Inputs an idle increase request signal to the PCM via CAN communication lines. • Controls the on-board diagnostic system and fail-safe function when an abnormality is detected in the EPS system.
PCM	Vehicle speed signal	• Inputs the vehicle speed signal to the EPS control module via CAN communication lines.
	Engine speed signal	• Inputs the engine speed signal to the EPS control module via CAN communication lines.
Instrument cluster	EPS warning light	• The light illuminates to inform the driver when a system malfunction is detected.

22140_MRX8_G0056

Fig. 107 Electronic Power Steering (EPS) control module

COIL SPRING

REMOVAL & INSTALLATION

See Figure 108.

1. Before servicing the vehicle, refer to the Precautions Section.

2. Remove or disconnect the following:
   - Shock absorber/coil spring
   - Place the shock/coil spring in a spring compressor and compress the spring
   - Piston rod nut
   - Upper retainer, bushing spring seat and rubber insulator
   - Bushing, spacer, bound stopper and casing
   - Coil spring

3. While pushing on the piston rod, be sure that the pull stroke is even and that there is no unusual noise or resistance. Also, inspect for any oil leakage around the piston rod.

4. Push the piston rod in, then release it. Be sure that the return rate is constant.

5. If the shock absorber does not operate as described, replace it.

To install:

6. Install or connect the following:
   - Strut assembly into a vise
   - Bound stopper and casing onto the piston rod
   - Temporarily install the lower spring seat, seat rubber and spring. Mark the seat, shock and spring assembly as illustrated for reassembly. Align the marks of the upper seat and coil spring. Protect the assembly with cloth and install the spring compressor.
   - Coil spring

7. Compress the coil spring with the spring compressor

8. Install or connect the following:
   - Bushing, spacer, bound stopper and casing
   - Upper retainer, bushing spring seat and rubber insulator
   - Piston rod upper nut

9. Be sure that the spring upper seat notched portion is facing inward and tighten the piston rod upper nut to 23–34 ft. lbs. (32–46 Nm).

10. Be sure that the spring is well seated in the upper seats.

11. Install the shock to the vehicle.

LOWER BALL JOINT

REMOVAL & INSTALLATION

1. Before servicing the vehicle, refer to the Precautions Section.

2. Remove or disconnect the following:
   - Wheel
   - Ball joint clip
   - Ball joint using a ball joint remover

To install:

3. Install or connect the following:
   - Ball joint to lower control arm using a ball joint installer
   - Ball joint clip
   - Wheel

4. Check and/or adjust the front wheel alignment.

LOWER CONTROL ARMS

REMOVAL & INSTALLATION

Performing the following procedures without first removing the ABS wheel-speed sensor may possibly cause an open circuit in the wiring harness if it is pulled by mistake. Before performing the following procedures, disconnect the ABS wheel-speed sensor harness connector (axle side) and fix it to an appropriate place where the sensor will not be pulled by mistake while servicing the vehicle.

1. Before servicing the vehicle, refer to the Precautions Section.

2. Remove the strut cross brace in the engine compartment.

3. Remove or disconnect the following:
   - Front wheel
   - Brake caliper and wire aside
   - Brake rotor
   - Tie rod end
   - Lower ball joint
   - Upper ball joint
   - Stabilizer link from the lower control arm
   - Lower control arm bolts and nuts
   - Lower control arm

To install:

4. Install or connect the following:
   - Lower control arm
   - Lower control arm bolt and nut and tighten to 62–72 ft. lbs. (84–98 Nm)
   - Stabilizer link to the lower control arm and tighten to 32–45 ft. lbs. (43–61 Nm)
   - Upper ball joint
   - Lower ball joint
   - Tie rod end

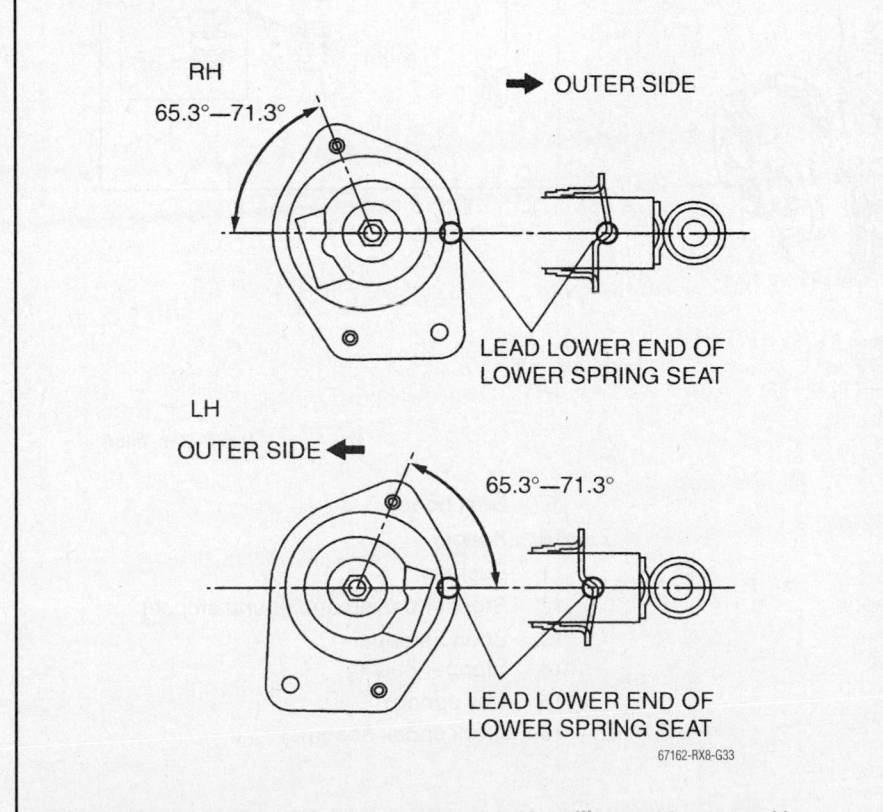

RH
65.3°—71.3°

OUTER SIDE

LEAD LOWER END OF
LOWER SPRING SEAT

LH
OUTER SIDE

65.3°—71.3°

LEAD LOWER END OF
LOWER SPRING SEAT

67162-RX8-G33

Fig. 108 Mark the lower seat, shock and spring assembly as illustrated for reassembly

- Brake rotor
- Brake caliper and wire aside
- Front wheel

5. Check and/or adjust the front wheel alignment.

SHOCK ABSORBERS

REMOVAL & INSTALLATION

See Figure 109.

1. Before servicing the vehicle, refer to the Precautions Section.

2. Remove the strut cross brace in the engine compartment and the upper shock mounting nuts.

3. Remove or disconnect the following:
- Front wheel
- Brake hose bracket
- Stabilizer bar nut
- Upper arm ball joint

4. Remove the upper and lower shock absorber mounting nuts and bolts and remove the shock absorber/coil spring assembly.

To install:

5. Installation is the reverse of removal. Tighten the upper shock nuts to 34–46 ft. lbs. (46–62 Nm) and the lower shock nut and bolt to 58–76 ft. lbs. (78–103 Nm)

6. Check and/or adjust the front end alignment.

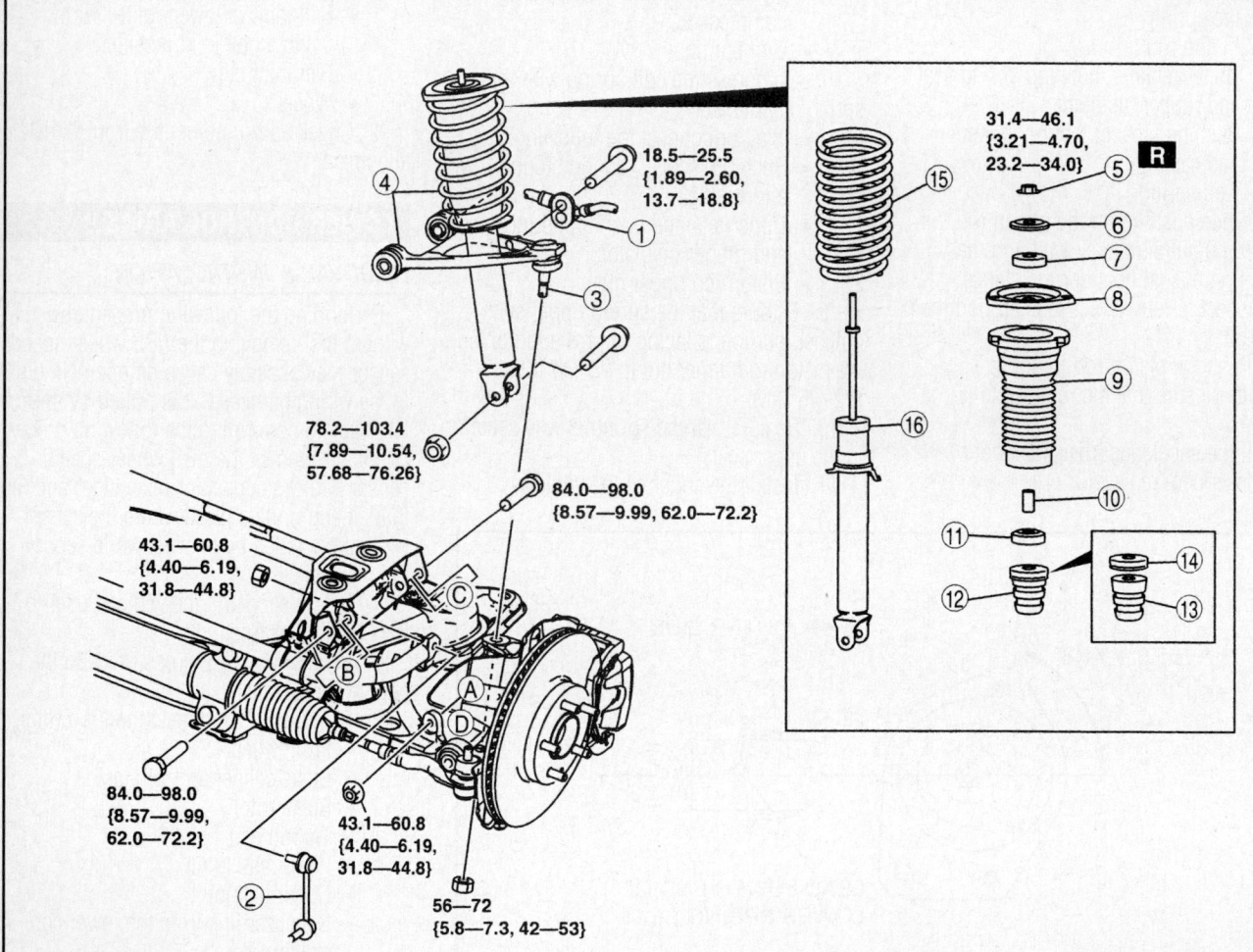

N·m {kgf·m, ft·lbf}

1	Brake hose bracket	9	Dust boot
2	Front stabilizer control link	10	Spacer
3	Front upper arm ball joint	11	Bushing
4	Front shock absorber and coil spring	12	Stopper casing and bound stopper
5	Piston rod nut	13	Bound stopper
6	Retainer	14	Stopper casing
7	Bushing	15	Coil spring
8	Upper spring seat	16	Front shock absorber

67162-RX8-G31

Fig. 109 Exploded view of front shock absorber mounting

STEERING KNUCKLE

REMOVAL & INSTALLATION

See Figures 110 and 111.

✳✳ WARNING

Performing the following procedures without first removing the ABS wheel-speed sensor may possibly cause an open circuit in the wiring harness if it is pulled by mistake. Before performing the following procedures, disconnect the ABS wheel-speed sensor harness connector (axle side) and fix it to an appropriate place where the sensor will not be pulled by mistake while servicing the vehicle.

1. Before servicing the vehicle, refer to the Precautions Section.

2. Remove or disconnect the following:
   - Front wheel
   - ABS wheel-speed sensor connector
   - Brake caliper component
   - Disc plate
   - Tie rod end
   - Stabilizer control link nut (lower)
   - Front upper arm ball joint
   - Front upper arm bolt
   - Front lower arm ball joint
   - Wheel hub, steering knuckle component
   - Steering knuckle
   - Dust cover
   - Wheel hub bolt
   - Wheel hub component

To install:

3. Install or connect the following:
   - Wheel hub component, steering knuckle component
   - Tighten the **UPPER** steering

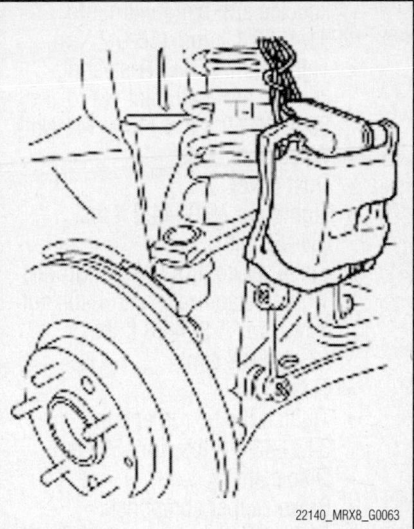

22140_MRX8_G0063

Fig. 111 Suspend the brake caliper component using a cable or equivalent

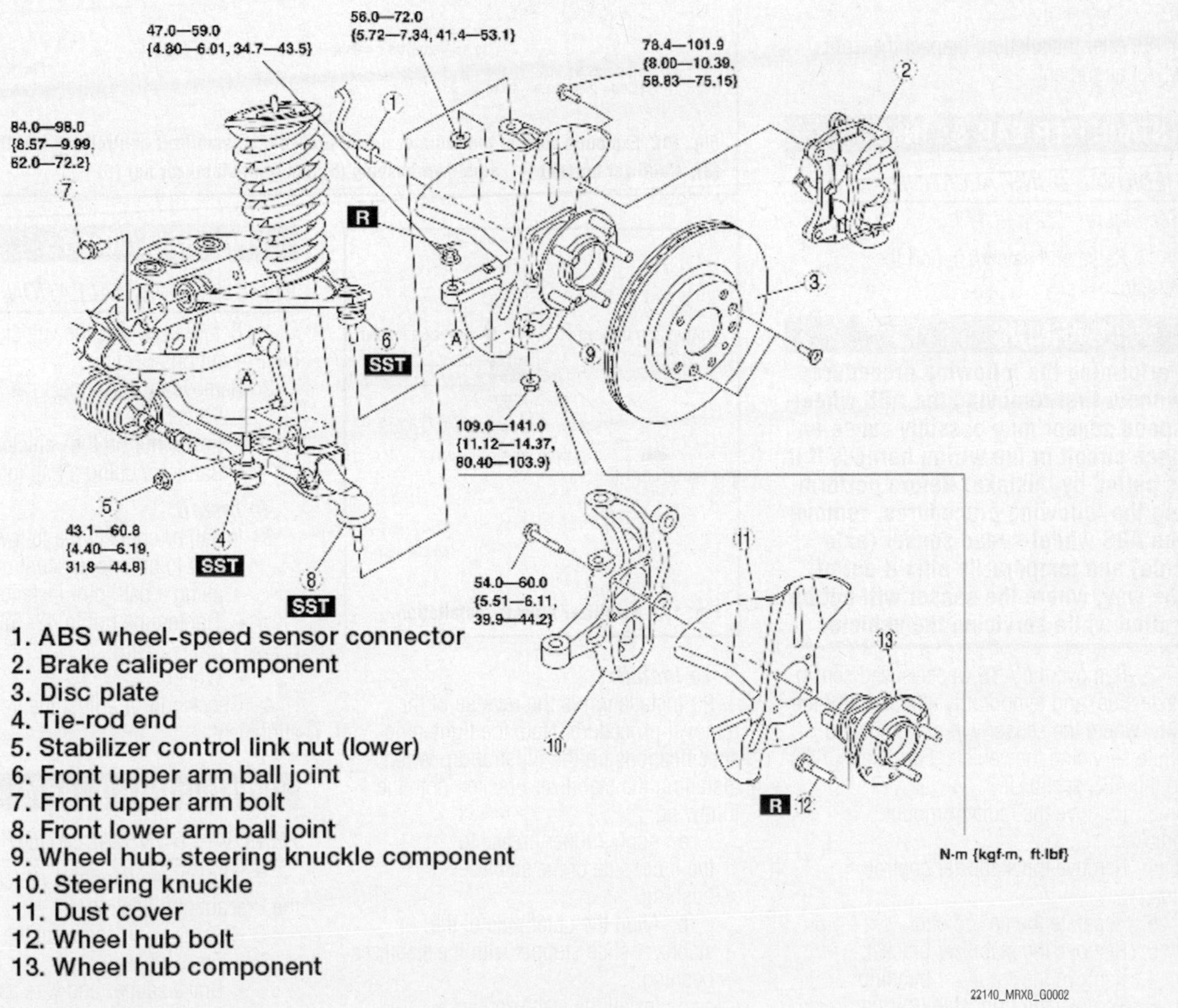

47.0—59.0
{4.80—6.01, 34.7—43.5}

56.0—72.0
{5.72—7.34, 41.4—53.1}

78.4—101.9
{8.00—10.39, 58.83—75.15}

84.0—98.0
{8.57—9.99, 62.0—72.2}

109.0—141.0
{11.12—14.37, 80.40—103.9}

43.1—60.8
{4.40—6.19, 31.8—44.8}

54.0—60.0
{5.51—6.11, 39.9—44.2}

1. ABS wheel-speed sensor connector
2. Brake caliper component
3. Disc plate
4. Tie-rod end
5. Stabilizer control link nut (lower)
6. Front upper arm ball joint
7. Front upper arm bolt
8. Front lower arm ball joint
9. Wheel hub, steering knuckle component
10. Steering knuckle
11. Dust cover
12. Wheel hub bolt
13. Wheel hub component

N·m {kgf·m, ft-lbf}

22140_MRX8_G0002

Fig. 110 Wheel hub, steering ,knuckle and associated components exploded view

knuckle arm ball joint nut to 41.4–53.1 ft lbs. (56–72 Nm).
- Tighten the **LOWER** steering knuckle arm ball joint nut to 80.4–103.9 ft lbs. (109–141 Nm).
- Wheel hub bolts
- Dust cover
- Tighten to 39.9–44.2 ft lbs. (54–60 Nm).
- Stabilizer control link nut (lower)
- Tighten stabilizer control link nut (lower)to 31.8–44.8 ft lbs. (43.1–60.8 Nm).
- Tie rod end
- Tighten the tie rod end to 34.7–43.5 ft lbs. (47–59 Nm).
- Disc plate
- Brake caliper component
- Tighten Brake caliper to 58.8–75.1 ft lbs. (78.4–101.9 Nm).
- ABS wheel-speed sensor connector
- Tighten Brake caliper to 58.8–75.1 ft lbs. (78.4–101.9 Nm).
- Tighten disc plate with wheel to 87 ft. lbs

4. After installation, inspect the front wheel alignment.

STABILIZER BAR & LINKS

REMOVAL & INSTALLATION

See Figures 112 and 113.

1. Raise and safely support the vehicle.

✴✴ WARNING

Performing the following procedures without first removing the ABS wheel-speed sensor may possibly cause an open circuit in the wiring harness if it is pulled by mistake. Before performing the following procedures, remove the ABS wheel-speed sensor (axle side) and temporarily affix it out of the way, where the sensor will not be pulled while servicing the vehicle.

2. Remove the ABS wheel-speed sensor (axle side) and temporarily affix it out of the way, where the sensor will not be pulled while servicing the vehicle. For details, refer to the ABS section.

3. Remove the radiator mount bracket.

4. Remove the stabilizer control link.

5. Separate the tie rod end.

6. Remove the stabilizer bracket.

7. Remove the stabilizer bushing.

8. Remove the front stabilizer bar from the vehicle.

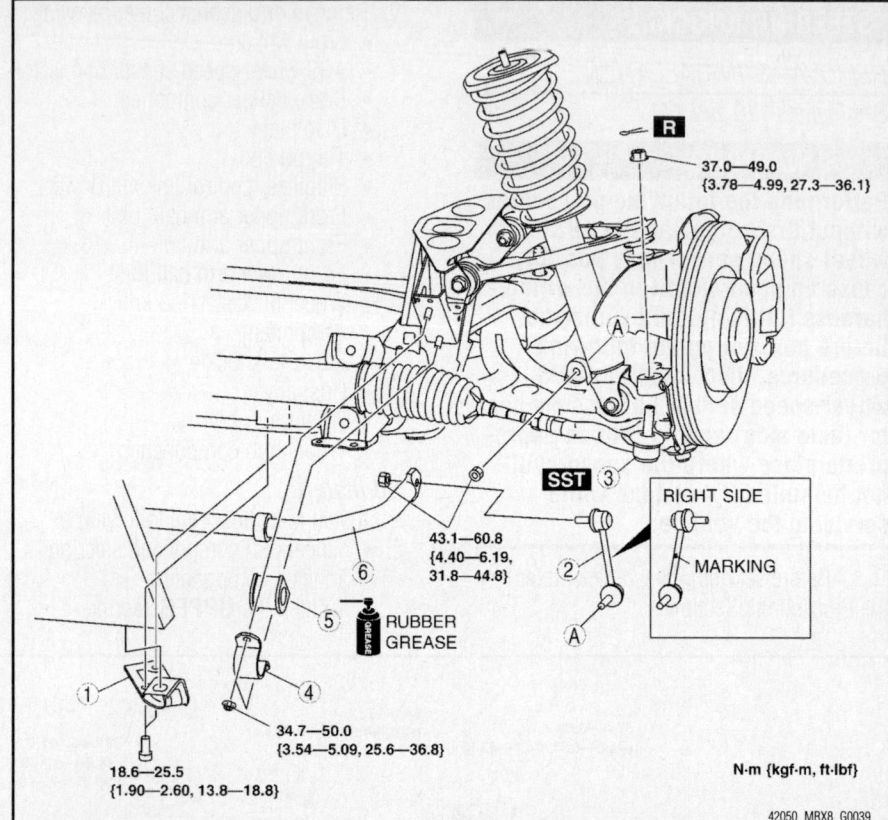

Fig. 112 Exploded view of the radiator mount bracket (1), stabilizer control link (2), tie-rod end (3), stabilizer bracket (4), stabilizer bushing (5) and front stabilizer bar (6)

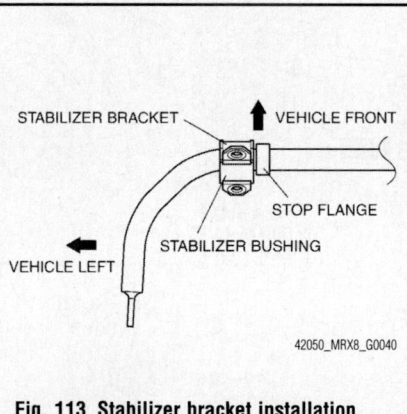

Fig. 113 Stabilizer bracket installation

To install:

9. Installation is the reverse of the removal procedure. Note the tightening specifications on the illustration. When installing the stabilizer bracket, note the following:

 a. Apply rubber grease to the inner side of the stabilizer bushing.

 b. Align the outer side of the stabilizer slide stopper with the stabilizer bushing.

 c. Install the stabilizer bracket.

UPPER BALL JOINT

REMOVAL & INSTALLATION

1. Before servicing the vehicle, refer to the Precautions Section.

2. Remove or disconnect the following:
- Wheel
- Lower nut on the vehicle side
- Ball joint using a ball joint remover

To install:

3. Install or connect the following:
- Ball joint to lower control arm using a ball joint installer
- Tighten the nut to 42–53 ft. lbs. (56–72 Nm).
- Wheel

4. Check and/or adjust the front wheel alignment.

UPPER CONTROL ARM

REMOVAL & INSTALLATION

1. Before servicing the vehicle, refer to the Precautions Section.

2. Remove or disconnect the following:
- Front wheel
- Brake caliper and wire aside
- Brake rotor
- Stabilizer bar nut

- Lower shock mounting bolt and nut
- Upper arm ball joint

3. Remove the front upper arm bolts, push down on the lower arm and remove the upper arm through the lower shock and lower arm gap.

To install:

4. Install the upper control arm.

5. Loosely tighten the bolt and nut.

6. Loosely install the lower strut mounting bolt.

7. Install the upper arm ball joint. Torque the nut 42–53 ft. lbs. (56–72 Nm).

8. Install the wheel.

9. Torque upper control arm bolt to 62–72 ft. lbs. (84–98 Nm) and the lower strut mounting bolt to 58–76 ft. lbs. (78–103 Nm).

10. Check and/or adjust the front wheel alignment.

WHEEL HUB & BEARING

REMOVAL & INSTALLATION

See Figure 114.

1. Before servicing the vehicle, refer to the Precautions Section.

2. Refer to the illustration for component location and torque specifications.

3. Remove or disconnect the following:
- Wheels
- ABS wheel speed sensor
- Brake caliper and rotor
- Tie rod end from the knuckle
- Stabilizer bar link
- Upper control arm ball joint from the knuckle
- Upper arm bolt
- Lower ball joint
- Wheel hub and knuckle assembly
- Separate the knuckle from the wheel hub and remove the backing plate.

4. Clean and inspect all parts but do not

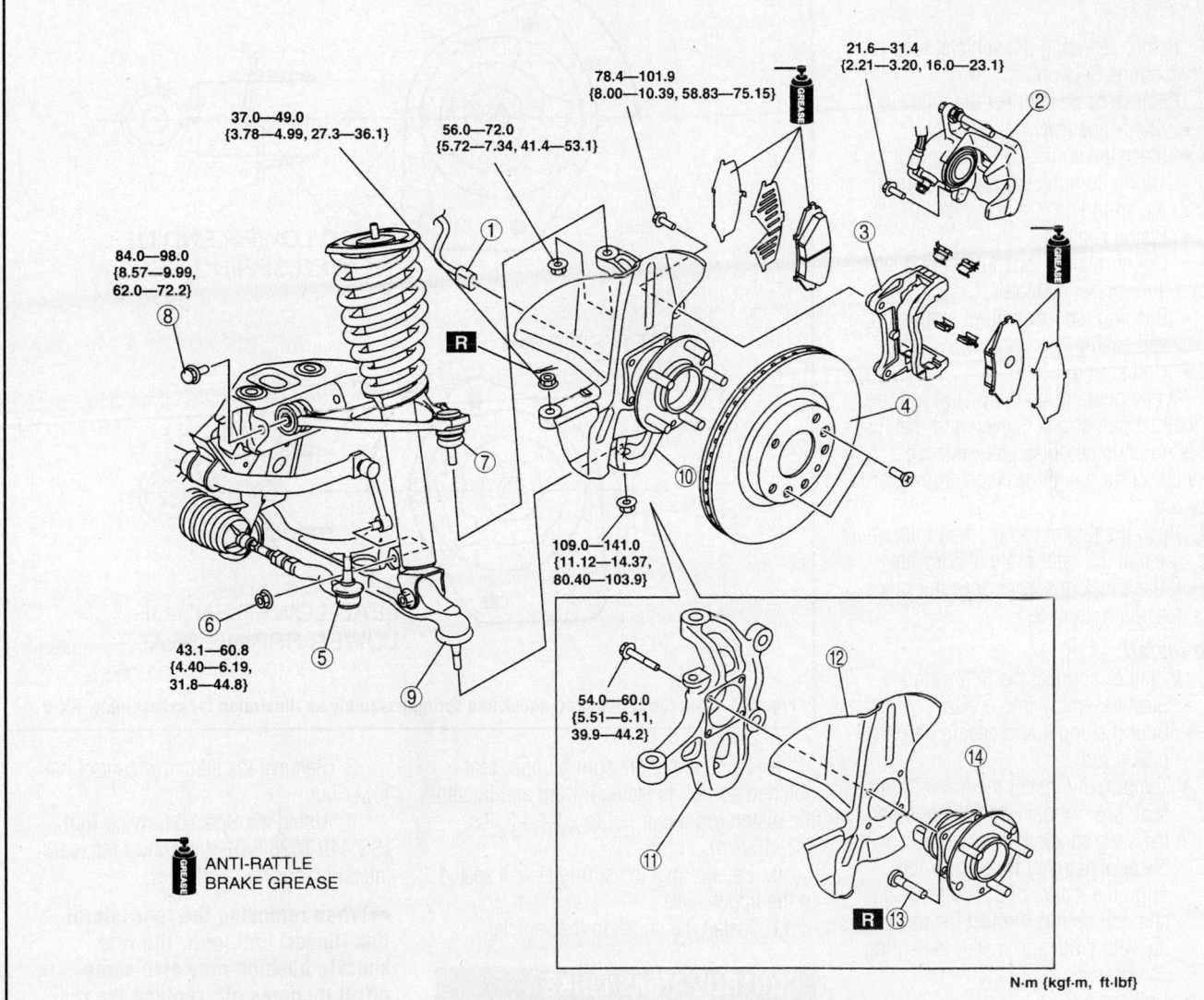

1	ABS wheel-speed sensor connector	5	Tie-rod end
2	Brake caliper component	6	Stabilizer control link (lower)
3	Mounting support	7	Front upper arm ball joint
4	Disc plate	8	Front upper arm bolt

67162-RX8-G39

Fig. 114 Exploded view of the front wheel bearing and knuckle assembly–RX-8

wash or clean the wheel bearing. The wheel hub/bearing must be replaced.

To install:

5. Install or connect the following:
- Install the backing plate to the hub, then install the knuckle to the hub and tighten the bolts to 40–44 ft. lbs. (54–60 Nm).

- Lower ball joint
- Upper arm bolt and tighten to 62–72 ft. lbs. (84–96 Nm).
- Upper control arm ball joint
- Stabilizer bar link and tighten to 32–45 ft. lbs. (43–61 Nm).
- Tie rod end
- ABS wheel speed sensor

- Brake caliper and rotor
- Wheels

ADJUSTMENT

The front wheel bearings are not adjustable. If the bearings become loose or make noise, they must be replaced.

SUSPENSION

REAR SUSPENSION

COIL SPRING

REMOVAL & INSTALLATION

See Figure 115.

1. Before servicing the vehicle, refer to the Precautions Section.
2. Remove or disconnect the following:
- Shock absorber/coil spring
- Place the shock/coil spring in a spring compressor and compress the spring
- Piston rod nut
- Upper retainer, bushing spring seat and rubber insulator
- Bushing, spacer, bound stopper and casing
- Coil spring

3. While pushing on the piston rod, be sure that the pull stroke is even and that there is no unusual noise or resistance. Also, inspect for any oil leakage around the piston rod.

4. Push the piston rod in, then release it. Be sure that the return rate is constant.

5. If the shock absorber does not operate as described, replace it.

To install:

6. Install or connect the following:
- Strut assembly into a vise
- Bound stopper and casing onto the piston rod
- Temporarily install the lower spring seat, seat rubber and spring. Mark the seat, shock and spring assembly as illustrated for reassembly. Align the marks of the upper seat and coil spring. Protect the assembly with cloth and install the spring compressor.
- Coil spring

7. Compress the coil spring with the spring compressor.

8. Install or connect the following:
- Bushing, spacer, bound stopper and casing
- Upper retainer, bushing spring seat and rubber insulator
- Piston rod upper nut

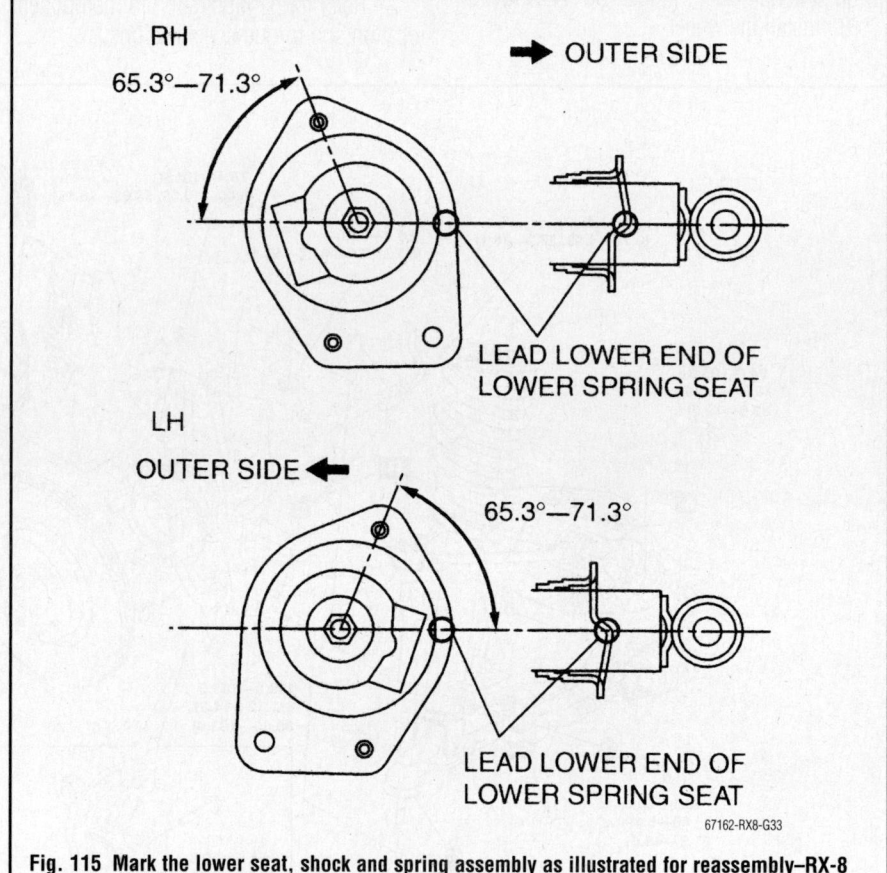

Fig. 115 Mark the lower seat, shock and spring assembly as illustrated for reassembly–RX-8

9. Be sure that the spring upper seat notched portion is facing inward and tighten the piston rod upper nut to 23–34 ft. lbs. (32–46 Nm).

10. Be sure that the spring is well seated in the upper seats.

11. Install the shock to the vehicle.

CONTROL ARMS/LINKS

REMOVAL & INSTALLATION

Lower Rear Lateral Link

See Figures 116 through 118.

1. Raise and safely support the vehicle.
2. Remove the rear wheels.

3. Remove the stabilizer control link lower nut.

4. Using the Special Service Tool (SST)49 T028 3A0, disconnect the rear lateral link (lower) ball joint.

➡**When removing the rear lateral link (lower) ball joint, the rear knuckle bushing may also come off. If it comes off, replace the rear knuckle.**

5. Remove the rear lateral link (lower)bolt.

6. Remove the rear dust boot.

To install:

7. Wipe the grease off the ball joint stud.

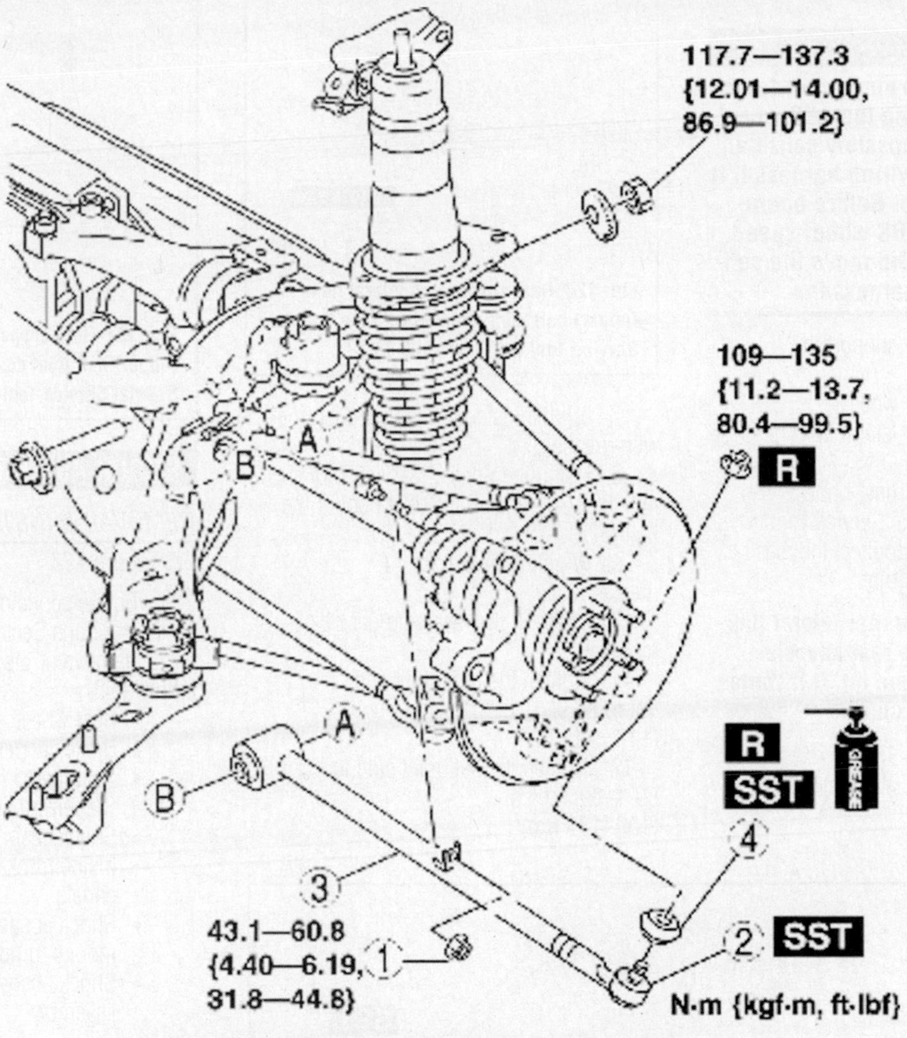

117.7—137.3
{12.01—14.00,
86.9—101.2}

109—135
{11.2—13.7,
80.4—99.5} **R**

R
SST

43.1—60.8
{4.40—6.19,
31.8—44.8} (1)

(3)

(4)

(2) **SST**

N·m {kgf·m, ft·lbf}

1. Stabilizer control link lower nut
2. Rear lateral link (lower) ball joint
3. Rear lateral link (lower)
4. Dust boot

22140_MRX8_G0060

Fig. 116 Rear suspension exploded view emphasizing the lower lateral link

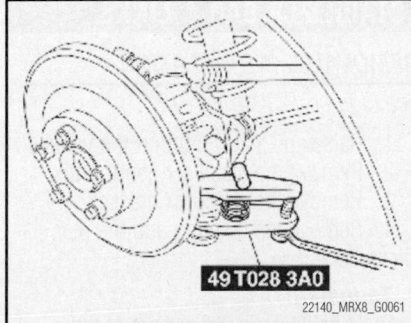

49 T028 3A0

22140_MRX8_G0061

**Fig. 117 Rear suspension lateral link
(lower) ball joint removal with the Special
Service Tool (SST)49 T028 3A0**

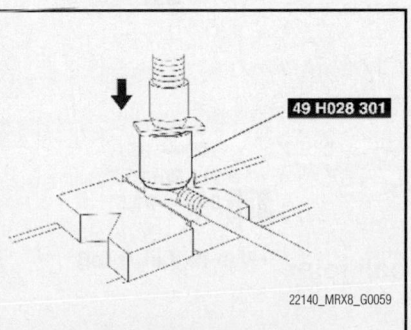

49 H028 301

22140_MRX8_G0059

**Fig. 118 Rear suspension lateral link
(lower) ball joint dust boot installation with
Special Service Tool (SST)49 H028 301**

8. Fill the inside of the new dust boot
with grease.

9. Using the Special Service Tool (SST)
49 H028 301, install the dust boot to the
ball joint.

10. Install in the reverse order of
removal.

11. Tighten the inner bolt and
nut to 86.9–101.2 ft.lbs.
(117.7–137.3 Nm).

12. Tighten the ball joint stud to
80.4–99.5 ft.lbs. (109–135 Nm).

13. Tighten the stabilizer control
link lower nut to 31.8–44.5 ft.lbs.
(43.1–60.8 Nm).

Upper Rear Lateral Link

See Figures 119 through 121.

✳✳ WARNING

Performing the following procedures without first removing the ABS wheel-speed sensor may possibly cause an open circuit in the wiring harness if it is pulled by mistake. Before operations, remove the ABS wheel-speed sensor (axle side) and move the sensor away from the harnesses.

1. Raise and safely support the vehicle.
2. Remove the rear wheels.
3. Remove the rear lateral link (lower)bolt.
4. Remove the rear dust boot.
5. Using the Special Service Tool (SST)49 T028 3A0, disconnect the rear lateral link (upper) ball joint.

➡**When removing the rear lateral link (upper) ball joint, the rear knuckle bushing may also come off. If it comes off, replace the rear knuckle.**

To install:

6. Wipe the grease off the ball joint stud.

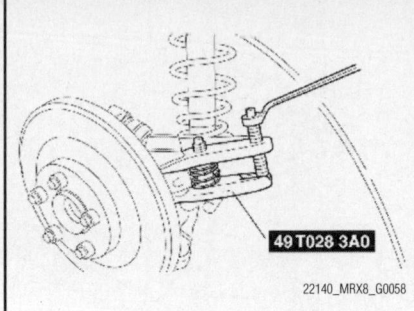

Fig. 120 Rear suspension lateral link (upper) ball joint removal with the Special Service Tool (SST) 49 T028 3A0

7. Fill the inside of the new dust boot with grease.
8. Using the Special Service Tool (SST) 49 H028 301, install the dust boot to the ball joint.
9. Wipe off the excess grease.
10. Install in the reverse order of removal.
11. Tighten the inner bolt and nut to 86.9–101.2 ft.lbs. (117.7–137.3 Nm).
12. Tighten the ball joint stud to 80.4–99.5 ft.lbs. (109–135 Nm).

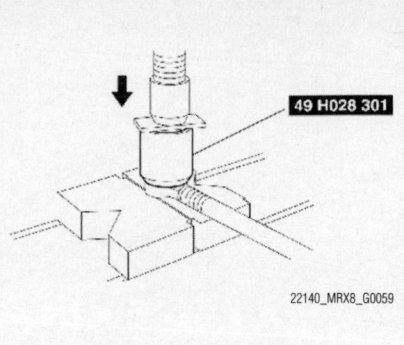

Fig. 121 Rear suspension lateral link (upper) ball joint dust boot installation with Special Service Tool (SST)49 H028 301

SHOCK ABSORBER

REMOVAL & INSTALLATION

See Figure 122.

1. Before servicing the vehicle, refer to the Precautions Section.
2. Remove or disconnect the following:
   - Rear wheel
   - Brake caliper and wire aside
   - Rear lateral link upper inner bolt
   - Stabilizer bar upper nut
   - Shock absorber lower bolt
   - Inside trunk end and side trim panels
   - Shock absorber upper bracket and mounting nuts
   - Shock absorber/coil spring assembly

To install:

3. Installation is the reverse of removal. Tighten the upper shock nuts to 34–46 ft. lbs. (46–62 Nm) and the lower shock nut and bolt to 65–88 ft. lbs. (88–119 Nm). Tighten the shock bracket bolts and nuts to 28–38 ft. lbs. (37–52 Nm).
4. Check and/or adjust the front end alignment.

STABILIZER BAR & LINKS

REMOVAL & INSTALLATION

See Figures 123 through 127.

1. Raise and safely support the vehicle.
2. Remove the stabilizer control link.
3. Remove the stabilizer bracket.
4. Remove the bushing and the rear stabilizer bar.

To install:

5. Installation is the reverse of the removal procedure, using the tightening specifications in the illustration and noting the following important steps:

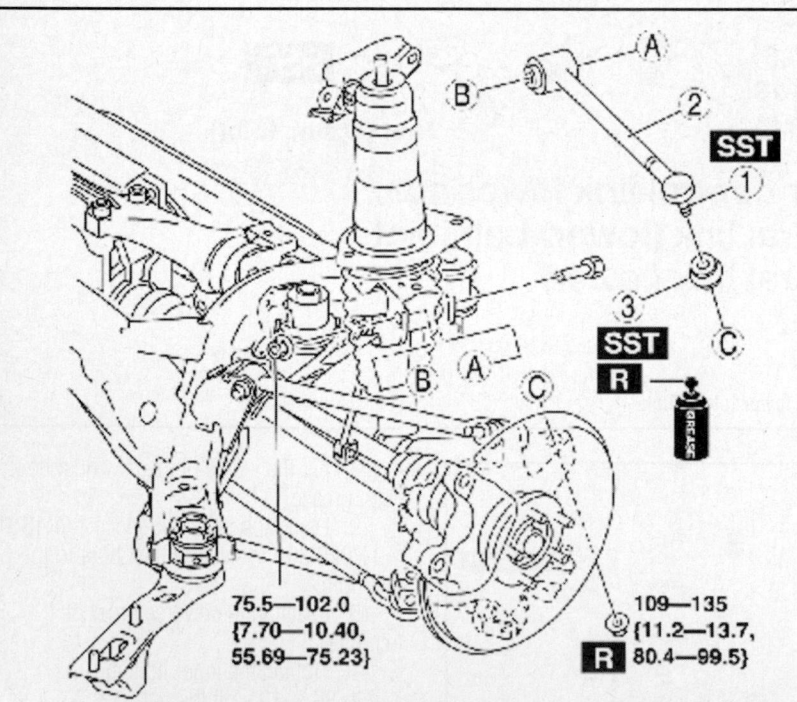

75.5—102.0
{7.70—10.40,
55.69—75.23}

109—135
{11.2—13.7,
80.4—99.5}

1. Rear lateral link (upper) ball joint
2. Rear lateral link (upper)
3. Dust boot

N·m {kgf·m, ft·lbf}

Fig. 119 Rear suspension exploded view emphasizing the upper lateral link

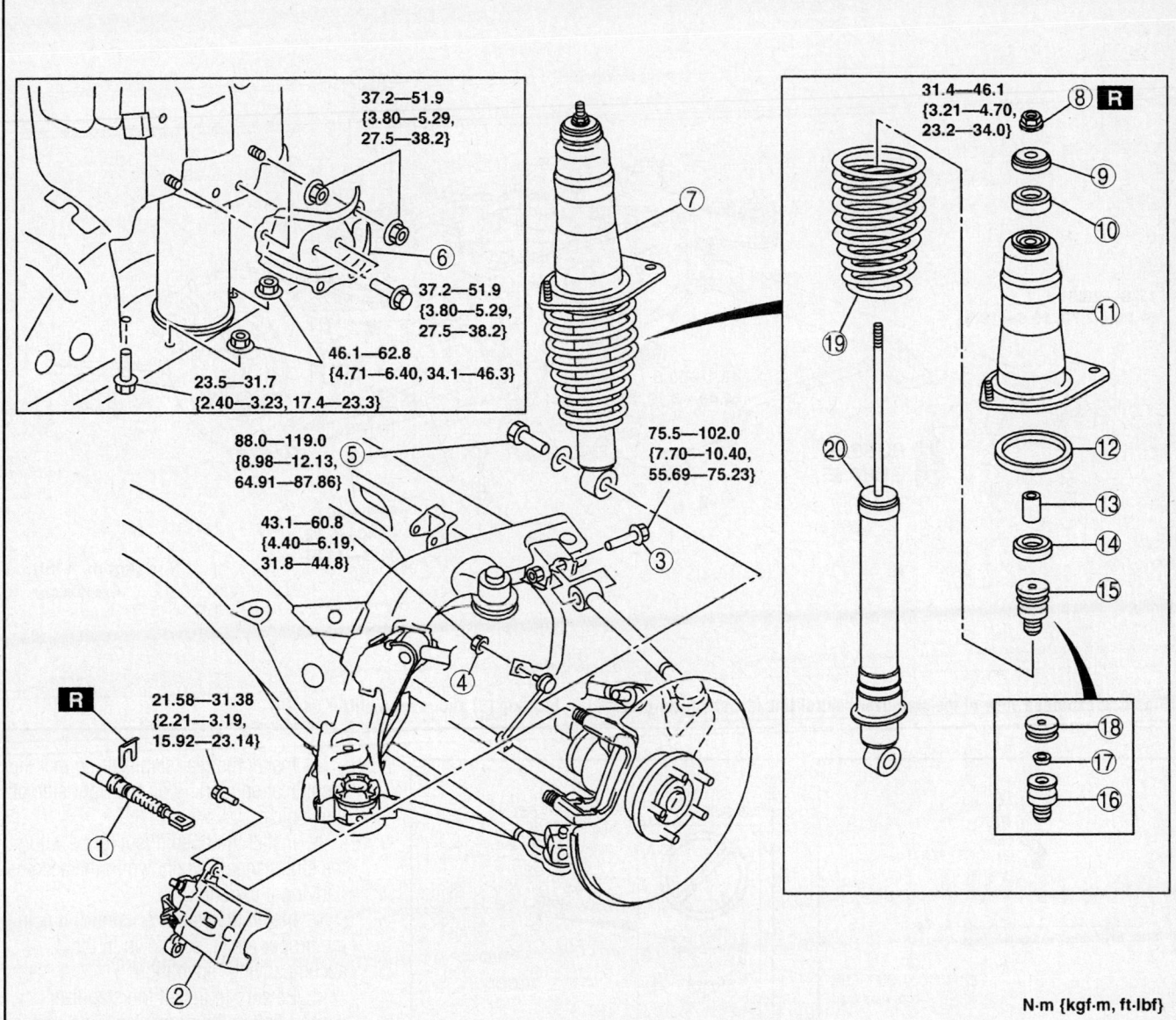

37.2—51.9
{3.80—5.29,
27.5—38.2}

37.2—51.9
{3.80—5.29,
27.5—38.2}

46.1—62.8
{4.71—6.40, 34.1—46.3}

23.5—31.7
{2.40—3.23, 17.4—23.3}

88.0—119.0
{8.98—12.13,
64.91—87.86}

43.1—60.8
{4.40—6.19,
31.8—44.8}

75.5—102.0
{7.70—10.40,
55.69—75.23}

31.4—46.1
{3.21—4.70,
23.2—34.0}

21.58—31.38
{2.21—3.19,
15.92—23.14}

N·m {kgf·m, ft·lbf}

1	Parking brake cable	11	Upper spring seat
2	Caliper	12	Spring seat rubber
3	Rear lateral link (upper) inner bolt	13	Bushing
4	Stabilizer control link upper nut	14	Spacer
5	Rear shock absorber lower bolt	15	Bound stopper and stopper casing
6	Rear shock absorber bracket	16	Bound stopper
7	Rear shock absorber and coil spring	17	Collar
8	Piston rod nut	18	Stopper casing
9	Retainer	19	Coil spring
10	Bushing	20	Rear shock absorber

67162-RX8-G32

Fig. 122 Exploded view of rear shock absorber mounting

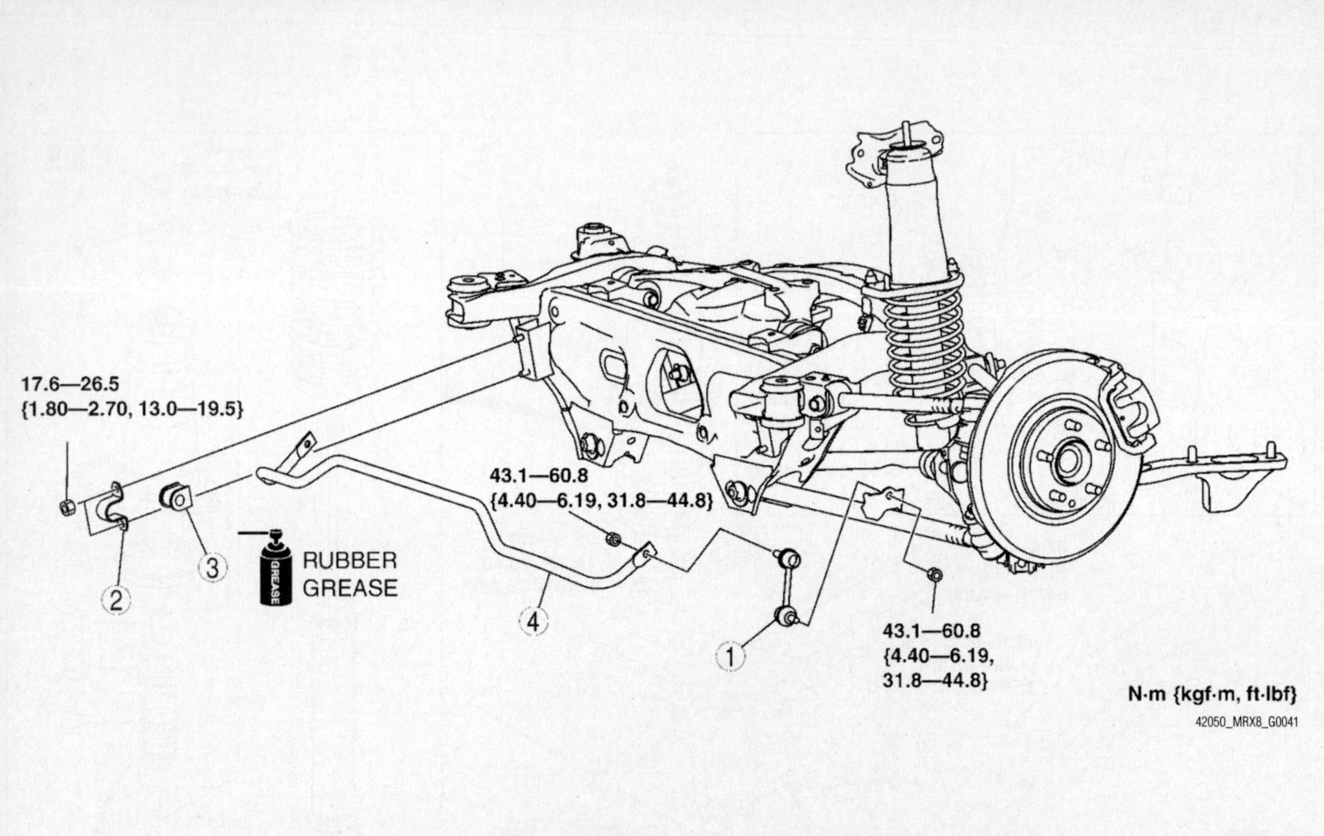

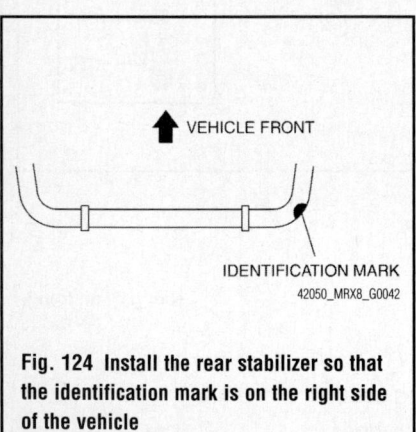

17.6—26.5
{1.80—2.70, 13.0—19.5}

RUBBER
GREASE

43.1—60.8
{4.40—6.19, 31.8—44.8}

43.1—60.8
{4.40—6.19,
31.8—44.8}

N·m {kgf·m, ft·lbf}

42050_MRX8_G0041

Fig. 123 Exploded view of the stabilizer control link (1), stabilizer bracket (2), bushing (3) and rear stabilizer bar (4)

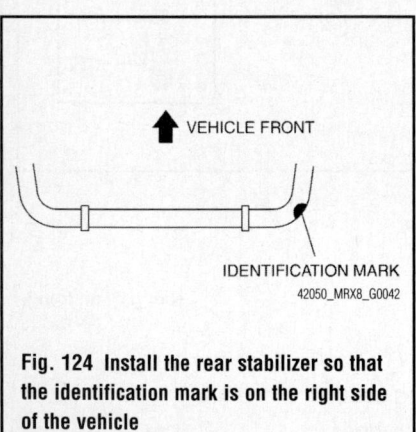

VEHICLE FRONT

IDENTIFICATION MARK

42050_MRX8_G0042

Fig. 124 Install the rear stabilizer so that the identification mark is on the right side of the vehicle

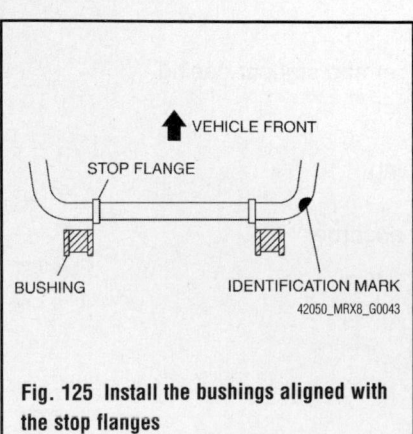

VEHICLE FRONT

STOP FLANGE

BUSHING   IDENTIFICATION MARK

42050_MRX8_G0043

Fig. 125 Install the bushings aligned with the stop flanges

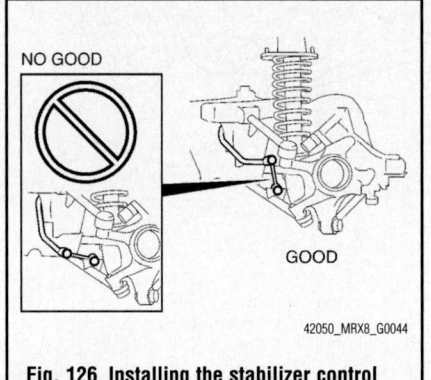

NO GOOD

GOOD

42050_MRX8_G0044

Fig. 126 Installing the stabilizer control link in the proper angle

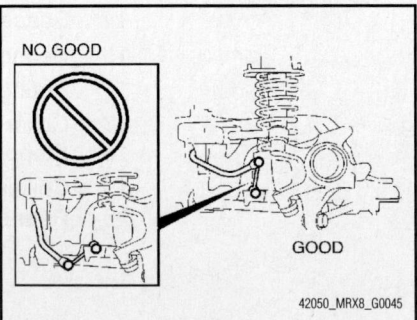

NO GOOD

GOOD

42050_MRX8_G0045

Fig. 127 Place the vehicle on the ground and verify that the stabilizer control link is installed in the angle shown

a. Install the rear stabilizer so that the identification mark is on the right side of the vehicle.

b. Install the bushings aligned with the stop flanges, as shown in the accompanying illustration.

c. Install the stabilizer control link in the proper angle, as shown in the accompanying illustration.

d. Be sure to install the stabilizer control link in the proper position. If it is not installed properly, the stabilizer control link may interfere with peripheral components when driving, causing damage to each other.

e. Place the vehicle on the ground and verify that the stabilizer control link is installed in the angle shown in the figure.

WHEEL HUB & BEARING

REMOVAL & INSTALLATION
See Figure 128.

1. Before servicing the vehicle, refer to the Precautions Section.

2. Refer to the illustration for component location and torque specifications.

➡The wheel bearings are not serviceable. If the bearings are bad, a new hub/bearing assembly must be installed.

3. Remove or disconnect the following:
- Rear wheels
- ABS wheel speed sensor

4. Raise the staked portion of the axle retaining nut with a hammer and chisel.

5. Remove or disconnect the following:
- Axle nut

- Parking brake cable
- Rear caliper and rotor assembly from the hub
- Rear lateral link upper ball joint
- Stabilizer bar lower link
- Rear lateral link lower ball joint
- Lower shock absorber bolt

- Rear trailing link lower outside bolt
- Rear trailing link upper ball joint
- Toe control link outside bolt
- Axle shaft
- Wheel hub/knuckle assembly

6. Press the wheel hub from the knuckle.

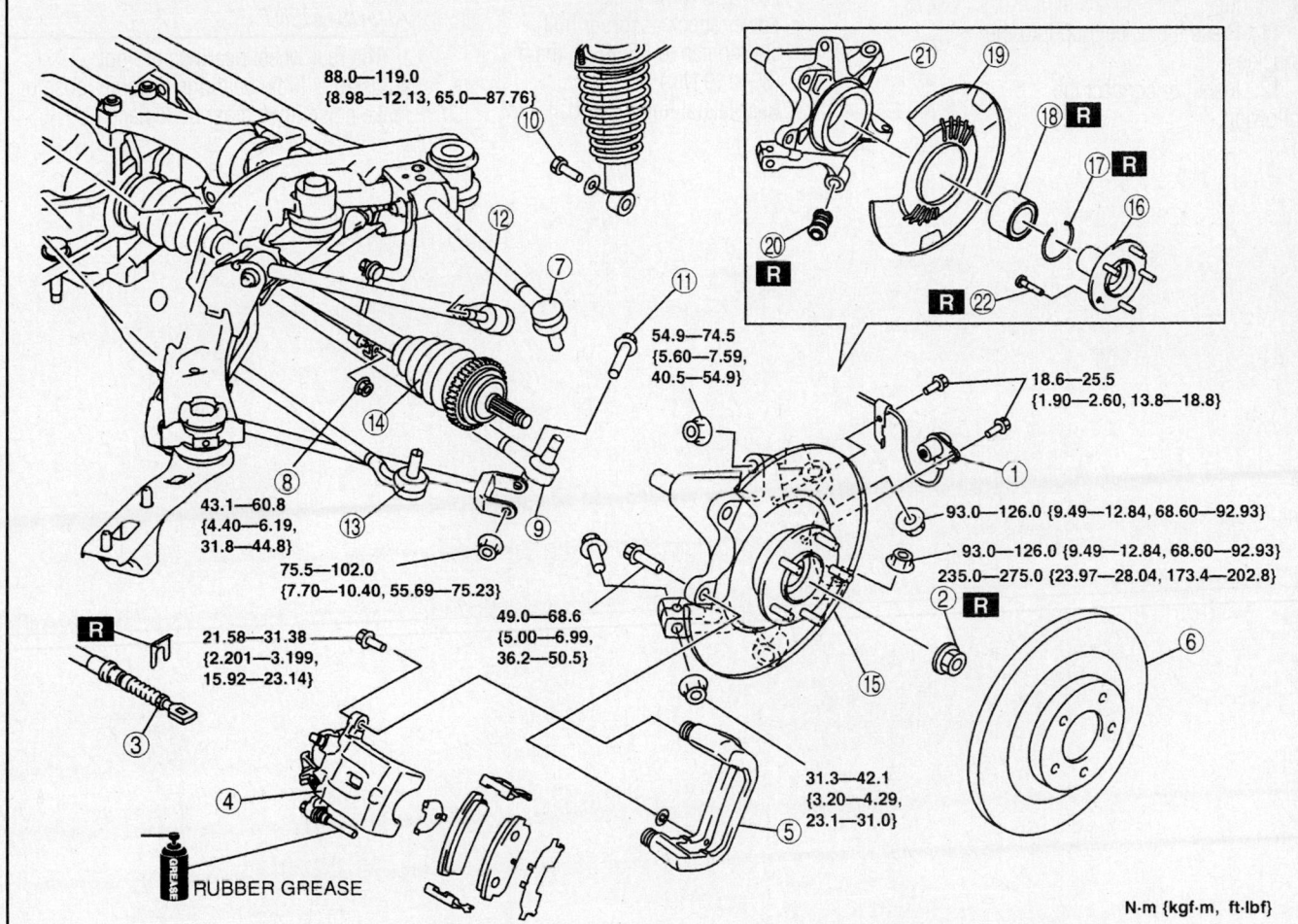

1	ABS wheel-speed sensor	12	Rear trailing link (upper) ball joint
2	Locknut	13	Toe control link outside bolt
3	Parking brake cable	14	Rear drive shaft
4	Brake caliper component	15	Rear knuckle component
5	Mounting support	16	Wheel hub component
6	Disc plate	17	Retaining ring
7	Rear lateral link (upper) ball joint	18	Wheel bearing
8	Stabilizer control link (lower)	19	Dust cover
9	Rear lateral link (lower) ball joint	20	Bushing
10	Shock absorber bolt (lower)	21	Rear knuckle
11	Rear trailing link (lower) outside bolt	22	Wheel hub bolt

Fig. 128 Exploded view of the rear wheel hub and bearing—RX-8

67162-RX8-G40

7. Press the bearing inner race from the wheel hub.

8. Press the wheel bearing from the rear knuckle.

To install:

9. Press the wheel bearing into the rear knuckle.

10. Press the bearing inner race into the wheel hub.

11. Press the wheel hub into the knuckle.

12. Install or connect the following:

- Wheel hub/knuckle assembly. Tighten the bolts to 23–31 ft. lbs. (31–42 Nm).
- Axle shaft
- Toe control link outside bolt
- Rear trailing link upper ball joint
- Rear trailing link lower outside bolt and tighten to 56–75 ft. lbs. (76–102 Nm).
- Lower shock absorber bolt and tighten to 65–88 ft. lbs. (86–119 Nm).
- Rear lateral link lower ball joint

- Stabilizer bar lower link and tighten to 32–45 ft. lbs. (43–61 Nm).
- Rear lateral link upper ball joint
- Rear caliper and rotor assembly
- Parking brake cable
- Axle nut and tighten to 174–203 ft. lbs. (235–275 Nm).

ADJUSTMENT

The rear wheel bearings are not adjustable. If the bearings become loose or make noise, they must be replaced.

MAZDA

Tribute

9

SPECIFICATIONS AND MAINTENANCE CHARTS

ENGINE AND VEHICLE IDENTIFICATION

		Engine					Model Year	
Code ①	Liters	Cu. In.	Cyl.	Fuel Sys.	Engine Type	Eng. Mfg.	Code ②	Year
B	2.5	152	4	MPI	DOHC	Ford	9	2009
1	3.0	183	6	MFI	DOHC	Ford	A	2010

MPI: Multi-port Fuel Injection

DOHC: Double Overhead Camshafts

① 8th digit of VIN

② 10th digit of VIN

37671_TRIB_C0001

GENERAL ENGINE SPECIFICATIONS

Year	Model	Engine Displacement Liters	Engine VIN	Net Horsepower @ rpm	Net Torque @ rpm (ft. lbs.)	Bore x Stroke (in.)	Com-pression Ratio	Oil Pressure @ rpm
2009	Tribute	2.5	B	171@6000	171@4500	3.50x3.94	9.7:1	11@1500
	Tribute	3.0	1	240@6550	223@4300	3.50x3.12	10.3:1	11@1500
2010	Tribute	2.5	B	171@6000	171@4500	3.44x3.70	9.7:1	11@1500
	Tribute	3.0	1	240@6550	223@4300	3.50x3.12	10.3:1	11@1500

37671_TRIB_C0002

ENGINE TUNE-UP SPECIFICATIONS

Year	Engine Displacement Liters	Engine VIN	Spark Plug Gap (in.)	Ignition Timing (deg.) MT	Ignition Timing (deg.) AT	Fuel Pump (psi)	Idle Speed (rpm) MT	Idle Speed (rpm) AT	Valve Clearance Intake	Valve Clearance Exhaust
2009	2.5	B	0.049-0.053	10 BTDC	10 BTDC	39	①	①	0.008-0.011	0.010-0.013
	3.0	1	0.045-0.049	10 BTDC	10 BTDC	39	①	①	HYD.	HYD.
2010	2.5	B	0.049-0.053	10 BTDC	10 BTDC	39	①	①	0.008-0.011	0.010-0.013
	3.0	1	0.045-0.049	10 BTDC	10 BTDC	39	①	①	HYD.	HYD.

BTDC: Before Top Dead Center

HYD: Hydraulic lash adjusters

37671_TRIB_C0003

CAPACITIES

Year	Model	Engine Displacement Liters	Engine VIN	Engine Oil with Filter (qts.)	Transmission (pts.) Manual	Transmission (pts.) Auto.	Transfer Case (pts.)	Drive Axle Front (pts.)	Drive Axle Rear (pts.)	Fuel Tank (gal.)	Cooling System (qts.)
2009	Tribute	2.5	B	4.5	5.0	18.0	2.95	—	2.43	17.5	①
	Tribute	3.0	1	6	—	18.0	2.95	—	2.43	17.5	10.5
2010	Tribute	2.5	B	4.5	5.0	18.0	2.95	—	2.43	17.5	①
	Tribute	3.0	1	6	—	18.0	2.95	—	2.43	17.5	10.5

NOTE: All capacities are approximate. Add fluid gradually and check to be sure a proper fluid level is obtained.

① With manual transaxle: 6.9 qts; with automatic transaxle: 8.0 qts.

37671_TRIB_C0004

FLUID SPECIFICATIONS

Year	Model	Engine Displacement Liters	Engine ID/VIN	Engine Oil ①	Auto. Trans.	Drive Axle	Power Steering Fluid	Brake Master Cylinder
2009	Tribute	2.5	B	5W-20	Mercon®LV	75W-90	NA	DOT 3
		3.0	1	5W-20	Mercon®LV	75W-90	NA	DOT 3
2010	Tribute	2.5	B	5W-20	Mercon®LV	75W-90	NA	DOT 3
		3.0	1	5W-20	Mercon®LV	75W-90	NA	DOT 3

NA - Not Available

DOT: Department Of Transpotation

®: Registerd Trademark

① SAE 5W-20 Premium Synthetic Blend Motor Oil (US only) SAE-5W-20 Super Premium Motor Oil (Canada)

37671_TRIB_C0005

VALVE SPECIFICATIONS

Year	Engine Displacement Liters	Engine VIN	Seat Angle (deg.)	Face Angle (deg.)	Spring Test Pressure (lbs. @ in.)	Spring Installed Height (in.)	Stem-to-Guide Clearance (in.) Intake	Stem-to-Guide Clearance (in.) Exhaust	Stem Diameter (in.) Intake	Stem Diameter (in.) Exhaust
2009	2.5	B	45	45	NA	1.492	0.0001	0.00011	0.2153-0.2159	0.2151-0.2157
	3.0	1	44.75	45.5	156@ 1.18	1.570	0.0017-0.0037	0.0017-0.0037	0.2350-0.2358	0.2343-0.2350
2010	2.5	B	45	45	NA	1.492	0.0001	0.00011	0.2153-0.2159	0.2151-0.2157
	3.0	1	44.75	45.5	156@ 1.18	1.570	0.0017-0.0037	0.0017-0.0037	0.2350-0.2358	0.2343-0.2350

37671_TRIB_C0006

CAMSHAFT SPECIFICATIONS CHART

All measurements are given in inches.

Year	Engine Displ. Liters	Engine VIN	Journal Dia.	Brg. Oil Clearance	Shaft End-play	Runout	Journal Bore	Lobe Height Intake	Exhaust
2009	2.5	B	0.9820-0.9830	0.001-0.003	NA	0.001 ①	NA	0.3240	0.3070
	3.0	1	1.0600-1.0610	0.001-0.00029	0.00748	NA	1.062-1.063	0.1890	0.1890
2010	2.5	B	0.9820-0.9830	0.001-0.003	NA	0.001 ①	NA	0.3240	0.3070
	3.0	1	1.0600-1.0610	0.001-0.00029	0.00748	NA	1.062-1.063	0.1890	0.1890

NA: Information not available

① Supported by No. 1 and No. 5 journals.

37671_TRIB_C0007

CRANKSHAFT AND CONNECTING ROD SPECIFICATIONS

All measurements are given in inches.

Year	Engine Displacement Liters	Engine VIN	Crankshaft Main Brg. Journal Dia.	Main Brg. Oil Clearance	Shaft End-play	Thrust on No.	Connecting Rod Journal Diameter	Oil Clearance	Side Clearance
2009	2.5	B	2.0460-2.0470	0.0006-0.0015	0.0070-0.0180	NA	2.0870-2.0880	0.001-0.002	0.0760-0.1200
	3.0	1	2.4670-2.4790	0.0009-0.0030	0.0050-0.0100	3	2.0872-2.0879	0.001-0.0025	0.0039-0.0118
2010	2.5	B	2.0460-2.0470	0.0006-0.0015	0.0070-0.0180	NA	2.0870-2.0880	0.001-0.002	0.0760-0.1200
	3.0	1	2.4670-2.4790	0.0009-0.0030	0.0050-0.0100	3	2.0872-2.0879	0.001-0.0025	0.0039-0.0118

NA: Information not available

37671_TRIB_C0008

PISTON AND RING SPECIFICATIONS

All measurements are given in inches.

Year	Engine Displacement Liters	Engine VIN	Piston Clearance	Ring Gap Top Compression	Bottom Compression	Oil Control	Ring Side Clearance Top Compression	Bottom Compression	Oil Control
2009	2.5	B	0.0009-0.0017	0.0060-0.0120	0.0120-0.0180	0.0070-0.0270	NA	NA	NA
	3.0	1	0.0005-0.0009	0.0039-0.0098	0.0106-0.0165	0.0059-0.0255	NA	NA	NA
2010	2.5	B	0.0009-0.0017	0.0060-0.0120	0.0120-0.0180	0.0070-0.0270	NA	NA	NA
	3.0	1	0.0005-0.0009	0.0039-0.0098	0.0106-0.0165	0.0059-0.0255	NA	NA	NA

NA: Information not available

37671_TRIB_C0009

TORQUE SPECIFICATIONS

All readings in ft. lbs.

Year	Engine Displacement Liters	Engine VIN	Cylinder Head Bolts	Main Bearing Bolts	Rod Bearing Bolts	Crankshaft Damper Bolts	Flywheel Bolts	Manifold Intake	Manifold Exhaust	Spark Plugs	Oil Pan Drain Plug
2009	2.5	B	①	②	③	④	⑤	13	35	9	21
	3.0	1	⑥	⑦	⑧	⑨	59	⑩	15	11	19
2010	2.5	B	①	②	③	④	⑤	13	35	9	21
	3.0	1	⑥	⑦	⑧	⑨	59	⑩	15	11	19

NA: Information not available

① Step 1: 44 inch lbs.

  Step 2: 11 ft. lbs.

  Step 3: 33 ft. lbs.

  Step 4: +90 degrees

  Step 5: + 90 degrees

② Step 1: 44 inch lbs.

  Step 2: 18 ft. lbs.

  Step3: plus 90 degrees

③ Step 1: 18 ft. lbs.

  Step 2: 59 ft. lbs

④ Step 1: 74 ft. lbs.

  Step 2: Tighten additional 90 degrees.

⑤ Step 1: 37 ft. lbs.

  Step 2: 59 ft. lbs.

  Step 3: 83 ft. lbs.

⑥ Step 1: 30 ft. lbs.

  Step 2: Tighten additional 90 degrees

  Step 3: Loosen 1 full turn

  Step 4: 30 ft. lbs.

  Step 5: Plus 90 degrees

  Step 6: Plus additional 90 degrees

⑦ Step 1: Tighten fasteners 1 through 8 tp 18 ft. lbs.

  Step 2: Tighten fasteners 9 through 16 tp 30 ft. lbs.

  Step 3: Tighten fasteners 1 through 16 and additional 90 degrees

  Step 4: Tighten fasteners 17 through 22 to 18 ft. lbs.

⑧ Step 1: 17 ft. lbs.

  Step 2: 32 ft. lbs.

⑨ Step 1: 89 ft. lbs.

  Step 2: Loosen the bolt one full turn

  Step 3: 37 ft. lbs.

  Step 4: Plus 90 degrees

⑩ Lower Intake Manifold: 89 inch lbs.

  Upper Intake Manifold:

  Step 1: 89 inch lbs.

  Step 2: Plus 45 degrees

37671_TRIB_C0010

WHEEL ALIGNMENT

Year	Model		Caster Range (+/-Deg.)	Caster Preferred Setting (Deg.)	Camber Range (+/-Deg.)	Camber Preferred Setting (Deg.)	Toe-in (in.)
2009	Tribute	F	0.50	+1.60	0.50	0.84	0.23+/-0.23
		R	NA	NA	0.70	0.00	0.18+/-0.20
2010	Tribute	F	0.50	+1.60	0.50	0.84	0.23+/-0.23
		R	NA	NA	0.70	0.00	0.18+/-0.20

NA: Information not available

37671_TRIB_C0011

TIRE, WHEEL AND BALL JOINT SPECIFICATIONS

Year	Model	OEM Tires		Tire Pressures (psi)		Wheel Size	Ball Joint Inspection	Lug Nuts (ft. lbs.)
		Standard	Optional	Front	Rear			
2009	Tribute	P235/70SR16	NA	①	①	7	0.080 in.	98
2010	Tribute	P235/70TR16	NA	①	①	7	0.080 in.	98

OEM: Original Equipment Manufacturer

PSI: Pounds Per Square Inch

NA: Not Available

① See certification label located on the inside driver's side front door jamb.

37671_TRIB_C0012

BRAKE SPECIFICATIONS

All measurements in inches unless noted

Year	Model		Brake Disc			Brake Drum		Minimum Lining Thickness	Brake Caliper	
			Original Thickness	Minimum Thickness	Maximum Run-out	Original Inside Diameter	Maximum Machine Diameter		Bracket Bolts (ft. lbs.)	Mounting Bolts (ft. lbs.)
2009	Tribute	F	NA	0.944	0.004	NA	NA	0.118	37	129
		R	NA	NA	0.004	—	10.09	0.039	NA	26
2010	Tribute	F	NA	0.944	0.004	NA	NA	0.118	37	129
		R	NA	NA	0.004	—	10.09	0.039	NA	26

NA: Information not available

37671_TRIB_C0013

SCHEDULED MAINTENANCE INTERVALS
MAZDA Tribute

Number of months or kilometers (miles), whichever comes first

Maintenance Interval	6	12	18	24	30	36	42	48	54	60	66	72
Months	6	12	18	24	30	36	42	48	54	60	66	72
×1000 km	12	24	36	48	60	72	84	96	108	120	132	144
×1000 miles	7.5	15	22.5	30	37.5	45	52.5	60	67.5	75	82.5	90
Engine valve clearance (2.5L)	Audible inspect every 120,000 km (75,000 miles), if noisy, adjust											
Engine oil	R	R	R	R	R	R	R	R	R	R	R	R
Oil filter	R	R	R	R	R	R	R	R	R	R	R	R
Drive belts (tension) (2.5L)			I						I			
Drive belts (tension) (3.0L)				I				I				I
PCV valve (3.0L)[1]	Replace every 160,000 (1000,000 miles)											
Spark plugs	Replace every 120,000 km (75,000 miles)											
Air cleaner filter		I		R		I		R		I		R
Fuel filter[1]				R				R				R
Fuel lines and hoses[1]				I				I				I
Hoses and tubes for emission[1]								I				
Cooling system and hoses				I			I		I		I	
Engine coolant (yellow)	Replace every 160,000 km (100,000 miles) or 60 months; after that, every 80,000 km (50,000											
Engine coolant level	I	I	I	I	I	I	I	I	I	I	I	I
Brake lines, hoses and connections				I				I				I
Brake fluid level	I	I	I	I	I	I	I	I	I	I	I	I
Disc brakes		I		I			I		I		I	
Drum brakes		I	I	I	I				I			I
Tire (rotation), check wheel lug nut torque[2]	Rotate every 12,000 km (7,500 miles)											
Tire inflation pressure and tire wear	I	I	I	I	I	I	I	I	I	I	I	I
Steering operation and linkages		I	I	I					I			I
Power steering fluid level	I	I	I	I	I	I	I	I	I	I	I	I
Manual transaxle oil												R
Automatic transaxle fluid	Replace every 240,000 km (150,000 miles)											
Rear differential fluid (4WD)[3]	Replace every 240,000 km (150,000 miles)											
Transfer case fluid (4WD)[3]	Replace every 240,000 km (150,000 miles)											
Front and rear suspension, ball joints								I				I
Halfshaft dust boots		I	I	I						I		I
Bolts and nuts on chassis and body				I						I		I
Exhaust system heat shields				I		I		I		I		I
All locks and hinges	L	L	L	L	L	L	L	L	L	L	L	L
Washer fluid level	I	I	I	I	I	I	I	I	I	I	I	I

Chart symbols:

I: Inspect and repair, clean, adjust, or replace if necessary (oil-permeated air filter cannot be cleaned using the air-blow method).

R: Replace

L: Lubricate

Remarks:

1: According to state and federal regulations, failure to perform maintenance on these items will not void your emissions warranties. However, Mazda recommends that all maintenance services be performed at the recommended time or mileage period to ensure long-term reliability.

2: The wheel lug nuts must be retightened to the correct specifications at 800 km (500 miles) of new vehicle operation, at any wheel change, or at any other time the wheel lug nuts have been loosened.

3: If this component has been submerged in water, the oil should be changed.

BRAKES — INFORMATION AND PRECAUTIONS

ANTI-LOCK SYSTEMS

• Certain components within the ABS system are not intended to be serviced or repaired individually.

• Do not use rubber hoses or other parts not specifically specified for and ABS system. When using repair kits, replace all parts included in the kit. Partial or incorrect repair may lead to functional problems and require the replacement of components.

• Lubricate rubber parts with clean, fresh brake fluid to ease assembly. Do not use shop air to clean parts; damage to rubber components may result.

• Use only DOT 3 brake fluid from an unopened container.

• If any hydraulic component or line is removed or replaced, it may be necessary to bleed the entire system.

• A clean repair area is essential. Always clean the reservoir and cap thoroughly before removing the cap. The slightest amount of dirt in the fluid may plug an orifice and impair the system function. Perform repairs after components have been thoroughly cleaned; use only denatured alcohol to clean components. Do not allow ABS components to come into contact with any substance containing mineral oil; this includes used shop rags.

• The Anti-Lock control unit is a microprocessor similar to other computer units in the vehicle. Ensure that the ignition switch is **OFF** before removing or installing controller harnesses. Avoid static electricity discharge at or near the controller.

• If any arc welding is to be done on the vehicle, the control unit should be unplugged before welding operations begin.

DISC AND DRUM SYSTEMS

✳✳ CAUTION

Dust and dirt accumulating on brake parts during normal use may contain asbestos fibers from production or aftermarket brake linings. Breathing excessive concentrations of asbestos fibers can cause serious bodily harm. Exercise care when servicing brake parts. Do not sand or grind brake lining unless equipment used is designed to contain the dust residue. Do not clean brake parts with compressed air or by dry brushing. Cleaning should be done by dampening the brake components with a fine mist of water, then wiping the brake components clean with a dampened cloth. Dispose of cloth and all residue containing asbestos fibers in an impermeable container with the appropriate label. Follow practices prescribed by the Occupational Safety and Health Administration (OSHA) and the Environmental Protection Agency (EPA) for the handling, processing, and disposing of dust or debris that may contain asbestos fibers.

BRAKES — BLEEDING THE BRAKE SYSTEM

BLEEDING PROCEDURE

BLEEDING PROCEDURE

Manual Bleeding

➡**Pressure bleeding the brake system is preferred to manual bleeding.**

1. Clean all dirt from around the brake fluid reservoir cap and remove the cap. Fill the brake master cylinder reservoir with clean, specified brake fluid.

2. Remove the RR bleeder screw cap and place a box-end wrench on the bleeder screw. Attach a rubber drain hose to the RR bleeder screw and submerge the free end of the hose in a container partially filled with clean, specified brake fluid.

3. Have an assistant pump and then hold firm pressure on the brake pedal.

4. Loosen the RR bleeder screw until a stream of brake fluid comes out. While an assistant maintains pressure on the brake pedal, tighten the RR bleeder screw.

   a. Repeat until clear, bubble-free fluid comes out.

   b. Refill the brake master cylinder reservoir as necessary.

5. Tighten the RR bleeder screw to specifications. Remove the rubber hose and install the bleeder screw cap.

6. Repeat Steps 2 through 5 for the LR bleeder screw.

7. Remove the RF bleeder cap and place a box-end wrench on the bleeder screw. Attach a rubber drain hose to the RF bleeder screw and submerge the free end of the hose in a container partially filled with clean, specified brake fluid.

8. Have an assistant pump and then hold firm pressure on the brake pedal.

9. Loosen the RF bleeder screw until a stream of brake fluid comes out. While an assistant maintains pressure on the brake pedal, tighten the RF bleeder screw.

   a. Repeat until clear, bubble-free fluid comes out.

   b. Refill the brake master cylinder reservoir as necessary.

10. Tighten the RF bleeder screw to specifications. Remove the rubber hose and install the bleeder screw cap.

11. Repeat Steps 7 through 10 for the LF bleeder screw.

Pressure Bleeding

✳✳ WARNING

Do not use any fluid other than clean brake fluid meeting manufacturer's specification. Additionally, do not use brake fluid that has been previously drained. Following these instructions will help prevent system contamination, brake component damage and the risk of serious personal injury.

✳✳ WARNING

Carefully read cautionary information on product label. For EMERGENCY MEDICAL INFORMATION seek medical advice. For additional information, consult the product Material Safety Data Sheet (MSDS) if available. Failure to follow these instructions may result in serious personal injury.

✳✳ WARNING

Do not spill brake fluid on painted or plastic surfaces or damage to the surface may occur. If brake fluid is spilled onto a painted or plastic surface, immediately wash the surface with water.

➡**The hydraulic control unit (HCU) bleeding procedure must be carried out if the HCU or any components upstream of the HCU are installed new.**

➡**Pressure bleed the brake system at 207-345 kPa (30-50 psi).**

1. Clean all dirt from and remove the brake master cylinder filler cap and fill the brake master cylinder reservoir with clean, specified brake fluid.

2. Install the bleeder adapter to the brake master cylinder reservoir, and attach the bleeder tank hose to the fitting on the adapter.

➡️Master cylinder pressure bleeder adapter tools are available from various manufacturers of pressure bleeding equipment. Follow the instructions of the manufacturer when installing the adapter.

3. Remove the RR bleeder screw cap and place a box-end wrench on the RR bleeder screw. Attach a rubber drain hose to the RR bleeder screw and submerge the free end of the hose in a container partially filled with clean, specified fluid.

➡️Make sure the bleeder tank contains enough clean, specified brake fluid to complete the bleeding operation.

4. Open the valve on the bleeder tank.
5. Apply 30–50 psi (207–345 kPa) to the brake system.
6. Loosen the RR bleeder screw. Leave open until clear, bubble-free brake fluid flows, then tighten the RR bleeder screw to specifications. Remove the rubber hose.
7. Continue bleeding the rest of the system, going in order from the LR bleeder screw to the RF bleeder screw, ending with the LF bleeder screw.

8. Close the bleeder tank valve. Remove the tank hose from the adapter and remove the adapter. Fill the reservoir with clean, specified brake fluid and install the reservoir cap.

BLEEDING THE ABS SYSTEM

※※ WARNING

Do not use any fluid other than clean brake fluid meeting manufacturer's specification. Additionally, do not use brake fluid that has been previously drained. Following these instructions will help prevent system contamination, brake component damage and the risk of serious personal injury.

※※ WARNING

Carefully read cautionary information on product label. For EMERGENCY MEDICAL INFORMATION seek medical advice. For additional information, consult the product Material Safety Data Sheet (MSDS) if available. Failure to follow these instructions may result in serious personal injury.

※※ WARNING

Do not allow the brake master cylinder to run dry during the bleeding operation. Master cylinder may be

damaged if operated without fluid, resulting in degraded braking performance. Failure to follow this instruction may result in serious personal injury.

➡️Bleeding the hydraulic control unit (HCU) is required only when removing or installing the HCU, master cylinder, or opening the tubes and hoses to the HCU.

※※ CAUTION

Do not spill brake fluid on painted or plastic surfaces or damage to the surface may occur. If brake fluid is spilled onto a painted or plastic surface, immediately wash the surface with water.

➡️Pressure bleeding the brake system is preferred to manual bleeding.

1. For all vehicles, follow the Pressure Bleeding or Manual Bleeding procedure steps to bleed the system.
2. Connect the scan tool, access the SYSTEM BLEED FUNCTION and follow the directions on the scan tool.
3. Access the SYSTEM BLEED FUNCTION and follow the directions on the scan tool.
4. Repeat the Pressure Bleeding or Manual Bleeding procedure steps to bleed the system.

BRAKES   ANTI-LOCK BRAKE SYSTEM (ABS)

SPEED SENSORS

REMOVAL & INSTALLATION

Front

See Figure 1.

1. Raise and safely support the vehicle.

➡️The harness connector is located in the engine compartment.

2. Disconnect the electrical connector.

※※ WARNING

Care must be taken during the removal of the plug to prevent damage. If the plug is damaged, a new sensor may need to be installed, even though the sensor is functional in all other aspects.

3. Remove the grommet from the body.

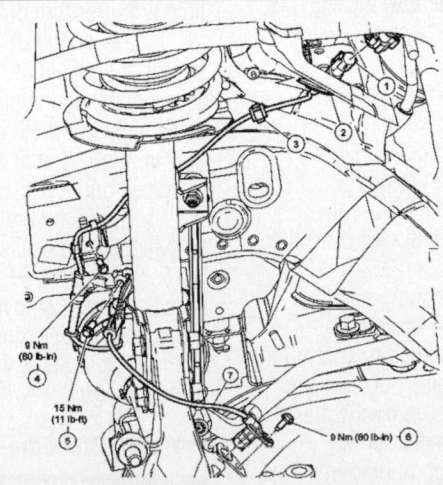

1. Front wheel speed sensor electrical connector
2. Grommet
3. Front wheel speed sensor harness retainer
4. Front wheel speed sensor harness-to-body bolt
5. Front wheel speed sensor harness-to-strut bolt
6. Front wheel speed sensor bolt
7. Front wheel speed sensor

32069_ESCA_G0056

Fig. 1 View of the front wheel speed sensor

4. When removing the body plug, rotate the plug into a position which allows the use of a small screwdriver to release the tabs on the underside of the body plug. These 2 tabs are located at right angles to the sensor wire.

5. Remove the front wheel speed sensor wire from the retainer.

6. Remove the front wheel speed sensor wire-to-body bolt.

7. Remove the front wheel speed sensor wire bolt.

8. Remove the front wheel speed sensor bolt from the wheel knuckle.

➡ **Clean off any foreign material that may have collected around the sensor before removal.**

9. Remove the front wheel speed sensor.

➡ **Thoroughly clean the mounting surface.**

10. Installation is the reverse of the removal procedure, noting the following tightening specifications:

   a. Front wheel speed sensor-to-knuckle bolt: 80 inch lbs. (9 Nm)

   b. Front wheel speed sensor wire bolt: 11 ft. lbs. (15 Nm)

   c. Front wheel speed sensor wire-to-body bolt: 80 inch lbs. (9 Nm)

Rear

See Figure 2.

1. Remove the wheel and tire.

✳✳ WARNING

Care must be taken during the removal of the plug to prevent damage. If the plug is damaged, a new sensor may need to be installed even though the sensor is functional in all other aspects.

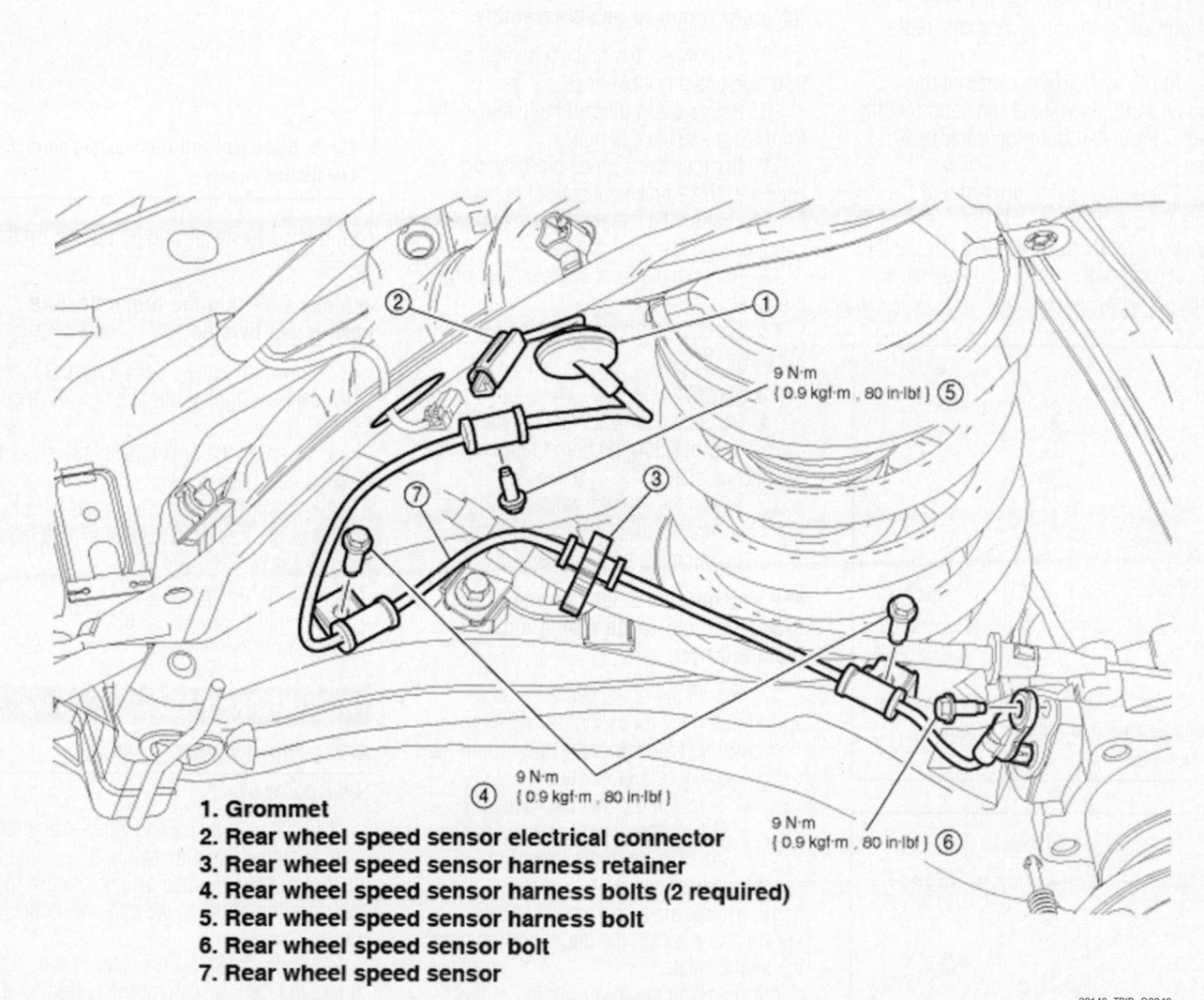

1. Grommet
2. Rear wheel speed sensor electrical connector
3. Rear wheel speed sensor harness retainer
4. Rear wheel speed sensor harness bolts (2 required)
5. Rear wheel speed sensor harness bolt
6. Rear wheel speed sensor bolt
7. Rear wheel speed sensor

22140_TRIB_G0049

Fig. 2 ABS rear wheel sensor full view

2. Remove the grommet from the body.

3. When removing the body plug, rotate the plug into a position which allows the use of a small screwdriver to release the tabs on the underside of the body plug. These 2 tabs are located at right angles to the sensor wire.

4. Disconnect the sensor wiring.

5. Detach the sensor wiring bolts.

➡**Clean off any dirt that may have collected around the sensor before removal.**

6. Remove the bolt and the sensor.

➡**Thoroughly clean the mounting surface.**

7. Installation is the reverse of the removal procedure. Tighten all retainers to 80 inch lbs. (9 Nm).

BRAKES

BRAKE CALIPER

REMOVAL & INSTALLATION

See Figures 3 through 5.

1. Remove the wheel and tire assembly.

2. Safely raise the vehicle.

3. For the LH brake caliper, release the lower portion of the brake pad anti-rattle spring.

4. Apply force to the center of the spring and pull outward at the bottom of the spring to remove it from the lower brake caliper cavity.

5. Rotate the spring upward and remove it from the brake caliper.

6. For the RH brake caliper, release the upper portion of the brake pad anti-rattle spring.

Fig. 3 For the LH brake caliper, rotate the spring upward and remove it from the brake caliper

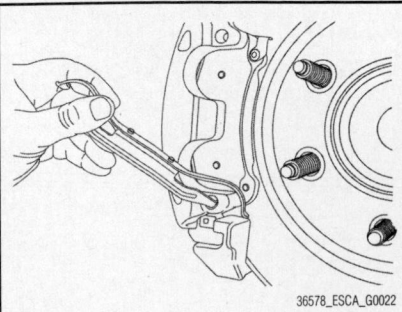

Fig. 4 For the RH brake caliper, rotate the spring downward and remove it from the brake caliper

7. Apply force to the center of the spring and pull outward at the top of the spring to remove it from the upper brake caliper cavity.

8. Rotate the spring downward and remove it from the brake caliper.

➡**The brake caliper and brake flexible hose are removed as an assembly.**

9. Disconnect the brake tube fitting from the brake flexible hose.

10. Remove and discard the retainer clip from the brake flexible hose.

11. Remove the 2 guide pin bushing caps and the 2 brake caliper guide pin bolts, position the caliper aside. Support the caliper using mechanic's wire.

12. Remove the front brake caliper from the vehicle.

13. Remove the brake flexible hose from the brake caliper.

To install:

14. Install the brake pads onto the caliper and position the brake caliper onto the anchor plate.

15. Install the 2 brake caliper guide pin bolts and tighten to 37 ft. lbs. (50 Nm).

16. Install the 2 bushing caps.

➡**If present, the 2-tabbed end of the brake pad anti-rattle spring must be installed first.**

17. Install the brake pad anti-rattle spring using the following procedure:

- Insert the tab of the spring into the brake caliper cavity.
- Twist the tab into the cavity (LH side in the upper brake caliper cavity, RH side in the lower brake caliper cavity).

18. Rotate the brake pad anti-rattle spring and position the upper portion onto the anchor plate.

19. Position the lower portion of the brake pad anti-rattle spring onto the anchor plate.

20. Push down and inward until the upper and lower ends of the brake pad anti-rattle spring are latched and seated in the brake caliper cavities.

21. Verify that the brake pad anti-rattle

FRONT DISC BRAKES

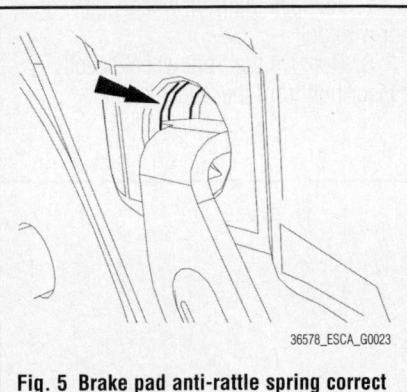

Fig. 5 Brake pad anti-rattle spring correct installation shown

spring is correctly latched by pulling on the spring.

➡**Make sure that the brake flexible hose is not twisted.**

22. Install the brake flexible hose to the brake caliper. Tighten the hose to 177 inch lbs. (20 Nm).

23. Position the brake flexible hose and install a new retainer clip.

24. Attach the brake tube fitting to the brake flexible hose and tighten to 159 inch lbs. (18 Nm).

25. Bleed the brake caliper.

26. Install the wheel and tire.

27. Lower the vehicle.

DISC BRAKE PADS

REMOVAL & INSTALLATION

See Figures 6 and 7.

1. Remove the tire and wheel assembly.

2. Safely raise the vehicle.

3. For the LH brake caliper, release the lower portion of the brake pad anti-rattle spring.

4. Apply force to the center of the spring and pull outward at the bottom of the spring to remove it from the lower brake caliper cavity.

5. Rotate the spring upward and remove it from the brake caliper.

6. For the RH brake caliper, release the upper portion of the brake pad anti-rattle spring.

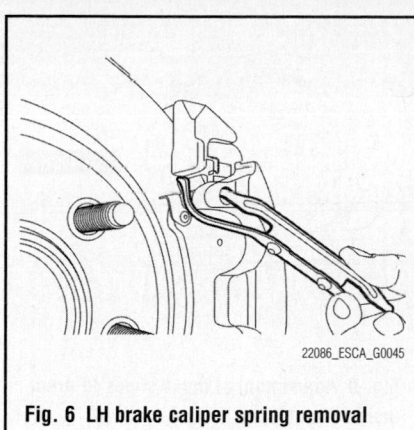

Fig. 6 LH brake caliper spring removal

7. Apply force to the center of the spring and pull outward at the top of the spring to remove it from the upper brake caliper cavity.

8. Rotate the spring downward and remove it from the brake caliper.

9. Remove the 2 guide pin bushing caps and the 2 brake caliper guide pin bolts, position the caliper aside.

10. Support the caliper using mechanic's wire.

11. Remove the 2 brake pads from the caliper.

12. Use a suitable tool to protect the brake caliper piston and compress the brake caliper piston into the brake caliper.

13. Inspect the brake disc and resurface or install new as necessary

To install:

14. Clean, dry and inspect the brake caliper anchor plate. Apply a light coat of specified lubricant to the 4 brake pad

➡**NOTE: Make sure that the brake flexible hose is not twisted.**

15. Install the brake pads onto the caliper and position the brake caliper onto the anchor plate.

16. Install the 2 brake caliper guide pin bolts and tighten to 37 ft. lbs. (50 Nm) install the 2 bushing caps.

➡**The 2-tabbed end of the brake pad anti-rattle spring must be installed first.**

17. Install the brake pad anti-rattle spring using the following procedure:

- Insert the tab of the spring into the brake caliper cavity.
- Twist the tab into the cavity (LH side in the upper brake caliper cavity, RH side in the lower brake caliper cavity).
- Rotate the brake pad anti-rattle spring and position the upper portion onto the anchor plate.
- Position the lower portion of the

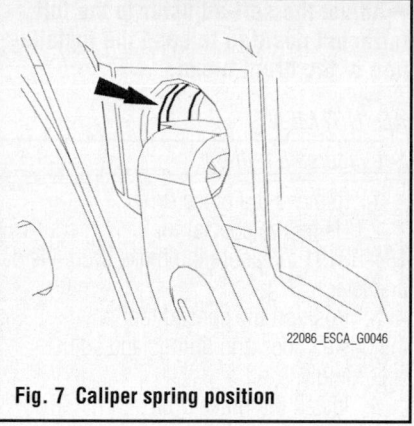

Fig. 7 Caliper spring position

brake pad anti-rattle spring onto the anchor plate.

- Push down and inward until the upper and lower ends of the brake pad anti-rattle spring are latched and seated in the brake caliper cavities.

⚹⚹ **WARNING**

The latch MUST be positioned as shown, or damage to component may occur.

➡**Verify that the brake pad anti-rattle spring is correctly latched by pulling on the spring.**

18. Install the wheel and tire.

BRAKES

BRAKE DRUM

REMOVAL & INSTALLATION

1. Remove the wheel and tire.
2. If the brake drum binds on the brake shoes, retract the brake shoes.

   a. Move the brake shoe adjuster actuator lever away from the adjuster.

   b. Rotate the brake shoe adjuster screw upward to retract the brake shoes.

⚹⚹ **CAUTION**

Use of a brake drum puller or a torch is not recommended. Brake drum distortion can result.

➡**If the brake drum is seized to the wheel hub pilot diameter, tap the center of the brake drum between the wheel studs.**

3. Remove the brake drum.

To install:

4. Using the Brake Drum Gauge, measure the inside diameter of the brake drum.

➡**Install a new brake drum if the inside diameter exceeds the specification stamped on the outside face of the brake drum.**

5. Adjust the rear drum brakes.
6. Position the brake drum of the vehicle.
7. Install the wheel and tire.

BRAKE SHOES

REMOVAL & INSTALLATION

⚹⚹ **WARNING**

Always install new brake shoes or pads at both ends of an axle to reduce the possibility of brake pulling vehicle to one side. Failure to follow this instruction may result in uneven braking and serious personal injury.

1. Remove the brake drum.

➡**If new rear brake shoes and linings are being installed, resurface the brake**

REAR DRUM BRAKES

drums to remove glazing and to provide an equal friction surface from side-to-side. Resurfacing also corrects out-of-round and bell conditions.

2. Remove the 2 brake shoe retaining springs and the 2 pins.
3. Remove the upper return spring.
4. Remove the self-adjuster and spring assembly.
5. Remove the lower return spring.
6. Remove the trailing brake shoe and parking brake actuator lever assembly.
7. Remove the leading brake shoe.

   a. Using specified brake parts cleaner, clean and dry the brake shoe contact points on the backing plate.

   b. Apply a thin coat of the specified silicone grease to the brake shoe contact points on the backing plate.

To install:

8. To install, reverse the removal procedure.
9. Adjust the rear brake shoes.

→Adjust the self-adjuster to the full retracted position to ease the installation of the drum brake.

ADJUSTMENT

See Figures 8 and 9.

1. Remove the brake drum.
2. Using the special tool, 134-R0191, measure the brake drum inside diameter.
3. Position the special tool on the brake shoes and linings and adjust accordingly.
4. Install the brake drum

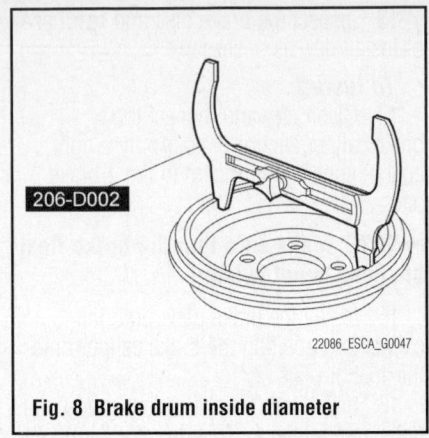

Fig. 8 Brake drum inside diameter

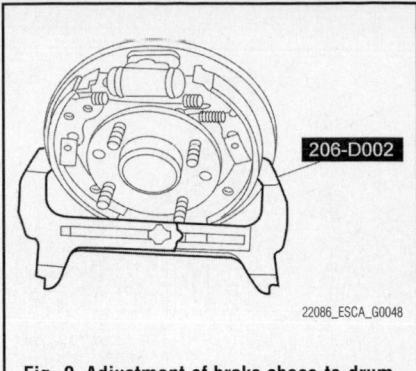

Fig. 9 Adjustment of brake shoes-to-drum inside diameter

CHASSIS ELECTRICAL   AIR BAG (SUPPLEMENTAL RESTRAINT SYSTEM)

GENERAL INFORMATION

✳✳ CAUTION

These vehicles are equipped with an air bag system. The system must be disarmed before performing service on, or around, system components, the steering column, instrument panel components, wiring and sensors. Failure to follow the safety precautions and the disarming procedure could result in accidental air bag deployment, possible injury and unnecessary system repairs.

SERVICE PRECAUTIONS

✳✳ CAUTION

Disconnect and isolate the battery negative cable before beginning any airbag system component diagnosis, testing, removal, or installation procedures. Wait at least 90 seconds after the ignition switch is turned off and the negative (-) terminal cable is disconnected from the battery before starting the operation. The SRS is equipped with a backup power source, so if work is started within 90 seconds after disconnecting the negative (-) terminal cable from the battery, the SRS may be deployed. Failure to disable the airbag system may result in accidental airbag deployment, personal injury, or death.

DISARMING THE SYSTEM

✳✳ WARNING

Never probe the electrical connectors on air bag, safety canopy or side air

curtain modules. Failure to follow this instruction may result in the accidental deployment of these modules, which increases the risk of serious personal injury or death.

✳✳ WARNING

Do not handle, move or change the original horizontal mounting position of the Restraints Control Module (RCM) while the RCM is connected and the ignition switch is ON. Failure to follow this instruction may result in the accidental deployment of the safety canopy and cause serious personal injury or death.

✳✳ WARNING

To reduce the risk of accidental deployment, do not use any memory saver devices. Failure to follow this instruction may result in serious personal injury or death.

→The air bag warning lamp illuminates when the RCM fuse is removed and the ignition switch is ON. This is normal operation and does not indicate a SRS fault.

→The SRS must be fully operational and free of faults before releasing the vehicle to the customer.

1. Turn all vehicle accessories OFF.
2. Turn the ignition switch to OFF.
3. At the Smart Power Distribution Junction Box (SPDJB), located at the RH side of the center console, remove the cover and the Restraints Control Module (RCM) fuse 32 (10A) from the SPDJB. For additional information, refer to the Wiring Diagram Manual.

4. Turn the ignition ON and visually monitor the air bag indicator for at least 30 seconds. The air bag indicator will remain lit continuously (no flashing) if the correct RCM fuse has been removed. If the air bag indicator does not remain lit continuously, remove the correct RCM fuse before proceeding.
5. Turn the ignition switch to OFF.
6. Disconnect the negative battery cable and wait at least one minute.

✳✳ WARNING

Always deplete the backup power supply before repairing or installing any new front or side air bag Supplemental Restraint System (SRS) component and before servicing, removing, installing, adjusting or striking components near the front or side impact sensors or the Restraints Control Module (RCM). Nearby components include doors, instrument panel, console, door latches, strikers, seats and hood latches. To deplete the backup power supply energy, disconnect the battery ground cable and wait at least 1 minute. Be sure to disconnect auxiliary batteries and power supplies (if equipped).Failure to follow these instructions may result in serious personal injury or death in the event of an accidental deployment.

ARMING THE SYSTEM

1. Turn the ignition switch from OFF to ON.
2. Install RCM fuses 32 (10A) into the SJB and install the cover.
3. Connect the negative battery cable.

Make sure no one is in the vehicle and there is nothing blocking or placed in front of any air bag module when the battery is connected. Failure to follow these instructions may result in serious personal injury in the event of an accidental deployment.

4. Prove out the SRS as follows:

a. Turn the ignition key from ON to OFF. Wait 10 seconds, then turn the key back to ON and visually monitor the air bag indicator with the air bag modules installed. The air bag indicator will light continuously for approximately 6 seconds and then turn off. If an air bag SRS fault is present, the air bag indicator will either:

• Fail to light
• Remain lit continuously
• Flash

b. The flashing might not occur until approximately 30 seconds after the ignition switch has been turned from the OFF to the ON position. This is the time required for the Restraints Control Module (RCM) to complete the testing of the SRS. If the air bag indicator is inoperative and a SRS fault exists, a chime will sound in a pattern of 5 sets of 5 beeps. If this occurs, the air bag indicator and any SRS fault discovered must be diagnosed and repaired.

5. Clear all continuous DTCs from the restraints control module using a scan tool.

CLOCKSPRING CENTERING

See Figures 10 through 12.

To reduce the risk of accidental deployment, do not use any memory saver devices. Failure to follow this instruction may result in serious personal injury or death.

➡The air bag warning lamp illuminates when the restraints control module (RCM) fuse is removed and the ignition switch is ON. This is normal operation and does not indicate a SRS fault.

➡The SRS must be fully operational and free of faults before releasing the vehicle to the customer.

➡Repair is made by installing a new part only. If the new part does not correct the condition, install the original

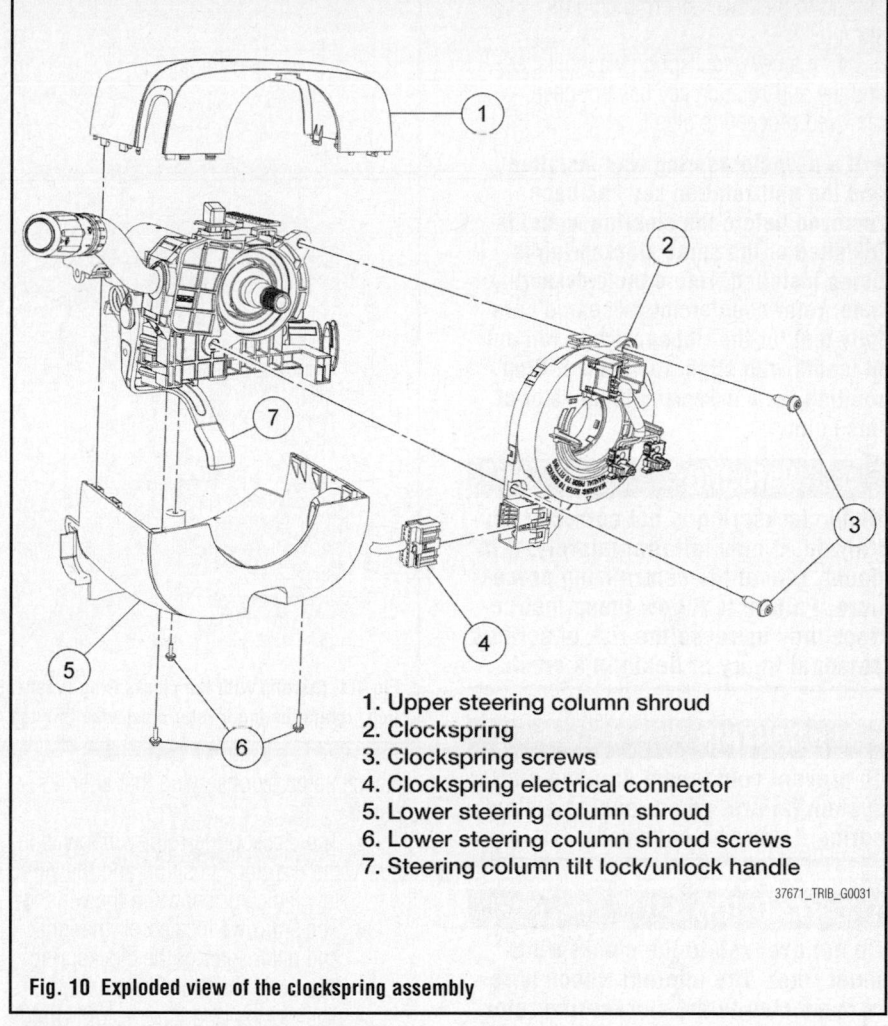

1. Upper steering column shroud
2. Clockspring
3. Clockspring screws
4. Clockspring electrical connector
5. Lower steering column shroud
6. Lower steering column shroud screws
7. Steering column tilt lock/unlock handle

37671_TRIB_G0031

Fig. 10 Exploded view of the clockspring assembly

part and carry out the diagnostic procedure again.

1. Disarm the system.
2. Remove the driver air bag module.
3. Tilt the steering wheel in the downward position and lock the tilt handle.
4. Remove the steering wheel.

To prevent damage to the clockspring make sure the road wheels are in the straight-ahead position.

5. Release the tabs and remove the upper steering column shroud.

➡The upper steering column shroud is a 2-piece design and must be removed as an assembly.

6. Remove the 3 lower steering column shroud screws. Then release the tilt column locking lever and remove the lower steering column shroud.

7. Disconnect the clockspring electrical connector.

8. Remove the 2 screws and the clockspring.

To install:

9. Install the clockspring and the 2 screws.

If installing a new clockspring, do not remove the clockspring anti-rotation key until the steering wheel is installed. If the anti-rotation key has been removed before installing the steering wheel, the clockspring must be centered. Failure to follow this instruction may result in component damage and/or system failure.

10. Connect the clockspring electrical connector.
11. Install the lower steering column shroud and the 3 screws.
12. Tilt the steering column down and lock the tilt column locking lever.
13. Attach the upper steering column

shroud to the lower steering column shroud.

14. If a new clockspring was installed and the anti-rotation key has not been removed proceed to step 8.

➡ If a new clockspring was installed and the anti-rotation key has been removed before the steering wheel is installed or the same clockspring is being installed, rotate the clockspring inner rotor counterclockwise and carefully feel for the ribbon wire to run out of length with slight resistance. Stop rotating the clockspring inner rotor at this point.

❋❋ WARNING

If the clockspring is not correctly centralized, it may fail prematurely. If in doubt, repeat the centralizing procedure. Failure to follow these instructions may increase the risk of serious personal injury or death in a crash.

❋❋ CAUTION

To prevent component damage and/or system failure, when reusing a clockspring it must be centered.

❋❋ CAUTION

Do not over-rotate the clockspring inner rotor. The internal ribbon wire is connected to the clockspring rotor. The internal ribbon wire acts as a stop and can be broken from its internal connection. Failure to follow this instruction may result in component damage and/or system failure.

15. Starting with the clockspring inner rotor, wiring and connector in the 12 o'clock position, rotate the inner rotor clockwise through 4 revolutions to center the clockspring.

16. Verify that the clockspring is cor-

37671_TRIB_G0032

Fig. 11 Starting with the clockspring inner rotor, wiring and connector in the 12 o'clock position, rotate the inner rotor clockwise through 4 revolutions to center the clockspring

rectly centered by observing that after 4 revolutions:

- The clockspring rotor window is in the 4 o'clock position and the yellow indicator shows in the window.
- The 2 arrows located on the inner and outer rotor of the clockspring line up in the 6 o'clock position.
- The clockspring inner rotor, wiring and connector are in the 12 o'clock position.

17. Install the steering wheel.

❋❋ CAUTION

To prevent damage to the clockspring make sure the road wheels are in the straight-ahead position.

➡ The clockspring inner rotor, wiring and connector must be in the 12 o'clock position to install the steering wheel.

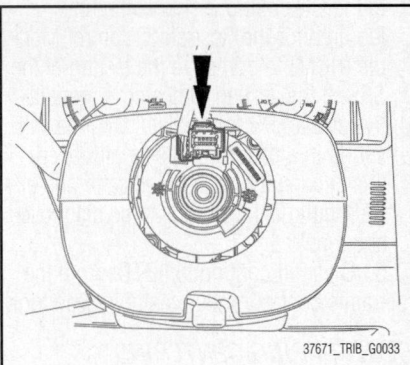

37671_TRIB_G0033

Fig. 12 The clockspring inner rotor, wiring and connector are in the 12 o'clock position

18. If a new clockspring was installed, remove the anti-rotation key.

19. Install the driver air bag module.

20. Arm the system.

DRIVE TRAIN

CLUTCH

REMOVAL & INSTALLATION

See Figures 13 through 15.

1. Disconnect the negative battery cable.
2. Remove the transaxle.
3. Using the flywheel holding tool, lock the flywheel to the engine.

➡**The clutch disc and clutch pressure plate are heavy and may fall if not held when the bolts are removed. Failure to follow this instruction may result in serious personal injury.**

4. Remove the 6 bolts, clutch pressure plate and clutch disc. Loosen the bolts evenly to prevent pressure plate damage.
5. Use a suitable cleaning solution to remove any oil film from the clutch pressure plate friction surface.
6. Inspect the clutch pressure plate surface for burn marks, scores, flatness or ridges.

➡**If the clutch disc is saturated with oil, inspect the rear engine crankshaft seal for leakage. If leakage is found, install a new seal prior to clutch disc installation.**

7. Use an emery cloth to remove minor imperfections in the clutch disc lining surface.
8. Install a new clutch disc if any of the following conditions are present:
   - Oil or grease saturation
   - Worn or loose facings
   - Warpage or loose rivets at the hub
   - Wear or rust on the splines
9. Check the flywheel runout and wear.

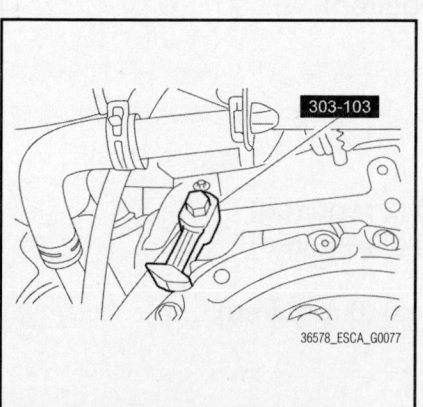

Fig. 13 Flywheel holding tool—303-103

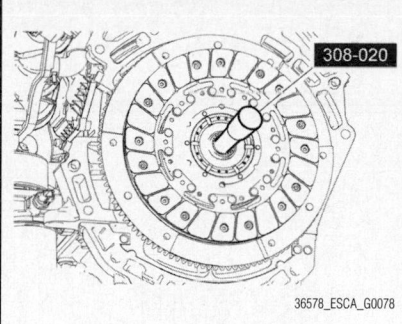

Fig. 14 Clutch disc installed with the alignment tool—308-020

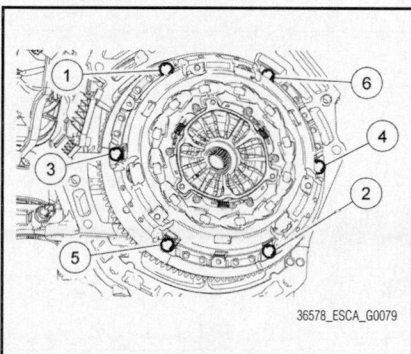

Fig. 15 Pressure plate tightening sequence

➡**The flywheel cannot be machined and must be replaced.**

To install:

10. Using the clutch aligner, position the clutch disc on the flywheel.
11. Position the clutch pressure plate on the flywheel and install the 6 clutch pressure plate bolts.
12. Tighten the bolts in the sequence shown to 21 ft. lbs. (29 Nm).
13. Install the transaxle.
14. Connect the negative battery cable.

TRANSFER CASE ASSEMBLY

REMOVAL & INSTALLATION

See Figure 16.

The transfer case is referred to as the Power Transfer Unit (PTU) for this model vehicle.

All vehicles

1. With the vehicle in NEUTRAL, position it on a hoist
2. Drain the Power Transfer Unit (PTU).
3. Remove the front RH intermediate shaft. Refer to Intermediate Shaft Removal & Installation in the Drive Train section.

4. Remove the driveshaft. Refer to Driveshaft Removal & Installation in the Drive Train section.
5. Remove the 4 bolts and the cross-member brace.

Vehicles equipped with 3.0L Engines

6. Remove the RH catalytic converter.

All vehicles

7. Remove the 3 PTU heat shield bolts and the PTU heat shield.
8. Remove the 2 exhaust bracket nuts.
9. Remove the 6 PTU -to-engine bracket bolts and the bracket.
10. Disconnect the PTU vent tube and position it aside.
11. Remove the 3 PTU -to-transaxle bolts.
12. Remove the PTU.

To install:
All vehicles

13. Position the Power Transfer Unit (PTU) to the transaxle.

❄❄ CAUTION

A new power transfer unit (PTU) intermediate shaft seal must be installed whenever the intermediate shaft or PTU is removed from the vehicle.

➡**If necessary, install a new RH differential fluid seal.**

14. Install the 3 PTU-to-transaxle bolts and tighten to 52 ft. lbs. (70 Nm).
15. Install the PTU-to-transaxle bolt (M10) and tighten to 35 ft. lbs. (48 Nm).
16. Connect the PTU vent tube.
17. Install the PTU-to-engine bracket and the 6 bracket bolts and 2 exhaust bracket bolts. Tighten the bolts to 35 ft. lbs. (48 Nm).
18. Install the PTU heat shield and the 3 PTU heat shield bolts. Tighten the bolts to 97 inch lbs. (11 Nm).

Vehicles equipped with 3.0L Engines

19. Install the RH catalytic converter.

All vehicles

20. Install the crossmember brace and the 4 bolts. Tighten the bolts to 30 ft. lbs. (40 Nm).
21. Install the driveshaft.
22. Install the front RH intermediate shaft.
23. Install the exhaust as required.
24. Fill the PTU with the correct fluid.

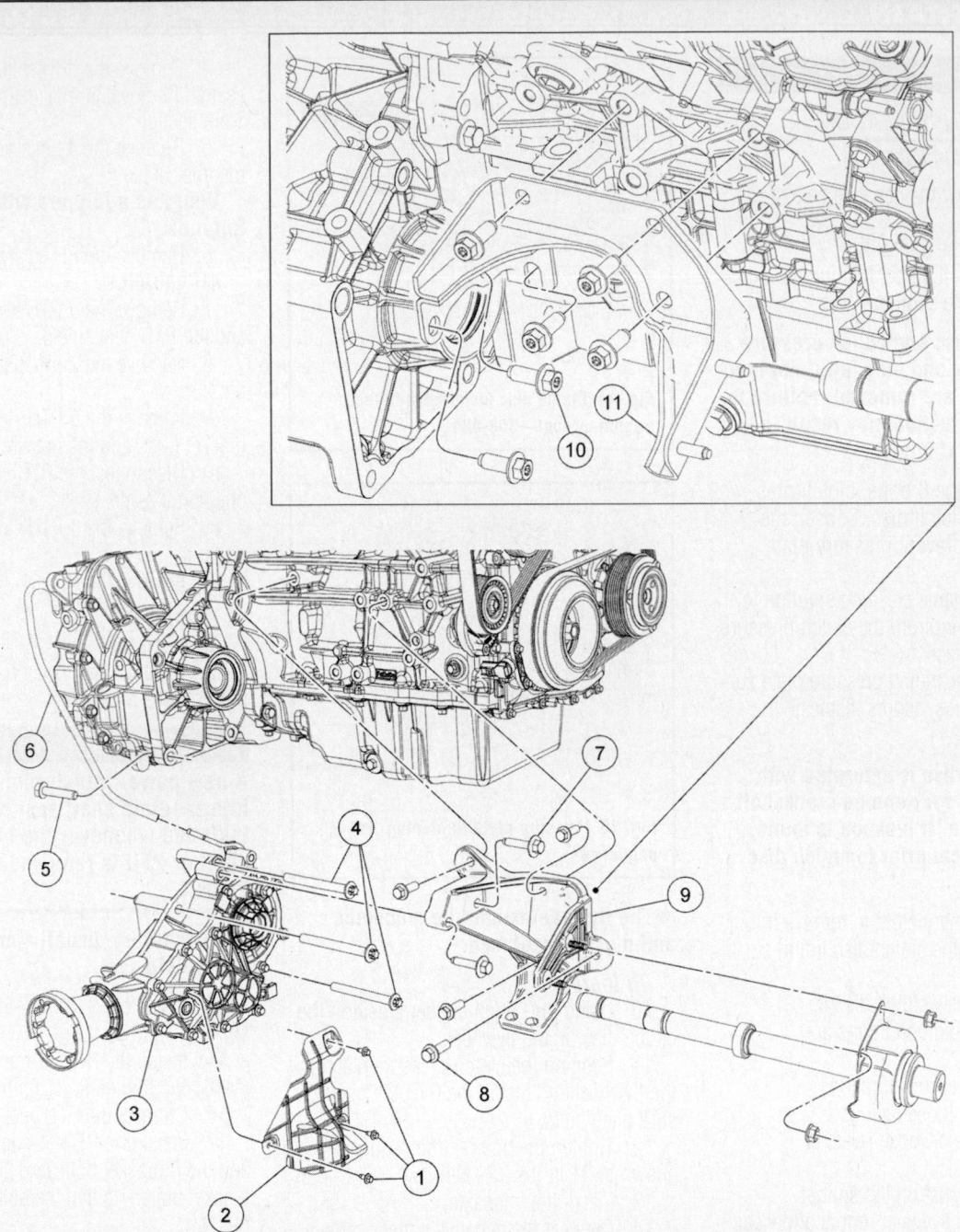

1. Power Transfer Unit (PTU) heat shield bolts (3 required)
2. PTU heat shield
3. PTU
4. PTU -to-transaxle bolts (3 required)
5. PTU -to-transaxle bolt (M10)
6. Vent tube
7. PTU -to-engine bracket bolt (4-cylinder engines) (2 required)
8. PTU -to-engine bracket bolt (4-cylinder engines) (4 required)
9. PTU -to-engine bracket (4-cylinder engines)
10. PTU -to-engine bracket bolts (6-cylinder engines) (6 required)
11. PTU -to-engine bracket (6-cylinder engines)

36578_ESCA_G0103

Fig. 16 Power Transfer Unit (PTU)

DRIVESHAFT

REMOVAL & INSTALLATION

See Figure 17.

❉❉ CAUTION

The normal operating temperature of the exhaust system is very high. Never attempt to remove any part of the system until it has cooled. Be especially careful when working around the catalytic converters. The temperature of the converter rises to a high level after only a few minutes of engine operation. Failure to follow these instructions may result in personal injury.

➡**Do not swap driveshaft assembles from different vehicles.**

1. With the vehicle in NEUTRAL, position it on a hoist.
2. Remove the ground strap bolt.

❉❉ WARNING

Do not reuse the CV-joint bolts and washers. Install new bolts and washers or damage to the vehicle may occur.

3. Remove and discard the 6 front driveshaft-to-transfer case bolts and washers.
4. Index-mark the front driveshaft to the center bearing.

❉❉ WARNING

Do not reuse the bolts and cap straps for the center U-joint. Install new bolts and cap straps or damage to the vehicle may occur.

➡**There is a difference in the length of the head of the replacement cap strap bolts from the production bolts. The longer head pinion bolts can be used in either location.**

5. Remove and discard the 4 universal joint cap strap bolts and 2 cap straps and remove the front driveshaft.
6. Index-mark the pinion and yoke to the driveshaft.

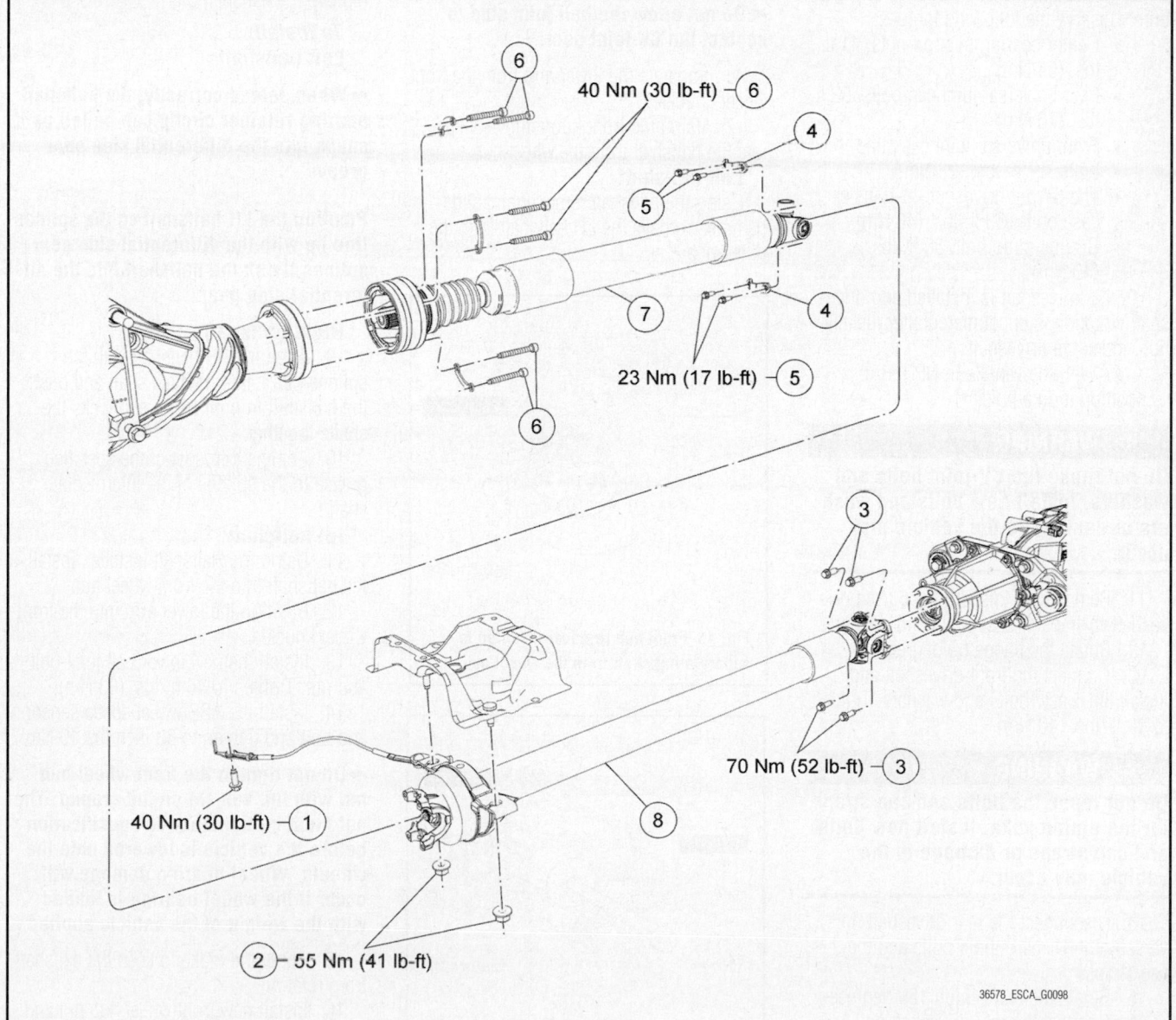

40 Nm (30 lb-ft)

23 Nm (17 lb-ft)

70 Nm (52 lb-ft)

40 Nm (30 lb-ft)

55 Nm (41 lb-ft)

36578_ESCA_G0098

Fig. 17 Rear driveshaft assembly

WARNING

Do not reuse the bolts and cap straps for the rear U-joint. Install new bolts and cap straps.

➡ **There is a difference in the length of the head of the replacement strap bolts from the production bolts. The longer head pinion bolts can be used in either location.**

7. Remove and discard the 4 universal joint cap bolts and 2 cap straps from the rear driveshaft universal joint.

8. With the help of an assistant, remove the center bearing support nuts and the driveshaft.

To install:

9. To install, reverse the removal procedure. Observe the following torques:
- Center bearing support nuts: 41 ft. lbs. (55 Nm)
- Rear universal joint cap bolts: 52 ft. lbs. (70 Nm)
- Front universal joint cap strap bolts: 17 ft. lbs. (23 Nm)
- The 6 front driveshaft-to-transfer case bolts: 30 ft. lbs. (40 Nm)
- Ground strap bolt: 30 ft. lbs. (40 Nm)

10. If a driveshaft is installed and driveshaft vibration is encountered after installation, index the driveshaft.

   a. With the vehicle in NEUTRAL, position it on a hoist.

WARNING

Do not reuse the CV-joint bolts and washers. Install new bolts and washers or damage to the vehicle may occur.

11. Remove and discard the 6 front driveshaft-to-transfer case bolts and washers.

12. Rotate the flange 60 degrees.

13. Connect the front driveshaft and install the 6 new bolts and washers. Tighten to 30 ft. lbs. (40 Nm).

WARNING

Do not reuse the bolts and cap straps for the pinion yoke. Install new bolts and cap straps or damage to the vehicle may occur.

14. Disconnect the rear driveshaft universal joint. Discard the 4 bolts and the 2 cap straps.

15. Rotate the rear pinion 180 degrees.

16. Connect the rear driveshaft and install 4 new bolts and 2 new cap straps. Tighten to 52 ft. lbs. (70 Nm).

17. Lower the vehicle and test drive.
18. Repeat the procedure if necessary.

FRONT HALFSHAFTS

REMOVAL & INSTALLATION
See Figures 18 through 20.

All halfshafts

1. With the vehicle in NEUTRAL, position it on a hoist.

2. Remove the front tire and wheel.

3. Remove and discard the front wheel hub nut.

4. Remove the ABS wheel speed sensor bolt and position the sensor aside.

5. Remove the lower arm pinch bolt and nut from the lower arm.

➡ **Do not allow the ball joint stud to contact the CV joint boot.**

6. Separate the lower arm from the front wheel knuckle.

7. Using the front hub remover, separate the halfshaft from the wheel hub.

Left halfshaft

Using the halfshaft remover and slide hammer, remove the LH halfshaft from the differential.

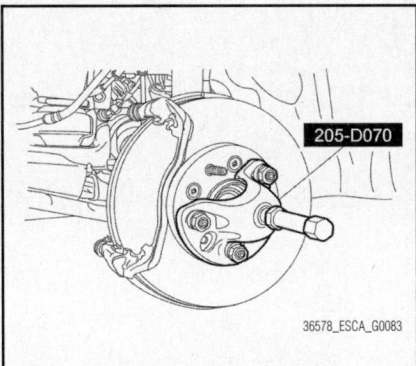

Fig. 18 Front hub remover installed to separate halfshaft from the wheel hub

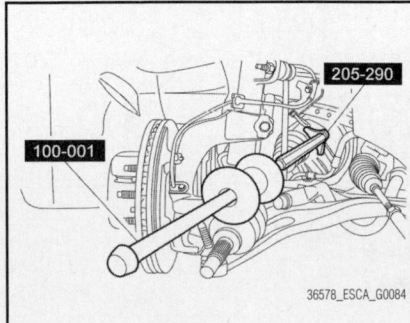

Fig. 19 Left halfshaft removal with slide hammer and adapter

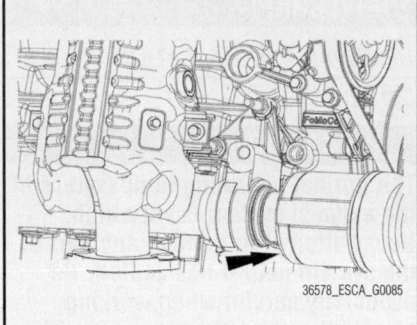

Fig. 20 Strike the RH in the area shown to remove the halfshaft

Right halfshaft

8. Using a brass drift to strike the RH halfshaft in the indicated area, separate and remove the halfshaft.

To install:
Left halfshaft

➡ **When seated correctly, the halfshaft bearing retainer circlip can be felt as it snaps into the differential side gear groove**

Position the LH halfshaft so the splines line up with the differential side gear splines. Push the halfshaft into the differential side gear.

Right halfshaft

9. Align the RH halfshaft with the splines of the intermediate shaft and push the halfshaft in until the circlip locks the shafts together.

10. Apply a thin coat of the specified grease to the splines of the intermediate shaft.

All halfshaft

11. Using the Halfshaft Installer, install the halfshaft into the front wheel hub.

12. Position the lower arm into the front wheel knuckle.

13. Install the new lower ball joint bolt and nut. Tighten to 46 ft. lbs. (63 Nm).

14. Install the ABS wheel speed sensor and bolt and tighten to 80 inch lbs. (9 Nm).

➡ **Do not tighten the front wheel hub nut with the vehicle on the ground. The nut must be tightened to specification before the vehicle is lowered onto the wheels. Wheel bearing damage will occur if the wheel bearing is loaded with the weight of the vehicle applied.**

15. Apply the brake to keep the halfshaft from rotating.

16. Install new front wheel hub nut and tighten to 222 ft. lbs. (300 Nm).

17. Install the front tires and wheels.

18. Check and fill the transaxle fluid as necessary.

CV-BOOTS INSPECTION

1. With the vehicle in NEUTRAL, position it on a hoist.

➡**Rotate the wheels to inspect the entire boot.**

2. Check the CV-boots for rips and tears.
3. Check for grease leaking from the CV-boots.
4. Check the boot bands for damage.
5. Replace any component if found to be defective.

FRONT INTERMEDIATE SHAFT

REMOVAL & INSTALLATION
See Figure 21.

➡**If removing the intermediate shaft in order to repair a separate component, it should only be removed as an assembly with the RH front drive half-shaft.**

1. Remove the RH halfshaft. Refer to halfshaft Removal & Installation in this section.
2. Remove the 2 intermediate shaft bearing retainer nuts.
3. Remove the intermediate shaft.

➡**On All-Wheel Drive (AWD) vehicles, the Power Transfer Unit (PTU) seal must be replaced every time the intermediate shaft is removed.**

To install:
4. Install a new PTU seal for AWD models.
5. Install the intermediate shaft.
6. Install the 2 intermediate shaft bearing retainer nuts. Tighten the retaining nuts to 20 ft. lbs. (27 Nm).
7. Apply a thin coat of the specified grease to the splines of the intermediate shaft.
8. Install the RH halfshaft.
9. Check and fill the transaxle fluid as necessary.

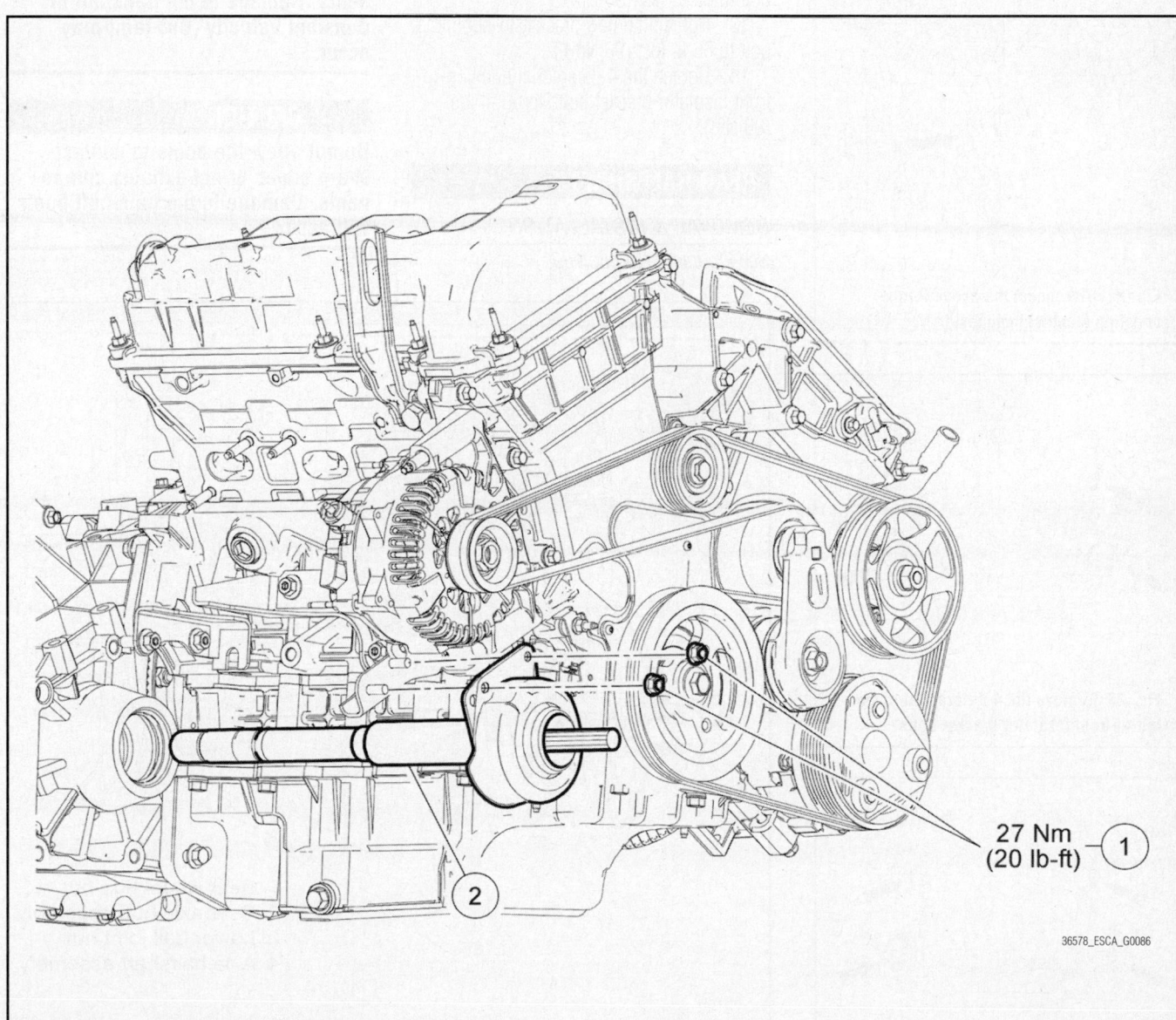

27 Nm
(20 lb-ft) ─①

②

36578_ESCA_G0086

Fig. 21 Intermediate shaft bearing retainer nuts (1) and shaft (2)

REAR DIFFERENTIAL

REMOVAL & INSTALLATION

See Figures 22 through 24.

1. Remove the rear driveshaft assembly.

2. Remove the rear halfshafts.

3. Position a suitable transmission hydraulic jack to the axle housing. Securely strap the jack to the housing.

4. Disconnect the active torque coupling electrical connector.

5. Remove the 4 differential housing-to-front insulator bracket bolts.

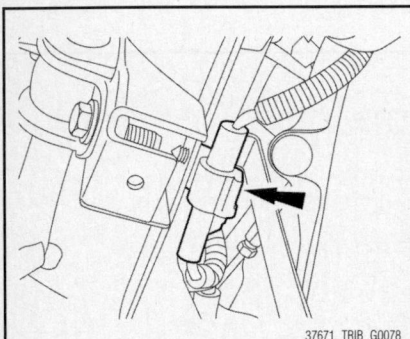

Fig. 22 Disconnect the active torque coupling electrical connector

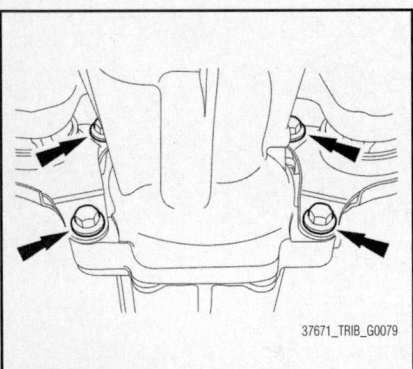

Fig. 23 Remove the 4 differential housing-to-front insulator bracket bolts

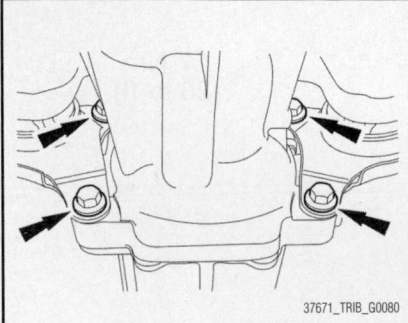

Fig. 24 Remove the 3 LH side insulator bracket-to-rear axle differential bolts

6. Remove and discard the LH front insulator bracket-to-subframe bolt and rotate the bracket aside.

7. Remove and discard the RH front insulator bracket-to-subframe bolt and the bracket.

8. Remove the 3 LH side insulator bracket-to-rear axle differential bolts.

9. Lower the rear axle assembly.

To install:

10. To install, reverse the removal procedure.

11. Tighten the 3 LH side insulator bracket-to-rear axle differential bolts to 66 ft. lbs. (90 Nm).

12. Tighten the new bracket-to-subframe bolt to 66 ft. lbs. (90 Nm).

13. Tighten the new bracket-to-subframe bolt to 66 ft. lbs. (90 Nm).

14. Tighten the 4 differential housing-to-front insulator bracket bolts to 66 ft. lbs. (90 Nm).

REAR HALFSHAFTS

REMOVAL & INSTALLATION

See Figures 25 through 29.

❋❋ CAUTION

Never pick up or hold the halfshaft by only the inner or outer Constant Velocity (CV) joint. Damage to the CV joint will occur.

❋❋ CAUTION

Never use a hammer to remove or install the halfshafts. Damage to the halfshaft may occur.

❋❋ CAUTION

Never use the halfshaft assembly as a lever to position other components. Damage to the halfshaft or Constant Velocity (CV) joint may occur.

❋❋ CAUTION

Do not allow the boots to contact sharp edges or hot exhaust components. Damage to the halfshaft boots will occur.

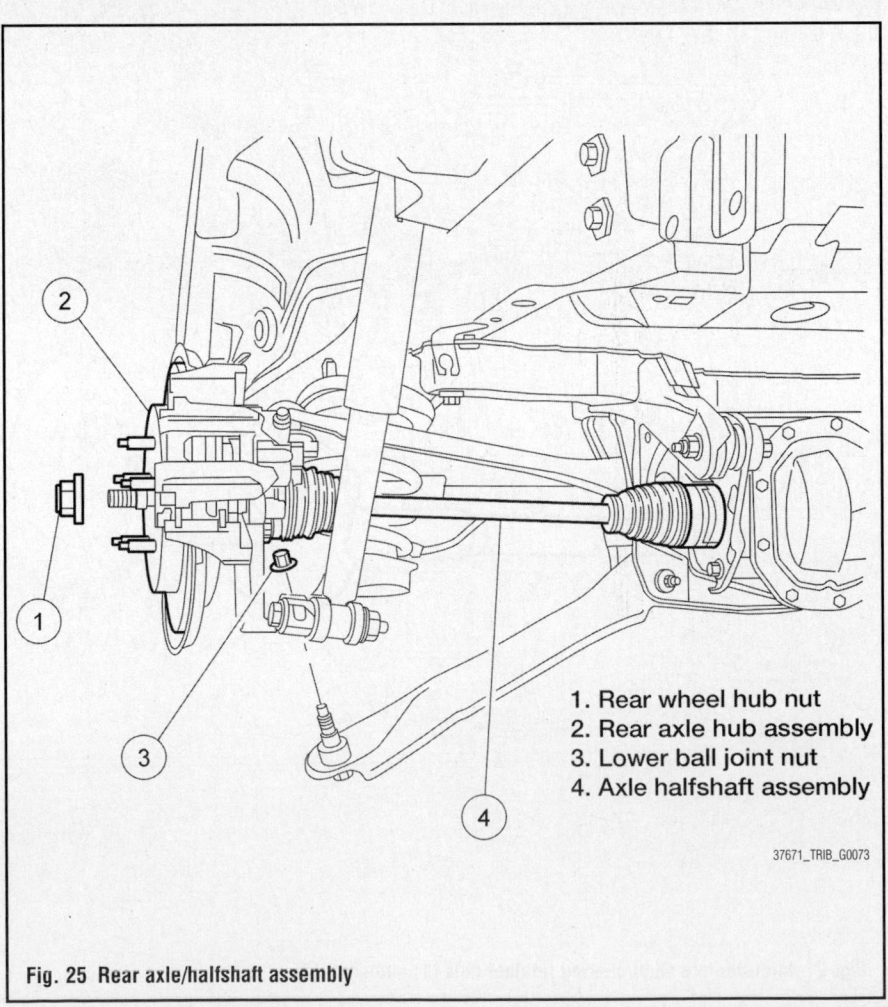

1. Rear wheel hub nut
2. Rear axle hub assembly
3. Lower ball joint nut
4. Axle halfshaft assembly

Fig. 25 Rear axle/halfshaft assembly

Do not drop assembled halfshafts. The impact may cut the boots from the inside without evidence of external damage.

1. Remove the rear coil spring.
2. Remove and discard the rear wheel hub nut.
3. Remove the anti-lock brake sensor harness-to-body retainer bolt.
4. Remove the nut and separate the lower ball joint.

➡ **Support the wheel knuckle.**

5. Remove the 4 differential housing-to-front insulator bracket bolts.
6. Remove the 3 side insulator bracket-to-rear axle differential bolts.
7. Remove the rear differential mount-to-subframe bolt and nut.
8. Using the Halfshaft Remover Tool, remove the halfshaft from the differential.

Do not damage the oil seal when removing the axle halfshaft from the differential.

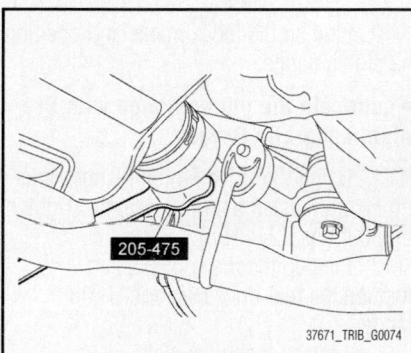

Fig. 26 Using the Halfshaft Remover Tool, remove the halfshaft from the differential

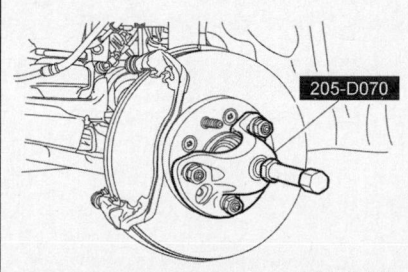

Fig. 27 Using the Front Hub Remover, separate the halfshaft from the rear axle hub assembly

➡ **Support the halfshaft inner joint.**

9. Using the Front Hub Remover, separate the halfshaft from the rear axle hub assembly.
10. Remove the halfshaft.

To install:

11. Using the Halfshaft Installer, install the outer halfshaft end into the hub assembly.
12. Using the Axle Seal Protector, install the halfshaft into the differential.

Make sure that the oil seal protector is correctly aligned with the differential oil seal during installation.

13. If the axle is equipped with the oil seal protector, make sure the oil seal lip and seal protector are correctly aligned.
14. Install the rear differential mount-to-subframe bolt and nut. Tighten to 98 ft. lbs. (133 Nm).

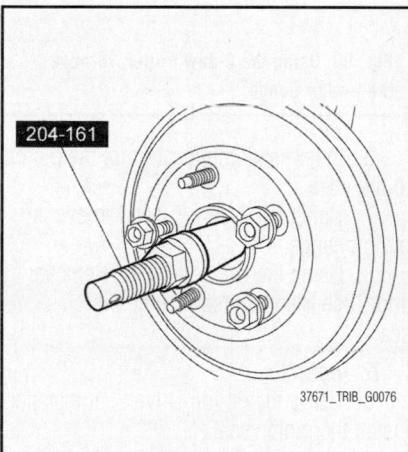

Fig. 28 Using the Halfshaft Installer, install the outer halfshaft end into the hub assembly

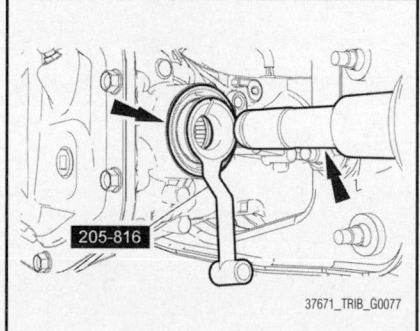

Fig. 29 Using the Axle Seal Protector, install the halfshaft into the differential

15. Install the 3 side insulator bracket-to-rear axle differential bolts. Tighten to 66 ft. lbs. (90 Nm).
16. Install the 4 differential housing-to-front insulator bracket bolts. Tighten to 66 ft. lbs. (90 Nm).
17. Position the lower ball joint and install the lower ball joint nut. Tighten to 76 ft. lbs. (103 Nm).
18. Install the rear coil spring.
19. Install the anti-lock brake sensor harness-to-body bolt.
20. Install the rear wheel hub nut. Tighten to 221 ft. lbs. (300 Nm).

Do not tighten the rear wheel hub nut with the vehicle on the ground. The nut must be tightened to specification before the vehicle is lowered onto the wheels. Wheel bearing damage will occur if the wheel bearing is loaded with the weight of the vehicle applied.

➡ **Apply the brake to keep the halfshaft from rotating.**

CV-BOOTS INSPECTION

1. With the vehicle in NEUTRAL, position it on a hoist.

➡ **Rotate the wheels to inspect the entire boot.**

2. Check the CV-boots for rips and tears.
3. Check for grease leaking from the CV-boots.
4. Check the boot bands for damage.
5. Replace any component if found to be defective.

REAR PINION SEAL

REMOVAL & INSTALLATION
See Figures 30 through 35.

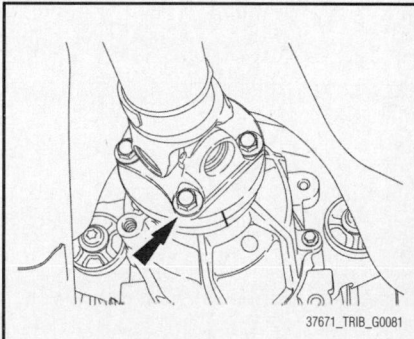

Fig. 30 Disconnect the rear driveshaft U-joint flange

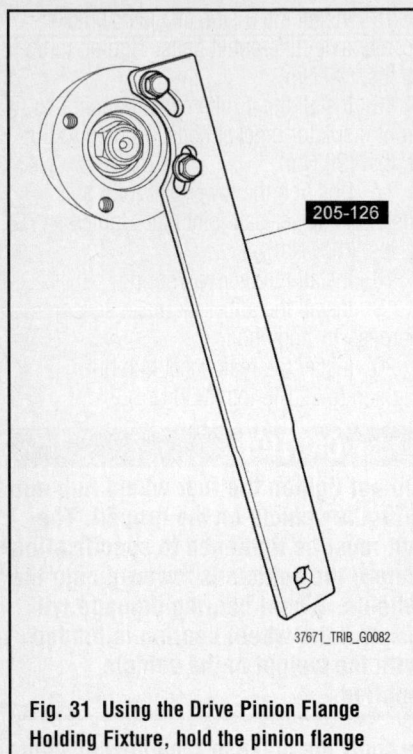

Fig. 31 Using the Drive Pinion Flange Holding Fixture, hold the pinion flange while removing the nut

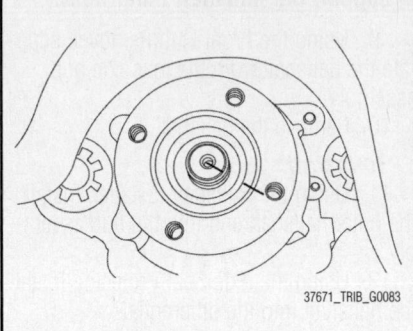

Fig. 32 Index-mark the location of the pinion to the yoke

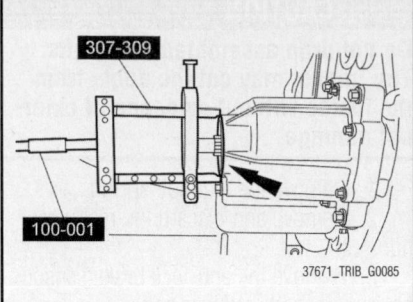

Fig. 34 Using the Converter Seal Remover and Slide Hammer, remove the pinion seal

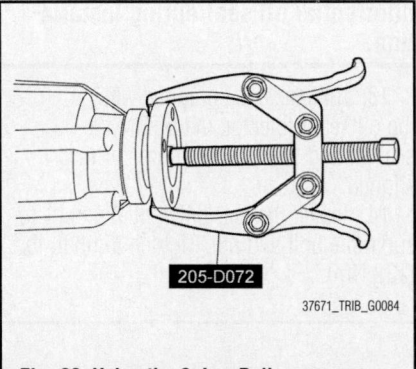

Fig. 33 Using the 2-Jaw Puller, remove the pinion flange

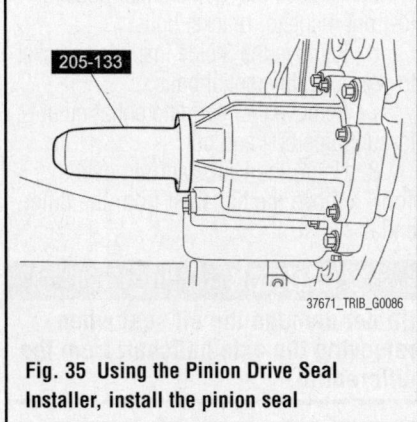

Fig. 35 Using the Pinion Drive Seal Installer, install the pinion seal

1. With the vehicle in NEUTRAL, position it on a hoist.

2. Index-mark the driveshaft flange and the pinion flange.

3. Disconnect the rear driveshaft U-joint flange.

  a. Remove and discard the 4 bolts.

  b. Position aside the driveshaft and flange.

➡**Support the driveshaft.**

4. Using the Drive Pinion Flange Holding Fixture, hold the pinion flange while removing the nut.

5. Remove and discard the pinion flange nut.

6. Index-mark the location of the pinion to the yoke.

7. Using the 2-Jaw Puller, remove the pinion flange.

8. Using the Converter Seal Remover and Slide Hammer, remove the pinion seal.

To install:

9. Using the Pinion Drive Seal Installer, install the pinion seal.

➡**Make sure the mating surface is clean before installing the new seal.**

10. Line up the index marks and position the pinion flange.

➡**Lubricate the pinion flange with Premium Long-Life Grease.**

11. Using the Drive Pinion Flange Holding Fixture, install a new pinion nut. Tighten to 180 ft. lbs. (244 Nm).

12. Line up the index marks and position the rear driveshaft and U-joint flange.

  a. Install the 4 new bolts.

  b. Tighten to 52 ft. lbs. (70 Nm).

ENGINE COOLING

ENGINE FAN

REMOVAL & INSTALLATION

See Figure 36.

All vehicles

1. With vehicle in NEUTRAL, position it on a hoist.

2.5L vehicles

2. Drain the cooling system.

All vehicles

3. Remove the front bumper cover.

4. Remove the 2 front impact severity sensors.

5. Remove the 2 pin-type retainers.

6. Remove the 4 bolts and the 2 radiator brackets.

7. Mark the hood latch position prior to removal of the bolts.

8. Loosen the nut, remove the 2 bolts and position aside the hood latch.

9. Remove the 2 wiring harness retainers from the radiator support.

10. Remove the center support bolt.

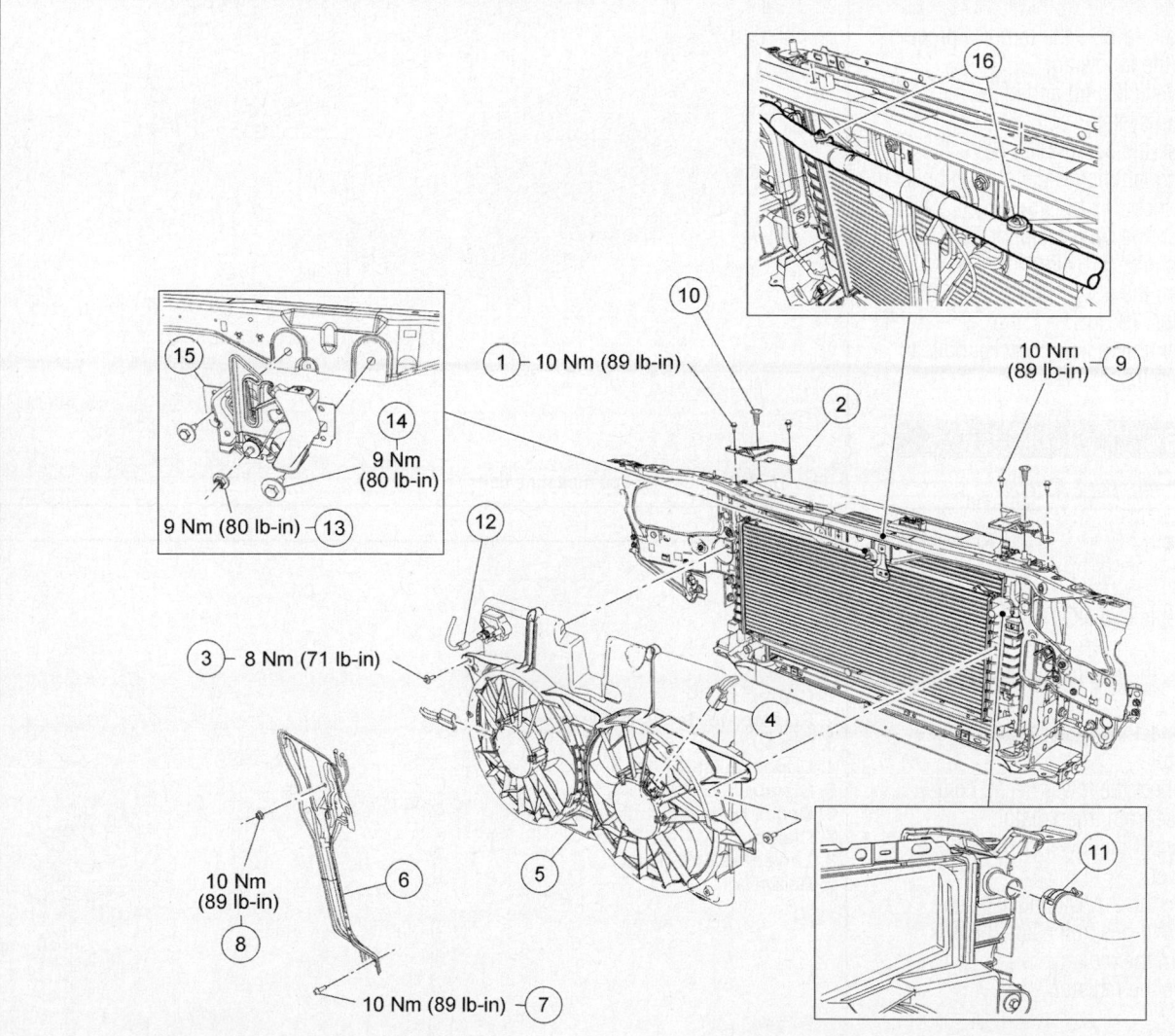

1. Radiator bracket bolt (4 required)
2. Radiator bracket (2 required)
3. Cooling fan bolt (2 required)
4. Cooling fan electrical connector (2 required)
5. Cooling fan motor and shroud
6. Center support
7. Center support lower bolt
8. Center support nut
9. Front grille bolt (2 required)
10. Pin-type retainer (2 required
11. Coolant recovery hose (2.3L only)
12. Cooling fan resistor electrical connector (2.3L only)
13. Hood latch nut
14. Hood latch bolt (2 required)
15. Hood latch
16. Wiring harness retainers (2 required)

22086_ESCA_G0021

Fig. 36 Cooling fan motor and shroud

2.5L vehicles

11. Disconnect the cooling fan resistor electrical connector.

12. Disconnect the coolant recovery hose from the radiator and position it aside.

All vehicles

13. Disconnect the 2 cooling fan electrical connectors and wire harness retainers.

14. Remove the 2 cooling fan bolts and the cooling fan motor and shroud.

To install:

15. To install, reverse the removal procedure and note the following:

- 2.5L vehicles, fill and bleed the cooling system.
- Tighten the cooling fan bolts, the cooling fan motor and shroud bolts to 71 inch. lbs (8 Nm).
- Tighten the center support bolt to 89 inch lbs. (10 Nm).
- Tighten the hood latch bolts to 80 inch lbs. (9 Nm).
- Tighten the radiator bracket bolts to 89 inch lbs. (10 Nm).

RADIATOR

REMOVAL & INSTALLATION

2.5L Engines

See Figure 37.

1. With vehicle in NEUTRAL, position it on a hoist.

2. Remove the cooling fan motor and shroud.

3. Disconnect the upper radiator hose from the radiator.

4. Disconnect the lower degas bottle-to-radiator hose from the radiator.

5. Disconnect the lower radiator hose from the radiator.

6. Remove the 2 A/C condenser-to-radiator bolts and position aside the A/C condenser from the radiator.

7. Remove the radiator.

To install:

8. To install, reverse the removal procedure.

9. Tighten the A/C condenser-to-radiator bolts to 71 inch lbs. (8 Nm).

10. Fill and bleed the cooling system.

3.0L Engines

See Figure 38.

1. With vehicle in NEUTRAL, position it on a hoist.

2. Drain the cooling system.

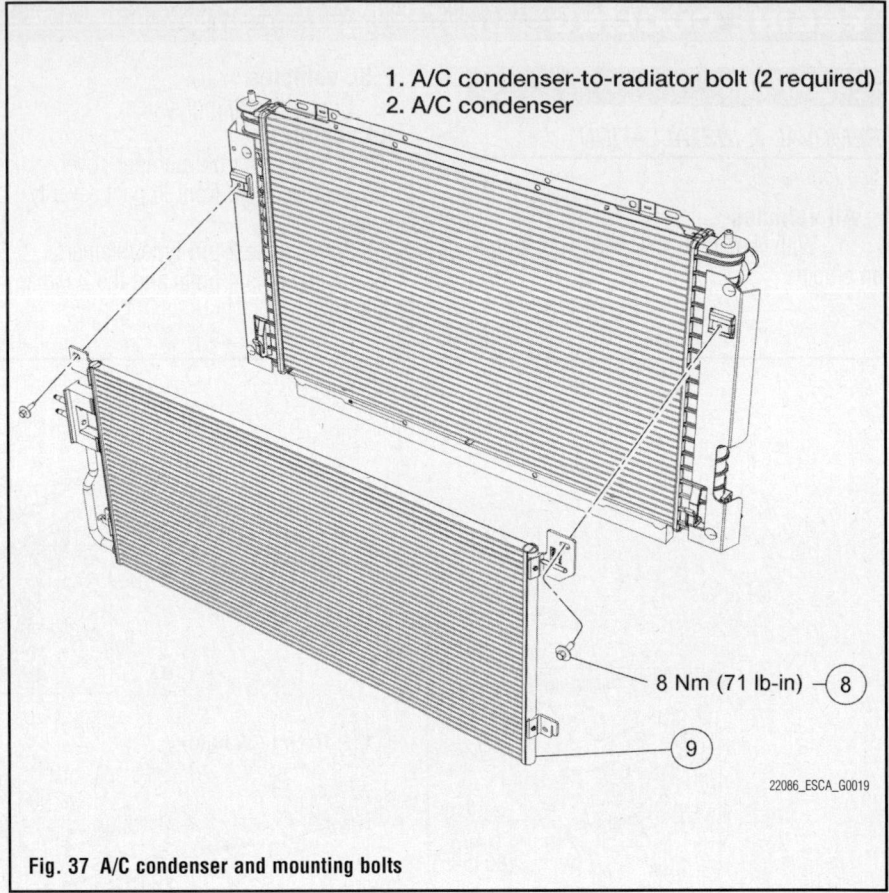

1. A/C condenser-to-radiator bolt (2 required)
2. A/C condenser

8 Nm (71 lb-in) — 8

9

22086_ESCA_G0019

Fig. 37 A/C condenser and mounting bolts

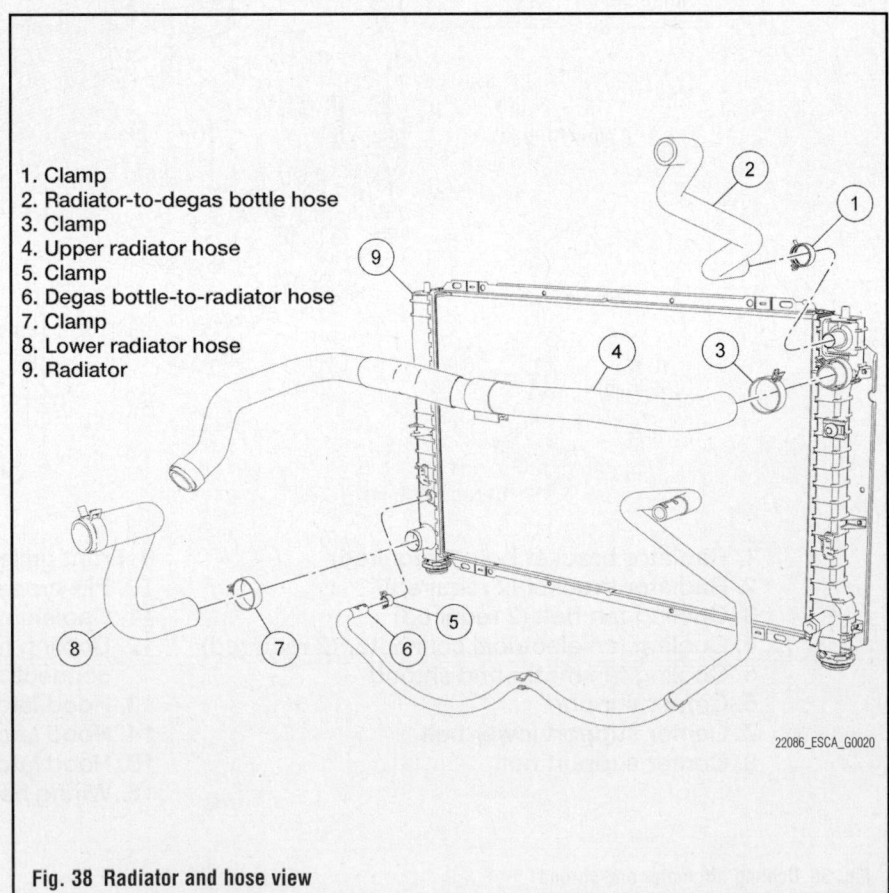

1. Clamp
2. Radiator-to-degas bottle hose
3. Clamp
4. Upper radiator hose
5. Clamp
6. Degas bottle-to-radiator hose
7. Clamp
8. Lower radiator hose
9. Radiator

22086_ESCA_G0020

Fig. 38 Radiator and hose view

3. Remove the cooling fan motor and shroud.

4. Disconnect the radiator-to-degas bottle hose and upper radiator hose from the radiator.

5. Disconnect the lower degas bottle-to-radiator hose from the radiator.

6. Disconnect the lower radiator hose from the radiator.

7. Remove the 2 A/C condenser-to-radiator bolts and position aside the A/C condenser from the radiator.

8. Remove the radiator.

To install:

9. To install, reverse the removal procedure.

10. Tighten the A/C condenser-to-radiator bolts to 71 inch lbs. (8 Nm).

11. Fill and bleed the cooling system

THERMOSTAT

REMOVAL & INSTALLATION

2.5L Engine

See Figure 39.

➡On the 2.5L engine, the thermostat and thermostat housing are serviced as an assembly.

1. Drain the cooling system.

2. Disconnect the heater hose at the thermostat housing.

3. Disconnect the lower radiator hose at the thermostat housing.

4. Remove the 3 bolts, thermostat housing and gasket.

➡The view of the thermostat housing bolts is obstructed by A/C and other

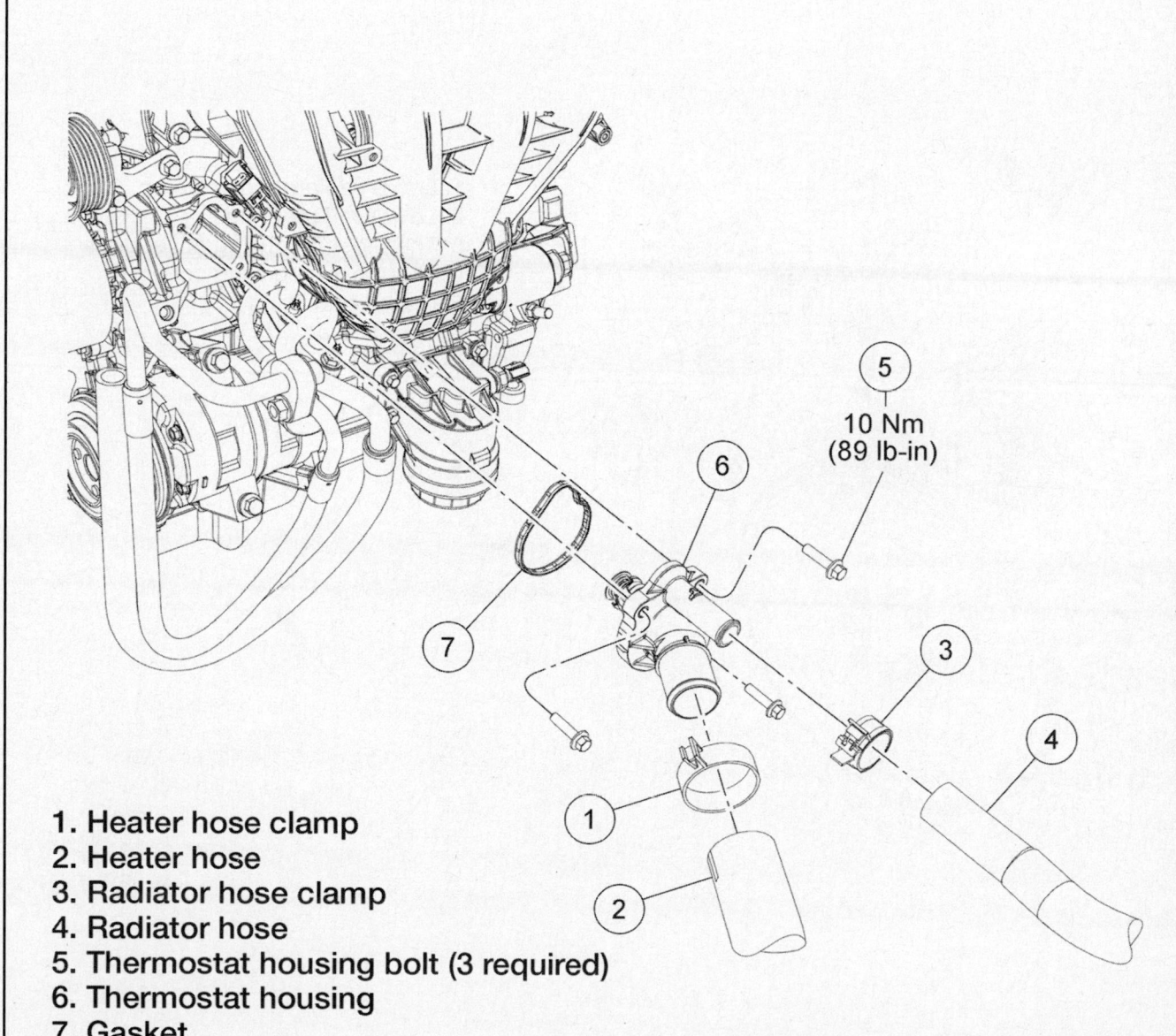

5
10 Nm
(89 lb-in)

1. **Heater hose clamp**
2. **Heater hose**
3. **Radiator hose clamp**
4. **Radiator hose**
5. **Thermostat housing bolt (3 required)**
6. **Thermostat housing**
7. **Gasket**

36578_ESCA_G0116

Fig. 39 2.5L engine thermostat mounting

engine components. However, the bolts can be removed by using ¼ inch drive hand tools.

5. Remove and discard the gasket.
6. To install, reverse the removal procedure.
7. Install a new gasket.
8. Tighten the thermostat housing bolts to 89 inch lbs. (10 Nm).

9. Fill and bleed the cooling system.

3.0L Engine

See Figure 40.

1. Drain the cooling system.
2. Disconnect the lower radiator hose from the thermostat housing.
3. Remove the 2 bolts, thermostat housing cover, O-ring seal and thermostat.

4. To install, reverse the removal procedure.
5. Clean and inspect the O-ring seal. Install a new seal if necessary.

➡To install, lubricate the thermostat housing O-ring seal with clean engine coolant.

6. Fill and bleed the cooling system.

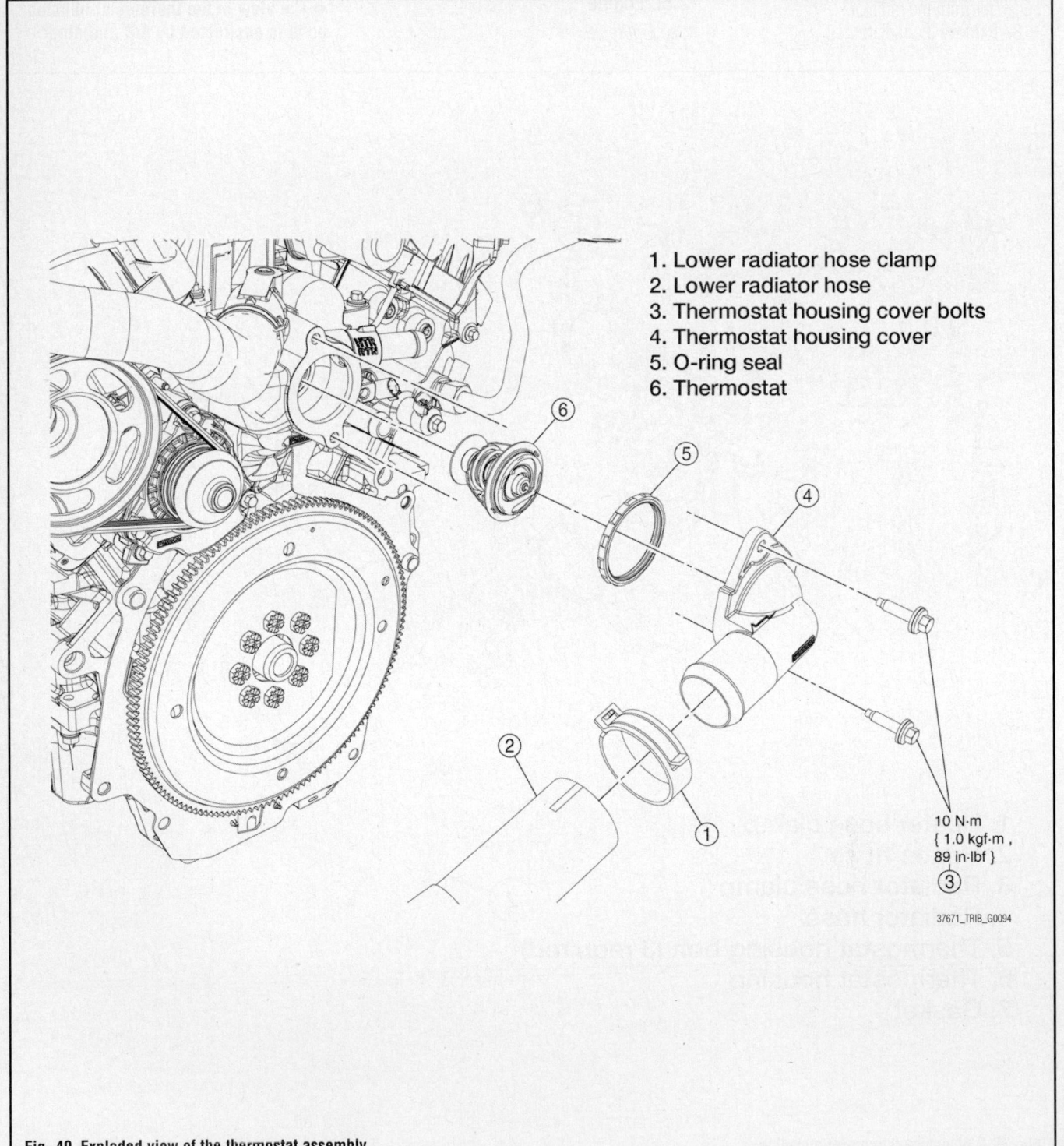

1. Lower radiator hose clamp
2. Lower radiator hose
3. Thermostat housing cover bolts
4. Thermostat housing cover
5. O-ring seal
6. Thermostat

10 N·m
{ 1.0 kgf·m ,
89 in·lbf }

37671_TRIB_G0094

Fig. 40 Exploded view of the thermostat assembly

WATER PUMP

REMOVAL & INSTALLATION

2.5L Engine

See Figure 41.

1. Drain the cooling system.
2. Loosen the 3 coolant pump pulley bolts.
3. Remove the accessory drive belt.
4. Remove the 3 coolant pump pulley bolts and the pulley.

5. Remove the 3 coolant pump bolts.
6. Remove the coolant pump.

✳✳ CAUTION

Make sure the coolant pump is correctly seated to the engine block before installing and tightening the fasteners, or damage to the coolant pump may occur.

7. Remove and discard the coolant pump O-ring seal.

To install:

8. To install, reverse the removal procedure and note the following:
   - Lubricate the new O-ring seal with clean engine coolant.
   - Tighten the water pump bolts to 89 inch lbs. (10 Nm).
   - Torque the pulley bolts to 15 ft. lbs. (20 Nm).
9. Fill and bleed the cooling system.

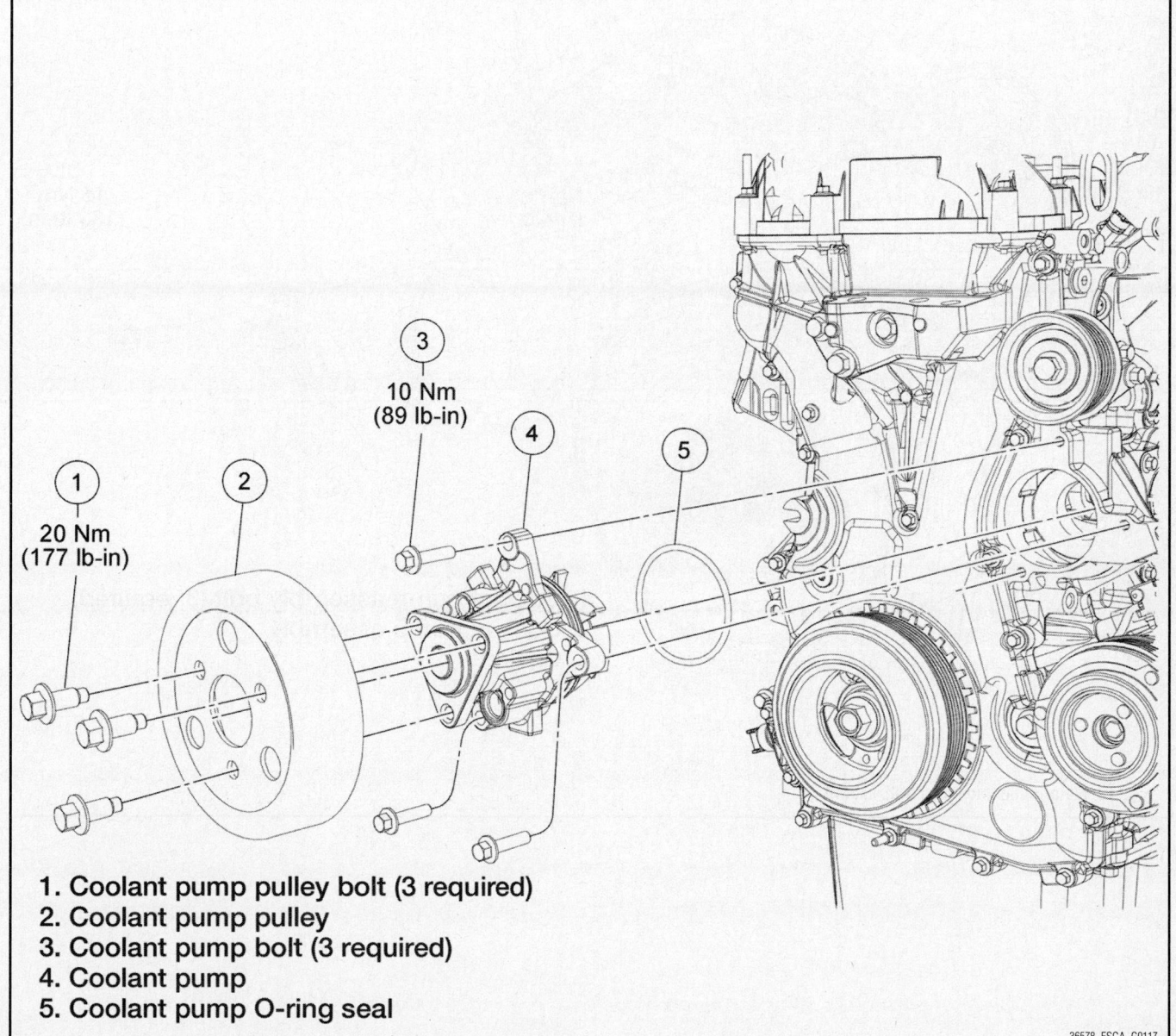

1. **Coolant pump pulley bolt (3 required)**
2. **Coolant pump pulley**
3. **Coolant pump bolt (3 required)**
4. **Coolant pump**
5. **Coolant pump O-ring seal**

36578_ESCA_G0117

Fig. 41 Water pump mounting—2.5L engine

3.0L Engine

See Figure 42.

1. Drain the cooling system.
2. Remove the Air Cleaner (ACL) outlet pipe.
3. Disconnect the lower radiator hose from the thermostat housing.
4. Remove the coolant pump belt. Refer to Accessory Drive Belts in the Engine Mechanical section.
5. Remove the 3 bolts and the coolant pump.
6. Remove and discard the gasket.

To install:

7. Using a new gasket, install the coolant pump and the 3 bolts.
8. Tighten the mounting bolts to 89 inch lbs. (10 Nm).
9. Install the coolant pump belt.
10. Connect the lower radiator hose to the thermostat housing.
11. Install the ACL outlet pipe.
12. Fill and bleed the cooling system.

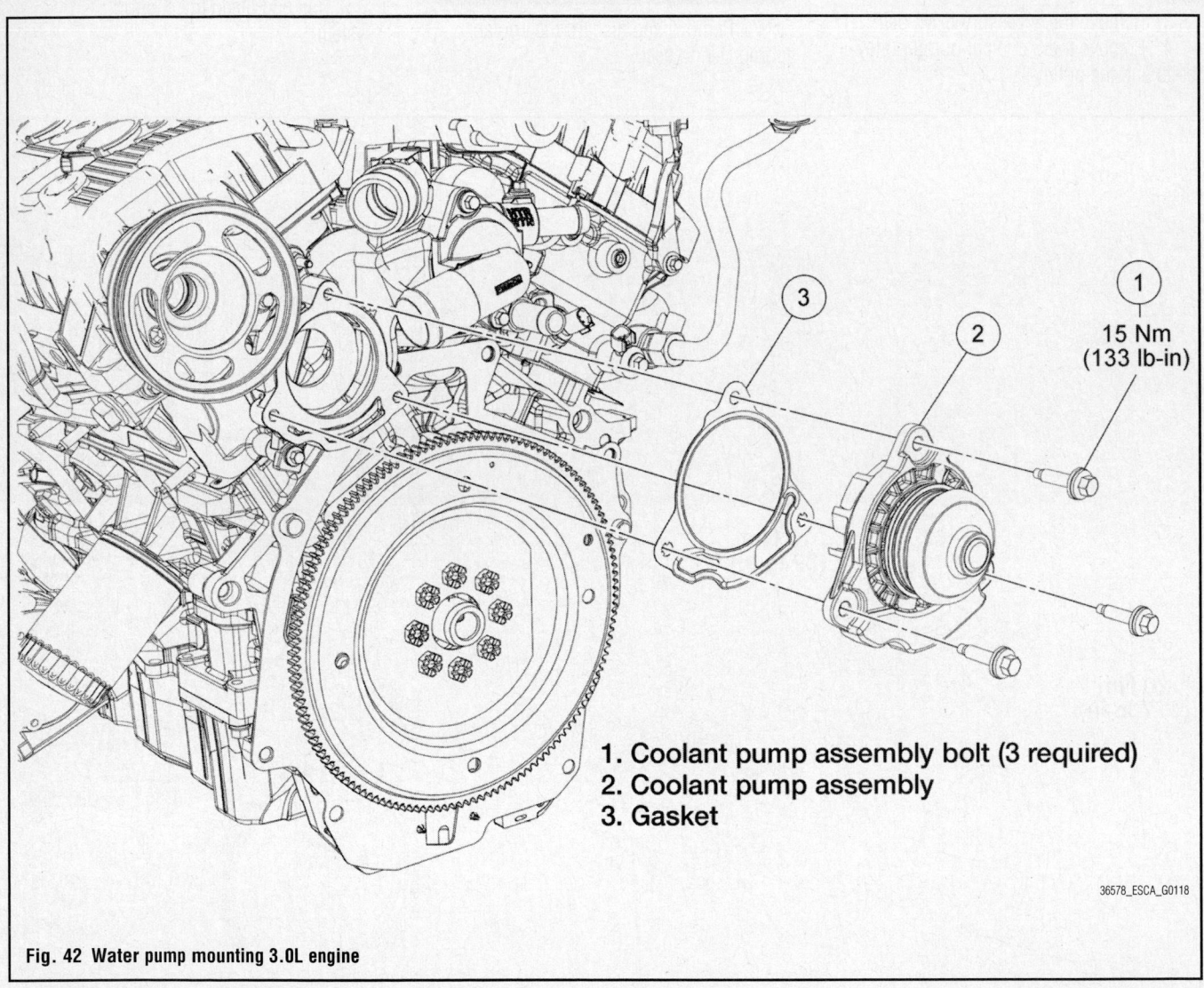

15 Nm
(133 lb-in)

1. Coolant pump assembly bolt (3 required)
2. Coolant pump assembly
3. Gasket

36578_ESCA_G0118

Fig. 42 Water pump mounting 3.0L engine

ENGINE ELECTRICAL   CHARGING SYSTEM

ALTERNATOR

REMOVAL & INSTALLATION

2.5L Engine
See Figure 43.

⁂ WARNING

Do not allow any metal object to come in contact with the alternator housing and internal diode cooling fins. A short circuit may result and burn out the diodes. Failure to follow this instruction may result in component damage.

1. With the vehicle in NEUTRAL, position it on a hoist.
2. Disconnect the battery.

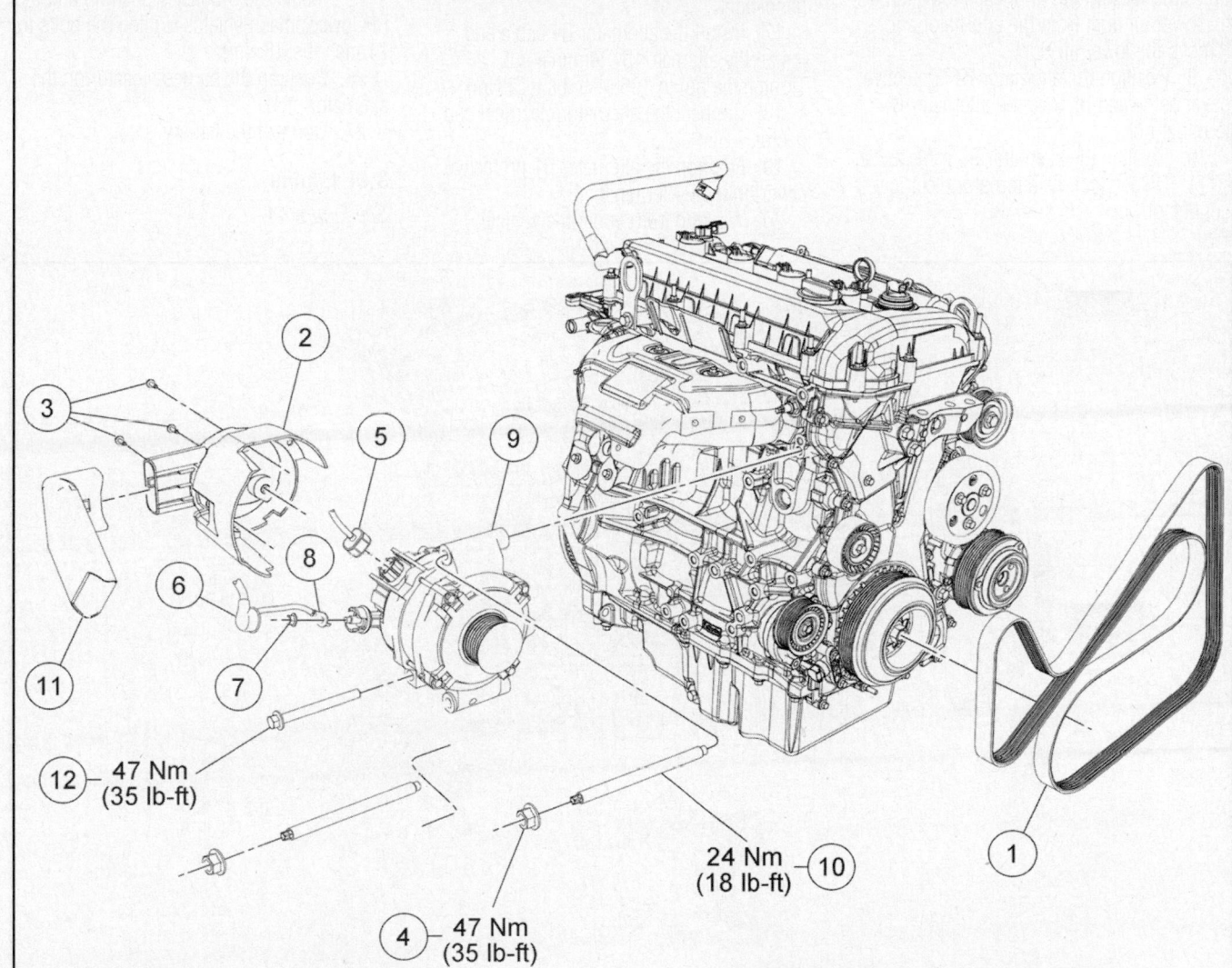

1. Front end accessory drive belt
2. Alternator upper air duct
3. Alternator upper air duct screws (3 required)
4. Alternator stud nut (2 required)
5. Alternator electrical connector
6. B+ protective cover
7. Alternator B+ terminal nut
8. Alternator B+ cable
9. Alternator
10. Alternator stud (2 required)
11. Alternator lower air duct
12. Alternator bolt

36578_ESCA_G0120

Fig. 43 Alternator mounting—2.5L engine

3. Remove the 5 bolts, 1 pushpin and the RH lower splash shield.

4. Rotate the front end accessory drive belt tensioner clockwise and position the accessory drive belt aside.

5. Remove the battery harness locator from the lower alternator stud.

6. Remove the alternator bolt.

7. Remove the 2 alternator stud nuts.

8. Working from the top of the vehicle, press the locking tab to release the alternator lower air duct from the alternator and remove the lower air duct.

9. Position the alternator B+ protective cover aside and remove the alternator B+ terminal nut.

10. Position the alternator B+ cable aside.

11. Disconnect the alternator electrical connector.

12. Remove the alternator.

13. Remove the 3 screws and the alternator upper air duct.

To install:

14. Install the 3 screws and the alternator upper air duct and tighten to 35 inch lbs. (4 Nm).

15. Working from the top of the vehicle, install the alternator and the alternator stud.

16. Install the 2 alternator stud nuts hand-tight.

17. Install the alternator B+ cable and install the alternator B+ terminal nut. Tighten the nut to 106 inch lbs. (12 Nm).

18. Connect the alternator electrical connector.

19. Position the alternator B+ protective cover on the B+ terminal.

20. Working from under the vehicle, install the lower alternator bolt hand-tight.

21. Tighten the 2 alternator stud nuts to 35 ft. lbs. (47 Nm).

22. Install the lower alternator air duct.

23. Tighten the alternator bolt to 35 ft. lbs. (47 Nm).

24. Rotate the front end accessory drive belt tensioner clockwise and position the accessory drive belt onto the pulleys.

25. Install the 5 bolts, 1 pushpin and the RH lower splash shield. Tighten the bolts to 71 inch lbs. (8 Nm).

26. Position the harness locator on the alternator stud.

27. Connect the battery.

3.0L Engine

See Figure 44.

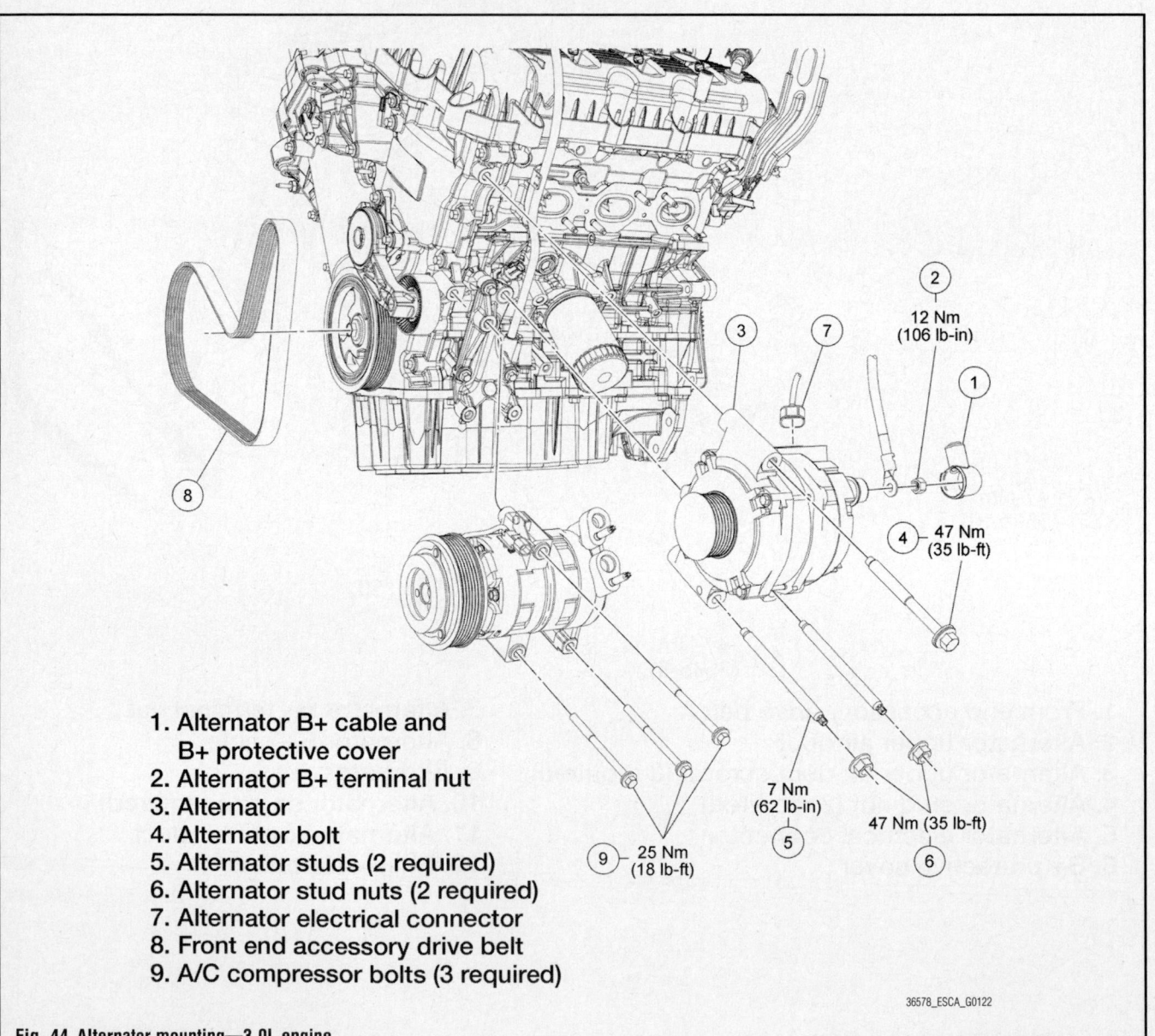

1. Alternator B+ cable and B+ protective cover
2. Alternator B+ terminal nut
3. Alternator
4. Alternator bolt
5. Alternator studs (2 required)
6. Alternator stud nuts (2 required)
7. Alternator electrical connector
8. Front end accessory drive belt
9. A/C compressor bolts (3 required)

36578_ESCA_G0122

Fig. 44 Alternator mounting—3.0L engine

✳ WARNING

Do not allow any metal object to come in contact with the alternator housing and internal diode cooling fins. A short circuit may result and burn out the diodes. Failure to follow this instruction may result in component damage.

1. Disconnect the negative battery cable.
2. Remove the 5 RH lower splash shield bolts and the 1 pin-type retainer.
3. Remove the RH lower splash shields.
4. Rotate the front end accessory drive tensioner counterclockwise and position the accessory drive belt aside

5. Disconnect the alternator electrical connector.
6. Position the alternator B+ protective cover aside and remove the alternator B+ terminal nut.
7. Remove the A/C compressor bolts. Use a tie-strap and position A/C compressor aside.
8. Loosen the 2 alternator nuts.
9. Remove the 2 lower alternator studs.
10. Remove the upper alternator bolt and the alternator.

To install:

11. Install alternator and upper bolt.
12. Tighten upper bolt to 35 ft. lbs. (47 Nm) with the vehicle on the ground.

13. Install the 2 lower alternator studs and tighten to 62 inch lbs. (7 Nm).
14. Install and tighten the 2 alternator nuts to 35 ft. lbs. (47 Nm).
15. Reposition and install A/C compressor bolts. Tighten to 18 ft. lbs. (25 Nm).
16. Reposition the alternator B+ protective cover and tighten terminal nut to 9 ft. lbs. (12 Nm).
17. Reconnect the alternator electrical connector.
18. Install accessory drive belt.
19. Install the splash shields and bolts and the 2 pin-type retainers.
20. Tighten the splash shield bolts to 71 inch lbs. (8 Nm).
21. Connect the negative battery cable.

ENGINE ELECTRICAL

IGNITION SYSTEM

FIRING ORDER

The firing order for the 2.5L engine is 1–3–4–2.
The firing order for the 3.0L engine is 1–4–2–5–3–6.

IGNITION COIL

REMOVAL & INSTALLATION

2.5L Engine

See Figure 45.

1. Disconnect the negative battery cable.

➡**When removing the ignition coil-on-plugs, a slight twisting motion will break the seal and ease removal.**

2. Disconnect the 4 ignition coil electrical connectors.
3. Remove the bolts and the ignition coils.

➡**Inspect the coil seals for rips, nicks or tears. Remove and discard any damaged coil seals.**

To install:

4. Install the ignition coils. Tighten the bolts to 71 inch lbs. (8 Nm).
5. Apply a small amount of dielectric grease to the inside of the ignition coil boots before attaching to the spark plugs.

3.0L Engine

Left Side

See Figure 46.

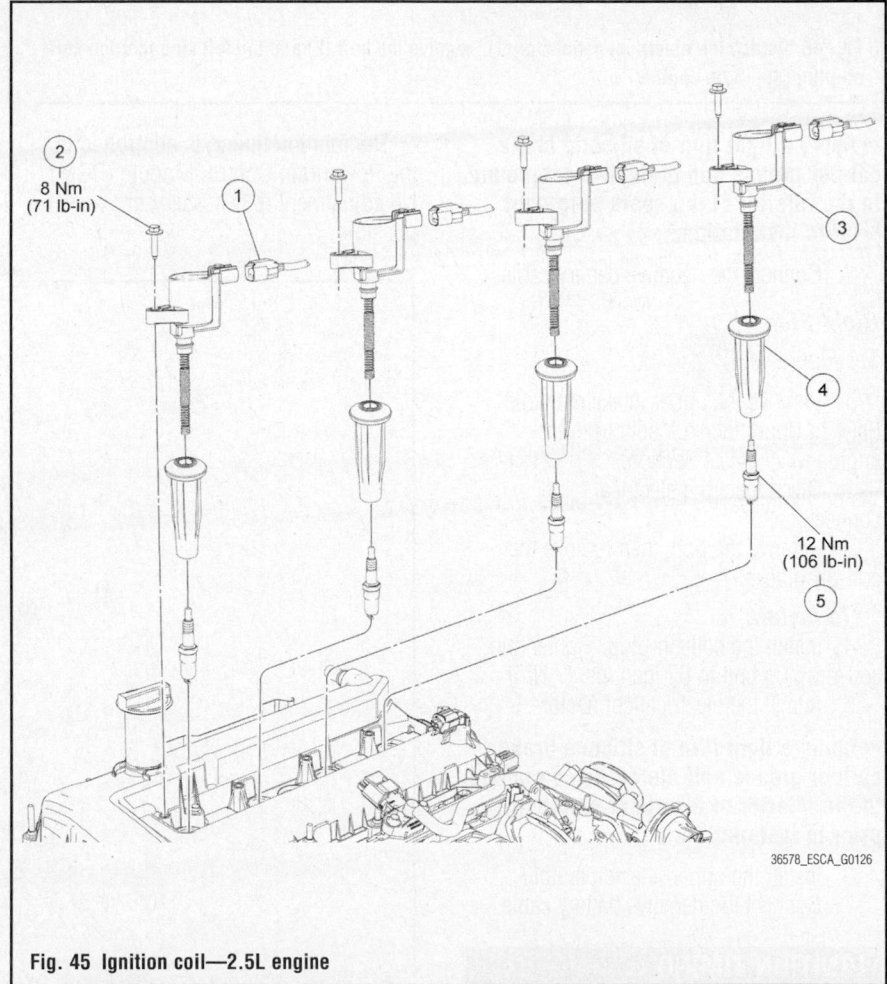

Fig. 45 Ignition coil—2.5L engine

36578_ESCA_G0126

1. Disconnect the negative battery cable.
2. Disconnect the ignition coil-on-plug electrical connector.
3. Remove the bolt, then remove the coil-on-plug.

To install:

4. Install the coil-on-plug. Tighten the coil-on-plug bolt to 62 inch lbs. (7 Nm).
5. Attach the electrical connector.

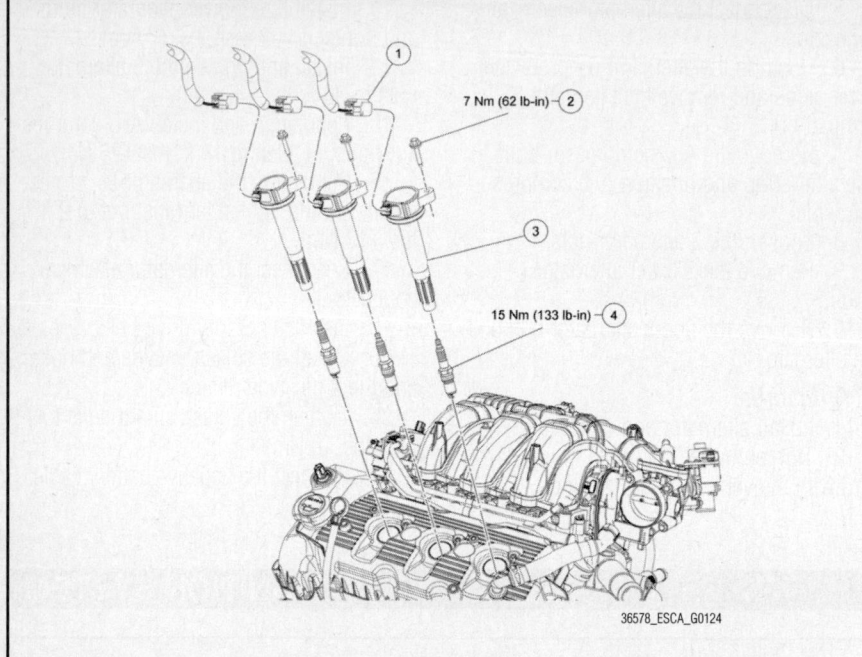

Fig. 46 Detach the electrical connector (1), remove the bolt (2) and the left side ignition coil-on-plug (3)—3.0L engine

➡Apply a light film of silicone brake caliper grease and dielectric compound to the interior of the spark plug boot prior to installation.

6. Connect the negative battery cable.

Right Side

See Figure 47.

1. Remove the upper intake manifold. Refer to Upper Intake Manifold in the Engine Mechanical Section.
2. Disconnect the electrical connector.
3. Remove the bolt, then remove the coil-on-plug.

To install:

4. Install the coil-on-plug. Tighten the coil-on-plug bolt to 62 inch lbs. (7 Nm).
5. Attach the electrical connector.

➡Apply a light film of silicone brake caliper grease and dielectric compound to the interior of the spark plug boot prior to installation.

6. Install the upper intake manifold.
7. Connect the negative battery cable.

IGNITION TIMING

ADJUSTMENT

The ignition timing is controlled by the Powertrain Control Module (PCM). No adjustment is necessary or possible.

SPARK PLUGS

REMOVAL & INSTALLATION

2.5L Engine

1. Remove the ignition coils.

✳✳ CAUTION

Only use hand tools when removing or installing the spark plugs, or damage can occur to the cylinder head or spark plug.

➡Use compressed air to remove any foreign material in the spark plug well before removing the spark plugs.

2. Remove the spark plugs.
3. Inspect spark plugs.
4. To install, reverse the removal procedure.
5. Adjust the spark plug gap as necessary.

a. 2.5L Engine: Specifications: 0.049–0.053 inches (1.25–1.35 mm). Tighten spark plugs to 9 ft. lbs. (12 Nm).
b. 3.0L Engine: Specifications: 0.045–0.049 inches (1.15–1.25 mm). Tighten spark plugs to 11 ft. lbs. (15 Nm).

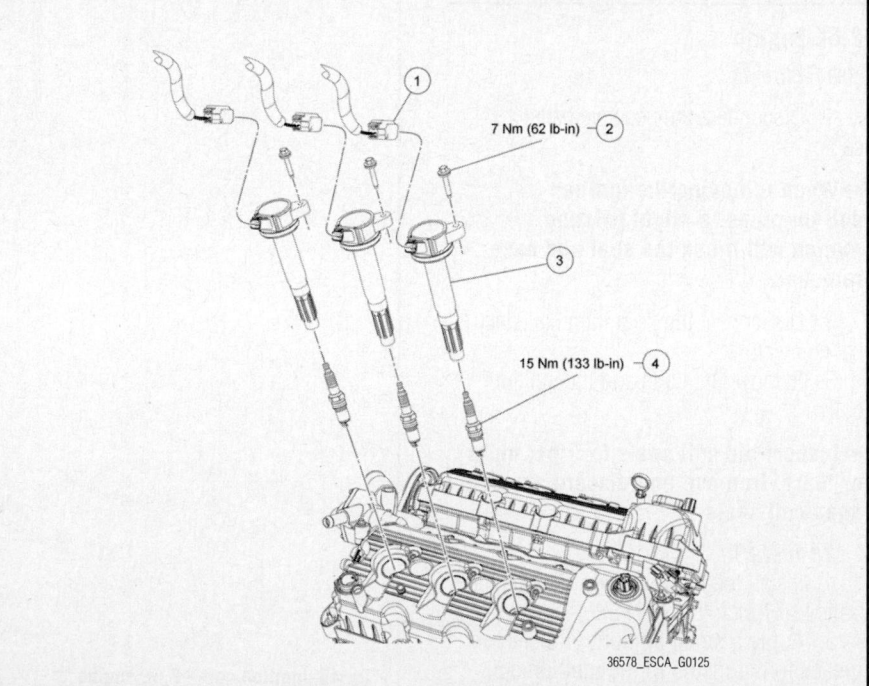

Fig. 47 View of the right side ignition coil-on-plug (3), electrical connector (1) and bolt (2)—3.0L engine

STARTER

REMOVAL & INSTALLATION

2.5L Engine

See Figure 48.

1. With the vehicle in NEUTRAL, position it on a hoist.

2. Disconnect the battery ground cable.

3. Remove the 5 bolts, the pin-type retainer and the RH splash shield.

4. Remove the starter solenoid wire nut.

5. Remove the starter solenoid battery cable nut and disconnect the starter motor solenoid terminal cover and cables.

6. Remove the ground wire nut and position aside the ground wire.

7. Remove the 2 stud bolts and the starter motor.

To install:

8. Install starter motor and mounting stud bolts, tighten bolts to 26 ft. lbs. (35 Nm).

9. Reposition ground strap and tighten ground strap nut to 159 inch lbs. (18 Nm).

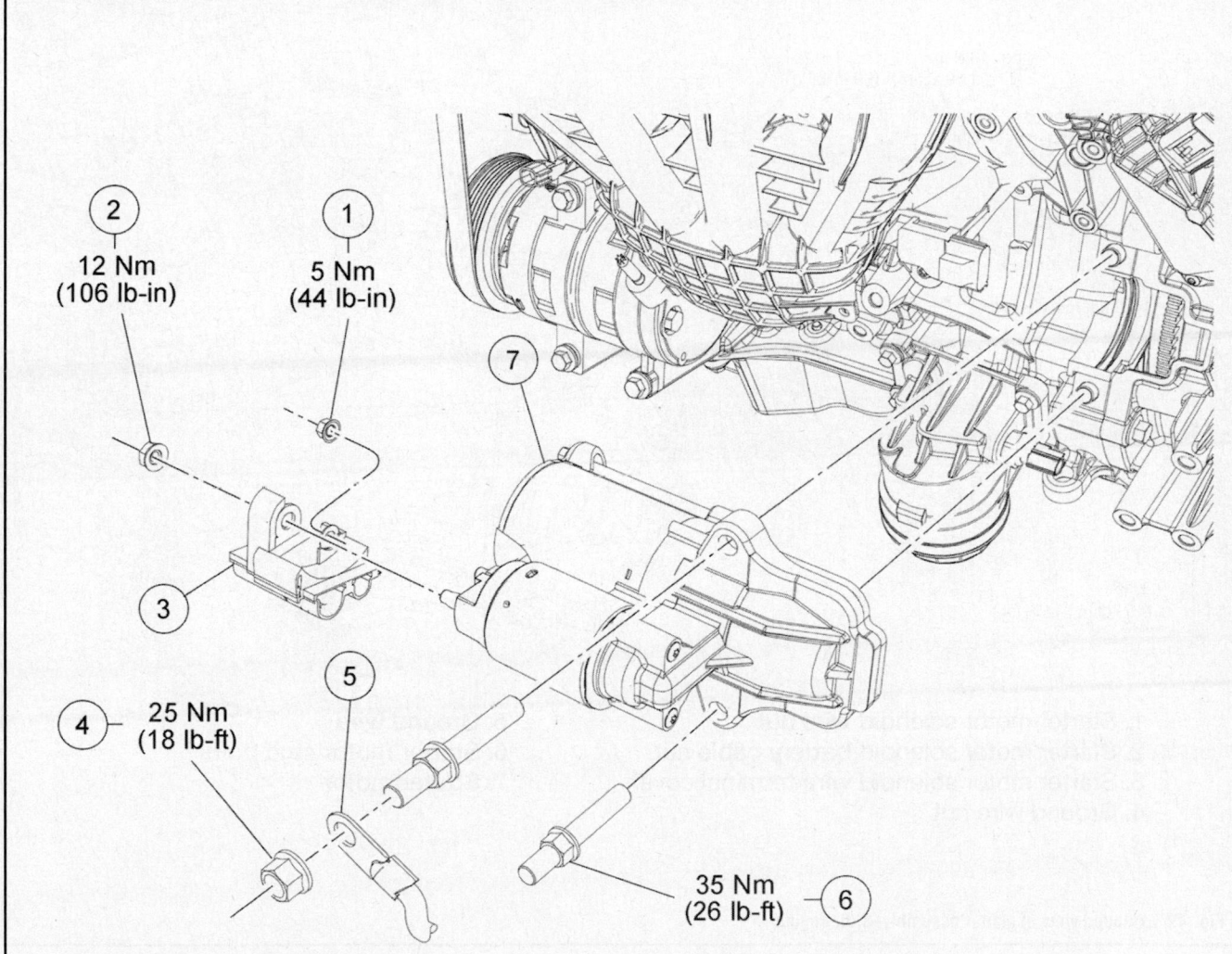

1. Starter motor solenoid wire nut
2. Starter motor solenoid battery cable nut
3. Starter motor solenoid terminal cover
4. Ground wire nut
5. Ground wire
6. Starter motor stud bolt (2 required)
7. Starter motor

36578_ESCA_G0134

Fig. 48 Starter motor mounting—2.5L engine

10. Install battery cable and tighten nut to 106 inch lbs. (12 Nm).

11. Install solenoid wire and tighten nut to 44 inch lbs. (5 Nm).

12. Secure all retainers, wires and cover.

13. Connect the negative battery cable.

3.0L Engine

See Figure 49.

1. Disconnect the battery ground cable.

2. Remove the starter motor solenoid wire nut.

3. Remove the starter motor solenoid battery cable nut and position aside the cables.

4. Remove the ground wire nut and position aside the ground wire.

5. Remove the 2 stud bolts and the starter motor.

6. To install, reverse the removal procedure.

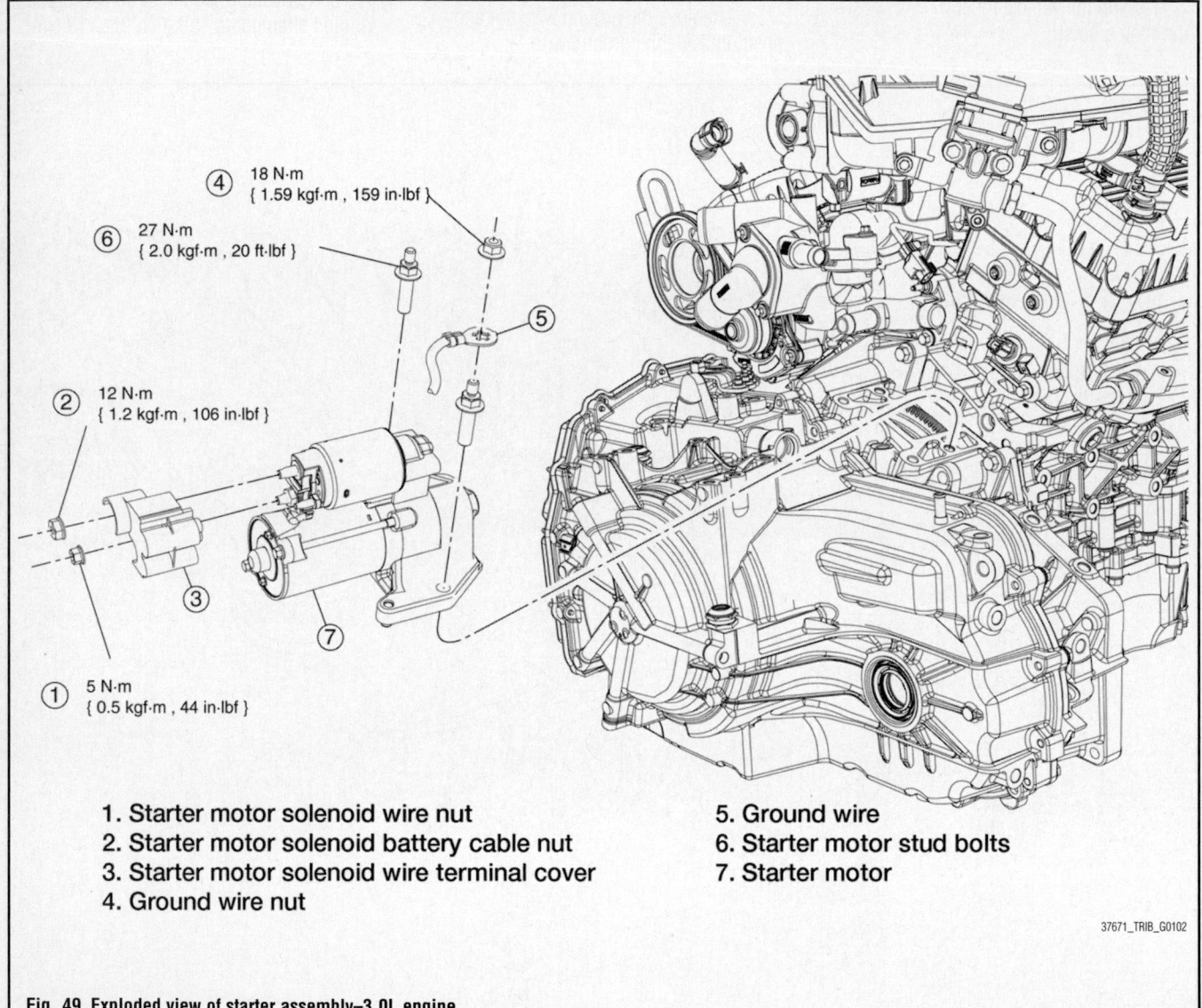

1. Starter motor solenoid wire nut
2. Starter motor solenoid battery cable nut
3. Starter motor solenoid wire terminal cover
4. Ground wire nut
5. Ground wire
6. Starter motor stud bolts
7. Starter motor

37671_TRIB_G0102

Fig. 49 Exploded view of starter assembly–3.0L engine

ENGINE MECHANICAL

➡️**Disconnecting the negative battery cable may interfere with the functions of the on board computer systems and may require the computer to undergo a relearning process, once the negative battery cable is reconnected.**

ACCESSORY DRIVE BELTS

ACCESSORY BELT ROUTING

See Figures 50 through 52.

INSPECTION

> ❊❊ **WARNING**
>
> **Under no circumstances should the accessory drive belt, tensioner or pulleys be lubricated as potential damage to the belt material and tensioner damping mechanism will occur. Do not apply any fluids or belt dressing to the accessory drive belt or pulleys.**

The water pump drive belt is on back of engine. It is driven off the rear cam pulley and doesn't have any adjustments.

Visual Inspection

Visually inspect the belt for obvious signs of mechanical damage:
- Drive belt cracking/chunking/wear
- Belt/pulley contamination
- Incorrectly routed belt
- Pulley misalignment or excessive pulley runout

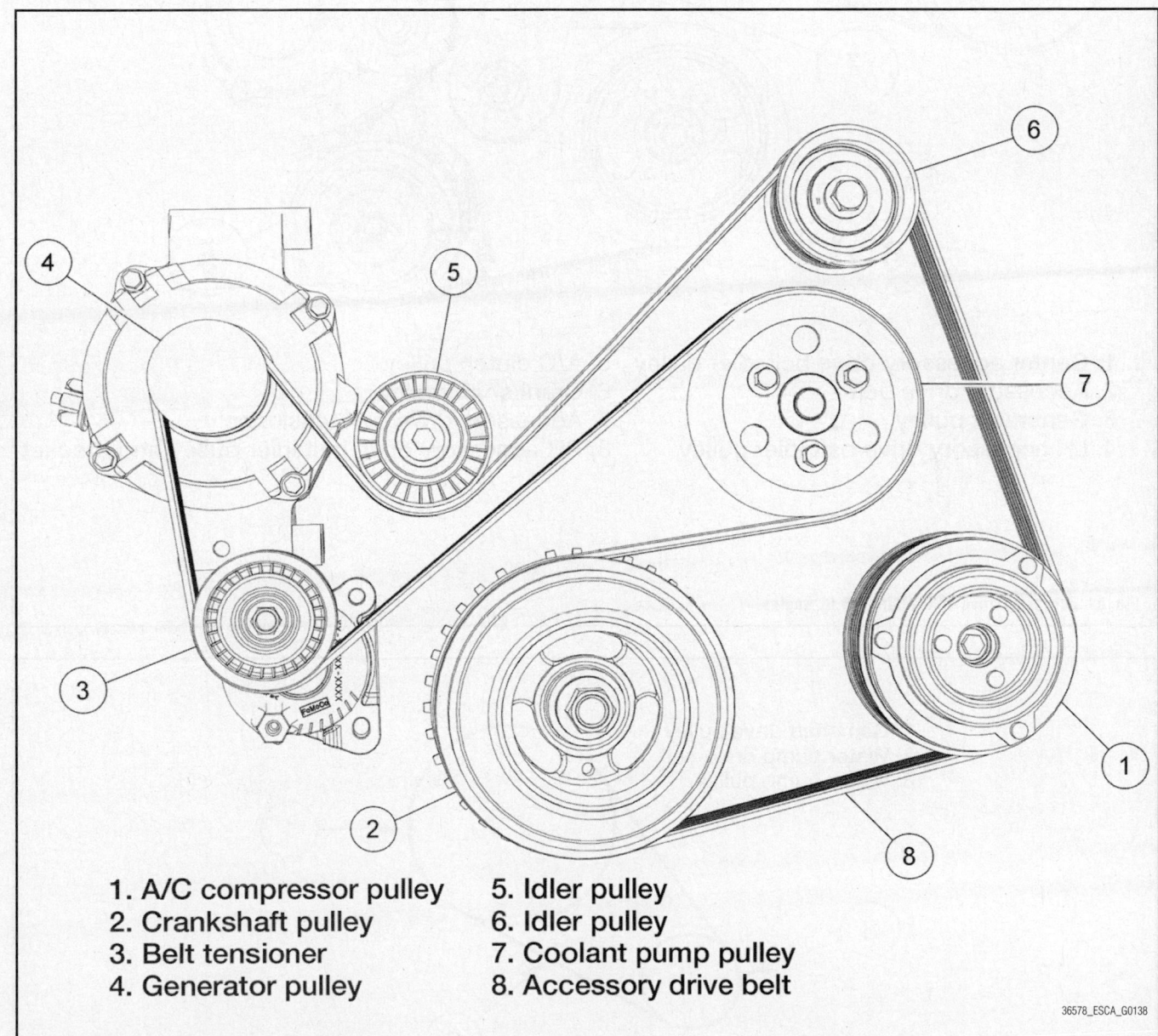

1. A/C compressor pulley
2. Crankshaft pulley
3. Belt tensioner
4. Generator pulley
5. Idler pulley
6. Idler pulley
7. Coolant pump pulley
8. Accessory drive belt

36578_ESCA_G0138

Fig. 50 Accessory drive belt routing—2.5L engine

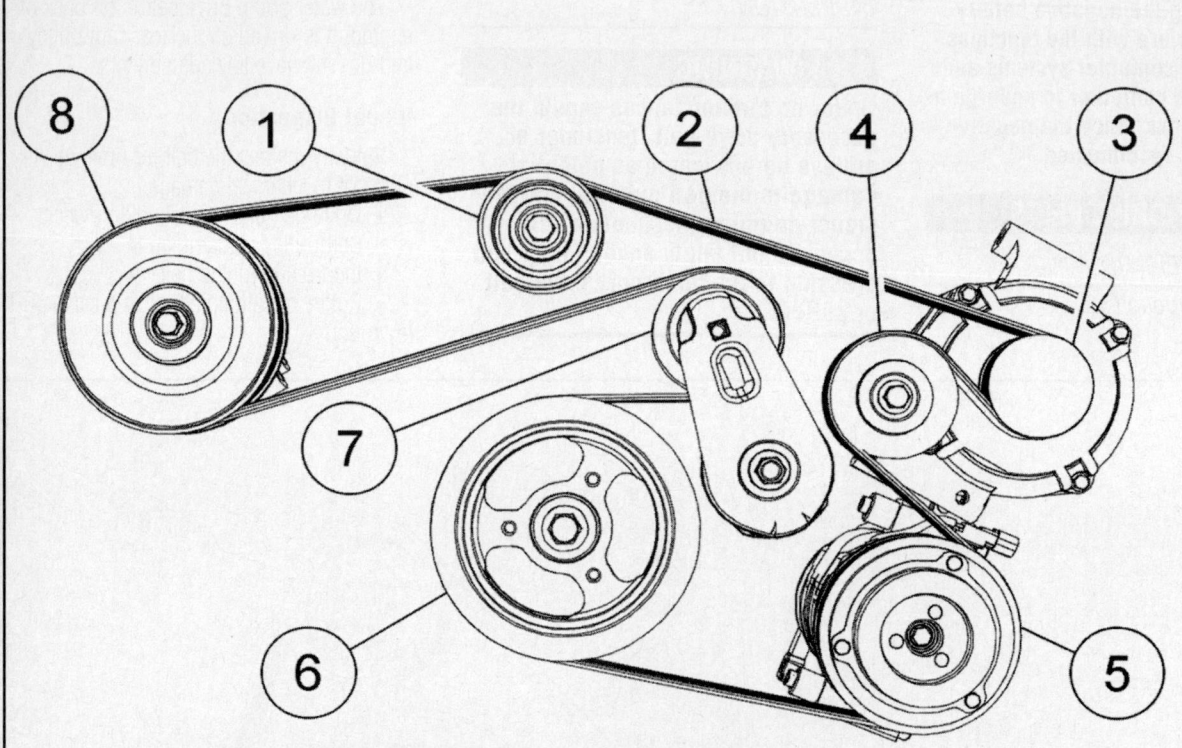

1. Center accessory drive belt idler pulley
2. Accessory drive belt
3. Generator pulley
4. LH accessory drive belt idler pulley
5. A/C clutch pulley
6. Crankshaft pulley
7. Accessory drive belt tensioner
8. RH accessory drive belt idler pulley and bracket

36578_ESCA_G0136

Fig. 51 Accessory drive belt routing—3.0L engine

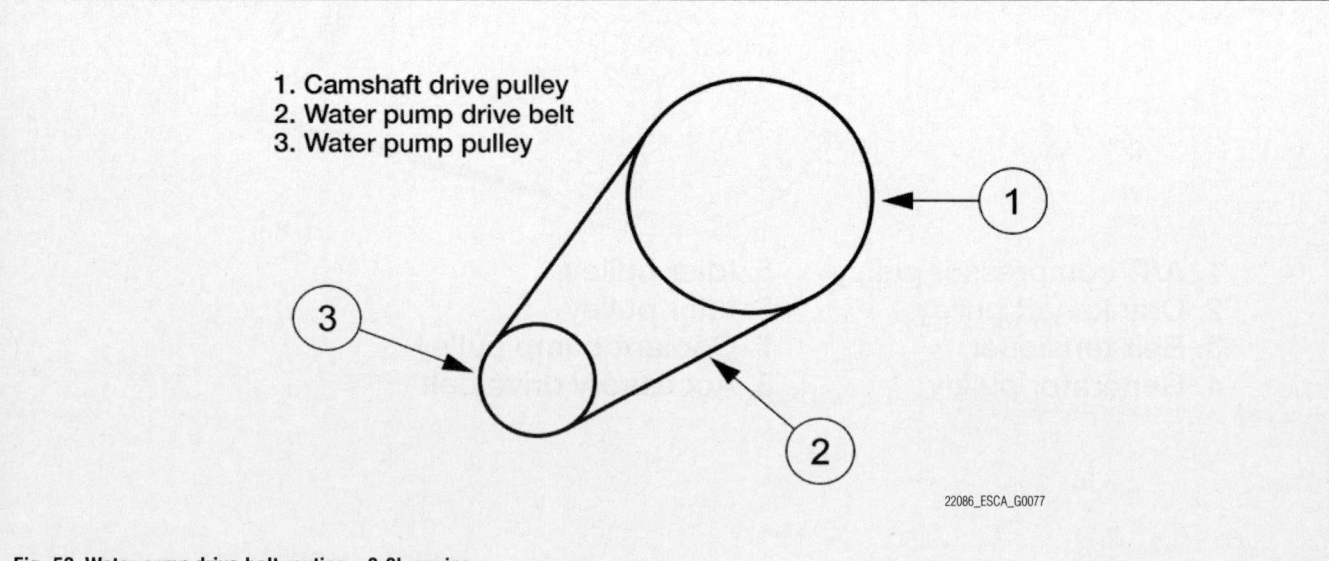

1. Camshaft drive pulley
2. Water pump drive belt
3. Water pump pulley

22086_ESCA_G0077

Fig. 52 Water pump drive belt routing—3.0L engine

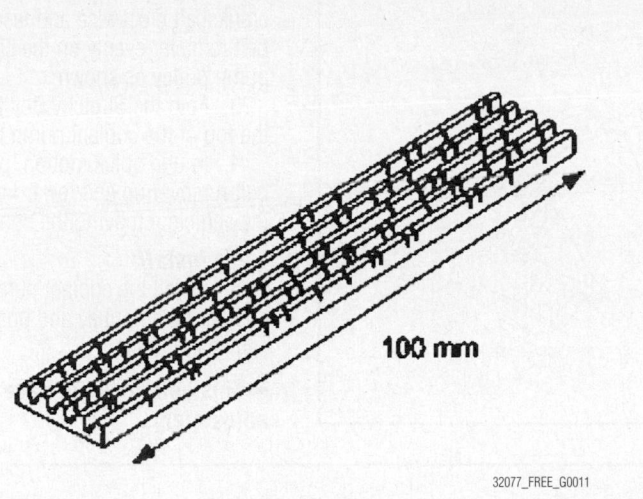

Fig. 53 Up to 15 cracks in a rib over a distance of 4 inches (100mm) can be considered accept-able. If cracks exceed this standard, install a new belt

- Loose or improperly located hardware
- Incorrectly routed power steering tubes (rubbing)

Eliminate all other non-belt related noises that could cause belt misdiagnosis, such as A/C compressor engagement chirp, power steering cavitations at low temperatures, variable camshaft timing (VCT) tick or alternator whine.

If a concern is found, correct the condition before proceeding to the next section.

V-Ribbed Serpentine Drive Belt With Cracks Across Ribs

See Figure 53.

➡**Up to 15 cracks in a rib over a distance of 4 inches (100mm) can be considered acceptable. If damage exceeds the acceptable limit or any chunks are found to be missing from the ribs, a new belt must be installed.**

1. Check the belt for cracks. Up to 15 cracks in a rib over a distance of 4 inches (100mm) can be considered acceptable. If cracks exceed this standard, install a new belt.

ADJUSTMENT

The belts used on these vehicle are equipped with automatic (spring load) tensioners which maintain tension. No adjustment is necessary or possible.

REMOVAL & INSTALLATION

2.5L Engine

See Figure 54.

1. With the vehicle in NEUTRAL, position it on a hoist.
2. Remove the pin-type retainer, 5 bolts and the RH splash shield.
3. Using the hex feature, rotate the accessory drive belt tensioner clockwise and remove the accessory drive belt from

the coolant pump pulley and remove the accessory drive belt from the pulley.

4. To install, reverse the removal procedure.

3.0L Engine

Accessory Drive Belt

See Figure 51.

1. With the vehicle in NEUTRAL, position it on a hoist.
2. Remove the pin-type retainer, 5 bolts and the RH lower splash shield.
3. Using a suitable belt tensioner release tool, rotate the accessory drive belt tensioner counterclockwise and remove the accessory drive belt.
4. Installation is the reverse of the removal procedure. Make sure the belt is properly routed.

Water Pump Belt

See Figures 55 through 57.

1. Position the stretchy belt remover under the coolant pump belt.

➡**Feed the stretchy belt remover on to the camshaft coolant pump pulley approximately 90 degrees.**

➡**If a stretchy belt remover is not available simply cut the belt to remove.**

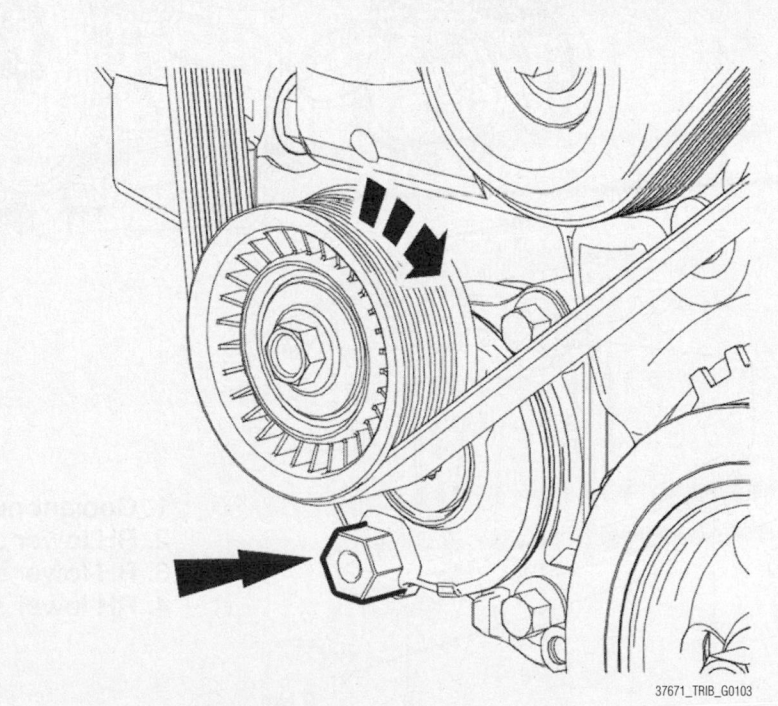

Fig. 54 Using the hex feature, rotate the accessory drive belt tensioner clockwise and remove the accessory drive belt from the coolant pump pulley and remove the accessory drive belt from the pulley

303-1419
90°
36578_ESCA_G0139

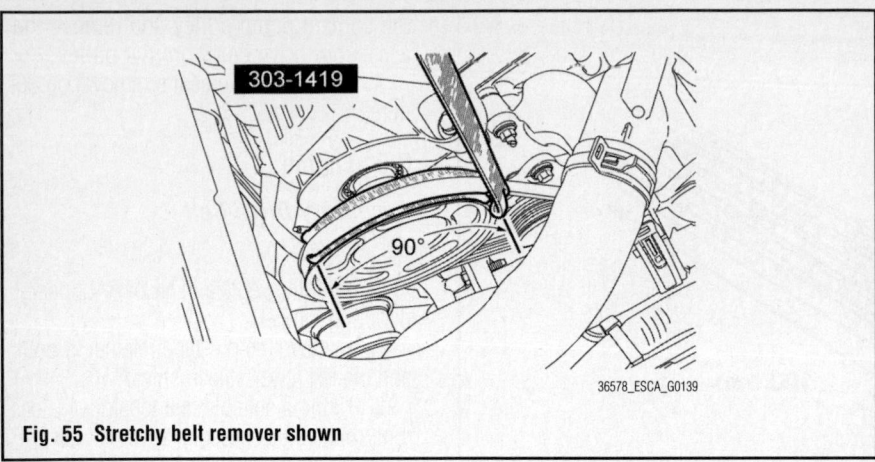

Fig. 55 Stretchy belt remover shown

2. With the help of an assistant, turn the crankshaft clockwise and feed the stretchy belt rcmover evenly on the camshaft coolant pump pulley as shown.

3. Fold the Stretchy Belt Remover over the top of the coolant pump belt.

4. In one quick motion, pull the stretchy belt remover up and toward the RH front of the vehicle removing the coolant pump belt.

To install:

5. Install the coolant pump belt on the coolant pump pulley and position it on the camshaft pulley.

➡This belt does not have any adjustments.

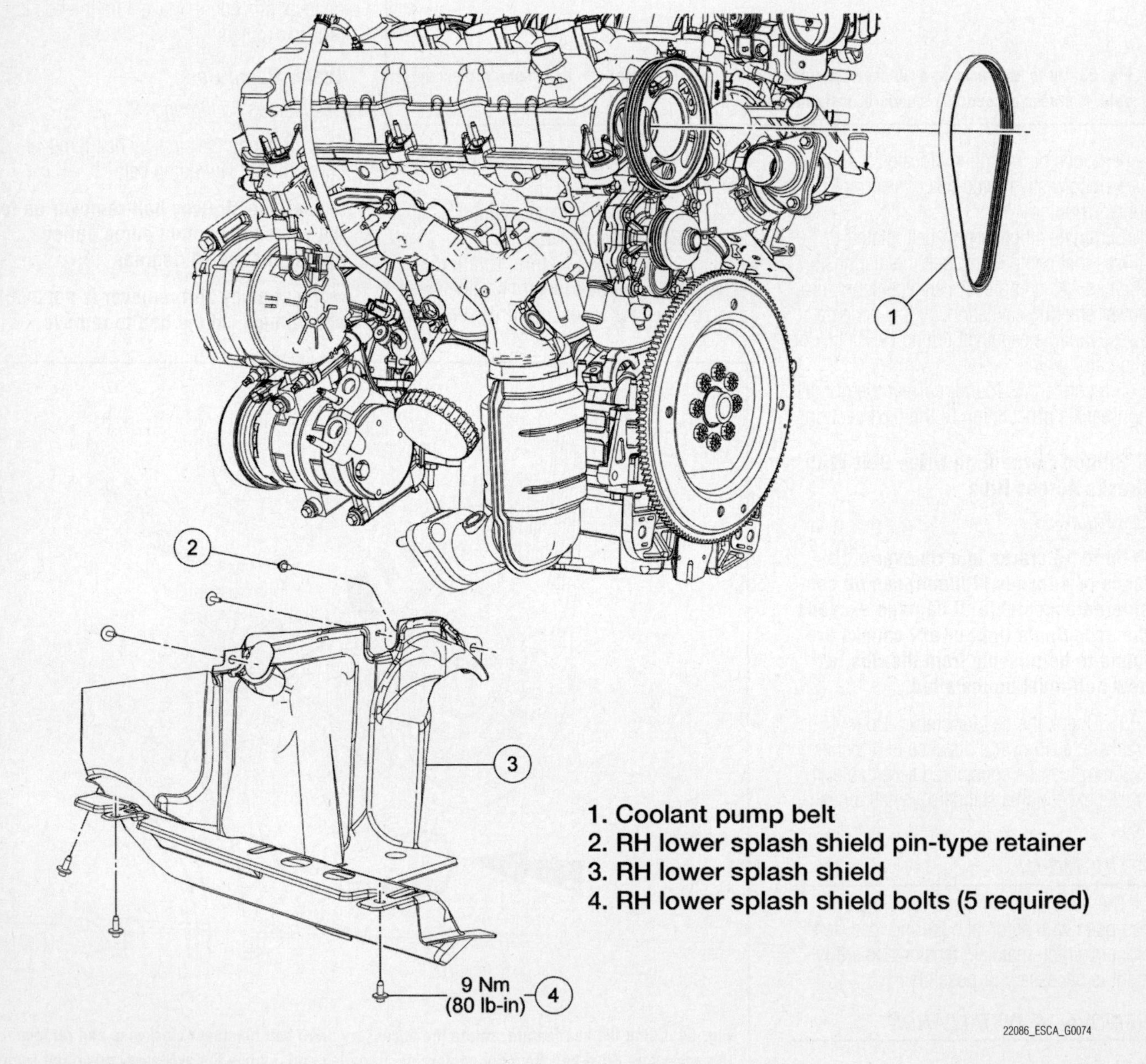

1
2
3
4
9 Nm
(80 lb-in)
1. Coolant pump belt
2. RH lower splash shield pin-type retainer
3. RH lower splash shield
4. RH lower splash shield bolts (5 required)
22086_ESCA_G0074

Fig. 56 Water pump drive belt

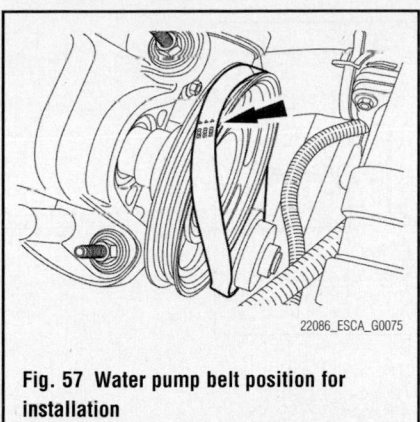

Fig. 57 Water pump belt position for installation

✳✳ WARNING

Do not use any screwdrivers, pliers or other metal objects that could cause damage to the belt or camshaft pulley while installing the belt.

6. Rotate the crankshaft clockwise to seat the coolant pump belt on the camshaft pulley.

CAMSHAFT AND VALVE LIFTERS

REMOVAL & INSTALLATION

2.5L Engine

See Figures 58 through 72.

✳✳ CAUTION

During engine repair procedures, cleanliness is extremely important. Any foreign material (including any material created while cleaning gasket surfaces) that enters the oil passages, coolant passages or the oil pan can cause engine failure.

✳✳ CAUTION

Do not rotate the camshafts unless instructed to in this procedure. Rotating the camshafts or crankshaft with timing components loosened or removed can cause serious damage to the valves or pistons.

1. With the vehicle in NEUTRAL, position it on a hoist.
2. Remove the accessory drive belt.
3. Remove the Variable Camshaft Timing (VCT) oil control solenoid.
   a. Remove the valve cover.
   b. Remove the bolt and the Variable Camshaft Timing (VCT) oil control solenoid.
4. Remove the front RH wheel and tire.

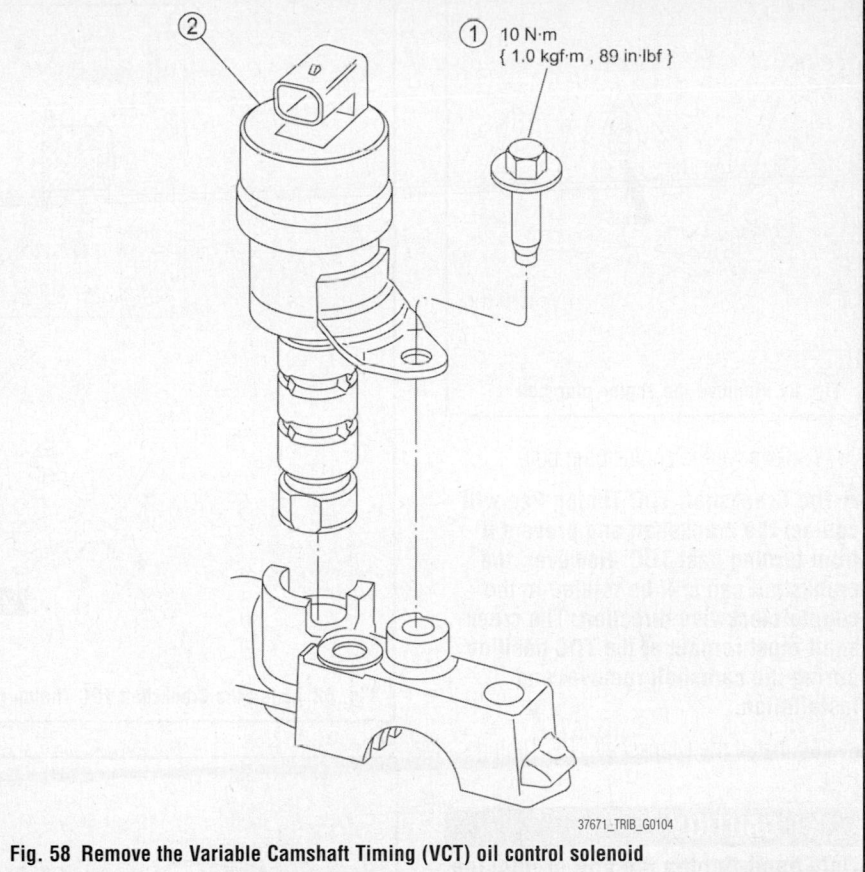

Fig. 58 Remove the Variable Camshaft Timing (VCT) oil control solenoid

✳✳ CAUTION

Failure to position the No. 1 piston at TDC can result in damage to the engine. Turn the engine in the normal direction of rotation only.

5. Using the crankshaft pulley bolt, turn the crankshaft clockwise to position the No. 1 piston at Top Dead Center (TDC).

➡The hole in the crankshaft pulley should be in the 6 o'clock position.

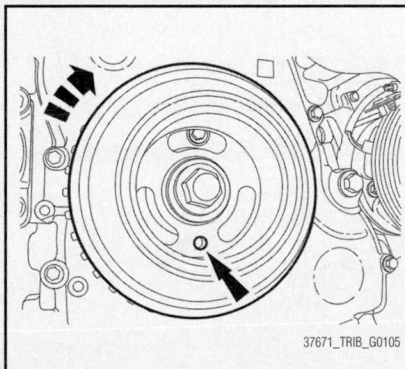

Fig. 59 The hole in the crankshaft pulley should be in the 6 o'clock position

✳✳ CAUTION

The Camshaft Alignment Plate is for camshaft alignment only. Using this tool to prevent engine rotation can result in engine damage.

➡The camshaft timing slots are offset. If the Camshaft Alignment Plate cannot be installed, rotate the crankshaft one complete revolution clockwise to correctly position the camshafts.

6. Install the Camshaft Alignment Plate in the slots on the rear of both camshafts.

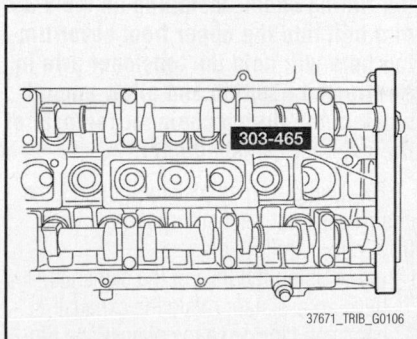

Fig. 60 Install the Camshaft Alignment Plate in the slots on the rear of both camshafts

Fig. 61 Remove the engine plug bolt

7. Remove the engine plug bolt.

➡**The Crankshaft TDC Timing Peg will contact the crankshaft and prevent it from turning past TDC. However, the crankshaft can still be rotated in the counterclockwise direction. The crankshaft must remain at the TDC position during the camshaft removal and installation.**

8. Install the Crankshaft TDC Timing Peg.

☆☆ CAUTION

Only hand-tighten the bolt or damage to the front cover can occur.

9. Install a 6 mm x 18 mm bolt through the crankshaft pulley and thread it into the front cover.

10. Remove the lower front cover timing hole plug from the engine front cover.

11. Remove the upper front cover timing hole plug from the engine front cover.

12. Reposition the Camshaft Alignment Plate to the slot on the rear of the intake camshaft only.

➡**Releasing the ratcheting mechanism in the timing chain tensioner allows the plunger to collapse and create slack in the timing chain. Installing an M6 x 30 mm bolt into the upper front cover timing hole will hold the tensioner arm in a retracted position and allow enough slack in the timing chain for removal of the exhaust camshaft gear.**

13. Using a small pick tool, unlock the chain tensioner ratchet through the lower front cover timing hole.

   a. Using the flats of the camshaft, have an assistant rotate the exhaust camshaft clockwise to collapse the timing chain tensioner plunger.

   b. Insert an M6 x 30 mm bolt into the upper front cover timing hole to hold the

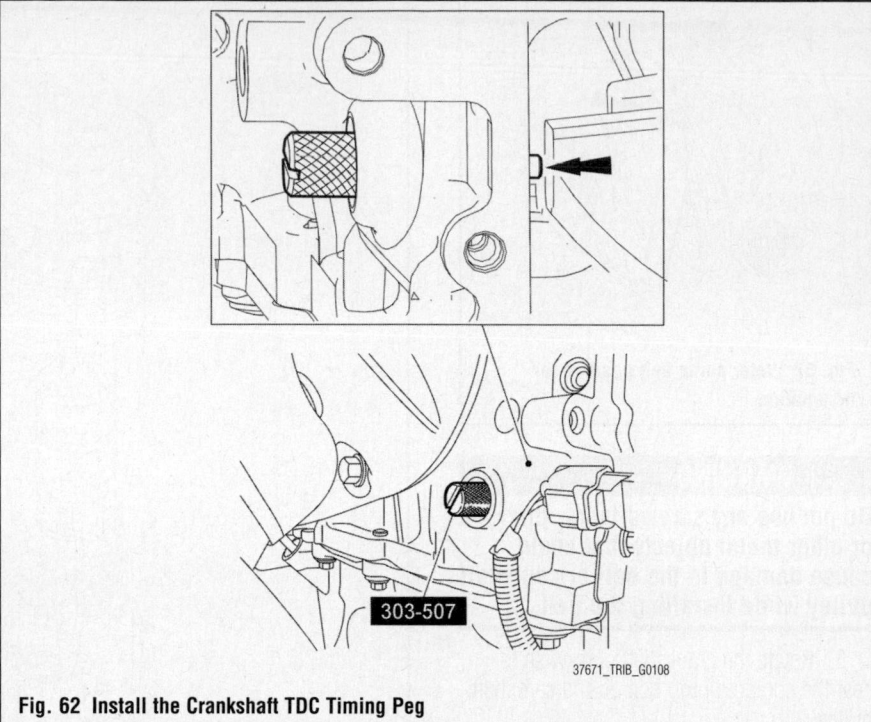

Fig. 62 Install the Crankshaft TDC Timing Peg

Fig. 63 Install a 6 mm x 18 mm bolt through the crankshaft pulley and thread it into the front cover

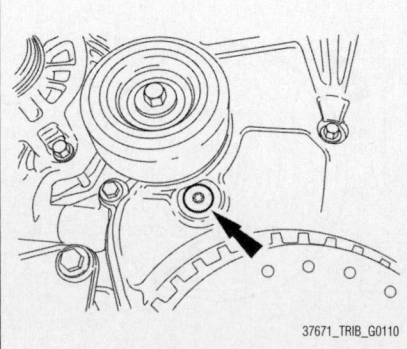

Fig. 64 Remove the lower front cover timing hole plug from the engine front cover

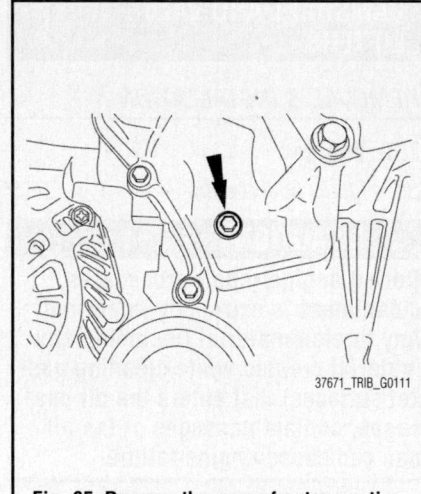

Fig. 65 Remove the upper front cover timing hole plug from the engine front cover

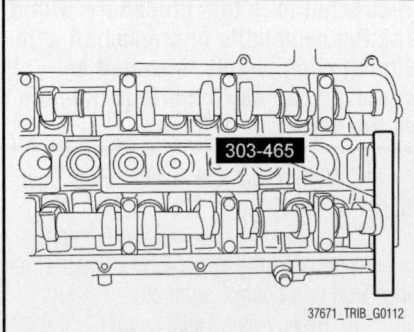

Fig. 66 Reposition the Camshaft Alignment Plate to the slot on the rear of the intake camshaft only

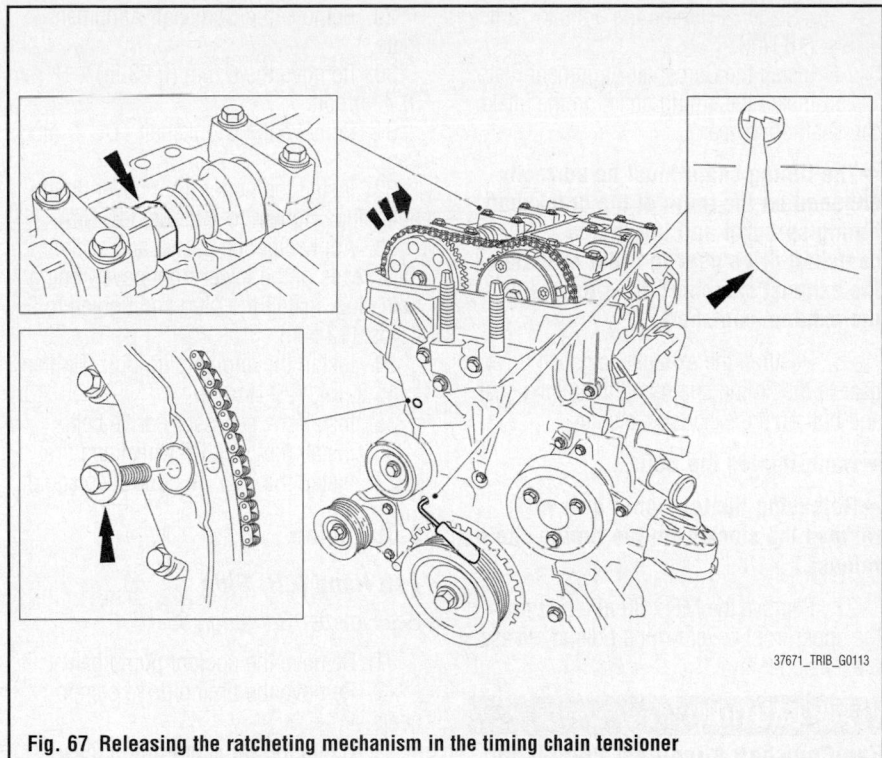

Fig. 67 Releasing the ratcheting mechanism in the timing chain tensioner

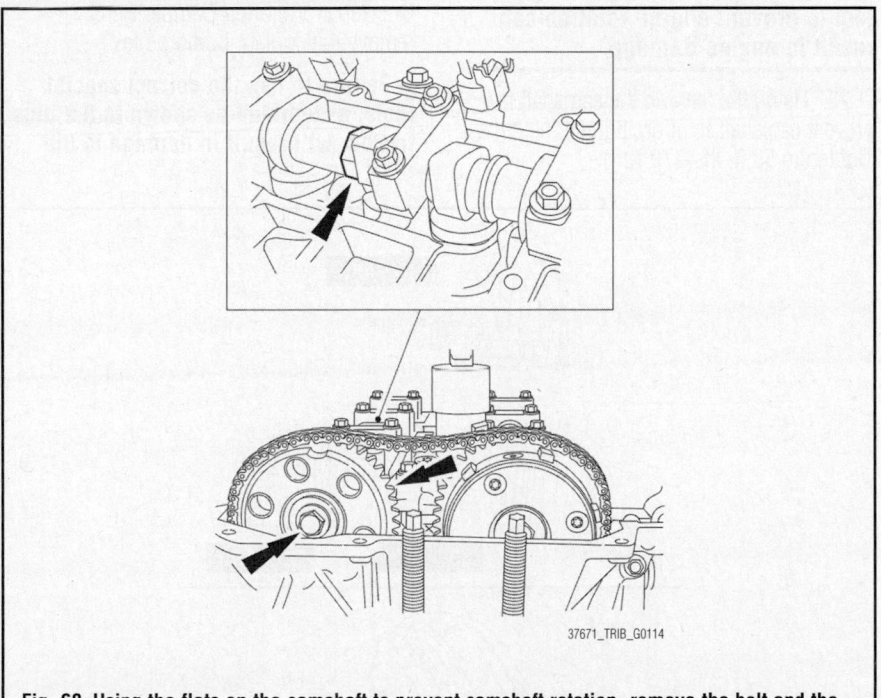

Fig. 68 Using the flats on the camshaft to prevent camshaft rotation, remove the bolt and the exhaust camshaft drive gear

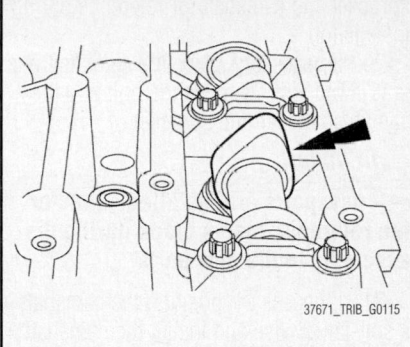

Fig. 69 Mark the position of the camshaft lobes on the No. 1 cylinder

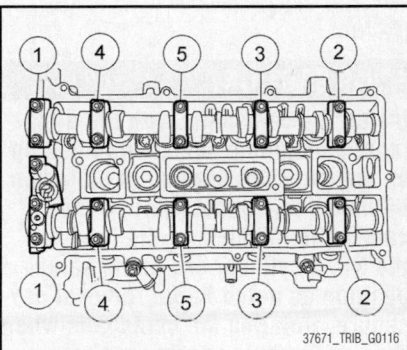

Fig. 70 Remove the camshafts from the engine

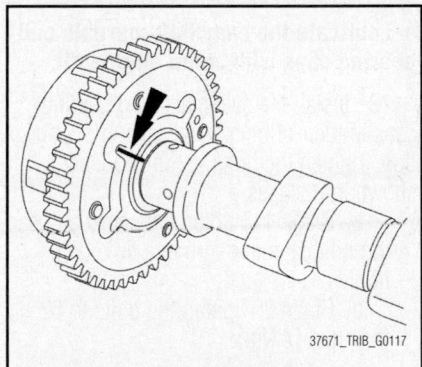

Fig. 71 If removal of the camshaft phaser and sprocket is necessary, mark the sprocket and camshaft for reference

tensioner arm in the retracted position.

14. Using the flats on the camshaft to prevent camshaft rotation, remove the bolt and the exhaust camshaft drive gear.

15. Remove the Camshaft Alignment Plate.

16. Remove the timing chain from the intake camshaft drive gear.

17. Mark the position of the camshaft lobes on the No. 1 cylinder for installation reference.

❋❋ CAUTION

Failure to follow the camshaft loosening procedure can result in damage to the camshafts.

➡ **Mark the location and orientation of each camshaft bearing cap.**

18. Remove the camshafts from the engine.

   a. Loosen the camshaft bearing cap bolts, in sequence, one turn at a time until all tension is released from the camshaft bearing caps.

   b. Remove the bolts and the camshaft bearing caps.

   c. Remove the camshafts.

19. If removal of the camshaft phaser and sprocket is necessary, mark the

sprocket and camshaft for reference during installation.

20. If necessary, place the camshaft in a soft-jawed vise. Remove the bolt and the camshaft phaser and sprocket.

To install:

➡**If new parts are installed, transfer the reference marks made during disassembly to the new parts.**

21. If necessary, position the camshaft in a soft-jawed vise and install the camshaft phaser and sprocket and the bolt.

22. Align the reference marks on the camshaft phaser and sprocket and the camshaft. Tighten the bolt to 53 ft. lbs. (72 Nm).

✲✲ CAUTION

Install the camshafts with the alignment slots in the camshafts lined up so the camshaft alignment plate can be installed without rotating the camshafts. Make sure the lobes on the No. 1 cylinder are in the same position as noted in the removal procedure. Rotating the camshafts when the timing chain is removed, or installing the camshafts 180 degrees out of position can cause severe damage to the valves and pistons.

➡**Lubricate the camshaft journals and bearing caps with clean engine oil.**

23. Install the camshafts and bearing caps in their original location and orientation. Tighten the bolts in the sequence shown in 3 stages.

    a. Stage 1: Tighten the camshaft bearing bolt caps one turn at a time until tight.

    b. Stage 2: Tighten the bolts to 62 inch lbs. (7 Nm).

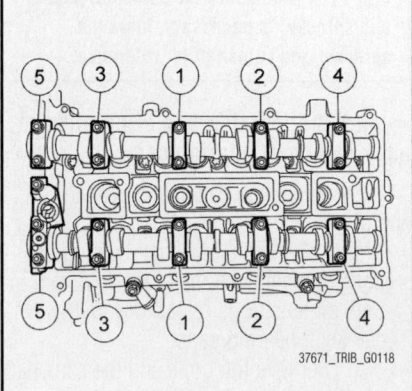

Fig. 72 Tighten the bolts in the sequence shown

    c. Stage 3: Tighten the bolts to 12 ft. lbs. (16 Nm).

24. Install the Camshaft Alignment Plate.

25. Install the timing chain on the intake camshaft drive gear.

➡**The timing chain must be correctly engaged on the teeth of the crankshaft timing sprocket and the intake camshaft drive gear in order to install the exhaust camshaft drive gear onto the exhaust camshaft.**

26. Position the exhaust camshaft drive gear in the timing chain and install the gear and bolt on the exhaust camshaft.

➡**Hand-tighten the bolt.**

➡**Releasing the tensioner arm will remove the slack from the timing chain release.**

27. Remove the M6 x 30 mm bolt from the upper front cover timing hole to release the tensioner arm.

✲✲ CAUTION

The Camshaft Alignment Plate is for camshaft alignment only. Using this tool to prevent engine rotation can result in engine damage.

28. Using the flats on the camshaft to prevent camshaft rotation, tighten the bolts. Tighten to 53 ft. lbs. (72 Nm).

29. Remove the Camshaft Alignment Plate.

30. Remove the 6 mm (0.23 in) x 18 mm (0.7 in) bolt.

31. Remove the Crankshaft TDC Timing Peg.

32. Install the upper front cover timing hole plug. Tighten to 7 ft. lbs. (10 Nm).

33. Apply silicone gasket and sealant to the threads of the lower front cover timing hole plug. Install the plug and tighten to 9 ft. lbs. (12 Nm).

34. Install the engine plug bolt. Tighten to 15 ft. lbs. (20 Nm).

35. Install the accessory drive belt.

36. Install the front RF wheel and tire.

37. Install the VCT oil control solenoid.

3.0L Engine

Left Hand (LH) Side

See Figures 73 through 86.

1. Remove the coolant pump belt.

2. Remove the timing drive components.

3. Using the Coolant Pump Pulley Plate, Coolant Pump Shaft Protector and the Crankshaft Vibration Damper Remover, remove the coolant pump pulley.

➡**Failure to use the correct special tools, assembled as shown in the illustration, will result in damage to the**

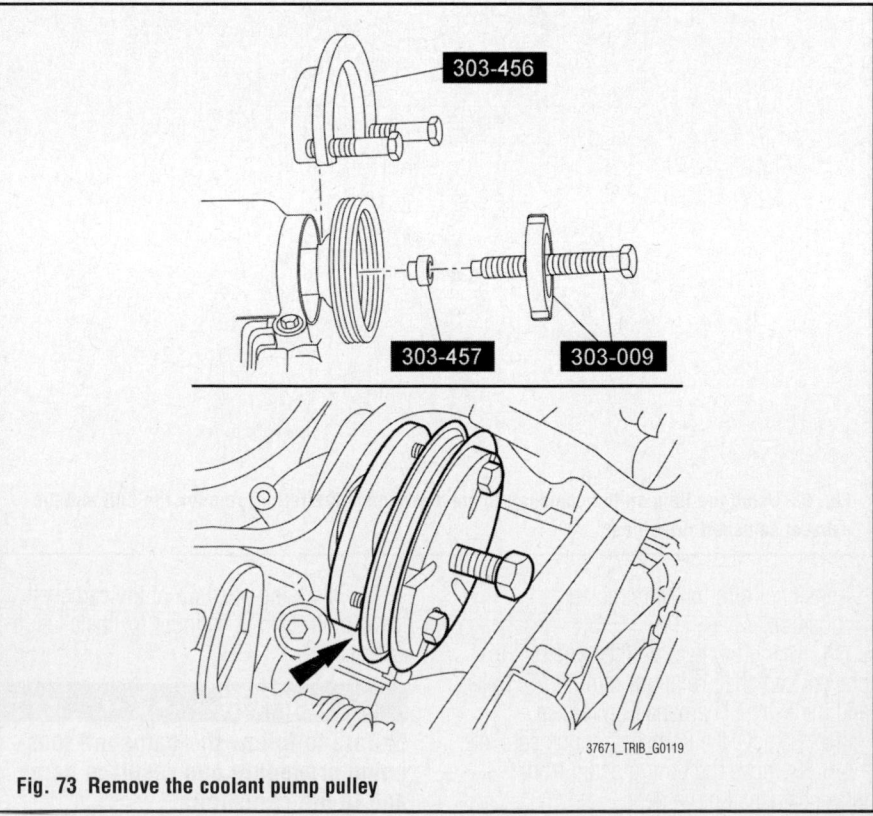

Fig. 73 Remove the coolant pump pulley

coolant pump pulley and/or special tools.

4. Using the Oil Seal Remover, remove and discard the camshaft oil seal.

⁂ CAUTION

Do not scratch the camshaft sealing surface while removing the camshaft oil seal. If scratched, camshaft oil seal leakage may occur.

5. Remove the 2 bolts and the camshaft oil seal retainer. Discard the press-in-place gasket.

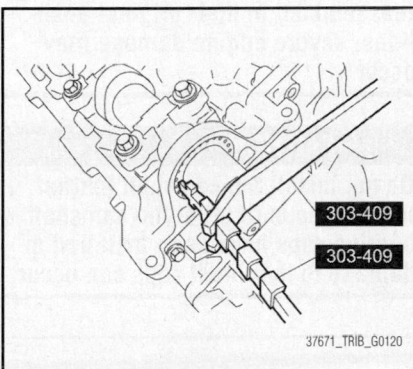

37671_TRIB_G0120

Fig. 74 Using the Oil Seal Remover, remove and discard the camshaft oil seal

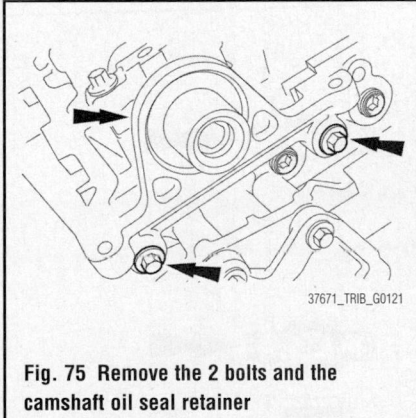

37671_TRIB_G0121

Fig. 75 Remove the 2 bolts and the camshaft oil seal retainer

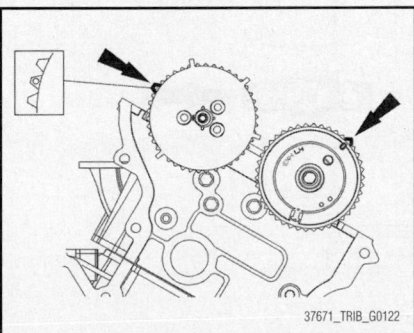

37671_TRIB_G0122

Fig. 76 Verify the LH camshafts are in the neutral position

6. Verify the LH camshafts are in the neutral position.

⁂ CAUTION

The camshafts must be in the neutral position before removing the bearing caps or damage to the engine may occur.

7. Remove the 3 bolts and the LH camshaft phaser and sprocket.

➡**Do not allow the camshaft to rotate from the neutral position while removing the camshaft phaser and sprocket or damage to the engine may occur.**

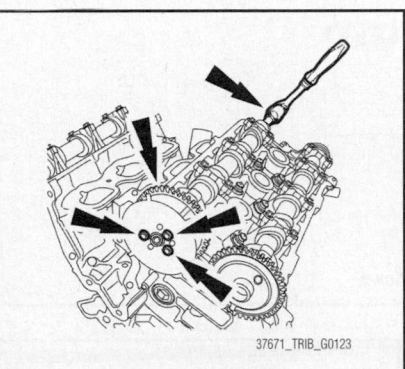

37671_TRIB_G0123

Fig. 77 Remove the 3 bolts and the LH camshaft phaser and sprocket

ing the camshaft phaser and sprocket or damage to the engine may occur.

➡**Install a 3/8-in ratchet and extension into the D-slot on the rear of the intake camshaft to hold the camshaft in place for removal of the camshaft phaser and sprocket bolts.**

8. If necessary, mark the camshaft bearing cap position and orientation as shown.

⁂ CAUTION

Cylinder head camshaft bearing caps must be assembled in their original positions. Some engines have factory markings on the camshaft bearing caps (as shown in illustration). Engines that do not have the factory markings must be marked for correct position and orientation prior to removal. Failure to install the camshaft bearing caps in their original positions may result in severe engine damage.

9. Loosen the bolts evenly in the sequence shown.
　　a. Remove the camshaft bearing thrust caps (1L and 5L).

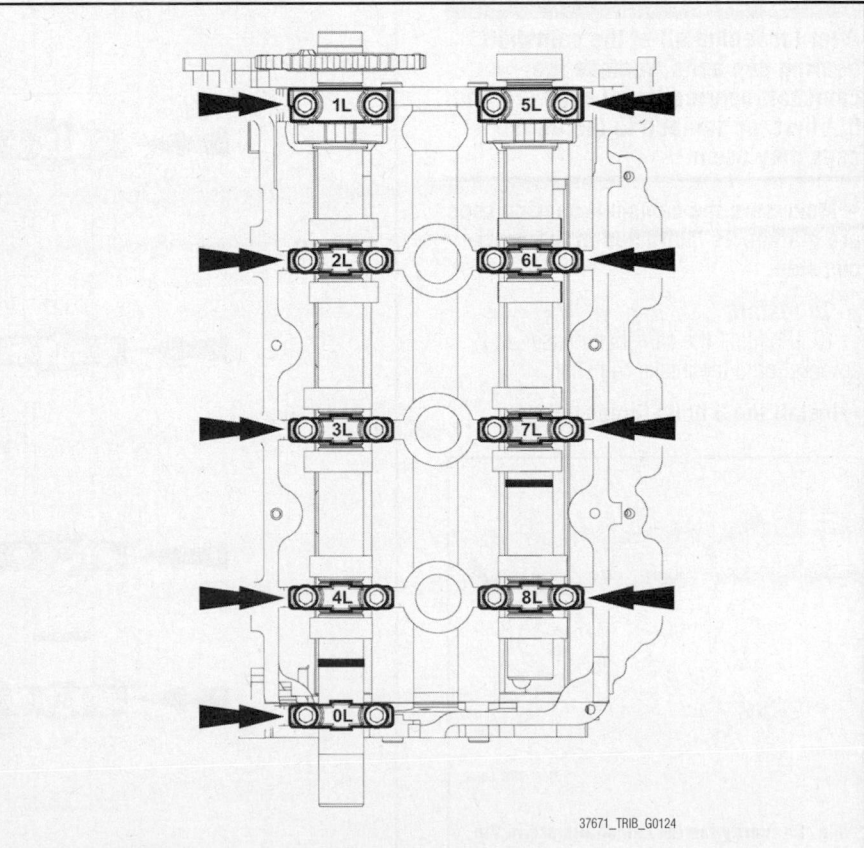

37671_TRIB_G0124

Fig. 78 If necessary, mark the camshaft bearing cap position and orientation as shown

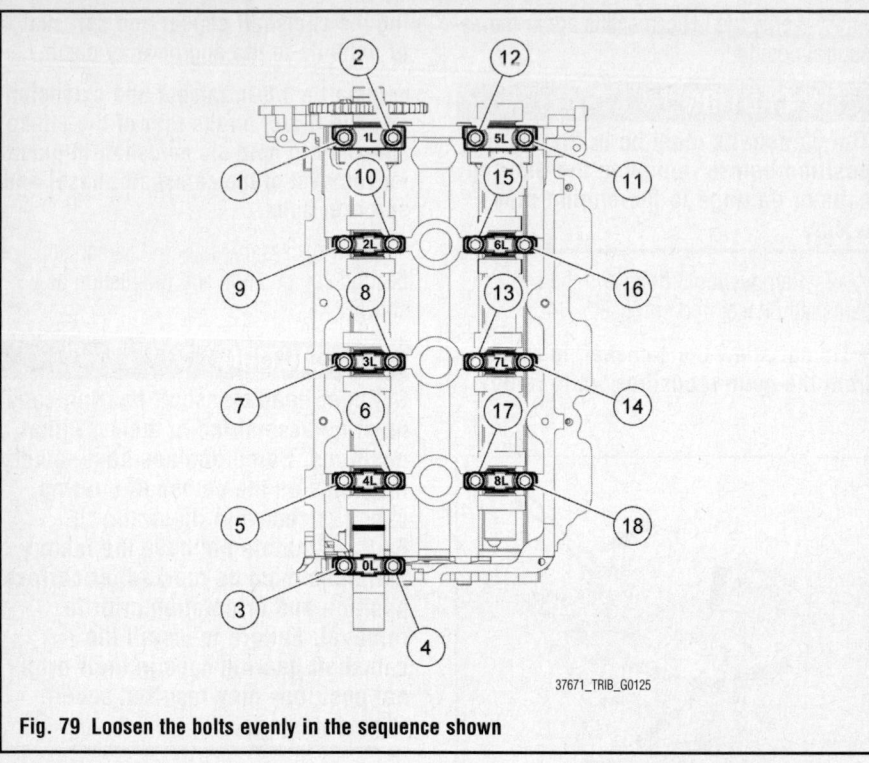

Fig. 79 Loosen the bolts evenly in the sequence shown

b. Remove the remaining camshaft bearing caps.

c. Remove the camshafts from the cylinder head.

✷✷ CAUTION

After loosening all of the camshaft bearing cap bolts, remove the camshaft bearing thrust caps (1L and 5L) first, or damage to the thrust caps may occur.

➡ Make sure the camshaft bearing caps are marked as instructed in the previous step.

To install:

10. Position the camshaft phaser and sprocket onto the intake camshaft.

➡ Install the 3 bolts finger-tight.

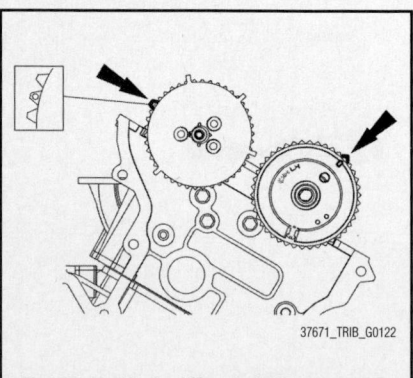

Fig. 80 Verify the LH camshafts are in the neutral position

11. Lubricate the LH camshafts with clean engine oil and carefully position the camshafts onto the cylinder head.

12. Align the LH camshafts as shown.

13. Lubricate the bearing surfaces of the camshaft bearing thrust caps with clean engine oil and install the bearing thrust caps.

14. Loosely install the bolts.

✷✷ CAUTION

Cylinder head camshaft journal caps and cylinder heads are numbered to verify that they are assembled in their original positions. If not reassembled in their original positions, severe engine damage may occur.

✷✷ CAUTION

Do not install the camshaft journal thrust caps until all of the camshaft bearing caps have been installed or damage to the thrust caps can occur.

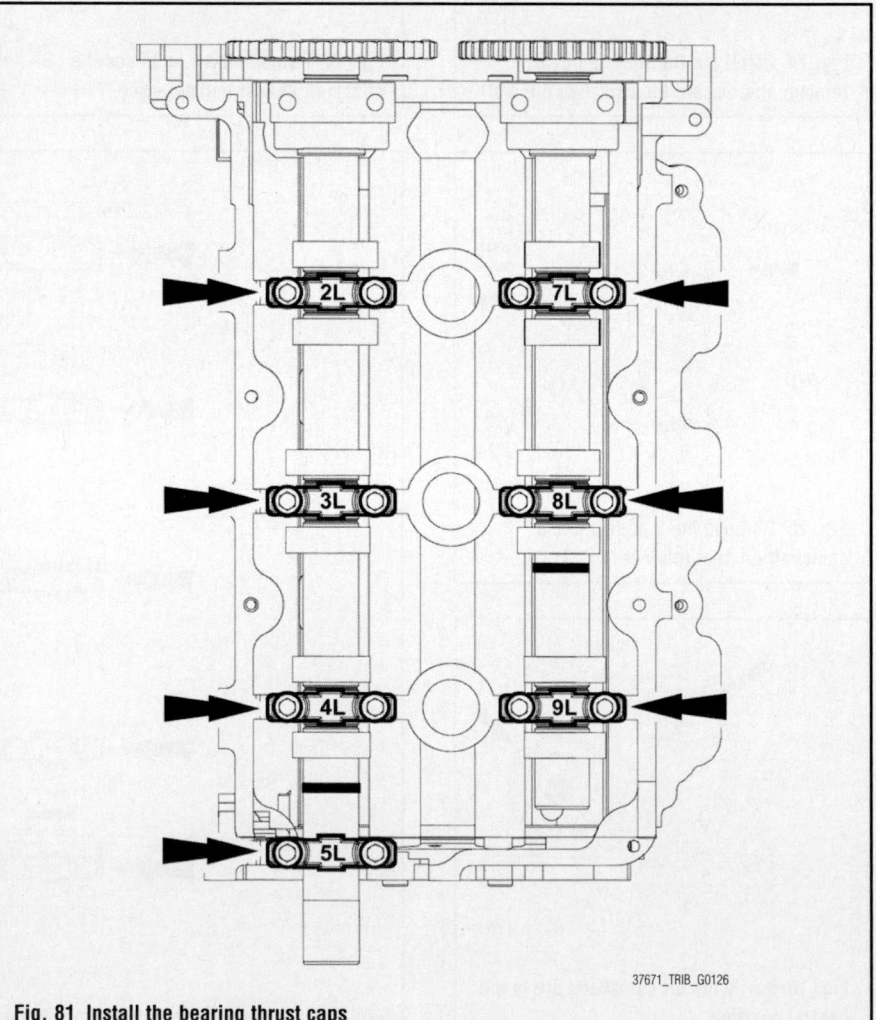

Fig. 81 Install the bearing thrust caps

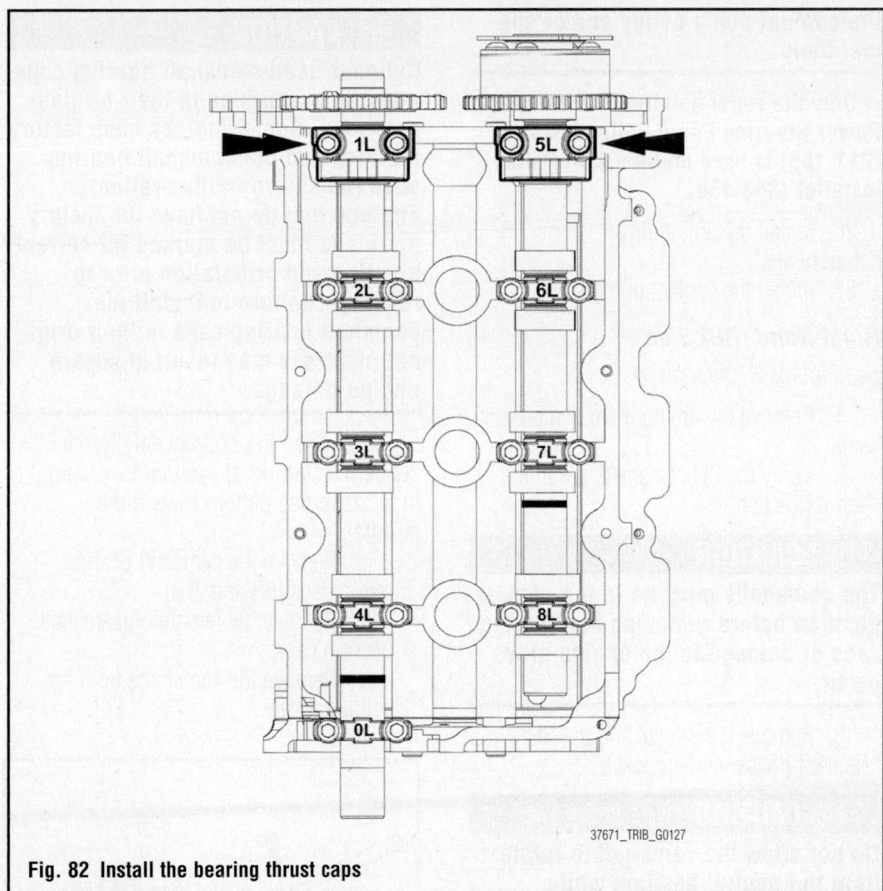

Fig. 82 Install the bearing thrust caps

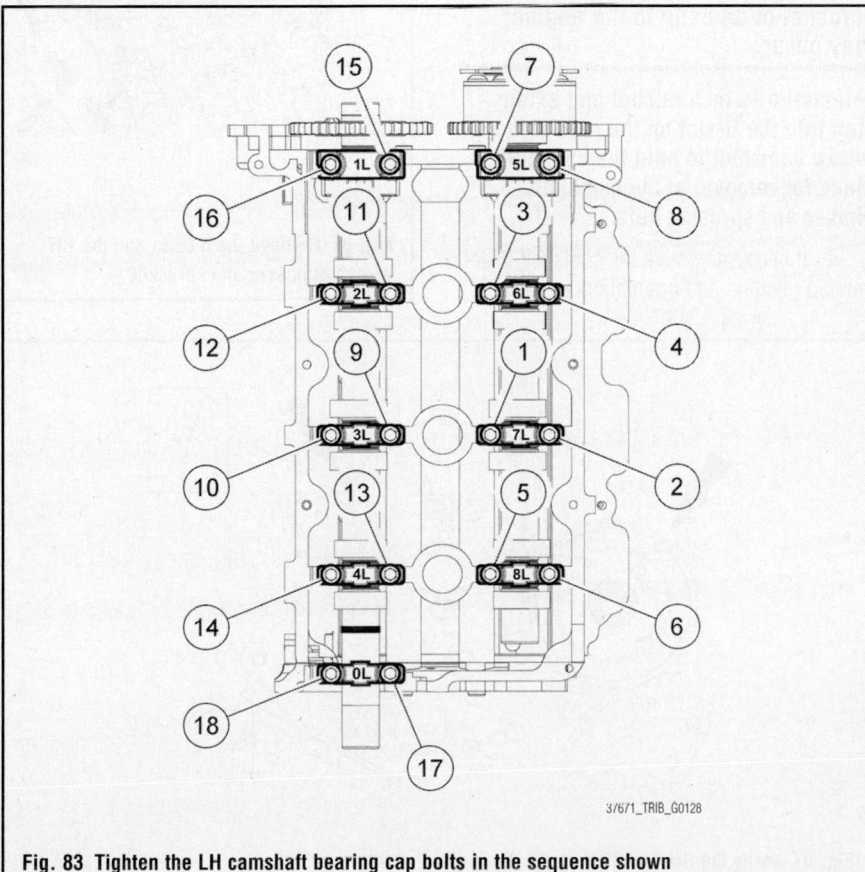

Fig. 83 Tighten the LH camshaft bearing cap bolts in the sequence shown

15. Lubricate the bearing surfaces of the LH camshaft bearing thrust caps with clean engine oil and install the bearing thrust caps.

16. Loosely install the bolts.

17. Tighten the LH camshaft bearing cap bolts in the sequence shown in 2 stages.

  a. Stage 1: Tighten to 89 inch lbs. (10 Nm).

  b. Stage 2: Individually loosen and then tighten each camshaft bearing cap to Tighten to 89 inch lbs. (10 Nm).

➡ **Make sure to tighten the camshaft bearing cap bolts in sequence in 2 stages.**

18. Tighten the 3 LH camshaft phaser and sprocket bolts to 13 ft. lbs. (18 Nm).

✳ CAUTION

Do not allow the camshaft to rotate from the neutral position while tightening the camshaft phaser and sprocket bolts or damage to the engine may occur.

➡ **Install a 3/8-in ratchet and extension into the D-slot on the rear of the intake camshaft to hold the camshaft in place for tightening of the camshaft phaser and sprocket bolts.**

19. Install the special tool in the camshaft as shown in the illustration.

20. Install the camshaft oil seal retainer and the 2 bolts. Tighten to 89 inch lbs. (10 Nm).

➡ **Clean the sealing surfaces with metal surface prep before installing a new press-in-place gasket.**

21. Using the Camshaft Oil Seal Installer, Camshaft Oil Seal Protector and the Power Steering Pump Pulley Installer, install the camshaft oil seal.

➡ **Lubricate the camshaft oil seal with clean engine oil.**

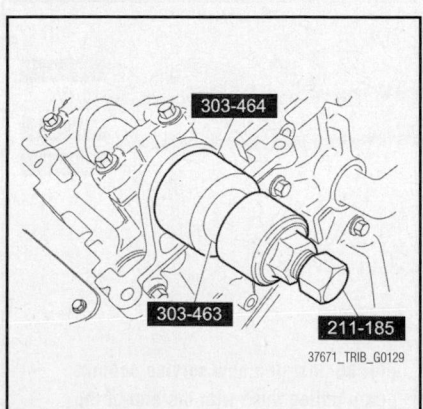

Fig. 84 Install the camshaft oil seal

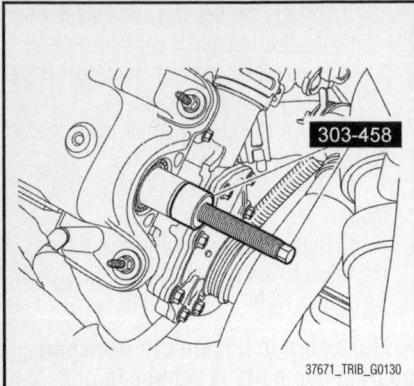

Fig. 85 Install the Camshaft Pulley Installer in the camshaft as shown

22. Install the Camshaft Pulley Installer in the camshaft as shown.

23. Adjust the collar on the Camshaft Pulley Installer screw to get the best thread engagement in the rear of the camshaft.

> **⊡ CAUTION**
>
> **Failure to use the correct special tools, assembled as shown, will result in damage to the coolant pump pulley and/or special tools.**

24. Position the coolant pump pulley over the previously installed Camshaft Pulley Installer and on the end of the camshaft. Install the Camshaft Pulley Installer, Power Steering Pump Pulley Installer and the Coolant Pump Pulley Spacer as shown.

25. Using the Camshaft Pulley Installer, Power Steering Pump Pulley Installer and the Coolant Pump Pulley Spacer, install a new service coolant pump pulley flush with the end of the camshaft.

> **⊡ CAUTION**
>
> **Failure to use the correct special tools, assembled as shown in the illustration, will result in damage to**

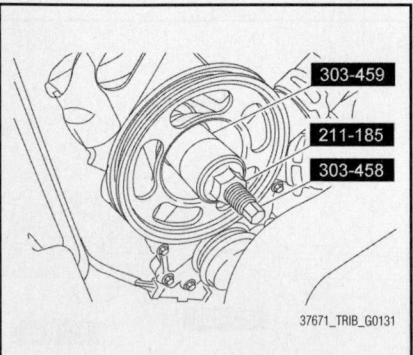

Fig. 86 Install a new service coolant pump pulley flush with the end of the camshaft

the coolant pump pulley and/or special tools.

➡Only the roller collared nut from the Power Steering Pump Pulley Installer (211-185) is used on Camshaft Pulley Installer (303-458).

26. Install the timing drive components.

27. Install the coolant pump belt.

Right Hand (RH) Side

See Figures 87 through 91.

1. Remove the timing drive components.

2. Verify the RH camshafts are in the neutral position.

> **✳✳ CAUTION**
>
> **The camshafts must be in the neutral position before removing the bearing caps or damage to the engine may occur.**

3. Remove the 3 bolts and the RH camshaft phaser and sprocket.

> **✳✳ CAUTION**
>
> **Do not allow the camshaft to rotate from the neutral position while removing the camshaft phaser and sprocket or damage to the engine may occur.**

➡Install a ⅜ inch ratchet and extension into the D-slot on the rear of the intake camshaft to hold the camshaft in place for removal of the camshaft phaser and sprocket bolts.

4. If necessary, mark the camshaft bearing cap position and orientation as shown.

> **✳✳ CAUTION**
>
> **Cylinder head camshaft bearing caps must be assembled in their original positions. Some engines have factory markings on the camshaft bearing caps (as shown in illustration). Engines that do not have the factory markings must be marked for correct position and orientation prior to removal. Failure to install the camshaft bearing caps in their original positions may result in severe engine damage.**

5. Loosen the bolts evenly in the sequence starting at each end working in a crisscross pattern toward the middle.

   a. Remove the camshaft bearing thrust caps (5R and 1R).

   b. Remove the remaining camshaft bearing caps.

   c. Remove the camshafts from the cylinder head.

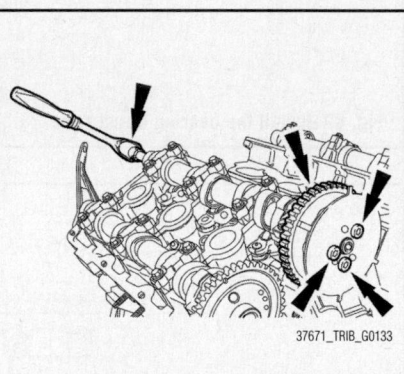

Fig. 88 Remove the 3 bolts and the RH camshaft phaser and sprocket

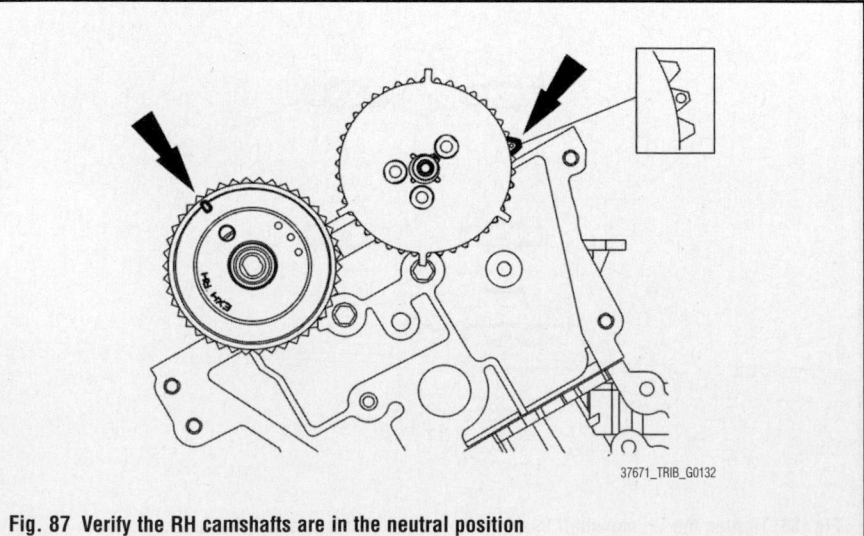

Fig. 87 Verify the RH camshafts are in the neutral position

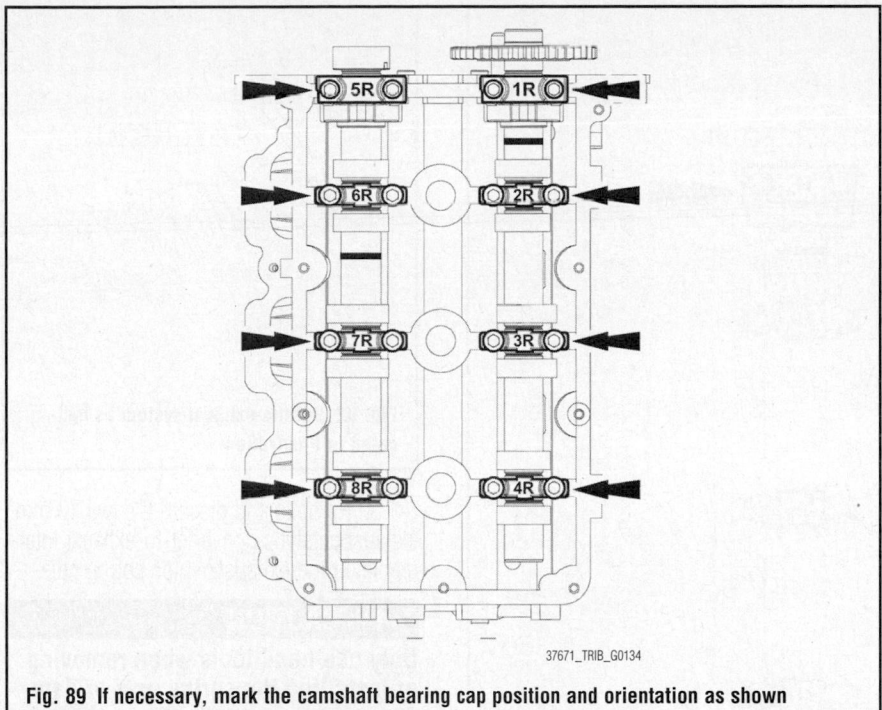

Fig. 89 If necessary, mark the camshaft bearing cap position and orientation as shown

12. Install the camshaft bearing thrust caps.

13. Loosely install the bolts.

➡ **Lubricate the bearing surfaces of the RH camshaft bearing thrust caps with clean engine oil.**

14. Tighten the RH camshaft bearing cap bolts in the sequence shown in 2 stages.

    a. Stage 1: Tighten to 89 inch lbs. (10 Nm).

    b. Stage 2: Individually loosen and then tighten each camshaft bearing cap to Tighten to 89 inch lbs. (10 Nm).

15. Tighten the 3 RH camshaft phaser and sprocket bolts to 13 ft. lbs. (18 Nm).

✳✳ CAUTION

Do not allow the camshaft to rotate from the neutral position while tightening the camshaft phaser and sprocket bolts or damage to the engine may occur.

➡ **Install a ⅜ inch ratchet and extension into the D-slot on the rear of the intake camshaft to hold the camshaft in place for tightening of the camshaft phaser and sprocket bolts.**

✳✳ CAUTION

After loosening all of the camshaft bearing cap bolts, remove the camshaft bearing thrust caps (5R and 1R) first, or damage to the thrust caps may occur.

➡ **Make sure the camshaft bearing caps are marked as instructed in the previous step.**

To install:

6. Position the RH camshaft phaser and sprocket onto the intake camshaft.

7. Install the 3 bolts finger-tight.

8. Lubricate the RH camshafts with clean engine oil and carefully position the camshafts onto the cylinder head.

9. Align the RH camshafts as shown.

10. Install the camshaft bearing caps.

11. Loosely install the bolts.

✳✳ CAUTION

Cylinder head camshaft journal caps and cylinder heads are numbered to verify that they are assembled in their original positions. If not reassembled in their original positions, severe engine damage may occur.

✳✳ CAUTION

Do not install the camshaft journal thrust caps until all of the camshaft bearing caps have been installed or damage to the thrust caps can occur.

➡ **Lubricate the bearing surfaces of the RH camshaft bearing caps with clean engine oil.**

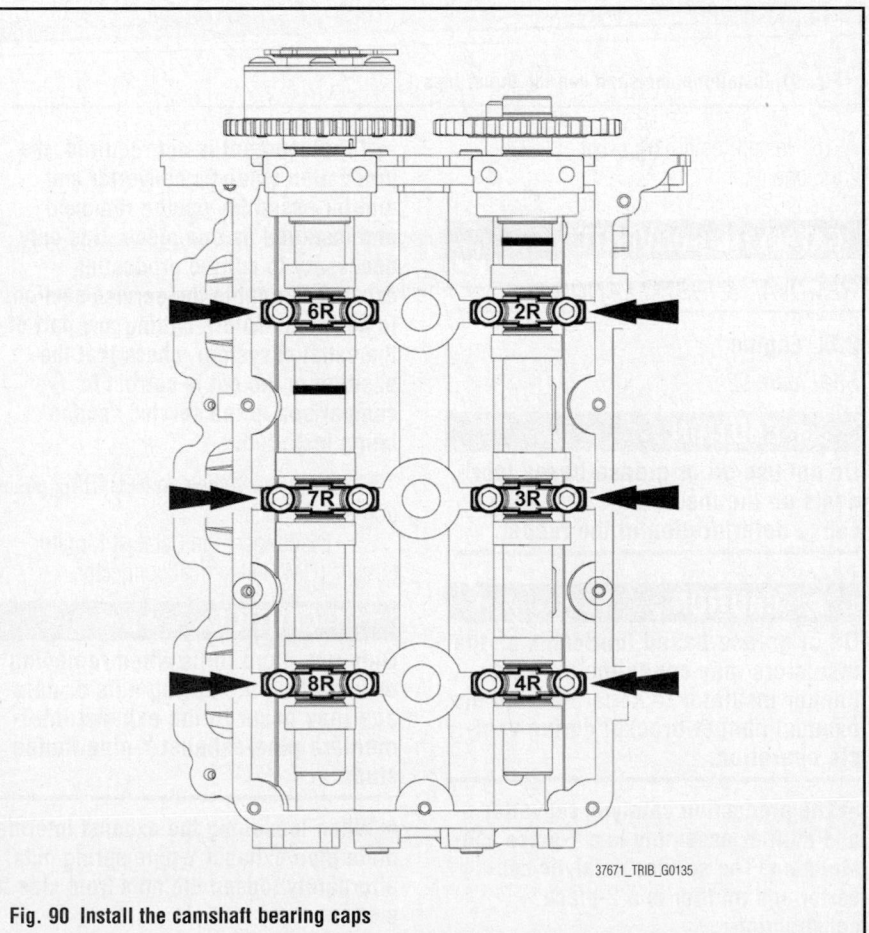

Fig. 90 Install the camshaft bearing caps

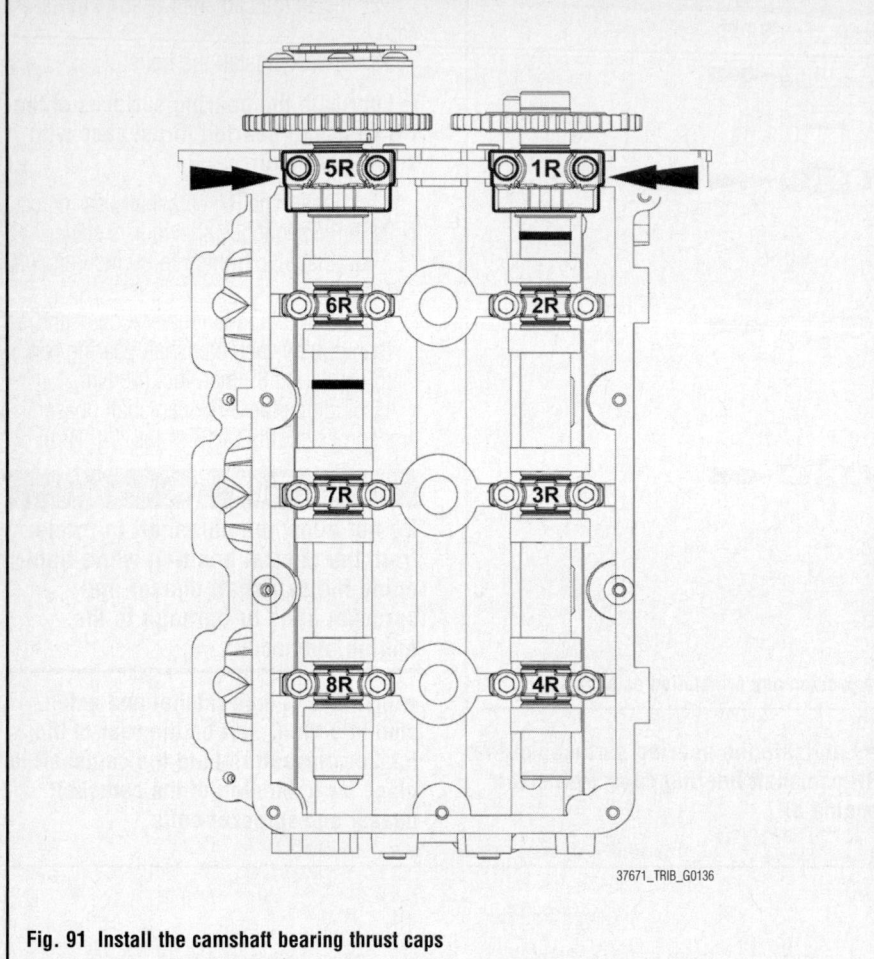

Fig. 91 Install the camshaft bearing thrust caps

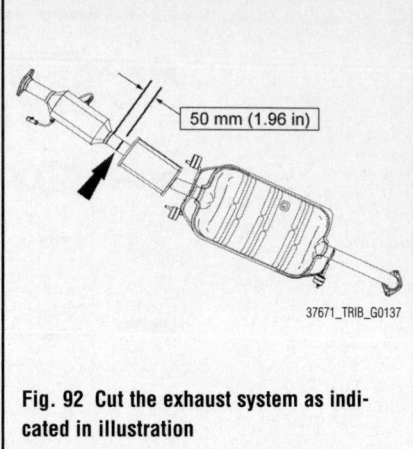

Fig. 92 Cut the exhaust system as indicated in illustration

16. Install the timing drive components.

CATALYTIC CONVERTER

REMOVAL & INSTALLATION

2.5L Engine

See Figure 92.

❋❋ CAUTION

Do not use oil or grease-based lubricants on the insulators. They may cause deterioration of the rubber.

❋❋ CAUTION

Oil or grease-based lubricants on the insulators may cause the exhaust hanger insulator to separate from the exhaust hanger bracket during vehicle operation.

➡The production catalytic converter and muffler assembly is a 1-piece construction. The service catalytic converter and muffler is a 2-piece construction.

➡If replacement is not required, the production catalytic converter and muffler assembly can be removed and installed as one piece. It is only necessary to cut the production exhaust to enable the service section to be fitted. Before cutting any part of the exhaust system, check that the position of the cut is correct in comparison to the service section being installed.

1. With the vehicle in NEUTRAL, position it on a hoist.
2. Disconnect the Catalyst Monitor Sensor (CMS) electrical connector.

❋❋ CAUTION

Only use hand tools when removing or installing the spring nuts or damage may occur to the exhaust intermediate pipe/exhaust Y-pipe flange studs.

➡When loosening the exhaust intermediate pipe/exhaust Y-pipe spring nuts, alternately loosen the nuts from side to side.

3. Remove and discard the two 10 mm exhaust catalytic converter-to-exhaust intermediate pipe/exhaust Y-pipe spring nuts.

❋❋ CAUTION

Only use hand tools when removing or installing the spring nuts or damage may occur to the resonator flange studs.

➡When loosening the resonator spring nuts, alternately loosen the nuts from side to side.

4. Remove and discard the two 8 mm resonator-to-muffler spring nuts.

❋❋ CAUTION

Remove the gaskets by simultaneously pulling up and twisting the gasket on the pipe. Do not pry under the gasket or damage to the flange may occur.

5. Detach the 3 exhaust hangers and remove the catalytic converter and muffler assembly.
6. Remove and discard the gaskets.
7. Cut the exhaust system as indicated in illustration.

To install:

❋❋ CAUTION

The exhaust hanger insulators are constructed of a special material. Use only the correct specification exhaust hanger insulator or damage to the exhaust system may occur.

➡Check the exhaust hanger insulators for damage and fatigue. Install new exhaust hanger insulators as required.

➡Thoroughly clean the sealing surfaces of the flanges using a finishing pad.

➡Do not tighten the service clamp at this time.

8. Install the service clamp onto the catalytic converter section.

9. Install the catalytic converter section and service clamp onto the muffler section.

10. Inspect the exhaust intermediate pipe/exhaust Y-pipe and resonator flange studs for damage.

➡**If damaged, replace stud(s), or if stud comes out when removing nut(s), replace the stud(s).**

➡**If replacement of the exhaust intermediate pipe/exhaust Y-pipe flange stud(s) is required, carry out the next two steps.**

11. Use the C-Frame and Screw Installer/Remover to push the 10 mm stud out of the flange.

➡**When positioning the new 10 mm stud in the exhaust intermediate/ exhaust Y-pipe flange, make sure to line up the new stud seat knurls with witness knurl grooves in the exhaust intermediate/exhaust Y-pipe flange.**

12. Use a C-clamp and a deep-well socket to push the 10 mm stud fully into the flange.

➡**Make sure the stud is fully and evenly seated into the flange.**

➡**If replacement of the resonator flange stud(s) is required, carry out the next two steps.**

13. Use the C-Frame and Screw Installer/Remover to push the 8 mm stud out of the flange.

➡**When positioning the new 8 mm stud in the resonator flange, make sure to line up the new stud seat knurls with witness knurl grooves in the resonator flange.**

14. Use a C-clamp and a deep-well socket to push the 8 mm stud fully into the flange.

➡**Make sure the stud is fully and evenly seated into the flange.**

➡**Do not tighten the service clamp at this time.**

15. Attach the catalytic converter and muffler assembly to the 3 exhaust hangers.

✳✳ CAUTION

Replace the gasket by hand. Do not rotate the gasket back and forth dur-

ing installation. Doing so may lessen the gaskets ability to seal and stay correctly positioned during assembly.

➡**Thoroughly clean the sealing surfaces of the flanges using a finishing pad. Inspect the cleaned sealing surface for nicks and scratches and replace as necessary.**

16. Install a new resonator-to-muffler gasket by hand.

17. Install the 2 new 8 mm resonator-to-muffler spring nuts and alternately tighten RH side to LH side in sequence in 3 stages:

    a. Stage 1: Tighten to 44 inch lbs. (5 Nm).

    b. Stage 2: Tighten to 89 inch lbs. (10 Nm).

    c. Stage 3: Tighten to 150 inch lbs. (17 Nm).

✳✳ CAUTION

Replace the gasket by hand. Do not rotate the gasket back and forth during installation. Doing so may lessen the gaskets ability to seal and stay correctly positioned during assembly.

➡**Thoroughly clean the sealing surfaces of the flanges using a finishing pad. Inspect the cleaned sealing surface for nicks and scratches and replace as necessary.**

18. Install a new converter-to-exhaust intermediate/exhaust Y-pipe gasket by hand.

19. Install the 2 new 10 mm catalytic converter-to-exhaust intermediate/exhaust Y-pipe spring nuts and alternately tighten RH side to LH side in sequence in 3 stages:

    a. Stage 1: Tighten to 44 inch lbs. (5 Nm).

    b. Stage 2: Tighten to 133 inch lbs. (15 Nm).

    c. Stage 3: Tighten to 18 ft. lbs. (25 Nm).

20. Tighten the 4 service clamp nuts to 35 ft. lbs. (47 Nm).

3.0L Engine

Under-Floor

See Figure 93.

✳✳ CAUTION

Do not use oil or grease-based lubricants on the insulators. They may cause deterioration of the rubber.

✳✳ CAUTION

Oil or grease-based lubricants on the insulators may cause the exhaust

hanger insulator to separate from the exhaust hanger bracket during vehicle operation.

✳✳ CAUTION

The catalytic converter and muffler assembly is 1-piece construction. The service catalytic converter and muffler is a 2-piece construction. It is necessary to cut the production exhaust to enable the service section to be fitted. Before cutting any part of the exhaust system, check that the position of the cut is correct in comparison to the service section being installed.

1. With the vehicle in NEUTRAL, position it on a hoist.

2. Remove and discard the 2 exhaust catalytic converter-to-intermediate pipe spring nuts.

3. Remove and discard the 2 resonator-to-muffler spring nuts.

4. Detach the 3 exhaust hangers and remove the catalytic converter and muffler assembly.

5. Discard the gaskets.

6. Cut the exhaust system as indicated in illustration.

To install:

✳✳ CAUTION

The exhaust hanger insulators are constructed of a special material. Use only the correct specification exhaust hanger insulator or damage to the exhaust system may occur.

✳✳ CAUTION

Check the exhaust hanger insulators for damage and fatigue. Install new exhaust hanger insulators as required.

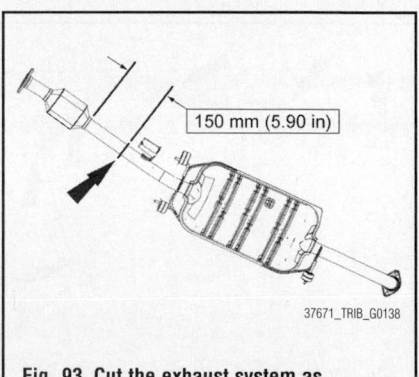

150 mm (5.90 in)

37671_TRIB_G0138

Fig. 93 Cut the exhaust system as indicated in illustration

✳✳ CAUTION

Thoroughly clean the sealing surfaces of the flanges using an abrasive pad.

7. Install the service clamp onto the catalytic converter section.

➡**Do not tighten the service clamp at this time.**

8. Install the catalytic converter section and service clamp onto the muffler section.

9. Attach the catalytic converter and muffler assembly to the 3 exhaust hangers.

➡**Do not tighten the service clamp at this time.**

10. Install a new gasket and 2 new resonator-to-muffler spring nuts. Tighten to 15 ft. lbs. (20 Nm).

11. Install a new gasket and 2 new catalytic converter-to-exhaust intermediate pipe spring nuts. Tighten to 18 ft. lbs. (25 Nm).

12. Tighten the 4 service clamp nuts to 35 ft. lbs. (47 Nm).

Left Hand (LH) Side

See Figures 94 and 95.

1. With the vehicle in NEUTRAL, position it on a hoist.

2. Remove the 6 retainers and the passenger side splash shield.

3. Disconnect the LH Heated Oxygen Sensor (HO2S).

4. Remove the 3 bolts and the LH catalytic converter heat shield.

5. Remove the exhaust crossover pipe.

6. Remove and discard the 6 LH catalytic converter manifold nuts.

7. Remove the LH catalytic converter manifold from the vehicle.

8. Discard the gasket.

9. Remove and discard the 6 LH catalytic converter manifold studs.

10. Clean and inspect the LH catalytic converter exhaust manifold.

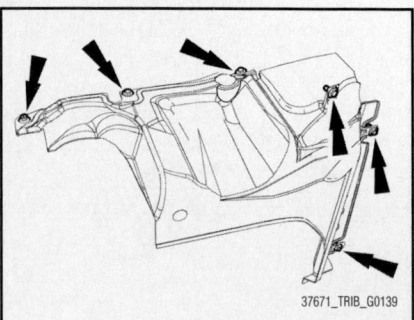

Fig. 94 Remove the 6 retainers and the passenger side splash shield

37671_TRIB_G0139

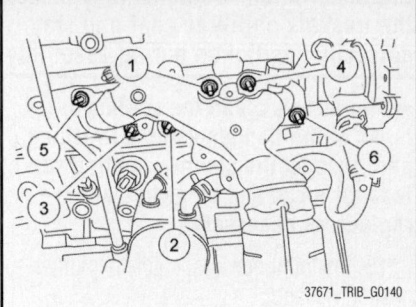

37671_TRIB_G0140

Fig. 95 Position the LH catalytic converter and tighten the 6 exhaust manifold nuts in 2 stages in the sequence shown

To install:

11. Install the 6 new LH catalytic converter manifold studs. Tighten to 8 ft. lbs. (11 Nm).

12. Position the LH catalytic converter and tighten the 6 exhaust manifold nuts in 2 stages in the sequence shown.

    a. Stage 1: Tighten to 18 ft. lbs. (25 Nm).

    b. Stage 2: Tighten to 18 ft. lbs. (25 Nm).

✳✳ CAUTION

Failure to tighten the catalytic converter nuts to specification before installing the converter bracket bolts will cause the converter to develop an exhaust leak.

✳✳ CAUTION

Failure to tighten the catalytic converter nuts to specification a second time will cause the converter to develop an exhaust leak.

13. Install the exhaust crossover pipe.

14. Install the LH catalytic converter heat shield and the 3 bolts. Tighten to 89 inch lbs. (10 Nm).

15. Install the LH heated oxygen sensor (HO2S).

16. Install the passenger side splash shield and the 6 retainers.

Right Hand (RH) Side

1. Remove the exhaust crossover pipe.

2. Disconnect the RH catalyst monitor sensor electrical connector.

3. Remove and discard the 3 RH catalytic converter nuts and remove the converter.

4. Discard the RH catalytic converter gasket.

To install:

5. To install, reverse the removal procedure.

6. Install a new gasket and nuts. Tighten to 30 ft. lbs. (40 Nm).

CRANKSHAFT FRONT SEAL

REMOVAL & INSTALLATION

2.5L Engine

See Figures 96 through 98.

✳✳ WARNING

Do not loosen or remove the crankshaft pulley bolt without first installing the special tools as instructed in this procedure. The crankshaft pulley and the crankshaft timing sprocket are not keyed to the crankshaft. The crankshaft, the crankshaft sprocket and the pulley are fitted together by friction, using diamond washers between the flange faces on each part. For that reason, the crankshaft sprocket is also unfastened if you loosen the pulley bolt. Before any repair requiring loosening or removal of the crankshaft pulley bolt, the crankshaft and camshafts must be locked in place by the special service tools, otherwise severe engine damage can occur.

✳✳ WARNING

During engine repair procedures, cleanliness is extremely important. Any foreign material (including any material created while cleaning gasket surfaces) that enters the oil passages, coolant passages or the oil pan can cause engine failure.

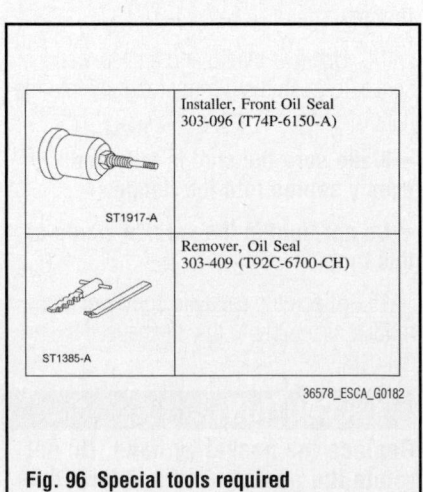

	Installer, Front Oil Seal 303-096 (T74P-6150-A)
ST1917-A	
	Remover, Oil Seal 303-409 (T92C-6700-CH)
ST1385-A	

36578_ESCA_G0182

Fig. 96 Special tools required

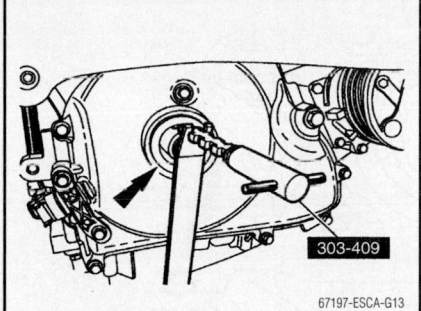

Fig. 97 Using the special tool, remove the crankshaft front oil seal

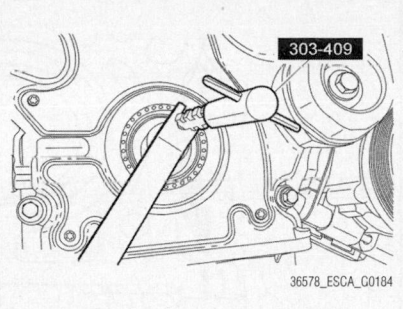

Fig. 99 Using the seal remover tool, remove and discard the crankshaft front seal

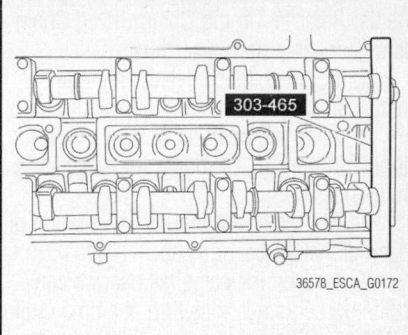

Fig. 101 Remove the camshaft alignment plate

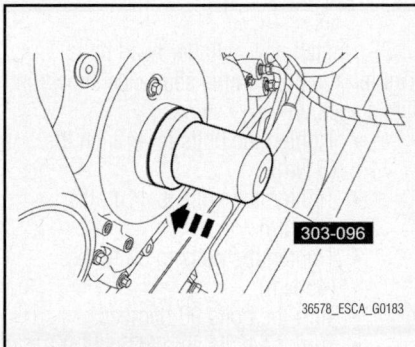

Fig. 98 Using the special tool, install the crankshaft front oil seal

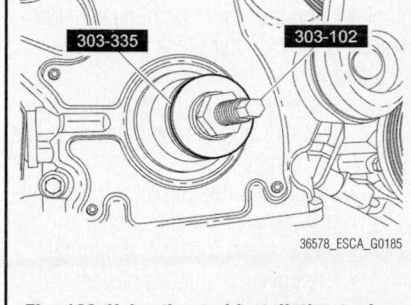

Fig. 100 Using the seal installation tools, install a new crankshaft front seal

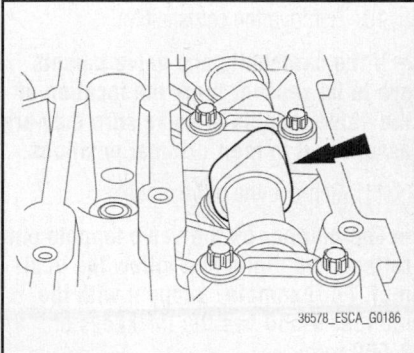

Fig. 102 Mark the position of the camshaft lobes on the No. 1 cylinder

1. Remove the crankshaft pulley.
2. Using the special tool, remove the crankshaft front oil seal. Use care not to damage the engine front cover or the crankshaft when removing the seal.

To install:

➡**Lubricate the oil seal with clean engine oil.**

3. Remove the through-bolt from the special tool.
4. Using the special tool, install the crankshaft front oil seal.
5. Install the crankshaft pulley.

3.0L Engine

See Figures 99 and 100.

1. With the vehicle in NEUTRAL, position it on a hoist.
2. Remove the crankshaft pulley.
3. Using the seal remover tool, remove and discard the crankshaft front seal.

To install:

4. Clean all sealing surfaces with metal surface prep.
5. Apply clean engine oil to the seal lip and seal bore before installing the seal.
6. Using the seal installation tools, install a new crankshaft front seal.

7. Install the crankshaft pulley.

CYLINDER HEAD

REMOVAL & INSTALLATION

2.5L Engine

See Figures 101 through 106.

✳✳ WARNING

Do not loosen or remove the crankshaft pulley bolt without first installing the special tools as instructed in this procedure. The crankshaft pulley and the crankshaft timing sprocket are not keyed to the crankshaft. The crankshaft, the crankshaft sprocket and the pulley are fitted together by friction, using diamond washers between the flange faces on each part. For that reason, the crankshaft sprocket is also unfastened if the pulley bolt is loosened. Before any repair requiring loosening or removal of the crankshaft pulley bolt, the crankshaft and camshafts must be locked in place by the special service tools, otherwise severe engine damage can occur.

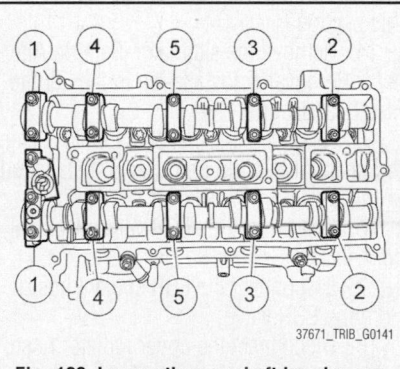

Fig. 103 Loosen the camshaft bearing cap bolts, in sequence

✳✳ WARNING

During engine repair procedures, cleanliness is extremely important. Any foreign material (including any material created while cleaning gasket surfaces) that enters the oil passages, coolant passages or the oil pan may cause engine failure.

1. With the vehicle in NEUTRAL, position it on a hoist.
2. Release the fuel system pressure. Refer to the Fuel System section.
3. Drain the engine cooling system.

4. Remove the timing drive components. Refer to Timing Chain Cover, Chain and Sprockets in this section.

5. Remove the camshaft alignment plate.

6. Mark the position of the camshaft lobes on the No. 1 cylinder for installation reference.

7. Mark the location and orientation of each camshaft bearing cap.

8. Loosen the camshaft bearing cap bolts, in sequence, one turn at a time until all tension is released from the camshaft bearing caps.

9. Remove the bolts and the camshaft bearing caps.

10. Remove the camshafts.

➡ **If the camshafts and valve tappets are to be reused, mark the location of the valve tappets to make sure they are assembled in their original positions.**

11. Remove the valve tappets.

➡ **The number on the valve tappets only reflects the digits that follow the decimal. For example, a tappet with the number 0.650 has the thickness of 3.650 mm.**

12. Inspect the valve tappets.

13. Remove the intake manifold. Refer to Intake Manifold Removal & Installation in this section.

14. Remove the alternator. Refer to Alternator Removal & Installation in the Engine Electrical section.

15. Remove the exhaust manifold. Refer to Exhaust Manifold Removal & Installation in this section.

16. Disconnect the EGR valve electrical connector.

17. Disconnect the EGR coolant hose from the EGR valve.

18. Disconnect the upper radiator hose, coolant bypass hose, heater hose and coolant vent hose from the engine coolant outlet.

19. Remove the 10 bolts and the cylinder head. Discard the bolts and the cylinder head gasket.

To install:

❊❊ WARNING

Do not use metal scrapers, wire brushes, power abrasive discs or other abrasive means to clean the sealing surfaces. These tools cause scratches and gouges that make leak paths. Use a plastic scraping tool to remove all traces of the head gasket.

Fig. 104 Apply silicone gasket and sealant to the locations shown—2.5L engine

20. Clean the cylinder head-to-cylinder block mating surface of both the cylinder head and the cylinder block in the following sequence:

- Remove any large deposits of silicone or gasket material with a plastic scraper.
- Apply silicone gasket remover, following package directions, and allow to set for several minutes.
- Remove the silicone gasket remover with a plastic scraper. A second application of silicone gasket remover may be required if residual traces of silicone or gasket material remain.
- Apply metal surface prep, following package directions, to remove any traces of oil or coolant, and to prepare the surfaces to bond with the new gasket. Do not attempt to make the metal shiny. Some staining of the metal surfaces is normal.

21. Support the cylinder head on a bench with the head gasket side up. Check the cylinder head distortion and the cylinder block distortion.

22. Clean the cylinder head bolt holes in the cylinder block. Make sure all coolant, oil or other foreign material is removed.

23. Apply silicone gasket and sealant to the locations shown.

24. Install a new head gasket.

❊❊ WARNING

The cylinder head bolts are torque-to-yield and must not be reused. New cylinder head bolts must be installed.

25. Lubricate the bolts with clean engine oil prior to installation.

26. Install the cylinder head and 10 new bolts. Tighten the bolts in the sequence shown in 5 stages:

27. Install new cylinder head bolts. Tighten the bolts in the sequence shown in five stages:

- Tighten the bolts to 44 inch lbs. (5 Nm).
- Tighten the bolts to 11 ft. lbs. (15 Nm).
- Tighten the bolts to 33 ft. lbs. (45 Nm).
- Turn the bolts 90 degrees.
- Turn the bolts an additional 90 degrees.

28. Connect the upper radiator hose, coolant bypass hose, heater hose and coolant vent hose to the engine coolant outlet.

29. Connect the EGR coolant hose to the EGR valve.

30. Connect the EGR valve electrical connector.

31. Install the exhaust manifold.

32. Install the alternator.

33. Install the intake manifold.

34. Lubricate the valve tappets with clean engine oil.

35. Install the valve tappets in their original positions.

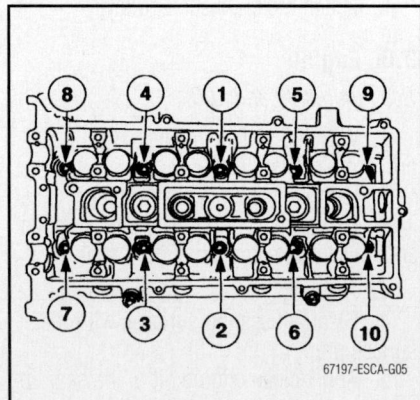

Fig. 105 Cylinder head bolt torque sequence—2.5L engine

8
4
1
5
9
7
3
2
6
10
36578_ESCA_G0187

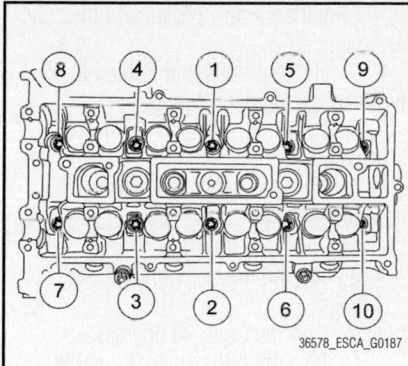

Fig. 106 Bearing cap tightening sequence

➡Install the camshafts with the alignment notches in the camshafts lined up so the camshaft alignment plate can be installed. Make sure the lobes on the No. 1 cylinder are in the same position as noted in the removal procedure. Failure to follow this procedure can cause severe damage to the valves and pistons.

36. Lubricate the camshaft journals and bearing caps with clean engine oil. Install the camshafts and bearing caps in their original location and orientation. Tighten the bearing caps in the sequence shown in 3 stages:

- Stage 1: Tighten the camshaft bearing cap bolts, one turn at a time, until finger tight.
- Stage 2: Tighten to (62 inch (7 Nm).
- Stage 3: Tighten to 142 inch lbs. (16 Nm).

37. Install the camshaft alignment plate.
38. Install the timing drive components.
39. Fill and bleed the engine cooling system.

3.0L Engine

See Figures 107 through 111.

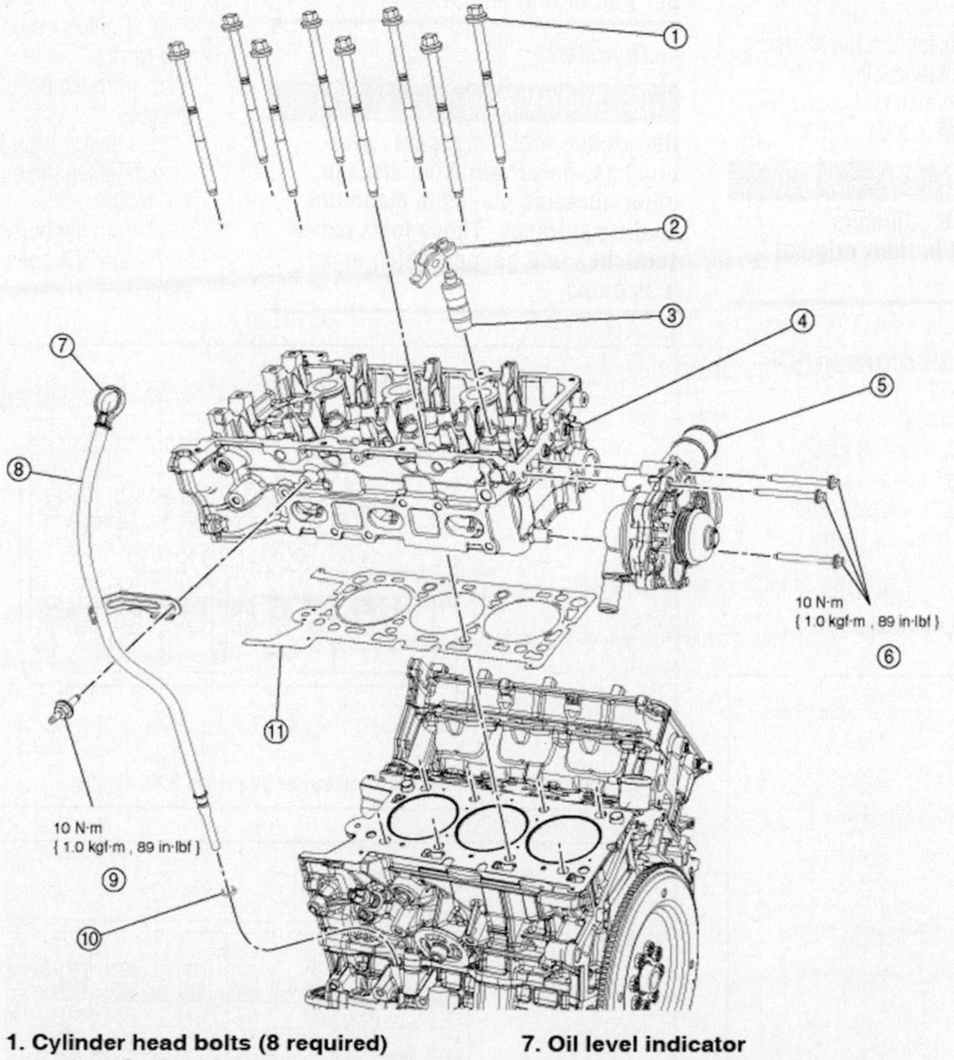

1
2
3
4
5
7
8
10 N·m
{ 1.0 kgf·m , 89 in·lbf }
6
11
10 N·m
{ 1.0 kgf·m , 89 in·lbf }
9
10
1. Cylinder head bolts (8 required)
2. Camshaft roller followers (12 required)
3. Hydraulic lash adjusters (12 required)
4. Cylinder head
5. Water pump
6. Water pump bolts (3 required)
7. Oil level indicator
8. Oil level indicator tube
9. Oil level indicator tube stud bolt
10. Oil level indicator tube O-ring
11. Cylinder head gasket
22140_TRIB_G0063

Fig. 107 Cylinder head— LH, 3.0L Engines exploded view

※ WARNING

During engine repair procedures, cleanliness is extremely important. Any foreign material, (including any material created while cleaning gasket surfaces) that enters the oil passages, coolant passages or the oil pan, can cause engine failure.

The procedure for the left side cylinder head and right side are similar. Changes in the procedure will be noted for either side cylinder head.

1. Properly relieve the fuel system pressure.
2. Drain the cooling system.
3. With the vehicle in **NEUTRAL**, position it on a hoist.
4. Remove or disconnect the following:
   - Negative battery cable
   - Lower intake manifold
   - Coolant bypass tube

※ WARNING

The hydraulic lash adjusters must be installed in their original positions.

- Camshaft
- Exhaust Gas Recirculation (EGR) tube, right side only
- Exhaust manifold
- oil level indicator and tube
- Camshaft followers
- Hydraulic lash adjusters and matchmark them for proper installation
- Remove the 3 bolts and position the water pump aside.

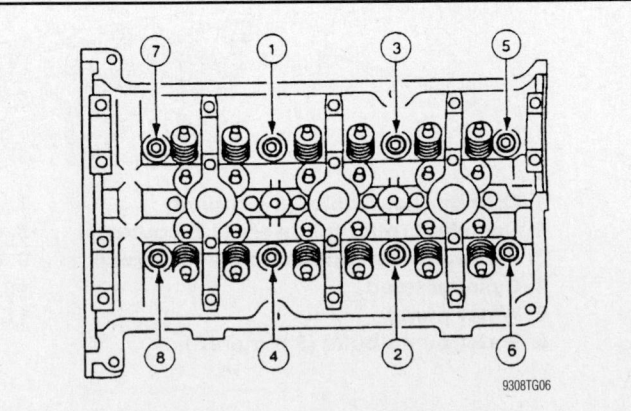

Fig. 108 Cylinder head bolt removal sequence— LH, 3.0L

- Cylinder head bolts in sequence and discard them
5. Remove the cylinder head
6. Remove the cylinder head and support the cylinder head on a bench with the head gasket up.
7. Discard the gasket and bolts.
8. Inspect all areas of the deck face with a straightedge and feeler gauge. The cylinder head must not have depressions deeper than 0.001 in (0.0254 mm) across a 1.5 inches(38.1 mm) square area, or scratches more than 0.001 in (0.0254 mm).

※ WARNING

The straightedge used must be flat within 0.0002 inches (0.0051 mm) per foot of tool length.

To install:

※ WARNING

Do not use metal scrapers, wire brushes, power abrasive discs or other abrasive means to clean the sealing surfaces. These tools cause scratches and gouges which make leak paths.

9. Position a new gasket and the cylinder head.
10. Install the bolts and tighten in 6 stages in the sequence shown.
   a. Tighten the bolts to 30 ft. lbs. (40 Nm).
   b. Turn the bolts an additional 90 degrees.
   c. Loosen 1 full turn.
   d. Tighten the bolts to 30 ft. lbs. (40 Nm).
   e. Turn the bolts 90 degrees.
   f. Turn the bolts an additional 90 degrees.
11. Lubricate the cylinder head bolt threads.
12. Torque the cylinder head bolts in the proper sequence as follows:
   a. Tighten the bolts to 30 ft. lbs. (40 Nm).
   b. Turn the bolts an additional 90 degrees.
   c. Loosen 1 full turn.
   d. Tighten the bolts to 30 ft. lbs. (40 Nm).
   e. Turn the bolts 90 degrees.
   f. Turn the bolts an additional 90 degrees.

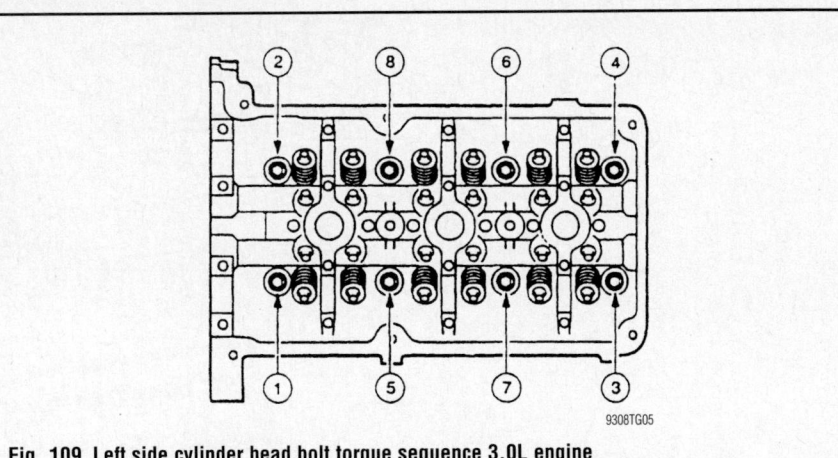

Fig. 109 Left side cylinder head bolt torque sequence 3.0L engine

Fig. 110 Right side cylinder head bolt torque sequence 3.0L engine

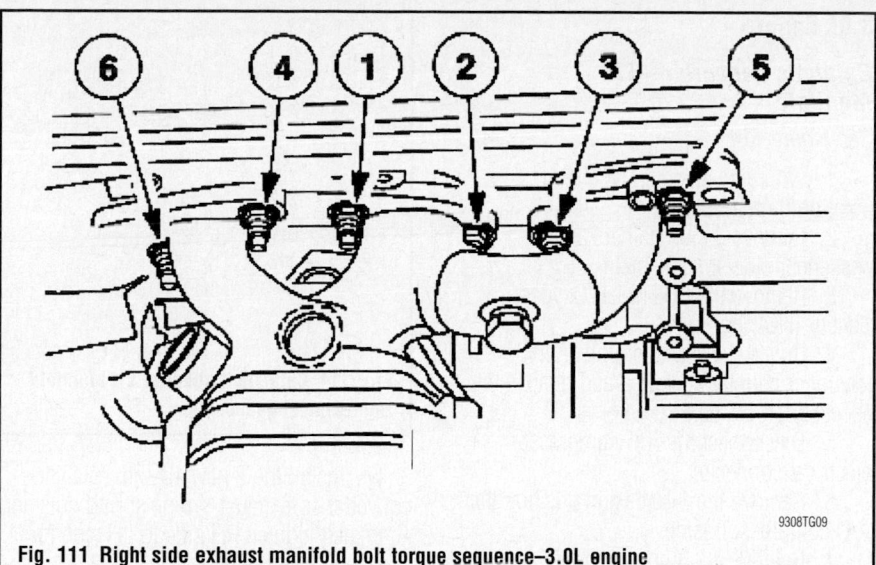

Fig. 111 Right side exhaust manifold bolt torque sequence–3.0L engine

9308TG09

5. Remove the 2 exhaust down pipe-to-exhaust manifold nuts and the exhaust down pipe.

6. Disconnect the Heated Oxygen Sensor (HO2S) electrical connector.

7. Remove the 4 exhaust manifold heat shield bolts and the heat shield.

8. Remove and discard the 7 exhaust manifold nuts.

9. Remove the exhaust manifold and discard the exhaust manifold gasket.

10. Remove and discard the 7 exhaust manifold studs.

11. Clean and inspect the exhaust manifold.

To install:

12. Install the 7 new exhaust manifold studs and tighten to 150 inch lbs. (17 Nm).

13. Install or connect the following:
- Position the water pump and install the bolts
- Tighten the bolts to 89 inch. lbs. (10 Nm).
- Install the oil level indicator tube and stud bolt.
- Install a new O-ring seal and lubricate with clean engine oil.
- Hydraulic lash adjusters
- Camshaft followers
- Camshaft
- Exhaust manifold. Torque the bolts in sequence to 15 ft. lbs. (20 Nm), right side only
- EGR tube, right side only
- Coolant bypass tube
- Negative battery cable

14. Fill the coolant to the proper level.

15. Start the vehicle and check for leaks, repair if necessary.

EXHAUST MANIFOLD

REMOVAL & INSTALLATION

2.5L Engine

See Figures 112 and 113.

1. With the vehicle in NEUTRAL, position it on a hoist.

2. Remove and discard the 3 exhaust intermediate pipe-to-exhaust down pipe nuts.

3. Remove and discard the two 10-mm catalytic converter-to-exhaust intermediate pipe spring nuts and remove the intermediate pipe.

4. Remove the 2 exhaust bracket bolts.

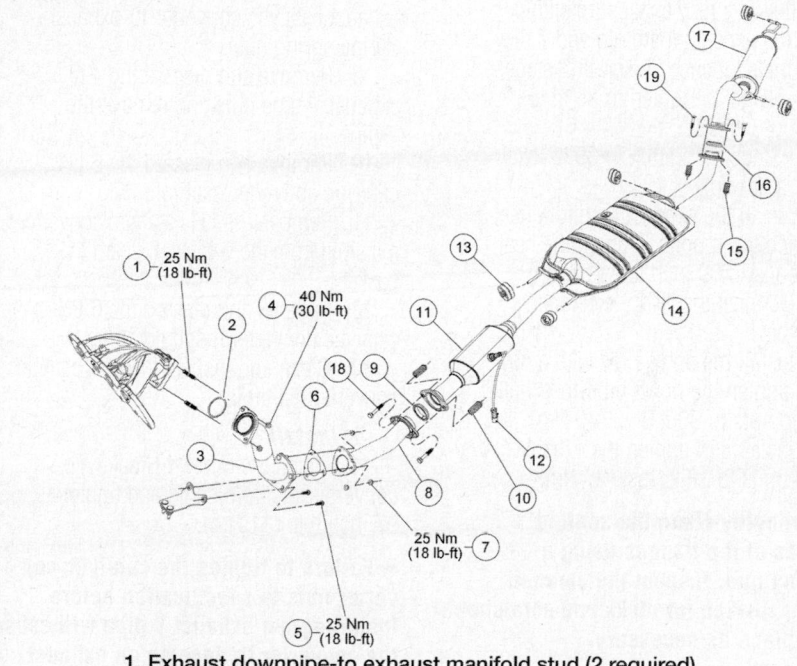

1. Exhaust downpipe-to exhaust manifold stud (2 required)
2. Gasket
3. Exhaust downpipe
4. Exhaust downpipe-to-exhaust manifold nut (2 required)
5. Exhaust downpipe bracket bolt (2 required)
6. Gasket
7. Exhaust intermediate pipe nut (3 required)
8. Exhaust intermediate pipe
9. Gasket
10. Catalytic converter-to-intermediate pipe spring nut (2 required)
11. Underbody catalytic converter
12. Catalyst Monitor Sensor (CMS) electrical connector
13. Exhaust hanger (5 required)
14. Muffler
15. Resonator-to-muffler spring nut (2 required)
16. Gasket
17. Resonator assembly
18. Intermediate pipe flange studs (2 required)
19. Resonator flange studs (2 required

Fig. 112 Exhaust system exploded view

36578_ESCA_G0191

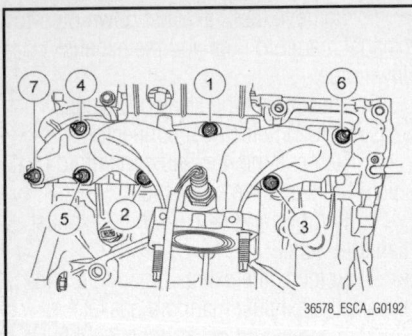

Fig. 113 Exhaust manifold tightening sequence

➡**Failure to tighten the catalytic converter nuts to specification before installing the converter bracket bolts will cause the converter to develop an exhaust leak.**

13. Install a new exhaust manifold gasket, the exhaust manifold and 7 new nuts in the sequence shown in 2 stages:
- Stage 1: Tighten to 35 ft. lbs. (48 Nm).
- Stage 2: Tighten to 35 ft. lbs. (48 Nm).

14. Install the exhaust manifold heat shield and the 4 bolts. Tighten the heat shield bolts to 89 inch lbs. (10 Nm).

15. Connect the HO2S electrical connector.

16. Install the down pipe with a new gasket, tighten the down pipe-to-exhaust manifold nuts to 30 ft. lbs. (40 Nm).

17. Install and tighten the exhaust bracket bolts to 18 ft. lbs. (25 Nm).

➡**Thoroughly clean the sealing surfaces of the flanges using a finishing pad. Inspect the cleaned sealing surface for nicks and scratches and replace as necessary.**

18. Install a new converter-to-exhaust intermediate pipe gasket by hand.

19. Install the 2 new 10-mm catalytic converter-to-exhaust intermediate pipe spring nuts and alternately tighten RH side to LH side in sequence in 3 stages:
- Stage 1: Tighten to 44 inch lbs. (5 Nm).
- Stage 2: Tighten to 133 inch lbs. (15 Nm).
- Stage 3: Tighten to 18 ft. lbs. (25 Nm).

20. Install 3 new exhaust intermediate pipe-to-exhaust down pipe nuts. Tighten the nuts to 18 ft. lbs. (25 Nm).

21. Start the engine and check for exhaust leaks, repair as needed.

3.0L Engine

Catalytic Converter—LH Manifold

See Figure 114.

1. With the vehicle in NEUTRAL, position it on a hoist.

2. Remove the 6 retainers and the passenger side splash shield.

3. Remove the LH Heated Oxygen Sensor (HO2S).

4. Rotate the accessory drive belt tensioner counterclockwise and remove the accessory drive belt.

5. Disconnect the A/C compressor electrical connector.

6. Remove the 3 bolts and position the A/C compressor aside.

7. Remove the 3 bolts and the LH catalytic converter heat shield.

8. Remove and discard the two 10-mm exhaust catalytic converter-to-exhaust Y-pipe spring nuts.

9. Remove and discard the 4 exhaust Y-pipe nuts and remove the Y-pipe.

10. Remove and discard the 6 LH catalytic converter manifold nuts.

11. Remove the LH catalytic converter manifold from the vehicle. Discard the gasket.

12. Remove and discard the 6 LH catalytic converter manifold studs.

13. Clean and inspect the LH catalytic converter manifold.

To install:

14. Install the 6 new LH catalytic converter manifold studs and tighten to 97 inch lbs. (11 Nm).

➡**Failure to tighten the catalytic converter nuts to specification before installing the exhaust Y-pipe will cause the converter to develop an exhaust leak.**

15. Using a new gasket, install the LH catalytic converter and the 6 nuts. Tighten in 2 stages in the sequence shown:
- Stage 1: Tighten to 18 ft. lbs. (25 Nm).
- Stage 2: Tighten to 18 ft. lbs. (25 Nm).

➡**Thoroughly clean the sealing surfaces of the flanges using a finishing pad. Inspect the cleaned sealing surface for nicks and scratches and replace as necessary.**

16. Install a new converter-to-exhaust Y-pipe gasket.

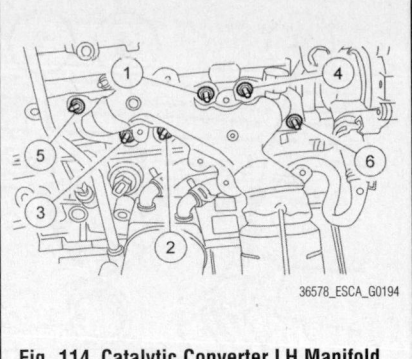

Fig. 114 Catalytic Converter LH Manifold tightening sequence

17. Install the 2 new 10-mm catalytic converter-to-exhaust Y-pipe spring nuts and alternately tighten RH side to LH side in sequence in 3 stages:
- Stage 1: Tighten to 44 inch lbs. (5 Nm).
- Stage 2: Tighten to 133 inch lbs. (15 Nm).
- Stage 3: Tighten to 18 ft. lbs. (25 Nm).

18. Install new gaskets, the exhaust Y-pipe and 4 new nuts. Tighten the 4 new exhaust Y-pipe nuts to 18 ft. lbs. (25 Nm).

19. Start the engine and check for exhaust leaks, repair as needed.

Catalytic Converter—RH Manifold

See Figure 115.

1. With the vehicle in NEUTRAL, position it on a hoist.

2. Remove the LH Heated Oxygen Sensor (HO2S).

3. Remove and discard the two 10-mm exhaust catalytic converter-to-exhaust Y-pipe spring nuts.

4. Remove and discard the 4 exhaust Y-pipe nuts and remove the Y-pipe.

5. Disconnect the RH Catalyst Monitor Sensor (CMS) electrical connector.

6. Remove the 3 bolts and the RH catalytic converter heat shield.

7. Remove and discard the 6 RH catalytic converter manifold nuts.

8. Remove the LH catalytic converter manifold from the vehicle. Discard the gasket.

9. Remove and discard the 6 RH catalytic converter manifold studs.

10. Clean and inspect the RH catalytic converter manifold.

To install:

11. Install the 6 new RH catalytic converter manifold studs and tighten to 97 inch lbs. (11 Nm).

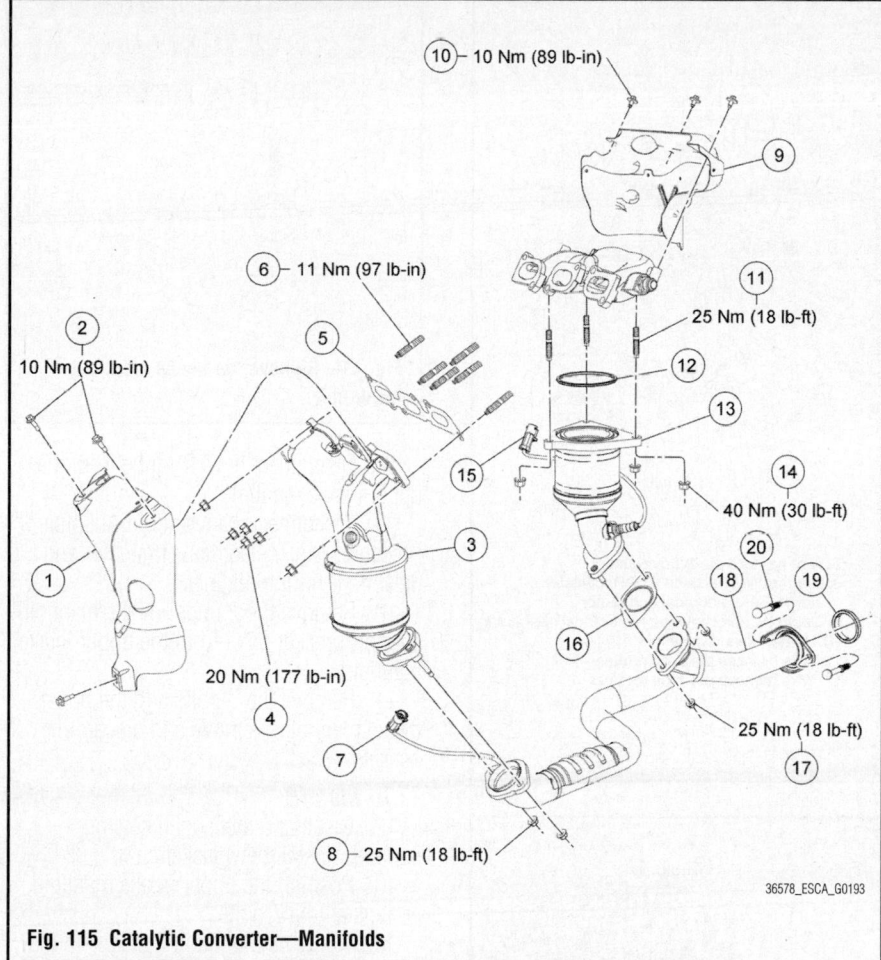

Fig. 115 Catalytic Converter—Manifolds

Labels on figure:
- 10 — 10 Nm (89 lb-in)
- 9
- 6 — 11 Nm (97 lb-in)
- 11
- 25 Nm (18 lb-ft)
- 2
- 10 Nm (89 lb-in)
- 5
- 12
- 13
- 15
- 14
- 40 Nm (30 lb-ft)
- 3
- 20
- 18
- 19
- 1
- 16
- 20 Nm (177 lb-in)
- 4
- 25 Nm (18 lb-ft)
- 17
- 7
- 8 — 25 Nm (18 lb-ft)
- 36578_ESCA_G0193

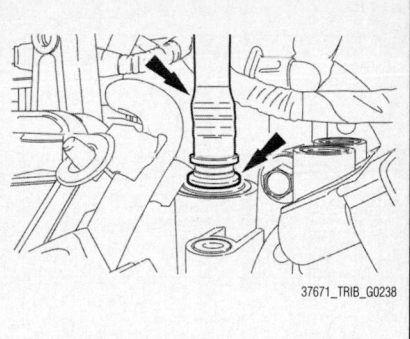

Fig. 116 Disconnect the vacuum supply hose

37671_TRIB_G0238

➡**Failure to tighten the catalytic converter nuts to specification before installing the exhaust Y-pipe will cause the converter to develop an exhaust leak.**

12. Using a new gasket, install the RH catalytic converter and the 6 nuts. Tighten the nuts in sequence in 2 stages:
- Stage 1: Tighten to 18 ft. lbs. (25 Nm).
- Stage 2: Tighten to 18 ft. lbs. (25 Nm).

➡**Thoroughly clean the sealing surfaces of the flanges using a finishing pad. Inspect the cleaned sealing surface for nicks and scratches and replace as necessary.**

13. Install a new converter-to-exhaust Y-pipe gasket.

14. Install the 2 new 10-mm catalytic converter-to-exhaust Y-pipe spring nuts and alternately tighten RH side to LH side in sequence in 3 stages:
- Stage 1: Tighten to 44 inch lbs. (5 Nm).
- Stage 2: Tighten to 133 inch lbs. (15 Nm).
- Stage 3: Tighten to 18 ft. lbs. (25 Nm).

15. Install new gaskets, the exhaust Y-pipe and 4 new nuts. Tighten the 4 new exhaust Y-pipe nuts to 18 ft. lbs. (25 Nm).

16. Start the engine and check for exhaust leaks, repair as needed.

INTAKE MANIFOLD

REMOVAL & INSTALLATION

2.5L Engine

See Figures 116 through 119.

1. With vehicle in NEUTRAL, position it on a hoist.
2. Remove the fuel rail.
3. Disconnect the vacuum supply hose.
   a. Depress the quick release locking ring.
   b. Pull the vacuum hose out of the quick release fitting.
4. Disconnect the fuel vapor return hose from the intake manifold.

5. Disconnect the Manifold Absolute Pressure (MAP) electrical connector.
6. Disconnect the Evaporative Emission (EVAP) canister purge valve electrical connector.
7. Disconnect the electronic throttle control electrical connector.
8. Disconnect the Knock Sensor (KS) electrical connector.
9. Detach the wire harness pin-type retainer.
10. Detach the heater hose pin-type retainer.
11. Detach all wiring harness pin-type retainers from the intake manifold and position the wiring harness aside.
12. Loosen the clamp and disconnect Air Cleaner (ACL) outlet pipe from the Throttle Body (TB).
13. Remove the intake manifold lower bolt.
14. Remove the 6 bolts and position the intake manifold aside to access the crankcase vent oil separator tube and the EGR tube.
15. Squeeze the 2 crankcase vent oil separator tube tabs and disconnect the tube from the intake manifold.
16. Remove the EGR tube.
17. Remove the intake manifold and gaskets.

✳✳ CAUTION

If the engine is repaired or replaced because of upper engine failure, typically including valve or piston damage, check the intake manifold for metal debris. If metal debris is found, install a new intake manifold. Failure to follow these instructions can result in engine damage.

To install:

18. To install, reverse the removal procedure.

19. Inspect and install new intake manifold gaskets if necessary.

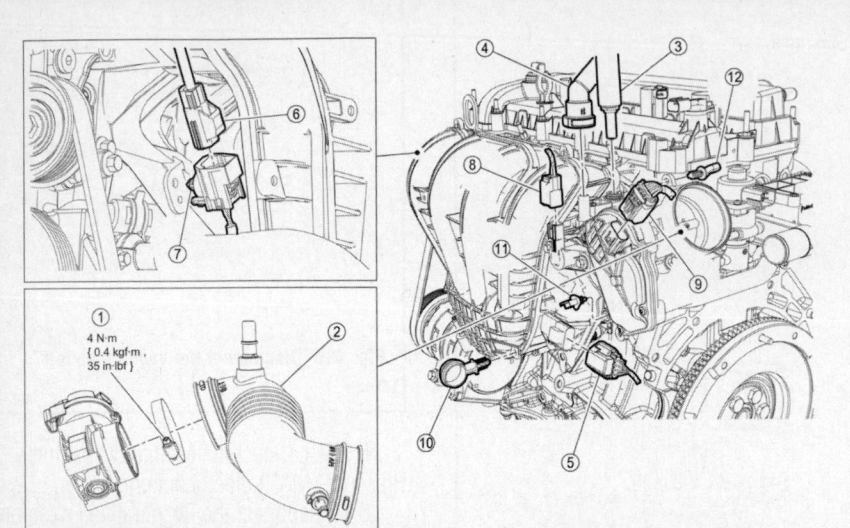

1. Air Cleaner (ACL) outlet pipe-to-Throttle Body (TB) clamp
2. ACL outlet pipe
3. Vacuum supply hose
4. Fuel vapor return hose
5. Knock Sensor (KS) electrical connector
6. Evaporative Emission (EVAP) canister purge valve electrical connector
7. Wire harness pin-type retainer
8. Evaporative Emission (EVAP) canister purge valve electrical connector
9. Electronic throttle control electrical connector
10. Heater hose retainer
11. Wire harness pin-type retainer
12. Wire harness pin-type retainer

37671_TRIB_G0239

Fig. 117 Intake manifold connections

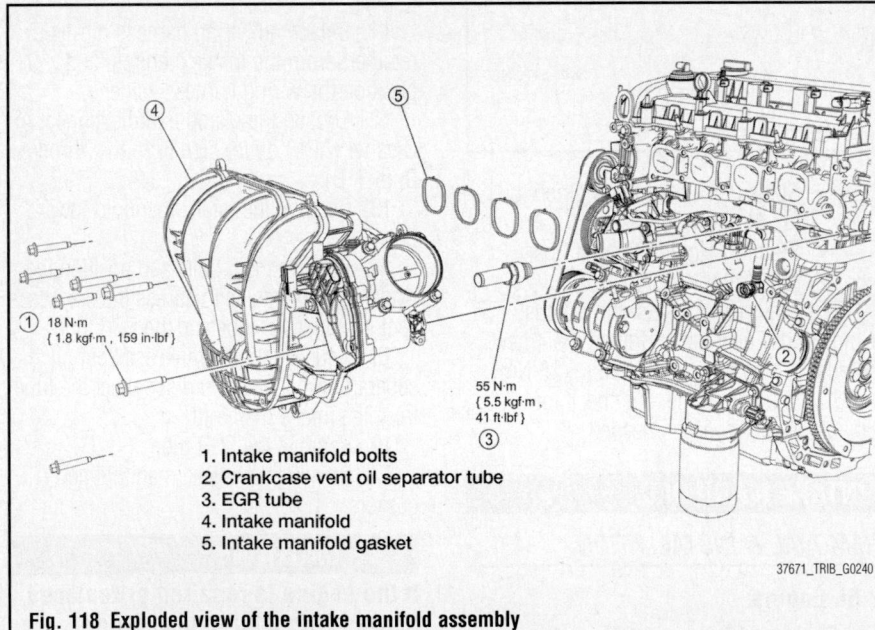

1. Intake manifold bolts
2. Crankcase vent oil separator tube
3. EGR tube
4. Intake manifold
5. Intake manifold gasket

37671_TRIB_G0240

Fig. 118 Exploded view of the intake manifold assembly

3.0L Engine

Upper

See Figures 120 and 121.

1. Remove the Air Cleaner (ACL) outlet pipe and the ACL.

2. Remove the EGR tube fitting from the EGR valve.

3. Disconnect the electronic throttle control electrical connector.

4. Disconnect the Evaporative Emission (EVAP) canister purge valve electrical connector.

5. Disconnect the EGR regulator electrical connector and detach the wiring retainer.

6. Disconnect the EVAP tube-to-EVAP canister purge valve quick connect coupling and the brake booster vacuum tube from the upper intake manifold.

7. Detach the engine control wiring harness retainer from the upper intake manifold.

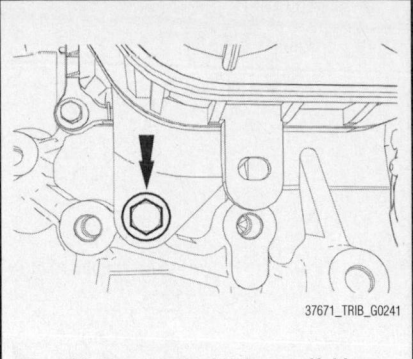

37671_TRIB_G0241

Fig. 119 Remove the intake manifold lower bolt

8. Disconnect the PCV tube from the upper intake manifold.

9. Disconnect the Manifold Absolute Pressure (MAP) electrical connector and detach the wiring retainer.

10. Remove the 2 upper intake manifold support bracket bolts from the upper intake manifold.

11. Remove the 7 bolts and the upper intake manifold. Remove and discard the gaskets.

To install:

12. Clean and inspect all sealing surfaces. Install new gaskets.

13. Position the upper intake manifold and install the 7 bolts.

14. Tighten the bolts in 2 stages in the sequence shown:
- Stage 1: Tighten to 89 inch lbs. (10 Nm).
- Stage 2: Tighten an additional 45 degrees.

15. Install the 2 upper intake manifold support bracket bolts and tighten to 89 inch lbs. (10 Nm).

16. Connect the MAP electrical connector and attach the wiring retainer.

17. Connect the PCV tube to the upper intake manifold.

18. Attach the engine control wiring harness retainer to the upper intake manifold.

19. Connect the EVAP tube-to-EVAP canister purge valve quick connect coupling and the brake booster vacuum tube to the upper intake manifold.

20. Connect the EGR regulator electrical connector and attach the wiring retainer.

21. Connect the EVAP canister purge valve electrical connector.

22. Connect the electronic throttle control electrical connector.

23. Install the EGR tube fitting to the EGR valve. Tighten the fitting to 30 ft. lbs. (40 Nm).

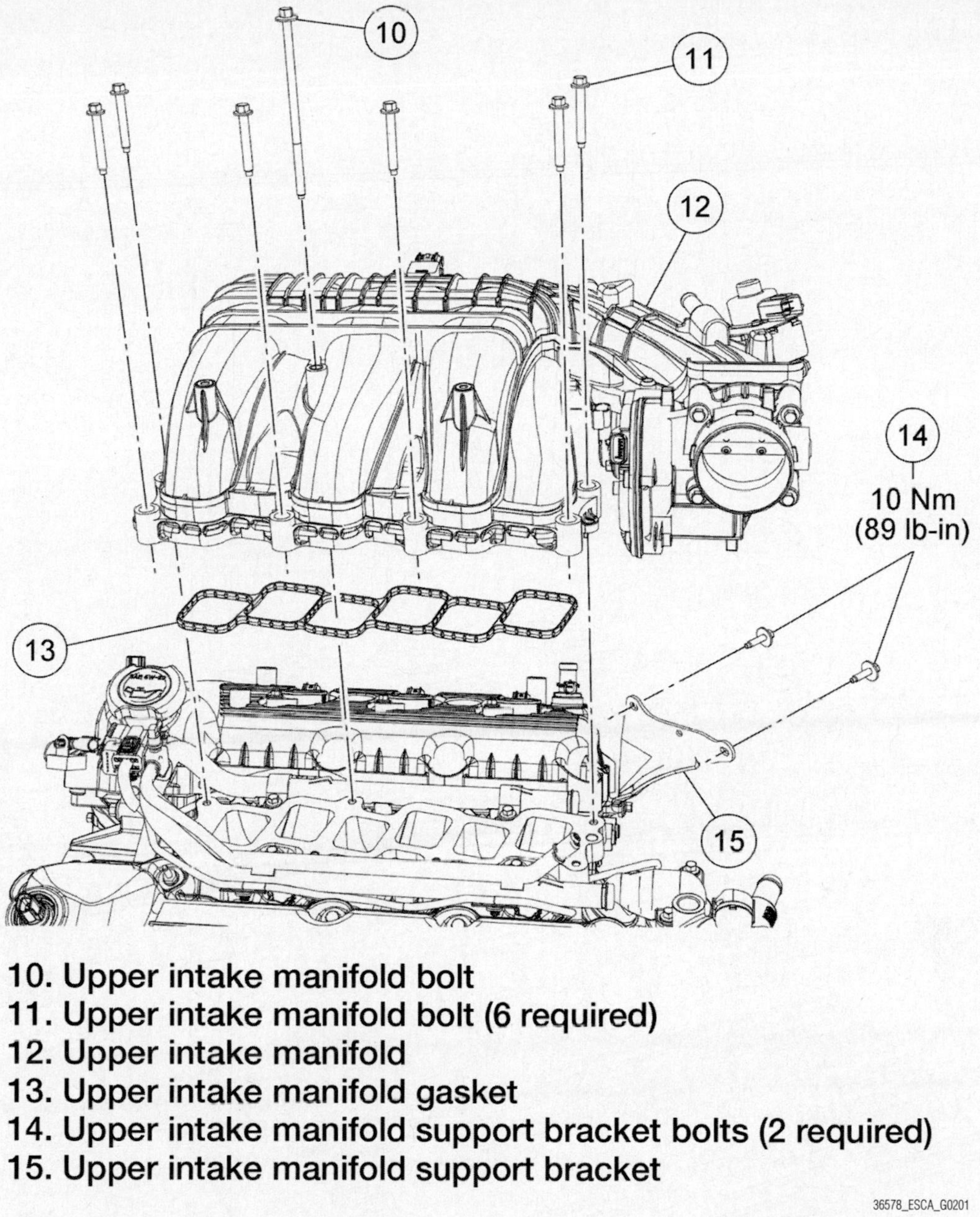

10. Upper intake manifold bolt
11. Upper intake manifold bolt (6 required)
12. Upper intake manifold
13. Upper intake manifold gasket
14. Upper intake manifold support bracket bolts (2 required)
15. Upper intake manifold support bracket

36578_ESCA_G0201

Fig. 120 Upper intake manifold

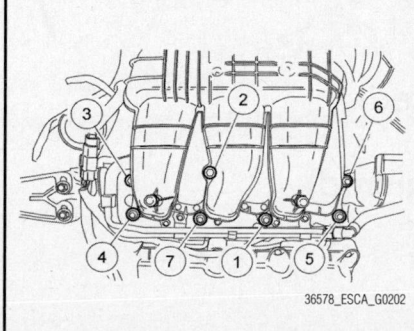

36578_ESCA_G0202

Fig. 121 Upper intake manifold tightening sequence

24. Install the ACL outlet pipe and the ACL.

Lower

See Figures 122 and 123.

1. Remove the upper intake manifold.
2. Release the fuel system pressure.
3. Disconnect the battery ground cable.
4. Remove the fuel rail.
5. Remove the 8 lower intake manifold bolts and the lower intake manifold. Remove and discard the gaskets.

To install:

6. Clean and inspect all sealing surfaces. Install new gaskets.
7. Position the lower intake manifold and install the 8 bolts.
8. Tighten in the sequence shown to 89 inch lbs. (10 Nm).
9. Install the fuel rail.
10. Install the upper intake manifold.
11. Connect the battery ground cable.

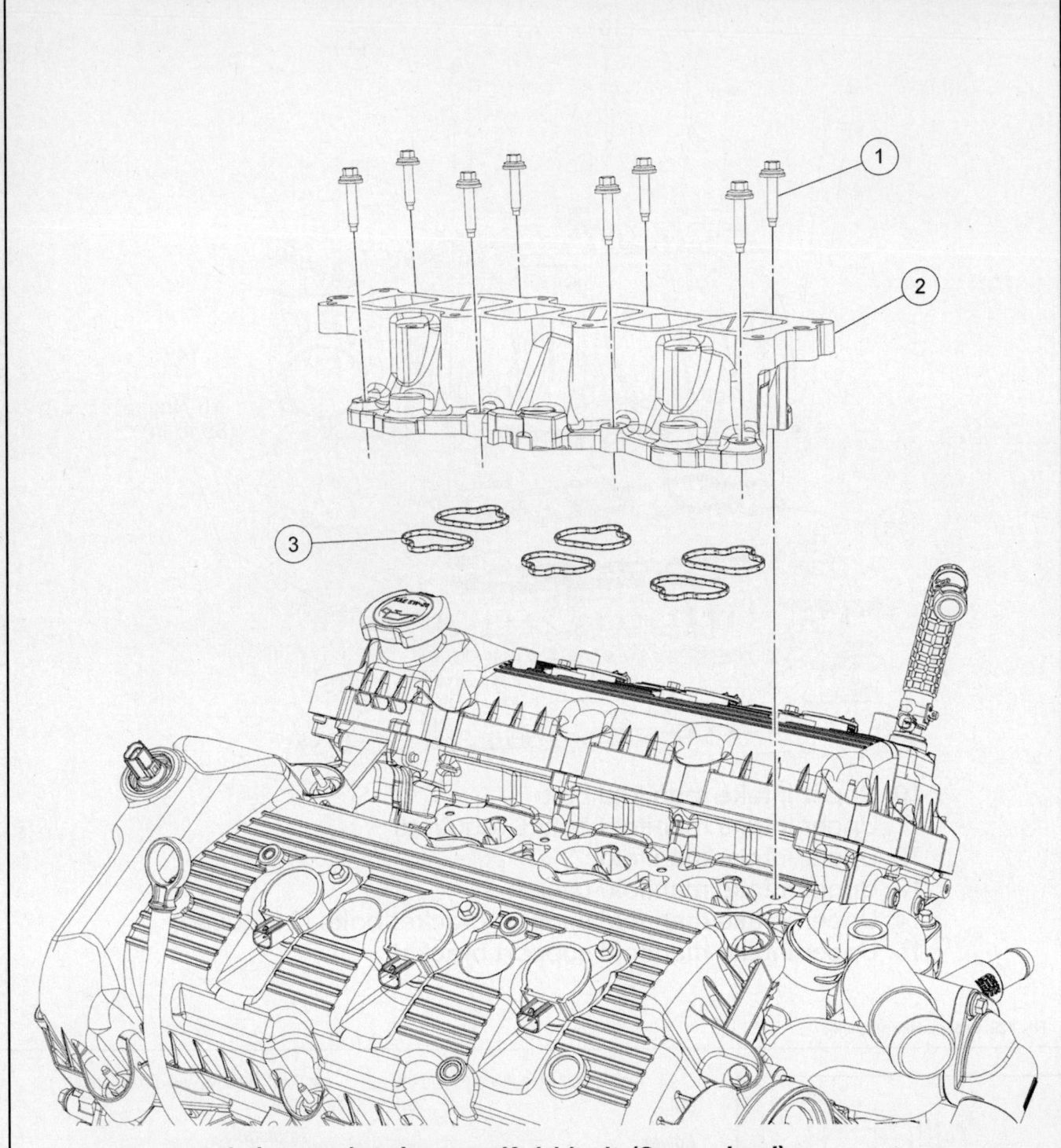

1. Lower intake manifold bolt (8 required)
2. Lower intake manifold
3. Lower intake manifold gasket (6 required)

36578_ESCA_G0203

Fig. 122 Lower intake manifold

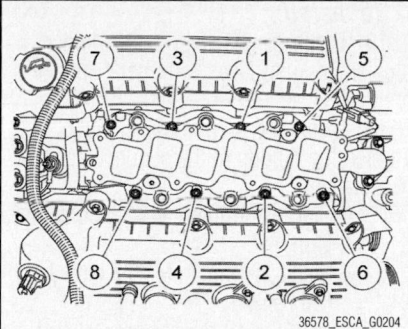

Fig. 123 Lower intake manifold tightening sequence

OIL PAN

REMOVAL & INSTALLATION

2.5L Engine

See Figures 124 and 125.

All vehicles:
1. Disconnect the negative battery cable.
2. With the vehicle in NEUTRAL, position it on a hoist.
3. Remove the air cleaner outlet pipe.
4. Drain the engine oil, then install the drain plug and tighten to 21 ft. lbs. (28 Nm).

> **✳✳ WARNING**
>
> **To prevent damage to the transaxle, do not loosen the transaxle-to-engine bolts more than 0.19 inch (5mm).**

5. Loosen the 2 top bell housing-to-engine bolts 0.19 inch (5mm).
AWD vehicles:
6. Working from the top of the vehicle, loosen the 2 rear lower engine-to-bell housing bolts 0.19 inch (5mm).
7. Working from under the vehicle, loosen the 2 upper engine bracket-to-Power Transfer Unit (PTU) bolts 0.19 inch (5mm).
All vehicles:
8. Remove the 7 retainers and the LH splash shield.
9. Remove the oil level indicator and tube
10. Loosen the 2 front lower bell housing-to-engine bolts 0.19 inch (5mm).
FWD vehicles:
11. Loosen the 1 (manual transaxle) and 2 (automatic transaxle) rear lower engine-to-bell housing bolt 0.19 inch (5mm).
All vehicles:
12. Remove the 2 oil pan-to-bell housing bolts.
13. Remove the 2 bell housing-to-oil pan bolt.

14. Slide the transaxle rearward 5 mm (0.19 in).
15. Drain the engine oil.
16. Remove the 4 engine front cover-to-oil pan bolts.
17. Remove the 13 bolts and the oil pan.

To install:
All vehicles:
18. Clean and inspect all mating surfaces.

➥**If the oil pan is not secured within 10 minutes of sealant application, the sealant must be removed and the sealing area cleaned with metal surface cleaner. Allow to dry until there is no sign of wetness, or 10 minutes, whichever is longer. Failure to follow this procedure can cause future oil leakage.**

19. Apply a 0.09 inch. (2.5mm) bead of silicone gasket and sealant to the oil pan-to-engine block and to the oil pan-to-engine front cover mating surface.
20. Position the oil pan onto the engine and install the oil pan bolts finger-tight.

> **✳✳ WARNING**
>
> **The engine front cover-to-oil pan bolts must be tightened first to align the front surface of the oil pan flush with the front surface of the engine block.**

21. Install the 4 engine front cover-to-oil pan bolts and tighten to 89 inch lbs. (10 Nm).
22. Tighten the oil pan bolts in sequence to 18 ft. lbs. (25 Nm).
FWD vehicles:
23. Alternate tightening the 1 front and 1 rear lower bolts to slide the transaxle and engine together. Tighten bolts to 35 ft. lbs. (48 Nm).
24. Tighten the remaining front lower bolt and rear lower bolt (automatic transaxle) to 35 ft. lbs. (48 Nm).

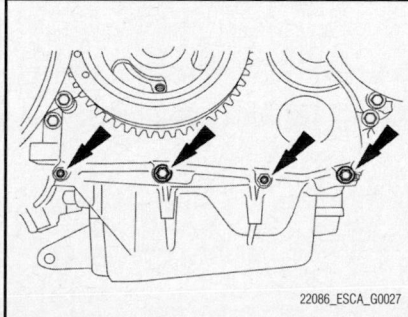

Fig. 124 Front cover-to-oil pan bolts shown

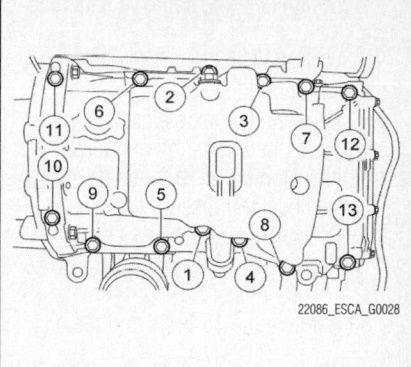

Fig. 125 Oil pan bolt tightening sequence

AWD vehicles:
25. Alternate tightening the 1 upper engine-to-PTU bracket bolt and 1 front lower bolt to slide transaxle and engine together.
26. Tighten the PTU bracket bolt to 33 ft. lbs. (45 Nm).
27. Tighten the front lower bolt to 33 ft. lbs. (45 Nm).
28. Tighten the remaining upper engine-to-PTU bracket bolt to 33 ft. lbs. (45 Nm).
29. Tighten the remaining front lower bolt to 33 ft. lbs. (45 Nm).
All vehicles:
30. Install the 2 bell housing-to-oil pan bolts to 33 ft. lbs. (45 Nm).
31. Install the 2 oil pan-to-bell housing bolts to 33 ft. lbs. (45 Nm).
32. Install the oil level indicator and tube.
33. Install the LH splash shield and the 7 retainers. Tighten to 80 inch. lbs (90 Nm).
AWD vehicles:
34. Working from the top of vehicle, tighten the 2 rear lower engine-to-bell housing bolts to 35 ft. lbs. (48 Nm).
All vehicles:
35. Tighten the 2 top bell housing-to-engine bolts to 35 ft. lbs. (48 Nm).
36. Install the air cleaner outlet pipe.
37. Connect the negative battery cable.
38. Fill the engine with clean engine oil.
39. Recheck for leaks.

3.0L Engine

See Figure 126.

> **✳✳ WARNING**
>
> **During engine repair procedures, cleanliness is extremely important. Any foreign material (including any material created while cleaning gasket surfaces) that enters the oil passages, coolant passages or the oil pan may cause engine failure.**

1. Disconnect the negative battery cable.

2. With the vehicle in NEUTRAL, position it on a hoist.

3. Remove the flexible exhaust pipe.

4. Drain the engine oil and install the drain plug. Tighten to 19 ft. lbs (26 Nm).

5. Remove and discard the oil filter.

6. Remove the access cover.

7. Remove the 2 oil pan-to-transaxle bolts.

8. Remove the 15 bolts and the oil pan.

9. Remove and discard the oil pan gasket.

To install:

❋❋ WARNING

Do not use metal scrapers, wire brushes, power abrasive discs or other abrasive means to clean the sealing surfaces. These tools cause scratches and gouges which make leak paths.

10. Use a plastic scraping tool to remove all traces of the oil pan gasket.

11. Clean all sealing surfaces with metal surface prep and install a new oil pan gasket.

➡ **The oil pan must be installed and the bolts tightened within 4 minutes of sealant application.**

12. Apply a 0.40 inch (10 mm) diameter dot of silicone sealant to the areas indicated.

13. Position the oil pan and loosely install the bolts and stud bolts.

14. Install the 2 oil pan-to-transaxle bolts and tighten to 30 ft. lbs. (40 Nm).

15. Tighten the oil pan-to-engine bolts and stud bolts in the sequence shown to 18 ft. lbs. (25 Nm).

16. Lubricate the engine oil filter gasket with clean engine oil prior to installing.

17. Install a new oil filter. Tighten to 44 inch lbs. (5 Nm) and then rotate an additional 180 degrees.

18. Install the exhaust Y-pipe.

19. Connect the negative battery cable.

20. Fill the engine with clean engine oil.

21. Start the engine and check for leaks.

OIL PUMP

REMOVAL & INSTALLATION

2.5L Engine

See Figures 127 through 130.

❋❋ WARNING

During engine repair procedures, cleanliness is extremely important. Any foreign material, including any material created while cleaning gasket surfaces that enters the oil passages, coolant passages or the oil pan, may cause engine failure.

1. With the engine in NEUTRAL, position it on a hoist.

2. Remove the engine front cover.

3. Drain the engine oil, then install the drain plug and tighten to 21 ft. lbs. (28 Nm).

4. Remove the 4 oil pan-to-bellhousing bolts.

5. Remove the 13 bolts and the oil pan.

6. Discard the gasket and clean and inspect the gasket mating surfaces.

7. Remove the 2 bolts and the oil pump screen and pickup tube.

8. Release the tension on the tensioner spring.

9. Remove the 2 shoulder bolts and the tensioner.

10. Remove the chain from the oil pump sprocket.

11. Remove the bolt and oil pump sprocket.

12. Remove the 4 bolts and the oil pump

To install:

13. Clean the oil pump and cylinder block mating surfaces with metal surface prep.

14. Install the oil pump assembly. Tighten the 4 bolts in the sequence shown in 2 stages:
- Stage 1: Tighten to 89 inch lbs. (10 Nm).
- Stage 2: Tighten to 177 inch lbs. (20 Nm).

15. Install the oil pump sprocket and bolt. Tighten the bolt to 18 ft. lbs. (25 Nm).

16. Install the chain onto the oil pump sprocket.

17. Install the oil pump drive chain tensioner shoulder bolt. Tighten the bolt to 89 inch lbs. (10 Nm).

18. Install the oil pump drive chain tensioner and bolt. Hook the tensioner spring around the shoulder bolt. Tighten to 89 inch lbs. (10 Nm).

19. Install the oil pump screen and pickup tube and the 2 bolts. Tighten the bolts to 89 inch lbs. (10 Nm).

20. Clean and inspect all mating surfaces.

➡ **If the oil pan is not secured within 10 minutes of sealant application, the sealant must be removed and the sealing area cleaned with metal surface cleaner. Allow to dry until there is no sign of wetness, or 10 minutes, whichever is longer. Failure to follow this procedure can cause future oil leakage.**

21. Apply a 0.09 inch. (2.5mm) bead of silicone gasket and sealant to the oil pan-to-engine block and to the oil pan-to-engine front cover mating surface.

22. Position the oil pan onto the engine and install the oil pan bolts finger-tight.

23. Using a suitable straight edge, align the front surface of the oil pan flush with the front surface of the engine block.

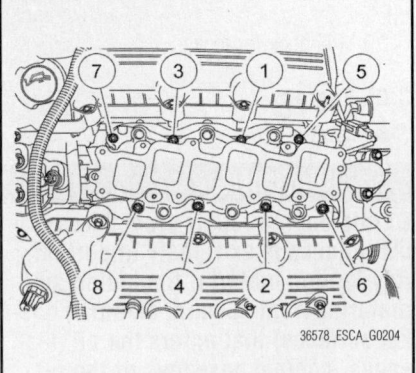

Fig. 126 Oil pan tightening sequence

36578_ESCA_G0204

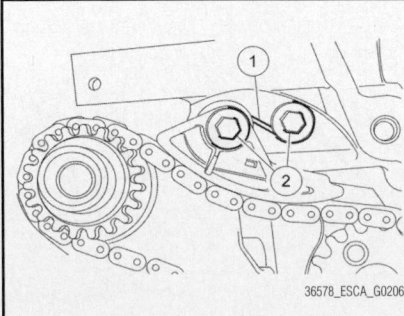

Fig. 127 Release the tension on the spring (1) and remove the shoulder bolts (2) to remove the tensioner

36578_ESCA_G0206

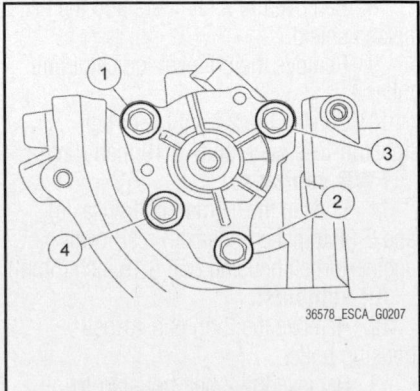

Fig. 128 Oil pump tightening sequence

36578_ESCA_G0207

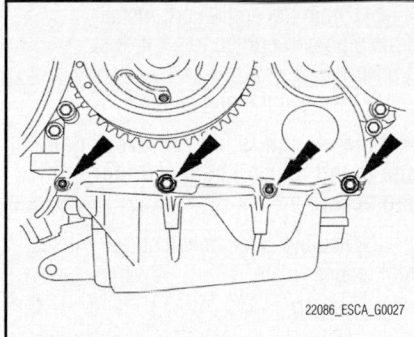

Fig. 129 Front cover-to-oil pan bolts shown

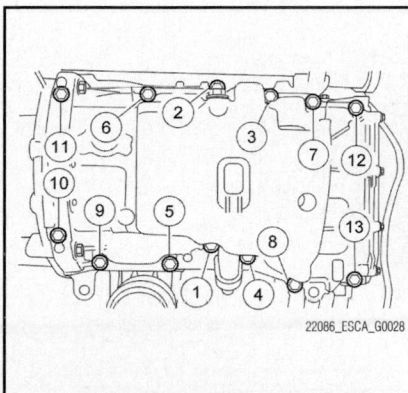

Fig. 130 Oil pan bolt tightening sequence

⊹ **WARNING**

The engine front cover-to-oil pan bolts must be tightened first to align the front surface of the oil pan flush with the front surface of the engine block.

24. Install the 4 engine front cover-to-oil pan bolts and tighten to 89 inch lbs. (10 Nm).

25. Tighten the oil pan bolts in sequence to 18 ft. lbs. (25 Nm).

26. Install the 4 oil pan-to-bellhousing bolts and tighten to 35 ft. lbs. (48 Nm).

27. Install the engine front cover.

28. Fill the engine with clean engine oil.

3.0L Engine

See Figures 131 through 133.

⊹ **WARNING**

During engine repair procedures, cleanliness is extremely important. Any foreign material, including any material created while cleaning gasket surfaces that enters the oil passages, coolant passages or the oil pan, may cause engine failure.

1. With the vehicle in NEUTRAL, position it on a hoist.

2. Drain the engine oil and install the drain plug. Tighten to 19 ft. lbs. (26 Nm).

3. Remove the timing drive components.

4. Remove the oil pan.

5. Remove the oil pump screen and pickup tube mounting bolts.

6. Remove the oil pump screen and pickup tube.

7. Remove the oil pump bolts in the sequence shown.

To install:

8. Position the oil pump and install the bolts.

9. Tighten in the sequence shown to 89 inch lbs. (10 Nm).

10. Position the oil pump screen and pickup tube.

11. Install the bolts and tighten to 89 inch lbs. (10 Nm).

12. Install the nut and tighten in 2 stages:

- Stage 1: Tighten to 44 inch lbs. (5 Nm).
- Stage 2: Tighten 45 degrees.

13. Install the oil pan.

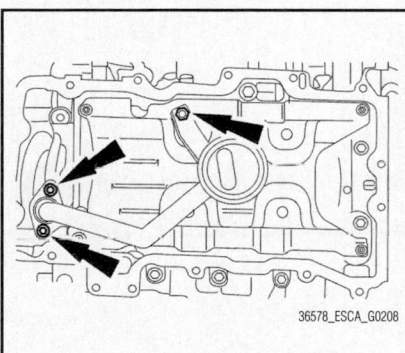

Fig. 131 Oil pump screen and tube mounting bolts

Fig. 132 Remove the oil pump bolts in the sequence

14. Install the timing drive components.

15. Fill the engine with clean engine oil.

PISTON AND RING

POSITIONING

See Figures 134 through 136.

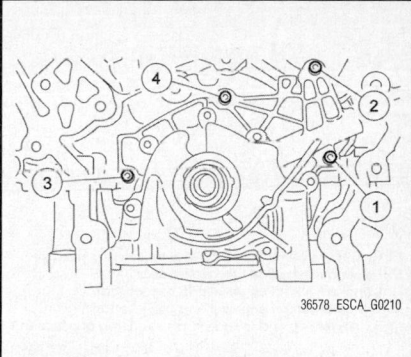

Fig. 133 Oil pump bolt tightening sequence

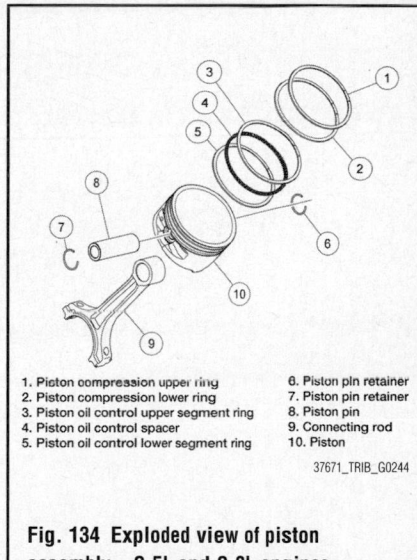

1. Piston compression upper ring
2. Piston compression lower ring
3. Piston oil control upper segment ring
4. Piston oil control spacer
5. Piston oil control lower segment ring
6. Piston pin retainer
7. Piston pin retainer
8. Piston pin
9. Connecting rod
10. Piston

Fig. 134 Exploded view of piston assembly—2.5L and 3.0L engines

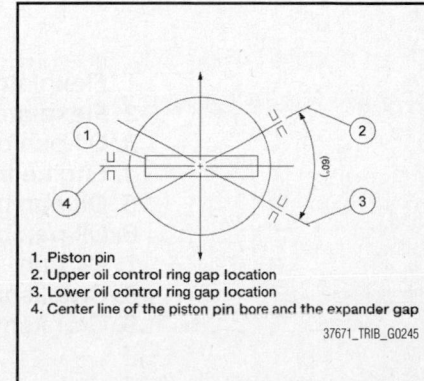

1. Piston pin
2. Upper oil control ring gap location
3. Lower oil control ring gap location
4. Center line of the piston pin bore and the expander gap

Fig. 135 Ring gap positions—2.5L engine

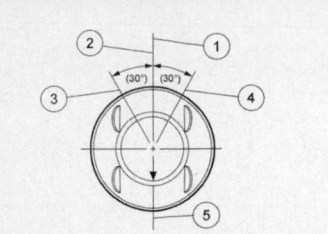

1. Center line of the piston parallel to the wrist pin bore
2. Upper compression ring gap location
3. Upper oil control segment ring gap location
4. Lower oil control segment ring gap location
5. Expander ring and lower compression ring gap location

37671_TRIB_G0246

Fig. 136 Ring gap positions—3.0L engine

REAR MAIN SEAL

REMOVAL & INSTALLATION

2.5L Engine

See Figures 137 through 140.

1. With the vehicle in NEUTRAL, position it on a hoist.
2. If equipped, remove the automatic transaxle.
3. If equipped, remove the manual transaxle and clutch.
4. Remove the flexplate or flywheel.
5. Drain the engine oil, install drain plug and tighten to 21 ft. lbs. (28 Nm).
6. Remove the oil pan.

➡**If the oil pan is not removed, damage to the rear oil seal retainer joint can occur.**

7. Remove the crankshaft rear oil seal with retainer plate.

To install:

8. Using a seal installer, position the crankshaft rear oil seal with retainer plate onto the crankshaft.

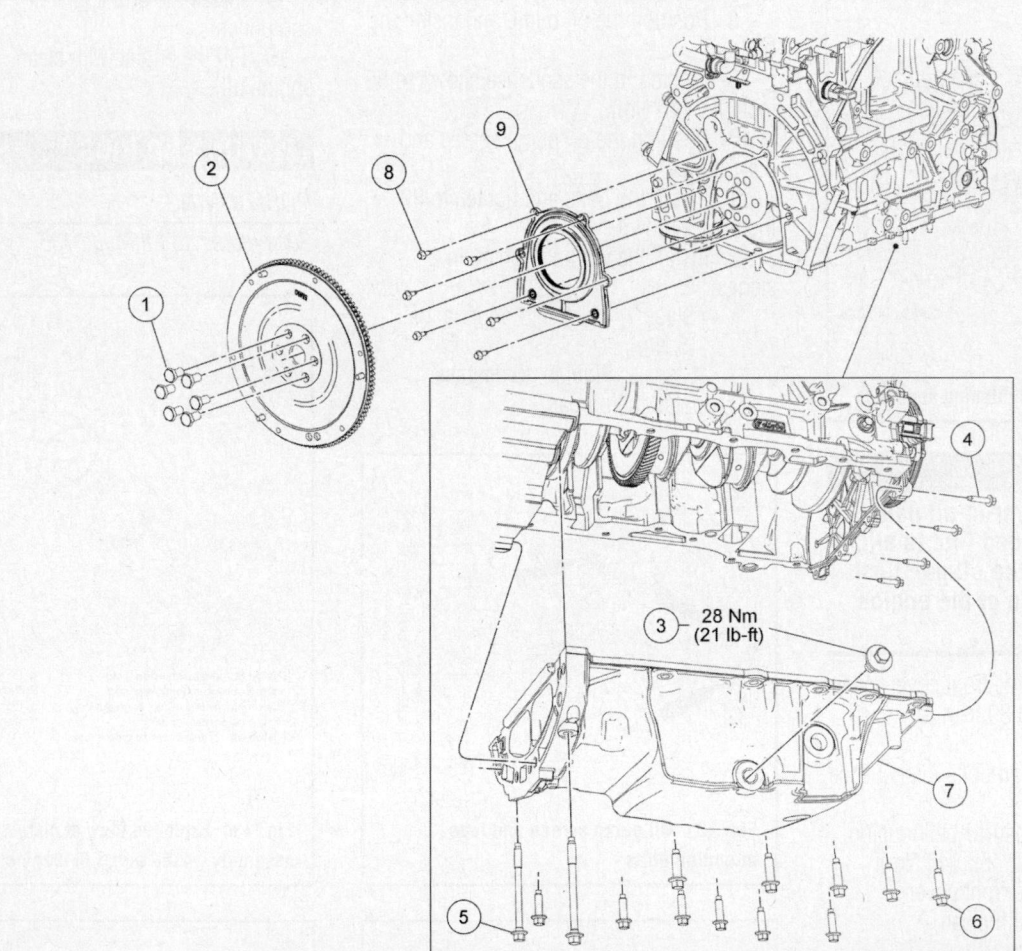

1. Flexplate or flywheel bolt (6 required)
2. Flexplate or flywheel
3. Oil pan drain plug
4. Engine front cover bolt (4 required)
5. Oil pan bolt (2 required)
6. Oil pan bolt (11 required)
7. Oil pan
8. Crankshaft rear oil seal with retainer plate bolt (6 required)
9. Crankshaft rear oil seal with retainer plate

36578_ESCA_G0212

Fig. 137 Exploded view—flexplate/flywheel and crankshaft rear seal

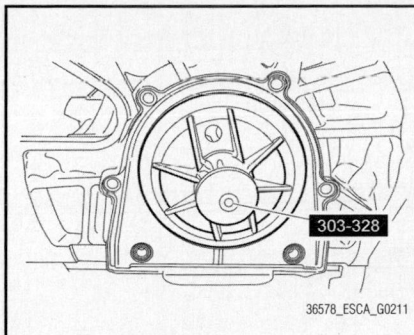

Fig. 138 Rear oil seal installation with installation tool

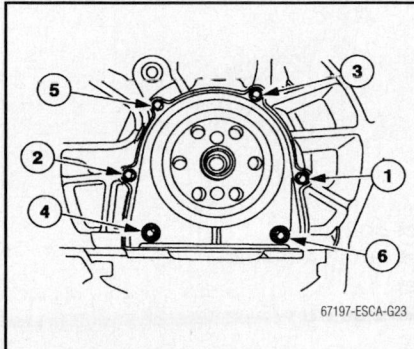

Fig. 139 Retainer plate torque sequence—I4 engine

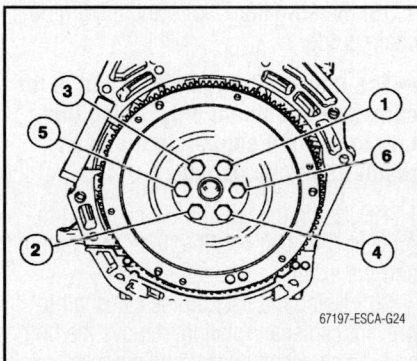

Fig. 140 Flywheel torque sequence—I4 engine

9. Install the crankshaft rear oil seal with retainer plate. Tighten the bolts in the sequence shown to 89 inch lbs. (10 Nm).

10. Install the oil pan.

➡**Special bolts are used for installation. Do not use standard bolts.**

11. Install the flywheel/flexplate.

12. Tighten the bolts in the sequence shown in three stages:
- Stage 1: Tighten to 37 ft. lbs. (50 Nm).
- Stage 2: Tighten to 50 ft. lbs. (80 Nm).

- Stage 3: Tighten to 83 ft. lbs. (112 Nm).

13. Fill the engine with clean engine oil.

3.0L Engine

See Figures 141 and 142.

1. Remove or disconnect the following:
- Negative battery cable
- Transaxle
- Flexplate

2. Using the Slide Hammer and the crankshaft rear oil seal remover, remove and discard the crankshaft rear oil seal.

To install:

3. Apply clean engine oil to the seal lip and seal bore before installing the seal.

4. Using the crankshaft rear main oil seal installer bolts and the crankshaft rear main oil seal installer, install the crankshaft rear oil seal.

5. Install or connect the following:
- Crankshaft rear oil seal
- Flywheel
- Transaxle
- Negative battery cable

TIMING CHAIN COMPONENTS

REMOVAL & INSTALLATION

2.5L Engine

See Figures 143 through 154.

✷✷ WARNING

Do not loosen or remove the crankshaft pulley bolt without first installing the special tools as instructed in this procedure. The crankshaft pulley and the crankshaft timing sprocket are not keyed to the crankshaft. The crankshaft, the crankshaft sprocket and the pulley are fitted together by friction, using diamond washers between the flange faces on each part. For that reason, the crankshaft sprocket is also unfastened if the pulley bolt is loosened. Before any repair requiring loosening or removal of the crankshaft pulley bolt, the crankshaft and camshafts must be locked in place by the special service tools, otherwise severe engine damage can occur. Refer to Crankshaft Damper in this section.

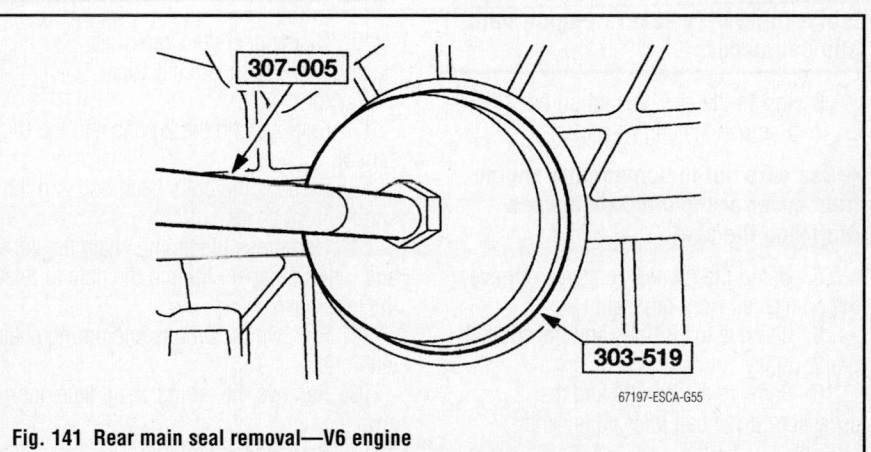

Fig. 141 Rear main seal removal—V6 engine

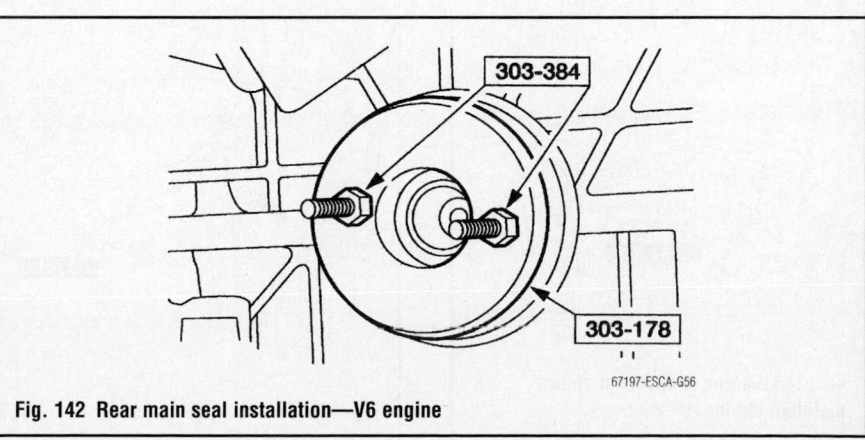

Fig. 142 Rear main seal installation—V6 engine

➡ During engine repair procedures, cleanliness is extremely important. Any foreign material, including any material created while cleaning gasket surfaces, which enters the oil passages, coolant passages or the oil pan can cause engine failure.

1. With the vehicle in NEUTRAL, position it on a hoist.

2. Remove the accessory drive belt and the smooth idler pulley.

3. Disconnect the Crankshaft Position (CKP) sensor electrical connector.

4. Remove the CKP sensor.

5. Remove the crankshaft pulley. Refer to Crankshaft Damper Removal & Installation in this section.

✳✳ WARNING

The crankshaft, the crankshaft sprocket and the pulley are fitted together by friction, using diamond washers between the flange faces on each part. For that reason, the crankshaft sprocket is also unfastened if the pulley bolt is loosened. Before any repair requiring loosening or removal of the crankshaft pulley bolt, the crankshaft and camshafts must be locked in place by the special service tools, otherwise severe engine damage can occur.

6. Install the engine support bar.

7. Remove the engine mount.

➡ Use care not to damage the engine front cover or the crankshaft when removing the seal.

8. Using the oil seal remover, remove the crankshaft front oil seal.

9. Remove the 3 bolts and the coolant pump pulley.

10. Remove the 2 bolts and the accessory drive belt idler pulley and bracket.

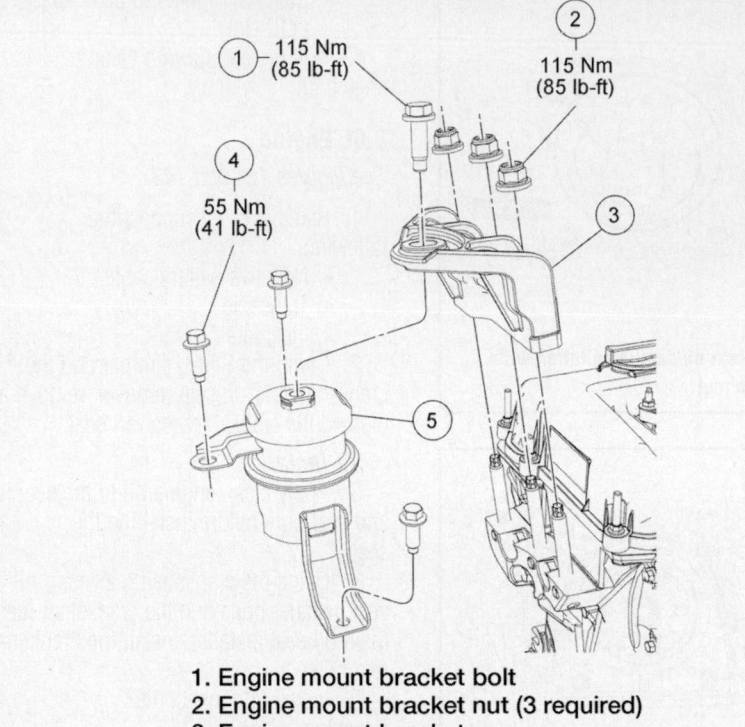

1. Engine mount bracket bolt
2. Engine mount bracket nut (3 required)
3. Engine mount bracket
4. Engine mount bolt (3 required)
5. Engine mount

1 115 Nm (85 lb-ft)
2 115 Nm (85 lb-ft)
4 55 Nm (41 lb-ft)

36578_ESCA_G0214

Fig. 144 Engine support mount

11. Disconnect the Crankshaft Position (CKP) sensor electrical connector.

12. Remove and the 2 bolts and the CKP sensor.

13. Remove the bolts, stud bolt and the engine front cover.

14. Compress the timing chain tensioner and insert a paper clip into the hole to retain the tensioner.

15. Remove the 2 bolts and timing chain tensioner.

16. Remove the timing chain tensioner arm.

17. Remove the timing chain.

18. Remove the 2 bolts and the timing chain guide.

➡ The Camshaft Alignment Plate is for camshaft alignment only. Using this tool to prevent engine rotation can result in engine damage.

19. Using the flats on the camshaft to prevent camshaft rotation, remove the bolt and the exhaust camshaft sprocket.

20. Using the flats on the camshaft to prevent camshaft rotation, remove the bolt and the camshaft phaser and sprocket.

To install:

21. Install the camshaft sprockets and the bolts. Do not tighten the bolts at this time.

22. Install the timing chain guide and the 2 bolts. Tighten to 89 inch lbs. (10 Nm).

23. Install the timing chain.

24. Install the timing chain tensioner arm.

25. Install the timing chain tensioner and the 2 bolts. Tighten the bolts to 89 inch lbs. (10 Nm). Remove the paper clip to release the piston.

➡ The Camshaft Alignment Plate is for camshaft alignment only. Using this tool to prevent engine rotation can result in engine damage.

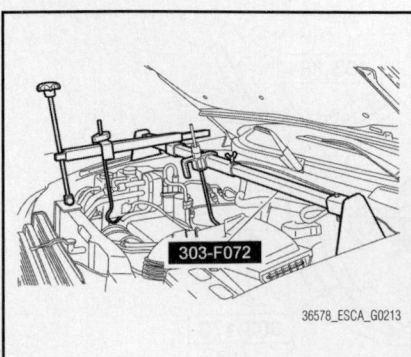

303-F072

36578_ESCA_G0213

Fig. 143 Engine support bar shown installed on

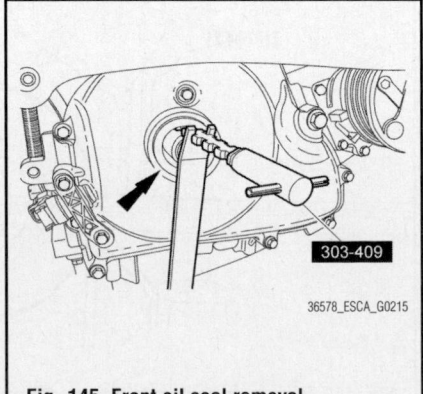

303-409

36578_ESCA_G0215

Fig. 145 Front oil seal removal

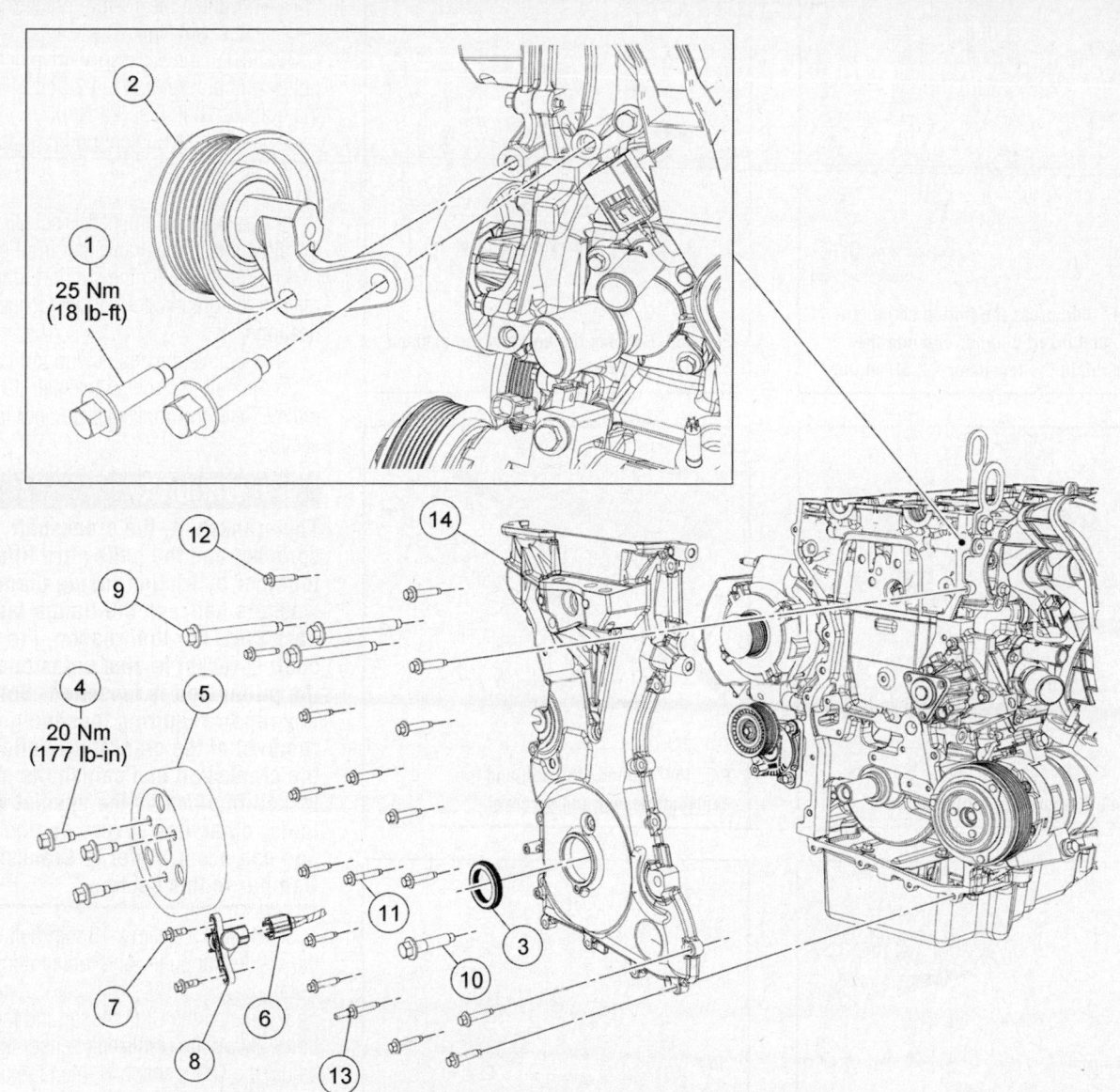

1. Accessory drive belt idler pulley bolt (2 required)
2. Accessory drive belt idler pulley and bracket
3. Crankshaft front seal
4. Coolant pump pulley bolt (3 required)
5. Coolant pump pulley
6. Crankshaft Position (CKP) sensor electrical connector
7. CKP sensor bolt (2 required)
8. CKP sensor
9. Engine front cover bolt (3 required)
10. Engine front cover bolt
11. Engine front cover bolt
12. Engine front cover bolt (16 required)
13. Engine front cover stud bolt
14. Engine front cover

36578_ESCA_G0216

Fig. 146 Engine front cover

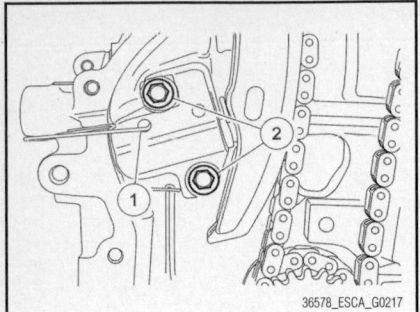

Fig. 147 Compress the timing chain tensioner, and insert a paper clip into the hole to retain the tensioner—2.5L engine

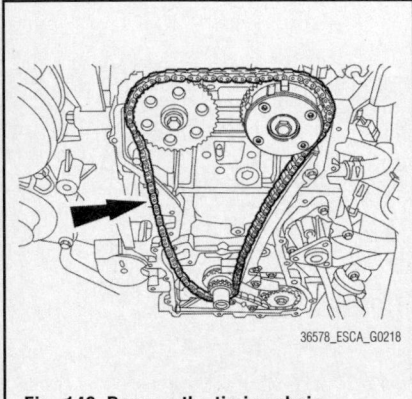

Fig. 148 Remove the timing chain

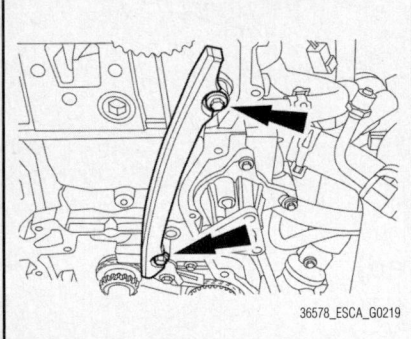

Fig. 149 Remove the 2 bolts and the timing chain guide

26. Using the flats on the camshafts to prevent camshaft rotation, tighten the bolts to 53 ft. lbs. (72 Nm).

27. Clean and inspect the mounting surfaces of the engine and the front cover.

➡ The engine front cover must be installed and the bolts tightened within 4 minutes of applying the silicone gasket and sealant.

28. Apply a 0.09 inch (2.5 mm) bead of silicone gasket and sealant to the cylinder head and oil pan joint areas. Apply a 0.09 inch (2.5 mm) bead of silicone gasket and sealant to the front cover.

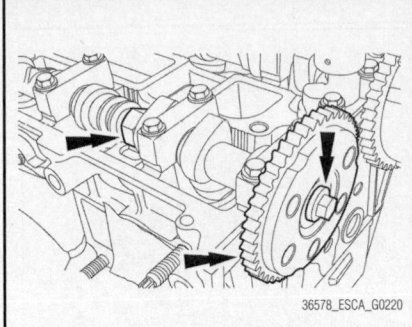

Fig. 150 Remove the bolt and the exhaust camshaft sprocket

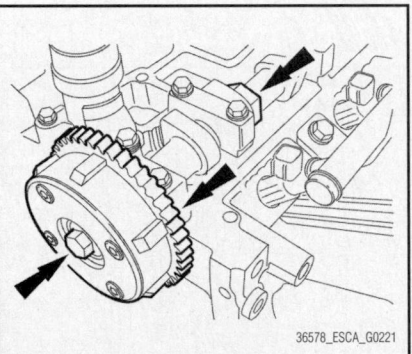

Fig. 151 Remove the bolt and the camshaft phaser and sprocket

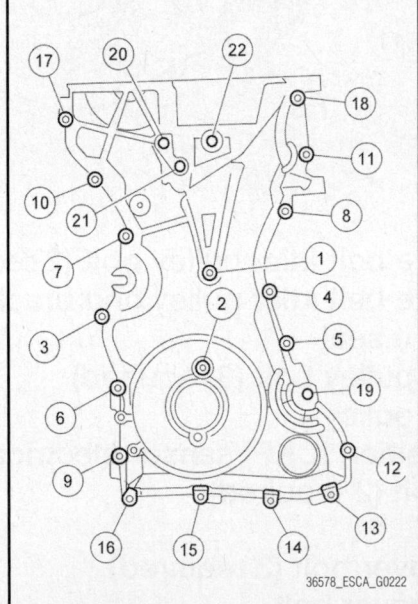

Fig. 152 Timing cover tightening sequence—2.5L engine

29. Install the engine front cover. Tighten the bolts in the sequence shown, to the following specifications:
- Tighten the 8-mm bolts and stud bolt to 89 inch lbs. (10 Nm).

- Tighten the 13-mm bolts to 35 ft. lbs. (48 Nm).

30. Install the accessory drive belt idler pulley and bracket and the 2 bolts. Tighten the bolts to 18 ft. lbs. (25 Nm).

31. Install the coolant pump pulley and bolts. Tighten the bolts to 177 inch lbs. (20 Nm).

32. Using the Camshaft Front Oil Seal Installer, install the crankshaft front oil seal.

33. Install the engine support mount, refer to the graphic for correct torque specifications.

34. Remove the engine support bar.

35. Install the front crankshaft damper pulley. Refer to Crankshaft Damper in this section.

✳✳ WARNING

The crankshaft, the crankshaft sprocket and the pulley are fitted together by friction, using diamond washers between the flange faces on each part. For that reason, the crankshaft sprocket is also unfastened if the pulley bolt is loosened. Before any repair requiring loosening or removal of the crankshaft pulley bolt, the crankshaft and camshafts must be locked in place by the special service tools, otherwise severe engine damage can occur. Refer to Crankshaft Damper in this section.

36. Install a 6 mm x 18 mm bolt through the crankshaft pulley and thread it into the front cover.

37. Install the CKP sensor and the 2 bolts. Using the crankshaft sensor aligner, adjust the CKP sensor. Tighten the bolts to 62 inch lbs. (7 Nm).

38. Connect the CKP sensor electrical connector.

39. Remove the 6 mm x 18 mm bolt.

40. Install the accessory drive belt and smooth idler pulley.

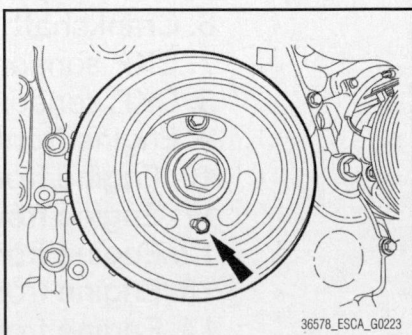

Fig. 153 Install a 6 mm x 18 mm bolt through the crankshaft pulley and thread it into the front cover

Fig. 154 Install and adjust the CKP sensor with the special tool

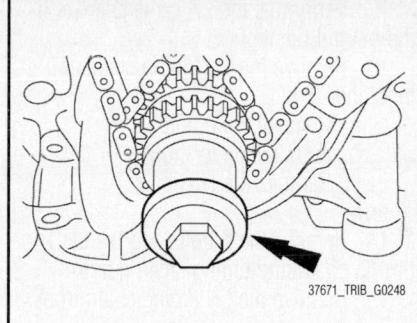

Fig. 156 Install the crankshaft pulley bolt and washer

3.0L Engine

See Figures 155 through 166.

> ✳✳ **CAUTION**
>
> **During engine repair procedures, cleanliness is extremely important. Any foreign material (including any material created while cleaning gasket surfaces) that enters the oil passages, coolant passages or the oil pan, may cause engine failure.**

> ✳✳ **CAUTION**
>
> **Failure to verify correct timing drive component alignment will result in severe engine damage.**

1. Remove the engine front cover.
2. Remove the LH and RH spark plugs.

➡**This pulse wheel is used in several different engines. Install the pulse wheel with the keyway in the slot stamped "30RFF" (orange in color).**

3. Remove the ignition pulse wheel.
4. Install the crankshaft pulley bolt and washer.
5. Rotate the crankshaft clockwise to position the crankshaft keyway in the 11

o'clock position and position the camshafts in the correct position. This will position the number one cylinder at Top Dead Center (TDC).

➡**Verify that the camshafts are correctly located. If not, rotate the crankshaft one additional turn and re-check.**

6. Rotate the crankshaft clockwise 120 degrees to the 3 o'clock position to locate the RH camshafts in the neutral position.
7. Verify that the RH camshafts are in the neutral position.

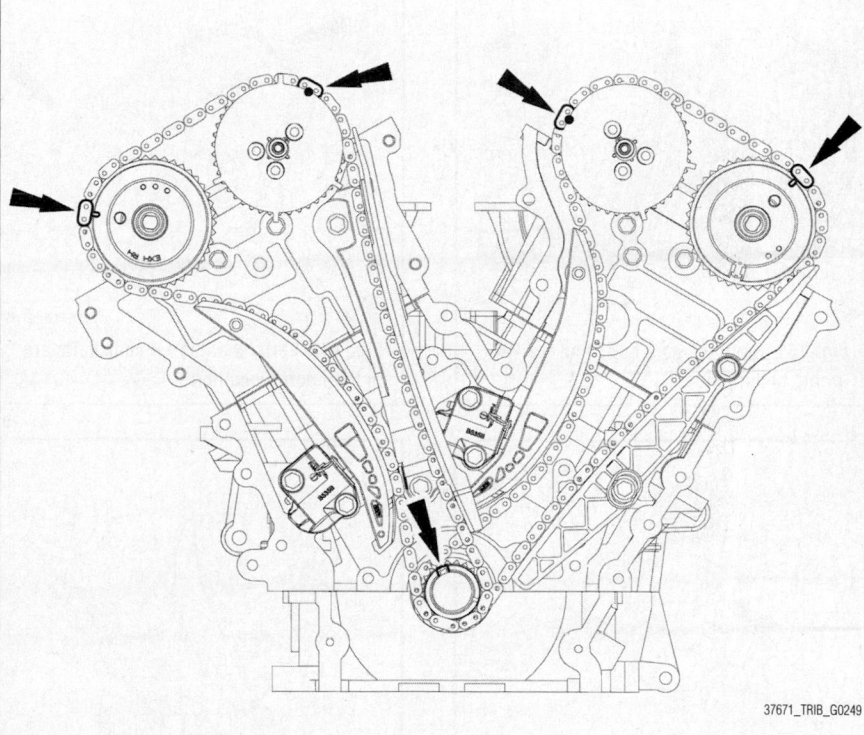

Fig. 157 Rotate the crankshaft clockwise to position the crankshaft keyway in the 11 o'clock position and position the camshafts in the correct position

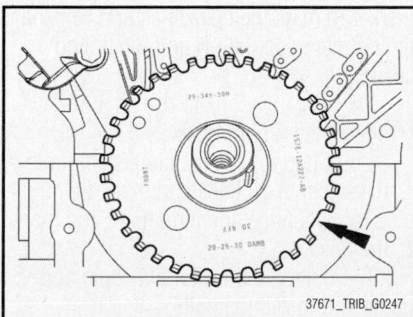

Fig. 155 Install the pulse wheel with the keyway in the slot stamped "30RFF" (orange in color)

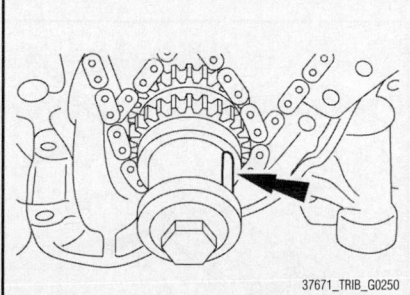

Fig. 158 Rotate the crankshaft clockwise 120 degrees to the 3 o'clock position to locate the RH camshafts in the neutral position

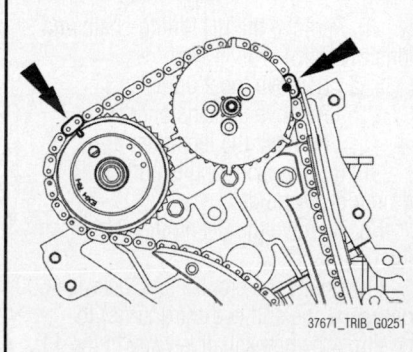

Fig. 159 Verify that the RH camshafts are in the neutral position

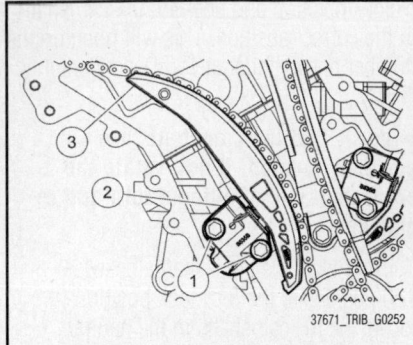

Fig. 160 Remove the RH timing chain and tensioner arm

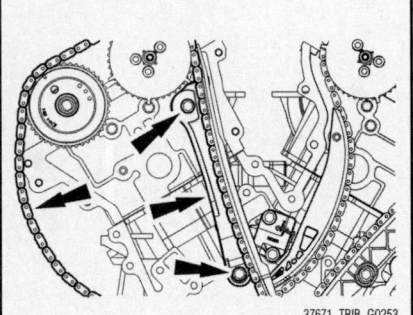

Fig. 161 Remove the 2 bolts and the RH timing chain guide

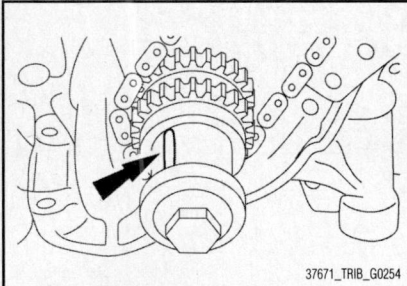

Fig. 162 Rotate the crankshaft clockwise 600 degrees (one and two-third turns) to position the crankshaft keyway in the 11 o'clock position

8. Remove the RH timing chain and tensioner arm.
  a. Remove the 2 bolts.
  b. Remove the tensioner.
  c. Remove the tensioner arm.
9. Remove the 2 bolts and the RH timing chain guide.
10. Remove the RH timing chain from the engine.
11. Rotate the crankshaft clockwise 600 degrees (one and two-third turns) to position the crankshaft keyway in the 11 o'clock position. This will position the LH camshafts in the neutral position.

12. Verify that the LH camshafts are in the neutral position.
13. Remove the LH timing chain and tensioner arm.
  a. Remove the 2 bolts.
  b. Remove the tensioner.
  c. Remove the tensioner arm.
14. Remove the 2 bolts and the LH timing chain and timing chain guide.
15. Remove the LH timing chain from the engine.
16. Remove the crankshaft pulley bolt and the crankshaft sprockets.

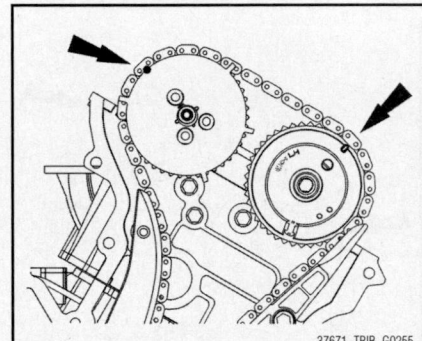

Fig. 163 Verify that the LH camshafts are in the neutral position

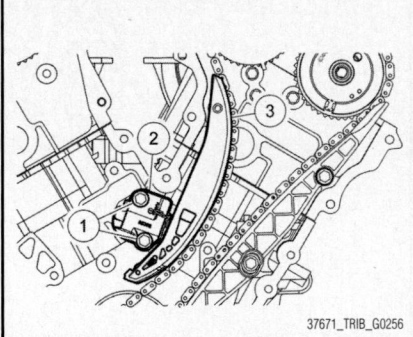

Fig. 164 Remove the LH timing chain and tensioner arm

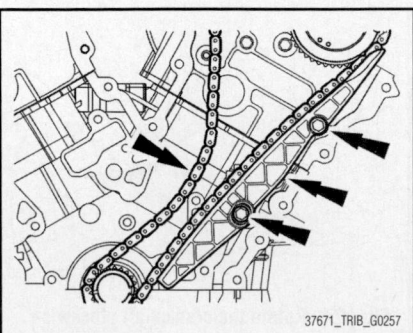

Fig. 165 Remove the 2 bolts and the LH timing chain and timing chain guide

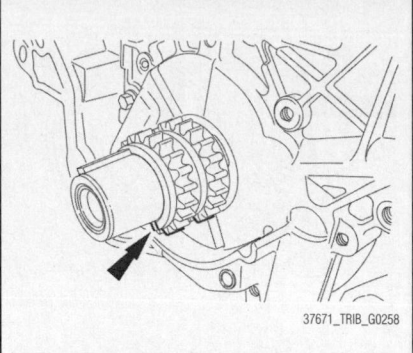

Fig. 166 Install the crankshaft sprocket with the timing mark facing out

To install:

✳✳ CAUTION

Failure to verify correct timing drive component alignment will result in severe engine damage.

17. Install the crankshaft sprocket with the timing mark facing out. The timing mark on the LH and RH timing chains will be aligned to this mark during assembly.
18. Position the chain tensioner in a soft-jawed vise.
19. Hold the chain tensioner ratchet lock mechanism away from the ratchet stem with a small pick.

✳✳ CAUTION

During tensioner compression, do not release the ratchet stem until the tensioner piston is fully bottomed in its bore or damage to the ratchet stem will result.

20. Slowly compress the timing chain tensioner.
21. Retain the tensioner piston with a 0.06 inches (1.5 mm) diameter wire or paper clip.
22. If timing marks on the timing chains are not evident, use a permanent-type marker to mark the crankshaft and camshaft timing marks on the LH and RH timing chains.
  a. Mark any link to use as the crankshaft timing mark.
  b. Starting with the crankshaft timing mark, count 29 links and mark the link.
  c. Continue counting to 42 and mark the link.
23. Position the LH timing chain and guide and install the bolts.
  a. Tighten to 18 ft. lbs. (25 Nm).
  b. Align the marks on the timing chain with the marks on the camshaft and crankshaft sprockets.

24. Install the LH timing chain and tensioner arm and the LH timing chain tensioner.

   a. Install the tensioner arm.

   b. Position the tensioner.

   c. Install the bolt. Tighten to 18 ft. lbs. (25 Nm).

25. Install the crankshaft pulley bolt and rotate the crankshaft clockwise 120 degrees until the crankshaft keyway is in the 3 o'clock position.

26. Verify that the RH camshafts are correctly positioned.

27. Position the RH timing chain and chain guide and install the bolts.

   a. Tighten to 18 ft. lbs. (25 Nm).

   b. Align the marks on the timing chain with the marks on the camshaft and crankshaft sprockets.

28. Install the RH timing chain tensioner and tensioner arm.

   a. Install the tensioner arm.

   b. Position the tensioner.

   c. Install the bolts. Tighten to 18 ft. lbs. (25 Nm).

29. Remove the LH and RH timing chain tensioner piston retaining wires.

30. Rotate the crankshaft counterclockwise 120 degrees to top dead center (TDC).

✳✳ CAUTION

Failure to verify correct timing drive component alignment will result in severe engine damage.

31. Verify the timing with the following steps.

   a. There should be 12 chain links between the camshaft timing marks.

   b. There should be 27 chain links between the camshaft and crankshaft timing marks.

   c. There should be 30 chain links between the camshaft and crankshaft timing marks.

32. Remove the crankshaft pulley bolt and washer.

➡**This pulse wheel is used in several different engines. Install the pulse wheel with the keyway in the slot stamped "30RFF" only (orange in color).**

33. Install the ignition pulse wheel.

34. Install the LH and RH spark plugs.

35. Install the engine front cover.

ENGINE PERFORMANCE & EMISSION CONTROLS

CAMSHAFT POSITION (CMP) SENSOR

LOCATION

2.5L Engine

See Figure 167.

The Camshaft Position (CMP) sensor is located on top the valve cover towards the front of the vehicle.

3.0L Engine

See Figure 168.

The Camshaft Position (CMP) sensors are located on the left and right cylinder head just below the valve cover.

REMOVAL & INSTALLATION

2.5L Engine

1. Disconnect the Camshaft Position (CMP) sensor electrical connector.

2. Remove the bolt and the CMP sensor.

To install:

➡**Lubricate the CMP sensor O-ring seal with clean engine oil.**

3. To install, reverse the removal procedure.

4. Tighten the mounting bolt to 62 inch lbs. (7 Nm)

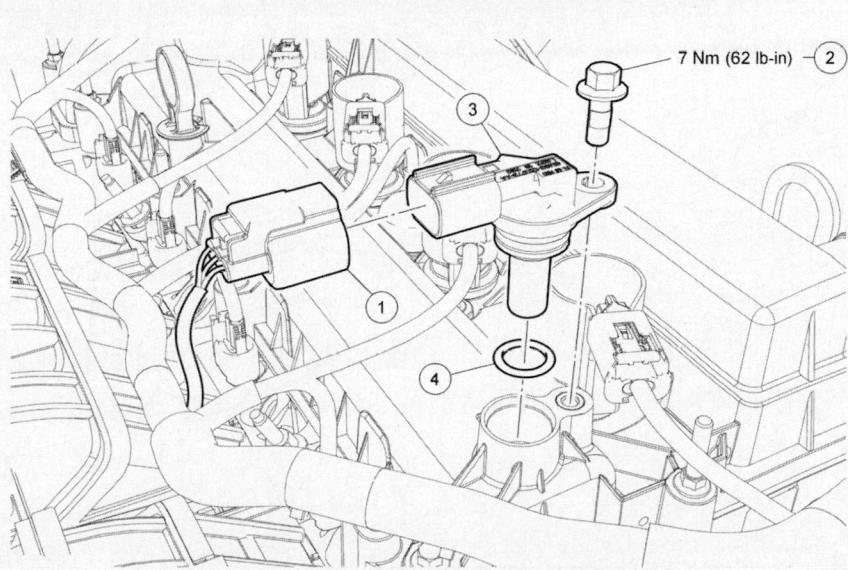

1. Camshaft Position (CMP) sensor electrical connector
2. CMP sensor bolt
3. CMP sensor
4. CMP sensor O-ring

36578_ESCA_G0262

Fig. 167 Camshaft Position (CMP) sensor location

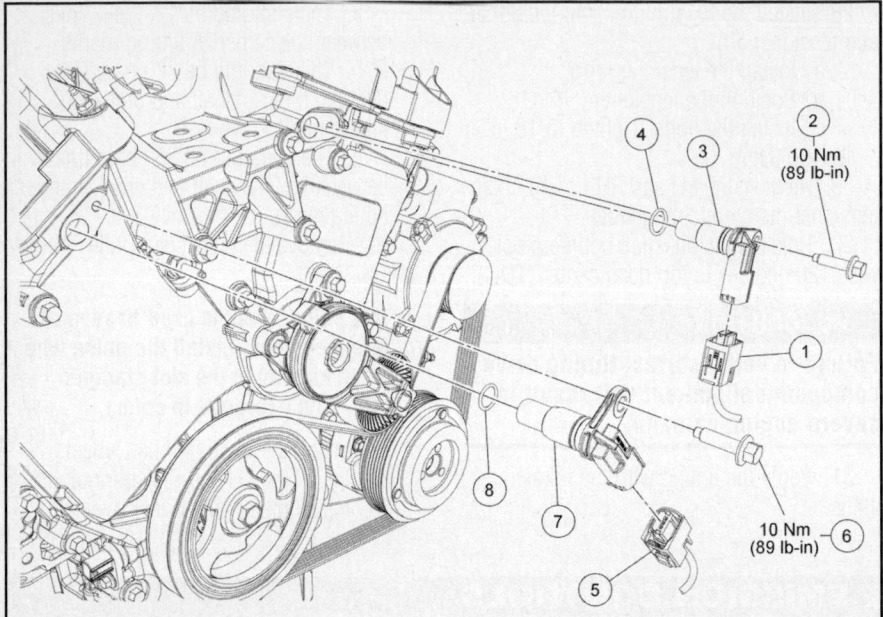

1. Camshaft Position (CMP) sensor electrical connector
2. CMP sensor bolt — LH
3. CMP sensor — LH
4. CMP sensor O-ring seal — LH
5. CMP sensor electrical connector
6. CMP sensor bolt — RH
7. CMP sensor — RH
8. CMP sensor O-ring seal — RH

36578_ESCA_G0261

Fig. 168 Camshaft Position (CMP) sensor location

3.0L Engine

1. Disconnect the Camshaft Position (CMP) sensor electrical connector.

2. Remove the bolt and the CMP sensor.

To install:

➡**Lubricate the CMP O-ring seal with clean engine oil.**

3. To install, reverse the removal procedure and tighten the mounting bolt to 89 inch lbs. (10 Nm).

CRANKSHAFT POSITION (CKP) SENSOR

LOCATION

The Crankshaft Position (CKP) sensor is located just behind the crankshaft pulley on engine block.

2.5L Engine

See Figure 169.

3.0L Engine

See Figure 170.

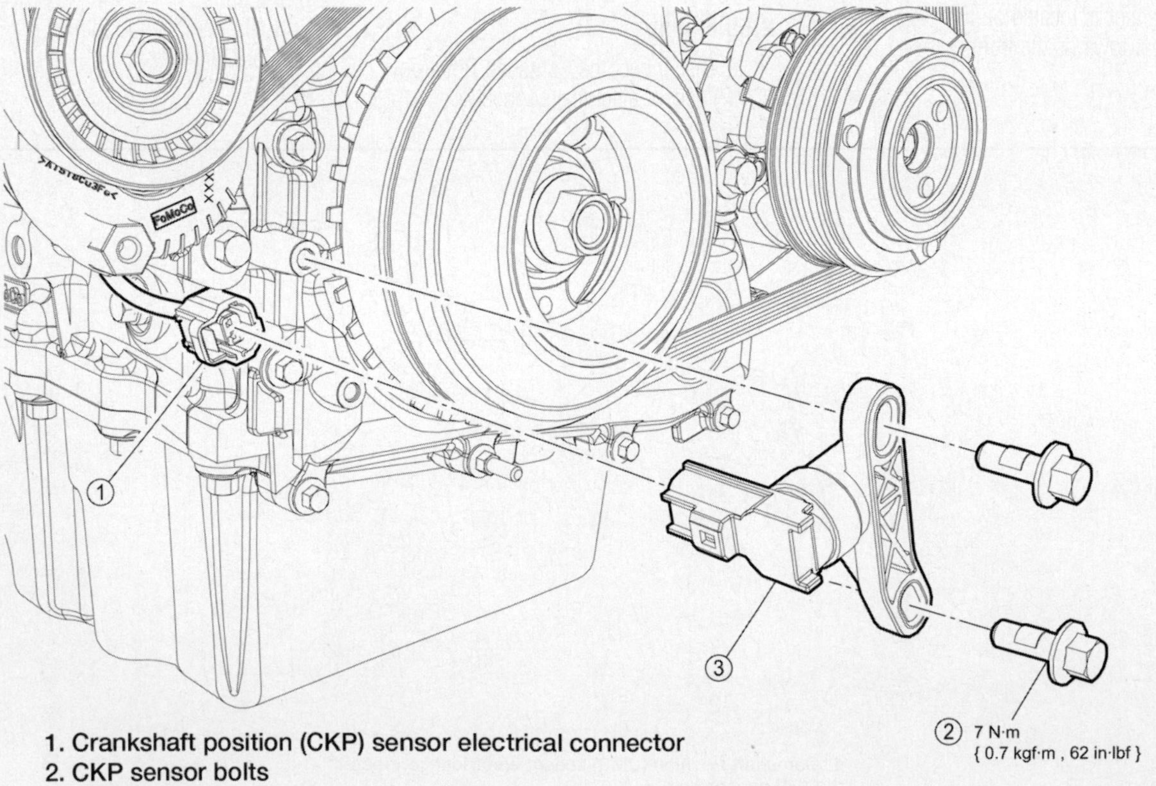

1. Crankshaft position (CKP) sensor electrical connector
2. CKP sensor bolts
3. CKP sensor

7 N·m
{ 0.7 kgf·m , 62 in·lbf }

37671_TRIB_G0260

Fig. 169 CKP sensor location—2.5L engine

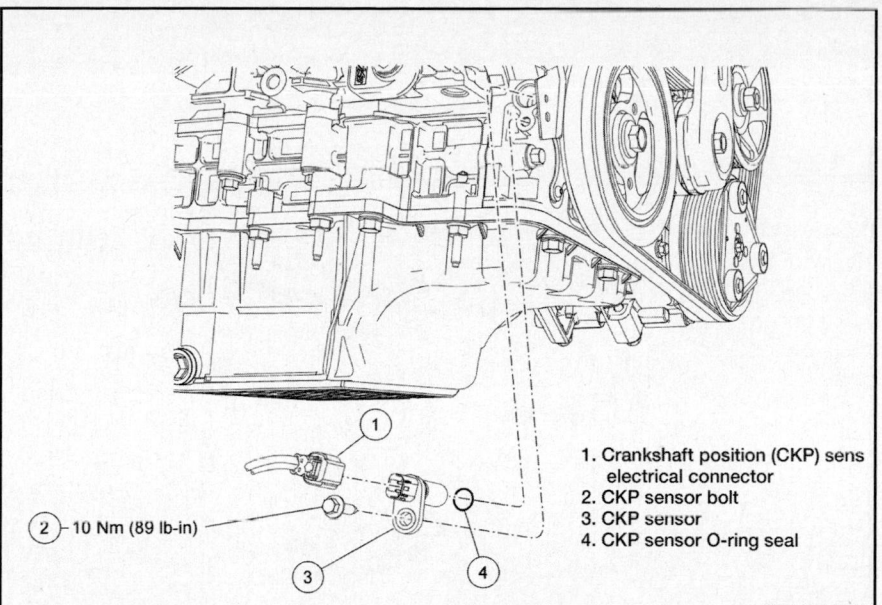

1. Crankshaft position (CKP) sens electrical connector
2. CKP sensor bolt
3. CKP sensor
4. CKP sensor O-ring seal

2 - 10 Nm (89 lb-in)

22086_ESCA_G0093

Fig. 170 Crankshaft Position (CKP) sensor location—3.0L engine

REMOVAL & INSTALLATION

2.5L Engine

See Figures 171 through 175.

1. With the vehicle in NEUTRAL, position on a hoist.

2. Remove the 5 bolts, the pin-type retainer (not shown) and the RH splash shield.

3. Turn the crankshaft clockwise until the hole in the crankshaft pulley is in the 3 o'clock position.

4. Remove the engine plug bolt.

5. Install the Crankshaft TDC Timing Peg and turn the crankshaft clockwise until the crankshaft contacts the Crankshaft TDC Timing Peg.

➡**When the crankshaft contacts the Crankshaft TDC Timing Peg, the No. 1 cylinder will be at Top Dead Center (TDC).**

6. Disconnect the Crankshaft Position (CKP) sensor electrical connector.

7. Remove the bolts and the CKP sensor.

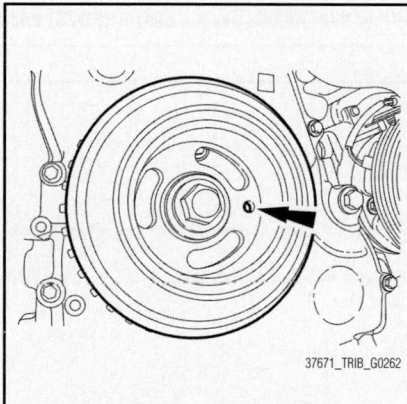

37671_TRIB_G0262

Fig. 172 Turn the crankshaft clockwise until the hole in the crankshaft pulley is in the 3 o'clock position

To install:

8. Install a 6 mm x 18 mm bolt through the crankshaft pulley and thread it into the front cover.

✳✳ CAUTION

Only hand-tighten the bolt or damage to the front cover can occur.

9. Install the CKP sensor and the 2 bolts.

➡**Do not tighten the bolts at this time.**

10. Adjust the CKP sensor with the Crankshaft Sensor Aligner.

11. Connect the CKP sensor electrical connector.

12. Remove the 6 mm x 18 mm bolt from the crankshaft pulley.

13. Install the engine plug bolt. Tighten to 15 ft. lbs. (20 Nm).

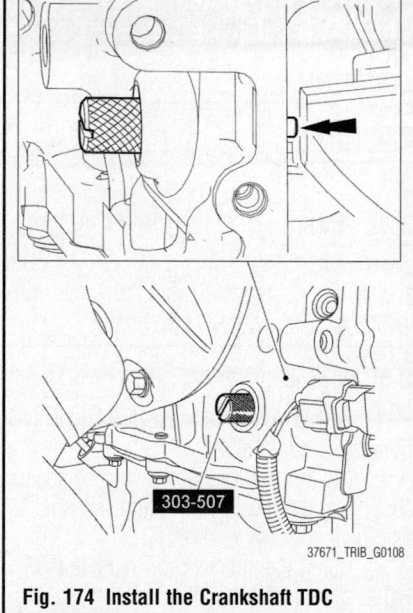

303-507

37671_TRIB_G0108

Fig. 174 Install the Crankshaft TDC Timing Peg

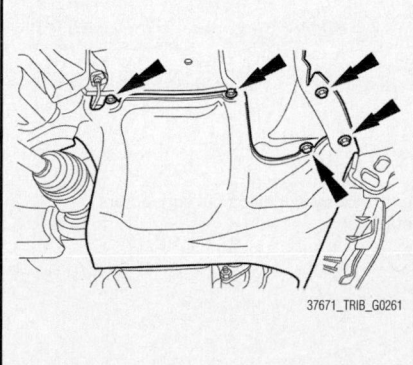

37671_TRIB_G0261

Fig. 171 Remove the RH splash Shield

37671_TRIB_G0263

Fig. 173 Remove the engine plug bolt

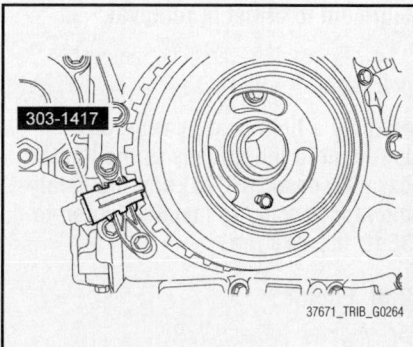

303-1417

37671_TRIB_G0264

Fig. 175 Adjust the CKP sensor with the Crankshaft Sensor Aligner

14. Install the RH splash shield, 5 bolts and the pin-type retainer (not shown).

15. Using the scan tool, carry out the Misfire Monitor Neutral Profile Correction procedure, following the on-screen instructions.

3.0L Engine

1. With the vehicle in NEUTRAL, position it on a hoist.

2. Remove the 5 bolts and the RH splash shield.

3. Disconnect the Crankshaft Position (CKP) sensor electrical connector.

4. Remove the bolt and the CKP sensor.

➡**Lubricate the CKP sensor O-ring seal with clean engine oil.**

5. To install, reverse the removal procedure and tighten sensor mounting bolt to 89 inch lbs. (10 Nm).

HEATED OXYGEN (HO2S) SENSOR

LOCATION

2.5L Engine

See Figure 176.

3.0L Engine

See Figure 177.

REMOVAL & INSTALLATION

2.5L Engine

1. With the vehicle in NEUTRAL, position it on a hoist.

2. Disconnect the heated oxygen sensor (HO2S) and the catalyst monitor sensor electrical connectors.

3. Using the special tool, remove the HO2S and catalyst monitor sensor.

➡**If necessary, lubricate the sensor threads with penetrating and lock lubricant to assist in removal.**

4. To install, reverse the removal procedure.

➡**Apply a light coat of anti-seize lubricant to the threads of the heated oxygen sensor (HO2S) and the catalyst monitor sensor. To install, tighten to 35 ft. lbs. (48 Nm).**

3.0L Engine

See Figures 178 and 179.

1. With the vehicle in NEUTRAL, position it on a hoist.

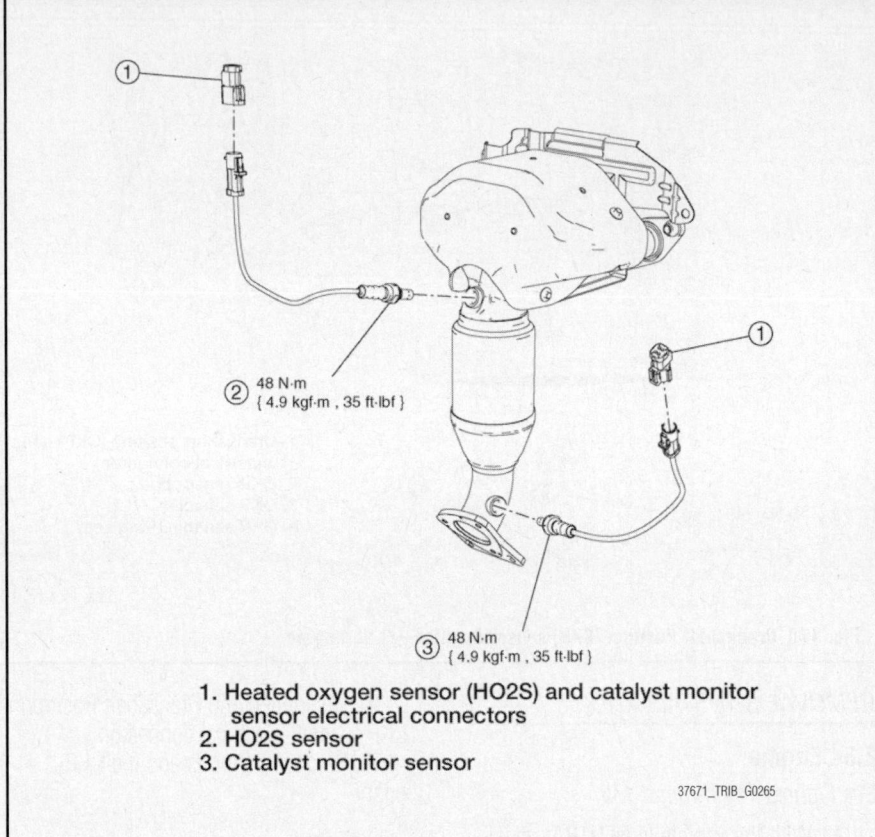

48 N·m
{ 4.9 kgf·m , 35 ft·lbf }

48 N·m
{ 4.9 kgf·m , 35 ft·lbf }

1. Heated oxygen sensor (HO2S) and catalyst monitor sensor electrical connectors
2. HO2S sensor
3. Catalyst monitor sensor

37671_TRIB_G0265

Fig. 176 Heated oxygen sensor (HO2S) and catalyst monitor sensor locations

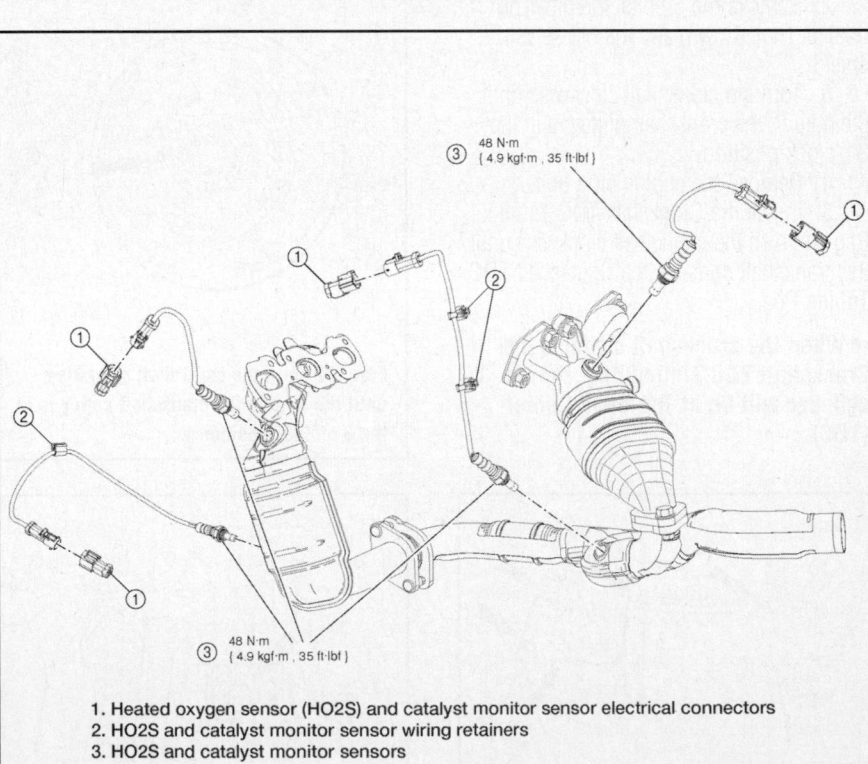

48 N·m
{ 4.9 kgf·m , 35 ft·lbf }

48 N·m
{ 4.9 kgf·m , 35 ft·lbf }

1. Heated oxygen sensor (HO2S) and catalyst monitor sensor electrical connectors
2. HO2S and catalyst monitor sensor wiring retainers
3. HO2S and catalyst monitor sensors

37671_TRIB_G0266

Fig. 177 Heated oxygen sensor (HO2S) and catalyst monitor sensor locations

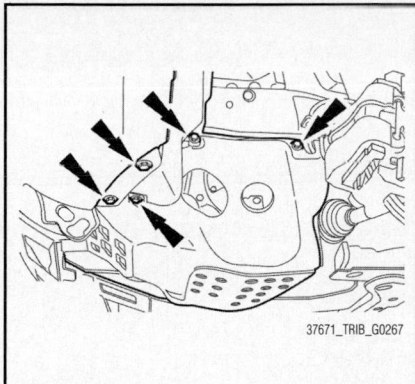

Fig. 178 Remove the LH splash shield

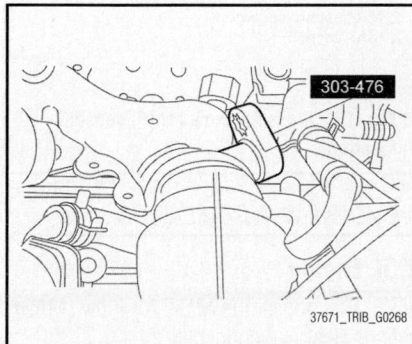

Fig. 179 Using the special tool, remove the HO2S and the catalyst monitor sensor

2. Remove the 7 bolts (5 shown) and the LH splash shield.

3. Disconnect the heated oxygen sensor (HO2S) and the catalyst monitor sensor electrical connectors.

4. Using the special tool, remove the HO2S and the catalyst monitor sensor.

➡️**If necessary, lubricate the HO2S and catalyst monitor sensor with penetrating and lock lubricant to aid in removal.**

5. To install, reverse the removal procedure.

➡️**Apply a light coat of anti-seize lubricant to the threads of the HO2S and the catalyst monitor sensor. To install, tighten to 35 ft. lbs. (48 Nm).**

INTAKE AIR TEMPERATURE (IAT) SENSOR

LOCATION

The Intake Air Temperature (IAT) sensor is an integral part of the Mass Air Flow (MAF) sensor/Intake Air Temperature (IAT) sensor assembly which is mounted on the air intake duct. Refer to the Mass Air Flow section for information regarding servicing this component.

KNOCK SENSOR (KS)

LOCATION

2.5L Engine

See Figure 180.

The Knock Sensor (KS) is located behind the intake manifold to the rear of engine block.

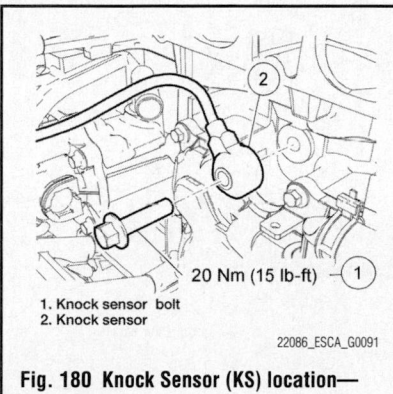

1. Knock sensor bolt
2. Knock sensor

Fig. 180 Knock Sensor (KS) location—2.5L engine

3.0L Engine

See Figures 181 and 182.

The 3.0L engine uses two Knock Sensors (KS), one is located at the cylinder head and one is located under the intake manifold mounted to the engine block.

REMOVAL & INSTALLATION

2.5L Engine

1. With the vehicle in NEUTRAL, position it on a hoist.

2. Remove the intake manifold. Refer to Intake Manifold Removal & Installation in the Engine Mechanical section.

3. Remove the bolt and the Knock Sensor (KS).

4. To install, reverse the removal procedure and tighten the sensor mounting bolt to 15ft. lbs. (20 Nm).

3.0L Engine

Cylinder head-mounted Knock Sensor (KS)

1. Disconnect the Knock Sensor (KS) electrical connector.

2. Remove the bolt and the KS.

Engine block-mounted KS

3. Remove the upper and lower intake manifold.

4. Disconnect the KS electrical connector.

5. Remove the bolt and the KS.

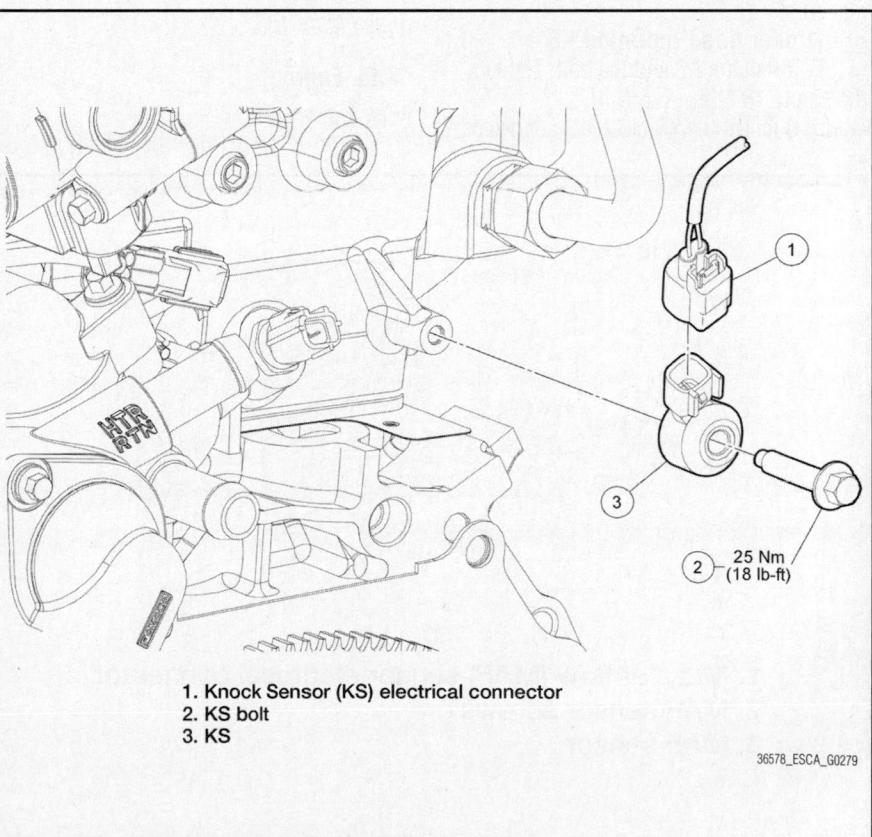

1. Knock Sensor (KS) electrical connector
2. KS bolt
3. KS

Fig. 181 Cylinder head mounted KS—3.0L engine

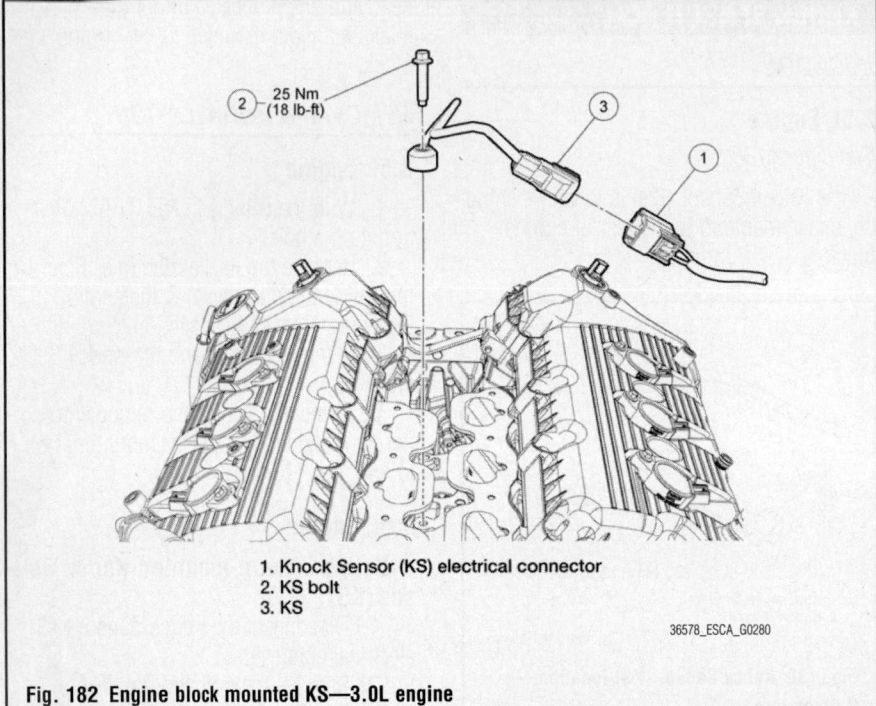

1. Knock Sensor (KS) electrical connector
2. KS bolt
3. KS

36578_ESCA_G0280

Fig. 182 Engine block mounted KS—3.0L engine

To install:
Engine block-mounted KS
6. Install the KS and the bolt. Tighten the bolt to 18 ft. lbs. (25 Nm).
7. Connect the KS electrical connector.
8. Install the upper and lower intake manifold.

Cylinder head-mounted KS
9. Install the KS and the bolt. Tighten the bolt to 18 ft. lbs. (25 Nm).
10. Connect the KS electrical connector.

MASS AIR FLOW (MAF) SENSOR

LOCATION

2.5L Engine
See Figure 183.

3.0L Engine
See Figure 184.

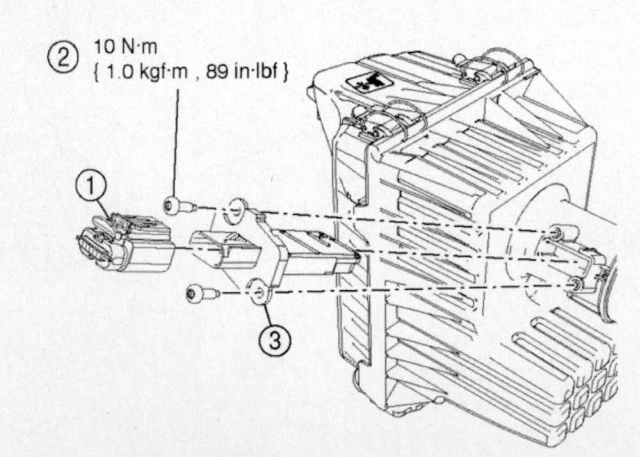

1. Mass airflow (MAF) sensor electrical connector
2. MAF sensor screws
3. MAF sensor

37671_TRIB_G0269

Fig. 183 Mass Air Flow (MAF) sensor location

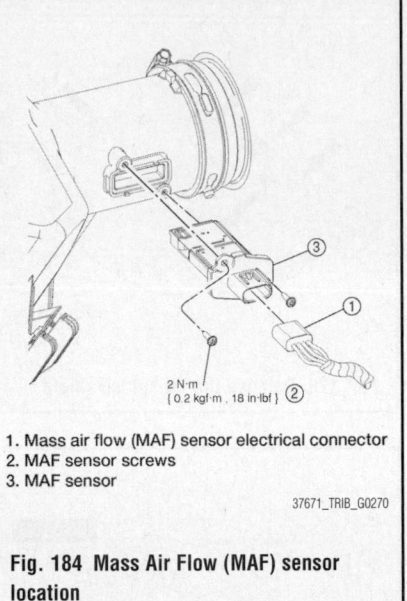

1. Mass air flow (MAF) sensor electrical connector
2. MAF sensor screws
3. MAF sensor

37671_TRIB_G0270

Fig. 184 Mass Air Flow (MAF) sensor location

REMOVAL & INSTALLATION

2.5L Engine
1. Disconnect the Mass Air Flow (MAF) sensor electrical connector.
2. Remove the 2 screws and the MAF sensor.
3. To install, reverse the removal procedure and tighten MAF sensor mounting screws to 89 inch lbs. (10 Nm).

3.0L Engine

> **✳✳ CAUTION**
>
> The mass air flow (MAF) sensor electronics module and body are calibrated as a unit and must be serviced as a unit. Do not tamper with the sensing elements located in the air flow bypass of the MAF body.

1. Disconnect the mass air flow (MAF) sensor electrical connector.
2. Remove the 2 screws and the MAF sensor.
3. To install, reverse the removal procedure.

MANIFOLD ABSOLUTE PRESSURE (MAP) SENSOR

LOCATION

2.5L Engine
See Figure 185.

3.0L Engine
See Figure 106.

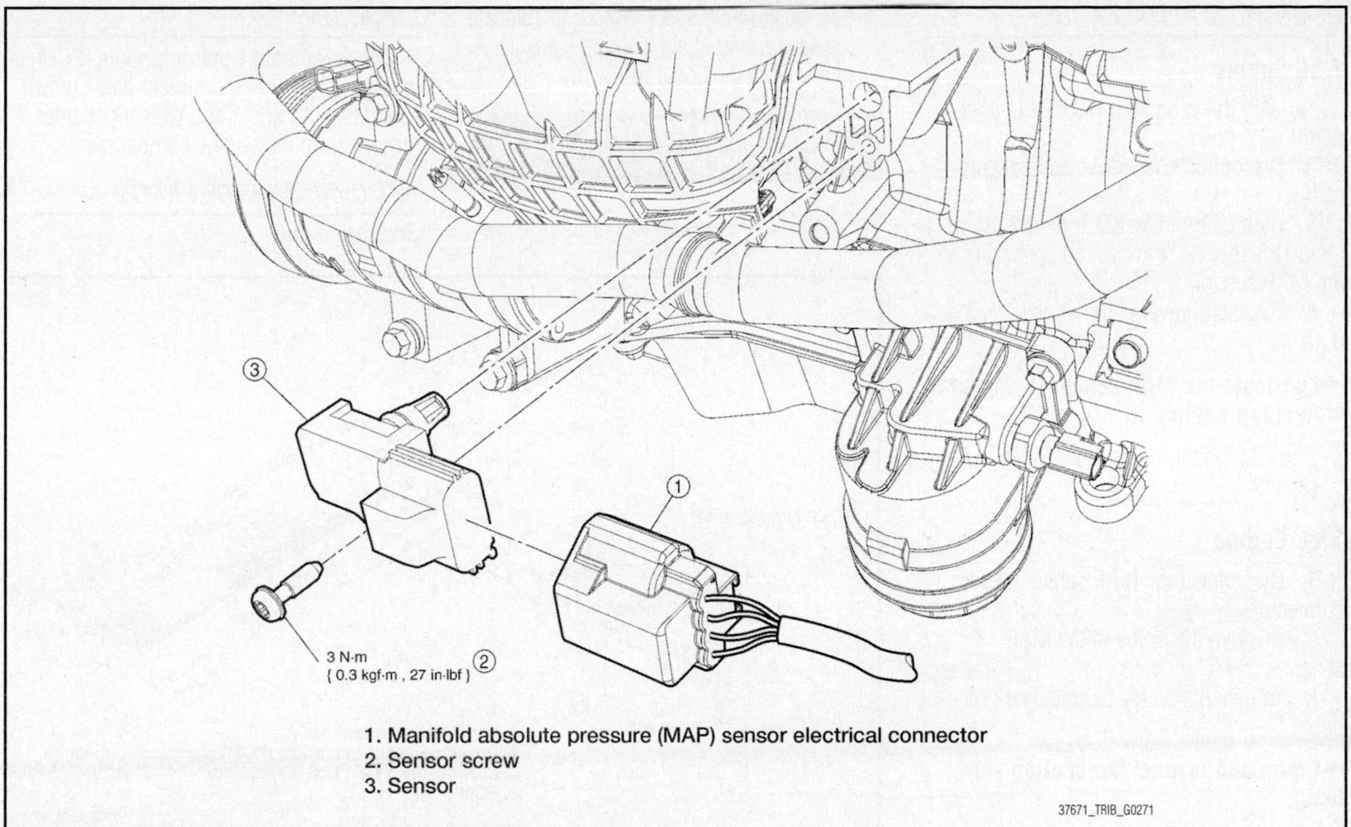

3 N·m
{ 0.3 kgf·m , 27 in·lbf }

1. Manifold absolute pressure (MAP) sensor electrical connector
2. Sensor screw
3. Sensor

37671_TRIB_G0271

Fig. 185 Manifold Absolute Pressure (MAP) sensor location

② 6 N·m
{ 0.6 kgf·m , 53 in·lbf }

1. Manifold absolute pressure (MAP) sensor electrical connector
2. Sensor bolt
3. Sensor

37671_TRIB_G0272

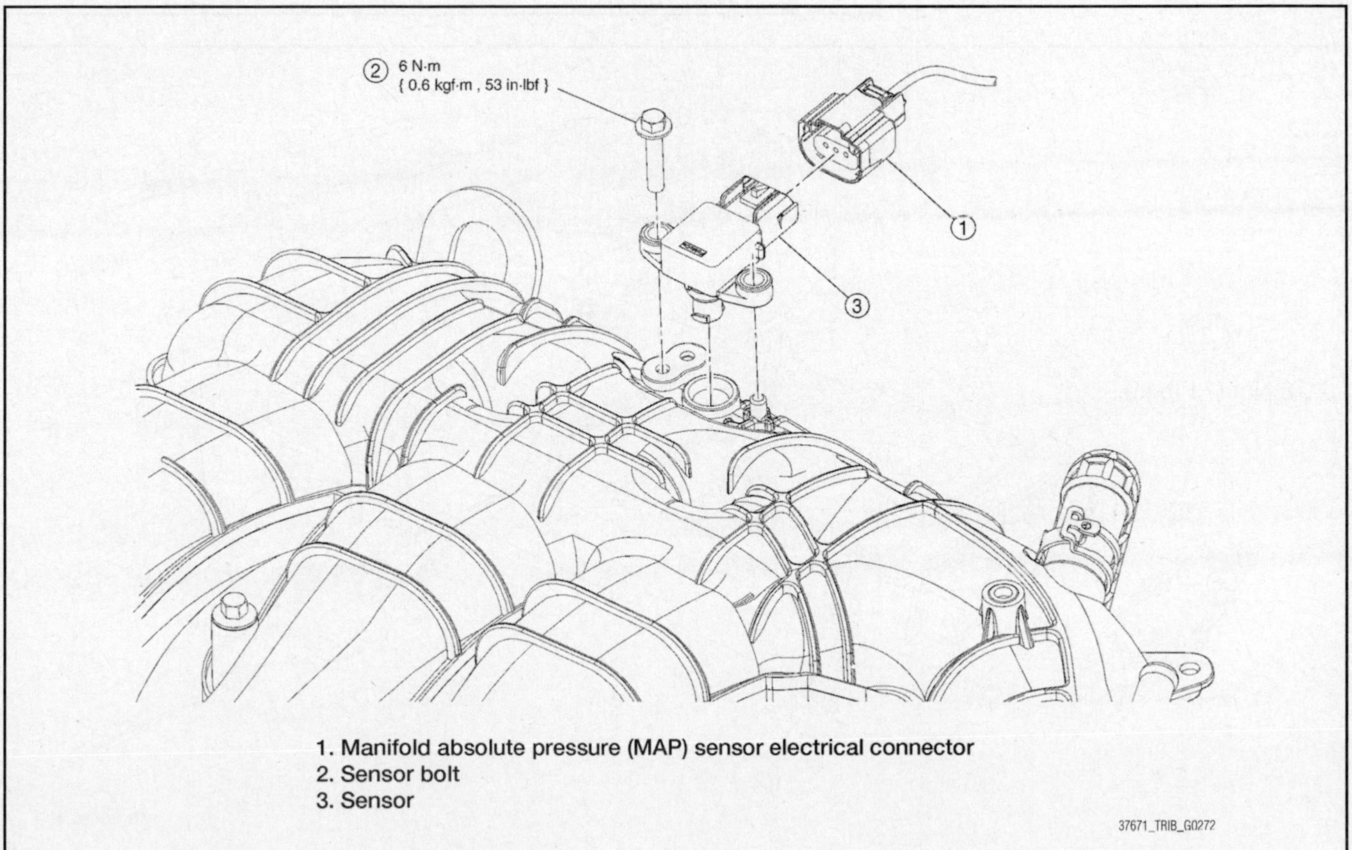

Fig. 186 Manifold Absolute Pressure (MAP) sensor location

REMOVAL & INSTALLATION

2.5L Engine

1. With the vehicle in NEUTRAL, position it on a hoist.

2. Disconnect the MAP electrical connector.

3. Using a tool like KD Tool #58181 or Lisle T25 Torx Bit, remove the screw and the MAP sensor.

4. To install, reverse the removal procedure.

➡**Lubricate the MAP sensor O-ring seal with clean engine oil.**

   a. To install, tighten to 27 inch lbs. (3 Nm).

3.0L Engine

1. Disconnect the MAP sensor electrical connector.

2. Remove the bolt and the MAP sensor.

3. To install, reverse the removal procedure.

➡**Clean and inspect the sealing surface.**

   a. To install, tighten to 53 inch lbs. (6 Nm).

   b. When installing the sensor, if the bolt fails to hold specified torque, relocate the sensor and retain, using the auxiliary bolt hole in the upper intake manifold.

POWERTRAIN CONTROL MODULE (PCM)

See Figure 187.

LOCATION

The Powertrain Control Module (PCM) is located behind the instrument panel (cowl), center to both driver and passenger sides (access from the engine compartment).

REMOVAL & INSTALLATION

See Figure 188.

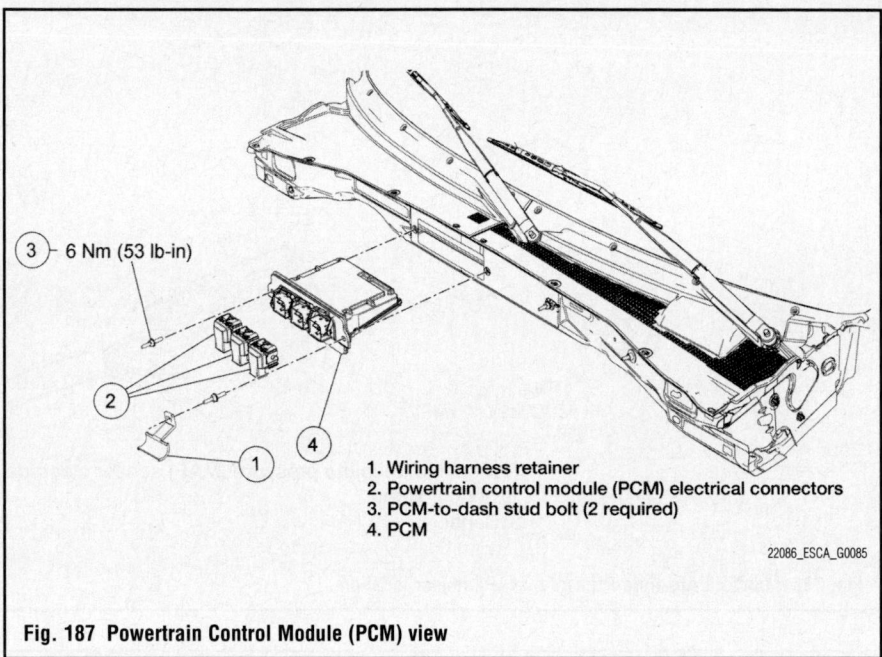

3 — 6 Nm (53 lb-in)

1. Wiring harness retainer
2. Powertrain control module (PCM) electrical connectors
3. PCM-to-dash stud bolt (2 required)
4. PCM

22086_ESCA_G0085

Fig. 187 Powertrain Control Module (PCM) view

8 Nm (71 lb-in)

36578_ESCA_G0299

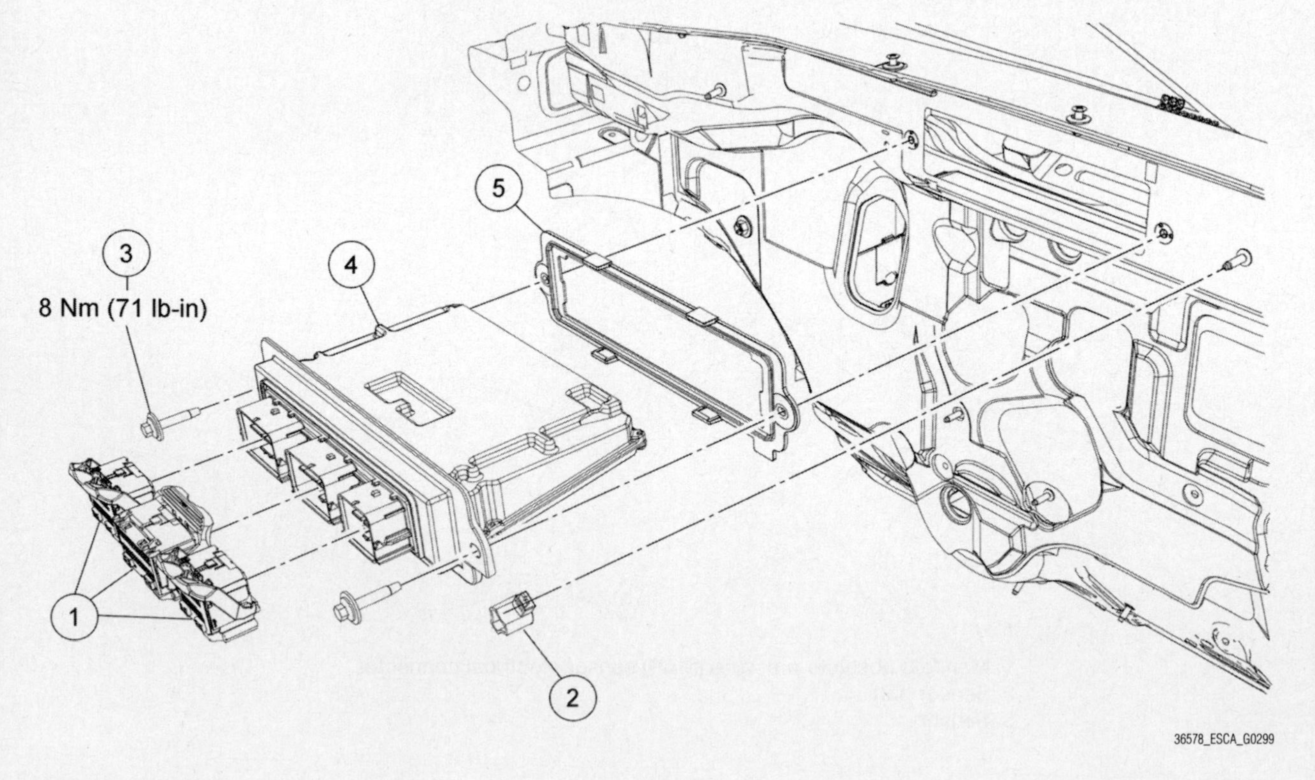

Fig. 188 Powertrain Control Module (PCM) view

➡ **PCM installation DOES NOT require new keys or programming of keys, only a Parameter Reset of the Passive Anti-Theft System (PATS).**

1. Retrieve the module configuration. Carry out the module configuration retrieval steps of the Programmable Module Installation (PMI) procedure. (Specialized equipment is required)

2. Detach the wiring harness retainer from the cowl stud bolt.

3. Disconnect the 3 PCM electrical connectors.

4. Remove the 2 bolts and the PCM.

5. Remove and inspect the PCM cowl seal; install new if damaged.

To install:

6. Install the PCM cowl seal.

7. Install the PCM and the 2 bolts. Tighten the retaining bolts to 71 inch lbs. (8 Nm).

8. Connect the 3 PCM electrical connectors.

9. Attach the wiring harness retainer to the cowl stud bolt.

10. Restore the module configuration. Carry out the module configuration restore steps of the

11. Programmable Module Installation (PMI) procedure.

12. Reprogram the PATS. Carry out the Parameter Reset procedure.

THROTTLE POSITION SENSOR (TPS)

LOCATION

3.0L Engine

See Figure 189.

The Throttle Position Sensor (TPS) is located on the throttle body just behind the EGR valve tube.

REMOVAL & INSTALLATION

3.0L Engine

1. Disconnect the Throttle Position Sensor (TPS) sensor electrical connector.

2. Remove the 2 screws and the TPS.

3. To install, reverse the removal procedure and tighten mounting screws to 27 inch lbs. (3 Nm).

VEHICLE SPEED SENSOR (VSS)

LOCATION

See Figure 190.

➡ **This sensor is used on the manual transaxle.**

1. Throttle position (TP) sensor electrical connector
2. TP sensor screws (2 required)
3. TP sensor

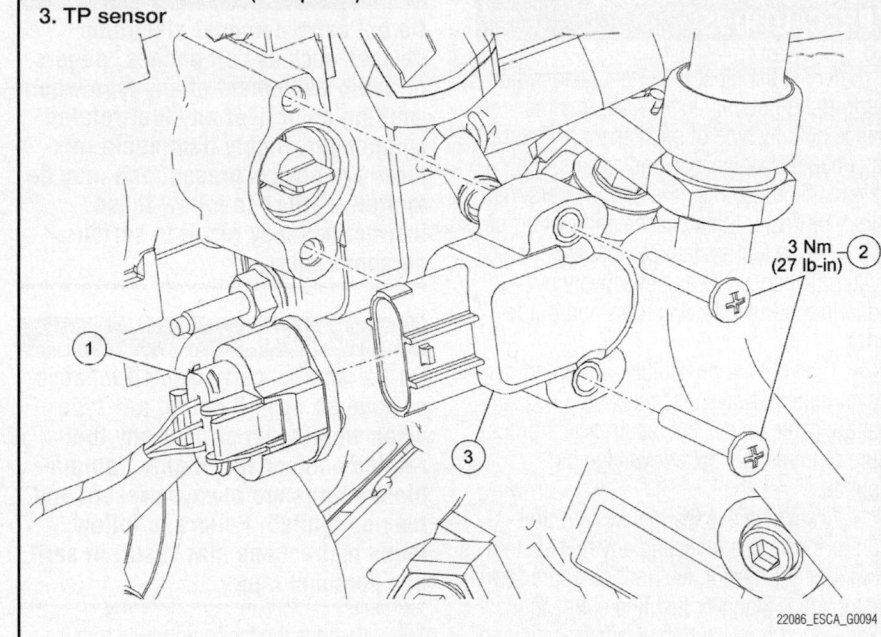

Fig. 189 Throttle Position Sensor (TPS) location—3.0L engine

1. Vehicle speed sensor (VSS) bolt
2. VSS
3. VSS electrical connector

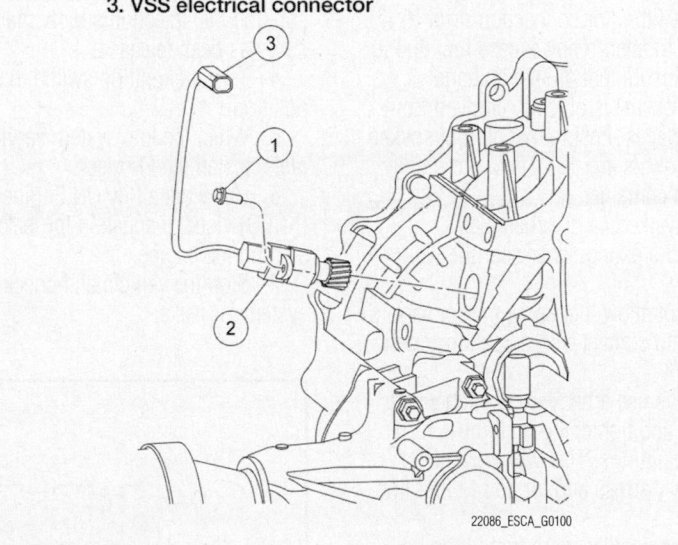

Fig. 190 Vehicle Speed Sensor (VSS) location—manual transaxle

REMOVAL & INSTALLATION

1. With the vehicle in NEUTRAL, position it on a hoist.

2. Remove the 7 retainers and the LH splash shield.

3. Disconnect the Vehicle Speed Sensor (VSS) electrical connector.

4. Remove the sensor bolt.

5. Remove the sensor.

To install:

➡ **When installing the sensor, lubricate the O-ring seal with clean transmission fluid.**

6. To install, reverse the removal procedure and note the following:

   • Tighten the sensor mounting bolt to 9 ft. lbs. (12 Nm).

FUEL GASOLINE FUEL INJECTION SYSTEM

FUEL SYSTEM SERVICE PRECAUTIONS

Safety is the most important factor when performing not only fuel system maintenance but any type of maintenance. Failure to conduct maintenance and repairs in a safe manner may result in serious personal injury or death. Maintenance and testing of the vehicle's fuel system components can be accomplished safely and effectively by adhering to the following rules and guidelines.

• To avoid the possibility of fire and personal injury, always disconnect the negative battery cable unless the repair or test procedure requires that battery voltage be applied.

• Always relieve the fuel system pressure prior to disconnecting any fuel system component (injector, fuel rail, pressure regulator, etc.), fitting or fuel line connection. Exercise extreme caution whenever relieving fuel system pressure to avoid exposing skin, face and eyes to fuel spray. Please be advised that fuel under pressure may penetrate the skin or any part of the body that it contacts.

• Always place a shop towel or cloth around the fitting or connection prior to loosening to absorb any excess fuel due to spillage. Ensure that all fuel spillage (should it occur) is quickly removed from engine surfaces. Ensure that all fuel soaked cloths or towels are deposited into a suitable waste container.

• Always keep a dry chemical (Class B) fire extinguisher near the work area.

• Do not allow fuel spray or fuel vapors to come into contact with a spark or open flame.

• Always use a back-up wrench when loosening and tightening fuel line connection fittings. This will prevent unnecessary stress and torsion to fuel line piping.

• Always replace worn fuel fitting O-rings with new Do not substitute fuel hose or equivalent where fuel pipe is installed.

Before servicing the vehicle, make sure to also refer to the precautions in the beginning of this section as well.

RELIEVING FUEL SYSTEM PRESSURE

See Figure 191.

✳✳ CAUTION

Do not carry personal electronic devices such as cell phones, pagers or audio equipment of any type when working on or near any fuel-related component. Highly flammable mixtures are always present and may be ignited. Failure to follow these instructions may result in serious personal injury.

✳✳ CAUTION

Do not smoke, carry lighted tobacco or have an open flame of any type when working on or near any fuel-related component. Highly flammable mixtures are always present and may be ignited. Failure to follow these instructions may result in serious personal injury.

1. Remove the fuel pump fuse that is located in the Battery Junction Box (BJB), location 22.
2. Start the engine and allow it to idle until it stalls.
3. After the engine stalls, crank the engine for approximately 5 seconds to make sure the fuel injection supply manifold pressure has been released.
4. Turn the ignition switch to the OFF position.
5. When the fuel system service is complete, install the FP fuse.
6. Carry out a Key ON Engine OFF (KOEO) visual inspection for leaks prior to starting the engine.
7. Start the vehicle and check the fuel system for leaks.

FUEL FILTER

REMOVAL & INSTALLATION

✳✳ WARNING

Do not smoke, or carry lighted tobacco or have an open flame of any type when working on or near any fuel-related component. Highly flammable mixtures are always present and may be ignited. Failure to follow these instructions may result in serious personal injury.

✳✳ WARNING

Do not carry personal electronic devices such as cell phones, pagers or audio equipment of any type when working on or near any fuel related component. Highly flammable mixtures are always present and may be ignited. Failure to follow these instructions may result in serious personal injury.

✳✳ WARNING

Before working on or disconnecting any of the fuel tubes or fuel system components, relieve the fuel system pressure to prevent accidental spraying of fuel. Fuel in the fuel system remains under pressure, even when the engine is not running. Failure to follow this instruction may result in serious personal injury.

1. With the vehicle in NEUTRAL, position it on a hoist.
2. Relieve the fuel pressure.

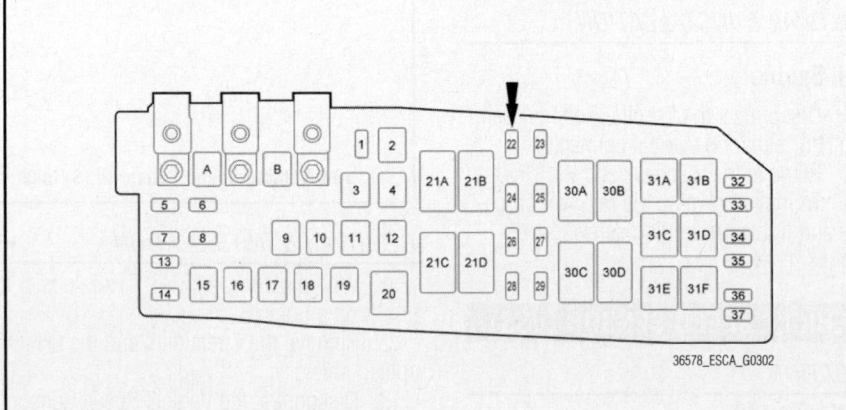

Fig. 191 Battery Junction Box (BJB), fuse location 22 shown

36578_ESCA_G0302

3. Disconnect the fuel supply tube-to-fuel filter inlet quick connect coupling.

➡Some residual fuel may remain in the fuel filter after releasing the fuel system pressure. Upon disconnecting or removing the fuel filter, carefully drain any residual fuel into a suitable container.

4. Disconnect the fuel supply tube-to-fuel filter outlet quick connect coupling.

5. Release the fuel filter clamp and remove the fuel filter.

6. To install, reverse the removal procedure.

FUEL PUMP MODULE

REMOVAL & INSTALLATION
See Figure 192.

✷✷ WARNING

Do not smoke, or carry lighted tobacco or have an open flame of any type when working on or near any fuel-related component. Highly flammable mixtures are always present and may be ignited. Failure to follow these instructions may result in serious personal injury.

✷✷ WARNING

Do not carry personal electronic devices such as cell phones, pagers or audio equipment of any type when working on or near any fuel related component. Highly flammable mixtures are always present and may be ignited. Failure to follow these instructions may result in serious personal injury.

✷✷ WARNING

When handling fuel, always observe fuel handling precautions and be prepared in the event of fuel spillage. Spilled fuel may be ignited by hot vehicle components or other ignition sources. Failure to follow these instructions may result in serious personal injury.

✷✷ WARNING

Always disconnect the battery ground cable at the battery when working on an Evaporative Emission (EVAP) system or fuel-related component.

Highly flammable mixtures are always present and may be ignited. Failure to follow these instructions may result in serious personal injury.

1. Release the fuel system pressure.
2. Disconnect the battery ground cable.
3. Remove the Fuel Pump (FP) module access cover.

➡Clean the FP module connections, couplings, flange surfaces and the immediate surrounding area of any dirt or foreign material.

4. Disconnect the FP module electrical connector.

➡Place absorbent toweling in the immediate surrounding area in case of fuel spillage.

5. Disconnect the fuel tube-to-FP module quick connect coupling.

6. Disconnect the fuel tank filler pipe recirculation tube-to-FP module quick connect coupling.

➡For initial draining on vehicles with a full tank of fuel, it is required to remove the fuel through the fuel tank filler pipe recirculation tube port on the FP

7. Connect the Fuel Storage Tanker hose to the fuel tank filler pipe recirculation tube port on the FP module and remove approximately one-fourth of the fuel from a completely full tank (approximately 4 gallons), lowering the fuel level below the FP module mounting flange.

✷✷ CAUTION

Carefully install the Fuel Tank Sender Unit Wrench to avoid damaging the Fuel Pump (FP) module when removing the lock ring.

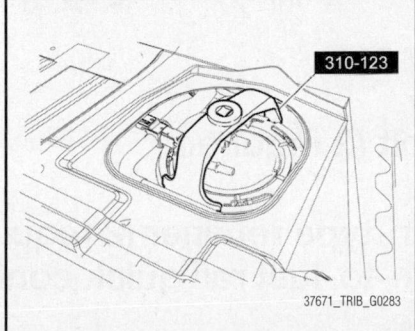

310-123

37671_TRIB_G0283

Fig. 192 Install the Fuel Tank Sender Unit Wrench and remove the FP module lock ring

➡When installing the FP module, install a new O-ring seal.

8. Install the Fuel Tank Sender Unit Wrench and remove the FP module lock ring.

✷✷ CAUTION

The Fuel Pump (FP) module must be handled carefully to avoid damage to the float arm.

➡Some residual fuel may remain in the FP module. Carefully drain into a suitable container.

9. Completely remove the FP module from the fuel tank.

➡Inspect the surfaces of the FP module flange and fuel tank seal contact surfaces. Do not polish or adjust the seal contact area of the FP module flange or fuel tank. Install a new FP module or fuel tank if the seal contact area is bent, scratched or corroded.

10. Remove and discard the FP module O-ring seal.

➡Make sure to install a new FP module O-ring seal.

To install:
11. To install, reverse the removal procedure.

a. Make sure the alignment tabs on the FP module and the fuel tank meet prior to tightening the FP module lock ring.

b. Tighten the FP module lock ring until it meets the stop tabs on the fuel tank.

FUEL RAIL AND INJECTOR

REMOVAL & INSTALLATION

2.5L Engine
See Figure 193.

✷✷ CAUTION

Do not carry personal electronic devices such as cell phones, pagers or audio equipment of any type when working on or near any fuel-related component. Highly flammable mixtures are always present and may be ignited. Failure to follow these instructions may result in serious personal injury.

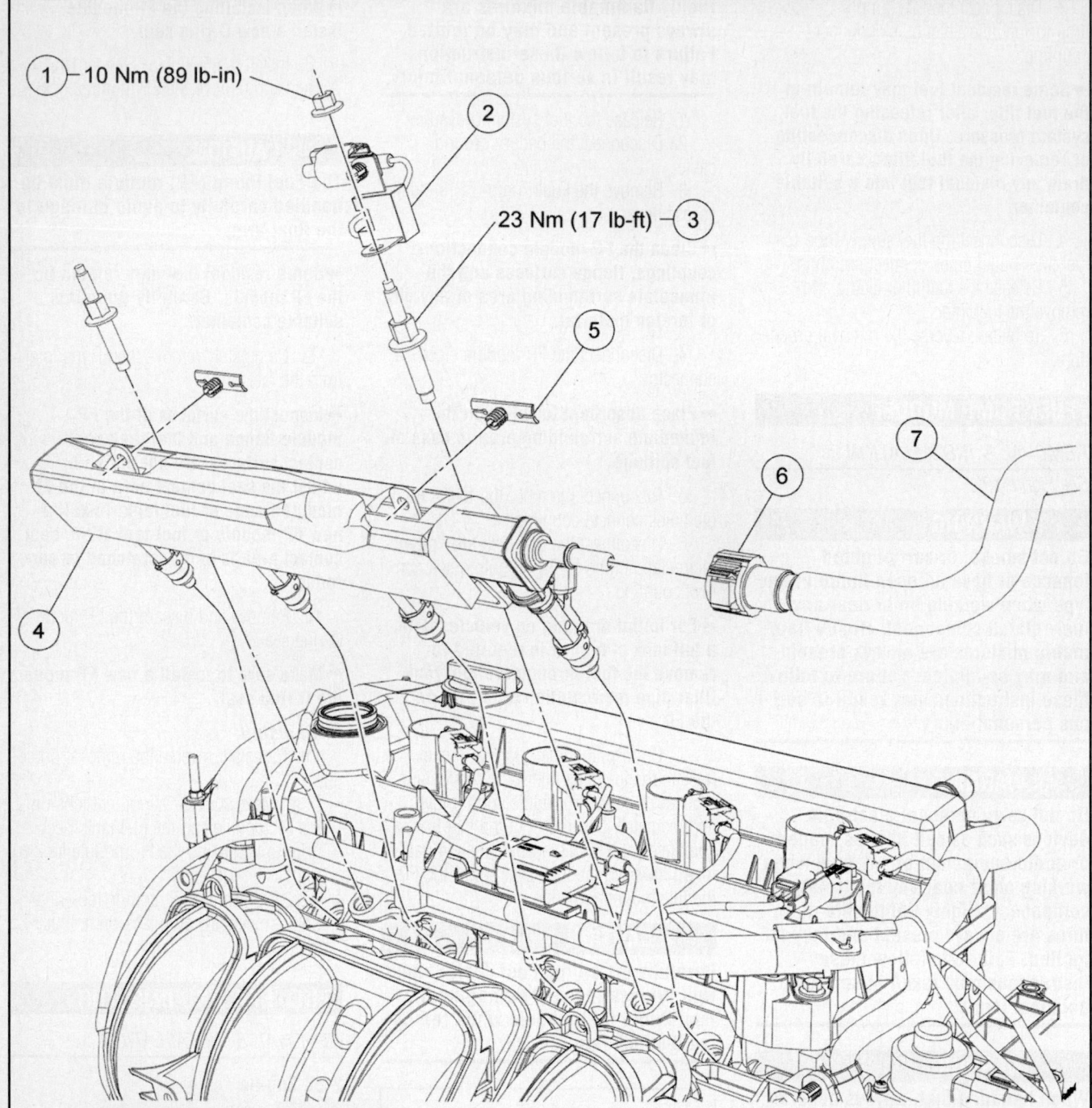

1 — 10 Nm (89 lb-in)

23 Nm (17 lb-ft) — 3

1. Radio capacitor nut
2. Radio capacitor
3. Fuel rail stud bolt (2 required)
4. Fuel rail
5. Wire harness pin-type retainer (2 required)
6. Fuel supply tube-to-fuel rail quick connect coupling
7. Fuel supply tube

36578_ESCA_G0308

Fig. 193 Fuel rail—2.5L engine

※※ **CAUTION**

Do not smoke or carry lighted tobacco or open flame of any type when working on or near any fuel-related components. Highly flammable mixtures are always present and may be ignited. Failure to follow these instructions may result in personal injury.

※※ **CAUTION**

Fuel in the fuel system remains under high pressure even when the engine is not running. Before working on or disconnecting any of the fuel lines or fuel system components, the fuel system pressure must be relieved. Failure to follow these instructions may result in personal injury.

1. Release the fuel pressure.
2. Disconnect the battery ground cable.
3. Disconnect the fuel supply tube-to-fuel quick connect coupling.
4. Disconnect the 4 fuel injector electrical connectors.
5. Remove the nut and position the radio capacitor aside.
6. Detach the 2 pin-type wire harness retainers from the fuel rail.
7. Remove the 2 fuel rail stud bolts.
8. Remove the fuel rail and injectors as an assembly.
9. Remove the 4 fuel injector retainer clips and the 4 fuel injectors.
10. Remove and discard the 8 fuel injector O-ring seals.

To install:

➡️Use O-ring seals that are made of special fuel-resistant material. Use of ordinary O-rings can cause the fuel system to leak. Do not reuse the O-ring seals.

11. Install 8 new fuel injector O-ring seals and lubricate them with clean engine oil.
12. Install the 4 fuel injectors and the 4 retainer clips on the fuel rail.
13. Install the fuel rail and injectors as an assembly.
14. Install the 2 fuel rail stud bolts and tighten to 17 ft lbs. (23 Nm).
15. Attach the 2 pin-type wire harness retainers to the fuel rail.
16. Position the radio capacitor and install and tighten the nut to 89 inch lbs. (10 Nm).
17. Connect the 4 fuel injector electrical connectors.

18. Connect the fuel supply tube-to-fuel rail quick connect coupling.
19. Connect the battery ground cable.

3.0L Engine

※※ **WARNING**

Do not smoke, carry lighted tobacco or have an open flame of any type when working on or near any fuel-related component. Highly flammable mixtures are always present and may be ignited. Failure to follow these instructions may result in serious personal injury.

※※ **WARNING**

Before working on or disconnecting any of the fuel tubes or fuel system components, relieve the fuel system pressure to prevent accidental spraying of fuel. Fuel in the fuel system remains under high pressure, even when the engine is not running. Failure to follow this instruction may result in serious personal injury.

※※ **WARNING**

Clean all fuel residue from the engine compartment. If not removed, fuel residue may ignite when the engine is returned to operation. Failure to follow this instruction may result in serious personal injury.

1. Release the fuel system pressure.
2. Disconnect the battery ground cable.
3. Remove the upper intake manifold.
4. Disconnect the fuel supply tube quick connect coupling at the fuel rail.
5. Release the 4 wire harness pin-type retainers from the fuel rail.
6. Release the engine wire harness electrical connector pin-type retainer from the fuel rail.
7. Disconnect the 6 fuel injector electrical connectors.
8. Remove the 4 fuel rail bolts.
9. Remove the fuel rail and injectors as an assembly.
10. Remove the 6 clips and the fuel injectors.
11. Remove and discard the 12 fuel injector O-ring seals.

To install:

12. Install the 6 fuel injectors and clips on the fuel rail.

※※ **CAUTION**

Use O-ring seals that are made of special fuel-resistant material. Use of ordinary O-rings can cause the fuel system to leak. Do not reuse the O-ring seals.

➡️The upper and lower fuel injector O-ring seals are similar in appearance, but are not interchangeable.

➡️Install new fuel injector O-ring seals and lubricate them with clean engine oil.

13. Position the fuel rail and install the bolts. Tighten to 89 inch lbs. (10 Nm).
14. Connect the engine wire harness electrical connector pin-type retainers to the fuel rail.
15. Connect the 4 wire harness pin-type retainers to the fuel rail.
16. Connect the fuel supply tube quick connect coupling at the fuel rail.
17. Connect the 6 fuel injector electrical connectors.
18. Install the upper intake manifold.
19. Connect the battery ground cable.

FUEL TANK

REMOVAL & INSTALLATION

See Figure 194.

※※ **WARNING**

Do not smoke, or carry lighted tobacco or have an open flame of any type when working on or near any fuel-related component. Highly flammable mixtures are always present and may be ignited. Failure to follow these instructions may result in serious personal injury.

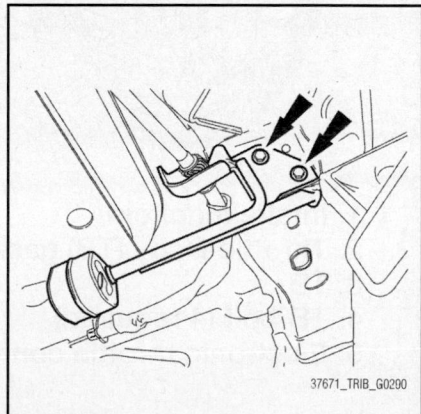

37671_TRIB_G0290

Fig. 194 Remove the 2 bolts and position the RR exhaust hanger aside

✳✳ WARNING

Do not carry personal electronic devices such as cell phones, pagers or audio equipment of any type when working on or near any fuel related component. Highly flammable mixtures are always present and may be ignited. Failure to follow these instructions may result in serious personal injury.

✳✳ WARNING

Before working on or disconnecting any of the fuel tubes or fuel system components, relieve the fuel system pressure to prevent accidental spraying of fuel. Fuel in the fuel system remains under high pressure, even when the engine is not running. Failure to follow this instruction may result in serious personal injury.

✳✳ WARNING

When handling fuel, always observe fuel handling precautions and be prepared in the event of fuel spillage. Spilled fuel may be ignited by hot vehicle components or other ignition sources. Failure to follow these instructions may result in serious personal injury.

✳✳ WARNING

Always disconnect the battery ground cable at the battery when working on an evaporative emission (EVAP) system or fuel-related component. Highly flammable mixtures are always present and may be ignited. Failure to follow these instructions may result in serious personal injury.

All Vehicles

1. With the vehicle in NEUTRAL, position it on a hoist.
2. Drain the fuel tank.
3. Remove the exhaust muffler and resonator.

AWD Vehicles

4. Remove the rear driveshaft.

All Vehicles

5. Release the clamp and remove the fuel tank filler pipe hose from the fuel tank.
6. Position a suitable lifting device under the fuel tank.

7. Detach the parking brake cable pin-type retainer from the LH fuel tank strap.
8. Detach the wire harness and fresh air tube pin-type retainers from the rear of LH fuel tank strap.
9. Remove the 2 bolts and position the 2 fuel tank straps aside. To install, tighten to 41 ft. lbs. (55 Nm).
10. Remove the 2 bolts and position the RR exhaust hanger aside. To install, tighten to 17 ft. lbs. (23 Nm).
11. Partially lower the fuel tank enough to disconnect the fuel vapor tube assembly-to-fuel tank quick connect coupling.
12. Completely lower and remove the fuel tank from the vehicle.
13. To install, reverse the removal procedure.

THROTTLE BODY

REMOVAL & INSTALLATION

2.5L Engine

See Figure 195.

1. Remove the Air Cleaner (ACL) outlet pipe.
2. Disconnect the electrical throttle control electrical connector.

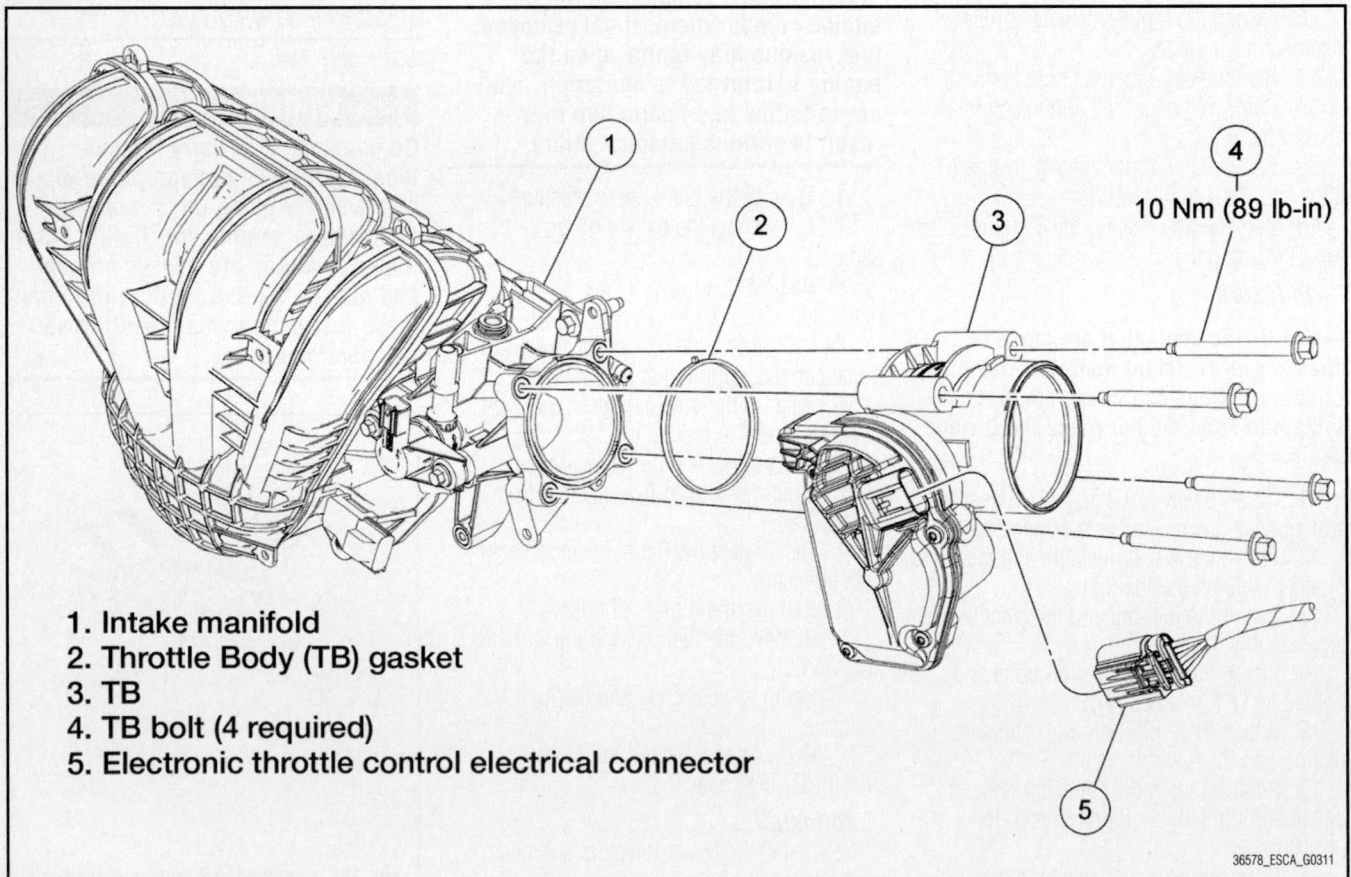

10 Nm (89 lb-in)

1. Intake manifold
2. Throttle Body (TB) gasket
3. TB
4. TB bolt (4 required)
5. Electronic throttle control electrical connector

36578_ESCA_G0311

Fig. 195 Throttle Body (TB)—2.5L engine

3. Remove the 4 bolts and the Throttle Body (TB).

4. Discard the TB gasket.

5. Installation is the reverse of the removal procedure, noting the following:

- Install a new gasket
- Tighten the throttle body retainers to 89 inch lbs. (10 Nm).

3.0L Engine

See Figure 196.

1. Disconnect the battery ground cable.

2. Remove the Air Cleaner (ACL) outlet pipe.

3. Disconnect the electronic throttle control electrical connector.

4. Remove the 4 bolts and the Throttle Body (TB).

5. Discard the TB gasket.

6. Installation is the reverse of the removal procedure, noting the following:

- Install a new gasket
- Tighten the throttle body retainers to 89 inch lbs. (10 Nm).

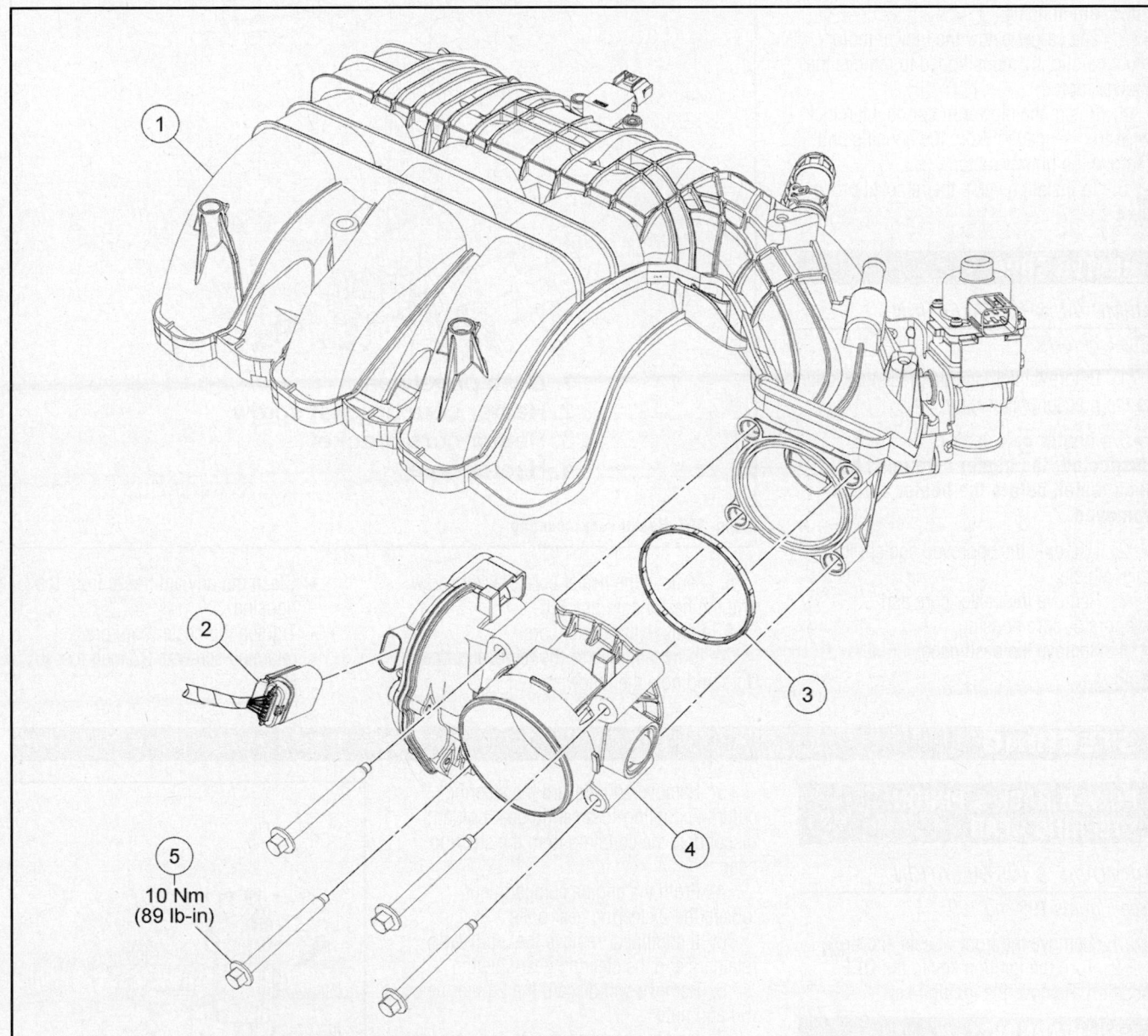

1. **Intake manifold**
2. **Electronic throttle control electrical connector**
3. **Throttle Body (TB) gasket**
4. **TB**
5. **TB bolt (4 required)**

5 — 10 Nm (89 lb-in)

36578_ESCA_G0312

Fig. 196 Throttle Body (TB)—3.0L engine

HEATING & AIR CONDITIONING COMPONENTS

BLOWER MOTOR

REMOVAL & INSTALLATION

1. Disconnect the blower motor electrical connector.

2. Release the 2 blower motor vent tube clips and pull the vent tube down until it is disengaged from the heater core and evaporator core housing.

3. The carpet below the blower motor must be slightly repositioned to remove the blower motor.

4. Rotate the blower motor counterclockwise to disengage it from the housing and remove the blower motor.

5. To install, reverse the removal procedure

HEATER CORE

REMOVAL & INSTALLATION

See Figure 197.

1. Before beginning this procedure, refer to the precautions section.

➡**If a heater core leak is suspected, the heater core must be leak tested before the heater core is removed.**

2. Use only the approved coolant for this vehicle.

3. Remove the heater core and evaporator core housing.

4. Remove the dash panel seal.

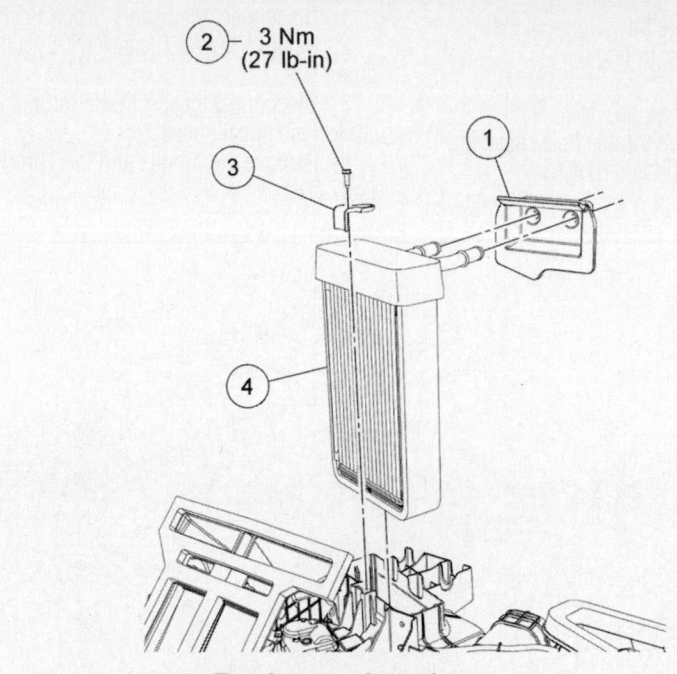

2. 3 Nm
(27 lb-in)

1. Dash panel seal
2. Heater core bracket screw
3. Heater core bracket
4. Heater core

36578_ESCA_G0321

Fig. 197 Heater core mounting

5. Remove the heater core bracket screw and the heater core bracket.

6. Remove the heater core.

7. To install, reverse the removal procedure and note the following:

• Clean out any antifreeze from the housing.

• Tighten the heater core bracket retaining screw to 27 inch lbs. (3 Nm).

STEERING

POWER RACK & PINION STEERING GEAR

REMOVAL & INSTALLATION

See Figures 198 and 199.

1. Remove the front wheels and tires.

2. Turn the ignition key to the **OFF** position. Remove the ignition key.

❊❊ WARNING

Do not allow the steering wheel to rotate while the intermediate shaft is disconnected or damage to the clockspring can result. If there is evidence that the shaft has rotated, the clockspring must be removed and re-centered

3. Remove and discard the steering column coupling-to-steering gear bolt and disconnect the coupling from the steering gear.

4. From the engine compartment, loosen the 2 steering gear bolts.

5. If equipped, remove the 3 pin-type retainers and the steering gear shield.

6. Remove and discard the 2 outer tie-rod end nuts.

7. Do not use a hammer to separate the tie-rod end from the wheel knuckle or damage to the wheel knuckle can result.

8. Using a suitable tool, separate the tie-rod ends from the wheel knuckles.

9. For AWD vehicles, remove the rear transaxle insulator through bolt.

10. Remove and discard the 2 steering gear bolts.

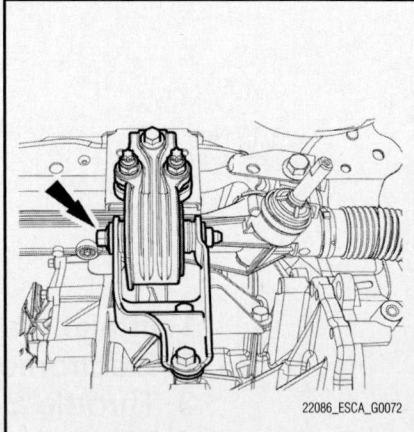

22086_ESCA_G0072

Fig. 198 Rear transaxle insulator and through bolt

➡**For All Wheel Drive (AWD) vehicles, it is necessary to grasp the driveshaft by hand and apply slight downward pressure to obtain clearance for the removal of the steering gear.**

11. Remove the steering gear from the LH side of the vehicle.

To install:

12. To install, reverse the removal procedure and note the following:

   a. Install a new steering column coupling-to-steering gear bolt and tighten to 41 ft. lbs. (55 Nm).

   b. Install and tighten the 2 steering gear bolts to 85 ft. lbs. (115 Nm).

   c. Install new outer tie-rod end nuts and tighten to 59 ft. lbs. (80 Nm).

   d. Tighten rear transaxle insulator through bolt to 66 ft. lbs. (90 Nm).

   e. Tighten transaxle damper bolts to 30 ft. lbs. (40 Nm).

13. Check and, if necessary, align the front end.

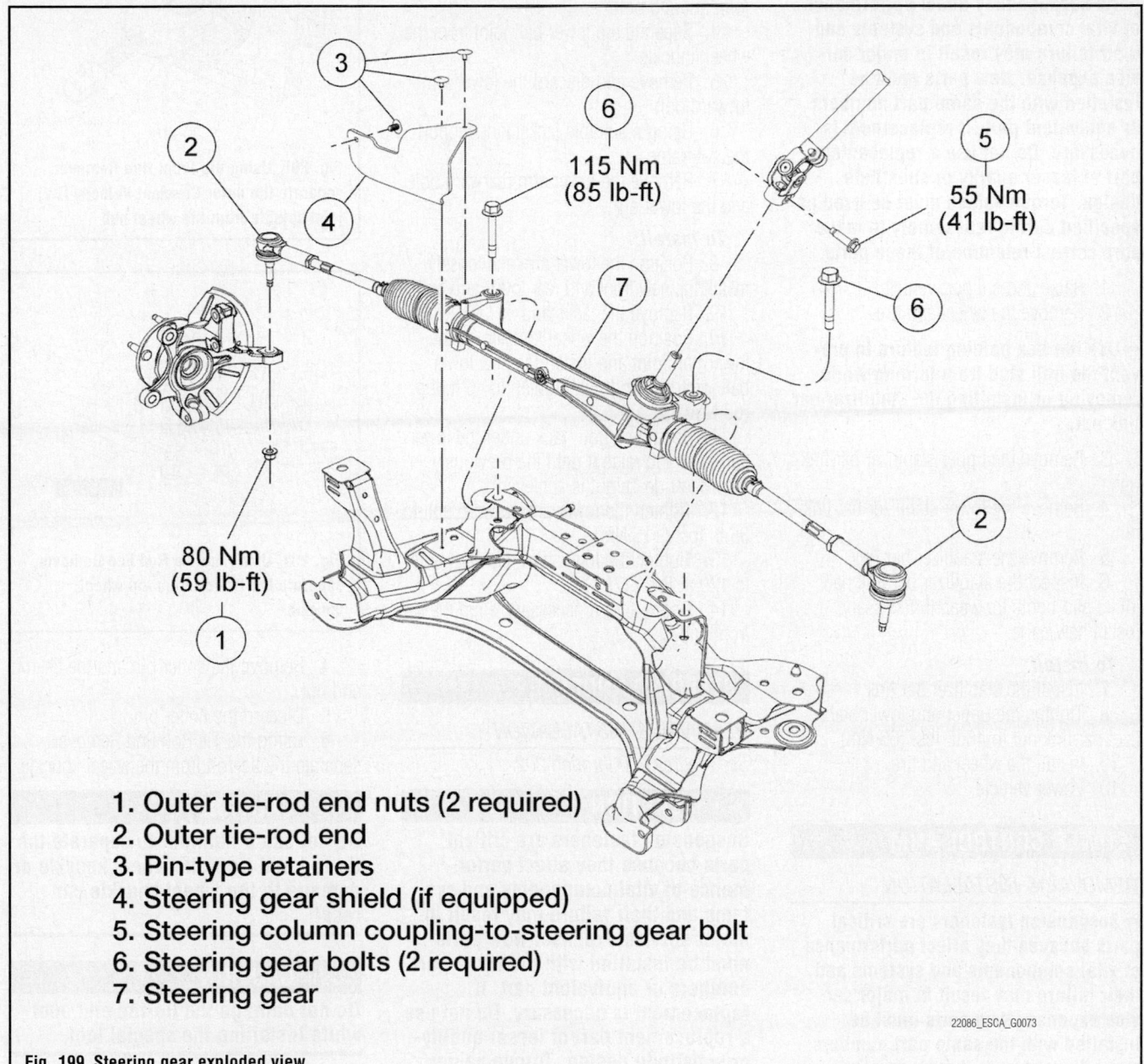

1. Outer tie-rod end nuts (2 required)
2. Outer tie-rod end
3. Pin-type retainers
4. Steering gear shield (if equipped)
5. Steering column coupling-to-steering gear bolt
6. Steering gear bolts (2 required)
7. Steering gear

22086_ESCA_G0073

Fig. 199 Steering gear exploded view

CONTROL LINKS

REMOVAL & INSTALLATION

Stabilizer Link

➡Suspension fasteners are critical parts because they affect performance of vital components and systems and their failure may result in major service expense. New parts must be installed with the same part numbers or equivalent part, if replacement is necessary. Do not use a replacement part of lesser quality or substitute design. Torque values must be used as specified during reassembly to make sure correct retention of these parts.

1. Raise and support vehicle
2. Remove the wheel and tire.

➡Use the hex holding feature to prevent the ball stud from turning while removing or installing the stabilizer bar link nut.

3. Remove the upper stabilizer bar link nut.
4. Remove the lower stabilizer bar link nut.
5. Remove the stabilizer bar link.
6. Inspect the stabilizer bar link ball joints and boots for wear. If necessary, install new parts.

To install:

7. Install the stabilizer bar link.
8. Tighten the upper and lower stabilizer bar link nut to 46 ft. lbs. (63 Nm).
9. Install the wheel and tire.
10. Lower vehicle

LOWER CONTROL ARM

REMOVAL & INSTALLATION

➡Suspension fasteners are critical parts because they affect performance of vital components and systems and their failure may result in major service expense. New parts must be installed with the same part numbers or equivalent part, if replacement is necessary. Do not use a replacement part of lesser quality or substitute design. Torque values must be used as specified during reassembly to make sure correct retention of these parts.

1. Record the ride height.

➡For reference during the installation of the lower arm, measure the distance between the center of the wheel hub and the lip of the fender with the weight of the vehicle resting on the wheel and tire assemblies.

2. Remove the wheel and tire.
3. Remove and discard the lower ball joint nut and bolt.
4. Separate the lower ball joint from the wheel knuckle.
5. Remove and discard the lower arm forward bolt.
6. Using a suitable jackstand, support the subframe.
7. Remove the lower arm rearward bolt and the lower arm.

To install:

8. Position the lower arm and loosely install the new front and rear lower arm bolts.
9. Remove the jackstand.
10. Position the wheel knuckle on the lower ball joint and install the new lower ball joint bolt and nut. Tighten to 46 ft. lbs. (63 Nm).
11. Position a floor jack under the lower ball joint and raise it until the previously recorded ride height is achieved.
12. Tighten the lower arm forward bolt to 85 ft. lbs. (115 Nm).
13. Tighten the lower arm rearward bolt to 129 ft. lbs. (175 Nm).
14. Check and, if necessary, align the front end.

STEERING KNUCKLE

REMOVAL & INSTALLATION

See Figures 200 through 202.

✳✳ CAUTION

Suspension fasteners are critical parts because they affect performance of vital components and systems and their failure may result in major service expense. New parts must be installed with the same part numbers or equivalent part, if replacement is necessary. Do not use a replacement part of lesser quality or substitute design. Torque values must be used as specified during reassembly to make sure correct retention of these parts.

1. Remove the brake disc.
2. Remove and discard the wheel hub nut.
3. Using the Front Hub Remover, separate the outer Constant Velocity (CV) joint spindle from the wheel hub.

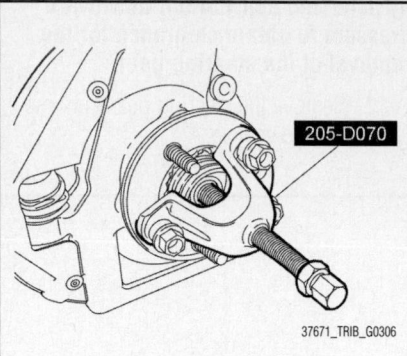

Fig. 200 Using the Front Hub Remover, separate the outer Constant Velocity (CV) joint spindle from the wheel hub

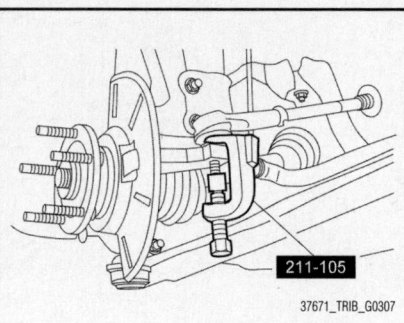

Fig. 201 Using the Tie Rod End Remover, separate the tie-rod from the wheel knuckle

4. Remove the cotter pin and the tie-rod end nut.
5. Discard the cotter pin.
6. Using the Tie Rod End Remover, separate the tie-rod from the wheel knuckle.

✳✳ CAUTION

Do not use a hammer to separate the tie-rod end from the wheel knuckle or damage to the wheel knuckle can result.

✳✳ CAUTION

Do not damage the tie-rod end boot while installing the special tool.

7. Remove and discard the lower ball joint bolt and nut.
8. Remove the wheel speed sensor bolt and position the sensor aside.
9. Separate the lower ball joint from the wheel knuckle.
10. Remove the 2 strut-to-knuckle nuts, the bolts and the wheel knuckle.
11. Discard the nuts and bolts.

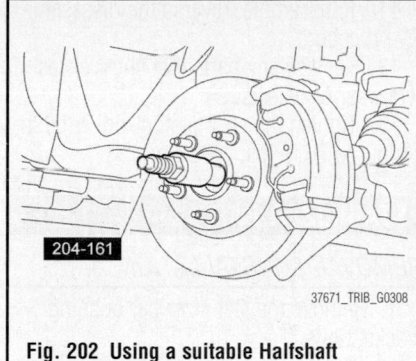

Fig. 202 Using a suitable Halfshaft Installer, insert the halfshaft into the wheel hub

To install:

12. Position the wheel knuckle and install the 2 new strut-to-knuckle bolts and nuts. Tighten to 85 ft. lbs. (115 Nm).

13. Position and align the ball joint stud into the wheel knuckle.

14. Install the new lower ball joint bolt and nut. Tighten to 46 ft. lbs. (63 Nm).

15. Install the wheel speed sensor and the bolt.

16. Position the tie rod-end into the wheel knuckle and install the new tie-rod end nut and a cotter pin. Tighten to 41 ft. lbs. (55 Nm).

17. Using a suitable Halfshaft Installer, insert the halfshaft into the wheel hub.

18. Install the brake disc.

19. Install the new front wheel hub nut. Tighten to 221 ft. lbs. (300 Nm).

✳✳ CAUTION

Do not tighten the front wheel hub nut with the vehicle on the ground. The nut must be tightened to specification before the vehicle is lowered onto the wheels. Wheel bearing damage will occur if the wheel bearing is loaded with the weight of the vehicle applied.

➡**Apply the brake to keep the halfshaft from rotating.**

20. Check and, if necessary, align the front end.

STRUT

REMOVAL & INSTALLATION

See Figure 203.

➡**Suspension fasteners are critical parts because they affect performance of vital components and systems and their failure may result in major service expense. New parts must be**

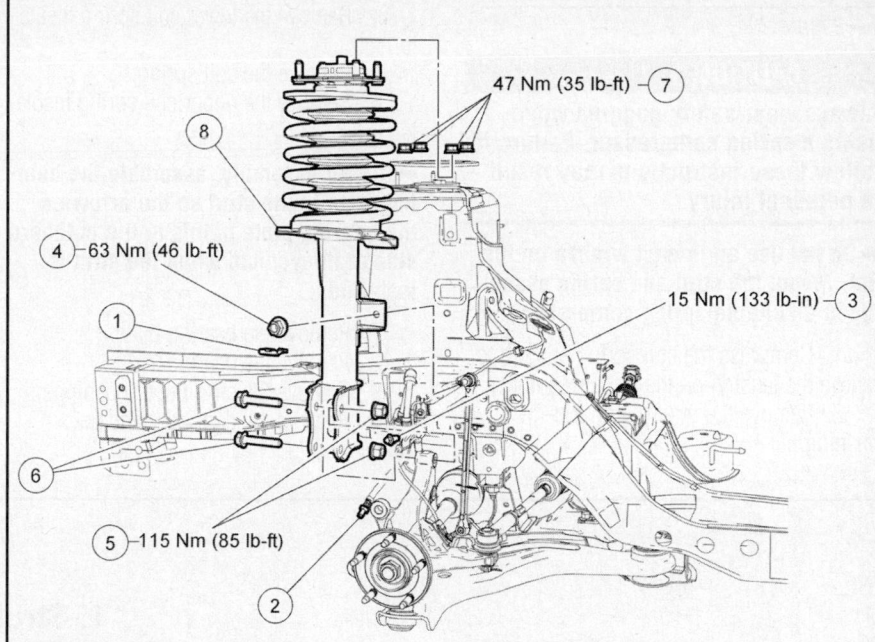

1. Brake jounce hose clip
2. Brake jounce hose (LH/RH)
3. Wheel speed sensor harness bolt
4. Upper stabilizer bar link nut
5. Strut-to-knuckle nuts (2 required)
6. Strut-to-knuckle bolts (2 required)
7. Strut upper bushing nuts (4 required)
8. Strut and spring assembly

Fig. 203 Strut and spring assembly

installed with the same part numbers or equivalent part, if replacement is necessary. Do not use a replacement part of lesser quality or substitute design. Torque values must be used as specified during reassembly to make sure correct retention of these parts.

1. Verify the steering wheel is in the unlocked position before removal.

2. Remove the wheel and tire.

3. Remove the brake jounce hose clip.

4. Pull the brake jounce hose downward slightly to remove the hose from the bracket and position the brake jounce hose aside.

5. Remove the wheel speed sensor harness bolt.

➡**Use the hex-holding feature to prevent the ball stud from turning while removing or installing the stabilizer bar link nut.**

6. Remove and discard the upper stabilizer bar link nut.

7. Remove and discard the 2 strut-to-knuckle nuts and bolts.

8. Reference mark the 4 strut upper bushing plate nuts.

9. Remove and discard the 4 strut upper bushing nuts.

➡**Do not allow the axle shaft to move outboard. Over-extension of the tripod Constant Velocity (CV) joint can result in the separation of internal parts, causing failure of the axle shaft.**

10. Remove the strut and spring assembly.

To install:

11. Position the strut and spring assembly upper mounting plate into the inner fender.

12. Align the 4 new strut upper bushing nuts to the reference marks and tighten to 35 ft. lbs. (47 Nm).

13. Install the 2 new strut-to-knuckle bolts and nuts. Tighten to 85 ft. lbs. (115 Nm).

14. Install the new upper stabilizer bar link nut and tighten to 46 ft. lbs. (63 Nm).

15. Install the wheel speed sensor harness bolt and tighten to 133 inch lbs. (15 Nm).

16. Position the brake jounce hose to the bracket and install the brake jounce hose clip.

17. Check the front end alignment and adjust as necessary.

OVERHAUL

See Figure 204.

⁂ CAUTION

Always wear safety goggles when using a spring compressor. Failure to follow these instructions may result in personal injury.

➡**Do not use an impact wrench on the nut. Mount the strut and spring assembly in a suitable spring compressor.**

1. Compress the coil spring enough to relieve the tension on the strut assembly.

2. Remove the strut piston rod-to-bushing nut.

3. Remove the strut.

4. Remove the lower coil spring insulator.

5. Remove the coil spring.

6. Remove the upper coil spring insulator.

➡**During assembly, assemble the bearing plate to the strut so the arrow on the bearing plate points to the outboard side of the vehicle when the strut is installed.**

7. Remove the bearing plate.

8. Remove the bearing.

9. Remove the strut upper bushing.

10. Remove the dust boot and the bumper.

11. To assemble, reverse the disassembly procedure.

12. Replace any parts that show excessive wear or defects.

13. Tighten the strut piston rod-to-bushing nut to 76 ft. lbs. (103 Nm).

STABILIZER BAR

REMOVAL & INSTALLATION

1. Remove the stabilizer bar bushing bracket bolts.

➡**Use the hex holding feature to prevent the ball stud from turning while removing or installing the stabilizer link nut.**

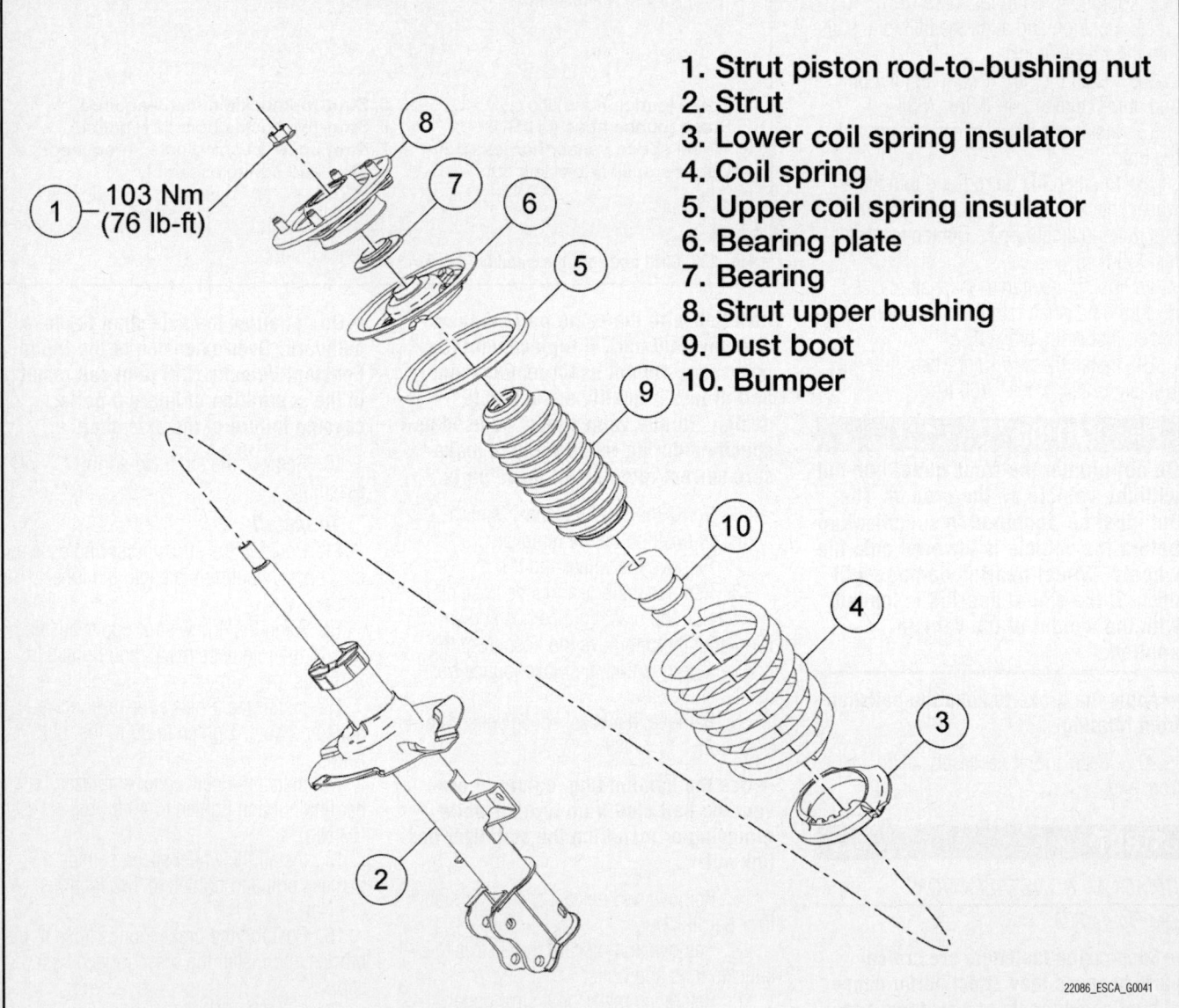

1. Strut piston rod-to-bushing nut
2. Strut
3. Lower coil spring insulator
4. Coil spring
5. Upper coil spring insulator
6. Bearing plate
7. Bearing
8. Strut upper bushing
9. Dust boot
10. Bumper

103 Nm (76 lb-ft)

22086_ESCA_G0041

Fig. 204 Front strut, spring and related parts

2. Remove the 2 lower stabilizer bar link nuts.

➡ **Access the stabilizer bar through the left wheel opening.**

3. Remove the stabilizer bar.
4. To install, reverse the removal procedure. Observe the following torques:
- Link nuts: 41 ft. lbs. (55 Nm)
- Bushing bracket bolts: 52 ft. lbs. (70 Nm)

WHEEL HUB & BEARING

REMOVAL & INSTALLATION

See Figures 205 through 209.

❋❋ CAUTION

Suspension fasteners are critical parts because they affect performance of vital components and systems and their failure may result in major service expense. New parts must be installed with the same part numbers or equivalent part, if replacement is necessary. Do not use a replacement part of lesser quality or substitute design. Torque values must be used as specified during reassembly to make sure correct retention of these parts.

➡ **If removing the wheel hub, a new wheel bearing must be installed.**

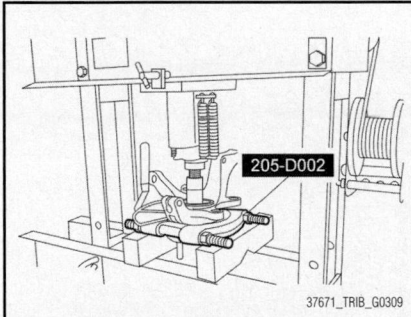

Fig. 205 Using the Pinion Bearing Cone Remover, press the wheel hub from the wheel bearing

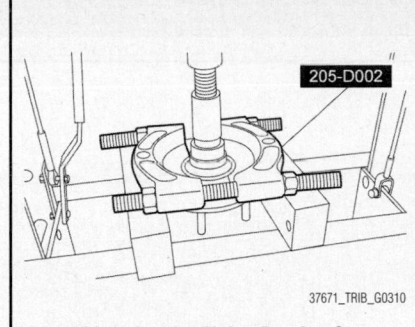

Fig. 206 Using the Pinion Bearing Cone Remover, press the inner wheel bearing race from the wheel hub

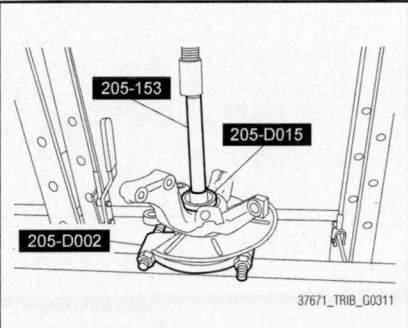

Fig. 207 Using the special tools, press the outer wheel bearing race from the wheel knuckle

1. Remove the wheel knuckle.
2. Using the Pinion Bearing Cone Remover, press the wheel hub from the wheel bearing.
3. Using the Pinion Bearing Cone Remover, press the inner wheel bearing race from the wheel hub.

➡ **This step may not be necessary if the inner wheel bearing race remains in the wheel knuckle after removing the wheel hub.**

4. Remove the snap ring.
5. Using the special tools, press the outer wheel bearing race from the wheel knuckle.

To install:

6. Position the wheel knuckle in a vise.

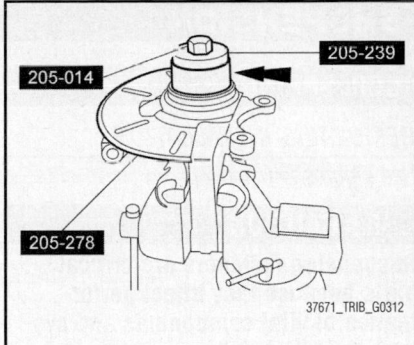

Fig. 208 Using the special tools, install the wheel bearing into the wheel knuckle

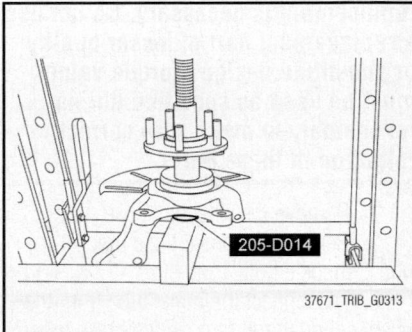

Fig. 209 Using the special tool, press the wheel hub into the wheel bearing

7. Using the special tools, install the wheel bearing into the wheel knuckle.

➡ **Special Tool 205-278 is not seen in place. It is located behind the wheel knuckle.**

8. Install the snap ring.
9. Using the special tool, press the wheel hub into the wheel bearing.
10. Install the wheel knuckle.

ADJUSTMENT

No adjustment is required or possible. If the tire and wheel (hub) is loose on the spindle, does not rotate freely, or has a rough feeling when spun, install a new wheel bearing.

COIL SPRING

REMOVAL & INSTALLATION

See Figures 210 and 211.

✳✳ CAUTION

Suspension fasteners are critical parts because they affect performance of vital components and systems and their failure may result in major service expense. New parts must be installed with the same part numbers or equivalent part, if replacement is necessary. Do not use a replacement part of lesser quality or substitute design. Torque values must be used as specified during reassembly to make sure correct retention of these parts.

1. Remove the wheel and tire.
2. Remove the brake hose bracket-to-wheel knuckle bolt.
3. Disconnect the brake tube from the wheel cylinder and remove the brake tube bracket.
4. Using a suitable jackstand, support the wheel knuckle.
5. Remove and discard the shock absorber lower nut, washer and bolt.
6. Remove the upper arm.
7. Loosen the lower arm inner bolt.
8. Using the jackstand, carefully lower the wheel knuckle.

➡Note the position of the coil spring insulator and coil spring for installation.

9. Remove the coil spring.

➡All Wheel Drive (AWD) vehicle shown, Front Wheel Drive (FWD) vehicle similar.

To install:

10. Align the coil spring and coil spring insulator to the previously noted position.
11. Using a suitable jackstand, carefully raise the wheel knuckle.
12. Install the new shock absorber lower bolt, washer and nut. Tighten to 121 ft. lbs. (175 Nm).
13. Install the upper arm.
14. Connect the brake tube fitting to the wheel cylinder. Tighten to 13 ft. lbs. (17 Nm).
15. Install the brake tube bracket bolt. Tighten to 16 ft. lbs. (22 Nm).

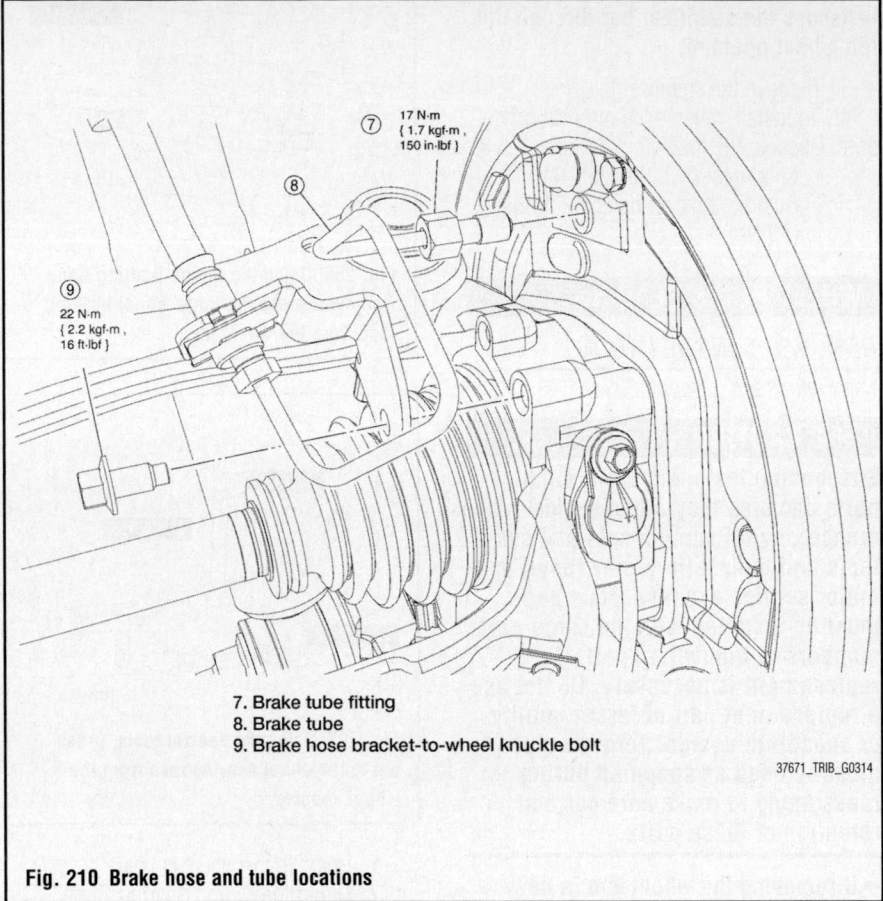

7. Brake tube fitting
8. Brake tube
9. Brake hose bracket-to-wheel knuckle bolt

37671_TRIB_G0314

Fig. 210 Brake hose and tube locations

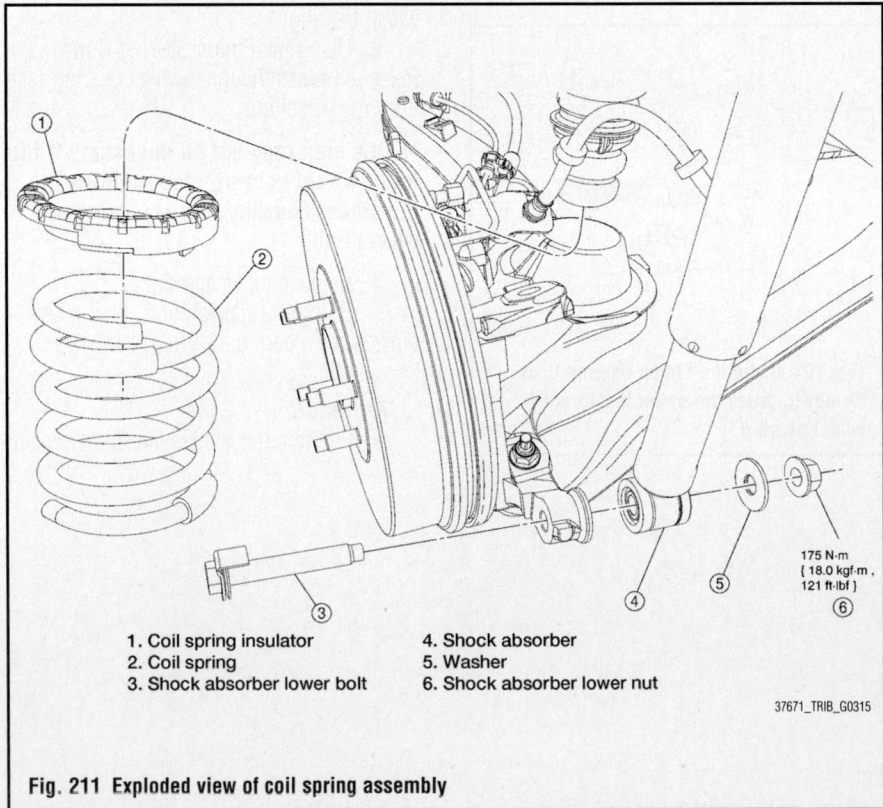

1. Coil spring insulator
2. Coil spring
3. Shock absorber lower bolt
4. Shock absorber
5. Washer
6. Shock absorber lower nut

37671_TRIB_G0315

Fig. 211 Exploded view of coil spring assembly

16. Install the brake hose bracket-to-wheel knuckle bolt. Tighten to 16 ft. lbs. (22 Nm).

17. Bleed the rear wheel cylinder.

CONTROL ARMS/LINKS

REMOVAL & INSTALLATION

Lower Arm

See Figure 212.

1. Remove the wheel and tire.

2. Remove and discard the lower ball joint nut and separate the lower ball joint from the wheel knuckle.

➡**Tighten the lower arm inner bolt with the weight of the vehicle on the wheels and tires.**

3. Remove the lower arm inner bolt and the lower arm.

To install:

4. To install, reverse the removal procedure and note the following:
- Ball joint nut: 46 ft. lbs. (63 Nm)
- Lower arm inner bolt: 111 ft. lbs. (150 Nm)

Upper Arm

1. Remove the wheel and tire..

2. Remove the upper ball joint nut.

3. Remove the upper arm inner bolt and the upper arm.

4. Discard the bolt.

➡**Tighten the upper arm inner bolt with the weight of the vehicle on the wheels and tires.**

To install:

5. To install, reverse the removal procedure.

6. Tighten the upper arm inner bolt to 85 ft. lbs. (115 Nm).

7. Tighten the upper ball joint nut to 46 ft. (63 Nm).

SHOCK ABSORBER

REMOVAL & INSTALLATION

1. Remove the rear quarter trim panel.

2. Remove the wheel and tire.

3. Using a suitable jackstand, support the rear suspension.

4. Remove and discard the upper shock absorber nuts, washer and bushing.

5. Remove and discard the lower shock absorber nut, washer and bolt.

6. Remove the shock absorber and the bushing.

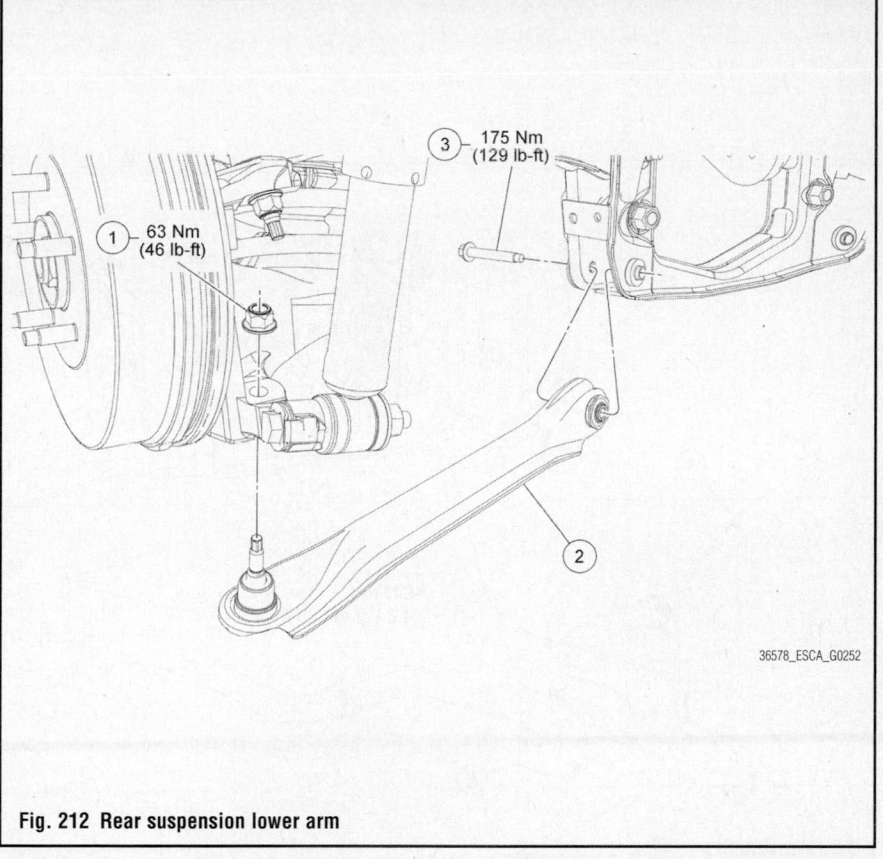

Fig. 212 Rear suspension lower arm

To install:

7. To install, reverse the removal procedure.

8. Tighten the lower shock absorber nut to 129 ft. lbs. (175 Nm).

9. Tighten the upper shock absorber nut to 30 ft. lbs. (40 Nm).

STABILIZER BAR

REMOVAL & INSTALLATION

See Figure 213.

✳✳ CAUTION

Suspension fasteners are critical parts because they affect performance of vital components and systems and their failure may result in major service expense. New parts must be installed with the same part numbers or equivalent part, if replacement is necessary. Do not use a replacement part of lesser quality or substitute design. Torque values must be used as specified during reassembly to make sure correct retention of these parts.

1. With the vehicle in NEUTRAL, position it on a hoist.

2. Remove and discard the stabilizer bar link upper nuts.

➡**Use the hex holding feature to prevent the stud from turning while removing the nut.**

3. Remove and discard the stabilizer bar link lower nut(s) and shield(s).

4. Remove the stabilizer bar links.

➡**Use the hex holding feature to prevent the stud from turning while removing the nut.**

5. Remove and discard the stabilizer bar bracket upper and lower bolts.

6. Remove the stabilizer bar brackets and the stabilizer bar.

To install:

7. To install, reverse the removal procedure.

➡**Inspect and, if necessary, install new stabilizer bar bushings.**

8. Tighten the new stabilizer bar bracket upper and lower bolts to 85 ft. lbs. (115 Nm).

9. Tighten the new stabilizer bar link lower nuts to 68 ft. lbs. (92 Nm).

10. Tighten the new stabilizer link upper nuts to 35 ft. lbs. (48 Nm).

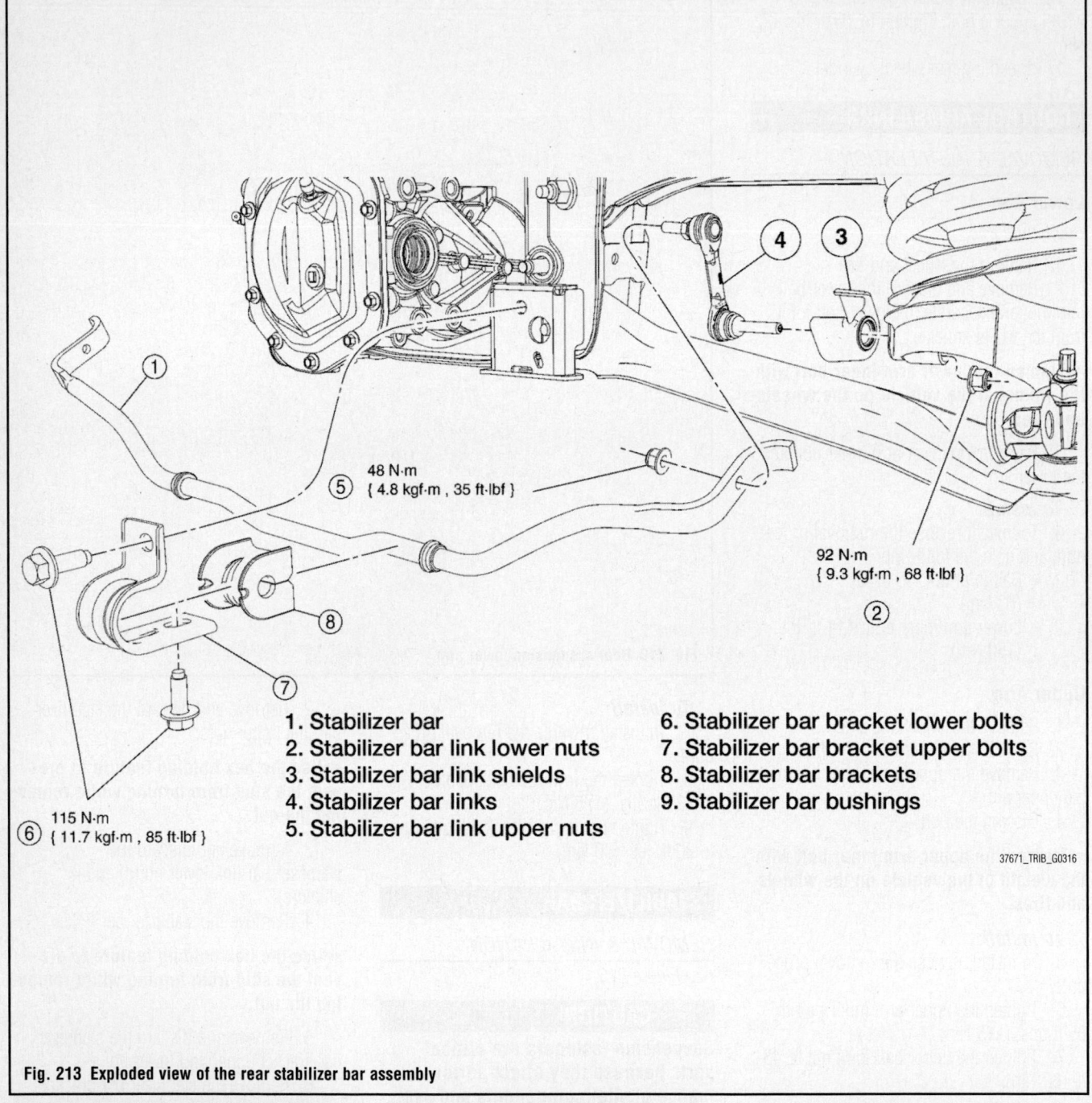

1. Stabilizer bar
2. Stabilizer bar link lower nuts
3. Stabilizer bar link shields
4. Stabilizer bar links
5. Stabilizer bar link upper nuts
6. Stabilizer bar bracket lower bolts
7. Stabilizer bar bracket upper bolts
8. Stabilizer bar brackets
9. Stabilizer bar bushings

48 N·m
{ 4.8 kgf·m , 35 ft·lbf }

92 N·m
{ 9.3 kgf·m , 68 ft·lbf }

115 N·m
{ 11.7 kgf·m , 85 ft·lbf }

37671_TRIB_G0316

Fig. 213 Exploded view of the rear stabilizer bar assembly

WHEEL HUB & BEARING

REMOVAL & INSTALLATION

Front Wheel Drive (FWD) Models
See Figures 214 through 219.

✷✷ CAUTION

Suspension fasteners are critical parts because they affect performance of vital components and systems and their failure may result in major service expense. New parts must be installed with the same part numbers or equivalent part, if replacement is necessary. Do not use a replacement part of lesser quality or substitute design. Torque values must be used as specified during reassembly to make sure correct retention of these parts.

1. Remove the wheel and tire.
2. Remove and discard the wheel hub nut.

➡**Apply the brake to keep the halfshaft from rotating.**

3. Remove the wheel speed sensor ring.
4. Remove the brake drum.
5. Using the special tools, remove the wheel hub.
6. Using a suitable press and the special tool or equivalent, press the inner bearing race from the wheel knuckle.

➡**This step may not be necessary if the inner wheel bearing race remains in the wheel knuckle after removing the wheel hub.**

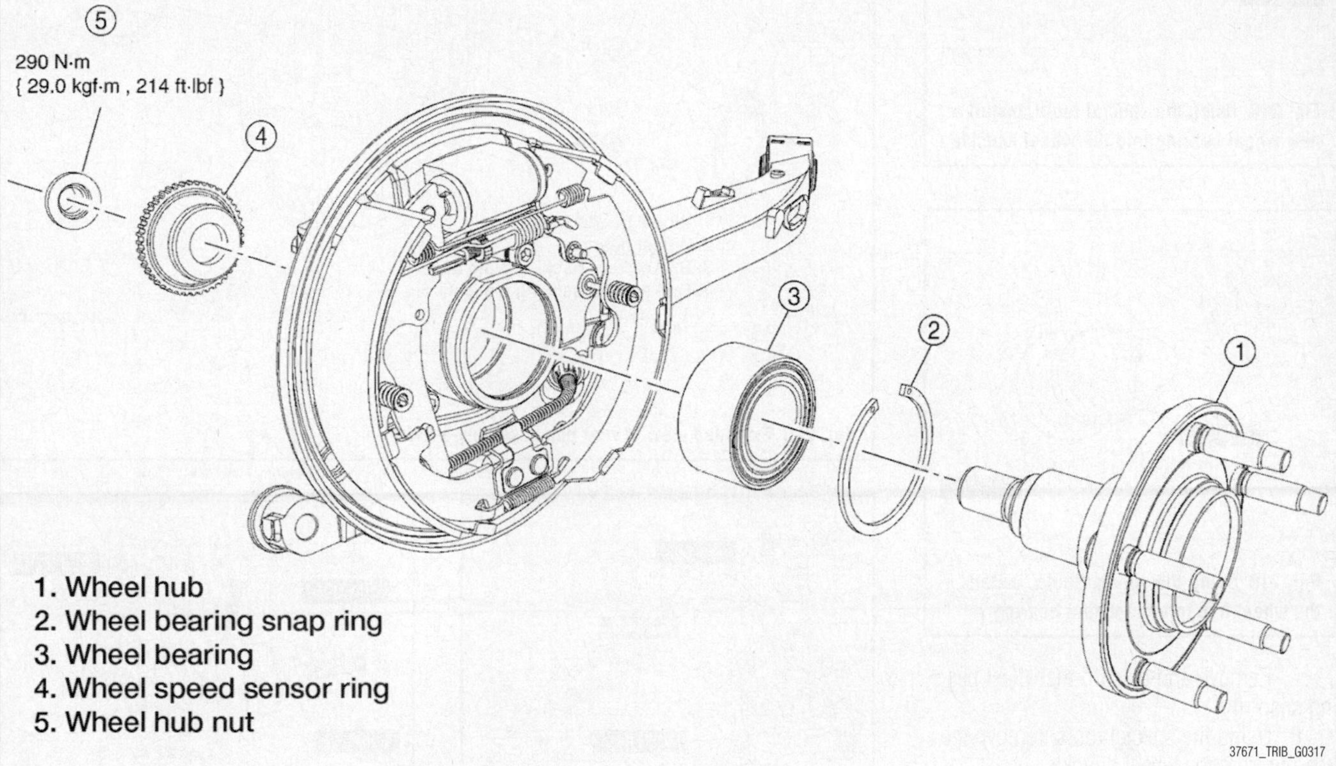

5
290 N·m
{ 29.0 kgf·m , 214 ft·lbf }
4
3
2
1
1. Wheel hub
2. Wheel bearing snap ring
3. Wheel bearing
4. Wheel speed sensor ring
5. Wheel hub nut
37671_TRIB_G0317

Fig. 214 Exploded view of rear hub assembly—FWD

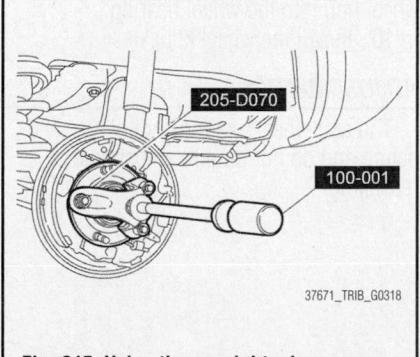

205-D070
100-001
37671_TRIB_G0318

Fig. 215 Using the special tools, remove the wheel hub

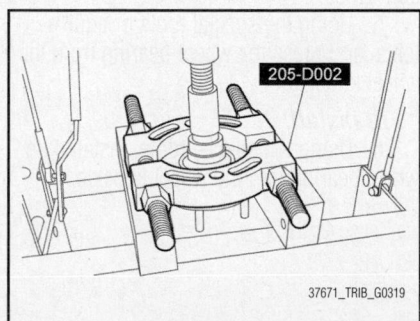

205-D002
37671_TRIB_G0319

Fig. 216 Using a suitable press and the special tool or equivalent, press the inner bearing race from the wheel knuckle

205-193
100-001
37671_TRIB_G0320

Fig. 217 Using the special tools, remove the bearing from the wheel knuckle

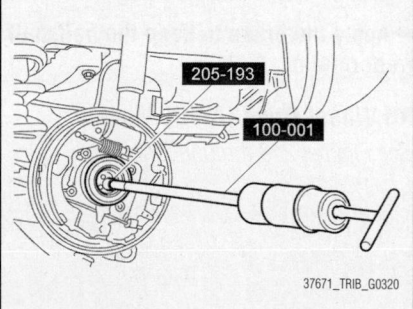

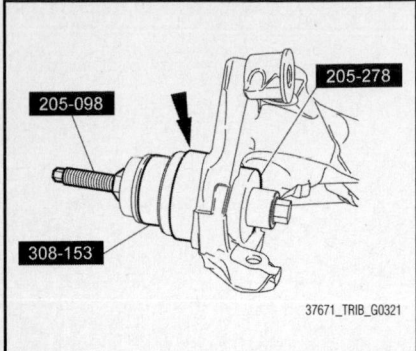

Fig. 218 Using the special tools, install a new wheel bearing into the wheel knuckle

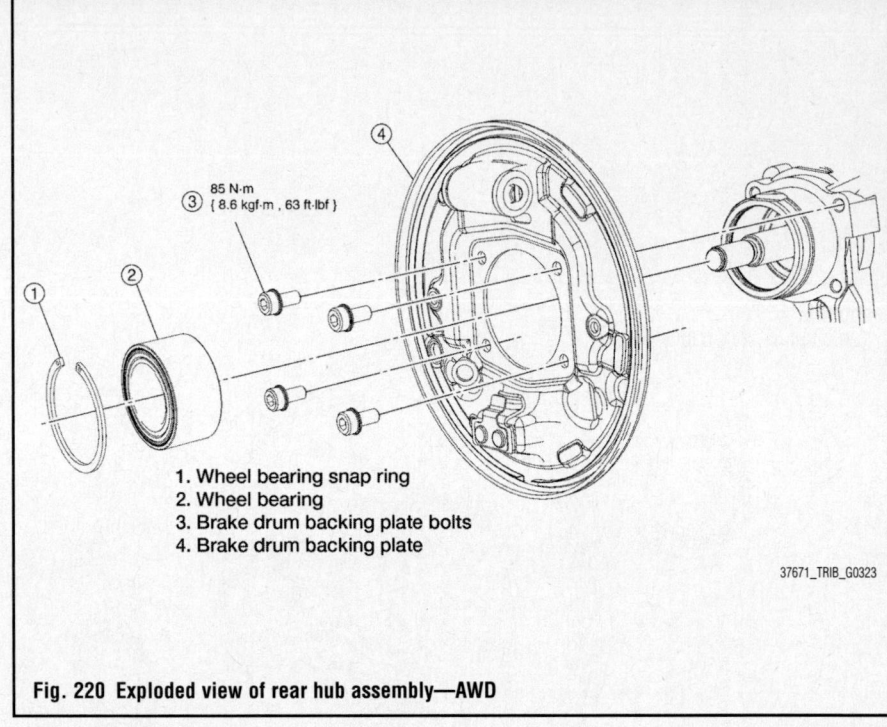

1. Wheel bearing snap ring
2. Wheel bearing
3. Brake drum backing plate bolts
4. Brake drum backing plate

Fig. 220 Exploded view of rear hub assembly—AWD

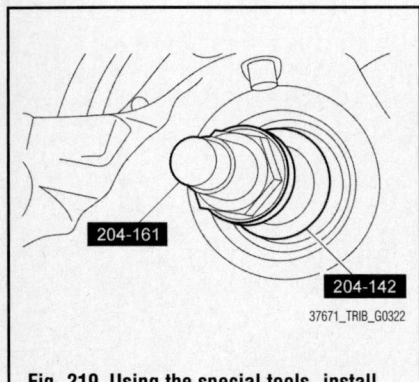

Fig. 219 Using the special tools, install the wheel hub into the wheel bearing

7. Remove and discard the wheel bearing snap ring.

8. Using the special tools, remove the bearing from the wheel knuckle.

To install:

9. Using the special tools, install a new wheel bearing into the wheel knuckle.

10. Install the new wheel bearing snap ring.

11. Using the special tools, install the wheel hub into the wheel bearing.

12. Install the wheel speed sensor ring.

13. Install the brake drum.

14. Install the new wheel hub nut. Tighten to 214 ft. lbs. (290 Nm).

➡**Apply the brake to keep the halfshaft from rotating.**

All Wheel Drive (AWD) Models

See Figures 219 through 222.

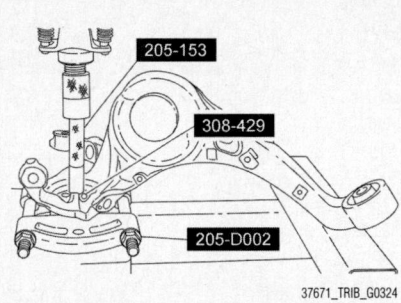

Fig. 221 Using the special tools or equivalents, and press the wheel bearing from the wheel knuckle

1. Remove the wheel knuckle.
2. Remove the wheel bearing snap ring.
3. Remove the 4 bolts and the brake drum backing plate.
4. Position the wheel knuckle on a suitable press.
5. Using the special tools or equivalents, and press the wheel bearing from the wheel knuckle.

To install:

6. Using the special tools, install a new wheel bearing into the wheel knuckle.

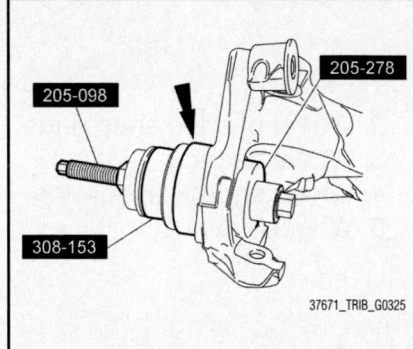

Fig. 222 Using the special tools, install a new wheel bearing into the wheel knuckle

7. Install the new wheel bearing snap ring.

8. Install the 4 bolts and the brake drum backing plate. Tighten to 63 ft. lbs. (85 Nm).

9. Using the special tools, install the wheel hub into the wheel bearing.

10. Install the wheel knuckle.

ADJUSTMENT

The rear wheel bearings are of a sealed nature and do not require adjustment or repacking.

DIAGNOSTIC TROUBLE CODES

OBD II VEHICLE APPLICATIONS

MAZDA

B-Series Truck
2009
- 2.3L I4 ....VIN D
- 4.0L V6.....................VIN E

CX-7
2009-2010
- 2.3L I4 ....VIN L3
  2.5L I4 ....VIN L5

CX-9
2009-2010
- 3.7L V6....................VIN V
- 3.7L V6....................VIN A

Mazda3 & MazdaSpeed3
2009-2010
- 2.0L I4 ....VIN F (Fed.)

- 2.0L I4 .... VIN G (Fed.)
- 2.3L I4 ....VIN 4 (CA)
- 2.3L I4 .... VIN 3 (Fed)
- 2.3L I4 .... VIN M (Turbo - CA)
- 2.3L I4 .... VIN L (Turbo – Fed.)
- 2.5L I4 ....VIN 5 (Fed.)
- 2.3L I4 ....VIN 6 (CA)

Mazda5
2009-2010
- 2.3L I4 ....VIN L (CA)
- 2.3L I4 .... VIN 3 (North America)

Mazda6
2009-2010
- 2.5L I4 ....VIN A
- 2.5L I4 ....VIN H
- 3.7L V6.....................VIN B

MX-5
2009-2010
- 2.0L I4 ....VIN F

RX-8
2009-2010
- 1.3L I4 ....VIN N (Std Power)
- 1.3L I4 .... VIN 3 (High Power)

Tribute
2009-2010
- 2.5L I4 ....VIN B
- 3.0L V6....................VIN 1

OBD II Trouble Code List (P0XXX Codes)

DTC	Trouble Code Title, Conditions & Possible Causes
DTC: P0010 **T PCM** **Year:** 2009, 2010 **Model:** 6, CX-9 **Engine:** 3.7L V6 VIN A, 3.7L V6 VIN B, 3.7L V6 VIN V **Transmission:** All	**Variable valve timing actuator circuit open (RH):** The PCM monitors the OCV (RH) circuit to the PCM for high and low voltage. The test fails if the voltage exceeds or falls below a calibrated limit for a calibrated amount of time. **Possible Causes:** • Connector or terminal malfunction • OCV (RH) malfunction • Open circuit between OCV (RH) terminal A and main relay terminal A • Short to power supply between OCV (RH) terminal B and PCM terminal 2BP • Short to ground circuit between OCV (RH) terminal B and PCM terminal 2BP • Open circuit between OCV (RH) terminal B and PCM terminal 2BP
DTC: P0011 **1T CCM, MIL: Yes** **Year:** 2009, 2010 **Model:** 3, 5, 6, CX-7, CX-9, MX-5 Miata **Engine:** 2.0L L4 VIN F, 2.0L L4 VIN G, 2.3L L4 VIN 3, 2.3L L4 VIN 4, 2.3L L4 VIN L, 2.3L L4 VIN M, 2.5L L4 VIN 5, 2.5L L4 VIN 6, 2.5L L4 VIN A, 2.5L L4 VIN M, 3.7L V6 VIN A, 3.7L V6 VIN B, 3.7L V6 VIN V **Transmission:** All	**Camshaft Position Timing Over-Advanced:** Engine running, and the PCM detected the Actual valve timing was more than 10 degrees advanced for 1 second from the Target valve timing with the engine running at maximum valve timing retard. **Possible Causes:** • OCV valve is damaged or has failed • OCV spool valve is stuck in the advanced position • Variable valve timing actuator is stuck in the retard position • Oil runners between oil pressure switch and the OCV, or between the OCV and the VVT actuator are dirty or clogged
DTC: P0012 **1T CCM, MIL: Yes** **Year:** 2009, 2010 **Model:** 3, 5, 6, CX-7, CX-9, MX-5 Miata **Engine:** 2.0L L4 VIN F, 2.0L L4 VIN G, 2.3L L4 VIN 3, 2.3L L4 VIN 4, 2.3L L4 VIN L, 2.3L L4 VIN M, 2.5L L4 VIN 5, 2.5L L4 VIN 6, 2.5L L4 VIN A, 2.5L L4 VIN M, 3.7L V6 VIN A, 3.7L V6 VIN B, 3.7L V6 VIN V **Transmission:** All	**Camshaft Position Timing Over-Retarded:** Engine running, and the PCM detected the Actual valve timing was more than 15 degrees retarded for 5 seconds from the Target valve timing with the Oil Control Valve system within feedback range. **Possible Causes:** • OCV valve is damaged or has failed • Low engine oil pressure condition • OCV spool valve is stuck in the retard position • Variable valve timing actuator is stuck in the advanced position • Timing belt is loose or improper valve timing due to a loose belt • PCM has failed
DTC: P0016 **2T CCM, MIL: Yes** **Year:** 2009, 2010 **Model:** 3, 5, 6, CX-7, CX-9, MX-5 Miata **Engine:** 2.0L L4 VIN F, 2.0L L4 VIN G, 2.3L L4 VIN 3, 2.3L L4 VIN 4, 2.3L L4 VIN L, 2.3L L4 VIN M, 2.5L L4 VIN 5, 2.5L L4 VIN 6, 2.5L L4 VIN A, 2.5L L4 VIN M, 3.7L V6 VIN A, 3.7L V6 VIN B, 3.7L V6 VIN V **Transmission:** All	**CKP-CPM correlation:** The PCM monitors the input pulses from the CKP sensor and CMP sensor. If the input pulse pick-up timing does not match each other, the PCM determines that the camshaft position does not coincide with the crankshaft position. **Possible Causes:** • Poor connection of connector • CMP sensor malfunction • CKP sensor malfunction • Damaged or foreign material on CKP or CMP sensor • Improper valve timing
DTC: P0018 **T** **Year:** 2009, 2010 **Model:** 6, CX-9 **Engine:** 3.7L V6 VIN A, 3.7L V6 VIN B, 3.7L V6 VIN V **Transmission:** All	**CKP-CMP sensor (LH) correlation:** The PCM monitors the variable valve timing position for a misalignment between the camshaft and crankshaft. The test fails when the misalignment is more than 1 tooth. **Possible Causes:** • OCV malfunction • Camshaft advanced mechanism binding • Improper valve timing

DTC	Trouble Code Title, Conditions & Possible Causes

DTC: P0020
T
Year: 2009, 2010
Model: 6, CX-9
Engine: 3.7L V6 VIN A, 3.7L V6 VIN B, 3.7L V6 VIN V
Transmission: All
	CMP actuator circuit open (LH):
	The PCM monitors the OCV (LH) circuit to the PCM for high and low voltage. The test fails if the voltage exceeds or falls below a calibrated limit for a calibrated amount of time.
	Possible Causes:
	  • Connector or terminal malfunction
	  • OCV (LH) malfunction
	  • Open circuit between OCV (LH) terminal A and main relay terminal A
	  • Short to power supply between OCV (LH) terminal B and PCM terminal 2BK
	  • Short to ground circuit between OCV (LH) terminal B and PCM terminal 2BK
	  • Open circuit between OCV (LH) terminal B and PCM terminal 2BK

DTC: P0021
T
Year: 2009, 2010
Model: 6, CX-9
Engine: 3.7L V6 VIN A, 3.7L V6 VIN B, 3.7L V6 VIN V
Transmission: All
	CMP timing over-advanced (LH):
	The PCM monitors the variable valve timing position for an over-advanced camshaft timing. The test fails when the camshaft timing exceeds a maximum calibrated value or remains in an advanced position.
	Possible Causes:
	  • Spool valve in OCV (LH) is stuck in advanced position
	  • Improper valve timing due to timing chain slippage
	  • OCV (LH) malfunction

DTC: P0022
T
Year: 2009, 2010
Model: 6, CX-9
Engine: 3.7L V6 VIN A, 3.7L V6 VIN B, 3.7L V6 VIN V
Transmission: All
	P timing over-retarded (LH):
	The PCM monitors the variable valve timing position for over-retarded camshaft timing. The test fails when the camshaft timing exceeds a maximum calibrated value or remains in a retarded position.
	Possible Causes:
	  • Low engine oil pressure
	  • Spool valve in the OCV (LH) is stuck in retard position
	  • Improper valve timing due to timing chain slippage
	  • OCV (LH) malfunction

DTC: P0030
2T CCM, MIL: Yes
Year: 2009, 2010
Model: 5, 6, B2300, B4000, CX-7, MX-5 Miata, RX-8
Engine: 1.3L R2 VIN 2, 1.3L R2 VIN 4, 1.3L R2 VIN M, 1.3L R2 VIN P, 2.0L L4 VIN F, 2.3L L4 VIN 3, 2.3L L4 VIN D, 2.3L L4 VIN L, 2.5L L4 VIN 5, 2.5L L4 VIN A, 2.5L L4 VIN M, 3.7L V6 VIN B, 4.0L V6 VIN E
Transmission: All
	HO2S Heater control circuit (Bank1, Sensor 1):
	The PCM monitors the front HO2S impedance when under the front HO2S heater control for 200 s. If the impedance is more than 44 ohms, the PCM determines that there is a front HO2S heater control circuit problem.
	Possible Causes:
	  • Front HO2S heater malfunction
	  • Connector or terminal damage
	  • PCM has failed

DTC: P0031
2T CCM, MIL: Yes
Year: 2009, 2010
Model: 3, 5, 6, B2300, B4000, CX-7, MX-5 Miata, RX-8
Engine: 1.3L R2 VIN 2, 1.3L R2 VIN 4, 1.3L R2 VIN M, 1.3L R2 VIN P, 2.0L L4 VIN F, 2.0L L4 VIN G, 2.3L L4 VIN 3, 2.3L L4 VIN 4, 2.3L L4 VIN D, 2.3L L4 VIN L, 2.3L L4 VIN M, 2.5L L4 VIN 5, 2.5L L4 VIN 6, 2.5L L4 VIN A, 2.5L L4 VIN M, 4.0L V6 VIN E
Transmission: All
	HO2S-11 (Bank 1 Sensor 1) Heater Circuit Low Input:
	Engine running, and after the front HO2S Heater Control duty cycle signal was commanded "off", the PCM detected a low input condition on the heater control circuit.
	Possible Causes:
	  • HO2S heater control circuit is shorted to chassis ground
	  • HO2S heater is damaged or has failed
	  • PCM has failed

DTC	Trouble Code Title, Conditions & Possible Causes
DTC: P0032 **2T CCM, MIL: Yes** **Year:** 2009, 2010 **Model:** 3, 5, 6, B2300, B4000, CX-7, MX-5 Miata, RX-8 **Engine:** 1.3L R2 VIN 2, 1.3L R2 VIN 4, 1.3L R2 VIN M, 1.3L R2 VIN P, 2.0L L4 VIN F, 2.0L L4 VIN G, 2.3L L4 VIN 3, 2.3L L4 VIN 4, 2.3L L4 VIN D, 2.3L L4 VIN L, 2.3L L4 VIN M, 2.5L L4 VIN 5, 2.5L L4 VIN 6, 2.5L L4 VIN A, 2.5L L4 VIN M, 4.0L V6 VIN E **Transmission:** All	**HO2S-11 (Bank 1 Sensor 1) Heater Circuit High Input:** Engine running, and after the front HO2S Heater Control duty cycle signal was commanded "on", the PCM detected a high input condition on the heater control circuit. **Possible Causes:** • HO2S heater control circuit is shorted to system power (B+) • HO2S heater is damaged or has failed • PCM has failed
DTC: P0037 **2T CCM, MIL: Yes** **Year:** 2009, 2010 **Model:** 3, 5, 6, B2300, B4000, CX-7, MX-5 Miata, RX-8, Tribute **Engine:** 1.3L R2 VIN 2, 1.3L R2 VIN 4, 1.3L R2 VIN M, 1.3L R2 VIN P, 2.0L L4 VIN F, 2.0L L4 VIN G, 2.3L L4 VIN 3, 2.3L L4 VIN 4, 2.3L L4 VIN D, 2.3L L4 VIN L, 2.3L L4 VIN M, 2.5L L4 VIN 3, 2.5L L4 VIN 5, 2.5L L4 VIN 6, 2.5L L4 VIN 7, 2.5L L4 VIN A, 2.5L L4 VIN M, 3.0L V6 VIN 1, 3.0L V6 VIN G, 4.0L V6 VIN E **Transmission:** All	**HO2S-12 (Bank 1 Sensor 2) Heater Circuit Low Input:** Engine running, and after the rear HO2S Heater Control duty cycle signal was commanded "off", the PCM detected a low input condition on the heater control circuit. **Possible Causes:** • HO2S heater control circuit is shorted to chassis ground • HO2S heater is damaged or has failed • PCM has failed
DTC: P0038 **2T CCM, MIL: Yes** **Year:** 2009, 2010 **Model:** 3, 5, 6, B2300, B4000, CX-7, MX-5 Miata, RX-8, Tribute **Engine:** 1.3L R2 VIN 2, 1.3L R2 VIN 4, 1.3L R2 VIN M, 1.3L R2 VIN P, 2.0L L4 VIN F, 2.0L L4 VIN G, 2.3L L4 VIN 3, 2.3L L4 VIN 4, 2.3L L4 VIN D, 2.3L L4 VIN L, 2.3L L4 VIN M, 2.5L L4 VIN 3, 2.5L L4 VIN 5, 2.5L L4 VIN 6, 2.5L L4 VIN 7, 2.5L L4 VIN A, 2.5L L4 VIN M, 3.0L V6 VIN 1, 3.0L V6 VIN G, 4.0L V6 VIN E **Transmission:** All	**HO2S-12 (Bank 1 Sensor 2) Heater Circuit High Input:** Engine running, and after the rear HO2S Heater Control duty cycle signal was commanded "on", the PCM detected a high input condition on the heater control circuit. **Possible Causes:** • HO2S heater control circuit is shorted to system power (B+) • HO2S heater is damaged or has failed • PCM has failed
DTC: P0040 **2T CCM** **Year:** 2009, 2010 **Model:** 6, CX-9, Tribute **Engine:** 2.5L L4 VIN 3, 2.5L L4 VIN 7, 3.0L V6 VIN 1, 3.0L V6 VIN G, 3.7L V6 VIN A, 3.7L V6 VIN B, 3.7L V6 VIN V **Transmission:** All	**HO2S-12 (Bank 1 Sensor 2) Heater Circuit High Input:** Engine running, and after the rear HO2S Heater Control duty cycle signal was commanded "on", the PCM detected a high input condition on the heater control circuit. **Possible Causes:** • HO2S heater control circuit is shorted to system power (B+) • HO2S heater is damaged or has failed • PCM has failed

DTC	Trouble Code Title, Conditions & Possible Causes
DTC: P0041 **2T CCM, MIL: Yes** **Year:** 2009, 2010 **Model:** 6, CX-9, Tribute **Engine:** 2.5L L4 VIN 3, 2.5L L4 VIN 7, 3.0L V6 VIN 1, 3.0L V6 VIN G, 3.7L V6 VIN A, 3.7L V6 VIN B, 3.7L V6 VIN V **Transmission:** All	**HO2S-12 (Bank 1 Sensor 2) Heater Circuit High Input:** Engine running, and after the rear HO2S Heater Control duty cycle signal was commanded "on", the PCM detected a high input condition on the heater control circuit. **Possible Causes:** • HO2S heater control circuit is shorted to system power (B+) • HO2S heater is damaged or has failed • PCM has failed
DTC: P0050 **T** **Year:** 2009, 2010 **Model:** 6 **Engine:** 3.7L V6 VIN B **Transmission:** All	**A/F sensor (LH) heater control circuit:** The PCM monitors the heater in the air fuel ratio (A/F) sensor (RH/LH) for correct operation. The PCM controls the heater on/off duty cycle to maintain a temperature of 780°C (1,436°F). The test fails when the sensor does not warm up to the required temperature in the calibrated amount of time. The test also fails when the PCM is not able to maintain the required temperature after the sensor is warm. **Possible Causes:** • Short to power supply in wiring harness between PCM and A/F sensor • Open circuit in wiring harness between PCM and A/F sensor • A/F sensor loose connection, and damaged or corroded terminals • Exhaust temperature significantly higher than expected • Damaged A/F sensor • PCM malfunction
DTC: P0051 **2T CCM, MIL: Yes** **Year:** 2009 **Model:** B2300, B4000 **Engine:** 2.3L L4 VIN D, 4.0L V6 VIN E **Transmission:** All	**HO2S-11 (Bank 1 Sensor 1) Heater Circuit Low Input:** Engine running, and after the front HO2S Heater Control duty cycle signal was commanded "off", the PCM detected a low input condition on the heater control circuit. **Possible Causes:** • HO2S heater control circuit is shorted to chassis ground • HO2S heater is damaged or has failed • PCM has failed
DTC: P0052 **2T CCM, MIL: Yes** **Year:** 2009 **Model:** B4000 **Engine:** 4.0L V6 VIN E **Transmission:** All	**HO2S-21 (Bank 2 Sensor 1) Heater Circuit High Input:** Engine running, and after the front HO2S Heater Control duty cycle signal was commanded "on", the PCM detected a high input condition on the heater control circuit. **Possible Causes:** • HO2S heater control circuit is shorted to system power (B+) • HO2S heater is damaged or has failed • PCM has failed
DTC: P0053 **2T CCM, MIL: Yes** **Year:** 2009, 2010 **Model:** 3, 5, 6, B2300, B4000, CX-7, CX-9, MX-5 Miata, RX-8, Tribute **Engine:** 1.3L R2 VIN 2, 1.3L R2 VIN 4, 1.3L R2 VIN M, 1.3L R2 VIN P, 2.0L L4 VIN F, 2.0L L4 VIN G, 2.3L L4 VIN 3, 2.3L L4 VIN 4, 2.3L L4 VIN D, 2.3L L4 VIN L, 2.3L L4 VIN M, 2.5L L4 VIN 3, 2.5L L4 VIN 5, 2.5L L4 VIN 6, 2.5L L4 VIN 7, 2.5L L4 VIN A, 2.5L L4 VIN M, 3.0L V6 VIN 1, 3.0L V6 VIN G, 3.7L V6 VIN A, 3.7L V6 VIN B, 3.7L V6 VIN V, 4.0L V6 VIN E **Transmission:** All	**HO2S-12 (Bank 1 Sensor 2) Heater Circuit High Input:** Engine running, and after the rear HO2S Heater Control duty cycle signal was commanded "on", the PCM detected a high input condition on the heater control circuit. **Possible Causes:** • HO2S heater control circuit is shorted to system power (B+) • HO2S heater is damaged or has failed • PCM has failed

DTC	Trouble Code Title, Conditions & Possible Causes
DTC: P0054 **2T CCM, MIL: Yes** **Year:** 2009, 2010 **Model:** 3, 5, 6, B2300, B4000, CX-7, CX-9, MX-5 Miata, RX-8, Tribute **Engine:** 1.3L R2 VIN 2, 1.3L R2 VIN 4, 1.3L R2 VIN M, 1.3L R2 VIN P, 2.0L L4 VIN F, 2.0L L4 VIN G, 2.3L L4 VIN 3, 2.3L L4 VIN 4, 2.3L L4 VIN D, 2.3L L4 VIN L, 2.3L L4 VIN M, 2.5L L4 VIN 3, 2.5L L4 VIN 5, 2.5L L4 VIN 6, 2.5L L4 VIN 7, 2.5L L4 VIN A, 2.5L L4 VIN M, 3.0L V6 VIN 1, 3.0L V6 VIN G, 3.7L V6 VIN A, 3.7L V6 VIN B, 3.7L V6 VIN V, 4.0L V6 VIN E **Transmission:** All	**HO2S-12 (Bank 1 Sensor 2) Heater Circuit High Input:** Engine running, and after the rear HO2S Heater Control duty cycle signal was commanded "on", the PCM detected a high input condition on the heater control circuit. **Possible Causes:** • HO2S heater control circuit is shorted to system power (B+) • HO2S heater is damaged or has failed • PCM has failed
DTC: P0059 **2T CCM, MIL: Yes** **Year:** 2009, 2010 **Model:** 6, B2300, B4000, CX-9, Tribute **Engine:** 2.3L L4 VIN D, 2.5L L4 VIN 3, 2.5L L4 VIN 7, 3.0L V6 VIN 1, 3.0L V6 VIN G, 3.7L V6 VIN A, 3.7L V6 VIN B, 3.7L V6 VIN V, 4.0L V6 VIN E **Transmission:** All	**HO2S-12 (Bank 1 Sensor 2) Heater Circuit High Input:** Engine running, and after the rear HO2S Heater Control duty cycle signal was commanded "on", the PCM detected a high input condition on the heater control circuit. **Possible Causes:** • HO2S heater control circuit is shorted to system power (B+) • HO2S heater is damaged or has failed • PCM has failed
DTC: P0060 **2T CCM, MIL: Yes** **Year:** 2009, 2010 **Model:** 6, B2300, B4000, CX-9, Tribute **Engine:** 2.3L L4 VIN D, 2.5L L4 VIN 3, 2.5L L4 VIN 7, 3.0L V6 VIN 1, 3.0L V6 VIN G, 3.7L V6 VIN A, 3.7L V6 VIN B, 3.7L V6 VIN V, 4.0L V6 VIN E **Transmission:** All	**HO2S-12 (Bank 1 Sensor 2) Heater Circuit High Input:** Engine running, and after the rear HO2S Heater Control duty cycle signal was commanded "on", the PCM detected a high input condition on the heater control circuit. **Possible Causes:** • HO2S heater control circuit is shorted to system power (B+) • HO2S heater is damaged or has failed • PCM has failed
DTC: P0068 **2T CCM, MIL: Yes** **Year:** 2009, 2010 **Model:** 6, B2300, B4000, CX-9, Tribute **Engine:** 2.3L L4 VIN D, 2.5L L4 VIN 3, 2.5L L4 VIN 7, 3.0L V6 VIN 1, 3.0L V6 VIN G, 3.7L V6 VIN A, 3.7L V6 VIN B, 3.7L V6 VIN V, 4.0L V6 VIN E **Transmission:** All	**PCM – Mass Air Flow (MAF/IAT) Sensor – Throttle Position (TP) Sensor:** TP signal went below 0.24 volts with a load greater than 55%, or TP signal went above 2.44 volts with a load less than 30%. **Possible Causes:** • TP Sensor has failed • PCM has failed
DTC: P0069 **2T CCM, MIL: Yes** **Year:** 2009, 2010 **Model:** 3, 5, 6, CX-7, MX-5 Miata **Engine:** 2.0L L4 VIN F, 2.0L L4 VIN G, 2.3L L4 VIN 3, 2.3L L4 VIN 4, 2.3L L4 VIN L, 2.3L L4 VIN M, 2.5L L4 VIN 5, 2.5L L4 VIN 6, 2.5L L4 VIN A, 2.5L L4 VIN M **Transmission:** All	**Manifold absolute pressure/atmospheric pressure correlation problem:** The PCM monitors differences between intake manifold vacuum and atmospheric pressure. If the difference is below -12 kPa {-90 mmHg, -3.5 in Hg} or above 12 kPa {90 mmHg, 3.5 in Hg} when the following conditions are met, the PCM determines that there is a MAP sensor performance problem. **Possible Causes:** • MAP sensor malfunction • BARO sensor (built-in PCM) malfunction • PCM malfunction

DTC	Trouble Code Title, Conditions & Possible Causes
DTC: P0076 **2T CCM** **Year:** 2009, 2010 **Model:** RX-8 **Engine:** 1.3L R2 VIN 2, 1.3L R2 VIN 4, 1.3L R2 VIN M, 1.3L R2 VIN P **Transmission:** All	**VDI solenoid valve control circuit low:** The PCM monitors the VDI solenoid valve control voltage when the PCM turns off the VDI solenoid valve. If the control voltage is low, the PCM determines that the VDI solenoid control circuit voltage is low. **Possible Causes:** • VDI solenoid valve malfunction • Connector or terminal malfunction • PCM has failed
DTC: P0077 **2T CCM** **Year:** 2009, 2010 **Model:** RX-8 **Engine:** 1.3L R2 VIN 2, 1.3L R2 VIN 4, 1.3L R2 VIN M, 1.3L R2 VIN P **Transmission:** All	**VDI solenoid valve control circuit high:** The PCM monitors the VDI solenoid valve control voltage when the PCM turns off the VDI solenoid valve. If the control voltage is high, the PCM determines that the VDI solenoid control circuit voltage is high. **Possible Causes:** • VDI solenoid valve malfunction • Connector or terminal malfunction • PCM has failed
DTC: P0089 **2T CCM, MIL: Yes** **Year:** 2009, 2010 **Model:** 3, CX-7 **Engine:** 2.3L L4 VIN 3, 2.3L L4 VIN 4, 2.3L L4 VIN L, 2.3L L4 VIN M **Transmission:** All	**Fuel pressure regulator performance problem:** The PCM monitors the actual fuel pressure (high-pressure side) using the fuel pressure sensor. The PCM compares the fuel pressure of the target that is calculated based on charging efficiency and engine speed and actual fuel pressure. The PCM detects that the difference of fuel pressures is above threshold or below threshold, the PCM determines that the spill valve control solenoid valve has a deterioration. **Possible Causes:** • Air suction in fuel line due to fuel runout • Improper operation of fuel pump speed control system • High pressure fuel pump connector or terminals malfunction • Fuel pressure sensor connector or terminals malfunction • PCM connector or terminals malfunction • Fuel pressure sensor malfunction • High pressure fuel pump malfunction • Spill valve control solenoid valve malfunction • Insufficient fuel pressure (low pressure line) • Fuel injector malfunction (fuel leakage) • Fuel pressure limiter malfunction • PCM malfunction
DTC: P0091 **2T PCM, MIL: Yes** **Year:** 2009, 2010 **Model:** 3, CX-7 **Engine:** 2.3L L4 VIN 3, 2.3L L4 VIN 4, 2.3L L4 VIN L, 2.3L L4 VIN M **Transmission:** All	**Fuel pressure regulator control circuit low input:** When the PCM turns the spill valve control solenoid valve off but the spill valve control solenoid valve control circuit voltage is low for 5 s, the PCM determines that the spill valve control solenoid valve control circuit has a malfunction. **Possible Causes:** • High pressure fuel pump connector or terminals malfunction • Short to ground in wiring harness between the High pressure fuel pump and PCM • Spill valve control solenoid valve malfunction • PCM connector or terminals malfunction • Open circuit in wiring harness between the High pressure fuel pump and PCM • PCM malfunction
DTC: P0092 **2T PCM, MIL: Yes** **Year:** 2009, 2010 **Model:** 3, CX-7 **Engine:** 2.3L L4 VIN 3, 2.3L L4 VIN 4, 2.3L L4 VIN L, 2.3L L4 VIN M **Transmission:** All	**Fuel pressure regulator control circuit high input:** When the PCM turns the spill valve control solenoid valve on but the spill valve control solenoid valve circuit voltage is high for 5 s, the PCM determines that the spill valve control solenoid valve control circuit has a malfunction. **Possible Causes:** • High pressure fuel pump connector or terminals malfunction • Spill valve control solenoid valve malfunction • PCM connector or terminals malfunction • Short to power supply in wiring harness between high pressure fuel pump and PCM • PCM malfunction

DTC	Trouble Code Title, Conditions & Possible Causes
DTC: P0096 **2T PCM, MIL: Yes** **Year:** 2009, 2010 **Model:** 3, CX-7 **Engine:** 2.3L L4 VIN 3, 2.3L L4 VIN 4, 2.3L L4 VIN L, 2.3L L4 VIN M **Transmission:** All	**Boost air temperature sensor circuit range/performance problem:** If the boost air temperature is higher than the engine coolant temperature by 23 °C {41.4 °F} for 1.2 s with the ignition switch is on*, the PCM determines that there is a boost air temperature sensor circuit range/performance problem. **Possible Causes:** • MAP sensor/boost air temperature sensor connector or terminals malfunction • Boost air temperature sensor malfunction • ECT sensor No.1 malfunction • PCM connector or terminals malfunction • PCM malfunction
DTC: P0097 **1T CCM, MIL: Yes** **Year:** 2009, 2010 **Model:** 3, CX-7 **Engine:** 2.3L L4 VIN 3, 2.3L L4 VIN 4, 2.3L L4 VIN L, 2.3L L4 VIN M **Transmission:** All	**Boost air temperature sensor circuit low input:** If the PCM detects that the boost air temperature sensor voltage is 0.1 V or less for 5 s, the PCM determines that the boost air temperature sensor circuit voltage is low. **Possible Causes:** • MAP sensor/boost air temperature sensor connector or terminals malfunction • Short to ground in wiring harness between MAP sensor/boost air temperature sensor and PCM • Boost air temperature sensor malfunction • PCM connector or terminals malfunction • Boost air temperature sensor signal circuit and ground circuit are shorted to each other. • PCM malfunction
DTC: P0098 **1T CCM, MIL: Yes** **Year:** 2009, 2010 **Model:** 3, CX-7 **Engine:** 2.3L L4 VIN 3, 2.3L L4 VIN 4, 2.3L L4 VIN L, 2.3L L4 VIN M **Transmission:** All	**Boost air temperature sensor circuit high input:** If the PCM detects that the boost air temperature sensor voltage is 4.96 V or more for 5 s, the PCM determines that the boost air temperature sensor circuit voltage is high. **Possible Causes:** • MAP sensor/boost air temperature sensor connector or terminals malfunction • Boost air temperature sensor malfunction • PCM connector or terminals malfunction • Short to power supply in wiring harness between the MAP sensor/boost air temperature sensor and PCM • Open circuit in wiring harness between the MAP sensor/boost air temperature sensor and PCM • PCM malfunction
DTC: P00B3 **1T PCM, MIL: Yes** **Year:** 2009, 2010 **Model:** 3 **Engine:** 2.0L L4 VIN F, 2.0L L4 VIN G, 2.3L L4 VIN 3, 2.3L L4 VIN 4, 2.3L L4 VIN L, 2.3L L4 VIN M **Transmission:** All	**ECT sensor No.2 circuit low input:** The PCM monitors the ECT sensor No.2 signal. If the PCM detects that the ECT sensor No.2 voltage is below 0.25 V for 5 s, the PCM determines that the ECT sensor No.2 circuit has a malfunction. **Possible Causes:** • ECT sensor No.2 connector or terminals malfunction • ECT sensor No.2 malfunction • Short to ground in wiring harness between ECT sensor No.2 terminal A and PCM terminal 2AP • PCM connector or terminals malfunction • ECT sensor No.2 signal circuit and ground circuit are shorted to each other. • PCM malfunction
DTC: P00B4 **1T PCM, MIL: Yes** **Year:** 2009, 2010 **Model:** 3 **Engine:** 2.0L L4 VIN F, 2.0L L4 VIN G, 2.3L L4 VIN 3, 2.3L L4 VIN 4, 2.3L L4 VIN L, 2.3L L4 VIN M **Transmission:** All	**ECT sensor No.2 circuit high input:** The PCM monitors the ECT sensor No.2 signal. If the PCM detects that the ECT sensor No.2 voltage is above 4.95 V for 5 s, the PCM determines that the ECT sensor No.2 circuit has a malfunction. **Possible Causes:** • ECT sensor No.2 connector or terminals malfunction • ECT sensor No.2 malfunction • PCM connector or terminals malfunction • Short to power supply in wiring harness between ECT sensor No.2 and PCM • Open circuit in wiring harness between the ECT sensor No.2 and PCM • PCM malfunction
DTC: P0100 **1T CCM, MIL: Yes** **Year:** 2009 **Model:** B2300, B4000 **Engine:** 2.3L L4 VIN D, 4.0L V6 VIN E **Transmission:** All	**MAF or VAF Sensor Circuit Malfunction:** Key on or engine running and the PCM detected the MAF or VAF sensor signal was less than 0.20v, or that it was more than 4.90v at any time during the CCM test. **Possible Causes:** • MAF sensor signal circuit is open or shorted to ground • MAF sensor power circuit is open or the ground circuit is open • MAF sensor is damaged or has failed • PCM has failed • TSB SSP056 (2/02) contains a MAF sensor warranty (Protégé)

DTC	Trouble Code Title, Conditions & Possible Causes
DTC: P0101 **2T CCM, MIL: Yes** **Year:** 2009, 2010 **Model:** 3, 5, 6, CX-7, MX-5 Miata, RX-8 **Engine:** 1.3L R2 VIN 2, 1.3L R2 VIN 4, 1.3L R2 VIN M, 1.3L R2 VIN P, 2.0L L4 VIN F, 2.0L L4 VIN G, 2.3L L4 VIN 3, 2.3L L4 VIN 4, 2.3L L4 VIN L, 2.3L L4 VIN M, 2.5L L4 VIN 5, 2.5L L4 VIN 6, 2.5L L4 VIN A, 2.5L L4 VIN M **Transmission:** All	**MAF Sensor inconsistent with TP Sensor:** The PCM compares the actual input signal from MAF sensor with expected input signal from MAF sensor which PCM calculates by input voltage from TP sensor or engine speed. With engine running and throttle opening angle at 50% for 5 seconds, if MAF amount is less than 8.93 g/sec., PCM determines that detected MAF amount is too low. With engine running at 2,000 rpm for 5 seconds, if MAF amount is over 103 g/sec., PSM determines that detected MAF amount is too high. **Possible Causes:** • MAF sensor malfunction • TP sensor malfunction • Electric corrosion in MAF signal circuit or MAF return circuit • Voltage drops in MAF signal circuit • Voltage drops in ground circuit
DTC: P0102 **1T CCM, MIL: Yes** **Year:** 2009, 2010 **Model:** 3, 5, 6, B2300, B4000, CX-7, CX-9, MX-5 Miata, RX-8, Tribute **Engine:** 1.3L R2 VIN 2, 1.3L R2 VIN 4, 1.3L R2 VIN M, 1.3L R2 VIN P, 2.0L L4 VIN F, 2.0L L4 VIN G, 2.3L L4 VIN 3, 2.3L L4 VIN 4, 2.3L L4 VIN D, 2.3L L4 VIN L, 2.3L L4 VIN M, 2.5L L4 VIN 3, 2.5L L4 VIN 5, 2.5L L4 VIN 6, 2.5L L4 VIN 7, 2.5L L4 VIN A, 2.5L L4 VIN M, 3.0L V6 VIN 1, 3.0L V6 VIN G, 3.7L V6 VIN A, 3.7L V6 VIN B, 3.7L V6 VIN V, 4.0L V6 VIN E **Transmission:** All	**MAF Sensor Circuit Low Input:** Key on or engine running and the PCM detected the MAF sensor signal was less than 0.36v at any time during the CCM test. **Possible Causes:** • MAF sensor signal circuit is open • MAF sensor signal is shorted to ground • MAF sensor power circuit is open • MAF sensor is damaged or has failed • PCM has failed
DTC: P0103 **1T CCM, MIL: Yes** **Year:** 2009, 2010 **Model:** 3, 5, 6, B2300, B4000, CX-7, CX-9, MX-5 Miata, RX-8, Tribute **Engine:** 1.3L R2 VIN 2, 1.3L R2 VIN 4, 1.3L R2 VIN M, 1.3L R2 VIN P, 2.0L L4 VIN F, 2.0L L4 VIN G, 2.3L L4 VIN 3, 2.3L L4 VIN 4, 2.3L L4 VIN D, 2.3L L4 VIN L, 2.3L L4 VIN M, 2.5L L4 VIN 3, 2.5L L4 VIN 5, 2.5L L4 VIN 6, 2.5L L4 VIN 7, 2.5L L4 VIN A, 2.5L L4 VIN M, 3.0L V6 VIN 1, 3.0L V6 VIN G, 3.7L V6 VIN A, 3.7L V6 VIN B, 3.7L V6 VIN V, 4.0L V6 VIN E **Transmission:** All	**MAF Sensor Circuit High Input:** Key on or engine running and the PCM detected the MAF sensor signal was more than 4.97v at any time during the CCM test. **Possible Causes:** • MAF sensor ground circuit is open • MAF sensor signal is shorted to VREF or system power (B+) • MAF sensor is damaged or has failed • PCM has failed
DTC: P0104 **T** **Year:** 2009, 2010 **Model:** 6, B2300, B4000, CX-9 **Engine:** 2.3L L4 VIN D, 3.7L V6 VIN A, 3.7L V6 VIN B, 3.7L V6 VIN V, 4.0L V6 VIN E **Transmission:** All	**MAF circuit intermittent/erratic:** A concern exists in the MAF sensor circuit, or the air tube containing the sensor, causing an incorrect air flow reading. **Possible Causes:** • Air leaks in the tube from the MAF to the throttle body • Connector or terminal malfunction • Short to power supply in wiring harness between MAF sensor and PCM • Short to ground circuit between MAF sensor and PCM • Open circuit wiring harness between MAF sensor and PCM

DTC	Trouble Code Title, Conditions & Possible Causes
DTC: P0106 **2T CCM, MIL: Yes** **Year:** 2009 **Model:** B2300, B4000 **Engine:** 2.3L L4 VIN D, 4.0L V6 VIN E **Transmission:** All	**BARO Sensor Circuit Performance:** Engine running, and the PCM detected the BARO sensor signal was less than 0.22v, or that it was more than 4.97v during the CCM test. **Possible Causes:** • BARO sensor signal circuit is open or shorted to ground • BARO sensor signal is shorted to VREF or system power • BARO sensor is damaged or has failed • PCM has failed
DTC: P0107 **1T CCM, MIL: Yes** **Year:** 2009, 2010 **Model:** 3, 5, 6, B2300, B4000, CX-7, MX-5 Miata, RX-8, Tribute **Engine:** 1.3L R2 VIN 2, 1.3L R2 VIN 4, 1.3L R2 VIN M, 1.3L R2 VIN P, 2.0L L4 VIN F, 2.0L L4 VIN G, 2.3L L4 VIN 3, 2.3L L4 VIN 4, 2.3L L4 VIN D, 2.3L L4 VIN L, 2.3L L4 VIN M, 2.5L L4 VIN 3, 2.5L L4 VIN 5, 2.5L L4 VIN 6, 2.5L L4 VIN 7, 2.5L L4 VIN A, 2.5L L4 VIN M, 3.0L V6 VIN 1, 3.0L V6 VIN G, 4.0L V6 VIN E **Transmission:** All	**BARO Sensor Circuit Low Input:** Engine running, EGR Boost Sensor Solenoid commanded "off" so that BARO pressure is applied to the sensor), IAT sensor signal more than 50°F, and the PCM detected the BARO sensor signal indicated less than 0.21v during the CCM test. **Possible Causes:** • EGR Boost sensor signal circuit is shorted to ground • EGR boost sensor VREF circuit is open or shorted to ground • EGR Boost sensor is damaged or has failed • PCM has failed
DTC: P0108 **1T CCM, MIL: Yes** **Year:** 2009, 2010 **Model:** 3, 5, 6, B2300, B4000, CX-7, RX-8, Tribute **Engine:** 1.3L R2 VIN 2, 1.3L R2 VIN 4, 1.3L R2 VIN M, 1.3L R2 VIN P, 2.0L L4 VIN F, 2.0L L4 VIN G, 2.3L L4 VIN 3, 2.3L L4 VIN 4, 2.3L L4 VIN D, 2.3L L4 VIN L, 2.3L L4 VIN M, 2.5L L4 VIN 3, 2.5L L4 VIN 5, 2.5L L4 VIN 6, 2.5L L4 VIN 7, 2.5L L4 VIN A, 2.5L L4 VIN M, 3.0L V6 VIN 1, 3.0L V6 VIN G, 4.0L V6 VIN E **Transmission:** All	**BARO Sensor Circuit High Input:** Engine running, EGR Boost Sensor Solenoid commanded "off" so that BARO pressure is applied to the sensor), IAT sensor signal more than 50°F, and the PCM detected the BARO sensor signal indicated more than 4.80v during the CCM test. **Possible Causes:** • EGR boost sensor signal circuit open from sensor to the PCM • EGR boost sensor ground circuit open from sensor to the PCM • EGR Boost sensor signal circuit is shorted to VREF or power • EGR Boost sensor is damaged or has failed • PCM has failed
DTC: P0109 **1T CCM, MIL: Yes** **Year:** 2009 **Model:** B2300, B4000 **Engine:** 2.3L L4 VIN D, 4.0L V6 VIN E **Transmission:** All	**Manifold Absolute Pressure (MAP) Sensor:** MAP sensor signal to the PCM is failing intermittently. **Possible Causes:** • MAP sensor is damaged or has failed • PCM has failed
DTC: P0110 **1T CCM, MIL: Yes** **Year:** 2009, 2010 **Model:** Tribute **Engine:** 2.5L L4 VIN 3, 2.5L L4 VIN 7, 3.0L V6 VIN 1, 3.0L V6 VIN G **Transmission:** All	**Intake Air Temperature Sensor Circuit Malfunction:** Key on or engine running, and the PCM detected the IAT sensor signal indicated less than 0.10v, or it detected the signal was more than 4.80v at any time during the CCM test period. **Possible Causes:** • IAT sensor signal circuit is open or shorted to ground • IAT sensor is damaged or has failed • PCM has failed

DTC	Trouble Code Title, Conditions & Possible Causes
DTC: P0111 **2T CCM, MIL: Yes** **Year:** 2009, 2010 **Model:** 3, 5, 6, B2300, B4000, CX-7, CX-9, MX-5 Miata, RX-8 **Engine:** 1.3L R2 VIN 2, 1.3L R2 VIN 4, 1.3L R2 VIN M, 1.3L R2 VIN P, 2.0L L4 VIN F, 2.0L L4 VIN G, 2.3L L4 VIN 3, 2.3L L4 VIN 4, 2.3L L4 VIN D, 2.3L L4 VIN L, 2.3L L4 VIN M, 2.5L L4 VIN 5, 2.5L L4 VIN 6, 2.5L L4 VIN A, 2.5L L4 VIN M, 3.7L V6 VIN A, 3.7L V6 VIN B, 3.7L V6 VIN V, 4.0L V6 VIN E **Transmission:** All	**Intake Air Temperature Sensor Circuit Range/Performance:** Key on or engine running, and the PCM detected the IAT sensor signal was more than 104°F higher than the ECT sensor signal. **Possible Causes:** • IAT sensor signal circuit has high resistance • IAT sensor has drifted out of calibration • IAT sensor is damaged or has failed • PCM has failed
DTC: P0112 **1T CCM, MIL: Yes** **Year:** 2009, 2010 **Model:** 3, 5, 6, B2300, B4000, CX-7, CX-9, MX-5 Miata, RX-8, Tribute **Engine:** 1.3L R2 VIN 2, 1.3L R2 VIN 4, 1.3L R2 VIN M, 1.3L R2 VIN P, 2.0L L4 VIN F, 2.0L L4 VIN G, 2.3L L4 VIN 3, 2.3L L4 VIN 4, 2.3L L4 VIN D, 2.3L L4 VIN L, 2.3L L4 VIN M, 2.5L L4 VIN 3, 2.5L L4 VIN 5, 2.5L L4 VIN 6, 2.5L L4 VIN 7, 2.5L L4 VIN A, 2.5L L4 VIN M, 3.0L V6 VIN 1, 3.0L V6 VIN G, 3.7L V6 VIN A, 3.7L V6 VIN B, 3.7L V6 VIN V, 4.0L V6 VIN E **Transmission:** All	**Intake Air Temperature Sensor Circuit Low Input:** Key on or engine running, and the PCM detected the IAT sensor signal indicated less than 0.20v (a Scan Tool PID near 250°F) at any time during the CCM test period. **Possible Causes:** • IAT sensor signal circuit is shorted to ground • IAT sensor is damaged or has failed • PCM has failed
DTC: P0113 **1T CCM, MIL: Yes** **Year:** 2009, 2010 **Model:** 3, 5, 6, B2300, B4000, CX-7, CX-9, MX-5 Miata, RX-8, Tribute **Engine:** 1.3L R2 VIN 2, 1.3L R2 VIN 4, 1.3L R2 VIN M, 1.3L R2 VIN P, 2.0L L4 VIN F, 2.0L L4 VIN G, 2.3L L4 VIN 3, 2.3L L4 VIN 4, 2.3L L4 VIN D, 2.3L L4 VIN L, 2.3L L4 VIN M, 2.5L L4 VIN 3, 2.5L L4 VIN 5, 2.5L L4 VIN 6, 2.5L L4 VIN 7, 2.5L L4 VIN A, 2.5L L4 VIN M, 3.0L V6 VIN 1, 3.0L V6 VIN G, 3.7L V6 VIN A, 3.7L V6 VIN B, 3.7L V6 VIN V, 4.0L V6 VIN E **Transmission:** All	**Intake Air Temperature Sensor Circuit High Input:** Key on or engine running, and the PCM detected the IAT sensor signal indicated more than 4.60v (a Scan Tool PID near -46°F) at any time during the CCM test period. **Possible Causes:** • IAT sensor signal circuit is open • IAT sensor signal circuit is shorted to VREF or system power • IAT sensor is damaged or has failed • PCM has failed
DTC: P0114 **1T CCM, MIL: Yes** **Year:** 2009, 2010 **Model:** 6, B2300, B4000, CX-9, Tribute **Engine:** 2.3L L4 VIN D, 2.5L L4 VIN 3, 2.5L L4 VIN 7, 3.0L V6 VIN 1, 3.0L V6 VIN G, 3.7L V6 VIN A, 3.7L V6 VIN B, 3.7L V6 VIN V, 4.0L V6 VIN E **Transmission:** All	**PCM - Intake Air Temperature (IAT) Sensor:** IAT sensor signal erratic. **Possible Causes:** • IAT sensor signal circuit is open • IAT sensor signal circuit is shorted to VREF or system power • IAT sensor is damaged or has failed • PCM has failed

DTC	Trouble Code Title, Conditions & Possible Causes
DTC: P0116 **2T CCM, MIL: Yes** **Year:** 2009, 2010 **Model:** 3, 5, 6, B2300, B4000, CX-7, CX-9, MX-5 Miata, RX-8, Tribute **Engine:** 1.3L R2 VIN 2, 1.3L R2 VIN 4, 1.3L R2 VIN M, 1.3L R2 VIN P, 2.0L L4 VIN F, 2.0L L4 VIN G, 2.3L L4 VIN 3, 2.3L L4 VIN 4, 2.3L L4 VIN D, 2.3L L4 VIN L, 2.3L L4 VIN M, 2.5L L4 VIN 3, 2.5L L4 VIN 5, 2.5L L4 VIN 6, 2.5L L4 VIN 7, 2.5L L4 VIN A, 2.5L L4 VIN M, 3.0L V6 VIN 1, 3.0L V6 VIN G, 3.7L V6 VIN A, 3.7L V6 VIN B, 3.7L V6 VIN V, 4.0L V6 VIN E **Transmission:** All	**Engine Coolant Temperature Sensor Circuit Performance Problem:** Key on or engine running, and the PCM detected the ECT sensor signal maximum value and minimum value is below 5.6 degrees C (4.2 degrees F), the PCM determines that ECT signal circuit has malfunctioned. **Possible Causes:** • ECT sensor malfunction • Poor connection at ECT sensor or PCM connector • PCM malfunction
DTC: P0117 **1T CCM, MIL: Yes** **Year:** 2009, 2010 **Model:** 3, 5, 6, B2300, B4000, CX-7, MX-5 Miata, RX-8, Tribute **Engine:** 1.3L R2 VIN 2, 1.3L R2 VIN 4, 1.3L R2 VIN M, 1.3L R2 VIN P, 2.0L L4 VIN F, 2.0L L4 VIN G, 2.3L L4 VIN 3, 2.3L L4 VIN 4, 2.3L L4 VIN D, 2.3L L4 VIN L, 2.3L L4 VIN M, 2.5L L4 VIN 3, 2.5L L4 VIN 5, 2.5L L4 VIN 6, 2.5L L4 VIN 7, 2.5L L4 VIN A, 2.5L L4 VIN M, 3.0L V6 VIN 1, 3.0L V6 VIN G, 4.0L V6 VIN E **Transmission:** All	**Engine Coolant Temperature Sensor Circuit Low Input:** Key on or engine running, and the PCM detected the ECT sensor signal indicated less than 0.20v (a Scan Tool PID near 250°F) at any time during the CCM test period. **Possible Causes:** • ECT sensor signal circuit is shorted to ground • ECT sensor is damaged or has failed • PCM has failed
DTC: P0118 **1T CCM, MIL: Yes** **Year:** 2009, 2010 **Model:** 3, 5, 6, B2300, B4000, CX-7, MX-5 Miata, RX-8, Tribute **Engine:** 1.3L R2 VIN 2, 1.3L R2 VIN 4, 1.3L R2 VIN M, 1.3L R2 VIN P, 2.0L L4 VIN F, 2.0L L4 VIN G, 2.3L L4 VIN 3, 2.3L L4 VIN 4, 2.3L L4 VIN D, 2.3L L4 VIN L, 2.3L L4 VIN M, 2.5L L4 VIN 3, 2.5L L4 VIN 5, 2.5L L4 VIN 6, 2.5L L4 VIN 7, 2.5L L4 VIN A, 2.5L L4 VIN M, 3.0L V6 VIN 1, 3.0L V6 VIN G, 4.0L V6 VIN E **Transmission:** All	**Engine Coolant Temperature Sensor Circuit High Input:** Key on or engine running, and the PCM detected the ECT sensor signal indicated more than 4.60v (a Scan Tool PID near -46°F) at any time during the CCM test period. **Possible Causes:** • ECT sensor signal circuit is open • ECT sensor signal circuit is shorted to VREF or system power • ECT sensor is damaged or has failed • PCM has failed
DTC: P0119 **1T CCM, MIL: Yes** **Year:** 2009, 2010 **Model:** 6, B2300, B4000, CX-9, Tribute **Engine:** 2.3L L4 VIN D, 2.5L L4 VIN 3, 2.5L L4 VIN 7, 3.0L V6 VIN 1, 3.0L V6 VIN G, 3.7L V6 VIN A, 3.7L V6 VIN B, 3.7L V6 VIN V, 4.0L V6 VIN E **Transmission:** All	**PCM - Engine Coolant Temperature (ECT) Sensor/Coolant Head Temperature (CHT) Sensor:** ECT / CHT sensor signal is erratic. **Possible Causes:** • ECT / CHT sensor signal circuit is open • ECT / CHT sensor signal circuit is shorted to VREF or system power • ECT / CHT sensor is damaged or has failed • PCM has failed

DTC	Trouble Code Title, Conditions & Possible Causes
DTC: P0121 **2T CCM, MIL: Yes** **Year:** 2009 **Model:** B2300, B4000 **Engine:** 2.3L L4 VIN D, 4.0L V6 VIN E **Transmission:** All	**TP Sensor In-Range Operating Circuit Malfunction:** Vehicle driven at light engine load at over 20 mph, and the PCM detected the TP sensor signal did not correlate when it was compared to the MAF sensor signal during the CCM test. **Possible Causes:** • TP sensor signal circuit is open to the PCM (intermittent fault) • TP sensor ground circuit is open • Throttle body is damaged or throttle linkage is bent or binding • TP sensor is damaged or has failed
DTC: P0122 **1T CCM, MIL: Yes** **Year:** 2009, 2010 **Model:** 3, 5, 6, B2300, B4000, CX-7, CX-9, MX-5 Miata, RX-8, Tribute **Engine:** 1.3L R2 VIN 2, 1.3L R2 VIN 4, 1.3L R2 VIN M, 1.3L R2 VIN P, 2.0L L4 VIN F, 2.0L L4 VIN G, 2.3L L4 VIN 3, 2.3L L4 VIN 4, 2.3L L4 VIN D, 2.3L L4 VIN L, 2.3L L4 VIN M, 2.5L L4 VIN 3, 2.5L L4 VIN 5, 2.5L L4 VIN 6, 2.5L L4 VIN 7, 2.5L L4 VIN A, 2.5L L4 VIN M, 3.0L V6 VIN 1, 3.0L V6 VIN G, 3.7L V6 VIN A, 3.7L V6 VIN B, 3.7L V6 VIN V, 4.0L V6 VIN E **Transmission:** All	**TP Sensor Circuit Low Input:** Key on or engine running, and the PCM detected the TP sensor signal indicated less than 0.17v (a throttle opening of 3.43%) during the CCM test. **Possible Causes:** • TP sensor signal circuit is shorted to ground • TP sensor VREF circuit is open or shorted to ground • Throttle body is damaged • Throttle linkage is bent or binding (sticking) • TP sensor is damaged or has failed
DTC: P0123 **1T CCM, MIL: Yes** **Year:** 2009, 2010 **Model:** 3, 5, 6, B2300, B4000, CX-7, CX-9, MX-5 Miata, RX-8, Tribute **Engine:** 1.3L R2 VIN 2, 1.3L R2 VIN 4, 1.3L R2 VIN M, 1.3L R2 VIN P, 2.0L L4 VIN F, 2.0L L4 VIN G, 2.3L L4 VIN 3, 2.3L L4 VIN 4, 2.3L L4 VIN D, 2.3L L4 VIN L, 2.3L L4 VIN M, 2.5L L4 VIN 3, 2.5L L4 VIN 5, 2.5L L4 VIN 6, 2.5L L4 VIN 7, 2.5L L4 VIN A, 2.5L L4 VIN M, 3.0L V6 VIN 1, 3.0L V6 VIN G, 3.7L V6 VIN A, 3.7L V6 VIN B, 3.7L V6 VIN V, 4.0L V6 VIN E **Transmission:** All	**TP Sensor Circuit High Input:** Key on or engine running, and the PCM detected the TP sensor signal indicated more than 4.60v (a throttle opening of 92.27%). **NOTE: This trouble code may set due to an intermittent fault.** **Possible Causes:** • TP sensor signal return circuit is open • TP sensor signal circuit is shorted to VREF or system power • TP sensor not mounted properly to the throttle body • TP sensor is damaged or has failed • PCM has failed
DTC: P0125 **2T CCM, MIL: Yes** **Year:** 2009, 2010 **Model:** 3, 5, 6, B2300, B4000, CX-7, MX-5 Miata, RX-8, Tribute **Engine:** 1.3L R2 VIN 2, 1.3L R2 VIN 4, 1.3L R2 VIN M, 1.3L R2 VIN P, 2.0L L4 VIN F, 2.0L L4 VIN G, 2.3L L4 VIN 3, 2.3L L4 VIN 4, 2.3L L4 VIN D, 2.3L L4 VIN L, 2.3L L4 VIN M, 2.5L L4 VIN 3, 2.5L L4 VIN 5, 2.5L L4 VIN 6, 2.5L L4 VIN 7, 2.5L L4 VIN A, 2.5L L4 VIN M, 3.0L V6 VIN 1, 3.0L V6 VIN G, 4.0L V6 VIN E **Transmission:** All	**Excessive Time To Enter Closed Loop:** DTC P0115, P0117 and P0118 not set, cold startup requirement met (ECT input less than 140°F, IAT sensor more than 50°F at startup), engine runtime over 10 minutes, and the PCM detected the ECT sensor signal did not reach 78°F after the engine warm-up period. **Possible Causes:** • Inspect for low coolant level or an incorrect coolant mixture • Check the operation of the thermostat (it may be stuck open) • ECT sensor signal circuit has high resistance • ECT sensor has failed

DTC	Trouble Code Title, Conditions & Possible Causes
DTC: P0126 **2T CCM, MIL: Yes** **Year:** 2009, 2010 **Model:** 3, 5, 6, B2300, B4000, CX-7, MX-5 Miata, RX-8 **Engine:** 1.3L R2 VIN 2, 1.3L R2 VIN 4, 1.3L R2 VIN M, 1.3L R2 VIN P, 2.0L L4 VIN F, 2.0L L4 VIN G, 2.3L L4 VIN 3, 2.3L L4 VIN 4, 2.3L L4 VIN D, 2.3L L4 VIN L, 2.3L L4 VIN M, 2.5L L4 VIN 5, 2.5L L4 VIN 6, 2.5L L4 VIN A, 2.5L L4 VIN M, 4.0L V6 VIN E **Transmission:** All	**Thermostat Malfunction:** Cold startup requirement met (IAT sensor more than 14(F and the difference between ECT and IAT sensor signals less than 43°F), engine runtime over 10 minutes, and the PCM detected the ECT sensor did not reach 160°F after the engine warm-up period. **Possible Causes:** • Inspect for low coolant level or an incorrect coolant mixture • Check the operation of the thermostat (it may be stuck open) • ECT sensor has drifted out of calibration or it has failed • PCM has failed
DTC: P0128 **1T CCM, MIL: Yes** **Year:** 2009, 2010 **Model:** 3, 5, 6, B2300, B4000, CX-7, CX-9, MX-5 Miata, Tribute **Engine:** 2.0L L4 VIN F, 2.0L L4 VIN G, 2.3L L4 VIN 3, 2.3L L4 VIN 4, 2.3L L4 VIN D, 2.3L L4 VIN L, 2.3L L4 VIN M, 2.5L L4 VIN 3, 2.5L L4 VIN 5, 2.5L L4 VIN 6, 2.5L L4 VIN 7, 2.5L L4 VIN A, 2.5L L4 VIN M, 3.0L V6 VIN 1, 3.0L V6 VIN G, 3.7L V6 VIN A, 3.7L V6 VIN B, 3.7L V6 VIN V, 4.0L V6 VIN E **Transmission:** All	**Thermostat Malfunction:** Cold startup requirement met (ECT input less than 95°F, IAT sensor more than 14°F and the difference between the ECT and IAT sensor signals less than 43°F), VSS input over 15 mph and the MAF, IAT and ECT sensor signals all within normal range, the PCM detected the radiator heat ratio exceeded its threshold after a warm-up period. **Possible Causes:** • Check the operation of the thermostat (it may be stuck open) • ECT sensor has failed • PCM has failed
DTC: P0130 **T** **Year:** 2009, 2010 **Model:** 3, 5, 6, B2300, B4000, CX-7, MX-5 Miata, RX-8 **Engine:** 1.3L R2 VIN 2, 1.3L R2 VIN 4, 1.3L R2 VIN M, 1.3L R2 VIN P, 2.0L L4 VIN F, 2.0L L4 VIN G, 2.3L L4 VIN 3, 2.3L L4 VIN 4, 2.3L L4 VIN D, 2.3L L4 VIN L, 2.3L L4 VIN M, 2.5L L4 VIN 5, 2.5L L4 VIN 6, 2.5L L4 VIN M, 3.7L V6 VIN B, 4.0L V6 VIN E **Transmission:** All	**HO2S-11 (Bank 1 Sensor 1) Circuit Malfunction:** Engine speed from 1500-3000 rpm while in closed loop, ECT sensor signal more than 14°F, engine load from 28-59%, vehicle speed over 3.5 mph, and after the PCM monitored the HO2S inversion cycle period, lean to rich response and rich to lean response time under these conditions, it determined that the HO2S failed one of the tests. **Possible Causes:** • Air leaks in intake manifold, exhaust pipes or exhaust manifold • Fuel delivery system component has failed (i.e., a clogged fuel filter, dirty or restricted fuel injectors, low or high fuel pressure) • Base engine "mechanical" problem (low cylinder compression, incorrect camshaft timing, intake or exhaust manifold leaks) • HO2S is damaged or has failed • HO2S heater element has deteriorated or failed • PRC solenoid or Purge solenoid is damaged or has failed • PCM has failed
DTC: P0131 **2T CCM, MIL: Yes** **Year:** 2009, 2010 **Model:** 3, 5, 6, B2300, B4000, CX-7, MX-5 Miata, RX-8, Tribute **Engine:** 1.3L R2 VIN 2, 1.3L R2 VIN 4, 1.3L R2 VIN M, 1.3L R2 VIN P, 2.0L L4 VIN F, 2.0L L4 VIN G, 2.3L L4 VIN 3, 2.3L L4 VIN 4, 2.3L L4 VIN D, 2.3L L4 VIN L, 2.3L L4 VIN M, 2.5L L4 VIN 3, 2.5L L4 VIN 5, 2.5L L4 VIN 6, 2.5L L4 VIN 7, 2.5L L4 VIN A, 2.5L L4 VIN M, 3.0L V6 VIN 1, 3.0L V6 VIN G, 3.7L V6 VIN B, 4.0L V6 VIN E **Transmission:** All	**HO2S-11 (Bank 1 Sensor 1) Circuit Malfunction:** Vehicle driven in closed loop at cruise speed and back to idle speed, and the PCM detected a negative voltage on the HO2S circuit. **Possible Causes:** • HO2S is water or fuel contaminated • HO2S is damaged or has failed • PCM has failed

DTC	Trouble Code Title, Conditions & Possible Causes
DTC: P0132 **2T CCM, MIL: Yes** **Year:** 2009, 2010 **Model:** 3, 5, 6, B2300, B4000, CX-7, CX-9, MX-5 Miata, RX-8, Tribute **Engine:** 1.3L R2 VIN 2, 1.3L R2 VIN 4, 1.3L R2 VIN M, 1.3L R2 VIN P, 2.0L L4 VIN F, 2.0L L4 VIN G, 2.3L L4 VIN 3, 2.3L L4 VIN 4, 2.3L L4 VIN D, 2.3L L4 VIN L, 2.3L L4 VIN M, 2.5L L4 VIN 3, 2.5L L4 VIN 5, 2.5L L4 VIN 6, 2.5L L4 VIN 7, 2.5L L4 VIN A, 2.5L L4 VIN M, 3.0L V6 VIN 1, 3.0L V6 VIN G, 3.7L V6 VIN A, 3.7L V6 VIN B, 3.7L V6 VIN V, 4.0L V6 VIN E **Transmission:** All	**HO2S-11 (Bank 1 Sensor 1) Circuit Malfunction:** Vehicle driven in closed loop at cruise speed and back to idle speed, and the PCM detected a negative voltage on the HO2S circuit. **Possible Causes:** • HO2S is water or fuel contaminated • HO2S is damaged or has failed • PCM has failed
DTC: P0133 **2T O2S1, MIL: Yes** **Year:** 2009, 2010 **Model:** 3, 5, 6, B2300, B4000, CX-7, CX-9, MX-5 Miata, RX-8 **Engine:** 1.3L R2 VIN 2, 1.3L R2 VIN 4, 1.3L R2 VIN M, 1.3L R2 VIN P, 2.0L L4 VIN F, 2.0L L4 VIN G, 2.3L L4 VIN 3, 2.3L L4 VIN 4, 2.3L L4 VIN D, 2.3L L4 VIN L, 2.3L L4 VIN M, 2.5L L4 VIN 5, 2.5L L4 VIN 6, 2.5L L4 VIN A, 2.5L L4 VIN M, 3.7L V6 VIN A, 3.7L V6 VIN B, 3.7L V6 VIN V, 4.0L V6 VIN E **Transmission:** All	**HO2S-11 (Bank 1 Sensor 1) Circuit Slow Response:** Vehicle driven in closed loop at cruise speed and back to idle speed, and the PCM detected the frequency and the amplitude response rate of the front HO2S was less than a calibrated window in memory. **Possible Causes:** • Air leaks in the intake manifold or exhaust manifold or pipes • IAT sensor or MAF sensor has deteriorated (out of calibration) • HO2S signal circuit is open or shorted to ground (intermittent) • HO2S contaminated with wrong fuel, has deteriorated or failed • PCM has failed
DTC: P0133 **T** **Year:** 2009, 2010 **Model:** Tribute **Engine:** 2.5L L4 VIN 3, 2.5L L4 VIN 7, 3.0L V6 VIN 1, 3.0L V6 VIN G **Transmission:** All	**HO2S-11 (Bank 1 Sensor 1) Circuit Slow Response:** Vehicle driven in closed loop at cruise speed and back to idle speed, and the PCM detected the frequency and the amplitude response rate of the front HO2S was less than a calibrated window in memory. **Possible Causes:** • Air leaks in the intake manifold or exhaust manifold or pipes • IAT sensor or MAF sensor has deteriorated (out of calibration) • HO2S signal circuit is open or shorted to ground (intermittent) • HO2S contaminated with wrong fuel, has deteriorated or failed • PCM has failed
DTC: P0134 **2T O2S1, MIL: Yes** **Year:** 2009, 2010 **Model:** 3, 5, 6, B2300, B4000, CX-7, MX-5 Miata, RX-8 **Engine:** 1.3L R2 VIN 2, 1.3L R2 VIN 4, 1.3L R2 VIN M, 1.3L R2 VIN P, 2.0L L4 VIN F, 2.0L L4 VIN G, 2.3L L4 VIN 3, 2.3L L4 VIN 4, 2.3L L4 VIN D, 2.3L L4 VIN L, 2.3L L4 VIN M, 2.5L L4 VIN 5, 2.5L L4 VIN 6, 2.5L L4 VIN A, 2.5L L4 VIN M, 4.0L V6 VIN E **Transmission:** All	**HO2S-11 (Bank 1 Sensor 1) Circuit No Activity Detected:** Engine speed from 1500-3000 rpm while in closed loop, ECT sensor signal more than 176°F, and the PCM detected the front HO2S signal did not exceed 550 mv for over 54.2 seconds during the test. **Possible Causes:** • Air leaks in the intake manifold or exhaust manifold or pipes • IAT sensor or MAF sensor has deteriorated (out of calibration) • HO2S signal circuit is open or shorted to ground (intermittent) • HO2S contaminated with wrong fuel, has deteriorated or failed • PCM has failed

DTC	Trouble Code Title, Conditions & Possible Causes
DTC: P0135 **2T O2S2, MIL: Yes** **Year:** 2009, 2010 **Model:** 6, B2300, B4000, CX-9 **Engine:** 2.3L L4 VIN D, 3.7L V6 VIN A, 3.7L V6 VIN B, 3.7L V6 VIN V, 4.0L V6 VIN E **Transmission:** All	**HO2S-11 (Bank 1 Sensor 1) Heater Circuit Malfunction:** Engine started, engine running, and the PCM detected the HO2S heater circuit was more than 11.5v with the heater turned "on", or it was less than 5.8v with the heater turned "off" for 331-361 seconds. **Possible Causes:** • HO2S heater control circuit is open or shorted to ground • HO2S heater control circuit is shorted to system power (B+) • HO2S heater power circuit open between the heater and PCM • HO2S heater is damaged or has failed • PCM has failed (it controls the heater with a duty cycle signal)
DTC: P0135 **T** **Year:** 2009, 2010 **Model:** Tribute **Engine:** 2.5L L4 VIN 3, 2.5L L4 VIN 7, 3.0L V6 VIN 1, 3.0L V6 VIN G **Transmission:** All	**HO2S-11 (Bank 1 Sensor 1) Heater Circuit Malfunction:** Engine started, engine running, and the PCM detected the HO2S heater circuit was more than 11.5v with the heater turned "on", or it was less than 5.8v with the heater turned "off" for 331-361 seconds. **Possible Causes:** • HO2S heater control circuit is open or shorted to ground • HO2S heater control circuit is shorted to system power (B+) • HO2S heater power circuit open between the heater and PCM • HO2S heater is damaged or has failed • PCM has failed (it controls the heater with a duty cycle signal)
DTC: P0136 **2T O2S1, MIL: Yes** **Year:** 2009 **Model:** B2300, B4000 **Engine:** 2.3L L4 VIN D, 4.0L V6 VIN E **Transmission:** All	**HO2S-12 (Bank 1 Sensor 2) No Activity Detected:** Engine running in closed loop for 5 minutes, and the PCM detected the middle HO2S signal was less than a calibrated value in memory. **Possible Causes:** • Air leaks in the intake manifold or exhaust manifold or pipes • IAT sensor or MAF sensor has deteriorated (out of calibration) • HO2S signal circuit is open or shorted to ground • HO2S contaminated with wrong fuel, has deteriorated or failed • PCM has failed
DTC: P0137 **2T CCM, MIL: Yes** **Year:** 2009, 2010 **Model:** 3, 5, 6, B2300, B4000, CX-7, MX-5 Miata, RX-8 **Engine:** 1.3L R2 VIN 2, 1.3L R2 VIN 4, 1.3L R2 VIN M, 1.3L R2 VIN P, 2.0L L4 VIN F, 2.0L L4 VIN G, 2.3L L4 VIN 3, 2.3L L4 VIN 4, 2.3L L4 VIN D, 2.3L L4 VIN L, 2.3L L4 VIN M, 2.5L L4 VIN 5, 2.5L L4 VIN 6, 2.5L L4 VIN A, 2.5L L4 VIN M, 4.0L V6 VIN E **Transmission:** All	**HO2S-12 (Bank 1 Sensor 2) Circuit Low Input:** Vehicle driven in closed loop at cruise speed at low engine load for over 5 minutes, and the PCM detected the HO2S signal remained at less than 500 mv during the CCM test for 55 seconds. **Possible Causes:** • HO2S signal circuit is open or shorted to ground • Air leaks at the exhaust manifold or exhaust pipes • HO2S contaminated with wrong fuel, has deteriorated or failed • PCM has failed
DTC: P0138 **2T CCM, MIL: Yes** **Year:** 2009, 2010 **Model:** 3, 5, 6, B2300, B4000, CX-7, CX-9, MX-5 Miata, RX-8, Tribute **Engine:** 1.3L R2 VIN 2, 1.3L R2 VIN 4, 1.3L R2 VIN M, 1.3L R2 VIN P, 2.0L L4 VIN F, 2.0L L4 VIN G, 2.3L L4 VIN 3, 2.3L L4 VIN 4, 2.3L L4 VIN D, 2.3L L4 VIN L, 2.3L L4 VIN M, 2.5L L4 VIN 3, 2.5L L4 VIN 5, 2.5L L4 VIN 6, 2.5L L4 VIN 7, 2.5L L4 VIN A, 2.5L L4 VIN M, 3.0L V6 VIN 1, 3.0L V6 VIN G, 3.7L V6 VIN A, 3.7L V6 VIN B, 3.7L V6 VIN V, 4.0L V6 VIN E **Transmission:** All	**HO2S-12 (Bank 1 Sensor 2) Circuit High Input:** Vehicle driven in closed loop at cruise speed at low engine load for over 5 minutes, and the PCM detected the HO2S signal remained at more than 500 mv during the CCM test for 55 seconds. **Possible Causes:** • HO2S signal tracking (wet/oily) in connector causing a short between the signal circuit and heater power circuit • HO2S signal circuit is shorted to system power HO2S contaminated with wrong fuel, has deteriorated or failed • PCM has failed

DTC	Trouble Code Title, Conditions & Possible Causes
DTC: P0139 **2T CCM, MIL: Yes** **Year:** 2009, 2010 **Model:** 3, 5, 6, B2300, B4000, CX-7, CX-9, MX-5 Miata, RX-8 **Engine:** 1.3L R2 VIN 2, 1.3L R2 VIN 4, 1.3L R2 VIN M, 1.3L R2 VIN P, 2.0L L4 VIN F, 2.0L L4 VIN G, 2.3L L4 VIN 3, 2.3L L4 VIN 4, 2.3L L4 VIN D, 2.3L L4 VIN L, 2.3L L4 VIN M, 2.5L L4 VIN 5, 2.5L L4 VIN 6, 2.5L L4 VIN A, 2.5L L4 VIN M, 3.7L V6 VIN A, 3.7L V6 VIN V, 4.0L V6 VIN E **Transmission:** All	**Middle HO2S Circuit problem:** The PCM monitors inversion cycle period, middle HO2S output voltage inclination. The PCM detects that the voltage inclinations are below threshold consecutive 5 times when following conditions are met, the PCM determines that circuit has malfunction. Under the following monitoring conditions, if 0.3 V or more is detected three times even if fuel cut is performed for 3 seconds or more, a circuit malfunction is determined. **Possible Causes:** • Middle HO2S circuit deterioration • Middle HO2S circuit malfunction • PCM has failed
DTC: P013A **T** **Year:** 2009, 2010 **Model:** 6 **Engine:** 3.7L V6 VIN B **Transmission:** All	**HO2S (RH) slow response-rich to lean:** During a deceleration fuel shut-off event, the PCM monitors how quickly the HO2S (RH) switches from rich to lean. The measured rate of the rich to lean switch is compared to a calibrated error threshold value. The threshold value takes into account the level of oxygen in the catalyst, which has an impact on how quickly the rich to lean switch occurs. The test fails if the measured value is slower than the threshold value. **Possible Causes:** • Exhaust leaks before or near the HO2S (RH) • HO2S (RH) malfunction
DTC: P013C **T** **Year:** 2009, 2010 **Model:** 6 **Engine:** 3.7L V6 VIN B **Transmission:** All	**HO2S (LH) slow response-rich to lean:** During a deceleration fuel shut-off event, the PCM monitors how quickly the HO2S (LH) switches from rich to lean. The measured rate of the rich to lean switch is compared to a calibrated error threshold value. The threshold value takes into account the level of oxygen in the catalyst, which has an impact on how quickly the rich to lean switch occurs. The test fails if the measured value is slower than the threshold value. **Possible Causes:** • Exhaust leaks before or near the HO2S (LH) • HO2S (LH) malfunction
DTC: P013E **T PCM** **Year:** 2009, 2010 **Model:** 6 **Engine:** 3.7L V6 VIN B **Transmission:** All	**HO2S (RH) delayed response-rich to lean:** During a deceleration fuel shut-off event, the PCM monitors the HO2S (RH) signal to determine if the signal is stuck in range. The PCM expects the signal to exceed a calibrated rich or lean value within a calibrated amount of time. If the signal voltage remains less than the rich value after a number of occurrences, the PCM forcefully controls the fuel system rich over increasing time periods in an attempt to force the signal to greater than the calibrated rich value. The test fails if after 3 consecutive forced attempts the signal cannot be forced greater than the calibrated rich value. Also, if the signal voltage remains greater than the lean value after a calibrated amount of time with the fuel injectors off, a counter is incremented. The test fails if after 3 consecutive occurrences the signal is not less than the calibrated lean value. **Possible Causes:** • Exhaust leaks before or near the HO2S (RH) • Aftermarket exhaust accessories or performance modifications • Ethanol content in the fuel • Circuit intermittent • HO2S (RH) malfunction
DTC: P0140 **2T O2S1, MIL: Yes** **Year:** 2009, 2010 **Model:** 3, 5, 6, B2300, B4000, CX-7, MX-5 Miata **Engine:** 2.0L L4 VIN F, 2.0L L4 VIN G, 2.3L L4 VIN 3, 2.3L L4 VIN 4, 2.3L L4 VIN D, 2.3L L4 VIN L, 2.3L L4 VIN M, 2.5L L4 VIN 5, 2.5L L4 VIN 6, 2.5L L4 VIN A, 2.5L L4 VIN M, 4.0L V6 VIN E **Transmission:** All	**HO2S-12 (Bank 1 Sensor 2) Circuit No Activity Detected:** Engine speed from 1500-3000 rpm while in closed loop, ECT sensor signal more than 176°F, and the PCM detected the HO2S signal voltage did not exceed 550 mv for over 54.2 seconds during the test. **Possible Causes:** • Air leaks in the intake manifold or exhaust manifold or pipes • IAT sensor or MAF sensor has deteriorated (out of calibration) • HO2S signal circuit is open or shorted to ground • HO2S contaminated with wrong fuel, has deteriorated or failed • TSB 0100700 (3/00) contains information for this code (MPV)

DTC	Trouble Code Title, Conditions & Possible Causes
DTC: P0141 **2T O2S1, MIL: Yes** **Year:** 2009, 2010 **Model:** 6, B2300, B4000, CX-9 **Engine:** 2.3L L4 VIN D, 3.7L V6 VIN A, 3.7L V6 VIN B, 3.7L V6 VIN V, 4.0L V6 VIN E **Transmission:** All	**HO2S-12 (Bank 1 Sensor 2) Heater Circuit Malfunction:** Engine started, engine running, and the PCM detected the HO2S heater circuit indicated more than 11.5v with the heater turned "on", or it indicated less than 5.8v with the heater turned "off" for 331-361 seconds during the CCM test. **Possible Causes:** • HO2S heater control circuit is open or shorted to ground • HO2S heater control circuit is shorted to system power (B+) • HO2S heater power circuit open between the heater and PCM • HO2S heater is damaged or has failed • PCM has failed (it controls the heater with a duty cycle signal)
DTC: P0141 **T** **Year:** 2009, 2010 **Model:** Tribute **Engine:** 2.5L L4 VIN 3, 2.5L L4 VIN 7, 3.0L V6 VIN 1, 3.0L V6 VIN G **Transmission:** All	**HO2S-12 (Bank 1 Sensor 2) Heater Circuit Malfunction:** Engine started, engine running, and the PCM detected the HO2S heater circuit indicated more than 11.5v with the heater turned "on", or it indicated less than 5.8v with the heater turned "off" for 331-361 seconds during the CCM test. **Possible Causes:** • HO2S heater control circuit is open or shorted to ground • HO2S heater control circuit is shorted to system power (B+) • HO2S heater power circuit open between the heater and PCM • HO2S heater is damaged or has failed • PCM has failed (it controls the heater with a duty cycle signal)
DTC: P0148 **T** **Year:** 2009, 2010 **Model:** 6, CX-9, Tribute **Engine:** 2.5L L4 VIN 3, 2.5L L4 VIN 7, 3.0L V6 VIN 1, 3.0L V6 VIN G, 3.7L V6 VIN A, 3.7L V6 VIN B, 3.7L V6 VIN V **Transmission:** All	**HO2S-13 (Bank 1 Sensor 3) Heater Circuit Malfunction:** At least one bank lean at wide open throttle. **Possible Causes:** • HO2S heater control circuit is open or shorted to ground • HO2S heater control circuit is shorted to system power (B+) • HO2S heater power circuit open between the heater and PCM • HO2S heater is damaged or has failed • PCM has failed (it controls the heater with a duty cycle signal)
DTC: P014A **T PCM** **Year:** 2009, 2010 **Model:** 6 **Engine:** 3.7L V6 VIN B **Transmission:** All	**HO2S (LH) delayed response-rich to lean:** During a deceleration fuel shut-off event, the PCM monitors the HO2S (LH) signal to determine if the signal is stuck in range. The PCM expects the signal to exceed a calibrated rich or lean value within a calibrated amount of time. If the signal voltage remains less than the rich value after a number of occurrences, the PCM forcefully controls the fuel system rich over increasing time periods in an attempt to force the signal to greater than the calibrated rich value. The test fails if after 3 consecutive forced attempts the signal cannot be forced greater than the calibrated rich value. Also, if the signal voltage remains greater than the lean value after a calibrated amount of time with the fuel injectors off, a counter is incremented. The test fails if after 3 consecutive occurrences the signal is not less than the calibrated lean value. **Possible Causes:** • Exhaust leaks before or near the HO2S (LH) • Aftermarket exhaust accessories or performance modifications • Ethanol content in the fuel • Circuit intermittent • HO2S (LH) malfunction
DTC: P0150 **2T CCM, MIL: Yes** **Year:** 2009, 2010 **Model:** Tribute **Engine:** 2.5L L4 VIN 3, 2.5L L4 VIN 7, 3.0L V6 VIN 1, 3.0L V6 VIN G **Transmission:** All	**HO2S-21 (Bank 2 Sensor 1) Circuit Malfunction:** Engine started, vehicle driven at over 3.5 mph at 1500-3000 rpm in closed loop, ECT sensor more than 14°F, engine load from 28-59%, and after the PCM monitored the HO2S inversion cycle period, lean to rich response and rich to lean response times under these conditions, it determined the HO2S-21 failed one of these tests. **Possible Causes:** • Air leaks in intake manifold, exhaust pipes or exhaust manifold • Fuel delivery system component has failed (i.e., a clogged fuel filter, dirty or restricted fuel injectors, low or high fuel pressure) • Base engine "mechanical" problem (low cylinder compression, incorrect camshaft timing, intake or exhaust manifold leaks) • HO2S is damaged or has failed • HO2S heater element has deteriorated or failed • PRC solenoid or Purge solenoid is damaged or has failed • PCM has failed

DTC	Trouble Code Title, Conditions & Possible Causes
DTC: P0150 **T** **Year:** 2009, 2010 **Model:** 6 **Engine:** 3.7L V6 VIN B **Transmission:** All	**A/F sensor (LH) circuit:** The PCM monitors the A/F sensor for a circuit concern. The test fails when the PCM detects a concern with one of the circuits used to determine the oxygen content in the exhaust gas. **Possible Causes:** • Open circuit between PCM and A/F sensor • Short to power supply or ground between PCM and A/F sensor • A/F sensor malfunction
DTC: P0151 **2T CCM, MIL: Yes** **Year:** 2009, 2010 **Model:** 6, B2300, B4000, Tribute **Engine:** 2.3L L4 VIN D, 2.5L L4 VIN 3, 2.5L L4 VIN 7, 3.0L V6 VIN 1, 3.0L V6 VIN G, 3.7L V6 VIN B, 4.0L V6 VIN E **Transmission:** All	**HO2S-21 (Bank 2 Sensor 1) Circuit Malfunction:** Engine started, vehicle driven in closed loop at cruise speed and then back to idle speed, and the PCM detected a negative voltage on the HO2S circuit during the CCM test. **Possible Causes:** • HO2S is water or fuel contaminated • HO2S is damaged or has failed • PCM has failed
DTC: P0152 **2T CCM, MIL: Yes** **Year:** 2009, 2010 **Model:** 6, B2300, B4000, CX-9, Tribute **Engine:** 2.3L L4 VIN D, 2.5L L4 VIN 3, 2.5L L4 VIN 7, 3.0L V6 VIN 1, 3.0L V6 VIN G, 3.7L V6 VIN A, 3.7L V6 VIN B, 3.7L V6 VIN V, 4.0L V6 VIN E **Transmission:** All	**HO2S-21 (Bank 2 Sensor 1) Circuit Malfunction:** Engine started, vehicle driven in closed loop at cruise speed and then back to idle speed, and the PCM detected a negative voltage on the HO2S circuit during the CCM test. **Possible Causes:** • HO2S is water or fuel contaminated • HO2S is damaged or has failed • PCM has failed
DTC: P0153 **2T CCM, MIL: Yes** **Year:** 2009, 2010 **Model:** 6, B2300, B4000, CX-9, Tribute **Engine:** 2.3L L4 VIN D, 2.5L L4 VIN 3, 2.5L L4 VIN 7, 3.0L V6 VIN 1, 3.0L V6 VIN G, 3.7L V6 VIN A, 3.7L V6 VIN B, 3.7L V6 VIN V, 4.0L V6 VIN E **Transmission:** All	**HO2S-21 (Bank 2 Sensor 1) Circuit Slow Response:** Vehicle driven in closed loop at cruise speed and back to idle speed, and the PCM detected the frequency and the amplitude response rate of the front HO2S was less than a calibrated window in memory. **Possible Causes:** • Air leaks in the intake manifold or exhaust manifold or pipes • IAT sensor or MAF sensor has deteriorated (out of calibration) • HO2S signal circuit is open or shorted to ground (intermittent) • HO2S contaminated with wrong fuel, has deteriorated or failed
DTC: P0154 **2T O2S HTR1, MIL: Yes** **Year:** 2009 **Model:** B4000 **Engine:** 4.0L V6 VIN E **Transmission:** All	**O2 Circuit No Activity Detected (Bank 2, Sensor 1):** The Powertrain Control Module (PCM) monitors the Heated Oxygen Sensor (HO2S) for a lack of movement concern. If the sensor signal value is not changing from the default value, the PCM commands an oscillating air/fuel ratio attempting to detect some movement in the signal value. The test fails when the PCM is unable to detect movement in the sensor signal while the air/fuel ratio is oscillating. **Possible Causes:** • Open UO2SPC circuit • Damaged universal HO2S • PCM has failed
DTC: P0154 **T** **Year:** 2009, 2010 **Model:** Tribute **Engine:** 2.5L L4 VIN 3, 2.5L L4 VIN 7, 3.0L V6 VIN 1, 3.0L V6 VIN G **Transmission:** All	**HO2S-21 (Bank 2 Sensor 1) Circuit No Activity Detected:** Engine speed from 1500-3000 rpm while in closed loop, ECT sensor signal more than 176°F, and the PCM detected the front HO2S signal did not exceed 0.55v for over 54.2 seconds during the test. **Possible Causes:** • Air leaks in the intake manifold or exhaust manifold or pipes • IAT sensor or MAF sensor has deteriorated (out of calibration) • HO2S signal circuit is open or shorted to ground • HO2S contaminated with wrong fuel, has deteriorated or failed

DTC	Trouble Code Title, Conditions & Possible Causes
DTC: P0155 **T** **Year:** 2009, 2010 **Model:** Tribute **Engine:** 2.5L L4 VIN 3, 2.5L L4 VIN 7, 3.0L V6 VIN 1, 3.0L V6 VIN G **Transmission:** All	**HO2S-21 (B2 S1) Heater Circuit Conditions:** Engine started, engine running, and the PCM detected the HO2S heater circuit indicated more than 11.5v with the heater turned "on", or it indicated less than 5.8v with the heater turned "off" for 331-361 seconds during the CCM test. **Possible Causes:** • HO2S heater control circuit is open or shorted to ground • HO2S heater control circuit is shorted to system power (B+) • HO2S heater power circuit open between the heater and PCM • HO2S heater is damaged or has failed • PCM has failed (it controls the heater with a duty cycle signal)
DTC: P0155 **2T O2S1, MIL: Yes** **Year:** 2009, 2010 **Model:** 6, CX-9 **Engine:** 3.7L V6 VIN A, 3.7L V6 VIN B, 3.7L V6 VIN V **Transmission:** All	**HO2S-21 (B2 S1) Heater Circuit Conditions:** Engine started, engine running, and the PCM detected the HO2S heater circuit indicated more than 11.5v with the heater turned "on", or it indicated less than 5.8v with the heater turned "off" for 331-361 seconds during the CCM test. **Possible Causes:** • HO2S heater control circuit is open or shorted to ground • HO2S heater control circuit is shorted to system power (B+) • HO2S heater power circuit open between the heater and PCM • HO2S heater is damaged or has failed • PCM has failed (it controls the heater with a duty cycle signal)
DTC: P0155 **2T , MIL: Yes** **Year:** 2009 **Model:** B2300, B4000 **Engine:** 2.3L L4 VIN D, 4.0L V6 VIN E **Transmission:** All	**HO2S-21 (B2 S1) Heater Circuit Conditions:** Engine started, engine running, and the PCM detected the HO2S heater circuit indicated more than 11.5v with the heater turned "on", or it indicated less than 5.8v with the heater turned "off" for 331-361 seconds during the CCM test. **Possible Causes:** • HO2S heater control circuit is open or shorted to ground • HO2S heater control circuit is shorted to system power (B+) • HO2S heater power circuit open between the heater and PCM • HO2S heater is damaged or has failed • PCM has failed (it controls the heater with a duty cycle signal)
DTC: P0158 **2T O2S1, MIL: Yes** **Year:** 2009, 2010 **Model:** 6, CX-9 **Engine:** 3.7L V6 VIN A, 3.7L V6 VIN B, 3.7L V6 VIN V **Transmission:** All	**HO2S-22 (Bank 2 Sensor 2) Circuit High Input:** Engine started, vehicle driven at cruise at low engine load for over 5 minutes, and the PCM detected the rear HO2S signal remained too high, the signal was fixed from 350-550 mv, or the HO2S signal switch time was too long during the CCM test. **Possible Causes:** • Air leaks at the exhaust manifold or exhaust pipes • HO2S signal tracking (wet/oily) in connector causing a short between the signal circuit and heater power circuit • HO2S contaminated with wrong fuel, has deteriorated or failed • PCM has failed
DTC: P0158 **T** **Year:** 2009, 2010 **Model:** Tribute **Engine:** 2.5L L4 VIN 3, 2.5L L4 VIN 7, 3.0L V6 VIN 1, 3.0L V6 VIN G **Transmission:** All	**HO2S-22 (Bank 2 Sensor 2) Circuit High Input:** Engine started, vehicle driven at cruise at low engine load for over 5 minutes, and the PCM detected the rear HO2S signal remained too high, the signal was fixed from 350-550 mv, or the HO2S signal switch time was too long during the CCM test. **Possible Causes:** • Air leaks at the exhaust manifold or exhaust pipes • HO2S signal tracking (wet/oily) in connector causing a short between the signal circuit and heater power circuit • HO2S contaminated with wrong fuel, has deteriorated or failed • PCM has failed
DTC: P0159 **T O2S** **Year:** 2009, 2010 **Model:** B2300, B4000, CX-9 **Engine:** 2.3L L4 VIN D, 3.7L V6 VIN A, 3.7L V6 VIN V, 4.0L V6 VIN E **Transmission:** All	**Rear HO2S (LH) circuit slow response:** The HO2S monitor checks the HO2S frequency and amplitude. The test fails if the frequency and amplitude fall below a calibrated limit during testing. **Possible Causes:** • Inlet air leaks • MAF sensor malfunction • Exhaust leaks • Open circuit in wiring harness between PCM (wiring harness-side) and rear HO2S (LH) (wiring harness-side) • Open circuit in wiring harness between PCM (wiring harness-side) and HO2S (wiring harness-side) • Short to ground in wiring harness between PCM (wiring harness-side) and PCM (wiring harness-side) • Contaminated HO2S • Incorrect fueling • Short to ground in wiring harness between HO2S (wiring harness-side) and battery negative terminal • Short to ground in wiring harness between HO2S (wiring harness-side) and HO2S (wiring harness-side) • Short to ground in wiring harness between PCM and rear HO2S (LH) • HO2S malfunction • Deteriorating HO2S • PCM malfunction

DTC	Trouble Code Title, Conditions & Possible Causes
DTC: P0161 **T** **Year:** 2009, 2010 **Model:** 6, B2300, B4000, CX-9, Tribute **Engine:** 2.3L L4 VIN D, 2.5L L4 VIN 3, 2.5L L4 VIN 7, 3.0L V6 VIN 1, 3.0L V6 VIN G, 3.7L V6 VIN A, 3.7L V6 VIN B, 3.7L V6 VIN V, 4.0L V6 VIN E **Transmission:** All	**HO2S-22 (Bank 2 Sensor 2) Heater Circuit Malfunction:** Engine started, engine running, and the PCM detected the HO2S heater circuit indicated more than 11.5v with the heater turned "on", or it indicated less than 5.8v with the heater turned "off" for 331-361 seconds during the CCM test. **Possible Causes:** • HO2S heater control circuit is open or shorted to ground • HO2S heater control circuit is shorted to system power (B+) • HO2S heater power circuit open between the heater and PCM • HO2S heater is damaged or has failed • PCM has failed (it controls the heater with a duty cycle signal)
DTC: P0171 **2T CCM, MIL: Yes** **Year:** 2009 **Model:** B2300, B4000 **Engine:** 2.3L L4 VIN D, 4.0L V6 VIN E **Transmission:** All	**Adaptive Fuel Trim Too Lean (Bank 1):** Engine running in closed loop at cruise speed for 2 minutes, and the PCM detected the A/F ratio remained leaner than the fuel correction limit in memory for 10 seconds during the Fuel System Monitor test. **Possible Causes:** • Vehicle driven low on fuel or until it ran out of fuel • One or more injectors restricted or pressure regulator has failed • Fuel delivery system supplying too much or too little fuel during cruise or idle periods (e.g., faulty fuel pump, or dirty fuel filter) • Air leaks after the MAF sensor, or air leaks in the PCV system • Air leaks at the EGR gasket, or at the EGR valve diaphragm • Exhaust leaks before or near where the front HO2S is mounted • HO2S is contaminated, deteriorated or it has failed • Fuel control sensor is out of calibration (i.e., ECT, IAT or MAP) • Base engine "mechanical" fault affecting one or more cylinders
DTC: P0171 **2T CCM, MIL: Yes** **Year:** 2009, 2010 **Model:** 3, 6, CX-7, CX-9, MX-5 Miata, RX-8, Tribute **Engine:** 1.3L R2 VIN 2, 1.3L R2 VIN 4, 1.3L R2 VIN M, 1.3L R2 VIN P, 2.0L L4 VIN F, 2.0L L4 VIN G, 2.3L L4 VIN 3, 2.3L L4 VIN 4, 2.3L L4 VIN L, 2.3L L4 VIN M, 2.5L L4 VIN 3, 2.5L L4 VIN 5, 2.5L L4 VIN 6, 2.5L L4 VIN 7, 2.5L L4 VIN A, 2.5L L4 VIN M, 3.0L V6 VIN 1, 3.0L V6 VIN G, 3.7L V6 VIN A, 3.7L V6 VIN B, 3.7L V6 VIN V **Transmission:** All	**Adaptive Fuel Trim Too Lean (Bank 1):** Engine running in closed loop at cruise speed for 2 minutes, and the PCM detected the A/F ratio remained leaner than the fuel correction limit in memory for 10 seconds during the Fuel System Monitor test. **Possible Causes:** • Air leaks after the MAF sensor, or in the PCV or EGR system • Base engine "mechanical" fault affecting one or more cylinders • Exhaust leaks before or near where the front HO2S is mounted • Fuel control sensor out of calibration (i.e., ECT, IAT or MAF) • Fuel delivery system supplying too much or too little fuel during idle or cruise speed (weak fuel pump, weak pressure regulator) • Fuel Injectors (one or more) is dirty, restricted or leaking fuel • HO2S is contaminated, deteriorated or it has failed • Vehicle driven low on fuel or driven until it ran out of fuel
DTC: P0172 **2T CCM, MIL: Yes** **Year:** 2009 **Model:** B2300, B4000 **Engine:** 2.3L L4 VIN D, 4.0L V6 VIN E **Transmission:** All	**Adaptive Fuel Trim Too Rich (Bank 1):** Engine running in closed loop at cruise speed for 2 minutes, and the PCM detected the A/F ratio remained richer than the fuel correction limit in memory for 10 seconds during the Fuel System Monitor test. **Possible Causes:** • One or more injectors leaking or pressure regulator is leaking • Fuel delivery system supplying too much fuel during cruise or idle periods (e.g., faulty fuel pump, or faulty pressure regulator) • Exhaust leaks before or near where the front HO2S is mounted • HO2S is contaminated, deteriorated or it has failed • EVAP system component has failed or canister fuel saturated • Fuel control sensor is out of calibration (i.e., ECT, IAT or MAP) • Base engine "mechanical" fault affecting one or more cylinders

DTC	Trouble Code Title, Conditions & Possible Causes
DTC: P0172 **2T CCM, MIL: Yes** **Year:** 2009, 2010 **Model:** 3, 6, CX-7, CX-9, MX-5 Miata, RX-8, Tribute **Engine:** 1.3L R2 VIN 2, 1.3L R2 VIN 4, 1.3L R2 VIN M, 1.3L R2 VIN P, 2.0L L4 VIN F, 2.0L L4 VIN G, 2.3L L4 VIN 3, 2.3L L4 VIN 4, 2.3L L4 VIN L, 2.3L L4 VIN M, 2.5L L4 VIN 3, 2.5L L4 VIN 5, 2.5L L4 VIN 6, 2.5L L4 VIN 7, 2.5L L4 VIN A, 2.5L L4 VIN M, 3.0L V6 VIN 1, 3.0L V6 VIN G, 3.7L V6 VIN A, 3.7L V6 VIN B, 3.7L V6 VIN V **Transmission:** All	**Adaptive Fuel Trim Too Rich (Bank 1):** Engine running in closed loop at cruise speed for 2 minutes, and the PCM detected the A/F ratio remained richer than the fuel correction limit in memory for 10 seconds during the Fuel System Monitor test. **Possible Causes:** • Base engine "mechanical" fault affecting one or more cylinders • EVAP system component has failed or canister fuel saturated • Exhaust leaks before or near where the front HO2S is mounted • Fuel control sensor out of calibration (i.e., IAT, MAP or MAF) • Fuel injectors (one or more) sticking or leaking fuel • Fuel delivery system supplying too much fuel during cruise or idle periods (e.g., faulty fuel pump, or faulty pressure regulator) • HO2S is contaminated, deteriorated or it has failed
DTC: P0174 **2T CCM, MIL: Yes** **Year:** 2009, 2010 **Model:** 6, B4000, CX-9, Tribute **Engine:** 3.0L V6 VIN 1, 3.0L V6 VIN G, 3.7L V6 VIN A, 3.7L V6 VIN B, 3.7L V6 VIN V, 4.0L V6 VIN E **Transmission:** All	**Adaptive Fuel Trim Too Lean (Bank 2):** Engine running in closed loop at cruise speed for 2 minutes, and the PCM detected the A/F ratio remained leaner than the fuel correction limit in memory for 10 seconds during the Fuel System Monitor test. **Possible Causes:** • Vehicle driven low on fuel or until it ran out of fuel • One or more injectors restricted or pressure regulator has failed • Fuel delivery system supplying too much or too little fuel during cruise or idle periods (e.g., faulty fuel pump, or dirty fuel filter) • Air leaks after the MAF sensor, or air leaks in the PCV system • Air leaks at the EGR gasket, or at the EGR valve diaphragm • Exhaust leaks before or near where the front HO2S is mounted • HO2S is contaminated, deteriorated or it has failed • Fuel control sensor is out of calibration (i.e., ECT, IAT or MAP) • Base engine "mechanical" fault affecting one or more cylinders
DTC: P0175 **2T CCM, MIL: Yes** **Year:** 2009, 2010 **Model:** 6, B4000, CX-9, Tribute **Engine:** 3.0L V6 VIN 1, 3.0L V6 VIN G, 3.7L V6 VIN A, 3.7L V6 VIN B, 3.7L V6 VIN V, 4.0L V6 VIN E **Transmission:** All	**Adaptive Fuel Trim Too Rich (Bank 2):** Engine running in closed loop at cruise speed for 2 minutes, and the PCM detected the A/F ratio remained richer than the fuel correction limit in memory for 10 seconds during the Fuel System Monitor test. **Possible Causes:** • One or more injectors leaking or pressure regulator is leaking • Fuel delivery system supplying too much fuel during cruise or idle periods (e.g., faulty fuel pump, or faulty pressure regulator) • Exhaust leaks before or near where the front HO2S is mounted • HO2S is contaminated, deteriorated or it has failed • EVAP system component has failed or canister fuel saturated • Fuel control sensor is out of calibration (i.e., ECT, IAT or MAP) • Base engine "mechanical" fault affecting one or more cylinders
DTC: P0191 **2T CCM, MIL: Yes** **Year:** 2009, 2010 **Model:** 3, CX-7 **Engine:** 2.3L L4 VIN 3, 2.3L L4 VIN 4, 2.3L L4 VIN L, 2.3L L4 VIN M **Transmission:** All	**Fuel pressure sensor circuit range/performance problem:** If the fluctuation range of the actual fuel pressure is below the set value when the fluctuation range of the target fuel pressure has exceeded the set value, the PCM determines that the fuel pressure sensor performance problem. **Possible Causes:** • Air suction in fuel line due to fuel is runout • High pressure fuel pump connector or terminals malfunction • Fuel pressure sensor connector or terminals malfunction • PCM connector or terminals malfunction • Fuel pressure sensor malfunction • Insufficient fuel pressure (low pressure line) • Fuel injector malfunction (leakage fuel) • Fuel pressure limiter malfunction • PCM malfunction

DTC	Trouble Code Title, Conditions & Possible Causes
DTC: P0192 **2T CCM, MIL: Yes** **Year:** 2009, 2010 **Model:** 3, CX-7 **Engine:** 2.3L L4 VIN 3, 2.3L L4 VIN 4, 2.3L L4 VIN L, 2.3L L4 VIN M **Transmission:** All	**Fuel pressure sensor circuit low input:** If the input voltage from the fuel pressure sensor is less than 0.19 V for 5.3 s, the PCM determines that the fuel pressure sensor circuit is low. **Possible Causes:** • Fuel pressure sensor connector or terminals malfunction • Short to ground in wiring harness between the Fuel pressure sensor and PCM • PCM connector or terminals malfunction • Fuel pressure sensor signal circuit and ground circuit are shorted to each other • Open circuit in wiring harness between fuel pressure sensor and PCM terminal • Fuel pressure sensor malfunction • PCM malfunction
DTC: P0193 **2T CCM, MIL: Yes** **Year:** 2009, 2010 **Model:** 3, CX-7 **Engine:** 2.3L L4 VIN 3, 2.3L L4 VIN 4, 2.3L L4 VIN L, 2.3L L4 VIN M **Transmission:** All	**Fuel pressure sensor circuit high input:** If the input voltage from the fuel pressure sensor is more than 4.8 V for 5.3 s, the PCM determines that the fuel pressure sensor circuit is high. **Possible Causes:** • Fuel pressure sensor connector or terminals malfunction • PCM connector or terminals malfunction • Short to power supply in wiring harness between fuel pressure sensor and PCM • Fuel pressure sensor power supply circuit and signal circuit are shorted to each other • Open circuit in wiring harness between the Fuel pressure sensor terminal and PCM • Fuel pressure sensor malfunction • PCM malfunction
DTC: P0201 **1T CCM, MIL: Yes** **Year:** 2009, 2010 **Model:** 3, 6, B2300, B4000, CX-7, CX-9, Tribute **Engine:** 2.3L L4 VIN 3, 2.3L L4 VIN 4, 2.3L L4 VIN D, 2.3L L4 VIN L, 2.3L L4 VIN M, 2.5L L4 VIN 3, 2.5L L4 VIN 7, 3.0L V6 VIN 1, 3.0L V6 VIN G, 3.7L V6 VIN A, 3.7L V6 VIN B, 3.7L V6 VIN V, 4.0L V6 VIN E **Transmission:** All	**Fuel Injector 1 Control Circuit Malfunction:** Engine running and the PCM detected an unexpected voltage condition on the fuel injector control circuit during the CCM test. **Possible Causes:** • Fuel injector 1 control circuit is open or grounded • Fuel injector 1 power circuit open between injector and VPWR • Fuel injector 1 has failed • PCM has failed (i.e., the PCM driver for Injector 1 has failed)
DTC: P0202 **1T CCM, MIL: Yes** **Year:** 2009, 2010 **Model:** 3, 6, B2300, B4000, CX-7, CX-9, Tribute **Engine:** 2.3L L4 VIN 3, 2.3L L4 VIN 4, 2.3L L4 VIN D, 2.3L L4 VIN L, 2.3L L4 VIN M, 2.5L L4 VIN 3, 2.5L L4 VIN 5, 2.5L L4 VIN 7, 3.0L V6 VIN 1, 3.0L V6 VIN G, 3.7L V6 VIN A, 3.7L V6 VIN B, 3.7L V6 VIN V, 4.0L V6 VIN E **Transmission:** All	**Fuel Injector 2 Control Circuit Malfunction:** Engine running and the PCM detected an unexpected voltage condition on the fuel injector control circuit during the CCM test. **Possible Causes:** • Fuel injector 2 control circuit is open or grounded • Fuel injector 2 power circuit open between injector and VPWR • Fuel injector 2 has failed • PCM has failed (i.e., the PCM driver for Injector 2 has failed)
DTC: P0203 **1T CCM, MIL: Yes** **Year:** 2009, 2010 **Model:** 3, 6, B2300, B4000, CX-7, CX-9, Tribute **Engine:** 2.3L L4 VIN 3, 2.3L L4 VIN 4, 2.3L L4 VIN D, 2.3L L4 VIN L, 2.3L L4 VIN M, 2.5L L4 VIN 3, 2.5L L4 VIN 7, 3.0L V6 VIN 1, 3.0L V6 VIN G, 3.7L V6 VIN A, 3.7L V6 VIN B, 3.7L V6 VIN V, 4.0L V6 VIN E **Transmission:** All	**Fuel Injector 3 Control Circuit Malfunction:** Engine running and the PCM detected an unexpected voltage condition on the fuel injector control circuit during the CCM test. **Possible Causes:** • Fuel injector 3 control circuit is open or grounded • Fuel injector 3 power circuit open between injector and VPWR • Fuel injector 3 has failed • PCM has failed (i.e., the PCM driver for Injector 3 has failed)

DTC	Trouble Code Title, Conditions & Possible Causes
DTC: P0204 **1T CCM, MIL: Yes** **Year:** 2009, 2010 **Model:** 3, 6, B2300, B4000, CX-7, CX-9, Tribute **Engine:** 2.3L L4 VIN 3, 2.3L L4 VIN 4, 2.3L L4 VIN D, 2.3L L4 VIN L, 2.3L L4 VIN M, 2.5L L4 VIN 3, 2.5L L4 VIN 7, 3.0L V6 VIN 1, 3.0L V6 VIN G, 3.7L V6 VIN A, 3.7L V6 VIN B, 3.7L V6 VIN V, 4.0L V6 VIN E **Transmission:** All	**Fuel Injector 4 Control Circuit Malfunction:** Engine running and the PCM detected an unexpected voltage condition on the fuel injector control circuit during the CCM test. **Possible Causes:** • Fuel injector 4 control circuit is open or grounded • Fuel injector 4 power circuit open between injector and VPWR • Fuel injector 4 has failed • PCM has failed (i.e., the PCM driver for Injector 4 has failed)
DTC: P0205 **1T CCM, MIL: Yes** **Year:** 2009, 2010 **Model:** 6, B4000, CX-9, Tribute **Engine:** 3.0L V6 VIN 1, 3.0L V6 VIN G, 3.7L V6 VIN A, 3.7L V6 VIN B, 3.7L V6 VIN V, 4.0L V6 VIN E **Transmission:** All	**Fuel Injector 5 Control Circuit Malfunction:** Engine running and the PCM detected an unexpected voltage condition on the fuel injector control circuit during the CCM test. **Possible Causes:** • Fuel injector 5 control circuit is open or grounded • Fuel injector 5 power circuit open between injector and VPWR • Fuel injector 5 has failed • PCM has failed (i.e., the PCM driver for Injector 5 has failed)
DTC: P0206 **1T CCM, MIL: Yes** **Year:** 2009, 2010 **Model:** 6, B4000, CX-9, Tribute **Engine:** 3.0L V6 VIN 1, 3.0L V6 VIN G, 3.7L V6 VIN A, 3.7L V6 VIN B, 3.7L V6 VIN V, 4.0L V6 VIN E **Transmission:** All	**Fuel Injector 6 Control Circuit Malfunction:** Engine running and the PCM detected an unexpected voltage condition on the fuel injector control circuit during the CCM test. **Possible Causes:** • Fuel injector 6 control circuit is open or grounded • Fuel injector 6 power circuit open between injector and VPWR • Fuel injector 6 has failed • PCM has failed (i.e., the PCM driver for Injector 6 has failed)
DTC: P0218 **2T CCM, MIL: Yes** **Year:** 2009 **Model:** B2300, B4000 **Engine:** 2.3L L4 VIN D, 4.0L V6 VIN E **Transmission:** All	**Transmission Fluid Temperature Over-Temperature Condition:** Indicates a transmission overheat condition was sensed by the Transmission Fluid Temperature (TFT) sensor. **NOTE: Monitor the transmission temperature PID TFT for an overheat condition.** **Possible Causes:** • Low transmission fluid level • Transmission cooling system concerns
DTC: P0219 **1T CCM, MIL: Yes** **Year:** 2009, 2010 **Model:** 6, B2300, B4000, Tribute **Engine:** 2.3L L4 VIN D, 2.5L L4 VIN 3, 2.5L L4 VIN 7, 3.0L V6 VIN 1, 3.0L V6 VIN G, 3.7L V6 VIN B, 4.0L V6 VIN E **Transmission:** All	**PCM – Engine RPM Limiter, PCM software:** Indicates the vehicle has been operated in a manner, which caused the engine speed to exceed a calibration limit. The engine speed is continuously monitored and evaluated by the PCM. The DTC is set when the rpm exceeds the calibrated limit set with in the PCM. **Possible Causes:** • TP Sensor malfunction • Connector or Terminal Malfunction • Open circuit between throttle body terminal A and PCM terminal 3M • Open circuit between throttle body terminal E and PCM terminal 3N
DTC: P0222 **1T CCM, MIL: Yes** **Year:** 2009, 2010 **Model:** 3, 5, 6, B2300, B4000, CX-7, CX-9, MX-5 Miata, RX-8 **Engine:** 1.3L R2 VIN 2, 1.3L R2 VIN 4, 1.3L R2 VIN M, 1.3L R2 VIN P, 2.0L L4 VIN F, 2.0L L4 VIN G, 2.3L L4 VIN 3, 2.3L L4 VIN 4, 2.3L L4 VIN D, 2.3L L4 VIN L, 2.3L L4 VIN M, 2.5L L4 VIN 5, 2.5L L4 VIN 6, 2.5L L4 VIN A, 2.5L L4 VIN M, 3.7L V6 VIN A, 3.7L V6 VIN B, 3.7L V6 VIN V, 4.0L V6 VIN E **Transmission:** All	**TP Sensor No. 2 circuit high input:** Key on or engine running, and the PCM detects TP Sensor No. 2 voltage at PCM terminal 3J is below 0.2 v after ignition key to ON, PCM determines that TP circuit has a malfunction. **Possible Causes:** • TP Sensor malfunction • Connector or Terminal Malfunction • Open circuit between throttle body terminal A and PCM terminal 3M • Open circuit between throttle body terminal E and PCM terminal 3N

DTC	Trouble Code Title, Conditions & Possible Causes
DTC: P0223 **1T CCM, MIL: Yes** **Year:** 2009, 2010 **Model:** 3, 5, 6, B2300, B4000, CX-7, CX-9, MX-5 Miata, RX-8 **Engine:** 1.3L R2 VIN 2, 1.3L R2 VIN 4, 1.3L R2 VIN M, 1.3L R2 VIN P, 2.0L L4 VIN F, 2.0L L4 VIN G, 2.3L L4 VIN 3, 2.3L L4 VIN 4, 2.3L L4 VIN D, 2.3L L4 VIN L, 2.3L L4 VIN M, 2.5L L4 VIN 5, 2.5L L4 VIN 6, 2.5L L4 VIN A, 2.5L L4 VIN M, 3.7L V6 VIN A, 3.7L V6 VIN B, 3.7L V6 VIN V, 4.0L V6 VIN E **Transmission:** All	**TP Sensor No. 2 circuit low input:** Key on or engine running, and the PCM detects TP Sensor No. 2 voltage at PCM terminal 3J is above 4.85 v after ignition key to ON, PCM determines that TP circuit has a malfunction. **Possible Causes:** • TP Sensor malfunction • Connector or Terminal Malfunction • Open circuit between throttle body terminal D and PCM terminal 3K • Open circuit between throttle body terminal C and PCM terminal 3J
DTC: P0230 **1T CCM, MIL: Yes** **Year:** 2009, 2010 **Model:** B2300, B4000, CX-9 **Engine:** 2.3L L4 VIN D, 3.7L V6 VIN A, 3.7L V6 VIN V, 4.0L V6 VIN E **Transmission:** All	**Fuel Pump Primary or Secondary Circuit Malfunction:** Key on or engine running and the PCM detected an unexpected voltage condition on the fuel pump primary circuit during the test. **NOTE: This trouble code may set due an intermittent fault!** **Possible Causes:** • Fuel pump primary circuit is open or shorted to ground • Fuel pump primary circuit is shorted to system power (VREF) • Fuel pump relay is damaged or has failed • PCM has failed
DTC: P0231 **1T CCM, MIL: Yes** **Year:** 2009, 2010 **Model:** B2300, B4000, CX-9 **Engine:** 2.3L L4 VIN D, 3.7L V6 VIN A, 3.7L V6 VIN V, 4.0L V6 VIN E **Transmission:** All	**Fuel Pump Primary Circuit or Secondary Malfunction:** Key on, then with the Fuel Pump commanded "on", the PCM did not detect system voltage (B+) on the fuel pump monitor circuit. **Possible Causes:** • Fuel pump feed circuit open between feed circuit and the pump • Fuel pump relay contacts "open" that provide B+ to the pump • FP PWR circuit open between relay and connection to the FPM • Fuel pump relay is damaged or has failed • PCM has failed (the engine will not start if the PCM circuit fails)
DTC: P0232 **1T CCM, MIL: Yes** **Year:** 2009, 2010 **Model:** B2300, B4000, CX-9 **Engine:** 2.3L L4 VIN D, 3.7L V6 VIN A, 3.7L V6 VIN V, 4.0L V6 VIN E **Transmission:** All	**Fuel Pump Primary Circuit or Secondary Malfunction:** Key on, then with the Fuel Pump commanded "on", the PCM did not detect system voltage (B+) on the fuel pump monitor circuit. **NOTE: the Fuel Pump Driver Module (FPM) modulates the voltage to the fuel pump to achieve the correct fuel pressure. Power to the fuel pump is supplied by the power relay or FPDM power supply relay.** **Possible Causes:** • Inertia switch needs resetting or its internal contacts are open • Fuel pump circuit open between the FPM connection and the FP PWWR circuit • Fuel pump ground circuit is open or has high resistance • Fuel pump relay is damaged or has failed • PCM has failed (the engine will not start and run) • Possible Causes (Engine Starts Condition) • Fuel pump secondary circuit is shorted to system power (B+) • Fuel pump relay contacts closed all of the time • Fuel pump circuit open between the PCM connection and the FP PWR circuit • PCM has failed (the engine will start and run)
DTC: P0234 **T PCM** **Year:** 2009, 2010 **Model:** 3, CX-7 **Engine:** 2.3L L4 VIN 3, 2.3L L4 VIN 4, 2.3L L4 VIN L, 2.3L L4 VIN M **Transmission:** All	**Turbo/supercharger overboost condition:** If the manifold absolute pressure or charging efficiency are more than the specification for the specified period of time, the PCM determines that the turbocharger is in an over boost condition. **Possible Causes:** • Wastegate control solenoid valve malfunction • Vacuum hose problem • Looseness • Damage • Improper installation • PCM malfunction

DTC	Trouble Code Title, Conditions & Possible Causes
DTC: P0245 **T** **Year:** 2009, 2010 **Model:** 3, CX-7 **Engine:** 2.3L L4 VIN 3, 2.3L L4 VIN 4, 2.3L L4 VIN L, 2.3L L4 VIN M **Transmission:** All	**Turbocharger wastegate solenoid low input:** When the PCM turns the wastegate control solenoid valve off but the wastegate control solenoid valve circuit voltage is low for 5 s, the PCM determines that the wastegate control solenoid valve control circuit voltage is low. **Possible Causes:** • Wastegate control solenoid valve connector or terminals malfunction • Short to ground or open circuit in wastegate control solenoid valve power supply circuit • Short to ground in wiring harness between main relay and wastegate control solenoid valve • fuse malfunction • Open circuit in wiring harness between main relay and wastegate control solenoid valve • Short to ground in wiring harness between wastegate control solenoid valve and PCM • Wastegate control solenoid valve malfunction • PCM connector or terminals malfunction • Open circuit in wiring harness between wastegate control solenoid valve and PCM • PCM malfunction
DTC: P0246 **T** **Year:** 2009, 2010 **Model:** 3, CX-7 **Engine:** 2.3L L4 VIN 3, 2.3L L4 VIN 4, 2.3L L4 VIN L, 2.3L L4 VIN M **Transmission:** All	**Turbocharger wastegate solenoid high input:** When the PCM turns the wastegate control solenoid valve on but the wastegate control solenoid valve circuit voltage is high for 5 s, the PCM determines that the wastegate control solenoid valve control circuit voltage is high. **Possible Causes:** • Wastegate control solenoid valve connector or terminals malfunction • Wastegate control solenoid valve malfunction • PCM connector or terminals malfunction • Short to power supply in wiring harness between wastegate control solenoid valve and PCM • PCM malfunction
DTC: P025A **T PCM** **Year:** 2009, 2010 **Model:** 6 **Engine:** 3.7L V6 VIN B **Transmission:** All	**Fuel pump control circuit-open:** The PCM monitors the fuel pump command (FPC) circuit for a concern. When the PCM commands the fuel pump (FP) ON, the PCM is able to detect a short to voltage on the FPC circuit. When the PCM commands the FP OFF, the PCM is able to detect an open circuit or a short to ground on the FPC circuit. The test fails if the voltage is less than or greater than a calibrated limit, for a calibrated amount of time. **Possible Causes:** • Open circuit between electronic fuel pump relay and PCM • Short to ground between electronic fuel pump relay and PCM • Short to power supply between electronic fuel pump relay and PCM • Electronic fuel pump relay malfunction
DTC: P025B **T PCM** **Year:** 2009, 2010 **Model:** 6 **Engine:** 3.7L V6 VIN B **Transmission:** All	**Fuel pump control circuit-range/performance:** The electronic fuel pump relay monitors the duty cycle and frequency of the signal it receives from the PCM. The fuel pump control module determines if the signal from the PCM on the fuel pump command (FPC) circuit is a valid duty cycle and frequency. If the duty cycle or frequency is invalid, the fuel pump control module sends a 20% duty cycle signal on the fuel pump monitor (FPM) circuit to report the concern to the PCM. The test fails if the fuel pump control module is still reporting that it is receiving an invalid duty cycle or frequency from the PCM after a calibrated amount of time. **Possible Causes:** • Open circuit between electronic fuel pump relay and PCM • Short to ground between electronic fuel pump relay and PCM • Short to power supply between electronic fuel pump relay and PCM • Radio frequency interference/electromagnetic interference (RFI/EMI) • Electronic fuel pump relay malfunction • PCM malfunction
DTC: P0297 **T PCM** **Year:** 2009, 2010 **Model:** 6, B2300, B4000, CX-9 **Engine:** 2.3L L4 VIN D, 3.7L V6 VIN A, 3.7L V6 VIN B, 3.7L V6 VIN V, 4.0L V6 VIN E **Transmission:** All	**Vehicle over speed condition:** P0297 indicates that the vehicle has been operated in a manner which caused the engine or vehicle to exceed a calibration limit. **Possible Causes:** • Wheel slippage (water, ice, mud and snow). • Excessive engine rpm in neutral. • Vehicle drive at a high rate of speed.

DTC	Trouble Code Title, Conditions & Possible Causes
DTC: P0298 **1T CCM, MIL: Yes** **Year:** 2009, 2010 **Model:** B2300, B4000, Tribute **Engine:** 2.3L L4 VIN D, 2.5L L4 VIN 3, 2.5L L4 VIN 7, 3.0L V6 VIN 1, 3.0L V6 VIN G, 4.0L V6 VIN E **Transmission:** All	**Engine Oil Over Temperature Condition:** Oil protection strategy in the PCM has been activated. PCM uses an oil algorithm to infer actual oil temperature. **Possible Causes:** • Overheating condition • Basic engine concerns • Engine cooling concerns • Open circuit or short in ECT harness • PCM has failed
DTC: P0300 **2T CCM, MIL: Yes** **Year:** 2009 **Model:** B2300, B4000 **Engine:** 2.3L L4 VIN D, 4.0L V6 VIN E **Transmission:** All	**Multiple Misfire Detected:** Engine started, vehicle speed over 3 mph at 400-4000 rpm, then the PCM detected irregular CKP signals indicating a random Misfire in two or more cylinders during the 200 or 1000-revolution Misfire test. **NOTE: If the misfire is severe, the MIL will flash on/off on the 1st trip!** **Possible Causes:** • Base engine mechanical fault affecting one or more cylinders • CMP sensor is damaged or failed (problem may be intermittent) • Fuel metering fault that affects more than one cylinder • Fuel pressure too low or too high, fuel supply contaminated • EVAP system problem or the EVAP canister is fuel saturated • EGR valve is stuck open or the PCV system has a vacuum leak • Ignition system fault that affects more than one cylinder • MAF sensor is contaminated (this can cause a lean condition)
DTC: P0300 **2T CCM, MIL: Yes** **Year:** 2009, 2010 **Model:** 3, 5, 6, CX-7, CX-9, MX-5 Miata, RX-8, Tribute **Engine:** 1.3L R2 VIN 2, 1.3L R2 VIN 4, 1.3L R2 VIN M, 1.3L R2 VIN P, 2.0L L4 VIN F, 2.0L L4 VIN G, 2.3L L4 VIN 3, 2.3L L4 VIN 4, 2.3L L4 VIN L, 2.3L L4 VIN M, 2.5L L4 VIN 3, 2.5L L4 VIN 5, 2.5L L4 VIN 6, 2.5L L4 VIN 7, 2.5L L4 VIN A, 2.5L L4 VIN M, 3.0L V6 VIN 1, 3.0L V6 VIN G, 3.7L V6 VIN A, 3.7L V6 VIN B, 3.7L V6 VIN V **Transmission:** All	**Random Misfire Detected:** Engine started, vehicle speed over 3 mph at 400-4000 rpm, and the PCM detected a random Misfire condition in two or more cylinders during the 200 or 1000-revolution Misfire test, or the PCM could not identify the misfiring cylinder due to a problem in the CMP sensor. **NOTE: If the misfire is severe, the MIL will flash on/off on the 1st trip!** **Possible Causes:** • Base engine mechanical fault affecting one or more cylinders • CMP sensor is damaged or failed (problem may be intermittent) • Fuel metering fault that affects more than one cylinder • Fuel pressure too low or too high, fuel supply contaminated • EVAP system problem or the EVAP canister is fuel saturated • EGR valve is stuck open or the PCV system has a vacuum leak • Ignition system fault that affects more than one cylinder • MAF sensor is contaminated (this can cause a lean condition)
DTC: P0301 **2T CCM, MIL: Yes** **Year:** 2009 **Model:** B2300, B4000 **Engine:** 2.3L L4 VIN D, 4.0L V6 VIN E **Transmission:** All	**Cylinder 1 Misfire Detected:** Engine started, vehicle speed over 3 mph at 400-4000 rpm, and the PCM detected irregular CKP signals indicating a Misfire condition present in one cylinder during the 200 or 1000-revolution Misfire test. **NOTE: If the misfire is severe, the MIL will flash on/off on the 1st trip!** **Possible Causes:** • Base engine mechanical fault affecting only Cylinder 1 • CKP sensor is damaged or failed (problem may be intermittent) • Fuel metering fault that affects only Cylinder 1 • Ignition system fault that affects only Cylinder 1

DTC	Trouble Code Title, Conditions & Possible Causes
DTC: P0301 **1T MISFIRE, MIL: Yes** **Year:** 2009, 2010 **Model:** RX-8 **Engine:** 1.3L R2 VIN 2, 1.3L R2 VIN 4, 1.3L R2 VIN M, 1.3L R2 VIN P **Transmission:** All	**Front rotor misfire detected:** The PCM monitors the eccentric shaft position sensor input signal interval time. The PCM calculates the change of the interval time for each rotor. If the change of interval time exceeds the preprogrammed criteria, the PCM detects a misfire in the corresponding rotor. While the engine is running, the PCM counts the number of misfires that occurred at 200 eccentric shaft revolutions and 1,000 eccentric shaft revolutions and calculates the misfire ratio for each eccentric shaft revolution. If the ratio exceeds the preprogrammed criteria, the PCM determines that a misfire, which can damage the catalytic converter or affect emission performance, has occurred. **Possible Causes:** • Erratic signal to PCM • APP sensor signal malfunction • ECT sensor signal malfunction • MAF sensor signal malfunction • TP sensor signal malfunction • VSS signal malfunction • Eccentric shaft position sensor malfunction • Spark plug malfunction • High-tension lead malfunction • Excess air suction in intake-air system • Fuel injector related wiring harness malfunction • Leakage engine coolant • Insufficient compression • Metering oil pump malfunction • Engine oil condition malfunction • Increased oil pressure • Oil passage malfunction • Engine malfunction • Fuel line pressure malfunction • Fuel pump unit malfunction • Pressure regulator (built-in fuel pump unit) malfunction • Fuel injector malfunction • PCM malfunction
DTC: P0301 **2T CCM, MIL: Yes** **Year:** 2009, 2010 **Model:** 3, 5, 6, CX-7, CX-9, MX-5 Miata, Tribute **Engine:** 2.0L L4 VIN F, 2.0L L4 VIN G, 2.3L L4 VIN 3, 2.3L L4 VIN 4, 2.3L L4 VIN L, 2.3L L4 VIN M, 2.5L L4 VIN 3, 2.5L L4 VIN 5, 2.5L L4 VIN 6, 2.5L L4 VIN 7, 2.5L L4 VIN A, 2.5L L4 VIN M, 3.0L V6 VIN 1, 3.0L V6 VIN G, 3.7L V6 VIN A, 3.7L V6 VIN B, 3.7L V6 VIN V **Transmission:** All	**Cylinder 1 Misfire Detected:** Engine started, vehicle speed over 3 mph at 400-4000 rpm, and the PCM detected a Misfire condition in a single cylinder during the 200 or 1000-revolution Misfire test under positive engine load conditions. **NOTE: If the misfire is severe, the MIL will flash on/off on the 1st trip!** **Possible Causes:** • Base engine mechanical fault affecting only Cylinder 1 • CKP sensor is damaged or failed (problem may be intermittent) • Fuel metering fault that affects only Cylinder 1 • Ignition system fault that affects only Cylinder 1
DTC: P0302 **2T CCM, MIL: Yes** **Year:** 2009 **Model:** B2300, B4000 **Engine:** 2.3L L4 VIN D, 4.0L V6 VIN E **Transmission:** All	**Cylinder 2 Misfire Detected:** Engine started, vehicle speed over 3 mph at 400-4000 rpm, and the PCM detected irregular CKP signals indicating a Misfire condition present in one cylinder during the 200 or 1000-revolution Misfire test. **NOTE: If the misfire is severe, the MIL will flash on/off on the 1st trip!** **Possible Causes:** • Base engine mechanical fault affecting only Cylinder 2 • CKP sensor is damaged or failed (problem may be intermittent) • Fuel metering fault that affects only Cylinder 2 • Ignition system fault that affects only Cylinder 2

DTC	Trouble Code Title, Conditions & Possible Causes
DTC: P0302 **2T CCM, MIL: Yes** **Year:** 2009, 2010 **Model:** 3, 5, 6, CX-7, CX-9, MX-5 Miata, Tribute **Engine:** 2.0L L4 VIN F, 2.0L L4 VIN G, 2.3L L4 VIN 3, 2.3L L4 VIN 4, 2.3L L4 VIN L, 2.3L L4 VIN M, 2.5L L4 VIN 3, 2.5L L4 VIN 5, 2.5L L4 VIN 6, 2.5L L4 VIN 7, 2.5L L4 VIN A, 2.5L L4 VIN M, 3.0L V6 VIN 1, 3.0L V6 VIN G, 3.7L V6 VIN A, 3.7L V6 VIN B, 3.7L V6 VIN V **Transmission:** All	**Cylinder 2 Misfire Detected:** Engine started, vehicle speed over 3 mph at 400-4000 rpm, and the PCM detected a Misfire condition in a single cylinder during the 200 or 1000-revolution Misfire test under positive engine load conditions. **NOTE: If the misfire is severe, the MIL will flash on/off on the 1st trip!** **Possible Causes:** • Base engine mechanical fault affecting only Cylinder 2 • CKP sensor is damaged or failed (problem may be intermittent) • Fuel metering fault that affects only Cylinder 2 • Ignition system fault that affects only Cylinder 2
DTC: P0302 **1T MISFIRE, MIL: Yes** **Year:** 2009, 2010 **Model:** RX-8 **Engine:** 1.3L R2 VIN 2, 1.3L R2 VIN 4, 1.3L R2 VIN M, 1.3L R2 VIN P **Transmission:** All	**Rear rotor misfire detected:** The PCM monitors the eccentric shaft position sensor input signal interval time. The PCM calculates the change of the interval time for each rotor. If the change of interval time exceeds the preprogrammed criteria, the PCM detects a misfire in the corresponding rotor. While the engine is running, the PCM counts the number of misfires that occurred at 200 eccentric shaft revolutions and 1,000 eccentric shaft revolutions and calculates the misfire ratio for each eccentric shaft revolution. If the ratio exceeds the preprogrammed criteria, the PCM determines that a misfire, which can damage the catalytic converter or affect emission performance, has occurred. **Possible Causes:** • Erratic signal to PCM • APP sensor signal malfunction • ECT sensor signal malfunction • MAF sensor signal malfunction • TP sensor signal malfunction • VSS signal malfunction • Eccentric shaft position sensor malfunction • Spark plug malfunction • High-tension lead malfunction • Excess air suction in intake-air system • Fuel injector related wiring harness malfunction • Leakage engine coolant • Insufficient compression • Metering oil pump malfunction • Engine oil condition malfunction • Increased oil pressure • Oil passage malfunction • Engine malfunction • Fuel line pressure malfunction • Fuel pump unit malfunction • Pressure regulator (built-in fuel pump unit) malfunction • Fuel injector malfunction • PCM malfunction
DTC: P0303 **2T CCM, MIL: Yes** **Year:** 2009, 2010 **Model:** 3, 5, 6, CX-7, CX-9, MX-5 Miata, Tribute **Engine:** 2.0L L4 VIN F, 2.0L L4 VIN G, 2.3L L4 VIN 3, 2.3L L4 VIN 4, 2.3L L4 VIN L, 2.3L L4 VIN M, 2.5L L4 VIN 3, 2.5L L4 VIN 5, 2.5L L4 VIN 6, 2.5L L4 VIN 7, 2.5L L4 VIN A, 2.5L L4 VIN M, 3.0L V6 VIN 1, 3.0L V6 VIN G, 3.7L V6 VIN A, 3.7L V6 VIN B, 3.7L V6 VIN V **Transmission:** All	**Cylinder 3 Misfire Detected:** Engine started, vehicle speed over 3 mph at 400-4000 rpm, and the PCM detected a Misfire condition in a single cylinder during the 200 or 1000-revolution Misfire test under positive engine load conditions. **NOTE: If the misfire is severe, the MIL will flash on/off on the 1st trip!** **Possible Causes:** • Base engine mechanical fault affecting only Cylinder 3 • CKP sensor is damaged or failed (problem may be intermittent) • Fuel metering fault that affects only Cylinder 3 • Ignition system fault that affects only Cylinder 3

DTC	Trouble Code Title, Conditions & Possible Causes

DTC: P0303
2T CCM, MIL: Yes
Year: 2009
Model: B2300, B4000
Engine: 2.3L L4 VIN D, 4.0L V6 VIN E
Transmission: All
	Cylinder 3 Misfire Detected:
	Engine started, vehicle speed over 3 mph at 400-4000 rpm, and the PCM detected irregular CKP signals indicating a Misfire condition present in one cylinder during the 200 or 1000-revolution Misfire test.
	NOTE: If the misfire is severe, the MIL will flash on/off on the 1st trip!
	Possible Causes:
	  • Base engine mechanical fault affecting only Cylinder 3
	  • CKP sensor is damaged or failed (problem may be intermittent)
	  • Fuel metering fault that affects only Cylinder 3
	  • Ignition system fault that affects only Cylinder 3

DTC: P0304
2T CCM, MIL: Yes
Year: 2009, 2010
Model: 3, 5, 6, CX-7, CX-9, MX-5 Miata, Tribute
Engine: 2.0L L4 VIN F, 2.0L L4 VIN G, 2.3L L4 VIN 3, 2.3L L4 VIN 4, 2.3L L4 VIN L, 2.3L L4 VIN M, 2.5L L4 VIN 3, 2.5L L4 VIN 5, 2.5L L4 VIN 6, 2.5L L4 VIN 7, 2.5L L4 VIN A, 2.5L L4 VIN M, 3.0L V6 VIN 1, 3.0L V6 VIN G, 3.7L V6 VIN A, 3.7L V6 VIN B, 3.7L V6 VIN V
Transmission: All
	Cylinder 4 Misfire Detected:
	Engine started, vehicle speed over 3 mph at 400-4000 rpm, and the PCM detected a Misfire condition in a single cylinder during the 200 or 1000-revolution Misfire test under positive engine load conditions.
	NOTE: If the misfire is severe, the MIL will flash on/off on the 1st trip!
	Possible Causes:
	  • Base engine mechanical fault affecting only Cylinder 4
	  • CKP sensor is damaged or failed (problem may be intermittent)
	  • Fuel metering fault that affects only Cylinder 4
	  • Ignition system fault that affects only Cylinder 4

DTC: P0304
2T CCM, MIL: Yes
Year: 2009
Model: B2300, B4000
Engine: 2.3L L4 VIN D, 4.0L V6 VIN E
Transmission: All
	Cylinder 4 Misfire Detected:
	Engine started, vehicle speed over 3 mph at 400-4000 rpm, and the PCM detected irregular CKP signals indicating a Misfire condition present in one cylinder during the 200 or 1000-revolution Misfire test.
	NOTE: If the misfire is severe, the MIL will flash on/off on the 1st trip!
	Possible Causes:
	  • Base engine mechanical fault affecting only Cylinder 4
	  • CKP sensor is damaged or failed (problem may be intermittent)
	  • Fuel metering fault that affects only Cylinder 4
	  • Ignition system fault that affects only Cylinder 4

DTC: P0305
2T CCM, MIL: Yes
Year: 2009
Model: B4000
Engine: 4.0L V6 VIN E
Transmission: All
	Cylinder 5 Misfire Detected:
	Engine started, vehicle speed over 3 mph at 400-4000 rpm, and the PCM detected irregular CKP signals indicating a Misfire condition present in one cylinder during the 200 or 1000-revolution Misfire test.
	NOTE: If the misfire is severe, the MIL will flash on/off on the 1st trip!
	Possible Causes:
	  • Base engine mechanical fault affecting only Cylinder 5
	  • CKP sensor is damaged or failed (problem may be intermittent)
	  • Fuel metering fault that affects only Cylinder 5
	  • Ignition system fault that affects only Cylinder 5

DTC: P0305
2T CCM, MIL: Yes
Year: 2009, 2010
Model: 6, CX-9, Tribute
Engine: 3.0L V6 VIN 1, 3.0L V6 VIN G, 3.7L V6 VIN A, 3.7L V6 VIN B, 3.7L V6 VIN V
Transmission: All
	Cylinder 5 Misfire Detected:
	Engine started, vehicle speed over 3 mph at 400-4000 rpm, and the PCM detected a Misfire condition in a single cylinder during the 200 or 1000-revolution Misfire test under positive engine load conditions.
	NOTE: If the misfire is severe, the MIL will flash on/off on the 1st trip!
	Possible Causes:
	  • Base engine mechanical fault affecting only Cylinder 5
	  • CKP sensor is damaged or failed (problem may be intermittent)
	  • Fuel metering fault that affects only Cylinder 5
	  • Ignition system fault that affects only Cylinder 5

DTC: P0306
2T CCM, MIL: Yes
Year: 2009, 2010
Model: Tribute
Engine: 3.0L V6 VIN 1, 3.0L V6 VIN G
Transmission: All
	Cylinder 6 Misfire Detected:
	Engine started, vehicle speed over 3 mph at 400-4000 rpm, and the PCM detected irregular CKP signals indicating a Misfire condition present in one cylinder during the 200 or 1000-revolution Misfire test.
	NOTE: If the misfire is severe, the MIL will flash on/off on the 1st trip!
	Possible Causes:
	  • Base engine mechanical fault affecting only Cylinder 6
	  • CKP sensor is damaged or failed (problem may be intermittent)
	  • Fuel metering fault that affects only Cylinder 6
	  • Ignition system fault that affects only Cylinder 6

DTC	Trouble Code Title, Conditions & Possible Causes
DTC: P0306 **2T CCM, MIL: Yes** **Year:** 2009, 2010 **Model:** 6, B4000, CX-9 **Engine:** 3.7L V6 VIN A, 3.7L V6 VIN B, 3.7L V6 VIN V, 4.0L V6 VIN E **Transmission:** All	**Cylinder 6 Misfire Detected:** Engine started, vehicle speed over 3 mph at 400-4000 rpm, and the PCM detected irregular CKP signals indicating a Misfire condition present in one cylinder during the 200 or 1000-revolution Misfire test. **NOTE: If the misfire is severe, the MIL will flash on/off on the 1st trip!** **Possible Causes:** • Base engine mechanical fault affecting only Cylinder 6 • CKP sensor is damaged or failed (problem may be intermittent) • Fuel metering fault that affects only Cylinder 6 • Ignition system fault that affects only Cylinder 6
DTC: P0315 **2T CCM, MIL: Yes** **Year:** 2009, 2010 **Model:** 6, B2300, B4000, CX-9, Tribute **Engine:** 2.3L L4 VIN D, 2.5L L4 VIN 3, 2.5L L4 VIN 7, 3.0L V6 VIN 1, 3.0L V6 VIN G, 3.7L V6 VIN A, 3.7L V6 VIN B, 3.7L V6 VIN V, 4.0L V6 VIN E **Transmission:** All	**Misfire Detected:** Engine started, vehicle speed over 3 mph at 400-4000 rpm, and the PCM detected irregular CKP signals indicating a Misfire condition present in one cylinder during the 200 or 1000-revolution Misfire test. **NOTE: If the misfire is severe, the MIL will flash on/off on the 1st trip!** **Possible Causes:** • Base engine mechanical fault • CKP sensor is damaged or failed (problem may be intermittent) • Fuel metering fault • Ignition system fault
DTC: P0316 **2T CCM, MIL: Yes** **Year:** 2009, 2010 **Model:** Tribute **Engine:** 2.5L L4 VIN 3, 2.5L L4 VIN 7, 3.0L V6 VIN 1, 3.0L V6 VIN G **Transmission:** All	**Misfire Detected:** Engine started, vehicle speed over 3 mph at 400-4000 rpm, and the PCM detected irregular CKP signals indicating a Misfire condition present in one cylinder during the 200 or 1000-revolution Misfire test. **NOTE: If the misfire is severe, the MIL will flash on/off on the 1st trip!** **Possible Causes:** • Base engine mechanical fault • CKP sensor is damaged or failed (problem may be intermittent) • Fuel metering fault • Ignition system fault
DTC: P0316 **2T CCM, MIL: Yes** **Year:** 2009 **Model:** B2300, B4000 **Engine:** 2.3L L4 VIN D, 4.0L V6 VIN E **Transmission:** All	**Misfire Detected:** Engine started, vehicle speed over 3 mph at 400-4000 rpm, and the PCM detected irregular CKP signals indicating a Misfire condition present in one cylinder during the 200 or 1000-revolution Misfire test. **NOTE: If the misfire is severe, the MIL will flash on/off on the 1st trip!** **Possible Causes:** • Base engine mechanical fault • CKP sensor is damaged or failed (problem may be intermittent) • Fuel metering fault • Ignition system fault
DTC: P0320 **1T CCM, MIL: Yes** **Year:** 2009, 2010 **Model:** 6, B2300, B4000, CX-9, Tribute **Engine:** 2.3L L4 VIN D, 2.5L L4 VIN 3, 2.5L L4 VIN 7, 3.0L V6 VIN 1, 3.0L V6 VIN G, 3.7L V6 VIN A, 3.7L V6 VIN B, 3.7L V6 VIN V, 4.0L V6 VIN E **Transmission:** All	**Ignition Engine Speed (PIP) Signal Error:** Engine running, and the PCM did not detect any engine speed (PIP) signals, or it detected that erratic speed signals were present. **Possible Causes:** • Engine speed or PIP signal circuit is open (an intermittent fault) • Engine speed signal circuit shorted to ground (intermittent) • Ignition system components arcing (ignition coils or wires) • Vehicle onboard transmitter interruption (i.e., 2-way radio)
DTC: P0325 **1T CCM, MIL: Yes** **Year:** 2009, 2010 **Model:** B2300, B4000, Tribute **Engine:** 2.3L L4 VIN D, 2.5L L4 VIN 3, 2.5L L4 VIN 7, 3.0L V6 VIN 1, 3.0L V6 VIN G, 4.0L V6 VIN E **Transmission:** All	**Knock Sensor Circuit Malfunction:** Engine running at idle speed, and the PCM detected an unexpected voltage condition on the Knock Sensor circuit during the CCM test. **NOTE: Check the Knock Sensor installation torque and connection.** **Possible Causes:** • Knock sensor signal circuit is open • Knock sensor signal circuit is shorted to ground • Knock sensor is damaged or has failed • PCM is damaged

DTC	Trouble Code Title, Conditions & Possible Causes
DTC: P0325 **1T CCM, MIL: Yes** **Year:** 2009, 2010 **Model:** 6, CX-9 **Engine:** 3.7L V6 VIN A, 3.7L V6 VIN B, 3.7L V6 VIN V **Transmission:** All	**Knock Sensor 1 Circuit Malfunction:** Engine running at idle speed, and the PCM detected an unexpected voltage condition on the Knock Sensor circuit during the CCM test. **NOTE: Check the Knock Sensor installation torque and connection.** **Possible Causes:** • Knock sensor signal circuit is open • Knock sensor signal circuit is shorted to ground • Knock sensor is damaged or has failed • PCM is damaged
DTC: P0326 **1T CCM, MIL: Yes** **Year:** 2009 **Model:** B2300, B4000 **Engine:** 2.3L L4 VIN D, 4.0L V6 VIN E **Transmission:** All	**Knock Sensor Circuit Range/Performance:** Engine running at idle speed, and the PCM detected the knock sensor (KS1) signals during engine acceleration and deceleration changes that were outside of a calibrated value stored in memory. **Possible Causes:** • Knock sensor signal circuit is open (intermittent fault) • Knock sensor signal circuit shorted to ground (intermittent fault) • Knock sensor is damaged or has failed • PCM is damaged
DTC: P0327 **1T CCM, MIL: Yes** **Year:** 2009, 2010 **Model:** 3, 5, 6, B2300, B4000, CX-7, MX-5 Miata, RX-8 **Engine:** 1.3L R2 VIN 2, 1.3L R2 VIN 4, 1.3L R2 VIN M, 1.3L R2 VIN P, 2.0L L4 VIN F, 2.0L L4 VIN G, 2.3L L4 VIN 3, 2.3L L4 VIN 4, 2.3L L4 VIN D, 2.3L L4 VIN L, 2.3L L4 VIN M, 2.5L L4 VIN 5, 2.5L L4 VIN 6, 2.5L L4 VIN A, 2.5L L4 VIN M, 4.0L V6 VIN E **Transmission:** All	**Knock Sensor Circuit Low Input:** Key on or engine running and the PCM detected the knock sensor (KS1) signal indicated less than 1.25v during the CCM test. **Possible Causes:** • Knock sensor signal circuit is shorted to ground • Knock sensor is damaged or has failed • PCM is damaged
DTC: P0328 **1T CCM, MIL: Yes** **Year:** 2009, 2010 **Model:** 3, 5, 6, B2300, B4000, CX-7, MX-5 Miata, RX-8 **Engine:** 1.3L R2 VIN 2, 1.3L R2 VIN 4, 1.3L R2 VIN M, 1.3L R2 VIN P, 2.0L L4 VIN F, 2.0L L4 VIN G, 2.3L L4 VIN 3, 2.3L L4 VIN 4, 2.3L L4 VIN D, 2.3L L4 VIN L, 2.3L L4 VIN M, 2.5L L4 VIN 5, 2.5L L4 VIN 6, 2.5L L4 VIN A, 2.5L L4 VIN M, 4.0L V6 VIN E **Transmission:** All	**Knock Sensor Circuit High Input:** Key on or engine running and the PCM detected the knock sensor (KS1) signal indicated more than 3.75v during the CCM test. **Possible Causes:** • Knock sensor signal circuit is open • Knock sensor signal circuit is shorted to VREF or system power • Knock sensor is damaged or has failed • PCM is damaged
DTC: P0330 **1T CCM, MIL: Yes** **Year:** 2009, 2010 **Model:** 6, CX-9, Tribute **Engine:** 2.5L L4 VIN 3, 2.5L L4 VIN 7, 3.0L V6 VIN 1, 3.0L V6 VIN G, 3.7L V6 VIN A, 3.7L V6 VIN B, 3.7L V6 VIN V **Transmission:** All	**Knock Sensor Circuit Malfunction:** Key on or engine running and the PCM detected the knock sensor (KS1) signal indicated a malfunction **Possible Causes:** • Knock sensor signal circuit is open • Knock sensor signal circuit is shorted to VREF or system power • Knock sensor is damaged or has failed • PCM is damaged

DTC	Trouble Code Title, Conditions & Possible Causes
DTC: P0332 **1T CCM, MIL: Yes** **Year:** 2009, 2010 **Model:** RX-8 **Engine:** 1.3L R2 VIN 2, 1.3L R2 VIN 4, 1.3L R2 VIN M, 1.3L R2 VIN P **Transmission:** All	**KS No.2 circuit low input:** The PCM monitors the input voltage from the KS No.2 when the engine is running. If the input voltage is less than 1.2 V, the PCM determines that the KS No.2 circuit input voltage is low. **Possible Causes:** • KS No.2 connector or terminals malfunction • Short to ground in wiring harness between KS No.2 terminal A and PCM terminal 2AC • Short to ground in wiring harness between KS No.2 terminal B and PCM terminal 2Y • KS No.2 malfunction • PCM connector or terminals malfunction • Open circuit in wiring harness between KS No.2 terminal A and PCM terminal 2AC • Open circuit in wiring harness between KS No.2 terminal B and PCM terminal 2Y • PCM malfunction
DTC: P0333 **1T CCM, MIL: Yes** **Year:** 2009, 2010 **Model:** RX-8 **Engine:** 1.3L R2 VIN 2, 1.3L R2 VIN 4, 1.3L R2 VIN M, 1.3L R2 VIN P **Transmission:** All	**KS No.2 circuit high input:** The PCM monitors the input voltage from the KS No.2 when the engine is running. If the input voltage is more than 4.0 V, the PCM determines that the KS No.2 circuit input voltage is high. **Possible Causes:** • KS No.2 connector or terminals malfunction • KS No.2 malfunction • PCM connector or terminals malfunction • Short to power supply in wiring harness between KS No.2 terminal A and PCM terminal 2AC • Short to power supply in wiring harness between KS No.2 terminal B and PCM terminal 2Y • PCM malfunction
DTC: P0335 **1T CCM, MIL: Yes** **Year:** 2009, 2010 **Model:** 3, 5, 6, CX-7, MX-5 Miata, RX-8 **Engine:** 1.3L R2 VIN 2, 1.3L R2 VIN 4, 1.3L R2 VIN M, 1.3L R2 VIN P, 2.0L L4 VIN F, 2.0L L4 VIN G, 2.3L L4 VIN 3, 2.3L L4 VIN 4, 2.3L L4 VIN L, 2.3L L4 VIN M, 2.5L L4 VIN 5, 2.5L L4 VIN 6, 2.5L L4 VIN A, 2.5L L4 VIN M **Transmission:** All	**Crankshaft Position (NE) Sensor Circuit Malfunction:** Engine running, MAF sensor signal indicating over 2.43 g/sec, and the PCM did not detect any CKP (NE) signals for 4.2 seconds. **Possible Causes:** • CKP sensor signal is open or shorted to ground • CKP sensor signal is shorted to VREF or system power (B+) • CKP sensor is damaged or has failed • Trigger wheel or tone wheel is damaged • PCM has failed
DTC: P0336 **1T CCM, MIL: Yes** **Year:** 2009, 2010 **Model:** RX-8 **Engine:** 1.3L R2 VIN 2, 1.3L R2 VIN 4, 1.3L R2 VIN M, 1.3L R2 VIN P **Transmission:** All	**Crankshaft Position (CKP) Sensor Circuit Malfunction:** Engine running, MAF sensor signal indicating over 2.43 g/sec, and the PCM did not detect any CKP (NE) signals for 4.2 seconds. **Possible Causes:** • CKP sensor signal is open or shorted to ground • CKP sensor signal is shorted to VREF or system power (B+) • CKP sensor is damaged or has failed • Trigger wheel or tone wheel is damaged • PCM has failed
DTC: P0340 **1T CCM, MIL: Yes** **Year:** 2009, 2010 **Model:** 3, 5, 6, B2300, B4000, CX-7, CX-9, MX-5 Miata, Tribute **Engine:** 2.0L L4 VIN F, 2.0L L4 VIN G, 2.3L L4 VIN 3, 2.3L L4 VIN 4, 2.3L L4 VIN D, 2.3L L4 VIN L, 2.3L L4 VIN M, 2.5L L4 VIN 3, 2.5L L4 VIN 5, 2.5L L4 VIN 6, 2.5L L4 VIN 7, 2.5L L4 VIN A, 2.5L L4 VIN M, 3.0L V6 VIN 1, 3.0L V6 VIN G, 3.7L V6 VIN A, 3.7L V6 VIN B, 3.7L V6 VIN V, 4.0L V6 VIN E **Transmission:** All	**Camshaft Position Sensor Circuit Malfunction:** Engine running, MAF sensor signal over 4.2 g/sec, and the PCM did not detect any CMP (SGT) sensor signals for 4.2 seconds during the CCM test. **Possible Causes:** • CMP sensor signal is open or shorted to ground • CMP sensor signal is shorted to VREF or system power (B+) • CMP sensor is damaged, or the sensor signal shielding is open • PCM has failed

DTC	Trouble Code Title, Conditions & Possible Causes
DTC: P0341 **T PCM** **Year:** 2009, 2010 **Model:** 6 **Engine:** 3.7L V6 VIN B **Transmission:** All	**CMP sensor (RH) circuit-range/performance:** The PCM monitors the CMP sensor (RH) for a noisy signal. **Possible Causes:** • Radio frequency interference/electromagnetic interference (RFI/EMI) • Camshaft phaser and sprocket malfunction
DTC: P0341 **1T CCM, MIL: Yes** **Year:** 2009, 2010 **Model:** Tribute **Engine:** 2.5L L4 VIN 3, 2.5L L4 VIN 7, 3.0L V6 VIN 1, 3.0L V6 VIN G **Transmission:** All	**Camshaft Position Sensor Circuit Malfunction:** Engine running, MAF sensor signal over 4.2 g/sec, and the PCM did not detect any CMP (SGT) sensor signals for 4.2 seconds during the CCM test. **Possible Causes:** • CMP sensor signal is open or shorted to ground • CMP sensor signal is shorted to VREF or system power (B+) • CMP sensor is damaged, or the sensor signal shielding is open • PCM has failed
DTC: P0344 **T PCM** **Year:** 2009, 2010 **Model:** 6, CX-9 **Engine:** 3.7L V6 VIN A, 3.7L V6 VIN B, 3.7L V6 VIN V **Transmission:** All	**CMP sensor (RH) circuit intermittent:** The test fails when the PCM detects an intermittent signal from the CMP sensor **Possible Causes:** • Connector or terminal malfunction • Short to power supply between CMP sensor (RH) and PCM • Short to ground circuit between CMP sensor (RH) and PCM • Open circuit between CMP sensor (RH) and PCM • CMP sensor (RH) malfunction • Sensor shielding malfunction
DTC: P0345 **T PCM** **Year:** 2009, 2010 **Model:** 6, CX-9 **Engine:** 3.7L V6 VIN A, 3.7L V6 VIN B, 3.7L V6 VIN V **Transmission:** All	**CMP sensor (LH) circuit:** The test fails when the PCM can no longer detect the signal from the CMP sensor (LH). **Possible Causes:** • Connector or terminal malfunction • Short to power supply between CMP sensor (LH) and PCM • Short to ground circuit between CMP sensor (LH) and PCM • Open circuit between CMP sensor (LH) and PCM • CMP sensor (LH) malfunction • PCM malfunction
DTC: P0346 **T PCM** **Year:** 2009, 2010 **Model:** 6 **Engine:** 3.7L V6 VIN B **Transmission:** All	**CMP sensor (LH) circuit-range/performance:** The PCM monitors the CMP sensor (LH) for a noisy signal. **Possible Causes:** • Radio frequency interference/electromagnetic interference (RFI/EMI) • Camshaft phaser and sprocket malfunction
DTC: P0349 **T PCM** **Year:** 2009, 2010 **Model:** 6, CX-9 **Engine:** 3.7L V6 VIN A, 3.7L V6 VIN B, 3.7L V6 VIN V **Transmission:** All	**CMP sensor (LH) circuit intermittent:** The test fails when the PCM detects an intermittent signal from the CMP sensor (LH). **Possible Causes:** • Connector or terminal malfunction • Short to power supply between CMP sensor (LH) and PCM • Short to ground circuit between CMP sensor (LH) and PCM • Open circuit between CMP sensor (LH) and PCM • CMP sensor (LH) malfunction • Sensor shielding malfunction
DTC: P0350 **2T CCM, MIL: Yes** **Year:** 2009, 2010 **Model:** B2300, B4000, Tribute **Engine:** 2.3L L4 VIN D, 2.5L L4 VIN 3, 2.5L L4 VIN 7, 3.0L V6 VIN 1, 3.0L V6 VIN G, 4.0L V6 VIN E **Transmission:** All	**Ignition Coil Primary Circuit Malfunction:** Engine running and the PCM detected an unexpected voltage condition on the ignition coil primary circuit during the CCM test. **Possible Causes:** • Ignition coil primary circuit is open or shorted to ground • Ignition coil primary circuit is shorted to system power (B+) • Ignition coil power circuit is open between coil and Start circuit • PCM has failed

DTC	Trouble Code Title, Conditions & Possible Causes
DTC: P0351 **2T CCM, MIL: Yes** **Year:** 2009, 2010 **Model:** B2300, Tribute **Engine:** 2.3L L4 VIN D, 2.5L L4 VIN 3, 2.5L L4 VIN 7, 3.0L V6 VIN 1, 3.0L V6 VIN G **Transmission:** All	**Ignition Coil 1 Primary Circuit Malfunction:** Engine running and the PCM detected an unexpected voltage condition on the Ignition Coil 1 primary circuit during the CCM test. **NOTE: DTC P0351 is related to the ignition coil on cylinder 1.** **Possible Causes:** • Ignition coil primary circuit is open, shorted to ground or power • Ignition coil power circuit is open between coil and Start circuit • Ignition coil has failed (arcing between primary and secondary) • PCM has failed
DTC: P0351 **T PCM** **Year:** 2009, 2010 **Model:** 6 **Engine:** 3.7L V6 VIN B **Transmission:** All	**Ignition coil No.1 primary/secondary circuit:** Each ignition primary circuit is continuously monitored. The test fails when the PCM does not receive a valid ignition diagnostic monitor pulse signal from the ignition module (integrated in the PCM). **Possible Causes:** • Connector or terminal malfunction • Open circuit between ignition coil and main relay • Short to power supply between ignition coil and PCM • Short to ground circuit between ignition coil and PCM • Open circuit between ignition coil and PCM • Ignition coil malfunction
DTC: P0352 **2T CCM, MIL: Yes** **Year:** 2009, 2010 **Model:** B2300, Tribute **Engine:** 2.3L L4 VIN D, 2.5L L4 VIN 3, 2.5L L4 VIN 7, 3.0L V6 VIN 1, 3.0L V6 VIN G **Transmission:** All	**Ignition Coil 2 Primary Circuit Malfunction:** Engine running and the PCM detected an unexpected voltage condition on the Ignition Coil 2 primary circuit during the CCM test. **NOTE: DTC P0352 is related to the ignition coil on cylinder 2.** **Possible Causes:** • Ignition coil primary circuit is open, shorted to ground or power • Ignition coil power circuit is open between coil and Start circuit • Ignition coil has failed (arcing between primary and secondary) • PCM has failed
DTC: P0352 **T PCM** **Year:** 2009, 2010 **Model:** 6, CX-9 **Engine:** 3.7L V6 VIN A, 3.7L V6 VIN B, 3.7L V6 VIN V **Transmission:** All	**Ignition coil No.2 primary/secondary circuit:** Each ignition primary circuit is continuously monitored. The test fails when the PCM does not receive a valid ignition diagnostic monitor pulse signal from the ignition module (integrated in the PCM). **Possible Causes:** • Connector or terminal malfunction • Open circuit between ignition coil and main relay • Short to power supply between ignition coil and PCM • Short to ground circuit between ignition coil and PCM • Open circuit between ignition coil and PCM • Ignition coil malfunction
DTC: P0353 **T PCM** **Year:** 2009, 2010 **Model:** 6, CX-9 **Engine:** 3.7L V6 VIN A, 3.7L V6 VIN B, 3.7L V6 VIN V **Transmission:** All	**Ignition coil No.3 primary/secondary circuit:** Each ignition primary circuit is continuously monitored. The test fails when the PCM does not receive a valid ignition diagnostic monitor pulse signal from the ignition module (integrated in the PCM). **Possible Causes:** • Connector or terminal malfunction • Open circuit between ignition coil and main relay • Short to power supply between ignition coil and PCM • Short to ground circuit between ignition coil and PCM • Open circuit between ignition coil and PCM • Ignition coil malfunction
DTC: P0353 **2T CCM, MIL: Yes** **Year:** 2009, 2010 **Model:** B2300, Tribute **Engine:** 2.3L L4 VIN D, 2.5L L4 VIN 3, 2.5L L4 VIN 7, 3.0L V6 VIN 1, 3.0L V6 VIN G **Transmission:** All	**Ignition Coil 3 Primary Circuit Malfunction:** Engine running and the PCM detected an unexpected voltage condition on the Ignition Coil 3 primary circuit during the CCM test. **NOTE: DTC P0353 is related to the ignition coil on cylinder 3.** **Possible Causes:** • Ignition coil primary circuit is open, shorted to ground or power • Ignition coil power circuit is open between coil and Start circuit • Ignition coil has failed (arcing between primary and secondary) • PCM has failed

DTC	Trouble Code Title, Conditions & Possible Causes
DTC: P0354 **T PCM** **Year:** 2009, 2010 **Model:** 6, CX-9 **Engine:** 3.7L V6 VIN A, 3.7L V6 VIN B, 3.7L V6 VIN V **Transmission:** All	**Ignition coil No.4 primary/secondary circuit:** Each ignition primary circuit is continuously monitored. The test fails when the PCM does not receive a valid ignition diagnostic monitor pulse signal from the ignition module (integrated in the PCM). **Possible Causes:** • Connector or terminal malfunction • Open circuit between ignition coil and main relay • Short to power supply between ignition coil and PCM • Short to ground circuit between ignition coil and PCM • Open circuit between ignition coil and PCM • Ignition coil malfunction
DTC: P0354 **2T CCM, MIL: Yes** **Year:** 2009, 2010 **Model:** B2300, Tribute **Engine:** 2.3L L4 VIN D, 2.5L L4 VIN 3, 2.5L L4 VIN 7, 3.0L V6 VIN 1, 3.0L V6 VIN G **Transmission:** All	**Ignition Coil 4 Primary Circuit Malfunction:** Engine running and the PCM detected an unexpected voltage condition on the Ignition Coil 4 primary circuit during the CCM test. **NOTE: DTC P0354 is related to the ignition coil on cylinder 4.** **Possible Causes:** • Ignition coil primary circuit is open, shorted to ground or power • Ignition coil power circuit is open between coil and Start circuit • Ignition coil has failed (arcing between primary and secondary) • PCM has failed
DTC: P0355 **2T CCM, MIL: Yes** **Year:** 2009, 2010 **Model:** Tribute **Engine:** 2.5L L4 VIN 3, 2.5L L4 VIN 7, 3.0L V6 VIN 1, 3.0L V6 VIN G **Transmission:** All	**Ignition Coil 3 Primary Circuit Malfunction:** Engine running and the PCM detected an unexpected voltage condition on the Ignition Coil 3 primary circuit during the CCM test. **NOTE: DTC P0353 is related to the ignition coils on cylinders 2 and 6.** **Possible Causes:** • Ignition coil primary circuit is open or shorted to ground • Ignition coil primary circuit is shorted to system power (B+) • Ignition coil power circuit is open between coil and Start circuit • Ignition coil has failed (arcing between primary and secondary) • PCM has failed
DTC: P0355 **T PCM** **Year:** 2009, 2010 **Model:** 6, CX-9 **Engine:** 3.7L V6 VIN A, 3.7L V6 VIN B, 3.7L V6 VIN V **Transmission:** All	**Ignition coil No.5 primary/secondary circuit:** Each ignition primary circuit is continuously monitored. The test fails when the PCM does not receive a valid ignition diagnostic monitor pulse signal from the ignition module (integrated in the PCM). **Possible Causes:** • Connector or terminal malfunction • Open circuit between ignition coil and main relay • Short to power supply between ignition coil and PCM • Short to ground circuit between ignition coil and PCM • Open circuit between ignition coil and PCM • Ignition coil malfunction
DTC: P0356 **T PCM** **Year:** 2009, 2010 **Model:** 6, CX-9 **Engine:** 3.7L V6 VIN A, 3.7L V6 VIN B, 3.7L V6 VIN V **Transmission:** All	**Ignition coil No.6 primary/secondary circuit:** Each ignition primary circuit is continuously monitored. The test fails when the PCM does not receive a valid ignition diagnostic monitor pulse signal from the ignition module (integrated in the PCM). **Possible Causes:** • Connector or terminal malfunction • Open circuit between ignition coil and main relay • Short to power supply between ignition coil and PCM • Short to ground circuit between ignition coil and PCM • Open circuit between ignition coil and PCM • Ignition coil malfunction
DTC: P0356 **2T CCM, MIL: Yes** **Year:** 2009, 2010 **Model:** Tribute **Engine:** 2.5L L4 VIN 3, 2.5L L4 VIN 7, 3.0L V6 VIN 1, 3.0L V6 VIN G **Transmission:** All	**Ignition Coil 3 Primary Circuit Malfunction:** Engine running and the PCM detected an unexpected voltage condition on the Ignition Coil 3 primary circuit during the CCM test. **NOTE: DTC P0353 is related to the ignition coils on cylinders 2 and 6.** **Possible Causes:** • Ignition coil primary circuit is open or shorted to ground • Ignition coil primary circuit is shorted to system power (B+) • Ignition coil power circuit is open between coil and Start circuit • Ignition coil has failed (arcing between primary and secondary) • PCM has failed

DTC	Trouble Code Title, Conditions & Possible Causes
DTC: P0400 **2T CCM, MIL: Yes** **Year:** 2009, 2010 **Model:** B2300, B4000, Tribute **Engine:** 2.3L L4 VIN D, 2.5L L4 VIN 3, 2.5L L4 VIN 7, 3.0L V6 VIN 1, 3.0L V6 VIN G, 4.0L V6 VIN E **Transmission:** All	**EGR System Flow Malfunction:** Vehicle driven to a speed of over 7 mph, then back to idle speed, ECT sensor signal more than 131°F, and the PCM detected little or no change in the EGR Differential Pressure sensor (DPFE) with the EGR vent commanded "on" and then "off" in the EGR Monitor test. **Possible Causes:** • EGR vent solenoid circuit open or shorted to ground • EGR vent solenoid power circuit is open • EGR vent solenoid is damaged or has failed • DPFE sensor signal circuit is open or shorted to ground • DPFE sensor is damaged or has failed • EGR valve assembly is leaking, sticking, damaged or has failed • EGR valve vacuum hose(s) lose, damaged or disconnected • Exhaust pipe to EGR valve is restricted or plugged • PCM has failed
DTC: P0401 **2T CCM, MIL: Yes** **Year:** 2009 **Model:** B2300, B4000 **Engine:** 2.3L L4 VIN D, 4.0L V6 VIN E **Transmission:** All	**EGR System Insufficient Flow Detected:** Vehicle driven to a speed of over 7 mph, then back to idle speed, ECT sensor signal more than 131°F, and the PCM detected little or no change in the EGR Differential Pressure sensor (DPFE) with the EGR vent commanded "on" and then "off" in the EGR Monitor test. **Possible Causes:** • EGR VR solenoid circuit open, shorted to ground or to power • EGR VR solenoid power circuit is open to the Main Relay • EGR VR solenoid is damaged or has failed • DPFE sensor signal circuit is open or shorted to ground • DPFE sensor hoses both off, reversed, or sensor is damaged • EGR valve assembly is leaking, stuck closed or is damaged • EGR valve vacuum hose(s) lose, damaged or disconnected • EGR orifice tube is restricted, plugged or damaged • PCM has failed
DTC: P0401 **2T CCM, MIL: Yes** **Year:** 2009, 2010 **Model:** 3, 5, 6, CX-7, MX-5 Miata **Engine:** 2.0L L4 VIN F, 2.0L L4 VIN G, 2.3L L4 VIN 3, 2.3L L4 VIN 4, 2.3L L4 VIN L, 2.3L L4 VIN M, 2.5L L4 VIN 5, 2.5L L4 VIN 6, 2.5L L4 VIN A, 2.5L L4 VIN M **Transmission:** All	**EGR System Insufficient Flow Detected:** DTC P0102, P0103, P0106, P0107, P0108, P0111, P0112, P0113, P0122, P0123, P0111, P0112, P0113, P0122, P0123, P1122, P1123, P1487, P1496, P1497, P1498 and P1499 not set, engine running in closed loop at 35-55 mph for two minutes, and the PCM detected insufficient EGR gas flow during the EGR Monitor test. **Possible Causes:** • EGR boost solenoid control circuit is open or shorted to ground • EGR boost solenoid power circuit is open to the Main Relay • EGR boost solenoid is damaged or has failed • EGR boost sensor is damaged or has failed • EGR valve position sensor is damaged, stuck or has failed • EGR valve assembly is leaking, damaged or has failed • EGR valve vacuum hose(s) lose, damaged or disconnected • IAT, MAF, TP and VSS (sensor) has drifted out-of-range • PCM has failed
DTC: P0402 **2T CCM, MIL: Yes** **Year:** 2009 **Model:** B2300, B4000 **Engine:** 2.3L L4 VIN D, 4.0L V6 VIN E **Transmission:** All	**EGR System Excessive Flow At Idle Speed Detected:** Vehicle driven to a speed of over 7 mph, then back to idle speed, ECT sensor signal more than 131°F, and the PCM detected little or no change in the EGR Differential Pressure sensor (DPFE) with the EGR vent commanded "on" and then "off" in the EGR Monitor test. **Possible Causes:** • EGR VR solenoid circuit open, shorted to ground or to power • EGR VR solenoid power circuit is open to the Main Relay • EGR VR solenoid is damaged or has failed • DPFE sensor signal circuit is open or shorted to ground • DPFE sensor hoses both off, reversed, or sensor is damaged • EGR valve assembly is leaking, stuck open or is damaged • EGR valve vacuum hose(s) lose, damaged or disconnected • PCM has failed

DTC	Trouble Code Title, Conditions & Possible Causes
DTC: P0403 **2T CCM, MIL: Yes** **Year:** 2009, 2010 **Model:** 3, 5, 6, B2300, B4000, CX-7, MX-5 Miata, Tribute **Engine:** 2.0L L4 VIN F, 2.0L L4 VIN G, 2.3L L4 VIN 3, 2.3L L4 VIN 4, 2.3L L4 VIN D, 2.3L L4 VIN L, 2.3L L4 VIN M, 2.5L L4 VIN 3, 2.5L L4 VIN 5, 2.5L L4 VIN 6, 2.5L L4 VIN 7, 2.5L L4 VIN A, 2.5L L4 VIN M, 3.0L V6 VIN 1, 3.0L V6 VIN G, 4.0L V6 VIN E **Transmission:** All	**EGR Valve (stepper motor) malfunction:** PCM monitors input voltage from EGR valve. If voltage at PCM terminals 4E, 4H, 4K and/or 4N remain low or high, PCM determines that ERG valve circuit has a malfunction. **Possible Causes:** • EGR valve malfunction • Connector or terminal malfunction • Short to power circuit in wiring to EGR valve terminals and PCM terminals • Open circuit in wiring between EGR valve terminals and PCM terminals • PCM has failed
DTC: P0410 **2T CCM, MIL: Yes** **Year:** 2009, 2010 **Model:** RX-8 **Engine:** 1.3L R2 VIN 2, 1.3L R2 VIN 4, 1.3L R2 VIN M, 1.3L R2 VIN P **Transmission:** All	**EGR Valve Position Sensor Circuit Range/Performance:** Engine speed from 1810-2190, TP angle from 3-13.4%, engine load 25-50%, and the PCM detected that the EGR solenoid accumulated on-time period exceeded a threshold with the EGR system enabled, or it detected an EGR vent solenoid problem existed during the test. **Possible Causes:** • EGR boost solenoid control circuit is open or shorted to ground • EGR boost solenoid power circuit is open to the Main Relay • EGR boost solenoid is damaged or has failed • EGR boost sensor is damaged or has failed • EGR valve position sensor is damaged, stuck or has failed • EGR valve assembly is leaking, damaged or has failed • EGR valve vacuum hose(s) lose, damaged or disconnected • Exhaust pipe to EGR boost sensor or solenoid leaking/plugged • PCM has failed
DTC: P0411 **2T AIR, MIL: Yes** **Year:** 2009, 2010 **Model:** RX-8 **Engine:** 1.3L R2 VIN 2, 1.3L R2 VIN 4, 1.3L R2 VIN M, 1.3L R2 VIN P **Transmission:** All	**Secondary air injection system incorrect upstream flow:** The PCM monitors the A/F sensor output current when the secondary air injection system is operating. If the output current is less than the specification, the PCM determines that there is a secondary air injection system problem. **Possible Causes:** • Secondary air injection system malfunction • Secondary air injection control valve malfunction • Secondary air injection solenoid valve malfunction • Secondary air injection pump malfunction • Restrict or damaged in exhaust system • Restrict in TWC • PCM malfunction
DTC: P0420 **2T CAT1, MIL: Yes** **Year:** 2009, 2010 **Model:** 6, B2300, B4000, CX-9 **Engine:** 2.3L L4 VIN D, 3.7L V6 VIN A, 3.7L V6 VIN B, 3.7L V6 VIN V, 4.0L V6 VIN E **Transmission:** All	**Catalyst Efficiency Below Normal (Bank 1):** Vehicle driven at a speed of 16-64 mph at 1090-3090 rpm with the calculated engine load from 16-55% for 2-3 minutes, and the PCM detected the switch rate of the rear HO2S was close to the switch rate of the front HO2S-11 during the Catalyst Monitor test. **Possible Causes:** • Air leaks at the exhaust manifold or in the exhaust pipes • Catalytic converter is contaminated, damaged or has failed • Front HO2S and/or the rear HO2S is loose in the mounting hole • Front HO2S older (aged) than the rear HO2S (HO2S is lazy) • Front HO2S or rear HO2S is contaminated with fuel or moisture
DTC: P0420 **T** **Year:** 2009, 2010 **Model:** CX-7, RX-8, Tribute **Engine:** 1.3L R2 VIN 2, 1.3L R2 VIN 4, 1.3L R2 VIN M, 1.3L R2 VIN P, 2.5L L4 VIN 3, 2.5L L4 VIN 5, 2.5L L4 VIN 7, 2.5L L4 VIN M, 3.0L V6 VIN 1, 3.0L V6 VIN G **Transmission:** All	**Catalyst Efficiency Below Normal (Bank 1):** Vehicle driven at a speed of 16-64 mph at 1090-3090 rpm with the calculated engine load from 16-55% for 2-3 minutes, and the PCM detected the inversion ratio of the rear HO2S and front HO2S was less than a stored threshold during the Catalyst Monitor test. **Possible Causes:** • Air leaks at the exhaust manifold or in the exhaust pipes • Catalytic converter is contaminated, damaged or has failed • Front HO2S and/or the rear HO2S is loose in the mounting hole • Front HO2S older (aged) than the rear HO2S (HO2S is lazy) • Front HO2S or rear HO2S is contaminated with fuel or moisture

DTC	Trouble Code Title, Conditions & Possible Causes
DTC: P0421 **T** **Year:** 2009, 2010 **Model:** 3, 5, 6, B2300, B4000, CX-7, MX-5 Miata **Engine:** 2.0L L4 VIN F, 2.0L L4 VIN G, 2.3L L4 VIN 3, 2.3L L4 VIN 4, 2.3L L4 VIN D, 2.3L L4 VIN L, 2.3L L4 VIN M, 2.5L L4 VIN 5, 2.5L L4 VIN 6, 2.5L L4 VIN A, 2.5L L4 VIN M, 4.0L V6 VIN E **Transmission:** All	**Warm-up Catalyst Efficiency Below Normal (Bank 1):** Vehicle driven at a speed of 17-74 mph at 1500-3000 rpm with the calculated engine load from 15-48% for 2-3 minutes, and the PCM detected the inversion ratio of the rear HO2S was close to the inversion ratio of the front HO2S during the Catalyst Monitor test. **Possible Causes:** • Air leaks at the exhaust manifold or in the exhaust pipes • Catalytic converter is contaminated, damaged or has failed • Front HO2S and/or the rear HO2S is loose in the mounting hole • Front HO2S older (aged) than the rear HO2S (HO2S is lazy) • Front HO2S or rear HO2S is contaminated with fuel or moisture
DTC: P0430 **2T CAT3, MIL: Yes** **Year:** 2009, 2010 **Model:** 6, CX-9 **Engine:** 3.7L V6 VIN A, 3.7L V6 VIN B, 3.7L V6 VIN V **Transmission:** All	**Catalyst Efficiency Below Normal (Bank 2):** Vehicle driven at a speed of 16-64 mph at 1090-3090 rpm with the calculated engine load from 16-55% for 2-3 minutes, and the PCM detected the inversion ratio of the rear HO2S and front HO2S was less than a stored threshold during the Catalyst Monitor test. **Possible Causes:** • Air leaks at the exhaust manifold or in the exhaust pipes • Catalytic converter is contaminated, damaged or has failed • Front HO2S and/or the rear HO2S is loose in the mounting hole • Front HO2S older (aged) than the rear HO2S (HO2S is lazy) • Front HO2S or rear HO2S is contaminated with fuel or moisture
DTC: P0430 **T** **Year:** 2009, 2010 **Model:** Tribute **Engine:** 2.5L L4 VIN 3, 2.5L L4 VIN 7, 3.0L V6 VIN 1, 3.0L V6 VIN G **Transmission:** All	**Catalyst Efficiency Below Normal (Bank 2):** Vehicle driven at a speed of 16-64 mph at 1090-3090 rpm with the calculated engine load from 16-55% for 2-3 minutes, and the PCM detected the inversion ratio of the rear HO2S and front HO2S was less than a stored threshold during the Catalyst Monitor test. **Possible Causes:** • Air leaks at the exhaust manifold or in the exhaust pipes • Catalytic converter is contaminated, damaged or has failed • Front HO2S and/or the rear HO2S is loose in the mounting hole • Front HO2S older (aged) than the rear HO2S (HO2S is lazy) • Front HO2S or rear HO2S is contaminated with fuel or moisture
DTC: P0430 **2T CAT3, MIL: Yes** **Year:** 2009 **Model:** B4000 **Engine:** 4.0L V6 VIN E **Transmission:** All	**Catalyst Efficiency Below Normal (Bank 2):** Vehicle driven at a speed of 16-64 mph at 1090-3090 rpm with the calculated engine load from 16-55% for 2-3 minutes, and the PCM detected the switch rate of the rear HO2S was close to the switch rate of the front HO2S-11 during the Catalyst Monitor test. **Possible Causes:** • Air leaks at the exhaust manifold or in the exhaust pipes • Catalytic converter is contaminated, damaged or has failed • Front HO2S and/or the rear HO2S is loose in the mounting hole • Front HO2S older (aged) than the rear HO2S (HO2S is lazy) • Front HO2S or rear HO2S is contaminated with fuel or moisture
DTC: P0441 **2T CCM, MIL: Yes** **Year:** 2009, 2010 **Model:** 5, B2300, B4000, MX-5 Miata, RX-8 **Engine:** 1.3L R2 VIN 2, 1.3L R2 VIN 4, 1.3L R2 VIN M, 1.3L R2 VIN P, 2.0L L4 VIN F, 2.3L L4 VIN 3, 2.3L L4 VIN D, 2.3L L4 VIN L, 4.0L V6 VIN E **Transmission:** All	**EVAP System Malfunction:** DTC P0443 not set, ECT sensor signal from 14-90°F and IAT sensor signal less than 14°F, engine running in closed loop at a cruise speed over 15 mph, then with the purge solenoid commanded "on", the PCM detected the fuel control "feedback" signal was less than a value stored in memory with "purge" enabled during the EVAP test. **Possible Causes:** • EVAP purge solenoid is damaged or has failed • EVAP charcoal canister is loaded with fuel or moisture • Vapor line is damaged or restricted between the purge solenoid and intake manifold, or between check valve and vapor valve • PCM has failed

DTC	Trouble Code Title, Conditions & Possible Causes
DTC: P0442 **2T CCM, MIL: Yes** **Year:** 2009, 2010 **Model:** B2300, B4000, Tribute **Engine:** 2.3L L4 VIN D, 2.5L L4 VIN 3, 2.5L L4 VIN 7, 3.0L V6 VIN 1, 3.0L V6 VIN G, 4.0L V6 VIN E **Transmission:** All	**EVAP System Small Leak (0.040") Detected:** DTC P0443 not set, ECT sensor from 14-90°F and IAT sensor less than 14°F at startup, vehicle driven at cruise speed, then with the purge solenoid enabled, the PCM detected a leak as small as 0.040" in the system during the EVAP Running Loss Monitor test. **NOTE: This code sets when there is less than 2.5" H2O bleed-up over a 15 second period with the fuel tank level more than 75% full. The bleed-up and evaluation time vary as a function of fuel level. Vapor generation is more than 2.5" H2O over a 120-second period of time.** **Possible Causes:** • Fuel filler cap loose, cross-threaded, incorrect part or damaged • EVAP purge solenoid valve is damaged or has failed • EVAP vent control solenoid is damaged or has failed • EVAP charcoal canister is loaded with fuel or moisture • Fuel tank pressure sensor damaged or has failed • Vapor line(s) damaged or leaking between the purge solenoid and the intake manifold, purge solenoid valve and the canister, fuel vapor control valve tube or fuel vapor vent valve assembly • PCM has failed
DTC: P0442 **2T CCM, MIL: Yes** **Year:** 2009, 2010 **Model:** 3, 5, 6, CX-7, CX-9, MX-5 Miata, RX-8 **Engine:** 1.3L R2 VIN 2, 1.3L R2 VIN 4, 1.3L R2 VIN M, 1.3L R2 VIN P, 2.0L L4 VIN F, 2.0L L4 VIN G, 2.3L L4 VIN 3, 2.3L L4 VIN 4, 2.3L L4 VIN L, 2.3L L4 VIN M, 2.5L L4 VIN 5, 2.5L L4 VIN 6, 2.5L L4 VIN A, 2.5L L4 VIN M, 3.7L V6 VIN A, 3.7L V6 VIN B, 3.7L V6 VIN V **Transmission:** All	**EVAP System Small Leak (0.040") Detected:** DTC P0443 not set, ECT sensor signal from 14-90°F, IAT sensor signal more than 14°F at startup, BARO sensor signal over 72 kPa, IAT sensor signal from 14-140°F and ECT sensor signal from 158-212°F during the EVAP leak test, fuel level from 15-85%, vehicle driven at 24-65 mph at an engine speed of 1000-4000 rpm, engine load from 9-65% with the throttle opening from 3.1-31.6%, then with the purge and CDCV valves both closed, the PCM detected a leak as small as 0.040" in the system during the EVAP Leak Monitor test. **NOTE: The fuel tank target pressure for this test is plus (+) 127 kPa.** **Possible Causes:** • Fuel filler cap loose, cross-threaded, incorrect part or damaged • Purge solenoid valve is damaged or has failed • Charcoal canister is loaded with fuel or moisture • Canister drain cut valve (CDCV) is damaged or has failed • Rollover valve, catch tank valve or fuel tank damaged/ leaking • Fuel tank pressure sensor is damaged or has failed • ECT, IAT, MAF, VSS or TP sensor signals out-of-calibration • Vapor line(s) damaged or leaking between the purge solenoid and the intake manifold, EVAP purge solenoid valve and the canister, canister drain cut valve and the rollover valve • PCM has failed
DTC: P0443 **1T CCM, MIL: Yes** **Year:** 2009, 2010 **Model:** B2300, B4000, Tribute **Engine:** 2.3L L4 VIN D, 2.5L L4 VIN 3, 2.5L L4 VIN 7, 3.0L V6 VIN 1, 3.0L V6 VIN G, 4.0L V6 VIN E **Transmission:** All	**EVAP Purge Solenoid Circuit Malfunction:** Key on or engine running and the PCM detected an unexpected voltage condition on the EVAP purge solenoid circuit during the test. **NOTE: This is a diagnostic support code only – not stored in the PCM!** **Possible Causes:** • VMV solenoid control circuit is open or shorted to ground • VMV solenoid control circuit is shorted to system power (B+) • VMV solenoid power circuit is open or shorted to ground • VMV solenoid is damaged or has failed • PCM has failed (the VMV solenoid driver may be damaged)
DTC: P0443 **1T CCM, MIL: Yes** **Year:** 2009, 2010 **Model:** 3, 5, 6, CX-7, CX-9, MX-5 Miata, RX-8 **Engine:** 1.3L R2 VIN 2, 1.3L R2 VIN 4, 1.3L R2 VIN M, 1.3L R2 VIN P, 2.0L L4 VIN F, 2.3L L4 VIN 3, 2.3L L4 VIN 4, 2.3L L4 VIN L, 2.3L L4 VIN M, 3.7L V6 VIN A, 3.7L V6 VIN B, 3.7L V6 VIN V **Transmission:** All	**EVAP Canister Purge Solenoid Circuit Malfunction:** Key on or engine running and the PCM detected an unexpected voltage condition on the EVAP Purge solenoid circuit during the test. **NOTE: This is a diagnostic support code only – not stored in the PCM!** **Possible Causes:** • Purge solenoid control circuit is open or shorted to ground • Purge solenoid control circuit is shorted to system power (B+) • Purge solenoid power circuit is open or shorted to ground • Purge solenoid is damaged or has failed • PCM has failed (the purge solenoid driver may be damaged)

DTC	Trouble Code Title, Conditions & Possible Causes
DTC: P0446 **1T CCM, MIL: Yes** **Year:** 2009, 2010 **Model:** 3, 5, 6, B2300, B4000, CX-7, CX-9, MX-5 Miata, RX-8, Tribute **Engine:** 1.3L R2 VIN 2, 1.3L R2 VIN 4, 1.3L R2 VIN M, 1.3L R2 VIN P, 2.0L L4 VIN F, 2.0L L4 VIN G, 2.3L L4 VIN 3, 2.3L L4 VIN 4, 2.3L L4 VIN D, 2.3L L4 VIN L, 2.3L L4 VIN M, 2.5L L4 VIN 3, 2.5L L4 VIN 5, 2.5L L4 VIN 6, 2.5L L4 VIN 7, 2.5L L4 VIN A, 2.5L L4 VIN M, 3.0L V6 VIN 1, 3.0L V6 VIN G, 3.7L V6 VIN A, 3.7L V6 VIN B, 3.7L V6 VIN V, 4.0L V6 VIN E **Transmission:** All	**EVAP System Vent Control Malfunction:** DTC P0443 not set, ECT sensor signal from 14-90°F and IAT sensor signal more than 14°F at startup, engine speed at 1000-3000 rpm, VSS input from 25-63 mph, calculated load at 9-70%, TP angle from 3-44%, then with the CDCV closed and then reopened, the PCM detected the change in fuel tank pressure was too small. **Possible Causes:** • Canister drain cut valve (CDCV) is damaged, sticking or failed • Tank pressure control valve (TPCV) is damaged or has failed • Charcoal canister is loaded with fuel or moisture • Air filter is severely restricted, or 2-way check valve is clogged • FTP sensor signal circuit is open, shorted to ground or to power • FTP sensor power circuit is open or shorted to ground • FTP sensor is damaged or has failed • BARO, ECT, IAT, MAF, VSS or TP sensor is out-of-calibration • Fuel tank level sensor is damaged or out-of-calibration • Fuel vapor line(s) kinked or blocked between the CDVC valve and intake manifold, or between the TPCV and the canister
DTC: P0451 **2T CCM, MIL: Yes** **Year:** 2009, 2010 **Model:** 3, 6, B2300, B4000, CX-7, CX-9, Tribute **Engine:** 2.0L L4 VIN F, 2.0L L4 VIN G, 2.3L L4 VIN 3, 2.3L L4 VIN 4, 2.3L L4 VIN D, 2.3L L4 VIN L, 2.3L L4 VIN M, 2.5L L4 VIN 3, 2.5L L4 VIN 5, 2.5L L4 VIN 6, 2.5L L4 VIN 7, 2.5L L4 VIN A, 2.5L L4 VIN M, 3.0L V6 VIN 1, 3.0L V6 VIN G, 3.7L V6 VIN A, 3.7L V6 VIN B, 3.7L V6 VIN V, 4.0L V6 VIN E **Transmission:** All	**EVAP Pressure Sensor Circuit Range/Performance:** Engine running, and the PCM detected the Fuel Tank Pressure (FTP) sensor changed more than 14" H2O within a 10 second period (indicating the FTP sensor circuit is noisy). **Possible Causes:** • FTP sensor signal circuit is open (intermittent fault) • FTP sensor signal circuit is shorted to ground (intermittent fault) • FTP sensor is damaged or has failed • PCM has failed
DTC: P0452 **2T CCM, MIL: Yes** **Year:** 2009, 2010 **Model:** 3, 6, B2300, B4000, CX-7, CX-9, Tribute **Engine:** 2.0L L4 VIN F, 2.0L L4 VIN G, 2.3L L4 VIN 3, 2.3L L4 VIN 4, 2.3L L4 VIN D, 2.3L L4 VIN L, 2.3L L4 VIN M, 2.5L L4 VIN 3, 2.5L L4 VIN 5, 2.5L L4 VIN 6, 2.5L L4 VIN 7, 2.5L L4 VIN A, 2.5L L4 VIN M, 3.0L V6 VIN 1, 3.0L V6 VIN G, 3.7L V6 VIN A, 3.7L V6 VIN B, 3.7L V6 VIN V, 4.0L V6 VIN E **Transmission:** All	**EVAP Pressure Sensor Circuit Low Input:** Engine running and the PCM detected the Fuel Tank Pressure (FTP) sensor signal indicated less than 0.22v during the CCM test. **Possible Causes:** • FTP sensor signal circuit is shorted to sensor ground • FTP sensor signal circuit is shorted to chassis ground • FTP sensor signal is shorted in the connector due to moisture • FTP sensor is damaged or has failed • PCM has failed
DTC: P0453 **2T CCM, MIL: Yes** **Year:** 2009, 2010 **Model:** 3, 6, B2300, B4000, CX-7, CX-9, Tribute **Engine:** 2.0L L4 VIN F, 2.0L L4 VIN G, 2.3L L4 VIN 3, 2.3L L4 VIN 4, 2.3L L4 VIN D, 2.3L L4 VIN L, 2.3L L4 VIN M, 2.5L L4 VIN 3, 2.5L L4 VIN 5, 2.5L L4 VIN 6, 2.5L L4 VIN 7, 2.5L L4 VIN A, 2.5L L4 VIN M, 3.0L V6 VIN 1, 3.0L V6 VIN G, 3.7L V6 VIN A, 3.7L V6 VIN B, 3.7L V6 VIN V, 4.0L V6 VIN E **Transmission:** All	**EVAP Pressure Sensor Circuit High Input:** Engine running and the PCM detected the Fuel Tank Pressure (FTP) sensor signal indicated less than 0.22v during the CCM test. **Possible Causes:** • FTP sensor signal circuit is shorted to VREF or power (B+) • FTP sensor ground circuit is open • FTP sensor is damaged or has failed • PCM has failed

DTC	Trouble Code Title, Conditions & Possible Causes
DTC: P0454 **2T CCM, MIL: Yes** **Year:** 2009, 2010 **Model:** 3, 6, B2300, B4000, CX-7, CX-9 **Engine:** 2.0L L4 VIN F, 2.0L L4 VIN G, 2.3L L4 VIN 3, 2.3L L4 VIN 4, 2.3L L4 VIN D, 2.3L L4 VIN L, 2.3L L4 VIN M, 2.5L L4 VIN 5, 2.5L L4 VIN 6, 2.5L L4 VIN A, 2.5L L4 VIN M, 3.7L V6 VIN A, 3.7L V6 VIN B, 3.7L V6 VIN V, 4.0L V6 VIN E **Transmission:** All	**Fuel Tank Pressure Sensor Intermittent:** The fuel tank pressure changes more than 1,993 Pa {8.0 in} of water in 0.1 s. **Possible Causes:** • Fuel tank pressure sensor malfunction • Intermittent open or short in the fuel tank pressure sensor or the fuel tank pressure sensor signal • PCM malfunction

DTC: P0455
2T CCM, MIL: Yes
Year: 2009, 2010
Model: B2300, B4000, Tribute
Engine: 2.3L L4 VIN D, 2.5L L4 VIN 3, 2.5L L4 VIN 7, 3.0L V6 VIN 1, 3.0L V6 VIN G, 4.0L V6 VIN E
Transmission: All
	EVAP System Large Leak (0.080") Detected:
	DTC P0443 not set, ECT sensor signal from 14-90°F, IAT sensor signal more than 14°F at startup, BARO sensor more than 75 kPa, fuel level from 15-85%, IAT sensor signal 14-131°F during testing, vehicle driven at 25-80 mph at an engine speed of 1100-3400 rpm, calculated engine load of 7-80% and a throttle angle from 3-12.5%, and with the purge and vent solenoids closed, the PCM detected the FTP value was minus (-) 7.0" H2O due to a leak (0.080") or blockage in the system for 30 seconds in the EVAP Flow/Running Loss test.
	NOTE: The target fuel tank pressure for this test is minus (-) 1.74 kPa.
	Possible Causes:
	    • Fuel filler cap loose, cross-threaded, incorrect part or damaged
	    • Charcoal canister is loaded with fuel or moisture
	    • Purge solenoid may be stuck mechanically in "closed" position
	    • Canister vent (CV) solenoid is stuck open or may be damaged
	    • Fuel vapor control valve tube assembly is restricted or blocked
	    • Fuel vapor vent valve assembly is restricted or blocked
	    • Fuel pressure sensor has failed (i.e., it failed mechanically)
	    • Vapor tube(s) loose, disconnected or blocked to the canister tube, EVAP canister purge outlet tube or the EVAP return tube
	    • PCM has failed

DTC: P0455
2T CCM, MIL: Yes
Year: 2009, 2010
Model: 3, 5, 6, CX-7, CX-9, MX-5 Miata, RX-8
Engine: 1.3L R2 VIN 2, 1.3L R2 VIN 4, 1.3L R2 VIN M, 1.3L R2 VIN P, 2.0L L4 VIN F, 2.0L L4 VIN G, 2.3L L4 VIN 3, 2.3L L4 VIN 4, 2.3L L4 VIN L, 2.3L L4 VIN M, 2.5L L4 VIN 5, 2.5L L4 VIN 6, 2.5L L4 VIN A, 2.5L L4 VIN M, 3.7L V6 VIN A, 3.7L V6 VIN B, 3.7L V6 VIN V
Transmission: All
	EVAP System Large Leak (0.080") Detected:
	DTC P0443 not set, ECT sensor signal from 14-90°F, IAT sensor signal more than 14°F at startup, BARO sensor more than 72 kPa, fuel level from 15-85%, ECT sensor signal from 158-212°F and IAT sensor signal from 14-131°F during testing, vehicle driven at 24-65 mph at an engine speed of 1000-4000 rpm, calculated engine load of 9-65% and a throttle angle from 3-31.6%, then with the CDCV (valve) closed, the PCM detected the fuel tank pressure was less than a threshold due to a blockage or a large leak (0.080") somewhere in the system during the EVAP Monitor test.
	NOTE: The target fuel tank pressure for this test is minus (-) 0.99 kPa.
	Possible Causes:
	    • Fuel filler cap loose, cross-threaded, incorrect part or damaged
	    • Purge solenoid valve is damaged or has failed
	    • Charcoal canister is loaded with fuel or moisture, or has failed
	    • Canister drain cut valve (CDCV) is damaged or leaking
	    • Fuel tank pressure (FTP) sensor is damaged or has failed
	    • Catch tank or rollover valve is damaged, or fuel tank is leaking
	    • Vapor line(s) damaged or leaking between the purge solenoid and intake manifold, purge solenoid and canister, or between the pressure control valve, check valve and rollover valve
	    • PCM has failed

DTC	Trouble Code Title, Conditions & Possible Causes
DTC: P0456 **2T CCM, MIL: Yes** **Year:** 2009, 2010 **Model:** 3, 5, 6, B2300, B4000, CX-7, CX-9, MX-5 Miata, RX-8, Tribute **Engine:** 1.3L R2 VIN 2, 1.3L R2 VIN 4, 1.3L R2 VIN M, 1.3L R2 VIN P, 2.0L L4 VIN F, 2.0L L4 VIN G, 2.3L L4 VIN 3, 2.3L L4 VIN 4, 2.3L L4 VIN D, 2.3L L4 VIN L, 2.3L L4 VIN M, 2.5L L4 VIN 3, 2.5L L4 VIN 5, 2.5L L4 VIN 6, 2.5L L4 VIN 7, 2.5L L4 VIN A, 2.5L L4 VIN M, 3.0L V6 VIN 1, 3.0L V6 VIN G, 3.7L V6 VIN A, 3.7L V6 VIN B, 3.7L V6 VIN V, 4.0L V6 VIN E **Transmission:** All	**EVAP System Small Leak (0.020") Detected:** DTC P0443 not set, ECT sensor signal from 14-90°F, IAT sensor signal more than 14°F at startup, engine speed at 1000-3000 rpm, VSS input from 25-63 mph, calculated load at 9-70%, fuel level over 75%, TP angle from 3-44%, and the PCM detected less than 2.5" H2O bleed-up over a 15 second period during the EVAP leak test. **NOTE: The vapor generation limit is over 2.5" H2O in 120 seconds.** **Possible Causes:** • Fuel filler cap loose, cross-threaded, incorrect part or damaged • Small holes or cuts in any of the vapor hoses and/or tubes • CV solenoid stuck part-way open while it is commanded closed • EVAP system component seals leaking at the purge valve, FTP sensor, CV solenoid, fuel vapor control valve or vapor vent tube • Fuel vapor hose or tube connections loose at the component • PCM has failed
DTC: P0457 **2T CCM, MIL: Yes** **Year:** 2009, 2010 **Model:** 3, 6, B2300, B4000, CX-7, CX-9, Tribute **Engine:** 2.0L L4 VIN F, 2.0L L4 VIN G, 2.3L L4 VIN 3, 2.3L L4 VIN 4, 2.3L L4 VIN D, 2.3L L4 VIN L, 2.3L L4 VIN M, 2.5L L4 VIN 3, 2.5L L4 VIN 5, 2.5L L4 VIN 6, 2.5L L4 VIN 7, 2.5L L4 VIN A, 2.5L L4 VIN M, 3.0L V6 VIN 1, 3.0L V6 VIN G, 3.7L V6 VIN A, 3.7L V6 VIN B, 3.7L V6 VIN V, 4.0L V6 VIN E **Transmission:** All	**EVAP System Gross Leak Detected:** Engine running, and immediately after a vehicle "refueling" event, the PCM detected it could not achieve any initial vacuum in the EVAP system with excessive vapor flow present (Gross EVAP leak). **Possible Causes:** • Fuel filler cap does not fit properly (it is the wrong part number) • Fuel filler cap is missing
DTC: P0460 **2T CCM, MIL: Yes** **Year:** 2009, 2010 **Model:** 3, 6, B2300, B4000, CX-7, CX-9, Tribute **Engine:** 2.0L L4 VIN F, 2.0L L4 VIN G, 2.3L L4 VIN 3, 2.3L L4 VIN 4, 2.3L L4 VIN D, 2.3L L4 VIN L, 2.3L L4 VIN M, 2.5L L4 VIN 3, 2.5L L4 VIN 5, 2.5L L4 VIN 6, 2.5L L4 VIN 7, 2.5L L4 VIN A, 2.5L L4 VIN M, 3.0L V6 VIN 1, 3.0L V6 VIN G, 3.7L V6 VIN A, 3.7L V6 VIN B, 3.7L V6 VIN V, 4.0L V6 VIN E **Transmission:** All	**Fuel Level Indicator Signal Circuit Malfunction:** Engine running, and the PCM detected the Fuel Level Indicator (FLI) signal indicated less than 0.1v, or that it indicated more than 3.48v during the CCM test period. **Possible Causes:** • FLI signal circuit is open, shorted to ground or to power (B+) • Fuel tank empty or overfull (FP module is stuck mechanically) • Wrong fuel gauge is installed, or instrument panel is damaged • Fuel gauge sender unit is damaged or has failed • PCM has failed
DTC: P0461 **2T CCM, MIL: Yes** **Year:** 2009, 2010 **Model:** 3, 5, 6, CX-7, CX-9, MX-5 Miata, RX-8 **Engine:** 1.3L R2 VIN 2, 1.3L R2 VIN 4, 1.3L R2 VIN M, 1.3L R2 VIN P, 2.0L L4 VIN F, 2.0L L4 VIN G, 2.3L L4 VIN 3, 2.3L L4 VIN 4, 2.3L L4 VIN L, 2.3L L4 VIN M, 2.5L L4 VIN 5, 2.5L L4 VIN 6, 2.5L L4 VIN A, 2.5L L4 VIN M, 3.7L V6 VIN A, 3.7L V6 VIN B, 3.7L V6 VIN V **Transmission:** All	**Fuel Tank Level Sensor Circuit Range/Performance:** Engine running and the PCM determined the fuel gauge sender unit signal was operating in too narrow a range during the CCM test. **Possible Causes:** • Fuel gauge sending unit is damaged or has failed • Fuel gauge sending unit signal is open or shorted to ground • Fuel gauge sending unit power circuit is open • Instrument cluster is damaged or has failed • PCM has failed

DTC	Trouble Code Title, Conditions & Possible Causes
DTC: P0461 **2T CCM, MIL: Yes** **Year:** 2009, 2010 **Model:** B2300, B4000, Tribute **Engine:** 2.3L I4 VIN D, 2.5L L4 VIN 3, 2.5L L4 VIN 7, 3.0L V6 VIN 1, 3.0L V6 VIN G, 4.0L V6 VIN E **Transmission:** All	**Fuel Level Indicator Signal Circuit Range/Performance:** Engine running, and the PCM detected the Fuel Level Indicator (FLI) signal indicated the circuit was noisy. **Possible Causes:** • FLI signal circuit is open or shorted to ground (intermittent fault) • FLI assembly is damaged or has failed • PCM has failed
DTC: P0462 **2T CCM, MIL: Yes** **Year:** 2009, 2010 **Model:** 3, 5, 6, B2300, B4000, CX-7, CX-9, MX-5 Miata, RX-8, Tribute **Engine:** 1.3L R2 VIN 2, 1.3L R2 VIN 4, 1.3L R2 VIN M, 1.3L R2 VIN P, 2.0L L4 VIN F, 2.0L L4 VIN G, 2.3L L4 VIN 3, 2.3L L4 VIN 4, 2.3L L4 VIN D, 2.3L L4 VIN L, 2.3L L4 VIN M, 2.5L L4 VIN 3, 2.5L L4 VIN 5, 2.5L L4 VIN 6, 2.5L L4 VIN 7, 2.5L L4 VIN A, 2.5L L4 VIN M, 3.0L V6 VIN 1, 3.0L V6 VIN G, 3.7L V6 VIN A, 3.7L V6 VIN B, 3.7L V6 VIN V, 4.0L V6 VIN E **Transmission:** All	**Fuel Level Sensor Circuit Low Input:** Engine running, system voltage from 11-16v, and the PCM detected the fuel level sensor signal was less than 0.08v during the CCM test. **Possible Causes:** • Fuel gauge sending unit signal is shorted to sensor ground • Fuel gauge sending unit signal is shorted to chassis ground • Fuel gauge sending unit is damaged or has failed • PCM has failed
DTC: P0463 **2T CCM, MIL: Yes** **Year:** 2009, 2010 **Model:** 3, 5, 6, B2300, B4000, CX-7, CX-9, MX-5 Miata, RX-8, Tribute **Engine:** 1.3L R2 VIN 2, 1.3L R2 VIN 4, 1.3L R2 VIN M, 1.3L R2 VIN P, 2.0L L4 VIN F, 2.0L L4 VIN G, 2.3L L4 VIN 3, 2.3L L4 VIN 4, 2.3L L4 VIN D, 2.3L L4 VIN L, 2.3L L4 VIN M, 2.5L L4 VIN 3, 2.5L L4 VIN 5, 2.5L L4 VIN 6, 2.5L L4 VIN 7, 2.5L L4 VIN A, 2.5L L4 VIN M, 3.0L V6 VIN 1, 3.0L V6 VIN G, 3.7L V6 VIN A, 3.7L V6 VIN B, 3.7L V6 VIN V, 4.0L V6 VIN E **Transmission:** All	**Fuel Level Sensor Circuit High Input:** Engine running, system voltage from 11-16v, and the PCM detected the fuel level sensor signal was more than 4.92v during the CCM test. **Possible Causes:** • Fuel gauge sending unit signal shorted to VREF or power • Fuel gauge sending unit ground circuit is open • Fuel gauge sender signal is damaged or has failed • PCM has failed
DTC: P0480 **2T CCM** **Year:** 2009, 2010 **Model:** 3, 5, 6, CX-7, CX-9, MX-5 Miata **Engine:** 2.0L L4 VIN F, 2.0L L4 VIN G, 2.3L L4 VIN 3, 2.3L L4 VIN 4, 2.3L L4 VIN L, 2.3L L4 VIN M, 2.5L L4 VIN 5, 2.5L L4 VIN 6, 2.5L L4 VIN A, 2.5L L4 VIN M, 3.7L V6 VIN A, 3.7L V6 VIN B, 3.7L V6 VIN V **Transmission:** All	**Condenser Fan Relay Control Circuit Malfunction:** Key on or engine running and the PCM detected an unexpected high or low voltage condition on the Condenser Fan Relay control circuit during the CCM test. **Possible Causes:** • Condenser fan relay control circuit open or shorted to ground • Condenser fan relay control power circuit is open • Condenser fan relay is damaged or has failed • PCM has failed

DTC	Trouble Code Title, Conditions & Possible Causes
DTC: P0481 **2T CCM** **Year:** 2009, 2010 **Model:** CX-7, MX-5 Miata, RX-8, Tribute **Engine:** 1.3L R2 VIN 2, 1.3L R2 VIN 4, 1.3L R2 VIN M, 1.3L R2 VIN P, 2.0L L4 VIN F, 2.3L L4 VIN 3, 2.3L L4 VIN L, 2.5L L4 VIN 3, 2.5L L4 VIN 7, 3.0L V6 VIN 1, 3.0L V6 VIN G **Transmission:** All	**Cooling Fan Relay 1 and Condenser Fan Relay 1 Control Circuit Malfunction:** Key on, and the PCM detected an unexpected high or low voltage on the Cooling Fan Relay 1 or Condenser Fan Relay 1 control circuit. **Possible Causes:** • Cooling or condenser fan relay 1 circuit open, shorted to ground • Cooling or condenser fan relay 1 control power circuit is open • Cooling or condenser fan relay 1 is damaged or has failed • PCM has failed
DTC: P0482 **2T CCM** **Year:** 2009, 2010 **Model:** MX-5 Miata, RX-8 **Engine:** 1.3L R2 VIN 2, 1.3L R2 VIN 4, 1.3L R2 VIN M, 1.3L R2 VIN P, 2.0L L4 VIN F **Transmission:** All	**Cooling Fan Relay and Condenser Fan Relay 2 Control Circuit Malfunction:** Key on, and the PCM detected an unexpected high or low voltage on the Cooling Fan Relay or Condenser Fan Relay 2 control circuit. **Possible Causes:** • Cooling or condenser fan relay 2 circuit open, shorted to ground • Cooling or condenser fan relay 2 control power circuit is open • Cooling or condenser fan relay 1 is damaged or has failed • PCM has failed
DTC: P0500 **2T CCM, MIL: Yes** **Year:** 2009, 2010 **Model:** Tribute **Engine:** 2.5L L4 VIN 3, 2.5L L4 VIN 7, 3.0L V6 VIN 1, 3.0L V6 VIN G **Transmission:** All	**Vehicle Speed Sensor Circuit Malfunction:** Vehicle driven in gear at more than 1000 rpm, ECT sensor signal more than 140°F, and the PCM detected an unexpected voltage condition (always high, low or missing) on the VSS circuit. **Possible Causes:** • VSS positive (+) signal is open or shorted to ground • VSS negative (-) signal is open or shorted to ground • VSS is damaged or has failed • PCM has failed
DTC: P0500 **2T CCM, MIL: Yes** **Year:** 2009, 2010 **Model:** 3, 5, 6, B2300, B4000, CX-7, MX-5 Miata, RX-8 **Engine:** 1.3L R2 VIN 2, 1.3L R2 VIN 4, 1.3L R2 VIN M, 1.3L R2 VIN P, 2.0L L4 VIN F, 2.0L L4 VIN G, 2.3L L4 VIN 3, 2.3L L4 VIN 4, 2.3L L4 VIN D, 2.3L L4 VIN L, 2.3L L4 VIN M, 2.5L L4 VIN 5, 2.5L L4 VIN 6, 2.5L L4 VIN A, 2.5L L4 VIN M, 4.0L V6 VIN E **Transmission:** All	**Vehicle Speed Sensor Circuit Malfunction:** Vehicle driven in Drive, 2nd or Low gear position, engine speed over 2000 rpm with the charging efficiency over 40%, and the PCM detected the VSS input indicated less than 2.34 mph for 33 seconds. **Possible Causes:** • VSS signal circuit is open, shorted to ground or to power • VSS power circuit is open or shorted to ground • VSS is damaged or has failed • PCM has failed
DTC: P0501 **2T CCM** **Year:** 2009 **Model:** B2300, B4000 **Engine:** 2.3L L4 VIN D, 4.0L V6 VIN E **Transmission:** All	**Vehicle Speed Sensor Circuit Malfunction:** Vehicle driven in gear at more than 1000 rpm, ECT sensor signal more than 140°F, and the PCM detected an invalid VSS signal from the ABS, CTM or GEM module during the CCM test. **Possible Causes:** • ABS wheel speed sensor circuit open or shorted to ground • ABS wheel speed sensor circuit shorted to VREF or power • ABS, CTM or GEM control module has failed • VSS positive (+) signal circuit is open or shorted to ground • VSS negative (-) signal circuit is open or shorted to ground • PCM has failed
DTC: P0503 **2T CCM, MIL: Yes** **Year:** 2009, 2010 **Model:** B2300, B4000, Tribute **Engine:** 2.3L L4 VIN D, 2.5L L4 VIN 3, 2.5L L4 VIN 7, 3.0L V6 VIN 1, 3.0L V6 VIN G, 4.0L V6 VIN E **Transmission:** All	**Vehicle Speed Sensor Circuit Malfunction:** Vehicle driven in gear at more than 1000 rpm, ECT sensor signal more than 140°F, and the PCM detected a poor or noisy VSS signal during the CCM test. **Possible Causes:** • Check for any devices that could cause RFI on the VSS circuit • VSS positive (+) signal open or shorted to ground (intermittent) • VSS negative (-) signal open or shorted to ground (intermittent) • VSS is damaged or has failed • PCM has failed

DTC	Trouble Code Title, Conditions & Possible Causes
DTC: P0505 **2T CCM, MIL: Yes** **Year:** 2009, 2010 **Model:** 3, 5, 6, B2300, B4000, CX-7, CX-9, MX-5 Miata, RX-8, Tribute **Engine:** 1.3L R2 VIN 2, 1.3L R2 VIN 4, 1.3L R2 VIN M, 1.3L R2 VIN P, 2.0L L4 VIN F, 2.3L L4 VIN 3, 2.3L L4 VIN 4, 2.3L L4 VIN D, 2.3L L4 VIN L, 2.3L L4 VIN M, 2.5L L4 VIN 3, 2.5L L4 VIN 5, 2.5L L4 VIN 6, 2.5L L4 VIN 7, 2.5L L4 VIN A, 2.5L L4 VIN M, 3.0L V6 VIN 1, 3.0L V6 VIN G, 3.7L V6 VIN A, 3.7L V6 VIN B, 3.7L V6 VIN V, 4.0L V6 VIN E **Transmission:** All	**Idle Speed Control System Malfunction:** Key on or engine running, system voltage over 11 volts, and the PCM detected an unexpected voltage condition on the IAC valve control circuit, condition met for 14 seconds, or the PCM could not control the IAC valve in order to reach the desired engine speed. **Possible Causes:** • Throttle body is damaged, dirty or idle speed out of adjustment • High resistance between PCM and IAC valve control circuits • IAC valve circuits open, shorted to ground or shorted to power • IAC valve air inlet is plugged, dirty or restricted • IAC valve is damaged or has failed
DTC: P0506 **2T CCM, MIL: Yes** **Year:** 2009, 2010 **Model:** 3, 5, 6, B2300, B4000, CX-7, CX-9, MX-5 Miata, RX-8, Tribute **Engine:** 1.3L R2 VIN 2, 1.3L R2 VIN 4, 1.3L R2 VIN M, 1.3L R2 VIN P, 2.0L L4 VIN F, 2.0L L4 VIN G, 2.3L L4 VIN 3, 2.3L L4 VIN 4, 2.3L L4 VIN D, 2.3L L4 VIN L, 2.3L L4 VIN M, 2.5L L4 VIN 3, 2.5L L4 VIN 5, 2.5L L4 VIN 6, 2.5L L4 VIN 7, 2.5L L4 VIN A, 2.5L L4 VIN M, 3.0L V6 VIN 1, 3.0L V6 VIN G, 3.7L V6 VIN A, 3.7L V6 VIN B, 3.7L V6 VIN V, 4.0L V6 VIN E **Transmission:** All	**Idle Control System RPM Lower Than Expected:** Engine running at idle, and the PCM detected the Actual idle speed was more than 100 rpm lower than the Target idle speed with the brake depressed and steering wheel straight ahead for 14 seconds. **NOTE: If the atmospheric pressure (BARO reading) is less than 72 kPa, or the IAT sensor signal is less than 14°F, the test is cancelled.** **Possible Causes:** • Air intake system leaks (in the intake manifold or PCV valve) • IAC valve power circuit is open between main relay and valve • High resistance between PCM and IAC valve control circuits • IAC valve circuits open, shorted to ground or shorted to power • IAC valve air inlet is plugged, dirty or restricted • IAC valve is damaged or has failed • Throttle body is damaged or dirty (it may need to be cleaned)
DTC: P0507 **2T CCM, MIL: Yes** **Year:** 2009, 2010 **Model:** 3, 5, 6, B2300, B4000, CX-7, CX-9, MX-5 Miata, RX-8, Tribute **Engine:** 1.3L R2 VIN 2, 1.3L R2 VIN 4, 1.3L R2 VIN M, 1.3L R2 VIN P, 2.0L L4 VIN F, 2.0L L4 VIN G, 2.3L L4 VIN 3, 2.3L L4 VIN 4, 2.3L L4 VIN D, 2.3L L4 VIN L, 2.3L L4 VIN M, 2.5L L4 VIN 3, 2.5L L4 VIN 5, 2.5L L4 VIN 6, 2.5L L4 VIN 7, 2.5L L4 VIN A, 2.5L L4 VIN M, 3.0L V6 VIN 1, 3.0L V6 VIN G, 3.7L V6 VIN A, 3.7L V6 VIN B, 3.7L V6 VIN V, 4.0L V6 VIN E **Transmission:** All	**Idle Control System RPM Higher Than Expected:** Engine running at idle, and the PCM detected the Actual idle speed was more than 200 rpm higher than the Target idle speed with the brake depressed and steering wheel straight ahead for 14 seconds. **NOTE: If the atmospheric pressure (BARO reading) is less than 72 kPa, or the IAT sensor signal is less than 14°F, the test is cancelled.** **Possible Causes:** • Air intake system leaks (in the intake manifold or PCV valve) • IAC valve power circuit is open between main relay and valve • High resistance between PCM and IAC valve control circuits • IAC valve circuits open, shorted to ground or shorted to power • IAC valve air inlet is plugged, dirty or restricted • IAC valve is damaged or has failed • Throttle body is damaged or dirty (it may need to be cleaned)

DTC	Trouble Code Title, Conditions & Possible Causes
DTC: P050A **2T CCM, MIL: Yes** **Year:** 2009, 2010 **Model:** 3, 5, 6, CX-7, MX-5 Miata, RX-8 **Engine:** 1.3L R2 VIN 2, 1.3L R2 VIN 4, 1.3L R2 VIN M, 1.3L R2 VIN P, 2.0L L4 VIN F, 2.0L L4 VIN G, 2.3L L4 VIN 3, 2.3L L4 VIN 4, 2.3L L4 VIN L, 2.3L L4 VIN M, 2.5L L4 VIN 5, 2.5L L4 VIN 6, 2.5L L4 VIN A, 2.5L L4 VIN M **Transmission:** All	**Cold start idle air control system performance problem:** The actual idle speed is lower than expected by specified for 8.4 s when the target idle speed correction value for cold start is above 0 rpm or ignition retard value is above as following: 9.0 ° (L5) 9.5 ° (LF/Except California emission regulation applicable model) 11.0 ° (LF/California emission regulation applicable model) **NOTE: If atmospheric pressure is less than 72.3 kPa {542 mmHg, 21.3 in Hg} or intake air temperature is below -10 °C {14 °F}, the PCM cancels diagnosis of P050A.** **Possible Causes:** • Air suction in intake air system • Electronic throttle control system malfunction • Throttle valve stuck or blockage • PCM malfunction
DTC: P050B **2T CCM, MIL: Yes** **Year:** 2009, 2010 **Model:** 3, 5, 6, CX-7, MX-5 Miata **Engine:** 2.0L L4 VIN F, 2.0L L4 VIN G, 2.3L L4 VIN 3, 2.3L L4 VIN 4, 2.3L L4 VIN L, 2.3L L4 VIN M, 2.5L L4 VIN 5, 2.5L L4 VIN 6, 2.5L L4 VIN A, 2.5L L4 VIN M **Transmission:** All	**Cold start ignition timing performance problem:** The PCM monitors actual ignition timing using the CKP sensor while electronic spark advance control fast idle correction operating. If the ignition timing is out of specified range, the PCM determines that the ignition timing at cold condition has performance problem. **Possible Causes:** • Damaged or chipped CKP sensor pulse wheel • CKP sensor malfunction • Damaged or chipped CMP sensor pulse wheel • CMP sensor malfunction • PCM malfunction
DTC: P050E **T PCM** **Year:** 2009, 2010 **Model:** 6, CX-9 **Engine:** 3.7L V6 VIN A, 3.7L V6 VIN B, 3.7L V6 VIN V **Transmission:** All	**Cold start engine exhaust temperature out of range:** The PCM calculates the actual catalyst warm up temperature during a cold start. The PCM then compares the actual temperature to the expected catalyst temperature model. The difference between the actual and expected temperatures is a ratio. When this ratio exceeds the calibrated value this DTC is set and the malfunction indicator lamp (MIL) illuminates. **Possible Causes:** • Intake air restriction • Vacuum leakage • Throttle body malfunction • Mechanical concern with engine • Exhaust restriction • PCM malfunction
DTC: P0511 **2T CCM, MIL: Yes** **Year:** 2009 **Model:** B2300, B4000 **Engine:** 2.3L L4 VIN D, 4.0L V6 VIN E **Transmission:** All	**PCM – Idle Air Control (IAC) valve assembly:** PCM detects electrical load failure on IAC output circuit **Possible Causes:** • IAC output circuit is open or shorted to ground • IAC output circuit is shorted to system power • IAC valve assembly is damaged or has failed • PCM has failed
DTC: P0522 **T CCM** **Year:** 2009, 2010 **Model:** RX-8 **Engine:** 1.3L R2 VIN 2, 1.3L R2 VIN 4, 1.3L R2 VIN M, 1.3L R2 VIN P **Transmission:** All	**Oil pressure sensor circuit low input:** The voltage of oil pressure sensor input terminal is less than 0.2 V for 0.5 s or more. **Possible Causes:** • Oil pressure sensor connector or terminals malfunction • Short to ground in wiring harness between oil pressure sensor and PCM • PCM connector or terminals malfunction • Open circuit in wiring harness between oil pressure sensor and PCM • Oil pressure sensor malfunction • PCM malfunction
DTC: P0523 **T CCM** **Year:** 2009, 2010 **Model:** RX-8 **Engine:** 1.3L R2 VIN 2, 1.3L R2 VIN 4, 1.3L R2 VIN M, 1.3L R2 VIN P **Transmission:** All	**Oil pressure sensor circuit high input:** The voltage of oil pressure sensor input terminal is more than 4.8 V for 0.5 s or more. **Possible Causes:** • Oil pressure sensor connector or terminals malfunction • PCM connector or terminals malfunction • Short to power supply in wiring harness between oil pressure sensor terminal B and PCM terminal 2R • Open circuit in wiring harness between oil pressure sensor terminal A and PCM terminal 2M • Oil pressure sensor malfunction • PCM malfunction

DTC	Trouble Code Title, Conditions & Possible Causes
DTC: P0532 **2T CCM, MIL: Yes** **Year:** 2009, 2010 **Model:** 6, B2300, B4000, Tribute **Engine:** 2.3L L4 VIN D, 2.5L L4 VIN 3, 2.5L L4 VIN 7, 3.0L V6 VIN 1, 3.0L V6 VIN G, 3.7L V6 VIN B, 4.0L V6 VIN E **Transmission:** All	**PCM – Air Conditioning Pressure (ACP) sensor:** ACP sensor voltage is above calibrated limit. **Possible Causes:** • ACP sensor is open or shorted to ground • ACP sensor is shorted to system power • ACP sensor is damaged or has failed • PCM has failed
DTC: P0533 **2T CCM, MIL: Yes** **Year:** 2009, 2010 **Model:** 6, B2300, B4000, Tribute **Engine:** 2.3L L4 VIN D, 2.5L L4 VIN 3, 2.5L L4 VIN 7, 3.0L V6 VIN 1, 3.0L V6 VIN G, 3.7L V6 VIN B, 4.0L V6 VIN E **Transmission:** All	**PCM – Air Conditioning Pressure (ACP) sensor:** ACP sensor voltage is below calibrated limit. **Possible Causes:** • ACP sensor is open or shorted to ground • ACP sensor is shorted to system power • ACP sensor is damaged or has failed • PCM has failed
DTC: P0534 **2T CCM, MIL: Yes** **Year:** 2009 **Model:** B2300, B4000 **Engine:** 2.3L L4 VIN D, 4.0L V6 VIN E **Transmission:** All	**PCM – Air Conditioning (A/C) system:** A/C compressor clutch cycling too frequently. **Possible Causes:** • A/C system is low on Freon • A/C system has too much Freon • A/C compressor clutch is damaged or failed • PCM has failed
DTC: P0550 **2T CCM, MIL: Yes** **Year:** 2009, 2010 **Model:** 6, B2300, B4000, CX-7, MX-5 Miata **Engine:** 2.0L L4 VIN F, 2.3L L4 VIN 3, 2.3L L4 VIN D, 2.3L L4 VIN L, 2.5L L4 VIN 5, 2.5L L4 VIN A, 2.5L L4 VIN M, 4.0L V6 VIN E **Transmission:** All	**Power Steering Pressure Switch Circuit Malfunction:** Engine running, ECT sensor signal more than 14°F, VSS more than 37.4 mph, and the PCM detected the PSP switch signal low signal (switch "on" signal), condition met for over one minute. **NOTE: This is a normally open (N.O.) type of pressure switch.** **Possible Causes:** • Power steering switch signal circuit shorted to sensor ground • Power steering switch signal circuit shorted to chassis ground • Power steering switch is damaged or has failed • Engine Speed System (power steering function) is damaged • PCM has failed
DTC: P0551 **2T CCM, MIL: Yes** **Year:** 2009, 2010 **Model:** Tribute **Engine:** 2.5L L4 VIN 3, 2.5L L4 VIN 7, 3.0L V6 VIN 1, 3.0L V6 VIN G **Transmission:** All	**Power Steering Pressure Switch Circuit Malfunction:** Engine running, ECT sensor signal more than 14°F, VSS more than 37.4 mph, and the PCM detected the PSP switch signal low signal (switch "on" signal), condition met for over one minute. **NOTE: This is a normally open (N.O.) type of pressure switch.** **Possible Causes:** • Power steering switch signal circuit shorted to sensor ground • Power steering switch signal circuit shorted to chassis ground • Power steering switch is damaged or has failed • Engine Speed System (power steering function) is damaged • PCM has failed
DTC: P0552 **2T CCM, MIL: Yes** **Year:** 2009, 2010 **Model:** Tribute **Engine:** 2.5L L4 VIN 3, 2.5L L4 VIN 7, 3.0L V6 VIN 1, 3.0L V6 VIN G **Transmission:** All	**Power Steering Pressure Switch Circuit Malfunction:** Engine running, ECT sensor signal more than 14°F, VSS more than 37.4 mph, and the PCM detected the PSP switch signal low signal (switch "on" signal), condition met for over one minute. **NOTE: This is a normally open (N.O.) type of pressure switch.** **Possible Causes:** • Power steering switch signal circuit shorted to sensor ground • Power steering switch signal circuit shorted to chassis ground • Power steering switch is damaged or has failed • Engine Speed System (power steering function) is damaged • PCM has failed

DTC	Trouble Code Title, Conditions & Possible Causes
DTC: P0553 **2T CCM, MIL: Yes** **Year:** 2009, 2010 **Model:** Tribute **Engine:** 2.5L L4 VIN 3, 2.5L L4 VIN 7, 3.0L V6 VIN 1, 3.0L V6 VIN G **Transmission:** All	**Power Steering Pressure Switch Circuit Malfunction:** Engine running, ECT sensor signal more than 14°F, VSS more than 37.4 mph, and the PCM detected the PSP switch signal low signal (switch "on" signal), condition met for over one minute. **NOTE: This is a normally open (N.O.) type of pressure switch.** **Possible Causes:** • Power steering switch signal circuit shorted to sensor ground • Power steering switch signal circuit shorted to chassis ground • Power steering switch is damaged or has failed • Engine Speed System (power steering function) is damaged • PCM has failed
DTC: P0562 **2T CCM, MIL: Yes** **Year:** 2009, 2010 **Model:** RX-8 **Engine:** 1.3L R2 VIN 2, 1.3L R2 VIN 4, 1.3L R2 VIN M, 1.3L R2 VIN P **Transmission:** All	**Cruise Control Switch Circuit Malfunction:** The PCM monitors the cruise control switch signal at the PCM terminal 3P. If the PCM detects that any one of the following switches (MAIN, CANCEL, SET/COAST, RESUME/ACCEL) remains on for 2 min., the PCM determines that the cruise control switch has a malfunction. **Possible Causes:** • Cruise control switch malfunction • Connector or terminal malfunction • Short to power circuit in wiring from cruise control terminal B and PCM terminal 3P. • Short to ground circuit in wiring from cruise control terminal B and PCM terminal 3P • PCM has failed
DTC: P0563 **T PCM** **Year:** 2009, 2010 **Model:** 6, CX-9 **Engine:** 3.7L V6 VIN A, 3.7L V6 VIN B, 3.7L V6 VIN V **Transmission:** All	**System voltage high:** This DTC is stored when the PCM detects high system voltage. **Possible Causes:** • Charging system concern.
DTC: P0564 **2T CCM, MIL: Yes** **Year:** 2009, 2010 **Model:** 5, MX-5 Miata, RX-8 **Engine:** 1.3L R2 VIN 2, 1.3L R2 VIN 4, 1.3L R2 VIN M, 1.3L R2 VIN P, 2.0L L4 VIN F, 2.3L L4 VIN 3, 2.3L L4 VIN L **Transmission:** All	**Cruise Control Switch Circuit Malfunction:** The PCM monitors the cruise control switch signal at the PCM terminal 3P. If the PCM detects that any one of the following switches (MAIN, CANCEL, SET/COAST, RESUME/ACCEL) remains on for 2 min., the PCM determines that the cruise control switch has a malfunction. **Possible Causes:** • Cruise control switch malfunction • Connector or terminal malfunction • Short to power circuit in wiring from cruise control terminal B and PCM terminal 3P. • Short to ground circuit in wiring from cruise control terminal B and PCM terminal 3P • PCM has failed
DTC: P0571 **2T CCM, MIL: Yes** **Year:** 2009, 2010 **Model:** 3, 5, 6, CX-7, MX-5 Miata, RX-8 **Engine:** 1.3L R2 VIN 2, 1.3L R2 VIN 4, 1.3L R2 VIN M, 1.3L R2 VIN P, 2.0L L4 VIN F, 2.0L L4 VIN G, 2.3L L4 VIN 3, 2.3L L4 VIN 4, 2.3L L4 VIN L, 2.3L L4 VIN M, 2.5L L4 VIN 5, 2.5L L4 VIN 6, 2.5L L4 VIN A, 2.5L L4 VIN M **Transmission:** All	**Cruise Control Switch Circuit Malfunction:** The PCM monitors the cruise control switch signal at the PCM terminal 3P. If the PCM detects that any one of the following switches (MAIN, CANCEL, SET/COAST, RESUME/ACCEL) remains on for 2 min., the PCM determines that the cruise control switch has a malfunction. **Possible Causes:** • Cruise control switch malfunction • Connector or terminal malfunction • Short to power circuit in wiring from cruise control terminal B and PCM terminal 3P. • Short to ground circuit in wiring from cruise control terminal B and PCM terminal 3P • PCM has failed
DTC: P0572 **T PCM** **Year:** 2009, 2010 **Model:** 6 **Engine:** 3.7L V6 VIN B **Transmission:** All	**Brake switch No.1 circuit low:** This DTC indicates the brake switch is stuck. **Possible Causes:** • Brake switch No.1 mis installation • Connector or terminal malfunction • Brake switch No.1 circuit malfunction • Brake switch No.1 malfunction

DTC	Trouble Code Title, Conditions & Possible Causes
DTC: P0573 **T PCM** **Year:** 2009, 2010 **Model:** 6 **Engine:** 3.7L V6 VIN B **Transmission:** All	**Brake switch No.1 circuit high:** This DTC indicates the brake switch is stuck. **Possible Causes:** • Brake switch No.1 mis installation • Connector or terminal malfunction • Brake switch No.1 circuit malfunction • Brake switch No.1 malfunction
DTC: P0579 **2T CCM, MIL: Yes** **Year:** 2009, 2010 **Model:** 3, 6, CX-7, CX-9, Tribute **Engine:** 2.0L L4 VIN F, 2.0L L4 VIN G, 2.3L L4 VIN 3, 2.3L L4 VIN 4, 2.3L L4 VIN L, 2.3L L4 VIN M, 2.5L L4 VIN 3, 2.5L L4 VIN 5, 2.5L L4 VIN 6, 2.5L L4 VIN 7, 2.5L L4 VIN A, 2.5L L4 VIN M, 3.0L V6 VIN 1, 3.0L V6 VIN G, 3.7L V6 VIN A, 3.7L V6 VIN B, 3.7L V6 VIN V **Transmission:** All	**PCM – Vehicle Speed Signal (VSS) Malfunction:** The PCM monitors the Vehicle speed signal (VSS) at the PCM terminal. If the PCM detects that the Turbine Shaft Speed (TSS) sensor or the Output Shaft Sensor (OSS) remains on for 2 min., the PCM determines that the VSS has a malfunction. **Possible Causes:** • VSS malfunction • Connector or terminal malfunction • Short to power circuit in wiring from VSS and PCM terminal. • Short to ground circuit in wiring from VSS and PCM terminal. • PCM has failed
DTC: P0581 **2T CCM, MIL: Yes** **Year:** 2009, 2010 **Model:** 3, 6, CX-7, CX-9, Tribute **Engine:** 2.0L L4 VIN F, 2.0L L4 VIN G, 2.3L L4 VIN 3, 2.3L L4 VIN 4, 2.3L L4 VIN L, 2.3L L4 VIN M, 2.5L L4 VIN 3, 2.5L L4 VIN 5, 2.5L L4 VIN 6, 2.5L L4 VIN 7, 2.5L L4 VIN A, 2.5L L4 VIN M, 3.0L V6 VIN 1, 3.0L V6 VIN G, 3.7L V6 VIN A, 3.7L V6 VIN B, 3.7L V6 VIN V **Transmission:** All	**PCM – Vehicle Speed Signal (VSS) malfunction:** The PCM monitors the Vehicle speed signal (VSS) at the PCM terminal. If the PCM detects that the Turbine Shaft Speed (TSS) sensor or the Output Shaft Sensor (OSS) remains on for 2 min., the PCM determines that the VSS has a malfunction. **Possible Causes:** • VSS malfunction • Connector or terminal malfunction • Short to power circuit in wiring from VSS and PCM terminal. • Short to ground circuit in wiring from VSS and PCM terminal. • PCM has failed
DTC: P0600 **1T CCM, MIL: Yes** **Year:** 2009, 2010 **Model:** 3, 5, 6, B2300, B4000, CX-7, CX-9 **Engine:** 2.0L L4 VIN F, 2.0L L4 VIN G, 2.3L L4 VIN 3, 2.3L L4 VIN 4, 2.3L L4 VIN D, 2.3L L4 VIN L, 2.3L L4 VIN M, 2.5L L4 VIN 5, 2.5L L4 VIN 6, 2.5L L4 VIN A, 2.5L L4 VIN M, 3.7L V6 VIN A, 3.7L V6 VIN B, 3.7L V6 VIN V, 4.0L V6 VIN E **Transmission:** All	**Serial communication link:** PCM internal malfunction. **Possible Causes:** • PCM connector or terminals malfunction • PCM malfunction
DTC: P0601 **1T CCM, MIL: Yes** **Year:** 2009, 2010 **Model:** 3, 5, 6, B2300, B4000, CX-7, MX-5 Miata, RX-8 **Engine:** 1.3L R2 VIN 2, 1.3L R2 VIN 4, 1.3L R2 VIN M, 1.3L R2 VIN P, 2.0L L4 VIN F, 2.0L L4 VIN G, 2.3L L4 VIN 3, 2.3L L4 VIN 4, 2.3L L4 VIN D, 2.3L L4 VIN L, 2.3L L4 VIN M, 2.5L L4 VIN 5, 2.5L L4 VIN 6, 2.5L L4 VIN A, 2.5L L4 VIN M, 4.0L V6 VIN E **Transmission:** All	**Internal control module memory check sum error:** PCM internal ROM malfunction **Possible Causes:** • Reprogramming has not been completed properly • PCM has failed

DTC	Trouble Code Title, Conditions & Possible Causes

DTC: P0602
1T CCM, MIL: Yes
Year: 2009, 2010
Model: 3, 5, 6, B2300, B4000, CX-7, CX-9, MX-5 Miata, RX-8, Tribute
Engine: 1.3L R2 VIN 2, 1.3L R2 VIN 4, 1.3L R2 VIN M, 1.3L R2 VIN P, 2.0L L4 VIN F, 2.0L L4 VIN G, 2.3L L4 VIN 3, 2.3L L4 VIN 4, 2.3L L4 VIN D, 2.3L L4 VIN L, 2.3L L4 VIN M, 2.5L L4 VIN 3, 2.5L L4 VIN 5, 2.5L L4 VIN 6, 2.5L L4 VIN 7, 2.5L L4 VIN A, 2.5L L4 VIN M, 3.0L V6 VIN 1, 3.0L V6 VIN G, 3.7L V6 VIN A, 3.7L V6 VIN B, 3.7L V6 VIN V, 4.0L V6 VIN E
Transmission: All
	PCM programming error:
	No configuration data in PCM
	Possible Causes:
	   • Reprogramming has not been completed properly
	   • PCM has failed

DTC: P0603
1T CCM
Year: 2009, 2010
Model: 6, B2300, B4000, CX-9, Tribute
Engine: 2.3L L4 VIN D, 2.5L L4 VIN 3, 2.5L L4 VIN 7, 3.0L V6 VIN 1, 3.0L V6 VIN G, 3.7L V6 VIN A, 3.7L V6 VIN B, 3.7L V6 VIN V, 4.0L V6 VIN E
Transmission: All
	PCM Keep Alive Memory Test Error:
	Key on, and the PCM detected an interruption to the Keep Alive Memory (KAM) circuit during the initial key "on" sequence.
	Possible Causes:
	   • KAM circuit is open (it may be an intermittent problem)
	   • KAM circuit is shorted to ground
	   • Battery terminals have high resistance due to corrosion or dirt
	   • PCM has failed

DTC: P0604
1T CCM, MIL: Yes
Year: 2009, 2010
Model: 3, 5, 6, B2300, B4000, CX-7, CX-9, MX-5 Miata, RX-8
Engine: 1.3L R2 VIN 2, 1.3L R2 VIN 4, 1.3L R2 VIN M, 1.3L R2 VIN P, 2.0L L4 VIN F, 2.0L L4 VIN G, 2.3L L4 VIN 3, 2.3L L4 VIN 4, 2.3L L4 VIN D, 2.3L L4 VIN L, 2.3L L4 VIN M, 2.5L L4 VIN 5, 2.5L L4 VIN 6, 2.5L L4 VIN A, 2.5L L4 VIN M, 3.7L V6 VIN A, 3.7L V6 VIN B, 3.7L V6 VIN V, 4.0L V6 VIN E
Transmission: All
	PCM RAM Error:
	PCM RAM Malfunction
	Possible Causes:
	   • Reprogramming has not been completed properly
	   • PCM has failed

DTC: P0605
1T CCM
Year: 2009, 2010
Model: B2300, B4000, CX-9, Tribute
Engine: 2.3L L4 VIN D, 2.5L L4 VIN 3, 2.5L L4 VIN 7, 3.0L V6 VIN 1, 3.0L V6 VIN G, 3.7L V6 VIN A, 3.7L V6 VIN V, 4.0L V6 VIN E
Transmission: All
	PCM Read Only Memory Test Error:
	Key on, and the PCM detected a Read Only Memory test error during its initial Self-Test key "on" sequence.
	Possible Causes:
	   • PCM has failed (an internal fault)
	   • PCM must be replaced when this code is set

DTC	Trouble Code Title, Conditions & Possible Causes
DTC: P0606 **1T CCM, MIL: Yes** **Year:** 2009, 2010 **Model:** 3, 5, 6, B2300, B4000, CX-7, CX-9, MX-5 Miata, RX-8, Tribute **Engine:** 1.3L R2 VIN 2, 1.3L R2 VIN 4, 1.3L R2 VIN M, 1.3L R2 VIN P, 2.0L L4 VIN F, 2.0L L4 VIN G, 2.3L L4 VIN 3, 2.3L L4 VIN 4, 2.3L L4 VIN D, 2.3L L4 VIN L, 2.3L L4 VIN M, 2.5L L4 VIN 3, 2.5L L4 VIN 5, 2.5L L4 VIN 6, 2.5L L4 VIN 7, 2.5L L4 VIN A, 2.5L L4 VIN M, 3.0L V6 VIN 1, 3.0L V6 VIN G, 3.7L V6 VIN A, 3.7L V6 VIN V, 4.0L V6 VIN E **Transmission:** All	**ECM/PCM Processor:** PCM internal malfunction. **Possible Causes:** • PCM internal CPU malfunction
DTC: P0607 **1T CCM, MIL: Yes** **Year:** 2009, 2010 **Model:** 5, 6, B2300, B4000, CX-9 **Engine:** 2.3L L4 VIN 3, 2.3L L4 VIN D, 2.3L L4 VIN L, 2.5L L4 VIN A, 3.7L V6 VIN A, 3.7L V6 VIN B, 3.7L V6 VIN V, 4.0L V6 VIN E **Transmission:** All	**Control Module vehicle operations error:** PCM data configuration error **Possible Causes:** • PCM internal malfunction
DTC: P060A **1T CCM, MIL: Yes** **Year:** 2009, 2010 **Model:** 6, CX-9 **Engine:** 2.5L L4 VIN A, 3.7L V6 VIN A, 3.7L V6 VIN B, 3.7L V6 VIN V **Transmission:** All	**Internal control module monitoring processor performance problem:** Indicates an error occurred in the PCM. **Possible Causes:** • PCM connector or terminal malfunction • Software incompatibility issue • PCM malfunction
DTC: P060B **T PCM** **Year:** 2009, 2010 **Model:** 6 **Engine:** 3.7L V6 VIN B **Transmission:** All	**Internal control module A/D processing performance:** Indicates that an error occurred in the PCM. **Possible Causes:** • PCM malfunction
DTC: P060B **1T CCM, MIL: Yes** **Year:** 2009, 2010 **Model:** 3, 6, CX-7, CX-9 **Engine:** 2.5L L4 VIN 5, 2.5L L4 VIN 6, 2.5L L4 VIN A, 2.5L L4 VIN M, 3.7L V6 VIN A, 3.7L V6 VIN V **Transmission:** All	**Internal control module A/D processing performance problem:** Indicates that an error occurred in the PCM. **Possible Causes:** • PCM connector or terminals malfunction • PCM malfunction
DTC: P060C **1T CCM, MIL: Yes** **Year:** 2009, 2010 **Model:** 3, 6, CX-7, CX-9 **Engine:** 2.5L L4 VIN 5, 2.5L L4 VIN 6, 2.5L L4 VIN A, 2.5L L4 VIN M, 3.7L V6 VIN A, 3.7L V6 VIN B, 3.7L V6 VIN V **Transmission:** All	**Internal control module main processor performance problem:** Indicates an error occurred in the PCM. **Possible Causes:** • PCM connector or terminals malfunction • Software incompatibility issue • PCM malfunction

DTC	Trouble Code Title, Conditions & Possible Causes
DTC: P0602 **1T CCM, MIL: Yes** **Year:** 2009, 2010 **Model:** 3, 5, 6, B2300, B4000, CX-7, CX-9, MX-5 Miata, RX-8, Tribute **Engine:** 1.3L R2 VIN 2, 1.3L R2 VIN 4, 1.3L R2 VIN M, 1.3L R2 VIN P, 2.0L L4 VIN F, 2.0L L4 VIN G, 2.3L L4 VIN 3, 2.3L L4 VIN 4, 2.3L L4 VIN D, 2.3L L4 VIN L, 2.3L L4 VIN M, 2.5L L4 VIN 3, 2.5L L4 VIN 5, 2.5L L4 VIN 6, 2.5L L4 VIN 7, 2.5L L4 VIN A, 2.5L L4 VIN M, 3.0L V6 VIN 1, 3.0L V6 VIN G, 3.7L V6 VIN A, 3.7L V6 VIN B, 3.7L V6 VIN V, 4.0L V6 VIN E **Transmission:** All	**PCM programming error:** No configuration data in PCM **Possible Causes:** • Reprogramming has not been completed properly • PCM has failed
DTC: P0603 **1T CCM** **Year:** 2009, 2010 **Model:** 6, B2300, B4000, CX-9, Tribute **Engine:** 2.3L L4 VIN D, 2.5L L4 VIN 3, 2.5L L4 VIN 7, 3.0L V6 VIN 1, 3.0L V6 VIN G, 3.7L V6 VIN A, 3.7L V6 VIN B, 3.7L V6 VIN V, 4.0L V6 VIN E **Transmission:** All	**PCM Keep Alive Memory Test Error:** Key on, and the PCM detected an interruption to the Keep Alive Memory (KAM) circuit during the initial key "on" sequence. **Possible Causes:** • KAM circuit is open (it may be an intermittent problem) • KAM circuit is shorted to ground • Battery terminals have high resistance due to corrosion or dirt • PCM has failed
DTC: P0604 **1T CCM, MIL: Yes** **Year:** 2009, 2010 **Model:** 3, 5, 6, B2300, B4000, CX-7, CX-9, MX-5 Miata, RX-8 **Engine:** 1.3L R2 VIN 2, 1.3L R2 VIN 4, 1.3L R2 VIN M, 1.3L R2 VIN P, 2.0L L4 VIN F, 2.0L L4 VIN G, 2.3L L4 VIN 3, 2.3L L4 VIN 4, 2.3L L4 VIN D, 2.3L L4 VIN L, 2.3L L4 VIN M, 2.5L L4 VIN 5, 2.5L L4 VIN 6, 2.5L L4 VIN A, 2.5L L4 VIN M, 3.7L V6 VIN A, 3.7L V6 VIN B, 3.7L V6 VIN V, 4.0L V6 VIN E **Transmission:** All	**PCM RAM Error:** PCM RAM Malfunction **Possible Causes:** • Reprogramming has not been completed properly • PCM has failed
DTC: P0605 **1T CCM** **Year:** 2009, 2010 **Model:** B2300, B4000, CX-9, Tribute **Engine:** 2.3L L4 VIN D, 2.5L L4 VIN 3, 2.5L L4 VIN 7, 3.0L V6 VIN 1, 3.0L V6 VIN G, 3.7L V6 VIN A, 3.7L V6 VIN V, 4.0L V6 VIN E **Transmission:** All	**PCM Read Only Memory Test Error:** Key on, and the PCM detected a Read Only Memory test error during its initial Self-Test key "on" sequence. **Possible Causes:** • PCM has failed (an internal fault) • PCM must be replaced when this code is set

DTC	Trouble Code Title, Conditions & Possible Causes
DTC: P0606 **1T CCM, MIL: Yes** **Year:** 2009, 2010 **Model:** 3, 5, 6, B2300, B4000, CX-7, CX-9, MX-5 Miata, RX-8, Tribute **Engine:** 1.3L R2 VIN 2, 1.3L R2 VIN 4, 1.3L R2 VIN M, 1.3L R2 VIN P, 2.0L L4 VIN F, 2.0L L4 VIN G, 2.3L L4 VIN 3, 2.3L L4 VIN 4, 2.3L L4 VIN D, 2.3L L4 VIN L, 2.3L L4 VIN M, 2.5L L4 VIN 3, 2.5L L4 VIN 5, 2.5L L4 VIN 6, 2.5L L4 VIN 7, 2.5L L4 VIN A, 2.5L L4 VIN M, 3.0L V6 VIN 1, 3.0L V6 VIN G, 3.7L V6 VIN A, 3.7L V6 VIN V, 4.0L V6 VIN E **Transmission:** All	**ECM/PCM Processor:** PCM internal malfunction. **Possible Causes:** • PCM internal CPU malfunction
DTC: P0607 **1T CCM, MIL: Yes** **Year:** 2009, 2010 **Model:** 5, 6, B2300, B4000, CX-9 **Engine:** 2.3L L4 VIN 3, 2.3L L4 VIN D, 2.3L L4 VIN L, 2.5L L4 VIN A, 3.7L V6 VIN A, 3.7L V6 VIN B, 3.7L V6 VIN V, 4.0L V6 VIN E **Transmission:** All	**Control Module vehicle operations error:** PCM data configuration error **Possible Causes:** • PCM internal malfunction
DTC: P060A **1T CCM, MIL: Yes** **Year:** 2009, 2010 **Model:** 6, CX-9 **Engine:** 2.5L L4 VIN A, 3.7L V6 VIN A, 3.7L V6 VIN B, 3.7L V6 VIN V **Transmission:** All	**Internal control module monitoring processor performance problem:** Indicates an error occurred in the PCM. **Possible Causes:** • PCM connector or terminal malfunction • Software incompatibility issue • PCM malfunction
DTC: P060B **T PCM** **Year:** 2009, 2010 **Model:** 6 **Engine:** 3.7L V6 VIN B **Transmission:** All	**Internal control module A/D processing performance:** Indicates that an error occurred in the PCM. **Possible Causes:** • PCM malfunction
DTC: P060B **1T CCM, MIL: Yes** **Year:** 2009, 2010 **Model:** 3, 6, CX-7, CX-9 **Engine:** 2.5L L4 VIN 5, 2.5L L4 VIN 6, 2.5L L4 VIN A, 2.5L L4 VIN M, 3.7L V6 VIN A, 3.7L V6 VIN V **Transmission:** All	**Internal control module A/D processing performance problem:** Indicates that an error occurred in the PCM. **Possible Causes:** • PCM connector or terminals malfunction • PCM malfunction
DTC: P060C **1T CCM, MIL: Yes** **Year:** 2009, 2010 **Model:** 3, 6, CX-7, CX-9 **Engine:** 2.5L L4 VIN 5, 2.5L L4 VIN 6, 2.5L L4 VIN A, 2.5L L4 VIN M, 3.7L V6 VIN A, 3.7L V6 VIN B, 3.7L V6 VIN V **Transmission:** All	**Internal control module main processor performance problem:** Indicates an error occurred in the PCM. **Possible Causes:** • PCM connector or terminals malfunction • Software incompatibility issue • PCM malfunction

DTC	Trouble Code Title, Conditions & Possible Causes
DTC: P060D **1T CCM, MIL:** Yes **Year:** 2009, 2010 **Model:** 3, 6, CX-7 **Engine:** 2.5L L4 VIN 5, 2.5L L4 VIN 6, 2.5L L4 VIN A, 2.5L L4 VIN M, 3.7L V6 VIN B **Transmission:** All	**Internal control module engine air mass performance problem:** Indicates an error occurred in the PCM. **Possible Causes:** • PCM connector or terminals malfunction • Software incompatibility issue • PCM malfunction
DTC: P0610 **1T CCM, MIL:** Yes **Year:** 2009, 2010 **Model:** 3, 5, 6, B2300, B4000, CX-7, CX-9, MX-5 Miata, RX-8 **Engine:** 1.3L R2 VIN 2, 1.3L R2 VIN 4, 1.3L R2 VIN M, 1.3L R2 VIN P, 2.0L L4 VIN F, 2.0L L4 VIN G, 2.3L L4 VIN 3, 2.3L L4 VIN 4, 2.3L L4 VIN D, 2.3L L4 VIN L, 2.3L L4 VIN M, 2.5L L4 VIN 5, 2.5L L4 VIN 6, 2.5L L4 VIN A, 2.5L L4 VIN M, 3.7L V6 VIN A, 3.7L V6 VIN B, 3.7L V6 VIN V, 4.0L V6 VIN E **Transmission:** All	**Control Module performance:** PCM internal CPU malfunction. **Possible Causes:** • PCM internal malfunction* Configuration has not been properly completed.
DTC: P061B **1T CCM, MIL:** Yes **Year:** 2009, 2010 **Model:** 3, 6, CX-7, CX-9 **Engine:** 2.5L L4 VIN 5, 2.5L L4 VIN 6, 2.5L L4 VIN A, 2.5L L4 VIN M, 3.7L V6 VIN A, 3.7L V6 VIN B, 3.7L V6 VIN V **Transmission:** All	**Internal control module torque calculation performance problem:** Indicates that a calculation error occurred in the PCM. **Possible Causes:** • NOTE: This DTC is an informational DTC and may be set in combination with a number of other DTCs which are causing the FMEM. Diagnose other DTCs first. • PCM connector or terminals malfunction • PCM malfunction
DTC: P061C **T PCM** **Year:** 2009, 2010 **Model:** 6, CX-9 **Engine:** 3.7L V6 VIN A, 3.7L V6 VIN B, 3.7L V6 VIN V **Transmission:** All	**Internal control module engine RPM performance:** Indicates that a calculation error occurred in the PCM. **Possible Causes:** • CKP sensor malfunction • CKP sensor circuit is open or short. • CKP sensor circuit intermittent • PCM malfunction • CMP sensor circuit is open or short. • CMP sensor circuit intermittent • CMP sensor malfunction
DTC: P061D **T PCM** **Year:** 2009, 2010 **Model:** 6 **Engine:** 3.7L V6 VIN B **Transmission:** All	**Internal control module engine air mass performance:** Indicates an error occurred in the PCM. **Possible Causes:** • Software incompatibility issue • PCM malfunction
DTC: P061D **T PCM** **Year:** 2009, 2010 **Model:** CX-9 **Engine:** 3.7L V6 VIN A, 3.7L V6 VIN V **Transmission:** All	**Internal control module engine air mass performance:** Indicates an error occurred in the PCM. **Possible Causes:** • Software incompatibility issue • PCM malfunction

DTC	Trouble Code Title, Conditions & Possible Causes
DTC: P061F **T** **Year:** 2009, 2010 **Model:** 3, 6, CX-7, CX-9 **Engine:** 2.5L L4 VIN 5, 2.5L L4 VIN 6, 2.5L L4 VIN A, 2.5L L4 VIN M, 3.7L V6 VIN A, 3.7L V6 VIN B, 3.7L V6 VIN V **Transmission:** All	**Internal control module throttle valve actuator controller performance problem:** Indicates that a calculation error occurred in the PCM. **Possible Causes:** • NOTE: This DTC is an informational DTC and may be set in combination with a number of other DTCs which are causing the FMEM. Diagnose other DTCs first. • PCM connector or terminals malfunction • PCM malfunction
DTC: P0620 **1T CCM, MIL: Yes** **Year:** 2009, 2010 **Model:** 6, CX-9, Tribute **Engine:** 2.5L L4 VIN 3, 2.5L L4 VIN 7, 3.0L V6 VIN 1, 3.0L V6 VIN G, 3.7L V6 VIN A, 3.7L V6 VIN B, 3.7L V6 VIN V **Transmission:** All	**Regulator / Generator malfunction:** PCM internal CPU malfunction. **Possible Causes:** • PCM internal malfunction* Configuration has not been properly completed.
DTC: P0625 **T PCM** **Year:** 2009, 2010 **Model:** 6, CX-9 **Engine:** 3.7L V6 VIN A, 3.7L V6 VIN B, 3.7L V6 VIN V **Transmission:** All	**Generator field terminal circuit low:** The PCM monitors generator load from the generator/regulator in the form of frequency. The concern indicates the input is lower than the load should be in normal operation. The load input could be low when no generator output exists. **Possible Causes:** • Drive belt damaged • Short to ground between generator and PCM • Short to ground between generator and PCM • Connector or terminal malfunction • Open circuit between terminal B and battery positive terminal • Low system voltage • Generator malfunction
DTC: P0626 **T PCM** **Year:** 2009, 2010 **Model:** 6, CX-9 **Engine:** 3.7L V6 VIN A, 3.7L V6 VIN B, 3.7L V6 VIN V **Transmission:** All	**Generator field terminal circuit high:** The PCM monitors generator load from the generator/regulator in the form of frequency. The concern indicates the input is higher than the load should be in normal operation. The load input could be high when a battery short to ground exists. **Possible Causes:** • Short to power supply between generator and PCM • Open circuit in wiring harness between generator and PCM • Connector or terminal malfunction
DTC: P0627 **T PCM** **Year:** 2009, 2010 **Model:** 6 **Engine:** 3.7L V6 VIN B **Transmission:** All	**Fuel pump circuit open:** The electronic fuel pump relay monitors the fuel pump unit and secondary circuits for a concern. If the electronic fuel pump relay detects a concern with the fuel pump unit or secondary circuits, the electronic fuel pump relay sends an 80% duty cycle signal on the fuel pump monitor (FPM) circuit to report the concern to the PCM. The test fails if the electronic fuel pump relay is still reporting a concern with the fuel pump unit or secondary circuits after a calibrated amount of time. **Possible Causes:** • Short to ground between electronic fuel pump relay and fuel pump unit • Open circuit between electronic fuel pump relay and fuel pump unit • Short to power supply between electronic fuel pump relay and fuel pump unit • Fuel pump unit malfunction • Electronic fuel pump relay malfunction
DTC: P0638 **1T CCM, MIL: Yes** **Year:** 2009, 2010 **Model:** 3, 5, 6, CX-7, MX-5 Miata, RX-8 **Engine:** 1.3L R2 VIN 2, 1.3L R2 VIN 4, 1.3L R2 VIN M, 1.3L R2 VIN P, 2.0L L4 VIN F, 2.0L L4 VIN G, 2.3L L4 VIN 3, 2.3L L4 VIN 4, 2.3L L4 VIN L, 2.3L L4 VIN M, 2.5L L4 VIN 5, 2.5L L4 VIN 6, 2.5L L4 VIN A, 2.5L L4 VIN M **Transmission:** All	**Throttle Actuator control range performance:** If the PCM detects that actual throttle angle opening is smaller or larger than the target throttle opening angle, the PCM determines that the throttle actuator control system has a malfunction **Possible Causes:** • PCM internal malfunction* Throttle Body malfunction

DTC	Trouble Code Title, Conditions & Possible Causes
DTC: P0642 **T PCM** **Year:** 2009, 2010 **Model:** 6, CX-9 **Engine:** 3.7L V6 VIN A, 3.7L V6 VIN B, 3.7L V6 VIN V **Transmission:** All	**Sensor reference voltage circuit low:** Indicates the reference voltage circuit is lower than reference voltage minimum. **Possible Causes:** • Short to ground between fuel tank pressure sensor and PCM • Connector or terminal malfunction • Fuel tank pressure sensor malfunction
DTC: P0643 **T PCM** **Year:** 2009, 2010 **Model:** 6, CX-9 **Engine:** 3.7L V6 VIN A, 3.7L V6 VIN B, 3.7L V6 VIN V **Transmission:** All	**Sensor reference voltage circuit high:** Indicates the reference voltage circuit is higher than reference voltage maximum. **Possible Causes:** • Short to power supply between fuel tank pressure sensor and PCM • Connector or terminal malfunction • Fuel tank pressure sensor malfunction
DTC: P0645 **1T CCM, MIL: Yes** **Year:** 2009, 2010 **Model:** 6, B2300, B4000, CX-9, Tribute **Engine:** 2.3L L4 VIN D, 2.5L L4 VIN 3, 2.5L L4 VIN 7, 3.0L V6 VIN 1, 3.0L V6 VIN G, 3.7L V6 VIN A, 3.7L V6 VIN B, 3.7L V6 VIN V, 4.0L V6 VIN E **Transmission:** All	**A/C clutch relay (ACCR) malfunction:** If the PCM detects excessive current draw when PCM grounds the circuit or voltage is not detected on the ACCR circuit when it is not grounded by the PCM, the PCM determines that the ACCR has a malfunction **Possible Causes:** • PCM internal malfunction* ACCR malfunction
DTC: P064D **1T CCM, MIL: Yes** **Year:** 2009, 2010 **Model:** 3, 5, 6, CX-7 **Engine:** 2.0L L4 VIN F, 2.0L L4 VIN G, 2.3L L4 VIN 3, 2.3L L4 VIN 4, 2.3L L4 VIN L, 2.3L L4 VIN M, 2.5L L4 VIN 5, 2.5L L4 VIN 6, 2.5L L4 VIN A, 2.5L L4 VIN M, 3.7L V6 VIN B **Transmission:** All	**Internal control module A/F sensor processor performance problem:** The A/F sensor IC integrated in PCM converts to voltage value for fuel control and for diagnosis based on the A/F sensor signal current and sends to CPU (integrated in PCM). If there is a transmission format error from the A/F sensor IC to the PCM, the PCM will detect a transmission error. **Possible Causes:** • PCM internal malfunction (between CPU and A/F sensor control IC communication line)
DTC: P064E **T PCM** **Year:** 2009, 2010 **Model:** 6 **Engine:** 3.7L V6 VIN B **Transmission:** All	**Internal control module exhaust gas sensor processor performance (LH):** The PCM monitors the application-specific integrated circuit that controls and monitors the A/F sensor and/or HO2S. The test fails when the PCM detects an internal circuit or communication concern. **Possible Causes:** • PCM malfunction • Influence of aftermarket product performance
DTC: P065B **T PCM** **Year:** 2009, 2010 **Model:** 6 **Engine:** 3.7L V6 VIN B **Transmission:** All	**Generator control circuit range/performance:** The PCM reads the generator and sends a DTC through the network when the generator indicates a concern. **Possible Causes:** • Connector or terminal malfunction • Short to power supply between generator and PCM • Short to ground between generator and PCM • Open circuit between generator and PCM • Generator malfunction • PCM malfunction

DTC	Trouble Code Title, Conditions & Possible Causes
DTC: P0661 **2T CCM** **Year:** 2009, 2010 **Model:** 3, 5, 6, CX-7, MX-5 Miata, RX-8 **Engine:** 1.3L R2 VIN 2, 1.3L R2 VIN 4, 1.3L R2 VIN M, 1.3L R2 VIN P, 2.0L L4 VIN F, 2.0L L4 VIN G, 2.3L L4 VIN 3, 2.3L L4 VIN L, 2.5L L4 VIN 5, 2.5L L4 VIN 6, 2.5L L4 VIN A, 2.5L L4 VIN M **Transmission:** All	**Variable intake-air system (VIS) control solenoid valve circuit low input:** PCM monitors the VIS control solenoid valve signal at PCM terminal 4R. If PCM turns VIS control solenoid valve OFF but voltage at PCM terminal 4R remains low, PCM determines that VIS control solenoid valve circuit has a malfunction. **Possible Causes:** • VIS control valve malfunction • Connector or terminal malfunction • Open circuit between VIS and PSM • PCM has failed
DTC: P0662 **2T CCM** **Year:** 2009, 2010 **Model:** 3, 5, 6, CX-7, MX-5 Miata, RX-8 **Engine:** 1.3L R2 VIN 2, 1.3L R2 VIN 4, 1.3L R2 VIN M, 1.3L R2 VIN P, 2.0L L4 VIN F, 2.0L L4 VIN G, 2.3L L4 VIN 3, 2.3L L4 VIN L, 2.5L L4 VIN 5, 2.5L L4 VIN 6, 2.5L L4 VIN A, 2.5L L4 VIN M **Transmission:** All	**Variable intake-air system (VIS) control solenoid valve circuit high input:** PCM monitors the VIS control solenoid valve signal at PCM terminal 4R. If PCM turns VIS control solenoid valve OFF but voltage at PCM terminal 4R remains high, PCM determines that VIS control solenoid valve circuit has a malfunction. **Possible Causes:** • VIS control valve malfunction • Connector or terminal malfunction • Open circuit between VIS and PSM • PCM has failed
DTC: P0685 **2T PCM, MIL: Yes** **Year:** 2009, 2010 **Model:** 3, 5, 6, CX-7, CX-9, MX-5 Miata **Engine:** 2.0L L4 VIN F, 2.0L L4 VIN G, 2.3L L4 VIN 3, 2.3L L4 VIN 4, 2.3L L4 VIN L, 2.3L L4 VIN M, 2.5L L4 VIN 5, 2.5L L4 VIN 6, 2.5L L4 VIN A, 2.5L L4 VIN M, 3.7L V6 VIN A, 3.7L V6 VIN B, 3.7L V6 VIN V **Transmission:** All	**Main relay control circuit open:** This DTC sets when the ignition switch position run circuit indicates the key is in the off, ACC, or LOCK position, and the amount of time the PCM remains powered through the PCM power relay exceeds a predetermined amount of time. **Possible Causes:** • Main relay malfunction • PCM malfunction
DTC: P0689 **T PCM** **Year:** 2009, 2010 **Model:** 6, CX-9 **Engine:** 3.7L V6 VIN A, 3.7L V6 VIN B, 3.7L V6 VIN V **Transmission:** All	**ECM/PCM power relay sense circuit low:** This DTC sets when the passive anti theft system (PATS) system indicates the key is in ON or START position and the ignition switch position run circuit indicates OFF, ACC, or LOCK position. **Possible Causes:** • PATS system malfunction • Ignition circuit fuse (MAIN, IG KEY 1, IG KEY 2) • Open circuit between battery positive terminal and PCM • Ignition switch malfunction • Short to ground circuit between battery positive terminal and PCM
DTC: P068A **T PCM** **Year:** 2009, 2010 **Model:** 6 **Engine:** 3.7L V6 VIN B **Transmission:** All	**ECM/PCM power relay de-engaged-too early:** P068A is set when the PCM powers down after a key off prior to completing the NVRAM write process. **Possible Causes:** • PCM malfunction
DTC: P0690 **T PCM** **Year:** 2009, 2010 **Model:** 6, CX-9 **Engine:** 3.7L V6 VIN A, 3.7L V6 VIN B, 3.7L V6 VIN V **Transmission:** All	**ECM/PCM power relay sense circuit high:** This DTC sets when the passive anti theft system (PATS) system indicates the key is in the OFF, ACC, or LOCK position and the ignition switch position run circuit indicates ON or START position. **Possible Causes:** • PATS system malfunction • Short to power supply between battery positive terminal and PCM • Ignition switch malfunction

DTC	Trouble Code Title, Conditions & Possible Causes
DTC: P06B8 **1T CCM, MIL: Yes** **Year:** 2009, 2010 **Model:** 3, 6, CX-7, MX-5 Miata **Engine:** 2.0L L4 VIN F, 2.0L L4 VIN G, 2.3L L4 VIN 3, 2.3L L4 VIN 4, 2.3L L4 VIN L, 2.3L L4 VIN M, 2.5L L4 VIN 5, 2.5L L4 VIN 6, 2.5L L4 VIN M, 3.7L V6 VIN B **Transmission:** All	**Internal control module non-volatile random access memory error:** PCM internal EEPROM malfunction. **Possible Causes:** • PCM internal EEPROM malfunction
DTC: P0703 **2T CCM** **Year:** 2009, 2010 **Model:** 3, 5, 6, CX-7, MX-5 Miata, RX-8 **Engine:** 1.3L R2 VIN 2, 1.3L R2 VIN 4, 1.3L R2 VIN M, 1.3L R2 VIN P, 2.0L L4 VIN F, 2.0L L4 VIN G, 2.3L L4 VIN 3, 2.3L L4 VIN 4, 2.3L L4 VIN L, 2.3L L4 VIN M, 2.5L L4 VIN 5, 2.5L L4 VIN 6, 2.5L L4 VIN A, 2.5L L4 VIN M **Transmission:** All	**Brake On/Off Switch Circuit Malfunction (Self-Test):** KOEO Self-Test: Key on, and the PCM did not detect any change in the Brake Switch status after the brake pedal was pressed and released during the self-test. KOER Self-Test: and the PCM did not detect any change in the Brake Switch status after the brake pedal was pressed and released during the self-test. **Possible Causes:** • Brake switch signal circuit is open or shorted ground • Brake switch power circuit (B+) open between switch and PCM • Brake switch is damaged, misadjusted or installed improperly • PCM has failed
DTC: P0703 **2T CCM** **Year:** 2009 **Model:** B2300, B4000 **Engine:** 2.3L L4 VIN D, 4.0L V6 VIN E **Transmission:** All	**Brake On/Off Switch Circuit Malfunction (Self-Test):** KOEO Self-Test: Key on, and the PCM did not detect any change in the Brake Switch status after the brake pedal was pressed and released during the self-test. KOER Self-Test: and the PCM did not detect any change in the Brake Switch status after the brake pedal was pressed and released during the self-test. **Possible Causes:** • Brake switch signal circuit is open or shorted ground • Brake switch power circuit (B+) is open or shorted to ground • Brake switch is damaged, misadjusted or installed improperly • PCM has failed
DTC: P0704 **2T CCM** **Year:** 2009, 2010 **Model:** Tribute **Engine:** 2.5L L4 VIN 3, 2.5L L4 VIN 7, 3.0L V6 VIN 1, 3.0L V6 VIN G **Transmission:** All	**Clutch Pedal Switch Circuit Malfunction (Self-Test):** Engine started, the after the vehicle was accelerated from 0-16 mph, the PCM detected the Clutch Pedal Position (CPP) switch input did not change status correctly, or it did not change during the KOEO Self-Test procedure (the clutch may not have been depressed). **Possible Causes:** • Starter relay disconnected during the Quick Test Procedure • Clutch switch signal circuit is open or shorted ground • Clutch switch power circuit (B+) open between switch and PCM • Clutch switch signal circuit is shorted to VREF or system power • Clutch switch is damaged, misadjusted or installed improperly • PCM has failed
DTC: P0704 **2T CCM** **Year:** 2009, 2010 **Model:** 3, 5, 6, B2300, B4000, MX-5 Miata, RX-8 **Engine:** 1.3L R2 VIN 2, 1.3L R2 VIN 4, 2.0L L4 VIN F, 2.0L L4 VIN G, 2.3L L4 VIN 3, 2.3L L4 VIN 4, 2.3L L4 VIN D, 2.3L L4 VIN L, 2.3L L4 VIN M, 2.5L L4 VIN 5, 2.5L L4 VIN 6, 2.5L L4 VIN A, 4.0L V6 VIN E **Transmission:** All	**Clutch Pedal Switch Circuit Malfunction:** Engine started, the after the vehicle was accelerated from 0-16 mph, the PCM detected the Clutch Pedal Position (CPP) switch input did not change status correctly under these conditions in the CCM test. **Possible Causes:** • Clutch switch signal circuit is open or shorted ground • Clutch switch power circuit (B+) open between switch and PCM • Clutch switch is damaged, misadjusted or installed improperly • PCM has failed

DTC	Trouble Code Title, Conditions & Possible Causes
DTC: P0705 **1T CCM, MIL: Yes** **Year:** 2009 **Model:** B2300, B4000 **Engine:** 2.3L L4 VIN D, 4.0L V6 VIN E **Transmission:** All	**Digital Transmission Range Switch Circuit Malfunction:** Key on or engine running and the PCM detected an unexpected low or high voltage condition on one of the Digital TR sensor circuits. **Possible Causes:** • TR4 signal circuit is open or shorted to ground (0v or 12v) • TR3A signal circuit is open or shorted to ground (0v or 12v) • TR2 signal circuit is open or shorted to ground (0v or 12v) • TR1 signal circuit is open or shorted to ground (0v or 12v) • DTR switch or its connector is damaged, or switch has failed • PCM has failed
DTC: P0706 **2T CCM, MIL: Yes** **Year:** 2009 **Model:** B2300, B4000 **Engine:** 2.3L L4 VIN D, 4.0L V6 VIN E **Transmission:** All	**Transmission Range Switch Circuit Malfunction:** Engine started, vehicle driven to a speed of over 37 mph, and the PCM did not detect any change in the TR Switch P-N-D-2-1 status. **Possible Causes:** • TR signal 'P' circuit open between the switch and the PCM (0v) • TR signal 'D' circuit open between the switch and the PCM (0v) • TR signal 'N' circuit open between the switch and the PCM (0v) • TR signal '2' circuit open between the switch and the PCM (0v) • TR signal '1' circuit open between the switch and the PCM (0v) • TR switch or its connector is damaged, or the switch has failed • PCM has failed
DTC: P0706 **2T CCM, MIL: Yes** **Year:** 2009, 2010 **Model:** Tribute **Engine:** 2.5L L4 VIN 3, 2.5L L4 VIN 7, 3.0L V6 VIN 1, 3.0L V6 VIN G **Transmission:** All	**Transmission Range Switch Circuit Malfunction:** Key on or engine running, and the PCM did not detect any TR switch inputs, or it detected 2 switch inputs simultaneously during the test. **Possible Causes:** • TR signal 'P' circuit open between the switch and the PCM (0v) • TR signal 'D' circuit open between the switch and the PCM (0v) • TR signal 'S' circuit open between the switch and the PCM (0v) • TR signal 'L' circuit open between the switch and the PCM (0v) • TR switch or its connector is damaged, or the switch has failed • PCM has failed
DTC: P0707 **2T CCM, MIL: Yes** **Year:** 2009, 2010 **Model:** B2300, B4000, Tribute **Engine:** 2.3L L4 VIN D, 2.5L L4 VIN 7, 3.0L V6 VIN 1, 3.0L V6 VIN G, 4.0L V6 VIN E **Transmission:** All	**Transmission Range Switch Range/Performance:** Engine running, and the PCM detected multiple T/R switch inputs, or the TR Switch signal did not change with the vehicle moving. **Possible Causes:** • TR switch signal circuit is shorted to another switch signal • TR switch signal circuit is open (problem may be intermittent) • TR switch is damaged, out of adjustment or it has failed • PCM has failed
DTC: P0708 **2T CCM, MIL: Yes** **Year:** 2009 **Model:** B2300, B4000 **Engine:** 2.3L L4 VIN D, 4.0L V6 VIN E **Transmission:** All	**Digital Transmission Range Switch Circuit Malfunction:** Key on or engine running and the PCM detected an unexpected "low" voltage condition on the TR switch circuit during the CCM test. **Possible Causes:** • TR switch signal circuit open between switch and the PCM (0v) • DTR switch or its connector is damaged, or switch has failed • PCM has failed
DTC: P0708 **2T CCM, MIL: Yes** **Year:** 2009, 2010 **Model:** Tribute **Engine:** 2.5L L4 VIN 7, 3.0L V6 VIN 1, 3.0L V6 VIN G **Transmission:** All	**Transmission Range Switch Circuit Malfunction:** Key on or engine running and the PCM detected an unexpected high or low voltage condition on the TR switch circuit during the test. **Possible Causes:** • TR switch signal circuit is open between the switch and PCM • TR switch signal circuit is shorted between switch and PCM • TR switch signal circuit is shorted to VREF or system power • TR switch ground circuit is open between switch and ground • TR switch or its connector is damaged, or the switch has failed • PCM has failed

DTC	Trouble Code Title, Conditions & Possible Causes
DTC: P0709 **2T CCM, MIL:** Yes **Year:** 2009 **Model:** B2300, B4000 **Engine:** 2.3L L4 VIN D, 4.0L V6 VIN E **Transmission:** All	**Transmission Range (TR) Sensor A Circuit Range/Performance:** TR sensor stuck in transition zone and possible no crank condition. Only PARK, REVERSE, NEUTRAL and 5th gear available. **NOTE: Will turn on wrench lamp.** **Possible Causes:** • DTR or TR sensor connector is damaged or open • DTR or TR sensor signal circuit is open • DTR or TR sensor is shorted to VREF (5v) • DTR or TR sensor is damaged • PCM has failed
DTC: P0710 **1T CCM, MIL:** Yes **Year:** 2009, 2010 **Model:** B2300, B4000, Tribute **Engine:** 2.3L L4 VIN D, 2.5L L4 VIN 7, 3.0L V6 VIN 1, 3.0L V6 VIN G, 4.0L V6 VIN E **Transmission:** All	**Transmission Fluid Temperature Sensor Circuit Malfunction:** Engine started, vehicle driven to a speed of over 12 mph, and the PCM detected the TFT sensor signal was less than 0.10v, or it was more than 4.90v, condition met for 100 seconds. **Possible Causes:** • TFT sensor signal circuit is open or shorted to ground • TFT sensor ground circuit is open between sensor and PCM • TFT sensor signal circuit shorted to VREF or system power • TFT sensor is damaged or has failed • PCM has failed
DTC: P0711 **2T CCM, MIL:** Yes **Year:** 2009, 2010 **Model:** B2300, B4000, Tribute **Engine:** 2.3L L4 VIN D, 2.5L L4 VIN 7, 3.0L V6 VIN 1, 3.0L V6 VIN G, 4.0L V6 VIN E **Transmission:** All	**Transmission Fluid Temperature Sensor Circuit Range/Performance:** DTC P0710 not set, engine started, vehicle driven to a speed of over 37 mph for 430 seconds, and the PCM detected the TFT sensor signal was less than 0.09v, or it was more than 4.99v in the test. **Possible Causes:** • TFT sensor signal circuit is open between sensor and the PCM • TFT sensor signal circuit is shorted to ground • TFT sensor is damaged or has failed • PCM has failed
DTC: P0712 **1T CCM** **Year:** 2009, 2010 **Model:** B2300, B4000, Tribute **Engine:** 2.3L L4 VIN D, 2.5L L4 VIN 7, 3.0L V6 VIN 1, 3.0L V6 VIN G, 4.0L V6 VIN E **Transmission:** All	**Transmission Fluid Temperature Sensor Low Input (High Temperature):** Key on or engine running and the PCM detected the Transmission Fluid Temperature (TFT) sensor signal indicated more than 315°F during the CCM test. **Possible Causes:** • TFT sensor signal circuit has a short to ground condition • TFT sensor is damaged or has failed (it may be shorted) • PCM has failed
DTC: P0713 **1T CCM** **Year:** 2009, 2010 **Model:** B2300, B4000, Tribute **Engine:** 2.3L L4 VIN D, 2.5L L4 VIN 7, 3.0L V6 VIN 1, 3.0L V6 VIN G, 4.0L V6 VIN E **Transmission:** All	**Transmission Fluid Temperature Sensor High Input (Low Temperature):** Key on or engine running and the PCM detected an unexpected "high" voltage condition on the Transaxle Fluid Temperature (TFT) sensor circuit during the CCM test. **Possible Causes:** • TFT sensor signal circuit is open between sensor and the PCM • TFT sensor signal circuit has a short to power condition • TFT sensor is damaged or has failed (it may be open) • PCM has failed
DTC: P0714 **2T CCM** **Year:** 2009 **Model:** B2300, B4000 **Engine:** 2.3L L4 VIN D, 4.0L V6 VIN E **Transmission:** All	**TFT Sensor Circuit Intermittent/Erratic:** Key on or engine running; and the PCM detected intermittent condition in the Transmission Fluid Temperature (TFT) sensor. **Possible Causes:** • TFT sensor signal circuit is open between the sensor and PCM • TFT sensor signal circuit is shorted to power • TFT sensor ground circuit is open between sensor and PCM • TFT sensor is damaged or has failed • PCM has failed
DTC: P0715 **1T CCM, MIL:** Yes **Year:** 2009, 2010 **Model:** B2300, B4000, Tribute **Engine:** 2.3L L4 VIN D, 2.5L L4 VIN 7, 3.0L V6 VIN 1, 3.0L V6 VIN G, 4.0L V6 VIN E **Transmission:** All	**Turbine Shaft Sensor Circuit Malfunction:** Engine started, vehicle driven to a speed of 31 mph, and the PCM detected the ISS/TSS signal dropped out for more than 1 second. **Possible Causes:** • TSS (+) signal circuit is open (an intermittent fault) • TSS (+) signal circuit shorted to ground (intermittent fault) • TSS (-) signal circuit is open (an intermittent fault) • TSS (-) signal circuit shorted to ground (intermittent fault) • TSS is damaged or has failed • PCM has failed

DTC	Trouble Code Title, Conditions & Possible Causes
DTC: P0716 **2T CCM** **Year:** 2009 **Model:** B2300, B4000 **Engine:** 2.3L L4 VIN D, 4.0L V6 VIN E **Transmission:** All	**Turbine Shaft Speed (TSS) Sensor Range/Performance:** Engine started, TSS signal more than 1 mph, and the PCM detected "noise" interference on the TSS sensor circuit. **NOTE: Will turn on TCIL, holds in 3rd gear.** **Possible Causes:** • TSS sensor signal circuit is shorted to ground • TSS sensor signal circuit is open • TSS sensor circuit is shorted to power • TSS sensor is damaged or it has failed • PCM has failed
DTC: P0717 **1T CCM, MIL: Yes** **Year:** 2009, 2010 **Model:** B2300, B4000, Tribute **Engine:** 2.3L L4 VIN D, 2.5L L4 VIN 7, 3.0L V6 VIN 1, 3.0L V6 VIN G, 4.0L V6 VIN E **Transmission:** All	**Turbine Shaft Sensor Circuit Malfunction:** Engine started, vehicle driven to a speed of 31 mph, and the PCM did not detect any TSS signals during the CCM test. **Possible Causes:** • TSS (+) signal circuit is open or shorted to ground • TSS (-) signal circuit is open or shorted to ground • TSS is damaged or has failed • PCM has failed
DTC: P0718 **1T CCM, MIL: Yes** **Year:** 2009, 2010 **Model:** B2300, B4000, Tribute **Engine:** 2.3L L4 VIN D, 2.5L L4 VIN 7, 3.0L V6 VIN 1, 3.0L V6 VIN G, 4.0L V6 VIN E **Transmission:** All	**Turbine Shaft Sensor Circuit Malfunction:** Engine started, vehicle driven to a speed of 31 mph, and the PCM detected an erratic or noisy TSS signal during the CCM test. **Possible Causes:** • TSS (+) signal circuit is open or shorted to ground • TSS (-) signal circuit is open or shorted to ground • TSS reluctor has metal chips or is damaged, or has failed • PCM has failed
DTC: P0720 **1T CCM, MIL: Yes** **Year:** 2009, 2010 **Model:** B2300, B4000, Tribute **Engine:** 2.3L L4 VIN D, 2.5L L4 VIN 7, 3.0L V6 VIN 1, 3.0L V6 VIN G, 4.0L V6 VIN E **Transmission:** All	**Output Shaft Sensor Circuit Malfunction:** Engine started, vehicle driven to a speed of 31 mph, and the PCM detected the OSS signal was lost (dropped out) for over 1 second. **Possible Causes:** • OSS (+) signal circuit is open (fault may be intermittent) • OSS (+) signal circuit shorted to ground (an intermittent fault) • OSS (-) signal circuit is open (fault may be intermittent) • OSS (-) signal circuit shorted to ground (an intermittent fault) • OSS is damaged or has failed • PCM has failed
DTC: P0721 **1T CCM, MIL: Yes** **Year:** 2009, 2010 **Model:** B2300, B4000, Tribute **Engine:** 2.3L L4 VIN D, 2.5L L4 VIN 7, 3.0L V6 VIN 1, 3.0L V6 VIN G, 4.0L V6 VIN E **Transmission:** All	**Output Shaft Sensor Circuit Signal Erratic:** Engine started, vehicle driven to a speed of over 12 mph, and the PCM detected the OSS signal was erratic during the CCM test. **Possible Causes:** • OSS (+) circuit is open or shorted to ground (noisy circuit) • OSS (-) circuit is open or shorted to ground (noisy circuit) • OSS reluctor contains metal chips or is damaged • OSS is damaged or has failed • PCM has failed
DTC: P0722 **1T CCM, MIL: Yes** **Year:** 2009, 2010 **Model:** B2300, B4000, Tribute **Engine:** 2.3L L4 VIN D, 2.5L L4 VIN 7, 3.0L V6 VIN 1, 3.0L V6 VIN G, 4.0L V6 VIN E **Transmission:** All	**Output Shaft Sensor Circuit Signal Intermittent:** Engine started, vehicle driven to a speed of over 12 mph, and the PCM detected the OSS signal was lost (dropped out) during the test. **NOTE: This code can set if the OSS signal drops "in" and "out".** **Possible Causes:** • OSS (+) circuit is open or shorted to ground (intermittent) • OSS (-) circuit is open or shorted to ground (intermittent) • OSS is damaged or has failed • PCM has failed
DTC: P0723 **1T CCM, MIL: Yes** **Year:** 2009, 2010 **Model:** B2300, B4000, Tribute **Engine:** 2.3L L4 VIN D, 2.5L L4 VIN 7, 3.0L V6 VIN 1, 3.0L V6 VIN G, 4.0L V6 VIN E **Transmission:** All	**Output Shaft Sensor Circuit Signal Intermittent:** Engine started, vehicle driven to a speed of over 12 mph, and the PCM detected the OSS signal was lost (dropped out) during the test. **NOTE: This code can set if the OSS signal drops "in" and "out".** **Possible Causes:** • OSS (+) circuit is open or shorted to ground (intermittent) • OSS (-) circuit is open or shorted to ground (intermittent) • OSS is damaged or has failed • PCM has failed

DTC	Trouble Code Title, Conditions & Possible Causes
DTC: P0731 **2T CCM, MIL: Yes** **Year:** 2009 **Model:** B2300, B4000 **Engine:** 2.3L L4 VIN D, 4.0L V6 VIN E **Transmission:** All	**Transmission 1st Gear Ratio Incorrect:** DTC P0500, P0710, P0715, P0755 and P0760 not set, engine started, then driven in Drive in 1st Gear (1GR) to a speed from 4-37 mph, engine speed over 500 rpm, TSS signal more than 75 rpm, TP angle over 6.25%, ATF sensor signal more than 68°F, brake switch indicating "off", and the PCM detected the ratio of the Turbine speed to the Vehicle speed was less than a preset value stored in memory. **Possible Causes:** • ATF level is too low, or the ATF is badly deteriorated • A/T line pressure is low, or the oil pump has failed • SSA, SSB, SSC or the PCS is stuck (mechanical fault) • A/T 1-2 clutch is slipping (mechanical fault) • A/T 1-2 shift valve is stuck (mechanical fault) • A/T Pressure regulator or the pressure modifier valve is stuck • A/T solenoid reducing valve is stuck (mechanical fault) • PCM has failed
DTC: P0732 **2T CCM, MIL: Yes** **Year:** 2009 **Model:** B2300, B4000 **Engine:** 2.3L L4 VIN D, 4.0L V6 VIN E **Transmission:** All	**Transmission 2nd Gear Ratio Incorrect:** DTC P0500, P0710, P0715, P0755 and P0760 not set, engine started, then driven in Drive in 2nd Gear (2GR) to a speed of 66 mph (± 6 mph), engine speed over 500 rpm, throttle opening over 6.25%, TSS signal more than 75 rpm, ATF sensor signal more than 68°F, brake switch indicating "off", and the PCM detected the ratio of the Turbine speed to the Vehicle speed was less than a preset value stored in memory. **Possible Causes:** • ATF level is too low, or the ATF is badly deteriorated • A/T line pressure is low, or the oil pump has failed • SSA, SSB, SSC or the PCS is stuck (mechanical fault) • A/T 2-3 clutch is slipping (mechanical fault) • A/T 1-2 or the 2-3 shift valve is stuck (mechanical fault) • A/T Pressure regulator or the pressure modifier valve is stuck • A/T solenoid reducing valve is stuck (mechanical fault) • PCM has failed
DTC: P0733 **2T CCM, MIL: Yes** **Year:** 2009, 2010 **Model:** B2300, B4000, Tribute **Engine:** 2.3L L4 VIN D, 2.5L L4 VIN 7, 3.0L V6 VIN 1, 3.0L V6 VIN G, 4.0L V6 VIN E **Transmission:** All	**Transmission 3rd Gear Ratio Incorrect:** DTC P0500, P0710, P0715, P0755 and P0760 not set, engine started, then driven in Drive in 3rd Gear (3GR) to a speed of 19-30 mph, engine speed over 500 rpm, TSS signal more than 75 rpm, ATF sensor signal more than 68°F, brake switch indicating "off", and the PCM detected the ratio of the Turbine speed to the Vehicle speed was less than a preset value stored in memory. **Possible Causes:** • ATF level is too low, or the ATF is badly deteriorated • A/T line pressure is low, or the oil pump has failed • SSA, SSB, SSC or the PCS is stuck (mechanical fault) • A/T 3-4 clutch is slipping (mechanical fault) • A/T 2-3 or the 3-4 shift valve is stuck (mechanical fault) • A/T Pressure regulator or the pressure modifier valve is stuck • A/T solenoid reducing valve is stuck (mechanical fault) • PCM has failed
DTC: P0734 **2T CCM, MIL: Yes** **Year:** 2009, 2010 **Model:** B2300, B4000, Tribute **Engine:** 2.3L L4 VIN D, 2.5L L4 VIN 7, 3.0L V6 VIN 1, 3.0L V6 VIN G, 4.0L V6 VIN E **Transmission:** All	**Transmission 4th Gear Ratio Incorrect:** DTC P0500, P0710, P0715, P0755 and P0760 not set, engine started, then driven in Drive in 4th Gear (4GR) to a speed of 47-64 mph, engine speed over 500 rpm, TSS signal more than 75 rpm, ATF sensor signal more than 68°F, brake switch indicating "off", and the PCM detected the ratio of the Turbine speed to the Vehicle speed was less than a preset value stored in memory. **Possible Causes:** • ATF level is too low, or the ATF is badly deteriorated • A/T line pressure is low, or the oil pump has failed • SSA, SSB, SSC or the PCS is stuck (mechanical fault) • A/T 2-4 brake bank or 3-4 clutch is slipping (mechanical fault) • A/T 1-2, 2-3 or 3-4 shift valve is stuck (mechanical fault) • A/T Pressure regulator, pressure modifier or solenoid reducing valve is stuck • PCM has failed
DTC: P0735 **2T CCM** **Year:** 2009 **Model:** B2300, B4000 **Engine:** 2.3L L4 VIN D, 4.0L V6 VIN E **Transmission:** All	**Incorrect Fifth Gear Ratio:** Engine started, vehicle operating with 5th Gear commanded "on", and the PCM detected an incorrect 5th gear ratio during the test. **Possible Causes:** • 5th Gear solenoid harness connector not properly seated • 5th Gear solenoid signal shorted to ground, or open • 5th Gear solenoid wiring harness connector is damaged • 5th Gear solenoid is damaged or not properly installed

DTC	Trouble Code Title, Conditions & Possible Causes
DTC: P0740 **2T CCM, MIL: Yes** **Year:** 2009, 2010 **Model:** B2300, B4000, Tribute **Engine:** 2.3L L4 VIN D, 2.5L L4 VIN 7, 3.0L V6 VIN 1, 3.0L V6 VIN G, 4.0L V6 VIN E **Transmission:** All	**Torque Converter Clutch System Malfunction:** DTC P0500, P0710, P0715, P0755 and P0760 not set, engine started, then driven in Drive in 4th Gear (4GR) to a speed of 47-64 mph with the throttle open, engine speed over 500 rpm, ATF sensor signal more than 68°F, Turbine speed over 75 rpm, brake switch indicating "off", TCC system operating, O/D OFF Switch is "off", and the PCM detected the difference between the engine speed and the Turbine speed was more than a preset value stored in memory. **Possible Causes:** • ATF level is too low, or the ATF is badly deteriorated • A/T line pressure is low, or the oil pump has failed • A/T control valve has failed (it may be stuck) • TCC solenoid valve or pressure control valve is stuck • TCC in the torque converter is slipping (mechanical fault) • TCC shift valve or converter relief valve is stuck • Pressure modifier or the pressure regulator valve is stuck • Solenoid reducing valve is stuck (mechanical fault) • PCM has failed
DTC: P0741 **1T CCM** **Year:** 2009, 2010 **Model:** B2300, B4000, Tribute **Engine:** 2.3L L4 VIN D, 2.5L L4 VIN 7, 3.0L V6 VIN 1, 3.0L V6 VIN G, 4.0L V6 VIN E **Transmission:** All	**Torque Converter Clutch Control Engagement Error:** Engine started, then driven to a speed of over 30 mph, and the PCM detected the TCC system did not operate (due to a mechanical fault. **Possible Causes:** • Brake switch is damaged, out of adjustment or has failed • Engine performance conditions are erratic or unstable • Engine or transmission temperature is too high • TCC control valve is stuck in the "off" position • An internal transaxle component is damaged or has failed • PCM has failed
DTC: P0743 **1T CCM, MIL: Yes** **Year:** 2009, 2010 **Model:** B2300, B4000, Tribute **Engine:** 2.3L L4 VIN D, 2.5L L4 VIN 7, 3.0L V6 VIN 1, 3.0L V6 VIN G, 4.0L V6 VIN E **Transmission:** All	**Torque Converter Clutch Solenoid Circuit Malfunction:** Engine running and the PCM detected an unexpected voltage condition on the Torque Converter Clutch (TCC) solenoid control circuit during the CCM test. **Possible Causes:** • TCC control circuit is open (continuous high signal) • TCC control circuit is shorted to ground (continuous low signal) • TCC control circuit is shorted to system power (B+) • TCC solenoid is damaged or has failed • PCM has failed
DTC: P0748 **1T CCM** **Year:** 2009, 2010 **Model:** B2300, B4000, Tribute **Engine:** 2.3L L4 VIN D, 2.5L L4 VIN 7, 3.0L V6 VIN 1, 3.0L V6 VIN G, 4.0L V6 VIN E **Transmission:** All	**Electronic Pressure Control Solenoid Circuit Malfunction:** Engine running vehicle driven to a speed of over 37 mph, and the PCM detected an unexpected voltage condition on the Electronic Pressure Control (EPC) solenoid control circuit during the CCM test. **Possible Causes:** • PCS control circuit is open (continuous high signal) • PCS control circuit is shorted to ground (continuous low signal) • PCS control circuit is shorted to system power (B+) • PCS is damaged or has failed • PCM has failed
DTC: P0750 **1T CCM, MIL: Yes** **Year:** 2009, 2010 **Model:** B2300, B4000, Tribute **Engine:** 2.3L L4 VIN D, 2.5L L4 VIN 7, 3.0L V6 VIN 1, 3.0L V6 VIN G, 4.0L V6 VIN E **Transmission:** All	**A/T Shift Solenoid 'A' Circuit Malfunction:** Engine started, and the PCM detected an unexpected voltage condition on the Shift Solenoid 'A' (SSA) circuit during the CCM test. **Possible Causes:** • SSA control circuit is open (continuous high signal) • SSA control circuit is shorted to ground (continuous low signal) • SSA control circuit is shorted to system power (B+) • SSA is damaged or has failed • PCM has failed
DTC: P0751 **2T CCM, MIL: Yes** **Year:** 2009, 2010 **Model:** B2300, B4000, Tribute **Engine:** 2.3L L4 VIN D, 2.5L L4 VIN 7, 3.0L V6 VIN 1, 3.0L V6 VIN G, 4.0L V6 VIN E **Transmission:** All	**A/T Shift Solenoid 'A' Performance (Mechanical):** Engine started, then driven to a speed of over 37 mph, and the PCM detected a mechanical shift failure in the Shift Solenoid 'A' (SSA) during the CCM test. **Possible Causes:** • SSA is stuck in "off" position (mechanical problem) • SSA is damaged is damaged or has failed (mechanical fault) • SSA has a hydraulic problem • PCM has failed

DTC	Trouble Code Title, Conditions & Possible Causes
DTC: P0752 **2T CCM, MIL: Yes** **Year:** 2009, 2010 **Model:** B2300, B4000, Tribute **Engine:** 2.3L L4 VIN D, 2.5L L4 VIN 7, 3.0L V6 VIN 1, 3.0L V6 VIN G, 4.0L V6 VIN E **Transmission:** All	**A/T Shift Solenoid 'A' Performance (Mechanical):** DTC P0500, P0705, P0706, P0710, P0715, P0734, P0751, P0753, P0756, P0758, P0761, P0762, P0763, P0766, P0767, P0768, P0771, P0772 and P0773 not set, engine started, then driven in Drive in 1st Gear (1GR) or 2nd Gear (2GR), Turbine speed at 4988 rpm ($\pm$ 225), ATF temperature over 68°F, engine speed over 450 rpm, throttle position indicating closed, brake pedal depressed, VSS at 0 mph, and the PCM detected the Input/Turbine speed sensor signal was 187.5 or higher. **Possible Causes:** • ATF level is too low, or the ATF is badly deteriorated • A/T line pressure is low, or the oil pump has failed • A/T control valve is stuck, or it is damaged (a mechanical fault) • SSA is stuck "on", or is damaged (a mechanical fault) • Transmission is damaged or has failed • TCM has failed
DTC: P0753 **1T CCM, MIL: Yes** **Year:** 2009, 2010 **Model:** B2300, B4000, Tribute **Engine:** 2.3L L4 VIN D, 2.5L L4 VIN 7, 3.0L V6 VIN 1, 3.0L V6 VIN G, 4.0L V6 VIN E **Transmission:** All	**A/T Shift Solenoid 'B' Circuit Malfunction:** Engine started, then driven to a speed of over 37 mph, and the PCM detected an unexpected voltage condition on the Shift Solenoid 'B' (SSB) control circuit during the CCM test. **Possible Causes:** • SSB control circuit is open (continuous high signal) • SSB control circuit is shorted to ground (continuous low signal) • SSB control circuit is shorted to system power (B+) • SSB is damaged or has failed • PCM has failed
DTC: P0755 **1T CCM, MIL: Yes** **Year:** 2009, 2010 **Model:** B2300, B4000, Tribute **Engine:** 2.3L L4 VIN D, 2.5L L4 VIN 7, 3.0L V6 VIN 1, 3.0L V6 VIN G, 4.0L V6 VIN E **Transmission:** All	**A/T Shift Solenoid 'B' Circuit Malfunction:** Engine started, and the PCM detected an unexpected voltage condition on the Shift Solenoid 'B' (SSB) circuit during the CCM test. **Possible Causes:** • SSB control circuit is open (continuous high signal) • SSB control circuit is shorted to ground (continuous low signal) • SSB control circuit is shorted to system power (B+) • SSB is damaged or has failed • PCM has failed
DTC: P0756 **2T CCM, MIL: Yes** **Year:** 2009, 2010 **Model:** B2300, B4000, Tribute **Engine:** 2.3L L4 VIN D, 2.5L L4 VIN 7, 3.0L V6 VIN 1, 3.0L V6 VIN G, 4.0L V6 VIN E **Transmission:** All	**A/T Shift Solenoid 'B' Performance (Mechanical):** Engine started, then driven to a speed of over 37 mph, and the PCM detected a mechanical shift failure in the Shift Solenoid 'B' (SSB) during the CCM test. **Possible Causes:** • SSB is stuck in "on" position (a mechanical fault) • SSB is damaged is damaged or has failed (a mechanical fault) • SSB has a hydraulic problem • PCM has failed
DTC: P0757 **2T CCM, MIL: Yes** **Year:** 2009, 2010 **Model:** B2300, B4000, Tribute **Engine:** 2.3L L4 VIN D, 2.5L L4 VIN 7, 3.0L V6 VIN 1, 3.0L V6 VIN G, 4.0L V6 VIN E **Transmission:** All	**A/T Shift Solenoid 'B' Performance (Mechanical):** 2nd Gear Test DTC P0731 and P0733 not set, engine started, then driven in Drive in 2nd Gear (2GR), Turbine speed at 4988 rpm ($\pm$ 225), engine speed over 450 rpm, ATF temperature over 68°F, differential gear output speed over 35 mph, and the PCM detected the revolution ratio of the forward clutch drum revolution to the differential gear case revolution was less than 1.249, or more than 2.157. 4th Gear Test With the conditions listed above for the 2nd Gear Test met, with the throttle closed, the PCM detected the revolution ratio of the forward clutch drum revolution to differential gear case revolution was less than 0.60, or it was 1.249 or higher while driving in 4th gear (4GR). **Possible Causes:** • ATF level is too low, or the ATF is badly deteriorated • A/T SSB is stuck "on", or it is damaged (a mechanical fault) • A/T control valve is stuck, or it is damaged (a mechanical fault) • PCM has failed
DTC: P0758 **1T CCM, MIL: Yes** **Year:** 2009, 2010 **Model:** B2300, B4000, Tribute **Engine:** 2.3L L4 VIN D, 2.5L L4 VIN 7, 3.0L V6 VIN 1, 3.0L V6 VIN G, 4.0L V6 VIN E **Transmission:** All	**A/T Shift Solenoid 'B' Circuit Malfunction:** Engine started, and the PCM detected an unexpected voltage condition on the Shift Solenoid 'B' (SSB) control circuit during the CCM test. **Possible Causes:** • SSB control circuit is open (continuous high signal) • SSB control circuit is shorted to ground (continuous low signal) • SSB control circuit is shorted to system power (B+) • SSB is damaged or has failed • PCM has failed

DTC	Trouble Code Title, Conditions & Possible Causes
DTC: P0760 **1T CCM, MIL: Yes** **Year:** 2009 **Model:** B2300, B4000 **Engine:** 2.3L L4 VIN D, 4.0L V6 VIN E **Transmission:** All	**A/T Shift Solenoid 'C' Circuit Malfunction:** Engine started, then driven to a speed of over 37 mph, and the PCM detected an unexpected voltage condition on the Shift Solenoid 'C' (SSC) circuit during the CCM test. **Possible Causes:** • SSC control circuit is open (continuous high signal) • SSC control circuit is shorted to ground (continuous low signal) • SSC control circuit is shorted to system power (B+) • SSC is damaged or has failed • PCM has failed
DTC: P0761 **1T CCM, MIL: Yes** **Year:** 2009 **Model:** B2300, B4000 **Engine:** 2.3L L4 VIN D, 4.0L V6 VIN E **Transmission:** All	**A/T Shift Solenoid 'C' Performance (Mechanical):** 1st Gear Test: DTC P0733 and P0734 not set, engine started, and then driven in Drive in 1st Gear (1GR), Turbine speed at 4988 rpm ($\pm$ 225), ATF temperature at 68°F or higher, engine speed over 450 rpm, differential gear case output speed over 35 mph, and the PCM detected the revolution ratio of the forward clutch drum revolution to the differential gear case revolution was less than 2.157. **Possible Causes:** • ATF level is too low, or the ATF is badly deteriorated • A/T SSC is stuck "off", or it is damaged (a mechanical fault) • A/T control valve is stuck, or it is damaged (a mechanical fault) • PCM has failed
DTC: P0762 **2T CCM, MIL: Yes** **Year:** 2009 **Model:** B2300, B4000 **Engine:** 2.3L L4 VIN D, 4.0L V6 VIN E **Transmission:** All	**A/T Shift Solenoid 'C' Performance (Mechanical):** 3rd Gear Test DTC P0731 and P0732 not set, engine started, and then driven in Drive in 3rd Gear (3GR), Turbine speed at 4988 rpm ($\pm$ 225), ATF temperature at 68°F or higher, engine speed over 450 rpm, differential gear case output speed over 35 mph, and the PCM detected the revolution ratio of the forward clutch drum revolution to the differential gear case revolution was less than 0.863, or it was more than 1.249 during the CCM rationality test. 4th Gear Test With the conditions listed above for the 3rd Gear Test met, the PCM detected the revolution ratio of the forward clutch drum revolution to differential gear case revolution was less than 0.6, or it was 1.249 or higher while driving in 4th gear (4GR) in the CCM rationality test. **Possible Causes:** • ATF level is too low, or the ATF is badly deteriorated • A/T SSC is stuck "on", or it is damaged (a mechanical fault) • A/T control valve stuck "on", or it is damaged (mechanical fault) • PCM has failed
DTC: P0763 **1T CCM, MIL: Yes** **Year:** 2009 **Model:** B2300, B4000 **Engine:** 2.3L L4 VIN D, 4.0L V6 VIN E **Transmission:** All	**A/T Shift Solenoid 'C' Circuit Malfunction:** Engine started, then driven in Drive in 4th Gear and the PCM detected an unexpected voltage condition on the Shift Solenoid 'C' (SSC) circuit during the CCM test. **Possible Causes:** • SSC control circuit is open (continuous high signal) • SSC control circuit is shorted to ground (continuous low signal) • SSC control circuit is shorted to system power (B+) • SSC is damaged or has failed • PCM has failed
DTC: P0766 **2T CCM, MIL: Yes** **Year:** 2009 **Model:** B2300, B4000 **Engine:** 2.3L L4 VIN D, 4.0L V6 VIN E **Transmission:** All	**A/T Shift Solenoid 'D' Performance (Mechanical):** 4th Gear Test DTC P0731, P0132 and P0733 not set, engine started, then driven in Drive in 4th Gear (4GR), Turbine speed at 4988 rpm ($\pm$ 225), ATF temperature over 68°F, engine speed over 450 rpm, differential gear case output speed over 35 mph, and the PCM detected the revolution ratio of the forward clutch drum revolution to the differential gear case revolution was less than 0.60, or it was 1.249 or higher during the CCM rationality test. **Possible Causes:** • ATF level is too low, or the ATF is badly deteriorated • A/T SSD is stuck "off", or it is damaged (a mechanical fault) • A/T control valve is stuck, or it is damaged (a mechanical fault) • PCM has failed

DTC	Trouble Code Title, Conditions & Possible Causes
DTC: P0767 **2T CCM, MIL: Yes** **Year:** 2009 **Model:** B2300, B4000 **Engine:** 2.3L L4 VIN D, 4.0L V6 VIN E **Transmission:** All	**A/T Shift Solenoid 'D' Performance (Mechanical):** 3rd Gear Test DTC P0731, P0132, P0734 and P0741 not set, engine started, then driven in Drive in 3rd Gear (3GR) Turbine speed at 4988 rpm (± 225), engine speed over 450 rpm, differential gear case output speed over 35 mph, ATF temperature over 68°F, and the PCM detected the revolution ratio of the forward clutch drum revolution to the differential gear case revolution was less than 0.863, or it was 1.249 or higher during the CCM rationality test. **Possible Causes:** • ATF level is too low, or the ATF is badly deteriorated • A/T SSD is stuck "on", or it is damaged (a mechanical fault) • A/T control valve is stuck, or it is damaged (a mechanical fault) • PCM has failed
DTC: P0768 **1T CCM, MIL: Yes** **Year:** 2009 **Model:** B2300, B4000 **Engine:** 2.3L L4 VIN D, 4.0L V6 VIN E **Transmission:** All	**A/T Shift Solenoid 'D' Circuit Malfunction:** Engine started, then driven in Drive in 4th Gear, and the PCM detected an unexpected voltage condition on the Shift Solenoid 'D' (SSD) control circuit during the CCM test. **Possible Causes:** • SSD control circuit is open (continuous high signal) • SSD control circuit is shorted to ground (continuous low signal) • SSD control circuit is shorted to system power (B+) • SSD is damaged or has failed • PCM has failed
DTC: P0791 **2T CCM, MIL: Yes** **Year:** 2009 **Model:** B2300, B4000 **Engine:** 2.3L L4 VIN D, 4.0L V6 VIN E **Transmission:** All	**Input Shaft Speed (ISS) Sensor:** PCM detected ISS sensor signal failure. **Possible Causes:** • Input Shaft Speed Sensor circuit is open • ISS circuit is shorted to ground • ISS circuit is shorted to system power (B+) • ISS Sensor is damaged or has failed • PCM has failed
DTC: P0794 **2T CCM, MIL: Yes** **Year:** 2009 **Model:** B2300, B4000 **Engine:** 2.3L L4 VIN D, 4.0L V6 VIN E **Transmission:** All	**Input Shaft Speed (ISS) Sensor:** PCM detected and intermittent failure with the ISS sensor signal. **Possible Causes:** • ISS circuit is open • ISS circuit is shorted to ground • ISS circuit is shorted to system power (B+) • ISS sensor is damaged or has failed • PCM has failed
DTC: P0850 **2T CCM, MIL: Yes** **Year:** 2009, 2010 **Model:** 3, 5, 6, MX-5 Miata, RX-8 **Engine:** 1.3L R2 VIN 2, 1.3L R2 VIN 4, 2.0L L4 VIN F, 2.0L L4 VIN G, 2.3L L4 VIN 3, 2.3L L4 VIN 4, 2.3L L4 VIN L, 2.3L L4 VIN M, 2.5L L4 VIN 5, 2.5L L4 VIN 6, 2.5L L4 VIN A **Transmission:** All	**Neutral switch input circuit problem:** The PCM monitors changes in input voltage from the neutral switch. If the PCM does not detect PCM terminal 1S voltage changes while running vehicle with vehicle speed above 30 km/h {19 mph} and clutch pedal turns press and depress 10 times repeatedly, the PCM determines that the neutral switch circuit has malfunction. **Possible Causes:** • Neutral switch malfunction • Poor connection of neutral switch connector or PCM connector • Short to ground in wiring harness between neutral switch terminal A and PCM terminal 1S • Open circuit in wiring harness between ground and neutral switch terminal B • PCM has failed
DTC: P0960 **2T CCM, MIL: Yes** **Year:** 2009 **Model:** B2300, B4000 **Engine:** 2.3L L4 VIN D, 4.0L V6 VIN E **Transmission:** All	**Electronic Pressure Control (EPC):** EPC solenoid circuit failure, open circuit. **Possible Causes:** • EPC solenoid circuit is open • PCM has failed

DTC	Trouble Code Title, Conditions & Possible Causes
DTC: P0962 **2T CCM, MIL: Yes** **Year:** 2009 **Model:** B2300, B4000 **Engine:** 2.3L L4 VIN D, 4.0L V6 VIN E **Transmission:** All	**Electronic Pressure Control (EPC):** EPC solenoid circuit failure, shorted to ground. **Possible Causes:** • EPC circuit is shorted to ground • PCM has failed
DTC: P0963 **2T CCM, MIL: Yes** **Year:** 2009 **Model:** B2300, B4000 **Engine:** 2.3L L4 VIN D, 4.0L V6 VIN E **Transmission:** All	**Electronic Pressure Control:** EPC solenoid circuit failure, short circuit to VBAT. **Possible Causes:** • EPC circuit is shorted to VBAT • PCM has failed

OBD II Trouble Code List (P1XXX Codes)

DTC	Trouble Code Title, Conditions & Possible Causes
DTC: P1000 **1T CCM, MIL: Yes** **Year:** 2009, 2010 **Model:** 6, B2300, B4000, CX-9 **Engine:** 2.3L L4 VIN D, 3.7L V6 VIN A, 3.7L V6 VIN B, 3.7L V6 VIN V, 4.0L V6 VIN E **Transmission:** All	**OBD II Monitor Testing No Completed:** Key on or engine running, and the PCM detected that DTC P1000 was set because one of the OBD II Main Monitors did not complete the I/M Readiness Test sequence. The recommended next step is to drive vehicle through the complete Mazda Drive Cycle pattern. **NOTE: This trouble code is deleted if the MIL is activated.** **Possible Causes:** • A PCM Reset step was performed with an OBD II Scan Tool • Battery keep alive power (KAPWR) was removed to the PCM • One or more OBD II Monitors did not complete during an official OBD II Drive Cycle
DTC: P1001 **1T CCM** **Year:** 2009, 2010 **Model:** 6, B2300, B4000, CX-9 **Engine:** 2.3L L4 VIN D, 3.7L V6 VIN A, 3.7L V6 VIN B, 3.7L V6 VIN V, 4.0L V6 VIN E **Transmission:** All	**Data Link Connector Circuit Malfunction:** Engine running and the PCM detected an unexpected voltage condition on the SCP communication circuit to the PCM. **NOTE: This code indicates the Scan Tool could not "talk" to the PCM.** **Possible Causes:** • DLC connector or the related terminals/pins are damaged • DLC Bus (+) circuit is open or shorted to ground • DLC Bus (+) circuit is shorted at an associated module • DLC ground circuit is open or has a high resistance condition • PCM power relay is damaged or has failed (no power to DLC) • Incorrect Self-Test procedure followed, or idle speed to low • PCM has failed (e.g., the PCM VREF may be out-of-range)
DTC: P1100 **2T CCM, MIL: Yes** **Year:** 2009 **Model:** B2300, B4000 **Engine:** 2.3L L4 VIN D, 4.0L V6 VIN E **Transmission:** All	**Mass Airflow Sensor Circuit Malfunction (Intermittent):** Engine running and the PCM detected the MAF signal was less than 0.39v, or it was more than 3.90v in the last 40 warm-up cycles. **NOTE: This code can set due to an interruption of the MAF signal.** **Possible Causes:** • MAF sensor continuity problems at the connector or harness • MAF sensor circuit is open (the fault may be intermittent) • MAF sensor is damaged or has failed • PCM has failed
DTC: P1101 **2T CCM, MIL: Yes** **Year:** 2009, 2010 **Model:** 6, B2300, B4000, CX-9, Tribute **Engine:** 2.3L L4 VIN D, 2.5L L4 VIN 3, 2.5L L4 VIN 7, 3.0L V6 VIN 1, 3.0L V6 VIN G, 3.7L V6 VIN A, 3.7L V6 VIN B, 3.7L V6 VIN V, 4.0L V6 VIN E **Transmission:** All	**Mass Airflow Sensor Malfunction (Self-Test):** KOEO Self-Test Enabled: Key on, and the PCM detected the MAF signal was more than 0.27v. KOER Self-Test Enabled: Engine running and the PCM detected the MAF signal was not in the normal operating range of 0.46-2.44v during the CCM test. **Possible Causes:** • MAF sensor is damaged, dirty or has failed • MAF sensor (inlet) screen is dirty or blocked • An intake air leak is present near the MAF sensor location • PCM has failed

DTC	Trouble Code Title, Conditions & Possible Causes
DTC: P1110 **1T CCM, MIL: Yes** **Year:** 2009, 2010 **Model:** Tribute **Engine:** 2.5L L4 VIN 3, 2.5L L4 VIN 7, 3.0L V6 VIN 1, 3.0L V6 VIN G **Transmission:** All	**Intake Air Temperature Sensor Circuit Malfunction (Dynamic Chamber):** Key on or engine running, IAT sensor signal more than 14°F, and the PCM detected the IAT Dynamic Chamber signal was less than 0.10v, or that it was more than 4.92v, or that the IAT sensor signal was higher than the ECT sensor signal by 104°F with the ECT sensor signal more than 158°F during the CCM Rationality test. **Possible Causes:** • IAT sensor signal circuit open or shorted to ground • IAT sensor ground circuit open between sensor and the PCM • IAT sensor is damaged or has failed • PCM has failed
DTC: P1112 **1T CCM** **Year:** 2009 **Model:** B2300, B4000 **Engine:** 2.3L L4 VIN D, 4.0L V6 VIN E **Transmission:** All	**Intake Air Temperature Sensor Signal Circuit Malfunction:** Key on or engine running, IAT sensor signal more than 50°F, and the PCM detected the IAT sensor signal was interrupted (dropped-out) during the CCM test. **Possible Causes:** • IAT sensor signal circuit is open (an intermittent fault) • IAT sensor signal circuit is shorted to ground (intermittent fault) • IAT sensor is damaged or has failed (an intermittent fault) • PCM has failed
DTC: P1116 **1T CCM, MIL: Yes** **Year:** 2009, 2010 **Model:** B2300, B4000, Tribute **Engine:** 2.3L L4 VIN D, 2.5L L4 VIN 3, 2.5L L4 VIN 7, 3.0L V6 VIN 1, 3.0L V6 VIN G, 4.0L V6 VIN E **Transmission:** All	**Engine Coolant Temperature Sensor Circuit Malfunction (Self-Test):** KOER Self-Test Enabled: Engine running, IAT sensor signal more than 50°F, and the PCM detected the ECT signal was less than 0.30v, or that it was more than 3.70v during the CCM test. **Possible Causes:** • ECT sensor signal circuit open or shorted to ground • ECT sensor signal shorted to VREF or system power • ECT sensor
DTC: P1117 **2T CCM, MIL: Yes** **Year:** 2009 **Model:** B2300, B4000 **Engine:** 2.3L L4 VIN D, 4.0L V6 VIN E **Transmission:** All	**Engine Coolant Temperature Sensor Circuit Malfunction:** Engine running, and the PCM detected the ECT sensor signal was interrupted (dropped-out) at some point during the CCM test. **Possible Causes:** • ECT sensor signal is open (an intermittent fault) • ECT sensor signal shorted to ground (an intermittent fault) • ECT sensor is damaged or has failed (an intermittent fault) • PCM has failed
DTC: P1120 **1T CCM, MIL: Yes** **Year:** 2009 **Model:** B2300, B4000 **Engine:** 2.3L L4 VIN D, 4.0L V6 VIN E **Transmission:** All	**TP Sensor Signal Out of Range Low:** Engine running at idle speed, and the PCM detected the TP sensor signal was less minimum operating range of 0.17-0.49v (Scan Tool indicates 3.43-98%) during the CCM Rationality test. **Possible Causes:** • TP sensor contacts are loose, broken or damaged • TP sensor pins are loose in the connector • TP sensor signal circuit shorted to ground (an intermittent fault) • TP sensor is damaged or has failed • PCM has failed
DTC: P1121 **1T CCM, MIL: Yes** **Year:** 2009 **Model:** B2300, B4000 **Engine:** 2.3L L4 VIN D, 4.0L V6 VIN E **Transmission:** All	**Throttle Position Signal Not Consistent With MAF Signal:** Engine running, and the PCM detected the TP sensor and MAF sensor signals were not consistent with the calibrated load values in memory (i.e., the PCM detected the mass airflow was too low). **Possible Causes:** • MAF sensor is damaged or has failed • TP sensor is damaged or has failed • Air leaks between the MAF sensor and the throttle body • PCM has failed
DTC: P1124 **1T CCM** **Year:** 2009, 2010 **Model:** 6, B2300, B4000 **Engine:** 2.3L L4 VIN D, 3.7L V6 VIN B, 4.0L V6 VIN E **Transmission:** All	**Throttle Position Sensor Signal Out Of Range (Self-Test):** KOER Self-Test Enabled: Engine running, and the PCM detected the TP sensor signal (the rotational setting) was not in the range of 0.66-1.20v (Scan Tool indicates 13.23-24.02%) during the CCM Rationality test. **Possible Causes:** • Throttle linkage is bent or binding • TP sensor is not seated properly • Throttle plate below closed throttle position, or throttle plate screw misadjusted • TP sensor is damaged • PCM has failed

DTC	Trouble Code Title, Conditions & Possible Causes
DTC: P1127 **1T CCM** **Year:** 2009, 2010 **Model:** 6, B2300, B4000, CX-9, Tribute **Engine:** 2.3L L4 VIN D, 2.5L L4 VIN 3, 2.5L L4 VIN 7, 3.0L V6 VIN 1, 3.0L V6 VIN G, 3.7L V6 VIN A, 3.7L V6 VIN B, 3.7L V6 VIN V, 4.0L V6 VIN E **Transmission:** All	**HO2S-12 (Bank 1 Sensor 2) Heater Not Enabled (Self-Test):** KOER Self-Test Enabled: Engine running, and the PCM detected the HO2S-12 heater was not hot enough for testing during the Self-Test. **Possible Causes:** • Engine not operated long enough prior to starting the KOER Self-Test • Exhaust system temperature is too cold to run the test
DTC: P1128 **1T CCM** **Year:** 2009 **Model:** B2300, B4000 **Engine:** 2.3L L4 VIN D, 4.0L V6 VIN E **Transmission:** All	**HO2S-11 (Bank 1 Sensor 1) Signals Swapped (Self-Test):** KOER Self-Test Enabled: Engine running and the PCM detected the front HO2S-11 signals indicated the wrong cylinder bank responded to a shift in fuel ratio. **Possible Causes:** • HO2S harness connectors are crossed • HO2S wiring is crossed at the harness connector • HO2S wiring is crossed at the PCM connector • PCM has failed
DTC: P1130 **2T CCM, MIL: Yes** **Year:** 2009 **Model:** B2300, B4000 **Engine:** 2.3L L4 VIN D, 4.0L V6 VIN E **Transmission:** All	**HO2S-11 (Bank 1 Sensor 1) Fuel Control Limit Reached:** Engine running in closed loop at cruise speed, and the PCM detected the front HO2S-11 signal indicated the fuel control had reached its maximum lean or rich limit (HO2S-11 is not switching). **Possible Causes:** • HO2S is damaged or has failed • HO2S circuits wet or oily, corroded, or poor terminal contact • HO2S signal circuit open, shorted to ground, short to power • Low fuel pressure or vehicle driven until it was out of fuel • Fuel pressure regulator leaking, or fuel injectors leaking • EVAP vapor recovery system has failed • Air leaks located after the MAF sensor • Exhaust leaks before or near the front HO2S • EGR valve stuck, EGR diaphragm leaking, or gasket leaking • Oil dipstick not seated or engine oil level too high (overfilled)
DTC: P1131 **2T CCM, MIL: Yes** **Year:** 2009 **Model:** B2300, B4000 **Engine:** 2.3L L4 VIN D, 4.0L V6 VIN E **Transmission:** All	**HO2S-11 (Bank 1 Sensor 1) Indicates A/F Ratio Too Lean:** Engine running in closed loop at cruise speed, and the PCM detected the front HO2S-11 signal indicated the Air Fuel (A/F) ratio was too lean (i.e., the PCM is correcting for an over-lean condition). **Possible Causes:** • Air leaks located after the MAF sensor • Base Engine: Check for vacuum leaks with a Smoke Tester • Intake air system is leaking, obstructed and damaged • Low fuel pressure or vehicle driven until it was out of fuel • PCV system has air leaks at the valve or related hoses
DTC: P1132 **2T CCM, MIL: Yes** **Year:** 2009 **Model:** B2300, B4000 **Engine:** 2.3L L4 VIN D, 4.0L V6 VIN E **Transmission:** All	**HO2S-11 (Bank 1 Sensor 1) Indicates A/F Ratio Too Lean:** Engine running in closed loop at cruise speed, and the PCM detected the front HO2S-11 signal indicated the Air Fuel (A/F) ratio was too lean (i.e., the PCM is correcting for an over-lean condition). **Possible Causes:** • Air leaks located after the MAF sensor • Base Engine: Check for vacuum leaks with a Smoke Tester • Intake air system is leaking, obstructed and damaged • Low fuel pressure or vehicle driven until it was out of fuel • PCV system has air leaks at the valve or related hoses

DTC	Trouble Code Title, Conditions & Possible Causes
DTC: P1137 **1T CCM** **Year:** 2009 **Model:** B2300, B4000 **Engine:** 2.3L L4 VIN D, 4.0L V6 VIN E **Transmission:** All	**HO2S-12 Lack of Switching, Indicates Lean (Self-Test):** KOER Self-Test Enabled: Engine running in closed loop, and the PCM detected the rear HO2S-12 signal indicated the fuel control had reached its maximum lean limit (i.e., the right bank HO2S-12 is not switching properly). **Possible Causes:** • Air leaks located after the MAF sensor • Base Engine: Check for vacuum leaks with a Smoke Tester • HO2S wiring may be crossed at the connector • Check intake air system for leaks, obstructions and damage • EGR system malfunction (the EGR valve may be stuck open) • PCV system has air leaks at the valve or related hoses • Fuel pressure is too low (fuel filter is dirty, or weak fuel pump) • Vehicle driven until it was out of fuel
DTC: P1138 **1T CCM** **Year:** 2009 **Model:** B2300, B4000 **Engine:** 2.3L L4 VIN D, 4.0L V6 VIN E **Transmission:** All	**HO2S-12 Lack of Switching, Indicates Rich (Self-Test):** KOER Self-Test Enabled: Engine running in closed loop, and the PCM detected the rear HO2S-12 signal indicated the fuel control had reached its maximum rich limit (i.e., the right bank HO2S-12 is not switching properly). **Possible Causes:** • Check air cleaner element, air cleaner housing for blockage • Fuel pressure regulator is leaking, or fuel pressure is too high • HO2S wiring may be crossed at the connector • Fuel injectors leaking or contaminated (they may be sticking) • EVAP vapor recovery system has failed (canister full of fuel)
DTC: P1150 **2T CCM, MIL: Yes** **Year:** 2009 **Model:** B2300, B4000 **Engine:** 2.3L L4 VIN D, 4.0L V6 VIN E **Transmission:** All	**HO2S-21 (Bank 2 Sensor 1) Fuel Control Limit Reached:** Engine running in closed loop at cruise speed, and the PCM detected the front HO2S-21 signal indicated the fuel control had reached its maximum lean or rich limit (HO2S-21 is not switching). **Possible Causes:** • HO2S is damaged or has failed • HO2S circuits wet or oily, corroded, or poor terminal contact • HO2S signal circuit open, shorted to ground, short to power • Low fuel pressure or vehicle driven until it was out of fuel • Excessive fuel pressure, leaking or contaminated fuel injectors • Leaking fuel pressure regulator • EVAP vapor recovery system has failed • Air leaks located after the MAF sensor • Exhaust leaks before or near the front HO2S • EGR valve stuck, EGR diaphragm leaking, or gasket leaking • Oil dipstick not seated or engine oil level too high (overfilled)
DTC: P1151 **2T CCM, MIL: Yes** **Year:** 2009 **Model:** B2300, B4000 **Engine:** 2.3L L4 VIN D, 4.0L V6 VIN E **Transmission:** All	**HO2S-21 (Bank 1 Sensor 1) Indicates A/F Ratio Too Lean:** Engine running in closed loop at cruise speed, and the PCM detected the front HO2S-21 signal indicated the Air Fuel (A/F) ratio was too lean (i.e., the PCM is correcting for an over-lean condition). **Possible Causes:** • Air leaks located after the MAF sensor • Base Engine: Check for vacuum leaks with a Smoke Tester • Intake air system is leaking, obstructed and damaged • Low fuel pressure or vehicle driven until it was out of fuel • PCV system has air leaks at the valve or related hoses
DTC: P1152 **2T CCM, MIL: Yes** **Year:** 2009 **Model:** B2300, B4000 **Engine:** 2.3L L4 VIN D, 4.0L V6 VIN E **Transmission:** All	**HO2S-21 (Bank 2 Sensor 1) Indicates A/F Ratio Too Rich:** Engine running in closed loop at cruise speed, and the PCM detected the front HO2S-21 signal indicated the Air Fuel (A/F) ratio was too rich (i.e., the PCM is correcting for an over-rich condition). **Possible Causes:** • Check air cleaner element, air cleaner housing for blockage • Fuel pressure regulator is leaking, or fuel pressure is too high • Fuel injectors leaking or contaminated (they may be sticking) • EVAP vapor recovery system has failed (canister full of fuel)

DTC	Trouble Code Title, Conditions & Possible Causes
DTC: P115E **T PCM** **Year:** 2009, 2010 **Model:** 6, CX-9 **Engine:** 3.7L V6 VIN A, 3.7L V6 VIN B, 3.7L V6 VIN V **Transmission:** All	**Throttle actuator control throttle body air flow trim at max limit:** During idle, the PCM monitors the throttle angle and air flow. If the air flow is determined to be less than expected, the PCM adjusts the throttle angle to compensate. The air flow reduction is typically the result of sludge buildup around the throttle plate. This DTC indicates the PCM has reached the maximum allowed compensation and is no longer able to compensate for the buildup. **Possible Causes:** • Sludge around the throttle plate
DTC: P120F **T CCM** **Year:** 2009, 2010 **Model:** 3, CX-7 **Engine:** 2.3L L4 3, 2.3L L4 VIN 4, 2.3L L4 VIN L, 2.3L L4 VIN M **Transmission:** All	**Fuel pressure regulator excessive variation:** If the fuel pressure average value measured by the PCM exceeds the specification when the camshaft is rotating at a specified rate, the PCM determines that there is a fuel pressure regulator performance problem. **Possible Causes:** • High pressure fuel pump connector or terminals malfunction • PCM connector or terminals malfunction • Spill valve control solenoid valve malfunction • PCM malfunction
DTC: P1260 **1T CCM, MIL: Yes** **Year:** 2009, 2010 **Model:** 3, 5, 6, CX-7, CX-9, MX-5 Miata, RX-8, Tribute **Engine:** 1.3L R2 VIN 2, 1.3L R2 VIN 4, 1.3L R2 VIN M, 1.3L R2 VIN P, 2.0L L4 VIN F, 2.0L L4 VIN G, 2.3L L4 VIN 3, 2.3L L4 VIN 4, 2.3L L4 VIN L, 2.3L L4 VIN M, 2.5L L4 VIN 3, 2.5L L4 VIN 5, 2.5L L4 VIN 6, 2.5L L4 VIN 7, 2.5L L4 VIN A, 2.5L L4 VIN M, 3.0L V6 VIN 1, 3.0L V6 VIN G, 3.7L V6 VIN A, 3.7L V6 VIN V **Transmission:** All	**Anti-Theft System Signal Detected, Engine Disabled:** Key on, and the PCM received a signal from the Anti-Theft System that a theft condition had occurred. The theft indicator on the dash will flash rapidly or remain on "solid" with the ignition switch in the "on" position. The engine may "start and stall", or may not crank if the vehicle is equipped with the PATS starter "disable" feature. **Possible Causes:** • Previous theft condition has occurred • Anti-Theft System is damaged or has failed
DTC: P1270 **1T CCM, MIL: Yes** **Year:** 2009 **Model:** B2300, B4000 **Engine:** 2.3L L4 VIN D, 4.0L V6 VIN E **Transmission:** All	**Engine RPM or Vehicle Speed Limit Reached:** Engine running and the PCM detected that the vehicle had been driven in a manner where the engine speed or the vehicle speed had exceeded a calibrated limit stored in memory. **Possible Causes:** • Excessive wheel slippage (due to water, ice, mud or snow) • Engine over-revved with the gear selector in Neutral position • Vehicle was driven at a very high rate of speed
DTC: P1285 **1T CCM, MIL: Yes** **Year:** 2009, 2010 **Model:** 6, B2300, B4000, CX-9, Tribute **Engine:** 2.3L L4 VIN D, 2.5L L4 VIN 3, 2.5L L4 VIN 7, 3.0L V6 VIN 1, 3.0L V6 VIN G, 3.7L V6 VIN A, 3.7L V6 VIN B, 3.7L V6 VIN V, 4.0L V6 VIN E **Transmission:** All	**Cylinder Head Temperature Sensor - Engine Overheated:** Engine running and the PCM detected a signal from the Cylinder Head Temperature (CHT) sensor that indicated an engine overheat condition existed during the CCM Rationality test. **Possible Causes:** • Engine coolant level is low, or wrong coolant mixture • Engine overheat condition exists • CHT sensor is out-of-calibration or damaged • PCM has failed
DTC: P1288 **1T CCM, MIL: Yes** **Year:** 2009, 2010 **Model:** 6, B2300, B4000, CX-9, Tribute **Engine:** 2.3L L4 VIN D, 2.5L L4 VIN 3, 2.5L L4 VIN 7, 3.0L V6 VIN 1, 3.0L V6 VIN G, 3.7L V6 VIN A, 3.7L V6 VIN B, 3.7L V6 VIN V, 4.0L V6 VIN E **Transmission:** All	**Cylinder Head Temperature Sensor Circuit Malfunction (Self-Test):** KOER Self-Test Enabled: Engine running IAT sensor signal more than 50°F, and the PCM detected a signal from the Cylinder Head Temperature (CHT) sensor that indicated an engine was not warm enough during the CCM test. **Possible Causes:** • Engine not warm enough to begin the KOER Self-Test • CHT sensor is out-of-calibration or damaged • PCM has failed

DTC	Trouble Code Title, Conditions & Possible Causes
DTC: P1289 **1T CCM, MIL: Yes** **Year:** 2009, 2010 **Model:** 6, B2300, B4000, CX-9, Tribute **Engine:** 2.3L L4 VIN D, 2.5L L4 VIN 3, 2.5L L4 VIN 7, 3.0L V6 VIN 1, 3.0L V6 VIN G, 3.7L V6 VIN A, 3.7L V6 VIN B, 3.7L V6 VIN V, 4.0L V6 VIN E **Transmission:** All	**Cylinder Head Temperature Sensor Overheat Condition:** Engine started, engine running and the PCM detected the Cylinder Head Temperature (CHT) sensor indicated an overheated engine condition, and the Fail-Safe Cooling (FMEM) strategy was enabled. **Possible Causes:** • Serious engine overheat condition exists • CHT sensor is out-of-calibration or damaged • PCM has failed
DTC: P1290 **1T CCM, MIL: Yes** **Year:** 2009, 2010 **Model:** 6, B2300, B4000, CX-9, Tribute **Engine:** 2.3L L4 VIN D, 2.5L L4 VIN 3, 2.5L L4 VIN 7, 3.0L V6 VIN 1, 3.0L V6 VIN G, 3.7L V6 VIN A, 3.7L V6 VIN B, 3.7L V6 VIN V, 4.0L V6 VIN E **Transmission:** All	**Cylinder Head Temperature Sensor Low Input:** Engine running and the PCM detected an unexpected "low" voltage condition on the Cylinder Head Temperature (CHT) sensor circuit. **Possible Causes:** • CHT sensor signal circuit is shorted to ground • CHT sensor is damaged or has failed • PCM has failed
DTC: P1299 **1T CCM, MIL: Yes** **Year:** 2009, 2010 **Model:** 6, B2300, B4000, CX-9, Tribute **Engine:** 2.3L L4 VIN D, 2.5L L4 VIN 3, 2.5L L4 VIN 7, 3.0L V6 VIN 1, 3.0L V6 VIN G, 3.7L V6 VIN A, 3.7L V6 VIN B, 3.7L V6 VIN V, 4.0L V6 VIN E **Transmission:** All	**Cylinder Head Temperature Sensor High Input:** Engine running and the PCM detected an unexpected "high" voltage condition on the Cylinder Head Temperature (CHT) sensor circuit. **Possible Causes:** • CHT sensor signal circuit is open • CHT sensor signal circuit is shorted to VRED • CHT sensor ground circuit is open • CHT sensor is damaged or has failed • PCM has failed
DTC: P1309 **2T CCM, MIL: Yes** **Year:** 2009 **Model:** B2300, B4000 **Engine:** 2.3L L4 VIN D, 4.0L V6 VIN E **Transmission:** All	**Misfire Detection Monitor:** Engine running for more from 2-6 complete engine cycles, and the PCM detected that the Misfire Monitor was not "enabled" due to an internal calculation error or problem. **Possible Causes:** • CMP sensor is not properly synchronized in the engine • CMP sensor is damaged or has failed • PCM has failed
DTC: P1336 **2T CCM, MIL: Yes** **Year:** 2009, 2010 **Model:** 6, CX-9, Tribute **Engine:** 2.5L L4 VIN 3, 2.5L L4 VIN 7, 3.0L V6 VIN 1, 3.0L V6 VIN G, 3.7L V6 VIN A, 3.7L V6 VIN B, 3.7L V6 VIN V **Transmission:** All	**Camshaft Position (SGC) Sensor Circuit Malfunction:** Engine cranking with 4 to 5 crankshaft rotations occurring and the PCM did not detect any CMP (SGC) sensor signals during the test. **Possible Causes:** • CMP sensor signal circuit is open or shorted to ground • CMP sensor signal circuit is shorted to VREF or system power • CMP sensor power circuit open between sensor and main relay • CMP sensor is damaged or has failed • PCM has failed
DTC: P1397 **T PCM** **Year:** 2009, 2010 **Model:** 6, CX-9 **Engine:** 3.7L V6 VIN A, 3.7L V6 VIN B, 3.7L V6 VIN V **Transmission:** All	**System voltage out of self-test range:** This DTC indicates that the 12-volt system voltage is too high or too low during the KOEO or KOER self-test. It sets if the system voltage falls below or exceeds the calibrated threshold at any time during the KOEO or KOER self-test. **Possible Causes:** • Battery or charging system concern.

DTC	Trouble Code Title, Conditions & Possible Causes
DTC: P1400 **1T CCM** **Year:** 2009 **Model:** B2300, B4000 **Engine:** 2.3L L4 VIN D, 4.0L V6 VIN E **Transmission:** All	**EGR Valve Position Sensor Circuit Low Input (Self-Test):** KOER Self-Test enabled: Key on, and the PCM detected the EGRP sensor signal was less than 0.20v during the CCM Rationality test. **Possible Causes:** • EGRP sensor signal circuit is shorted to ground • EGRP sensor power circuit is open • EGRP sensor is damaged or has failed • PCM has failed
DTC: P1401 **1T CCM** **Year:** 2009 **Model:** B2300, B4000 **Engine:** 2.3L L4 VIN D, 4.0L V6 VIN E **Transmission:** All	**EGR Valve Position Sensor Circuit High Input (Self-Test):** KOER Self-Test enabled: Key on, and the PCM detected the EGRP sensor signal was more than 4.50v during the CCM test. **Possible Causes:** • EGRP sensor signal circuit is open • EGRP sensor ground circuit is open • EGRP sensor is damaged or has failed • PCM has failed
DTC: P1408 **1T CCM** **Year:** 2009 **Model:** B2300, B4000 **Engine:** 2.3L L4 VIN D, 4.0L V6 VIN E **Transmission:** All	**EGR System Flow Out Of Range (Self-Test):** KOER Self-Test enabled: Vehicle driven to a speed of over 7 mph, then back to idle speed, ECT sensor signal more than 131°F, and the PCM detected little or no change in the EGR Differential Pressure sensor (DPFE) with the EGR vent commanded "on" and then "off" in the EGR Monitor test. **Possible Causes:** • EGR valve stuck closed or iced up, or the flow path is restricted • EGR valve diaphragm leaking, hose is off, plugged or leaking • EGR system vacuum lines are loose or damaged • EGRA solenoid is damaged or has failed • EGRV solenoid is damaged or has failed • PCM has failed
DTC: P1409 **2T CCM, MIL: Yes** **Year:** 2009 **Model:** B2300, B4000 **Engine:** 2.3L L4 VIN D, 4.0L V6 VIN E **Transmission:** All	**Exhaust Gas Recirculation (EGR) Vacuum Regulator Solenoid Circuit:** Engine started, and the PCM detected a fault in the EGR VR solenoid circuit (i.e., the VR circuit was too high or low when compared to the expected range with the solenoid enabled). **NOTE: The EGR vacuum regulator solenoid resistance is between 26 and 40 ohms.** **Possible Causes:** • VPWR circuit is open to EGR Vacuum Regulator (VR)solenoid • EGR VR solenoid circuit is open, or shorted to ground • EGR vacuum regulator solenoid is damaged
DTC: P1443 **2T CCM, MIL: Yes** **Year:** 2009 **Model:** B2300, B4000 **Engine:** 2.3L L4 VIN D, 4.0L V6 VIN E **Transmission:** All	**EVAP System Purge Flow Malfunction:** Engine running in closed loop at cruise speed, TP Angle steady at medium load, VSS at 35-55 mph, and the PCM detected a leak or blockage between the intake manifold, canister purge valve and the charcoal canister during the EVAP Monitor test. **Possible Causes:** • Fuel vapor hose blocked between EVAP purge valve and PF sensor, or blocked between purge valve and intake manifold, or vacuum hose blocked between purge valve and intake manifold • EVAP canister purge solenoid stuck closed (mechanically)
DTC: P144A **2T EVAP, MIL: Yes** **Year:** 2009, 2010 **Model:** 3, 6, CX-7 **Engine:** 2.0L L4 VIN F, 2.0L L4 VIN G, 2.3L L4 VIN 3, 2.3L L4 VIN 4, 2.3L L4 VIN L, 2.3L L4 VIN M, 2.5L L4 VIN 5, 2.5L L4 VIN 6, 2.5L L4 VIN A, 2.5L L4 VIN M, 3.7L V6 VIN B **Transmission:** All	**EVAP system purge vapor line restricted/blocked:** P144A:00 indicates that the P144A:00 blocked line diagnostic is designed to detect a full blockage in the vapor lines between the fuel tank pressure sensor and the fuel tank. **Possible Causes:** • Blockage(s) between fuel tank pressure sensor and fuel tank • PCM malfunction

DTC	Trouble Code Title, Conditions & Possible Causes
DTC: P1450 **2T CCM, MIL: Yes** **Year:** 2009, 2010 **Model:** 3, 6, B2300, B4000, CX-7, CX-9, Tribute **Engine:** 2.0L L4 VIN F, 2.0L L4 VIN G, 2.3L L4 VIN 3, 2.3L L4 VIN 4, 2.3L L4 VIN D, 2.3L L4 VIN L, 2.3L L4 VIN M, 2.5L L4 VIN 3, 2.5L L4 VIN 5, 2.5L L4 VIN 6, 2.5L L4 VIN 7, 2.5L L4 VIN A, 2.5L L4 VIN M, 3.0L V6 VIN 1, 3.0L V6 VIN G, 3.7L V6 VIN A, 3.7L V6 VIN B, 3.7L V6 VIN V, 4.0L V6 VIN E **Transmission:** All	**EVAP Control System Excessive Vacuum Detected:** Engine startcd, ECT sensor signal more than 95°F and IAT sensor signal more than 14°F at startup, vehicle driven at a speed of over 61.9 mph, engine load from 15-80%, fuel level between 15-85%, fuel tank pressure from minus (-) 1.9-10.0 kPa, BARO sensor signal over 76.2 kPa, then with the EVAP control solenoid commanded "on", the PCM detected the FTP sensor signal indicated higher than minus (-) 3.91 kPa for 35 seconds in the EVAP Monitor Leak test. **Possible Causes:** • Air filter element is extremely dirty or clogged • Canister drain cut valve (CDCV) is damaged or has failed • Charcoal canister is full of fuel or moisture, or is damaged • EVAP system drain passage is clogged • Fuel tank pressure sensor is damaged or out-of-calibration • Purge solenoid valve is damaged or has failed
DTC: P1451 **2T CCM, MIL: Yes** **Year:** 2009, 2010 **Model:** B2300, B4000, Tribute **Engine:** 2.3L L4 VIN D, 2.5L L4 VIN 3, 2.5L L4 VIN 7, 3.0L V6 VIN 1, 3.0L V6 VIN G, 4.0L V6 VIN E **Transmission:** All	**Canister Vent Solenoid Circuit Malfunction:** Key on, and the PCM detected an unexpected voltage condition on the Canister Vent Solenoid control circuit during the CCM test. **Possible Causes:** • CV solenoid control circuit is open or shorted to ground • CV solenoid control circuit is shorted to system power (B+) • CV solenoid is damaged or has failed • PCM has failed
DTC: P1460 **1T CCM, MIL: Yes** **Year:** 2009 **Model:** B2300, B4000 **Engine:** 2.3L L4 VIN D, 4.0L V6 VIN E **Transmission:** All	**Wide Open Throttle A/C Cut-Off Relay Circuit Malfunction:** Engine running and the PCM detected the A/C Relay control signal did not change between an A/C "on" request and wide open throttle event, or it detected "high" current draw on the circuit during testing. **Possible Causes:** • A/C relay is damaged or has failed • A/C relay power circuit is open between relay and ENGINE fuse • A/C relay control circuit is open between the relay and the PCM • A/C relay control circuit is shorted to ground • A/C relay control circuit is shorted to system power (B+) • PCM has failed
DTC: P1461 **2T CCM, MIL: Yes** **Year:** 2009 **Model:** B2300, B4000 **Engine:** 2.3L L4 VIN D, 4.0L V6 VIN E **Transmlssion:** All	**A/C Pressure Sensor Circuit High Input:** Engine started, and the PCM detected the A/C Pressure sensor signal was over the test limit. **NOTE: Verify a VREF voltage between 4 and 6 volts.** **Possible Causes:** • ACP sensor circuit shorted to VREF or to power (VPWR) • ACP sensor circuit is open, or the ground circuit is open • ACP sensor is damaged or has failed
DTC: P1462 **2T CCM, MIL: Yes** **Year:** 2009 **Model:** B2300, B4000 **Engine:** 2.3L L4 VIN D, 4.0L V6 VIN E **Transmission:** All	**Air Conditioning Pressure (A/CP) Sensor Low Voltage Detected:** Engine started, and the PCM detected the A/C Pressure sensor signal was under the test limit. **NOTE: Verify a VREF voltage between 4 and 6 volts.** **Possible Causes:** • ACP circuit short to GND or SIGRTN • VREF circuit open • Open ACP circuit • Damaged ACP sensor
DTC: P1463 **2T CCM, MIL: Yes** **Year:** 2009 **Model:** B2300, B4000 **Engine:** 2.3L L4 VIN D, 4.0L V6 VIN E **Transmission:** All	**A/C Pressure Sensor Insufficient Pressure Change:** Engine started, and with the A/C compressor operating, the PCM detected the A/C refrigerant pressure did not change as the compressor cycled during the self-test period. **NOTE: Verify the A/C system function, including refrigerant charge.** **Possible Causes:** • A/C system mechanical failure, or A/C clutch always engaged • ACP sensor signal open, or sensor ground circuit open • A/C sensor is damaged or the PCM has failed

DTC	Trouble Code Title, Conditions & Possible Causes
DTC: P1464 **1T CCM** **Year:** 2009, 2010 **Model:** 6, B2300, B4000, CX-9 **Engine:** 2.3L L4 VIN D, 3.7L V6 VIN A, 3.7L V6 VIN B, 3.7L V6 VIN V, 4.0L V6 VIN E **Transmission:** All	**A/C Control Signal Circuit Malfunction (Self-Test):** KOER Self-Test enabled: Engine running and the PCM detected an unexpected "high" voltage condition on the A/C Relay control signal during the CCM test. **Possible Causes:** • A/C relay control circuit is shorted to system power (B+) • A/C relay is damaged or has failed • A/C WOT relay is damaged or has failed • A/C clutch is damaged or has failed • PCM has failed
DTC: P1474 **1T CCM** **Year:** 2009 **Model:** B2300, B4000 **Engine:** 2.3L L4 VIN D, 4.0L V6 VIN E **Transmission:** All	**Low Fan Relay Control Circuit Malfunction:** Key on or engine running, ECT sensor signal more than 212°F, and the PCM detected excessive current draw on the Low Fan Control Relay control circuit during the CCM test. **Possible Causes:** • LFC relay circuit is open or shorted to ground • LFC relay circuit is shorted to system power (B+) • LFC relay is shorted or damaged • PCM has failed
DTC: P1480 **1T CCM** **Year:** 2009, 2010 **Model:** Tribute **Engine:** 2.5L L4 VIN 3, 2.5L L4 VIN 7, 3.0L V6 VIN 1, 3.0L V6 VIN G **Transmission:** All	**Condenser Fan Relay Control Circuit Malfunction:** Engine running, ECT sensor signal over 226°F, and the PCM detected the Condenser Fan Relay control circuit signal did not change, or that the Condenser Fan Relay control signal did not change during the KOER Self-Test. **Possible Causes:** • Condenser fan relay control circuit is open or shorted to ground • Condenser fan relay control circuit is shorted to system power • Condenser fan relay power circuit is open or shorted to ground • Condenser fan relay is damaged or has failed • PCM has failed
DTC: P1500 **1T CCM, MIL: Yes** **Year:** 2009, 2010 **Model:** CX-9 **Engine:** 3.7L V6 VIN A, 3.7L V6 VIN V **Transmission:** All	**Vehicle Speed Sensor Intermittent Signal:** Engine running in gear, VSS inputs indicate the vehicle is moving, and the PCM detected an erratic, intermittent or noisy VSS signal. **Possible Causes:** • VSS signal circuit is open (an intermittent fault) • VSS signal circuit is shorted to ground (an intermittent fault) • VSS signal circuit is shorted to VREF (an intermittent fault) • VSS is damaged or has failed • PCM has failed
DTC: P1500 **1T CCM, MIL: Yes** **Year:** 2009, 2010 **Model:** Tribute **Engine:** 2.5L L4 VIN 3, 2.5L L4 VIN 7, 3.0L V6 VIN 1, 3.0L V6 VIN G **Transmission:** All	**Vehicle Speed Sensor Intermittent Signal:** Engine running in gear, VSS inputs indicate the vehicle is moving, and the PCM detected the VSS signal was interrupted in the test. **Possible Causes:** • VSS signal circuit is open (an intermittent fault) • VSS signal circuit is shorted to ground (an intermittent fault) • VSS signal circuit is shorted to VREF (an intermittent fault) • VSS is damaged or has failed • PCM has failed
DTC: P1500 **2T CCM, MIL: Yes** **Year:** 2009 **Model:** B2300, B4000 **Engine:** 2.3L L4 VIN D, 4.0L V6 VIN E **Transmission:** All	**Vehicle Speed Sensor (VSS) Signal Intermittent:** Engine running in gear with a VSS signal present, and the PCM detected that the VSS signal was intermittent. **NOTE: Check the wiring, connector, and sensor for damage.** **Possible Causes:** • VSS pins damaged, loose or pushed in at the connector • VSS circuit open or shorted in the wiring harness (insulation) • VSS wiring harness routing incorrect or VSS mounting incorrect • Damaged VSS
DTC: P1501 **1T CCM** **Year:** 2009 **Model:** B2300, B4000 **Engine:** 2.3L L4 VIN D, 4.0L V6 VIN E **Transmission:** All	**VSS Signal Out Of Self-Test Range:** Indicates the VSS input signal is out of self-test range. If the Powertrain Control Module (PCM) detects a VSS input signal any time during the self-test, DTC P1501 is set and the test aborts. **NOTE: Verify the VSS input is 0 km/h (0 mph) when the vehicle transmission is in PARK.** **Possible Causes:** • VSS signal is noisy due to Radio Frequency Interference/ Electro-Magnetic Interference (RFI/EMI) from outside devices (ignition wires, charging circuit or aftermarket devices)

DTC	Trouble Code Title, Conditions & Possible Causes
DTC: P1501 **1T CCM, MIL: Yes** **Year:** 2009, 2010 **Model:** 6, CX-9, Tribute **Engine:** 2.5L L4 VIN 3, 2.5L L4 VIN 7, 3.0L V6 VIN 1, 3.0L V6 VIN G, 3.7L V6 VIN A, 3.7L V6 VIN B, 3.7L V6 VIN V **Transmission:** All	**Vehicle Speed Sensor Circuit Malfunction (Self-Test):** KOER Self-Test Enabled: Engine running, gear selector in Park or Neutral, and the PCM detected a VSS signal with the vehicle not moving during the test. **Possible Causes:** • Charging system problem (check alternator for AC leakage) • Ignition system component has failed (i.e., check for signs of primary or secondary voltage leaking from the coil, coil wire, spark plug wires or boots, or the spark plug insulators • PCM has failed
DTC: P1502 **1T CCM, MIL: Yes** **Year:** 2009 **Model:** B2300, B4000 **Engine:** 2.3L L4 VIN D, 4.0L V6 VIN E **Transmission:** All	**VSS Signal Intermittent:** Engine started, and the PCM detected an intermittent VSS signal. The TCIL will flash on the first trip that this code is set. The VSS signal is received from the VSS, transfer case speed sensor, ABS Control module, GEM or the Central Timer module (depends upon the vehicle). **Possible Causes:** • VSS+ or VSS- harness circuit is open • TCSS signal or TCSS signal return harness circuit is open • VSS harness circuit, TCSS harness circuit is shorted to ground • VSS harness circuit, CSS harness circuit is shorted to power • VSS circuit open between the PCM and related control module • VSS or TCSS, or wheel speed sensors circuits are damaged • Modules connected to VSC/VSS harness circuits are damaged • Mechanical drive mechanism for the VSS or TCSS is damaged
DTC: P1504 **1T CCM, MIL: Yes** **Year:** 2009 **Model:** B2300, B4000 **Engine:** 2.3L L4 VIN D, 4.0L V6 VIN E **Transmission:** All	**Idle Air Control Solenoid Circuit Malfunction:** Engine running and the PCM detected an electrical "load" failure on the IAC solenoid circuit during the CCM Rationality test. **NOTE: This solenoid is controlled by an On or Off signal from the PCM. The resistance (range) of the IAC solenoid is 7.7-9.3 ohms.** **Possible Causes:** • IAC solenoid control circuit shorted to ground • IAC solenoid is damaged or has failed • PCM has failed
DTC: P1506 **2T CCM, MIL: Yes** **Year:** 2009 **Model:** B2300, B4000 **Engine:** 2.3L L4 VIN D, 4.0L V6 VIN E **Transmission:** All	**Idle Air Control System RPM Higher Than Expected:** Engine running at hot idle in closed loop, and the PCM detected the Actual idle speed was more than the Desired idle speed in the test. **Possible Causes:** • EGR valve is stuck open or leaking, or the gasket is leaking • EVAP purge valve is leaking or stuck in open position • IAC solenoid control circuit is shorted to ground (intermittent) • IAC valve is stuck open (it may be dirty or have failed) • PCV valve is leaking, or a leak in the intake manifold exists • PCM has failed
DTC: P1507 **2T CCM, MIL: Yes** **Year:** 2009 **Model:** B2300, B4000 **Engine:** 2.3L L4 VIN D, 4.0L V6 VIN E **Transmission:** All	**Idle Air Control System RPM Lower Than Expected:** Engine running at hot idle in closed loop, and the PCM detected the Actual idle speed was less than the Desired idle speed in the test. **Possible Causes:** • Air filter element is severely restricted, or air inlet is plugged • Base engine problems: cylinder compression or valve timing • IAC valve is stuck in one position (it may be dirty or has failed) • IAC solenoid or valve is damaged or has failed • Throttle body is dirty, damaged or contaminated • PCM has failed
DTC: P1525 **2T CCM, MIL: Yes** **Year:** 2009, 2010 **Model:** Tribute **Engine:** 2.5L L4 VIN 3, 2.5L L4 VIN 7, 3.0L V6 VIN 1, 3.0L V6 VIN G **Transmission:** All	**ABV Vacuum Solenoid Circuit Malfunction:** Key on or engine running and the PCM detected an unexpected voltage condition on the Air Bypass Valve Vacuum solenoid circuit during the CCM test. **Possible Causes:** • ABV vacuum solenoid control circuit open or shorted to ground • ABV vacuum solenoid control circuit is shorted to system power • ABV vacuum solenoid is damaged or has failed • PCM has failed

DTC	Trouble Code Title, Conditions & Possible Causes
DTC: P1526 **2T CCM, MIL: Yes** **Year:** 2009, 2010 **Model:** Tribute **Engine:** 2.5L L4 VIN 3, 2.5L L4 VIN 7, 3.0L V6 VIN 1, 3.0L V6 VIN G **Transmission:** All	**ABV Vent Solenoid Circuit Malfunction:** Key on or engine running and the PCM detected an unexpected voltage condition on the Air Bypass Valve Vent solenoid circuit during the CCM test. **Possible Causes:** • ABV vent solenoid control circuit open or shorted to ground • ABV vent solenoid control circuit is shorted to system power • ABV vent solenoid is damaged or has failed • PCM has failed
DTC: P1527 **2T CCM, MIL: Yes** **Year:** 2009, 2010 **Model:** Tribute **Engine:** 2.5L L4 VIN 3, 2.5L L4 VIN 7, 3.0L V6 VIN 1, 3.0L V6 VIN G **Transmission:** All	**Bypass Air Solenoid Circuit Malfunction:** Engine running, system voltage over 11.5v, and PCM detected the Bypass Air solenoid signal was less than 5.8v with the solenoid "off", or the difference in Intake Airflow was less than 2.3 g/sec before and after the Bypass Air Solenoid valve was turned "on" during the test. **Possible Causes:** • Bypass air solenoid control circuit is open or shorted to ground • Bypass air solenoid power circuit is open • Bypass air solenoid is damaged or has failed • PCM has failed
DTC: P1540 **2T CCM, MIL: Yes** **Year:** 2009, 2010 **Model:** Tribute **Engine:** 2.5L L4 VIN 3, 2.5L L4 VIN 7, 3.0L V6 VIN 1, 3.0L V6 VIN G **Transmission:** All	**Air Bypass Valve System:** Key on or engine running and the PCM detected a problem in the Air Bypass Valve system during normal operation in the CCM test. **Possible Causes:** • Engine control sensor is out-of-range (i.e., ECT, IAT or MAF) • ABV vacuum or vent solenoid is damaged or has failed • Air intake leak in the engine or related vacuum hose or pipes • Air intake restriction in a component related to the ABV system • PCM has failed
DTC: P1550 **1T CCM** **Year:** 2009 **Model:** B2300, B4000 **Engine:** 2.3L L4 VIN D, 4.0L V6 VIN E **Transmission:** All	**Power Steering Pressure (PSP) Sensor Out of Self-Test Range:** The PSP sensor input signal to the Powertrain Control Module (PCM) is continuously monitored. The test fails when the signal falls out of a maximum or minimum calibrated range. **NOTE: The DTC indicates the PSP sensor is out of self-test range.** **Possible Causes:** • Steering wheel not turned during self-test. • PSP sensor or circuit damaged • Power steering concern • PSP sensor
DTC: P1572 **2T CCM, MIL: Yes** **Year:** 2009 **Model:** B2300, B4000 **Engine:** 2.3L L4 VIN D, 4.0L V6 VIN E **Transmission:** All	**Brake Pedal Switch Circuit Malfunction:** KOER Self-Test enabled, the brake input rationality test for Brake Pedal Position (BPP) and Brake Pressure Switch (BPS) has detected a concern. One or both inputs to the Powertrain Control Module (PCM) did not change state when expected. On some vehicles with stability assist, the BPP switch is connected to the Anti lock Brake System (ABS) module and the ABS generates a driver brake application signal, which is then sent to the PCM. **NOTE: DTC P1572 sets when the PCM does not sense the correct sequence of the brake pedal input signal from both the BPP and BPS switches when the brake pedal is pressed and released.** **Possible Causes:** • Misadjusted brake switches, BPP or BPS • Blown fuse • Damaged BPP switch • Damaged BPS switch • Open or short in the BPP circuit • Open or short in the DBA circuit • Open or short in the BPS circuit • PCM has failed
DTC: P1633 **1T PCM, MIL: Yes** **Year:** 2009 **Model:** B2300, B4000 **Engine:** 2.3L L4 VIN D, 4.0L V6 VIN E **Transmission:** All	**Keep Alive Power (KAPWR) Voltage Too Low:** Key on, and the PCM detected that the KAPWR circuit has experienced a voltage interrupt. **NOTE: Loss of KAPWR to the Powertrain Control Module (PCM) results in immediate Malfunction Indicator Lamp (MIL) illumination and DTC P1633.** **Possible Causes:** • KAPWR circuit has been interrupted (this problem may be an intermittent condition) • PCM has failed

DTC	Trouble Code Title, Conditions & Possible Causes
DTC: P1633 **1T CCM, MIL: Yes** **Year:** 2009, 2010 **Model:** 6, CX-9, Tribute **Engine:** 2.5L L4 VIN 3, 2.5L L4 VIN 7, 3.0L V6 VIN 1, 3.0L V6 VIN G, 3.7L V6 VIN A, 3.7L V6 VIN B, 3.7L V6 VIN V **Transmission:** All	**PCM Long Term Memory Interruption:** Key on or engine running and the PCM detected an interruption in the Keep Alive Memory circuit during the CCM test. **Possible Causes:** • KAM circuit is open between battery and the PCM connector • KAM circuit is shorted to ground (check for a blown fuse) • KAM circuit has high resistance at the battery connections • PCM has failed
DTC: P1635 **1T CCM** **Year:** 2009, 2010 **Model:** 6, CX-9, Tribute **Engine:** 2.5L L4 VIN 3, 2.5L L4 VIN 7, 3.0L V6 VIN 1, 3.0L V6 VIN G, 3.7L V6 VIN A, 3.7L V6 VIN B, 3.7L V6 VIN V **Transmission:** All	**Generator Battery Terminal Open Circuit Malfunction:** Engine running for 1 minute, and the PCM detected the Generator Battery Terminal Charging voltage was more than 17v, while the battery voltage was less than 11.0v for 5 seconds in the CCM test. **Possible Causes:** • Generator Battery Terminal circuit is open • Generator is damaged or has failed (i.e., no Generator output) • Battery has failed (it is unable to accept a charge) • PCM has failed
DTC: P1636 **1T CCM** **Year:** 2009, 2010 **Model:** Tribute **Engine:** 2.5L L4 VIN 3, 2.5L L4 VIN 7, 3.0L V6 VIN 1, 3.0L V6 VIN G **Transmission:** All	**Generator Battery Terminal Open Circuit Malfunction:** Engine running for 1 minute, and the PCM detected the Generator Battery Terminal Charging voltage was more than 17v, while the battery voltage was less than 11.0v for 5 seconds in the CCM test. **Possible Causes:** • Generator Battery Terminal circuit is open • Generator is damaged or has failed (i.e., no Generator output) • Battery has failed (it is unable to accept a charge) • PCM has failed
DTC: P1639 **1T CCM** **Year:** 2009, 2010 **Model:** 6, CX-9, Tribute **Engine:** 2.5L L4 VIN 3, 2.5L L4 VIN 7, 3.0L V6 VIN 1, 3.0L V6 VIN G, 3.7L V6 VIN A, 3.7L V6 VIN B, 3.7L V6 VIN V **Transmission:** All	**Generator Battery Terminal Open Circuit Malfunction:** Engine running for 1 minute, and the PCM detected the Generator Battery Terminal Charging voltage was more than 17v, while the battery voltage was less than 11.0v for 5 seconds in the CCM test. **Possible Causes:** • Generator Battery Terminal circuit is open • Generator is damaged or has failed (i.e., no Generator output) • Battery has failed (it is unable to accept a charge) • PCM has failed
DTC: P1639 **1T PCM, MIL: Yes** **Year:** 2009 **Model:** B2300, B4000 **Engine:** 2.3L L4 VIN D, 4.0L V6 VIN E **Transmission:** All	**Vehicle ID (VID) Block Corrupted, Not Programmed:** Key on, and the PCM determined the Vehicle ID Block information was incorrect. **NOTE: Program the PCM to the most recent calibration available.** **Possible Causes:** • PCM may not be the correct application • PCM may need to be reprogrammed • VID configuration may not be correct
DTC: P1641 **1T CCM** **Year:** 2009, 2010 **Model:** Tribute **Engine:** 2.5L L4 VIN 3, 2.5L L4 VIN 7, 3.0L V6 VIN 1, 3.0L V6 VIN G **Transmission:** All	**Generator Battery Terminal Open Circuit Malfunction:** Engine running for 1 minute, and the PCM detected the Generator Battery Terminal Charging voltage was more than 17v, while the battery voltage was less than 11.0v for 5 seconds in the CCM test. **Possible Causes:** • Generator Battery Terminal circuit is open • Generator is damaged or has failed (i.e., no Generator output) • Battery has failed (it is unable to accept a charge) • PCM has failed
DTC: P1646 **T PCM** **Year:** 2009, 2010 **Model:** 6 **Engine:** 3.7L V6 VIN B **Transmission:** All	**A/F sensor (RH) control chip:** The PCM monitors the application-specific integrated circuit that controls and monitors the HO2S. The test fails when the PCM detects an internal circuit or communication concern. **Possible Causes:** • PCM malfunction

DTC	Trouble Code Title, Conditions & Possible Causes
DTC: P1647 **T PCM** **Year:** 2009, 2010 **Model:** 6 **Engine:** 3.7L V6 VIN B **Transmission:** All	**A/F sensor (LH) control chip :** The PCM monitors the application-specific integrated circuit that controls and monitors the HO2S. The test fails when the PCM detects an internal circuit or communication concern. **Possible Causes:** • PCM malfunction
DTC: P164A **T PCM** **Year:** 2009, 2010 **Model:** 6 **Engine:** 3.7L V6 VIN B **Transmission:** All	**A/F sensor (RH) positive current trim circuit performance:** A resistor is installed in the A/F sensor (RH) connector for part to part variance. The PCM determines the value of this resistor by taking multiple measurements of the resistor during each key on event. The PCM uses this value in order to compensate for the variance in the pumping current signal. The test fails if the PCM receives an inconsistent or erratic measurement of the resistor **Possible Causes:** • Connector or terminal malfunction • A/F sensor (RH) malfunction
DTC: P164B **T PCM** **Year:** 2009, 2010 **Model:** 6 **Engine:** 3.7L V6 VIN B **Transmission:** All	**A/F sensor (LH) positive current trim circuit performance:** A resistor is installed in the A/F sensor (LH) connector for part to part variance. The PCM determines the value of this resistor by taking multiple measurements of the resistor during each key on event. The PCM uses this value in order to compensate for the variance in the pumping current signal. The test fails if the PCM receives an inconsistent or erratic measurement of the resistor. **Possible Causes:** • Connector or terminal malfunction • A/F sensor (LH) malfunction
DTC: P1650 **1T CCM** **Year:** 2009, 2010 **Model:** 6, B2300, B4000, CX-9 **Engine:** 2.3L L4 VIN D, 3.7L V6 VIN A, 3.7L V6 VIN B, 3.7L V6 VIN V, 4.0L V6 VIN E **Transmission:** All	**Power Steering Pressure Switch Circuit Malfunction (Self-Test):** KOER Self-Test Enabled: Engine running and the PCM detected an unexpected voltage condition on the Power Steering Pressure Switch (PSPS) circuit. **NOTE: If the steering wheel is not turned during at the correct time during the Self-Test step, the PCM will set this trouble code.** **Possible Causes:** • PSP signal circuit is open or shorted to ground • PSPS ground circuit is open (at the switch mounting point) • PSPS is damaged or has failed • PCM has failed
DTC: P1651 **1T CCM** **Year:** 2009 **Model:** B2300, B4000 **Engine:** 2.3L L4 VIN D, 4.0L V6 VIN E **Transmission:** All	**Power Steering Pressure Switch Circuit Malfunction:** Engine started, then vehicle driven to a speed over 37.4 mph for 1 minute, ECT sensor signal more than 140°F, and the PCM did not detect any change in the Power Steering Pressure Switch (PSPS) signal under these conditions during the CCM Rationality test. **Possible Causes:** • PSP signal circuit is open or shorted to ground • PSPS ground circuit is open (at the switch mounting point) • PSPS is damaged or has failed • PCM has failed
DTC: P1674 **1T CCM, MIL: Yes** **Year:** 2009, 2010 **Model:** 6, CX-9 **Engine:** 2.5L L4 VIN A, 3.7L V6 VIN A, 3.7L V6 VIN B, 3.7L V6 VIN V **Transmission:** All	**Control module software corrupted:** Indicates that an error occurred in the PCM. This DTC is set in combination with P2105:00. **Possible Causes:** • PCM malfunction • Software incompatibility issue
DTC: P1676 **T PCM** **Year:** 2009, 2010 **Model:** 6 **Engine:** 3.7L V6 VIN B **Transmission:** All	**Pedal position out of self test range:** During KOEO self-test, the PCM monitors the APP sensor inputs to determine if the APP1 and APP2 signals are less than an expected value. If either APP1 or APP2 is greater than the expected value, the DTC is set. **Possible Causes:** • Accelerator pedal applied during KOEO self-test

DTC	Trouble Code Title, Conditions & Possible Causes
DTC: P1680 **T CCM** **Year:** 2009, 2010 **Model:** RX-8 **Engine:** 1.3L R2 VIN 2, 1.3L R2 VIN 4, 1.3L R2 VIN M, 1.3L R2 VIN P **Transmission:** All	**OCV circuit low input:** The PCM monitors input voltage from the OCV. If the voltage of the OCV input terminal is less than the specification for 1 s when the battery voltage is more than 10 V, the PCM determines the OCV circuit voltage is low. **Possible Causes:** • OCV connector or terminals malfunction • Short to ground or open circuit in OCV power supply circuit • Short to ground in wiring harness between main relay and OCV • OCV related fuse malfunction • Open circuit in wiring harness between main relay and OCV • Short to ground in wiring harness between OCV and PCM • PCM connector or terminals malfunction • Open circuit in wiring harness between OCV and PCM • OCV malfunction • PCM malfunction
DTC: P1681 **T CCM** **Year:** 2009, 2010 **Model:** RX-8 **Engine:** 1.3L R2 VIN 2, 1.3L R2 VIN 4, 1.3L R2 VIN M, 1.3L R2 VIN P **Transmission:** All	**OCV circuit high input:** The PCM monitors input voltage from the OCV. If the OCV current is more than 3.5 A for 2 s when the battery voltage is more than 10 V, the PCM determines the OCV circuit has a malfunction. **Possible Causes:** • OCV connector or terminals malfunction • PCM connector or terminals malfunction • Short to power supply in wiring harness between OCV and PCM • OCV malfunction • PCM malfunction
DTC: P1682 **T CCM** **Year:** 2009, 2010 **Model:** RX-8 **Engine:** 1.3L R2 VIN 2, 1.3L R2 VIN 4, 1.3L R2 VIN M, 1.3L R2 VIN P **Transmission:** All	**Metering oil pump No.1 circuit low input:** The PCM monitors the input voltage from the metering oil pump No.1 when the battery voltage is more than 8 V and the metering oil pump No.1 control signal turned from ON to OFF. If the input voltage is less than the specification, the PCM determines that the metering oil pump No.1 circuit has a malfunction. **Possible Causes:** • Metering oil pump No.1 connector or terminals malfunction • Metering oil pump driver connector or terminals malfunction • Short to ground or open circuit in metering oil pump No.1 power supply circuit • Short to ground in wiring harness between main relay and metering oil pump No.1 • Metering oil pump No.1 related fuse malfunction • Open circuit in wiring harness between main relay and metering oil pump No.1 • Short to ground or open circuit in metering oil pump driver power supply circuit • Short to ground in wiring harness between main relay and metering oil pump driver • Metering oil pump driver related fuse malfunction • Open circuit in wiring harness between main relay and metering oil pump driver • PCM connector or terminals malfunction • Short to power supply in wiring harness between metering oil pump driver and PCM • Short to ground in wiring harness between metering oil pump No.1 and PCM • Short to ground in wiring harness between metering oil pump driver and PCM • Metering oil pump No.1 malfunction • Metering oil pump driver malfunction • Open circuit in wiring harness between metering oil pump No.1 and PCM • Open circuit in wiring harness between metering oil pump driver and PCM • PCM malfunction

DTC	Trouble Code Title, Conditions & Possible Causes
DTC: P1683 **T CCM** **Year:** 2009, 2010 **Model:** RX-8 **Engine:** 1.3L R2 VIN 2, 1.3L R2 VIN 4, 1.3L R2 VIN M, 1.3L R2 VIN P **Transmission:** All	**Metering oil pump No.1 circuit high input:** The PCM monitors the input voltage from the metering oil pump No.1 when the battery voltage Is more than 8 V and the metering oil pump No.1 control signal turned from ON to OFF. If the input voltage is more than the specification, the PCM determines that the metering oil pump No.1 circuit has a malfunction. **Possible Causes:** • Metering oil pump No.1 connector or terminals malfunction • Metering oil pump driver connector or terminals malfunction • PCM connector or terminals malfunction • Short to ground in wiring harness between metering oil pump driver and PCM • Short to power supply in wiring harness between metering oil pump No.1 and PCM • Short to power supply in wiring harness between metering oil pump driver and PCM • Metering oil pump No.1 malfunction • Metering oil pump driver malfunction • Open circuit in wiring harness between metering oil pump No.1 and metering oil pump driver • Open circuit in wiring harness between metering oil pump driver and PCM • PCM malfunction
DTC: P1684 **T CCM** **Year:** 2009, 2010 **Model:** RX-8 **Engine:** 1.3L R2 VIN 2, 1.3L R2 VIN 4, 1.3L R2 VIN M, 1.3L R2 VIN P **Transmission:** All	**Metering oil pump oil pressure sensor—oil pressure is low:** It is that the oil pressure at the metering oil pump system is less than 40 kPa {0.41 kgf/cm2, 5.8 psi} continues for 10 s, after specified period passes after the engine starts. **Possible Causes:** • Low engine oil level • Engine oil leakage (oil hose and/or drain hose looseness) • Oil pressure sensor malfunction (abnormal characteristic) • Oil pump malfunction (oil pressure regulator stuck open) • Engine oil leakage between OCV and metering oil pump drain line (quick connector poor connection or comes off) • Looseness of oil pressure sensor • Looseness of OCV • OCV malfunction (stuck close) • PCM malfunction
DTC: P1685 **T CCM** **Year:** 2009, 2010 **Model:** RX-8 **Engine:** 1.3L R2 VIN 2, 1.3L R2 VIN 4, 1.3L R2 VIN M, 1.3L R2 VIN P **Transmission:** All	**Metering oil pump oil pressure sensor—oil pressure is high:** It is that the oil pressure at the metering oil pump system is more than 180 kPa {1.84 kgf/cm2, 26.1 psi} continues for 10 s, after specified period passes after the engine starts. **Possible Causes:** • Oil pressure sensor malfunction (abnormal characteristic) • Oil pump malfunction (oil pressure regulator stuck close) • Metering oil pump drain passage clogged • OCV malfunction (stuck open) • PCM malfunction
DTC: P1686 **2T CCM** **Year:** 2009, 2010 **Model:** RX-8 **Engine:** 1.3L R2 VIN 2, 1.3L R2 VIN 4, 1.3L R2 VIN M, 1.3L R2 VIN P **Transmission:** All	**Metering oil pump control circuit low flow side problem:** The PCM monitors the input signal from the metering oil pump switch when the metering oil pump stepping motor is more than the standard step. If the input signal is off, the PCM determines that the metering oil pump control circuit has a problem on the low flow side. **Possible Causes:** • Metering oil pump malfunction • Metering oil pump switch malfunction • Connector or terminal malfunction, short or open circuit • PCM has failed
DTC: P1687 **2T CCM** **Year:** 2009, 2010 **Model:** RX-8 **Engine:** 1.3L R2 VIN 2, 1.3L R2 VIN 4, 1.3L R2 VIN M, 1.3L R2 VIN P **Transmission:** All	**Metering oil pump control circuit high flow side problem:** The PCM monitors the input signal from the metering oil pump switch when the metering oil pump stepping motor is less than the standard step. If the input signal is on, the PCM determines that the metering oil pump control circuit has a problem on the high flow side. **Possible Causes:** • Metering oil pump malfunction • Metering oil pump switch malfunction • Connector or terminal malfunction, short or open circuit • PCM has failed

DTC	Trouble Code Title, Conditions & Possible Causes
DTC: P1702 **1T CCM, MIL: Yes** **Year:** 2009, 2010 **Model:** B2300, B4000, Tribute **Engine:** 2.3L L4 VIN D, 2.5L L4 VIN 7, 3.0L V6 VIN 1, 3.0L V6 VIN G, 4.0L V6 VIN E **Transmission:** All	**Transmission Range Sensor Circuit Malfunction:** Engine started, vehicle driven to a speed over 31 mph, and the PCM detected an interruption in the Transmission Range (TR) sensor signal during the CCM test (DTC P0707 or P0708 may also be set). **Possible Causes:** • TR sensor signal circuit is open (an intermittent fault) • TR sensor signal circuit is shorted to ground (intermittent fault) • TR sensor signal circuit is shorted to VREF (intermittent fault) • TR sensor is damaged or has failed (an intermittent fault) • PCM has failed
DTC: P1703 **1T CCM** **Year:** 2009, 2010 **Model:** 6, B2300, B4000, CX-9, Tribute **Engine:** 2.3L L4 VIN D, 2.5L L4 VIN 3, 2.5L L4 VIN 7, 3.0L V6 VIN 1, 3.0L V6 VIN G, 3.7L V6 VIN A, 3.7L V6 VIN B, 3.7L V6 VIN V, 4.0L V6 VIN E **Transmission:** All	**Brake Switch Circuit Malfunction (Self-Test):** KOEO Self-Test Enabled: Key on, and the PCM detected an unexpected "high" voltage condition on the Brake Switch circuit during the CCM test. KOER Self-Test Enabled: Engine running and the PCM detected an unexpected voltage "low" or "high" voltage condition on the Brake Switch circuit in the test. **Possible Causes:** • Brake not depressed at the proper time during KOEO Self-Test. • Brake was depressed at all times during the KOEO Self-Test. • Brake was not cycled "on" and "off" during the KOER Self-Test. • Brake switch signal circuit shorted to system power (B+) • Brake switch is damaged or has failed, or is out of adjustment • PCM has failed
DTC: P1704 **1T CCM** **Year:** 2009 **Model:** B2300, B4000 **Engine:** 2.3L L4 VIN D, 4.0L V6 VIN E **Transmission:** All	**Transmission Range Sensor Circuit Malfunction (Self-Test):** KOEO Self-Test Enabled: Key on, and the PCM detected an invalid TR switch signal occurred. KOER Self-Test Enabled: Engine running, and the PCM detected an invalid TR switch signal occurred during the CCM test. **Possible Causes:** • TR switch signal circuit is open or shorted to ground • TR switch is damaged, has failed or is out of adjustment • Transmission shift cable is damaged or out-of-adjustment • PCM had failed
DTC: P1705 **1T CCM** **Year:** 2009, 2010 **Model:** 6, B2300, B4000, Tribute **Engine:** 2.3L L4 VIN D, 2.5L L4 VIN 7, 3.0L V6 VIN 1, 3.0L V6 VIN G, 3.7L V6 VIN B, 4.0L V6 VIN E **Transmission:** All	**Digital Transmission Range Sensor Circuit Malfunction (Self-Test):** KOEO Self-Test Enabled: Key on, and the PCM detected a TR switch signal that indicated the gear selector was not in Park during the CCM test. **Possible Causes:** • The gear selector was not In Park during the Self-Test • TR switch is damaged, has failed or is out of adjustment • Starter or starter relay is damaged or has failed • Backup lamp circuit is open or shorted • PCM had failed
DTC: P1709 **2T CCM** **Year:** 2009 **Model:** B2300, B4000 **Engine:** 2.3L L4 VIN D, 4.0L V6 VIN E **Transmission:** All	**Clutch Pedal Position Switch Circuit Malfunction (Self-Test):** KOEO Self-Test Enabled: Key on, clutch engaged (in), and the PCM detected the CPP Switch input was high (switch open) when it should have been low (switch closed) during the CCM test. **Possible Causes:** • CPP switch circuit is open between the switch and the PCM • CPP switch circuit is shorted between the switch and the PCM • TR switch is damaged or has failed • PCM has failed
DTC: P1711 **1T CCM** **Year:** 2009, 2010 **Model:** B2300, B4000, Tribute **Engine:** 2.3L L4 VIN D, 2.5L L4 VIN 7, 3.0L V6 VIN 1, 3.0L V6 VIN G, 4.0L V6 VIN E **Transmission:** All	**Transaxle Fluid Temperature Sensor Circuit Malfunction (Self-Test):** KOEO Self-Test Enabled: Key on, and the PCM detected the TFT sensor signal was less than the allowable temperature during the CCM test. KOER Self-Test Enabled: Engine running and the PCM detected the TFT sensor signal was less than the allowable temperature during the CCM test. **Possible Causes:** • ATF level is too low • ATF temperature not at operating temperature during the test • TFT sensor is damaged or has failed • PCM has failed

DTC	Trouble Code Title, Conditions & Possible Causes
DTC: P1713 **1T CCM** **Year:** 2009 **Model:** B2300, B4000 **Engine:** 2.3L L4 VIN D, 4.0L V6 VIN E **Transmission:** All	**Transaxle Fluid Temperature Sensor Circuit Malfunction:** Engine running and the PCM did not detect any change in the TFT sensor signal during in-range low transaxle operation during the test. **Possible Causes:** • ATF level is low, or the fluid is contaminated • TFT sensor is damaged (it may be out-of-calibration) • TFT sensor signal circuit shorted to ground (intermittent fault) • PCM has failed
DTC: P1714 **2T CCM** **Year:** 2009 **Model:** B2300, B4000 **Engine:** 2.3L L4 VIN D, 4.0L V6 VIN E **Transmission:** All	**Transmission Shift Solenoid 'A' Mechanical Malfunction:** DTC P1715, P1716, P1717 and P1743 not set, engine running at cruise speed, and the PCM detected a mechanical fault related to the operation of the Shift Solenoid 'A' (SSA) in the CCM test. **Possible Causes:** • ATF level is low, or the fluid is contaminated • SSA has failed (mechanically) or is damaged • PCM has failed
DTC: P1715 **2T CCM, MIL: Yes** **Year:** 2009 **Model:** B2300, B4000 **Engine:** 2.3L L4 VIN D, 4.0L V6 VIN E **Transmission:** All	**Transmission Shift Solenoid 'B' Mechanical Malfunction:** DTC P1714, P1716, P1717 and P1743 not set, engine running at cruise speed, and the PCM detected a mechanical fault related to the operation of the Shift Solenoid 'B' (SSB) in the CCM test. **Possible Causes:** • ATF level is low, or the fluid is contaminated • SSB has failed (mechanically) or is damaged • PCM has failed
DTC: P1716 **2T CCM** **Year:** 2009 **Model:** B2300, B4000 **Engine:** 2.3L L4 VIN D, 4.0L V6 VIN E **Transmission:** All	**Transmission Shift Solenoid 'C' Mechanical Malfunction:** DTC P1714, P1715, P1717 and P1743 not set, engine running at cruise speed, and the PCM detected a mechanical fault related to the operation of the Shift Solenoid 'C' (SSB) in the CCM test. **Possible Causes:** • ATF level is low, or the fluid is contaminated • SSC has failed (mechanically) or is damaged • PCM has failed
DTC: P1717 **2T CCM** **Year:** 2009 **Model:** B2300, B4000 **Engine:** 2.3L L4 VIN D, 4.0L V6 VIN E **Transmission:** All	**Transmission Shift Solenoid 'D' Mechanical Malfunction:** DTC P1714, P1715, P1716 and P1743 not set, engine running at cruise speed, and the PCM detected a mechanical fault related to the operation of the Shift Solenoid 'D' (SSD) in the CCM test. **Possible Causes:** • ATF level is low, or the fluid is contaminated • SSD has failed (mechanically) or is damaged • PCM has failed
DTC: P1718 **1T CCM** **Year:** 2009 **Model:** B2300, B4000 **Engine:** 2.3L L4 VIN D, 4.0L V6 VIN E **Transmission:** All	**Transaxle Fluid Temperature Sensor Circuit Malfunction:** Engine running and the PCM did not detect any change in the TFT sensor signal during in-range high transaxle operation during the CCM Rationality test. **Possible Causes:** • ATF level is low • TFT sensor is damaged (it may be out-of-calibration) • TFT sensor signal circuit shorted to power (intermittent fault) • PCM has failed
DTC: P1740 **1T CCM, MIL: Yes** **Year:** 2009 **Model:** B2300, B4000 **Engine:** 2.3L L4 VIN D, 4.0L V6 VIN E **Transmission:** All	**Torque Converter Clutch Solenoid Malfunction (Mechanical):** Engine running at cruise speed, and the PCM detected a symptom that indicated the A/T Shift Solenoid had a mechanical problem. **Possible Causes:** • ATF level is low, or the fluid is contaminated • A/T shift solenoid is damaged (mechanical fault) or has failed • PCM has failed
DTC: P1741 **1T CCM, MIL: Yes** **Year:** 2009 **Model:** B2300, B4000 **Engine:** 2.3L L4 VIN D, 4.0L V6 VIN E **Transmission:** All	**Torque Converter Clutch Control Circuit Malfunction:** Key on or engine running and the PCM detected an unexpected voltage condition on the TCC solenoid control circuit during the test. **Possible Causes:** • TCC solenoid control circuit is shorted to system power (B+) • TCC solenoid control circuit is open • TCC solenoid is damaged or has failed • TCM has failed

DTC	Trouble Code Title, Conditions & Possible Causes
DTC: P1742 **1T CCM** **Year:** 2009 **Model:** B2300, B4000 **Engine:** 2.3L L4 VIN D, 4.0L V6 VIN E **Transmission:** All	**Torque Converter Clutch Solenoid Circuit Malfunction:** DTP P0443 not set, engine running at cruise speed at light engine load for 2-3 minutes, and the PCM detected the a mechanical (low-slip) in the TCC system during the CCM Rationality test. **NOTE: This trouble code sets due to a mechanical problem in unit.** **Possible Causes:** • ATF is low, or the fluid is burnt, contaminated or dirty • TCC control valve is stuck or not working properly • An internal transaxle component is damaged or has failed • PCM has failed
DTC: P1743 **1T CCM** **Year:** 2009 **Model:** B2300, B4000 **Engine:** 2.3L L4 VIN D, 4.0L V6 VIN E **Transmission:** All	**Torque Converter Clutch Solenoid Circuit Malfunction:** Key on or engine running and the PCM detected an unexpected low voltage condition on the TCC solenoid control circuit. **NOTE: The O/D Lamp is "on" when this trouble code is set.** **Possible Causes:** • TCC solenoid control circuit has failed in the "on" position • TCC solenoid control circuit is shorted to ground • TCC solenoid is damaged or has failed • PCM is damaged
DTC: P1744 **2T CCM, MIL: Yes** **Year:** 2009, 2010 **Model:** B2300, B4000, Tribute **Engine:** 2.3L L4 VIN D, 2.5L L4 VIN 7, 3.0L V6 VIN 1, 3.0L V6 VIN G, 4.0L V6 VIN E **Transmission:** All	**Torque Converter Clutch Mechanical Malfunction:** DTP P0443 not set, engine running at cruise speed at light engine load for 2-3 minutes, and the PCM detected the a mechanical (high-slippage) in the TCC system during the CCM Rationality test. **NOTE: This trouble code sets due to a mechanical problem in unit.** **Possible Causes:** • Brake switch is damaged, out of adjustment or has failed • Engine or transmission temperature is too high • TCC control valve is stuck in the "off" position • An internal transaxle component is damaged or has failed • PCM has failed
DTC: P1745 **2T CCM, MIL: Yes** **Year:** 2009 **Model:** B2300, B4000 **Engine:** 2.3L L4 VIN D, 4.0L V6 VIN E **Transmission:** All	**Electronic Pressure Control Solenoid Circuit Malfunction:** Key on or engine running and the PCM detected an unexpected high voltage condition on the EPC solenoid circuit in the CCM test. **Possible Causes:** • EPC solenoid control circuit is open • EPC solenoid control circuit is shorted to system power (B+) • EPC solenoid is damaged or has failed • PCM has failed
DTC: P1746 **2T CCM** **Year:** 2009 **Model:** B2300, B4000 **Engine:** 2.3L L4 VIN D, 4.0L V6 VIN E **Transmission:** All	**Electronic Pressure Control Solenoid Circuit Malfunction:** Key on or engine running and the PCM detected an unexpected high voltage condition on the EPC solenoid circuit in the CCM test. **NOTE: The EPC is this application is a Variable Force Style solenoid that combines an electro-hydraulic actuator and a regulating valve.** **Possible Causes:** • EPC solenoid control circuit is open • EPC solenoid control circuit is shorted to system power (B+) • EPC solenoid is damaged or has failed • PCM has failed
DTC: P1747 **2T CCM** **Year:** 2009 **Model:** B2300, B4000 **Engine:** 2.3L L4 VIN D, 4.0L V6 VIN E **Transmission:** All	**Electronic Pressure Control Solenoid Circuit Malfunction:** Key on or engine running and the PCM detected an unexpected "low" voltage condition on the EPC solenoid circuit in the CCM test. **Possible Causes:** • EPC solenoid control circuit open between solenoid and PCM • EPC solenoid control circuit is shorted to ground • EPC solenoid is damaged or has failed • PCM has failed
DTC: P1748 **2T CCM, MIL: Yes** **Year:** 2009 **Model:** B2300, B4000 **Engine:** 2.3L L4 VIN D, 4.0L V6 VIN E **Transmission:** All	**Electronic Pressure Control Solenoid Circuit Malfunction:** Key on or engine running and the PCM detected an unexpected "low" voltage condition on the EPC solenoid circuit in the CCM test. **NOTE: The EPC is this application is a Variable Force Style solenoid that combines an electro-hydraulic actuator and a regulating valve.** **Possible Causes:** • EPC solenoid control circuit is shorted to ground • EPC solenoid is damaged or has failed • PCM has failed

DTC	Trouble Code Title, Conditions & Possible Causes
DTC: P1752 **2T CCM, MIL: Yes** **Year:** 2009 **Model:** B2300, B4000 **Engine:** 2.3L L4 VIN D, 4.0L V6 VIN E **Transmission:** All	**Transmission Shift Solenoid 'A' Circuit Malfunction:** Key on or engine running, and the PCM detected an unexpected voltage "low" voltage condition (near 0.0v) on the Shift Solenoid 'A' (SSA) during the CCM test. **Possible Causes:** • SSA solenoid control circuit is shorted to ground • SSA solenoid is damaged or has failed • TCM has failed
DTC: P1756 **2T CCM, MIL: Yes** **Year:** 2009 **Model:** B2300, B4000 **Engine:** 2.3L L4 VIN D, 4.0L V6 VIN E **Transmission:** All	**Transmission Shift Solenoid 'B' Mechanical Malfunction:** Engine started, then driven to a speed of 30-35 mph, and the PCM detected a mechanical shift problem related to the Shift Solenoid 'B' (SSB) during the CCM Rationality test. **Possible Causes:** • Problems in the internal transmission components • SSB is damaged or has failed • PCM has failed
DTC: P1757 **2T CCM, MIL: Yes** **Year:** 2009 **Model:** B2300, B4000 **Engine:** 2.3L L4 VIN D, 4.0L V6 VIN E **Transmission:** All	**Transmission Shift Solenoid 'B' Circuit Malfunction:** Key on or engine running and the PCM detected an unexpected voltage "high" voltage condition (near 0.0v) on the Shift Solenoid 'B' (SSB) during the CCM test. **Possible Causes:** • SSA solenoid control circuit is shorted to ground • SSA solenoid is damaged or has failed • TCM has failed
DTC: P1780 **2T CCM** **Year:** 2009, 2010 **Model:** B2300, B4000, Tribute **Engine:** 2.3L L4 VIN D, 2.5L L4 VIN 7, 3.0L V6 VIN 1, 3.0L V6 VIN G, 4.0L V6 VIN E **Transmission:** All	**Transmission Control Switch Circuit Malfunction (Self-Test:** KOER Self-Test Enabled: Engine running and the PCM detected the Transmission Control switch signal did not change its status not cycled in the CCM test. **Possible Causes:** • TCS switch not cycled at the proper time in the KOER Self-Test • TCS switch signal circuit is open or shorted to ground • TCS switch signal circuit is shorted to VREF or system power • TCS switch is damaged, out-of-adjustment or has failed • PCM has failed
DTC: P1783 **1T CCM** **Year:** 2009, 2010 **Model:** B2300, B4000, Tribute **Engine:** 2.3L L4 VIN D, 2.5L L4 VIN 7, 3.0L V6 VIN 1, 3.0L V6 VIN G, 4.0L V6 VIN E **Transmission:** All	**Transmission Fluid Temperature Sensor Circuit High Input:** Key on or engine running and the PCM detected the Transmission Fluid Temperature (TFT) sensor signal exceeded 260°F in the test. **Possible Causes:** • ATF level is low, or the fluid is contaminated or burnt • Vehicle driven with excessive loads, or trailer towing conditions • TCC clutch is damaged or has failed (it may be slipping) • TFT sensor signal circuit is open (an intermittent fault) • Transaxle cooling system is damaged or has failed • PCM has failed
DTC: P1900 **2T CCM** **Year:** 2009 **Model:** B2300, B4000 **Engine:** 2.3L L4 VIN D, 4.0L V6 VIN E **Transmission:** All	**Output Speed Sensor Circuit Malfunction:** Engine started, vehicle driven at cruise speed, and the PCM detected an interruption (or drop out) of the Output Speed Sensor (OSS) signal during the CCM Rationality test. **Possible Causes:** • OSS signal circuit is open (an intermittent fault) • OSS signal circuit is shorted to ground (an intermittent fault) • OSS signal circuit is shorted to system power (intermittent fault) • OSS is damaged or has failed (an intermittent fault) • PCM has failed

OBD II Trouble Code List (P2XXX Codes)

DTC	Trouble Code Title, Conditions & Possible Causes
DTC: P2004 **2T CCM, MIL: Yes** **Year:** 2009, 2010 **Model:** 3, 5, 6, CX-7, MX-5 Miata, RX-8 **Engine:** 1.3L R2 VIN 2, 1.3L R2 VIN 4, 1.3L R2 VIN M, 1.3L R2 VIN P, 2.0L L4 VIN F, 2.0L L4 VIN G, 2.3L L4 VIN 3, 2.3L L4 VIN 4, 2.3L L4 VIN L, 2.3L L4 VIN M, 2.5L L4 VIN 5, 2.5L L4 VIN 6, 2.5L L4 VIN A, 2.5L L4 VIN M **Transmission:** All	**Variable Tumble Control System (VTCS) shutter valve stuck open:** PCM monitors mass VTCS shutter valve position using VTCS position sensor. If PCM turns VTCS solenoid valve ON but the VTCS position still remains open (VTCS shutter valve switch output: approx. 5.0 V), PCM determines VTCS shutter valve has been stuck open. **Possible Causes:** • VTCS shutter valve malfunction (stuck open) • Misconnecting or pull out the vacuum hose • Variable tumble control valve malfunction • PCM has failed
DTC: P2005 **2T CCM, MIL: Yes** **Year:** 2009, 2010 **Model:** RX-8 **Engine:** 1.3L R2 VIN 2, 1.3L R2 VIN 4, 1.3L R2 VIN M, 1.3L R2 VIN P **Transmission:** All	**APV stuck open (No.2):** The PCM monitors the input voltage from the APV position sensor No.2 when the APV is closed. If the input voltage is more than 1.0 V, the PCM determines that the APV is stuck open. **Possible Causes:** • APV control malfunction • APV motor malfunction • APV position sensor No.2 connector or terminals malfunction • APV position sensor No.2 malfunction • PCM connector or terminals malfunction • Open circuit in wiring harness between APV position sensor No.2 and PCM • PCM malfunction
DTC: P2006 **2T CCM, MIL: Yes** **Year:** 2009, 2010 **Model:** 3, 5, 6, CX-7, MX-5 Miata, RX-8 **Engine:** 1.3L R2 VIN 2, 1.3L R2 VIN 4, 1.3L R2 VIN M, 1.3L R2 VIN P, 2.0L L4 VIN F, 2.0L L4 VIN G, 2.3L L4 VIN 3, 2.3L L4 VIN 4, 2.3L L4 VIN L, 2.3L L4 VIN M, 2.5L L4 VIN 5, 2.5L L4 VIN 6, 2.5L L4 VIN A, 2.5L L4 VIN M **Transmission:** All	**Variable Tumble Control System (VTCS) shutter valve stuck closed:** PCM monitors mass VTCS shutter valve position using VTCS position sensor. If PCM turns VTCS solenoid valve ON but the VTCS position still remains closed (VTCS shutter valve switch output: approx. 5.0 V), PCM determines VTCS shutter valve has been stuck closed. **Possible Causes:** • VTCS shutter valve malfunction (stuck closed) • Misconnecting or pull out the vacuum hose • Variable tumble control valve malfunction • PCM has failed
DTC: P2007 **2T CCM, MIL: Yes** **Year:** 2009, 2010 **Model:** RX-8 **Engine:** 1.3L R2 VIN 2, 1.3L R2 VIN 4, 1.3L R2 VIN M, 1.3L R2 VIN P **Transmission:** All	**APV stuck closed (No.2):** The PCM monitors the input voltage from the APV position sensor No.2 when the APV is opened. If the input voltage is less than 1.0 V, the PCM determines that the APV is stuck closed. **Possible Causes:** • APV control malfunction • APV motor malfunction • APV position sensor No.2 connector or terminals malfunction • APV position sensor No.2 malfunction • Short to ground in wiring harness between APV position sensor No.2 and PCM • PCM connector or terminals malfunction • PCM malfunction

DTC	Trouble Code Title, Conditions & Possible Causes
DTC: P2009 **2T CCM, MIL: Yes** **Year:** 2009, 2010 **Model:** 3, 5, 6, CX-7, MX-5 Miata, RX-8 **Engine:** 1.3L R2 VIN 2, 1.3L R2 VIN 4, 1.3L R2 VIN M, 1.3L R2 VIN P, 2.0L L4 VIN F, 2.0L L4 VIN G, 2.3L L4 VIN 3, 2.3L L4 VIN 4, 2.3L L4 VIN L, 2.3L L4 VIN M, 2.5L L4 VIN 5, 2.5L L4 VIN 6, 2.5L L4 VIN A, 2.5L L4 VIN M **Transmission:** All	**Variable Tumble Control Solenoid Valve circuit input low:** PCM monitors Variable Tumble Control Solenoid Valve signal at PCM terminal 4T. If PCM turns variable tumble control solenoid valve OFF but voltage at PCM terminal 4T remains low, PCM determines VTCS solenoid valve circuit has a malfunction. **Possible Causes:** • Poor connection of connectors for PCM and/or VTCS • Open circuit or short between VTCS terminals and PCM terminals • Variable tumble control valve malfunction • PCM has failed
DTC: P2010 **2T CCM, MIL: Yes** **Year:** 2009, 2010 **Model:** 3, 5, 6, CX-7, MX-5 Miata, RX-8 **Engine:** 1.3L R2 VIN 2, 1.3L R2 VIN 4, 1.3L R2 VIN M, 1.3L R2 VIN P, 2.0L L4 VIN F, 2.0L L4 VIN G, 2.3L L4 VIN 3, 2.3L L4 VIN 4, 2.3L L4 VIN L, 2.3L L4 VIN M, 2.5L L4 VIN 5, 2.5L L4 VIN 6, 2.5L L4 VIN A, 2.5L L4 VIN M **Transmission:** All	**Variable Tumble Control Solenoid Valve circuit input high:** PCM monitors Variable Tumble Control Solenoid Valve signal at PCM terminal 4T. If PCM turns variable tumble control solenoid valve OFF but voltage at PCM terminal 4T remains high, PCM determines VTCS solenoid valve circuit has a malfunction. **Possible Causes:** • Poor connection of connectors for PCM and/or VTCS • Open circuit or short between VTCS terminals and PCM terminals • Variable tumble control valve malfunction • PCM has failed
DTC: P2016 **2T CCM, MIL: Yes** **Year:** 2009, 2010 **Model:** 3, 5, 6, CX-7 **Engine:** 2.0L L4 VIN F, 2.0L L4 VIN G, 2.3L L4 VIN 3, 2.3L L4 VIN L, 2.5L L4 VIN 5, 2.5L L4 VIN 6, 2.5L L4 VIN A, 2.5L L4 VIN M **Transmission:** All	**AVP position sensor circuit low input:** The PCM monitors the input voltage from the APV position sensor when the engine is running. If the input voltage is less than 0.2 V, the PCM determines that the APV position sensor circuit input voltage is low. **Possible Causes:** • AVP position sensor malfunction • Open circuit or short between AVP terminals and PCM terminals • PCM has failed
DTC: P2017 **2T CCM, MIL: Yes** **Year:** 2009, 2010 **Model:** 3, 5, 6, CX-7 **Engine:** 2.0L L4 VIN F, 2.0L L4 VIN G, 2.3L L4 VIN 3, 2.3L L4 VIN L, 2.5L L4 VIN 5, 2.5L L4 VIN 6, 2.5L L4 VIN A, 2.5L L4 VIN M **Transmission:** All	**AVP position sensor circuit high input:** The PCM monitors the input voltage from the APV position sensor when the engine is running. If the input voltage is greater than 4.2 V, the PCM determines that the APV position sensor circuit input voltage is high. **Possible Causes:** • AVP position sensor malfunction • Open circuit or short between AVP terminals and PCM terminals • PCM has failed
DTC: P2067 **2T CCM, MIL: Yes** **Year:** 2009, 2010 **Model:** B2300, B4000, RX-8 **Engine:** 1.3L R2 VIN 2, 1.3L R2 VIN 4, 1.3L R2 VIN M, 1.3L R2 VIN P, 2.3L L4 VIN D, 4.0L V6 VIN E **Transmission:** All	**Fuel gauge sender unit (sub) circuit low input:** The PCM monitors the fuel tank level and input voltage from the fuel gauge sender unit (sub) when the engine is running. If the input voltage is less than 0.78 V and fuel tank level is full, the PCM determines that the fuel gauge sender unit (sub) circuit input voltage is low. **Possible Causes:** • Fuel gauge sender unit (sub) malfunction • Instrument cluster malfunction • PCM has failed
DTC: P2068 **2T CCM, MIL: Yes** **Year:** 2009, 2010 **Model:** B2300, B4000, RX-8 **Engine:** 1.3L R2 VIN 2, 1.3L R2 VIN 4, 1.3L R2 VIN M, 1.3L R2 VIN P, 2.3L L4 VIN D, 4.0L V6 VIN E **Transmission:** All	**Fuel gauge sender unit (sub) circuit high input:** The PCM monitors the fuel tank level and input voltage from the fuel gauge sender unit (sub) when the engine is running. If the input voltage is greater than 4.9 V and fuel tank level is full, the PCM determines that the fuel gauge sender unit (sub) circuit input voltage is low. **Possible Causes:** • Fuel gauge sender unit (sub) malfunction • Instrument cluster malfunction • PCM has failed

DTC	Trouble Code Title, Conditions & Possible Causes
DTC: P2070 **2T CCM, MIL: Yes** **Year:** 2009, 2010 **Model:** RX-8, Tribute **Engine:** 1.3L R2 VIN 2, 1.3L R2 VIN 4, 1.3L R2 VIN M, 1.3L R2 VIN P, 2.5L L4 VIN 3, 2.5L L4 VIN 7, 3.0L V6 VIN 1, 3.0L V6 VIN G **Transmission:** All	**SSV malfunction:** The PCM monitors the input signal from the SSV switch when the PCM turns the SSV solenoid valve off. If the input signal is on, the PCM determines that the SSV is stuck open. **Possible Causes:** • SSV stuck open • SSV control malfunction • SSV actuator malfunction • PCM has failed
DTC: P2071 **2T CCM, MIL: Yes** **Year:** 2009, 2010 **Model:** RX-8, Tribute **Engine:** 1.3L R2 VIN 2, 1.3L R2 VIN 4, 1.3L R2 VIN M, 1.3L R2 VIN P, 2.5L L4 VIN 3, 2.5L L4 VIN 7, 3.0L V6 VIN 1, 3.0L V6 VIN G **Transmission:** All	**SSV malfunction:** The PCM monitors the input signal from the SSV switch when the PCM turns the SSV solenoid valve off. If the input signal is on, the PCM determines that the SSV is stuck open. **Possible Causes:** • SSV stuck open • SSV control malfunction • SSV actuator malfunction • PCM has failed
DTC: P2072 **T PCM** **Year:** 2009, 2010 **Model:** 6, B2300, B4000 **Engine:** 2.3L L4 VIN D, 3.7L V6 VIN B, 4.0L V6 VIN E **Transmission:** All	**Throttle actuator control system-ice blockage:** This DTC only identifies that the strategy has carried out several open and close cycles to remove potential ice buildup. This DTC does not imply any system concerns, only that the mode has occurred, and that mode may be causing a long start time. **Possible Causes:** • Ice or oil in the intake-air system could be the result of a PCV system concern
DTC: P2088 **2T CCM, MIL: Yes** **Year:** 2009, 2010 **Model:** 3, 5, 6, CX-7, MX-5 Miata **Engine:** 2.0L L4 VIN F, 2.0L L4 VIN G, 2.3L L4 VIN 3, 2.3L L4 VIN 4, 2.3L L4 VIN L, 2.3L L4 VIN M, 2.5L L4 VIN 5, 2.5L L4 VIN 6, 2.5L L4 VIN A, 2.5L L4 VIN M **Transmission:** All	**CMP actuator circuit low:** PCM monitors OCV voltage. If PCM detects OCV voltage (calculated from OCV) is below the threshold voltage (calculated from battery positive voltage), PCM determines that OCV circuit has a malfunction. **Possible Causes:** • Poor connection of connectors for PCM and/or OCV • Open circuit or short between OCV terminals and PCM terminals • OCV malfunction • PCM has failed
DTC: P2089 **2T CCM, MIL: Yes** **Year:** 2009, 2010 **Model:** 3, 5, 6, CX-7, MX-5 Miata **Engine:** 2.0L L4 VIN F, 2.0L L4 VIN G, 2.3L L4 VIN 3, 2.3L L4 VIN 4, 2.3L L4 VIN L, 2.3L L4 VIN M, 2.5L L4 VIN 5, 2.5L L4 VIN 6, 2.5L L4 VIN A, 2.5L L4 VIN M **Transmission:** All	**CMP actuator circuit high:** PCM monitors OCV voltage. If PCM detects OCV voltage (calculated from OCV) is above the threshold voltage (calculated from battery positive voltage), PCM determines that OCV circuit has a malfunction. **Possible Causes:** • Poor connection of connectors for PCM and/or OCV • Open circuit or short between OCV terminals and PCM terminals • OCV malfunction • PCM has failed
DTC: P2096 **2T CCM, MIL: Yes** **Year:** 2009, 2010 **Model:** 3, 5, 6, CX-7, MX-5 Miata, RX-8 **Engine:** 1.3L R2 VIN 2, 1.3L R2 VIN 4, 1.3L R2 VIN M, 1.3L R2 VIN P, 2.0L L4 VIN F, 2.0L L4 VIN G, 2.3L L4 VIN 3, 2.3L L4 VIN 4, 2.3L L4 VIN L, 2.3L L4 VIN M, 2.5L L4 VIN 5, 2.5L L4 VIN 6, 2.5L L4 VIN A, 2.5L L4 VIN M, 3.7L V6 VIN B **Transmission:** All	**Target A/F feedback system too lean (right bank):** The PCM monitors the target A/F fuel trim when under the target A/F feedback control. If the fuel trim is more than the specification, the PCM determines that the target A/F feedback system too lean. **Possible Causes:** • Leaking exhaust gas • HO2S (RR, LR) malfunction • MAF malfunction • PCM has failed

DTC	Trouble Code Title, Conditions & Possible Causes
DTC: P2097 **2T CCM, MIL: Yes** **Year:** 2009, 2010 **Model:** 3, 5, 6, CX-7, MX-5 Miata, RX-8 **Engine:** 1.3L R2 VIN 2, 1.3L R2 VIN 4, 1.3L R2 VIN M, 1.3L R2 VIN P, 2.0L L4 VIN F, 2.0L L4 VIN G, 2.3L L4 VIN 3, 2.3L L4 VIN 4, 2.3L L4 VIN L, 2.3L L4 VIN M, 2.5L L4 VIN 5, 2.5L L4 VIN 6, 2.5L L4 VIN A, 2.5L L4 VIN M, 3.7L V6 VIN B **Transmission:** All	**Target A/F feedback system too rich (right bank):** The PCM monitors the target A/F fuel trim when under the target A/F feedback control. If the fuel trim is more than the specification, the PCM determines that the target A/F feedback system too rich. **Possible Causes:** • Leaking exhaust gas • HO2S (RR, LR) malfunction • MAF malfunction • PCM has failed
DTC: P2098 **2T CCM, MIL: Yes** **Year:** 2009, 2010 **Model:** 6 **Engine:** 3.7L V6 VIN B **Transmission:** All	**Target A/F feedback system too lean (left bank):** The PCM monitors the target A/F fuel trim when under the target A/F feedback control. If the fuel trim is more than the specification, the PCM determines that the target A/F feedback system too lean. **Possible Causes:** • Leaking exhaust gas • HO2S (RR, LR) malfunction • MAF malfunction • PCM has failed
DTC: P2099 **2T CCM, MIL: Yes** **Year:** 2009, 2010 **Model:** 6 **Engine:** 3.7L V6 VIN B **Transmission:** All	**Target A/F feedback system too rich (left bank):** The PCM monitors the target A/F fuel trim when under the target A/F feedback control. If the fuel trim is more than the specification, the PCM determines that the target A/F feedback system too rich. **Possible Causes:** • Leaking exhaust gas • HO2S (RR, LR) malfunction • MAF malfunction • PCM has failed
DTC: P2100 **1T CCM, MIL: Yes** **Year:** 2009, 2010 **Model:** 3, 5, 6, CX-7, CX-9, MX-5 Miata **Engine:** 2.0L L4 VIN F, 2.0L L4 VIN G, 2.3L L4 VIN 3, 2.3L L4 VIN 4, 2.3L L4 VIN L, 2.3L L4 VIN M, 2.5L L4 VIN 5, 2.5L L4 VIN 6, 2.5L L4 VIN A, 2.5L L4 VIN M, 3.7L V6 VIN A, 3.7L V6 VIN V **Transmission:** All	**Throttle Actuator circuit open:** PCM monitors electronic throttle valve motor current. If PCM detects that the electronic throttle valve motor current is below the threshold current, PCM determines that the electronic throttle valve motor current has a malfunction. **Possible Causes:** • Poor connection of connectors for PCM and/or OCV • Open circuit or short between OCV terminals and PCM terminals • OCV malfunction • PCM has failed
DTC: P2101 **1T CCM, MIL: Yes** **Year:** 2009, 2010 **Model:** 3, 5, 6, CX-7, CX-9, RX-8 **Engine:** 1.3L R2 VIN 2, 1.3L R2 VIN 4, 1.3L R2 VIN M, 1.3L R2 VIN P, 2.0L L4 VIN F, 2.0L L4 VIN G, 2.3L L4 VIN 3, 2.3L L4 VIN 4, 2.3L L4 VIN L, 2.3L L4 VIN M, 2.5L L4 VIN 5, 2.5L L4 VIN 6, 2.5L L4 VIN A, 2.5L L4 VIN M, 3.7L V6 VIN A, 3.7L V6 VIN B, 3.7L V6 VIN V **Transmission:** All	**Throttle Actuator circuit range/performance:** If the PCM detects any of the following conditions, the PCM determines that the electronic throttle valve motor current has a malfunction: Default throttle angle that PCM memorized and the throttle angle with ET control relay OFF is not much; Voltage from ET relay is too high or too low; PCM detects a big voltage difference between the ET control relay and from the main relay; or PCM internal malfunction. **Possible Causes:** • ET control relay and related circuit malfunction • Main relay and related circuit malfunction • PCM has failed
DTC: P2104 **T PCM** **Year:** 2009, 2010 **Model:** CX-9 **Engine:** 3.7L V6 VIN A, 3.7L V6 VIN V **Transmission:** All	**Throttle actuator control system - forced idle:** The TAC system is in the failure mode effects management (FMEM) mode. **Possible Causes:** • This DTC is an informational DTC and may be set in combination with a number of other DTCs which are causing the FMEM. Diagnose other DTCs first.

DTC	Trouble Code Title, Conditions & Possible Causes
DTC: P2105 **1T CCM, MIL: Yes** **Year:** 2009, 2010 **Model:** 3, 6, CX-7, CX-9 **Engine:** 2.0L L4 VIN F, 2.0L L4 VIN G, 2.3L L4 VIN 3, 2.3L L4 VIN 4, 2.3L L4 VIN L, 2.3L L4 VIN M, 2.5L L4 VIN 5, 2.5L L4 VIN 6, 2.5L L4 VIN A, 2.5L L4 VIN M, 3.7L V6 VIN A, 3.7L V6 VIN V **Transmission:** All	**Throttle valve actuator control system-forced engine shutdown:** The throttle valve actuator control system is in the failure mode effects management mode. **Possible Causes:** • Damaged throttle body and/or PCM • Throttle valve actuator control module internal processor malfunction • PCM malfunction
DTC: P2107 **1T CCM, MIL: Yes** **Year:** 2009, 2010 **Model:** 3, 5, 6, CX-7, CX-9, MX-5 Miata, RX-8 **Engine:** 1.3L R2 VIN 2, 1.3L R2 VIN 4, 1.3L R2 VIN M, 1.3L R2 VIN P, 2.0L L4 VIN F, 2.0L L4 VIN G, 2.3L L4 VIN 3, 2.3L L4 VIN 4, 2.3L L4 VIN L, 2.3L L4 VIN M, 2.5L L4 VIN 5, 2.5L L4 VIN 6, 2.5L L4 VIN A, 2.5L L4 VIN M, 3.7L V6 VIN A, 3.7L V6 VIN B, 3.7L V6 VIN V **Transmission:** All	**Throttle Actuator control module processor:** If the PCM detects that either the electronic control module has malfunctioned or the target throttle opening angle is more than the actual throttle opening angle, the PCM determines that the throttle actuator control module has a malfunction. **Possible Causes:** • PCM has failed
DTC: P2108 **1T CCM, MIL: Yes** **Year:** 2009, 2010 **Model:** 3, 5, 6, CX-7, MX-5 Miata, RX-8 **Engine:** 1.3L R2 VIN 2, 1.3L R2 VIN 4, 1.3L R2 VIN M, 1.3L R2 VIN P, 2.0L L4 VIN F, 2.0L L4 VIN G, 2.3L L4 VIN 3, 2.3L L4 VIN 4, 2.3L L4 VIN L, 2.3L L4 VIN M, 2.5L L4 VIN 5, 2.5L L4 VIN 6, 2.5L L4 VIN A, 2.5L L4 VIN M **Transmission:** All	**Throttle Actuator control module performance:** If the PCM detects any of the following conditions, the PCM determines that the throttle actuator control system has a malfunction: TP sensor power supply voltage below 4.4 V; TP sensor No. 1 output voltage is below 0.20 V or above 4.85 V; TP sensor No. 2 output voltage is below 0.20 V or above 4.85 V; PCM internal circuit for TP sensor No. 1 input circuit malfunction; or the wrong communication between main CPU and throttle control system CPU in PCM internal. **Possible Causes:** • Open circuits between throttle body terminals and PCM terminals • Short circuits between throttle body terminals and PCM terminals* Poor connection of throttle body or PCM connector • TP sensor No. 1 or TP sensor No. 2 malfunction • PCM has failed
DTC: P2109 **1T CCM, MIL: Yes** **Year:** 2009, 2010 **Model:** MX-5 Miata **Engine:** 2.0L L4 VIN F **Transmission:** All	**TP sensor minimum stop range/performance problem:** The PCM monitors the minimum TP when the closed TP learning is completed. If the TP is less than 11.5% or more than 24.3%, the PCM determines that there is a TP sensor minimum stop range/performance problem. **Possible Causes:** • Drive-by-wire control system malfunction • Throttle actuator malfunction* Throttle valve malfunction • PCM has failed
DTC: P2110 **1T CCM, MIL: Yes** **Year:** 2009, 2010 **Model:** 3, 6, CX-7, CX-9 **Engine:** 2.0L L4 VIN F, 2.0L L4 VIN G, 2.5L L4 VIN 5, 2.5L L4 VIN 6, 2.5L L4 VIN A, 2.5L L4 VIN M, 3.7L V6 VIN A, 3.7L V6 VIN V **Transmission:** All	**Throttle valve actuator control system-forced limited RPM:** The throttle valve actuator control system is in the failure mode effects management mode. **Possible Causes:** • Throttle valve actuator control module internal processor error • APP sensor malfunction • PCM malfunction

DTC	Trouble Code Title, Conditions & Possible Causes
DTC: P2111 **T PCM** **Year:** 2009, 2010 **Model:** 6, CX-9 **Engine:** 3.7L V6 VIN A, 3.7L V6 VIN B, 3.7L V6 VIN V **Transmission:** All	**Throttle actuator control system - stuck open:** This PCM fault status indicates the throttle plate is at a more angle than commanded. **Possible Causes:** • Binding throttle body, stuck open • Throttle actuator malfunction • Open circuit between throttle body and PCM • Short to ground between throttle body and PCM • Short to TP sensor power supply between throttle body and PCM • Throttle actuator control circuits are shorted to each other • Throttle actuator control circuits are cross-wired • TP sensor related circuit intermittent malfunction • PCM malfunction
DTC: P2112 **1T CCM, MIL: Yes** **Year:** 2009, 2010 **Model:** 6, CX-9, MX-5 Miata, RX-8 **Engine:** 1.3L R2 VIN 2, 1.3L R2 VIN 4, 1.3L R2 VIN M, 1.3L R2 VIN P, 2.0L L4 VIN F, 3.7L V6 VIN A, 3.7L V6 VIN B, 3.7L V6 VIN V **Transmission:** All	**TP sensor minimum stop range/performance problem:** The PCM monitors the minimum TP when the closed TP learning is completed. If the TP is less than 11.5% or more than 24.3%, the PCM determines that there is a TP sensor minimum stop range/performance problem. **Possible Causes:** • Drive-by-wire control system malfunction • Throttle actuator malfunction* Throttle valve malfunction • PCM has failed
DTC: P2119 **1T CCM, MIL: Yes** **Year:** 2009, 2010 **Model:** 3, 5, 6, CX-7, MX-5 Miata, RX-8 **Engine:** 1.3L R2 VIN 2, 1.3L R2 VIN 4, 1.3L R2 VIN M, 1.3L R2 VIN P, 2.0L L4 VIN F, 2.0L L4 VIN G, 2.3L L4 VIN 3, 2.3L L4 VIN 4, 2.3L L4 VIN L, 2.3L L4 VIN M, 2.5L L4 VIN 5, 2.5L L4 VIN 6, 2.5L L4 VIN A, 2.5L L4 VIN M **Transmission:** All	**TP sensor minimum stop range/performance problem:** The PCM monitors the minimum TP when the closed TP learning is completed. If the TP is less than 11.5% or more than 24.3%, the PCM determines that there is a TP sensor minimum stop range/performance problem. **Possible Causes:** • Drive-by-wire control system malfunction • Throttle actuator malfunction* Throttle valve malfunction • PCM has failed
DTC: P2121 **T PCM** **Year:** 2009, 2010 **Model:** CX-9 **Engine:** 3.7L V6 VIN A, 3.7L V6 VIN V **Transmission:** All	**APP sensor No.1 circuit range/performance:** APP sensor fault flag is set for sensor No.1, No.2 and No.3 by the PCM, indicating the signal is out of the normal self-test operating range. **Possible Causes:** • APP sensor signal circuits are shorted each other • APP sensor malfunction • PCM malfunction • Open circuit wiring harness between APP sensor and PCM • Short to ground circuit between APP sensor and PCM • Short to power supply in wiring harness between APP sensor and PCM
DTC: P2122 **1T CCM, MIL: Yes** **Year:** 2009, 2010 **Model:** 3, 5, 6, CX-7, CX-9, MX-5 Miata, RX-8 **Engine:** 1.3L R2 VIN 2, 1.3L R2 VIN 4, 1.3L R2 VIN M, 1.3L R2 VIN P, 2.0L L4 VIN F, 2.0L L4 VIN G, 2.3L L4 VIN 3, 2.3L L4 VIN 4, 2.3L L4 VIN L, 2.3L L4 VIN M, 2.5L L4 VIN 5, 2.5L L4 VIN 6, 2.5L L4 VIN A, 2.5L L4 VIN M, 3.7L V6 VIN A, 3.7L V6 VIN B, 3.7L V6 VIN V **Transmission:** All	**APP sensor No. 1 circuit low input:** The PCM monitors the input voltage from the APP sensor No.1 when the engine is running. If the input voltage is less than 0.35 V, the PCM determines that the APP sensor No.1 circuit input voltage is low. **Possible Causes:** • Open circuits between APP sensor terminals and PCM terminals • Short circuits between APP sensor terminals and PCM terminals* Poor connection of APP sensor or PCM connector • APP sensor No. 1 malfunction • PCM has failed

DTC	Trouble Code Title, Conditions & Possible Causes
DTC: P2123 **1T CCM, MIL: Yes** **Year:** 2009, 2010 **Model:** 3, 5, 6, CX-7, CX-9, MX-5 Miata, RX-8 **Engine:** 1.3L R2 VIN 2, 1.3L R2 VIN 4, 1.3L R2 VIN M, 1.3L R2 VIN P, 2.0L L4 VIN F, 2.0L L4 VIN G, 2.3L L4 VIN 3, 2.3L L4 VIN 4, 2.3L L4 VIN L, 2.3L L4 VIN M, 2.5L L4 VIN 5, 2.5L L4 VIN 6, 2.5L L4 VIN A, 2.5L L4 VIN M, 3.7L V6 VIN A, 3.7L V6 VIN B, 3.7L V6 VIN V **Transmission:** All	**APP sensor No. 1 circuit high input:** The PCM monitors the input voltage from the APP sensor No.1 when the engine is running. If the input voltage is more than 4.8 V, the PCM determines that the APP sensor No.1 circuit input voltage is high. **Possible Causes:** • Open circuits between APP sensor terminals and PCM terminals • Short circuits between APP sensor terminals and PCM terminals* Poor connection of APP sensor or PCM connector • APP sensor No. 1 malfunction • PCM has failed
DTC: P2126 **1T CCM, MIL: Yes** **Year:** 2009, 2010 **Model:** 3, 5 **Engine:** 2.0L L4 VIN F, 2.0L L4 VIN G, 2.3L L4 VIN 3, 2.3L L4 VIN 4, 2.3L L4 VIN L, 2.3L L4 VIN M **Transmission:** All	**Throttle Actuator control module performance:** If the PCM detects any of the following conditions, the PCM determines that the throttle actuator control system has a malfunction: TP sensor power supply voltage below 4.4 V; TP sensor No. 1 output voltage is below 0.20 V or above 4.85 V; TP sensor No. 2 output voltage is below 0.20 V or above 4.85 V; PCM internal circuit for TP sensor No. 1 input circuit malfunction; or the wrong communication between main CPU and throttle control system CPU in PCM internal. **Possible Causes:** • Open circuits between throttle body terminals and PCM terminals • Short circuits between throttle body terminals and PCM terminals* Poor connection of throttle body or PCM connector • TP sensor No. 1 or TP sensor No. 2 malfunction • PCM has failed
DTC: P2126 **T PCM** **Year:** 2009, 2010 **Model:** CX-9 **Engine:** 3.7L V6 VIN A, 3.7L V6 VIN V **Transmission:** All	**APP sensor No.2 circuit range/performance:** APP sensor fault flag is set for sensor No.1, No.2 and No.3 by the PCM, indicating the signal is out of the normal self-test operating range. **Possible Causes:** • APP sensor signal circuits are shorted each other • APP sensor malfunction • PCM malfunction • Open circuit wiring harness between APP sensor and PCM • Short to ground circuit between APP sensor and PCM • Short to power supply in wiring harness between APP sensor and PCM
DTC: P2127 **1T CCM, MIL: Yes** **Year:** 2009, 2010 **Model:** 3, 5, 6, CX-7, CX-9, MX-5 Miata, RX-8 **Engine:** 1.3L R2 VIN 2, 1.3L R2 VIN 4, 1.3L R2 VIN M, 1.3L R2 VIN P, 2.0L L4 VIN F, 2.0L L4 VIN G, 2.3L L4 VIN 3, 2.3L L4 VIN 4, 2.3L L4 VIN L, 2.3L L4 VIN M, 2.5L L4 VIN 5, 2.5L L4 VIN 6, 2.5L L4 VIN A, 2.5L L4 VIN M, 3.7L V6 VIN A, 3.7L V6 VIN B, 3.7L V6 VIN V **Transmission:** All	**APP sensor No. 2 circuit low input:** The PCM monitors the input voltage from the APP sensor No.1 when the engine is running. If the input voltage is less than 0.35 V, the PCM determines that the APP sensor No.1 circuit input voltage is low. **Possible Causes:** • Open circuits between APP sensor terminals and PCM terminals • Short circuits between APP sensor terminals and PCM terminals* Poor connection of APP sensor or PCM connector • APP sensor No. 1 malfunction • PCM has failed
DTC: P2128 **1T CCM, MIL: Yes** **Year:** 2009, 2010 **Model:** 3, 5, 6, CX-7, CX-9, MX-5 Miata, RX-8 **Engine:** 1.3L R2 VIN 2, 1.3L R2 VIN 4, 1.3L R2 VIN M, 1.3L R2 VIN P, 2.0L L4 VIN F, 2.0L L4 VIN G, 2.3L L4 VIN 3, 2.3L L4 VIN 4, 2.3L L4 VIN L, 2.3L L4 VIN M, 2.5L L4 VIN 5, 2.5L L4 VIN 6, 2.5L L4 VIN A, 2.5L L4 VIN M, 3.7L V6 VIN A, 3.7L V6 VIN B, 3.7L V6 VIN V **Transmission:** All	**APP sensor No. 2 circuit high input:** The PCM monitors the input voltage from the APP sensor No.1 when the engine is running. If the input voltage is more than 4.8 V, the PCM determines that the APP sensor No.1 circuit input voltage is high. **Possible Causes:** • Open circuits between APP sensor terminals and PCM terminals • Short circuits between APP sensor terminals and PCM terminals* Poor connection of APP sensor or PCM connector • APP sensor No. 1 malfunction • PCM has failed

DTC	Trouble Code Title, Conditions & Possible Causes
DTC: P2131 **T PCM** **Year:** 2009, 2010 **Model:** CX-9 **Engine:** 3.7L V6 VIN A, 3.7L V6 VIN V **Transmission:** All	**APP sensor No.3 circuit range/performance:** APP sensor fault flag is set for sensor No.1, No.2 and No.3 by the PCM, indicating the signal is out of the normal self-test operating range. **Possible Causes:** • APP sensor signal circuits are shorted each other • APP sensor malfunction • PCM malfunction • Open circuit wiring harness between APP sensor and PCM • Short to ground circuit between APP sensor and PCM • Short to power supply in wiring harness between APP sensor and PCM
DTC: P2132 **T PCM** **Year:** 2009, 2010 **Model:** CX-9 **Engine:** 3.7L V6 VIN A, 3.7L V6 VIN V **Transmission:** All	**APP sensor No.3 circuit low input:** APP sensor No.3 is out of self-test range low. **Possible Causes:** • Connector or terminal malfunction • Short to ground circuit between APP sensor and PCM • APP sensor malfunction
DTC: P2133 **T PCM** **Year:** 2009, 2010 **Model:** CX-9 **Engine:** 3.7L V6 VIN A, 3.7L V6 VIN V **Transmission:** All	**APP sensor No.3 circuit high input:** APP sensor No.3 is out of self-test range high. **Possible Causes:** • Connector or terminal malfunction • APP sensor malfunction • Open circuit wiring harness between APP sensor and PCM • Short to power supply in wiring harness between APP sensor and PCM
DTC: P2135 **1T CCM, MIL: Yes** **Year:** 2009, 2010 **Model:** 3, 5, 6, CX-7, CX-9, MX-5 Miata, RX-8 **Engine:** 1.3L R2 VIN 2, 1.3L R2 VIN 4, 1.3L R2 VIN M, 1.3L R2 VIN P, 2.0L L4 VIN F, 2.0L L4 VIN G, 2.3L L4 VIN 3, 2.3L L4 VIN 4, 2.3L L4 VIN L, 2.3L L4 VIN M, 2.5L L4 VIN 5, 2.5L L4 VIN 6, 2.5L L4 VIN A, 2.5L L4 VIN M, 3.7L V6 VIN A, 3.7L V6 VIN B, 3.7L V6 VIN V **Transmission:** All	**TP sensor No. 1/No. 2 voltage correlation problem:** The PCM compares the input voltage from TP sensor No.1 with the input voltage from TP sensor No.2 when the engine is running. If the difference is more than the specification, the PCM determines that there is a TP sensor No.1/No.2 voltage correlation problem. **Possible Causes:** • TP sensor No. 1 malfunction • TP sensor No. 2 malfunction* Poor connection of throttle body or PCM connector • PCM has failed
DTC: P2138 **1T CCM, MIL: Yes** **Year:** 2009, 2010 **Model:** 3, 5, 6, CX-7, MX-5 Miata, RX-8 **Engine:** 1.3L R2 VIN 2, 1.3L R2 VIN 4, 1.3L R2 VIN M, 1.3L R2 VIN P, 2.0L L4 VIN F, 2.0L L4 VIN G, 2.3L L4 VIN 3, 2.3L L4 VIN 4, 2.3L L4 VIN L, 2.3L L4 VIN M, 2.5L L4 VIN 5, 2.5L L4 VIN 6, 2.5L L4 VIN A, 2.5L L4 VIN M, 3.7L V6 VIN B **Transmission:** All	**TP sensor No. 3/No. 4 voltage correlation problem:** The PCM compares the input voltage from TP sensor No.3 with the input voltage from TP sensor No.4 when the engine is running. If the difference is more than the specification, the PCM determines that there is a TP sensor No.3/No.4 voltage correlation problem. **Possible Causes:** • TP sensor No. 3 malfunction • TP sensor No. 4 malfunction* Poor connection of throttle body or PCM connector • PCM has failed
DTC: P2177 **1T CCM, MIL: Yes** **Year:** 2009, 2010 **Model:** 5, 6 **Engine:** 2.3L L4 VIN 3, 2.3L L4 VIN L, 2.5L L4 VIN A **Transmission:** All	**Fuel system too lean at off idle (right bank):** PCM monitors short term fuel trim (SHRTFT), long term fuel trim (LONGFT) during closed loop fuel control at off-idle. If the LONGFT and the sum total of these fuel trims exceed preprogrammed criteria. PCM determines that fuel system is too lean at off-idle. **Possible Causes:** • Misfire • Exhaust system leak* Fuel pump malfunction • Fuel filter clogged • PCM has failed

DTC	Trouble Code Title, Conditions & Possible Causes
DTC: P2178 **1T CCM, MIL: Yes** **Year:** 2009, 2010 **Model:** 5, 6 **Engine:** 2.3L L4 VIN 3, 2.3L L4 VIN L, 2.5L L4 VIN A **Transmission:** All	**Fuel system too rich at off idle (right bank):** PCM monitors short term fuel trim (SHRTFT), long term fuel trim (LONGFT) during closed loop fuel control at off-idle. If the LONGFT and the sum total of these fuel trims exceed preprogrammed criteria. PCM determines that fuel system is too rich at off-idle. **Possible Causes:** • Misfire • Exhaust system leak* Fuel pump malfunction • Fuel filter clogged • PCM has failed
DTC: P2183 **2T CCM, MIL: Yes** **Year:** 2009, 2010 **Model:** 3 **Engine:** 2.0L L4 VIN F, 2.0L L4 VIN G, 2.3L L4 VIN 3, 2.3L L4 VIN 4, 2.3L L4 VIN L, 2.3L L4 VIN M, 2.5L L4 VIN 5, 2.5L L4 VIN 6 **Transmission:** All	**ECT sensor No.2 circuit range/performance problem:** When the ECT sensor No.2 detected temperature is approx. 8.1 °C higher than the IAT, or the ECT sensor No.2 detected temperature is approx. 13.8 °C less than the IAT with the ignition switch is ON*. *: Ignition switch is ON when 6s or more have passed since the ignition was switched to off. **Possible Causes:** • ECT sensor No.2 connector or terminals malfunction • PCM connector or terminals malfunction • ECT sensor No.2 malfunction • Thermostat malfunction • PCM malfunction
DTC: P2187 **1T CCM, MIL: Yes** **Year:** 2009, 2010 **Model:** 5, 6 **Engine:** 2.3L L4 VIN 3, 2.3L L4 VIN L, 2.5L L4 VIN A **Transmission:** All	**Fuel system too lean at off idle (right bank):** PCM monitors short term fuel trim (SHRTFT), long term fuel trim (LONGFT) during closed loop fuel control at off-idle. If the LONGFT and the sum total of these fuel trims exceed preprogrammed criteria. PCM determines that fuel system is too lean at off-idle. **Possible Causes:** • Misfire • Exhaust system leak* Fuel pump malfunction • Fuel filter clogged • PCM has failed
DTC: P2188 **1T CCM, MIL: Yes** **Year:** 2009, 2010 **Model:** 5, 6 **Engine:** 2.3L L4 VIN 3, 2.3L L4 VIN L, 2.5L L4 VIN A **Transmission:** All	**Fuel system too lean at off idle (left bank):** PCM monitors short term fuel trim (SHRTFT), long term fuel trim (LONGFT) during closed loop fuel control at off-idle. If the LONGFT and the sum total of these fuel trims exceed preprogrammed criteria. PCM determines that fuel system is too lean at off-idle. **Possible Causes:** • Misfire • Exhaust system leak* Fuel pump malfunction • Fuel filter clogged • PCM has failed
DTC: P2195 **1T CCM, MIL: Yes** **Year:** 2009, 2010 **Model:** 3, 5, 6, B2300, B4000, CX-7, CX-9, MX-5 Miata, RX-8, Tribute **Engine:** 1.3L R2 VIN 2, 1.3L R2 VIN 4, 1.3L R2 VIN M, 1.3L R2 VIN P, 2.0L L4 VIN F, 2.0L L4 VIN G, 2.3L L4 VIN 3, 2.3L L4 VIN 4, 2.3L L4 VIN D, 2.3L L4 VIN L, 2.3L L4 VIN M, 2.5L L4 VIN 3, 2.5L L4 VIN 5, 2.5L L4 VIN 6, 2.5L L4 VIN 7, 2.5L L4 VIN A, 2.5L L4 VIN M, 3.0L V6 VIN 1, 3.0L V6 VIN G, 3.7L V6 VIN A, 3.7L V6 VIN B, 3.7L V6 VIN V, 4.0L V6 VIN E **Transmission:** All	**HO2S (RF) signal stuck lean:** The PCM monitors the HO2S (RF, LF) output voltage when the following conditions are met. If output voltage is less than 0.45 V for 25.6 s, the PCM determines that the HO2S (RF, LF) signal remains lean. **Possible Causes:** • HO2S (RF, LF) malfunction • Exhaust gas leaking* MAF sensor malfunction • ECT sensor malfunction • PCM has failed

DTC	Trouble Code Title, Conditions & Possible Causes
DTC: P2196 **1T CCM, MIL: Yes** **Year:** 2009, 2010 **Model:** 3, 5, 6, B2300, B4000, CX-7, CX-9, MX-5 Miata, RX-8, Tribute **Engine:** 1.3L R2 VIN 2, 1.3L R2 VIN 4, 1.3L R2 VIN M, 1.3L R2 VIN P, 2.0L L4 VIN F, 2.0L L4 VIN G, 2.3L L4 VIN 3, 2.3L L4 VIN 4, 2.3L L4 VIN D, 2.3L L4 VIN L, 2.3L L4 VIN M, 2.5L L4 VIN 3, 2.5L L4 VIN 5, 2.5L L4 VIN 6, 2.5L L4 VIN 7, 2.5L L4 VIN A, 2.5L L4 VIN M, 3.0L V6 VIN 1, 3.0L V6 VIN G, 3.7L V6 VIN A, 3.7L V6 VIN B, 3.7L V6 VIN V, 4.0L V6 VIN E **Transmission:** All	**HO2S (RF) signal stuck rich:** The PCM monitors the HO2S (RF, LF) output voltage when the following conditions are met. If output voltage is more than 0.45 V for 25.6 s, the PCM determines that the HO2S (RF, LF) signal remains rich. **Possible Causes:** • HO2S (RF, LF) malfunction • Exhaust gas leaking* MAF sensor malfunction • ECT sensor malfunction • PCM has failed
DTC: P2197 **1T CCM, MIL: Yes** **Year:** 2009, 2010 **Model:** 6, B4000, CX-9, Tribute **Engine:** 2.5L L4 VIN 3, 2.5L L4 VIN 7, 3.0L V6 VIN 1, 3.0L V6 VIN G, 3.7L V6 VIN A, 3.7L V6 VIN B, 3.7L V6 VIN V, 4.0L V6 VIN E **Transmission:** All	**HO2S (LF) signal stuck lean:** The PCM monitors the HO2S (RF, LF) output voltage when the following conditions are met. If output voltage is less than 0.45 V for 25.6 s, the PCM determines that the HO2S (RF, LF) signal remains lean. **Possible Causes:** • HO2S (RF, LF) malfunction • Exhaust gas leaking* MAF sensor malfunction • ECT sensor malfunction • PCM has failed
DTC: P2198 **1T CCM, MIL: Yes** **Year:** 2009, 2010 **Model:** 6, B4000, CX-9, Tribute **Engine:** 2.5L L4 VIN 3, 2.5L L4 VIN 7, 3.0L V6 VIN 1, 3.0L V6 VIN G, 3.7L V6 VIN A, 3.7L V6 VIN B, 3.7L V6 VIN V, 4.0L V6 VIN E **Transmission:** All	**HO2S (LF) signal stuck rich:** The PCM monitors the HO2S (RF, LF) output voltage when the following conditions are met. If output voltage is more than 0.45 V for 25.6 s, the PCM determines that the HO2S (RF, LF) signal remains rich. **Possible Causes:** • HO2S (RF, LF) malfunction • Exhaust gas leaking* MAF sensor malfunction • ECT sensor malfunction • PCM has failed
DTC: P2228 **1T CCM, MIL: Yes** **Year:** 2009, 2010 **Model:** 3, 5, 6, CX-7, MX-5 Miata **Engine:** 2.0L L4 VIN F, 2.0L L4 VIN G, 2.3L L4 VIN 3, 2.3L L4 VIN 4, 2.3L L4 VIN L, 2.3L L4 VIN M, 2.5L L4 VIN 5, 2.5L L4 VIN 6, 2.5L L4 VIN A, 2.5L L4 VIN M **Transmission:** All	**EGR boost sensor solenoid valve circuit low input:** The PCM monitors the EGR boost sensor solenoid valve control signal. If PCM turns the EGR boost sensor solenoid valve off but voltage still remains low, the PCM determines that the EGR boost sensor solenoid valve circuit has malfunction. **Possible Causes:** • EGR boost sensor solenoid malfunction • Connector or terminal malfunction* PCM has failed
DTC: P2229 **1T CCM, MIL: Yes** **Year:** 2009, 2010 **Model:** 3, 5, 6, CX-7, MX-5 Miata **Engine:** 2.0L L4 VIN F, 2.0L L4 VIN G, 2.3L L4 VIN 3, 2.3L L4 VIN 4, 2.3L L4 VIN L, 2.3L L4 VIN M, 2.5L L4 VIN 5, 2.5L L4 VIN 6, 2.5L L4 VIN A, 2.5L L4 VIN M **Transmission:** All	**EGR boost sensor solenoid valve circuit high input:** The PCM monitors the EGR boost sensor solenoid valve control signal. If PCM turns the EGR boost sensor solenoid valve off but voltage still remains high, the PCM determines that the EGR boost sensor solenoid valve circuit has malfunction. **Possible Causes:** • EGR boost sensor solenoid malfunction • Connector or terminal malfunction* PCM has failed

DTC	Trouble Code Title, Conditions & Possible Causes
DTC: P2237 **2T PCM, MIL: Yes** **Year:** 2009, 2010 **Model:** 3, 5, 6, CX-7 **Engine:** 2.0L L4 VIN F, 2.0L L4 VIN G, 2.3L L4 VIN 3, 2.3L L4 VIN 4, 2.3L L4 VIN L, 2.3L L4 VIN M, 2.5L L4 VIN 5, 2.5L L4 VIN 6, 2.5L L4 VIN A, 2.5L L4 VIN M **Transmission:** All	**A/F sensor positive current control circuit open:** The PCM monitors the A/F sensor terminal D current. If the current approx. 0 A while the A/F sensor active, the PCM determines that the A/F sensor positive current control circuit is open. **Possible Causes:** • A/F sensor connector or terminals malfunction • PCM connector or terminals malfunction • Open circuit in wiring harness between A/F sensor terminal B and PCM terminal 2AD • A/F sensor malfunction • PCM malfunction
DTC: P2243 **2T PCM, MIL: Yes** **Year:** 2009, 2010 **Model:** 3, 5, 6, CX-7 **Engine:** 2.0L L4 VIN F, 2.0L L4 VIN G, 2.3L L4 VIN 3, 2.3L L4 VIN 4, 2.3L L4 VIN L, 2.3L L4 VIN M, 2.5L L4 VIN 5, 2.5L L4 VIN 6, 2.5L L4 VIN A, 2.5L L4 VIN M **Transmission:** All	**A/F sensor reference voltage circuit open:** The PCM monitors the A/F sensor terminal F voltage. If the either of the following condition is met, the PCM determines that the A/F sensor reference voltage circuit is open. The A/F sensor terminal F voltage is specified voltage or more for 30 s. The PCM detects the DTC P0134 while the pending code P2243 is stored. **Possible Causes:** • A/F sensor connector or terminals malfunction • PCM connector or terminals malfunction • Open circuit in wiring harness between A/F sensor terminal F and PCM terminal 2Z • A/F sensor malfunction • PCM malfunction
DTC: P2245 **2T O2S, MIL: Yes** **Year:** 2009, 2010 **Model:** 3, CX-7 **Engine:** 2.3L L4 VIN 3, 2.3L L4 VIN 4, 2.3L L4 VIN L, 2.3L L4 VIN M, 2.5L L4 VIN 5, 2.5L L4 VIN M **Transmission:** All	**A/F sensor reference voltage circuit low input:** The PCM monitors the A/F sensor reference voltage circuit voltage. If the voltage below the threshold value while the engine is running, the PCM determines that the A/F sensor reference voltage circuit is low. **Possible Causes:** • A/F sensor connector or terminals malfunction • A/F sensor malfunction • Short to ground in reference voltage circuit at A/F sensor • Short to ground in wiring harness between A/F sensor and PCM • PCM connector or terminals malfunction • PCM malfunction
DTC: P2246 **2T O2S, MIL: Yes** **Year:** 2009, 2010 **Model:** 3, CX-7 **Engine:** 2.3L L4 VIN 3, 2.3L L4 VIN 4, 2.3L L4 VIN L, 2.3L L4 VIN M, 2.5L L4 VIN 5, 2.5L L4 VIN M **Transmission:** All	**A/F sensor reference voltage circuit high input:** The PCM monitors the A/F sensor reference voltage circuit voltage. If the voltage is above the threshold value while the engine is running, the PCM determines that the A/F sensor reference voltage circuit is high. **Possible Causes:** • A/F sensor connector or terminals malfunction • A/F sensor malfunction • Short to power supply in reference voltage circuit at A/F sensor • PCM connector or terminals malfunction • Short to power supply in wiring harness between A/F sensor and PCM • Open circuit in wiring harness between the A/F sensor and PCM • PCM malfunction
DTC: P2251 **2T O2S, MIL: Yes** **Year:** 2009, 2010 **Model:** 3, 5, 6, CX-7 **Engine:** 2.0L L4 VIN F, 2.0L L4 VIN G, 2.3L L4 VIN 3, 2.3L L4 VIN 4, 2.3L L4 VIN L, 2.3L L4 VIN M, 2.5L L4 VIN 5, 2.5L L4 VIN 6, 2.5L L4 VIN A, 2.5L L4 VIN M **Transmission:** All	**A/F sensor negative current control circuit open:** The PCM monitors the A/F sensor terminal F voltage. If either of the following conditions are met, the PCM determines that the A/F sensor negative current control circuit is open. MONITORING CONDITIONS The A/F sensor element impedance specified or more. Circuit voltage oscillation or A/F sensor terminal F voltage is below the specified. The PCM detects the DTC P0134:00 while the pending code P2251:00 is stored. **Possible Causes:** • A/F sensor connector or terminals malfunction • PCM connector or terminals malfunction • Open circuit in wiring harness between A/F sensor terminal D and PCM terminal 2AC • A/F sensor malfunction • PCM malfunction

DTC	Trouble Code Title, Conditions & Possible Causes
DTC: P2257 **1T CCM, MIL: Yes** **Year:** 2009, 2010 **Model:** RX-8 **Engine:** 1.3L R2 VIN 2, 1.3L R2 VIN 4, 1.3L R2 VIN M, 1.3L R2 VIN P **Transmission:** All	**AIR pump relay circuit low:** The PCM monitors the AIR pump relay control voltage when the AIR pump is not operating. If the control voltage is less than 5.8 V, the PCM determines that the AIR pump relay control circuit voltage is low. **Possible Causes:** • AIR pump relay malfunction • Connector or terminal malfunction* PCM has failed
DTC: P2258 **1T CCM, MIL: Yes** **Year:** 2009, 2010 **Model:** RX-8 **Engine:** 1.3L R2 VIN 2, 1.3L R2 VIN 4, 1.3L R2 VIN M, 1.3L R2 VIN P **Transmission:** All	**AIR pump relay circuit high:** The PCM monitors the AIR pump relay control voltage when the AIR pump is not operating. If the control voltage is more than 11.5 V, the PCM determines that the AIR pump relay control circuit voltage is low. **Possible Causes:** • AIR pump relay malfunction • Connector or terminal malfunction* PCM has failed
DTC: P2259 **1T CCM, MIL: Yes** **Year:** 2009, 2010 **Model:** RX-8 **Engine:** 1.3L R2 VIN 2, 1.3L R2 VIN 4, 1.3L R2 VIN M, 1.3L R2 VIN P **Transmission:** All	**AIR solenoid valve control circuit low:** The PCM monitors the AIR solenoid valve control voltage when the AIR pump is not operating. If the control voltage is less than 5.8 V, the PCM determines that the AIR solenoid valve control circuit voltage is low. **Possible Causes:** • AIR solenoid valve malfunction • Connector or terminal malfunction* PCM has failed
DTC: P2260 **1T CCM, MIL: Yes** **Year:** 2009, 2010 **Model:** RX-8 **Engine:** 1.3L R2 VIN 2, 1.3L R2 VIN 4, 1.3L R2 VIN M, 1.3L R2 VIN P **Transmission:** All	**AIR solenoid valve control circuit low:** The PCM monitors the AIR solenoid valve control voltage when the AIR pump is not operating. If the control voltage is more than 11.5 V, the PCM determines that the AIR solenoid valve control circuit voltage is low. **Possible Causes:** • AIR solenoid valve malfunction • Connector or terminal malfunction* PCM has failed
DTC: P2270 **1T CCM, MIL: Yes** **Year:** 2009, 2010 **Model:** 6, B2300, B4000, CX-9, RX-8, Tribute **Engine:** 1.3L R2 VIN 2, 1.3L R2 VIN 4, 1.3L R2 VIN M, 1.3L R2 VIN P, 2.3L L4 VIN D, 2.5L L4 VIN 3, 2.5L L4 VIN 7, 3.0L V6 VIN 1, 3.0L V6 VIN G, 3.7L V6 VIN A, 3.7L V6 VIN B, 3.7L V6 VIN V, 4.0L V6 VIN E **Transmission:** All	**Rear HO2S signal stuck lean:** The PCM monitors the input voltage from the rear HO2S when the following conditions are met. If the input voltage is less than 0.4 V for 40 seconds, the PCM determines that the rear HO2S signal remains lean. **Possible Causes:** • Front or Rear HO2S malfunction • Front or Rear HO2S heater malfunction* Fuel pressure malfunction • Fuel injector malfunction • PCM has failed
DTC: P2271 **1T CCM, MIL: Yes** **Year:** 2009, 2010 **Model:** 6, B2300, B4000, CX-9, RX-8, Tribute **Engine:** 1.3L R2 VIN 2, 1.3L R2 VIN 4, 1.3L R2 VIN M, 1.3L R2 VIN P, 2.3L L4 VIN D, 2.5L L4 VIN 3, 2.5L L4 VIN 7, 3.0L V6 VIN 1, 3.0L V6 VIN G, 3.7L V6 VIN A, 3.7L V6 VIN B, 3.7L V6 VIN V, 4.0L V6 VIN E **Transmission:** All	**Rear HO2S signal stuck rich:** The PCM monitors the input voltage from the rear HO2S when the following conditions are met. If the input voltage is more than 0.85 V for 40 seconds, the PCM determines that the rear HO2S signal remains lean. **Possible Causes:** • Front or Rear HO2S malfunction • Front or Rear HO2S heater malfunction* Fuel pressure malfunction • Fuel injector malfunction • PCM has failed
DTC: P2272 **1T CCM, MIL: Yes** **Year:** 2009, 2010 **Model:** 6, CX-9, Tribute **Engine:** 2.5L L4 VIN 3, 2.5L L4 VIN 7, 3.0L V6 VIN 1, 3.0L V6 VIN G, 3.7L V6 VIN A, 3.7L V6 VIN B, 3.7L V6 VIN V **Transmission:** All	**Rear HO2S signal stuck lean:** The PCM monitors the input voltage from the rear HO2S when the following conditions are met. If the input voltage is less than 0.4 V for 40 seconds, the PCM determines that the rear HO2S signal remains lean. **Possible Causes:** • Front or Rear HO2S malfunction • Front or Rear HO2S heater malfunction* Fuel pressure malfunction • Fuel injector malfunction • PCM has failed

DTC	Trouble Code Title, Conditions & Possible Causes
DTC: P2273 **1T CCM, MIL: Yes** **Year:** 2009, 2010 **Model:** 6, CX-9, Tribute **Engine:** 2.5L L4 VIN 3, 2.5L L4 VIN 7, 3.0L V6 VIN 1, 3.0L V6 VIN G, 3.7L V6 VIN A, 3.7L V6 VIN B, 3.7L V6 VIN V **Transmission:** All	**Rear HO2S signal stuck rich:** The PCM monitors the input voltage from the rear HO2S when the following conditions are met. If the input voltage is more than 0.85 V for 40 seconds, the PCM determines that the rear HO2S signal remains lean. **Possible Causes:** • Front or Rear HO2S malfunction • Front or Rear HO2S heater malfunction* Fuel pressure malfunction • Fuel injector malfunction • PCM has failed
DTC: P2401 **1T CCM, MIL: Yes** **Year:** 2009, 2010 **Model:** 5, MX-5 Miata, RX-8 **Engine:** 1.3L R2 VIN 2, 1.3L R2 VIN 4, 1.3L R2 VIN M, 1.3L R2 VIN P, 2.0L L4 VIN F, 2.3L L4 VIN 3, 2.3L L4 VIN L **Transmission:** All	**EVAP system leak detection pump motor circuit low:** The PCM monitors pump load current (EVAP line pressure), while evaporative leak monitor is operating. If the pump load current is lower than specified, the PCM determines EVAP system leak detection pump motor circuit has a malfunction. **Possible Causes:** • EVAP system leak detection pump malfunction • Poor connection of EVAP system or PCM connector • PCM has failed
DTC: P2402 **1T CCM, MIL: Yes** **Year:** 2009, 2010 **Model:** 5, MX-5 Miata, RX-8 **Engine:** 1.3L R2 VIN 2, 1.3L R2 VIN 4, 1.3L R2 VIN M, 1.3L R2 VIN P, 2.0L L4 VIN F, 2.3L L4 VIN 3, 2.3L L4 VIN L **Transmission:** All	**EVAP system leak detection pump motor circuit low:** The PCM monitors pump load current (EVAP line pressure), while evaporative leak monitor is operating. If the pump load current is higher than specified, the PCM determines EVAP system leak detection pump motor circuit has a malfunction. **Possible Causes:** • EVAP system leak detection pump malfunction • Poor connection of EVAP system or PCM connector • PCM has failed
DTC: P2404 **1T CCM, MIL: Yes** **Year:** 2009, 2010 **Model:** 5, MX-5 Miata, RX-8 **Engine:** 1.3L R2 VIN 2, 1.3L R2 VIN 4, 1.3L R2 VIN M, 1.3L R2 VIN P, 2.0L L4 VIN F, 2.3L L4 VIN 3, 2.3L L4 VIN L **Transmission:** All	**EVAP system leak detection pump sense circuit problem:** The PCM monitors pump load current (EVAP line pressure), while evaporative leak monitor is operating. After obtaining the reference current value, if the time in which the pump load current reaches the reference current value is less than the specification, the PCM determines air filter has a malfunction. **Possible Causes:** • Air filter clogging • EVAP hose bending • PCM has failed
DTC: P2405 **1T CCM, MIL: Yes** **Year:** 2009, 2010 **Model:** 5, MX-5 Miata, RX-8 **Engine:** 1.3L R2 VIN 2, 1.3L R2 VIN 4, 1.3L R2 VIN M, 1.3L R2 VIN P, 2.0L L4 VIN F, 2.3L L4 VIN 3, 2.3L L4 VIN L **Transmission:** All	**EVAP system leak detection pump sense circuit low input:** The PCM monitors pump load current (EVAP line pressure), while evaporative leak monitor is operating. If the current is lower than the specification while the PCM obtains the reference current value, the PCM determines EVAP system leak detection pump orifice has a malfunction. **Possible Causes:** • EVAP system leak detection orifice has fallen off • EVAP system leak detection pump motor malfunction • PCM has failed
DTC: P2407 **1T CCM, MIL: Yes** **Year:** 2009, 2010 **Model:** RX-8 **Engine:** 1.3L R2 VIN 2, 1.3L R2 VIN 4, 1.3L R2 VIN M, 1.3L R2 VIN P **Transmission:** All	**EVAP system leak detection pump sense circuit intermittent:** The PCM monitors pump load current (EVAP line pressure), while evaporative leak monitor is operating. When either of the following is detected 6 times or more successively, the PCM determines EVAP system leak detection pump heater has a malfunction. **Possible Causes:** • EVAP system leak detection pump heater malfunction • PCM has failed

DTC	Trouble Code Title, Conditions & Possible Causes
DTC: P2502 **1T CCM, MIL: Yes** **Year:** 2009, 2010 **Model:** 3, 5, 6, CX-7, MX-5 Miata, RX-8 **Engine:** 1.3L R2 VIN 2, 1.3L R2 VIN 4, 1.3L R2 VIN M, 1.3L R2 VIN P, 2.0L L4 VIN F, 2.0L L4 VIN G, 2.3L L4 VIN 3, 2.3L L4 VIN 4, 2.3L L4 VIN L, 2.3L L4 VIN M, 2.5L L4 VIN 5, 2.5L L4 VIN 6, 2.5L L4 VIN A, 2.5L L4 VIN M **Transmission:** All	**Charging system voltage problem:** PCM judges' generator output voltage is above 17 V or battery voltage is below 11 V during engine running. **Possible Causes:** • Open circuits between battery terminals • Battery is malfunctioning* Poor connection of PCM connectors • Generator malfunction • PCM has failed
DTC: P2503 **1T CCM, MIL: Yes** **Year:** 2009, 2010 **Model:** 3, 5, 6, CX-7, MX-5 Miata, RX-8 **Engine:** 1.3L R2 VIN 2, 1.3L R2 VIN 4, 1.3L R2 VIN M, 1.3L R2 VIN P, 2.0L L4 VIN F, 2.0L L4 VIN G, 2.3L L4 VIN 3, 2.3L L4 VIN 4, 2.3L L4 VIN L, 2.3L L4 VIN M, 2.5L L4 VIN 5, 2.5L L4 VIN 6, 2.5L L4 VIN A, 2.5L L4 VIN M **Transmission:** All	**Charging system voltage low:** PCM needs more than 20 A from generator, and judges' generator output voltage to be below 8.5 V during engine running. **Possible Causes:** • Open circuits between battery terminals • Battery is malfunctioning* Poor connection of PCM connectors • Generator malfunction • PCM has failed
DTC: P2504 **1T CCM, MIL: Yes** **Year:** 2009, 2010 **Model:** 3, 5, 6, CX-7, MX-5 Miata, RX-8 **Engine:** 1.3L R2 VIN 2, 1.3L R2 VIN 4, 1.3L R2 VIN M, 1.3L R2 VIN P, 2.0L L4 VIN F, 2.0L L4 VIN G, 2.3L L4 VIN 3, 2.3L L4 VIN 4, 2.3L L4 VIN L, 2.3L L4 VIN M, 2.5L L4 VIN 5, 2.5L L4 VIN 6, 2.5L L4 VIN A, 2.5L L4 VIN M **Transmission:** All	**Charging system voltage high:** PCM judges generator output voltage is above 18.5 V or battery voltage is above 16.0 V during engine running. **Possible Causes:** • Short to power circuit between generator and PCM • Generator malfunction • PCM has failed
DTC: P2507 **1T CCM, MIL: Yes** **Year:** 2009, 2010 **Model:** 3, 5, 6, CX-7, MX-5 Miata **Engine:** 2.0L L4 VIN F, 2.0L L4 VIN G, 2.3L L4 VIN 3, 2.3L L4 VIN 4, 2.3L L4 VIN L, 2.3L L4 VIN M, 2.5L L4 VIN 5, 2.5L L4 VIN 6, 2.5L L4 VIN A, 2.5L L4 VIN M **Transmission:** All	**PCM B+ voltage low:** The PCM monitors the voltage of back-up battery positive terminal at PCM terminal 4AG. If the PCM detected battery positive terminal voltage below 2.5 V for 2 seconds, the PCM determines that the backup voltage circuit has malfunction. **Possible Causes:** • Melt down fuse* Poor connection of PCM • PCM has failed
DTC: P260F **2T CCM, MIL: Yes** **Year:** 2009, 2010 **Model:** 6, CX-7, CX-9 **Engine:** 2.5L L4 VIN 5, 2.5L L4 VIN A, 2.5L L4 VIN M, 3.7L V6 VIN A, 3.7L V6 VIN B, 3.7L V6 VIN V **Transmission:** All	**EVAP system monitoring processor performance problem:** This DTC sets when a concern is detected internal to the PCM. The microprocessor that controls the engine off natural vacuum (EONV) leak check monitor is separate from the main processor within the PCM. **Possible Causes:** • Module communications network concerns • PCM connector or terminals malfunction • PCM malfunction

DTC	Trouble Code Title, Conditions & Possible Causes
DTC: P2610 **2T PCM, MIL: Yes** **Year:** 2009, 2010 **Model:** 3, 5, 6, CX-7, MX-5 Miata **Engine:** 2.0L L4 VIN F, 2.0L L4 VIN G, 2.3L L4 VIN 3, 2.3L L4 VIN 4, 2.3L L4 VIN L, 2.3L L4 VIN M, 2.5L L4 VIN 5, 2.5L L4 VIN 6, 2.5L L4 VIN A, 2.5L L4 VIN M, 3.7L V6 VIN B **Transmission:** All	**PCM internal engine off timer performance problem:** The PCM internal engine off timer is damaged. **Possible Causes:** • PCM internal engine off timer malfunction
DTC: P2627 **T PCM** **Year:** 2009, 2010 **Model:** 6 **Engine:** 3.7L V6 VIN B **Transmission:** All	**A/F sensor (RH) positive current trim circuit low:** A resistor is installed in the A/F sensor (RH) connector for part to part variance. The PCM determines the value of this resistor by taking multiple measurements of the resistor during each key on event. The PCM uses this value in order to compensate for the variance in the pumping current signal. The test fails if the PCM determines the resistance value is too high. **Possible Causes:** • Open circuit between A/F sensor (RH) and PCM • Short to ground between A/F sensor (RH) and PCM • A/F sensor (RH) malfunction
DTC: P2630 **T** **Year:** 2009, 2010 **Model:** 6 **Engine:** 3.7L V6 VIN B **Transmission:** All	**A/F sensor (LH) positive current trim circuit low:** A resistor is installed in the A/F sensor (LH) connector for part to part variance. The PCM determines the value of this resistor by taking multiple measurements of the resistor during each key on event. The PCM uses this value in order to compensate for the variance in the pumping current signal. The test fails if the PCM determines the resistance value is too high. **Possible Causes:** • Open circuit between A/F sensor (LH) and PCM • Short to ground between A/F sensor (LH) and PCM • A/F sensor (LH) malfunction

OBD II Trouble Code List (U0XXX Codes)

DTC	Trouble Code Title, Conditions & Possible Causes
DTC: U0109 **T PCM** **Year:** 2009, 2010 **Model:** 6 **Engine:** 3.7L V6 VIN B **Transmission:** All	**Lost communication with fuel pump control module:** The PCM monitors the fuel pump monitor (FPM) circuit for the presence of a duty cycled signal. If the FPM circuit is fixed at a low or high voltage, the PCM begins to increment a counter. The test fails if the PCM is still not detecting a duty cycled signal on the FPM circuit after a calibrated amount of time. **Possible Causes:** • Open circuit between electronic fuel pump relay and PCM • Short to ground between electronic fuel pump relay and PCM • Short to power supply between electronic fuel pump relay and PCM • Open circuit between electronic fuel pump relay and fuel pump relay • Open circuit between electronic fuel pump relay and body ground • Fuel pump relay malfunction
DTC: U0300 **T PCM** **Year:** 2009, 2010 **Model:** 6 **Engine:** 3.7L V6 VIN B **Transmission:** All	**Internal control module software incompatibility:** This DTC indicates there are incompatible software levels within the PCM that control the drive-by wire control system. The drive-by wire control system uses multiple microprocessors within the PCM, each having its own software level and function. The microprocessors must have the correct level of software in order to communicate and function together. **Possible Causes:** • PCM calibration level

OBD II Trouble Code List (U1XXX Codes)

DTC	Trouble Code Title, Conditions & Possible Causes
DTC: U1020 **1T CCM** **Year:** 2009 **Model:** B2300, B4000 **Engine:** 2.3L L4 VIN D, 4.0L V6 VIN E **Transmission:** All	**Data Circuit Malfunction:** Key on, and the PCM detected the SCP was invalid, or that it was missing data from the A/C System during the CCM test. **Possible Causes:** • SCP data bus to other controller(s) is open or shorted to ground • A/C System controller is damaged or has failed • PCM has failed

DTC	Trouble Code Title, Conditions & Possible Causes
DTC: U1039 **1T CCM** **Year:** 2009, 2010 **Model:** B2300, B4000, Tribute **Engine:** 2.3L L4 VIN D, 2.5L L4 VIN 3, 2.5L L4 VIN 7, 3.0L V6 VIN 1, 3.0L V6 VIN G, 4.0L V6 VIN E **Transmission:** All	**Data Circuit Malfunction:** Key on, and the PCM detected the SCP was invalid, or that it was missing data from the Vehicle Speed System during the CCM test. **Possible Causes:** • SCP data bus to other controller(s) is open or shorted to ground • Vehicle Speed System controller is damaged or has failed • PCM has failed
DTC: U1051 **1T CCM** **Year:** 2009 **Model:** B2300, B4000 **Engine:** 2.3L L4 VIN D, 4.0L V6 VIN E **Transmission:** All	**Data Circuit Malfunction:** Key on, and the PCM detected the SCP was invalid, or that it was missing data from the Antilock Brake System during the CCM test. **Possible Causes:** • SCP data bus to other controller(s) is open or shorted to ground • ABS System controller is damaged or has failed • PCM has failed
DTC: U1131 **1T CCM** **Year:** 2009 **Model:** B2300, B4000 **Engine:** 2.3L L4 VIN D, 4.0L V6 VIN E **Transmission:** All	**Data Circuit Malfunction:** Key on, and the PCM detected the SCP was invalid, or that it was missing data from the Fuel System during the CCM test. **Possible Causes:** • SCP data bus to other controller(s) is open or shorted to ground • Fuel System controller is damaged or has failed • PCM has failed
DTC: U1262 **1T CCM** **Year:** 2009 **Model:** B2300, B4000 **Engine:** 2.3L L4 VIN D, 4.0L V6 VIN E **Transmission:** All	**Data Circuit Malfunction:** Key on, and the PCM detected a problem in the SCP communications data bus circuit during the CCM test. **NOTE: Perform a network communications test of the whole system.** **Possible Causes:** • SCP data bus to other controller(s) is open or shorted to ground • Anti-Theft System controller is damaged or has failed • PCM has failed
DTC: U1451 **1T CCM** **Year:** 2009 **Model:** B2300, B4000 **Engine:** 2.3L L4 VIN D, 4.0L V6 VIN E **Transmission:** All	**Data Circuit Malfunction:** Key on, and the PCM detected the SCP was invalid, or that it was missing data from the Anti-Theft System during the CCM test. **Possible Causes:** • SCP data bus to other controller(s) is open or shorted to ground • Anti-Theft System controller is damaged or has failed • PCM has failed

OBD II Trouble Code List (U2XXX Codes)

DTC	Trouble Code Title, Conditions & Possible Causes
DTC: U2243 **1T CCM** **Year:** 2009 **Model:** B2300, B4000 **Engine:** 2.3L L4 VIN D, 4.0L V6 VIN E **Transmission:** All	**Data Circuit Malfunction:** Key on, and the PCM detected the SCP was invalid, or that it was missing data from the Exterior Environment System in the CCM test. **Possible Causes:** • SCP data bus to other controller(s) is open or shorted to ground • Vehicle Security System controller is damaged or has failed • PCM has failed

OBD II Trouble Code List (U3XXX Codes)

DTC	Trouble Code Title, Conditions & Possible Causes
DTC: U3000 **T** **Year:** 2009, 2010 **Model:** 3, 6, CX-7 **Engine:** 2.0L L4 VIN F, 2.0L L4 VIN G, 2.3L L4 VIN 3, 2.3L L4 VIN 4, 2.3L L4 VIN L, 2.3L L4 VIN M, 2.5L L4 VIN 5, 2.5L L4 VIN 6, 2.5L L4 VIN A, 2.5L L4 VIN M **Transmission:** All	**PCM processor error:** The PCM internal EEPROM malfunction. **Possible Causes:** • PCM internal malfunction

SPECIFICATIONS AND MAINTENANCE CHARTS

ENGINE AND VEHICLE IDENTIFICATION

Engine							Model Year	
Code ①	Liters (cc)	Cu. In.	Cyl.	Fuel Sys.	Engine Type	Eng. Mfg.	Code ②	Year
4G69/F	2.4 (2,378)	145.1	4	MPFI	SOHC	Mitsubishi	9	2009
6G75/T	3.8 (3,828)	233.6	6	MPFI	SOHC	Mitsubishi	A	2010

MPFI: Multi-Point Fuel Injection

DOHC: Double Overhead Camshafts

① Engine Code/8th digit of VIN

② 10th digit of VIN

37671_ECLI_C0001

GENERAL ENGINE SPECIFICATIONS

All measurements are given in inches.

Year	Model	Engine Displacement Liters	Engine Series ID/VIN	Net Horsepower @ rpm	Net Torque @ rpm (ft. lbs.)	Bore x Stroke (in.)	Compression Ratio	Oil Pressure @ rpm
2009	Eclipse/Spyder	2.4	4G69/F	162@6000	162@4000	3.43 x 3.94	9.5:1	43-100@3500
	Eclipse/Spyder	3.8	6G75/T	265@5750	262@4500	3.74 x 3.54	10.5:1	43-100@3500
2010	Eclipse/Spyder	2.4	4G69/F	162@6000	162@4000	3.43 x 3.94	9.5:1	43-100@3500
	Eclipse/Spyder	3.8	6G75/T	265@5750	262@4500	3.74 x 3.54	10.5:1	43-100@3500

37671_ECLI_C0002

GASOLINE ENGINE TUNE-UP SPECIFICATIONS

Year	Engine Displacement Liters	Engine Series ID/VIN	Spark Plug Gap (in.)	Ignition Timing (deg.) MT	Ignition Timing (deg.) AT	Fuel Pump (psi)	Idle Speed (rpm) MT	Idle Speed (rpm) AT	Valve Clearance (in.) In.	Valve Clearance (in.) Ex.
2009	2.4	4G69/F	0.028-0.031	①	①	47	②	②	0.0040	0.0800
	3.8	6G75/T	0.028-0.031	①	①	47	②	②	HYD	HYD
2010	2.4	4G69/F	0.028-0.031	①	①	47	②	②	0.0040	0.0800
	3.8	6G75/T	0.028-0.031	①	①	47	②	②	HYD	HYD

NOTE: The Vehicle Emission Control Information label reflects specification changes made during production.

Follow the figures on the label if they differ from those in this chart.

HYD: Hydraulic

① Ignition timing is preset and cannot be adjusted

② Idle speed is maintained by the Electronic Control Module (ECM)

37671_ECLI_C0003

CAPACITIES

Year	Model	Engine Displacement Liters	Engine Series ID/VIN	Engine Oil with Filter (qts.)	Transmission (pts.) Manual	Auto. ①	Fuel Tank (gal.)	Cooling System (qts.)
2009	Eclipse/Spyder	2.4	4G69/F	②	4.6	16.2	17.7	③
	Eclipse/Spyder	3.8	6G75/T	②	4.6	17.8	17.7	④
2010	Eclipse/Spyder	2.4	4G69/F	②	4.6	16.2	17.7	③
	Eclipse/Spyder	3.8	6G75/T	②	4.6	17.8	17.7	④

NOTE: All capacities are approximate. Add fluid gradually and check to be sure a proper fluid level is obtained.

① Drain and refill

② Oil pan (4.2 qts.), Oil filter (0.32 qt.), except MIVEC. MEVIC Oil pan (4.5 qts.), Oil filter (0.32 qt.

③ Automatic: 9.2. Manual: 9.3

④ Automatic: 8.5. Manual: 8.6

37671_ECLI_C0004

FLUID SPECIFICATIONS

Year	Model	Engine Displacement Liters	Engine Series ID/VIN	Engine Oil	Manual Trans.	Auto. Trans.	Power Steering Fluid	Brake Master Cylinder	Cooling System
2009	Eclipse/Spyder	2.4	4G69/F	5W-20	①	②	③	④	⑤
	Eclipse/Spyder	3.8	6G75/T	5W-20	①	②	③	④	⑤
2010	Eclipse/Spyder	2.4	4G69/F	5W-20	①	②	③	④	⑤
	Eclipse/Spyder	3.8	6G75/T	5W-20	①	②	③	④	⑤

DOT: Department Of Transportation

① DiaQueen NEW MULTI GEAR OIL (GL-3)

② DIAMOND ATF SP-3 or equivalent

③ Genuine MITSUBISHI power steering fluid

④ DOT 3, DOT 4, or equivalent

⑤ Long life antifreeze/coolant or an equivalent

37671_ECLI_C0005

VALVE SPECIFICATIONS

Year	Engine Displacement Liters	Engine Series ID/VIN	Seat Angle (deg.)	Face Angle (deg.)	Spring Test Pressure (lbs. @ in.)	Spring Installed Height (in.)	Stem-to-Guide Clearance (in.) Intake	Exhaust	Stem Diameter (in.) Intake	Exhaust
2009	2.4	4G69/F	NA	43.5-44.0	60.0 @1.74	1.74	0.0008-0.0019	0.0014-0.0024	0.240	0.240
	3.8	6G75/T	NA	43.5-44.0	①	1.74	0.0008-0.0019	0.0016-0.0023	0.240	0.240
2010	2.4	4G69/F	NA	43.5-44.0	60.0 @1.74	1.74	0.0008-0.0019	0.0014-0.0024	0.240	0.240
	3.8	6G75/T	NA	43.5-44.0	①	1.74	0.0008-0.0019	0.0016-0.0023	0.240	0.240

NA: Not Available

① Intake: 59.0 lbs. @ 1.74 in.

Exhaust: 53.0 @ 1.74 in.

37671_ECLI_C0006

CAMSHAFT AND BEARING SPECIFICATIONS CHART

All measurements are given in inches.

Year	Engine Displ. Liters	Engine Series ID/VIN	Journal Dia.	Brg. Oil Clearance	Shaft End-play	Runout	Journal Bore	Lobe Height Intake	Lobe Height Exhaust
2009	2.4	4G69/F	1.800	NA	NA	NA	NA	①	1.471-1.491
	3.8	6G75/T	1.800	NA	NA	NA	NA	②	1.471-1.491
2010	2.4	4G69/F	1.800	NA	NA	NA	NA	①	1.471-1.491
	3.8	6G75/T	1.800	NA	NA	NA	NA	②	1.471-1.491

NA: Not Available

① Intake Low speed cam A: 1.335-1.355 in.

   Intake Low speed cam B: 1.456-1.475 in.

   Intake High speed cam: 1.445-1.465 in.

② Intake Low speed cam A: 1.301-1.321 in.

   Intake Low speed cam B: 1.451-1.4705 in.

   Intake High speed cam: 1.445-1.465 in.

37671_ECLI_C0007

CRANKSHAFT AND CONNECTING ROD SPECIFICATIONS

All measurements are given in inches.

Year	Engine Displacement Liters	Engine Series ID/VIN	Crankshaft Main Brg. Journal Dia.	Crankshaft Main Brg. Oil Clearance	Crankshaft Shaft End-play	Crankshaft Thrust on No.	Connecting Rod Journal Diameter	Connecting Rod Oil Clearance	Connecting Rod Side Clearance
2009	2.4	4G69/F	2.240	①	0.0020-0.0090	3	1.770	②	0.0040-0.0090
	3.8	6G75/T	2.520	③	0.0020-0.0090	3	2.165	0.0008-0.0015	0.0040-0.0090
2010	2.4	4G69/F	2.240	①	0.0020-0.0090	3	1.770	②	0.0040-0.0090
	3.8	6G75/T	2.520	③	0.0020-0.0090	3	2.165	0.0008-0.0015	0.0040-0.0090

NA: Not Available

① Number 1, 2, 4, 5: 0.0008-0.0012 in.

   Number 3: 0.0012-0.0016 in.

② ID Mark or color - I or Yellow: 0.0007-0.0089 in.

   II or None: 0.0006-0.0019 in.

   III or Blue: 0.0007-0.0018 in.

③ Number 1 and 4: 0.0008-0.0012 in.

   Number 2 and 3: 0.0008-0.0015 in.

37671_ECLI_C0008

PISTON AND RING SPECIFICATIONS

All measurements are given in inches.

Year	Engine Displ. Liters	Engine Series ID/VIN	Piston Clearance	Ring Gap			Ring Side Clearance		
				Top Compression	Bottom Compression	Oil Control	Top Compression	Bottom Compression	Oil Control
2009	2.4	4G69/F	0.0008-0.0015	0.0060-0.0120	0.0110-0.0170	0.0030-0.0150	0.0012-0.0028	0.0008-0.0024	NA
	3.8	6G75/T	0.0012-0.0019	0.0100-0.0160	0.0140-0.0190	0.0040-0.0140	0.0012-0.0027	0.0008-0.0023	NA
2010	2.4	4G69/F	0.0008-0.0015	0.0060-0.0120	0.0110-0.0170	0.0030-0.0150	0.0012-0.0028	0.0008-0.0024	NA
	3.8	6G75/T	0.0012-0.0019	0.0100-0.0160	0.0140-0.0190	0.0040-0.0140	0.0012-0.0027	0.0008-0.0023	NA

NA: Not Available

37671_ECLI_C0009

TORQUE SPECIFICATIONS

All readings in ft. lbs.

Year	Engine Displacement Liters	Engine Series ID/VIN	Cylinder Head Bolts	Main Bearing Bolts	Rod Bearing Bolts	Crankshaft Damper Bolts	Flywheel Bolts	Manifold		Spark Plugs	Oil Pan Drain Plug
								Intake	Exhaust		
2009	2.4	4G69/F	①	②	③	123	95-101	④	⑤	15-22	25-33
	3.8	6G75/T	⑥	51-57	⑦	134-140	53-56	⑩	⑨	15-22	25-33
2010	2.4	4G69/F	①	②	③	123	95-101	④	⑤	15-22	25-33
	3.8	6G75/T	⑥	51-57	⑦	134-140	53-56	⑩	⑨	15-22	25-33

NOTE: Dip main bearing bolts and crankshaft damper bolt in clean engine oil prior to tightening.

① Step 1: Tighten all bolts to 57-59 ft. lbs.
  Step 2: Loosen all bolts to 0 ft. lbs.
  Step 3: Tighten all bolts to 14-16 ft. lbs.
  Step 4: Plus 90 degrees
  Step 4: Plus another 90 degrees

② Step 1: 17-19 ft. lbs.
  Step 2: Plus 90 degrees

③ Step 1: 14-16 ft. lbs.
  Step 2: Plus 90-94 degrees

④ Intake manifold bolt 17-19 ft. lbs.
  Intake manifold nut 14-16 ft. lbs.
  Stay bolt 21-25 ft. lbs.

⑤ Exhaust manifold nut 33-39 ft. lbs.
  Cover bolt 116-132 inch lbs.

⑥ Step 1: Tighten all bolts to 76-84 ft. lbs.
  Step 2: Loosen all bolts to 0 ft. lbs.
  Step 3: Tighten all bolts to 76-84 ft. lbs.

⑦ Step 1: 19-21 ft. lbs.
  Step 2: Plus 90-94 degrees

⑧ Intake manifold bolt 18-24 ft. lbs.
  Stay bolts: M8 (12-14 ft. lbs.), M10 (23-31 ft. lbs.)

⑨ Exhaust manifold nut 29-37 ft. lbs.
  Stay bolts: M8 (12-16 ft. lbs.), M10 (28-38 ft. lbs.)
  Stay bolts: M12 (48-63 ft. lbs.)
  Engine hanger bolt 22-30 ft. lbs.

⑩ Intake manifold bolts 15-17 ft. lbs.
  Stay bolts: M8 (15-17 ft. lbs.), M10 (31-39 ft. lbs.)

37671_ECLI_C0010

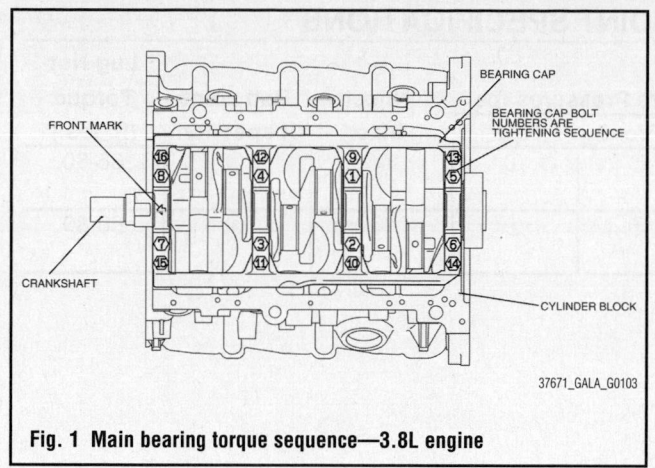

Fig. 1 Main bearing torque sequence—3.8L engine

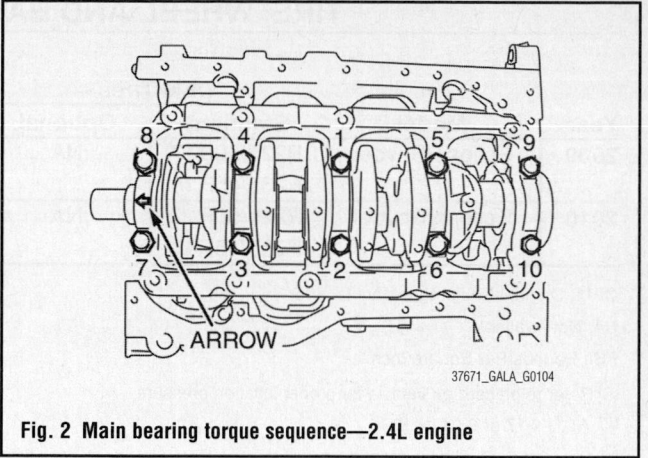

Fig. 2 Main bearing torque sequence—2.4L engine

WHEEL ALIGNMENT

Year	Model		Caster Range (+/-Deg.)	Caster Preferred Setting (Deg.)	Camber Range (+/-Deg.)	Camber Preferred Setting (Deg.)	Toe-in (Deg.)
2009	Eclipse/Spyder	Front	0.30	+3.00	0.30	0.00	0 +/- 0.12
		Rear	—	—	0.30	-0.50	0.12 +/- 0.12
2010	Eclipse/Spyder	Front	0.30	+3.00	0.30	0.00	0 +/- 0.12
		Rear	—	—	0.30	-0.50	0.12 +/- 0.12

37671_ECLI_C0011

TIRE, WHEEL AND BALL JOINT SPECIFICATIONS

| Year | Model | OEM Tires | | Tire Pressures (psi) | | Wheel Size | Ball Joint Inspection | Lug Nut Torque (ft. lbs.) |
		Standard	Optional	Front	Rear			
2009	Eclipse/Spyder	P225/50R17 P235/45R18	NA	①	①	②	③	66-80
2010	Eclipse/Spyder	P225/50R17 P235/45R18	NA	①	①	②	③	66-80

OEM: Original Equipment Manufacturer

NA: Not Available

PSI: Pounds Per Square Inch

① Refer to placard on vehicle for proper inflation pressure.

② 7.5JJ x 17 or 8.0JJ x 18

③ Replace the ball joint if too loose or if rotating torque exceeds specification: 31-61 inch lbs.

37671_ECLI_C0012

BRAKE SPECIFICATIONS
All measurements in inches unless noted

| Year | Model | | Brake Disc | | | Minimum Lining Thickness | Brake Caliper | |
			Original Thickness	Minimum Thickness	Maximum Runout		Bracket Bolts (ft. lbs.)	Mounting Bolts (ft. lbs.)
2009	Eclipse/Spyder	F	1.020	0.960	0.0039	0.080	25-31	67-81
		R	0.390	0.330	0.0016	0.080	28-36	42-48
2010	Eclipse/Spyder	F	1.020	0.960	0.0039	0.080	25-31	67-81
		R	0.390	0.330	0.0016	0.080	28-36	42-48

NA: Not Available

F: Front

R: Rear

37671_ECLI_C0013

SCHEDULED MAINTENANCE INTERVALS
Eclipse/Spyder

TO BE SERVICED	TYPE OF	VEHICLE MILEAGE INTERVAL (x1000)														
		7.5	15	22.5	30	37.5	45	52.5	60	67.5	75	82.5	90	97.5	105	120
Accessory drive belts	S/I				✔				✔				✔			✔
Air cleaner element (engine)	R				✔				✔				✔			✔
Air conditioner system	S/I	Inspect the system operation annually														
Automatic transaxle fluid	S/I				✔				✔				✔			✔
Ball joint and steering linkage seals	S/I				✔				✔				✔			✔
Brake lines, hoses, and connections	S/I		✔		✔		✔		✔		✔		✔		✔	✔
Brake pads, calipers, & rotors	S/I		✔		✔		✔		✔		✔		✔		✔	✔
Cooling system hoses and coolant level	S/I				✔				✔				✔			✔
Driveshafts and CV-boots	S/I		✔		✔		✔		✔		✔		✔			✔
Engine coolant	R								✔				✔			✔
Engine oil and filter	R	✔	✔	✔	✔	✔	✔	✔	✔	✔	✔	✔	✔	✔	✔	✔
Exhaust pipe connections, muffler, and suspension bolts	S/I				✔				✔				✔			✔
Evaporative emission control system (except canister)	S/I								✔							✔
Fuel hoses	S/I				✔				✔				✔			✔
Fuel system (tank, connections, gas cap)	S/I								✔							✔
Manual Transaxle	S/I				✔				✔				✔			✔
Spark plugs (standard)	R				✔				✔				✔			✔
Spark plugs (Iridium coated)	R	Every 84 months or 105,000 miles (under normal usage)														
Spark plugs (platinum coated)	R								✔							✔
Suspension system	S/I				✔				✔				✔			✔
Timing belt	R														✔	
Tires (rotate)	S/I	✔	✔	✔	✔	✔	✔	✔	✔	✔	✔	✔	✔	✔	✔	✔
Valve clearance	S/I				✔				✔				✔			✔

R: Replace   S/I: Service or Inspect

FREQUENT OPERATION MAINTENANCE (SEVERE SERVICE)

If a vehicle is operated under any of the following conditions it is considered severe service:

- Extremely dusty areas.

- 50% or more of the vehicle operation is in 90°F (32°C) or higher temperatures, or constant operation in temperatures below 32°F (0°C).

- Prolonged idling (vehicle operation in stop and go traffic).

- Frequent short running periods (engine does not warm to normal operating temperatures).

- Police, taxi, delivery usage, or trailer towing usage.

Air cleaner filter: replace every 15,000 miles.

Automatic transaxle fluid & filter: check every 15,000 miles, replace every 24,000 miles.

Brake pads, calipers & rotors: service or inspect every 6 months or 7,500 miles.

Manual transaxle oil change every 30,000 miles

Oil & oil filter: change every 3,750 miles.

Spark plugs: service or inspect every 15,000 miles (standard plugs only)

Suspension system inspect for looseness and damage every 6 months or 7,500 miles

37671_ECLI_C0014

BRAKES — INFORMATION AND PRECAUTIONS

ANTI-LOCK SYSTEMS

• Certain components within the ABS system are not intended to be serviced or repaired individually.

• Do not use rubber hoses or other parts not specifically specified for and ABS system. When using repair kits, replace all parts included in the kit. Partial or incorrect repair may lead to functional problems and require the replacement of components.

• Lubricate rubber parts with clean, fresh brake fluid to ease assembly. Do not use shop air to clean parts; damage to rubber components may result.

• Use only DOT 3 brake fluid from an unopened container.

• If any hydraulic component or line is removed or replaced, it may be necessary to bleed the entire system.

• A clean repair area is essential. Always clean the reservoir and cap thoroughly before removing the cap. The slightest amount of dirt in the fluid may plug an orifice and impair the system function. Perform repairs after components have been thoroughly cleaned; use only denatured alcohol to clean components. Do not allow ABS components to come into contact with any substance containing mineral oil; this includes used shop rags.

• The Anti-Lock control unit is a microprocessor similar to other computer units in the vehicle. Ensure that the ignition switch is **OFF** before removing or installing controller harnesses. Avoid static electricity discharge at or near the controller.

• If any arc welding is to be done on the vehicle, the control unit should be unplugged before welding operations begin.

DISC AND DRUM SYSTEMS

✳✳ CAUTION

Dust and dirt accumulating on brake parts during normal use may contain asbestos fibers from production or aftermarket brake linings. Breathing excessive concentrations of asbestos fibers can cause serious bodily harm. Exercise care when servicing brake parts. Do not sand or grind brake lining unless equipment used is designed to contain the dust residue. Do not clean brake parts with compressed air or by dry brushing. Cleaning should be done by dampening the brake components with a fine mist of water, then wiping the brake components clean with a dampened cloth. Dispose of cloth and all residue containing asbestos fibers in an impermeable container with the appropriate label. Follow practices prescribed by the Occupational Safety and Health Administration (OSHA) and the Environmental Protection Agency (EPA) for the handling, processing, and disposing of dust or debris that may contain asbestos fibers.

BRAKES — BLEEDING THE BRAKE SYSTEM

BLEEDING PROCEDURE

BLEEDING PROCEDURE

See Figure 3.

When any part of the hydraulic system has been disconnected for repair or replacement, air may get into the lines and cause spongy pedal action (because air can be compressed and brake fluid cannot). To correct this condition, it is necessary to bleed the hydraulic system so to be sure all air is purged.

When bleeding the brake system, bleed one brake cylinder at a time, beginning at the cylinder with the longest hydraulic line (farthest from the master cylinder) first. ALWAYS keep the master cylinder reservoir filled with brake fluid during the bleeding operation. Never use brake fluid that has been drained from the hydraulic system, no matter how clean it is.

The primary and secondary hydraulic brake systems are separate and are bled independently. During the bleeding operation, do not allow the reservoir to run dry. Keep the master cylinder reservoir filled with brake fluid.

1. Clean all dirt from around the master cylinder fill cap, remove the cap and fill the master cylinder with brake fluid until the level is within ¼ inch (6mm) of the top edge of the reservoir.

2. Clean the bleeder screws at all 4 wheels. The bleeder screws are located on the back of the brake backing plate (drum brakes) and on the top of the brake calipers (disc brakes).

3. Attach a length of rubber hose over the bleeder screw and place the other end of the hose in a glass jar, submerged in brake fluid.

4. Open the bleeder screw ½–¾ turn. Have an assistant slowly depress the brake pedal.

✳✳ CAUTION

Brake fluid contains polyglycol ethers and polyglycols. Avoid contact with the eyes and wash your hands thoroughly after handling brake fluid. If you do get brake fluid in your eyes, flush your eyes with clean, running water for 15 minutes. If eye irritation persists, or if you have taken brake fluid internally, IMMEDIATELY seek medical assistance.

5. Close the bleeder screw and tell your assistant to allow the brake pedal to return slowly. Continue this process to purge all air from the system.

6. When bubbles cease to appear at the end of the bleeder hose, close the bleeder screw and remove the hose.

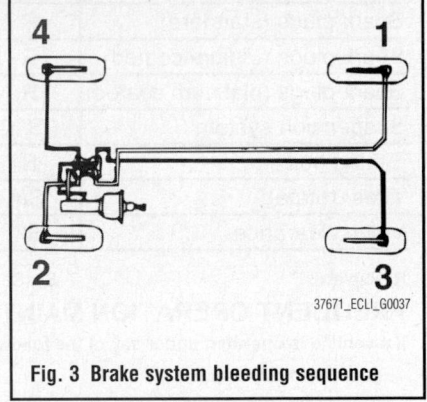

Fig. 3 Brake system bleeding sequence

Tighten the bleeder screw to the proper torque.

7. Check the master cylinder fluid level and add fluid accordingly. Do this after bleeding each wheel.

8. Repeat the bleeding operation at the remaining 3 wheels, ending with the one closet to the master cylinder.

9. Fill the master cylinder reservoir to the proper level.

BLEEDING THE ABS SYSTEM

There are no special procedures for bleeding the ABS system. Refer to the conventional bleeding procedures.

BRAKES

ANTI-LOCK BRAKE SYSTEM (ABS)

WHEEL SPEED SENSORS

REMOVAL & INSTALLATION

See Figures 4 and 5.

1. Before servicing the vehicle, refer to the Precautions Section.

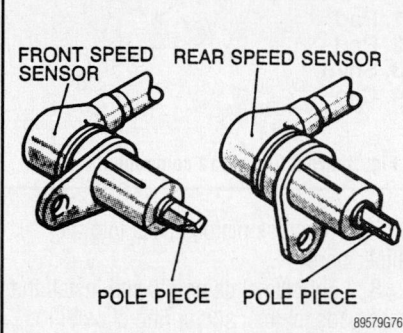

89579G76

Fig. 4 Inspect removed speed sensor for damaged pole piece. Also check pole piece for foreign material or metal adhesion

➡**If working near and/or around the SRS system and components, be sure to disable the SRS system. Tape the negative battery cable with insulating tape. Always disconnect the negative battery cable first.**

✳✳ CAUTION

Wait for 1 minute after disconnecting the negative battery cable before working inside the vehicle. The air bag system is set to deploy for a short period of time after the battery is disconnected.

2. Raise and safely support the vehicle.
3. Remove the wheel and tire assembly.
4. Remove the inner fender or splash shield.
5. Beginning at the sensor end, carefully disconnect or release each clip and retainer along the sensor wire.

➡**Take careful note of the exact position of each clip; they must be reinstalled in the identical position. Rear wheel sensor harnesses may be held by plastic wire ties; these may be cut away but must be replaced at reassembly.**

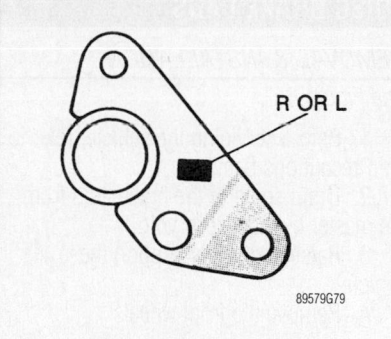

89579G79

Fig. 5 Rear speed sensor bracket identification marking

6. Detach the sensor connector at the end of the harness.
7. Remove the 2 bolts holding the speed sensor bracket to the knuckle and remove the assembly from the vehicle.

✳✳ WARNING

The speed sensor has a pole piece projecting from it. This exposed tip must be protected from impact or scratches. Do not allow the pole piece to contact the toothed wheel during removal or installation.

8. Remove the wheel speed sensor from the bracket.

To install:

➡**Be sure to use new fasteners, as required.**

9. Assemble the sensor into the bracket. Note that the brackets are different for the left and right front wheels, as well as both side rear wheels. Each bracket has identifying letters stamped on it.
10. Identify the front speed sensor brackets as follows:
   a. FR: Indicates that the bracket is for the front speed sensor.
   b. R: Indicates that the bracket is for the right wheel.
   c. L: Indicates that the bracket is for the left wheel.
11. Identify the rear speed sensor brackets as follows:
   a. R: Indicates that the bracket is for the right wheel.

b. L: Indicates that the bracket is for the left wheel.
12. Temporarily install the speed sensor to the spindle or knuckle; tighten the bolts only finger tight.

✳✳ WARNING

During speed sensor installation to the mounting bracket, make sure the letters FR are visible. Be careful when installing the speed sensor on the vehicle, that the pole piece at the tip of the sensor does not strike the toothed edge of the rotor, and damage them.

13. Route the cable correctly and loosely install the clips and retainers. All clips must be in their original position and the sensor cable must not be twisted. Improper installation may cause cable damage and system failure.
14. Use a brass or other non-magnetic feeler gauge to check the air gap between the tip of the pole piece and the toothed wheel. The correct gap is 0.012–0.035 inch (0.3–0.9mm). Tighten the 2 sensor bracket bolts to 10 ft. lbs. (14 Nm) with the sensor located so that the gap is the same at several points on the toothed wheel. If the gap is incorrect, it is likely that the toothed wheel is worn or improperly installed.
15. Tighten the screws and bolts for the cable retaining clips.
16. Install the inner fender or splash shield.
17. Install the wheel and tire assembly. Lower the vehicle to the ground.
18. Inspect the brake system for proper operation.

WHEEL SPEED SENSOR RINGS (TOOTHED RINGS)

REMOVAL & INSTALLATION

Front

➡**The front wheel speed rotors are integrated with the BJ assembly of the halfshaft and cannot be disassembled.**

Rear

➡**The rear wheel speed rotors are integrated with the rear hub assembly and cannot be disassembled.**

BRAKES

FRONT DISC BRAKES

BRAKE CALIPER

REMOVAL & INSTALLATION

1. As required, partially drain the master cylinder.
2. Raise and support the vehicle safely.
3. Remove the tire and wheel assembly.
4. Disconnect and plug the brake hose connection. Discard the gasket.
5. Remove the caliper retaining bolts. Remove the caliper from the vehicle.

To install:

➡**Be sure to use new fasteners, as required.**

6. Position the caliper to its mounting on the vehicle.
7. Install the caliper mounting bolts and tighten to 67–81 ft. lbs. (91–110 Nm).
8. Connect the brake line to the caliper with new gaskets. Torque the brake line union bolt to 18–22 ft. lbs. (25–30 Nm).
9. Bleed the brake system.
10. Install the wheel and tire.
11. Before attempting to move the vehicle, pump the brake pedal to seat the pads against the rotors. Make sure the vehicle has a firm brake pedal. Check the level of the brake fluid and add fluid if necessary.

DISC BRAKE PADS

REMOVAL & INSTALLATION

See Figure 6.

1. Before servicing the vehicle, refer to the Precautions Section.
2. Drain some of the brake fluid from the master cylinder reservoir.
3. Raise and safely support the vehicle.
4. Remove the front wheels.

➡**Do not allow the caliper to hang by the brake hose. Properly support the caliper with mechanics wire, as required.**

5. Remove the caliper guide and lock pins and lift the caliper assembly from the caliper support.

➡**On some vehicles, the caliper can be flipped up by leaving the upper pin in place and using it as a pivot point.**

6. Remove the brake pads, spring clip, and shims.

To install:

➡**Be sure to use new fasteners, as required.**

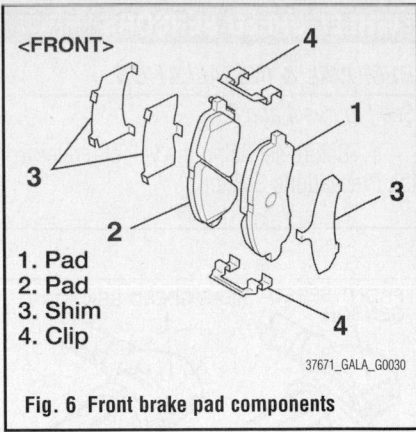

1. Pad
2. Pad
3. Shim
4. Clip

37671_GALA_G0030

Fig. 6 Front brake pad components

7. Compress pistons back into the caliper bore.
8. Lubricate slide points and install the brake pads, shims, and spring clips onto the caliper support.
9. Install the caliper over the brake pads.
10. Lubricate and install the caliper guide and lock pins in their original positions.
11. Install the wheels.
12. Before attempting to move the vehicle, pump the brake pedal to seat the pads against the rotors. Make sure the vehicle has a firm brake pedal. Check the level of the brake fluid and add fluid if necessary.

BRAKES

REAR DISC BRAKES

BRAKE CALIPER

REMOVAL & INSTALLATION

See Figure 7.

1. Before servicing the vehicle, refer to the Precautions Section.

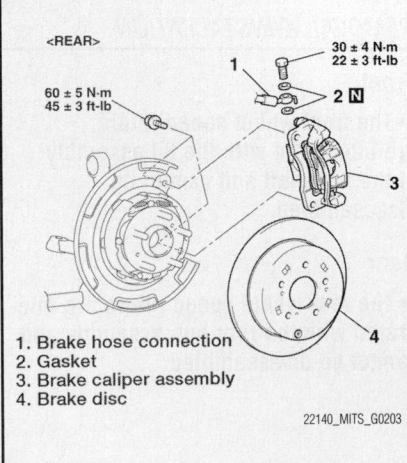

1. Brake hose connection
2. Gasket
3. Brake caliper assembly
4. Brake disc

22140_MITS_G0203

Fig. 7 View of typical rear disc brakes

2. As required, partially drain the master cylinder.
3. Remove or disconnect the following:
   • Wheels
   • Brake hose from the caliper
   • Caliper guide and lock pins and lift the caliper assembly from the caliper support

To install:

➡**Be sure to use new fasteners, as required.**

4. Install or connect the following:
   • Caliper onto the caliper support
   • Guide pin and lock pin and tighten to specification
   • Brake hose or banjo bolt with new washers
5. Bleed the brake system.
6. Install the wheels.

DISC BRAKE PADS

REMOVAL & INSTALLATION

See Figure 8.

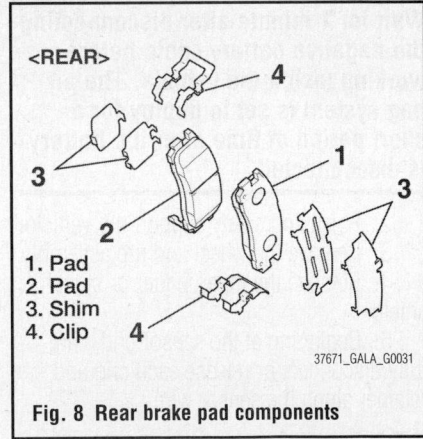

1. Pad
2. Pad
3. Shim
4. Clip

37671_GALA_G0031

Fig. 8 Rear brake pad components

1. Before servicing the vehicle, refer to the Precautions Section.
2. Raise and support the vehicle safely.
3. Remove or disconnect the following:
   • Rear wheels
   • Lower caliper mounting bolt and rotate the caliper upward
   • Pads from the caliper support
   • Pad retainers, if necessary

➡ **Do not allow the caliper to hang by the brake hose. Properly support the caliper with mechanics wire, as required.**

To install:

➡ **Be sure to use new fasteners, as required.**

4. Install or connect the following:

- Pad retainers, if removed
- Pads onto the pad retainers

5. Compress the caliper piston using a C-clamp.

6. Rotate the caliper downward and install the mounting bolt.

7. Install the wheel.

8. Pump the brake pedal until the brake pads are seated and a firm pedal is achieved before attempting to move the vehicle.

✳✳ CAUTION

Do not move the vehicle until a firm pedal is obtained.

9. Road test the vehicle to check for proper brake operation.

BRAKES

PARKING BRAKE CABLES

ADJUSTMENT

See Figures 9 and 10.

1. Before servicing the vehicle, refer to the Precautions Section.

2. Pull the parking brake lever with a force of 45 lbs. (200 N) and count the number of notches. Standard value is: 5–7 notches.

3. If the parking brake lever is not the standard value, adjust in the following manner.

4. Release the parking brake.

5. Remove the inner compartment mat of the floor console.

6. Loosen the adjusting nut at the end of the cable rod, freeing the parking brake.

7. Raise and safely support the vehicle.

8. Remove the tire and wheel assemblies.

9. Remove the adjustment hole plug on the brake disc.

10. Use a flat bladed tool to turn the adjuster in the direction of the arrow (see illustration) which expands the shoe. Return the adjuster five notches in the direction opposite to the direction of the arrow.

11. Install the tire and wheel assemblies.

PARKING BRAKE

➡ **Be careful that the parking brake lever stroke is within the standard value. If the stroke is too short, brake drag can occur.**

12. Turn the adjusting nut to adjust the parking brake lever stroke to specification.

13. After adjustment check to see that the nut and cable rod are not loose.

PARKING BRAKE SHOES

REMOVAL & INSTALLATION

See Figure 11.

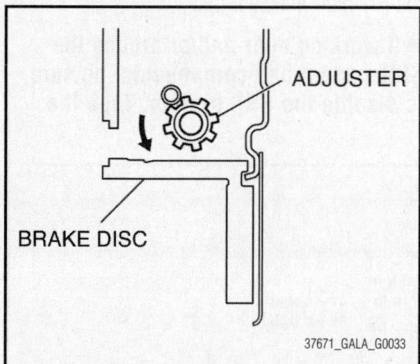

Fig. 9 Rear parking brake adjuster positioning

ADJUSTER

BRAKE DISC

37671_GALA_G0033

◁INSIDE THE FLOOR CONSOLE▷

CABLE ROD

ADJUSTING NUT

37671_GALA_G0032

Fig. 10 Rear parking brake adjusting nut location

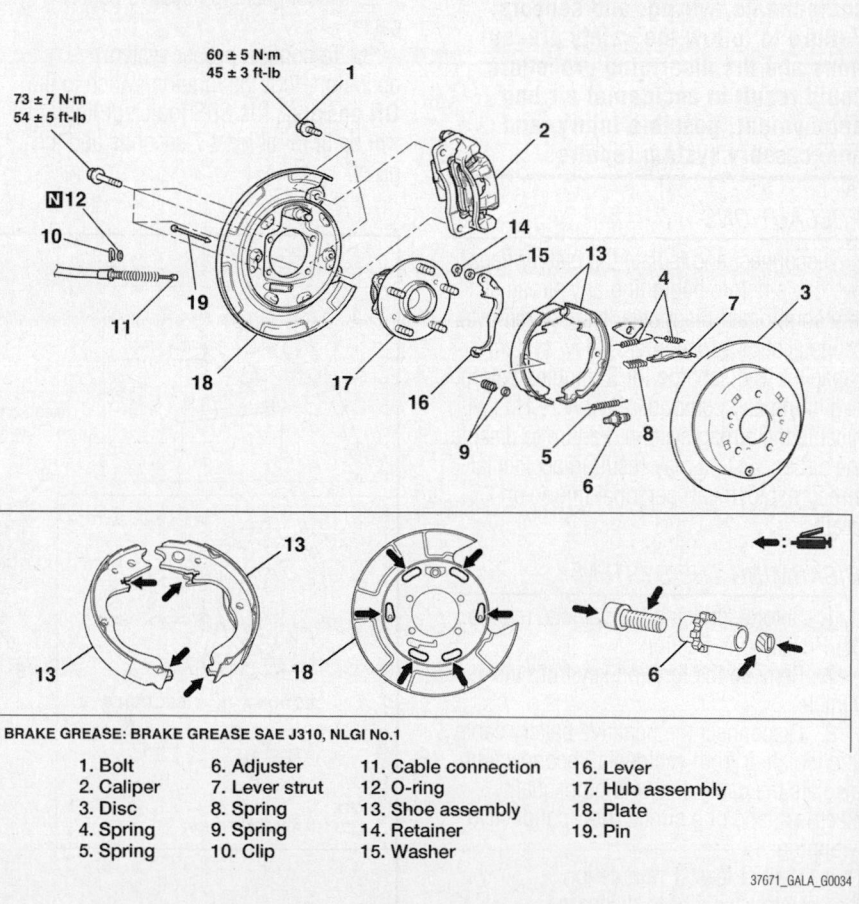

BRAKE GREASE: BRAKE GREASE SAE J310, NLGI No.1

1. Bolt	6. Adjuster	11. Cable connection	16. Lever
2. Caliper	7. Lever strut	12. O-ring	17. Hub assembly
3. Disc	8. Spring	13. Shoe assembly	18. Plate
4. Spring	9. Spring	14. Retainer	19. Pin
5. Spring	10. Clip	15. Washer	

37671_GALA_G0034

Fig. 11 Rear parking brake shoes and related components

1. Before servicing the vehicle, refer to the Precautions Section.

2. Raise and safely support the vehicle.

3. Remove the caliper assembly.

4. Remove the rear brake rotor.

➡ **When servicing, only dissemble and assemble one side at a time, leaving the remaining side intact for reference.**

5. Remove the front and rear shoe-to-anchor springs.

6. Remove the adjusting wheel spring and the adjuster.

7. Remove the strut and the strut return spring.

8. Remove the shoe hold-down cup, spring, and pin.

9. Remove the shoe and lining assembly.

10. Unfasten the clips and the retaining bolts.

11. Remove the parking brake cable.

To install:

➡ **Be sure to use new fasteners, as required.**

12. Installation is the reverse of the removal procedure.

ADJUSTMENT

Brake Lining Seating

1. Adjust the parking brake lever stroke.

2. Hook a spring scale onto the center of the parking brake lever grip and pull it with a force of about 22–23 ft. lbs. in the direction perpendicular to the handle.

3. Drive the vehicle at a constant speed of 22–31 MPH for about 328 feet.

4. Release the parking brake and let the brakes cool for 5–10 minutes.

5. Repeat the above about 4–5 times.

CHASSIS ELECTRICAL　　AIR BAG (SUPPLEMENTAL RESTRAINT SYSTEM)

GENERAL INFORMATION

✳✳ CAUTION

This vehicle is equipped with an air bag system. The system must be disarmed before performing service on, or around, system components, the steering column, instrument panel components, wiring, and sensors. Failure to follow the safety precautions and the disarming procedure could result in accidental air bag deployment, possible injury, and unnecessary system repairs.

PRECAUTIONS

Disconnect and isolate the battery negative cable before beginning any airbag system component diagnosis, testing, removal, or installation procedures. Allow system capacitor to discharge for 3 minutes before beginning any component service. This will disable the airbag system. Failure to disable the airbag system may result in accidental airbag deployment, personal injury, or death.

DISARMING THE SYSTEM

1. Before servicing the vehicle, refer to the Precautions Section.

2. Remove the ignition key from the vehicle.

3. Disconnect the negative battery cable and isolate it from accidental reconnection. Insulate the cable end with high-quality electrical tape or a similar non-conductive wrapping.

4. Wait at least 1 minute for the system capacitor to discharge before performing any service. The air

bag system is designed to retain enough voltage to deploy the air bag for a short period of time after the battery has been disconnected.

ARMING THE SYSTEM

1. Before servicing the vehicle, refer to the Precautions Section.

2. Reconnect the negative battery cable.

3. To confirm proper system operation, turn the ignition switch to the **ON** position. The SRS indicator light will be lit for at least 7 seconds and then go off.

CLOCKSPRING CENTERING

See Figures 12 and 13.

At this time the manufacturer does not provide removal and installation procedures for this component, refer to the illustration as required. The following procedure is a guideline and may differ from the vehicle you are servicing.

1. Before servicing the vehicle, refer to the Precautions Section.

➡ **If working near and/or around the SRS system and components, be sure to disable the SRS system. Tape the**

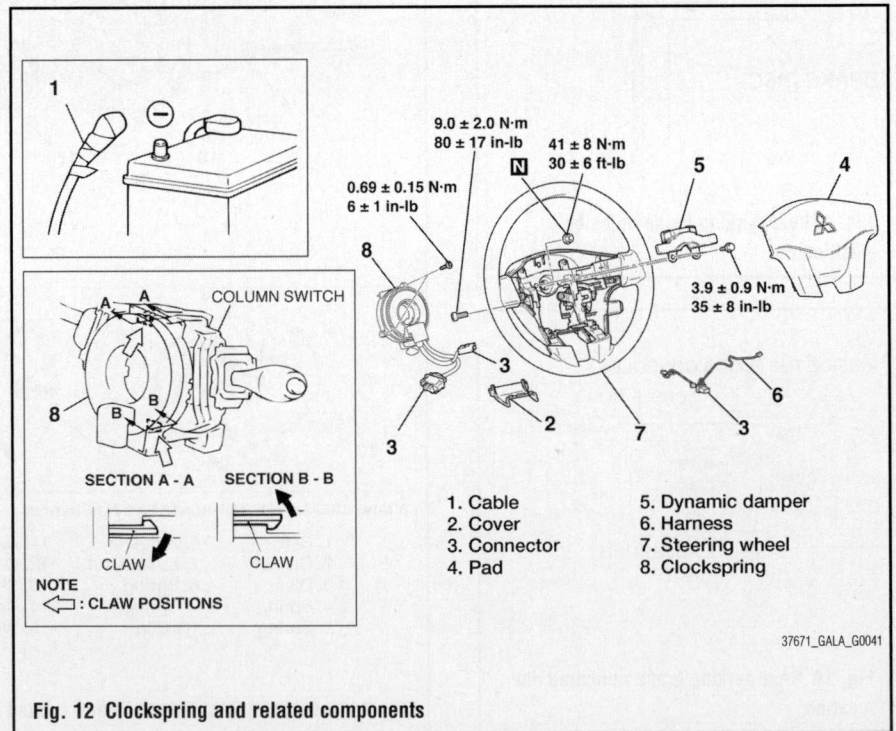

1. Cable
2. Cover
3. Connector
4. Pad
5. Dynamic damper
6. Harness
7. Steering wheel
8. Clockspring

9.0 ± 2.0 N·m
80 ± 17 in-lb

41 ± 8 N·m
30 ± 6 ft-lb

0.69 ± 0.15 N·m
6 ± 1 in-lb

3.9 ± 0.9 N·m
35 ± 8 in-lb

SECTION A - A　SECTION B - B

CLAW　CLAW

NOTE
⬅ : CLAW POSITIONS

37671_GALA_G0041

Fig. 12 Clockspring and related components

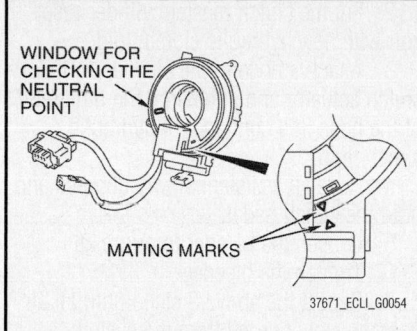

Fig. 13 Align the mating marks of the clockspring module

negative battery cable with insulating tape. Always disconnect the negative battery cable first.

☀ CAUTION
Wait for 1 minute after disconnecting the negative battery cable before working inside the vehicle. The air

bag system is set to deploy for a short period of time after the battery is disconnected.

2. Disconnect and isolate the battery negative cable. Allow the system capacitor to discharge for 1 minute before beginning any component service. This will disable the airbag system.

☀ CAUTION
Failure to disable the airbag system may result in accidental airbag deployment, personal injury, or death.

3. Remove the ignition key from the vehicle.

☀ WARNING
Ensure that the clockspring's mating marks are properly aligned. If not, the steering wheel may not rotate completely during a turn, or the flat

cable in the clockspring could be damaged. This would prevent normal SRS operation and possibly cause serious injury to the driver.

4. Align the mating marks of the clockspring.
5. Turn the clockspring clockwise fully. Then, turn it back approximately 3 ¾ turns counterclockwise to align the mating marks.
6. Turn the wheels to the straight ahead position. Then install the clock spring to the column switch.
7. Connect the battery negative cable.
8. After installing the airbag, confirm proper system operation:
   a. Turn the ignition switch ON; the SRS indicator light should be turned on for about 7 seconds and then go off.
   b. Make sure the horn functions properly.

DRIVE TRAIN

CLUTCH DRIVEN DISC & PRESSURE PLATE

REMOVAL & INSTALLATION
See Figures 14 and 15.

☀ CAUTION
The clutch driven disc may contain asbestos, which has been determined to be a cancer causing agent. Never clean clutch surfaces with compressed air! Avoid inhaling any dust from any clutch surface! When cleaning clutch surfaces, use a commercially available brake cleaning fluid.

1. Before servicing the vehicle, refer to the Precautions Section.

➡**If working near and/or around the SRS system and components, be sure to disable the SRS system. Tape the negative battery cable with insulating tape. Always disconnect the negative battery cable first.**

☀ CAUTION
Wait for 1 minute after disconnecting the negative battery cable before working inside the vehicle. The air bag system is set to deploy for a short period of time after the battery is disconnected.

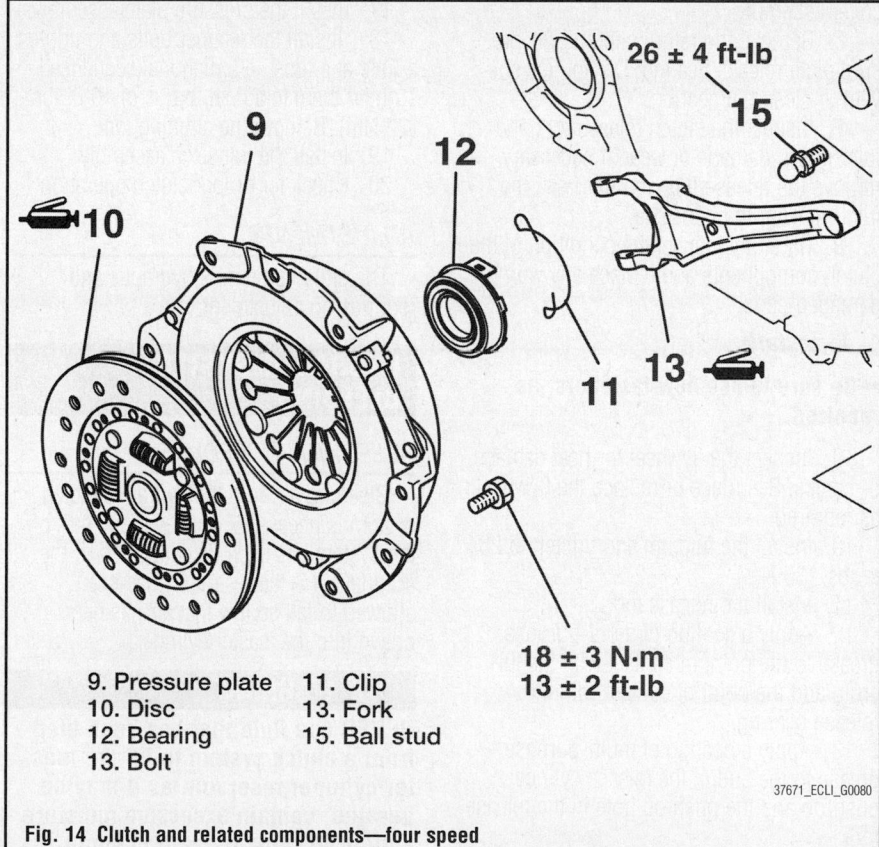

9. Pressure plate	11. Clip
10. Disc	13. Fork
12. Bearing	15. Ball stud
13. Bolt	

26 ± 4 ft-lb

18 ± 3 N·m
13 ± 2 ft-lb

Fig. 14 Clutch and related components—four speed

2. Disconnect the negative battery cable.
3. Raise and safely support the vehicle.

4. Remove the transaxle assembly from the vehicle.
5. Remove the pressure plate attaching

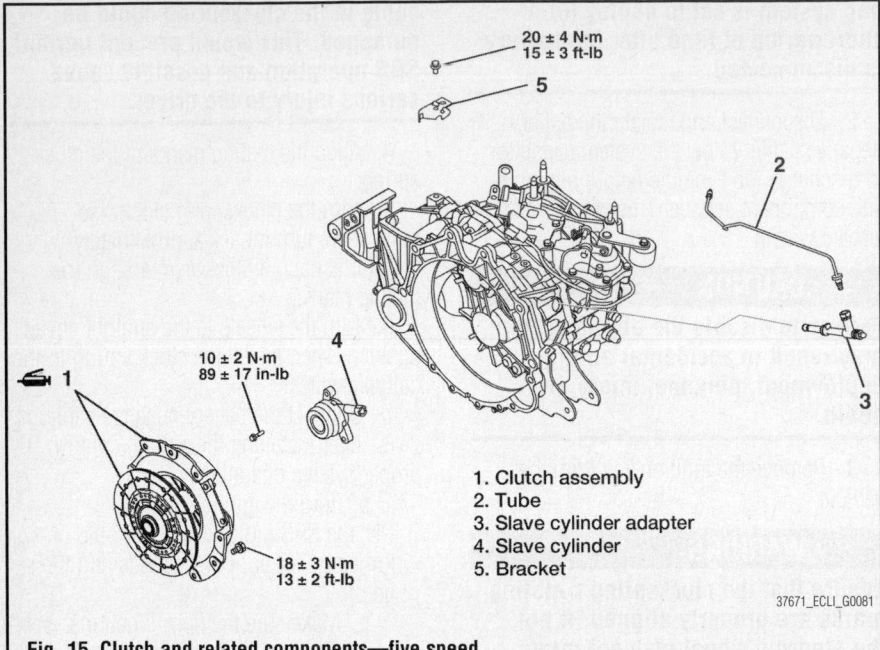

20 ± 4 N·m
15 ± 3 ft-lb

10 ± 2 N·m
89 ± 17 in-lb

18 ± 3 N·m
13 ± 2 ft-lb

1. Clutch assembly
2. Tube
3. Slave cylinder adapter
4. Slave cylinder
5. Bracket

37671_ECLI_G0081

Fig. 15 Clutch and related components—five speed

bolts, pressure plate and clutch disc. If the pressure plate is to be reused, loosen the bolts in a diagonal pattern, 1 or 2 turns at a time. This will prevent warping the clutch cover assembly.

6. Remove the return clip and the pressure plate release bearing. Do not use solvent to clean the bearing.

7. Inspect the clutch release fork and fulcrum for damage or wear. If necessary, remove the release fork and unthread the fulcrum from the transaxle.

8. Carefully inspect the condition of the clutch components and replace any worn or damaged parts.

To install:

➡**Be sure to use new fasteners, as required.**

9. Inspect the flywheel for heat damage or cracks. Resurface or replace the flywheel as required.

10. Install the fulcrum and tighten to 25 ft. lbs. (35 Nm).

11. Install the release fork.

12. Apply a coating of multi-purpose grease to the point of contact with the fulcrum and the point of contact with the release bearing.

13. Apply a coating of multi-purpose grease to the end of the release cylinder pushrod and the pushrod hole in the release fork.

14. Apply multi-purpose grease to the clutch release bearing. Pack the bearing inner surface and the groove with grease. Do not apply grease to the resin portion of the bearing.

15. Place the bearing in position and install the return clip.

16. Using the proper alignment tool, install the clutch disc to the flywheel.

17. Install the pressure plate assembly.

18. Install the retainer bolts and tighten a little at a time, in a diagonal sequence. Tighten them to a final torque of 16 ft. lbs. (22 Nm). Remove the aligning tool.

19. Install the transaxle assembly.

20. Check for proper clutch operation.

ADJUSTMENTS

The clutch system is hydraulic and requires no adjustment.

HYDRAULIC SYSTEM BLEEDING

BLEEDING PROCEDURE

Bleeding air from the hydraulic clutch system is necessary whenever any part of the system has been disconnected or the fluid level (in the reservoir) has been allowed to fall so low that air has been drawn into the master cylinder.

✳✳ WARNING

NEVER use fluid that has been bled from a clutch system to fill the master cylinder reservoir, as it may be aerated, contain excessive moisture and/or be contaminated in some other way.

1. Before servicing the vehicle, refer to the Precautions Section.

2. Fill the clutch master cylinder reservoir with new hydraulic clutch fluid.

3. Attach a hose to the bleeder on the clutch actuator and submerge the other end of the hose in a container of hydraulic clutch fluid.

4. Have an assistant slowly depress and hold the clutch pedal.

5. Loosen the bleeder to purge air.

6. Tighten the bleeder.

7. Repeat the above 3 steps until all air is completely purged from the system.

➡**Be sure that the fluid reservoir level stays between MAX and MIN throughout the bleeding procedure.**

8. Refill the clutch master cylinder reservoir.

FRONT HALFSHAFT

REMOVAL & INSTALLATION

See Figure 16.

1. Before servicing the vehicle, refer to the Precautions Section.

2. Raise and support the vehicle safely.

3. Remove the front undercover. Remove the side undercover.

4. Drain the transaxle fluid.

5. On vehicles equipped with the 3.8L engine, disconnect the front exhaust pipe if working on the right side halfshaft.

6. Disconnect the speed sensor connection. Remove the wheel speed sensor. Remove the brake hose clip.

7. Remove the cotter pin. Install tool MB990767 to the hub and remove the halfshaft nut. Remove the washer.

8. Remove the lower ball joint cotter pin.

➡**Do not remove the nut from the ball joint. Loosen it and use special tool MB991897 to avoid possible damage to the ball joint threads. Hang the special tool in place with wire or string to prevent it from falling.**

9. Install the special tool. Turn the bolt and knob as necessary to make the jaws of the tool parallel. Tighten the bolt by hand and confirm that the jaws are still parallel.

➡**When adjusting the jaws in parallel, make sure the knob is in the vertical (upward) position.**

10. Tighten the bolt with a wrench to disconnect the lower arm ball joint connection.

11. Remove the tie rod end cotter pin. Install tool MB990767 to the hub and remove the halfshaft nut. Remove the washer.

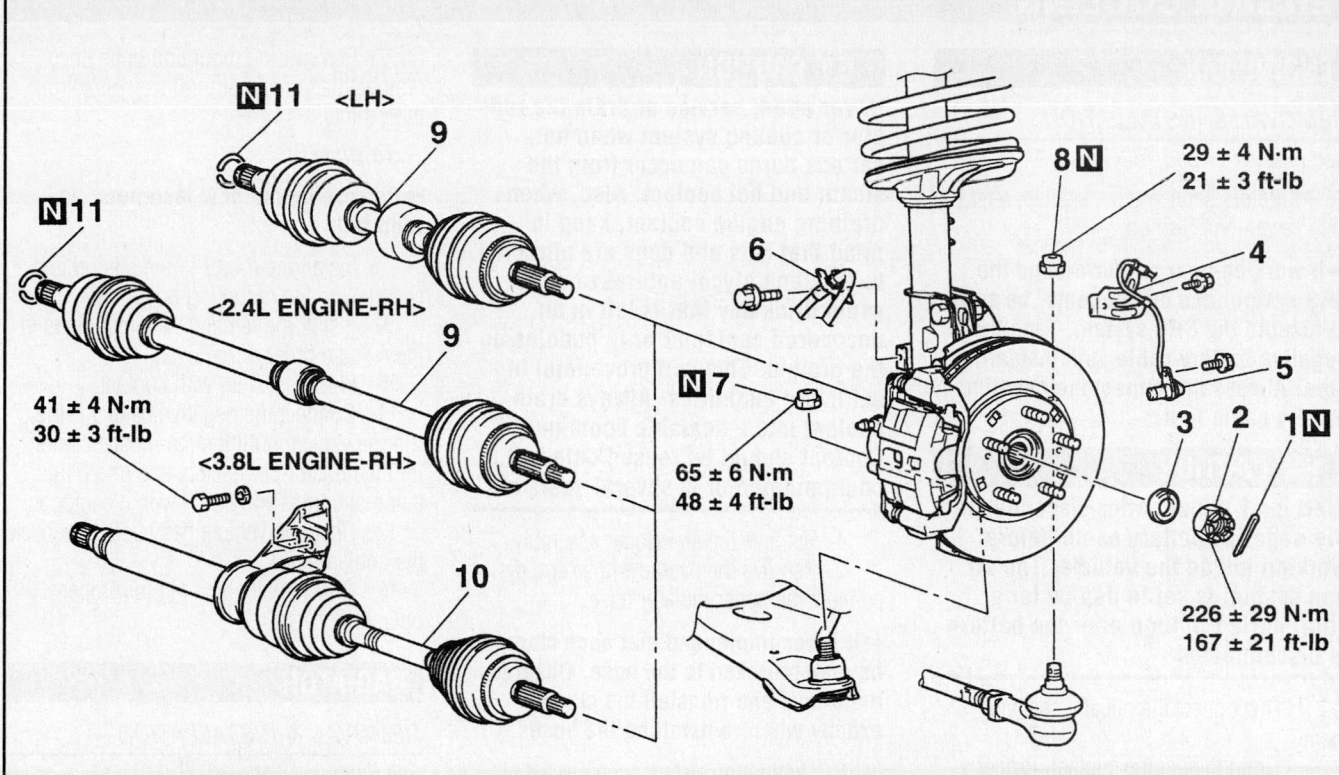

1. SPLIT PIN
2. DRIVE SHAFT NUT
3. WASHER
4. FRONT WHEEL SPEED SENSOR
   BRACKET
5. FRONT WHEEL SPEED SENSOR
6. BRAKE HOSE BRACKET
7. SELF LOCKING NUT (LOWER
   ARM BALL JOINT CONNECTION)

8. SELF LOCKING NUT (TIE ROD
   END CONNECTION)
9. DRIVE SHAFT
10. DRIVE SHAFT AND INNER
    SHAFT ASSEMBLY<3.8L
    ENGINE-RH>
11. CIRCLIP

09482_GALA_G0169

Fig. 16 Halfshaft and related components

➡Do not remove the nut from the tie rod end. Loosen it and use special tool MB991897 to avoid possible damage to the ball joint threads. Hang the special tool in place with wire or string to prevent it from falling.

12. Install the special tool. Turn the bolt and knob as necessary to make the jaws of the tool parallel. Tighten the bolt by hand and confirm that the jaws are still parallel.

➡When adjusting the jaws in parallel, make sure the knob is in the vertical (upward) position.

13. Tighten the bolt with a wrench to disconnect the tie rod end.
14. Disconnect the stabilizer link connection.

➡Do not strike the ABS rotor attached to the BJ or EBJ outer race, of the halfshaft against other parts when removing the halfshaft as damage to the rotors will result.

15. Use special tools MB991354, MB990242, and MB990767 to push the halfshaft out from the hub.

➡Do not pull on the halfshaft, doing so will damage the TJ or PTJ. Be sure to use a prybar. Do not insert the prybar so deep as to damage the oil seal.

16. Remove the halfshaft from the hub by pulling the bottom of the brake disc towards you.
17. Insert a prybar between the transaxle case and the halfshaft, and then

pry and remove the halfshaft from the transaxle.

➡Insert a prybar, taking care not to damage the protrusion of the transaxle case when removing the halfshaft (left side).

18. If the inner shaft and transaxle are tightly joined, tap the center bearing bracket with a plastic hammer to remove the halfshaft and inner shaft from the transaxle.

To install:

➡Be sure to use new fasteners, as required.

19. Installation is the reverse of the removal procedure.
20. Check and adjust the front end alignment, as necessary.

ENGINE COOLING

ENGINE FAN

REMOVAL & INSTALLATION

See Figures 17 and 18.

1. Before servicing the vehicle, refer to the Precautions Section.

➡ **If working near and/or around the SRS system and components, be sure to disable the SRS system. Tape the negative battery cable with insulating tape. Always disconnect the negative battery cable first.**

❊❊ **CAUTION**

Wait for 1 minute after disconnecting the negative battery cable before working inside the vehicle. The air bag system is set to deploy for a short period of time after the battery is disconnected.

2. Disconnect the negative battery cable.
3. Drain the engine coolant. Properly dispose of used coolant.

❊❊ **CAUTION**

Never open, service or drain the radiator or cooling system when hot; serious burns can occur from the steam and hot coolant. Also, when draining engine coolant, keep in mind that cats and dogs are attracted to ethylene glycol antifreeze and could drink any that is left in an uncovered container or in puddles on the ground. This will prove fatal in sufficient quantities. Always drain coolant into a sealable container. Coolant should be reused unless it is contaminated or is several years old.

4. Remove the air cleaner assembly.
5. Remove the overflow hose and disconnect the upper radiator hose.

➡ **It is recommended that each clamp be matchmarked to the hose. Observe the marks and reinstall the clamps exactly when reinstalling the hoses.**

6. Unplug the electrical connector(s) from the coolant fan motor(s).

7. Remove the mounting bolts, then remove the fan and shroud assembly from the vehicle.

To install:

➡ **Be sure to use new fasteners, as required.**

8. Install the motor to the shroud and secure with the mounting bolts.
9. Install the remaining components in the reverse order of removal.
10. Fill the system with coolant.
11. Connect the negative battery cable, run the vehicle until the thermostat opens, fill the radiator completely and check the automatic transaxle fluid level, if equipped.
12. Once the vehicle has cooled, recheck the coolant level.
13. Check the engine fan for proper operation.

RADIATOR

REMOVAL & INSTALLATION

See Figures 17 and 18.

1. Before servicing the vehicle, refer to the Precautions Section.

➡ **If working near and/or around the SRS system and components, be sure to disable the SRS system. Tape the negative battery cable with insulating tape. Always disconnect the negative battery cable first.**

❊❊ **CAUTION**

Wait for 1 minute after disconnecting the negative battery cable before working inside the vehicle. The air bag system is set to deploy for a short period of time after the battery is disconnected.

2. Drain the engine cooling system. Remove the radiator cap and radiator drain plug to ensure the coolant is completely drained.

❊❊ **CAUTION**

Never open, service or drain the radiator or cooling system when hot; serious burns can occur from the steam and hot coolant. Also, when draining engine coolant, keep in mind that cats and dogs are attracted to ethylene glycol antifreeze and could drink any that is left in an uncovered container or in puddles on the ground. This will prove fatal in

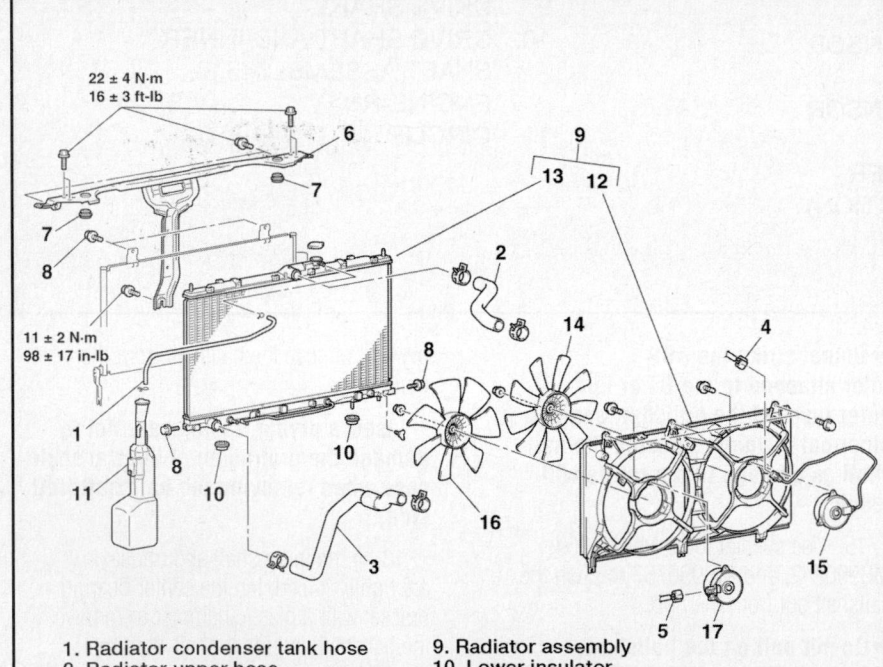

22 ± 4 N·m
16 ± 3 ft-lb

11 ± 2 N·m
98 ± 17 in-lb

1. Radiator condenser tank hose
2. Radiator upper hose
3. Radiator lower hose
4. Cooling fan motor connector (RH)
5. Cooling fan motor connector (LH)
6. Front end structure bar
7. Upper insulator
8. Condenser bolts
9. Radiator assembly
10. Lower insulator
11. Radiator condenser tank assembly
12. Cooling fan shroud assembly
13. Radiator
14. Cooling fan (RH)
15. Cooling fan motor (RH)
16. Cooling fan (LH)
17. Cooling fan motor (LH)

22140_MITS_G0070

Fig. 17 Radiator, fan and related components—2.4L engine

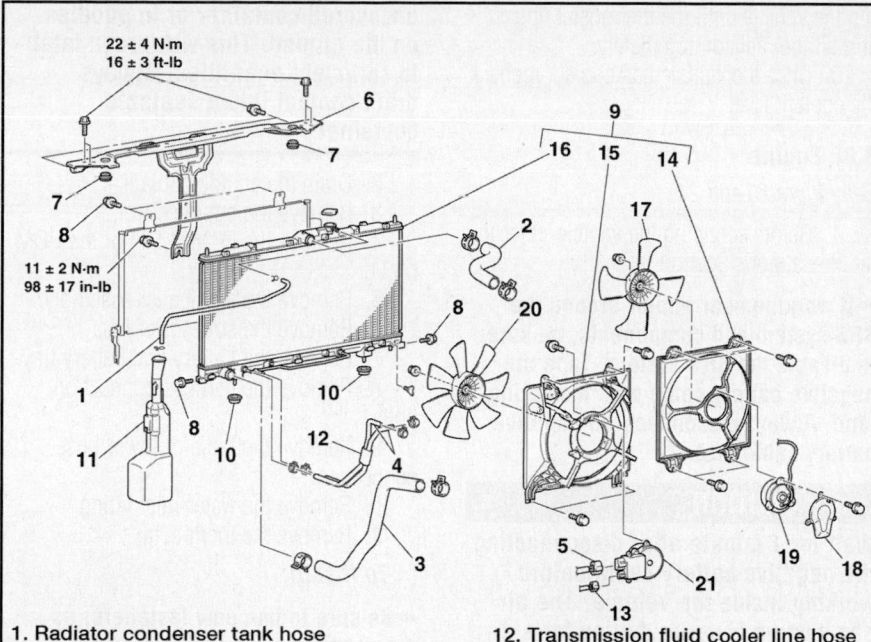

1. Radiator condenser tank hose
2. Radiator upper hose
3. Radiator lower hose
4. Transmission fluid cooler line hose connection
5. Fan control module connector
6. Front end structure bar
7. Upper insulator
8. Condenser bolts
9. Radiator assembly
10. Lower insulator
11. Radiator condenser tank assembly
12. Transmission fluid cooler line hose
13. Condenser fan motor connector
14. Condenser fan shroud assembly
15. Cooling fan shroud assembly
16. Radiator
17. Condenser fan
18. Heat protector
19. Condenser fan motor
20. Cooling fan
21. Cooling fan motor

22140_MITS_G0071

Fig. 18 Radiator, fan and related components—3.8L engine

sufficient quantities. Always drain coolant into a sealable container.

3. Remove the air cleaner assembly.

4. Remove the overflow hose and overflow tank.

5. Disconnect the upper and lower radiator hoses. Matchmark the upper radiator hose to assure installation in the proper orientation.

6. Remove the hood latch, as required.

7. If equipped with an automatic transaxle, disconnect the transaxle oil cooler hoses. Place matchmarks on the hose before removal to assure proper installation.

8. Remove the upper radiator supports.

9. Disconnect the required electrical connectors.

10. Remove the radiator assembly.

To install:

➡**Be sure to use new fasteners, as required.**

11. Install the radiator assembly and upper radiator supports. Tighten the mounting bolts to 13–19 ft. lbs. (18–26 Nm).

12. Connect the automatic transaxle oil

cooler lines. Tighten the bracket mounting bolt to 44 inch lbs. (5 Nm).

13. Connect the upper and lower radiator hoses.

14. Install the overflow tank. Tighten the mounting bolt to 100 inch lbs. (12 Nm). Connect the overflow hose.

15. Fill the system with coolant.

16. Connect the negative battery cable, run the vehicle until the thermostat opens, fill the radiator completely and check the automatic transaxle fluid level, if equipped.

17. Once the vehicle has cooled, recheck the coolant level.

THERMOSTAT

REMOVAL & INSTALLATION

2.4L Engine

See Figures 19 and 20.

1. Before servicing the vehicle, refer to the Precautions Section.

➡**If working near and/or around the SRS system and components, be sure to disable the SRS system. Tape the negative battery cable with insulating tape. Always disconnect the negative battery cable first.**

✸✸ CAUTION

Wait for 1 minute after disconnecting the negative battery cable before working inside the vehicle. The air bag system is set to deploy for a short period of time after the battery is disconnected.

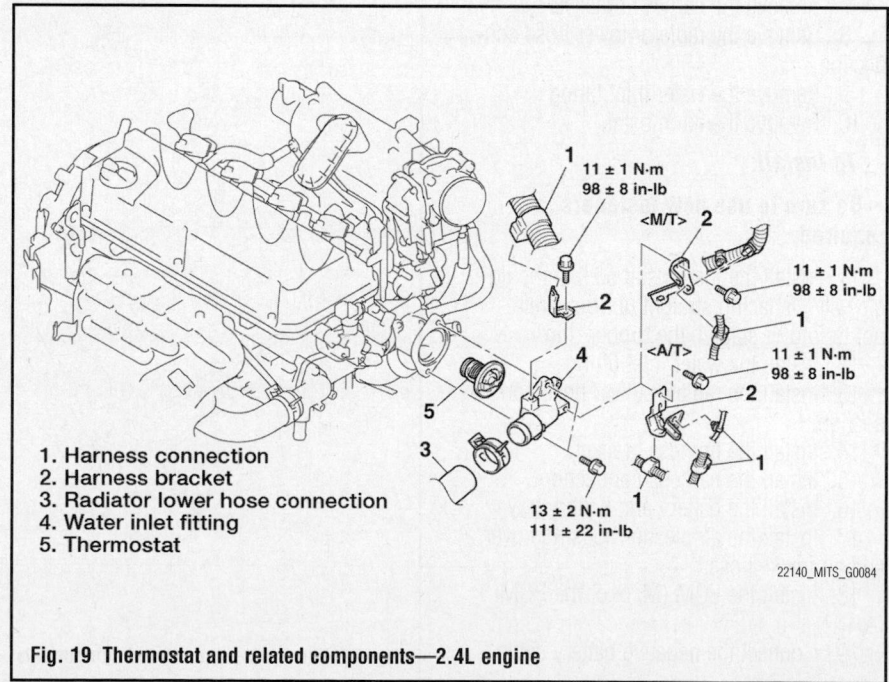

1. Harness connection
2. Harness bracket
3. Radiator lower hose connection
4. Water inlet fitting
5. Thermostat

11 ± 1 N·m
98 ± 8 in-lb

11 ± 1 N·m
98 ± 8 in-lb

11 ± 1 N·m
98 ± 8 in-lb

13 ± 2 N·m
111 ± 22 in-lb

22140_MITS_G0084

Fig. 19 Thermostat and related components—2.4L engine

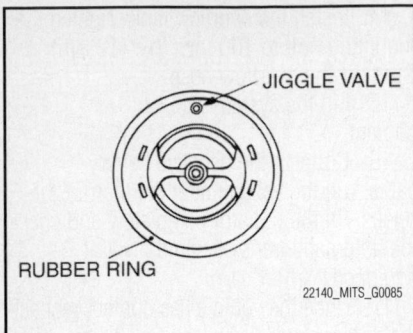

Fig. 20 Install the thermostat so that the jiggle valve is facing straight up

⁂ CAUTION

Never open, service or drain the radiator or cooling system when hot; serious burns can occur from the steam and hot coolant. Also, when draining engine coolant, keep in mind that cats and dogs are attracted to ethylene glycol antifreeze and could drink any that is left in an uncovered container or in puddles on the ground. This will prove fatal in sufficient quantities. Always drain coolant into a sealable container.

2. Drain the engine coolant.
3. Remove the ECM (M/T) or the PCM (A/T).
4. Remove the air cleaner housing cover and air intake hose.
5. Remove the battery and battery tray.
6. Remove the harness connection.
7. Remove the harness bracket.
8. Remove the radiator lower hose connection.
9. Remove the water inlet fitting.
10. Remove the thermostat.

To install:

➡**Be sure to use new fasteners, as required.**

11. Install the thermostat so that the jiggle valve is facing straight up. Be careful not to fold or scratch the rubber ring.
12. Install the water inlet fitting.
13. Install the radiator lower hose connection.
14. Install the harness bracket.
15. Install the harness connection.
16. Install the battery and battery tray.
17. Install the air cleaner housing cover and air intake hose.
18. Install the ECM (M/T) or the PCM (A/T).
19. Connect the negative battery cable,

run the vehicle until the thermostat opens and fill the radiator completely.
20. Once the vehicle has cooled, recheck the coolant level.

3.8L Engine

See Figure 20 and 21.

1. Before servicing the vehicle, refer to the Precautions Section.

➡**If working near and/or around the SRS system and components, be sure to disable the SRS system. Tape the negative battery cable with insulating tape. Always disconnect the negative battery cable first.**

⁂ CAUTION

Wait for 1 minute after disconnecting the negative battery cable before working inside the vehicle. The air bag system is set to deploy for a short period of time after the battery is disconnected.

⁂ CAUTION

Never open, service or drain the radiator or cooling system when hot; serious burns can occur from the steam and hot coolant. Also, when draining engine coolant, keep in mind that cats and dogs are attracted to ethylene glycol antifreeze and could drink any that is left in an uncovered container or in puddles on the ground. This will prove fatal in sufficient quantities. Always drain coolant into a sealable container.

2. Drain the engine coolant.
3. Remove the engine cover.
4. Remove the ECM (M/T) or the PCM (A/T).
5. Remove the air cleaner assembly.
6. Remove the strut tower bar.
7. Remove the battery and battery tray.
8. Remove the harness connection bolts.
9. Remove the radiator lower hose connection.
10. Remove the water inlet fitting.
11. Remove the thermostat.

To install:

➡**Be sure to use new fasteners, as required.**

12. Install the thermostat so that the jiggle valve is facing straight up. Be careful not to fold or scratch the rubber ring.
13. Install the water inlet fitting.
14. Install the radiator lower hose connection.
15. Install the harness bracket.
16. Install the harness connection.
17. Install the battery and battery tray.
18. Install the strut tower bar.

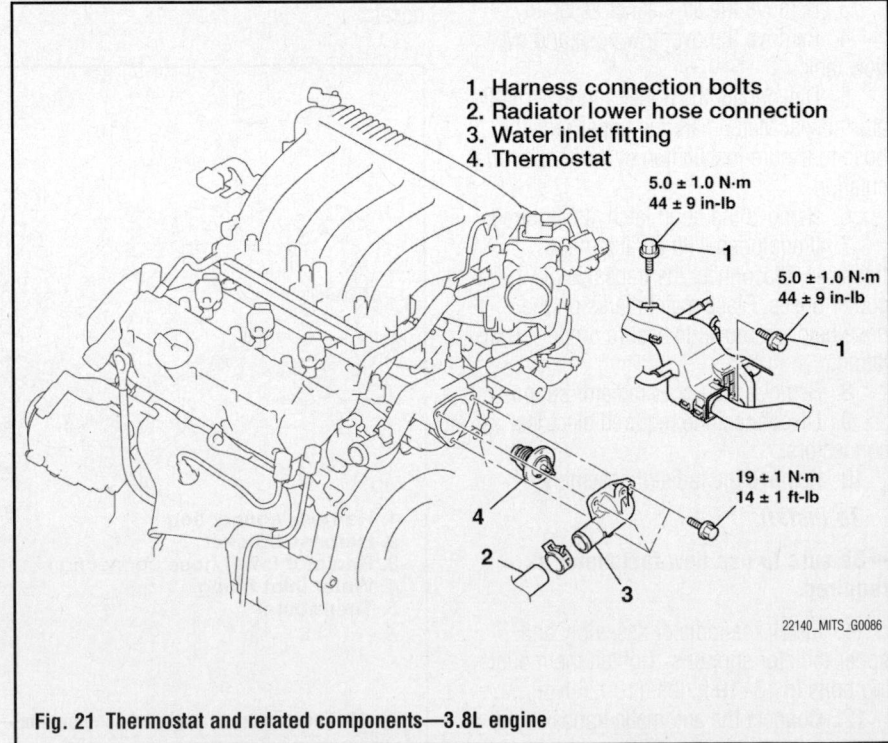

1. Harness connection bolts
2. Radiator lower hose connection
3. Water inlet fitting
4. Thermostat

5.0 ± 1.0 N·m
44 ± 9 in-lb

5.0 ± 1.0 N·m
44 ± 9 in-lb

19 ± 1 N·m
14 ± 1 ft-lb

Fig. 21 Thermostat and related components—3.8L engine

19. Install the air cleaner housing cover and air intake hose.

20. Install the ECM (M/T) or the PCM (A/T).

21. Install the engine cover.

22. Connect the negative battery cable, run the vehicle until the thermostat opens and fill the radiator completely.

23. Once the vehicle has cooled, recheck the coolant level.

WATER PUMP

REMOVAL & INSTALLATION

2.4L Engine

See Figure 22.

1. Before servicing the vehicle, refer to the Precautions Section.

➡**If working near and/or around the SRS system and components, be sure to disable the SRS system. Tape the negative battery cable with insulating tape. Always disconnect the negative battery cable first.**

✴✴ CAUTION

Wait for 1 minute after disconnecting the negative battery cable before working inside the vehicle. The air bag system is set to deploy for a short period of time after the battery is disconnected.

2. Disconnect the negative battery cable.

3. Drain the engine coolant.

4. Remove the timing belt.

5. Remove the water pump retaining bolts.

6. Remove the water pump from the engine. Discard the water pump gasket and O-ring.

➡**To install:**

➡**Be sure to use new fasteners, as required.**

7. Install or connect the following:

• New O-ring on the water inlet pipe. Coat the O-ring with water or coolant. Do not allow oil or other grease to contact the O-ring.

• Water pump to the engine block, with new gasket. Tighten the mounting bolts as illustrated.

8. Continue the installation in the reverse order of the removal procedure.

9. Fill the engine with the proper grade and type engine coolant. Start the engine and check for leaks.

3.8L Engine

See Figure 23.

1. Before servicing the vehicle, refer to the Precautions Section.

➡**If working near and/or around the SRS system and components, be sure to disable the SRS system. Tape the negative battery cable with insulating tape. Always disconnect the negative battery cable first.**

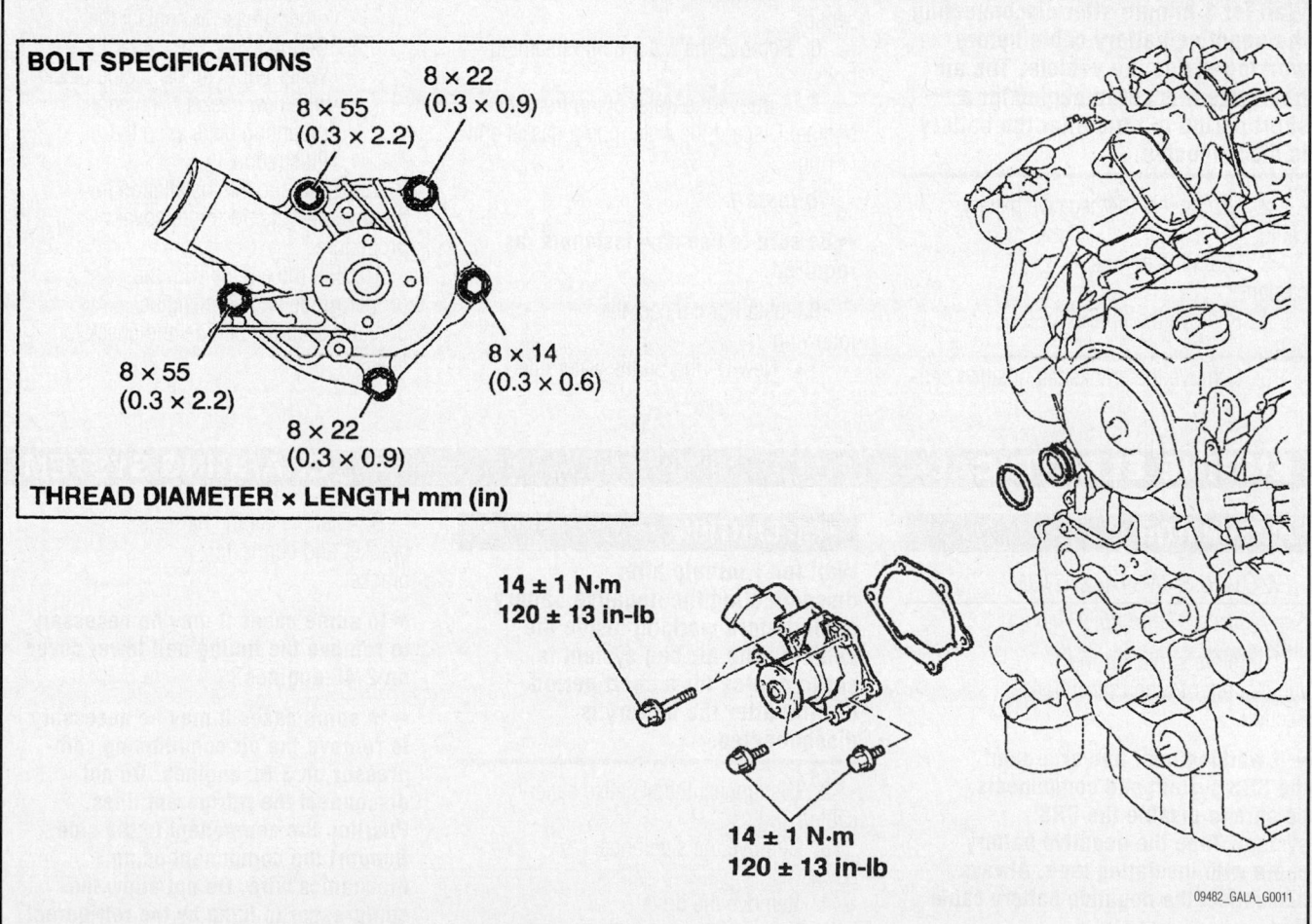

BOLT SPECIFICATIONS

8 × 55 (0.3 × 2.2)
8 × 22 (0.3 × 0.9)
8 × 55 (0.3 × 2.2)
8 × 14 (0.3 × 0.6)
8 × 22 (0.3 × 0.9)

THREAD DIAMETER × LENGTH mm (in)

14 ± 1 N·m
120 ± 13 in-lb

14 ± 1 N·m
120 ± 13 in-lb

09482_GALA_G0011

Fig. 22 Water pump and related components—2.4L engine

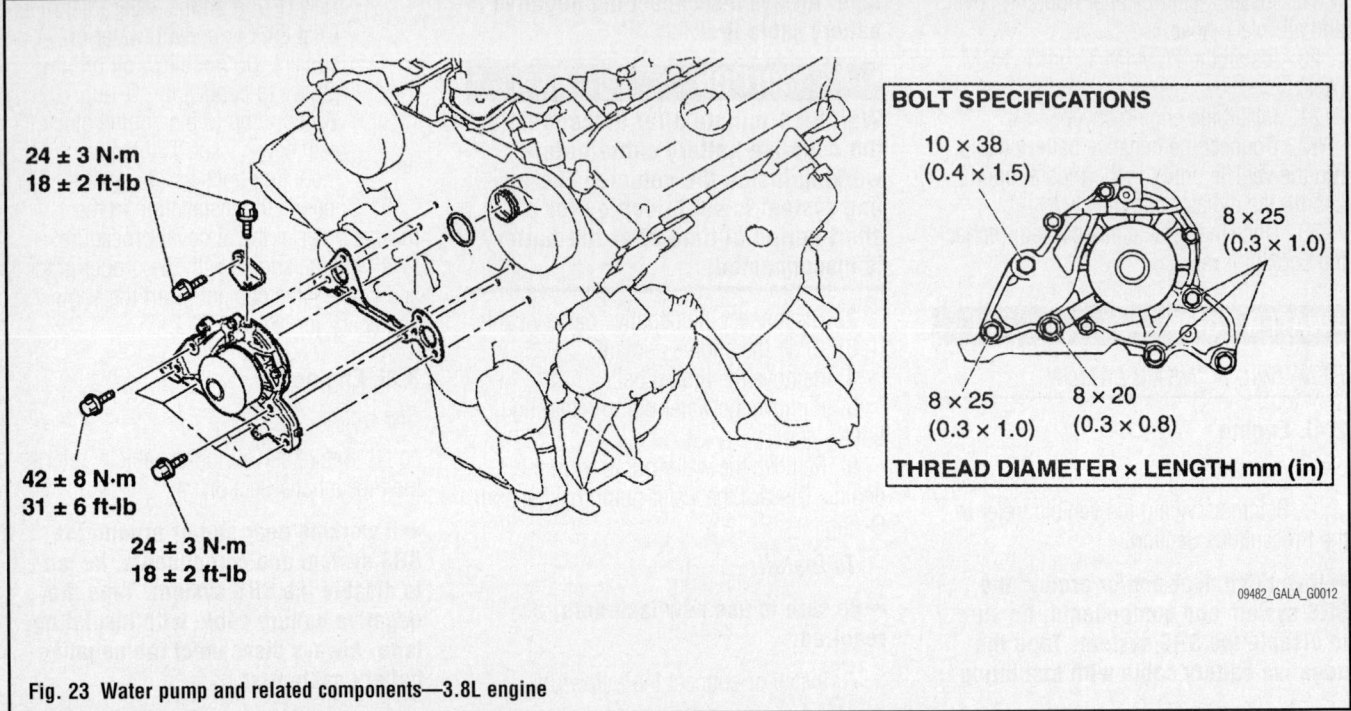

24 ± 3 N·m
18 ± 2 ft-lb

42 ± 8 N·m
31 ± 6 ft-lb

24 ± 3 N·m
18 ± 2 ft-lb

BOLT SPECIFICATIONS

10 × 38
(0.4 × 1.5)

8 × 25
(0.3 × 1.0)

8 × 25
(0.3 × 1.0)

8 × 20
(0.3 × 0.8)

THREAD DIAMETER × LENGTH mm (in)

09482_GALA_G0012

Fig. 23 Water pump and related components—3.8L engine

✷ CAUTION

Wait for 1 minute after disconnecting the negative battery cable before working inside the vehicle. The air bag system is set to deploy for a short period of time after the battery is disconnected.

2. Disconnect the negative battery cable.

3. Drain the engine coolant.

4. Remove the timing belt.

5. Remove the crankshaft position sen-sor connector clip. Remove the crankshaft position sensor mounting bolt. Remove the sensor.

6. Remove the water pump retaining bolts.

7. Remove the water pump from the engine. Discard the water pump gasket and O-ring.

To install:

➡**Be sure to use new fasteners, as required.**

8. Install or connect the following:
   • New O-ring on the water inlet pipe. Coat the O-ring with water or coolant. Do not allow oil or other grease to contact the O-ring.
   • Water pump to the engine block, with new gasket. Tighten the mounting bolts as in the illustration.

9. Continue the installation in the reverse order of the removal procedure.

10. Fill the engine with the proper grade and type engine coolant. Start the engine and check for leaks.

ENGINE ELECTRICAL

ALTERNATOR

REMOVAL & INSTALLATION

See Figures 24 and 25.

1. Before servicing the vehicle, refer to the Precautions Section.

➡**If working near and/or around the SRS system and components, be sure to disable the SRS system. Tape the negative battery cable with insulating tape. Always disconnect the negative battery cable first.**

✷✷ CAUTION

Wait for 1 minute after disconnecting the negative battery cable before working inside the vehicle. The air bag system is set to deploy for a short period of time after the battery is disconnected.

2. Disconnect the negative battery cable.

3. Remove the side under cover.

4. Remove the drive belts.

CHARGING SYSTEM

5. Disconnect the harness bracket and connector bracket.

➡**In some cases it may be necessary to remove the timing belt lower cover on 2.4L engines.**

➡**In some cases it may be necessary to remove the air conditioning compressor on 3.8L engines. Do not disconnect the refrigerant lines. Position the component to the side. Support the component using mechanics wire. Do not allow the compressor to hang by the refrigerant hoses.**

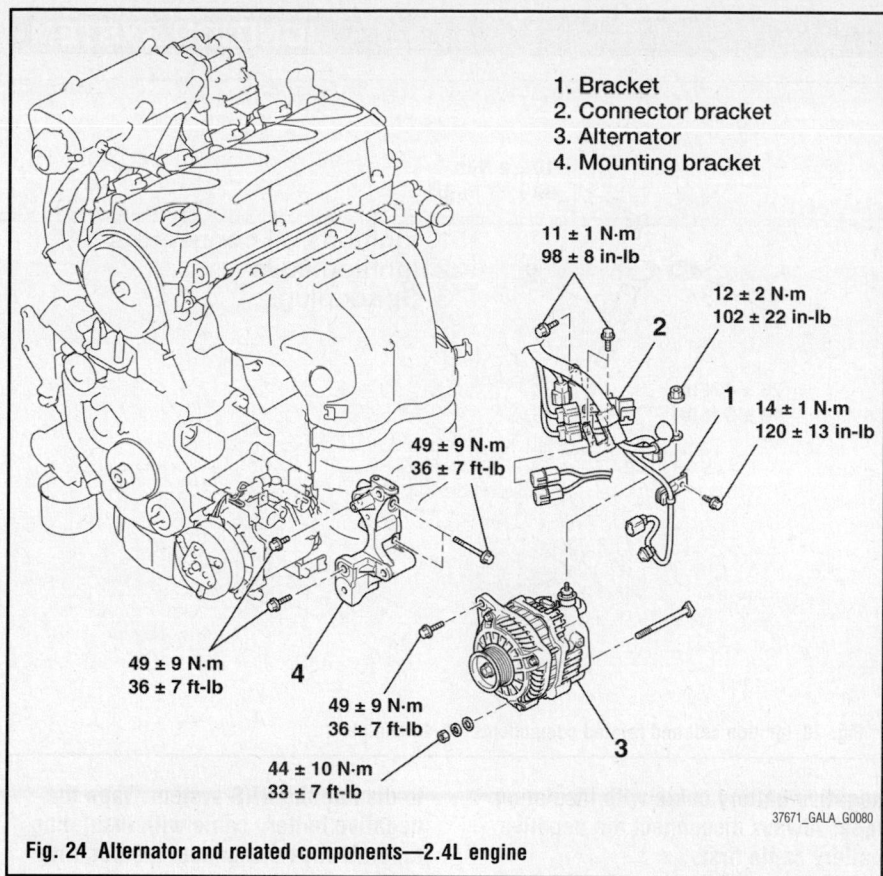

1. Bracket
2. Connector bracket
3. Alternator
4. Mounting bracket

11 ± 1 N·m
98 ± 8 in-lb

12 ± 2 N·m
102 ± 22 in-lb

14 ± 1 N·m
120 ± 13 in-lb

49 ± 9 N·m
36 ± 7 ft-lb

49 ± 9 N·m
36 ± 7 ft-lb

49 ± 9 N·m
36 ± 7 ft-lb

44 ± 10 N·m
33 ± 7 ft-lb

2

1

4

3

37671_GALA_G0080

Fig. 24 Alternator and related components—2.4L engine

➡In some cases it may be necessary to remove the engine oil dipstick and guide on 3.8L engines.

6. Remove the alternator retaining bolts.
7. Disconnect the electrical connectors.
8. Remove the component from the vehicle.

To install:

➡Be sure to use new fasteners, as required.

9. Installation is the reverse of the removal procedure.

VOLTAGE REGULATOR

ADJUSTMENT

The voltage regulator is an internal component of the alternator. In order to replace the voltage regulator, the entire alternator assembly must be replaced.

REMOVAL & INSTALLATION

The voltage regulator is an internal component of the alternator. In order to replace the voltage regulator, the entire alternator assembly must be replaced.

5

14 ± 1 N·m
120 ± 13 in-lb

49 ± 9 N·m
36 ± 7 ft-lb

1

12 ± 2 N·m
102 ± 22 in-lb

7

🛢 N 6
(ENGINE OIL)

3

4

11 ± 1 N·m
98 ± 8 in-lb

49 ± 9 N·m
36 ± 7 ft-lb

1. Connector
2. Compressor
3. Alternator
4. Bracket
5. Dipstick and guide
6. O-ring
7. Bracket

2

37671_ECLI_G0094

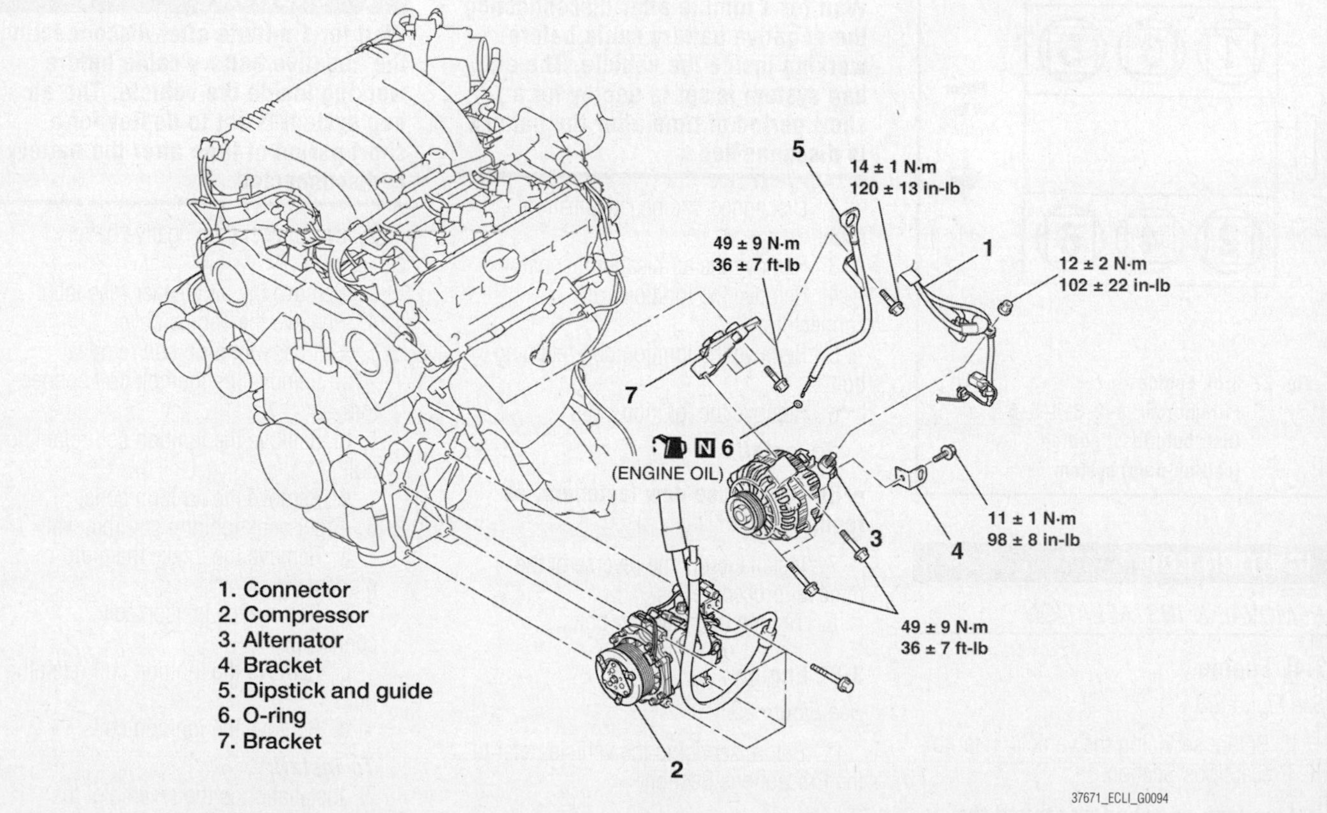

Fig. 25 Alternator and related components—3.8L engine

ENGINE ELECTRICAL   **IGNITION SYSTEM**

FIRING ORDERS

See Figures 26 and 27.

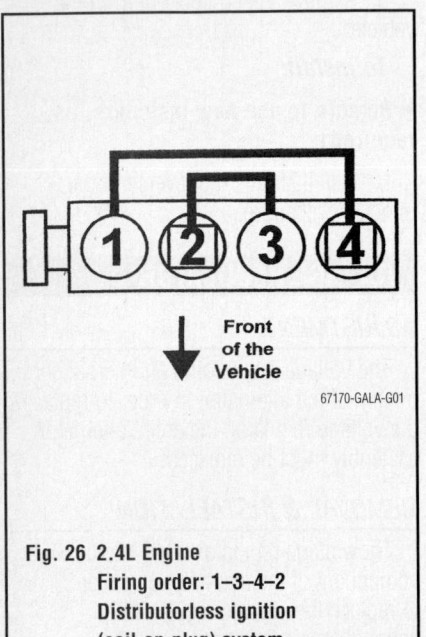

Fig. 26 2.4L Engine
Firing order: 1–3–4–2
Distributorless ignition
(coil-on-plug) system

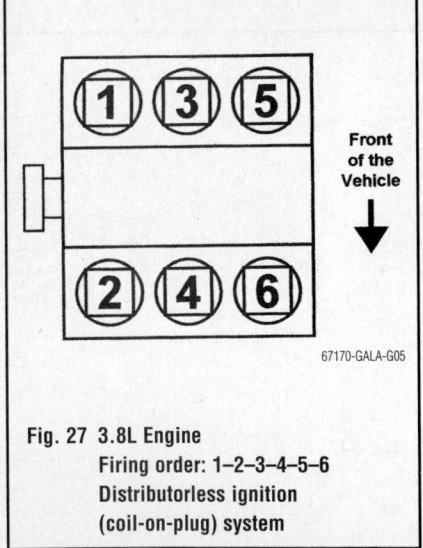

Fig. 27 3.8L Engine
Firing order: 1–2–3–4–5–6
Distributorless ignition
(coil-on-plug) system

IGNITION COIL

REMOVAL & INSTALLATION

2.4L Engine

See Figure 28.

1. Before servicing the vehicle, refer to the Precautions Section.

➥If working near and/or around the SRS system and components, be sure to disable the SRS system. Tape the

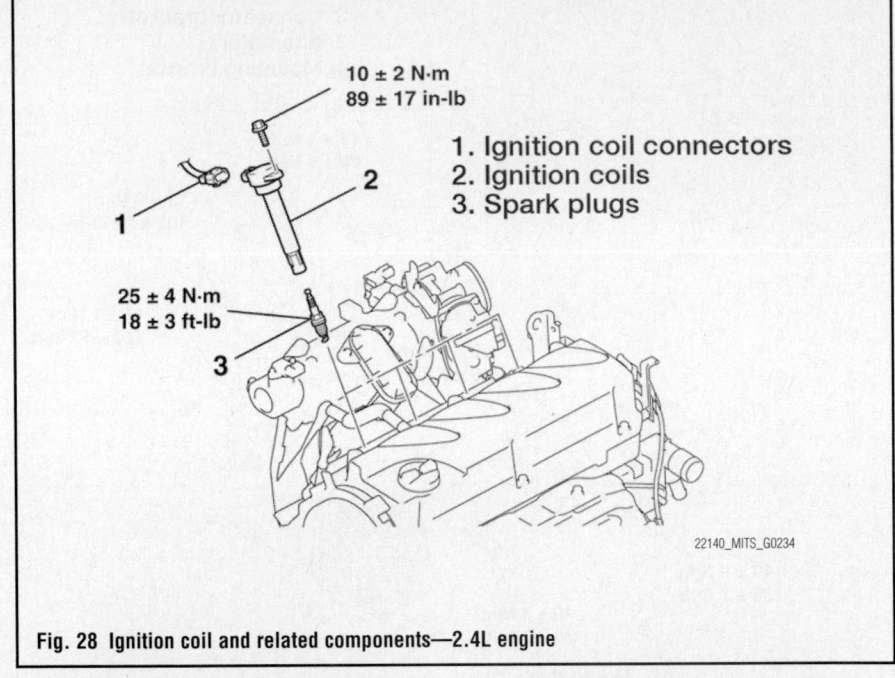

1. Ignition coil connectors
2. Ignition coils
3. Spark plugs

10 ± 2 N·m
89 ± 17 in-lb

25 ± 4 N·m
18 ± 3 ft-lb

22140_MITS_G0234

Fig. 28 Ignition coil and related components—2.4L engine

negative battery cable with insulating tape. Always disconnect the negative battery cable first.

✷✷ CAUTION

Wait for 1 minute after disconnecting the negative battery cable before working inside the vehicle. The air bag system is set to deploy for a short period of time after the battery is disconnected.

2. Disconnect the negative battery cable.
3. Remove the air cleaner resonator.
4. Remove the ignition coil connectors.
5. Remove the ignition coil retaining bolt.
6. Remove the ignition coils.

To install:

➥Be sure to use new fasteners, as required.

7. Installation is the reverse of the removal procedure.
8. Tighten the retaining bolts.

3.8L Engine

See Figure 29.

1. Before servicing the vehicle, refer to the Precautions Section.

➥If working near and/or around the SRS system and components, be sure

to disable the SRS system. Tape the negative battery cable with insulating tape. Always disconnect the negative battery cable first.

✷✷ CAUTION

Wait for 1 minute after disconnecting the negative battery cable before working inside the vehicle. The air bag system is set to deploy for a short period of time after the battery is disconnected.

2. Disconnect the negative battery cable.
3. Remove the air cleaner resonator.
4. Remove the engine cover.
5. Left bank ignition coil removal:
   a. Remove the ignition coil connectors.
   b. Remove the ignition coil retaining bolt.
   c. Remove the ignition coils.
6. Right bank ignition coil removal:
   a. Remove the intake manifold plenum.
   b. Remove the ignition coil connectors.
   c. Remove the ignition coil retaining bolt.
   d. Remove the ignition coils.

To install:
7. Installation is the reverse of the removal procedure.
8. Tighten the retaining bolts.

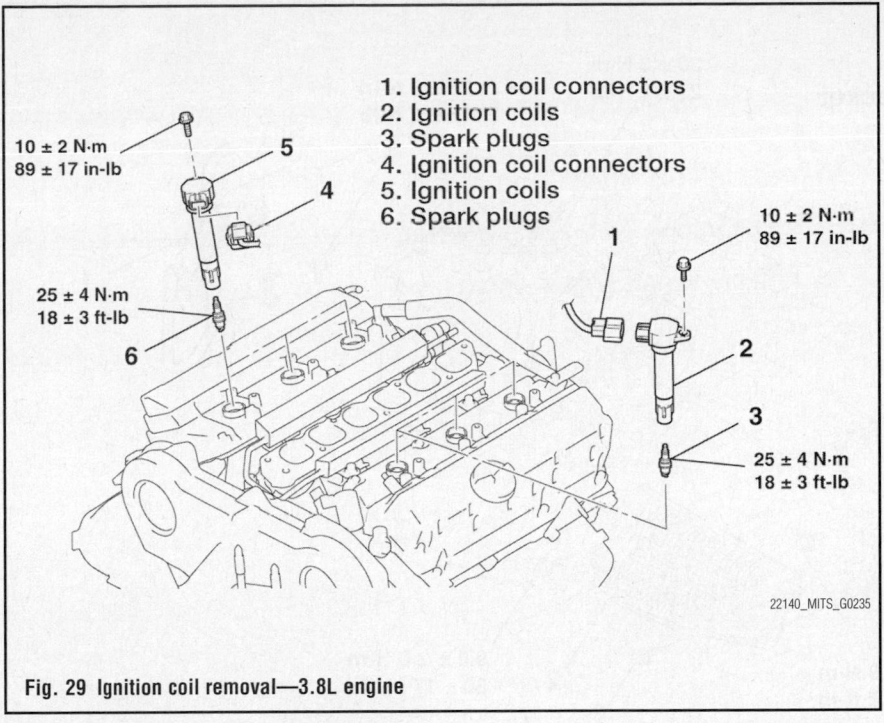

1. Ignition coil connectors
2. Ignition coils
3. Spark plugs
4. Ignition coil connectors
5. Ignition coils
6. Spark plugs

10 ± 2 N·m
89 ± 17 in-lb

25 ± 4 N·m
18 ± 3 ft-lb

10 ± 2 N·m
89 ± 17 in-lb

25 ± 4 N·m
18 ± 3 ft-lb

22140_MITS_G0235

Fig. 29 Ignition coil removal—3.8L engine

IGNITION TIMING

INSPECTION

➡**Mitsubishi's Multi-Use Tester-III (MUT III) scan tool or an equivalent OBD-II scan tool must be used for this procedure.**

1. Set the transaxle in Park for automatic transaxles or Neutral for manual transaxles.

2. Connect an OBD-II compliant scan tool to the Data Link Connector, located under the instrument panel on the driver's side.

3. Set the timing light to the power supply line of ignition coil No. 1.

4. Start the engine and allow it to run at idle.

5. Using the scan tool (Item No. 17 of the actuator test), check that the ignition timing is within specification.

ADJUSTMENT

The ignition timing is controlled by the Powertrain Control Module (Or the Electronic Control Module (ECM). No adjustment is necessary or possible.

SPARK PLUGS

REMOVAL & INSTALLATION

1. Before servicing the vehicle, refer to the Precautions Section.

➡**If working near and/or around the SRS system and components, be sure to disable the SRS system. Tape the negative battery cable with insulating tape. Always disconnect the negative battery cable first.**

✳✳ CAUTION

Wait for 1 minute after disconnecting the negative battery cable before working inside the vehicle. The air bag system is set to deploy for a short period of time after the battery is disconnected.

2. Disconnect the negative battery cable.

3. Remove the engine cover (as necessary).

4. Disconnect the ignition coil connector.

5. Remove the ignition coil.

6. Use a spark plug socket and wrench to remove the spark plugs.

✳✳ WARNING

Be careful that no contaminates enter through the spark plug holes.

➡**Check the electrode gap on the spark plugs before installation.**

7. To install, reverse the removal procedure. Tighten the spark plugs to 15–22 ft. lbs. (20–30 Nm).

ENGINE ELECTRICAL

STARTER

REMOVAL & INSTALLATION
See Figures 30 through 32.

1. Before servicing the vehicle, refer to the Precautions Section.

➡**If working near and/or around the SRS system and components, be sure to disable the SRS system. Tape the negative battery cable with insulating tape. Always disconnect the negative battery cable first.**

✳✳ CAUTION

Wait for 1 minute after disconnecting the negative battery cable before working inside the vehicle. The air bag system is set to deploy for a short period of time after the battery is disconnected.

2. Disconnect the negative battery cable.

➡**Raise and support the vehicle as required to gain access to the starter and its attaching points.**

➡**On 3.8L engines the starter is removed from underneath the vehicle.**

3. Remove the air cleaner assembly, 2.4L engines.

4. Remove the engine under cover, 3.8L engines.

5. Remove the transmission warmer fluid cooler line and bracket assembly from

STARTING SYSTEM

the transmission case front roll stopper bracket, 2.4L engines.

6. Remove the starter cover.

7. Disconnect the electrical connectors.

8. Remove the retaining bolts.

9. Remove the component from the vehicle.

To install:

➡**Be sure to use new fasteners, as required.**

10. Installation is the reverse of the removal procedure.

SOLENOID OR RELAY REPLACEMENT

1. Before servicing the vehicle, refer to the Precautions Section.

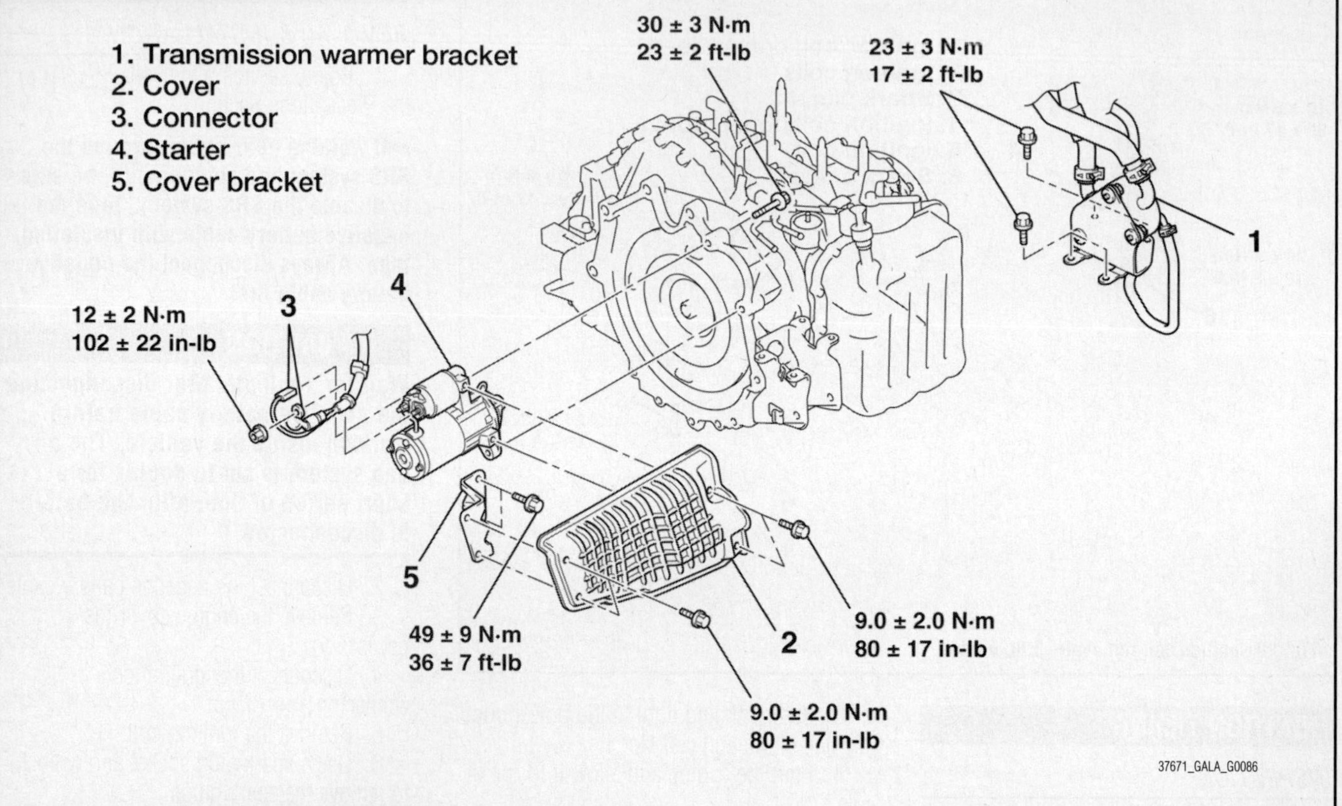

1. Transmission warmer bracket
2. Cover
3. Connector
4. Starter
5. Cover bracket

30 ± 3 N·m
23 ± 2 ft-lb

23 ± 3 N·m
17 ± 2 ft-lb

12 ± 2 N·m
102 ± 22 in-lb

49 ± 9 N·m
36 ± 7 ft-lb

9.0 ± 2.0 N·m
80 ± 17 in-lb

9.0 ± 2.0 N·m
80 ± 17 in-lb

37671_GALA_G0086

Fig. 30 Starter and related components automatic transaxle—2.4L engine

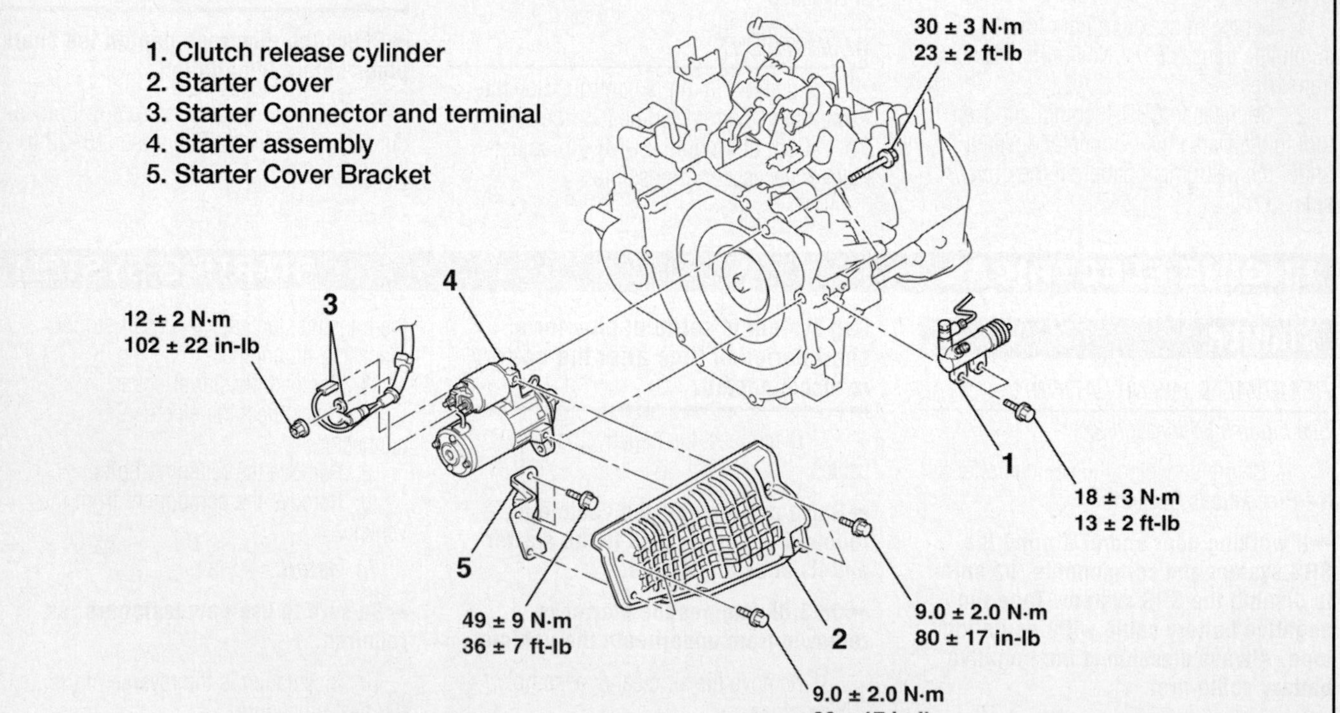

1. Clutch release cylinder
2. Starter Cover
3. Starter Connector and terminal
4. Starter assembly
5. Starter Cover Bracket

30 ± 3 N·m
23 ± 2 ft-lb

12 ± 2 N·m
102 ± 22 in-lb

18 ± 3 N·m
13 ± 2 ft-lb

49 ± 9 N·m
36 ± 7 ft-lb

9.0 ± 2.0 N·m
80 ± 17 in-lb

9.0 ± 2.0 N·m
80 ± 17 in-lb

37671_ECLI_G0095

Fig. 31 Starter and related components manual transaxle—2.4L engine

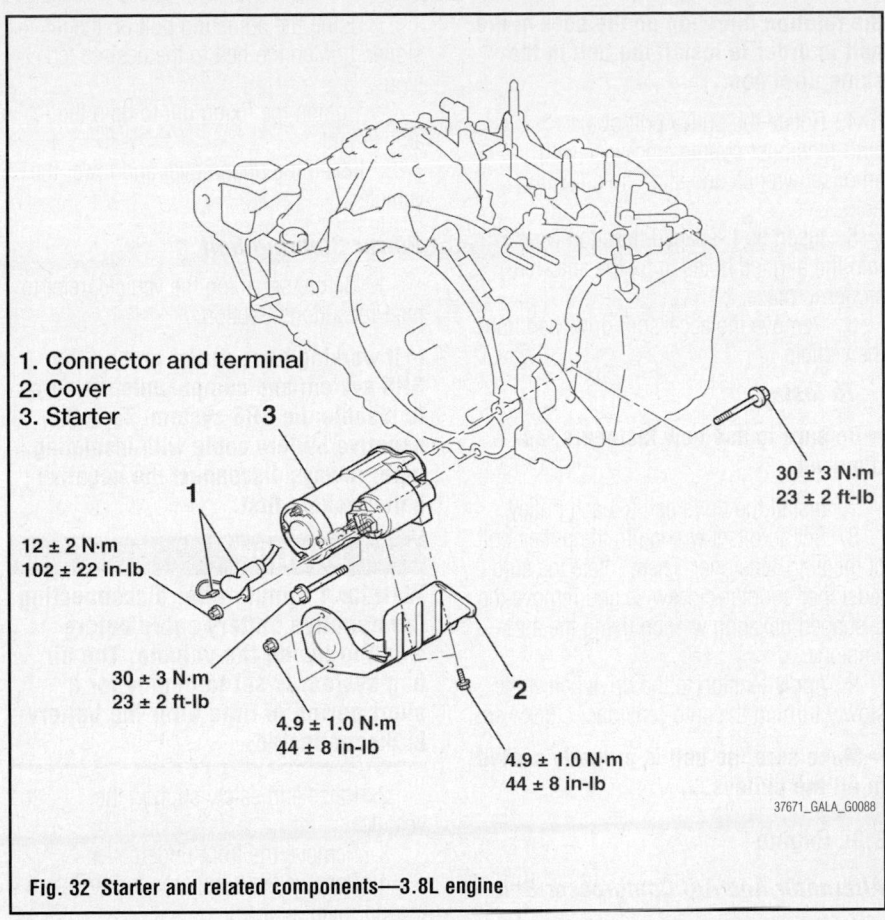

1. Connector and terminal
2. Cover
3. Starter

30 ± 3 N·m
23 ± 2 ft-lb

12 ± 2 N·m
102 ± 22 in-lb

30 ± 3 N·m
23 ± 2 ft-lb

4.9 ± 1.0 N·m
44 ± 8 in-lb

4.9 ± 1.0 N·m
44 ± 8 in-lb

37671_GALA_G0088

Fig. 32 Starter and related components—3.8L engine

➡If working near and/or around the SRS system and components, be sure to disable the SRS system. Tape the negative battery cable with insulating tape. Always disconnect the negative battery cable first.

❋❋ CAUTION

Wait for 1 minute after disconnecting the negative battery cable before working inside the vehicle. The air bag system is set to deploy for a short period of time after the battery is disconnected.

2. Disconnect the negative battery cable.
3. Remove the starter.
4. Remove the solenoid attaching screws.
5. Remove the solenoid.

To install:

➡Be sure to use new fasteners, as required.

6. Install the solenoid.
7. Install the solenoid attaching screws.
8. Install the starter.
9. Connect the negative battery cable.

ENGINE MECHANICAL

ACCESSORY DRIVE BELTS

ACCESSORY BELT ROUTING

See Figures 33 and 34.

Refer to the accompanying illustrations.

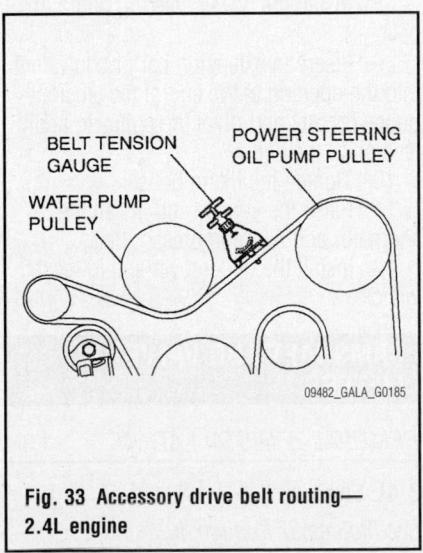

BELT TENSION GAUGE

POWER STEERING OIL PUMP PULLEY

WATER PUMP PULLEY

09482_GALA_G0185

Fig. 33 Accessory drive belt routing— 2.4L engine

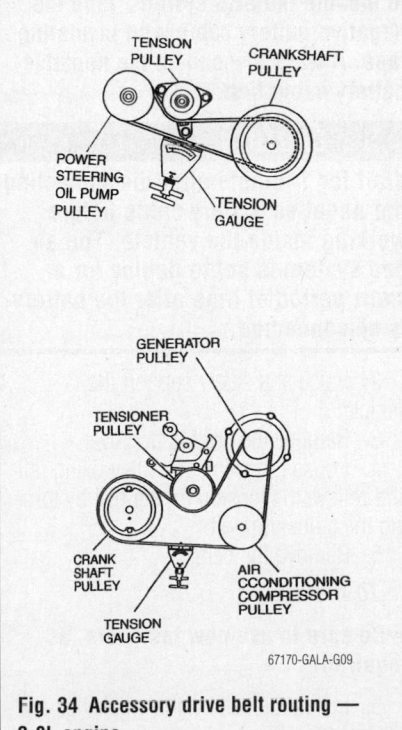

TENSION PULLEY

CRANKSHAFT PULLEY

POWER STEERING OIL PUMP PULLEY

TENSION GAUGE

GENERATOR PULLEY

TENSIONER PULLEY

CRANK SHAFT PULLEY

AIR CCONDITIONING COMPRESSOR PULLEY

TENSION GAUGE

67170-GALA-G09

Fig. 34 Accessory drive belt routing — 3.8L engine

INSPECTION

Inspect the drive belt for signs of glazing or cracking. A glazed belt will be perfectly smooth from slippage, while a good belt will have a slight texture of fabric visible. Cracks will usually start at the inner edge of the belt and run outward. All worn or damaged drive belts should be replaced immediately.

ADJUSTMENT

Excessive belt tension will cause damage to the alternator and water pump pulley bearings, while, on the other hand, loose belt tension will produce slip and premature wear on the belt. Therefore, be sure to adjust the belt tension to the proper level.

If the engine is not equipped with an auto-tensioner, loosen the adjusting bolt or fixing bolt locknut on the alternator, alternator bracket or tension pulley. Then, move the alternator or turn the adjusting bolt to adjust belt tension. Once the desired value is reached, secure the bolt or locknut and recheck tension.

REMOVAL & INSTALLATION

2.4L Engine

See Figures 35 and 36.

1. Before servicing the vehicle, refer to the Precautions Section.

➡**If working near and/or around the SRS system and components, be sure to disable the SRS system. Tape the negative battery cable with insulating tape. Always disconnect the negative battery cable first.**

❋❋ CAUTION
Wait for 1 minute after disconnecting the negative battery cable before working inside the vehicle. The air bag system is set to deploy for a short period of time after the battery is disconnected.

2. Remove the engine undercover.
3. As required, remove the radiator condenser tank assembly mounting bolt, and move the radiator condenser tank assembly to a place where it does not interfere with the drive belt removal and installation.

➡**To reuse the accessory drive belt, use chalk to draw an arrow indicating**

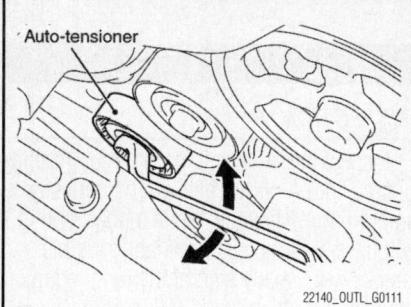

Fig. 35 Rotate the pulley bolt of the auto-tensioner counterclockwise with an offset wrench—2.4L engine

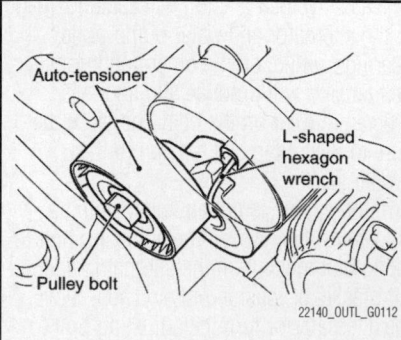

Fig. 36 Insert the hexagon wrench to fix the auto-tensioner—2.4L engine

the rotation direction on the back of the belt in order to install the belt in the same direction.

4. Rotate the pulley bolt of the auto-tensioner counterclockwise with an offset wrench and align hole A with hole B.
5. Insert an L-shaped hexagon wrench into the aligned holes to fix the auto-tensioner in place.
6. Remove the accessory drive belt from the vehicle.

To install:

➡**Be sure to use new fasteners, as required.**

7. Install the drive belt to each pulley.
8. Set an offset wrench to the pulley bolt of the auto-tensioner. Then, rotate the auto-tensioner counter-clockwise and remove the L-shaped hexagon wrench fixing the auto-tensioner.
9. Apply tension to the drive belt while slowly turning the auto-tensioner clockwise.

➡**Make sure the belt is properly seated in all the pulleys.**

3.8L Engine

Alternator And A/C Compressor Belt

1. Before servicing the vehicle, refer to the Precautions Section.

➡**If working near and/or around the SRS system and components, be sure to disable the SRS system. Tape the negative battery cable with insulating tape. Always disconnect the negative battery cable first.**

❋❋ CAUTION
Wait for 1 minute after disconnecting the negative battery cable before working inside the vehicle. The air bag system is set to deploy for a short period of time after the battery is disconnected.

2. Raise and safely support the vehicle.
3. Remove the front undercover.
4. Loosen the tension pulley fixing nut and relieve the tension on the belt by turning the adjusting bolt.
5. Remove the belt.

To install:

➡**Be sure to use new fasteners, as required.**

6. Install the belt on the crankshaft and alternator pulleys.

7. Using the adjusting bolt on the tensioner, tighten the belt to the desired tension.
8. Tighten the fixing nut to hold the adjustment.
9. Install the undercover and lower the vehicle.

Power Steering Belt

1. Before servicing the vehicle, refer to the Precautions Section.

➡**If working near and/or around the SRS system and components, be sure to disable the SRS system. Tape the negative battery cable with insulating tape. Always disconnect the negative battery cable first.**

❋❋ CAUTION
Wait for 1 minute after disconnecting the negative battery cable before working inside the vehicle. The air bag system is set to deploy for a short period of time after the battery is disconnected.

2. Raise and safely support vehicle.
3. Remove the front undercover.
4. Remove the alternator and A/C compressor belt.
5. Lower the vehicle and remove the cruise control pump link assembly.
6. Place the power steering hose under the oil reservoir.
7. Loosen the tension pulley fixing bolts and remove the power steering pump drive belt.

To install:

➡**Be sure to use new fasteners, as required.**

8. Install the power steering pump drive belt.
9. Insert an extension bar or equivalent into the opening at the end of the tension pulley bracket and pivot the pulley to apply tension to the belt.
10. Tighten the fixing bolts.
11. Raise the vehicle and install the alternator and A/C compressor belt.
12. Install the undercover and lower the vehicle.

CAMSHAFT AND VALVE LIFTERS

REMOVAL & INSTALLATION

2.4L Engine

See Figures 37 through 40.

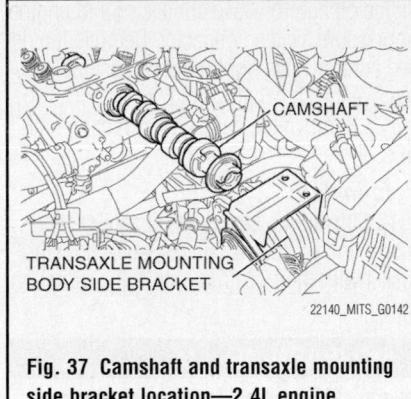

Fig. 37 Camshaft and transaxle mounting side bracket location—2.4L engine

1. Before servicing the vehicle, refer to the Precautions Section.

➡If working near and/or around the SRS system and components, be sure to disable the SRS system. Tape the negative battery cable with insulating tape. Always disconnect the negative battery cable first.

※※ CAUTION

Wait for 1 minute after disconnecting the negative battery cable before working inside the vehicle. The air bag system is set to deploy for a short period of time after the battery is disconnected.

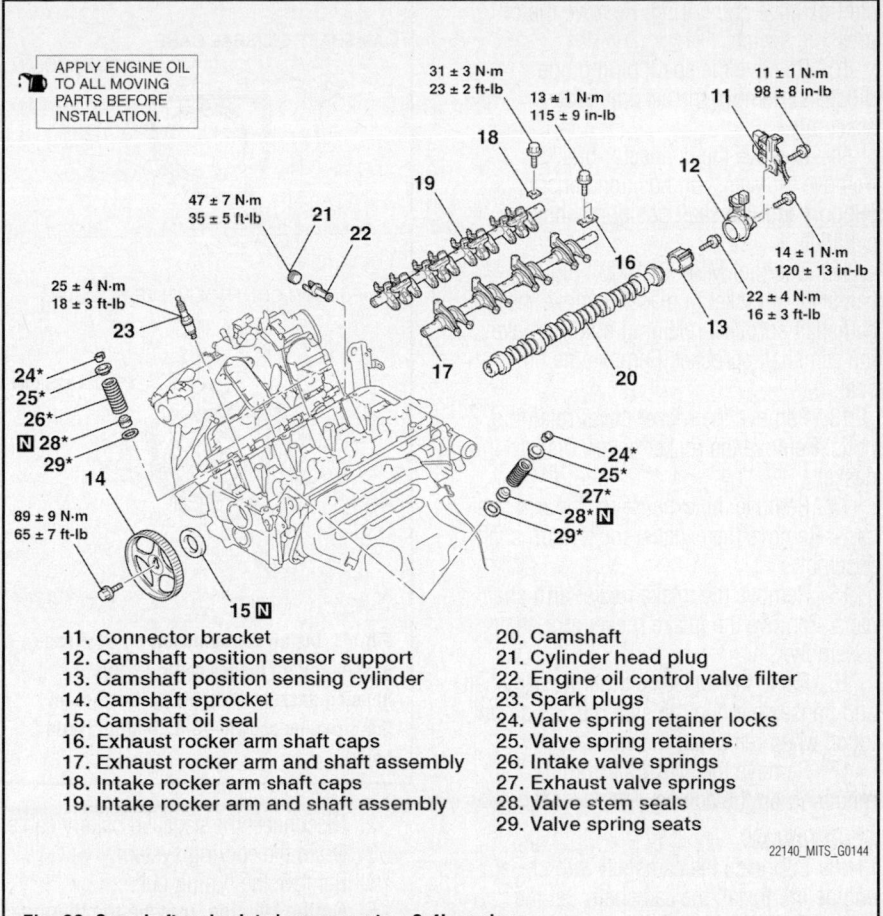

11. Connector bracket
12. Camshaft position sensor support
13. Camshaft position sensing cylinder
14. Camshaft sprocket
15. Camshaft oil seal
16. Exhaust rocker arm shaft caps
17. Exhaust rocker arm and shaft assembly
18. Intake rocker arm shaft caps
19. Intake rocker arm and shaft assembly
20. Camshaft
21. Cylinder head plug
22. Engine oil control valve filter
23. Spark plugs
24. Valve spring retainer locks
25. Valve spring retainers
26. Intake valve springs
27. Exhaust valve springs
28. Valve stem seals
29. Valve spring seats

Fig. 39 Camshaft and related components—2.4L engine

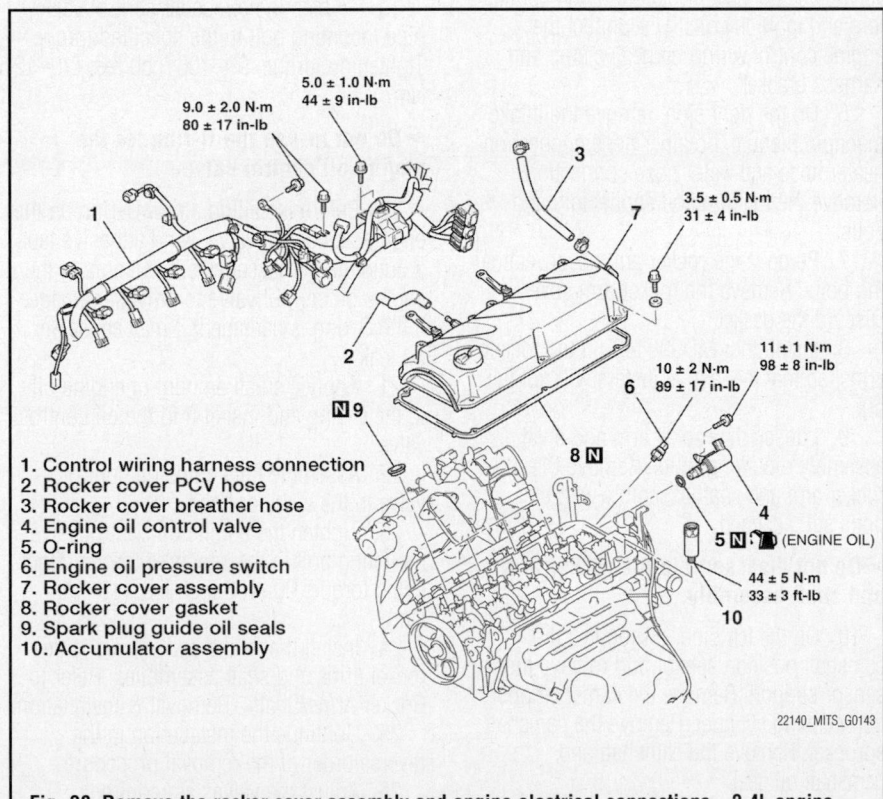

1. Control wiring harness connection
2. Rocker cover PCV hose
3. Rocker cover breather hose
4. Engine oil control valve
5. O-ring
6. Engine oil pressure switch
7. Rocker cover assembly
8. Rocker cover gasket
9. Spark plug guide oil seals
10. Accumulator assembly

Fig. 38 Remove the rocker cover assembly and engine electrical connections—2.4L engine

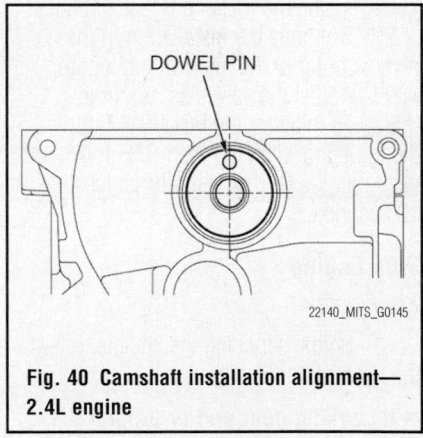

Fig. 40 Camshaft installation alignment—2.4L engine

2. Disconnect the negative battery cable.

3. Remove the ECM/PCM.

4. Remove the air cleaner assembly.

5. Remove the battery.

6. Remove the ignition coils.

7. Remove the timing belt.

8. Disconnect the control wiring harness connection. Remove the rocker cover PCV connection.

9. Remove the rocker cover breather hose connection. Remove the engine oil

control valve and O-ring. Remove the oil pressure switch.

10. Remove the spark plug guide oil seals. Remove the accumulator assembly.

11. Remove the connector bracket. Remove the camshaft position sensor support and camshaft position sensing cylinder.

12. Use tool MB998719 to hold the camshaft sprocket in place. Remove the camshaft sprocket retaining bolt. Remove the camshaft sprocket. Remove the oil seal.

13. Remove the rocker cover retaining bolts. Remove the rocker cover. Discard the gasket.

14. Remove the exhaust rocker arm shaft caps. Remove the exhaust rocker arm shaft assembly.

15. Remove the intake rocker arm shaft caps. Remove the intake rocker arm shaft assembly.

16. Raise the transaxle until the camshaft and transaxle mounting side bracket do not touch when removing the camshaft.

17. Remove the camshaft from its mounting on the engine.

To install:

18. Lubricate the camshaft with clean engine oil. Install the camshaft. Set the dowel pin of the camshaft in the position as shown in the illustration.

19. Install the rocker arm assemblies.

20. Continue the installation in the reverse order of the removal procedure.

21. Adjust the valves as required.

22. To initialize the PCM/ECM, turn the ignition switch ON then OFF and keep it in the OFF position for at least 10 seconds.

3.8L Engine

See Figure 41.

1. Before servicing the vehicle, refer to the Precautions Section.

➡If working near and/or around the SRS system and components, be sure to disable the SRS system. Tape the negative battery cable with insulating tape. Always disconnect the negative battery cable first.

✳✳ CAUTION

Wait for 1 minute after disconnecting the negative battery cable before working inside the vehicle. The air bag system is set to deploy for a short period of time after the battery is disconnected.

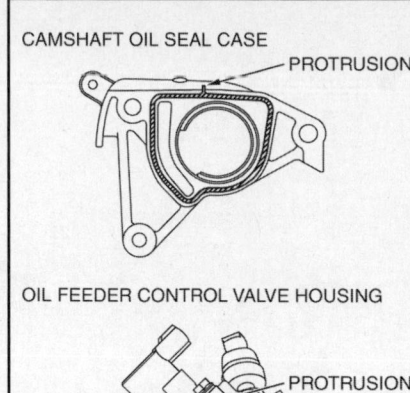

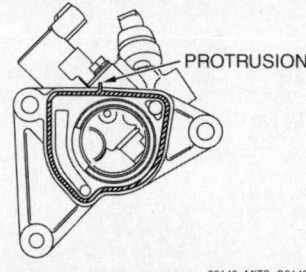

22140_MITS_G0148

Fig. 41 Install the camshaft oil seal case gasket and oil feeder control valve housing gasket with their protrusions in the direction shown—3.8L engine (right bank)

2. Disconnect the negative battery cable.

3. Drain the cooling system.

4. Remove the timing belt.

5. On the left side, remove the thermostat housing assembly. Remove the PCV valve hose connection, ignition coil connectors and ignition coils. Disconnect the engine control wiring harness clamp and harness bracket.

6. On the right side, remove the intake manifold plenum, breather hose connection, heater hose and water hose connection. Remove the ignition coil connectors and coils.

7. Remove the rocker arm cover retaining bolts. Remove the rocker arm cover. Discard the gasket.

8. Install tool MD998443 on the rocker arms, so that the lash adjusters will not fall out.

9. Loosen the rocker arm and shaft assembly mounting bolts. Remove the rocker arm and shaft assembly with the bolts still attached.

➡Do not disassemble the rocker arm and shaft assembly.

10. On the left side, disconnect the camshaft position sensor and remove the sensor support. Remove the camshaft position sensing cylinder. Remove the camshaft sprocket. Remove the camshaft and camshaft oil seal.

11. On the right side, disconnect the engine oil control valve connector and engine oil pressure switch connector. Remove the oil feed control valve right housing assembly. Remove the camshaft sprocket.

12. Remove the camshaft and camshaft oil seal.

To install:

13. Install the camshaft oil seal case gasket and oil feeder control valve housing gasket with their protrusions in the direction shown.

✳✳ WARNING

Be careful that no foreign material gets into the oil passages. Make sure the mating surfaces are clean.

14. Install the camshaft oil seal case and oil feeder control valve housing to the cylinder head.

15. Tighten the mounting bolt to: 16–20 ft. lbs. (21–27 Nm).

16. Install a gasket to one of oil feeder control valve pipes and tighten the eye bolt by hand.

17. Install a gasket to the other oil feeder control valve pipe and tighten the eye bolt to the specified torque. Tightening torque: 20–24 ft. lbs. (27–33 Nm).

18. Tighten the eye bolt which is temporarily tightened to the specified torque. Tightening torque: 20–24 ft. lbs. (27–33 Nm).

19. Tighten the oil feeder control valve pipe mounting bolt to the specified torque. Tightening torque: 90–106 inch lbs. (10–12 Nm).

➡Do not re-use the O-ring for the engine oil control valve.

20. Before installing a new O-ring on the engine oil control valve, wind adhesive tape around the oil passages cut-out area of the engine oil control valve to prevent damage. If the O-ring is damaged, it may cause an oil leak.

21. Apply a small amount of engine oil to the O-ring and install it to the oil control valve.

22. Assemble the engine oil control valve to the cylinder head.

23. Tighten the engine oil control valve mounting bolt to the specified torque. Tightening torque: 90–106 inch lbs. (10–12 Nm).

24. Install the intake and exhaust side rocker arms and shaft assemblies. Refer to Rocker Arms/Shafts, Removal & Installation.

25. Continue the installation in the reverse order of the removal procedure.

26. Adjust the valves as required.

27. To initialize the PCM, turn the ignition switch ON then OFF and keep it in the OFF position for at least 10 seconds.

CATALYTIC CONVERTER

REMOVAL & INSTALLATION

The catalytic converter(s) are removed with the exhaust manifold(s).

CRANKSHAFT DAMPER

REMOVAL & INSTALLATION

See Figures 42 and 43.

1. Before servicing the vehicle, refer to the Precautions Section.

➡**If working near and/or around the SRS system and components, be sure**

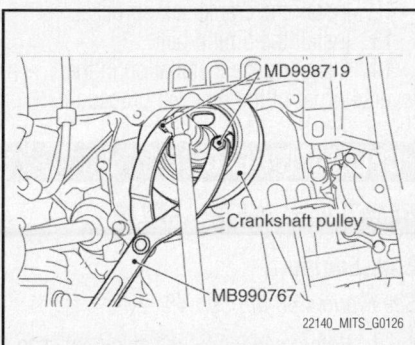

Fig. 42 Hold the crankshaft damper sprocket with special tools MB990767 and MD998719 for removal

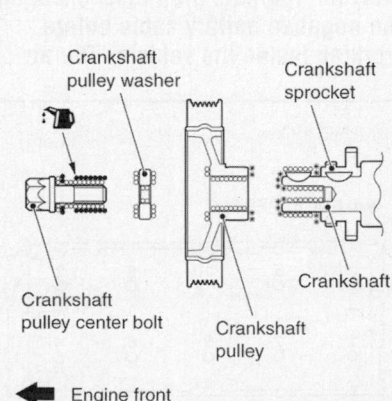

Fig. 43 Wipe off any dirt on the crankshaft and the crankshaft damper and degrease the parts before assembly

to disable the SRS system. Tape the negative battery cable with insulating tape. Always disconnect the negative battery cable first.

❊❊ CAUTION

Wait for 1 minute after disconnecting the negative battery cable before working inside the vehicle. The air bag system is set to deploy for a short period of time after the battery is disconnected.

2. Disconnect the negative battery cable.
3. Remove the accessory drive belts from around the crankshaft pulley.
4. Raise and support the vehicle, as required.
5. Remove the passenger side front wheel, as required.
6. Remove the passenger side inner fender splash shield to gain access to the crankshaft damper, as necessary.
7. Hold the crankshaft damper sprocket with special tools MB990767 and MD998719.
8. Remove the crankshaft damper center bolt and washer.
9. Remove the crankshaft damper.

To install:

➡**Be sure to use new fasteners, as required.**

10. Wipe off any dirt on the crankshaft and the crankshaft damper. Degrease the parts before assembly.

➡Degrease the crankshaft damper and crankshaft to prevent a drop in the friction coefficient of the pressed area, which is caused by oil adhesion.

11. Install the crankshaft pulley.
12. Apply an adequate and minimum amount of engine oil to the threads of the crankshaft pulley center bolt and the lower area of the flange.
13. Hold the crankshaft pulley with special tools MB990767 and MD998719 in the same manner as removal.
14. Tighten the crankshaft pulley center bolt to specification.
15. Install the splash shield.
16. Install the wheel, then carefully lower the vehicle.
17. Install the accessory drive belts.
18. Connect the negative battery cable.

CRANKSHAFT FRONT SEAL

REMOVAL & INSTALLATION

2.4L Engine

See Figure 44.

1. Before servicing the vehicle, refer to the Precautions Section.

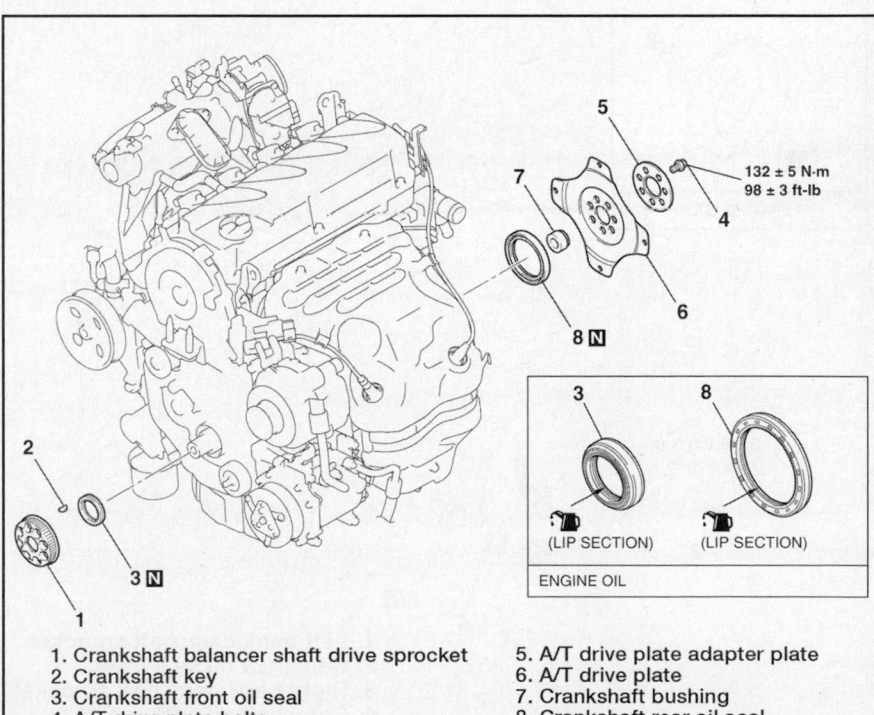

1. Crankshaft balancer shaft drive sprocket
2. Crankshaft key
3. Crankshaft front oil seal
4. A/T drive plate bolts
5. A/T drive plate adapter plate
6. A/T drive plate
7. Crankshaft bushing
8. Crankshaft rear oil seal

Fig. 44 Expanded view of the crankshaft oil seals (front and rear)—2.4L engine

→If working near and/or around the SRS system and components, be sure to disable the SRS system. Tape the negative battery cable with insulating tape. Always disconnect the negative battery cable first.

✳✳ CAUTION

Wait for 1 minute after disconnecting the negative battery cable before working inside the vehicle. The air bag system is set to deploy for a short period of time after the battery is disconnected.

2. Disconnect the negative battery cable.
3. Remove the timing belt.
4. Remove the crankshaft sprocket.
5. Carefully pry the oil seal out of the front case. Be careful not to damage the oil seal bore or the crankshaft sealing surface.

To install:

→Be sure to use new fasteners, as required.

6. Apply clean engine oil to the oil seal lip. Using a seal driver, install the oil seal.

7. Install the crankshaft sprocket.
8. Install the timing belt.
9. Install the negative battery cable.

3.8L Engine

See Figure 45.

1. Before servicing the vehicle, refer to the Precautions Section.

→If working near and/or around the SRS system and components, be sure to disable the SRS system. Tape the negative battery cable with insulating tape. Always disconnect the negative battery cable first.

✳✳ CAUTION

Wait for 1 minute after disconnecting the negative battery cable before working inside the vehicle. The air bag system is set to deploy for a short period of time after the battery is disconnected.

2. Remove the timing belt.
3. Remove the Crankshaft Position (CKP) sensor.
4. Remove the crankshaft sprocket.
5. Remove the crankshaft sensing blade.

6. Remove the crankshaft spacer and key.
7. Remove the front oil seal.

To install:

→Be sure to use new fasteners, as required.

8. Install the front oil seal. Apply oil to the seal and install using a Crankshaft Front Oil Seal Installer tool MD998717.
9. Install the crankshaft key and spacer.
10. Install the crankshaft sensing blade.
11. Install the CKP sensor.

→To be sure the crankshaft pulley bolt does not loosen, make sure to clean the mating areas of the crankshaft, spacer, sensing blade, and sprocket.

12. Install the crankshaft sprocket.
13. Install the timing belt.
14. Continue the installation in the reverse order of the removal procedure.

CYLINDER HEAD

REMOVAL & INSTALLATION

2.4L Engine

See Figures 46 through 48.

1. Before servicing the vehicle, refer to the Precautions Section.

→If working near and/or around the SRS system and components, be sure to disable the SRS system. Tape the negative battery cable with insulating tape. Always disconnect the negative battery cable first.

✳✳ CAUTION

Wait for 1 minute after disconnecting the negative battery cable before working inside the vehicle. The air

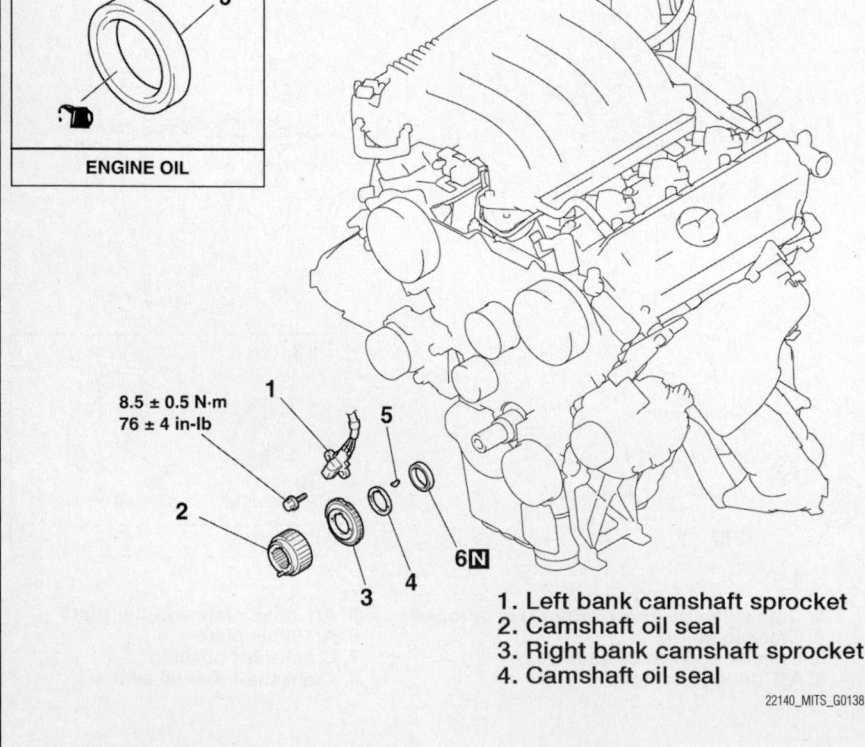

8.5 ± 0.5 N·m
76 ± 4 in-lb

ENGINE OIL

1. Left bank camshaft sprocket
2. Camshaft oil seal
3. Right bank camshaft sprocket
4. Camshaft oil seal

22140_MITS_G0138

Fig. 45 Expanded view of the crankshaft oil front seal—3.8L engine

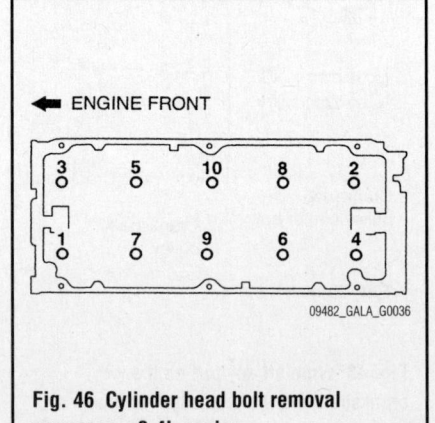

← ENGINE FRONT

09482_GALA_G0036

Fig. 46 Cylinder head bolt removal sequence—2.4L engine

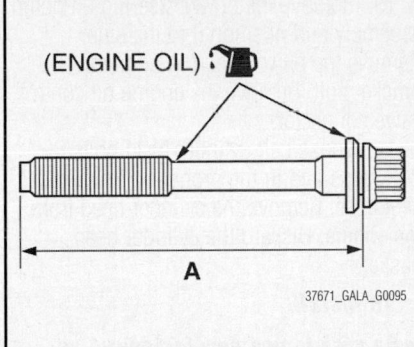

Fig. 47 Cylinder head bolt length check—2.4L engine

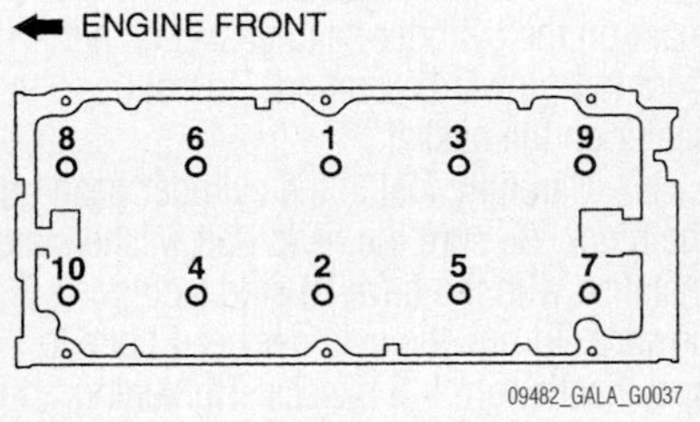

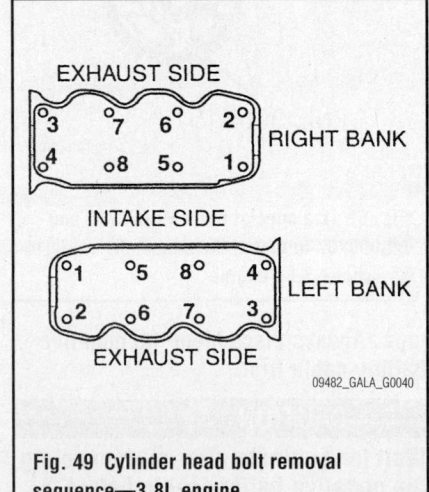

Fig. 48 Cylinder head bolt torque sequence—2.4L engine

bag system is set to deploy for a short period of time after the battery is disconnected.

2. Relieve the fuel system pressure.
3. Disconnect the negative battery cable.
4. Drain the cooling system.
5. Remove the ECM/PCM.
6. Remove the air cleaner assembly.
7. Remove the battery and battery tray.
8. Disconnect the control wiring harness, radiator hose lower clamp, and the water hose clamp.
9. Disconnect the EVAP hose connection. Disconnect the brake booster vacuum hose, pressure hose clamp, and knock sensor connector.
10. Remove the engine oil dipstick and dipstick guide.
11. Remove the intake manifold stay.
12. Remove the exhaust manifold.
13. Remove the timing belt upper cover.
14. Remove the engine front mounting bracket.
15. Turn the crankshaft clockwise, align the timing marks on the camshaft sprocket to position the number 1 piston to Top Dead Center (TDC) of its compression stroke.

☀ WARNING

Never rotate the crankshaft in the counterclockwise direction.

16. Remove the timing belt undercover rubber plug, using tool MD998738. Screw the tool in until it contacts the timing belt tensioner arm.
17. Secure the camshaft and valve timing belt with wire to prevent slippage between the camshaft sprocket and valve timing belt.
18. Remove the camshaft sprocket. Do not turn the crankshaft after the camshaft sprocket is removed.
19. Remove the upper radiator hose connection, water cooler hose connection (automatic transaxle), and water hose connection. Disconnect the high pressure hose connection.
20. Remove the valve cover.
21. Using special tool MB991654, loosen the cylinder head bolts. Loosen each bolt evenly, little by little, in 2–3 steps and in the proper bolt removal sequence.
22. Remove the cylinder head from the engine. Discard the cylinder head gasket.

➡If the cylinder head bolts cannot be pulled out due to the washer being trapped in the valve spring, raise the bolt slightly, then remove it while holding it with a magnet.

To install:

➡Be sure to use new fasteners, as required.

23. Thoroughly clean the mating surfaces of the head and block.
24. Place a new head gasket on the cylinder block. Position the gasket on the block so that the identification mark "381" is at the top surface and on the exhaust side.
25. Inspect the cylinder head bolt length prior to installation. If the length (A) exceeds 3.91 inches (99.4mm), the bolt must be replaced. Apply a small amount of engine oil to the thread section and washer of the bolt.
26. Position the cylinder head on the engine. Install and torque the retaining bolts to specification and in the proper sequence.
   a. Step 1: 57–59 ft. lbs. (76–80 Nm).
   b. Step 2: Loosen all bolts fully in the reverse order of tightening.
   c. Step 3: Retighten the loosened bolts to the specified torque in the tightening sequence shown. Tightening torque: 14–16 ft. lbs. (18–22 Nm).
   d. Step 4: Make a paint mark across each bolt head and cylinder head. Tighten the cylinder head bolts 90° in the specified order.
   e. Step 5: Tighten the bolts another 90° degrees in the same order and check that the paint marks on the cylinder head bolt are aligned with the paint marks on the cylinder head.

☀ WARNING

If the bolt is turned less than 90°, proper fastening performance may not be achieved. Be careful to turn each bolt exactly 90°. If the bolt is over-tightened, loosen the bolt completely and then retighten it by repeating the tightening procedure from step 1.

27. Torque the camshaft bolt to 58—72 ft. lbs. (80—98 Nm).
28. Continue the installation in the reverse order of the removal procedure.
29. Fill the engine with the proper grade and type engine oil.
30. Fill the cooling system with the proper grade and type engine coolant.
31. Initialize the ECM/PCM, by turning the ignition switch ON then OFF. Keep it off for at least 10 seconds.
32. Start the engine and check for leaks.

3.8L Engine

See Figures 49 through 51.

1. Before servicing the vehicle, refer to the Precautions Section.

➡If working near and/or around the SRS system and components, be sure to disable the SRS system. Tape the negative battery cable with insulating

Fig. 49 Cylinder head bolt removal sequence—3.8L engine

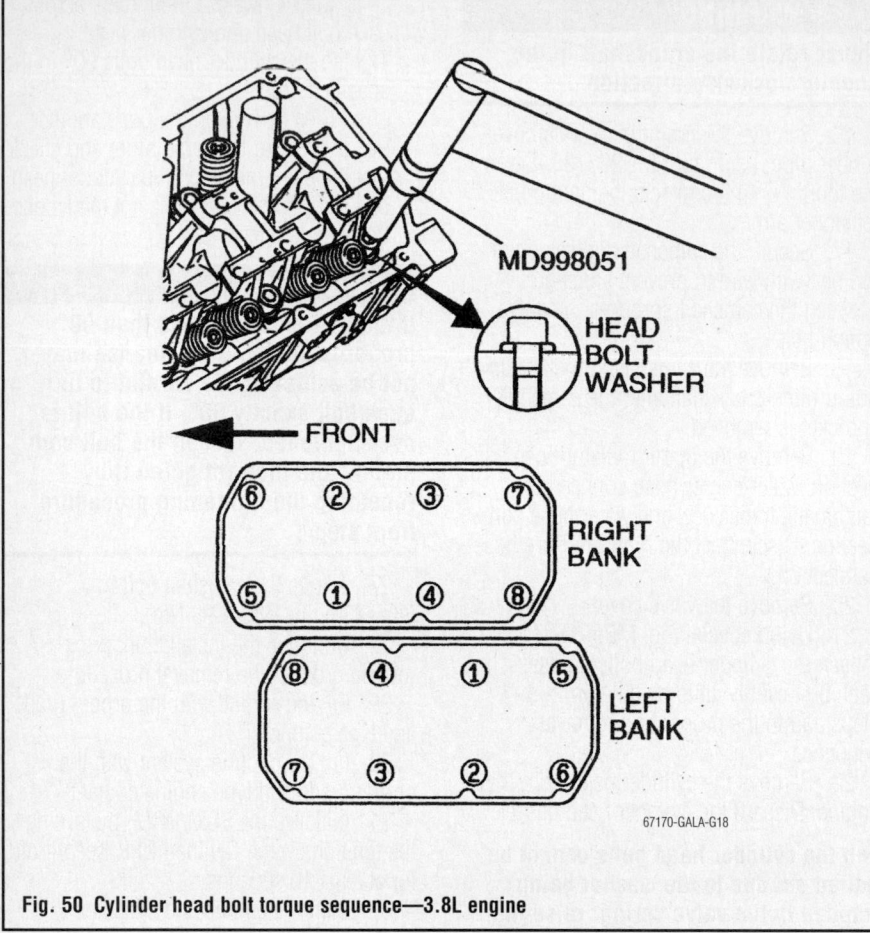

Fig. 50 Cylinder head bolt torque sequence—3.8L engine

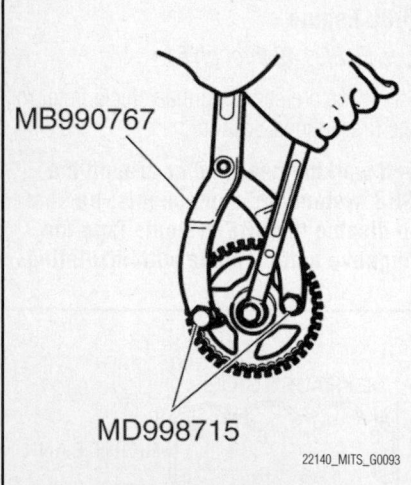

Fig. 51 Use special tools MD998715 and MB990767 to install the camshaft sprocket—3.8L engine

tape. Always disconnect the negative battery cable first.

✴✴ CAUTION

Wait for 1 minute after disconnecting the negative battery cable before

working inside the vehicle. The air bag system is set to deploy for a short period of time after the battery is disconnected.

2. Relieve the fuel system pressure.
3. Disconnect the negative battery cable.
4. Drain the cooling system.
5. Remove the intake manifold. Remove the exhaust manifolds.
6. Remove the timing belt.
7. Remove the thermostat housing. Disconnect the PCV hose connection. Remove the PCV valve.
8. Remove the ignition coils.
9. Disconnect the engine control wiring harness clamp. Remove the harness bracket.
10. Remove the rocker arm covers and gaskets. Remove the camshaft position sensor connector.
11. Remove the engine oil control valve connector. Remove the grounding bolt. Remove the engine oil dipstick assembly.
12. Remove the camshaft sprockets, using tool MB990767 or equivalent. Remove the timing belt rear center cover.

13. Remove the power steering oil pump assembly and position it to the side. Remove the power steering oil pump bracket bolt. Remove the engine oil control valve connector.
14. Loosen all cylinder head bolts in 2–3 steps and in the proper bolt removal sequence. Remove the cylinder head from the engine. Discard the cylinder head gasket.

To install:

➡ **Be sure to use new fasteners, as required.**

15. Thoroughly clean the sealing surfaces of the head and block.
16. Place a new head gasket on the cylinder block making sure the identification mark on the cylinder head gasket is in the front top (upward) location. Do not use sealer on the gasket.
17. Carefully install the cylinder head on the block. Be sure the head bolt washers are installed with the beveled side facing upward. Torque the cylinder head bolts to specification in 2–3 passes. Tightening torque:
   a. Step 1: 77–83 ft. lbs. (103–113 Nm).
   b. Step 2: 0 inch lbs. (0 Nm).
   c. Step 3: 77–83 ft. lbs. (103–113 Nm).
18. Use special tools MD998715 and MB990767 in the same way as during removal to install the camshaft sprocket.
19. Tighten the camshaft sprocket mounting bolt to 58–72 ft. lbs. (78–98 Nm).
20. Torque the camshaft bolt to 58—72 ft. lbs. (80—98 Nm).
21. Continue the installation in the reverse order of the removal procedure.
22. Fill the engine with the proper grade and type engine oil.
23. Fill the cooling system with the proper grade and type engine coolant.
24. Initialize the ECM/PCM, by turning the ignition switch ON then OFF. Keep it off for at least 10 seconds.
25. Start the engine and check for leaks.

EXHAUST MANIFOLD

REMOVAL & INSTALLATION

2.4L Engine

1. Before servicing the vehicle, refer to the Precautions Section.

➡ If working near and/or around the SRS system and components, be sure to disable the SRS system. Tape the negative battery cable with insulating tape. Always disconnect the negative battery cable first.

✳✳ CAUTION

Wait for 1 minute after disconnecting the negative battery cable before working inside the vehicle. The air bag system is set to deploy for a short period of time after the battery is disconnected.

2. Disconnect the negative battery cable.
3. Disconnect the heated oxygen sensor electrical connection. Remove the heated oxygen sensor.
4. Remove the exhaust manifold cover retaining screws. Remove the cover.
5. Remove the exhaust manifold bracket. Remove the lower heat shield.
6. Remove the exhaust manifold retaining bolts.
7. Remove the exhaust manifold from the engine.
8. Discard the old gasket.

To install:

➡ Be sure to use new fasteners, as required.

9. Position a new exhaust manifold gasket to the mating surface of the cylinder.
10. Install the exhaust manifold to the engine. Torque all bolts to specification according to the illustration.
11. Continue the installation in the reverse order of the removal procedure.

3.8L Engine

See Figures 52 and 53.

1. Before servicing the vehicle, refer to the Precautions Section.
2. Disconnect the negative battery cable.
3. Remove the engine undercover. Remove the air cleaner intake duct assembly.
4. Remove the right side exhaust manifold:
    a. Remove the right bank heated oxygen sensor (front).
    b. Remove the right bank heated oxygen sensor (rear).
    c. Remove the heat protector (right) and the exhaust manifold stay (right "B").
    d. Remove the exhaust manifold stay (right "A").
    e. Remove the heat protector, lower right.
    f. Remove the right side exhaust manifold and discard the gasket.

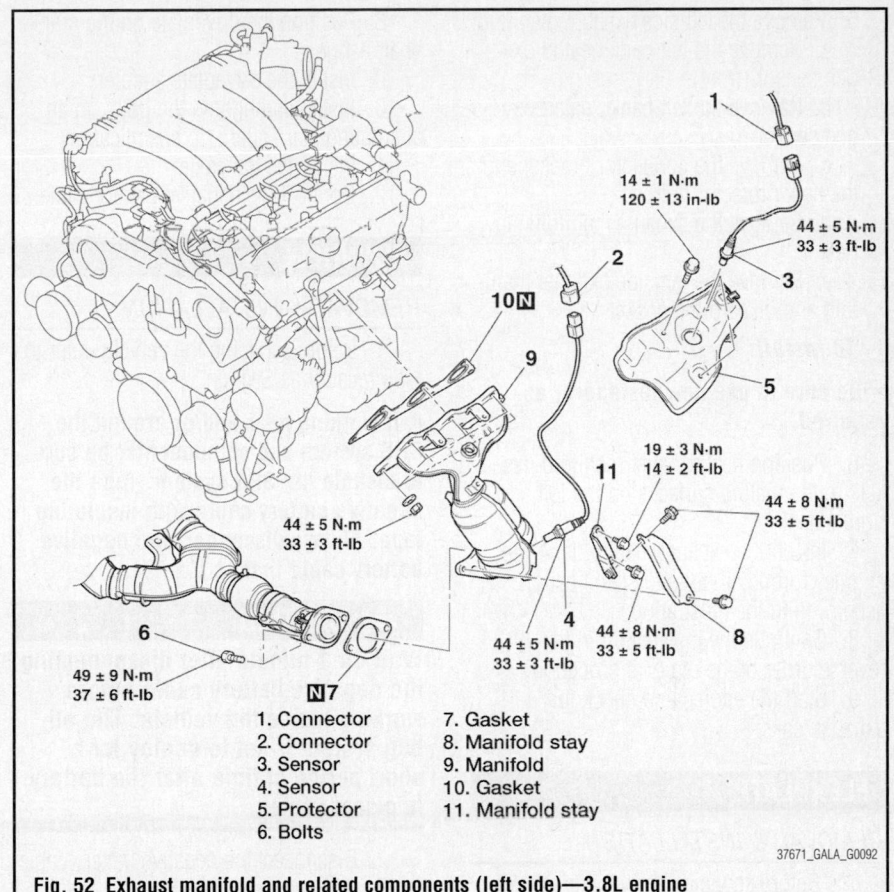

1. Connector
2. Connector
3. Sensor
4. Sensor
5. Protector
6. Bolts
7. Gasket
8. Manifold stay
9. Manifold
10. Gasket
11. Manifold stay

37671_GALA_G0092

Fig. 52 Exhaust manifold and related components (left side)—3.8L engine

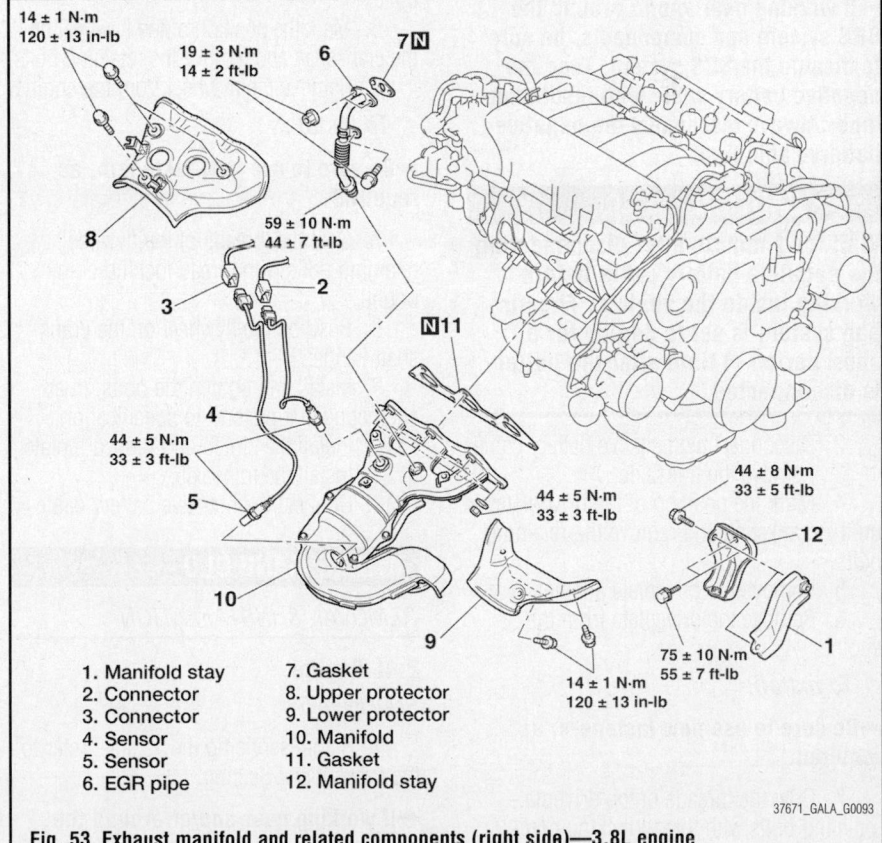

1. Manifold stay
2. Connector
3. Connector
4. Sensor
5. Sensor
6. EGR pipe
7. Gasket
8. Upper protector
9. Lower protector
10. Manifold
11. Gasket
12. Manifold stay

37671_GALA_G0093

Fig. 53 Exhaust manifold and related components (right side)—3.8L engine

5. Remove the left side exhaust manifold:
   a. Remove the left bank heated oxygen sensor (front).
   b. Remove the left bank heated oxygen sensor (rear).
   c. Remove the connector bracket and the heat protector (left).
   d. Remove the exhaust manifold stay (left "B").
   e. Remove the left side exhaust manifold and discard the gasket.

To install:

➡ **Be sure to use new fasteners, as required.**

6. Position new exhaust manifold gaskets to the mating surfaces on the left and right sides.

7. Install the exhaust manifolds to the engine. Torque all bolts to specification according to the illustration.

8. Continue the installation in the reverse order of the removal procedure.

9. Start the engine and check for exhaust leaks.

FLEXPLATE

REMOVAL & INSTALLATION

1. Before servicing the vehicle, refer to the Precautions Section.

➡ **If working near and/or around the SRS system and components, be sure to disable the SRS system. Tape the negative battery cable with insulating tape. Always disconnect the negative battery cable first.**

✳ CAUTION

Wait for 1 minute after disconnecting the negative battery cable before working inside the vehicle. The air bag system is set to deploy for a short period of time after the battery is disconnected.

2. Disconnect the negative battery cable.
3. Remove the transaxle.
4. Mark the position of the driveplate on the crankshaft and remove the retaining bolts.
5. Remove the driveplate adapter.
6. Remove the driveplate from the engine.

To install:

➡ **Be sure to use new fasteners, as required.**

7. Coat the threads of the driveplate retaining bolts with thread locking compound.

8. Position the driveplate on the crankshaft flange.
9. Install the driveplate adapter.
10. Install and tighten the bolts, in an alternating star pattern, to specification.
11. Install the transaxle.
12. Connect the negative battery cable.

FLYWHEEL

REMOVAL & INSTALLATION

1. Before servicing the vehicle, refer to the Precautions Section.

➡ **If working near and/or around the SRS system and components, be sure to disable the SRS system. Tape the negative battery cable with insulating tape. Always disconnect the negative battery cable first.**

✳ CAUTION

Wait for 1 minute after disconnecting the negative battery cable before working inside the vehicle. The air bag system is set to deploy for a short period of time after the battery is disconnected.

2. Disconnect the negative battery cable.
3. Remove the transaxle.
4. Remove the clutch disc and pressure plate.
5. Mark the position of the flywheel on the crankshaft and remove the retaining bolts.
6. Remove the flywheel from the engine.

To install:

➡ **Be sure to use new fasteners, as required.**

7. Coat the threads of the flywheel retaining bolts with thread locking compound.

8. Position the flywheel on the crankshaft flange.
9. Install and tighten the bolts, in an alternating star pattern, to specification.
10. Install the clutch and pressure plate.
11. Install the transaxle.
12. Connect the negative battery cable.

INTAKE MANIFOLD

REMOVAL & INSTALLATION

2.4L Engine
See Figure 54.

1. Before servicing the vehicle, refer to the Precautions Section.

➡ **If working near and/or around the SRS system and components, be sure**

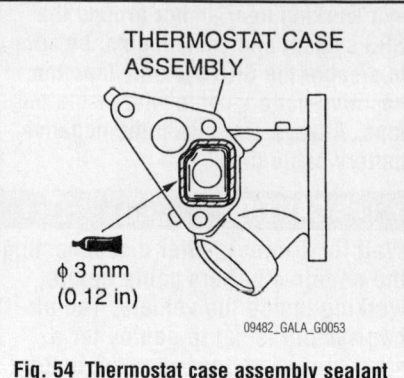

THERMOSTAT CASE ASSEMBLY

φ 3 mm
(0.12 in)

09482_GALA_G0053

Fig. 54 Thermostat case assembly sealant application—2.4L engine

to disable the SRS system. Tape the negative battery cable with insulating tape. Always disconnect the negative battery cable first.

✳ CAUTION

Wait for 1 minute after disconnecting the negative battery cable before working inside the vehicle. The air bag system is set to deploy for a short period of time after the battery is disconnected.

2. Relieve the fuel system pressure. Disconnect the negative battery cable.

3. Remove the air cleaner assembly. Drain the cooling system. Remove the battery.

➡ **Do not loosen the retaining screws for the resin cover of the throttle body assembly. If these screws are loosened, the sensor incorporated in the resin cover becomes misaligned and the throttle body will not function properly.**

4. Disconnect the accelerator cable connection, throttle position sensor connector and water return hose connection

5. Remove the throttle body bracket. Remove the throttle body retaining bolts. Remove the throttle body. Discard the gasket.

6. Remove the delivery pipe and injector assembly.

7. Disconnect the EVAP purge solenoid valve connector, the evaporative emission purge hose connection and the EGR valve connector.

8. Disconnect the MAP sensor connector, brake booster vacuum hose connection, pressure hose clamp, and knock sensor connector bracket.

9. Disconnect the harness clamp bracket and harness clamp. Remove the engine oil dipstick and tube assembly.

10. Disconnect the lower radiator hose. Remove the thermostat case assembly, resonator, and gasket.

11. Remove the EGR valve and gasket. Remove the harness clamp bracket, EVAP purge solenoid valve, and brake booster vacuum pipe.

12. Remove the brake booster vacuum hose and MAP sensor.

13. Remove the intake manifold bracket.

14. Remove the intake manifold retaining bolts. Remove the intake manifold from the engine. Discard the gasket.

To install:

➡ **Be sure to use new fasteners, as required.**

15. Be sure to use a new intake manifold gasket. Position the gasket on the engine.

16. Install the intake manifold. Torque the retaining bolts to specification.

17. When installing the thermostat case assembly, be sure to apply a bead of sealant, part number 8672 or equivalent, to the cylinder head mating surface of the case assembly.

➡ **Be sure to assemble the components with 15 minutes of applying the sealant to the surfaces.**

18. Continue the installation in the reverse order of the removal procedure.

➡ **Be sure to install the throttle body gasket in the proper direction. If this gasket is not installed correctly, poor engine idling may result.**

19. Complete the vehicle initialization procedure.

➡ **To complete the initialization procedure the following tools are needed: MB991958 scan tool, MB991824 VCI, MB991827 MUT III USB cable, MB991910 MUT III Main harness "A."**

20. Connect the scan tool to the data link connector. To prevent damage to the scan tool, be sure that the ignition switch is in the **LOC** position before connecting the scan tool.

21. Turn the ignition switch to the **ON** position.

22. Select "check mode" from the menu screen.

23. Select "erase memory" from the menu screen.

24. Initialize the learning value.

25. After initialization, complete the idle learning procedure.

➡ **This procedure must be performed when the ECM/PCM is replaced, or when the learning value is initialized, as the idling is not stabilized because the learning value in the MFI engine is not completed.**

26. Start the engine. Allow the coolant temperature to reach 176°F (80°C) or more.

27. Stop the engine and place the ignition switch in the **LOCK** position.

28. After 10 seconds, restart the engine.

29. For 10 minutes, carry out the idling procedure below to confirm that the engine has normal idling.

- Position the transaxle selector lever in the "P" range or "N" range
- The engine fan is not to be operated
- The engine coolant temperature should be 176°F (80°C) or more

➡ **If the engine stalls during idling, check the throttle valve of the throttle body for dirt. Correct and perform the procedure again.**

3.8L Engine

1. Before servicing the vehicle, refer to the Precautions Section.

➡ **If working near and/or around the SRS system and components, be sure to disable the SRS system. Tape the negative battery cable with insulating tape. Always disconnect the negative battery cable first.**

✳✳ CAUTION

Wait for 1 minute after disconnecting the negative battery cable before working inside the vehicle. The air bag system is set to deploy for a short period of time after the battery is disconnected.

2. Relieve the fuel system pressure.

3. Disconnect the negative battery cable.

4. Remove the air cleaner assembly.

5. Drain the cooling system.

6. Remove the intake manifold plenum assembly.

7. Disconnect the injector connector wiring assembly, high pressure hose connection, O-ring, and fuel return hose.

8. Disconnect the vacuum hose connection. Remove the fuel rail, injector, and fuel pressure regulator. Remove the harness bracket.

9. Remove the PCV hose connection. Remove the left timing belt front upper cover. Remove the right timing belt front upper cover. Remove the water pump bracket.

10. Remove the intake manifold retaining bolts. Remove the intake manifold from the engine. Discard the gaskets.

To install:

➡ **Be sure to use new fasteners, as required.**

11. Be sure to use a new intake manifold gasket. Position the gasket on the engine.

12. Install the intake manifold. Coat the retaining bolts in clean engine oil. Torque the retaining bolts to specification.

13. Continue the installation in the reverse order of the removal procedure.

14. Complete the vehicle initialization procedure.

➡ **To complete the initialization procedure the following tools are needed: MB991958 scan tool, MB991824 VCI, MB991827 MUT III USB cable, MB991910 MUT III Main harness "A."**

15. Connect the scan tool to the data link connector. To prevent damage to the scan tool, be sure that the ignition switch is in the **LOC** position before connecting the scan tool.

16. Turn the ignition switch to the **ON** position.

17. Select "check mode" from the menu screen.

18. Select "erase memory" from the menu screen.

19. Initialize the learning value.

20. After initialization, complete the idle learning procedure.

➡ **This procedure must be performed when the ECM/PCM is replaced, or when the learning value is initialized, as the idling is not stabilized because the learning value in the MFI engine is not completed.**

21. Start the engine. Allow the coolant temperature to reach 176°F (80°C) or more.

22. Stop the engine and place the ignition switch in the **LOCK** position.

23. After 10 seconds, restart the engine.

24. For 10 minutes, carry out the idling procedure below to confirm that the engine has normal idling.

- Position the transaxle selector lever in the "P" range or "N" range
- The engine fan is not to be operated
- The engine coolant temperature should be 176°F (80°C) or more

➡ **If the engine stalls during idling, check the throttle valve of the throttle body for dirt. Correct and perform the procedure again.**

OIL PAN

REMOVAL & INSTALLATION

2.4L Engine
See Figures 55 and 56.

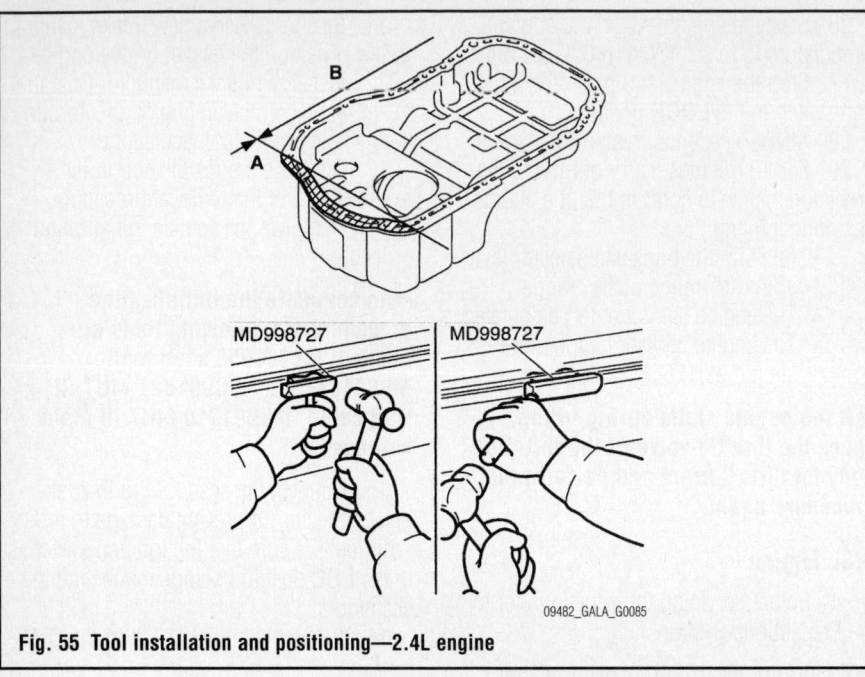

Fig. 55 Tool installation and positioning—2.4L engine

1. Before servicing the vehicle, refer to the Precautions Section.

➡ **If working near and/or around the SRS system and components, be sure to disable the SRS system. Tape the negative battery cable with insulating tape. Always disconnect the negative battery cable first.**

✳✳ CAUTION

Wait for 1 minute after disconnecting the negative battery cable before working inside the vehicle. The air bag system is set to deploy for a short period of time after the battery is disconnected.

2. Remove the negative battery cable.

3. Drain the engine oil.

4. Remove the engine undercover. Remove the front exhaust pipe.

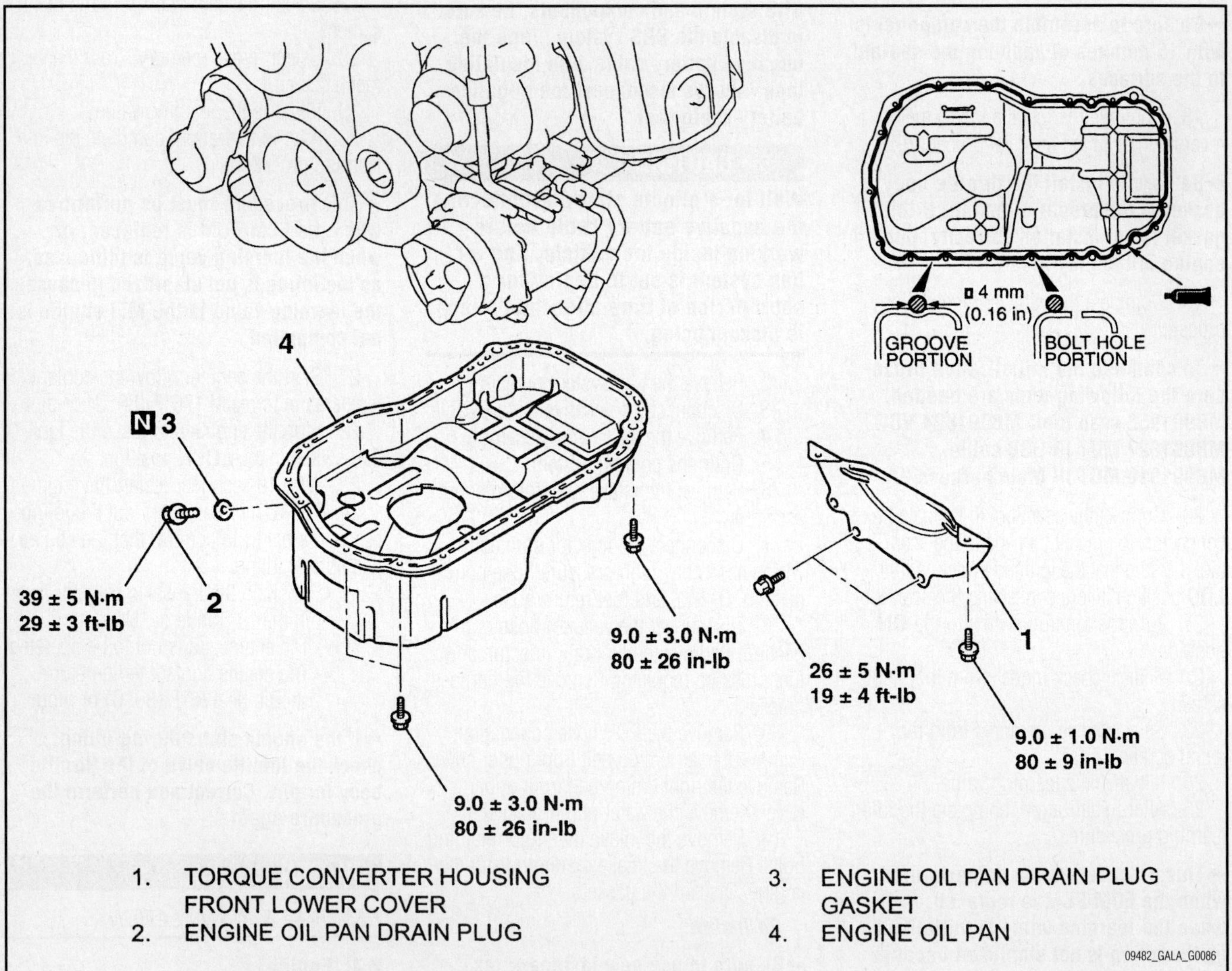

4 mm (0.16 in)
GROOVE PORTION
BOLT HOLE PORTION

N 3

39 ± 5 N·m
29 ± 3 ft-lb

2

9.0 ± 3.0 N·m
80 ± 26 in-lb

9.0 ± 3.0 N·m
80 ± 26 in-lb

26 ± 5 N·m
19 ± 4 ft-lb

1

9.0 ± 1.0 N·m
80 ± 9 in-lb

1. TORQUE CONVERTER HOUSING FRONT LOWER COVER
2. ENGINE OIL PAN DRAIN PLUG
3. ENGINE OIL PAN DRAIN PLUG GASKET
4. ENGINE OIL PAN

Fig. 56 Oil pan and related components—2.4L engine

5. Remove the torque converter housing front lower cover.

6. Remove the oil pan retaining bolts. Remove the oil pan. Discard the gasket.

➡Use tool MD998727 to remove the oil pan. Tap the tool into range B between the cylinder block and the engine oil pan, and then slide the tool sideway. Do not position the tool in area A of the engine oil pan as this may cause deformation of the front case because the front case is made of aluminum.

To install:

➡Be sure to use new fasteners, as required.

7. Clean the sealing surface on the oil pan and engine block. Apply a continuous bead of sealant to the oil pan.

8. Install the oil pan. Torque the bolts to specification.

➡Install the pan within 15 minutes after applying sealant.

9. Continue the installation in the reverse order of the removal procedure.

➡Allow the engine to sit for at least one hour before starting. Never start the engine or allow oil or coolant to come in contact with the sealant.

10. Be sure to fill the engine with the proper grade and type engine oil.

11. After the waiting period start the engine and check for leaks. Correct, as required.

3.8L Engine

See Figures 57 through 60.

1. Before servicing the vehicle, refer to the Precautions Section.

➡If working near and/or around the SRS system and components, be sure

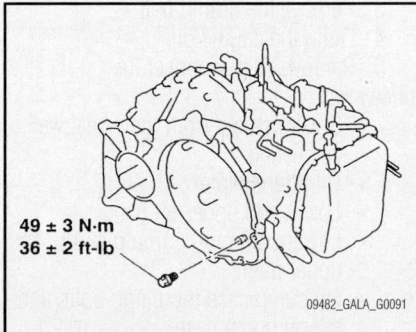

49 ± 3 N·m
36 ± 2 ft-lb

09482_GALA_G0091

Fig. 57 Automatic transaxle torque converter connecting bolt location—3.8L engine

FLYWHEEL

FLYWHEEL

09482_GALA_G0092

Fig. 58 Manual transaxle alignment location—3.8L engine

to disable the SRS system. Tape the negative battery cable with insulating tape. Always disconnect the negative battery cable first.

❊❊ CAUTION

Wait for 1 minute after disconnecting the negative battery cable before working inside the vehicle. The air bag system is set to deploy for a short period of time after the battery is disconnected.

2. Remove the negative battery cable.

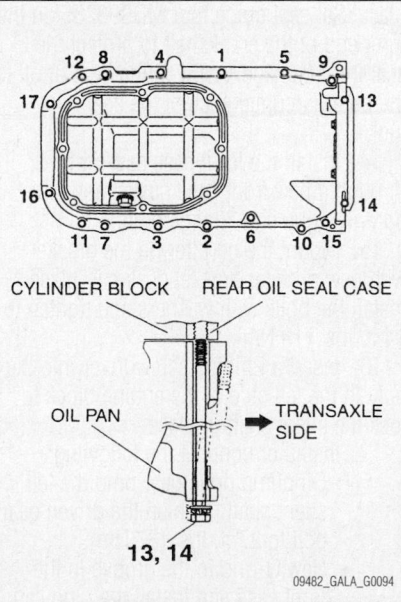

CYLINDER BLOCK   REAR OIL SEAL CASE

OIL PAN

TRANSAXLE SIDE

13, 14

09482_GALA_G0094

Fig. 59 Upper oil pan torque sequence—3.8L engine

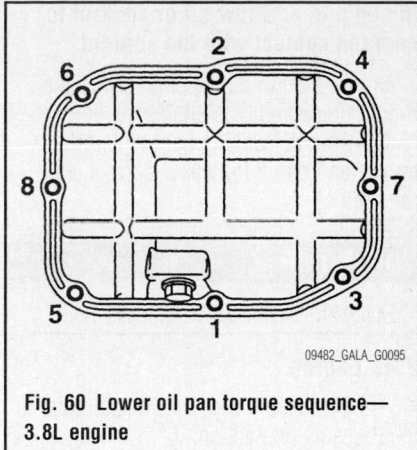

09482_GALA_G0095

Fig. 60 Lower oil pan torque sequence—3.8L engine

3. Remove the engine undercover. Remove the starter. Remove the engine oil dipstick and guide.

4. If removing the upper oil pan, remove the front exhaust pipe. Remove the cover and connecting bolt.

➡If the vehicle is equipped with a manual transaxle, align the recessed area in the flywheel with the location shown in the illustration. Mark the flywheel. Turn the crankshaft so that the alignment mark is positioned as shown in the illustration.

5. Remove the lower engine oil pan retaining bolts. Remove the engine lower oil pan. Discard the gasket.

6. Remove the upper oil pan retaining bolts. Remove the upper oil pan. Discard the gasket.

To install:

➡Be sure to use new fasteners, as required.

7. Clean the sealing surface on the oil pan and engine block. Apply a continuous bead of sealant to the oil pan.

➡Install the pan within 15 minutes after applying sealant.

8. Install the upper oil pan. Torque the bolts to specification.

➡The bolt holes for bolts 13 and 14 are cut away on the transaxle side. Be careful not to insert these bolts at an angle.

➡Install the pan within 15 minutes after applying sealant.

9. Install the lower oil pan. Torque the bolts to specification.

10. Continue the installation in the reverse order of the removal procedure.

➡Allow the engine to sit for at least one hour before starting. Never start the engine or allow oil or coolant to come in contact with the sealant.

11. Be sure to fill the engine with the proper grade and type engine oil.

12. After the waiting period start the engine and check for leaks. Correct, as required.

OIL PUMP

REMOVAL & INSTALLATION

2.4L Engine

1. Before servicing the vehicle, refer to the Precautions Section.

➡ **If working near and/or around the SRS system and components, be sure to disable the SRS system. Tape the negative battery cable with insulating tape. Always disconnect the negative battery cable first.**

✳✳ CAUTION

Wait for 1 minute after disconnecting the negative battery cable before working inside the vehicle. The air bag system is set to deploy for a short period of time after the battery is disconnected.

➡ **Whenever the oil pump is disassembled or the cover is removed, the gear cavity must be filled with petroleum jelly to seal the pump and act as a prime. Do not use grease.**

2. Disconnect the negative battery cable. Rotate the engine so number 1 cylinder is on Top Dead Center (TDC) of its compression stroke.

3. Drain the engine oil.

4. Using the proper equipment, support the weight of the engine. Remove the front engine mount bracket and accessory drive belts.

5. Remove or disconnect the following:
   - Timing belt upper and lower covers
   - Timing belt and crankshaft sprocket
   - Electrical connector from the oil pressure sending unit
   - Oil pressure sensor
   - Oil filter and the oil filter bracket
   - Oil pan, oil screen and gasket

6. Using special tool MD998162, remove the plug cap in the engine front cover.

7. Remove or disconnect the following:
   - Plug on the side of the engine block. Insert a steel rod with a shank diameter of 0.32 in. (8mm) into the plug hole. This will hold the silent shaft

   - Driven gear bolt that secures the oil pump driven gear to the silent shaft
   - Front cover mounting bolts. Note the lengths of the mounting bolts as they are removed for proper installation
   - Front case cover and oil pump assembly. If necessary, the silent shaft can come out with the cover assembly
   - Oil pump cover, located on the back of the engine front cover. Remove the oil pump drive and driven gears

8. After disassembling the oil pump, clean all components and remove any gasket material from mating surfaces

9. Assemble the oil pump gears into the front case and rotate it to ensure smooth rotation and no looseness. Be sure there is no ridge wear on the contact surface between the front case and the gear surface of the oil pump front cover.

To install:

➡ **Be sure to use new fasteners, as required.**

10. Align the timing mark on the oil pump drive gear with that on the driven gear and install them into the engine front case. Apply engine oil to the gears.

11. Install the oil pump cover and tighten the retainer bolts to 17 ft. lbs. (24 Nm).

12. Using the appropriate driver, install a new crankshaft seal into the front case.

13. Position new front case gasket in place. Set seal guide tool MD998285 on the front end of the crankshaft to protect the seal from damage. Apply a thin coat of oil to the outer circumference of the seal pilot tool.

14. Install the front case assembly through a new front case gasket and temporarily tighten the flange bolts.

15. Mount the oil filter on the bracket with new oil filter bracket gasket in place. Install the bolts with washers and tighten to 14 ft. lbs. (19 Nm).

16. Insert a Phillips screwdriver into the hole in the left side of the engine block to lock the silent shaft in place.

17. Install or connect the following:
   - Oil pump drive gear onto the left silent shaft. Tighten the driven gear bolt to 27 ft. lbs. (37 Nm).
   - New O-ring to the groove in the front case and install the plug cap. Tighten the cap to 17 ft. lbs. (24 Nm).

   - Oil screen in position with new gasket in place

18. Clean both mating surfaces of the oil pan and the cylinder block. Apply sealant in the groove in the oil pan flange.

➡ **After applying sealant to the oil pan, do not exceed 15 minutes before installing the oil pan.**

19. Install or connect the following:
   - Oil pan to the engine and secure with the retainers. Tighten bolts to 60 inch lbs. (7 Nm).
   - Oil pressure gauge unit and the oil pressure switch
   - Electrical harness connector
   - Oil cooler. Oil cooler bolt to 31 ft. lbs. (43 Nm).

20. Refill the crankcase with the correct amount and type of oil. Install new oil filter.

21. Connect the negative battery cable and start the engine.

22. Verify there is correct oil pressure. Inspect for leaks.

3.8L Engine

1. Before servicing the vehicle, refer to the Precautions Section.

➡ **If working near and/or around the SRS system and components, be sure to disable the SRS system. Tape the negative battery cable with insulating tape. Always disconnect the negative battery cable first.**

✳✳ CAUTION

Wait for 1 minute after disconnecting the negative battery cable before working inside the vehicle. The air bag system is set to deploy for a short period of time after the battery is disconnected.

2. Disconnect the negative battery cable.

3. Remove the timing belt.

4. Drain the engine oil.

5. Remove or disconnect the following:
   - Splash shield from the wheel well, as needed
   - Oil filter adapter
   - Lower and upper oil pans
   - Lower baffle, oil pump pick-up, and upper baffle
   - Oil pump case mounting bolts and the oil pump case
   - Oil pump gear cover

6. Make matchmarks on the oil pump rotors before removing them.

7. Remove the crankshaft seal from the oil pump case.

To install:

➡ **Be sure to use new fasteners, as required.**

8. Install a new crankshaft seal in the oil pump cover.

9. Apply engine oil to the rotors, then align the matchmarks and install the rotors in the oil pump case.

10. Install the rotor cover and tighten the bolts to 84 inch lbs. (10 Nm).

11. Apply a 0.113 in. (3mm) bead of sealant to the back of the oil pump case. Install the case on the engine and tighten the bolts to 10 ft. lbs. (13 Nm).

12. Install the upper baffle plate and oil pump pick-up using a new gasket.

13. Tighten the baffle bolts to 84 inch lbs. (10 Nm) and the pick-up bolts to 13 ft. lbs. (18 Nm).

14. Lower the baffle in the upper oil pan. Tighten the bolts to 96 inch lbs. (11 Nm).

15. Install the oil pans.

16. Install the oil filter adapter using a new gasket. Tighten the larger bolt to 30 ft. lbs. (41 Nm) and the smaller bolt to 17 ft. lbs. (23 Nm).

17. Install the timing belt and remaining the components from the removal procedure.

18. Fill the engine with the correct amount of oil.

19. Connect the negative battery cable.

20. Start the engine and check for leaks.

PISTON AND RING

POSITIONING

See Figures 61 through 65.

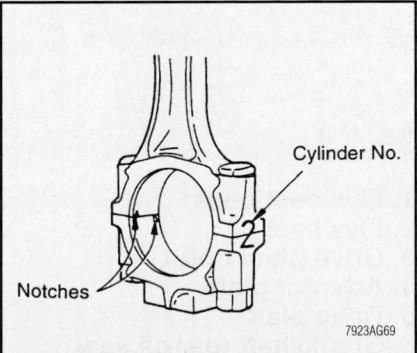

Fig. 61 Before removing the caps from the connecting rods, be sure to matchmark them as shown

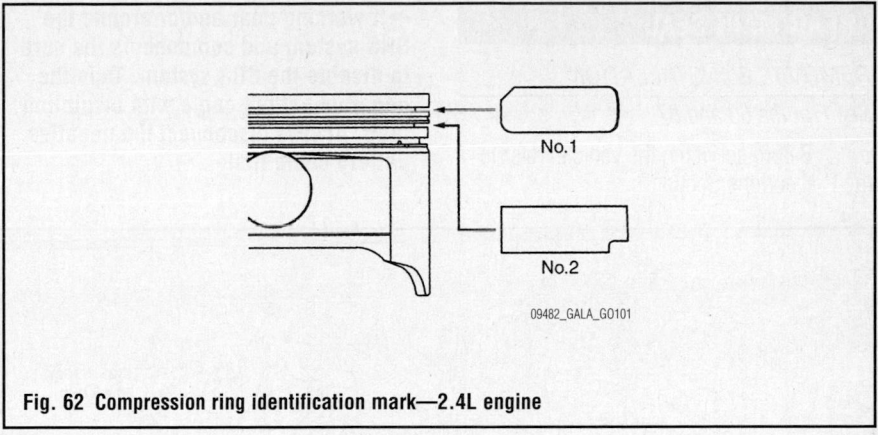

Fig. 62 Compression ring identification mark—2.4L engine

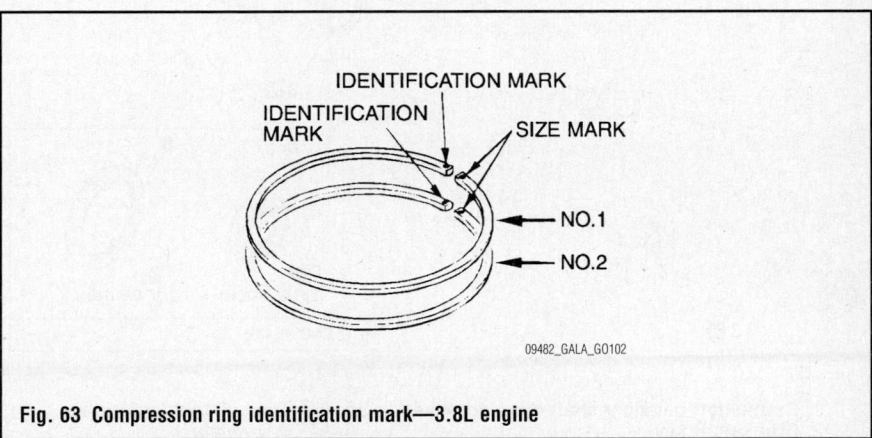

Fig. 63 Compression ring identification mark—3.8L engine

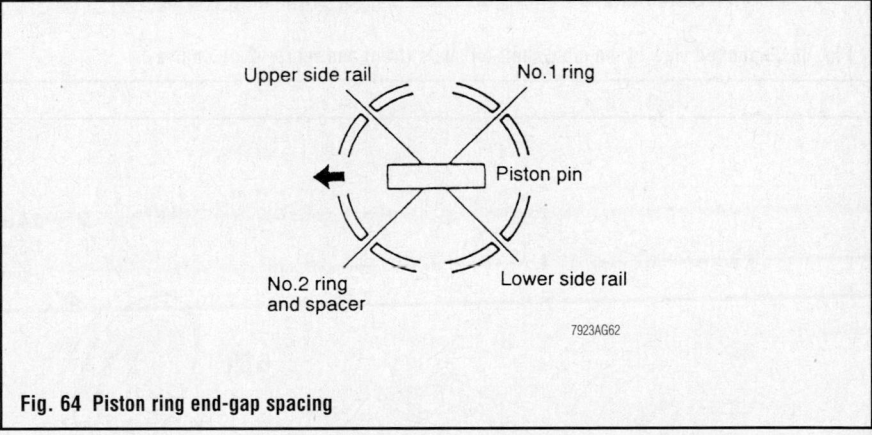

Fig. 64 Piston ring end-gap spacing

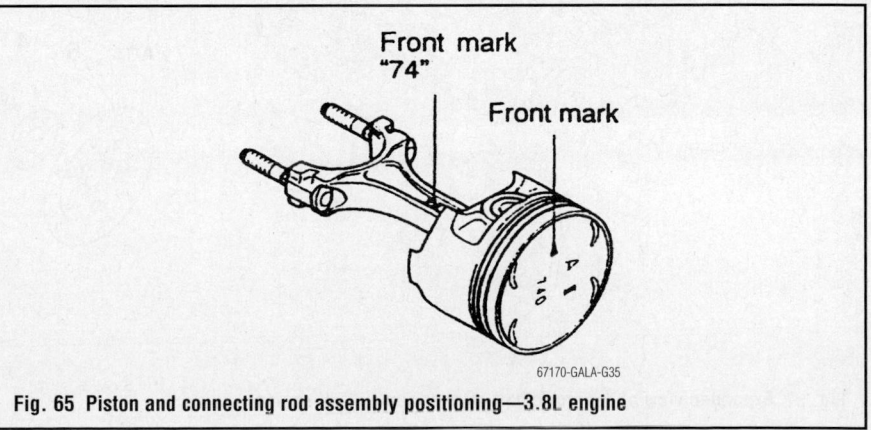

Fig. 65 Piston and connecting rod assembly positioning—3.8L engine

REAR MAIN SEAL

REMOVAL & INSTALLATION

See Figures 66 and 67.

1. Before servicing the vehicle, refer to the Precautions Section.

➡ If working near and/or around the SRS system and components, be sure to disable the SRS system. Tape the negative battery cable with insulating tape. Always disconnect the negative battery cable first.

✳✳ CAUTION

Wait for 1 minute after disconnecting the negative battery cable before working inside the vehicle. The air bag system is set to deploy for a short period of time after the battery is disconnected.

2. Disconnect the negative battery cable.
3. Remove the transaxle assembly.
4. Remove the flywheel/driveplate.
5. Remove the crankshaft bushing.
6. Remove the rear main seal. Be careful not to damage the oil seal bore or the crankshaft sealing surface.

To install:

➡ Be sure to use new fasteners, as required.

7. Apply clean engine oil to the oil seal lip. Using a seal driver, install the oil seal.
8. Use special tools MB990938 and MD998776 to press-fit the oil seal.
9. Installation continues in the reverse of the removal procedure.

ROCKER ARMS

REMOVAL & INSTALLATION

2.4L Engine

See Figures 68 and 69.

1. Before servicing the vehicle, refer to the Precautions Section.

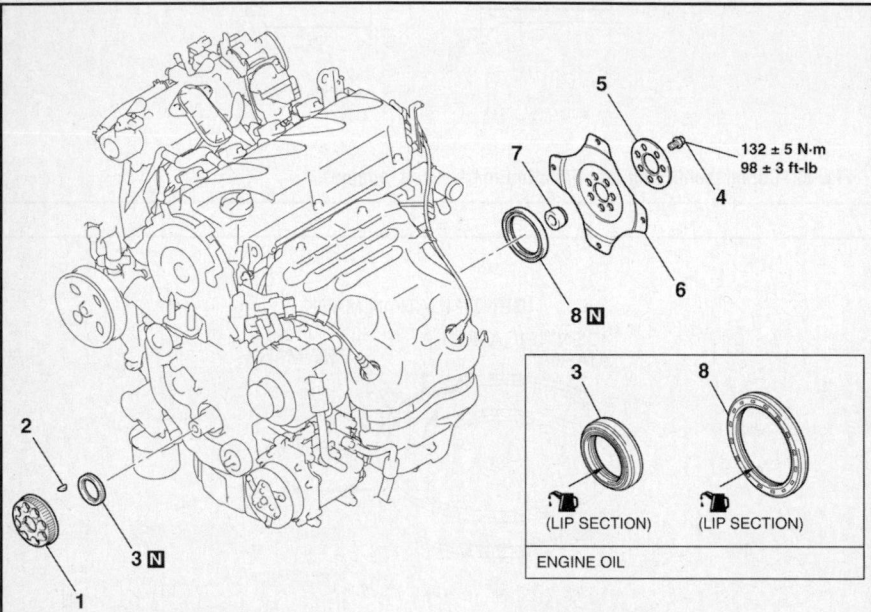

1. Crankshaft balancer shaft drive sprocket
2. Crankshaft key
3. Crankshaft front oil seal
4. A/T drive plate bolts
5. A/T drive plate adapter plate
6. A/T drive plate
7. Crankshaft bushing
8. Crankshaft rear oil seal

132 ± 5 N·m
98 ± 3 ft-lb

(LIP SECTION)   (LIP SECTION)

ENGINE OIL

22140_MITS_G0137

Fig. 66 Expanded view of the crankshaft oil seals (front and rear)—2.4L engine

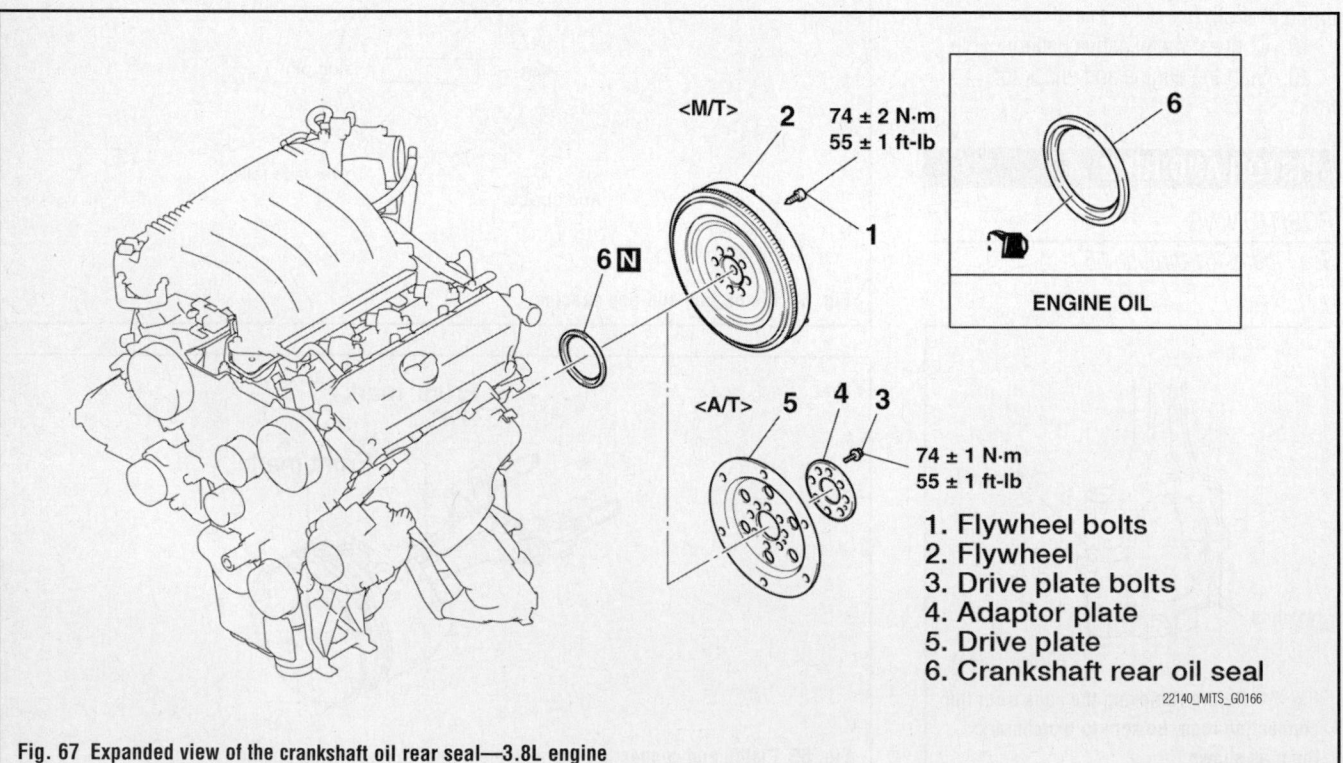

<M/T> 74 ± 2 N·m
55 ± 1 ft-lb

<A/T> 74 ± 1 N·m
55 ± 1 ft-lb

ENGINE OIL

1. Flywheel bolts
2. Flywheel
3. Drive plate bolts
4. Adaptor plate
5. Drive plate
6. Crankshaft rear oil seal

22140_MITS_G0166

Fig. 67 Expanded view of the crankshaft oil rear seal—3.8L engine

➡If working near and/or around the SRS system and components, be sure to disable the SRS system. Tape the negative battery cable with insulating tape. Always disconnect the negative battery cable first.

✳✳ CAUTION

Wait for 1 minute after disconnecting the negative battery cable before working inside the vehicle. The air bag system is set to deploy for a short period of time after the battery is disconnected.

2. Disconnect the negative battery cable.
3. Remove the ECM/PCM.
4. Remove the air cleaner assembly.
5. Remove the battery.
6. Remove the ignition coils.
7. Remove the timing belt upper cover.
8. Disconnect the control wiring harness connection.
9. Remove the rocker cover PCV connection.
10. Remove the rocker cover breather hose connection.
11. Remove the engine oil control valve and O-ring. Remove the oil pressure switch.
12. Remove the rocker cover retaining bolts. Remove the rocker cover. Discard the gasket.

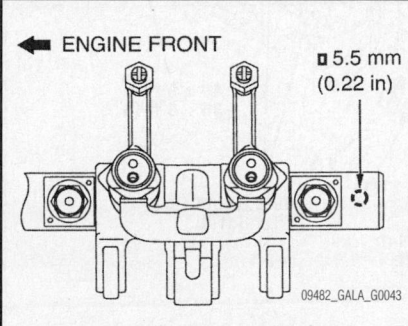

Fig. 68 Intake rocker arm shaft installation locating point—2.4L engine

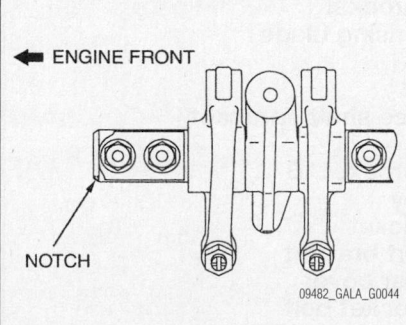

Fig. 69 Exhaust rocker arm shaft installation locating point—2.4L engine

13. Remove the spark plug guide oil seals. Remove the accumulator assembly.
14. Remove the exhaust rocker arm shaft caps. Remove the exhaust rocker arm shaft assembly.
15. Remove the intake rocker arm shaft caps. Remove the intake rocker arm shaft assembly.

➡Do not disassemble the exhaust or intake rocker arm assembly.

To install:

➡Be sure to use new fasteners, as required.

16. Position the intake rocker arm assembly so that its 0.22 inch hole faces toward the cylinder head.
17. Install the rocker shaft mounting bolts. Torque to 21—25 ft. lbs. (29—34 Nm).
18. Position the exhaust rocker arm assembly so that its notch is positioned as shown in the illustration.
19. Install the rocker shaft mounting bolts. Torque to 106—124 inch lbs. (12—14 Nm).
20. Continue the installation in the reverse order of the removal procedure.
21. To initialize the ECM/PCM, turn the ignition switch ON then OFF and keep it in the OFF position for at least 10 seconds.

3.8L Engine

See Figure 70.

1. Before servicing the vehicle, refer to the Precautions Section.

➡If working near and/or around the SRS system and components, be sure to disable the SRS system. Tape the negative battery cable with insulating tape. Always disconnect the negative battery cable first.

✳✳ CAUTION

Wait for 1 minute after disconnecting the negative battery cable before working inside the vehicle. The air bag system is set to deploy for a short period of time after the battery is disconnected.

2. Disconnect the negative battery cable.
3. Drain the cooling system.
4. On the left side:
  a. Remove the thermostat housing assembly.
  b. Remove the PCV valve hose connection, ignition coil connector, and ignition coil.
  c. Disconnect the engine control wiring harness clamp and harness bracket.
5. On the right side:

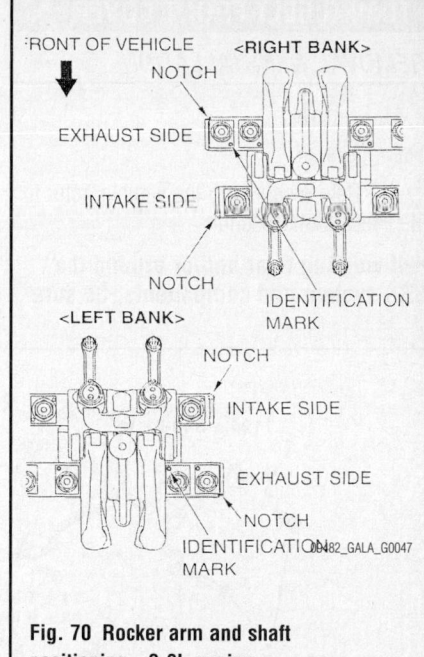

Fig. 70 Rocker arm and shaft positioning—3.8L engine

  a. Remove the intake manifold plenum, breather hose connection, heater hose, and water hose connection.
  b. Remove the ignition coil connector and coil.
6. Remove the rocker arm cover retaining bolts. Remove the rocker arm cover. Discard the gasket.
7. Install tool MD998443 on the rocker arms, so that the lash adjusters will not fall out.
8. Loosen the rocker arm and shaft assembly mounting bolts. Remove the rocker arm and shaft assembly with the bolts still attached.

➡Do not disassemble the rocker arm and shaft assembly.

To install:

➡Be sure to use new fasteners, as required.

9. Install the rocker arm, shaft, and lash adjuster assembly.
10. Check that the notches in the rocker shaft are facing the direction, shown in the illustration. Install the rocker shaft cap with its identification mark as shown in the illustration.
11. Tighten the intake side rocker arm and shaft assembly mounting bolts to 21—25 ft. lbs. (28—34 Nm).
12. Tighten the exhaust side rocker arm and shaft assembly mounting bolts to 106—124 inch lbs. (12—14 Nm).
13. Remove special tool MD998443.
14. Continue the installation in the reverse order of the removal procedure. Be sure to use new valve cover gaskets.

TIMING BELT FRONT COVER

REMOVAL & INSTALLATION

2.4L Engine

See Figure 71.

1. Before servicing the vehicle, refer to the Precautions Section.

➡ If working near and/or around the SRS system and components, be sure to disable the SRS system. Tape the negative battery cable with insulating tape. Always disconnect the negative battery cable first.

❊❊ CAUTION

Wait for 1 minute after disconnecting the negative battery cable before working inside the vehicle. The air bag system is set to deploy for a short period of time after the battery is disconnected.

2. Disconnect the negative battery cable.

3. Remove the engine undercover.

4. Remove the crankshaft damper pulley.

5. Disconnect the control wiring harness connector, battery wiring harness connector, and connector bracket to engine mounting insulator.

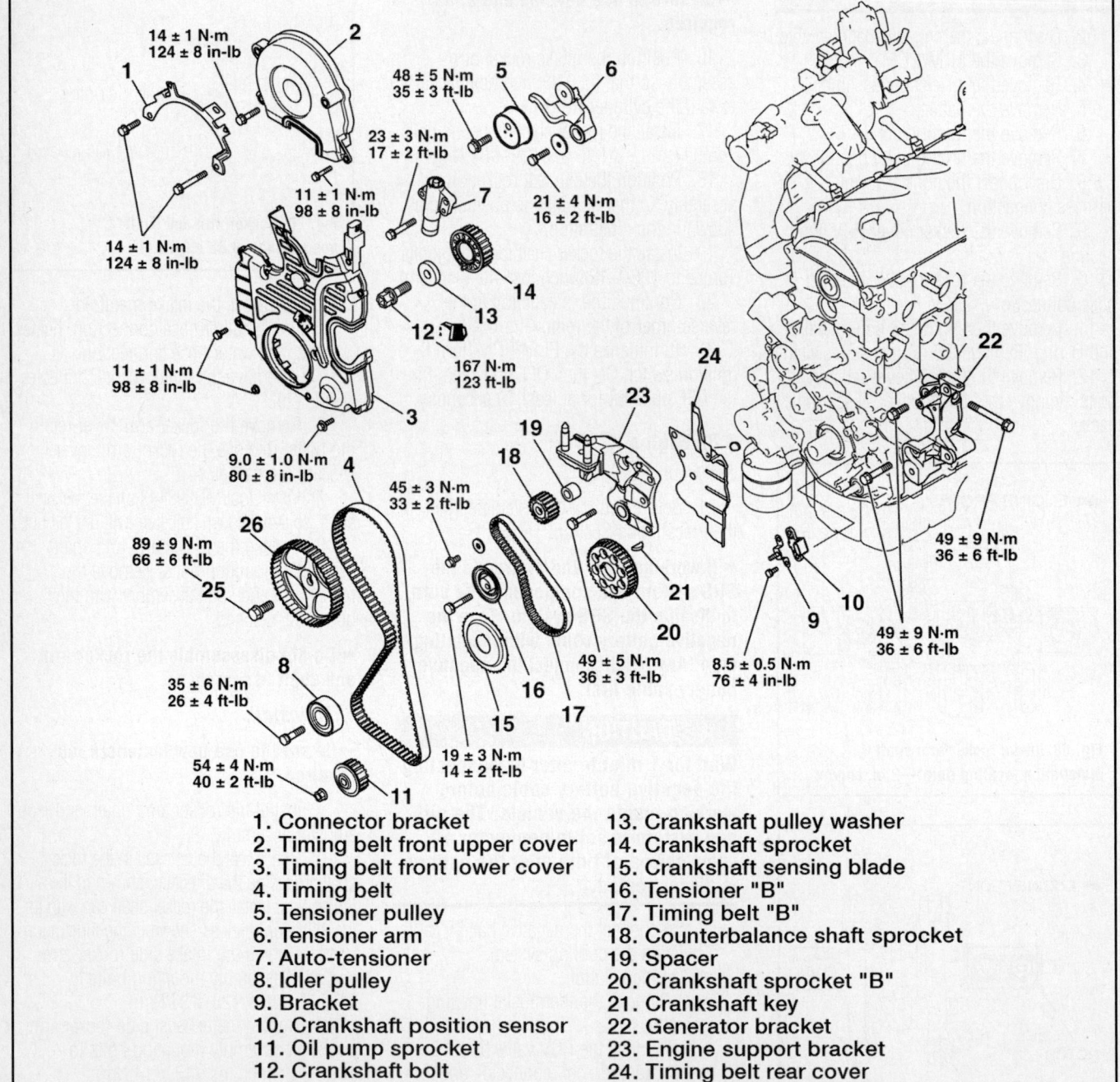

1. Connector bracket
2. Timing belt front upper cover
3. Timing belt front lower cover
4. Timing belt
5. Tensioner pulley
6. Tensioner arm
7. Auto-tensioner
8. Idler pulley
9. Bracket
10. Crankshaft position sensor
11. Oil pump sprocket
12. Crankshaft bolt
13. Crankshaft pulley washer
14. Crankshaft sprocket
15. Crankshaft sensing blade
16. Tensioner "B"
17. Timing belt "B"
18. Counterbalance shaft sprocket
19. Spacer
20. Crankshaft sprocket "B"
21. Crankshaft key
22. Generator bracket
23. Engine support bracket
24. Timing belt rear cover
25. Camshaft sprocket bolt
26. Camshaft sprocket

Fig. 71 Exploded view of timing belt and related components—2.4L engine

22140_MITS_G0128

6. Remove the harness bracket.
7. Remove the front upper timing belt cover.
8. Remove the water pump pulley.
9. Remove the idler pulley.
10. Remove the auto tensioner.
11. Remove the front lower timing belt cover.

To install:

➡**Be sure to use new fasteners, as required.**

12. Install the front lower timing belt cover.
13. Install the auto tensioner.
14. Install the idler pulley.
15. Install the water pump pulley.
16. Install the front upper timing belt cover.
17. Tighten all bolts to specification as illustrated.
18. Install the harness bracket.
19. Connect the control wiring harness connector, battery wiring harness connector, and connector bracket to engine mounting insulator.
20. Install the crankshaft damper pulley.
21. Install the engine undercover.
22. Connect the negative battery cable.

3.8L Engine

See Figure 72.

1. Before servicing the vehicle, refer to the Precautions Section.

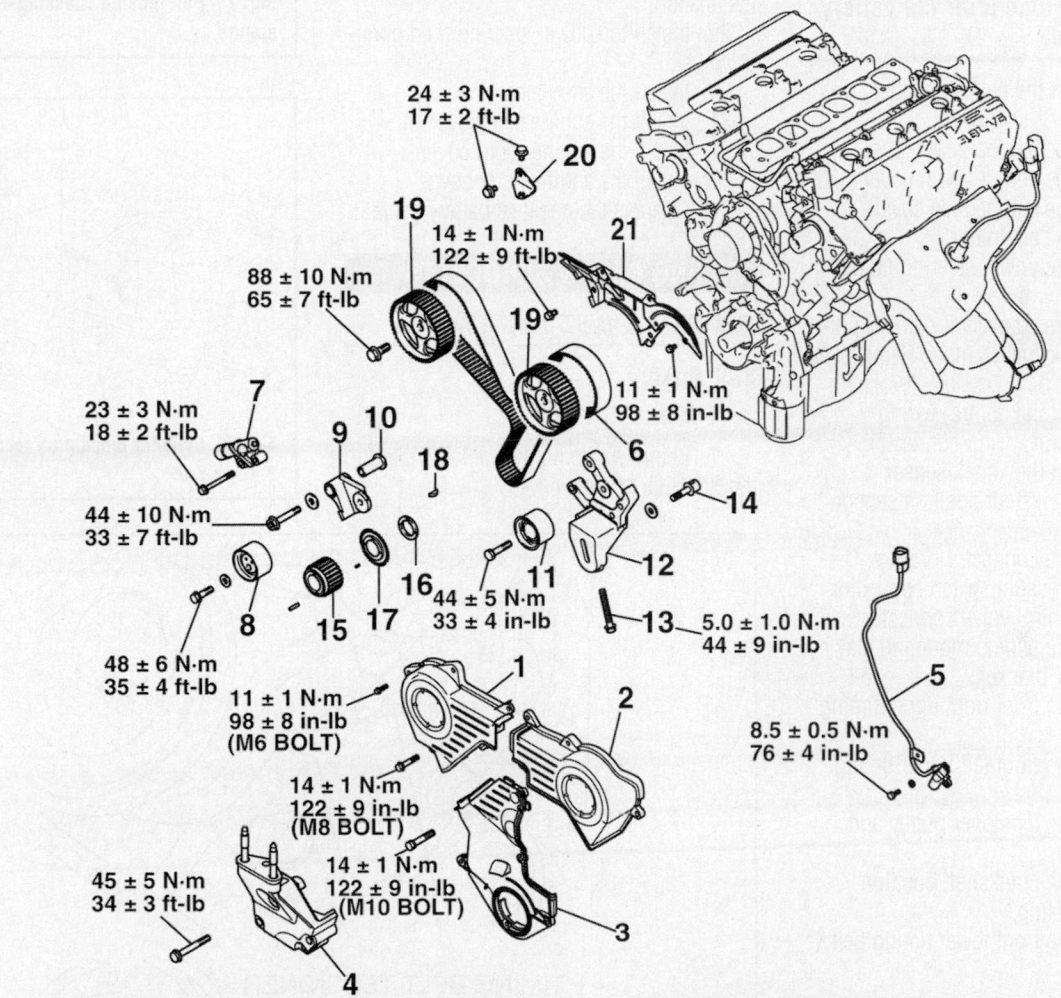

1. Timing belt front upper cover, right
2. Timing belt front upper cover, left
3. Timing belt front lower cover
4. Engine support bracket, right
5. Crankshaft position sensor
6. Timing belt
7. Auto-tensioner
8. Tensioner pulley
9. Tensioner arm
10. Shaft
11. Idler pulley
12. Tensioner bracket
13. Adjusting bolt
14. Adjusting stud
15. Crankshaft sprocket
16. Crankshaft spacer
17. Crankshaft sensing blade
18. Key
19. Camshaft sprocket
20. Bracket
21. Timing belt rear cover

Fig. 72 Exploded view of timing belt and related components—3.8L engine

22140_MITS_G0129

➡ If working near and/or around the SRS system and components, be sure to disable the SRS system. Tape the negative battery cable with insulating tape. Always disconnect the negative battery cable first.

✳✳ CAUTION

Wait for 1 minute after disconnecting the negative battery cable before working inside the vehicle. The air bag system is set to deploy for a short period of time after the battery is disconnected.

2. Disconnect the negative battery cable.

3. Remove the engine undercover.

4. Remove the engine room cover.

5. Remove the engine side cover.

6. Remove the alternator.

7. Remove the power steering fluid pump drive belt.

8. Remove the crankshaft damper pulley. Refer to Crankshaft Damper, Removal & Installation.

9. Disconnect the control wiring harness and injector wiring connector.

10. Disconnect the knock sensor connector and crankshaft position sensor connector.

11. Remove the connector bracket. Disconnect the oil pressure switch connector and engine oil control valve connector.

12. Remove the engine mounting stay and the connector bracket.

13. Remove the front right upper timing belt cover.

14. Remove the front left upper timing belt cover.

15. Remove the tensioner pulley and tensioner bracket.

16. Remove the crankshaft position sensor harness clamp.

17. Remove the front lower timing belt cover.

To install:

➡ Be sure to use new fasteners, as required.

18. Install the front lower timing belt cover.

19. Install the crankshaft position sensor harness clamp.

20. Install the tensioner pulley and tensioner bracket.

21. Install the front left upper timing belt cover.

22. Install the front right upper timing belt cover.

23. Install the engine mounting stay and the connector bracket.

24. Tighten all bolts to specification as illustrated.

25. Install the connector bracket. Connect the oil pressure switch connector and engine oil control valve connector.

26. Connect the knock sensor connector and crankshaft position sensor connector.

27. Connect the control wiring harness and injector wiring connector.

28. Install the crankshaft damper pulley. Refer to Crankshaft Damper, Removal & Installation.

29. Install the power steering fluid pump drive belt.

30. Install the alternator.

31. Install the engine side cover.

32. Install the engine room cover.

33. Install the engine undercover.

34. Connect the negative battery cable.

TIMING BELT & SPROCKETS

REMOVAL & INSTALLATION

2.4L Engine

See Figures 73 through 76.

1. Before servicing the vehicle, refer to the Precautions Section.

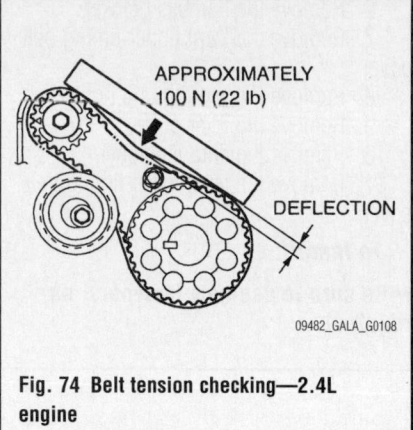

Fig. 74 Belt tension checking—2.4L engine

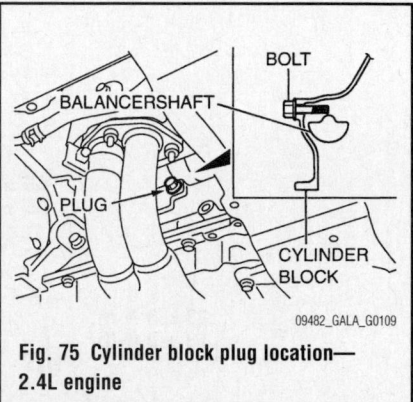

Fig. 75 Cylinder block plug location— 2.4L engine

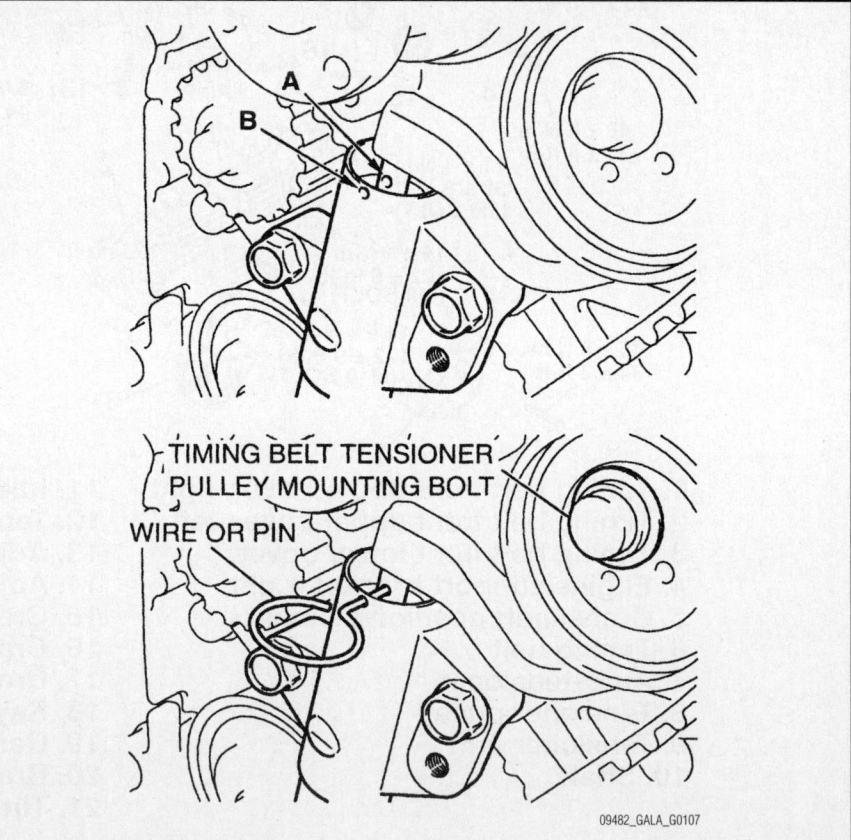

Fig. 73 Timing belt tensioner adjuster rod set hole "A" and cylinder set hole "B"—2.4L engine

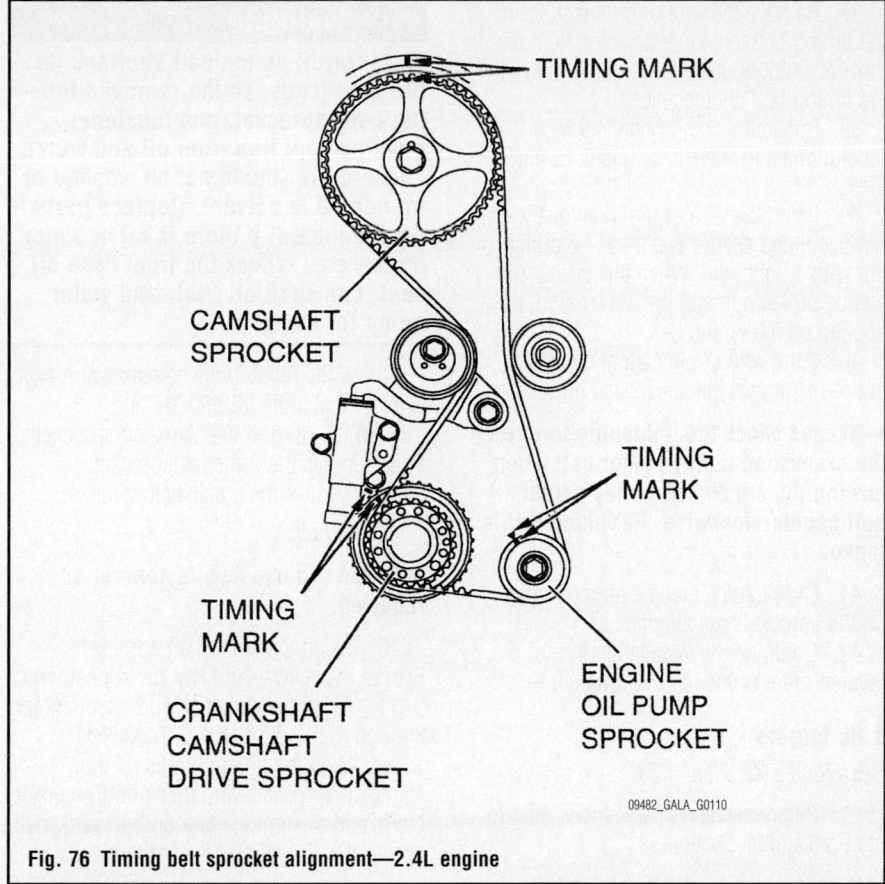

TIMING MARK

CAMSHAFT
SPROCKET

TIMING
MARK

TIMING
MARK

ENGINE
OIL PUMP
SPROCKET

CRANKSHAFT
CAMSHAFT
DRIVE SPROCKET

09482_GALA_G0110

Fig. 76 Timing belt sprocket alignment—2.4L engine

➡If working near and/or around the SRS system and components, be sure to disable the SRS system. Tape the negative battery cable with insulating tape. Always disconnect the negative battery cable first.

✳ CAUTION

Wait for 1 minute after disconnecting the negative battery cable before working inside the vehicle. The air bag system is set to deploy for a short period of time after the battery is disconnected.

2. Disconnect the negative battery cable.
3. Remove the timing belt front cover(s).
4. Turn the crankshaft clockwise and align each timing mark to set the number 1 piston to Top Dead Center (TDC) of its compression stroke.
5. Remove the timing belt undercover rubber plug and then install special tool MD998738. Screw the special tool until it contacts the timing belt tensioner arm.

✳ WARNING

The special tool must be screwed in gradually at the rate of a 30° turn per

second. If it is turned in all at once, the timing belt tensioner adjuster rod will not easily retract and the tool may bend.

6. Gradually screw in the special tool and then align the timing belt tensioner adjuster rod set hole "A" with the timing belt tensioner adjuster cylinder set hole "B".
7. Insert a wire or pin in the set holes to lock the assembly in place. After removing the special tool, loosen the timing belt tensioner pulley mounting bolts and remove the timing belt.

➡If the belt is being reused be sure to mark the direction of rotation (clockwise) on the belt.

8. Remove the timing belt tensioner pulley, tensioner arm, and adjuster.
9. Remove the timing belt idler pulley.
10. Remove the timing belt lower cover bracket.
11. Remove the crankshaft position sensor.
12. Remove the crankshaft pulley center bolt and washer. Remove the drive sprocket. Remove the crankshaft angle sensing blade.
13. Remove the balancer timing belt tensioner.
14. Remove the timing belt "B" (balancer

belt) from its mounting. Remove the sprocket.

To install:

➡**Be sure to use new fasteners, as required.**

15. Be sure that the crankshaft balancer shaft drive sprocket timing marks and balancer shaft sprocket timing marks are aligned.
16. Install the timing belt "B" (balancer belt) on the crankshaft balancer drive sprocket and balancer shaft sprocket. There should be no slack on the tension side.
17. Assemble an temporarily fix the center of the pulley of the balancer timing belt tensioner so that it is at the top left from the center of the assembling bolt, and the pulley flange is at the front side of the engine.

➡**When tightening the mounting bolts, ensure that the tensioner does not rotate with the bolts. Allowing it to rotate can cause excessive tension of the belt.**

18. Lift your fingers and the balancer timing belt tensioner in the counterclockwise position. Apply minimal torque to the balancer timing belt so the belt is tense without looseness. Tighten the assembling bolt to 12—16 ft. lbs. (16—22 Nm).
19. Turn the crankshaft clockwise 2 turns to set the number 1 piston to TDC on the compression stroke. Check that the sprocket timing marks are aligned.
20. Apply slight pressure at the center of the belt between both sprockets. Inspect whether the belt deflection is within specification 0.31–0.47 inch (8–12mm). Correct as required.
21. Install the crankshaft angle sensing blade, crankshaft drive sprocket, pulley washer and pulley center bolt. Apply clean engine oil to the retaining bolt prior to installation. Tighten the assembling bolt to 123 ft. lbs. (167 Nm).
22. Slowly compress the timing belt tensioner adjuster rod using a press or vise. Align the set hole "A" of the rod with set hole "B" of the timing belt tensioner adjuster cylinder.

➡**Do not compress the assembly too fast as damage to the rod may occur.**

23. Insert a wire or pin in the aligned set holes to lock the assembly in place.
24. Install the timing belt tensioner adjuster to the engine and tighten the mounting bolt to 15—19 ft. lbs. (20—26 Nm).
25. Temporarily tighten the timing belt tensioner pulley. Align the timing marks on

the camshaft sprocket, crankshaft drive sprocket and engine oil pump sprocket.

26. Adjust the timing mark of the engine oil pump sprocket, by removing the cylinder block plug. Insert a bolt (M6, section width 10MM, nominal length 45MM) from the plug hole.

➡ **If the bolt comes in contact with the balancer shaft, turn the engine oil sprocket one rotation. Re-adjust the timing mark and check to see that the bolt fits. Do not remove the bolt until the valve timing belt is assembled.**

27. Position the timing belt on the timing belt tensioner pulley and crankshaft driver sprocket, support it with your hand so that it does not slide.

28. Position the belt on the engine oil pump sprocket while pulling it with your other hand.

29. Position the timing belt on the timing belt idler pulley.

➡ **Incorporate the timing belt. Then apply reverse rotation (counterclockwise) pressure to the camshaft sprocket. Recheck to see that each timing mark is aligned while the tension side of the belt is right.**

30. Position the timing mark on the camshaft sprocket.

31. Turn the timing belt tensioner pulley upward using tool MD998767 to apply tension to the belt. Temporarily tighten and fix the belt tensioner pulley mounting bolt.

32. Check that the timing marks are aligned.

33. Remove the bolt that was inserted in the cylinder block plug hole. Replace the cylinder block plug hole bolt and torque to 21—25 ft. lbs. (27—33 Nm).

➡ **Install tool MD998738 by hand. Do not use any other tools to install the special tool as damage to the wire or pin inserted in the timing belt tensioner may occur.**

34. Gradually screw the tool into position so that the wire or pin inserted in the timing belt tensioner adjuster moves slightly.

35. Turn the crankshaft in the clockwise direction; align each timing mark to set number one piston to TDC on the compression stroke.

36. Loosen the timing belt tensioner pulley mounting bolt. With tool MD998767 and a torque wrench, apply tension torque 31 inch lbs. (42 Nm) to the timing belt. Tighten the timing belt tensioner pulley mounting bolt to 33—39 ft. lbs. (43—53 Nm).

37. Remove the wire or pin inserted in the timing belt tensioner adjuster. Remove tool MD998738. Install the rubber plug of the timing belt undercover.

38. Rotate the crankshaft clockwise 2 revolutions and leave it for about 15 minutes.

39. Insert the wire or pin, previously removed, and ensure that it can be pulled out with a light load. When the wire or pin can be pulled out, appropriate tension is applied on the timing belt.

40. If the wire or pin cannot be pulled out easily, repeat the above step until it can.

➡ **Always check the tightening torque of the crankshaft pulley center bolt when turning the crankshaft pulley center bolt counterclockwise. Retighten if it is loose.**

41. Check again that the timing marks on the sprockets are aligned.

42. Continue the installation in the reverse order of the removal procedure.

3.8L Engine
See Figures 72, 77 and 78.

1. Before servicing the vehicle, refer to the Precautions Section.

➡ **If working near and/or around the SRS system and components, be sure to disable the SRS system. Tape the negative battery cable with insulating tape. Always disconnect the negative battery cable first.**

> ✳✳ **CAUTION**
>
> **Wait for 1 minute after disconnecting the negative battery cable before working inside the vehicle. The air bag system is set to deploy for a short period of time after the battery is disconnected.**

2. Disconnect the negative battery cable.

3. Remove the timing belt front cover(s).

4. Remove the engine support bracket.

5. Turn the crankshaft clockwise to align each timing mark and set the number 1 cylinder to Top Dead Center (TDC) on the compression stroke.

➡ **If the belt is going to be reused, mark the direction of rotation.**

6. Remove the timing belt.

7. Remove the auto tensioner, tensioner pulley, and tensioner arm.

> ✳✳ **WARNING**
>
> **Water or oil on the belt shortens its life drastically, so the removed timing belt, sprocket, and tensioner must be kept free from oil and water. These parts should not be washed or immersed in solvent. Replace parts if contaminated. If there is oil or water on any part, check the front case oil seal, camshaft oil seal, and water pump for leaks.**

8. While holding the camshaft sprocket with special tools MB990767 and MD998715, loosen the camshaft sprocket bolt. Remove the camshaft sprocket.

9. Remove the crankshaft sprocket.

To install:

➡ **Be sure to use new fasteners, as required.**

10. Install the crankshaft sprocket. Ensure the keyway and key are in position.

11. Install the camshaft sprocket. Tighten the bolt to 58–72 ft. lbs. (78–98 Nm).

12. Align the timing marks on the camshaft sprockets with those on the rocker cover and the timing mark on the crankshaft sprocket with that on the engine block.

> ✳✳ **CAUTION**
>
> **The right side camshaft sprocket can turn easily due to the spring force applied. Be careful of hands and fingers when turning the camshaft as they could get caught in an uncomfortable position.**

13. Install the timing belt first on the crankshaft sprocket, than on the idler pulley, then on the left camshaft sprocket, then on the water pump pulley, then on the right camshaft sprocket, and finally on the tensioner pulley.

14. Turn the right camshaft sprocket counterclockwise until the tension side of the timing belt is firmly stretched. Check all timing marks, again.

15. Use special tool MD998767 to push the tensioner pulley into the timing belt, and then temporarily tighten the center bolt.

16. Use special tool MD998769 to turn the crankshaft ¼ turn counterclockwise, then turn it again clockwise until the timing marks are aligned.

17. Holding the auto tensioner, with your hand, press the end of the pushrod against a metal surface with a force of about 32 lbs. Measure how far in the push rod is pushed. Specification should be 0.189–0.236 inch

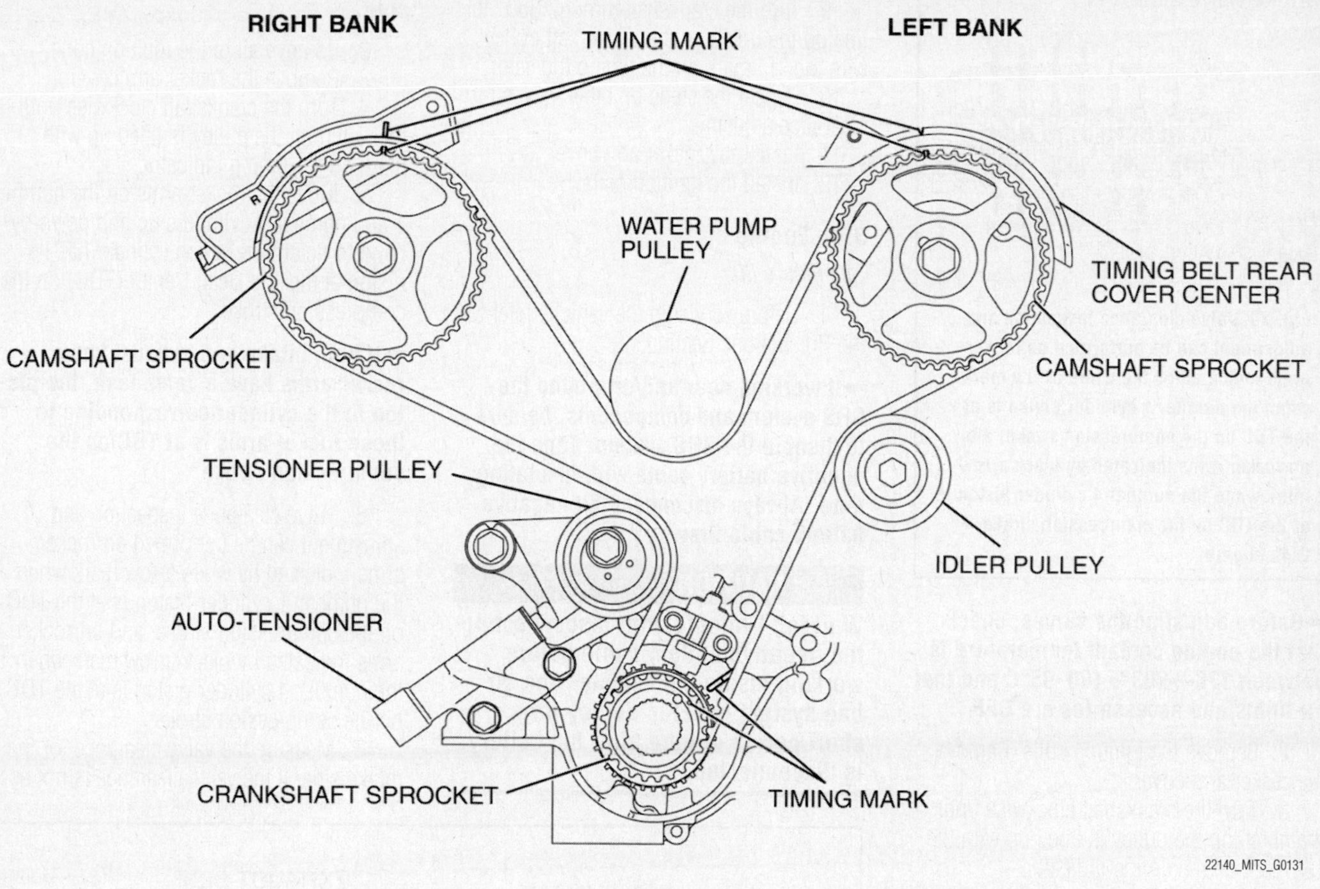

RIGHT BANK

TIMING MARK

LEFT BANK

WATER PUMP
PULLEY

TIMING BELT REAR
COVER CENTER

CAMSHAFT SPROCKET

CAMSHAFT SPROCKET

TENSIONER PULLEY

IDLER PULLEY

AUTO-TENSIONER

CRANKSHAFT SPROCKET

TIMING MARK

22140_MITS_G0131

Fig. 77 Check to see that the timing marks of all the sprockets are in alignment—3.8L engine

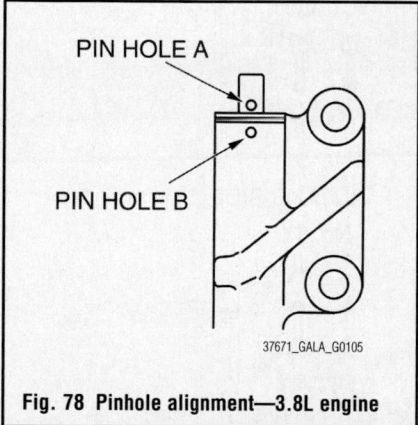

PIN HOLE A

PIN HOLE B

37671_GALA_G0105

Fig. 78 Pinhole alignment—3.8L engine

(4.8–6.0mm). If not within specification, replace the auto tensioner assembly.

18. Position the auto tensioner perpendicular in a vise. If the tensioner has a plug at the base, be sure to use a washer to protect the plug.

19. Slowly compress the pushrod of the auto tensioner until the pin hole "A" is aligned the pin hole "B" in the cylinder.

Insert a pin into both holes once they are aligned. Install the auto tensioner to the engine.

➡**When tightening the center bolt, be careful that the tensioner pulley does not turn with the bolt.**

20. Loosen the center bolt of the tensioner pulley. Use tool MD998767 and a torque wrench and apply tension torque to the timing belt in a downward motion. Tighten the center bolt to 32—40 ft. lbs. (42—54 Nm).

21. Remove the tensioner set pin. Turn the crankshaft clockwise twice to align the timing marks. Wait at least 5 minutes, then check that the auto tensioner pushrod extends within the standard value range "A" of 0.19—0.24 inch. If not repeat the operation again.

22. Check again that the timing marks of the sprockets are aligned.

23. Continue the installation in the reverse order of the removal procedure.

VALVE LASH

ADJUSTMENT

2.4L Engine

See Figure 79.

1. Before servicing the vehicle, refer to the Precautions Section.

➡**If working near and/or around the SRS system and components, be sure to disable the SRS system. Tape the negative battery cable with insulating tape. Always disconnect the negative battery cable first.**

✳✳ CAUTION

Wait for 1 minute after disconnecting the negative battery cable before working inside the vehicle. The air bag system is set to deploy for a short period of time after the battery is disconnected.

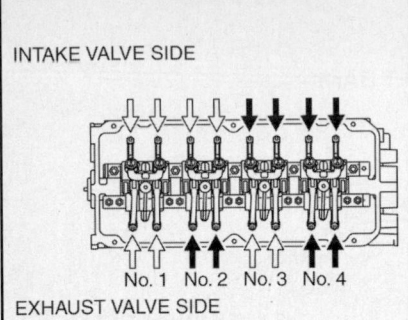

INTAKE VALVE SIDE

No. 1 No. 2 No. 3 No. 4

EXHAUST VALVE SIDE

22140_MITS_G0050

Fig. 79 Valve clearance inspection and adjustment can be performed on rocker arms indicated by the white arrow mark when the number 1 cylinder piston is at the TDC on the compression stroke, and on rocker arms indicated by black arrow mark when the number 4 cylinder piston is at the TDC on the compression stroke— 2.4L engine

➡️**Before adjusting the valves, check that the engine coolant temperature is between 176—203°F (80–95°C and that all lights and accessories are OFF.**

2. Remove the ignition coils. Remove the rocker arm cover.

3. Turn the crankshaft clockwise until the notch on the pulley is lined up with the "T" mark on the timing indicator.

4. Move the rocker arms on the number 1 and number 4 cylinders up and down by hand to determine which cylinder has its piston at the Top Dead Center (TDC) position on the compression stroke.

➡️**If both intake and exhaust valve rocker arms have a valve lash, the piston in the cylinder corresponding to these rocker arms is at TDC on the compression stroke.**

5. Valve clearance inspection and adjustment can be performed on rocker arms indicated by the white arrow mark when the number 1 cylinder piston is at the TDC on the compression stroke, and on rocker arms indicated by black arrow mark when the number 4 cylinder piston is at the TDC on the compression stroke.

6. Measure the valve clearance. If the valve clearance is not as specified, loosen the rocker arm lock nut and adjust the clearance using a thickness gauge while turning the adjusting screw.

7. Standard value (hot engine):
   a. Intake valve: 0.008 inch (0.20mm).
   b. Exhaust valve: 0.012 inch (0.30mm).

8. While holding the adjusting screw with a screwdriver to prevent it from turning,

tighten the lock nut to the specified torque: 71–89 inch lbs. (8–10 Nm).

9. Turn the crankshaft through 360° to line up the notch on the crankshaft pulley with the "T" mark on the timing indicator.

10. Repeat the steps on other valves for clearance adjustment.

11. Install the rocker arm cover.

12. Install the ignition coils.

3.8L Engine

See Figure 80.

1. Before servicing the vehicle, refer to the Precautions Section.

➡️**If working near and/or around the SRS system and components, be sure to disable the SRS system. Tape the negative battery cable with insulating tape. Always disconnect the negative battery cable first.**

✲✲ CAUTION

Wait for 1 minute after disconnecting the negative battery cable before working inside the vehicle. The air bag system is set to deploy for a short period of time after the battery is disconnected.

➡️**Perform the valve clearance check and adjustment when the engine is cold.**

2. Remove all of the ignition coils.

3. Remove the rocker arm covers.

4. Turn the crankshaft clockwise until the notch on the pulley is lined up with "T" mark on the timing indicator.

5. Move the rocker arms on the number 1 and number 4 cylinders up and down by hand to determine which cylinder has its piston at the Top Dead Center (TDC) on the compression stroke.

➡️**If both intake and exhaust valve rocker arms have a valve lash, the piston in the cylinder corresponding to these rocker arms is at TDC on the compression stroke.**

6. Valve clearance inspection and adjustment can be performed on rocker arms indicated by white arrow mark when the number 1 cylinder piston is at the TDC on the compression stroke, and on rocker arms indicated by black arrow mark when the number 4 cylinder piston is at the TDC on the compression stroke.

7. Measure the valve clearance for the intake side. If the valve clearance is not as

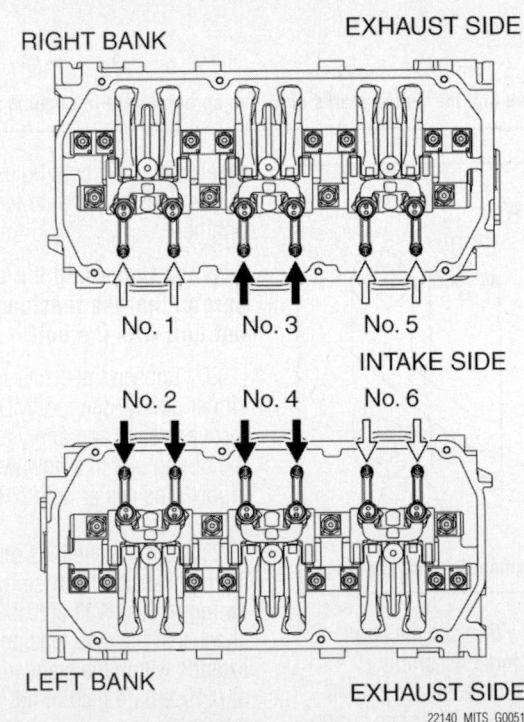

RIGHT BANK   EXHAUST SIDE

No. 1   No. 3   No. 5

INTAKE SIDE

No. 2   No. 4   No. 6

LEFT BANK   EXHAUST SIDE

22140_MITS_G0051

Fig. 80 Valve clearance inspection and adjustment can be performed on rocker arms indicated by the white arrow mark when the number 1 cylinder piston is at the TDC on the compression stroke, and on rocker arms indicated by black arrow mark when the number 4 cylinder piston is at the TDC on the compression stroke—3.8L engine

specified, loosen the rocker arm lock nut and adjust the clearance using a thickness gauge while turning the adjusting screw. Standard value (cold engine): 0.004 inch (0.10mm).

➥Valve clearance check and adjustment is unnecessary for the exhaust side due to auto lash adjusters installed.

8. While holding the adjusting screw with a screwdriver to prevent it from turning, tighten the lock nut to the specified torque: 71–89 inch lbs. (8–10 Nm).

9. Turn the crankshaft through 360° to line up the notch on the crankshaft pulley with the "T" mark on the timing indicator.

10. Repeat the steps on other valves for clearance adjustment.

11. Install the rocker arm covers.

12. Install the ignition coils.

ENGINE PERFORMANCE & EMISSION CONTROLS

CAMSHAFT POSITION (CMP) SENSOR

LOCATION

See Figures 81 and 82.

Refer to the accompanying illustrations.

REMOVAL & INSTALLATION

See Figures 81 and 82.

1. Before servicing the vehicle, refer to the Precautions Section.

➥If working near and/or around the SRS system and components, be sure to disable the SRS system. Tape the negative battery cable with insulating tape. Always disconnect the negative battery cable first.

❄❄ CAUTION

Wait for 1 minute after disconnecting the negative battery cable before working inside the vehicle. The air bag system is set to deploy for a short period of time after the battery is disconnected.

2. Disconnect the negative battery cable.

3. Remove the necessary components in order to gain access to the sensor retaining bolts.

4. Disconnect the connector from the CMP sensor.

5. Remove the bolt that retains the CMP sensor.

6. Remove the CMP sensor.

To install:

➥Be sure to use new fasteners, as required.

7. Installation is the reverse of the removal procedure.

8. Tighten the bolt that retains the CMP sensor to: 98 inch lbs. (11 Nm).

9. Connect the sensor connector.

CRANKSHAFT POSITION (CKP) SENSOR

LOCATION

See Figures 83 and 84.

Refer to the accompanying illustrations.

REMOVAL & INSTALLATION

See Figures 83 and 84.

1. Before servicing the vehicle, refer to the Precautions Section.

➥If working near and/or around the SRS system and components, be sure to disable the SRS system. Tape the negative battery cable with insulating tape. Always disconnect the negative battery cable first.

❄❄ CAUTION

Wait for 1 minute after disconnecting the negative battery cable before working inside the vehicle. The air bag system is set to deploy for a short period of time after the battery is disconnected.

2. Disconnect the negative battery cable.

3. Remove the necessary components in order to gain access to the sensor retaining bolts.

4. Disconnect the connector from the sensor.

5. Remove the bolt that retains the sensor in place.

6. Remove the sensor from its mounting.

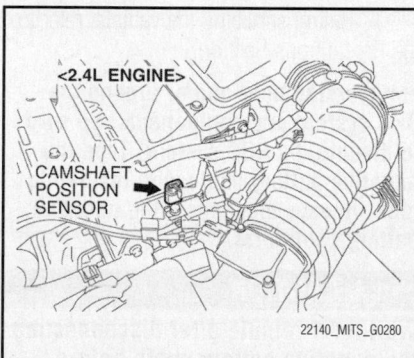

Fig. 81 Camshaft Position (CMP) sensor location—2.4L engine

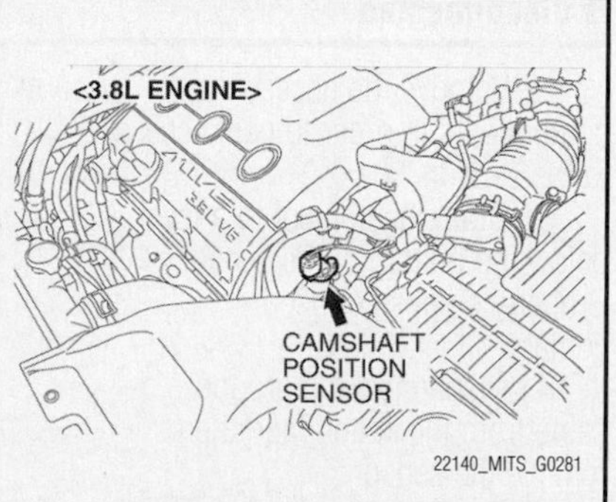

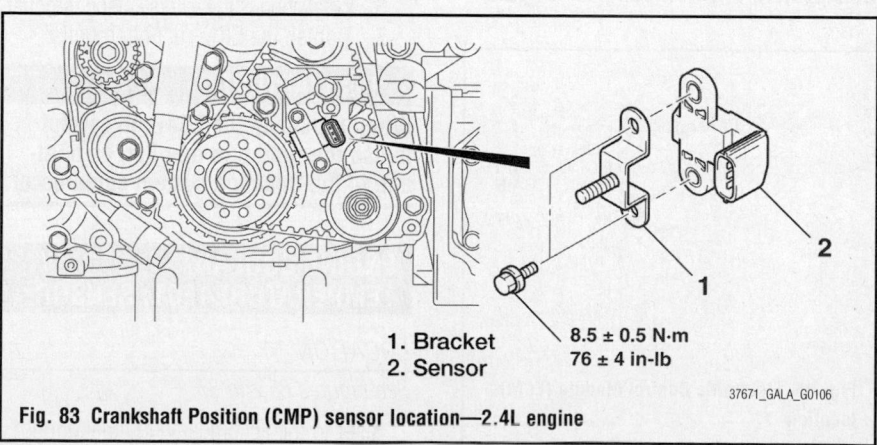

Fig. 82 Camshaft Position (CMP) sensor location—3.8L engine

1. Bracket
2. Sensor

8.5 ± 0.5 N·m
76 ± 4 in-lb

37671_GALA_G0106

Fig. 83 Crankshaft Position (CMP) sensor location—2.4L engine

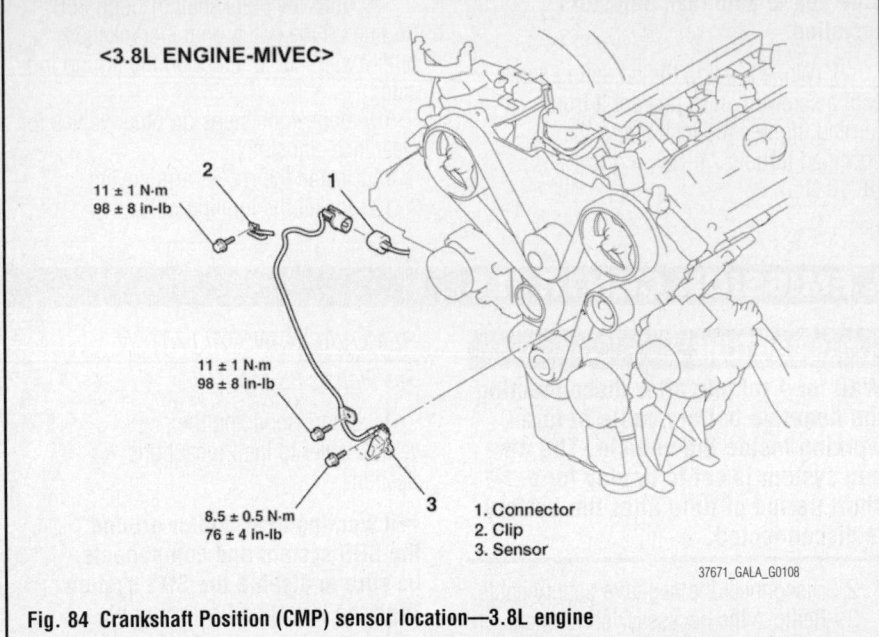

Fig. 84 Crankshaft Position (CMP) sensor location—3.8L engine

1. Connector
2. Clip
3. Sensor

11 ± 1 N·m
98 ± 8 in-lb

11 ± 1 N·m
98 ± 8 in-lb

8.5 ± 0.5 N·m
76 ± 4 in-lb

<3.8L ENGINE-MIVEC>

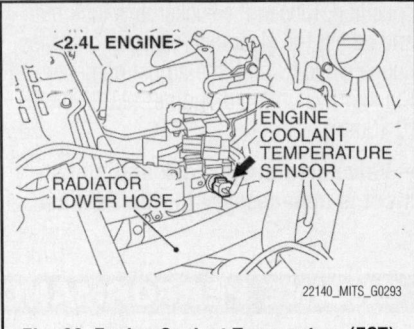

Fig. 86 Engine Coolant Temperature (ECT) sensor location—2.4L engine

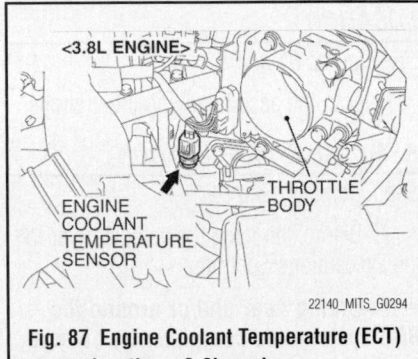

Fig. 87 Engine Coolant Temperature (ECT) sensor location—3.8L engine

To install:

➡ Be sure to use new fasteners, as required.

7. Installation is the reverse of the removal procedure.

8. Tighten the sensor retaining bolt to: 78 inch lbs. (9 Nm).

ELECTRONIC CONTROL MODULE (ECM)

LOCATION

See Figure 85.

Refer to the accompanying illustration.

REMOVAL & INSTALLATION

See Figure 85.

1. Before servicing the vehicle, refer to the Precautions Section.

➡ If working near and/or around the SRS system and components, be sure

to disable the SRS system. Tape the negative battery cable with insulating tape. Always disconnect the negative battery cable first.

✳✳ CAUTION

Wait for 1 minute after disconnecting the negative battery cable before working inside the vehicle. The air bag system is set to deploy for a short period of time after the battery is disconnected.

2. Turn ignition switch off.
3. Disconnect the negative battery cable from the battery.
4. Disconnect the ECM connectors.
5. Remove the ECM mounting bolts and remove the ECM from the vehicle.

To install:

➡ Be sure to use new fasteners, as required.

6. Installation is the reverse of the removal.
7. Tighten the ECM mounting bolts.

✳✳ WARNING

When replacing the ECM, be careful to use the right part number, as damage to the injection system could occur.

ENGINE COOLANT TEMPERATURE (ECT) SENSOR

LOCATION

See Figures 86 and 87.

Refer to the accompanying illustrations.

REMOVAL & INSTALLATION

See Figures 86 and 87

1. Before servicing the vehicle, refer to the Precautions Section.

➡ If working near and/or around the SRS system and components, be sure to disable the SRS system. Tape the negative battery cable with insulating tape. Always disconnect the negative battery cable first.

✳✳ CAUTION

Wait for 1 minute after disconnecting the negative battery cable before working inside the vehicle. The air bag system is set to deploy for a short period of time after the battery is disconnected.

2. Remove the necessary components in order to gain access to the sensor.

3. Drain the coolant to a level below the bottom of the sensor. Properly dispose of used coolant.

4. Disconnect the ground cable from the battery and then remove the sensor connector.

5. Remove the coolant temperature sensor.

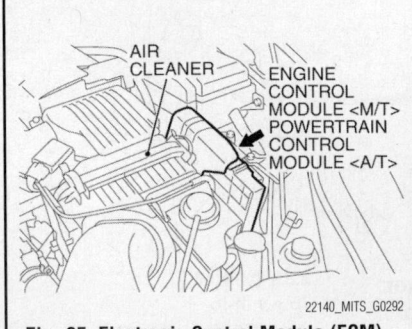

Fig. 85 Electronic Control Module (ECM) location

AIR CLEANER

ENGINE CONTROL MODULE <M/T> POWERTRAIN CONTROL MODULE <A/T>

To install:

➡**Be sure to use new fasteners, as required.**

6. Coat the threads of the sensor with a suitable sealant and thread into the housing.

7. Tighten the sensor to 22 ft. lbs. (30 Nm).

8. Refill the cooling system to the proper level.

9. Attach the electrical connector to the sensor securely.

10. Connect the negative battery cable.

3.8L Engine

See Figure 88.

1. Before servicing the vehicle, refer to the Precautions Section.

➡**If working near and/or around the SRS system and components, be sure to disable the SRS system. Tape the negative battery cable with insulating tape. Always disconnect the negative battery cable first.**

❉❉ CAUTION

Wait for 1 minute after disconnecting the negative battery cable before working inside the vehicle. The air bag system is set to deploy for a short period of time after the battery is disconnected.

2. Remove the strut tower bar.
3. Disconnect the connector.
4. Remove the retaining bolt.
5. Remove the EGR valve. Discard the gasket.

To install:

➡**Be sure to use new fasteners, as required.**

6. Installation is the reverse of the removal procedure.
7. Be sure to use a new gasket.

HEATED OXYGEN (HO2S) SENSOR

LOCATION

The sensors are located in the exhaust system. On some vehicles, one sensor is located up at the exhaust manifold(s) and the other sensor is located down at the catalytic converter.

REMOVAL & INSTALLATION

See Figures 89 through 91.

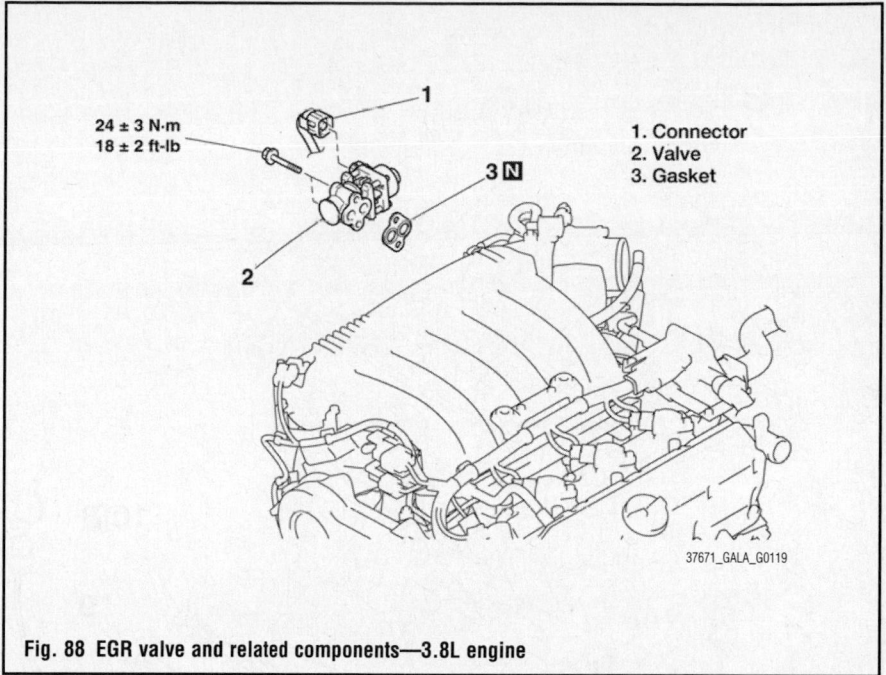

1. Connector
2. Valve
3. Gasket

Fig. 88 EGR valve and related components—3.8L engine

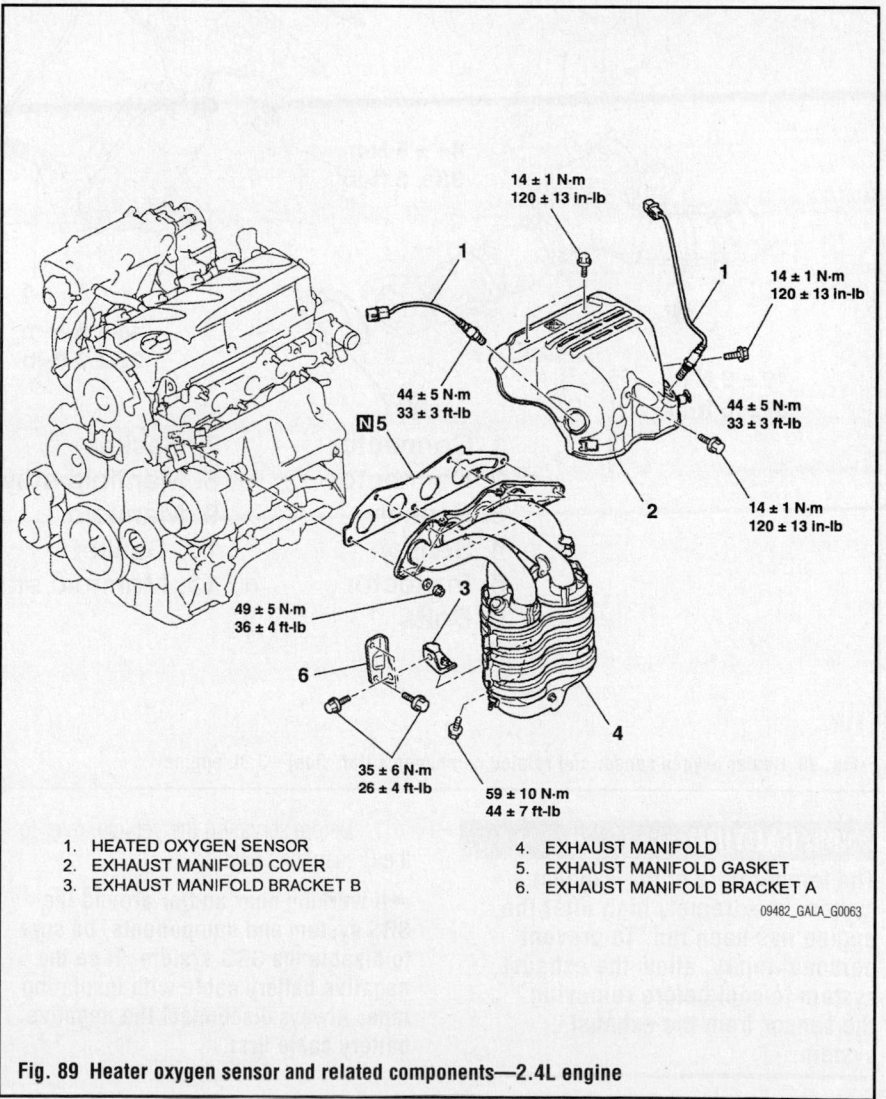

1. HEATED OXYGEN SENSOR
2. EXHAUST MANIFOLD COVER
3. EXHAUST MANIFOLD BRACKET B
4. EXHAUST MANIFOLD
5. EXHAUST MANIFOLD GASKET
6. EXHAUST MANIFOLD BRACKET A

Fig. 89 Heater oxygen sensor and related components—2.4L engine

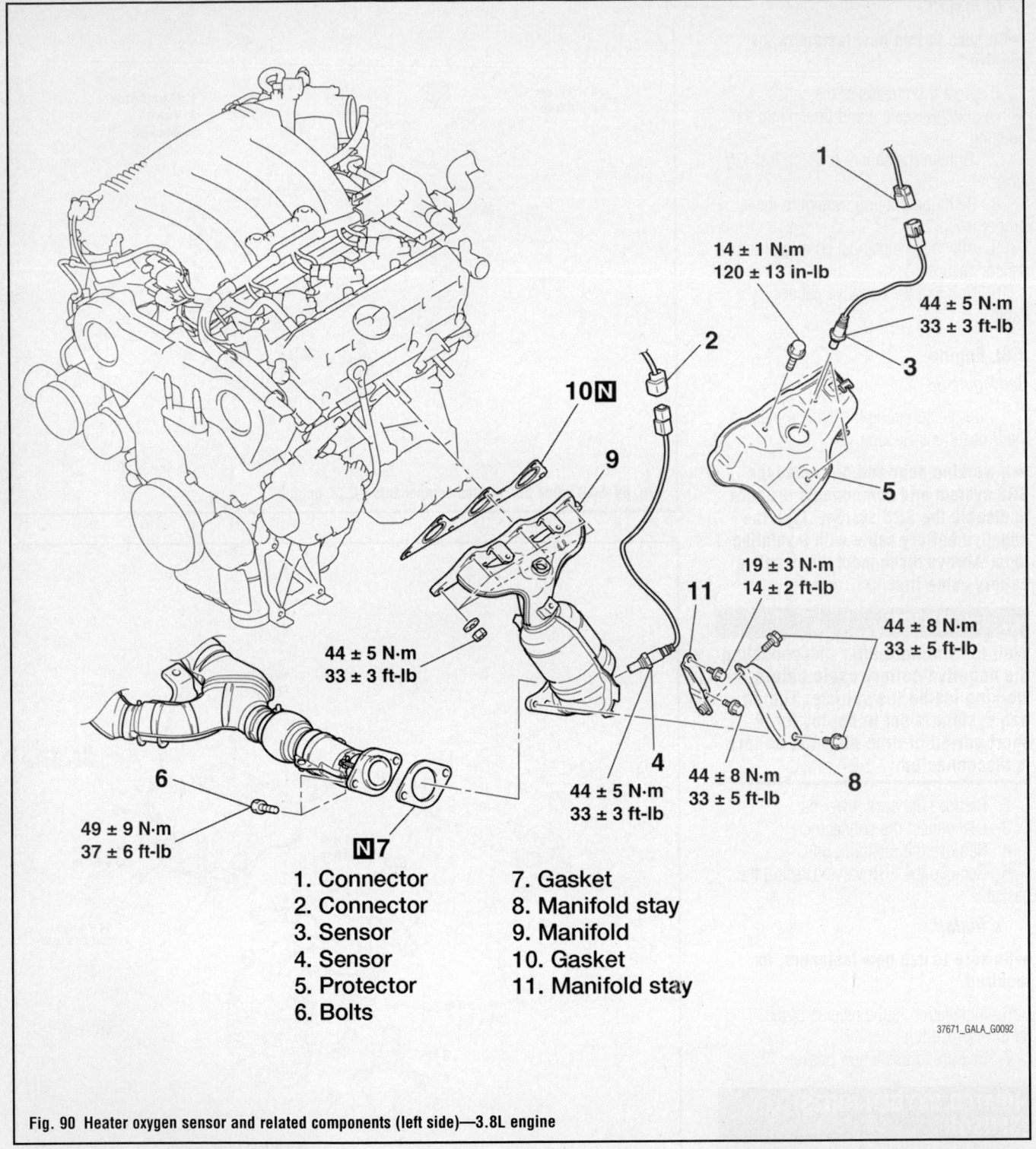

1. Connector
2. Connector
3. Sensor
4. Sensor
5. Protector
6. Bolts
7. Gasket
8. Manifold stay
9. Manifold
10. Gasket
11. Manifold stay

37671_GALA_G0092

Fig. 90 Heater oxygen sensor and related components (left side)—3.8L engine

---

⁂ **CAUTION**

The temperature of the exhaust system is extremely high after the engine has been run. To prevent personal injury, allow the exhaust system to cool before removing the sensor from the exhaust system.

1. Before servicing the vehicle, refer to the Precautions Section.

➡ If working near and/or around the SRS system and components, be sure to disable the SRS system. Tape the negative battery cable with insulating tape. Always disconnect the negative battery cable first.

⁂ **CAUTION**

Wait for **1** minute after disconnecting the negative battery cable before working inside the vehicle. The air bag system is set to deploy for a short period of time after the battery is disconnected.

2. Disconnect the negative battery cable.

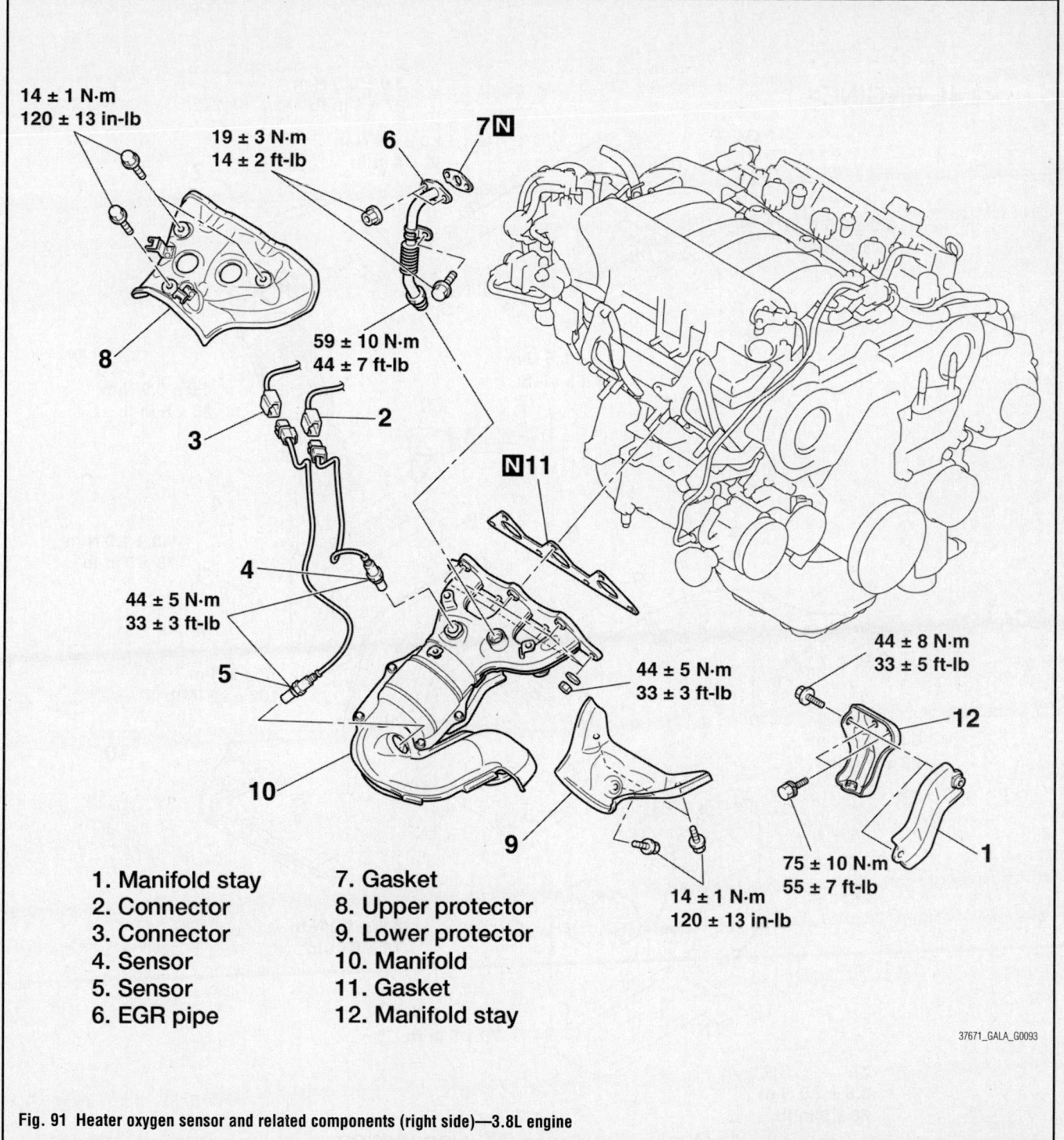

1. Manifold stay
2. Connector
3. Connector
4. Sensor
5. Sensor
6. EGR pipe
7. Gasket
8. Upper protector
9. Lower protector
10. Manifold
11. Gasket
12. Manifold stay

37671_GALA_G0093

Fig. 91 Heater oxygen sensor and related components (right side)—3.8L engine

3. Raise and safely support the vehicle, as needed.

4. Detach the electrical connector from the oxygen sensor.

5. Using socket MD998770, or equivalent oxygen sensor socket, remove the heated oxygen sensor.

To install:

➡**Be sure to use new fasteners, as required.**

6. If installing the old oxygen sensor, coat the threads with anti-seize compound. New sensors are already coated. Take care not to contaminate the oxygen sensor probe with the anti-seize compound.

7. Install the oxygen sensor. Using the correct tool, tighten the sensor to 33 ft. lbs. (45 Nm).

8. Attach the wiring to the sensor.

9. Connect the negative battery cable.

INTAKE AIR TEMPERATURE (IAT) SENSOR

LOCATION

See Figures 92.

The Intake Air Temperature (IAT) sensor is mounted in the intake air hose of the air cleaner assembly.

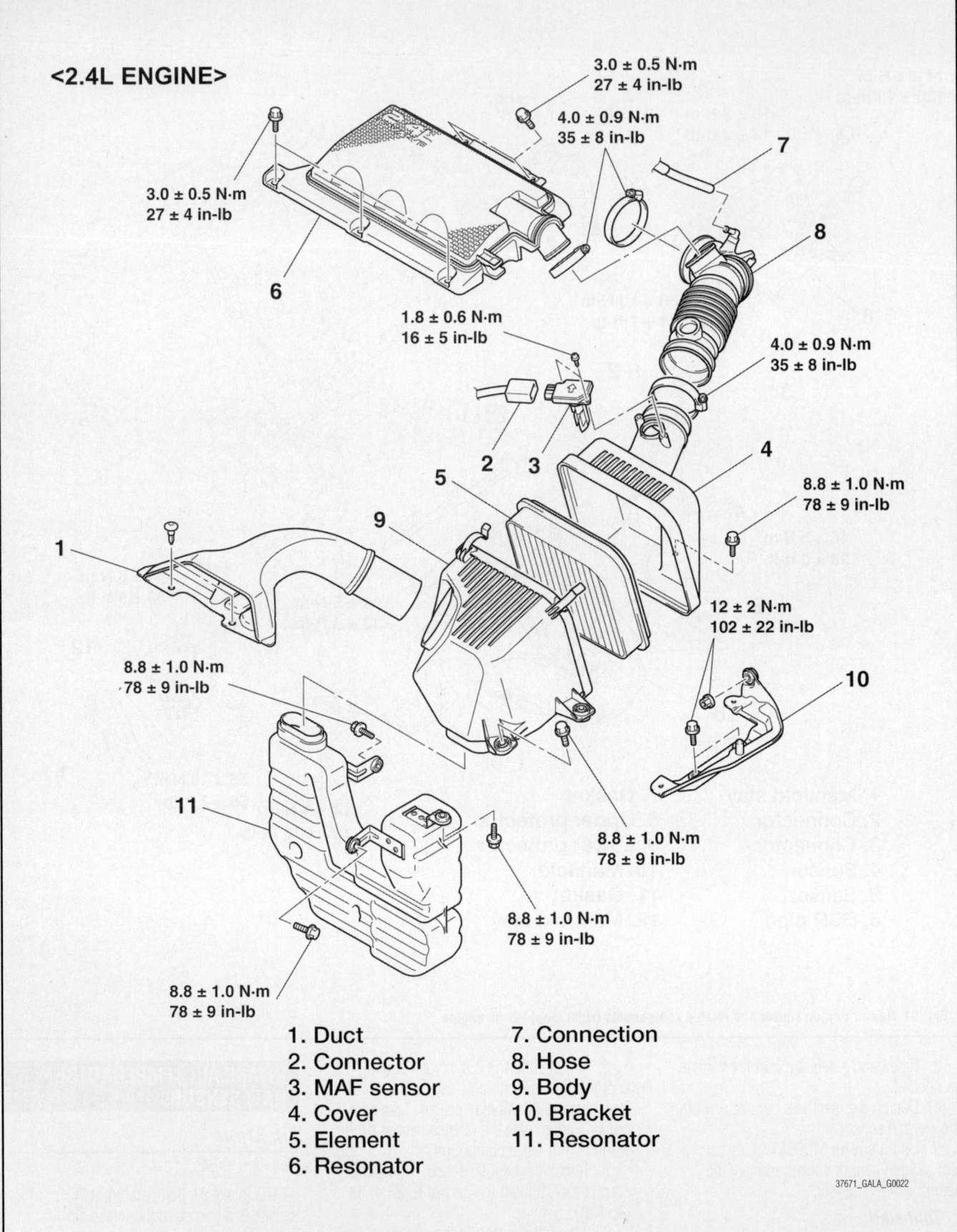

<2.4L ENGINE>

3.0 ± 0.5 N·m
27 ± 4 in-lb

4.0 ± 0.9 N·m
35 ± 8 in-lb

3.0 ± 0.5 N·m
27 ± 4 in-lb

1.8 ± 0.6 N·m
16 ± 5 in-lb

4.0 ± 0.9 N·m
35 ± 8 in-lb

8.8 ± 1.0 N·m
78 ± 9 in-lb

12 ± 2 N·m
102 ± 22 in-lb

8.8 ± 1.0 N·m
78 ± 9 in-lb

8.8 ± 1.0 N·m
78 ± 9 in-lb

8.8 ± 1.0 N·m
78 ± 9 in-lb

1. Duct
2. Connector
3. MAF sensor
4. Cover
5. Element
6. Resonator
7. Connection
8. Hose
9. Body
10. Bracket
11. Resonator

37671_GALA_G0022

Fig. 92 Sensor location—2.4L engine shown, 3.8L engine similar

REMOVAL & INSTALLATION

See Figures 92.

1. Before servicing the vehicle, refer to the Precautions Section.

➡ **If working near and/or around the SRS system and components, be sure to disable the SRS system. Tape the negative battery cable with insulating tape. Always disconnect the negative battery cable first.**

❋❋ CAUTION

Wait for 1 minute after disconnecting the negative battery cable before working inside the vehicle. The air bag system is set to deploy for a short period of time after the battery is disconnected.

2. Disconnect the negative battery cable.
3. Disconnect the connector from the sensor.
4. Remove the sensor retaining screws.
5. Remove the sensor from its mounting.

To install:

➡ **Be sure to use new fasteners, as required.**

6. Installation is the reverse of the removal procedure.
7. Handle the sensor assembly carefully, protecting it from impact, extremes of temperature and/or exposure to shop chemicals.

KNOCK SENSOR (KS)

LOCATION

See Figures 93 and 94.

Refer to the accompanying illustrations.

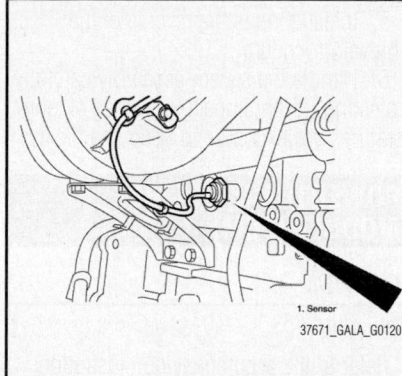

Fig. 93 Knock sensor location—2.4L engine

REMOVAL & INSTALLATION

See Figures 93 and 94.

➡ **After replacing this sensor use the Mitsubishi diagnostic scan tool, or equivalent and perform the initialization procedure for learning value in MFI engine.**

1. Before servicing the vehicle, refer to the Precautions Section.

➡ **If working near and/or around the SRS system and components, be sure**

to disable the SRS system. Tape the negative battery cable with insulating tape. Always disconnect the negative battery cable first.

❋❋ CAUTION

Wait for 1 minute after disconnecting the negative battery cable before working inside the vehicle. The air bag system is set to deploy for a short period of time after the battery is disconnected.

2. Disconnect the negative battery cable.
3. Disconnect the sensor connector.
4. Remove the sensor from its mounting.

To install:

5. Installation is the reverse of the removal procedure.
6. Tighten the sensor to 12–17 ft. lbs. (16–24 Nm).

MASS AIR FLOW (MAF) SENSOR

LOCATION

See Figures 92.

The Mass Air Flow (MAF) sensor is mounted in the intake air hose of the air cleaner assembly.

REMOVAL & INSTALLATION

See Figures 92.

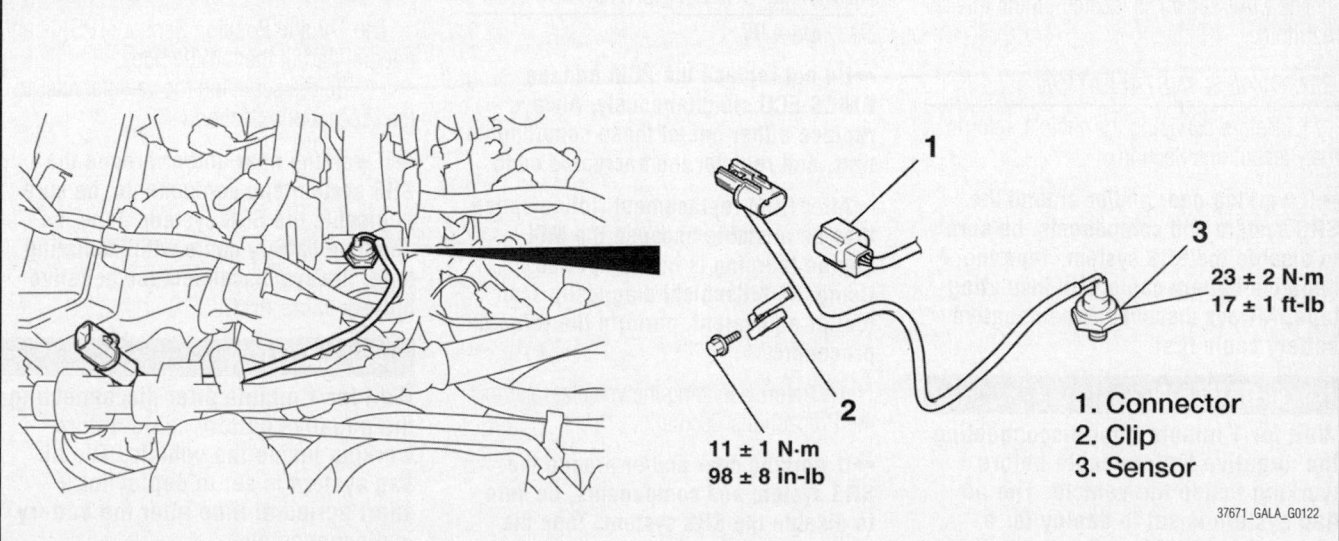

11 ± 1 N·m
98 ± 8 in-lb

23 ± 2 N·m
17 ± 1 ft-lb

1. Connector
2. Clip
3. Sensor

Fig. 94 Knock sensor location—3.8L engine

1. Before servicing the vehicle, refer to the Precautions Section.

➡ If working near and/or around the SRS system and components, be sure to disable the SRS system. Tape the negative battery cable with insulating tape. Always disconnect the negative battery cable first.

✳✳ CAUTION

Wait for 1 minute after disconnecting the negative battery cable before working inside the vehicle. The air bag system is set to deploy for a short period of time after the battery is disconnected.

2. Disconnect the negative battery cable.
3. Disconnect the connector from the sensor.
4. Remove the air cleaner and air intake assembly, as required.
5. Remove the sensor from its mounting.

To install:

➡ Be sure to use new fasteners, as required.

6. Installation is the reverse of the removal procedure.
7. Handle the sensor assembly carefully, protecting it from impact, extremes of temperature, and exposure to shop chemicals.

MANIFOLD ABSOLUTE PRESSURE (MAP) SENSOR

LOCATION

The MAP sensor is located on the intake manifold.

REMOVAL & INSTALLATION

1. Before servicing the vehicle, refer to the Precautions Section.

➡ If working near and/or around the SRS system and components, be sure to disable the SRS system. Tape the negative battery cable with insulating tape. Always disconnect the negative battery cable first.

✳✳ CAUTION

Wait for 1 minute after disconnecting the negative battery cable before working inside the vehicle. The air bag system is set to deploy for a short period of time after the battery is disconnected.

2. Disconnect the negative battery cable.

3. Disconnect the connector from the sensor.
4. Remove the sensor from its mounting.

To install:

➡ Be sure to use new fasteners, as required.

5. Installation is the reverse of the removal procedure.
6. Handle the sensor assembly carefully, protecting it from impact, extremes of temperature, and exposure to shop chemicals.

POWERTRAIN CONTROL MODULE (PCM)

LOCATION

See Figure 95.

Refer to the accompanying illustration.

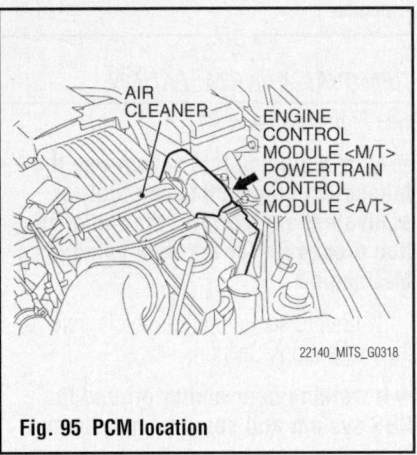

22140_MITS_G0318

Fig. 95 PCM location

REMOVAL & INSTALLATION

See Figure 95.

➡ Do not replace the PCM and the ETACS-ECU simultaneously. Always replace either one of these components first, and register the encrypted code.

➡ After PCM replacement, idling speed may be unstable because the MFI engine learning is not completed. Using the Mitsubishi diagnostic scan tool or equivalent, perform the relearn procedure.

1. Before servicing the vehicle, refer to the Precautions Section.

➡ If working near and/or around the SRS system and components, be sure to disable the SRS system. Tape the negative battery cable with insulating tape. Always disconnect the negative battery cable first.

✳✳ CAUTION

Wait for 1 minute after disconnecting the negative battery cable before working inside the vehicle. The air bag system is set to deploy for a short period of time after the battery is disconnected.

2. Turn the ignition switch off.
3. Disconnect the negative battery cable from the battery.
4. Disconnect the PCM connector(s).
5. Remove the PCM mounting bolts and remove the PCM.

To install:

➡ Be sure to use new fasteners, as required.

6. Installation is the reverse of the removal.
7. Tighten the PCM mounting bolts to: 86–104 inch lbs. (10–12 Nm).
8. Using the Mitsubishi diagnostic scan tool, or equivalent, perform the relearning procedures. Follow the directions on the screen of the tool.

THROTTLE POSITION SENSOR (TPS)

LOCATION

See Figures 96 and 97.

The Throttle Position Sensor (TPS) is mounted on the throttle body and is incorporated into the throttle body assembly.

REMOVAL & INSTALLATION

See Figures 96 and 97.

The Throttle Position Sensor (TPS) is an integral part of the throttle body.
1. Before servicing the vehicle, refer to the Precautions Section.

➡ If working near and/or around the SRS system and components, be sure to disable the SRS system. Tape the negative battery cable with insulating tape. Always disconnect the negative battery cable first.

✳✳ CAUTION

Wait for 1 minute after disconnecting the negative battery cable before working inside the vehicle. The air bag system is set to deploy for a short period of time after the battery is disconnected.

2. Properly relieve the fuel system pressure.

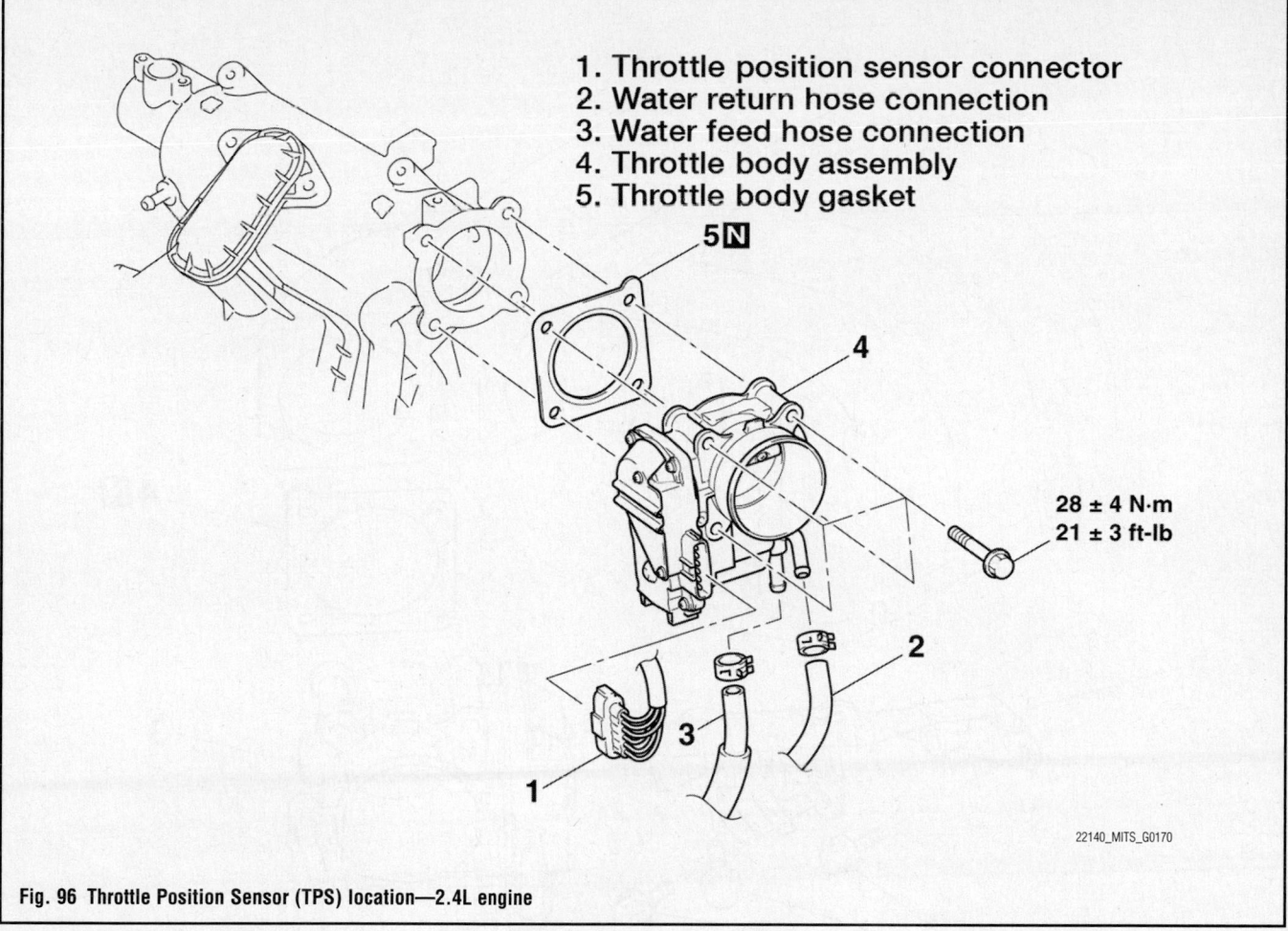

1. Throttle position sensor connector
2. Water return hose connection
3. Water feed hose connection
4. Throttle body assembly
5. Throttle body gasket

28 ± 4 N·m
21 ± 3 ft-lb

22140_MITS_G0170

Fig. 96 Throttle Position Sensor (TPS) location—2.4L engine

3. Drain the engine coolant.

4. Remove the air intake hose.

5. Remove the battery.

6. Disconnect the throttle position sensor connector.

7. Disconnect the water hose connection.

8. Remove the throttle body retaining bolts.

9. Remove the throttle body from the engine.

10. Discard the gasket.

To install:

➡Be sure to use new fasteners, as required.

✳✳ WARNING

Do not loosen the retaining screws for the resin cover of the throttle body assembly. If the screws are loosened, the sensor incorporated in the resin cover becomes misaligned and the throttle body may not work properly.

11. Align the recess on the intake manifold plenum with the projection of the throttle body gasket.

12. Install the gasket. Install the throttle body to the engine and tighten the retaining bolts to 18–24 ft. lbs. (24–32 Nm).

➡**Poor idling may result if the throttle body gasket is not installed properly.**

13. Continue the installation in the reverse order of the removal procedure.

14. Connect the negative battery cable.

15. Turn the ignition **ON** and then **OFF**, and keep it off for at least 10 seconds.

16. Complete the vehicle initialization procedure.

➡**To complete the initialization procedure the following tools are needed. MB991958 scan tool, MB991824 VCI, MB991827 MUT III USB cable, MB991910 MUT III Main harness "A."**

a. Connect the scan tool to the Data Link Connector (DLC). To prevent damage to the scan tool, be sure that the ignition switch is in the **LOCK** position before connecting the scan tool.

b. Turn the ignition switch to the **ON** position.

c. Select "check mode" from the menu screen.

d. Select "erase memory" from the menu screen.

e. Initialize the learning value.

f. After initialization, complete the idle learning procedure.

➡**This procedure must be performed when the ECM/PCM is replaced, or when the learning value is initialized, as the idling is not stabilized when the learning value in the MFI engine is not completed.**

g. Start the engine. Allow the coolant temperature to reach 176°F (80°C) or more.

h. Stop the engine and place the ignition switch in the **LOCK** position.

i. After 10 seconds, restart the engine.

j. Carry out the following idling procedure for 10 minutes to confirm that the engine has normal idling:

• Position the transaxle selector lever in the **P** or neutral range

• The engine fan is not to be operated

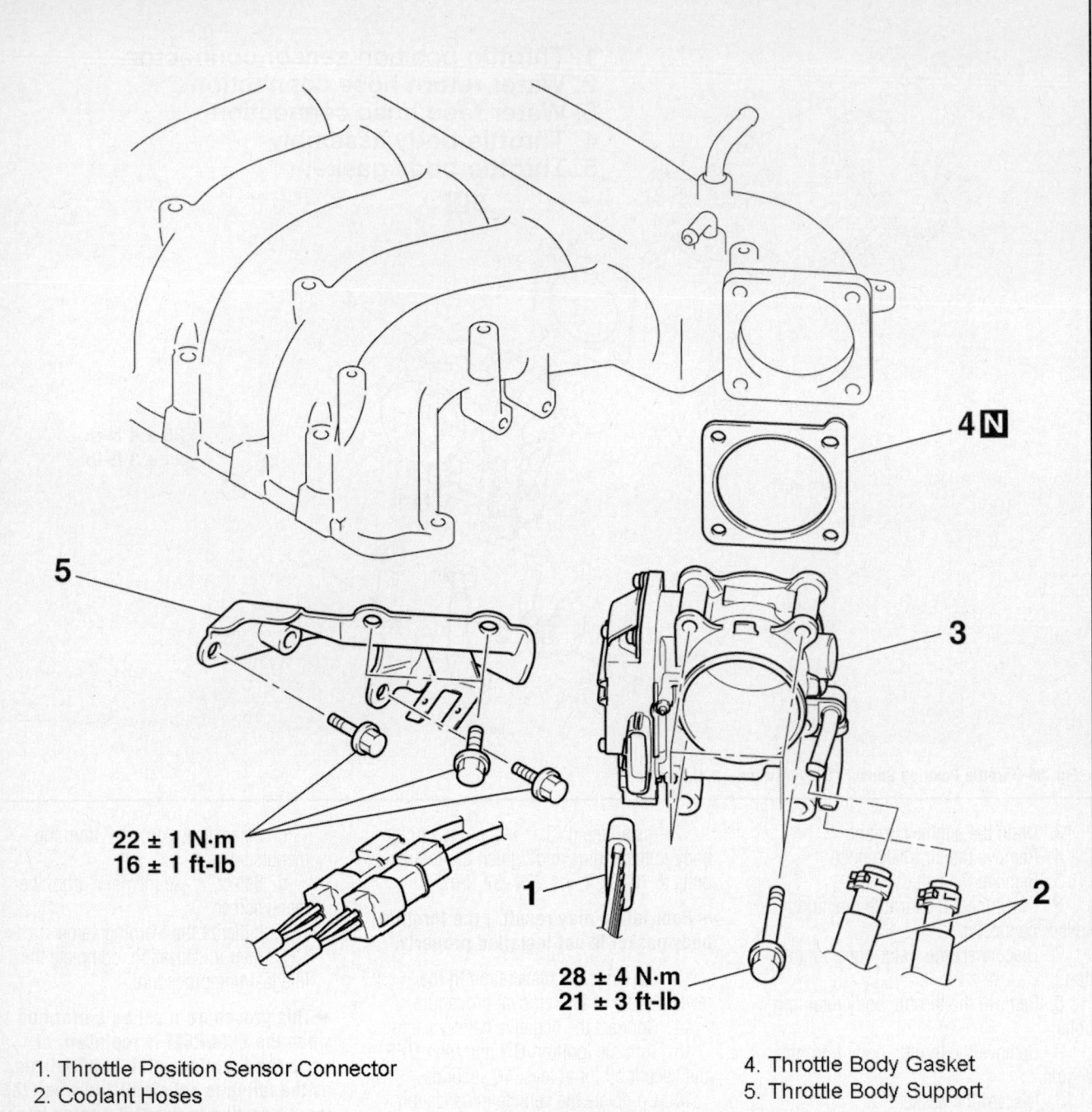

22 ± 1 N·m
16 ± 1 ft-lb

28 ± 4 N·m
21 ± 3 ft-lb

1. Throttle Position Sensor Connector
2. Coolant Hoses
3. Throttle Body Assembly
4. Throttle Body Gasket
5. Throttle Body Support

42050_MITS_G0010

Fig. 97 Throttle Position Sensor (TPS) location—3.8L engine

- The engine coolant temperature should be 176°F (80°C) or more

➡**If the engine stalls during idling, check the throttle valve of the throttle body for dirt. Correct and perform the procedure again.**

VARIABLE CAMSHAFT TIMING OIL CONTROL SOLENOID

LOCATION

See Figure 98.

REMOVAL & INSTALLATION

See Figure 99.

1. Before servicing the vehicle, refer to the Precautions Section.

➡**If working near and/or around the SRS system and components, be sure**

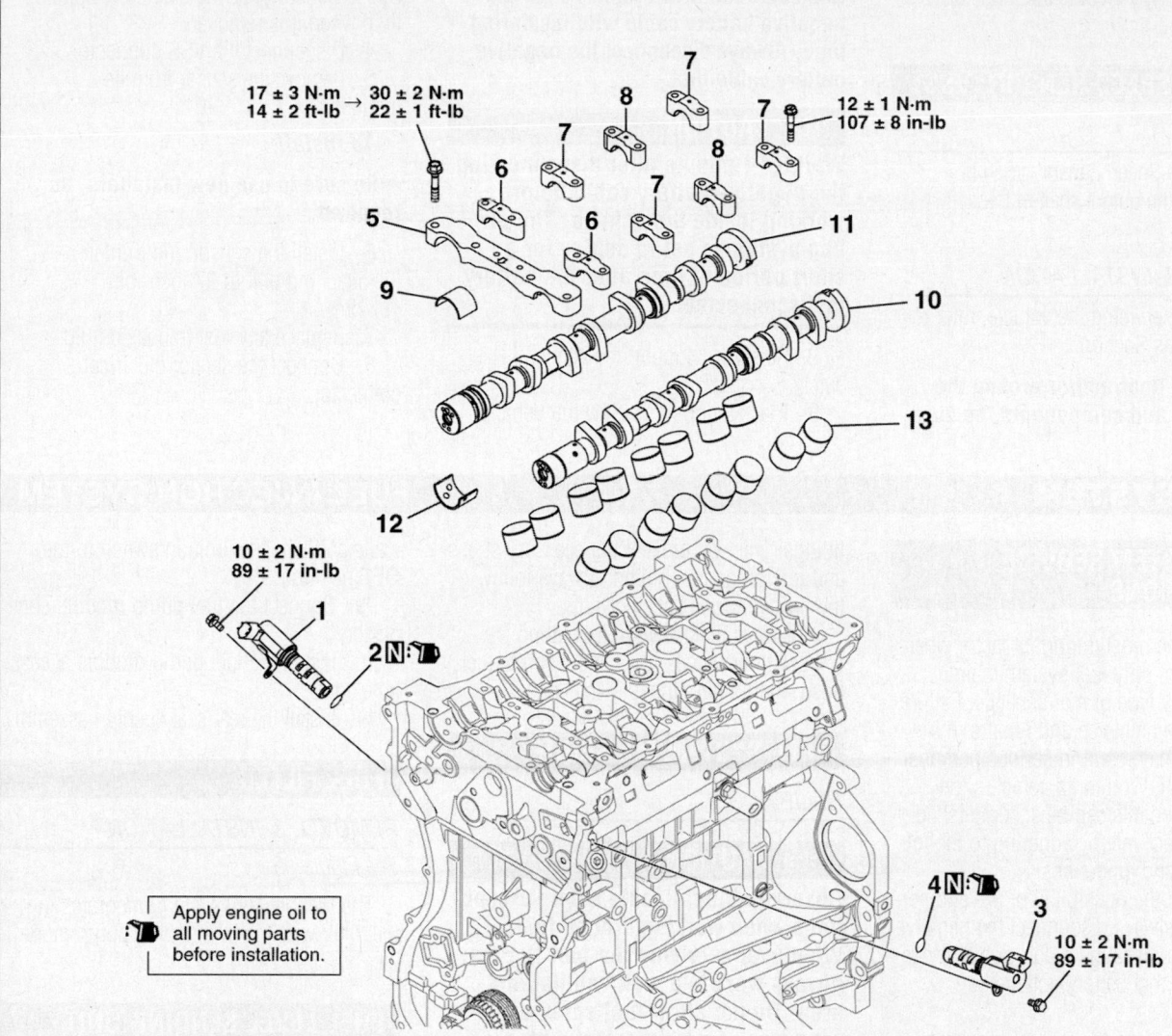

17 ± 3 N·m → 30 ± 2 N·m
14 ± 2 ft-lb → 22 ± 1 ft-lb

12 ± 1 N·m
107 ± 8 in-lb

10 ± 2 N·m
89 ± 17 in-lb

Apply engine oil to all moving parts before installation.

10 ± 2 N·m
89 ± 17 in-lb

1. Engine Oil Control Valve (OCV) exhaust
2. O-ring
3. Engine Oil Control Valve (OCV) intake
4. O-ring
5. Front camshaft bearing cap
6. Oil feeding camshaft bearing cap
7. Camshaft bearing cap
8. Thrust camshaft bearing cap
9. Bearing
10. Camshaft intake
11. Camshaft exhaust
12. Bearing
13. Valve tappet

22140_MITS_G0251

Fig. 98 Location of the variable camshaft timing oil control solenoid (1), (3), also called the engine Oil Control Valve (OCV)

to disable the SRS system. Tape the negative battery cable with insulating tape. Always disconnect the negative battery cable first.

✳✳ CAUTION

Wait for 1 minute after disconnecting the negative battery cable before working inside the vehicle. The air bag system is set to deploy for a short period of time after the battery is disconnected.

2. Disconnect the ground cable from the battery.

3. Disconnect the connector from Variable Camshaft Timing Oil Control Solenoid (VCTOCS) on the right-hand bank and/or left-hand bank.

4. Remove the bolt retaining the VCTOCS.

5. Remove the VCTOCS.

To install:

➡ Be sure to use new fasteners, as required.

➡ Always use a new gasket/O-ring.

6. Apply a small amount of engine oil to the new O-ring of the VCTOCS.

7. Install the VCTOCS on the cylinder head.

8. Tighten the VCTOCS mounting bolt to 72–106 inch lbs. (8–12 Nm).

9. Connect the electrical connector to the VCTOCS.

VEHICLE SPEED SENSOR (VSS)

LOCATION

The Vehicle Speed Sensor (VSS) is attached near the output shaft of the transaxle.

REMOVAL & INSTALLATION

1. Before servicing the vehicle, refer to the Precautions Section.

➡ **If working near and/or around the SRS system and components, be sure** to disable the SRS system. Tape the negative battery cable with insulating tape. Always disconnect the negative battery cable first.

✳✳ CAUTION

Wait for 1 minute after disconnecting the negative battery cable before working inside the vehicle. The air bag system is set to deploy for a short period of time after the battery is disconnected.

2. Raise and support the vehicle safely.
3. Place a drip pan below the Vehicle Speed Sensor (VSS) to catch any spilled fluid when it is removed.
4. Disconnect the VSS connector.
5. Remove the sensor from its mounting.

To install:

➡ **Be sure to use new fasteners, as required.**

6. Install the sensor and tighten the attaching bolt to 97 inch lbs. (11 Nm).
7. Replace any lost transaxle fluid.
8. Connect the sensor electrical connector.

FUEL SYSTEM

FUEL SYSTEM SERVICE PRECAUTIONS

Safety is the most important factor when performing not only fuel system maintenance, but any type of maintenance. Failure to conduct maintenance and repairs in a safe manner may result in serious personal injury or death. Work on a vehicle's fuel system components can be accomplished safely and effectively by adhering to the following rules and guidelines.

• To avoid the possibility of fire and personal injury, always disconnect the negative battery cable unless the repair or test procedure requires that battery voltage be applied.

• Always relieve the fuel system pressure prior to disconnecting any fuel system component (injector, fuel rail, pressure regulator, etc.) fitting or fuel line connection. Exercise extreme caution whenever relieving fuel system pressure to avoid exposing skin, face and eyes to fuel spray. Please be advised that fuel under pressure may penetrate the skin or any part of the body that it contacts.

• Always place a shop towel or cloth around the fitting or connection prior to loosening to absorb any excess fuel due to spillage. Ensure that all fuel spillage is quickly removed from engine surfaces. Ensure that all fuel-soaked cloths or towels are deposited into a flame-proof waste container with a lid.

• Always keep a dry chemical (Class B) fire extinguisher near the work area.

• Do not allow fuel spray or fuel vapors to come into contact with a spark or open flame.

• Always use a second wrench when loosening or tightening fuel line connection fittings. This will prevent unnecessary stress and torsion on fuel piping. Always follow the proper torque specifications.

• Always replace worn fuel fitting O-rings with new ones. Do not substitute fuel hose where rigid pipe is installed.

FUEL SYSTEM PRESSURE

RELIEVING

✳✳ CAUTION

Observe all applicable safety precautions when working around fuel. Whenever servicing the fuel system, always work in a well ventilated area. Do not allow fuel spray or vapors to come in contact with a spark or open flame. Keep a dry chemical fire extinguisher near the work area. Always keep fuel in a container specifically designed for fuel storage; also, always properly seal fuel containers to avoid the possibility of fire or explosion.

1. Before servicing the vehicle, refer to the Precautions Section.
2. Remove the rear seat cushion assembly.
3. Remove the fuel pump module access hole cover.
4. Disconnect the fuel pump module connector.
5. After starting the engine and letting it run until it stops, turn the ignition switch to the **OFF** position.
6. Crank the engine for 2 seconds or more.
7. If the engine does not start, turn the ignition switch to the **OFF** position.
8. If the engine starts, let it run until it

FUEL INJECTION SYSTEM

stalls and turn the ignition switch to the **OFF** position.
9. Connect the fuel pump module connector.
10. Install the fuel pump module access hole cover.
11. Install the rear seat cushion assembly.

FUEL FILTER

REMOVAL & INSTALLATION
See Figure 99.

The fuel delivery system integrates the fuel filter with the in-tank fuel pump module.

FUEL LEVEL SENDING UNIT

LOCATION

The fuel delivery system integrates the fuel level sensing unit with the in-tank fuel pump module.

REMOVAL & INSTALLATION
See Figure 99.

✳✳ CAUTION

Observe all applicable safety precautions when working around fuel. Whenever servicing the fuel system, always work in a well ventilated area. Do not allow fuel spray or vapors to come in contact with a spark or open flame. Keep a dry chemical fire extinguisher near the work area. Always keep fuel in a container specifically designed for fuel storage; also, always properly seal fuel containers to avoid the possibility of fire or explosion.

1. Before servicing the vehicle, refer to the Precautions Section.

➡️**If working near and/or around the SRS system and components, be sure to disable the SRS system. Tape the negative battery cable with insulating tape. Always disconnect the negative battery cable first.**

✳✳ CAUTION

Wait for 1 minute after disconnecting the negative battery cable before working inside the vehicle. The air bag system is set to deploy for a short period of time after the battery is disconnected.

2. Properly relieve the fuel system pressure.

3. Disconnect the negative battery cable.

4. Remove the rear seat cushion assembly.

5. Remove the access hole cover.

6. Disconnect the fuel level sensor connector and fuel tank differential pressure sensor connector.

✳ CAUTION

When removing and installing the fuel level sensor from the fuel tank, be careful not damage the sensor unit and the float.

7. Remove the fuel level sensor mounting bolts and remove the fuel tank gauge unit from service hole.

To install:

➡️**Be sure to use new fasteners, as required.**

8. Install the fuel level sensor to the fuel tank through the service hole.

9. Connect the fuel level sensor (sub) connector and fuel tank differential pressure sensor connector.

10. Install the access hole cover.

11. Install the rear seat cushion assembly.

FUEL PUMP MODULE

REMOVAL & INSTALLATION
See Figure 99.

✳✳ CAUTION

Observe all applicable safety precautions when working around fuel. Whenever servicing the fuel system, always work in a well

ventilated area. Do not allow fuel spray or vapors to come in contact with a spark or open flame. Keep a dry chemical fire extinguisher near the work area. Always keep fuel in a container specifically designed for fuel storage; also, always properly seal fuel containers to avoid the possibility of fire or explosion.

1. Before servicing the vehicle, refer to the Precautions Section.

➡️**If working near and/or around the SRS system and components, be sure to disable the SRS system. Tape the negative battery cable with insulating tape. Always disconnect the negative battery cable first.**

✳✳ CAUTION

Wait for 1 minute after disconnecting the negative battery cable before working inside the vehicle. The air bag system is set to deploy for a short period of time after the battery is disconnected.

2. Relieve the fuel system pressure.

3. Disconnect the negative battery cable.

4. Remove or disconnect the following:
  • Rear seat cushion by pulling the seat stopper near the floor and lifting the cushion up
  • Inspection cover on the right side of the vehicle
  • Harness connector and the fuel lines
  • Fuel pump assemble from the tank.

To install:

5. Install or connect the following:
  • Fuel pump in the tank
  • Hoses and the harness connector
  • Inspection cover
  • Rear seat
  • Negative battery cable

FUEL PRESSURE REGULATOR

REMOVAL & INSTALLATION
See Figure 99.

A fuel pressure regulator is built into the fuel pump.

✳✳ CAUTION

Observe all applicable safety precautions when working around fuel.

Whenever servicing the fuel system, always work in a well ventilated area. Do not allow fuel spray or vapors to come in contact with a spark or open flame. Keep a dry chemical fire extinguisher near the work area. Always keep fuel in a container specifically designed for fuel storage; also, always properly seal fuel containers to avoid the possibility of fire or explosion.

1. Before servicing the vehicle, refer to the Precautions Section.

➡️**If working near and/or around the SRS system and components, be sure to disable the SRS system. Tape the negative battery cable with insulating tape. Always disconnect the negative battery cable first.**

✳✳ CAUTION

Wait for 1 minute after disconnecting the negative battery cable before working inside the vehicle. The air bag system is set to deploy for a short period of time after the battery is disconnected.

2. Relieve the fuel system pressure.

3. Disconnect the negative battery cable.

4. Remove or disconnect the following:
  • Rear seat cushion by pulling the seat stopper near the floor and lifting the cushion up
  • Inspection cover on the right side of the vehicle
  • Harness connector and the fuel lines
  • Fuel pump assemble from the tank.

To install:

5. Install or connect the following:
  • Fuel pump in the tank
  • Hoses and the harness connector
  • Inspection cover
  • Rear seat
  • Negative battery cable

FUEL RAIL AND INJECTOR

REMOVAL & INSTALLATION

2.4L Engine

1. Before servicing the vehicle, refer to the Precautions Section.

➡️**If working near and/or around the SRS system and components, be sure**

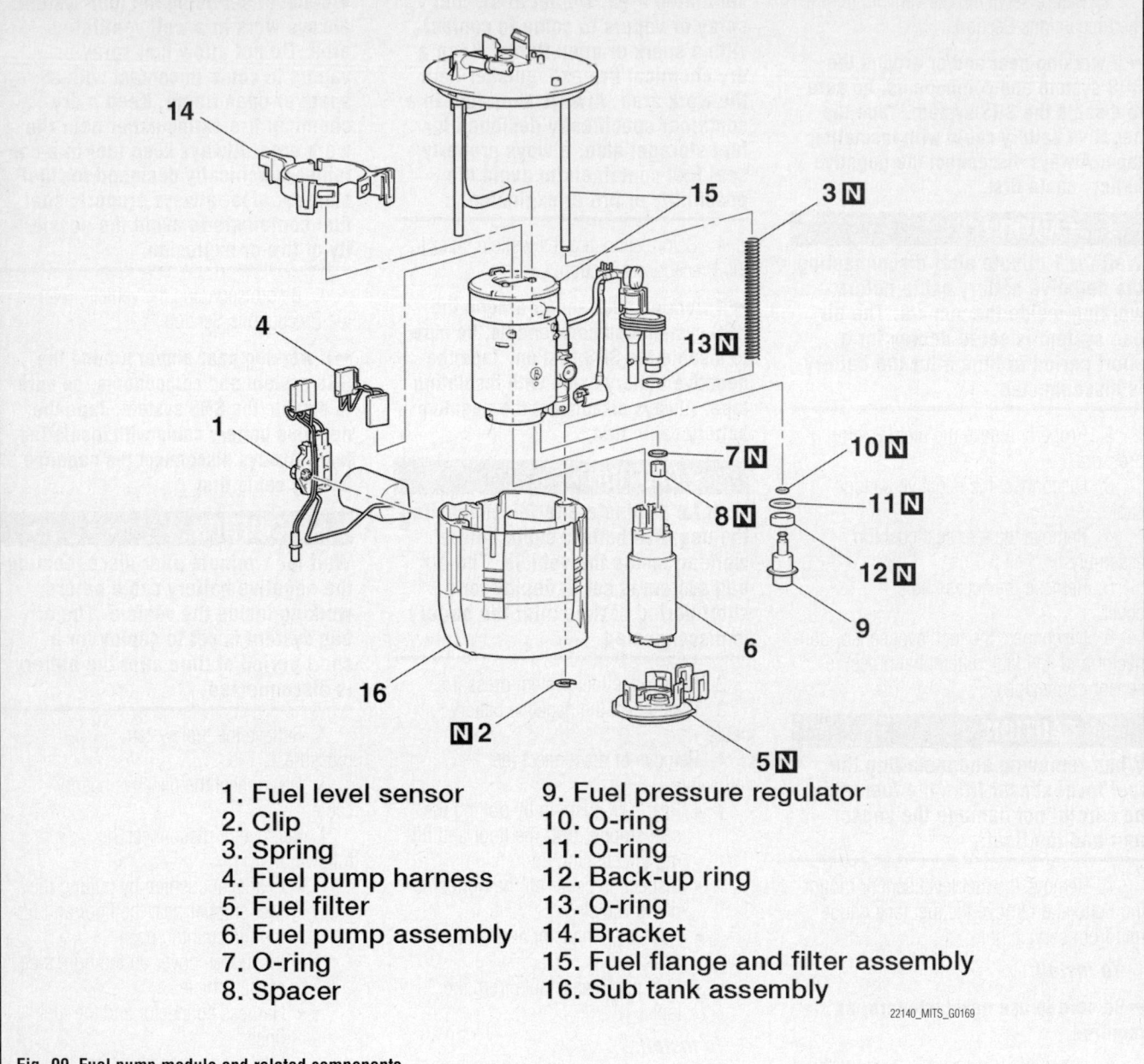

1. Fuel level sensor
2. Clip
3. Spring
4. Fuel pump harness
5. Fuel filter
6. Fuel pump assembly
7. O-ring
8. Spacer
9. Fuel pressure regulator
10. O-ring
11. O-ring
12. Back-up ring
13. O-ring
14. Bracket
15. Fuel flange and filter assembly
16. Sub tank assembly

22140_MITS_G0169

Fig. 99 Fuel pump module and related components

to disable the SRS system. Tape the negative battery cable with insulating tape. Always disconnect the negative battery cable first.

❊❊ CAUTION
Wait for 1 minute after disconnecting the negative battery cable before working inside the vehicle. The air bag system is set to deploy for a short period of time after the battery is disconnected.

2. Relieve the fuel system pressure.

3. Disconnect the negative battery cable.

4. Remove the air cleaner cover and air intake hose assembly.

5. Disconnect the PCV valve hose connection, ignition coil connectors, EGR valve connector and fuel injector connectors.

6. Disconnect the throttle position sensor connector, manifold absolute pressure sensor connector, EVAP emission purge solenoid connector and knock sensor connector.

7. Disconnect the power steering pressure switch connector.

8. Remove the rocker cover bracket bolts.

9. Disconnect the high pressure hose connection. Remove the fuel rail, insulators, grommets, and injectors. Discard the O-rings.

To install:

➡**Be sure to use new fasteners, as required.**

10. Installation is the reverse of the removal procedure. Be sure to use new O-rings and gaskets as required.

11. Connect the scan tool to the data link connector. To prevent damage to the scan tool be sure that the ignition switch is in the LOCK position before connecting the scan tool.

12. Turn the ignition switch to the ON position.

13. Select "check mode" from the menu screen.

14. Select "erase memory" from the menu screen.

15. Initialize the learning value.

16. Start the engine and check for leaks, correct as required.

3.8L Engine

See Figure 100.

1. Before servicing the vehicle, refer to the Precautions Section.

➡**If working near and/or around the SRS system and components, be sure to disable the SRS system. Tape the negative battery cable with insulating tape. Always disconnect the negative battery cable first.**

❋❋ CAUTION

Wait for 1 minute after disconnecting the negative battery cable before working inside the vehicle. The air bag system is set to deploy for a short period of time after the battery is disconnected.

2. Relieve the fuel system pressure.

3. Disconnect the negative battery cable.

4. Remove the intake manifold plenum assembly.

5. Disconnect the fuel injector connectors.

6. Remove the control wiring harness bracket mounting bolts.

7. Remove the engine mounting stay.

8. Disconnect the high pressure fuel hose connection. Remove the fuel rail and fuel injector assembly.

9. Remove the insulators, O-rings, and fuel injectors. Discard the gaskets and O-rings.

To install:

➡**Be sure to use new fasteners, as required.**

10. Installation is the reverse of the removal procedure. Be sure to use new O-rings and gaskets as required.

11. Connect the scan tool to the data link connector. To prevent damage to the scan tool be sure that the ignition switch is in the LOCK position before connecting the scan tool.

12. Turn the ignition switch to the ON position.

13. Select "check mode" from the menu screen.

14. Select "erase memory" from the menu screen.

15. Initialize the learning value.

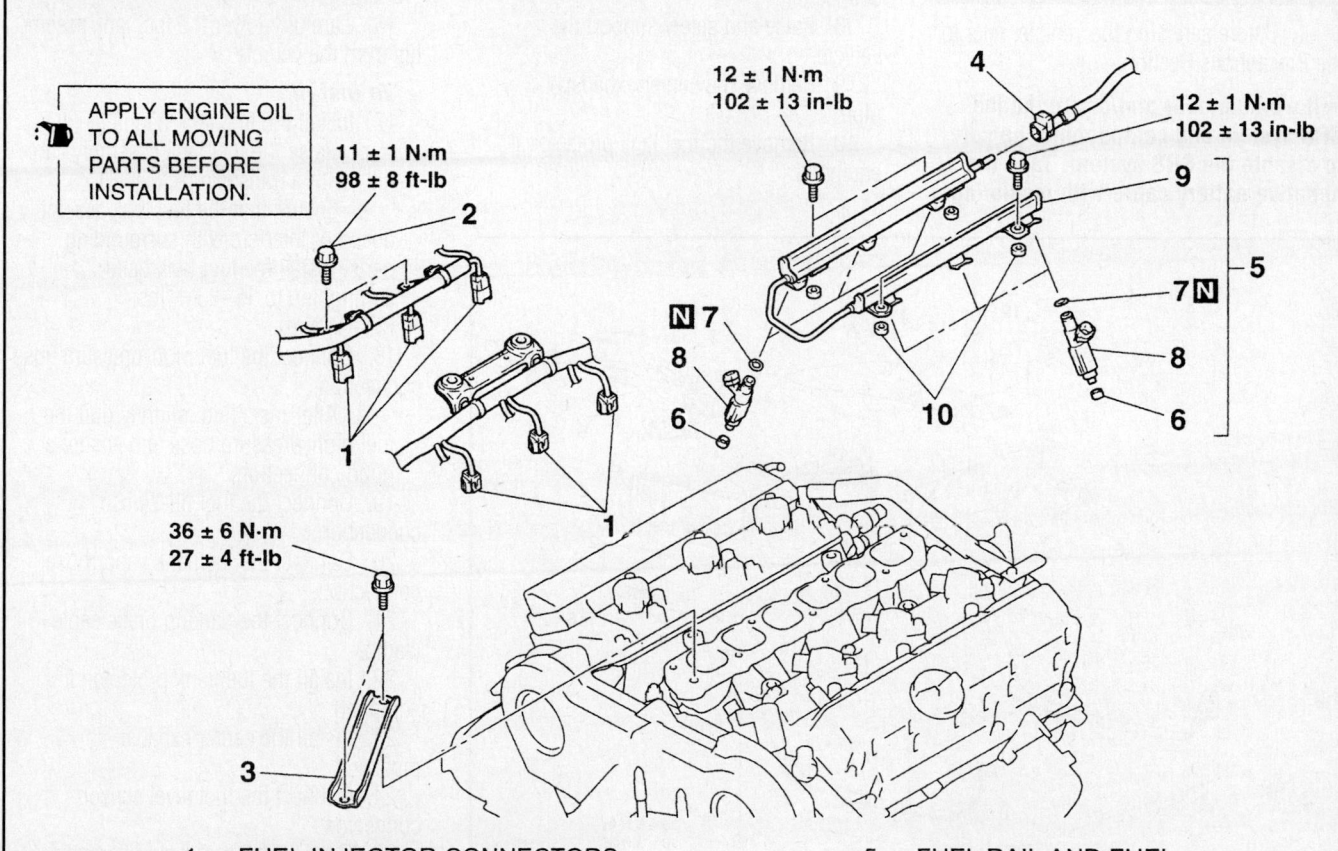

APPLY ENGINE OIL TO ALL MOVING PARTS BEFORE INSTALLATION.

12 ± 1 N·m
102 ± 13 in-lb

12 ± 1 N·m
102 ± 13 in-lb

11 ± 1 N·m
98 ± 8 ft-lb

36 ± 6 N·m
27 ± 4 ft-lb

1. FUEL INJECTOR CONNECTORS
2. CONTROL WIRING HARNESS BRACKET MOUNTING BOLTS
3. ENGINE MOUNTING STAY
4. FUEL HIGH-PRESSURE HOSE CONNECTION (FUEL RAIL SIDE)
5. FUEL RAIL AND FUEL INJECTOR ASSEMBLY
6. INSULATORS
7. O-RINGS
8. FUEL INJECTORS
9. FUEL RAIL
10. INSULATORS

09482_GALA_G0126

Fig. 100 Fuel injectors and related components—3.8L engine

16. Start the engine and check for leaks, correct as required.

FUEL TANK

REMOVAL & INSTALLATION

See Figures 101 through 103.

�֍֍ CAUTION

The fuel injection system remains under pressure even after the engine has been turned OFF. Properly relieve fuel pressure before disconnecting any fuel lines. Failure to do so may result in fire or personal injury. Do not allow fuel spray or fuel vapors to come in contact with a spark or an open flame. Keep a dry chemical fire extinguisher nearby. Never store fuel in an open container due to risk of fire or explosion.

1. Before servicing the vehicle, refer to the Precautions Section.

➡**If working near and/or around the SRS system and components, be sure to disable the SRS system. Tape the negative battery cable with insulating tape. Always disconnect the negative battery cable first.**

✶✶ CAUTION

Wait for 1 minute after disconnecting the negative battery cable before working inside the vehicle. The air bag system is set to deploy for a short period of time after the battery is disconnected.

2. Relieve the fuel system pressure.
3. Drain the fuel from the tank.
4. Remove the negative battery cable.
5. Remove the rear seat cushion assembly.
6. Remove the fuel pump module hole cover.
7. Disconnect the fuel pump module connector and fuel tank differential pressure sensor connector connection.
8. Raise and safely support the vehicle.
9. Remove the center exhaust pipe.
10. Remove the fuel tank protector, if equipped.

11. Disconnect the parking brake cable clamp.
12. Disconnect the fuel tank vapor hose connection.
13. Disconnect the fuel filler hose connection.
14. Disconnect the fuel high-pressure hose connection.

✶✶ CAUTION

As there will be some pressure remaining in the fuel pipe line, cover it with a shop towel to prevent fuel from spraying out.

15. Remove the fuel tank band.
  a. Support the fuel tank with a transaxle jack.
  b. Remove the front securing nut of the fuel tank band. Tilt the fuel tank assembly forward and lower it gradually to remove it. Remove the fuel tank band.
16. Carefully lower the fuel tank assembly from the vehicle.

To install:

17. Install the fuel tank bands.
  a. Raise the fuel tank assembly carefully with a transaxle jack.
  b. Ensure that the fuel tank assembly does not interfere with surrounding parts. Install the fuel tank band and tighten to 15–23 ft. lbs. (21–31 Nm).
18. Connect the fuel high-pressure hose connection.
  a. After installing, slightly pull the fuel high-pressure hose and ensure a good connection.
19. Connect the fuel filler hose connection.
20. Connect the fuel tank vapor hose connection.
21. Connect the parking brake cable clamp.
22. Install the fuel tank protector, if equipped.
23. Install the center exhaust pipe.
24. Connect the fuel level sensor connector.
25. Connect the fuel tank differential pressure sensor connector.
26. Connect the fuel pump module connector.
27. Install the service hole cover.
28. Install the rear seat cushion assembly.
29. Fill the fuel tank.
30. Start the engine and check for leaks.

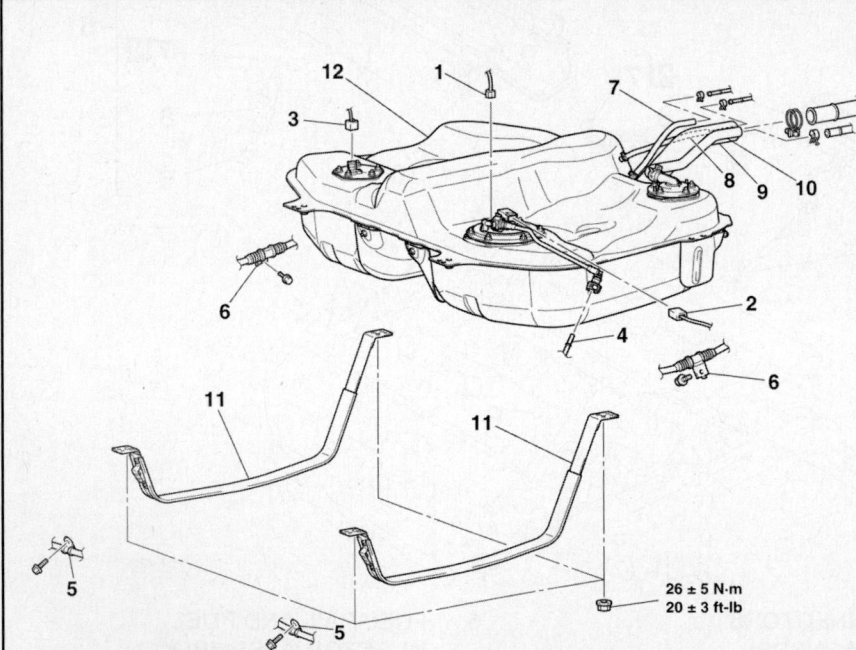

26 ± 5 N·m
20 ± 3 ft-lb

1. Fuel pump module connector connection
2. Fuel tank differential pressure sensor connector connection
4. Fuel high pressure hose connection
5. Parking brake cable clamp connection
6. Parking brake cable clamp connection
7. Fuel tank vapor hose b connection
8. Fuel tank vapor hose connection
9. Fuel filler hose connection
10. Fuel tank vapor hose a connection
11. Fuel tank band
12. Fuel tank assembly

22140_MITS_G0252

Fig. 101 Fuel tank and related components (1 of 3)

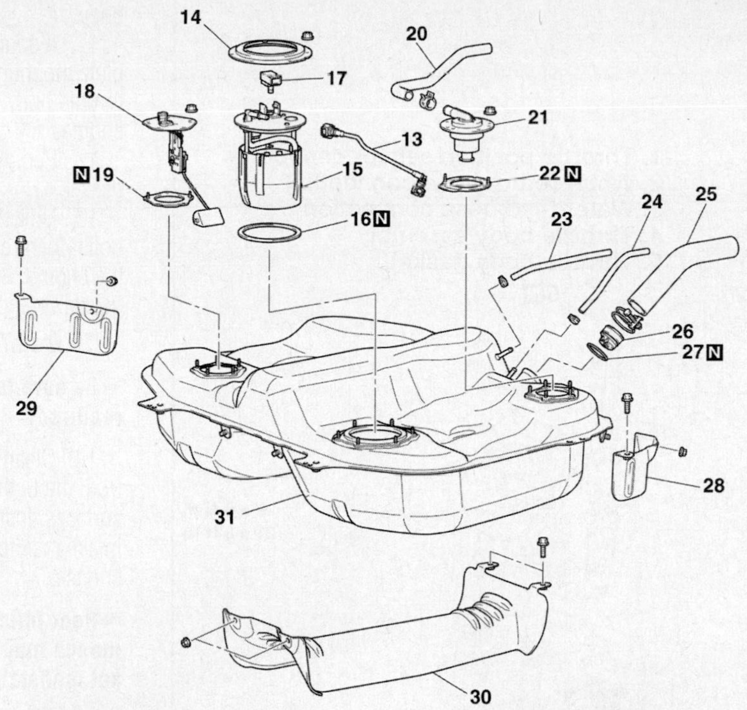

13. Fuel highpressure hose
14. Plate
15. Fuel pump module
16. Packing
17. Fuel tank differential pressure sensor
18. Fuel level sensor (sub)
19. Packing
20. Fuel tank vapor hose A
21. Fuel tank leveling valve assembly
22. Packing

23. Fuel tank vapor hose C
24. Fuel tank vapor hose B
25. Fuel filler hose
26. Fuel tank shut-off valve
27. O-ring
28. Fuel tank protector (A)
29. Fuel tank protector (B)
30. Fuel tank center protector
31. Fuel tank

22140_MITS_G0253

Fig. 102 Fuel tank and related components (1 of 3)

32. Fuel cap
33. Vapor hose connection
34. Fuel filler hose and fuel
    tank vapor hose connection
35. Fuel tank filler tube protector
36. Fuel tank filler tube
37. Fuel tank filler tube packing
38. Fuel tank filler tube vapor hose
39. Check valve

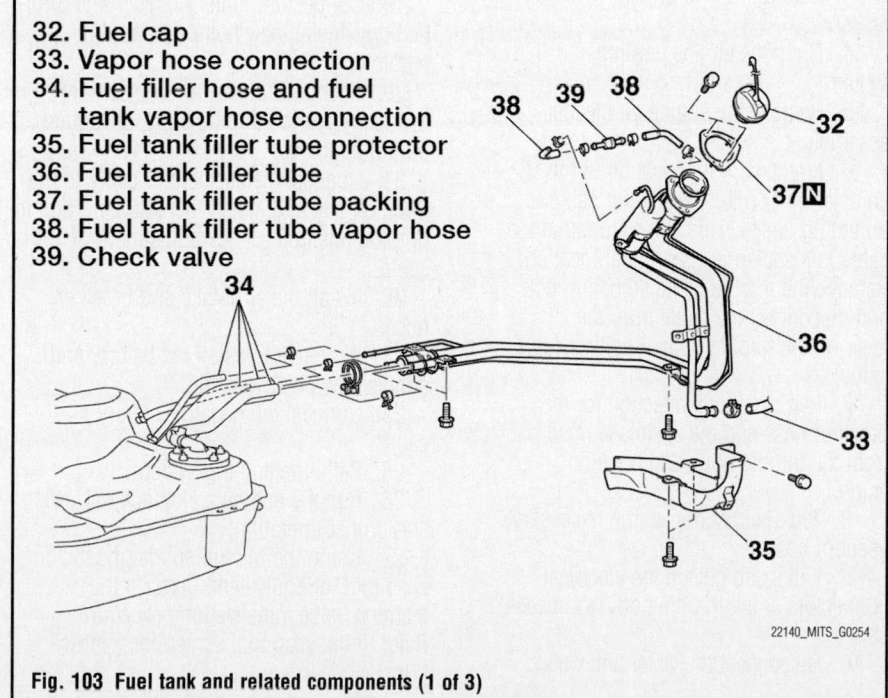

22140_MITS_G0254

Fig. 103 Fuel tank and related components (1 of 3)

IDLE SPEED

ADJUSTMENT

Idle speed is maintained by the Engine Control Module (ECM) or Powertrain Control Module (PCM). No adjustment is necessary or possible.

THROTTLE BODY

REMOVAL & INSTALLATION
See Figures 104 and 105.

✳✳ CAUTION

The fuel injection system remains under pressure even after the engine has been turned OFF. Properly relieve fuel pressure before disconnecting any fuel lines. Failure to do so may result in fire or personal injury. Do not allow fuel spray or fuel vapors to come in contact with a spark or an open flame. Keep a dry chemical fire extinguisher nearby. Never store fuel in an open container due to risk of fire or explosion.

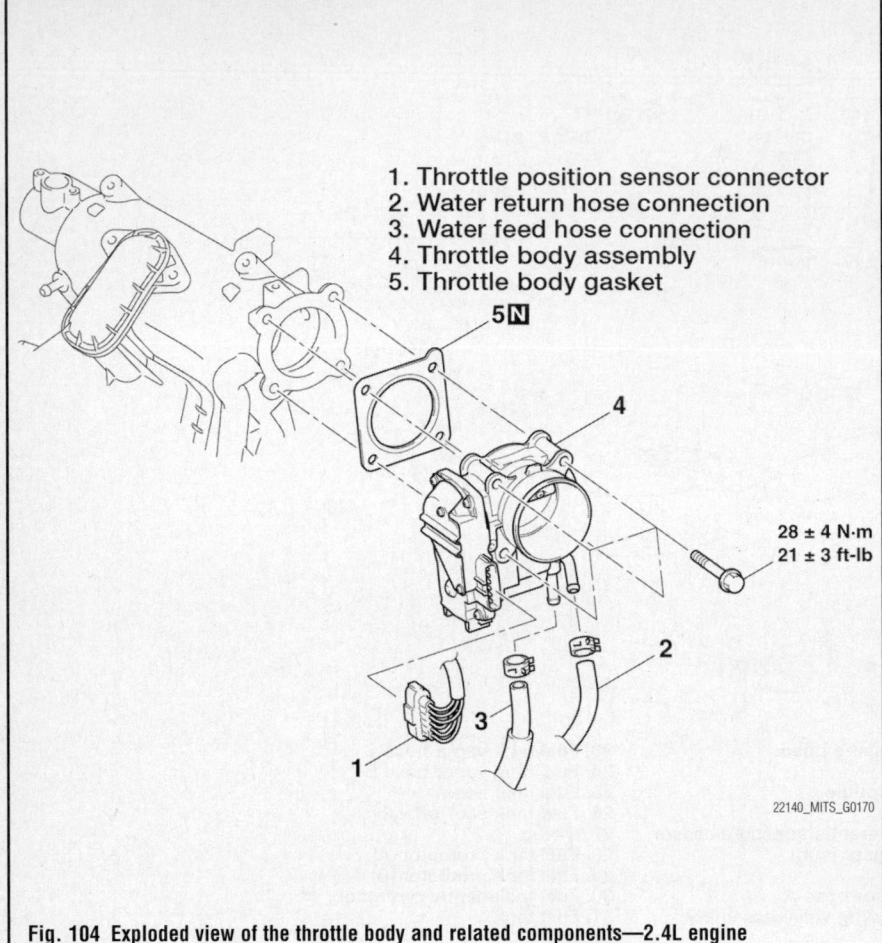

1. Throttle position sensor connector
2. Water return hose connection
3. Water feed hose connection
4. Throttle body assembly
5. Throttle body gasket

28 ± 4 N·m
21 ± 3 ft-lb

22140_MITS_G0170

Fig. 104 Exploded view of the throttle body and related components—2.4L engine

➡**When the throttle body assembly replacement is performed, use scan tool MB991958 to initialize the learning value. Do not loosen the fixing screws for the resin cover of throttle body assembly. If the screws are loosened, the sensor incorporated in the resin cover becomes misaligned and the throttle body may not work normally.**

1. Before servicing the vehicle, refer to the Precautions Section.

➡**If working near and/or around the SRS system and components, be sure to disable the SRS system. Tape the negative battery cable with insulating tape. Always disconnect the negative battery cable first.**

❊❊ CAUTION

Wait for 1 minute after disconnecting the negative battery cable before working inside the vehicle. The air bag system is set to deploy for a short period of time after the battery is disconnected.

2. Disconnect the negative battery cable. As necessary, remove the battery.
3. Relieve the fuel system pressure.
4. Drain the engine cooling system.
5. Remove the engine air cleaner assembly.
6. Matchmark the location of the adjuster bolt on the accelerator cable mounting flange. This will assure that the cable is installed in its original location. Remove the throttle cable adjusting bolt and disconnect the cable from the lever on the throttle body. Position cable aside.
7. Remove the connection for the breather hose and the air intake hose from the throttle body and position aside.
8. Tag and disconnect the necessary vacuum hoses.
9. Label and detach the electrical connectors at the throttle body, as necessary.
10. Disconnect the water and water

by-pass hoses at the base of the throttle body.
11. If equipped, unfasten the ground plate mounting screws, then remove the throttle body stay and ground plate from the engine.
12. Remove the air fitting and gasket.
13. Unfasten the throttle body mounting bolts, then remove the throttle body from the engine. Remove and discard the gasket.

To install:

➡**Be sure to use new fasteners, as required.**

14. Clean all old gasket material from the both throttle body mounting surfaces. Install new gasket onto the intake manifold plenum mounting surface.

➡**Poor idling quality and poor performance may be experienced if the gasket is installed incorrectly.**

15. Install the throttle body to the intake manifold plenum and tighten the mounting bolts.
16. Install the air fitting, if equipped, making sure new gasket is in place.
17. If equipped, install the throttle body stay and ground plate. Secure with retainers tightened to 11–16 ft. lbs. (15–22 Nm).
18. Install the ground plate mounting screw.
19. Connect the water hoses to the throttle body. Install new hose clamps if required.
20. Attach the electrical and vacuum connectors to the throttle body, as tagged during removal.
21. Connect the accelerator cable to the throttle body and install the adjusting bolt in original position. Check adjustment of cable.
22. Install the air intake and breather hoses.
23. If removed, install the battery and connect the positive cable.
24. Connect the negative battery cable.
25. Refill the cooling system.
26. Run the engine and check for leaks and proper operation.
27. Using the Mitsubishi diagnostic scan tool, or equivalent, perform the learning valve initialization procedure. Refer to the scan tool screen for instructions.

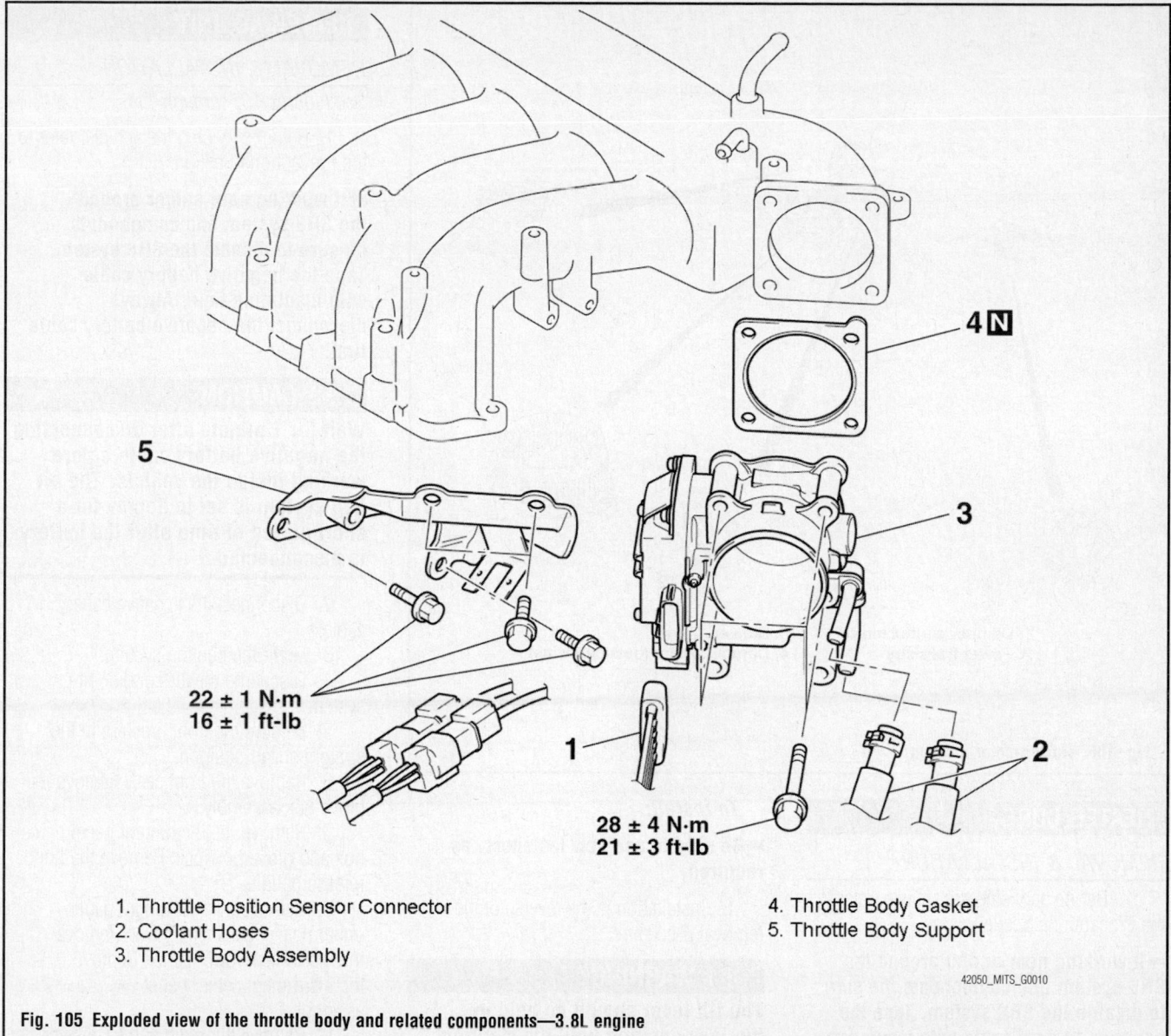

22 ± 1 N·m
16 ± 1 ft-lb

28 ± 4 N·m
21 ± 3 ft-lb

1. Throttle Position Sensor Connector
2. Coolant Hoses
3. Throttle Body Assembly
4. Throttle Body Gasket
5. Throttle Body Support

42050_MITS_G0010

Fig. 105 Exploded view of the throttle body and related components—3.8L engine

HEATING & AIR CONDITIONING SYSTEM

BLOWER MOTOR

REMOVAL & INSTALLATION

See Figure 106.

1. Before servicing the vehicle, refer to the Precautions Section.

➡If working near and/or around the SRS system and components, be sure to disable the SRS system. Tape the negative battery cable with insulating tape. Always disconnect the negative battery cable first.

❄❄ CAUTION

Wait for 1 minute after disconnecting the negative battery cable before working inside the vehicle. The air bag system is set to deploy for a short period of time after the battery is disconnected.

2. Disconnect the negative battery cable.
3. Remove the three instrument panel undercover mounting screws and remove the cover.
4. Detach the electrical connector from the fan motor.
5. Remove the three small bolts holding the motor to the housing and remove the motor and fan.

To install:

➡Be sure to use new fasteners, as required.

6. Check the inside of the case carefully; any debris can snag the fan and cause noise or poor airflow.
7. Install the blower motor, in the blower case and secure with the three mounting bolts.
8. Attach the blower motor electrical connector.
9. Install the undercover, taking care to insure it is in place and all the fasteners are secure.
10. Connect the negative battery cable.

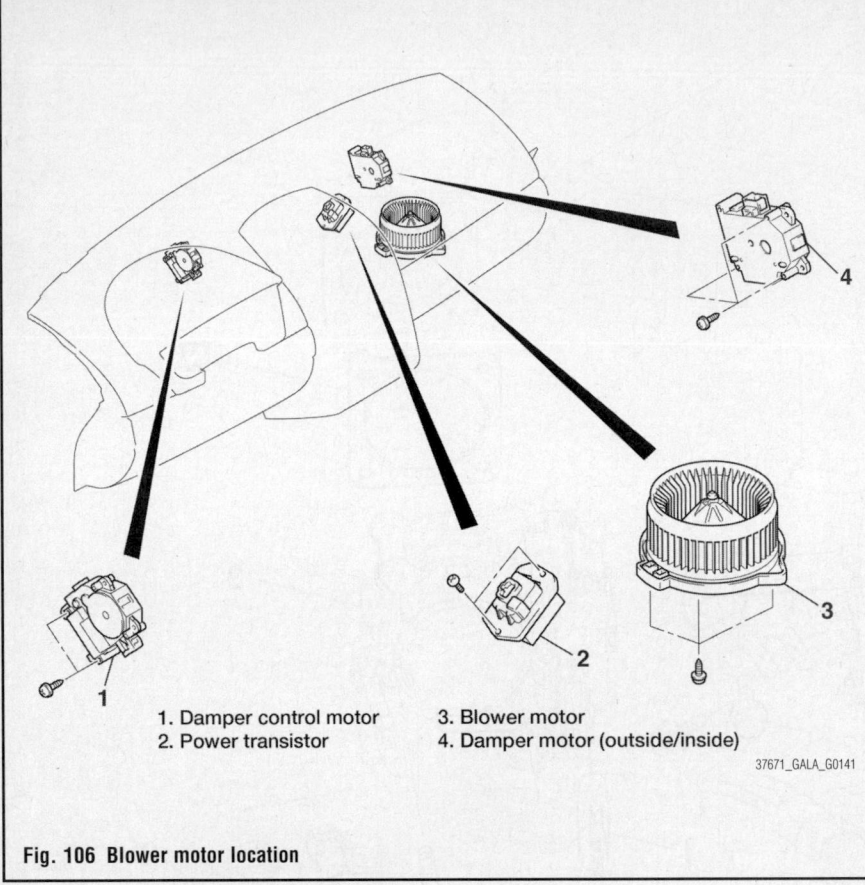

1. Damper control motor
2. Power transistor
3. Blower motor
4. Damper motor (outside/inside)

37671_GALA_G0141

Fig. 106 Blower motor location

HEATER CORE

REMOVAL & INSTALLATION

1. Before servicing the vehicle, refer to the Precautions Section.

➡If working near and/or around the SRS system and components, be sure to disable the SRS system. Tape the negative battery cable with insulating tape. Always disconnect the negative battery cable first.

✳✳ CAUTION

Wait for 1 minute after disconnecting the negative battery cable before working inside the vehicle. The air bag system is set to deploy for a short period of time after the battery is disconnected.

2. Disconnect the negative battery cable.
3. Drain the cooling system.
4. Discharge the air conditioning system.
5. Be sure the front tires are in the straight ahead position.
6. Remove the Heater/AC unit, as outlined in this section
7. Remove the heater core from the assembly.

To install:

➡Be sure to use new fasteners, as required.

8. Installation is the reverse of the removal procedure.

✳✳ WARNING

The tilt lever should be held in the lock position until the steering column is reinstalled in the vehicle. If the column is removed with the lever released, or the lever released after the column is removed from the vehicle, the steering column cannot be reinstalled correctly. If the steering column is installed incorrectly, the collision energy absorbing mechanism may be damaged.

➡After the seats have been reinstalled be sure to perform the accuracy check of the occupant classification sensor, using the Mitsubishi Diagnostic Scan tool, or equivalent. Follow the directions on the tool screen.

9. Be sure to fill the cooling system with the proper grade and type coolant.
10. Recharge the air conditioning system.

HEATER/AC UNIT

REMOVAL & INSTALLATION

See Figures 107 through 111.

1. Before servicing the vehicle, refer to the Precautions Section.

➡If working near and/or around the SRS system and components, be sure to disable the SRS system. Tape the negative battery cable with insulating tape. Always disconnect the negative battery cable first.

✳✳ CAUTION

Wait for 1 minute after disconnecting the negative battery cable before working inside the vehicle. The air bag system is set to deploy for a short period of time after the battery is disconnected.

2. Disconnect the negative battery cable.
3. Drain the cooling system.
4. Discharge the air conditioning system.
5. Be sure the front tires are in the straight ahead position.
6. Remove the front seat. Remove the glove box assembly.
7. Remove the instrument panel parcel box and glove box lock. Remove the hood release handle.
8. Remove the instrument panel under passenger's side cover. Remove the instrument lower panel. Remove the instrument center panel assembly.
9. Remove the radio and CD player. Remove the CD changer, if equipped. Remove the accessory box.
10. Remove the console meter hood. Remove the passenger's side air bag indicator light and passenger's seat belt warning light.
11. If equipped, remove the multi center display. Remove the center console assembly.
12. Remove the instrument panel cover. Remove the combination meter assembly. Remove the front pillar trim.
13. Remove the front speaker covers. Remove the speakers. Remove the instrument panel side cover. Remove the instrument panel side air outlet.
14. Remove the driver's side air bag module. Remove the steering wheel. Remove the steering column cover. Remove the clockspring connector and clockspring and column switch assembly.

CLIP AND CLAW POSITIONS

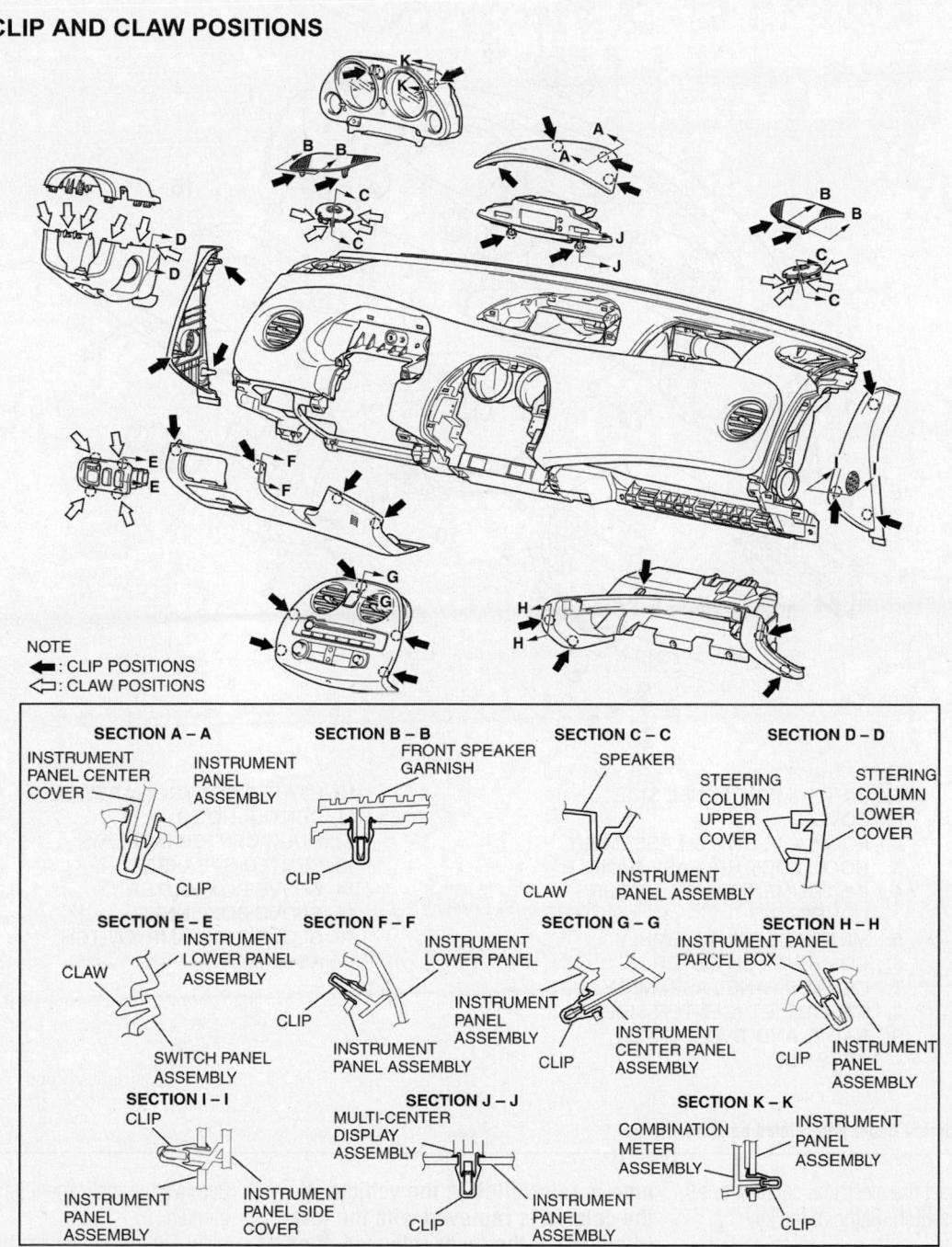

NOTE
← : CLIP POSITIONS
◁ : CLAW POSITIONS

SECTION A – A
INSTRUMENT PANEL CENTER COVER
INSTRUMENT PANEL ASSEMBLY
CLIP

SECTION B – B
FRONT SPEAKER GARNISH
CLIP

SECTION C – C
SPEAKER
CLAW
INSTRUMENT PANEL ASSEMBLY

SECTION D – D
STEERING COLUMN UPPER COVER
STTERING COLUMN LOWER COVER

SECTION E – E
CLAW
INSTRUMENT LOWER PANEL ASSEMBLY
SWITCH PANEL ASSEMBLY

SECTION F – F
INSTRUMENT LOWER PANEL
CLIP
INSTRUMENT PANEL ASSEMBLY

SECTION G – G
INSTRUMENT PANEL ASSEMBLY
INSTRUMENT CENTER PANEL ASSEMBLY
CLIP

SECTION H – H
INSTRUMENT PANEL PARCEL BOX
CLIP
INSTRUMENT PANEL ASSEMBLY

SECTION I – I
CLIP
INSTRUMENT PANEL ASSEMBLY
INSTRUMENT PANEL SIDE COVER

SECTION J – J
MULTI-CENTER DISPLAY ASSEMBLY
CLIP
INSTRUMENT PANEL ASSEMBLY

SECTION K – K
COMBINATION METER ASSEMBLY
INSTRUMENT PANEL ASSEMBLY
CLIP

37671_ECLI_G0009

Fig. 107 Instrument panel fastener locations

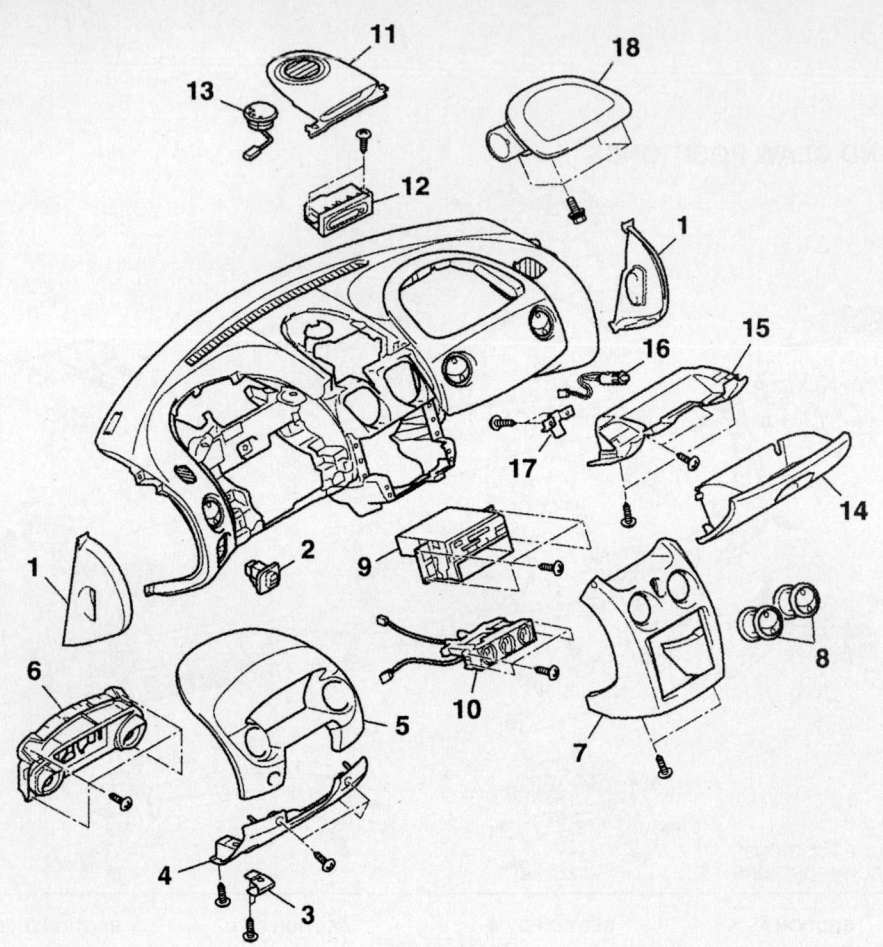

1. INSTRUMENT PANEL SIDE COVER
2. FOG LAMP SWITCH ASSEMBLY
3. HOOD LOCK RELEASE HANDLE
4. INSTRUMENT PANEL UNDER COVER
5. METER BEZEL ASSEMBLY
6. COMBINATION METER
7. CENTER PANEL ASSEMBLY
8. AIR OUTLET (CENTER SIDE)
9. RADIO AND TAPE PLAYER ASSEMBLY
10. HEATER CONTROL ASSEMBLY
11. CENTER HOOD
12. MULTI CENTER DISPLAY
13. CENTER SPEAKER
14. GLOVE BOX, OUTER
15. GLOVE BOX, INNER
16. GLOVE BOX LAMP SWITCH ASSEMBLY

09482_GALA_G0017

Fig. 108 Instrument panel and related components

15. Disconnect the electrical connections for the ignition switch. Remove the key interlock cable.

16. Remove the steering shaft pad. Remove the steering column retaining bolts. Remove the steering column assembly. Pinch the steering column shaft clip with a pliers, and pull up the shaft to disengage the steering column assembly.

❊❊ WARNING

The tilt lever should be held in the lock position until the steering col-umn is reinstalled in the vehicle. If the column is removed with the lever released, or the lever released after the column is removed from the vehicle, the steering column cannot be reinstalled correctly. If the steering column is installed incorrectly, the collision energy absorbing mecha-nism may be damaged.

17. If equipped with automatic temperature control, remove the interior temperature sensor and photo sensor.

Remove the instrument panel front end garnish.

18. Remove the gearshift lever handle, heated seat switch, accessory socket, socket cover, front plate, and front box.

19. Remove the lid lock lever, hinge, lid, liner box, accessory socket and cover.

20. Remove the floor console retaining screws. Remove the floor console. Remove the accessory socket harness and rear bracket.

21. Remove the cowl side trim. Remove the passenger's side air bag module.

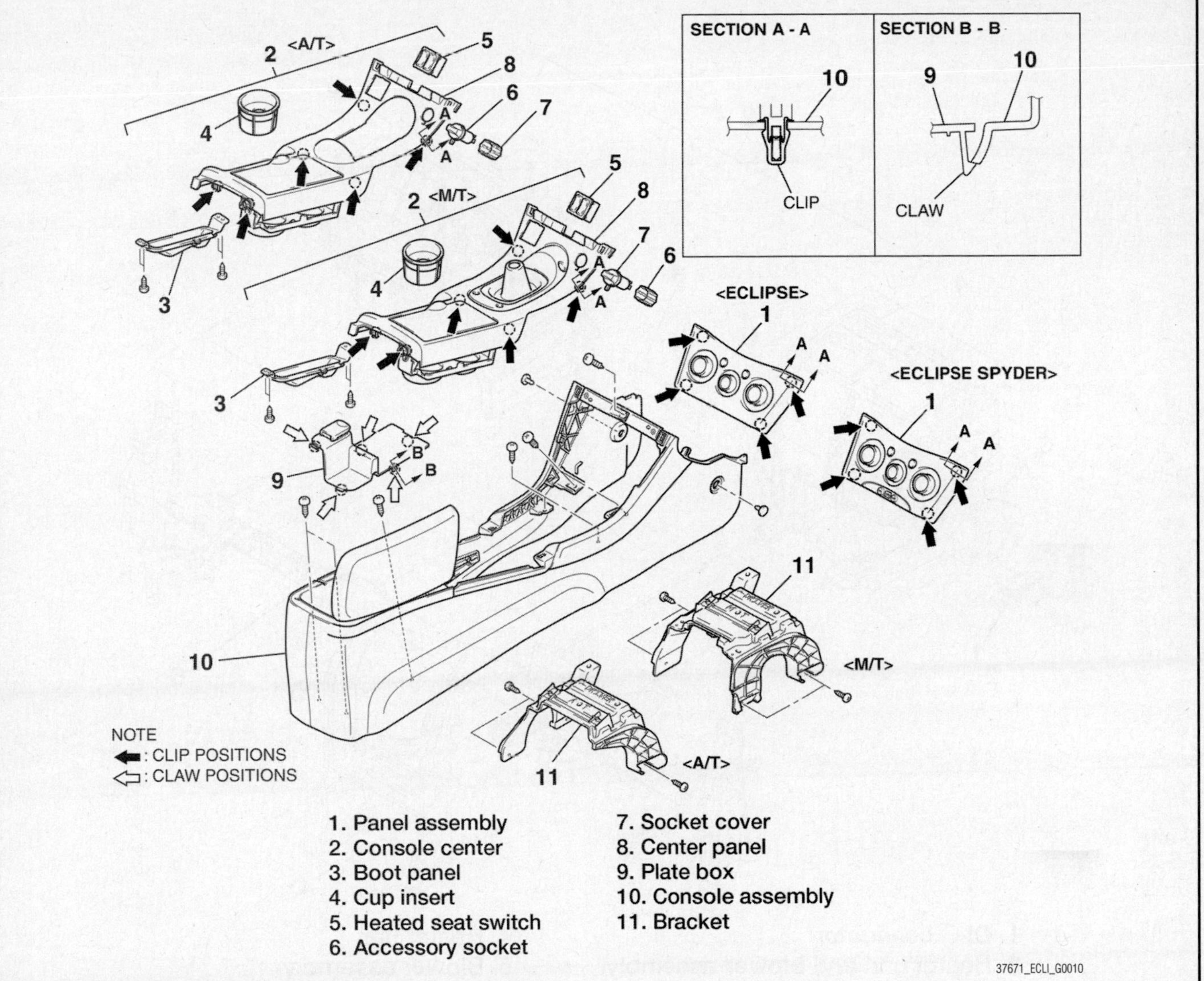

SECTION A - A | SECTION B - B

10 CLIP

9 10 CLAW

2 <A/T>
5
8
4
6
7
3

2 <M/T>
5
8
4
6
7
3
9

<ECLIPSE>
1

<ECLIPSE SPYDER>
1

11 <M/T>

11 <A/T>

10

NOTE
← : CLIP POSITIONS
◁ : CLAW POSITIONS

1. Panel assembly
2. Console center
3. Boot panel
4. Cup insert
5. Heated seat switch
6. Accessory socket
7. Socket cover
8. Center panel
9. Plate box
10. Console assembly
11. Bracket

37671_ECLI_G0010

Fig. 109 Floor console and related components

22. Remove the instrument panel assembly from the vehicle.

23. If equipped remove the strut tower retaining bolts and remove the strut tower bar. Remove the battery. Remove the air cleaner body.

24. Remove the heater hoses.

25. Remove the air conditioning lines from the evaporator.

26. Remove the rear heat duct. Remove the heater unit and deck crossmember assembly.

To install:

➡**Be sure to use new fasteners, as required.**

27. Installation is the reverse of the removal procedure.

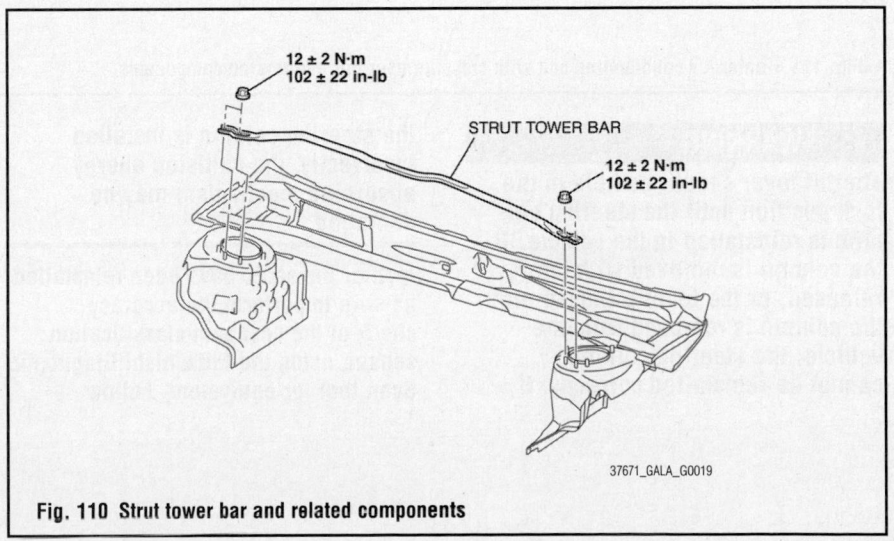

12 ± 2 N·m
102 ± 22 in-lb

STRUT TOWER BAR

12 ± 2 N·m
102 ± 22 in-lb

37671_GALA_G0019

Fig. 110 Strut tower bar and related components

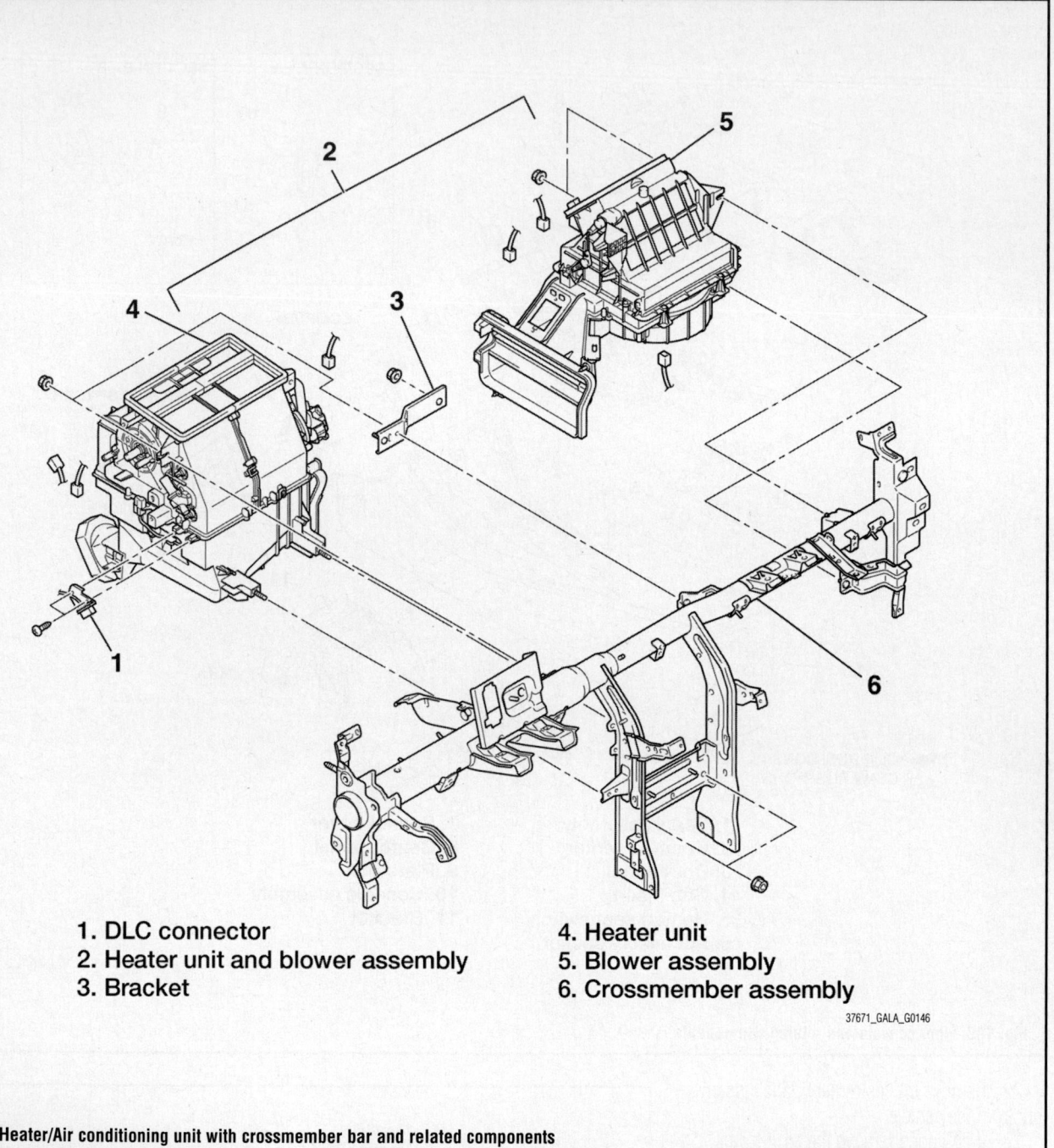

1. DLC connector
2. Heater unit and blower assembly
3. Bracket
4. Heater unit
5. Blower assembly
6. Crossmember assembly

37671_GALA_G0146

Fig. 111 Heater/Air conditioning unit with crossmember bar and related components

☀☀ WARNING

The tilt lever should be held in the lock position until the steering column is reinstalled in the vehicle. If the column is removed with the lever released, or the lever released after the column is removed from the vehicle, the steering column cannot be reinstalled correctly. If the steering column is installed incorrectly, the collision energy absorbing mechanism may be damaged.

➡After the seats have been reinstalled be sure to perform the accuracy check of the occupant classification sensor, using the Mitsubishi Diagnostic Scan tool, or equivalent. Follow the directions on the tool screen.

28. Be sure to fill the cooling system with the proper grade and type coolant.

29. Recharge the air conditioning system.

30. Using the Mitsubishi diagnostic scan tool, or equivalent, reprogram the required components.

STEERING

POWER RACK & PINION STEERING GEAR

REMOVAL & INSTALLATION

See Figures 112 through 120.

At this time the manufacturer does not provide removal and installation procedures for this component, refer to the illustration as required. The following procedure is a guideline and may differ from the vehicle you are servicing.

➡**Prior to removal of the steering rack, center the front wheels and remove the ignition key. Failure to do so may damage the SRS clockspring and render SRS system inoperative.**

1. Before servicing the vehicle, refer to the Precautions Section.

➡**If working near and/or around the SRS system and components, be sure to disable the SRS system. Tape the negative battery cable with insulating tape. Always disconnect the negative battery cable first.**

※ CAUTION

Wait for 1 minute after disconnecting the negative battery cable before working inside the vehicle. The air bag system is set to deploy for a short period of time after the battery is disconnected.

2. Disconnect the negative battery cable.
3. Remove the driver's side air bag module.
4. Remove the steering wheel. Remove the column lower cover. Remove the clockspring.
5. Drain the power steering fluid. Properly dispose of used fluid.
6. Remove the centermember.
7. Remove the lower control arm.
8. Remove the stabilizer bar.
9. Remove the steering shaft pad.
10. Remove the required components in order to remove the steering gear from the vehicle.

To install:

➡**Be sure to use new fasteners, as required.**

11. Position the steering rack in the vehicle and install the clamp and the mounting bolts. Be sure the rack is centered before connecting it to the joint

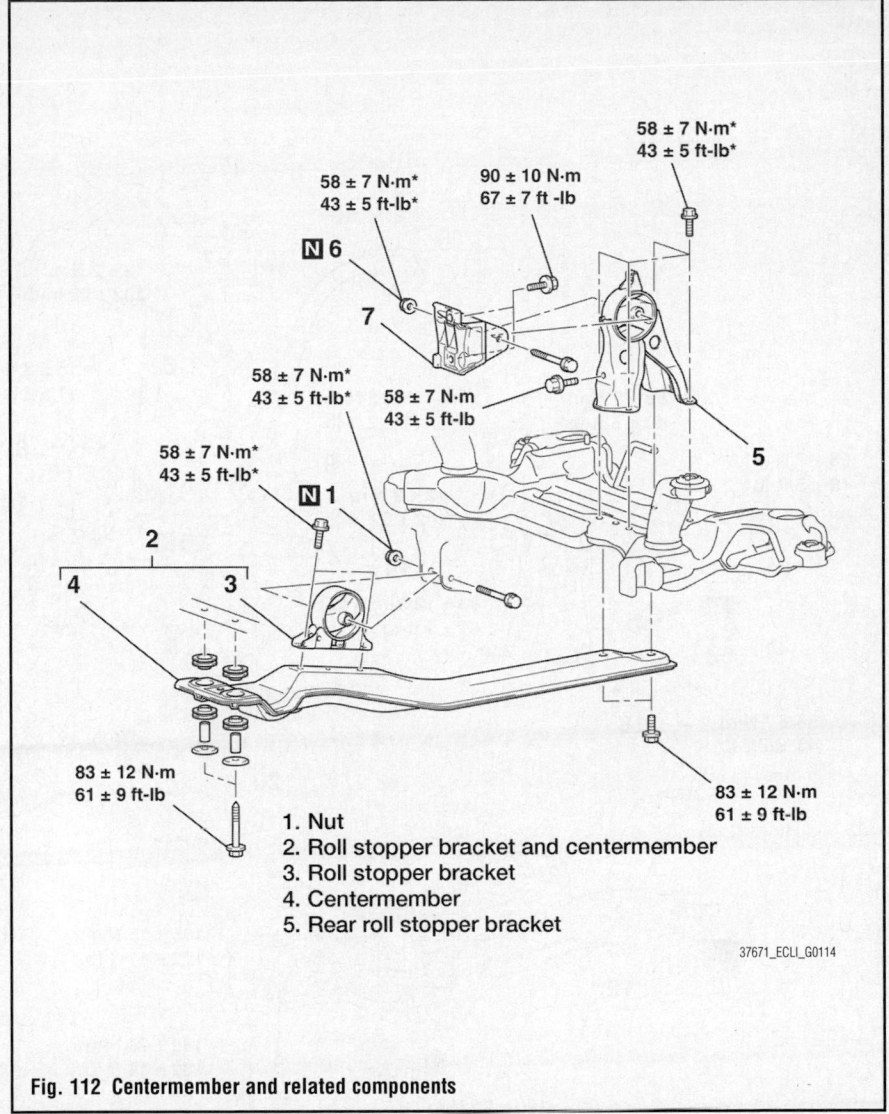

1. Nut
2. Roll stopper bracket and centermember
3. Roll stopper bracket
4. Centermember
5. Rear roll stopper bracket

37671_ECLI_G0114

Fig. 112 Centermember and related components

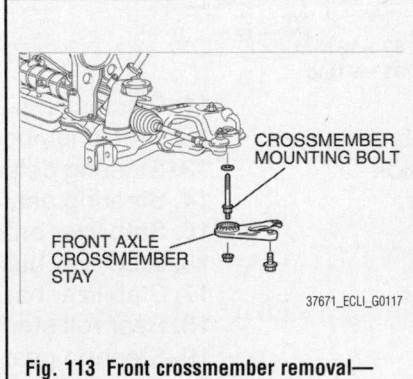

37671_ECLI_G0117

Fig. 113 Front crossmember removal— 1 of 2

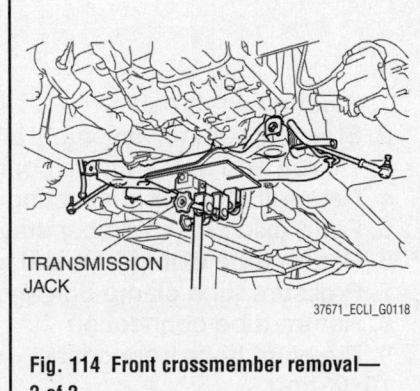

37671_ECLI_G0118

Fig. 114 Front crossmember removal— 2 of 2

assembly. Note the following for clamp installation:

a. Install the pinch clam and align the steering column dash panel cover notch (arrow A) with the steering gear lug (arrow

B), and then install the steering column dash panel cover to the steering gear.
12. Use the end pliers to tighten the pinch clamp. The pinched dimension (A) should be 0.08–0.12 in. (2.0–3.0mm).

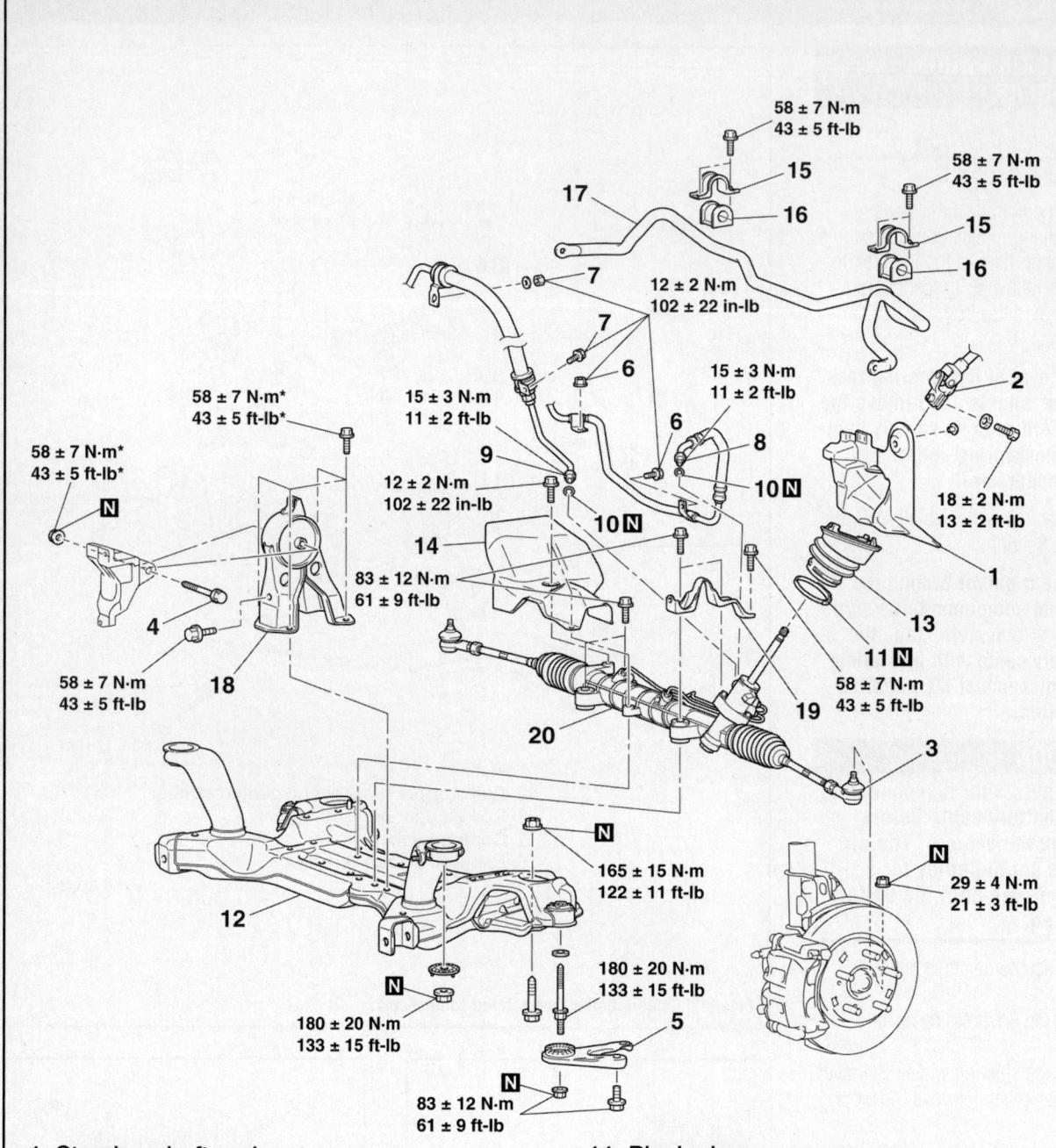

1. Steering shaft pad
2. Steering column assembly and steering gear
3. Tie rod end and knuckle connection
4. Rear roll stopper connecting bolt
5. Front axle crossmember stay
6. Return tube clamp bolt and nut
7. Pressure tube clamp bolt and nut
8. Return tube connection
9. Pressure hose connection
10. O-ring
11. Pinch clamp
12. Crossmember assembly
13. Steering column dash panel cover
14. Steering gear and linkage protector, if equipped
15. Stabilizer bracket
16. Stabilizer bushing
17. Stabilizer bar
18. Rear roll stopper
19. Steering gear bracket
20. Gear and linkage

37671_ECLI_G0115

Fig. 115 Power steering gear and related components

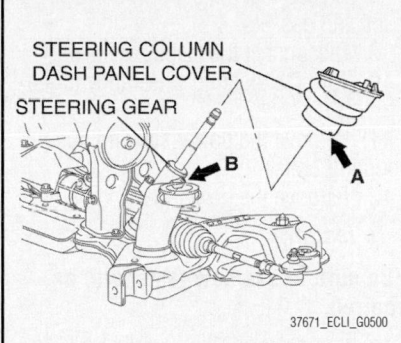

Fig. 116 Install the pinch clam and align the steering column dash panel cover notch (arrow A) with the steering gear lug (arrow B), and then install the steering column dash panel cover to the steering gear

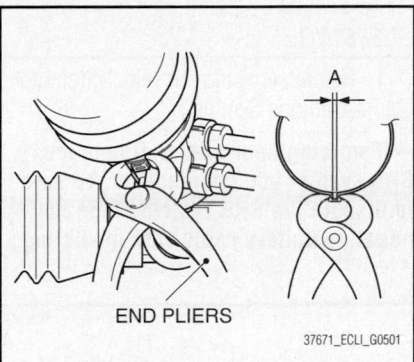

Fig. 117 Use the end pliers to tighten the pinch clamp. The pinched dimension (A) should be 0.08–0.12 in. (2.0–3.0mm)

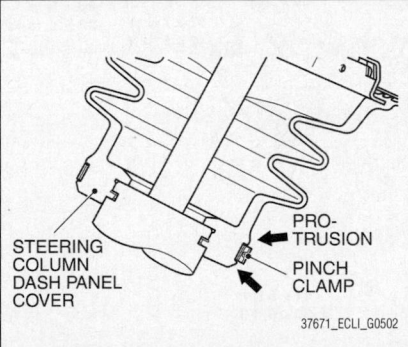

Fig. 118 The pinch clamp must be located securely between protrusions of the steering column dash panel cover

a. The pinch clamp must be located securely between protrusions of the steering column dash panel cover.

b. The pinching location must be in the area shown in the illustration.

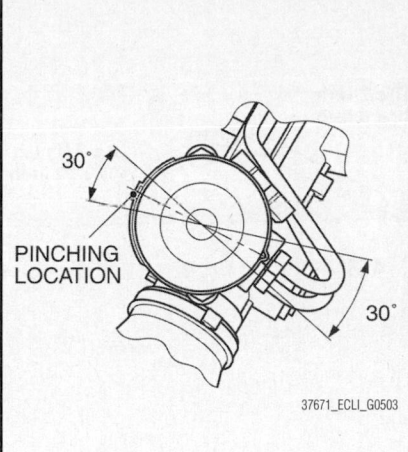

Fig. 119 The pinching location must be in the area shown

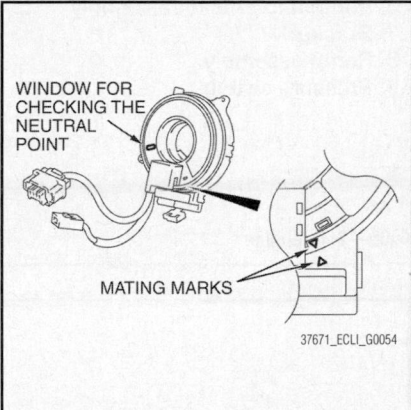

Fig. 120 Align the mating marks of the clockspring module

➡ When installing the clockspring be sure that the springs mating marks are properly aligned. If not the steering wheel may not rotate completely during a turn, or the flat cable in the clockspring could be damaged. This would prevent normal SRS operation and possibly cause serious injury to the driver.

13. After aligning the mating surfaces of the clockspring, turn the front wheels to the straight ahead position. Install the clockspring to the column switch.

➡ Turn the clockspring clockwise fully. Then turn it back approximately 3 ¾ turns counterclockwise to align the mating marks.

14. Continue the installation in the reverse order of the removal procedure.

15. Check and adjust the front end alignment, as required.

POWER STEERING PUMP

REMOVAL & INSTALLATION

2.4L Engine

See Figure 121.

1. Before servicing the vehicle, refer to the Precautions Section.

➡ If working near and/or around the SRS system and components, be sure to disable the SRS system. Tape the negative battery cable with insulating tape. Always disconnect the negative battery cable first.

✳✳ CAUTION

Wait for 1 minute after disconnecting the negative battery cable before working inside the vehicle. The air bag system is set to deploy for a short period of time after the battery is disconnected.

2. Disconnect the battery negative cable.

3. Loosen and remove the power steering pump drive belt.

4. Remove the pressure switch connector from the side of the pump.

5. Disconnect the return fluid line. Remove the reservoir cap and allow the return line to drain the fluid from the reservoir.

6. Disconnect the pressure line.

7. Unbolt and remove the pump from the mounting bracket.

To install:

➡ Be sure to use new fasteners, as required.

8. Install the pump, wrap the belt around the pulley and lightly tighten the mounting bolts.

9. Replace the O-rings and connect the pressure line. Connect the pressure line so the notch in the fitting aligns and contacts the pump's guide bracket. Tighten the fitting to specification.

10. Connect the return line and secure with the clamp.

11. Attach the pressure switch connector.

12. Adjust the power steering belt for proper tension and tighten the adjusting bolts.

13. Reconnect the negative battery cable.

14. Refill the reservoir and bleed the system.

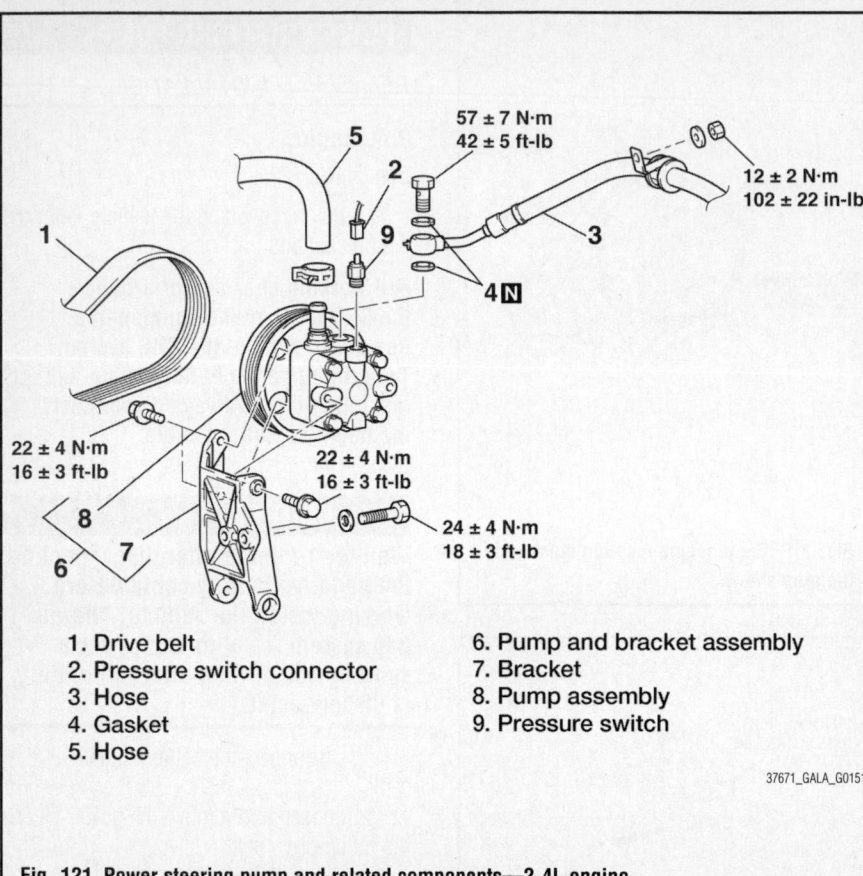

1. Drive belt
2. Pressure switch connector
3. Hose
4. Gasket
5. Hose
6. Pump and bracket assembly
7. Bracket
8. Pump assembly
9. Pressure switch

37671_GALA_G0151

Fig. 121 Power steering pump and related components—2.4L engine

3.8L Engine

See Figure 122.

1. Before servicing the vehicle, refer to the Precautions Section.

➡**If working near and/or around the SRS system and components, be sure to disable the SRS system. Tape the negative battery cable with insulating tape. Always disconnect the negative battery cable first.**

⁂ **CAUTION**

Wait for 1 minute after disconnecting the negative battery cable before working inside the vehicle. The air bag system is set to deploy for a short period of time after the battery is disconnected.

2. Disconnect the negative battery cable.

3. Remove the undercover.

4. Remove the strut tower bar.

5. Remove the power steering drive belt.

6. Drain the fluid. Be sure to properly dispose of used fluid.

7. Disconnect the pressure switch connector electrical connector.

8. Disconnect the supply and return hoses and gaskets.

9. Disconnect the stabilizer links.

10. Turn the steering wheel fully to the left.

11. Remove the power steering pump mounting bolts.

12. Remove the power steering pump.

To install:

➡**Be sure to use new fasteners, as required.**

13. Installation is the reverse order of removal.

14. Torque the power steering pump mounting bolts to specification.

15. Be sure to fill the system with new fluid. Be sure to use the proper grade and type fluid.

16. Start the engine and check for leaks.

17. Bleed the system.

BLEEDING

1. Before servicing the vehicle, refer to the Precautions Section.

➡**If working near and/or around the SRS system and components, be sure to disable the SRS system. Tape the negative battery cable with insulating**

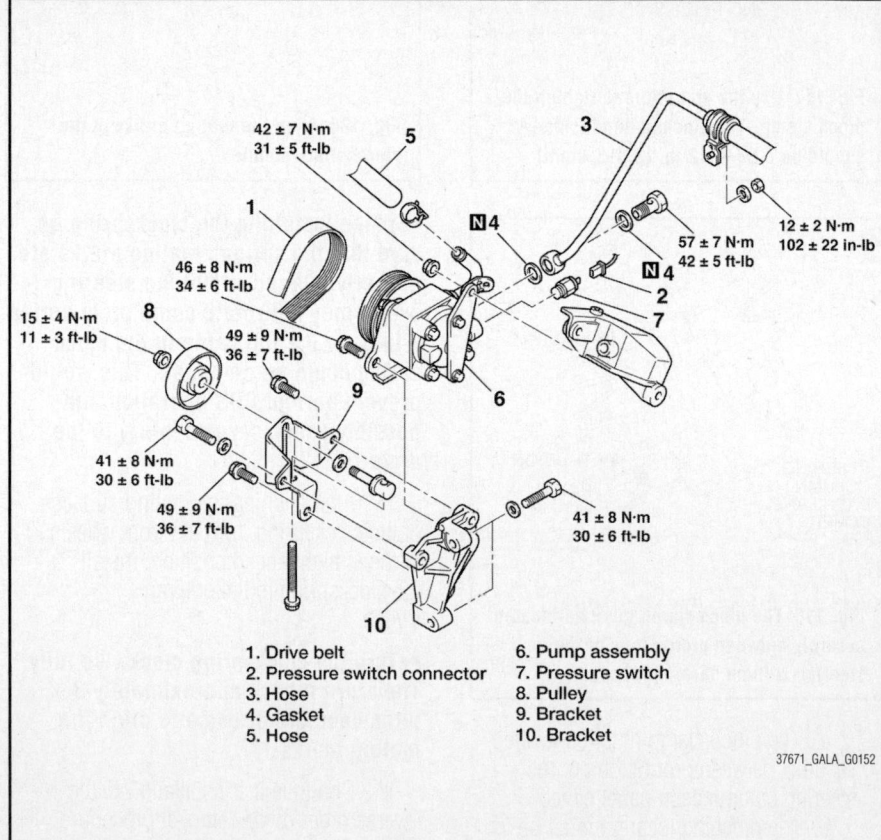

1. Drive belt
2. Pressure switch connector
3. Hose
4. Gasket
5. Hose
6. Pump assembly
7. Pressure switch
8. Pulley
9. Bracket
10. Bracket

37671_GALA_G0152

Fig. 122 Power steering pump and related components—3.8L engine

tape. Always disconnect the negative battery cable first.

2. Raise and safely support the vehicle.

3. Disconnect the crankshaft position sensor connector.

➡ **Perform air bleeding only, while cranking the engine. Do not perform air bleeding while the engine is running. If you do so, air in the fluid will be increased and air bleeding will become more difficult.**

➡ **During air bleeding refill the steering fluid so that the level never falls below the low mark on the dipstick.**

4. Turn the steering wheel all the way to the left and to the right 5 or 6 times, while using the starter motor to crank the engine.

5. Connect the crankshaft position sensor connector.

6. Start engine and allow to idle.

7. Turn the steering wheel left and right until there are no air bubbles in the reservoir. Confirm that the fluid is not milky and the level is up to the specified position on the gauge. Confirm that there is very little change in the fluid level when the steering wheel is turned.

SUSPENSION | FRONT SUSPENSION

LOWER BALL JOINT

REMOVAL & INSTALLATION

See Figure 123.

The lower ball joint is an integral part of the lower control arm assembly and cannot be serviced separately. A worn or damaged ball joint requires replacement of lower control arm assembly.

LOWER CONTROL ARM

REMOVAL & INSTALLATION

See Figure 123.

1. Before servicing the vehicle, refer to the Precautions Section.

➡ **If working near and/or around the SRS system and components, be sure to disable the SRS system. Tape the negative battery cable with insulating tape. Always disconnect the negative battery cable first.**

2. Raise and support the vehicle safely.

3. Remove the tire and wheel assembly.

➡ **Do not remove the nut from the ball joint. Loosen it and use special tool MB991897 to avoid possible damage to the ball joint threads. Hang the special tool in place with wire or string to prevent it from falling.**

4. To disconnect the lower arm from the knuckle, replace the self-locking nut for the lower arm ball joint with a regular nut. This is done because the original one is too large to install the special tool. Install special tool MB991897.

5. Turn the bolt and knob as necessary to make the jaws of the tool parallel. Tighten the bolt by hand and confirm that the jaws are still parallel.

➡ **When adjusting the jaws in parallel, make sure the knob is in the vertical (upward) position.**

6. Tighten the bolt with a wrench to disconnect the ball joint.

7. Remove the lower control arm retaining bolts.

8. Properly support the lower control arm assembly, using a suitable jack.

9. Remove the lower control arm from the vehicle.

To install:

➡ **Be sure to use new fasteners, as required.**

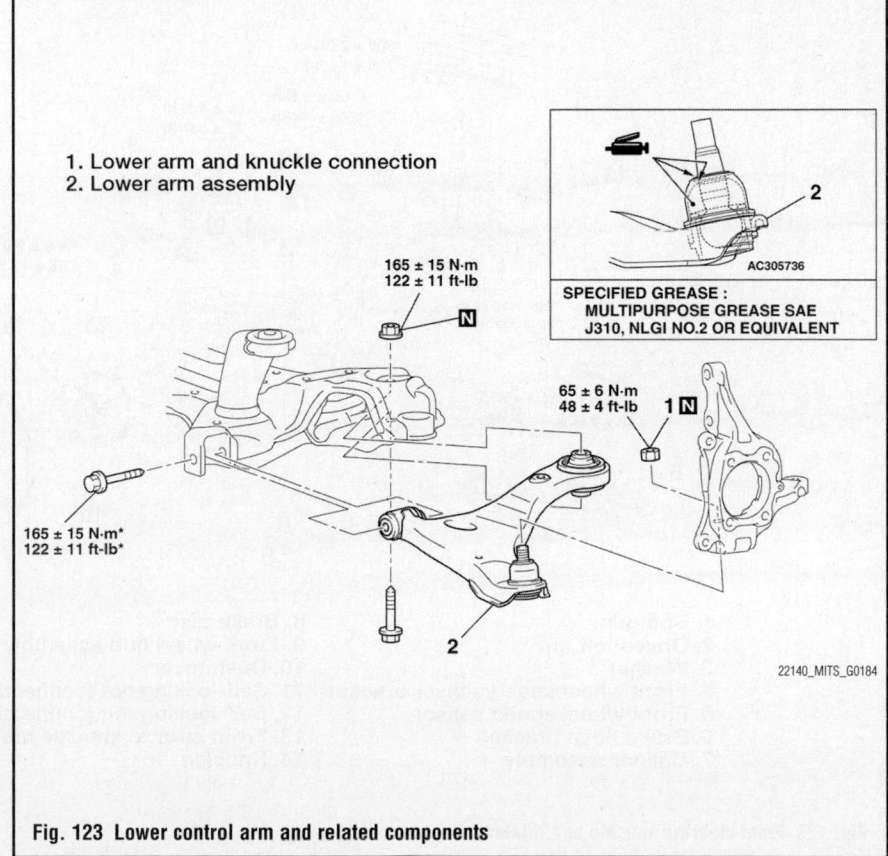

1. Lower arm and knuckle connection
2. Lower arm assembly

165 ± 15 N·m
122 ± 11 ft-lb

165 ± 15 N·m*
122 ± 11 ft-lb*

65 ± 6 N·m
48 ± 4 ft-lb

AC305736

SPECIFIED GREASE :
MULTIPURPOSE GREASE SAE
J310, NLGI NO.2 OR EQUIVALENT

22140_MITS_G0184

Fig. 123 Lower control arm and related components

10. Installation is the reverse of the removal procedure.

11. Tighten bolts to specification.

➡ **Refer to illustration and fully tighten bolts. Bolts with should be temporarily tightened, and fully tightened with the vehicle on the ground in an unladen condition.**

12. Check and adjust the front end alignment, as required.

MACPHERSON STRUT

REMOVAL & INSTALLATION

See Figures 124 and 127.

1. Before servicing the vehicle, refer to the Precautions Section.

➡ **If working near and/or around the SRS system and components, be sure to disable the SRS system. Tape the negative battery cable with insulating tape. Always disconnect the negative battery cable first.**

✳✳ CAUTION

Wait for 1 minute after disconnecting the negative battery cable before working inside the vehicle. The air

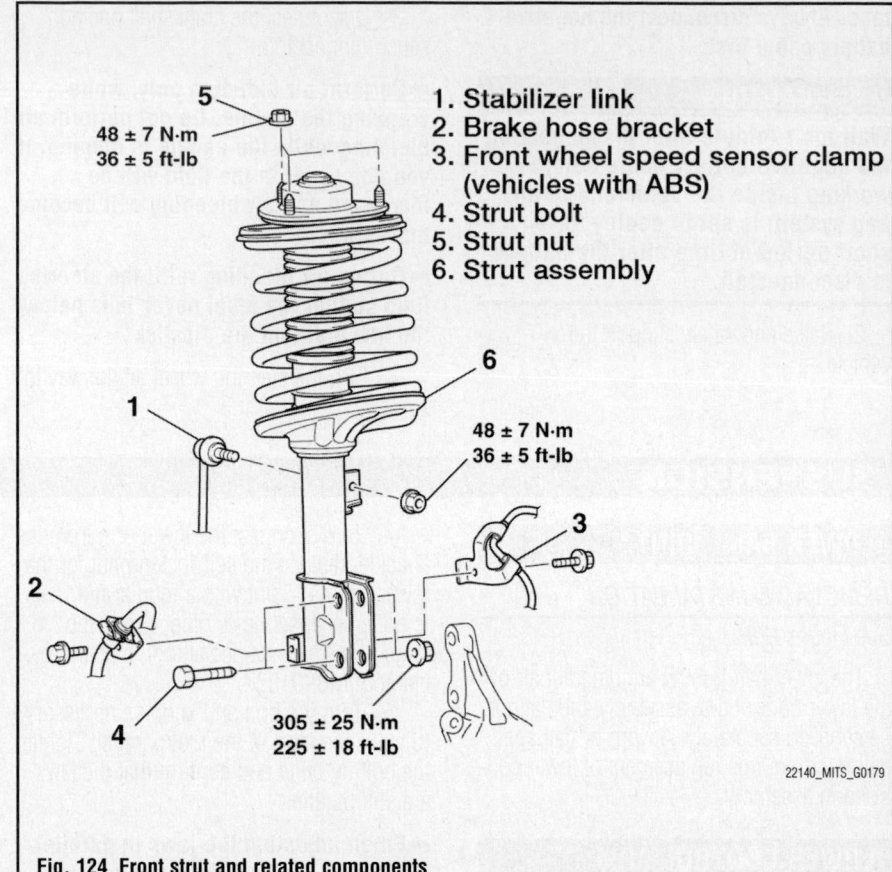

5
48 ± 7 N·m
36 ± 5 ft-lb

1. Stabilizer link
2. Brake hose bracket
3. Front wheel speed sensor clamp (vehicles with ABS)
4. Strut bolt
5. Strut nut
6. Strut assembly

6

48 ± 7 N·m
36 ± 5 ft-lb

1

3

2

4

305 ± 25 N·m
225 ± 18 ft-lb

22140_MITS_G0179

Fig. 124 Front strut and related components

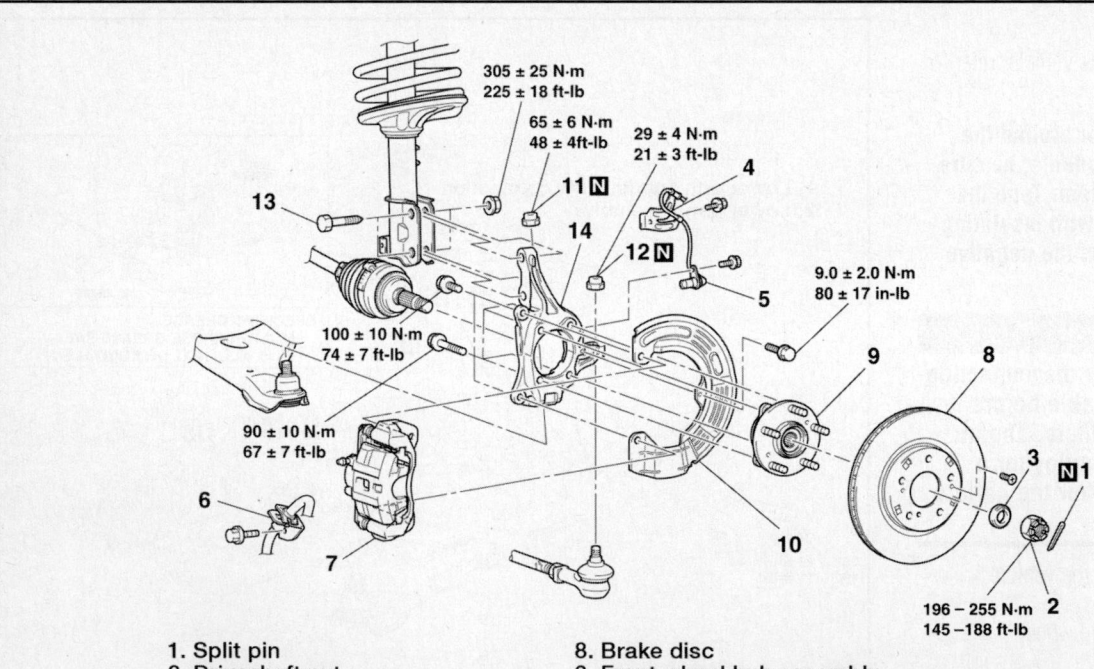

305 ± 25 N·m
225 ± 18 ft-lb

65 ± 6 N·m
48 ± 4ft-lb

29 ± 4 N·m
21 ± 3 ft-lb

13

11 N

4

14

12 N

5

9.0 ± 2.0 N·m
80 ± 17 in-lb

100 ± 10 N·m
74 ± 7 ft-lb

9

8

90 ± 10 N·m
67 ± 7 ft-lb

6

3

N 1

7

10

2

196 – 255 N·m
145 –188 ft-lb

1. Split pin
2. Driveshaft nut
3. Washer
4. Front wheel speed sensor bracket
5. Front wheel speed sensor
6. Brake hose bracket
7. Caliper assembly
8. Brake disc
9. Front wheel hub assembly
10. Dust cover
11. Self-locking nut (connection for lower arm ball joint)
12. Self-locking nut (connection for tie rod end)
13. Front strut to knuckle mounting bolt and nut
14. Knuckle

22140_MITS_G0193

Fig. 125 Front steering knuckle and related components

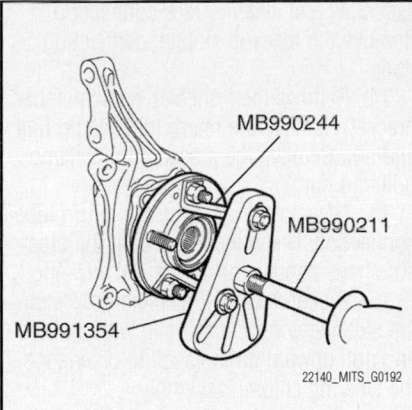

Fig. 126 Use special tools MB990244, MB991354, and MB990211 to pull out the front wheel hub from the steering knuckle

bag system is set to deploy for a short period of time after the battery is disconnected.

2. Raise and support the vehicle safely.

3. Remove the tire and wheel assembly, as necessary.

4. Matchmark alignment markings on the camber adjusting bolt and strut for approximate installation alignment later.

5. Disconnect the stabilizer link. Remove the brake hose bracket.

6. On vehicles equipped with ABS, remove the wheel speed sensor clamp.

7. Remove the lower strut bolts and nuts. Remove the upper strut nuts.

8. Remove the strut assembly from the vehicle.

To install:

➡️Be sure to use new fasteners, as required.

9. Installation is the reverse of the removal procedure.

10. Use the alignment marks made before removal to align the strut in position.

11. Tighten the bolts to specification.

12. Be sure to check and adjust the front end alignment, as required.

13. Be sure to check and adjust the front end alignment, as required.

STEERING KNUCKLE

REMOVAL & INSTALLATION
See Figures 125 and 126.

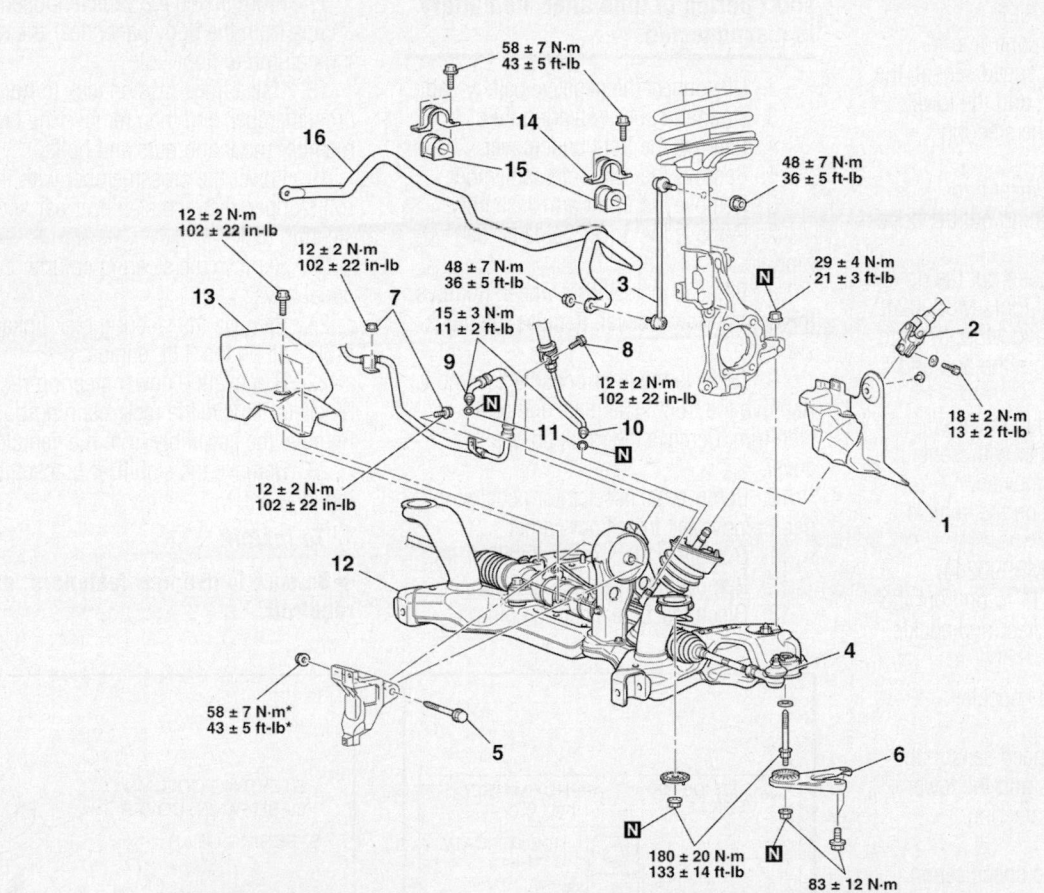

1. Steering shaft pad
2. Steering gear and steering column assembly connection
3. Stabilizer link
4. Tie rod end and knuckle connection
5. Rear roll stopper connection bolt
6. Front axle crossmember stay
7. Return tube clamp nut and bolt
8. Pressure tube clamp bolt
9. Steering gear and return tube connection
10. Steering gear and pressure tube connection
11. O-ring
12. Front axle crossmember, rear roll stopper and steering gear assembly
13. Steering gear and linkage protector
14. Stabilizer bracket
15. Stabilizer bushing
16. Stabilizer bar

Fig. 127 Stabilizer bar and related components

1. Before servicing the vehicle, refer to the Precautions Section.

➡ **If working near and/or around the SRS system and components, be sure to disable the SRS system. Tape the negative battery cable with insulating tape. Always disconnect the negative battery cable first.**

※※ CAUTION

Wait for 1 minute after disconnecting the negative battery cable before working inside the vehicle. The air bag system is set to deploy for a short period of time after the battery is disconnected.

2. Raise and safely support the vehicle.

3. Remove the front wheels and tires.

4. Remove the brake rotor.

5. Remove the lower control arm.

6. Remove the wheel speed sensor, the strut lower mounting bolt, and the lower arm mounting bolt from the steering knuckle.

7. Remove the front wheel hub mounting bolts while pushing out the driveshaft by hand.

8. If it is difficult to push out the driveshaft by hand, use special tools MB990242, MB990244, MB991354, and MB990767 to push out the driveshaft from the hub and knuckle.

9. If the front wheel hub is seized, remove the knuckle together with front wheel hub and fix them in a vise.

10. Hang the halfshaft on the vehicle body with a rope.

11. Use special tools MB990244, MB991354, and MB990211 to pull out the front wheel hub from the steering knuckle.

To install:

12. Install the hub and knuckle assembly.

13. Install the wheel speed sensor, the strut lower mounting bolt, and the lower arm mounting bolt to the steering knuckle.

14. Tighten the bolts to specification.

15. Install the lower ball joint.

16. Install the brake rotor.

17. Install the front wheels and tires.

18. Check alignment and adjust as necessary.

STABILIZER BAR

REMOVAL & INSTALLATION

See Figures 127 through 129.

➡ **Prior to removal of the stabilizer bar, center the front wheels and remove the ignition key. Failure to do so may damage the SRS clockspring and render the SRS system inoperative.**

1. Before servicing the vehicle, refer to the Precautions Section.

➡ **If working near and/or around the SRS system and components, be sure to disable the SRS system. Tape the negative battery cable with insulating tape. Always disconnect the negative battery cable first.**

※※ CAUTION

Wait for 1 minute after disconnecting the negative battery cable before working inside the vehicle. The air bag system is set to deploy for a short period of time after the battery is disconnected.

2. Disconnect the negative battery cable.

3. Drain the power steering fluid.

4. Remove the front undercover.

5. Remove the center crossmember.

6. Remove the lower arm assembly.

7. Remove the driver's side air bag module.

8. Remove the steering wheel. Remove the column lower cover. Remove the clockspring.

9. Remove the floor console assembly. Remove the front scuff plate and cowl side trim. Remove the trunk lid opener cover.

10. Remove the accelerator pedal stopper. Remove the front floor carpet.

11. Disconnect the upper stabilizer link.

12. Remove the steering shaft pad.

13. Disconnect the steering column

assembly and steering rack connection. Remove the rear roll stopper connecting bolt.

14. Remove the front axle crossmember bracket. Remove the return tube clamp bolt and nut. Remove the pressure tube clamp bolt and nut.

15. Disconnect and plug the return tube connection. Disconnect and plug the pressure hose connection. Discard the O-ring.

16. Disconnect the pinch clamp. Pinch the steering column shaft clip with a pliers and pull upward on the shaft to disengage the steering column assembly.

➡ **If the steering column shaft is removed accidentally, remove the steering column assembly and be sure to insert the steering column shaft into the steering column.**

17. From inside the vehicle loosen the 3 clips from the body panel near the steering column to floor hole.

18. Use a transmission jack to hold the crossmember, and then remove the crossmember mounting nuts and bolts.

19. Lower the crossmember with the rear roll stopper, then the stabilizer bar and steering rack.

20. Remove the steering column dash panel cover.

21. Remove the steering rack linkage protector, on the 3.8L engine.

22. Remove the power steering rack bracket. Remove the rack retaining bolts. Remove the assembly from the vehicle.

23. Remove the stabilizer bracket, bushing, and bar.

To install:

➡ **Be sure to use new fasteners, as required.**

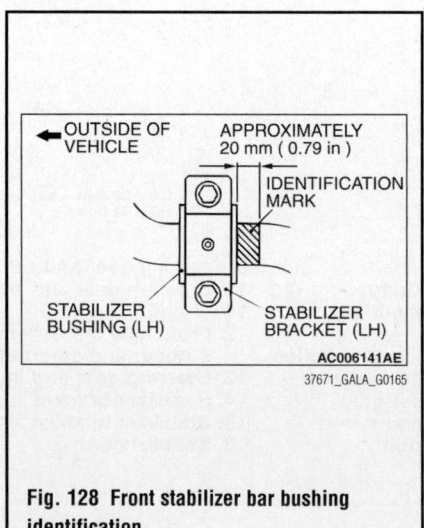

Fig. 128 Front stabilizer bar bushing identification

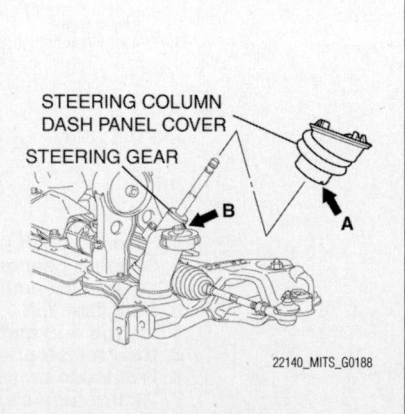

Fig. 129 Align the steering column dash panel cover notch "A" with the steering gear lug "B"

24. Align the stabilizer bar identification mark with the right end of the bushing. Install the stabilizer bar.

25. Position the steering rack in the vehicle and install the clamp and the mounting bolts. Be sure the rack is centered before connecting it to the joint assembly.

26. When installing the steering column dash panel cover pinch clamp, align the steering column dash panel cover notch "A" with the steering gear lug "B" and then install the steering column dash panel cover to the steering gear.

✳✳ CAUTION

When installing the clockspring be sure that the springs mating marks are properly aligned. If not the steering wheel may not rotate completely during a turn, or the flat cable in the clockspring could be damaged. This would prevent normal SRS operation and possibly cause serious injury to the driver.

27. After aligning the mating surfaces of the clockspring, turn the front wheels to the straight ahead position. Install the clockspring to the column switch.

➡**Turn the clockspring clockwise fully. Then turn it back approximately 3 ¾ turns counterclockwise to align the mating marks.**

28. Continue the installation in the reverse order of the removal procedure.

29. Check and adjust the front end alignment, as required.

WHEEL HUB & BEARING

REMOVAL & INSTALLATION

See Figures 125 and 126.

1. Before servicing the vehicle, refer to the Precautions Section.

➡**If working near and/or around the SRS system and components, be sure to disable the SRS system. Tape the negative battery cable with insulating tape. Always disconnect the negative battery cable first.**

✳✳ CAUTION

Wait for 1 minute after disconnecting the negative battery cable before working inside the vehicle. The air bag system is set to deploy for a short period of time after the battery is disconnected.

2. Remove the steering knuckle, as required.

3. Use special tools MB990244, MB991354, and MB990211 to pull out the front wheel hub from the steering knuckle.

To install:

4. Install the hub bearing into the knuckle assembly.

5. Install the knuckle, as required.

6. Tighten the bolts to specification.

7. Install the brake rotor.

8. Install the front wheels and tires.

9. Check alignment and adjust as necessary.

ADJUSTMENT

The front wheel bearings are not adjustable. If the bearings are noisy or become loose, they must be replaced.

SUSPENSION

CONTROL ARMS/LINKS

REMOVAL & INSTALLATION

Lower Control Arm
See Figure 130.

1. Before servicing the vehicle, refer to the Precautions Section.

➡**If working near and/or around the SRS system and components, be sure to disable the SRS system. Tape the negative battery cable with insulating tape. Always disconnect the negative battery cable first.**

✳✳ CAUTION

Wait for 1 minute after disconnecting the negative battery cable before working inside the vehicle. The air bag system is set to deploy for a short period of time after the battery is disconnected.

2. Raise and support the vehicle safely.

3. Disconnect the stabilizer link connection.

4. Disconnect the wheel speed sensor mounting bolts.

REAR SUSPENSION

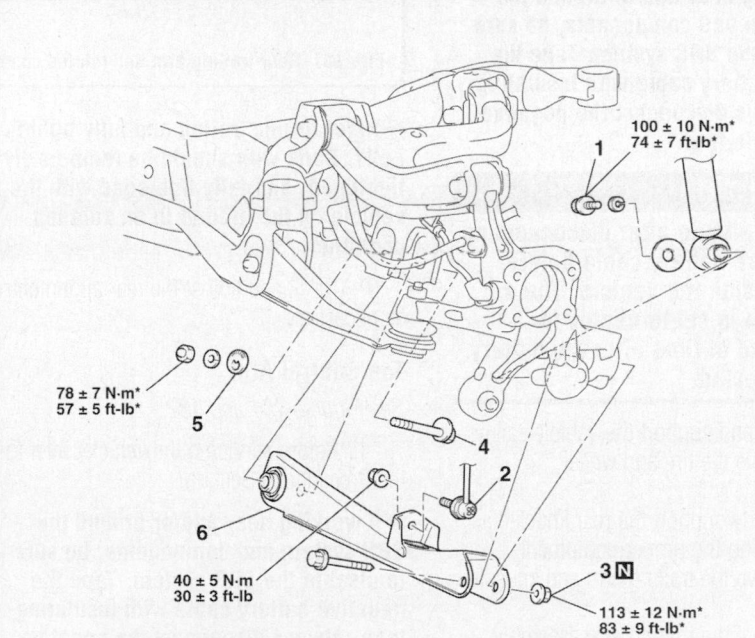

100 ± 10 N·m*
74 ± 7 ft-lb*

78 ± 7 N·m*
57 ± 5 ft-lb*

40 ± 5 N·m
30 ± 3 ft-lb

113 ± 12 N·m*
83 ± 9 ft-lb*

1. Shock Absorber assembly and knuckle connection
2. Lower arm assembly and stabilizer bar link assembly connection
3. Lower arm assembly and knuckle connection
4. Lower arm bolt
5. Lower arm plate
6. Lower arm assembly

22140_MITS_G0197

Fig. 130 View of the lower control arm

5. Properly support the lower control arm assembly.

6. Remove the lower arm assembly and knuckle connecting bolt.

7. Remove the lower arm mounting bolt.

8. Remove the lower control arm assembly from the vehicle.

To install:

➡**Be sure to use new fasteners, as required.**

9. Install the lower control arm to its mounting.

10. Continue the installation in the reverse order of the removal procedure.

➡**Refer to illustration and fully tighten bolts. Bolts with should be temporarily tightened, and fully tightened with the vehicle on the ground in an unladen condition.**

11. Check and adjust the rear alignment, as necessary.

Rear Trailing Arm

See Figure 131.

1. Before servicing the vehicle, refer to the Precautions Section.

➡**If working near and/or around the SRS system and components, be sure to disable the SRS system. Tape the negative battery cable with insulating tape. Always disconnect the negative battery cable first.**

❋❋ CAUTION

Wait for 1 minute after disconnecting the negative battery cable before working inside the vehicle. The air bag system is set to deploy for a short period of time after the battery is disconnected.

2. Raise and support the vehicle safely.
3. Remove the tire and wheel assembly.
4. Properly support the rear knuckle as required, using the proper equipment.
5. Remove the trailing arm and knuckle bolt.
6. Remove the trailing arm assembly.
7. Remove the trailing arm bracket.

To install:

➡**Be sure to use new fasteners, as required.**

8. Installation is the reverse of the removal procedure.

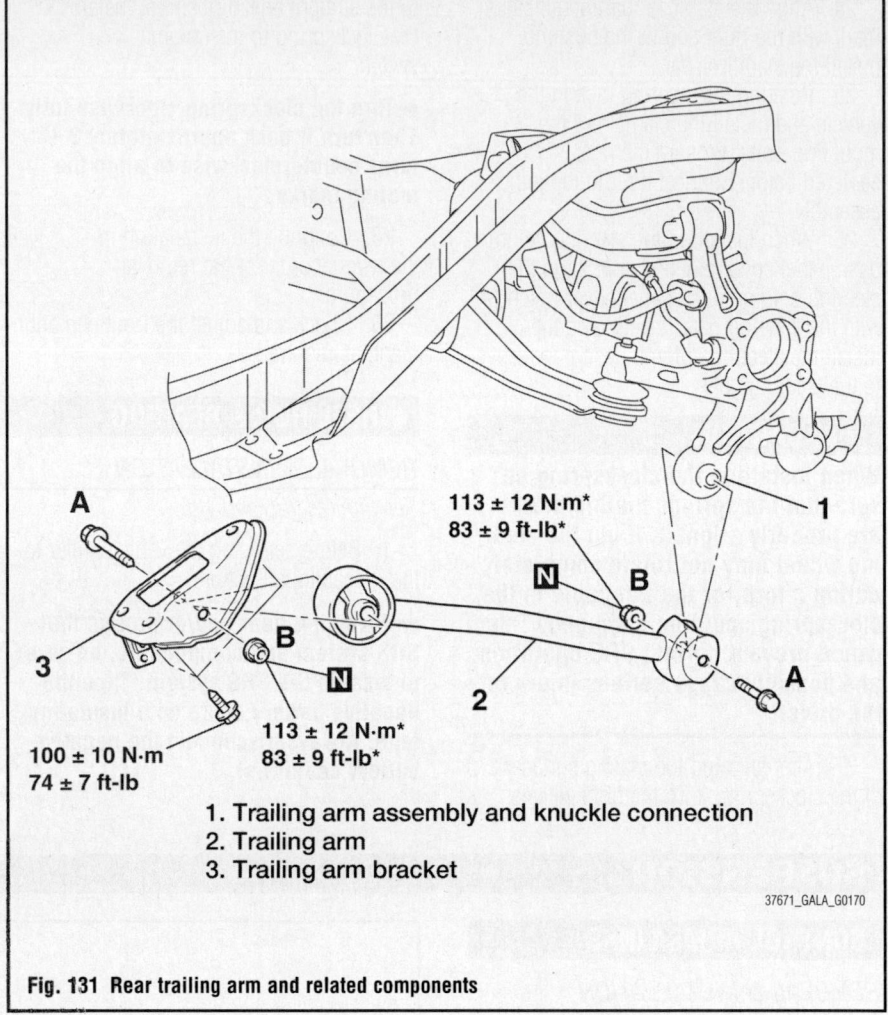

113 ± 12 N·m*
83 ± 9 ft-lb*

100 ± 10 N·m
74 ± 7 ft-lb

113 ± 12 N·m*
83 ± 9 ft-lb*

1. Trailing arm assembly and knuckle connection
2. Trailing arm
3. Trailing arm bracket

37671_GALA_G0170

Fig. 131 Rear trailing arm and related components

➡**Refer to illustration and fully tighten bolts. Bolts with should be temporarily tightened, and fully tightened with the vehicle on the ground in an unladen condition.**

9. Check and adjust the rear alignment, as necessary.

Toe Control Arm

See Figures 131 and 132.

1. Before servicing the vehicle, refer to the Precautions Section.

➡**If working near and/or around the SRS system and components, be sure to disable the SRS system. Tape the negative battery cable with insulating tape. Always disconnect the negative battery cable first.**

❋❋ CAUTION

Wait for 1 minute after disconnecting the negative battery cable before working inside the vehicle. The air bag system is set to deploy for a short period of time after the battery is disconnected.

2. Raise and support the vehicle safely.
3. Remove the tire and wheel assembly.
4. Properly support the rear knuckle as required, using the proper equipment.
5. Remove the trailing arm assembly.
6. Remove the toe control arm assembly and knuckle connection.
7. Remove the assist link bolt.
8. Remove the assist link plate.
9. Remove the toe control arm assembly.

To install:

➡**Be sure to use new fasteners, as required.**

10. Installation is the reverse of the removal procedure.

➡**Refer to illustration and fully tighten bolts. Bolts with should be temporarily**

port the knuckle before removing the upper control arm.

6. Remove the upper arm assembly and knuckle retaining bolt and nut.

7. Remove the upper arm assembly retaining bolts and nuts.

8. Remove the upper control arm from the vehicle.

To install:

➡ **Be sure to use new fasteners, as required.**

9. Install the upper control arm to its mounting.

10. Continue the installation in the reverse order of the removal procedure.

➡ **Refer to illustration and fully tighten bolts. Bolts with should be temporarily tightened, and fully tightened with the vehicle on the ground in an unladen condition.**

11. Check and adjust the rear alignment, as necessary.

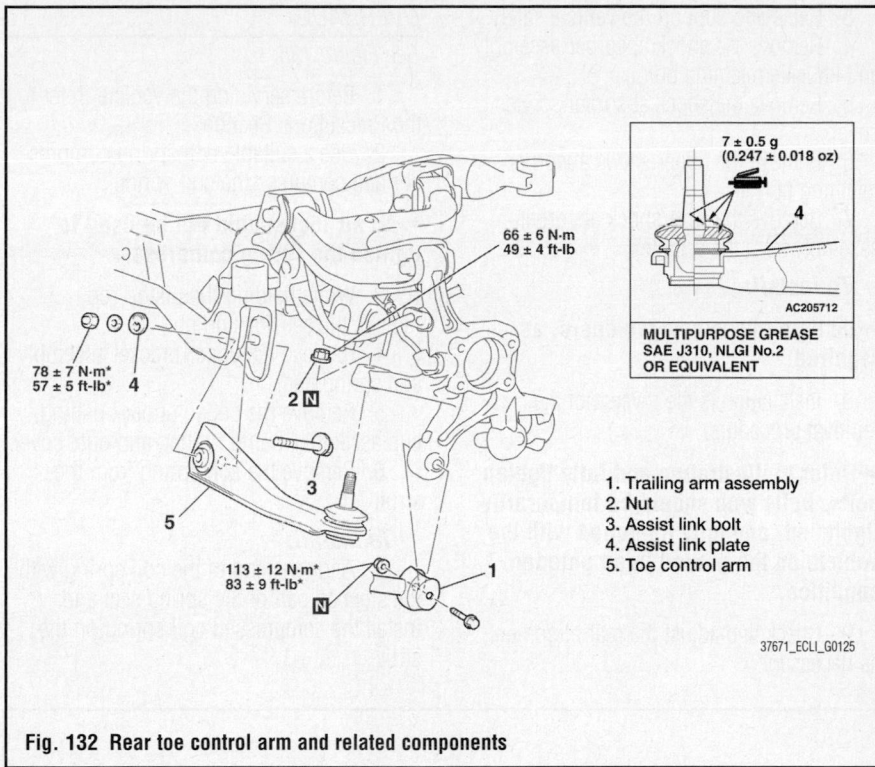

1. Trailing arm assembly
2. Nut
3. Assist link bolt
4. Assist link plate
5. Toe control arm

37671_ECLI_G0125

Fig. 132 Rear toe control arm and related components

tightened, and fully tightened with the vehicle on the ground in an unladen condition.

11. Check and adjust the rear alignment, as necessary.

Upper Control Arm

See Figure 133.

1. Before servicing the vehicle, refer to the Precautions Section.

➡ **If working near and/or around the SRS system and components, be sure to disable the SRS system. Tape the negative battery cable with insulating tape. Always disconnect the negative battery cable first.**

✳✳ CAUTION

Wait for 1 minute after disconnecting the negative battery cable before working inside the vehicle. The air bag system is set to deploy for a short period of time after the battery is disconnected.

2. Raise and support the vehicle safely.

3. Remove the tire and wheel assembly.

4. If equipped with ABS, remove the wheel speed sensor retaining bolts.

5. Position a floor jack to properly sup-

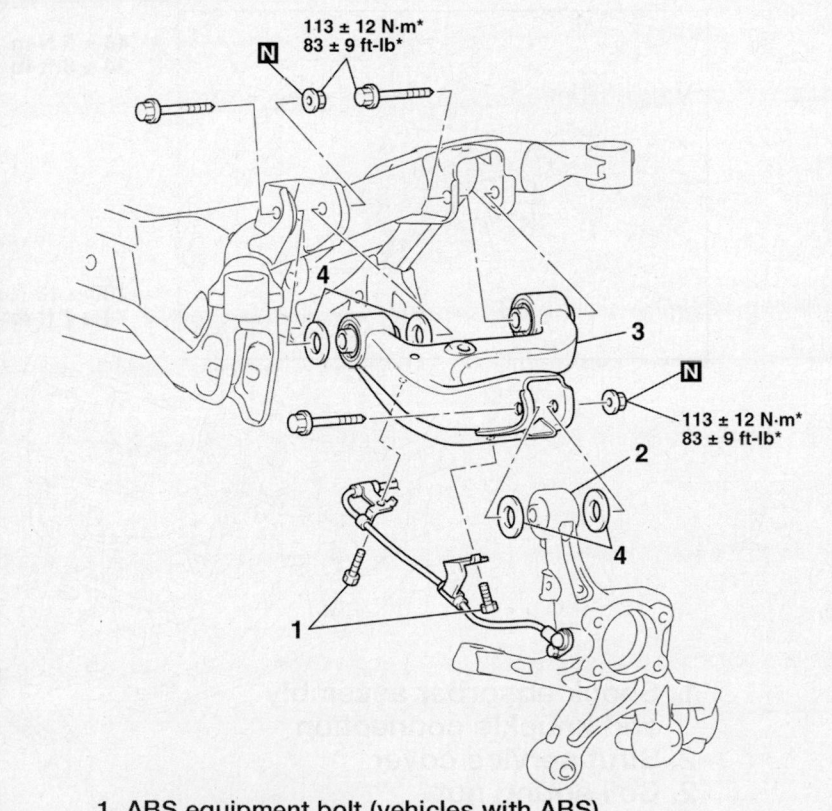

1. ABS equipment bolt (vehicles with ABS)
2. Upper arm assembly and knuckle connection
3. Upper arm assembly
4. Upper arm stopper

22140_MITS_G0201

Fig. 133 Rear upper control arm and related components

STRUTS

REMOVAL & INSTALLATION

See Figure 134.

1. Before servicing the vehicle, refer to the Precautions Section.

➡**If working near and/or around the SRS system and components, be sure to disable the SRS system. Tape the negative battery cable with insulating tape. Always disconnect the negative battery cable first.**

✳✳ CAUTION

Wait for 1 minute after disconnecting the negative battery cable before working inside the vehicle. The air bag system is set to deploy for a short period of time after the battery is disconnected.

2. Remove the rear sub woofer speaker.

3. Raise and support the vehicle safely.
4. Remove the shock absorber assembly and knuckle retaining bolt.
5. Remove the shock absorber service cover.
6. Remove the upper shock absorber retaining nuts.
7. Remove the strut/shock absorber assembly from the vehicle.

To install:

➡**Be sure to use new fasteners, as required.**

8. Installation is the reverse of the removal procedure.

➡**Refer to illustration and fully tighten bolts. Bolts with should be temporarily tightened, and fully tightened with the vehicle on the ground in an unladen condition.**

9. Check and adjust the rear alignment, as necessary.

OVERHAUL

See Figure 135.

1. Before servicing the vehicle, refer to the Precautions Section.
2. Use a suitable coil spring compressor and compress the coil spring.

➡**An air tool should not be used to tighten the spring compressor.**

3. While holding the piston rod, remove the self-locking nut.
4. Remove the upper bracket assembly and spring pad.
5. Remove the collar, upper bushing, cup assembly, bump rubber, and dust cover.
6. Remove the coil spring from the strut.

To install:
7. Align the end of the coil spring with the stepped part of the spring seat and install the compressed coil spring on the strut.

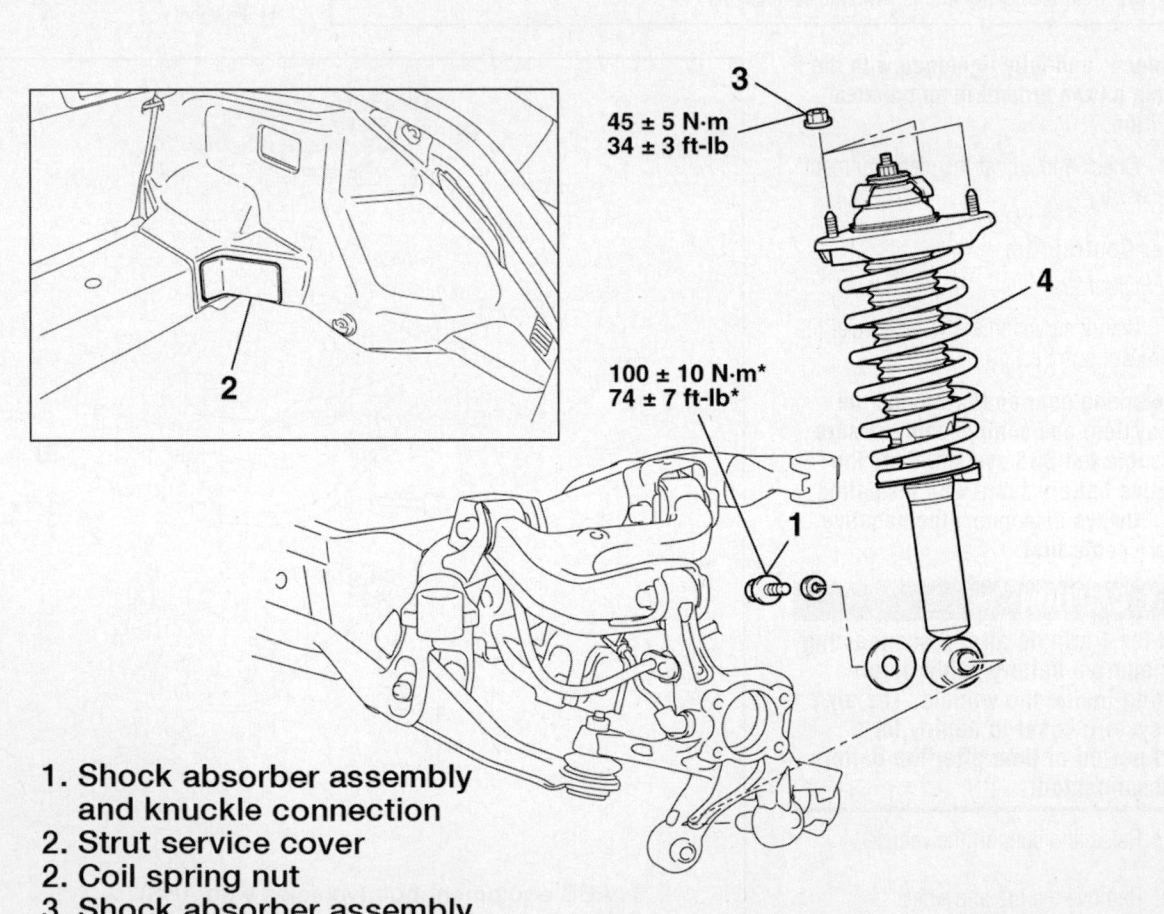

3
45 ± 5 N·m
34 ± 3 ft-lb

4

100 ± 10 N·m*
74 ± 7 ft-lb*

1

1. Shock absorber assembly and knuckle connection
2. Strut service cover
2. Coil spring nut
3. Shock absorber assembly

22140_MITS_G0194

Fig. 134 Rear strut/shock and related components

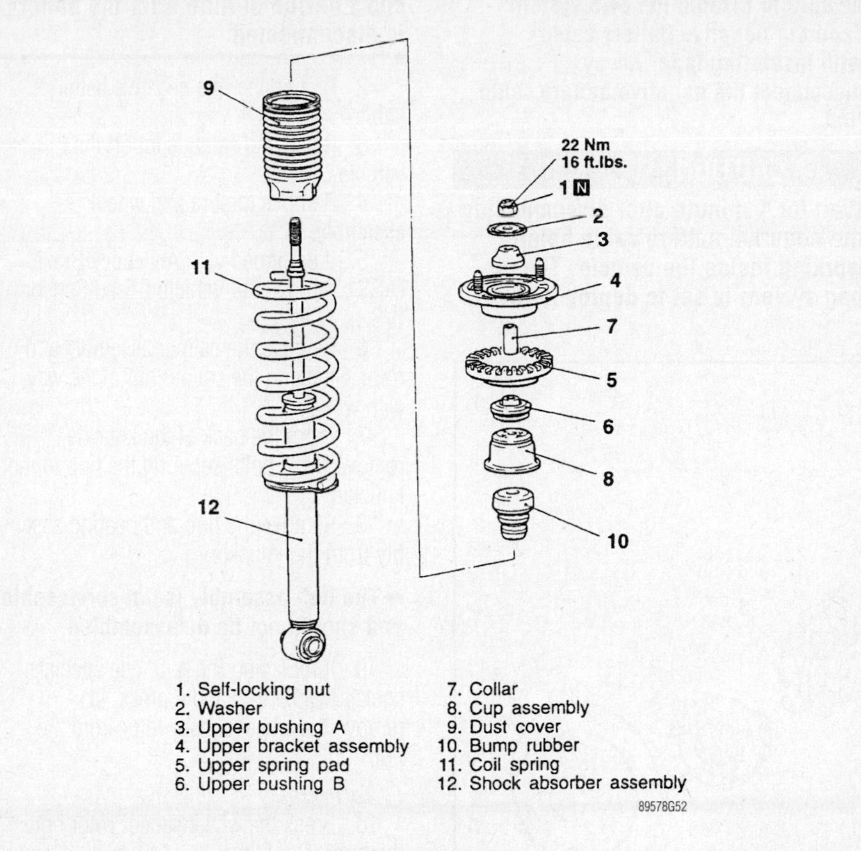

Fig. 135 Exploded view of the strut/shock assembly

22 Nm
16 ft.lbs.
1 N

1. Self-locking nut
2. Washer
3. Upper bushing A
4. Upper bracket assembly
5. Upper spring pad
6. Upper bushing B
7. Collar
8. Cup assembly
9. Dust cover
10. Bump rubber
11. Coil spring
12. Shock absorber assembly

89578G52

8. Install the dust cover, bump rubber, cup assembly, upper bushing, collar, upper spring pad, and bracket assembly on the strut.

9. Install the upper bushing and washer on the piston rod.

10. Install a new self-locking nut on the piston rod. Temporarily tighten the nut.

11. Carefully remove the spring compressor from the spring. Tighten the self-locking nut to 16 ft. lbs. (25 Nm).

STABILIZER BAR

REMOVAL & INSTALLATION

See Figure 136.

1. Before servicing the vehicle, refer to the Precautions Section.

➡**If working near and/or around the SRS system and components, be sure to disable the SRS system. Tape the negative battery cable with insulating tape. Always disconnect the negative battery cable first.**

✻✻ CAUTION

Wait for 1 minute after disconnecting the negative battery cable before working inside the vehicle. The air bag system is set to deploy for a short period of time after the battery is disconnected.

2. Raise and support the vehicle safely.

3. Remove the tire and wheel assemblies.

4. Remove the stabilizer bar link assembly.

5. Remove the stabilizer bar bracket.

6. Remove the bushings.

7. Remove the stabilizer bar.

To install:

➡**Be sure to use new fasteners, as required.**

8. Installation is the reverse of the removal procedure.

➡**Refer to illustration and fully tighten bolts. Bolts with should be temporarily tightened, and fully tightened with the vehicle on the ground in an unladen condition.**

9. Check and adjust the rear alignment, as necessary.

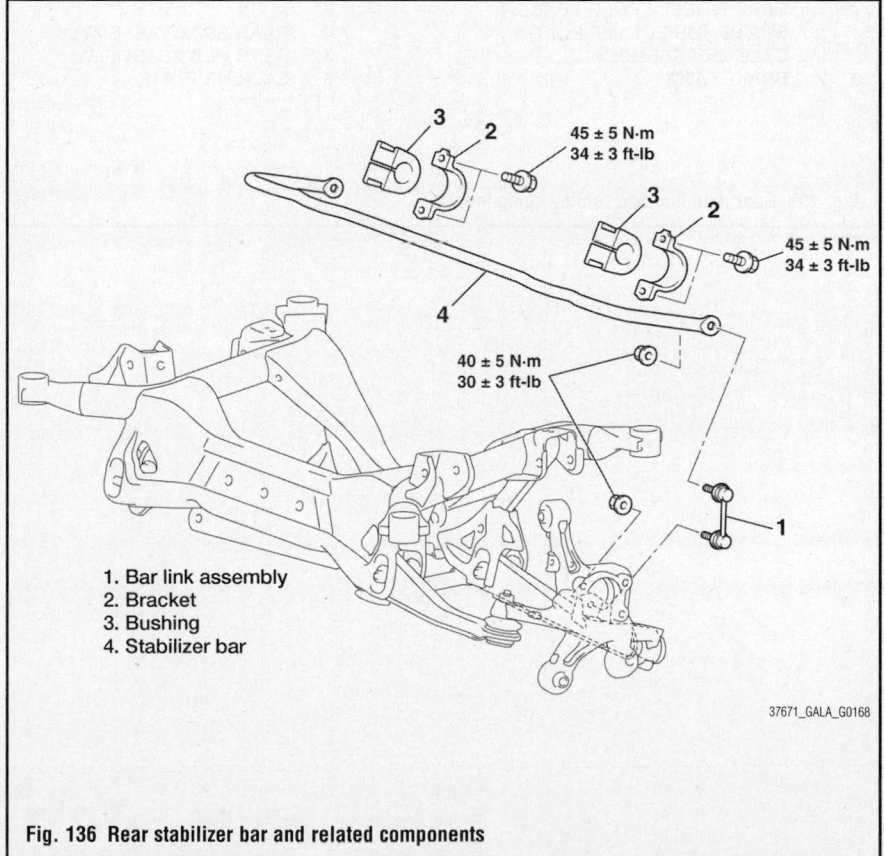

45 ± 5 N·m
34 ± 3 ft-lb

45 ± 5 N·m
34 ± 3 ft-lb

40 ± 5 N·m
30 ± 3 ft-lb

1. Bar link assembly
2. Bracket
3. Bushing
4. Stabilizer bar

37671_GALA_G0168

Fig. 136 Rear stabilizer bar and related components

WHEEL HUB & BEARING

REMOVAL & INSTALLATION
See Figure 137.

➡**The hub and bearing assembly is serviced as a unit.**

1. Before servicing the vehicle, refer to the Precautions Section.

➡**If working near and/or around the SRS system and components,** be sure to disable the SRS system. Tape the negative battery cable with insulating tape. Always disconnect the negative battery cable first.

✸✸ CAUTION
Wait for 1 minute after disconnecting the negative battery cable before working inside the vehicle. The air bag system is set to deploy for a short period of time after the battery is disconnected.

2. Disconnect the negative battery cable.

3. Raise and safely support the vehicle.

4. Remove the tire and wheel assembly.

5. If equipped with Anti-Lock Brake (ABS), remove the Vehicle Speed Sensor (VSS).

6. Remove the caliper assembly and rotor. Suspend the caliper out of the way with wire.

7. From the back of the knuckle, remove the 4 bolts securing the hub to the knuckle.

8. Remove the hub and bearing assembly from the knuckle.

➡**The hub assembly is not serviceable and should not be disassembled.**

9. If replacing the hub, use special socket MB991248 and a press, to remove the wheel sensor rotor from the hub.

To install:

10. Press the wheel sensor rotor onto the hub.

11. Install or connect the following:
- Hub to the knuckle and tighten the mounting bolts to 54–65 ft. lbs. (74–88 Nm)
- Brake rotor on the hub
- VSS, if equipped with ABS
- Wheel assembly and lower the vehicle

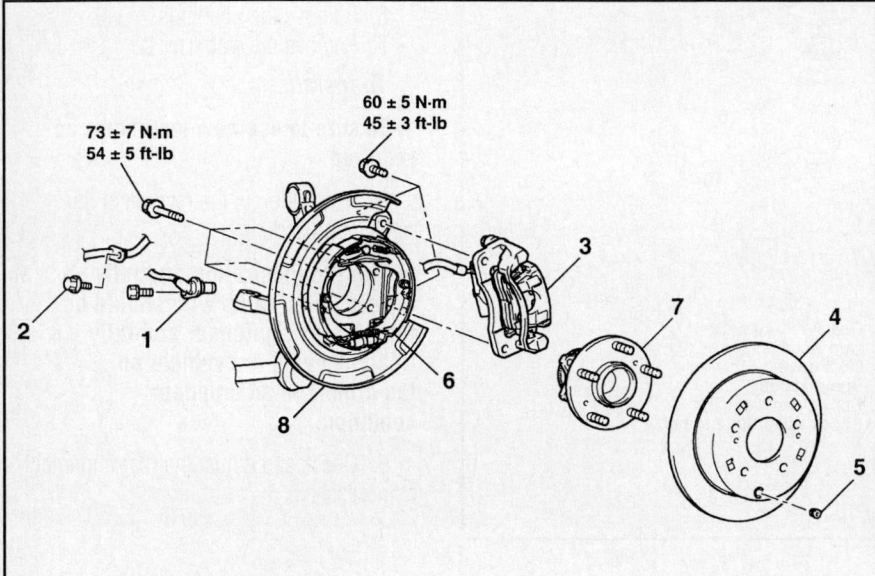

73 ± 7 N·m
54 ± 5 ft-lb

60 ± 5 N·m
45 ± 3 ft-lb

1.	REAR WHEEL SPEED SENSOR	5.	PLUG
2.	BRAKE HOSE CLAMP BOLT	6.	REAR BRAKE ASSEMBLY
3.	CALIPER ASSEMBLY	7.	REAR HUB ASSEMBLY
4.	BRAKE DISC	8.	BACKING PLATE

09482_GALA_G0157

Fig. 137 Rear axle hub and related components

MITSUBISHI

Endeavor

12

SPECIFICATIONS AND MAINTENANCE CHARTS

ENGINE AND VEHICLE IDENTIFICATION CHART

Engine							Model Year	
Code	Liters (cc)	Cu. In.	Cyl.	Fuel Sys.	Engine Type	Eng. Mfg.	Code	Year
6G75/S	3.8 (3828)	233.6	6	MFI	SOHC	Mitsubishi	9	2009
							A	2010

MFI: Multi-port Fuel Injection

37671_ENDE_C0001

GENERAL ENGINE SPECIFICATIONS

Year	Engine Displacement Liters	Engine ID/VIN	Net Horsepower @ rpm	Net Torque @ rpm (ft. lbs.)	Bore x Stroke (in.)	Com-pression Ratio	Oil Pressure @ rpm
2009	3.8	6G75/S	225@5000	255@3750	3.74x3.54	10.0:1	①
2010	3.8	6G75/S	225@5000	255@3750	3.74x3.54	10.0:1	①

① 4.2 or more psi @ idle
  43-100 @ 3500 rpm

37671_ENDE_C0002

ENGINE TUNE-UP SPECIFICATIONS

Year	Engine Displacement Liters	Engine ID/VIN	Spark Plugs Gap (in.)	Ignition Timing (deg.) MT	Ignition Timing (deg.) AT	Fuel Pump (psi)	Idle Speed (rpm) MT	Idle Speed (rpm) AT	Valve Clearance In.	Valve Clearance Ex.
2009	3.8	6G75/S	0.028-0.031	—	①	47	—	②	HYD	HYD
2010	3.8	6G75/S	0.028-0.031	—	①	47	—	②	HYD	HYD

HYD: Hydraulic
① Coil on plug 2-8 degrees BTDC
  10 degrees BTDC actual
② 580-780 rpm's

37671_ENDE_C0003

CAPACITIES

Year	Model	Engine Displacement Liters	Engine ID/VIN	Engine Oil with Filter (qts.)	Transmission (pts.) 5-Spd	Auto.	Transfer Case (pts.)	Drive Axle Front (pts.)	Rear (pts.)	Fuel Tank (gal.)	Cooling System (qts.)
2009	Endeavor	3.8	6G75/S	4.5	—	①	1.12	—	2.4	21.4	②
2010	Endeavor	3.8	6G75/S	4.5	—	①	1.12	—	2.4	21.4	②

① FWD: 17.8 pts.

 AWD: 18.6 pts.

② FWD & AWD without tow kit: 9.5 qts.

 AWD with tow kit: 10.1 qts.

37671_ENDE_C0004

FLUID SPECIFICATIONS

Year	Model	Engine Displ. Liters (VIN)	Engine Oil	Auto. Trans.	Drive Axle Front	Rear	Transfer Case	Power Steering Fluid	Brake Master Cylinder	Cooling System
2009	Endeavor	3.8 (S)	5W-20	ATF SPIII	—	80W-90	SAE 90	Mitsubishi	DOT 3-4	Long Life
2010	Endeavor	3.8 (S)	5W-20	ATF SPIII	—	80W-90	SAE 90	Mitsubishi	DOT 3-4	Long Life

DOT: Department Of Transpotation

37671_ENDE_C0014

VALVE SPECIFICATIONS

Year	Engine Displacement Liters	Engine ID/VIN	Seat Angle (deg.)	Face Angle (deg.)	Spring Test Pressure (lbs. @ in.)	Spring Installed Height (in.)	Stem-to-Guide Clearance (in.) Intake	Exhaust	Stem Diameter (in.) Intake	Exhaust
2009	3.8	6G75/S	NS	43.5-44	60@1.74	1.740	0.0008-0.0019	0.0014-0.0024	0.240	0.240
2010	3.8	6G75/S	NS	43.5-44	60@1.74	1.740	0.0008-0.0019	0.0014-0.0024	0.240	0.240

NS - Not specified by manufacturer

37671_ENDE_C0005

CAMSHAFT SPECIFICATIONS CHART

All measurements are given in inches.

Year	Engine Displacement Liters	Engine VIN	Journal Dia.	Brg. Oil Clearance	Shaft End-play	Runout	Lobe Height Intake	Lobe Height Exhaust
2009	3.8	6G75/S	1.8000	NS	NS	NS	①	②
2010	3.8	6G75/S	1.8000	NS	NS	NS	①	②

NS - Not specified by manufacturer

① Standard value: 1.472
   Minimum value: 1.452

② Standard value: 1.462
   Minimum value: 1.443

37671_ENDE_C0006

CRANKSHAFT AND CONNECTING ROD SPECIFICATIONS

All measurements are given in inches.

Year	Engine Displacement Liters	Engine ID/VIN	Crankshaft Main Brg. Journal Dia.	Crankshaft Main Brg. Oil Clearance	Crankshaft Shaft End-play	Crankshaft Thrust on No.	Connecting Rod Journal Diameter	Connecting Rod Oil Clearance	Connecting Rod Side Clearance
2009	3.8	6G75/S	2.5200	①	0.0020-0.0090	NS	2.1650	0.0008-0.0015	0.0030-0.0090
2010	3.8	6G75/S	2.5200	①	0.0020-0.0090	NS	2.1650	0.0008-0.0015	0.0030-0.0090

NS - Not specified by manufacturer

① Nos. 1 & 4: 0.0007-0.0014 inch
   Nos. 2 & 3: 0.0009-0.0017 inch

37671_ENDE_C0007

PISTON AND RING SPECIFICATIONS

All measurements are given in inches.

Year	Engine Displacement Liters	Engine ID/VIN	Piston Clearance	Ring Gap Top Compression	Ring Gap Bottom Compression	Ring Gap Oil Control	Ring Side Clearance Top Compression	Ring Side Clearance Bottom Compression	Ring Side Clearance Oil Control
2009	3.8	6G75/S	0.0008-0.0015	0.0100-0.0160	0.0140-0.0200	0.0030-0.0140	0.0012-0.0027	0.0008-0.0023	Snug
2010	3.8	6G75/S	0.0008-0.0015	0.0100-0.0160	0.0140-0.0200	0.0030-0.0140	0.0012-0.0027	0.0008-0.0023	Snug

37671_ENDE_C0008

TORQUE SPECIFICATIONS
All readings in ft. lbs.

Year	Engine Displacement Liters	Engine ID/VIN	Cylinder Head Bolts	Main Bearing Bolts	Rod Bearing Bolts	Crankshaft Damper Bolts	Flywheel Bolts	Manifold Intake	Manifold Exhaust	Spark Plugs	Oil Pan Drain Plug
2009	3.8	6G75/S	①	②	③	137	54	④	⑤	⑥	⑦
2010	3.8	6G75/S	①	②	③	137	54	④	⑤	⑥	⑦

① Step 1: 76-84 ft. lbs.
　Step 2: Loosen completely
　Step 3: 76-84 ft. lbs.
② 51-57 ft. lbs.
③ 20 ft. lbs., plus an additional 90-94 degrees
④ Step 1: 45-71 inch lbs.
　Step 2: 15-17 ft. lbs.

⑤ Exhaust manifold nut: 29-37 ft. lbs.
　Exhaust manifold stay bolt (M8): 12-16 ft. lbs.
　Exhaust manifold stay bolt (M10): 28-38 ft. lbs.
　Exhaust manifold stay bolt (M12): 49-63 ft. lbs.
⑥ 14-22 ft. lbs.
⑦ 25-33 ft. lbs.

37671_ENDE_C0009

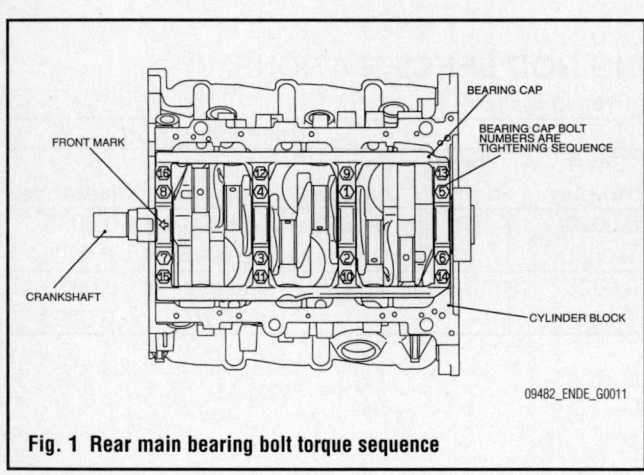

09482_ENDE_G0011

Fig. 1 Rear main bearing bolt torque sequence

WHEEL ALIGNMENT

Year	Model		Caster Range (+/-Deg.)	Caster Preferred Setting (Deg.)	Camber Range (+/-Deg.)	Camber Preferred Setting (Deg.)	Toe-in (in.)
2009	Endeavor	F	0.30	+3.00	0.30	0	0.0+/-0.12
		R	—	—	0.30	-0.50	0.12+/-0.12
2010	Endeavor	F	0.30	+3.00	0.30	0	0.0+/-0.12
		R	—	—	0.30	-0.50	0.12+/-0.12

37671_ENDE_C0010

TIRE, WHEEL AND BALL JOINT SPECIFICATIONS

Year	Model	OEM Tires		Tire Pressures (psi)		Wheel Size	Ball Joint Inspection	Lug Nut (ft. lbs.)
		Standard	Optional	Front	Rear			
2009	Endeavor	P235/65R17	P255/55R18	29	29	17 7-JJ or 18 8 JJ	U: NA L: 13-43 in. ①	②
2010	Endeavor	P235/65R17	P255/55R18	29	29	17 7-JJ or 18 8 JJ	U: NA L: 13-43 in. ①	②

OEM: Original Equipment Manufacturer

PSI: Pounds Per Square Inch

① Torque required in inch lbs. to rotate ball joint when removed from the knuckle

② 66-80 ft. lbs.

37671_ENDE_C0011

BRAKE SPECIFICATIONS

All measurements in inches unless noted

Year	Model		Brake Disc			Brake Drum Diameter			Minimum Lining Thickness		Brake Caliper	
			Original Thickness	Minimum Thickness	Maximum Runout	Original Inside Diameter	Max. Wear Limit	Maximum Machine Diameter	Front	Rear	Bracket Bolts (ft. lbs.)	Mounting Bolts (ft. lbs.)
2009	Endeavor	F	1.020	0.960	0.0012	—	—	—	0.080	—	74	34
		R	0.390	0.330	0.0031	—	—	—	—	0.080	45	32
2010	Endeavor	F	1.020	0.960	0.0012	—	—	—	0.080	—	74	34
		R	0.390	0.330	0.0031	—	—	—	—	0.080	45	32

37671_ENDE_C0012

SCHEDULED MAINTENANCE INTERVALS
Mitsubishi Endeavor

TO BE SERVICED	TYPE OF SERVICE	VEHICLE MILEAGE INTERVAL (x1000)												
		7.5	15	22.5	30	37.5	45	52.5	60	67.5	75	82.5	90	97.5
Air cleaner filter	R				✓				✓				✓	
Automatic transaxle & transfer case oil (AWD)	S/I				✓				✓				✓	
Automatic transaxle & transfer case oil (AWD)	R								✓					
Ball joints & steering linkage seals	S/I				✓				✓				✓	
Brake hoses	S/I		✓		✓		✓		✓		✓		✓	
Disc brake pads & rotors	S/I		✓		✓		✓		✓		✓		✓	
Drive belt(s)	S/I				✓				✓				✓	
Drive shaft boots	S/I		✓		✓		✓		✓		✓		✓	
Engine coolant	R	Initially at 60,000, then every 30,000												
Engine coolant hoses	S/I				✓				✓				✓	
Engine oil & filter	R	✓	✓	✓	✓	✓	✓	✓	✓	✓	✓	✓	✓	✓
EVAP canister	S/I	Every 100,000												
EVAP system (except EVAP canister)	S/I								✓					
Exhaust system	S/I				✓				✓				✓	
Front & rear axle	S/I				✓				✓				✓	
Fuel hoses	S/I				✓				✓				✓	
Fuel system	S/I								✓					
PCV system	S/I	✓	✓	✓	✓	✓	✓	✓	✓	✓	✓	✓	✓	✓
Rear differential	S/I				✓				✓				✓	
Spark plugs	R	Every 105,000												
Suspension system	S/I				✓				✓				✓	
Timing belt	R	Every 105,000												
Tires (rotate)	R	✓	✓	✓	✓	✓	✓	✓	✓	✓	✓	✓	✓	✓
Transfer case oil	R								✓					
Transfer case oil	S/I				✓				✓				✓	

R: Replace   S/I: Service or Inspect

FREQUENT OPERATION MAINTENANCE (SEVERE SERVICE)

If a vehicle is operated under any of the following conditions it is considered severe service:

- Extremely dusty areas.

- 50% or more of the vehicle operation is in 32°C (90°F) or higher temperatures, or constant operation in temperatures below 0°C (32°F).

- Prolonged idling (vehicle operation in stop and go traffic).

- Frequent short running periods (engine does not warm to normal operating temperatures).

- Police, taxi, delivery usage or trailer towing usage.

Air cleaner filter: service or inspect every 15,000 miles.

Front disc brake pads: service or inspect every 7500 miles.

Oil & oil filter: replace every 3750 miles.

PCV system: service or inspect every 60,000 miles.

Suspension system: service or inspect every 7500 miles.

Transmission fluid: check every 15,000 miles. Replace every 30,000 miles.

37671_ENDE_C0013

BRAKES — INFORMATION AND PRECAUTIONS

ANTI-LOCK SYSTEMS

- Certain components within the ABS system are not intended to be serviced or repaired individually.
- Do not use rubber hoses or other parts not specifically specified for and ABS system. When using repair kits, replace all parts included in the kit. Partial or incorrect repair may lead to functional problems and require the replacement of components.
- Lubricate rubber parts with clean, fresh brake fluid to ease assembly. Do not use shop air to clean parts; damage to rubber components may result.
- Use only DOT 3 brake fluid from an unopened container.
- If any hydraulic component or line is removed or replaced, it may be necessary to bleed the entire system.
- A clean repair area is essential. Always clean the reservoir and cap thoroughly before removing the cap. The slightest amount of dirt in the fluid may plug an orifice and impair the system function. Perform repairs after components have been thoroughly cleaned; use only denatured alcohol to clean components. Do not allow ABS components to come into contact with any substance containing mineral oil; this includes used shop rags.
- The Anti-Lock control unit is a microprocessor similar to other computer units in the vehicle. Ensure that the ignition switch is **OFF** before removing or installing controller harnesses. Avoid static electricity discharge at or near the controller.
- If any arc welding is to be done on the vehicle, the control unit should be unplugged before welding operations begin.

DISC AND DRUM SYSTEMS

✳✳ CAUTION

Dust and dirt accumulating on brake parts during normal use may contain asbestos fibers from production or aftermarket brake linings. Breathing excessive concentrations of asbestos fibers can cause serious bodily harm. Exercise care when servicing brake parts. Do not sand or grind brake lining unless equipment used is designed to contain the dust residue. Do not clean brake parts with compressed air or by dry brushing. Cleaning should be done by dampening the brake components with a fine mist of water, then wiping the brake components clean with a dampened cloth. Dispose of cloth and all residue containing asbestos fibers in an impermeable container with the appropriate label. Follow practices prescribed by the Occupational Safety and Health Administration (OSHA) and the Environmental Protection Agency (EPA) for the handling, processing, and disposing of dust or debris that may contain asbestos fibers.

BRAKES — BLEEDING THE BRAKE SYSTEM

BLEEDING PROCEDURE

BLEEDING PROCEDURE

Brake Lines

See Figure 2.

➡**Use only DOT 3 or DOT 4 brake fluid. Never mix the specified brake fluid with other fluid as it will influence the braking performance significantly.**

1. Make sure the master cylinder reserve tank is filled the fluid.
2. Start the engine.

➡**Bleed the air in the sequence shown in the figure.**

3. Raise and safely support the vehicle.
4. Connect a piece of vinyl tubing to the brake caliper.
5. Have an assistant depress the brake pedal several times, then loosen the bleeder plug with the pedal held down.
6. When fluid stops coming out, tighten the bleeder plug, then release the brake pedal.
7. Repeat steps 2. and 3. until all the air in the fluid has been bled out.
8. Tighten the brake bleeder plug to 74 inch lbs. (8.3 Nm).
9. Repeat the above steps to bleed the air out of the brake line for each wheel.

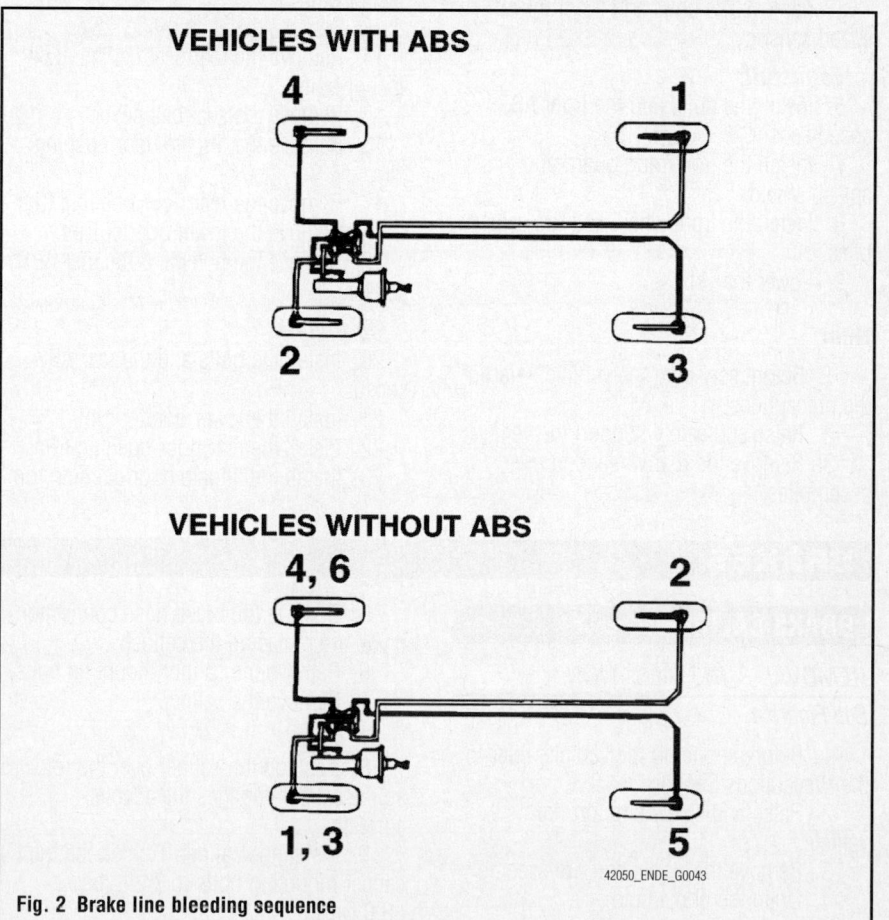

VEHICLES WITH ABS

VEHICLES WITHOUT ABS

42050_ENDE_G0043

Fig. 2 Brake line bleeding sequence

Master Cylinder

See Figure 3.

➡**Use only DOT 3 or DOT 4 brake fluid. Never mix the specified brake fluid with other fluid as it will influence the braking performance significantly.**

The master cylinder used has no check valve, so if bleeding is carried out by the following procedure, bleeding of air from the brake pipeline will become easier. (When brake fluid is not contained in the master cylinder).

1. Fill the reserve tank with brake fluid.

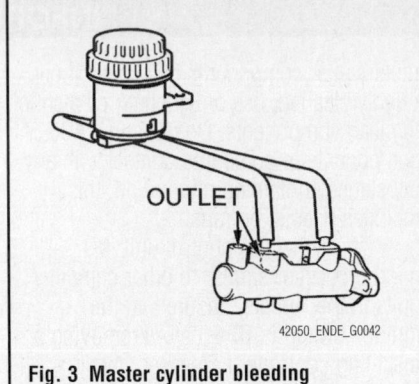

Fig. 3 Master cylinder bleeding

2. Keep the brake pedal depressed.
3. Have another person cover the master cylinder outlet with a finger.
4. With the outlet still closed, release the brake pedal.
5. Repeat steps 2–4 three or four times to fill the inside of the master cylinder with brake fluid.

BLEEDING THE ABS SYSTEM

There are no special procedures for bleeding the ABS system. Refer to the conventional bleeding procedures.

BRAKES

ANTI-LOCK BRAKE SYSTEM (ABS)

WHEEL SPEED SENSORS

REMOVAL & INSTALLATION

Front

1. Before servicing the vehicle, refer to the precautions.
2. Raise and safely support the vehicle.
3. Remove the front wheel and tire assemblies.
4. Remove the front mud guard and splash shield.
5. Remove the bolts and the front ABS speed sensor.

To install:

6. Install the bolts and the front ABS speed sensor.
7. Install the front mud guard and splash shield.
8. Install the front wheel and tire assemblies.
9. Lower the vehicle.

Rear

1. Before servicing the vehicle, refer to the precautions.
2. Raise and safely support the vehicle.
3. Remove the rear wheel and tire assemblies.

4. Remove the luggage floor rear board.
5. Remove the tonneau cover.
6. Remove the luggage floor front board.
7. Remove the parcel strap hook.
8. Remove the luggage floor side board.
9. Remove the rear end trim.
10. Remove the rear seat assembly.
11. Remove the scuff plate cover and rear scuff plate.
12. Remove the rear seat belt lower anchor bolt.
13. Remove the sash guide cover.
14. Remove the rear seat belt shoulder anchor bolt.
15. Remove the seat belt cover.
16. Remove the liftgate door opening trim.
17. Remove the rear door opening trim.
18. Remove the lower quarter trim.
19. Remove the bolts and the rear ABS sensor.

To install:

20. Install the bolts and the rear ABS sensor
21. Install the lower quarter trim.
22. Install the rear door opening trim.
23. Install the liftgate door opening trim.

24. Install the seat belt cover.
25. Install the rear seat belt shoulder anchor bolt.
26. Install the sash guide cover.
27. Install the rear seat belt lower anchor bolt.
28. Install the scuff plate cover and rear scuff plate.
29. Install the rear seat assembly.
30. Install the rear end trim.
31. Install the luggage floor side board.
32. Install the parcel strap hook.
33. Install the luggage floor front board.
34. Install the tonneau cover.
35. Install the luggage floor rear board.
36. Install the rear wheel and tire assemblies.
37. Lower the vehicle.

WHEEL SPEED SENSOR RINGS (TOOTHED RINGS)

REMOVAL & INSTALLATION

The front ABS rotor and the rear ABS rotor on AWD models are integrated with the BJ assembly of the Halfshaft and cannot be disassembled. The rear ABS sensor rotor on FWD models is integrated with the rear hub assembly and cannot be disassembled.

BRAKES

FRONT DISC BRAKES

BRAKE CALIPER

REMOVAL & INSTALLATION

See Figure 4.

1. Before servicing the vehicle, refer to the Precautions Section.
2. Raise and safely support the vehicle.
3. Remove the wheel and tire assembly.
4. Drain the brake fluid.

5. Remove the brake hose connection, if you are replacing the caliper.
6. Remove the caliper mounting bolts.
7. Remove the caliper.

To install:

8. Position the caliper over the rotor so the caliper engages the adapter correctly.
9. Install the caliper. Tighten the front caliper mounting bolts to 74 ft. lbs. (100 Nm).

10. Install the brake hose, if removed. Tighten banjo bolt to 22 ft. lbs. (30 Nm).
11. Install the wheel and tire assembly.
12. Fill and bleed the brake system.
13. Lower the vehicle.

DISC BRAKE PADS

REMOVAL & INSTALLATION

See Figure 5.

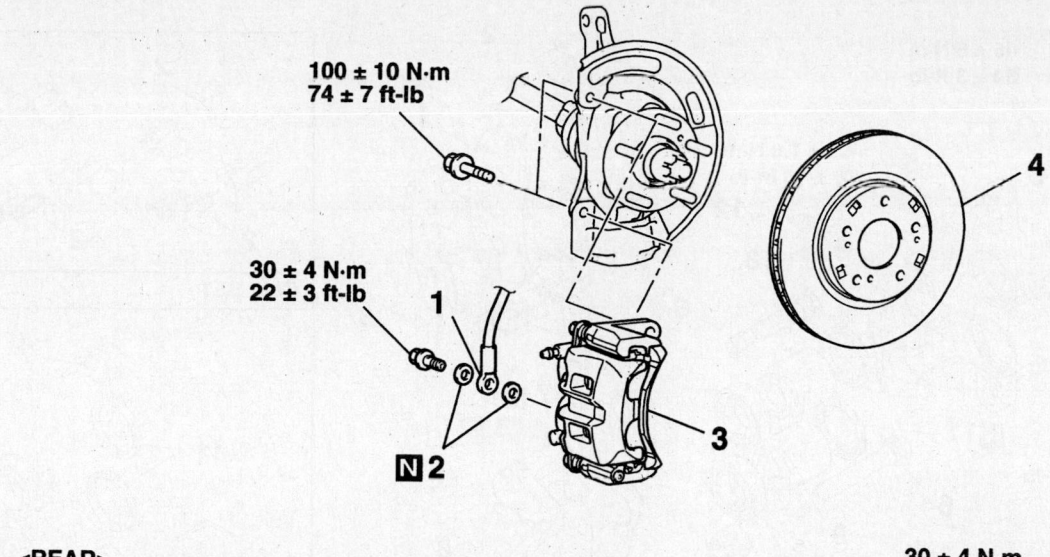

100 ± 10 N·m
74 ± 7 ft-lb

30 ± 4 N·m
22 ± 3 ft-lb

1

3

N 2

4

<REAR>

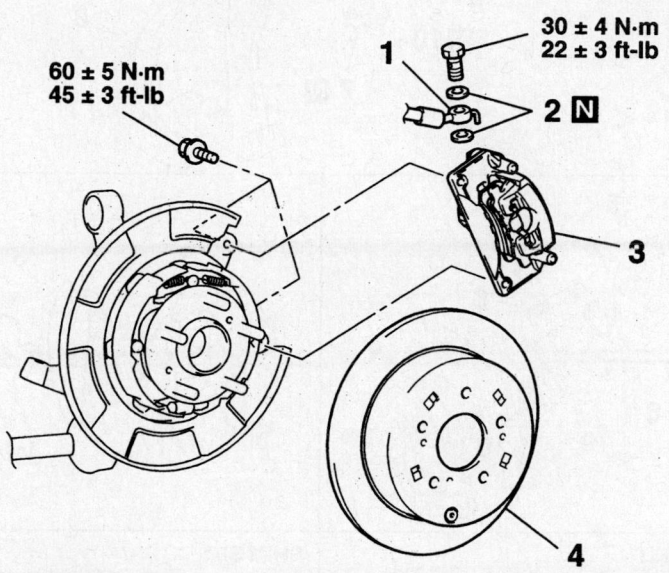

30 ± 4 N·m
22 ± 3 ft-lb

60 ± 5 N·m
45 ± 3 ft-lb

1

2 N

3

4

1. BRAKE HOSE CONNECTION
2. GASKET
3. BRAKE CALIPER ASSEMBLY
4. BRAKE DISC

67170-ENDE-G44

Fig. 4 Front and rear calipers and related components

1. Before servicing the vehicle, refer to the Precautions Section.

2. Remove ½ of the brake fluid from the master cylinder.

3. Raise and safely support the vehicle.

4. Remove the wheel and tire assembly.

5. Remove the lower caliper guide pin bolt.

6. Remove the caliper from the caliper support but do NOT disconnect the fluid line. Suspend the caliper from the suspension with a piece of wire.

7. Remove the disc brake pads, shims, and the clips from the caliper support.

To install:

8. Clean the exposed portion of the caliper piston, then press the piston back into the caliper bore using the old inner brake pad and a C-clamp.

9. Install the disc brake pads, shims, and the clips. make sure the shims and clips are properly positioned.

10. Install the caliper over the rotor so the caliper engages the adapter correctly.

11. Install the mounting pin(s).

12. Install the wheel and tire assembly and lower the vehicle.

13. Apply the brake pedal several times until a firm pedal is obtained. Check the fluid level in the master cylinder and add fluid, as necessary.

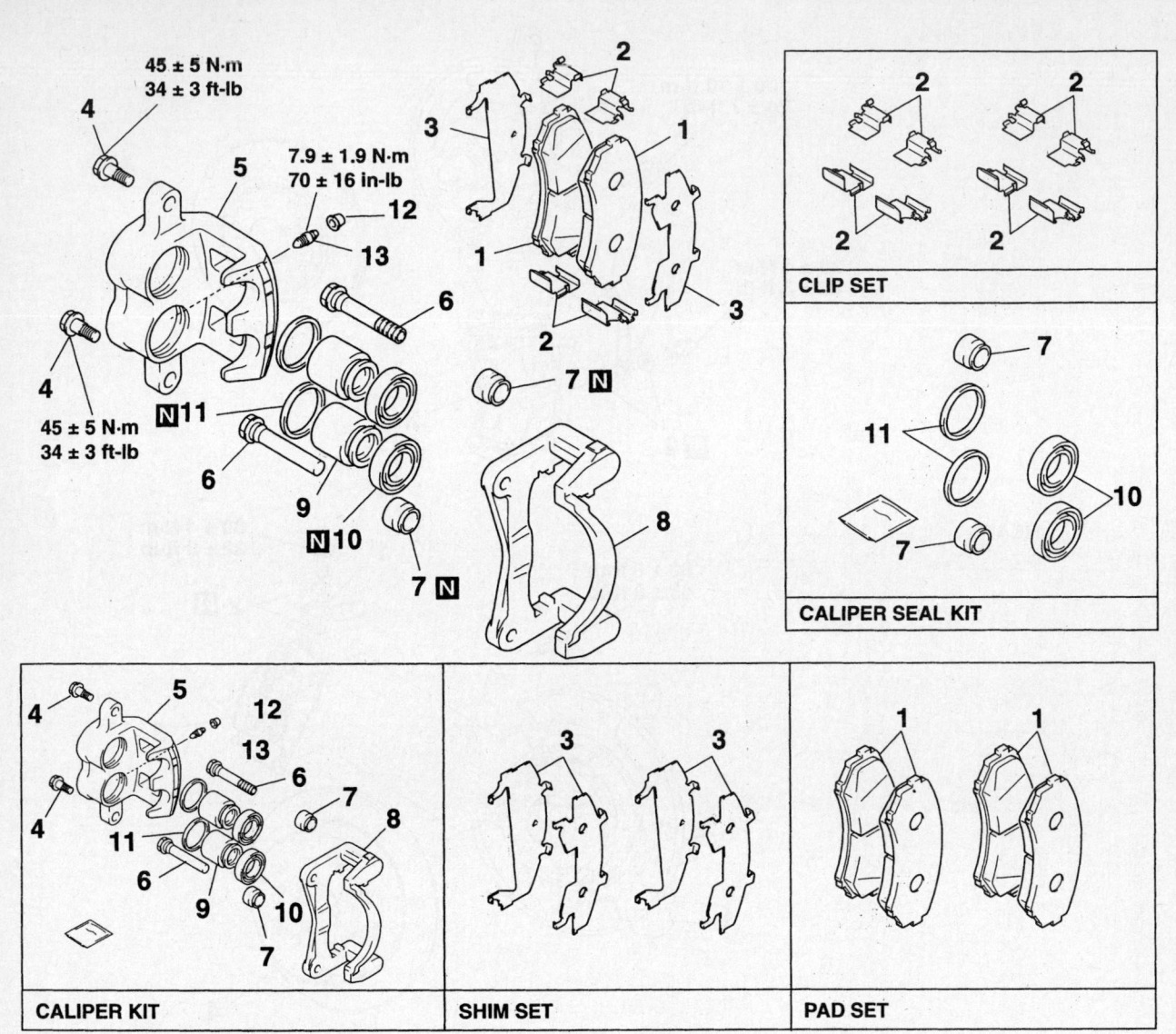

45 ± 5 N·m
34 ± 3 ft-lb

7.9 ± 1.9 N·m
70 ± 16 in-lb

45 ± 5 N·m
34 ± 3 ft-lb

CLIP SET

CALIPER SEAL KIT

CALIPER KIT

SHIM SET

PAD SET

1. PAD (AND WEAR INDICATOR) ASSEMBLY
2. CLIP
3. SHIM
4. FRONT BRAKE BOLT
5. CALIPER BODY
6. FRONT BRAKE PIN

7. BOOT
8. CALIPER SUPPORT
9. CALIPER PISTON
10. PISTON BOOT
11. PISTON SEAL
12. CALIPER BLEEDER CAP
13. CALIPER BLEEDER

67170-ENDE-G45

Fig. 5 Front brake pads and related components

BRAKES

REAR DISC BRAKES

BRAKE CALIPER

REMOVAL & INSTALLATION

See Figure 4.

1. Before servicing the vehicle, refer to the Precautions Section.

2. Raise and safely support the vehicle.

3. Remove the wheel and tire assembly.

4. Remove the brake hose connection, if you are replacing the caliper.

5. Remove the caliper mounting bolts.
6. Remove the caliper.

To install:

7. Position the caliper over the rotor so the caliper engages the adapter correctly

8. Install the caliper. tighten the rear caliper mounting bolts to 45 ft. lbs. (60 Nm).

9. Install the brake hose. Tighten to 22 ft. lbs. (30 Nm).

10. Install the wheel and tire assembly.

11. Fill and bleed the brake system.

DISC BRAKE PADS

REMOVAL & INSTALLATION

See Figure 6.

1. Before servicing the vehicle, refer to the Precautions Section.

2. Remove ½ of the brake fluid from the master cylinder.

3. Raise and safely support the vehicle.

4. Remove the wheel and tire assembly.

5. Remove the lower caliper guide pin bolt.

6. Remove the caliper from the caliper support.

7. Remove the disc brake pads, shims, and the clips from the caliper support.

To install:

8. Clean the exposed portion of the caliper piston, then press the piston back

into the caliper bore using the old inner brake pad and a C-clamp.

9. Install the disc brake pads, shims, and the clips. make sure the shims and clips are properly positioned.

10. Install the caliper over the rotor so the caliper engages the adapter correctly.

11. Install the mounting pin(s).

12. Install the wheel and tire assembly and lower the vehicle.

13. Apply the brake pedal several times until a firm pedal is obtained. Check the fluid level in the master cylinder and add fluid, as necessary.

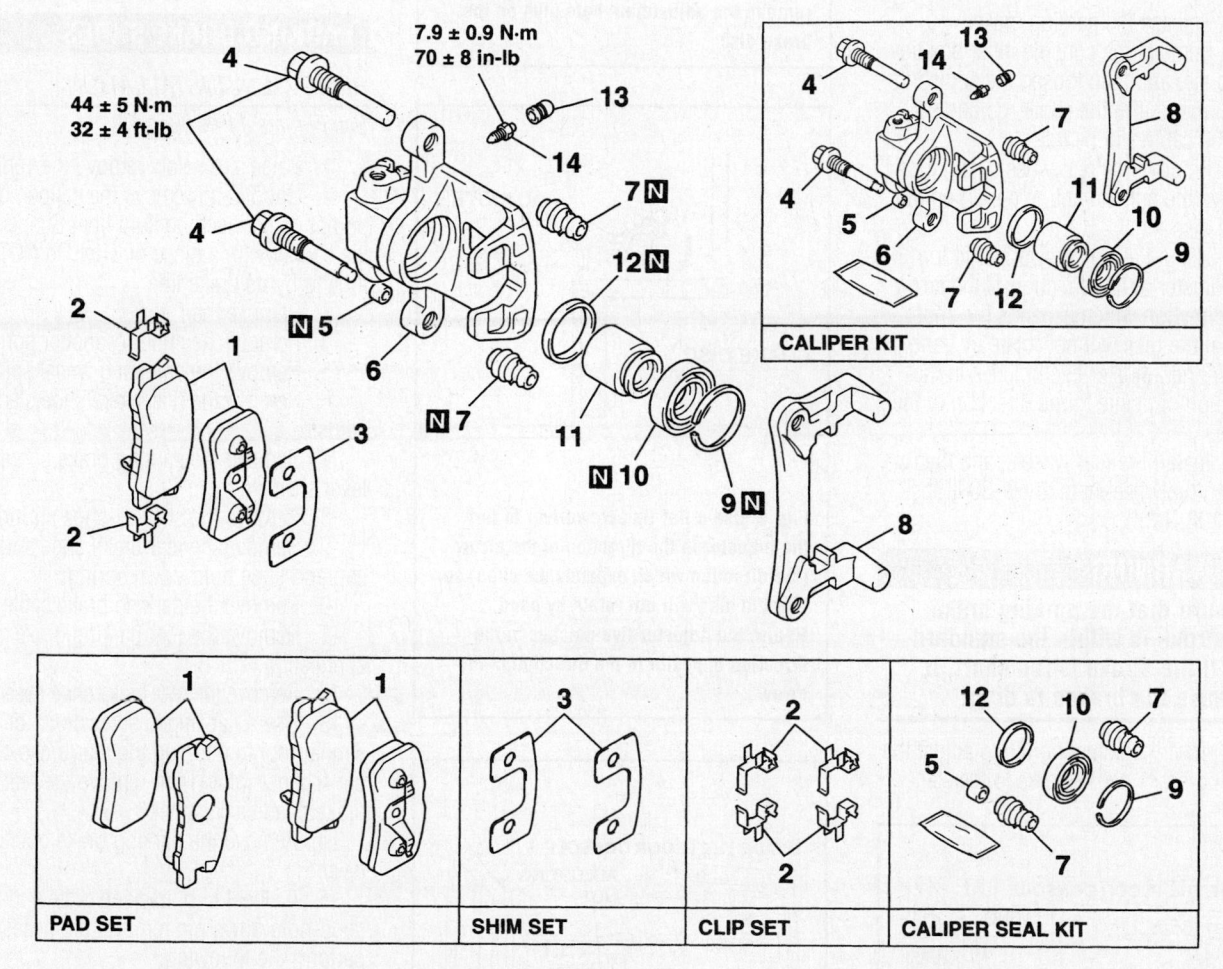

1. PAD (AND WEAR INDICATOR) ASSEMBLY
2. CLIP
3. SHIM
4. REAR BRAKE PIN
5. REAR BRAKE BUSHING
6. CALIPER BODY
7. PIN BOOT
8. CALIPER SUPPORT
9. BOOT RING
10. PISTON BOOT
11. CALIPER PISTON
12. PISTON SEAL
13. REAR BRAKE CAP
14. CALIPER BLEEDER

67170-ENDE-G46

Fig. 6 Rear brake pads and related components

PARKING BRAKE CABLES

ADJUSTMENT

Parking Brake Pedal Stroke Check And Adjustment

See Figures 7 through 10.

1. Depress the parking brake pedal with a force of 44–55 lbs. (196–245 N) and count the number of notches. The standard value is 1 notch

2. If the parking brake pedal stroke is not within the standard value, adjust as follows:

   a. Release the parking brake.

   b. Remove the console inner box tray and plate, and then loosen the adjusting nut to move it to the cable rod end so that the cable will be free.

   c. Remove the rear wheels, and then remove the adjustment hole plug on the brake disc.

   d. Use a flat-tip screwdriver to turn the adjuster in the direction of the arrow (the direction which expands the shoe) so that the disc will not rotate by hand. Return the adjuster five notches in the direction opposite to the direction of the arrow.

   e. Install the rear wheels, and then tighten the wheel nuts to 66–80 ft. lbs. (88–108 Nm).

✳✳ WARNING

Make sure that the parking brake pedal stroke is within the standard value. If the stroke is too short, it may cause the brakes to drag.

   f. Turn the adjusting nut to adjust the parking brake pedal stroke to the stan-

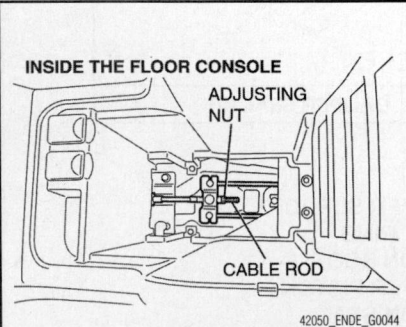

Fig. 7 Remove the console inner box tray and plate, and then loosen the adjusting nut to move it to the cable rod end so that the cable will be free

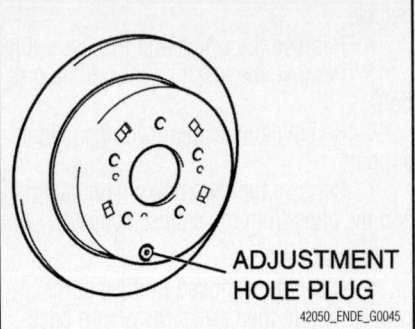

Fig. 8 Remove the rear wheels, and then remove the adjustment hole plug on the brake disc

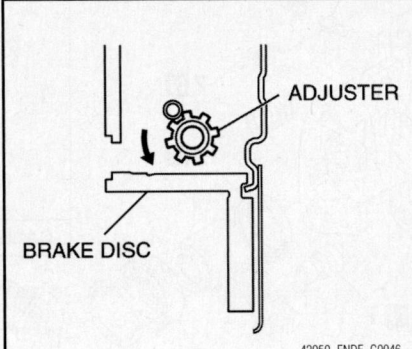

Fig. 9 Use a flat-tip screwdriver to turn the adjuster in the direction of the arrow (the direction which expands the shoe) so that the disc will not rotate by hand. Return the adjuster five notches in the direction opposite to the direction of the arrow

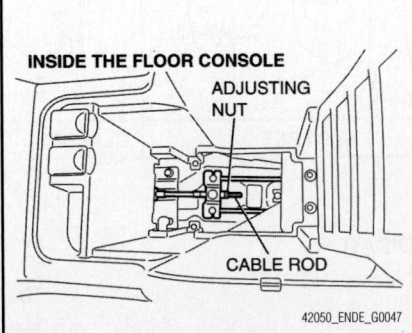

Fig. 10 Turn the adjusting nut to adjust the parking brake pedal stroke to the standard value. After adjustment, check that the adjust nut and the cable rod is not loose

dard value. After adjustment, check that the adjust nut and the cable rod is not loose.

   g. Release the parking brake and turn the rear wheels to check that the rear brakes are not dragging.

3. If either of the parking brake cables is replaced, adjust the parking brake pedal stroke as described previously, depress the parking brake pedal 10 times with approximately 135 lbs. (600 N) to eliminate the initial slack of the cable. Then adjust the parking brake pedal stroke as outlined above again.

PARKING BRAKE SHOES

REMOVAL & INSTALLATION

See Figures 11 through 15.

1. Raise and safely remove the vehicle.

2. Unbolt and remove the caliper, but do not disconnect the fluid line. Suspend the caliper with a piece of wire. Do NOT let it hang by its brake line.

3. Remove the rotor.

4. Remove the shoe-to-anchor spring.

5. Remove the adjusting wheel spring.

6. Remove the rear brake shoe slack adjuster.

7. Remove the parking brake operating lever strut.

8. Remove the strut-to-shoe spring.

9. Remove the rear brake shoe spring cup and shoe hold-down spring.

10. Remove the parking brake cable clip.

11. Remove the rear parking brake cable connection.

12. Remove the rear brake shoe assembly.

13. Use a flat-tipped screwdriver or a similar tool to open up the rear brake chamber retainer joint. Then remove the rear brake chamber retainer.

14. Remove the parking brake operating lever.

15. If wheel hub, backing plate and shoe-hold down pin removal are necessary, perform the following:

   a. Remove the driveshaft nut (AWD models) and rear ABS sensor.

   b. Remove the rear wheel hub assembly.

   c. Remove the backing plate.

   d. Remove the shoe hold-down pin.

16. Inspect the parking brake lining, as follows:

   a. Measure the thickness of the brake lining at several places. The standard value is 2.8 mm (0.11 inch) and the minimum limit is 1.0 mm (0.04 inch)

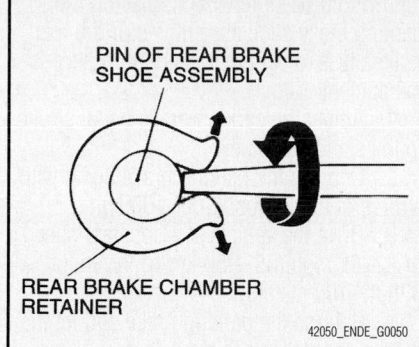

Fig. 11 Use a prytool to open up the rear brake chamber retainer joint in order to remove the chamber retainer

b. If the thickness of the brake lining is below the limit, replace the shoe assemblies on both sides of the vehicle. Never replace only one side.

To install:

17. Installation is the reverse of the removal procedure, noting the following:

a. Refer to the tightening specifications shown on the accompanying illustration.

b. Use pliers or a similar tool to close the rear brake chamber retainer end onto the pin.

c. Install the rear brake shoe slack adjuster as shown in the accompanying illustration.

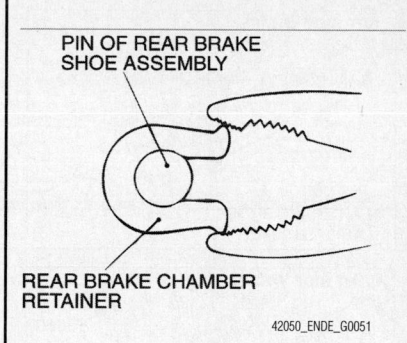

Fig. 13 Use pliers or a similar tool to close the rear brake chamber retainer end onto the pin

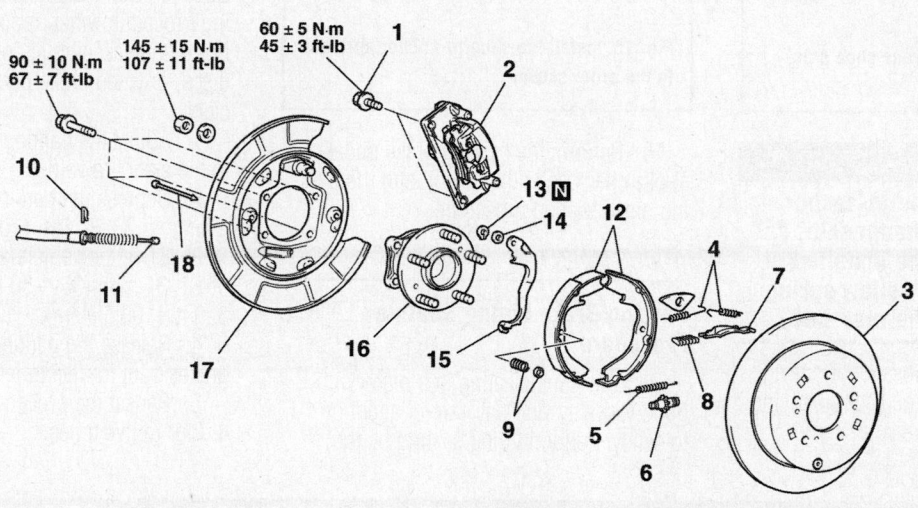

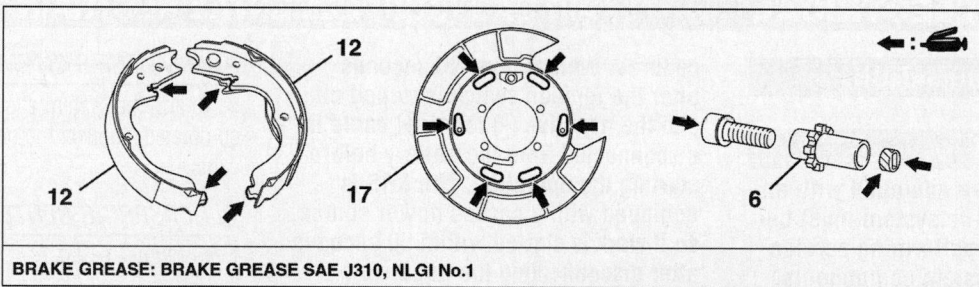

BRAKE GREASE: BRAKE GREASE SAE J310, NLGI No.1

1. REAR BRAKE BOLT
2. REAR BRAKE CALIPER ASSEMBLY
3. REAR BRAKE DISC
4. SHOE-TO-ANCHOR SPRING
5. ADJUSTING WHEEL SPRING
6. REAR BRAKE SHOE SLACK ADJUSTER
7. PARKING BRAKE OPERATING LEVER STRUT
8. STRUT-TO- SHOE SPRING
9. REAR BRAKE SHOE SPRING CUP AND SHOE HOLD-DOWN SPRING
10. PARKING BRAKE CABLE CLIP

11. REAR PARKING BRAKE CABLE CONNECTION
12. REAR BRAKE SHOE ASSEMBLY
13. REAR BRAKE CHAMBER RETAINER
14. REAR BRAKE WASHER
15. PARKING BRAKE OPERATING LEVER
• DRIVE SHAFT NUT <AWD>
• REAR ABS SENSOR <VEHICLES WITH ABS>
16. REAR WHEEL HUB ASSEMBLY
17. BACKING PLATE
18. SHOE HOLD-DOWN PIN

42050_ENDE_G0048

Fig. 12 Exploded view of the parking brake shoes (linings) and related components

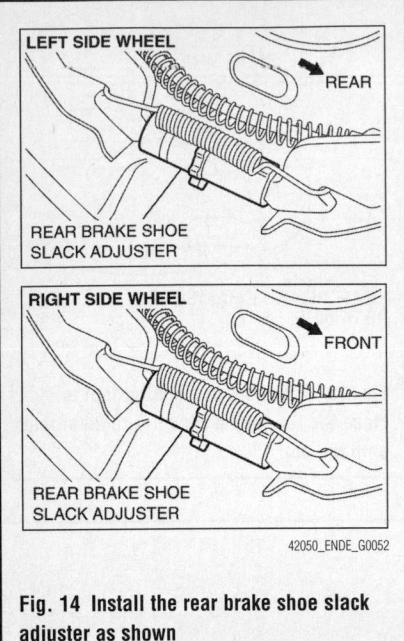

Fig. 14 Install the rear brake shoe slack adjuster as shown

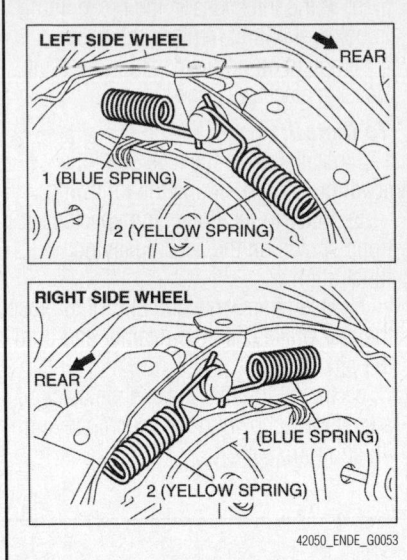

Fig. 15 Install the shoe-to-anchor springs in the order shown

❋ WARNING

The front and rear shoe-to-anchor springs are not interchangeable, so the blue spring must be installed at the front side and the yellow spring must be installed at the rear side.

    d. Install the shoe-to-anchor springs in the order shown in the accompanying illustration.

  18. Perform the parking brake pedal stroke check and adjustment, and the parking brake seating procedure.

ADJUSTMENT

Parking Brake Lining Seating Procedure

  Perform lining seating in a place with good visibility, and pay careful attention to safety. Perform lining seating by the following procedure when replacing the parking brake shoe assemblies or the rear brake discs, or when brake performance is insufficient.

  1. Adjust the parking brake pedal stroke to the standard value.

  2. Depress the parking brake pedal with a force of 23–34 lbs. (100–150 N).

  3. Drive the vehicle at a constant speed of 22–31 mph (35—50 km/h) for about 328 ft. (100 meters).

  4. Release the parking brake and let the brakes cool for five to ten minutes.

  5. Repeat the procedure four to five times.

Lining Break-In

  Perform lining seating in a place with good visibility, and pay careful attention to safety. Perform lining seating by the following procedure when replacing the parking brake shoe assemblies or the rear brake discs, or when brake performance is insufficient.

  1. Adjust the parking brake pedal stroke to the standard value.

  2. Depress the parking brake pedal with a force of 23–34 lbs. (100–150 N).

  3. Drive the vehicle at a constant speed of 22–31 mph (35—50 km/h) for about 328 ft. (100 meters).

  4. Release the parking brake and let the brakes cool for five to ten minutes.

  5. Repeat the procedure in steps 2. to 4. four to five times.

CHASSIS ELECTRICAL

GENERAL INFORMATION

❋ CAUTION

These vehicles are equipped with an air bag system. The system must be disarmed before performing service on, or around, system components, the steering column, instrument panel components, wiring and sensors. Failure to follow the safety precautions and the disarming procedure could result in accidental air bag deployment, possible injury and unnecessary system repairs.

SERVICE PRECAUTIONS

❋ CAUTION

Disconnect and isolate the battery negative cable before beginning any airbag system component diagnosis, testing, removal, or installation pro-

cedures. Wait at least 90 seconds after the ignition switch is turned off and the negative (-) terminal cable is disconnected from the battery before starting the operation. The SRS is equipped with a backup power source, so if work is started within 90 seconds after disconnecting the negative (-) terminal cable from the battery, the SRS may be deployed. Failure to disable the airbag system may result in accidental airbag deployment, personal injury, or death.

DISARMING THE SYSTEM

  To avoid personal injury when working on vehicles equipped with an air bag, the negative battery cable must be disconnected and at least 90 seconds must elapse before working on the system. Failure to do so may result in deployment of the air bag. You should also wrap or isolate the negative battery cable with electrical or other non-conductive tape.

AIR BAG (SUPPLEMENTAL RESTRAINT SYSTEM)

ARMING THE SYSTEM

  To arm the system after service is completed, connect the negative battery cable.

CLOCKSPRING CENTERING

See Figures 16 and 17.

❋❋ WARNING

Ensure that the clock spring's mating marks are properly aligned. If not, the steering wheel may not rotate completely during a turn, or the flat cable in the clock spring could be damaged, This would prevent normal SRS operation and possibly cause serious injury to the driver.

  1. Align the mating marks and install the clock spring.

    a. Turn the clock spring clockwise fully. Then turn it back approximately

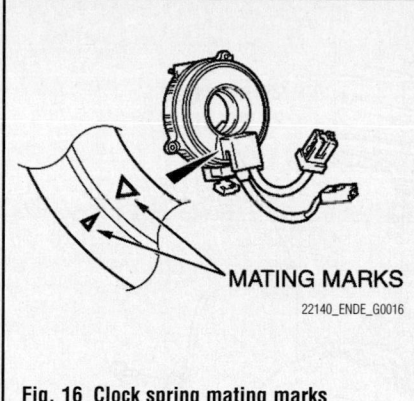

Fig. 16 Clock spring mating marks

3-3/4 turns counterclockwise to align the mating marks.

b. Turn the front wheels to the straight-ahead position. Then install

the clock spring to the column switch.

❋❋ **WARNING**

Ensure that the steering wheel sensor's mating marks are properly aligned. If not, the steering wheel sensor could be damaged.

2. Align the mating marks and install the steering wheel sensor.

a. Turn the steering wheel sensor to the arrow direction shown to align the mating marks.

b. Align the mating marks on the clock spring and the steering wheel sensor, and install the steering wheel sensor to the column switch assembly.

c. Connect the steering wheel sensor connector.

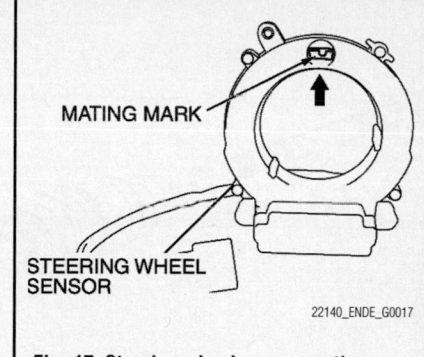

Fig. 17 Steering wheel sensor mating marks

3. Install the column switch, clock spring and steering wheel assembly.

4. Connect a scan tool and calibrate the steering wheel sensor.

DRIVE TRAIN

FRONT HALFSHAFT

REMOVAL & INSTALLATION
See Figures 18 and 19.

➥If the vehicle is equipped with ABS, do not strike the ABS rotors installed to

the BJ outer race of the halfshaft against other parts when removing or installing the halfshaft. Otherwise the ABS rotors will be damaged.

1. Before servicing the vehicle, refer to the Precautions Section.

2. Drain the transaxle and transfer case fluid, as necessary.

3. Remove the negative battery cable.

4. Remove the front and side under covers.

5. Remove the front exhaust pipe.

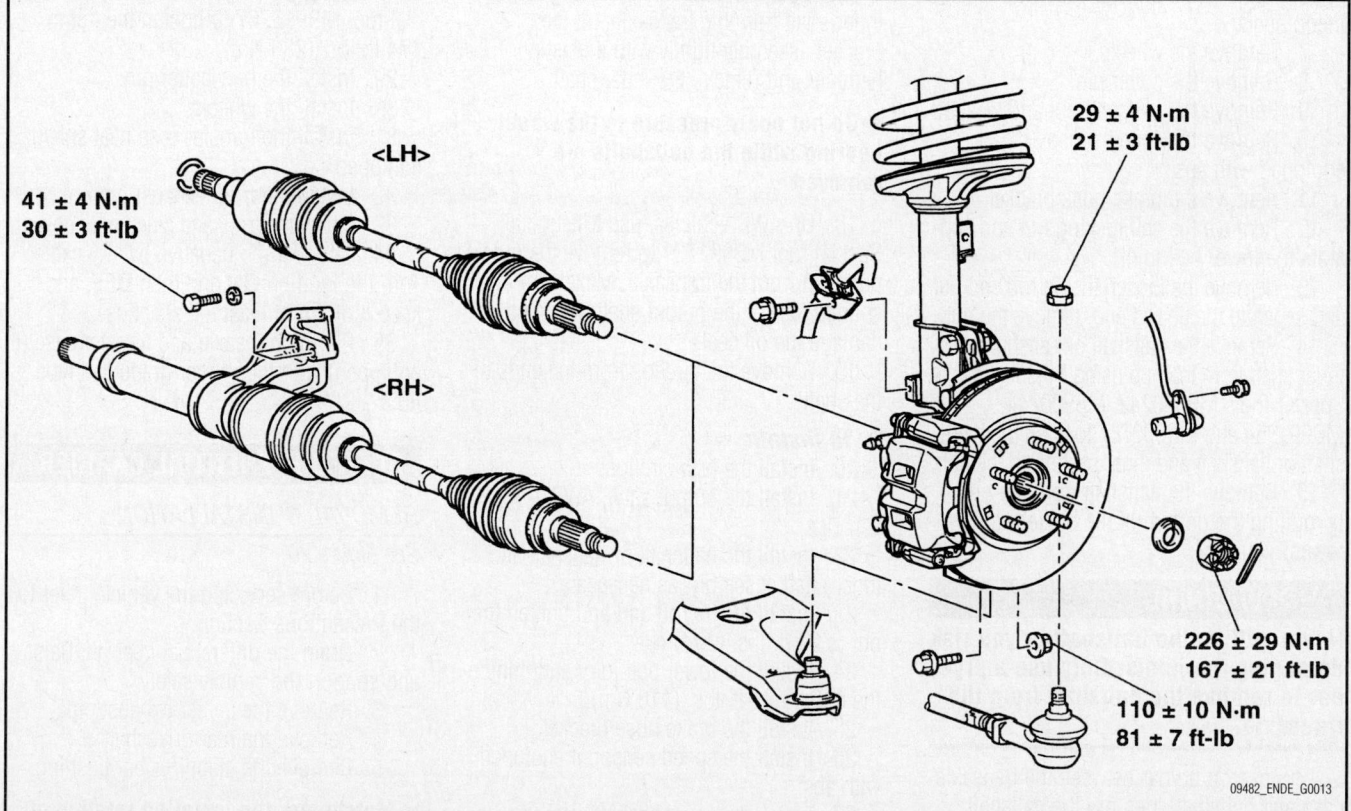

Fig. 18 Front halfshaft and related components—FWD vehicles

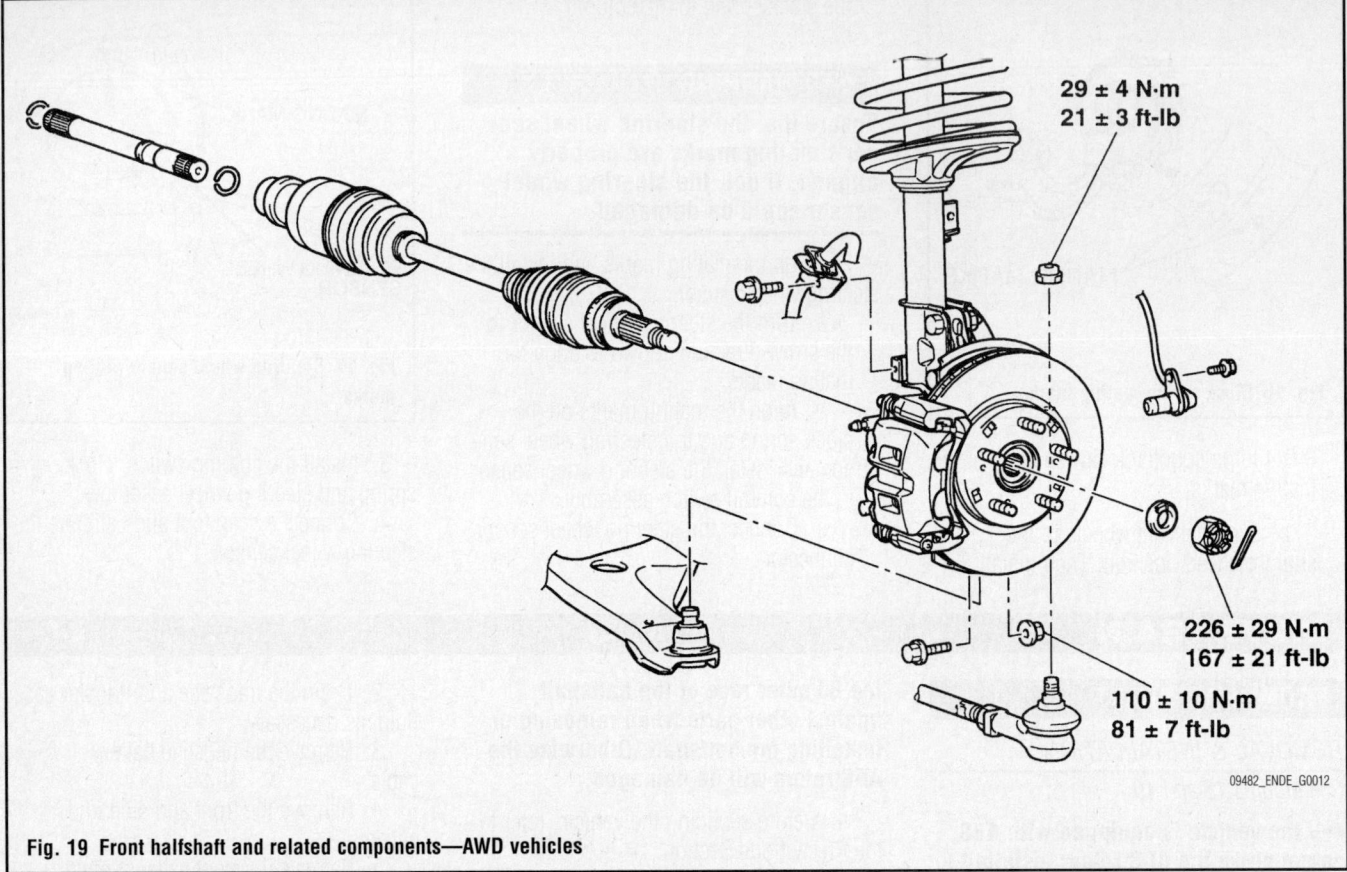

29 ± 4 N·m
21 ± 3 ft-lb

226 ± 29 N·m
167 ± 21 ft-lb

110 ± 10 N·m
81 ± 7 ft-lb

09482_ENDE_G0012

Fig. 19 Front halfshaft and related components—AWD vehicles

6. Remove the transfer case heat shield, if equipped.

7. Remove the wheels.

8. Remove the cotter pin.

9. Remove the halfshaft nut and washer.

10. Remove the speed sensor, if equipped with abs.

11. Remove the brake hose bracket.

12. Remove the self locking nut and separate the lower ball joint.

13. Remove the loosen the tie rod end nut and separate the tie rod end. remove the nut.

14. Remove the halfshaft or halfshaft and inner shaft from the hub using Mitsubishi Special Tools MB990242, MB990244, MB991354 and MB990767 to push the halfshaft or halfshaft and inner shaft from the hub.

15. Remove the halfshaft from the hub by pulling the bottom of the brake rotor toward you.

※※ WARNING

Never pull on the halfshaft or you risk damaging the joints. Only use a prybar to remove the halfshaft from the transaxle.

16. Insert a prybar between the transaxle case and halfshaft, then pry the halfshaft from the transaxle. Make sure the spline of the halfshaft does not damage the oil seal.

17. If you have trouble removing the inner shaft from the transaxle, hit the bracket assembly lightly with a plastic hammer and remove the inner shaft.

➡**Do not apply pressure to the wheel bearing while the halfshafts are removed.**

18. On AWD vehicles, use Mitsubishi Special tool MB991721 to remove the output shaft from the transaxle. Make sure the splined part of the output shaft does not damage the oil seal.

19. Remove the circlips from the ends of the shafts.

To install:

20. Install the new circlips.

21. Install the output shaft, AWD vehicles.

22. Install the halfshaft or halfshaft and inner shaft assembly, as necessary.

23. Install the tie rod end and tighten the nut to 21 ft. lbs. (29 Nm).

24. Install the lower ball joint and tighten the nuts to 81 ft. lbs. (110 Nm).

25. Install the brake hose bracket.

26. Install the speed sensor, if equipped with abs.

27. Install the new halfshaft washer, with the beveled edge facing out.

28. Install the halfshaft nut. with no load

on the wheel bearings, use Mitsubishi special tool MB990767 to tighten the nut to 174 ft. lbs. (236 Nm).

29. Install the new cotter pin.

30. Install the wheels.

31. Install the transfer case heat shield, if equipped.

32. Install the front exhaust pipe.

33. Install the front and side under covers.

34. Connect the negative battery cable. Turn the ignition **ON** and then **OFF**, and keep it off for at least ten seconds.

35. Fill the transaxle and transfer case (if equipped) with the proper grade and type fluid

REAR DIFFERENTIAL CARRIER

REMOVAL & INSTALLATION

See Figure 20.

1. Before servicing the vehicle, refer to the Precautions Section.

2. Drain the differential gear oil. Raise and support the vehicle safely.

3. Remove the center exhaust pipe.

4. Remove the rear driveshaft.

5. Remove the stabilizer bar bushing.

➡**Matchmark the installed relation of the propeller shaft to the differential carrier before removal.**

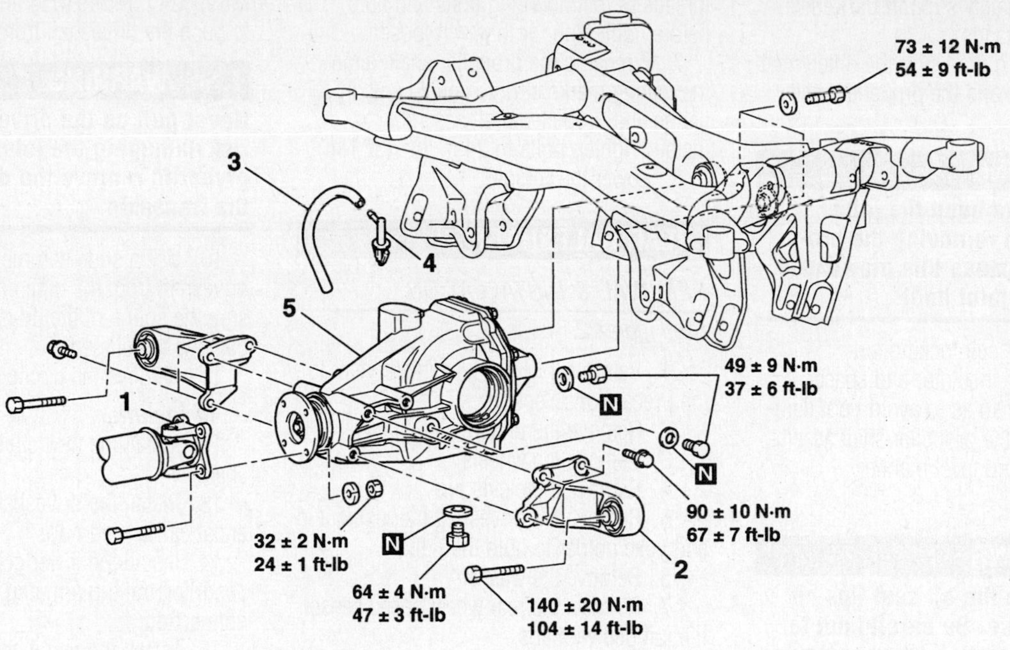

73 ± 12 N·m
54 ± 9 ft-lb

3

4

5

49 ± 9 N·m
37 ± 6 ft-lb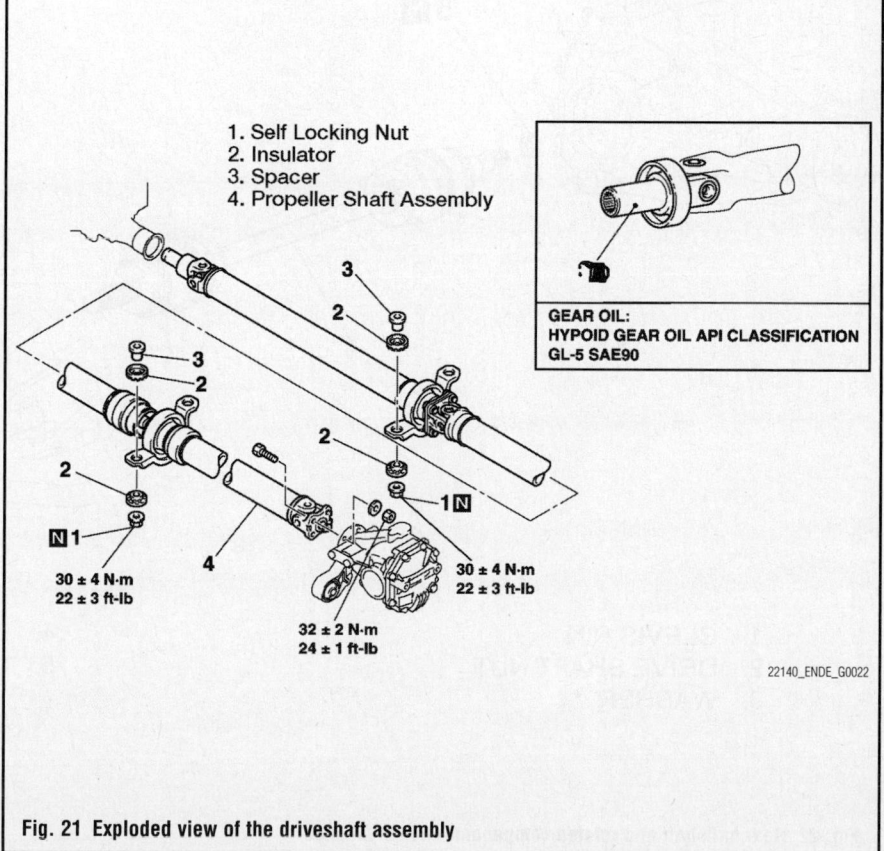

1

90 ± 10 N·m
67 ± 7 ft-lb

32 ± 2 N·m
24 ± 1 ft-lb

2

64 ± 4 N·m
47 ± 3 ft-lb

140 ± 20 N·m
104 ± 14 ft-lb

1. PROPELLER SHAFT CONNECTION
2. DIFFERENTIAL MOUNT BRACKET
3. HOSE

4. NIPPLE
5. DIFFERENTIAL CARRIER
   ASSEMBLY

67170-ENDE-G33

Fig. 20 Rear differential carrier and related components

6. Remove the propeller shaft from the differential carrier. suspend the propeller shaft from the body to avoid damaging or bending the shaft.

7. Remove the differential mount bracket.

8. Remove the hose and nipple.

9. Remove the retainers and differential carrier. Discard the washers.

To install:

10. Install the differential mount bracket. Tighten the retainers, with new washers, to 67 ft. lbs. (90 Nm).

11. Install the differential carrier. Tighten the through bolts, with new washers, to 104 ft. lbs. (140 Nm).

12. Install the nipple and hose.

13. Install the propeller shaft, aligning the matchmarks made during removal. Tighten bolts to 24 ft. lbs. (32 Nm).

14. Fill the differential with the proper grade and type gear oil.

15. Check and adjust the rear wheel alignment, as necessary.

REAR DRIVESHAFT

REMOVAL & INSTALLATION

See Figure 21.

1. Self Locking Nut
2. Insulator
3. Spacer
4. Propeller Shaft Assembly

GEAR OIL:
HYPOID GEAR OIL API CLASSIFICATION
GL-5 SAE90

3

2

3
2

2

2

N 1

4

30 ± 4 N·m
22 ± 3 ft-lb

30 ± 4 N·m
22 ± 3 ft-lb

32 ± 2 N·m
24 ± 1 ft-lb

22140_ENDE_G0022

Fig. 21 Exploded view of the driveshaft assembly

1. Raise and safely support the vehicle securely on jackstands.

2. Make mating marks on the differential companion flange and the propeller shaft assembly.

> **⁂ WARNING**
>
> **Be careful not to bend the joint assembly when removing the propeller shaft because this may cause damage to the joint boot.**

3. Remove the self locking nut.

4. Remove the insulator and spacer.

5. Insert a rag so as to avoid boot damage, and remove the propeller shaft assembly in a straight and level manner.

To install:

> **⁂ WARNING**
>
> **Do not damage the oil seal lips on the transfer case. Be careful not to bend the joint portion when removing the propeller shaft, because this will damage the joint boot.**

6. Remove oil and grease from the threads of the mounting bolts and nuts before tightening, or they will loosen.

7. If reusing the propeller shaft, align the mating marks of differential companion flange and propeller shaft assembly to install. Tighten bolts to 24 ft. lbs. (32 Nm).

8. Lower the vehicle.

REAR HALFSHAFT

REMOVAL & INSTALLATION

See Figure 22.

1. Before servicing the vehicle, refer to the Precautions Section.

2. Remove the undercover.

3. Remove the wheels.

4. Remove the clevis pin.

5. Remove the driveshaft nut, using and end yoke holder to hold the hub.

6. Remove the washer.

7. Remove the rear wheel speed sensor, if equipped with abs.

8. Remove the lower control arm, shock absorber, trailing arm and toe control arm connection.

9. Remove the driveshaft. use Mitsubishi special tools MB990242, MB990244, MB991354 and MB990767 to push the driveshaft from the hub.

> **⁂ WARNING**
>
> **Never pull on the driveshaft or you risk damaging the joints. Only use a prybar to remove the driveshaft from the transaxle.**

10. Use a slide hammer to remove the driveshaft from the differential carrier. Make sure the spline of the driveshaft does not damage the oil seal.

11. Remove the circlip.

To install:

12. Install the new circlip on the driveshaft.

13. Install the driveshaft into the differential carrier and hub.

14. Install the lower control arm, shock absorber, trailing arm and toe control arm connection.

15. Install the rear wheel speed sensor, if equipped with abs.

16. Install the new driveshaft washer, with the beveled edge facing out.

17. Install the driveshaft nut. with no

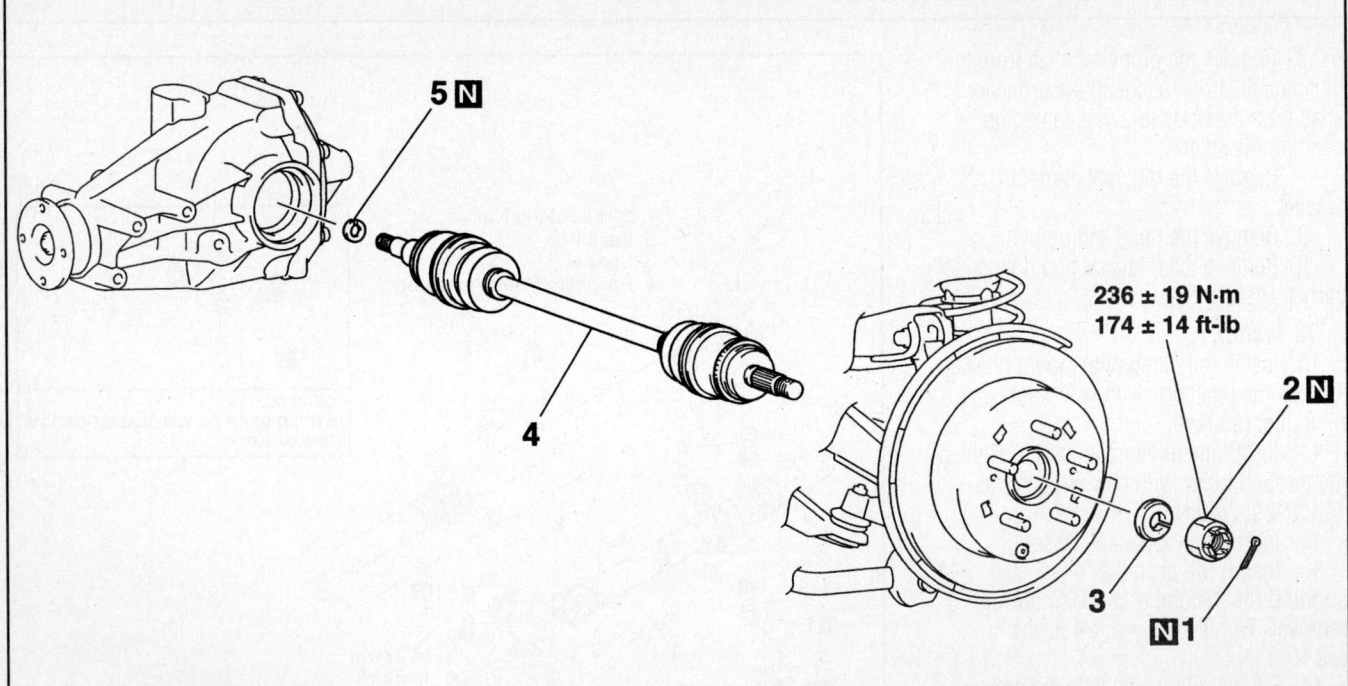

1. CLEVIS PIN
2. DRIVE SHAFT NUT
3. WASHER
4. DRIVE SHAFT
5. CIRCLIP

236 ± 19 N·m
174 ± 14 ft-lb

67170-ENDE-G32

Fig. 22 Rear halfshaft and related components—AWD vehicles

load on the wheel bearings, use Mitsubishi special tool MB990767 to tighten the nut to 174 ft. lbs. (236 Nm).

18. Install the new cotter pin.

19. Install the wheels.

20. Check and adjust the rear wheel alignment, as necessary.

REAR PINION SEAL

REMOVAL & INSTALLATION

See Figures 23 through 26.

1. Before servicing the vehicle, refer to the Precautions Section.

2. Remove the differential assembly from the vehicle.

3. Use special tool MB990850 to

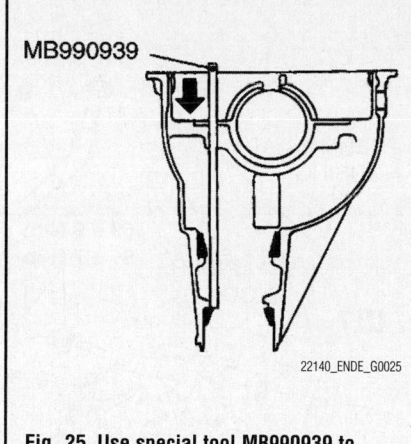

Fig. 25 Use special tool MB990939 to remove the oil seal

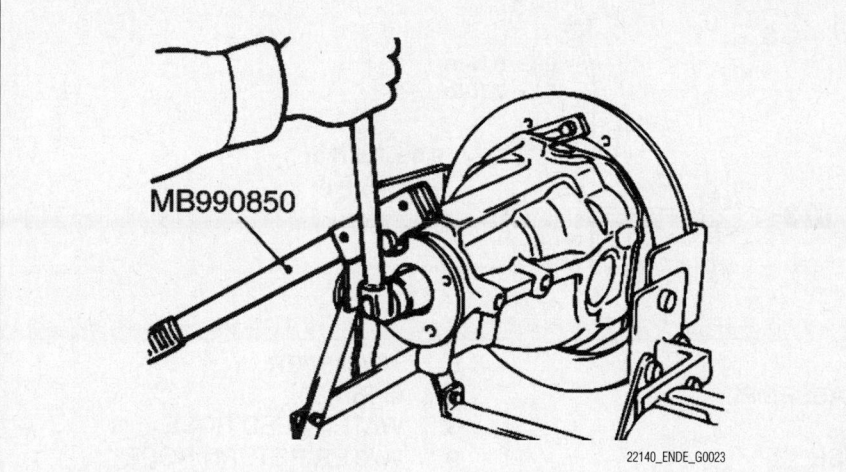

Fig. 23 Use special tool MB990850 to hold the companion flange, and then remove the companion flange self-locking nut

hold the companion flange, and then remove the companion flange self-locking nut.

4. Make mating marks on the drive pinion and companion flange.

5. Use special tool MB990810 to pull out the companion flange.

6. Use special tool MB990939 to remove the oil seal.

To install:

7. Use special tools MB990938 and MB991380 to press-fit the oil seal.

8. Adjust the drive pinion turning torque by the following procedures:

a. Install the drive pinion assembly and companion flange with the mating marks properly aligned. Tighten the companion flange self-locking nut to 137 ft. lbs. (186 Nm) while holding the companion flange with special tool MB990850.

b. Use special tools MB990685 and MB990326 to measure the drive pinion turning torque (with drive pinion oil seal) to verify that the drive pinion turning torque is 12.22–15.66 inch lbs. (1.38–1.77 Nm).

c. If the turning torque is not within the standard value, check the tightening torque of the companion flange self-locking nut, and the installation of the oil seal.

9. Once proper turning torque has been achieved, install the differential assembly in the vehicle.

TRANSFER CASE ASSEMBLY

REMOVAL & INSTALLATION

See Figure 27.

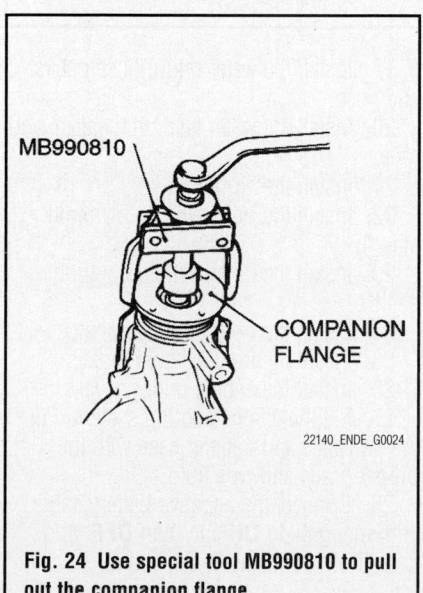

Fig. 24 Use special tool MB990810 to pull out the companion flange

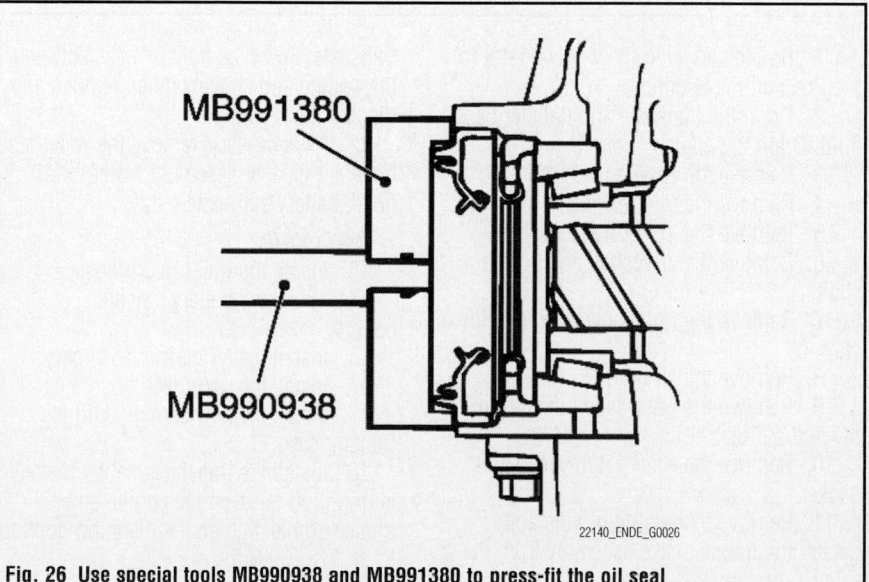

Fig. 26 Use special tools MB990938 and MB991380 to press-fit the oil seal

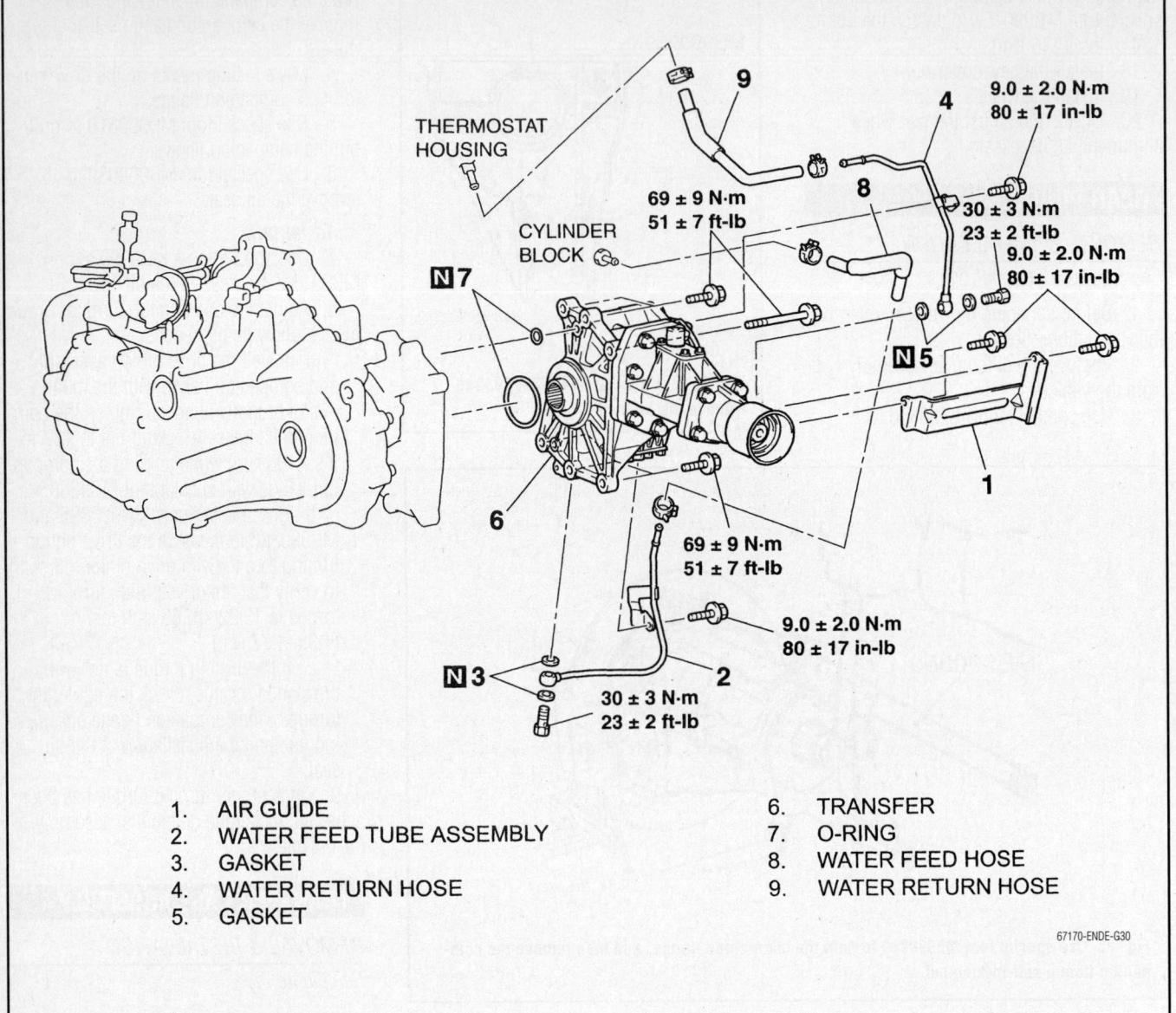

THERMOSTAT
HOUSING

9

4

9.0 ± 2.0 N·m
80 ± 17 in-lb

69 ± 9 N·m
51 ± 7 ft-lb

CYLINDER
BLOCK

8

30 ± 3 N·m
23 ± 2 ft-lb

9.0 ± 2.0 N·m
80 ± 17 in-lb

N 7

N 5

1

6

69 ± 9 N·m
51 ± 7 ft-lb

9.0 ± 2.0 N·m
80 ± 17 in-lb

N 3

2

30 ± 3 N·m
23 ± 2 ft-lb

1.	AIR GUIDE	6.	TRANSFER
2.	WATER FEED TUBE ASSEMBLY	7.	O-RING
3.	GASKET	8.	WATER FEED HOSE
4.	WATER RETURN HOSE	9.	WATER RETURN HOSE
5.	GASKET		

67170-ENDE-G30

Fig. 27 Transfer case and related components

1. Before servicing the vehicle, refer to the Precautions Section.
2. Drain the transaxle and transfer case fluid. Drain the engine coolant.
3. Remove the engine under cover.
4. Remove the front exhaust pipe.
5. Remove the front propeller shaft.
6. Remove the driveshaft and output shaft.
7. Remove the right exhaust manifold stay "b".
8. Remove the air guide.
9. Remove the water feed tube assembly and gasket.
10. Remove the water return hose and gasket.
11. Remove the retaining bolts and detach the transfer case assembly from the

transaxle. lower the transfer case between the engine and crossmember. remove and discard the o-rings.
12. As necessary, remove the water feed hose, air cleaner assembly, water return hose, battery and battery tray.

To install:
13. Install the water return hose.
14. Install the battery tray and battery.
15. Install the air cleaner assembly.
16. Install the water feed hose.
17. Install the new o-rings onto the transfer case.
18. Install the transfer case, by maneuvering it up between the engine and crossmember. tighten the retaining bolts to 51 ft. lbs. (69 Nm).

19. Install the water return hose gasket and hose.
20. Install the water feed tube gasket and tube.
21. Install the air guide.
22. Install the right exhaust manifold stay "b".
23. Install the driveshaft and output shaft.
24. Install the front propeller shaft.
25. Install the front exhaust pipe.
26. Install the engine under cover.
27. Fill the engine cooling system. Fill the transaxle and transfer case with the proper grade and type fluid.
28. Connect the negative battery cable. Turn the ignition **ON** and then **OFF**, and keep it off for at least ten seconds.

ENGINE COOLING

ENGINE FAN

REMOVAL & INSTALLATION

Engine Fan

See Figures 28 and 29.

1. Before servicing the vehicle, refer to the Precautions Section.
2. Drain the engine coolant.
3. Remove the air cleaner assembly.
4. Remove the radiator grille.
5. Remove the rubber hose.
6. Matchmark the installed position of the hose and clamp, then disconnect the upper radiator hose.
7. Detach the fan motor connector.

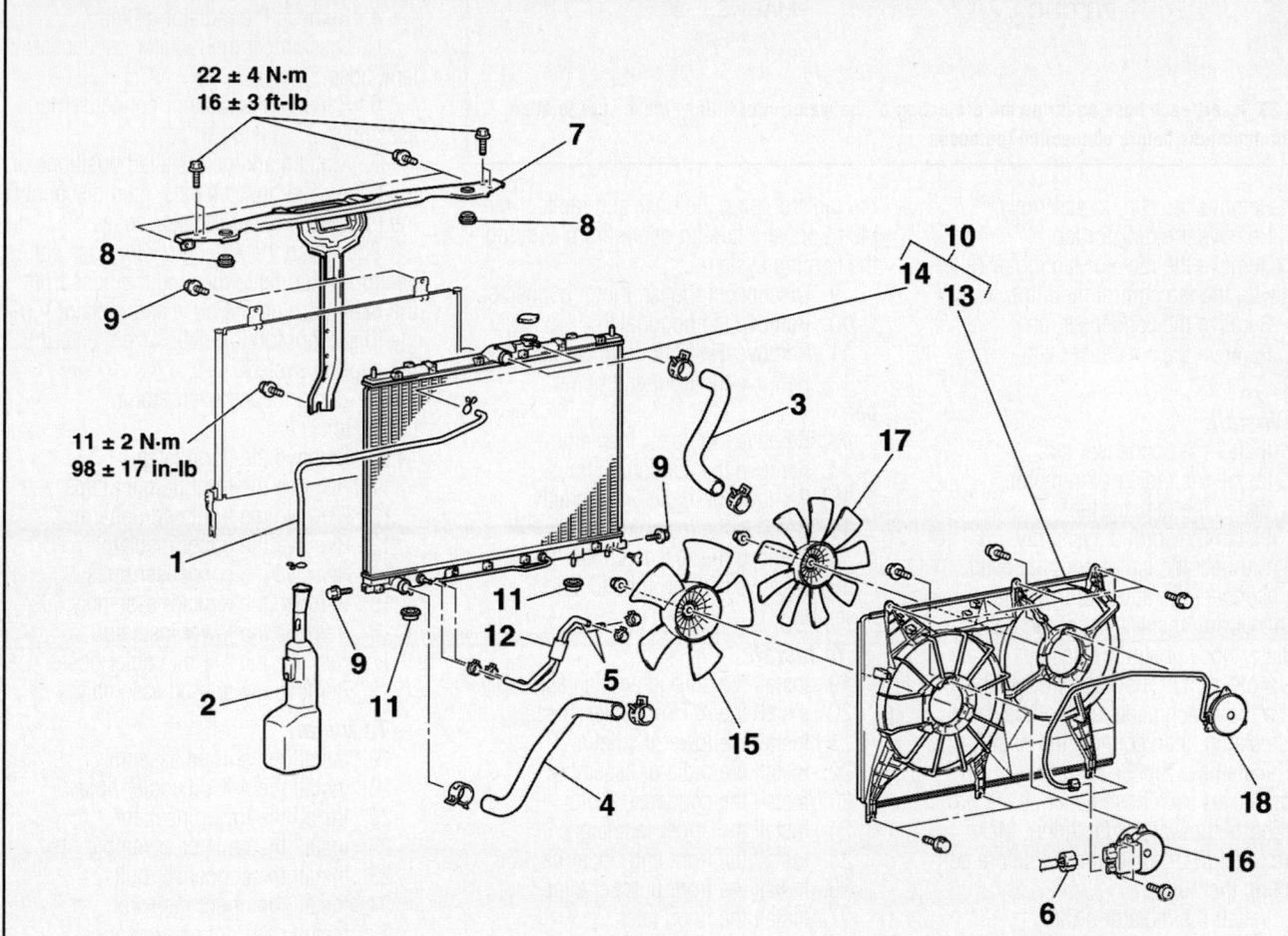

22 ± 4 N·m
16 ± 3 ft-lb

11 ± 2 N·m
98 ± 17 in-lb

1. RADIATOR CONDENSER TANK HOSE
2. RADIATOR CONDENSER TANK ASSEMBLY
3. RADIATOR UPPER HOSE
4. RADIATOR LOWER HOSE
5. A/T OIL COOLER HOSE CONNECTION
6. FAN MOTOR CONNECTOR
7. FRONT END UPPER BAR
8. UPPER INSULATOR
9. CONDENSER BOLTS
10. RADIATOR ASSEMBLY
11. LOWER INSULATOR
12. A/T OIL COOLER HOSE
13. SHROUD ASSEMBLY
14. RADIATOR
15. RADIATOR FAN
16. RADIATOR FAN MOTOR (INTEGRATED WITH FAN CONTROL MODULE)
17. CONDENSER FAN
18. CONDENSER FAN MOTOR

42050_ENDE_G0027

Fig. 28 Exploded view of the radiator, fan and related components

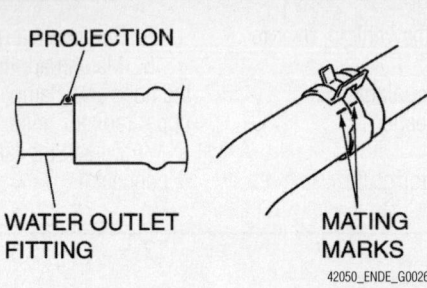

PROJECTION

WATER OUTLET FITTING

MATING MARKS

42050_ENDE_G0026

Fig. 29 Insert each hose as far as the projection of the water inlet fitting. Make sure to align the matchmarks before connecting the hoses

8. Remove the shroud assembly.

9. Remove the radiator fan.

10. Remove the radiator fan motor (integrated with the fan control module).

11. Remove the condenser fan.

12. Remove the condenser fan motor.

To install:

13. Install the condenser fan.

14. Install the radiator fan motor.

15. Install the radiator fan.

16. Install the shroud assembly.

17. Connect the fan motor connector.

18. Connect the upper radiator hose.

a. During installation of the upper radiator hose, insert the hose as far as the projection of the water inlet fitting. Align the matchmarks on the hose and hose clamp, then connect the hose.

19. Install the rubber hose.

a. Insert each hose as far as the projection of the water inlet fitting. Make sure to align the matchmarks before connecting the hoses.

20. Install the radiator grille.

21. Install the air cleaner assembly.

22. Fill the engine with coolant.

23. Start the engine and check for proper fan operation.

Condenser Fan

1. Before servicing the vehicle, refer to the Precautions Section.

2. Drain the engine coolant.

3. Remove the air cleaner assembly.

4. Remove the radiator grille.

5. Disconnect the radiator condenser tank hose.

6. Remove the radiator condenser tank assembly.

7. Matchmark the installed positions of the hoses and hose clamps, then disconnect the upper and lower radiator hoses.

8. Detach the A/T oil cooler hose connection. After disconnecting the hose from

the radiator, plug the hose and radiator nipple to prevent foreign debris from entering the cooling system.

9. Disconnect the fan motor connector.

10. Remove the hood latch.

11. Remove the front bumper clips.

12. Remove the front end upper bar.

13. Remove the upper insulator.

14. Remove the condenser bolts.

15. Remove the radiator assembly.

16. Remove the lower insulator.

17. Remove the A/T oil cooler hose.

18. Remove the shroud and fan assembly.

To install:

19. Install the shroud and fan assembly.

20. Install the A/T oil cooler hose.

21. Install the lower insulator.

22. Install the radiator assembly.

23. Install the condenser bolts.

24. Install the upper insulator.

25. Install the front end upper bar.

26. Install the front bumper clips.

27. Install the hood latch.

28. Connect the fan motor connector.

29. Connect the A/T oil cooler hose connection.

30. Connect the upper and lower radiator hoses.

a. During installation of the upper and lower radiator hoses, insert the hoses as far as the projection of the water inlet fitting. Align the matchmarks on the hoses and hose clamps, then connect the hoses.

31. Install the radiator condenser tank assembly.

32. Disconnect the radiator condenser tank hose.

33. Install the radiator grille.

34. Install the air cleaner assembly.

35. Fill the engine with coolant.

36. Start the engine and check for proper fan operation.

RADIATOR

REMOVAL & INSTALLATION

See Figures 28 and 29.

1. Before servicing the vehicle, refer to the Precautions Section.

2. Drain the engine coolant. Refer to Engine Coolant

3. Remove the air cleaner assembly.

4. Remove the radiator grille.

5. Disconnect the radiator condenser tank hose.

6. Remove the radiator condenser tank assembly.

7. Matchmark the installed positions of the hoses and hose clamps, then disconnect the upper and lower radiator hoses.

8. Detach the A/T oil cooler hose connection. After disconnecting the hose from the radiator, plug the hose and radiator nipple to prevent foreign debris from entering the cooling system.

9. Disconnect the fan motor connector.

10. Remove the hood latch.

11. Remove the front bumper clips.

12. Remove the front end upper bar.

13. Remove the upper insulator.

14. Remove the condenser bolts.

15. Remove the radiator assembly.

16. Remove the lower insulator.

17. Remove the A/T oil cooler hose.

18. Remove the shroud assembly.

To install:

19. Install the shroud assembly.

20. Install the A/T oil cooler hose.

21. Install the lower insulator.

22. Install the radiator assembly.

23. Install the condenser bolts.

24. Install the upper insulator.

25. Install the front end upper bar.

26. Install the front bumper clips.

27. Install the hood latch.

28. Connect the fan motor connector.

29. Connect the A/T oil cooler hose connection.

30. Connect the upper and lower radiator hoses.

a. During installation of the upper and lower radiator hoses, insert the hoses as far as the projection of the water inlet fitting. Align the matchmarks on the hoses and hose clamps, then connect the hoses.

31. Install the radiator condenser tank assembly.

32. Disconnect the radiator condenser tank hose.

33. Install the radiator grille.

34. Install the air cleaner assembly.

35. Fill the engine with coolant.

36. Start the engine and check for proper fan operation.

THERMOSTAT

REMOVAL & INSTALLATION

See Figures 30 through 32.

1. Before servicing the vehicle, refer to the Precautions Section.

2. Drain the cooling system.

3. Remove the air cleaner.

4. Disconnect the manifold differential pressure (MDP) sensor connector.

5. Disconnect the knock sensor (KS) connector.

6. Disconnect the crankshaft position (CKP) sensor connector.

7. Disconnect the control wiring harness and wiring harness combination connector.

8. Disconnect the right bank (rear) heated oxygen sensor (HO2S) connector.

9. Disconnect the right bank (front) HO2S connector.

10. Disconnect the power steering pressure (PSP) switch connector.

11. Disconnect the evaporative emission purge solenoid connector.

12. Disconnect the injector connector.

13. Disconnect the left bank (front) heated oxygen sensor (HO2S) connector.

14. Disconnect the left bank (rear) HO2S connector.

15. Disconnect the throttle position (TP) sensor connector.

16. Disconnect the engine coolant temperature (ECT) connector.

17. Disconnect the exhaust gas recirculation (EGR) valve connector.

18. Disconnect the capacitor connector.

19. Disconnect the output shaft speed sensor connector.

20. Disconnect the ground strap.

21. Matchmark the installed position of the hose and clamp, then disconnect the radiator lower hose connection.

22. Remove the water inlet fitting, then remove the thermostat.

23. Thoroughly clean the gasket mating surfaces.

To install:

24. Install the thermostat so the jiggle valve is facing straight up. Be careful not to fold or otherwise damage the rubber ring.

➡**Make absolutely sure that no oil adheres to the rubber ring of the thermostat. Also do not fold or scratch the rubber ring during installation.**

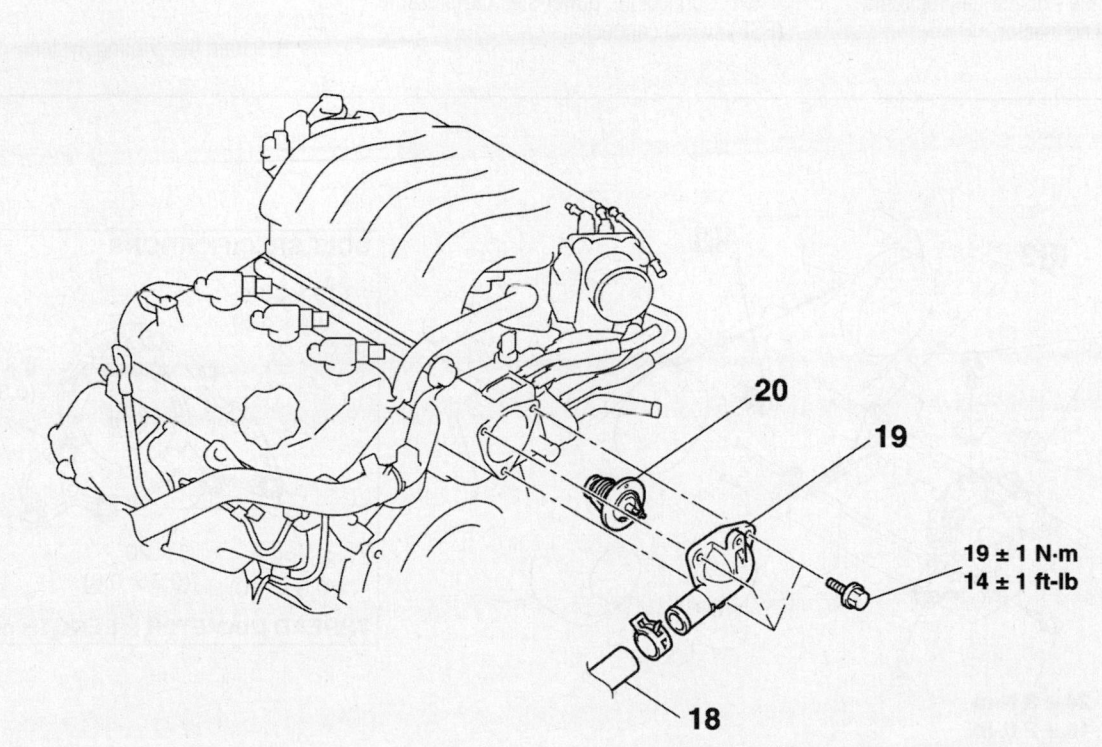

18. RADIATOR LOWER HOSE CONNECTION

19. WATER INLET FITTING
20. THERMOSTAT

19 ± 1 N·m
14 ± 1 ft-lb

42050_ENDE_G0030

Fig. 30 Exploded view of the thermostat, water inlet and radiator hose

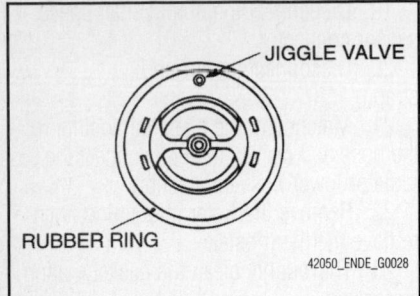

Fig. 31 The thermostat's jiggle valve must be facing up, at the 12 o'clock position during installation

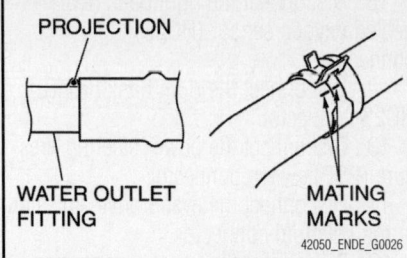

Fig. 32 Insert each hose as far as the projection of the water inlet fitting. Make sure to align the matchmarks before connecting the hoses

25. Install the water inlet fitting. Tighten the bolts to 14 ft. lbs. (19 Nm).

26. Install the lower radiator hose. Insert the hose as far as the projection of the water inlet fitting. Align the matchmarks on the hose and hose clamp, then connect the hose.

27. Connect the ground strap.

28. Connect the output shaft speed sensor connector.

29. Connect the capacitor connector.

30. Connect the exhaust gas recirculation (EGR) valve connector.

31. Connect the engine coolant temperature (ECT) connector.

32. Connect the throttle position (TP) sensor connector.

33. Connect the left bank (rear) HO2S connector.

34. Connect the left bank (front) heated oxygen sensor (HO2S) connector.

35. Connect the injector connector.

36. Connect the evaporative emission purge solenoid connector.

37. Connect the power steering pressure (PSP) switch connector.

38. Connect the right bank (front) HO2S connector.

39. Connect the right bank (rear) heated oxygen sensor (HO2S) connector.

40. Connect the control wiring harness and wiring harness combination connector.

41. Connect the crankshaft position (CKP) sensor connector.

42. Connect the knock sensor (KS) connector.

43. Connect the manifold differential pressure (MDP) sensor connector.

44. Install the air cleaner.

45. Fill the engine with coolant.

46. Start the engine and check for proper fan operation.

WATER PUMP

REMOVAL & INSTALLATION

See Figure 33.

1. Before servicing the vehicle, refer to the Precautions Section.

2. If necessary, properly release the fuel pressure.

3. Drain the cooling system.

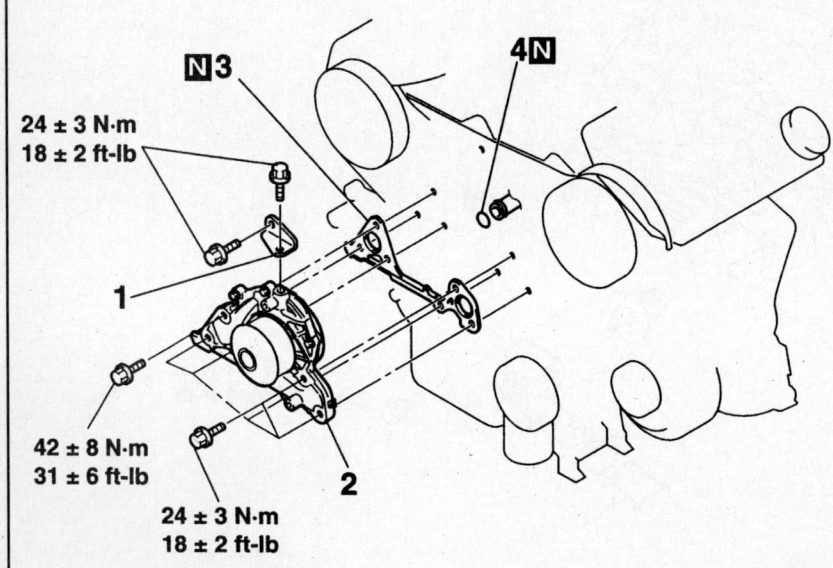

1. WATER PUMP BRACKET
2. WATER PUMP
3. WATER PUMP GASKET
4. O-RING

Fig. 33 Water pump and bolt torque specifications

4. Disconnect the negative battery cable.

> ✳✳ **CAUTION**
>
> **Wait at least 90 seconds after the negative battery cable is disconnected to prevent possible deployment of the air bag.**

5. Remove the timing belt.
6. Remove the crankshaft position (CKP) sensor.

7. Remove the water pump bracket.
8. Remove the retaining bolts, water pump assembly, gasket and o-ring.

To install:

9. Clean and dry the mating surfaces of the block and water pump
10. Install the water pump assembly, using a new gasket and O-ring. Torque the water pump bolts to the specifications shown in the accompanying figure.
11. Install the water outlet fitting bracket.

12. Install the CKP sensor.
13. Install the timing belt.
14. Refill the radiator with coolant. This cooling system has a self-bleeding thermostat, so system bleeding is not required.
15. Run the vehicle until the thermostat opens and fill the overflow tank. Check for leaks.
16. Once the vehicle has cooled, recheck the coolant level.

ENGINE ELECTRICAL

ALTERNATOR

REMOVAL & INSTALLATION
See Figure 34.

1. Before servicing the vehicle, refer to the Precautions Section.
2. Disconnect the negative battery cable.

CHARGING SYSTEM

3. Remove the undercover.
4. Remove the side under cover.
5. Remove the drive belt.
6. Remove the alternator connector.

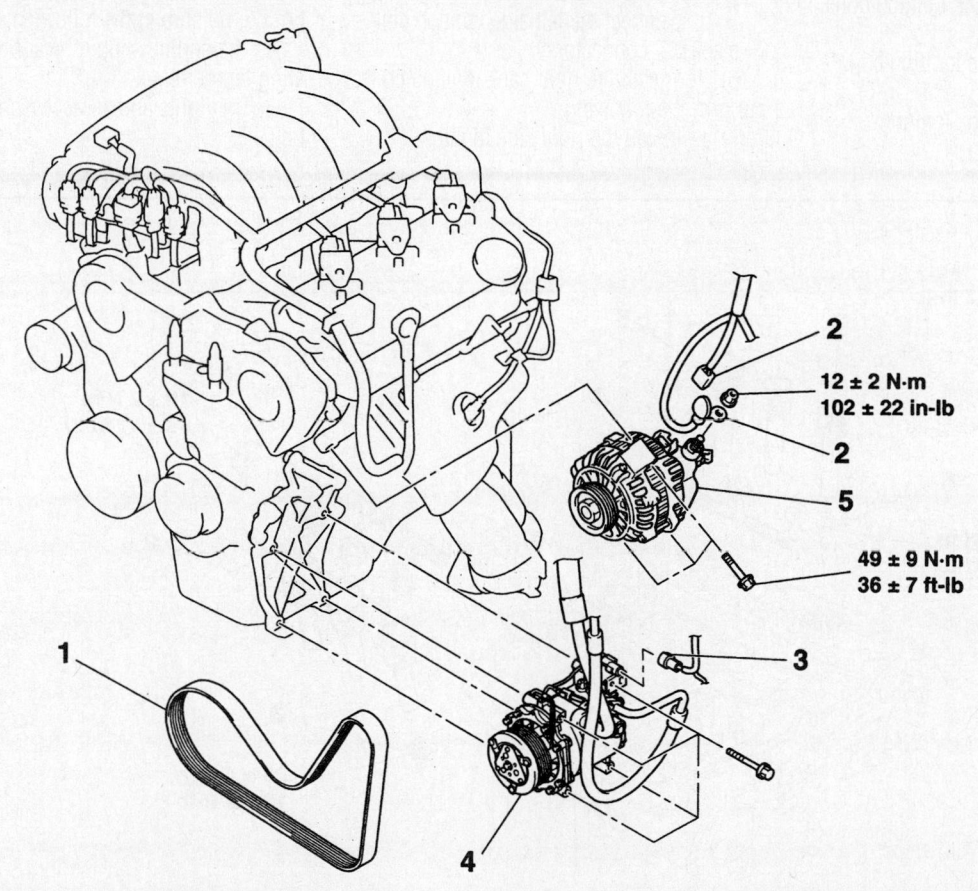

12 ± 2 N·m
102 ± 22 in-lb

49 ± 9 N·m
36 ± 7 ft-lb

1. GENERATOR DRIVE BELT	4. A/C COMPRESSOR ASSEMBLY
2. GENERATOR CONNECTOR	5. GENERATOR
3. A/C COMPRESSOR ASSEMBLY CONNECTOR	

67170-ENDE-G02

Fig. 34 Alternator mounting and related components

7. Disconnect the A/C compressor connector.

8. Remove the A/C compressor and position aside. do not disconnect the refrigerant lines.

9. Remove the alternator.

To install:

10. Install the alternator. Torque the bolts to 36 ft. lbs. (49 Nm).

11. Install the A/C compressor.

12. Install the A/C compressor connector.

13. Connect the alternator connector. Torque the nut to 102 inch lbs. (12 Nm).

14. Install the drive belt.

15. Install the side under cover.

16. Install the undercover.

17. Disconnect the negative battery cable.

ENGINE ELECTRICAL

FIRING ORDER

The distributorless ignition system has no spark plug wires that require routing.

IGNITION COIL

REMOVAL & INSTALLATION

See Figure 35.

1. Before servicing the vehicle, refer to the Precautions Section.

2. To remove the left bank ignition coils, perform the following:

a. Detach the left side ignition coil electrical connectors.

b. Remove the left side ignition coils.

3. Remove the intake manifold plenum.

4. To remove the right bank ignition coils, perform the following:

a. Detach the right side ignition coil electrical connectors.

b. Remove the right side ignition coils.

To install:

5. To install the left bank ignition coils, perform the following:

a. Install the left side ignition coils. Tighten screws to 89 inch lbs. (10 Nm).

b. Connect the left side ignition coil electrical connectors.

6. To install the right bank ignition coils, perform the following:

a. Install the right side ignition

IGNITION SYSTEM

coils. Tighten screws to 89 inch lbs. (10 Nm).

b. Connect the right side ignition coil electrical connectors.

c. Install the intake manifold plenum.

IGNITION TIMING

INSPECTION

1. Before servicing the vehicle, refer to the Precautions Section.

Before attempting to adjust the ignition timing, be sure of the following:

• The engine should be at normal operating temperature.

• The lights and all accessories should be OFF.

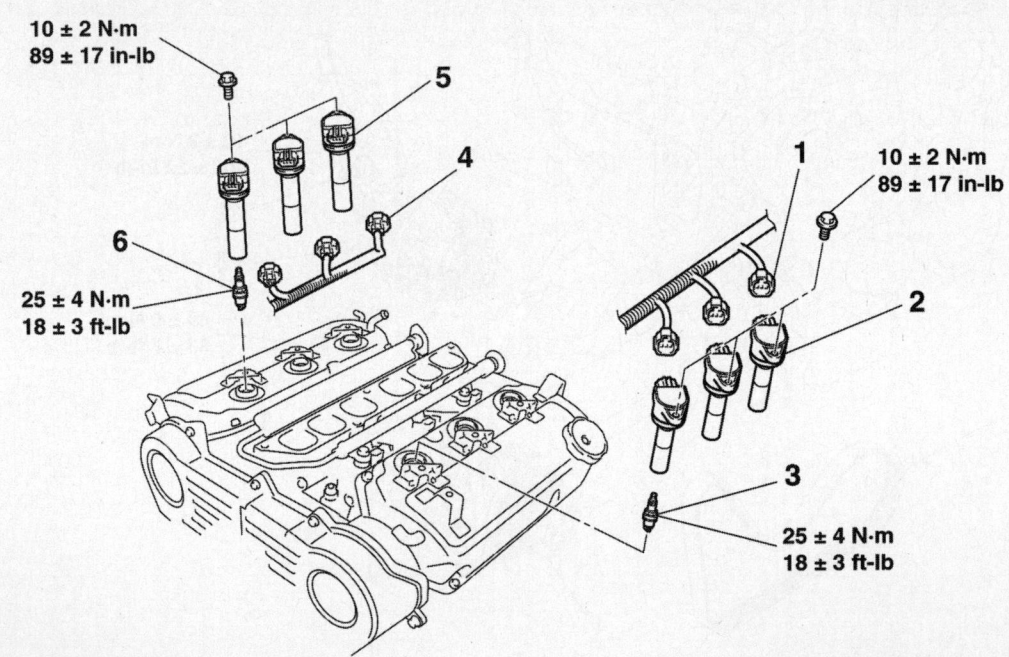

```
10 ± 2 N·m
89 ± 17 in-lb
```
```
25 ± 4 N·m
18 ± 3 ft-lb
```
```
10 ± 2 N·m
89 ± 17 in-lb
```
```
25 ± 4 N·m
18 ± 3 ft-lb
```

1. IGNITION COIL CONNECTORS (LH)
2. IGNITION COILS (LH)
3. SPARK PLUGS (LH)
• INTAKE MANIFOLD PLENUM

4. IGNITION COIL CONNECTORS (RH)
5. IGNITION COILS (RH)
6. SPARK PLUGS (RH)

42050_ENDE_G0009

Fig. 35 Exploded view of the ignition coils and related components

- The transaxle should be in **P** or **N**.
  2. Connect scan tool MB991958 to the data link connector
  3. Set up the timing light.
  4. Start the engine and run at idle.
  5. Verify that the idle speed is about 680 rpm.
  6. Select scan tool MB991958 actuator test "item number 17".
  7. Check that basic timing is within standard, it should be 2–8° BTDC.
  8. If the base timing is out of specification:
  a. Check to see if the distributor is aligned properly
  b. Check to see if the timing belt cover and Crankshaft Position (CKP) sensor installation is conditions.
  c. Crankshaft sensing blade conditions.
  9. Press the clear key on the scan tool, select forced drive stop mode and cancel the actuator test.

❋❋ CAUTION

If the actuator test is not canceled, the forced drive will continue for 27 minutes. Driving in this state could lead to engine failure.

10. Check that the actual ignition timing is approximately 10°BTDC.

➡**Keep in mind that the ignition timing may fluctuate by as much as +/-7°BTDC even under normal operation conditions. It is also further advanced by 5–10°BTDC at higher altitudes.**

ADJUSTMENT

The ignition timing is controlled by the Powertrain Control Module (PCM) and is not adjustable. The PCM determines the timing based on input from the crankshaft position sensor.

SPARK PLUGS

REMOVAL & INSTALLATION

See Figures 36 and 37.

Endeavor models use iridium spark plugs. Be careful not to damage the iridium tips of the plugs. Do not adjust the spark plug gap. Spark plugs must spark properly to assure proper engine performance and reduce exhaust emission level. Therefore,

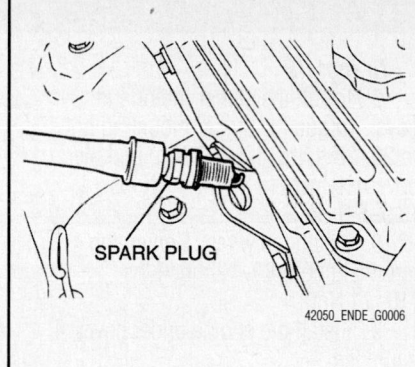

Fig. 36 Remove the spark plug and connect to the ignition coil

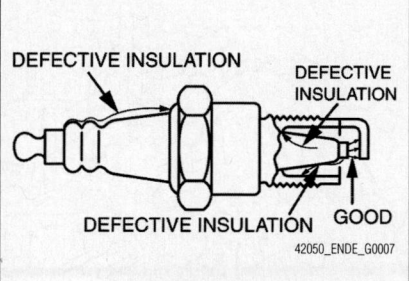

Fig. 37 Ground the spark plug outer electrode (body), and crank the engine. Check that there is an electrical discharge between the electrodes

they should be replaced periodically with new ones.
1. Disconnect the negative battery cable.
2. Remove the ignition coil(s).
3. Remove the spark plug(s).
4. Check, but do not adjust, the plugs to make sure they have the proper gap.
5. Once the spark plugs is removed, you can test them, as follows:
a. Remove the spark plug and connect to the ignition coil.
b. Ground the spark plug outer electrode (body), and crank the engine. Check that there is an electrical discharge between the electrodes. If not, replace the plug.

To install:
6. Install the spark plugs and tighten to 14–22 ft. lbs. (21–29 Nm).
7. Install the ignition coil(s).
8. Connect the negative battery cable.

Turn the ignition **ON** and then **OFF**, and keep it off for at least ten seconds.

INSPECTION

See Figure 38.

Endeavor models use iridium spark plugs. Be careful not to damage the iridium tips of the plugs. Do not adjust the spark plug gap. Spark plugs must spark properly to assure proper engine performance and reduce exhaust emission level. Therefore, they should be replaced periodically with new ones.

❋❋ WARNING

Cleaning of the iridium plug may result in damage to the iridium tips. Therefore, if carbon deposits must be removed, use a plug cleaner and complete cleaning within 20 seconds to protect the electrode. Do not use a wire brush.

1. Remove the spark plug(s).
2. Check, but do not adjust, the plugs to make sure they have the proper gap.
3. Once the spark plugs is removed, you can test them, as follows:
a. Remove the spark plug and connect to the ignition coil.
b. Ground the spark plug outer electrode (body), and crank the engine. Check that there is an electrical discharge between the electrodes. If not, replace the plug.

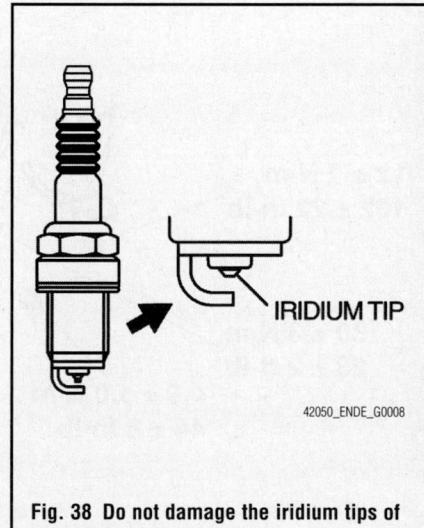

Fig. 38 Do not damage the iridium tips of the spark plugs

STARTER

REMOVAL & INSTALLATION

See Figure 39.

1. Before servicing the vehicle, refer to the precautions section.
2. Disconnect the negative battery cable.
3. Remove the engine under cover, if equipped.
4. Remove the wires.
5. Remove the starter cover.
6. Remove the starter motor.

To install:

7. Install the starter motor and cover. Torque the starter mounting bolts to 20–25 ft. lbs. (26–33 Nm) and the cover retainers to 36–42 inch lbs. (3.9–5.9 Nm).
8. Install the wires. Tighten the terminal nut to 80–124 inch lbs. (10–14 Nm).
9. Install the engine under cover, if equipped.
10. Disconnect the negative battery cable.

SOLENOID REPLACEMENT

1. Before servicing the vehicle, refer to the precautions section.
2. Disconnect the negative battery cable.
3. Remove the starter.
4. Remove the solenoid attaching screws.
5. Remove the solenoid.

To install:

6. Install the solenoid.
7. Install the solenoid attaching screws.
8. Install the starter.
9. Connect the negative battery cable.

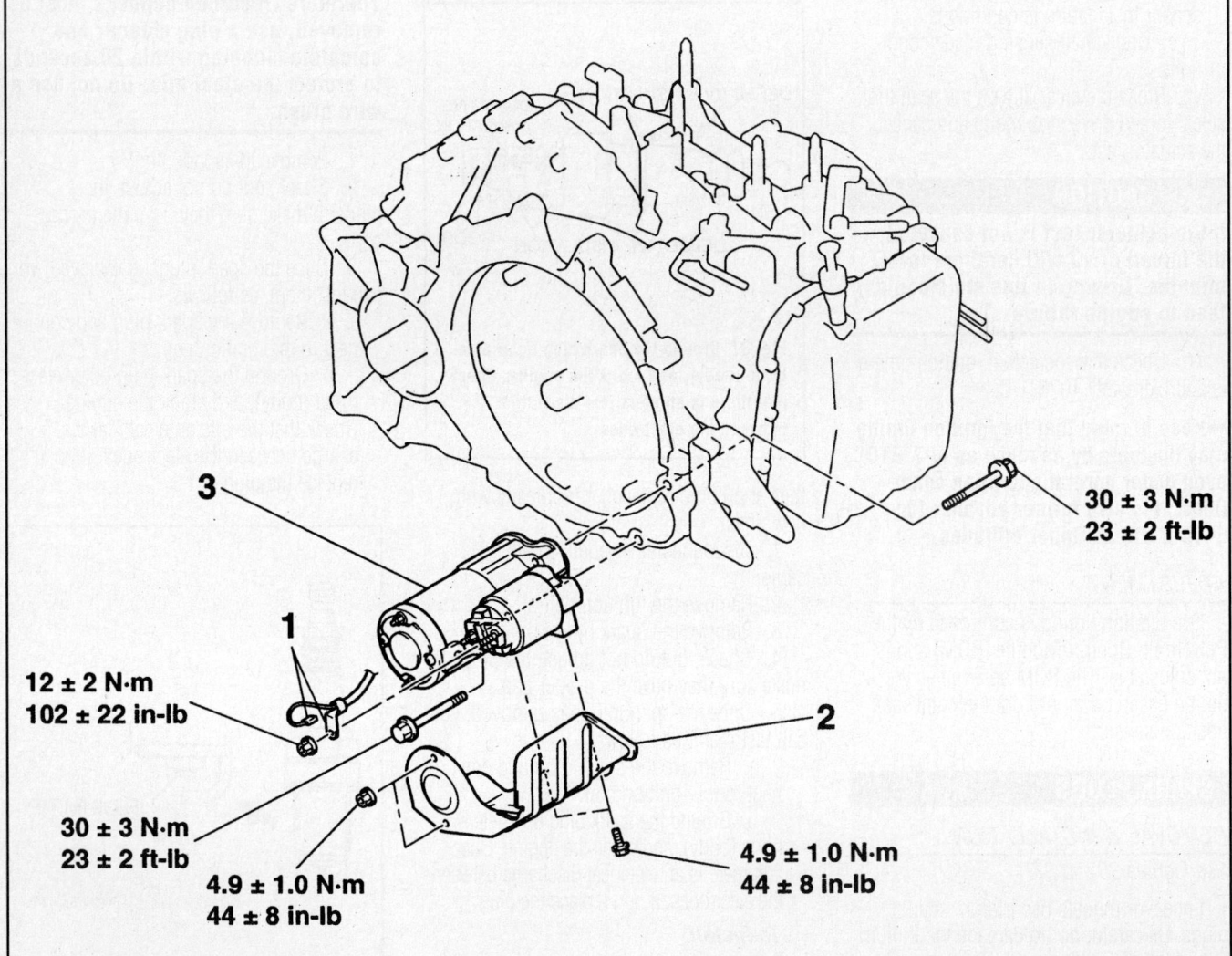

3

1

**12 ± 2 N·m
102 ± 22 in-lb**

**30 ± 3 N·m
23 ± 2 ft-lb**

**30 ± 3 N·m
23 ± 2 ft-lb**

2

**4.9 ± 1.0 N·m
44 ± 8 in-lb**

**4.9 ± 1.0 N·m
44 ± 8 in-lb**

1. **STARTER CONNECTOR**	2. **STARTER COVER**
	3. **STARTER ASSEMBLY**

67170-ENDE-G16

Fig. 39 Starter motor mounting

ENGINE MECHANICAL

ACCESSORY DRIVE BELTS

ACCESSORY BELT ROUTING

See Figure 40.

Refer to the accompanying illustration.

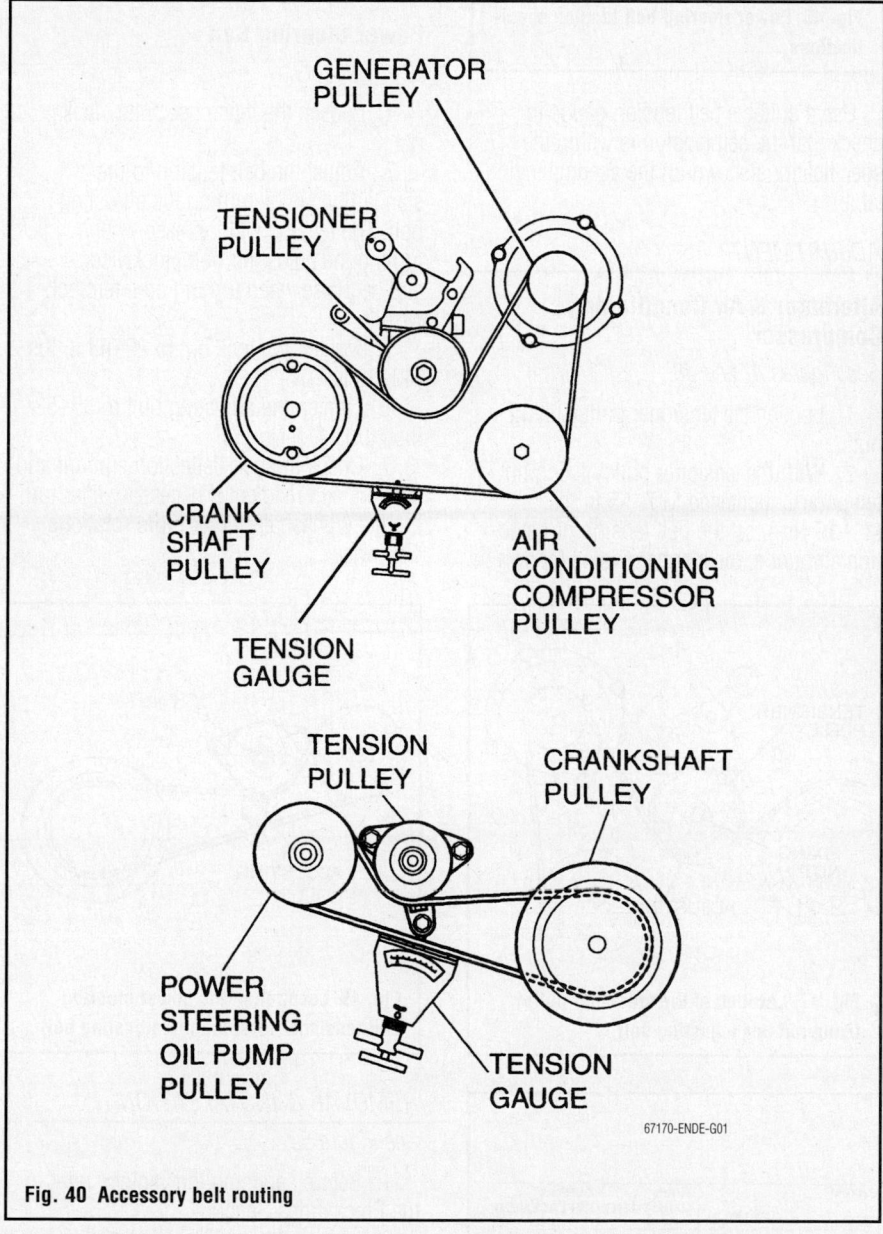

Fig. 40 Accessory belt routing

INSPECTION

Inspect the drive belt for signs of glazing or cracking. A glazed belt will be perfectly smooth from slippage, while a good belt will have a slight texture of fabric visible. Cracks will usually start at the inner edge of the belt and run outward. All worn or damaged drive belts should be replaced immediately.

Alternator-A/C Compressor Belt

There are 3 methods to check the belt tension: using a scan tool, using a belt tension gauge or the measuring the belt deflection. Each method is outlined below.

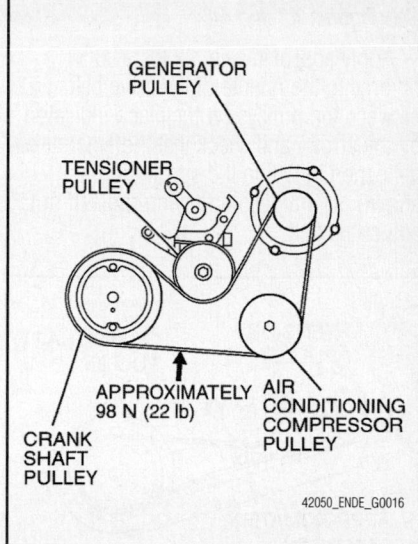

Fig. 41 Apply approximately 22 lbs. (98 N) of force to the middle of the drive belt between the pulleys (at the place indicated by the arrow) and check that the amount of deflection in within specifications

Belt Tension Gauge Method

See Figure 42.

Use a suitable belt tension gauge to check that the belt tension is within 110–154 lbs. (490–686 N). If not, adjust the belt.

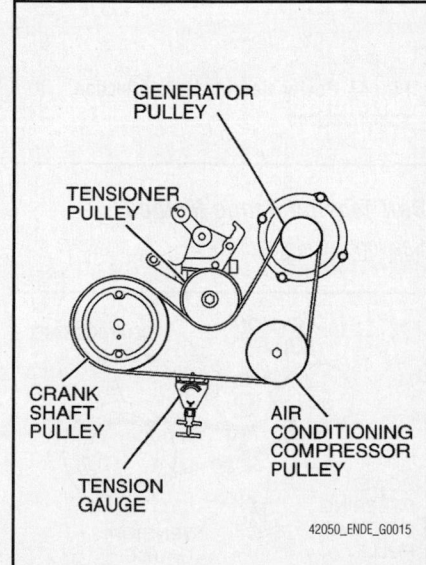

Fig. 42 Use a suitable belt tension gauge to check that the belt tension is within specifications

Belt Deflection Method

See Figure 41.

Apply approximately 22 lbs. (98 N) of force to the middle of the drive belt between the pulleys (at the place indicated by the arrow) and check that the amount of deflection in within 0.33–0.42 in. (8.4–10.7mm). If not, adjust the belt.

Power Steering Pump Belt

Belt Deflection Method

See Figures 43 and 44.

Apply approximately 22 lbs. (98 N) of force to the middle of the drive belt between the pulleys (at the place indicated by the arrow) and check that the amount of deflection in within the specifications shown on the accompanying table. If not, adjust the belt.

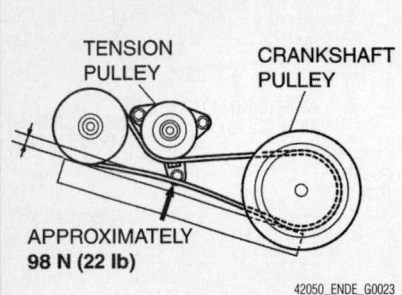

Fig. 43 Apply approximately 22 lbs. (98 N) of force to the middle of the drive belt between the pulleys (arrow) and check that the amount of deflection in within specifications

ITEM	WHEN CHECKED	DURING ADJUSTMENT	DURING REPLACEMENT
Deflection (Reference value) (in)	12.3 – 16.2 (0.48 – 0.64)	13.2 – 15.1 (0.52 – 0.59)	9.6 – 12.3 (0.38 – 0.48)

42050_ENDE_G0024

Fig. 44 Power steering belt deflection specifications

Belt Tension Gauge Method

See Figures 45 and 46.

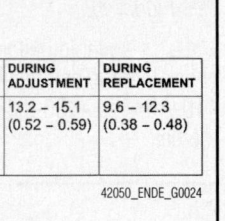

42050_ENDE_G0021

Fig. 45 Use a suitable belt tension gauge to check that the belt tension is within specifications

ITEM	WHEN CHECKED	DURING ADJUSTMENT	DURING REPLACEMENT
Tension N (lb)	294 – 490 (66 – 110)	343 – 441 (77 – 99)	490 – 686 (110 – 154)

42050_ENDE_G0022

Fig. 46 Power steering belt tension specifications

Use a suitable belt tension gauge to check that the belt tension is within the specifications shown on the accompanying table.

ADJUSTMENT

Alternator & Air Conditioning Compressor

See Figures 47 and 48.

1. Loosen the tensioner pulley fixing nut.

2. With the tensioner pulley fixing nut temporarily tightened to 7–15 ft. lbs. (11–19 Nm), set the belt tension or deflection amount to the standard value (shown in the accompanying table) using the adjusting bolt.

➡ **Because the frequency depends on the belt material, confirm Part No. shown on the reverse of the belt.**

3. After the proper specification is achieved, tighten the tension pulley fixing nut to 29–43 ft. lbs. (39–59 Nm).

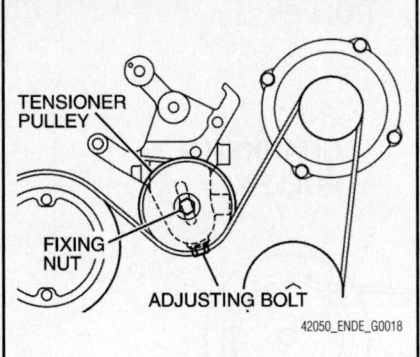

Fig. 47 Location of the tensioner pulley fixing nut and adjusting bolt

ITEM	Part No.	DURING ADJUSTMENT	DURING REPLACEMENT
Vibration frequency Hz	MD368275	140 – 152	168 – 188
	MN158101, MN187016	150 – 163	180 – 202
Tension N (lb)		539 – 637 (121 – 143)	785 – 981 (176 – 221)
Deflection (Reference value) mm (in)		8.9 – 10.1 (0.35 – 0.40)	6.2 – 7.5 (0.24 – 0.30)

42050_ENDE_G0017

Fig. 48 Belt tension specification table

Power Steering Belt

See Figure 49.

1. Loosen the tensioner pulley lock nut.

2. Adjust the belt tension to the standard value by turning the adjusting bolt. The tension will increase when turning the adjusting bolt clockwise, and decrease when turning counterclockwise.

3. Tighten the lock nut to 29–43 ft. lbs. (40–58 Nm).

4. Tighten the adjusting bolt to 35–53 inch lbs. (4–6 Nm).

5. Check the belt deflection amount and tension, and readjust if necessary after turning the crankshaft one or more rotations clockwise

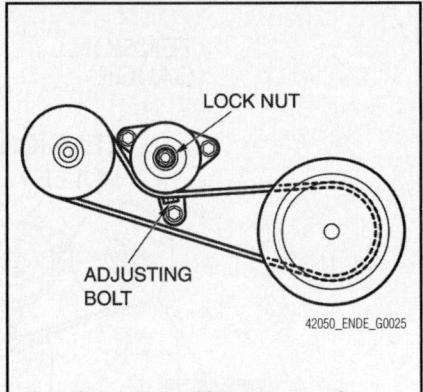

Fig. 49 Location of the power steering belt tensioner lock nut and adjusting bolt

REMOVAL & INSTALLATION

See Figure 50.

1. Before servicing the vehicle, refer to the Precautions Section.

2. Loosen the tensioner pulley fixing nut or locknut, then remove the belt.

To install:

3. Install the belt, carefully routing it as shown in the accompanying illustration.

4. Adjust the belt tension and tighten the fixing bolt or lock nut and adjusting bolt as specified in the adjustment procedures.

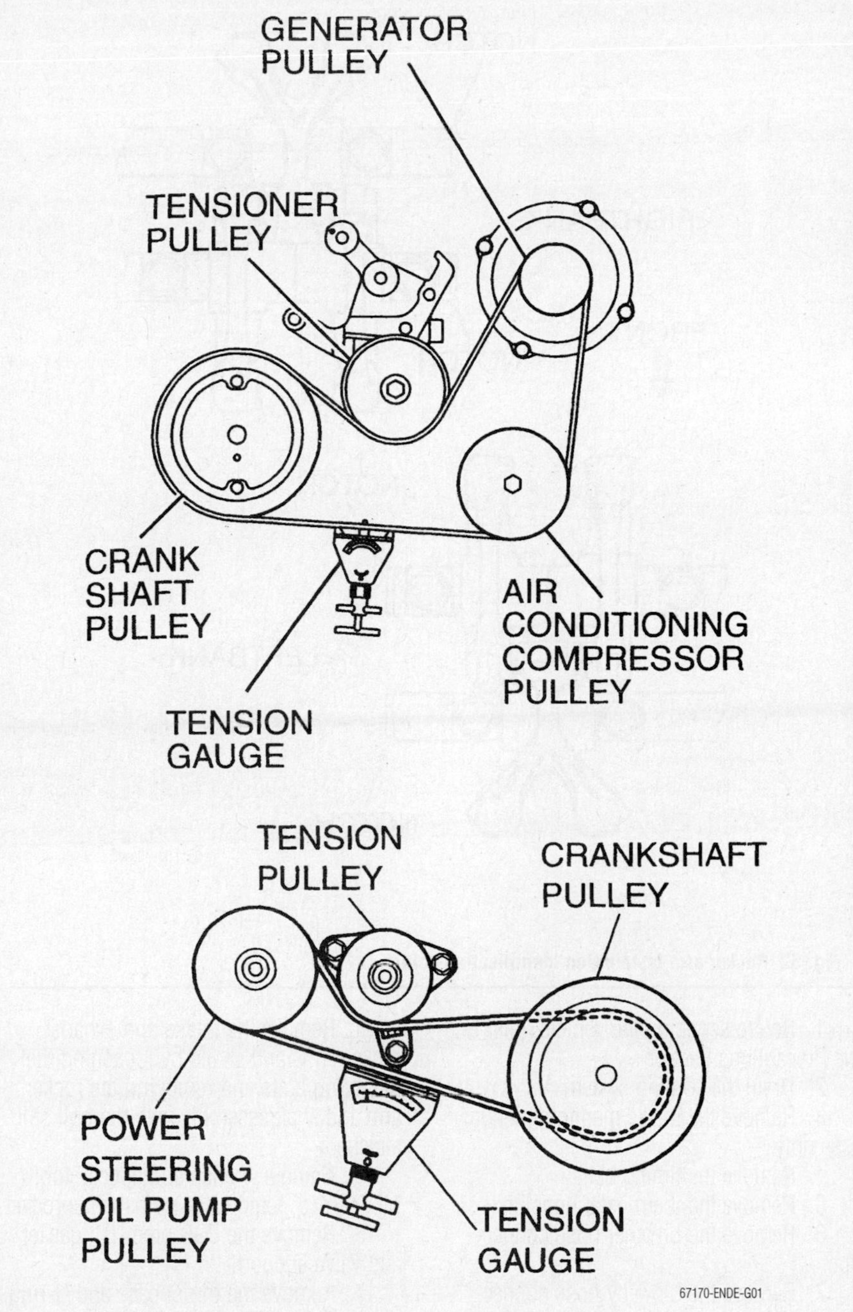

GENERATOR PULLEY

TENSIONER PULLEY

CRANK SHAFT PULLEY

TENSION GAUGE

AIR CONDITIONING COMPRESSOR PULLEY

TENSION PULLEY

CRANKSHAFT PULLEY

POWER STEERING OIL PUMP PULLEY

TENSION GAUGE

67170-ENDE-G01

Fig. 50 Accessory belt routing

CAMSHAFT AND VALVE LIFTERS

INSPECTION

1. Remove the camshaft from the engine.
2. Check the camshaft bearing journals for damage and binding.
3. If the journals are binding, check the cylinder head for damage.

4. Check the cylinder head for clogged oil holes.
5. As required, check the tooth surface of the distributor drive gear teeth of the camshaft. Replace the camshaft if wear is evident.
6. Check the camshaft surface for abnormal wear and damage. Replace the camshaft, as required.
7. Measure the cam height and

replace the camshaft if not within specification.

REMOVAL & INSTALLATION

Left Side

See Figure 51.

1. Before servicing the vehicle, refer to the Precautions Section.
2. Drain the cooling system.
3. Remove the timing belt.
4. Remove the thermostat housing.
5. Remove the blow-by hose connection.
6. Remove the PCV valve connection.
7. Remove the PCV valve.
8. Disconnect the ignition coil connector.
9. Remove the ignition coil.
10. Disconnect the control wiring harness clamp.
11. Remove the valve cover retaining bolts.
12. Remove the valve cover.
13. Remove the spark plug guide oil seal.

➡️**Install auto lash adjuster retainers SST MD998443 on the rocker arms.**

14. Remove the intake and exhaust rocker arms and shafts by loosening the mounting bolts and removing the rocker arm and shaft assembly with the bolt still attached.
15. Disconnect the camshaft position sensor connector.
16. Remove the camshaft position sensor.
17. Remove the camshaft position sensor support.
18. Remove the camshaft position sensing cylinder.

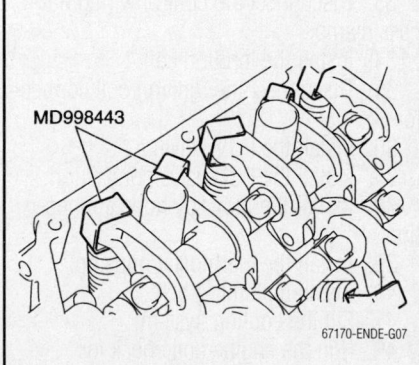

MD998443

67170-ENDE-G07

Fig. 51 Install the special tool to prevent the lash adjusters from falling to the floor during rocker arm removal

19. Using a wrench and special tool MD998715, remove the camshaft sprocket.

20. Remove the camshaft.

21. Remove the camshaft oil seal.

    a. Make a notch in the oil seal lip section with a knife.

    b. Cover the end of a flat-tipped screwdriver with a shop towel and insert into the notched section of the oil seal, and pry out the oil seal to remove it.

To install:

22. Install the camshaft oil seal.

    a. Apply grease to the camshaft oil seal, prior to installation. Use special tools MD998713 and MB991559 to press fit the camshaft oil seal into position.

23. Install the camshaft.

24. Using a wrench and special tool MD998715, install the camshaft sprocket. Tighten to 65 ft. lbs. (88 Nm).

25. Install the camshaft position sensing cylinder.

26. Install the camshaft position sensor support.

27. Install the camshaft position sensor. Tighten to 120 inch lbs. (14 Nm).

28. Connect the camshaft position sensor connector.

29. Install the intake and exhaust rocker arms and shafts. Tighten bolts to 23 ft. lbs. (31 Nm).

➡**Remove auto lash adjuster retainers SST MD998443 from the rocker arms.**

30. Install the spark plug guide oil seal.

31. Install the valve cover.

32. Install the valve cover retaining bolts. Tighten to 31 inch lbs. (4 Nm).

33. Disconnect the control wiring harness clamp.

34. Install the ignition coil.

35. Disconnect the ignition coil connector.

36. Install the PCV valve.

37. Install the PCV valve connection.

38. Install the blow-by hose connection.

39. Install the thermostat housing.

40. Install the timing belt.

41. Fill the cooling system.

42. Run the engine and check for leaks.

Right Side

See Figures 51 and 52.

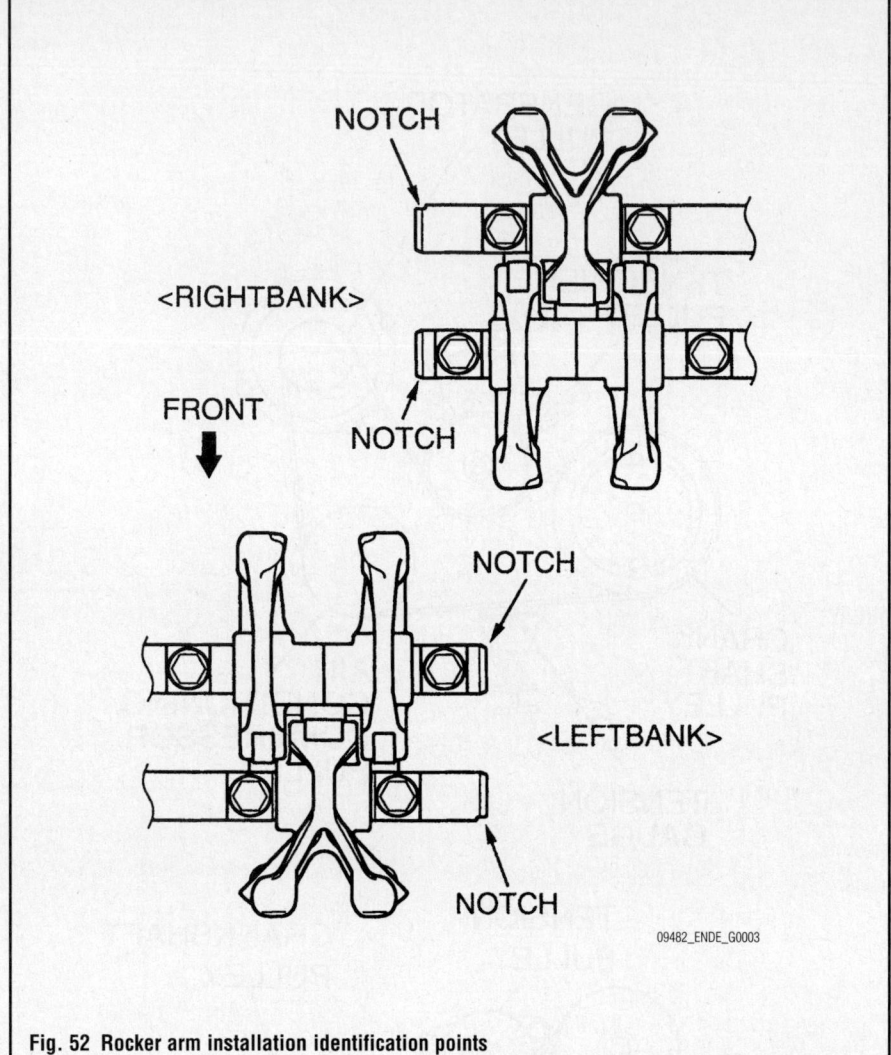

Fig. 52 Rocker arm installation identification points

1. Before servicing the vehicle, refer to the Precautions Section.

2. Drain the cooling system.

3. Remove the intake manifold plenum assembly.

4. Remove the timing belt.

5. Remove the thermostat housing.

6. Remove the breather hose connection.

7. Remove the blow-by hose connection.

8. Remove the ignition coil connector.

9. Remove the ignition coil.

10. Remove the control wiring harness clamp.

11. Remove the valve cover retaining bolts.

12. Remove the valve cover.

13. Remove the spark plug guide oil seal.

➡**Remove the auto lash adjuster retainers SST MD998443 on the rocker arms.**

14. Remove the intake and exhaust rocker arms and shafts by loosening the mounting bolts and removing the rocker arm and shaft assembly with the bolt still attached.

15. Using a wrench and special tool MD998715, remove the camshaft sprocket.

16. Remove the EGR pipe "B", gasket and valve support.

17. Remove the thrust case and O-ring.

18. Remove the camshaft.

    a. Make a notch in the oil seal lip section with a knife.

    b. Cover the end of a flat-tipped screwdriver with a shop towel and insert into the notched section of the oil seal, and pry out the oil seal to remove it.

To install:

19. Install the camshaft oil seal.

    a. Apply grease to the camshaft oil seal, prior to installation. Use special tools MD998713 and MB991559 to press fit the camshaft oil seal into position.

20. Install the camshaft.
21. Install the thrust case and O-ring. Tighten bolts to 14 ft. lbs. (19 Nm).
22. Install the EGR pipe "B", gasket and valve support. Tighten to 120 inch lbs. (14 Nm).
23. Using a wrench and special tool MD998715, install the camshaft sprocket.
24. Install the intake and exhaust rocker arms and shafts. Tighten bolts to 23 ft. lbs. (31 Nm).

➡**Remove the auto lash adjuster retainers SST MD998443 on the rocker arms.**

25. Install the spark plug guide oil seal.
26. Install the valve cover.
27. Install the valve cover retaining bolts. Tighten to 31 inch lbs. (4 Nm).
28. Install the control wiring harness clamp.
29. Install the ignition coil.
30. Install the ignition coil connector.
31. Install the blow-by hose connection.
32. Install the breather hose connection.
33. Install the thermostat housing.
34. Install the timing belt.
35. Install the intake manifold plenum assembly.
36. Fill the cooling system.
37. Run the engine and check for leaks, correct as required.

CRANKSHAFT DAMPER

REMOVAL & INSTALLATION
See Figure 53.

1. Before servicing the vehicle, refer to the Precautions Section.
2. Remove the engine under cover and side under cover.
3. Remove the drive belts.
4. Remove the crankshaft pulley, using special tools MB991800 and MB991802.

To install:
5. Install the crankshaft pulley using pulley holder MB991800 and the 2 crankshaft pulley holder pin tools MB991802, to hold the crankshaft pulley. Tighten the crankshaft pulley bolt to 136 ft. lbs. (185 nm).
6. Install the drive belts.
7. Install the side and under cover.

CRANKSHAFT FRONT SEAL

REMOVAL & INSTALLATION
See Figures 54 and 55.

1. Before servicing the vehicle, refer to the Precautions Section.
2. Remove the timing belt.
3. Remove the crankshaft position (CKP) sensor.
4. Remove the crankshaft sprocket.
5. Remove the crankshaft sensing blade.
6. Remove the crankshaft spacer and key.
7. Remove the front oil seal.

To install:
8. Install the front oil seal. Apply oil to the seal, and install using Crankshaft Front Oil Seal Installer tool no. MD998717.

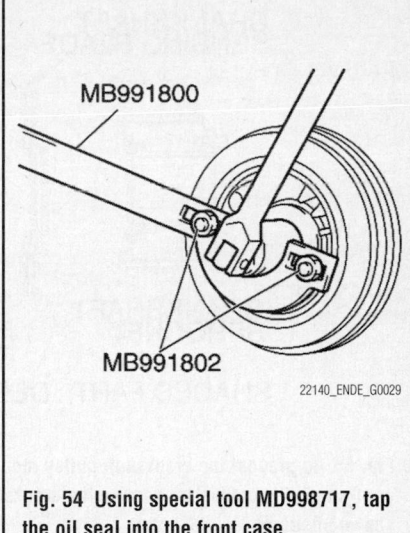

Fig. 54 Using special tool MD998717, tap the oil seal into the front case

9. Install the crankshaft key and spacer.
10. Install the crankshaft sensing blade.
11. Install the CKP sensor.

➡**To be sure the crankshaft pulley bolt does not loosen, make sure the clean the mating areas of the crankshaft, spacer, sensing blade and sprocket.**

12. Install the crankshaft sprocket.
13. Install the timing belt.

CYLINDER HEAD

REMOVAL & INSTALLATION
See Figures 56 and 57.

1. Before servicing the vehicle, refer to the Precautions Section.

✻✻ CAUTION

The fuel injection system remains under pressure after the engine has beenOFF. Properly relieve fuel pressure before disconnecting any fuel lines. Failure to do so may result in fire or personal injury.

2. Relieve fuel system pressure.
3. Drain the cooling system.
4. Disconnect the negative battery cable.

➡**Wait at least 90 seconds after disconnecting the cable from the negative (-) battery terminal to prevent airbag activation.**

5. Remove the intake manifold.
6. Remove the exhaust manifold.

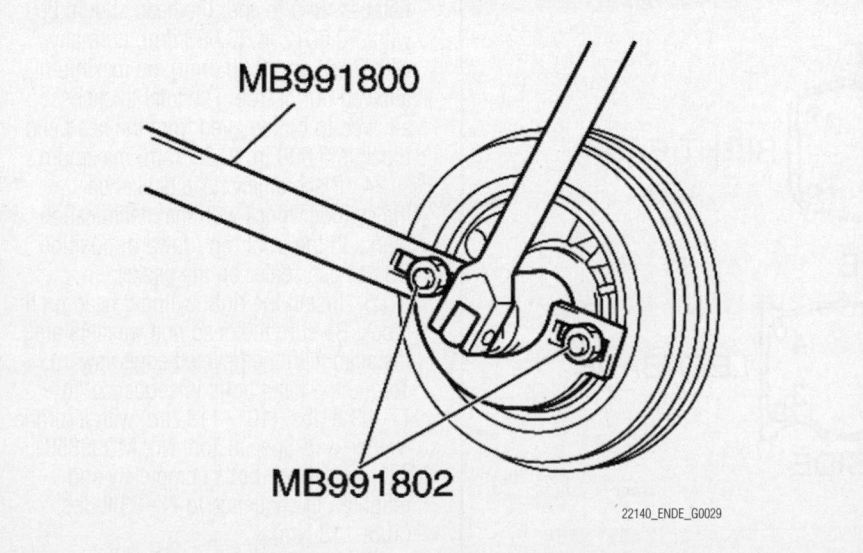

Fig. 53 Removing the crankshaft pulley, using special tools MB991800 and MB991802

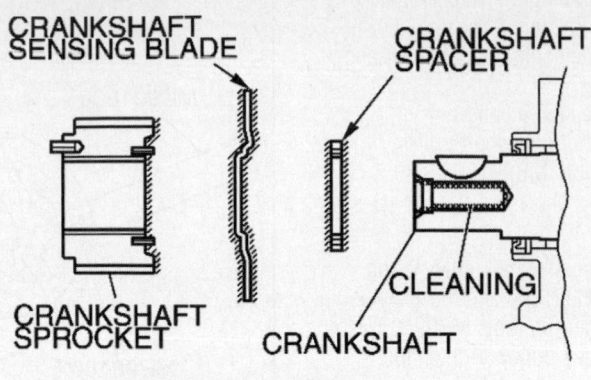

SHADED PART : DEGREASE

22140_ENDE_G0031

Fig. 55 To prevent the crankshaft pulley mounting bolt from loosening, degrease or clean the crankshaft, the crankshaft spacer, the crankshaft sensing blade and the crankshaft at the shown positions

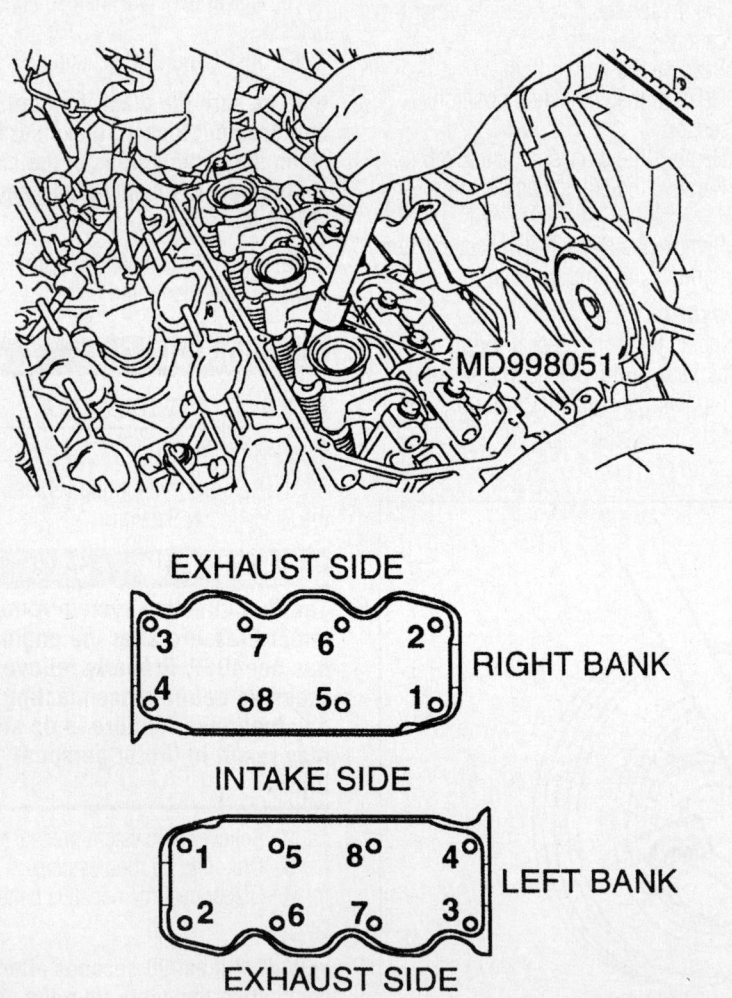

09482_ENDE_G0001

Fig. 56 Cylinder head bolt removal sequence

7. Remove the timing belt.

8. Remove the thermostat housing.

9. Remove the alternator.

10. Remove the blow-by hose from the left and right valve covers.

11. Remove the positive crankcase ventilation (PCV) hose from the left and right valve covers.

12. Remove the spark plug wires; tag before disconnecting.

13. Remove the ignition coil connectors and ignition coils.

14. Remove the engine control wiring harness clamp.

15. Remove the rocker covers and gaskets.

16. Disconnect the camshaft position (CMP) sensor connector.

17. Remove the ground strap.

18. Remove the timing belt rear center cover.

19. Loosen the left cylinder head mounting bolts in 3 steps, in the sequence shown. Lift off the left cylinder head assembly and remove the head gasket.

20. Remove the power steering oil pump bracket.

21. Remove the Exhaust Gas Recirculation (EGR) pipe B and gasket.

22. Loosen the right cylinder head mounting bolts in 3 steps, in the sequence shown. Lift off the right cylinder head assembly and remove the head gasket.

To install:

23. Thoroughly clean and dry the mating surfaces of the head and block. Check the cylinder head for cracks, damage or engine coolant leakage. Remove scale, sealing compound and carbon. Clean oil passages thoroughly. Check the head for flatness. End to end, the head should be within 0.0012 in. (0.030mm), normally with 0.008 in. (0.203mm) the maximum allowed out of true. The total thickness allowed to be removed from the head and block is 0.008 in. (0.203mm) maximum.

24. Place a new head gasket on the cylinder block with the identification marks in the front top (upward) position. Do not use sealer on the gasket.

25. Install the right cylinder head on the block. Be sure the head bolt washers are installed with the beveled edge upward. Torque the head bolts in sequence, to 77–83 ft. lbs. (105–113 Nm) with a torque wrench and Special Tool No. MD998501, then loosen the bolts completely and retighten in sequence to 77–83 ft. lbs. (105–113 Nm).

26. Install the exhaust gas recirculation (EGR) pipe "B" and gasket

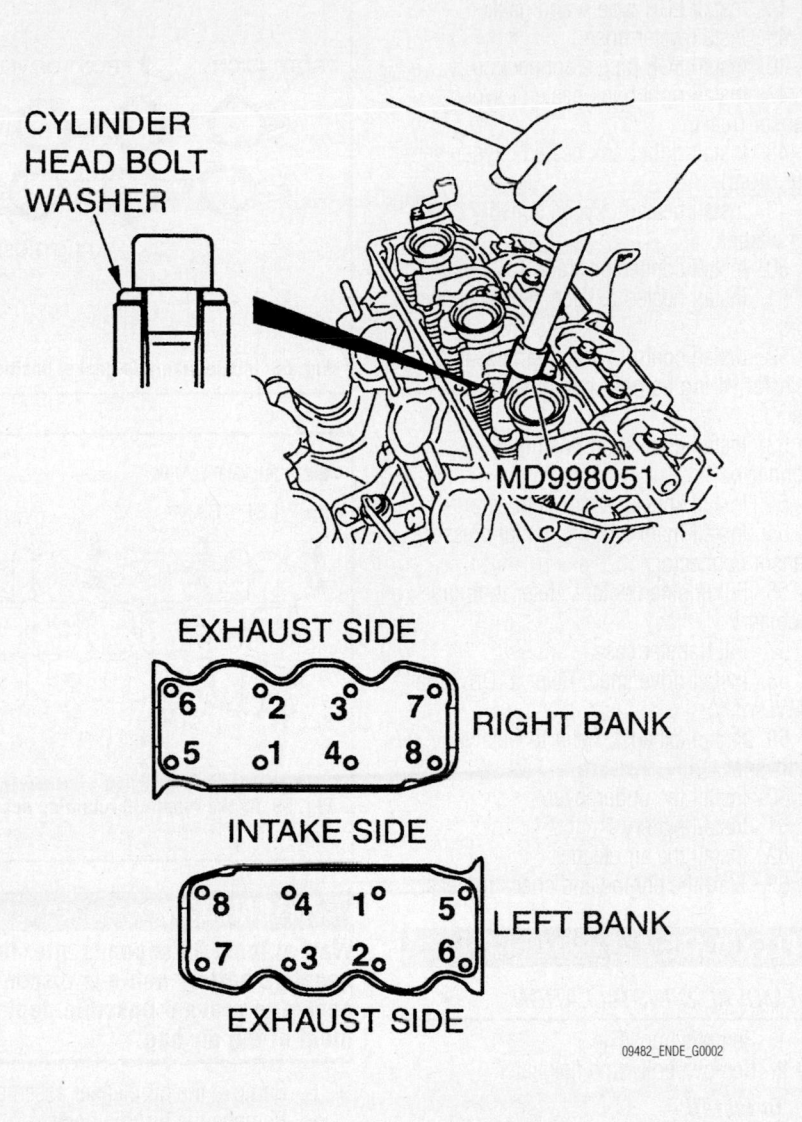

CYLINDER
HEAD BOLT
WASHER

MD998051

EXHAUST SIDE

6	2	3	7
5	1	4	8

RIGHT BANK

INTAKE SIDE

8	4	1	5
7	3	2	6

LEFT BANK

EXHAUST SIDE

09482_ENDE_G0002

Fig. 57 Cylinder head bolt torque sequence

27. Install the power steering oil pump bracket

28. Install the left cylinder head on the block. Be sure the head bolt washers are installed with the beveled edge upward. Torque the head bolts in sequence, to 77–83 ft. lbs. (105–113 Nm) with a torque wrench and Special Tool No. MD998501, then loosen the bolts completely and retighten in sequence to 77–83 ft. lbs. (105–113 Nm).

29. Install the timing belt rear center cover.

30. Install the ground strap.

31. Install the CMP sensor connector.

32. Install the rocker covers with new gaskets.

33. Install the engine control wiring harness clamp.

34. Install the ignition coils and their connectors.

35. Install the spark plug wires.

36. Install the positive crankcase ventilation (PCV) hoses.

37. Install the blow-by hoses .

38. Install the alternator.

39. Install the thermostat housing.

40. Install the timing belt.

41. Install the exhaust manifold.

42. Install the intake manifold.

43. Connect the negative battery cable.

44. Turn the ignition **ON** and then **OFF**, and keep it off for at least ten seconds.

45. Change the engine oil and oil filter.

46. Refill the system with coolant.

47. Run the engine until the thermostat opens.

48. Once the engine has cooled, recheck the coolant level.

EXHAUST MANIFOLD

REMOVAL & INSTALLATION

Left Side

1. Before servicing the vehicle, refer to the Precautions Section.

2. Remove the engine under cover

3. Remove the air duct.

4. Remove the left Heated Oxygen Sensor (HO2S) connectors and sensors.

5. Remove the engine oil dipstick, guide and O-ring.

6. Remove the heat shield.

7. Remove the front exhaust pipe from the manifold. Remove and discard the gasket.

8. Remove the exhaust manifold stay (left B).

9. Remove the left side exhaust manifold and gasket.

10. Remove the exhaust manifold stay (left A).

11. Clean the gasket mounting surfaces. Inspect the manifolds for cracks, flatness and/or damage.

To install:

12. Install the exhaust manifold stay (left A).

13. Install the new gasket and left side exhaust manifold. Torque the nuts to specification.

14. Install the exhaust manifold stay (left B).

15. Install the exhaust pipe to the manifold with a new gasket. Torque the bolts to 31–43 ft. lbs. (40–58 Nm).

16. Install the left side heat shield.

17. Install the oil dipstick, guide and new O-ring.

18. Install the left HO2S and connectors.

19. Install the air duct.

20. Install the undercover.

21. Start the engine and check for exhaust leaks.

Right Side

1. Before servicing the vehicle, refer to the Precautions Section.

2. Remove air cleaner.

3. Remove battery.

4. Remove engine under cover.

5. Remove exhaust. Refer to Catalytic Converter.

6. Remove drive shaft. Refer to Drive-shaft (AWD only).

7. Remove transfer case oil.

8. Remove engine coolant. Refer to Engine Coolant.

9. Remove manifold differential pressure sensor connector.

10. Remove knock sensor connector.

11. Remove crankshaft position sensor connector.

12. Remove control wiring harness and injector wiring harness combination connector.

13. Remove right bank heated oxygen sensor (front) connector.

14. Remove right bank heated oxygen sensor (rear) connector.

15. Remove connector bracket.

16. Remove right bank heated oxygen sensor (front) connector clamp.

17. Remove right bank heated oxygen sensor (rear) connector clamp.

18. Use special tool MD998770 to remove right bank heated oxygen sensor (front).

19. Use special tool MD998770 to remove right bank heated oxygen sensor (rear).

20. Remove EGR pipe a connection.

21. Remove water hose.

22. Remove water hose.

23. Remove EGR pipe A.

24. Remove EGR pipe A gasket.

25. Remove exhaust manifold stay, right b.

26. Remove steering gear and linkage protector.

27. Remove front floor backbone brace.

28. Remove front heat protector panel.

29. Remove transfer extension housing (AWD only).

30. Remove O-ring (AWD only).

31. Remove lower heat protector.

32. Remove upper heat protector.

33. Remove power steering return pipe clamp connecting bolt, nut (AWD only).

34. Remove exhaust manifold and gasket.

To install:

35. Install exhaust manifold and gasket.

36. Install power steering return pipe clamp connecting bolt, nut (AWD only).

37. Install upper heat protector.

38. Install lower heat protector.

39. Install O-ring and transfer extension housing (AWD only).

40. Install front heat protector panel.

41. Install front floor backbone brace.

42. Install steering gear and linkage protector.

43. Install exhaust manifold stay, right b.

44. Install EGR pipe A and gasket.

45. Install water hose.

46. Install EGR pipe a connection.

47. Install right bank heated oxygen sensor (rear).

48. Install right bank heated oxygen sensor (front).

49. Install heated oxygen sensor connector clamps.

50. Install connector bracket.

51. Install heated oxygen sensor connectors.

52. Install control wiring harness and injector wiring harness combination connector.

53. Install crankshaft position sensor connector.

54. Install knock sensor connector.

55. Install manifold differential pressure sensor connector.

56. Fill engine coolant. Refer to Engine Coolant.

57. Fill transfer case.

58. Install drive shaft. Refer to Driveshaft (AWD only).

59. Install exhaust. Refer to Catalytic Converter.

60. Install the undercover.

61. Install battery.

62. Install the air cleaner.

63. Start the engine and check for leaks.

FLEXPLATE

REMOVAL & INSTALLATION

1. Remove transaxle.
2. Remove bolts and flexplate.

To install:

3. Install flexplate and bolts. Tighten to 33–39 ft. lbs. (46–52Nm).
4. Install transaxle.

INTAKE MANIFOLD

REMOVAL & INSTALLATION

See Figures 58 and 59.

1. Before servicing the vehicle, refer to the Precautions Section.

❉❉ CAUTION

The fuel injection system remains under pressure after the engine has beenOFF. Properly relieve fuel pressure before disconnecting any fuel lines. Failure to do so may result in fire or personal injury.

2. Relieve the fuel pressure.
3. Partially drain the cooling system.
4. Disconnect the negative battery cable.

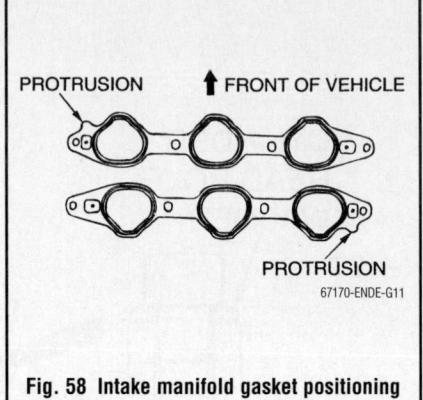

Fig. 58 Intake manifold gasket positioning

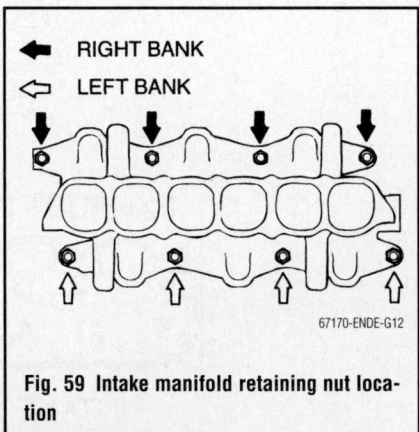

Fig. 59 Intake manifold retaining nut location

❉❉ CAUTION

Wait at least 90 seconds after the negative battery cable is disconnected to prevent possible deployment of the air bag.

5. Remove the air cleaner assembly.

6. Remove the throttle body.

7. Remove the manifold differential pressure (MDP) sensor connector.

8. Remove the knock sensor (KS) connector.

9. Remove the crankshaft position (CKP) sensor connector.

10. Remove the control wiring harness and injector wiring harness combination connector.

11. Remove the right bank heated oxygen sensor (ho2s) connector connections and clamps.

12. Remove the connector bracket.

13. Remove the evaporative emission (EVAP) purge solenoid connector.

14. Remove the purge hose.

15. Remove the purge hose connection.

16. Remove the EVAP purge solenoid.

17. Remove the front intake manifold plenum stay.

18. Remove the exhaust gas recirculation (EGR) pipe a clamp.

19. Remove the power steering pressure hose clamp.

20. Remove the power steering pressure hose clamp bracket.

21. Remove the rear intake manifold plenum stay.

22. Remove the EGR pipe b connection.

23. Remove the retaining bolts and intake manifold plenum.

24. Remove the intake manifold plenum gasket.

25. Remove the EGR adapter and gasket.

26. Remove the MDP sensor and o-ring.

27. Remove the ignition coil connector and coil.

28. Remove the injector connector.

29. Remove the engine mount stay.

30. Remove the high pressure fuel hose and o-ring.

31. Remove the blow-by hose.

32. Remove the fuel rail and injectors. also remove the damper assembly, if equipped.

33. Remove the positive crankcase ventilation (PCV) hose.

34. Remove the right and left front upper timing belt cover.

35. Remove the water pump bracket.

36. Remove the intake manifold retainers, manifold and gasket. thoroughly clean and dry the mating surfaces of the manifold and heads.

To install:

37. Install the new intake manifold gasket. Make sure the gaskets are installed with the protrusions as shown in the illustration.

38. Install the intake manifold.

39. Coat the intake manifold retaining studs with clean engine oil. Install the intake manifold bolts and tighten as follows:

   a. 1st step: Right bank nuts to 45–71 inch lbs. (5–8 Nm).

   b. 2nd step: Left bank nuts to 15–17 ft. lbs. (21–23 Nm).

   c. 3rd step: Right bank nuts to 15–17 ft. lbs. (21–23 Nm).

   d. 4th step: Left bank nuts to 15–17 ft. lbs. (21–23 Nm).

   e. 5th step: Right bank nuts to 15–17 ft. lbs. (21–23 Nm).

40. Install the water pump bracket.

41. Install the right and left front upper timing belt cover.

42. Install the PCV hose.

43. Install the fuel rail and injectors. also remove the damper assembly, if equipped.

44. Install the blow-by hose.

45. Install the high pressure fuel hose with a new o-ring.

46. Install the engine mount stay.

47. Install the injector connectors.

48. Install the ignition coil connectors and coils.

49. Install the MDP sensor and o-ring.

50. Install the EGR adapter and gasket.

51. Install the new intake manifold plenum gasket.

52. Install the intake manifold plenum. tighten the plenum mounting bolts to 20 ft. lbs. (28 Nm).

53. Install the EGR pipe b connection.

54. Install the rear intake manifold plenum stay.

55. Install the power steering pressure hose clamp bracket.

56. Install the power steering pressure hose clamp.

57. Install the EGR pipe a clamp.

58. Install the front intake manifold plenum stay.

59. Install the EVAP purge solenoid, hose and connector.

60. Install the connector bracket.

61. Install the right bank ho2s clamps and connectors.

62. Install the control wiring harness and injector wiring harness combination connector.

63. Install the CKP sensor connector.

64. Install the KS connector.

65. Install the MDP sensor connector.

66. Install the throttle body.

67. Install the air cleaner assembly.

68. Refill the radiator with coolant.

69. Connect the negative battery cable.

70. Turn the ignition **ON** and then **OFF**, and keep it off for at least ten seconds.

71. Run the engine and check for fuel leaks, correct as required.

72. Correct and adjust all fluid levels, as required.

OIL PAN

REMOVAL & INSTALLATION

See Figures 60 through 62.

1. Before servicing the vehicle, refer to the Precautions Section.

2. Drain the engine oil.

3. Remove the engine undercover.

4. Remove the front exhaust pipe.

5. Remove the oil pan drain plug and gasket.

6. Remove the starter motor.

7. Remove the engine oil dipstick and o-ring.

8. Remove the lower oil pan. if necessary, use a block of wood and

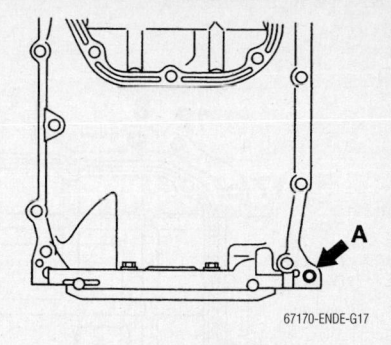

Fig. 60 Oil pan removal bolt installation location

hammer to carefully dislodge the lower oil pan.

9. Remove the cover.

10. Remove the 2 lower torque converter connecting bolts.

11. Remove the upper oil pan. screw the m10 bolts holding the oil pan to the transaxle assembly into the bolt hole shown in the illustration to remove the pan.

12. Remove the oil screen.

13. Remove the oil pan gasket.

To install:

14. Before installing, thoroughly clean the oil pan and cylinder block mating surfaces.

15. Apply liquid gasket around the surface of the oil pan.

➡**Assemble the oil pan to the cylinder block within 30 minutes after applying the liquid gasket.**

16. Install the oil screen. Torque the bolts to 12–16 ft. lbs. (16–22 Nm).

17. Install the upper oil pan. torque the bolts to specification and, in the proper sequence. the bolt holes for bolts 13 and 14 are cut away on the transaxle side. be sure you do not insert the bolts at an angle.

18. Install the two lower torque converter connecting bolts. tighten the bolts to 36 ft. lbs. (49 Nm).

19. Install the cover.

20. Install the lower oil pan. torque the bolts to specification and in the proper sequence.

21. Install the engine oil dipstick and o-ring.

22. Install the starter motor.

23. Install the oil pan drain plug with a new gasket. tighten to specification.

24. Install the front exhaust pipe.

25. Install the engine undercover.

26. Fill the crankcase with oil.

27. Start the engine and check for leaks.

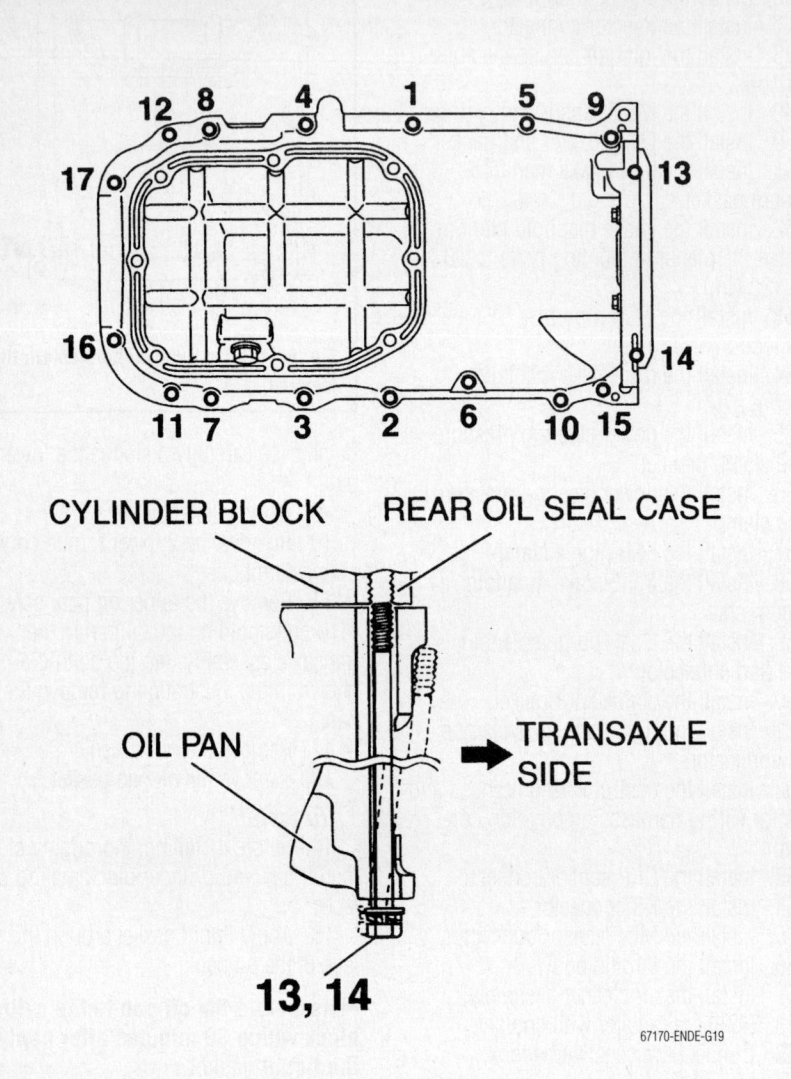

Fig. 61 Upper oil pan bolt torque sequence

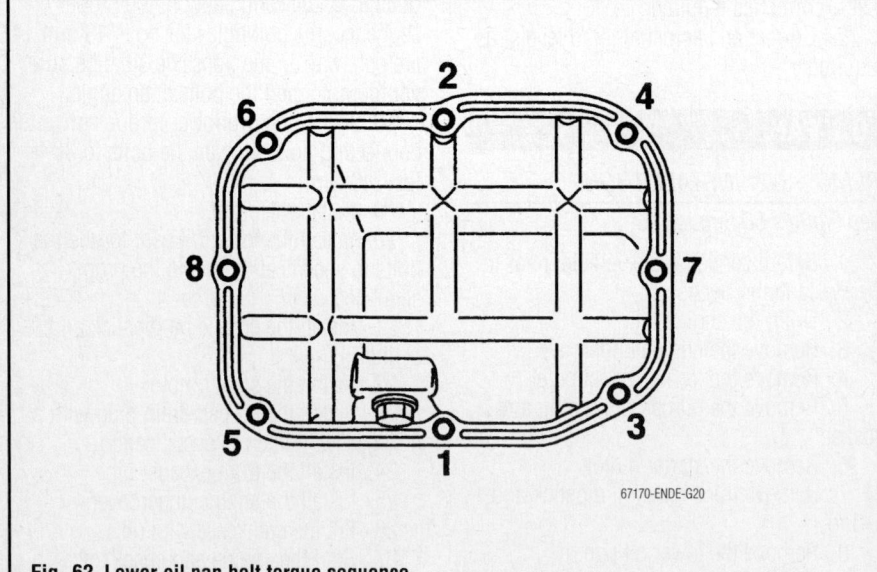

Fig. 62 Lower oil pan bolt torque sequence

OIL PUMP

REMOVAL & INSTALLATION

See Figure 63.

1. Before servicing the vehicle, refer to the Precautions Section.
2. Drain the engine oil.
3. Remove the timing belt.
4. Remove the oil pressure switch.
5. Remove the oil dipstick.
6. Remove the oil pans from the engine.
7. Remove the oil baffle and screen.
8. Remove the oil pump mounting bolts and the pump from the front of the engine.

➡Note the position of each oil pump case retaining bolts to facilitate installation. The bolts are of different length.

To install:

9. Clean the gasket mounting surfaces of the pump and engine block.
10. Prime the pump by pouring fresh oil into the inlet and turning the rotors or by packing pump with petroleum jelly. Using a new gasket, install the oil pump on the engine and tighten all bolts to 10 ft. lbs. (14 Nm).
11. Clean the gasket mounting surfaces of the pump and engine block.
12. Apply a 0.1 inch diameter bead of sealant (part number MD970389 or equivalent) to the oil pump case.

➡Apply sealant as indicated by the broken line in the illustration. The grooves must be traced and the bolt holes must be surrounded with a bead of sealant.

13. Install the oil pump case assembly to the front of the cylinder block.

➡Be sure to install the oil pump case quickly, while the sealant is wet. There is a fifteen minute working window.

14. Torque the oil pump case mounting bolts to 113–131 inch lbs. (13–15 Nm).

➡After installation keep the sealed area free from oil and coolant for at least one hour.

15. Clean out the oil pick-up or replace as required. Replace the oil pick-up gasket ring and install the pick-up to the pump.

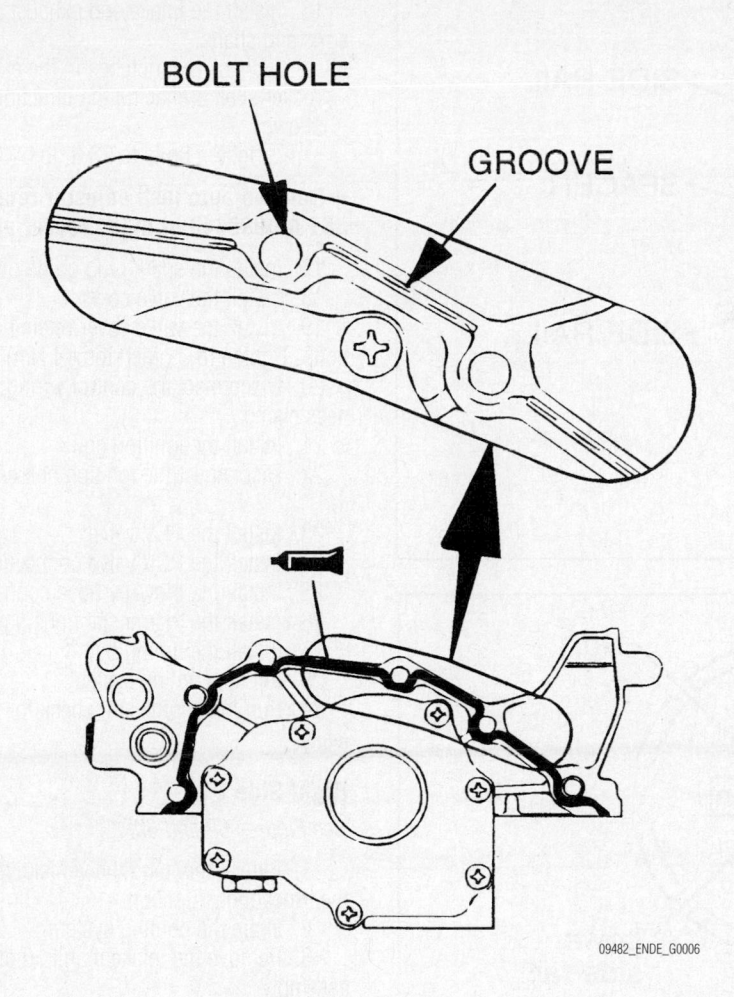

Fig. 63 Oil pump to case sealant application

REAR MAIN SEAL

REMOVAL & INSTALLATION

See Figure 67.

1. Before servicing the vehicle, refer to the Precautions Section.
2. Remove the transaxle assembly.
3. Remove the transfer case, if AWD.
4. Remove the driveplate and adapter plate, matchmark for reassembly. Use Mitsubishi tool (md998781) to hold the driveplate in position to remove the bolts.
5. Remove the rear oil seal as follows:
   a. Cut out a portion in the crankshaft oil seal lip.
   b. Cover the tip of a small prytool with a cloth and apply it to the cutout in the oil seal to pry the oil seal out.

✱✱ CAUTION

Take care not to damage the crankshaft and oil seal case.

To install:
6. Inspect the sealing surface at the rear of the crankshaft. If a deep groove is worn into the surface, the crankshaft will have to be replaced. Coat the sealing lip of the seal with fresh, clean engine oil. Press the new seal into the case with a seal installing tool. The seal must be pressed in squarely until it bottoms in the case. It is necessary to use the proper tool (MD998718-01) to fit the seal into place.
7. Install the drive plate and adapter. Use Mitsubishi tool (md998781) to hold the driveplate in position while tightening the bolts to 55 ft. lbs. (75 Nm).

16. Install the oil filter and the bracket. torque the bolts to 17 ft. lbs. (23 Nm).
17. Install the oil baffle and screen. torque the bolts to 13 ft. lbs. (18 Nm).
18. Install the oil pans.
19. Install the oil pressure switch. torque the switch to 87 inch lbs. (9.8 Nm).
20. Install the timing belt.
21. Install the dipstick.
22. Refill the engine with the proper amount of oil.
23. Start the engine and check for proper oil pressure.
24. Check for and correct leaks.

PISTON AND RING

POSITIONING

See Figures 64 through 66.

Fig. 64 Piston ring identification

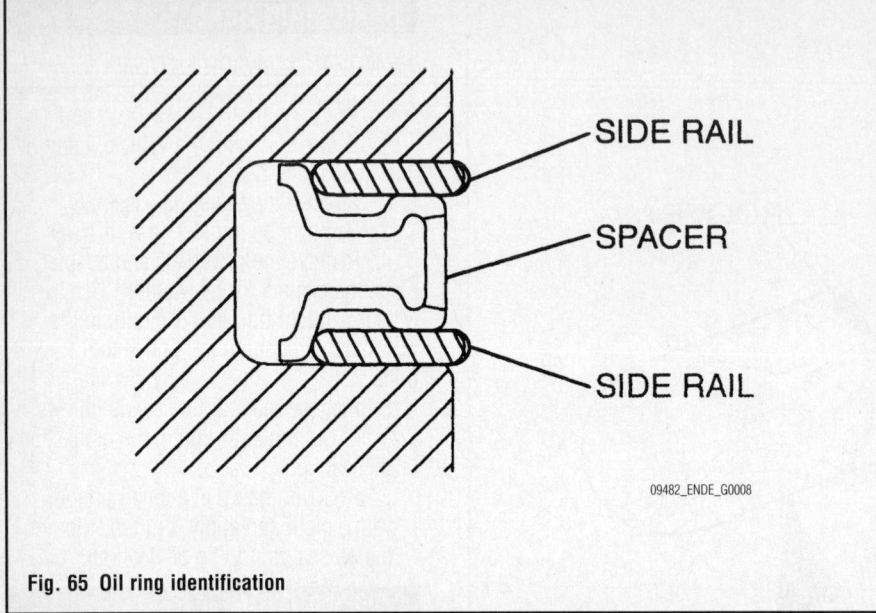

Fig. 65 Oil ring identification

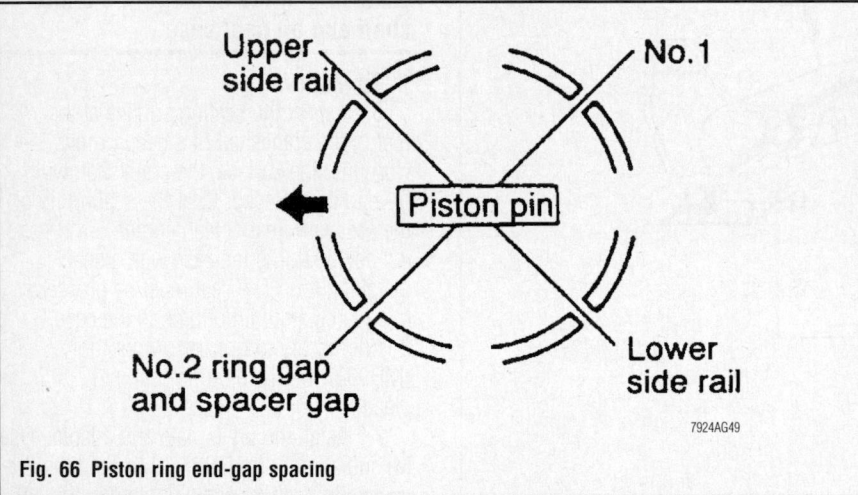

Fig. 66 Piston ring end-gap spacing

8. Install the transfer case, if AWD.

9. Install the transaxle and related components as necessary.

ROCKER ARMS

REMOVAL & INSTALLATION

Left Side

See Figures 68 and 69.

1. Before servicing the vehicle, refer to the Precautions Section.
2. Drain the cooling system.
3. Remove the timing belt.
4. Remove the thermostat housing.
5. Remove the blow-by hose connection.
6. Remove the PCV valve connection.
7. Remove the PCV valve.
8. Disconnect the ignition coil connector.

9. Remove the ignition coil.
10. Disconnect the control wiring harness clamp.
11. Remove the valve cover retaining bolts.
12. Remove the valve cover.
13. Remove the spark plug guide oil seal.

➡**Install auto lash adjuster retainers SST MD998443 on the rocker arms.**

14. Remove the intake and exhaust rocker arms and shafts by loosening he mounting bolts and removing the rocker arm and shaft assembly with the bolt still attached.

To install:

➡**Lubricate the valve train components with clean engine oil.**

15. Bleed and install the lash adjusters

in their original bores in the cylinder head.

16. Install the intake and exhaust rocker arms and shafts.

  a. Check that notches in the each rocker shaft are facing the direction shown.

  b. Tighten bolts to 23 ft. lbs. (31 Nm).

➡**Remove auto lash adjuster retainers SST MD998443 from the rocker arms.**

17. Install the spark plug guide oil seal.
18. Install the valve cover.
19. Install the valve cover retaining bolts. Tighten to 31 inch lbs. (4 Nm).
20. Disconnect the control wiring harness clamp.
21. Install the ignition coil.
22. Disconnect the ignition coil connector.
23. Install the PCV valve.
24. Install the PCV valve connection.
25. Install the blow-by hose connection.
26. Install the thermostat housing.
27. Install the timing belt.
28. Fill the cooling system.
29. Run the engine and check for leaks.

Right Side

See Figures 68 and 69.

1. Before servicing the vehicle, refer to the Precautions Section.
2. Drain the cooling system.
3. Remove the intake manifold plenum assembly.
4. Remove the timing belt.
5. Remove the thermostat housing.
6. Remove the breather hose connection.
7. Remove the blow-by hose connection.
8. Remove the ignition coil connector.
9. Remove the ignition coil.
10. Remove the control wiring harness clamp.
11. Remove the valve cover retaining bolts.
12. Remove the valve cover.
13. Remove the spark plug guide oil seal.

➡**Install the auto lash adjuster retainers SST MD998443 on the rocker arms.**

14. Remove the intake and exhaust rocker arms and shafts by loosening the mounting bolts and removing the rocker arm and shaft assembly with the bolt still attached.

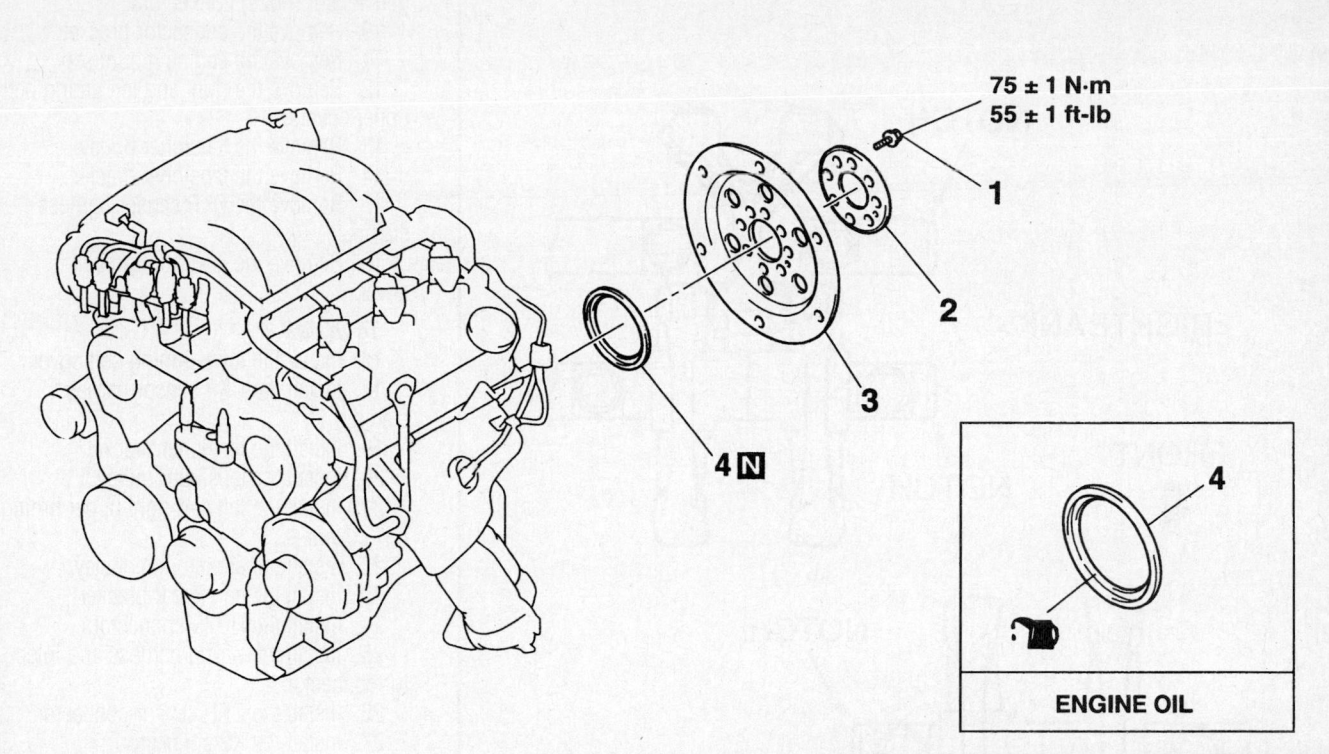

75 ± 1 N·m
55 ± 1 ft-lb

ENGINE OIL

1. DRIVE PLATE BOLTS
2. ADAPTOR PLATE
3. DRIVE PLATE
4. CRANKSHAFT REAR OIL SEAL

67170-ENDE-G24

Fig. 67 Rear main seal and related components

To install:

➡️**Lubricate the valve train components with clean engine oil.**

15. Bleed and install the lash adjusters in their original bores in the cylinder head.

16. Install the intake and exhaust rocker arms and shafts.

　a. Check that notches in the each rocker shaft are facing the direction shown.

　b. Tighten bolts to 23 ft. lbs. (31 Nm).

➡️**Remove the auto lash adjuster retainers SST MD998443 on the rocker arms.**

17. Install the spark plug guide oil seal.

18. Install the valve cover.

19. Install the valve cover retaining bolts. Tighten to 31 inch lbs. (4 Nm).

20. Install the control wiring harness clamp.

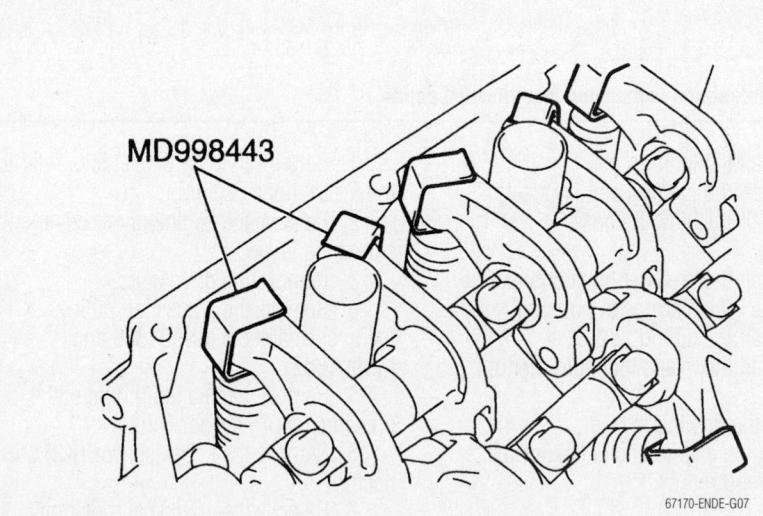

MD998443

67170-ENDE-G07

Fig. 68 Install the special tool to prevent the lash adjusters from falling to the floor during rocker arm removal

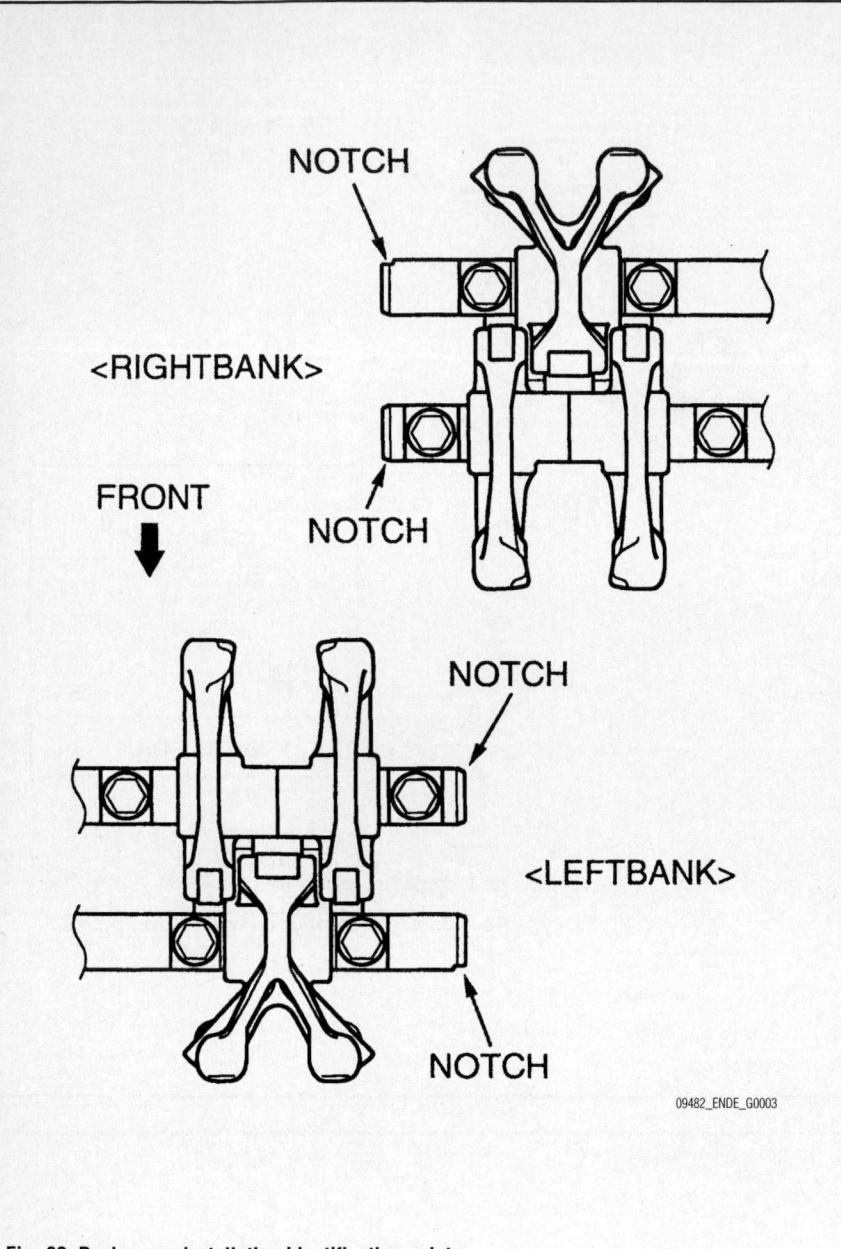

NOTCH

<RIGHTBANK>

FRONT

NOTCH

NOTCH

<LEFTBANK>

NOTCH

09482_ENDE_G0003

Fig. 69 Rocker arm installation identification points

21. Install the ignition coil.
22. Install the ignition coil connector.
23. Install the blow-by hose connection.
24. Install the breather hose connection.
25. Install the thermostat housing.
26. Install the timing belt.
27. Install the intake manifold plenum assembly.
28. Fill the cooling system.
29. Run the engine and check for leaks, correct as required.

TIMING BELT COVER

REMOVAL & INSTALLATION

See Figure 70.

1. Before servicing the vehicle, refer to the Precautions Section.
2. Remove the engine under cover and side under cover.
3. Remove the drive belts.
4. Remove the crankshaft pulley, using special tools MB991800 and MB991802.
5. Remove the manifold differential pressure (MDP) sensor connector.
6. Remove the knock sensor (KS) connector.
7. Remove the crankshaft position (CKP) sensor connector.
8. Remove the control wiring harness and injector wiring harness combination connector.

9. Remove the right bank heated oxygen sensor (ho2s) connectors.
10. Remove the connector bracket.
11. Remove the engine mount stay.
12. Remove the right and left timing belt upper covers.
13. Remove the tensioner pulley.
14. Remove the tensioner bracket.
15. Remove the CKP sensor harness clamp.
16. Remove the lower timing belt cover.

To install:
17. Install the lower timing belt cover.
18. Install the CKP sensor harness clamp.
19. Install the tensioner bracket.
20. Install the tensioner pulley.
21. Install the left and right upper timing belt covers.
22. Install the engine mount stay.
23. Install the connector bracket.
24. Install the HO2S connectors.
25. Install the wiring harness and injector connector.
26. Install the CKP sensor connector.
27. Install the KS connector.
28. Install the MDP sensor connector.
29. Install the crankshaft pulley. Using pulley holder MB991800 and the 2 crankshaft pulley holder pin tools MD991802 to hold the crankshaft pulley, tighten the crankshaft pulley bolt to 136 ft. Lbs. (185 nm).
30. Install the drive belts.
31. Install the side and under covers.

TIMING BELT & SPROCKETS

REMOVAL & INSTALLATION

See Figures 70 through 72.

1. Before servicing the vehicle, refer to the Precautions Section.
2. Remove the engine under cover and side under cover.
3. Remove the drive belts.
4. Remove the crankshaft pulley, using special tools MB991800 and MB991802.
5. Remove the manifold differential pressure (MDP) sensor connector.
6. Remove the knock sensor (KS) connector.
7. Remove the crankshaft position (CKP) sensor connector.
8. Remove the control wiring harness and injector wiring harness combination connector.
9. Remove the right bank heated oxygen sensor (ho2s) connectors.

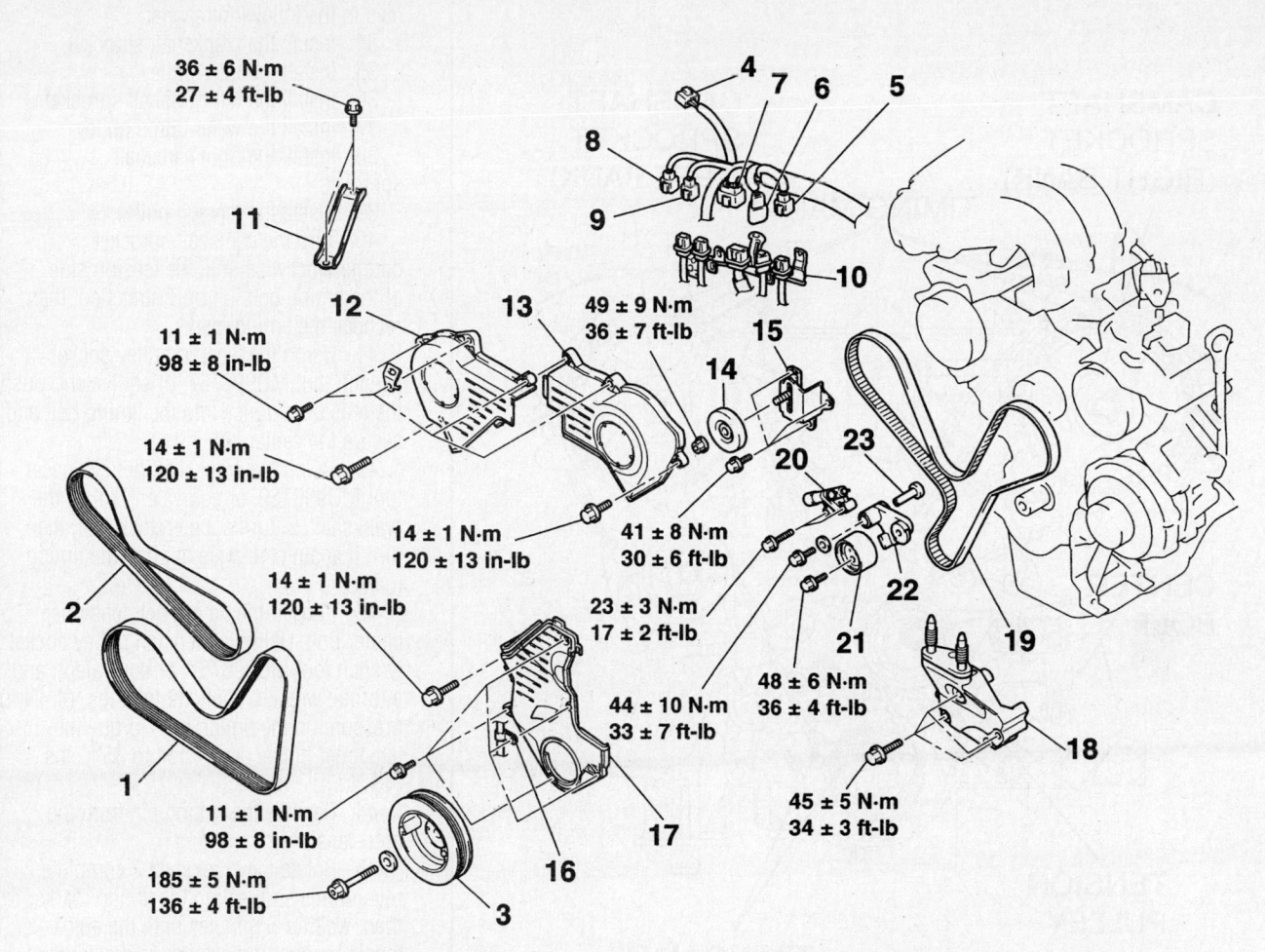

36 ± 6 N·m
27 ± 4 ft-lb

11

4 7 6
8   5
9

10

12   13   49 ± 9 N·m
              36 ± 7 ft-lb   15

11 ± 1 N·m   14
98 ± 8 in-lb

14 ± 1 N·m   20   23
120 ± 13 in-lb

2   22
                14 ± 1 N·m
                120 ± 13 in-lb   41 ± 8 N·m   21   19
                                  30 ± 6 ft-lb
        14 ± 1 N·m
        120 ± 13 in-lb   23 ± 3 N·m   18
                          17 ± 2 ft-lb
1   48 ± 6 N·m
                44 ± 10 N·m   36 ± 4 ft-lb
                33 ± 7 ft-lb
        11 ± 1 N·m   17   45 ± 5 N·m
        98 ± 8 in-lb   34 ± 3 ft-lb
                          16
185 ± 5 N·m
136 ± 4 ft-lb   3

1. GENERATOR DRIVE BELT
2. POWER STEERING OIL PUMP
   DRIVE BELT
3. CRANKSHAFT PULLEY
4. MANIFOLD DIFFERENTIAL
   PRESSURE SENSOR
   CONNECTOR
5. KNOCK SENSOR CONNECTOR
6. CRANKSHAFT POSITION
   SENSOR CONNECTOR
7. CONTROL WIRING HARNESS
   AND INJECTOR WIRING
   HARNESS COMBINATION
   CONNECTOR

8. RIGHT BANK HEATED OXYGEN
   SENSOR (REAR) CONNECTOR
9. RIGHT BANK HEATED OXYGEN
   SENSOR (FRONT) CONNECTOR
10. CONNECTOR BRACKET
11. ENGINE MOUNT STAY
12. TIMING BELT FRONT UPPER
    COVER, RIGHT
13. TIMING BELT FRONT UPPER
    COVER, LEFT
14. TENSIONER PULLEY
15. TENSIONER BRACKET
16. CRANKSHAFT POSITION
    SENSOR HARNESS CLAMP
17. TIMING BELT LOWER COVER

67170-ENDE-G22

Fig. 70 Timing belt and related components

10. Remove the connector bracket.
11. Remove the engine mount stay.
12. Remove the right and left timing belt
upper covers.

13. Remove the tensioner pulley.
14. Remove the tensioner bracket.
15. Remove the CKP sensor harness
clamp.

16. Remove the lower timing belt cover.
17. Remove the engine mount.
18. Remove the engine support
bracket.

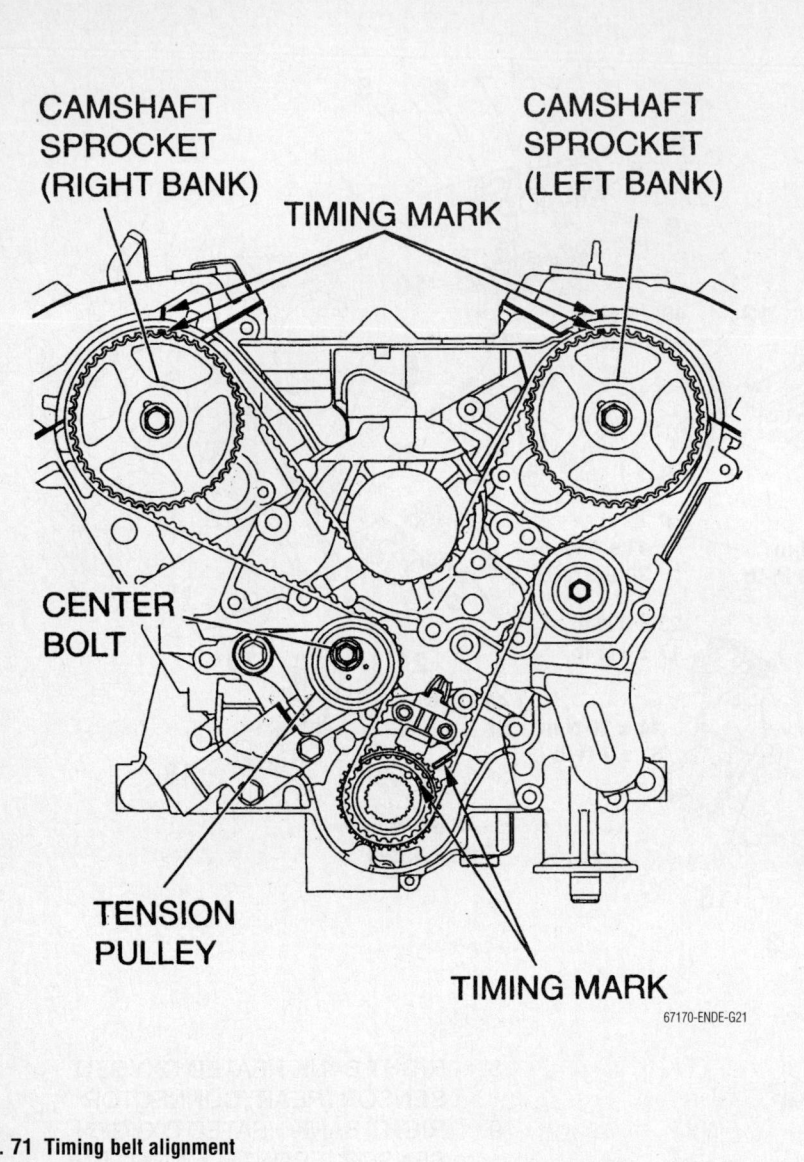

CAMSHAFT
SPROCKET
(RIGHT BANK)

CAMSHAFT
SPROCKET
(LEFT BANK)

TIMING MARK

CENTER
BOLT

TENSION
PULLEY

TIMING MARK

67170-ENDE-G21

Fig. 71 Timing belt alignment

19. Turn the crankshaft clockwise to align the timing marks and set the No. 1 cylinder at Top Dead Center (TDC). If you are reusing the timing belt, mark the flat side of the belt with an arrow showing the clockwise direction.

20. Loosen the center bolt of the tension pulley, and then remove the timing belt.

21. Remove the auto tensioner.
22. Remove the tensioner pulley.
23. Remove the tensioner arm.
24. Remove the shaft.

To install:

25. Install the idler pulley.
26. Install the shaft.
27. Install the tensioner arm assembly.
28. Install the tension pulley.
29. Press the end of the auto-tensioner inward with 72–145 ft. lbs. (98–196 nm) of force and measure the distance that the pushrod is pushed in. If the standard distance is not 0.04 in. (1mm), replace the auto-tensioner.

30. Position the auto tensioner in a soft-jawed vise and slowly compress the pushrod until the pushrod and housing holes align; then, install a setting pin to secure the auto-tensioner in the retracted position.

➡If you are installing a new auto-tensioner, the pin will already be inserted into the pin holes of the new tensioner.

31. Install the auto-tensioner.
32. Align the camshaft and crankshaft TDC timing marks.
33. Install the timing belt (noting its rotational direction) so that there is no deflection between the sprockets and pulleys in the following manner:

34. Install the crankshaft sprocket.
35. Install the idler pulley.
36. Install the left camshaft sprocket.
37. Install the water pump pulley.
38. Install the right camshaft sprocket.
39. Install the tension pulley.
40. Turn the camshaft sprocket counterclockwise until the tension side of the timing belt is firmly stretched, then, recheck the timing marks.

41. Using the tension pulley socket wrench tool MD998767, or equivalent, push the tensioner pulley into the timing belt and secure the center bolt.

42. Using the crankshaft pulley spacer tool MD998769, or equivalent, rotate the crankshaft ¼ turn counterclockwise, then, turn it again clockwise to align the timing marks.

43. Loosen the timing belt tensioner center bolt. Using the tension pulley socket wrench tool MD998767, or equivalent, and a torque wrench, apply 39 inch lbs. (4.4Nm) pressure on the timing belt. torque the tensioner pulley center bolt to 35 ft. lbs. (48 Nm).

44. Remove the setting pin from the auto-tensioner.

45. Rotate the crankshaft 2 complete revolutions and realign the timing marks. then, wait for 5 minutes until the auto-tensioner pushrod extends to its standard value. If the standard value is not 0.19–0.24 in. (4.8–6.0 mm), repeat the adjustment procedure. If the standard value is still not achieved, replace the auto-tensioner.

46. Install the engine support bracket.
47. Install the engine mount.
48. Install the lower timing belt cover.
49. Install the CKP sensor harness clamp.
50. Install the tensioner bracket.
51. Install the tensioner pulley.
52. Install the left and right upper timing belt covers.
53. Install the engine mount stay.
54. Install the connector bracket.
55. Install the HO2S connectors.
56. Install the wiring harness and injector connector.
57. Install the CKP sensor connector.
58. Install the KS connector.
59. Install the MDP sensor connector.
60. Install the crankshaft pulley. Using pulley holder MB991800 and the 2 crankshaft pulley holder pin tools MD991802 to hold the crankshaft pulley, tighten the crankshaft pulley bolt to 136 ft. Lbs. (185 nm).

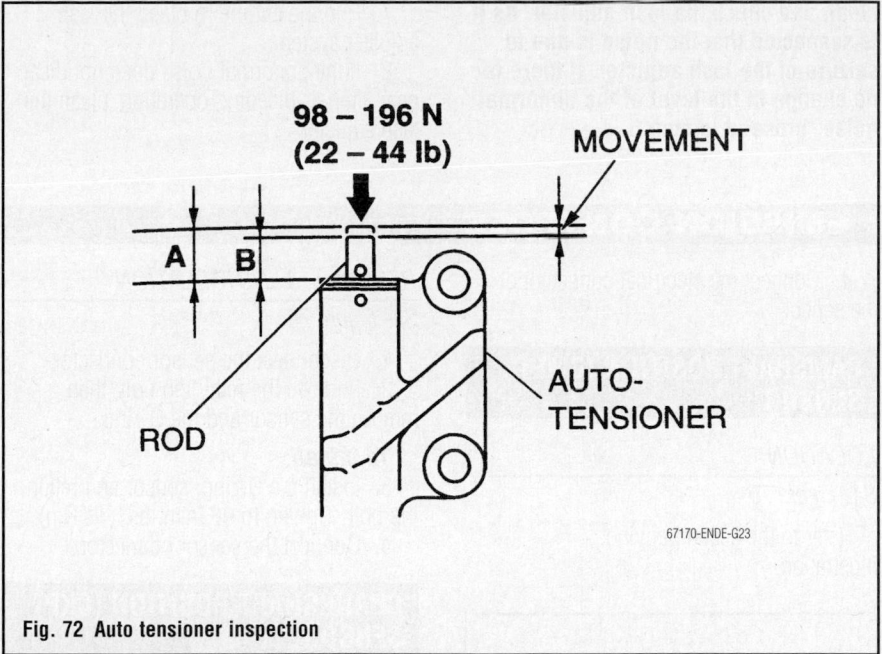

**98 – 196 N
(22 – 44 lb)**

MOVEMENT

A B

ROD

AUTO-
TENSIONER

67170-ENDE-G23

Fig. 72 Auto tensioner inspection

61. Install the drive belts.
62. Install the side under cover.
63. Install the undercover.
64. Connect the negative battery cable.
65. Turn the ignition **ON** and then **OFF**, and keep it off for at least ten seconds.
66. Refill the cooling system.

VALVE LASH

INSPECTION

Bleeding the Lash Adjuster
See Figure 73.

1. Check engine oil and add or change oil if required.
   a. If the engine oil level is low, air is sucked from the oil screen, causing air to enter the oil passage.
   b. If the engine oil level is higher than specification, oil may be stirred by the crankshaft, causing oil to be mixed with a large quantity of air.
   c. If oil is deteriorated, air is not easily separated from oil, increasing the quantity of air contained in oil.
   d. If air mixed with oil enters the high pressure chamber inside the lash adjuster from the above causes, air in the high pressure chamber is compressed excessively while the valve is opened, resulting in an abnormal noise when the valve closes. This is the same phenomenon as that observed when the valve clearance has become excessive. The lash adjuster can resume normal function when air entered the lash adjuster is removed.

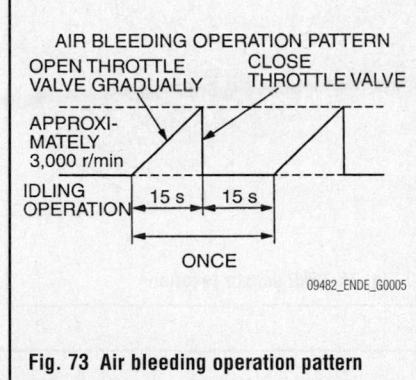

AIR BLEEDING OPERATION PATTERN
OPEN THROTTLE   CLOSE
VALVE GRADUALLY   THROTTLE VALVE

APPROXI-
MATELY
3,000 r/min

IDLING
OPERATION   15 s   15 s

ONCE

09482_ENDE_G0005

Fig. 73 Air bleeding operation pattern

2. Idle the engine for one to three minutes to warm it up.
3. Repeat the operation pattern shown, at no load to check for abnormal noise. Usually the abnormal noise is eliminated after repetition of the operation 10 to 30 times. If, however, no change is observed in the level of abnormal noise after repeating the operation more than 30 times, suspect that the abnormal noise is due to some other factors.
4. After elimination of abnormal noise, repeat the operation shown in left figure five more times.
5. Run the engine at idle for one to three minutes to make sure that the abnormal noise has been eliminated.

Lash Adjuster Check

1. The engine used in the Endeavor uses hydraulic lash adjusters that do not require adjustment. However, if a valve tap

(abnormal noise) is noticed, perform the following procedure to attempt to fix the problem.
   a. If an abnormal noise (chattering noise) suspected to be caused by malfunction of the lash adjuster is produced immediately after starting the engine and does not disappear, perform the following check.
   b. Parking the vehicle on a grade for a long time may decrease oil in the lash adjuster, causing air to enter the high pressure chamber when starting the engine. After parking for many hours, oil may run out from the oil passage and take time before oil is supplied to the lash adjuster, causing air to enter the high pressure chamber. Abnormal noise can be eliminated by bleeding the lash adjuster system.
   c. An abnormal noise due to malfunction of the lash adjuster is produced immediately after starting the engine and changes with the engine speed, irrespective of the engine load. If, the abnormal noise is not produced immediately after starting the engine or does not change with the engine speed, or it changes with the engine load, the lash adjuster is not the cause for the abnormal noise.
   d. When the lash adjuster is malfunctioning, the abnormal noise is rarely eliminated by continuing the warming-up of the engine at idle speed.
   e. The abnormal noise may disappear only when seizure is caused by oil sludge in the engine whose oil is not maintained properly.
2. Start the engine.
3. Check if the abnormal noise produced immediately after starting the engine, changes with the change in the engine speed.
4. If the abnormal noise is not produced immediately after starting the engine or it does not change with the engine speed, the lash adjuster is not the cause for the noise. Therefore, investigate other causes. The abnormal noise is probably caused by some other parts than the engine proper if it does not change with the engine speed. (In this case, the lash adjuster is in good condition.)
5. With the engine idling, change the engine load (shift from N to D range, for example) to make sure that there is no change in the level of abnormal noise.

If there is a change in the level of abnormal noise, suspect a tapping noise due to worn crankshaft bearing or connecting rod bearing (In this case, the lash adjuster is in good condition.).

6. After completion of warm-up, run the engine at idle to check for abnormal noise.

→If the noise is reduced or disappears,

clean and check the lash adjuster. As it is suspected that the noise is due to seizure of the lash adjuster. If there is no change in the level of the abnormal noise, proceed to step 5.

7. Run the engine to bleed the lash adjuster system.

8. If the abnormal noise does not disappear after air bleeding operation, clean the lash adjuster.

ENGINE PERFORMANCE & EMISSION CONTROLS

ACCELERATOR PEDAL POSITION (APP) SENSOR

LOCATION

See Figure 74.

Refer to the accompanying illustration.

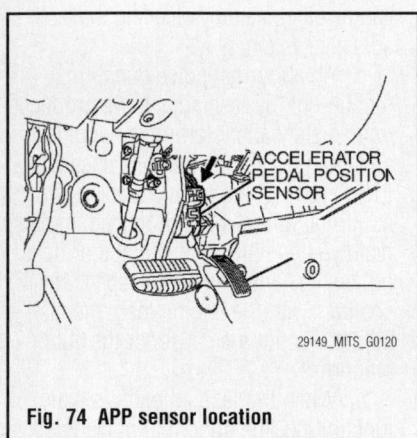

Fig. 74 APP sensor location

REMOVAL & INSTALLATION

The accelerator pedal position sensor is an integral part of the accelerator pedal assembly. It should not be serviced unnecessarily; it has been precisely adjusted by the manufacturer.

❋❋ WARNING

Never loosen the screw fixing the accelerator pedal assembly resin cover. If the screw is loosened, the sensor position which exists inside the resin cover is off and the accelerator pedal position sensor does not work normally. Do not remove the accelerator pedal pad. If the pad is removed and installed, excessive force may damage accelerator pedal position sensor.

1. Disconnect the sensor electrical connector.

2. Remove the accelerator pedal.

To install:

3. Install the accelerator pedal assembly. Tighten the pedal mounting nuts to 111 inch lbs. (13 Nm).

4. Connect the electrical connector at the sensor.

CAMSHAFT POSITION (CMP) SENSOR

LOCATION

See Figure 75.

Refer to the accompanying illustration.

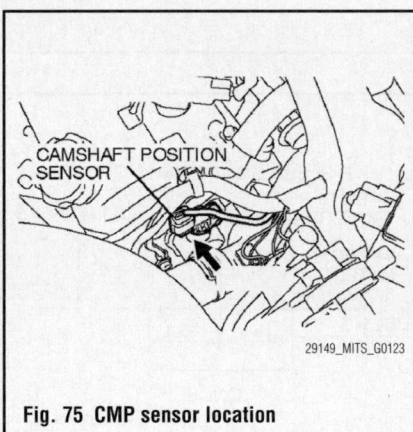

Fig. 75 CMP sensor location

REMOVAL & INSTALLATION

See Figure 76.

1. Disconnect the sensor connector.

2. Remove the retaining bolt, then remove the sensor and the O-ring.

To install:

3. Install the O-ring, sensor and retaining bolt. Tighten to 98 inch lbs. (11 Nm).

4. Connect the sensor connector.

CRANKSHAFT POSITION (CKP) SENSOR

LOCATION

See Figure 77.

Refer to the accompanying illustration.

REMOVAL & INSTALLATION

See Figure 78.

1. Detach the crankshaft position sensor electrical connector.

2. Remove the retaining bolt, then remove the sensor.

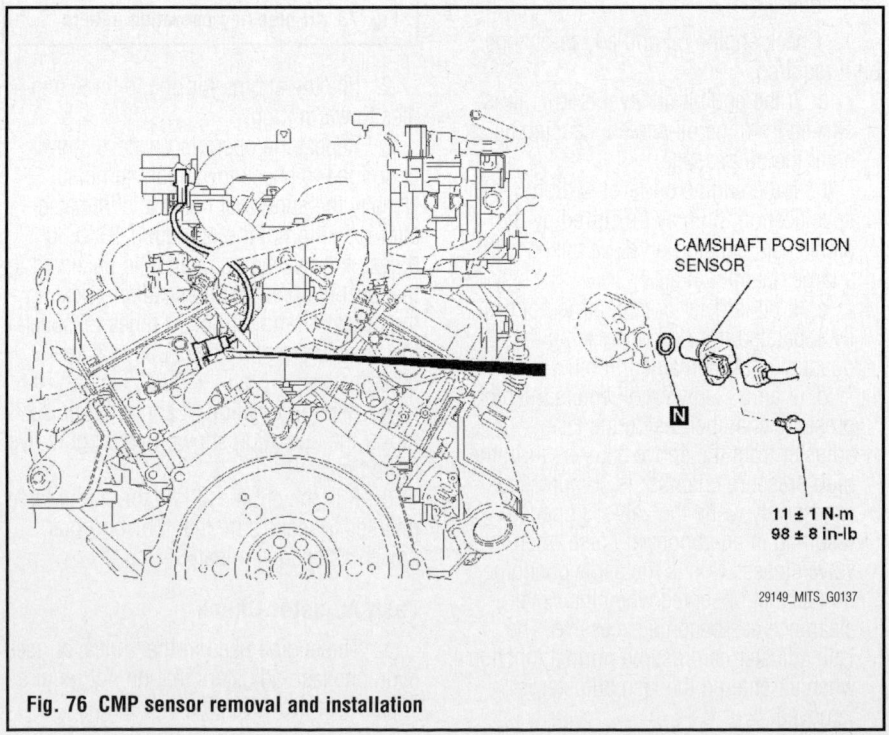

Fig. 76 CMP sensor removal and installation

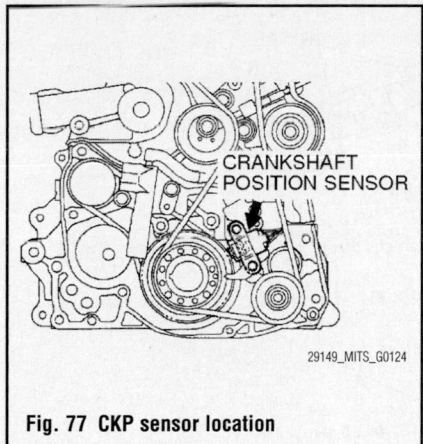

Fig. 77 CKP sensor location

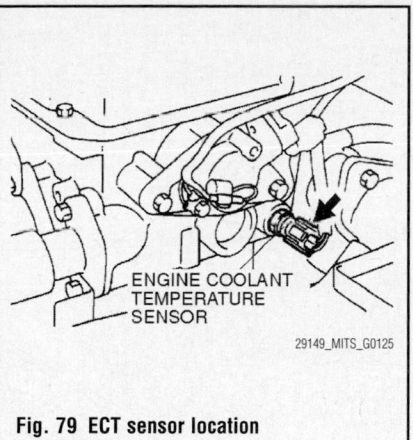

Fig. 79 ECT sensor location

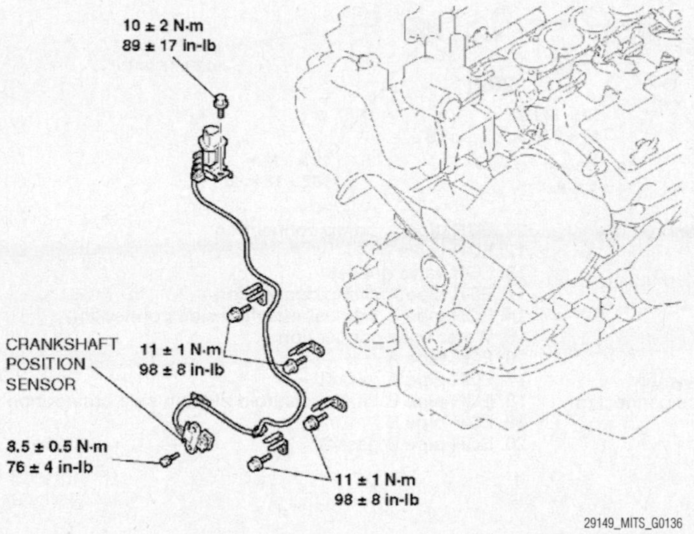

3. Use a deep socket and an extension to reach the ECT sensor, remove the ECT sensor from the thermostat housing.

To install:

4. Coat the threads of the sensor with a suitable sealant and thread into the housing. Tighten the sensor to 22 ft. lbs. (30 Nm).

5. Refill the cooling system to the proper level.

6. Attach the electrical connector to the sensor securely.

EVAPORATIVE EMISSIONS (EVAP) CANISTER

LOCATION

See Figure 80.

Refer to the accompanying illustration.

REMOVAL & INSTALLATION

1. Raise and support vehicle.
2. Remove harness connectors and hoses.
3. Remove bracket bolts and bracket.
4. Remove EVAP canister.

To install:

5. Install EVAP canister.
6. Install bracket bolts and bracket.
7. Install harness connectors and hoses.
8. Lower vehicle.

EXHAUST GAS RECIRCULATION (EGR) VALVE

LOCATION

See Figure 81.

Refer to the accompanying illustration.

Fig. 78 CKP sensor removal and installation

To install:

3. Position the sensor and install the retaining bolt. Tighten to 78 inch lbs. (9 Nm).

4. Connect the crankshaft position sensor electrical connector.

ENGINE COOLANT TEMPERATURE (ECT) SENSOR

LOCATION

See Figure 79.

Refer to the accompanying illustration.

REMOVAL & INSTALLATION

1. Drain the engine coolant to a level below the intake manifold.

2. Unplug the ECT sensor electrical connector

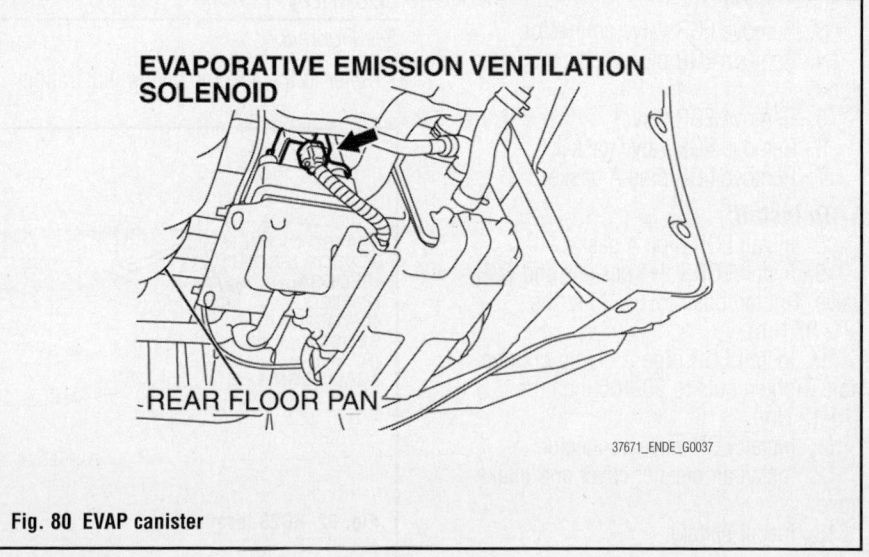

Fig. 80 EVAP canister

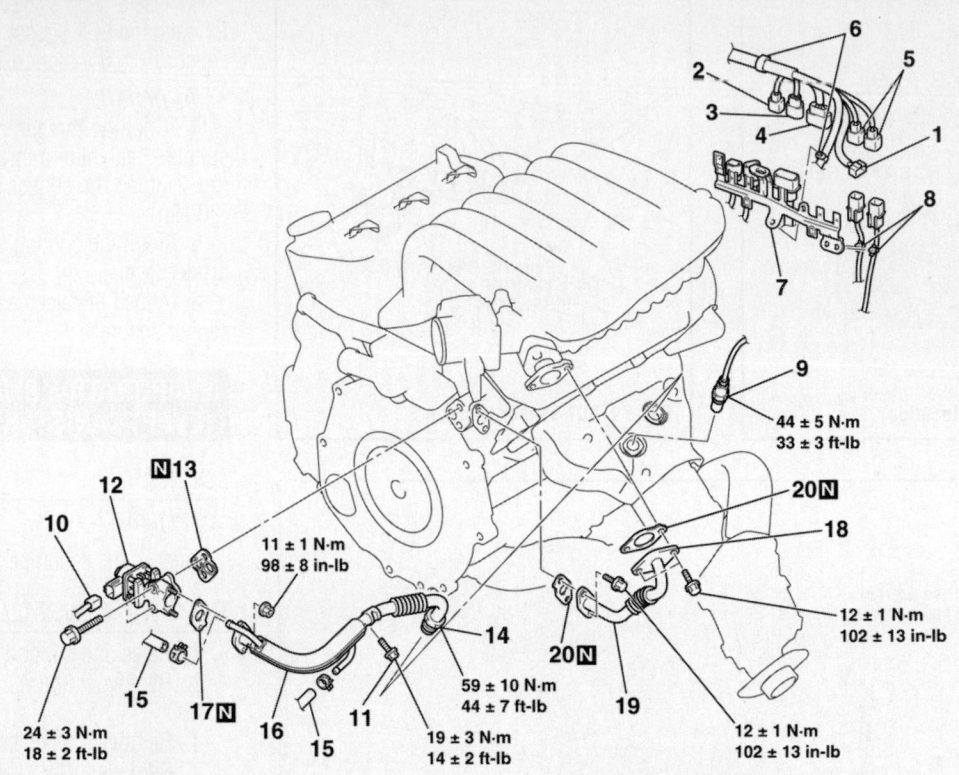

1. Manifold differential pressure sensor connector
2. Knock sensor connector
3. Crankshaft position sensor connector
4. Control wiring harness and injector wiring harness combination connector
5. Right bank heated oxygen sensor connectors
6. Harness clips connection
7. Connector bracket and harness connector assembly
8. Right bank heated oxygen sensor harness clips connection
9. Right bank heated oxygen sensor (front)
10. EGR valve connector
11. EGR pipe A clamp connection
12. EGR valve
13. EGR valve gasket
11. EGR pipe A clamp connection
14. EGR pipe A exhaust manifold side connection
15. Water hose connection
16. EGR pipe A
17. EGR pipe A gasket
18. EGR pipe B intake manifold plenum side connection
19. EGR pipe B
20. EGR pipe B gasket

37671_ENDE_G0036

Fig. 81 EGR valve components

REMOVAL & INSTALLATION

1. Remove battery.
2. Remove air cleaner cover and intake hose.
3. Remove EGR valve connector.
4. Remove EGR pipe A clamp connection.
5. Remove EGR valve.
6. Remove EGR valve gasket.
7. Remove EGR pipe A gasket.

To install:

8. Install EGR pipe A gasket.
9. Install EGR valve gaskets and EGR valve. Tighten bolts to 16–20 ft. lbs. (21–27 Nm).
10. Install EGR pipe A clamp connection. Tighten nuts to 90-106 inch lbs. (10-12 Nm).
11. Install EGR valve connector.
12. Install air cleaner cover and intake hose.
13. Install battery.

HEATED OXYGEN SENSOR (HO2S)

LOCATION

See Figure 82.

Refer to the accompanying illustration.

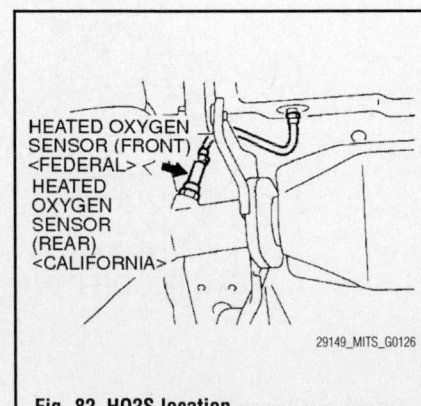

HEATED OXYGEN SENSOR (FRONT) <FEDERAL>
HEATED OXYGEN SENSOR (REAR) <CALIFORNIA>

29149_MITS_G0126

Fig. 82 HO2S location

REMOVAL & INSTALLATION

❊❊ CAUTION

The temperature of the exhaust system is extremely high after the engine has been run. To prevent personal injury, allow the exhaust system to cool completely before removing sensor from the exhaust system.

1. Raise and safely support the vehicle.
2. Disconnect the sensor electrical connector.
3. Using oxygen sensor socket, remove the oxygen sensor.

To install:

4. If installing old oxygen sensor, coat the threads with anti-seize compound. New sensors are already coated. Take care not to contaminate the oxygen sensor probe with the anti-seize compound.

5. Install the oxygen sensor into the exhaust manifold. Tighten the sensor, using the correct tool, to 33 ft. lbs. (45 Nm)
6. Connect the sensor electrical connector.
7. Lower the vehicle.

INTAKE AIR TEMPERATURE (IAT) SENSOR

LOCATION

See Figure 83.

Refer to the accompanying illustration.

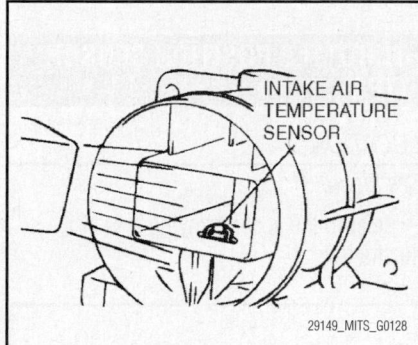

29149_MITS_G0128

Fig. 83 Intake Air Temperature (IAT) Sensor

REMOVAL & INSTALLATION

1. Disconnect the sensor electrical connector.
2. Remove the sensor from the air intake duct.

To install:
3. Install the sensor on the air intake duct.
4. Connect the electrical connector at the sensor.

KNOCK SENSOR (KS)

LOCATION

See Figure 84.

Refer to the accompanying illustration.

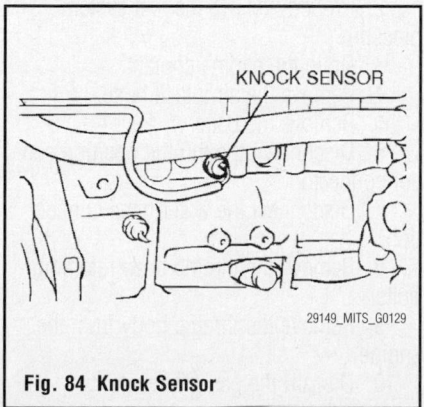

29149_MITS_G0129

Fig. 84 Knock Sensor

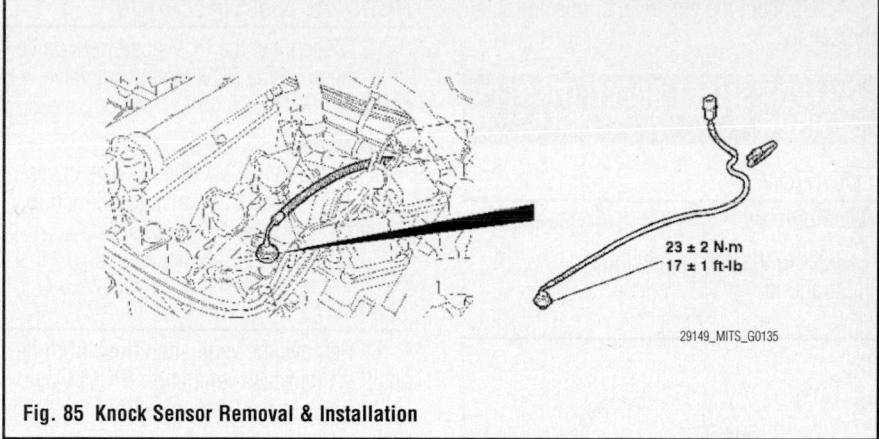

23 ± 2 N·m
17 ± 1 ft-lb

29149_MITS_G0135

Fig. 85 Knock Sensor Removal & Installation

REMOVAL & INSTALLATION

See Figure 85.

1. Disconnect the electrical connector at the sensor.
2. Remove the sensor from the engine block.

To install:
3. Install the knock sensor in the opening in the engine block and tighten the sensor retainer to 17 ft. lbs. (23 Nm).
4. Attach the electrical connector to the sensor.

MALFUNCTION INDICATOR LIGHT (MIL)

RESET PROCEDURE

1. Connect a scan tool to the diagnostic connector.
2. Clear DTCs.
3. The MIL should turn **OFF**.

MASS AIR FLOW (MAF) SENSOR

LOCATION

See Figure 86.

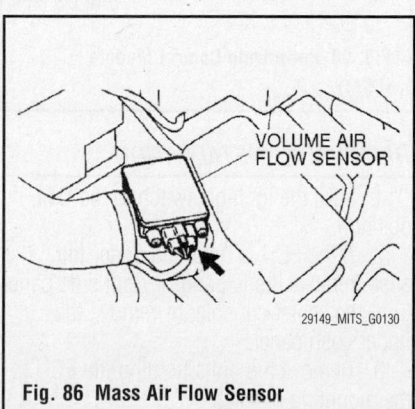

VOLUME AIR FLOW SENSOR

29149_MITS_G0130

Fig. 86 Mass Air Flow Sensor

Refer to the accompanying illustration.

REMOVAL & INSTALLATION

1. Disconnect the sensor electrical connector.
2. Remove the sensor from the air intake duct.

To install:
3. Install the sensor on the air intake duct.
4. Connect the electrical connector at the sensor.

MANIFOLD ABSOLUTE PRESSURE (MAP) SENSOR

LOCATION

See Figure 87.

Refer to the accompanying illustration.

REMOVAL & INSTALLATION

1. Disconnect the sensor electrical connector.
2. Remove the barometric sensor from the air intake duct.

To install:
3. Install the barometric sensor on the air intake duct.

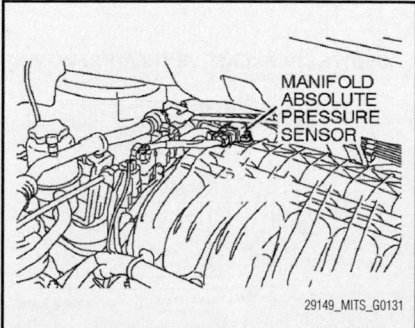

MANIFOLD ABSOLUTE PRESSURE SENSOR

29149_MITS_G0131

Fig. 87 Manifold Absolute Pressure (MAP) Sensor

4. Connect the electrical connector at the sensor.

OUTPUT SHAFT SPEED (OSS) SENSOR

LOCATION

See Figure 88.

Refer to the accompanying illustration.

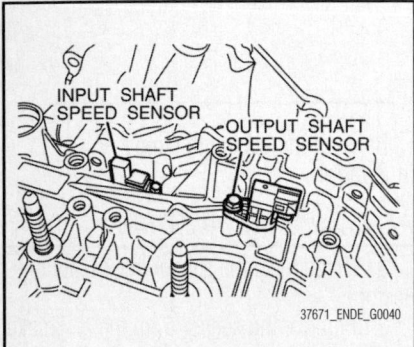

Fig. 88 Input and output shaft speed sensors

REMOVAL & INSTALLATION

1. Disconnect the sensor electrical connector.
2. Remove the output shaft speed sensor from the transaxle.

To install:

3. Install the sensor on the transaxle.
4. Connect the electrical connector at the sensor.

POSITIVE CRANKCASE VENTILATION (PCV) VALVE

LOCATION

See Figure 89.

Refer to the accompanying illustration.

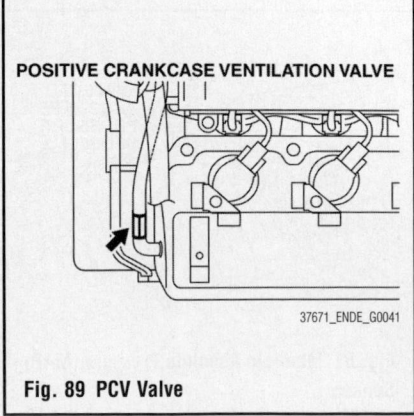

Fig. 89 PCV Valve

REMOVAL & INSTALLATION

1. Disconnect the PCV hose from valve.
2. Remove the PCV valve from valve cover.

To install:

3. Install the PCV valve in valve cover.
4. Tighten the PCV valve to 87 inch lbs. (10 Nm)
5. Connect the PCV hose to valve.

TESTING

1. Remove the ventilation hose from he positive crankcase ventilation (PCV) valve.
2. Remove the PCV valve from the rocker cover.
3. Hold the PCV valve with the vacuum side down. Insert a thin rod, and using light pressure, depress the end of the PCV valve spring by 5 - 10 mm (0.2 - 0.3 inch). Release pressure on the rod to see if the PCV valve spring will lift the rod to its original position.
4. If the rod returns quickly to its original position, the PCV valve is OK. If the stick does not return quickly, clean or replace the PCV valve.
5. Tighten the PCV valve to 87 inch lbs. (10 Nm)

POWERTRAIN CONTROL MODULE (PCM)

LOCATION

See Figure 90.

Refer to the accompanying illustration.

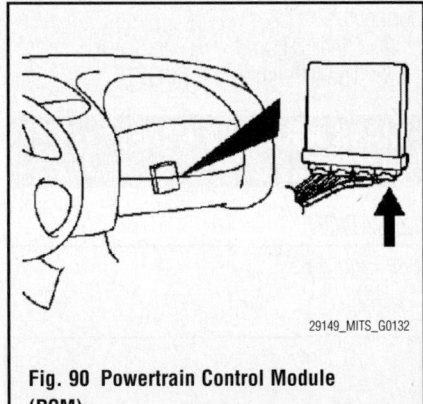

Fig. 90 Powertrain Control Module (PCM)

REMOVAL & INSTALLATION

1. Turn the ignition switch to the **OFF** position.
2. If the ECU is mounted under the dash, remove the left and/or right side panel from the center console, or remove the under dash panel.
3. Remove the bolts holding the ECU to the mounting bracket.

4. Disconnect the wiring harness from the ECU and remove ECU from the vehicle.

To install:

5. Connect the electrical harness to the ECU. Make certain the multi–pin connector is firmly and squarely seated to the ECU.
6. Install the ECU in the mounting bracket and secure in position.
7. If necessary, install the side panels to the center console or dash panel.
8. Connect the negative battery cable.

THROTTLE POSITION SENSOR (TPS)

LOCATION

See Figure 91.

Refer to the accompanying illustration.

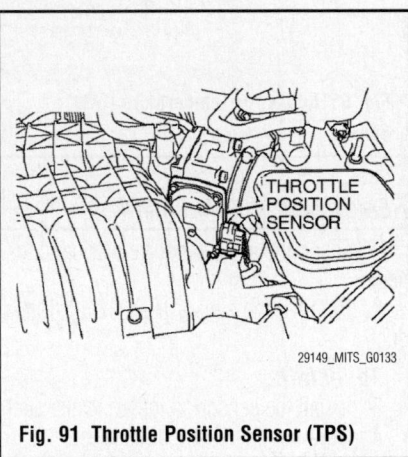

Fig. 91 Throttle Position Sensor (TPS)

REMOVAL & INSTALLATION

See Figure 92.

The Throttle Position Sensor (TPS) is a integral part of the throttle body.

1. Before servicing the vehicle, refer to the Precautions Section.
2. Properly relieve the fuel system pressure.
3. Drain the engine coolant.
4. Remove the air intake hose.
5. Remove the battery.
6. Disconnect the throttle position sensor connector.
7. Disconnect the water hose connection.
8. Remove the throttle body retaining bolts.
9. Remove the throttle body from the engine.
10. Discard the gasket.

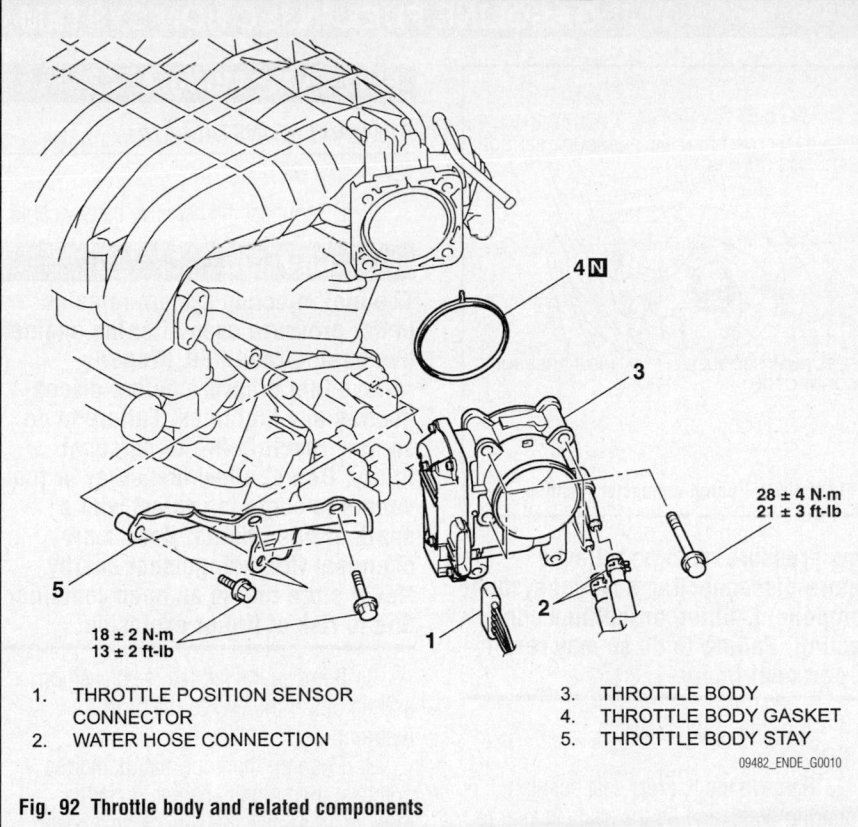

1. THROTTLE POSITION SENSOR
   CONNECTOR
2. WATER HOSE CONNECTION

3. THROTTLE BODY
4. THROTTLE BODY GASKET
5. THROTTLE BODY STAY

18 ± 2 N·m
13 ± 2 ft-lb

28 ± 4 N·m
21 ± 3 ft-lb

09482_ENDE_G0010

Fig. 92 Throttle body and related components

To install:

✷✷ WARNING

Do not loosen the retaining screws for the resin cover of the throttle body assembly. If the screws are loosened, the sensor incorporated in the resin cover becomes misaligned and the throttle body cannot work properly.

11. Align the recess on the intake manifold plenum with the projection of the throttle body gasket.

12. Install the gasket. Install the throttle body to the engine and tighten the retaining bolts to 18–24 ft. lbs. (24–32 Nm).

✷✷ WARNING

Poor idling etc. may result if the throttle body gasket is not installed properly.

13. Continue the installation in the reverse order of the removal procedure.

14. Connect the negative battery cable.

15. Turn the ignition **ON** and then **OFF**, and keep it off for at least ten seconds.

16. Complete the vehicle initialization procedure.

➡**To complete the initialization procedure the following tools are needed. MB991958 scan tool, MB991824 VCI, MB991827 MUT III USB cable, MB991910 MUT III Main harness "A".**

a. Connect the scan tool to the data link connector. To prevent damage to the scan tool be sure that the ignition switch is in the **LOCK** position before connecting the scan tool.

b. Turn the ignition switch to the **ON** position.

c. Select "check mode" from the menu screen.

d. Select "erase memory" from the menu screen.

e. Initialize the learning value.

f. After initialization complete the idle learning procedure.

➡**This procedure must be performed when the PCM is replaced, or when the learning value is initialized, as the idling is not stabilized because the learning value in the MFI engine is not completed.**

g. Start the engine. Allow the coolant temperature to reach 176 degrees or more.

h. Stop the engine and place the ignition switch in the **LOCK** position.

i. After ten seconds, restart the engine.

j. For ten minutes carry out the idling procedure below to confirm that the engine has normal idling.

- Position the transaxle selector lever in the **P** range
- The engine fan is not to be operated
- The engine coolant temperature should be 176 degrees or more

➡**If the engine stalls during idling, check the throttle valve of the throttle body for dirt. Correct and perform the procedure again.**

VEHICLE SPEED SENSOR (VSS)

LOCATION

See Figure 93.

Refer to the accompanying illustration.

22140_ENDE_G0045

Fig. 93 Vehicle Speed Sensor (VSS)

REMOVAL & INSTALLATION

1. Before servicing the vehicle, refer to the precautions section.

2. Disconnect the sensor electrical connector.

3. Remove the sensor attaching bolt/screw.

4. Remove the sensor.

To install:

5. Install the sensor and tighten the attaching bolt to 97 inch lbs. (11 Nm).

6. Connect the sensor electrical connector.

FUEL SYSTEM SERVICE PRECAUTIONS

Safety is the most important factor when performing not only fuel system maintenance, but any type of maintenance. Failure to conduct maintenance and repairs in a safe manner may result in serious personal injury or death. Work on a vehicle's fuel system components can be accomplished safely and effectively by adhering to the following rules and guidelines.

• To avoid the possibility of fire and personal injury, always disconnect the negative battery cable unless the repair or test procedure requires that battery voltage be applied.

• Always relieve the fuel system pressure prior to disconnecting any fuel system component (injector, fuel rail, pressure regulator, etc.) fitting or fuel line connection. Exercise extreme caution whenever relieving fuel system pressure to avoid exposing skin, face and eyes to fuel spray. Please be advised that fuel under pressure may penetrate the skin or any part of the body that it contacts.

• Always place a shop towel or cloth around the fitting or connection prior to loosening to absorb any excess fuel due to spillage. Ensure that all fuel spillage is quickly removed from engine surfaces. Ensure that all fuel-soaked cloths or towels are deposited into a flame-proof waste container with a lid.

• Always keep a dry chemical (Class B) fire extinguisher near the work area.

• Do not allow fuel spray or fuel vapors to come into contact with a spark or open flame.

• Always use a second wrench when loosening or tightening fuel line connection fittings. This will prevent unnecessary stress and torsion on fuel piping. Always follow the proper torque specifications.

• Always replace worn fuel fitting O-rings with new ones. Do not substitute fuel hose where rigid pipe is installed.

FUEL SYSTEM PRESSURE

RELIEVING

See Figure 94.

✳✳ CAUTION

The fuel system is under constant pressure, even with the engine off.

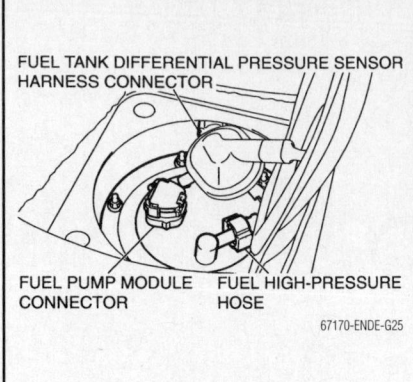

FUEL TANK DIFFERENTIAL PRESSURE SENSOR HARNESS CONNECTOR

FUEL PUMP MODULE CONNECTOR   FUEL HIGH-PRESSURE HOSE

67170-ENDE-G25

Fig. 94 Fuel pump connector location

This pressure must be relieved before disconnecting any fuel system component, fitting or fuel line connection. Failure to do so may result in personal injury.

1. Turn the ignition switch to **LOCK**.
2. Remove the left rear seat cushion mounting bolts, and lift the seat cushion to tie it to the head restraint.
3. Push on the floor mat to find the notches, then cut the carpet along the notches to access the service hole cover. Remove the service hole cover.
4. Detach the fuel pump module connector.
5. Start the engine and let it run out of fuel.
6. Attach the fuel pump module connector.
7. Install the service hole cover, floor mat and left rear seat cushion.

FUEL FILTER

REMOVAL & INSTALLATION

The fuel filter used on this vehicle is an integral part of the fuel pump module and is not normally serviced separately.

FUEL LEVEL SENDING UNIT

LOCATION

The fuel level sending unit is part of the fuel pump module, located inside the fuel tank.

REMOVAL & INSTALLATION

Refer to Fuel Pump Module.

FUEL PUMP MODULE

REMOVAL & INSTALLATION

See Figure 95.

1. Disconnect the negative battery cable.

✳✳ CAUTION

The fuel injection system remains under pressure even after the engine has been turned OFF. Properly relieve fuel pressure before disconnecting any fuel lines. Failure to do so may result in fire or personal injury. Do not allow fuel spray or fuel vapors to come in contact with a spark or open flame. Keep a dry chemical fire extinguisher nearby. Never store fuel in an open container due to risk of fire or explosion.

2. Remove the left rear seat cushion and tie out of the way to the head restraint.
3. Push on the floor mat to find the notches, then cut the carpet along the notches to access the service hole cover. Remove the service hole cover.
4. Remove the fuel pump module connector.
5. Remove the fuel tank differential pressure sensor connector.
6. Remove the fuel high-pressure hose.
7. Remove the mounting nuts and plate and fuel pump module assembly.

✳✳ WARNING

When removing the fuel pump module from the tank, be careful not to damage the module unit and float.

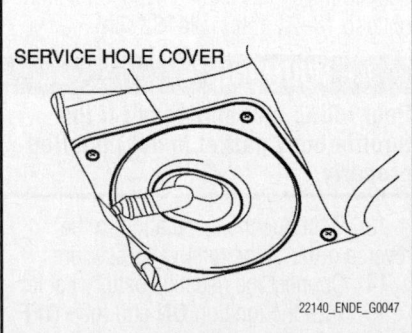

SERVICE HOLE COVER

22140_ENDE_G0047

Fig. 95 Push on the floor mat to find the notches, then cut the carpet along the notches to access the service hole cover. Remove the service hole cover

To install:

8. Install the fuel pump assembly into the fuel tank. torque the nuts to 24 inch lbs. (2.5 Nm).

9. Install the fuel lines.

10. Install the fuel tank differential pressure sensor connector.

11. Install the fuel pump module connector.

12. Install the fuel pump cover. torque the bolts to 14 inch lbs. (1.5 Nm).

13. Install the rear floor carpeting.

14. Connect the negative battery cable. Turn the ignition **ON** and then **OFF**, and keep it off for at least ten seconds.

15. Start the vehicle, check for leaks and proper operation.

FUEL RAIL AND INJECTOR

REMOVAL & INSTALLATION

See Figure 96.

1. Before servicing the vehicle, refer to the Precautions Section.

2. Properly relieve the fuel system pressure.

3. Remove the intake manifold lenum.

4. Remove the ignition coil connectors and coils.

5. Remove the injector connectors.

6. Remove the high pressure fuel hose and o-ring.

7. Remove the blow-by hose connection.

8. Remove the fuel rail, injector and damper assembly, as necessary.

9. Remove the insulators from the fuel rail.

To install:

10. Install the fuel rail and insulators. Tighten the bolts to 102 inch lbs. (12 Nm).

11. Install the injector connectors.

12. Install the blow-by hose connection.

13. Install the high pressure fuel hose and o-ring. torque the bolts to 44 inch lbs. (5.0 Nm). .

14. Install the ignition coils and connectors.

15. Install the intake manifold plenum.

FUEL TANK

REMOVAL & INSTALLATION

1. Before servicing the vehicle, refer to the Precautions Section.

2. Properly relieve the fuel system pressure.

3. Drain the fuel from the tank.

4. Remove the left rear seat cushion and tie out of the way to the head restraint.

5. Push on the floor mat to find the notches, then cut the carpet along the notches to access the service hole cover. Remove the service hole cover.

6. Disconnect the fuel pump module connector.

7. Disconnect the fuel tank differential pressure sensor connector.

8. Disconnect the fuel level sensor connector.

9. Raise and safely support the vehicle securely on jackstands.

10. Remove the center exhaust pipe.

11. Remove the propeller shaft on AWD vehicles

12. Remove the fuel tank protector.

13. Disconnect the parking brake cable clamp.

14. Disconnect the fuel tank vapor hose connection.

15. Disconnect the fuel filler hose connection.

16. Disconnect the fuel high-pressure hose connection.

17. Remove the fuel tank band.

   a. Support the fuel tank with a transaxle jack.

   b. On FWD vehicles, remove the fuel tank band and lower the fuel tank assembly gradually to remove it.

   c. On AWD vehicles, remove the front securing nut of the fuel tank band. Tilt the fuel tank assembly forward and lower it gradually to remove it. Remove the fuel tank band.

18. Carefully lower the fuel tank assembly from the vehicle.

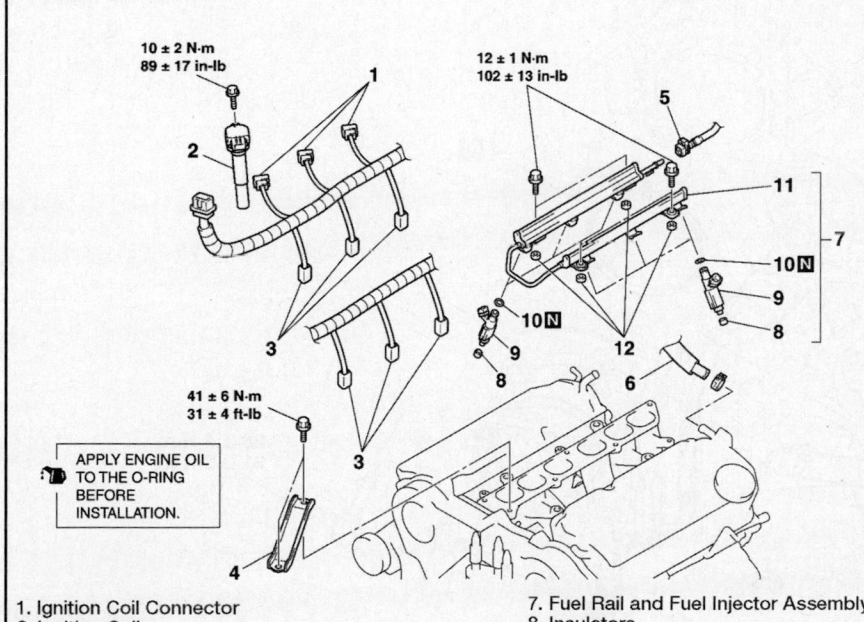

10 ± 2 N·m
89 ± 17 in-lb

12 ± 1 N·m
102 ± 13 in-lb

41 ± 6 N·m
31 ± 4 ft-lb

APPLY ENGINE OIL TO THE O-RING BEFORE INSTALLATION.

1. Ignition Coil Connector
2. Ignition Coil
3. Fuel Injector Connector
4. Engine Mounting Stay
5. Fuel High-Pressure Hose Connection (Fuel Rail Side)
6. Blow-By Hose Connection
7. Fuel Rail and Fuel Injector Assembly
8. Insulators
9. Fuel Injectors
10. O-Ring
11. Fuel Rail
12. Insulators

22140_ENDE_G0050

Fig. 96 Fuel rail and related components

To install:

19. Install the fuel tank bands

    a. Raise the fuel tank assembly carefully with a transaxle jack.

    b. Ensure that the fuel tank assembly does not interfere with surrounding parts. Then install the fuel tank band and tighten to 19 ft. lbs. (26 Nm).

20. Connect the fuel high-pressure hose connection.

    a. After installing, slightly pull the fuel high-pressure hose and ensure that there is no disengaged fuel high-pressure hose. Also confirm that there is approximately 0.12 inch (3 mm) play.

21. Connect the fuel filler hose connection.

22. Connect the fuel tank vapor hose connection.

23. Connect the parking brake cable clamp.

24. Install the fuel tank protector.

25. Install the propeller shaft on AWD vehicles

26. Install the center exhaust pipe.

27. Raise and safely support the vehicle securely on jackstands.

28. Connect the fuel level sensor connector.

29. Connect the fuel tank differential pressure sensor connector.

30. Connect the fuel pump module connector.

31. Install the service hole cover.

32. Install the left rear seat cushion.

33. Fill the fuel tank.

34. Start the engine and check for leaks.

IDLE SPEED

ADJUSTMENT

Idle speed is maintained by the Powertrain Control Module (PCM). No adjustment is possible.

THROTTLE BODY

REMOVAL & INSTALLATION

See Figure 97.

The Throttle Position (TPS) sensor is a integral part of the throttle body.

1. Before servicing the vehicle, refer to the Precautions Section.

2. Properly relieve the fuel system pressure.

3. Drain the engine coolant.

4. Remove the air intake hose.

5. Remove the battery.

6. Disconnect the throttle position sensor connector.

7. Disconnect the water hose connection.

8. Remove the throttle body retaining bolts.

9. Remove the throttle body from the engine.

10. Discard the gasket.

To install:

❊❊ WARNING

Do not loosen the retaining screws for the resin cover of the throttle body assembly. If the screws are loosened, the sensor incorporated in the

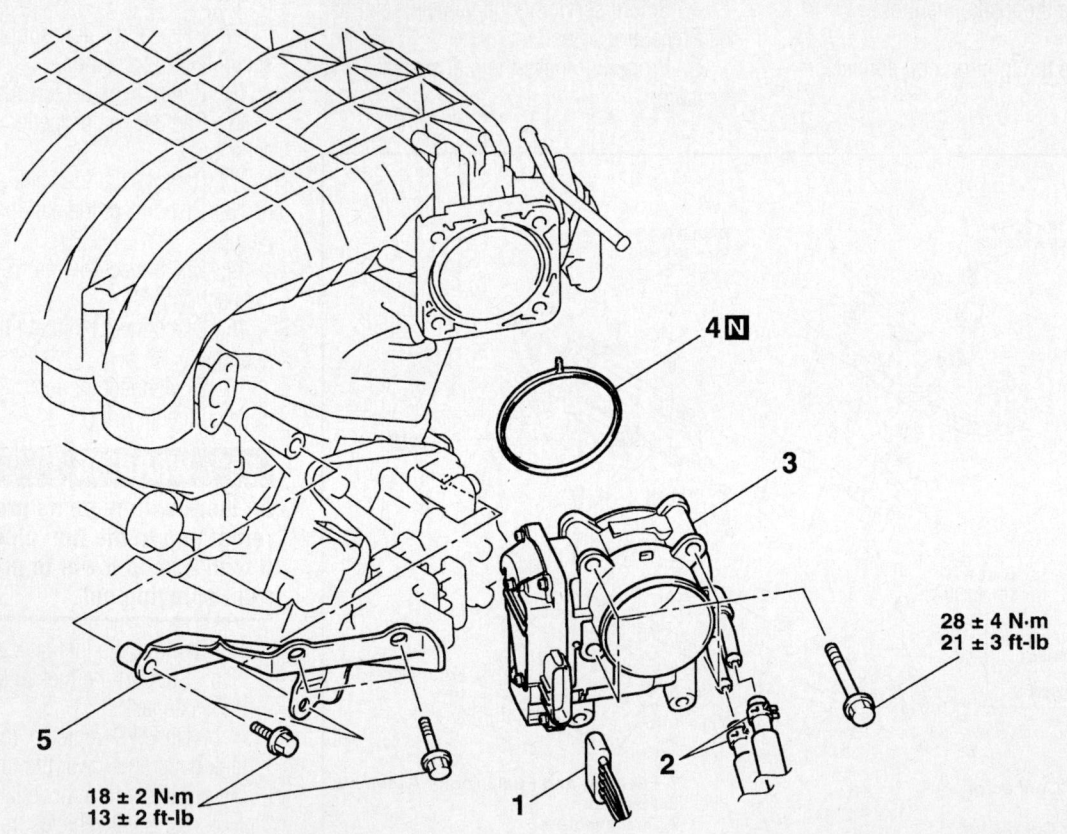

28 ± 4 N·m
21 ± 3 ft-lb

18 ± 2 N·m
13 ± 2 ft-lb

1. THROTTLE POSITION SENSOR CONNECTOR
2. WATER HOSE CONNECTION
3. THROTTLE BODY
4. THROTTLE BODY GASKET
5. THROTTLE BODY STAY

09482_ENDE_G0010

Fig. 97 Throttle body and related components

resin cover becomes misaligned and the throttle body cannot work properly.

11. Align the recess on the intake manifold plenum with the projection of the throttle body gasket.

12. Install the gasket. Install the throttle body to the engine and tighten the retaining bolts to 18–24 ft. lbs. (24–32 Nm).

❋❋ WARNING

Poor idling etc. may result if the throttle body gasket is not installed properly.

13. Continue the installation in the reverse order of the removal procedure.

14. Connect the negative battery cable.

15. Turn the ignition **ON** and then **OFF**, and keep it off for at least ten seconds.

16. Complete the vehicle initialization procedure.

➡To complete the initialization procedure the following tools are needed. MB991958 scan tool, MB991824 VCI, MB991827 MUT III USB cable, MB991910 MUT III Main harness "A".

a. Connect the scan tool to the data link connector. To prevent damage to the scan tool be sure that the ignition switch is in the **LOCK**position before connecting the scan tool.

b. Turn the ignition switch to the **ON** position.

c. Select "check mode" from the menu screen.

d. Select "erase memory" from the menu screen.

e. Initialize the learning value.

f. After initialization complete the idle learning procedure.

➡**This procedure must be performed when the PCM is replaced, or when the learning value is initialized, as the**

idling is not stabilized because the learning value in the MFI engine is not completed.

g. Start the engine. Allow the coolant temperature to reach 176 degrees or more.

h. Stop the engine and place the ignition switch in the **LOCK** position.

i. After ten seconds, restart the engine.

j. For ten minutes carry out the idling procedure below to confirm that the engine has normal idling.

- Position the transaxle selector lever in the **P** range
- The engine fan is not to be operated
- The engine coolant temperature should be 176 degrees or more

➡**If the engine stalls during idling, check the throttle valve of the throttle body for dirt. Correct and perform the procedure again.**

HEATING & AIR CONDITIONING SYSTEM

BLOWER MOTOR

REMOVAL & INSTALLATION

Front
See Figure 98.

1. Before servicing the vehicle, refer to the Precautions Section.

2. Disconnect the blower motor connector.

3. Remove the blower motor attaching screws.

4. Carefully lower the blower motor from the heater unit assembly.

To install:

5. Installation is the reverse of the removal procedure.

Rear
See Figure 99.

1. Before servicing the vehicle, refer to the Precautions Section.

2. Remove the accessory socket panel.

3. Remove the rear A/C floor console switch assembly.

4. Remove the air outlet.

5. Remove the rear blower unit.

6. Remove the resistor.

To install:

7. Installation is the reverse of the removal procedure.

HEATER CORE

REMOVAL AND INSTALLATION

See Figures 100 through 103.

1. Before servicing the vehicle, refer to the Precautions Section.

2. Discharge the A/C system and recycle the refrigerant.

3. Drain the engine coolant.

4. Remove the instrument panel.

5. Remove the steering column as follows:

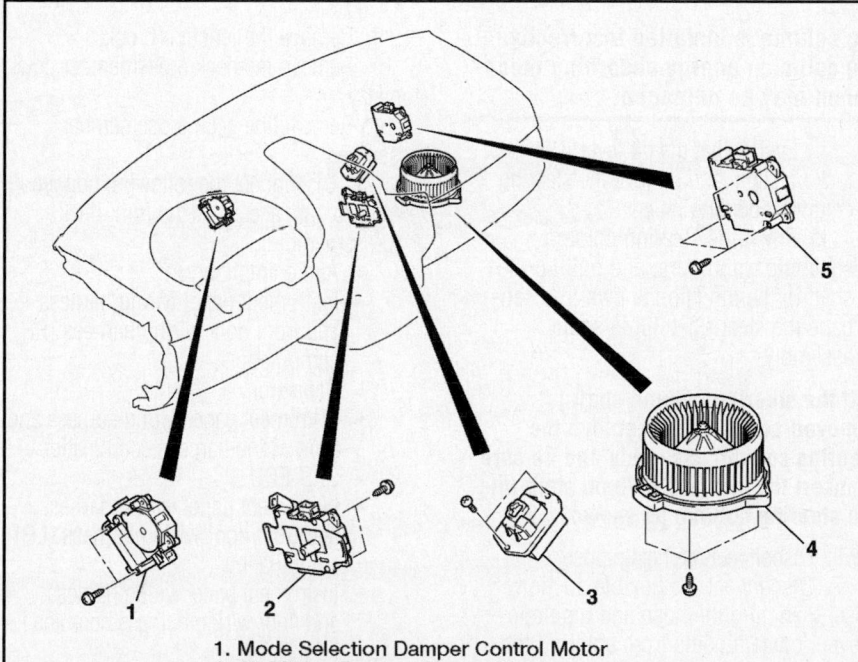

1. Mode Selection Damper Control Motor
2. Air Mixing Damper Control Motor
3. Power Transistor
4. Front Blower Motor
5. Outside/Inside Air Selection Damper Motor

22140_ENDE_G0051

Fig. 98 Blower motor and actuator locations

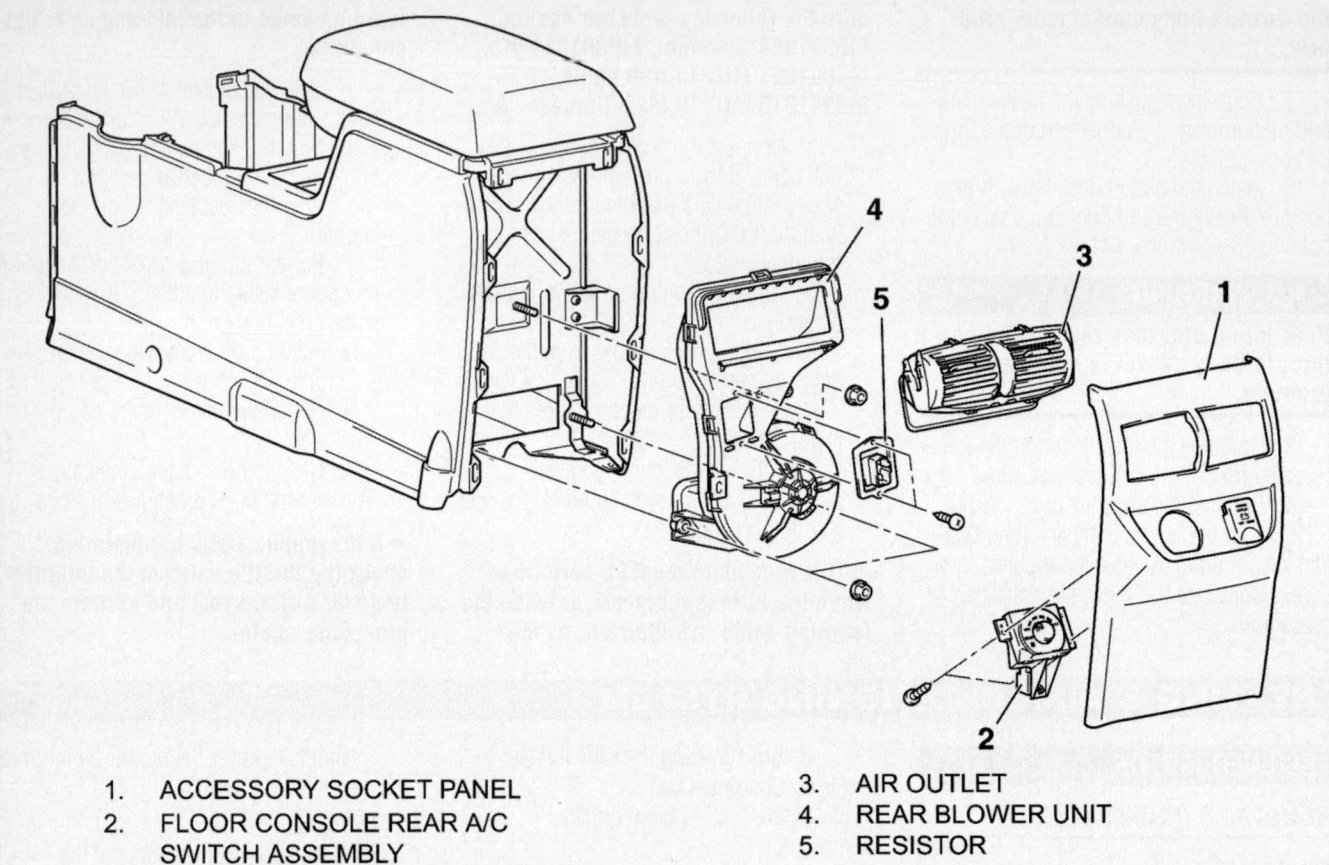

1. ACCESSORY SOCKET PANEL
2. FLOOR CONSOLE REAR A/C
   SWITCH ASSEMBLY

3. AIR OUTLET
4. REAR BLOWER UNIT
5. RESISTOR

42050_ENDE_G0063

Fig. 99 Exploded view of the rear blower unit

a. Remove the front scuff plate and cowl side trim.

b. Remove the steering wheel and air bag module assembly.

c. Remove the instrument panel side cover and under cover.

d. Remove the floor console side cover.

e. Remove the front floor carpet

f. Remove the lower and upper column covers.

g. Remove the steering wheel sensor, clock spring and column switch.

h. Disconnect the key interlock cable.

i. Remove the steering shaft pad.

⁑ WARNING

The tilt lever should be held in the lock position until the steering column shaft is installed to the vehicle. If the steering column is removed with the tilt lever released, or the tilt lever is released after the steering column shaft was removed from the vehicle, the steering column cannot be reinstalled correctly. If the steer-

ing column is installed incorrectly, the collision energy absorbing mechanism may be damaged.

j. Ensure that the tilt lever is in the lock position, and remove the steering column mounting bolts.

k. Pinch the steering column shaft clip with pliers, and pull up the shaft in the direction shown to disengage the steering column shaft assembly.

➡If the steering column shaft is removed accidentally, remove the steering column assembly and be sure to insert the steering column shaft into the steering column as shown.

6. Disconnect the heater hose.

7. Disconnect the flexible suction hose, then plug the hose and nipple to prevent contaminants from entering the system. Remove and discard the O-ring.

8. Disconnect the flexible discharge hose, then plug the hose and nipple to prevent contaminants from entering the system. Remove and discard the O-ring.

9. Remove the junction block.

10. Remove the deck crossmember cowl top stay.

11. Remove the deck crossmember assembly.

a. Disconnect the following connectors to gain access to the front deck crossmember:

- Audio amplifier
- Instrument panel wiring harness and front door wiring harness (RH) combination
- Capacitor
- Instrument panel wiring harness and front wiring harness combination
- SRS-ECU
- Instrument panel wiring harness and front door wiring harness (LH) combination
- Instrument panel wiring harness and floor wiring harness combination
- Instrument panel wiring harness and front wiring harness combination
- Front wiring harness and floor wiring harness combination

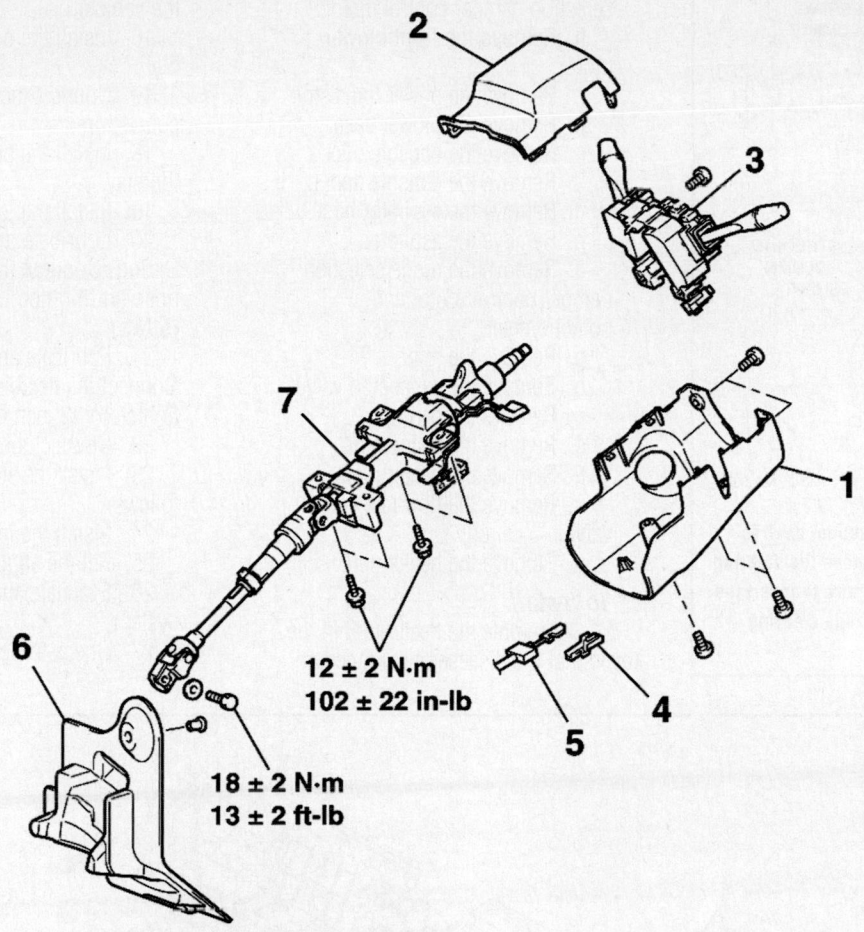

1. Lower Column Cover
2. Upper Column Cover
3. Steering Wheel Sensor,
   Clock Spring and Column Switch Assembly
4. Cover
5. Key Interlock Cable
6. Steering Shaft Pad
7. Steering Column Shaft Assembly

22140_ENDE_G0056

Fig. 100 Exploded view of the steering column assembly

- Instrument panel wiring harness and floor wiring harness combination
- Instrument panel wiring harness and console wiring harness combination
- Floor wiring harness and junction block combination
- Roof wiring harness and junction block combination
- Front wiring harness and junction block combination
- Front wiring harness and junction block combination

- Front wiring harness and junction block combination
- Ignition switch
- Key reminder switch
12. Remove the heater unit assembly.
13. Disassemble the heater unit as follows:
   a. Remove the foot light bracket.
   b. Remove the air mixing damper control motor and potentiometer.
   c. Remove the antenna cable.
   d. Remove the console duct c.
   e. Remove the deck crossmember.
   f. Remove the joint bracket.

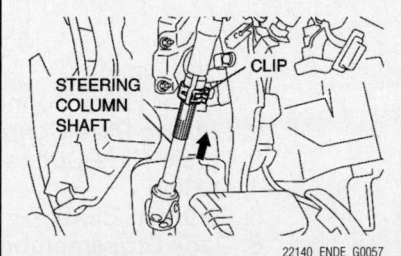

22140_ENDE_G0057

Fig. 101 Pinch the steering column shaft clip with pliers, and pull up the shaft in the direction shown to disengage the steering column shaft assembly

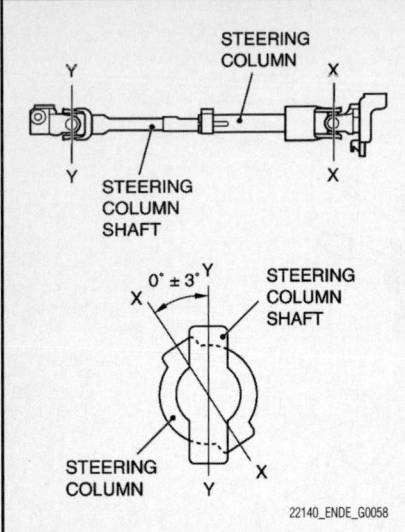

Fig. 102 If the steering column shaft is removed accidentally, remove the steering column assembly and be sure to insert the steering column shaft into the steering column as shown

g. Remove the outside/inside air selection damper control motor.

h. Remove the front blower motor.

i. Remove the power transistor.

j. Remove the blower case.

k. Remove the console duct a.

l. Remove the console duct b.

m. Remove the aspirator hose.

n. Remove the aspirator.

o. Remove the mode selection damper control motor and potentiometer.

p. Remove the joint.

q. Remove the expansion valve.

r. Remove the o-ring.

s. Remove the evaporator.

t. Remove the heater pipe.

u. Remove the flow rate control valve.

v. Remove the heater core.

To install:

14. Assemble the heater unit in the reverse of the disassembly procedure.

15. Install the heater unit assembly in the vehicle.

16. Install the deck crossmember assembly.

17. Connect the previously disconnected connectors.

18. Install the deck crossmember cowl top stay.

19. Install the junction block.

20. Lubricate and install a new O-ring. Connect the flexible discharge hose and tighten to 43 inch lbs. (5 Nm).

21. Lubricate and install a new O-ring. Connect the flexible suction hose and tighten to 43 inch lbs. (5 Nm).

22. Connect the heater hose.

23. Install the steering column as follows:

24. Install the instrument panel.

25. Fill the engine with coolant.

26. Evacuate and charge the A/C system.

27. Check for leaks.

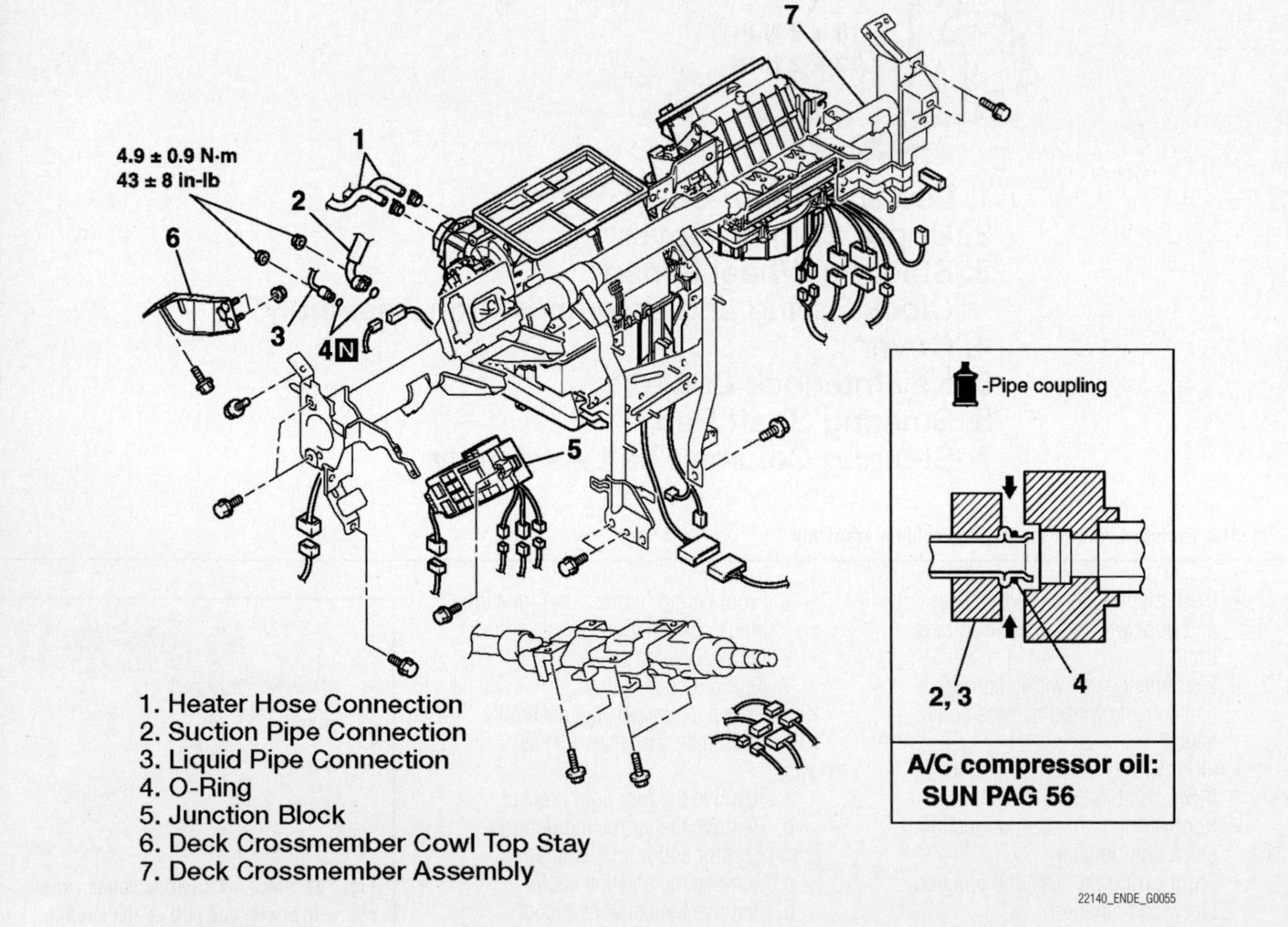

4.9 ± 0.9 N·m
43 ± 8 in-lb

-Pipe coupling

A/C compressor oil:
SUN PAG 56

1. Heater Hose Connection
2. Suction Pipe Connection
3. Liquid Pipe Connection
4. O-Ring
5. Junction Block
6. Deck Crossmember Cowl Top Stay
7. Deck Crossmember Assembly

Fig. 103 Heater unit assembly

STEERING

POWER RACK & PINION STEERING GEAR

REMOVAL & INSTALLATION

See Figures 104 through 109.

1. Before servicing the vehicle, refer to the precautions.

2. Disconnect the negative battery cable.

➡**Wait at least 90 seconds after disconnecting the cable from the negative (-) battery terminal to prevent airbag activation.**

3. Drain the power steering system.
4. Remove the steering wheel.

5. Remove the clock spring and put aside in a safe place.
6. Remove the front scuff plate and cowl side trim.
7. Remove the front floor left side console cover.
8. Remove the front floor carpet.

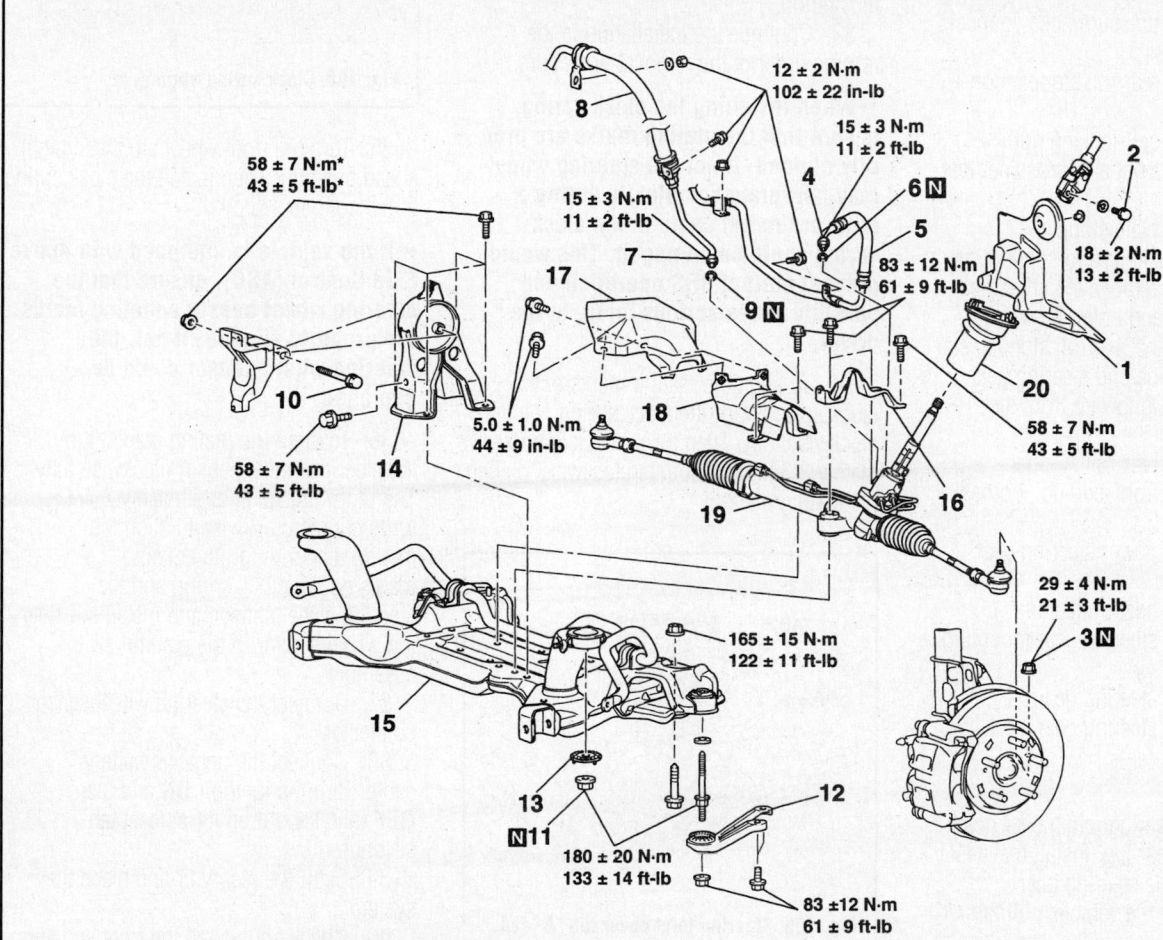

1. STEERING SHAFT PAD
2. STEERING COLUMN SHAFT ASSEMBLY AND STEERING GEAR CONNECTING BOLT
3. SELF LOCKING NUT (TIE ROD END AND KNUCKLE CONNECTION)
4. RETURN TUBE CONNECTION
5. RETURN TUBE
6. O-RING
7. PRESSURE TUBE CONNECTION
8. PRESSURE TUBE/PRESSURE HOSE
9. O-RING
10. REAR ROLL STOPPER CONNECTING BOLT
11. SELF LOCKING NUT

12. FRONT AXLE CROSSMEMBER STAY
13. PLATE STOPPER
14. REAR ROLL STOPPER
15. FRONT AXLE NO.1 CROSSMEMBER
16. POWER STEERING GEAR BRACKET
17. HEAT PROTECTOR
18. STEERING GEAR MOUNTING GEAR SIDE BRACKET
19. STEERING GEAR
20. STEERING JOINT COVER

67170-ENDE-G34

Fig. 104 Power steering gear and related components

9. Remove the stabilizer link and lower control arm assembly.

10. Remove the center member.

11. Remove the shaft pad.

12. Remove the steering column shaft assembly and steering gear connecting bolt.

13. Remove the tie rod end from the steering knuckle, using a suitable puller.

14. Remove the fluid return hose and return tube. plug the hoses. remove and discard the o-ring.

15. Remove the pressure hose connection and hose gasket.

16. Remove the rear roll stopper connecting bolt.

17. Remove the self-locking nut.

18. Remove the front axle crossmember stay.

19. Remove the plate stopper.

20. Use a transaxle jack to support the crossmember, then remove the crossmember mounting nuts and bolts. Lower the crossmember with the rear roll stopper, stabilizer bar, return tube and steering gear.

21. Remove the following from the crossmember.

22. Remove the rear roll stopper.

23. Remove the front axle no. 1 crossmember.

24. Remove the power steering gear bracket.

25. Remove the heat protector.

26. Remove the steering gear mounting gear side bracket.

27. Remove the steering gear.

28. Remove the steering gear joint cover.

To install:

29. Align the steering joint cover notch with the steering gear lug. Install the steering joint cover to the steering gear.

30. From inside the vehicle, pull tab "A" and then tab "B" to secure the three clips to the body panel.

➡**When securing the joint cover to the body panel, be careful that the cover seal lip does not move backwards.**

31. After installing the joint cover, check that the joint cover rubber is not disengaged from the retainer. If there is any doubt, release the clips from the body and start the procedure again.

32. Tighten the power steering gear bracket bolts in the order shown in the illustration.

33. Continue the installation in the reverse order of the removal procedure.

➡**When installing the clock spring, ensure that the mating marks are properly aligned. If not the steering wheel may not rotate completely during a turn, or the flat cable in the clock spring could be damaged. This would prevent normal SRS operation and possibly cause serious injury to the driver.**

34. To align the mating marks of the clock spring, turn the clock spring counterclockwise fully. Then turn it back approximately 3 3/4 turns counterclockwise to align the mating marks.

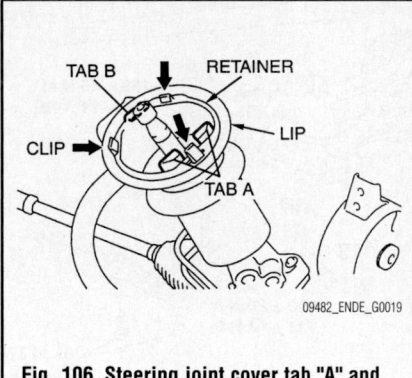

Fig. 106 Steering joint cover tab "A" and tab "B" location

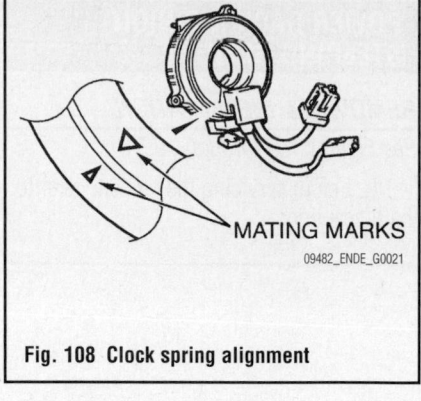

Fig. 108 Clock spring alignment

35. Turn the front wheels to the straight ahead position. Then install the clock spring to the column switch.

➡**If the vehicle is equipped with Active Skid Control (ASC), ensure that the steering wheel sensor's mating marks are properly aligned. If not, the steering wheel sensor could be damaged.**

36. To align the mating marks, turn the steering wheel sensor clockwise fully. Then turn it back approximately 2 3/4 turns counterclockwise to align the mating marks. Align the mating marks on the clock spring and the steering wheel sensor, and install the steering wheel sensor to the column switch assembly.

37. Connect the steering wheel sensor connector.

38. Connect the negative battery cable. Turn the ignition **ON** and then **OFF**, and keep it off for at least ten seconds.

39. Refill the reservoir and bleed the system.

40. Check and adjust the front end alignment, as required.

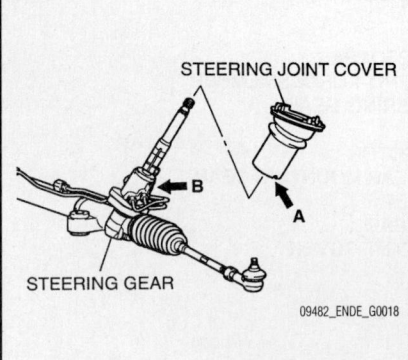

Fig. 105 Steering joint cover notch alignment

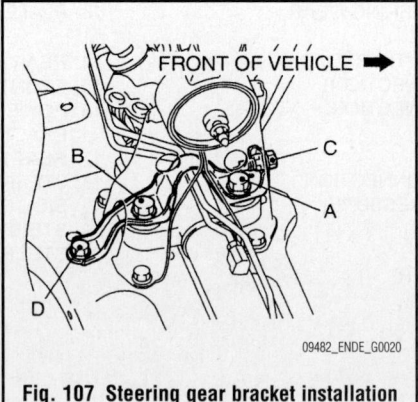

Fig. 107 Steering gear bracket installation points

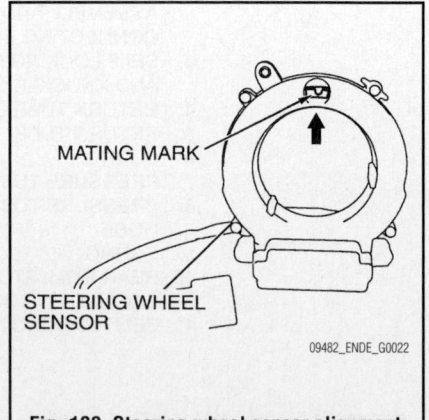

Fig. 109 Steering wheel sensor alignment

POWER STEERING PUMP

REMOVAL & INSTALLATION

See Figures 110 through 113.

1. Raise and safely support the front of the vehicle.

2. Remove the front and side under covers.

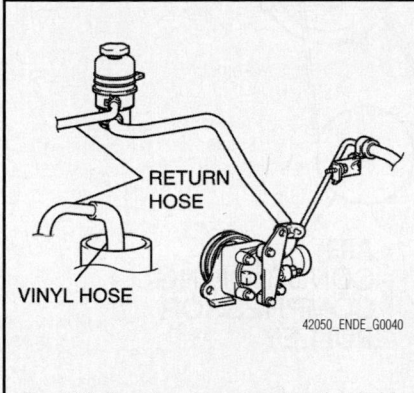

Fig. 110 Draining the power steering fluid

3. Drain the power steering fluid, as follows:

 a. Disconnect the return hose connection, and then connect a vinyl hose to the return hose, and drain the fluid into a container.

 b. Disconnect the ignition coil connectors.

 c. While operating the starter motor intermittently, turn the steering wheel all the way to the left and right several times to drain all of the fluid.

 d. Connect the return hose securely, and then secure with the clip.

4. Remove the power steering pump drive belt, by loosen the tensioner pulley fixing nut or locknut, then remove the belt.

5. Remove the belt tensioner pulley and bracket.

6. Detach the pressure switch connector.

7. Disconnect the power steering pressure hose and gasket.

8. Disconnect the power steering suction hose.

9. Remove the power steering pump bracket.

10. Remove the power steering pump assembly.

➡ **Before removing the power steering pressure switch, wipe the power steering pump clean to prevent foreign debris from entering the pump.**

11. If necessary, remove the power steering pressure switch, as follows:

 a. Detach the harness connector from the male terminal of the pressure switch.

 b. Use a socket wrench to loosen and remove the switch.

To install:

✳✳ WARNING

Do not touch the terminal part of the switch during installation as this may damage the switch.

12. If removed, install the power steering switch, as follows:

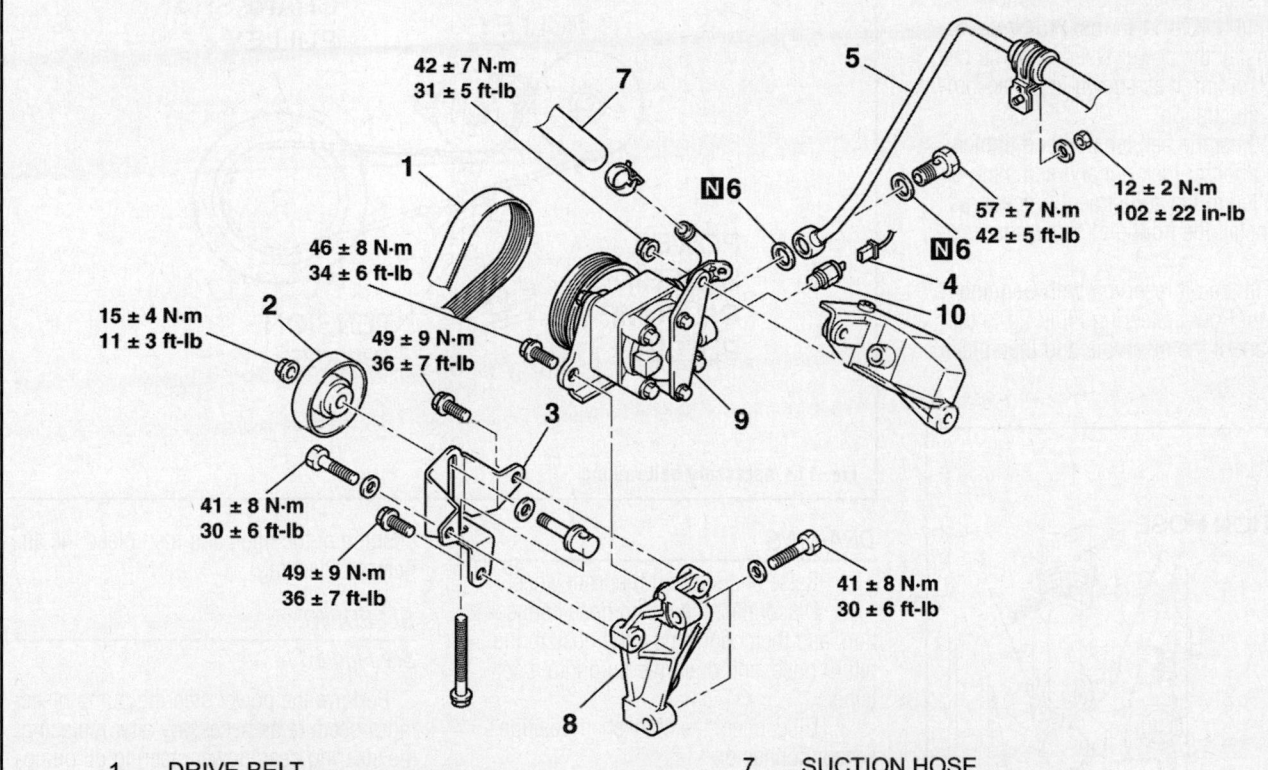

1. DRIVE BELT	7. SUCTION HOSE
2. BELT TENSIONER PULLEY	8. POWER STEERING PUMP
3. BELT TENSIONER BRACKET	BRACKET
4. PRESSURE SWITCH CONNECTOR	9. OIL PUMP ASSEMBLY
5. PRESSURE HOSE	10. POWER STEERING PRESSURE
6. GASKET	SWITCH

42050_ENDE_G0038

Fig. 111 Exploded view of the power steering pump and fastener tightening specifications

a. Insert the new switch and tighten hand-tight.

b. Use a socket to tighten the pressure switch to 14.5 ft. lbs. (19.6 Nm).

c. Attach the harness connector to the pressure switch.

13. Install the power steering pump and tighten the retainers to the specifications shown in the accompanying illustration.

14. Install the power steering pump bracket and tighten the retainers to the specifications shown in the accompanying illustration.

15. Install the suction hose. Make sure the marks on the hose and nipple are positioned as shown in the accompanying illustration.

16. Install the power steering pressure hose and gasket. Tighten the eye bolt to 42 ft. lbs. (57 Nm).

17. Install the belt tensioner bracket and tighten the retainers to the specifications shown in the accompanying illustration.

18. Install the belt tensioner pulley. Tighten the bolt to 11 ft. lbs. (15 Nm).

19. Install the power steering pump belt, carefully routing it as shown in the accompanying illustration.

20. Adjust the belt tension and tighten the fixing bolt or lock nut and adjusting bolt as specified in the adjustment procedures.

21. Install the front and side under covers.

22. Fill the oil reservoir with Genuine Mitsubishi Power Steering Fluid up to the lower mark of the reservoir, and then bleed the air.

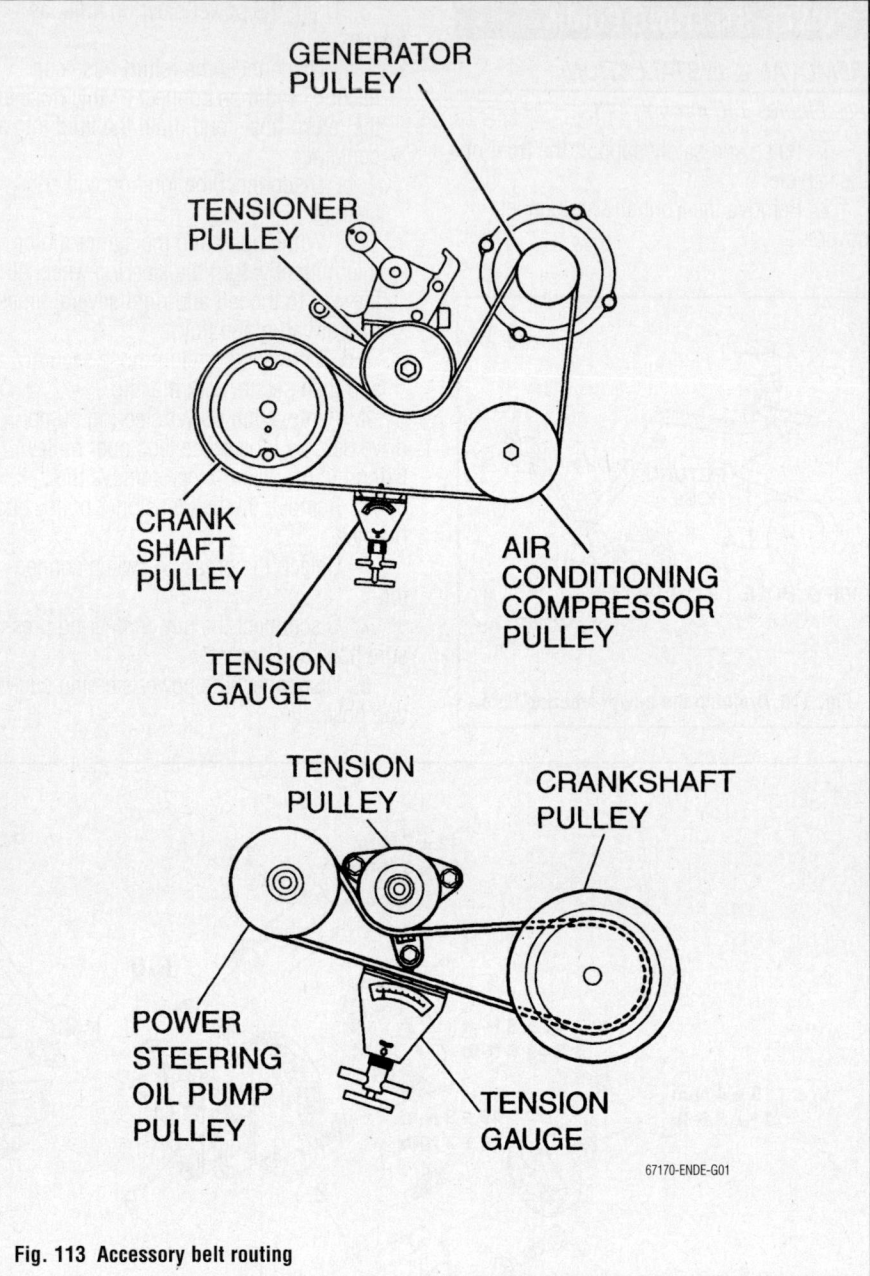

Fig. 113 Accessory belt routing

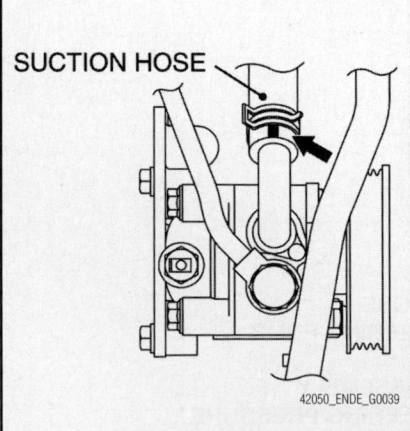

Fig. 112 When installing the suction hose, make sure the marks are aligned as shown

DRAINING

1. Raise and support the front wheels.

2. Disconnect the return hose connection, and then connect a vinyl hose to the return hose, and drain the fluid into a container.

3. Disconnect the crankshaft position sensor connector.

4. While operating the starter motor intermittently, turn the steering wheel all the way to the left and right several times to drain all of the fluid.

5. Connect the return hose securely, and then secure with the clip.

6. Fill the oil reservoir with GENUINE MITSUBISHI POWER STEERING FLUID or ATF DEXRON III/DEXRON II up to the lower position of the filler, and then bleed the air. Refer to Bleeding.

BLEEDING

See Figure 114.

Perform the power steering pump bleeding procedure as necessary after replacing the steering gear, power steering oil pump or the steering fluid lines.

1. Raise and safely support the front of the vehicle.

2. Disconnect the ignition coil connectors.

✷✷ WARNING

Perform air bleeding only while cranking the engine. Do not

perform air bleeding while the engine is running. If you do so, air in the fluid will be increased and air bleeding will become more difficult. During air bleeding, refill the steering fluid so that the level never falls below the lower mark on the dipstick.

3. Turn the steering wheel all the way to the left and right five or six times while using the starter motor to crank the engine intermittently several times (for 15 to 20 seconds).

4. Connect the ignition coil connectors.

5. Start the engine and allow to idle.

6. Turn the steering wheel to the left and right until there are no air bubbles in the oil reservoir. Make sure that the fluid is not

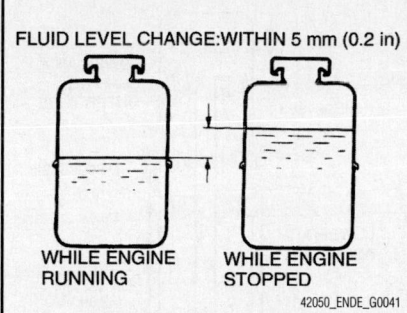

FLUID LEVEL CHANGE:WITHIN 5 mm (0.2 in)

WHILE ENGINE RUNNING

WHILE ENGINE STOPPED

42050_ENDE_G0041

Fig. 114 Make sure that the change in the fluid level is no more than 5 mm (0.2 inch) when the engine is stopped

milky, and that the level is between the high and low dipstick marks.

7. Check that there is minimal change in

the fluid level when the steering wheel is turned left and right.

If the fluid level rises suddenly after the engine is stopped, the air has not been completely bled. If air bleeding is not complete, there will be abnormal noises from the pump and the flow-control valve, and this condition could reduce the life of the power steering components.

8. Make sure that the change in the fluid level is no more than 5 mm (0.2 inch) when the engine is stopped.

9. If the change of the fluid level is 5 mm (0.2 inch) or more, the air has not been completely bled from the system. The air bleeding procedure must be repeated.

SUSPENSION

FRONT SUSPENSION

LOWER BALL JOINT

REMOVAL & INSTALLATION

The lower ball joint is an integral part of the lower control arm and is not serviced separately.

LOWER CONTROL ARM

REMOVAL AND & INSTALLATION
See Figure 115.

1. Before servicing the vehicle, refer to the Precautions Section.

2. Raise and support the vehicle safely.
3. Remove the wheel.
4. Remove the lower control arm mounting bolts and nuts.
5. Remove the lower ball joint from knuckle.
6. Remove the lower control arm.

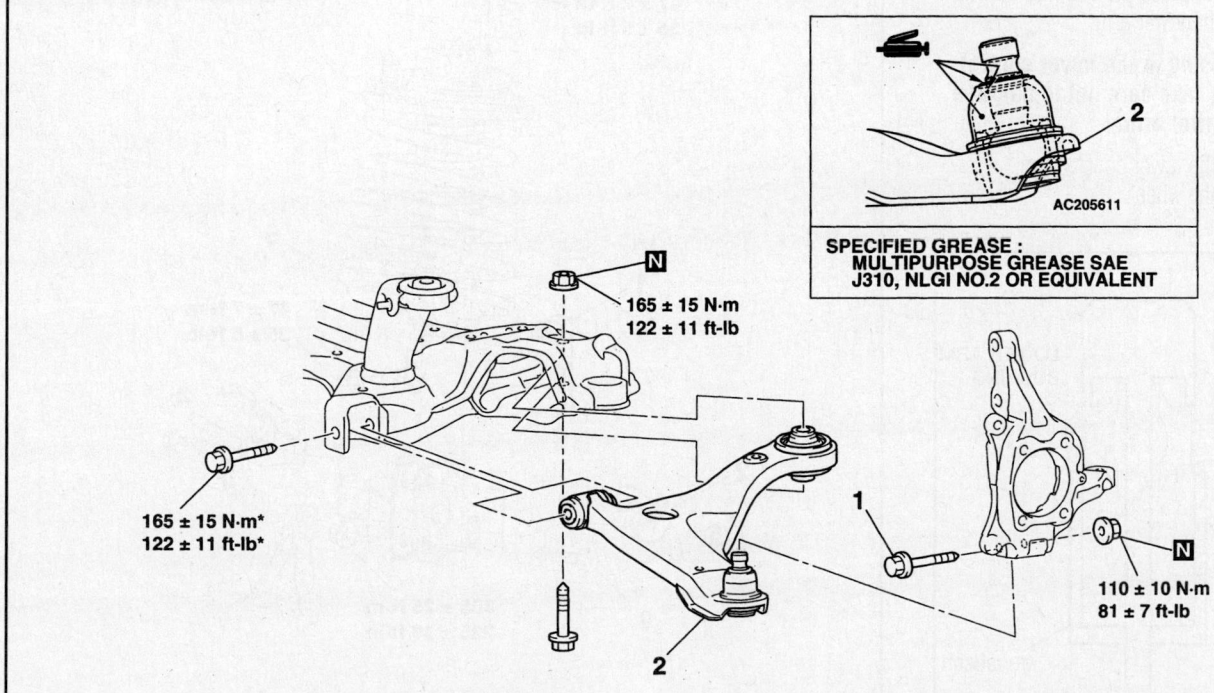

165 ± 15 N·m
122 ± 11 ft-lb

165 ± 15 N·m*
122 ± 11 ft-lb*

110 ± 10 N·m
81 ± 7 ft-lb

AC205611

SPECIFIED GREASE : MULTIPURPOSE GREASE SAE J310, NLGI NO.2 OR EQUIVALENT

1. LOWER ARM BOLT
2. LOWER ARM ASSEMBLY

67170-ENDE-G40

Fig. 115 Front lower control arm and related components

To install:

7. Install the lower control arm.

8. Install the lower control arm retainers. Tighten the knuckle nut and bolt to 81 ft. lbs. (110 Nm) and the other retainers hand-tight only at this time.

9. Install the wheel.

10. Once the weight of the vehicle is resting on the suspension, torque the lower control arm mounting bolts and nuts 122 ft. lbs. (165 Nm).

11. Check and adjust the front wheel alignment.

BUSHING REPLACEMENT

See Figures 116 and 117.

1. Before servicing the vehicle, refer to the Precautions Section.

2. Raise and support the vehicle safely.

3. Remove the wheel.

4. Remove the lower control arm and place in a vise.

5. Using tools MB991963 and MB990890 remove the bushing.

To install:

6. Position the bushing with the larger end facing the front of the vehicle.

➥**Coat the bushing with a soap solution and take care not to twist.**

7. Using tools MB991963, MB990889 and MB990890, press the bushing into its mounting on the lower arm.

➥**When pressing in the lower control arm bushing, take care not to damage the lower control arm.**

8. Install the lower control arm.

9. Install the wheel.

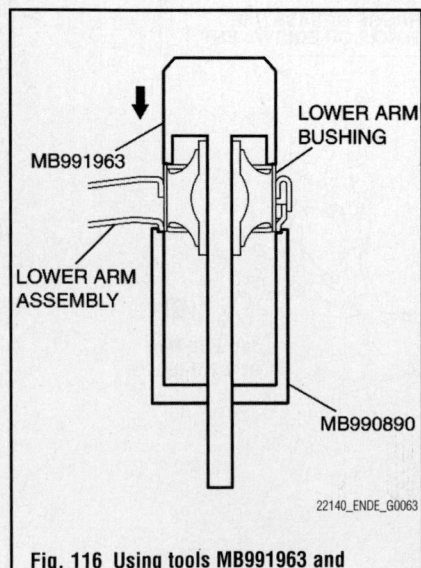

Fig. 116 Using tools MB991963 and MB990890 remove the bushing

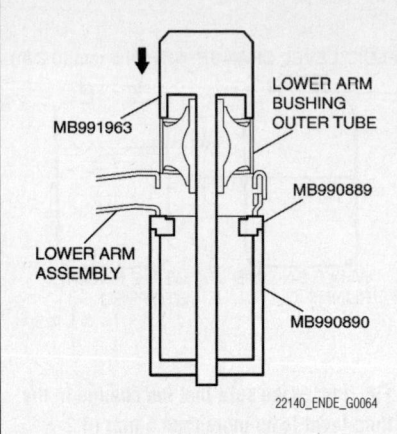

Fig. 117 Using tools MB991963, MB990889 and MB990890, press the bushing into its mounting on the lower arm

10. Check and adjust the front end alignment.

MACPHERSON STRUT

REMOVAL & INSTALLATION

See Figure 118.

1. Before servicing the vehicle, refer to the Precautions Section.

2. Remove the windshield wiper arm assemblies and front deck garnish.

3. Raise and support the vehicle safely.

4. Remove the tire and wheel.

5. Remove the stabilizer link.

6. Remove the brake hose bracket.

7. Remove the lower strut mounting bolt and nut.

8. Remove the front abs sensor clamp, if equipped.

9. Remove the upper strut mounting nuts.

10. Remove the strut from the vehicle.

To install:

11. Install the strut into the vehicle.

12. Install the upper strut mounting nuts and tighten to 35 ft. lbs. (47 Nm).

13. Install the front abs sensor clamp, if equipped.

14. Install the lower strut mounting bolt and nut. torque the nut to 225 ft. lbs. (305 Nm).

15. Install the brake hose bracket.

16. Install the stabilizer link.

17. Install the tire and wheel.

18. Install the front deck garnish and wiper arms.

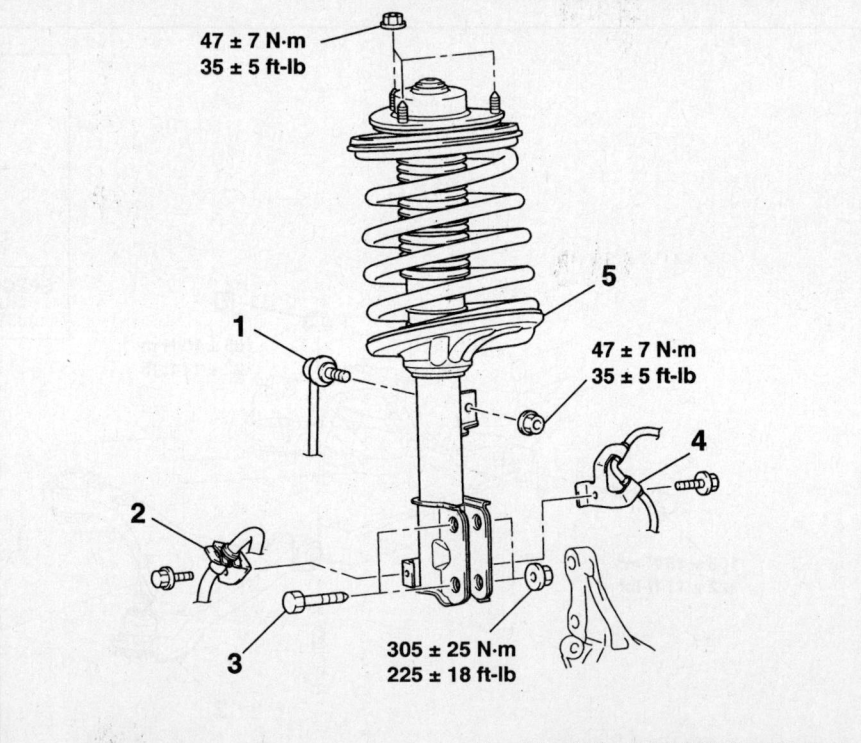

1. STABILIZER LINK	4. FRONT ABS SENSOR CLAMP <VEHICLES WITH ABS>
2. BRAKE HOSE BRACKET	5. STRUT ASSEMBLY
3. STRUT BOLT	

Fig. 118 Front strut and related components

STEERING KNUCKLE

REMOVAL & INSTALLATION

See Figures 119 through 121.

1. Before servicing the vehicle, refer to the Precautions Section.
2. Raise and support the vehicle safely.
3. Remove the split pin.
4. Remove the driveshaft nut and washer.
5. Remove the front ABS sensor.

✻✻ WARNING

Do not strike the ABS rotors installed to the BJ outer race of driveshaft against other parts when removing or installing the driveshaft. Otherwise the ABS rotors will be damaged. Be careful not to strike the pole piece at the tip of the front ABS sensor with tools during servicing work.

6. Remove the brake hose bracket.
7. Remove the caliper assembly.

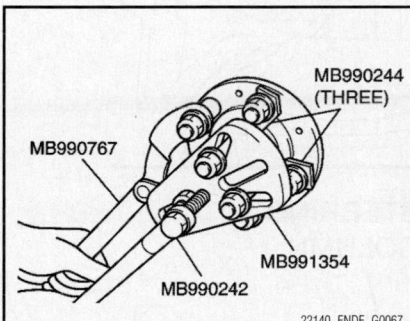

Fig. 119 Use special tools MB990242, MB990244, MB991354 and MB990767 to push out the driveshaft from the hub and knuckle

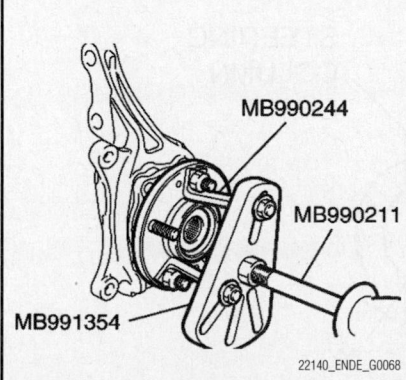

Fig. 120 Use special tools MB990244, MB991354 and MB990211 to pull out the front wheel hub from the knuckle

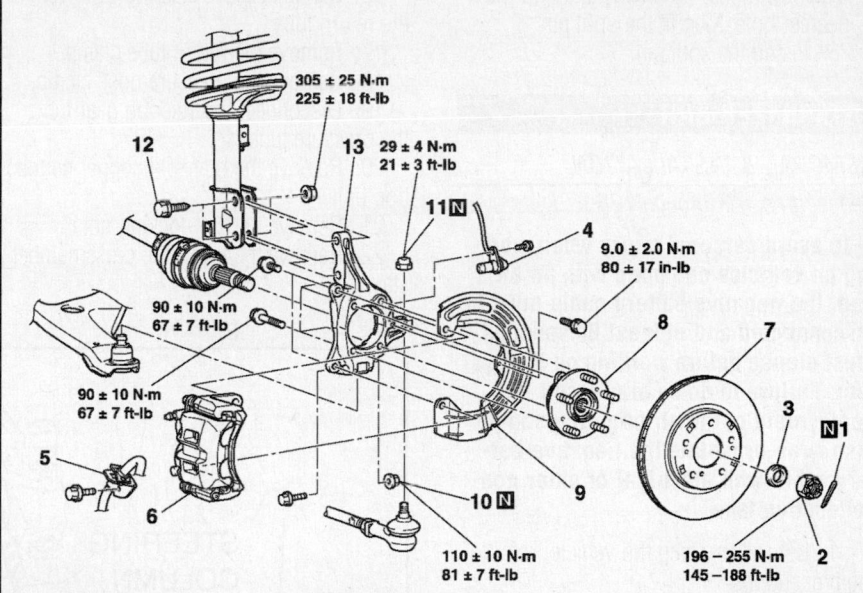

1. Split Pin
2. Driveshaft Nut
3. Washer
4. Front ABS Sensor
5. Brake Hose Bracket
6. Caliper Assembly
7. Brake Disc
8. Front Wheel Hub Assembly
9. Dust Cover
10. Self Locking Nut (Connection For Lower Arm Ball Joint)
11. Self Locking Nut (Connection For Tie Rod End)
12. Front Strut To Hub and Knuckle Mounting Bolt and Nut
13. Knuckle

Fig. 121 Exploded view of the front steering knuckle

a. Remove the caliper assembly with brake hose.

b. Retain the removed caliper assembly with a wire to prevent from falling.

8. Remove the brake disc.

a. If the brake disc is seized, install a M8 × 1.25-mm bolts in the provided holes, and remove the disc by tightening the bolts evenly and gradually.

9. Remove the front wheel hub assembly.

a. Use special tools MB990242, MB990244, MB991354 and MB990767 to push out the driveshaft from the hub and knuckle.

b. If the front wheel hub is seized, remove the knuckle together with front wheel hub and fix them with a vise.

c. Hang the driveshaft on the vehicle body with a rope.

d. Use special tools MB990244, MB991354 and MB990211 to pull out the front wheel hub from the knuckle.

10. Remove the dust cover.

11. Remove the self locking nut for lower arm ball joint.

12. Remove the self locking nut for tie rod end.

13. Remove the front strut to hub and knuckle mounting bolt and nut.

14. Remove the knuckle.

To install:

15. Install the knuckle.

16. Install the front strut to hub and knuckle mounting bolt and nut.

17. Install the self locking nut for tie rod end.

18. Install the self locking nut for lower arm ball joint.

19. Install the dust cover.

20. Install the front wheel hub assembly.

21. Install the brake disc.

22. Install the caliper assembly.

23. Install the brake hose bracket.

24. Install the front ABS sensor.

25. Install the driveshaft nut and washer.

a. Before securely tightening the driveshaft nuts, make sure there is no load on the wheel bearings. Otherwise the wheel bearings will be damaged.

b. Be sure to install the driveshaft washer in the specified direction.

c. Using special tool MB990767, tighten the driveshaft nut. At this time, tighten the nut to 145–188 ft. lbs. (196–255 Nm) torque so that the pin hole may align with split pin.

d. If the pin hole does not align with the pin, tighten the driveshaft nut to less than 188 ft. lbs. (255 Nm) and find the nearest hole, then fit the split pin.

26. Install the split pin.

STABILIZER BAR

REMOVAL & INSTALLATION

See Figures 122 through 124.

➡ **To avoid personal injury when working on vehicles equipped with an air bag, the negative battery cable must be disconnected and at least 60 seconds must elapse before working on the system. Failure to do so may result in deployment of the air bag. You should also wrap or isolate the negative battery cable with electrical or other non-conductive tape.**

1. Before servicing the vehicle, refer to the precautions.
2. Disconnect the negative battery cable.

➡ **Wait at least 90 seconds after disconnecting the cable from the negative (-) battery terminal to prevent airbag activation.**

3. Raise and safely support the vehicle securely on jackstands.
4. Remove the wheels.
5. Remove the front under cover.
6. Drain the power steering fluid.
7. Remove the center member.
8. Remove the steering wheel.
9. Remove the floor console side cover.
10. Remove the front scruff plate and cowl side trim.
11. Pull back the floor carpeting.
12. Remove the stabilizer nut.
13. Disconnect the tie rod end front the knuckle.
14. Disconnect the lower arm assembly from the knuckle.
15. Disconnect the steering gear from the steering column shaft.
   a. Remove the steering gear and steering column shaft connecting bolt.
   b. Pinch the steering column shaft clip with pliers, and pull up the shaft towards the direction shown to disengage the steering column shaft from the steering gear.

✳✳ WARNING

If the steering column shaft is removed accidentally, remove the steering column assembly and be sure to insert the steering column shaft into the steering column as shown in the figure.

16. Disconnect the steering gear from the return tube.
17. Remove the return tube clamp.
18. Remove the pressure hose clamp.
19. Disconnect the steering gear from the pressure tube.
20. Remove the rear roll stopper connection bolt.
21. Remove the self-locking nut.
22. Remove the front axle crossmember stay.
23. Remove the front axle No.1 crossmember, rear roll stopper and steering gear assembly.
24. Remove the steering gear and linkage protector.
25. Remove the stabilizer bracket.
26. Remove the stabilizer bushing.
27. Remove the stabilizer bar.

To install:

28. Install the stabilizer bar.
   a. Align the stabilizer bar identifica-

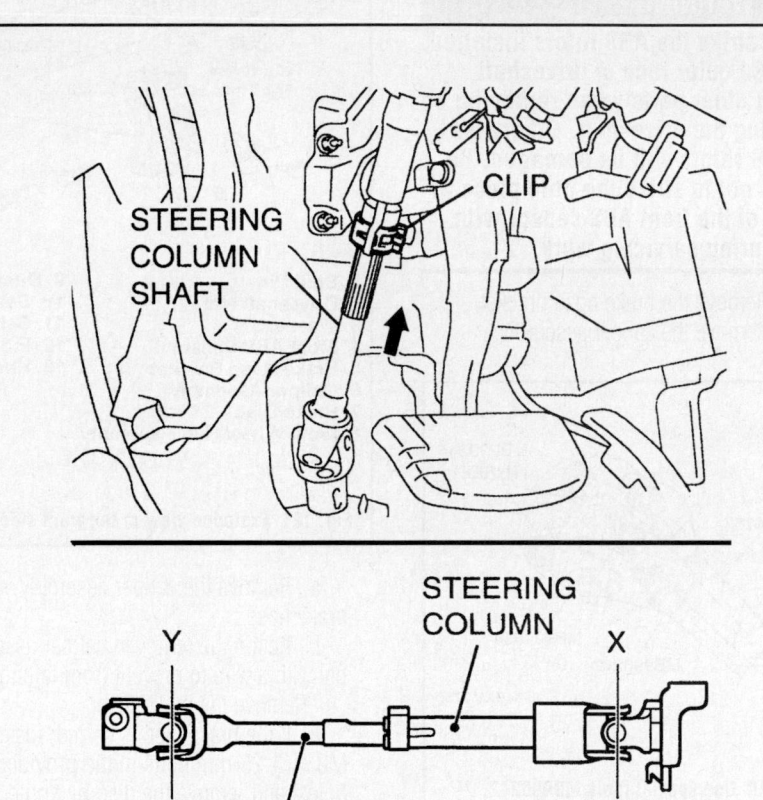

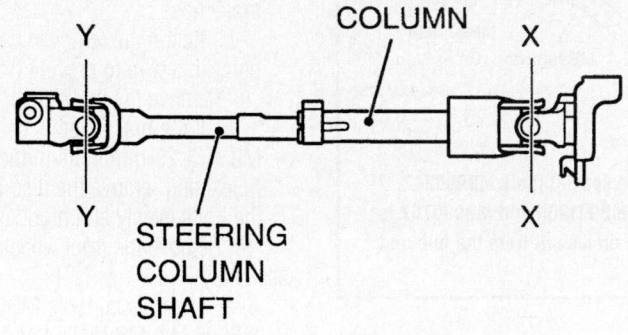

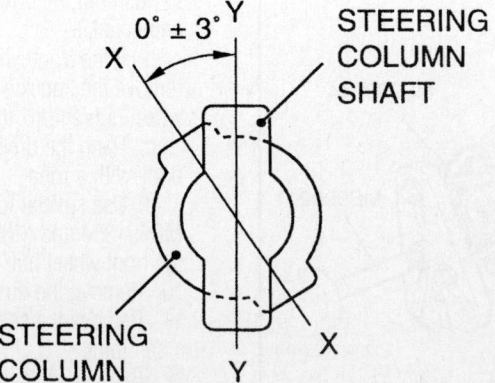

09482_ENDE_G0023

Fig. 122 Steering gear to steering column disconnection

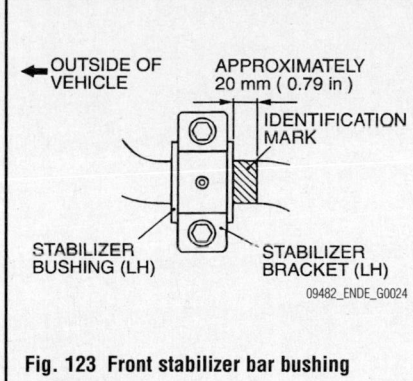

Fig. 123 Front stabilizer bar bushing alignment

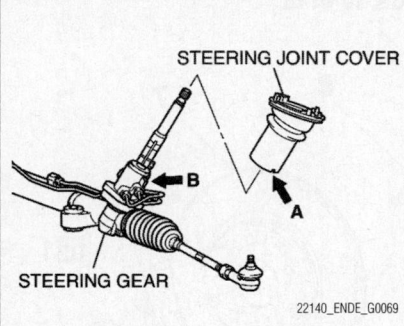

Fig. 124 Align the steering joint cover notch (arrow A) with the steering gear lug (arrow B), and install the front axle number 1 crossmember, the rear roll stopper and steering gear assembly

tion mark with the right end of the bushing.

29. Install the stabilizer bushing.

30. Install the stabilizer bracket. Tighten to 43 ft. lbs. (58 Nm).

31. Install the steering gear and linkage protector.

32. Install the front axle No.1 crossmember, rear roll stopper and steering gear assembly.

    a. Align the steering joint cover notch (arrow A) with the steering gear lug (arrow B), and install the front axle number 1 crossmember, the rear roll stopper and steering gear assembly.

33. Install the front axle crossmember stay.

34. Install the self-locking nut.

35. Install the rear roll stopper connection bolt.

36. Connect the steering gear from the pressure tube.

37. Install the pressure hose clamp.

38. Install the return tube clamp.

39. Connect the steering gear from the return tube.

40. Connect the steering gear from the steering column shaft.

41. Connect the lower arm assembly from the knuckle.

42. Connect the tie rod end front the knuckle.

43. Install the stabilizer link nut. Tighten to 35 ft. lbs. (47 Nm).

44. Pull back the floor carpeting.

45. Install the front scruff plate and cowl side trim.

46. Install the floor console side cover.

47. Install the steering wheel.

48. Install the center member.

49. Drain the power steering fluid.

50. Install the front under cover.

51. Install the wheels and lower the vehicle.

52. Connect the negative battery cable. Turn the ignition **ON** and then **OFF**, and keep it off for at least ten seconds.

53. Check and adjust the power steering fluid level,

54. Check and adjust the front end alignment, as required.

WHEEL HUB & BEARING

REMOVAL & INSTALLATION

See Figures 125 through 127.

1. Before servicing the vehicle, refer to the Precautions Section.

2. Raise and support the vehicle safely.

3. Remove the split pin.

4. Remove the driveshaft nut and washer.

5. Remove the front ABS sensor.

✻✻ WARNING

Do not strike the ABS rotors installed to the BJ outer race of driveshaft against other parts when removing or installing the driveshaft. Otherwise the ABS rotors will be damaged. Be careful not to strike the pole piece at the tip of the front ABS sensor with tools during servicing work.

6. Remove the brake hose bracket.

7. Remove the caliper assembly.

    a. Remove the caliper assembly with brake hose.

    b. Retain the removed caliper assembly with a wire to prevent from falling.

8. Remove the brake disc.

    a. If the brake disc is seized, install a M8 × 1.25-mm bolts in the provided holes, and remove the disc by tightening the bolts evenly and gradually.

9. Remove the front wheel hub assembly.

    a. Use special tools MB990242, MB990244, MB991354 and MB990767

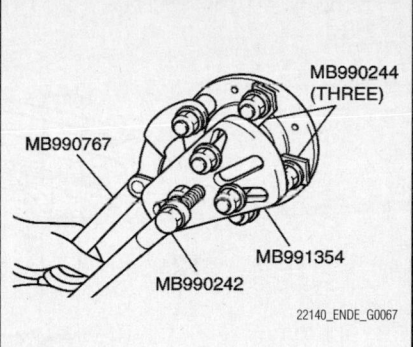

Fig. 125 Use special tools MB990242, MB990244, MB991354 and MB990767 to push out the driveshaft from the hub and knuckle

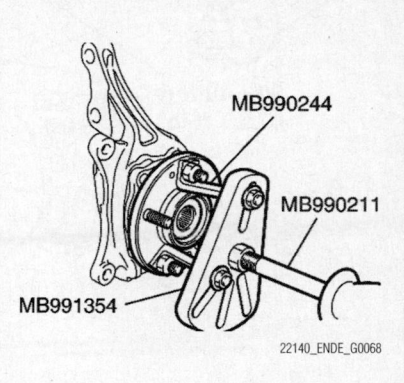

Fig. 126 Use special tools MB990244, MB991354 and MB990211 to pull out the front wheel hub from the knuckle

to push out the driveshaft from the hub and knuckle.

    b. If the front wheel hub is seized, remove the knuckle together with front wheel hub and fix them with a vise.

    c. Hang the driveshaft on the vehicle body with a rope.

    d. Use special tools MB990244, MB991354 and MB990211 to pull out the front wheel hub from the knuckle.

10. Remove the dust cover.

11. Remove the self locking nut for lower arm ball joint.

12. Remove the self locking nut for tie rod end.

13. Remove the front strut to hub and knuckle mounting bolt and nut.

14. Remove the steering knuckle.

15. Remove the hub from the steering knuckle.

To install:

16. Install the hub on the steering knuckle. Tighten the attaching bolts to 67 ft. lbs. (90 Nm).

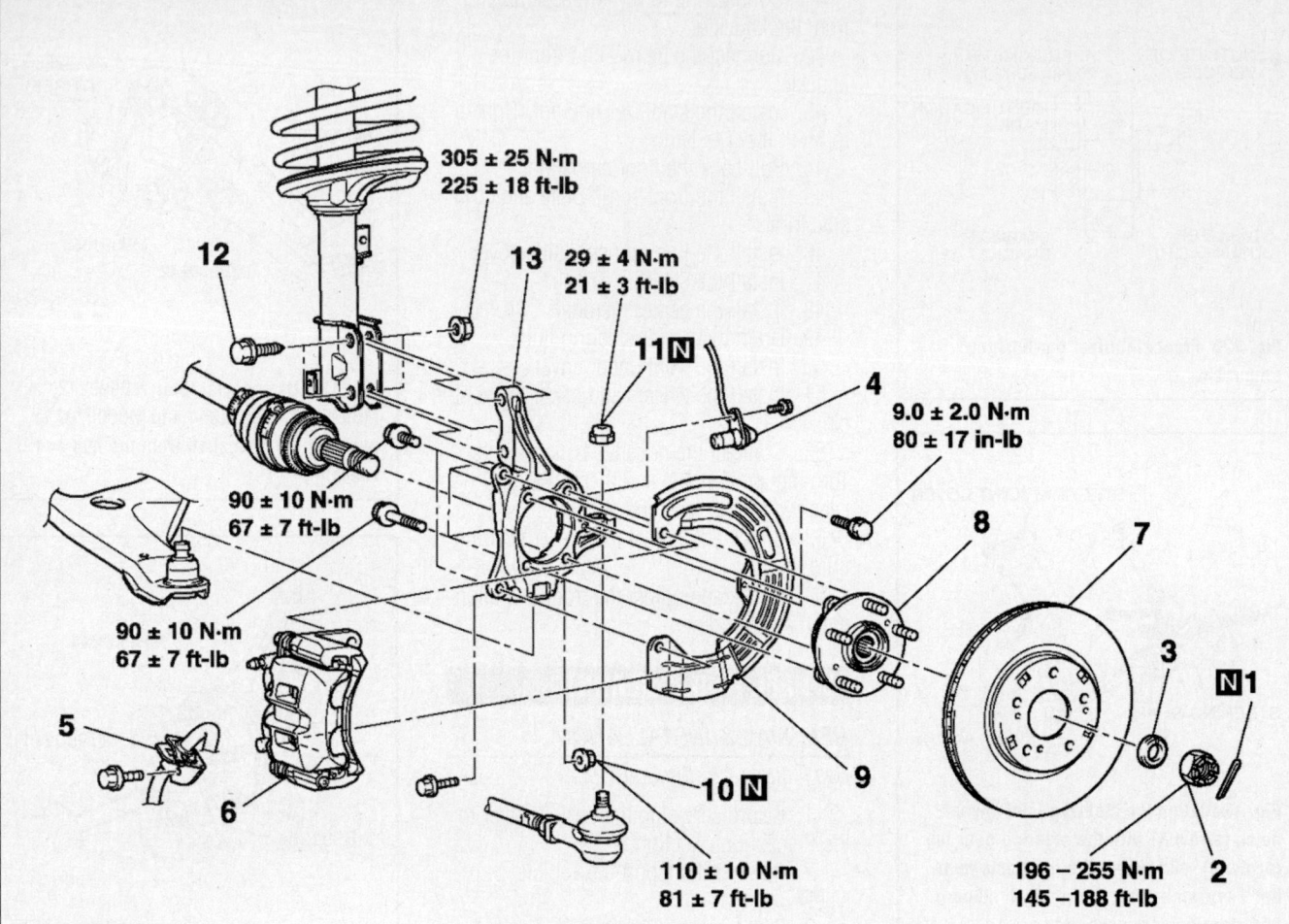

305 ± 25 N·m
225 ± 18 ft-lb

29 ± 4 N·m
21 ± 3 ft-lb

9.0 ± 2.0 N·m
80 ± 17 in-lb

90 ± 10 N·m
67 ± 7 ft-lb

90 ± 10 N·m
67 ± 7 ft-lb

110 ± 10 N·m
81 ± 7 ft-lb

196 – 255 N·m
145 –188 ft-lb

1. Split Pin
2. Driveshaft Nut
3. Washer
4. Front ABS Sensor
5. Brake Hose Bracket
6. Caliper Assembly
7. Brake Disc
8. Front Wheel Hub Assembly
9. Dust Cover
10. Self Locking Nut (Connection For Lower Arm Ball Joint)
11. Self Locking Nut (Connection For Tie Rod End)
12. Front Strut To Hub and Knuckle Mounting Bolt and Nut
13. Knuckle

22140_ENDE_G0066

Fig. 127 Exploded view of the front steering knuckle

17. Install the knuckle.
18. Install the front strut to hub and knuckle mounting bolt and nut.
19. Install the self locking nut for tie rod end.
20. Install the self locking nut for lower arm ball joint.
21. Install the dust cover.
22. Install the front wheel hub assembly.
23. Install the brake disc.
24. Install the caliper assembly.
25. Install the brake hose bracket.

26. Install the front ABS sensor.
27. Install the driveshaft nut and washer.
  a. Before securely tightening the driveshaft nuts, make sure there is no load on the wheel bearings. Otherwise the wheel bearings will be damaged.
  b. Be sure to install the driveshaft washer in the specified direction.
  c. Using special tool MB990767, tighten the driveshaft nut. At this time, tighten the nut to 145–188 ft. lbs.

(196–255 Nm) torque so that the pin hole may align with split pin.
  d. If the pin hole does not align with the pin, tighten the driveshaft nut to less than 188 ft. lbs. (255 Nm) and find the nearest hole, then fit the split pin.
28. Install the split pin.

ADJUSTMENT

The wheel bearings are not adjustable. If the bearings are noisy or become loose, they must be replaced.

SUSPENSION

COIL SPRING

REMOVAL & INSTALLATION

See Figure 128.

1. Before servicing the vehicle, refer to the Precautions Section.

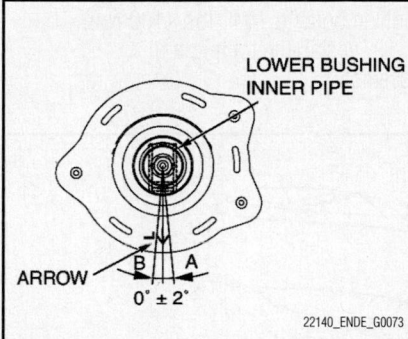

Fig. 128 Position a center line (A) of the shock absorber lower bushing inner pipe as shown from the arrow (B) on the shock absorber insulator

2. Raise and support the vehicle safely.
3. Remove the rear wheel.
4. Remove the rear shock absorber assembly.
5. Compress the spring using a suitable spring compressor and remove the coil spring nut.
6. Remove the washer.
7. Remove the shock absorber insulator assembly.
8. Remove the spring upper pad.
9. Remove the shock absorber cover.
10. Remove the shock absorber damper.
11. Remove the coil spring.

To install:

12. Install the coil spring.
13. Install the shock absorber damper.
14. Install the shock absorber cover.
15. Install the spring upper pad.
16. Install the shock absorber insulator assembly.
   a. Position a center line (A) of the shock absorber lower bushing

inner pipe as shown from the arrow (B) on the shock absorber insulator. Then install the shock absorber insulator.
17. Install the washer.
18. Compress the spring using a suitable spring compressor and install the coil spring nut.
19. Install the rear shock absorber assembly.
20. Install the rear wheel.
21. With the weight of the vehicle resting on the suspension, tighten the lower shock-to-knuckle bolt to 74 ft. lbs. (100 Nm).

CONTROL ARMS/LINKS

REMOVAL & INSTALLATION

Toe Control Arm

See Figure 129.

1. Before servicing the vehicle, refer to the Precautions Section.
2. Raise and support the vehicle safely.

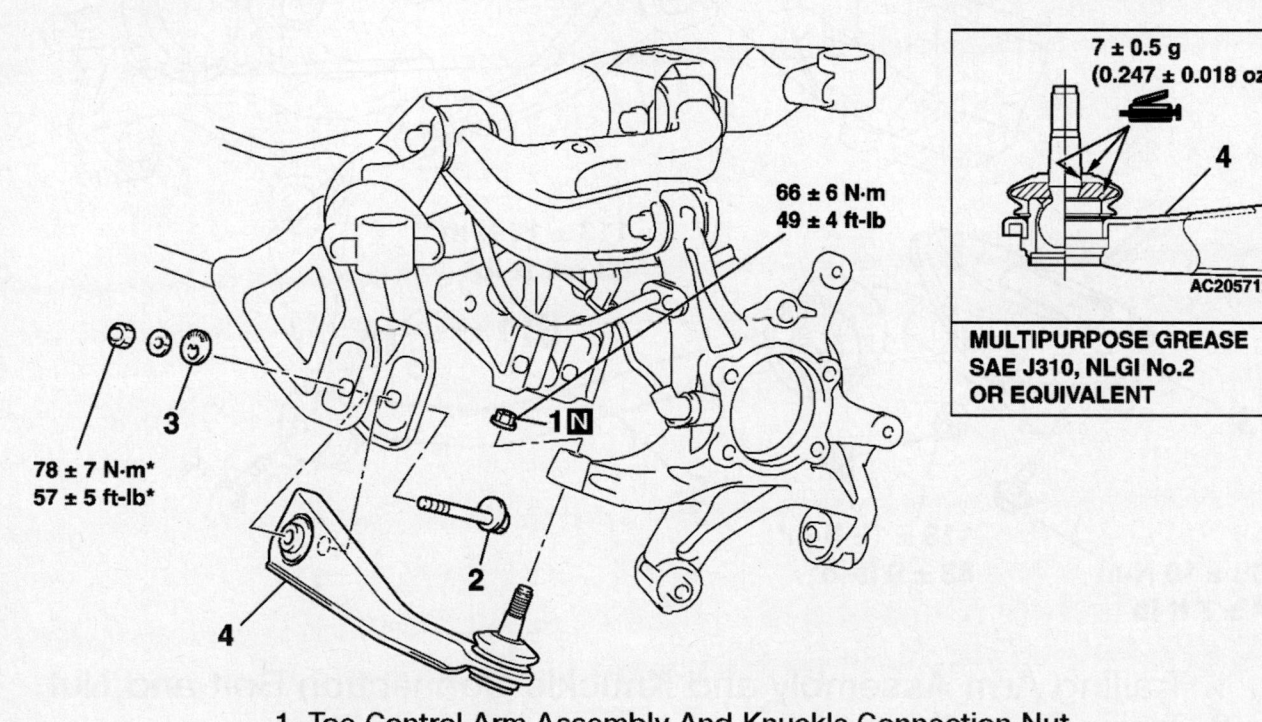

1. Toe Control Arm Assembly And Knuckle Connection Nut
2. Assist Link Bolt
3. Assist Link Plate
4. Toe Control Arm Assembly

Fig. 129 Toe control arm and related components

3. Remove the wheel.

4. Remove the toe control arm assembly to knuckle bolt and nut.

5. Remove the assist link bolt.

6. Remove the assist link plate.

7. Remove the toe control arm assembly.

To install:

8. Install the toe control arm assembly.

9. Install the assist link plate.

10. Install the assist link bolt.

11. Install the toe control arm assembly to knuckle bolt and nut.

12. Install the wheel.

13. Once the weight of the vehicle is on the suspension, tighten the toe control arm assembly to knuckle bolt and nut 57 ft. lbs. (78 Nm) and the assist link bolt to 49 ft. lbs. (66 Nm).

14. Check and adjust the rear wheel alignment.

Trailing Arm

See Figure 130.

1. Before servicing the vehicle, refer to the Precautions Section.

2. Raise and support the vehicle safely.

3. Remove the wheel.

4. Remove the trailing arm assembly to knuckle bolt and nut.

5. Remove the trailing arm assembly.

6. Remove the trailing arm bracket.

To install:

7. Install the trailing arm bracket. Tighten bolts to 74 ft. lbs. (100 Nm).

8. Install the trailing arm assembly.

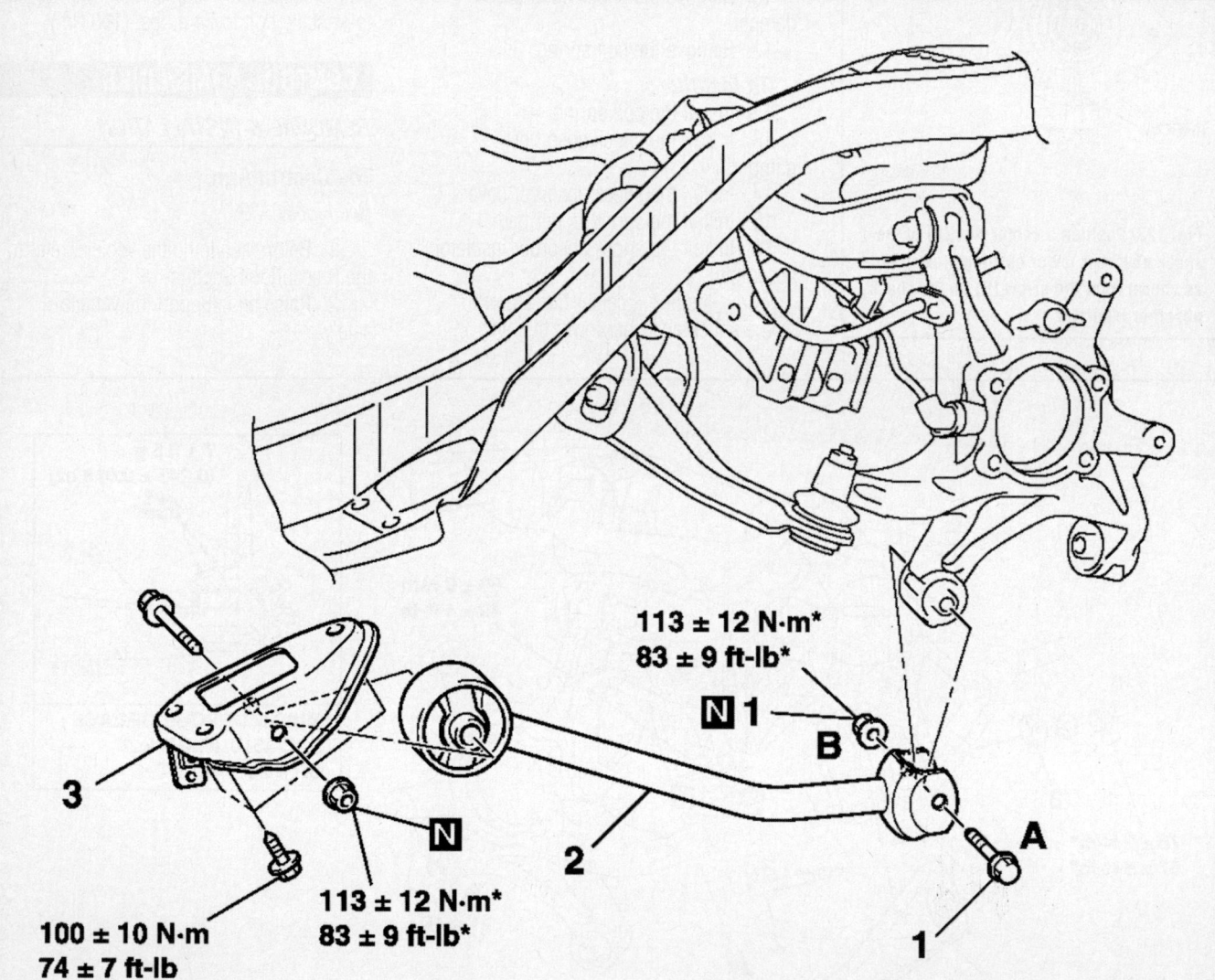

113 ± 12 N·m*
83 ± 9 ft-lb*

113 ± 12 N·m*
83 ± 9 ft-lb*

100 ± 10 N·m
74 ± 7 ft-lb

1. Trailing Arm Assembly and Knuckle Connection Bolt and Nut
2. Trailing Arm Assembly
3. Trailing Arm Bracket

22140_ENDE_G0070

Fig. 130 Trailing arm and related components

9. Install the trailing arm assembly to knuckle bolt and nut.

10. Install the wheel.

11. Once the weight of the vehicle is on the suspension, tighten the trailing arm assembly to knuckle bolt and nut 83 ft. lbs. (113 Nm).

12. Check and adjust the rear wheel alignment.

Lower Control Arm

See Figure 131.

1. Before servicing the vehicle, refer to the Precautions Section.

2. Raise and support the vehicle safely.

3. Remove the rear wheel.

4. Remove the lower control arm from the knuckle.

5. Remove the retaining nut, then separate the lower control arm from the stabilizer bar link.

➡ **Matchmark the crossmember and the plate before removing the lower control arm bolt.**

6. Remove the lower control arm bolt, plate and lower control arm assembly.

To install:

7. Install the lower control arm.

8. Install the lower control arm plate.

9. Install the lower control arm bolt, hand-tight only at this time.

10. Install the lower control arm to the stabilizer bar link. Tighten the retaining to 30 ft. lbs. (40 Nm).

11. Install the lower control arm to the knuckle. Use a new nut and secure hand-tight.

12. Install the rear wheel.

13. Once the weight of the vehicle is resting on the suspension, torque the

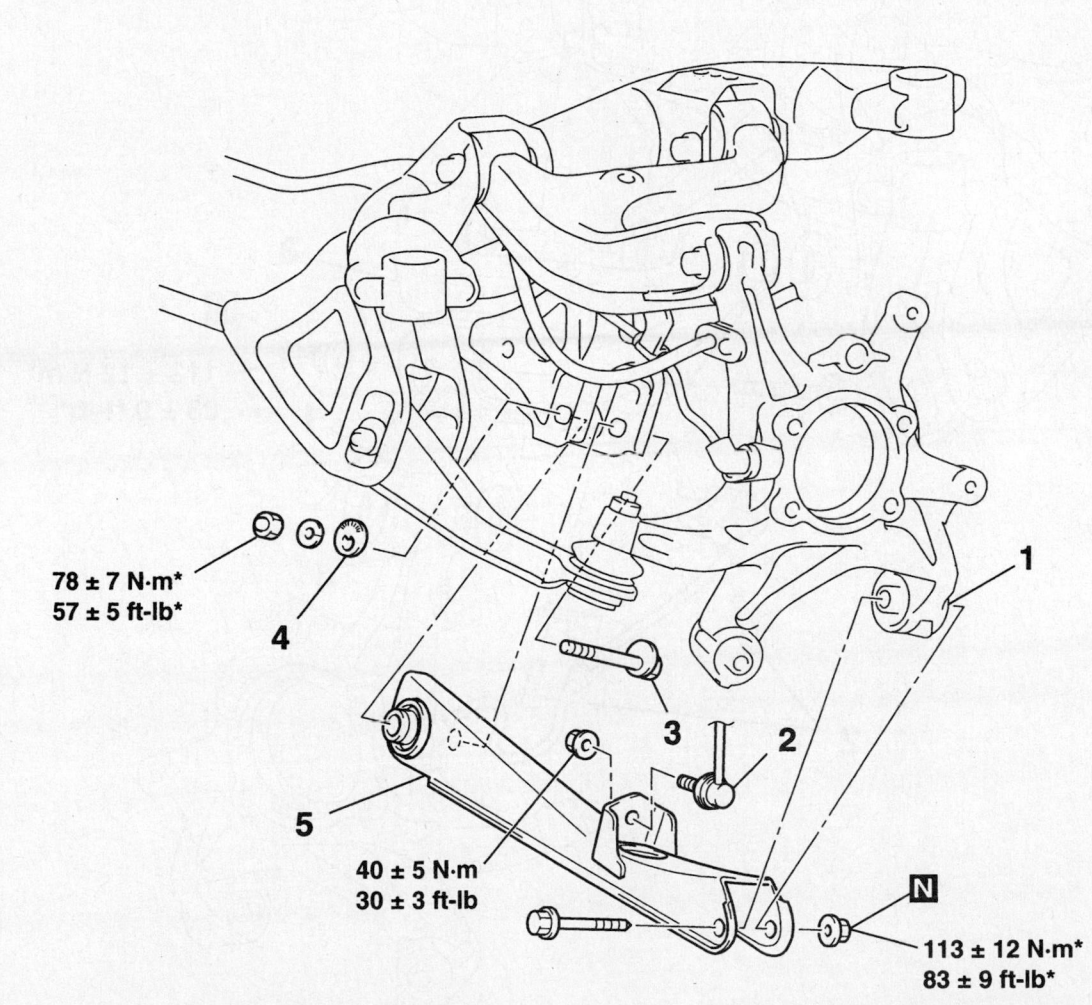

78 ± 7 N·m*
57 ± 5 ft-lb*

40 ± 5 N·m
30 ± 3 ft-lb

113 ± 12 N·m*
83 ± 9 ft-lb*

1. LOWER ARM ASSEMBLY AND KNUCKLE CONNECTION
2. LOWER ARM ASSEMBLY AND STABILIZER BAR LINK ASSEMBLY CONNECTION
3. LOWER ARM BOLT
4. LOWER ARM PLATE
5. LOWER ARM ASSEMBLY

67170-ENDE-G39

Fig. 131 Rear lower control arm and related components

lower control arm mounting bolt to 57 ft. lbs. (78 Nm), and the lower control arm to knuckle nut to 83 ft. lbs. (113 Nm).

14. Check and adjust the rear wheel alignment.

Upper Control Arm
See Figure 132.

1. Before servicing the vehicle, refer to the Precautions Section.

2. Raise and support the vehicle safely.

3. Remove the wheel assembly.

4. Remove the upper control arm from the knuckle.

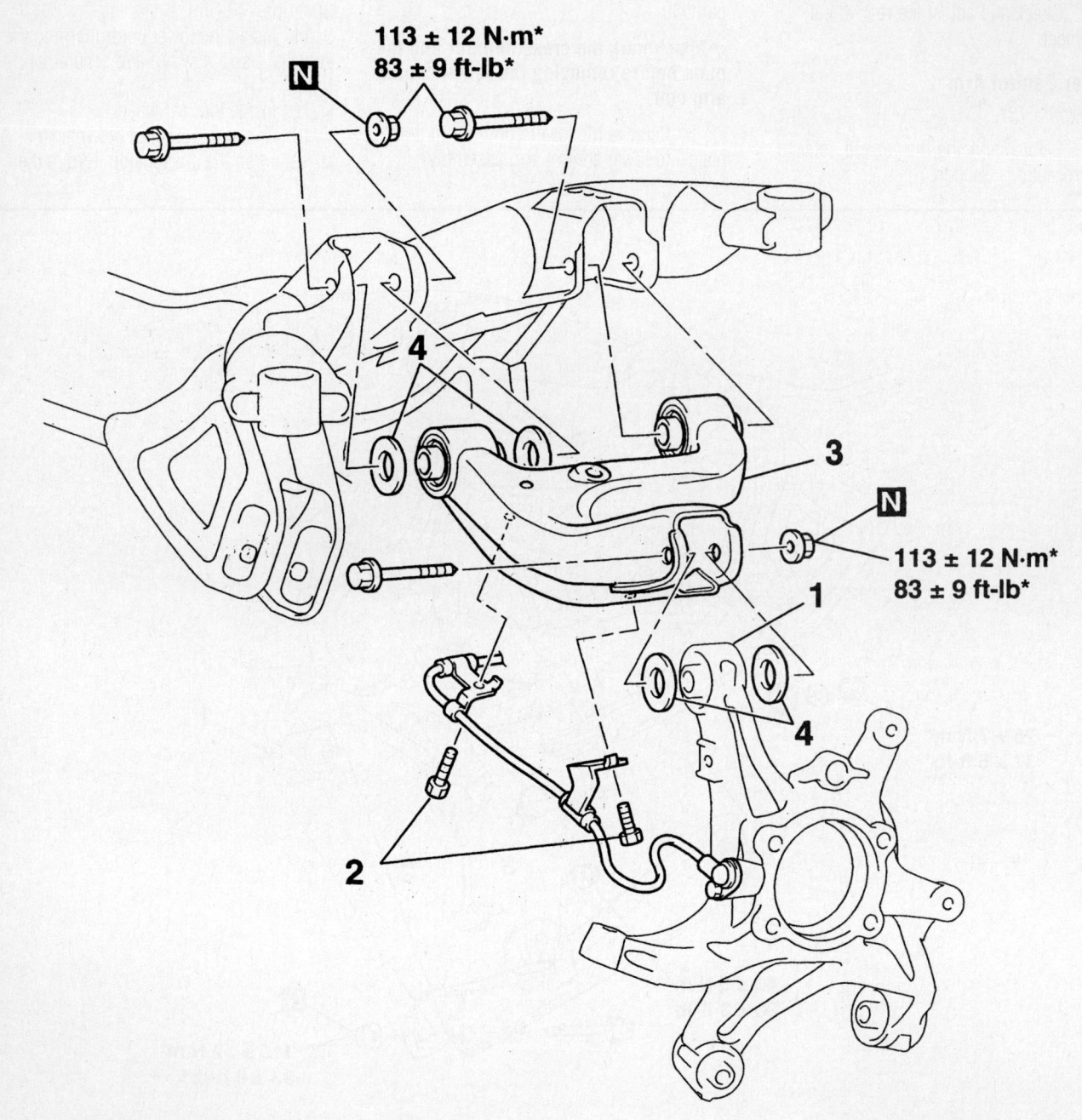

113 ± 12 N·m*
83 ± 9 ft-lb*

113 ± 12 N·m*
83 ± 9 ft-lb*

1. **UPPER ARM ASSEMBLY AND KNUCKLE CONNECTION**	3. **UPPER ARM ASSEMBLY**
2. **ABS EQUIPMENT BOLT**	4. **UPPER ARM STOPPER**

67170-ENDE-G38

Fig. 132 Rear upper control arm and related components

5. Remove the ABS sensor clamp bolts, if equipped.

6. Remove the upper control arm.

7. Remove the upper arm stopper.

To install:

8. Install the upper arm stoppers.

9. Install the upper arm into the vehicle and hand-tighten the retainers.

10. Install the upper control arm-to-knuckle bolt and nut and tighten hand-tight.

11. Install the ABS sensor clamp bolts.

12. Install the wheel assembly.

13. Once the weight of the vehicle is on the suspension, tighten the control arm-to-knuckle nut and the upper control

arm mounting nuts to 83 ft. lbs. (113 Nm).

14. Check and adjust the rear wheel alignment.

SHOCK ABSORBER

REMOVAL & INSTALLATION
See Figure 133.

1. Before servicing the vehicle, refer to the Precautions Section.

2. Remove the luggage floor rear board.

3. Remove the tonneau cover.

4. Remove the luggage floor front board.

5. Remove the parcel strap hook.

6. Remove the luggage floor side board.

7. Remove the rear end trim.

8. Remove the luggage floor carpet bracket.

9. Raise and support the vehicle safely.

10. Remove the wheel and tire.

11. Remove the lower shock absorber bolt and washer, and separate the shock from the knuckle.

12. Remove the upper shock mounting nuts.

13. Remove the shock absorber from the vehicle.

To install:

14. Install the shock absorber. Tighten the upper nuts to 34 ft. lbs. (45 Nm) and the lower shock-to-knuckle bolt hand-tight.

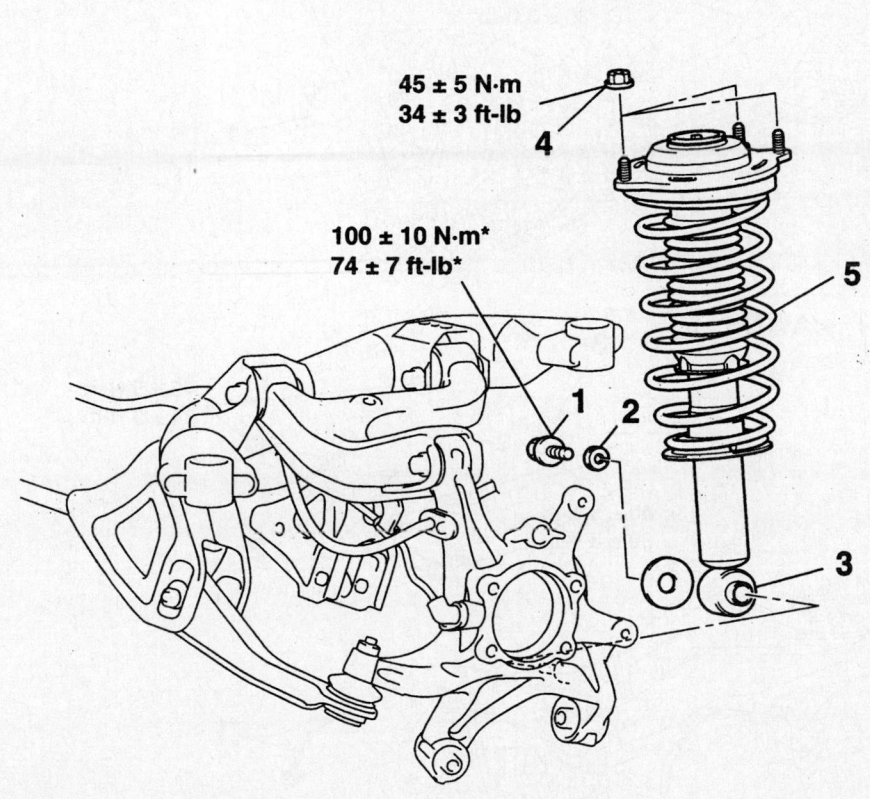

45 ± 5 N·m
34 ± 3 ft-lb

100 ± 10 N·m*
74 ± 7 ft-lb*

1. **COIL SPRING BOLT**
2. **COIL SPRING WASHER**
3. **SHOCK ABSORBER ASSEMBLY AND KNUCKLE CONNECTION**

4. **COIL SPRING NUT**
5. **SHOCK ABSORBER ASSEMBLY**

67170-ENDE-G37

Fig. 133 Rear shock absorber and related components

15. Install the wheel and tire
16. Install the luggage floor carpet bracket
17. Install the rear end trim
18. Install the luggage floor side board
19. Install the parcel strap hook
20. Install the luggage floor front board
21. Install the tonneau cover
22. Install the luggage floor rear board

23. With the weight of the vehicle resting on the suspension, tighten the lower shock-to-knuckle bolt to 74 ft. lbs. (100 Nm).

TESTING

1. Check the rubber parts for cracks and wear.
2. Check the shock absorber for malfunctions, oil leakage, or abnormal noise.

3. If shock absorber does not perform properly, replace the shock absorber.

STABILIZER BAR

REMOVAL & INSTALLATION

See Figure 134.

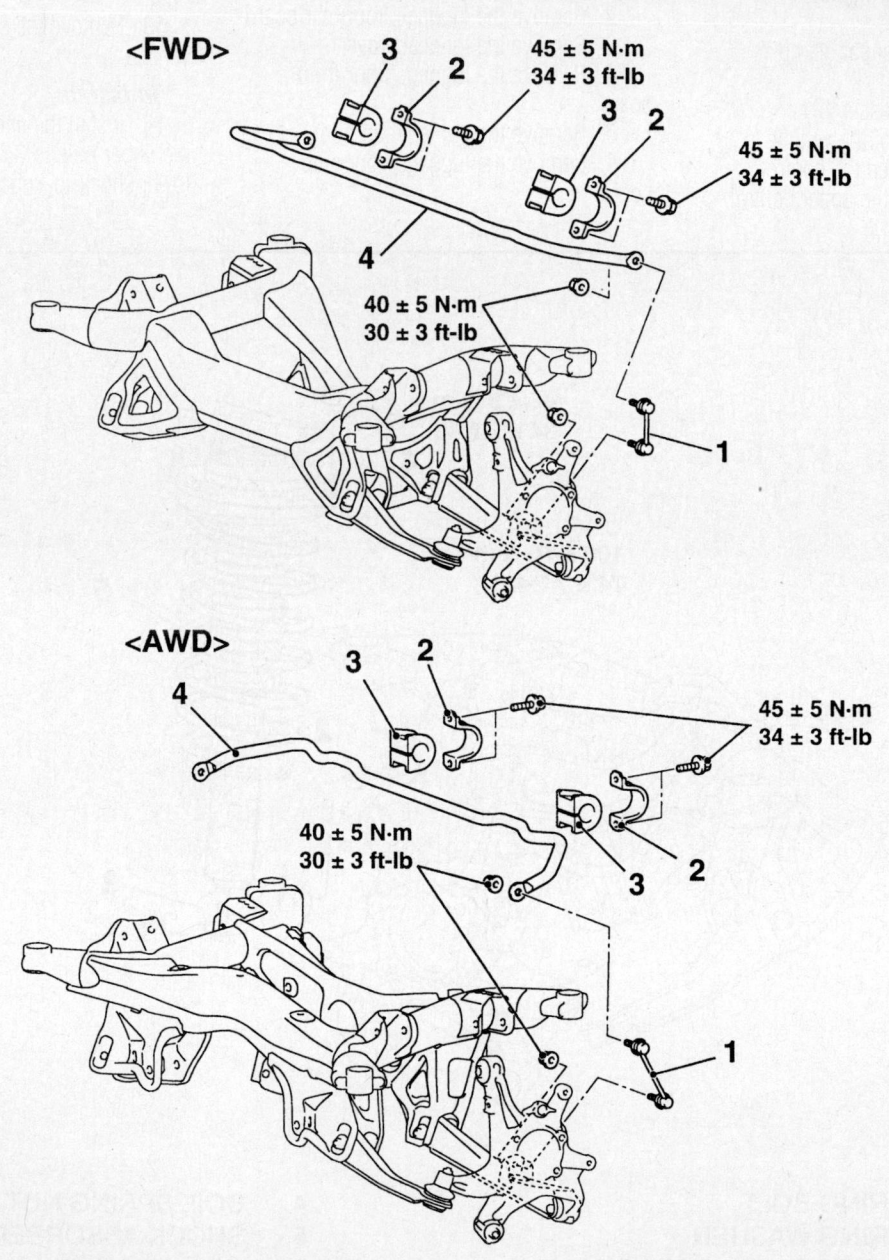

1. STABILIZER BAR LINK ASSEMBLY
2. STABILIZER BAR BRACKET
3. STABILIZER BUSHING
4. STABILIZER BAR

09482_ENDE_G0026

Fig. 134 Rear stabilizer bar and related components

1. Before servicing the vehicle, refer to the Precautions Section.

2. Raise and support the vehicle safely.

3. Remove the tire and wheel assembly.

4. Remove the stabilizer link retaining bolts.

5. Remove the stabilizer link assembly.

6. Remove the stabilizer bar bracket retaining bolts.

7. Remove the bushing.

8. Remove the stabilizer from the vehicle.

To install:

9. Install the stabilizer from the vehicle.

10. Install the bushing.

11. Install the stabilizer bar bracket retaining bolts. Tighten to 34 ft. lbs. (45 Nm).

12. Install the stabilizer link assembly.

13. Install the stabilizer link retaining bolts. Tighten to 30 ft. lbs. (40 Nm).

14. Install the tire and wheel assembly.

15. Lower the vehicle.

WHEEL HUB & BEARING

REMOVAL & INSTALLATION

See Figures 135 and 136.

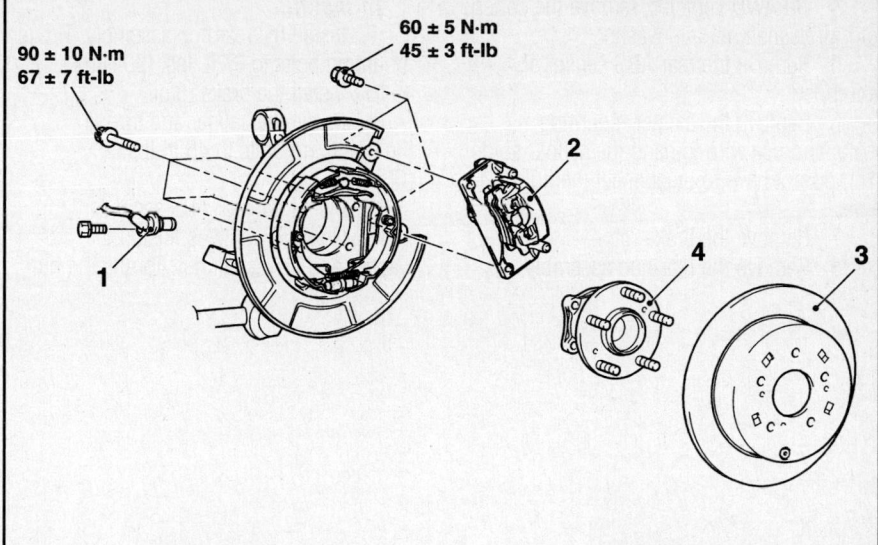

1. REAR ABS SENSOR<VEHICLES WITH ABS>	3. BRAKE DISC
2. CALIPER ASSEMBLY	4. REAR HUB ASSEMBLY

67170-ENDE-G42

Fig. 135 Rear hub and related components—FWD vehicles

➡**If the vehicle is equipped with ABS, do not to strike the pole piece at the tip of the rear ABS sensor, as damage may result. The hub assembly should not be disassembled.**

1. Before servicing the vehicle, refer to the Precautions Section.

2. Raise and support the vehicle safely.

3. Remove the wheels.

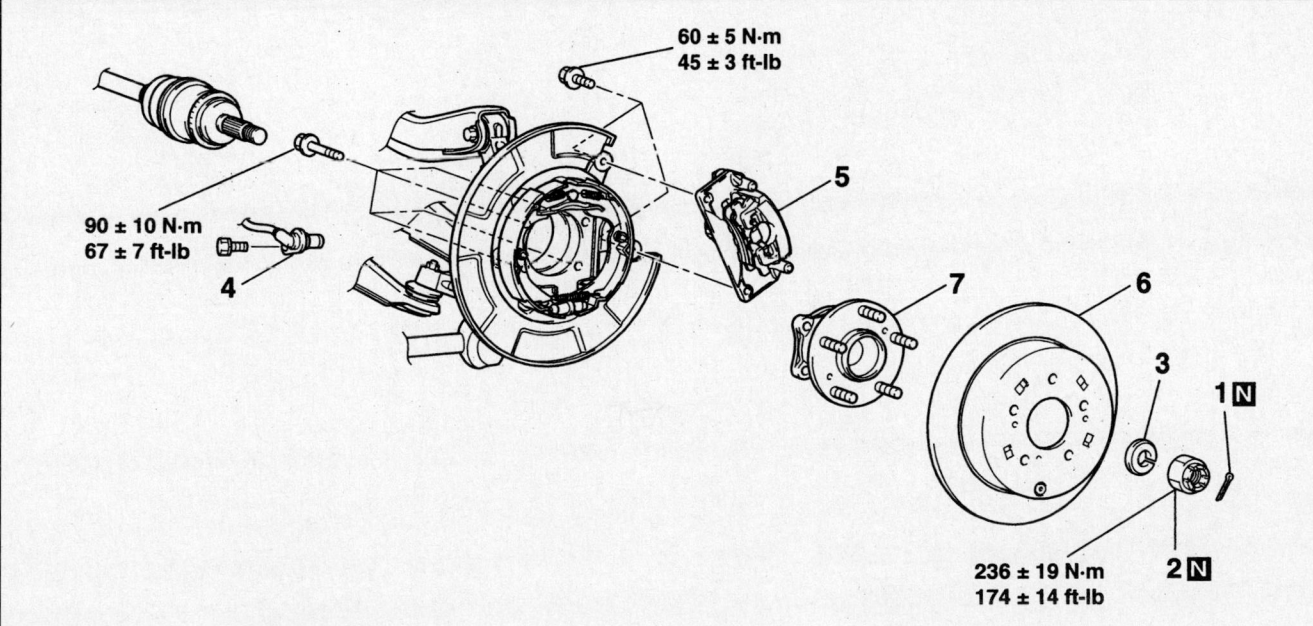

1. SPLIT PIN	5. CALIPER ASSEMBLY
2. DRIVE SHAFT NUT	6. BRAKE DISC
3. WASHER	7. REAR WHEEL HUB ASSEMBLY
4. REAR ABS SENSOR<VEHICLES WITH ABS>	

67170-ENDE-G43

Fig. 136 Rear hub and related components—AWD vehicles

4. On AWD vehicles, remove the cotter pin, driveshaft nut and washer.

5. Remove the rear ABS sensor, if equipped.

6. Remove the caliper mounting bolts and use wire to hang the caliper aside. You do not have to disconnect the fluid line.

7. Remove the brake rotor.

8. Remove the rear hub assembly.

To install:

9. Install the rear hub assembly. Tighten mounting bolts to 67 ft. lbs. (90 Nm).

10. Install the brake rotor.

11. Install the caliper and torque the mounting bolts to 45 ft. lbs. (60 Nm).

12. Install the rear ABS sensor.

13. On AWD vehicles, install the washer and driveshaft nut. Torque the nut to 174 ft. lbs. (236 Nm). Install a new cotter pin.

14. Install the wheels.

15. Road test the vehicle and check for leaks.

ADJUSTMENT

The wheel bearings are not adjustable. If the bearings are noisy or become loose, they must be replaced.

SPECIFICATIONS AND MAINTENANCE CHARTS

ENGINE AND VEHICLE IDENTIFICATION

Engine							Model Year	
Code ①	Liters (cc)	Cu. In.	Cyl.	Fuel Sys.	Engine Type	Eng. Mfg.	Code ②	Year
4G69/F	2.4 (2,378)	145.1	4	MPFI	SOHC	Mitsubishi	9	2009
6G75/S	3.8 (3,828)	233.6	6	MPFI	SOHC	Mitsubishi	A	2010
6G75/T	3.8 (3,828)	233.6	6	MPFI	SOHC	Mitsubishi		

MPFI: Multi-Point Fuel Injection

DOHC: Double Overhead Camshafts

① Engine Code/8th digit of VIN

② 10th digit of VIN

37671_GALA_C0001

GENERAL ENGINE SPECIFICATIONS

All measurements are given in inches.

Year	Model	Engine Displacement Liters	Engine Series ID/VIN	Net Horsepower @ rpm	Net Torque @ rpm (ft. lbs.)	Bore x Stroke (in.)	Compression Ratio	Oil Pressure @ rpm
2009	Galant	2.4	4G69/F	160@5500	155@4000	3.43 x 3.94	9.5:1	43-100@3500
	Galant	3.8	6G75/S	230@5250	250@4000	3.74 x 3.54	10.0:1	43-100@3500
	Galant	3.8	6G75/T	258@5750	258@4500	3.74 x 3.54	10.5:1	43-100@3500
2010	Galant	2.4	4G69/F	160@5500	155@4000	3.43 x 3.94	9.5:1	43-100@3500

37671_GALA_C0002

GASOLINE ENGINE TUNE-UP SPECIFICATIONS

Year	Engine Displacement Liters	Engine Series ID/VIN	Spark Plug Gap (in.)	Ignition Timing (deg.) MT	Ignition Timing (deg.) AT	Fuel Pump (psi)	Idle Speed (rpm) MT	Idle Speed (rpm) AT	Valve Clearance (in.) In.	Valve Clearance (in.) Ex.
2009	2.4	4G69/F	0.028-0.031	—	①	47	—	②	0.0040	0.0800
	3.8	6G75/S	0.028-0.031	—	①	47	—	②	HYD	HYD
	3.8	6G75/T	0.028-0.031	—	①	47	—	②	HYD	HYD
2010	2.4	4G69/F	0.028-0.031	—	①	47	—	②	0.0040	0.0080

NOTE: The Vehicle Emission Control Information label reflects specification changes made during production.

Follow the figures on the label if they differ from those in this chart.

HYD: Hydraulic

① Ignition timing is preset and cannot be adjusted

② Idle speed is maintained by the Electronic Control Module (ECM)

37671_GALA_C0003

CAPACITIES

Year	Model	Engine Displacement Liters	Engine Series ID/VIN	Engine Oil with Filter (qts.)	Transmission (pts.) Manual	Transmission (pts.) Auto. ①	Transfer Case (pts.)	Fuel Tank (gal.)	Cooling System (qts.)
2009	Galant	2.4	4G69/F	②	—	16.2	—	17.7	8.5
	Galant	3.8	6G75/S	②	—	17.8	—	17.7	9.2
	Galant	3.8	6G75/T	②	—	17.8	—	17.7	9.2
2010	Galant	2.4	4G69/F	②	—	16.2	—	17.7	8.5

NOTE: All capacities are approximate. Add fluid gradually and check to be sure a proper fluid level is obtained.

① Drain and refill

② Oil pan (4.2 qts.), Oil filter (0.32 qt.), except MIVEC. MEVIC Oil pan (4.5 qts.), Oil filter (0.32 qt.

37671_GALA_C0004

FLUID SPECIFICATIONS

Year	Model	Engine Displacement Liters	Engine Series ID/VIN	Engine Oil	Auto. Trans.	Power Steering Fluid	Brake Master Cylinder	Cooling System
2009	Galant	2.4	4G69/F	5W-20	①	②	③	④
	Galant	3.8	6G75/S	5W-20	①	②	③	④
	Galant	3.8	6G75/T	5W-20	①	②	③	④
2010	Galant	2.4	4G69/F	5W-20	①	②	③	④

DOT: Department Of Transportation

① DIAMOND ATF SP-3 or equivalent

② Genuine MITSUBISHI power steering fluid

③ DOT 3, DOT 4, or equivalent

④ Long life antifreeze/coolant or an equivalent

37671_GALA_C0005

VALVE SPECIFICATIONS

Year	Engine Displacement Liters	Engine Series ID/VIN	Seat Angle (deg.)	Face Angle (deg.)	Spring Test Pressure (lbs. @ in.)	Spring Installed Height (in.)	Stem-to-Guide Clearance (in.) Intake	Stem-to-Guide Clearance (in.) Exhaust	Stem Diameter (in.) Intake	Stem Diameter (in.) Exhaust
2009	2.4	4G69/F	NA	43.5-44.0	60.0 @1.74	1.74	0.0008-0.0019	0.0014-0.0024	0.240	0.240
	3.8	6G75/S	NA	43.5-44.0	60.0 @1.74	1.74	0.0008-0.0019	0.0014-0.0024	0.240	0.240
	3.8	6G75/T	NA	43.5-44.0	①	1.74	0.0008-0.0019	0.0016-0.0023	0.240	0.240
2010	2.4	4G69/F	NA	43.5-44.0	60.0 @1.74	1.74	0.0008-0.0019	0.0014-0.0024	0.240	0.240

NA: Not Available

① Intake: 59.0 lbs. @ 1.74 in.

Exhaust: 53.0 @ 1.74 in.

37671_GALA_C0006

CAMSHAFT AND BEARING SPECIFICATIONS CHART

All measurements are given in inches.

Year	Engine Displ. Liters	Engine Series ID/VIN	Journal Dia.	Brg. Oil Clearance	Shaft End-play	Runout	Journal Bore	Lobe Height	
								Intake	Exhaust
2009	2.4	4G69/F	1.800	NA	NA	NA	NA	①	1.471-1.491
	3.8	6G75/S	1.800	NA	NA	NA	NA	1.465-1.485	1.443-1.462
	3.8	6G75/T	1.800	NA	NA	NA	NA	②	1.471-1.491
2010	2.4	4G69/F	1.800	NA	NA	NA	NA	①	1.471-1.491

NA: Not Available

① Intake Low speed cam A: 1.335-1.355 in.
Intake Low speed cam B: 1.456-1.475 in.
Intake High speed cam: 1.445-1.465 in.

② Intake Low speed cam A: 1.301-1.321 in.
Intake Low speed cam B: 1.451-1.4705 in.
Intake High speed cam: 1.445-1.465 in.

37671_GALA_C0007

CRANKSHAFT AND CONNECTING ROD SPECIFICATIONS

All measurements are given in inches.

Year	Engine Displacement Liters	Engine Series ID/VIN	Crankshaft				Connecting Rod		
			Main Brg. Journal Dia.	Main Brg. Oil Clearance	Shaft End-play	Thrust on No.	Journal Diameter	Oil Clearance	Side Clearance
2009	2.4	4G69/F	2.240	①	0.0020-0.0090	3	1.770	②	0.0040-0.0090
	3.8	6G75/S	2.520	③	0.0020-0.0090	3	2.165	0.0008-0.0019	0.0030-0.0090
	3.8	6G75/T	2.520	④	0.0020-0.0090	3	2.165	0.0008-0.0015	0.0040-0.0090
2010	2.4	4G69/F	2.240	①	0.0020-0.0090	3	1.770	②	0.0040-0.0090

NA: Not Available

① Number 1, 2, 4, 5: 0.0008-0.0012 in.
Number 3: 0.0012-0.0016 in.

② ID Mark or color - I or Yellow: 0.0007-0.0089 in.
II or None: 0.0006-0.0019 in.
III or Blue: 0.0007-0.0018 in.

③ Number 1 and 4: 0.0007-0.0014 in.
Number 2 and 3: 0.0009-0.0017 in.

④ Number 1 and 4: 0.0008-0.0012 in.
Number 2 and 3: 0.0008-0.0015 in.

37671_GALA_C0008

PISTON AND RING SPECIFICATIONS
All measurements are given in inches.

| Year | Engine Displ. Liters | Engine Series ID/VIN | Piston Clearance | Ring Gap | | | Ring Side Clearance | | |
				Top Compression	Bottom Compression	Oil Control	Top Compression	Bottom Compression	Oil Control
2009	2.4	4G69/F	0.0008- 0.0015	0.0060- 0.0120	0.0110- 0.0170	0.0030- 0.0150	0.0012- 0.0028	0.0008- 0.0024	NA
	3.8	6G75/S	0.0008- 0.0016	0.0100- 0.0160	0.0140- 0.0190	0.0030 0.0140	0.0120 0.0270	0.0008- 0.0023	NA
	3.8	6G75/T	0.0012- 0.0019	0.0100- 0.0160	0.0140- 0.0190	0.0040- 0.0140	0.0012- 0.0027	0.0008- 0.0023	NA
2010	2.4	4G69/F	0.0008- 0.0015	0.0060- 0.0120	0.0110- 0.0170	0.0030- 0.0150	0.0012- 0.0028	0.0008- 0.0024	NA

NA: Not Available

37671_GALA_C0009

TORQUE SPECIFICATIONS
All readings in ft. lbs.

| Year | Engine Displacement Liters | Engine Series ID/VIN | Cylinder Head Bolts | Main Bearing Bolts | Rod Bearing Bolts | Crankshaft Damper Bolts | Flywheel Bolts | Manifold | | Spark Plugs | Oil Pan Drain Plug |
								Intake	Exhaust		
2009	2.4	4G69/F	①	②	③	123	95-101	④	⑤	15-22	25-33
	3.8	6G75/S	⑥	51-57	⑦	134-140	53-56	⑧	⑨	15-22	25-33
	3.8	6G75/T	⑥	51-57	⑦	134-140	53-56	⑩	⑨	15-22	25-33
2010	2.4	4G69/F	①	②	③	123	95-101	④	⑤	15-22	25-33

NOTE: Dip main bearing bolts and crankshaft damper bolt in clean engine oil prior to tightening.

① Step 1: Tighten all bolts to 57-59 ft. lbs.
Step 2: Loosen all bolts to 0 ft. lbs.
Step 3: Tighten all bolts to 14-16 ft. lbs.
Step 4: Plus 90 degrees
Step 4: Plus another 90 degrees

② Step 1: 17-19 ft. lbs.
Step 2: Plus 90 degrees

③ Step 1: 14-16 ft. lbs.
Step 2: Plus 90-94 degrees

④ Intake manifold bolt 17-19 ft. lbs.
Intake manifold nut 14-16 ft. lbs.
Stay bolt 21-25 ft. lbs.

⑤ Exhaust manifold nut 33-39 ft. lbs.
Cover bolt 116-132 inch lbs.

⑥ Step 1: Tighten all bolts to 76-84 ft. lbs.
Step 2: Loosen all bolts to 0 ft. lbs.
Step 3: Tighten all bolts to 76-84 ft. lbs.

⑦ Step 1: 19-21 ft. lbs.
Step 2: Plus 90-94 degrees

⑧ Intake manifold bolt 18-24 ft. lbs.
Stay bolts: M8 (12-14 ft. lbs.), M10 (23-31 ft. lbs.)

⑨ Exhaust manifold nut 29-37 ft. lbs.
Stay bolts: M8 (12-16 ft. lbs.), M10 (28-38 ft. lbs.)
Stay bolts: M12 (48-63 ft. lbs.)
Engine hanger bolt 22-30 ft. lbs.

⑩ Intake manifold bolts 15-17 ft. lbs.
Stay bolts: M8 (15-17 ft. lbs.), M10 (31-39 ft. lbs.)

37671_GALA_C0010

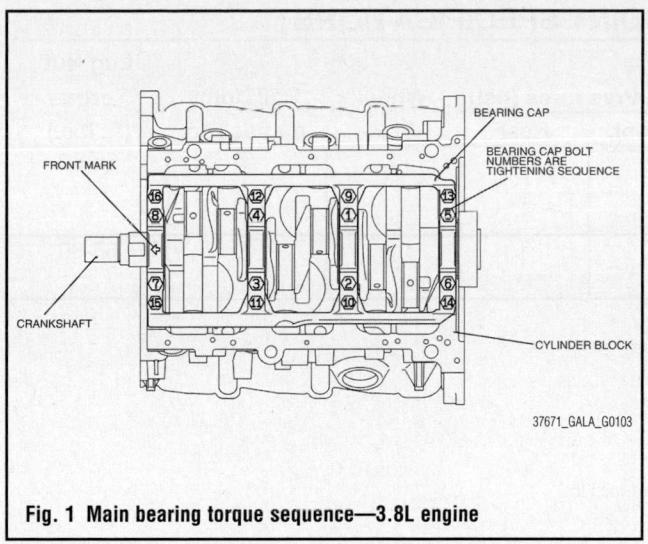

Fig. 1 Main bearing torque sequence—3.8L engine

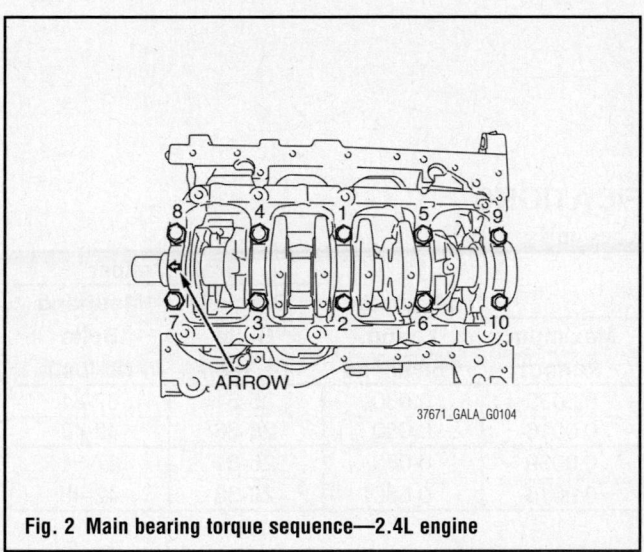

Fig. 2 Main bearing torque sequence—2.4L engine

WHEEL ALIGNMENT

Year	Model		Caster		Camber		Toe-in (Deg.)
			Range (+/-Deg.)	Preferred Setting (Deg.)	Range (+/-Deg.)	Preferred Setting (Deg.)	
2009	Galant	Front	0.30	+3.00	0.30	0.00	0 +/- 0.12
		Rear	—	—	0.30	-0.50	0.12 +/- 0.12
2010	Galant	Front	0.30	+3.00	0.30	0.00	0 +/- 0.12
		Rear	—	—	0.30	-0.50	0.12 +/- 0.12

37671_GALA_C0011

TIRE, WHEEL AND BALL JOINT SPECIFICATIONS

| Year | Model | OEM Tires | | Tire Pressures (psi) | | Wheel Size | Ball Joint Inspection | Lug Nut Torque (ft. lbs.) |
		Standard	Optional	Front	Rear			
2009	Galant	P215/60R16 P215/55R17 P235/45R18	NA	①	①	②	③	66-80
2010	Galant	P215/60R16 P235/45R18	NA	①	①	④	③	66-80

OEM: Original Equipment Manufacturer

NA: Not Available

PSI: Pounds Per Square Inch

① Refer to placard on vehicle for proper inflation pressure.

② 2.4L: 6.5JJ x 16 or 7.0JJ x 17. 3.8L: 8.0JJ x 18

③ Replace the ball joint if too loose or if rotating torque exceeds specification: 31-61 inch lbs.

④ 2.4L: 6.5JJ x 16 or 8.0JJ x 18. 3.8L: 8.0JJ x 18

37671_GALA_C0012

BRAKE SPECIFICATIONS
All measurements in inches unless noted

| Year | Model | | Brake Disc | | | Minimum Lining Thickness | Brake Caliper | |
			Original Thickness	Minimum Thickness	Maximum Runout		Bracket Bolts (ft. lbs.)	Mounting Bolts (ft. lbs.)
2009	Galant	F	1.020	0.960	0.0039	0.080	25-31	67-81
		R	0.390	0.330	0.0016	0.080	28-36	42-48
2010	Galant	F	1.020	0.960	0.0039	0.080	25-31	67-81
		R	①	②	0.0016	0.080	28-36	42-48

NA: Not Available

F: Front

R: Rear

① 2.4L 0.390. 3.8L 0.790.

② 2.4L 0.330. 3.8L 0.720.

37671_GALA_C0013

SCHEDULED MAINTENANCE INTERVALS
MITSUBISHI GALANT

TO BE SERVICED	TYPE OF	VEHICLE MILEAGE INTERVAL (x1000)														
		7.5	15	22.5	30	37.5	45	52.5	60	67.5	75	82.5	90	97.5	105	120
Accessory drive belts	S/I				✔				✔				✔			✔
Air cleaner element (engine)	R				✔				✔				✔			✔
Air conditioner system	S/I	Inspect the system operation annually														
Automatic transaxle fluid	S/I				✔				✔				✔			✔
Ball joint and steering linkage seals	S/I				✔				✔				✔			✔
Brake lines, hoses, and connections	S/I		✔		✔		✔		✔		✔		✔		✔	✔
Brake pads, calipers, & rotors	S/I		✔		✔		✔		✔		✔		✔		✔	✔
Cooling system hoses and coolant level	S/I				✔				✔				✔			✔
Driveshafts and CV-boots	S/I		✔		✔				✔		✔		✔		✔	✔
Engine coolant	R								✔				✔			✔
Engine oil and filter	R	✔	✔	✔	✔	✔	✔	✔	✔	✔	✔	✔	✔	✔	✔	✔
Exhaust pipe connections, muffler, and suspension bolts	S/I				✔				✔				✔			✔
Evaporative emission control system (except canister)	S/I								✔							✔
Fuel hoses	S/I				✔				✔				✔			✔
Fuel System (tank, connections, gas cap)	S/I								✔							✔
Spark plugs (standard)	R				✔				✔				✔			✔
Spark plugs (Iridium coated)	R	Every 84 months or 105,000 miles (under normal usage)														
Spark plugs (Platinum	R								✔							✔
Suspension system	S/I				✔				✔				✔			✔
Timing Belt	R														✔	
Tires (rotate)	S/I	✔	✔	✔	✔	✔	✔	✔	✔	✔	✔	✔	✔	✔	✔	✔
Valve clearance	S/I				✔				✔				✔			✔

R: Replace   S/I: Service or Inspect

FREQUENT OPERATION MAINTENANCE (SEVERE SERVICE)

If a vehicle is operated under any of the following conditions it is considered severe service:

- Extremely dusty areas.

- 50% or more of the vehicle operation is in 90°F (32°C) or higher temperatures, or constant operation in temperatures below 32°F (0°C).

- Prolonged idling (vehicle operation in stop and go traffic).

- Frequent short running periods (engine does not warm to normal operating temperatures).

- Police, taxi, delivery usage, or trailer towing usage.

Air cleaner filter: replace every 15,000 miles.

Automatic transaxle fluid & filter: check every 15,000 miles, replace every 30,000 miles.

Brake pads, calipers & rotors: service or inspect every 6 months or 7,500 miles.

Oil & oil filter: change every 3,750 miles.

Spark plugs: service or inspect every 15,000 miles (standard plugs only)

Suspension system inspect for looseness and damage every 6 months or 7,500 miles

ANTI-LOCK SYSTEMS

• Certain components within the ABS system are not intended to be serviced or repaired individually.

• Do not use rubber hoses or other parts not specifically specified for and ABS system. When using repair kits, replace all parts included in the kit. Partial or incorrect repair may lead to functional problems and require the replacement of components.

• Lubricate rubber parts with clean, fresh brake fluid to ease assembly. Do not use shop air to clean parts; damage to rubber components may result.

• Use only DOT 3 brake fluid from an unopened container.

• If any hydraulic component or line is removed or replaced, it may be necessary to bleed the entire system.

• A clean repair area is essential. Always clean the reservoir and cap thoroughly before removing the cap. The slightest amount of dirt in the fluid may plug an orifice and impair the system function. Perform repairs after components have been thoroughly cleaned; use only denatured alcohol to clean components. Do not allow ABS components to come into contact with any substance containing mineral oil; this includes used shop rags.

• The Anti-Lock control unit is a microprocessor similar to other computer units in the vehicle. Ensure that the ignition switch is **OFF** before removing or installing controller harnesses. Avoid static electricity discharge at or near the controller.

• If any arc welding is to be done on the vehicle, the control unit should be unplugged before welding operations begin.

DISC AND DRUM SYSTEMS

> ✳✳ **CAUTION**
>
> **Dust and dirt accumulating on brake parts during normal use may contain asbestos fibers from production or aftermarket brake linings. Breathing excessive concentrations of asbestos fibers can cause serious bodily harm. Exercise care when servicing brake parts. Do not sand or grind brake lining unless equipment used is designed to contain the dust residue. Do not clean brake parts with compressed air or by dry brushing. Cleaning should be done by dampening the brake components with a fine mist of water, then wiping the brake components clean with a dampened cloth. Dispose of cloth and all residue containing asbestos fibers in an impermeable container with the appropriate label. Follow practices prescribed by the Occupational Safety and Health Administration (OSHA) and the Environmental Protection Agency (EPA) for the handling, processing, and disposing of dust or debris that may contain asbestos fibers.**

BLEEDING PROCEDURE

BLEEDING PROCEDURE

See Figure 3.

When any part of the hydraulic system has been disconnected for repair or replacement, air may get into the lines and cause spongy pedal action (because air can be compressed and brake fluid cannot). To correct this condition, it is necessary to bleed the hydraulic system so to be sure all air is purged.

When bleeding the brake system, bleed one brake cylinder at a time, beginning at the cylinder with the longest hydraulic line (farthest from the master cylinder) first. ALWAYS keep the master cylinder reservoir filled with brake fluid during the bleeding operation. Never use brake fluid that has been drained from the hydraulic system, no matter how clean it is.

The primary and secondary hydraulic brake systems are separate and are bled independently. During the bleeding operation, do not allow the reservoir to run dry. Keep the master cylinder reservoir filled with brake fluid.

1. Clean all dirt from around the master cylinder fill cap, remove the cap and fill the master cylinder with brake fluid until the level is within ¼ inch (6mm) of the top edge of the reservoir.

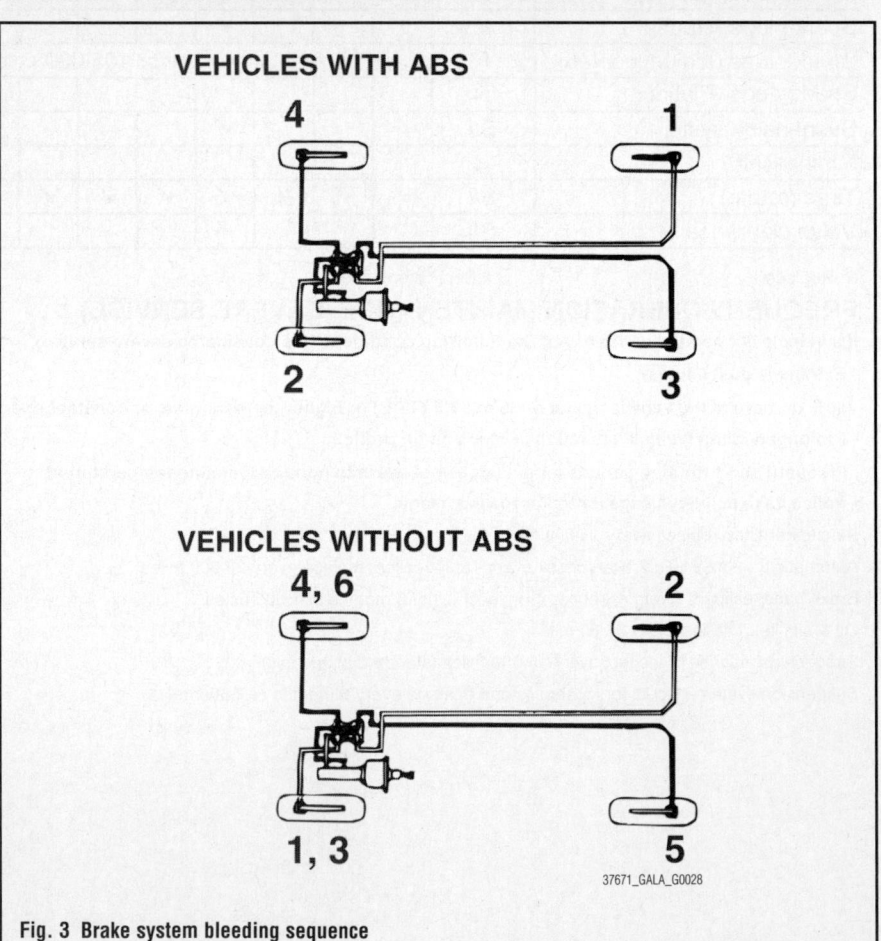

VEHICLES WITH ABS

VEHICLES WITHOUT ABS

37671_GALA_G0028

Fig. 3 Brake system bleeding sequence

2. Clean the bleeder screws at all 4 wheels. The bleeder screws are located on the back of the brake backing plate (drum brakes) and on the top of the brake calipers (disc brakes).

3. Attach a length of rubber hose over the bleeder screw and place the other end of the hose in a glass jar, submerged in brake fluid.

4. Open the bleeder screw ½–¾ turn. Have an assistant slowly depress the brake pedal.

✳✳ CAUTION

Brake fluid contains polyglycol ethers and polyglycols. Avoid contact with

the eyes and wash your hands thoroughly after handling brake fluid. If you do get brake fluid in your eyes, flush your eyes with clean, running water for 15 minutes. If eye irritation persists, or if you have taken brake fluid internally, IMMEDIATELY seek medical assistance.

5. Close the bleeder screw and tell your assistant to allow the brake pedal to return slowly. Continue this process to purge all air from the system.

6. When bubbles cease to appear at the end of the bleeder hose, close the bleeder screw and remove the

hose. Tighten the bleeder screw to the proper torque.

7. Check the master cylinder fluid level and add fluid accordingly. Do this after bleeding each wheel.

8. Repeat the bleeding operation at the remaining 3 wheels, ending with the one closet to the master cylinder.

9. Fill the master cylinder reservoir to the proper level.

BLEEDING THE ABS SYSTEM

There are no special procedures for bleeding the ABS system. Refer to the conventional bleeding procedures.

BRAKES

ANTI-LOCK BRAKE SYSTEM (ABS)

WHEEL SPEED SENSORS

REMOVAL & INSTALLATION

See Figures 4 and 5.

➡**The front wheel speed rotors are integrated with the BJ assembly of the halfshaft and cannot be disassembled. The rear wheel speed rotors are integrated with the rear hub assembly and cannot be disassembled.**

1. Before servicing the vehicle, refer to the Precautions Section.

➡**If working near and/or around the SRS system and components, be sure to disable the SRS system. Tape the negative battery cable with insulating tape. Always disconnect the negative battery cable first.**

✳✳ CAUTION

Wait for 1 minute after disconnecting the negative battery cable before working inside the vehicle. The air bag system is set to deploy for a short period of time after the battery is disconnected.

2. Raise and safely support the vehicle.

3. Remove the wheel and tire assembly.

4. Remove the inner fender splash shield.

5. Beginning at the sensor end, carefully disconnect or release each clip and retainer along the sensor wire.

➡**Take careful note of the exact position of each clip; they must be reinstalled in the identical position. Rear wheel sensor harnesses may be held**

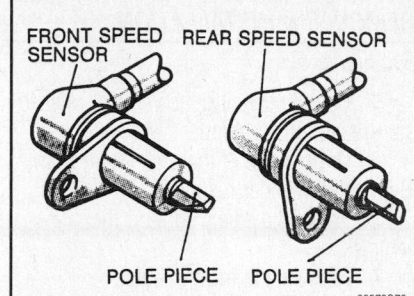

Fig. 4 Inspect removed speed sensor for damaged pole piece. Also check pole piece for foreign material or metal adhesion

by plastic wire ties; these may be cut away but must be replaced at reassembly.

6. Detach the sensor connector at the end of the harness.

7. Remove the 2 bolts holding the speed sensor bracket to the knuckle and remove the assembly from the vehicle.

✳✳ WARNING

The speed sensor has a pole piece projecting from it. This exposed tip must be protected from impact or scratches. Do not allow the pole piece to contact the toothed wheel during removal or installation.

8. Remove the wheel speed sensor from the bracket.

To install:

9. Assemble the sensor into the bracket.

10. On some vehicles note that the brackets are different for the left and right front

wheels, as well as both rear wheels. Each bracket has identifying letters stamped on it.

11. Identify the front speed sensor brackets as follows:

  a. FR: Indicates that the bracket is for the front speed sensor.

  b. R: Indicates that the bracket is for the right wheel.

  c. L: Indicates that the bracket is for the left wheel.

12. Identify the rear speed sensor brackets as follows:

  a. R: Indicates that the bracket is for the right wheel.

  b. L: Indicates that the bracket is for the left wheel.

13. Temporarily install the speed sensor to the spindle or knuckle; tighten the bolts only finger tight.

✳✳ WARNING

During speed sensor installation to the mounting bracket, make sure the letters FR are visible. Be careful when installing the speed sensor on the vehicle, that the pole piece at the tip of the sensor does not strike the toothed edge of the rotor, and damage them.

14. Route the cable correctly and loosely install the clips and retainers. All clips must be in their original position and the sensor cable must not be twisted. Improper installation may cause cable damage and system failure.

15. Use a brass or other non-magnetic feeler gauge to check the air gap between the tip of the pole piece and the toothed wheel. The correct gap is 0.012–0.035 inch (0.3–0.9mm). Tighten the 2 sensor bracket

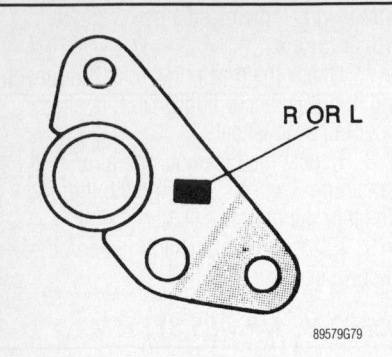

Fig. 5 Rear speed sensor bracket identification marking

bolts to 10 ft. lbs. (14 Nm) with the sensor located so that the gap is the same at several points on the toothed wheel. If the gap is incorrect, it is likely that the toothed wheel is worn or improperly installed.

16. Tighten the screws and bolts for the cable retaining clips.

17. Install the inner fender splash shield.

18. Install the wheel and tire assembly. Lower the vehicle to the ground.

19. Inspect the brake system for proper operation.

WHEEL SPEED SENSOR RINGS (TOOTHED RINGS)

REMOVAL & INSTALLATION

Front

➡The front wheel speed rotors are integrated with the BJ assembly of the halfshaft and cannot be disassembled.

Rear

➡The rear wheel speed rotors are integrated with the rear hub assembly and cannot be disassembled.

BRAKES

BRAKE CALIPER

REMOVAL & INSTALLATION

1. As required, partially drain the master cylinder.

2. Raise and support the vehicle safely.

3. Remove the tire and wheel assembly.

4. Disconnect and plug the brake hose connection. Discard the gasket.

5. Remove the caliper retaining bolts. Remove the caliper from the vehicle.

To install:

➡Be sure to use new fasteners, as required.

6. Position the caliper to its mounting on the vehicle.

7. Install the caliper mounting bolts and tighten to 67–81 ft. lbs. (91–110 Nm).

8. Connect the brake line to the caliper with new gaskets. Torque the brake line union bolt to 18–22 ft. lbs. (25–30 Nm).

9. Bleed the brake system.

10. Install the wheel and tire.

11. Before attempting to move the vehicle, pump the brake pedal to seat the pads against the rotors. Make sure the vehicle has a firm brake pedal. Check the level of the brake fluid and add fluid if necessary.

DISC BRAKE PADS

REMOVAL & INSTALLATION

See Figure 6.

1. Before servicing the vehicle, refer to the Precautions Section.

2. Drain some of the brake fluid from the master cylinder reservoir.

3. Raise and safely support the vehicle.

4. Remove the front wheels.

➡Do not allow the caliper to hang by the brake hose. Properly support the caliper with mechanics wire, as required.

5. Remove the caliper guide and lock pins and lift the caliper assembly from the caliper support.

➡On some vehicles, the caliper can be flipped up by leaving the upper pin in place and using it as a pivot point.

6. Remove the brake pads, spring clip, and shims.

To install:

➡Be sure to use new fasteners, as required.

7. Compress pistons back into the caliper bore.

8. Lubricate slide points and install the

FRONT DISC BRAKES

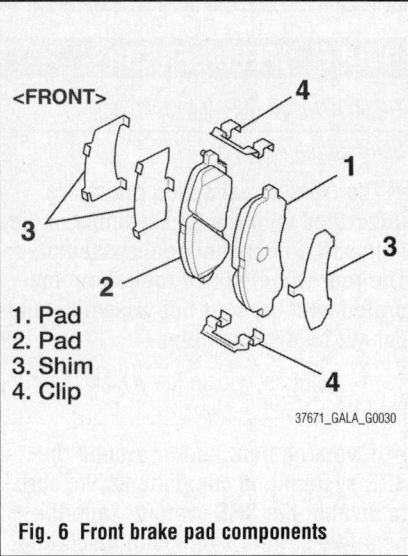

1. Pad
2. Pad
3. Shim
4. Clip

37671_GALA_G0030

Fig. 6 Front brake pad components

brake pads, shims, and spring clips onto the caliper support.

9. Install the caliper over the brake pads.

10. Lubricate and install the caliper guide and lock pins in their original positions.

11. Install the wheels.

12. Before attempting to move the vehicle, pump the brake pedal to seat the pads against the rotors. Make sure the vehicle has a firm brake pedal. Check the level of the brake fluid and add fluid if necessary.

BRAKES

BRAKE CALIPER

REMOVAL & INSTALLATION

1. Before servicing the vehicle, refer to the Precautions Section.

2. As required, partially drain the master cylinder.

3. Raise and safely support the vehicle.

4. Remove or disconnect the following:
- Wheels
- Brake hose from the caliper
- Caliper guide and lock pins and lift the caliper assembly from the caliper support

REAR DISC BRAKES

To install:

➡Be sure to use new fasteners, as required.

5. Install or connect the following:
- Caliper onto the caliper support
- Guide pin and lock pin and tighten to specification

- Brake hose or banjo bolt with new washers
6. Bleed the brake system.
7. Install the wheels.

DISC BRAKE PADS

REMOVAL & INSTALLATION
See Figure 7.

1. Before servicing the vehicle, refer to the Precautions Section.
2. Raise and support the vehicle safely.
3. Remove or disconnect the following:
- Rear wheels
- Lower caliper mounting bolt and rotate the caliper upward
- Pads from the caliper support
- Pad retainers, if necessary

➡**Do not allow the caliper to hang by the brake hose. Properly support the caliper with mechanics wire, as required.**

To install:

➡**Be sure to use new fasteners, as required.**

4. Install or connect the following:
- Pad retainers, if removed
- Pads onto the pad retainers
5. Compress the caliper piston using a C-clamp.
6. Rotate the caliper downward and install the mounting bolt.
7. Install the wheel.
8. Pump the brake pedal until the brake pads are seated and a firm pedal is achieved before attempting to move the vehicle.

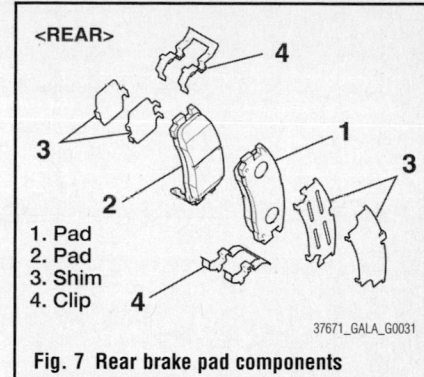

1. Pad
2. Pad
3. Shim
4. Clip

37671_GALA_G0031

Fig. 7 Rear brake pad components

✳✳ CAUTION

Do not move the vehicle until a firm pedal is obtained.

9. Road test the vehicle to check for proper brake operation.

BRAKES

PARKING BRAKE CABLES

ADJUSTMENT
See Figures 8 and 9.

1. Before servicing the vehicle, refer to the Precautions Section.
2. Pull the parking brake lever with a force of 45 lbs. (200 N) and count the number of notches. Standard value is: 5–7 notches.
3. If the parking brake lever is not the standard value, adjust in the following manner.
4. Release the parking brake.
5. Remove the inner compartment mat of the floor console.
6. Loosen the adjusting nut at the end of the cable rod, freeing the parking brake.
7. Raise and safely support the vehicle.

8. Remove the tire and wheel assemblies.
9. Remove the adjustment hole plug on the brake disc.
10. Use a flat bladed tool to turn the adjuster in the direction of the arrow (see illustration) which expands the shoe. Return the adjuster five notches in the direction opposite to the direction of the arrow.
11. Install the tire and wheel assemblies.

➡**Be careful that the parking brake lever stroke is within the standard value. If the stroke is too short, brake drag can occur.**

12. Turn the adjusting nut to adjust the parking brake lever stroke to specification.
13. After adjustment check to see that the nut and cable rod are not loose.

PARKING BRAKE

PARKING BRAKE SHOES

REMOVAL & INSTALLATION
See Figure 10.

1. Before servicing the vehicle, refer to the Precautions Section.
2. Raise and safely support the vehicle.
3. Remove the caliper assembly.
4. Remove the rear brake rotor.

➡**When servicing, only dissemble and assemble one side at a time, leaving the remaining side intact for reference.**

5. Remove the front and rear shoe-to-anchor springs.
6. Remove the adjusting wheel spring and the adjuster.
7. Remove the strut and the strut return spring.
8. Remove the shoe hold-down cup, spring, and pin.
9. Remove the shoe and lining assembly.
10. Unfasten the clips and the retaining bolts.
11. Remove the parking brake cable.

To install:

➡**Be sure to use new fasteners, as required.**

12. Installation is the reverse of the removal procedure.

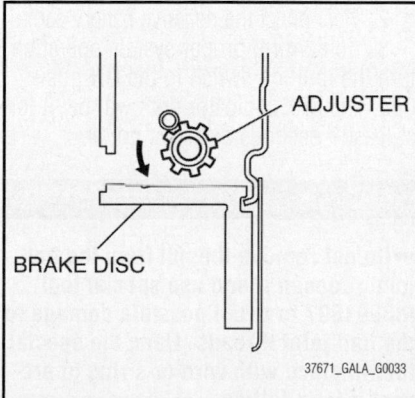

37671_GALA_G0033

Fig. 8 Rear parking brake adjuster positioning

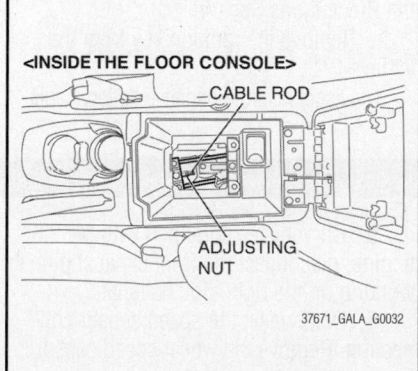

37671_GALA_G0032

Fig. 9 Rear parking brake adjusting nut location

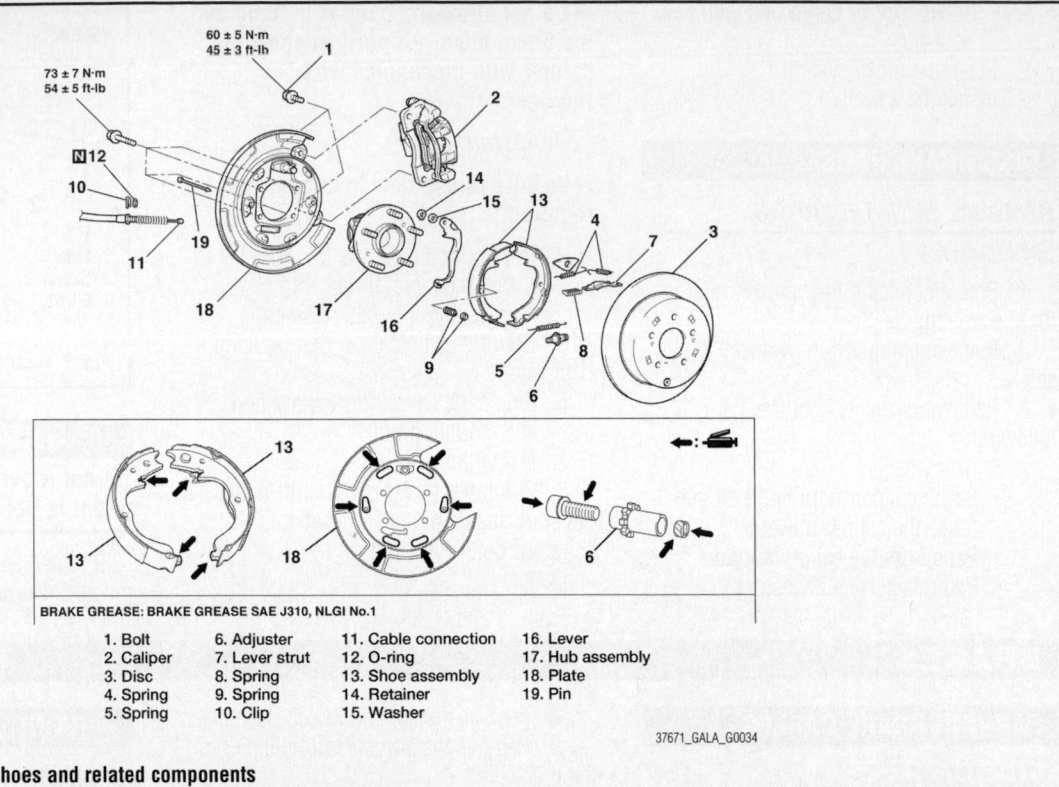

BRAKE GREASE: BRAKE GREASE SAE J310, NLGI No.1

1. Bolt	6. Adjuster	11. Cable connection	16. Lever
2. Caliper	7. Lever strut	12. O-ring	17. Hub assembly
3. Disc	8. Spring	13. Shoe assembly	18. Plate
4. Spring	9. Spring	14. Retainer	19. Pin
5. Spring	10. Clip	15. Washer	

37671_GALA_G0034

Fig. 10 Rear parking brake shoes and related components

CHASSIS ELECTRICAL   AIR BAG (SUPPLEMENTAL RESTRAINT SYSTEM)

GENERAL INFORMATION

✳✳ CAUTION

This vehicle is equipped with an air bag system. The system must be disarmed before performing service on, or around, system components, the steering column, instrument panel components, wiring, and sensors. Failure to follow the safety precautions and the disarming procedure could result in accidental air bag deployment, possible injury, and unnecessary system repairs.

PRECAUTIONS

Disconnect and isolate the battery negative cable before beginning any airbag system component diagnosis, testing, removal, or installation procedures. Allow system capacitor to discharge before beginning any component service. This will disable the airbag system. Failure to disable the airbag system may result in accidental airbag deployment, personal injury, or death.

DISARMING THE SYSTEM

1. Before servicing the vehicle, refer to the Precautions Section.
2. Remove the ignition key from the vehicle.
3. Disconnect the negative battery cable and isolate it from accidental reconnection. Insulate the cable end with high-quality electrical tape or a similar non-conductive wrapping.
4. Wait at least 1 minute for the system capacitor to discharge before performing any service. The air bag system is designed to retain enough voltage to deploy the air bag for a short period of time after the battery has been disconnected.

ARMING THE SYSTEM

1. Before servicing the vehicle, refer to the Precautions Section.
2. Reconnect the negative battery cable.
3. To confirm proper system operation, turn the ignition switch to the **ON** position. The SRS indicator light will be lit for at least 7 seconds and then go off.

DRIVE TRAIN

FRONT HALFSHAFT

REMOVAL & INSTALLATION

See Figures 11 and 128.

1. Before servicing the vehicle, refer to the Precautions Section.
2. Raise and support the vehicle safely.
3. Remove the front undercover. Remove the side undercover.
4. Drain the transaxle fluid.

5. On vehicles equipped with the 3.8L engine, disconnect the front exhaust pipe if working on the right side halfshaft.
6. Disconnect the speed sensor connection. Remove the wheel speed sensor. Remove the brake hose clip.
7. Remove the cotter pin. Install tool MB990767 to the hub and remove the halfshaft nut. Remove the washer.
8. Remove the lower ball joint cotter pin.

➡Do not remove the nut from the ball joint. Loosen it and use special tool MB991897 to avoid possible damage to the ball joint threads. Hang the special tool in place with wire or string to prevent it from falling.

9. Install the special tool. Turn the bolt and knob as necessary to make the jaws of the tool parallel. Tighten the bolt by hand and confirm that the jaws are still parallel.

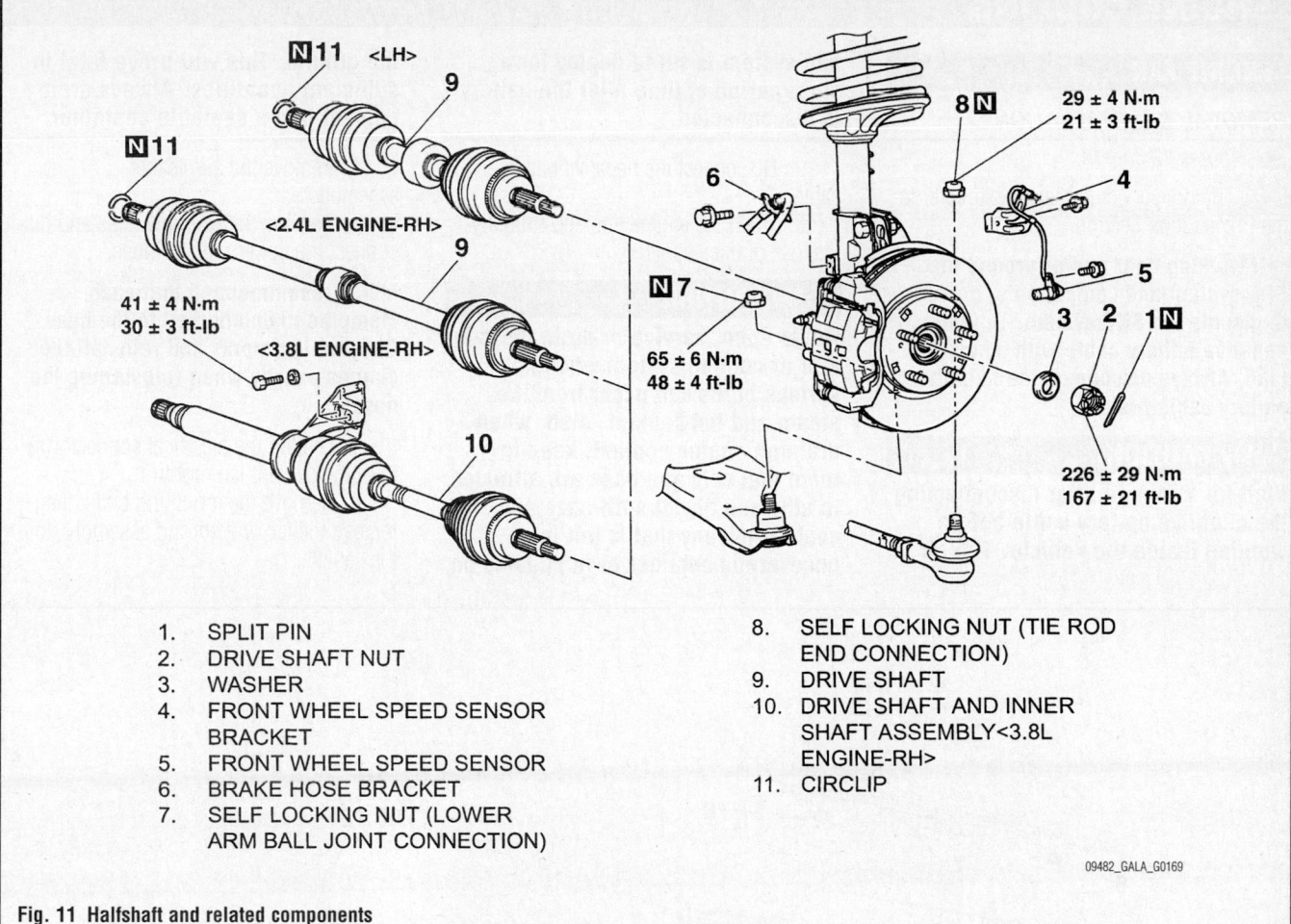

1. SPLIT PIN
2. DRIVE SHAFT NUT
3. WASHER
4. FRONT WHEEL SPEED SENSOR BRACKET
5. FRONT WHEEL SPEED SENSOR
6. BRAKE HOSE BRACKET
7. SELF LOCKING NUT (LOWER ARM BALL JOINT CONNECTION)
8. SELF LOCKING NUT (TIE ROD END CONNECTION)
9. DRIVE SHAFT
10. DRIVE SHAFT AND INNER SHAFT ASSEMBLY<3.8L ENGINE-RH>
11. CIRCLIP

09482_GALA_G0169

Fig. 11 Halfshaft and related components

➡**When adjusting the jaws in parallel, make sure the knob is in the vertical (upward) position.**

10. Tighten the bolt with a wrench to disconnect the lower arm ball joint connection.

11. Remove the tie rod end cotter pin. Install tool MB990767 to the hub and remove the halfshaft nut. Remove the washer.

➡**Do not remove the nut from the tie rod end. Loosen it and use special tool MB991897 to avoid possible damage to the ball joint threads. Hang the special tool in place with wire or string to prevent it from falling.**

12. Install the special tool. Turn the bolt and knob as necessary to make the jaws of the tool parallel. Tighten the bolt by hand and confirm that the jaws are still parallel.

➡**When adjusting the jaws in parallel, make sure the knob is in the vertical (upward) position.**

13. Tighten the bolt with a wrench to disconnect the tie rod end.

14. Disconnect the stabilizer link connection.

➡**Do not strike the ABS rotor attached to the BJ or EBJ outer race, of the halfshaft against other parts when removing the halfshaft as damage to the rotors will result.**

15. Use special tools MB991354, MB990242, and MB990767 to push the halfshaft out from the hub.

➡**Do not pull on the halfshaft, doing so will damage the TJ or PTJ. Be sure to use a prybar. Do not insert the prybar so deep as to damage the oil seal.**

16. Remove the halfshaft from the hub by pulling the bottom of the brake disc towards you.

17. Insert a prybar between the transaxle case and the halfshaft, and then pry and remove the halfshaft from the transaxle.

➡**Insert a prybar, taking care not to damage the protrusion of the transaxle case when removing the halfshaft (left side).**

18. If the inner shaft and transaxle are tightly joined, tap the center bearing

bracket with a plastic hammer to remove the halfshaft and inner shaft from the transaxle.

To install:

➡**Be sure to use new fasteners, as required.**

19. Installation is the reverse of the removal procedure.

20. Check and adjust the front end alignment, as necessary.

CV-BOOTS INSPECTION

1. Before servicing the vehicle, refer to the Precautions Section.

2. Check the halfshaft boots for damage and deterioration.

  - Raise the vehicle
  - Rotate the axle and inspect for cracked or ripped CV-boot material on the inner and outer CV joints on both sides of the vehicle
  - Inspect for excessive grease deposits on or around the CV-boot

3. Replace the boot if it is damaged or deteriorated.

ENGINE COOLING

ENGINE FAN

REMOVAL & INSTALLATION

See Figures 12 and 13.

1. Before servicing the vehicle, refer to the Precautions Section.

➡ If working near and/or around the SRS system and components, be sure to disable the SRS system. Tape the negative battery cable with insulating tape. Always disconnect the negative battery cable first.

✳✳ CAUTION

Wait for 1 minute after disconnecting the negative battery cable before working inside the vehicle. The air bag system is set to deploy for a short period of time after the battery is disconnected.

2. Disconnect the negative battery cable.

3. Drain the engine coolant. Properly dispose of used coolant.

✳✳ CAUTION

Never open, service or drain the radiator or cooling system when hot; serious burns can occur from the steam and hot coolant. Also, when draining engine coolant, keep in mind that cats and dogs are attracted to ethylene glycol antifreeze and could drink any that is left in an uncovered container or in puddles on the ground. This will prove fatal in sufficient quantities. Always drain coolant into a sealable container.

4. Remove the air cleaner assembly.

5. Remove the overflow hose and disconnect the upper radiator hose.

➡ It is recommended that each clamp be matchmarked to the hose. Observe the marks and reinstall the clamps exactly when reinstalling the hoses.

6. Unplug the electrical connector(s) from the coolant fan motor(s).

7. Remove the mounting bolts, then remove the fan and shroud assembly from the vehicle.

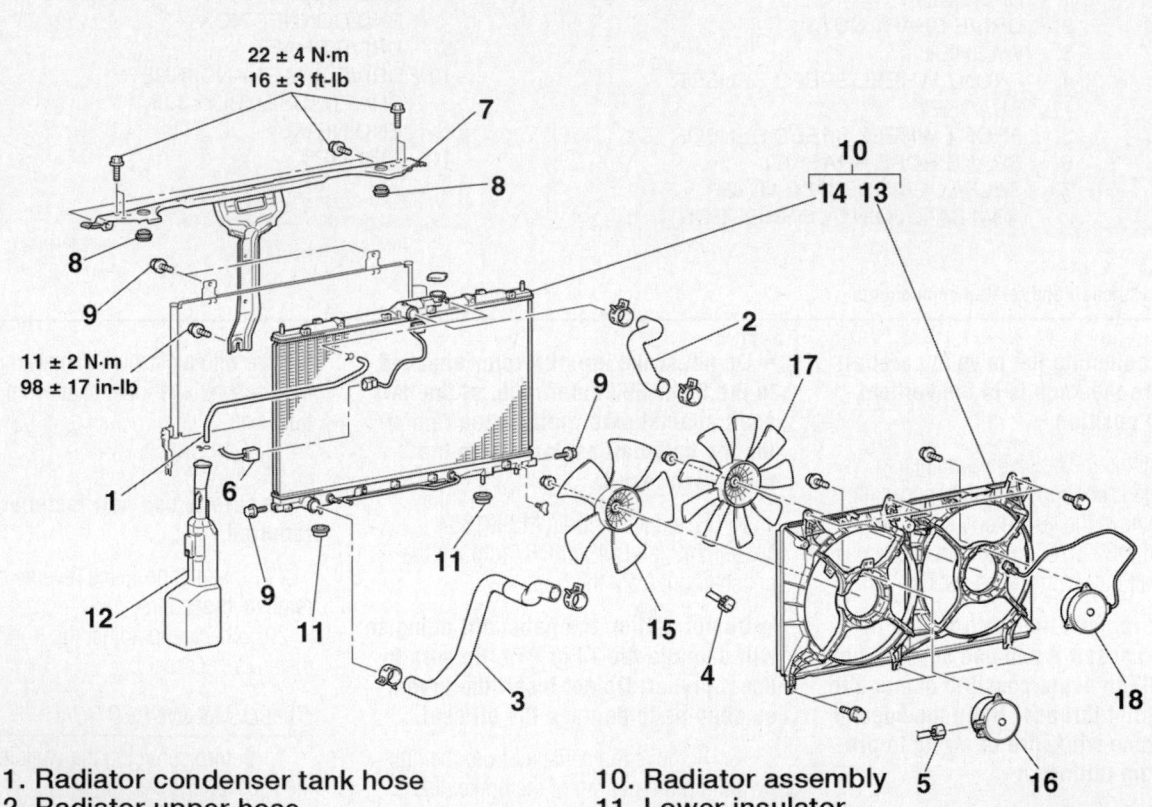

22 ± 4 N·m
16 ± 3 ft-lb

11 ± 2 N·m
98 ± 17 in-lb

1. Radiator condenser tank hose
2. Radiator upper hose
3. Radiator lower hose
4. Condenser fan motor connector
5. Radiator fan motor connector
6. Radiator sensor connector (dor radiator>)
7. Front end structure bar
8. Upper insulator
9. Condenser bolts
10. Radiator assembly
11. Lower insulator
12. Radiator condenser tank assembly
13. Shroud assembly
14. Radiator
15. Radiator fan
16. Radiator fan motor
17. Condenser fan
18. Condenser fan motor

22140_MITS_G0072

Fig. 12 Radiator, fan and related components—2.4L engine

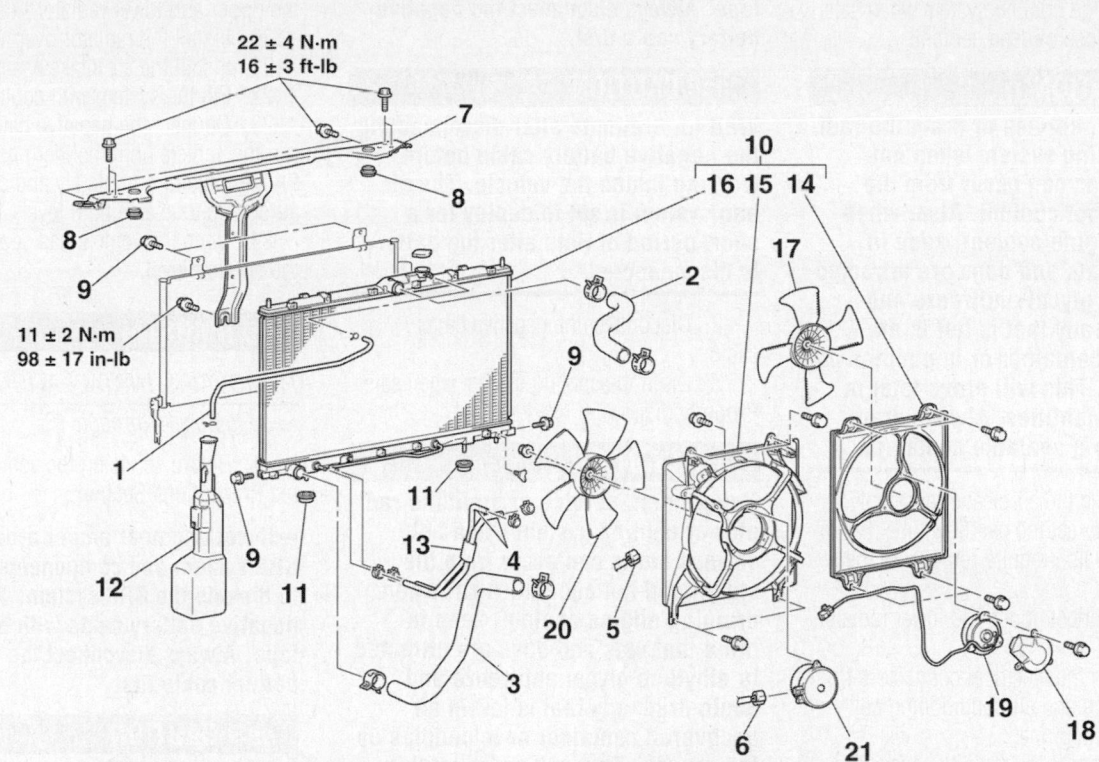

1. Radiator condenser tank hose
2. Radiator upper hose
3. Radiator lower hose
4. A/T oil cooler hose connection
5. Condenser fan motor connector
6. Radiator fan motor connector
7. Front end structure bar
8. Upper insulator
9. Condenser bolts
10. Radiator assembly
11. Lower insulator
12. Radiator condenser tank assembly
13. A/T oil cooler hose
14. Condenser fan shroud assembly
15. Cooling fan shroud assembly
16. Radiator
17. Condenser fan
18. Heat protector
19. Condenser fan motor
20. Radiator fan
21. Radiator fan motor

22140_MITS_G0073

Fig. 13 Radiator, fan and related components—3.8L engine

To install:

➡**Be sure to use new fasteners, as required.**

8. Install the motor to the shroud and secure with the mounting bolts.

9. Install the remaining components in the reverse order of removal.

10. Fill the system with coolant.

11. Connect the negative battery cable, run the vehicle until the thermostat opens, fill the radiator completely and check the automatic transaxle fluid level, if equipped.

12. Once the vehicle has cooled, recheck the coolant level.

13. Check the engine fan for proper operation.

RADIATOR

REMOVAL & INSTALLATION

2.4L Engine

See Figure 12.

For California vehicle emission regulation: Never remove the radiator sensor from the radiator because it cannot be disassembled. When replacing the radiator sensor, replace it with the radiator as a set. When the radiator sensor is removed from the radiator, the Powertrain Control Module (PCM) detects an error and sets a diagnostic trouble code. Never replace the DOR radiator with the NON-DOR radiator. Never clean the DOR radiator with a high concentration of alkaline cleaner.

1. Before servicing the vehicle, refer to the Precautions Section.

➡**If working near and/or around the SRS system and components, be sure to disable the SRS system. Tape the negative battery cable with insulating tape. Always disconnect the negative battery cable first.**

✳✳ CAUTION

Wait for 1 minute after disconnecting the negative battery cable before working inside the vehicle. The air bag system is set to deploy for a short period of time after the battery is disconnected.

2. Disconnect the negative battery cable.

3. Drain the cooling system when safe. Properly dispose of used coolant.

✳✳ CAUTION

Never open, service or drain the radiator or cooling system when hot; serious burns can occur from the steam and hot coolant. Also, when draining engine coolant, keep in mind that cats and dogs are attracted to ethylene glycol antifreeze and could drink any that is left in an uncovered container or in puddles on the ground. This will prove fatal in sufficient quantities. Always drain coolant into a sealable container.

4. Remove the air cleaner assembly.

5. Disconnect the overflow tube. Some vehicles may also require removal of the overflow tank.

6. Disconnect upper and lower radiator hoses.

7. Detach the electrical connectors from the cooling fan and air conditioning condenser fan, if equipped.

8. Disconnect the radiator sensor connector (DOR radiator), see illustration.

9. Disconnect and plug the automatic transaxle cooler lines.

10. Remove the upper radiator mounts and lift out the radiator assembly.

To install:

11. Install the radiator or radiator and fan, if removed as an assembly.

12. Connect the automatic transaxle cooler lines, if disconnected.

13. Connect the radiator sensor connector (DOR radiator), see illustration.

14. Install the fan, if removed separately.

15. Install the radiator hoses.

16. Install the air cleaner support bracket.

17. Install the overflow tube and reservoir.

18. Fill the system with coolant.

19. Connect the negative battery cable, run the vehicle until the thermostat opens, fill the radiator completely and check the automatic transaxle fluid level, if equipped.

20. Once the vehicle has cooled, recheck the coolant level.

3.8L Engine

See Figure 13.

1. Before servicing the vehicle, refer to the Precautions Section.

➡ **If working near and/or around the SRS system and components, be sure to disable the SRS system. Tape the**

negative battery cable with insulating tape. Always disconnect the negative battery cable first.

✳✳ CAUTION

Wait for 1 minute after disconnecting the negative battery cable before working inside the vehicle. The air bag system is set to deploy for a short period of time after the battery is disconnected.

2. Disconnect the negative battery cable.

3. Drain the cooling system when safe. Properly dispose of used coolant.

✳✳ CAUTION

Never open, service or drain the radiator or cooling system when hot; serious burns can occur from the steam and hot coolant. Also, when draining engine coolant, keep in mind that cats and dogs are attracted to ethylene glycol antifreeze and could drink any that is left in an uncovered container or in puddles on the ground. This will prove fatal in sufficient quantities. Always drain coolant into a sealable container.

4. Remove the air intake assembly.

5. Remove the radiator overflow hose.

6. Disconnect the upper and lower radiator hoses. Matchmark the hoses to ensure proper installation.

7. Disconnect the automatic transaxle oil cooler hose. After removing the hose from the radiator, plug the hose and the radiator nipple to prevent dust or foreign particles from getting in.

8. Disconnect the fan motor electrical connectors.

9. Remove the hood latch.

10. Remove the front end grille structure bar.

11. Remove the condenser mounting bolts with the upper insulators.

12. Remove the radiator assembly.

To install:

13. Install the radiator assembly.

14. Install the condenser mounting bolts and upper radiator insulators.

15. Install the front end grille structure bar. Tighten the mounting bolts to 16 ft. lbs. (22 Nm).

16. Install the hood latch.

17. Connect the fan motor electrical connectors.

18. Connect the automatic transaxle oil cooler hoses.

19. Aligning the matchmarks, connect the upper and lower radiator hoses.

20. Install the radiator overflow hose.

21. Install the air intake assembly.

22. Fill the system with coolant.

23. Connect the negative battery cable, run the vehicle until the thermostat opens, fill the radiator completely and check the automatic transaxle fluid level, if equipped.

24. Once the vehicle has cooled, recheck the coolant level.

THERMOSTAT

REMOVAL & INSTALLATION

See Figures 14 through 16.

1. Before servicing the vehicle, refer to the Precautions Section.

➡ **If working near and/or around the SRS system and components, be sure to disable the SRS system. Tape the negative battery cable with insulating tape. Always disconnect the negative battery cable first.**

✳✳ CAUTION

Wait for 1 minute after disconnecting the negative battery cable before working inside the vehicle. The air bag system is set to deploy for a short period of time after the battery is disconnected.

2. Disconnect the negative battery cable.

3. Drain the cooling system when safe. Properly dispose of used coolant.

4. Remove the PCM.

5. Remove the engine cover, 3.8L engine.

6. Remove the air cleaner housing cover and air intake hose.

7. Remove the strut tower bar, 3.8L engine.

8. Remove the battery and battery tray.

9. Remove the harness connection.

10. Remove the harness bracket.

11. Remove the radiator lower hose connection.

12. Remove the water inlet fitting.

13. Remove the thermostat. Discard the gasket.

To install:

➡ **Be sure to use new fasteners, as required.**

14. Install the thermostat so that the jiggle valve is facing straight up. Be careful not to fold or scratch the rubber ring. Be sure to use a new gasket.

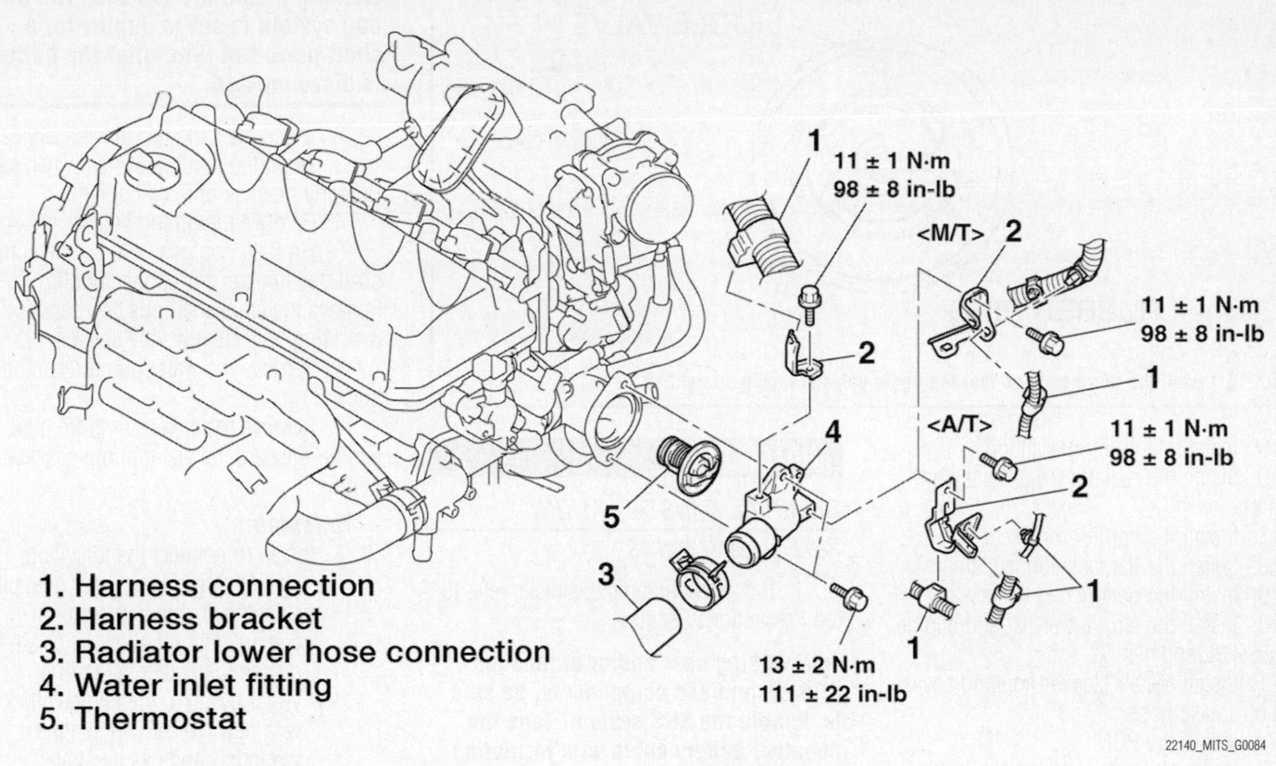

1. Harness connection
2. Harness bracket
3. Radiator lower hose connection
4. Water inlet fitting
5. Thermostat

11 ± 1 N·m
98 ± 8 in-lb

<M/T>

11 ± 1 N·m
98 ± 8 in-lb

<A/T>

11 ± 1 N·m
98 ± 8 in-lb

13 ± 2 N·m
111 ± 22 in-lb

22140_MITS_G0084

Fig. 14 Thermostat and related parts—2.4L engine

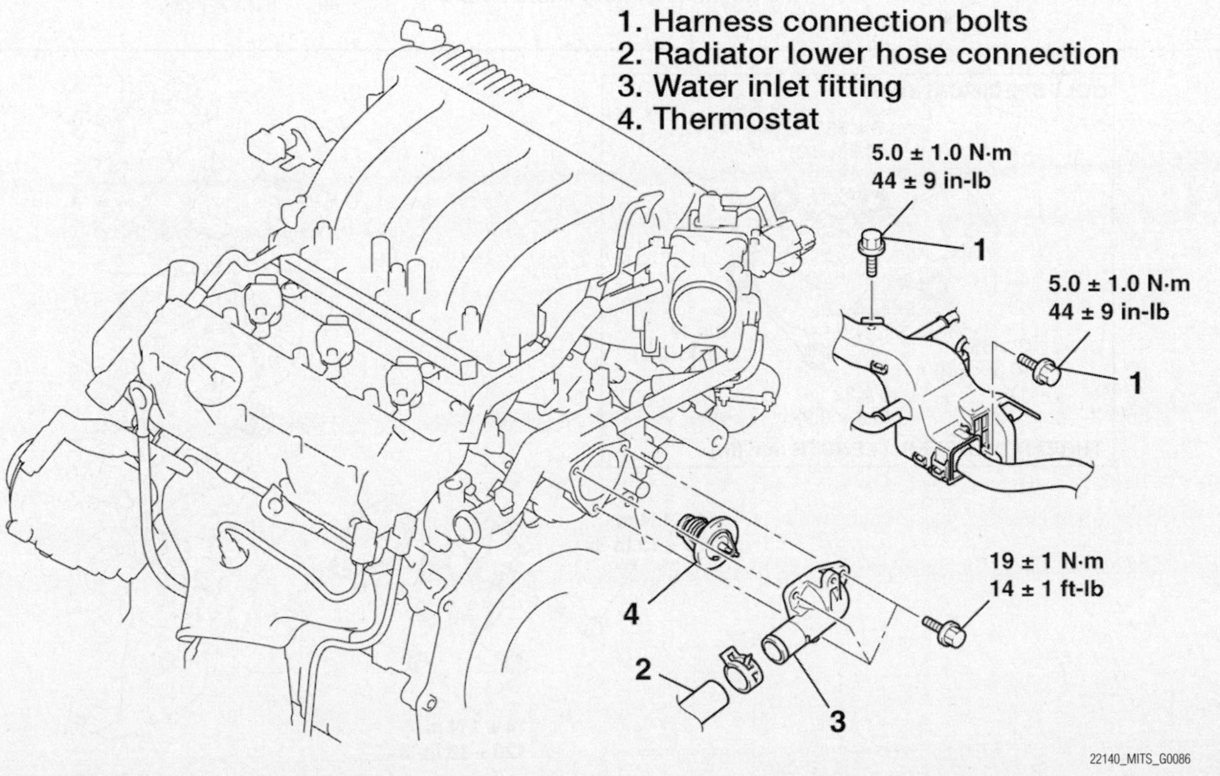

1. Harness connection bolts
2. Radiator lower hose connection
3. Water inlet fitting
4. Thermostat

5.0 ± 1.0 N·m
44 ± 9 in-lb

5.0 ± 1.0 N·m
44 ± 9 in-lb

19 ± 1 N·m
14 ± 1 ft-lb

22140_MITS_G0086

Fig. 15 Thermostat and related components—3.8L engine

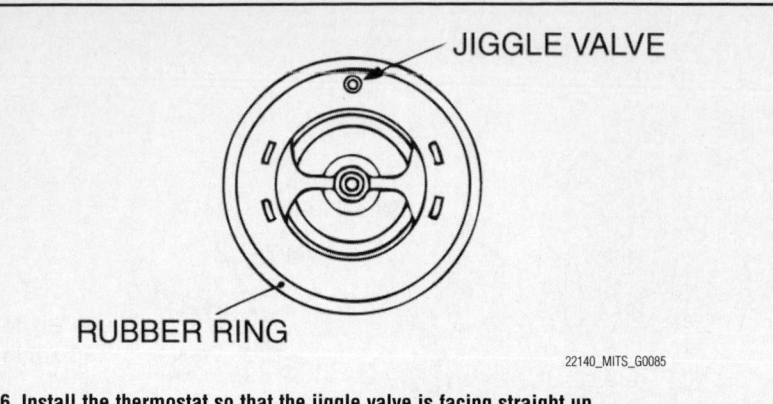

Fig. 16 Install the thermostat so that the jiggle valve is facing straight up

15. Install the water inlet fitting.

16. Install the radiator lower hose connection.

17. Install the harness bracket.

18. Install the harness connection.

19. Install the battery and battery tray.

20. Install the strut tower bar and engine cover, 3.8L engines.

21. Install the air cleaner housing cover and air intake hose.

22. Install the PCM.

23. Connect the negative battery cable, run the vehicle until the thermostat opens and fill the radiator completely.

24. Once the vehicle has cooled, recheck the coolant level.

WATER PUMP

REMOVAL & INSTALLATION

See Figures 17 and 18.

1. Before servicing the vehicle, refer to the Precautions Section.

➡ If working near and/or around the SRS system and components, be sure to disable the SRS system. Tape the negative battery cable with insulating tape. Always disconnect the negative battery cable first.

❊❊ CAUTION

Wait for 1 minute after disconnecting the negative battery cable before working inside the vehicle. The air bag system is set to deploy for a short period of time after the battery is disconnected.

2. Disconnect the negative battery cable.

3. Drain the cooling system when safe. Properly dispose of used coolant.

4. Remove the timing belt.

5. On 3.8L engines, remove the crankshaft position sensor connector clip. Remove the crankshaft position sensor mounting bolt. Remove the sensor.

6. Remove the water pump retaining bolts.

7. Remove the water pump from the engine. Discard the water pump gasket and O-ring.

To install:

8. Install or connect the following:
- New O-ring on the water inlet pipe. Coat the O-ring with water or coolant. Do not allow oil or other grease to contact the O-ring.
- Water pump to the engine block, with new gasket. Tighten the mounting bolts as illustrated.

9. Continue the installation in the reverse order of the removal procedure.

10. Fill the engine with the proper grade and type engine coolant. Start the engine and check for leaks.

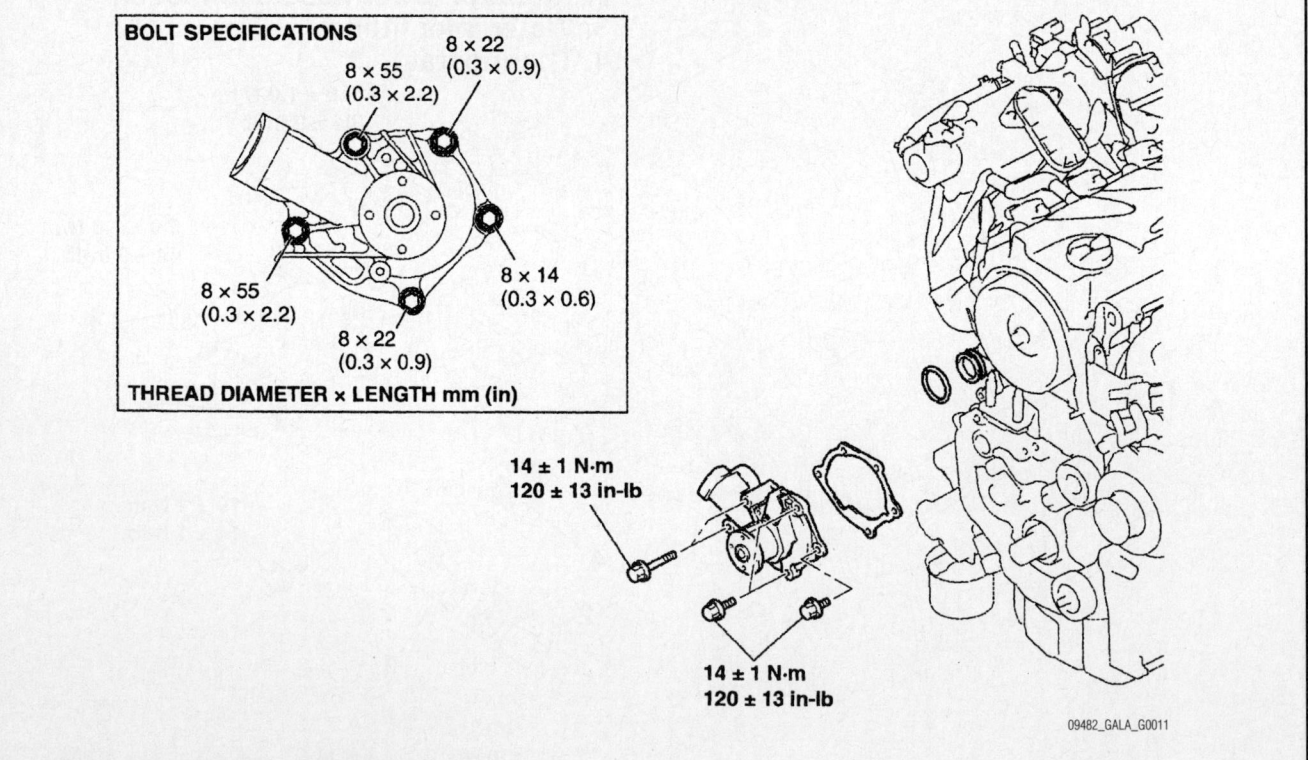

Fig. 17 Water pump and related components—2.4L engine

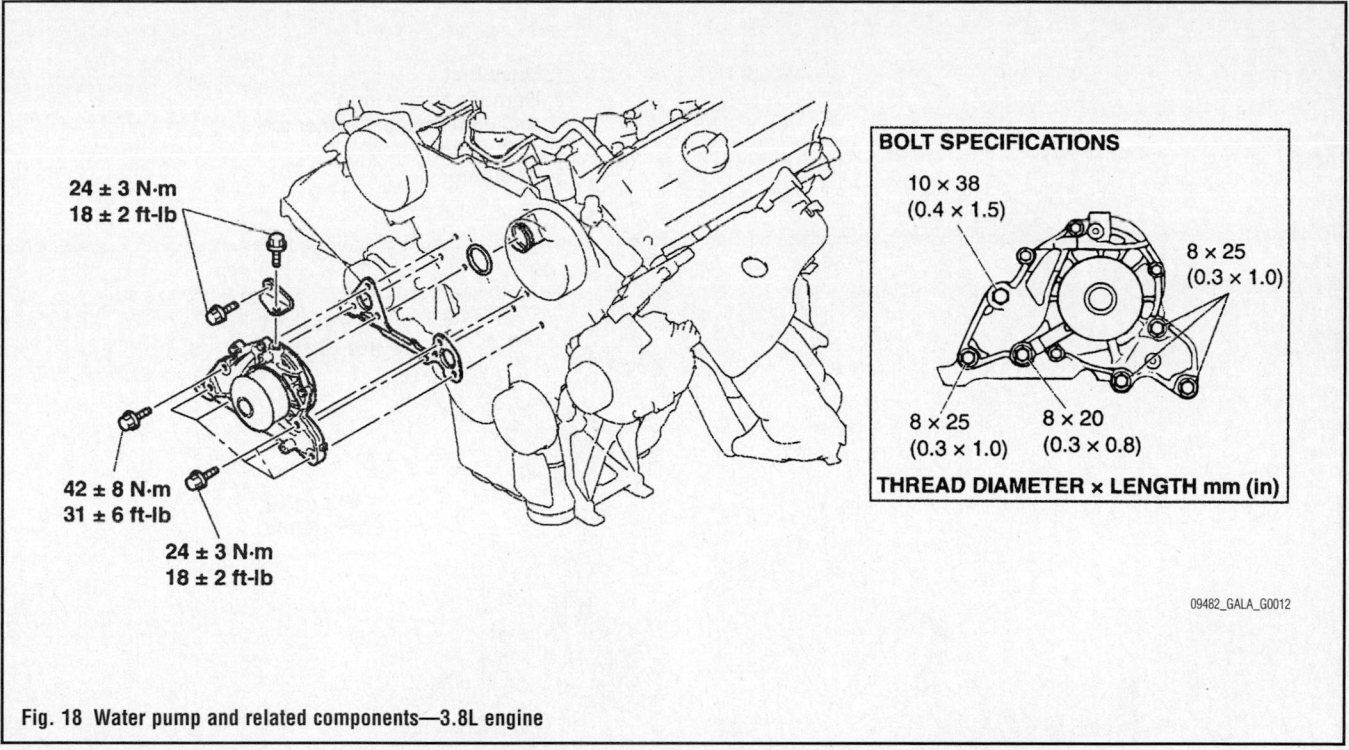

24 ± 3 N·m
18 ± 2 ft-lb

42 ± 8 N·m
31 ± 6 ft-lb

24 ± 3 N·m
18 ± 2 ft-lb

BOLT SPECIFICATIONS

10 × 38
(0.4 × 1.5)

8 × 25
(0.3 × 1.0)

8 × 25
(0.3 × 1.0)

8 × 20
(0.3 × 0.8)

THREAD DIAMETER × LENGTH mm (in)

09482_GALA_G0012

Fig. 18 Water pump and related components—3.8L engine

ENGINE ELECTRICAL   CHARGING SYSTEM

ALTERNATOR

REMOVAL & INSTALLATION
See Figures 19 through 21.

1. Before servicing the vehicle, refer to the Precautions Section.

➡**If working near and/or around the SRS system and components, be sure to disable the SRS system. Tape the negative battery cable with insulating tape. Always disconnect the negative battery cable first.**

⁑ CAUTION

Wait for 1 minute after disconnecting the negative battery cable before working inside the vehicle. The air bag system is set to deploy for a short period of time after the battery is disconnected.

2. Disconnect the negative battery cable.
3. Remove the side under cover.
4. Remove the drive belts.
5. Disconnect the harness bracket and connector bracket.

➡**In some cases it may be necessary to remove the timing belt lower cover on 2.4L engines.**

➡**In some cases it may be necessary to remove the air conditioning compressor**

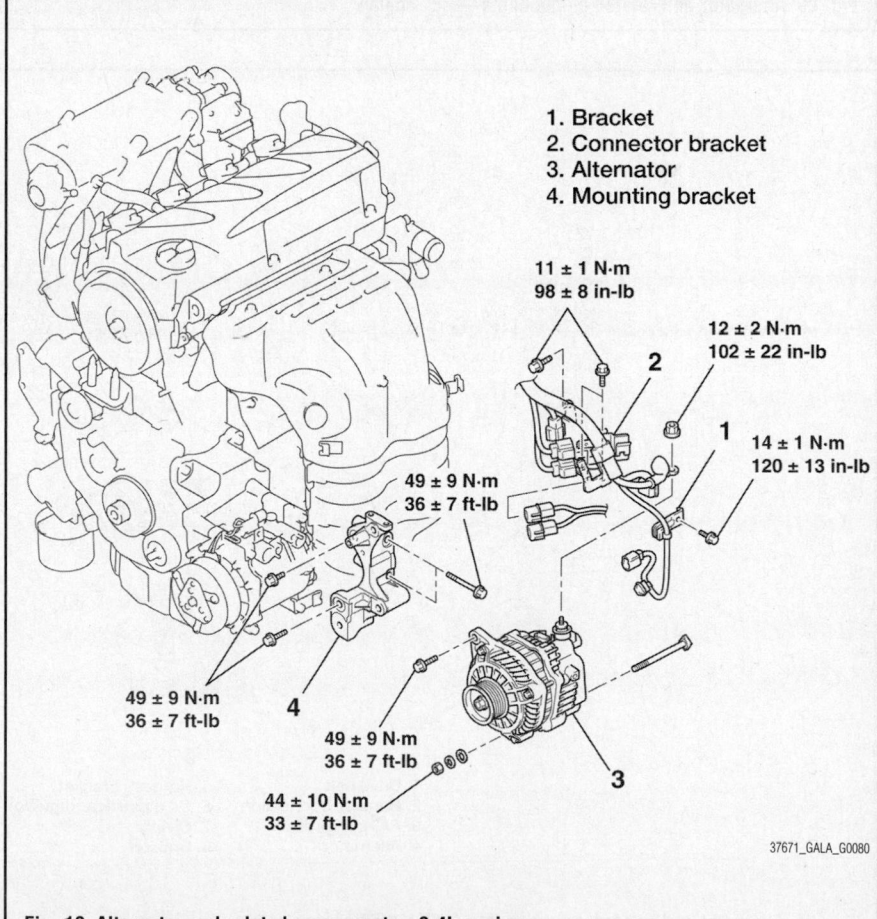

1. Bracket
2. Connector bracket
3. Alternator
4. Mounting bracket

11 ± 1 N·m
98 ± 8 in-lb

12 ± 2 N·m
102 ± 22 in-lb

14 ± 1 N·m
120 ± 13 in-lb

49 ± 9 N·m
36 ± 7 ft-lb

49 ± 9 N·m
36 ± 7 ft-lb

49 ± 9 N·m
36 ± 7 ft-lb

44 ± 10 N·m
33 ± 7 ft-lb

37671_GALA_G0080

Fig. 19 Alternator and related components—2.4L engine

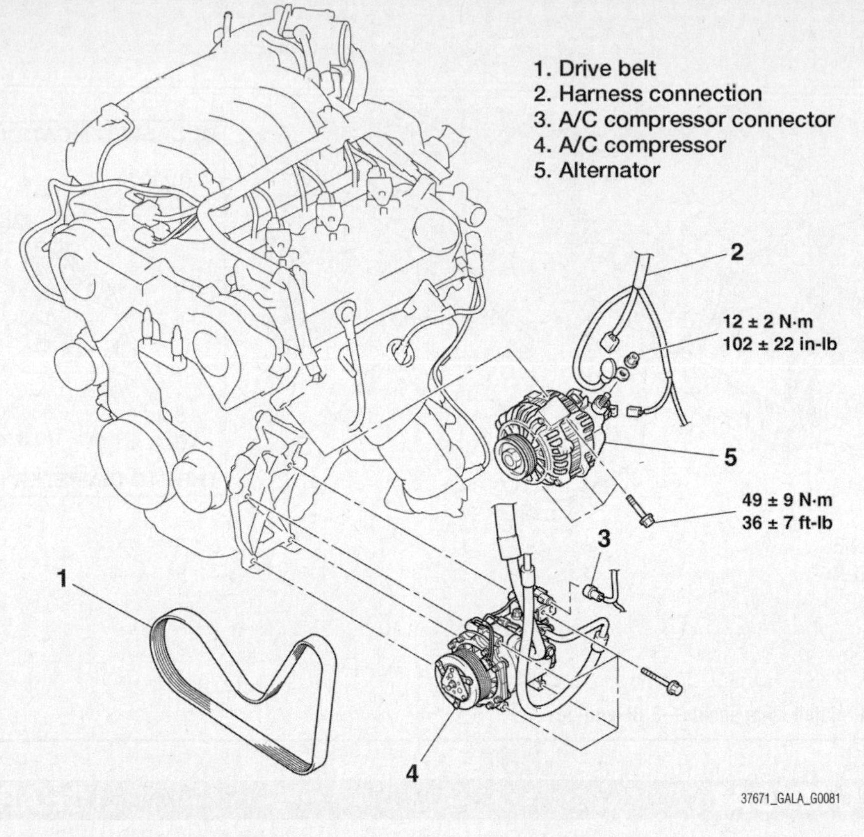

1. Drive belt
2. Harness connection
3. A/C compressor connector
4. A/C compressor
5. Alternator

2

12 ± 2 N·m
102 ± 22 in-lb

5

49 ± 9 N·m
36 ± 7 ft-lb

3

1

4

37671_GALA_G0081

Fig. 20 Alternator and related components—3.8L engine

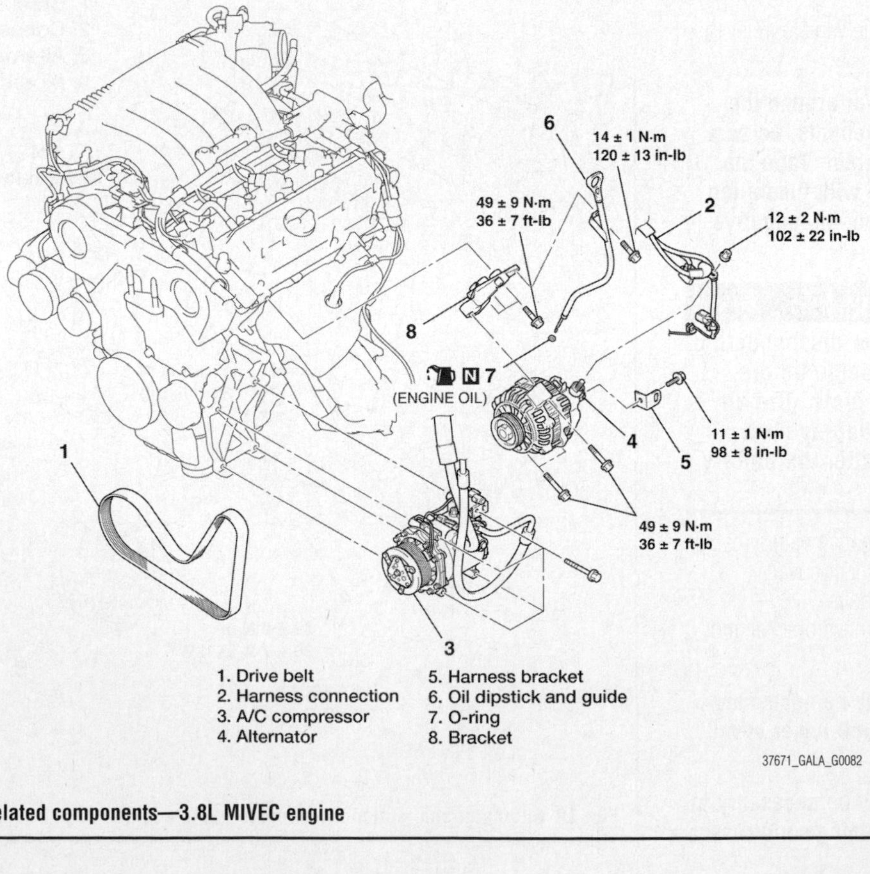

6

14 ± 1 N·m
120 ± 13 in-lb

49 ± 9 N·m
36 ± 7 ft-lb

2

12 ± 2 N·m
102 ± 22 in-lb

8

N 7
(ENGINE OIL)

4

5

11 ± 1 N·m
98 ± 8 in-lb

1

49 ± 9 N·m
36 ± 7 ft-lb

3

1. Drive belt   5. Harness bracket
2. Harness connection   6. Oil dipstick and guide
3. A/C compressor   7. O-ring
4. Alternator   8. Bracket

37671_GALA_G0082

Fig. 21 Alternator and related components—3.8L MIVEC engine

on 3.8L engines. Do not disconnect the refrigerant lines. Position the component to the side. Support the component using mechanics wire. Do not allow the compressor to hang by the refrigerant hoses.

➡In some cases it may be necessary to remove the engine oil dipstick and guide on 3.8L MIVEC engines.

6. Remove the alternator retaining bolts.
7. Disconnect the electrical connectors.
8. Remove the component from the vehicle.

To install:

➡Be sure to use new fasteners, as required.

9. Installation is the reverse of the removal procedure.

VOLTAGE REGULATOR

REMOVAL & INSTALLATION

The voltage regulator is an internal component of the alternator. In order to replace the voltage regulator, the entire alternator assembly must be replaced.

ENGINE ELECTRICAL

FIRING ORDER

See Figures 22 and 23.

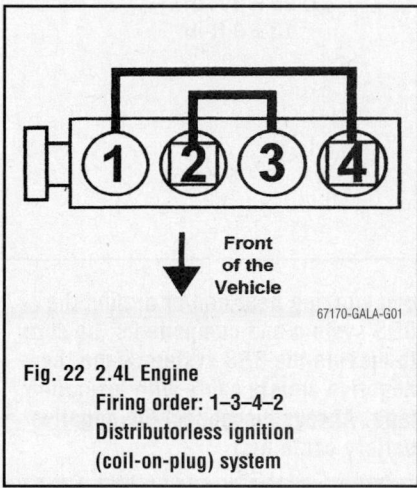

Fig. 22 2.4L Engine
Firing order: 1–3–4–2
Distributorless ignition
(coil-on-plug) system

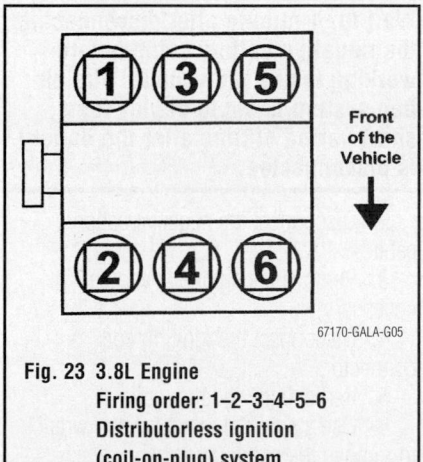

Fig. 23 3.8L Engine
Firing order: 1–2–3–4–5–6
Distributorless ignition
(coil-on-plug) system

IGNITION COIL

REMOVAL & INSTALLATION

2.4L Engine
See Figure 24.

1. Before servicing the vehicle, refer to the Precautions Section.

IGNITION SYSTEM

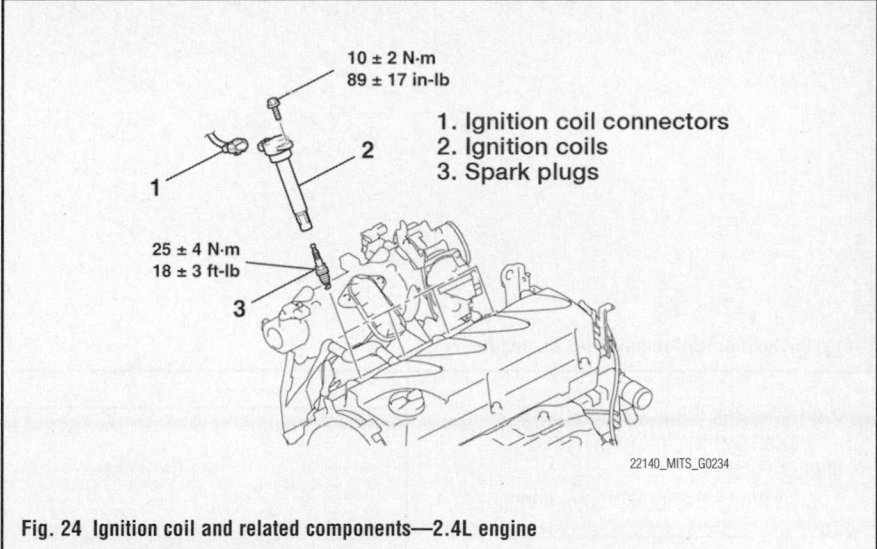

1. Ignition coil connectors
2. Ignition coils
3. Spark plugs

10 ± 2 N·m
89 ± 17 in-lb

25 ± 4 N·m
18 ± 3 ft-lb

Fig. 24 Ignition coil and related components—2.4L engine

➡If working near and/or around the SRS system and components, be sure to disable the SRS system. Tape the negative battery cable with insulating tape. Always disconnect the negative battery cable first.

❊❊ **CAUTION**

Wait for 1 minute after disconnecting the negative battery cable before working inside the vehicle. The air bag system is set to deploy for a short period of time after the battery is disconnected.

2. Disconnect the negative battery cable.
3. Remove the air cleaner resonator.
4. Remove the ignition coil connectors.
5. Remove the ignition coil retaining bolt.
6. Remove the ignition coils.

To install:

➡Be sure to use new fasteners, as required.

7. Installation is the reverse of the removal procedure.
8. Tighten the retaining bolts as illustrated.

3.8L Engine
See Figure 25.

1. Before servicing the vehicle, refer to the Precautions Section.

➡If working near and/or around the SRS system and components, be sure to disable the SRS system. Tape the negative battery cable with insulating tape. Always disconnect the negative battery cable first.

❊❊ **CAUTION**

Wait for 1 minute after disconnecting the negative battery cable before working inside the vehicle. The air bag system is set to deploy for a short period of time after the battery is disconnected.

2. Disconnect the negative battery cable.
3. Remove the air cleaner resonator.
4. Remove the engine cover.
5. Left bank ignition coil removal:
   a. Remove the ignition coil connectors.
   b. Remove the ignition coil retaining bolt.
   c. Remove the ignition coils.

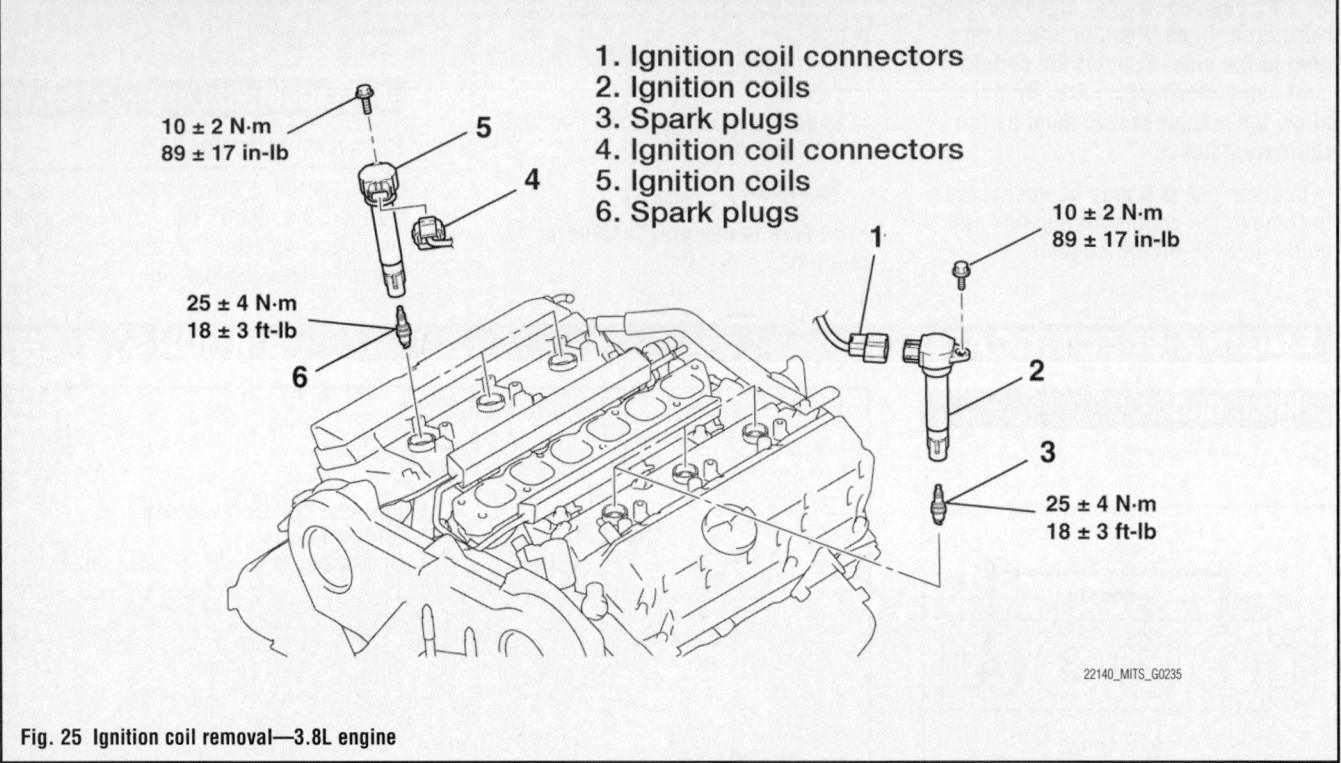

1. Ignition coil connectors
2. Ignition coils
3. Spark plugs
4. Ignition coil connectors
5. Ignition coils
6. Spark plugs

10 ± 2 N·m
89 ± 17 in-lb

25 ± 4 N·m
18 ± 3 ft-lb

10 ± 2 N·m
89 ± 17 in-lb

25 ± 4 N·m
18 ± 3 ft-lb

22140_MITS_G0235

Fig. 25 Ignition coil removal—3.8L engine

6. Right bank ignition coil removal:
   a. Remove the intake manifold plenum.
   b. Remove the ignition coil connectors.
   c. Remove the ignition coil retaining bolt.
   d. Remove the ignition coils.

To install:

7. Installation is the reverse of the removal procedure.

8. Tighten the retaining bolts as illustrated.

IGNITION TIMING

INSPECTION

See Figure 26.

➡ **Mitsubishi's Multi-Use Tester-III (MUT III) scan tool or an equivalent OBD-II scan tool must be used for this procedure.**

1. Set the transaxle in Park.
2. Connect an OBD-II compliant scan tool to the Data Link Connector, located under the instrument panel on the driver's side.
3. Set the timing light to the power supply line of ignition coil No. 1.
4. Start the engine and allow it to run at idle.
5. Using the scan tool (Item No. 17 of the actuator test), check that the ignition timing is within specification.

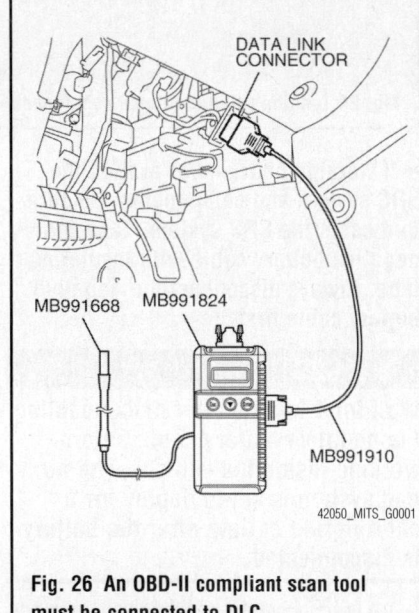

DATA LINK CONNECTOR

MB991668   MB991824

MB991910

42050_MITS_G0001

Fig. 26 An OBD-II compliant scan tool must be connected to DLC

ADJUSTMENT

The ignition timing is controlled by the Powertrain Control Module (PCM). No adjustment is necessary or possible.

SPARK PLUGS

REMOVAL & INSTALLATION

1. Before servicing the vehicle, refer to the Precautions Section.

➡ **If working near and/or around the SRS system and components, be sure to disable the SRS system. Tape the negative battery cable with insulating tape. Always disconnect the negative battery cable first.**

✳✳ CAUTION

Wait for 1 minute after disconnecting the negative battery cable before working inside the vehicle. The air bag system is set to deploy for a short period of time after the battery is disconnected.

2. Disconnect the negative battery cable.
3. Remove the engine cover (as necessary).
4. Disconnect the ignition coil connector.
5. Remove the ignition coil.
6. Use a spark plug socket and wrench to remove the spark plugs.

✳✳ WARNING

Be careful that no contaminates enter through the spark plug holes.

➡ **Check the electrode gap on the spark plugs before installation.**

7. To install, reverse the removal procedure. Tighten the spark plugs to 15–22 ft. lbs. (20–30 Nm).

STARTER

REMOVAL & INSTALLATION

See Figures 27 and 28.

1. Before servicing the vehicle, refer to the Precautions Section.

➡**If working near and/or around the SRS system and components, be sure to disable the SRS system. Tape the negative battery cable with insulating tape. Always disconnect the negative battery cable first.**

✳✳ **CAUTION**

Wait for 1 minute after disconnecting the negative battery cable before working inside the vehicle. The air bag system is set to deploy for a short period of time after the battery is disconnected.

2. Disconnect the negative battery cable.

➡**Raise and support the vehicle as required to gain access to the starter and its attaching points.**

➡**On 3.8L engines the starter is removed from underneath the vehicle.**

3. Remove the air cleaner assembly, 2.4L engines.
4. Remove the engine under cover, 3.8L engines.
5. Remove the exhaust manifold stay, 3.8L MIVEC engines.
6. Remove the transmission warmer fluid cooler line and bracket assembly from the transmission case front roll stopper bracket, 2.4L engines.
7. Remove the starter cover.
8. Disconnect the electrical connectors.
9. Remove the retaining bolts.
10. Remove the component from the vehicle.

To install:

➡**Be sure to use new fasteners, as required.**

11. Installation is the reverse of the removal procedure.

SOLENOID OR RELAY REPLACEMENT

1. Before servicing the vehicle, refer to the Precautions Section.

➡**If working near and/or around the SRS system and components, be sure to disable the SRS system. Tape the negative battery cable with insulating tape. Always disconnect the negative battery cable first.**

✳✳ **CAUTION**

Wait for 1 minute after disconnecting the negative battery cable before working inside the vehicle. The air bag system is set to deploy for a short period of time after the battery is disconnected.

2. Disconnect the negative battery cable.
3. Remove the starter.
4. Remove the solenoid attaching screws.
5. Remove the solenoid.

To install:

➡**Be sure to use new fasteners, as required.**

6. Install the solenoid.
7. Install the solenoid attaching screws.
8. Install the starter.
9. Connect the negative battery cable.

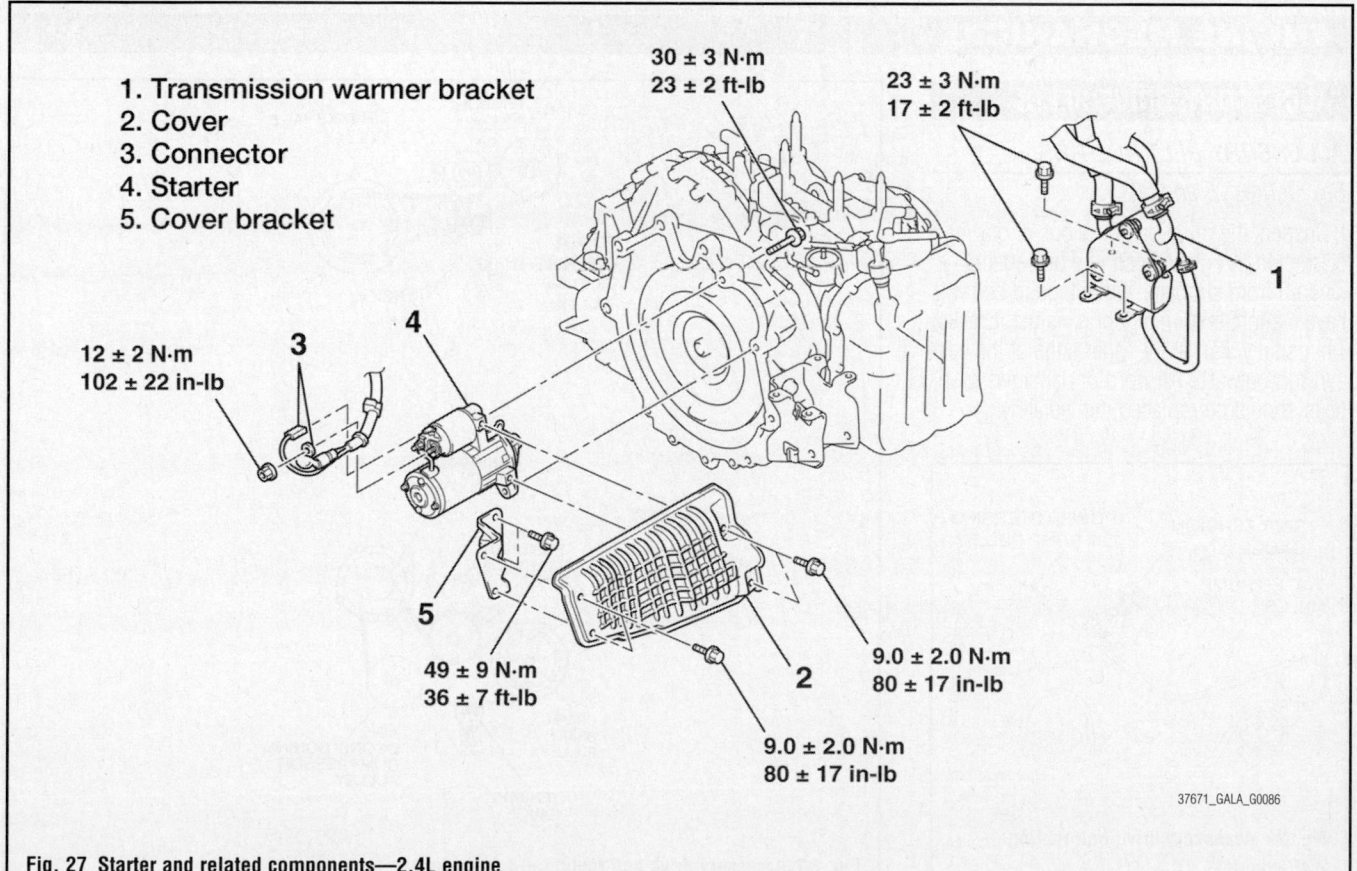

1. Transmission warmer bracket
2. Cover
3. Connector
4. Starter
5. Cover bracket

30 ± 3 N·m
23 ± 2 ft-lb

23 ± 3 N·m
17 ± 2 ft-lb

12 ± 2 N·m
102 ± 22 in-lb

49 ± 9 N·m
36 ± 7 ft-lb

9.0 ± 2.0 N·m
80 ± 17 in-lb

9.0 ± 2.0 N·m
80 ± 17 in-lb

37671_GALA_G0086

Fig. 27 Starter and related components—2.4L engine

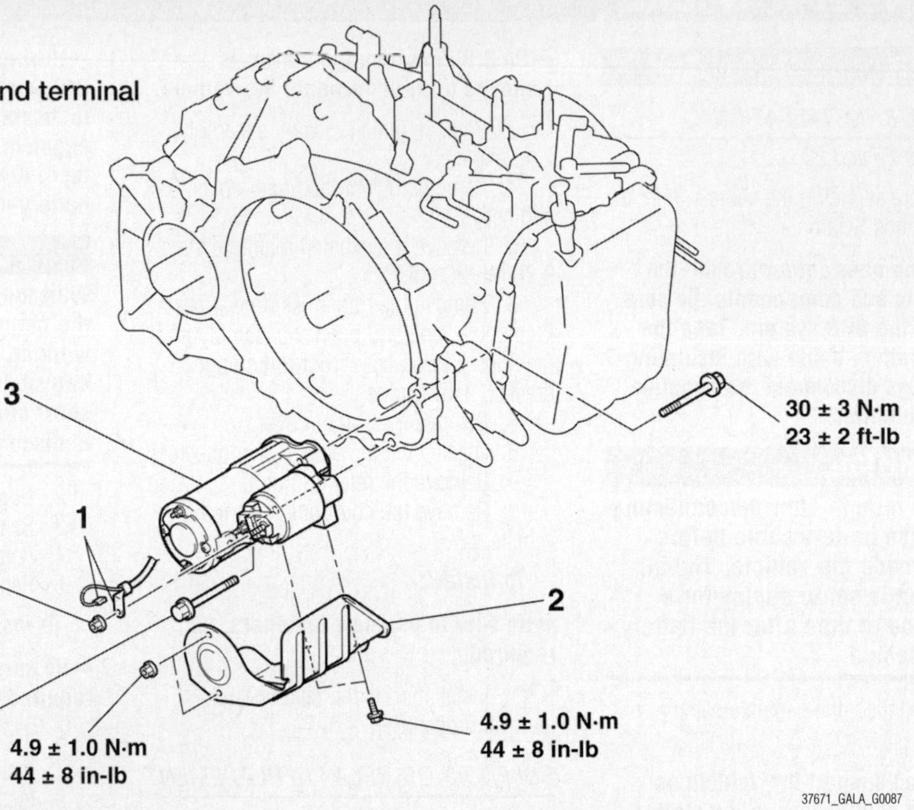

1. Connector and terminal
2. Cover
3. Starter

30 ± 3 N·m
23 ± 2 ft-lb

12 ± 2 N·m
102 ± 22 in-lb

30 ± 3 N·m
23 ± 2 ft-lb

4.9 ± 1.0 N·m
44 ± 8 in-lb

4.9 ± 1.0 N·m
44 ± 8 in-lb

37671_GALA_G0087

Fig. 28 Starter and related components—3.8L engine shown (MIVEC engine similar)

ENGINE MECHANICAL

ACCESSORY DRIVE BELTS

ACCESSORY BELT ROUTING

See Figures 29 and 30.

Inspect the drive belt for signs of glazing or cracking. A glazed belt will be perfectly smooth from slippage, while a good belt will have a slight texture of fabric visible. Cracks will usually start at the inner edge of the belt and run outward. All worn or damaged drive belts should be replaced immediately.

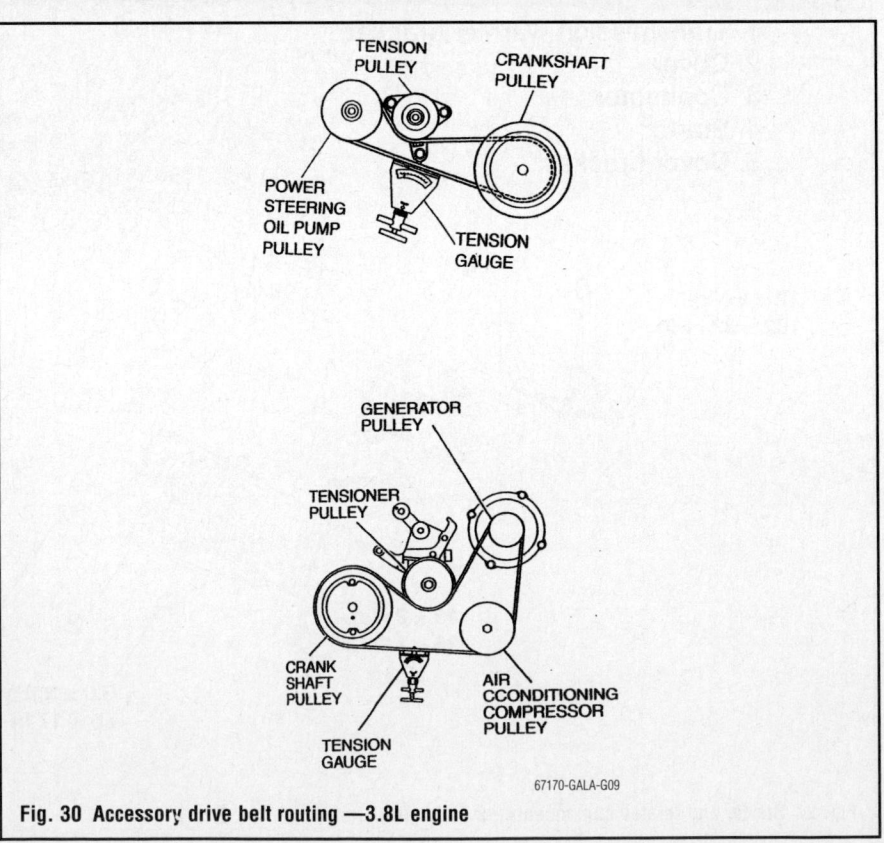

TENSION PULLEY
CRANKSHAFT PULLEY
POWER STEERING OIL PUMP PULLEY
TENSION GAUGE

GENERATOR PULLEY
TENSIONER PULLEY
CRANK SHAFT PULLEY
AIR CCONDITIONING COMPRESSOR PULLEY
TENSION GAUGE

67170-GALA-G09

Fig. 30 Accessory drive belt routing —3.8L engine

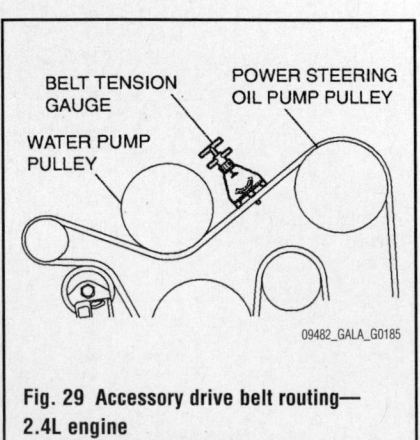

BELT TENSION GAUGE
POWER STEERING OIL PUMP PULLEY
WATER PUMP PULLEY

09482_GALA_G0185

Fig. 29 Accessory drive belt routing— 2.4L engine

ADJUSTMENT

Excessive belt tension will cause damage to the alternator and water pump pulley bearings, while, on the other hand, loose belt tension will produce slip and premature wear on the belt. Therefore, be sure to adjust the belt tension to the proper level.

If the engine is not equipped with an auto-tensioner, loosen the adjusting bolt or fixing bolt locknut on the alternator, alternator bracket or tension pulley. Then, move the alternator or turn the adjusting bolt to adjust belt tension. Once the desired value is reached, secure the bolt or locknut and recheck tension.

REMOVAL & INSTALLATION

2.4L Engine

See Figures 31 and 32.

1. Before servicing the vehicle, refer to the Precautions Section.

➡ **If working near and/or around the SRS system and components, be sure to disable the SRS system. Tape the negative battery cable with insulating tape. Always disconnect the negative battery cable first.**

❄❄ **CAUTION**

Wait for 1 minute after disconnecting the negative battery cable before working inside the vehicle. The air bag system is set to deploy for a short period of time after the battery is disconnected.

2. Remove the engine undercover.
3. As required, remove the radiator condenser tank assembly mounting bolt, and move the radiator condenser tank assembly to a place where it does not interfere with the drive belt removal and installation.

➡ **To reuse the accessory drive belt, use chalk to draw an arrow indicating the rotation direction on the back of the belt in order to install the belt in the same direction.**

4. Rotate the pulley bolt of the auto-tensioner counterclockwise with an offset wrench and align hole A with hole B.
5. Insert an L-shaped hexagon wrench into the aligned holes to fix the auto-tensioner in place.
6. Remove the accessory drive belt from the vehicle.

To install:

➡ **Be sure to use new fasteners, as required.**

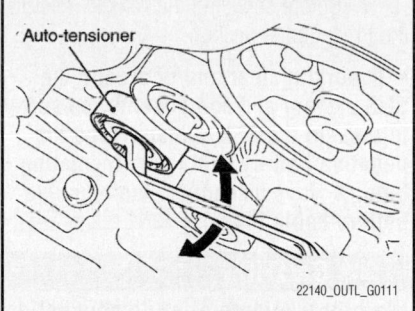

Fig. 31 Rotate the pulley bolt of the auto-tensioner counterclockwise with an offset wrench—2.4L engine

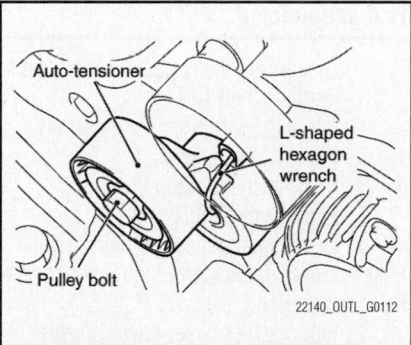

Fig. 32 Insert the hexagon wrench to fix the auto-tensioner—2.4L engine

7. Install the drive belt to each pulley.
8. Set an offset wrench to the pulley bolt of the auto-tensioner. Then, rotate the auto-tensioner counter-clockwise and remove the L-shaped hexagon wrench fixing the auto-tensioner.
9. Apply tension to the drive belt while slowly turning the auto-tensioner clockwise.

➡ **Make sure the belt is properly seated in all the pulleys.**

3.8L Engine

Alternator And A/C Compressor Belt

1. Before servicing the vehicle, refer to the Precautions Section.

➡ **If working near and/or around the SRS system and components, be sure to disable the SRS system. Tape the negative battery cable with insulating tape. Always disconnect the negative battery cable first.**

❄❄ **CAUTION**

Wait for 1 minute after disconnecting the negative battery cable before working inside the vehicle. The air bag system is set to deploy for a short period of time after the battery is disconnected.

2. Raise and safely support the vehicle.
3. Remove the front undercover.
4. Loosen the tension pulley fixing nut and relieve the tension on the belt by turning the adjusting bolt.
5. Remove the belt.

To install:

➡ **Be sure to use new fasteners, as required.**

6. Install the belt on the crankshaft and alternator pulleys.
7. Using the adjusting bolt on the tensioner, tighten the belt to the desired tension.
8. Tighten the fixing nut to hold the adjustment.
9. Install the undercover and lower the vehicle.

Power Steering Belt

1. Before servicing the vehicle, refer to the Precautions Section.

➡ **If working near and/or around the SRS system and components, be sure to disable the SRS system. Tape the negative battery cable with insulating tape. Always disconnect the negative battery cable first.**

❄❄ **CAUTION**

Wait for 1 minute after disconnecting the negative battery cable before working inside the vehicle. The air bag system is set to deploy for a short period of time after the battery is disconnected.

2. Raise and safely support the vehicle.
3. Remove the front undercover.
4. Remove the alternator and A/C compressor belt.
5. Lower the vehicle and remove the cruise control pump link assembly.
6. Place the power steering hose under the oil reservoir.
7. Loosen the tension pulley fixing bolts and remove the power steering pump drive belt.

To install:

➡ **Be sure to use new fasteners, as required.**

8. Install the power steering pump drive belt.
9. Insert an extension bar or equivalent into the opening at the end of the tension pulley bracket and pivot the pulley to apply tension to the belt.
10. Tighten the fixing bolts.

11. Raise the vehicle and install the alternator and A/C compressor belt.

12. Install the undercover and lower the vehicle.

CAMSHAFT AND VALVE LIFTERS

REMOVAL & INSTALLATION

2.4L Engine

See Figures 33 through 35.

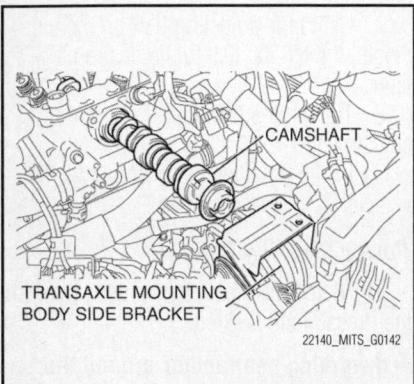

Fig. 33 Camshaft and transaxle mounting side bracket location—2.4L engine

1. Before servicing the vehicle, refer to the Precautions Section.

➡ **If working near and/or around the SRS system and components, be sure to disable the SRS system. Tape the negative battery cable with insulating tape. Always disconnect the negative battery cable first.**

❄❄ CAUTION

Wait for 1 minute after disconnecting the negative battery cable before working inside the vehicle. The air bag system is set to deploy for a short period of time after the battery is disconnected.

2. Disconnect the negative battery cable.
3. Remove the PCM.
4. Remove the air cleaner assembly.
5. Remove the battery.
6. Remove the ignition coils.
7. Remove the timing belt.
8. Disconnect the control wiring harness connection. Remove the rocker cover PCV connection.
9. Remove the rocker cover breather hose connection. Remove the engine oil

control valve and O-ring. Remove the oil pressure switch.

10. Remove the spark plug guide oil seals. Remove the accumulator assembly.

11. Remove the connector bracket. Remove the camshaft position sensor support and camshaft position sensing cylinder.

12. Use tool MB998719 to hold the camshaft sprocket in place. Remove the camshaft sprocket retaining bolt. Remove the camshaft sprocket. Remove the oil seal.

13. Remove the rocker cover retaining bolts. Remove the rocker cover. Discard the gasket.

14. Remove the exhaust rocker arm shaft caps. Remove the exhaust rocker arm shaft assembly.

15. Remove the intake rocker arm shaft caps. Remove the intake rocker arm shaft assembly.

16. Raise the transaxle until the camshaft and transaxle mounting side bracket do not touch when removing the camshaft.

17. Remove the camshaft from its mounting on the engine.

To install:

18. Lubricate the camshaft with clean engine oil. Install the camshaft. Set the

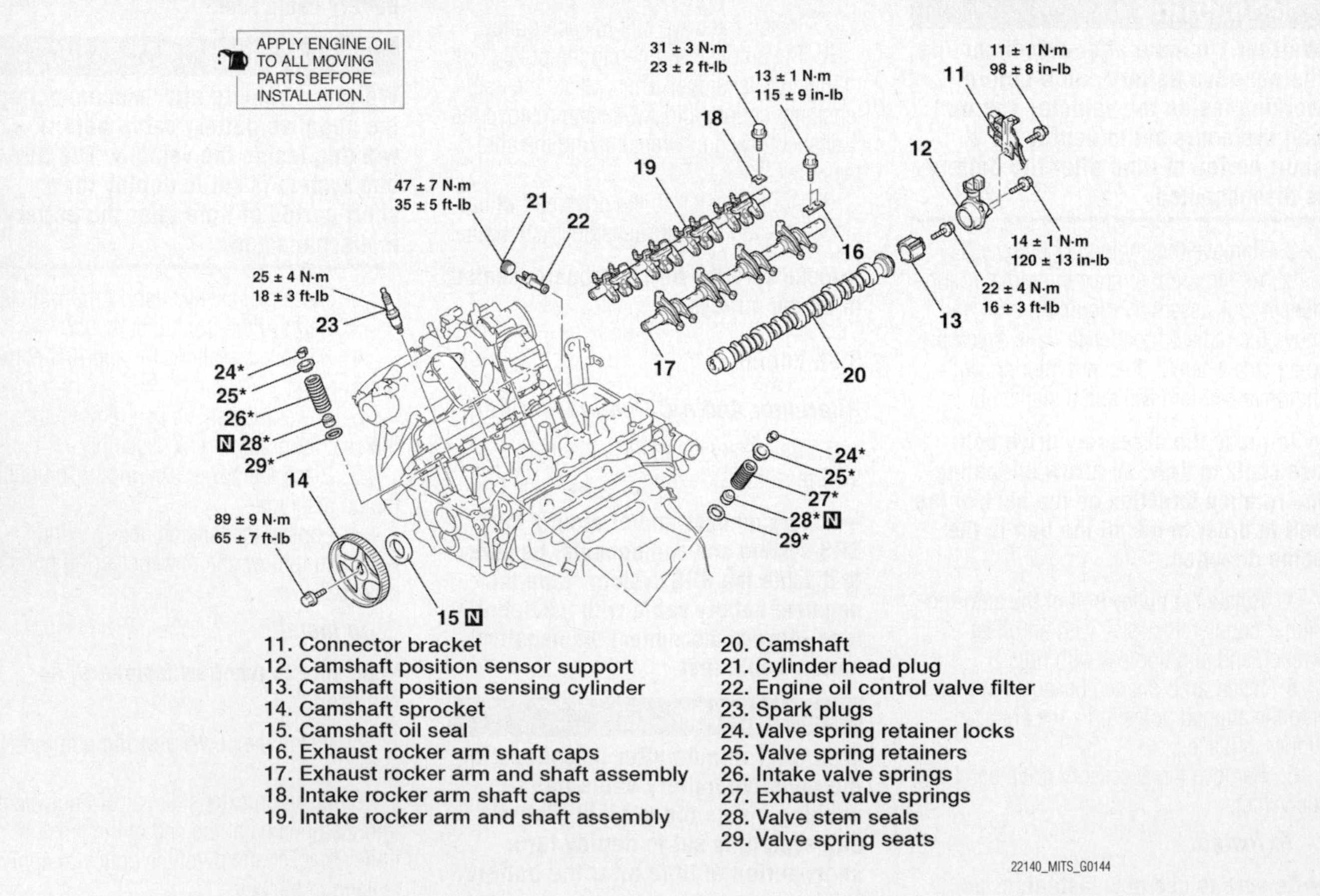

APPLY ENGINE OIL TO ALL MOVING PARTS BEFORE INSTALLATION.

31 ± 3 N·m
23 ± 2 ft-lb

13 ± 1 N·m
115 ± 9 in-lb

11 ± 1 N·m
98 ± 8 in-lb

47 ± 7 N·m
35 ± 5 ft-lb

14 ± 1 N·m
120 ± 13 in-lb

25 ± 4 N·m
18 ± 3 ft-lb

22 ± 4 N·m
16 ± 3 ft-lb

89 ± 9 N·m
65 ± 7 ft-lb

11. Connector bracket
12. Camshaft position sensor support
13. Camshaft position sensing cylinder
14. Camshaft sprocket
15. Camshaft oil seal
16. Exhaust rocker arm shaft caps
17. Exhaust rocker arm and shaft assembly
18. Intake rocker arm shaft caps
19. Intake rocker arm and shaft assembly

20. Camshaft
21. Cylinder head plug
22. Engine oil control valve filter
23. Spark plugs
24. Valve spring retainer locks
25. Valve spring retainers
26. Intake valve springs
27. Exhaust valve springs
28. Valve stem seals
29. Valve spring seats

Fig. 34 Camshaft and related components—2.4L engine

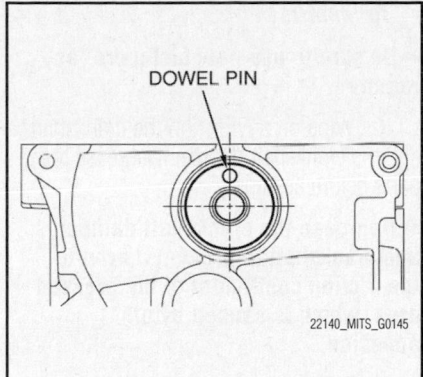

Fig. 35 Camshaft installation alignment—2.4L engine

dowel pin of the camshaft in the position as shown in the illustration.

19. Install the rocker arm assemblies.

20. Continue the installation in the reverse order of the removal procedure.

21. Adjust the valves as required.

22. To initialize the PCM, turn the ignition switch ON then OFF and keep it in the OFF position for at least 10 seconds.

3.8L Engine

See Figure 36.

1. Before servicing the vehicle, refer to the Precautions Section.

➡**If working near and/or around the SRS system and components, be sure to disable the SRS system. Tape the negative battery cable with insulating tape. Always disconnect the negative battery cable first.**

✳✳ CAUTION

Wait for 1 minute after disconnecting the negative battery cable before working inside the vehicle. The air bag system is set to deploy for a short period of time after the battery is disconnected.

2. Disconnect the negative battery cable.

3. Drain the cooling system.

4. Remove the timing belt.

5. On the left side, remove the thermostat housing assembly. Remove the PCV valve hose connection, ignition coil connectors and ignition coils. Disconnect the engine control wiring harness clamp and harness bracket.

6. On the right side, remove the intake manifold plenum, breather hose connection, heater hose and water hose connection. Remove the ignition coil connectors and coils.

7. Remove the rocker arm cover retain-

ing bolts. Remove the rocker arm cover. Discard the gasket.

8. Install tool MD998443 on the rocker arms, so that the lash adjusters will not fall out.

9. Loosen the rocker arm and shaft assembly mounting bolts. Remove the rocker arm and shaft assembly with the bolts still attached.

➡**Do not disassemble the rocker arm and shaft assembly.**

10. On the left side, disconnect the camshaft position sensor and remove the sensor support. Remove the camshaft position sensing cylinder. Remove the camshaft sprocket. Remove the camshaft and camshaft oil seal.

11. On the right side, disconnect the engine oil control valve connector and engine oil pressure switch connector. Remove the oil feed control valve right housing assembly. Remove the camshaft sprocket.

12. Remove the camshaft and camshaft oil seal.

To install:

13. Install the camshaft oil seal case gasket and oil feeder control valve housing gasket with their protrusions in the direction shown.

✳✳ WARNING

Be careful that no foreign material gets into the oil passages. Make sure the mating surfaces are clean.

14. Install the camshaft oil seal case and oil feeder control valve housing to the cylinder head.

15. Tighten the mounting bolt to: 16–20 ft. lbs. (21–27 Nm).

16. Install a gasket to one of oil feeder control valve pipes and tighten the eye bolt by hand.

17. Install a gasket to the other oil feeder control valve pipe and tighten the eye bolt to the specified torque. Tightening torque: 20–24 ft. lbs. (27–33 Nm).

18. Tighten the eye bolt which is temporarily tightened to the specified torque. Tightening torque: 20–24 ft. lbs. (27–33 Nm).

19. Tighten the oil feeder control valve pipe mounting bolt to the specified torque. Tightening torque: 90–106 inch lbs. (10–12 Nm).

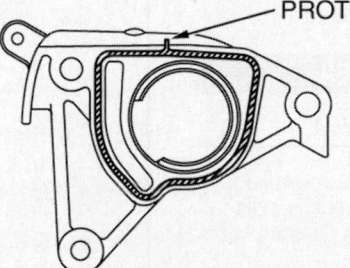
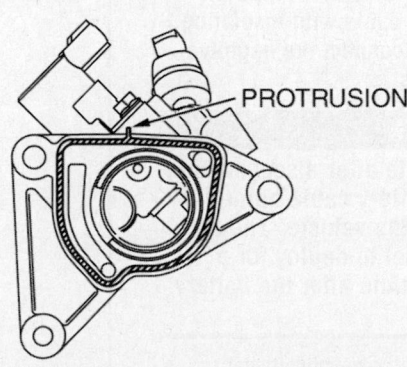

Fig. 36 Install the camshaft oil seal case gasket and oil feeder control valve housing gasket with their protrusions in the direction shown—3.8L engine (right bank)

➥Do not re-use the O-ring for the engine oil control valve.

20. Before installing a new O-ring on the engine oil control valve, wind adhesive tape around the oil passages cut-out area of the engine oil control valve to prevent damage. If the O-ring is damaged, it may cause an oil leak.

21. Apply a small amount of engine oil to the O-ring and install it to the oil control valve.

22. Assemble the engine oil control valve to the cylinder head.

23. Tighten the engine oil control valve mounting bolt to the specified torque. Tightening torque: 90–106 inch lbs. (10–12 Nm).

24. Install the intake and exhaust side rocker arms and shaft assemblies. Refer to Rocker Arms/Shafts, Removal & Installation.

25. Continue the installation in the reverse order of the removal procedure.

26. Adjust the valves as required.

27. To initialize the PCM, automatic transaxle equipped vehicles, turn the ignition switch ON then OFF and keep it in the OFF position for at least 10 seconds.

CATALYTIC CONVERTER

REMOVAL & INSTALLATION

The catalytic converter(s) are removed with the exhaust manifold(s).

CRANKSHAFT DAMPER

REMOVAL & INSTALLATION

See Figures 37 and 38.

1. Before servicing the vehicle, refer to the Precautions Section.

➥If working near and/or around the SRS system and components, be sure to disable the SRS system. Tape the negative battery cable with insulating tape. Always disconnect the negative battery cable first.

❋ CAUTION

Wait for 1 minute after disconnecting the negative battery cable before working inside the vehicle. The air bag system is set to deploy for a short period of time after the battery is disconnected.

2. Disconnect the negative battery cable.

3. Remove the accessory drive belts from around the crankshaft pulley.

4. Raise and support the vehicle.

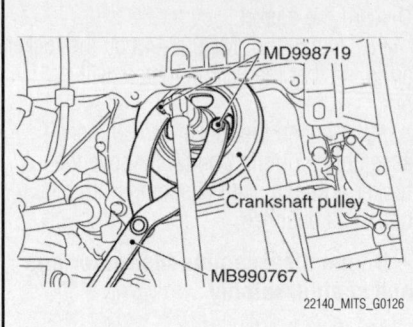

Fig. 37 Hold the crankshaft damper sprocket with special tools MB990767 and MD998719 for removal

5. Remove the passenger side front wheel.

6. Remove the passenger side inner fender splash shield to gain access to the crankshaft damper.

7. Hold the crankshaft damper sprocket with special tools MB990767 and MD998719.

8. Remove the crankshaft damper center bolt and washer.

9. Remove the crankshaft damper.

To install:

➥Be sure to use new fasteners, as required.

10. Wipe off any dirt on the crankshaft and the crankshaft damper. Degrease the parts before assembly.

➥Degrease the crankshaft damper and crankshaft to prevent a drop in the friction coefficient of the pressed area, which is caused by oil adhesion.

11. Install the crankshaft pulley.

12. Apply an adequate and minimum amount of engine oil to the threads of the crankshaft pulley center bolt and the lower area of the flange.

13. Hold the crankshaft pulley with special tools MB990767 and MD998719 in the same manner as removal.

14. Tighten the crankshaft pulley center bolt to specification.

15. Install the splash shield.

16. Install the wheel, then carefully lower the vehicle.

17. Install the accessory drive belts.

18. Connect the negative battery cable.

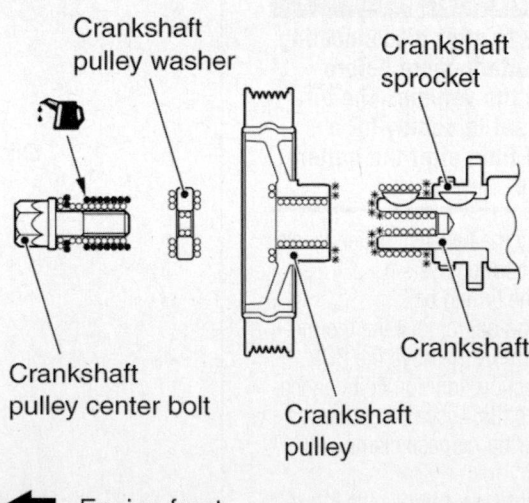

○ : Wipe clean with a rag.
✱ : Wipe clean with a rag and degrease.
● : Apply a small amount of engine oil.

Fig. 38 Wipe off any dirt on the crankshaft and the crankshaft damper and degrease the parts before assembly

CRANKSHAFT FRONT SEAL

REMOVAL & INSTALLATION

2.4L Engine

See Figure 39.

1. Before servicing the vehicle, refer to the Precautions Section.

➡**If working near and/or around the SRS system and components, be sure to disable the SRS system. Tape the negative battery cable with insulating tape. Always disconnect the negative battery cable first.**

❋❋ CAUTION

Wait for 1 minute after disconnecting the negative battery cable before working inside the vehicle. The air bag system is set to deploy for a short period of time after the battery is disconnected.

2. Disconnect the negative battery cable.
3. Remove the timing belt.
4. Remove the crankshaft sprocket.
5. Carefully pry the oil seal out of the front case. Be careful not to damage the oil seal bore or the crankshaft sealing surface.

To install:

➡**Be sure to use new fasteners, as required.**

6. Apply clean engine oil to the oil seal lip. Using a seal driver, install the oil seal.
7. Install the crankshaft sprocket.
8. Install the timing belt.
9. Install the negative battery cable.

3.8L Engine

See Figure 40.

1. Before servicing the vehicle, refer to the Precautions Section.

➡**If working near and/or around the SRS system and components, be sure to disable the SRS system. Tape the negative battery cable with insulating tape. Always disconnect the negative battery cable first.**

❋❋ CAUTION

Wait for 1 minute after disconnecting the negative battery cable before working inside the vehicle. The air bag system is set to deploy for a short period of time after the battery is disconnected.

2. Remove the timing belt.
3. Remove the Crankshaft Position (CKP) sensor.
4. Remove the crankshaft sprocket.
5. Remove the crankshaft sensing blade.
6. Remove the crankshaft spacer and key.
7. Remove the front oil seal.

To install:

➡**Be sure to use new fasteners, as required.**

8. Install the front oil seal. Apply oil to the seal and install using a Crankshaft Front Oil Seal Installer tool MD998717.
9. Install the crankshaft key and spacer.
10. Install the crankshaft sensing blade.
11. Install the CKP sensor.

➡**To be sure the crankshaft pulley bolt does not loosen, make sure to clean the mating areas of the crankshaft, spacer, sensing blade, and sprocket.**

12. Install the crankshaft sprocket.
13. Install the timing belt.

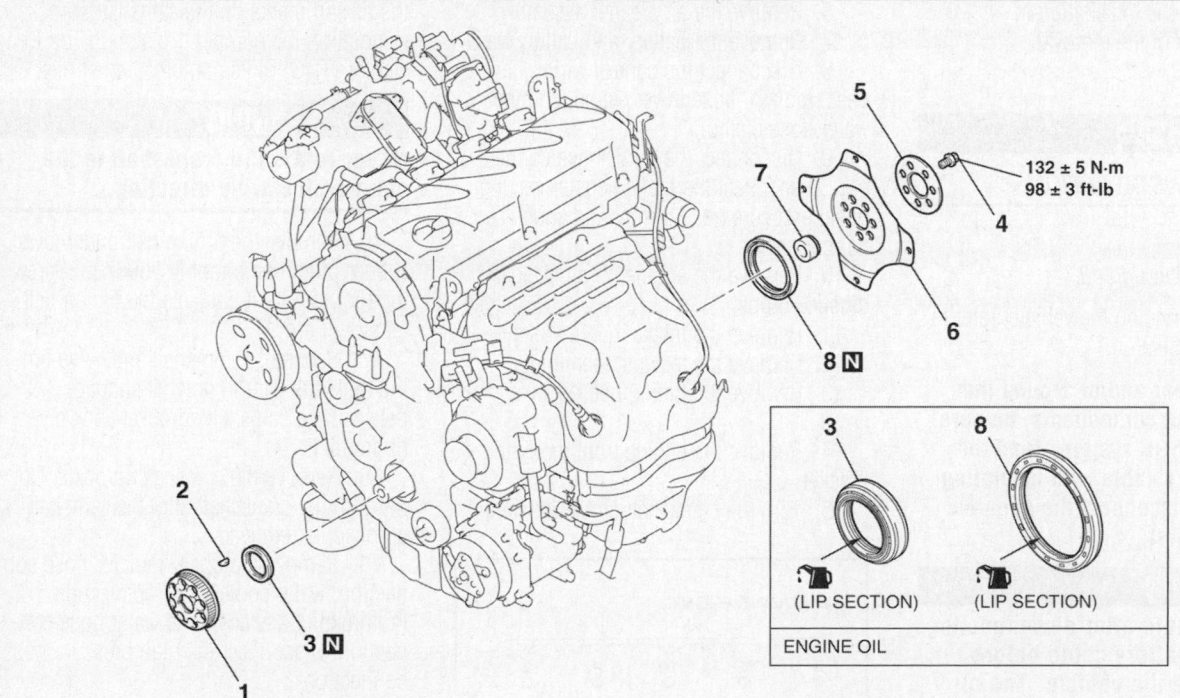

1. Crankshaft balancer shaft drive sprocket
2. Crankshaft key
3. Crankshaft front oil seal
4. A/T drive plate bolts
5. A/T drive plate adapter plate
6. A/T drive plate
7. Crankshaft bushing
8. Crankshaft rear oil seal

22140_MITS_G0137

Fig. 39 Expanded view of the crankshaft oil seals (front and rear)—2.4L engine

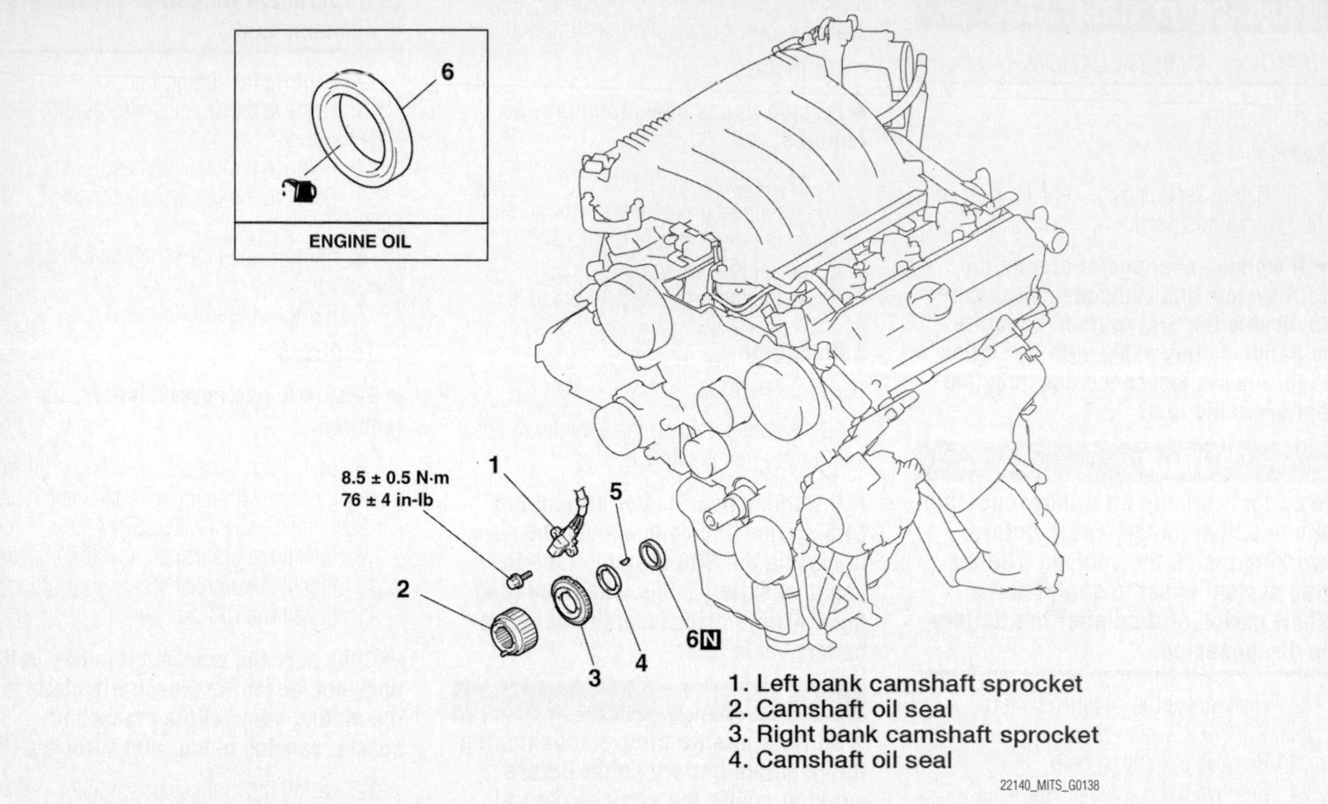

8.5 ± 0.5 N·m
76 ± 4 in-lb

1. Left bank camshaft sprocket
2. Camshaft oil seal
3. Right bank camshaft sprocket
4. Camshaft oil seal

ENGINE OIL

22140_MITS_G0138

Fig. 40 Expanded view of the crankshaft oil front seal—3.8L engine

14. Continue the installation in the reverse order of the removal procedure.

CYLINDER HEAD

REMOVAL & INSTALLATION

2.4L Engine

See Figures 41 through 43.

1. Before servicing the vehicle, refer to the Precautions Section.

➡ **If working near and/or around the SRS system and components, be sure to disable the SRS system. Tape the negative battery cable with insulating tape. Always disconnect the negative battery cable first.**

✳ CAUTION

Wait for 1 minute after disconnecting the negative battery cable before working inside the vehicle. The air bag system is set to deploy for a short period of time after the battery is disconnected.

2. Relieve the fuel system pressure.
3. Disconnect the negative battery cable.
4. Drain the cooling system.
5. Remove the PCM.

6. Remove the air cleaner assembly.
7. Remove the battery and battery tray.
8. Disconnect the control wiring harness, radiator hose lower clamp, and the water hose clamp.
9. Disconnect the EVAP hose connection. Disconnect the brake booster vacuum hose, pressure hose clamp, and knock sensor connector.
10. Remove the engine oil dipstick and dipstick guide.
11. Remove the intake manifold stay.
12. Remove the exhaust manifold.
13. Remove the timing belt upper cover.
14. Remove the engine front mounting bracket.
15. Turn the crankshaft clockwise, align

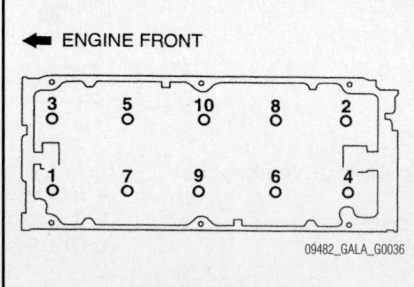

⬅ ENGINE FRONT

| 3 | 5 | 10 | 8 | 2 |
| 1 | 7 | 9 | 6 | 4 |

09482_GALA_G0036

Fig. 41 Cylinder head bolt removal sequence—2.4L engine

the timing marks on the camshaft sprocket to position the number 1 piston to Top Dead Center (TDC) of its compression stroke.

✳ WARNING

Never rotate the crankshaft in the counterclockwise direction.

16. Remove the timing belt undercover rubber plug, using tool MD998738. Screw the tool in until it contacts the timing belt tensioner arm.
17. Secure the camshaft and valve timing belt with wire to prevent slippage between the camshaft sprocket and valve timing belt.
18. Remove the camshaft sprocket. Do not turn the crankshaft after the camshaft sprocket is removed.
19. Remove the upper radiator hose connection, water cooler hose connection (automatic transaxle), and water hose connection. Disconnect the high pressure hose connection.
20. Remove the valve cover.
21. Using special tool MB991654, loosen the cylinder head bolts. Loosen each bolt evenly, little by little, in 2–3 steps and in the proper bolt removal sequence.
22. Remove the cylinder head from the engine. Discard the cylinder head gasket.

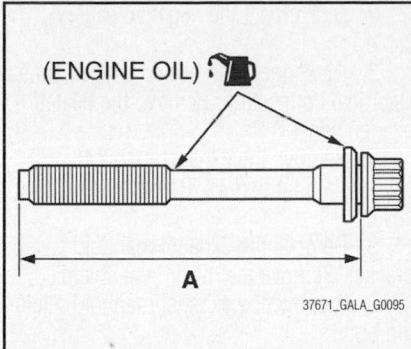

Fig. 42 Cylinder head bolt length check— 2.4L engine

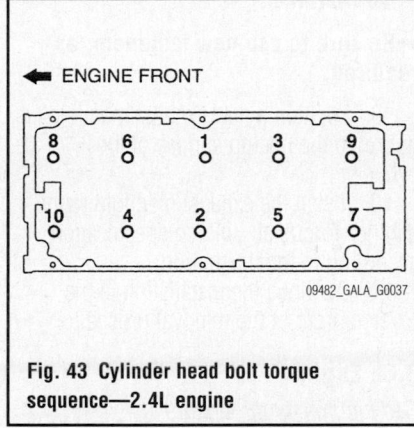

Fig. 43 Cylinder head bolt torque sequence—2.4L engine

➡If the cylinder head bolts cannot be pulled out due to the washer being trapped in the valve spring, raise the bolt slightly, then remove it while holding it with a magnet.

To install:

➡Be sure to use new fasteners, as required.

23. Thoroughly clean the mating surfaces of the head and block.

24. Place a new head gasket on the cylinder block. Position the gasket on the block so that the identification mark "381" is at the top surface and on the exhaust side.

25. Inspect the cylinder head bolt length prior to installation. If the length (A) exceeds 3.91 inches (99.4mm), the bolt must be replaced. Apply a small amount of engine oil to the thread section and washer of the bolt.

26. Position the cylinder head on the engine. Install and torque the retaining bolts to specification and in the proper sequence.

   a. Step 1: 57–59 ft. lbs. (76–80 Nm).

   b. Step 2: Loosen all bolts fully in the reverse order of tightening.

   c. Step 3: Retighten the loosened bolts to the specified torque in the tight-

ening sequence shown. Tightening torque: 14–16 ft. lbs. (18–22 Nm).

   d. Step 4: Make a paint mark across each bolt head and cylinder head. Tighten the cylinder head bolts 90° in the specified order.

   e. Step 5: Tighten the bolts another 90° degrees in the same order and check that the paint marks on the cylinder head bolt are aligned with the paint marks on the cylinder head.

✳✳ WARNING

If the bolt is turned less than 90°, proper fastening performance may not be achieved. Be careful to turn each bolt exactly 90°. If the bolt is over-tightened, loosen the bolt completely and then retighten it by repeating the tightening procedure from torque step 1.

27. Torque the camshaft bolt to 58—72 ft. lbs. (80—98 Nm).

28. Continue the installation in the reverse order of the removal procedure.

29. Fill the engine with the proper grade and type engine oil.

30. Fill the cooling system with the proper grade and type engine coolant.

31. Initialize the PCM, by turning the ignition switch ON then OFF. Keep it off for at least 10 seconds.

32. Start the engine and check for leaks.

3.8L Engine
See Figures 44 through 46.

1. Before servicing the vehicle, refer to the Precautions Section.

➡If working near and/or around the SRS system and components, be sure to disable the SRS system. Tape the negative battery cable with insulating tape. Always disconnect the negative battery cable first.

✳✳ CAUTION

Wait for 1 minute after disconnecting the negative battery cable before working inside the vehicle. The air bag system is set to deploy for a short period of time after the battery is disconnected.

2. Relieve the fuel system pressure.
3. Disconnect the negative battery cable.
4. Drain the cooling system.
5. Remove the intake manifold. Remove the exhaust manifolds.
6. Remove the timing belt.
7. Remove the thermostat housing. Dis-

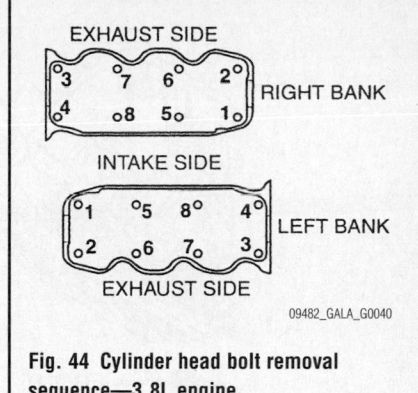

Fig. 44 Cylinder head bolt removal sequence—3.8L engine

connect the PCV hose connection. Remove the PCV valve.

8. Remove the ignition coils.

9. Disconnect the engine control wiring harness clamp. Remove the harness bracket.

10. Remove the rocker arm covers and gaskets. Remove the camshaft position sensor connector.

11. Remove the engine oil control valve connector. Remove the grounding bolt. Remove the engine oil dipstick assembly.

12. Remove the camshaft sprockets, using tool MB990767 or equivalent. Remove the timing belt rear center cover.

13. Remove the power steering oil pump assembly and position it to the side. Remove the power steering oil pump bracket bolt. Remove the engine oil control valve connector.

14. Loosen all cylinder head bolts in 2–3 steps and in the proper bolt removal sequence. Remove the cylinder head from the engine. Discard the cylinder head gasket.

To install:

➡Be sure to use new fasteners, as required.

15. Thoroughly clean the sealing surfaces of the head and block.

16. Place a new head gasket on the cylinder block making sure the identification mark on the cylinder head gasket is in the front top (upward) location. Do not use sealer on the gasket.

17. Carefully install the cylinder head on the block. Be sure the head bolt washers are installed with the beveled side facing upward. Torque the cylinder head bolts to specification in 2–3 passes. Tightening torque:

   a. Step 1: 77–83 ft. lbs. (103–113 Nm).

   b. Step 2: 0 inch lbs. (0 Nm).

   c. Step 3: 77–83 ft. lbs. (103–113 Nm).

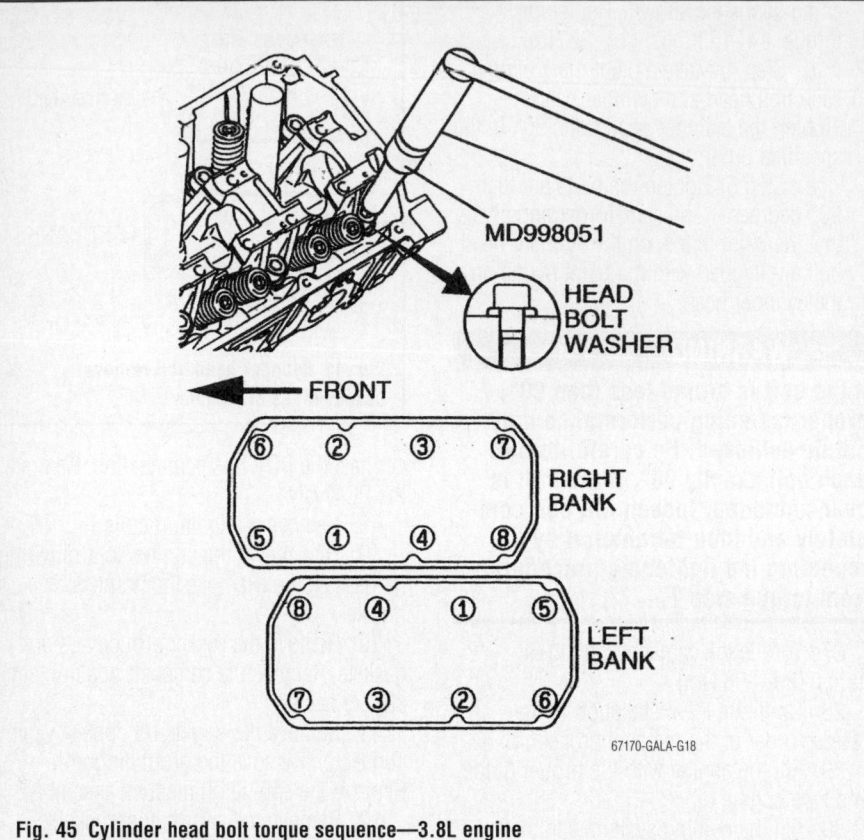

Fig. 45 Cylinder head bolt torque sequence—3.8L engine

18. Use special tools MD998715 and MB990767 in the same way as during removal to install the camshaft sprocket.

19. Tighten the camshaft sprocket mounting bolt to 58–72 ft. lbs. (78–98 Nm).

20. Torque the camshaft bolt to 58—72 ft. lbs. (80—98 Nm).

21. Continue the installation in the reverse order of the removal procedure.

22. Fill the engine with the proper grade and type engine oil.

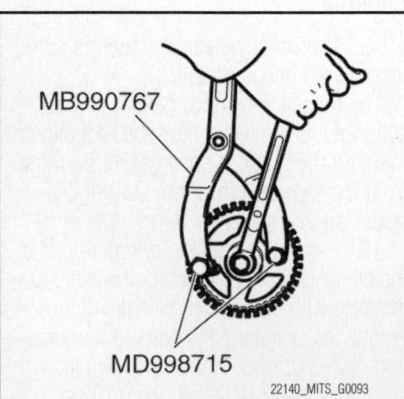

Fig. 46 Use special tools MD998715 and MB990767 to install the camshaft sprocket—3.8L engine

23. Fill the cooling system with the proper grade and type engine coolant.

24. Initialize the PCM, by turning the ignition switch ON then OFF. Keep it off for at least 10 seconds.

25. Start the engine and check for leaks.

EXHAUST MANIFOLD

REMOVAL & INSTALLATION

2.4L Engine

See Figure 47.

1. Before servicing the vehicle, refer to the Precautions Section.

➡ **If working near and/or around the SRS system and components, be sure to disable the SRS system. Tape the negative battery cable with insulating tape. Always disconnect the negative battery cable first.**

❖❖ CAUTION

Wait for 1 minute after disconnecting the negative battery cable before working inside the vehicle. The air bag system is set to deploy for a short period of time after the battery is disconnected.

2. Disconnect the negative battery cable.

3. Disconnect the heated oxygen sensor electrical connection. Remove the heated oxygen sensor.

4. Remove the exhaust manifold cover retaining screws. Remove the cover.

5. Remove the exhaust manifold bracket. Remove the lower heat shield.

6. Remove the exhaust manifold retaining bolts.

7. Remove the exhaust manifold from the engine.

8. Discard the old gasket.

To install:

➡ **Be sure to use new fasteners, as required.**

9. Position a new exhaust manifold gasket to the mating surface of the cylinder.

10. Install the exhaust manifold to the engine. Torque all bolts to specification according to the illustration.

11. Continue the installation in the reverse order of the removal procedure.

3.8L Engine

See Figures 48 through 49.

1. Before servicing the vehicle, refer to the Precautions Section.

2. Disconnect the negative battery cable.

3. Remove the engine undercover. Remove the air cleaner intake duct assembly.

4. Remove the right side exhaust manifold:

    a. Remove the right bank heated oxygen sensor (front).

    b. Remove the right bank heated oxygen sensor (rear).

    c. Remove the heat protector (right) and the exhaust manifold stay (right "B").

    d. Remove the exhaust manifold stay (right "A").

    e. Remove the heat protector, lower right.

    f. Remove the right side exhaust manifold and discard the gasket.

5. Remove the left side exhaust manifold:

    a. Remove the left bank heated oxygen sensor (front).

    b. Remove the left bank heated oxygen sensor (rear).

    c. Remove the connector bracket and the heat protector (left).

    d. Remove the exhaust manifold stay (left "B").

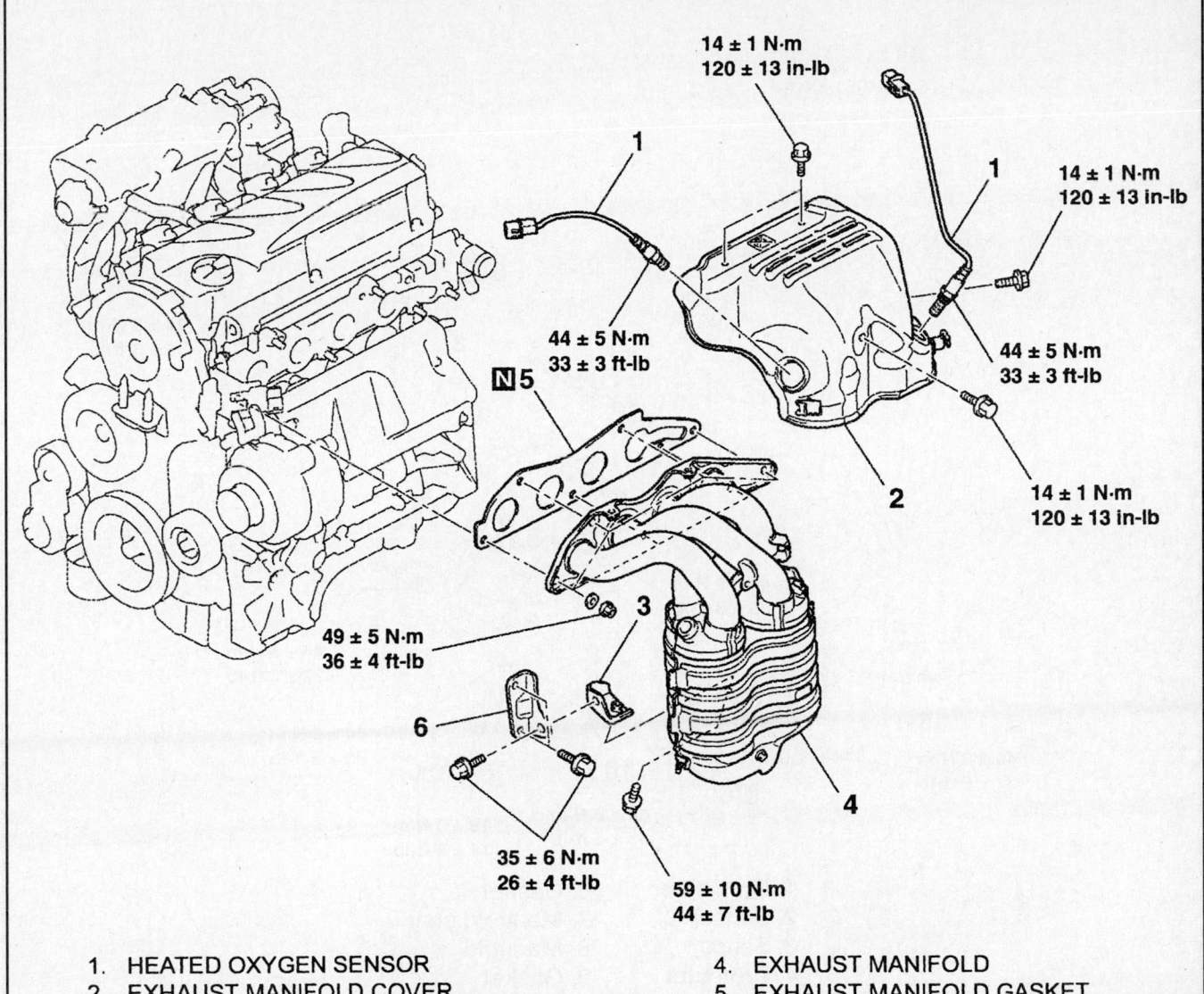

14 ± 1 N·m
120 ± 13 in-lb

14 ± 1 N·m
120 ± 13 in-lb

44 ± 5 N·m
33 ± 3 ft-lb

44 ± 5 N·m
33 ± 3 ft-lb

14 ± 1 N·m
120 ± 13 in-lb

49 ± 5 N·m
36 ± 4 ft-lb

35 ± 6 N·m
26 ± 4 ft-lb

59 ± 10 N·m
44 ± 7 ft-lb

1. HEATED OXYGEN SENSOR
2. EXHAUST MANIFOLD COVER
3. EXHAUST MANIFOLD BRACKET B
4. EXHAUST MANIFOLD
5. EXHAUST MANIFOLD GASKET
6. EXHAUST MANIFOLD BRACKET A

00482_GALA_G0063

Fig. 47 Exhaust manifold and related components—2.4L engine

e. Remove the left side exhaust manifold and discard the gasket.

To install:

➡**Be sure to use new fasteners, as required.**

6. Position new exhaust manifold gaskets to the mating surfaces on the left and right sides.

7. Install the exhaust manifolds to the engine. Torque all bolts to specification according to the illustration.

8. Continue the installation in the reverse order of the removal procedure.

9. Start the engine and check for exhaust leaks.

FLEXPLATE

REMOVAL & INSTALLATION

1. Before servicing the vehicle, refer to the Precautions Section.

➡**If working near and/or around the SRS system and components, be sure to disable the SRS system. Tape the negative battery cable with insulating tape. Always disconnect the negative battery cable first.**

✳✳ CAUTION

Wait for 1 minute after disconnecting the negative battery cable before working inside the vehicle. The air bag system is set to deploy for a short period of time after the battery is disconnected.

2. Disconnect the negative battery cable.

3. Remove the transaxle.

4. Mark the position of the driveplate on the crankshaft and remove the retaining bolts.

5. Remove the driveplate adapter.

6. Remove the driveplate from the engine.

To install:

➡**Be sure to use new fasteners, as required.**

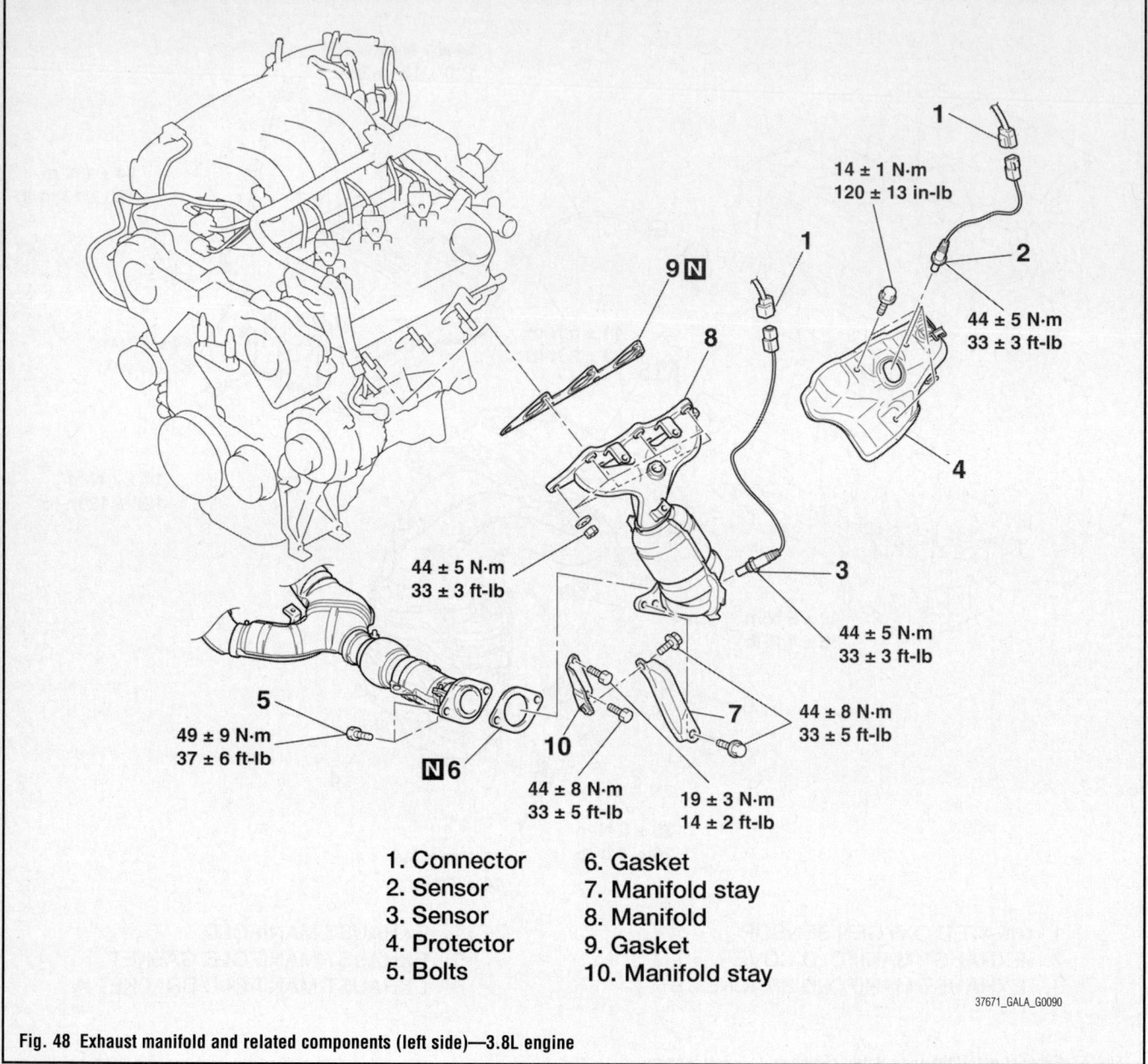

1. Connector
2. Sensor
3. Sensor
4. Protector
5. Bolts
6. Gasket
7. Manifold stay
8. Manifold
9. Gasket
10. Manifold stay

37671_GALA_G0090

Fig. 48 Exhaust manifold and related components (left side)—3.8L engine

7. Coat the threads of the driveplate retaining bolts with thread locking compound.

8. Position the driveplate on the crankshaft flange.

9. Install the driveplate adapter.

10. Install and tighten the bolts, in an alternating star pattern, to specification.

11. Install the transaxle.

12. Connect the negative battery cable.

INTAKE MANIFOLD

REMOVAL & INSTALLATION

2.4L Engine

See Figures 50 and 51.

1. Before servicing the vehicle, refer to the Precautions Section.

➡If working near and/or around the SRS system and components, be sure to disable the SRS system. Tape the negative battery cable with insulating tape. Always disconnect the negative battery cable first.

❄❄ CAUTION

Wait for 1 minute after disconnecting the negative battery cable before working inside the vehicle. The air bag system is set to deploy for a short period of time after the battery is disconnected.

2. Relieve the fuel system pressure. Disconnect the negative battery cable.

3. Remove the air cleaner assembly. Drain the cooling system. Remove the battery.

➡Do not loosen the retaining screws for the resin cover of the throttle body assembly. If these screws are loosened, the sensor incorporated in the resin cover becomes misaligned and the throttle body will not function properly.

4. Disconnect the accelerator cable connection, throttle position sensor connector and water return hose connection

5. Remove the throttle body bracket. Remove the throttle body retaining bolts. Remove the throttle body. Discard the gasket.

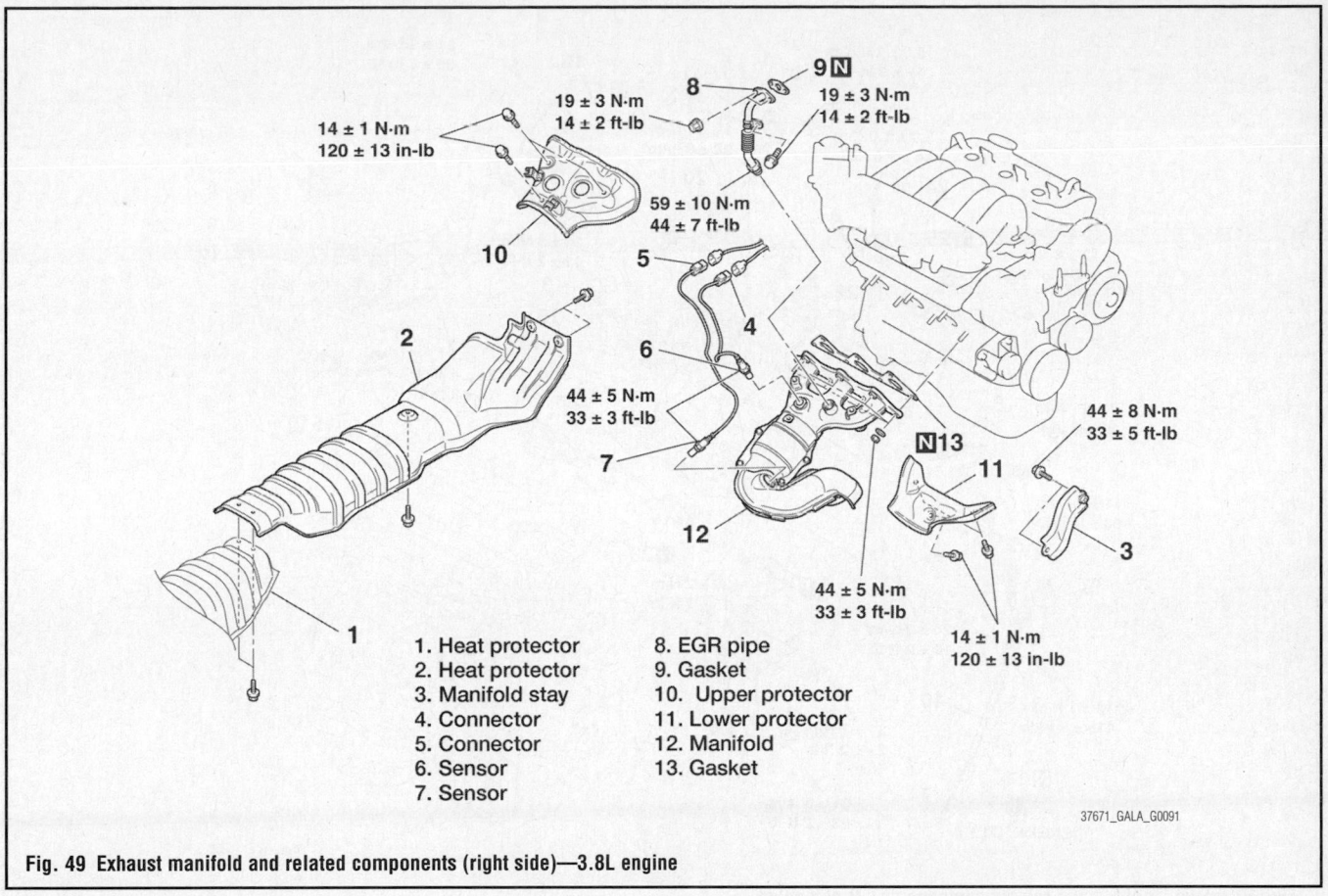

1. Heat protector
2. Heat protector
3. Manifold stay
4. Connector
5. Connector
6. Sensor
7. Sensor
8. EGR pipe
9. Gasket
10. Upper protector
11. Lower protector
12. Manifold
13. Gasket

14 ± 1 N·m
120 ± 13 in-lb

19 ± 3 N·m
14 ± 2 ft-lb

19 ± 3 N·m
14 ± 2 ft-lb

59 ± 10 N·m
44 ± 7 ft-lb

44 ± 5 N·m
33 ± 3 ft-lb

44 ± 8 N·m
33 ± 5 ft-lb

44 ± 5 N·m
33 ± 3 ft-lb

14 ± 1 N·m
120 ± 13 in-lb

37671_GALA_G0091

Fig. 49 Exhaust manifold and related components (right side)—3.8L engine

6. Remove the delivery pipe and injector assembly.

7. Disconnect the EVAP purge solenoid valve connector, the evaporative emission purge hose connection and the EGR valve connector.

8. Disconnect the MAP sensor connector, brake booster vacuum hose connection, pressure hose clamp, and knock sensor connector bracket.

9. Disconnect the harness clamp bracket and harness clamp. Remove the engine oil dipstick and tube assembly.

10. Disconnect the lower radiator hose. Remove the thermostat case assembly, resonator, and gasket.

11. Remove the EGR valve and gasket. Remove the harness clamp bracket, EVAP purge solenoid valve, and brake booster vacuum pipe.

12. Remove the brake booster vacuum hose and MAP sensor.

13. Remove the intake manifold bracket.

14. Remove the intake manifold retaining bolts. Remove the intake manifold from the engine. Discard the gasket.

To install:

➡**Be sure to use new fasteners, as required.**

15. Be sure to use a new intake manifold gasket. Position the gasket on the engine.

16. Install the intake manifold. Torque the retaining bolts to specification.

17. When installing the thermostat case assembly, be sure to apply a bead of sealant, part number 8672 or equivalent, to the cylinder head mating surface of the case assembly.

➡**Be sure to assemble the components with 15 minutes of applying the sealant to the surfaces.**

18. Continue the installation in the reverse order of the removal procedure.

➡**Be sure to install the throttle body gasket in the proper direction. If this gasket is not installed correctly, poor engine idling may result.**

19. Complete the vehicle initialization procedure.

➡**To complete the initialization procedure the following tools are needed: MB991958 scan tool, MB991824 VCI, MB991827 MUT III USB cable, MB991910 MUT III Main harness "A."**

20. Connect the scan tool to the data link connector. To prevent damage to the scan tool, be sure that the ignition switch is

in the **LOC** position before connecting the scan tool.

21. Turn the ignition switch to the **ON** position.

22. Select "check mode" from the menu screen.

23. Select "erase memory" from the menu screen.

24. Initialize the learning value.

25. After initialization, complete the idle learning procedure.

➡**This procedure must be performed when the PCM is replaced, or when the learning value is initialized, as the idling is not stabilized because the learning value in the MFI engine is not completed.**

26. Start the engine. Allow the coolant temperature to reach 176°F (80°C) or more.

27. Stop the engine and place the ignition switch in the **LOCK** position.

28. After 10 seconds, restart the engine.

29. For 10 minutes, carry out the idling procedure below to confirm that the engine has normal idling.

- Position the transaxle selector lever in the "P" range
- The engine fan is not to be operated

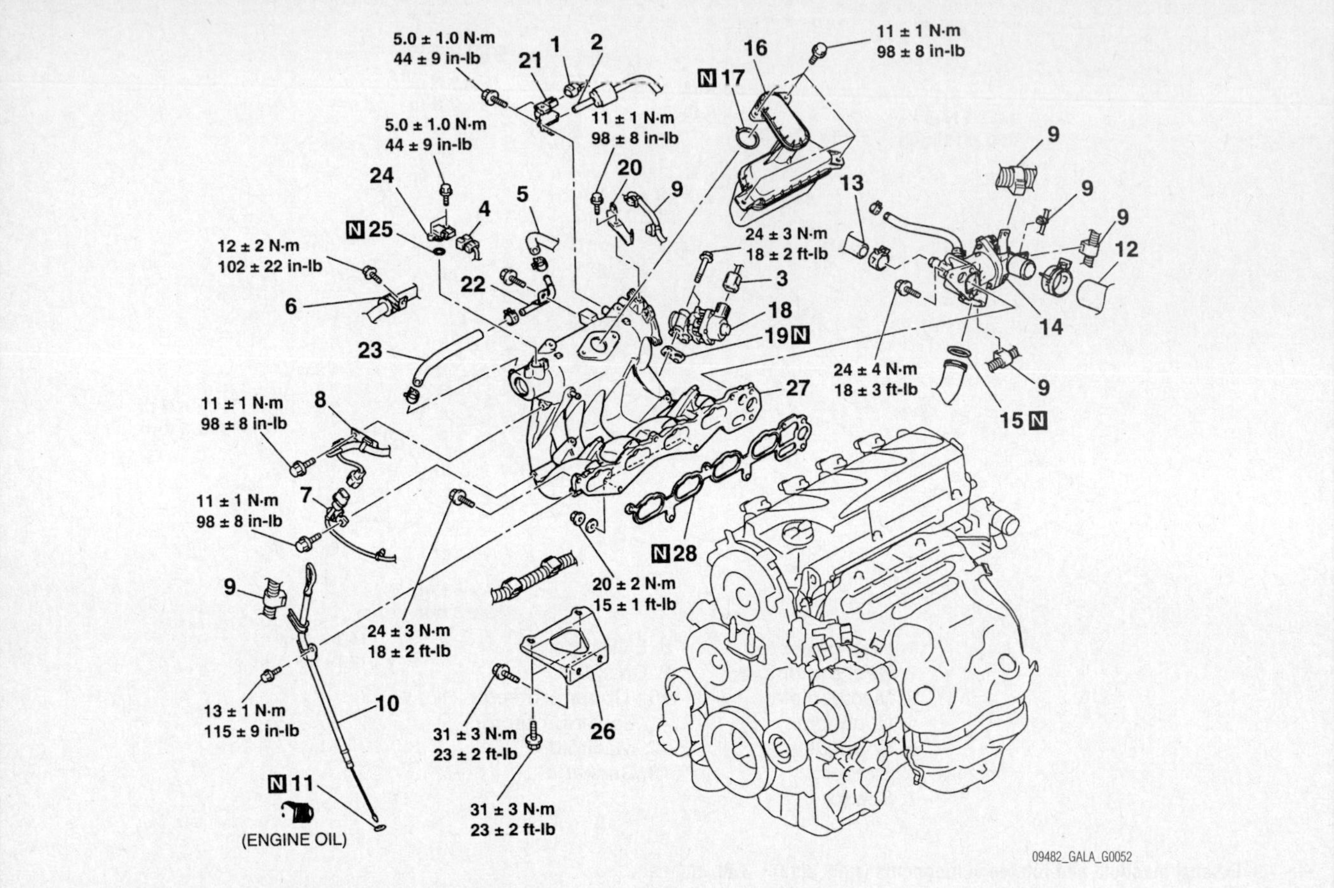

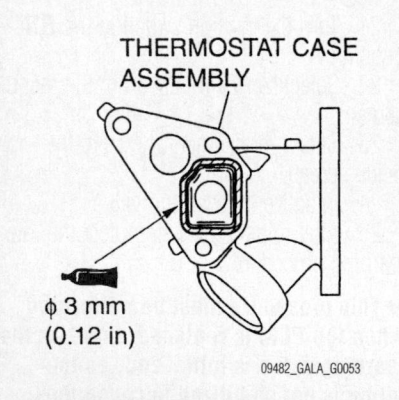

Fig. 50 Intake manifold and related components—2.4L engine

THERMOSTAT CASE ASSEMBLY

φ 3 mm
(0.12 in)

09482_GALA_G0053

Fig. 51 Thermostat case assembly sealant application—2.4L engine

- The engine coolant temperature should be 176°F (80°C) or more

➡**If the engine stalls during idling, check the throttle valve of the throttle body for dirt. Correct and perform the procedure again.**

3.8L Engine

See Figures 52 and 53.

1. Before servicing the vehicle, refer to the Precautions Section.

➡**If working near and/or around the SRS system and components, be sure to disable the SRS system. Tape the negative battery cable with insulating tape. Always disconnect the negative battery cable first.**

❋❋ CAUTION

Wait for 1 minute after disconnecting the negative battery cable before working inside the vehicle. The air bag system is set to deploy for a short period of time after the battery is disconnected.

2. Relieve the fuel system pressure.
3. Disconnect the negative battery cable.
4. Remove the air cleaner assembly.
5. Drain the cooling system.
6. Remove the intake manifold plenum assembly.
7. Disconnect the injector connector wiring assembly, high pressure hose connection, O-ring, and fuel return hose.
8. Disconnect the vacuum hose connection. Remove the fuel rail, injector, and fuel pressure regulator. Remove the harness bracket.
9. Remove the PCV hose connection.

Remove the left timing belt front upper cover. Remove the right timing belt front upper cover. Remove the water pump bracket.
10. Remove the intake manifold retaining bolts. Remove the intake manifold from the engine. Discard the gaskets.

To install:

➡**Be sure to use new fasteners, as required.**

11. Be sure to use a new intake manifold gasket. Position the gasket on the engine.
12. Install the intake manifold. Coat the retaining bolts in clean engine oil. Torque the retaining bolts to specification.
13. Continue the installation in the reverse order of the removal procedure.
14. Complete the vehicle initialization procedure.

➡**To complete the initialization procedure the following tools are needed: MB991958 scan tool, MB991824 VCI, MB991827 MUT III USB cable, MB991910 MUT III Main harness "A."**

15. Connect the scan tool to the data link connector. To prevent damage to the

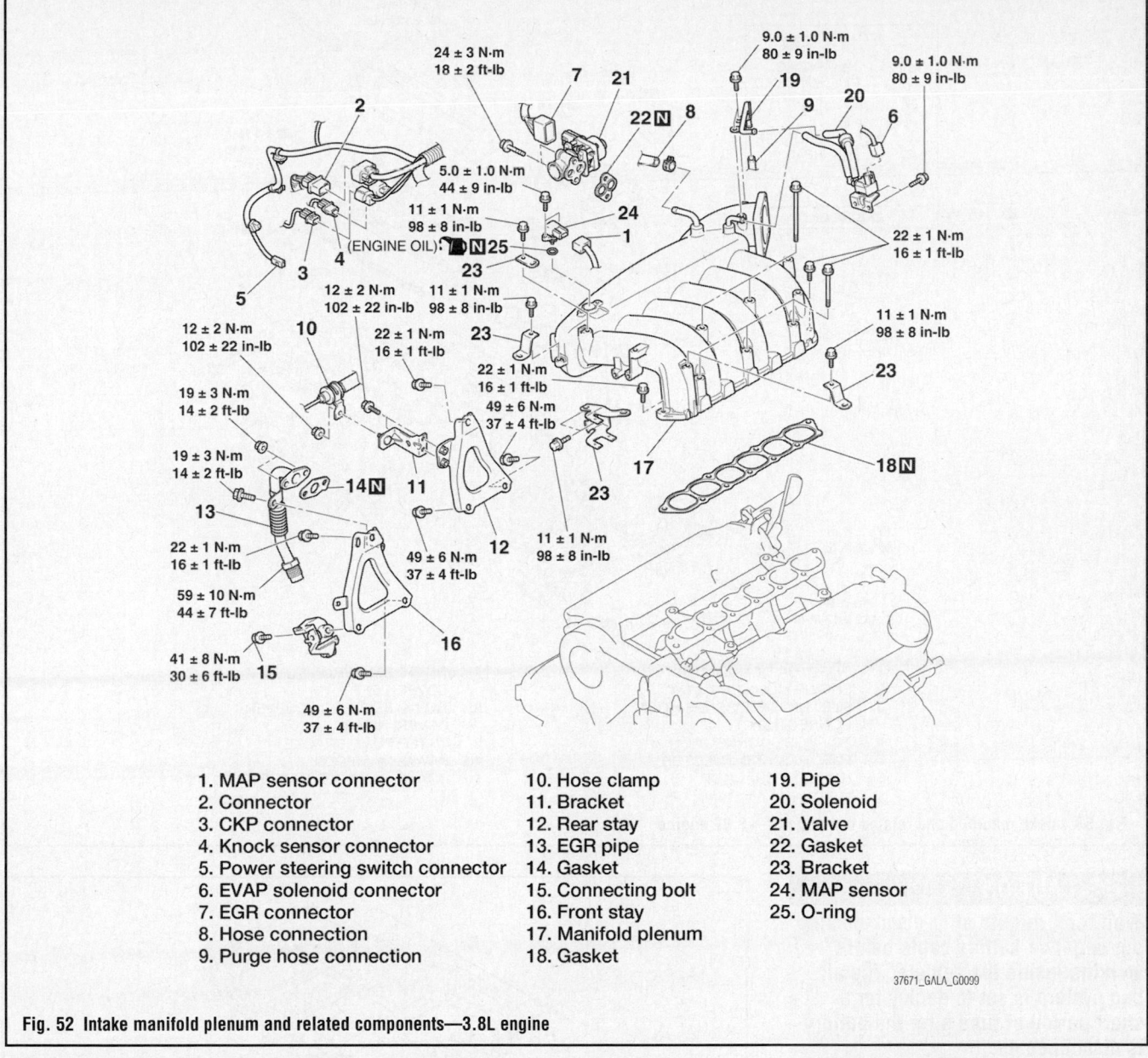

1. MAP sensor connector
2. Connector
3. CKP connector
4. Knock sensor connector
5. Power steering switch connector
6. EVAP solenoid connector
7. EGR connector
8. Hose connection
9. Purge hose connection
10. Hose clamp
11. Bracket
12. Rear stay
13. EGR pipe
14. Gasket
15. Connecting bolt
16. Front stay
17. Manifold plenum
18. Gasket
19. Pipe
20. Solenoid
21. Valve
22. Gasket
23. Bracket
24. MAP sensor
25. O-ring

37671_GALA_C0099

Fig. 52 Intake manifold plenum and related components—3.8L engine

scan tool, be sure that the ignition switch is in the **LOC** position before connecting the scan tool.

16. Turn the ignition switch to the **ON** position.

17. Select "check mode" from the menu screen.

18. Select "erase memory" from the menu screen.

19. Initialize the learning value.

20. After initialization, complete the idle learning procedure.

➡**This procedure must be performed when the PCM is replaced, or when the learning value is initialized, as the idling is not stabilized because the learning value in the MFI engine is not completed.**

21. Start the engine. Allow the coolant temperature to reach 176°F (80°C) or more.

22. Stop the engine and place the ignition switch in the **LOCK** position.

23. After 10 seconds, restart the engine.

24. For 10 minutes, carry out the idling procedure below to confirm that the engine has normal idling.

- Position the transaxle selector lever in the "P" range
- The engine fan is not to be operated
- The engine coolant temperature should be 176°F (80°C) or more

➡If the engine stalls during idling, check the throttle valve of the throttle body for dirt. Correct and perform the procedure again.

OIL PAN

REMOVAL & INSTALLATION

2.4L Engine

See Figures 54 and 55.

1. Before servicing the vehicle, refer to the Precautions Section.

➡**If working near and/or around the SRS system and components, be sure to disable the SRS system. Tape the negative battery cable with insulating tape. Always disconnect the negative battery cable first.**

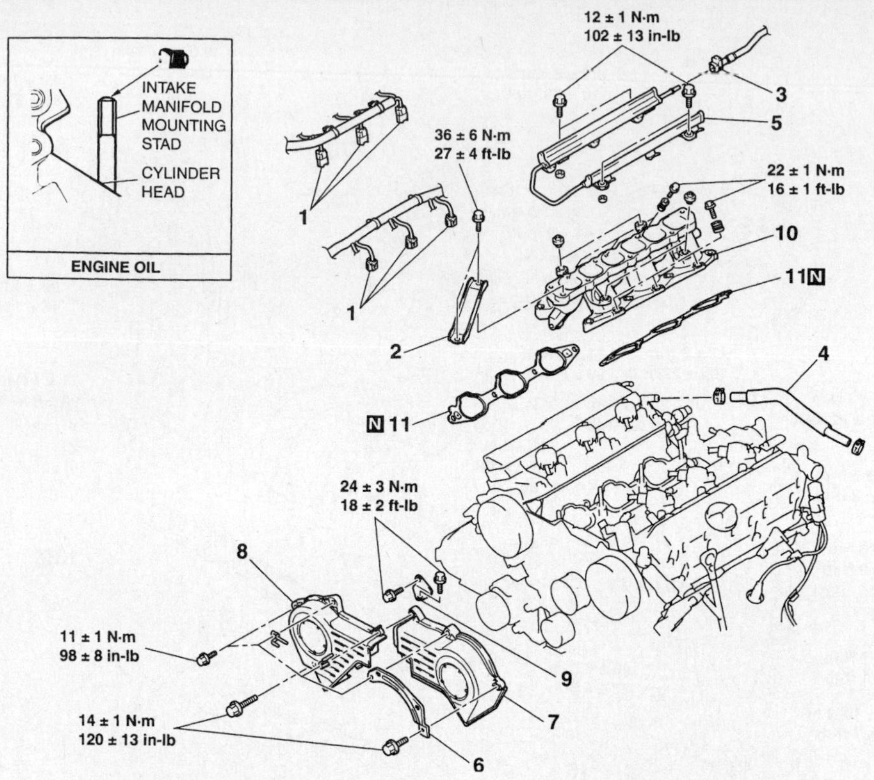

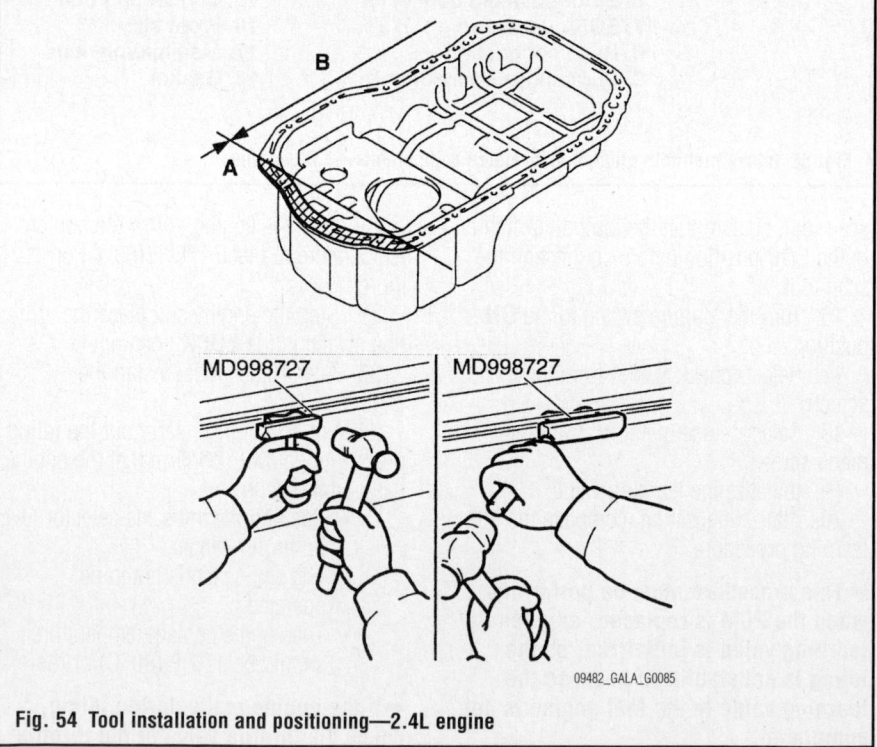

INTAKE MANIFOLD MOUNTING STAD

CYLINDER HEAD

ENGINE OIL

12 ± 1 N·m
102 ± 13 in-lb

36 ± 6 N·m
27 ± 4 ft-lb

22 ± 1 N·m
16 ± 1 ft-lb

24 ± 3 N·m
18 ± 2 ft-lb

11 ± 1 N·m
98 ± 8 in-lb

14 ± 1 N·m
120 ± 13 in-lb

1. INJECTOR CONNECTOR
2. ENGINE MOUNTING STAY
3. FUEL HIGH-PRESSURE HOSE CONNECTION
4. PCV HOSE CONNECTION
5. FUEL RAIL AND INJECTOR
6. HARNESS BRACKET
7. TIMING BELT FRONT UPPER COVER, LEFT
8. TIMING BELT FRONT UPPER COVER, RIGHT
9. WATER PUMP BRACKET
10. INTAKE MANIFOLD
11. INTAKE MANIFOLD GASKET

09482_GALA_G0056

Fig. 53 Intake manifold and related components—3.8L engine

✳✳ CAUTION

Wait for 1 minute after disconnecting the negative battery cable before working inside the vehicle. The air bag system is set to deploy for a short period of time after the battery is disconnected.

2. Remove the negative battery cable.
3. Drain the engine oil.
4. Remove the engine undercover. Remove the front exhaust pipe.
5. Remove the torque converter housing front lower cover.
6. Remove the oil pan retaining bolts. Remove the oil pan. Discard the gasket.

➡Use tool MD998727 to remove the oil pan. Tap the tool into range B between the cylinder block and the engine oil pan, and then slide the tool sideway. Do not position the tool in area A of the engine oil pan as this may cause deformation of the front case because the front case is made of aluminum.

B

A

MD998727   MD998727

09482_GALA_G0085

Fig. 54 Tool installation and positioning—2.4L engine

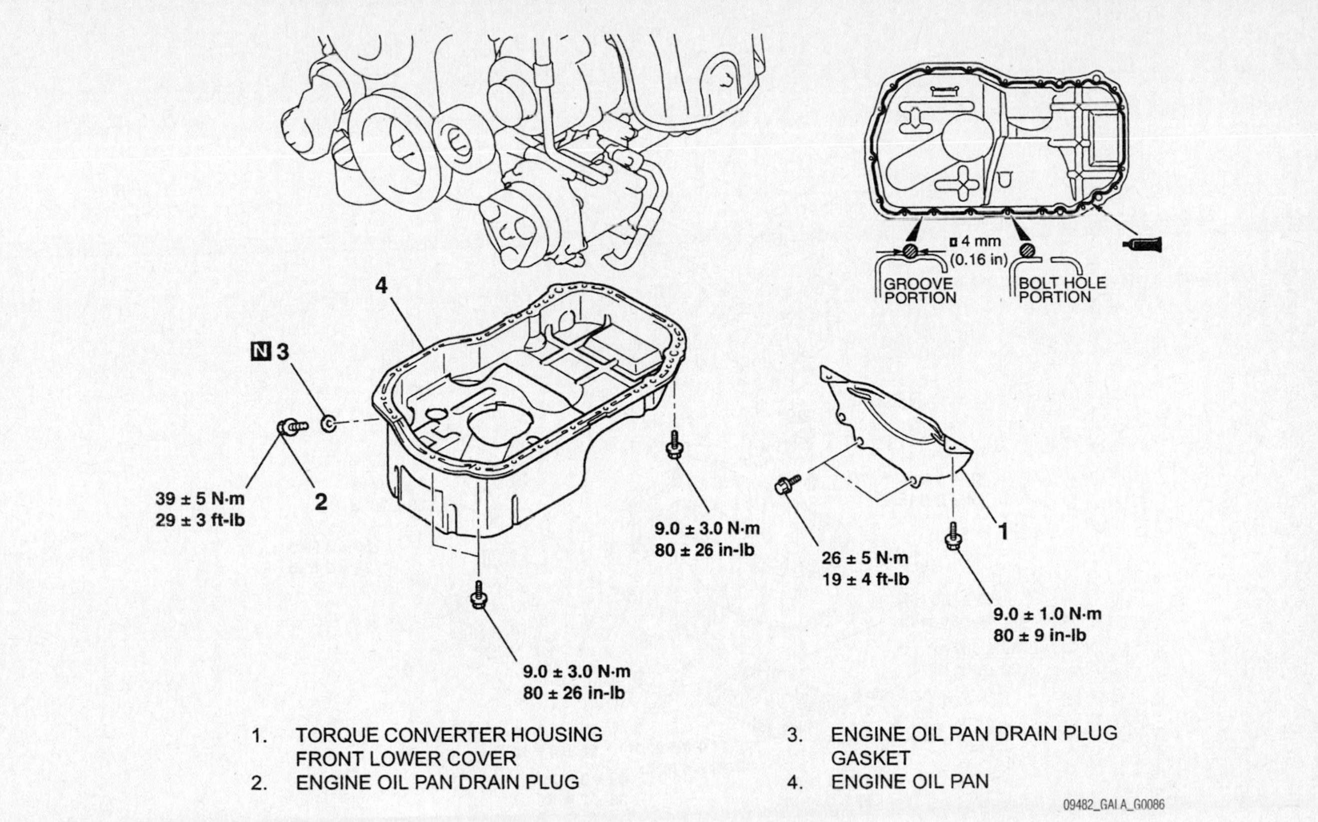

39 ± 5 N·m
29 ± 3 ft-lb

□ 4 mm
(0.16 in)

GROOVE
PORTION

BOLT HOLE
PORTION

9.0 ± 3.0 N·m
80 ± 26 in-lb

26 ± 5 N·m
19 ± 4 ft-lb

9.0 ± 1.0 N·m
80 ± 9 in-lb

9.0 ± 3.0 N·m
80 ± 26 in-lb

1. TORQUE CONVERTER HOUSING
   FRONT LOWER COVER
2. ENGINE OIL PAN DRAIN PLUG

3. ENGINE OIL PAN DRAIN PLUG
   GASKET
4. ENGINE OIL PAN

09482_GAI A_G0086

Fig. 55 Oil pan and related components—2.4L engine

To install:

➡Be sure to use new fasteners, as required.

7. Clean the sealing surface on the oil pan and engine block. Apply a continuous bead of sealant to the oil pan.

8. Install the oil pan. Torque the bolts to specification.

➡Install the pan within 15 minutes after applying sealant.

9. Continue the installation in the reverse order of the removal procedure.

➡Allow the engine to sit for at least one hour before starting. Never start the engine or allow oil or coolant to come in contact with the sealant.

10. Be sure to fill the engine with the proper grade and type engine oil.

11. After the waiting period start the engine and check for leaks. Correct, as required.

3.8L Engine

See Figures 56 through 59.

1. Before servicing the vehicle, refer to the Precautions Section.

➡If working near and/or around the SRS system and components, be sure to disable the SRS system. Tape the negative battery cable with insulating tape. Always disconnect the negative battery cable first.

❋ CAUTION

Wait for 1 minute after disconnecting the negative battery cable before working inside the vehicle. The air bag system is set to deploy for a short period of time after the battery is disconnected.

2. Remove the negative battery cable.

3. Remove the engine undercover. Remove the starter. Remove the engine oil dipstick and guide.

4. If removing the upper oil pan, remove the front exhaust pipe. Remove the cover and connecting bolt.

5. Remove the lower engine oil pan retaining bolts. Remove the engine lower oil pan. Discard the gasket.

6. Remove the upper oil pan retaining bolts. Remove the upper oil pan. Discard the gasket.

To install:

➡Be sure to use new fasteners, as required.

7. Clean the sealing surface on the oil pan and engine block. Apply a continuous bead of sealant to the oil pan.

➡Install the pan within 15 minutes after applying sealant.

8. Install the upper oil pan. Torque the bolts to specification.

➡The bolt holes for bolts 13 and 14 are cut away on the transaxle side. Be careful not to insert these bolts at an angle.

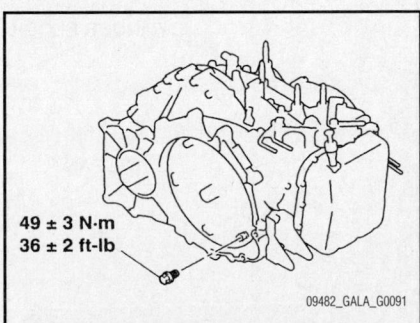

49 ± 3 N·m
36 ± 2 ft-lb

09482_GALA_G0091

Fig. 56 Automatic transaxle torque converter connecting bolt location—3.8L engine

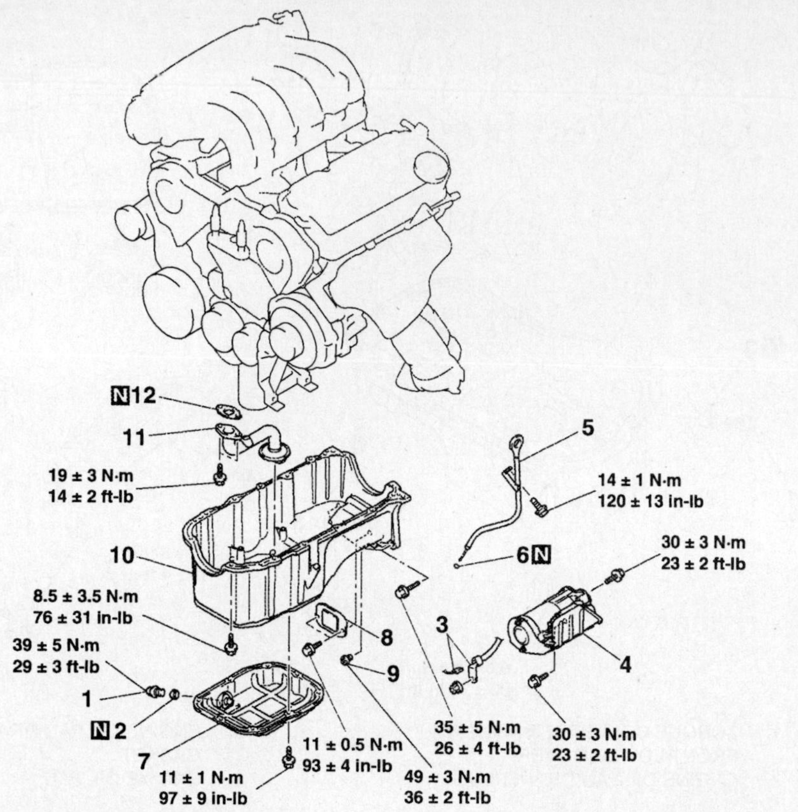

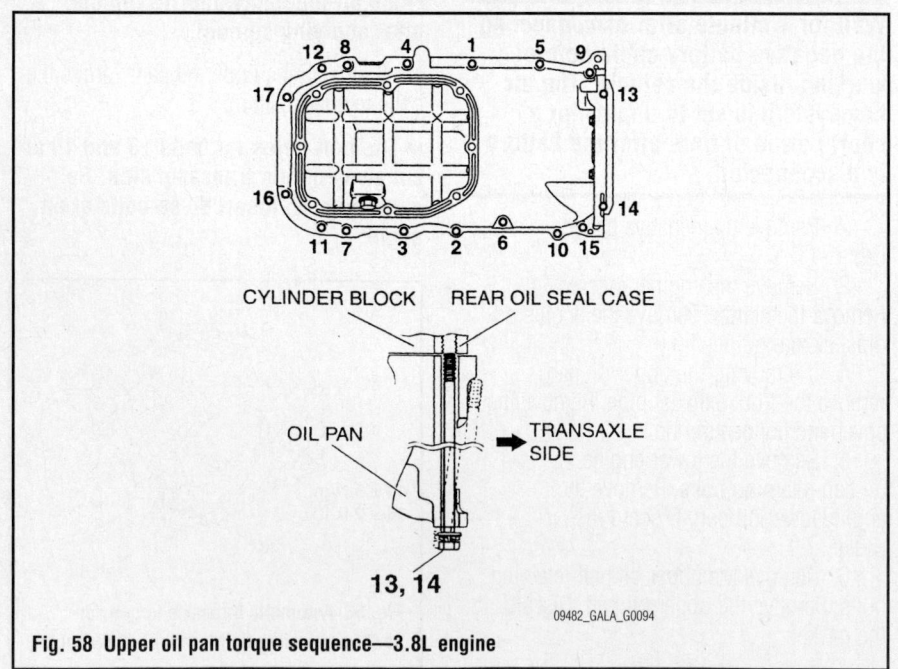

N12

11

19 ± 3 N·m
14 ± 2 ft-lb

10

8.5 ± 3.5 N·m
76 ± 31 in-lb

39 ± 5 N·m
29 ± 3 ft-lb

1

N 2

7

11 ± 1 N·m
97 ± 9 in-lb

11 ± 0.5 N·m
93 ± 4 in-lb

8

9

3

35 ± 5 N·m
26 ± 4 ft-lb

49 ± 3 N·m
36 ± 2 ft-lb

5

14 ± 1 N·m
120 ± 13 in-lb

6 N

30 ± 3 N·m
23 ± 2 ft-lb

4

30 ± 3 N·m
23 ± 2 ft-lb

1. ENGINE OIL PAN DRAIN PLUG
2. ENGINE OIL PAN DRAIN PLUG
   GASKET
3. STARTER CONNECTOR
4. STARTER ASSEMBLY
5. ENGINE OIL DIPSTICK
   ASSEMBLY
6. O-RING
7. ENGINE LOWER OIL PAN

• FRONT NO.1 EXHAUST PIPE
8. COVER
9. TORQUE CONVERTER
   CONNECTING BOLT
10. ENGINE UPPER OIL PAN
11. OIL SCREEN
12. GASKET

09482_GALA_G0093

Fig. 57 Oil pan and related components—3.8L engine

CYLINDER BLOCK   REAR OIL SEAL CASE

OIL PAN

TRANSAXLE
SIDE

13, 14

09482_GALA_G0094

Fig. 58 Upper oil pan torque sequence—3.8L engine

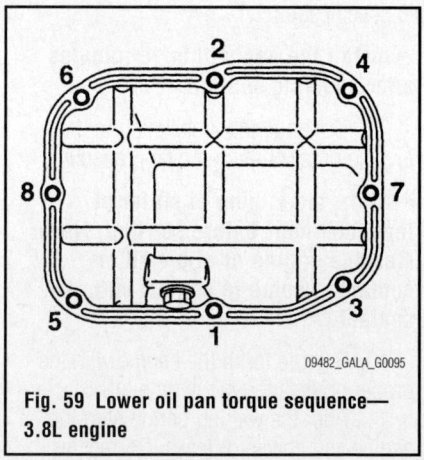

09482_GALA_G0095

**Fig. 59 Lower oil pan torque sequence—
3.8L engine**

➥**Install the pan within 15 minutes
after applying sealant.**

9. Install the lower oil pan. Torque the
bolts to specification.
10. Continue the installation in the
reverse order of the removal procedure.

➡ **Allow the engine to sit for at least one hour before starting. Never start the engine or allow oil or coolant to come in contact with the sealant.**

11. Be sure to fill the engine with the proper grade and type engine oil.

12. After the waiting period start the engine and check for leaks. Correct, as required.

OIL PUMP

REMOVAL & INSTALLATION

2.4L Engine

1. Before servicing the vehicle, refer to the Precautions Section.

➡ **If working near and/or around the SRS system and components, be sure to disable the SRS system. Tape the negative battery cable with insulating tape. Always disconnect the negative battery cable first.**

❋❋ CAUTION

Wait for 1 minute after disconnecting the negative battery cable before working inside the vehicle. The air bag system is set to deploy for a short period of time after the battery is disconnected.

➡ **Whenever the oil pump is disassembled or the cover is removed, the gear cavity must be filled with petroleum jelly to seal the pump and act as a prime. Do not use grease.**

2. Disconnect the negative battery cable. Rotate the engine so number 1 cylinder is on Top Dead Center (TDC) of its compression stroke.

3. Drain the engine oil.

4. Using the proper equipment, support the weight of the engine. Remove the front engine mount bracket and accessory drive belts.

5. Remove or disconnect the following:
- Timing belt upper and lower covers
- Timing belt and crankshaft sprocket
- Electrical connector from the oil pressure sending unit
- Oil pressure sensor
- Oil filter and the oil filter bracket
- Oil pan, oil screen and gasket

6. Using special tool MD998162, remove the plug cap in the engine front cover.

7. Remove or disconnect the following:
- Plug on the side of the engine

block. Insert a steel rod with a shank diameter of 0.32 in. (8mm) into the plug hole. This will hold the silent shaft

- Driven gear bolt that secures the oil pump driven gear to the silent shaft
- Front cover mounting bolts. Note the lengths of the mounting bolts as they are removed for proper installation
- Front case cover and oil pump assembly. If necessary, the silent shaft can come out with the cover assembly
- Oil pump cover, located on the back of the engine front cover. Remove the oil pump drive and driven gears

8. After disassembling the oil pump, clean all components and remove any gasket material from mating surfaces

9. Assemble the oil pump gears into the front case and rotate it to ensure smooth rotation and no looseness. Be sure there is no ridge wear on the contact surface between the front case and the gear surface of the oil pump front cover.

To install:

➡ **Be sure to use new fasteners, as required.**

10. Align the timing mark on the oil pump drive gear with that on the driven gear and install them into the engine front case. Apply engine oil to the gears.

11. Install the oil pump cover and tighten the retainer bolts to 17 ft. lbs. (24 Nm).

12. Using the appropriate driver, install a new crankshaft seal into the front case.

13. Position new front case gasket in place. Set seal guide tool MD998285 on the front end of the crankshaft to protect the seal from damage. Apply a thin coat of oil to the outer circumference of the seal pilot tool.

14. Install the front case assembly through a new front case gasket and temporarily tighten the flange bolts.

15. Mount the oil filter on the bracket with new oil filter bracket gasket in place. Install the bolts with washers and tighten to 14 ft. lbs. (19 Nm).

16. Insert a Phillips screwdriver into the hole in the left side of the engine block to lock the silent shaft in place.

17. Install or connect the following:
- Oil pump drive gear onto the left silent shaft. Tighten the driven gear bolt to 27 ft. lbs. (37 Nm).
- New O-ring to the groove in the front case and install the plug cap.

Tighten the cap to 17 ft. lbs. (24 Nm).
- Oil screen in position with new gasket in place

18. Clean both mating surfaces of the oil pan and the cylinder block. Apply sealant in the groove in the oil pan flange.

➡ **After applying sealant to the oil pan, do not exceed 15 minutes before installing the oil pan.**

19. Install or connect the following:
- Oil pan to the engine and secure with the retainers. Tighten bolts to 60 inch lbs. (7 Nm).
- Oil pressure gauge unit and the oil pressure switch
- Electrical harness connector
- Oil cooler. Oil cooler bolt to 31 ft. lbs. (43 Nm).

20. Refill the crankcase with the correct amount and type of oil. Install new oil filter.

21. Connect the negative battery cable and start the engine.

22. Verify there is correct oil pressure. Inspect for leaks.

3.8L Engine

1. Before servicing the vehicle, refer to the Precautions Section.

➡ **If working near and/or around the SRS system and components, be sure to disable the SRS system. Tape the negative battery cable with insulating tape. Always disconnect the negative battery cable first.**

❋❋ CAUTION

Wait for 1 minute after disconnecting the negative battery cable before working inside the vehicle. The air bag system is set to deploy for a short period of time after the battery is disconnected.

2. Disconnect the negative battery cable.

3. Remove the timing belt.

4. Drain the engine oil.

5. Remove or disconnect the following:
- Splash shield from the wheel well, as needed
- Oil filter adapter
- Lower and upper oil pans
- Lower baffle, oil pump pick-up, and upper baffle
- Oil pump case mounting bolts and the oil pump case
- Oil pump gear cover

6. Make matchmarks on the oil pump rotors before removing them.

7. Remove the crankshaft seal from the oil pump case.

To install:

➡**Be sure to use new fasteners, as required.**

8. Install a new crankshaft seal in the oil pump cover.

9. Apply engine oil to the rotors, then align the matchmarks and install the rotors in the oil pump case.

10. Install the rotor cover and tighten the bolts to 84 inch lbs. (10 Nm).

11. Apply a 0.113 in. (3mm) bead of sealant to the back of the oil pump case. Install the case on the engine and tighten the bolts to 10 ft. lbs. (13 Nm).

12. Install the upper baffle plate and oil pump pick-up using a new gasket.

13. Tighten the baffle bolts to 84 inch lbs. (10 Nm) and the pick-up bolts to 13 ft. lbs. (18 Nm).

14. Lower the baffle in the upper oil pan. Tighten the bolts to 96 inch lbs. (11 Nm).

15. Install the oil pans.

16. Install the oil filter adapter using a new gasket. Tighten the larger bolt to 30 ft. lbs. (41 Nm) and the smaller bolt to 17 ft. lbs. (23 Nm).

17. Install the timing belt and remaining the components from the removal procedure.

18. Fill the engine with the correct amount of oil.

19. Connect the negative battery cable.

20. Start the engine and check for leaks.

PISTON AND RING

POSITIONING

See Figures 60 through 64.

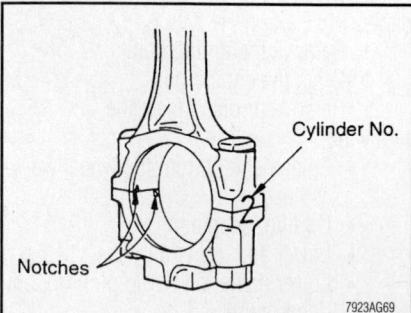

Fig. 60 Before removing the caps from the connecting rods, be sure to matchmark them as shown

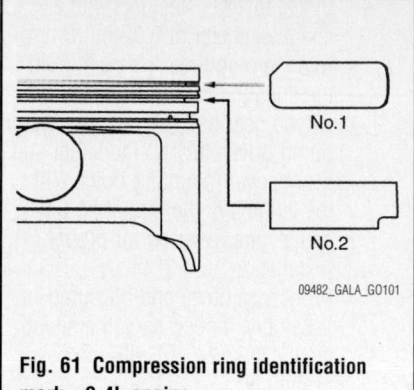

Fig. 61 Compression ring identification mark—2.4L engine

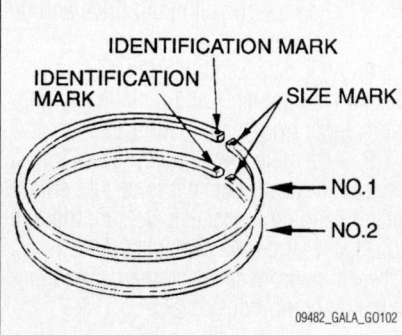

Fig. 62 Compression ring identification mark—3.8L engine

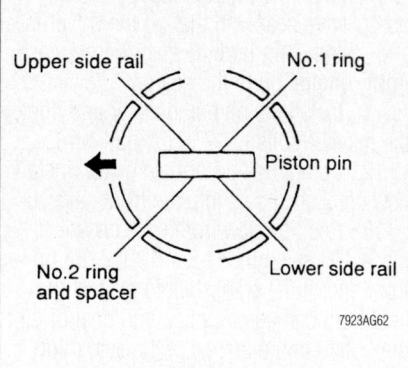

Fig. 63 Piston ring end-gap spacing

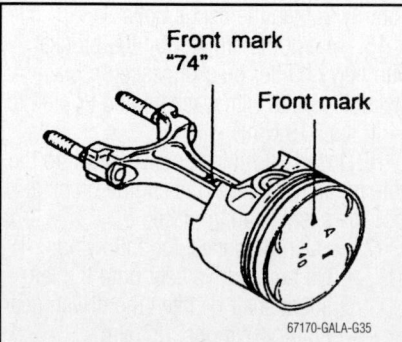

Fig. 64 Piston and connecting rod assembly positioning—3.8L engine

REAR MAIN SEAL

REMOVAL & INSTALLATION

See Figures 65 and 66.

1. Before servicing the vehicle, refer to the Precautions Section.

➡**If working near and/or around the SRS system and components, be sure to disable the SRS system. Tape the negative battery cable with insulating tape. Always disconnect the negative battery cable first.**

✶✶ CAUTION

Wait for 1 minute after disconnecting the negative battery cable before working inside the vehicle. The air bag system is set to deploy for a short period of time after the battery is disconnected.

2. Disconnect the negative battery cable.

3. Remove the transaxle assembly.

4. Remove the driveplate.

5. Remove the crankshaft bushing.

6. Remove the rear main seal. Be careful not to damage the oil seal bore or the crankshaft sealing surface.

To install:

➡**Be sure to use new fasteners, as required.**

7. Apply clean engine oil to the oil seal lip. Using a seal driver, install the oil seal.

8. Use special tools MB990938 and MD998776 to press-fit the oil seal.

9. Installation continues in the reverse of the removal procedure.

ROCKER ARMS

REMOVAL & INSTALLATION

2.4L Engine

See Figures 67 and 68.

1. Before servicing the vehicle, refer to the Precautions Section.

➡**If working near and/or around the SRS system and components, be sure to disable the SRS system. Tape the negative battery cable with insulating tape. Always disconnect the negative battery cable first.**

✶✶ CAUTION

Wait for 1 minute after disconnecting the negative battery cable before working inside the vehicle. The air

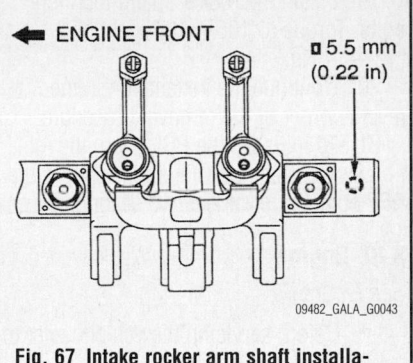

Fig. 67 Intake rocker arm shaft installation locating point—2.4L engine

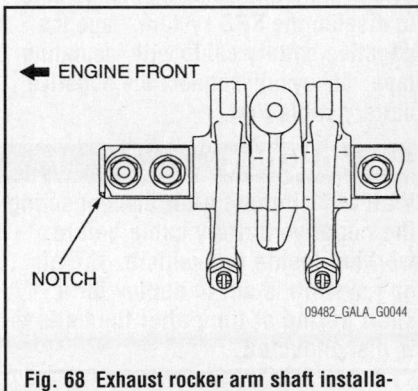

Fig. 68 Exhaust rocker arm shaft installation locating point—2.4L engine

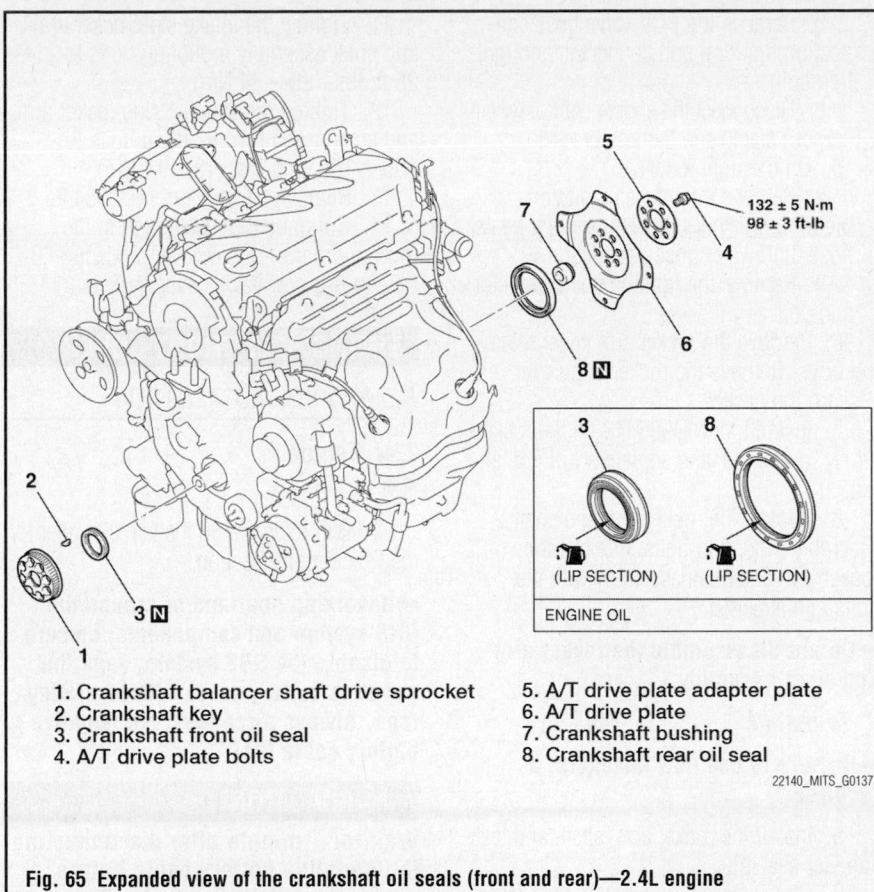

1. Crankshaft balancer shaft drive sprocket
2. Crankshaft key
3. Crankshaft front oil seal
4. A/T drive plate bolts
5. A/T drive plate adapter plate
6. A/T drive plate
7. Crankshaft bushing
8. Crankshaft rear oil seal

Fig. 65 Expanded view of the crankshaft oil seals (front and rear)—2.4L engine

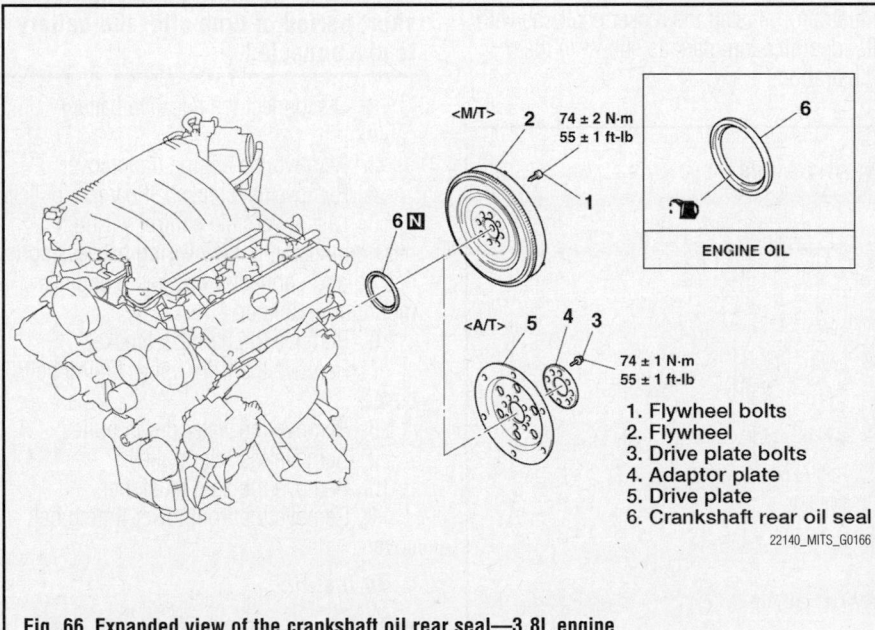

1. Flywheel bolts
2. Flywheel
3. Drive plate bolts
4. Adaptor plate
5. Drive plate
6. Crankshaft rear oil seal

Fig. 66 Expanded view of the crankshaft oil rear seal—3.8L engine

bag system is set to deploy for a short period of time after the battery is disconnected.

2. Disconnect the negative battery cable.
3. Remove the PCM.
4. Remove the air cleaner assembly.
5. Remove the battery.
6. Remove the ignition coils.
7. Remove the timing belt upper cover.
8. Disconnect the control wiring harness connection.
9. Remove the rocker cover PCV connection.

10. Remove the rocker cover breather hose connection.
11. Remove the engine oil control valve and O-ring. Remove the oil pressure switch.
12. Remove the rocker cover retaining bolts. Remove the rocker cover. Discard the gasket.
13. Remove the spark plug guide oil seals. Remove the accumulator assembly.
14. Remove the exhaust rocker arm shaft caps. Remove the exhaust rocker arm shaft assembly.
15. Remove the intake rocker arm shaft caps. Remove the intake rocker arm shaft assembly.

➥**Do not disassemble the exhaust or intake rocker arm assembly.**

To install:

➥**Be sure to use new fasteners, as required.**

16. Position the intake rocker arm assembly so that its 0.22 inch hole faces toward the cylinder head.
17. Install the rocker shaft mounting bolts. Torque to 21—25 ft. lbs. (29—34 Nm).
18. Position the exhaust rocker arm assembly so that its notch is positioned as shown in the illustration.

19. Install the rocker shaft mounting bolts. Torque to 106—124 inch lbs. (12—14 Nm).

20. Continue the installation in the reverse order of the removal procedure.

21. To initialize the PCM, turn the ignition switch ON then OFF and keep it in the OFF position for at least 10 seconds.

3.8L Engine

See Figure 69.

1. Before servicing the vehicle, refer to the Precautions Section.

➡**If working near and/or around the SRS system and components, be sure to disable the SRS system. Tape the negative battery cable with insulating tape. Always disconnect the negative battery cable first.**

❋❋ CAUTION

Wait for 1 minute after disconnecting the negative battery cable before working inside the vehicle. The air bag system is set to deploy for a short period of time after the battery is disconnected.

2. Disconnect the negative battery cable.

3. Drain the cooling system.

4. On the left side:

   a. Remove the thermostat housing assembly.

b. Remove the PCV valve hose connection, ignition coil connector, and ignition coil.

   c. Disconnect the engine control wiring harness clamp and harness bracket.

5. On the right side:

   a. Remove the intake manifold plenum, breather hose connection, heater hose, and water hose connection.

   b. Remove the ignition coil connector and coil.

6. Remove the rocker arm cover retaining bolts. Remove the rocker arm cover. Discard the gasket.

7. Install tool MD998443 on the rocker arms, so that the lash adjusters will not fall out.

8. Loosen the rocker arm and shaft assembly mounting bolts. Remove the rocker arm and shaft assembly with the bolts still attached.

➡**Do not disassemble the rocker arm and shaft assembly.**

To install:

➡**Be sure to use new fasteners, as required.**

9. Install the rocker arm, shaft, and lash adjuster assembly.

10. Check that the notches in the rocker shaft are facing the direction, shown in the illustration. Install the rocker shaft cap with its identification mark as shown in the illustration.

11. Tighten the intake side rocker arm and shaft assembly mounting bolts to 21—25 ft. lbs. (28—34 Nm).

12. Tighten the exhaust side rocker arm and shaft assembly mounting bolts to 106—124 inch lbs. (12—14 Nm).

13. Remove special tool MD998443.

14. Continue the installation in the reverse order of the removal procedure. Be sure to use new valve cover gaskets.

TIMING BELT FRONT COVER

REMOVAL & INSTALLATION

2.4L Engine

See Figure 70.

1. Before servicing the vehicle, refer to the Precautions Section.

➡**If working near and/or around the SRS system and components, be sure to disable the SRS system. Tape the negative battery cable with insulating tape. Always disconnect the negative battery cable first.**

❋❋ CAUTION

Wait for 1 minute after disconnecting the negative battery cable before working inside the vehicle. The air bag system is set to deploy for a short period of time after the battery is disconnected.

2. Disconnect the negative battery cable.

3. Remove the engine undercover.

4. Remove the crankshaft damper pulley.

5. Disconnect the control wiring harness connector, battery wiring harness connector, and connector bracket to engine mounting insulator.

6. Remove the harness bracket.

7. Remove the front upper timing belt cover.

8. Remove the water pump pulley.

9. Remove the idler pulley.

10. Remove the auto tensioner.

11. Remove the front lower timing belt cover.

To install:

➡**Be sure to use new fasteners, as required.**

12. Install the front lower timing belt cover.

13. Install the auto tensioner.

14. Install the idler pulley.

15. Install the water pump pulley.

16. Install the front upper timing belt cover.

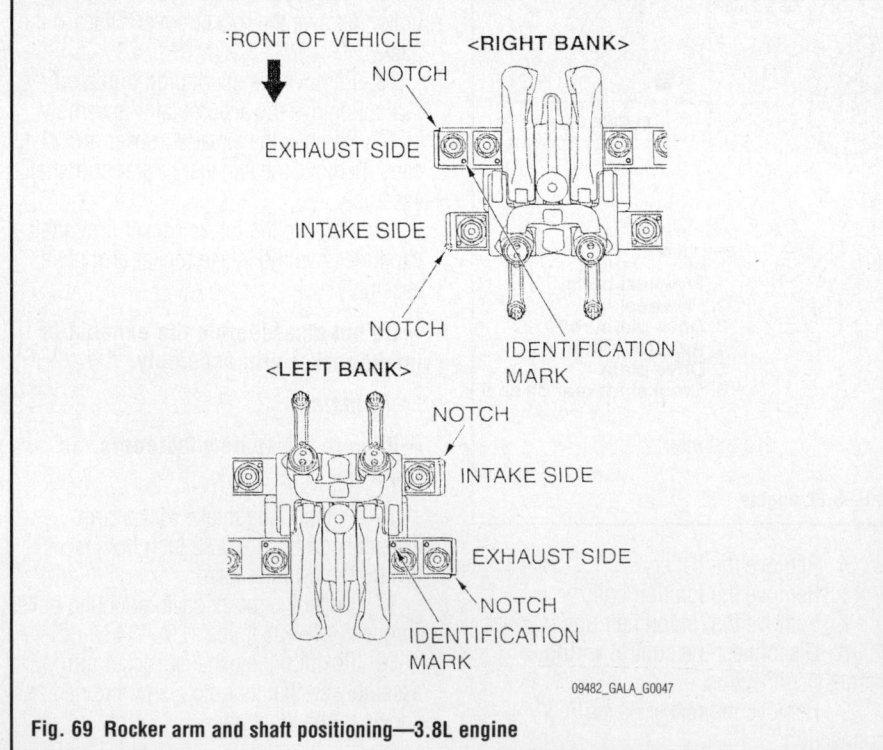

Fig. 69 Rocker arm and shaft positioning—3.8L engine

09482_GALA_G0047

(In figure: FRONT OF VEHICLE, <RIGHT BANK>, NOTCH, EXHAUST SIDE, INTAKE SIDE, NOTCH, IDENTIFICATION MARK, <LEFT BANK>, NOTCH, INTAKE SIDE, EXHAUST SIDE, NOTCH, IDENTIFICATION MARK)

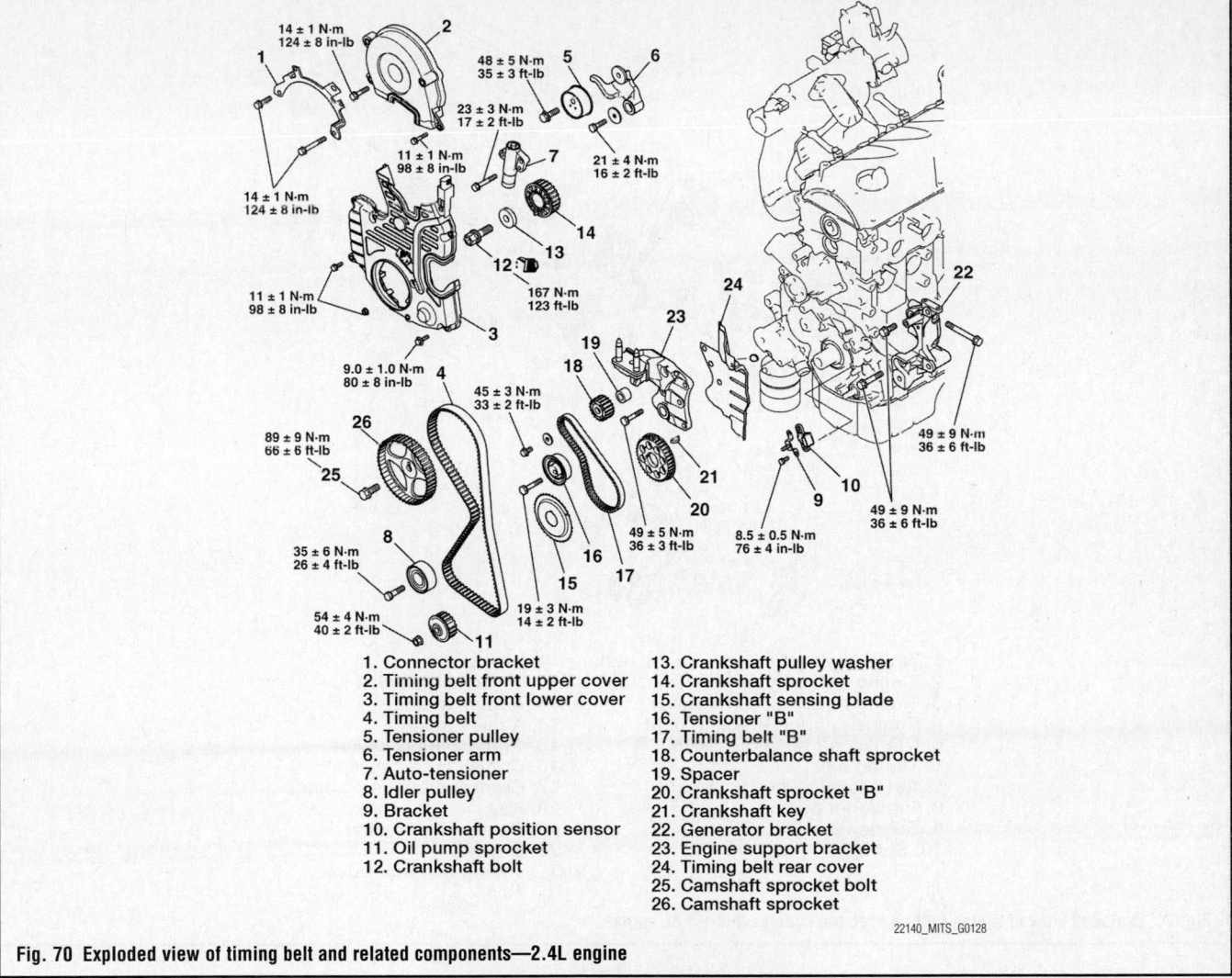

1. Connector bracket
2. Timing belt front upper cover
3. Timing belt front lower cover
4. Timing belt
5. Tensioner pulley
6. Tensioner arm
7. Auto-tensioner
8. Idler pulley
9. Bracket
10. Crankshaft position sensor
11. Oil pump sprocket
12. Crankshaft bolt
13. Crankshaft pulley washer
14. Crankshaft sprocket
15. Crankshaft sensing blade
16. Tensioner "B"
17. Timing belt "B"
18. Counterbalance shaft sprocket
19. Spacer
20. Crankshaft sprocket "B"
21. Crankshaft key
22. Generator bracket
23. Engine support bracket
24. Timing belt rear cover
25. Camshaft sprocket bolt
26. Camshaft sprocket

22140_MITS_G0128

Fig. 70 Exploded view of timing belt and related components—2.4L engine

17. Tighten all bolts to specification as illustrated.

18. Install the harness bracket.

19. Connect the control wiring harness connector, battery wiring harness connector, and connector bracket to engine mounting insulator.

20. Install the crankshaft damper pulley.

21. Install the engine undercover.

22. Connect the negative battery cable.

3.8L Engine

See Figure 71.

1. Before servicing the vehicle, refer to the Precautions Section.

➡ **If working near and/or around the SRS system and components, be sure to disable the SRS system. Tape the negative battery cable with insulating tape. Always disconnect the negative battery cable first.**

✳✳ **CAUTION**

Wait for 1 minute after disconnecting the negative battery cable before working inside the vehicle. The air bag system is set to deploy for a short period of time after the battery is disconnected.

2. Disconnect the negative battery cable.

3. Remove the engine undercover.

4. Remove the engine room cover.

5. Remove the engine side cover.

6. Remove the alternator.

7. Remove the power steering fluid pump drive belt.

8. Remove the crankshaft damper pulley. Refer to Crankshaft Damper, Removal & Installation.

9. Disconnect the control wiring harness and injector wiring connector.

10. Disconnect the knock sensor connector and crankshaft position sensor connector.

11. Remove the connector bracket. Disconnect the oil pressure switch connector and engine oil control valve connector.

12. Remove the engine mounting stay and the connector bracket.

13. Remove the front right upper timing belt cover.

14. Remove the front left upper timing belt cover.

15. Remove the tensioner pulley and tensioner bracket.

16. Remove the crankshaft position sensor harness clamp.

17. Remove the front lower timing belt cover.

To install:

➡ **Be sure to use new fasteners, as required.**

18. Install the front lower timing belt cover.

19. Install the crankshaft position sensor harness clamp.

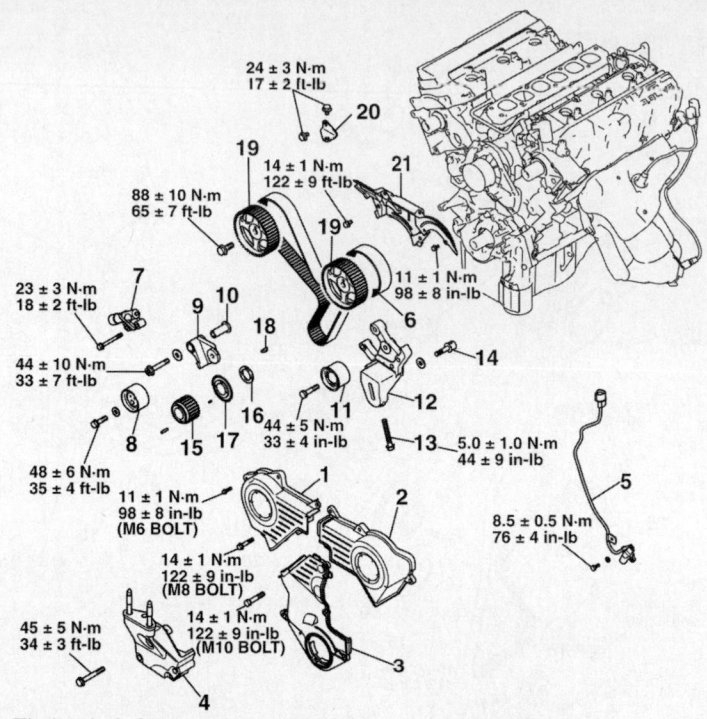

24 ± 3 N·m
17 ± 2 ft-lb

14 ± 1 N·m
122 ± 9 ft-lb

88 ± 10 N·m
65 ± 7 ft-lb

23 ± 3 N·m
18 ± 2 ft-lb

44 ± 10 N·m
33 ± 7 ft-lb

11 ± 1 N·m
98 ± 8 in-lb

44 ± 5 N·m
33 ± 4 in-lb

5.0 ± 1.0 N·m
44 ± 9 in-lb

8.5 ± 0.5 N·m
76 ± 4 in-lb

48 ± 6 N·m
35 ± 4 ft-lb

11 ± 1 N·m
98 ± 8 in-lb
(M6 BOLT)

14 ± 1 N·m
122 ± 9 in-lb
(M8 BOLT)

14 ± 1 N·m
122 ± 9 in-lb
(M10 BOLT)

45 ± 5 N·m
34 ± 3 ft-lb

1. Timing belt front upper cover, right
2. Timing belt front upper cover, left
3. Timing belt front lower cover
4. Engine support bracket, right
5. Crankshaft position sensor
6. Timing belt
7. Auto-tensioner
8. Tensioner pulley
9. Tensioner arm
10. Shaft
11. Idler pulley
12. Tensioner bracket
13. Adjusting bolt
14. Adjusting stud
15. Crankshaft sprocket
16. Crankshaft spacer
17. Crankshaft sensing blade
18. Key
19. Camshaft sprocket
20. Bracket
21. Timing belt rear cover

22140_MITS_G0129

Fig. 71 Exploded view of timing belt and related components—3.8L engine

20. Install the tensioner pulley and tensioner bracket.

21. Install the front left upper timing belt cover.

22. Install the front right upper timing belt cover.

23. Install the engine mounting stay and the connector bracket.

24. Tighten all bolts to specification as illustrated.

25. Install the connector bracket. Connect the oil pressure switch connector and engine oil control valve connector.

26. Connect the knock sensor connector and crankshaft position sensor connector.

27. Connect the control wiring harness and injector wiring connector.

28. Install the crankshaft damper pulley. Refer to Crankshaft Damper, Removal & Installation.

29. Install the power steering fluid pump drive belt.

30. Install the alternator.

31. Install the engine side cover.

32. Install the engine room cover.

33. Install the engine undercover.

34. Connect the negative battery cable.

TIMING BELT & SPROCKETS

REMOVAL & INSTALLATION

2.4L Engine

See Figures 72 through 75.

1. Before servicing the vehicle, refer to the Precautions Section.

➡️If working near and/or around the SRS system and components, be sure to disable the SRS system. Tape the negative battery cable with insulating tape. Always disconnect the negative battery cable first.

✳️ CAUTION

Wait for 1 minute after disconnecting the negative battery cable before working inside the vehicle. The air bag system is set to deploy for a short period of time after the battery is disconnected.

2. Disconnect the negative battery cable.

3. Remove the timing belt front cover(s).

4. Turn the crankshaft clockwise and align each timing mark to set the number 1 piston to Top Dead Center (TDC) of its compression stroke.

5. Remove the timing belt undercover rubber plug and then install special tool MD998738. Screw the special tool until it contacts the timing belt tensioner arm.

✳️ WARNING

The special tool must be screwed in gradually at the rate of a 30° turn per second. If it is turned in all at once, the timing belt tensioner adjuster rod will not easily retract and the tool may bend.

6. Gradually screw in the special tool and then align the timing belt tensioner

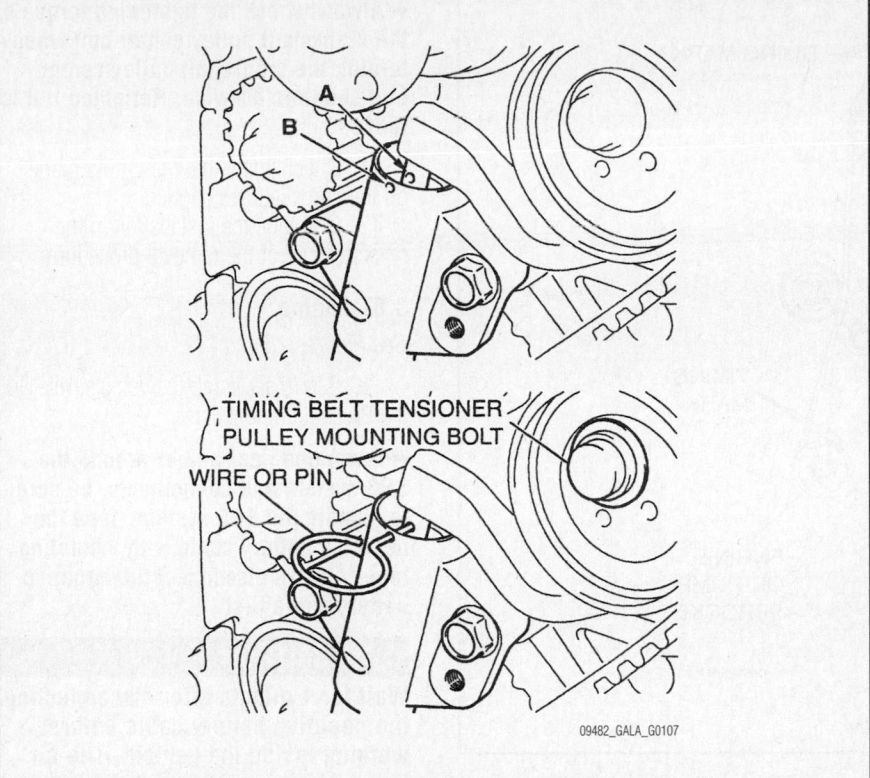

Fig. 72 Timing belt tensioner adjuster rod set hole "A" and cylinder set hole "B"—2.4L engine

adjuster rod set hole "A" with the timing belt tensioner adjuster cylinder set hole "B".

7. Insert a wire or pin in the set holes to lock the assembly in place. After removing the special tool, loosen the timing belt tensioner pulley mounting bolts and remove the timing belt.

➡**If the belt is being reused be sure to mark the direction of rotation (clockwise) on the belt.**

8. Remove the timing belt tensioner pulley, tensioner arm, and adjuster.
9. Remove the timing belt idler pulley.
10. Remove the timing belt lower cover bracket.
11. Remove the crankshaft position sensor.
12. Remove the crankshaft pulley center bolt and washer. Remove the drive sprocket. Remove the crankshaft angle sensing blade.
13. Remove the balancer timing belt tensioner.
14. Remove the timing belt "B" (balancer belt) from its mounting. Remove the sprocket.

To install:

➡**Be sure to use new fasteners, as required.**

15. Be sure that the crankshaft balancer shaft drive sprocket timing marks and·

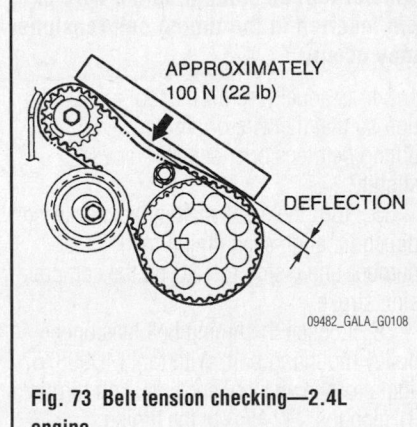

Fig. 73 Belt tension checking—2.4L engine

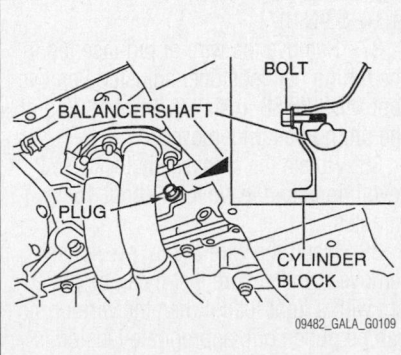

Fig. 74 Cylinder block plug location—2.4L engine

balancer shaft sprocket timing marks are aligned.

16. Install the timing belt "B" (balancer belt) on the crankshaft balancer drive sprocket and balancer shaft sprocket. There should be no slack on the tension side.

17. Assemble an temporarily fix the center of the pulley of the balancer timing belt tensioner so that it is at the top left from the center of the assembling bolt, and the pulley flange is at the front side of the engine.

➡**When tightening the mounting bolts, ensure that the tensioner does not rotate with the bolts. Allowing it to rotate can cause excessive tension of the belt.**

18. Lift your fingers and the balancer timing belt tensioner in the counterclockwise position. Apply minimal torque to the balancer timing belt so the belt is tense without looseness. Tighten the assembling bolt to 12—16 ft. lbs. (16—22 Nm).

19. Turn the crankshaft clockwise 2 turns to set the number 1 piston to TDC on the compression stroke. Check that the sprocket timing marks are aligned.

20. Apply slight pressure at the center of the belt between both sprockets. Inspect whether the belt deflection is within specification 0.31–0.47 inch (8–12mm). Correct as required.

21. Install the crankshaft angle sensing blade, crankshaft drive sprocket, pulley washer and pulley center bolt. Apply clean engine oil to the retaining bolt prior to installation. Tighten the assembling bolt to 123 ft. lbs. (167 Nm).

22. Slowly compress the timing belt tensioner adjuster rod using a press or vise. Align the set hole "A" of the rod with set hole "B" of the timing belt tensioner adjuster cylinder.

➡**Do not compress the assembly too fast as damage to the rod may occur.**

23. Insert a wire or pin in the aligned set holes to lock the assembly in place.
24. Install the timing belt tensioner adjuster to the engine and tighten the mounting bolt to 15—19 ft. lbs. (20—26 Nm).
25. Temporarily tighten the timing belt tensioner pulley. Align the timing marks on the camshaft sprocket, crankshaft drive sprocket and engine oil pump sprocket.
26. Adjust the timing mark of the engine oil pump sprocket, by removing the cylinder block plug. Insert a bolt (M6, section width 10MM, nominal length 45MM) from the plug hole.

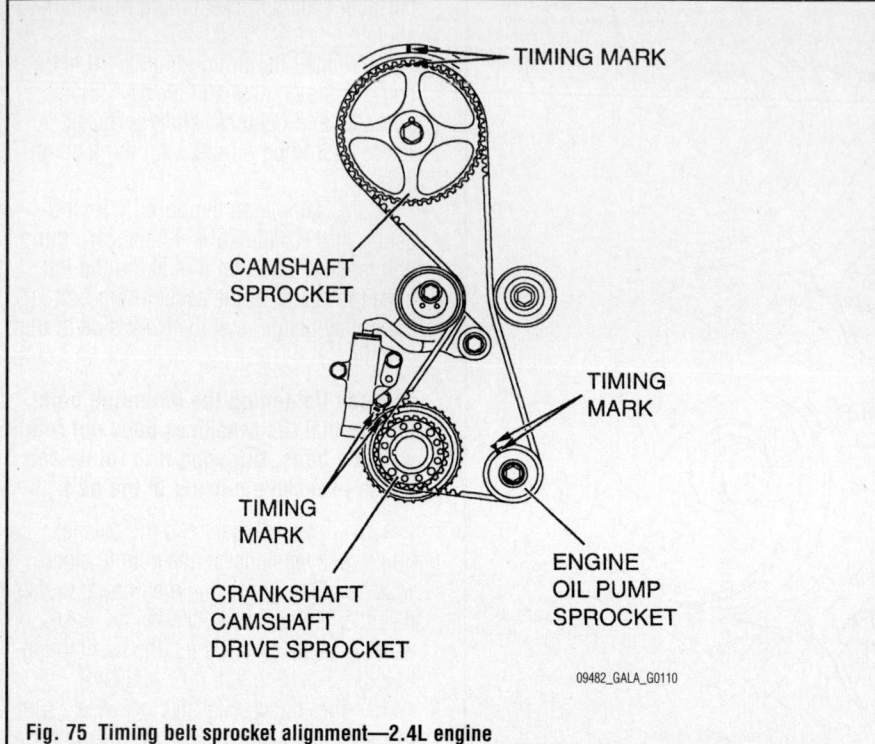

TIMING MARK

CAMSHAFT
SPROCKET

TIMING
MARK

TIMING
MARK

CRANKSHAFT
CAMSHAFT
DRIVE SPROCKET

ENGINE
OIL PUMP
SPROCKET

09482_GALA_G0110

Fig. 75 Timing belt sprocket alignment—2.4L engine

➡ If the bolt comes in contact with the balancer shaft, turn the engine oil sprocket one rotation. Re-adjust the timing mark and check to see that the bolt fits. Do not remove the bolt until the valve timing belt is assembled.

27. Position the timing belt on the timing belt tensioner pulley and crankshaft driver sprocket, support it with your hand so that it does not slide.

28. Position the belt on the engine oil pump sprocket while pulling it with your other hand.

29. Position the timing belt on the timing belt idler pulley.

➡ Incorporate the timing belt. Then apply reverse rotation (counterclockwise) pressure to the camshaft sprocket. Recheck to see that each timing mark is aligned while the tension side of the belt is right.

30. Position the timing mark on the camshaft sprocket.

31. Turn the timing belt tensioner pulley upward using tool MD998767 to apply tension to the belt. Temporarily tighten and fix the belt tensioner pulley mounting bolt.

32. Check that the timing marks are aligned.

33. Remove the bolt that was inserted in the cylinder block plug hole. Replace the cylinder block plug hole bolt and torque to 21—25 ft. lbs. (27—33 Nm).

➡ Install tool MD998738 by hand. Do not use any other tools to install the special tool as damage to the wire or pin inserted in the timing belt tensioner may occur.

34. Gradually screw the tool into position so that the wire or pin inserted in the timing belt tensioner adjuster moves slightly.

35. Turn the crankshaft in the clockwise direction; align each timing mark to set number one piston to TDC on the compression stroke.

36. Loosen the timing belt tensioner pulley mounting bolt. With tool MD998767 and a torque wrench, apply tension torque 31 inch lbs. (42 Nm) to the timing belt. Tighten the timing belt tensioner pulley mounting bolt to 33—39 ft. lbs. (43—53 Nm).

37. Remove the wire or pin inserted in the timing belt tensioner adjuster. Remove tool MD998738. Install the rubber plug of the timing belt undercover.

38. Rotate the crankshaft clockwise 2 revolutions and leave it for about 15 minutes.

39. Insert the wire or pin, previously removed, and ensure that it can be pulled out with a light load. When the wire or pin can be pulled out, appropriate tension is applied on the timing belt.

40. If the wire or pin cannot be pulled out easily, repeat the above step until it can.

➡ Always check the tightening torque of the crankshaft pulley center bolt when turning the crankshaft pulley center bolt counterclockwise. Retighten if it is loose.

41. Check again that the timing marks on the sprockets are aligned.

42. Continue the installation in the reverse order of the removal procedure.

3.8L Engine

See Figures 76 and 77.

1. Before servicing the vehicle, refer to the Precautions Section.

➡ If working near and/or around the SRS system and components, be sure to disable the SRS system. Tape the negative battery cable with insulating tape. Always disconnect the negative battery cable first.

✳✳ CAUTION

Wait for 1 minute after disconnecting the negative battery cable before working inside the vehicle. The air bag system is set to deploy for a short period of time after the battery is disconnected.

2. Disconnect the negative battery cable.
3. Remove the timing belt front cover(s).
4. Remove the engine support bracket.
5. Turn the crankshaft clockwise to align each timing mark and set the number 1 cylinder to Top Dead Center (TDC) on the compression stroke.

➡ If the belt is going to be reused, mark the direction of rotation.

6. Remove the timing belt.
7. Remove the auto tensioner, tensioner pulley, and tensioner arm.

✳✳ WARNING

Water or oil on the belt shortens its life drastically, so the removed timing belt, sprocket, and tensioner must be kept free from oil and water. These parts should not be washed or immersed in solvent. Replace parts if contaminated. If there is oil or water on any part, check the front case oil seal, camshaft oil seal, and water pump for leaks.

8. While holding the camshaft sprocket with special tools MB990767 and MD998715, loosen the camshaft sprocket bolt. Remove the camshaft sprocket.
9. Remove the crankshaft sprocket.

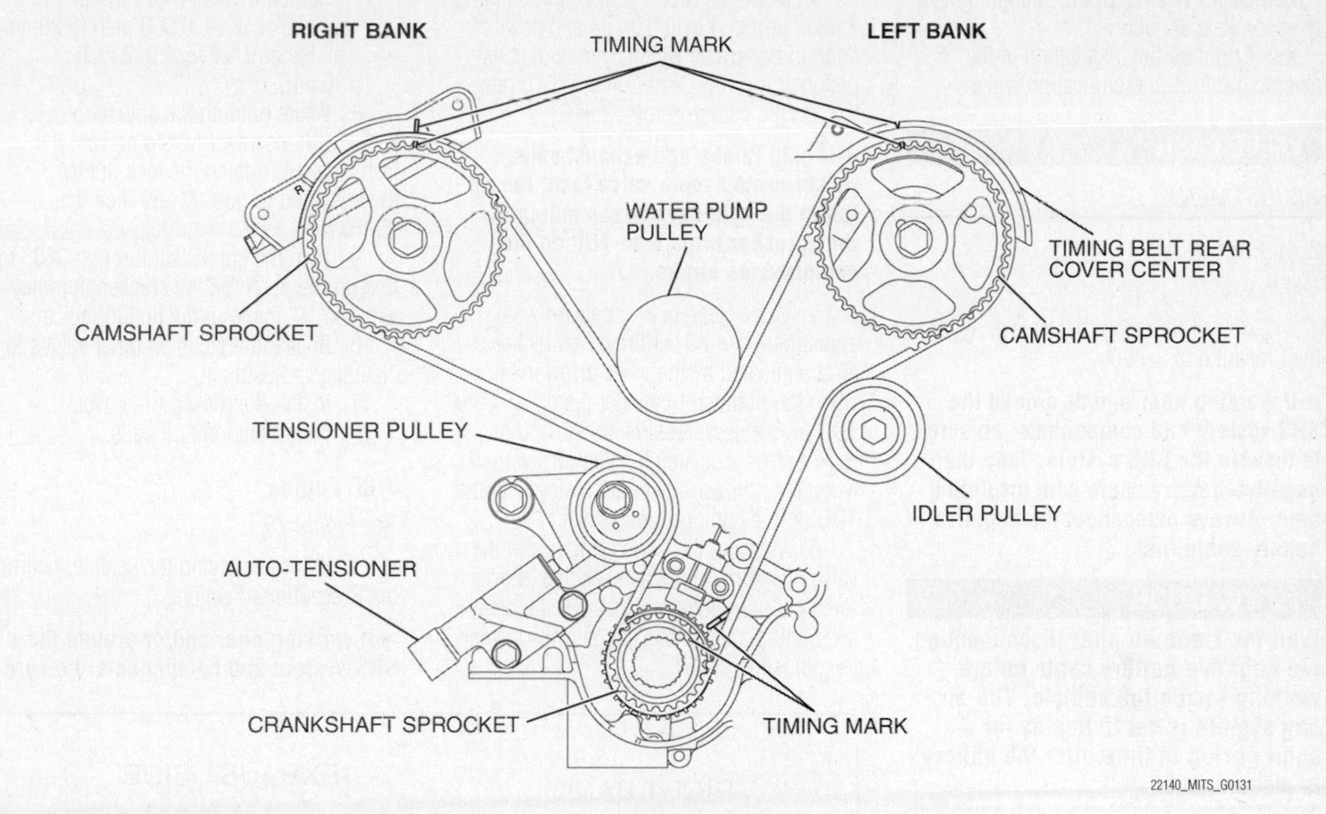

RIGHT BANK   TIMING MARK   LEFT BANK

WATER PUMP PULLEY

TIMING BELT REAR COVER CENTER

CAMSHAFT SPROCKET

CAMSHAFT SPROCKET

TENSIONER PULLEY

IDLER PULLEY

AUTO-TENSIONER

CRANKSHAFT SPROCKET

TIMING MARK

22140_MITS_G0131

Fig. 76 Check to see that the timing marks of all the sprockets are in alignment—3.8L engine

To install:

➡**Be sure to use new fasteners, as required.**

10. Install the crankshaft sprocket. Ensure the keyway and key are in position.

11. Install the camshaft sprocket. Tighten the bolt to 58–72 ft. lbs. (78–98 Nm).

12. Align the timing marks on the camshaft sprockets with those on the rocker cover and the timing mark on the crankshaft sprocket with that on the engine block.

✷✷ CAUTION

The right side camshaft sprocket can turn easily due to the spring force applied. Be careful of hands and fingers when turning the camshaft as they could get caught in an uncomfortable position.

13. Install the timing belt first on the crankshaft sprocket, than on the idler pulley, then on the left camshaft sprocket, then on the water pump pulley, then on the right camshaft sprocket, and finally on the tensioner pulley.

14. Turn the right camshaft sprocket counterclockwise until the tension side of the timing belt is firmly stretched. Check all timing marks, again.

15. Use special tool MD998767 to push the tensioner pulley into the timing belt, and then temporarily tighten the center bolt.

16. Use special tool MD998769 to turn the crankshaft ¼ turn counterclockwise, then turn it again clockwise until the timing marks are aligned.

17. Holding the auto tensioner, with your hand, press the end of the pushrod against a metal surface with a force of about 32 lbs. Measure how far in the push rod is

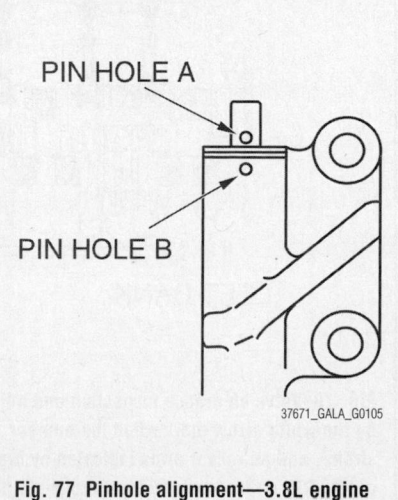

PIN HOLE A

PIN HOLE B

37671_GALA_G0105

Fig. 77 Pinhole alignment—3.8L engine

pushed. Specification should be 0.189–0.236 inch (4.8–6.0mm). If not within specification, replace the auto tensioner assembly.

18. Position the auto tensioner perpendicular in a vise. If the tensioner has a plug at the base, be sure to use a washer to protect the plug.

19. Slowly compress the pushrod of the auto tensioner until the pin hole "A" is aligned the pin hole "B" in the cylinder. Insert a pin into both holes once they are aligned. Install the auto tensioner to the engine.

➡**When tightening the center bolt, be careful that the tensioner pulley does not turn with the bolt.**

20. Loosen the center bolt of the tensioner pulley. Use tool MD998767 and a torque wrench and apply tension torque to the timing belt in a downward motion. Tighten the center bolt to 32—40 ft. lbs. (42—54 Nm).

21. Remove the tensioner set pin. Turn the crankshaft clockwise twice to align the timing marks. Wait at least 5 minutes, then check that the auto tensioner pushrod extends within the standard value range "A" of 0.19—0.24 inch. If not repeat the operation again.

22. Check again that the timing marks of the sprockets are aligned.

23. Continue the installation in the reverse order of the removal procedure.

VALVE LASH

ADJUSTMENT

2.4L Engine

See Figure 78.

1. Before servicing the vehicle, refer to the Precautions Section.

➡ **If working near and/or around the SRS system and components, be sure to disable the SRS system. Tape the negative battery cable with insulating tape. Always disconnect the negative battery cable first.**

✳✳ CAUTION

Wait for 1 minute after disconnecting the negative battery cable before working inside the vehicle. The air bag system is set to deploy for a short period of time after the battery is disconnected.

➡ **Before adjusting the valves, check that the engine coolant temperature is between 176—203°F (80—95°C and that all lights and accessories are OFF.**

2. Remove the ignition coils. Remove the rocker arm cover.

3. Turn the crankshaft clockwise until the notch on the pulley is lined up with the "T" mark on the timing indicator.

4. Move the rocker arms on the number 1 and number 4 cylinders up and down by hand to determine which cylinder has its piston at the Top Dead Center (TDC) position on the compression stroke.

➡ **If both intake and exhaust valve rocker arms have a valve lash, the piston in the cylinder corresponding to these rocker arms is at TDC on the compression stroke.**

5. Valve clearance inspection and adjustment can be performed on rocker arms indicated by the white arrow mark when the number 1 cylinder piston is at the TDC on the compression stroke, and on rocker arms indicated by black arrow mark when the number 4 cylinder piston is at the TDC on the compression stroke.

6. Measure the valve clearance. If the valve clearance is not as specified, loosen the rocker arm lock nut and adjust the clearance using a thickness gauge while turning the adjusting screw.

7. Standard value (hot engine):
   a. Intake valve: 0.008 inch (0.20mm).
   b. Exhaust valve: 0.012 inch (0.30mm).

8. While holding the adjusting screw with a screwdriver to prevent it from turning, tighten the lock nut to the specified torque: 71–89 inch lbs. (8–10 Nm).

9. Turn the crankshaft through 360° to line up the notch on the crankshaft pulley with the "T" mark on the timing indicator.

10. Repeat the steps on other valves for clearance adjustment.

11. Install the rocker arm cover.

12. Install the ignition coils.

3.8L Engine

See Figure 79.

1. Before servicing the vehicle, refer to the Precautions Section.

➡ **If working near and/or around the SRS system and components, be sure**

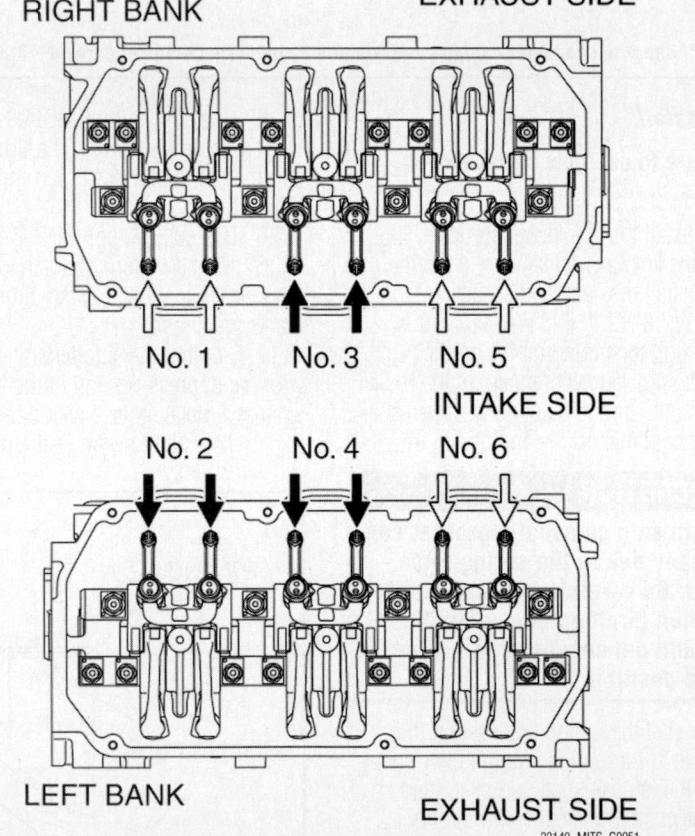

Fig. 79 Valve clearance inspection and adjustment can be performed on rocker arms indicated by the white arrow mark when the number 1 cylinder piston is at the TDC on the compression stroke, and on rocker arms indicated by black arrow mark when the number 4 cylinder piston is at the TDC on the compression stroke—3.8L engine

---

INTAKE VALVE SIDE

No. 1 No. 2 No. 3 No. 4

EXHAUST VALVE SIDE

22140_MITS_G0050

Fig. 78 Valve clearance inspection and adjustment can be performed on rocker arms indicated by the white arrow mark when the number 1 cylinder piston is at the TDC on the compression stroke, and on rocker arms indicated by black arrow mark when the number 4 cylinder piston is at the TDC on the compression stroke— 2.4L engine

to disable the SRS system. Tape the negative battery cable with insulating tape. Always disconnect the negative battery cable first.

➡ **Perform the valve clearance check and adjustment when the engine is cold.**

2. Remove all of the ignition coils.
3. Remove the rocker arm covers.
4. Turn the crankshaft clockwise until the notch on the pulley is lined up with "T" mark on the timing indicator.

5. Move the rocker arms on the number 1 and number 4 cylinders up and down by hand to determine which cylinder has its piston at the Top Dead Center (TDC) on the compression stroke.

➡ **If both intake and exhaust valve rocker arms have a valve lash, the piston in the cylinder corresponding to these rocker arms is at TDC on the compression stroke.**

6. Valve clearance inspection and adjustment can be performed on rocker arms indicated by white arrow mark when the number 1 cylinder piston is at the TDC on the compression stroke, and on rocker arms indicated by black arrow mark when the number 4 cylinder piston is at the TDC on the compression stroke.

7. Measure the valve clearance for the intake side. If the valve clearance is not as

specified, loosen the rocker arm lock nut and adjust the clearance using a thickness gauge while turning the adjusting screw. Standard value (cold engine): 0.004 inch (0.10mm).

➡ **Valve clearance check and adjustment is unnecessary for the exhaust side due to auto lash adjusters installed.**

8. While holding the adjusting screw with a screwdriver to prevent it from turning, tighten the lock nut to the specified torque: 71–89 inch lbs. (8–10 Nm).
9. Turn the crankshaft through 360° to line up the notch on the crankshaft pulley with the "T" mark on the timing indicator.
10. Repeat the steps on other valves for clearance adjustment.
11. Install the rocker arm covers.
12. Install the ignition coils.

ENGINE PERFORMANCE & EMISSION CONTROLS

CAMSHAFT POSITION (CMP) SENSOR

LOCATION

See Figures 80 through 82.

Refer to the accompanying illustrations.

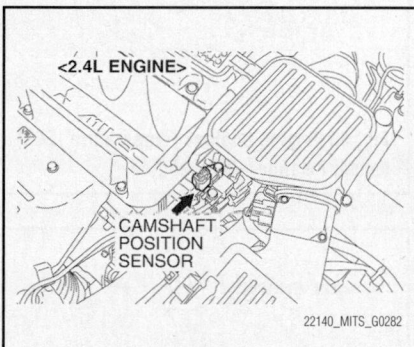

Fig. 80 Camshaft Position (CMP) sensor location—2.4L engine

REMOVAL & INSTALLATION

See Figures 80 through 82.

1. Before servicing the vehicle, refer to the Precautions Section.

➡ **If working near and/or around the SRS system and components, be sure to disable the SRS system. Tape the negative battery cable with insulating tape. Always disconnect the negative battery cable first.**

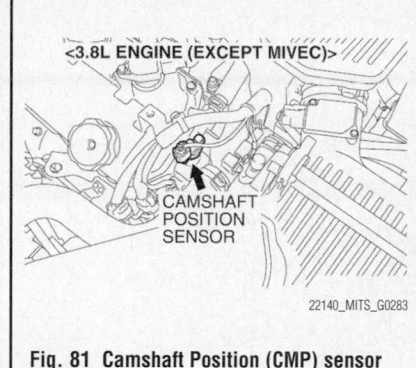

Fig. 81 Camshaft Position (CMP) sensor location—3.8L engine

2. Disconnect the negative battery cable.
3. Remove the necessary components in order to gain access to the sensor retaining bolts.
4. Disconnect the connector from the CMP sensor.
5. Remove the bolt that retains the CMP sensor.
6. Remove the CMP sensor.

To install:

➡ **Be sure to use new fasteners, as required.**

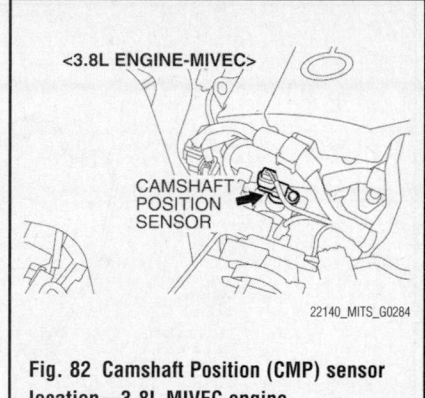

Fig. 82 Camshaft Position (CMP) sensor location—3.8L MIVEC engine

7. Installation is the reverse of the removal procedure.
8. Tighten the bolt that retains the CMP sensor to: 98 inch lbs. (11 Nm).
9. Connect the sensor connector.

CRANKSHAFT POSITION (CKP) SENSOR

LOCATION

See Figures 83 through 85.

Refer to the accompanying illustrations.

REMOVAL & INSTALLATION

See Figures 83 through 85.

1. Before servicing the vehicle, refer to the Precautions Section.

➡ **If working near and/or around the SRS system and components, be sure**

to disable the SRS system. Tape the negative battery cable with insulating tape. Always disconnect the negative battery cable first.

⁂ CAUTION

Wait for 1 minute after disconnecting the negative battery cable before

working inside the vehicle. The air bag system is set to deploy for a short period of time after the battery is disconnected.

2. Disconnect the negative battery cable.

3. Remove the necessary components in order to gain access to the sensor retaining bolts.

4. Disconnect the connector from the sensor.

5. Remove the bolt that retains the sensor in place.

6. Remove the sensor from its mounting.

To install:

➡**Be sure to use new fasteners, as required.**

7. Installation is the reverse of the removal procedure.

8. Tighten the sensor retaining bolt to: 78 inch lbs. (9 Nm).

ENGINE COOLANT TEMPERATURE (ECT) SENSOR

LOCATION

See Figures 86 and 87.

Refer to the accompanying illustrations.

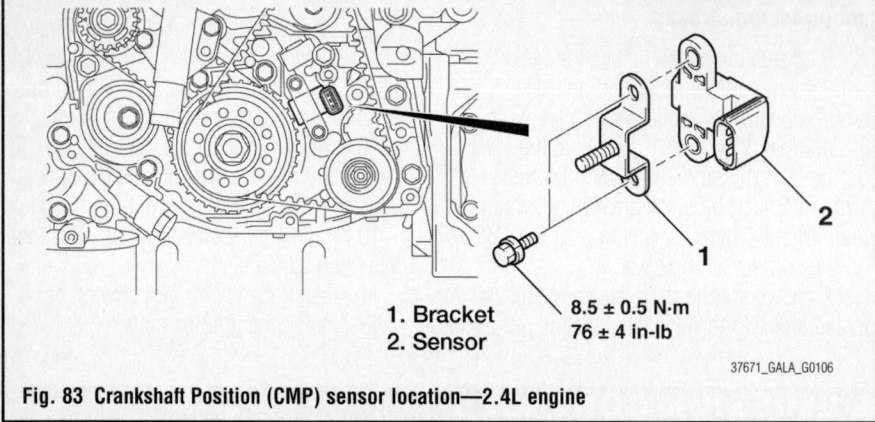

1. Bracket
2. Sensor

8.5 ± 0.5 N·m
76 ± 4 in-lb

37671_GALA_G0106

Fig. 83 Crankshaft Position (CMP) sensor location—2.4L engine

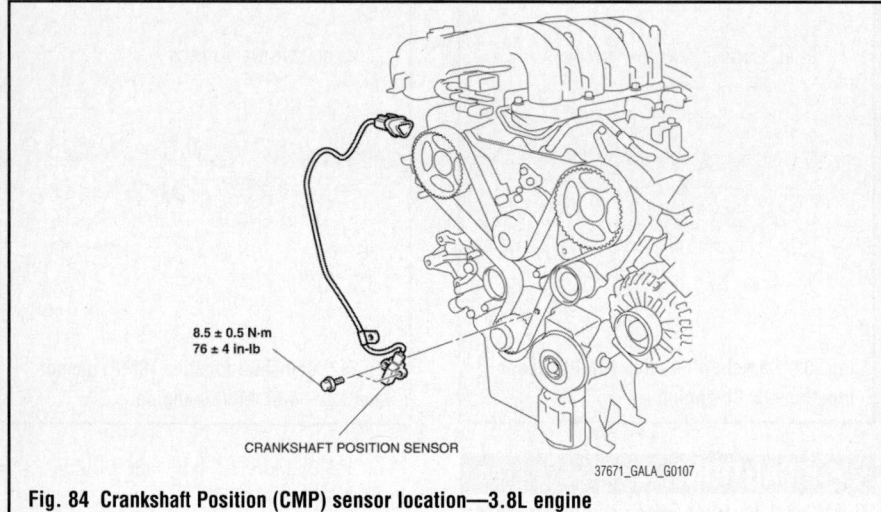

8.5 ± 0.5 N·m
76 ± 4 in-lb

CRANKSHAFT POSITION SENSOR

37671_GALA_G0107

Fig. 84 Crankshaft Position (CMP) sensor location—3.8L engine

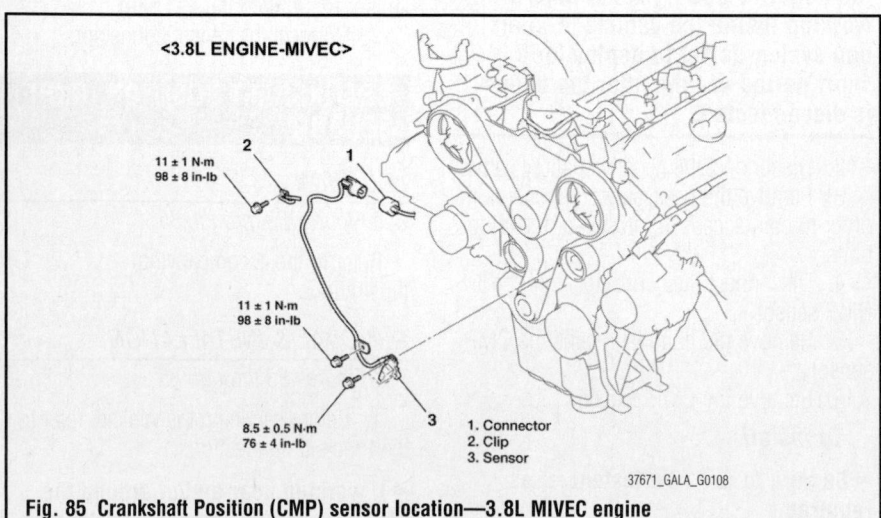

<3.8L ENGINE-MIVEC>

11 ± 1 N·m
98 ± 8 in-lb

2

1

11 ± 1 N·m
98 ± 8 in-lb

8.5 ± 0.5 N·m
76 ± 4 in-lb

3

1. Connector
2. Clip
3. Sensor

37671_GALA_G0108

Fig. 85 Crankshaft Position (CMP) sensor location—3.8L MIVEC engine

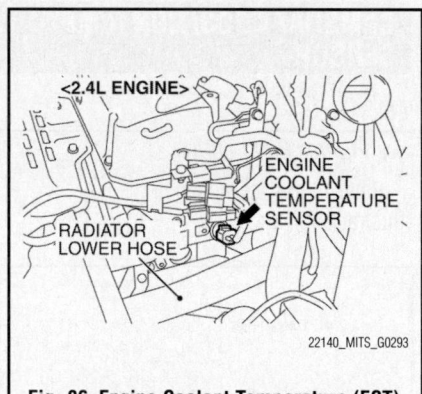

<2.4L ENGINE>

ENGINE COOLANT TEMPERATURE SENSOR

RADIATOR LOWER HOSE

22140_MITS_G0293

Fig. 86 Engine Coolant Temperature (ECT) sensor location—2.4L engine

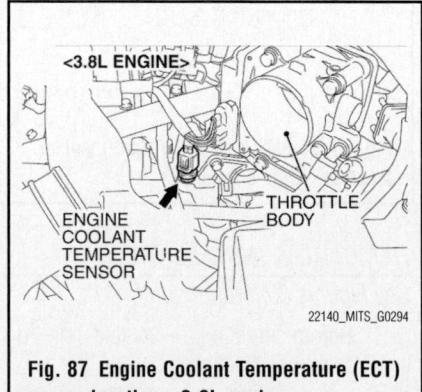

<3.8L ENGINE>

THROTTLE BODY

ENGINE COOLANT TEMPERATURE SENSOR

22140_MITS_G0294

Fig. 87 Engine Coolant Temperature (ECT) sensor location—3.8L engine

REMOVAL & INSTALLATION

See Figures 86 and 87.

1. Before servicing the vehicle, refer to the Precautions Section.

➡If working near and/or around the SRS system and components, be sure to disable the SRS system. Tape the negative battery cable with insulating tape. Always disconnect the negative battery cable first.

✳✳ CAUTION

Wait for 1 minute after disconnecting the negative battery cable before working inside the vehicle. The air bag system is set to deploy for a short period of time after the battery is disconnected.

2. Remove the necessary components in order to gain access to the sensor.

3. Drain the coolant to a level below the bottom of the sensor. Properly dispose of used coolant.

4. Disconnect the ground cable from the battery and then remove the sensor connector.

5. Remove the coolant temperature sensor.

To install:

➡Be sure to use new fasteners, as required.

6. Coat the threads of the sensor with a suitable sealant and thread into the housing.

7. Tighten the sensor to 22 ft. lbs. (30 Nm).

8. Refill the cooling system to the proper level.

9. Attach the electrical connector to the sensor securely.

10. Connect the negative battery cable.

HEATED OXYGEN SENSOR (HO2S)

LOCATION

The sensors are located in the exhaust system. On some vehicles, one sensor is located up at the exhaust manifold(s) and the other sensor is located down at the catalytic converter.

REMOVAL & INSTALLATION

See Figures 88 through 90.

✳✳ CAUTION

The temperature of the exhaust system is extremely high after the engine has been run. To prevent personal injury, allow the exhaust system to cool before removing the sensor from the exhaust system.

1. Before servicing the vehicle, refer to the Precautions Section.

➡If working near and/or around the SRS system and components, be sure to disable the SRS system. Tape the negative battery cable with insulating tape. Always disconnect the negative battery cable first.

✳✳ CAUTION

Wait for 1 minute after disconnecting the negative battery cable before working inside the vehicle. The air bag system is set to deploy for a short period of time after the battery is disconnected.

2. Disconnect the negative battery cable.

3. Raise and safely support the vehicle, as needed.

4. Detach the electrical connector from the oxygen sensor.

5. Using socket MD998770, or equivalent oxygen sensor socket, remove the heated oxygen sensor.

To install:

➡Be sure to use new fasteners, as required.

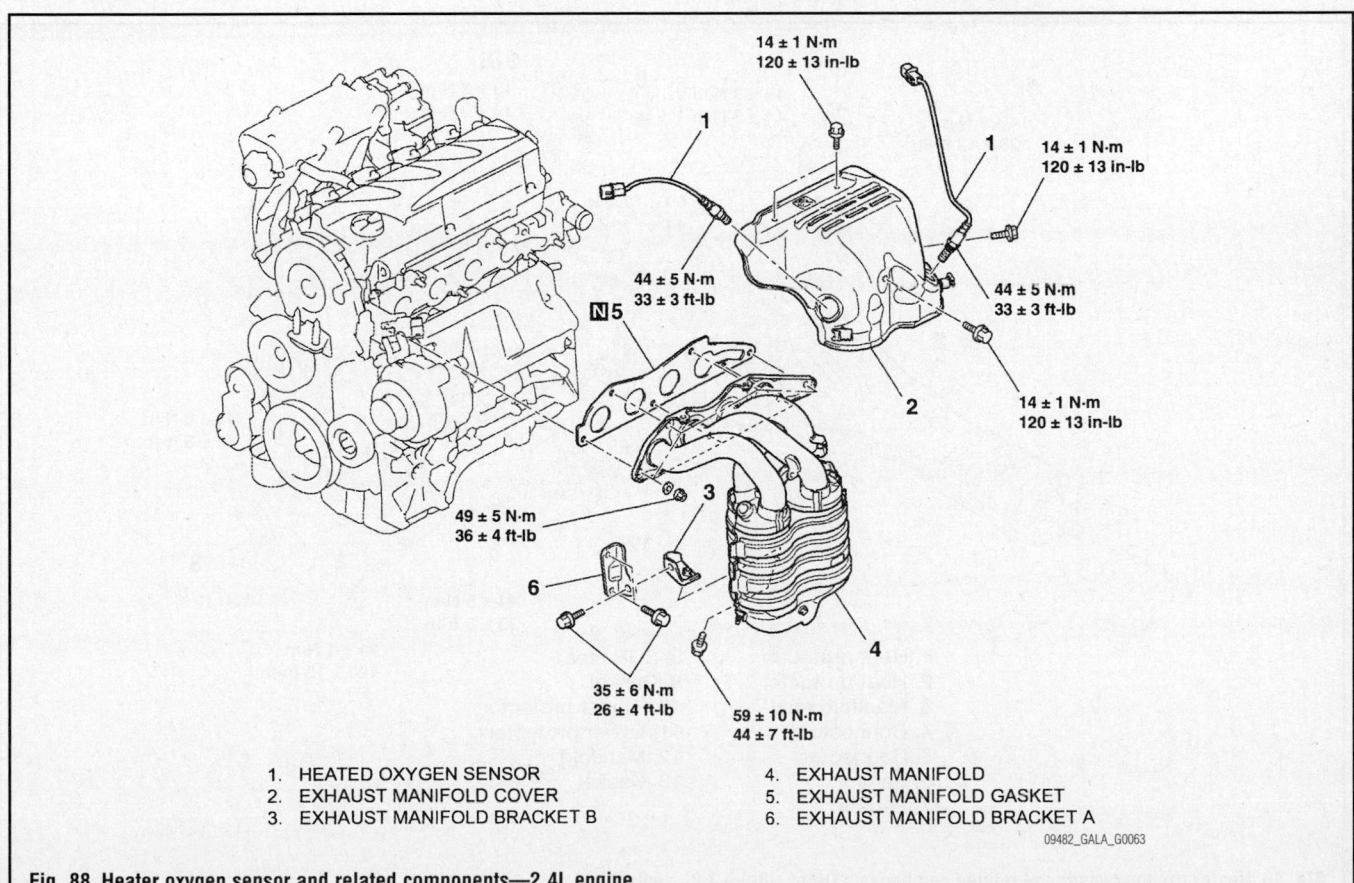

1.	HEATED OXYGEN SENSOR
2.	EXHAUST MANIFOLD COVER
3.	EXHAUST MANIFOLD BRACKET B
4.	EXHAUST MANIFOLD
5.	EXHAUST MANIFOLD GASKET
6.	EXHAUST MANIFOLD BRACKET A

09482_GALA_G0063

Fig. 88 Heater oxygen sensor and related components—2.4L engine

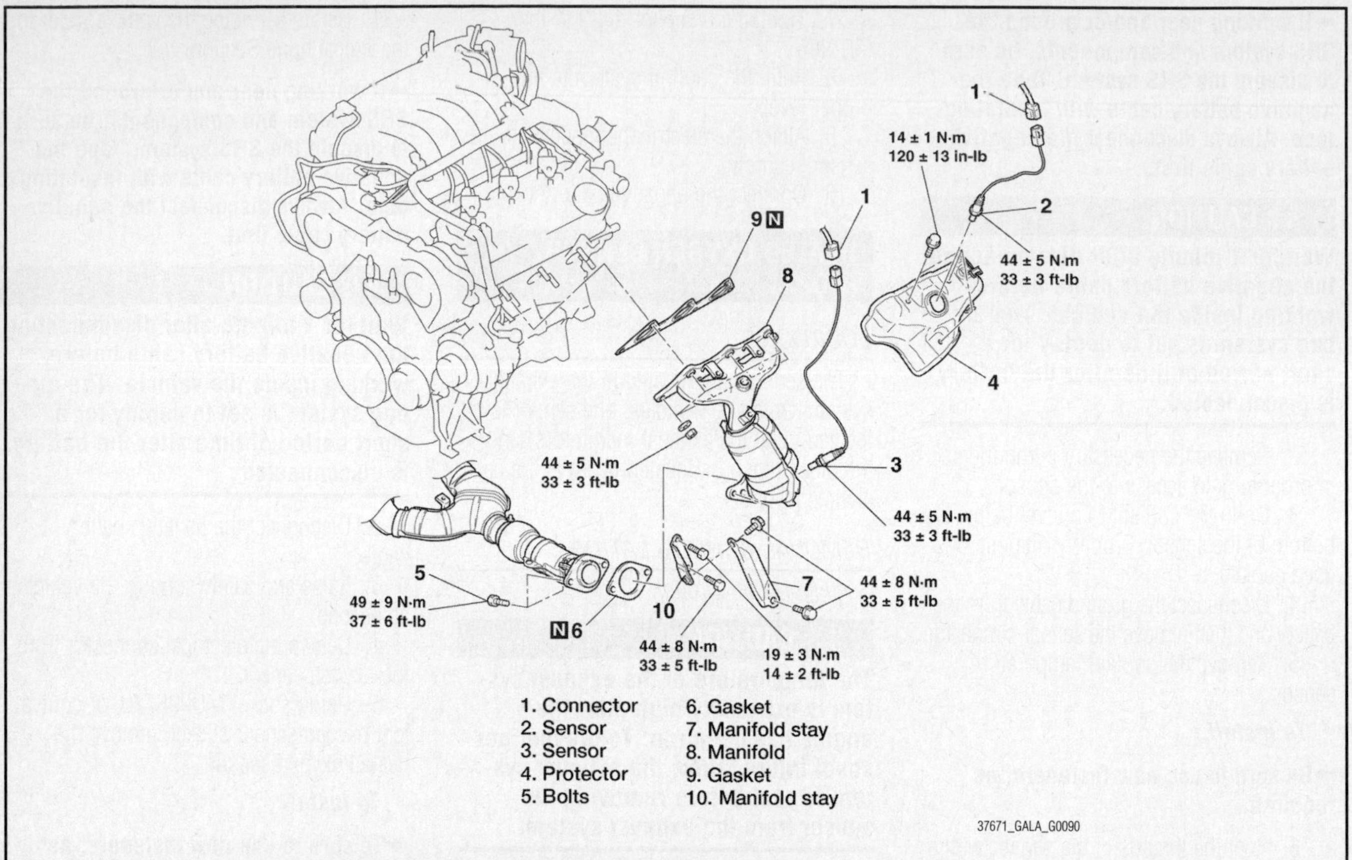

14 ± 1 N·m
120 ± 13 in-lb

44 ± 5 N·m
33 ± 3 ft-lb

9 **N**

44 ± 5 N·m
33 ± 3 ft-lb

44 ± 5 N·m
33 ± 3 ft-lb

49 ± 9 N·m
37 ± 6 ft-lb

44 ± 8 N·m
33 ± 5 ft-lb

N6

44 ± 8 N·m
33 ± 5 ft-lb

19 ± 3 N·m
14 ± 2 ft-lb

1. Connector
2. Sensor
3. Sensor
4. Protector
5. Bolts
6. Gasket
7. Manifold stay
8. Manifold
9. Gasket
10. Manifold stay

37671_GALA_G0090

Fig. 89 Heater oxygen sensor and related components (left side)—3.8L engine

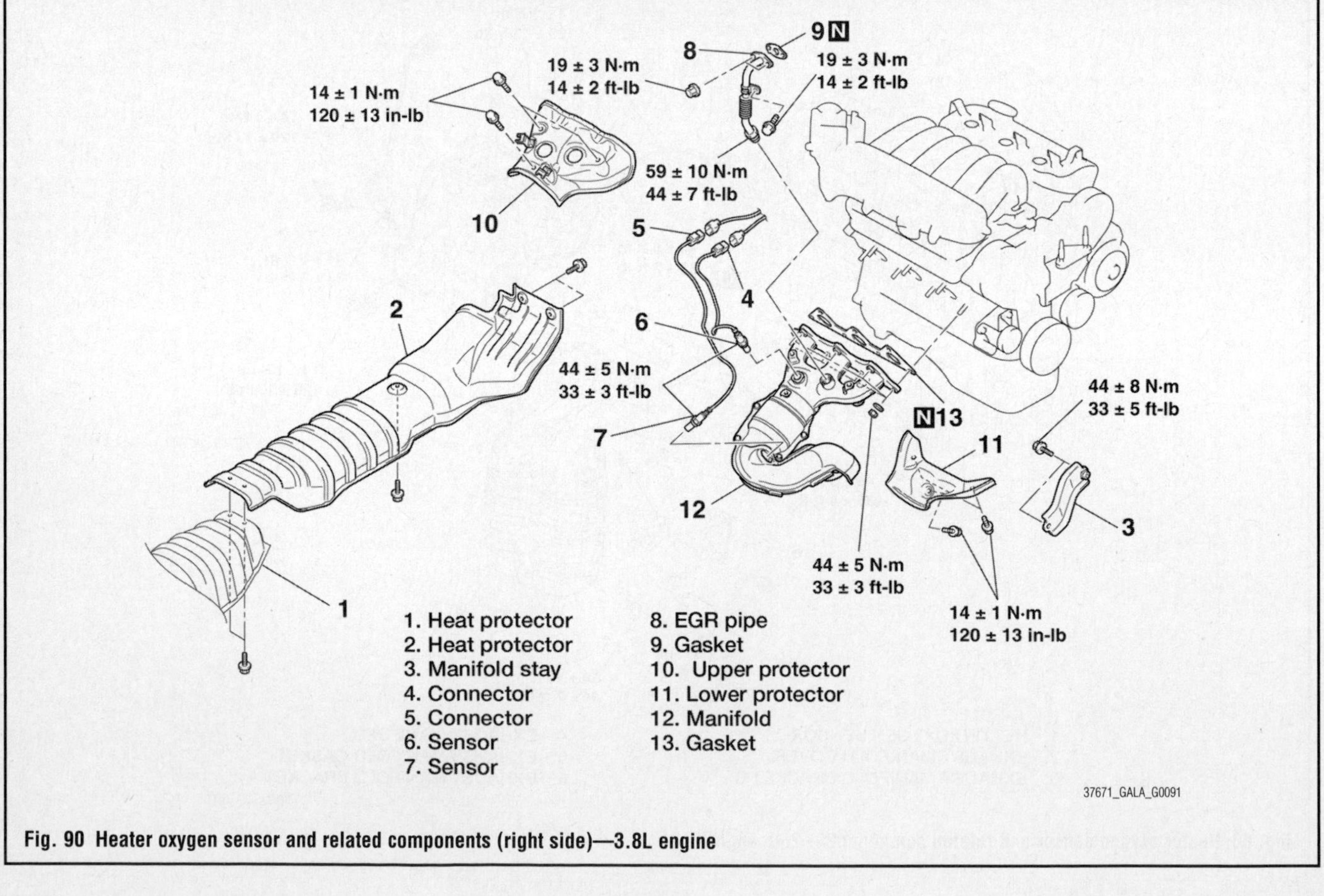

14 ± 1 N·m
120 ± 13 in-lb

19 ± 3 N·m
14 ± 2 ft-lb

8

9 **N**

19 ± 3 N·m
14 ± 2 ft-lb

59 ± 10 N·m
44 ± 7 ft-lb

44 ± 5 N·m
33 ± 3 ft-lb

44 ± 8 N·m
33 ± 5 ft-lb

N13

44 ± 5 N·m
33 ± 3 ft-lb

14 ± 1 N·m
120 ± 13 in-lb

1. Heat protector
2. Heat protector
3. Manifold stay
4. Connector
5. Connector
6. Sensor
7. Sensor
8. EGR pipe
9. Gasket
10. Upper protector
11. Lower protector
12. Manifold
13. Gasket

37671_GALA_G0091

Fig. 90 Heater oxygen sensor and related components (right side)—3.8L engine

6. If installing the old oxygen sensor, coat the threads with anti-seize compound. New sensors are already coated. Take care not to contaminate the oxygen sensor probe with the anti-seize compound.

7. Install the oxygen sensor. Using the correct tool, tighten the sensor to 33 ft. lbs. (45 Nm).

8. Attach the wiring to the sensor.

9. Connect the negative battery cable.

INTAKE AIR TEMPERATURE (IAT) SENSOR

LOCATION

The Intake Air Temperature (IAT) sensor is mounted in the intake air hose of the air cleaner assembly.

REMOVAL & INSTALLATION

1. Before servicing the vehicle, refer to the Precautions Section.

➡️**If working near and/or around the SRS system and components, be sure to disable the SRS system. Tape the negative battery cable with insulating tape. Always disconnect the negative battery cable first.**

✳✳ CAUTION

Wait for 1 minute after disconnecting the negative battery cable before working inside the vehicle. The air bag system is set to deploy for a short period of time after the battery is disconnected.

2. Disconnect the negative battery cable.

3. Disconnect the connector from the sensor.

4. Remove the sensor retaining screws.

5. Remove the sensor from its mounting.

To install:

➡️**Be sure to use new fasteners, as required.**

6. Installation is the reverse of the removal procedure.

7. Handle the sensor assembly carefully, protecting it from impact, extremes of temperature and/or exposure to shop chemicals.

KNOCK SENSOR (KS)

LOCATION

See Figures 91 through 93.

Refer to the accompanying illustrations.

REMOVAL & INSTALLATION

See Figures 91 through 93.

➡️**After replacing this sensor use the Mitsubishi diagnostic scan tool, or equivalent and perform the initialization procedure for learning value in MFI engine.**

1. Before servicing the vehicle, refer to the Precautions Section.

➡️**If working near and/or around the SRS system and components, be sure to disable the SRS system. Tape the**

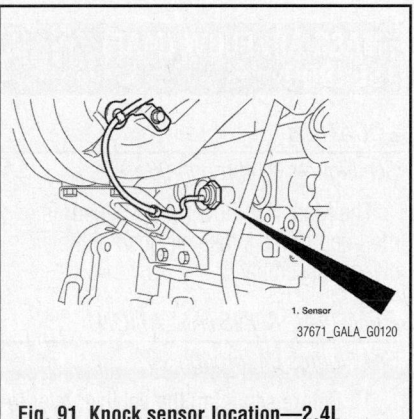

Fig. 91 Knock sensor location—2.4L engine

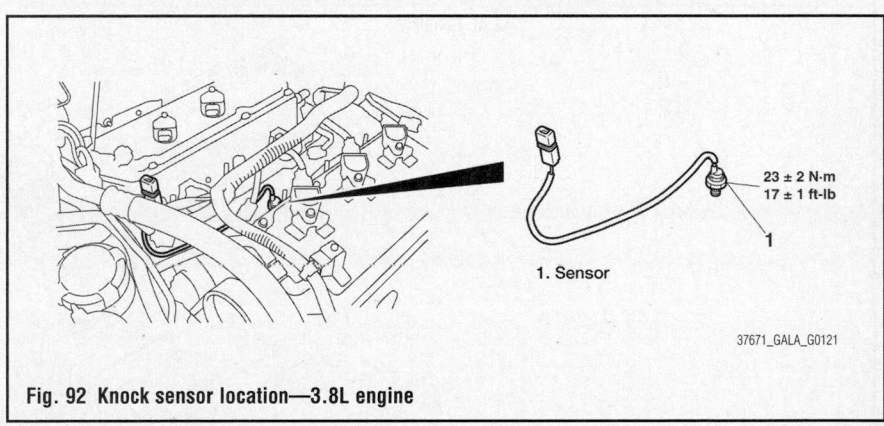

Fig. 92 Knock sensor location—3.8L engine

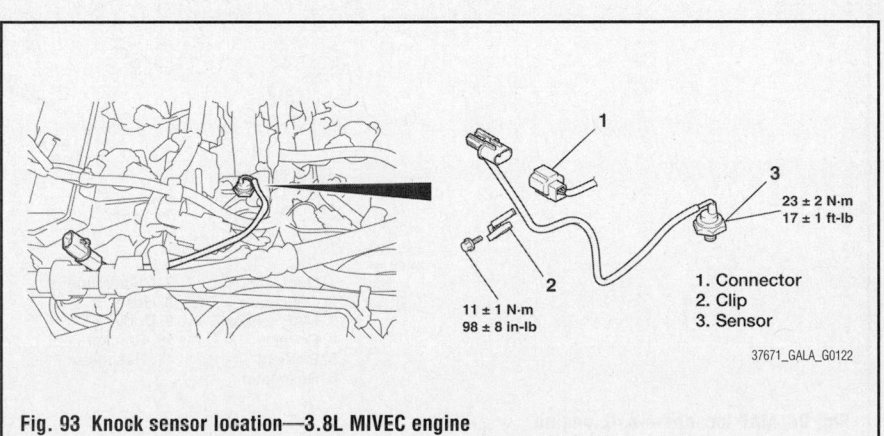

Fig. 93 Knock sensor location—3.8L MIVEC engine

negative battery cable with insulating tape. Always disconnect the negative battery cable first.

✳✳ CAUTION

Wait for 1 minute after disconnecting the negative battery cable before working inside the vehicle. The air bag system is set to deploy for a short period of time after the battery is disconnected.

2. Disconnect the negative battery cable.

3. On 3.8L engines, remove the intake manifold.

4. Disconnect the sensor connector.

5. Remove the sensor from its mounting.

To install:

6. Installation is the reverse of the removal procedure.

7. Tighten the sensor to 12–17 ft. lbs. (16–24 Nm).

MALFUNCTION INDICATOR LIGHT (MIL)

The Malfunction Indicator Light (MIL) operates as follows:

• The MIL will illuminate with the ignition switch ON and the engine OFF
• The MIL will turn OFF when the engine is started
• The MIL will remain ON if the self-diagnostic system has detected a malfunction
• The MIL may turn OFF if the malfunction is no longer present
• If the MIL is illuminated and then the engine stalls, the MIL will remain illuminated as long as the ignition switch is ON
• If the MIL is not illuminated and the engine stalls, the MIL will not illuminate until the ignition switch is cycled OFF, then ON

RESET PROCEDURE

Reset as follows:
• The control module turns OFF the MIL after 3 consecutive ignition cycles that the diagnostic system runs and does not fail
• The control module turns OFF the MIL after a current Diagnostic Trouble Code (DTC) clears when the diagnostic cycle runs and passes
• There may still be a history of DTC's stored in the system. These will clear after

40 consecutive warm-up cycles, if no failures are reported by any other related diagnostic system
• Manual resetting of the MIL and any DTC stored in the system, requires the use of an OBD2 scan tool connected to the Data Link Connector (DLC) for communication with the vehicle. Follow the instructions of the scan tool for both retrieval and resetting of DTC's.

➡ If the error symptoms causing the MIL to illuminate have been corrected, the MIL will return to normal operation.

MASS AIR FLOW (MAF) SENSOR

LOCATION
See Figures 94 through 96.

The Mass Air Flow (MAF) sensor is mounted in the intake air hose of the air cleaner assembly.

REMOVAL & INSTALLATION
See Figures 94 through 96.

1. Before servicing the vehicle, refer to the Precautions Section.

➡ If working near and/or around the SRS system and components, be sure to disable the SRS system. Tape the negative battery cable with insulating tape. Always disconnect the negative battery cable first.

✳✳ CAUTION
Wait for 1 minute after disconnecting the negative battery cable before working inside the vehicle. The air bag system is set to deploy for a short period of time after the battery is disconnected.

2. Disconnect the negative battery cable.
3. Disconnect the connector from the sensor.
4. Remove the air cleaner and air intake assembly, as required.
5. Remove the sensor from its mounting.

To install:

➡ Be sure to use new fasteners, as required.

6. Installation is the reverse of the removal procedure.
7. Handle the sensor assembly carefully,

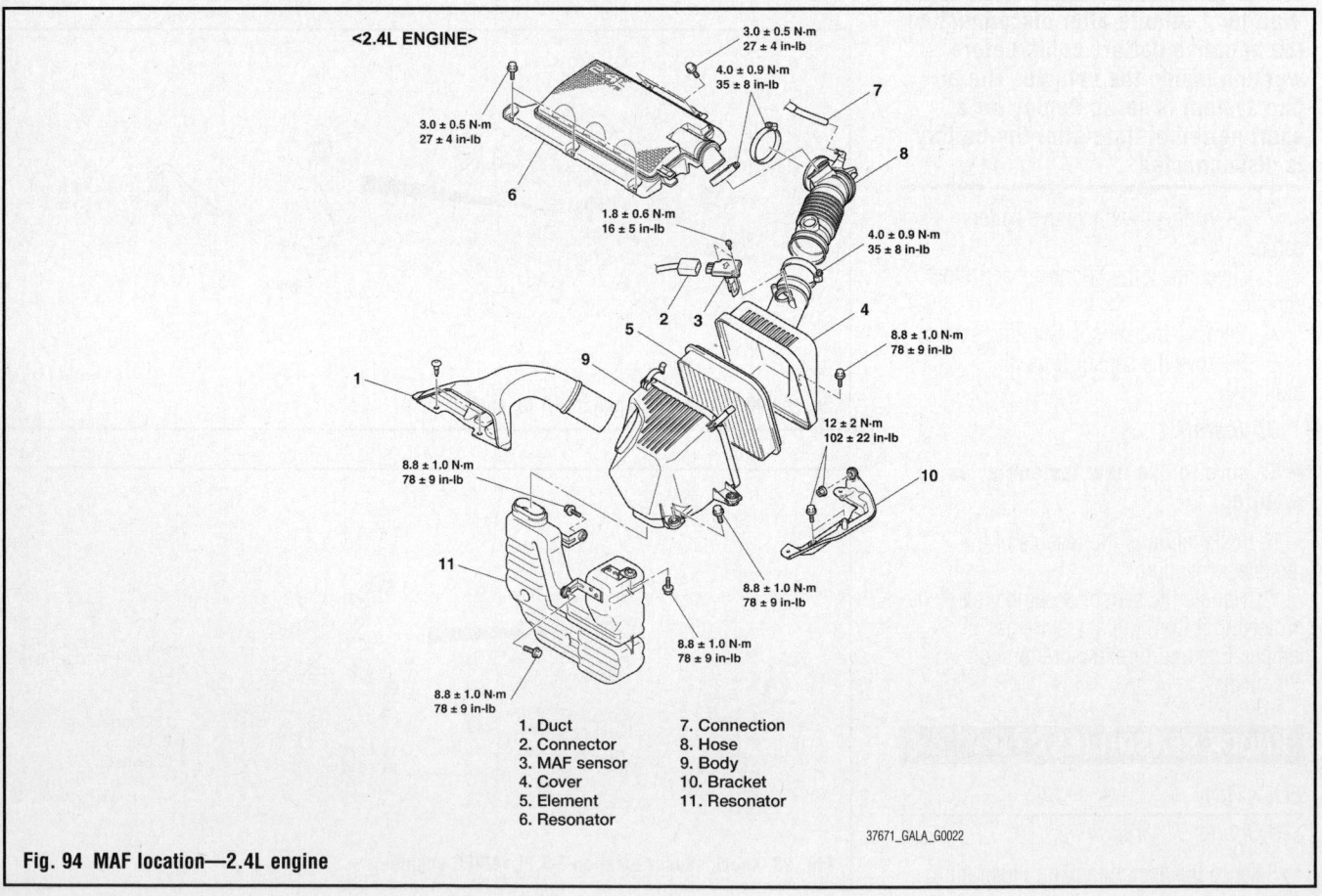

<2.4L ENGINE>

3.0 ± 0.5 N·m
27 ± 4 in-lb

4.0 ± 0.9 N·m
35 ± 8 in-lb

3.0 ± 0.5 N·m
27 ± 4 in-lb

1.8 ± 0.6 N·m
16 ± 5 in-lb

4.0 ± 0.9 N·m
35 ± 8 in-lb

8.8 ± 1.0 N·m
78 ± 9 in-lb

12 ± 2 N·m
102 ± 22 in-lb

8.8 ± 1.0 N·m
78 ± 9 in-lb

8.8 ± 1.0 N·m
78 ± 9 in-lb

8.8 ± 1.0 N·m
78 ± 9 in-lb

8.8 ± 1.0 N·m
78 ± 9 in-lb

1. Duct
2. Connector
3. MAF sensor
4. Cover
5. Element
6. Resonator
7. Connection
8. Hose
9. Body
10. Bracket
11. Resonator

37671_GALA_G0022

Fig. 94 MAF location—2.4L engine

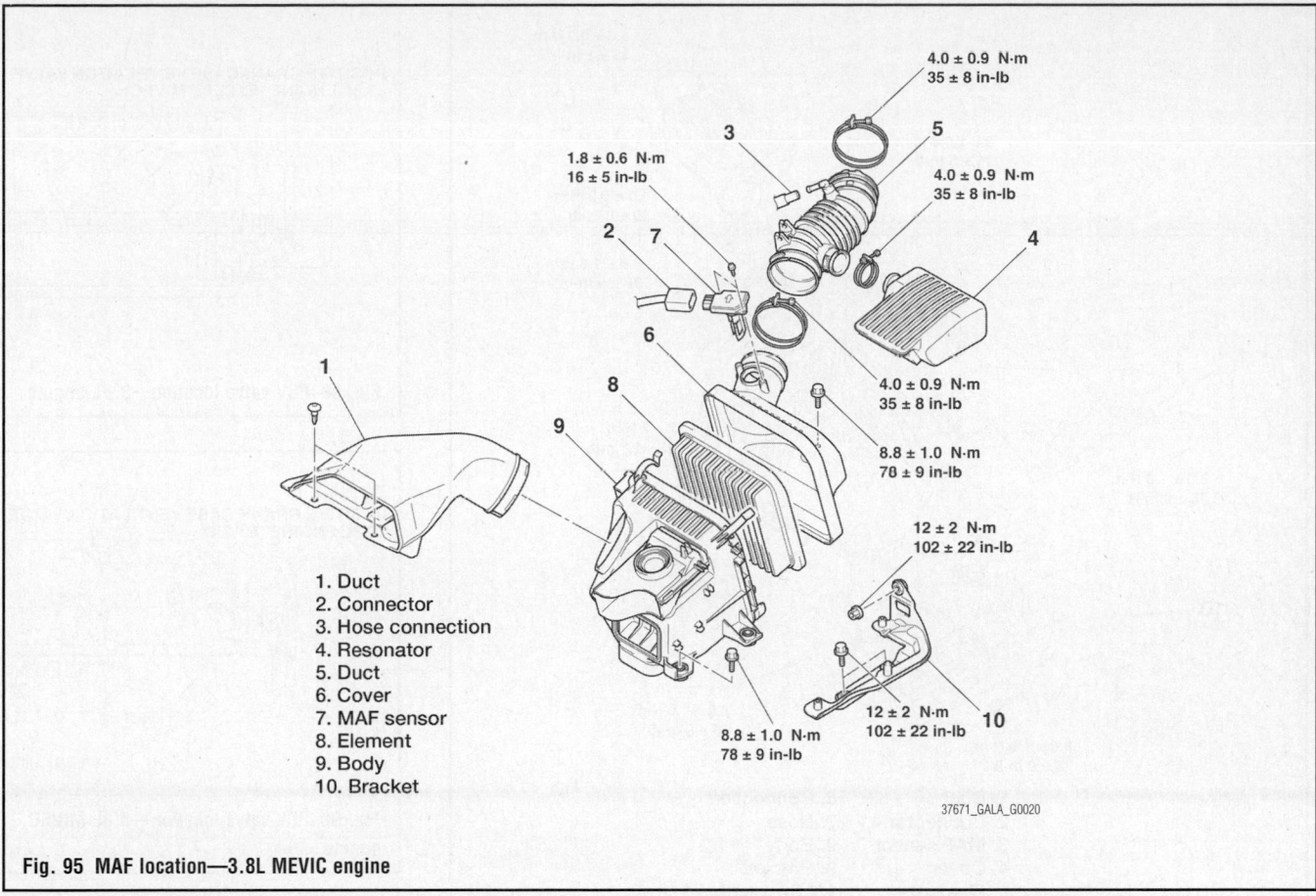

4.0 ± 0.9 N·m
35 ± 8 in-lb

1.8 ± 0.6 N·m
16 ± 5 in-lb

4.0 ± 0.9 N·m
35 ± 8 in-lb

4.0 ± 0.9 N·m
35 ± 8 in-lb

8.8 ± 1.0 N·m
78 ± 9 in-lb

12 ± 2 N·m
102 ± 22 in-lb

12 ± 2 N·m
102 ± 22 in-lb

8.8 ± 1.0 N·m
78 ± 9 in-lb

1. Duct
2. Connector
3. Hose connection
4. Resonator
5. Duct
6. Cover
7. MAF sensor
8. Element
9. Body
10. Bracket

37671_GALA_G0020

Fig. 95 MAF location—3.8L MEVIC engine

protecting it from impact, extremes of temperature, and exposure to shop chemicals.

MANIFOLD ABSOLUTE PRESSURE (MAP) SENSOR

LOCATION

The MAP sensor is located on the intake manifold.

REMOVAL & INSTALLATION

1. Before servicing the vehicle, refer to the Precautions Section.

➡**If working near and/or around the SRS system and components, be sure to disable the SRS system. Tape the negative battery cable with insulating tape. Always disconnect the negative battery cable first.**

✳✳ CAUTION

Wait for 1 minute after disconnecting the negative battery cable before working inside the vehicle. The air bag system is set to deploy for a short period of time after the battery is disconnected.

2. Disconnect the negative battery cable.

3. Disconnect the connector from the sensor.

4. Remove the sensor from its mounting.

To install:

➡**Be sure to use new fasteners, as required.**

5. Installation is the reverse of the removal procedure.

6. Handle the sensor assembly carefully, protecting it from impact, extremes of temperature, and exposure to shop chemicals.

POSITIVE CRANKCASE VENTILATION (PCV) VALVE

LOCATION

See Figures 96 through 99.

Refer to the accompanying illustrations.

REMOVAL & INSTALLATION

See Figures 96 through 99.

At this time the manufacturer does not provide removal and installation procedures

for this component. The following procedure is a guideline and may differ from the vehicle you are servicing.

1. Before servicing the vehicle, refer to the Precautions Section.

➡**If working near and/or around the SRS system and components, be sure to disable the SRS system. Tape the negative battery cable with insulating tape. Always disconnect the negative battery cable first.**

✳✳ CAUTION

Wait for 1 minute after disconnecting the negative battery cable before working inside the vehicle. The air bag system is set to deploy for a short period of time after the battery is disconnected.

2. Remove the necessary components in order to gain access to the component.

3. Disconnect the PCV hose.

4. Remove the valve from its mounting.

To install:

➡**Be sure to use new fasteners, as required.**

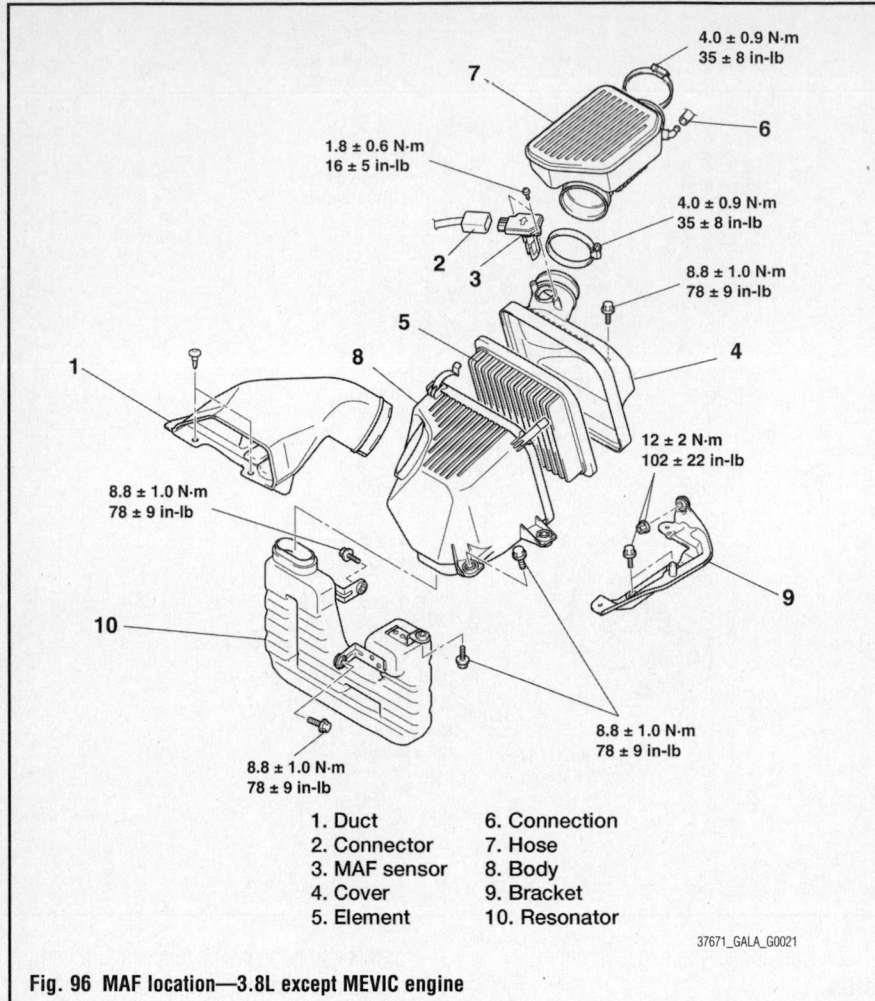

4.0 ± 0.9 N·m
35 ± 8 in-lb

1.8 ± 0.6 N·m
16 ± 5 in-lb

4.0 ± 0.9 N·m
35 ± 8 in-lb

8.8 ± 1.0 N·m
78 ± 9 in-lb

8.8 ± 1.0 N·m
78 ± 9 in-lb

12 ± 2 N·m
102 ± 22 in-lb

8.8 ± 1.0 N·m
78 ± 9 in-lb

8.8 ± 1.0 N·m
78 ± 9 in-lb

1. Duct	6. Connection
2. Connector	7. Hose
3. MAF sensor	8. Body
4. Cover	9. Bracket
5. Element	10. Resonator

37671_GALA_G0021

Fig. 96 MAF location—3.8L except MEVIC engine

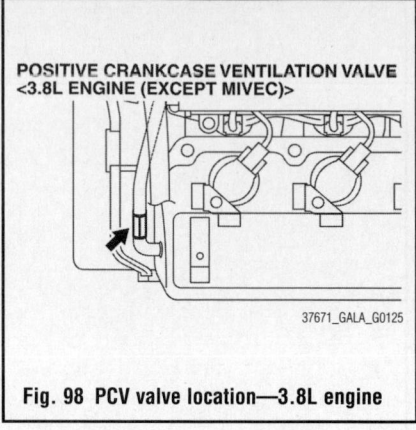

37671_GALA_G0125

Fig. 98 PCV valve location—3.8L engine

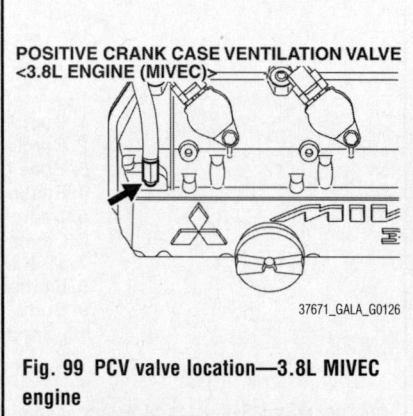

37671_GALA_G0126

Fig. 99 PCV valve location—3.8L MIVEC engine

engine learning is not completed. Using the Mitsubishi diagnostic scan tool or equivalent, perform the relearn procedure.

1. Before servicing the vehicle, refer to the Precautions Section.

➡If working near and/or around the SRS system and components, be sure to disable the SRS system. Tape the negative battery cable with insulating tape. Always disconnect the negative battery cable first.

❋❋ CAUTION

Wait for 1 minute after disconnecting the negative battery cable before working inside the vehicle. The air bag system is set to deploy for a short period of time after the battery is disconnected.

2. Turn the ignition switch off.
3. Disconnect the negative battery cable from the battery.
4. Disconnect the PCM connector(s).
5. Remove the PCM mounting bolts and remove the PCM from the air cleaner assembly.

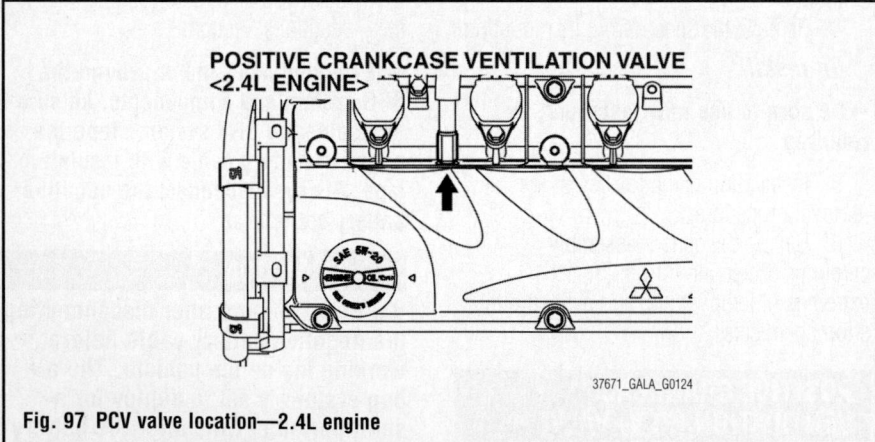

37671_GALA_G0124

Fig. 97 PCV valve location—2.4L engine

5. Installation is the reverse of the removal procedure.

POWERTRAIN CONTROL MODULE (PCM)

LOCATION

See Figure 100.

Refer to the accompanying illustration.

REMOVAL & INSTALLATION

See Figure 100.

➡Do not replace the PCM and the ETACS-ECU simultaneously. Always replace either one of these components first, and register the encrypted code.

➡After PCM replacement, idling speed may be unstable because the MFI

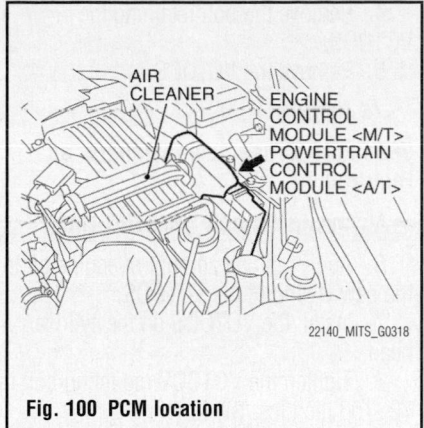

AIR CLEANER

ENGINE CONTROL MODULE <M/T> POWERTRAIN CONTROL MODULE <A/T>

22140_MITS_G0318

Fig. 100 PCM location

To install:

➡**Be sure to use new fasteners, as required.**

6. Installation is the reverse of the removal.

7. Tighten the PCM mounting bolts to: 86–104 inch lbs. (10–12 Nm).

8. Using the Mitsubishi diagnostic scan tool, or equivalent, perform the relearning procedures. Follow the directions on the screen of the tool.

THROTTLE POSITION SENSOR (TPS)

LOCATION

See Figures 101 and 102.

The Throttle Position Sensor (TPS) is mounted on the throttle body and is incorporated into the throttle body assembly.

REMOVAL & INSTALLATION

See Figures 101 and 102.

The throttle position sensor is an integral part of the throttle body.

1. Before servicing the vehicle, refer to the Precautions Section.

➡**If working near and/or around the SRS system and components, be sure to disable the SRS system. Tape the negative battery cable with insulating tape. Always disconnect the negative battery cable first.**

❊❊ CAUTION

Wait for 1 minute after disconnecting the negative battery cable before working inside the vehicle. The air bag system is set to deploy for a short period of time after the battery is disconnected.

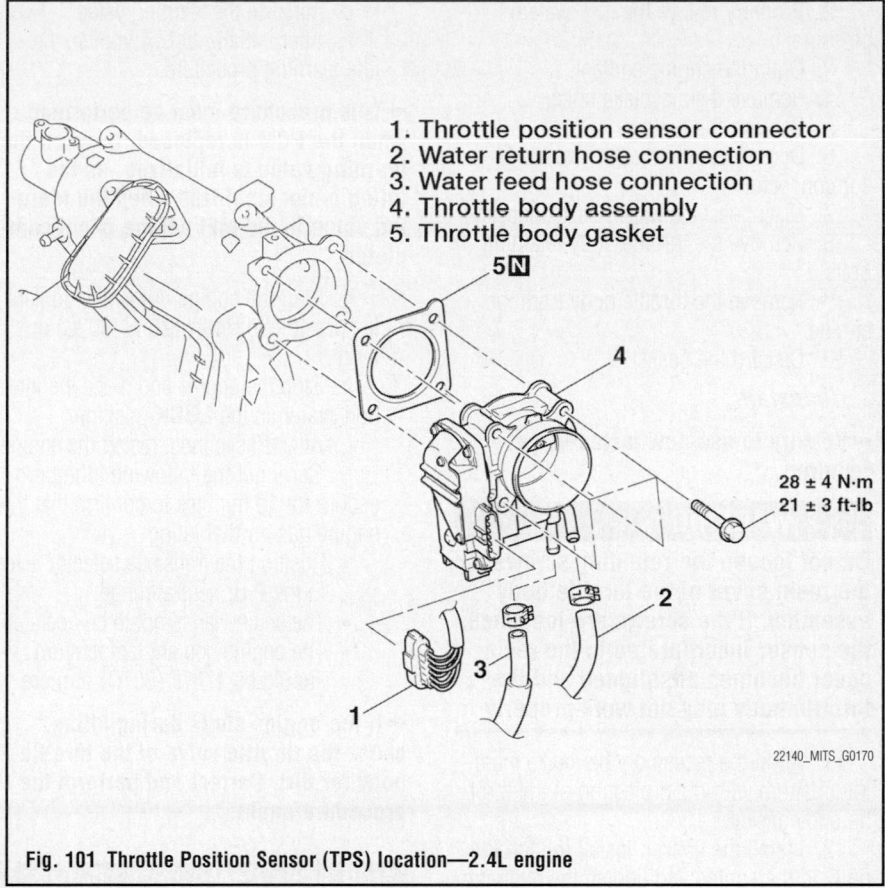

1. Throttle position sensor connector
2. Water return hose connection
3. Water feed hose connection
4. Throttle body assembly
5. Throttle body gasket

28 ± 4 N·m
21 ± 3 ft-lb

22140_MITS_G0170

Fig. 101 Throttle Position Sensor (TPS) location—2.4L engine

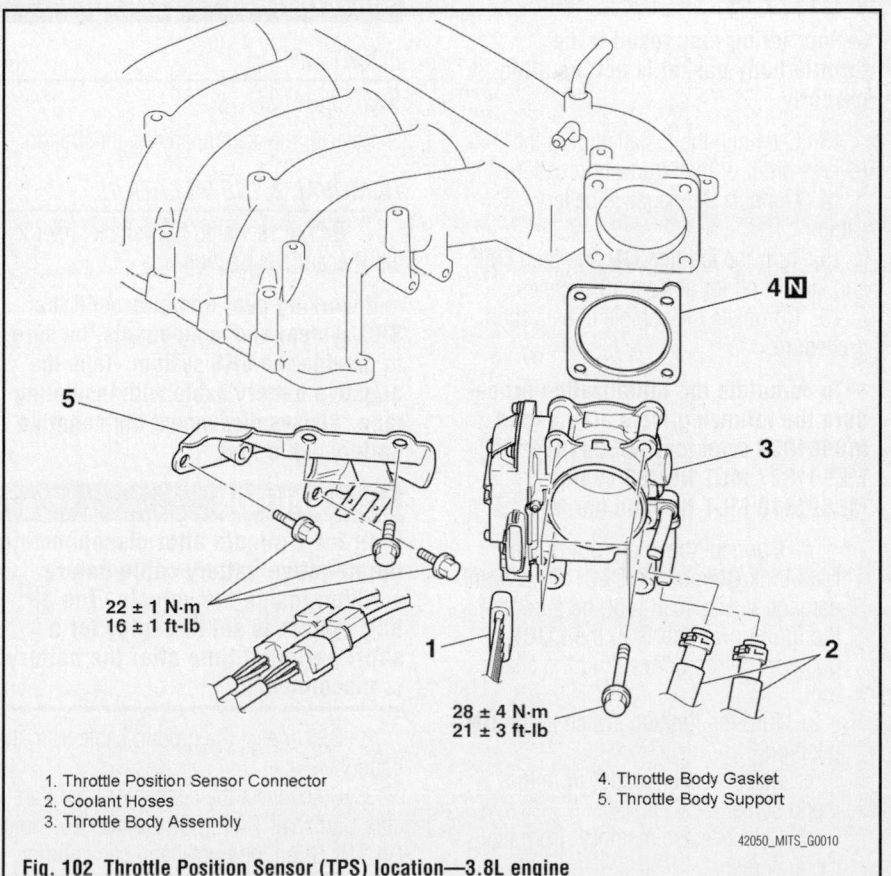

22 ± 1 N·m
16 ± 1 ft-lb

28 ± 4 N·m
21 ± 3 ft-lb

1. Throttle Position Sensor Connector
2. Coolant Hoses
3. Throttle Body Assembly
4. Throttle Body Gasket
5. Throttle Body Support

42050_MITS_G0010

Fig. 102 Throttle Position Sensor (TPS) location—3.8L engine

2. Properly relieve the fuel system pressure.

3. Drain the engine coolant.

4. Remove the air intake hose.

5. Remove the battery.

6. Disconnect the throttle position sensor connector.

7. Disconnect the water hose connection.

8. Remove the throttle body retaining bolts.

9. Remove the throttle body from the engine.

10. Discard the gasket.

To install:

➡ Be sure to use new fasteners, as required.

✳✳ WARNING

Do not loosen the retaining screws for the resin cover of the throttle body assembly. If the screws are loosened, the sensor incorporated in the resin cover becomes misaligned and the throttle body may not work properly.

11. Align the recess on the intake manifold plenum with the projection of the throttle body gasket.

12. Install the gasket. Install the throttle body to the engine and tighten the retaining bolts to 18–24 ft. lbs. (24–32 Nm).

➡ Poor idling may result if the throttle body gasket is not installed properly.

13. Continue the installation in the reverse order of the removal procedure.

14. Connect the negative battery cable.

15. Turn the ignition **ON** and then **OFF**, and keep it off for at least 10 seconds.

16. Complete the vehicle initialization procedure.

➡ To complete the initialization procedure the following tools are needed. MB991958 scan tool, MB991824 VCI, MB991827 MUT III USB cable, MB991910 MUT III Main harness "A."

a. Connect the scan tool to the Data Link Connector (DLC). To prevent damage to the scan tool, be sure that the ignition switch is in the **LOCK** position before connecting the scan tool.

b. Turn the ignition switch to the **ON** position.

c. Select "check mode" from the menu screen.

d. Select "erase memory" from the menu screen.

e. Initialize the learning value.

f. After initialization, complete the idle learning procedure.

➡ This procedure must be performed when the PCM is replaced, or when the learning value is initialized, as the idling is not stabilized when the learning value in the MFI engine is not completed.

g. Start the engine. Allow the coolant temperature to reach 176°F (80°C) or more.

h. Stop the engine and place the ignition switch in the **LOCK** position.

i. After 10 seconds, restart the engine.

j. Carry out the following idling procedure for 10 minutes to confirm that the engine has normal idling:

- Position the transaxle selector lever in the **P** or neutral range
- The engine fan is not to be operated
- The engine coolant temperature should be 176°F (80°C) or more

➡ If the engine stalls during idling, check the throttle valve of the throttle body for dirt. Correct and perform the procedure again.

VARIABLE CAMSHAFT TIMING OIL CONTROL SOLENOID

LOCATION
See Figure 103.

Refer to the accompanying illustration.

REMOVAL & INSTALLATION

1. Before servicing the vehicle, refer to the Precautions Section.

➡ If working near and/or around the SRS system and components, be sure to disable the SRS system. Tape the negative battery cable with insulating tape. Always disconnect the negative battery cable first.

✳✳ CAUTION

Wait for 1 minute after disconnecting the negative battery cable before working inside the vehicle. The air bag system is set to deploy for a short period of time after the battery is disconnected.

2. Disconnect the ground cable from the battery.

3. Disconnect the connector from Variable Camshaft Timing Oil Control Solenoid (VCTOCS) on the right-hand bank and/or left-hand bank.

4. Remove the bolt retaining the VCTOCS.

5. Remove the VCTOCS.

To install:

➡ Be sure to use new fasteners, as required.

➡ Always use a new gasket/O-ring.

6. Apply a small amount of engine oil to the new O-ring of the VCTOCS.

7. Install the VCTOCS on the cylinder head.

8. Tighten the VCTOCS mounting bolt to 72–106 inch lbs. (8–12 Nm).

9. Connect the electrical connector to the VCTOCS.

VEHICLE SPEED SENSOR (VSS)

LOCATION

The Vehicle Speed Sensor (VSS) is attached near the output shaft of the transaxle.

REMOVAL & INSTALLATION

1. Before servicing the vehicle, refer to the Precautions Section.

➡ If working near and/or around the SRS system and components, be sure to disable the SRS system. Tape the negative battery cable with insulating tape. Always disconnect the negative battery cable first.

✳✳ CAUTION

Wait for 1 minute after disconnecting the negative battery cable before working inside the vehicle. The air bag system is set to deploy for a short period of time after the battery is disconnected.

2. Raise and support the vehicle safely.

3. Place a drip pan below the Vehicle Speed Sensor (VSS) to catch any spilled fluid when it is removed.

4. Disconnect the VSS connector.

5. Remove the sensor from its mounting.

To install:

➡ Be sure to use new fasteners, as required.

6. Install the sensor and tighten the attaching bolt to 97 inch lbs. (11 Nm).

7. Replace any lost transaxle fluid.

8. Connect the sensor electrical connector.

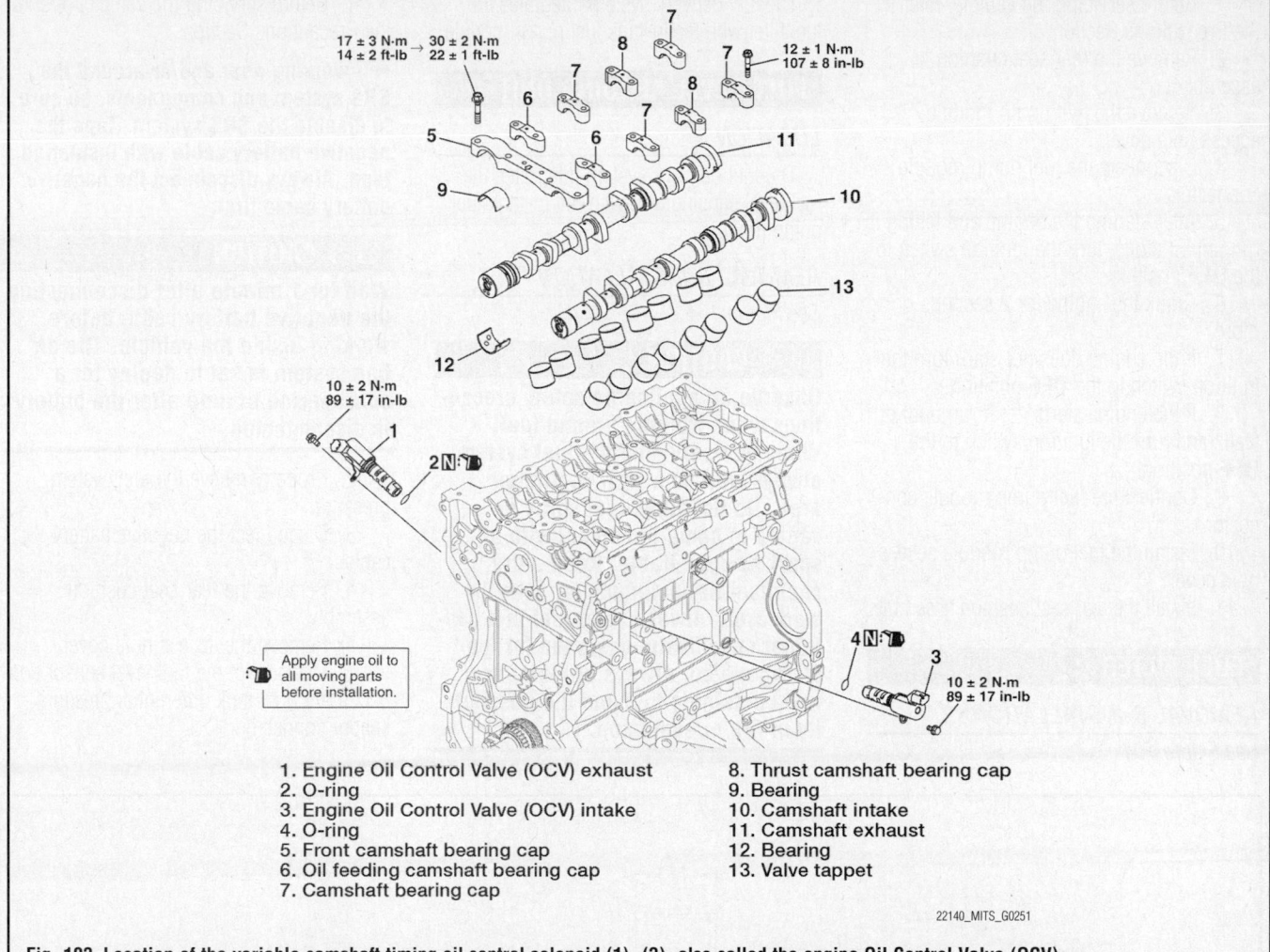

17 ± 3 N·m
14 ± 2 ft-lb → 30 ± 2 N·m
22 ± 1 ft-lb

12 ± 1 N·m
107 ± 8 in-lb

10 ± 2 N·m
89 ± 17 in-lb

10 ± 2 N·m
89 ± 17 in-lb

Apply engine oil to all moving parts before installation.

1. Engine Oil Control Valve (OCV) exhaust
2. O-ring
3. Engine Oil Control Valve (OCV) intake
4. O-ring
5. Front camshaft bearing cap
6. Oil feeding camshaft bearing cap
7. Camshaft bearing cap
8. Thrust camshaft bearing cap
9. Bearing
10. Camshaft intake
11. Camshaft exhaust
12. Bearing
13. Valve tappet

22140_MITS_G0251

Fig. 103 Location of the variable camshaft timing oil control solenoid (1), (3), also called the engine Oil Control Valve (OCV)

| FUEL | GASOLINE FUEL INJECTION SYSTEM |

FUEL SYSTEM SERVICE PRECAUTIONS

Safety is the most important factor when performing not only fuel system maintenance, but any type of maintenance. Failure to conduct maintenance and repairs in a safe manner may result in serious personal injury or death. Work on a vehicle's fuel system components can be accomplished safely and effectively by adhering to the following rules and guidelines.

• To avoid the possibility of fire and personal injury, always disconnect the negative battery cable unless the repair or test procedure requires that battery voltage be applied.

• Always relieve the fuel system pressure prior to disconnecting any fuel system component (injector, fuel rail, pressure regulator, etc.) fitting or fuel line connection. Exercise extreme caution whenever relieving fuel sys-

tem pressure to avoid exposing skin, face and eyes to fuel spray. Please be advised that fuel under pressure may penetrate the skin or any part of the body that it contacts.

• Always place a shop towel or cloth around the fitting or connection prior to loosening to absorb any excess fuel due to spillage. Ensure that all fuel spillage is quickly removed from engine surfaces. Ensure that all fuel-soaked cloths or towels are deposited into a flame-proof waste container with a lid.

• Always keep a dry chemical (Class B) fire extinguisher near the work area.

• Do not allow fuel spray or fuel vapors to come into contact with a spark or open flame.

• Always use a second wrench when loosening or tightening fuel line connection fittings. This will prevent unnecessary stress and torsion on fuel piping. Always follow the proper torque specifications.

• Always replace worn fuel fitting

O-rings with new ones. Do not substitute fuel hose where rigid pipe is installed.

FUEL SYSTEM PRESSURE

RELIEVING

❄ CAUTION

Observe all applicable safety precautions when working around fuel. Whenever servicing the fuel system, always work in a well ventilated area. Do not allow fuel spray or vapors to come in contact with a spark or open flame. Keep a dry chemical fire extinguisher near the work area. Always keep fuel in a container specifically designed for fuel storage; also, always properly seal fuel containers to avoid the possibility of fire or explosion.

1. Before servicing the vehicle, refer to the Precautions Section.

2. Remove the rear seat cushion assembly.

3. Remove the fuel pump module access hole cover.

4. Disconnect the fuel pump module connector.

5. After starting the engine and letting it run until it stops, turn the ignition switch to the **OFF** position.

6. Crank the engine for 2 seconds or more.

7. If the engine does not start, turn the ignition switch to the **OFF** position.

8. If the engine starts, let it run until it stalls and turn the ignition switch to the **OFF** position.

9. Connect the fuel pump module connector.

10. Install the fuel pump module access hole cover.

11. Install the rear seat cushion assembly.

FUEL FILTER

REMOVAL & INSTALLATION

See Figure 104.

The fuel delivery system integrates the fuel filter with the in-tank fuel pump module.

FUEL LEVEL SENDING UNIT

LOCATION

The fuel delivery system integrates the fuel level sensing unit with the in-tank fuel pump module.

REMOVAL & INSTALLATION

See Figure 104.

✳✳ CAUTION

Observe all applicable safety precautions when working around fuel. Whenever servicing the fuel system, always work in a well ventilated area. Do not allow fuel spray or vapors to come in contact with a spark or open flame. Keep a dry chemical fire extinguisher near the work area. Always keep fuel in a container specifically designed for fuel storage; also, always properly seal fuel containers to avoid the possibility of fire or explosion.

1. Before servicing the vehicle, refer to the Precautions Section.

➡ If working near and/or around the SRS system and components, be sure to disable the SRS system. Tape the negative battery cable with insulating tape. Always disconnect the negative battery cable first.

✳✳ CAUTION

Wait for 1 minute after disconnecting the negative battery cable before working inside the vehicle. The air bag system is set to deploy for a short period of time after the battery is disconnected.

2. Properly relieve the fuel system pressure.

3. Disconnect the negative battery cable.

4. Remove the rear seat cushion assembly.

5. Remove the access hole cover.

6. Disconnect the fuel level sensor connector and fuel tank differential pressure sensor connector.

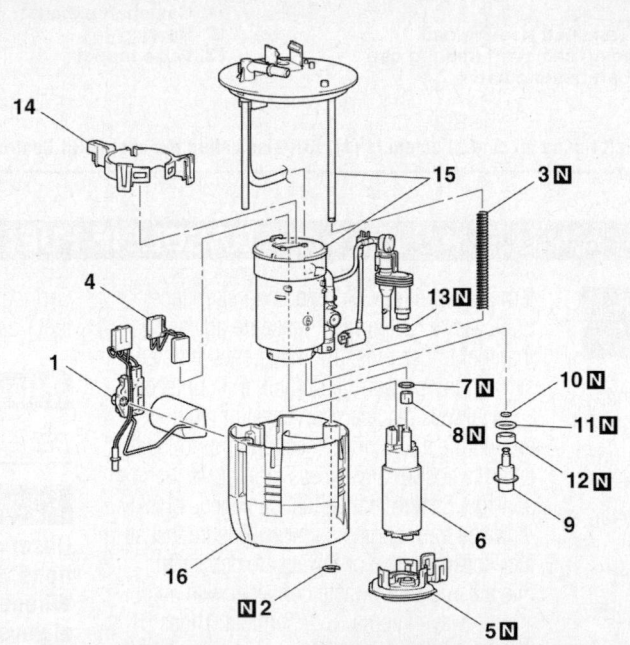

1. Fuel level sensor
2. Clip
3. Spring
4. Fuel pump harness
5. Fuel filter
6. Fuel pump assembly
7. O-ring
8. Spacer
9. Fuel pressure regulator
10. O-ring
11. O-ring
12. Back-up ring
13. O-ring
14. Bracket
15. Fuel flange and filter assembly
16. Sub tank assembly

22140_MITS_G0169

Fig. 104 Fuel pump module and related components

✳✳ CAUTION

When removing and installing the fuel level sensor from the fuel tank, be careful not damage the sensor unit and the float.

7. Remove the fuel level sensor mounting bolts and remove the fuel tank gauge unit from service hole.

To install:

➡ **Be sure to use new fasteners, as required.**

8. Install the fuel level sensor to the fuel tank through the service hole.

9. Connect the fuel level sensor (sub) connector and fuel tank differential pressure sensor connector.

10. Install the access hole cover.

11. Install the rear seat cushion assembly.

FUEL PUMP MODULE

REMOVAL & INSTALLATION
See Figure 104.

✳✳ CAUTION

Observe all applicable safety precautions when working around fuel. Whenever servicing the fuel system, always work in a well ventilated area. Do not allow fuel spray or vapors to come in contact with a spark or open flame. Keep a dry chemical fire extinguisher near the work area. Always keep fuel in a container specifically designed for fuel storage; also, always properly seal fuel containers to avoid the possibility of fire or explosion.

1. Before servicing the vehicle, refer to the Precautions Section.

➡ **If working near and/or around the SRS system and components, be sure to disable the SRS system. Tape the negative battery cable with insulating tape. Always disconnect the negative battery cable first.**

✳✳ CAUTION

Wait for 1 minute after disconnecting the negative battery cable before working inside the vehicle. The air bag system is set to deploy for a short period of time after the battery is disconnected.

2. Relieve the fuel system pressure.

3. Disconnect the negative battery cable.

4. Remove or disconnect the following:
- Rear seat cushion by pulling the seat stopper near the floor and lifting the cushion up
- Inspection cover on the right side of the vehicle
- Harness connector and the fuel lines
- Fuel pump assemble from the tank.

To install:

5. Install or connect the following:
- Fuel pump in the tank
- Hoses and the harness connector
- Inspection cover
- Rear seat
- Negative battery cable

FUEL RAIL AND INJECTOR

REMOVAL & INSTALLATION

2.4L Engine

1. Before servicing the vehicle, refer to the Precautions Section.

➡ **If working near and/or around the SRS system and components, be sure to disable the SRS system. Tape the negative battery cable with insulating tape. Always disconnect the negative battery cable first.**

✳✳ CAUTION

Wait for 1 minute after disconnecting the negative battery cable before working inside the vehicle. The air bag system is set to deploy for a short period of time after the battery is disconnected.

2. Relieve the fuel system pressure.

3. Disconnect the negative battery cable.

4. Remove the air cleaner cover and air intake hose assembly.

5. Disconnect the PCV valve hose connection, ignition coil connectors, EGR valve connector and fuel injector connectors.

6. Disconnect the throttle position sensor connector, manifold absolute pressure sensor connector, EVAP emission purge solenoid connector and knock sensor connector.

7. Disconnect the power steering pressure switch connector.

8. Remove the rocker cover bracket bolts.

9. Disconnect the high pressure hose connection. Remove the fuel rail, insulators, grommets, and injectors. Discard the O-rings.

To install:

➡ **Be sure to use new fasteners, as required.**

10. Installation is the reverse of the removal procedure. Be sure to use new O-rings and gaskets as required.

11. Connect the scan tool to the data link connector. To prevent damage to the scan tool be sure that the ignition switch is in the LOCK position before connecting the scan tool.

12. Turn the ignition switch to the ON position.

13. Select "check mode" from the menu screen.

14. Select "erase memory" from the menu screen.

15. Initialize the learning value.

16. Start the engine and check for leaks, correct as required.

3.8L Engine

See Figure 105.

1. Before servicing the vehicle, refer to the Precautions Section.

➡ **If working near and/or around the SRS system and components, be sure to disable the SRS system. Tape the negative battery cable with insulating tape. Always disconnect the negative battery cable first.**

✳✳ CAUTION

Wait for 1 minute after disconnecting the negative battery cable before working inside the vehicle. The air bag system is set to deploy for a short period of time after the battery is disconnected.

2. Relieve the fuel system pressure.

3. Disconnect the negative battery cable.

4. Remove the intake manifold plenum assembly.

5. Disconnect the fuel injector connectors.

6. Remove the control wiring harness bracket mounting bolts.

7. Remove the engine mounting stay.

8. Disconnect the high pressure fuel hose connection. Remove the fuel rail and fuel injector assembly.

9. Remove the insulators, O-rings, and fuel injectors. Discard the gaskets and O-rings.

To install:

➡ **Be sure to use new fasteners, as required.**

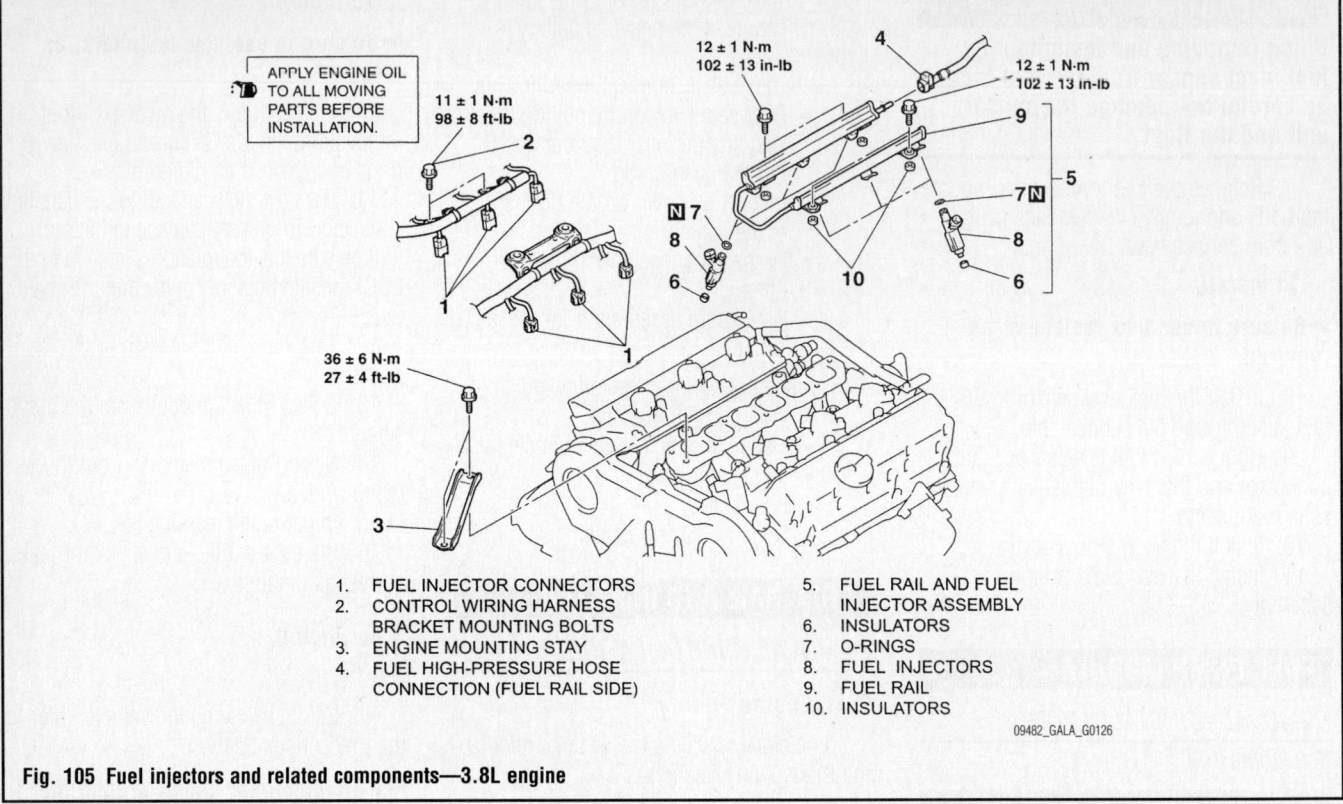

APPLY ENGINE OIL TO ALL MOVING PARTS BEFORE INSTALLATION.

11 ± 1 N·m
98 ± 8 ft-lb

12 ± 1 N·m
102 ± 13 in-lb

12 ± 1 N·m
102 ± 13 in-lb

36 ± 6 N·m
27 ± 4 ft-lb

1. FUEL INJECTOR CONNECTORS
2. CONTROL WIRING HARNESS BRACKET MOUNTING BOLTS
3. ENGINE MOUNTING STAY
4. FUEL HIGH-PRESSURE HOSE CONNECTION (FUEL RAIL SIDE)
5. FUEL RAIL AND FUEL INJECTOR ASSEMBLY
6. INSULATORS
7. O-RINGS
8. FUEL INJECTORS
9. FUEL RAIL
10. INSULATORS

09482_GALA_G0126

Fig. 105 Fuel injectors and related components—3.8L engine

10. Installation is the reverse of the removal procedure. Be sure to use new O-rings and gaskets as required.

11. Connect the scan tool to the data link connector. To prevent damage to the scan tool be sure that the ignition switch is in the LOCK position before connecting the scan tool.

12. Turn the ignition switch to the ON position.

13. Select "check mode" from the menu screen.

14. Select "erase memory" from the menu screen.

15. Initialize the learning value.

16. Start the engine and check for leaks, correct as required.

FUEL TANK

REMOVAL & INSTALLATION

✳✳ CAUTION

The fuel injection system remains under pressure even after the engine has been turned OFF. Properly relieve fuel pressure before disconnecting any fuel lines. Failure to do so may result in fire or personal injury. Do not allow fuel spray or fuel vapors to come in contact with a spark or an open flame. Keep a dry chemical fire extinguisher nearby. Never store fuel in an open container due to risk of fire or explosion.

1. Before servicing the vehicle, refer to the Precautions Section.

➡**If working near and/or around the SRS system and components, be sure to disable the SRS system. Tape the negative battery cable with insulating tape. Always disconnect the negative battery cable first.**

✳✳ CAUTION

Wait for 1 minute after disconnecting the negative battery cable before working inside the vehicle. The air bag system is set to deploy for a short period of time after the battery is disconnected.

2. Relieve the fuel system pressure.
3. Drain the fuel from the tank.
4. Remove the negative battery cable.
5. Remove the rear seat cushion assembly.
6. Remove the fuel pump module hole cover.
7. Disconnect the fuel pump module connector and fuel tank differential pressure sensor connector connection.
8. Raise and safely support the vehicle.
9. Remove the center exhaust pipe.

10. Remove the fuel tank protector, if equipped.
11. Disconnect the parking brake cable clamp.
12. Disconnect the fuel tank vapor hose connection.
13. Disconnect the fuel filler hose connection.
14. Disconnect the fuel high-pressure hose connection.

✳✳ CAUTION

As there will be some pressure remaining in the fuel pipe line, cover it with a shop towel to prevent fuel from spraying out.

15. Remove the fuel tank band.
   a. Support the fuel tank with a transaxle jack.
   b. Remove the front securing nut of the fuel tank band. Tilt the fuel tank assembly forward and lower it gradually to remove it. Remove the fuel tank band.
16. Carefully lower the fuel tank assembly from the vehicle.

To install:
17. Install the fuel tank bands.
   a. Raise the fuel tank assembly carefully with a transaxle jack.
   b. Ensure that the fuel tank assembly does not interfere with surrounding parts.

Install the fuel tank band and tighten to 15–23 ft. lbs. (21–31 Nm).

18. Connect the fuel high-pressure hose connection.

a. After installing, slightly pull the fuel high-pressure hose and ensure a good connection.

19. Connect the fuel filler hose connection.

20. Connect the fuel tank vapor hose connection.

21. Connect the parking brake cable clamp.

22. Install the fuel tank protector, if equipped.

23. Install the center exhaust pipe.

24. Connect the fuel level sensor connector.

25. Connect the fuel tank differential pressure sensor connector.

26. Connect the fuel pump module connector.

27. Install the service hole cover.

28. Install the rear seat cushion assembly.

29. Fill the fuel tank.

30. Start the engine and check for leaks.

IDLE SPEED

ADJUSTMENT

Idle speed is maintained by the Powertrain Control Module (PCM). No adjustment is necessary or possible.

THROTTLE BODY

REMOVAL & INSTALLATION

See Figures 106 and 107.

❋❋ CAUTION

The fuel injection system remains under pressure even after the engine has been turned OFF. Properly relieve fuel pressure before disconnecting any fuel lines. Failure to do so may result in fire or personal injury. Do not allow fuel spray or fuel vapors to come in contact with a spark or an open flame. Keep a dry chemical fire extinguisher nearby. Never store fuel in an open container due to risk of fire or explosion.

➡When the throttle body assembly replacement is performed, use scan tool MB991958 to initialize the learning value. Do not loosen the fixing screws for the resin cover of throttle body assembly. If the screws are loosened, the sensor incorporated in the resin

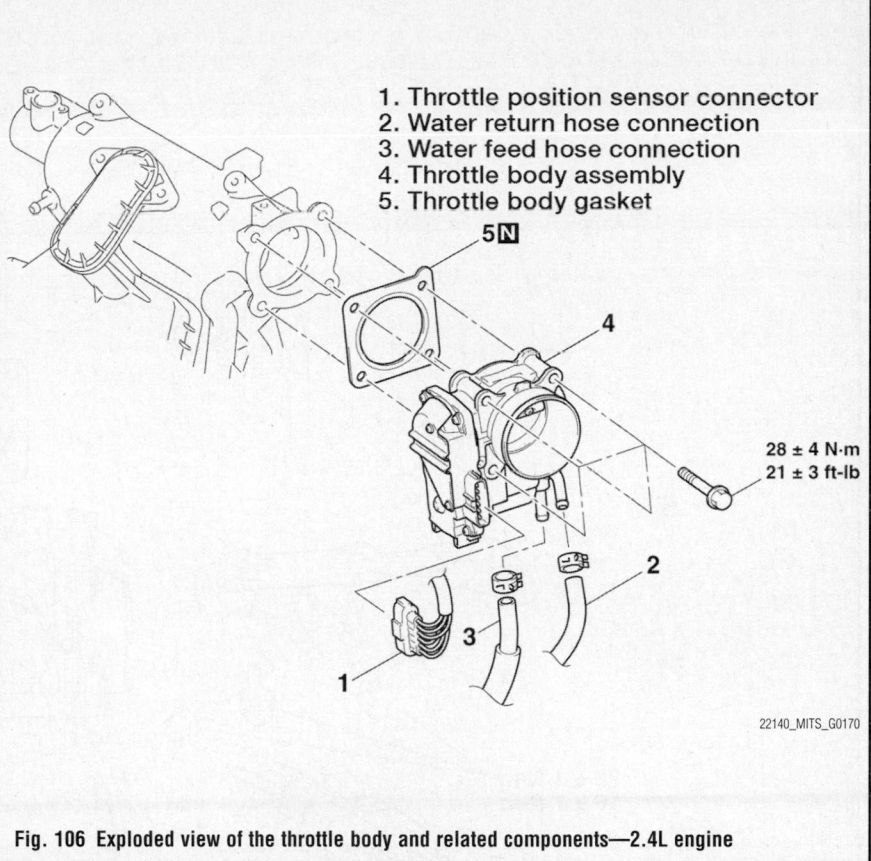

1. Throttle position sensor connector
2. Water return hose connection
3. Water feed hose connection
4. Throttle body assembly
5. Throttle body gasket

28 ± 4 N·m
21 ± 3 ft-lb

22140_MITS_G0170

Fig. 106 Exploded view of the throttle body and related components—2.4L engine

cover becomes misaligned and the throttle body may not work normally.

1. Before servicing the vehicle, refer to the Precautions Section.

➡**If working near and/or around the SRS system and components, be sure to disable the SRS system. Tape the negative battery cable with insulating tape. Always disconnect the negative battery cable first.**

❋❋ CAUTION

Wait for 1 minute after disconnecting the negative battery cable before working inside the vehicle. The air bag system is set to deploy for a short period of time after the battery is disconnected.

2. Relieve the fuel system pressure.

3. Drain the engine cooling system.

4. Remove the engine air cleaner assembly.

5. Matchmark the location of the adjuster bolt on the accelerator cable mounting flange. This will assure that the cable is installed in its original location. Remove the throttle cable adjusting bolt and disconnect the cable from the lever on the throttle body. Position cable aside.

6. Remove the connection for the breather hose and the air intake hose from the throttle body and position aside.

7. Tag and disconnect the necessary vacuum hoses.

8. Label and detach the electrical connectors at the throttle body, as necessary.

9. Disconnect the water and water bypass hoses at the base of the throttle body.

10. If equipped, unfasten the ground plate mounting screws, then remove the throttle body stay and ground plate from the engine.

11. Remove the air fitting and gasket.

12. Unfasten the throttle body mounting bolts, then remove the throttle body from the engine. Remove and discard the gasket.

To install:

➡**Be sure to use new fasteners, as required.**

13. Clean all old gasket material from the both throttle body mounting surfaces. Install new gasket onto the intake manifold plenum mounting surface.

➡**Poor idling quality and poor performance may be experienced if the gasket is installed incorrectly.**

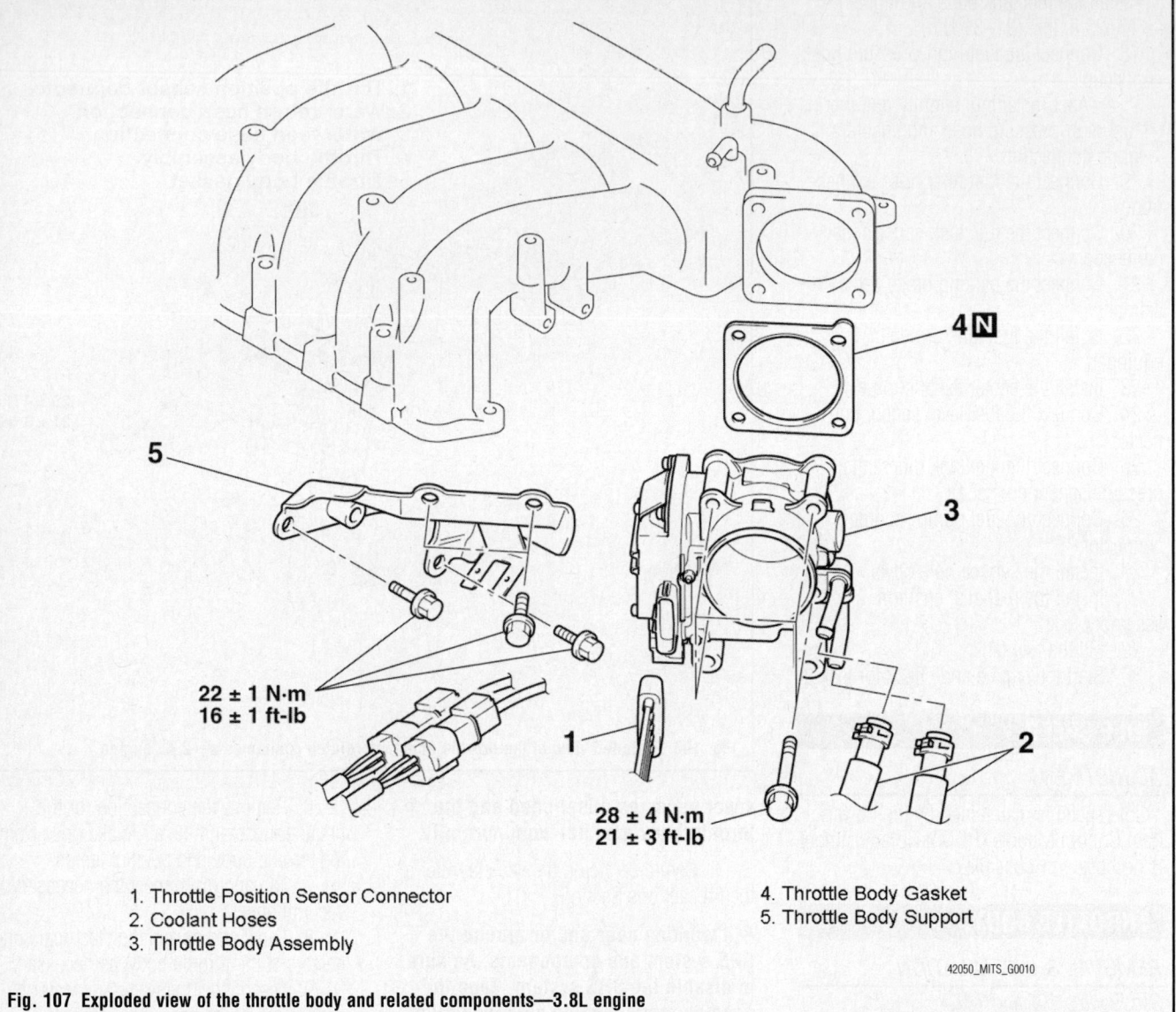

1. Throttle Position Sensor Connector
2. Coolant Hoses
3. Throttle Body Assembly
4. Throttle Body Gasket
5. Throttle Body Support

22 ± 1 N·m
16 ± 1 ft-lb

28 ± 4 N·m
21 ± 3 ft-lb

42050_MITS_G0010

Fig. 107 Exploded view of the throttle body and related components—3.8L engine

14. Install the throttle body to the intake manifold plenum and tighten the mounting bolts.

15. Install the air fitting, if equipped, making sure new gasket is in place.

16. If equipped, install the throttle body stay and ground plate. Secure with retainers tightened to 11–16 ft. lbs. (15–22 Nm).

17. Install the ground plate mounting screw.

18. Connect the water hoses to the throttle body. Install new hose clamps if required.

19. Attach the electrical and vacuum connectors to the throttle body, as tagged during removal.

20. Connect the accelerator cable to the throttle body and install the adjusting bolt in original position. Check adjustment of cable.

21. Install the air intake and breather hoses.

22. If removed, install the battery and connect the positive cable.

23. Connect the negative battery cable.

24. Refill the cooling system.

25. Run the engine and check for leaks and proper operation.

HEATING & AIR CONDITIONING SYSTEM

BLOWER MOTOR

REMOVAL & INSTALLATION

1. Before servicing the vehicle, refer to the Precautions Section.

➥If working near and/or around the SRS system and components, be sure to disable the SRS system. Tape the negative battery cable with insulating tape. Always disconnect the negative battery cable first.

✳✳ CAUTION

Wait for 1 minute after disconnecting the negative battery cable before working inside the vehicle. The air bag system is set to deploy for a short period of time after the battery is disconnected.

2. Disconnect the negative battery cable.

3. Remove the three instrument panel

undercover mounting screws and remove the cover.

4. Detach the electrical connector from the fan motor.

5. Remove the three small bolts holding the motor to the housing and remove the motor and fan.

To install:

➡ **Be sure to use new fasteners, as required.**

6. Check the inside of the case carefully; any debris can snag the fan and cause noise or poor airflow.

7. Install the blower motor, in the blower case and secure with the three mounting bolts.

8. Attach the blower motor electrical connector.

9. Install the undercover, taking care to insure it is in place and all the fasteners are secure.

10. Connect the negative battery cable.

HEATER CORE

REMOVAL & INSTALLATION

1. Before servicing the vehicle, refer to the Precautions Section.

➡ **If working near and/or around the SRS system and components, be sure to disable the SRS system. Tape the negative battery cable with insulating tape. Always disconnect the negative battery cable first.**

✳✳ CAUTION

Wait for 1 minute after disconnecting the negative battery cable before working inside the vehicle. The air bag system is set to deploy for a short period of time after the battery is disconnected.

2. Disconnect the negative battery cable.
3. Drain the cooling system.
4. Discharge the air conditioning system.
5. Be sure the front tires are in the straight ahead position.
6. Remove the Heater/Air Conditioning unit, as outlined in this section.
7. Remove the heater core from the assembly.

To install:

➡ **Be sure to use new fasteners, as required.**

8. Installation is the reverse of the removal procedure.

✳✳ WARNING

The tilt lever should be held in the lock position until the steering column is reinstalled in the vehicle. If the column is removed with the lever released, or the lever released after the column is removed from the vehicle, the steering column cannot be reinstalled correctly. If the steering column is installed incorrectly, the collision energy absorbing mechanism may be damaged.

➡ **After the seats have been reinstalled be sure to perform the accuracy check of the occupant classification sensor, using the Mitsubishi Diagnostic Scan tool, or equivalent. Follow the directions on the tool screen.**

9. Be sure to fill the cooling system with the proper grade and type coolant.

10. Recharge the air conditioning system.

HEATER/AIR CONDITIONING UNIT

REMOVAL & INSTALLATION

See Figures 108 through 111.

1. Before servicing the vehicle, refer to the Precautions Section.

➡ **If working near and/or around the SRS system and components, be sure to disable the SRS system. Tape the negative battery cable with insulating tape. Always disconnect the negative battery cable first.**

✳✳ CAUTION

Wait for 1 minute after disconnecting the negative battery cable before working inside the vehicle. The air bag system is set to deploy for a short period of time after the battery is disconnected.

2. Disconnect the negative battery cable.
3. Drain the cooling system.
4. Discharge the air conditioning system.
5. Be sure the front tires are in the straight ahead position.
6. Remove the front seat. Remove the glovebox assembly.
7. Remove the instrument panel parcel box and glovebox lock. Remove the hood release handle.
8. Remove the instrument panel under passenger's side cover. Remove the instrument lower panel. Remove the instrument center panel assembly.

9. Remove the radio and CD player. Remove the CD changer, if equipped. Remove the accessory box.

10. Remove the console meter hood. Remove the passenger's side air bag indicator light and passenger's seat belt warning light.

11. If equipped, remove the multi center display. Remove the center console assembly.

12. Remove the instrument panel cover. Remove the combination meter assembly. Remove the front pillar trim.

13. Remove the front speaker covers. Remove the speakers. Remove the instrument panel side cover. Remove the instrument panel side air outlet.

14. Remove the driver's side air bag module. Remove the steering wheel. Remove the steering column cover. Remove the clockspring connector and clockspring and column switch assembly.

15. Disconnect the electrical connections for the ignition switch. Remove the key interlock cable.

16. Remove the steering shaft pad. Remove the steering column retaining bolts. Remove the steering column assembly. Pinch the steering column shaft clip with a pliers, and pull up the shaft to disengage the steering column assembly.

✳✳ WARNING

The tilt lever should be held in the lock position until the steering column is reinstalled in the vehicle. If the column is removed with the lever released, or the lever released after the column is removed from the vehicle, the steering column cannot be reinstalled correctly. If the steering column is installed incorrectly, the collision energy absorbing mechanism may be damaged.

17. If equipped with automatic temperature control, remove the interior temperature sensor and photo sensor. Remove the instrument panel front end garnish.

18. Remove the gearshift lever handle, heated seat switch, accessory socket, socket cover, front plate, and front box.

19. Remove the lid lock lever, hinge, lid, liner box, accessory socket and cover.

20. Remove the floor console retaining screws. Remove the floor console. Remove the accessory socket harness and rear bracket.

21. Remove the cowl side trim. Remove the passenger's side air bag module.

22. Remove the instrument panel assembly from the vehicle.

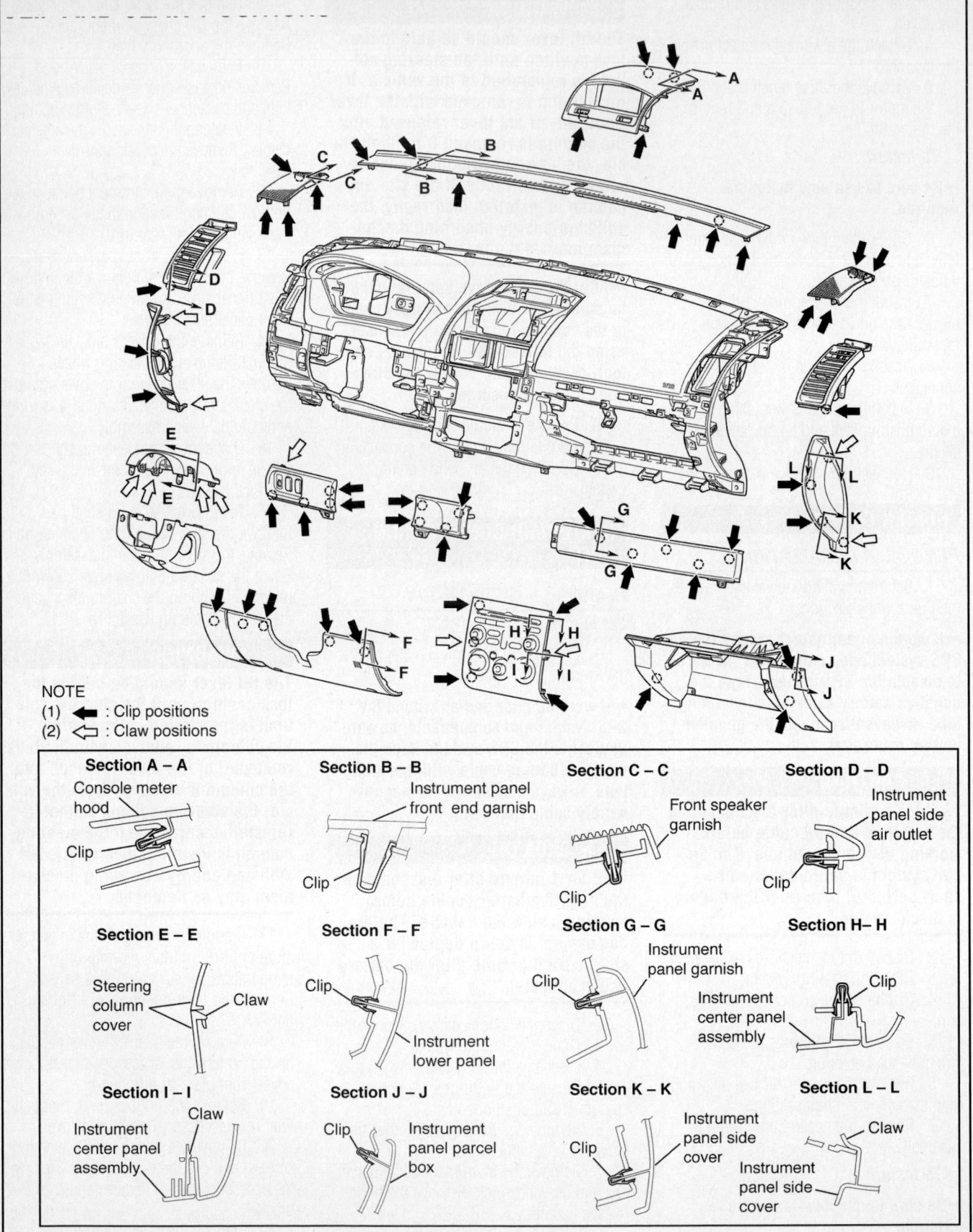

NOTE
(1) ◄━ : Clip positions
(2) ◄▭ : Claw positions

Section A – A
Console meter hood
Clip

Section B – B
Instrument panel front end garnish
Clip

Section C – C
Front speaker garnish
Clip

Section D – D
Instrument panel side air outlet
Clip

Section E – E
Steering column cover
Claw

Section F – F
Clip
Instrument lower panel

Section G – G
Instrument panel garnish
Clip

Section H– H
Clip
Instrument center panel assembly

Section I – I
Claw
Instrument center panel assembly

Section J – J
Clip
Instrument panel parcel box

Section K – K
Instrument panel side cover
Clip

Section L – L
Claw
Instrument panel side cover

37671_GALA_G0009

Fig. 108 Instrument panel retaining clip locations

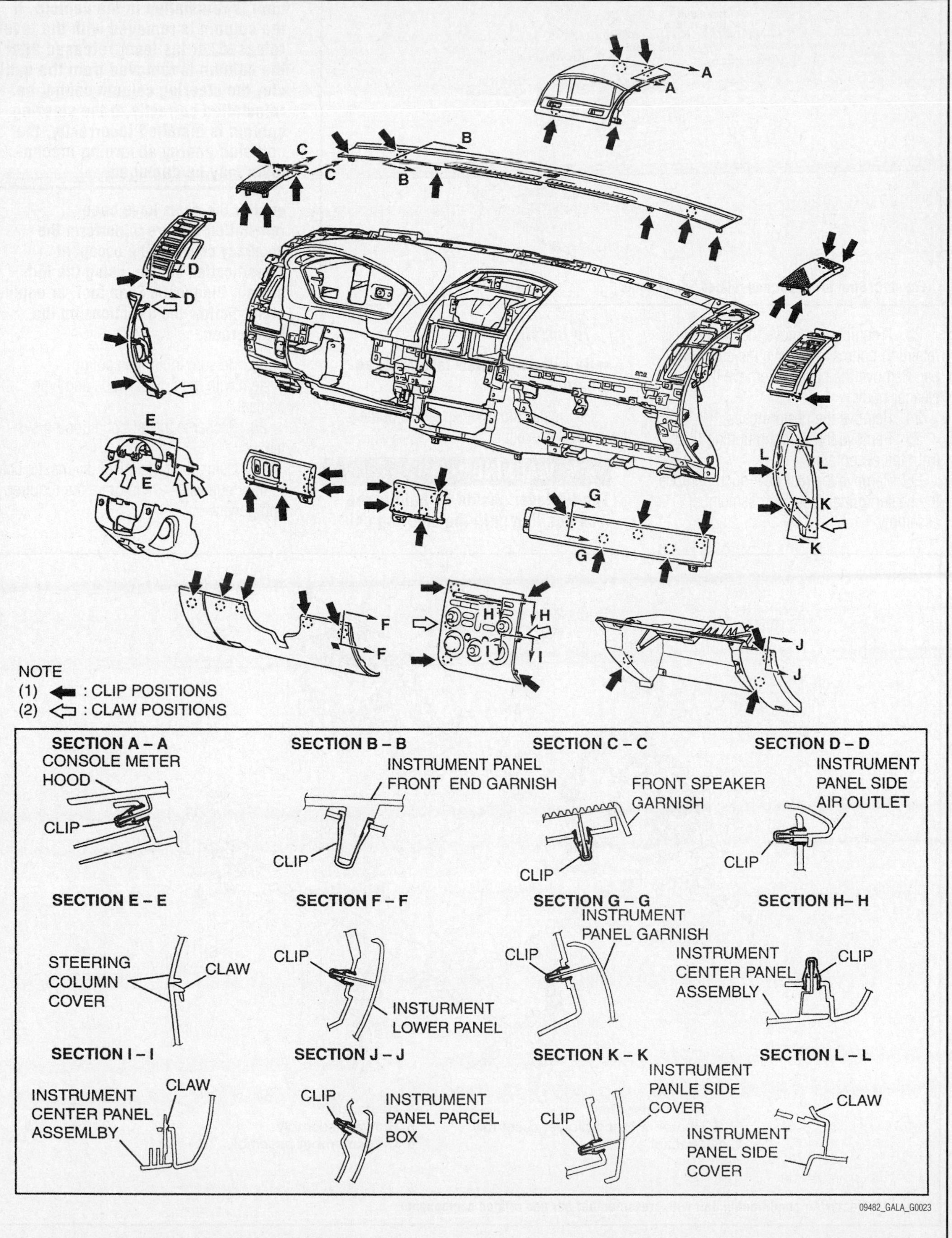

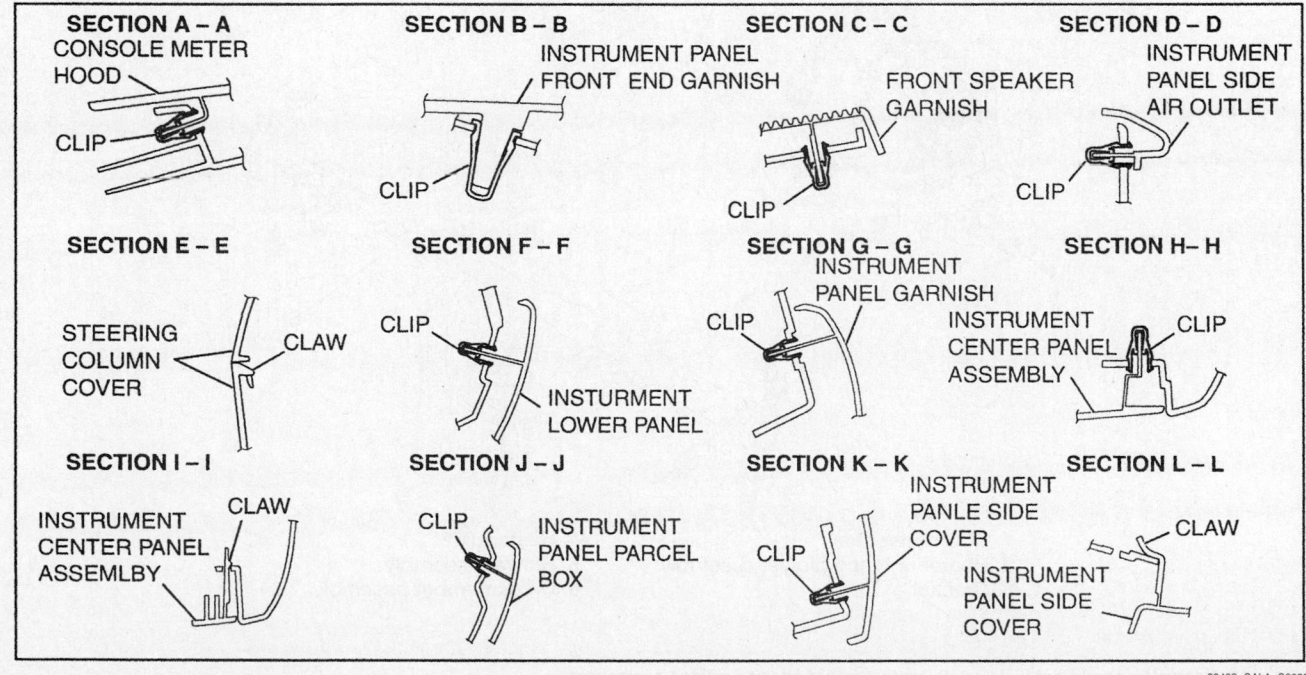

NOTE
(1) ← : CLIP POSITIONS
(2) ⇐ : CLAW POSITIONS

SECTION A – A
CONSOLE METER
HOOD
CLIP

SECTION B – B
INSTRUMENT PANEL
FRONT END GARNISH
CLIP

SECTION C – C
FRONT SPEAKER
GARNISH
CLIP

SECTION D – D
INSTRUMENT
PANEL SIDE
AIR OUTLET
CLIP

SECTION E – E
STEERING
COLUMN
COVER
CLAW

SECTION F – F
CLIP
INSTURMENT
LOWER PANEL

SECTION G – G
INSTRUMENT
PANEL GARNISH
CLIP

SECTION H– H
INSTRUMENT
CENTER PANEL
ASSEMBLY
CLIP

SECTION I – I
INSTRUMENT
CENTER PANEL
ASSEMLBY
CLAW

SECTION J – J
CLIP
INSTRUMENT
PANEL PARCEL
BOX

SECTION K – K
INSTRUMENT
PANLE SIDE
COVER
CLIP

SECTION L – L
CLAW
INSTRUMENT
PANEL SIDE
COVER

09482_GALA_G0023

Fig. 109 Instrument panel and related components

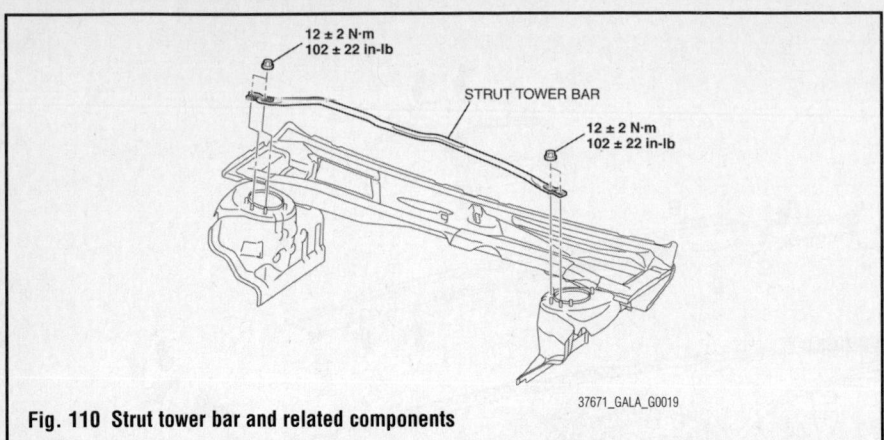

12 ± 2 N·m
102 ± 22 in-lb

STRUT TOWER BAR

12 ± 2 N·m
102 ± 22 in-lb

37671_GALA_G0019

Fig. 110 Strut tower bar and related components

23. If equipped remove the strut tower retaining bolts and remove the strut tower bar. Remove the battery. Remove the air cleaner body.

24. Remove the heater hoses.

25. Remove the air conditioning lines from the evaporator.

26. Remove the rear heat duct. Remove the heater unit and deck crossmember assembly.

To install:

➡**Be sure to use new fasteners, as required.**

27. Installation is the reverse of the removal procedure.

✳✳ WARNING

The tilt lever should be held in the lock position until the steering col- umn is reinstalled in the vehicle. If the column is removed with the lever released, or the lever released after the column is removed from the vehicle, the steering column cannot be reinstalled correctly. If the steering column is installed incorrectly, the collision energy absorbing mechanism may be damaged.

➡**After the seats have been reinstalled be sure to perform the accuracy check of the occupant classification sensor, using the Mitsubishi Diagnostic Scan tool, or equivalent. Follow the directions on the tool screen.**

28. Be sure to fill the cooling system with the proper grade and type coolant.

29. Recharge the air conditioning system.

30. Using the Mitsubishi diagnostic scan tool, or equivalent, reprogram the required components.

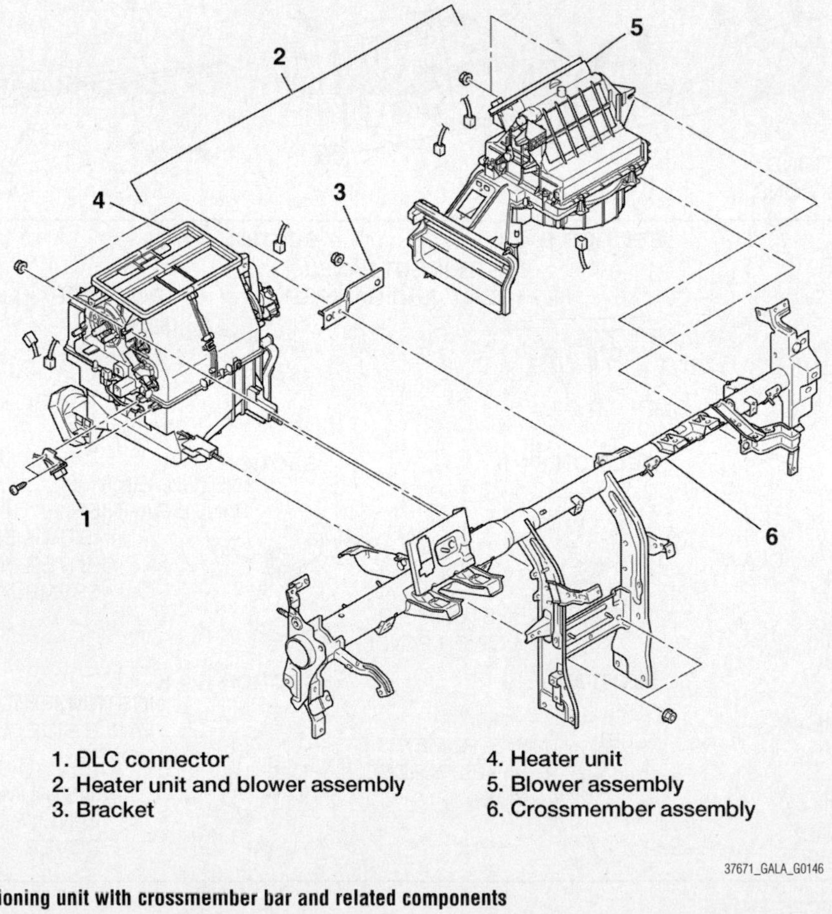

2
5
4
3
1
6

1. DLC connector
2. Heater unit and blower assembly
3. Bracket
4. Heater unit
5. Blower assembly
6. Crossmember assembly

37671_GALA_G0146

Fig. 111 Heater/Air conditioning unit with crossmember bar and related components

STEERING

POWER RACK & PINION STEERING GEAR

REMOVAL & INSTALLATION

See Figures 112 and 113.

➡ **Prior to removal of the steering rack, center the front wheels and remove the ignition key. Failure to do so may damage the SRS clockspring and render SRS system inoperative.**

1. Before servicing the vehicle, refer to the Precautions Section.

➡ **If working near and/or around the SRS system and components, be sure to disable the SRS system. Tape the negative battery cable with insulating tape. Always disconnect the negative battery cable first.**

❋❋ CAUTION

Wait for 1 minute after disconnecting the negative battery cable before working inside the vehicle. The air bag system is set to deploy for a short period of time after the battery is disconnected.

2. Disconnect the negative battery cable.

3. Drain the power steering fluid.

4. Remove the front undercover.

5. Remove the center crossmember.

6. Remove the lower arm assembly.

7. Remove the driver's side air bag module. Remove the steering wheel. Remove the column lower cover. Remove the clockspring.

8. Remove the floor console assembly. Remove the front scuff plate and cowl side trim. Remove the trunk lid opener cover.

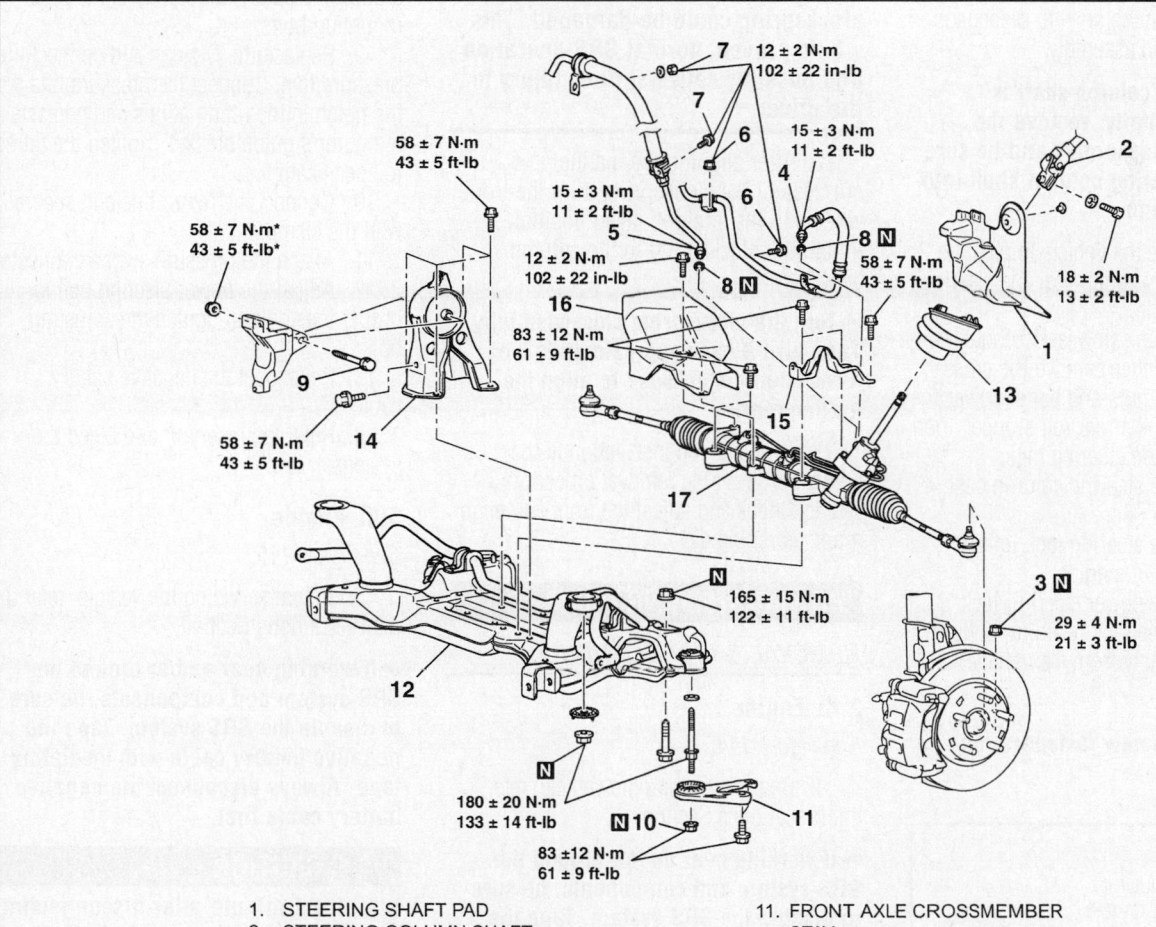

1. STEERING SHAFT PAD
2. STEERING COLUMN SHAFT ASSEMBLY AND STEERING GEAR CONNECTION
3. JAM NUT (TIE ROD END AND KNUCKLE CONNECTION)
4. RETURN TUBE CONNECTION
5. PRESSURE HOSE CONNECTION
6. RETURN TUBE CLAMP
7. PRESSURE HOSE CLAMP
8. O-RING
9. REAR ROLL STOPPER CONNECTING BOLT
10. JAM NUT
11. FRONT AXLE CROSSMEMBER STAY
12. CROSSMEMBER ASSEMBLY
13. STEERING COLUMN DASH PANEL COVER
14. REAR ROLL STOPPER
15. POWER STEERING GEAR BRACKET
16. STEERING GEAR AND LINKAGE PROTECTOR <3.8L ENGINE>
17. POWER STEERING GEAR AND LINKAGE

09482_GALA_G0135

Fig. 112 Power steering rack and related components

9. Remove the accelerator pedal stopper. Remove the front floor carpet.

10. Disconnect the stabilizer link and stabilizer bar. Remove the steering shaft pad.

11. Disconnect the steering column assembly and steering rack connection. Remove the rear roll stopper connecting bolt.

12. Remove the front axle crossmember bracket. Remove the return tube clamp bolt and nut. Remove the pressure tube clamp bolt and nut.

13. Disconnect and plug the return tube connection. Disconnect and plug the pressure hose connection. Discard the O-ring.

14. Disconnect the pinch clamp. Pinch the steering column shaft clip with a pliers, and pull upward on the shaft to disengage the steering column assembly.

➡If the steering column shaft is removed accidentally, remove the steering column assembly and be sure to insert the steering column shaft into the steering column.

15. From inside the vehicle loosen the 3 clips from the body panel, near the steering column to floor hole.

16. Use a transmission jack to hold the crossmember, and then remove the crossmember mounting nuts and bolts. Lower the crossmember with the rear roll stopper, then the stabilizer bar and steering rack.

17. Remove the steering column dash panel cover.

18. Remove the steering rack linkage protector, on the 3.8L engine.

19. Remove the power steering rack bracket. Remove the rack retaining bolts. Remove the assembly from the vehicle.

To install:

➡Be sure to use new fasteners, as required.

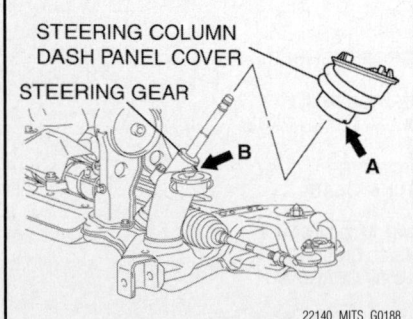

Fig. 113 Align the steering column dash panel cover notch "A" with the steering gear lug "B"

22140_MITS_G0188

20. Position the steering rack in the vehicle and install the clamp and the mounting bolts. Be sure the rack is centered before connecting it to the joint assembly.

21. When installing the steering column dash panel cover pinch clamp, align the steering column dash panel cover notch "A" with the steering gear lug arrow "B" and then install the steering column dash panel cover to the steering gear.

✳✳ CAUTION

When installing the clockspring be sure that the springs mating marks are properly aligned. If not the steering wheel may not rotate completely during a turn, or the flat cable in the clockspring could be damaged. This would prevent normal SRS operation and possibly cause serious injury to the driver.

22. After aligning the mating surfaces of the clockspring, turn the front wheels to the straight ahead position. Install the clockspring to the column switch.

➡Turn the clockspring clockwise fully. Then turn it back approximately 3 ¾ turns counterclockwise to align the mating marks.

23. Continue the installation in the reverse order of the removal procedure.

24. Check and adjust the front end alignment, as required.

POWER STEERING PUMP

REMOVAL & INSTALLATION

2.4L Engine

See Figure 114.

1. Before servicing the vehicle, refer to the Precautions Section.

➡If working near and/or around the SRS system and components, be sure to disable the SRS system. Tape the negative battery cable with insulating tape. Always disconnect the negative battery cable first.

✳✳ CAUTION

Wait for 1 minute after disconnecting the negative battery cable before working inside the vehicle. The air bag system is set to deploy for a short period of time after the battery is disconnected.

2. Disconnect the battery negative cable.

3. Loosen and remove the power steering pump drive belt.

4. Remove the pressure switch connector from the side of the pump.

5. Disconnect the return fluid line. Remove the reservoir cap and allow the return line to drain the fluid from the reservoir.

6. Disconnect the pressure line.

7. Unbolt and remove the pump from the mounting bracket.

To install:

➡Be sure to use new fasteners, as required.

8. Install the pump, wrap the belt around the pulley and lightly tighten the mounting bolts.

9. Replace the O-rings and connect the pressure line. Connect the pressure line so the notch in the fitting aligns and contacts the pump's guide bracket. Tighten the fitting to specification.

10. Connect the return line and secure with the clamp.

11. Attach the pressure switch connector.

12. Adjust the power steering belt for proper tension and tighten the adjusting bolts.

13. Reconnect the negative battery cable.

14. Refill the reservoir and bleed the system.

3.8L Engine

See Figure 115.

1. Before servicing the vehicle, refer to the Precautions Section.

➡If working near and/or around the SRS system and components, be sure to disable the SRS system. Tape the negative battery cable with insulating tape. Always disconnect the negative battery cable first.

✳✳ CAUTION

Wait for 1 minute after disconnecting the negative battery cable before working inside the vehicle. The air bag system is set to deploy for a short period of time after the battery is disconnected.

2. Disconnect the negative battery cable.

3. Remove the undercover.

4. Remove the strut tower bar, on RALLIART.

5. Remove the power steering drive belt.

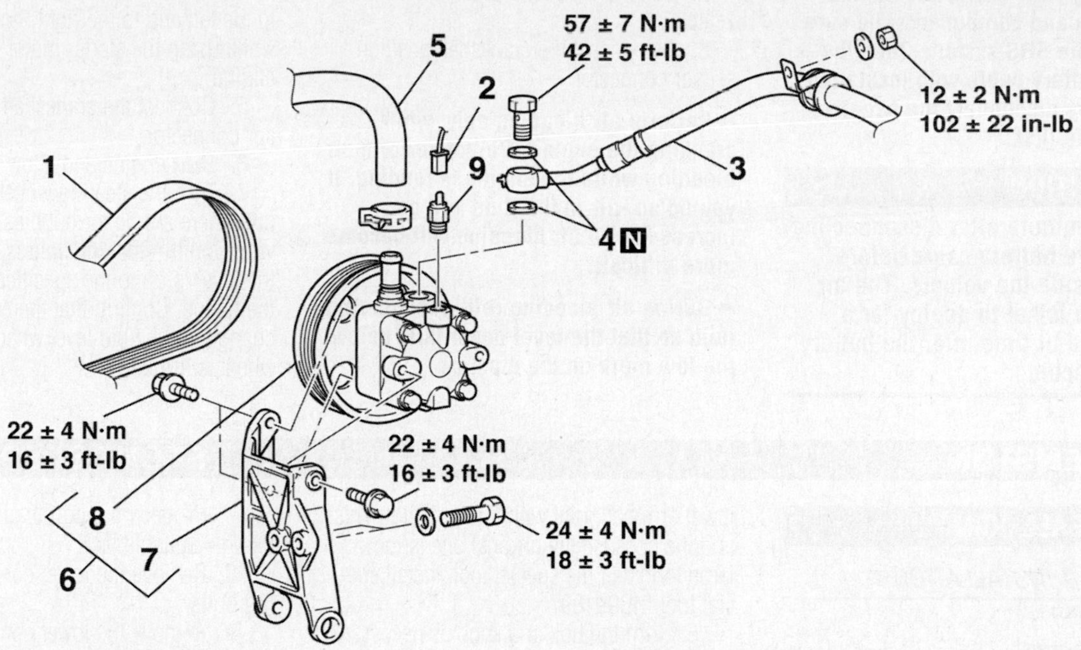

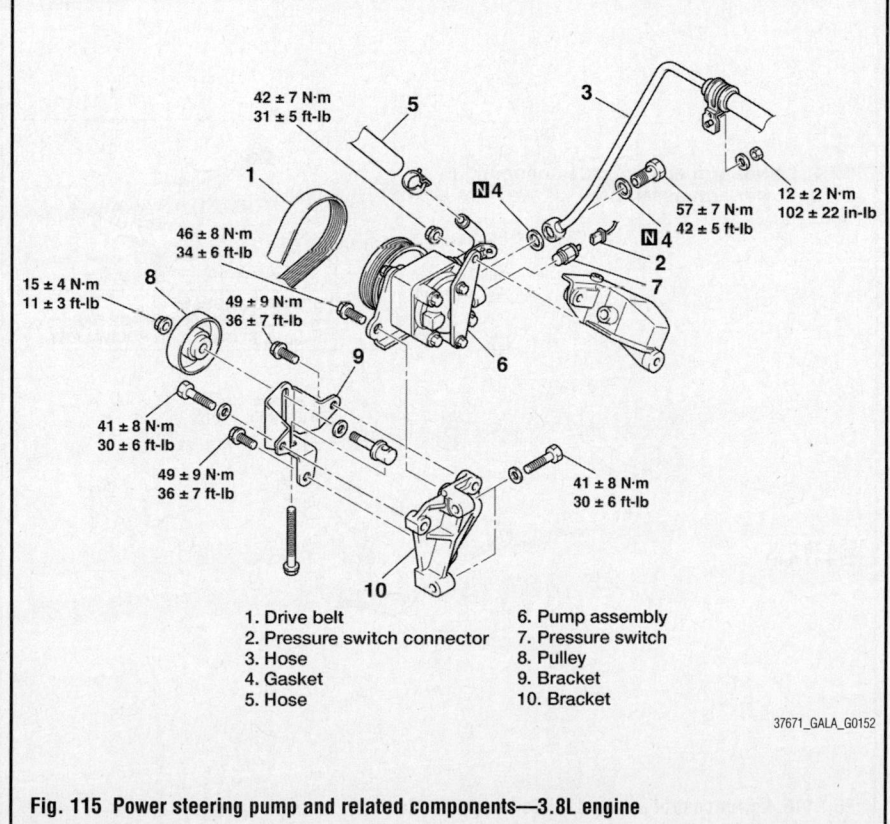

57 ± 7 N·m
42 ± 5 ft-lb

12 ± 2 N·m
102 ± 22 in-lb

22 ± 4 N·m
16 ± 3 ft-lb

22 ± 4 N·m
16 ± 3 ft-lb

24 ± 4 N·m
18 ± 3 ft-lb

1. Drive belt	6. Pump and bracket assembly
2. Pressure switch connector	7. Bracket
3. Hose	8. Pump assembly
4. Gasket	9. Pressure switch
5. Hose	

37671_GALA_G0151

Fig. 114 Power steering pump and related components—2.4L engine

42 ± 7 N·m
31 ± 5 ft-lb

46 ± 8 N·m
34 ± 6 ft-lb

15 ± 4 N·m
11 ± 3 ft-lb

57 ± 7 N·m
42 ± 5 ft-lb

12 ± 2 N·m
102 ± 22 in-lb

49 ± 9 N·m
36 ± 7 ft-lb

41 ± 8 N·m
30 ± 6 ft-lb

49 ± 9 N·m
36 ± 7 ft-lb

41 ± 8 N·m
30 ± 6 ft-lb

1. Drive belt	6. Pump assembly
2. Pressure switch connector	7. Pressure switch
3. Hose	8. Pulley
4. Gasket	9. Bracket
5. Hose	10. Bracket

37671_GALA_G0152

Fig. 115 Power steering pump and related components—3.8L engine

6. Drain the fluid. Be sure to properly dispose of used fluid.

7. Disconnect the pressure switch connector electrical connector.

8. Disconnect the supply and return hoses and gaskets.

9. Disconnect the stabilizer links.

10. Turn the steering wheel fully to the left.

11. Remove the power steering pump mounting bolts.

12. Remove the power steering pump.

To install:

➡**Be sure to use new fasteners, as required.**

13. Installation is the reverse order of removal.

14. Torque the power steering pump mounting bolts to specification.

15. Be sure to fill the system with new fluid. Be sure to use the proper grade and type fluid.

16. Start the engine and check for leaks.

17. Bleed the system.

BLEEDING

1. Before servicing the vehicle, refer to the Precautions Section.

➡If working near and/or around the SRS system and components, be sure to disable the SRS system. Tape the negative battery cable with insulating tape. Always disconnect the negative battery cable first.

✵✵ CAUTION

Wait for 1 minute after disconnecting the negative battery cable before working inside the vehicle. The air bag system is set to deploy for a short period of time after the battery is disconnected.

SUSPENSION

LOWER BALL JOINT

REMOVAL & INSTALLATION

See Figure 116.

The lower ball joint is an integral part of the lower control arm assembly and cannot be serviced separately. A worn or damaged ball joint requires replacement of lower control arm assembly.

LOWER CONTROL ARM

REMOVAL & INSTALLATION

See Figure 116.

1. Before servicing the vehicle, refer to the Precautions Section.

➡If working near and/or around the SRS system and components, be sure to disable the SRS system. Tape the negative battery cable with insulating tape. Always disconnect the negative battery cable first.

✵✵ CAUTION

Wait for 1 minute after disconnecting the negative battery cable before working inside the vehicle. The air bag system is set to deploy for a short period of time after the battery is disconnected.

2. Raise and support the vehicle safely.
3. Remove the tire and wheel assembly.

➡Do not remove the nut from the ball joint. Loosen it and use special tool MB991897 to avoid possible damage to the ball joint threads. Hang the special tool in place with wire or string to prevent it from falling.

4. To disconnect the lower arm from the knuckle, replace the self-locking nut for the

lower arm ball joint with a regular nut. This is done because the original one is too large to install the special tool. Install special tool MB991897.

5. Turn the bolt and knob as necessary to make the jaws of the tool parallel. Tighten the bolt by hand and confirm that the jaws are still parallel.

➡When adjusting the jaws in parallel, make sure the knob is in the vertical (upward) position.

6. Tighten the bolt with a wrench to disconnect the ball joint.

---

2. Raise and safely support the vehicle.
3. Disconnect the crankshaft position sensor connector.

➡Perform air bleeding only, while cranking the engine. Do not perform air bleeding while the engine is running. If you do so, air in the fluid will be increased and air bleeding will become more difficult.

➡During air bleeding refill the steering fluid so that the level never falls below the low mark on the dipstick.

4. Turn the steering wheel all the way to the left and to the right 5 or 6 times, while using the starter motor to crank the engine.
5. Connect the crankshaft position sensor connector.
6. Start engine and allow to idle.
7. Turn the steering wheel left and right until there are no air bubbles in the reservoir. Confirm that the fluid is not milky and the level is up to the specified position on the gauge. Confirm that there is very little change in the fluid level when the steering wheel is turned.

FRONT SUSPENSION

7. Properly support the lower control using a suitable jack.
8. Remove the lower control arm retaining bolts.
9. Remove the lower control arm from the vehicle.

To install:

➡Be sure to use new fasteners, as required.

10. Installation is the reverse of the removal procedure.
11. Tighten bolts to specification.

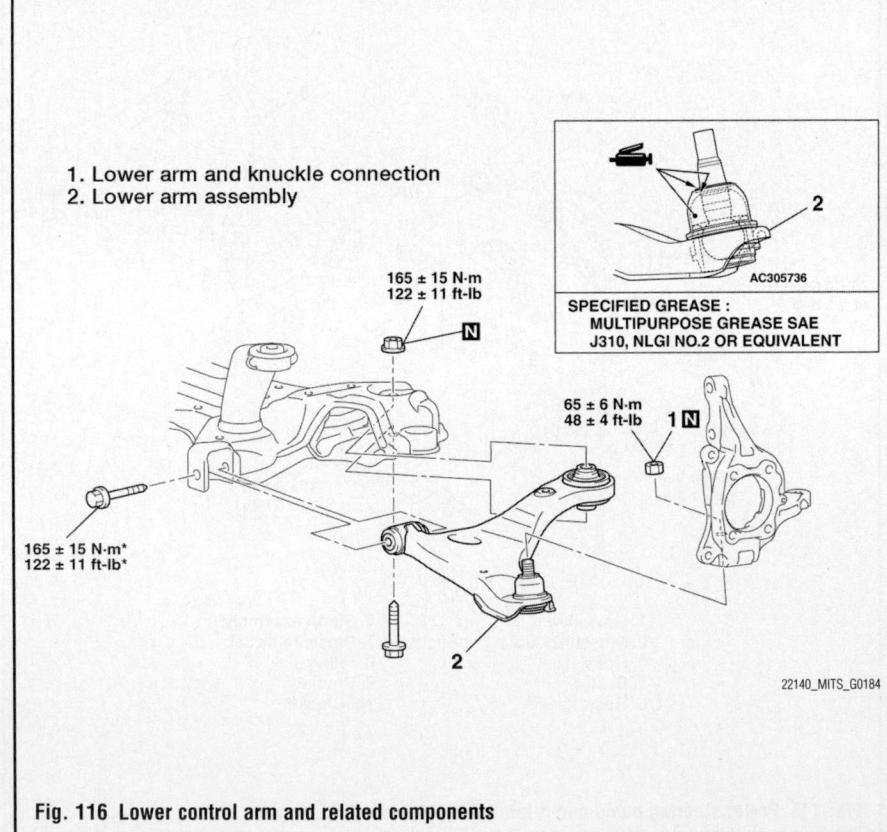

1. Lower arm and knuckle connection
2. Lower arm assembly

165 ± 15 N·m
122 ± 11 ft-lb

65 ± 6 N·m
48 ± 4 ft-lb

165 ± 15 N·m*
122 ± 11 ft-lb*

AC305736

SPECIFIED GREASE :
MULTIPURPOSE GREASE SAE
J310, NLGI NO.2 OR EQUIVALENT

22140_MITS_G0184

Fig. 116 Lower control arm and related components

➡️Refer to illustration and fully tighten bolts. Bolts with should be temporarily tightened, and fully tightened with the vehicle on the ground in an unladen condition.

12. Check and adjust the front end alignment, as required.

MACPHERSON STRUT

REMOVAL & INSTALLATION
See Figure 117.

1. Before servicing the vehicle, refer to the Precautions Section.

➡️If working near and/or around the SRS system and components, be sure to disable the SRS system. Tape the negative battery cable with insulating tape. Always disconnect the negative battery cable first.

✳️✳️ CAUTION

Wait for 1 minute after disconnecting the negative battery cable before working inside the vehicle. The air

bag system is set to deploy for a short period of time after the battery is disconnected.

2. Raise and support the vehicle safely.
3. Remove the tire and wheel assembly, as necessary.
4. Matchmark alignment markings on the camber adjusting bolt and strut for approximate installation alignment later.
5. Remove the strut tower bar, on RALLIART.
6. Disconnect the stabilizer link. Remove the brake hose bracket.
7. On vehicles equipped with ABS, remove the wheel speed sensor clamp.
8. Remove the lower strut bolts and nuts. Remove the upper strut nuts.
9. Remove the strut assembly from the vehicle.

To install:

➡️Be sure to use new fasteners, as required.

10. Installation is the reverse of the removal procedure.

11. Use the alignment marks made before removal to align the strut in position.
12. Tighten the bolts to specification.
13. Be sure to check and adjust the front end alignment, as required.

STABILIZER BAR

REMOVAL & INSTALLATION
See Figures 118 through 120.

➡️Prior to removal of the stabilizer bar, center the front wheels and remove the ignition key. Failure to do so may damage the SRS clockspring and render the SRS system inoperative.

1. Before servicing the vehicle, refer to the Precautions Section.

➡️If working near and/or around the SRS system and components, be sure to disable the SRS system. Tape the negative battery cable with insulating tape. Always disconnect the negative battery cable first.

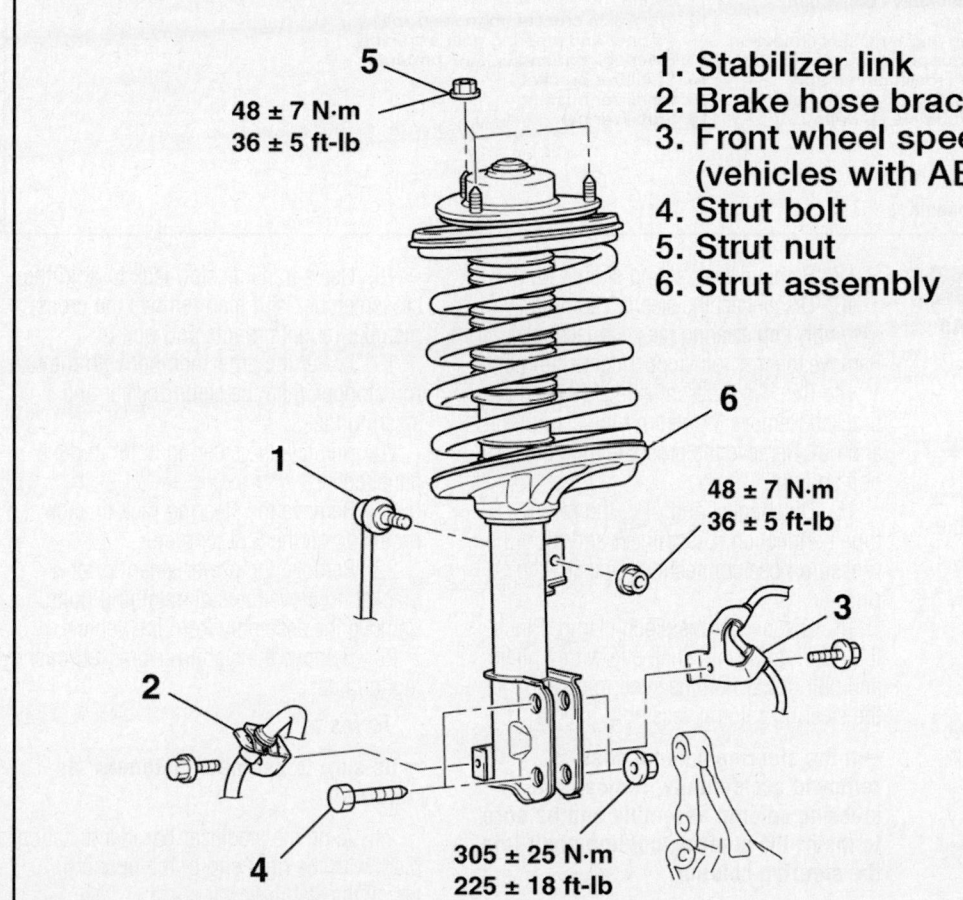

5
48 ± 7 N·m
36 ± 5 ft-lb

1. Stabilizer link
2. Brake hose bracket
3. Front wheel speed sensor clamp (vehicles with ABS)
4. Strut bolt
5. Strut nut
6. Strut assembly

48 ± 7 N·m
36 ± 5 ft-lb

305 ± 25 N·m
225 ± 18 ft-lb

22140_MITS_G0179

Fig. 117 Front strut and related components

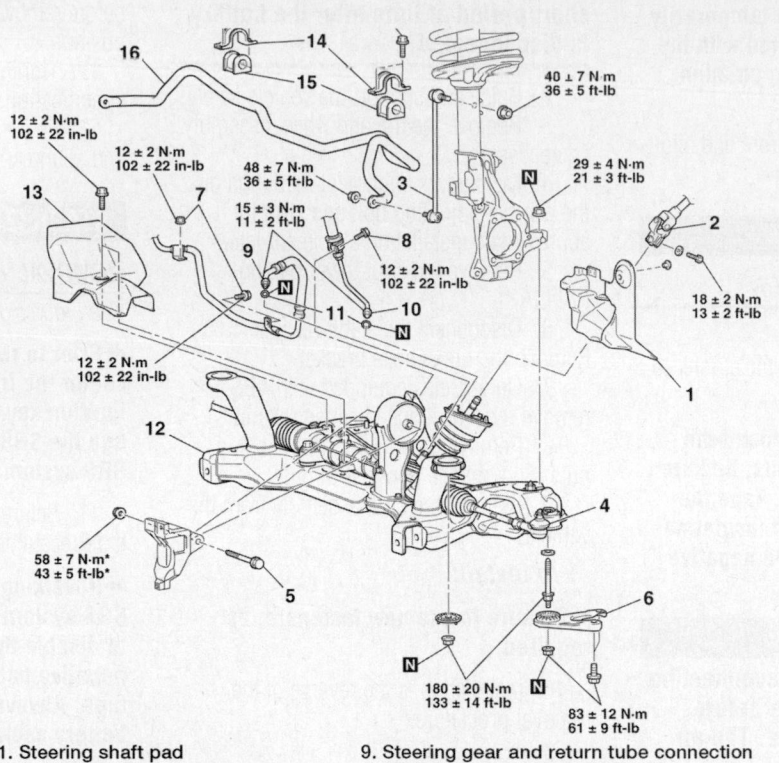

12 ± 2 N·m
102 ± 22 in-lb

12 ± 2 N·m
102 ± 22 in-lb

48 ± 7 N·m
36 ± 5 ft-lb

40 ± 7 N·m
36 ± 5 ft-lb

29 ± 4 N·m
21 ± 3 ft-lb

15 ± 3 N·m
11 ± 2 ft-lb

12 ± 2 N·m
102 ± 22 in-lb

18 ± 2 N·m
13 ± 2 ft-lb

12 ± 2 N·m
102 ± 22 in-lb

58 ± 7 N·m*
43 ± 5 ft-lb*

180 ± 20 N·m
133 ± 14 ft-lb

83 ± 12 N·m
61 ± 9 ft-lb

1. Steering shaft pad
2. Steering gear and steering
   column assembly connection
3. Stabilizer link
4. Tie rod end and knuckle connection
5. Rear roll stopper connection bolt
6. Front axle crossmember stay
7. Return tube clamp nut and bolt
8. Pressure tube clamp bolt

9. Steering gear and return tube connection
10. Steering gear and pressure tube connection
11. O-ring
12. Front axle crossmember, rear roll
    stopper and steering gear assembly
13. Steering gear and linkage protector
14. Stabilizer bracket
15. Stabilizer bushing
16. Stabilizer bar

22140_MITS_G0187

Fig. 118 Stabilizer bar and related components

⁂ CAUTION

Wait for 1 minute after disconnecting the negative battery cable before working inside the vehicle. The air bag system is set to deploy for a short period of time after the battery is disconnected.

2. Disconnect the negative battery cable.

3. Drain the power steering fluid.

4. Remove the front undercover.

5. Remove the center crossmember.

6. Remove the lower arm assembly.

7. Remove the driver's side air bag module.

8. Remove the steering wheel. Remove the column lower cover. Remove the clockspring.

9. Remove the floor console assembly. Remove the front scuff plate and cowl side trim. Remove the trunk lid opener cover.

10. Remove the accelerator pedal stopper. Remove the front floor carpet.

11. Disconnect the upper stabilizer link.

12. Remove the steering shaft pad.

13. Disconnect the steering column assembly and steering rack connection. Remove the rear roll stopper connecting bolt.

14. Remove the front axle crossmember bracket. Remove the return tube clamp bolt and nut. Remove the pressure tube clamp bolt and nut.

15. Disconnect and plug the return tube connection. Disconnect and plug the pressure hose connection. Discard the O-ring.

16. Disconnect the pinch clamp. Pinch the steering column shaft clip with a pliers and pull upward on the shaft to disengage the steering column assembly.

➡**If the steering column shaft is removed accidentally, remove the steering column assembly and be sure to insert the steering column shaft into the steering column.**

17. From inside the vehicle loosen the 3 clips from the body panel near the steering column to floor hole.

18. Use a transmission jack to hold the crossmember, and then remove the crossmember mounting nuts and bolts.

19. Lower the crossmember with the rear roll stopper, then the stabilizer bar and steering rack.

20. Remove the steering column dash panel cover.

21. Remove the steering rack linkage protector, on the 3.8L engine.

22. Remove the power steering rack bracket. Remove the rack retaining bolts. Remove the assembly from the vehicle.

23. Remove the stabilizer bracket, bushing, and bar.

To install:

➡**Be sure to use new fasteners, as required.**

24. Align the stabilizer bar identification mark with the right end of the bushing. Install the stabilizer bar.

25. Position the steering rack in the vehicle and install the clamp and the

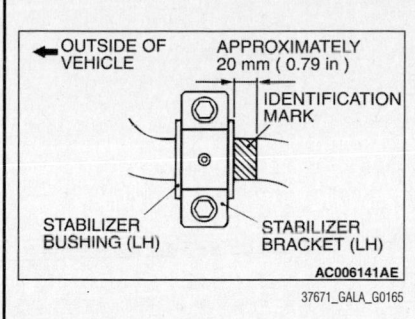

Fig. 119 Front stabilizer bar bushing identification

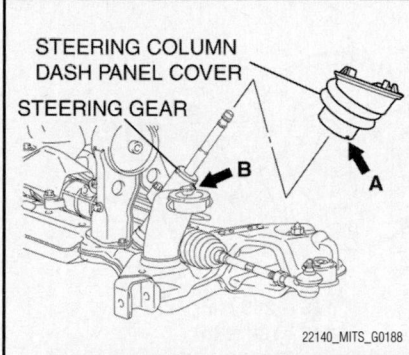

Fig. 120 Align the steering column dash panel cover notch "A" with the steering gear lug "B"

mounting bolts. Be sure the rack is centered before connecting it to the joint assembly.

26. When installing the steering column dash panel cover pinch clamp, align the steering column dash panel cover notch "A" with the steering gear lug "B" and then install the steering column dash panel cover to the steering gear.

❋❋ CAUTION

When installing the clockspring be sure that the springs mating marks are properly aligned. If not the steering wheel may not rotate completely during a turn, or the flat cable in the clockspring could be damaged. This would prevent normal SRS operation and possibly cause serious injury to the driver.

27. After aligning the mating surfaces of the clockspring, turn the front wheels to the straight ahead position. Install the clockspring to the column switch.

➡**Turn the clockspring clockwise fully. Then turn it back approximately 3 ¾ turns counterclockwise to align the mating marks.**

28. Continue the installation in the reverse order of the removal procedure.
29. Check and adjust the front end alignment, as required.

STEERING KNUCKLE

REMOVAL & INSTALLATION
See Figures 121 through 123.

1. Before servicing the vehicle, refer to the Precautions Section.

➡**If working near and/or around the SRS system and components, be sure to disable the SRS system. Tape the negative battery cable with insulating tape. Always disconnect the negative battery cable first.**

❋❋ CAUTION

Wait for 1 minute after disconnecting the negative battery cable before working inside the vehicle. The air bag system is set to deploy for a short period of time after the battery is disconnected.

2. Raise and safely support the vehicle.
3. Remove the front wheels and tires.
4. Remove the brake rotor.
5. Remove the lower control arm.
6. Remove the wheel speed sensor, the strut lower mounting bolt, and the lower arm mounting bolt from the steering knuckle.
7. Remove the front wheel hub mounting bolts while pushing out the driveshaft by hand.
8. If it is difficult to push out the driveshaft by hand, use special tools MB990242, MB990244, MB991354, and MB990767 to push out the driveshaft from the hub and knuckle.
9. If the front wheel hub is seized, remove the knuckle together with front wheel hub and fix them in a vise.
10. Hang the halfshaft on the vehicle body with a rope.
11. Use special tools MB990244, MB991354, and MB990211 to pull out the front wheel hub from the steering knuckle.

To install:
12. Install the hub and knuckle assembly.
13. Install the wheel speed sensor, the strut lower mounting bolt, and the lower arm mounting bolt to the steering knuckle.
14. Tighten the bolts to specification.
15. Install the lower ball joint.
16. Install the brake rotor.
17. Install the front wheels and tires.
18. Check alignment and adjust as necessary.

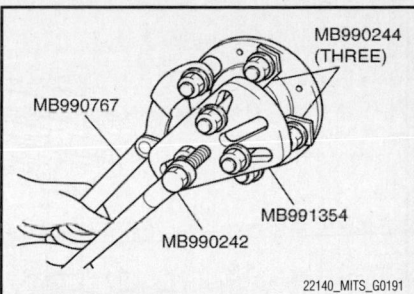

Fig. 121 Use special tools MB990242, MB990244, MB991354, and MB990767 to push out the driveshaft from the hub and knuckle

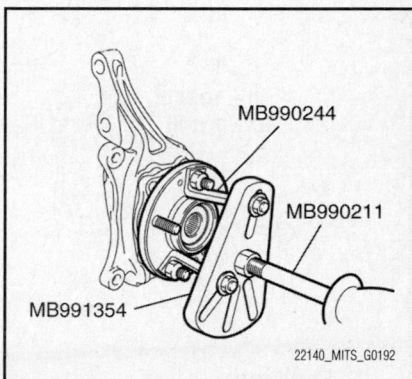

Fig. 122 Use special tools MB990244, MB991354, and MB990211 to pull out the front wheel hub from the steering knuckle

WHEEL HUB & BEARING

REMOVAL & INSTALLATION
See Figures 122 and 123.

1. Before servicing the vehicle, refer to the Precautions Section.

➡**If working near and/or around the SRS system and components, be sure to disable the SRS system. Tape the negative battery cable with insulating tape. Always disconnect the negative battery cable first.**

❋❋ CAUTION

Wait for 1 minute after disconnecting the negative battery cable before working inside the vehicle. The air bag system is set to deploy for a short period of time after the battery is disconnected.

2. Remove the steering knuckle, as required.
3. Use special tools MB990244, MB991354, and MB990211 to pull out the front wheel hub from the steering knuckle.

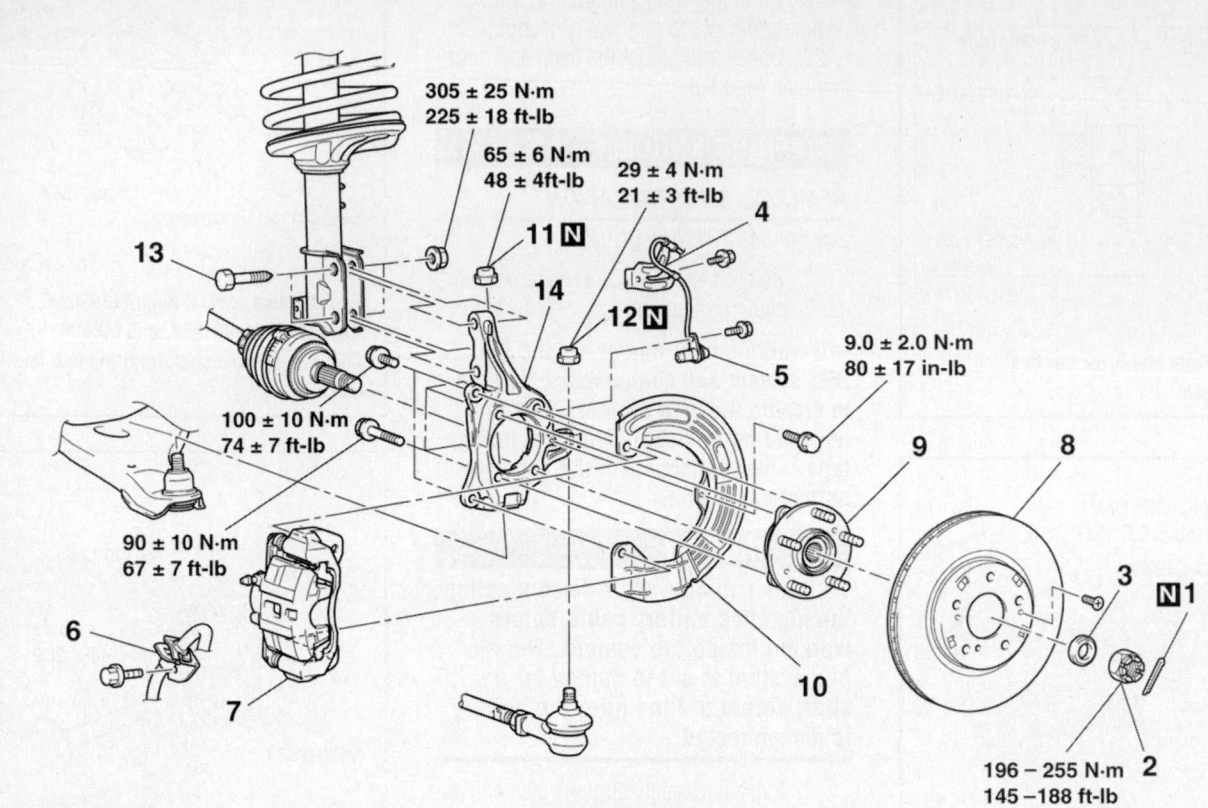

1. Split pin
2. Driveshaft nut
3. Washer
4. Front wheel speed sensor bracket
5. Front wheel speed sensor
6. Brake hose bracket
7. Caliper assembly
8. Brake disc
9. Front wheel hub assembly
10. Dust cover
11. Self-locking nut (connection for lower arm ball joint)
12. Self-locking nut (connection for tie rod end)
13. Front strut to knuckle mounting bolt and nut
14. Knuckle

22140_MITS_G0193

Fig. 123 Front steering knuckle, wheel bearing and related components

To install:

4. Install the hub bearing into the knuckle assembly.

5. Install the knuckle, as required.

6. Tighten the bolts to specification.

7. Install the brake rotor.

8. Install the front wheels and tires.

9. Check alignment and adjust as necessary.

ADJUSTMENT

The front wheel bearings are not adjustable. If the bearings are noisy or become loose, they must be replaced.

SUSPENSION

CONTROL ARMS/LINKS

REMOVAL & INSTALLATION

Lower Control Arm

See Figure 124.

1. Before servicing the vehicle, refer to the Precautions Section.

➡**If working near and/or around the SRS system and components, be sure to disable the SRS system. Tape the** negative battery cable with insulating tape. Always disconnect the negative battery cable first.

✳ CAUTION

Wait for 1 minute after disconnecting the negative battery cable before working inside the vehicle. The air bag system is set to deploy for a short period of time after the battery is disconnected.

REAR SUSPENSION

2. Raise and support the vehicle safely.

3. Remove the tire and wheel assembly.

4. Remove the shock absorber assembly and knuckle connection retaining bolt.

5. Properly support the lower control arm assembly.

6. Remove the lower control arm assembly and stabilizer link assembly connection.

7. Remove the lower control arm assembly and knuckle connection bolt.

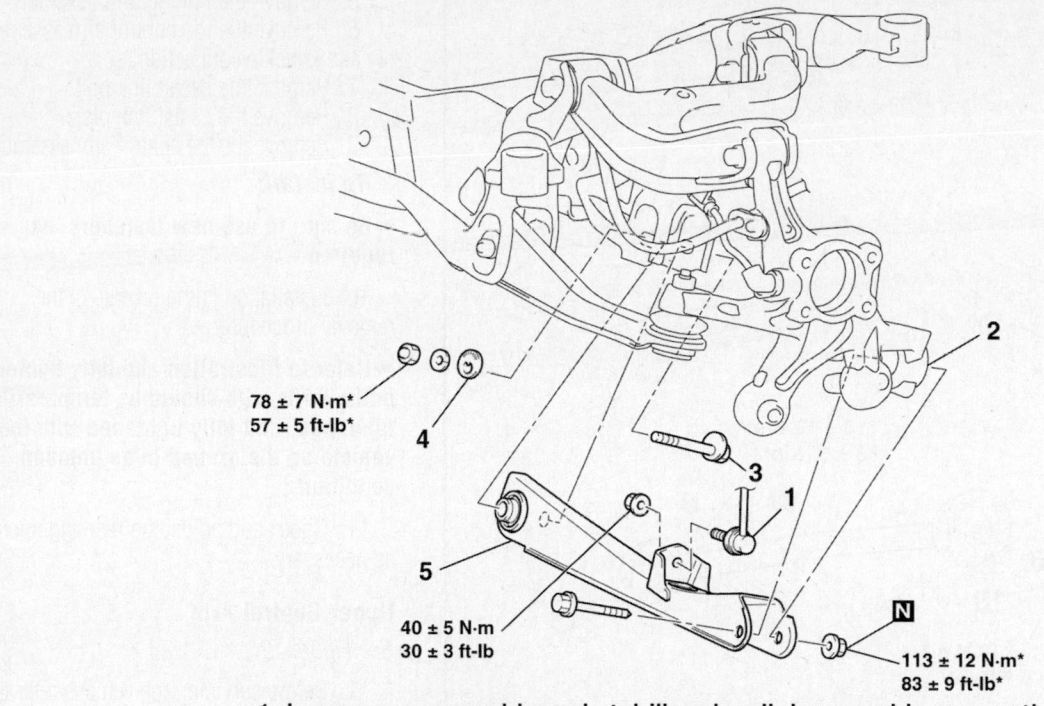

78 ± 7 N·m*
57 ± 5 ft-lb*

40 ± 5 N·m
30 ± 3 ft-lb

113 ± 12 N·m*
83 ± 9 ft-lb*

1. Lower arm assembly and stabilizer bar link assembly connection
2. Lower arm assembly and knuckle connection
3. Lower arm bolt
4. Lower arm plate
5. Lower arm assembly

22140_MITS_G0198

Fig. 124 Rear lower control arm and related components

8. Remove the lower arm bolt and arm plate.

9. Remove the lower control arm from the vehicle.

To install:

➡**Be sure to use new fasteners, as required.**

10. Install the lower control arm to its mounting.

11. Continue the installation in the reverse order of the removal procedure.

➡**Refer to illustration and fully tighten bolts. Bolts with should be temporarily tightened, and fully tightened with the vehicle on the ground in an unladen condition.**

12. Check and adjust the rear alignment, as necessary.

Rear Trailing Arm

See Figure 157.

1. Before servicing the vehicle, refer to the Precautions Section.

➡**If working near and/or around the SRS system and components, be sure to disable the SRS system. Tape the**

negative battery cable with insulating tape. Always disconnect the negative battery cable first.

❊❊ CAUTION

Wait for 1 minute after disconnecting the negative battery cable before working inside the vehicle. The air bag system is set to deploy for a short period of time after the battery is disconnected.

2. Raise and support the vehicle safely.
3. Remove the tire and wheel assembly.
4. Properly support the rear knuckle as required, using the proper equipment.
5. Remove the trailing arm and knuckle bolt.
6. Remove the trailing arm assembly.
7. Remove the trailing arm bracket.

To install:

➡**Be sure to use new fasteners, as required.**

8. Installation is the reverse of the removal procedure.

➡**Refer to illustration and fully tighten bolts. Bolts with should be temporarily tightened, and fully tightened with the**

vehicle on the ground in an unladen condition.

9. Check and adjust the rear alignment, as necessary.

Toe Control Arm

See Figures 125 and 126.

1. Before servicing the vehicle, refer to the Precautions Section.

➡**If working near and/or around the SRS system and components, be sure to disable the SRS system. Tape the negative battery cable with insulating tape. Always disconnect the negative battery cable first.**

❊❊ CAUTION

Wait for 1 minute after disconnecting the negative battery cable before working inside the vehicle. The air bag system is set to deploy for a short period of time after the battery is disconnected.

2. Raise and support the vehicle safely.
3. Remove the tire and wheel assembly.
4. Properly support the rear knuckle as required, using the proper equipment.

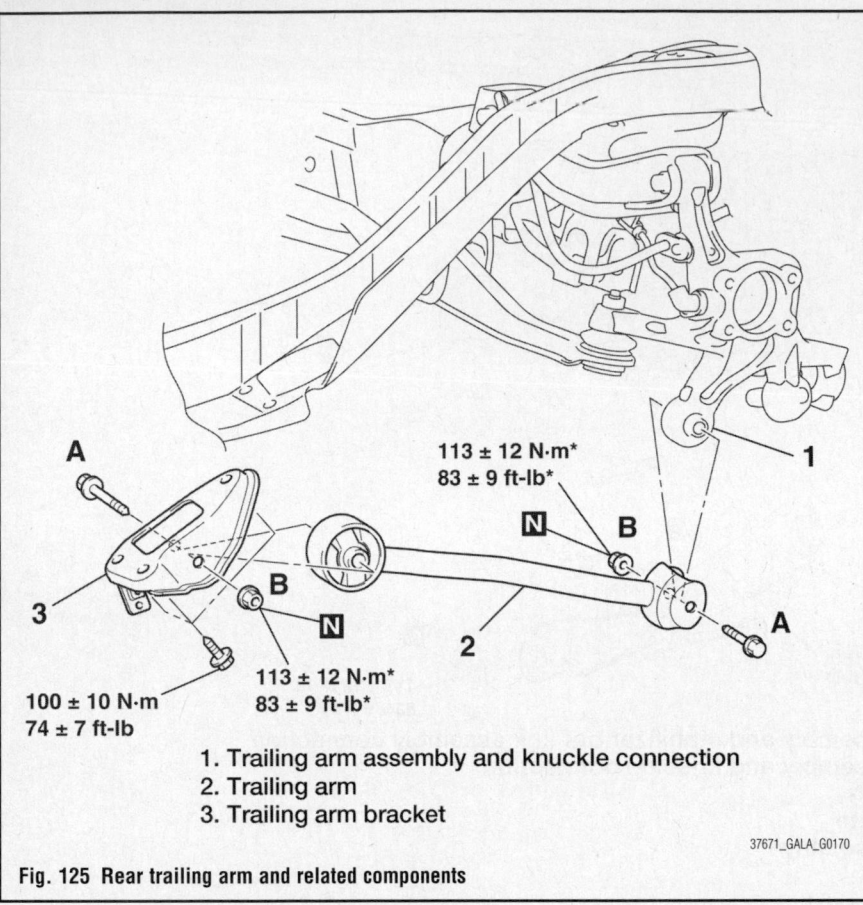

113 ± 12 N·m*
83 ± 9 ft-lb*

A

3

B

B

N

N

2

A

100 ± 10 N·m
74 ± 7 ft-lb

113 ± 12 N·m*
83 ± 9 ft-lb*

1. Trailing arm assembly and knuckle connection
2. Trailing arm
3. Trailing arm bracket

37671_GALA_G0170

Fig. 125 Rear trailing arm and related components

5. Remove the trailing arm assembly.
6. Remove the toe control arm assembly and knuckle connection.
7. Remove the assist link bolt.
8. Remove the assist link plate.
9. Remove the toe control arm assembly.

To install:

➡ **Be sure to use new fasteners, as required.**

10. Installation is the reverse of the removal procedure.

➡ **Refer to illustration and fully tighten bolts. Bolts with should be temporarily tightened, and fully tightened with the vehicle on the ground in an unladen condition.**

11. Check and adjust the rear alignment, as necessary.

Upper Control Arm

See Figure 127.

1. Before servicing the vehicle, refer to the Precautions Section.

➡ **If working near and/or around the SRS system and components, be sure to disable the SRS system. Tape the negative battery cable with insulating**

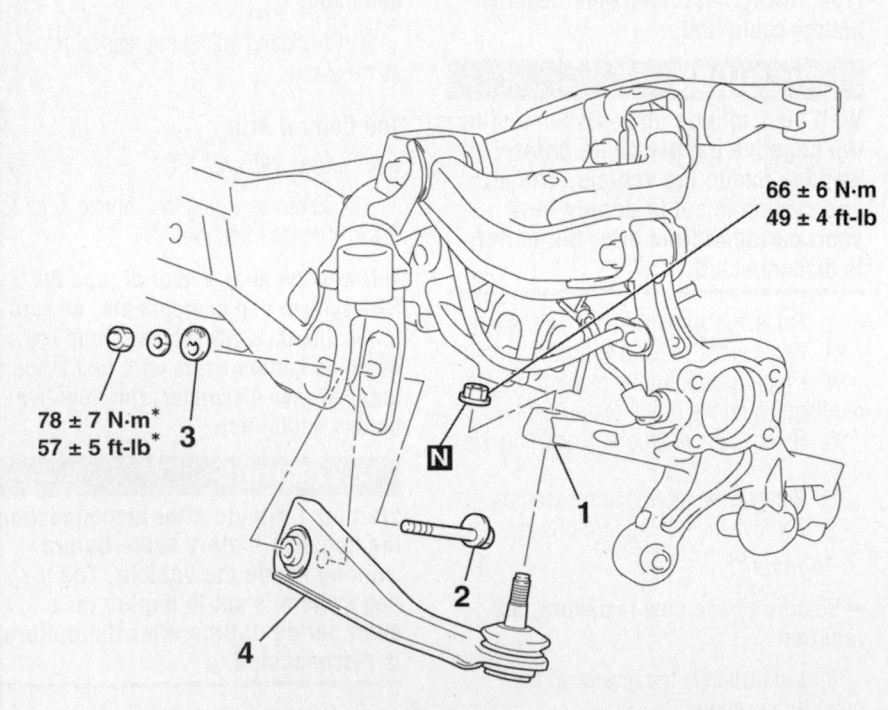

66 ± 6 N·m
49 ± 4 ft-lb

78 ± 7 N·m*
57 ± 5 ft-lb*

3

N

1

2

4

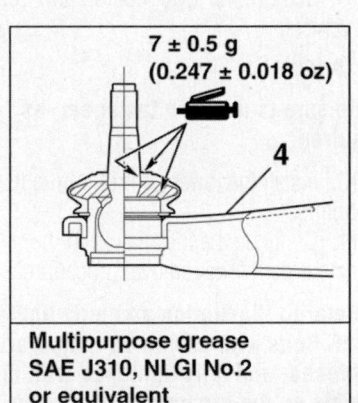

7 ± 0.5 g
(0.247 ± 0.018 oz)

4

**Multipurpose grease
SAE J310, NLGI No.2
or equivalent**

1. Trailing arm assembly
2. Assist link bolt
3. Assist link plate
4. Toe control arm assembly

37671_GALA_G0171

Fig. 126 Rear toe control arm and related components

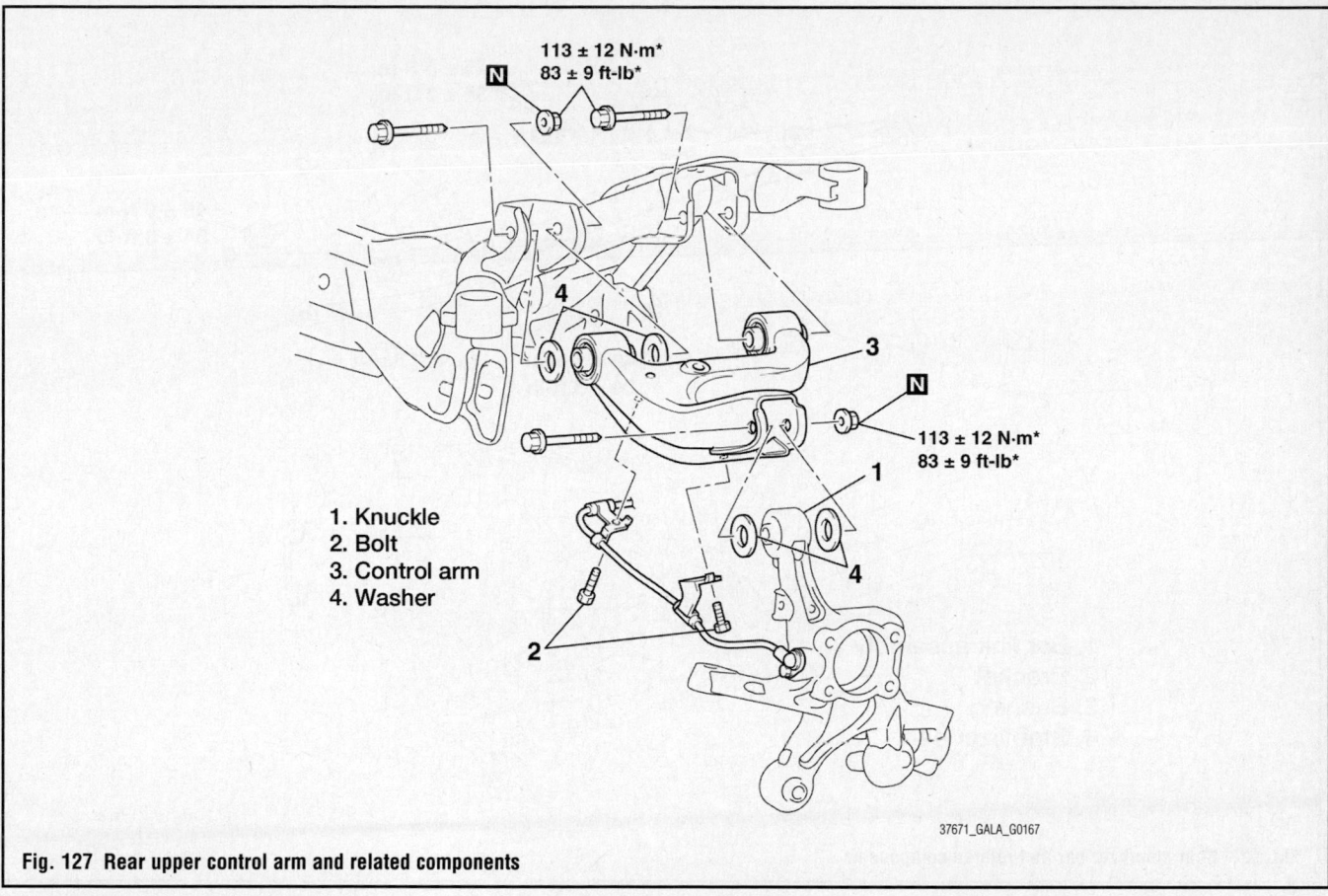

113 ± 12 N·m*
83 ± 9 ft-lb*

113 ± 12 N·m*
83 ± 9 ft-lb*

1. Knuckle
2. Bolt
3. Control arm
4. Washer

37671_GALA_G0167

Fig. 127 Rear upper control arm and related components

tape. Always disconnect the negative battery cable first.

❋❋ CAUTION

Wait for 1 minute after disconnecting the negative battery cable before working inside the vehicle. The air bag system is set to deploy for a short period of time after the battery is disconnected.

2. Raise and support the vehicle safely.

3. Remove the tire and wheel assembly.

4. If equipped with ABS, remove the wheel speed sensor retaining bolts.

5. Position a floor jack to properly support the knuckle before removing the upper control arm.

6. Remove the upper arm assembly and knuckle retaining bolt and nut.

7. Remove the upper arm assembly retaining bolts and nuts.

8. Remove the upper control arm from the vehicle.

To install:

➡Be sure to use new fasteners, as required.

9. Install the upper control arm to its mounting.

10. Continue the installation in the reverse order of the removal procedure.

➡Refer to illustration and fully tighten bolts. Bolts with should be temporarily tightened, and fully tightened with the vehicle on the ground in an unladen condition.

11. Check and adjust the rear alignment, as necessary.

STABILIZER BAR

REMOVAL & INSTALLATION
See Figures 128 and 129.

1. Before servicing the vehicle, refer to the Precautions Section.

➡If working near and/or around the SRS system and components, be sure to disable the SRS system. Tape the negative battery cable with insulating tape. Always disconnect the negative battery cable first.

❋❋ CAUTION

Wait for 1 minute after disconnecting the negative battery cable before working inside the vehicle. The air bag system is set to deploy for a short period of time after the battery is disconnected.

2. Raise and support the vehicle safely.

3. Remove the tire and wheel assemblies.

4. Remove the stabilizer bar link assembly.

5. Remove the stabilizer bar bracket.

6. Remove the bushings.

7. Remove the stabilizer bar.

To install:

➡Be sure to use new fasteners, as required.

8. Installation is the reverse of the removal procedure.

9. Install the bushings and align as shown in the illustration.

➡Refer to illustration and fully tighten bolts. Bolts with should be temporarily tightened, and fully tightened with the vehicle on the ground in an unladen condition.

10. Check and adjust the rear alignment, as necessary.

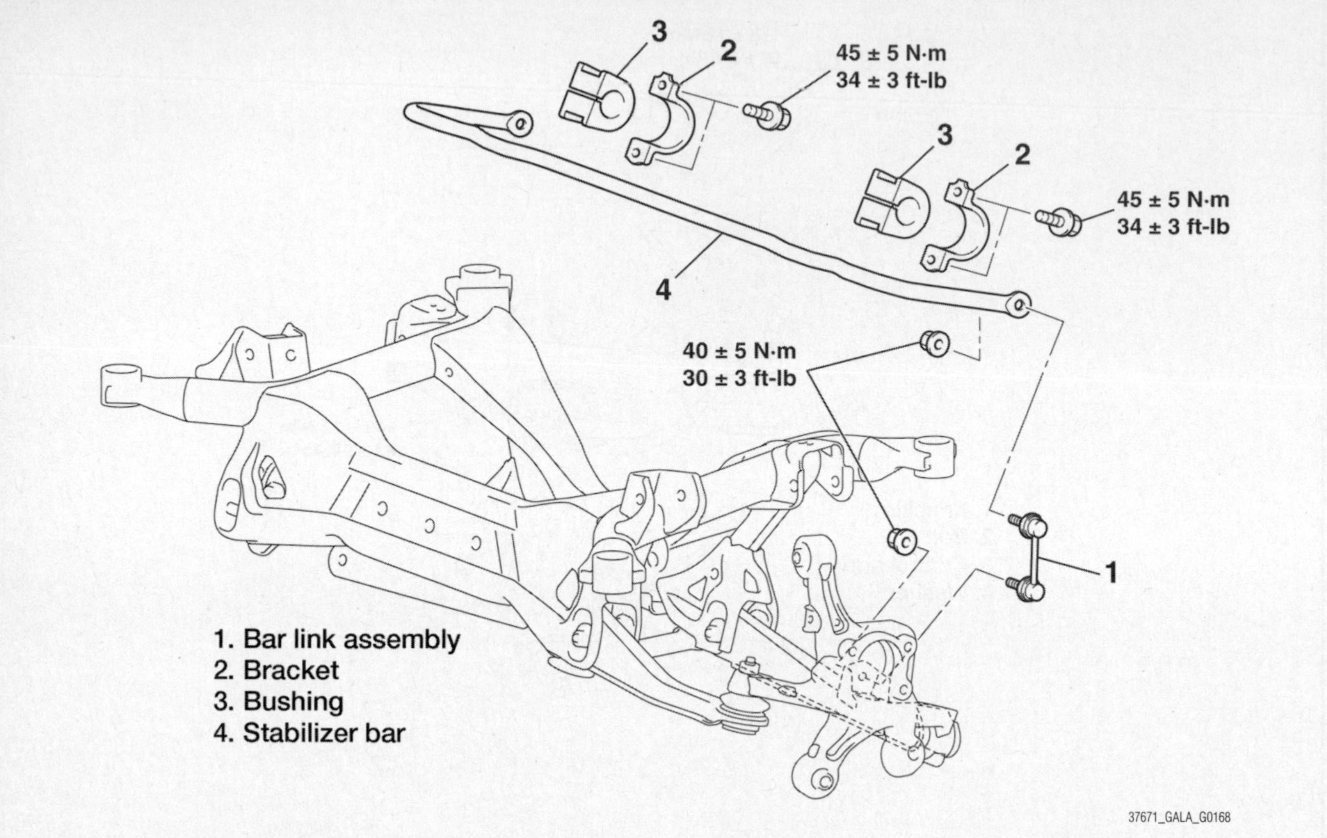

Fig. 128 Rear stabilizer bar and related components

1. Bar link assembly
2. Bracket
3. Bushing
4. Stabilizer bar

45 ± 5 N·m
34 ± 3 ft-lb

45 ± 5 N·m
34 ± 3 ft-lb

40 ± 5 N·m
30 ± 3 ft-lb

37671_GALA_G0168

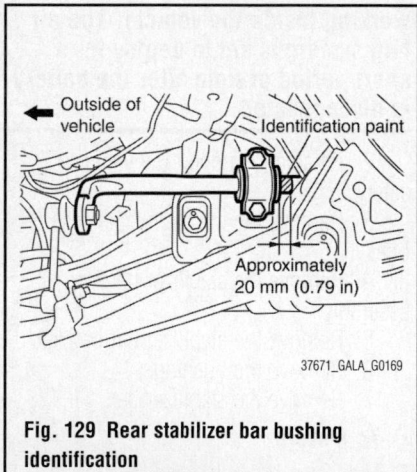

Fig. 129 Rear stabilizer bar bushing identification

Outside of vehicle

Identification paint

Approximately 20 mm (0.79 in)

37671_GALA_G0169

STRUTS

REMOVAL & INSTALLATION

See Figure 130.

1. Before servicing the vehicle, refer to the Precautions Section.

➡**If working near and/or around the SRS system and components, be sure to disable the SRS system. Tape the negative battery cable with insulating tape. Always disconnect the negative battery cable first.**

✳✳ CAUTION

Wait for 1 minute after disconnecting the negative battery cable before working inside the vehicle. The air bag system is set to deploy for a short period of time after the battery is disconnected.

2. Remove the rear sub woofer speaker.
3. Remove the trunk trim (front side).
4. Raise and support the vehicle safely.
5. Remove the shock absorber assembly and knuckle retaining bolt.
6. Remove the shock absorber service cover.
7. Remove the upper shock absorber retaining nuts.
8. Remove the strut/shock absorber assembly from the vehicle.

To install:

➡**Be sure to use new fasteners, as required.**

9. Installation is the reverse of the removal procedure.

➡**Refer to illustration and fully tighten bolts. Bolts with should be temporarily**

tightened, and fully tightened with the vehicle on the ground in an unladen condition.

10. Check and adjust the rear alignment, as necessary.

WHEEL HUB & BEARING

REMOVAL & INSTALLATION

See Figure 131.

➡**The hub and bearing assembly is serviced as a unit.**

1. Before servicing the vehicle, refer to the Precautions Section.

➡**If working near and/or around the SRS system and components, be sure to disable the SRS system. Tape the negative battery cable with insulating tape. Always disconnect the negative battery cable first.**

✳✳ CAUTION

Wait for 1 minute after disconnecting the negative battery cable before working inside the vehicle. The air bag system is set to deploy for a short period of time after the battery is disconnected.

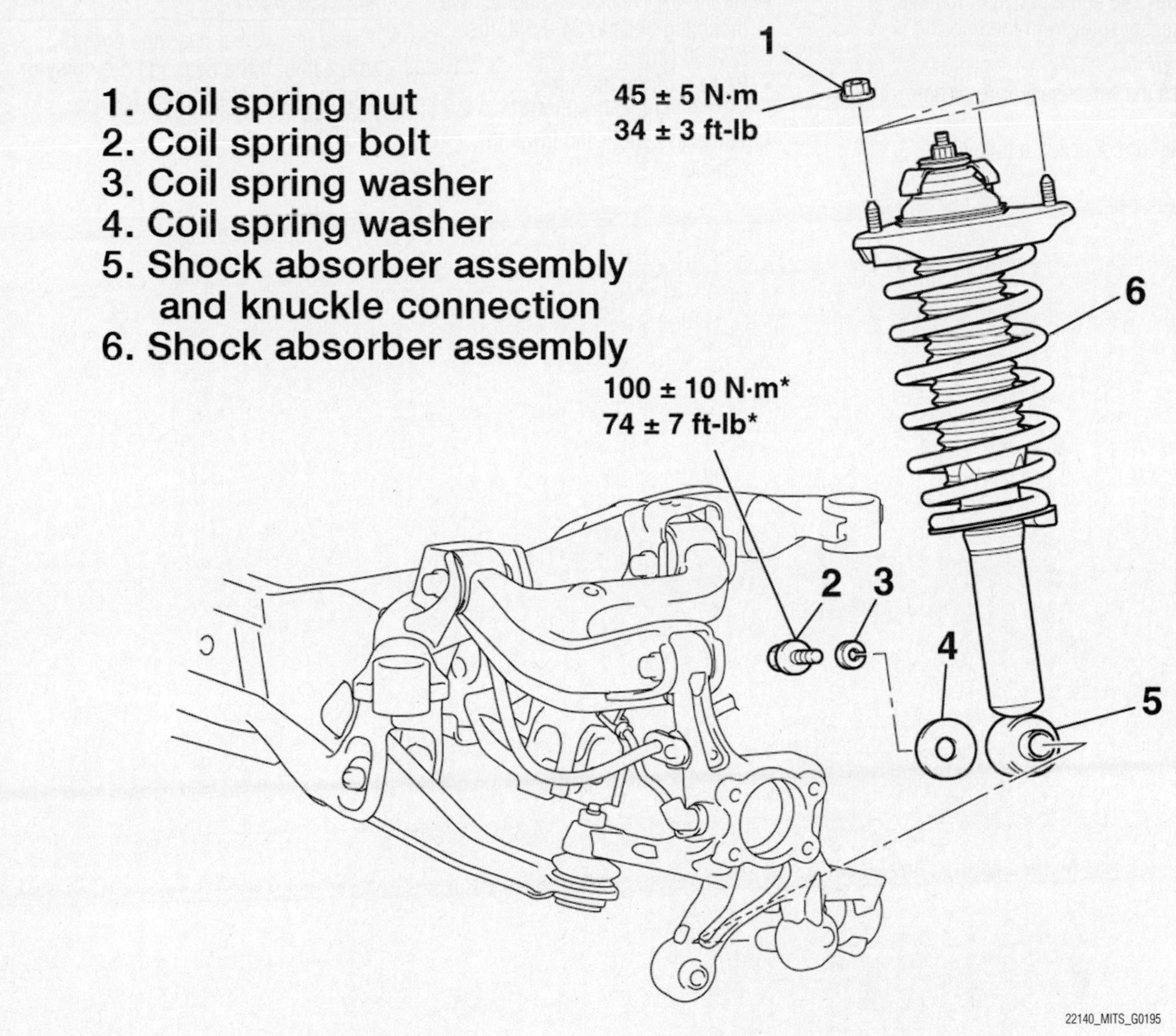

1. Coil spring nut
2. Coil spring bolt
3. Coil spring washer
4. Coil spring washer
5. Shock absorber assembly
   and knuckle connection
6. Shock absorber assembly

45 ± 5 N·m
34 ± 3 ft-lb

100 ± 10 N·m*
74 ± 7 ft-lb*

22140_MITS_G0195

Fig. 130 Rear strut/shock and related components

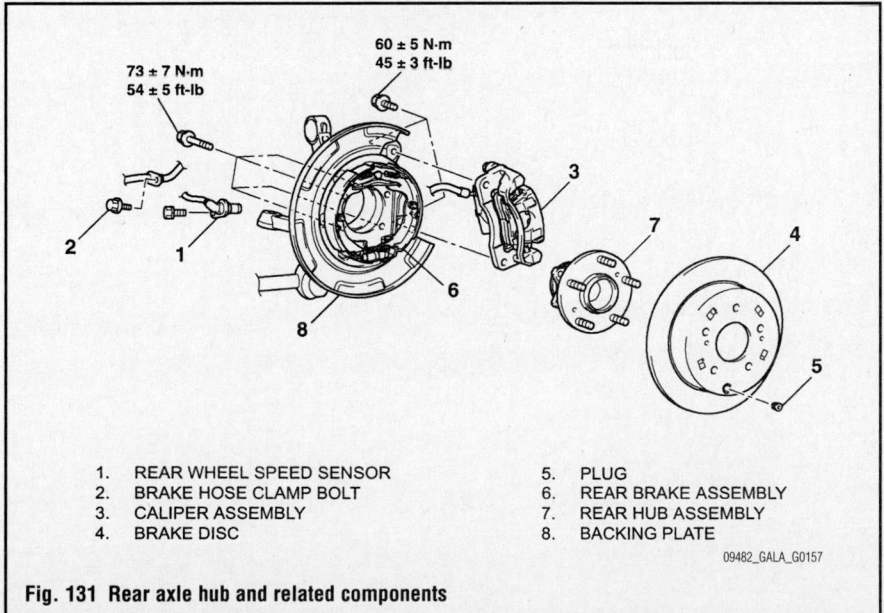

73 ± 7 N·m
54 ± 5 ft-lb

60 ± 5 N·m
45 ± 3 ft-lb

1. REAR WHEEL SPEED SENSOR	5. PLUG
2. BRAKE HOSE CLAMP BOLT	6. REAR BRAKE ASSEMBLY
3. CALIPER ASSEMBLY	7. REAR HUB ASSEMBLY
4. BRAKE DISC	8. BACKING PLATE

09482_GALA_G0157

Fig. 131 Rear axle hub and related components

2. Disconnect the negative battery cable.

3. Raise and safely support the vehicle.

4. Remove the tire and wheel assembly.

5. If equipped with Anti-Lock Brake (ABS), remove the Vehicle Speed Sensor (VSS).

6. Remove the caliper assembly and rotor. Suspend the caliper out of the way with wire.

7. From the back of the knuckle, remove the 4 bolts securing the hub to the knuckle.

8. Remove the hub and bearing assembly from the knuckle.

➡ **The hub assembly is not serviceable and should not be disassembled.**

9. If replacing the hub, use special

socket MB991248 and a press, to remove the wheel sensor rotor from the hub.

To Install:

10. Press the wheel sensor rotor onto the hub.

11. Install or connect the following:

- Hub to the knuckle and tighten the mounting bolts to 54–65 ft. lbs. (74–88 Nm)
- Brake rotor on the hub
- VSS, if equipped with ABS
- Wheel assembly and lower the vehicle

ADJUSTMENT

The rear wheel bearings are not adjustable. If the bearings are noisy or become loose, they must be replaced.

MITSUBISHI

14

Lancer

SPECIFICATIONS AND MAINTENANCE CHARTS

ENGINE AND VEHICLE IDENTIFICATION

Code ①	Liters (cc)	Cu. In.	Cyl.	Fuel Sys.	Engine Type	Eng. Mfg.	Code ②	Year
			Engine				**Model Year**	
4B11/U	2.0 (1,998)	121.9	4	MPFI	DOHC	Mitsubishi	9	2009
4B11/V	2.0 (1,998)	121.9	4	MPFI-Turbo	DOHC	Mitsubishi	A	2010
4B12/W	2.4 (2,359)	1441.0	4	MPFI	DOHC	Mitsubishi		

MPFI: Multi-Point Fuel Injection

DOHC: Double Overhead Camshafts

① Engine Code/8th digit of VIN

② 10th digit of VIN

37671_LANC_C0001

GENERAL ENGINE SPECIFICATIONS

All measurements are given in inches.

Year	Model	Engine Disp. Liters	Engine Series ID/VIN	Net Horsepower @ rpm	Net Torque @ rpm (ft. lbs.)	Bore x Stroke (in.)	Com-pression Ratio	Oil Pressure @ rpm
2009	Lancer	2.0	4B11/U	152@6000	146@4250	3.39 x 3.39	10.0:1	43-100@3500
	Lancer (exc Evolution)	2.0	4B11/V	237@6000	253@4750	3.39 x 3.39	9.0:1	43-100@3500
	Lancer (Evolution)	2.0	4B11/V	291@6500	300@4400	3.39 x 3.39	9.0:1	43-100@3500
	Lancer	2.4	4B12/W	168@6000	167@4100	3.50 x 3.80	10.5:1	43-100@3500
2010	Lancer	2.0	4B11/U	152@6000	146@4250	3.39 x 3.39	10.0:1	43-100@3500
	Lancer (exc Evolution)	2.0	4B11/V	237@6000	253@4750	3.39 x 3.39	9.0:1	43-100@3500
	Lancer (Evolution)	2.0	4B11/V	291@6500	300@4400	3.39 x 3.39	9.0:1	43-100@3500
	Lancer	2.4	4B12/W	168@6000	167@4100	3.50 x 3.80	10.5:1	43-100@3500

37671_LANC_C0002

GASOLINE ENGINE TUNE-UP SPECIFICATIONS

Year	Engine Displacement Liters	Engine Series ID/VIN	Spark Plug Gap (in.)	Ignition Timing (deg.) MT	Ignition Timing (deg.) AT	Fuel Pump (psi)	Idle Speed (rpm) MT	Idle Speed (rpm) AT	Valve Clearance (in.) Intake	Valve Clearance (in.) Exhaust
2009	2.0	4B11/U	0.028-0.031	①	①	47	②	②	④	⑤
	2.0	4B11/V	0.020-0.023	①	①	③	②	②	④	⑤
	2.4	4B12/W	0.028-0.031	①	①	47	②	②	④	⑤
2010	2.0	4B11/U	0.028-0.031	①	①	47	②	②	④	⑤
	2.0	4B11/V	0.020-0.023	①	①	③	②	②	④	⑤
	2.4	4B12/W	0.028-0.031	①	①	47	②	②	④	⑤

NOTE: The Vehicle Emission Control Information label reflects specification changes made during production.

Follow the figures on the label if they differ from those in this chart.

① Ignition timing is preset and cannot be adjusted

② Idle speed is maintained by the Electronic Control Module (ECM)

③ 45-50, vacuum hose disconnected.

④ Intake valve clearance: 0.008 +/- 0.0012 inch

⑤ Exhaust valve clearance: 0.012 +/- 0.0012 inch

37671_LANC_C0003

CAPACITIES

Year	Model	Engine Disp. Liters	Engine Series ID/VIN	Engine Oil with Filter (qts.)	Transmission (pts.) Manual	Transmission (pts.) Auto. ①	Transfer Case (pts.)	Drive Axle Front (pts.)	Drive Axle Rear (pts.)	Fuel Tank (gal.)	Cooling System (qts.)
2009	Lancer	2.0	4B11/U	②	5.2	③	1.7	NA	④	14.5	7.9
	Lancer (exc Evolution)	2.0	4B11/V	②	5.2	③	1.7	NA	④	14.5	7.9
	Lancer (Evolution)	2.0	4B11/V	②	5.2	③	1.7	NA	④	14.5	7.9
	Lancer	2.4	4B12/W	②	5.2	③	1.7	NA	④	14.5	7.9
2010	Lancer	2.0	4B11/U	②	5.2	③	1.7	NA	④	14.5	7.9
	Lancer (exc Evolution)	2.0	4B11/V	②	5.2	③	1.7	NA	④	14.5	7.9
	Lancer (Evolution)	2.0	4B11/V	②	5.2	③	1.7	NA	④	14.5	7.9
	Lancer	2.4	4B12/W	②	5.2	③	1.7	NA	④	14.5	7.9

NOTE: All capacities are approximate. Add fluid gradually and check to be sure a proper fluid level is obtained.

NA: Not Available

① Drain and refill

② Oil pan (5.07 qts.), Oil cooler (0.52 qt.), Oil filter (0.32 qt.)

③ W5M6A (5.2 pts.). W6DGA - Twin Clutch Sport Shift Transaxle (TC-SST) (16.5 pts.). AWD (15.0 pts.).

④ AWD: 1.16

37671_LANC_C0004

FLUID SPECIFICATIONS

Year	Model	Engine Disp. Liters	Engine Series ID/VIN	Engine Oil	Manual Trans.	Auto. Trans.	Drive Axle	Transfer Case	Power Steering Fluid	Brake Master Cylinder	Cooling System
2009	Lancer	2.0	4B11/U	5W-30	①	②	③	④	⑤	⑥	⑦
	Lancer (exc Evolution)	2.0	4B11/V	5W-30	①	②	③	④	⑤	⑥	⑦
	Lancer (Evolution)	2.0	4B11/V	5W-30	①	②	③	④	⑤	⑥	⑦
	Lancer	2.4	4B12/W	5W-30	①	②	③	④	⑤	⑥	⑦
2010	Lancer	2.0	4B11/U	5W-30	①	②	③	④	⑤	⑥	⑦
	Lancer (exc Evolution)	2.0	4B11/V	5W-30	①	②	③	④	⑤	⑥	⑦
	Lancer (Evolution)	2.0	4B11/V	5W-30	①	②	③	④	⑤	⑥	⑦
	Lancer	2.4	4B12/W	5W-30	①	②	③	④	⑤	⑥	⑦

① DiaQueen NEW MULTI GEAR OIL (GL-3), SAE 75W-80.

  Twin Clutch Sport Shift Transaxle (TC-SST): Use DiaQueen SSTF-I

② DiaQueen CVTF-J1

③ Active Center Differential (ACD) fluid: DIAMOND ATF SP-3

  Rear differential gear oil: MITSUBISHI Limited Slip Differential Oil (LSD) or equivalent

④ MITSUBISHI Limited Slip Differential Oil (LSD) or equivalent

⑤ Genuine MITSUBISHI power steering fluid

⑥ DOT 3, DOT 4, or equivalent

⑦ Long life antifreeze/coolant or an equivalent

37671_LANC_C0005

VALVE SPECIFICATIONS

Year	Engine Displacement Liters	Engine Series ID/VIN	Seat Angle (deg.)	Face Angle (deg.)	Spring Test Pressure (lbs. @ in.)	Spring Installed Height (in.)	Stem-to-Guide Clearance (in.)		Stem Diameter (in.)	
							Intake	Exhaust	Intake	Exhaust
2009	2.0	4B11/U	①	①	NA	②	0.0008-0.0019	0.0012-0.0021	NA	NA
	2.0	4B11/V	①	①	NA	③	0.0008-0.0019	0.0012-0.0022	NA	NA
	2.4	4B12/W	①	①	NA	④	0.0008-0.0019	0.0012-0.0021	NA	NA
2010	2.0	4B11/U	①	①	NA	②	0.0008-0.0019	0.0012-0.0021	NA	NA
	2.0	4B11/V	①	①	NA	③	0.0008-0.0019	0.0012-0.0022	NA	NA
	2.4	4B12/W	①	①	NA	④	0.0008-0.0019	0.0012-0.0021	NA	NA

NA: Not Available

① Valve seat contact width - Intake: 0.046 - 0.058 in.

   Exhaust: 0.053 - 0.065 in.

② Free height of valve spring: 1.868 inches

③ Free height of valve spring: 2.030 inches

④ Free height of valve spring: 1.858 inches

37671_LANC_C0006

CAMSHAFT AND BEARING SPECIFICATIONS CHART

All measurements are given in inches.

Year	Engine Displ. Liters	Engine Series ID/VIN	Journal Dia.	Brg. Oil Clearance	Shaft End-play	Runout	Journal Bore	Lobe Height	
								Intake	Exhaust
2009	2.0	4B11/U	NA	0.0000-0.0013	NA	NA	NA	1.683-1.703	1.752-1.772
	2.0	4B11/V	NA	0.0014-0.0028	NA	NA	NA	1.720-1.740	1.750-1.770
	2.4	4B12/W	NA	0.0013	NA	NA	NA	1.717 1.736	1.752 1.772
2010	2.0	4B11/U	NA	0.0000-0.0013	NA	NA	NA	1.683-1.703	1.752-1.772
	2.0	4B11/V	NA	0.0014-0.0028	NA	NA	NA	1.720-1.740	1.750-1.770
	2.4	4B12/W	NA	0.0013	NA	NA	NA	1.717 1.736	1.752 1.772

NA: Not Available

37671_LANC_C0007

CRANKSHAFT AND CONNECTING ROD SPECIFICATIONS

All measurements are given in inches.

Year	Engine Displacement Liters	Engine Series ID/VIN	Crankshaft				Connecting Rod		
			Main Brg. Journal Dia.	Main Brg. Oil Clearance	Shaft End-play	Thrust on No.	Journal Diameter	Oil Clearance	Side Clearance
2009	2.0	4B11/U	NA	0.0005-0.0012	0.0020-0.0010	3	NA	0.0007-0.0018	0.0040-0.0010
	2.0	4B11/V	NA	①	0.0020-0.0100	3	NA	0.0015-0.0027	0.0040-0.0010
	2.4	4B12/W	NA	0.0005-0.0012	0.0020-0.0010	3	NA	0.0007-0.0018	0.0040-0.0090
2010	2.0	4B11/U	NA	0.0005-0.0012	0.0020-0.0010	3	NA	0.0007-0.0018	0.0040-0.0010
	2.0	4B11/V	NA	①	0.0020-0.0100	3	NA	0.0015-0.0027	0.0040-0.0010
	2.4	4B12/W	NA	0.0005-0.0012	0.0020-0.0010	3	NA	0.0007-0.0018	0.0040-0.0090

NA: Not Available

① Number 1, 2, 4, 5: 0.0015-0.0026 in.

   Number 3: 0.0020-0.0030 in.

37671_LANC_C0008

PISTON AND RING SPECIFICATIONS

All measurements are given in inches.

Year	Engine Displ. Liters	Engine Series ID/VIN	Piston Clearance	Ring Gap			Ring Side Clearance		
				Top Compression	Bottom Compression	Oil Control	Top Compression	Bottom Compression	Oil Control
2009	2.0	4B11/U	NA	0.0060-0.0110	0.0100-0.0160	0.0040-0.0140	0.0010-0.0030	0.0010-0.0030	NA
	2.0	4B11/V	NA	0.0070-0.0110	0.0110-0.0170	0.0040-0.0130	0.0010-0.0020	0.0010-0.0020	NA
	2.4	4B14/W	NA	0.0060-0.0110	0.0100-0.0160	0.0040-0.0140	0.0010-0.0030	0.0010-0.0030	NA
2010	2.0	4B11/U	NA	0.0060-0.0110	0.0100-0.0160	0.0040-0.0140	0.0010-0.0030	0.0010-0.0030	NA
	2.0	4B11/V	NA	0.0070-0.0110	0.0110-0.0170	0.0040-0.0130	0.0010-0.0020	0.0010-0.0020	NA
	2.4	4B14/W	NA	0.0060-0.0110	0.0100-0.0160	0.0040-0.0140	0.0010-0.0030	0.0010-0.0030	NA

NA: Not Available

37671_LANC_C0009

TORQUE SPECIFICATIONS
All readings in ft. lbs.

Year	Engine Disp. Liters	Engine Series ID/VIN	Cylinder Head Bolts	Main Bearing Bolts	Rod Bearing Bolts	Crankshaft Damper Bolts	Flywheel Bolts	Manifold		Spark Plugs	Oil Pan Drain Plug
								Intake	Exhaust		
2009	2.0	4B11/U	①	②	③	155	④	⑤	⑥	15-21	25-33
	2.0	4B11/V	①	⑦	③	155	④	14-16	33-39	12-14	37-43
	2.4	4B12/W	①	②	③	155	④	⑧	33-39	15-21	26-32
2010	2.0	4B11/U	①	②	③	155	④	⑤	⑥	15-21	25-33
	2.0	4B11/V	①	⑦	③	155	④	14-16	33-39	12-14	37-43
	2.4	4B12/W	①	②	③	155	④	⑧	33-39	15-21	26-32

NOTE: Dip main bearing bolts and crankshaft damper bolt in clean engine oil prior to tightening.

① Step 1: 25-27 ft. lbs.
　Step 2: Plus 90 degrees
　Step 3: Plus another 90 degrees

② Step 1: 19-21 ft. lbs.
　Step 2: Plus 45 degrees

③ Step 1: 44 inch lbs.
　Step 2: 15 ft. lbs.
　Step 3: Plus 90 degrees

④ Step 1: 30 ft. lbs.
　Step 2: 96 ft. lbs.

⑤ Step 1: 2-28 inch lbs.
　Step 2: 14-16 ft. lbs.

⑥ Exhaust manifold nut 33-39 ft. lbs.
　Upper and lower covers: 116-132 inch lbs.

⑦ Step 1: M8 bolts (72-106 inch lbs.)
　and M10 bolts (27-29 ft. lbs.)
　Step 2: Plus 90 degrees

⑧ 18-44 inch lbs.
　14-16 ft. lbs.

37671_LANC_C0010

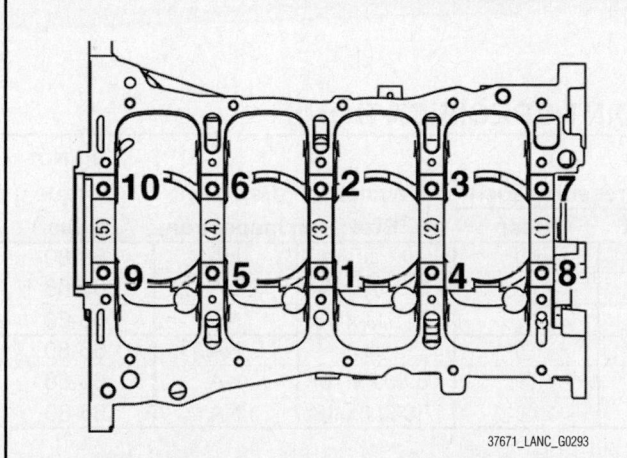

Fig. 1 Main bearing torque sequence—2.0L non turbocharged engine and 2.4L engine

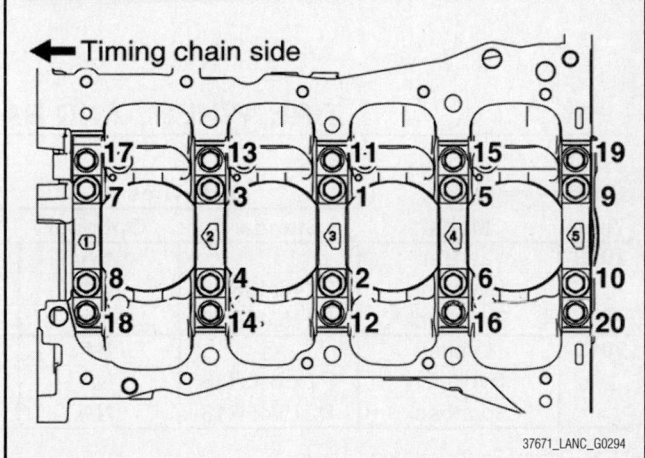

Fig. 2 Main bearing torque sequence—2.0L turbocharged engine

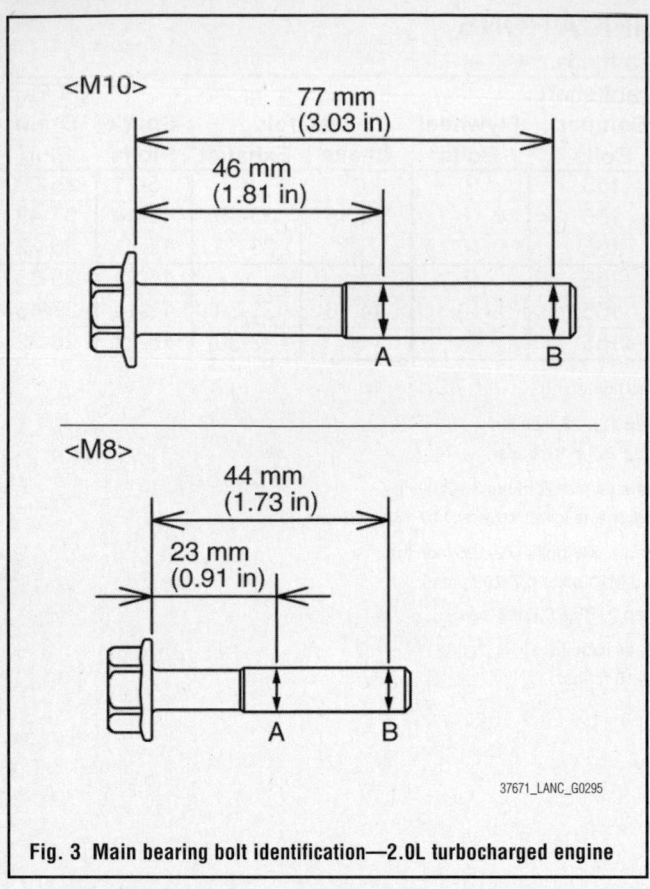

<M10>

77 mm
(3.03 in)

46 mm
(1.81 in)

A   B

<M8>

44 mm
(1.73 in)

23 mm
(0.91 in)

A   B

37671_LANC_G0295

Fig. 3 Main bearing bolt identification—2.0L turbocharged engine

TIRE, WHEEL AND BALL JOINT SPECIFICATIONS

| Year | Model | OEM Tires | | Tire Pressures (psi) | | Wheel Size | Ball Joint Inspection | Lug Nut Torque (ft. lbs.) |
		Standard	Optional	Front	Rear			
2009	Lancer	①	NA	②	②	③	NA	66-80
	Evolution	P245/40R18	NA	②	②	8.5JJ x 18	NA	66-68
	Sportback	P215/45R18	NA	②	②	7.05JJ x 18	NA	66-80
2010	Lancer	①	NA	②	②	③	NA	66-80
	Evolution	P245/40R18	NA	②	②	8.5JJ x 18	NA	66-68
	Sportback	P215/45R18	NA	②	②	7.05JJ x 18	NA	66-80

OEM: Original Equipment Manufacturer

PSI: Pounds Per Square Inch

NA: Not Available

① P205/60R16, P215/45R18

② Refer to placard on vehicle for proper inflation pressure.

③ 6.5JJ x 16, 7.0JJ x 18

37671_LANC_C0012

WHEEL ALIGNMENT

Year	Model		Caster Range (+/-Deg.)	Caster Preferred Setting (Deg.)	Camber Range (+/-Deg.)	Camber Preferred Setting (Deg.)	Toe-in (Deg.)
2009	Lancer	Front	0.30	+2.40	0.30	-0.50	0.04 +/- 0.08
		Rear	—	—	0.30	-0.55	0.12 +/- 0.08
2010	Lancer	Front	0.30	+2.40	0.30	-0.50	0.04 +/- 0.08
		Rear	—	—	0.30	-0.55	0.12 +/- 0.08

37671_LANC_C0011

BRAKE SPECIFICATIONS

All measurements in inches unless noted

Year	Model		Brake Disc Original Thickness	Brake Disc Minimum Thickness	Brake Disc Maximum Run-out	Lining Thickness Standard	Lining Thickness Limit	Brake Caliper Bracket Bolts (ft. lbs.)	Brake Caliper Mounting Bolts (ft. lbs.)
2009	Lancer	F	①	NA	0.0024	0.039	0.080	NA	②
		R	0.390	NA	0.0032	0.039	0.080	NA	28-36
	Evolution	F	1.260	NA	0.0024	0.039	0.080	NA	NA
		R	0.870	NA	0.0032	0.035	0.080	NA	NA
	Sportback	F	③	NA	0.0024	0.039	0.080	NA	④
		R	0.390	NA	0.0032	0.039	0.080	NA	28-36
2010	Lancer	F	①	NA	0.0024	0.039	0.080	NA	②
		R	0.390	NA	0.0032	0.039	0.080	NA	28-36
	Evolution	F	1.260	NA	0.0024	0.039	0.080	NA	NA
		R	0.870	NA	0.0032	0.035	0.080	NA	NA
	Sportback	F	③	NA	0.0024	0.039	0.080	NA	④
		R	0.390	NA	0.0032	0.039	0.080	NA	28-36

NA: Not Available

① RALLIART: 1.020. RAILLART: 0.940.

② 15" 28-36. 16" except RALLIART 28-36. 16" RALLIART 49-62.

③ 2.0L engine: 0.940. 2.4L engine: 1.020.

④ 2.0L engine: 49-62. 2.4L engine: 28-36.

37671_LANC_C0013

SCHEDULED MAINTENANCE INTERVALS
LANCER, EVOLUTION and SPORTBACK

TO BE SERVICED	TYPE OF SERVICE	VEHICLE MILEAGE INTERVAL (x1000)														
		7.5	15	22.5	30	37.5	45	52.5	60	67.5	75	82.5	90	97.5	105	120
Accessory drive belts	S/I				✓				✓				✓			✓
Air cleaner element (engine)	R				✓				✓				✓			✓
Air conditioner system	S/I	Inspect the system operation annually														
Automatic transaxle fluid	S/I				✓				✓				✓			✓
Ball joint and steering linkage seals	S/I				✓				✓				✓			✓
Brake lines, hoses, and connections	S/I		✓		✓		✓		✓		✓		✓		✓	✓
Brake pads, calipers, & rotors	S/I		✓		✓		✓		✓		✓		✓		✓	✓
Cooling system hoses and coolant level	S/I				✓				✓				✓			✓
Driveshafts and CV-boots	S/I		✓		✓		✓				✓		✓		✓	✓
Engine coolant	R								✓				✓			✓
Transfer Case Fluid w/ACD	S/I		✓	✓	✓	✓	✓	✓	✓	✓	✓	✓	✓	✓	✓	✓
Transfer Case Fluid w/ACD	R				✓				✓				✓			✓
Engine oil and filter, exc 2.0L turbo	R	✓	✓	✓	✓	✓	✓	✓	✓	✓	✓	✓	✓	✓	✓	✓
Engine oil and filter, 2.0L turbo	R	Every 5 months or 5,000 miles														
Exhaust pipe connections, muffler, and suspension bolts	S/I				✓				✓				✓			✓
Evaporative emission control system (except canister)	S/I								✓							✓
Fuel hoses	S/I				✓				✓				✓			✓
Fuel System (tank, connections, gas cap)	S/I								✓							✓
Manual transaxle oil	S/I				✓				✓				✓			✓
Twin Clurch Sportmatic Shift	S/I		✓	✓	✓	✓	✓	✓	✓	✓	✓	✓	✓	✓	✓	✓
Twin Clurch Sportmatic Shift	R								✓							✓
Spark plugs (standard)	R				✓				✓				✓			✓
Spark plugs exc 2.0L turbo (Iridium coated)	R	Every 84 months or 105,000 miles (under normal usage)														
Spark plugs exc 2.0L turbo (Iridium coated)	R								✓							✓
Spark plugs (Platinum coated)	R								✓							✓
Suspension system	S/I				✓				✓				✓			✓
Rear Axle with LSD	R				✓				✓				✓			✓
Timing Belt	R														✓	
Tires (rotate), 2.0L turbo	S/I	Every 5 months or 5,000 miles														
Tires (rotate), exc. 2.0L turbo	S/I	✓	✓	✓	✓	✓	✓	✓	✓	✓	✓	✓	✓	✓	✓	✓
Valve clearance	S/I								✓							✓

37671_LANC_C0014

SCHEDULED MAINTENANCE INTERVALS
LANCER, EVOLUTION and SPORTBACK (footnotes cont.)

R: Replace   S/I: Service or Inspect

FREQUENT OPERATION MAINTENANCE (SEVERE SERVICE)

If a vehicle is operated under any of the following conditions it is considered severe service:

- **Extremely dusty areas.**

- **50% or more of the vehicle operation is in 90°F (32°C) or higher temperatures, or constant operation in temperatures below 32°F (0°C).**

- **Prolonged idling (vehicle operation in stop and go traffic).**

- **Frequent short running periods (engine does not warm to normal operating temperatures).**

- **Police, taxi, delivery usage, or trailer towing usage.**

Air cleaner element: replace every 15,000 miles, 2.4L engine

Automatic transaxle fluid & filter: check every 15,000 miles, replace every 30,000 miles.

Twin clutch sportmatic shift transmission: change fluid every 30,000 miles

Brake pads, calipers & rotors: service or inspect every 7,500 miles, except 2.0L turbo. 2.0L turbo every 3,000 miles or 3 months..

Manual transaxle oil change every 30,000 miles

Oil & oil filter: change every 3,750 miles, except 2.0L turbo. 2.0L turbo every 3,000 miles or 3 months..

Spark plugs: service or inspect every 15,000 miles (standard plugs only)

Suspension system inspect for looseness and damage every 7,500 miles, except 2.0L turbo. 2.0L turbo every 3,000 miles or 3 months.

Halfshaft boots: inspect every 3,000 miles or 3 months.

Rotate tires every 3,750 miles, vehicles with 2.4L engine. Every 3,000 miles or 3 months, vehicles with 2.0L turbo engine.

Air filter: 2.4L engine inspect for clogging every 3,750 miles or 3 months. Replace every 7,500 miles or 6 months.

Air filter: 2.0L turbo engine inspect for clogging every 3,000 miles or 3 months. Replace every 6,000 miles or 6 months.

37671_LANC_C0015

BRAKES · INFORMATION AND PRECAUTIONS

ANTI-LOCK SYSTEMS

- Certain components within the ABS system are not intended to be serviced or repaired individually.
- Do not use rubber hoses or other parts not specifically specified for and ABS system. When using repair kits, replace all parts included in the kit. Partial or incorrect repair may lead to functional problems and require the replacement of components.
- Lubricate rubber parts with clean, fresh brake fluid to ease assembly. Do not use shop air to clean parts; damage to rubber components may result.
- Use only DOT 3 brake fluid from an unopened container.
- If any hydraulic component or line is removed or replaced, it may be necessary to bleed the entire system.
- A clean repair area is essential. Always clean the reservoir and cap thoroughly before removing the cap. The slightest amount of dirt in the fluid may plug an orifice and impair the system function. Perform repairs after components have been thoroughly cleaned; use only denatured alcohol to clean components. Do not allow ABS components to come into contact with any substance containing mineral oil; this includes used shop rags.
- The Anti-Lock control unit is a microprocessor similar to other computer units in the vehicle. Ensure that the ignition switch is **OFF** before removing or installing controller harnesses. Avoid static electricity discharge at or near the controller.
- If any arc welding is to be done on the vehicle, the control unit should be unplugged before welding operations begin.

DISC AND DRUM SYSTEMS

> ※ **CAUTION**
>
> Dust and dirt accumulating on brake parts during normal use may contain asbestos fibers from production or aftermarket brake linings.

Breathing excessive concentrations of asbestos fibers can cause serious bodily harm. Exercise care when servicing brake parts. Do not sand or grind brake lining unless equipment used is designed to contain the dust residue. Do not clean brake parts with compressed air or by dry brushing. Cleaning should be done by dampening the brake components with a fine mist of water, then wiping the brake components clean with a dampened cloth. Dispose of cloth and all residue containing asbestos fibers in an impermeable container with the appropriate label. Follow practices prescribed by the Occupational Safety and Health Administration (OSHA) and the Environmental Protection Agency (EPA) for the handling, processing, and disposing of dust or debris that may contain asbestos fibers.

BRAKES · BLEEDING THE BRAKE SYSTEM

BLEEDING PROCEDURE

BLEEDING PROCEDURE

See Figure 4.

When any part of the hydraulic system has been disconnected for repair or replacement, air may get into the lines and cause spongy pedal action (because air can be compressed and brake fluid cannot). To correct this condition, it is necessary to bleed the hydraulic system so to be sure all air is purged.

When bleeding the brake system, bleed one brake cylinder at a time, beginning at the cylinder with the longest hydraulic line (farthest from the master cylinder) first. ALWAYS keep the master cylinder reservoir filled with brake fluid during the bleeding operation. Never use brake fluid that has been drained from the hydraulic system, no matter how clean it is.

The primary and secondary hydraulic brake systems are separate and are bled independently. During the bleeding operation, do not allow the reservoir to run dry. Keep the master cylinder reservoir filled with brake fluid.

1. Clean all dirt from around the master cylinder fill cap, remove the cap and fill the master cylinder with brake fluid until the level is within ¼ inch (6mm) of the top edge of the reservoir.

2. Clean the bleeder screws at all 4 wheels. The bleeder screws are located on the back of the brake backing plate (drum brakes) and on the top of the brake calipers (disc brakes).

3. Attach a length of rubber hose over the bleeder screw and place the other end of the hose in a glass jar, submerged in brake fluid.

4. Open the bleeder screw ½–¾ turn. Have an assistant slowly depress the brake pedal.

> ※ **CAUTION**
>
> Brake fluid contains polyglycol ethers and polyglycols. Avoid contact with the eyes and wash your hands thoroughly after handling brake fluid. If you do get brake fluid in your eyes, flush your eyes with clean, running water for 15 minutes. If eye irritation persists, or if you have taken brake fluid internally, IMMEDIATELY seek medical assistance.

5. Close the bleeder screw and tell your assistant to allow the brake pedal to return slowly. Continue this process to purge all air from the system.

6. When bubbles cease to appear at the end of the bleeder hose, close the bleeder screw and remove the hose. Tighten the bleeder screw to the proper torque.

7. Check the master cylinder fluid level and add fluid accordingly. Do this after bleeding each wheel.

8. Repeat the bleeding operation at the remaining 3 wheels, ending with the one closet to the master cylinder.

9. Fill the master cylinder reservoir to the proper level.

BLEEDING THE ABS SYSTEM

There are no special procedures for bleeding the ABS system. Refer to the conventional bleeding procedures.

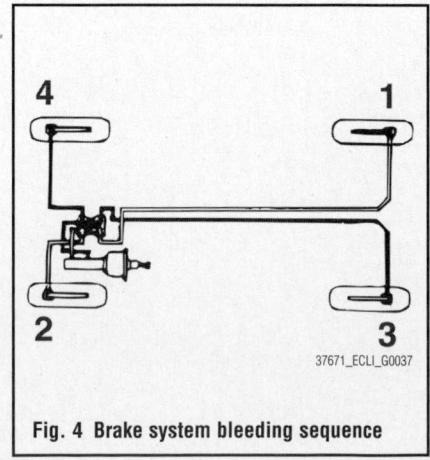

37671_ECLI_G0037

Fig. 4 Brake system bleeding sequence

BRAKES   **ANTI-LOCK BRAKE SYSTEM (ABS)**

WHEEL SPEED SENSORS

REMOVAL & INSTALLATION

1. Before servicing the vehicle, refer to the Precautions Section.

➡If working near and/or around the SRS system and components, be sure to disable the SRS system. Tape the negative battery cable with insulating tape. Always disconnect the negative battery cable first.

✳✳ CAUTION

Wait for 1 minute after disconnecting the negative battery cable before working inside the vehicle. The air bag system is set to deploy for a short period of time after the battery is disconnected.

2. Remove the air cleaner assembly, if removing the front sensor.
3. Remove the rear scuff plate, rear door opening trim, cargo box, trunk room rear end trim, flex floor plug if removing the rear sensor.
4. Raise and safely support the vehicle.
5. Remove the wheel and tire assembly, as necessary.
6. Disconnect the electrical connector.
7. Remove the harness clips, as required.
8. Remove the sensor from its mounting.

➡Take careful note of the exact position of each clip; they must be reinstalled in the identical position. Rear wheel sensor harnesses may be held by plastic wire ties; these may be cut away but must be replaced at reassembly.

✳✳ WARNING

The speed sensor has a pole piece projecting from it. This exposed tip must be protected from impact or scratches. Do not allow the pole piece to contact the toothed wheel during removal or installation.

To install:

➡Be sure to use new fasteners, as required.

9. Installation is the reverse of the removal procedure.

WHEEL SPEED SENSOR RINGS (TOOTHED RINGS)

REMOVAL & INSTALLATION

Front

➡The front wheel speed rotors are integrated with the BJ assembly of the halfshaft and cannot be disassembled.

Rear

➡The rear wheel speed rotors are integrated with the rear hub assembly and cannot be disassembled.

BRAKES   **FRONT DISC BRAKES**

BRAKE CALIPER

REMOVAL & INSTALLATION

See Figure 5.

1. Before servicing the vehicle, refer to the Precautions Section.
2. As required, partially drain the master cylinder.
3. Raise and support the vehicle safely.
4. Remove the tire and wheel assembly.

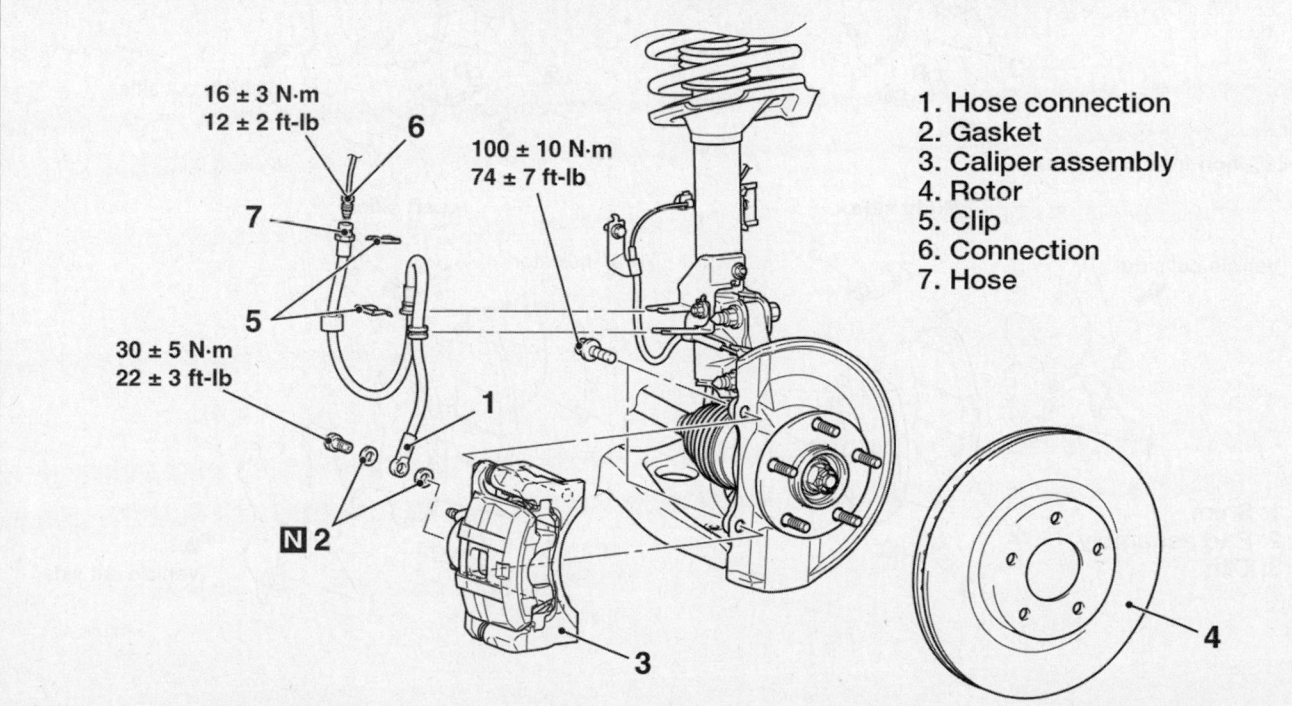

16 ± 3 N·m
12 ± 2 ft-lb

100 ± 10 N·m
74 ± 7 ft-lb

30 ± 5 N·m
22 ± 3 ft-lb

1. Hose connection
2. Gasket
3. Caliper assembly
4. Rotor
5. Clip
6. Connection
7. Hose

37671_LANC_G0204

Fig. 5 Front caliper and related components

5. Disconnect and plug the brake hose connection. Discard the gasket.

6. Remove the caliper retaining bolts. Remove the caliper from the vehicle.

To install

7. Position the caliper to its mounting on the vehicle.

8. Install the caliper mounting bolts and tighten to specification.

9. Connect the brake line to the caliper with new gaskets.

10. Bleed the brake system.

11. Install the wheel and tire.

12. Before attempting to move the vehicle, pump the brake pedal to seat the pads against the rotors. Make sure the vehicle has a firm brake pedal. Check the level of the brake fluid and add fluid if necessary.

DISC BRAKE PADS

REMOVAL & INSTALLATION

See Figures 6 through 10.

1. Before servicing the vehicle, refer to the Precautions Section.

2. Drain some of the brake fluid from the master cylinder reservoir.

3. Remove the front wheels.

4. Remove the caliper guide and lock pins and lift the caliper assembly from the caliper support.

➡**On some vehicles, the caliper can be flipped up by leaving the upper pin in place and using it as a pivot point.**

5. Remove the brake pads, spring clip, and shims.

To install:

6. Compress pistons back into the caliper bore.

7. Lubricate slide points and install the brake pads, shims, and spring clips onto the caliper support.

8. Install the caliper over the brake pads.

9. Lubricate and install the caliper guide and lock pins in their original positions.

10. Install the wheels.

11. Before attempting to move the vehicle, pump the brake pedal to seat the pads against the rotors. Make sure the vehicle has a firm brake pedal. Check the level of the brake fluid and add fluid if necessary.

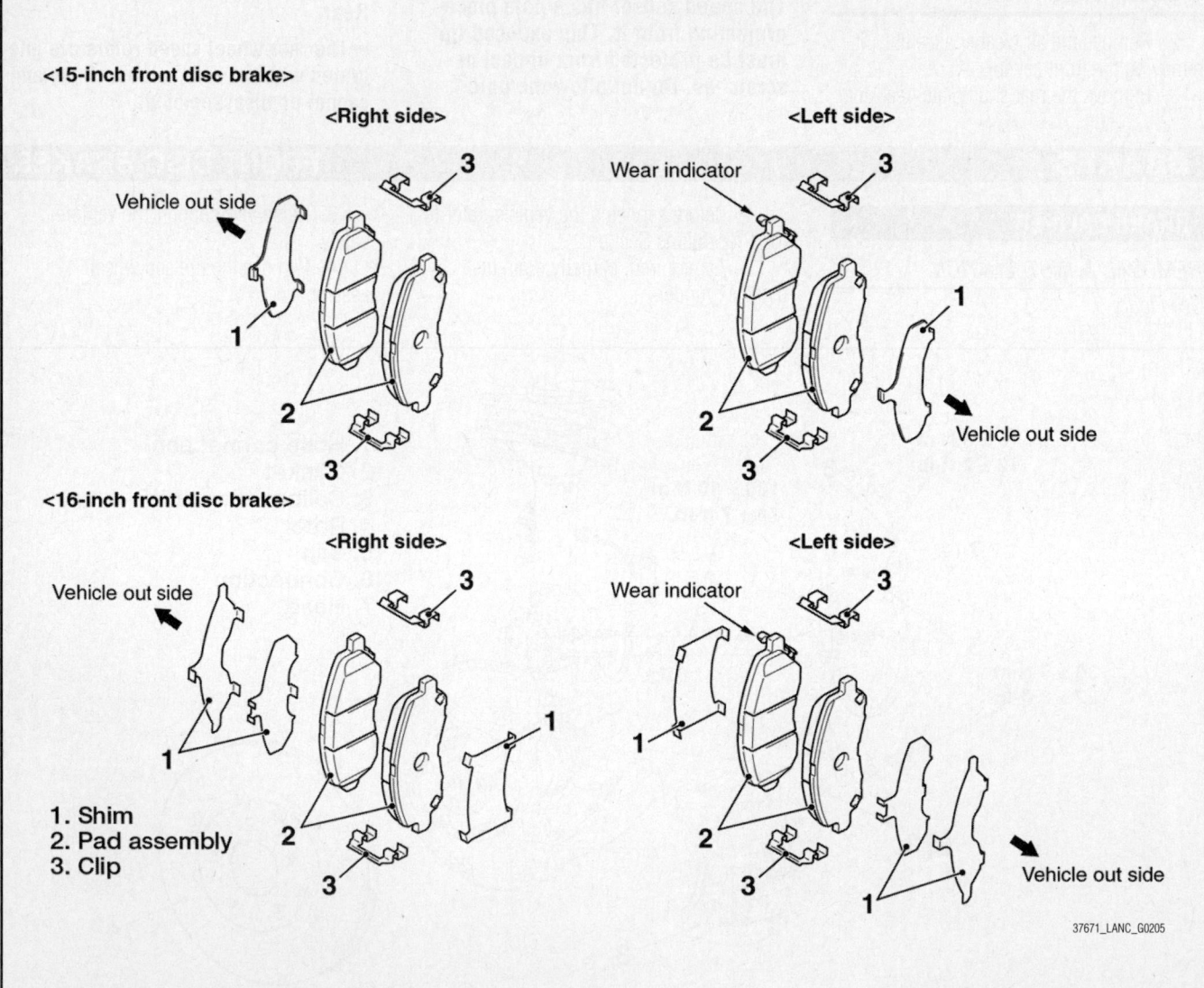

Fig. 6 Front pads and related components—Lancer except Ralliart

37671_LANC_G0205

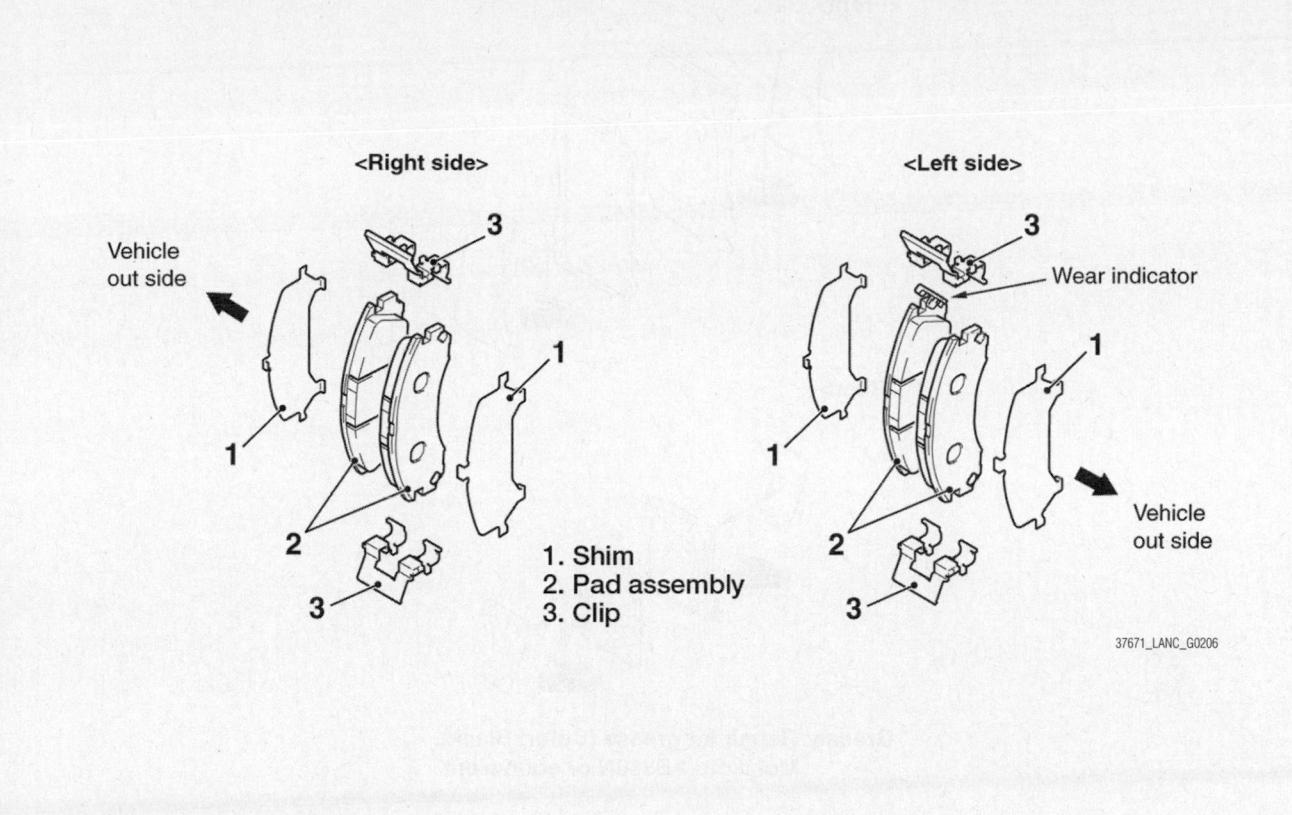

Fig. 7 Front pads and related components—Ralliart

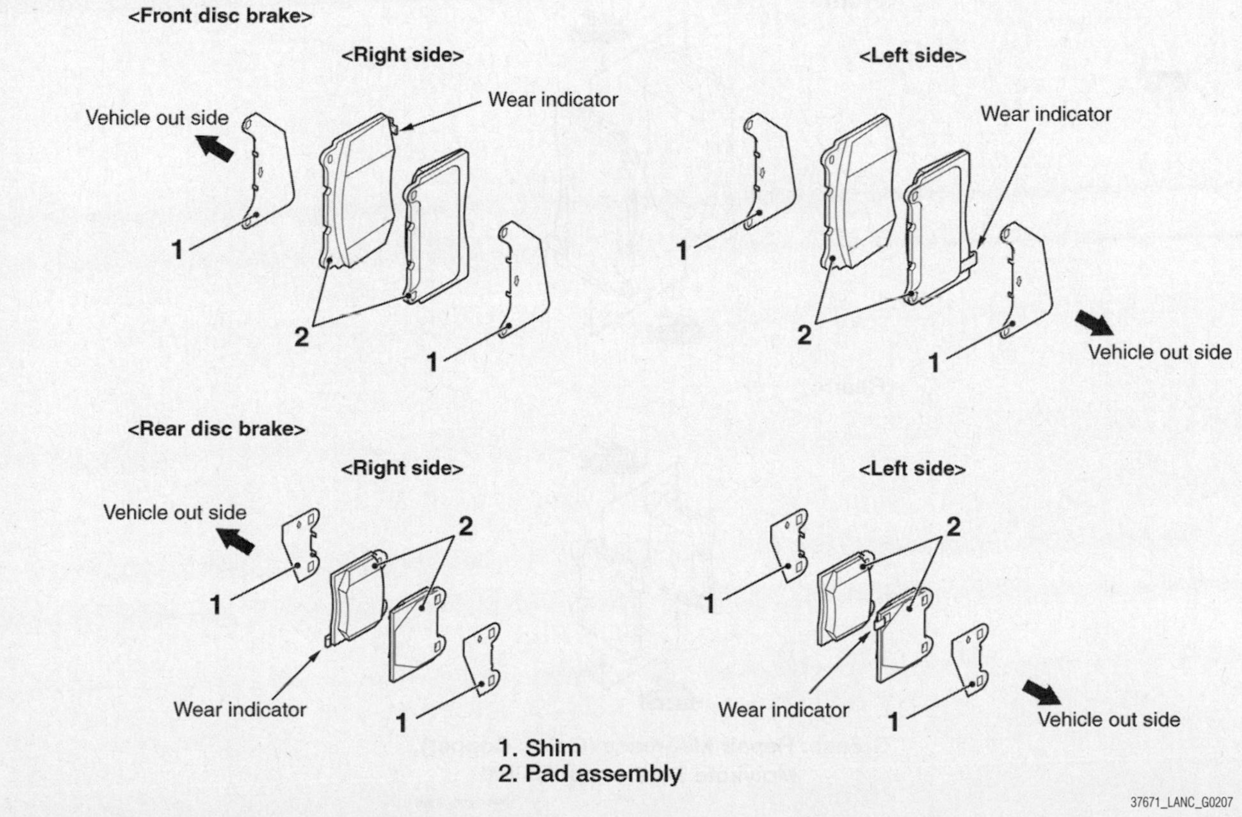

Fig. 8 Front and rear pads and related components—Evolution

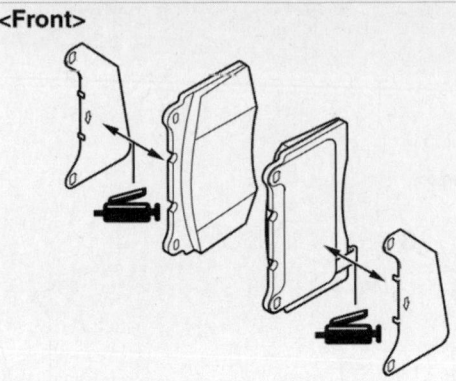

<Front>

<Rear>

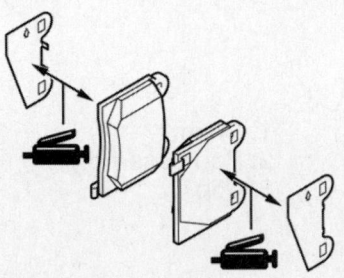

**Grease: Repair kit grease (Color: Black),
Molykote AS880N or equivalent**

37671_LANC_G0208

Fig. 9 Front and rear pad grease application points 1 of 2—Evolution

<Front>

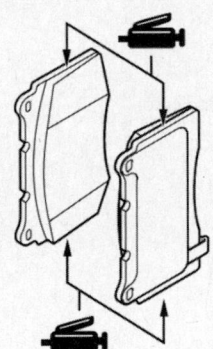

<Rear>

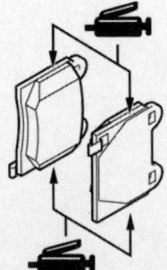

**Grease: Repair kit grease (Color: Copper),
Molykote 7439 or equivalent**

37671_LANC_G0209

Fig. 10 Front and rear pad grease application points 2 of 2—Evolution

BRAKES   **REAR DISC BRAKES**

BRAKE CALIPER

REMOVAL & INSTALLATION

See Figures 11 and 12.

1. Before servicing the vehicle, refer to the Precautions Section.

2. As required, partially drain the master cylinder.

3. Remove or disconnect the following:
   - Wheels
   - Brake hose from the caliper
   - Caliper guide and lock pins and lift the caliper assembly from the caliper support

To install

4. Install or connect the following:
   - Caliper onto the caliper support
   - Guide pin and lock pin and tighten to specification
   - Brake hose or banjo bolt with new washers
5. Bleed the brake system.
6. Install the wheels.

DISC BRAKE PADS

REMOVAL & INSTALLATION

See Figures 13 through 17.

1. Before servicing the vehicle, refer to the Precautions Section.

2. Remove or disconnect the following:
   - Rear wheels

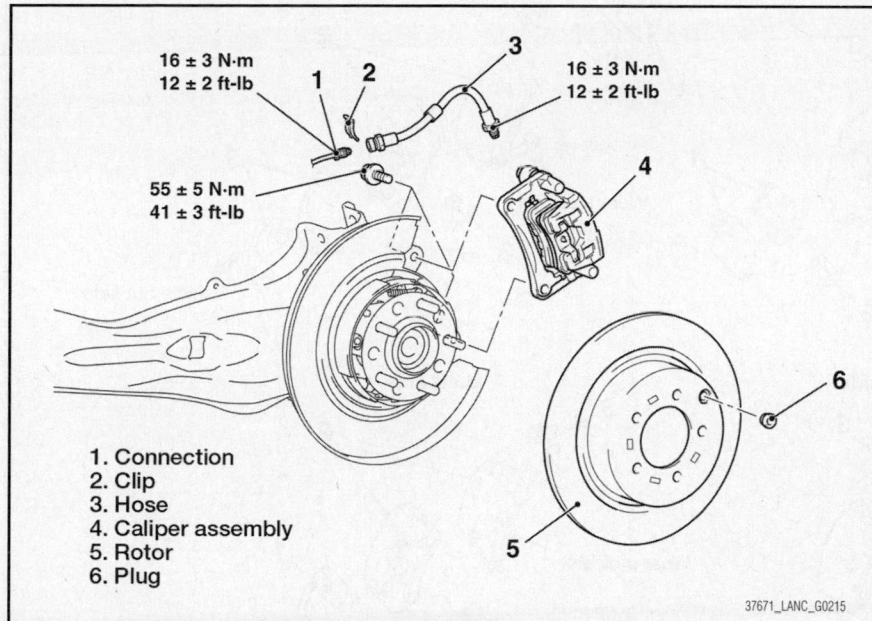

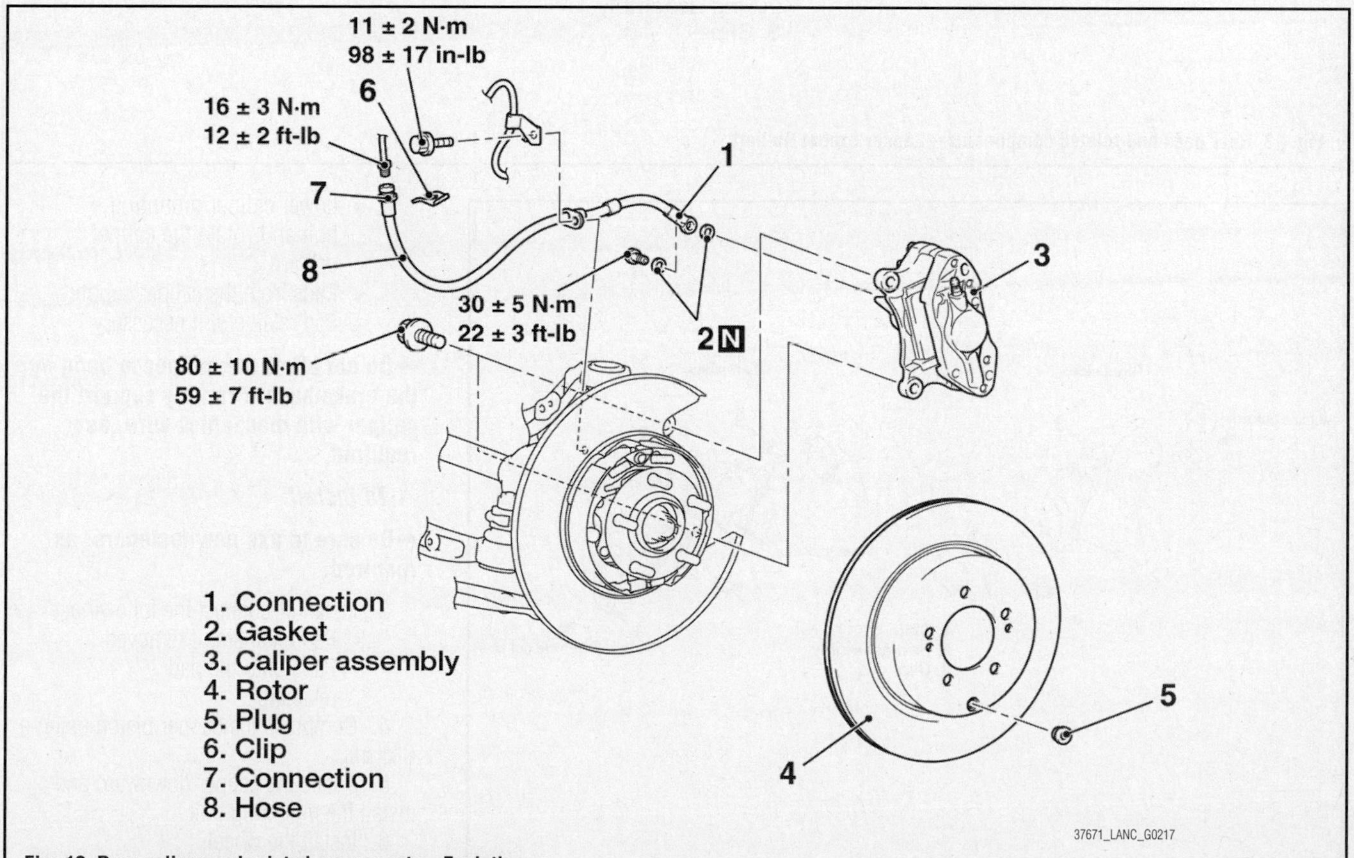

16 ± 3 N·m
12 ± 2 ft-lb

16 ± 3 N·m
12 ± 2 ft-lb

55 ± 5 N·m
41 ± 3 ft-lb

1. Connection
2. Clip
3. Hose
4. Caliper assembly
5. Rotor
6. Plug

37671_LANC_G0215

Fig. 11 Rear caliper and related components—Lancer

11 ± 2 N·m
98 ± 17 in-lb

16 ± 3 N·m
12 ± 2 ft-lb

30 ± 5 N·m
22 ± 3 ft-lb

80 ± 10 N·m
59 ± 7 ft-lb

2 N

1. Connection
2. Gasket
3. Caliper assembly
4. Rotor
5. Plug
6. Clip
7. Connection
8. Hose

37671_LANC_G0217

Fig. 12 Rear caliper and related components—Evolution

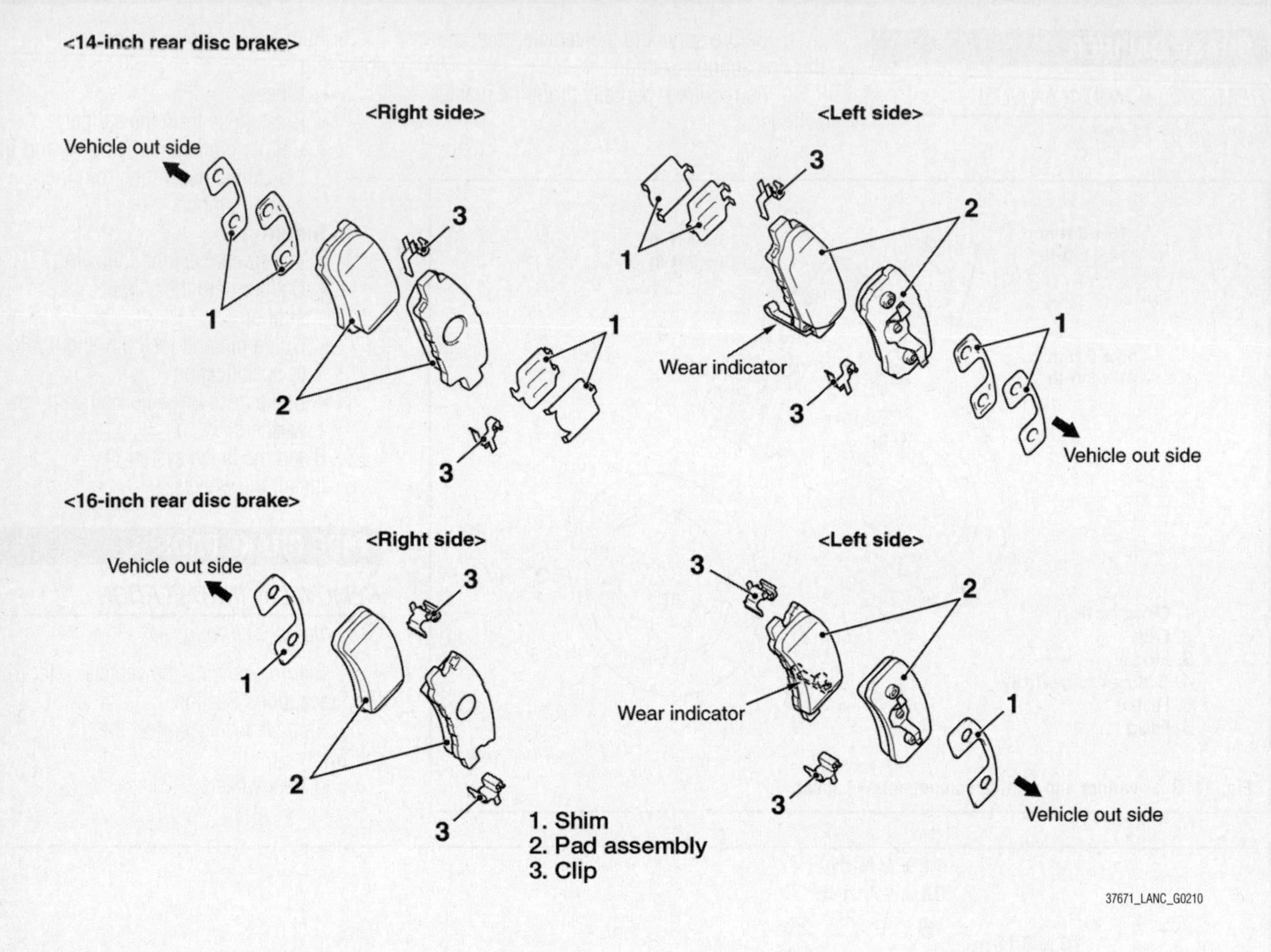

<14-inch rear disc brake>

<Right side>

Vehicle out side

3

1

1

2

3

<Left side>

3

2

1

1

Wear indicator

3

Vehicle out side

<16-inch rear disc brake>

<Right side>

Vehicle out side

3

1

2

3

<Left side>

3

2

Wear indicator

3

1

Vehicle out side

1. Shim
2. Pad assembly
3. Clip

37671_LANC_G0210

Fig. 13 Rear pads and related components—Lancer except Ralliart

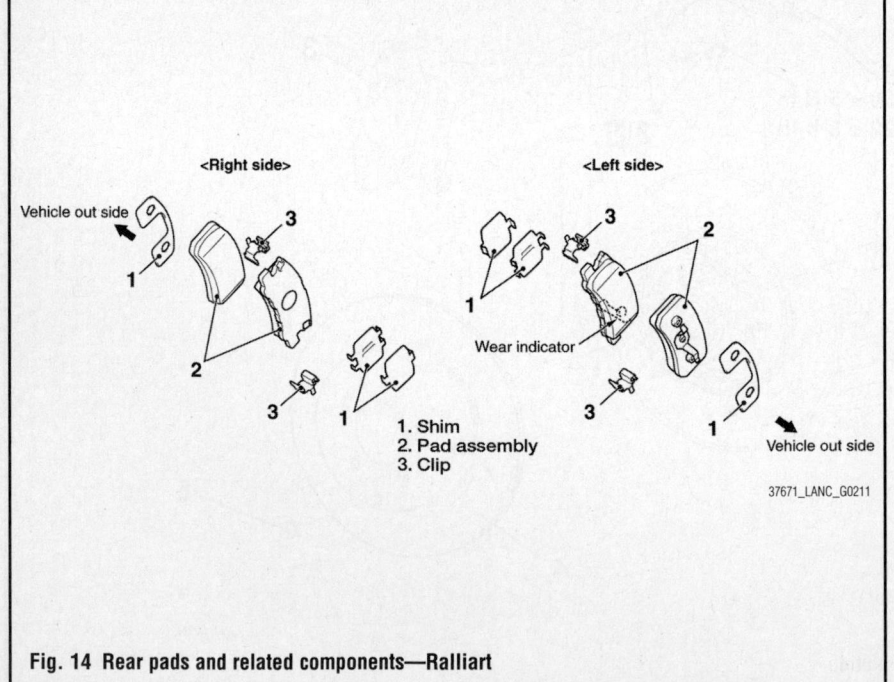

<Right side>

Vehicle out side

1

3

2

3

1

<Left side>

3

2

1

Wear indicator

3

1

Vehicle out side

1. Shim
2. Pad assembly
3. Clip

37671_LANC_G0211

Fig. 14 Rear pads and related components—Ralliart

- Lower caliper mounting bolt and rotate the caliper upward
- Pads from the caliper support
- Pad retainers, if necessary

➡**Do not allow the caliper to hang by the brake hose. Properly support the caliper with mechanics wire, as required.**

To install:

➡**Be sure to use new fasteners, as required.**

3. Install or connect the following:
- Pad retainers, if removed
- Pads onto the pad retainers

4. Compress the caliper piston using a C-clamp.

5. Rotate the caliper downward and install the mounting bolt.

6. Install the wheel.

7. Pump the brake pedal until the

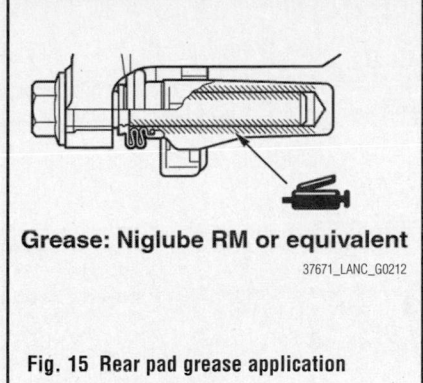

Grease: Niglube RM or equivalent

37671_LANC_G0212

Fig. 15 Rear pad grease application points 1 of 2—Ralliart

Grease: Repair kit grease (Color: Yellow)

37671_LANC_G0213

Fig. 16 Rear pad grease application points 2 of 2—Ralliart

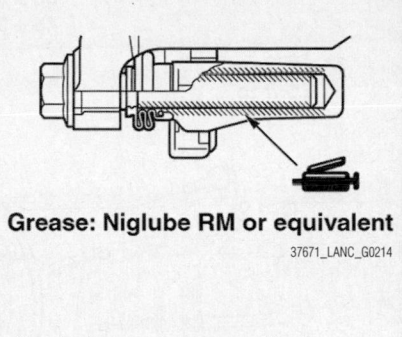

Grease: Niglube RM or equivalent

37671_LANC_G0214

Fig. 17 Rear pad grease application point— Lancer except Ralliart

brake pads are seated and a firm pedal is achieved before attempting to move the vehicle.

※※ **CAUTION**

Do not move the vehicle until a firm pedal is obtained.

8. Road test the vehicle to check for proper brake operation.

BRAKES

PARKING BRAKE CABLES

ADJUSTMENT

See Figure 18.

1. Before servicing the vehicle, refer to the Precautions Section.
2. Pull the parking brake lever with a force of 44 lbs. (196 N) and count the number of notches. Standard value is: 3–5 notches.
3. If the parking brake lever is not the standard value, adjust in the following manner:

    a. Remove the inner compartment mat of the floor console.

    b. Loosen the adjusting nut at the end of the cable rod, freeing the parking brake.

    c. With the engine idling, forcefully depress the brake pedal 5–6 times and confirm that the pedal stroke stops changing. If the pedal stroke stops changing, the automatic-adjustment mechanism is functioning normally, and the clearance between the shoe and the drum is correct.

    d. After adjusting the parking brake lever stroke, safely raise and support the rear of the vehicle and with the parking brake lever in the released position, turn the rear wheels to confirm that there is no brake drag.

➡**Be careful that the parking brake lever notch number is within the standard range. If the notch num-**

PARKING BRAKE

ber is too low, rear brake dragging may result.

PARKING BRAKE SHOES

REMOVAL & INSTALLATION

See Figure 19.

1. Before servicing the vehicle, refer to the Precautions Section.
2. Raise and safely support the vehicle.
3. Remove the caliper assembly.
4. Remove the rear brake rotor.

➡**When servicing, only dissemble and assemble one side at a time, leaving the remaining side intact for reference.**

5. Remove the front and rear shoe-to-anchor springs.
6. Remove the adjusting wheel spring and the adjuster.
7. Remove the strut and the strut return spring.
8. Remove the shoe hold-down cup, spring, and pin.
9. Remove the shoe and lining assembly.
10. Unfasten the clips and the retaining bolts.
11. Remove the parking brake cable.

    To install:

➡**Be sure to use new fasteners, as required.**

12. Installation is the reverse of the removal procedure.

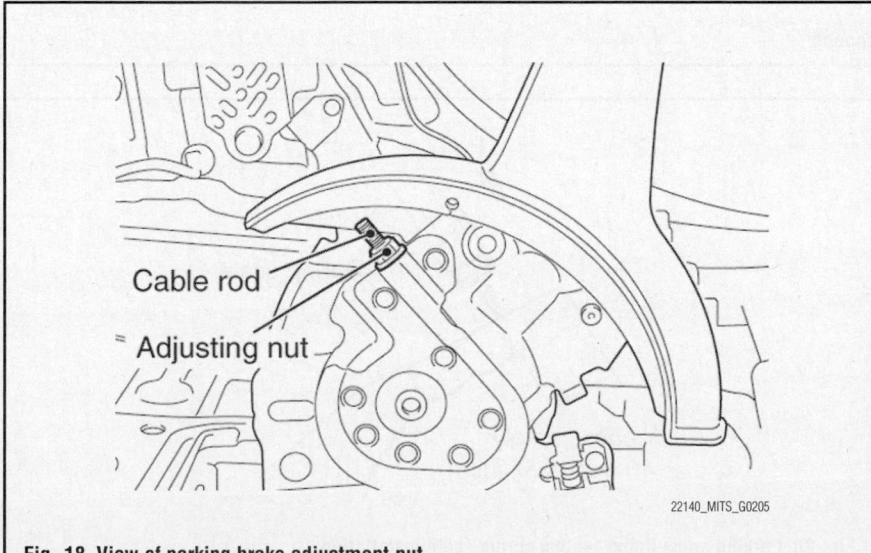

Cable rod

Adjusting nut

22140_MITS_G0205

Fig. 18 View of parking brake adjustment nut

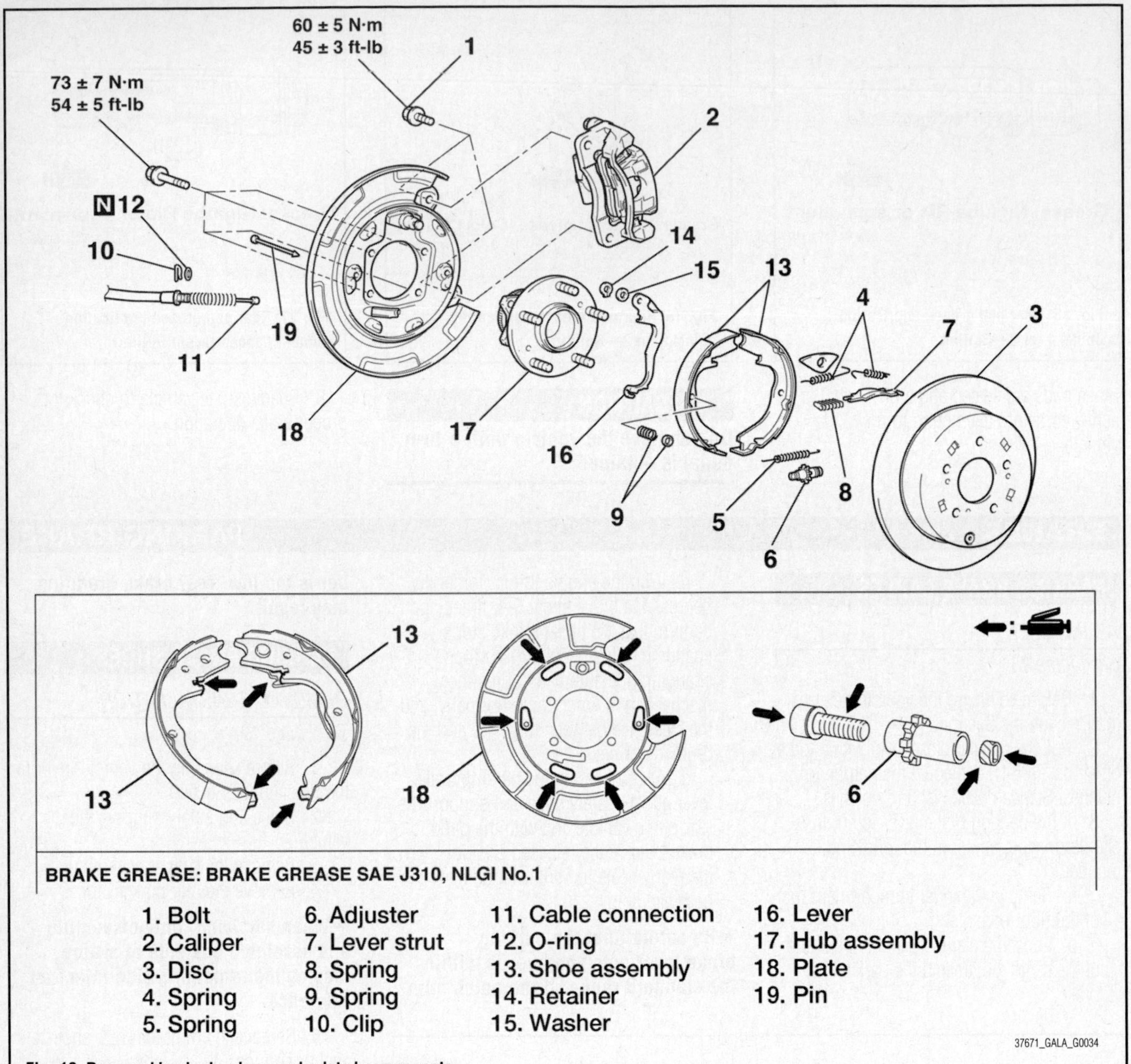

60 ± 5 N·m
45 ± 3 ft-lb — 1

73 ± 7 N·m
54 ± 5 ft-lb

N12

10

11

19

18

17

16

9

5

6

2

14

15

13

4

7

3

8

13

13

18

18

6

BRAKE GREASE: BRAKE GREASE SAE J310, NLGI No.1

1. Bolt	6. Adjuster	11. Cable connection	16. Lever
2. Caliper	7. Lever strut	12. O-ring	17. Hub assembly
3. Disc	8. Spring	13. Shoe assembly	18. Plate
4. Spring	9. Spring	14. Retainer	19. Pin
5. Spring	10. Clip	15. Washer	

37671_GALA_G0034

Fig. 19 Rear parking brake shoes and related components

ADJUSTMENT

Brake Lining Seating

See Figure 20.

1. Adjust the parking brake lever stroke.

2. Hook a spring scale onto the center of the parking brake lever grip and pull it with a force of about 22–23 ft. lbs. in the direction perpendicular to the handle.

3. Drive the vehicle at a constant speed of 22–31 MPH for about 328 feet.

4. Release the parking brake and let the brakes cool for 5–10 minutes.

5. Repeat the above about 4–5 times.

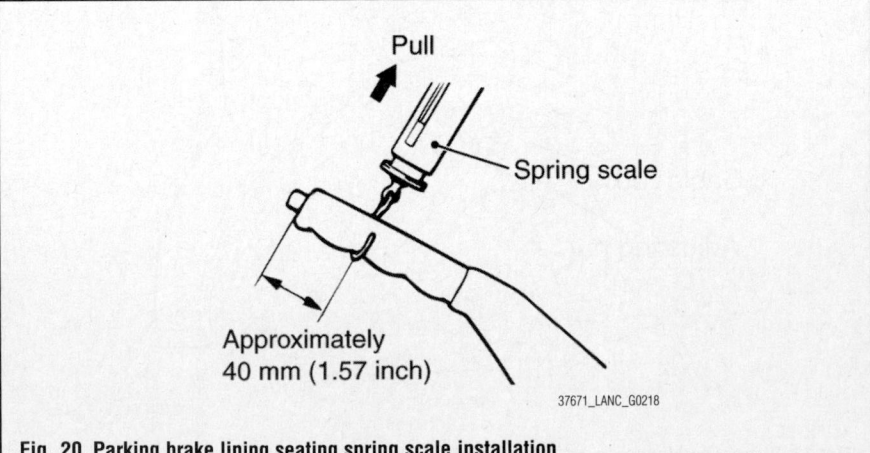

Pull

Spring scale

Approximately
40 mm (1.57 inch)

37671_LANC_G0218

Fig. 20 Parking brake lining seating spring scale installation

CHASSIS ELECTRICAL — AIR BAG (SUPPLEMENTAL RESTRAINT SYSTEM)

GENERAL INFORMATION

❊ CAUTION

This vehicle is equipped with an air bag system. The system must be disarmed before performing service on, or around, system components, the steering column, instrument panel components, wiring, and sensors. Failure to follow the safety precautions and the disarming procedure could result in accidental air bag deployment, possible injury, and unnecessary system repairs.

PRECAUTIONS

Disconnect and isolate the battery negative cable before beginning any airbag system component diagnosis, testing, removal, or installation procedures. Allow system capacitor to discharge for 3 minutes before beginning any component service. This will disable the airbag system. Failure to disable the airbag system may result in accidental airbag deployment, personal injury, or death.

DISARMING THE SYSTEM

1. Before servicing the vehicle, refer to the Precautions Section.
2. Remove the ignition key from the vehicle.
3. Disconnect the negative battery cable and isolate it from accidental reconnection. Insulate the cable end with high-quality electrical tape or a similar non-conductive wrapping.

4. Wait at least 1 minute for the system capacitor to discharge before performing any service. The air bag system is designed to retain enough voltage to deploy the air bag for a short period of time after the battery has been disconnected.

ARMING THE SYSTEM

1. Before servicing the vehicle, refer to the Precautions Section.
2. Reconnect the negative battery cable.
3. To confirm proper system operation, turn the ignition switch to the **ON** position. The SRS indicator light will be lit for at least 7 seconds and then go off.

DRIVE TRAIN

CLUTCH DRIVEN DISC & PRESSURE PLATE

REMOVAL & INSTALLATION
See Figures 21 and 22.

❊❊ CAUTION

The clutch driven disc may contain asbestos, which has been determined to be a cancer causing agent. Never clean clutch surfaces with compressed air! Avoid inhaling any dust from any clutch surface! When cleaning clutch surfaces, use a commercially available brake cleaning fluid.

1. Before servicing the vehicle, refer to the Precautions Section.

➡ **If working near and/or around the SRS system and components, be sure to disable the SRS system. Tape the negative battery cable with insulating tape. Always disconnect the negative battery cable first.**

❊❊ CAUTION

Wait for 1 minute after disconnecting the negative battery cable before working inside the vehicle. The air bag system is set to deploy for a short period of time after the battery is disconnected.

2. Disconnect the negative battery cable.
3. Raise and safely support the vehicle.
4. Remove the transaxle assembly from the vehicle.
5. Remove the pressure plate attaching bolts, pressure plate and clutch disc. If the pressure plate is to be reused, loosen the bolts in a diagonal pattern, 1 or 2 turns at a time. This will prevent warping the clutch cover assembly.
6. Remove the return clip and the pressure plate release bearing. Do not use solvent to clean the bearing.

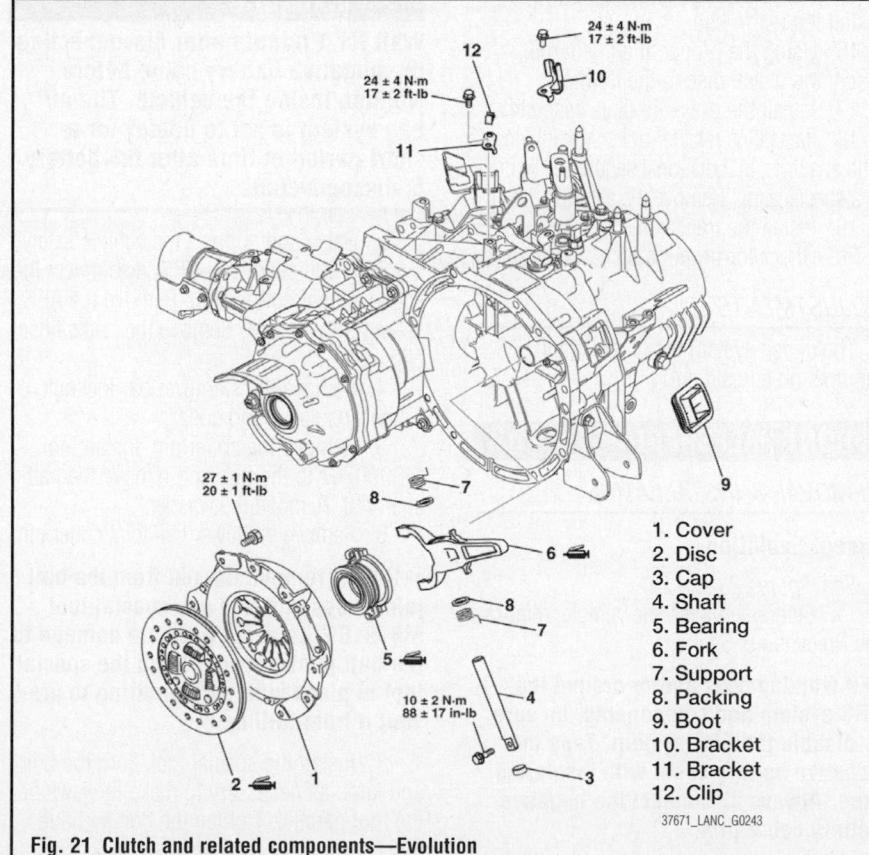

24 ± 4 N·m
17 ± 2 ft-lb

24 ± 4 N·m
17 ± 2 ft-lb

27 ± 1 N·m
20 ± 1 ft-lb

10 ± 2 N·m
88 ± 17 in-lb

1. Cover
2. Disc
3. Cap
4. Shaft
5. Bearing
6. Fork
7. Support
8. Packing
9. Boot
10. Bracket
11. Bracket
12. Clip

37671_LANC_G0243

Fig. 21 Clutch and related components—Evolution

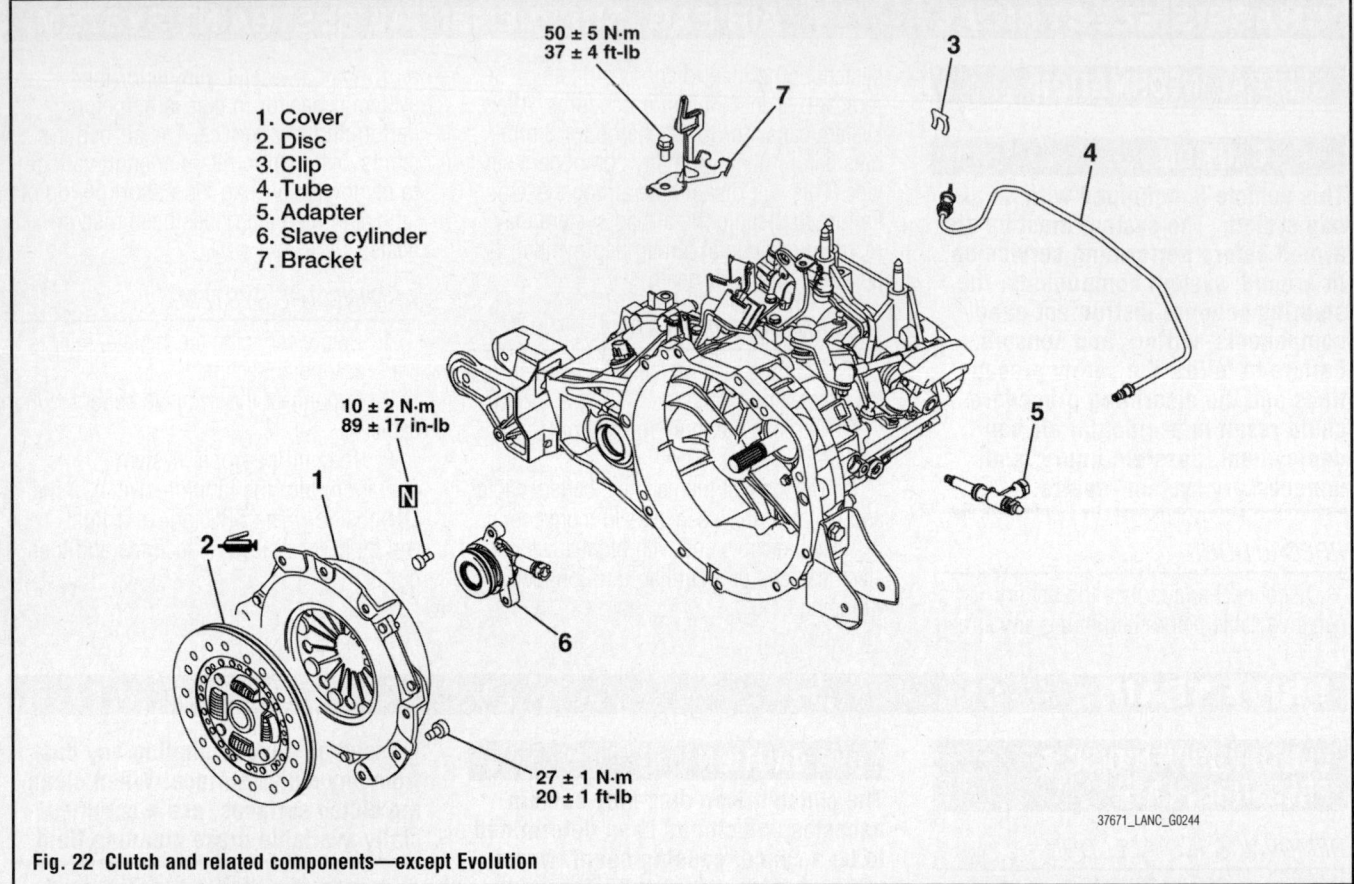

1. Cover
2. Disc
3. Clip
4. Tube
5. Adapter
6. Slave cylinder
7. Bracket

50 ± 5 N·m
37 ± 4 ft-lb

10 ± 2 N·m
89 ± 17 in-lb

27 ± 1 N·m
20 ± 1 ft-lb

37671_LANC_G0244

Fig. 22 Clutch and related components—except Evolution

7. Inspect the clutch release fork and fulcrum for damage or wear. If necessary, remove the release fork and unthread the fulcrum from the transaxle.

8. Carefully inspect the condition of the clutch components and replace any worn or damaged parts.

To install:

→**Be sure to use new fasteners, as required.**

9. Inspect the flywheel for heat damage or cracks. Resurface or replace the flywheel as required.

10. Install the fulcrum and tighten to 25 ft. lbs. (35 Nm).

11. Install the release fork.

12. Apply a coating of multi-purpose grease to the point of contact with the fulcrum and the point of contact with the release bearing.

13. Apply a coating of multi-purpose grease to the end of the release cylinder pushrod and the pushrod hole in the release fork.

14. Apply multi-purpose grease to the clutch release bearing. Pack the bearing inner surface and the groove with grease. Do not apply grease to the resin portion of the bearing.

15. Place the bearing in position and install the return clip.

16. Using the proper alignment tool, install the clutch disc to the flywheel.

17. Install the pressure plate assembly.

18. Install the retainer bolts and tighten a little at a time, in a diagonal sequence. Tighten to a specification. Remove the aligning tool.

19. Install the transaxle assembly.

20. Check for proper clutch operation.

ADJUSTMENTS

The clutch system is hydraulic and requires no adjustment.

FRONT HALFSHAFT

REMOVAL & INSTALLATION

Except Evolution

See Figures 23 and 24.

1. Before servicing the vehicle, refer to the Precautions Section.

→**If working near and/or around the SRS system and components, be sure to disable the SRS system. Tape the negative battery cable with insulating tape. Always disconnect the negative battery cable first.**

✴✴ CAUTION

Wait for 1 minute after disconnecting the negative battery cable before working inside the vehicle. The air bag system is set to deploy for a short period of time after the battery is disconnected.

2. Raise and support the vehicle safely.

3. If equipped with ABS, disconnect the speed sensor connection. Remove the ABS sensor and bracket. Remove the brake hose clip.

4. Remove the stabilizer bar locknut, rubber insulator, and collar.

5. Remove the cotter pin. Install tool MB990767 to the hub and remove the halfshaft nut. Remove the washer.

6. Remove the lower ball joint cotter pin.

→**Do not remove the nut from the ball joint. Loosen it and use special tool MB991897 to avoid possible damage to the ball joint threads. Hang the special tool in place with wire or string to prevent it from falling.**

7. Install the special tool. Turn the bolt and knob as necessary to make the jaws of the tool parallel. Tighten the bolt by hand and confirm that the jaws are still parallel.

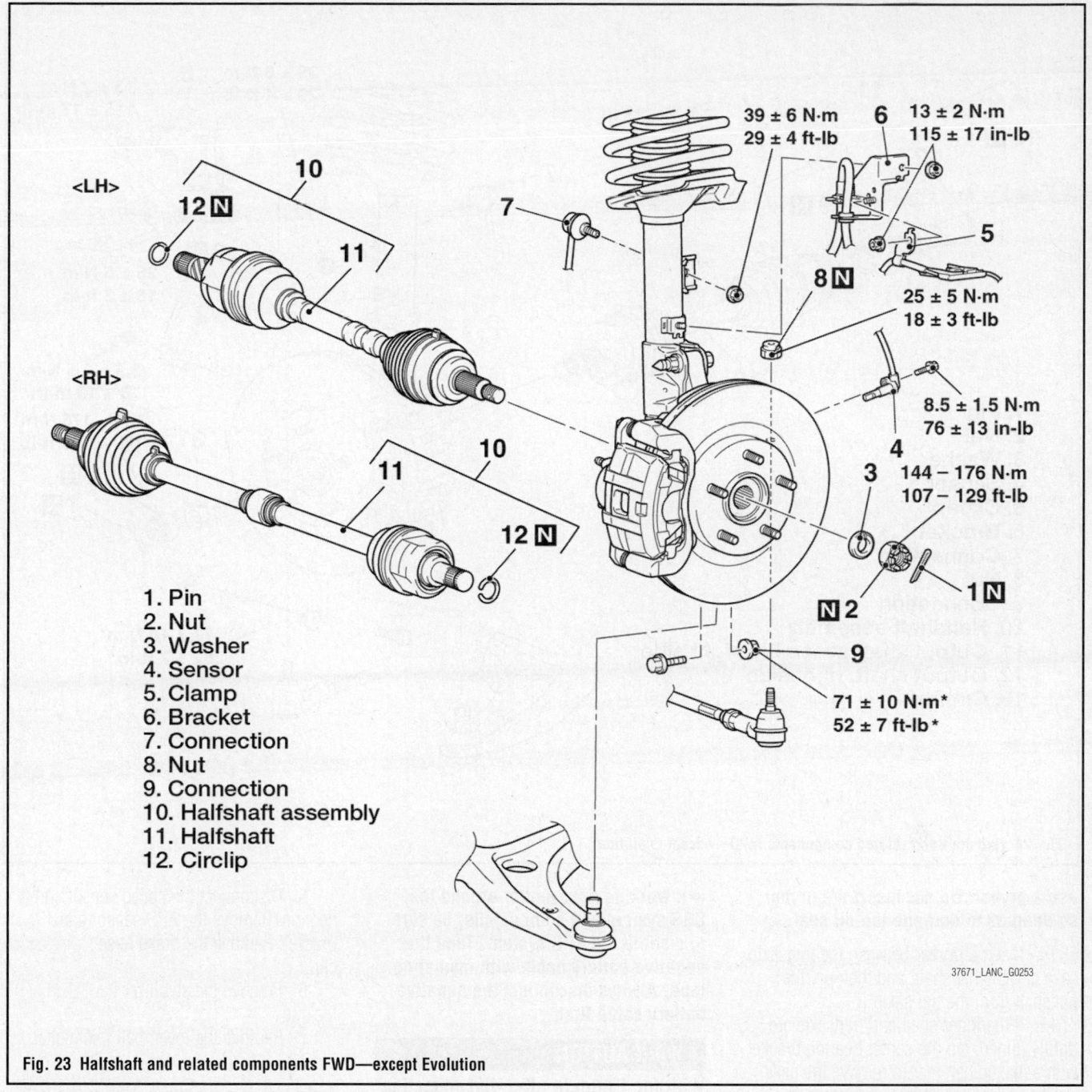

1. Pin
2. Nut
3. Washer
4. Sensor
5. Clamp
6. Bracket
7. Connection
8. Nut
9. Connection
10. Halfshaft assembly
11. Halfshaft
12. Circlip

Fig. 23 Halfshaft and related components FWD—except Evolution

➡**When adjusting the jaws in parallel, make sure the knob is in the vertical (upward) position.**

8. Tighten the bolt with a wrench to disconnect the lower arm ball joint connection.

9. Remove the tie rod end cotter pin. Install tool MB990767 to the hub and remove the halfshaft nut. Remove the washer.

➡**Do not remove the nut from the tie rod end. Loosen it and use special tool MB991897 to avoid possible damage to the ball joint threads. Hang the special**

tool in place with wire or string to prevent it from falling.

10. Install the special tool. Turn the bolt and knob as necessary to make the jaws of the tool parallel. Tighten the bolt by hand and confirm that the jaws are still parallel.

➡**When adjusting the jaws in parallel, make sure the knob is in the vertical (upward) position.**

11. Tighten the bolt with a wrench to disconnect the tie rod end.

➡**Do not damage the ABS rotor**

attached to the BJ outer race, on vehicle equipped with ABS.

12. Use special tools MB990241 and MB990767 (vehicles without center bearing) or tools MB991354, MB990242 and MB990244 (vehicles equipped with center bearing) to push the halfshaft out from the hub and knuckle.

13. Remove the halfshaft from the hub by pulling the bottom of the brake disc toward you, and then remove the hub retaining bolts.

➡**Do not pull on the halfshaft, doing so will damage the TJ or ETJ. Be sure to**

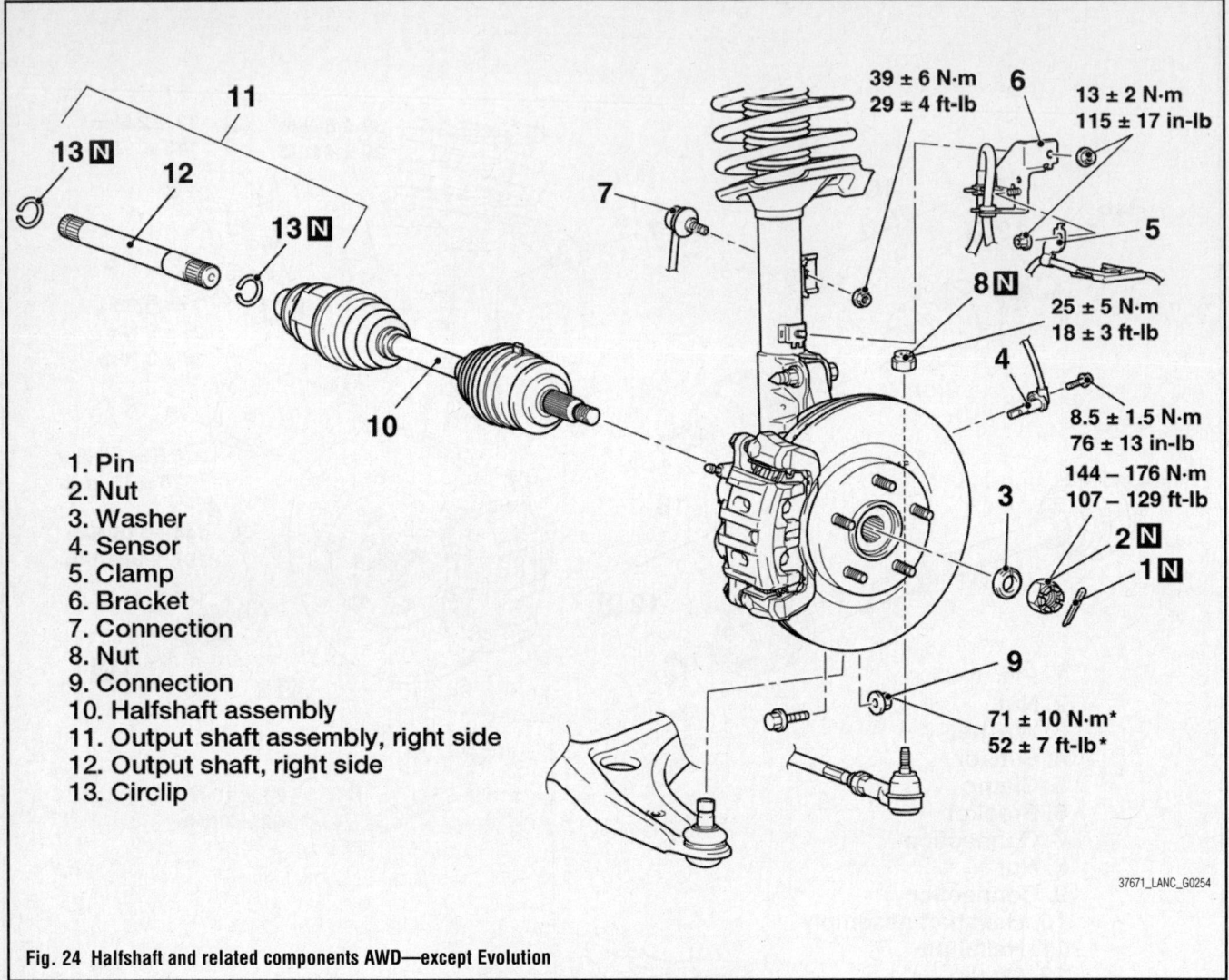

39 ± 6 N·m
29 ± 4 ft-lb

13 ± 2 N·m
115 ± 17 in-lb

25 ± 5 N·m
18 ± 3 ft-lb

8.5 ± 1.5 N·m
76 ± 13 in-lb

144 – 176 N·m
107 – 129 ft-lb

71 ± 10 N·m*
52 ± 7 ft-lb*

1. Pin
2. Nut
3. Washer
4. Sensor
5. Clamp
6. Bracket
7. Connection
8. Nut
9. Connection
10. Halfshaft assembly
11. Output shaft assembly, right side
12. Output shaft, right side
13. Circlip

37671_LANC_G0254

Fig. 24 Halfshaft and related components AWD—except Evolution

use a prybar. Do not insert the prybar so deep as to damage the oil seal.

14. Insert a prybar between the transaxle case and the halfshaft, and then pry the halfshaft from the transaxle.

15. If the inner shaft and transaxle are tightly joined, tap the center bearing bracket with a plastic hammer to remove the halfshaft and inner shaft from the transaxle.

To install:

➡Be sure to use new fasteners, as required.

16. Installation is the reverse of the removal procedure.

17. Check and adjust the front end alignment, as necessary.

Evolution

See Figure 25.

1. Before servicing the vehicle, refer to the Precautions Section.

➡If working near and/or around the SRS system and components, be sure to disable the SRS system. Tape the negative battery cable with insulating tape. Always disconnect the negative battery cable first.

✳✳ CAUTION

Wait for 1 minute after disconnecting the negative battery cable before working inside the vehicle. The air bag system is set to deploy for a short period of time after the battery is disconnected.

2. Raise and support the vehicle safely.

3. Remove the undercover. Remove the side cover. Drain the transaxle fluid. Drain the transfer case.

4. Remove the cotter pin. Install tool MB990767 to the hub and remove the driveshaft nut. Remove the washer.

5. Disconnect the speed sensor connection. Remove the ABS sensor and bracket. Remove the brake hose clip.

6. Remove the stabilizer bar locknut.

7. Remove the lower ball joint cotter pin.

➡Do not remove the nut from the ball joint. Loosen it and use special tool MB991897 to avoid possible damage to the ball joint threads. Hang the special tool in place with wire or string to prevent it from falling.

8. Install the special tool. Turn the bolt and knob as necessary to make the jaws of the tool parallel. Tighten the bolt by hand and confirm that the jaws are still parallel.

➡When adjusting the jaws in parallel, make sure the knob is in the vertical (upward) position.

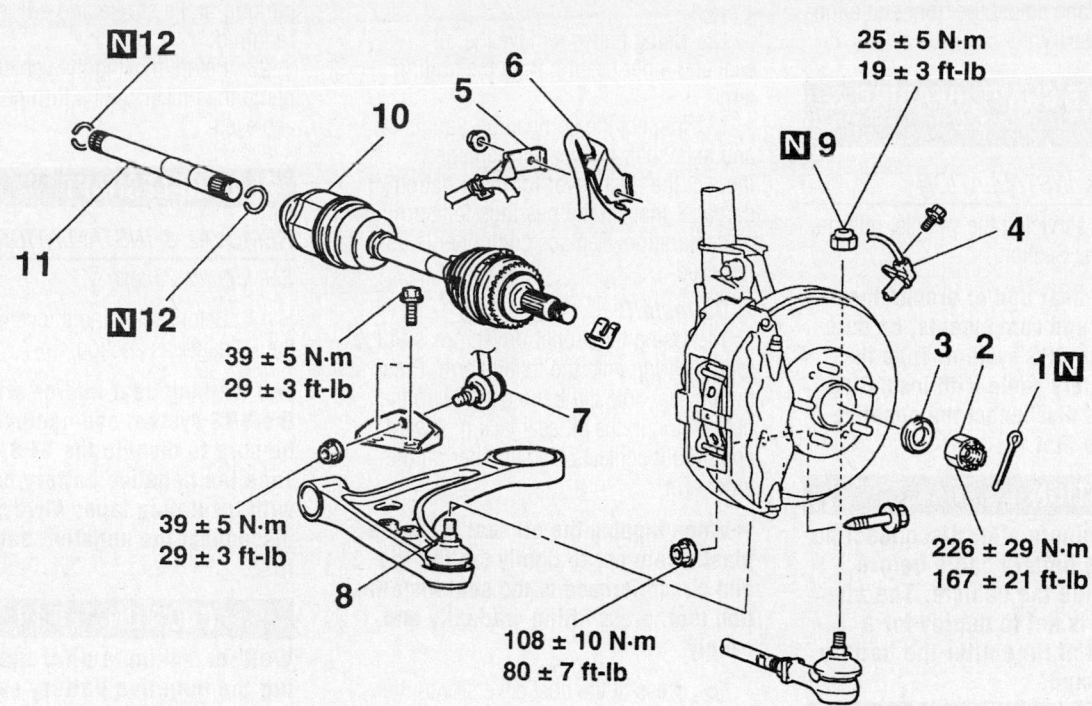

25 ± 5 N·m
19 ± 3 ft-lb

39 ± 5 N·m
29 ± 3 ft-lb

39 ± 5 N·m
29 ± 3 ft-lb

108 ± 10 N·m
80 ± 7 ft-lb

226 ± 29 N·m
167 ± 21 ft-lb

1. COTTER PIN
2. DRIVESHAFT NUT
3. WASHER
4. FRONT ABS SENSOR
5. FRONT ABS SENSOR HARNESS BRACKET
6. BRAKE HOSE BRACKET
7. STABILIZER BAR LINK CONNECTION
8. LOWER ARM BALL JOINT CONNECTION
9. SELF LOCKING NUT (TIE ROD END CONNECTION)
10. DRIVESHAFT
11. OUTPUT SHAFT
12. CIRCLIP

09482_GALA_G0171

Fig. 25 Halfshaft and related components—Evolution

9. Tighten the bolt with a wrench to disconnect the lower arm ball joint connection.

10. Remove the tie rod end cotter pin. Install tool MB990767 to the hub and remove the halfshaft nut. Remove the washer.

➡**Do not remove the nut from the tie rod end. Loosen it and use special tool MB991897 to avoid possible damage to the ball joint threads. Hang the special tool in place with wire or string to prevent it from falling.**

11. Install the special tool. Turn the bolt and knob as necessary to make the jaws of the tool parallel. Tighten the bolt by hand and confirm that the jaws are still parallel.

➡**When adjusting the jaws in parallel, make sure the knob is in the vertical (upward) position.**

12. Tighten the bolt with a wrench to disconnect the tie rod end.

➡**Do not strike the ABS rotor attached to the EBJ outer race, of the halfshaft against other parts when removing the halfshaft as damage to the rotors will result.**

13. Use special tools MB990241 (MB990242 and MB990244), MB991354 and MB990767 to push the halfshaft out from the hub.

14. Remove the halfshaft from the hub by pulling the bottom of the brake disc toward you, and then remove the hub retaining bolts.

➡**Do not pull on the halfshaft, doing so will damage the TJ. Be sure to use a prybar. Do not insert the prybar so deep as to damage the oil seal.**

15. Insert a prybar between the transaxle case and the halfshaft, and then pry and remove the driveshaft from the transaxle.

To install:

➡**Be sure to use new fasteners, as required.**

16. Installation is the reverse of the removal procedure.

17. Check and adjust the front end alignment, as necessary.

REAR AXLE SHAFT, BEARING & SEAL

REMOVAL & INSTALLATION

1. Before servicing the vehicle, refer to the Precautions Section.

➡ **If working near and/or around the SRS system and components, be sure to disable the SRS system. Tape the negative battery cable with insulating tape. Always disconnect the negative battery cable first.**

※※ CAUTION

Wait for 1 minute after disconnecting the negative battery cable before working inside the vehicle. The air bag system is set to deploy for a short period of time after the battery is disconnected.

2. Disconnect the negative battery cable.

3. Raise and support the vehicle safely.

4. Remove the tire and wheel assembly from the vehicle.

5. If equipped with ABS, remove the rear wheel speed sensor.

➡ **Be cautious to ensure that the tip of the pole piece on the rear speed sensor does not come in contact with other parts during removal. Sensor damage could occur.**

6. Remove the rear caliper and support assembly out of the way. Remove the brake disc.

7. Remove the driveshaft and companion flange installation bolts, nuts and washers. Move the end of shaft slightly to access the self-locking nut.

8. Using axle holding tool MB990211-01 or equivalent, secure the rear axle shaft in position, then remove the self-locking nut.

9. Using puller and adapter MB990211-01 and MB990241-01 or equivalents, remove the rear axle shaft from the trailing arm.

10. If equipped with ABS, remove the rear rotor from the axle assembly using collar and press. The rotor is a press fit.

11. Remove the outer bearing and dust cover concurrently from the axle shaft using a press.

12. Using puller, remove the oil seal and inner bearing from the trailing arm.

13. Inspect the companion flange and axle shaft for wear or damage. Inspect the dust cover for deformation or damage. Inspect the bearings for burning or declaration. Replace components as required.

To install:

14. Using the proper driver, press fit the inner bearing onto the trailing arm. Press fit the oil seal onto the trailing arm with the depression in the oil seal facing upward, and until it contacts the shoulder on the inner arm.

➡ **When tapping the oil seal in, use a plastic hammer to lightly tap the top and circumference of the seal installation tool, press fitting gradually and evenly.**

15. Press fit the dust covers onto the axle until it contacts the axle shaft shoulder. Install the innermost cover so the depression is facing upward.

➡ **When tapping the oil seal in, use a plastic hammer to lightly tap the top and circumference of the seal installation tool, press fitting gradually and evenly.**

16. Apply multi-purpose grease around the entire circumference of the inner side of the outer bearing seal lip. Press fit the outer bearing to the axle shaft so that the bearing seal lip surface is facing towards the axle shaft flange.

17. Press fit the rear rotor to the axle shaft with the rear rotor groove surface towards the axle shaft flange.

18. Install the rear axle shaft to the trailing arm temporarily. Install the companion flange to the rear axle shaft, then install a new self-locking nut.

19. While holding the rear axle shaft in position using holding fixture tool MB990767-01 or equivalent, tighten a new self-locking nut to 159 ft. lbs. (220 Nm).

20. Install the drive shaft nuts, washers and bolts. Tighten to 40–47 ft. lbs. (55–65 Nm).

21. Install the rear brake disc, caliper assembly and parking brake.

22. Install the tire and wheel assembly and lower the vehicle. Check the parking brake stroke and adjust as required.

23. Before moving the vehicle, pump the brakes until a firm pedal is achieved.

REAR HALFSHAFT

REMOVAL & INSTALLATION

See Figures 26 and 27.

1. Before servicing the vehicle, refer to the Precautions Section.

➡ **If working near and/or around the SRS system and components, be sure to disable the SRS system. Tape the negative battery cable with insulating tape. Always disconnect the negative battery cable first.**

※※ CAUTION

Wait for 1 minute after disconnecting the negative battery cable before working inside the vehicle. The air bag system is set to deploy for a short period of time after the battery is disconnected.

➡ **Be cautious to ensure that the tip of the pole piece on the rear speed sensor does not come in contact with other parts during removal. Sensor damage could occur.**

2. Drain the gear oil. Properly dispose of used oil.

3. On Evolution, remove the center exhaust pipe.

4. Disconnect the joints between the lower arm, trailing arm, shock and stabilizer link.

5. Disconnect the joint between the control link and trailing arm.

6. Remove the cotter pin. Remove the halfshaft nut. Remove the washer.

7. Remove the speed sensor.

8. Remove the halfshaft.

To install:

➡ **Be sure to use new fasteners, as required.**

9. Installation is the reverse of the removal procedure.

10. Be sure to fill the differential using the proper grade and type oil.

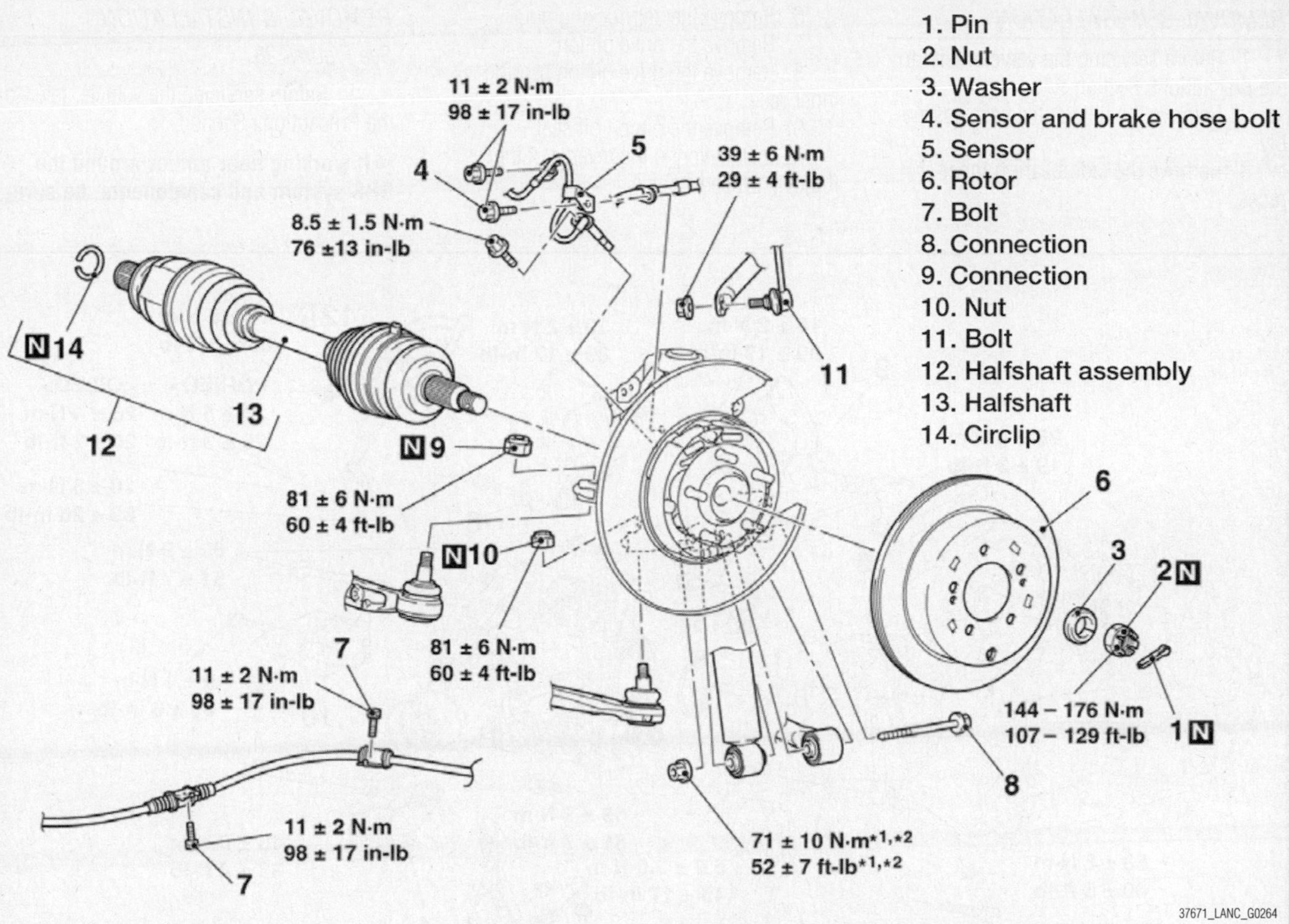

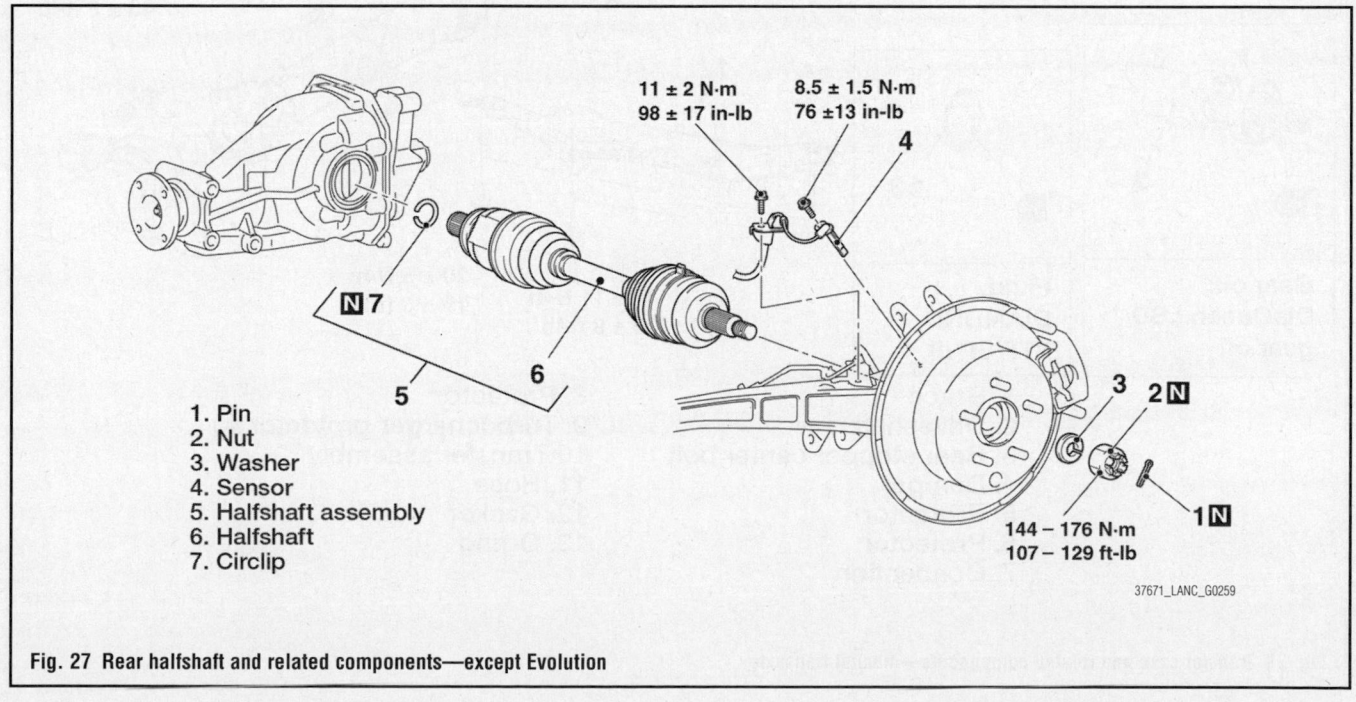

1. Pin
2. Nut
3. Washer
4. Sensor and brake hose bolt
5. Sensor
6. Rotor
7. Bolt
8. Connection
9. Connection
10. Nut
11. Bolt
12. Halfshaft assembly
13. Halfshaft
14. Circlip

11 ± 2 N·m
98 ± 17 in-lb

8.5 ± 1.5 N·m
76 ± 13 in-lb

39 ± 6 N·m
29 ± 4 ft-lb

N14

13

12

N9

81 ± 6 N·m
60 ± 4 ft-lb

N10

7

11 ± 2 N·m
98 ± 17 in-lb

81 ± 6 N·m
60 ± 4 ft-lb

7

11 ± 2 N·m
98 ± 17 in-lb

71 ± 10 N·m*1,*2
52 ± 7 ft-lb*1,*2

6

3

2N

144 – 176 N·m
107 – 129 ft-lb

8

1N

37671_LANC_G0264

Fig. 26 Rear differential and related components—except Evolution

11 ± 2 N·m
98 ± 17 in-lb

8.5 ± 1.5 N·m
76 ± 13 in-lb

4

N7

5

6

3

2N

1N

144 – 176 N·m
107 – 129 ft-lb

1. Pin
2. Nut
3. Washer
4. Sensor
5. Halfshaft assembly
6. Halfshaft
7. Circlip

37671_LANC_G0259

Fig. 27 Rear halfshaft and related components—except Evolution

REAR PINION SEAL

REMOVAL & INSTALLATION

1. Before servicing the vehicle, refer to the Precautions Section.
2. Remove the differential case assembly.
3. Remove the side bearing inner races.

4. Remove the drive gear.
5. Remove the lock pin.
6. Remove the self-locking nut.
7. Remove the drive pinion.
8. Remove the drive pinion bearing inner race.
9. Remove the pinion oil seal.
10. Installation is the reverse of the removal procedure.

TRANSFER CASE ASSEMBLY

REMOVAL & INSTALLATION

See Figures 28.

1. Before servicing the vehicle, refer to the Precautions Section.

→If working near and/or around the SRS system and components, be sure

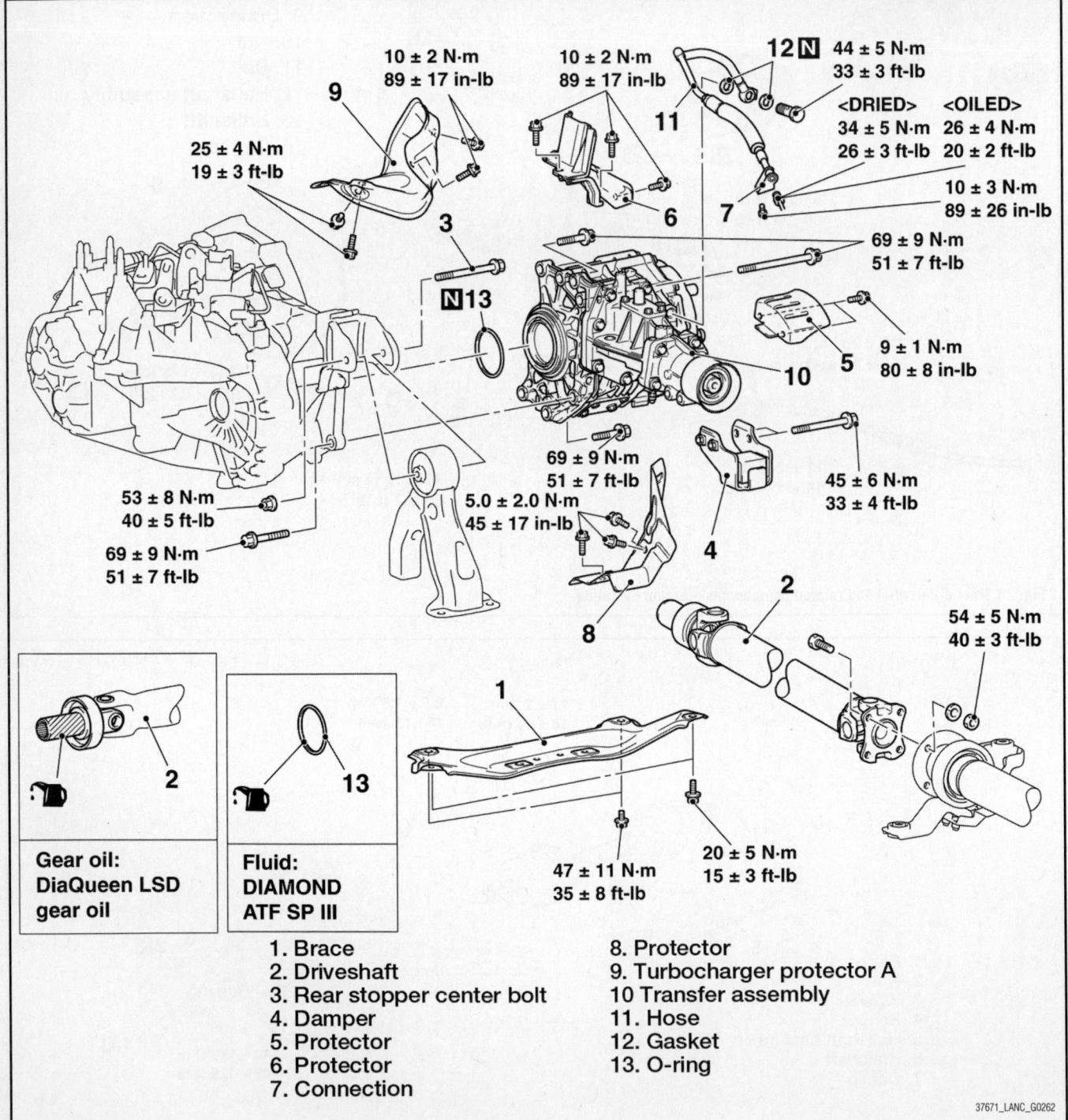

Gear oil:
DiaQueen LSD
gear oil

Fluid:
DIAMOND
ATF SP III

1. Brace
2. Driveshaft
3. Rear stopper center bolt
4. Damper
5. Protector
6. Protector
7. Connection
8. Protector
9. Turbocharger protector A
10 Transfer assembly
11. Hose
12. Gasket
13. O-ring

37671_LANC_G0262

Fig. 28 Transfer case and related components—manual transaxle

to disable the SRS system. Tape the negative battery cable with insulating tape. Always disconnect the negative battery cable first.

> ⁕ **CAUTION**
>
> **Wait for 1 minute after disconnecting the negative battery cable before working inside the vehicle. The air bag system is set to deploy for a short period of time after the battery is disconnected.**

2. Disconnect the negative battery cable.

3. Remove the undercover assembly.

4. Drain the transaxle fluid.
5. Drain the transfer case fluid.
6. Drain the engine coolant.
7. Remove the front axle crossmember assembly.
8. Remove the front exhaust pipe.
9. Remove the battery and the battery tray.
10. Remove the air cleaner and air intake hose assembly.
11. Remove the strut tower bar assembly.
12. Remove the air hose, air by-pass hose, and air by-pass valve.
13. Remove the radiator.
14. Remove the output shaft. Remove the driveshaft.
15. Remove the rear roll stopper connection bolt.

16. Remove the crossmember assembly.
17. Disconnect the pressure hose connection and discard the gasket.
18. Remove the dust seal guard.
19. Remove the transfer case retaining bolts. Remove the transfer case from the vehicle. Discard the O-ring.

To install:

20. Position the transfer case on a suitable holding fixture. Install the transfer case to its mounting in the vehicle.
21. Continue the installation in the reverse order of the removal procedure.
22. Be sure to check and adjust all fluid levels, as necessary.

ENGINE COOLING

ENGINE FAN

REMOVAL & INSTALLATION

See Figures 29 through 31.

At this time the manufacturer does not provide removal and installation procedures for this component. The following procedure is a guideline and may differ from the vehicle you are servicing.

➡ **It may first be necessary to remove the radiator from the vehicle before separation the required components.**

1. Before servicing the vehicle, refer to the Precautions Section.

➡ **If working near and/or around the SRS system and components, be sure to disable the SRS system. Tape the negative battery cable with insulating tape. Always disconnect the negative battery cable first.**

> ⁕ **CAUTION**
>
> **Wait for 1 minute after disconnecting the negative battery cable before working inside the vehicle. The air bag system is set to deploy for a short period of time after the battery is disconnected.**

2. Disconnect the negative battery cable.
3. Drain and recycle the engine coolant.

> ⁕ **CAUTION**
>
> **Never open, service or drain the radiator or cooling system when hot; serious burns can occur from the steam and hot coolant. Also,**

when draining engine coolant, keep in mind that cats and dogs are attracted to ethylene glycol antifreeze and could drink any that is left in an uncovered container or in puddles on the ground. This will prove fatal in sufficient quantities. Always drain coolant into a sealable container.

4. Remove the overflow hose and disconnect the upper radiator hose.

➡ **It is recommended that each clamp be matchmarked to the hose. Observe the marks and reinstall the clamps exactly when reinstalling the hoses.**

5. Unplug the electrical connector(s) from the coolant fan motor(s).
6. Remove the mounting bolts, then remove the fan and shroud assembly from the vehicle.

To install:

➡ **Be sure to use new fasteners, as required.**

7. Install the motor to the shroud and secure with the mounting bolts.
8. Install the remaining components in the reverse order of removal.
9. Fill the system with coolant.
10. Connect the negative battery cable, run the vehicle until the thermostat opens, fill the radiator completely and check the automatic transaxle fluid level, if equipped.
11. Once the vehicle has cooled, recheck the coolant level.
12. Check the engine fan for proper operation.

RADIATOR

REMOVAL & INSTALLATION

2.0L Non Turbocharged Engine and 2.4L Engine

See Figure 29.

1. Before servicing the vehicle, refer to the Precautions Section.

➡ **If working near and/or around the SRS system and components, be sure to disable the SRS system. Tape the negative battery cable with insulating tape. Always disconnect the negative battery cable first.**

> ⁕ **CAUTION**
>
> **Wait for 1 minute after disconnecting the negative battery cable before working inside the vehicle. The air bag system is set to deploy for a short period of time after the battery is disconnected.**

> ⁕ **CAUTION**
>
> **Never open, service or drain the radiator or cooling system when hot; serious burns can occur from the steam and hot coolant. Also, when draining engine coolant, keep in mind that cats and dogs are attracted to ethylene glycol antifreeze and could drink any that is left in an uncovered container or in puddles on the ground. This will prove fatal in sufficient quantities. Always drain coolant into a sealable container.**

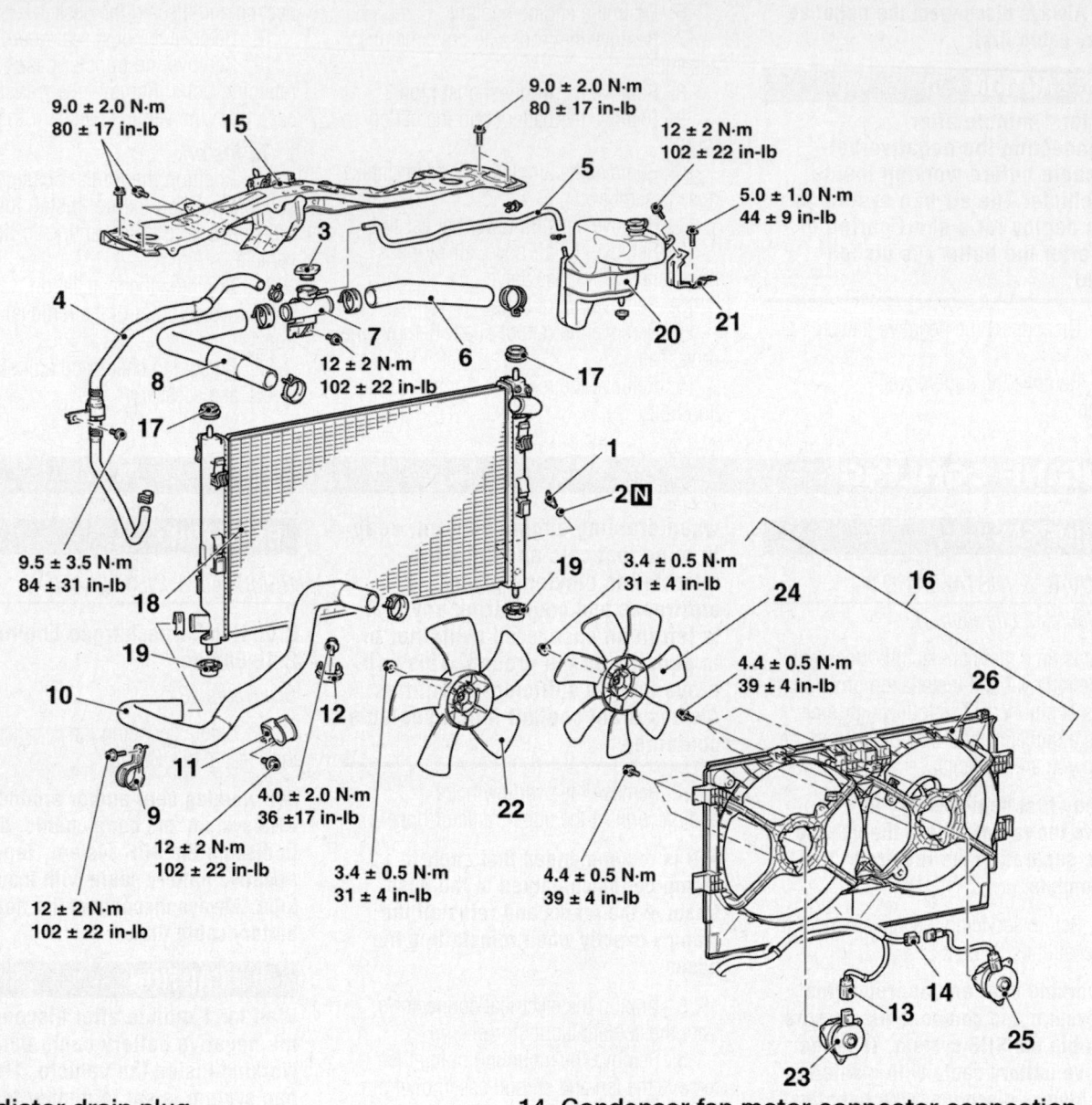

1. Radiator drain plug
2. O-ring
3. Radiator cap
4. CVT fluid cooler water hose assembly (CVT)
5. Radiator condenser tank hose
6. Radiator upper hose
7. Radiator cap assembly
8. Radiator upper hose
9. Radiator hose clamp
10. Radiator lower hose
11. Radiator hose bracket (M/T)
12. Radiator hose support (CVT)
13. Radiator fan motor connector connection
14. Condenser fan motor connector connection
15. Front end upper bar assembly
16. Fan, fan motor, and cooling fan shroud assembly
17. Support upper insulator
18. Radiator assembly
19. Support lower insulator
20. Radiator condenser tank
21. Radiator condenser tank bracket
22. Radiator fan
23. Radiator fan motor
24. Condenser fan
25. Condenser fan motor
26. Cooling fan shroud

22140_MITS_G0076

Fig. 29 Cooling fan, radiator and related components—2.0L non turbocharged engine and 2.4L engine

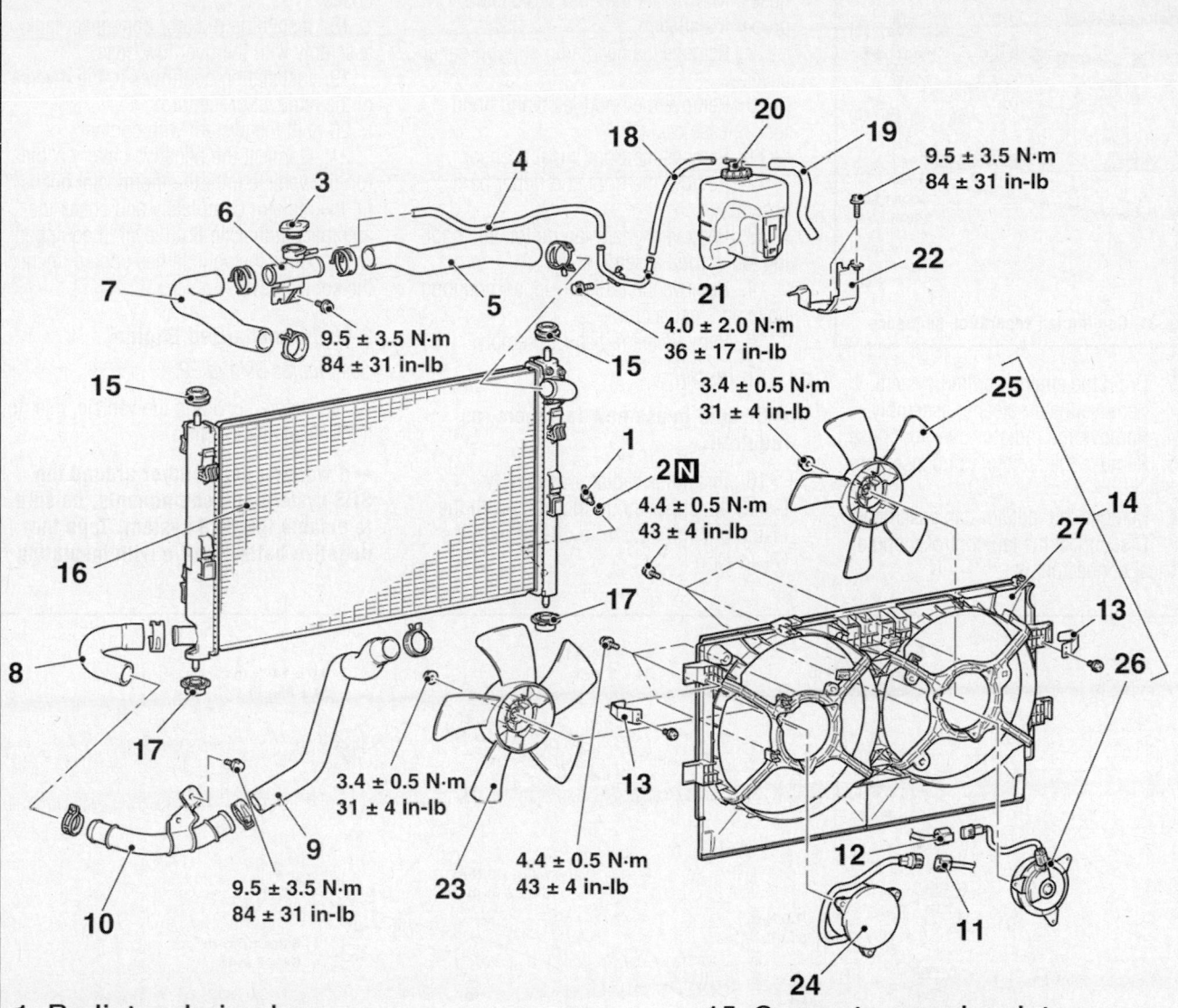

1. Radiator drain plug
2. O-ring
3. Radiator cap
4. Radiator condenser tank hose
5. Radiator upper hose
6. Radiator cap assembly
7. Radiator upper hose
8. Radiator lower hose
9. Radiator lower hose
10. Radiator lower pipe assembly
11. Radiator fan motor connector connection
12. Condenser fan motor connector connection
13. Bracket
14. Fan, fan motor, and fan shroud assembly
15. Support upper insulator
16. Radiator assembly
17. Support lower insulator
18. Radiator condenser tank hose
19. Radiator condenser tank hose
20. Radiator condenser tank
21. Radiator condenser tank pipe
22. Radiator condenser tank bracket
23. Radiator fan
24. Radiator fan motor
25. Condenser fan
26. Condenser fan motor
27. Fan shroud

22140_MITS_G0080

Fig. 30 Cooling fan, radiator and related components—2.0L turbocharged engine

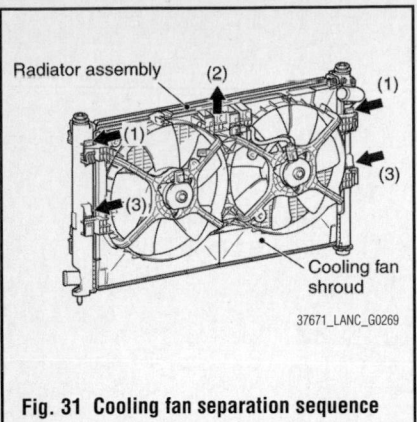

Fig. 31 Cooling fan separation sequence

2. Drain the engine cooling system.

3. Remove the air cleaner assembly.

4. Remove the radiator overflow hose.

5. Remove the radiator condenser tank assembly.

6. Remove the radiator cap assembly.

7. Disconnect the fan control module electrical connector.

8. Remove the upper and lower radiator hoses. Matchmark the hoses to ensure proper installation.

9. Remove the headlamp support panel cover.

10. Remove the hood latch and hood lock release cable.

11. Remove the front impact sensor.

12. Remove the front end upper bar assembly.

13. Remove the fan, fan motor, and cooling fan shroud assembly.

14. Remove the radiator upper mounting bolts and insulators.

15. Remove the radiator assembly.

To install:

➡Be sure to use new fasteners, as required.

16. Install the radiator assembly. Install the upper mounting bolts with the insulators and tighten to 63–97 inch lbs. (7–11 Nm).

17. Install the upper and lower radiator hoses.

18. Install the radiator condenser tank assembly with the overflow hose.

19. Installation continues in the reverse of the removal procedure.

20. Fill the system with coolant.

21. Connect the negative battery cable, run the vehicle until the thermostat opens, fill the radiator completely and check the automatic transaxle fluid level, if equipped.

22. Once the vehicle has cooled, recheck the coolant level.

2.0L Turbocharged Engine

See Figures 30 and 32.

1. Before servicing the vehicle, refer to the Precautions Section.

➡If working near and/or around the SRS system and components, be sure to disable the SRS system. Tape the negative battery cable with insulating

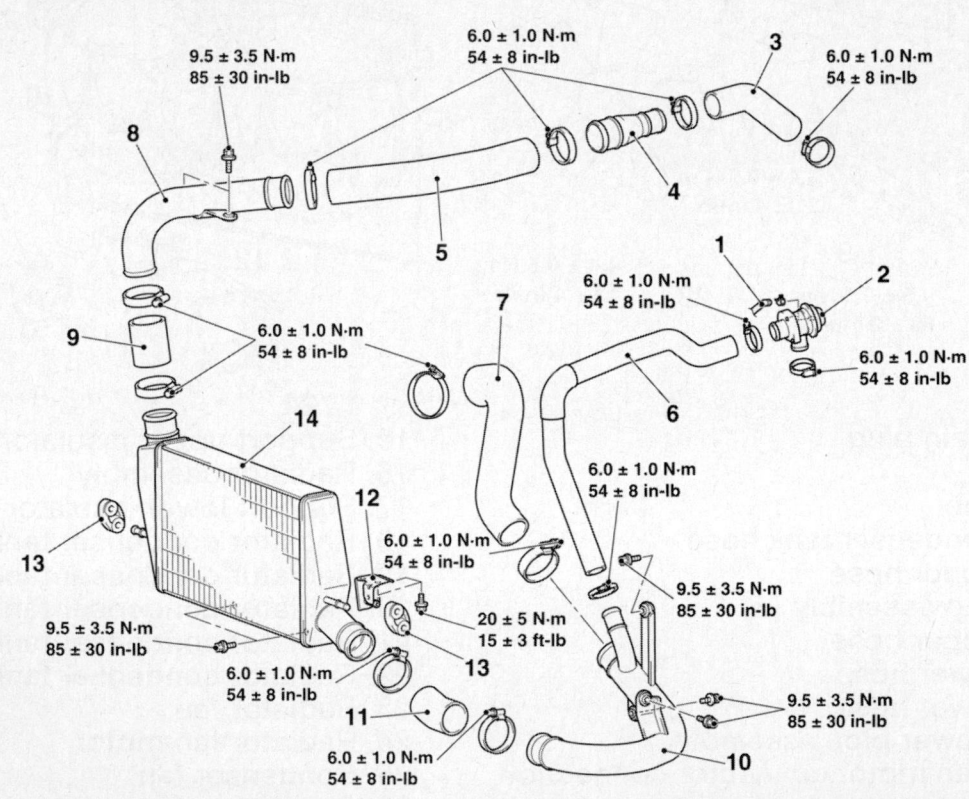

1. Turbocharger by-pass valve vacuum hose connection
2. Turbocharger by-pass valve
3. Charge air cooler intake hose A
4. Charge air cooler intake pipe A
5. Charge air cooler intake hose B
6. Turbocharger by-pass valve hose
7. Charge air cooler outlet hose E
8. Charge air cooler intake pipe B
9. Charge air cooler intake hose D
10. Charge air cooler outlet pipe C
11. Charge air cooler outlet hose C
12. Charge air cooler hanger bracket (LH)
13. Charge air cooler hanger
14. Charge air cooler assembly

Fig. 32 Air charge cooler and related components—2.0L turbocharged engine

tape. Always disconnect the negative battery cable first.

> ※※ **CAUTION**
>
> **Wait for 1 minute after disconnecting the negative battery cable before working inside the vehicle. The air bag system is set to deploy for a short period of time after the battery is disconnected.**

> ※※ **CAUTION**
>
> **Never open, service or drain the radiator or cooling system when hot; serious burns can occur from the steam and hot coolant. Also, when draining engine coolant, keep in mind that cats and dogs are attracted to ethylene glycol antifreeze and could drink any that is left in an uncovered container or in puddles on the ground. This will prove fatal in sufficient quantities. Always drain coolant into a sealable container.**

2. Remove the engine room undercover front A and engine room undercover center.

3. Drain the engine coolant.

4. Remove the charge air cooler intake hose B and charge air cooler intake pipe B.

5. Remove the radiator condenser tank hose.

6. Remove the radiator upper hoses, radiator cap assembly, and radiator lower hoses. Matchmark the hoses to ensure proper installation.

7. Remove the radiator lower pipe assembly.

8. Remove the radiator fan motor connector connection.

9. Remove the condenser fan motor connector connection.

10. Remove the front end upper bar.

11. Remove the fan assembly brackets.

12. Remove the fan, fan motor, and fan shroud assembly.

13. Remove the upper support insulator.

14. Remove the radiator assembly.

To install:

➡**Be sure to use new fasteners, as required.**

15. Install the radiator assembly. Tighten all bolts as illustrated in the exploded view.

16. Install the upper support insulator.

17. Install the fan, fan motor, and fan shroud assembly.

18. Install the fan assembly brackets.

19. Install the front end upper bar.

20. Install the condenser fan motor connector connection.

21. Install the radiator fan motor connector connection.

22. Install the radiator lower pipe assembly.

23. Install the radiator upper hoses, radiator cap assembly, and radiator lower hoses.

24. Install the radiator condenser tank hose.

25. Install the charge air cooler intake hose B and charge air cooler intake pipe B.

26. Fill the system with coolant.

27. Install the engine room undercover front A and engine room undercover center panel.

28. Connect the negative battery cable, run the vehicle until the thermostat opens, fill the radiator completely and check the automatic transaxle fluid level, if equipped.

29. Once the vehicle has cooled, recheck the coolant level.

THERMOSTAT

REMOVAL & INSTALLATION

See Figures 33 and 34.

1. Before servicing the vehicle, refer to the Precautions Section.

➡**If working near and/or around the SRS system and components, be sure to disable the SRS system. Tape the negative battery cable with insulating tape. Always disconnect the negative battery cable first.**

> ※※ **CAUTION**
>
> **Wait for 1 minute after disconnecting the negative battery cable before working inside the vehicle. The air bag system is set to deploy for a short period of time after the battery is disconnected.**

> ※※ **CAUTION**
>
> **Never open, service or drain the radiator or cooling system when hot; serious burns can occur from the steam and hot coolant. Also, when draining engine coolant, keep in mind that cats and dogs are attracted to ethylene glycol antifreeze and could drink any that is left in an uncovered container or in puddles on the ground. This will prove fatal in sufficient quantities. Always drain coolant into a sealable container.**

2. Remove the engine upper cover.

3. Remove the air cleaner assembly.

4. On 2.0L turbocharged engine, remove the battery tray.

5. Drain the engine coolant.

6. Remove the radiator lower hose connection.

7. Remove the control wiring harness clamp connection.

8. Remove the harness bracket.

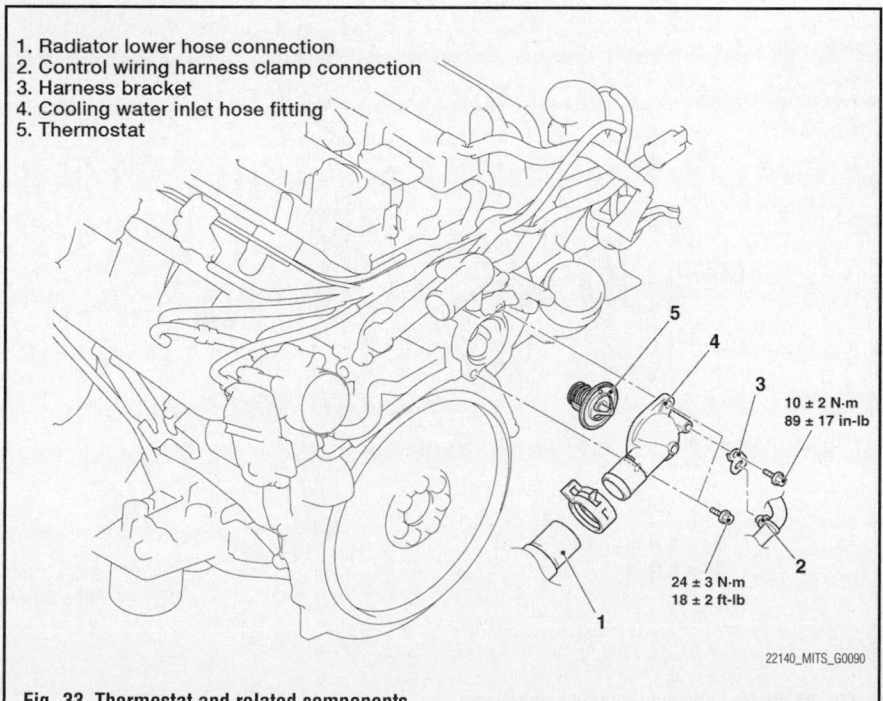

1. Radiator lower hose connection
2. Control wiring harness clamp connection
3. Harness bracket
4. Cooling water inlet hose fitting
5. Thermostat

10 ± 2 N·m
89 ± 17 in-lb

24 ± 3 N·m
18 ± 2 ft-lb

22140_MITS_G0090

Fig. 33 Thermostat and related components

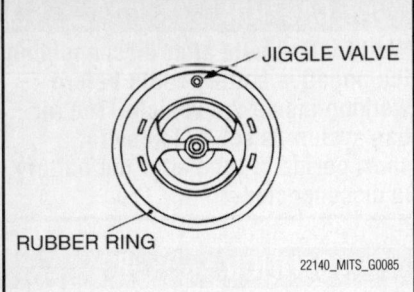

JIGGLE VALVE

RUBBER RING

22140_MITS_G0085

Fig. 34 Install the thermostat so that the jiggle valve is facing straight up

9. Remove the cooling water inlet hose fitting.

10. Remove the thermostat.

To install:

➡ Be sure to use new fasteners, as required.

11. Install the thermostat so that the jiggle valve is facing straight up. Be careful not to fold or scratch the rubber ring.

12. Install the cooling water inlet hose fitting.

13. Install the harness bracket.

14. Install the control wiring harness clamp connection.

15. Install the radiator lower hose connection.

16. Install the air cleaner intake hose.

17. Install the engine upper cover.

18. Fill the system with coolant.

19. Install the engine undercover.

20. Connect the negative battery cable, run the vehicle until the thermostat opens and fill the radiator completely.

21. Once the vehicle has cooled, recheck the coolant level.

WATER PUMP

REMOVAL & INSTALLATION

See Figure 35.

1. Before servicing the vehicle, refer to the Precautions Section.

➡ If working near and/or around the SRS system and components, be sure to disable the SRS system. Tape the negative battery cable with insulating tape. Always disconnect the negative battery cable first.

✳✳ CAUTION

Wait for 1 minute after disconnecting the negative battery cable before working inside the vehicle. The air bag system is set to deploy for a short period of time after the battery is disconnected.

2. Disconnect the negative battery cable.

3. Drain the engine coolant.

4. Remove the accessory drive belt.

5. On 2.0L turbocharged engine, remove the strut tower bar and engine upper cover.

6. On 2.0L turbocharged engine, except Evolution, remove the exhaust manifold cover.

7. Remove the water pump pulley.

8. Remove the water pump inlet pipe mounting nuts.

9. Remove the water pump.

To install:

➡ Be sure to use new fasteners, as required.

10. Completely remove the water pump gasket material from the mating surfaces.

11. Install new gaskets and the water pump into position. Tighten the bolts to specification.

12. Install the water pump inlet pipe mounting nuts.

13. Install the water pump pulley.

14. Install the accessory drive belt.

15. Refill the cooling system and connect the negative battery cable.

---

24 ± 3 N·m
18 ± 2 ft-lb

2

N4

24 ± 3 N·m
18 ± 2 ft-lb

24 ± 3 N·m
18 ± 2 ft-lb

3

5N

1

9.0 ± 1.0 N·m
80 ± 9 in-lb

1. Water pump pulley
2. Water pump inlet pipe mounting nuts
3. Water pump
4. Cooling water line gasket
5. Water pump gasket

22140_MITS_G0082

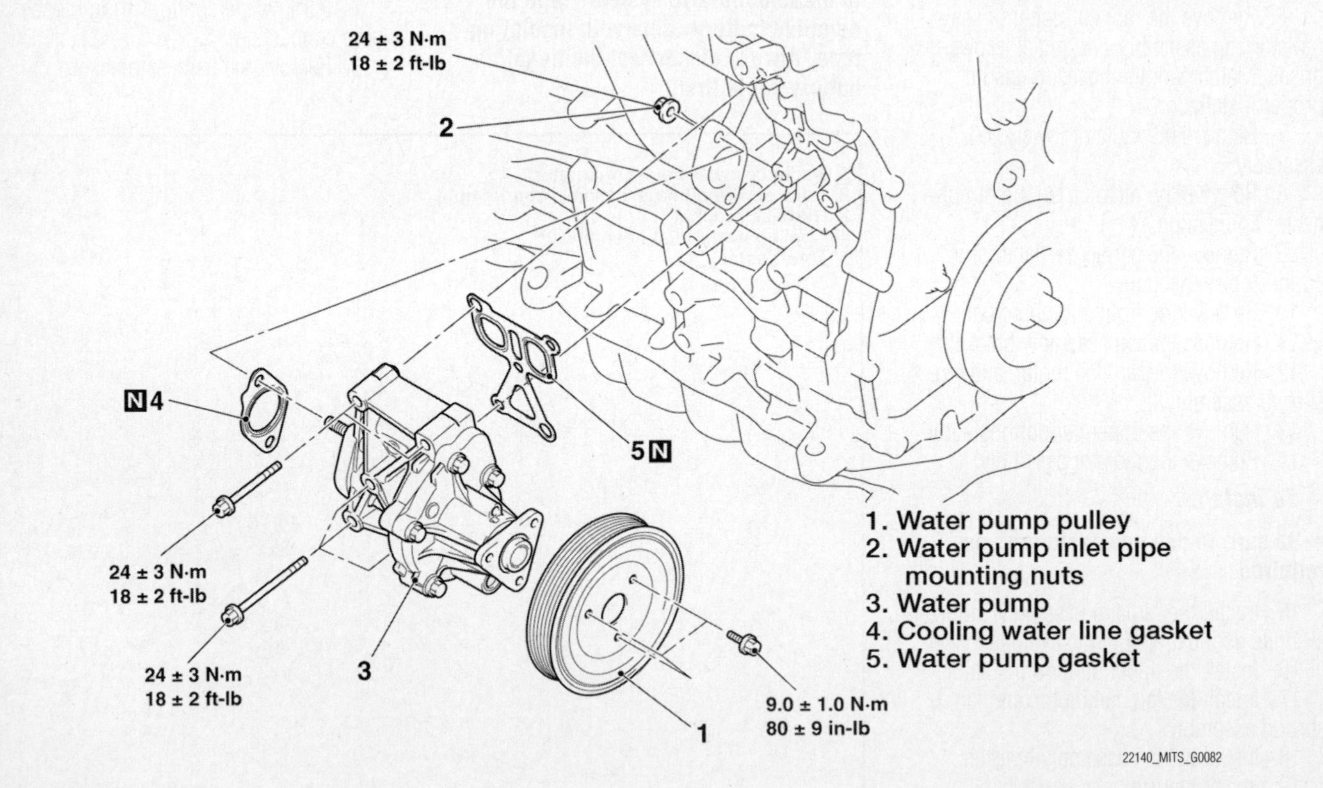

Fig. 35 Water pump and related components

ENGINE ELECTRICAL   **CHARGING SYSTEM**

ALTERNATOR

REMOVAL & INSTALLATION

See Figures 36 through 38.

1. Before servicing the vehicle, refer to the Precautions Section.

➡**If working near and/or around the SRS system and components, be sure to disable the SRS system. Tape the negative battery cable with insulating tape. Always disconnect the negative battery cable first.**

✳✳ CAUTION

Wait for 1 minute after disconnecting the negative battery cable before working inside the vehicle. The air bag system is set to deploy for a short period of time after the battery is disconnected.

➡**It may be necessary to remove and reposition the air condition compressor assembly. Do not allow the compressor to hang by the refrigerant lines. Do not discharge the system.**

2. Remove the drive belt.
3. On 2.0L turbocharged engines,

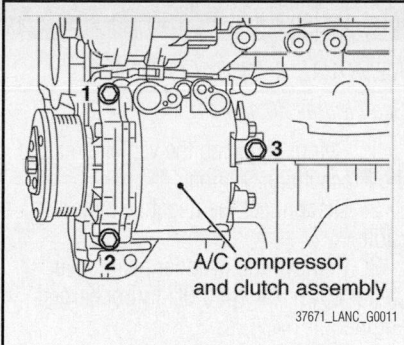

37671_LANC_G0011

Fig. 37 Air conditioning compressor mounting bolt tightening sequence

remove the fan, fan motor and cooling fan shroud, as required.

4. Remove the idler pulley.
5. Disconnect the electrical connections.
6. Remove the retaining bolts.
7. Remove the component from its mounting.

To install:

➡**Be sure to use new fasteners, as required.**

8. Installation is the reverse of the removal procedure.
9. If removed, be sure to tighten the air

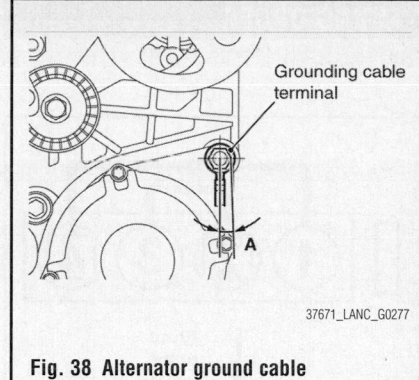

Grounding cable terminal

37671_LANC_G0277

Fig. 38 Alternator ground cable installation

conditioning compressor in the proper sequence. Tighten bolts to 13–21 ft. lbs.

10. Tighten the alternator grounding bolt as shown in the illustration. Tightening specification is 26–40 ft. lbs.

VOLTAGE REGULATOR

REMOVAL & INSTALLATION

The voltage regulator is an internal component of the alternator. In order to replace the voltage regulator, the entire alternator assembly must be replaced.

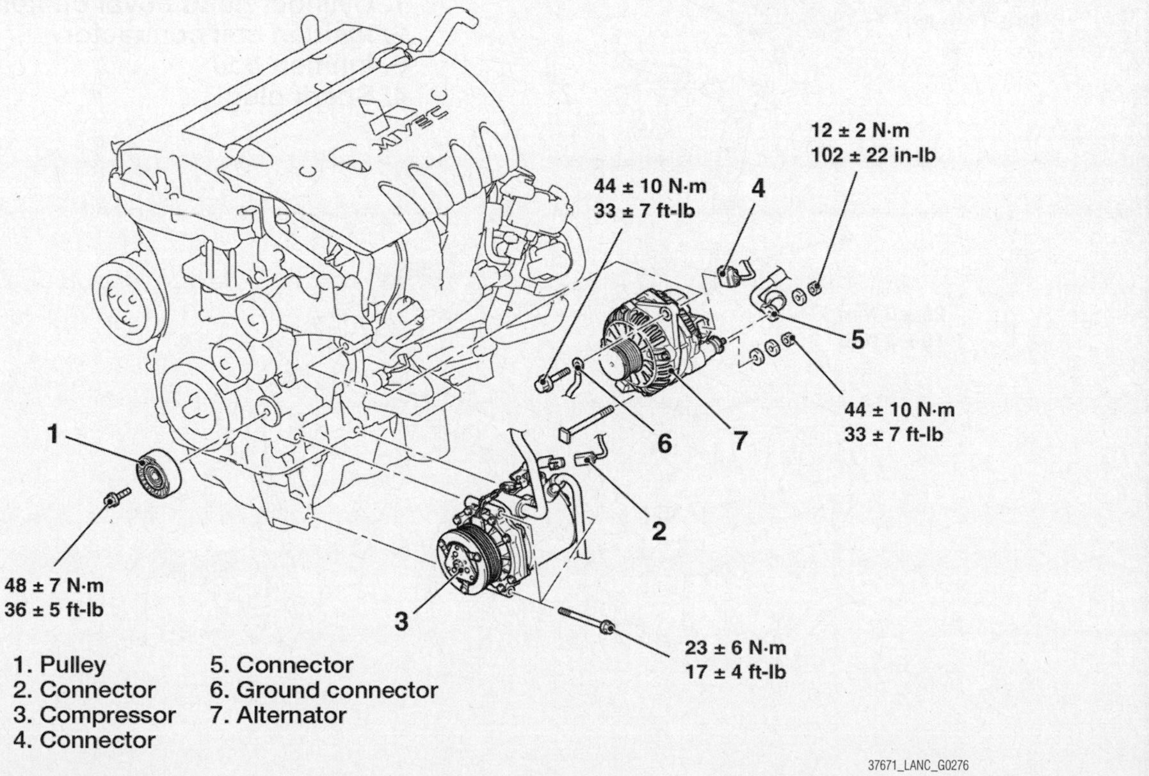

12 ± 2 N·m
102 ± 22 in-lb

44 ± 10 N·m
33 ± 7 ft-lb

44 ± 10 N·m
33 ± 7 ft-lb

48 ± 7 N·m
36 ± 5 ft-lb

23 ± 6 N·m
17 ± 4 ft-lb

1. Pulley
2. Connector
3. Compressor
4. Connector
5. Connector
6. Ground connector
7. Alternator

37671_LANC_G0276

Fig. 36 Alternator and related components

ENGINE ELECTRICAL

IGNITION SYSTEM

FIRING ORDER

See Figure 39.

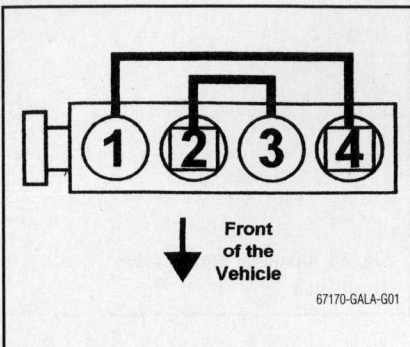

Front
of the
Vehicle

67170-GALA-G01

**Fig. 39 2.0L Engine and 2.4L Engine
Firing order: 1–3–4–2
Distributorless ignition
(coil-on-plug) system**

IGNITION COIL

REMOVAL & INSTALLATION

See Figures 40 and 41.

1. Before servicing the vehicle, refer to the Precautions Section.
2. Disconnect the negative battery cable.
3. Remove the cylinder head cover center cover, except 2.0L Turbocharged engine.
4. Remove the engine upper cover, 2.0L Turbocharged engine.
5. Remove the ignition coil connector.
6. Remove the ignition coil retaining bolt.
7. Remove the ignition coil.

To install:

8. Installation is the reverse of the removal procedure.

9. Tighten the retaining bolts to specification.

IGNITION TIMING

INSPECTION

➡**Mitsubishi's Multi-Use Tester-III (MUT III) scan tool or an equivalent OBD-II scan tool must be used for this procedure.**

1. Set the transaxle in Park for automatic transaxles or Neutral for manual transaxles.
2. Connect an OBD-II compliant scan tool to the Data Link Connector, located under the instrument panel on the driver's side.
3. Set the timing light to the power supply line of ignition coil No. 1.
4. Start the engine and allow it to run at idle.

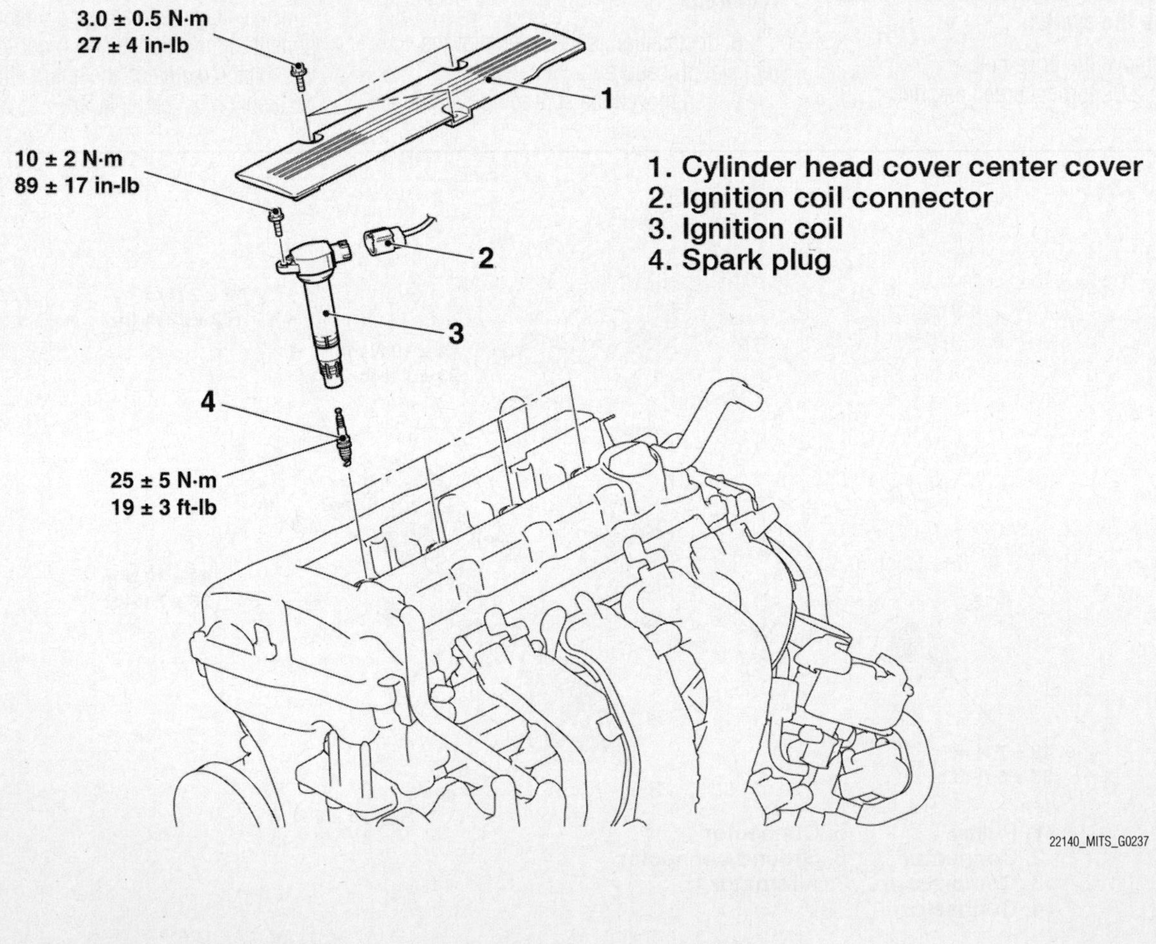

3.0 ± 0.5 N·m
27 ± 4 in-lb

10 ± 2 N·m
89 ± 17 in-lb

25 ± 5 N·m
19 ± 3 ft-lb

1

2

3

4

1. Cylinder head cover center cover
2. Ignition coil connector
3. Ignition coil
4. Spark plug

22140_MITS_G0237

Fig. 40 Ignition coil and related components—2.0L non turbocharged engine and 2.4L engine

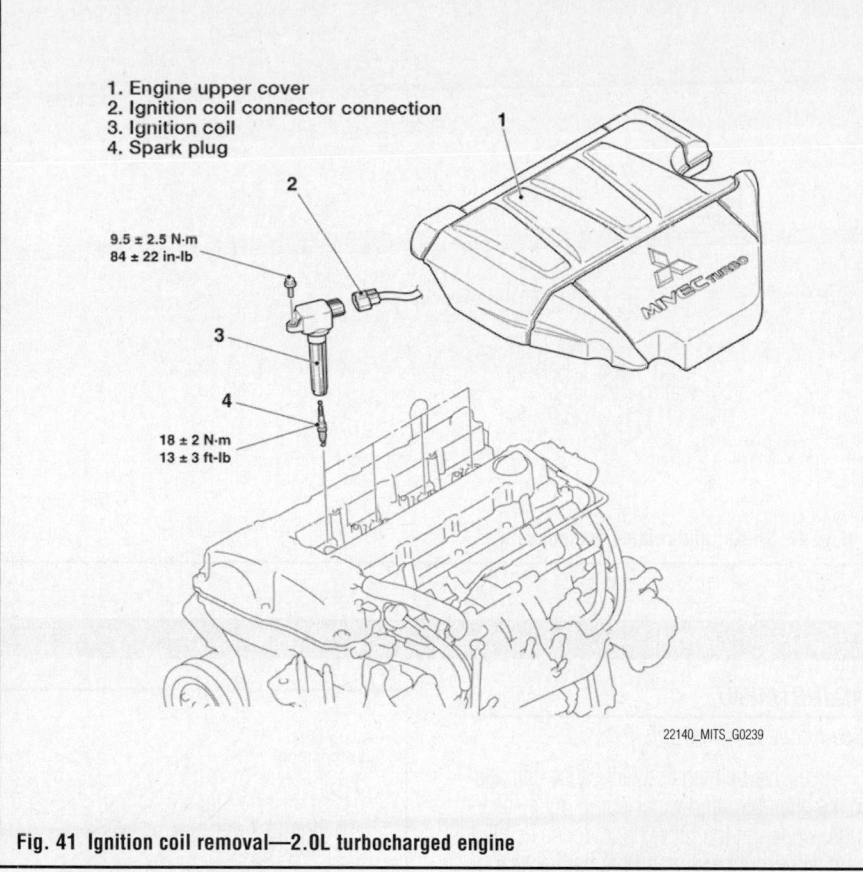

1. Engine upper cover
2. Ignition coil connector connection
3. Ignition coil
4. Spark plug

9.5 ± 2.5 N·m
84 ± 22 in-lb

18 ± 2 N·m
13 ± 3 ft-lb

22140_MITS_G0239

Fig. 41 Ignition coil removal—2.0L turbocharged engine

5. Using the scan tool (Item No. 17 of the actuator test), check that the ignition timing is within specification.

ADJUSTMENT

The ignition timing is controlled by the Powertrain Control Module (Or the Electronic Control Module (ECM). No adjustment is necessary or possible.

ENGINE ELECTRICAL

STARTER

REMOVAL & INSTALLATION
See Figure 42.

1. Before servicing the vehicle, refer to the Precautions Section.

➡If working near and/or around the SRS system and components, be sure to disable the SRS system. Tape the negative battery cable with insulating tape. Always disconnect the negative battery cable first.

✳✳ CAUTION

Wait for 1 minute after disconnecting the negative battery cable before

SPARK PLUGS

REMOVAL & INSTALLATION

1. Before servicing the vehicle, refer to the Precautions Section.

➡If working near and/or around the SRS system and components, be sure to disable the SRS system. Tape the

working inside the vehicle. The air bag system is set to deploy for a short period of time after the battery is disconnected.

2. Remove the air cleaner assembly, as required.
3. Remove the battery and battery tray.
4. Raise and safely support the vehicle.
5. Remove the engine undercover.
6. On 2.0L non turbocharged engine and 2.4L engines, remove the brake booster vacuum hose connection. Remove the emission vacuum hose connection. Remove the throttle body stay.
7. On 2.0L turbocharged engines remove the air cooler outlet hose.

negative battery cable with insulating tape. Always disconnect the negative battery cable first.

✳✳ CAUTION

Wait for 1 minute after disconnecting the negative battery cable before working inside the vehicle. The air bag system is set to deploy for a short period of time after the battery is disconnected.

2. Disconnect the negative battery cable.
3. Remove the engine cover (as necessary). Remove the cylinder head cover (as necessary).
4. Disconnect the ignition coil connector.
5. Remove the ignition coil.
6. Use a spark plug socket and wrench to remove the spark plugs.

✳✳ WARNING

Be careful that no contaminates enter through the spark plug holes.

➡Check the electrode gap on the spark plugs before installation.

To install:

➡Be sure to use new fasteners, as required.

7. Installation is the reverse of the removal procedure.
8. Tighten the spark plugs to 15–22 ft. lbs. (20–30 Nm).

STARTING SYSTEM

8. Detach the starter motor electrical connections.
9. Remove the starter motor mounting bolts and remove the starter.

To install:
10. Installation is the reverse of removal. Tighten starter motor bolts to specification.

SOLENOID OR RELAY REPLACEMENT

1. Before servicing the vehicle, refer to the Precautions Section.

➡If working near and/or around the SRS system and components, be sure to disable the SRS system. Tape the negative battery cable with insulating tape. Always disconnect the negative battery cable first.

Wait for 1 minute after disconnecting the negative battery cable before working inside the vehicle. The air bag system is set to deploy for a short period of time after the battery is disconnected.

2. Disconnect the negative battery cable.
3. Remove the starter.
4. Remove the solenoid attaching screws.
5. Remove the solenoid.

To install:

➡ Be sure to use new fasteners, as required.

6. Install the solenoid.
7. Install the solenoid attaching screws.
8. Install the starter.
9. Connect the negative battery cable.

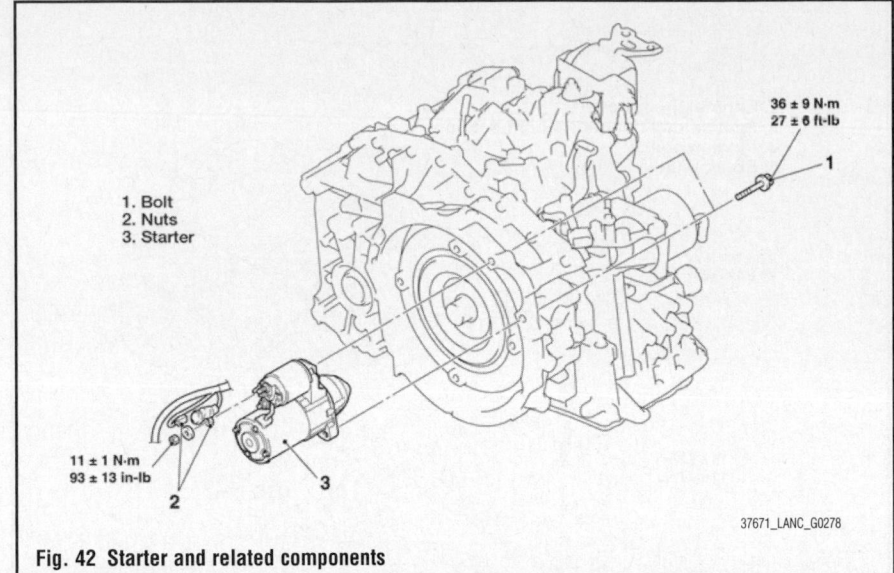

1. Bolt
2. Nuts
3. Starter

36 ± 9 N·m
27 ± 6 ft-lb

11 ± 1 N·m
93 ± 13 in-lb

37671_LANC_G0278

Fig. 42 Starter and related components

ENGINE MECHANICAL

ACCESSORY DRIVE BELTS

ACCESSORY BELT ROUTING

See Figure 43.

Refer to the accompanying illustration for belt routing.

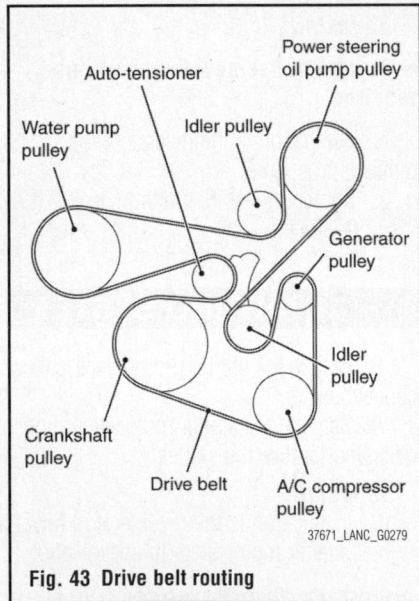

Power steering oil pump pulley
Auto-tensioner
Idler pulley
Water pump pulley
Generator pulley
Idler pulley
Crankshaft pulley
Drive belt
A/C compressor pulley

37671_LANC_G0279

Fig. 43 Drive belt routing

INSPECTION

Inspect the drive belt for signs of glazing or cracking. A glazed belt will be perfectly smooth from slippage, while a good belt will have a slight texture of fabric visible. Cracks will usually start at the inner edge of the belt and run outward. All worn or damaged drive belts should be replaced immediately.

ADJUSTMENT

See Figures 44 through 46.

Excessive belt tension will cause damage to the alternator and water pump pulley bearings, while, on the other hand, loose belt tension will produce slip and premature wear on

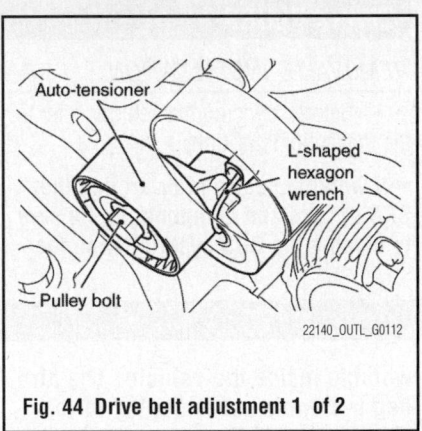

Auto-tensioner
L-shaped hexagon wrench
Pulley bolt

22140_OUTL_G0112

Fig. 44 Drive belt adjustment 1 of 2

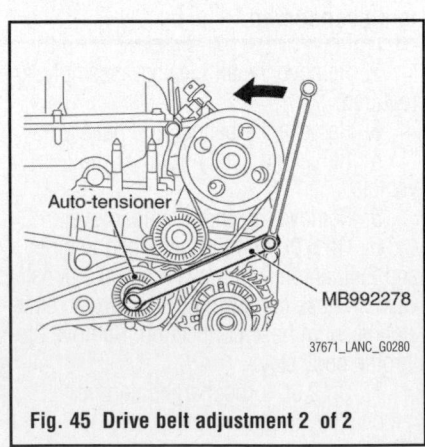

Auto-tensioner

MB992278

37671_LANC_G0280

Fig. 45 Drive belt adjustment 2 of 2

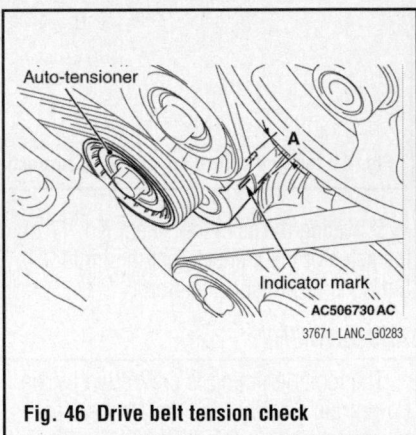

Auto-tensioner
A
Indicator mark

AC506730 AC
37671_LANC_G0283

Fig. 46 Drive belt tension check

the belt. Therefore, be sure to adjust the belt tension to the proper level. If the mark is not within area A (see illustration) replace the belt.

REMOVAL & INSTALLATION

See Figures 43, 47 and 48.

1. Before servicing the vehicle, refer to the Precautions Section.

➡ If working near and/or around the SRS system and components, be sure to disable the SRS system. Tape the negative battery cable with insulating tape. Always disconnect the negative battery cable first.

Wait for 1 minute after disconnecting the negative battery cable before working inside the vehicle. The air bag system is set to deploy for a

short period of time after the battery is disconnected.

2. Remove the engine undercover.
3. Remove the radiator condenser tank assembly mounting bolt, and move the radiator condenser tank assembly to a place where it does not interfere with the drive belt removal and installation.

➡**To reuse the accessory drive belt, use chalk to draw an arrow indicating the rotation direction on the back of the belt in order to install the belt in the same direction.**

4. Rotate the pulley bolt of the auto-tensioner counterclockwise with an offset wrench.
5. Insert an L-shaped hexagon wrench into the aligned holes to fix the auto-tensioner in place.
6. Remove the accessory drive belt from the vehicle.

Water pump pulley
Generator
Crankshaft pulley
37671_LANC_G0282

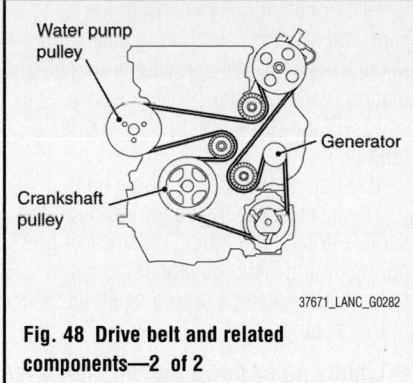

Fig. 48 Drive belt and related components—2 of 2

To install:

➡**Be sure to use new fasteners, as required.**

7. Install the drive belt to each pulley.
8. Set an offset wrench to the pulley bolt of the auto-tensioner. Then, rotate the auto-tensioner counter-clockwise and remove the L-shaped hexagon wrench fixing the auto-tensioner.

➡**Make sure the belt is properly seated in all the pulleys.**

9. Apply tension to the drive belt while slowly turning the auto-tensioner clockwise.

CAMSHAFT AND VALVE LIFTERS

REMOVAL & INSTALLATION

See Figures 49 through 55.

1. Before servicing the vehicle, refer to the Precautions Section.

➡**If working near and/or around the SRS system and components, be sure to disable the SRS system. Tape the negative battery cable with insulating tape. Always disconnect the negative battery cable first.**

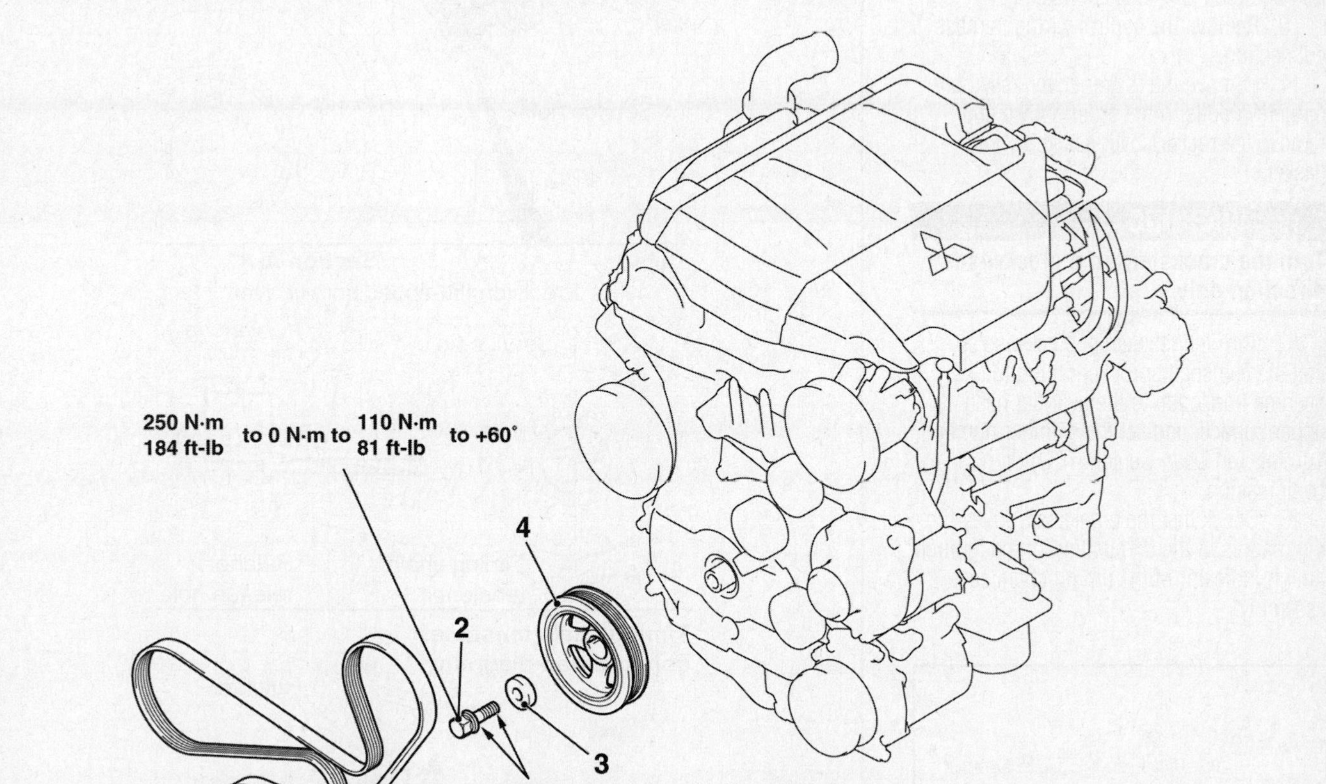

250 N·m
184 ft-lb
to 0 N·m to
110 N·m
81 ft-lb
to +60°
4
2
3
(Engine oil)
1
1. Belt
2. Bolt
3. Washer
4. Pulley
37671_LANC_G0281

Fig. 47 Drive belt and related components—1 of 2

✳✳ CAUTION

Wait for 1 minute after disconnecting the negative battery cable before working inside the vehicle. The air bag system is set to deploy for a short period of time after the battery is disconnected.

2. Disconnect the negative battery cable.

3. On 2.4L engines, remove the strut tower bar.

4. Remove the engine undercover. Remove the side cover (RH).

5. Remove the air cleaner assembly.

6. If equipped with turbocharger, remove the charge air cooler intake hose A, B and charge air cooler intake pipe A. Remove the strut tower bar.

7. Remove the ignition coils.

8. Remove the engine upper cover.

9. Remove the breather hose connection and the PCV hose connection.

10. Remove the control wiring harness connection.

11. Remove the rocker cover assembly mounting bolts in the order shown and remove the rocker cover assembly and gasket.

✳✳ WARNING

Turn the crankshaft in a clockwise direction only.

12. Turn the crankshaft clockwise so that the camshaft sprocket timing marks become horizontal to the cylinder head upper surface, and set the cylinder number 1 to the Top Dead Center (TDC) of compression.

13. Check that the crankshaft pulley timing mark is in the 0° position of the ignition timing indicator of the timing chain case assembly.

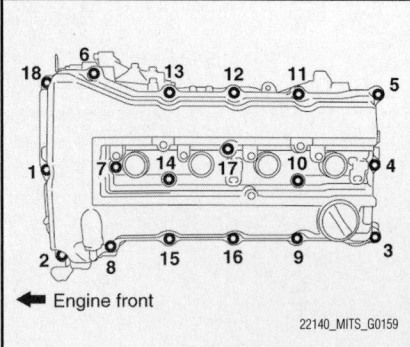

Fig. 49 Remove the rocker cover assembly mounting bolts in the order shown

22140_MITS_G0159

14. Put paint marks on both the camshaft sprocket and timing chain at the position of camshaft sprocket timing chain mating mark (circular hole).

15. Remove the timing chain upper guide.

16. Remove the service hole bolt.

17. Insert a flat-tipped tool through the service hole of the timing chain case, press up the timing chain tensioner ratchet to unlock, and keep the timing chain tensioner in this position.

➡**Lightly press down the tail end of the flat-tipped tool to press up the tip of the flat-tipped tool inserted into the timing chain tensioner in order to unlock it.**

18. With the timing chain tensioner unlocked, insert special tool MB992103 inside the timing chain case assembly along the tension side of the timing chain until the insertion guide line aligns with the upper surface of the timing chain case assembly.

19. With the special tool inserted up to the insertion guide line, press the special tool against the intake side camshaft sprocket and spread and hold the timing chain tension side guide.

20. Remove the flat-tipped precision tool unlocking the timing chain tensioner.

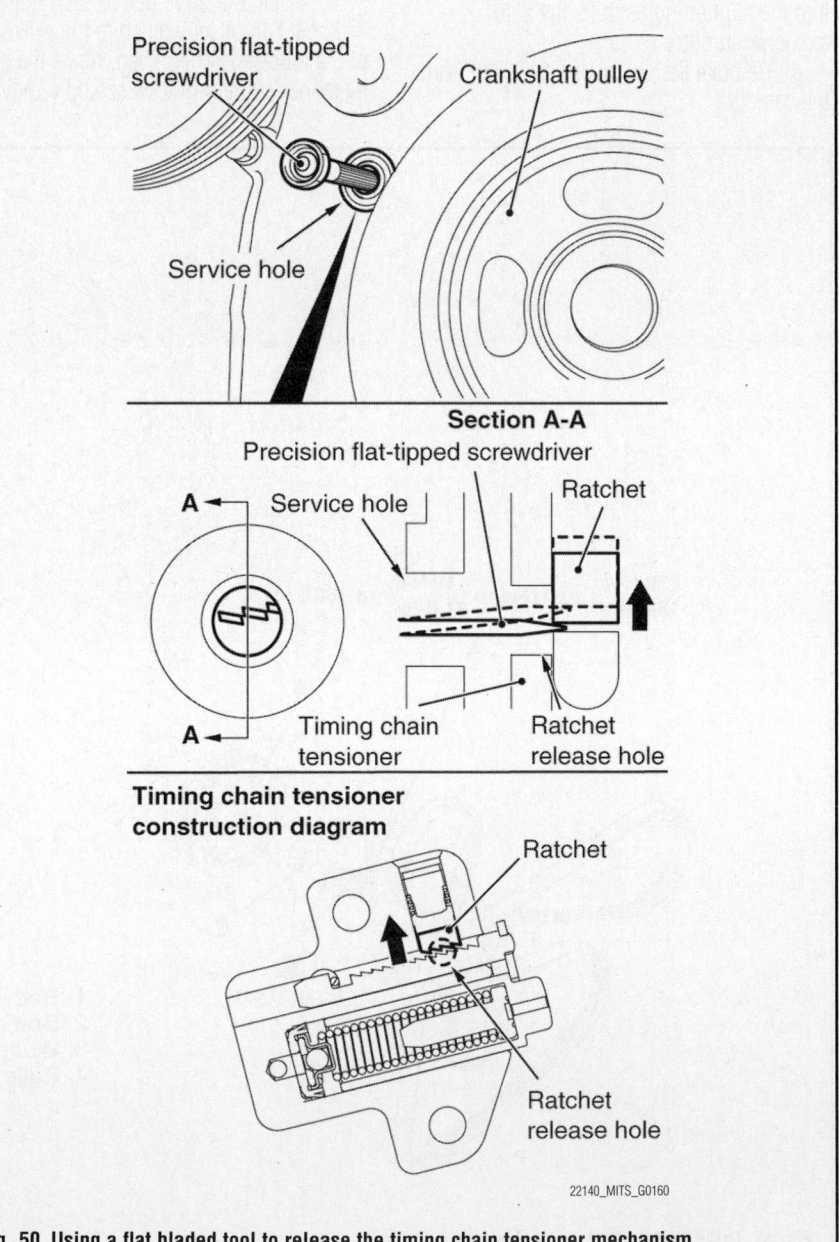

22140_MITS_G0160

Fig. 50 Using a flat bladed tool to release the timing chain tensioner mechanism

The timing chain may snag on other parts. After removing the tension from the timing chain, never rotate the crankshaft.

21. With the timing chain tension side guide spread, hook the special tool over the hexagon part of the camshaft on the exhaust side, and turn the camshaft clockwise to apply slack to the timing chain between the camshaft sprockets.

22. Remove the mounting bolts of front camshaft bearing cap in the order shown in the figure and remove the front camshaft bearing cap assembly.

When the camshaft bearing cap mounting bolts are loosened all at once, the mounting bolts jump out by spring force and the threads are damaged. Always loosen the mounting bolts in 4–5 steps.

23. Loosen the mounting bolts of the camshaft bearing caps in the order shown in the figure in 4–5 steps. Remove the camshaft bearing caps.

24. Slightly raise the transaxle side of the camshaft and camshaft sprocket assembly (exhaust side) by using the slack of the timing chain, and remove from the cam bearing.

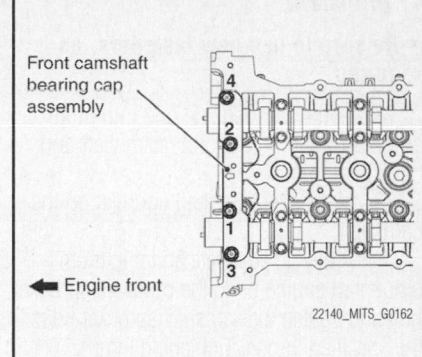

Front camshaft bearing cap assembly

← Engine front

22140_MITS_G0162

Fig. 52 Remove the mounting bolts of front camshaft bearing cap in the order shown

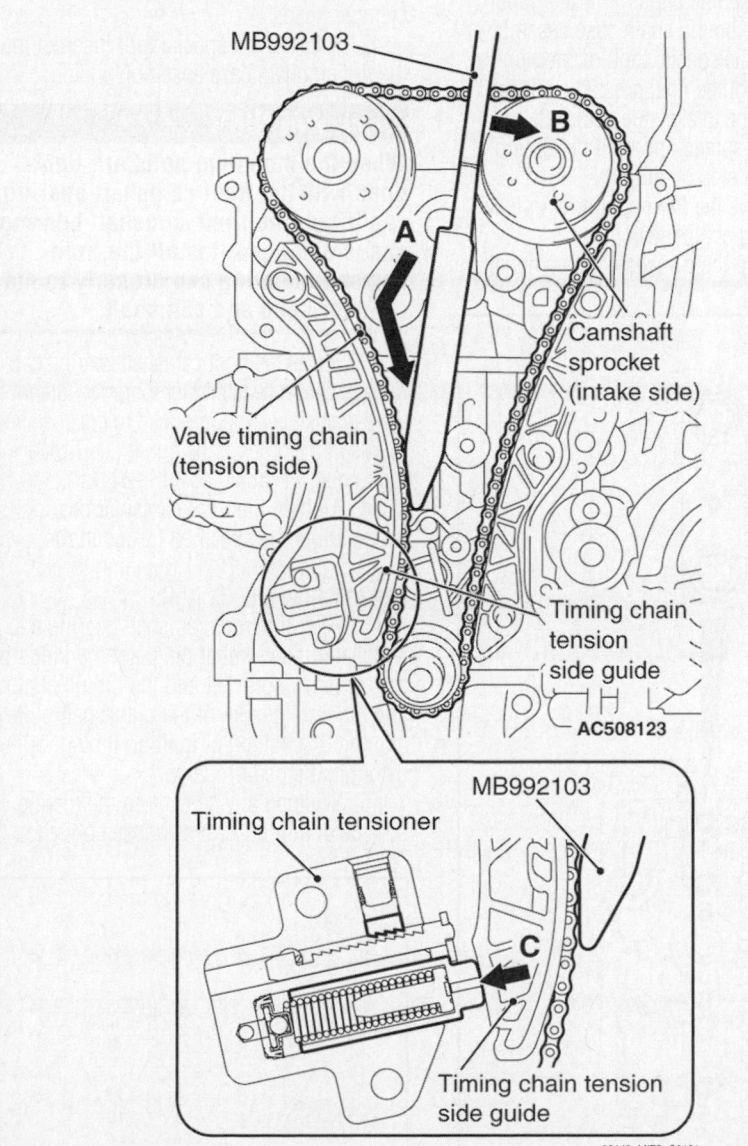

MB992103

A

B

Valve timing chain (tension side)

Camshaft sprocket (intake side)

Timing chain tension side guide

AC508123

MB992103

Timing chain tensioner

C

Timing chain tension side guide

22140_MITS_G0161

Fig. 51 With the timing chain tensioner unlocked, insert special tool MB992103 inside the timing chain case assembly along the tension side of the timing chain

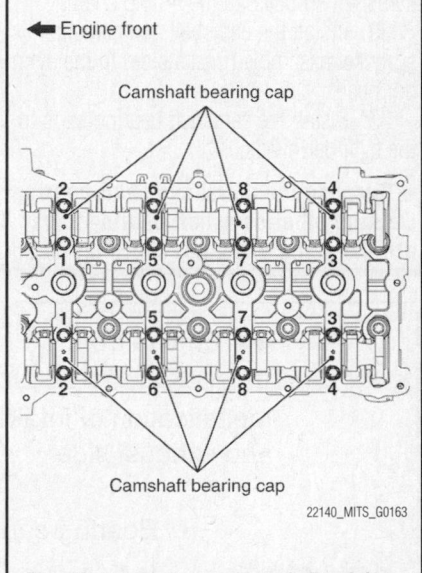

← Engine front

Camshaft bearing cap

Camshaft bearing cap

22140_MITS_G0163

Fig. 53 Loosen the mounting bolts of the camshaft bearing caps in the order shown

25. Remove the timing chain from the camshaft and camshaft sprocket assembly (exhaust side) toward the timing chain case assembly, and remove the camshaft and camshaft sprocket assembly (exhaust side) toward the transaxle.

26. Remove special tool MB992103 inserted into the timing chain case assembly.

The timing chain may snag on other parts. After removing the camshaft and camshaft sprocket assembly, never rotate the crankshaft.

27. After removing the camshaft and camshaft sprocket assembly (exhaust side), hang up the timing chain with a rope to prevent the timing chain from falling into the timing chain case assembly.

To install:

➤Be sure to use new fasteners, as required.

28. Apply an adequate and minimum amount of engine oil to the camshaft and camshaft sprocket.

29. Install the camshaft sprocket to the camshaft.

30. Apply an adequate and minimum amount of engine oil to the camshaft sprocket bolt and tighten the camshaft sprocket bolts to the specified torque. Tightening torque: 41–47 ft. lbs. (54–64 Nm).

31. Align the intake side paint mark of the timing chain, which was made at removal, with the paint mark of the intake side camshaft sprocket and install the camshaft sprocket to the timing chain.

32. Install the camshaft and camshaft sprocket assembly (intake side) to the cylinder head.

33. Install the camshaft bearing caps to the cylinder heads.

➤Because the thrust camshaft bearing cap and other camshaft bearing caps are the same in shape, check the bearing cap number and additionally its symbol to identify the intake and exhaust sides for correct installation.

34. Tighten each camshaft bearing cap mounting bolt to the specified torque in the order shown in the figure in 2–3 steps. Tightening torque: 99–115 inch lbs. (11–13 Nm).

35. In the same manner as removal, insert the flat-tipped tool through the service hole of the timing chain case, press up the ratchet of timing chain tensioner to unlock, and hold the unlocked timing chain tensioner in place.

36. With the timing chain tensioner unlocked, insert special tool MB992103 inside the timing chain case assembly along the tension side of the timing chain until the insertion guide line aligns with the upper surface of the timing chain case assembly.

37. With the special tool inserted up to the insertion guide line, press the special tool against the intake side camshaft sprocket and spread and hold the timing chain tension side guide.

38. Remove the flat-tipped tool unlocking the timing chain tensioner.

39. Pull up the camshaft and camshaft sprocket assembly (exhaust side) mounting area of the timing chain to provide allowance for easy installation of the camshaft and camshaft sprocket assembly (exhaust side) to the timing chain.

➤When installing the camshaft and camshaft sprocket assembly (exhaust side), be careful not to let the camshaft bearing which is installed to the front cam bearing deviate from its position.

40. Align the exhaust side paint mark of the timing chain, which was made during removal, with the paint mark of the exhaust side camshaft sprocket and install the timing chain to the camshaft sprocket.

41. Install the camshaft and camshaft sprocket assembly (exhaust side) to the cylinder head.

42. Remove the special tool inserted into the timing chain case assembly inside.

❊❊ WARNING

When the mounting bolts are tightened with the front camshaft bearing cap tilted, the front camshaft bearing cap is damaged. Install the front camshaft bearing cap properly to the cylinder head and camshaft.

43. Install the front camshaft bearing cap to the cylinder head and temporarily tighten the camshaft bearing front cap to the specified torque in the order of the figure (1). Tightening torque: 11–15 ft. lbs. (14–20 Nm).

44. Tighten the front camshaft bearing cap again to the specified torque in the order of the figure (2). Tightening torque: 22–24 ft. lbs. (28–32 Nm).

45. After the front camshaft bearing cap installation, check that the paint markings of the camshaft sprocket and the timing chain and the timing mark of the crankshaft pulley and the 0° position of ignition timing indicator are aligned properly.

46. Wipe off any sealant on the mating surface of the rocker cover assembly and

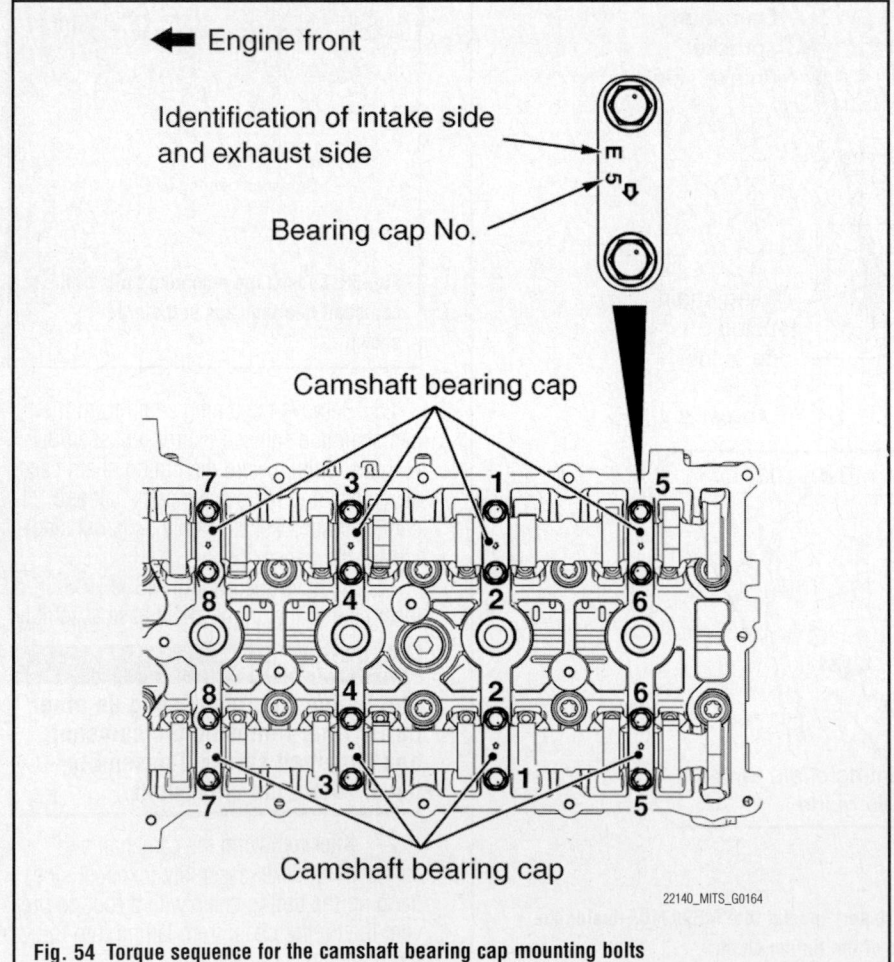

Fig. 54 Torque sequence for the camshaft bearing cap mounting bolts

22140_MITS_G0164

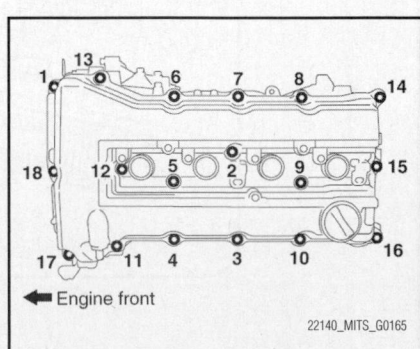

22140_MITS_G0165

Fig. 55 Torque sequence for the rocker cover assembly mounting bolts

the cylinder head and timing chain case assembly, and degrease the surface where the sealant is applied with white gasoline or an equivalent degreaser.

47. Apply sealant to the joint between the cylinder head and timing chain case assembly and install the rocker cover assembly to the cylinder head. Specified sealant: Three Bond® 1217G or equivalent.

➡**Install the rocker cover assembly within 3 minutes after the application of sealant.**

48. Tighten the rocker cover assembly mounting bolts to the specified torque in the order shown. Tightening torque: 19–35 inch lbs. (2–4 Nm).

49. Tighten the rocker cover assembly mounting bolts again to the specified torque in the order shown. Tightening torque: 45–53 inch lbs. (5–6 Nm).

50. Installation continues in the reverse order of the removal procedure.

CATALYTIC CONVERTER

REMOVAL & INSTALLATION

At this time the manufacturer does not provide removal and installation procedures for this component. The following procedure is a guideline and may differ from the vehicle you are servicing.

1. Before servicing the vehicle, refer to the Precautions Section.

➡**If working near and/or around the SRS system and components, be sure to disable the SRS system. Tape the negative battery cable with insulating tape. Always disconnect the negative battery cable first.**

✳✳ CAUTION

Wait for 1 minute after disconnecting the negative battery cable before working inside the vehicle. The air bag system is set to deploy for a short period of time after the battery is disconnected.

➡**If the converter is protected with a heat shield, the shield may be fastened using rivets. These rivets will have to be removed. Be careful when removing these rivets.**

2. Raise and safely support the vehicle.
3. Remove the harness cover.
4. Disconnect the oxygen sensor connector.
5. Remove the sensor.
6. Remove the converter retaining bolts.

7. Remove the converter from its mounting. Discard the gaskets.

To install:

➡**Be sure to use new fasteners, as required.**

8. Installation is the reverse of the removal procedure.
9. Tighten the sensor to 30—36 ft. lbs.
10. If reusing the old sensor be sure to coat the threads using the proper antiseize compound.

CRANKSHAFT DAMPER

REMOVAL & INSTALLATION

See Figures 56 and 57.

1. Before servicing the vehicle, refer to the Precautions Section.

➡**If working near and/or around the SRS system and components, be sure to disable the SRS system. Tape the negative battery cable with insulating tape. Always disconnect the negative battery cable first.**

✳✳ CAUTION

Wait for 1 minute after disconnecting the negative battery cable before working inside the vehicle. The air bag system is set to deploy for a short period of time after the battery is disconnected.

2. Remove the accessory drive belts from around the crankshaft pulley.
3. Remove the engine undercover.
4. Raise and support the vehicle.
5. As necessary, remove the passenger side front wheel. Remove the passenger side inner fender splash shield to gain access to the crankshaft damper.
6. Hold the crankshaft damper sprocket with special tools MB990767 and MD998719.
7. Remove the crankshaft damper center bolt and washer.
8. Remove the crankshaft damper.

To install:

➡**Be sure to use new fasteners, as required.**

9. Wipe off any dirt on the crankshaft and the crankshaft damper. Degrease the parts before assembly.

➡**Degrease the crankshaft damper and crankshaft to prevent a drop in the friction coefficient of the pressed area, which is caused by oil adhesion.**

10. Install the crankshaft pulley.

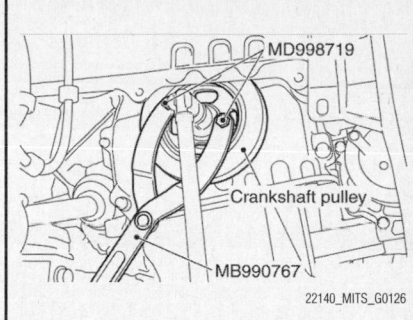

Fig. 56 Hold the crankshaft damper sprocket with special tools MB990767 and MD998719 for removal

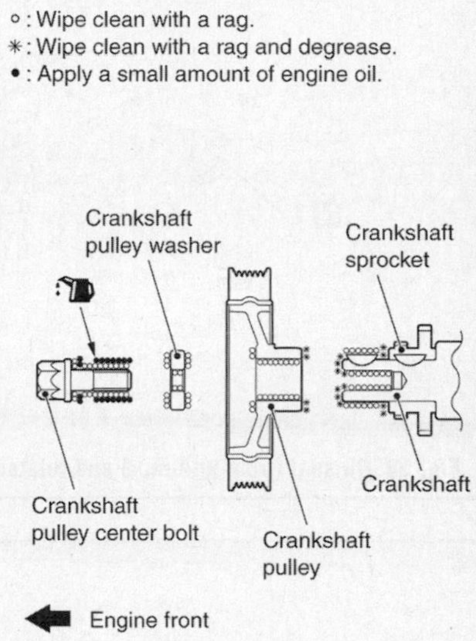

○ : Wipe clean with a rag.
✳ : Wipe clean with a rag and degrease.
● : Apply a small amount of engine oil.

Crankshaft pulley washer
Crankshaft sprocket
Crankshaft pulley center bolt
Crankshaft pulley
Crankshaft

◀ Engine front

22140_MITS_G0127

Fig. 57 Wipe off any dirt on the crankshaft and the crankshaft damper and degrease the parts before assembly

11. Apply an adequate and minimum amount of engine oil to the threads of the crankshaft pulley center bolt and the lower area of the flange.

12. Hold the crankshaft pulley with special tools MB990767 and MD998719 in the same manner as removal.

13. Tighten the crankshaft pulley center bolt to 155 ft. lbs. (210 Nm).

14. Install the splash shield.

15. Install the wheel, then carefully lower the vehicle.

16. Install the accessory drive belts.

17. Connect the negative battery cable.

CRANKSHAFT FRONT SEAL

REMOVAL & INSTALLATION

See Figures 58 and 59.

1. Before servicing the vehicle, refer to the Precautions Section.

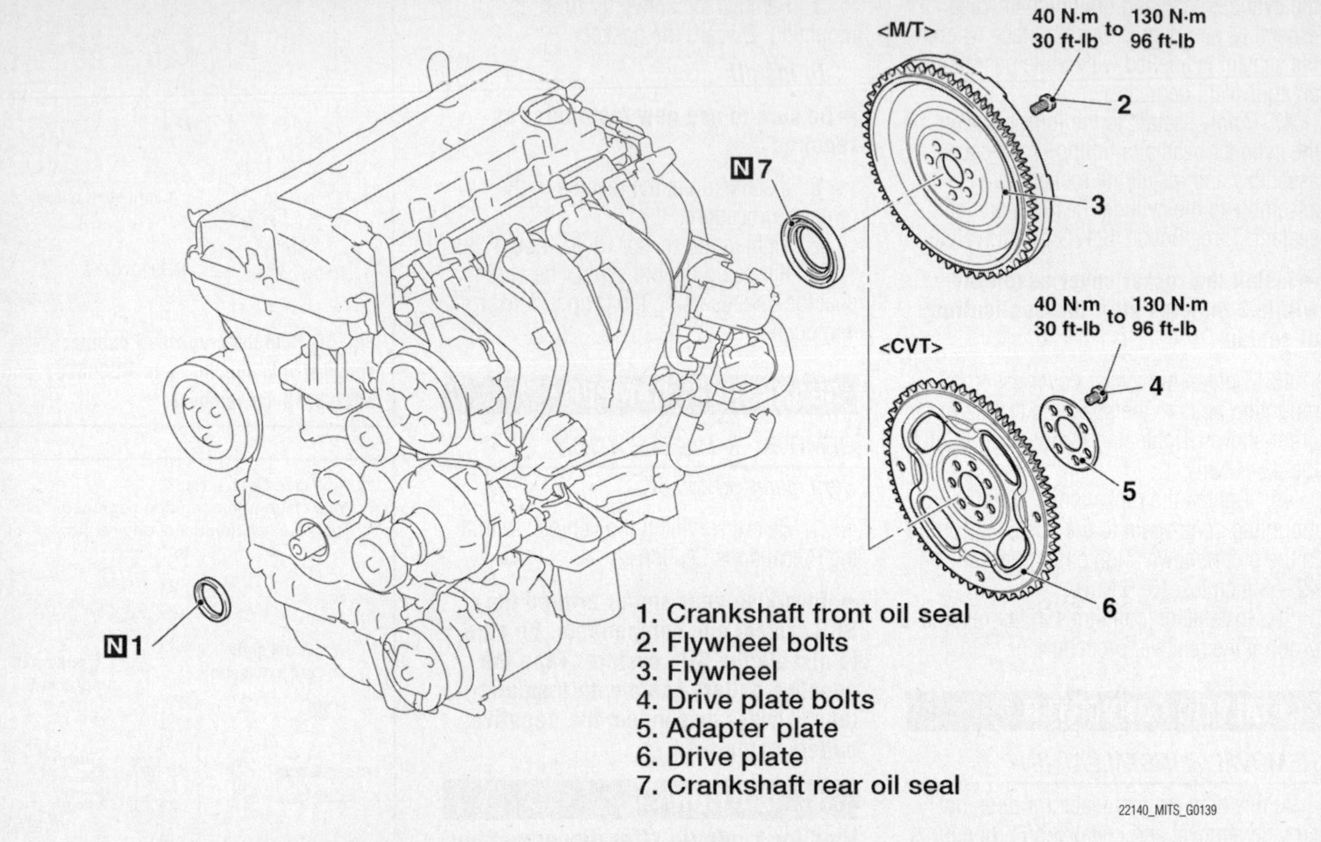

<M/T>

40 N·m to 130 N·m
30 ft-lb   96 ft-lb

2

N 7

3

40 N·m to 130 N·m
30 ft-lb   96 ft-lb

<CVT>

4

5

6

1. Crankshaft front oil seal
2. Flywheel bolts
3. Flywheel
4. Drive plate bolts
5. Adapter plate
6. Drive plate
7. Crankshaft rear oil seal

N 1

22140_MITS_G0139

Fig. 58 Oil seal (front and rear) and related components—except turbocharged engine

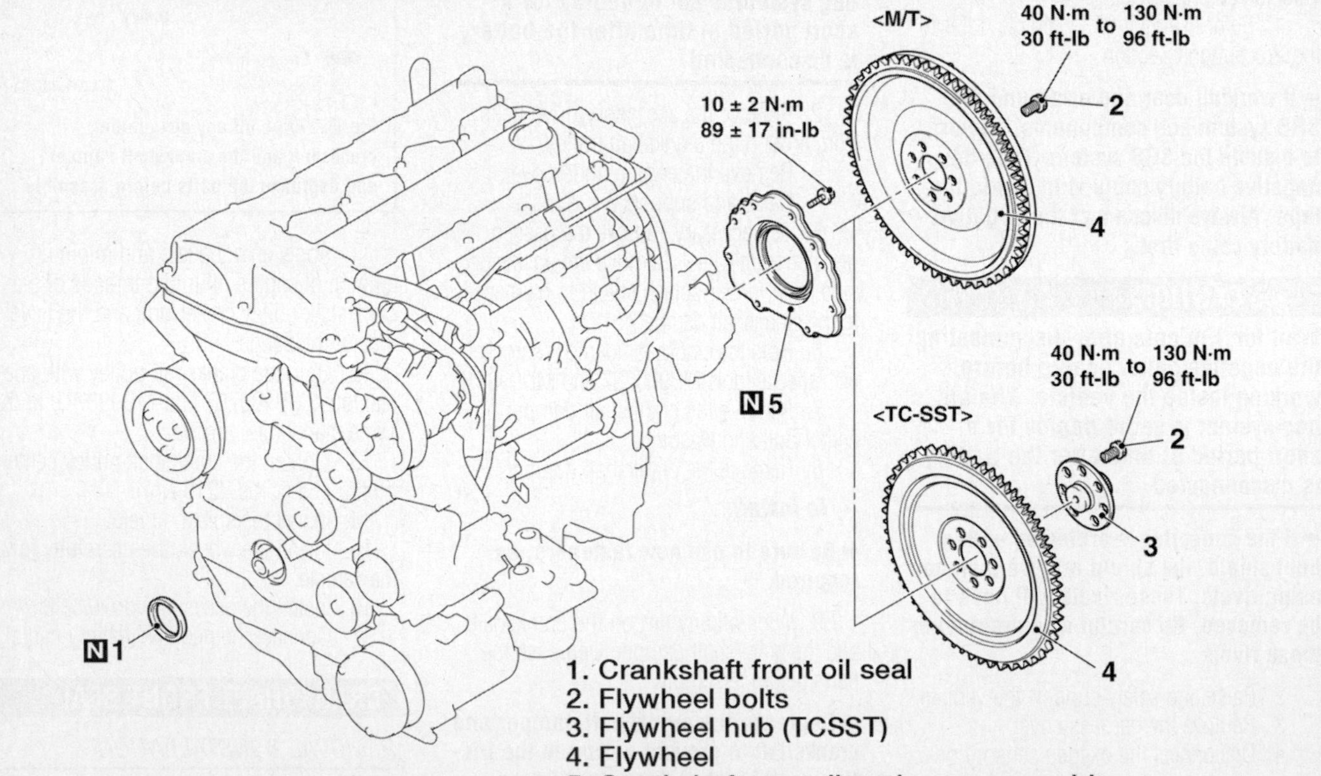

<M/T>

40 N·m to 130 N·m
30 ft-lb   96 ft-lb

2

10 ± 2 N·m
89 ± 17 in-lb

4

N 5

<TC-SST>

40 N·m to 130 N·m
30 ft-lb   96 ft-lb

2

3

4

1. Crankshaft front oil seal
2. Flywheel bolts
3. Flywheel hub (TCSST)
4. Flywheel
5. Crankshaft rear oil seal case assembly

N 1

22140_MITS_G0141

Fig. 59 Oil seal (front and rear) and related components—turbocharged engine

→If working near and/or around the SRS system and components, be sure to disable the SRS system. Tape the negative battery cable with insulating tape. Always disconnect the negative battery cable first.

※※ CAUTION

Wait for 1 minute after disconnecting the negative battery cable before working inside the vehicle. The air bag system is set to deploy for a short period of time after the battery is disconnected.

2. Disconnect the negative battery cable.
3. Remove the crankshaft pulley.
4. Carefully remove the crankshaft front oil seal.

To install:

5. Install the front oil seal. Apply oil to the seal and install using a crankshaft front oil seal installer tool install the seal.
6. Continue the installation in the reverse order of the removal procedure.

CYLINDER HEAD

REMOVAL & INSTALLATION

2.0L Non Turbocharged Engine and 2.4L Engine

See Figures 60 through 65.

1. Before servicing the vehicle, refer to the Precautions Section.

→If working near and/or around the SRS system and components, be sure to disable the SRS system. Tape the negative battery cable with insulating tape. Always disconnect the negative battery cable first.

※※ CAUTION

Wait for 1 minute after disconnecting the negative battery cable before working inside the vehicle. The air bag system is set to deploy for a short period of time after the battery is disconnected.

2. Relieve the fuel system pressure.
3. Disconnect the negative battery cable.
4. On 2.4L engines, remove the strut bar.
5. Remove the engine room undercover and side cover (RH).
6. Drain the cooling system.

7. Remove the air cleaner assembly.
8. Remove the ignition coils.
9. Remove the exhaust manifold.
10. Remove the throttle body assembly.
11. Remove the EGR valve and EGR valve stay (California Vehicles).
12. Remove the water pump.
13. Remove the control wiring harness connection.
14. Remove the radiator upper and lower hose connection.
15. Remove the heater hose connection.
16. Remove the cooling water line hose (M/T) or CVT fluid cooler water return hose B (CVT) connection.
17. Remove the water pump inlet pipe.
18. Remove the cooling water line gasket and O-ring.
19. Remove the emission vacuum hose connection and brake booster vacuum hose connection.
20. Remove the engine oil level gauge.
21. Remove the intake manifold stay.
22. Remove the cylinder head cover PCV hose connection.
23. Remove the fuel high pressure hose connection.
24. Remove the valve timing chain.

※※ WARNING

When the camshaft bearing cap mounting bolts are loosened all at once, the mounting bolts jump out by the spring force and the threads are damaged. Always loosen the mounting bolts in 4–5 steps.

25. Loosen the mounting bolts of the camshaft bearing caps in the order of number shown in the figure in 4–5 steps, and remove the camshaft bearing caps.
26. Remove the camshaft and camshaft sprocket assembly.
27. Remove the camshaft bearing.
28. Remove the intake manifold stay B (except for California).
29. Temporarily install the engine oil pan which was removed at the valve timing chain removal.
30. Place a garage jack against the engine oil pan with a piece of wood in between to support the engine and transaxle assembly.
31. Remove special tool MB991928 or MB991895 which was installed for supporting the engine and transaxle assembly when the valve timing chain was removed.
32. Loosen and remove the cylinder head bolts in 2–3 steps in the order of number shown in the figure.

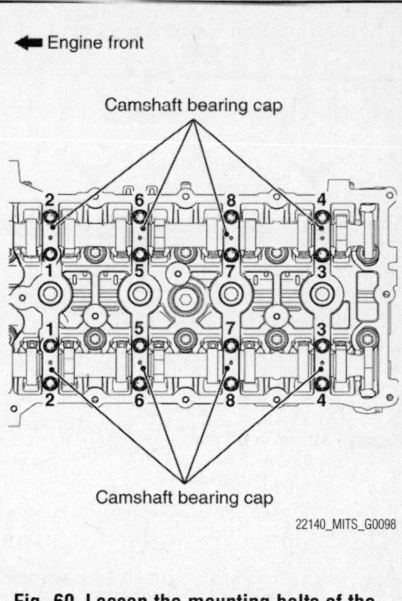

Fig. 60 Loosen the mounting bolts of the camshaft bearing caps in the order shown

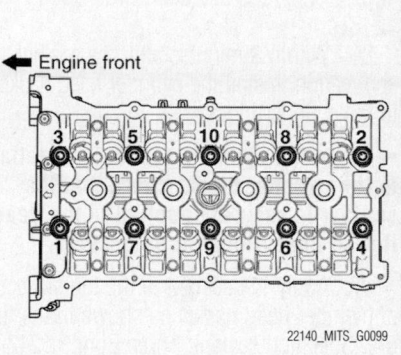

Fig. 61 Loosen and remove the cylinder head bolts in 2–3 steps in the order shown

33. Remove the cylinder head assembly and the gasket.

To install:

→Be sure to use new fasteners, as required.

※※ WARNING

Do not allow any foreign materials get into the coolant passages, oil passages, or cylinder.

34. Remove the sealant and grease on the top surface of cylinder block and on the bottom surface of the cylinder head. Then, use a quick-drying degreasing agent (white gasoline) to degrease the sealant application surface.
35. Apply sealant to the top surface of cylinder block as shown in the

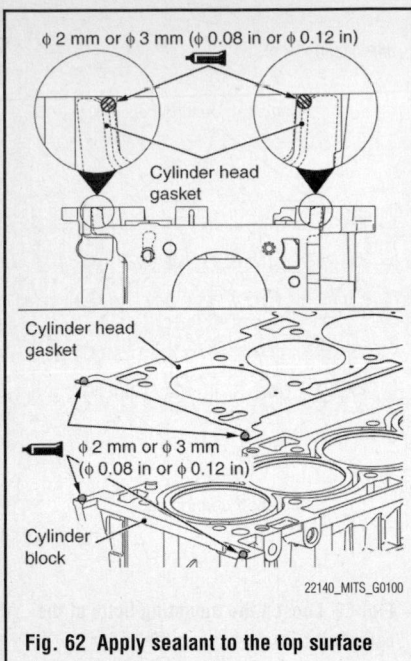

Fig. 62 Apply sealant to the top surface of cylinder block as shown

figure. Specified sealant: Three Bond® 1217G.

36. Within 3 minutes after the sealant application, install the cylinder head gasket to the cylinder block.

➡️**When the cylinder gasket is installed to the cylinder block, check that the sealant is securely applied to the bead line of the cylinder head gasket.**

37. Apply the sealant to the top surface of cylinder head gasket as shown in the figure. Specified sealant: Three Bond® 1217G.

※※ **WARNING**

Within 2 hours after the cylinder head assembly installation, do not apply oil or water to the sealant application area or start the engine.

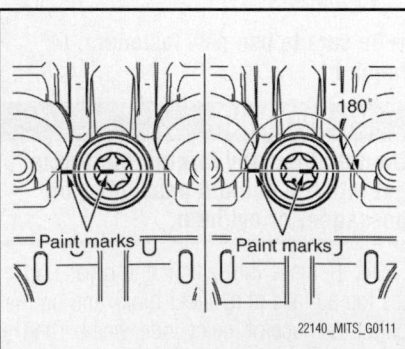

Fig. 63 Check that the paint mark on the cylinder head bolt head aligns with the paint mark on the cylinder head when tightening

38. Within 3 minutes after the sealant application, install the cylinder head assembly.

39. Replace cylinder head bolts with a new ones.

40. For the 2 bolts of the timing chain side, the washer can be removed from the bolt. Install the washer, with its sag facing upward, to the bolts.

41. Apply a small amount of engine oil to the cylinder head bolt threads and the washers.

42. Tighten the bolts by the following procedure (plastic region angular tightening method).

   a. Step 1: tighten the bolts to 25–27 ft. lbs. (33–37 Nm) in the order shown in the figure.

※※ **WARNING**

The bolt is not tightened sufficiently if the tightening angle is less than a 180° angle. If the tightening angle exceeds the standard specification,

remove the bolt and repeat the installation steps.

   b. Step 2: Put a paint mark on the cylinder head bolt head and cylinder head, tighten to 178–182° in the order shown in the figure, and check that the paint mark on the cylinder head bolt head aligns with the paint mark on the cylinder head.

43. Install special tool MB991928 or MB991895 which was installed for supporting the engine and transaxle assembly when the valve timing chain was removed.

44. Remove the garage jack which supports the engine and transaxle assembly.

45. Remove the engine oil pan installed temporarily.

※※ **WARNING**

Be careful not to drop the camshaft bearing. When installing the

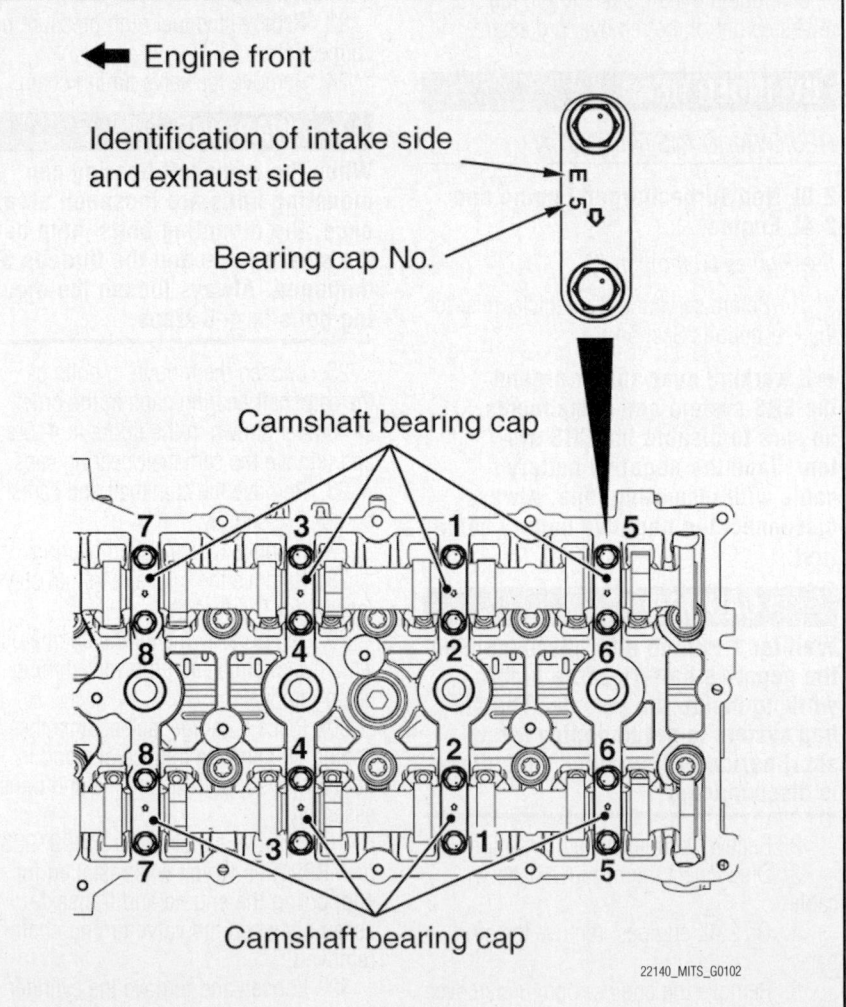

Fig. 64 Tighten each camshaft bearing cap mounting bolt to the specified torque in 2–3 passes in the sequence shown

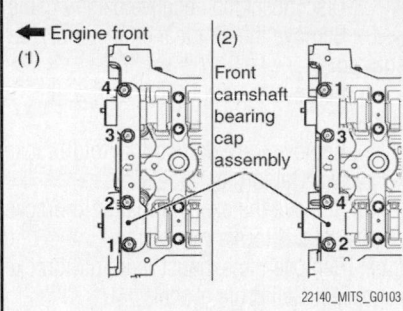

Fig. 65 Install the front camshaft bearing cap to the cylinder head and tighten the front camshaft bearing cap to the specified torque in the order shown

camshaft and camshaft sprocket assembly (exhaust side), be careful not to let the camshaft bearing, which is installed to the front cam bearing, deviate from its position.

46. Install the camshaft bearing caps to the cylinder heads.

➡ **Because the thrust camshaft bearing cap and the other camshaft bearing caps are the same in shape, check the bearing cap number and additionally its symbol to identify the intake and exhaust sides for correct installation.**

47. Tighten each camshaft bearing cap mounting bolt to the specified torque in 2–3 passes according to the figure shown. Tightening torque: 99–115 inch lbs. (11–13 Nm).

✳✳ WARNING

When the mounting bolts are tightened with the front camshaft bearing cap tilted, the front camshaft bearing cap is damaged. Install the front camshaft bearing cap properly to the cylinder head and camshaft.

48. Install the front camshaft bearing cap to the cylinder head and temporarily tighten the front camshaft bearing cap to the specified torque in the order shown (1). Tightening torque: 11–15 ft. lbs. (14–20 Nm).

49. Tighten the front camshaft bearing cap again to the specified torque in the order shown (2). Tightening torque: 22–24 ft. lbs. (28–32 Nm).

50. Installation continues in the reverse of the removal procedure.

51. Run the engine and check for fluid leaks and proper operation.

2.0L Turbocharged Engine

See Figures 60 through 65.

1. Before servicing the vehicle, refer to the Precautions Section.

➡ **If working near and/or around the SRS system and components, be sure to disable the SRS system. Tape the negative battery cable with insulating tape. Always disconnect the negative battery cable first.**

✳✳ CAUTION

Wait for 1 minute after disconnecting the negative battery cable before working inside the vehicle. The air bag system is set to deploy for a short period of time after the battery is disconnected.

2. Relieve the fuel system pressure. Disconnect the negative battery cable. Drain the cooling system.

3. Remove the engine undercover. Remove the side cover.

4. Drain the engine coolant.

5. Remove the air cleaner assembly.

6. Remove the charge air cooler intake hose A and B and charge air cooler intake pipe A.

7. Remove the ignition coils.

8. Remove the strut tower bar.

9. Remove the exhaust manifold and turbocharger assembly.

10. Remove the throttle body assembly.

11. Remove the water pump.

12. Remove the control wiring harness connection.

13. Remove the radiator upper and lower hose connections.

14. Remove the heater hose connection.

15. Remove the fuel high pressure hose connection and fuel return hose connection.

16. Remove the turbocharger by-pass valve purge hose connection.

17. Remove the canister vacuum hose connection.

18. Remove the brake booster vacuum hose connection.

19. Remove the water pump intake pipe and O-ring.

20. Remove the engine oil level gauge and O-ring.

21. Remove the starter wiring harness clamp.

22. Remove the intake manifold stay (front).

23. Remove the starter wiring harness clamp.

24. Remove the intake manifold stay (rear).

25. Remove the valve timing chain.

✳✳ WARNING

When the camshaft bearing cap mounting bolts are loosened all at once, the mounting bolts jump out by the spring force and the threads are damaged. Always loosen the mounting bolts in 4–5 steps.

26. Loosen the mounting bolts of the camshaft bearing caps in the order of number shown in the figure in 4–5 steps, and remove the camshaft bearing caps.

27. Remove the camshaft and camshaft sprocket assembly.

28. Remove the camshaft bearing.

29. Remove the intake manifold stay B (except for California).

30. Temporarily install the engine oil pan which was removed at the valve timing chain removal.

31. Place a garage jack against the engine oil pan with a piece of wood in between to support the engine and transaxle assembly.

32. Remove special tool MB991928 or MB991895 which was installed for supporting the engine and transaxle assembly when the valve timing chain was removed.

33. Loosen and remove the cylinder head bolts in 2–3 steps in the order of number shown in the figure.

34. Remove the cylinder head assembly and the gasket.

To install:

➡ **Be sure to use new fasteners, as required.**

✳✳ WARNING

Do not allow any foreign materials get into the coolant passages, oil passages, or cylinder.

35. Remove the sealant and grease on the top surface of cylinder block and on the bottom surface of the cylinder head. Then, use a quick-drying degreasing agent (white gasoline) to degrease the sealant application surface.

36. Apply sealant to the top surface of cylinder block as shown in the figure. Specified sealant: Three Bond® 1217G.

37. Within 3 minutes after the sealant application, install the cylinder head gasket to the cylinder block.

➡ **When the cylinder gasket is installed to the cylinder block, check that the sealant is securely applied to the bead line of the cylinder head gasket.**

38. Apply the sealant to the top surface of cylinder head gasket as shown in the figure. Specified sealant: Three Bond® 1217G.

※※ WARNING

Within 2 hours after the cylinder head assembly installation, do not apply oil or water to the sealant application area or start the engine.

39. Within 3 minutes after the sealant application, install the cylinder head assembly.

40. Replace cylinder head bolts with a new ones.

41. For the 2 bolts of the timing chain side, the washer can be removed from the bolt. Install the washer, with its sag facing upward, to the bolts.

42. Apply a small amount of engine oil to the cylinder head bolt threads and the washers.

43. Tighten the bolts by the following procedure (plastic region angular tightening method).

　　a. Step 1: tighten the bolts to 25–27 ft. lbs. (33–37 Nm) in the order shown in the figure.

※※ WARNING

The bolt is not tightened sufficiently if the tightening angle is less than a 180° angle. If the tightening angle exceeds the standard specification, remove the bolt and repeat the installation steps.

　　b. Step 2: Put a paint mark on the cylinder head bolt head and cylinder head, tighten to 178–182° in the order shown in the figure, and check that the paint mark on the cylinder head bolt head aligns with the paint mark on the cylinder head.

44. Install special tool MB991928 or MB991895 which was installed for supporting the engine and transaxle assembly when the valve timing chain was removed.

45. Remove the garage jack which supports the engine and transaxle assembly.

46. Remove the engine oil pan installed temporarily.

※※ WARNING

Be careful not to drop the camshaft bearing. When installing the camshaft and camshaft sprocket assembly (exhaust side), be careful not to let the camshaft bearing, which is installed to the front cam bearing, deviate from its position.

47. Install the camshaft bearing caps to the cylinder heads.

➡**Because the thrust camshaft bearing cap and the other camshaft bearing caps are the same in shape, check the bearing cap number and additionally its symbol to identify the intake and exhaust sides for correct installation.**

48. Tighten each camshaft bearing cap mounting bolt to the specified torque in 2–3 passes according to the figure shown. Tightening torque: 99–115 inch lbs. (11–13 Nm).

※※ WARNING

When the mounting bolts are tightened with the front camshaft bearing cap tilted, the front camshaft bearing cap is damaged. Install the front camshaft bearing cap properly to the cylinder head and camshaft.

49. Install the front camshaft bearing cap to the cylinder head and temporarily tighten the front camshaft bearing cap to the specified torque in the order shown (1). Tightening torque: 11–15 ft. lbs. (14–20 Nm).

50. Tighten the front camshaft bearing cap again to the specified torque in the order shown (2). Tightening torque: 22–24 ft. lbs. (28–32 Nm).

51. Installation continues in the reverse of the removal procedure.

52. Run the engine and check for fluid leaks and proper operation.

EXHAUST MANIFOLD

REMOVAL & INSTALLATION

2.0L Non Turbocharged Engine and 2.4L Engine

See Figure 66.

1. Before servicing the vehicle, refer to the Precautions Section.

➡**If working near and/or around the SRS system and components, be sure to disable the SRS system. Tape the negative battery cable with insulating tape. Always disconnect the negative battery cable first.**

※※ CAUTION

Wait for 1 minute after disconnecting the negative battery cable before working inside the vehicle. The air bag system is set to deploy for a short period of time after the battery is disconnected.

2. Disconnect the negative battery cable.

3. Remove the air cleaner assembly, as required.

4. Remove the strut tower bar, 2.4L engines.

5. Remove the exhaust manifold bracket D—except California models.

6. Remove the exhaust manifold bracket A—except California models.

7. Remove the exhaust manifold bracket B—except California models.

8. Remove the Crankshaft Position (CKP) sensor cover.

9. Remove the Crankshaft Position (CKP) sensor and O-ring.

10. Remove the exhaust manifold upper cover and lower cover.

11. Remove the exhaust manifold and discard the gasket.

To install:

12. Position a new exhaust manifold gasket to the mating surface of the cylinder.

➡**The exhaust manifold gasket, washers and nuts must not be reused.**

13. Install the exhaust manifold to the engine. Torque all exhaust manifold nuts to specification.

14. Continue the installation in the reverse order of the removal procedure. Tighten the all bolts to specification. See illustration.

2.0L Turbocharged Engine

1. Before servicing the vehicle, refer to the Precautions Section.

➡**If working near and/or around the SRS system and components, be sure to disable the SRS system. Tape the negative battery cable with insulating tape. Always disconnect the negative battery cable first.**

※※ CAUTION

Wait for 1 minute after disconnecting the negative battery cable before working inside the vehicle. The air bag system is set to deploy for a short period of time after the battery is disconnected.

2. Disconnect the negative battery cable.

3. Remove the air cleaner assembly.

4. Remove the exhaust manifold cover.

5. Remove the turbocharger compressor bracket.

6. Remove the exhaust fitting bracket.

7. Remove the oil return pipe and gasket.

8. Remove the turbocharger bracket, turbocharger, pipe assembly, and gasket.

9. Remove the exhaust manifold and discard the gasket.

To install:

10. Clean all gasket material from the mating surfaces.

11. Install a new gasket and the exhaust manifold. Torque the manifold nuts to: 33–39 ft. lbs. (44–54 Nm).

12. Install the turbocharger assembly.

13. Install the exhaust manifold cover.

14. Install the air cleaner assembly.

15. Connect the negative battery cable.

16. Start the engine and check for exhaust leaks.

FLEXPLATE

REMOVAL & INSTALLATION

1. Before servicing the vehicle, refer to the Precautions Section.

➡ **If working near and/or around the SRS system and components, be sure to disable the SRS system. Tape the** negative battery cable with insulating tape. Always disconnect the negative battery cable first.

✳✳ CAUTION

Wait for 1 minute after disconnecting the negative battery cable before working inside the vehicle. The air bag system is set to deploy for a short period of time after the battery is disconnected.

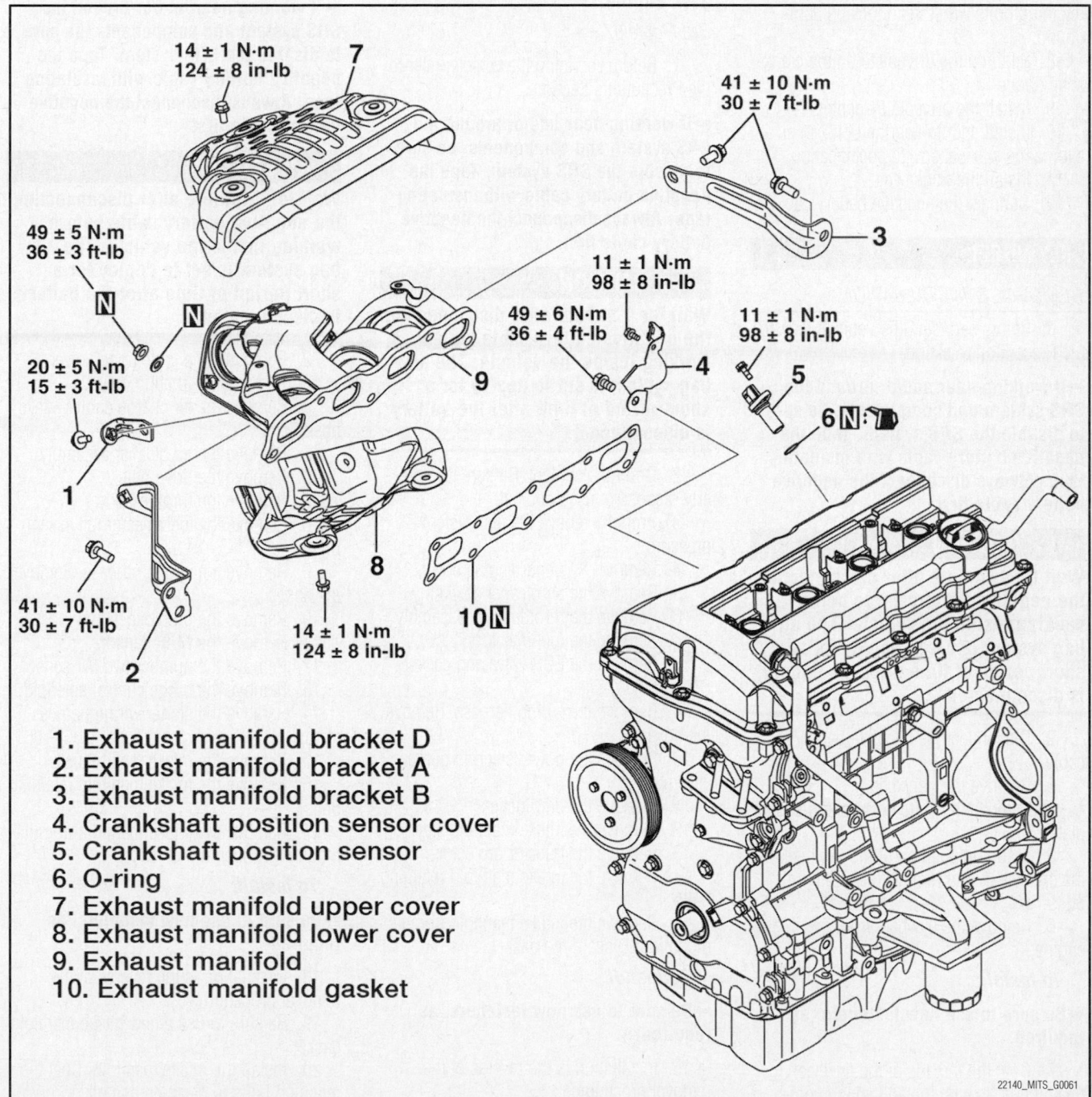

1. Exhaust manifold bracket D
2. Exhaust manifold bracket A
3. Exhaust manifold bracket B
4. Crankshaft position sensor cover
5. Crankshaft position sensor
6. O-ring
7. Exhaust manifold upper cover
8. Exhaust manifold lower cover
9. Exhaust manifold
10. Exhaust manifold gasket

22140_MITS_G0061

Fig. 66 Exhaust manifold and related components—2.0L non turbocharged engine and 2.4L engine (except California)

2. Disconnect the negative battery cable.
3. Remove the transaxle.
4. Mark the position of the driveplate on the crankshaft and remove the retaining bolts.
5. Remove the driveplate adapter.
6. Remove the driveplate from the engine.

To install:

➡ **Be sure to use new fasteners, as required.**

7. Coat the threads of the driveplate retaining bolts with thread locking compound.
8. Position the driveplate on the crankshaft flange.
9. Install the driveplate adapter.
10. Install and tighten the bolts, in an alternating star pattern, to specification.
11. Install the transaxle.
12. Connect the negative battery cable.

FLYWHEEL

REMOVAL & INSTALLATION

1. Before servicing the vehicle, refer to the Precautions Section.

➡ **If working near and/or around the SRS system and components, be sure to disable the SRS system. Tape the negative battery cable with insulating tape. Always disconnect the negative battery cable first.**

✳✳ CAUTION

Wait for 1 minute after disconnecting the negative battery cable before working inside the vehicle. The air bag system is set to deploy for a short period of time after the battery is disconnected.

2. Disconnect the negative battery cable.
3. Remove the transaxle.
4. Remove the clutch disc and pressure plate.
5. Mark the position of the flywheel on the crankshaft and remove the retaining bolts.
6. Remove the flywheel from the engine.

To install:

➡ **Be sure to use new fasteners, as required.**

7. Coat the threads of the flywheel retaining bolts with thread locking compound.

8. Position the flywheel on the crankshaft flange.
9. Install and tighten the bolts, in an alternating star pattern, to specification.
10. Install the clutch and pressure plate.
11. Install the transaxle.
12. Connect the negative battery cable.

INTAKE MANIFOLD

REMOVAL & INSTALLATION

2.0L Non Turbocharged Engine and 2.4L Engine

See Figure 67.

1. Before servicing the vehicle, refer to the Precautions Section.

➡ **If working near and/or around the SRS system and components, be sure to disable the SRS system. Tape the negative battery cable with insulating tape. Always disconnect the negative battery cable first.**

✳✳ CAUTION

Wait for 1 minute after disconnecting the negative battery cable before working inside the vehicle. The air bag system is set to deploy for a short period of time after the battery is disconnected.

2. Drain the coolant. Be sure to properly dispose of used coolant.
3. Properly relieve the fuel system pressure.
4. Remove the upper engine cover.
5. Remove the air cleaner assembly.
6. Remove the throttle body assembly.
7. Remove the fuel injectors.
8. Remove the EGR valve and gasket. Discard the gasket.
9. Remove the wiring harness. Remove the hoses.
10. Remove the power steering pump assembly.
11. Remove the oil dipstick.
12. Remove the intake manifold stay.
13. Remove the injector protector.
14. Remove the intake manifold retaining bolts.
15. Remove the intake manifold from its mounting. Discard the gasket.

To install:

➡ **Be sure to use new fasteners, as required.**

16. Installation is the reverse of the removal procedure.
17. Be sure to use a new intake manifold gasket.

18. Install the intake manifold. Coat the retaining bolts in clean engine oil. Torque the retaining bolts to specification.
19. Fill the cooling system with the proper grade and type engine coolant.
20. Start the engine and check for leaks, correct as required.

2.0L Turbocharged Engine

See Figure 68.

1. Before servicing the vehicle, refer to the Precautions Section.

➡ **If working near and/or around the SRS system and components, be sure to disable the SRS system. Tape the negative battery cable with insulating tape. Always disconnect the negative battery cable first.**

✳✳ CAUTION

Wait for 1 minute after disconnecting the negative battery cable before working inside the vehicle. The air bag system is set to deploy for a short period of time after the battery is disconnected.

2. Remove the engine upper cover.
3. Remove the ignition coil.
4. Remove the air charge cooler intake hose .
5. Remove the air cleaner assembly.
6. Remove the drive belt.
7. Remove the throttle body.
8. Remove the fuel injector rail assembly.
9. Remove the wiring harness. Remove the hoses.
10. Remove the oil dipstick and tube.
11. Remove the MAP sensor.
12. Remove the number two IAT sensor.
13. Remove the purge control solenoid.
14. Remove the starter wiring harness clamp.
15. Remove the intake manifold stay.
16. Remove the intake manifold retaining bolts.
17. Remove the intake manifold from its mounting. Discard the gasket.

To install:

➡ **Be sure to use new fasteners, as required.**

18. Installation is the reverse of the removal procedure.
19. Be sure to use a new intake manifold gasket.
20. Install the intake manifold. Coat the retaining bolts in clean engine oil. Torque the retaining bolts to specification.

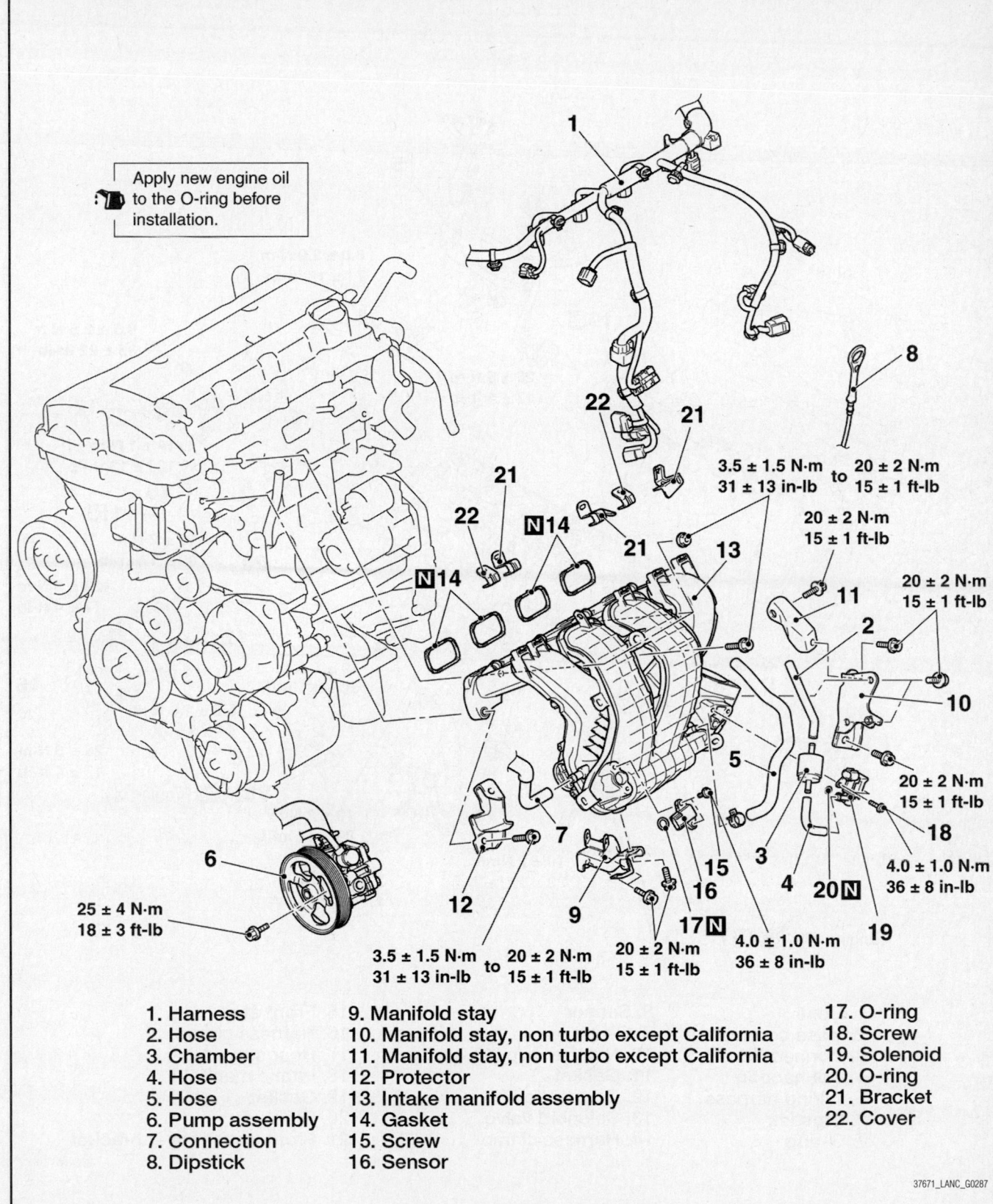

Apply new engine oil to the O-ring before installation.

3.5 ± 1.5 N·m
31 ± 13 in-lb to 20 ± 2 N·m
15 ± 1 ft-lb

20 ± 2 N·m
15 ± 1 ft-lb

20 ± 2 N·m
15 ± 1 ft-lb

20 ± 2 N·m
15 ± 1 ft-lb

4.0 ± 1.0 N·m
36 ± 8 in-lb

25 ± 4 N·m
18 ± 3 ft-lb

3.5 ± 1.5 N·m
31 ± 13 in-lb to 20 ± 2 N·m
15 ± 1 ft-lb

20 ± 2 N·m
15 ± 1 ft-lb

4.0 ± 1.0 N·m
36 ± 8 in-lb

1. Harness	9. Manifold stay	17. O-ring
2. Hose	10. Manifold stay, non turbo except California	18. Screw
3. Chamber	11. Manifold stay, non turbo except California	19. Solenoid
4. Hose	12. Protector	20. O-ring
5. Hose	13. Intake manifold assembly	21. Bracket
6. Pump assembly	14. Gasket	22. Cover
7. Connection	15. Screw	
8. Dipstick	16. Sensor	

Fig. 67 Intake manifold and related components—2.0L non turbocharged engine and 2.4L engine

37671_LANC_G0287

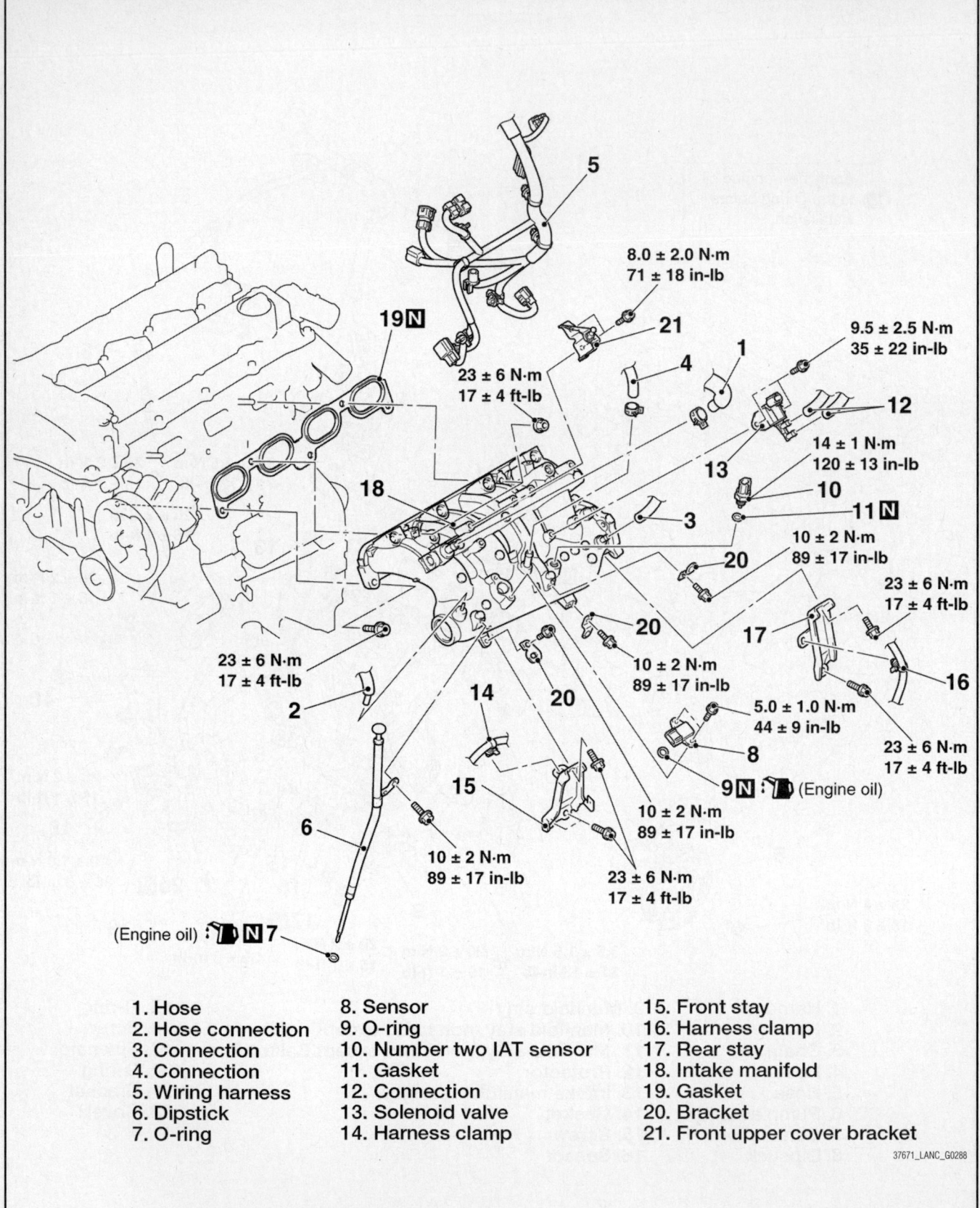

8.0 ± 2.0 N·m
71 ± 18 in-lb

9.5 ± 2.5 N·m
35 ± 22 in-lb

23 ± 6 N·m
17 ± 4 ft-lb

14 ± 1 N·m
120 ± 13 in-lb

10 ± 2 N·m
89 ± 17 in-lb

23 ± 6 N·m
17 ± 4 ft-lb

23 ± 6 N·m
17 ± 4 ft-lb

10 ± 2 N·m
89 ± 17 in-lb

5.0 ± 1.0 N·m
44 ± 9 in-lb

23 ± 6 N·m
17 ± 4 ft-lb

(Engine oil)

10 ± 2 N·m
89 ± 17 in-lb

23 ± 6 N·m
17 ± 4 ft-lb

10 ± 2 N·m
89 ± 17 in-lb

(Engine oil)

1. Hose
2. Hose connection
3. Connection
4. Connection
5. Wiring harness
6. Dipstick
7. O-ring

8. Sensor
9. O-ring
10. Number two IAT sensor
11. Gasket
12. Connection
13. Solenoid valve
14. Harness clamp

15. Front stay
16. Harness clamp
17. Rear stay
18. Intake manifold
19. Gasket
20. Bracket
21. Front upper cover bracket

37671_LANC_G0288

Fig. 68 Intake manifold and related components—2.0L turbocharged engine

21. Fill the cooling system with the proper grade and type engine coolant.

22. Start the engine and check for leaks, correct as required.

OIL PAN

REMOVAL & INSTALLATION

See Figures 69 through 71.

1. Before servicing the vehicle, refer to the Precautions Section.

➡**If working near and/or around the SRS system and components, be sure to disable the SRS system. Tape the negative battery cable with insulating tape. Always disconnect the negative battery cable first.**

✸✸ CAUTION

Wait for 1 minute after disconnecting the negative battery cable before working inside the vehicle. The air bag system is set to deploy for a short period of time after the battery is disconnected.

2. Remove the negative battery cable.

3. Remove the engine undercover and side cover.

4. Drain the engine oil.

5. Remove the accessory drive belt.

6. Remove the A/C compressor and clutch assembly together with the hose from the bracket.

7. Tie the removed A/C compressor and clutch assembly with a string at a position where they will not interfere with the removal and installation of engine oil pan.

8. Remove the engine oil pan mounting bolts.

✸✸ WARNING

Do not forcibly drive in special tool MD998727 to avoid damage to the engine oil pan seal surface of cylinder block assembly.

9. Insert special tool MD998727 into the engine oil pan removal groove of the cylinder block assembly.

10. Lightly tap the special tool with a hammer to slide the oil pan seal surface, cut off the liquid gasket, and remove the engine oil pan.

To install:

11. Remove all the traces of sealant adhering to the engine oil pan and cylinder block assembly using a remover. Then, degrease them using a quick-drying degreasing agent (white gasoline).

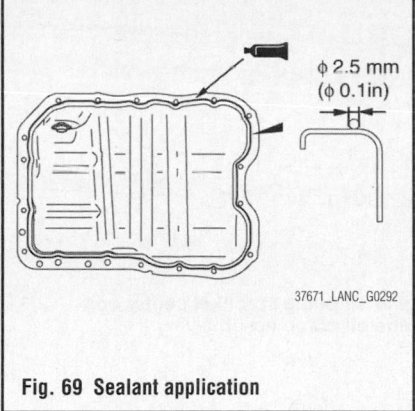

Fig. 69 Sealant application

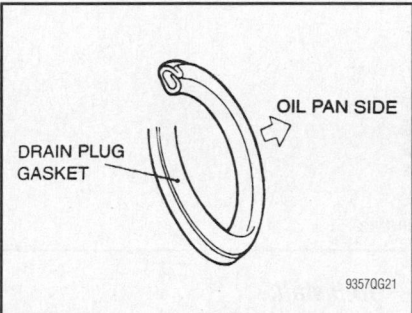

Fig. 70 Make sure to the install the new drain plug gasket as shown to avoid oil leakage

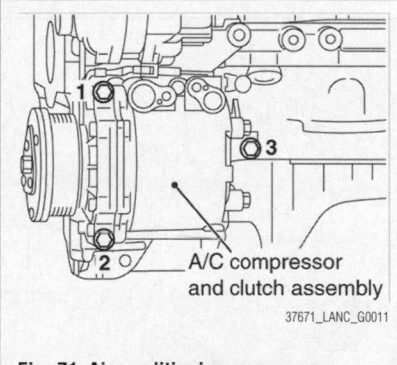

Fig. 71 Air conditioning compressor mounting bolt tightening sequence

12. Apply the sealant without any gap to the mating surface of engine oil pan as shown in the figure. Specified sealant: Three Bond® 1217G or equivalent.

13. Within 3 minutes, install the engine oil pan to the cylinder block assembly.

✸✸ WARNING

Do not apply oil or water to the sealant area or start up the engine within 2 hours after the installation of the engine oil pan.

14. Tighten the engine oil pan mounting bolts to specification.

15. Install the engine undercover.

16. Install the oil drain plug with a new gasket and tighten to specification

17. Tighten A/C compressor and clutch assembly mounting bolts to the specified torque in the proper sequence.

18. Lower the vehicle and fill the crankcase to the proper level with clean engine oil.

19. Connect the negative battery cable. Start the engine and check for leaks.

OIL PUMP

REMOVAL & INSTALLATION

2.0L Engine

See Figures 72 and 73.

1. Before servicing the vehicle, refer to the Precautions Section.

➡**If working near and/or around the SRS system and components, be sure to disable the SRS system. Tape the negative battery cable with insulating tape. Always disconnect the negative battery cable first.**

✸✸ CAUTION

Wait for 1 minute after disconnecting the negative battery cable before working inside the vehicle. The air bag system is set to deploy for a short period of time after the battery is disconnected.

2. Disconnect the negative battery cable.

3. raise and safely support the vehicle.

4. Remove the engine oil pan.

5. Secure the engine oil pump sprocket and the engine oil pump drive chain with tie-wrap to prevent slippage between the engine oil pump sprocket and the engine oil pump drive chain.

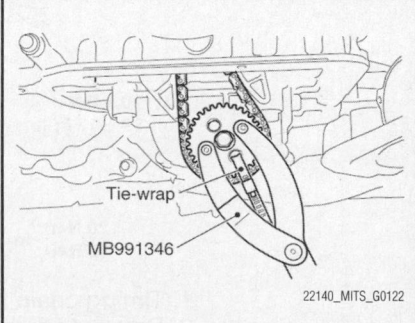

Fig. 72 Hold the engine oil pump sprocket with special tool MB991346 for removal— 2.0L engine

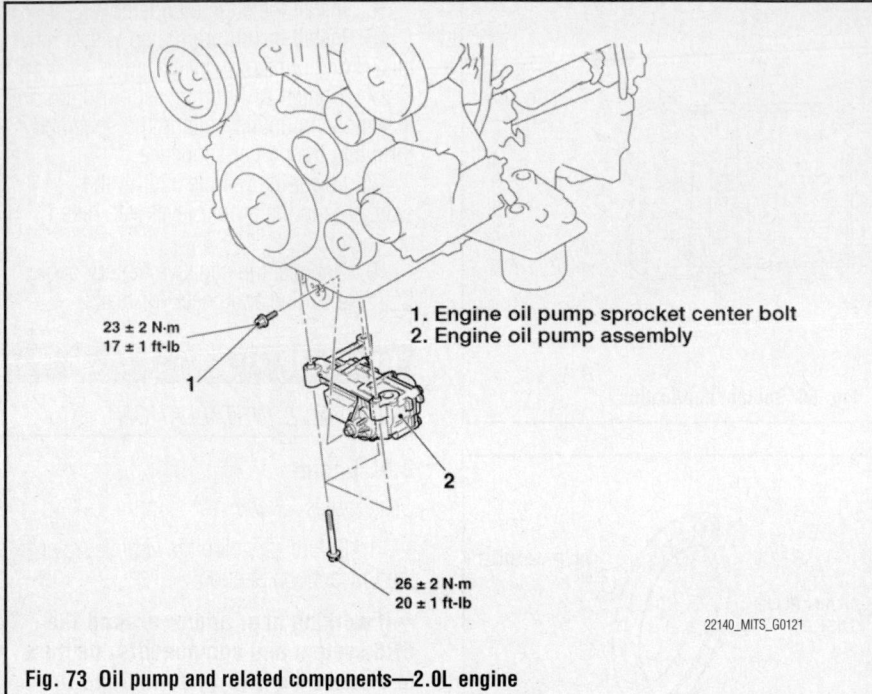

1. Engine oil pump sprocket center bolt
2. Engine oil pump assembly

23 ± 2 N·m
17 ± 1 ft-lb

26 ± 2 N·m
20 ± 1 ft-lb

22140_MITS_G0121

Fig. 73 Oil pump and related components—2.0L engine

6. Hold the engine oil pump sprocket with special tool MB991346.

7. Remove the engine oil pump sprocket with the engine oil pump drive chain attached.

To install:

➡Be sure to use new fasteners, as required.

8. Installation is the reverse of the removal procedure.

9. Tighten the engine oil pump sprocket center bolt to 16–18 ft. lbs. (21–25 Nm).

10. Tighten the engine oil pump assembly attaching bolts to 19–21 ft. lbs. (24–28 Nm).

11. Fill the engine with the correct amount and type of oil.

12. Connect the negative battery cable.

13. Start the engine and check for leaks.

2.4L Engine

See Figures 74 through 77.

1. Before servicing the vehicle, refer to the Precautions Section.

➡If working near and/or around the SRS system and components, be sure to disable the SRS system. Tape the negative battery cable with insulating tape. Always disconnect the negative battery cable first.

✳ CAUTION

Wait for 1 minute after disconnecting the negative battery cable before working inside the vehicle. The air bag system is set to deploy for a short period of time after the battery is disconnected.

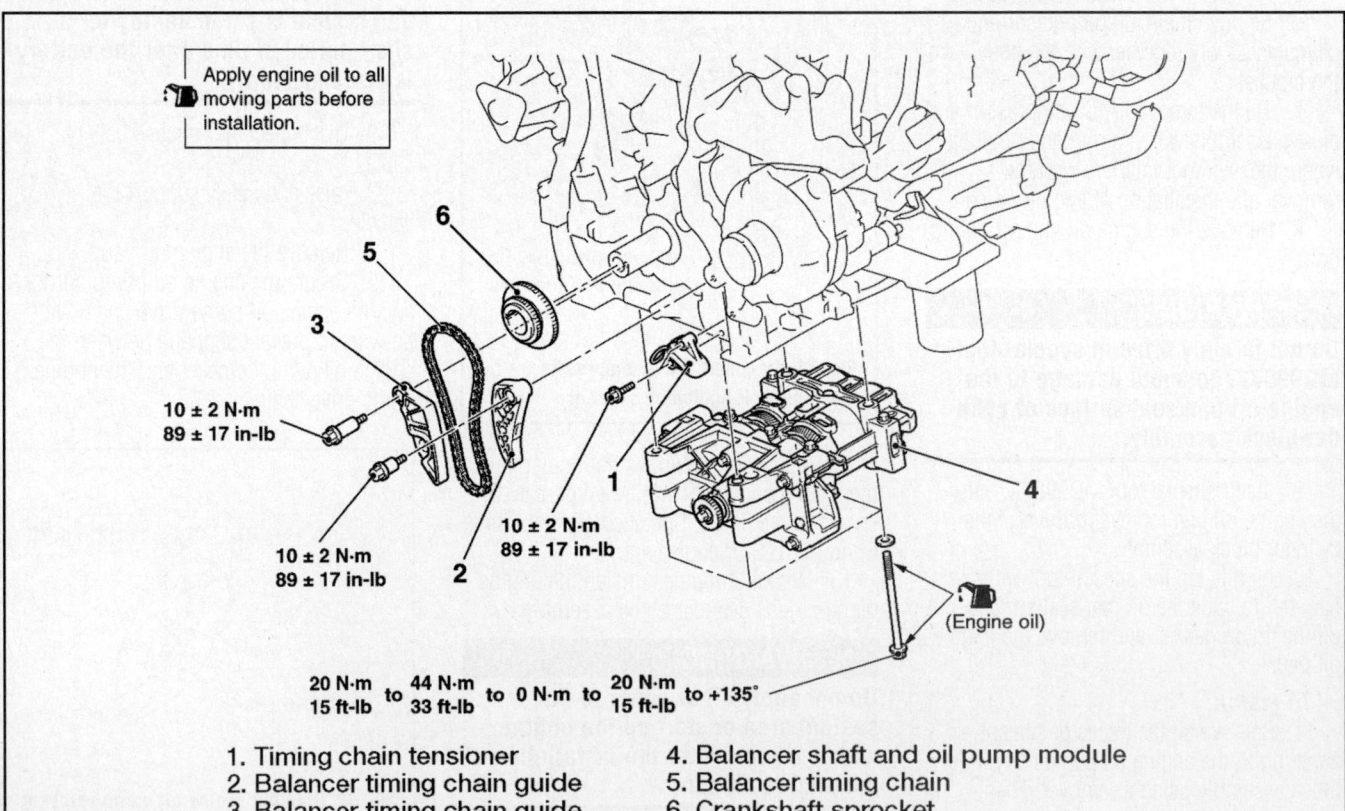

Apply engine oil to all moving parts before installation.

10 ± 2 N·m
89 ± 17 in-lb

10 ± 2 N·m
89 ± 17 in-lb

10 ± 2 N·m
89 ± 17 in-lb

(Engine oil)

20 N·m   to 44 N·m   to 0 N·m to 20 N·m   to +135°
15 ft-lb   33 ft-lb   15 ft-lb

1. Timing chain tensioner
2. Balancer timing chain guide
3. Balancer timing chain guide
4. Balancer shaft and oil pump module
5. Balancer timing chain
6. Crankshaft sprocket

22140_OUTL_G0136

Fig. 74 Balance shaft timing chain and module

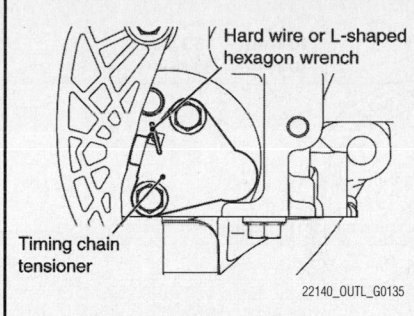

Fig. 75 Compress the plunger of the timing chain tensioner and insert hard wire or L-shaped hexagon wrench to fix the plunger of the timing chain tensioner

2. Remove the timing chain.

3. Remove the balance shaft timing chain tensioner.

a. Securely install the plunger of the timing chain tensioner. Otherwise, it may pop out.

b. Press the balancer timing chain against the timing chain tensioner, compress the plunger of the timing chain tensioner and insert hard wire (piano wire, etc.) or L-shaped hexagon wrench 0.05 inch (1.5 mm) to fix the plunger of the timing chain tensioner.

c. Remove the balance shaft timing chain tensioner.

4. Remove the balance shaft timing chain guides.

5. Remove the balance shaft and oil pump module.

6. Remove the balance shaft timing chain.

To install:

➡ **Be sure to use new fasteners, as required.**

7. Install the balance shaft and oil pump module.

a. When installing the new balancer shaft and oil pump module, apply oil to the oil pump in the balancer shaft and oil pump module and the balancer shaft bearing as follows:

b. Clean the inside of the removed engine oil pan, and put the balancer shaft and oil pump module into the engine oil pan with its oil inlet port facing up.

c. Pour new engine oil until two-thirds of the balancer shaft and oil pump module is soaked.

d. Fill the engine oil (approximately 3.05 cu. in. [50 cm3]) into the balancer shaft and oil pump module from the oil inlet port.

e. Turn the balancer shaft sprocket of

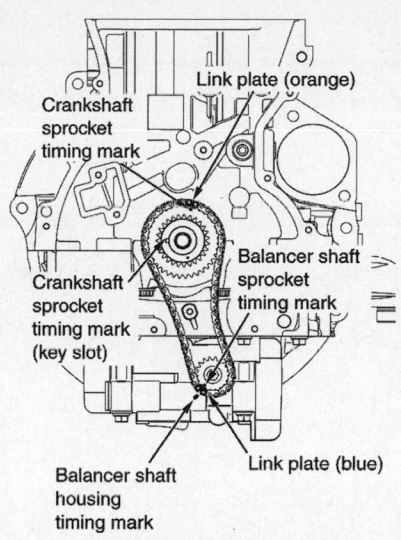

Fig. 76 With the link marks (orange or blue) of balancer timing chain aligned with the timing marks of balancer sprocket and crankshaft sprocket, install the balancer shaft and oil pump module together with the balancer timing chain and crankshaft sprocket as one unit to the cylinder block

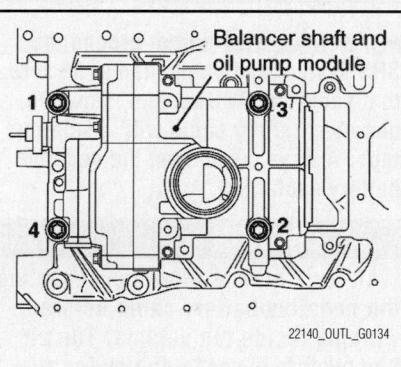

Fig. 77 Tighten the balancer shaft and oil pump module bolts to the specified torque in the order shown

the balancer shaft and oil pump module clockwise four rotations or more to apply the engine oil to the entire area of the oil pump and the balancer shaft bearing.

f. With the link marks (orange or blue) of balancer timing chain aligned with the timing marks of balancer sprocket and crankshaft sprocket, install the balancer shaft and oil pump module together with the balancer timing chain and crankshaft sprocket as one unit to the cylinder block. At this time, securely bring the balancer shaft and oil pump module into contact with the rudder frame mounting area.

g. Apply an adcquatc and minimum amount of engine oil to the threads and bearing surfaces of the balancer shaft and oil pump module bolts.

h. Tighten the balancer shaft and oil pump module bolts to the specified torque in the order shown:

• Tighten to 15 ft. lbs. (20 Nm)
• Retighten to 15 ft. lbs. (20 Nm)
• Loosen all bolts in the reverse sequence fully
• Tighten to 15 ft. lbs. (20 Nm)
• Tighten to 15 ft. lbs. (20 Nm)

8. Install the balance shaft timing chain guides.

9. Install the timing chain tensioner.

a. Install the timing chain tensioner to the cylinder block.

b. Remove the hard wire or L-shaped hexagon wrench fixing the plunger of the timing chain tensioner to apply tension to the balancer timing chain.

10. Install the timing chain.

PISTON AND RING

POSITIONING

See Figure 78.

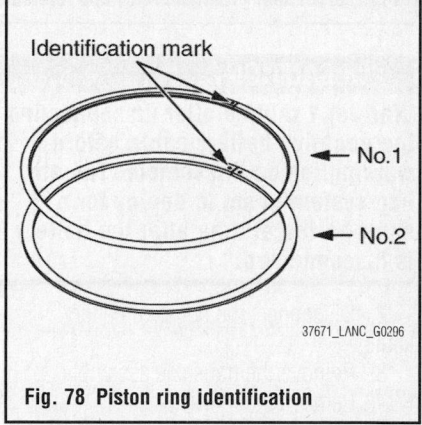

Fig. 78 Piston ring identification

REAR MAIN SEAL

REMOVAL & INSTALLATION

2.0L Non Turbocharged Engine and 2.4L Engine

See Figure 79.

1. Before servicing the vehicle, refer to the Precautions Section.

➡ **If working near and/or around the SRS system and components, be sure to disable the SRS system. Tape the negative battery cable with insulating tape. Always disconnect the negative battery cable first.**

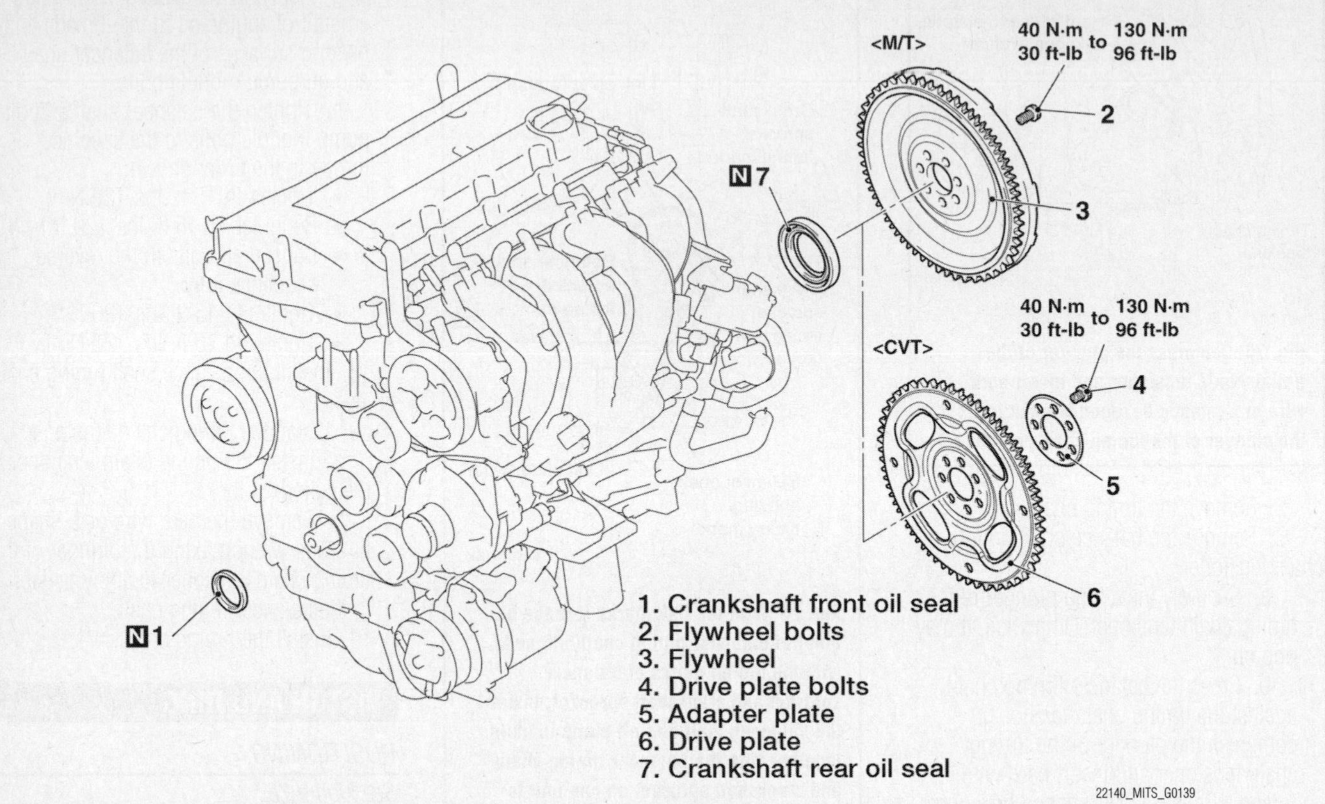

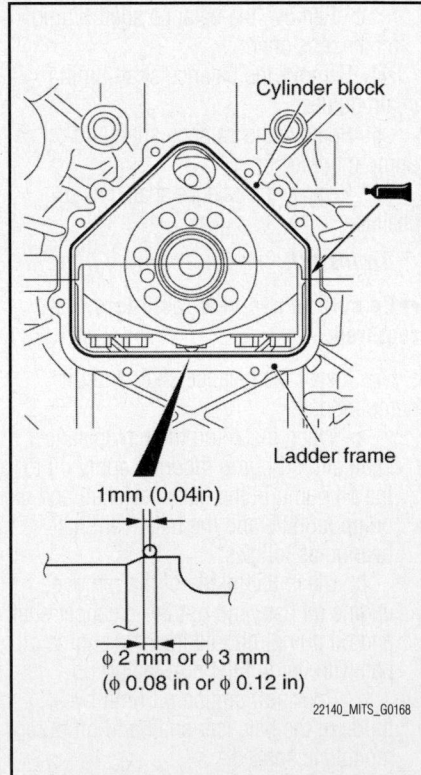

1. Crankshaft front oil seal
2. Flywheel bolts
3. Flywheel
4. Drive plate bolts
5. Adapter plate
6. Drive plate
7. Crankshaft rear oil seal

22140_MITS_G0139

Fig. 79 Oil seal (front and rear) and related components—except turbocharged engine

⁂ CAUTION

Wait for 1 minute after disconnecting the negative battery cable before working inside the vehicle. The air bag system is set to deploy for a short period of time after the battery is disconnected.

2. Disconnect the negative battery cable.
3. Remove the transaxle assembly.
4. Remove the flywheel.
5. Remove the crankshaft rear oil seal.

To install:

➡**Be sure to use new fasteners, as required.**

6. Apply a small amount of engine oil to the entire circumference of the rear oil seal lip.
7. Use special tool MD998718 to tap in the oil seal.
8. Installation continues in the reverse of the removal procedure.

2.0L Turbocharged Engine

See Figure 80.

1. Before servicing the vehicle, refer to the Precautions Section.

➡If working near and/or around the SRS system and components, be sure to disable the SRS system. Tape the negative battery cable with insulating tape. Always disconnect the negative battery cable first.

⁂ CAUTION

Wait for 1 minute after disconnecting the negative battery cable before working inside the vehicle. The air bag system is set to deploy for a short period of time after the battery is disconnected.

2. Disconnect the negative battery cable.
3. Remove the transfer assembly.
4. Remove the transaxle assembly.
5. Remove the flywheel.
6. Remove the crankshaft rear main seal case assembly.

To install:

➡**Be sure to use new fasteners, as required.**

7. Remove all the traces of sealant adhering to the cylinder block and ladder frame using a remover. Degrease using quick-drying degreasing agent (white gasoline).

Cylinder block

Ladder frame

1mm (0.04in)

φ 2 mm or φ 3 mm
(φ 0.08 in or φ 0.12 in)

22140_MITS_G0168

Fig. 80 Apply the sealant without any gap to the cylinder block and ladder frame—2.0L turbocharged engine

8. Apply the sealant without any gap to the cylinder block and ladder frame as shown. Within 3 minutes, install the crankshaft rear oil seal case assembly. Specified sealant: Three Bond® 1227D or equivalent.

9. Apply a small amount of engine oil to the entire inner diameter of the oil seal lip, and install the crankshaft rear oil seal case assembly.

⁑ WARNING

Do not apply oil or water to the sealant-applied area or start the engine within 2 hours after the installation of the crankshaft rear oil seal.

10. Tighten the crankshaft rear oil seal case assembly mounting bolts to the specified torque. Tightening torque: 72– 106 inch lbs. (8–12 Nm).

11. Installation continues in the reverse of the removal procedure.

ROCKER ARMS

REMOVAL & INSTALLATION

These engines are not equipped with rocker arms. The camshafts act directly on the valves through the valve tappets.

TIMING CHAIN FRONT COVER

REMOVAL & INSTALLATION

2.0L Non Turbocharged Engine and 2.4L Engine

See Figures 81 through 83.

1. Before servicing the vehicle, refer to the Precautions Section.

➥ If working near and/or around the SRS system and components, be sure to disable the SRS system. Tape the negative battery cable with insulating tape. Always disconnect the negative battery cable first.

⁑ CAUTION

Wait for 1 minute after disconnecting the negative battery cable before working inside the vehicle. The air bag system is set to deploy for a short period of time after the battery is disconnected.

2. Remove the front engine undercover and side covers (RH).
3. Drain the engine oil.
4. Remove the rocker cover assembly.
5. Remove the oil pan.

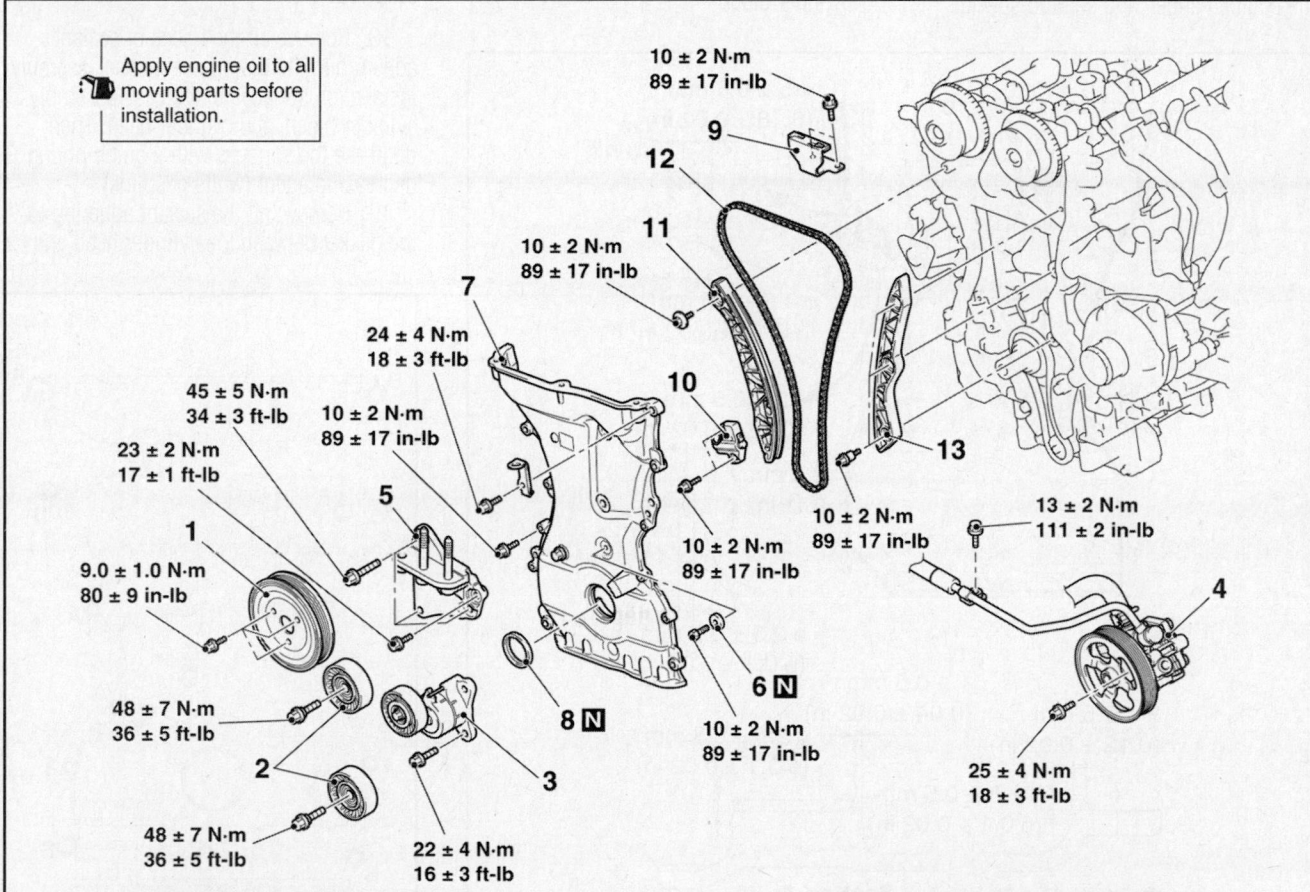

Apply engine oil to all moving parts before installation.

1. Water pump pulley
2. Idler pulley
3. Autotensioner
4. Power steering oil pump assembly
5. Cylinder block engine front mounting bracket
6. Gasket
7. Timing chain case assembly
8. Crankshaft front oil seal
9. Timing chain upper guide
10. Timing chain tensioner
11. Timing chain tension side guide
12. Timing chain
13. Timing chain loose side guide

22140_MITS_G0240

Fig. 81 Timing chain cover and related components

6. Remove the crankshaft damper pulley.

7. Remove the water pump pulley.

8. Remove the idler pulley.

9. Remove the auto tensioner.

10. Remove the power steering oil pump assembly.

11. Install a special tool, engine hanger MB991928, for holding the engine and transaxle assembly.

   a. Assemble the engine hanger. Set the following parts on the base hanger.

- Slide bracket (HI)
- Foot x 4 (standard) (MB991932)
- Joint x 2 (90) (MB991930)

   b. Set the foot of the special tool to the strut mounts and front bumper brace.

   c. Move the slide bracket (HI) to adjust the engine hanger balance.

   d. Mount special tool MB991454 to the power steering oil pump bracket and the engine hanger, and set it to special

tool MB991928 to support the engine and transaxle assembly.

12. Remove the engine mounting insulator.

13. Remove the cylinder block engine front mounting bracket and gasket.

14. Remove the timing chain case assembly.

✳✳ WARNING

The adhesive strength of the sealant on the timing chain case assembly may be so strong that the boss may be damaged by peeling it off. Do not peel it off forcibly.

   a. After removing the timing chain case assembly mounting bolts, slightly pry the boss of the timing chain case assembly using a flat-tipped tool

   b. Remove the timing chain case assembly from the cylinder head and cylinder block.

   c. If the sealant cannot be peeled off easily, insert a wooden hammer shank into the timing chain case assembly, pry slightly, and remove the timing chain case assembly from the cylinder head and cylinder block.

15. Remove the crankshaft front oil seal.

To install:

➡ **Be sure to use new fasteners, as required.**

16. Apply a small amount of engine oil to the entire inner diameter of the crankshaft front oil seal lip.

17. Using special tool MB991448, press in the crankshaft front oil seal up to the chamfered surface of timing chain case.

➡ **Be sure to remove the sealant remaining in the mounting hole, O-ring groove, and the gap between parts. Do not touch the degreased area once cleaned.**

18. Remove all the traces of sealant adhering to the timing chain case assembly installation surfaces of the case assembly, cylinder block, and cylinder head. Then, degrease the surfaces with a quick-drying degreasing agent (white gasoline).

19. Remove all the sealant adhering to the gasket between the cylinder head and

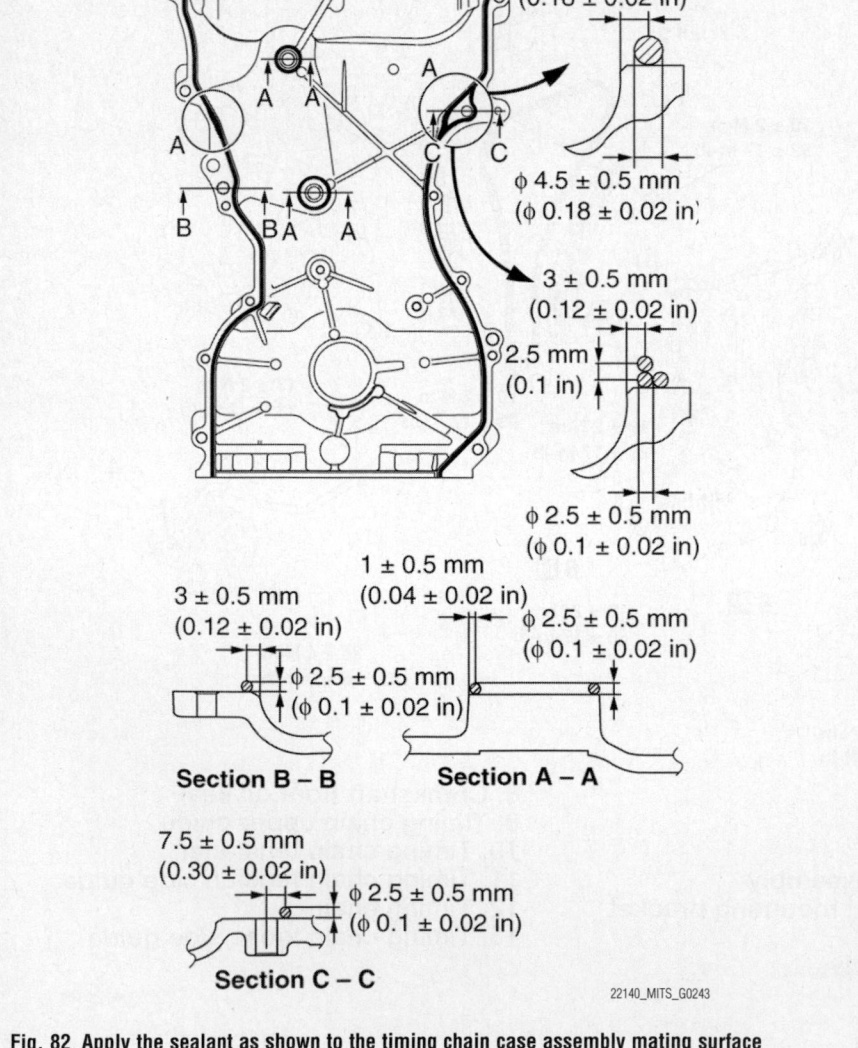

Fig. 82 Apply the sealant as shown to the timing chain case assembly mating surface

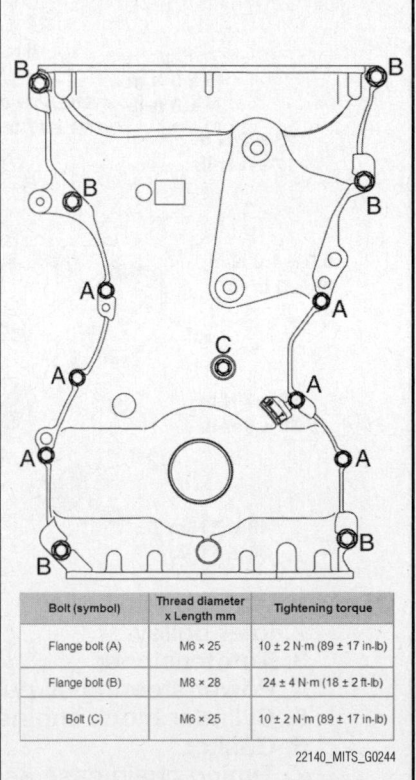

Bolt (symbol)	Thread diameter x Length mm	Tightening torque
Flange bolt (A)	M6 × 25	10 ± 2 N·m (89 ± 17 in-lb)
Flange bolt (B)	M8 × 28	24 ± 4 N·m (18 ± 2 ft-lb)
Bolt (C)	M6 × 25	10 ± 2 N·m (89 ± 17 in-lb)

22140_MITS_G0244

Fig. 83 Timing chain torque sequence and bolt tightening specifications

cylinder block (3-surface aligned part). Then, degrease the surfaces with the quick-drying degreasing agent (white gasoline).

➡ **The engine oil may leak from the cylinder head gasket. Thus, quickly apply the sealant to it after degreasing.**

20. Apply a bead of the sealant to the timing chain case assembly mounting surface. The bead diameter should be 0.08–0.12 inch (2–3mm). Overlap part "A" with the diameter of 0.16–0.20 inch (4–5mm) or 0.08–0.12 inch (2–3mm) as shown in the figure, and apply the sealant. Specified sealant: Three Bond® 1217G or equivalent.

21. If the sealant contacts any other part during installation of the timing chain case assembly, apply sealant again before installing the timing chain case assembly.

➡ **Do not apply oil or water to the sealant applied area or start up the engine within 2 hours after the installation of the timing chain case assembly.**

22. Install the timing chain case assembly to the cylinder block and cylinder head so that the sealant does not contact other parts. Install the timing chain case assembly within 3 minutes after the application of sealant.

23. Insert the bolts to the timing chain case assembly as shown, and tighten them to the specified torque.

24. Continue the installation in the reverse order of the removal procedure.

2.0L Turbocharged Engine
See Figures 106 through 108.

1. Before servicing the vehicle, refer to the Precautions Section.

➡ **If working near and/or around the SRS system and components, be sure to disable the SRS system. Tape the negative battery cable with insulating tape. Always disconnect the negative battery cable first.**

✳✳ **CAUTION**

Wait for 1 minute after disconnecting the negative battery cable before working inside the vehicle. The air bag system is set to deploy for a short period of time after the battery is disconnected.

2. Remove the front engine undercover and side covers (RH).

3. Remove the strut tower bar.
4. Drain the engine oil.
5. Remove the rocker cover assembly.
6. Remove the oil pan.
7. Remove the power steering oil pump assembly.
8. Remove the headlight support panel cover.
9. Remove the engine mounting bracket.
10. Remove the crankshaft damper pulley.
11. Remove the water pump pulley.
12. Remove the idler pulley.
13. Remove the auto tensioner.
14. Install a special tool, engine hanger MB991928, for holding the engine and transaxle assembly.

   a. Assemble the engine hanger. Set the following parts on the base hanger.
   • Slide bracket (HI)
   • Foot x 4 (standard) (MB991932)
   • Joint x 2 (90) (MB991930)
   b. Set the foot of the special tool to the strut mounts and front bumper brace.
   c. Move the slide bracket (HI) to adjust the engine hanger balance.
   d. Mount special tool MB991454 to the power steering oil pump bracket and the engine hanger, and set it to special tool MB991928 to support the engine and transaxle assembly.

15. Remove the engine mounting insulator.
16. Remove the cylinder block engine front mounting bracket and gasket.
17. Remove the timing chain case assembly.

✳✳ **WARNING**

The adhesive strength of the sealant on the timing chain case assembly may be so strong that the boss may be damaged by peeling it off. Do not peel it off forcibly.

   a. After removing the timing chain case assembly mounting bolts, slightly pry the boss of the timing chain case assembly using a flat-tipped tool
   b. Remove the timing chain case assembly from the cylinder head and cylinder block.
   c. If the sealant cannot be peeled off easily, insert a wooden hammer shank into the timing chain case assembly, pry slightly, and remove the timing chain case assembly from the cylinder head and cylinder block.
18. Remove the crankshaft front oil seal.

To install:

➡ **Be sure to use new fasteners, as required.**

19. Apply a small amount of engine oil to the entire inner diameter of the crankshaft front oil seal lip.

20. Using special tool MB991448, press in the crankshaft front oil seal up to the chamfered surface of timing chain case.

➡ **Be sure to remove the sealant remaining in the mounting hole, O-ring groove, and the gap between parts. Do not touch the degreased area once cleaned.**

21. Remove all the traces of sealant adhering to the timing chain case assembly installation surfaces of the case assembly, cylinder block, and cylinder head. Then, degrease the surfaces with a quick-drying degreasing agent (white gasoline).

22. Remove all the sealant adhering to the gasket between the cylinder head and cylinder block (3-surface aligned part). Then, degrease the surfaces with the quick-drying degreasing agent (white gasoline).

➡ **The engine oil may leak from the cylinder head gasket. Thus, quickly apply the sealant to it after degreasing.**

23. Apply a bead of the sealant to the timing chain case assembly mounting surface. The bead diameter should be 0.08–0.12 inch (2–3mm). Overlap part "A" with the diameter of 0.16–0.20 inch (4–5mm) or 0.08–0.12 inch (2–3mm) as shown in the figure, and apply the sealant. Specified sealant: Three Bond® 1217G or equivalent.

24. If the sealant contacts any other part during installation of the timing chain case assembly, apply sealant again before installing the timing chain case assembly.

➡ **Do not apply oil or water to the sealant applied area or start up the engine within 2 hours after the installation of the timing chain case assembly.**

25. Install the timing chain case assembly to the cylinder block and cylinder head so that the sealant does not contact other parts. Install the timing chain case assembly within 3 minutes after the application of sealant.

26. Insert the bolts to the timing chain case assembly, and tighten them to the specified torque.

27. Continue the installation in the reverse order of the removal procedure.

TIMING CHAIN & SPROCKETS

REMOVAL & INSTALLATION
See Figures 84 through 92.

1. Before servicing the vehicle, refer to the Precautions Section.

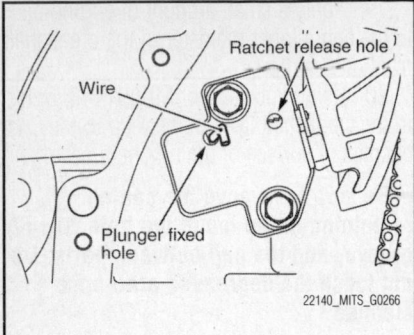

Fig. 84 Insert a hard wire or hexagonal bar wrench into the fixing hole of the plunger to fix the tensioner in place

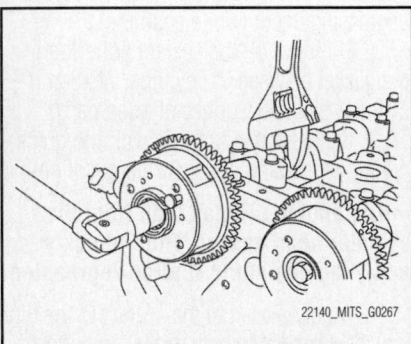

Fig. 85 Remove the exhaust Variable Valve Timing (VVT) Sprocket bolt

➡If working near and/or around the SRS system and components, be sure to disable the SRS system. Tape the negative battery cable with insulating tape. Always disconnect the negative battery cable first.

❊❊ CAUTION

Wait for 1 minute after disconnecting the negative battery cable before working inside the vehicle. The air bag system is set to deploy for a short period of time after the battery is disconnected.

2. Remove the timing chain cover and crankshaft oil seal.

3. Remove the timing chain upper guide (1).

4. Insert a flat blade tool into the release hole of the timing chain tensioner to release the latch.

5. Push the tensioner lever by hand and push in the plunger of the timing chain tensioner until it hits the bottom. Then, insert a hard wire or hexagonal bar wrench 0.05 inch (1.5mm) into the fixing hole of the plunger.

6. Remove the timing chain tensioner (2).

7. Remove the timing chain lever (4).

8. Remove the timing chain (5).

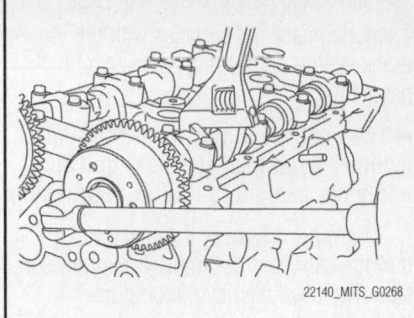

Fig. 86 Remove the intake Variable Valve Timing (VVT) Sprocket bolt

9. Remove the chain oil jet (6).

10. Remove the exhaust Variable Valve Timing (VVT) sprocket assembly.

a. Hold the hexagonal portion of the exhaust camshaft with a wrench and loosen the exhaust VVT sprocket bolt.

b. Remove the exhaust VVT sprocket assembly.

11. Remove the intake VVT sprocket assembly.

a. Hold the hexagonal portion of the intake camshaft with a wrench and loosen the intake VVT sprocket bolt.

b. Remove the intake VVT sprocket assembly.

Apply engine oil to all moving parts before installation.

85 ± 5 N·m
63 ± 3 ft-lb

10 ± 2 N·m
89 ± 17 in-lb

10 ± 2 N·m
89 ± 17 in-lb

10 ± 2 N·m
89 ± 17 in-lb

10 ± 2 N·m
89 ± 17 in-lb

85 ± 5 N·m
63 ± 3 ft-lb

1. Chain upper guide
2. Timing chain tensioner
3. Tensioner lever
4. Timing chain guide
5. Timing chain
6. Chain oil jet
7. Exhaust V.V.T. sprocket bolt
8. Exhaust V.V.T. sprocket assembly
9. Intake V.V.T. sprocket bolt
10. Intake V.V.T. sprocket assembly

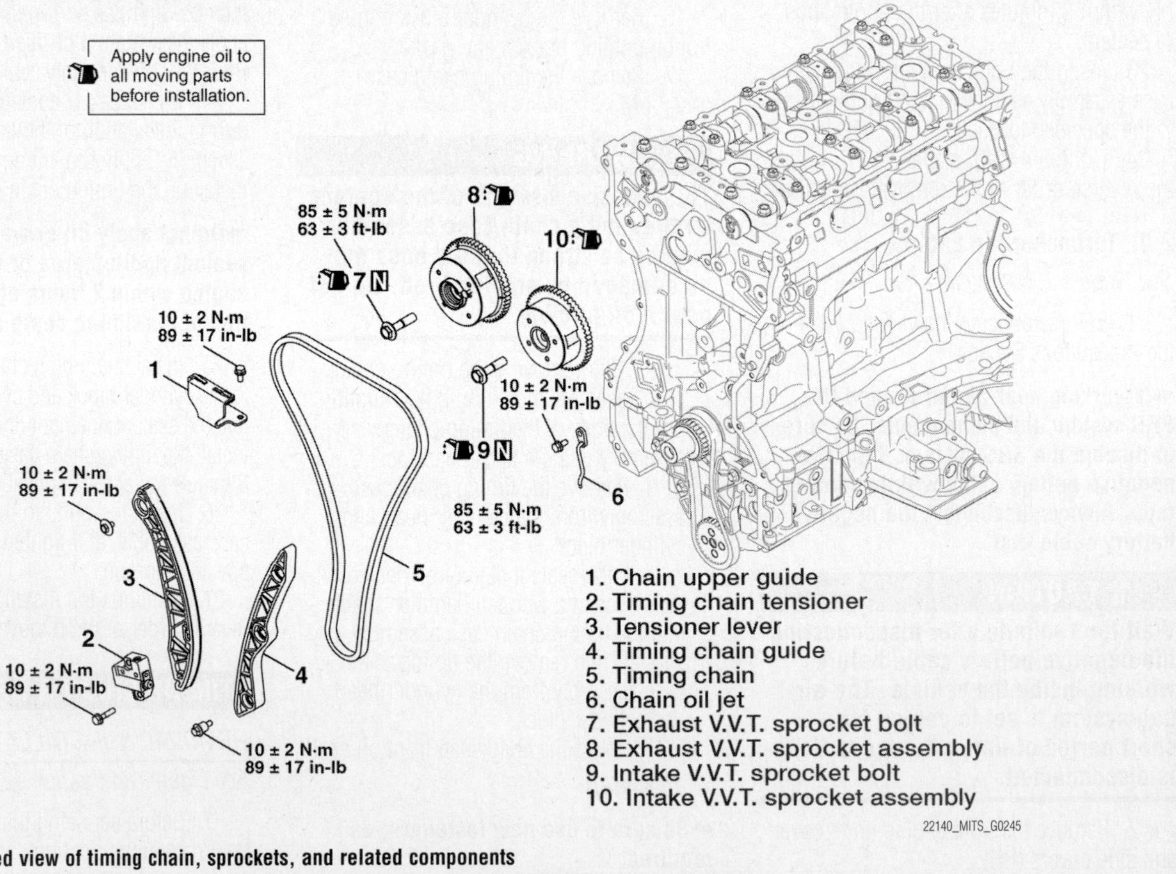

Fig. 87 Exploded view of timing chain, sprockets, and related components

To install:

➡ **Be sure to use new fasteners, as required.**

12. Assemble the intake VVT sprocket assembly in the following procedure.

a. Make sure that the knock pin of the inlet camshaft assembly is positioned facing straight upward.

b. Apply an appropriate and minimum amount of engine oil to the circumference of the tip of the intake VVT sprocket assembly and the entire circumference of the area into which the intake VVT sprocket assembly is inserted.

c. Slowly insert the intake VVT sprocket assembly into the normal position of the inlet camshaft assembly with its knock pin hole facing straight upward.

d. Install the VVT sprocket.

e. Make sure that the VVT sprocket is securely inserted into the bottom and that the VVT sprocket does not rotate with the hexagonal portion of the camshaft secured with a wrench.

f. Hold the hexagonal portion of the camshaft with a wrench and tighten the intake VVT sprocket bolt to 60–66 ft. lbs. (80–90 Nm).

13. Assemble the exhaust VVT sprocket assembly in the following procedure.

a. Make sure that the knock pin of the exhaust camshaft assembly is positioned facing straight upward.

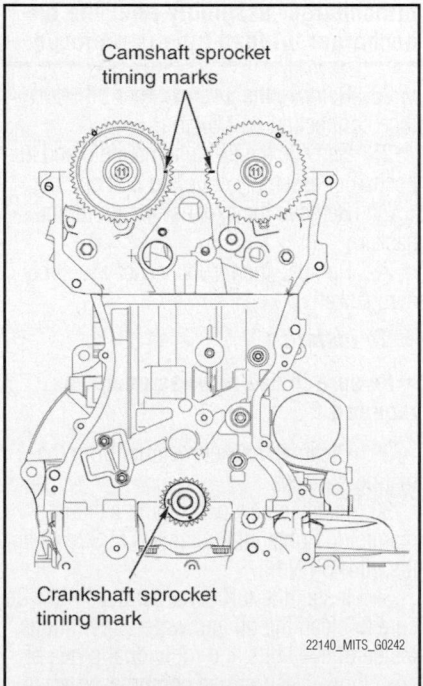

Fig. 88 Set the timing marks of the camshaft sprockets and the crankshaft sprocket as shown

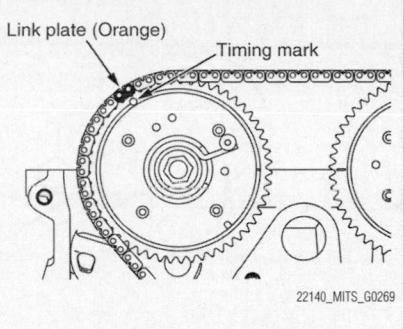

Fig. 89 Align the link plate (Orange) with the timing mark of the exhaust VVT sprocket and loop the timing chain

b. Apply an appropriate and minimum amount of engine oil to the circumference of the tip of the exhaust VVT sprocket assembly and the entire circumference of the area into which the exhaust VVT sprocket assembly is inserted.

c. Slowly insert the exhaust VVT sprocket assembly into the normal position of the exhaust camshaft assembly with its knock pin hole facing straight upward.

d. Install the VVT sprocket.

e. Make sure that the VVT sprocket is securely inserted into the bottom and that the VVT sprocket does not rotate with the hexagonal portion of the camshaft secured with a wrench.

f. Hold the hexagonal portion of the camshaft with a wrench and tighten the camshaft sprocket bolt to 60–66 ft. lbs. (80–90 Nm).

14. Align the timing marks of the VVT sprockets and the crankshaft sprocket key as illustrated.

15. Align the link plate (Orange) with the timing mark of the exhaust VVT sprocket and loop the timing chain.

16. Align the link plate (Blue) with the timing mark of the intake VVT sprocket to loop the timing chain. Rotate the intake VVT

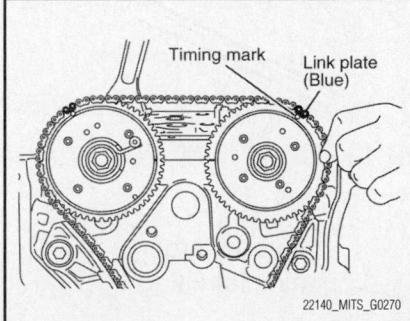

Fig. 90 Align the link plate (Blue) with the timing mark of the intake VVT sprocket to loop the timing chain

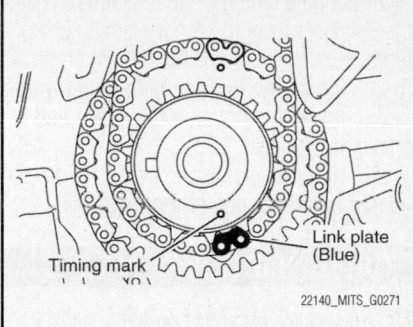

Fig. 91 Align the timing mark of the crankshaft sprocket with the link plate (Blue) to loop the timing chain

sprocket by 1 or 2 teeth to align with the timing mark.

17. Align the timing mark of the crankshaft sprocket with the link plate (Blue) to loop the timing chain. Because of the timing chain slacks, hold it to prevent the timing mark from coming off the link plate (Blue).

18. Make sure that the timing mark of each sprocket is aligned with the link plate (blue) of the timing chain at all 3 locations.

19. Install the timing chain guide.

20. Install the timing chain tensioner on the cylinder block and tighten it to the specified torque. Specified torque: 72–106 inch lbs. (8–12 Nm).

21. Remove the hard wire or hexagonal bar wrench 0.05 inch (1.5mm) from the timing chain tensioner. This enables the

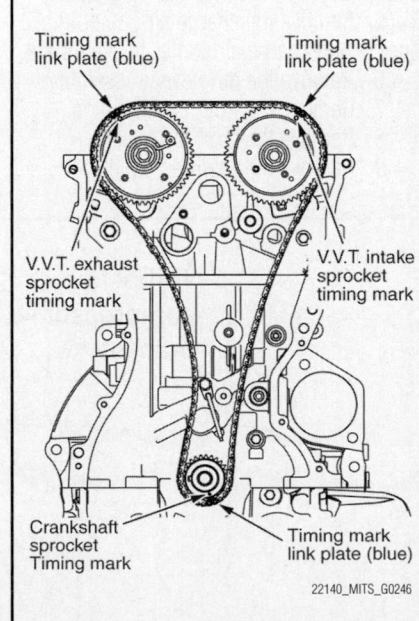

Fig. 92 Make sure that the timing mark of each sprocket is aligned with the link plate (blue) of the timing chain at all 3 locations

plunger of the timing chain tensioner to push the tensioner lever to keep the timing chain tight.

22. Install the timing chain upper guide.

23. Install the timing chain cover and crankshaft oil seal.

24. Run the engine and check for proper operation and for any oil leakage.

TURBOCHARGER

REMOVAL & INSTALLATION

See Figures 93 through 98.

1. Before servicing the vehicle, refer to the Precautions Section.

➡**If working near and/or around the SRS system and components, be sure to disable the SRS system. Tape the negative battery cable with insulating tape. Always disconnect the negative battery cable first.**

❊❊ CAUTION

Wait for 1 minute after disconnecting the negative battery cable before working inside the vehicle. The air bag system is set to deploy for a short period of time after the battery is disconnected.

2. Remove the engine room undercover and side cover.

3. Drain the engine oil and coolant.

4. Remove the engine upper cover and the ignition coil assemblies.

5. Remove the charge air cooler intake hose A and charge air cooler intake pipe A.

6. Remove the air cleaner assembly.

7. Remove the front exhaust pipe.

8. Remove the strut tower bar.

9. Remove the cowl top panel.

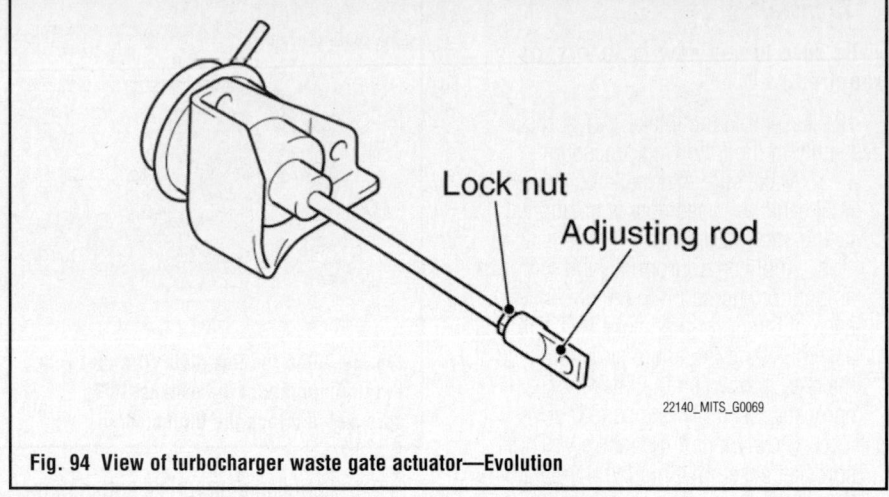

Fig. 94 View of turbocharger waste gate actuator—Evolution

10. Remove the dash panel heat protector.

11. Remove the exhaust manifold cover.

12. Remove the emission vacuum control hose connection.

13. Remove the turbocharger air outlet fitting and gasket.

14. Remove the turbocharger compressor bracket.

➡**Use special MB992274 to remove the turbocharger compressor bracket mounting bolt (cylinder block side).**

15. Remove the turbocharger air inlet fitting and gasket.

16. Remove the turbocharger water feed hose, water feed pipe, and gasket.

17. Remove the turbocharger water return hose.

18. Remove the turbocharger exhaust outlet fitting bracket.

19. Remove the transfer heat protector and the drive shaft heat protector.

20. Remove the steering gear and linkage heat protector, the turbocharger protectors A and B.

21. Remove the turbocharger pin and the waste gate actuator.

➡**Do not loosen the locking nuts and adjusting rod of the waste gate actuator.**

22. Remove the turbocharger exhaust outlet fitting and gasket.

23. Remove the turbocharger bracket and turbocharger assembly coupling bolt.

24. Remove the turbocharger oil feed tube connection and gasket.

❊❊ WARNING

Take care not to allow foreign objects to get into the oil passage hole of the turbocharger assembly after the turbocharger oil feed tube is removed.

25. Remove the turbocharger oil return tube connection and O-ring.

26. Remove the exhaust manifold and turbocharger assembly coupling bolt and nut.

27. Remove the exhaust manifold and gasket.

28. Remove the turbocharger assembly and gasket.

To install:

➡**Be sure to use new fasteners, as required.**

29. Clean all gasket material from the mating surfaces.

30. Install a new gasket and exhaust manifold. Tighten the retainers to 32–40 ft. lbs. (44–54 Nm).

31. Install the turbocharger assembly. Be sure to clean the oil and water pipe fittings, inside of eye bolts and individual pipes of any clogs. Clean or use compressed air to remove any carbon matter stuck in the oil passage of the turbocharger. Refill new engine oil at the oil feed pipe fitting hole.

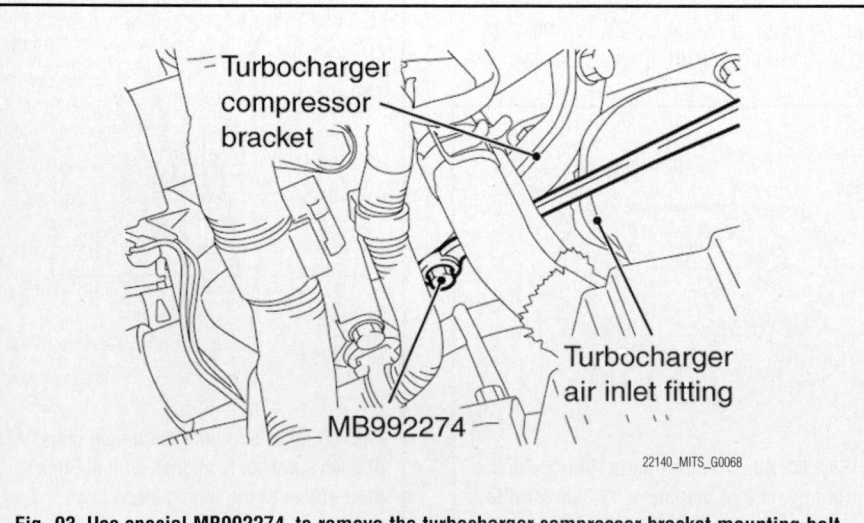

Fig. 93 Use special MB992274 to remove the turbocharger compressor bracket mounting bolt

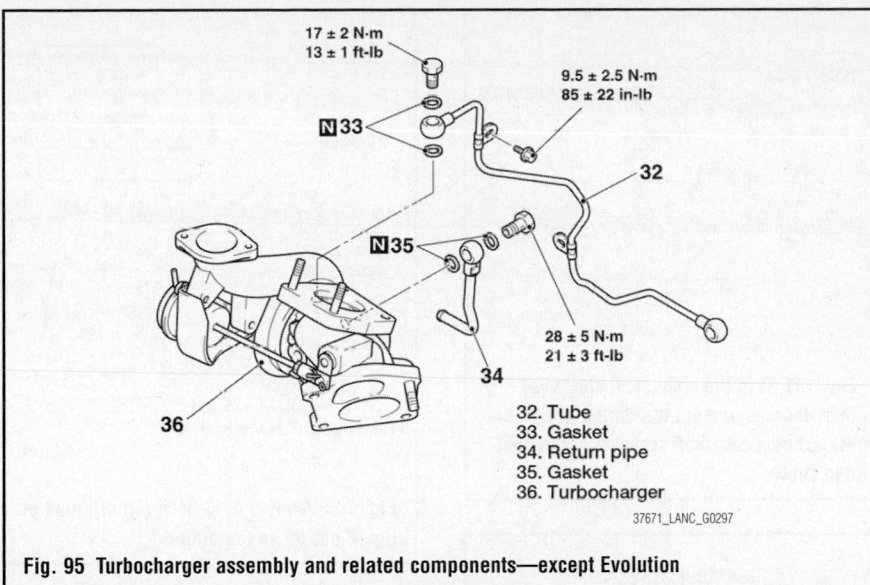

Fig. 95 Turbocharger assembly and related components—except Evolution

32. Tube
33. Gasket
34. Return pipe
35. Gasket
36. Turbocharger

37671_LANC_G0297

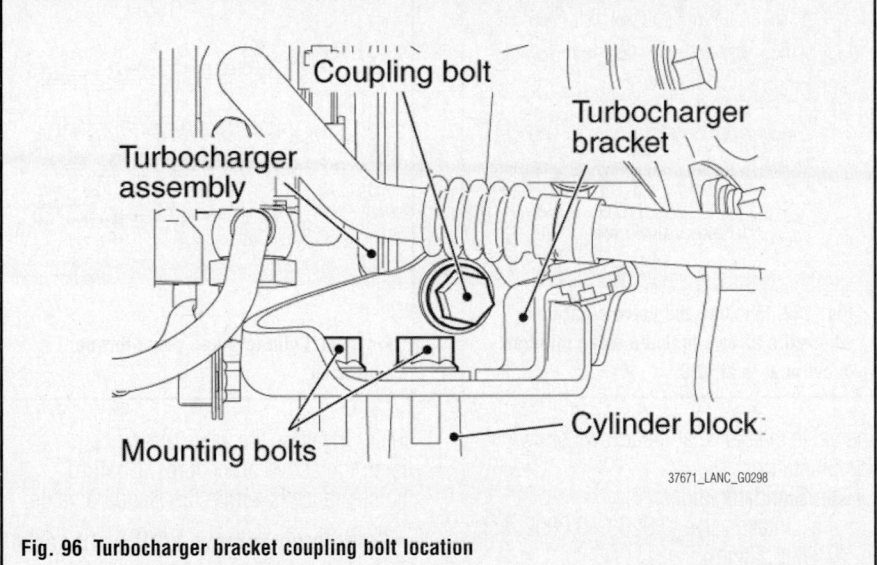

Fig. 96 Turbocharger bracket coupling bolt location

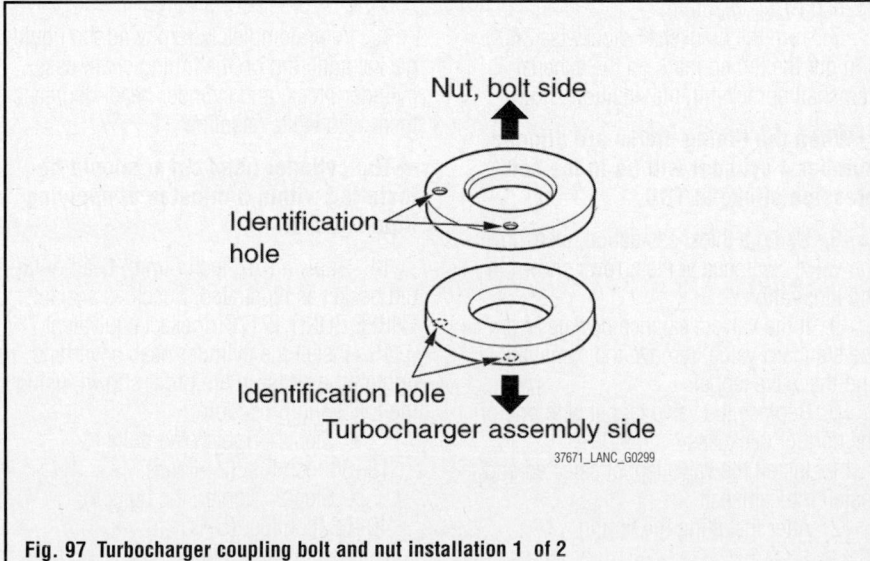

Fig. 97 Turbocharger coupling bolt and nut installation 1 of 2

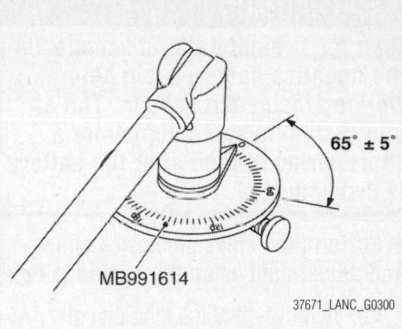

Fig. 98 Turbocharger coupling bolt and nut installation 2 of 2

32. Install the exhaust fitting assembly, along with a new gasket.

33. Install the exhaust fitting bracket.

34. Install the air outlet fitting and gasket.

35. Connect the vacuum hose connections.

36. Connect the turbocharger oil feed and oil return pipes, along with new gaskets. Tighten the oil feed pipe fastener to 10–14 ft. lbs. (15–19 Nm). Tighten the oil return pipe fasteners to 71–89 inch lbs. (8–10 Nm).

37. Connect the turbocharger water feed pipe and return hoses, along with new gaskets. Tighten the pipe fasteners to 26–36 ft. lbs. (35–49 Nm).

38. Install the turbocharger heat protector.

39. Install the front Heated Oxygen (HO2S) sensor.

40. Install the exhaust manifold cover.

41. Install the front exhaust pipe to the exhaust manifold.

42. Install the air pipes and hoses to the charge air cooler assembly.

43. Install the air intake hose to the air cleaner assembly.

44. Install the front undercover and secure with the retaining clips.

45. Refill the cooling system and the engine oil with the correct amount and type of fluid.

46. Connect the negative battery cable.

VALVE LASH

ADJUSTMENT

See Figures 99 through 104.

1. Before servicing the vehicle, refer to the Precautions Section.

➡ If working near and/or around the SRS system and components, be sure to disable the SRS system. Tape the negative battery cable with insulating tape. Always disconnect the negative battery cable first.

✳✳ CAUTION

Wait for 1 minute after disconnecting the negative battery cable before working inside the vehicle. The air bag system is set to deploy for a short period of time after the battery is disconnected.

➡**Perform the valve clearance check and adjustment when the engine is cold.**

2. Remove all of the ignition coils.
3. Remove the cylinder head cover.

✳✳ WARNING

Always turn the crankshaft in a clockwise direction.

4. Turn the crankshaft clockwise, and align the timing mark on the exhaust camshaft sprocket against the upper face of the cylinder head as shown in the illustration.

➡**When the marks are aligned, number 1 cylinder will move to the compression stroke at Top Dead Center (TDC).**

5. Using a thickness gauge, measure

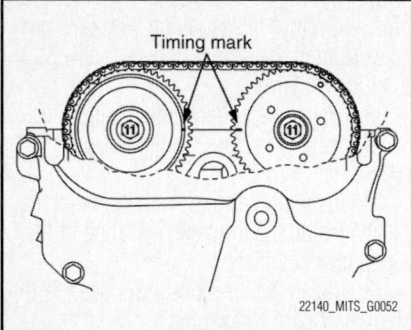

Fig. 99 Turn the crankshaft clockwise, and align the timing mark on the exhaust camshaft sprocket against the upper face of the cylinder head

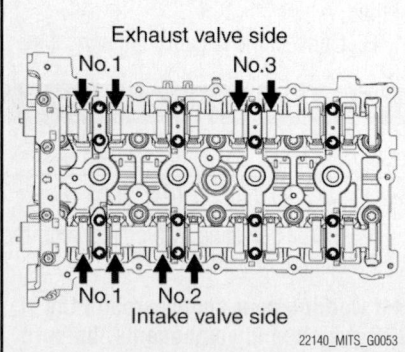

Fig. 100 Measure the valve clearance where the arrows indicate when number 1 cylinder is at TDC

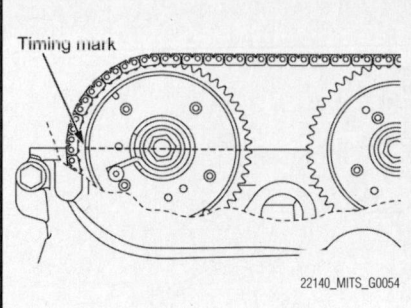

Fig. 101 Turn the crankshaft clockwise 360 degrees and put the timing mark on the exhaust camshaft sprocket in the position shown

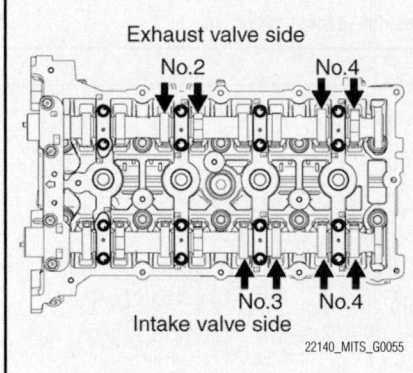

Fig. 102 Measure the valve clearance where the arrows indicate when number 4 cylinder is at TDC

the valve clearance at the arrows shown in the illustration.

6. Standard values:
    a. Intake valve: 0.0068– 0.0092 inch (0.17–0.23mm).
    b. Exhaust valve: 0.0108– 0.0132 inch (0.27–0.33mm).

7. Turn the crankshaft clockwise 360° and put the timing mark on the exhaust camshaft sprocket in the position shown.

➡**When the timing marks are aligned, number 4 cylinder will be in the compression stroke at TDC.**

8. Using a thickness gauge, measure the valve clearance at the arrows shown in the illustration.

9. If the valve clearance deviates from the standard value, remove the camshaft and the valve tappet.

10. Replace the valve tappet with one of the correct thickness.

11. Install the valve tappet selected and install the camshaft

12. After installing the timing

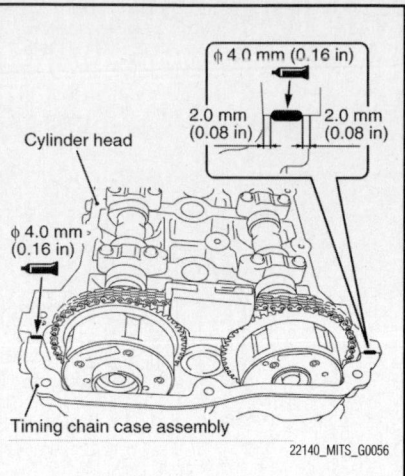

Fig. 103 Apply a 0.16 inch (4mm) bead of liquid gasket as illustrated

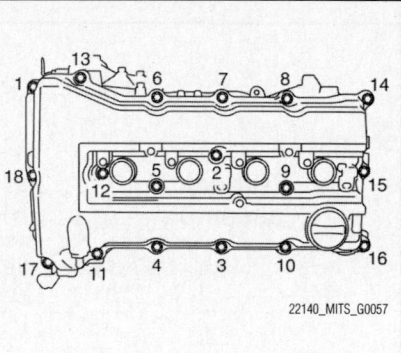

Fig. 104 Cylinder head cover torque sequence

chain, measure the valve clearance using the above procedures. Confirm the clearance is within the standard value.

➡**Completely remove all the old gasket material, which might be remaining among the components.**

13. After completely removing the liquid gasket adhering on the timing chain case, cylinder block, and cylinder head, degrease them with white gasoline.

➡**The cylinder head cover should be installed within 3 minutes of applying liquid gasket.**

14. Apply a 0.16 inch (4mm) bead of liquid gasket as illustrated. Specified sealant: THREE BOND 1217G or exact equivalent.

15. Install the cylinder head cover and tighten the bolts, in the order shown, using the following procedure:
    a. Step 1: Tighten the bolts to 18–36 inch lbs. (2–4 Nm).
    b. Step 2: Tighten the bolts to 40–58 inch lbs. (5–6 Nm).

16. Install the ignition coils.

ENGINE PERFORMANCE & EMISSION CONTROLS

CAMSHAFT POSITION (CMP) SENSOR

LOCATION

See Figures 105 and 106.

REMOVAL & INSTALLATION

See Figures 105 and 106.

1. Before servicing the vehicle, refer to the Precautions Section.

➡**If working near and/or around the SRS system and components, be sure to disable the SRS system. Tape the negative battery cable with insulating tape. Always disconnect the negative battery cable first.**

❄ CAUTION

Wait for 1 minute after disconnecting the negative battery cable before working inside the vehicle. The air bag system is set to deploy for a short period of time after the battery is disconnected.

2. Disconnect the negative battery cable.

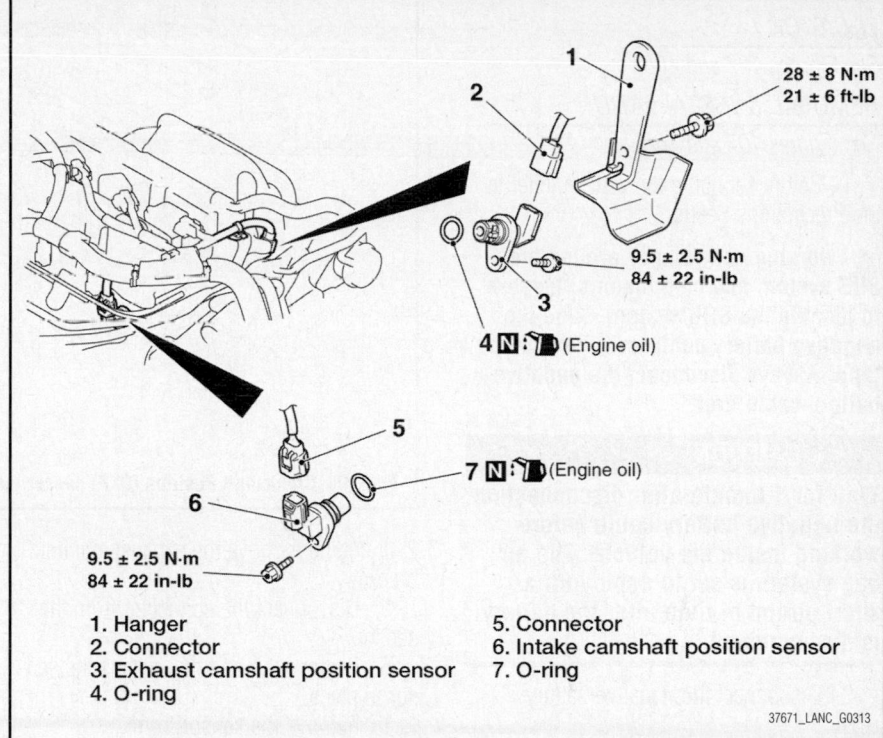

1. Hanger
2. Connector
3. Exhaust camshaft position sensor
4. O-ring
5. Connector
6. Intake camshaft position sensor
7. O-ring

37671_LANC_G0313

Fig. 106 Camshaft Position (CMP) sensor location—2.0L turbocharged engine

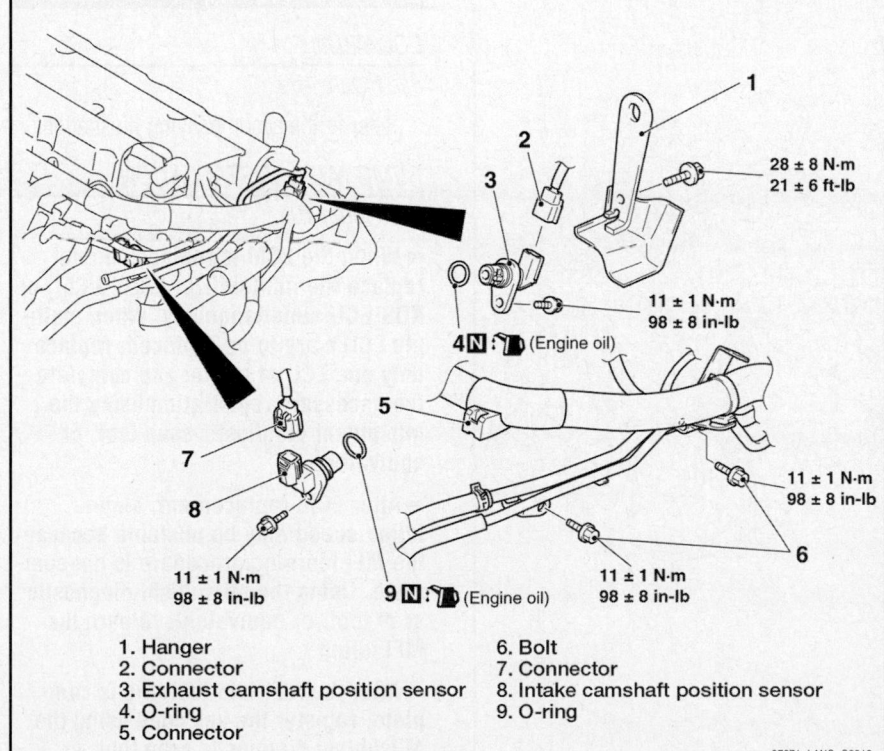

1. Hanger
2. Connector
3. Exhaust camshaft position sensor
4. O-ring
5. Connector
6. Bolt
7. Connector
8. Intake camshaft position sensor
9. O-ring

37671_LANC_G0312

Fig. 105 Camshaft Position (CMP) sensor location—2.0L non turbocharged engine and 2.4L engine

3. Remove the engine cover.

4. To remove the exhaust sensor, remove the engine hanger.

5. On 2.0L non turbocharged engine and 2.4L engine to remove the intake sensor, remove the air cleaner assembly. Disconnect the vacuum hose and pipe assembly. Remove the injector protection cover.

6. Remove the necessary components in order to gain access to the sensor retaining bolts.

7. Disconnect the connector from the CMP sensor.

8. Remove the bolt that retains the CMP sensor.

9. Remove the CMP sensor. Discard the O-ring.

To install:

➡**Be sure to use new fasteners, as required.**

10. Installation is the reverse of the removal procedure.

11. Tighten the bolt that retains the CMP sensor to: 98 inch lbs. (11 Nm).

12. Connect the sensor connector.

CRANKSHAFT POSITION (CKP) SENSOR

LOCATION

See Figures 107 and 108.

REMOVAL & INSTALLATION

See Figures 107 and 108.

1. Before servicing the vehicle, refer to the Precautions Section.

➡**If working near and/or around the SRS system and components, be sure to disable the SRS system. Tape the negative battery cable with insulating tape. Always disconnect the negative battery cable first.**

> ✳✳ **CAUTION**
>
> **Wait for 1 minute after disconnecting the negative battery cable before working inside the vehicle. The air bag system is set to deploy for a short period of time after the battery is disconnected.**

2. Disconnect the negative battery cable.
3. Remove the air cleaner assembly.
4. On 2.0L non turbocharged engine and

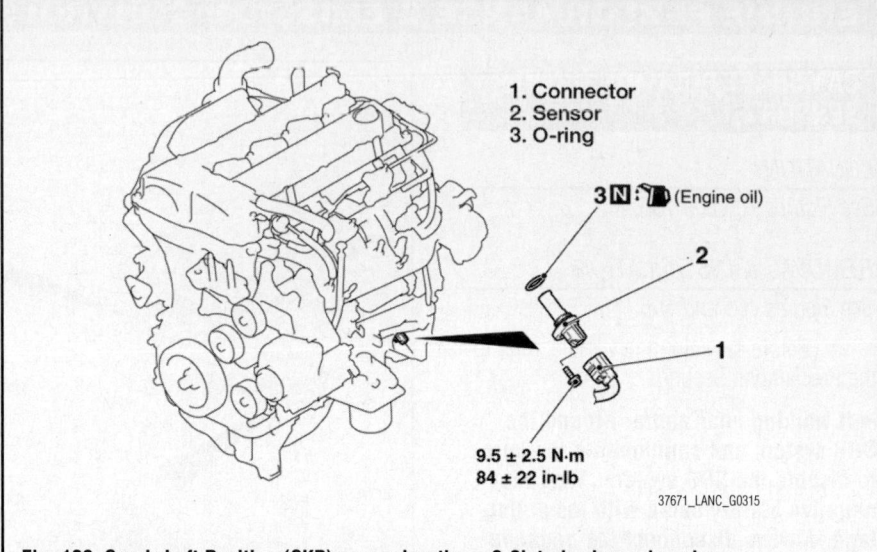

Fig. 108 Crankshaft Position (CKP) sensor location—2.0L turbocharged engine

2.4L engine, remove the exhaust manifold bracket.

5. Disconnect the connector from the sensor.

6. Remove the bolt that retains the sensor in place.

7. Remove the sensor from its mounting. Discard the O-ring.

To install:

➡**Be sure to use new fasteners, as required.**

8. Installation is the reverse of the removal procedure.

9. Tighten the sensor retaining bolt to specification.

ELECTRONIC CONTROL MODULE (ECM)

LOCATION

See Figure 109.

Refer to the accompanying illustration.

REMOVAL & INSTALLATION

See Figure 109.

➡**When the ECM is replaced do not replace the immobilizer ECU (WCM) or KOS-ECU simultaneously. When multiple ECU's are to be replaced, replace only one ECU at a time and complete the necessary registration using the Mitsubishi diagnostic scan tool, or equivalent.**

➡**After ECM replacement, engine idling speed may be unstable because the MFI learning procedure is not complete. Using the Mitsubishi diagnostic scan tool, or equivalent, relearn the MFI idling.**

➡**After the MFI idle learning is complete, register the key code using the Mitsubishi diagnostic scan tool, or equivalent.**

➡**When the ECM is replaced, save the vehicle identification number and per-**

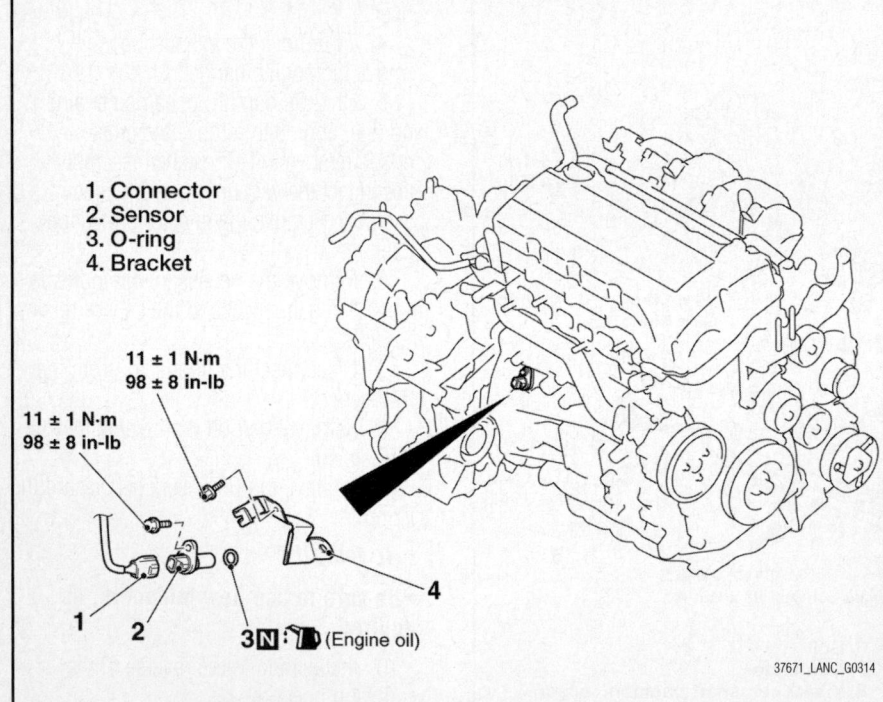

Fig. 107 Crankshaft Position (CKP) sensor location—2.0L non turbocharged engine and 2.4L engine

form the variant coding procedure, using the Mitsubishi diagnostic scan tool, or equivalent.

1. Before servicing the vehicle, refer to the Precautions Section.

➡ **If working near and/or around the SRS system and components, be sure to disable the SRS system. Tape the negative battery cable with insulating tape. Always disconnect the negative battery cable first.**

❊❊ CAUTION
Wait for 1 minute after disconnecting the negative battery cable before working inside the vehicle. The air bag system is set to deploy for a short period of time after the battery is disconnected.

2. Turn ignition switch off.
3. Disconnect the negative battery cable from the battery.
4. On Canadian vehicles remove the break off bolts. Remove the cover.
5. Disconnect the ECM connectors.
6. Remove the ECM mounting bolts and remove the ECM from the vehicle.

To install:
7. Installation is the reverse of the removal.
8. Tighten the ECM mounting bolts.

❊❊ WARNING
When replacing the ECM, be careful to use the right part number, as damage to the injection system could occur.

9. Turn the ignition switch to the ON position and then to the LOCK (OFF) position and hold it for at least ten seconds, to perform the initialization operation.

ENGINE COOLANT TEMPERATURE (ECT) SENSOR

LOCATION
See Figures 110 and 111.

Refer to the accompanying illustrations.

REMOVAL & INSTALLATION
See Figures 110 and 111.

1. Before servicing the vehicle, refer to the Precautions Section.

➡ **If working near and/or around the SRS system and components, be sure to disable the SRS system. Tape the**

negative battery cable with insulating tape. Always disconnect the negative battery cable first.

❊❊ CAUTION
Wait for 1 minute after disconnecting the negative battery cable before working inside the vehicle. The air bag system is set to deploy for a short period of time after the battery is disconnected.

2. Drain the coolant to a level below the bottom of the sensor.

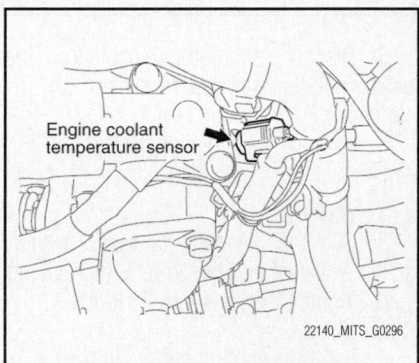

Engine coolant temperature sensor

22140_MITS_G0296

Fig. 110 Engine Coolant Temperature (ECT) sensor location—2.0L non turbocharged engine and 2.4L engine

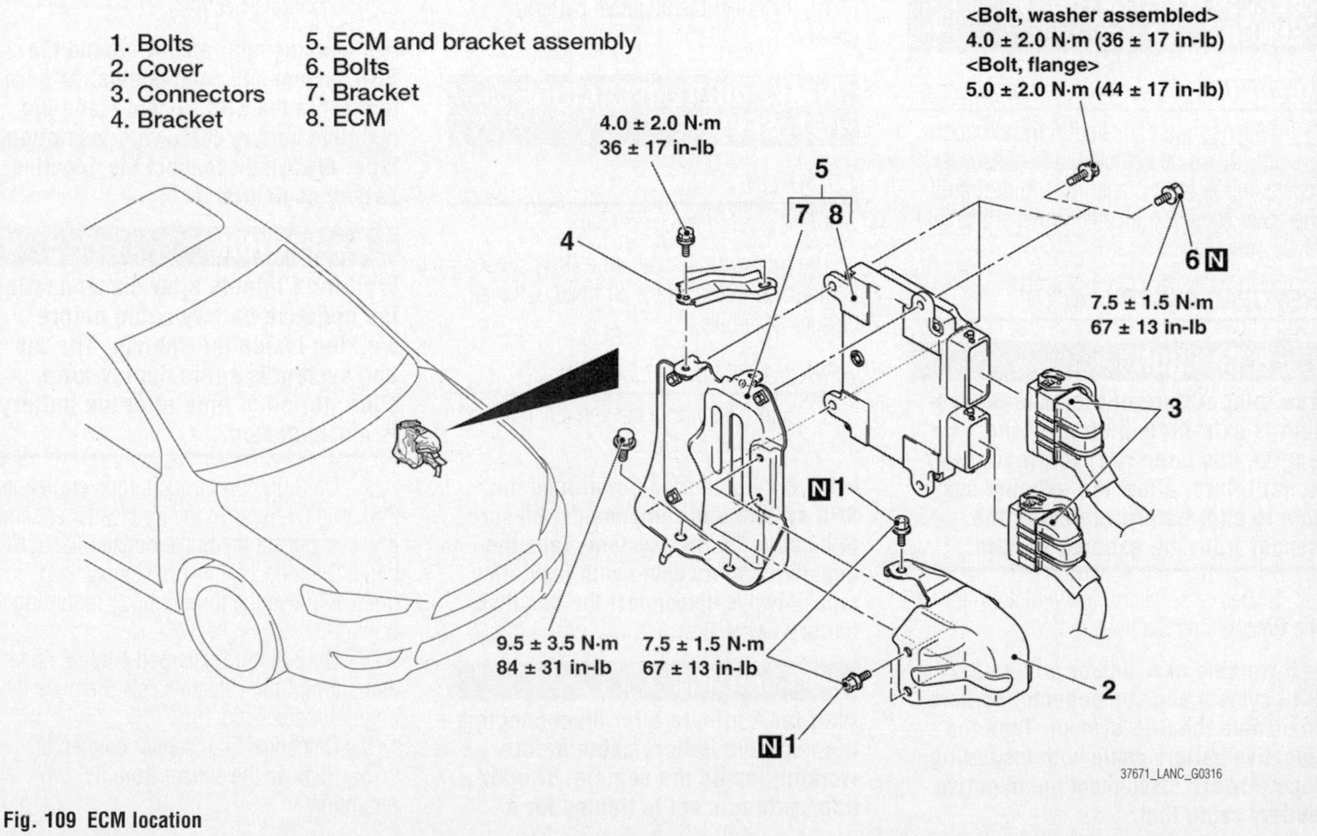

---

1. Bolts
2. Cover
3. Connectors
4. Bracket
5. ECM and bracket assembly
6. Bolts
7. Bracket
8. ECM

<Bolt, washer assembled>
4.0 ± 2.0 N·m (36 ± 17 in-lb)
<Bolt, flange>
5.0 ± 2.0 N·m (44 ± 17 in-lb)

4.0 ± 2.0 N·m
36 ± 17 in-lb

7.5 ± 1.5 N·m
67 ± 13 in-lb

9.5 ± 3.5 N·m
84 ± 31 in-lb

7.5 ± 1.5 N·m
67 ± 13 in-lb

37671_LANC_G0316

Fig. 109 ECM location

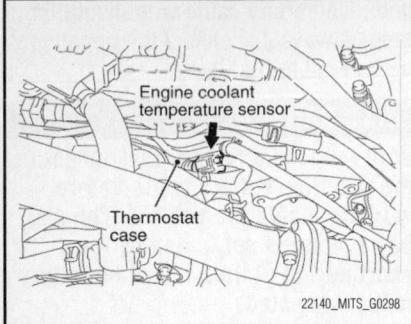

Engine coolant temperature sensor

Thermostat case

22140_MITS_G0298

Fig. 111 Engine Coolant Temperature (ECT) sensor location—2.0L turbocharged engine

3. Disconnect the ground cable from the battery and then remove the sensor connector.

4. Remove the coolant temperature sensor.

To install:

5. Coat the threads of the sensor with a suitable sealant and thread into the housing.

6. Tighten the sensor to 22 ft. lbs. (30 Nm).

7. Refill the cooling system to the proper level.

8. Attach the electrical connector to the sensor securely.

9. Connect the negative battery cable.

HEATED OXYGEN (HO2S) SENSOR

LOCATION

The sensors are located in the exhaust system. On some vehicles, one sensor is located up at the exhaust manifold(s) and the other sensor is located down at the catalytic converter.

REMOVAL & INSTALLATION

✳✳ CAUTION

The temperature of the exhaust system is extremely high after the engine has been run. To prevent personal injury, allow the exhaust system to cool before removing the sensor from the exhaust system.

1. Before servicing the vehicle, refer to the Precautions Section.

➡**If working near and/or around the SRS system and components, be sure to disable the SRS system. Tape the negative battery cable with insulating tape. Always disconnect the negative battery cable first.**

✳✳ CAUTION

Wait for 1 minute after disconnecting the negative battery cable before working inside the vehicle. The air bag system is set to deploy for a short period of time after the battery is disconnected.

2. Disconnect the negative battery cable.

3. Raise and safely support the vehicle, as needed.

4. Remove the engine undercover, as required.

5. Detach the electrical connector from the oxygen sensor.

6. Using socket MD998770, or equivalent oxygen sensor socket, remove the heated oxygen sensor.

To install:

➡**Be sure to use new fasteners, as required.**

7. If installing the old oxygen sensor, coat the threads with anti-seize compound. New sensors are already coated. Take care not to contaminate the oxygen sensor probe with the anti-seize compound.

8. Install the oxygen sensor. Using the correct tool, tighten the sensor to 33 ft. lbs. (45 Nm).

9. Attach the wiring to the sensor.

10. Connect the negative battery cable.

INTAKE AIR TEMPERATURE (IAT) SENSOR

LOCATION

See Figures 151 through 153.

The Intake Air Temperature (IAT) sensor is mounted in the intake air hose of the air cleaner assembly.

REMOVAL & INSTALLATION

1. Before servicing the vehicle, refer to the Precautions Section.

➡**If working near and/or around the SRS system and components, be sure to disable the SRS system. Tape the negative battery cable with insulating tape. Always disconnect the negative battery cable first.**

✳✳ CAUTION

Wait for 1 minute after disconnecting the negative battery cable before working inside the vehicle. The air bag system is set to deploy for a

short period of time after the battery is disconnected.

2. Disconnect the negative battery cable.

3. Disconnect the connector from the sensor.

4. Remove the sensor retaining screws.

5. Remove the sensor from its mounting.

To install:

6. Installation is the reverse of the removal procedure.

7. Handle the sensor assembly carefully, protecting it from impact, extremes of temperature and/or exposure to shop chemicals.

KNOCK SENSOR (KS)

LOCATION

See Figures 112 and 113.

Refer to the accompanying illustrations.

REMOVAL & INSTALLATION

See Figures 114 and 115.

➡**After replacing this sensor use the Mitsubishi diagnostic scan tool, or equivalent and perform the initialization procedure for learning value in MFI engine.**

1. Before servicing the vehicle, refer to the Precautions Section.

➡**If working near and/or around the SRS system and components, be sure to disable the SRS system. Tape the negative battery cable with insulating tape. Always disconnect the negative battery cable first.**

✳✳ CAUTION

Wait for 1 minute after disconnecting the negative battery cable before working inside the vehicle. The air bag system is set to deploy for a short period of time after the battery is disconnected.

2. On 2.0L non turbocharged engine and 2.4L engine, remove the air cleaner assembly. Disconnect the brake booster vacuum hose. Disconnect the emission vacuum hose. Disconnect the throttle body mounting bolts.

3. On 2.0L turbocharged engine, raise and support the vehicle safely. Remove the engine undercover.

4. Disconnect the sensor connector.

5. Remove the sensor from its mounting.

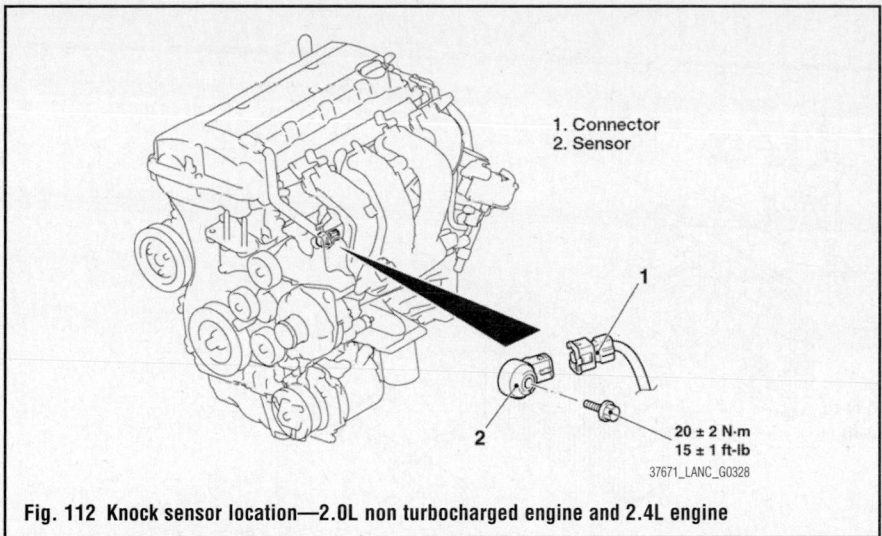

1. Connector
2. Sensor

20 ± 2 N·m
15 ± 1 ft-lb

37671_LANC_G0328

Fig. 112 Knock sensor location—2.0L non turbocharged engine and 2.4L engine

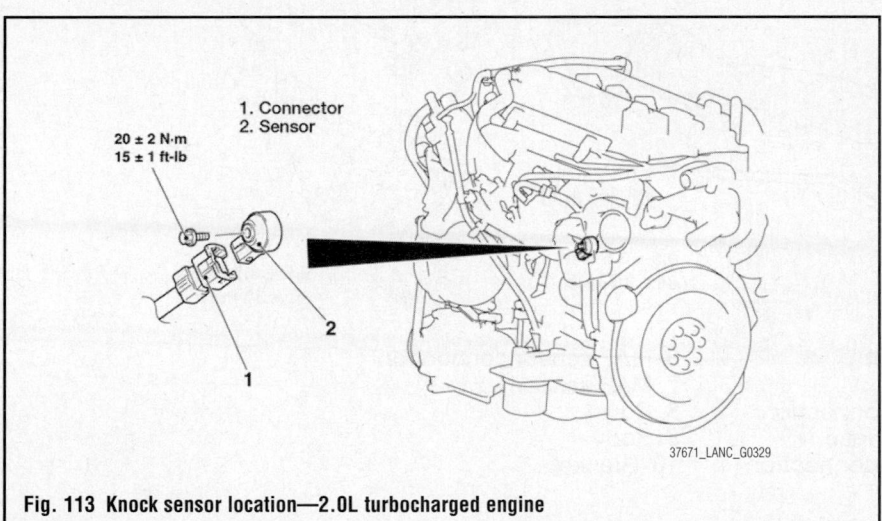

1. Connector
2. Sensor

20 ± 2 N·m
15 ± 1 ft-lb

37671_LANC_G0329

Fig. 113 Knock sensor location—2.0L turbocharged engine

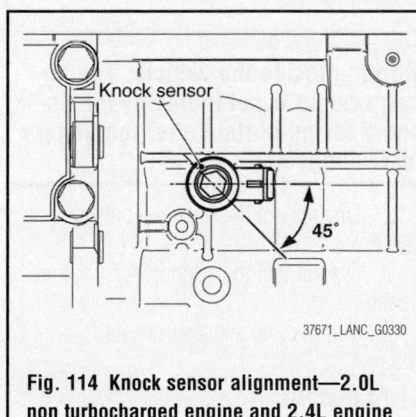

Knock sensor

45°

37671_LANC_G0330

Fig. 114 Knock sensor alignment—2.0L non turbocharged engine and 2.4L engine

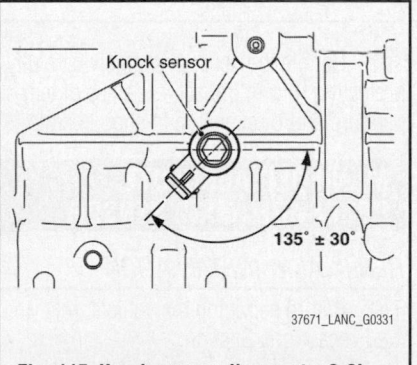

Knock sensor

135° ± 30°

37671_LANC_G0331

Fig. 115 Knock sensor alignment—2.0L turbocharged engine

To install:

➡**Be sure to use new fasteners, as required.**

6. Installation is the reverse of the removal procedure.

7. Tighten the sensor retaining bolt to specification.

MALFUNCTION INDICATOR LIGHT (MIL)

RESET PROCEDURE

1. Proper operation of the Malfunction Indicator Light (MIL):
   - The MIL will illuminate with the ignition switch ON and the engine OFF
   - The MIL will turn OFF when the engine is started
   - The MIL will remain ON if the self-diagnostic system has detected a malfunction
   - The MIL may turn OFF if the malfunction is no longer present
   - If the MIL is illuminated and then the engine stalls, the MIL will remain illuminated as long as the ignition switch is ON
   - If the MIL is not illuminated and the engine stalls, the MIL will not illuminate until the ignition switch is cycled OFF, then ON

2. Resetting the MIL:
   - The control module turns OFF the MIL after 3 consecutive ignition cycles that the diagnostic system runs and does not fail
   - The control module turns OFF the MIL after a current Diagnostic Trouble Code (DTC) clears when the diagnostic cycle runs and passes
   - There may still be a history of DTC's stored in the system. These will clear after 40 consecutive warm-up cycles, if no failures are reported by any other related diagnostic system
   - Manual resetting of the MIL and any DTC stored in the system, requires the use of an OBD2 scan tool connected to the Data Link Connector (DLC) for communication with the vehicle. Follow the instructions of the scan tool for both retrieval and resetting of DTC's.

➡**If the error symptoms causing the MIL to illuminate have been corrected, the MIL will return to normal operation.**

MASS AIR FLOW (MAF) SENSOR

LOCATION
See Figure 116.

The Mass Air Flow (MAF) sensor is mounted in the intake air hose of the air cleaner assembly.

REMOVAL & INSTALLATION
See Figure 116.

1. Before servicing the vehicle, refer to the Precautions Section.

➡**If working near and/or around the SRS system and components, be sure**

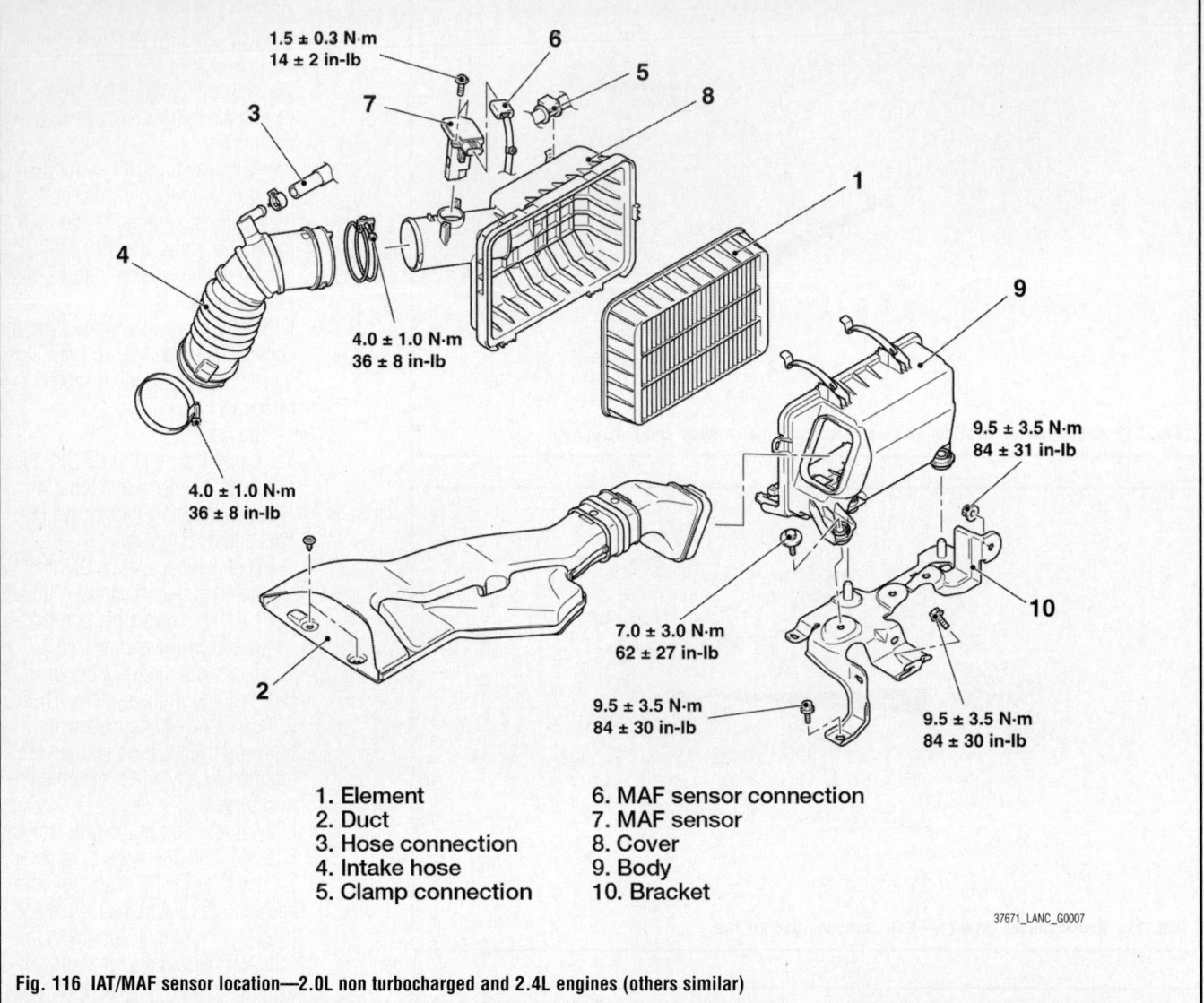

1.5 ± 0.3 N·m
14 ± 2 in-lb

4.0 ± 1.0 N·m
36 ± 8 in-lb

4.0 ± 1.0 N·m
36 ± 8 in-lb

9.5 ± 3.5 N·m
84 ± 31 in-lb

7.0 ± 3.0 N·m
62 ± 27 in-lb

9.5 ± 3.5 N·m
84 ± 30 in-lb

9.5 ± 3.5 N·m
84 ± 30 in-lb

1. Element
2. Duct
3. Hose connection
4. Intake hose
5. Clamp connection
6. MAF sensor connection
7. MAF sensor
8. Cover
9. Body
10. Bracket

37671_LANC_G0007

Fig. 116 IAT/MAF sensor location—2.0L non turbocharged and 2.4L engines (others similar)

to disable the SRS system. Tape the negative battery cable with insulating tape. Always disconnect the negative battery cable first.

✳✳ CAUTION

Wait for 1 minute after disconnecting the negative battery cable before working inside the vehicle. The air bag system is set to deploy for a short period of time after the battery is disconnected.

2. Disconnect the negative battery cable.
3. Disconnect the connector from the sensor.
4. Remove the sensor retaining screws.
5. Remove the sensor from its mounting.

To install:
6. Installation is the reverse of the removal procedure.

7. Handle the sensor assembly carefully, protecting it from impact, extremes of temperature and/or exposure to shop chemicals.

MANIFOLD ABSOLUTE PRESSURE (MAP) SENSOR

REMOVAL & INSTALLATION

1. Before servicing the vehicle, refer to the Precautions Section.

➡ If working near and/or around the SRS system and components, be sure to disable the SRS system. Tape the negative battery cable with insulating tape. Always disconnect the negative battery cable first.

✳✳ CAUTION

Wait for 1 minute after disconnecting the negative battery cable before

working inside the vehicle. The air bag system is set to deploy for a short period of time after the battery is disconnected.

2. Disconnect the negative battery cable.
3. Disconnect the connector from the sensor.
4. Remove the sensor from its mounting.

To install:

➡ Be sure to use new fasteners, as required.

5. Installation is the reverse of the removal procedure.
6. Handle the sensor assembly carefully, protecting it from impact, extremes of temperature, and exposure to shop chemicals.

POWERTRAIN CONTROL MODULE (PCM)

LOCATION

See Figure 117.

Refer to the accompanying illustration.

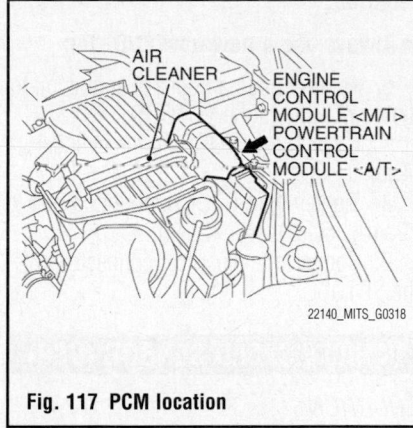

Fig. 117 PCM location

REMOVAL & INSTALLATION

See Figure 117.

➡ Do not replace the PCM and the ETACS-ECU simultaneously. Always replace either one of these components first, and register the encrypted code.

➡ After PCM replacement, idling speed may be unstable because the MFI engine learning is not completed. Using the Mitsubishi diagnostic scan tool or equivalent, perform the relearn procedure.

1. Before servicing the vehicle, refer to the Precautions Section.

➡ If working near and/or around the SRS system and components, be sure to disable the SRS system. Tape the negative battery cable with insulating tape. Always disconnect the negative battery cable first.

✳✳ CAUTION

Wait for 1 minute after disconnecting the negative battery cable before working inside the vehicle. The air bag system is set to deploy for a short period of time after the battery is disconnected.

2. Turn the ignition switch off.
3. Disconnect the negative battery cable from the battery.
4. Disconnect the PCM connector(s).
5. Remove the PCM mounting bolts and remove the PCM from the air cleaner assembly.

To install:

6. Installation is the reverse of the removal.
7. Tighten the PCM mounting bolts to: 86–104 inch lbs. (10–12 Nm).

THROTTLE POSITION SENSOR (TPS)

LOCATION

See Figures 118 and 119.

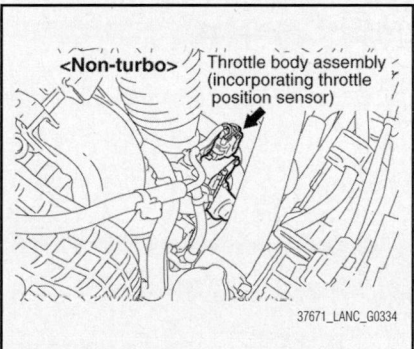

Fig. 118 TPS location—2.0L non turbocharged engine and 2.4L engine

The Throttle Position Sensor (TPS) is mounted on the throttle body and is incorporated into the throttle body assembly.

REMOVAL & INSTALLATION

See Figures 118 and 119.

At this time the manufacturer does not provide removal and installation procedures for this component.

The throttle position sensor is an integral part of the throttle body.

➡ When the throttle body assembly replacement is performed, use scan tool MB991958 to initialize the learning value. Do not loosen the fixing screws for the resin cover of throttle body assembly. If the screws are loosened, the sensor incorporated in the resin cover becomes misaligned and the throttle body may not work normally.

1. Before servicing the vehicle, refer to the Precautions Section.

➡ If working near and/or around the SRS system and components, be sure to disable the SRS system. Tape the negative battery cable with insulating tape. Always disconnect the negative battery cable first.

✳✳ CAUTION

Wait for 1 minute after disconnecting the negative battery cable before working inside the vehicle. The air bag system is set to deploy for a short period of time after the battery is disconnected.

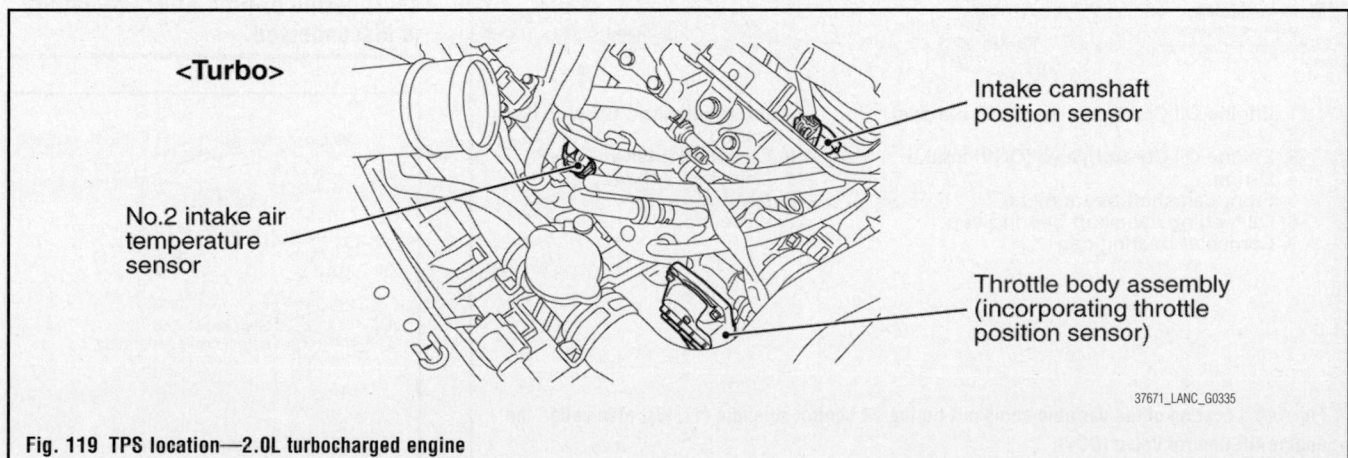

Fig. 119 TPS location—2.0L turbocharged engine

VARIABLE CAMSHAFT TIMING OIL CONTROL SOLENOID

LOCATION

See Figure 120.

Refer to the accompanying illustration.

REMOVAL & INSTALLATION

1. Before servicing the vehicle, refer to the Precautions Section.

➡ **If working near and/or around the SRS system and components, be sure to disable the SRS system. Tape the negative battery cable with insulating tape. Always disconnect the negative battery cable first.**

❋❋ CAUTION

Wait for 1 minute after disconnecting the negative battery cable before working inside the vehicle. The air bag system is set to deploy for a short period of time after the battery is disconnected.

2. Disconnect the ground cable from the battery.

3. Disconnect the connector from Variable Camshaft Timing Oil Control Solenoid (VCTOCS) on the right-hand bank and/or left-hand bank.

4. Remove the bolt retaining the VCTOCS.

5. Remove the VCTOCS.

To install:

➡ **Be sure to use new fasteners, as required.**

➡ **Always use a new gasket/O-ring.**

6. Apply a small amount of engine oil to the new O-ring of the VCTOCS.

7. Install the VCTOCS on the cylinder head.

8. Tighten the VCTOCS mounting bolt to 72–106 inch lbs. (8–12 Nm).

9. Connect the electrical connector to the VCTOCS.

VEHICLE SPEED SENSOR (VSS)

LOCATION

See Figure 121.

The Vehicle Speed Sensor (VSS) is attached near the output shaft of the transaxle.

REMOVAL & INSTALLATION

See Figure 121.

1. Before servicing the vehicle, refer to the Precautions Section.

➡ **If working near and/or around the SRS system and components, be sure to disable the SRS system. Tape the negative battery cable with insulating tape. Always disconnect the negative battery cable first.**

❋❋ CAUTION

Wait for 1 minute after disconnecting the negative battery cable before working inside the vehicle. The air bag system is set to deploy for a short period of time after the battery is disconnected.

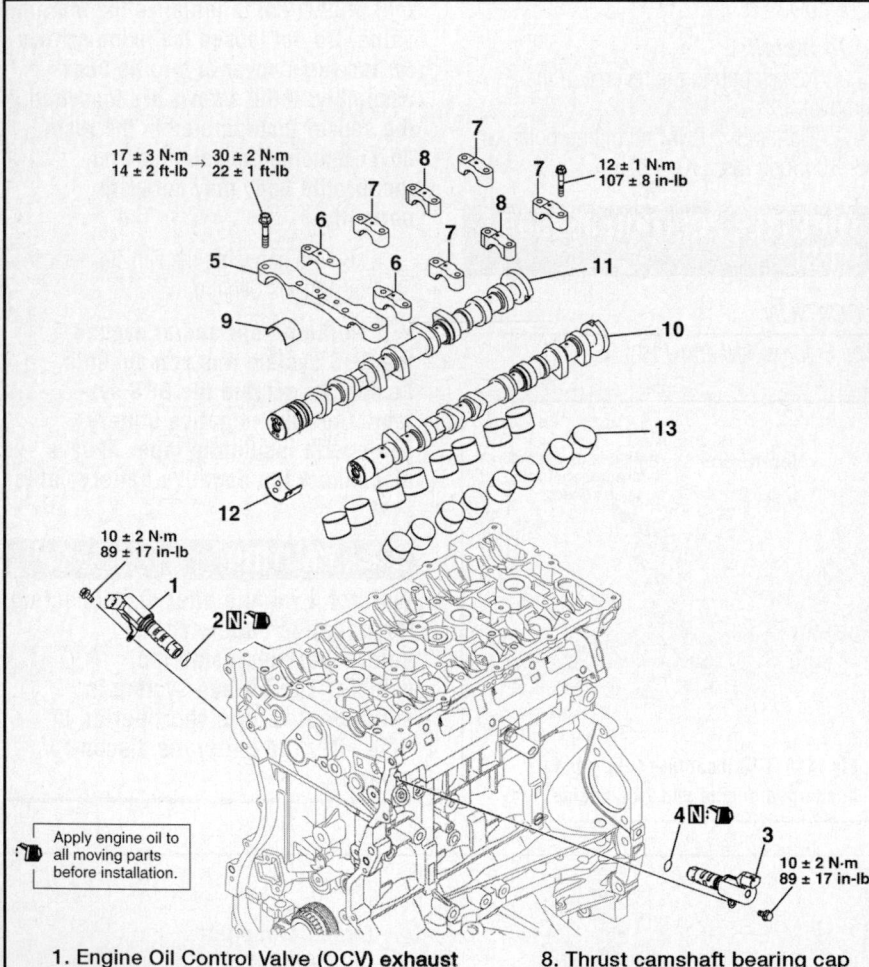

Apply engine oil to all moving parts before installation.

1. Engine Oil Control Valve (OCV) **exhaust**
2. O-ring
3. Engine Oil Control Valve (OCV) **intake**
4. O-ring
5. Front camshaft bearing cap
6. Oil feeding camshaft bearing cap
7. Camshaft bearing cap
8. Thrust camshaft bearing cap
9. Bearing
10. Camshaft intake
11. Camshaft exhaust
12. Bearing
13. Valve tappet

22140_MITS_G0251

Fig. 120 Location of the variable camshaft timing oil control solenoid (1), (3), also called the engine Oil Control Valve (OCV)

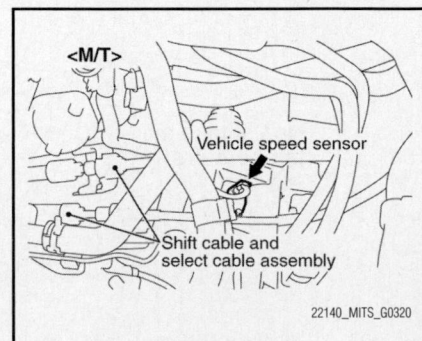

<M/T>

Vehicle speed sensor

Shift cable and select cable assembly

22140_MITS_G0320

Fig. 121 VSS sensor location

2. Raise and support the vehicle safely.

3. Place a drip pan below the Vehicle Speed Sensor (VSS) to catch any spilled fluid when it is removed.

4. Disconnect the VSS connector.

5. Remove the sensor from its mounting.

To install:

➡**Be sure to use new fasteners, as required.**

6. Install the sensor and tighten the attaching bolt to 97 inch lbs. (11 Nm).

7. Replace any lost transaxlc fluid.

8. Connect the sensor electrical connector.

FUEL | GASOLINE FUEL INJECTION SYSTEM

FUEL SYSTEM SERVICE PRECAUTIONS

Safety is the most important factor when performing not only fuel system maintenance, but any type of maintenance. Failure to conduct maintenance and repairs in a safe manner may result in serious personal injury or death. Work on a vehicle's fuel system components can be accomplished safely and effectively by adhering to the following rules and guidelines.

• To avoid the possibility of fire and personal injury, always disconnect the negative battery cable unless the repair or test procedure requires that battery voltage be applied.

• Always relieve the fuel system pressure prior to disconnecting any fuel system component (injector, fuel rail, pressure regulator, etc.) fitting or fuel line connection. Exercise extreme caution whenever relieving fuel system pressure to avoid exposing skin, face and eyes to fuel spray. Please be advised that fuel under pressure may penetrate the skin or any part of the body that it contacts.

• Always place a shop towel or cloth around the fitting or connection prior to loosening to absorb any excess fuel due to spillage. Ensure that all fuel spillage is quickly removed from engine surfaces. Ensure that all fuel-soaked cloths or towels are deposited into a flame-proof waste container with a lid.

• Always keep a dry chemical (Class B) fire extinguisher near the work area.

• Do not allow fuel spray or fuel vapors to come into contact with a spark or open flame.

• Always use a second wrench when loosening or tightening fuel line connection fittings. This will prevent unnecessary stress and torsion on fuel piping. Always follow the proper torque specifications.

• Always replace worn fuel fitting O-rings with new ones. Do not substitute fuel hose where rigid pipe is installed.

FUEL SYSTEM PRESSURE

RELIEVING

See Figures 122 and 123.

✳ CAUTION

Observe all applicable safety precautions when working around fuel. Whenever servicing the fuel system, always work in a well ventilated area. Do not allow fuel spray or vapors to come in contact with a spark or open flame. Keep a dry chemical fire extinguisher near the work area. Always keep fuel in a container specifically designed for fuel storage; also, always properly seal fuel containers to avoid the possibility of fire or explosion.

➡**When disconnecting fuel lines, refer to the illustrations as needed.**

1. Before servicing the vehicle, refer to the Precautions Section.

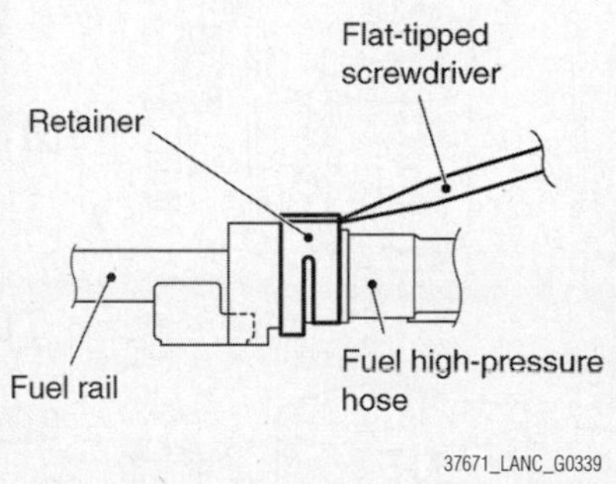

Flat-tipped screwdriver

Retainer

Fuel rail

Fuel high-pressure hose

37671_LANC_G0339

Fig. 122 Fuel line disconnection procedure—1 of 2

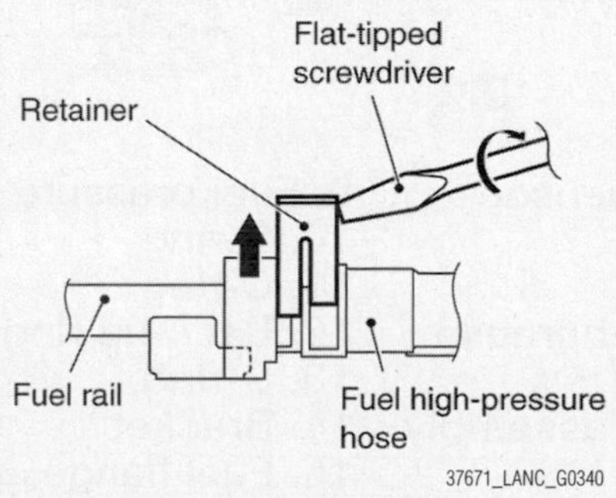

Flat-tipped screwdriver

Retainer

Fuel rail

Fuel high-pressure hose

37671_LANC_G0340

Fig. 123 Fuel line disconnection procedure—2 of 2

2. Remove the rear seat cushion assembly.

3. Remove the fuel pump module access hole cover.

4. Disconnect the fuel pump module connector.

5. After starting the engine and letting it run until it stops, turn the ignition switch to the **OFF** position.

6. Crank the engine for 2 seconds or more.

7. If the engine does not start, turn the ignition switch to the **OFF** position.

8. If the engine starts, let it run until it stalls and turn the ignition switch to the **OFF** position.

9. Connect the fuel pump module connector.

10. Install the fuel pump module access hole cover.

11. Install the rear seat cushion assembly.

FUEL FILTER

REMOVAL & INSTALLATION

See Figure 125.

The fuel delivery system integrates the fuel filter with the in-tank fuel pump module.

FUEL LEVEL SENDING UNIT

LOCATION

See Figure 124.

The fuel delivery system integrates the fuel level sensing unit with the in-tank fuel pump

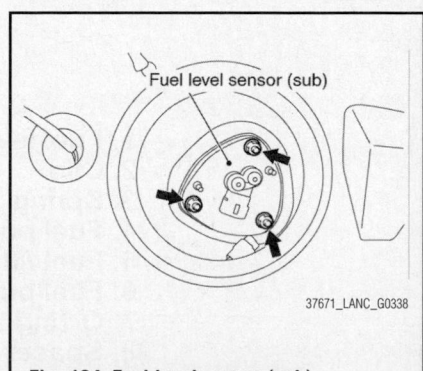

Fuel level sensor (sub)

37671_LANC_G0338

Fig. 124 Fuel level sensor (sub) location—2.0L turbocharged engine

module. However the 2.0L turbocharged engine uses a fuel level sensor (sub).

REMOVAL & INSTALLATION

1. Before servicing the vehicle, refer to the Precautions Section.

➡ If working near and/or around the SRS system and components, be sure to disable the SRS system. Tape the negative battery cable with insulating tape. Always disconnect the negative battery cable first.

✳✳ CAUTION

Wait for 1 minute after disconnecting the negative battery cable before working inside the vehicle. The air bag system is set to deploy for a short period of time after the battery is disconnected.

2. Remove the rear seat cushion assembly.
3. Remove the access hole cover.
4. Disconnect the fuel level sensor connector and fuel tank differential pressure sensor connector.

✳✳ CAUTION

When removing and installing the fuel level sensor from the fuel tank, be careful not damage the sensor unit and the float.

5. Remove the fuel level sensor mounting bolts and remove the fuel tank gauge unit from service hole.

To install:

➡ Be sure to use new fasteners, as required.

6. Install the fuel level sensor to the fuel tank through the service hole.

7. Connect the fuel level sensor (sub) connector and fuel tank differential pressure sensor connector.
8. Install the access hole cover.
9. Install the rear seat cushion assembly.

FUEL PUMP MODULE

REMOVAL & INSTALLATION

See Figure 125.

1. Before servicing the vehicle, refer to the Precautions Section.

➡ If working near and/or around the SRS system and components, be sure to disable the SRS system. Tape the negative battery cable with insulating tape. Always disconnect the negative battery cable first.

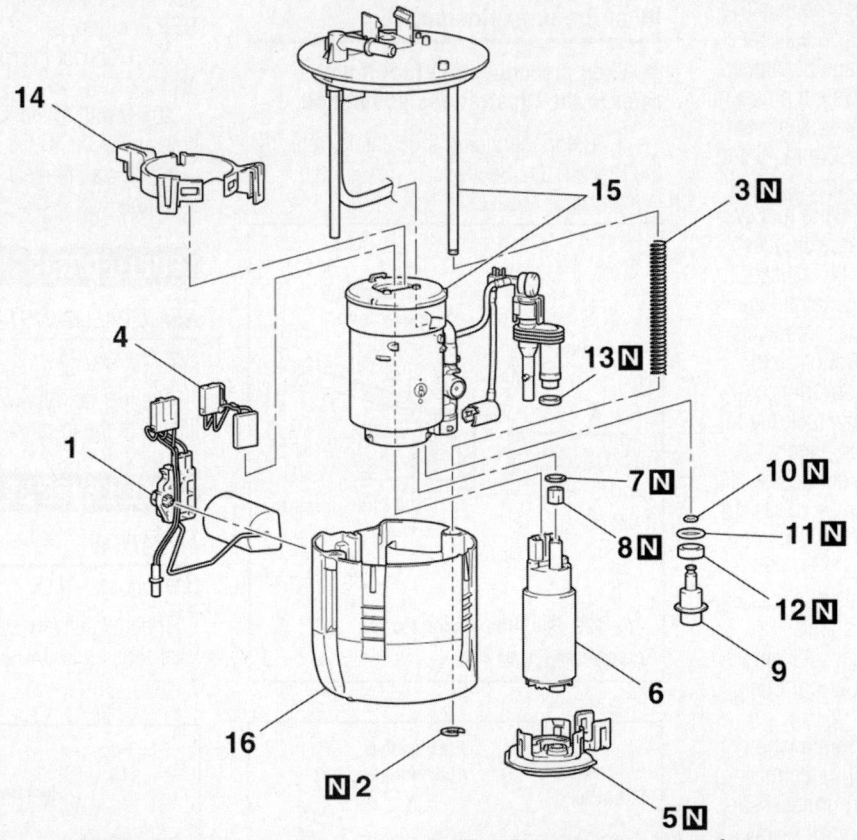

1. Fuel level sensor
2. Clip
3. Spring
4. Fuel pump harness
5. Fuel filter
6. Fuel pump assembly
7. O-ring
8. Spacer
9. Fuel pressure regulator
10. O-ring
11. O-ring
12. Back-up ring
13. O-ring
14. Bracket
15. Fuel flange and filter assembly
16. Sub tank assembly

22140_MITS_G0169

Fig. 125 Fuel pump module and related components

※※ **CAUTION**

Wait for 1 minute after disconnecting the negative battery cable before working inside the vehicle. The air bag system is set to deploy for a short period of time after the battery is disconnected.

2. Relieve the fuel system pressure.
3. Disconnect the negative battery cable.
4. Remove or disconnect the following:
- Rear seat cushion by pulling the seat stopper near the floor and lifting the cushion up
- Inspection cover on the right side of the vehicle
- Harness connector and the fuel lines
- Fuel pump assemble from the tank.

To install:

5. Install or connect the following:
- Fuel pump in the tank
- Hoses and the harness connector
- Inspection cover
- Rear seat
- Negative battery cable

FUEL RAIL AND INJECTOR

REMOVAL & INSTALLATION

2.0L Non Turbocharged Engine and 2.4L Engine
See Figures 122 and 123, 126 and 127.

➡**When the fuel injectors are replaced, initialize the learned value using the** Mitsubishi diagnostic scan tool, or equivalent.

1. Before servicing the vehicle, refer to the Precautions Section.

➡**If working near and/or around the SRS system and components, be sure to disable the SRS system. Tape the negative battery cable with insulating tape. Always disconnect the negative battery cable first.**

※※ **CAUTION**

Wait for 1 minute after disconnecting the negative battery cable before working inside the vehicle. The air bag system is set to deploy for a short period of time after the battery is disconnected.

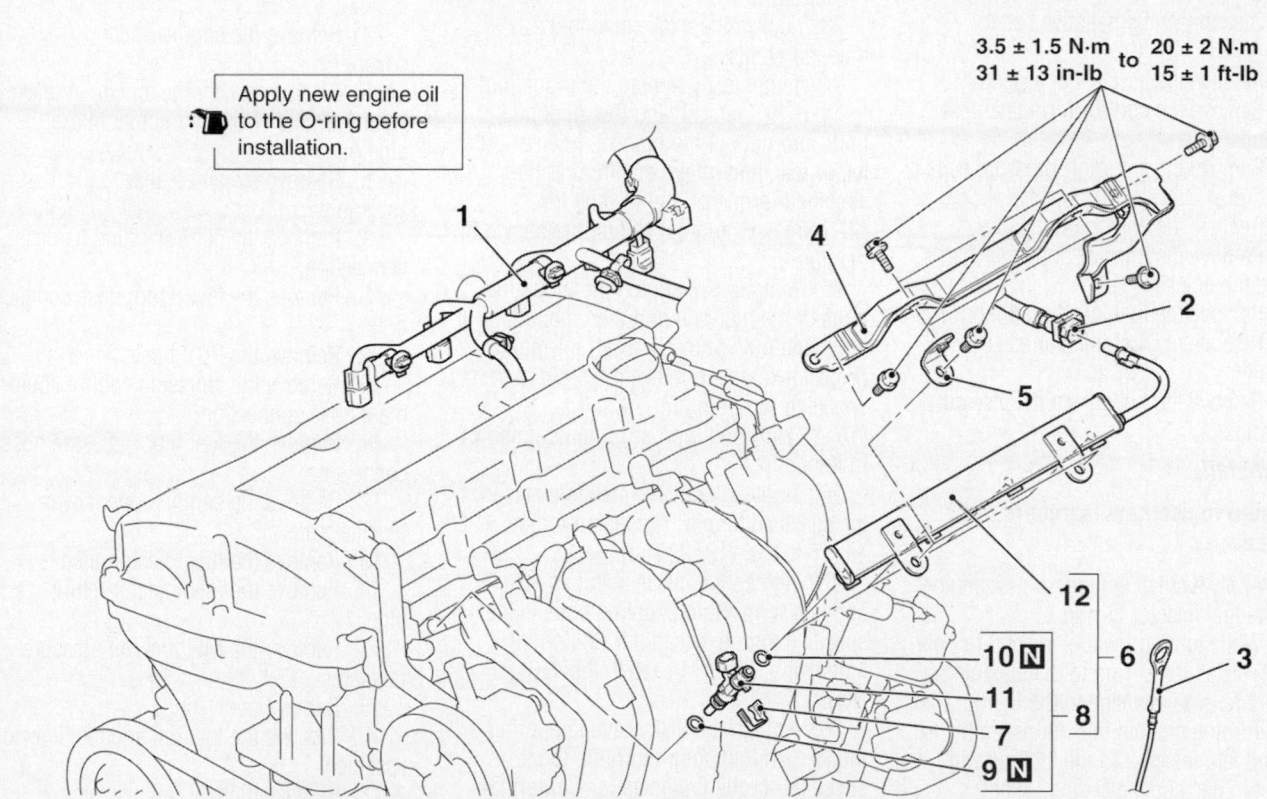

1. Control wiring harness connection
2. Fuel high-pressure hose connection
3. Engine oil dipstick
4. Injector protector rear
5. Bracket
6. Fuel rail and fuel injector assembly
7. Fuel injector support
8. Fuel injector assembly
9. O-ring
10. O-ring
11. Fuel injector
12. Fuel rail

22140_MITS_G0172

Fig. 126 Fuel injectors and related components—2.0L non turbocharged engine and 2.4L engine

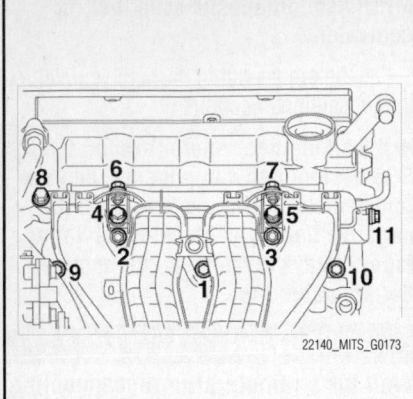

Fig. 127 Loosen the intake manifold mounting bolts and nuts 1, 2, 3, and 9— 2.0L non turbocharged engine and 2.4L engine

2. Relieve the fuel system pressure.

3. Disconnect the negative battery cable.

4. Remove the air cleaner assembly.

5. Remove the control wiring harness connection.

6. Remove the fuel high-pressure hose connection.

7. Remove the engine oil dipstick.

8. Remove the injector protector rear. Remove the bracket.

9. Remove the fuel injector support.

10. Remove the fuel injector as an assembly.

11. Remove the fuel injectors. Discard the O-rings.

To install:

➡**Be sure to use new fasteners, as required.**

12. Apply a small amount of new engine oil to the fuel injector O-ring.

13. While turning the fuel injector to right and left, install the O-ring to the fuel injector with care to avoid damage to the O-ring.

14. Turning the fuel injector assembly to right and left, install it to the fuel rail with care not to damage the O-ring. After the installation, check for its smooth rotation. At this time, check that the projection of the fuel injector assembly is in the center.

15. If the rotation is not smooth, the O-ring may be caught. Remove the fuel injector assembly and check the O-ring for damage. Re-insert it into the fuel rail and check for its smooth rotation.

16. Apply a small amount of new engine oil to the O-ring at the end of fuel injector assembly.

⁂ WARNING

When installing the fuel rail and fuel injector assembly to the intake manifold, pay attention to avoid damage to the O-ring at the end of the fuel injector assembly.

17. Install the fuel rail and fuel injector assembly to the intake manifold.

18. Install the bracket and injector protector rear.

19. Loosen the intake manifold mounting bolts and nuts. (Bolts and nuts 1, 2, 3, and 9 shown in the figure).

20. Remove the EGR valve (vehicles for California).

21. Loosen the intake manifold stay B and intake manifold coupling bolts (except for California).

22. Loosen the EGR valve support and intake manifold coupling bolts (vehicles for California).

23. Loosen the intake manifold stay mounting bolts.

24. Temporarily tighten the mounting bolts of the fuel rail, and the mounting bolts and nuts of the bracket, injector protector rear, and intake manifold to the specified torque of 18–44 inch lbs. (2–5 Nm) in the order of number shown in the figure.

25. Temporarily tighten the mounting bolts of the fuel rail, and then mounting bolts and nuts of the bracket, injector protector rear, and intake manifold to the specified torque of 14–16 ft. lbs. (18–22 Nm) in the order of number shown in the figure.

26. Tighten the intake manifold stay to the specified torque. Tightening torque: 14–16 ft. lbs. (18–22 Nm).

27. Tighten the intake manifold stay B and intake manifold coupling bolts to the specified torque (except for California). Tightening torque: 14–16 ft. lbs. (18–22 Nm).

28. Tighten the EGR valve support and intake manifold coupling bolts to the specified torque (vehicles for California). Tightening torque: 14–16 ft. lbs. (18–22 Nm).

29. Install the EGR valve (vehicles for California).

30. Installation continues in the reverse of the removal procedure.

2.0L Turbocharged Engine

See Figures 122, 123 and 128.

➡**When the fuel injectors are replaced, initialize the learned value using the**

Mitsubishi diagnostic scan tool, or equivalent.

1. Before servicing the vehicle, refer to the Precautions Section.

➡**If working near and/or around the SRS system and components, be sure to disable the SRS system. Tape the negative battery cable with insulating tape. Always disconnect the negative battery cable first.**

⁂ CAUTION

Wait for 1 minute after disconnecting the negative battery cable before working inside the vehicle. The air bag system is set to deploy for a short period of time after the battery is disconnected.

2. Relieve the fuel system pressure.

3. Disconnect the negative battery cable.

4. Remove the engine upper cover.

5. Remove the charge air cooler intake hose A, B and charge air cooler intake pipe A.

6. Remove the air cleaner assembly.

7. Remove the control wiring harness connection.

8. Remove the fuel return hose connection.

9. Remove the PCV hose.

10. Remove the emission control equipment hose connection.

11. Remove the fuel high-pressure hose connection.

12. Remove the engine upper cover bracket rear.

13. Remove the fuel injector hose.

14. Remove the fuel injector return pipe.

15. Remove the MFI fuel rail pressure regulator.

16. Remove the O-ring.

17. Remove the fuel rail and fuel injector assembly.

18. Remove the fuel rail insulator.

19. Remove the fuel injector support.

20. Remove the fuel injector assembly.

21. Remove the fuel injector.

To install:

22. Apply a small amount of new engine oil to the fuel injector O-ring.

23. While turning the fuel injector to right and left, install the O-ring to the fuel injector with care to avoid damage to the O-ring.

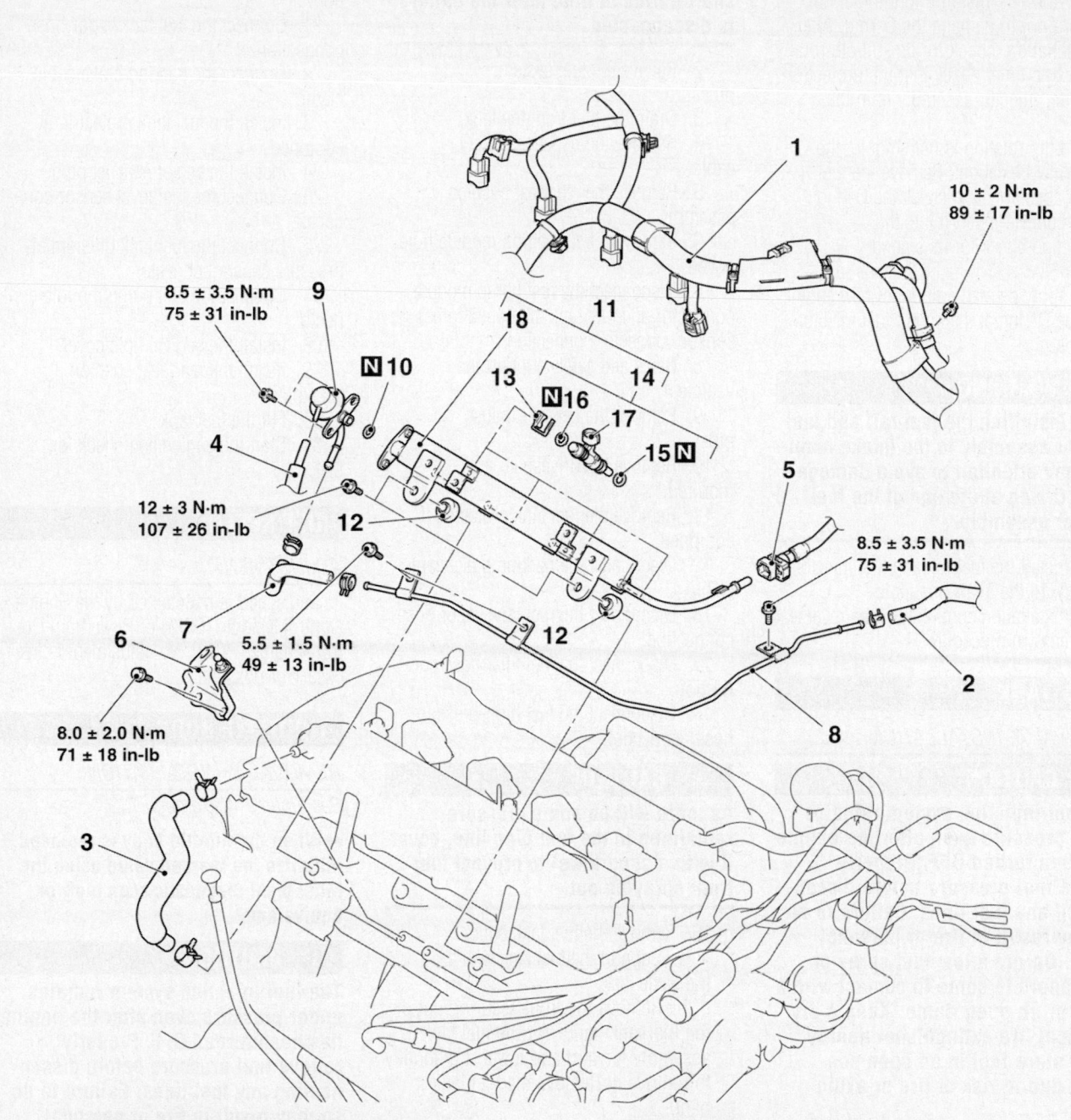

10 ± 2 N·m
89 ± 17 in-lb

8.5 ± 3.5 N·m
75 ± 31 in-lb

12 ± 3 N·m
107 ± 26 in-lb

5.5 ± 1.5 N·m
49 ± 13 in-lb

8.0 ± 2.0 N·m
71 ± 18 in-lb

8.5 ± 3.5 N·m
75 ± 31 in-lb

1. Control wiring harness connection
2. Fuel return hose connection
3. PCV hose
4. Emission control equipment hose connection
5. Fuel high-pressure hose connection
6. Engine upper cover bracket rear
7. Fuel injector hose
8. Fuel injector return pipe
9. MFI fuel rail pressure regulator
10. O-ring
11. Fuel rail and fuel injector assembly
12. Fuel rail insulator
13. Fuel injector support
14. Fuel injector assembly
15. O-ring
16. O-ring
17. Fuel injector
18. Fuel rail

22140_MITS_G0174

Fig. 128 Fuel injectors and related components—2.0L turbocharged engine

24. Turning the fuel injector assembly to right and left, install it to the fuel rail with care not to damage the O-ring. After the installation, check for its smooth rotation. At this time, check that the projection of the fuel injector assembly is in the center.

25. If the rotation is not smooth, the O-ring may be caught. Remove the fuel injector assembly and check the O-ring for damage. Re-insert it into the fuel rail and check for its smooth rotation.

26. Apply a small amount of new engine oil to the O-ring at the end of fuel injector assembly.

❋❋ WARNING

When installing the fuel rail and fuel injector assembly to the intake manifold, pay attention to avoid damage to the O-ring at the end of the fuel injector assembly.

27. Install the fuel rail and fuel injector assembly to the intake manifold.

28. Installation continues in the reverse of the removal procedure.

FUEL TANK

REMOVAL & INSTALLATION

❋❋ CAUTION

The fuel injection system remains under pressure even after the engine has been turned OFF. Properly relieve fuel pressure before disconnecting any fuel lines. Failure to do so may result in fire or personal injury. Do not allow fuel spray or fuel vapors to come in contact with a spark or an open flame. Keep a dry chemical fire extinguisher nearby. Never store fuel in an open container due to risk of fire or explosion.

1. Before servicing the vehicle, refer to the Precautions Section.

➥If working near and/or around the SRS system and components, be sure to disable the SRS system. Tape the negative battery cable with insulating tape. Always disconnect the negative battery cable first.

❋❋ CAUTION

Wait for 1 minute after disconnecting the negative battery cable before working inside the vehicle. The air

bag system is set to deploy for a short period of time after the battery is disconnected.

2. Relieve the fuel system pressure.

3. Drain the fuel from the tank.

4. Remove the negative battery cable.

5. Remove the rear seat cushion assembly.

6. Remove the fuel pump module hole cover.

7. Disconnect the fuel pump module connector and fuel tank differential pressure sensor connector connection.

8. Raise and safely support the vehicle.

9. Remove the center exhaust pipe.

10. Remove the driveshaft, as required.

11. Remove the fuel tank protector, if equipped.

12. Disconnect the parking brake cable clamp.

13. Disconnect the fuel tank vapor hose connection.

14. Disconnect the fuel filler hose connection.

15. Disconnect the fuel high-pressure hose connection.

❋❋ CAUTION

As there will be some pressure remaining in the fuel pipe line, cover it with a shop towel to prevent fuel from spraying out.

16. Remove the fuel tank band.
   a. Support the fuel tank with a transaxle jack.
   b. Remove the front securing nut of the fuel tank band. Tilt the fuel tank assembly forward and lower it gradually to remove it. Remove the fuel tank band.

17. Carefully lower the fuel tank assembly from the vehicle.

To install:

18. Install the fuel tank bands.
   a. Raise the fuel tank assembly carefully with a transaxle jack.
   b. Ensure that the fuel tank assembly does not interfere with surrounding parts. Install the fuel tank band and tighten to 15–23 ft. lbs. (21–31 Nm).

19. Connect the fuel high-pressure hose connection.
   a. After installing, slightly pull the fuel high-pressure hose and ensure a good connection.

20. Connect the fuel filler hose connection.

21. Connect the fuel tank vapor hose connection.

22. Connect the parking brake cable clamp.

23. Install the fuel tank protector, if equipped.

24. Install the center exhaust pipe.

25. Connect the fuel level sensor connector.

26. Connect the fuel tank differential pressure sensor connector.

27. Connect the fuel pump module connector.

28. Install the service hole cover.

29. Install the rear seat cushion assembly.

30. Fill the fuel tank.

31. Start the engine and check for leaks.

IDLE SPEED

ADJUSTMENT

Idle speed is maintained by the Engine Control Module (ECM) or Powertrain Control Module (PCM). No adjustment is necessary or possible.

THROTTLE BODY

REMOVAL & INSTALLATION
See Figures 129 through 132.

➥When the throttle body is replaced, initialize the learned value using the Mitsubishi diagnostic scan tool, or equivalent.

❋❋ CAUTION

The fuel injection system remains under pressure even after the engine has been turned OFF. Properly relieve fuel pressure before disconnecting any fuel lines. Failure to do so may result in fire or personal injury. Do not allow fuel spray or fuel vapors to come in contact with a spark or an open flame. Keep a dry chemical fire extinguisher nearby. Never store fuel in an open container due to risk of fire or explosion.

➥When the throttle body assembly replacement is performed, use scan tool MB991958 to initialize the learning value. Do not loosen the fixing screws for the resin cover of throttle body assembly. If the screws are loosened, the sensor incorporated in the resin cover becomes misaligned and

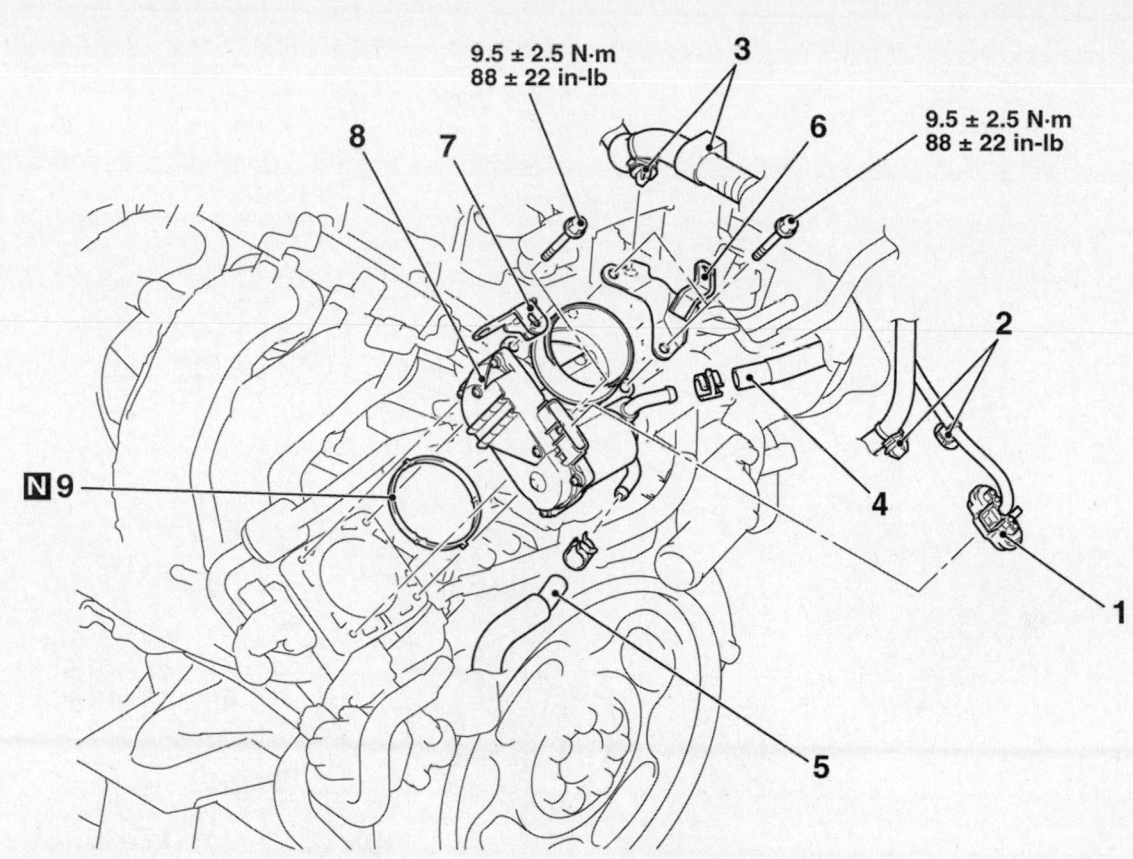

Fig. 129 Throttle body and related components—2.0L non turbocharged engine and 2.4L engine

9.5 ± 2.5 N·m
88 ± 22 in-lb

9.5 ± 2.5 N·m
88 ± 22 in-lb

1. Connector
2. Clamp connection
3. Clamp connection
4. Hose connection
5. Hose connection
6. Stay
7. Connector bracket
8. Throttle body assembly
9. Gasket

37671_LANC_G0341

the throttle body may not work normally.

1. Before servicing the vehicle, refer to the Precautions Section.

➡If working near and/or around the SRS system and components, be sure to disable the SRS system. Tape the negative battery cable with insulating tape. Always disconnect the negative battery cable first.

✳✳ CAUTION
Wait for 1 minute after disconnecting the negative battery cable before working inside the vehicle. The air bag system is set to deploy for a short period of time after the battery is disconnected.

2. Remove the engine upper cover.
3. On the 2.0L turbocharged engine, remove the battery and battery tray. Remove the air charge cooler outlet hose.
4. Drain the engine coolant. Be sure to properly dispose of used coolant.
5. Remove the air cleaner assembly.
6. Disconnect the electrical connectors.
7. Disconnect the hoses.
8. Remove the retaining bolts.
9. Remove the component from its mounting. Discard the gasket.

To install:

➡Be sure to use new fasteners, as required.

10. Installation is the reverse of the removal procedure.
11. Be sure to use a new gasket.

➡Poor idling quality and poor performance may be experienced if the gasket is installed incorrectly.

12. Perform the initialization operation (at installation). Turn the ignition switch to the ON position and then to the LOCK (OFF) position and hold for at least ten seconds.

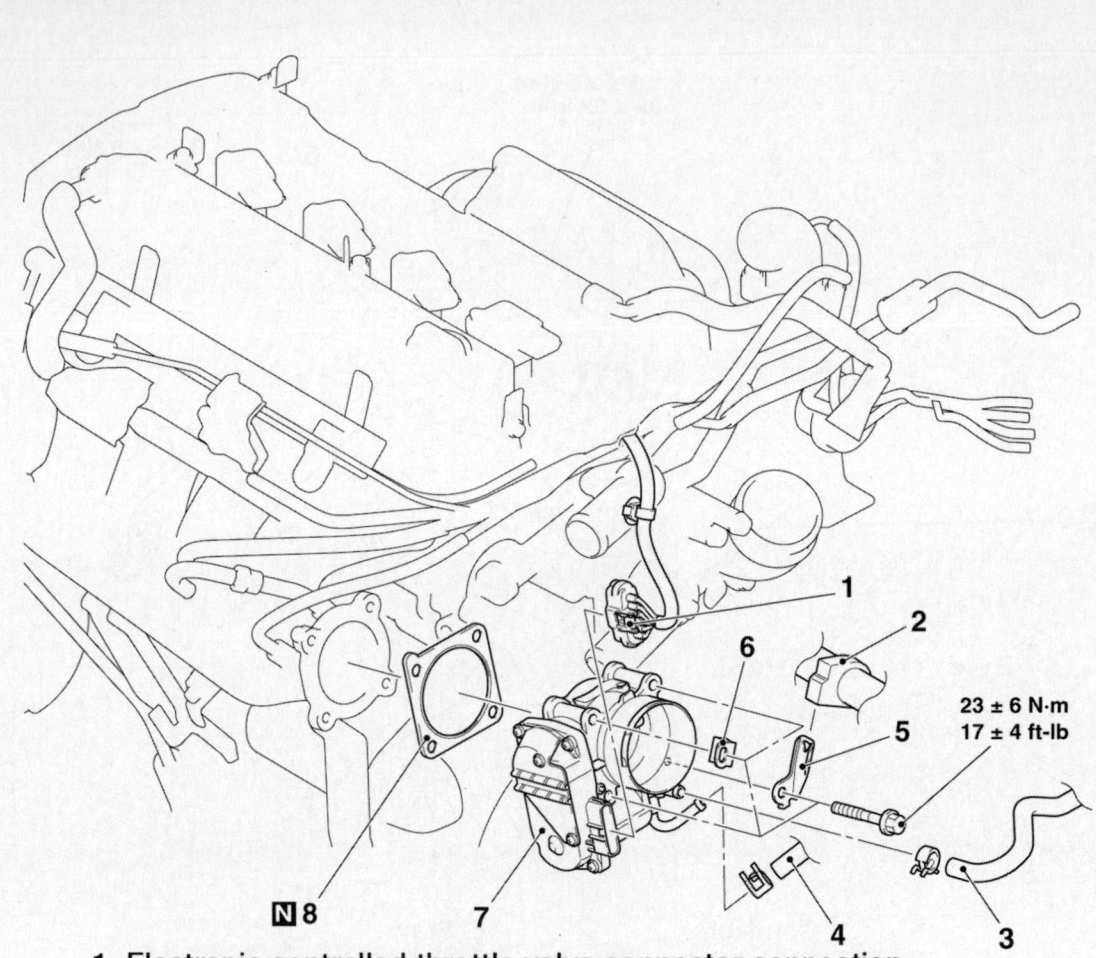

23 ± 6 N·m
17 ± 4 ft-lb

N8

1. Electronic controlled throttle valve connector connection
2. Battery wiring harness clamp
3. Cooling water line hose connection
4. Cooling water line hose connection
5. Harness bracket
6. Harness bracket
7. Throttle body assembly
8. Throttle body gasket

22140_MITS_G0171

Fig. 130 Throttle body and related components—2.0L turbocharged engine

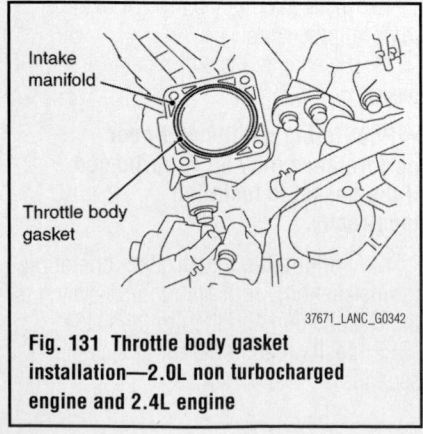

Intake manifold

Throttle body gasket

37671_LANC_G0342

Fig. 131 Throttle body gasket installation—2.0L non turbocharged engine and 2.4L engine

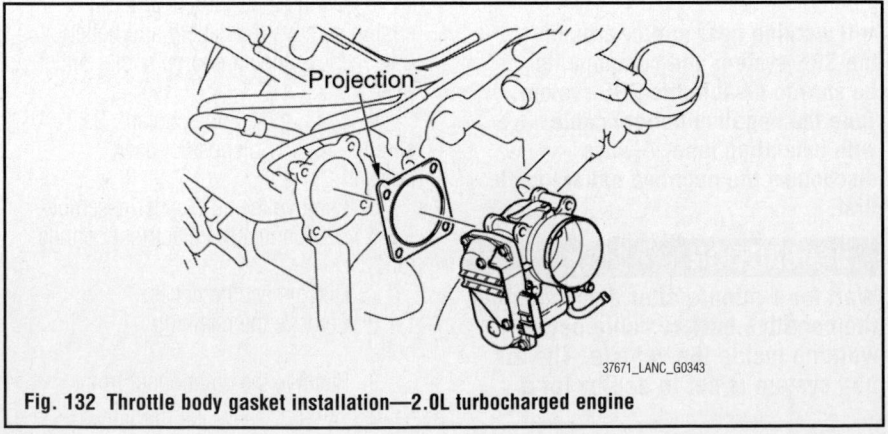

Projection

37671_LANC_G0343

Fig. 132 Throttle body gasket installation—2.0L turbocharged engine

HEATING & AIR CONDITIONING SYSTEM

BLOWER MOTOR

REMOVAL & INSTALLATION

See Figure 133.

1. Before servicing the vehicle, refer to the Precautions Section.

➡**If working near and/or around the SRS system and components, be sure to disable the SRS system. Tape the negative battery cable with insulating tape. Always disconnect the negative battery cable first.**

❋❋ CAUTION

Wait for 1 minute after disconnecting the negative battery cable before working inside the vehicle. The air bag system is set to deploy for a short period of time after the battery is disconnected.

2. Remove the bottom cover instrument panel assembly, left side.
3. Remove the power transistor.
4. Remove the blower motor hose.
5. Remove the blower motor retaining screws.

6. Remove the blower motor from its mounting.

To install:

➡**Be sure to use new fasteners, as required.**

7. Installation is the reverse of the removal procedure.

HEATER CORE

REMOVAL & INSTALLATION

➡**Depending on the particular model you are servicing, the front seats may have to be removed. After the seats have been reinstalled be sure to perform the accuracy check of the occupant classification sensor, using the Mitsubishi Diagnostic Scan tool, or equivalent. Follow the directions on the tool screen.**

1. Before servicing the vehicle, refer to the Precautions Section.

➡**If working near and/or around the SRS system and components, be sure to disable the SRS system. Tape the negative battery cable with insulating**

tape. Always disconnect the negative battery cable first.

❋❋ CAUTION

Wait for 1 minute after disconnecting the negative battery cable before working inside the vehicle. The air bag system is set to deploy for a short period of time after the battery is disconnected.

2. Remove the heater/AC unit from the vehicle.
3. Remove the heater core from the assembly.

To install:

➡**Be sure to use new fasteners, as required.**

4. Installation is the reverse of the removal procedure.
5. Be sure to properly recharge the system.
6. Be sure to fill the cooling system with the proper grade and type engine coolant.
7. Start the engine and check for leaks, correct as required.

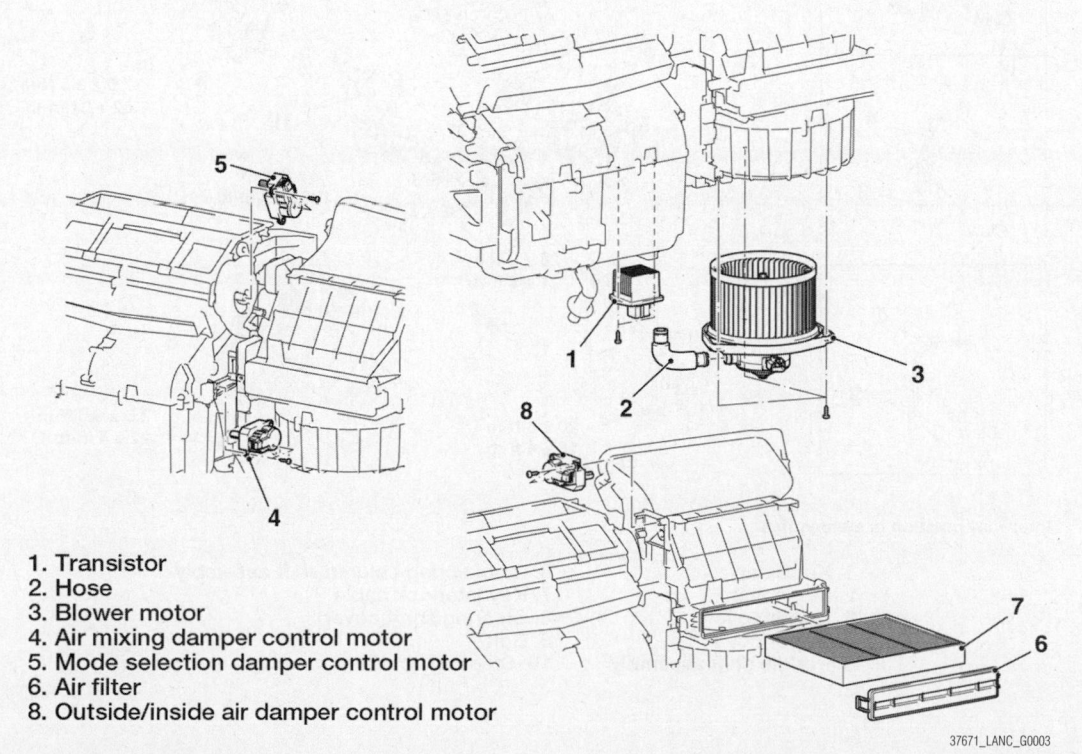

1. Transistor
2. Hose
3. Blower motor
4. Air mixing damper control motor
5. Mode selection damper control motor
6. Air filter
8. Outside/inside air damper control motor

37671_LANC_G0003

Fig. 133 Blower motor and related components

HEATER/AC UNIT

REMOVAL & INSTALLATION

See Figures 134 through 137.

➡Depending on the particular model you are servicing, the front seats may have to be removed. After the seats have been reinstalled be sure to perform the accuracy check of the occupant classification sensor, using the Mitsubishi Diagnostic Scan tool, or equivalent. Follow the directions on the tool screen.

1. Before servicing the vehicle, refer to the Precautions Section.

➡If working near and/or around the SRS system and components, be sure to disable the SRS system. Tape the negative battery cable with insulating tape. Always disconnect the negative battery cable first.

⁕⁕ CAUTION

Wait for 1 minute after disconnecting the negative battery cable before working inside the vehicle. The air bag system is set to deploy for a short period of time after the battery is disconnected.

2. Disconnect the negative battery cable.
3. Properly discharge the air conditioning system.
4. Drain the cooling system. Be sure to properly dispose of used coolant.
5. Remove the steering wheel.
6. Remove the steering column.
7. Remove the floor console.
8. Remove the instrument panel assembly.
9. Remove the heat protector and hose connection.

10. Disconnect the refrigerant lines. Discard the O-rings. Plug the lines.
11. Disconnect and plug the heater hoses.
12. Remove the crossmember
13. Remove the drain hose.
14. Remove the heater/AC unit from the vehicle.

To install:

➡Be sure to use new fasteners, as required.

15. Installation is the reverse of the removal procedure.
16. Be sure to properly recharge the system.
17. Be sure to fill the cooling system with the proper grade and type engine coolant.
18. Start the engine and check for leaks, correct as required.

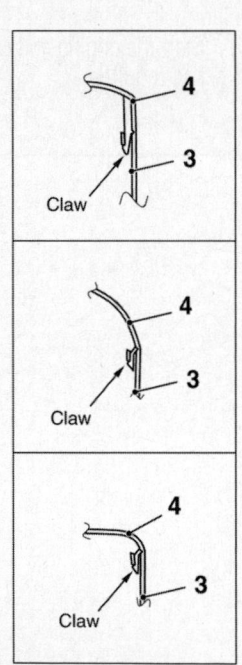

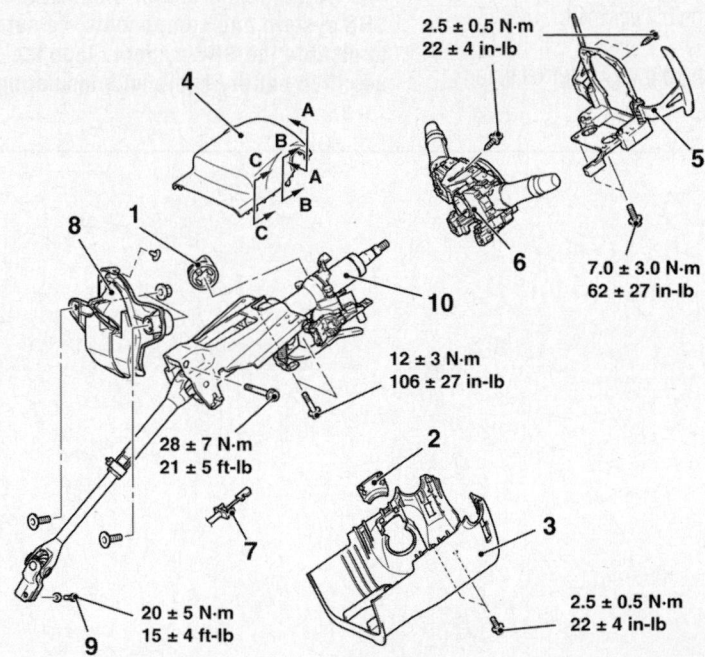

Note: claw position is symmetrical

1. Knob cap
2. Key cover
3. Lower cover
4. Upper cover
5. Paddle shift assembly
6. Clockspring column shift assembly
7. Key interlock cable
8. Steering shaft cover
9. Bolt
10. Steering column assembly

37671_LANC_G0025

Fig. 134 Steering column and related components.

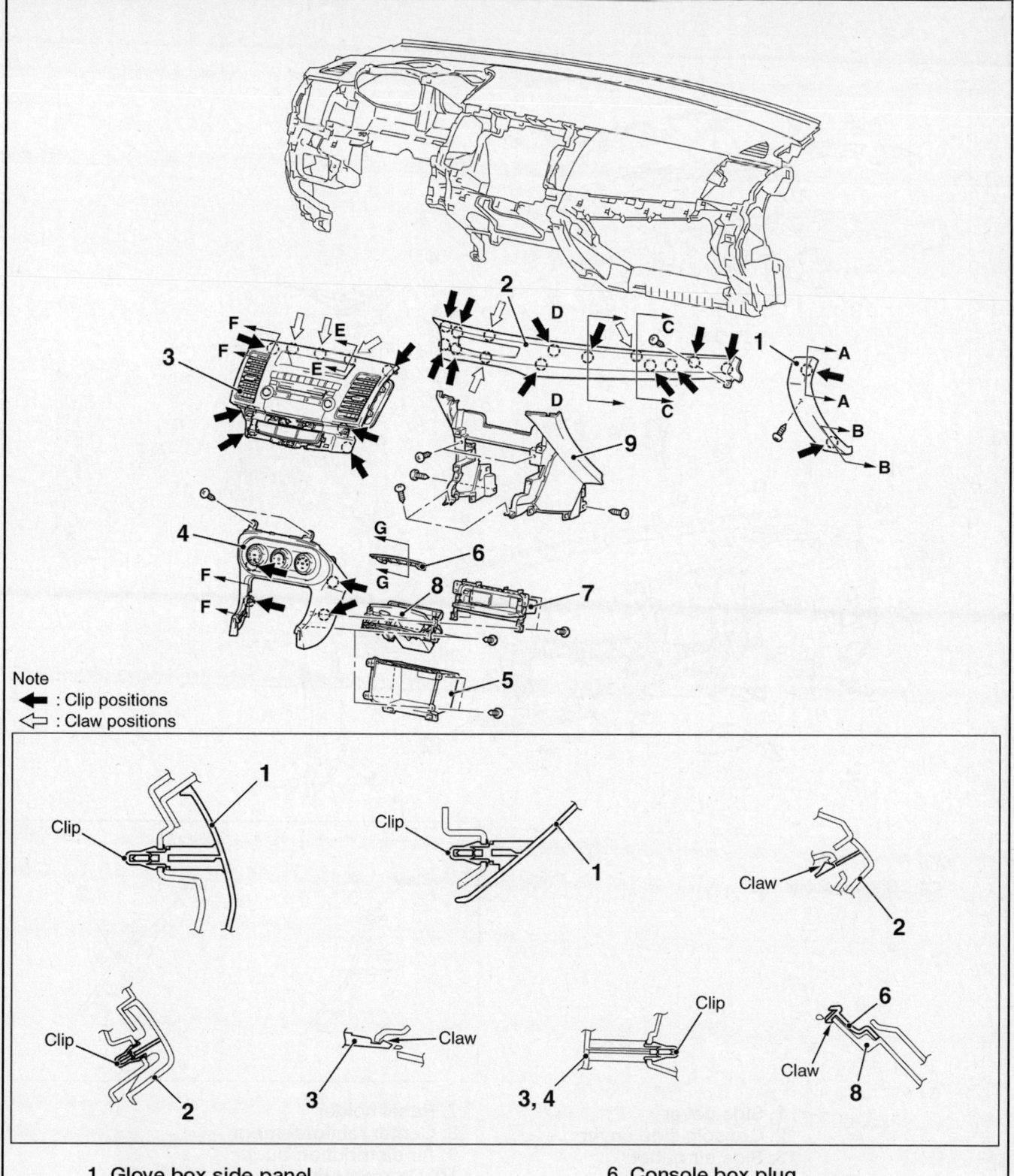

Note
← : Clip positions
⇐ : Claw positions

1. Glove box side panel
2. Air outlet garnish lower (right)
3. Instrument panel center panel
4. Instrument panel center panel lower
5. Console box
6. Console box plug
7. Tray, if equipped
8. Extra box, if equipped
9. Console side cover
10. Undercover

37671_LANC_G0023

Fig. 135 Instrument panel pad and related components

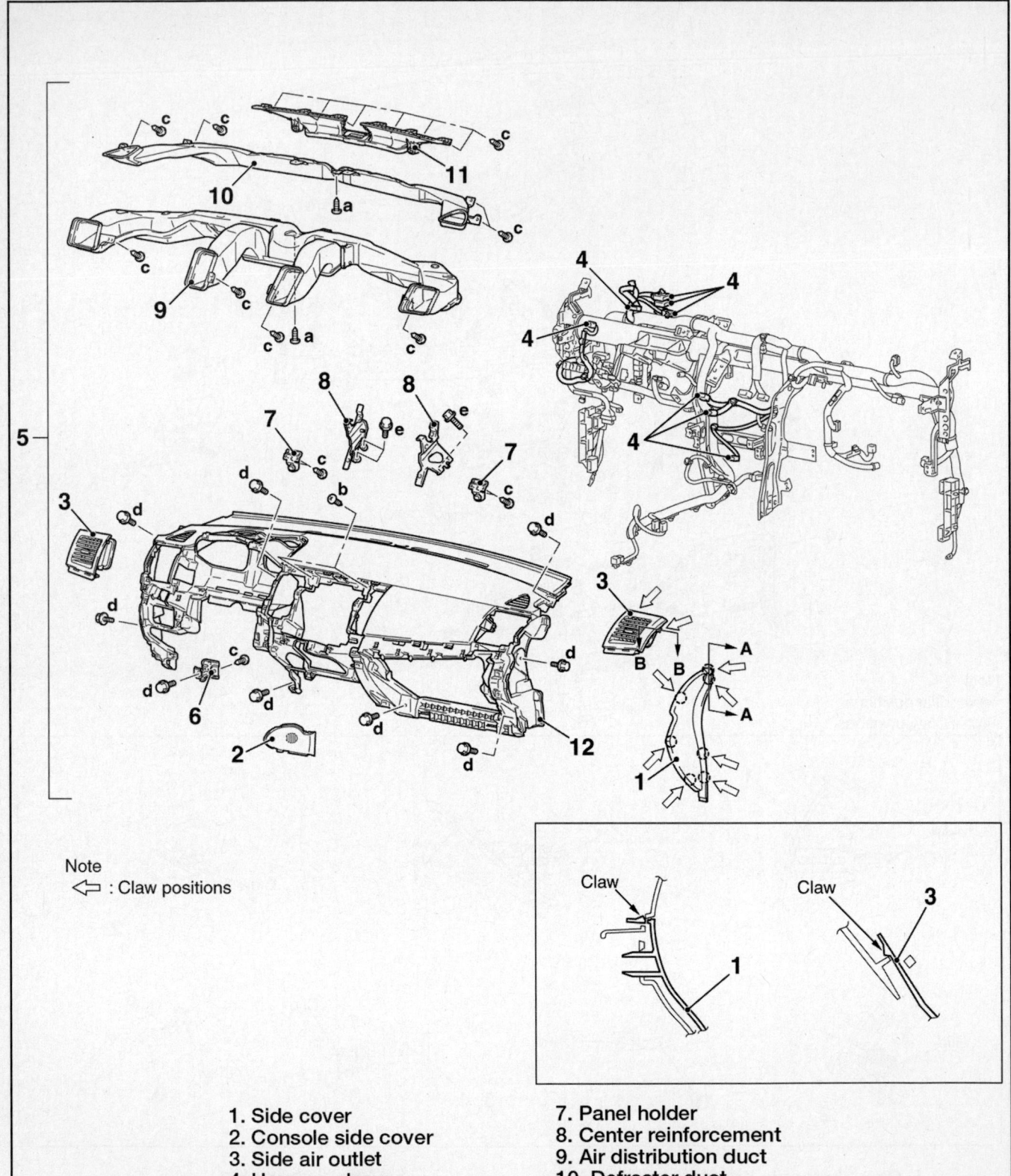

Note
⇐ : Claw positions

Claw

Claw

1. Side cover
2. Console side cover
3. Side air outlet
4. Harness clamp
5. Instrument panel assembly
6. Side lower bracket

7. Panel holder
8. Center reinforcement
9. Air distribution duct
10. Defroster duct
11. Defroster nozzle
12. Instrument panel

37671_LANC_G0026

Fig. 136 Instrument panel and related components

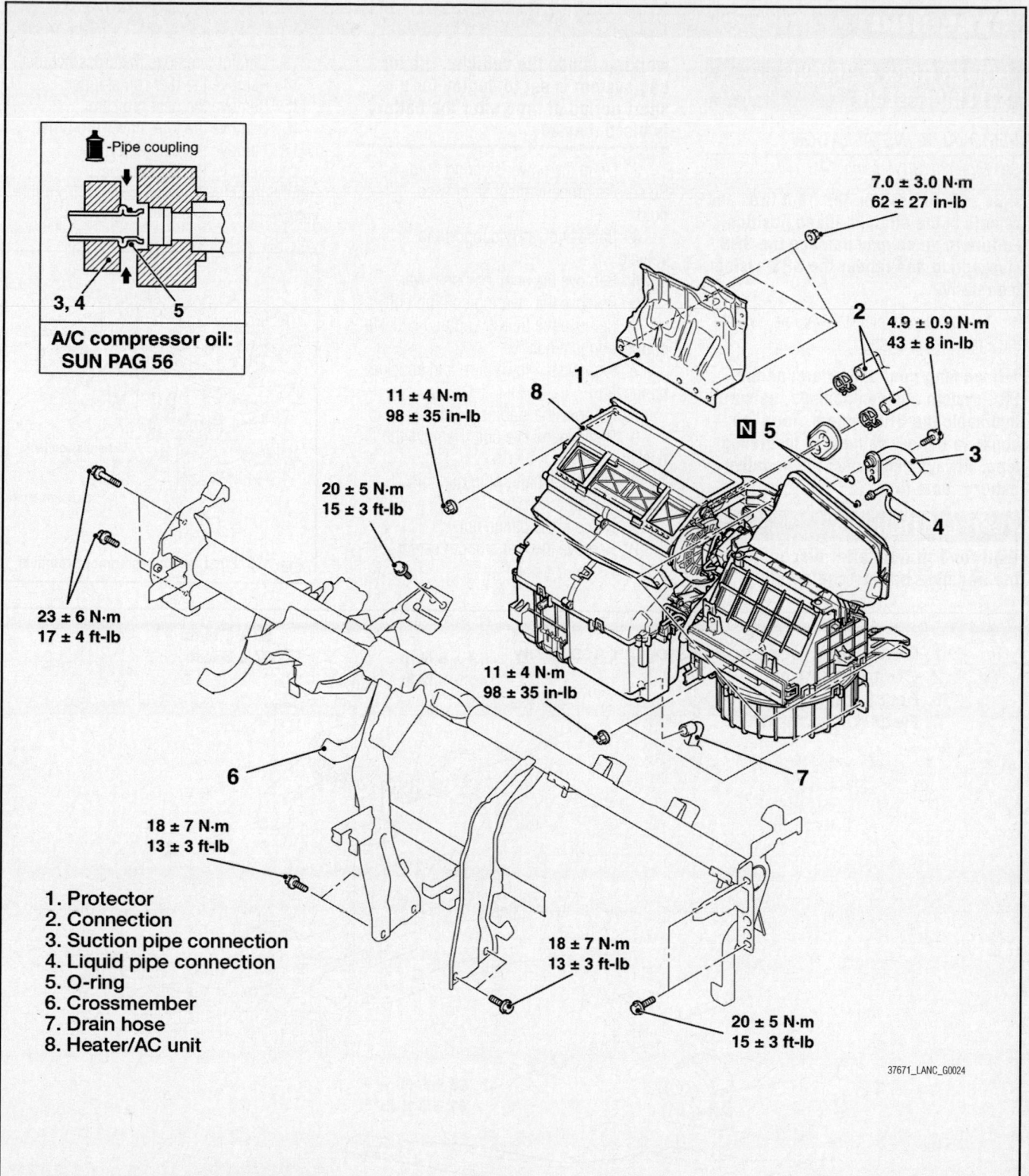

-Pipe coupling

3, 4   5

A/C compressor oil:
SUN PAG 56

7.0 ± 3.0 N·m
62 ± 27 in-lb

4.9 ± 0.9 N·m
43 ± 8 in-lb

2

N 5

3

4

11 ± 4 N·m
98 ± 35 in-lb

8

1

20 ± 5 N·m
15 ± 3 ft-lb

23 ± 6 N·m
17 ± 4 ft-lb

11 ± 4 N·m
98 ± 35 in-lb

7

6

18 ± 7 N·m
13 ± 3 ft-lb

18 ± 7 N·m
13 ± 3 ft-lb

20 ± 5 N·m
15 ± 3 ft-lb

1. Protector
2. Connection
3. Suction pipe connection
4. Liquid pipe connection
5. O-ring
6. Crossmember
7. Drain hose
8. Heater/AC unit

37671_LANC_G0024

Fig. 137 Heater/AC unit and related components

STEERING

POWER RACK & PINION STEERING GEAR

REMOVAL & INSTALLATION

See Figures 138 through 141.

➡ **Be sure to position the front tires and wheels in the straight ahead position. Failure to do so may damage the SRS clockspring and render the SRS system inoperative.**

1. Before servicing the vehicle, refer to the Precautions Section.

➡ **If working near and/or around the SRS system and components, be sure to disable the SRS system. Tape the negative battery cable with insulating tape. Always disconnect the negative battery cable first.**

✳✳ CAUTION

Wait for 1 minute after disconnecting the negative battery cable before working inside the vehicle. The air bag system is set to deploy for a short period of time after the battery is disconnected.

2. Drain the power steering fluid. Be sure to properly dispose of used fluid.
3. Raise and safely support the vehicle.
4. Remove the engine undercover.
5. Remove the steering column bolt.
6. Remove the tie rod end and knuckle connection jam nut.
7. Remove the lower arm and knuckle connection.
8. Remove the stabilizer link.
9. Remove the eye bolt. Remove the gasket.
10. Disconnect and plug the fluid lines.
11. Remove the flange nut.
12. Remove the rear stopper center bolt.
13. Carefully remove the crossmember.
14. Remove the rear roll stopper.
15. Remove the heat protector.
16. Remove the joint cover grommet.
17. Remove the flanged bolt.
18. Remove the steering gear mounting bolts. Remove the gear from the vehicle.

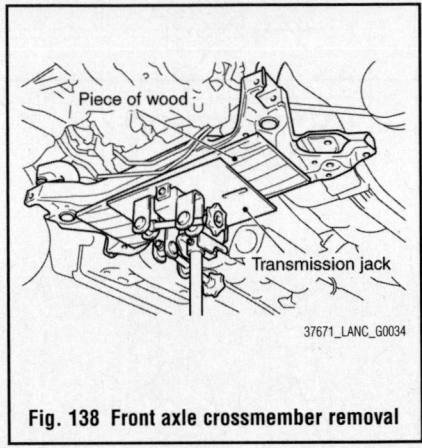

37671_LANC_G0034

Fig. 138 Front axle crossmember removal

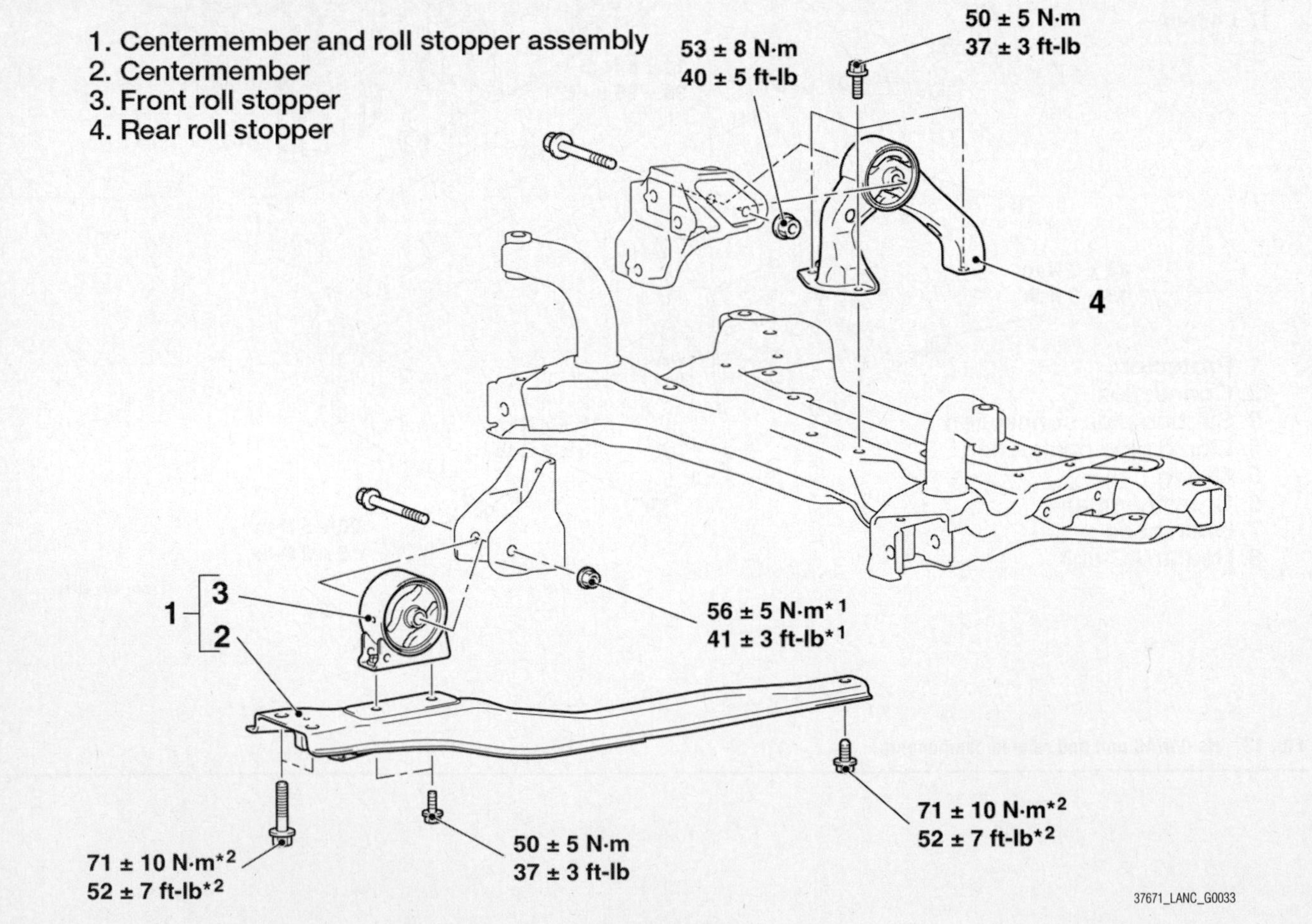

1. Centermember and roll stopper assembly
2. Centermember
3. Front roll stopper
4. Rear roll stopper

50 ± 5 N·m
37 ± 3 ft-lb

53 ± 8 N·m
40 ± 5 ft-lb

56 ± 5 N·m*1
41 ± 3 ft-lb*1

71 ± 10 N·m*2
52 ± 7 ft-lb*2

50 ± 5 N·m
37 ± 3 ft-lb

71 ± 10 N·m*2
52 ± 7 ft-lb*2

37671_LANC_G0033

Fig. 139 Centermember and front/rear roll stopper assembly and related components

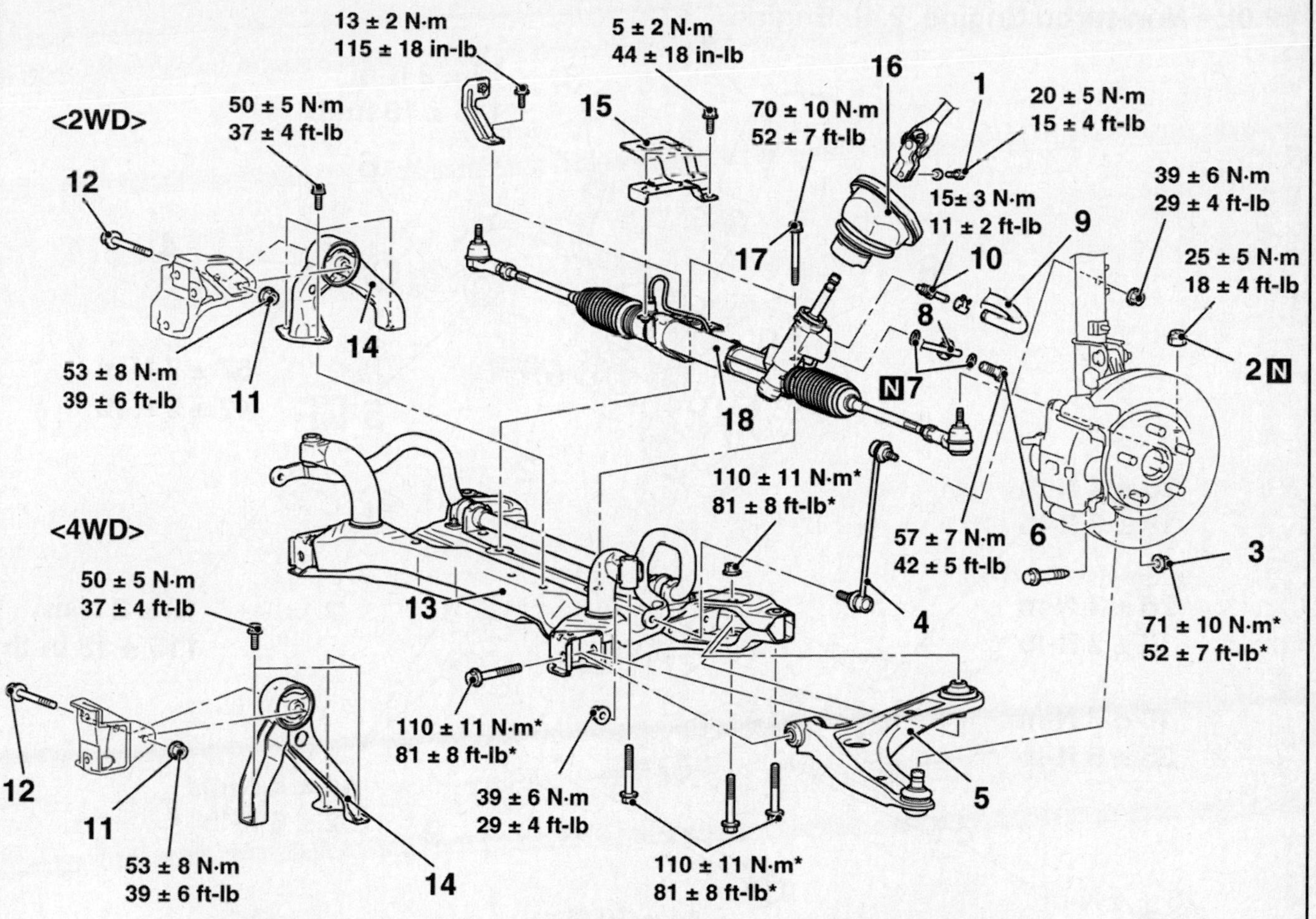

13 ± 2 N·m
115 ± 18 in-lb
5 ± 2 N·m
44 ± 18 in-lb
16
1
<2WD>
50 ± 5 N·m
37 ± 4 ft-lb
15
70 ± 10 N·m
52 ± 7 ft-lb
20 ± 5 N·m
15 ± 4 ft-lb
12
15± 3 N·m
11 ± 2 ft-lb
9
39 ± 6 N·m
29 ± 4 ft-lb
17
10
25 ± 5 N·m
18 ± 4 ft-lb
14
8
53 ± 8 N·m
39 ± 6 ft-lb
11
N 7
2 N
18
110 ± 11 N·m*
81 ± 8 ft-lb*
<4WD>
57 ± 7 N·m
42 ± 5 ft-lb
6
3
50 ± 5 N·m
37 ± 4 ft-lb
13
4
71 ± 10 N·m*
52 ± 7 ft-lb*
110 ± 11 N·m*
81 ± 8 ft-lb*
12
39 ± 6 N·m
29 ± 4 ft-lb
5
11
53 ± 8 N·m
39 ± 6 ft-lb
14
110 ± 11 N·m*
81 ± 8 ft-lb*

1. Bolt
2. Jam nut
3. Flange nut
4. Stabilizer link
5. Lower arm assembly
6. Eye bolt
7. Gasket
8. Hose connection
9. Hose connection
10. Tube assembly
11. Nut
12. Center bolt
13. Crossmember
14. Roll stopper
15. Protector
16. Grommet
17. Bolt
18. Gear and linkage

37671_LANC_G0031

Fig. 140 Steering gear and related components—Except Evolution

Steering joint
cover assembly
Protrusion part
Joint cover
grommet
Steering gear

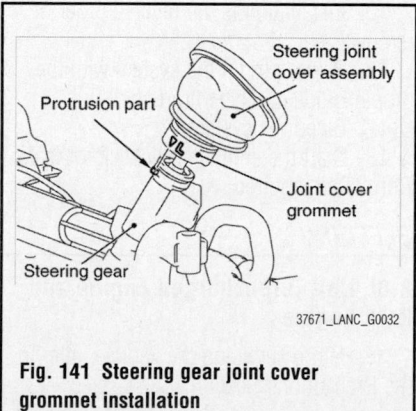

37671_LANC_G0032

Fig. 141 Steering gear joint cover grommet installation

To install:

➡Be sure to use new fasteners, as required.

19. Installation is the reverse of the removal procedure.
20. Install the joint cover grommet to the steering gear and linkage by aligning the L mark and the protrusion part of the gearbox as shown in the illustration.
21. Using the Mitsubishi diagnostic scan tool, or equivalent, perform the ASC-ECU relearn procedure to learn the steering wheel sensor neutral point, if the vehicle is equipped with ASC.

POWER STEERING PUMP

REMOVAL & INSTALLATION

See Figure 142.

1. Before servicing the vehicle, refer to the Precautions Section.

➡If working near and/or around the SRS system and components, be sure to disable the SRS system. Tape the negative battery cable with insulating tape. Always disconnect the negative battery cable first.

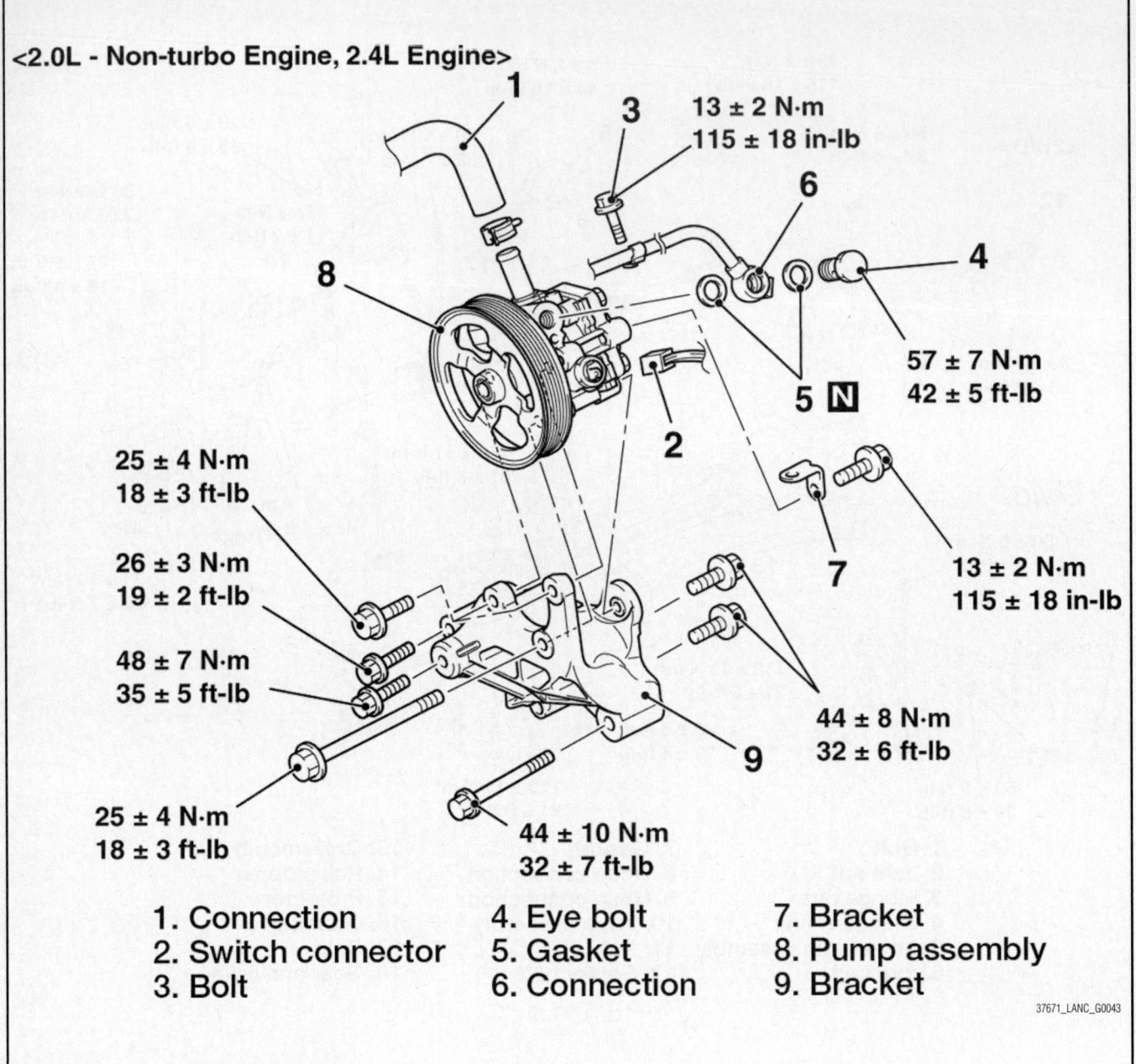

<2.0L - Non-turbo Engine, 2.4L Engine>

3   13 ± 2 N·m
    115 ± 18 in-lb

57 ± 7 N·m
42 ± 5 ft-lb

25 ± 4 N·m
18 ± 3 ft-lb

26 ± 3 N·m
19 ± 2 ft-lb

48 ± 7 N·m
35 ± 5 ft-lb

13 ± 2 N·m
115 ± 18 in-lb

44 ± 8 N·m
32 ± 6 ft-lb

25 ± 4 N·m
18 ± 3 ft-lb

44 ± 10 N·m
32 ± 7 ft-lb

1. Connection   4. Eye bolt   7. Bracket
2. Switch connector 5. Gasket   8. Pump assembly
3. Bolt   6. Connection   9. Bracket

37671_LANC_G0043

Fig. 142 Power steering pump and related components—2.0L non-turbo and 2.4L engines shown, Turbo engine similar

※※ CAUTION

Wait for 1 minute after disconnecting the negative battery cable before working inside the vehicle. The air bag system is set to deploy for a short period of time after the battery is disconnected.

2. Disconnect the negative battery cable.

3. Remove accessory drive belt.

4. Remove the engine cover.

5. Remove the radiator condenser tank.

6. Drain the power steering fluid. Be sure to dispose of used fluid.

7. Disconnect the supply and return hoses. Discard the gaskets. Plug the lines.

8. Remove the power steering pump mounting bolts.

9. Remove the power steering pump.

To install:

➡**Be sure to use new fasteners, as required.**

10. Installation is the reverse order of removal.

11. Be sure to fill the system with the proper grade and type fluid.

12. Bleed the system.

13. Start the engine and check for leaks. Correct as required.

BLEEDING

2.0L Non Turbocharged Engine and 2.4L Engine

1. Before servicing the vehicle, refer to the Precautions Section.

➡ If working near and/or around the SRS system and components, be sure to disable the SRS system. Tape the negative battery cable with insulating tape. Always disconnect the negative battery cable first.

✳✳ CAUTION

Wait for 1 minute after disconnecting the negative battery cable before working inside the vehicle. The air bag system is set to deploy for a short period of time after the battery is disconnected.

2. Raise and safely support the vehicle.
3. Disconnect the injector connectors.

➡ Perform air bleeding only, while cranking the engine. Do not perform air bleeding while the engine is running. If you do so, air in the fluid will be increased and air bleeding will become more difficult.

➡ During air bleeding refill the steering fluid so that the level never falls below the low mark on the dipstick.

4. Turn the steering wheel all the way to the left and to the right 5 or 6 times, while using the starter motor to crank the engine.
5. Connect the injector connectors.

6. Start engine and allow to idle.
7. Turn the steering wheel left and right until there are no air bubbles in the reservoir. Confirm that the fluid is not milky and the level is up to the specified position on the gauge. Confirm that there is very little change in the fluid level when the steering wheel is turned.
8. Using the Mitsubishi diagnostic scan tool, or equivalent, check for diagnostic codes and clear.

2.0L Turbocharged Engine

1. Before servicing the vehicle, refer to the Precautions Section.

➡ If working near and/or around the SRS system and components, be sure to disable the SRS system. Tape the negative battery cable with insulating tape. Always disconnect the negative battery cable first.

✳✳ CAUTION

Wait for 1 minute after disconnecting the negative battery cable before working inside the vehicle. The air bag system is set to deploy for a short period of time after the battery is disconnected.

2. Raise and safely support the vehicle.

3. Disconnect the Crankshaft Position (CKP) sensor connector.

➡ Perform air bleeding only, while cranking the engine. Do not perform air bleeding while the engine is running. If you do so, air in the fluid will be increased and air bleeding will become more difficult.

➡ During air bleeding refill the steering fluid so that the level never falls below the low mark on the dipstick.

4. Turn the steering wheel all the way to the left and to the right 5 or 6 times, while using the starter motor to crank the engine.
5. Connect the Crankshaft Position (CKP) sensor connector.
6. Start engine and allow to idle.
7. Turn the steering wheel left and right until there are no air bubbles in the reservoir. Confirm that the fluid is not milky and the level is up to the specified position on the gauge. Confirm that there is very little change in the fluid level when the steering wheel is turned.
8. Using the Mitsubishi diagnostic scan tool, or equivalent, check for diagnostic codes and clear.

SUSPENSION

LOWER BALL JOINT

REMOVAL & INSTALLATION

The lower ball joint is an integral part of the lower control arm assembly and cannot be serviced separately. A worn or damaged ball joint requires replacement of lower control arm assembly.

LOWER CONTROL ARM

REMOVAL & INSTALLATION

Except Evolution

See Figure 143.

1. Before servicing the vehicle, refer to the Precautions Section.

➡ If working near and/or around the SRS system and components, be sure to disable the SRS system. Tape the negative battery cable with insulating tape. Always disconnect

the negative battery cable first.

✳✳ CAUTION

Wait for 1 minute after disconnecting the negative battery cable before working inside the vehicle. The air bag system is set to deploy for a short period of time after the battery is disconnected.

2. Raise the vehicle and support safely.
3. Remove the wheel and tire assembly.
4. Remove the engine undercover.
5. Remove the stabilizer bar self-locking nut, rubber bushings, stabilizer bar, and collar. Discard the nut.
6. Remove the lower control arm-to-knuckle bolt and nut.
7. Lift the transaxle with a jack, then remove the front control arm-to-crossmember bolt.
8. Remove the lower control arm.

FRONT SUSPENSION

To install:

➡ Be sure to use new fasteners, as required.

9. Install or connect the following:
   • Lower control arm into the vehicle
   • Lower control arm-to-crossmember bolts. Torque the bottom bolt snug until the vehicle is lowered.
   • Lower control arm-to-steering knuckle bolt. Torque to specification.
   • Stabilizer bar collar, stabilizer bar, bushings and new self-locking nut.
10. Lower the vehicle and install the wheels. Then, with the weight of the vehicle on the wheels, torque the side control arm-to-crossmember bolt to specification.
11. Check and adjust the front end alignment, as required.

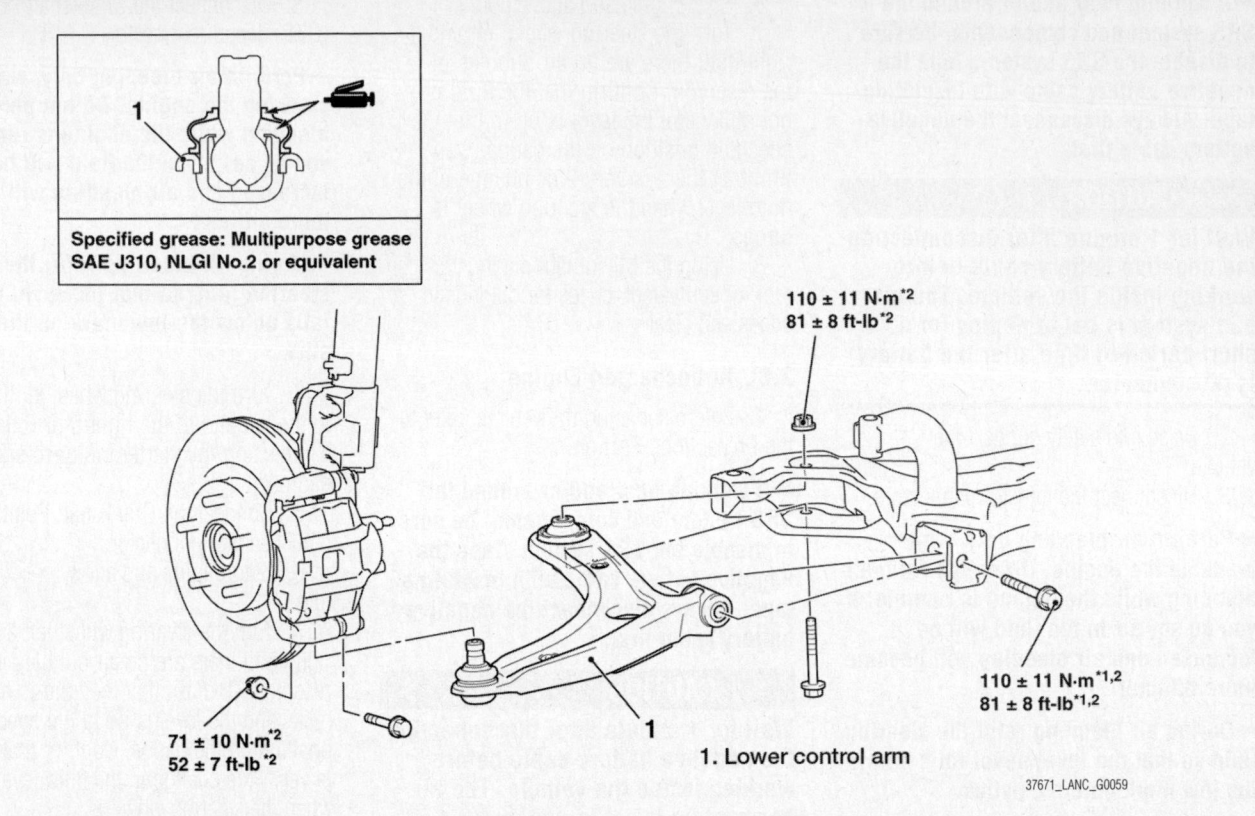

Specified grease: Multipurpose grease
SAE J310, NLGI No.2 or equivalent

110 ± 11 N·m[*2]
81 ± 8 ft-lb[*2]

110 ± 11 N·m[*1,2]
81 ± 8 ft-lb[*1,2]

71 ± 10 N·m[*2]
52 ± 7 ft-lb[*2]

1. Lower control arm

37671_LANC_G0059

Fig. 143 Lower control arm and related components —except Evolution

Evolution

See Figure 144.

1. Before servicing the vehicle, refer to the Precautions Section.

➡**If working near and/or around the SRS system and components, be sure to disable the SRS system. Tape the negative battery cable with insulating tape. Always disconnect the negative battery cable first.**

※※ CAUTION

Wait for 1 minute after disconnecting the negative battery cable before working inside the vehicle. The air bag system is set to deploy for a short period of time after the battery is disconnected.

2. Raise and support the vehicle safely.
3. Remove the undercover and side cover.

➡**Do not remove the nut from the ball joint. Loosen it and use special tool MB991897 to avoid possible damage to the ball joint threads. Hang the special tool in place with wire or string to prevent it from falling.**

4. To disconnect the lower arm from the knuckle, replace the self-locking nut for the lower arm ball joint with a regular nut. This is done because the original one is too large to install the special tool. Install special tool MB991897.

5. Turn the bolt and knob as necessary to make the jaws of the tool parallel. Tighten the bolt by hand and confirm that the jaws are still parallel.

➡**When adjusting the jaws in parallel, make sure the knob is in the vertical (upward) position.**

6. Tighten the bolt with a wrench to disconnect the ball joint.
7. Disconnect the stabilizer control link from the lower control arm.
8. Remove the lower control arm retaining bolts.
9. Remove the lower control arm from the vehicle.

To install:

➡**Be sure to use new fasteners, as required.**

10. Installation is the reverse of the removal procedure.
11. Tighten the bolts as illustrated in the figure.
12. Check and adjust the front end alignment, as required.

MACPHERSON STRUT

REMOVAL & INSTALLATION

Except Evolution

See Figure 145.

1. Before servicing the vehicle, refer to the Precautions Section.

➡**If working near and/or around the SRS system and components, be sure to disable the SRS system. Tape the negative battery cable with insulating tape. Always disconnect the negative battery cable first.**

※※ CAUTION

Wait for 1 minute after disconnecting the negative battery cable before working inside the vehicle. The air bag system is set to deploy for a short period of time after the battery is disconnected.

2. Raise and safely support the vehicle.
3. Matchmark alignment markings on the camber adjusting bolt and strut for approximate installation alignment later.
4. Remove the brake hose bracket.
5. On vehicles equipped with ABS, remove the wheel speed sensor clamp.

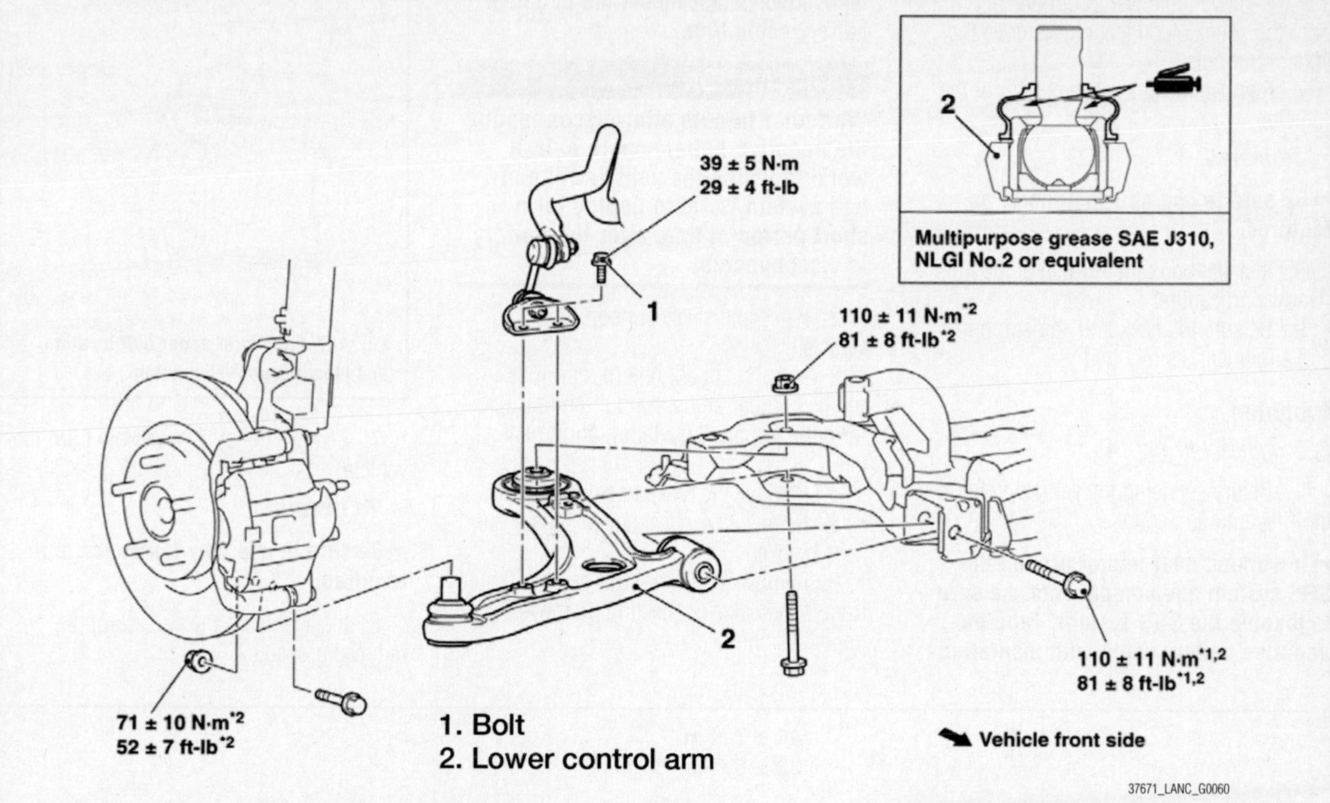

39 ± 5 N·m
29 ± 4 ft-lb

Multipurpose grease SAE J310, NLGI No.2 or equivalent

110 ± 11 N·m*2
81 ± 8 ft-lb*2

71 ± 10 N·m*2
52 ± 7 ft-lb*2

110 ± 11 N·m*1,2
81 ± 8 ft-lb*1,2

1. Bolt
2. Lower control arm

◄ Vehicle front side

37671_LANC_G0060

Fig. 144 Lower control arm and related components —Evolution

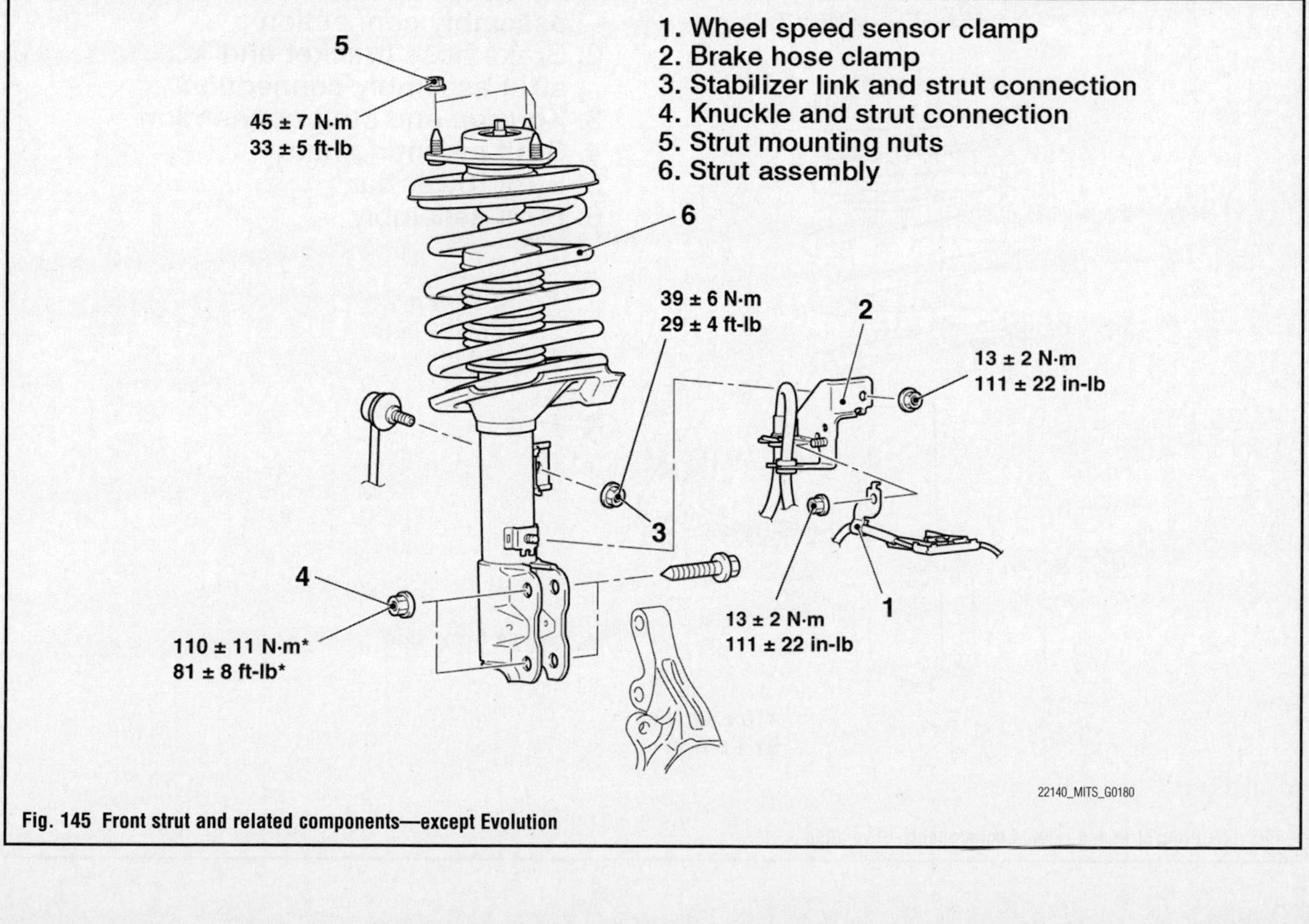

1. Wheel speed sensor clamp
2. Brake hose clamp
3. Stabilizer link and strut connection
4. Knuckle and strut connection
5. Strut mounting nuts
6. Strut assembly

45 ± 7 N·m
33 ± 5 ft-lb

39 ± 6 N·m
29 ± 4 ft-lb

13 ± 2 N·m
111 ± 22 in-lb

13 ± 2 N·m
111 ± 22 in-lb

110 ± 11 N·m*
81 ± 8 ft-lb*

22140_MITS_G0180

Fig. 145 Front strut and related components—except Evolution

6. Remove the lower strut bolts and nuts from the knuckle connection. Remove the upper strut nuts.

7. Remove the strut assembly from the vehicle.

To install:

➡ Be sure to use new fasteners, as required.

8. Installation is the reverse of the removal procedure.

9. Be sure to check and adjust the front end alignment, as required.

Evolution

See Figures 146 and 147.

1. Before servicing the vehicle, refer to the Precautions Section.

➡ If working near and/or around the SRS system and components, be sure to disable the SRS system. Tape the negative battery cable with insulating tape. Always disconnect the negative battery cable first.

✴✴ CAUTION

Wait for 1 minute after disconnecting the negative battery cable before working inside the vehicle. The air bag system is set to deploy for a short period of time after the battery is disconnected.

2. Raise and safely support the vehicle.

3. Matchmark alignment markings on the camber adjusting bolt and strut for approximate installation alignment later.

4. Remove the brake hose bracket.

5. Remove the wheel speed sensor harness bracket.

6. Remove the lower strut bolts and nuts from the knuckle connection. Remove the upper strut nuts.

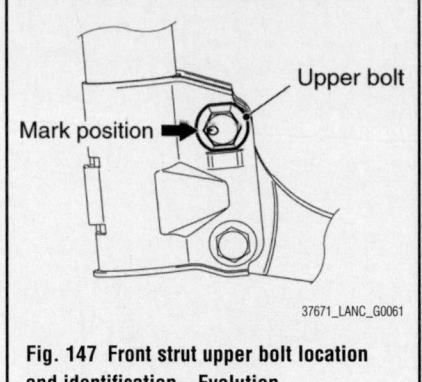

Fig. 147 Front strut upper bolt location and identification—Evolution

7. Remove the strut assembly from the vehicle.

To install:

➡ Be sure to use new fasteners, as required.

8. Installation is the reverse of the removal procedure.

**45 ± 7 N·m
33 ± 5 ft-lb**

1. Harness clip and strut assembly connection
2. Brake hose bracket and strut assembly connection
3. Knuckle and strut connection
4. Strut mounting nuts
5. Strut tower bar
6. Strut assembly

**13 ± 2 N·m
111 ± 22 in-lb**

◀ Vehicle front side

**110 ± 11 N·m
81 ± 8 ft-lb**

22140_MITS_G0181

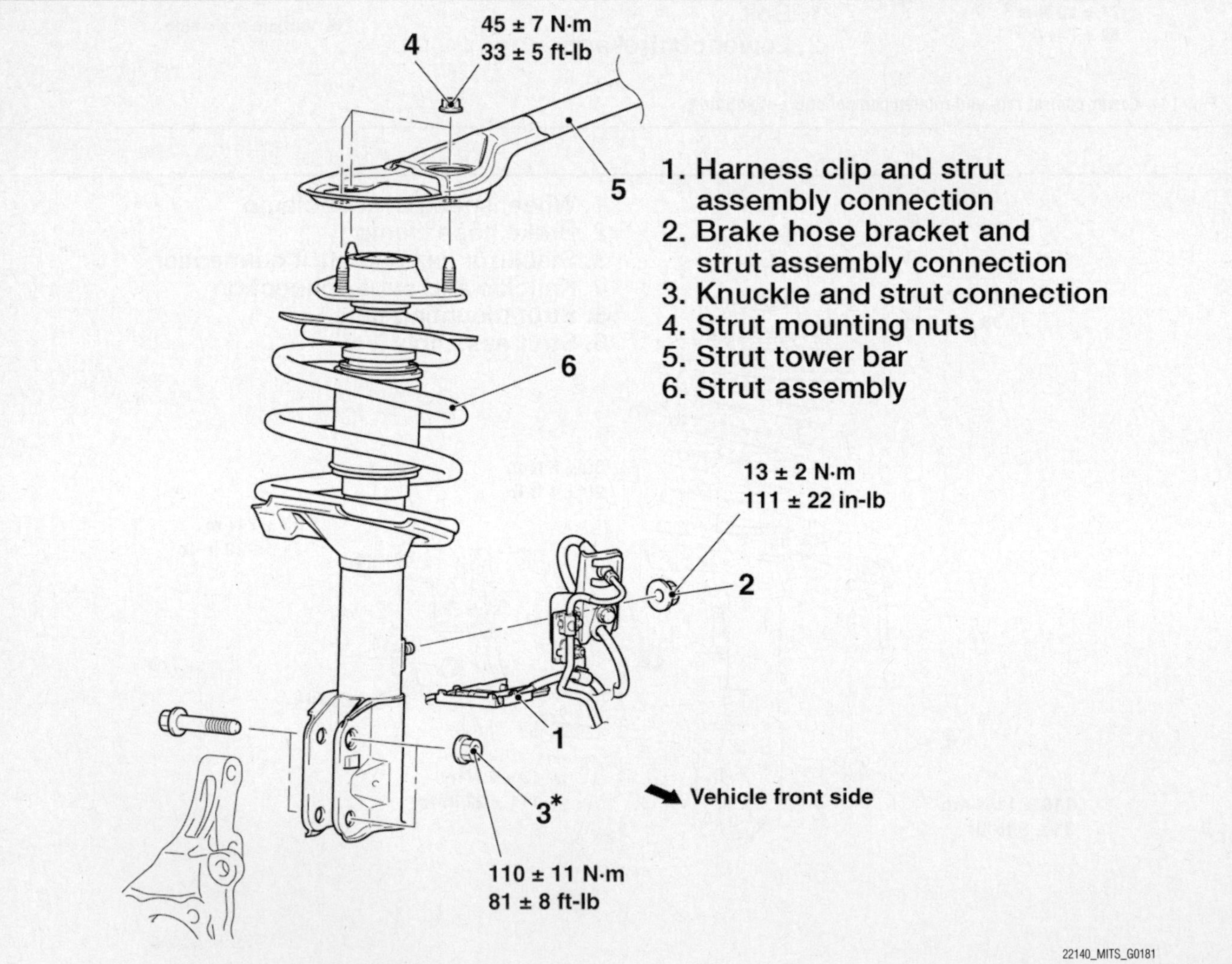

Fig. 146 Front strut and related components—Evolution

9. Be sure to check and adjust the front end alignment, as required.

STEERING KNUCKLE

REMOVAL & INSTALLATION

See Figures 148 through 151.

1. Before servicing the vehicle, refer to the Precautions Section.

➡**If working near and/or around the SRS system and components, be sure to disable the SRS system. Tape the negative battery cable with insulating tape. Always disconnect the negative battery cable first.**

✳✳ CAUTION

Wait for 1 minute after disconnecting the negative battery cable before working inside the vehicle. The air bag system is set to deploy for a short period of time after the battery is disconnected.

2. Raise and safely support the vehicle.

3. Remove the front wheels and tires.

4. Remove the brake rotor.

5. Remove the lower ball joint.

6. Remove the wheel speed sensor, the strut lower mounting bolt, and the lower arm mounting bolt from the steering knuckle.

7. Remove the front wheel hub mounting bolts while pushing out the driveshaft by hand.

8. If it is difficult to push out the driveshaft by hand, use special tools MB990242, MB990244, MB991354, and MB990767 to push out the driveshaft from the hub and knuckle.

9. If the front wheel hub is seized,

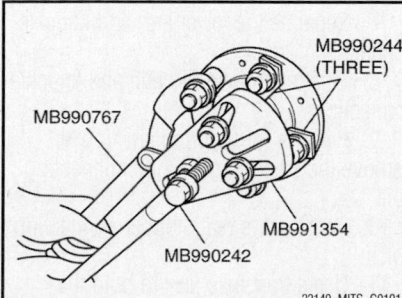

Fig. 148 Use special tools MB990242, MB990244, MB991354, and MB990767 to push out the driveshaft from the hub and knuckle

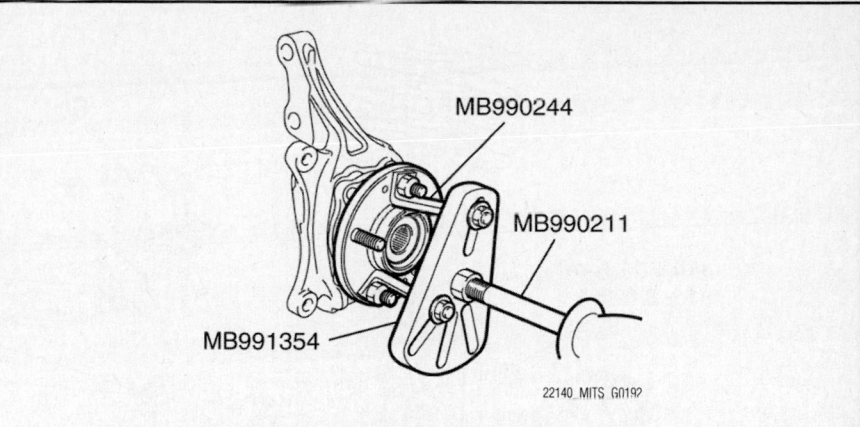

Fig. 149 Use special tools MB990244, MB991354, and MB990211 to pull out the front wheel hub from the steering knuckle

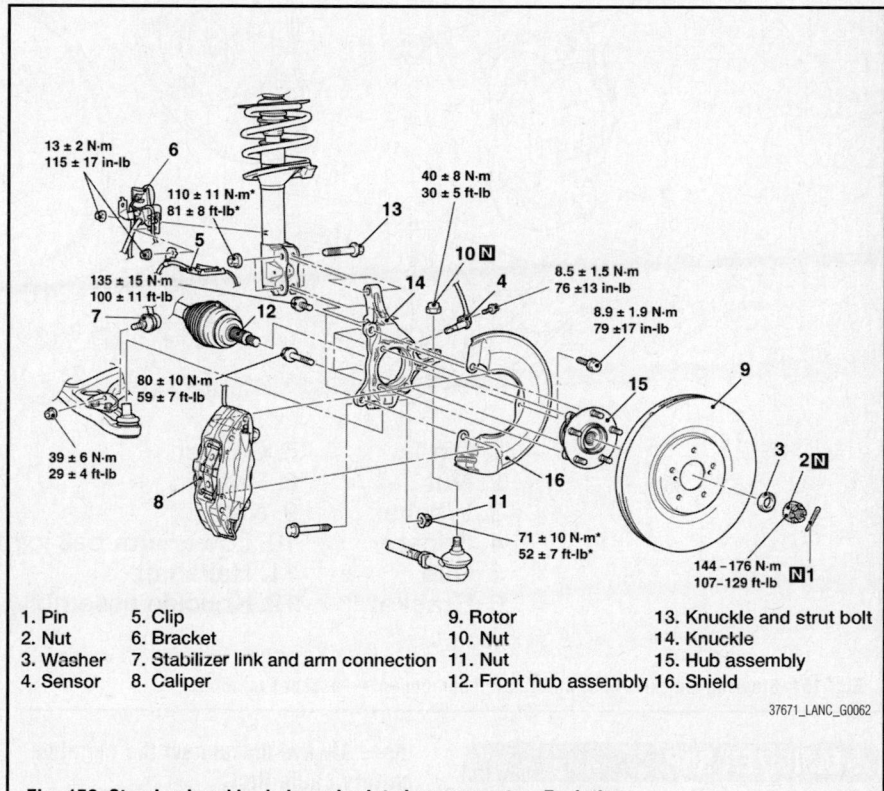

Fig. 150 Steering knuckle, hub, and related components—Evolution

1. Pin
2. Nut
3. Washer
4. Sensor
5. Clip
6. Bracket
7. Stabilizer link and arm connection
8. Caliper
9. Rotor
10. Nut
11. Nut
12. Front hub assembly
13. Knuckle and strut bolt
14. Knuckle
15. Hub assembly
16. Shield

remove the knuckle together with front wheel hub and fix them in a vise.

10. Hang the driveshaft on the vehicle body with a rope.

11. Use special tools MB990244, MB991354, and MB990211 to pull out the front wheel hub from the steering knuckle.

To install:

➡**Be sure to use new fasteners, as required.**

12. Install the hub and knuckle assembly.

13. Install the wheel speed sensor, the strut lower mounting bolt, and the lower arm mounting bolt to the steering knuckle.

14. Tighten the bolts according to the illustration.

15. Install the lower ball joint.

16. Install the brake rotor.

17. Install the front wheels and tires.

18. Check alignment and adjust as necessary.

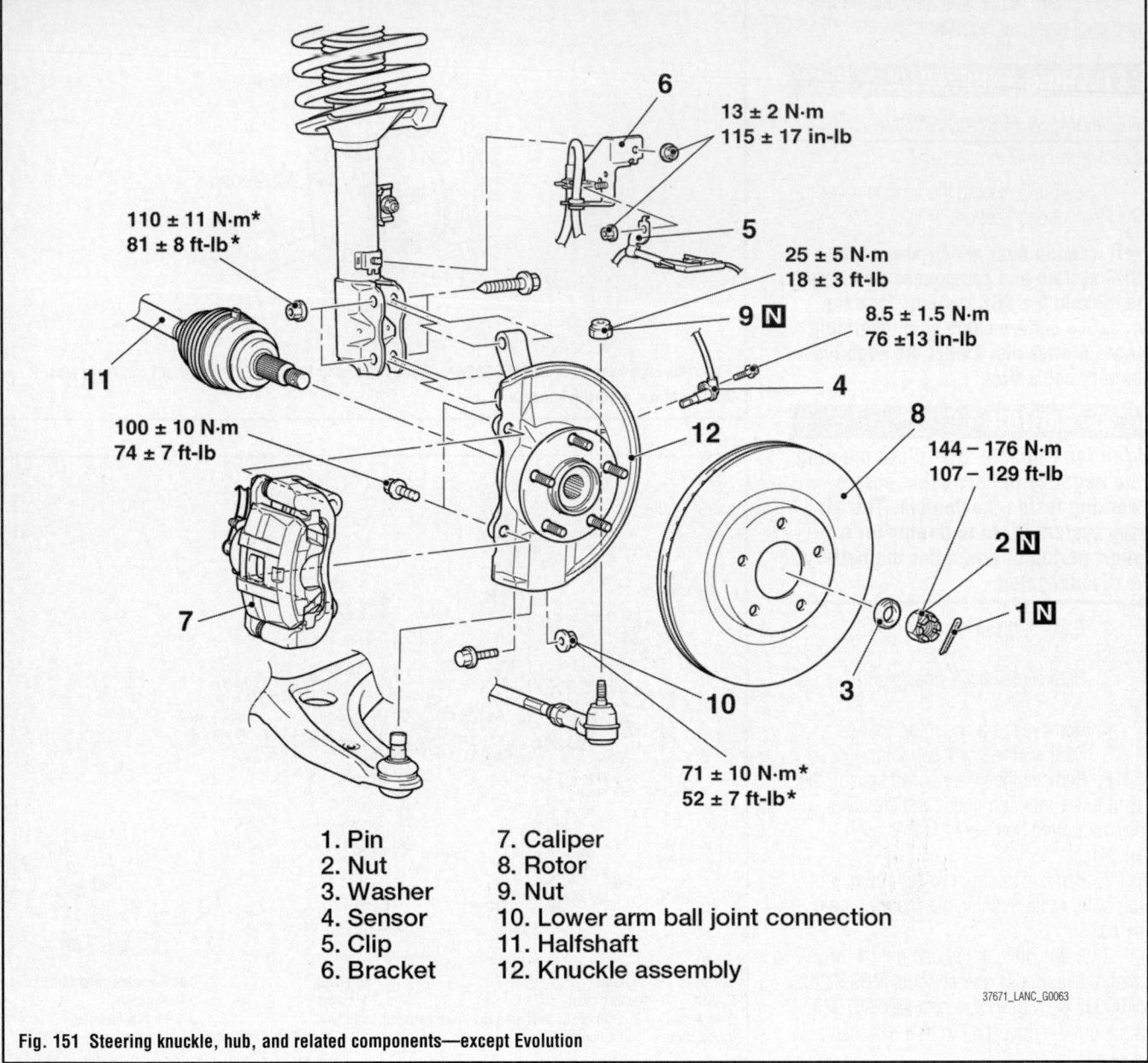

1. Pin
2. Nut
3. Washer
4. Sensor
5. Clip
6. Bracket
7. Caliper
8. Rotor
9. Nut
10. Lower arm ball joint connection
11. Halfshaft
12. Knuckle assembly

37671_LANC_G0063

Fig. 151 Steering knuckle, hub, and related components—except Evolution

STABILIZER BAR

REMOVAL & INSTALLATION

Except Evolution

See Figures 152 through 155.

➡ Prior to removal of the stabilizer bar, center the front wheels and remove the ignition key. Failure to do so may damage the SRS clockspring and render SRS system inoperative.

1. Before servicing the vehicle, refer to the Precautions Section.

➡ If working near and/or around the SRS system and components, be sure to disable the SRS system. Tape the negative battery cable with insulating tape. Always disconnect the negative battery cable first.

✳✳ CAUTION

Wait for 1 minute after disconnecting the negative battery cable before working inside the vehicle. The air bag system is set to deploy for a short period of time after the battery is disconnected.

2. Disconnect the negative battery cable.

3. Remove the driver's side air bag module. Remove the steering wheel. Remove the column lower cover. Remove the clockspring.

4. Remove the engine undercover.

5. Remove the center member.

6. Remove the stabilizer bar locknut. Remove the stabilizer bar rubber and collar.

7. Separate the lower arm and knuckle connection

8. Separate the tie rod end and knuckle connection.

9. Remove the steering shaft cover. Remove the steering gear joint connecting bolt.

10. Remove the rear stopper connecting bolt.

11. Use a transaxle jack to hold the crossmember in place. Remove the crossmember mounting bolts and nuts.

➡ Be careful not to lower the crossmember too much, otherwise the power steering return hose bracket may deform.

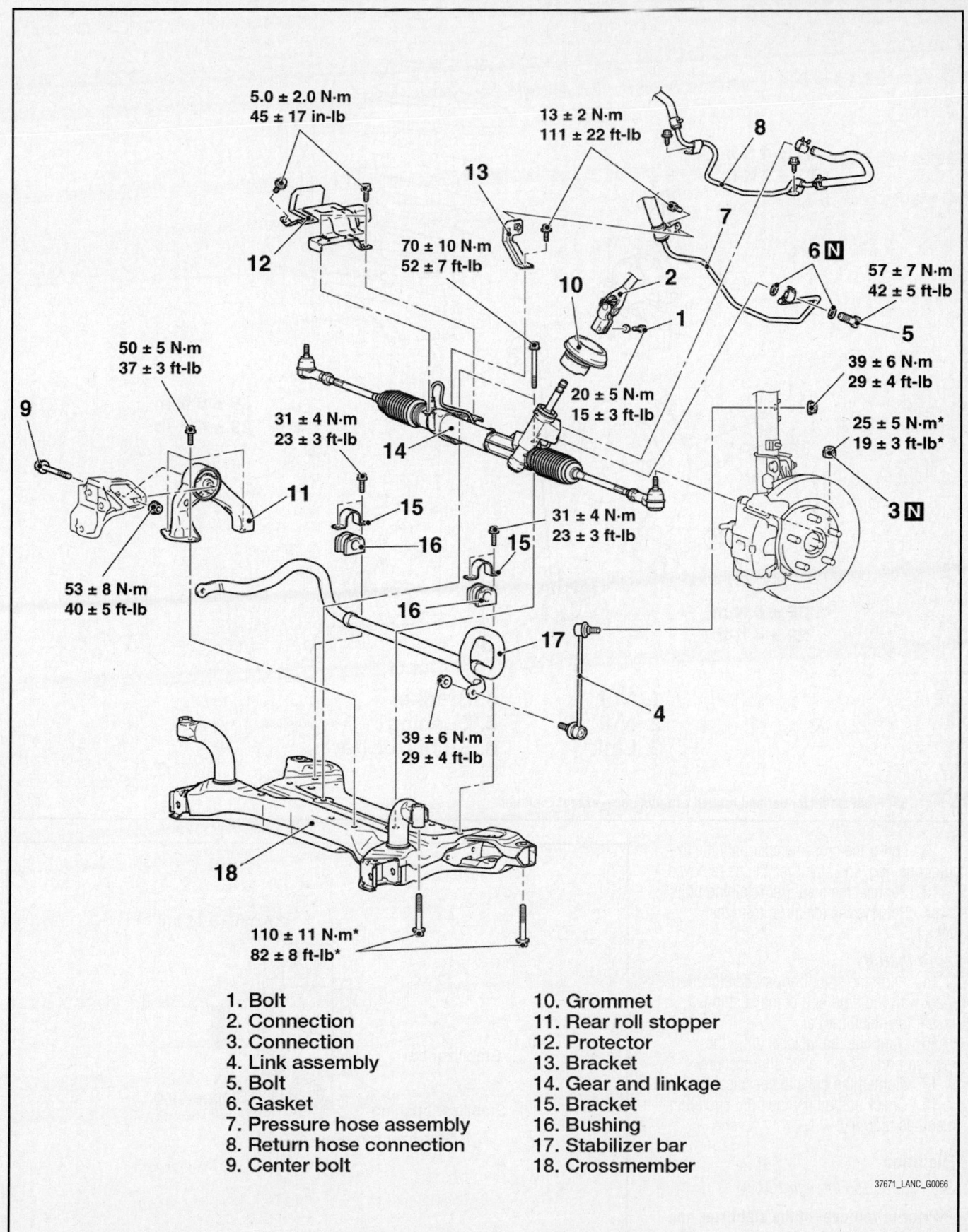

5.0 ± 2.0 N·m
45 ± 17 in-lb

13 ± 2 N·m
111 ± 22 ft-lb

70 ± 10 N·m
52 ± 7 ft-lb

50 ± 5 N·m
37 ± 3 ft-lb

57 ± 7 N·m
42 ± 5 ft-lb

39 ± 6 N·m
29 ± 4 ft-lb

25 ± 5 N·m*
19 ± 3 ft-lb*

20 ± 5 N·m
15 ± 3 ft-lb

31 ± 4 N·m
23 ± 3 ft-lb

53 ± 8 N·m
40 ± 5 ft-lb

31 ± 4 N·m
23 ± 3 ft-lb

39 ± 6 N·m
29 ± 4 ft-lb

110 ± 11 N·m*
82 ± 8 ft-lb*

1. Bolt
2. Connection
3. Connection
4. Link assembly
5. Bolt
6. Gasket
7. Pressure hose assembly
8. Return hose connection
9. Center bolt
10. Grommet
11. Rear roll stopper
12. Protector
13. Bracket
14. Gear and linkage
15. Bracket
16. Bushing
17. Stabilizer bar
18. Crossmember

37671_LANC_G0066

Fig. 152 Crossmember and related components—except Evolution

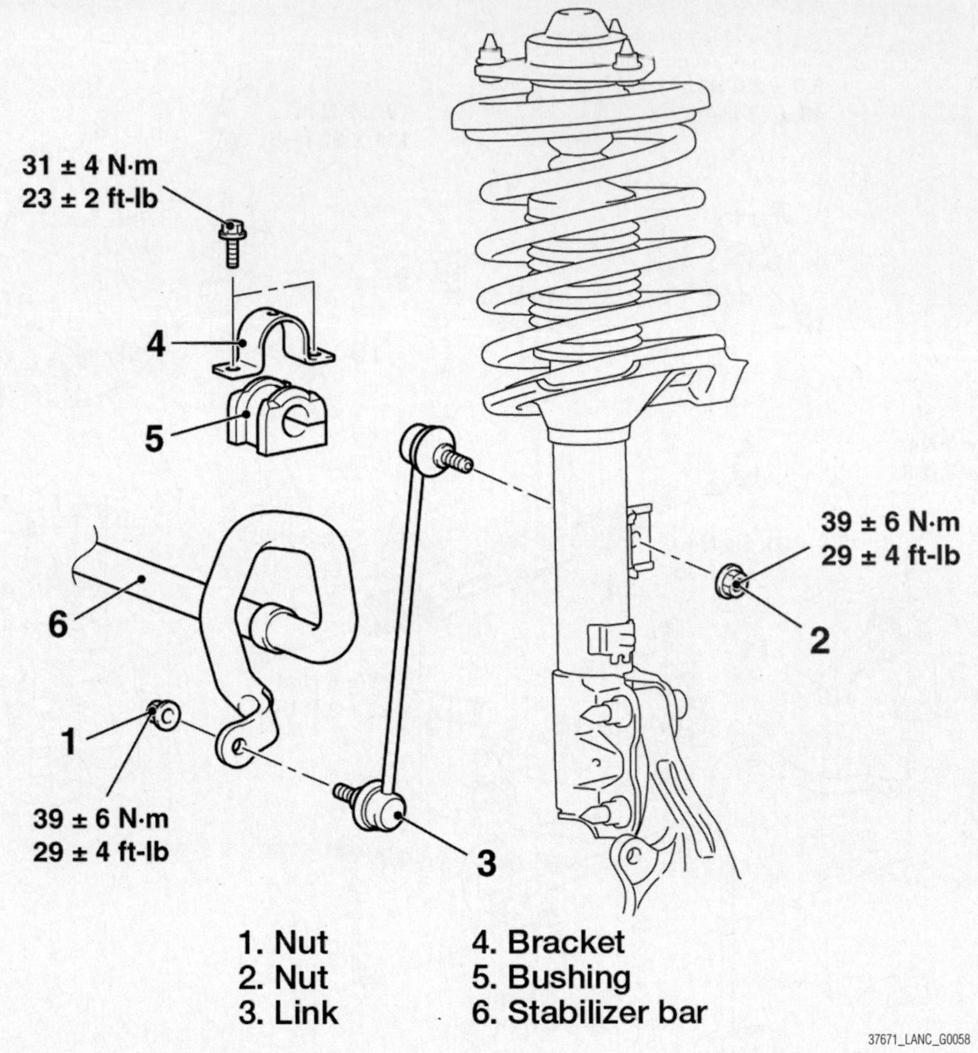

31 ± 4 N·m
23 ± 2 ft-lb

4

5

6

1

39 ± 6 N·m
29 ± 4 ft-lb

3

39 ± 6 N·m
29 ± 4 ft-lb

2

1. Nut
2. Nut
3. Link
4. Bracket
5. Bushing
6. Stabilizer bar

37671_LANC_G0058

Fig. 153 Front stabilizer bar and related components—except Evolution

12. Lower the crossmember until the fixture, bushing, and stabilizer can be removed.

13. Remove the stabilizer retaining bolts.

14. Remove the stabilizer from the vehicle.

To install:

15. Align the stabilizer bar identification mark with the right end of the bushing. Install the stabilizer bar.

16. Continue the installation in the reverse order of the removal procedure.

17. Tighten the bolts to specification.

18. Check and adjust the front end alignment, as required.

Evolution

See Figures 156 through 160.

→**Prior to removal of the stabilizer bar, center the front wheels and remove the ignition key. Failure to do so may**

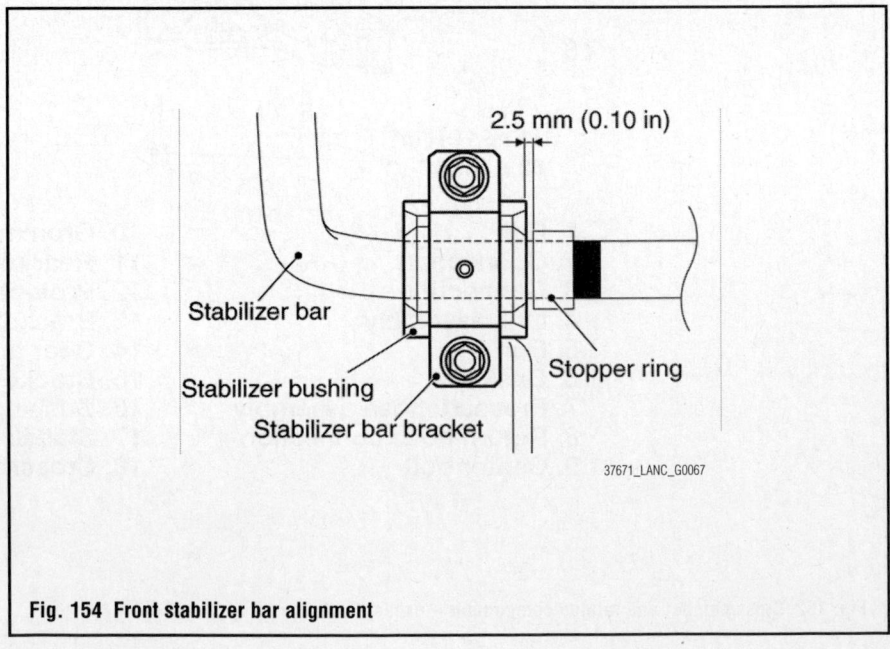

2.5 mm (0.10 in)

Stabilizer bar

Stabilizer bushing

Stabilizer bar bracket

Stopper ring

37671_LANC_G0067

Fig. 154 Front stabilizer bar alignment

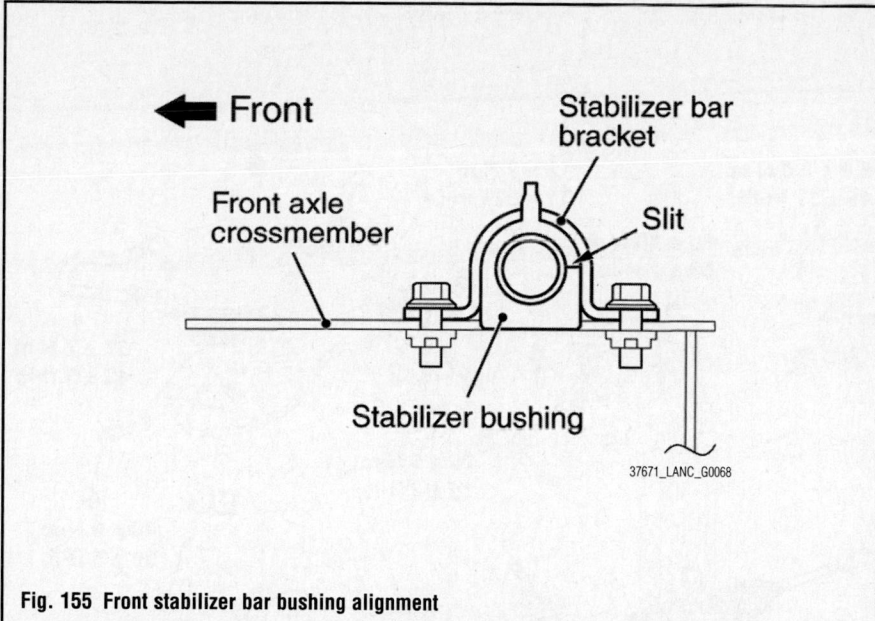

Front

Stabilizer bar bracket

Slit

Front axle crossmember

Stabilizer bushing

37671_LANC_G0068

Fig. 155 Front stabilizer bar bushing alignment

damage the SRS clockspring and render SRS system inoperative.

1. Before servicing the vehicle, refer to the Precautions Section.

➡**If working near and/or around the SRS system and components, be sure to disable the SRS system. Tape the negative battery cable with insulating tape. Always disconnect the negative battery cable first.**

❊❊ CAUTION

Wait for 1 minute after disconnecting the negative battery cable before working inside the vehicle. The air bag system is set to deploy for a short period of time after the battery is disconnected.

2. Disconnect the negative battery cable.

1. Centermember and roll stopper assembly
2. Centermember
3. Front roll stopper
4. Rear roll stopper

53 ± 8 N·m
40 ± 5 ft-lb

50 ± 5 N·m
37 ± 3 ft-lb

56 ± 5 N·m*1
41 ± 3 ft-lb*1

71 ± 10 N·m*2
52 ± 7 ft-lb*2

50 ± 5 N·m
37 ± 3 ft-lb

71 ± 10 N·m*2
52 ± 7 ft-lb*2

37671_LANC_G0033

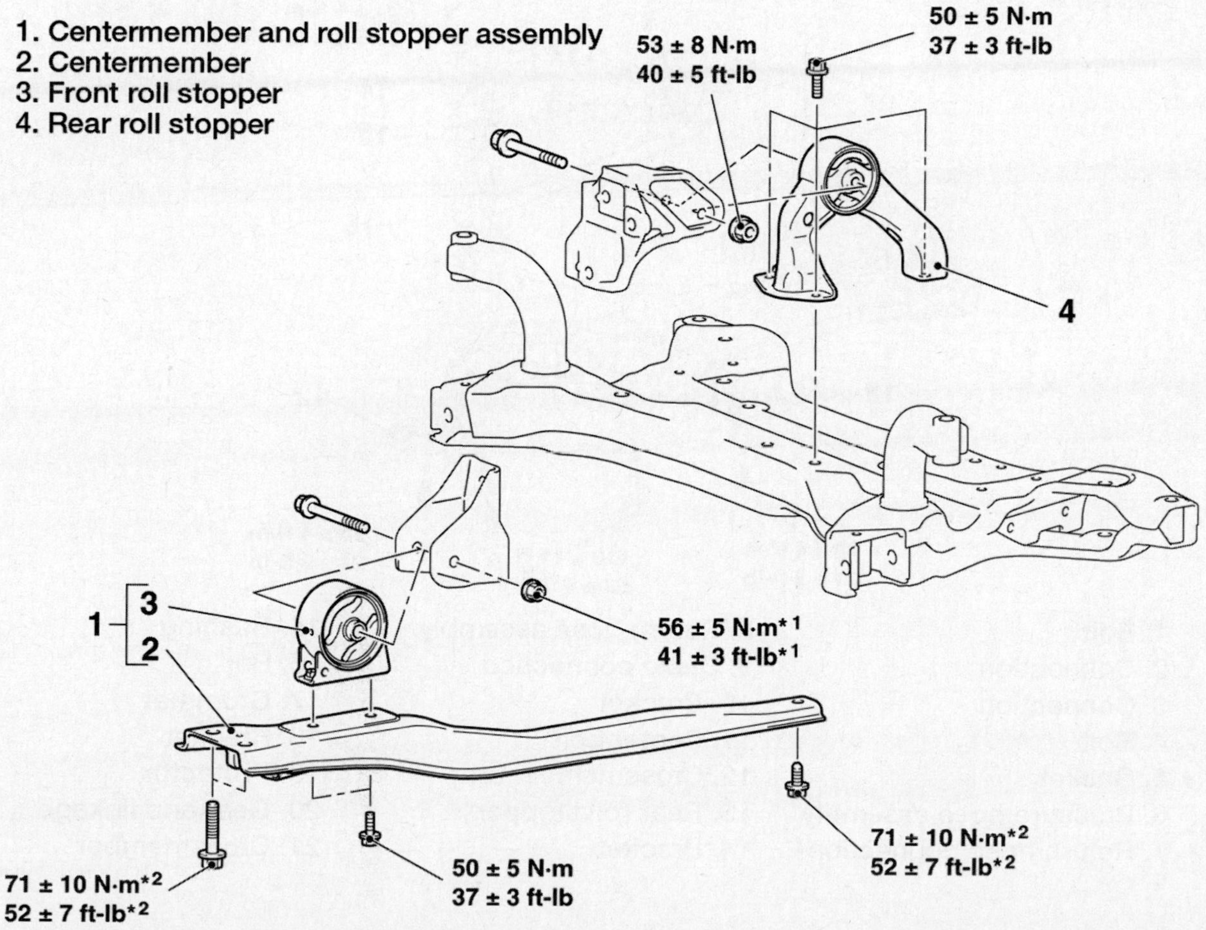

Fig. 156 Centermember and front/rear roll stopper assembly and related components

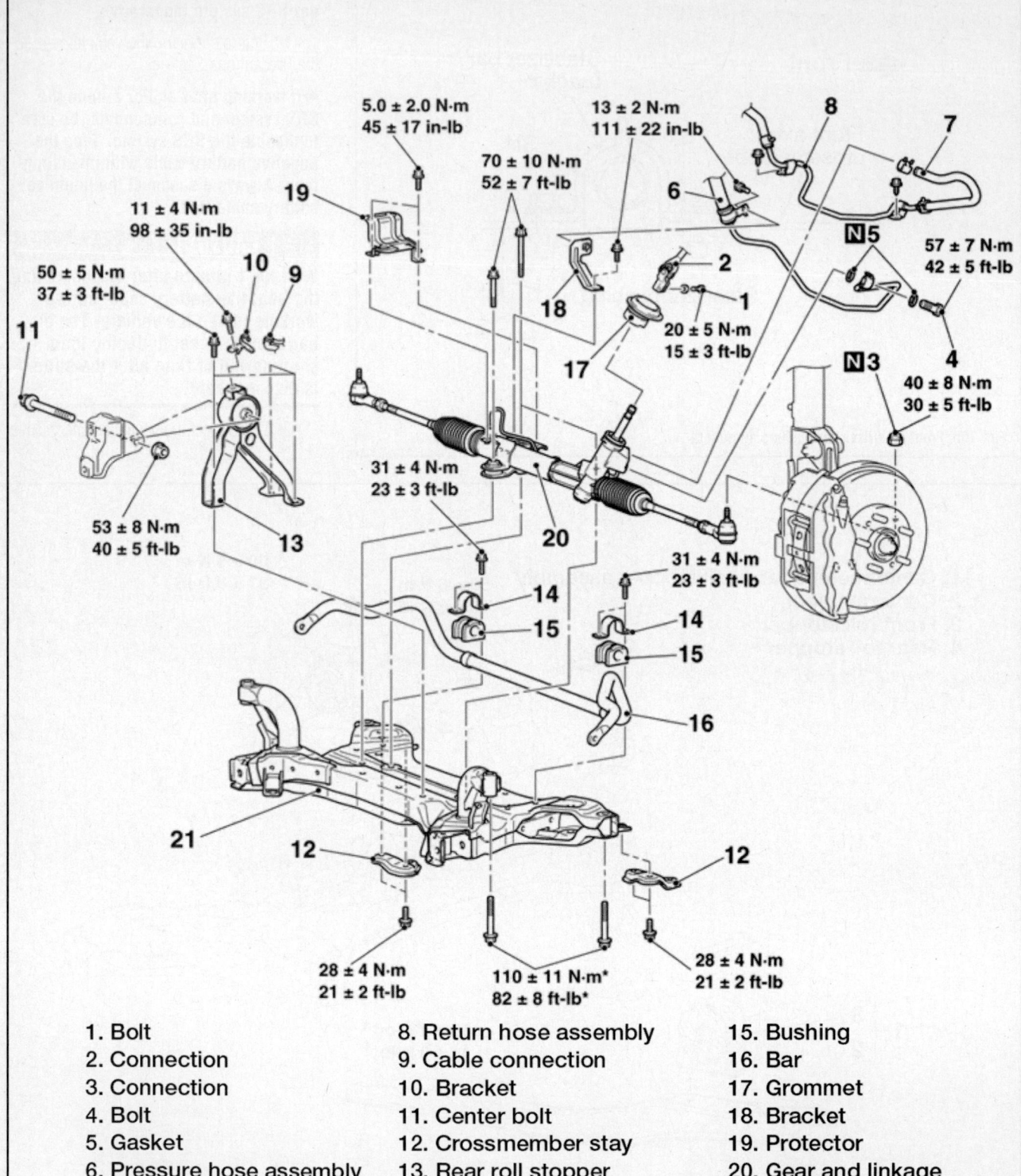

5.0 ± 2.0 N·m
45 ± 17 in-lb

13 ± 2 N·m
111 ± 22 in-lb

70 ± 10 N·m
52 ± 7 ft-lb

11 ± 4 N·m
98 ± 35 in-lb

50 ± 5 N·m
37 ± 3 ft-lb

57 ± 7 N·m
42 ± 5 ft-lb

20 ± 5 N·m
15 ± 3 ft-lb

40 ± 8 N·m
30 ± 5 ft-lb

53 ± 8 N·m
40 ± 5 ft-lb

31 ± 4 N·m
23 ± 3 ft-lb

31 ± 4 N·m
23 ± 3 ft-lb

28 ± 4 N·m
21 ± 2 ft-lb

110 ± 11 N·m*
82 ± 8 ft-lb*

28 ± 4 N·m
21 ± 2 ft-lb

1. Bolt
2. Connection
3. Connection
4. Bolt
5. Gasket
6. Pressure hose assembly
7. Return hose connection
8. Return hose assembly
9. Cable connection
10. Bracket
11. Center bolt
12. Crossmember stay
13. Rear roll stopper
14. Bracket
15. Bushing
16. Bar
17. Grommet
18. Bracket
19. Protector
20. Gear and linkage
21. Crossmember

37671_LANC_G0065

Fig. 157 Crossmember and related components—Evolution

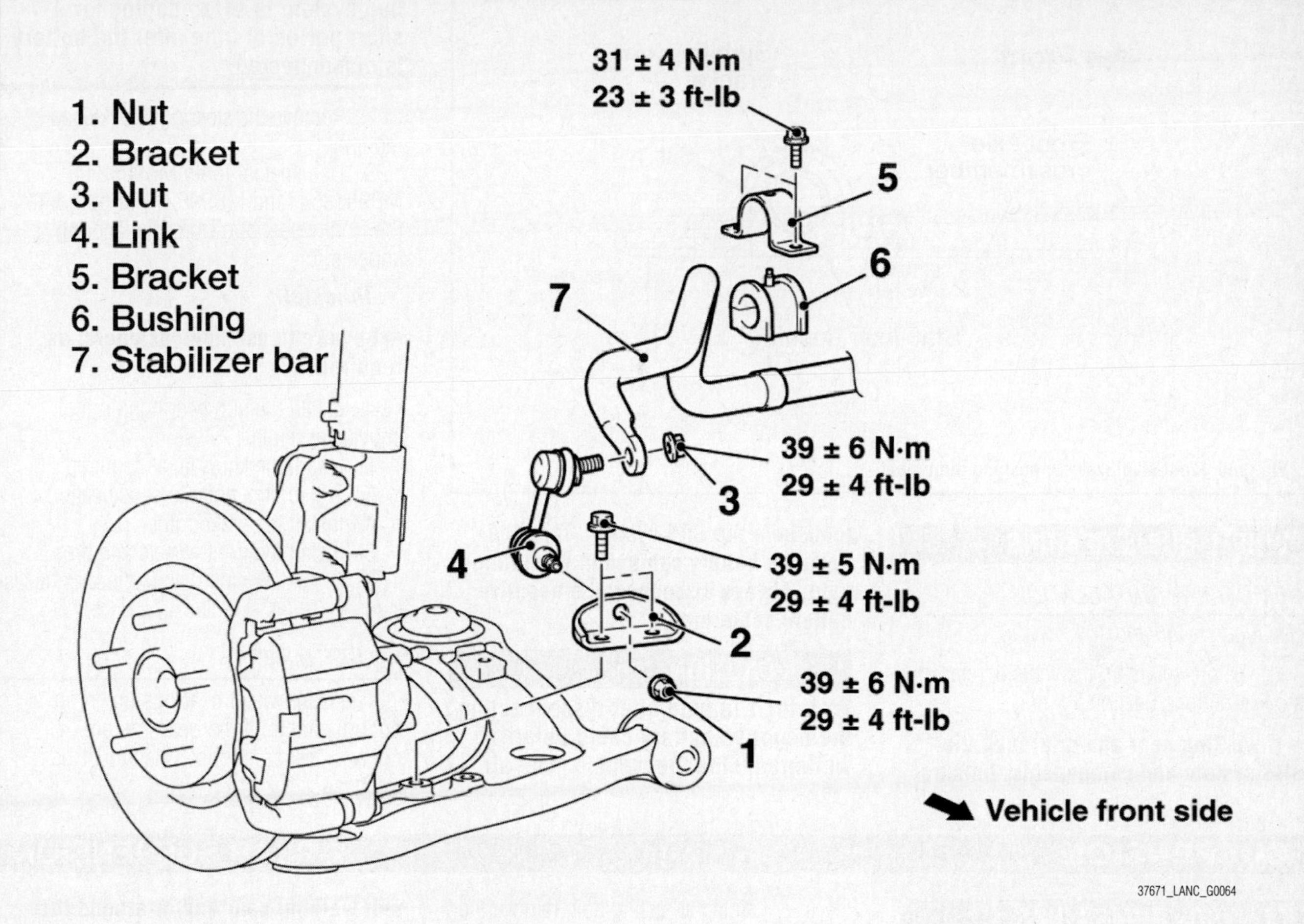

1. Nut
2. Bracket
3. Nut
4. Link
5. Bracket
6. Bushing
7. Stabilizer bar

31 ± 4 N·m
23 ± 3 ft-lb

5

6

7

39 ± 6 N·m
29 ± 4 ft-lb

3

39 ± 5 N·m
29 ± 4 ft-lb

4

2

39 ± 6 N·m
29 ± 4 ft-lb

1

◥ Vehicle front side

37671_LANC_G0064

Fig. 158 Front stabilizer bar and related components—Evolution

3. Remove the driver's side air bag module. Remove the steering wheel. Remove the column lower cover. Remove the clockspring.

4. Remove the engine undercover.

5. Remove the steering column shaft.

6. Remove the centermember.

7. Remove the stabilizer bar locknut. Remove the stabilizer bar rubber and collar.

8. Separate the lower arm and knuckle connection

9. Separate the tie rod end and knuckle connection.

10. Remove the steering shaft cover.

Remove the steering gear joint connecting bolt.

11. Remove the rear stopper connecting bolt.

12. Use a transaxle jack to hold the crossmember in place. Remove the crossmember mounting bolts and nuts.

➥**Be careful not to lower the crossmember too much, otherwise the power steering return hose bracket may deform.**

13. Lower the crossmember until the fixture, bushing and stabilizer can be removed.

14. Remove the stabilizer retaining bolts. Remove the stabilizer from the vehicle.

To install:

15. Align the stabilizer bar identification mark with the right end of the bushing. Install the stabilizer bar.

16. Continue the installation in the reverse order of the removal procedure.

17. Tighten the bolts to specification.

18. Check and adjust the front end alignment, as required.

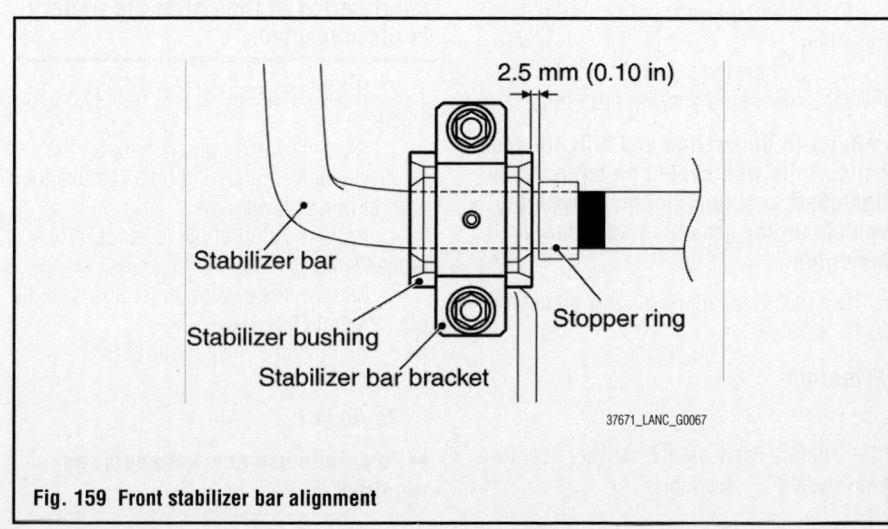

2.5 mm (0.10 in)

Stabilizer bar

Stabilizer bushing

Stabilizer bar bracket

Stopper ring

37671_LANC_G0067

Fig. 159 Front stabilizer bar alignment

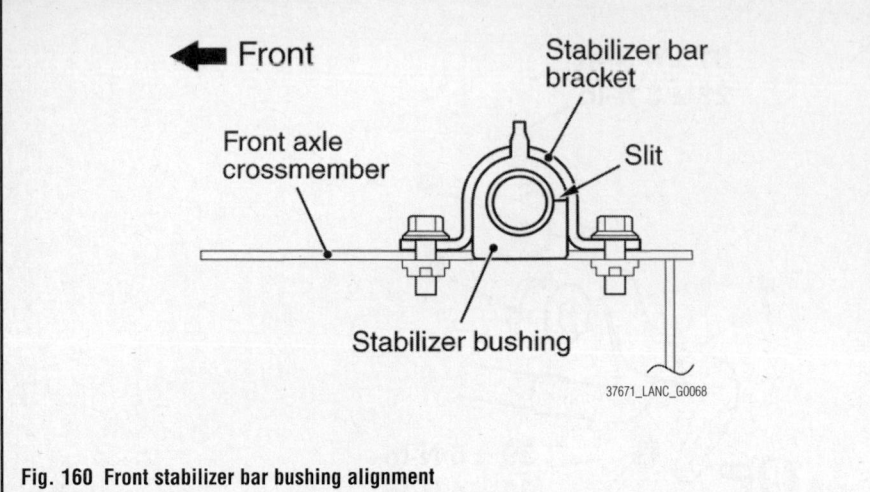

Fig. 160 Front stabilizer bar bushing alignment

WHEEL HUB & BEARING

REMOVAL & INSTALLATION

See Figures 150 and 151.

1. Before servicing the vehicle, refer to the Precautions Section.

➡ **If working near and/or around the SRS system and components, be sure** to disable the SRS system. Tape the negative battery cable with insulating tape. Always disconnect the negative battery cable first.

❋❋ CAUTION

Wait for 1 minute after disconnecting the negative battery cable before working inside the vehicle. The air bag system is set to deploy for a short period of time after the battery is disconnected.

2. Remove the steering knuckle, as required.
3. Use special tools MB990244, MB991354, and MB990211 to pull out the front wheel hub from the steering knuckle.

To install:

➡ **Be sure to use new fasteners, as required.**

4. Install the hub bearing into the knuckle assembly.
5. Install the knuckle, as required.
6. Tighten the bolts to specification.
7. Install the brake rotor.
8. Install the front wheels and tires.
9. Check alignment and adjust as necessary.

ADJUSTMENT

The front wheel bearings are not adjustable. If the bearings are noisy or become loose, they must be replaced.

SUSPENSION

CONTROL ARMS/LINKS

REMOVAL & INSTALLATION

Lower Control Arm

Except Evolution

See Figures 162 and 163.

1. Before servicing the vehicle, refer to the Precautions Section.

➡ **If working near and/or around the SRS system and components, be sure to disable the SRS system. Tape the negative battery cable with insulating tape. Always disconnect the negative battery cable first.**

❋❋ CAUTION

Wait for 1 minute after disconnecting the negative battery cable before working inside the vehicle. The air bag system is set to deploy for a short period of time after the battery is disconnected.

2. Raise and support the vehicle safely.
3. Support the lower arm assembly, using a floor jack.
4. After making a mating mark on the toe-in or camber adjusting bolt, remove the control link or lower arm.

5. Remove the lower arm and trailing arm bolt and nut.
6. Remove the lower shock absorber retaining bolt and nut.
7. Remove the lower control arm from the vehicle.

To install:

➡ **Be sure to use new fasteners, as required.**

8. Install the lower control arm to its mounting.
9. Continue the installation in the reverse order of the removal procedure.

➡ **Refer to illustration and fully tighten bolts. Bolts with should be temporarily tightened, and fully tightened with the vehicle on the ground in an unladen condition.**

10. Check and adjust the rear alignment, as necessary.

Evolution

See Figure 163.

1. Before servicing the vehicle, refer to the Precautions Section.

REAR SUSPENSION

➡ **If working near and/or around the SRS system and components, be sure to disable the SRS system. Tape the negative battery cable with insulating tape. Always disconnect the negative battery cable first.**

❋❋ CAUTION

Wait for 1 minute after disconnecting the negative battery cable before working inside the vehicle. The air bag system is set to deploy for a short period of time after the battery is disconnected.

2. Raise and support the vehicle safely.
3. After making a mating mark on the toe-in or camber adjusting bolt, remove the control link or lower arm.
4. Properly support the lower control arm assembly.
5. Remove the lower control arm assembly mounting bolt.
6. Remove the lower control arm assembly.

To install:

➡ **Be sure to use new fasteners, as required.**

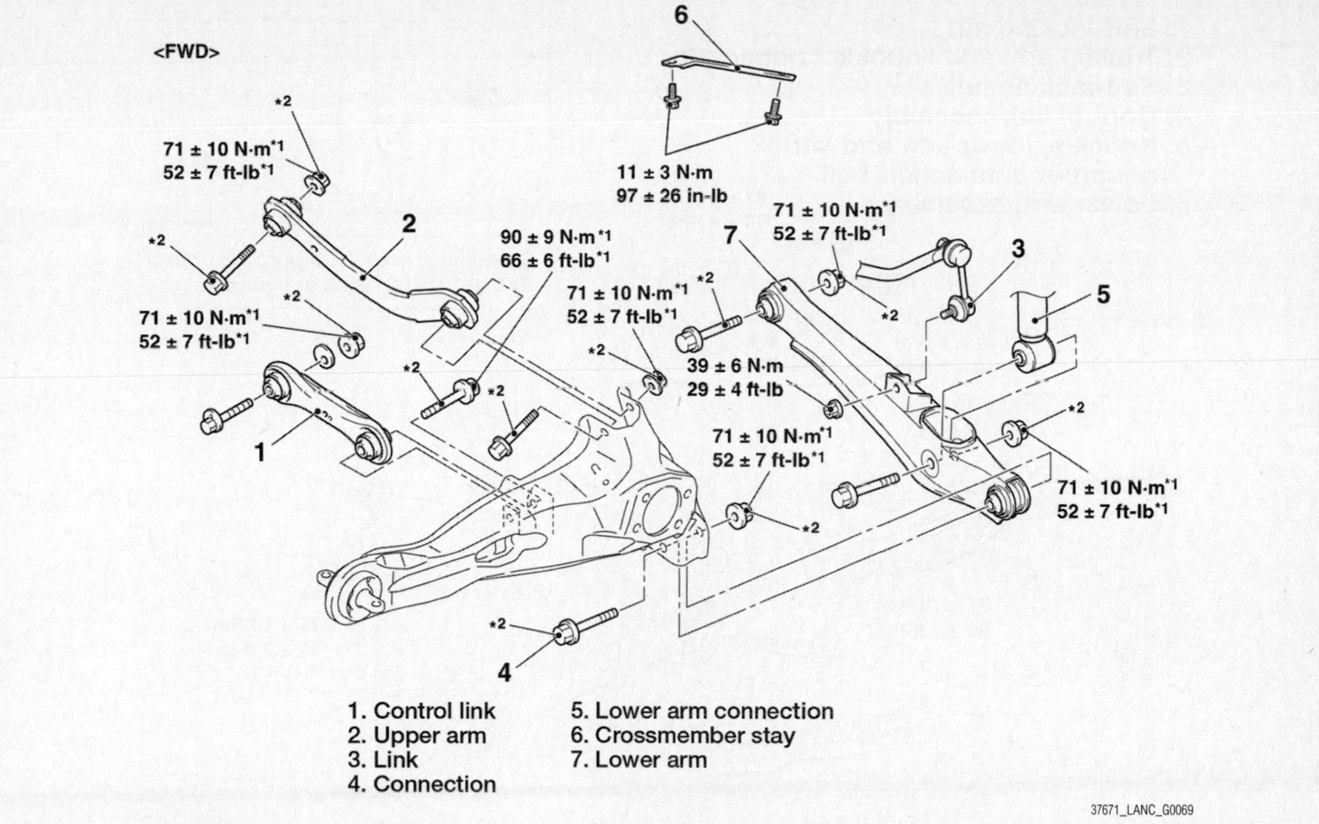

<FWD>

71 ± 10 N·m*1
52 ± 7 ft-lb*1

*2

*2

71 ± 10 N·m*1
52 ± 7 ft-lb*1

2

90 ± 9 N·m*1
66 ± 6 ft-lb*1

71 ± 10 N·m*1
52 ± 7 ft-lb*1

*2

*2

*2

*2

1

6

11 ± 3 N·m
97 ± 26 in-lb

7

71 ± 10 N·m*1
52 ± 7 ft-lb*1

*2

3

5

39 ± 6 N·m
29 ± 4 ft-lb

71 ± 10 N·m*1
52 ± 7 ft-lb*1

*2

71 ± 10 N·m*1
52 ± 7 ft-lb*1

*2

4

1. Control link
2. Upper arm
3. Link
4. Connection

5. Lower arm connection
6. Crossmember stay
7. Lower arm

37671_LANC_G0069

Fig. 161 Rear suspension control arms and related components (FWD)—except Evolution

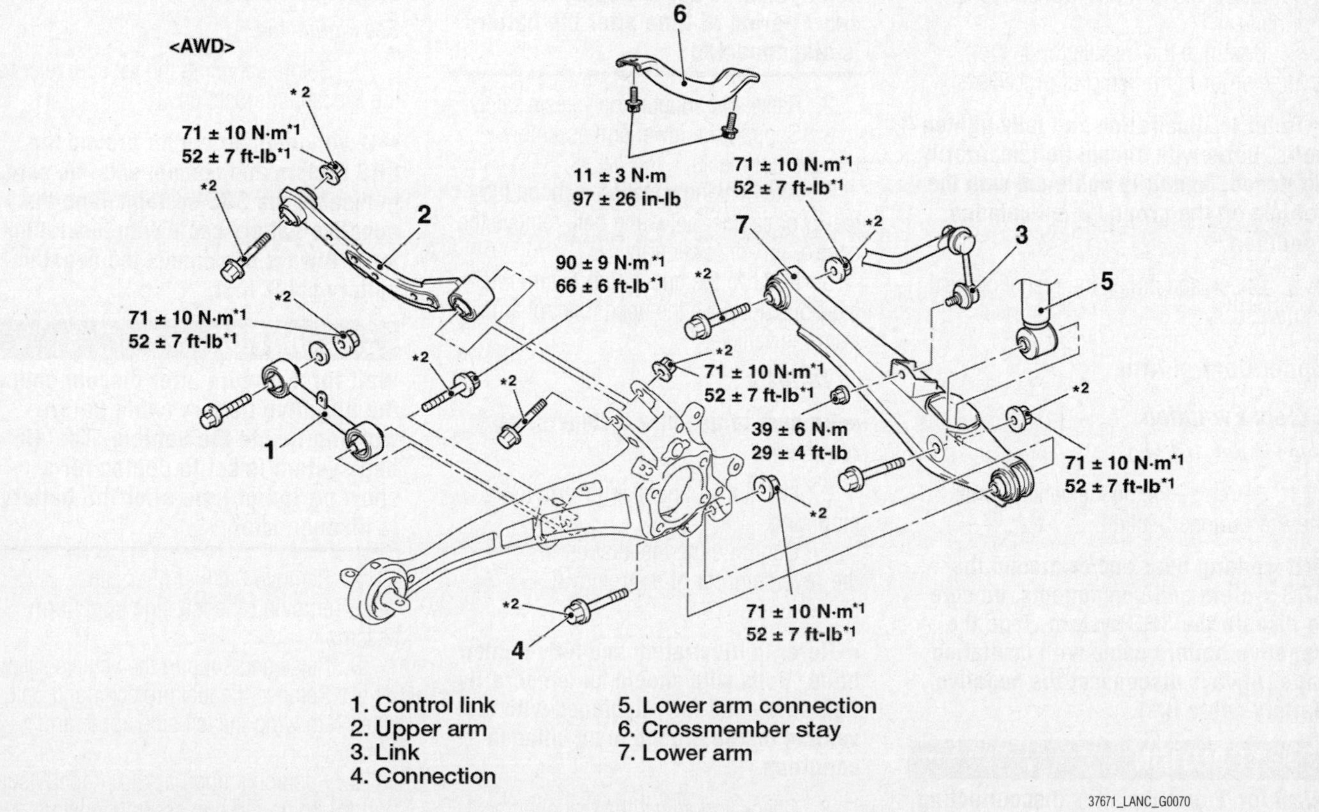

<AWD>

71 ± 10 N·m*1
52 ± 7 ft-lb*1

*2

*2

71 ± 10 N·m*1
52 ± 7 ft-lb*1

2

90 ± 9 N·m*1
66 ± 6 ft-lb*1

*2

*2

*2

1

6

11 ± 3 N·m
97 ± 26 in-lb

71 ± 10 N·m*1
52 ± 7 ft-lb*1

7

*2

3

5

71 ± 10 N·m*1
52 ± 7 ft-lb*1

*2

39 ± 6 N·m
29 ± 4 ft-lb

*2

71 ± 10 N·m*1
52 ± 7 ft-lb*1

*2

71 ± 10 N·m*1
52 ± 7 ft-lb*1

4

1. Control link
2. Upper arm
3. Link
4. Connection

5. Lower arm connection
6. Crossmember stay
7. Lower arm

37671_LANC_G0070

Fig. 162 Rear suspension control arms and related components (AWD)—except Evolution

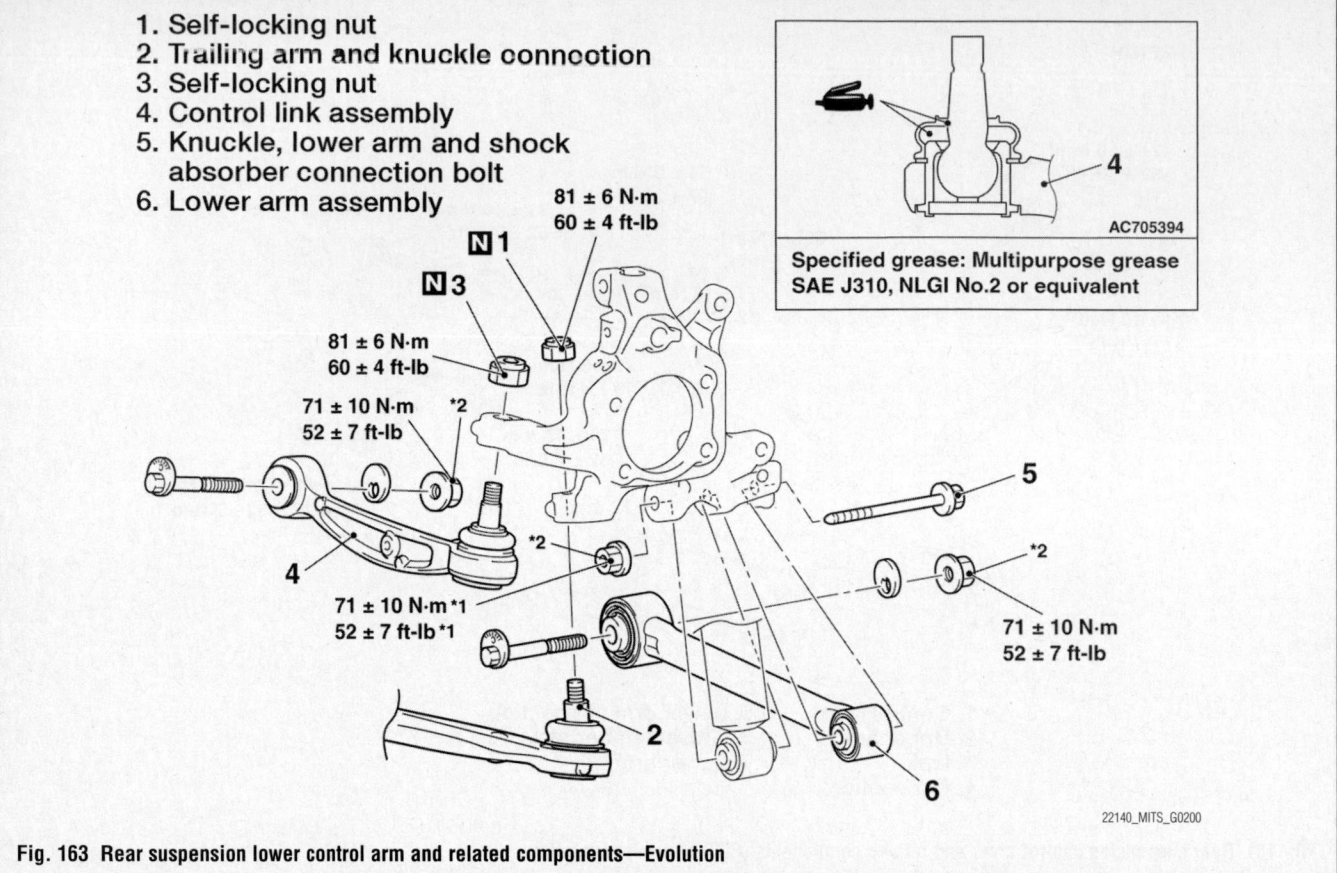

1. Self-locking nut
2. Trailing arm and knuckle connection
3. Self-locking nut
4. Control link assembly
5. Knuckle, lower arm and shock absorber connection bolt
6. Lower arm assembly

81 ± 6 N·m
60 ± 4 ft-lb

81 ± 6 N·m
60 ± 4 ft-lb

71 ± 10 N·m
52 ± 7 ft-lb

71 ± 10 N·m *1
52 ± 7 ft-lb *1

71 ± 10 N·m
52 ± 7 ft-lb

AC705394

Specified grease: Multipurpose grease
SAE J310, NLGI No.2 or equivalent

22140_MITS_G0200

Fig. 163 Rear suspension lower control arm and related components—Evolution

7. Install the lower control arm to its mounting.

8. Continue the installation in the reverse order of the removal procedure.

➡**Refer to illustration and fully tighten bolts. Bolts with should be temporarily tightened, and fully tightened with the vehicle on the ground in an unladen condition.**

9. Check and adjust the rear alignment, as necessary.

Upper Control Arm

Except Evolution

See Figures 161 and 162.

1. Before servicing the vehicle, refer to the Precautions Section.

➡**If working near and/or around the SRS system and components, be sure to disable the SRS system. Tape the negative battery cable with insulating tape. Always disconnect the negative battery cable first.**

✳✳ CAUTION

Wait for 1 minute after disconnecting the negative battery cable before working inside the vehicle. The air

bag system is set to deploy for a short period of time after the battery is disconnected.

2. Raise and support the vehicle safely.

3. Support the lower arm assembly, using a floor jack.

4. After making a mating mark on the toe-in or camber adjusting bolt, remove the control link or lower arm.

5. Remove the upper control arm retaining bolts. Remove the upper control arm from the vehicle.

To install:

➡**Be sure to use new fasteners, as required.**

6. Install the upper control arm to its mounting.

7. Continue the installation in the reverse order of the removal procedure.

➡**Refer to illustration and fully tighten bolts. Bolts with should be temporarily tightened, and fully tightened with the vehicle on the ground in an unladen condition.**

8. Check and adjust the rear alignment, as necessary.

Lancer Evolution

See Figure 164.

1. Before servicing the vehicle, refer to the Precautions Section.

➡**If working near and/or around the SRS system and components, be sure to disable the SRS system. Tape the negative battery cable with insulating tape. Always disconnect the negative battery cable first.**

✳✳ CAUTION

Wait for 1 minute after disconnecting the negative battery cable before working inside the vehicle. The air bag system is set to deploy for a short period of time after the battery is disconnected.

2. Remove the fuel filler cap, when removing the left side upper arm assembly.

3. Raise and support the vehicle safely.

4. Remove the fuel protector and bolt, when removing the left side upper arm assembly.

5. Position a floor jack to properly support the rotor and hub assembly before removing the upper control

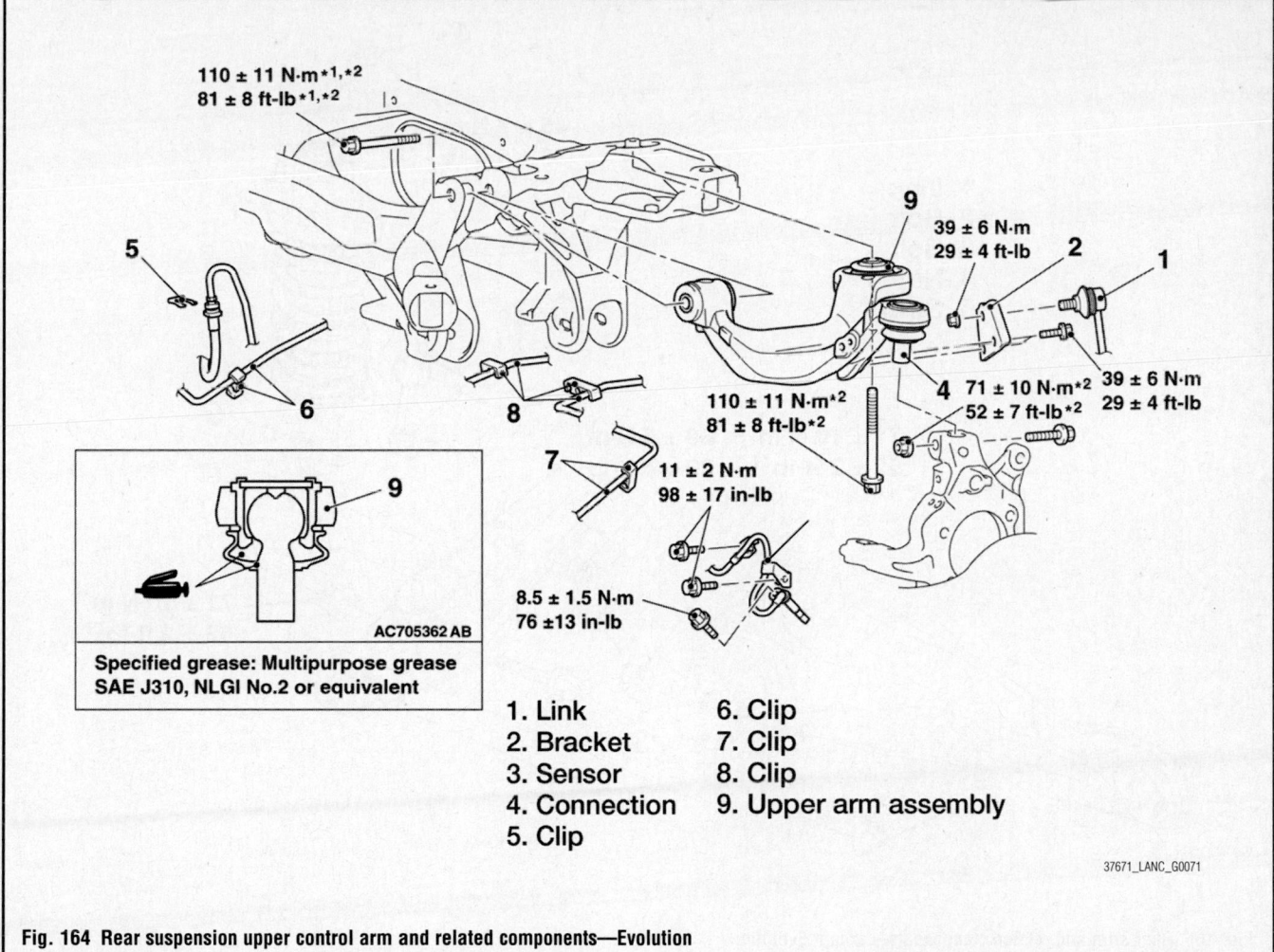

110 ± 11 N·m *1, *2
81 ± 8 ft-lb *1, *2

39 ± 6 N·m
29 ± 4 ft-lb

110 ± 11 N·m *2
81 ± 8 ft-lb *2

11 ± 2 N·m
98 ± 17 in-lb

8.5 ± 1.5 N·m
76 ±13 in-lb

71 ± 10 N·m *2
52 ± 7 ft-lb *2

39 ± 6 N·m
29 ± 4 ft-lb

AC705362 AB

**Specified grease: Multipurpose grease
SAE J310, NLGI No.2 or equivalent**

1. Link
2. Bracket
3. Sensor
4. Connection
5. Clip
6. Clip
7. Clip
8. Clip
9. Upper arm assembly

37671_LANC_G0071

Fig. 164 Rear suspension upper control arm and related components—Evolution

➥Do not remove the nut from the ball joint. Loosen it and use special tool MB991897 to avoid possible damage to the ball joint threads. Hang the special tool in place with wire or string to prevent it from falling.

6. Install the special tool. Turn the bolt and knob as necessary to make the jaws of the tool parallel. Tighten the bolt by hand and confirm that the jaws are still parallel.

➥When adjusting the jaws in parallel, make sure the knob is in the vertical (upward) position.

7. Tighten the bolt with a wrench to disconnect the upper arm assembly and the knuckle.

8. Remove the upper arm assembly mounting bolts. Remove the upper arm stopper.

9. Remove the upper control arm from the vehicle.

To install:

➥Be sure to use new fasteners, as required.

10. Install the upper control arm to its mounting.

11. Continue the installation in the reverse order of the removal procedure.

➥Refer to illustration and fully tighten bolts. Bolts with should be temporarily tightened, and fully tightened with the vehicle on the ground in an unladen condition.

12. Check and adjust the rear alignment, as necessary.

MACPHERSON STRUTS

REMOVAL & INSTALLATION

Except Evolution

See Figure 165.

1. Before servicing the vehicle, refer to the Precautions Section.

➥If working near and/or around the SRS system and components, be sure to disable the SRS system. Tape the negative battery cable with insulating

tape. Always disconnect the negative battery cable first.

✳✳ CAUTION

Wait for 1 minute after disconnecting the negative battery cable before working inside the vehicle. The air bag system is set to deploy for a short period of time after the battery is disconnected.

2. Remove the stabilizer link connection.

3. Support the lower control arm with a jack.

4. Remove or disconnect the following:
- Lower control arm and trailing arm bolt
- Upper shock absorber mounting nut
- Shock absorber-to-lower control arm attaching bolt
- Shock absorber/strut assembly

To install:

➥Be sure to use new fasteners, as required.

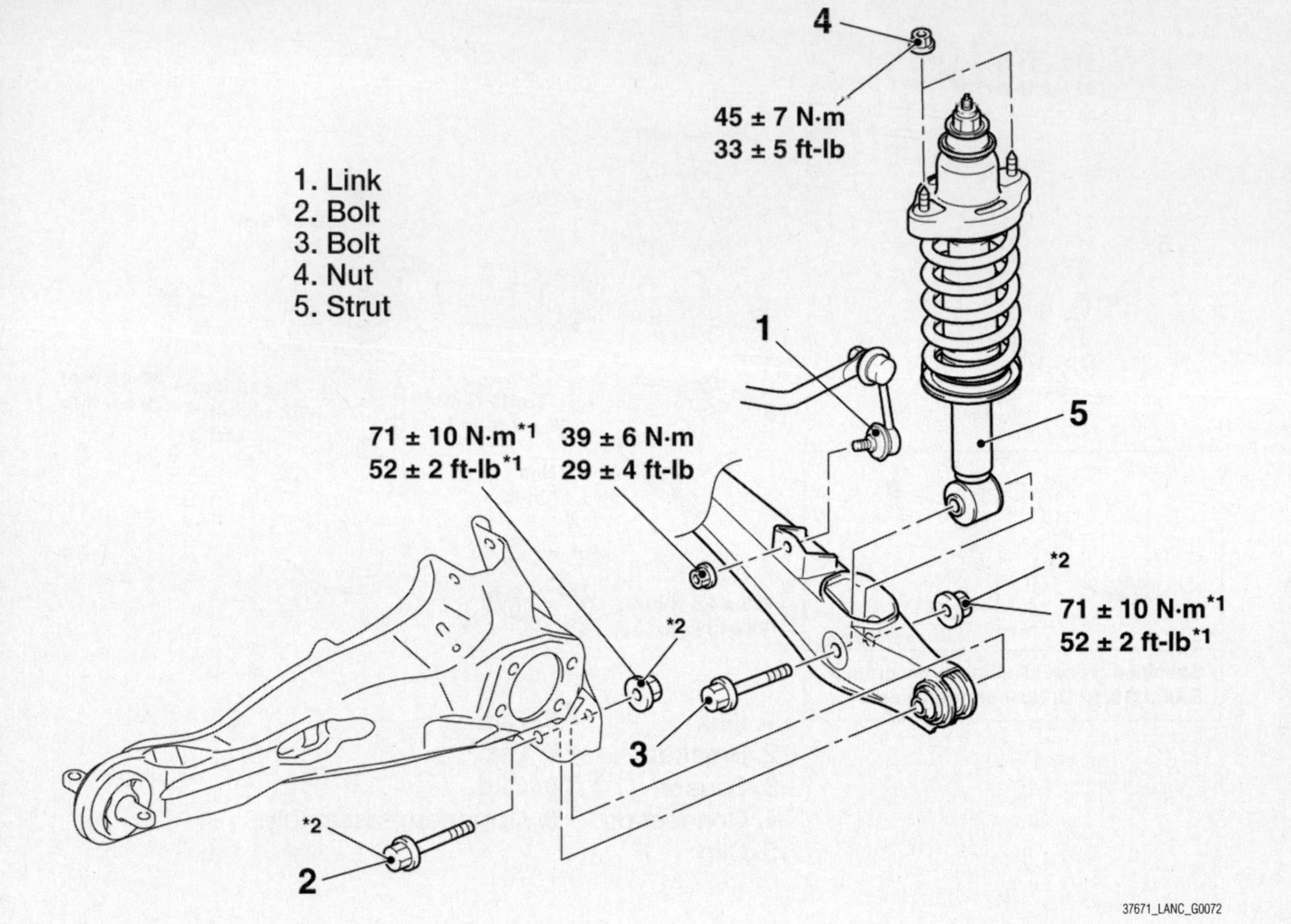

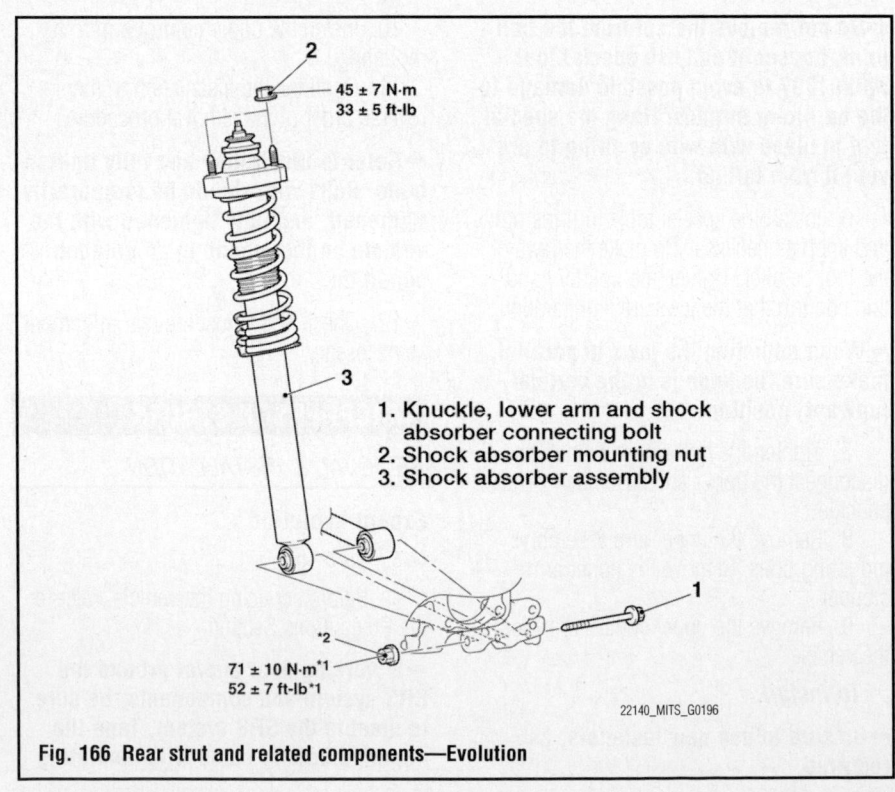

1. Link
2. Bolt
3. Bolt
4. Nut
5. Strut

45 ± 7 N·m
33 ± 5 ft-lb

71 ± 10 N·m^{*1} 39 ± 6 N·m
52 ± 2 ft-lb^{*1} 29 ± 4 ft-lb

*2
71 ± 10 N·m^{*1}
52 ± 2 ft-lb^{*1}

*2

*2

37671_LANC_G0072

Fig. 165 Rear strut and related components—except Evolution

5. Position the shock absorber into the vehicle. Install the spring seat stepped section so that it points toward the rear side of the vehicle.

6. Continue the installation in the reverse order of the removal procedure.

➡**Refer to illustration and fully tighten bolts. Bolts with should be temporarily tightened, and fully tightened with the vehicle on the ground in an unladen condition.**

7. Check and adjust the rear alignment, as necessary.

Evolution

See Figure 166.

1. Before servicing the vehicle, refer to the Precautions Section.

➡**If working near and/or around the SRS system and components, be sure to disable the SRS system. Tape the negative battery cable with insulating tape. Always disconnect the negative battery cable first.**

2

45 ± 7 N·m
33 ± 5 ft-lb

3

1. Knuckle, lower arm and shock absorber connecting bolt
2. Shock absorber mounting nut
3. Shock absorber assembly

1

*2
71 ± 10 N·m^{*1}
52 ± 7 ft-lb^{*1}

22140_MITS_G0196

Fig. 166 Rear strut and related components—Evolution

Wait for 1 minute after disconnecting the negative battery cable before working inside the vehicle. The air bag system is set to deploy for a short period of time after the battery is disconnected.

2. Raise and support the vehicle safely.

3. Remove the upper shock absorber retaining nuts.

4. Remove the lower shock absorber retaining nuts.

5. Remove the shock absorber from the vehicle.

To install:

➡Be sure to use new fasteners, as required.

6. Installation is the reverse of the removal procedure.

➡Refer to illustration and fully tighten bolts. Bolts with should be temporarily tightened, and fully tightened with the vehicle on the ground in an unladen condition.

7. Check and adjust the rear alignment, as necessary.

REAR SUSPENSION CROSSMEMBER

REMOVAL & INSTALLATION

See Figures 167 and 168.

At this time the manufacturer does not provide removal and installation procedures for this component, refer to the illustration as required.

STABILIZER BAR

REMOVAL & INSTALLATION

See Figures 169 through 172.

1. Before servicing the vehicle, refer to the Precautions Section.

➡If working near and/or around the SRS system and components, be sure to disable the SRS system. Tape the negative battery cable with insulating tape. Always disconnect the negative battery cable first.

Wait for 1 minute after disconnecting the negative battery cable before working inside the vehicle. The air bag system is set to deploy for a short period of time after the battery is disconnected.

2. Raise and support the vehicle safely.

3. Remove the tire and wheel assemblies.

4. On Lancer and Ralliart, remove the rear differential.

5. On Lancer and Ralliart, remove the rear suspension crossmember.

6. Remove the stabilizer bar link assembly.

7. Remove the stabilizer bar bracket.

8. Remove the bushings.

9. Remove the stabilizer bar.

To install:

➡Be sure to use new fasteners, as required.

10. Installation is the reverse of the removal procedure.

➡Refer to illustration and fully tighten bolts. Bolts with should be temporarily tightened, and fully tightened with the vehicle on the ground in an unladen condition.

11. Check and adjust the rear alignment, as necessary.

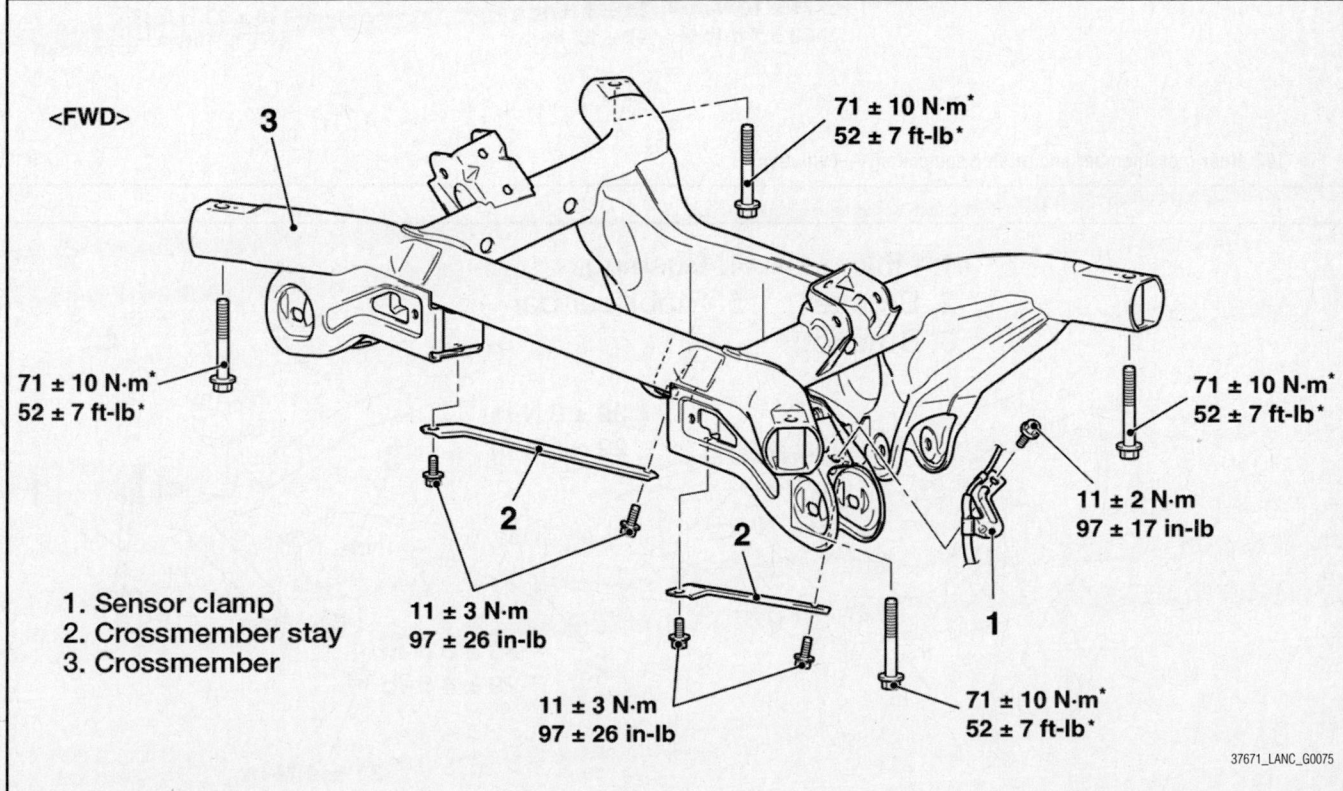

<FWD>

71 ± 10 N·m*
52 ± 7 ft-lb*

71 ± 10 N·m*
52 ± 7 ft-lb*

71 ± 10 N·m*
52 ± 7 ft-lb*

11 ± 2 N·m
97 ± 17 in-lb

11 ± 3 N·m
97 ± 26 in-lb

11 ± 3 N·m
97 ± 26 in-lb

71 ± 10 N·m*
52 ± 7 ft-lb*

1. Sensor clamp
2. Crossmember stay
3. Crossmember

37671_LANC_G0075

Fig. 167 Rear crossmember and related components (FWD)—except Evolution

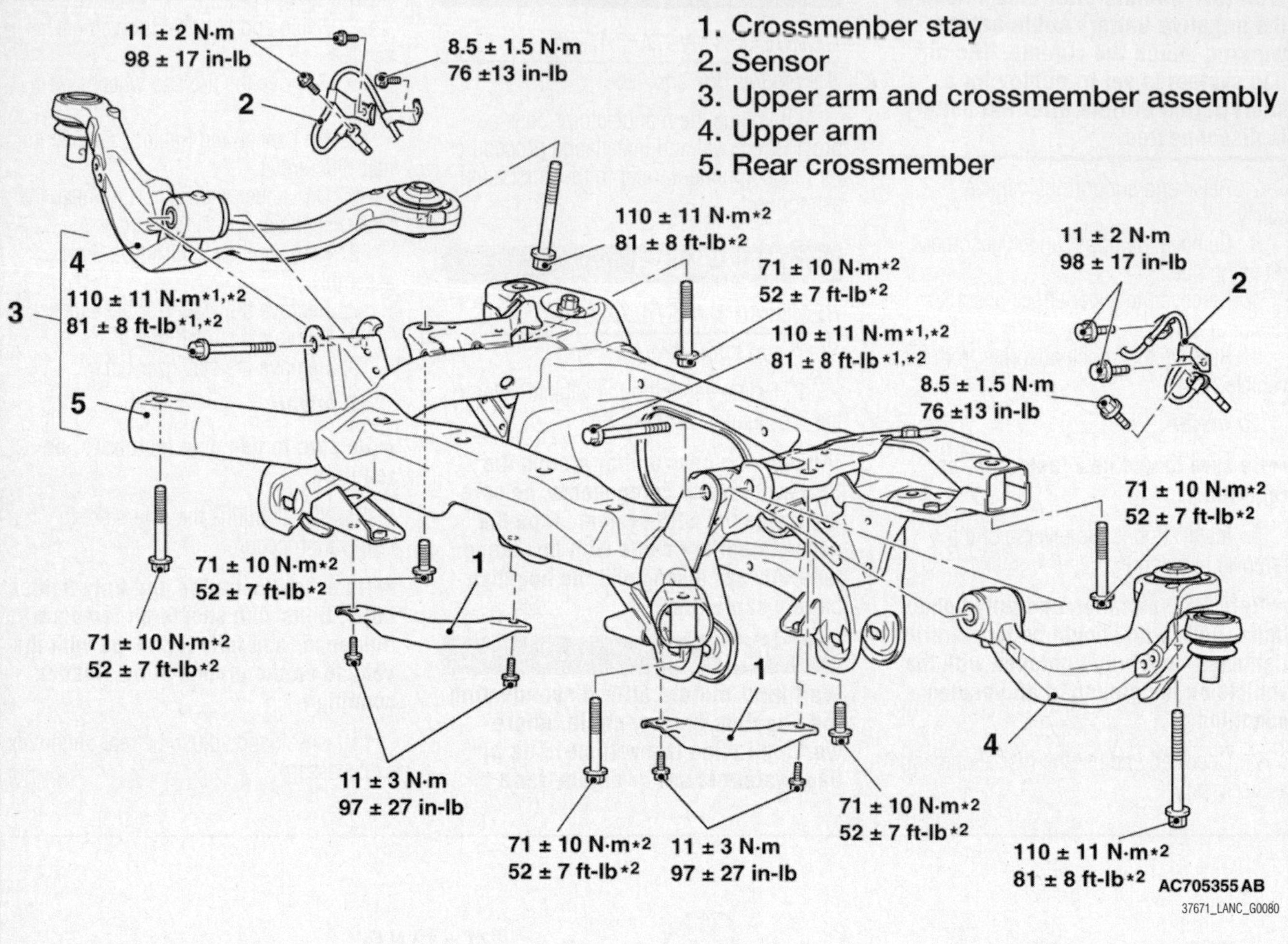

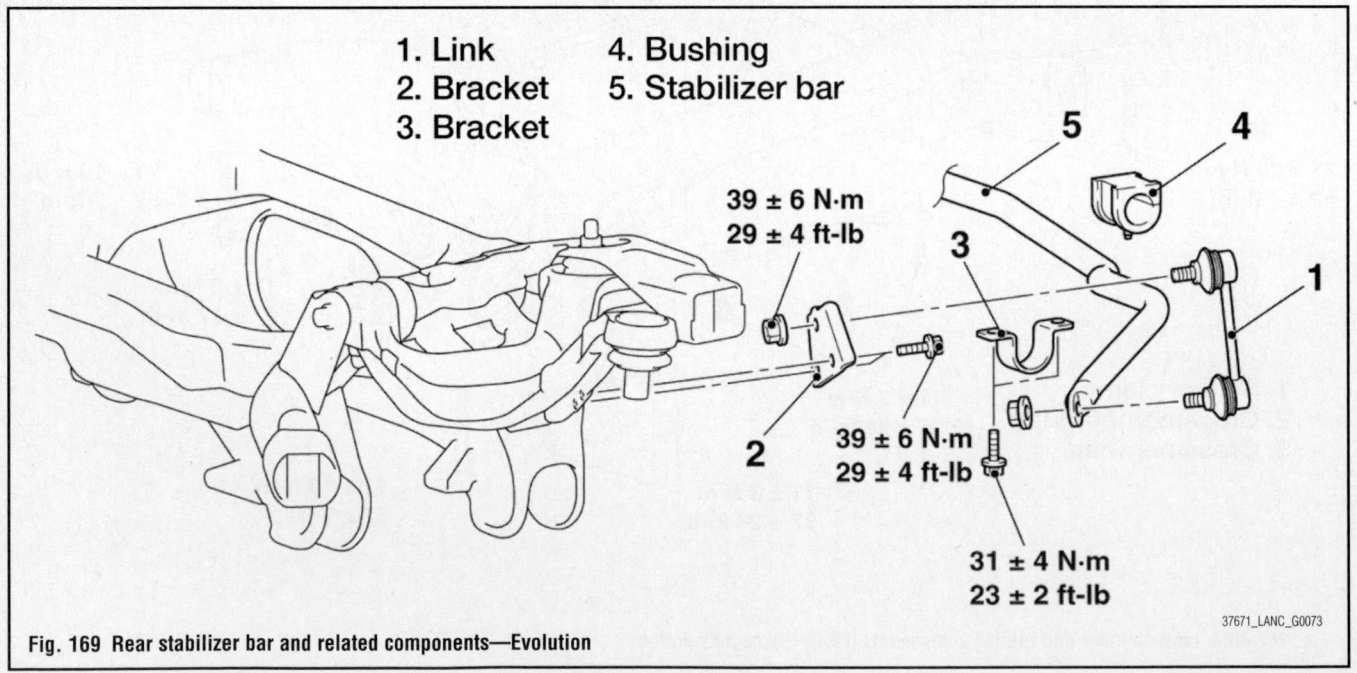

11 ± 2 N·m
98 ± 17 in-lb

8.5 ± 1.5 N·m
76 ±13 in-lb

2

1. Crossmember stay
2. Sensor
3. Upper arm and crossmember assembly
4. Upper arm
5. Rear crossmember

4

110 ± 11 N·m*2
81 ± 8 ft-lb*2

71 ± 10 N·m*2
52 ± 7 ft-lb*2

11 ± 2 N·m
98 ± 17 in-lb

2

3

110 ± 11 N·m*1,*2
81 ± 8 ft-lb*1,*2

110 ± 11 N·m*1,*2
81 ± 8 ft-lb*1,*2

8.5 ± 1.5 N·m
76 ±13 in-lb

5

71 ± 10 N·m*2
52 ± 7 ft-lb*2

71 ± 10 N·m*2
52 ± 7 ft-lb*2

1

1

71 ± 10 N·m*2
52 ± 7 ft-lb*2

11 ± 3 N·m
97 ± 27 in-lb

4

11 ± 3 N·m
97 ± 27 in-lb

71 ± 10 N·m*2
52 ± 7 ft-lb*2

71 ± 10 N·m*2
52 ± 7 ft-lb*2

11 ± 3 N·m
97 ± 27 in-lb

71 ± 10 N·m*2
52 ± 7 ft-lb*2

110 ± 11 N·m*2
81 ± 8 ft-lb*2

AC705355 AB

37671_LANC_G0080

Fig. 168 Rear crossmember and related components—Evolution

1. Link   4. Bushing
2. Bracket   5. Stabilizer bar
3. Bracket

5   **4**

39 ± 6 N·m
29 ± 4 ft-lb

3

1

2

39 ± 6 N·m
29 ± 4 ft-lb

31 ± 4 N·m
23 ± 2 ft-lb

Fig. 169 Rear stabilizer bar and related components—Evolution

37671_LANC_G0073

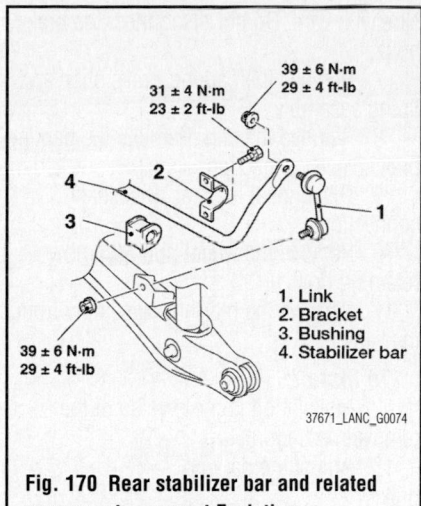

Fig. 170 Rear stabilizer bar and related components—except Evolution

1. Link
2. Bracket
3. Bushing
4. Stabilizer bar

31 ± 4 N·m
23 ± 2 ft-lb

39 ± 6 N·m
29 ± 4 ft-lb

39 ± 6 N·m
29 ± 4 ft-lb

37671_LANC_G0074

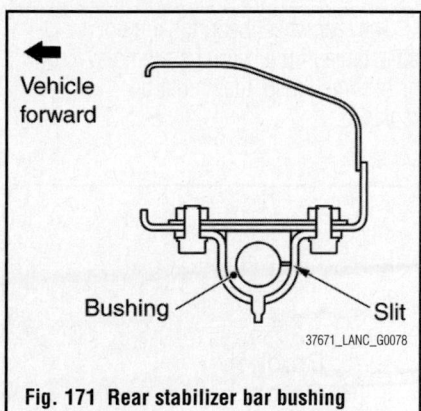

Vehicle forward

Bushing   Slit

Fig. 171 Rear stabilizer bar bushing installation—Evolution

37671_LANC_G0078

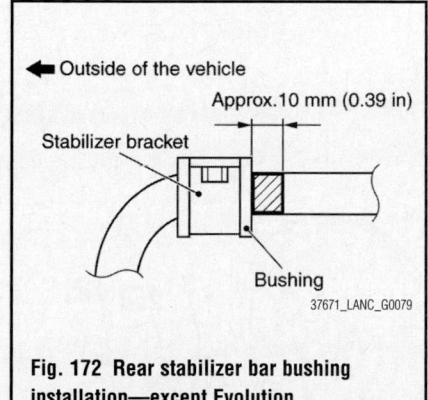

Outside of the vehicle

Approx.10 mm (0.39 in)

Stabilizer bracket

Bushing

37671_LANC_G0079

Fig. 172 Rear stabilizer bar bushing installation—except Evolution

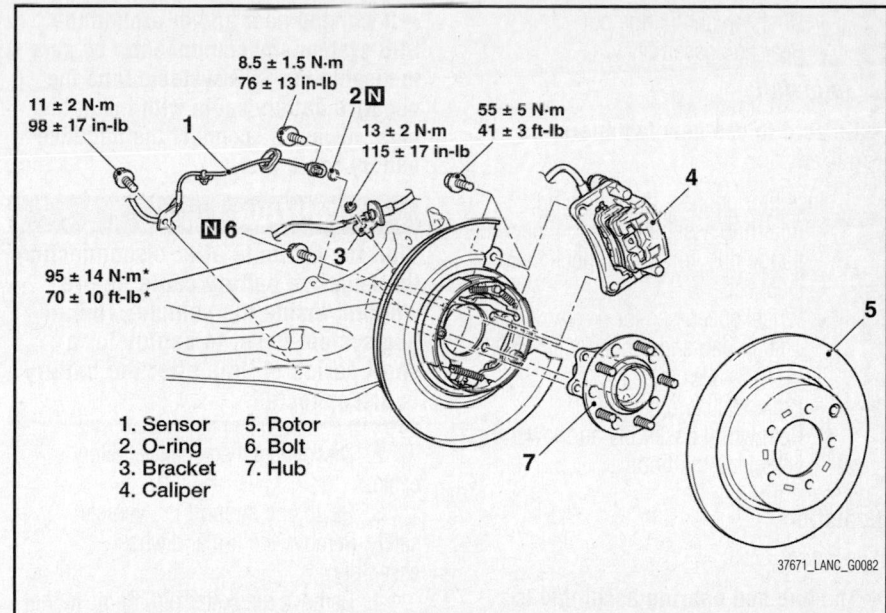

8.5 ± 1.5 N·m
76 ± 13 in-lb

11 ± 2 N·m
98 ± 17 in-lb

13 ± 2 N·m
115 ± 17 in-lb

55 ± 5 N·m
41 ± 3 ft-lb

95 ± 14 N·m*
70 ± 10 ft-lb*

1. Sensor
2. O-ring
3. Bracket
4. Caliper
5. Rotor
6. Bolt
7. Hub

37671_LANC_G0082

Fig. 173 Rear hub and bearing assembly and related components (FWD)—except Evolution

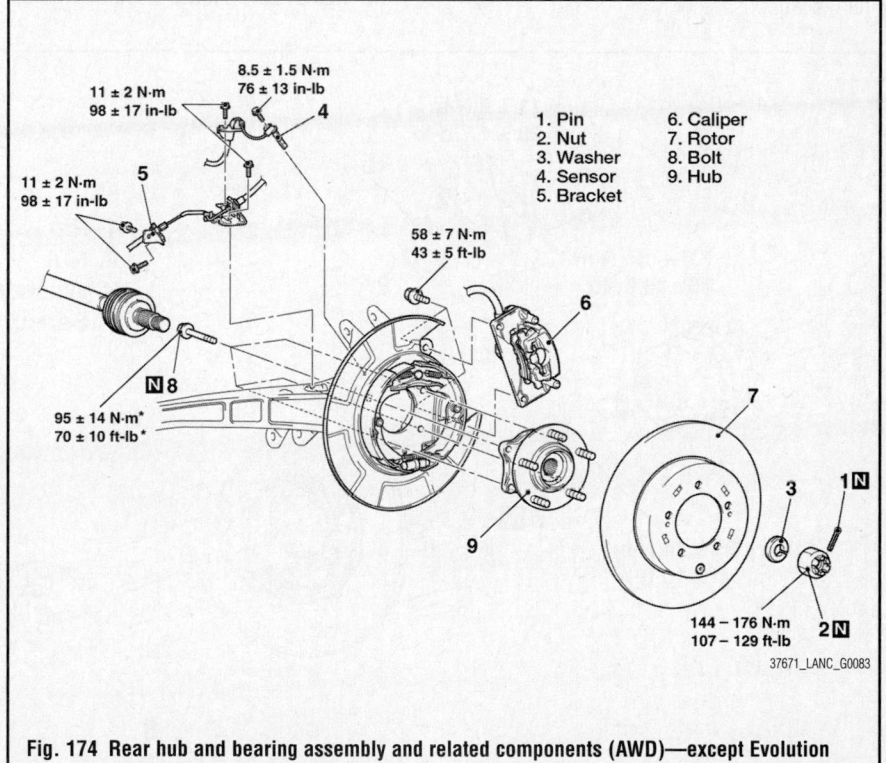

8.5 ± 1.5 N·m
76 ± 13 in-lb

11 ± 2 N·m
98 ± 17 in-lb

11 ± 2 N·m
98 ± 17 in-lb

58 ± 7 N·m
43 ± 5 ft-lb

95 ± 14 N·m*
70 ± 10 ft-lb*

144 – 176 N·m
107 – 129 ft-lb

1. Pin
2. Nut
3. Washer
4. Sensor
5. Bracket
6. Caliper
7. Rotor
8. Bolt
9. Hub

37671_LANC_G0083

Fig. 174 Rear hub and bearing assembly and related components (AWD)—except Evolution

WHEEL HUB & BEARING

REMOVAL & INSTALLATION

Except Evolution

See Figures 173 and 174.

➡**The hub and bearing assembly is serviced as a unit.**

1. Before servicing the vehicle, refer to the Precautions Section.

➡If working near and/or around the SRS system and components, be sure to disable the SRS system. Tape the negative battery cable with insulating tape. Always disconnect the negative battery cable first.

❄ CAUTION

Wait for 1 minute after disconnecting the negative battery cable before working inside the vehicle. The air bag system is set to deploy for a short period of time after the battery is disconnected.

2. Disconnect the negative battery cable.
3. If equipped with Anti-Lock Brake (ABS), remove the wheel speed sensor.
4. Raise and safely support the vehicle.
5. Remove or disconnect the following:
   • Rear wheel
   • Caliper and brake disc

- Dust cap and flange nut
- Rear hub assembly

To install:

➡ **Be sure to use new fasteners, as required.**

6. Install or connect the following:
- Rear hub assembly using a new flange nut. Torque to specification.
- Dust cap
- Wheel speed sensor if removed. The air gap should be 0.012–0.035 in. (0.3–0.9mm).
- Brake disc and caliper
- Rear wheel assembly and lower the vehicle to the floor

Evolution

See Figure 175.

➡ **The hub and bearing assembly is serviced as a unit.**

1. Before servicing the vehicle, refer to the Precautions Section.

➡ **If working near and/or around the SRS system and components, be sure to disable the SRS system. Tape the negative battery cable with insulating tape. Always disconnect the negative battery cable first.**

✳✳ CAUTION

Wait for 1 minute after disconnecting the negative battery cable before working inside the vehicle. The air bag system is set to deploy for a short period of time after the battery is disconnected.

2. Disconnect the negative battery cable.

3. Raise and support the vehicle safely. Remove the tire and wheel assembly.

4. Remove the cotter pin. Remove the driveshaft nut and washer.

5. Remove the rear ABS sensor.

6. Remove the caliper. Position it to the side with wire. Do not disconnect the brake hose.

7. Remove the parking brake shoe and lining assembly.

8. Remove the clip. Remove the parking brake cable connection.

9. Disconnect the rear driveshaft connection.

10. Remove the wheel hub assembly retaining bolts.

11. Remove the rear hub assembly from the vehicle.

To install:

12. Installation is the reverse of the removal procedure.

13. Adjust the parking brake.

ADJUSTMENT

The rear wheel bearings are not adjustable. If the bearings are noisy or become loose, they must be replaced.

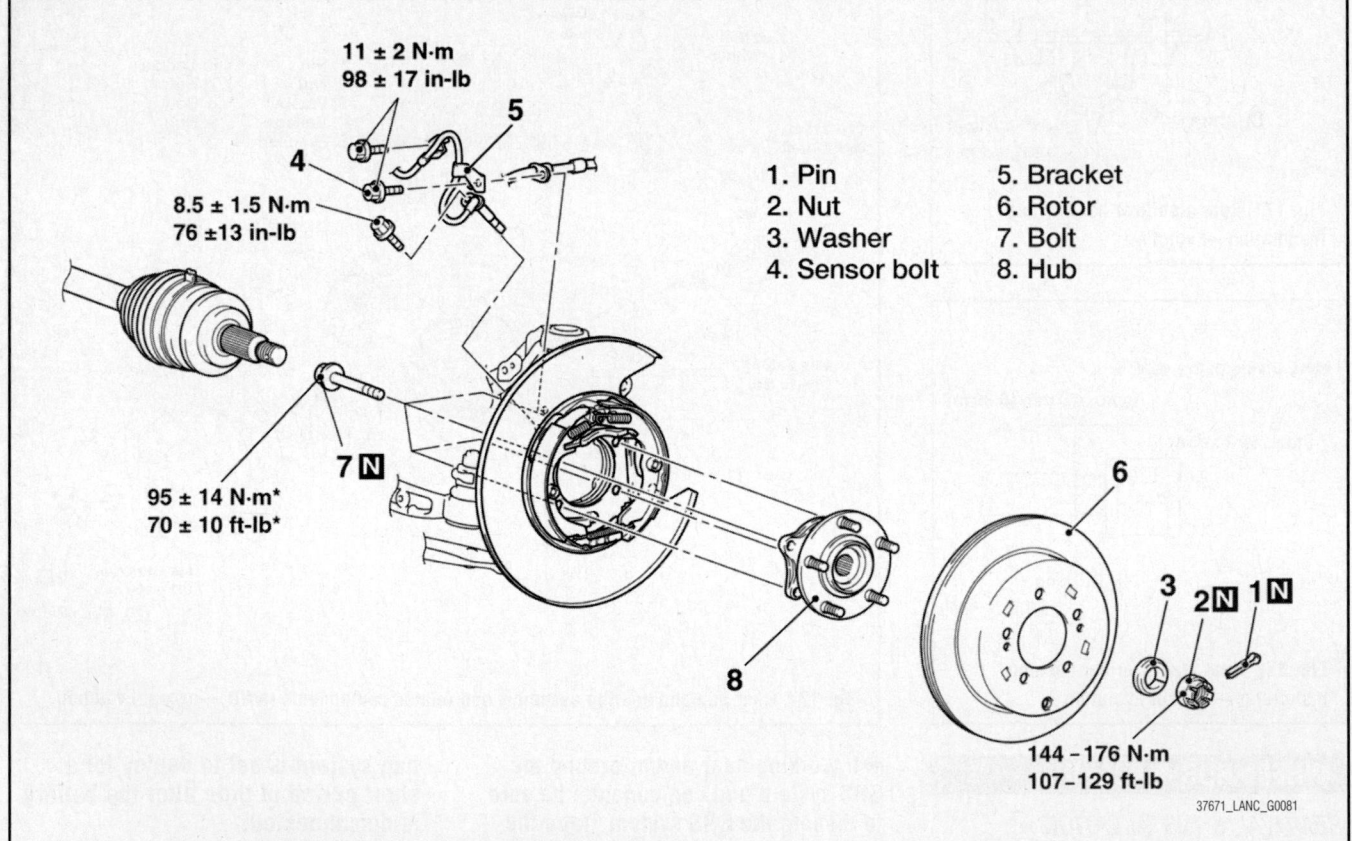

11 ± 2 N·m
98 ± 17 in-lb

8.5 ± 1.5 N·m
76 ±13 in-lb

95 ± 14 N·m*
70 ± 10 ft-lb*

144 – 176 N·m
107–129 ft-lb

1. Pin
2. Nut
3. Washer
4. Sensor bolt
5. Bracket
6. Rotor
7. Bolt
8. Hub

37671_LANC_G0081

Fig. 175 Rear hub and bearing assembly and related components—Evolution

SPECIFICATIONS AND MAINTENANCE CHARTS

ENGINE AND VEHICLE IDENTIFICATION

Engine							Model Year	
Code ①	Liters (cc)	Cu. In	Cyl.	Fuel Sys.	Type	Eng. Mfg.	Code ②	Year
4B12/W	2.4 (2359)	144	4	MFI	DOHC	Mitsubishi	9	2009
6B31/X	3.0 (2998)	183	6	MFI	SOHC	Mitsubishi		

MFI: Multiport fuel injection

SOHC: Single overhead camshaft

① Engine ID / 8th digit of the VIN

② 10th digit of the VIN

37671_OUTL_C0001

GENERAL ENGINE SPECIFICATIONS

Year	Model	Engine Displacement Liters	Engine ID/VIN	Net Horsepower @ rpm	Net Torque @ rpm (ft. lbs.)	Bore x Stroke (in.)	Compression Ratio	Oil Pressure @ rpm
2009	Outlander	2.4	4B12/W	168@6000	167@4100	3.46x3.82	10.5:1	①
	Outlander	3.0	6B31/X	220@6250	204@4000	3.45x3.26	9.5:1	①

① 4.2 psi or more at curb idle speed

37671_OUTL_C0002

ENGINE TUNE-UP SPECIFICATIONS

Year	Engine Displacement Liters	Engine ID/VIN	Spark Plugs Gap (in.)	Ignition Timing (deg.) MT	Ignition Timing (deg.) AT	Fuel Pump (psi)	Idle Speed (rpm) MT	Idle Speed (rpm) AT	Valve Clearance In.	Valve Clearance Ex.
2009	2.4	4B12/W	0.028-0.031	—	2-8B	47	—	550-750	0.007-0.009	0.011-0.013
	3.0	6B31/X	0.028-0.031	—	2-8B	47	—	550-750	0.004	HYD

NOTE: The Vehicle Emission Control Information label often reflects specification changes made during production.

The label figures must be used if they differ from those in this chart.

B: Before top dead center

37671_OUTL_C0003

CAPACITIES

Year	Model	Engine Displacement Liters	Engine ID/VIN	Engine Oil with Filter	Transmission (pts.) Man.	Transmission (pts.) Auto.	Transfer Case (pts.)	Drive Axle Front (pts.)	Drive Axle Rear (pts.)	Fuel Tank (gal.)	Cooling System (qts.)
2009	Outlander	2.4	4B12/W	4.8	—	①	1.12	—	1.06	16.6	②
	Outlander	3.0	6B31/X	4.5	—	①	1.12	—	1.06	16.6	②

NOTE: All capacities are approximate. Add fluid gradually and ensure a proper fluid level is obtained.

① Automatic transaxle: 17.4 pts.

CVT tranaxle: 16.4 pts.

② California: 10.6 qts.

N on-California: 10.0 qts.

37671_OUTL_C0004

FLUID SPECIFICATIONS

Year	Model	Engine Displacement Liters	Engine ID/VIN	Engine Oil	Auto. Trans.	Drive Axle	Transfer Case	Power Steering Fluid	Brake Master Cylinder
2009	Outlander	2.4	4B12/W	5W-20	①	85W-90	85W-90	Genuine Mitsubishi	DOT 3 or 4
	Outlander	3.0	6B31/X	5W-20	①	85W-90	85W-90	Genuine Mitsubishi	DOT 3 or 4

DOT: Department Of Transpotation

① Automatic transaxle: DIA Queen J2

CVT transaxle: DIA Queen CVTF-J1

37671_OUTL_C0014

VALVE SPECIFICATIONS

Year	Engine Displacement Liters	Engine ID/VIN	Seat Angle (deg.)	Face Angle (deg.)	Spring Test Pressure (lbs. @ in.)	Spring Installed Height (in.)	Stem-to-Guide Clearance (in.) Intake	Stem-to-Guide Clearance (in.) Exhaust	Stem Diameter (in.) Intake	Stem Diameter (in.) Exhaust
2009	2.4	4B12/W	NA	45	NA	NA	0.0008-0.0019	0.0012-0.0021	NA	NA
	3.0	6B31/X	NA	43.5-44	55.13@1.940	1.9400	0.0008-0.0018	0.0014-0.0024	0.240	0.240

NA - Not Available

37671_OUTL_C0007

CAMSHAFT SPECIFICATIONS CHART
All measurements are given in inches.

Year	Engine Displacement Liters	Engine VIN	Journal Dia.	Brg. Oil Clearance	Shaft End-play	Runout	Lobe Height Intake	Lobe Height Exhaust
2009	2.4	4B12/W	1.8000	0.0013	NS	NS	1.736	1.772
	3.0	6B31/X	1.8000	NS	NS	NS	①	②

NS - Not specified by manufacturer

① Standard value low speed cam : 1.468

   Standard value high speed cam: 1.426

② Standard value: 1.490

37671_OUTL_C0013

CRANKSHAFT AND CONNECTING ROD SPECIFICATIONS
All measurements are given in inches.

Year	Engine Displacement Liters	Engine ID/VIN	Crankshaft Main Brg. Journal Dia.	Crankshaft Main Brg. Oil Clearance	Crankshaft Shaft End-play	Crankshaft Thrust on No.	Connecting Rod Journal Diameter	Connecting Rod Oil Clearance	Connecting Rod Side Clearance
2009	2.4	4B12/W	NA	0.0005-0.0012	0.0020-0.0010	3	NA	0.0007-0.0018	0.0040-0.0100
	3.0	6B31/X	2.700	①	0.0020-0.0090	3	NA	0.0005-0.0015	0.0040-0.0090

NA - Not Available

① Journal 1 and 4: 0.0008-0.0014 in..

   Journal 2 and 3: 0.0010-0.0017 in..

37671_OUTL_C0005

PISTON AND RING SPECIFICATIONS
All measurements are given in inches.

Year	Engine Displacement Liters	Engine ID/VIN	Piston Clearance	Ring Gap Top Compression	Ring Gap Bottom Compression	Ring Gap Oil Control	Ring Side Clearance Top Compression	Ring Side Clearance Bottom Compression	Ring Side Clearance Oil Control
2009	2.4	4B12/W	NA	0.0060-0.0110	0.0100-0.0160	0.0040-0.0140	0.0010-0.0030	0.0010-0.0030	NA
	3.0	6B31/X	NA	0.0080-0.0120	0.0120-0.0180	0.0040-0.0230	0.0016-0.0031	0.0012-0.0027	0.0040-0.023

NA - Not Available

37671_OUTL_C0006

TORQUE SPECIFICATIONS
All readings in ft. lbs.

Year	Engine Displacement Liters	Engine ID/VIN	Cylinder Head Bolts	Main Bearing Bolts	Rod Bearing Bolts	Crankshaft Damper Bolts	Flexplate Bolts	Manifold		Spark Plugs	Oil Pan Drain Plug
								Intake	Exhaust		
2009	2.4	4B12/W	①	20 ②	14 ③	155	34-38	15	36	19	21-23
	3.0	6B31/X	④	18 ③	15 ③	④	55-59	16	33-39	13	29

① Step 1: Tighten all bolts to 26 ft. lbs.
  Step 2: Tighten all bolts 90 degrees.
  Step 3: Tighten all bolts an additional 90 degrees.

② Torque to specification plus
  an additional 45 degrees.

③ Torque to specification plus
  an additional 90 degrees.

④ Step 1: Tighten bolt to 148 ft. lbs.
  Step 2: Loosen bolt to 0 ft. lbs.
  Step 3: Tighten bolt to 81 ft. lbs.
  Step 4: Tighten bolt 60 degrees.

37671_OUTL_C0008

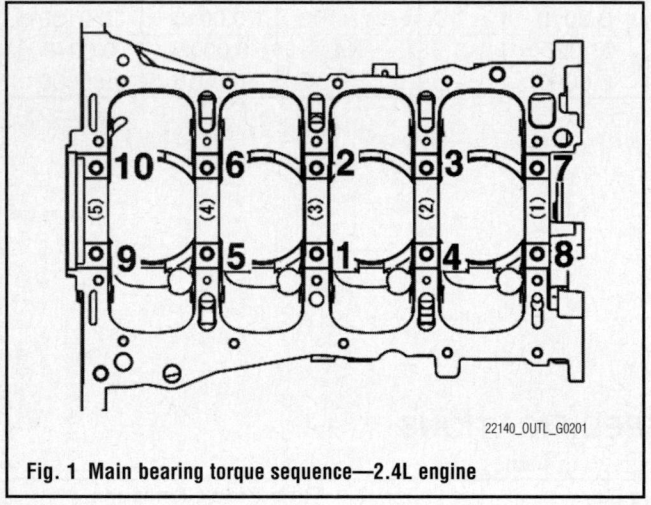

Fig. 1 Main bearing torque sequence—2.4L engine

22140_OUTL_G0201

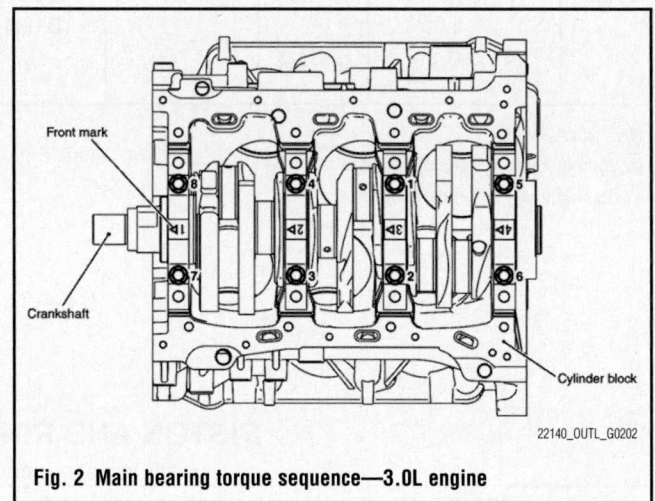

Fig. 2 Main bearing torque sequence—3.0L engine

22140_OUTL_G0202

WHEEL ALIGNMENT

Year	Model		Caster		Camber		Toe-in (in.)
			Range (+/-Deg.)	Preferred Setting (Deg.)	Range (+/-Deg.)	Preferred Setting (Deg.)	
2009	Outlander	F	0.30	+2.35	0.30	0.20	0.04 +/- 0.09
		R	—	—	0.30	-0.25	0.12 +/- 0.08

37671_OUTL_C0009

TIRE, WHEEL AND BALL JOINT SPECIFICATIONS

Year	Model	OEM Tires		Tire Pressures (psi)		Wheel Size	Ball Joint Inspection ①	Lug Nut (ft. lbs.)
		Standard	Optional	Front	Rear			
2009	Outlander	P215/70R16	P225/55R18	②	②	③	U: 4-26 in. L: 0-35 in.	66-80

OEM: Original Equipment Manufacturer

PSI: Pounds Per Square Inch. If specification differs from the one located on driver's door, use the driver's door specification

L: Lower

U: Upper

① Torque required in inch lbs. to rotate ball joint when removed from the knuckle

② Refer to the label attached to the center pillar on the Driver's side of the vehicle

③ Standard: 16X6.5JJ. Optional: 18X7JJ

37671_OUTL_C0010

BRAKE SPECIFICATIONS
All measurements in inches unless noted

Year	Model		Brake Disc			Minimum Lining Thickness		Brake Caliper	
			Original Thickness	Minimum Thickness	Maximum Runout	Front	Rear	Bracket Bolts (ft. lbs.)	Mounting Bolts (ft. lbs.)
2009	Outlander	F	1.020	0.960	0.0024	0.080	—	55	74
		R	0.390	0.330	0.0032	—	0.080	33	33

NA: Not Available

F: Front

R: Rear

37671_OUTL_C0011

SCHEDULED MAINTENANCE INTERVALS
Mitsubishi—Outlander

TO BE SERVICED	TYPE OF SERVICE	VEHICLE MILEAGE INTERVAL (x1000)												
		7.5	15	22.5	30	37.5	45	52.5	60	67.5	75	82.5	90	97.5	102.5
Engine oil & filter	R	✓	✓	✓	✓	✓	✓	✓	✓	✓	✓	✓	✓	✓	✓
Air cleaner element	R		✓		✓		✓		✓		✓		✓		✓
Automatic transaxle fluid	S/I		✓		✓		✓		✓		✓		✓		✓
Manual transaxle fluid	S/I		✓		✓		✓		✓		✓		✓		✓
Brake hoses	S/I		✓		✓		✓		✓		✓		✓		✓
Disc brake pads	S/I		✓		✓		✓		✓		✓		✓		✓
Driveshaft boots	S/I		✓		✓		✓		✓		✓		✓		✓
Valve clearance	S/I				✓				✓				✓		
Engine coolant	R								✓				✓		
Coolant Hoses	S/I				✓				✓				✓		
Spark plugs (iridium)	R														✓
Ball joints & steering linkage seals	S/I				✓				✓				✓		
Drive belt(s)	S/I				✓				✓				✓		
Exhaust system	S/I				✓				✓				✓		
Fuel hoses	S/I				✓				✓				✓		
Transfer case fluid	S/I				✓				✓				✓		
Transfer case fluid	R								✓						
Rear Axle Oil	S/I				✓				✓				✓		
Rear drum brake linings & rear wheel cylinders	S/I				✓				✓				✓		
Timing belt(s) 3.0L	R														✓
EVAP system (except canister)	S/I								✓						
Fuel system (tank, pipe line, connection & fuel tank filler tube cap)	S/I				✓				✓				✓		
Tires (rotate)	S/I	✓	✓	✓	✓	✓	✓	✓	✓	✓	✓	✓	✓	✓	✓

R: Replace   S/I: Service or Inspect

FREQUENT OPERATION MAINTENANCE (SEVERE SERVICE)

If a vehicle is operated under any of the following conditions it is considered severe service:

- **Extremely dusty areas.**

- **50% or more of the vehicle operation is in 32°C (90°F) or higher temperatures, or constant operation in temperatures below 0°C (32°F).**

- **Prolonged idling (vehicle operation in stop and go traffic).**

- **Frequent short running periods (engine does not warm to normal operating temperatures).**

- **Police, taxi, delivery usage or trailer towing usage.**

Oil & oil filter: change every 3750 miles.

Disc brake pads: service or inspect every 7500 miles.

Suspension system service or inspect every 7500 miles

Air filter element: service or inspect every 15,000 miles.

Automatic transaxle fluid & filter: replace every 30,000 miles.

Transfer case fluid: replace every 30,000 miles.

ANTI-LOCK SYSTEMS

- Certain components within the ABS system are not intended to be serviced or repaired individually.

- Do not use rubber hoses or other parts not specifically specified for and ABS system. When using repair kits, replace all parts included in the kit. Partial or incorrect repair may lead to functional problems and require the replacement of components.

- Lubricate rubber parts with clean, fresh brake fluid to ease assembly. Do not use shop air to clean parts; damage to rubber components may result.

- Use only DOT 3 brake fluid from an unopened container.

- If any hydraulic component or line is removed or replaced, it may be necessary to bleed the entire system.

- A clean repair area is essential. Always clean the reservoir and cap thoroughly before removing the cap. The slightest amount of dirt in the fluid may plug an ori-fice and impair the system function. Perform repairs after components have been thoroughly cleaned; use only denatured alcohol to clean components. Do not allow ABS components to come into contact with any substance containing mineral oil; this includes used shop rags.

- The Anti-Lock control unit is a microprocessor similar to other computer units in the vehicle. Ensure that the ignition switch is **OFF** before removing or installing controller harnesses. Avoid static electricity discharge at or near the controller.

- If any arc welding is to be done on the vehicle, the control unit should be unplugged before welding operations begin.

DISC AND DRUM SYSTEMS

> ※※ **CAUTION**
>
> **Dust and dirt accumulating on brake parts during normal use may contain asbestos fibers from production or aftermarket brake linings. Breathing excessive concentrations of asbestos fibers can cause serious bodily harm. Exercise care when servicing brake parts. Do not sand or grind brake lining unless equipment used is designed to contain the dust residue. Do not clean brake parts with compressed air or by dry brushing. Cleaning should be done by dampening the brake components with a fine mist of water, then wiping the brake components clean with a dampened cloth. Dispose of cloth and all residue containing asbestos fibers in an impermeable container with the appropriate label. Follow practices prescribed by the Occupational Safety and Health Administration (OSHA) and the Environmental Protection Agency (EPA) for the handling, processing, and disposing of dust or debris that may contain asbestos fibers.**

BLEEDING PROCEDURE

MASTER CYLINDER BLEEDING

See Figure 3.

1. Before servicing the vehicle, refer to the precautions.

➡ **Use only DOT 3 or DOT 4 brake fluid. Never mix the specified brake fluid with other fluid as it will influence the braking performance significantly.**

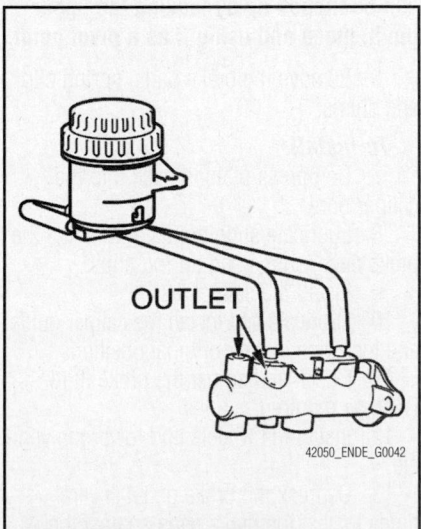

Fig. 3 Master cylinder bleeding

The master cylinder used has no check valve, so if bleeding is carried out by the following procedure, bleeding of air from the brake pipeline will become easier. (When brake fluid is not contained in the master cylinder).

2. Fill the reserve tank with brake fluid.

3. Keep the brake pedal depressed.

4. Have another person cover the master cylinder outlet with a finger.

5. With the outlet still closed, release the brake pedal.

6. Repeat steps 2–4 three or four times to fill the inside of the master cylinder with brake fluid.

BRAKE LINE BLEEDING

See Figure 4.

1. Before servicing the vehicle, refer to the precautions.

➡ **Use only DOT 3 or DOT 4 brake fluid. Never mix the specified brake fluid with other fluid as it will influence the braking performance significantly.**

2. Make sure the master cylinder reserve tank is filled the fluid.

3. Start the engine.

➡ **Bleed the air in the sequence shown in the figure.**

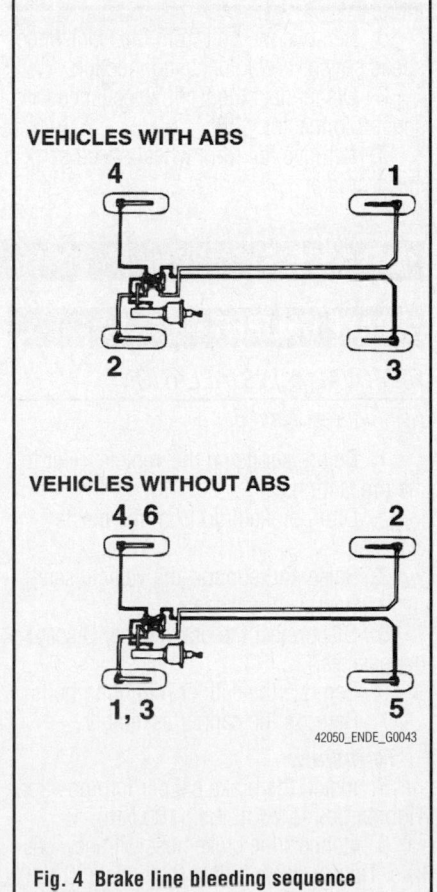

VEHICLES WITH ABS

VEHICLES WITHOUT ABS

42050_ENDE_G0043

Fig. 4 Brake line bleeding sequence

4. Raise and safely support the vehicle.

5. Connect a piece of vinyl tubing to the brake caliper.

6. Have an assistant depress the brake pedal several times, then loosen the bleeder plug with the pedal held down.

7. When fluid stops coming out, tighten the bleeder plug, then release the brake pedal.

8. Repeat steps until all the air in the fluid has been bled out.

9. Tighten the brake bleeder plug to 74 inch lbs. (8.3 Nm).

10. Repeat the above steps to bleed the air out of the brake line for each wheel.

BLEEDING THE ABS SYSTEM

There are no special procedures for bleeding the ABS system. Refer to the bleeding procedures listed above.

BRAKES   ANTI-LOCK BRAKE SYSTEM (ABS)

WHEEL SPEED SENSORS

REMOVAL & INSTALLATION

Front

1. Before servicing the vehicle, refer to the precautions.

2. Remove the air cleaner assembly.

3. Raise and safely support the vehicle.

4. Remove the front wheel and tire assemblies.

☀ WARNING

The wheel speed sensor collects any metallic particles easily, because it is magnetized. Make sure that the sensor does not collect any metallic particles. Check that there is no trouble prior to reassembling it.

5. Remove the bolt from the front wheel speed sensor and knuckle connection.

6. Disconnect the front wheel speed sensor connector.

7. Remove the front wheel speed sensor grommet

8. Remove the front wheel speed sensor.

To install:

9. Install the front wheel speed sensor. Tighten to 76 inch lbs. (9 Nm).

10. Install the front wheel speed sensor grommet

11. Connect the front wheel speed sensor connector.

12. Install the bolt from the front wheel speed sensor and knuckle connection. Tighten to 98 inch lbs. (11 Nm).

13. Install the front wheel and tire assemblies.

14. Lower the vehicle.

15. Install the air cleaner assembly.

Rear

1. Before servicing the vehicle, refer to the precautions.

2. Raise and safely support the vehicle.

3. Remove the rear wheel and tire assemblies.

4. Remove the rear wheel speed sensor and trailing arm assembly connection.

5. Disconnect the rear wheel speed sensor connector.

6. Remove the rear wheel speed sensor grommet.

7. Remove the rear wheel speed sensor.

To install:

8. Install the rear wheel speed sensor. Tighten to 76 inch lbs. (9 Nm).

9. Install the rear wheel speed sensor grommet.

10. Connect the rear wheel speed sensor connector.

11. Install the rear wheel speed sensor and trailing arm assembly connection. Tighten to 98 inch lbs. (11 Nm).

12. Install the rear wheel and tire assemblies.

13. Lower the vehicle.

WHEEL SPEED SENSOR RINGS (TOOTHED RINGS)

REMOVAL & INSTALLATION

The ABS wheel speed sensor ring is an integral part of either the BJ assembly of the halfshaft or the hub and cannot be disassembled.

BRAKES   FRONT DISC BRAKES

BRAKE CALIPER

REMOVAL & INSTALLATION

See Figures 5 and 6.

1. Before servicing the vehicle, refer to the precautions.

2. Drain brake fluid from the master cylinder.

3. Raise and support the vehicle safely.

4. Remove the wheels.

5. Disconnect the brake hose. Discard the gaskets.

6. Remove the caliper mounting bolts.

7. Remove the caliper assembly.

To install:

8. Install the brake caliper into position. Tighten bolt to 74 ft. lbs. (100 Nm).

9. Connect the brake hose with new gaskets. Tighten banjo bolt to 22 ft. lbs. (30 Nm).

10. Bleed the brake system.

11. Install the wheels and lower the vehicle.

12. Depress the brake pedal several times to seat the brake pads against the rotors.

DISC BRAKE PADS

REMOVAL & INSTALLATION

See Figures 5 and 6.

1. Before servicing the vehicle, refer to the precautions.

2. Raise and support the vehicle safely.

3. Remove some of the brake fluid from the master cylinder reservoir.

4. Remove the wheels.

5. Remove the caliper guide and lock pins and lift the caliper assembly from the caliper support.

➡On some vehicles, the front caliper can be flipped up by leaving the upper pin in place and using it as a pivot point.

6. Remove the brake pads, spring clip and shims.

To install:

7. Compress pistons back into the caliper bore.

8. Lubricate slide points and install the brake pads, shims and spring clips.

9. Install the caliper.

10. Lubricate and install the caliper guide and lock pins in their original positions.

11. Check and adjust the brake fluid level, as required.

12. Install the wheels and lower the vehicle.

13. Depress the brake pedal several times to seat the brake pads against the rotors.

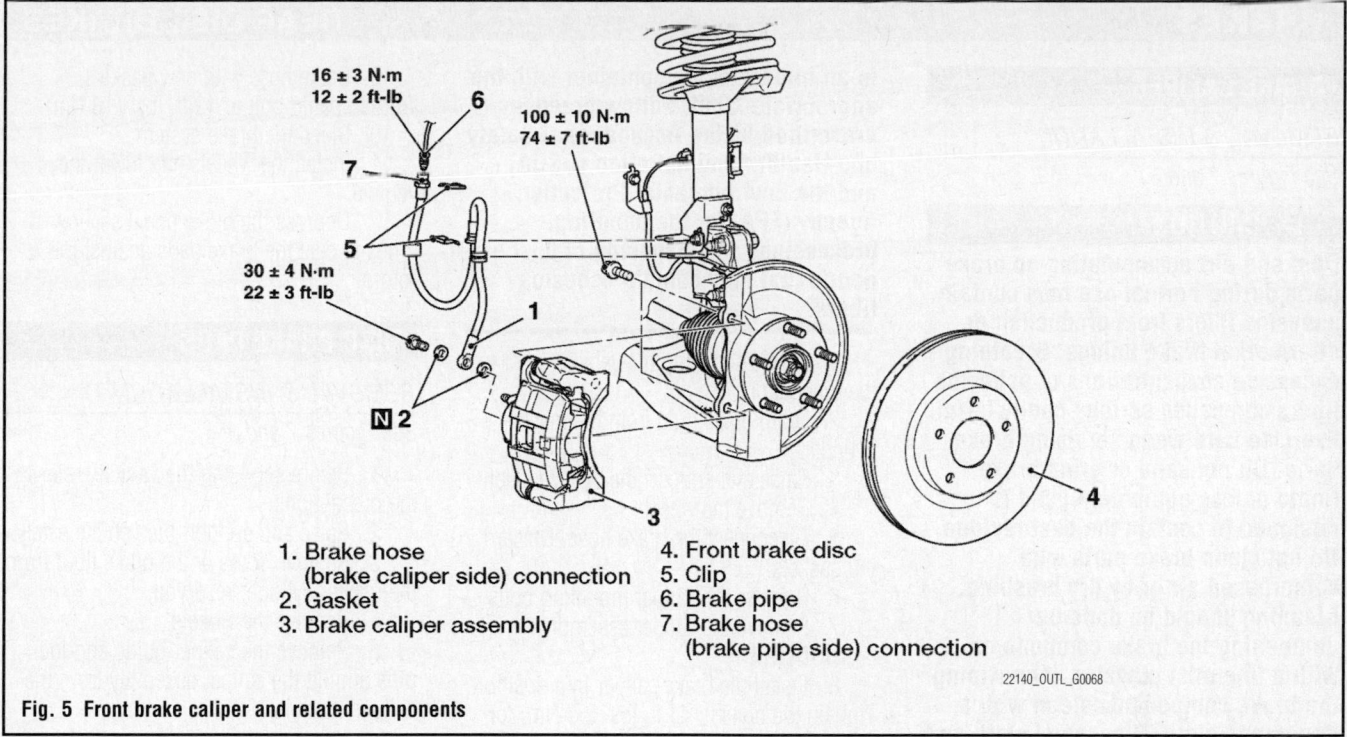

16 ± 3 N·m
12 ± 2 ft-lb

100 ± 10 N·m
74 ± 7 ft-lb

30 ± 4 N·m
22 ± 3 ft-lb

1. Brake hose
   (brake caliper side) connection
2. Gasket
3. Brake caliper assembly

4. Front brake disc
5. Clip
6. Brake pipe
7. Brake hose
   (brake pipe side) connection

22140_OUTL_G0068

Fig. 5 Front brake caliper and related components

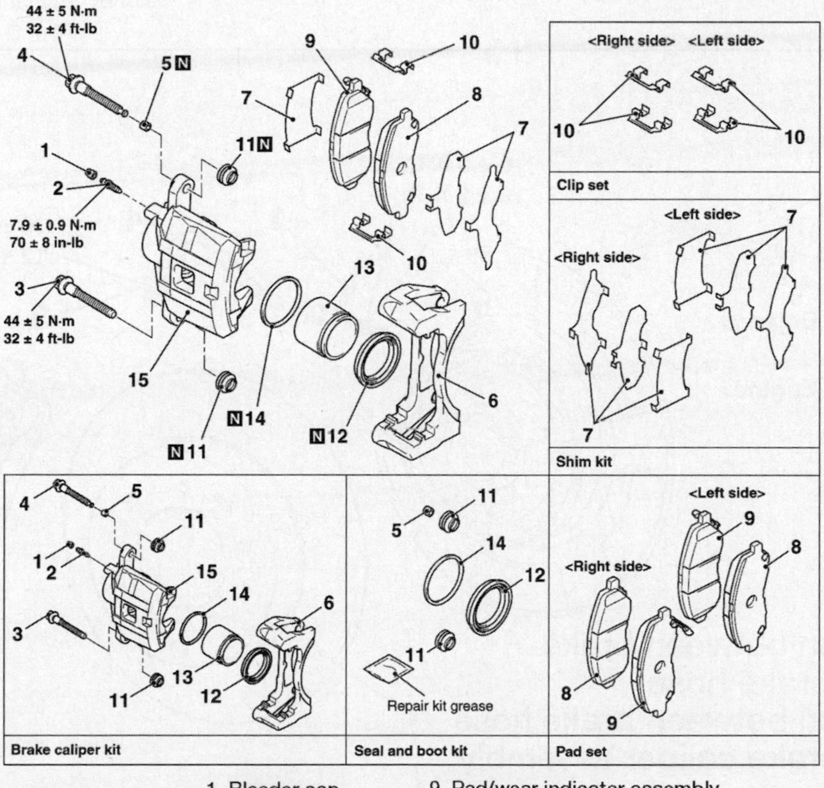

44 ± 5 N·m
32 ± 4 ft-lb

7.9 ± 0.9 N·m
70 ± 8 in-lb

44 ± 5 N·m
32 ± 4 ft-lb

<Right side> <Left side>

Clip set

<Left side>

<Right side>

Shim kit

Repair kit grease

Brake caliper kit

Seal and boot kit

<Left side>

<Right side>

Pad set

1. Bleeder cap
2. Bleeder
3. Guide pin
4. Lock pin
5. Bushing
6. Caliper support
7. Shim
8. Pad assembly

9. Pad/wear indicator assembly
10. Clip
11. Pin boot
12. Piston boot
13. Piston 14. Piston seal
15. Caliper body

22140_OUTL_G0069

Fig. 6 Exploded view of the front brake caliper

BRAKE CALIPER

REMOVAL & INSTALLATION

See Figures 7 and 8.

❋❋ CAUTION

Dust and dirt accumulating on brake parts during normal use may contain asbestos fibers from production or aftermarket brake linings. Breathing excessive concentrations of asbestos fibers can cause serious bodily harm. Exercise care when servicing brake parts. Do not sand or grind brake lining unless equipment used is designed to contain the dust residue. Do not clean brake parts with compressed air or by dry brushing. Cleaning should be done by dampening the brake components with a fine mist of water, then wiping the brake components clean with a dampened cloth. Dispose of cloth and all residue containing asbestos fibers

in an impermeable container with the appropriate label. Follow practices prescribed by the Occupational Safety and Health Administration (OSHA) and the Environmental Protection Agency (EPA) for the handling, processing, and disposing of dust or debris that may contain asbestos fibers.

1. Before servicing the vehicle, refer to the precautions.
2. Drain brake fluid from the master cylinder.
3. Raise and support the vehicle safely.
4. Remove the wheels.
5. Disconnect the brake hose. Discard the gaskets.
6. Remove the caliper mounting bolts
7. Remove the caliper assembly.

To install

8. Install the brake caliper into position. Tighten the bolts to 41 ft. lbs. (55 Nm) for 2.4L engine and 43 ft. lbs. (58 Nm) for 3.0L engine.

9. Brake hose with new gaskets. Tighten banjo bolt to 12 ft. lbs. (18 Nm).
10. Bleed the brake system.
11. Install the wheels and lower the vehicle.
12. Depress the brake pedal several times to seat the brake pads against the rotors.

DISC BRAKE PADS

REMOVAL & INSTALLATION

See Figures 7 and 8.

1. Before servicing the vehicle, refer to the precautions.
2. Raise and support the vehicle safely.
3. Remove some of the brake fluid from the master cylinder reservoir.
4. Remove the wheels.
5. Remove the caliper guide and lock pins and lift the caliper assembly from the caliper support.

➡On some vehicles, the front caliper can be flipped up by leaving the upper

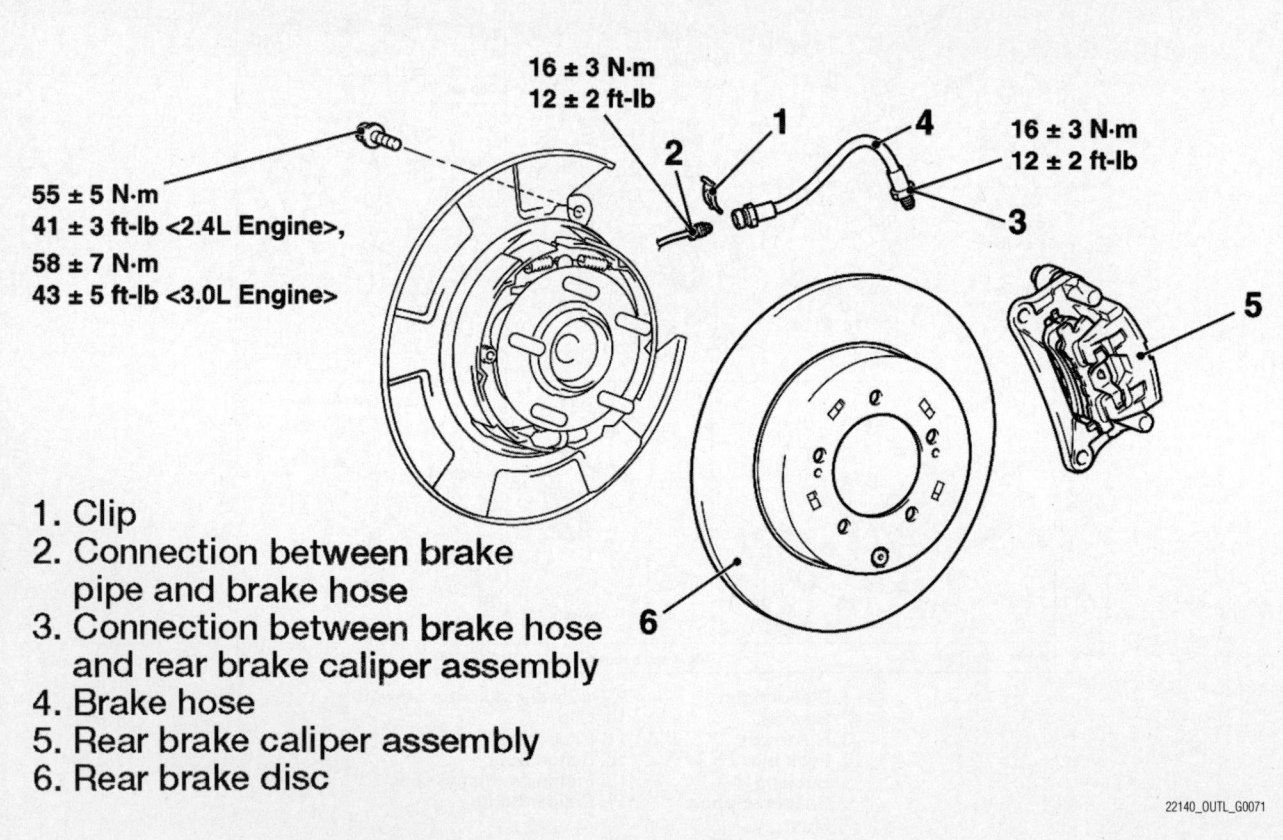

1. Clip
2. Connection between brake pipe and brake hose
3. Connection between brake hose and rear brake caliper assembly
4. Brake hose
5. Rear brake caliper assembly
6. Rear brake disc

22140_OUTL_G0071

Fig. 7 Rear brake caliper and related components

pin in place and using it as a pivot point.

6. Remove the brake pads, spring clip and shims.

To install:

7. Compress pistons back into the caliper bore.

8. Lubricate slide points and install the brake pads, shims and spring clips.

9. Install the caliper.

10. Lubricate and install the caliper guide and lock pins in their original positions.

11. Check and adjust the brake fluid level, as required.

12. Install the wheels and lower the vehicle.

13. Depress the brake pedal several times to seat the brake pads against the rotors.

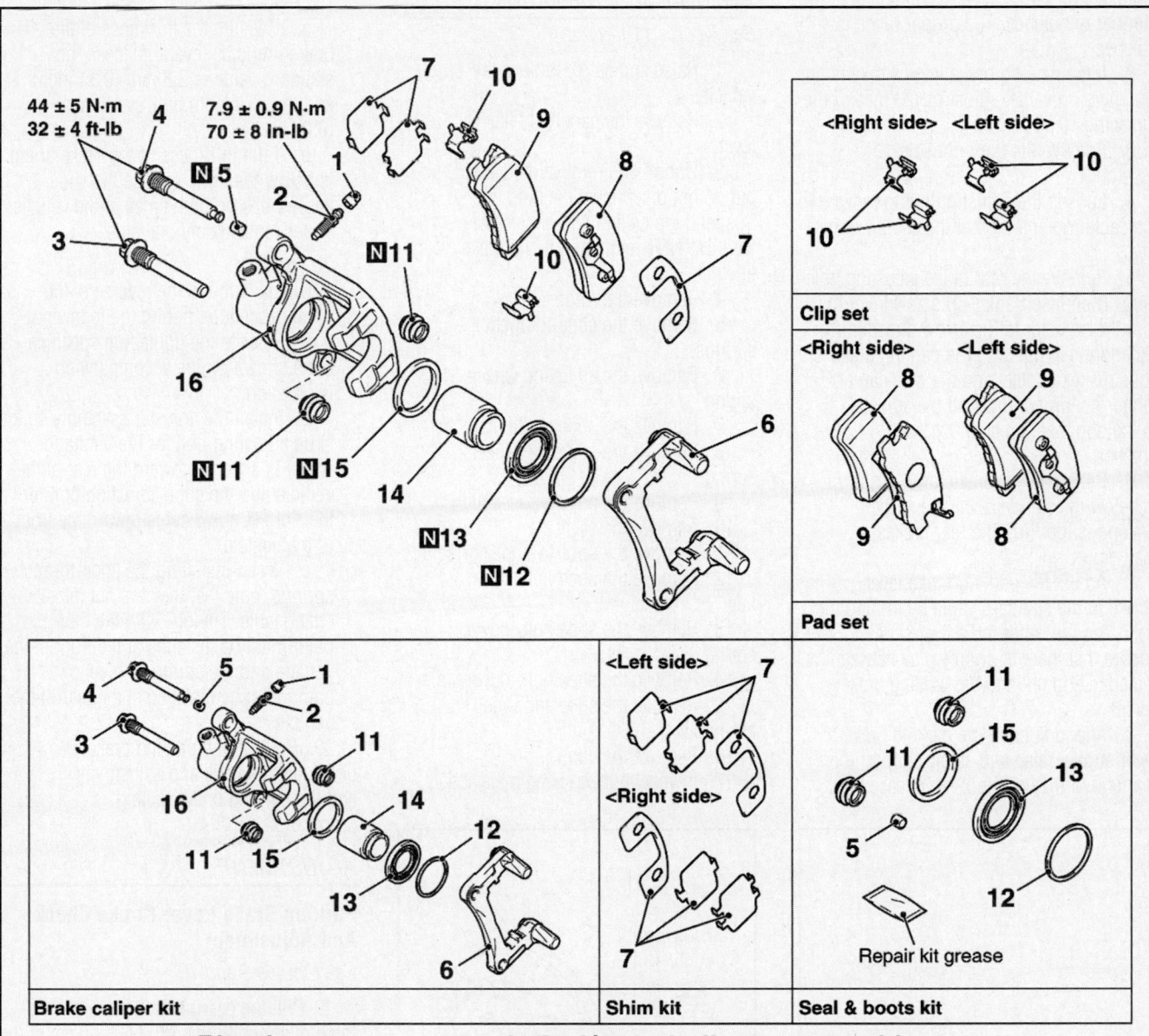

1. Bleeder cap
2. Bleeder
3. Guide pin
4. Lock pin
5. Bushing
6. Caliper support
7. Shim
8. Pad/clip assembly
9. Pad/wear indicator assembly
10. Clip
11. Pin boot
12. Boot ring
13. Piston boot
14. Piston
15. Piston seal
16. Caliper body

22140_OUTL_G0072

Fig. 8 Exploded view of the front rear brake caliper

BRAKES   **PARKING BRAKE**

PARKING BRAKE CABLES

ADJUSTMENT

See Figures 9 and 10.

1. Pull the parking brake lever with a force of about 45 lbs. (200 N) and count the number of notches. The proper range is 4–5 notches.

2. If the parking brake lever stroke is not the standard value, adjust as outlined in the following steps.

3. Remove the floor console assembly.

4. Loosen the adjusting nut to move it to the cable rod end so that the cable will be free.

5. Remove the rear brake adjusting hole plug. Then insert a flat-tip screwdriver to turn the adjuster to the arrow direction (to expand the shoe) until the parking brake shoe makes contact and the disc can no longer be turned. Back off the adjuster to the opposite direction by five notches.

6. If the parking brake lever stroke is below the standard value and the braking is too firm, the rear brakes may drag.

7. Adjust the parking brake lever stroke to the standard value by turning the adjusting nut. After the adjustment, ensure that there is no free play between the adjusting nut and the parking brake lever.

8. After adjusting the parking brake lever stroke, raise and safely support the rear end of the vehicle, then release the parking brake and turn the rear wheels to make sure the rear brakes are not dragging.

PARKING BRAKE SHOES

REMOVAL & INSTALLATION

See Figures 11 through 13.

1. Raise and safely remove the vehicle.

2. Release the parking brake lever.

3. Unbolt and remove the caliper, but do not disconnect the fluid line. Suspend the caliper with a piece of wire. Do NOT let it hang by its brake line.

4. Remove the rotor.

5. Remove the shoe-to-anchor springs.

6. Remove the adjusting wheel spring.

7. Remove the shoe guide plate.

8. Remove the adjusting wheel spring.

9. Remove the adjuster assembly.

10. Remove the strut.

11. Remove the strut-to-shoe spring.

12. Remove the shoe hold-down cup.

13. Remove the shoe hold-down spring.

14. Remove the shoe hold-down pin.

15. Remove the shoe and lining assembly.

16. Remove the clip.

17. If wheel hub, backing plate and shoe-hold down pin removal are necessary, perform the following:

   a. Remove the rear wheel hub assembly.

   b. Remove the backing plate.

   c. Remove the shoe hold-down pin.

18. Inspect the parking brake lining.

   a. Measure the thickness of the brake lining at several places. The standard value is 2.8 mm (0.11 inch) and the minimum limit is 1.0 mm (0.04 inch)

   b. If the thickness of the brake lining is below the limit, replace the shoe assemblies on both sides of the vehicle. Never replace only one side.

To install:

19. Installation is the reverse of the removal procedure, noting the following:

   a. Refer to the tightening specifications shown on the accompanying illustration.

   b. Install the adjuster assembly so the shoe adjusting bolt for the left hand wheel is attached toward the rear of the vehicle and the shoe adjusting bolt for the right hand wheel is toward the front of the vehicle.

   c. When installing the shoe-to-anchor springs, note that they are not interchangeable. The one with the blue paint must be installed at the front of the vehicle and the spring with yellow paint must be installed at the rear of the vehicle.

20. Perform the parking brake pedal stroke check and adjustment, and the parking brake seating procedure.

ADJUSTMENT

Parking Brake Lever Stroke Check And Adjustment

See Figures 9 and 10.

1. Pull the parking brake lever with a force of about 45 lbs. (200 N) and count the number of notches. The proper range is 4–5 notches.

2. If the parking brake lever stroke is not the standard value, adjust as outlined in the following steps.

3. Remove the floor console assembly.

4. Loosen the adjusting nut to move it to the cable rod end so that the cable will be free.

5. Remove the rear brake adjusting hole plug. Then insert a flat-tip screwdriver to turn the adjuster to the arrow direction (to

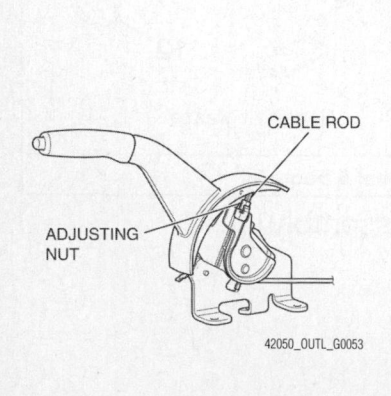

Fig. 9 Loosen the adjusting nut to move it to the cable rod end so that the cable will be free

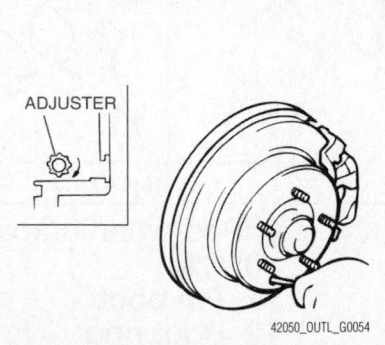

Fig. 10 Insert a flat-tip screwdriver to turn the adjuster to the arrow direction (to expand the shoe) until the parking brake shoe makes contact and the disc can no longer be turned

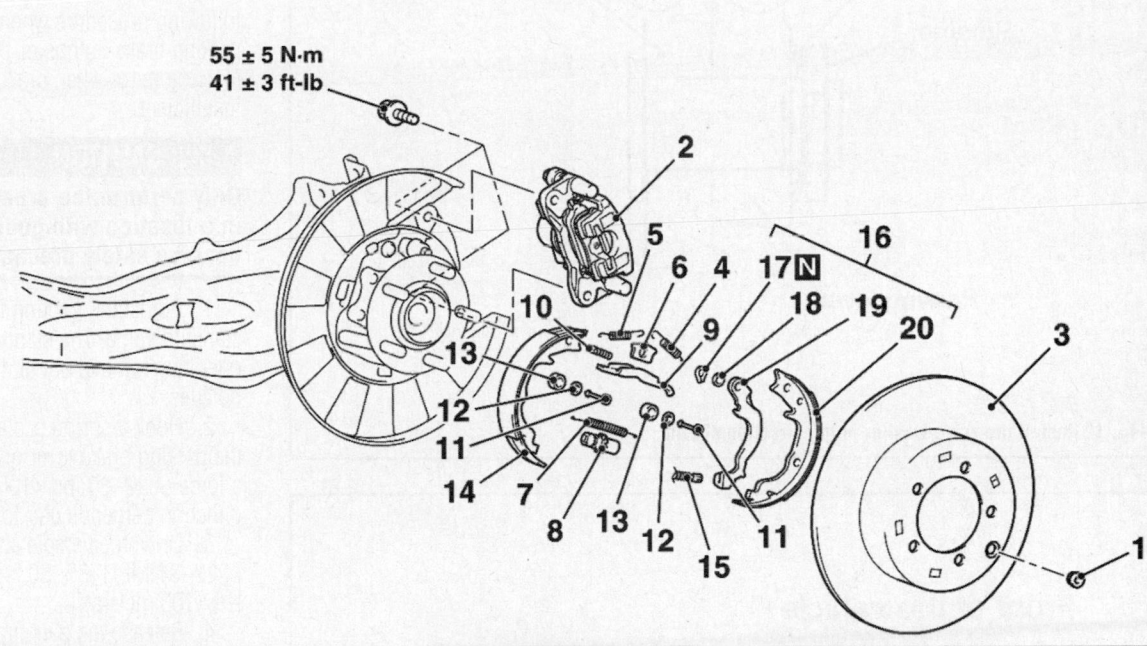

55 ± 5 N·m
41 ± 3 ft-lb

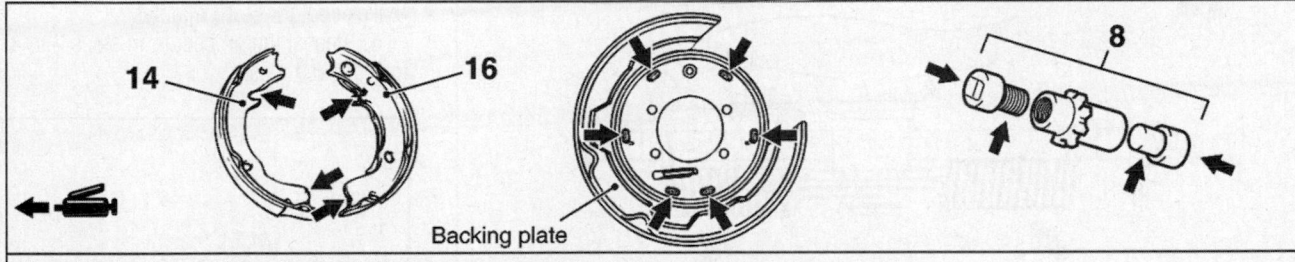

Backing plate

Specified grease: Chuo Yuka AKB 100 or equivalent

1. Plug
2. Rear brake caliper assembly
3. Rear brake disc
4. Shoe-to-anchor spring
5. Shoe-to-anchor spring
6. Shoe guide plate
7. Adjusting wheel spring
8. Adjuster assembly
9. Strut
10. Strut shoe-to-spring
11. Shoe hold down pin
12. Shoe hold down cup
13. Shoe hold down spring
14. Shoe and lining assembly
15. Parking brake rear cable assembly connection
16. Shoe and lever assembly
17. Retainer
18. Wave washer
19. Parking lever
20. Shoe and lining assembly

22140_OUTL_G0074

Fig. 11 Exploded view of the parking brake shoes (linings) and related components

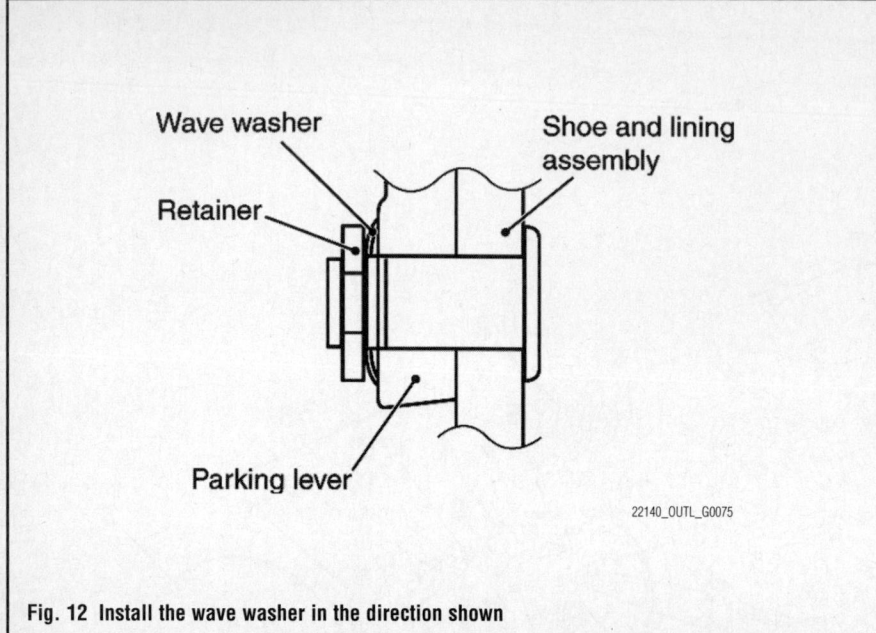

Fig. 12 Install the wave washer in the direction shown

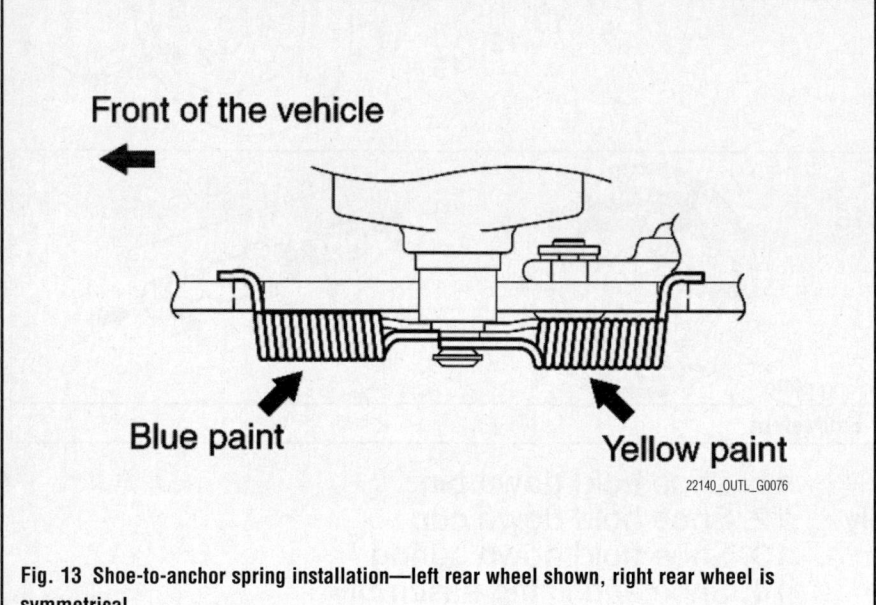

Front of the vehicle

Blue paint

Yellow paint

Fig. 13 Shoe-to-anchor spring installation—left rear wheel shown, right rear wheel is symmetrical

brake and turn the rear wheels to make sure the rear brakes are not dragging.

Lining Break-In

See Figure 14.

Perform the break-in using the following procedure when replacing the parking brake linings or the rear brake disc rotors, or when brake performance is insufficient.

✳✳ CAUTION

Only perform the break-in procedure in a location with good visibility, and observe safety precautions.

1. Adjust the parking brake stroke to the standard value. The standard value (with an operation force of about 45 lbs.) is 4–5 notches.

2. Hook a spring scale onto the center of the parking brake lever grip and pull it with a force of 22–34 lbs. (100–150 N) in a direction perpendicular to the handle.

3. Drive the vehicle at a constant speed of 22–31 MPH (35–50 km/h) for about 328 feet (100 meters).

4. Release the parking brake and let the brakes cool for 5–10 minutes.

5. Repeat the procedure in steps 2 to 4, four or five times.

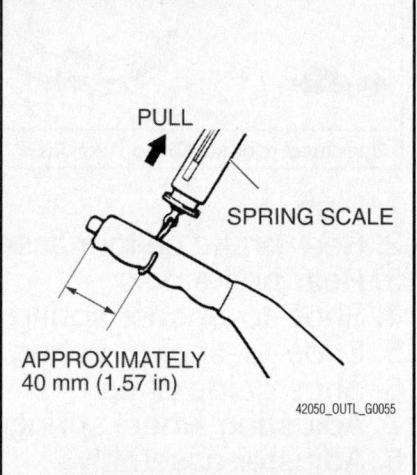

PULL

SPRING SCALE

APPROXIMATELY
40 mm (1.57 in)

Fig. 14 Hook a spring scale onto the center of the parking brake lever grip and pull it with a force of 22–34 lbs. (100–150 N) in a direction perpendicular to the handle

expand the shoe) until the parking brake shoe makes contact and the disc can no longer be turned. Back off the adjuster to the opposite direction by five notches.

6. If the parking brake lever stroke is below the standard value and the braking is too firm, the rear brakes may drag.

7. Adjust the parking brake lever stroke to the standard value by turning the adjusting nut. After the adjustment, ensure that there is no free play between the adjusting nut and the parking brake lever.

8. After adjusting the parking brake lever stroke, raise and safely support the rear end of the vehicle, then release the parking

CHASSIS ELECTRICAL | AIR BAG (SUPPLEMENTAL RESTRAINT SYSTEM)

GENERAL INFORMATION

✳✳ CAUTION

Some vehicles are equipped with an air bag system. The system must be disarmed before performing service on, or around, system components, the steering column, instrument panel components, wiring and sensors. Failure to follow the safety precautions and the disarming procedure could result in accidental air bag deployment, possible injury and unnecessary system repairs.

PRECAUTIONS

Disconnect and isolate the battery negative cable before beginning any airbag system component diagnosis, testing, removal, or installation procedures. Allow system capacitor to discharge for two minutes before beginning any component

service. This will disable the airbag system. Failure to disable the airbag system may result in accidental airbag deployment, personal injury, or death.

DISARMING THE SYSTEM

See Figure 15.

1. Before servicing the vehicle, refer to the precautions.
2. Position the front wheels in the straight-ahead position and place the key in the **LOCK** position. Remove the key from the ignition lock cylinder.
3. Disconnect the negative battery cable and insulate the cable end with high-quality electrical tape or similar non-conductive wrapping.
4. Wait at least 60 seconds minute before working on the vehicle. The air bag system is designed to retain enough voltage to deploy the air bag for a short period of time after the battery has been disconnected.

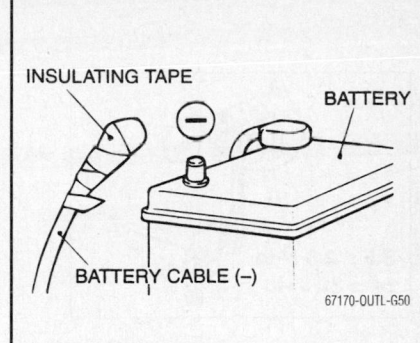

INSULATING TAPE   BATTERY

BATTERY CABLE (–)

67170-OUTL-G50

Fig. 15 Insulate the negative battery cable to prevent accidental deployment of the air bag

ARMING THE SYSTEM

1. Connect the negative battery cable, turn the ignition switch to the **ON** position and check the Supplemental Restraint System (SRS) warning light for proper operation.

DRIVE TRAIN

FRONT HALFSHAFT

REMOVAL & INSTALLATION

See Figures 16 and 17.

✳✳ WARNING

The wheel speed sensor/toothed ring collect metallic particles easily, because it is magnetized. Make sure that the sensor does not collect metallic particles. Check that there is not any trouble prior to reassembling it. When removing and installing the driveshaft assembly, make sure that the sensor and ring (integrated with the inner oil seal) do not contact with surrounding parts to avoid damage. When removing and installing the front wheel speed sensor, make sure that the sensor head at the end does not contact with surrounding parts to avoid damage.

1. Before servicing the vehicle, refer to the Precautions Section.
2. Raise the vehicle and support it safely.
3. Drain the transaxle fluid.
4. Drain the transfer case, if equipped with AWD.
5. Remove the wheel and tire assembly.

6. Remove the halfshaft cotter pin, nut and washer.
7. Remove the front wheel speed sensor and harness bracket.
8. Disconnect the front height sensor from the lower arm.
9. Remove the stabilizer link.
10. Remove the lower control arm ball joint and tie rod end from the steering knuckle.
11. Remove the halfshaft or halfshaft and inner shaft as follows:
    a. Use Mitsubishi Special tool Nos. MB990241, MB991354 and MB990767, or suitable puller, to push the halfshaft or halfshaft and inner shaft assembly from the hub.
    b. Remove the halfshaft from the hub by pulling the bottom of the rotor toward you.

✳✳ WARNING

When pulling the halfshaft from the transaxle, be careful that the spline of the halfshaft does not damage the oil seal.

    c. Insert a pry bar between the transaxle case and halfshaft, then pry the halfshaft out of the transaxle.
    d. If the inner shaft is difficult to remove, tap the bracket assembly lightly with a plastic hammer, then remove the inner shaft.

    e. Cover the halfshaft opening in the transaxle case to prevent foreign debris from entering.

➡**Do not pull on the shaft; doing so damages the inboard joint.**

12. For AWD vehicles, Use Mitsubishi Special tool No. MB991721 to remove the output shaft.

To install:

13. Replace the circlip on the ends of the halfshafts.
14. If AWD, install the output shaft.

➡**When installing the output shaft, halfshaft or halfshaft and inner shaft assembly, make sure the splines do not damage the oil seal.**

15. Insert the halfshaft or halfshaft and inner shaft into the transaxle. Be sure it is fully seated.
16. Push out on the knuckle assembly and install the halfshaft through the hub.
17. Install the self-locking nut. Tighten the nut so the protruding length of the stabilizer link is 0.35–0.39 inches (9.0–9.8mm).
18. Connect the tie rod end to the steering knuckle. Torque the retaining nut to 21 ft. lbs. (29 Nm) and secure with a new cotter pin.

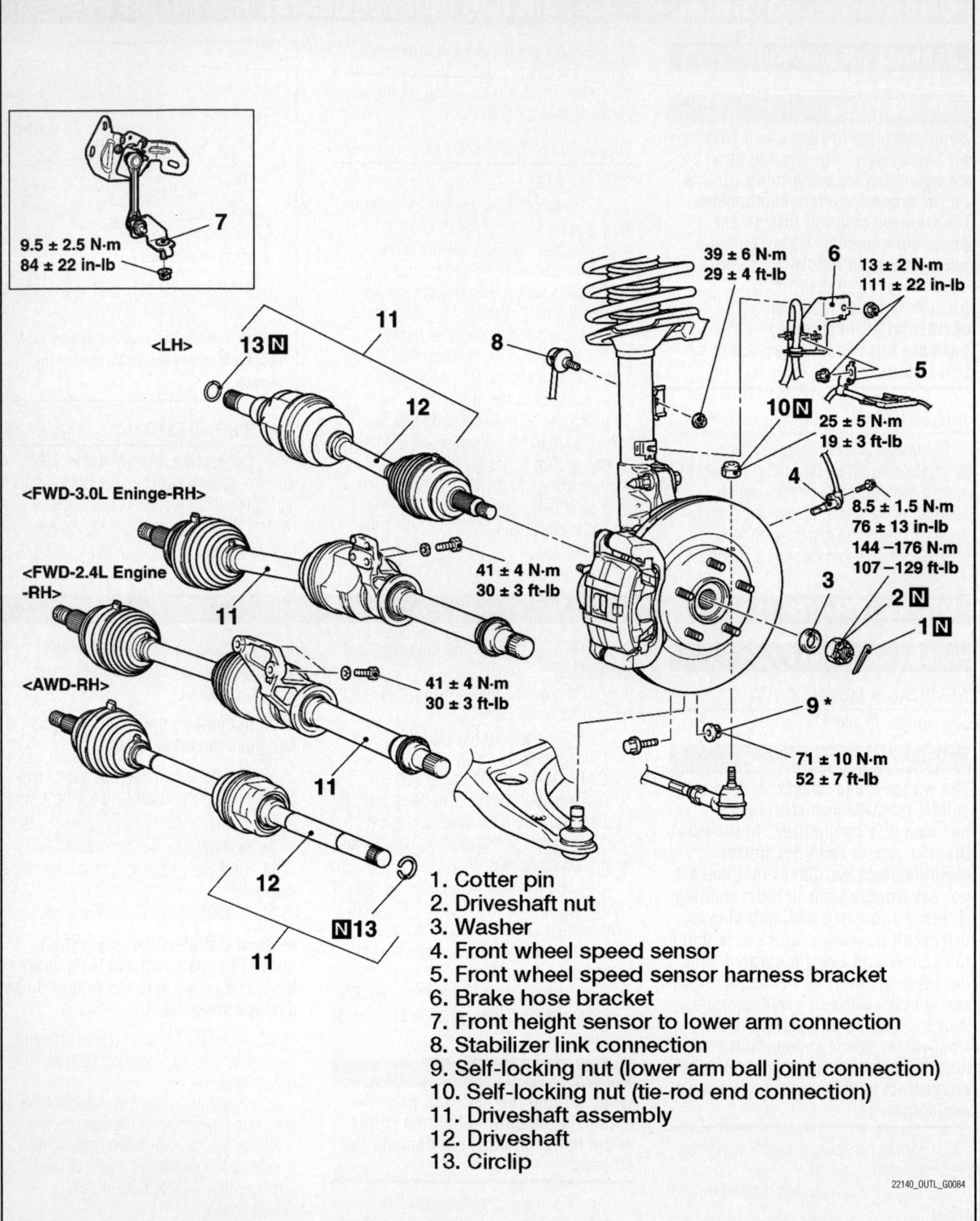

9.5 ± 2.5 N·m
84 ± 22 in-lb

7

39 ± 6 N·m
29 ± 4 ft-lb

6

13 ± 2 N·m
111 ± 22 in-lb

8

5

<LH>   13 N   11

12

10 N

25 ± 5 N·m
19 ± 3 ft-lb

4

<FWD-3.0L Eninge-RH>

8.5 ± 1.5 N·m
76 ± 13 in-lb
144 −176 N·m
107 −129 ft-lb

3

2 N

1 N

<FWD-2.4L Engine -RH>

41 ± 4 N·m
30 ± 3 ft-lb

11

<AWD-RH>

41 ± 4 N·m
30 ± 3 ft-lb

11

9 *

71 ± 10 N·m
52 ± 7 ft-lb

12

N 13

11

1. Cotter pin
2. Driveshaft nut
3. Washer
4. Front wheel speed sensor
5. Front wheel speed sensor harness bracket
6. Brake hose bracket
7. Front height sensor to lower arm connection
8. Stabilizer link connection
9. Self-locking nut (lower arm ball joint connection)
10. Self-locking nut (tie-rod end connection)
11. Driveshaft assembly
12. Driveshaft
13. Circlip

22140_OUTL_G0084

Fig. 16 Front halfshaft and related components

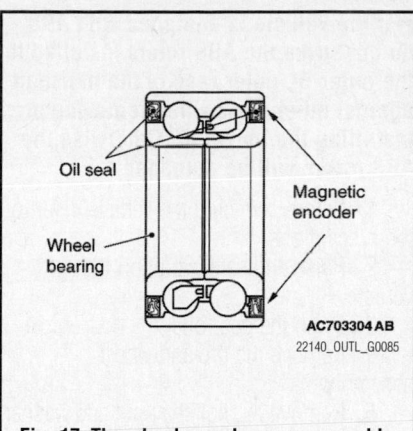

Fig. 17 The wheel speed sensor assembly is an integral part of the wheel bearing

19. Connect the ball joint to the steering knuckle. Torque the new retaining nut to 43–52 ft. lbs. (60–72 Nm) and secure with a new cotter pin.

20. Install the stabilizer link and rubber insulator.

21. Connect the front height sensor from the lower arm.

22. Install the ABS sensor harness bracket and ABS sensor.

23. Install the washer so the chamfered edge faces outward. Install the halfshaft nut and tighten to 107–129 ft. lbs. (144–176 Nm). Secure with a new cotter pin.

24. Install the wheel and lower the vehicle to the floor.

25. Fill the transaxle with fluid.

26. On AWD models, fill the transfer case.

27. Test drive the vehicle and check for leaks.

28. Check and adjust the alignment, as required.

REAR DRIVESHAFT

REMOVAL & INSTALLATION

See Figure 18.

1. Before servicing the vehicle, refer to the precautions.

2. Raise and safely support the vehicle.

3. Remove the flange yoke and electronic control coupling connecting nut.

1. Flange yoke and electronic control coupling connecting nut
2. Hear protector
3. Bolt
4. Insulator
5. Spacer
6. Propeller shaft assembly

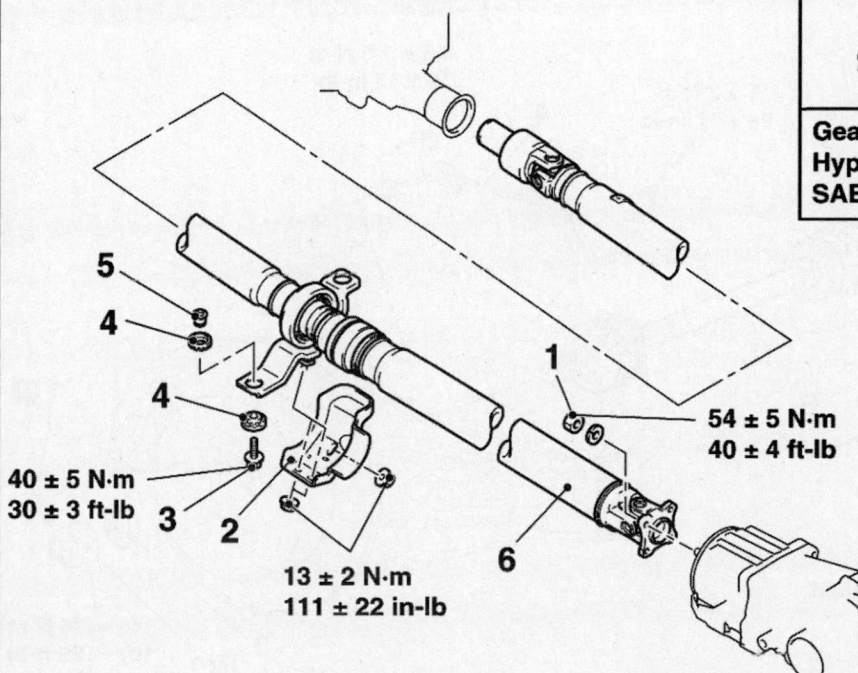

Gear oil:
Hypoid gear oil API classification GL-5 SAE90

54 ± 5 N·m
40 ± 4 ft-lb

40 ± 5 N·m
30 ± 3 ft-lb

13 ± 2 N·m
111 ± 22 in-lb

Fig. 18 Driveshaft assembly

4. Remove the hear protector.
5. Remove the bolt.
6. Remove the insulator.
7. Remove the spacer.

   a. Put mating marks on the flange yoke and the electronic control coupling.

   b. If the joint assembly is bent, it may be damaged when pinching the joint boots.

   c. Insert a rag or similar materials into the joint boots, and remove the propeller shaft assembly by aligning the front propeller shaft with the rear shaft.

   d. Cover the transfer case to prevent the entry of foreign materials.

8. Remove the driveshaft shaft assembly.

To install:

✻✻ WARNING

Do not damage the oil seal lip of the transfer. The mounting bolt and nut may be loosened if oil or grease is stuck on the threads of the bolt and nut. Tighten them after degreasing the threads. If the joint assembly is bent, it may be damaged when pinching the joint boots.

9. Align the mating marks on the flange yoke and the electronic control coupling.
10. Install the driveshaft shaft assembly.
11. Install the spacer.
12. Install the insulator.
13. Install the bolt. Tighten to 30 ft. lbs. (40 Nm).
14. Install the hear protector.
15. Install the flange yoke and electronic control coupling connecting nut. Tighten to 40 ft. lbs. (54 Nm).
16. Lower the vehicle.

REAR HALFSHAFTS

REMOVAL & INSTALLATION

See Figure 19.

➡**If the vehicle is equipped with ABS, do not strike the ABS rotors installed to the outer BL outer race of the halfshaft against other parts when removing or installing the halfshaft. Otherwise the ABS rotors will be damaged.**

1. Before servicing the vehicle, refer to the precautions.
2. Raise and safely support the vehicle.
3. Drain the gear oil.
4. Remove the tire and wheel assembly.
5. Remove the halfshaft nut and washer.

➡**Prevent the hub assembly from turning by using a tool such as MB990767 and remove the halfshaft nut and washer.**

✻✻ WARNING

Be careful not to strike the pole piece at the tip of the rear wheel speed sensor, as damage may result.

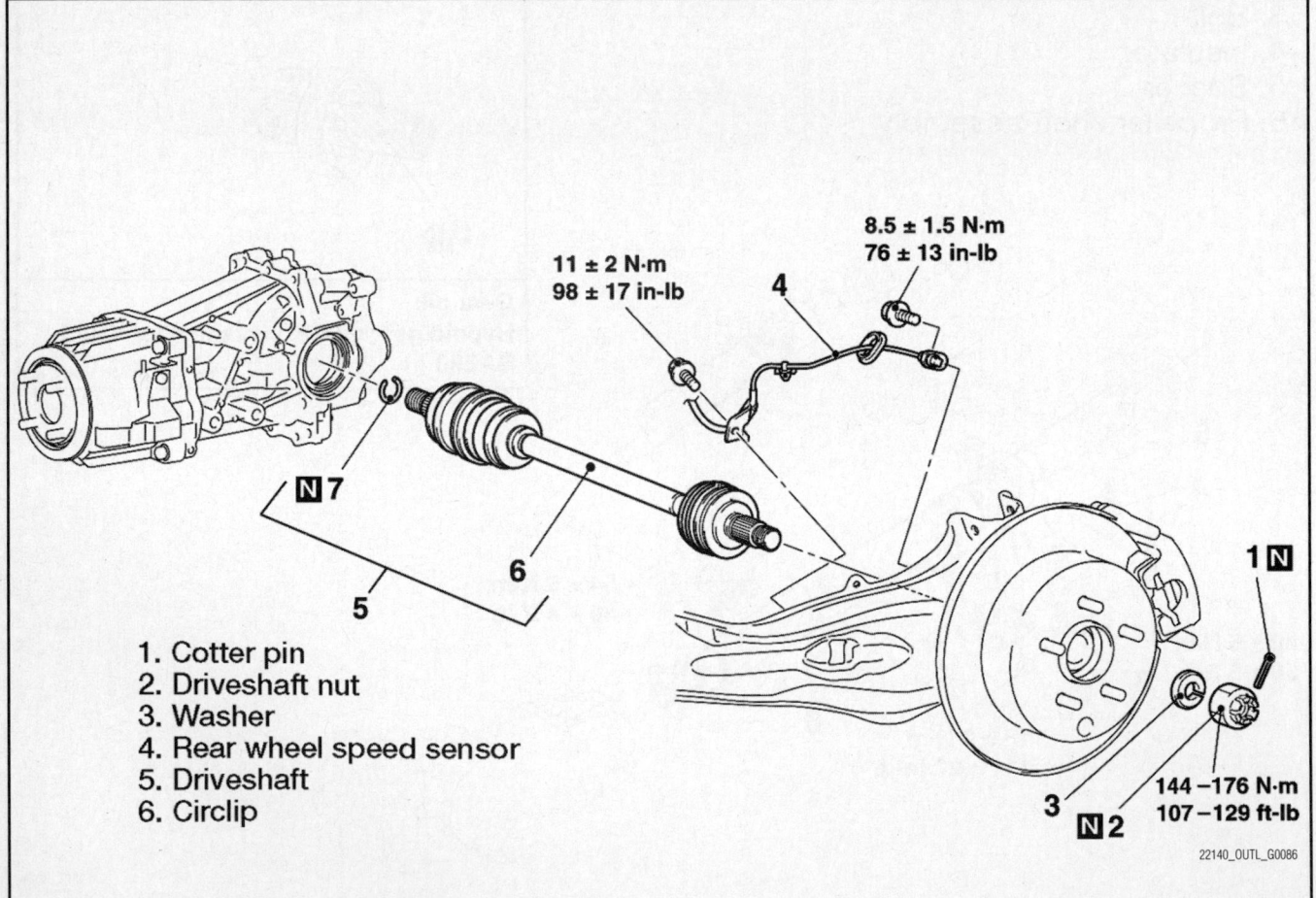

1. Cotter pin
2. Driveshaft nut
3. Washer
4. Rear wheel speed sensor
5. Driveshaft
6. Circlip

11 ± 2 N·m
98 ± 17 in-lb

8.5 ± 1.5 N·m
76 ± 13 in-lb

144 –176 N·m
107 –129 ft-lb

22140_OUTL_G0086

Fig. 19 Rear halfshaft assembly—AWD vehicle

6. Disconnect the lower arm from the trailing arm, the shock absorber and the stabilizer link.

7. Disconnect the upper arm from the trailing arm.

8. Remove the rear wheel speed sensor.

9. Remove the halfshaft and circlip.

To install:

10. Replace the circlip on the end of the halfshaft.

11. Install the halfshaft.

12. Install the rear wheel speed sensor.

13. Connect the upper arm to the trailing arm.

14. Connect the lower arm to the trailing arm, the shock absorber and the stabilizer link.

15. Install the halfshaft nut and washer. Tighten to 107–129 ft. lbs. (144–176 Nm).

16. Install the tire and wheel assembly.

17. Fill the differential with oil.

18. Test drive the vehicle and check for leaks.

19. Check and adjust the rear alignment, as required.

REAR PINION SEAL

REMOVAL & INSTALLATION

See Figure 20.

1. Raise and safely support the vehicle securely on jackstands.

2. Remove the halfshaft from the differential carrier.

3. Remove the differential carrier oil seal.

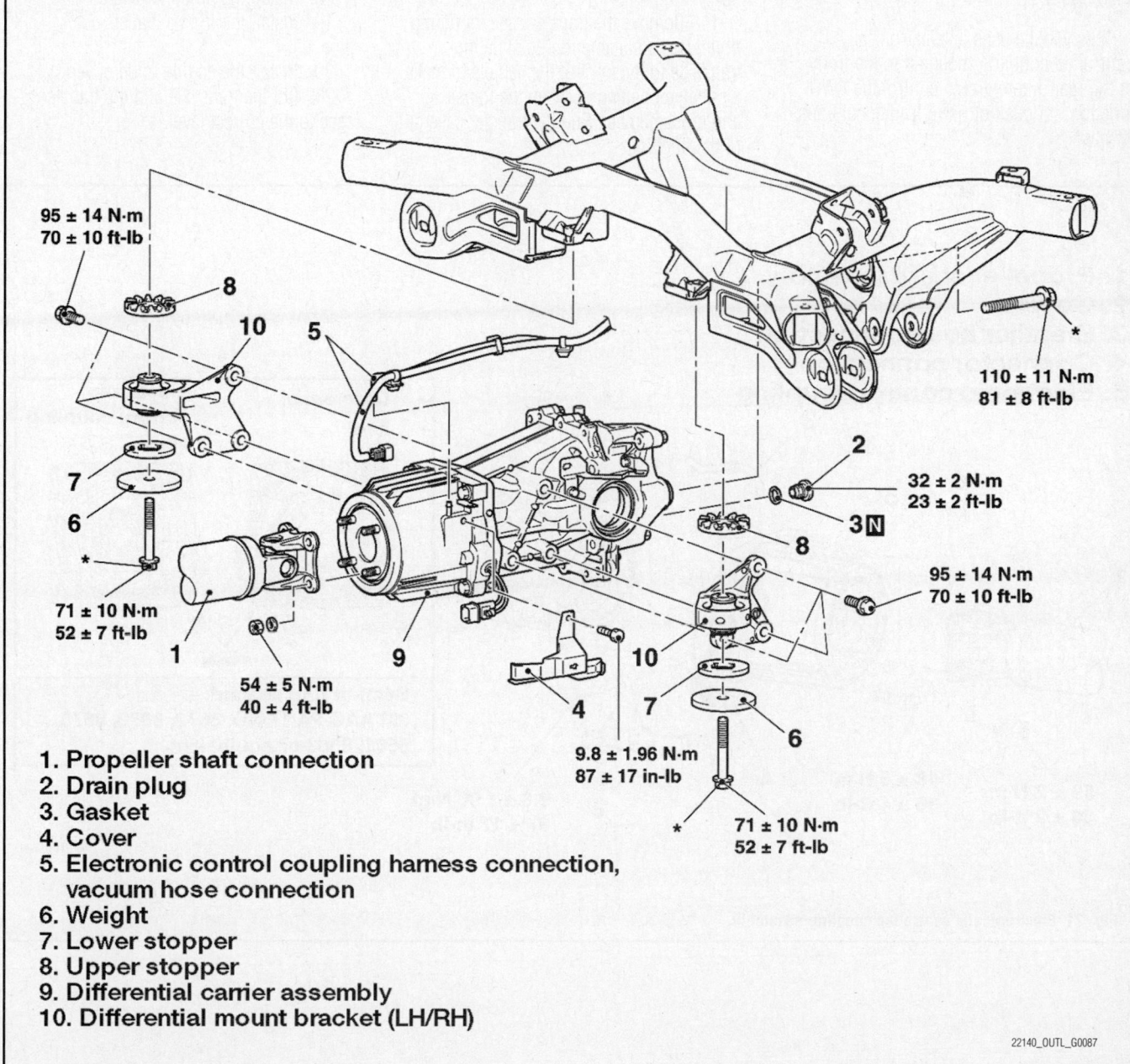

95 ± 14 N·m
70 ± 10 ft-lb

110 ± 11 N·m
81 ± 8 ft-lb

32 ± 2 N·m
23 ± 2 ft-lb

95 ± 14 N·m
70 ± 10 ft-lb

71 ± 10 N·m
52 ± 7 ft-lb

54 ± 5 N·m
40 ± 4 ft-lb

9.8 ± 1.96 N·m
87 ± 17 in-lb

71 ± 10 N·m
52 ± 7 ft-lb

1. **Propeller shaft connection**
2. **Drain plug**
3. **Gasket**
4. **Cover**
5. **Electronic control coupling harness connection, vacuum hose connection**
6. **Weight**
7. **Lower stopper**
8. **Upper stopper**
9. **Differential carrier assembly**
10. **Differential mount bracket (LH/RH)**

22140_OUTL_G0087

Fig. 20 Differential carrier assembly

To install:

4. Using the special tools MB990938 and MB991115, press fit a new oil seal.

5. Apply multi-purpose grease to the oil seal lip and driveshaft oil seal seating area.

6. Replace the halfshaft circlip with a new one, and install the halfshaft to the differential carrier.

7. Lower the vehicle.

TRANSFER CASE ASSEMBLY

REMOVAL & INSTALLATION

See Figure 21.

This vehicle uses an electronically controlled coupling mounted to the front of the rear differential to control the AWD function. It does not use a traditional transfer case.

1. Before servicing the vehicle, refer to the precautions.

2. Drain the transaxle fluid and transfer case fluid.

3. Remove the engine undercover(s).

4. Remove the front exhaust pipe.

5. Remove the propeller shaft.

6. Remove the center member.

7. Remove the air guide.

8. Remove the dust seal guard, manual transaxle.

9. Remove the halfshaft and output shaft.

10. Remove the rear roll stopper bolt and nut.

11. Remove the transfer case mounting bolts. Use a suitable tool to slide the transaxle to the front of the vehicle to make a suitable opening between the transaxle and crossmember. Pull the transfer case out of the opening.

To install:

12. Install the transfer case to the transaxle. Torque the bolts to 44–58 ft. lbs. (60–78 Nm).

13. Install the rear roll stopper bolt and nut hand-tight. After the full weight of the vehicle is on the ground, torque the nut to 34–44 ft. lbs. (45–59 Nm).

14. Install the halfshaft and output shaft.

15. Install the air guide.

16. Install the dust seal guard, manual transaxle.

17. Install the center member.

18. Install the propeller shaft.

19. Install the front exhaust pipe.

20. Install the engine undercover(s).

21. Fill the transaxle and the transfer case to the correct level.

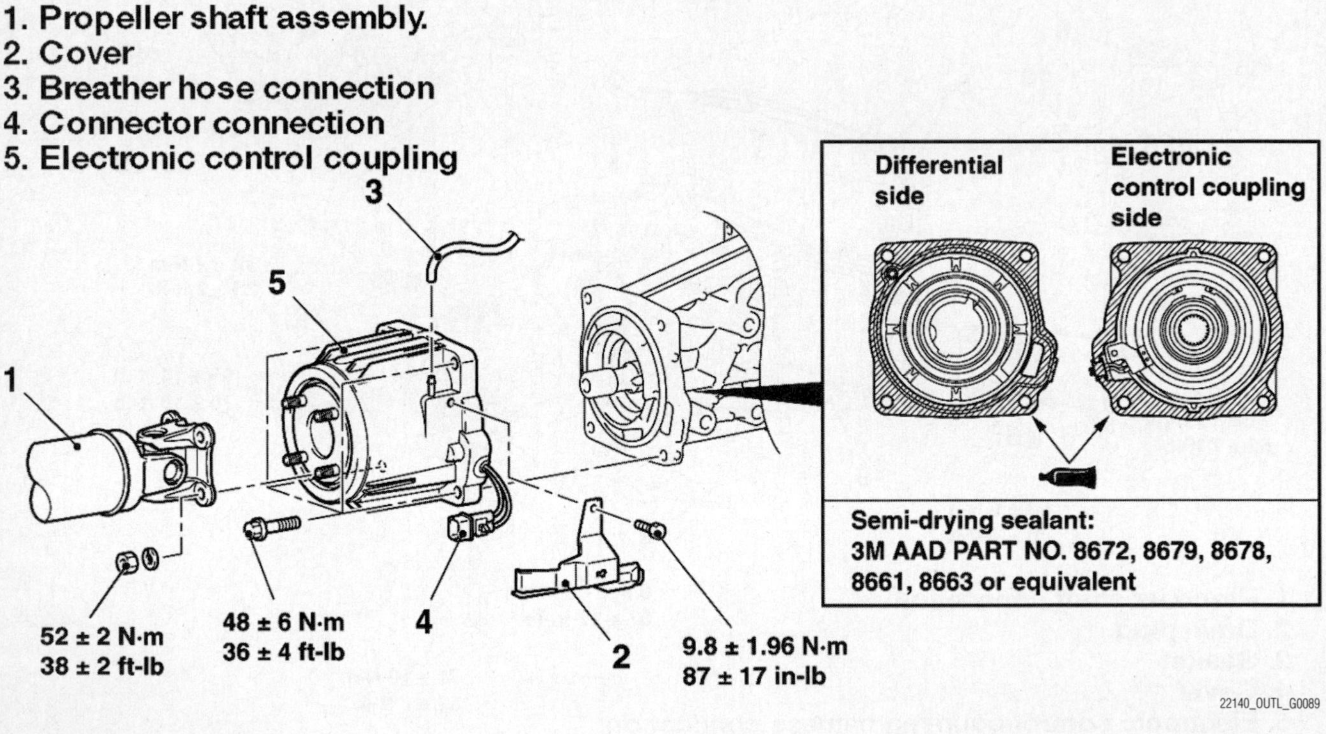

1. Propeller shaft assembly.
2. Cover
3. Breather hose connection
4. Connector connection
5. Electronic control coupling

52 ± 2 N·m
38 ± 2 ft-lb

48 ± 6 N·m
36 ± 4 ft-lb

9.8 ± 1.96 N·m
87 ± 17 in-lb

Differential side

Electronic control coupling side

Semi-drying sealant:
3M AAD PART NO. 8672, 8679, 8678, 8661, 8663 or equivalent

22140_OUTL_G0089

Fig. 21 Electronically controlled coupling assembly

ENGINE COOLING

ENGINE FAN

REMOVAL & INSTALLATION

See Figures 22 through 24.

1. Before servicing the vehicle, refer to the precautions.

2. Disconnect the negative battery cable.

❈❈ CAUTION

Wait at least 90 seconds after the negative battery cable is discon-nected to prevent possible deploy-ment of the air bag.

3. Remove the transaxle fluid cooler hose.

4. Remove the water feed hose.

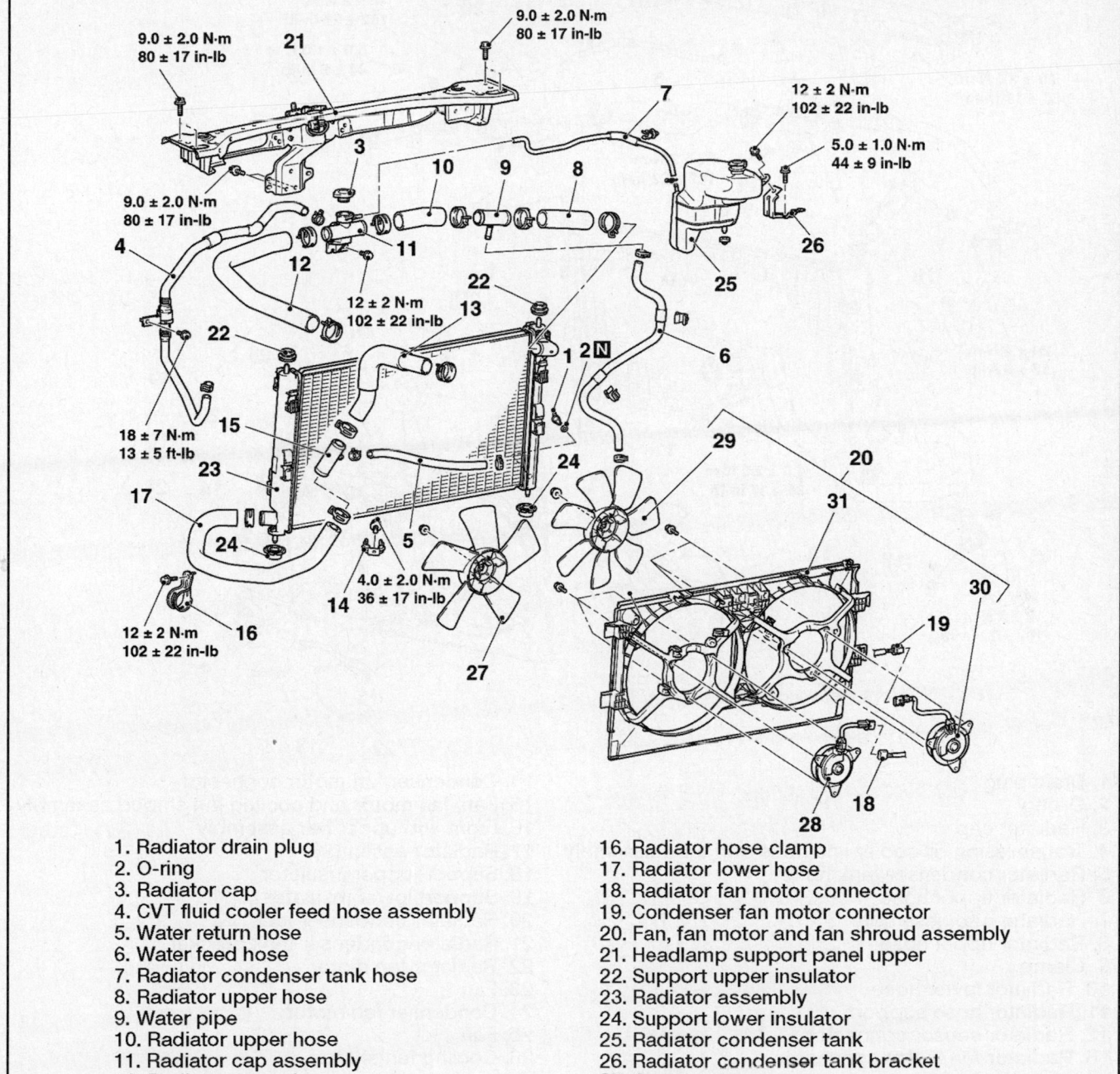

1. Radiator drain plug
2. O-ring
3. Radiator cap
4. CVT fluid cooler feed hose assembly
5. Water return hose
6. Water feed hose
7. Radiator condenser tank hose
8. Radiator upper hose
9. Water pipe
10. Radiator upper hose
11. Radiator cap assembly
12. Radiator upper hose
13. Radiator lower hose
14. Radiator hose support
15. Water pipe

16. Radiator hose clamp
17. Radiator lower hose
18. Radiator fan motor connector
19. Condenser fan motor connector
20. Fan, fan motor and fan shroud assembly
21. Headlamp support panel upper
22. Support upper insulator
23. Radiator assembly
24. Support lower insulator
25. Radiator condenser tank
26. Radiator condenser tank bracket
27. Radiator fan
28. Radiator fan motor
29. Condenser fan
30. Condenser fan motor
31. Fan shroud

22140_OUTL_G0092

Fig. 22 Exploded view of the radiator assembly—2.4L engine

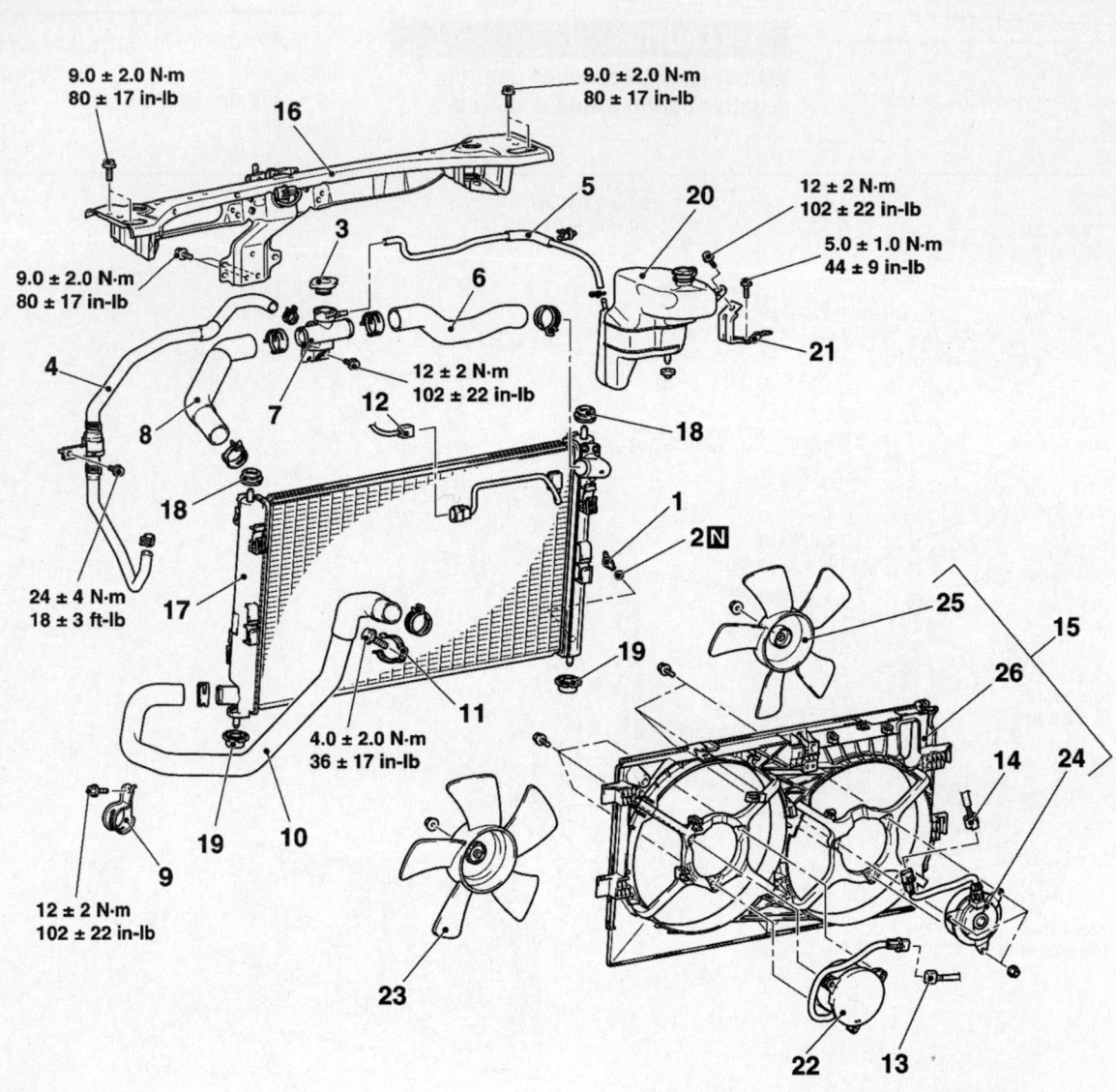

9.0 ± 2.0 N·m
80 ± 17 in-lb

9.0 ± 2.0 N·m
80 ± 17 in-lb

9.0 ± 2.0 N·m
80 ± 17 in-lb

12 ± 2 N·m
102 ± 22 in-lb

5.0 ± 1.0 N·m
44 ± 9 in-lb

12 ± 2 N·m
102 ± 22 in-lb

24 ± 4 N·m
18 ± 3 ft-lb

4.0 ± 2.0 N·m
36 ± 17 in-lb

12 ± 2 N·m
102 ± 22 in-lb

1. Drain plug
2. O-ring
3. Radiator cap
4. Transmission oil cooler line hose and tube assembly
5. Radiator condenser tank hose
6. Radiator upper hose
7. Radiator cap assembly
8. Radiator upper hose
9. Clamp
10. Radiator lower hose
11. Radiator hose support
12. Radiator sensor connector
13. Radiator fan motor connector
14. Condenser fan motor connector
15. Fan, fan motor and cooling fan shroud assembly
16. Front end upper bar assembly
17. Radiator assembly
18. Support upper insulator
19. Support lower insulator
20. Radiator condenser tank
21. Radiator condenser tank bracket
22. Radiator fan motor
23. Fan
24. Condenser fan motor
25. Fan
26. Cooling fan shroud

22140_OUTL_G0093

Fig. 23 Exploded view of the radiator assembly—3.0L engine

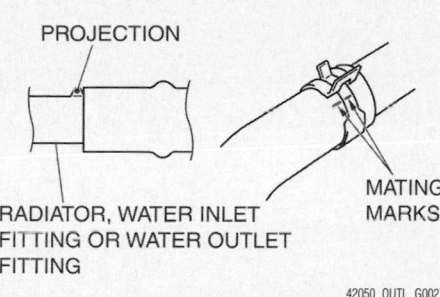

PROJECTION

RADIATOR, WATER INLET
FITTING OR WATER OUTLET
FITTING

MATING
MARKS

42050_OUTL_G0027

Fig. 24 Insert the hose as far as the projection of the water inlet or outlet fitting or radiator. Make sure to align the matchmarks before connecting the hose

5. Remove the radiator condenser tank hose.

6. Make mating marks on the radiator hoses and the hose clamps.

7. Remove the radiator hoses.

8. Remove the water pipes.

9. Remove the radiator cap.

10. Disconnect the radiator fan motor connector.

11. Disconnect the condenser fan motor connector.

12. Remove the fan, fan motor and fan shroud.

13. Remove the radiator fan.

14. Remove the radiator fan motor.

15. Remove the condenser fan.

16. Remove the condenser fan motor.

17. Remove the fan shroud.

To install:

18. Install the fan shroud.

19. Install the condenser fan motor.

20. Install the condenser fan.

21. Install the radiator fan motor.

22. Install the radiator fan.

23. Install the fan, fan motor and fan shroud.

24. Connect the condenser fan motor connector.

25. Connect the radiator fan motor connector.

26. Install the radiator cap.

27. Install the water pipes.

28. Install the radiator hoses.

29. Make mating marks on the radiator hoses and the hose clamps.

30. Install the radiator condenser tank hose.

31. Install the water feed hose.

32. Install the transaxle fluid cooler hose.

33. Connect the negative battery cable.

34. Start the engine and check for proper fan operation.

RADIATOR

REMOVAL & INSTALLATION

See Figures 22 through 24.

1. Before servicing the vehicle, refer to the precautions.

2. Disconnect the negative battery cable.

❄ CAUTION

Wait at least 90 seconds after the negative battery cable is disconnected to prevent possible deployment of the air bag.

3. Remove the air cleaner assembly.

4. Remove the engine undercover.

5. Remove the radiator drain plug and drain the engine coolant.

6. Remove the O-ring.

7. Remove the radiator cap.

8. Remove the transaxle fluid cooler hose.

9. Remove the water feed and return hoses.

10. Remove the radiator condenser tank hose.

11. Make mating marks on the radiator hoses and the hose clamps.

12. Remove the radiator hoses.

13. Remove the water pipes.

14. Remove the radiator cap.

15. Remove the radiator hose support.

16. Disconnect the radiator fan motor connector.

17. Disconnect the condenser fan motor connector.

18. Remove the fan, fan motor and fan shroud assembly.

19. Remove the radiator grille, headlamp support upper panel cover.

20. Remove the front impact sensor.

21. Remove the hood lock release cable, hood switch.

22. Remove the upper headlamp support panel.

23. Remove the support upper insulator.

24. Remove the radiator assembly.

To install:

25. Install the radiator assembly.

26. Install the support upper insulator.

27. Install the upper headlamp support panel.

28. Install the hood lock release cable, hood switch.

29. Install the front impact sensor.

30. Install the radiator grille, headlamp support upper panel cover.

31. Install the fan, fan motor and fan shroud assembly.

32. Connect the condenser fan motor connector.

33. Connect the radiator fan motor connector.

34. Install the radiator hose support.

35. Install the radiator cap.

36. Install the upper and lower radiator hoses.

a. Insert radiator hose as far as the projection of the water inlet fitting, water outlet fitting or radiator.

b. Align the mating marks on the radiator hose and hose clamp, and then connect the radiator hose.

37. Install the water pipes.

38. Install the radiator condenser tank hose.

39. Install the water feed and return hoses.

40. Install the CVT fluid cooler feed hose.

41. Install the radiator cap.

42. Install the O-ring.

43. Install the radiator drain plug and drain the engine coolant.

44. Install the engine undercover.

45. Install the air cleaner assembly.

46. Connect the negative battery cable.

47. Fill the engine with coolant.

48. Start the engine and check for leaks.

THERMOSTAT

REMOVAL & INSTALLATION

See Figures 25 through 27.

1. Disconnect the negative battery cable, then the positive battery cable.

❄ CAUTION

Wait at least 90 seconds after the negative battery cable is disconnected to prevent possible deployment of the air bag.

2. Drain the engine coolant.

3. Remove the air cleaner assembly.

4. Remove the battery and battery tray, as necessary.

5. Matchmark the installed positions of

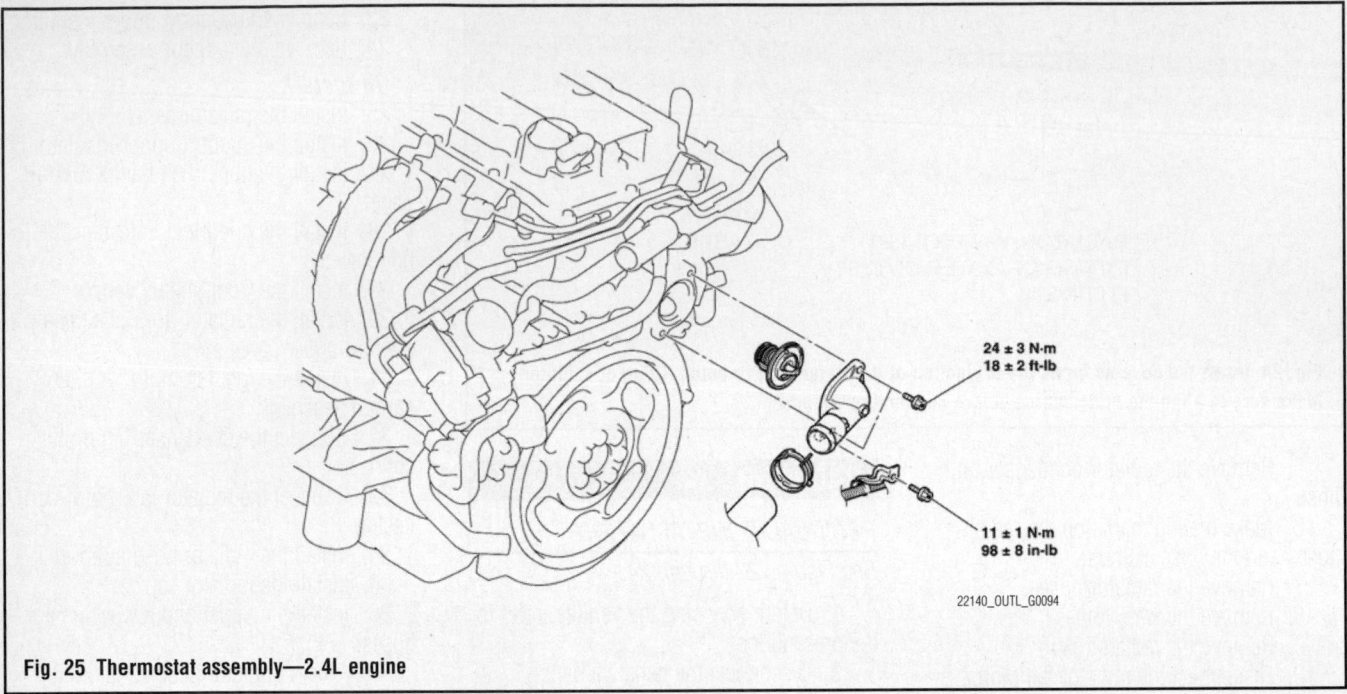

Fig. 25 Thermostat assembly—2.4L engine

24 ± 3 N·m
18 ± 2 ft-lb

11 ± 1 N·m
98 ± 8 in-lb

22140_OUTL_G0094

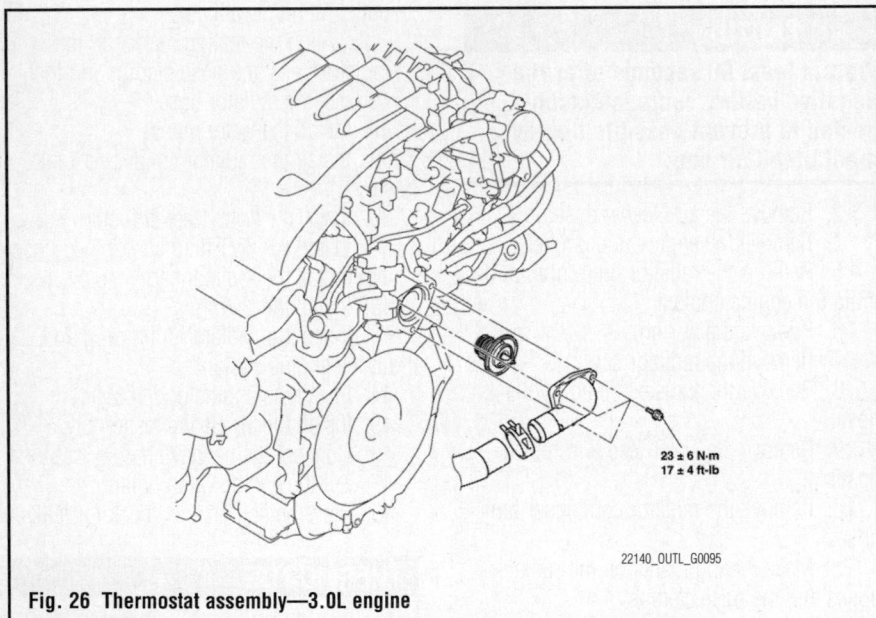

Fig. 26 Thermostat assembly—3.0L engine

23 ± 6 N·m
17 ± 4 ft-lb

22140_OUTL_G0095

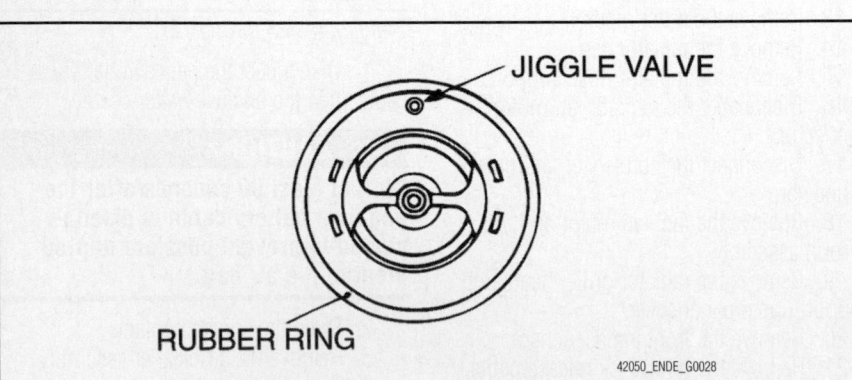

JIGGLE VALVE

RUBBER RING

42050_ENDE_G0028

Fig. 27 The thermostat's jiggle valve must be facing up, at the 12 o'clock position during installation

the hose and hose clamp, then disconnect the lower radiator hose from the water inlet fitting.

6. Detach the control hose electrical connection.

7. Remove the control wiring harness connection bracket, as necessary.

8. Unfasten the retainer(s), then remove the water inlet fitting.

9. Remove the thermostat.

To install:

10. Installation is the reverse of the removal procedure, noting the following:

- Install the thermostat so the jiggle valve is facing straight up. Be careful not to fold or otherwise damage the rubber ring.
- Tighten the thermostat housing bolts to 18 ft. lbs. (24 Nm) on 2.4L engine and 17 ft. lbs. (23 Nm) on 3.0L engine
- During installation of the lower radiator hose connection, insert the hose as far as the projection of the water inlet fitting. Align the matchmarks on the hose and hose clamp, then connect the hose.
- Fill the engine with the proper type and amount of engine coolant.

11. Connect the positive, then the negative battery cable.

12. Start the engine and check for leaks.

WATER PUMP

REMOVAL & INSTALLATION

See Figures 28 and 29.

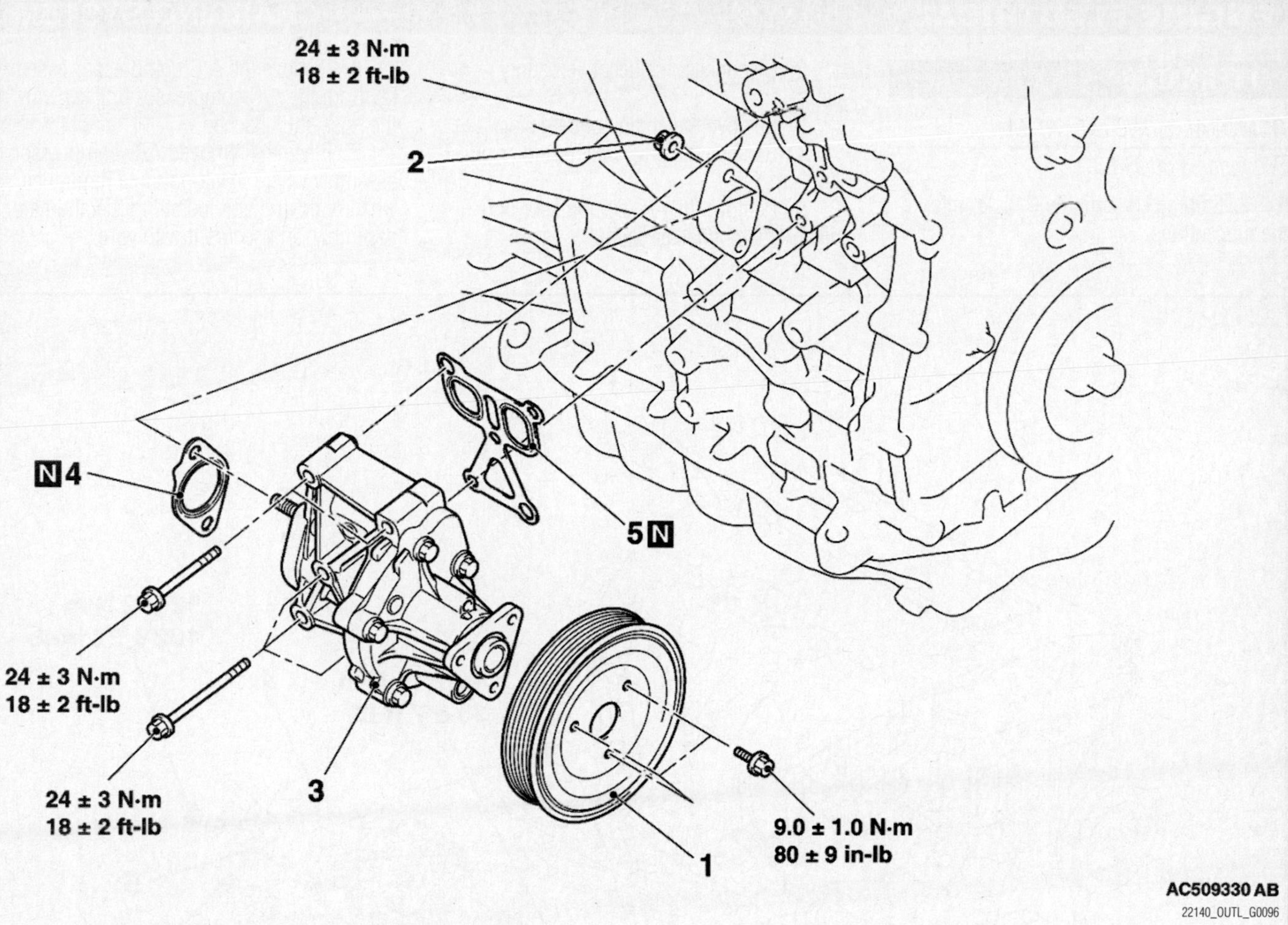

24 ± 3 N·m
18 ± 2 ft-lb

2

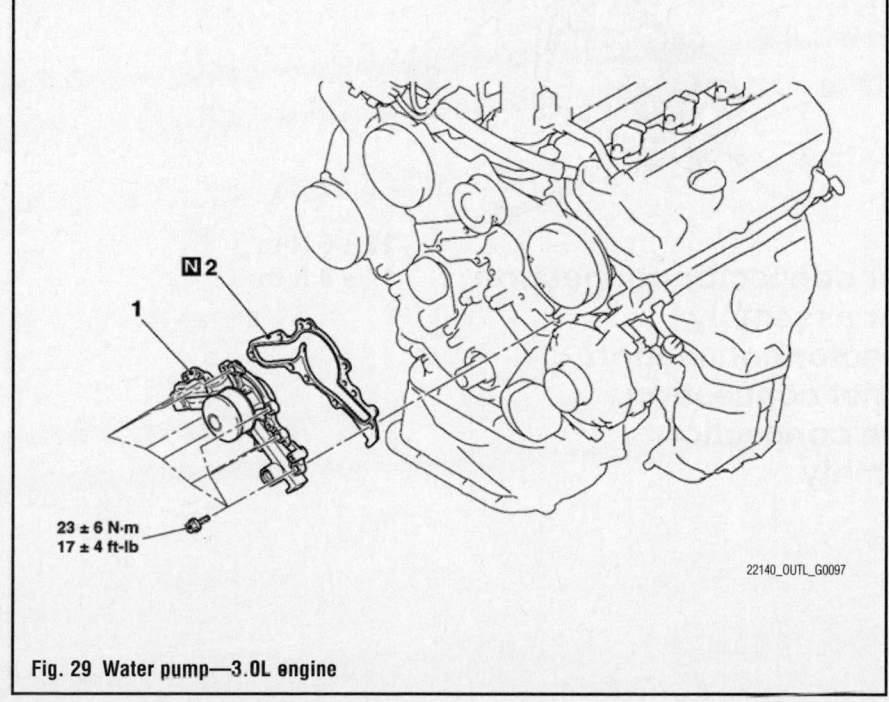

4

5

24 ± 3 N·m
18 ± 2 ft-lb

24 ± 3 N·m
18 ± 2 ft-lb

3

1

9.0 ± 1.0 N·m
80 ± 9 in-lb

AC509330 AB
22140_OUTL_G0096

Fig. 28 Water pump—2.4L engine

2

1

23 ± 6 N·m
17 ± 4 ft-lb

22140_OUTL_G0097

Fig. 29 Water pump—3.0L engine

1. Before servicing the vehicle, refer to the precautions.

2. Disconnect the negative battery cable.

3. Drain the engine coolant.

4. On 2.4L engines, remove the accessory drive belt(s).

5. On 3.0L engines, remove the timing belt.

6. Remove the water pump mounting bolts.

7. Remove the water pump, gasket.

To install:

8. Install the water pump to the engine block, with new gasket.

9. Tighten the mounting bolts to 18 ft. lbs. (24 Nm) on 2.4L engine and 17 ft. lbs. (23 Nm) on 3.0L engine.

10. On 2.4L engines, install the accessory drive belt(s).

11. On 3.0L engines, install the timing belt.

12. Refill the engine with coolant.

13. Connect the negative battery cable.

14. Start the engine and check for leaks.

ALTERNATOR

REMOVAL & INSTALLATION

See Figures 30 and 31.

1. Before servicing the vehicle, refer to the precautions.

2. Disconnect the negative battery cable.

3. Remove the engine undercover.

4. Remove the drive belt and idler pulley.

5. Remove the A/C compressor connector and connector clamp.

6. Remove the A/C compressor assembly from the A/C compressor bracket with the hose still attached.

7. Place the removed A/C compressor assembly where it will not be a hindrance when removing and installing the alternator assembly, and secure it with wire.

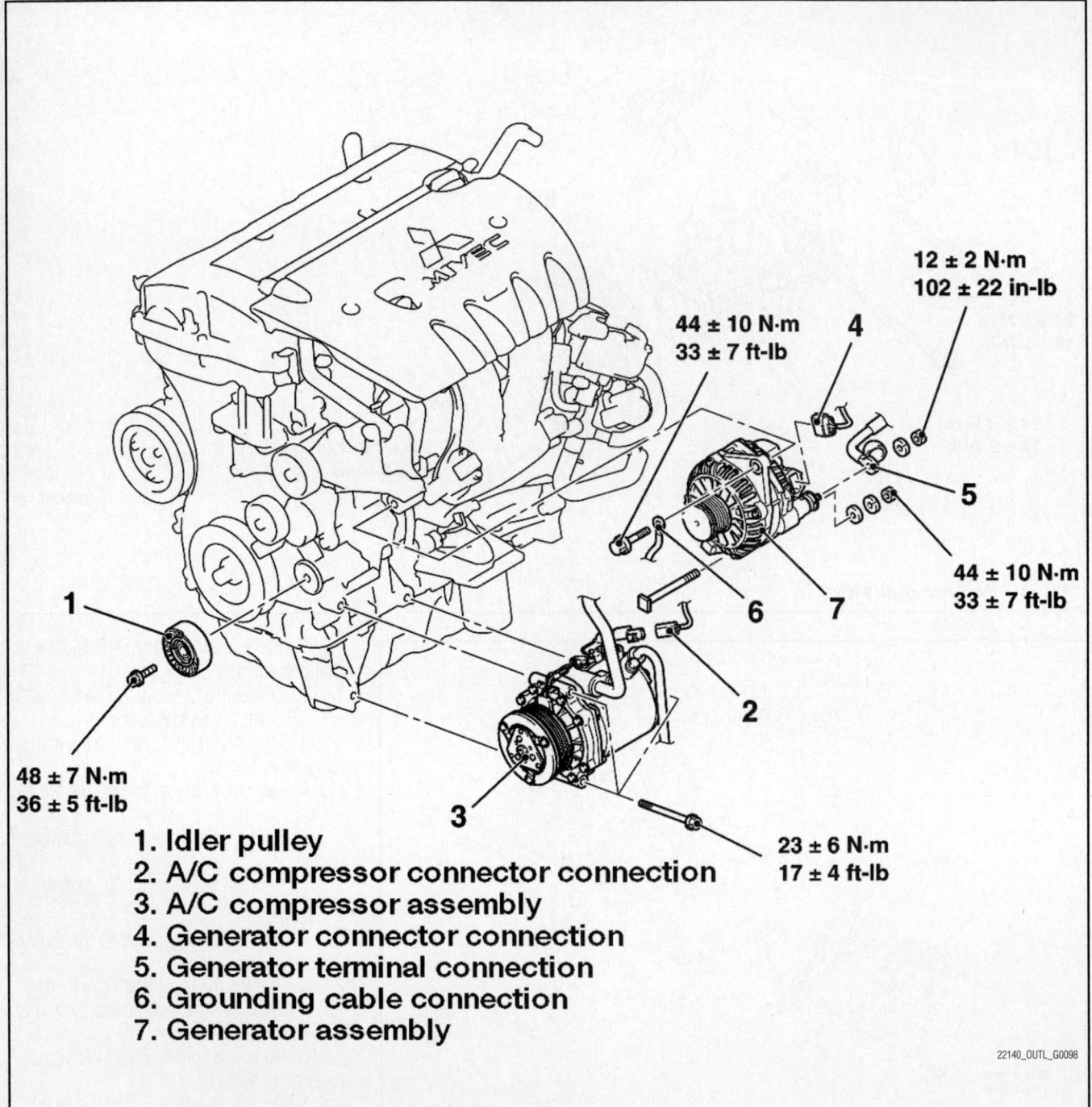

1. Idler pulley
2. A/C compressor connector connection
3. A/C compressor assembly
4. Generator connector connection
5. Generator terminal connection
6. Grounding cable connection
7. Generator assembly

22140_OUTL_G0098

Fig. 30 Alternator assembly—2.4L engine

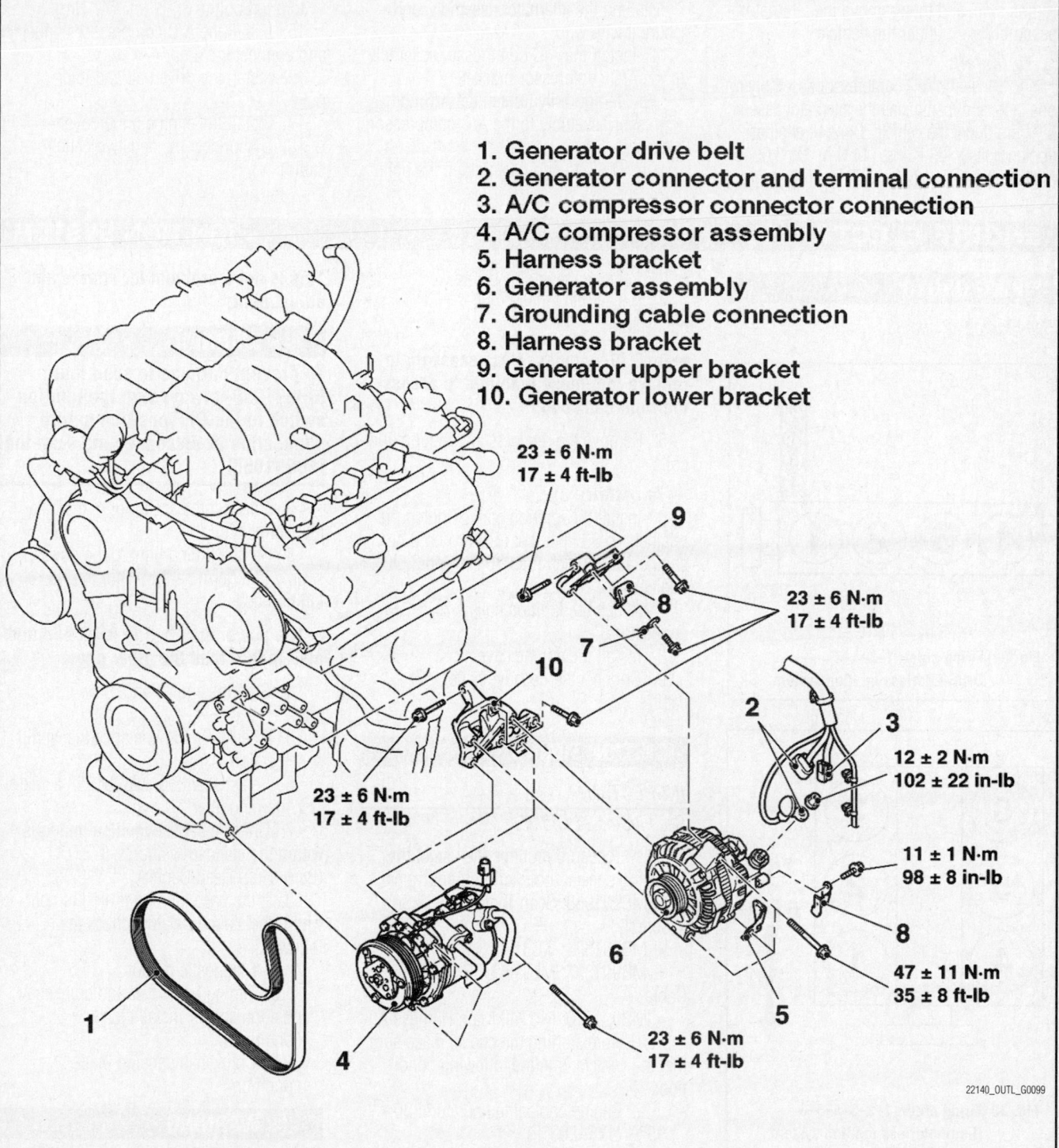

1. Generator drive belt
2. Generator connector and terminal connection
3. A/C compressor connector connection
4. A/C compressor assembly
5. Harness bracket
6. Generator assembly
7. Grounding cable connection
8. Harness bracket
9. Generator upper bracket
10. Generator lower bracket

23 ± 6 N·m
17 ± 4 ft-lb

23 ± 6 N·m
17 ± 4 ft-lb

23 ± 6 N·m
17 ± 4 ft-lb

12 ± 2 N·m
102 ± 22 in-lb

11 ± 1 N·m
98 ± 8 in-lb

47 ± 11 N·m
35 ± 8 ft-lb

23 ± 6 N·m
17 ± 4 ft-lb

22140_OUTL_G0099

Fig. 31 Alternator assembly—3.0L engine

8. Remove the alternator connector and terminal.

9. Remove the ground cable.

10. Move the A/C compressor assembly to one side, and then remove the alternator assembly from under the vehicle.

To install:

11. Move the A/C compressor assembly to one side, and then install the alternator assembly from under the vehicle. On 2.4L engines, tighten bolts to 33 ft. lbs. (44 Nm). On 3.0L engines, tighten bolts to 35 ft. lbs. (47 Nm).

12. Install the ground cable with the cable positioned straight down.

13. Install the alternator connector and terminal.

installing the alternator assembly, and secure it with wire.

14. Install the A/C compressor assembly on the A/C compressor bracket.

a. Temporarily tighten the A/C compressor assembly to the A/C compressor bracket.

b. On 2.4L engines, tighten the bolts in a counterclockwise order starting with the top bolt to 17 ft. lbs. (23 Nm).

c. On 3.0L engines, tighten the bolts in a clockwise order starting with the bottom left bolt to 35 ft. lbs. (47 Nm).

15. Install the A/C compressor connector and connector clamp.

16. Install the drive belt and idler pulley.

17. Install the engine undercover.

18. Disconnect the negative battery cable.

ENGINE ELECTRICAL

FIRING ORDER

See Figures 32 and 33.

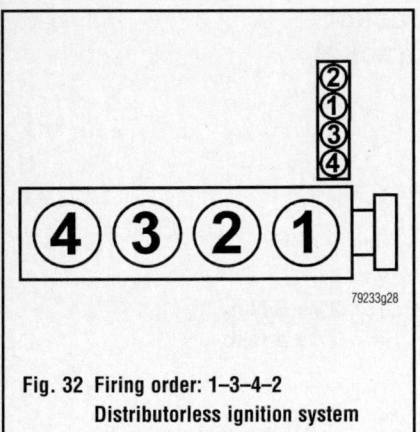

Fig. 32 Firing order: 1–3–4–2 Distributorless ignition system

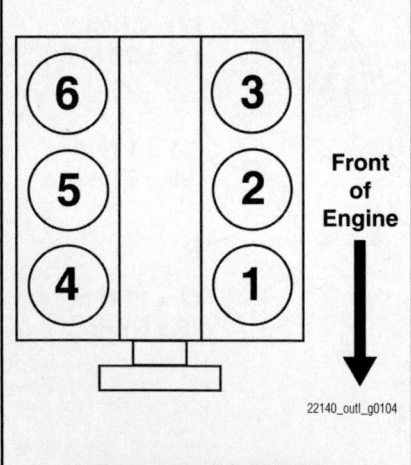

Fig. 33 Firing order: 1–2–3–4–5–6 Distributorless ignition system

IGNITION COIL

REMOVAL & INSTALLATION

1. Before servicing the vehicle, refer to the precautions.

2. Disconnect the negative battery cable.

3. Remove the engine cover.

4. Detach the ignition coil connectors.

➡**On 3.0L engines, it is necessary to remove the intake manifold to access the right bank coils.**

5. Remove the retainers and the ignition coils.

To install:

6. Install the ignition coils. Tighten the retainers to 89 inch lbs. (10 Nm) on 2.4L engine and 84 inch lbs. (10 Nm) on 3.0L engine.

7. Attach the ignition coil connectors.

8. Install the engine cover.

9. Connect the negative battery cable.

IGNITION TIMING

INSPECTION

See Figures 34 and 35.

This procedure requires the use of the following special tools, or their equivalents:
- MB991958: Scan Tool (MUT-III Sub Assembly)
- MB991824: V.C.I.
- MB991827: MUT-III USB Cable
- MB991910: MUT-III Main Harness A

1. Before starting the check, make sure that the vehicle is in the following condition:

a. Engine coolant temperature: 80–95°C (176–203°F)

b. Lights and all accessories: OFF

c. Transaxle: Neutral (or P range on vehicles with A/T)

➡**On Canadian vehicles, the headlight, taillight, etc. remain lit even when the light switch is in the "OFF" position.**

IGNITION SYSTEM

This is not a problem for checks and adjustment.

✸✸ WARNING

To prevent damage to scan tool MB991958, always turn the ignition switch to the OFF position before connecting or disconnecting scan tool MB991958.

2. Connect the scan tool to the data link connector.

3. Set the timing light to the power supply line (terminal No. 1) of the No. 1 ignition coil.

➡**The power supply line is looped and also longer than the other ones.**

4. Start the engine and allow it to idle.

5. Check that the idle speed is within specification.

6. Select scan tool MB991958 actuator test "item number 17".

7. Check that base ignition timing is within the standard value: 2–8°BTDC (Before Top Dead Center).

8. If the base ignition timing is not within the standard value, check the following:

a. Diagnostic output

b. Timing belt cover and crankshaft position sensor installation conditions

c. Crankshaft sensing blade condition

✸✸ WARNING

If the actuator test is not canceled, the forced drive will continue for 27 minutes. Driving in this state could lead to engine failure.

9. Press the clear key on scan tool MB991958 (select forced drive stop mode), and cancel the actuator test.

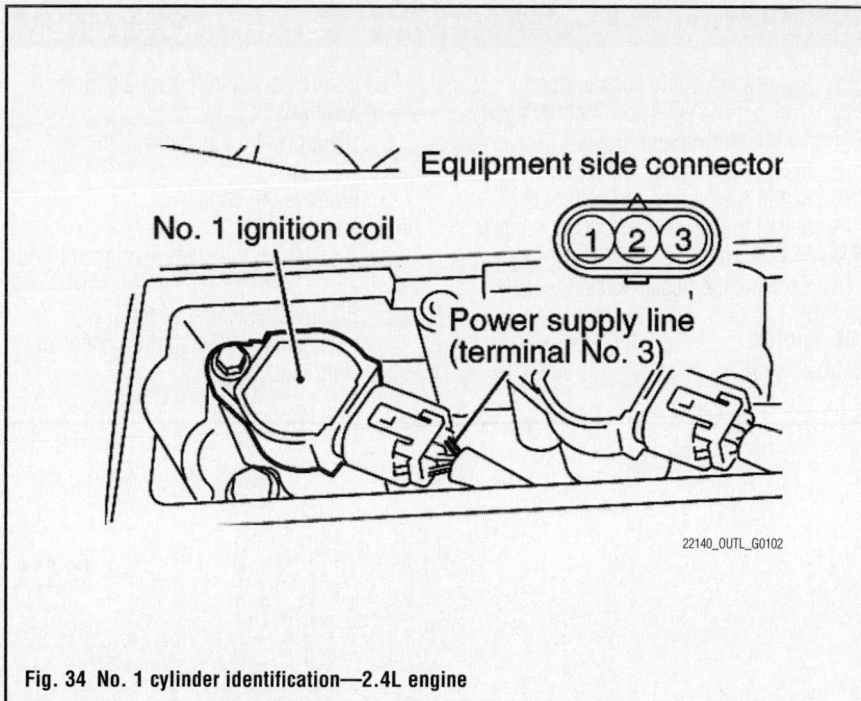

Fig. 34 No. 1 cylinder identification—2.4L engine

Equipment side connector

No. 1 ignition coil

Power supply line (terminal No. 3)

1 2 3

22140_OUTL_G0102

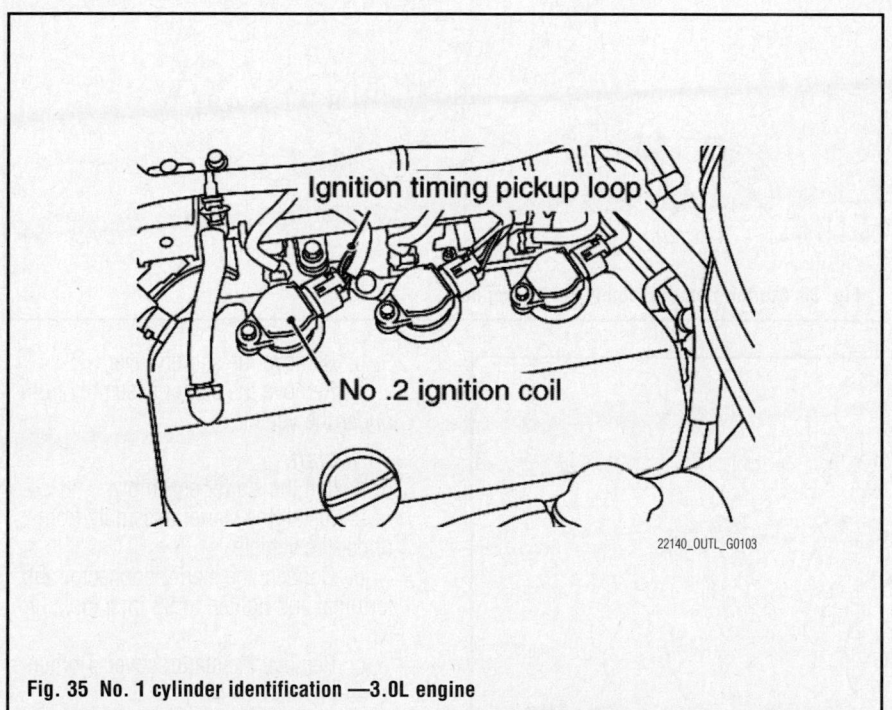

Fig. 35 No. 1 cylinder identification —3.0L engine

Ignition timing pickup loop

No .2 ignition coil

22140_OUTL_G0103

10. Check that the actual ignition timing is at the standard value: about 10°BTDC.

➡**Ignition timing fluctuates approximately +/- 7°Before Top Dead Center, even under normal operating conditions.**

➡**At higher altitude, ignition timing is automatically further advanced by about 5–10°BTDC.**

ADJUSTMENT

The ignition timing is controlled by the Electronic Control Module (ECM) and is not adjustable.

SPARK PLUGS

REMOVAL & INSTALLATION

See Figure 36.

The Outlander models use iridium spark plugs. Be careful not to damage the iridium tips of the plugs. Do not adjust the spark plug gap. Spark plugs must spark properly to assure proper engine performance and reduce exhaust emission level. Therefore, they should be replaced periodically with new ones.

1. Before servicing the vehicle, refer to the precautions.

2. Disconnect the negative battery cable.

3. Remove the ignition coil(s).

4. Remove the spark plug(s).

5. Check, but do not adjust, the plugs to make sure they have the specified gap.

6. Once the spark plugs is removed, you can test them.

   a. Remove the spark plug and connect to the ignition coil.

   b. Ground the spark plug outer electrode (body), and crank the engine. Check that there is an electrical discharge between the electrodes. If not, replace the plug.

To install:

7. Install the spark plugs and tighten to 19 ft. lbs. (25 Nm) on 2.4L engines and 13 ft. lbs. (18 Nm) on 3.0L engines.

8. Install the ignition coil(s).

9. Connect the negative battery cable.

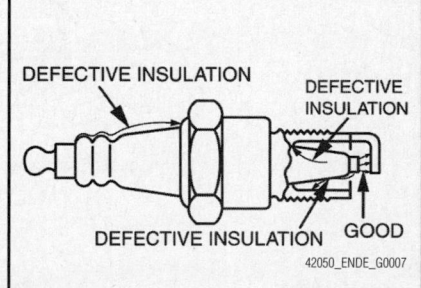

DEFECTIVE INSULATION

DEFECTIVE INSULATION

DEFECTIVE INSULATION

GOOD

42050_ENDE_G0007

Fig. 36 Ground the spark plug outer electrode (body), and crank the engine. Check that there is an electrical discharge between the electrodes

STARTER

REMOVAL & INSTALLATION

2.4L Engine

See Figure 37.

1. Before servicing the vehicle, refer to the precautions.
2. Disconnect the negative battery cable.
3. Remove the air cleaner duct and intake hose.
4. Remove the battery and battery tray.
5. Remove the engine undercover.
6. Disconnect the emission vacuum hose and brake booster vacuum hose.
7. Remove the throttle body support.
8. Disconnect the starter connector and terminal.
9. Remove the starter assembly.
   a. Slide the starter assembly, and disconnect the starter connector and terminal.
   b. Remove the starter assembly from the lower front of the engine.

To install:
10. Install the starter assembly.
   a. Install the starter assembly from the lower front of the engine.
   b. Connect the starter connector and terminal and tighten to 93 inch lbs. (11 Nm).
   c. Slide the starter assembly into place. Tighten starter mounting bolts to 27 ft. lbs. (35 Nm).

11. Install the throttle body support.
12. Connect the emission vacuum hose and brake booster vacuum hose.
13. Install the engine undercover.
14. Install the battery and battery tray.
15. Install the air cleaner duct and intake hose.
16. Connect the negative battery cable.

3.0L Engine

See Figure 38.

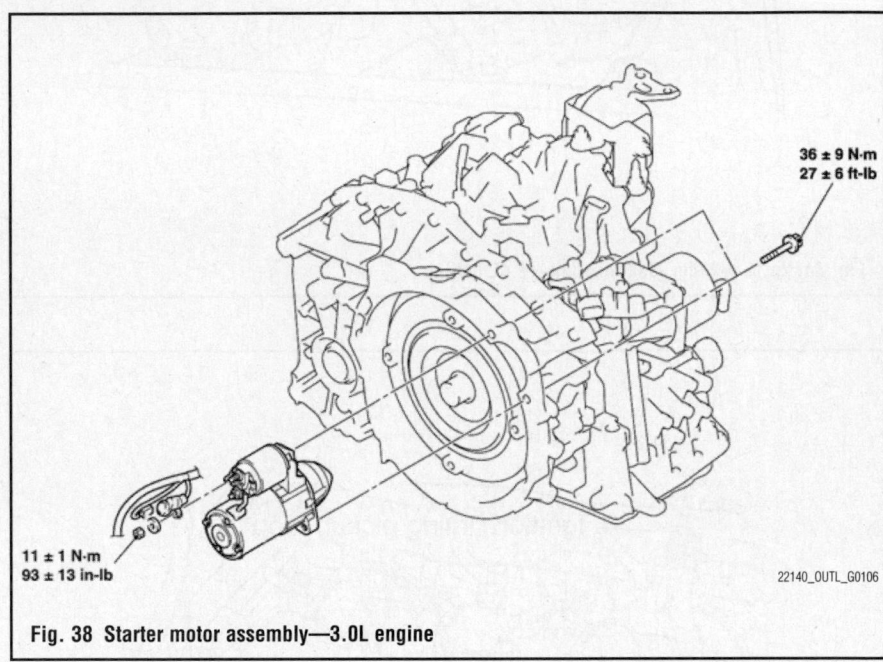

36 ± 9 N·m
27 ± 6 ft-lb

11 ± 1 N·m
93 ± 13 in-lb

22140_OUTL_G0106

Fig. 38 Starter motor assembly—3.0L engine

1. Before servicing the vehicle, refer to the precautions.
2. Disconnect the negative battery cable.
3. Remove the engine undercover.
4. Disconnect the starter connector and terminal.
5. Remove the starter assembly.
   a. Disconnect the starter connector and terminal.
   b. Remove the starter cover.
   c. Remove the starter assembly from under the vehicle..

To install:
6. Install the starter assembly.
   a. Install the starter assembly from under the vehicle.
   b. Connect the starter connector and terminal and tighten to 93 inch lbs. (11 Nm).
   c. Remove the starter cover. Tighten to 44 inch lbs. (5 Nm).
   d. Slide the starter assembly into place. Tighten starter mounting bolts to 27 ft. lbs. (35 Nm).
7. Install the engine undercover.
8. Connect the negative battery cable.

SOLENOID OR RELAY REPLACEMENT

See Figure 39.

1. Remove the starter from the vehicle.

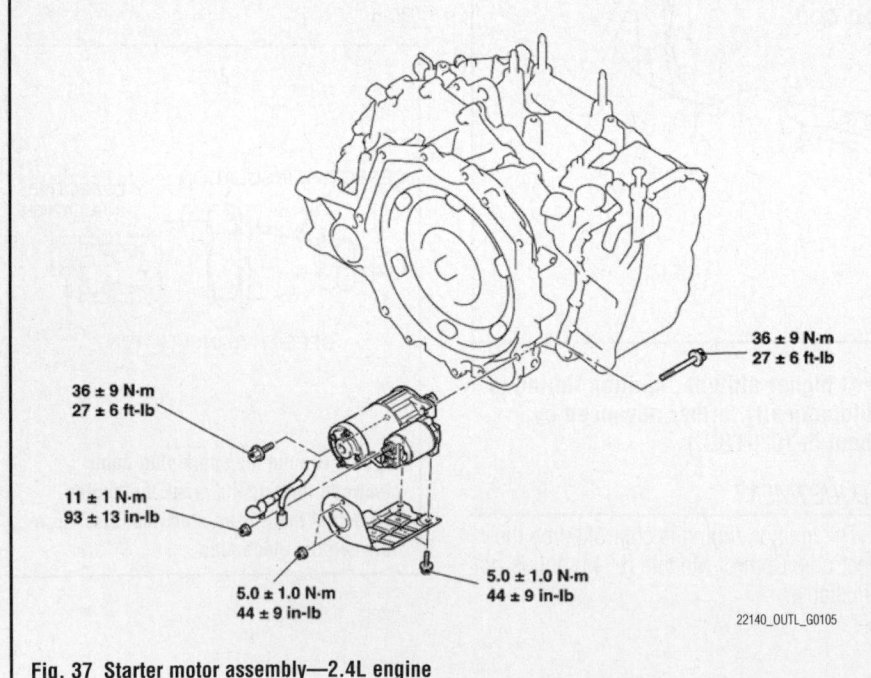

36 ± 9 N·m
27 ± 6 ft-lb

11 ± 1 N·m
93 ± 13 in-lb

36 ± 9 N·m
27 ± 6 ft-lb

5.0 ± 1.0 N·m
44 ± 9 in-lb

5.0 ± 1.0 N·m
44 ± 9 in-lb

22140_OUTL_G0105

Fig. 37 Starter motor assembly—2.4L engine

➡**Do not clamp the yoke assembly with a vise.**

2. Disconnect the lead from the M terminal of the magnetic switch.

3. Remove the screw and bracket from the solenoid.

4. Remove the solenoid from the starter.

To install:

5. Installation is the reverse of removal.

6. Tighten the solenoid screws to 51 inch lbs. (6 Nm).

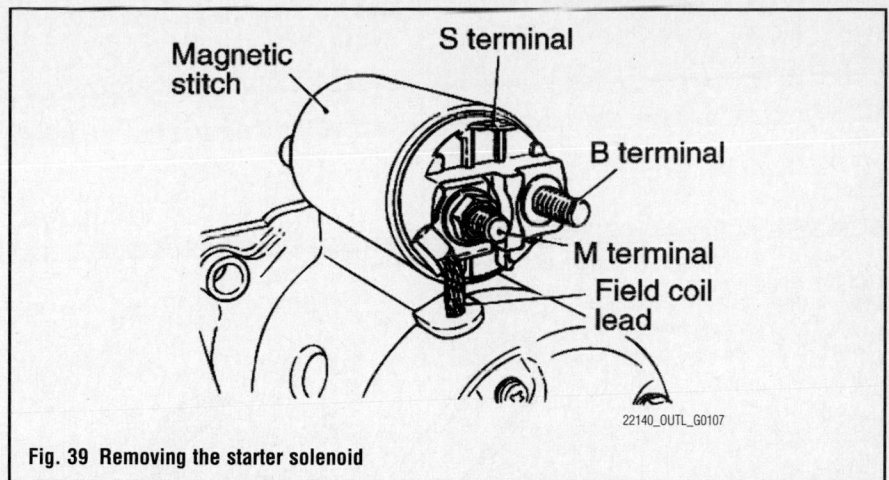

Fig. 39 Removing the starter solenoid

ENGINE MECHANICAL

ACCESSORY DRIVE BELTS

ACCESSORY BELT ROUTING

See Figures 40 and 41.

Refer to the accompanying illustrations for belt routing.

INSPECTION

See Figures 42 and 43.

1. Before servicing the vehicle, refer to the precautions.

2. On 2.4L engine, remove the radiator condenser tank mounting bolts.

   a. Move the radiator condenser

tank to a place where it will not be a hindrance when checking the drive belt tension.

➡**Check the drive belt tension after turning the crankshaft clockwise one turn or more.**

3. Make sure that the indicator mark on the auto-tensioner is within the area marked with "A".

4. If the mark is out of the area A, replace the drive belt.

5. On 2.4L engine, tighten the radiator condenser tank mounting bolts to 102 inch lbs. (12 Nm).

ADJUSTMENT

Belt tension is maintained by a automatic tensioner. No adjustments are necessary.

REMOVAL & INSTALLATION

See Figures 44 through 46.

1. Before servicing the vehicle, refer to the precautions.

2. Remove the engine undercover.

3. Remove the radiator condenser tank assembly mounting bolt, and move the radiator condenser tank assembly to a place where it does not interfere with the drive belt removal and installation.

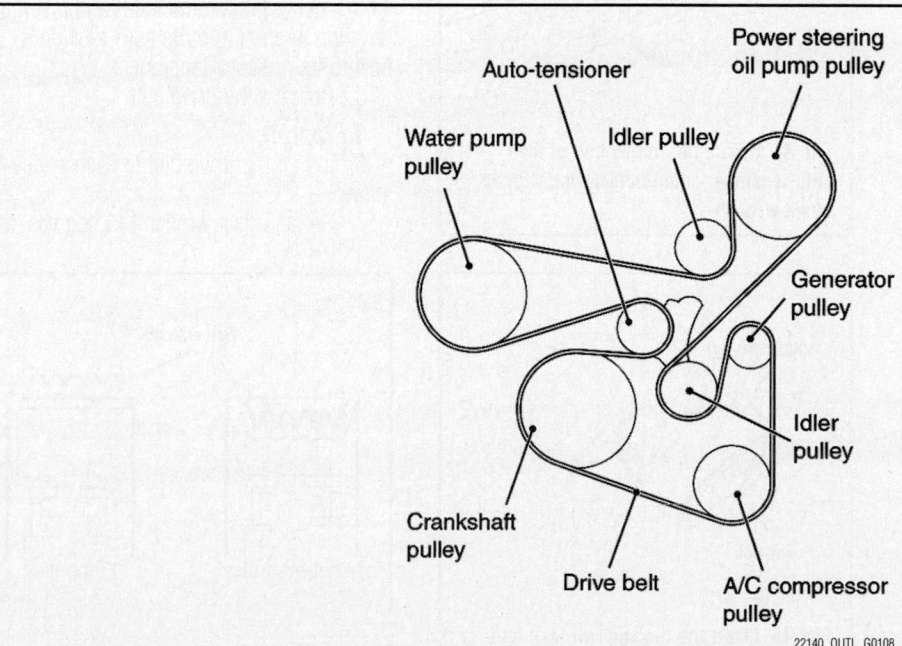

Fig. 40 Accessory drive belt routing—2.4L engine

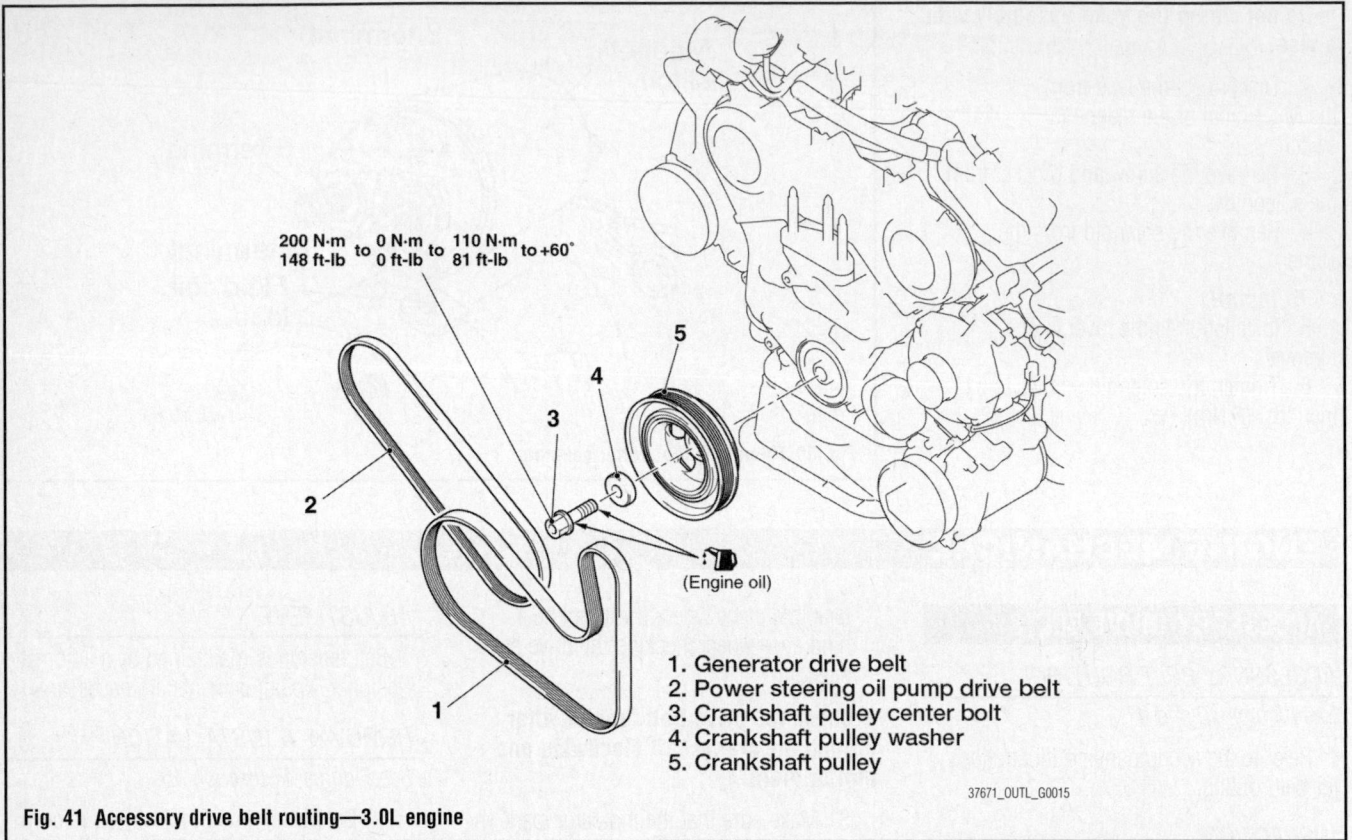

200 N·m to 0 N·m to 110 N·m to +60°
148 ft-lb to 0 ft-lb to 81 ft-lb

5

4

3

2

1

(Engine oil)

1. Generator drive belt
2. Power steering oil pump drive belt
3. Crankshaft pulley center bolt
4. Crankshaft pulley washer
5. Crankshaft pulley

37671_OUTL_G0015

Fig. 41 Accessory drive belt routing—3.0L engine

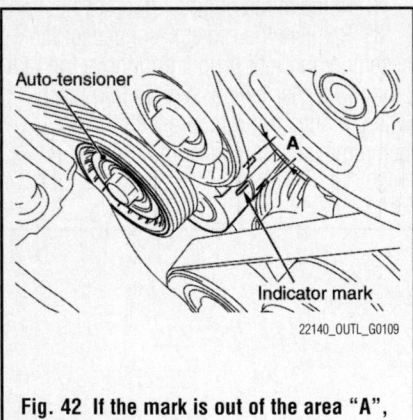

Auto-tensioner

A

Indicator mark

22140_OUTL_G0109

Fig. 42 If the mark is out of the area "A", replace the drive belt —2.4L engine

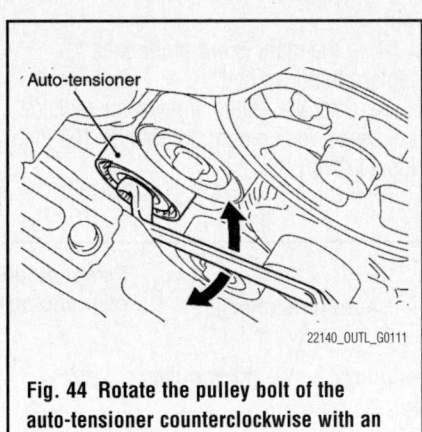

Auto-tensioner

22140_OUTL_G0111

Fig. 44 Rotate the pulley bolt of the auto-tensioner counterclockwise with an offset wrench

➡ To reuse the drive belt, draw an arrow indicating the rotating direction on the back of the belt using chalk to install the same direction.

4. Rotate the pulley bolt of the auto-tensioner counterclockwise with an offset wrench [45°, a long offset wrench (5/8 x 11/16 inches) recommended] and insert the hexagon wrench into the auto-tensioner hole to fix the auto-tensioner.
5. Remove the drive belt.

To install:

6. Install the drive belt to each pulley as shown.
7. Set an offset wrench [45 degrees, a

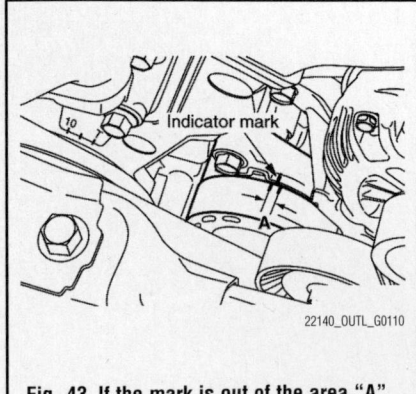

Indicator mark

10

A

22140_OUTL_G0110

Fig. 43 If the mark is out of the area "A", replace the drive belt —3.0L engine

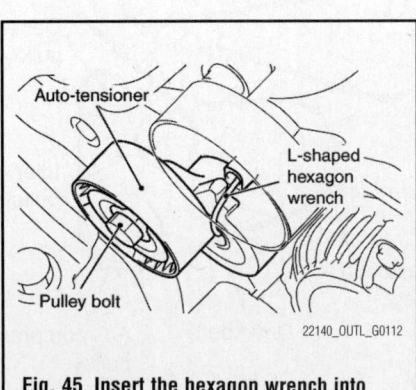

Auto-tensioner

L-shaped hexagon wrench

Pulley bolt

22140_OUTL_G0112

Fig. 45 Insert the hexagon wrench into the auto-tensioner hole to fix the auto-tensioner

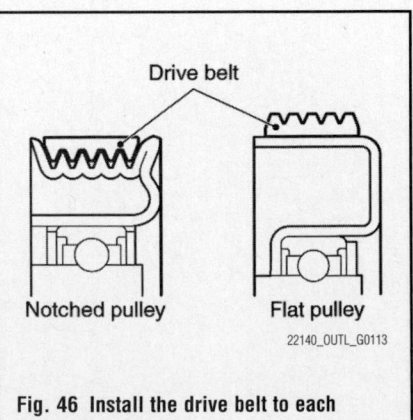

Drive belt

Notched pulley   Flat pulley

22140_OUTL_G0113

Fig. 46 Install the drive belt to each pulley as shown

long offset wrench (5/8 x 11/16 inches) recommended] to the pulley bolt of the auto-tensioner. Then, rotate the auto-tensioner counter-clockwise and remove the L-shaped hexagon wrench fixing the auto-tensioner.

➡**Apply tension to the drive belt while slowly turning the auto-tensioner clockwise.**

CAMSHAFT AND VALVE LIFTERS

REMOVAL & INSTALLATION

2.4L Engine

See Figures 47 through 63.

1. Before servicing the vehicle, refer to the precautions.

2. Remove the engine under cover and engine room side cover.

3. Remove the air cleaner.

4. Remove the strut tower bar.

5. Remove the engine upper cover.

6. Remove the ignition coils.

7. Disconnect the breather hose connection.

8. Disconnect the PCV hose connection.

9. Disconnect the control wiring harness connection.

10. Remove the valve cover assembly.

   a. Loosen the valve cover assembly mounting bolts in the order of number shown, and remove the valve cover assembly.

   b. Remove the valve cover gasket.

11. Position cylinder No. 1 top dead center on the compression stroke.

   a. Turn the crankshaft clockwise so that the camshaft sprocket timing marks become horizontal to the cylinder head upper surface, and set cylinder No. 1 to Top Dead Center (TDC) on the compression stroke. At this time, check that the crankshaft pulley timing mark is in the 0-degree position of the ignition timing

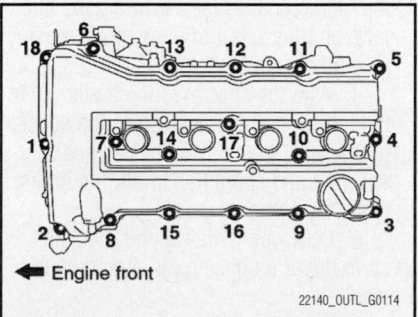

Fig. 47 Loosen the valve cover assembly mounting bolts in the order of number shown

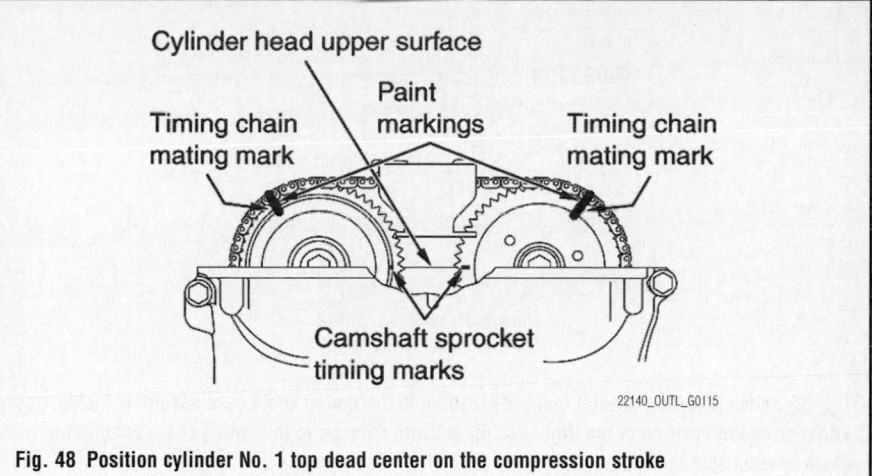

Fig. 48 Position cylinder No. 1 top dead center on the compression stroke

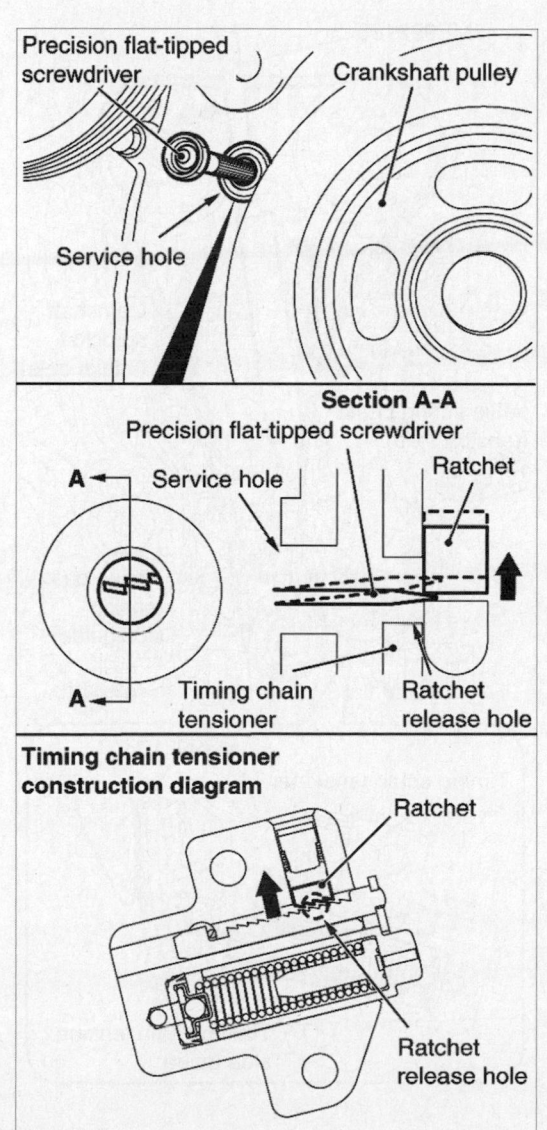

Fig. 49 Insert a precision flat-tipped screwdriver through the service hole of the timing chain case, press up the timing chain tensioner ratchet to unlock

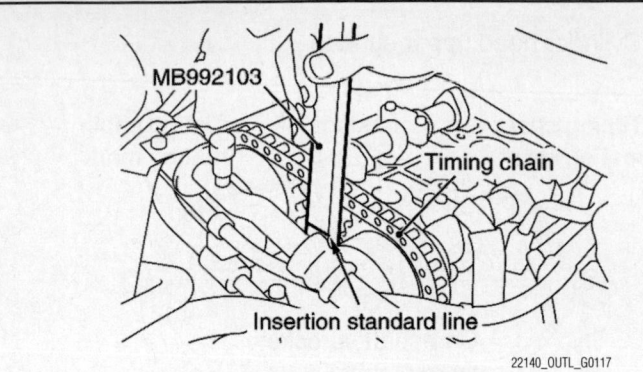

Fig. 50 When inserting special tool MB992103 into the timing chain case assembly inside, pay attention to the position of the timing chain to avoid damage to the timing chain and timing chain tension side guide

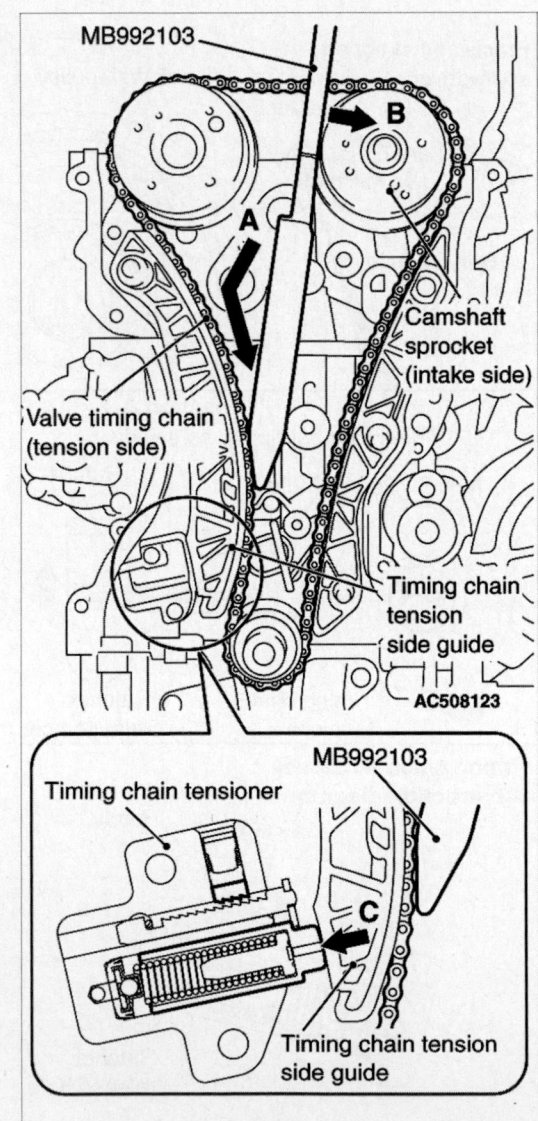

Fig. 51 With the timing chain tensioner unlocked, insert special tool MB992103 inside the timing chain case assembly along the tension side of the timing chain until the insertion guide line aligns with the upper surface of the timing chain case assembly

indicator of the timing chain case assembly.

b. Put paint marks on both the camshaft sprocket and timing chain at the position of camshaft sprocket timing chain mating mark (circular hole).

12. Remove the timing chain upper guide.

13. Remove the service hole bolt.

14. Remove the camshaft and camshaft sprocket assembly (exhaust side).

a. Insert a precision flat-tipped screwdriver through the service hole of the timing chain case, press up the timing chain tensioner ratchet to unlock, and keep the timing chain tensioner in that state.

b. Lightly press down the tail end of the precision flat-tipped screwdriver to press up the tip of the precision flat-tipped screwdriver inserted to the timing chain tensioner to unlock.

c. When inserting special tool MB992103 into the timing chain case assembly inside, pay attention to the position of the timing chain to avoid damage to the timing chain and timing chain tension side guide. Do not insert the special tool beyond its insertion guideline.

d. If unlocking the timing chain tensioner is insufficient, the special tool cannot be inserted to the insertion guideline. Do not insert the special tool forcibly, follow steps again to unlock the timing chain tensioner and insert the special tool.

e. With the timing chain tensioner unlocked, insert special tool MB992103 inside the timing chain case assembly along the tension side of the timing chain until the insertion guide line aligns with the upper surface of the timing chain case assembly.

f. With the timing chain tensioner unlocked, insert the special tool along the tension side of the timing chain, according to the special tool top shape. The special tool can be inserted smoothly to the position where the special tool insertion guide line aligns with the timing chain case assembly top surface, and the spread timing chain tension side guide can be held.

g. With the special tool inserted up to the insertion guide line, press the special tool against the intake side camshaft sprocket and spread and hold the timing chain tension side guide.

h. Remove the flat-tipped precision screwdriver unlocking the timing chain tensioner.

i. The timing chain may snag on by other parts. After sagging the timing chain, never rotate the crankshaft.

j. With the timing chain tension side

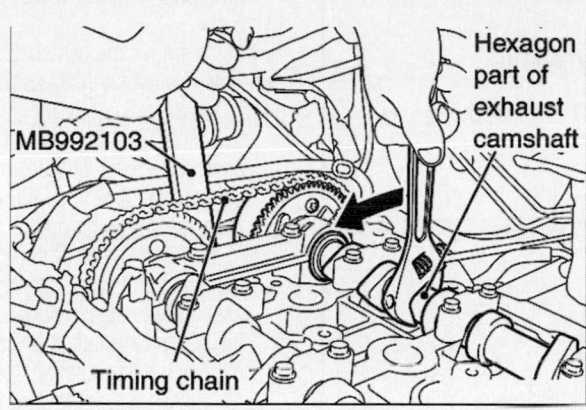

Fig. 52 With the timing chain tension side guide spread, hook the special tool over the hexagon part of the camshaft on the exhaust side, and turn the camshaft clockwise to apply slack to the timing chain between the camshaft sprockets

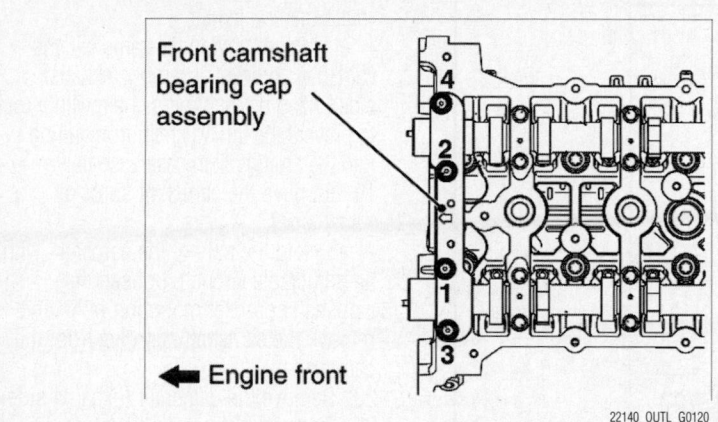

Fig. 53 Loosen the mounting bolts of front camshaft bearing cap in the order shown

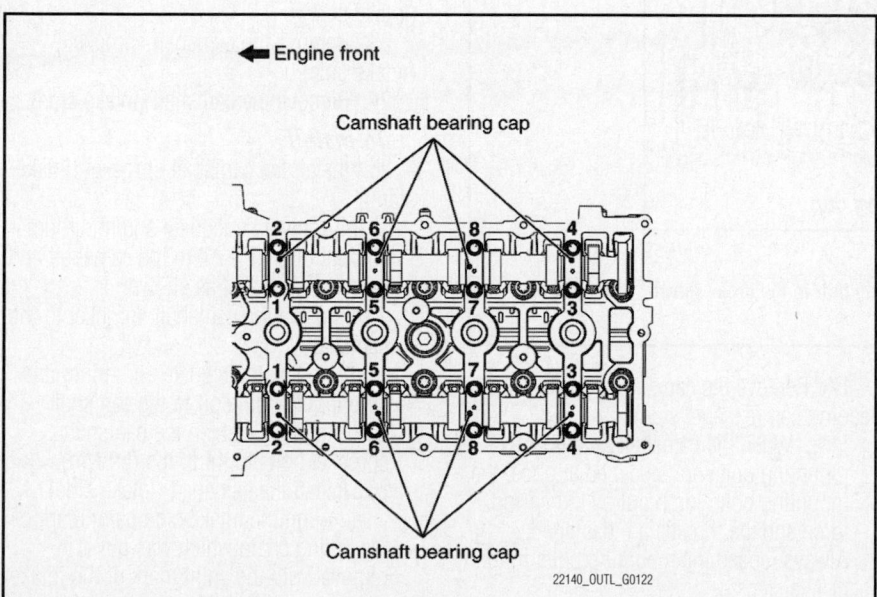

Fig. 54 Loosen the mounting bolts of the camshaft bearing caps in the order shown in four or five steps

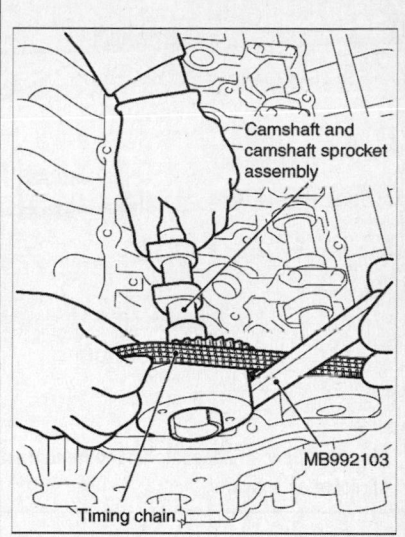

Fig. 55 Remove the timing chain from the camshaft and camshaft sprocket assembly (exhaust side) toward the timing chain case assembly, and remove the camshaft and camshaft sprocket assembly (exhaust side) toward the transaxle

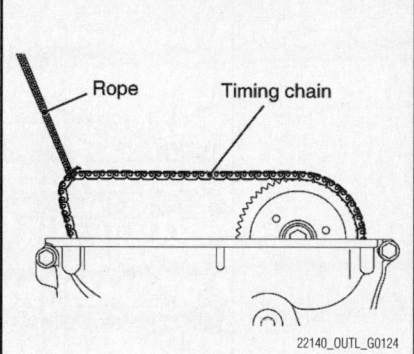

Fig. 56 After removing the camshaft and camshaft sprocket assembly (exhaust side), hang up the timing chain with a rope to prevent the timing chain from falling into the timing chain case assembly

guide spread, hook the special tool over the hexagon part of the camshaft on the exhaust side, and turn the camshaft clockwise to apply slack to the timing chain between the camshaft sprockets.

15. Remove the camshaft bearing front caps.

  a. Loosen the mounting bolts of front camshaft bearing cap in the order shown, and remove the front camshaft bearing cap assembly.

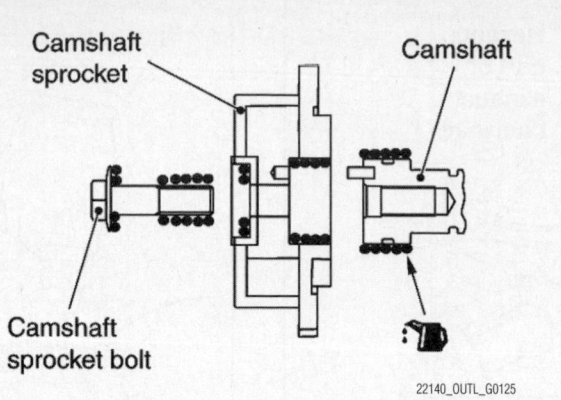

Fig. 57 Apply an adequate and minimum amount of engine oil to the camshaft and camshaft sprocket as shown

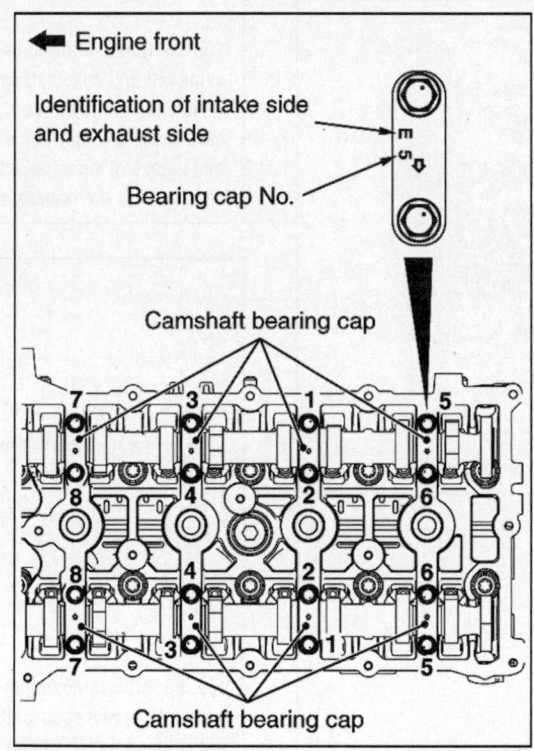

Fig. 58 Tighten each camshaft bearing cap mounting bolt in the order shown, using two or three steps

16. Remove the camshaft bearing.

a. When the camshaft bearing cap mounting bolts are loosened at once, the mounting bolts jump out by the spring force and the threads are damaged. Always loosen the mounting bolts in four or five steps.

b. Loosen the mounting bolts of the camshaft bearing caps in the order shown in four or five steps, and remove the camshaft bearing caps.

17. Remove the camshaft bearing oil feeding cap (exhaust side).

a. When the camshaft bearing cap mounting bolts are loosened at once, the mounting bolts jump out by the spring force and the threads are damaged. Always loosen the mounting bolts in four or five steps.

b. Loosen the mounting bolts of the camshaft bearing caps in the order shown in four or five steps,

and remove the camshaft bearing caps.

18. Remove the camshaft and camshaft sprocket assembly (exhaust side).

a. Raise slightly the transaxle side of the camshaft and camshaft sprocket assembly (exhaust side) by using the slack of the timing chain, and remove from the cam bearing.

b. Remove the timing chain from the camshaft and camshaft sprocket assembly (exhaust side) toward the timing chain case assembly, and remove the camshaft and camshaft sprocket assembly (exhaust side) toward the transaxle.

c. Remove special tool MB992103 inserted into the timing chain case assembly.

d. The timing chain may snag on other parts. After removing the camshaft and camshaft sprocket assembly, never rotate the crankshaft.

e. After removing the camshaft and camshaft sprocket assembly (exhaust side), hang up the timing chain with a rope to prevent the timing chain from falling into the timing chain case assembly.

19. Remove the camshaft sprocket (exhaust side).

a. Hold the flats of the camshaft with an adjustable wrench. Loosen the camshaft sprocket mounting bolts and remove the camshaft sprocket from the camshaft.

20. Remove the camshaft (exhaust side).

21. Remove the camshaft bearing oil feeding cap (intake side).

22. Remove the camshaft bearing cap (intake side).

23. Remove the camshaft sprocket (intake side).

24. Remove the camshaft (intake side).

To install:

25. Install the camshaft sprocket (intake side).

a. Apply an adequate and minimum amount of engine oil to the camshaft and camshaft sprocket as shown.

b. Install the camshaft sprocket to the camshaft.

c. Apply an adequate and minimum amount of engine oil to the camshaft sprocket bolt. Tighten the camshaft sprocket bolts to 44 ft. lbs (59 Nm).

26. Install the camshaft (intake side).

a. Align the intake side paint mark of the timing chain which was put at removal with the paint mark of the intake side camshaft sprocket, and install the camshaft sprocket to the timing chain.

b. Install the camshaft and camshaft

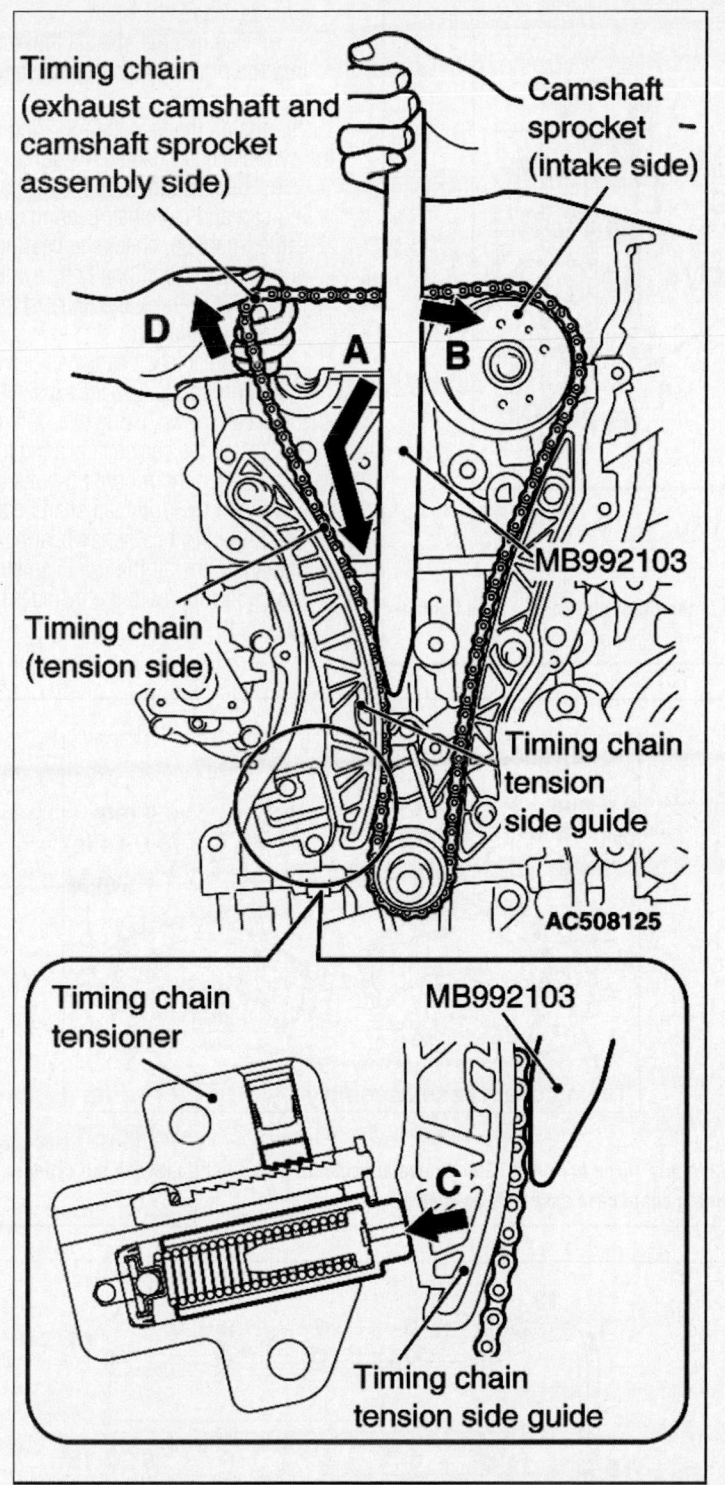

Fig. 59 Insert the special tool inside the timing chain case assembly along the tension side of the timing chain until the insertion guide line aligns with the upper surface of the timing chain case assembly "A", press the special tool against the intake side camshaft sprocket, "B", and spread and hold the timing chain tension side guide "C". Pull up the camshaft and camshaft sprocket assembly (exhaust side) mounting area of the timing chain "D"

sprocket assembly (intake side) to the cylinder head.

27. Install the camshaft bearing caps to the cylinder head (intake side).

   a. Because the thrust camshaft bearing cap and camshaft bearing cap are the same in shape, check the bearing cap number and additionally its symbol to identify the intake and exhaust sides for correct installation.

   b. Tighten each camshaft bearing cap mounting bolt to 107 inch lbs. 912 Nm) in the order shown, using two or three steps.

28. Install the camshaft (exhaust side).

   a. Insert the precision flat-tipped screwdriver through the service hole of the timing chain case, press up the ratchet of timing chain tensioner to unlock, and hold the unlocked timing chain tensioner.

   b. Lightly press down the tail end of the precision flat-tipped screwdriver to press up the tip of the precision flat-tipped screwdriver inserted to the timing chain tensioner to unlock.

   c. When inserting special tool MB992103 into the timing chain case assembly, pay attention to the position of the timing chain to avoid damage to the timing chain and timing chain tension side guide. Do not insert the special tool beyond its insertion guideline.

   d. If unlocking the timing chain tensioner is insufficient, the special tool cannot be inserted to the insertion guideline. Do not insert the special tool forcibly, follow Step 2 again to unlock the timing chain tensioner and insert the special tool.

   e. With the timing chain tensioner unlocked, insert special tool MB992103 inside the timing chain case assembly along the tension side of the timing chain until the insertion guide line aligns with the upper surface of the timing chain case assembly, as shown "A".

   f. With the timing chain tensioner unlocked, insert the special tool along the tension side of the timing chain, according to the special tool top shape. The special tool can be inserted smoothly to the position where the special tool insertion guideline aligns with the timing chain case assembly top surface, and the spread timing chain tension side guide can be hold.

   g. With the special tool inserted up to the insertion guide line, press the special tool against the intake side camshaft sprocket, as shown "B", and spread and hold the timing chain tension side guide, as shown "C".

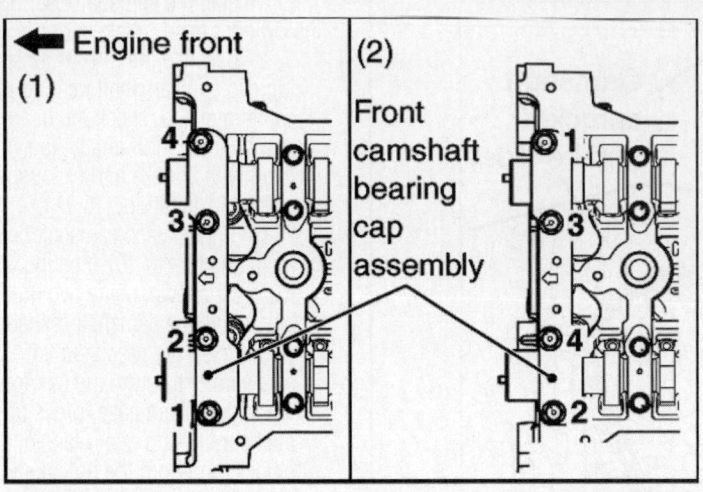

Fig. 60 Temporarily tighten the camshaft bearing front cap in the order shown (1). Then tighten again in the order (2)

h. Remove the flat-tipped precision screwdriver unlocking the timing chain tensioner.

i. Pull up the camshaft and camshaft sprocket assembly (exhaust side) mounting area of the timing chain, as shown "D", to provide allowance for easy installation of the camshaft and camshaft sprocket assembly (exhaust side) to the timing chain.

j. When installing the camshaft and camshaft sprocket assembly (exhaust side), be careful not to let the camshaft bearing which is installed to the front cam bearing deviate from its position.

k. Align the exhaust side paint mark of the timing chain which was put at removal with the paint mark of the exhaust side camshaft sprocket, and install the timing chain to the camshaft sprocket.

l. Install the camshaft and camshaft sprocket assembly (exhaust side) to the cylinder head.

m. Remove the special tool inserted into the timing chain case assembly inside.

29. Install the camshaft bearing caps to the cylinder head (exhaust side).

a. Because the thrust camshaft bearing cap and camshaft bearing cap are the same in shape, check the bearing cap number and additionally its symbol to identify the intake and exhaust sides for correct installation.

b. Tighten each camshaft bearing cap mounting bolt to 107 inch lbs. 912 Nm) in the order shown, using two or three steps.

30. Install the camshaft bearing front caps.

a. When the mounting bolts are tightened with the front camshaft bearing cap tilted, the front camshaft bearing cap is damaged. Install the front camshaft bearing cap properly to the cylinder head and camshaft.

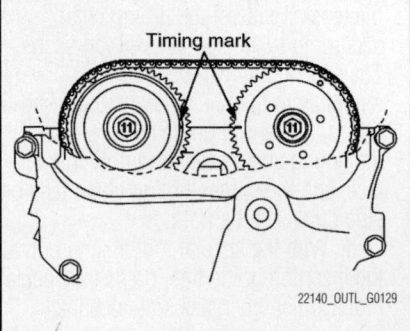

Fig. 61 Timing sprockets with No. 1 cylinder at TDC

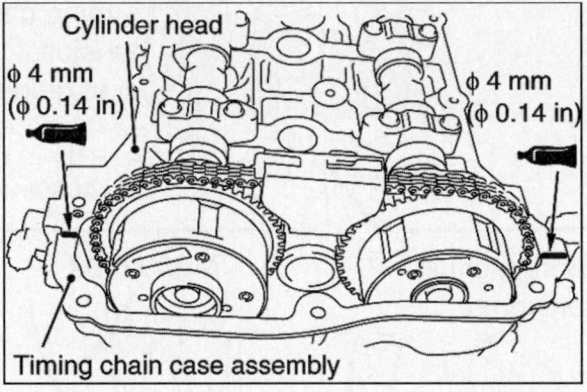

Fig. 62 Apply Three bond 1217G or equivalent sealant to the joint between the cylinder head and timing chain case assembly as shown

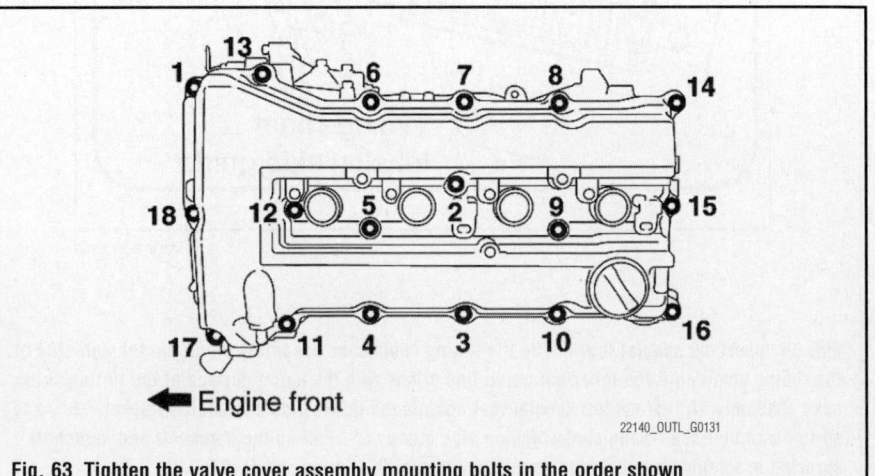

Fig. 63 Tighten the valve cover assembly mounting bolts in the order shown

b. Install the front camshaft bearing cap to the cylinder head, and temporarily tighten the camshaft bearing front cap to 13 ft. lbs. (17 Nm) in the order shown of the figure (1). Then tighten again in the order of the figure (2).

c. After the front camshaft bearing cap installation, check that the paint markings of the camshaft sprocket and the timing chain and the timing mark of the crankshaft pulley and the "T" mark position of ignition timing indicator are aligned respectively.

31. Install the service hole bolt.
32. Install the timing chain upper guide.
33. Adjust the valve clearance.

a. Perform the valve clearance check and adjustment at the engine cold state.

b. If the engine has been turned, turn the crankshaft clockwise, and align the timing mark on the exhaust camshaft sprocket against the upper face of the cylinder head. The No.1 cylinder should now be at TDC on the compression stroke.

c. Using a thickness gauge, measure the valve clearance of cylinders No. 1 and 2 on the intake side and cylinders No. 1 and 3 on the exhaust side. If not within specification, make note for the valve clearance adjustment.

d. Turn the crankshaft clockwise 360 degrees, and put the timing mark on the exhaust camshaft sprocket as illustrated. The No.1 cylinder should now be at TDC on the compression stroke.

e. Check the valve clearance of cylinders No. 3 and 4 on the intake side and cylinders No. 2 and 4 on the exhaust side.

f. If the valve clearance is not within specification, remove the camshaft and the valve tappet.

g. Using a micrometer, measure the thickness of the removed valve tappet.

h. Calculate the thickness of the newly installed valve tappet through the following equation.
- A: thickness of newly installed valve tappet
- B: thickness of removed valve tappet
- C: measured valve clearance

i. Calculate the thickness of the valve tappet as follows:
- Intake valve: A = B + [C - 0.20 mm (0.008 inch)]
- Exhaust valve: A = B + [C - 0.30 mm (0.012 inch)]

j. Install the valve tappet selected and put the camshaft in position.

k. After installing the timing chain, measure the valve clearance again.

34. Install the valve cover assembly

a. Wipe off the sealant on the mating surface of the valve cover assembly and the cylinder head and timing chain case assembly, and degrease the surface where the sealant is applied by white gasoline or the like.

b. Apply Three bond 1217G or equivalent sealant to the joint between the cylinder head and timing chain case assembly as shown and install the valve cover assembly to the cylinder head. note Install the valve cover assembly within 3 minutes after the application of sealant.

c. Tighten the valve cover assembly mounting bolts to 27 inch lbs. (3 Nm) in the order shown. Then tighten again to specification in the order shown.

35. Connect the control wiring harness connection.
36. Connect the PCV hose connection.
37. Connect the breather hose connection.
38. Install the ignition coils.
39. Install the engine upper cover.
40. Install the strut tower bar.
41. Install the air cleaner.
42. Install the engine under cover and engine room side cover.

3.0L Engine

Left Bank

See Figures 64 through 70.

1. Before servicing the vehicle, refer to the precautions.
2. Remove the engine cover.
3. Remove the timing belt.
4. Remove the thermostat housing.
5. Remove the engine oil level gauge and O-ring.
6. Disconnect the blow-by hose.
7. Disconnect the PCV hose.
8. Disconnect the ignition coil connector.
9. Remove the ignition coil.

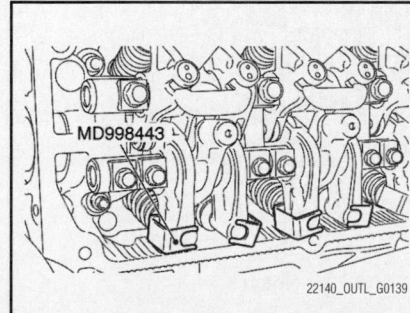

Fig. 64 Install special tool MD998443 as shown so that the lash adjusters will not fall out

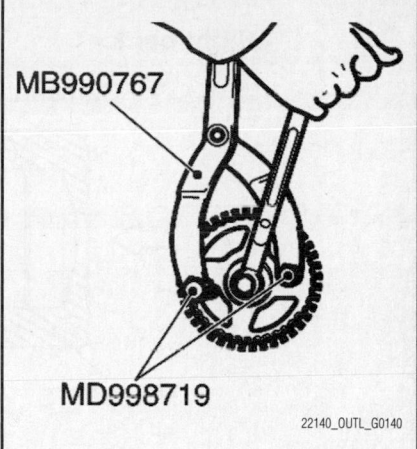

Fig. 65 Using special tools MB990767 and MD998719, remove the camshaft sprocket

10. Remove the valve cover and gasket.
11. Remove the rocker arm, shaft and the lash adjuster assembly (exhaust side).

a. Install special tool MD998443 as shown so that the lash adjusters will not fall out.

☀☀ WARNING

Never disassemble the rocker arm and shaft assembly.

b. Loosen the rocker arm and shaft assembly mounting bolt, and then remove the rocker arm and shaft assembly with the bolt still attached.

12. Remove the rocker arm and shaft assembly (intake side).
13. Remove the oil pipe assembly and gasket.
14. Remove the harness bracket.
15. Disconnect the camshaft position sensor connector.
16. Remove the camshaft position sensor and O-ring.
17. Remove the camshaft position sensor support and gasket.

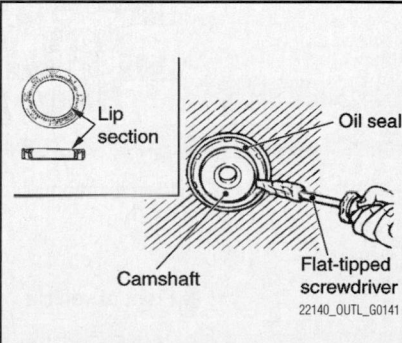

Fig. 66 Make a notch in the oil seal lip section with a knife

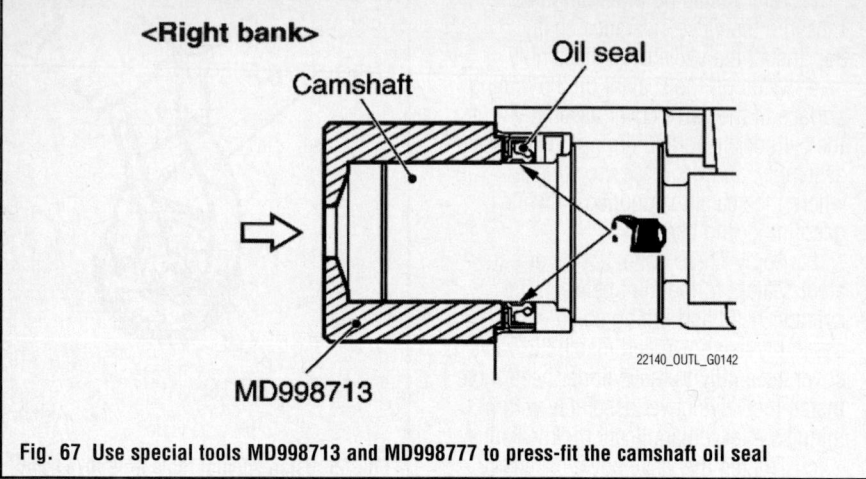

Fig. 67 Use special tools MD998713 and MD998777 to press-fit the camshaft oil seal

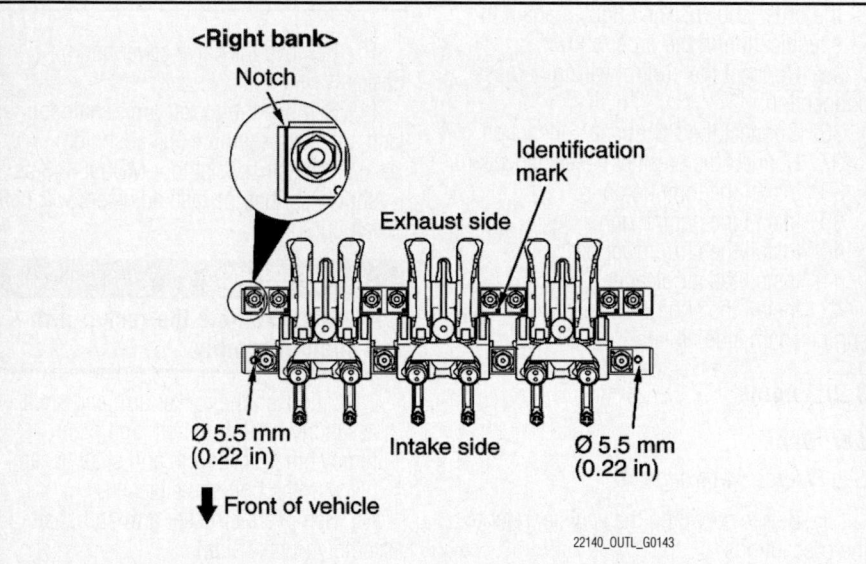

Fig. 68 Install the intake side rocker arm and shaft assembly so that the holes of rocker arm shaft face the cylinder head side

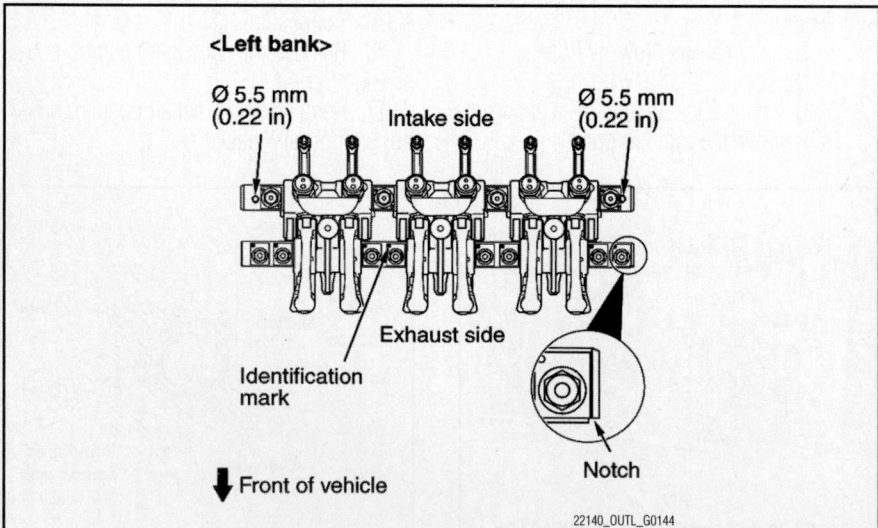

Fig. 69 Install the exhaust side rocker arm, shaft and lash adjuster assembly so that the notch of rocker arm shaft is located as shown

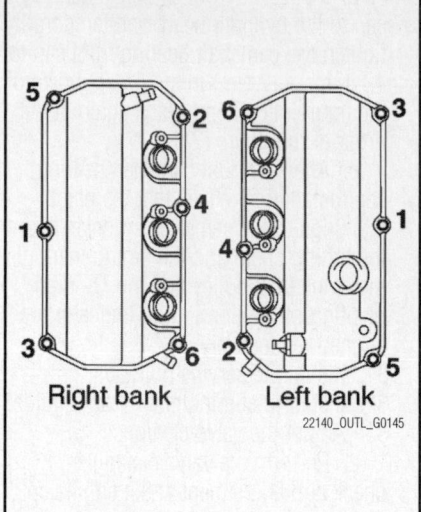

Fig. 70 Tighten the valve cover bolts in the order shown

18. Remove the camshaft position sensing cylinder.

19. Using special tools MB990767 and MD998719, remove the camshaft sprocket.

20. Remove the camshaft.

21. Remove the camshaft oil seal.

a. Make a notch in the oil seal lip section with a knife.

b. Be careful not to damage the camshaft and the cylinder head.

c. Cover the end of a flat-tipped screwdriver with a shop towel and insert into the notched section of the oil seal, and pry out the oil seal to remove it.

To install:

22. Install the camshaft oil seal.

a. Apply engine oil to the camshaft oil seal lip.

b. Use special tools MD998713 and MD998777 to press-fit the camshaft oil seal.

23. Install the camshaft.

24. Using special tools MB990767 and MD998719, install the camshaft sprocket.

25. Install the camshaft position sensing cylinder.

26. Install the camshaft position sensor support and gasket.

27. Install the camshaft position sensor and O-ring.

28. Connect the camshaft position sensor connector.

29. Install the harness bracket.

30. Install the oil pipe assembly and gasket.

31. Install the intake side rocker arm and shaft assembly so that the 0.22 inch (5.5 mm) holes of rocker arm shaft face the cylinder head side. Tighten the intake side

rocker arm shaft mounting bolts to 23 ft. lbs. (31 Nm).

32. Install the exhaust side rocker arm, shaft and lash adjuster assembly so that the notch of rocker arm shaft is located as shown.

    a. Check that the identification mark of exhaust side rocker shaft cap is located as shown.

    b. Tighten the exhaust side rocker arm shaft mounting bolts to 9 ft. lbs. (13 Nm).

    c. Remove special tool MD998443.

33. Install the valve cover and gasket.

    a. Tighten the bolts in the order shown to 74 inch lbs. (8 Nm).

34. Install the ignition coil.

35. Connect the ignition coil connector.

36. Connect the PCV hose.

37. Connect the blow-by hose.

38. Install the engine oil level gauge and O-ring.

39. Install the thermostat housing.

40. Install the timing belt.

41. Install the engine cover.

42. Change the engine oil.

43. Start the engine and check for leaks.

Right Bank

See Figures 71 through 77.

1. Before servicing the vehicle, refer to the precautions.

2. Remove the intake manifold plenum.

3. Remove the timing belt.

4. Remove the thermostat housing.

5. Disconnect the blow-by hose.

6. Disconnect the breather hose.

7. Disconnect the ignition coil connector.

8. Remove the ignition coil.

9. Remove the valve cover and gasket.

10. Remove the rocker arm, shaft and the lash adjuster assembly (exhaust side).

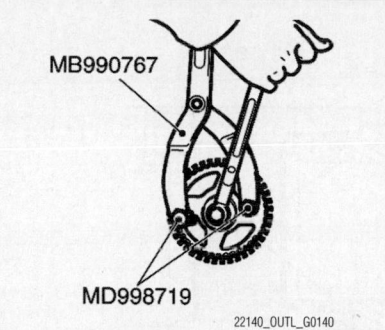

MB990767

MD998719

22140_OUTL_G0140

Fig. 72 Using special tools MB990767 and MD998719, remove the camshaft sprocket

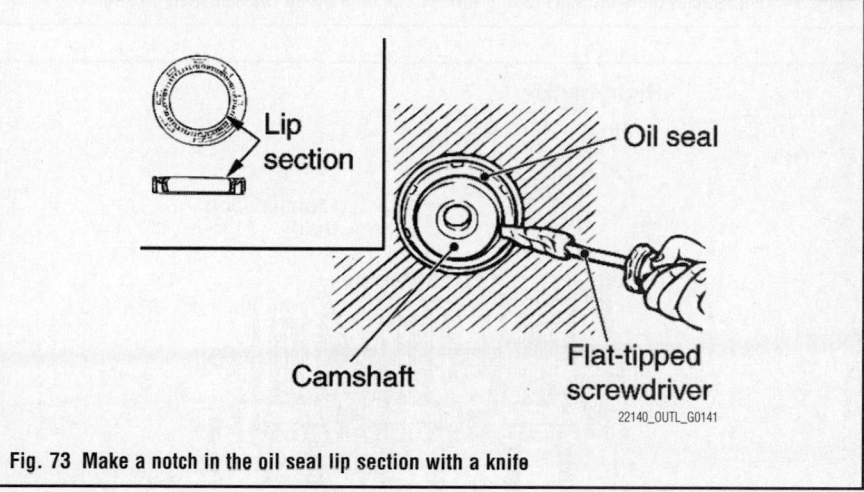

Lip section

Oil seal

Camshaft

Flat-tipped screwdriver

22140_OUTL_G0141

Fig. 73 Make a notch in the oil seal lip section with a knife

    a. Install special tool MD998443 as shown so that the lash adjusters will not fall out.

❄❄ WARNING

Never disassemble the rocker arm and shaft assembly.

    b. Loosen the rocker arm and shaft assembly mounting bolt, and then

remove the rocker arm and shaft assembly with the bolt still attached.

11. Remove the rocker arm and shaft assembly (intake side).

12. Disconnect the engine oil control valve connector.

13. Remove the engine oil pressure switch connector.

14. Remove the engine oil control valve and O-ring.

15. Remove the engine oil pressure switch.

16. Remove the engine oil control valve filter.

17. Remove the engine oil control valve housing and gasket.

18. Remove the engine oil control valve housing gasket

19. Using special tools MB990767 and MD998719, remove the camshaft sprocket.

20. Remove the engine control module.

21. Remove the air cleaner bracket.

22. Remove the camshaft.

23. Remove the camshaft oil seal.

    a. Make a notch in the oil seal lip section with a knife.

    b. Be careful not to damage the camshaft and the cylinder head.

    c. Cover the end of a flat-tipped screwdriver with a shop towel and insert into the notched section of the oil seal, and pry out the oil seal to remove it.

To install:

24. Install the camshaft oil seal.

    a. Apply engine oil to the camshaft oil seal lip.

    b. Use special tools MD998713 and MD998777 to press-fit the camshaft oil seal

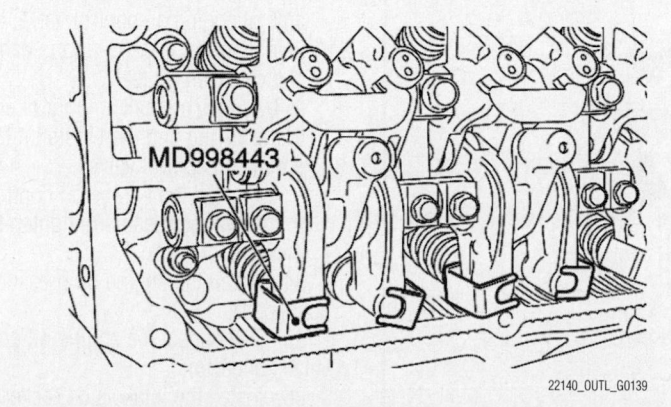

MD998443

22140_OUTL_G0139

Fig. 71 Install special tool MD998443 as shown so that the lash adjusters will not fall out

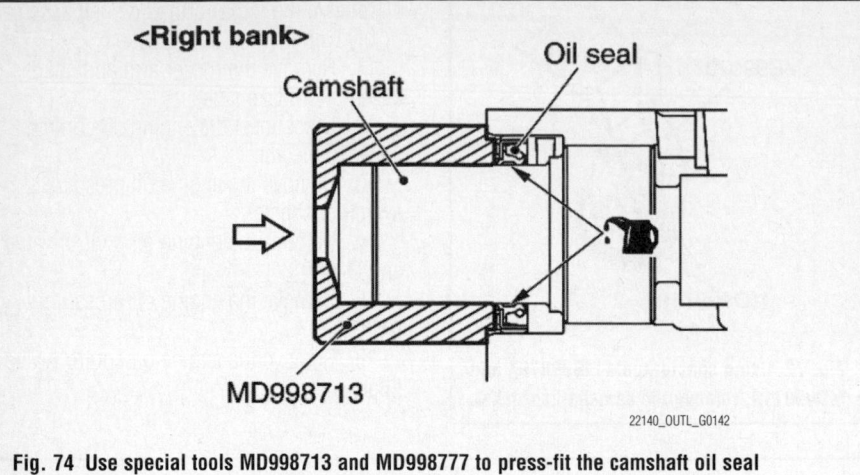

Fig. 74 Use special tools MD998713 and MD998777 to press-fit the camshaft oil seal

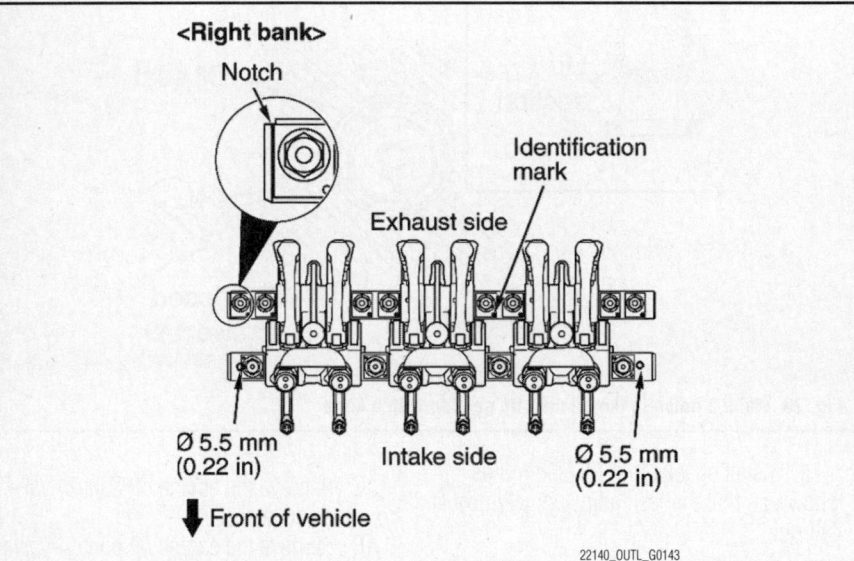

Fig. 75 Install the intake side rocker arm and shaft assembly so that the holes of rocker arm shaft face the cylinder head side

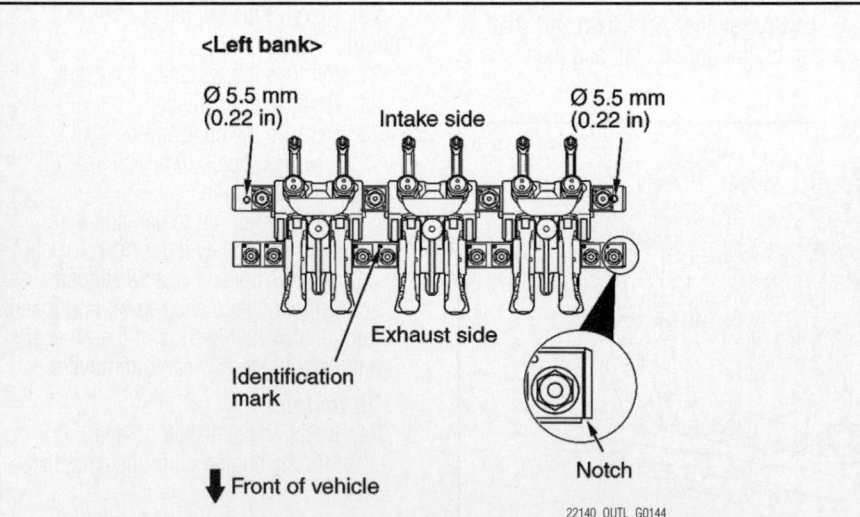

Fig. 76 Install the exhaust side rocker arm, shaft and lash adjuster assembly so that the notch of rocker arm shaft is located as shown

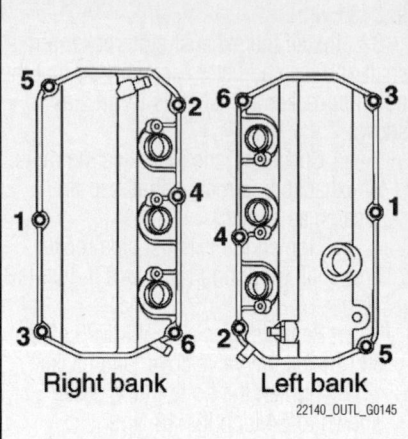

Fig. 77 Tighten the valve cover bolts in the order shown

25. Install the camshaft.
26. Install the air cleaner bracket.
27. Install the engine control module.
28. Using special tools MB990767 and MD998719, install the camshaft sprocket.
29. Install the engine oil control valve housing gasket
30. Install the engine oil control valve housing and gasket.
31. Install the engine oil control valve filter.
32. Install the engine oil pressure switch.
   a. Apply Three bond 1215 or equivalent sealant to the thread of the engine oil pressure switch.
   b. Tighten to 89 inch lbs. (10 Nm).
33. Install the engine oil control valve and O-ring.

❋❋ WARNING
Never re-use the O-ring.

   a. Before installing O-ring, wind sealing tape around the oil passages cut-out area of engine oil control valve, to prevent damage. If the O-ring is damaged, it can cause an oil leak.
   b. Apply a small amount of engine oil to the O-ring and then install it to the engine oil control valve.
   c. Install the engine oil control valve to the cylinder head and tighten to 98 inch lbs. (11 Nm).
34. Install the engine oil pressure switch connector.
35. Disconnect the engine oil control valve connector.
36. Install the intake side rocker arm and shaft assembly so that the 0.22 inch (5.5 mm) holes of rocker arm shaft face the

cylinder head side. Tighten the intake side rocker arm shaft mounting bolts to 23 ft. lbs. (31 Nm).

37. Install the exhaust side rocker arm, shaft and lash adjuster assembly so that the notch of rocker arm shaft is located as shown.

   a. Check that the identification mark of exhaust side rocker shaft cap is located as shown.

   b. Tighten the exhaust side rocker arm shaft mounting bolts to 9 ft. lbs. (13 Nm).

   c. Remove special tool MD998443.

38. Install the valve cover and gasket.

   a. Tighten the valve cover bolts in the order shown to 74 inch lbs. (8 Nm).

39. Install the ignition coil.

40. Disconnect the ignition coil connector.

41. Disconnect the breather hose.

42. Disconnect the blow-by hose.

43. Install the thermostat housing.

44. Install the timing belt.

45. Install the intake manifold plenum.

46. Change the engine oil.

47. Start the engine and check for leaks.

CRANKSHAFT DAMPER

REMOVAL & INSTALLATION

2.4L Engine

See Figures 78 and 79.

1. Before servicing the vehicle, refer to the precautions.

2. Remove the radiator condenser tank assembly.

3. Remove the drive belt.

4. Remove the crankshaft pulley center bolt.

   a. Hold the crankshaft pulley with special tools MB990767 and MD998719.

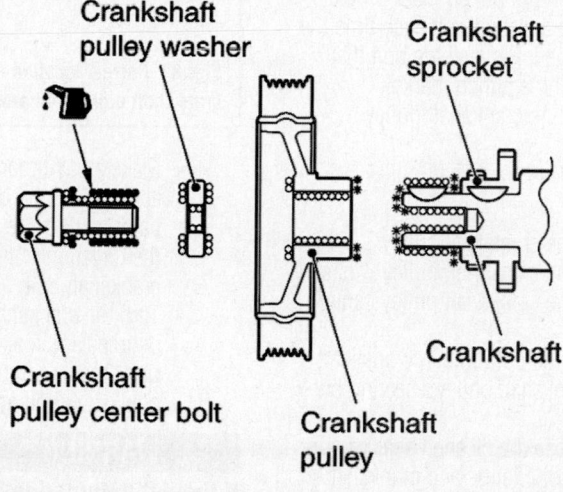

○ : Wipe clean with a rag.
✳ : Wipe clean with a rag and degrease.
● : Apply a small amount of engine oil.

Crankshaft pulley washer

Crankshaft sprocket

Crankshaft pulley center bolt

Crankshaft pulley

Crankshaft

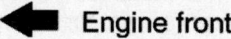 Engine front

22140_OUTL_G0146

Fig. 79 Clean and lubricate the crankshaft and pulley bolt as shown

5. Remove the crankshaft pulley washer.

6. Remove the crankshaft pulley.

To install:

7. Install the crankshaft pulley.

   a. Wipe off the dirt on the crankshaft and the crankshaft pulley as shown in the figure using a rag.

   b. Wipe off the dirt on the crankshaft sprocket, the crankshaft and the crankshaft pulley as shown in the figure using a rag, and then degrease them. note Degrease them to prevent drop in the friction coefficient of the pressed area, which is caused by oil adhesion.

   c. Install the crankshaft pulley.

   d. Wipe off the dirt on the crankshaft pulley washer and the crankshaft pulley center bolt as shown in the figure using a rag.

   e. Apply an adequate and minimum amount of engine oil to the threads of the crankshaft pulley center bolt and the lower area of the flange.

   f. Hold the crankshaft pulley with special tools MB990767 and MD998719.

   g. Tighten the crankshaft pulley center bolt to 155 ft. lbs. (210 Nm).

8. Install the drive belt.

9. Install the radiator condenser tank assembly.

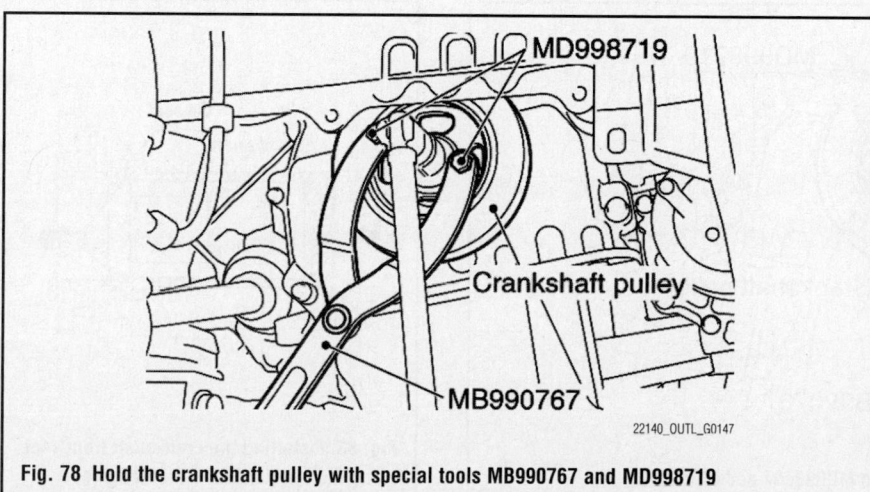

MD998719

Crankshaft pulley

MB990767

22140_OUTL_G0147

Fig. 78 Hold the crankshaft pulley with special tools MB990767 and MD998719

3.0L Engine

See Figures 79 through 81.

1. Remove the alternator drive belt.
2. Remove the power steering tensioner pulley bracket.
3. Remove the power steering oil pump drive belt.
4. Remove the crankshaft pulley center bolt.

 a. Hold the crankshaft pulley with special tools MB990767 and MD998719.

 b. Provide one punch mark on the head of the crankshaft bolt each time the bolt is removed. Replace the bolt that already has three punch marks.

5. Remove the crankshaft pulley washer.
6. Remove the crankshaft pulley.

To install:

7. Install the crankshaft pulley.
8. Install the crankshaft pulley washer.
9. Install the crankshaft pulley center bolt.

 a. Clean the bolt hole in crankshaft bolt and crankshaft pulley's seating surface.

 b. Degrease the cleaned seating surface of the front flange and crankshaft pulley.

 c. Install the front flange and crankshaft pulley.

 d. Apply oil to the threads of crankshaft bolt and the outer surface of washer.

 e. Use special tools MB990767 and MD998719 to install the crankshaft pulley.

 f. Tighten the bolt as follows:
 - Tighten to 148 ft. lbs. (200 Nm)
 - Loosen the crankshaft bolt fully
 - Tighten to 81 ft. lbs. (110 Nm)
 - Make a paint mark on the crankshaft bolt.

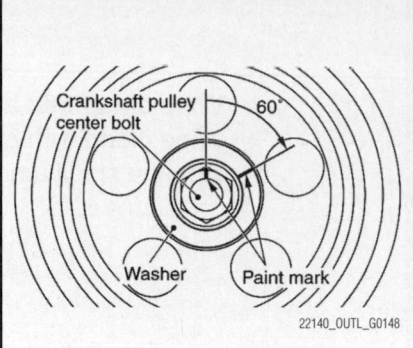

Fig. 81 Correct location of paint marks for crankshaft pulley bolt and washer

- Make a paint mark on the bolt end at a position 60 degrees from the paint mark made on the washer in the direction of tightening the crankshaft bolt
- Turn the crankshaft bolt another 60 degrees and make sure that the paint marks on the washer and crankshaft bolt are aligned.

✳✳ WARNING

If the nut is turned less than 60 degrees, proper fastening performance may not be achieved. Be careful to tighten the nut exactly 60 degrees. If the nut is over tightened (exceeding 60 degrees), loosen the nut completely and then retighten it by repeating the tightening procedure.

10. Install the power steering oil pump drive belt.
11. Install the power steering tensioner pulley bracket.
12. Install the alternator drive belt.

CRANKSHAFT FRONT SEAL

REMOVAL & INSTALLATION

See Figures 82 and 83.

1. Before servicing the vehicle, refer to the precautions.
2. On 3.0L engines, remove the timing belt.
3. Remove the crankshaft pulley.
4. Carefully pry the oil seal out of the front case.

✳✳ WARNING

Be careful not to damage the oil seal bore or the crankshaft sealing surface.

To install:

5. Install the crankshaft seal.

 a. Apply a small amount of engine oil to the entire inner diameter of the oil seal lip.

 b. Using special tool MD998718, press in the crankshaft rear oil seal up to the cylinder block assembly end surface.

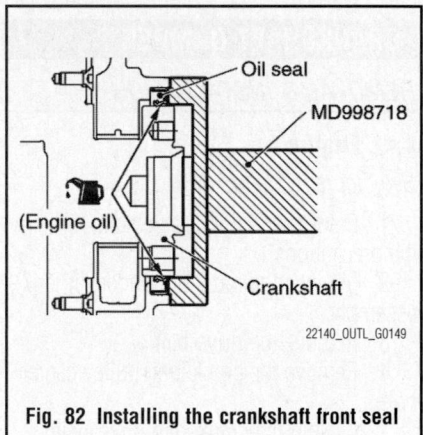

Fig. 82 Installing the crankshaft front seal using the special tool—2.4L engine

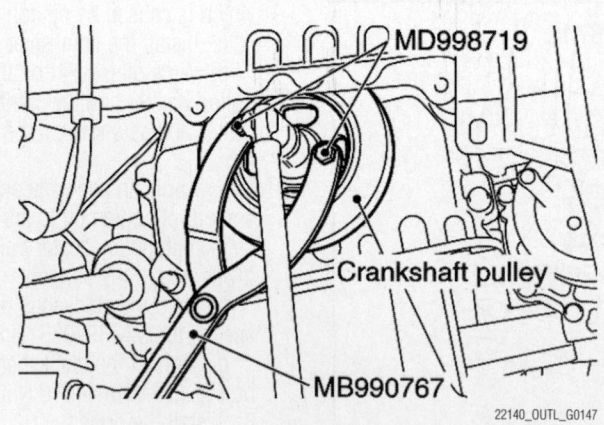

Fig. 80 Hold the crankshaft pulley with special tools MB990767 and MD998719

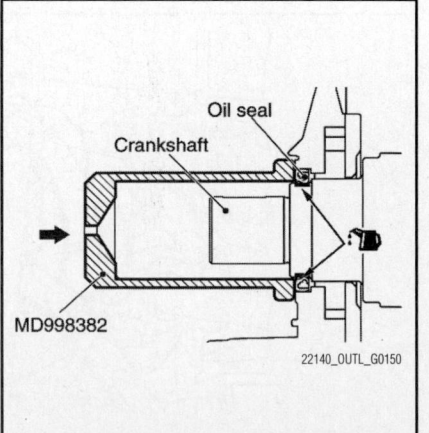

Fig. 83 Installing the crankshaft front seal using the special tool—3.0L engine

6. Install the crankshaft pulley.

7. On 3.0L engines, install the timing belt.

CYLINDER HEAD

REMOVAL & INSTALLATION

2.4L Engine

See Figures 84 through 89.

1. Before servicing the vehicle, refer to the precautions.

2. Relieve fuel system pressure.

3. Remove the engine under cover and engine side cover.

4. Drain the engine coolant.

5. Remove the air cleaner assembly.

6. Remove the ignition coil.

7. Remove the strut tower bar.

8. Remove the exhaust manifold. Refer to "Engine Mechanical Components, Exhaust Manifold, Removal & Installation."

9. Remove the throttle body assembly.

10. Remove the EGR valve and EGR valve bracket.

11. Remove the water pump.

12. Disconnect the control wiring harness.

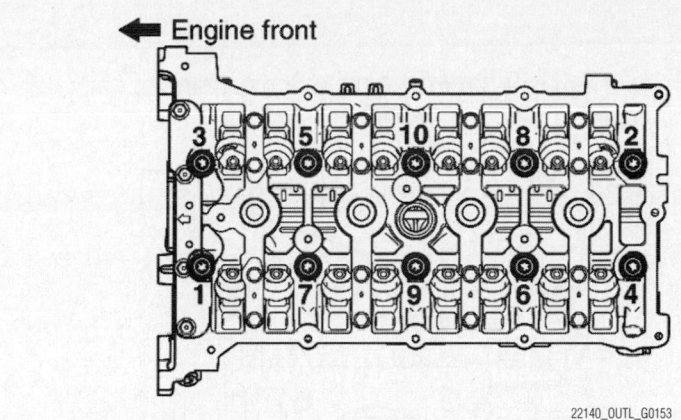

Fig. 86 Loosen and remove the bolts in two or three steps in the order shown

13. Disconnect the upper and lower radiator hoses

14. Disconnect the heater hose.

15. Disconnect the CVT fluid cooler water return hose.

16. Disconnect the water pump intake pipe.

17. Remove the cooling water line gasket.

18. Remove the O-ring.

19. Disconnect the canister vacuum hose.

20. Disconnect the brake booster vacuum hose.

21. Disconnect the engine oil level gauge.

22. Remove the intake manifold bracket

23. Disconnect the valve cover PCV hose.

24. Disconnect the fuel high-pressure hose.

   a. Remove the stopper of the fuel high-pressure hose.

   b. Raise the retainer of the fuel high-pressure hose and pull out the fuel high-pressure hose in the direction shown.

   c. If the retainer is released, install it securely after removing the fuel high-pressure hose.

25. Remove the timing chain. Refer to "Engine Mechanical Components, Timing

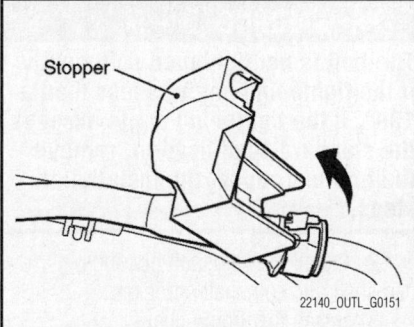

Fig. 84 Remove the stopper of the fuel high-pressure hose

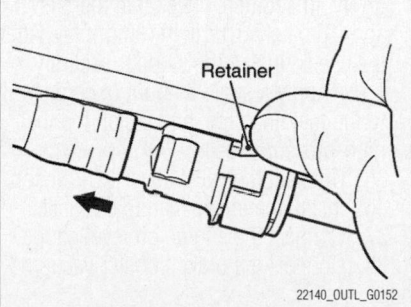

Fig. 85 Raise the retainer of the fuel high-pressure hose and pull out the fuel high-pressure hose in the direction shown

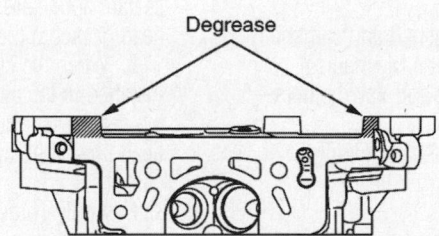

Bottom view of cylinder head

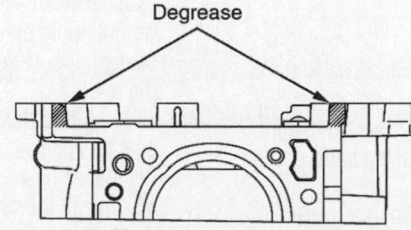

Top view of cylinder block

Fig. 87 Remove the sealant and grease on the top surface of cylinder block and on the bottom surface of the cylinder head. Then degrease the sealant application surface

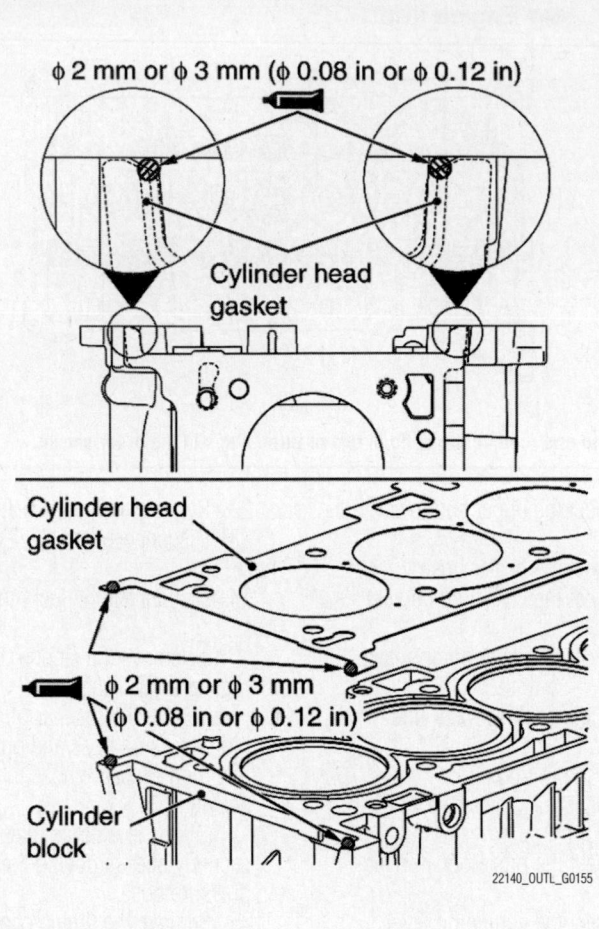

Fig. 88 Apply the sealant to the top surface of cylinder block as shown

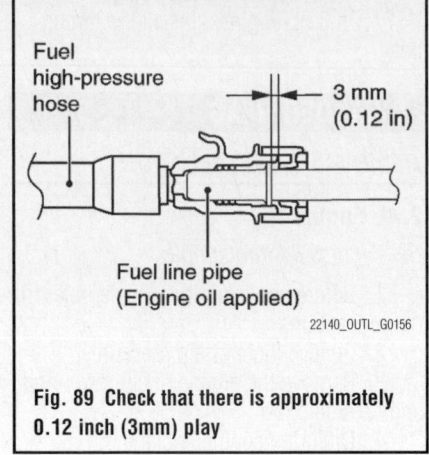

Fig. 89 Check that there is approximately 0.12 inch (3mm) play

Chain and Sprockets, Removal & Installation."

26. Remove the camshaft and camshaft sprocket. Refer to "Engine Mechanical Components, Camshaft and Valve Lifters, Removal & Installation."

27. Remove the camshaft bearings.

28. Remove the cylinder head bolts.

   a. Loosen and remove the bolts in two or three steps in the order shown.

29. Remove the cylinder head and gasket.

To install:

30. Install the cylinder head gasket and cylinder head.

   a. Do not allow any foreign materials get into the coolant passages, oil passages and cylinder.

   b. Remove the sealant and grease on the top surface of cylinder block and on the bottom surface of the cylinder head. Then degrease the sealant application surface.

   c. Apply Three bond 1217G sealant to the top surface of cylinder block as shown.

   d. Within three minutes after the sealant application, install the cylinder head gasket to the cylinder block.

   e. When the cylinder gasket is installed to the cylinder block, check that the sealant is securely applied to the bead line of the cylinder head gasket.

   f. Apply Three bond 1217G sealant to the top surface of cylinder head gasket as shown.

   g. Within two hours after the cylinder head assembly installation, do not apply oil or water to the sealant application area or start the engine.

   h. Within three minutes after the sealant application, install the cylinder head assembly.

31. Install the cylinder head bolts.

   a. Replace cylinder head bolts with a new ones.

   b. For two bolts of the timing chain side, the washer can be removed from the bolt. Install the washer, with its sag facing upward, to the bolts.

   c. Apply a small amount of engine oil to the cylinder head bolt threads and the washers.

   d. Tighten the bolts as follows:

- Tighten bolts to 26 ft. lbs. (35 Nm).
- Put a paint mark on the cylinder head bolt head and cylinder head
- Tighten an additional 180° in the order shown
- Check that the paint mark on the cylinder head bolt head aligns with the paint mark on the cylinder head.

✳✳ WARNING

The bolt is not tightened sufficiently if the tightening angle is less than a 180°. If the tightening angle exceeds the standard specification, remove the bolt and repeat the installation steps.

32. Install the camshaft bearings, camshaft and camshaft sprocket.

33. Install the timing chain.

34. Connect the fuel high-pressure hose.

   a. After connecting the fuel high-pressure hose, slightly pull it in the pull-out direction to check that it is installed firmly. In addition, check that there is approximately 0.12 inch (3mm)play. After the check, install the stopper securely.

   b. Apply a small amount of engine oil to the fuel line pipe, and install the fuel high-pressure hose.

35. Connect the valve cover PCV hose.

36. Install the intake manifold bracket

37. Connect the engine oil level gauge.

38. Connect the brake booster vacuum hose.

39. Connect the canister vacuum hose.

40. Install the O-ring.

41. Install the cooling water line gasket.

42. Connect the water pump intake pipe.

43. Connect the CVT fluid cooler water return hose.

44. Connect the heater hose.

45. Connect the upper and lower radiator hoses

46. Connect the control wiring harness.

47. Install the water pump.

48. Install the EGR valve and EGR valve bracket.

49. Install the throttle body assembly.

50. Install the exhaust manifold.

51. Install the strut tower bar.

52. Install the ignition coil.

53. Install the air cleaner assembly.

54. Fill the engine with coolant.

55. Install the engine under cover and engine side cover.

56. Start the engine and check for leaks.

3.0L Engine

See Figures 90 through 92.

1. Before servicing the vehicle, refer to the precautions.

2. Remove the intake manifold. Refer to "Engine Mechanical Components, Exhaust Manifold, Removal & Installation."

3. Remove the exhaust manifold. Refer to "Engine Mechanical Components, Intake Manifold, Removal & Installation."

4. Remove the timing belt. Refer to "Engine Mechanical Components, Timing Belt and Sprockets, Removal & Installation."

5. Remove the thermostat housing.

6. Remove the alternator.

7. Disconnect the blow-by hose.

8. On the left bank, disconnect the PCV hose.

9. On the left bank, remove the engine oil dipstick.

10. On the right bank, disconnect the breather hose.

11. On the right bank, disconnect the engine oil control valve connector.

12. On the right bank, disconnect the engine oil pressure switch connector.

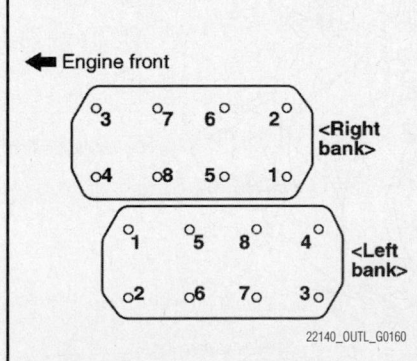

Fig. 90 Loosen the bolts in two or three steps in the order shown

13. On the right bank, disconnect the solenoid valve connector.

14. Disconnect the ignition coil connector.

15. Remove the ignition coil.

16. Remove the valve cover and gasket.

17. Remove the oil pipe and gasket.

18. Disconnect the camshaft position sensor connector.

19. Remove the camshaft sprocket. Refer to "Engine Mechanical Components, Timing Belt and Sprockets, Removal & Installation."

20. Remove the idler pulley

21. Remove the rear timing belt cover.

22. Remove the cylinder head and gasket assembly.

   a. Loosen the bolts in two or three steps in the order shown.

To install:

23. Install the cylinder head and gasket assembly.

   a. Be careful that no foreign material gets into the cylinder, coolant passages or oil passages. Engine damage may result.

   b. Use a scraper to clean the gasket surface of the cylinder head assembly.

   c. Check the cylinder head bolt as follows:

   * Measure the outside diameter "A"
   * Measure the smallest outside diameter "B" within the range "X" shown
   * If the difference of outside diameter of thread exceeds 0.0039 inch (0.1mm), replace the cylinder head bolt.

   d. Tighten the cylinder head bolts as follows:

   * Tighten in the order shown in two or three steps to 33 ft. lbs. (45 Nm).
   * Tighten an additional 150–154°

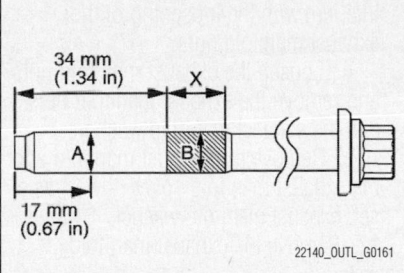

Fig. 91 Measure the outside diameter "A", then measure the smallest outside diameter "B" within the range "X". If the difference of outside diameter of thread exceeds 0.0039 inch (0.1mm), replace the cylinder head bolt

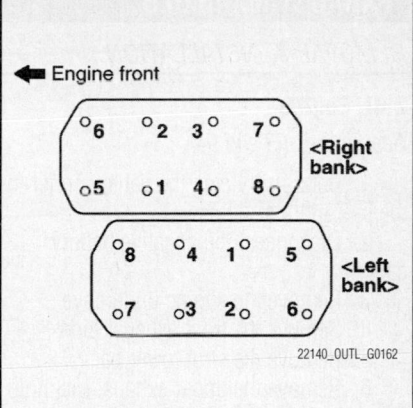

Fig. 92 Tighten the bolts in two or three steps in the order shown

✱✱ WARNING

If the bolt is turned less than 150–154°, proper fastening performance may not be achieved. Be sure to turn the bolt exactly 150 to 154 degrees. If the bolts is over tightened, loosen the bolt completely and then retighten it by repeating the tightening procedure.

24. Install the rear timing belt cover.

25. Install the idler pulley

26. Install the camshaft sprocket.

27. Connect the camshaft position sensor connector.

28. Install the oil pipe and gasket.

29. Install the valve cover and gasket.

30. Install the ignition coil.

31. Connect the ignition coil connector.

32. On the right bank, connect the solenoid valve connector.

33. On the right bank, connect the engine oil pressure switch connector.

34. On the right bank, connect the engine oil control valve connector.

35. On the right bank, connect the breather hose.

36. On the left bank, Install the engine oil dipstick.

37. On the left bank, connect the PCV hose.

38. Connect the blow-by hose.

39. Install the alternator.

40. Install the thermostat housing.

41. Install the timing belt.

42. Install the exhaust manifold.

43. Install the intake manifold.

44. Fill the engine with coolant.

45. Change the engine oil.

46. Start the engine and check for leaks.

EXHAUST MANIFOLD

REMOVAL & INSTALLATION

2.4L Engine

See Figures 93 and 94.

1. Before servicing the vehicle, refer to the precautions.
2. Disconnect the negative battery cable.
3. Remove the engine undercover.
4. Remove the front exhaust pipe.
5. Remove the strut tower bar.
6. Remove the upper exhaust manifold cover.
7. Remove the engine hanger.
8. Remove the exhaust manifold bracket (D).
9. Remove the exhaust manifold bracket (B).
10. Remove the exhaust manifold bracket (A).
11. Remove the exhaust manifold.
   a. Remove the mounting bolts of the exhaust manifold cover (lower) and move them to the position where they will not interfere with the loosening of the exhaust manifold nuts.
   b. Loosen the exhaust manifold nuts, and remove the exhaust manifold nuts and the exhaust manifold washers.
   c. Remove the exhaust manifold and the exhaust manifold cover (lower) as a set. exhaust manifold washer
12. Remove the exhaust manifold gasket.

To install:

13. Install the exhaust manifold and gasket.
   a. Install the exhaust manifold cover (lower) to the exhaust manifold (with mounting bolts removed), install the exhaust manifold and the exhaust manifold cover (lower) to the engine as a set.
   b. Move the exhaust manifold cover (lower) to the position where it will not interfere with the installation of the exhaust manifold nuts.
   c. Install the new exhaust manifold nuts and exhaust manifold washers, and tighten the exhaust manifold nuts to 36 ft. lbs. (49 Nm).
   d. Install the exhaust manifold cover (lower) to the exhaust manifold, and tighten the exhaust manifold cover (lower) mounting bolts to 10 ft. lbs. (14 Nm).
14. On AWD vehicles, install the exhaust manifold bracket (B).
   a. Make sure that the exhaust manifold bracket is closely contacted with the exhaust manifold and transfer assembly, and then temporarily tighten the bolts.
   b. First, tighten bolt A on the exhaust manifold side to 15 ft. lbs. (20 Nm).
   c. Then, tighten bolt B on the transfer assembly side to 15 ft. lbs. (20 Nm).
15. Install the exhaust manifold bracket (A). Tighten to 30 ft. lbs. (41 Nm).
16. Install the exhaust manifold bracket (D). Tighten to 30 ft. lbs. (41 Nm).
17. Install the engine hanger. Tighten to 21 ft. lbs. (28 Nm).

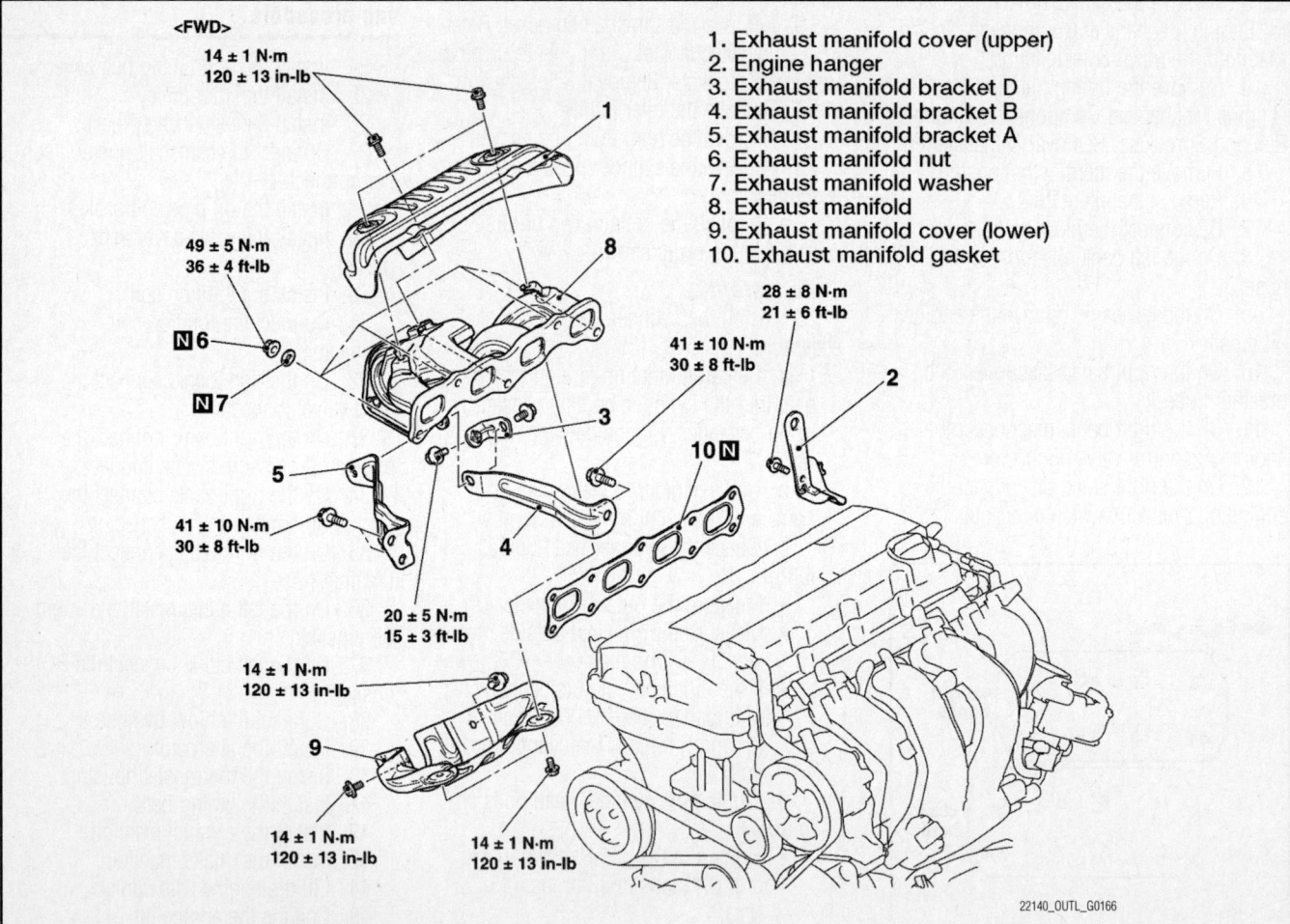

1. Exhaust manifold cover (upper)
2. Engine hanger
3. Exhaust manifold bracket D
4. Exhaust manifold bracket B
5. Exhaust manifold bracket A
6. Exhaust manifold nut
7. Exhaust manifold washer
8. Exhaust manifold
9. Exhaust manifold cover (lower)
10. Exhaust manifold gasket

22140_OUTL_G0166

Fig. 93 Exploded view of the exhaust manifold—2.4L engine, FWD shown, AWD similar

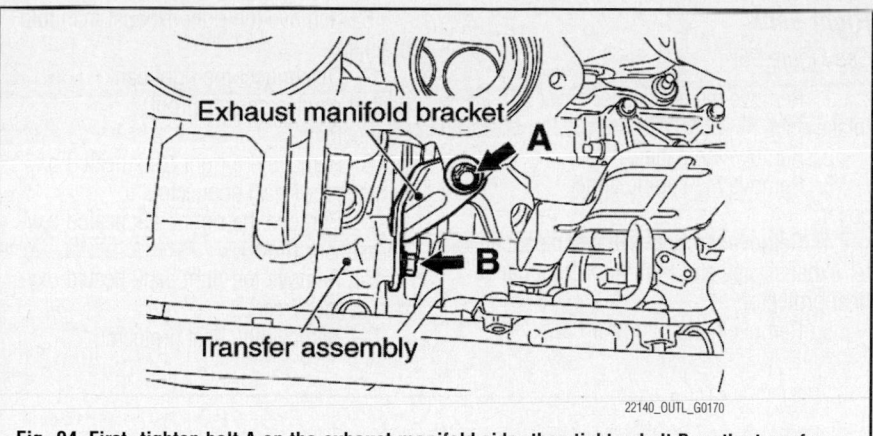

Fig. 94 First, tighten bolt A on the exhaust manifold side, then tighten bolt B on the transfer assembly side

18. Install the upper exhaust manifold cover. Tighten to 120 inch lbs. (14 Nm).
19. Install the strut tower bar.
20. Install the front exhaust pipe. Tighten to 30 ft. lbs. (41 Nm).
21. Install the engine undercover.
22. Connect the negative battery cable.

3.0L Engine

Left Bank

See Figure 95.

1. Remove the engine under cover.
2. Remove the air duct.

3. Remove the front exhaust pipe.
4. Disconnect the left bank heated oxygen sensor (front) connector.
5. Disconnect the left bank heated oxygen sensor (rear) connector.
6. Remove the left bank heated oxygen sensor (front).
7. Remove the left bank heated oxygen sensor (rear).
8. Remove the heat protector.
9. Remove the left exhaust manifold bracket.
10. Remove the exhaust manifold and gasket.

To install:

11. Install the exhaust manifold and gasket. Tighten to 36 ft. lbs. (49 Nm).
12. Install the left exhaust manifold bracket. Tighten to 35 ft. lbs. (47 Nm).
13. Install the heat protector. Tighten to 12 ft. lbs. (16 Nm).
14. Install the left bank heated oxygen sensor (rear). Tighten to 33 ft. lbs. (44 Nm).

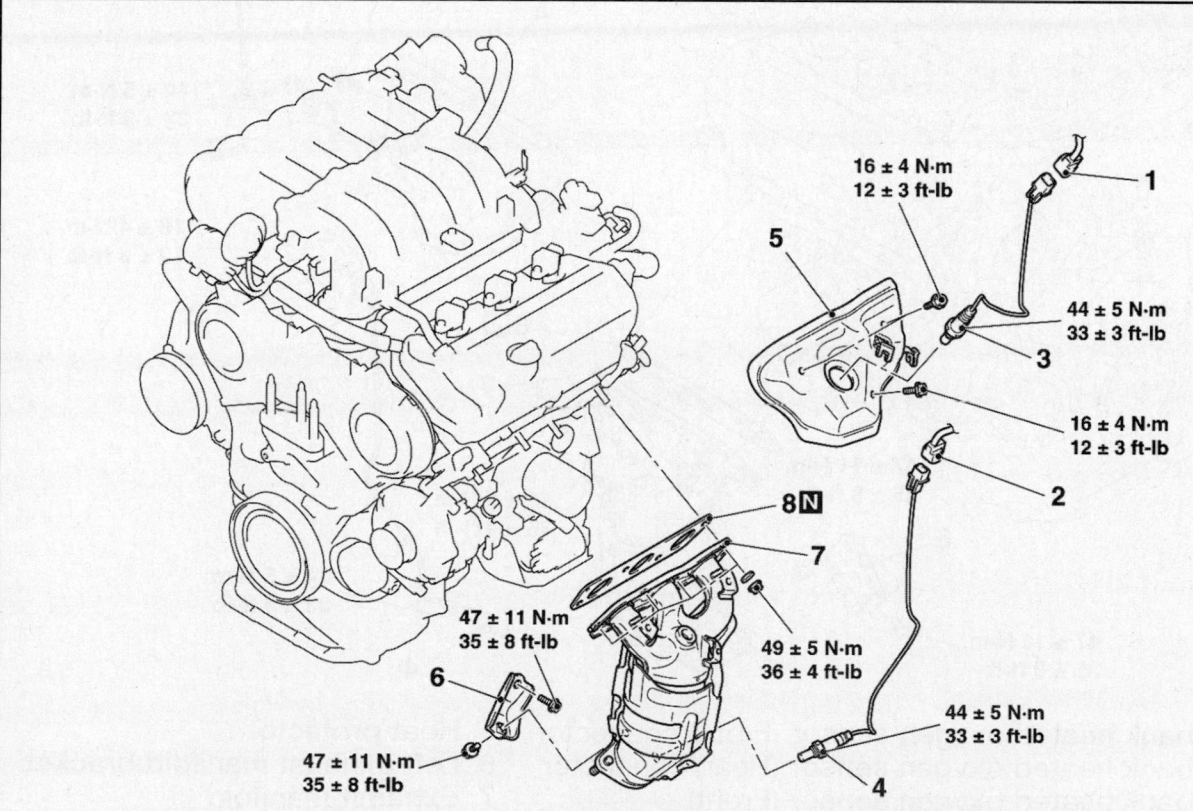

1. Left bank heated oxygen sensor (Front) connector
2. Left bank heated oxygen sensor (Rear) connector
3. Left bank heated oxygen sensor (Front)
4. Left bank heated oxygen sensor (Rear)
5. Heat protector
6. Left exhaust manifold bracket
7. Exhaust manifold
8. Exhaust manifold gasket

Fig. 95 Exploded view of the left bank exhaust manifold—3.0L engine

15. Install the left bank heated oxygen sensor (front). Tighten to 33 ft. lbs. (44 Nm).

16. Connect the left bank heated oxygen sensor (rear) connector.

17. Connect the left bank heated oxygen sensor (front) connector.

18. Install the front exhaust pipe. Tighten to 30 ft. lbs. (41 Nm).

19. Install the air duct.

20. Install the engine under cover.

Right Bank
See Figure 96.

1. Remove the air cleaner cover and air intake duct.

2. Remove the battery.

3. Remove the engine under cover.

4. Remove the front exhaust pipe, center exhaust pipe and front floor front panel heat protector.

5. Remove the strut tower bar.

6. Remove the right exhaust manifold bracket.

7. Disconnect the right bank heated oxygen sensor (front) connector.

8. Remove the right bank heated oxygen sensor (rear) connector.

9. Remove the right bank heated oxygen sensor (front).

10. Remove the right bank heated oxygen sensor (rear).

11. Remove the heat protector.

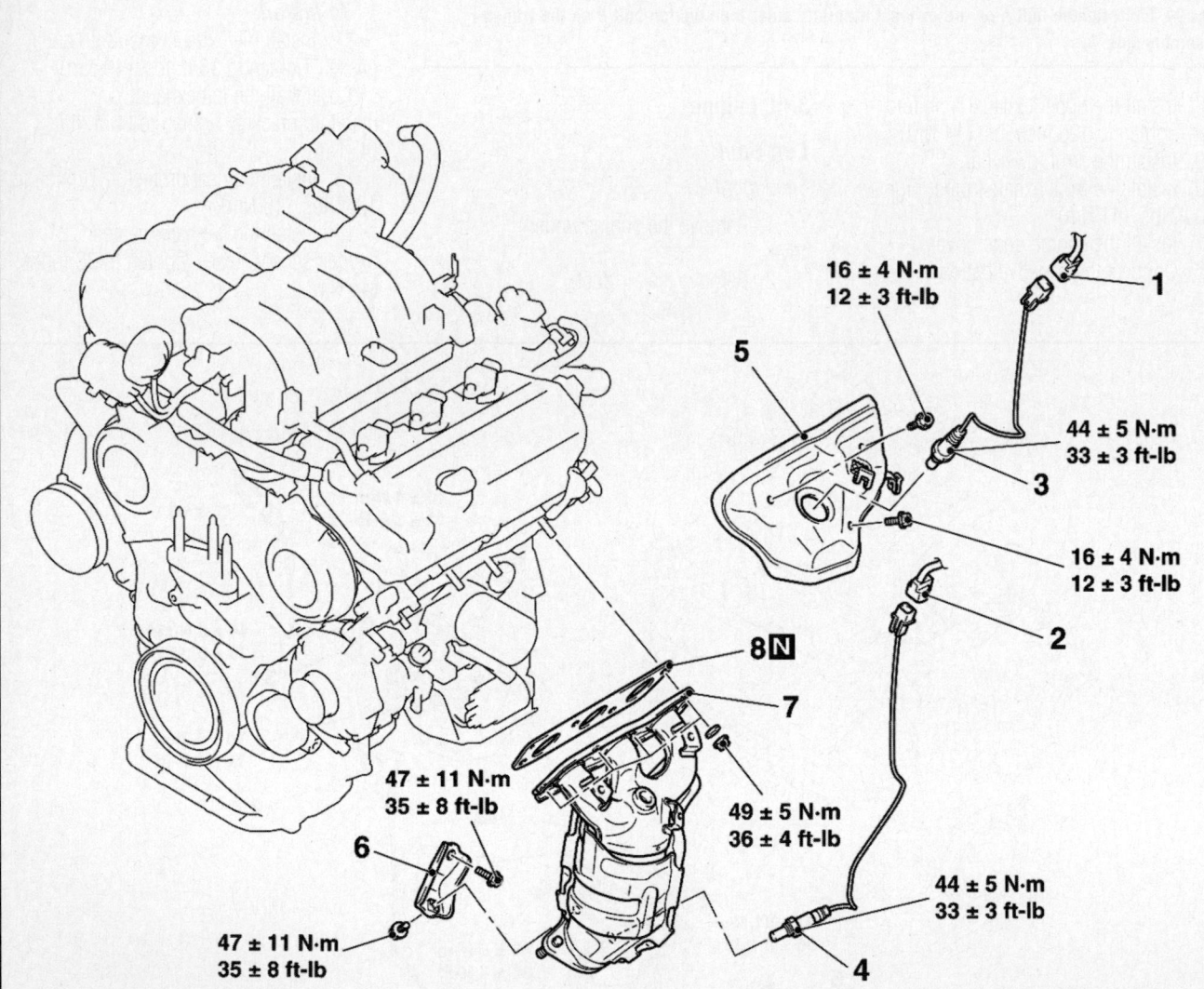

16 ± 4 N·m
12 ± 3 ft-lb

44 ± 5 N·m
33 ± 3 ft-lb

16 ± 4 N·m
12 ± 3 ft-lb

47 ± 11 N·m
35 ± 8 ft-lb

49 ± 5 N·m
36 ± 4 ft-lb

47 ± 11 N·m
35 ± 8 ft-lb

44 ± 5 N·m
33 ± 3 ft-lb

1. Left bank heated oxygen sensor (Front) connector
2. Left bank heated oxygen sensor (Rear) connector
3. Left bank heated oxygen sensor (Front)
4. Left bank heated oxygen sensor (Rear)
5. Heat protector
6. Left exhaust manifold bracket
7. Exhaust manifold
8. Exhaust manifold gasket

22140_OUTL_G0171

Fig. 96 Exploded view of the right bank exhaust manifold—3.0L engine

12. Remove the exhaust manifold and gasket.

To install:

13. Install the exhaust manifold and gasket. Tighten to 36 ft. lbs. (49 Nm).

14. Install the heat protector. Tighten to 12 ft. lbs. (16 Nm).

15. Install the right bank heated oxygen sensor (rear). Tighten to 33 ft. lbs. (44 Nm).

16. Install the right bank heated oxygen sensor (front). Tighten to 33 ft. lbs. (44 Nm).

17. Install the right bank heated oxygen sensor (rear) connector.

18. Connect the right bank heated oxygen sensor (front) connector.

19. Install the right exhaust manifold bracket. Tighten to 35 ft. lbs. (47 Nm).

20. Install the strut tower bar.

21. Install the front exhaust pipe and center exhaust pipe. Tighten to 37 ft. lbs. (50 Nm).

22. Install the floor panel heat protector.

23. Install the engine under cover.

24. Install the battery.

25. Install the air cleaner cover and air intake duct.

FLEXPLATE

REMOVAL & INSTALLATION

See Figures 97 through 99.

1. Remove the transaxle.

2. Use special tool MB991883 on 2.4L engine and MD998781 on 3.0L engine to secure the flywheel and adapter plate.

3. Remove the flywheel bolts, then remove the flywheel.

To install:

4. Remove the sealant, the engine oil, and other adhering materials from the flywheel and adapter plate installation face, the crankshaft screw hole and flywheel bolts.

5. Install the flywheel and adapter plate to the crankshaft.

6. Use special tool MB991883 on 2.4L engine and MD998781 on 3.0L engine to secure the flywheel and adapter plate in the same manner as removal.

7. Apply a small amount of engine oil to the screw holes of the crankshaft and the bearing surface of the flywheel bolts and the adapter plate bolts.

8. Apply Three bond 1324 or equivalent

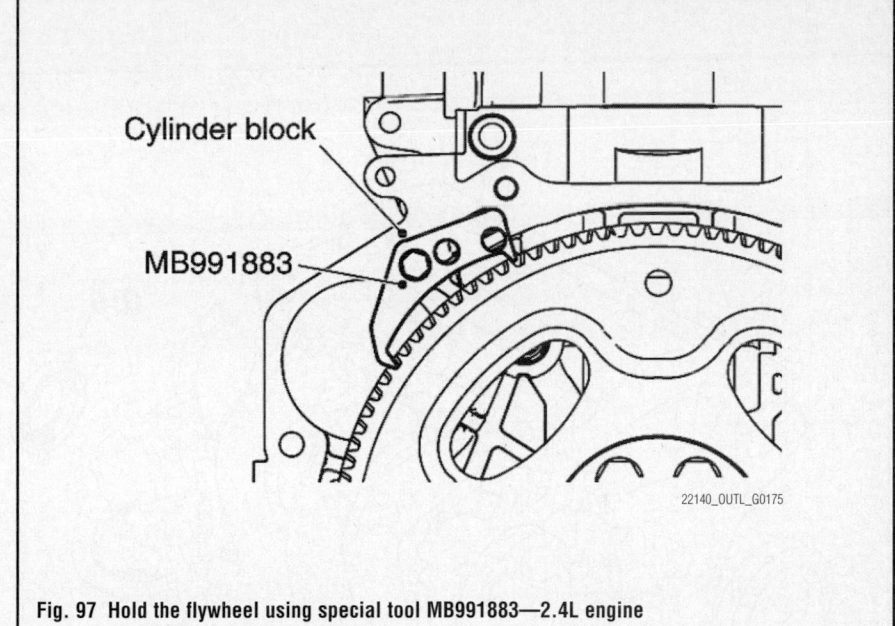

Fig. 97 Hold the flywheel using special tool MB991883—2.4L engine

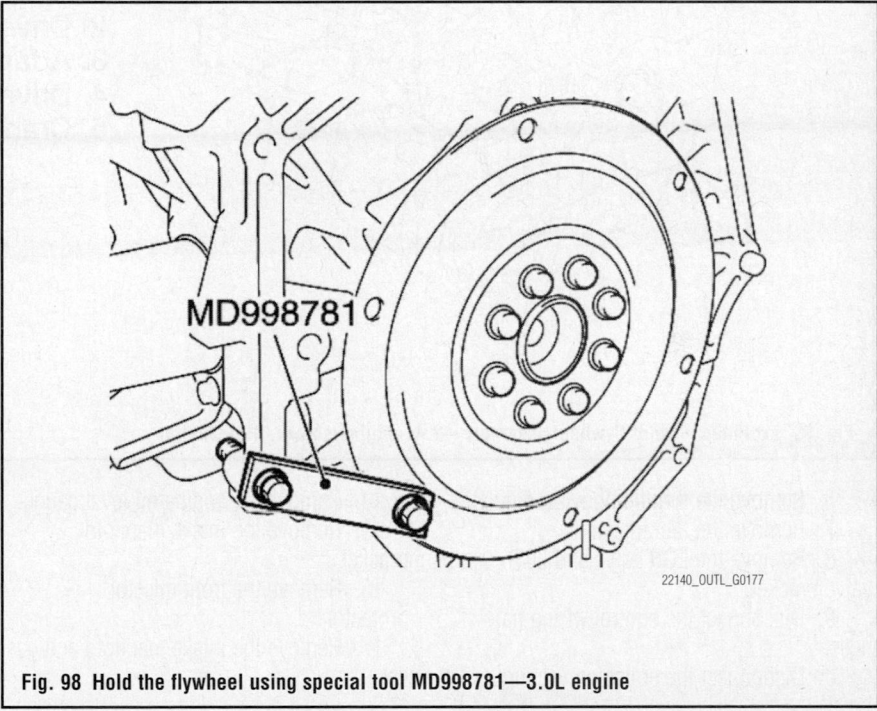

Fig. 98 Hold the flywheel using special tool MD998781—3.0L engine

sealant to the thread of the flywheel bolts and the flywheel bolts.

9. On 2.4L engines, tighten flywheel bolts as follows:

a. Tighten flywheel bolts to temporary torque 30 ft. lbs. (40 Nm) in a star pattern.

b. Tighten flywheel bolts to 96 ft. lbs. (130 Nm) in a star pattern.

10. On 3.0L engines, tighten flywheel bolts to 56 ft. lbs. (76 Nm) in a star pattern.

11. Install the transaxle.

INTAKE MANIFOLD

REMOVAL & INSTALLATION

2.4L Engine

See Figure 100.

1. Before servicing the vehicle, refer to the precautions.

2. Drain the engine coolant.

3. Remove the upper engine cover.

4. Remove the drive belt.

5. Remove the air cleaner intake hose and air cleaner assembly.

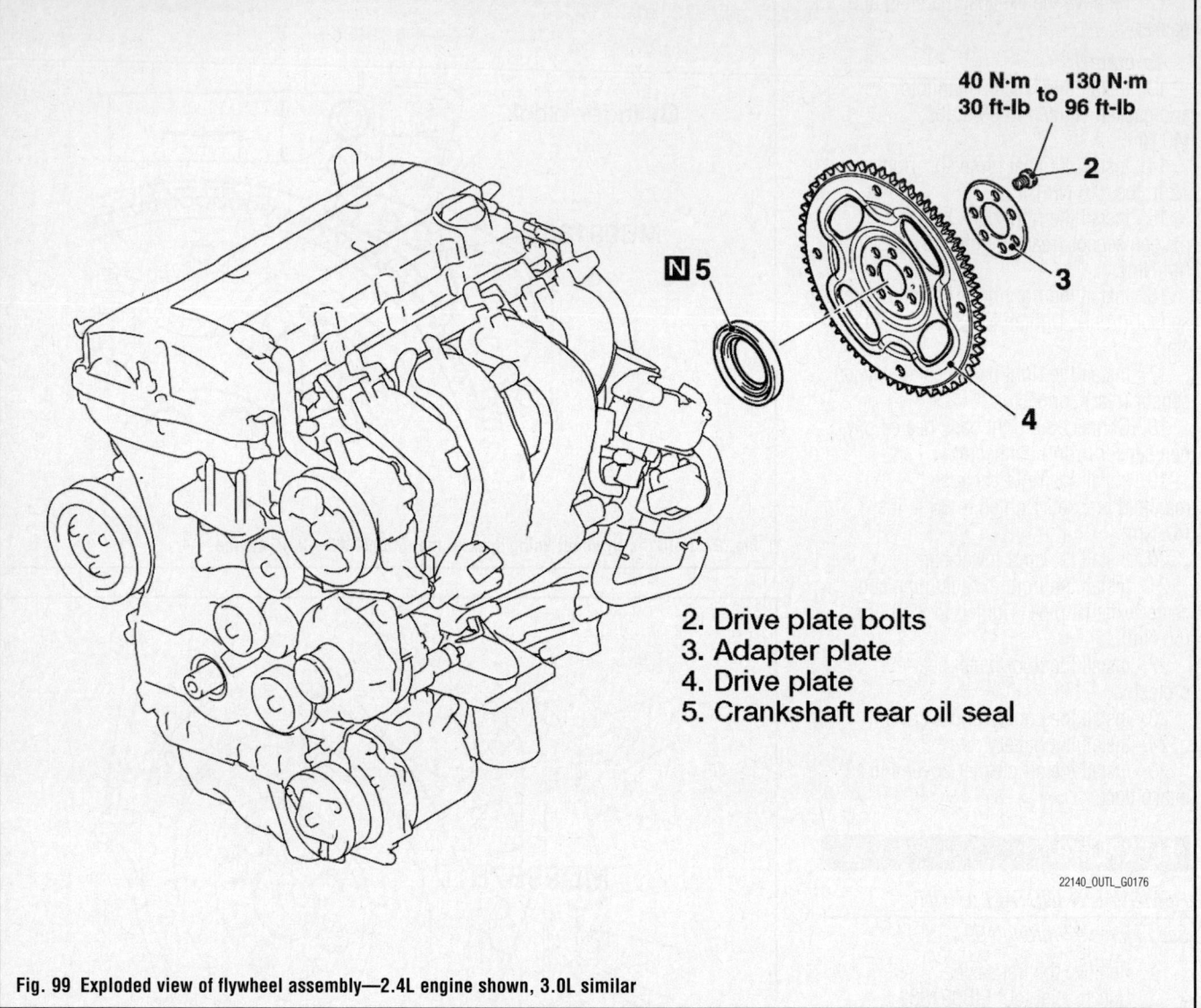

40 N·m to 130 N·m
30 ft-lb to 96 ft-lb

2. Drive plate bolts
3. Adapter plate
4. Drive plate
5. Crankshaft rear oil seal

22140_OUTL_G0176

Fig. 99 Exploded view of flywheel assembly—2.4L engine shown, 3.0L similar

6. Remove the throttle body assembly.
7. Remove the fuel injectors.
8. Remove the EGR valve and EGR valve bracket.
9. Disconnect the control wiring harness.
10. Disconnect the emission vacuum hose.
11. Disconnect the brake booster vacuum hose
12. Remove the power steering oil pump.
   a. With the hose installed, remove the power steering oil pump assembly from the bracket.
   b. Tie the removed power steering oil pump assembly with wire at a position where it will not interfere with the removal and installation of the intake manifold.
13. Disconnect the valve cover PCV hose.

14. Remove the engine oil level gauge.
15. Remove the intake manifold bracket.
16. Remove the front injector protector.
17. Remove the intake manifold and gasket.

To install:
18. Install the intake manifold and gasket.
   a. Install the intake manifold assembly and the injector protector front, and tighten mounting bolts and nuts temporarily.

➡**The tightening of the fuel delivery pipe, the intake manifold assembly and the injector protector front and rear has the specified order. Temporarily tighten the intake manifold assembly and injector protector front mounting bolts and nuts.**

   b. Loosen the intake manifold mounting bolts and nuts 1, 2, 3, and 9 as shown.
   c. Remove the EGR valve.
   d. Loosen the EGR valve support and intake manifold coupling bolts.
   e. Loosen the intake manifold bracket mounting bolts.
   f. Temporarily tighten the mounting bolts and nuts of the intake manifold, bracket, fuel rail and injector protector rear to 31 inch lbs. (4 Nm) in the order shown.
   g. Tighten the mounting bolts and nuts of the intake manifold, bracket, fuel rail and injector protector rear to 15 ft. lbs. (20 Nm) in the order shown.
   h. Tighten the intake manifold bracket mounting bolts to 15 ft. lbs. (20 Nm).
   i. Tighten the EGR valve support and

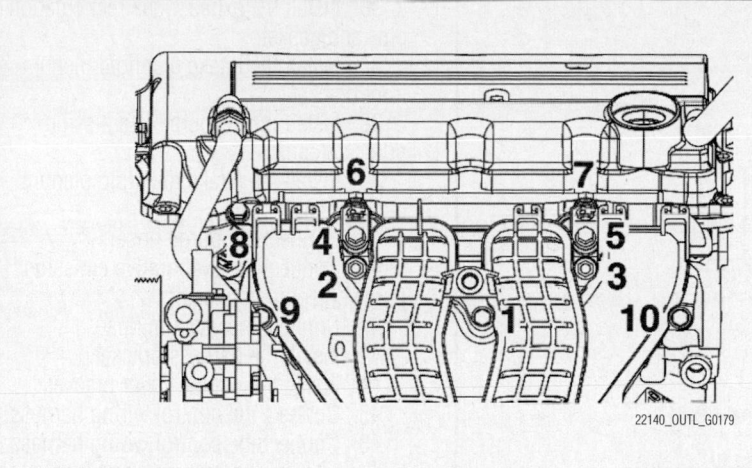

Fig. 100 Loosen the intake manifold mounting bolts and nuts 1, 2, 3, and 9 as shown

intake manifold coupling bolts to 15 ft. lbs. (20 Nm).

   j. Install the EGR valve to 18 ft. lbs. (24 Nm).

19. Install the engine oil level gauge.

20. Connect the valve cover PCV hose.

21. Install the power steering oil pump.

22. Connect the brake booster vacuum hose

23. Connect the emission vacuum hose.

24. Connect the control wiring harness.

25. Install the EGR valve and EGR valve bracket.

26. Install the fuel injectors.

27. Install the throttle body assembly.

28. Install the air cleaner intake hose and air cleaner assembly.

29. Install the drive belt.

30. Install the upper engine cover.

31. Fill the engine coolant.

3.0L Engine

Lower

See Figures 101 through 103.

1. Before servicing the vehicle, refer to the precautions.

2. Discharge the fuel system pressure.

3. Remove the intake manifold plenum.

4. Disconnect the injector connector.

5. Disconnect the fuel high-pressure hose.

> ❊❊ **WARNING**
>
> **So not kink the fuel high-pressure hose as it is made of plastic.**

   a. Insert a flat-tipped screwdriver [width 0.24 inch (6 mm), thickness 0.04 inch (1 mm)] to the retainer.

   b. Turn the flat-tipped screwdriver approximately 90 degrees to the right, and lift the retainer to unlock.

6. Remove the fuel rail and fuel injector assemblies.

> ❊❊ **WARNING**
>
> **Do not drop the fuel injectors.**

➡ **Remove the fuel rail with the fuel injectors still attached.**

7. Disconnect the harness bracket.

8. Remove the intake manifold and gasket.

To install:

9. Install the intake manifold and gasket.

   a. Coat the intake manifold mounting studs with engine oil.

   b. Tighten the intake manifold mounting nuts in the following order:

- Left bank nuts to 58 inch lbs. (7 Nm)

- Right bank nuts to 16 ft. lbs. (22 Nm)
- Left bank nuts to 16 ft. lbs. (22 Nm)
- Right bank nuts to 16 ft. lbs. (22 Nm)
- Left bank nuts to 16 ft. lbs. (22 Nm)

10. Connect the harness bracket.

11. Install the fuel rail and fuel injector assemblies.

12. Connect the fuel high-pressure hose.

   a. Pull up the lock of fuel high-pressure hose to unlock before installing.

   b. Install the fuel high-pressure hose to the fuel rail securely and push the lock of fuel high-pressure hose downward and lock thoroughly.

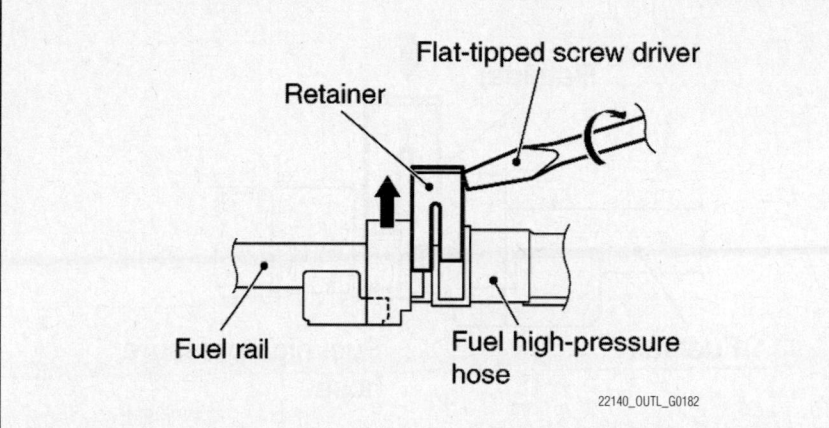

Fig. 101 Turn the flat-tipped screwdriver approximately 90 degrees to the right, and lift the retainer to unlock

   c. After installing, slightly pull the fuel high-pressure hose and ensure that there is no disengaged fuel high-pressure hose. Also confirm that there is approximately 0.04 inch (1 mm) play at this time.

13. Connect the injector connector.

14. Install the intake manifold plenum.

Intake Manifold Plenum

1. Before servicing the vehicle, refer to the precautions.

2. Drain the engine coolant

3. Remove the engine cover.

4. Remove the strut tower bar.

5. Remove the air cleaner assembly.

6. Remove the throttle body.

7. Disconnect the manifold absolute pressure sensor connector.

8. Remove the manifold absolute pressure sensor.

9. Remove the O-ring.

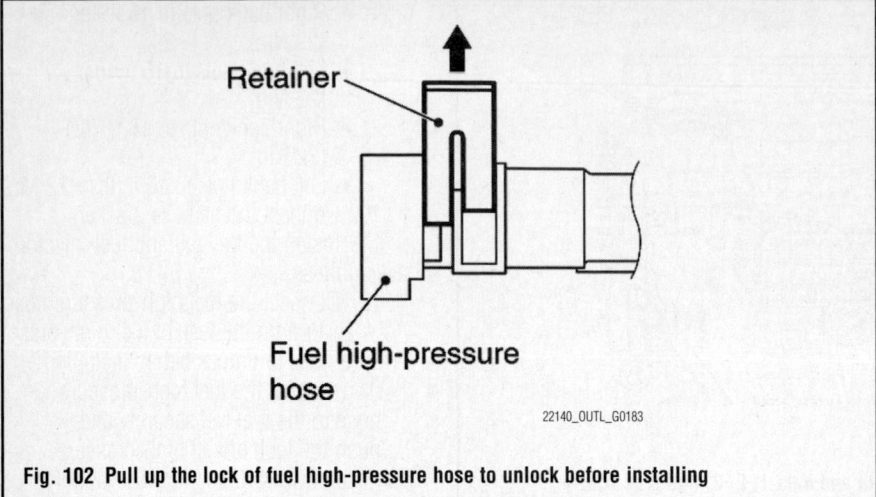

Fig. 102 Pull up the lock of fuel high-pressure hose to unlock before installing

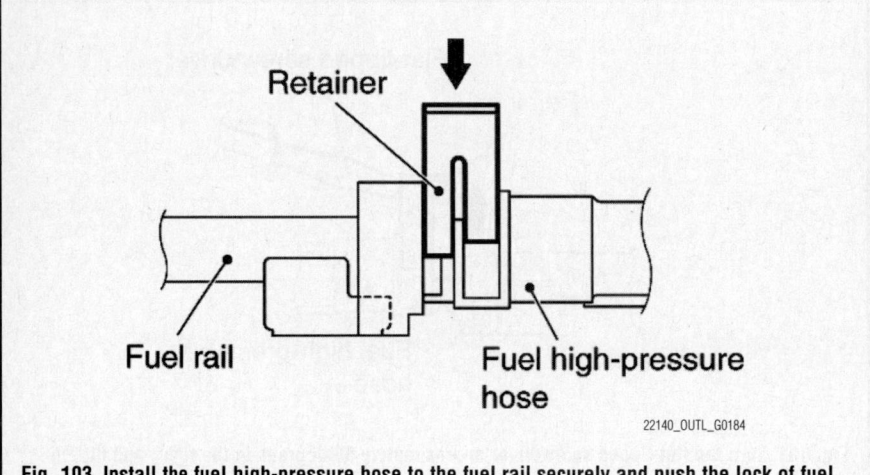

Fig. 103 Install the fuel high-pressure hose to the fuel rail securely and push the lock of fuel high-pressure hose downward and lock thoroughly

10. Disconnect the variable intake air control solenoid valve connector.

11. Disconnect the right bank heated oxygen sensor connectors.

12. Remove the right bank heated oxygen sensors.

13. Disconnect the purge hose.

14. Disconnect the vacuum hose.

15. Remove the vacuum tank and bracket.

16. Remove the variable intake air control solenoid valve.

17. Disconnect the left bank knock sensor connector

18. Remove the left bank knock sensor.

19. Disconnect the right bank knock sensor connector

20. Remove the right bank knock sensor.

21. Disconnect the control wiring harness and injector wiring harness combination connector.

22. Disconnect the control wiring harness.

23. Remove the engine cover bracket.

24. Remove the harness bracket.

25. Disconnect the vacuum hose.

26. Disconnect the evaporative emission purge solenoid connector

27. Disconnect the purge hose.

28. Remove the intake manifold plenum rear bracket.

29. Remove the evaporative emission purge solenoid.

30. Remove the intake manifold plenum front bracket.

31. Remove the exhaust gas recirculation pipe and gasket.

32. Disconnect the blow-by hose.

33. Remove the intake manifold plenum and gasket.

To install:

34. Install the intake manifold plenum and gasket.

35. Connect the blow-by hose.

36. Install the exhaust gas recirculation pipe and gasket.

37. Install the intake manifold plenum front bracket.

38. Install the evaporative emission purge solenoid.

39. Install the intake manifold plenum rear bracket.

40. Connect the purge hose.

41. Connect the evaporative emission purge solenoid connector

42. Connect the vacuum hose.

43. Install the harness bracket.

44. Install the engine cover bracket.

45. Connect the control wiring harness.

46. Connect the control wiring harness and injector wiring harness combination connector.

47. Install the right bank knock sensor.

48. Connect the right bank knock sensor connector

49. Install the left bank knock sensor.

50. Connect the left bank knock sensor connector

51. Install the variable intake air control solenoid valve.

52. Install the vacuum tank and bracket.

53. Connect the vacuum hose.

54. Connect the purge hose.

55. Install the right bank heated oxygen sensors.

56. Connect the right bank heated oxygen sensor connectors.

57. Connect the variable intake air control solenoid valve connector.

58. Install the O-ring.

59. Install the manifold absolute pressure sensor.

60. Connect the manifold absolute pressure sensor connector.

61. Install the throttle body.

62. Install the air cleaner assembly.

63. Install the strut tower bar.

64. Install the engine cover.

65. Fill the engine coolant.

66. Start the engine and check for leaks.

OIL PAN

REMOVAL & INSTALLATION

2.4L Engine

See Figures 104 through 106.

1. Before servicing the vehicle, refer to the precautions.

2. Remove the engine under cover and engine side cover.

3. Drain the engine oil.

4. Remove the drive belt.

5. Remove the A/C compressor and clutch assembly

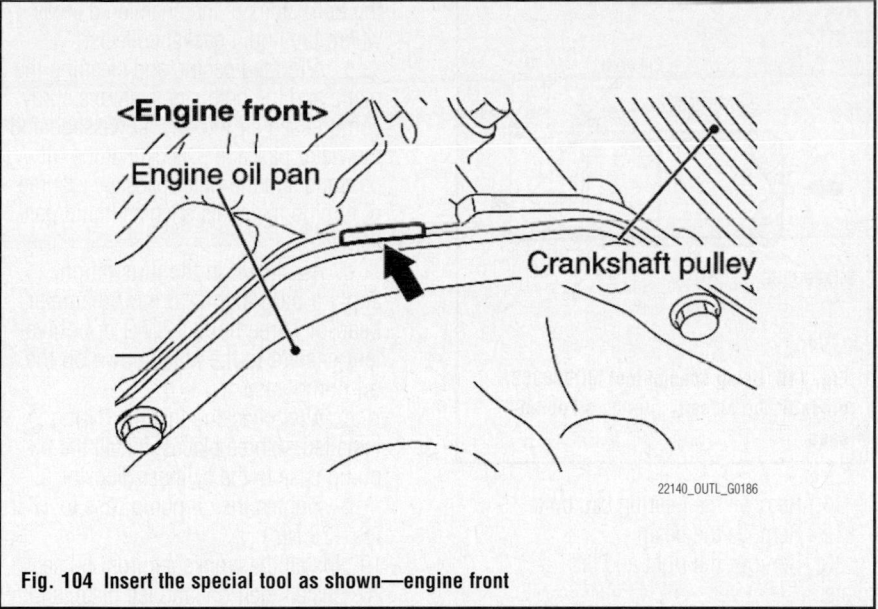

Fig. 104 Insert the special tool as shown—engine front

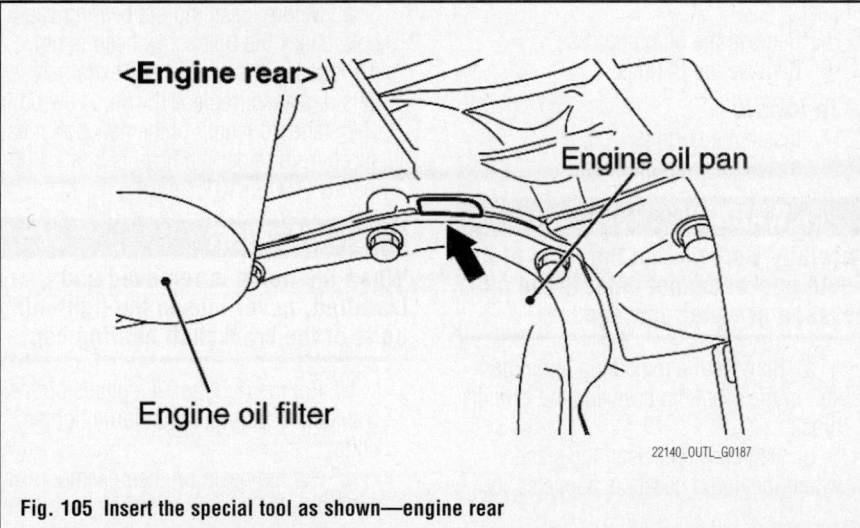

Fig. 105 Insert the special tool as shown—engine rear

a. Remove the A/C compressor and clutch assembly together with the hose from the bracket.

b. Tie the removed A/C compressor and clutch assembly with a string at a position where they will not interfere with the removal and installation of engine oil pan.

6. Remove the A/C compressor bracket.

7. Remove the engine oil pan.

a. Remove the engine oil pan mounting bolts.

❊❊ WARNING

Do not forcibly drive in special tool MD998727 to avoid damage to the engine oil pan seal surface of cylinder block assembly.

b. Insert special tool MD998727 from the engine oil pan removal groove of the cylinder block assembly.

c. Lightly tap the special tool with a hammer to slide the oil pan seal surface, cut off the liquid gasket, and remove the engine oil pan.

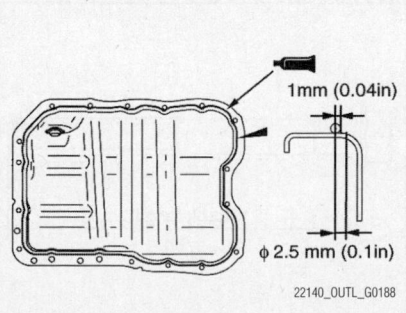

Fig. 106 Apply the sealant without any gap to the mating surface of engine oil pan as shown

To install:

8. Install the engine oil pan.

a. Remove all the traces of sealant adhering to the engine oil pan and cylinder block assembly using a remover or others. Then, degrease them using white gasoline.

b. Apply the sealant without any gap to the mating surface of engine oil pan as shown in the figure. Within three minutes, install the engine oil pan to the cylinder block assembly. Specified sealant: Three bond 1217G or equivalent

c. Do not apply oil or water to the sealant-applied area or start up the engine within 2 hours after the installation of the engine oil pan.

d. Tighten the M6 89 inch lbs. (10 Nm) and the M8 to 22 ft. lbs. (29 Nm).

9. Install the A/C compressor bracket.

10. Install the A/C compressor and clutch assembly. Tighten bolts to 17 ft. lbs. (23 Nm) in a counter clockwise direction from the top.

11. Install the drive belt.

12. Replace the engine oil pan drain plug gasket with a new one. Install the new gasket with the chamfer toward the engine oil pan.

13. Fill the engine with oil.

14. Install the engine under cover and engine side cover.

15. Start the engine and check for leaks.

3.0L Engine

See Figures 107 through 117.

1. Before servicing the vehicle, refer to the precautions.

2. Remove the engine under cover and engine side cover.

3. Drain the engine oil.

4. Remove the oil filter and cover.

5. Remove the heat protector.

6. Remove the engine oil pressure switch.

7. Remove the lower oil pan.

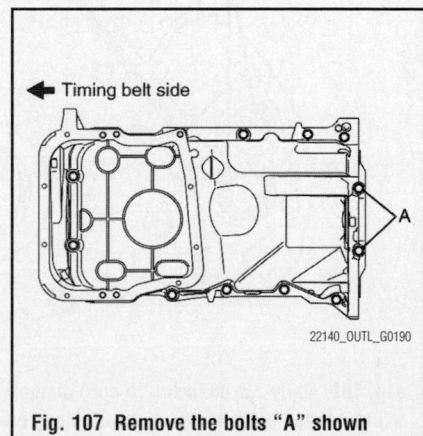

Fig. 107 Remove the bolts "A" shown

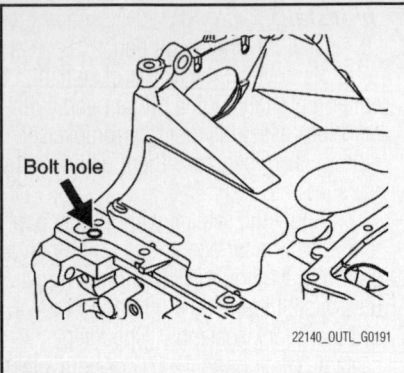

Fig. 108 Thread the M10 × 1.25 pitch bolt into the illustrated bolt hole to remove the oil pan upper

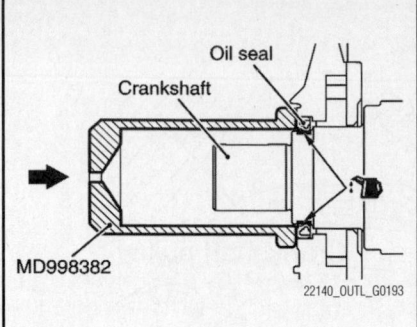

Fig. 110 Using special tool MD998382, press-fit the oil seal into the oil pump case

---

✳✳ WARNING

To prevent the sealed area from being damaged, carefully insert the special tool.

a. Insert the special tool MD998727, into the groove. Strike and slide it and then cut the liquid gasket.

8. Remove the oil pan cover.
9. Remove the upper oil pan.

a. Remove the bolts A shown in the illustration first.

b. Remove all other bolts.

✳✳ WARNING

Do not use a scraper or special tool to remove the oil pan upper.

c. Thread the M10 _ 1.25 pitch bolt into the illustrated bolt hole to remove the oil pan upper.

10. Remove the oil screen and gasket.

---

11. Remove the bearing cap bolt.
12. Remove the beam.
13. Remove the right and left plate.
14. Remove the crankshaft front oil seal.
15. Remove the oil pump case.
16. Remove the O-rings.

To install:

17. Install the O-rings.
18. Install the oil pump case.

✳✳ WARNING

Carefully work so that the rests of the liquid gasket cannot enter to the oil passage or water passage.

a. Remove the remaining liquid gasket on the oil pump case and the cylinder block.

b. Degrease and clean the plane where the liquid gasket is applied and

---

the both sides of the chamfered areas where the liquid gasket collects.

c. After degreasing and cleaning the plane and the both sides, always spray compressed air into the oil passage and the water passage. Check that the oil passage and the water passage are free of foreign materials such as liquid gasket.

d. As shown in the illustration, apply a 0.08 inch (2.0 mm) diameter bead of Three bond 1217G or equivalent sealant to the area shown on the oil pump case.

e. After checking that the O-ring is installed at three places, install the oil pump case to the cylinder block.

f. Tighten the oil pump case to 17 ft. lbs. (23 Nm).

19. Install the crankshaft front oil seal.
20. Install the right and left plate, and the beam.

a. Before installing the bearing cap bolt, check the bolt screw head is not damaged. If the screw head extremely gets damaged, replace it with a new bolt. The standard length of the new bolt measured from under the head is 3.97–4.00 inches (100.7–101.7 mm).

✳✳ WARNING

When the beam is removed and installed, never loosen the tightening bolts of the crankshaft bearing cap.

b. Put the right and left plates on the crankshaft bearing, temporarily tighten bolts.

c. Put the beam on them, Temporarily tighten bolt.

d. Tighten the beam bolts to 18 ft. lbs. (24 Nm) in the sequence shown.

✳✳ WARNING

When the tightening angle is smaller than the specified

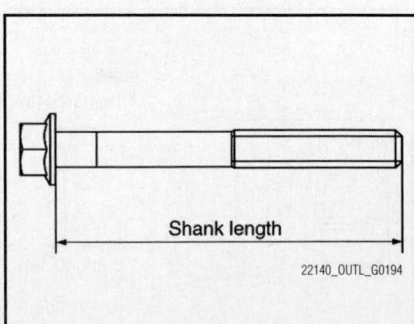

Fig. 111 The standard length of the new bolt measured from under the head is 3.97–4.00 inches (100.7–101.7 mm)

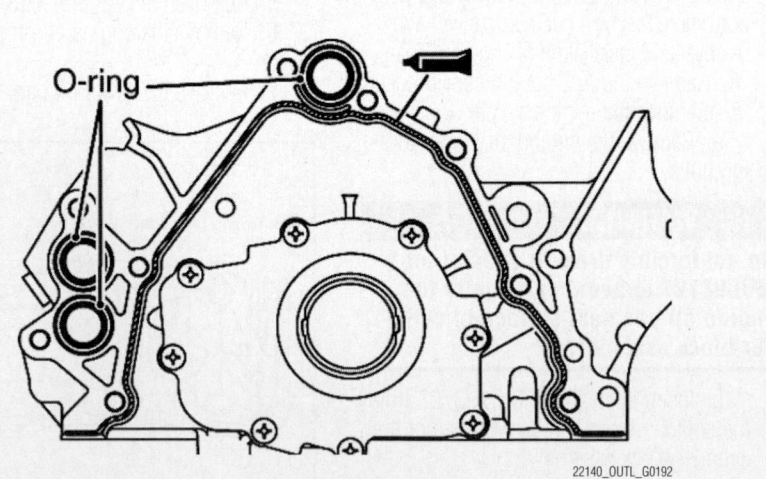

Fig. 109 Apply a 0.08 inch (2.0 mm) diameter bead of Three bond 1217G or equivalent sealant to the area shown on the oil pump case

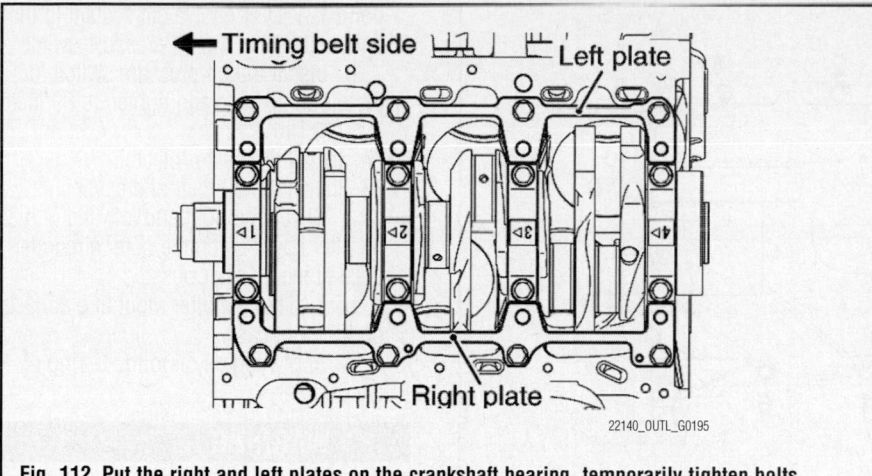

Fig. 112 Put the right and left plates on the crankshaft bearing, temporarily tighten bolts

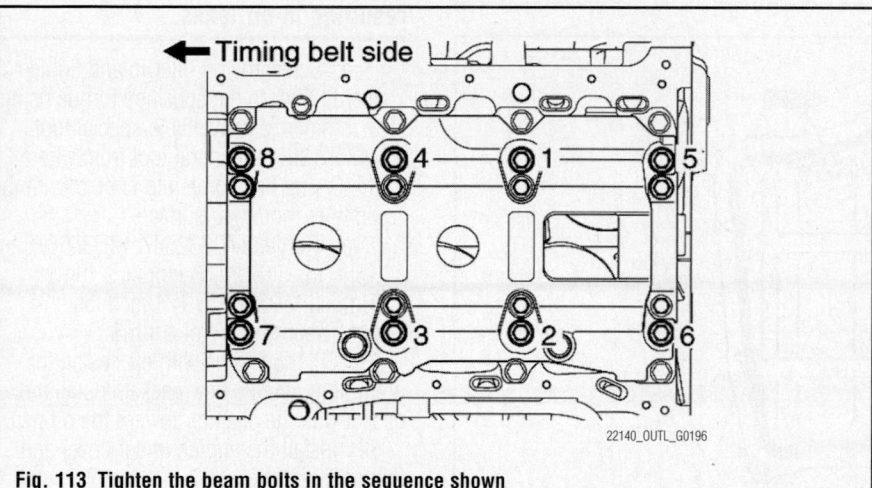

Fig. 113 Tighten the beam bolts in the sequence shown

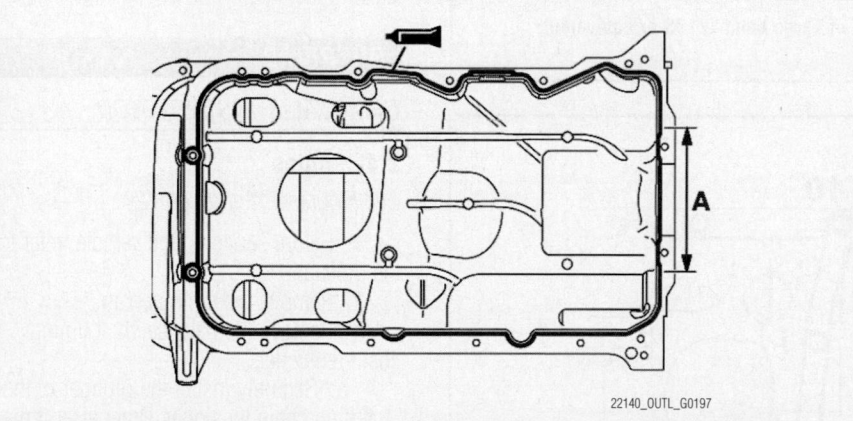

Fig. 114 Apply a 0.10 inch (2.5 mm) diameter bead of Three bond 1217G or equivalent sealant to the oil pan

tightening angle, the appropriate tightening capacity cannot be secured. When the tightening angle is larger than the specified tightening angle, remove the bolt to start from the beginning again according to the procedure.

✳✳ WARNING

Using a torque angle meter, tighten the beam bolt (M9) 60 degrees in the specified order.

    e. Tighten the plate bolts to 17 ft. lbs. (23 Nm) in the sequence shown.

21. Install the bearing cap bolt.
22. Install the oil screen and gasket.
23. Install the upper oil pan.
    a. Clean the gasket surfaces of the oil pan upper and oil pan lower.

✳✳ WARNING

When installing the upper oil pan, be sure not to expel the sealant from the oil pan flange at portion A.

    b. Apply a 0.10 inch (2.5 mm) diameter bead of Three bond 1217G or equivalent sealant to the oil pan.
    c. Be sure to install the oil pan quickly while the sealant is wet. Install the oil pan within 15 minutes after applying liquid gasket.
    d. Then wait at least one hour. Never start the engine or let engine oil or coolant touch the adhesion surface during that time.
    e. Tighten the upper oil pan bolts to 89 ft. lbs. (10 Nm) in the sequence shown.
    f. After installation, keep the sealed area away from the oil and coolant for approximately one hour.
24. Install the oil pan cover.
25. Install the lower oil pan.
    a. Clean the gasket surfaces of the cylinder block and oil pan lower.
    b. Apply a 2.5 ± 0.5 mm (0.10 ± 0.01 inch) diameter bead of sealant (Three bond 1217G or exact equivalent) to the oil pan. Be sure to install the oil pan quickly while the sealant is wet.

✳✳ WARNING

Install the oil pan within 15 minutes after applying liquid gasket. Then wait at least one hour. Never start the engine or let engine oil or coolant touch the adhesion surface during that time.

    c. Tighten the upper oil pan bolts to 89 inch lbs. (10 Nm) in the sequence shown.
    d. After installation, keep the sealed area away from the oil and coolant for approximately one hour.
26. Install the engine oil pressure switch.
    a. Completely remove existing sealant from the oil pressure switch and the switch mounting hole on the oil pump case.
    b. Apply Three bond 1215, Three

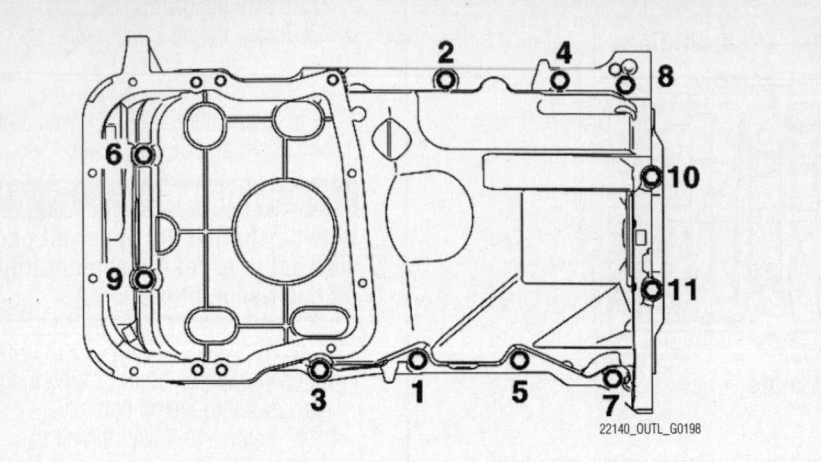

Fig. 115 Tighten the upper oil pan bolts in the sequence shown

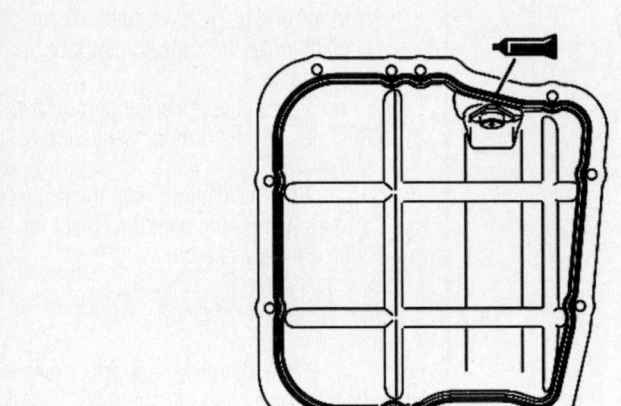

Fig. 116 Apply a 0.10 inch (2.5 mm) diameter bead of Three bond 1217G or equivalent sealant to the oil pan

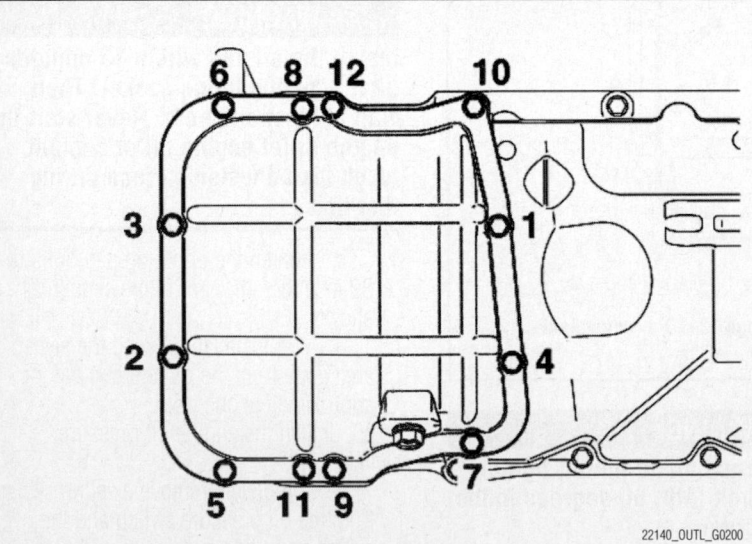

Fig. 117 Tighten the upper oil pan bolts in the sequence shown

bond 1212D or equivalent sealant to the threaded part of the oil pressure switch.

c. Install the oil pressure switch to the oil pump case and tighten to 89 inch lbs. (10 Nm).

27. Install the heat protector.

28. Install the oil filter and cover.

a. Check the right and left ribs of the oil filter cover touch the outer circumference of the oil filter boss.

b. 1.Clean the filter mounting surface on the front case.

c. Apply engine oil to the O-ring of the oil filter.

✳✳ WARNING

If the filter is tightened by hand only, it will be insufficiently torqued, resulting in oil leaks.

d. Screw the oil filter in and tighten the oil filter to the specified torque using a commercially available special tool MB991396 or general tool from where the O-ring has come into contact with the oil filter mounting surface.

e. Tighten MD332687, MD365876 to 13 ft. lbs. (16Nm) &approx. 1 turn$. Tighten MD360935 124 inch lbs. (14 Nm) &approx. 3/4 of a turn$.

29. Fill the engine with oil. Install the drain plug using a new gasket. Install the gasket with the chamfer toward the oil pan.

30. Install the engine under cover and engine side cover.

31. Start the engine and check for leaks.

OIL PUMP

REMOVAL & INSTALLATION

2.4L Engine

See Figures 118 through 121.

1. Before servicing the vehicle, refer to the precautions.

2. Remove the timing chain.

3. Remove the balance shaft timing chain tensioner.

a. Securely install the plunger of the timing chain tensioner. Otherwise, it may pop out.

b. Press the balancer timing chain against the timing chain tensioner, compress the plunger of the timing chain tensioner and insert hard wire (piano wire, etc.) or L-shaped hexagon wrench 0.05 inch (1.5 mm) to fix the plunger of the timing chain tensioner.

c. Remove the balance shaft timing chain tensioner.

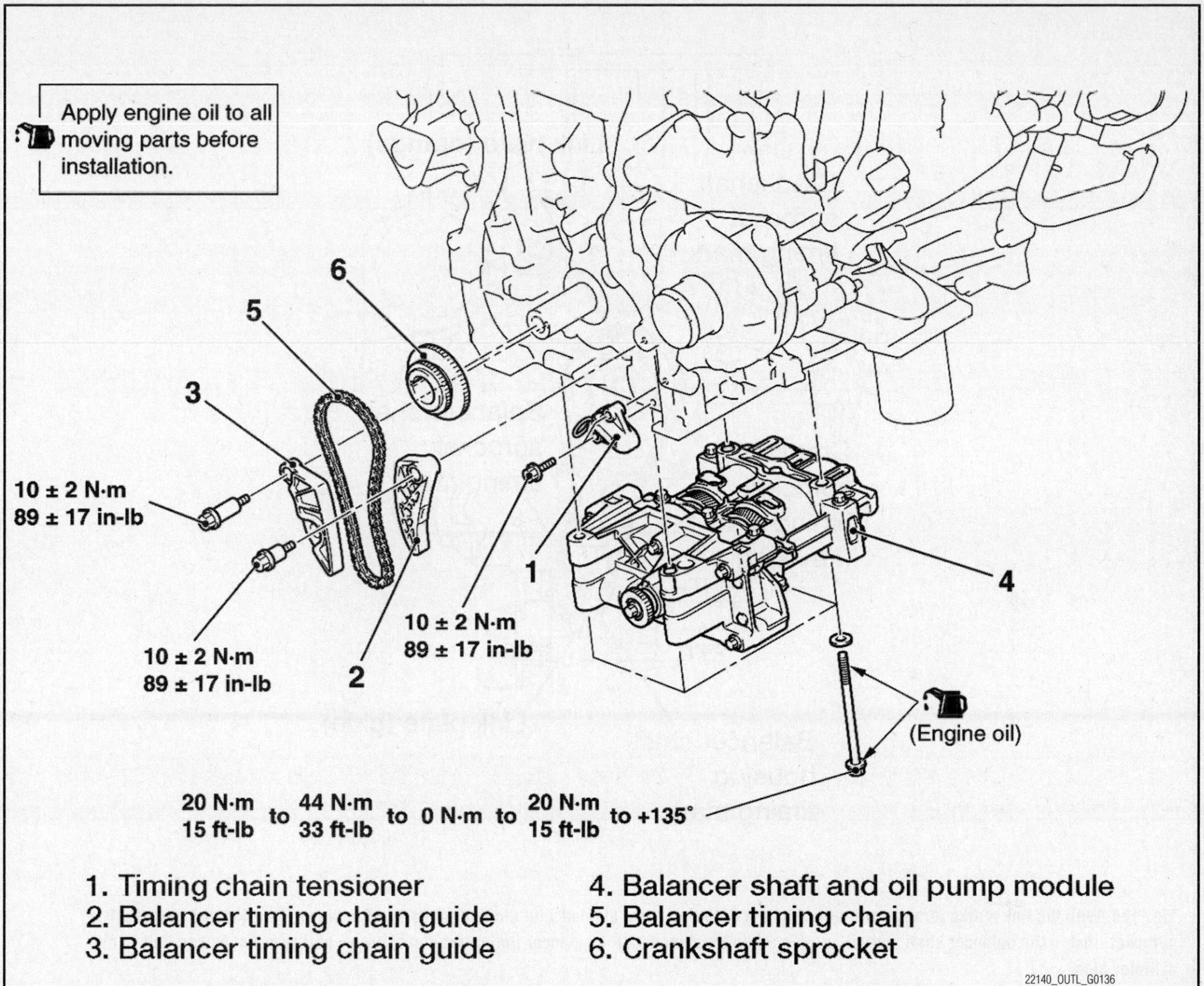

Apply engine oil to all moving parts before installation.

6

5

3

10 ± 2 N·m
89 ± 17 in·lb

10 ± 2 N·m
89 ± 17 in·lb

2

1

10 ± 2 N·m
89 ± 17 in·lb

4

(Engine oil)

| 20 N·m | | 44 N·m | | | 20 N·m | |
| 15 ft·lb | to | 33 ft·lb | to 0 N·m to | | 15 ft·lb | to +135° |

1. Timing chain tensioner
2. Balancer timing chain guide
3. Balancer timing chain guide
4. Balancer shaft and oil pump module
5. Balancer timing chain
6. Crankshaft sprocket

22140_OUTL_G0136

Fig. 118 Balance shaft timing chain and module

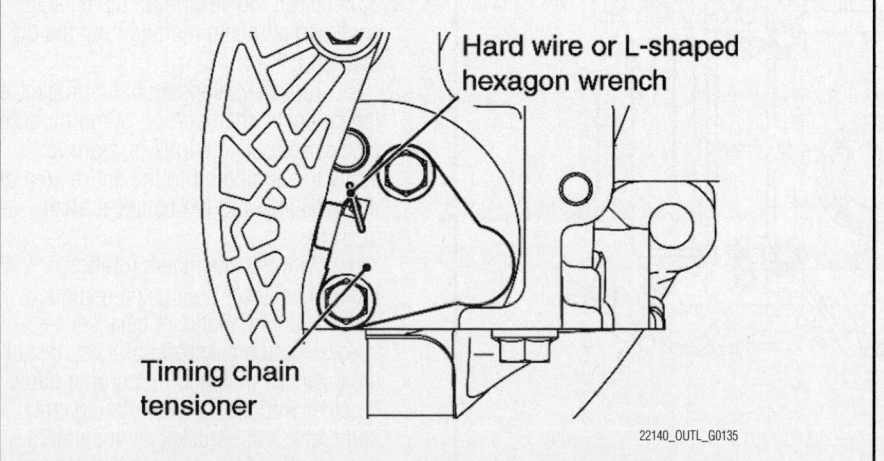

Hard wire or L-shaped hexagon wrench

Timing chain tensioner

22140_OUTL_G0135

Fig. 119 Compress the plunger of the timing chain tensioner and insert hard wire or L-shaped hexagon wrench to fix the plunger of the timing chain tensioner

4. Remove the balance shaft timing chain guides.

5. Remove the balance shaft and oil pump module.

6. Remove the balance shaft timing chain.

To install:

7. Install the balance shaft and oil pump module.

a. When installing the new balancer shaft and oil pump module, apply oil to the oil pump in the balancer shaft and oil pump module and the balancer shaft bearing as follows:

b. Clean the inside of the removed engine oil pan, and put the balancer shaft and oil pump module into the engine oil pan with its oil inlet port facing up.

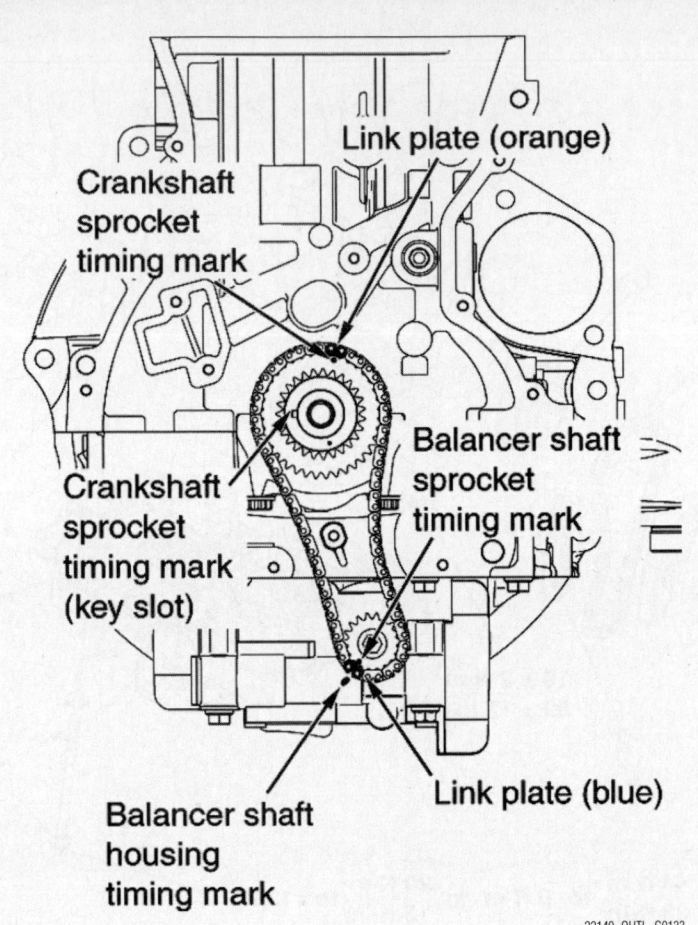

Link plate (orange)

Crankshaft sprocket timing mark

Crankshaft sprocket timing mark (key slot)

Balancer shaft sprocket timing mark

Balancer shaft housing timing mark

Link plate (blue)

22140_OUTL_G0133

Fig. 120 With the link marks (orange or blue) of balancer timing chain aligned with the timing marks of balancer sprocket and crankshaft sprocket, install the balancer shaft and oil pump module together with the balancer timing chain and crankshaft sprocket as one unit to the cylinder block

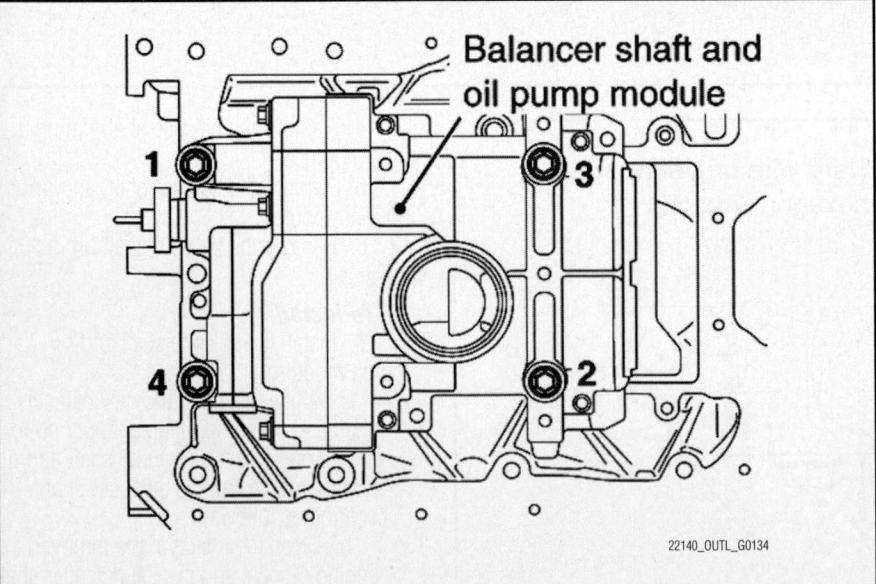

Balancer shaft and oil pump module

22140_OUTL_G0134

Fig. 121 Tighten the balancer shaft and oil pump module bolts to the specified torque in the order shown

c. Pour new engine oil until two-thirds of the balancer shaft and oil pump module is soaked.

d. Fill the engine oil (approximately 3.05 cu. in. [50 cm3]) into the balancer shaft and oil pump module from the oil inlet port.

e. Turn the balancer shaft sprocket of the balancer shaft and oil pump module clockwise four rotations or more to apply the engine oil to the entire area of the oil pump and the balancer shaft bearing.

f. With the link marks (orange or blue) of balancer timing chain aligned with the timing marks of balancer sprocket and crankshaft sprocket, install the balancer shaft and oil pump module together with the balancer timing chain and crankshaft sprocket as one unit to the cylinder block. At this time, securely bring the balancer shaft and oil pump module into contact with the rudder frame mounting area.

g. Apply an adequate and minimum amount of engine oil to the threads and bearing surfaces of the balancer shaft and oil pump module bolts.

h. Tighten the balancer shaft and oil pump module bolts to the specified torque in the order shown:

- Tighten to 15 ft. lbs. (20 Nm)
- Retighten to 15 ft. lbs. (20 Nm)
- Loosen all bolts in the reverse sequence fully
- Tighten to 15 ft. lbs. (20 Nm)
- Tighten to 15 ft. lbs. (20 Nm)

8. Install the balance shaft timing chain guides.

9. Install the timing chain tensioner.

a. Install the timing chain tensioner to the cylinder block.

b. Remove the hard wire or L-shaped hexagon wrench fixing the plunger of the timing chain tensioner to apply tension to the balancer timing chain.

10. Install the timing chain.

3.0L Engine

See Figures 122 through 132.

1. Before servicing the vehicle, refer to the precautions.

2. Remove the engine under cover and engine side cover.

3. Drain the engine oil.

4. Remove the oil filter and cover.

5. Remove the heat protector.

6. Remove the engine oil pressure switch.

7. Remove the lower oil pan.

✳✳ WARNING

To prevent the sealed area from being damaged, carefully insert the special tool.

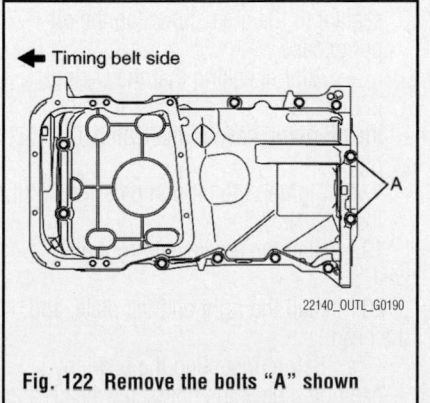

Fig. 122 Remove the bolts "A" shown

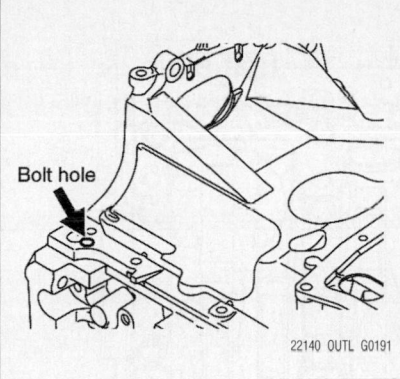

Fig. 123 Thread the M10 × 1.25 pitch bolt into the illustrated bolt hole to remove the oil pan upper

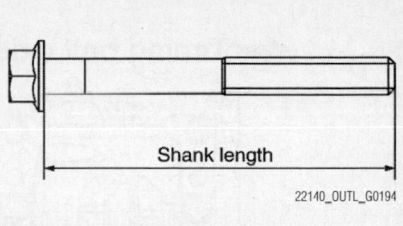

Fig. 126 The standard length of the new bolt measured from under the head is 3.97–4.00 inches (100.7–101.7 mm)

a. Insert the special tool MD998727, into the groove. Strike and slide it and then cut the liquid gasket.

8. Remove the oil pan cover.

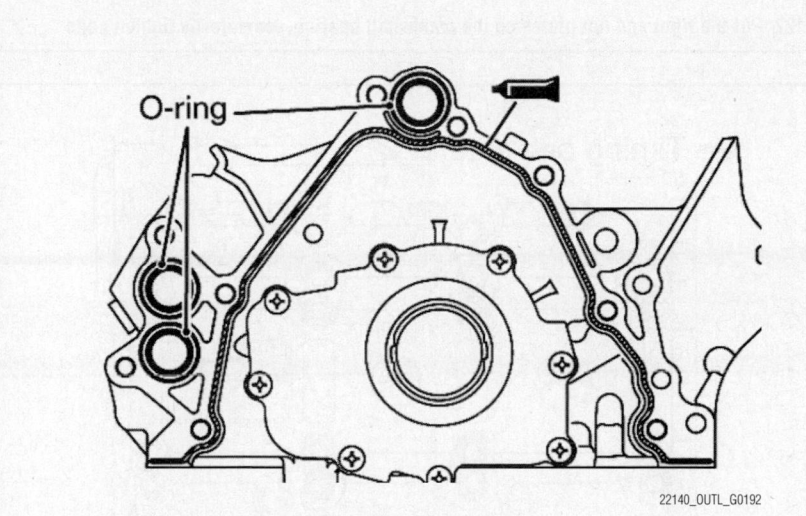

Fig. 124 Apply a 0.08 inch (2.0 mm) diameter bead of Three bond 1217G or equivalent sealant to the area shown on the oil pump case

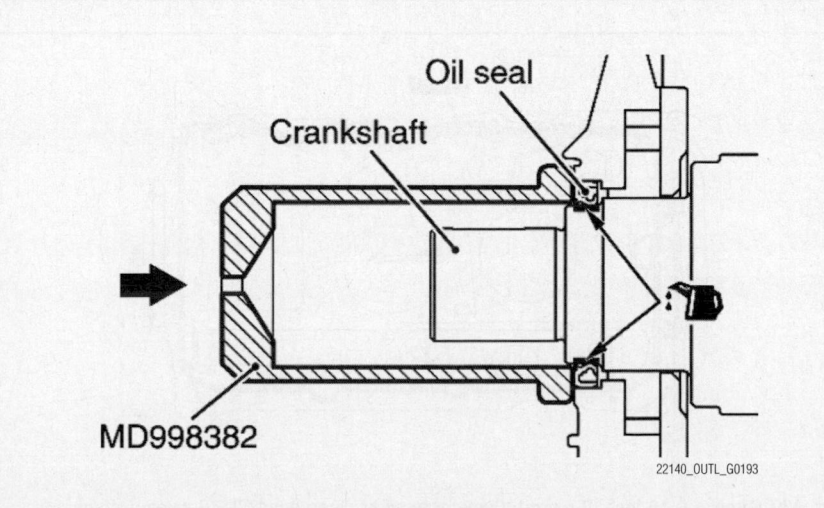

Fig. 125 Using special tool MD998382, press-fit the oil seal into the oil pump case

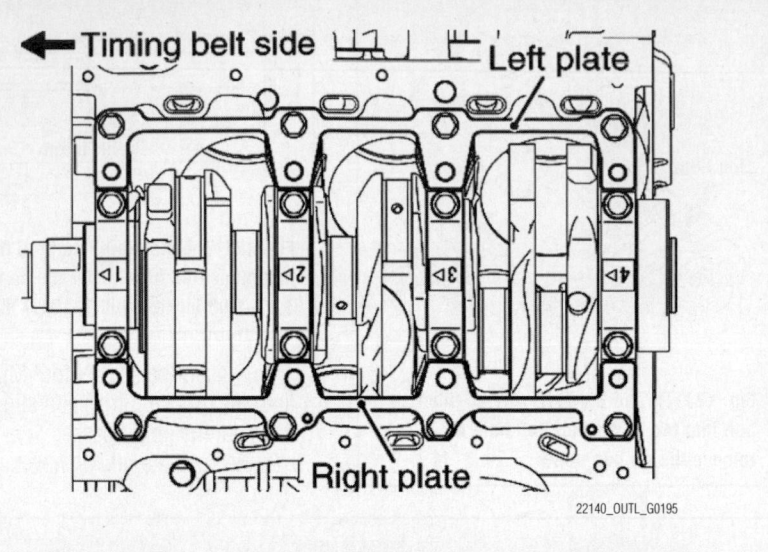

Fig. 127 Put the right and left plates on the crankshaft bearing, temporarily tighten bolts

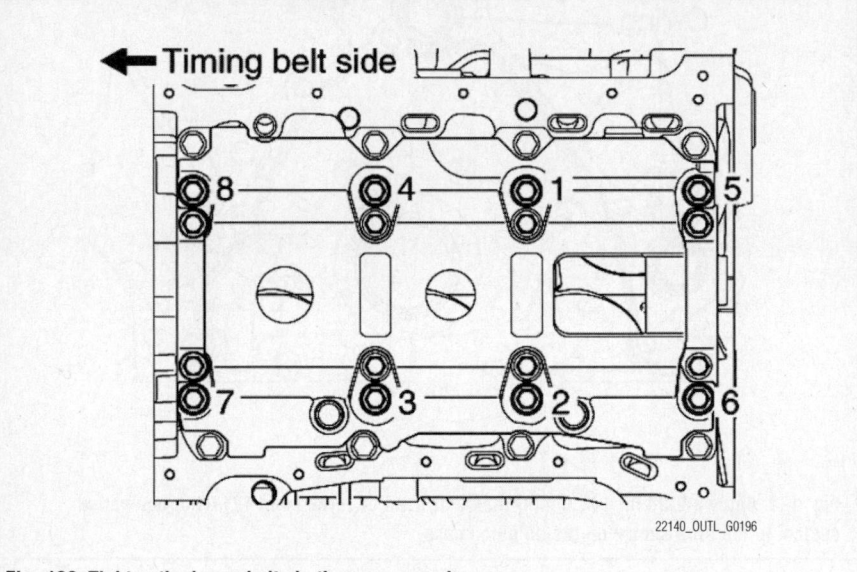

Fig. 128 Tighten the beam bolts in the sequence shown

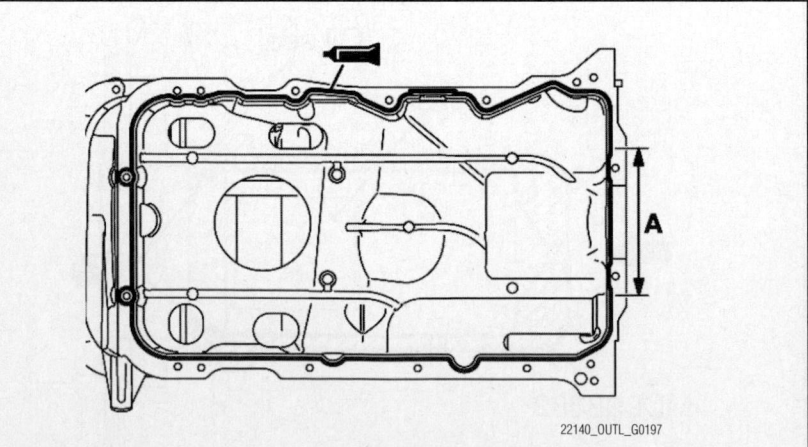

Fig. 129 Apply a 0.10 inch (2.5 mm) diameter bead of Three bond 1217G or equivalent sealant to the oil pan

9. Remove the upper oil pan.
   a. Remove the bolts A shown in the illustration first.
   b. Remove all other bolts.

✳✳ WARNING
Do not use a scraper or special tool to remove the oil pan upper.

   c. Thread the M10 _ 1.25 pitch bolt into the illustrated bolt hole to remove the oil pan upper.
10. Remove the oil screen and gasket.
11. Remove the bearing cap bolt.
12. Remove the beam.
13. Remove the right and left plate.
14. Remove the crankshaft front oil seal.
15. Remove the oil pump case.
16. Remove the O-rings.

To install:
17. Install the O-rings.
18. Install the oil pump case.

✳✳ WARNING
Carefully work so that the rests of the liquid gasket cannot enter to the oil passage or water passage.

   a. Remove the remaining liquid gasket on the oil pump case and the cylinder block.
   b. Degrease and clean the plane where the liquid gasket is applied and the both sides of the chamfered areas where the liquid gasket collects.
   c. After degreasing and cleaning the plane and the both sides, always spray compressed air into the oil passage and the water passage. Check that the oil passage and the water passage are free of foreign materials such as liquid gasket.
   d. As shown in the illustration, apply a 0.08 inch (2.0 mm) diameter bead of Three bond 1217G or equivalent sealant to the area shown on the oil pump case.
   e. After checking that the O-ring is installed at three places, install the oil pump case to the cylinder block.
   f. Tighten the oil pump case to 17 ft. lbs. (23 Nm).
19. Install the crankshaft front oil seal.
20. Install the right and left plate, and the beam.
   a. Before installing the bearing cap bolt, check the bolt screw head

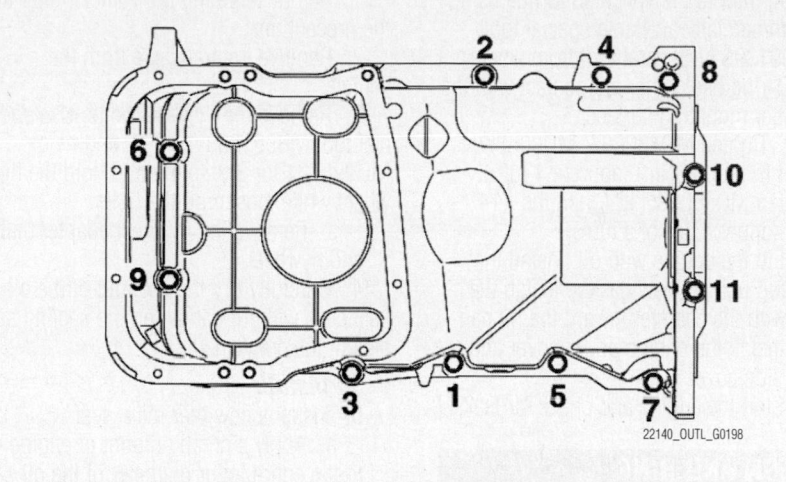

Fig. 130 Tighten the upper oil pan bolts in the sequence shown

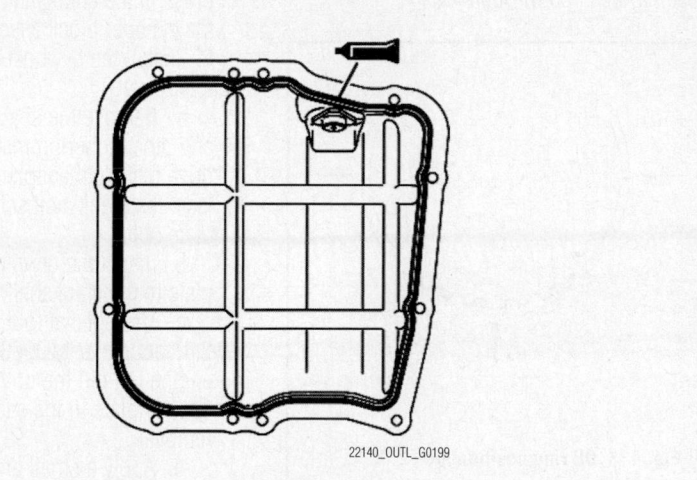

Fig. 131 Apply a 0.10 inch (2.5 mm) diameter bead of Three bond 1217G or equivalent sealant to the oil pan

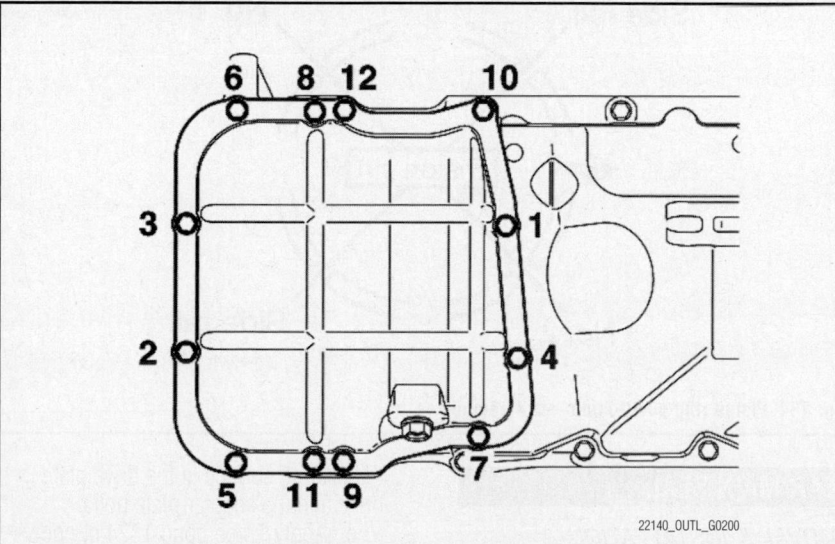

Fig. 132 Tighten the upper oil pan bolts in the sequence shown

is not damaged. If the screw head extremely gets damaged, replace it with a new bolt. The standard length of the new bolt measured from under the head is 3.97–4.00 inches (100.7–101.7 mm).

✳✳ WARNING

When the beam is removed and installed, never loosen the tightening bolts of the crankshaft bearing cap.

    b. Put the right and left plates on the crankshaft bearing, temporarily tighten bolts.

    c. Put the beam on them, Temporarily tighten bolt.

    d. Tighten the beam bolts to 18 ft. lbs. (24 Nm) in the sequence shown.

✳✳ WARNING

When the tightening angle is smaller than the specified tightening angle, the appropriate tightening capacity cannot be secured. When the tightening angle is larger than the specified tightening angle, remove the bolt to start from the beginning again according to the procedure.

✳✳ WARNING

Using a torque angle meter, tighten the beam bolt (M9) 60 degrees in the specified order.

    e. Tighten the plate bolts to 17 ft. lbs. (23 Nm) in the sequence shown.

21. Install the bearing cap bolt.
22. Install the oil screen and gasket.
23. Install the upper oil pan.

    a. Clean the gasket surfaces of the oil pan upper and oil pan lower.

✳✳ WARNING

When installing the upper oil pan, be sure not to expel the sealant from the oil pan flange at portion A.

    b. Apply a 0.10 inch (2.5 mm) diameter bead of Three bond 1217G or equivalent sealant to the oil pan.

    c. Be sure to install the oil pan quickly while the sealant is wet. Install the oil pan within 15 minutes after applying liquid gasket.

    d. Then wait at least one hour. Never start the engine or let engine oil or coolant touch the adhesion surface during that time.

e. Tighten the upper oil pan bolts to 89 ft. lbs. (10 Nm) in the sequence shown.

f. After installation, keep the sealed area away from the oil and coolant for approximately one hour.

24. Install the oil pan cover.

25. Install the lower oil pan.

a. Clean the gasket surfaces of the cylinder block and oil pan lower.

b. Apply a 2.5 ± 0.5 mm (0.10 ± 0.01 inch) diameter bead of sealant (Three bond 1217G or exact equivalent) to the oil pan. Be sure to install the oil pan quickly while the sealant is wet.

> ### ✳✳ WARNING
>
> **Install the oil pan within 15 minutes after applying liquid gasket. Then wait at least one hour. Never start the engine or let engine oil or coolant touch the adhesion surface during that time.**

c. Tighten the upper oil pan bolts to 89 inch lbs. (10 Nm) in the sequence shown.

d. After installation, keep the sealed area away from the oil and coolant for approximately one hour.

26. Install the engine oil pressure switch.

a. Completely remove existing sealant from the oil pressure switch and the switch mounting hole on the oil pump case.

b. Apply Three bond 1215, Three bond 1212D or equivalent sealant to the threaded part of the oil pressure switch.

c. Install the oil pressure switch to the oil pump case and tighten to 89 inch lbs. (10 Nm).

27. Install the heat protector.

28. Install the oil filter and cover.

a. Check the right and left ribs of the oil filter cover touch the outer circumference of the oil filter boss.

b. 1.Clean the filter mounting surface on the front case.

c. Apply engine oil to the O-ring of the oil filter.

> ### ✳✳ WARNING
>
> **If the filter is tightened by hand only, it will be insufficiently torqued, resulting in oil leaks.**

d. Screw the oil filter in and tighten

the oil filter to the specified torque using a commercially available special tool MB991396 or general tool from where the O-ring has come into contact with the oil filter mounting surface.

e. Tighten MD332687, MD365876 to 13 ft. lbs. (16Nm) (approx. 1 turn). Tighten MD360935 124 inch lbs. (14 Nm) (approx. 3/4 of a turn).

29. Fill the engine with oil. Install the drain plug using a new gasket. Install the gasket with the chamfer toward the oil pan.

30. Install the engine under cover and engine side cover.

31. Start the engine and check for leaks.

PISTON AND RING

POSITIONING

See Figures 133 through 135.

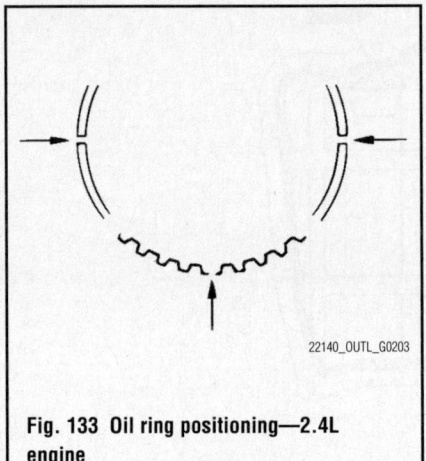

Fig. 133 Oil ring positioning—2.4L engine

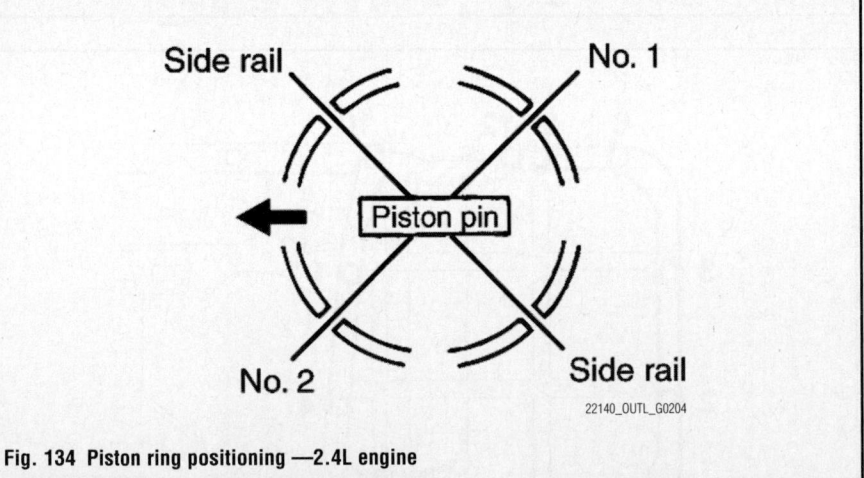

Fig. 134 Piston ring positioning —2.4L engine

REAR MAIN SEAL

REMOVAL & INSTALLATION

See Figures 136 and 137.

1. Before servicing the vehicle, refer to the precautions.

2. Remove the transaxle from the vehicle.

3. Remove the flywheel bolts. Use special tool MB991883 for 2.4L engine or MD998781 for 3.0L engine to hold the flywheel while loosening the bolts.

a. Remove the flywheel adapter plate and flywheel.

4. Carefully pry the seal out of the oil seal case without damaging the sealing surface of the crankshaft.

To install:

5. Install a new rear main seal.

a. Apply a small amount of engine oil to the entire inner diameter of the oil seal lip.

b. Using special tool MD998718, press in the crankshaft rear oil seal up to the cylinder block assembly end surface.

6. Install the flywheel and flywheel adapter plate.

a. Remove the sealant, the engine oil, and other adhering materials from the drive plate and adapter plate installation face, the crankshaft screw hole and drive plate bolts.

b. Install the drive plate and adapter plate to the crankshaft.

c. Use special tool MB991883 for 2.4L engine or MD998781 for 3.0L engine to hold the drive plate and adapter plate in the same manner as removal.

d. Apply a small amount of engine oil to the screw holes of the crankshaft and

the bearing surface of the drive plate bolts and the adapter plate bolts.

e. Apply Three bond 1324 or equivalent sealant to the thread of the drive plate bolts and the flywheel bolts.

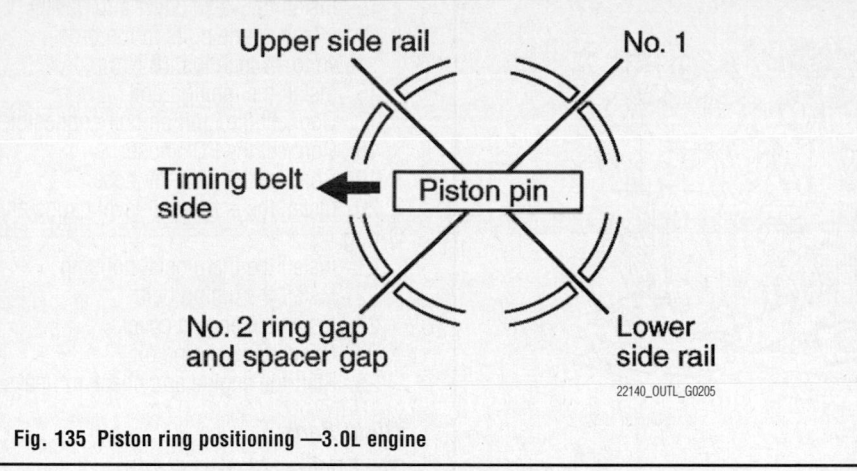

Fig. 135 Piston ring positioning —3.0L engine

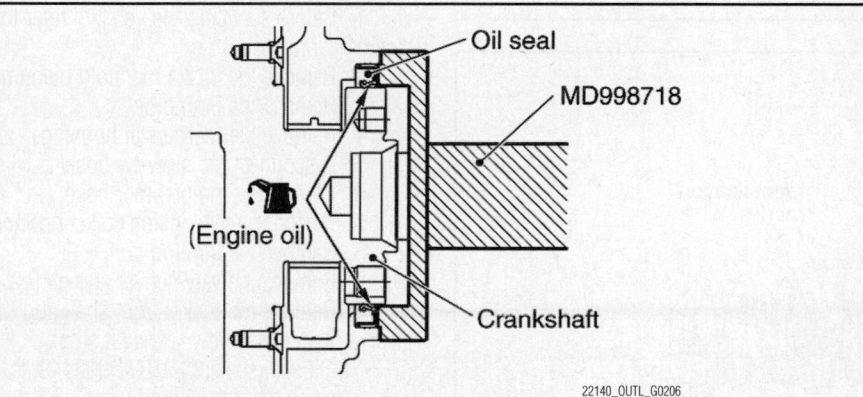

Fig. 136 Using special tool MD998718, press in the crankshaft rear oil seal up to the cylinder block assembly end surface—2.4L engine

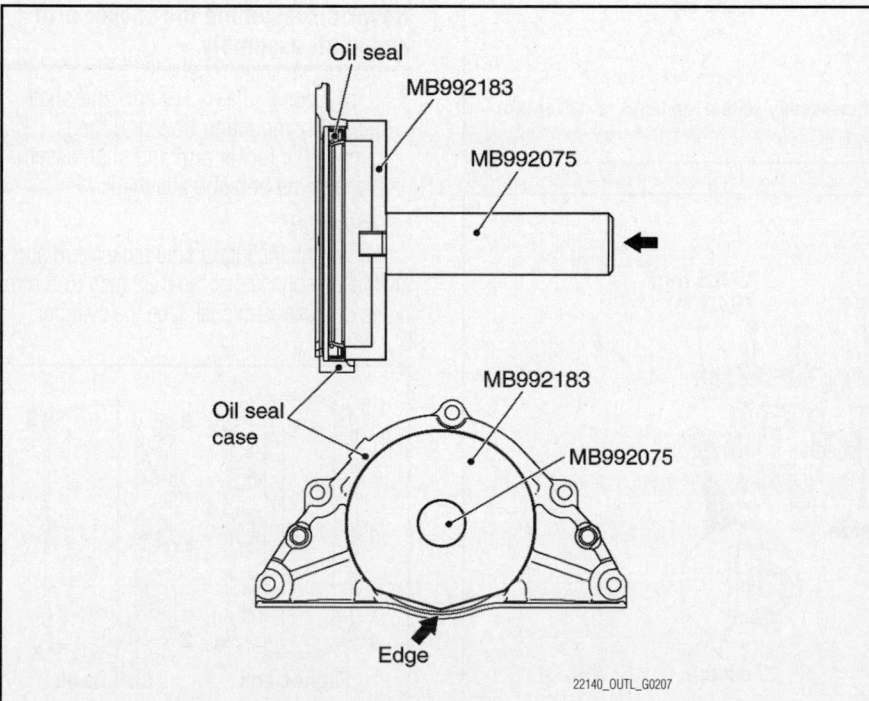

Fig. 137 Using special tool MB992075 and MB992183, press in the crankshaft rear oil seal up to the cylinder block assembly end surface—3.0L engine

f. Tighten drive plate bolts to temporary torque 30 ft. lbs. (40 Nm) in a star pattern.

g. Tighten drive plate bolts to 96 ft. lbs. (130 Nm) in a star pattern.

7. Install the transaxle assembly.

ROCKER ARMS

REMOVAL & INSTALLATION

3.0L Engine

Left Bank

See Figures 138 through 140.

1. Before servicing the vehicle, refer to the precautions.

2. Remove the engine cover.

3. Remove the timing belt.

4. Remove the thermostat housing.

5. Remove the engine oil level gauge and O-ring.

6. Disconnect the blow-by hose.

7. Disconnect the PCV hose.

8. Disconnect the ignition coil connector.

9. Remove the ignition coil.

10. Remove the valve cover and gasket.

11. Remove the rocker arm, shaft and the lash adjuster assembly (exhaust side).

a. Install special tool MD998443 as shown so that the lash adjusters will not fall out.

✳✳ WARNING

Never disassemble the rocker arm and shaft assembly.

b. Loosen the rocker arm and shaft assembly mounting bolt, and then remove the rocker arm and shaft assembly with the bolt still attached.

12. Remove the rocker arm and shaft assembly (intake side).

To install:

13. Install the intake side rocker arm and shaft assembly so that the 0.22 inch (5.5 mm) holes of rocker arm shaft face the cylinder head side. Tighten the intake side rocker arm shaft mounting bolts to 23 ft. lbs. (31 Nm).

14. Install the exhaust side rocker arm, shaft and lash adjuster assembly so that the notch of rocker arm shaft is located as shown.

a. Check that the identification mark of exhaust side rocker shaft cap is located as shown.

b. Tighten the exhaust side rocker arm shaft mounting bolts to 9 ft. lbs. (13 Nm).

c. Remove special tool MD998443.

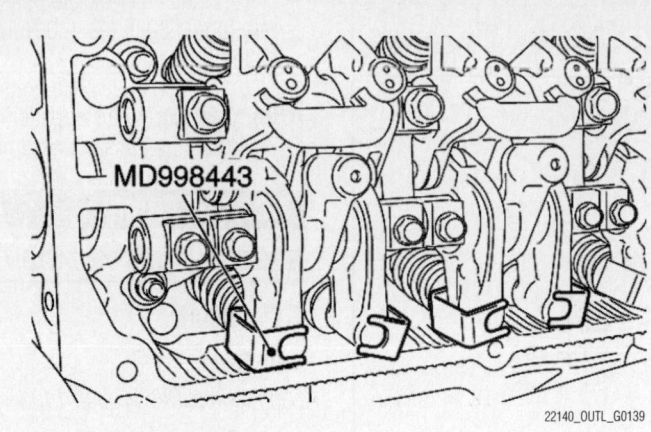

Fig. 138 Install special tool MD998443 as shown so that the lash adjusters will not fall out

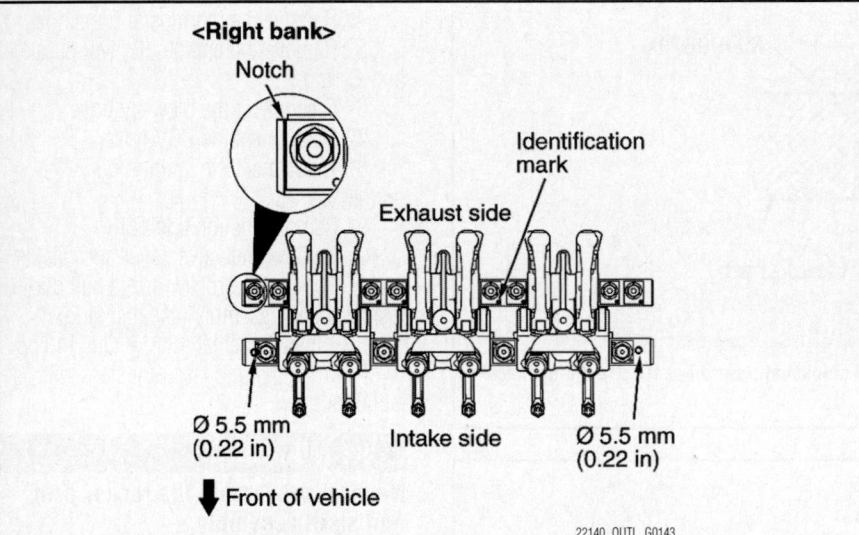

Fig. 139 Install the intake side rocker arm and shaft assembly so that the holes of rocker arm shaft face the cylinder head side

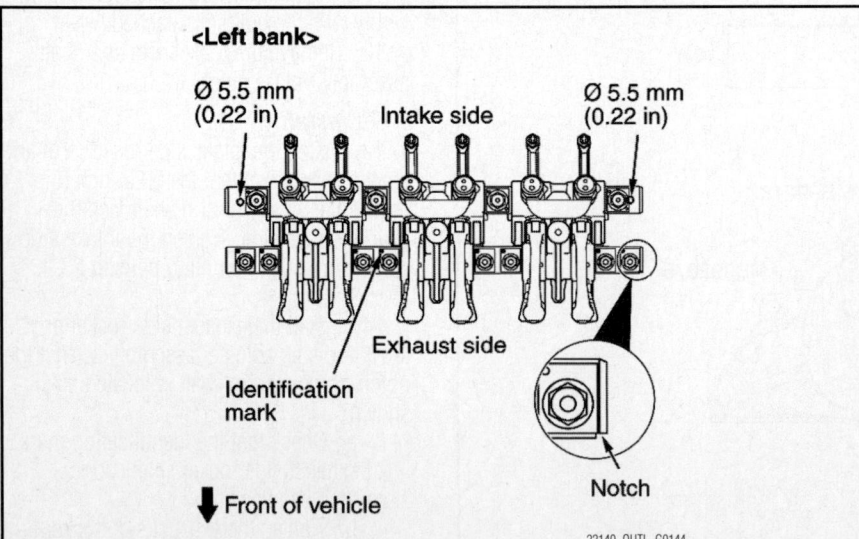

Fig. 140 Install the exhaust side rocker arm, shaft and lash adjuster assembly so that the notch of rocker arm shaft is located as shown

15. Install the valve cover and gasket.
   a. Tighten the bolts in the order shown to 74 inch lbs. (8 Nm).
16. Install the ignition coil.
17. Connect the ignition coil connector.
18. Connect the PCV hose.
19. Connect the blow-by hose.
20. Install the engine oil level gauge and O-ring.
21. Install the thermostat housing.
22. Install the timing belt.
23. Install the engine cover.
24. Change the engine oil.
25. Start the engine and check for leaks.

Right Bank

See Figures 138 through 141.

1. Before servicing the vehicle, refer to the precautions.
2. Remove the intake manifold plenum.
3. Remove the timing belt.
4. Remove the thermostat housing.
5. Disconnect the blow-by hose.
6. Disconnect the breather hose.
7. Disconnect the ignition coil connector.
8. Remove the ignition coil.
9. Remove the valve cover and gasket.
10. Remove the rocker arm, shaft and the lash adjuster assembly (exhaust side).
   a. Install special tool MD998443 as shown so that the lash adjusters will not fall out.

✳✳ WARNING

Never disassemble the rocker arm and shaft assembly.

   b. Loosen the rocker arm and shaft assembly mounting bolt, and then remove the rocker arm and shaft assembly with the bolt still attached.

To install:

11. Install the intake side rocker arm and shaft assembly so that the 0.22 inch (5.5 mm) holes of rocker arm shaft face the cylinder

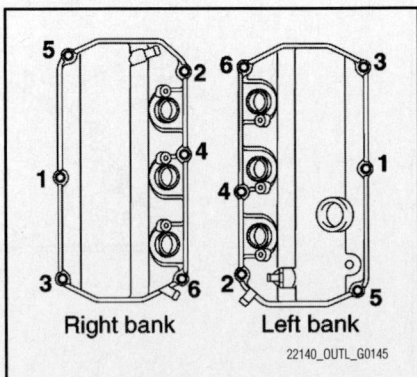

Fig. 141 Tighten the valve cover bolts in the order shown

head side. Tighten the intake side rocker arm shaft mounting bolts to 23 ft. lbs. (31 Nm).

12. Install the exhaust side rocker arm, shaft and lash adjuster assembly so that the notch of rocker arm shaft is located as shown.

    a. Check that the identification mark of exhaust side rocker shaft cap is located as shown.

    b. Tighten the exhaust side rocker arm shaft mounting bolts to 9 ft. lbs. (13 Nm).

    c. Remove special tool MD998443.

13. Install the valve cover and gasket

    a. Tighten the bolts in the order shown to 74 inch lbs. (8 Nm).

14. Install the ignition coil.

15. Disconnect the ignition coil connector.

16. Disconnect the breather hose.

17. Disconnect the blow-by hose.

18. Install the thermostat housing.

19. Install the timing belt.

20. Install the intake manifold plenum.

21. Change the engine oil.

22. Start the engine and check for leaks.

TIMING BELT, & SPROCKETS

REMOVAL & INSTALLATION

3.0L Engine

See Figures 142 through 148.

1. Before servicing the vehicle, refer to the precautions.

2. Remove the engine cover.

3. Remove the engine under cover and side cover.

4. Remove the drive belt auto-tensioner.

5. Remove the alternator drive belt.

✳✳ WARNING

When the alternator drive belt is reused, draw an arrow indicating the rotating direction on the back of the belt using chalk to install the same direction.

    a. Turn the drive belt auto-tensioner to counterclockwise, and insert the L-shaped hexagon wrench to the auto-tensioner hole in order to fix the auto-tensioner.

    b. Remove the alternator drive belt.

    c. Remove the drive belt auto-tensioner.

6. Remove the power steering tensioner pulley bracket.

7. Remove the power steering oil pump drive belt.

8. Remove the crankshaft pulley.

    a. Use special tools MB990767 and MD998719 to remove the crankshaft pulley from the crankshaft.

9. Remove the timing belt front upper covers.

10. Remove the timing belt lower cover

11. Remove the engine mounting bracket.

12. Remove the engine support bracket.

13. Remove the front flange.

14. Remove the timing belt auto-tensioner

    a. Remove the upper tightening bolt of the auto-tensioner.

✳✳ WARNING

The auto-tensioner rotates centering on the flange bolt due to the rod thrust, so please make sure your finger is not trapped.

    b. Loosen the lower tightening bolt of the auto-tensioner slowly and slide the auto-tensioner slightly. Remove the rod from the tensioner arm.

    c. Remove the lower tightening bolt of the auto-tensioner.

15. Remove the timing belt.

✳✳ WARNING

Never turn the crankshaft counterclockwise.

    a. Turn the crankshaft clockwise to align each timing mark and to set the number 1 cylinder to compression top dead center.

    b. If the timing belt is to be reused, chalk an arrow on the flat side of the belt, indicating the clockwise direction.

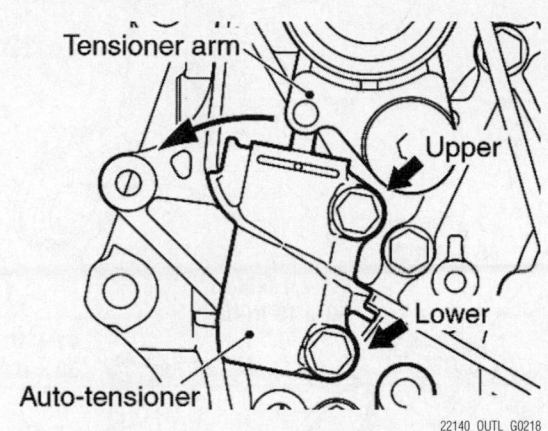

22140_OUTL_G0218

Fig. 142 Loosen the lower tightening bolt of the auto-tensioner slowly and slide the auto-tensioner slightly. Remove the rod from the tensioner arm. Remove the lower tightening bolt of the auto-tensioner

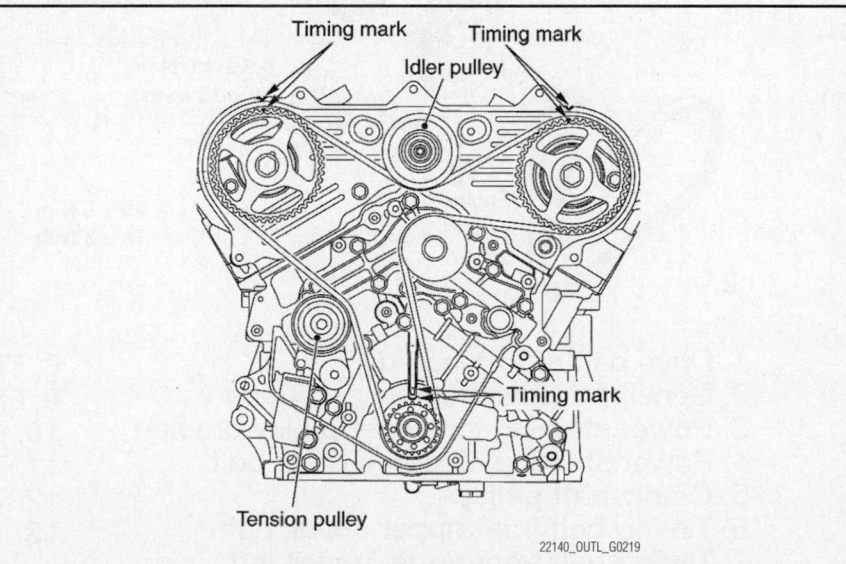

22140_OUTL_G0219

Fig. 143 Loosen the lower tightening bolt of the auto-tensioner slowly and slide the auto-tensioner slightly. Remove the rod from the tensioner arm. Remove the lower tightening bolt of the auto-tensioner

c. Loosen the center bolt of the tensioner pulley, then remove the timing belt.

16. Remove the tensioner pulley.

To install:

※ **WARNING**

Always bleed the auto-tensioner of air before installing the auto-tensioner.

17. Bleed the auto-tensioner.
   a. Always use the vertical press and put the auto-tensioner vertically.

※ **WARNING**

Do not apply the load of 1,124 pound (5,000 N) or more to the rod. Do not press the rod beyond Dimension "A" shown.

   b. Set the auto-tensioner as shown.
   c. Press the rod slowly down to the lowest point "A" shown.
   d. Repeat the procedure 2 three times.
   e. While the rod is projected at the point "B" shown in the illustration, push the rod with 22–44 pound (100–200 N).

Check the enough stiffness. If the stiffness is not enough, replace the auto-tensioner.
   f. Press down the rod slowly. Put the pin through the hole and secure it.

18. Insert the pin into the rod of the auto-tensioner under the following procedures.
   a. Put the auto-tensioner vertically to the vertical press not to be in the sideway direction.
   b. Slowly close the vice to force the rod in until the hole (A) of the rod is lined up with set hole (B) of the cylinder.

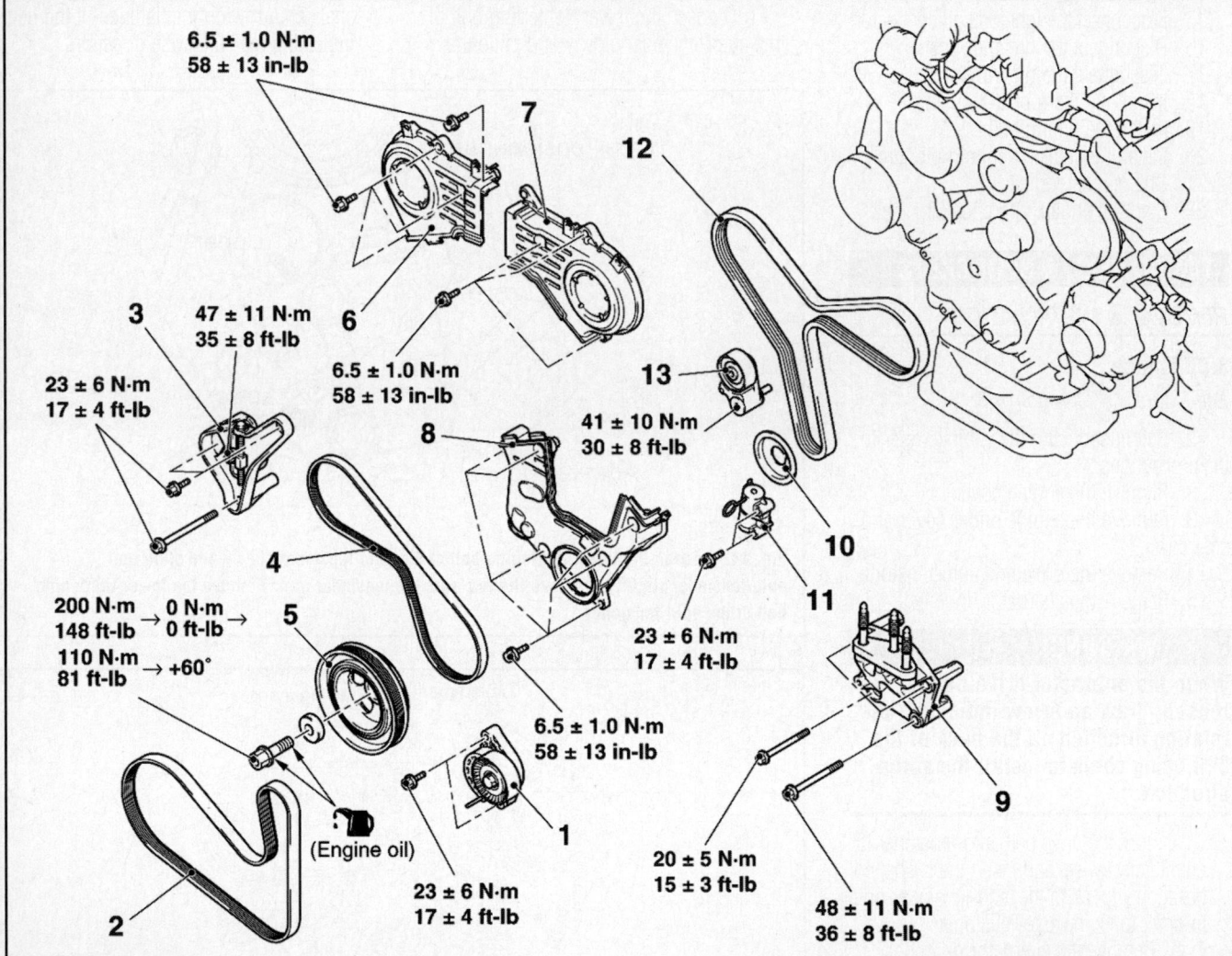

1. Drive belt auto-tensioner
2. Generator drive belt
3. Power steering tensioner pulley bracket
4. Power steering oil pump drive belt
5. Crankshaft pulley
6. Timing belt front upper cover, right
7. Timing belt front upper cover, left
8. Timing belt lower cover
9. Engine support bracket
10. Front flange
11. Timing belt auto-tensioner
12. Timing belt
13. Tensioner pulley assembly

Fig. 144 Exploded view of the timing belt and covers—3.0L engine

22140_OUTL_G0224

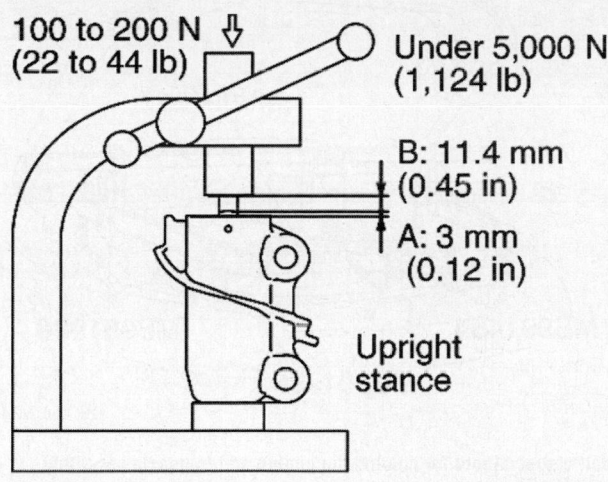

Fig. 145 Set the auto-tensioner as shown and press the rod slowly down to the lowest point "A"

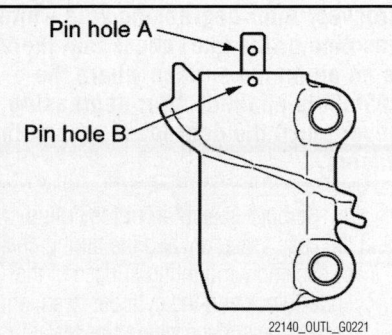

Fig. 146 Slowly close the vice to force the rod in until the hole (A) of the rod is lined up with set hole (B) of the cylinder

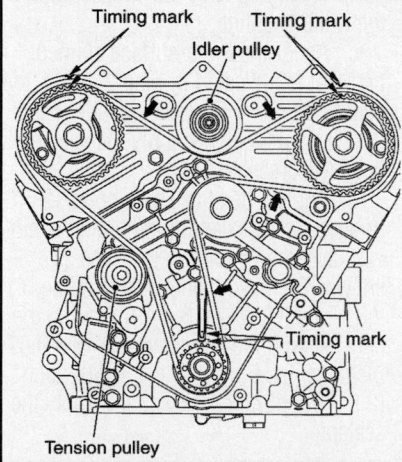

Fig. 147 Align the timing marks on the camshaft sprockets with those on the timing belt rear cover and the timing mark on the crankshaft sprocket with that on the engine block as shown

c. Insert a pin into the set holes.

d. Remove the auto-tensioner from the vice.

19. Install the timing belt and auto-tensioner.

a. Align the timing marks on the camshaft sprockets with those on the timing belt rear cover and the timing mark on the crankshaft sprocket with that on the engine block as shown.

✳✳ WARNING

The camshaft sprocket (right bank) can turn easily due to the spring force applied, so be careful not to get your fingers caught.

b. Install the timing belt in the following order so that there is no deflection in the timing belt between each sprocket and pulley.

- Crankshaft sprocket
- Water pump pulley
- Camshaft sprocket (Left bank)
- Idler pulley
- Camshaft sprocket (Right bank)
- Tensioner pulley

c. Turn the camshaft sprocket (Right bank) counterclockwise until the tension side of the timing belt is firmly stretched. Check all the timing marks again.

d. Use special tool MD998716 to turn the crankshaft 1/4 turn counterclockwise, then turn it again clockwise until the timing marks are aligned.

e. Remove the setting pin that has been inserted into the auto-tensioner.

f. Turn the crankshaft clockwise twice to align the timing marks.

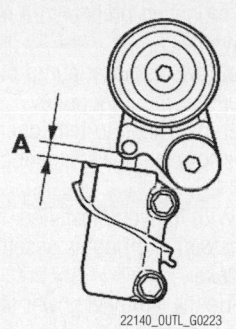

Fig. 148 Wait for at least five minutes, then check that the auto-tensioner pushrod extends within 0.36–0.52 inch (9.1–13.4 mm)

g. Wait for at least five minutes, then check that the auto-tensioner pushrod extends within 0.36–0.52 inch (9.–13.4 mm).

h. If not, repeat the operation.

i. Check again that the timing marks of the sprockets are aligned.

20. Install the front flange.

21. Install the engine support bracket.

22. Install the engine mounting bracket.

23. Install the timing belt lower cover. Tighten to 58 inch lbs. (7 Nm).

24. Install the timing belt front upper covers. Tighten to 58 inch lbs. (7 Nm).

25. Install the crankshaft pulley.

a. Use special tools MB990767 and MD998719 to install the crankshaft pulley.

26. Install the power steering oil pump drive belt.

27. Install the power steering tensioner pulley bracket.

28. Install the alternator drive belt.

29. Install the drive belt auto-tensioner.

30. Install the engine under cover and side cover.

31. Install the engine cover.

TIMING CHAIN FRONT COVER

REMOVAL & INSTALLATION

2.4L Engine

See Figures 149 through 154.

1. Before servicing the vehicle, refer to the precautions.

2. Remove the engine room under cover and engine side.

3. Drain the engine oil.

4. Remove the valve cover.

5. Remove the engine oil pan.

6. Remove the crankshaft pulley.

a. When removing the crankshaft

pulley, slightly loosen the water pump pulley mounting bolts before removal of the drive belt.

7. Remove the water pump pulley.

8. Remove the idler pulley.

9. Remove the auto-tensioner.

10. Remove the power steering oil pump.

a. With the hose installed, remove the power steering oil pump assembly from the bracket.

b. Tie the removed power steering oil pump assembly with a string at a position where it will not interfere with the removal and installation of valve timing chain.

11. Install the engine and transaxle assembly holding fixture.

a. If special tool engine hanger MB991928 is used:

b. Set the foot of the special tools (MB991930, MB991932 and MB991933) as shown.

c. Slide the slide bracket (HI) to adjust the engine hanger balance.

d. Install special tool MB991956 to the cylinder head, and set special tool MB991527 and the chains of special tool MB991454 to the engine assembly to hold the engine and transaxle assembly.

e. If special tool engine hanger MB991895 is used:

f. Place a rag between special tool MB991895 and the windshield to prevent the special tool MB991895 from interfering with the windshield.

g. Set the foot of special tool MB991895 as shown.

h. Slide the foot to adjust the engine hanger balance.

i. Install special tool MB991956 to the cylinder head, and set special tool MB991527 and the chains of special tool MB991454 to the engine assembly to hold the engine and transaxle assembly.

12. Remove the engine mounting insulator.

13. Remove the cylinder block engine front mounting bracket

14. Remove the gasket.

15. Remove the timing chain case.

a. If the adhesive strength of sealant on the timing chain case assembly is so strong that the boss may be damaged by peeling off, do not peel it off forcibly.

b. After removing the timing chain case assembly mounting bolts, slightly pry the boss of the timing chain case assembly shown in the figure using a flat-tipped screwdriver (-), and remove the timing chain case assembly from the cylinder head and cylinder block.

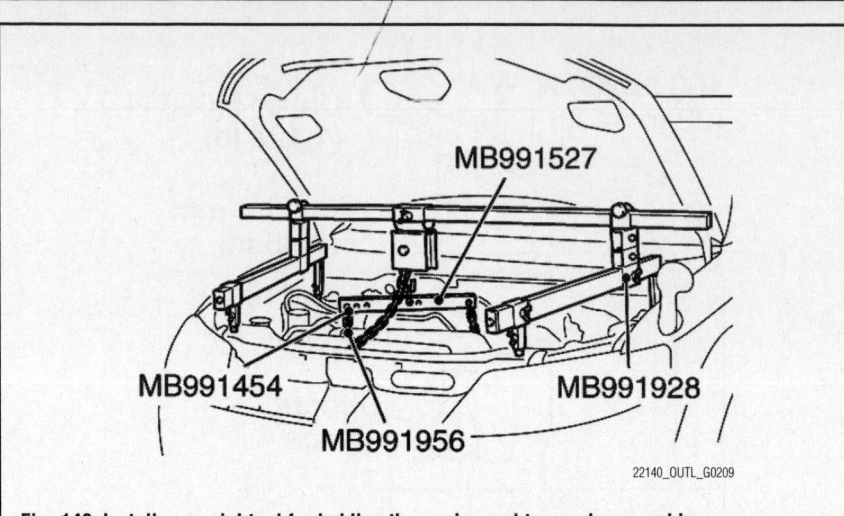

Fig. 149 Install a special tool for holding the engine and transaxle assembly

22140_OUTL_G0209

c. If the sealant cannot be peeled off easily, insert a wooden hammer shank into the timing chain case assembly inside as shown in the figure, pry slightly, and remove the timing chain case assembly from the cylinder head and cylinder block.

16. Remove the crankshaft front oil seal.

To install:

17. Install the crankshaft front oil seal.

a. Apply a small amount of engine oil to the entire inner diameter of the crankshaft front oil seal lip.

✴✴ WARNING

When installing the crankshaft front oil seal, be careful to avoid damage to the crankshaft front oil seal.

b. Using a seal installer, press in the crankshaft front oil seal up to the chamfered surface of timing chain case.

18. Install the timing chain case.

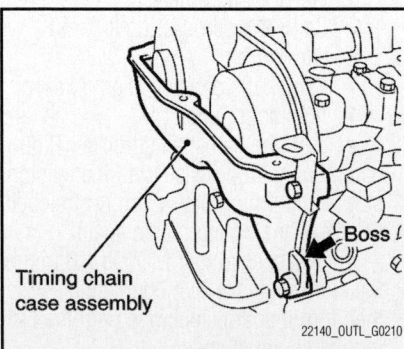

Timing chain case assembly

Boss

22140_OUTL_G0210

Fig. 150 Slightly pry the boss of the timing chain case assembly using a flat-tipped screwdriver (-), and remove the timing chain case assembly

✴✴ WARNING

Be sure to remove the sealant inside the mounting holes and the O-ring grooves. After degreasing with white gasoline or the like, check that there is no oil on the surface where the sealant is applied. After degreasing, never touch the degreased area with fingers.

a. Remove sealant from the timing chain case assembly and the timing chain case assembly mounting surface of the cylinder block and the cylinder head, and degrease the surface where the sealant is applied by white gasoline or the like.

b. Remove all the sealant adhering to the gasket between the cylinder head and cylinder block (three-surface aligned part). Then, degrease the surfaces with the white gasoline.

c. As for the three-surface aligned part that is indicated in Step 2 above, the engine oil oozes from the cylinder head gasket. Thus, quickly apply the sealant to it after degreasing.

d. Apply a bead of Three bond 1217G or equivalent sealant to the timing chain case assembly mounting surface as shown. The bead diameter should be 0.1 ± 0.02 inch (2.5 ± 0.5 mm). Overlap the part "A" with the diameter of 0.18 ± 0.02 inch (4.5 ± 0.5 mm) or 0.1 ± 0.02 inch (2.5 ± 0.5 mm) as shown, and apply the sealant.

✴✴ WARNING

If the sealant contacts any other part during installation of the timing chain case assembly, apply sealant again before installing the timing chain case

Hammer
Timing chain case assembly
22140_OUTL_G0211

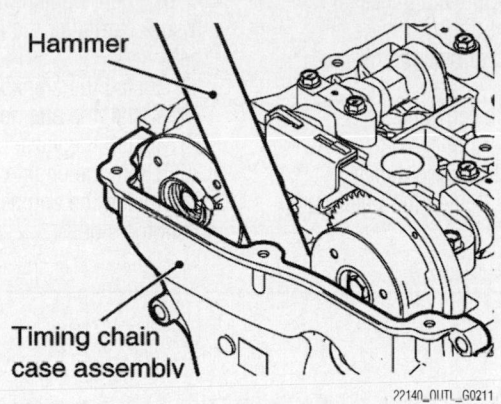

Fig. 151 If the sealant cannot be peeled off easily, insert a wooden hammer shank into the timing chain case assembly inside as shown in the figure, pry slightly, and remove the timing chain case assembly

assembly. Do not apply oil or water to the sealant-applied area or start up the engine within 2 hours after the installation of the timing chain case assembly.

e. Install the timing chain case assembly to the cylinder block and cylinder head so that the sealant does not contact other parts.

➡Install the timing chain case assembly within 3 minutes after the application of sealant.

f. Insert the bolts to the timing chain case assembly as shown, and tighten them as follows:

- Flange bolt (A)–M6 _ 25–Tighten to 89 inch lbs. (10 Nm)

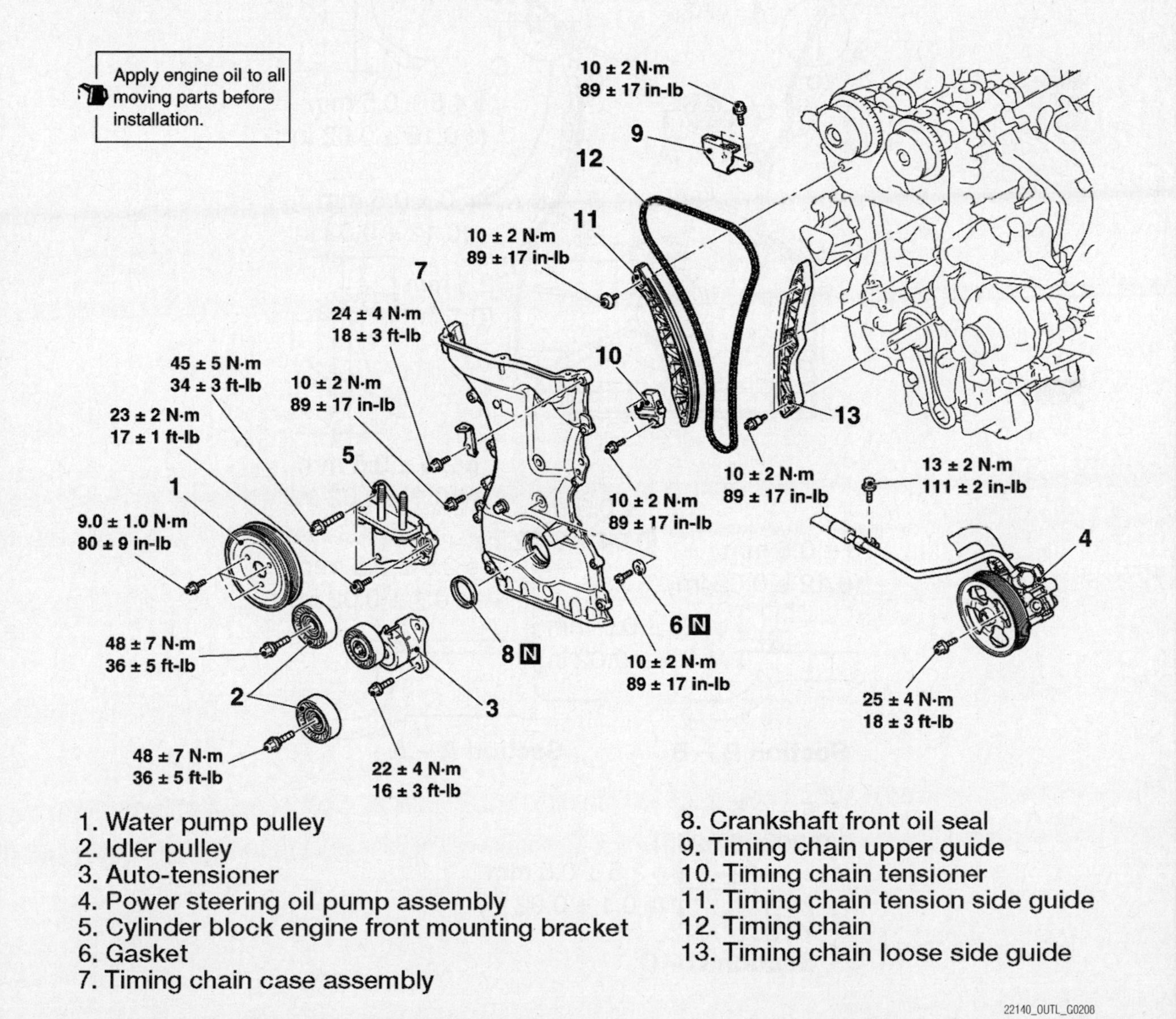

Apply engine oil to all moving parts before installation.
10 ± 2 N·m
89 ± 17 in-lb
9
12
11
10 ± 2 N·m
89 ± 17 in-lb
7
24 ± 4 N·m
18 ± 3 ft-lb
10
45 ± 5 N·m
34 ± 3 ft-lb
10 ± 2 N·m
89 ± 17 in-lb
23 ± 2 N·m
17 ± 1 ft-lb
13
5
10 ± 2 N·m
89 ± 17 in-lb
13 ± 2 N·m
111 ± 2 in-lb
1
10 ± 2 N·m
89 ± 17 in-lb
9.0 ± 1.0 N·m
80 ± 9 in-lb
4
6 N
48 ± 7 N·m
36 ± 5 ft-lb
8 N
10 ± 2 N·m
89 ± 17 in-lb
2
3
25 ± 4 N·m
18 ± 3 ft-lb
48 ± 7 N·m
36 ± 5 ft-lb
22 ± 4 N·m
16 ± 3 ft-lb
1. Water pump pulley
2. Idler pulley
3. Auto-tensioner
4. Power steering oil pump assembly
5. Cylinder block engine front mounting bracket
6. Gasket
7. Timing chain case assembly
8. Crankshaft front oil seal
9. Timing chain upper guide
10. Timing chain tensioner
11. Timing chain tension side guide
12. Timing chain
13. Timing chain loose side guide
22140_OUTL_G0208

Fig. 152 Exploded view of timing chain and cover assembly—2.4L engine

- Flange bolt (B)–M8 _ 28–Tighten to 18 ft. lbs. (24 Nm)
- Bolt (C) M6 _ 25–Tighten to 89 inch lbs. (10 Nm)

19. Install the gasket.
20. Install the cylinder block engine front mounting bracket
21. Install the engine mounting insulator.

22. Install the engine and transaxle assembly holding fixture.
23. Install the power steering oil pump.
24. Install the auto-tensioner.
25. Install the idler pulley.
26. Install the water pump pulley.
   a. Temporarily tighten the water pump pulley mounting bolts.

b. Then, tighten them to 80 inch lbs. (9 Nm) after the installation of drive belt.
27. Install the crankshaft pulley.
28. Install the engine oil pan.
29. Install the valve cover.
30. Drain the engine oil.
31. Install the engine room under cover and engine side.

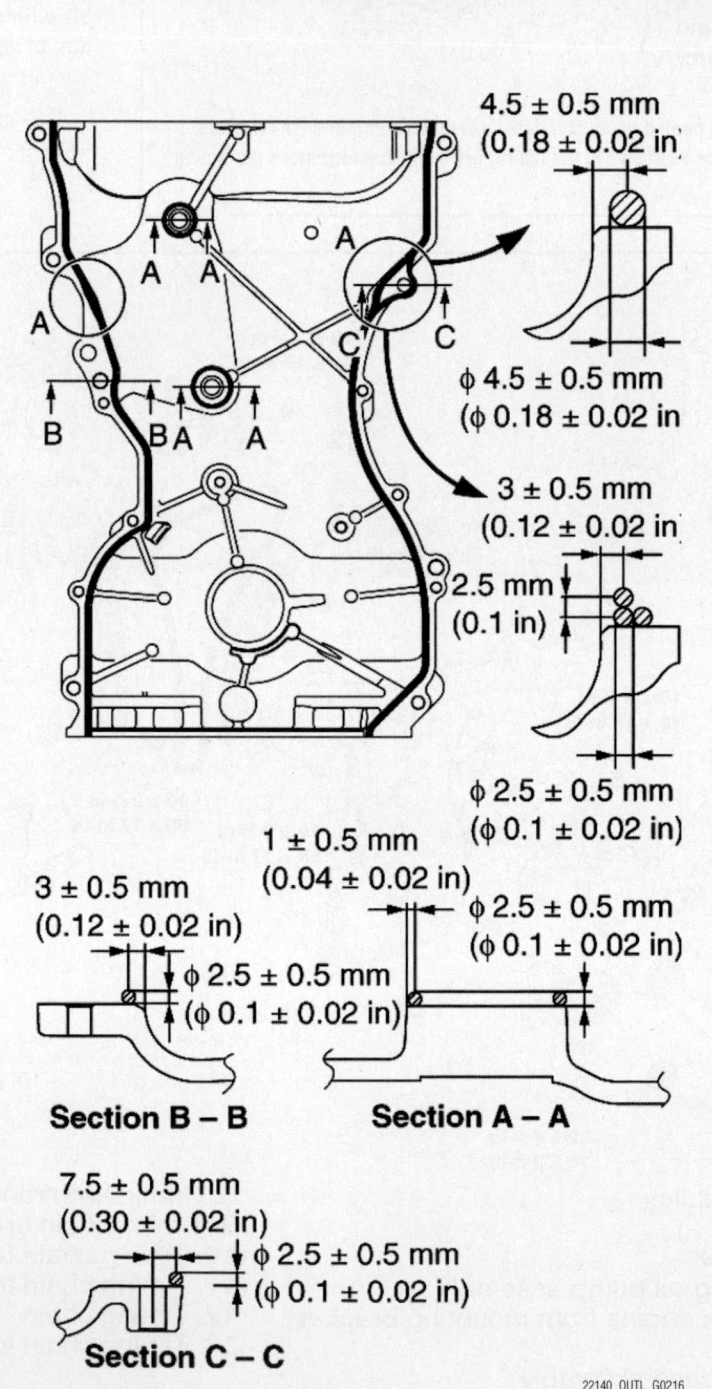

Fig. 153 Apply a bead of the sealant to the timing chain case assembly mounting surface as shown

22140_OUTL_G0216

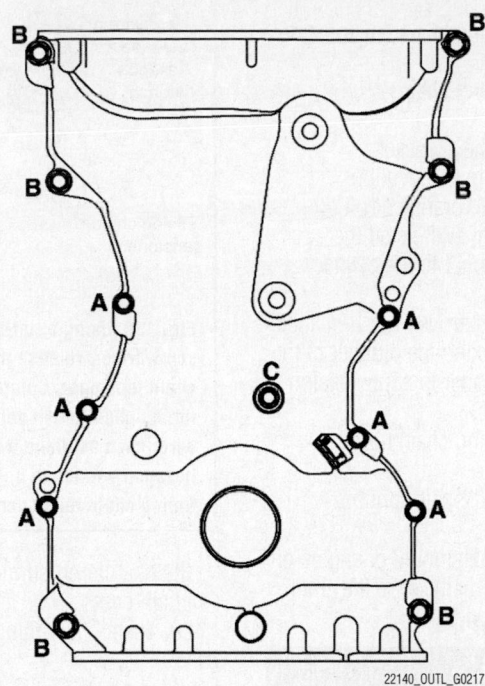

Fig. 154 Insert the bolts to the timing chain case assembly as shown, and tighten them to the specified torque

TIMING CHAIN & SPROCKETS

REMOVAL & INSTALLATION

2.4L Engine

See Figures 152, 155 through 163.

1. Before servicing the vehicle, refer to the precautions.

2. Remove the engine room under cover and engine side.

3. Drain the engine oil.

4. Remove the valve cover.

5. Remove the engine oil pan.

6. Remove the crankshaft pulley.

   a. When removing the crankshaft pulley, slightly loosen the water pump pulley mounting bolts before removal of the drive belt.

7. Remove the water pump pulley.

8. Remove the idler pulley.

9. Remove the auto-tensioner.

10. Remove the power steering oil pump.

   a. With the hose installed, remove the power steering oil pump assembly from the bracket.

   b. Tie the removed power steering oil pump assembly with a string at a position where it will not interfere with the removal and installation of valve timing chain.

11. Install the engine and transaxle assembly holding fixture.

   a. If special tool engine hanger MB991928 is used:

   b. Set the foot of the special tools (MB991930, MB991932 and MB991933) as shown.

   c. Slide the slide bracket (HI) to adjust the engine hanger balance.

   d. Install special tool MB991956 to the cylinder head, and set special tool MB991527 and the chains of special tool MB991454 to the engine assembly to hold the engine and transaxle assembly.

   e. If special tool engine hanger MB991895 is used:

   f. Place a rag between special tool MB991895 and the windshield to prevent the special tool MB991895 from interfering with the windshield.

   g. Set the foot of special tool MB991895 as shown.

   h. Slide the foot to adjust the engine hanger balance.

   i. Install special tool MB991956 to the cylinder head, and set special tool MB991527 and the chains of special tool MB991454 to the engine assembly to hold the engine and transaxle assembly.

12. Remove the engine mounting insulator.

13. Remove the cylinder block engine front mounting bracket

14. Remove the gasket.

15. Remove the timing chain case.

   a. If the adhesive strength of sealant on the timing chain case assembly is so strong that the boss may be damaged by peeling off, do not peel it off forcibly.

   b. After removing the timing chain case assembly mounting bolts, slightly pry the boss of the timing chain case assembly shown in the figure using a flat-tipped screwdriver (-), and remove the timing chain case assembly from the cylinder head and cylinder block.

   c. If the sealant cannot be peeled off easily, insert a wooden hammer shank into the timing chain case assembly inside as shown in the figure, pry slightly, and remove the timing chain case assembly from the cylinder head and cylinder block.

16. Remove the crankshaft front oil seal.

17. Remove the timing chain upper guide.

18. Remove the timing chain tensioner.

   a. Temporarily install the crankshaft pulley to the crankshaft.

   b. Turn the crankshaft clockwise.

   c. Turn the crankshaft clockwise to align the sprocket timing marks as shown in the figure and set the cylinder No. 1 to the top dead center of compression stroke.

   d. At this time, it is not necessary that the link plate (orange or blue) of the timing chain always aligns with each sprocket timing mark.

   e. Remove the crankshaft pulley installed temporarily.

   f. Using a flat-tipped precision screwdriver, release the ratchet of timing chain tensioner.

   g. Compress the plunger of timing chain tensioner and insert hard wire (such as piano wire) or the L-shaped hexagon wrench (1.5 mm[0.05 inch]) to fix the plunger of the timing chain tensioner.

   h. Remove the timing chain tensioner.

19. Remove the timing chain tension side guide.

20. Remove the timing chain.

21. Remove the timing chain loose side guide.

To install:

22. Install the timing chain loose side guide.

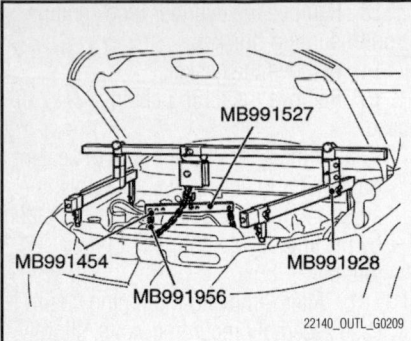

Fig. 155 Install a special tool for holding the engine and transaxle assembly

23. Install the timing chain.
   a. Set the timing marks of the camshaft sprockets and the crankshaft sprocket as shown in the figure.
   b. Align each sprocket timing chain mating mark with the link plate (orange or blue) of timing chain to avoid slack of the timing chain tension side,

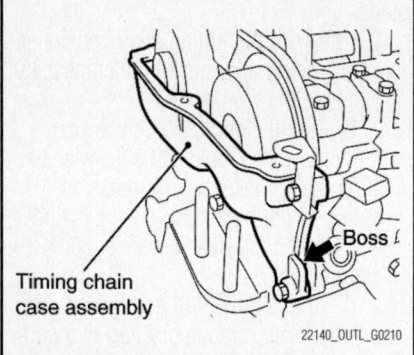

Fig. 156 Slightly pry the boss of the timing chain case assembly using a flat-tipped screwdriver (-), and remove the timing chain case assembly

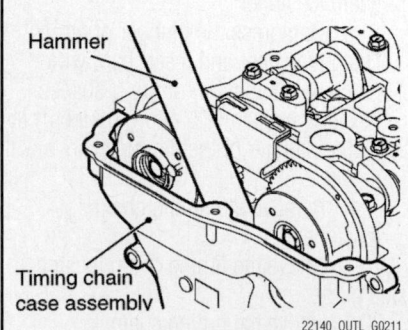

Fig. 157 If the sealant cannot be peeled off easily, insert a wooden hammer shank into the timing chain case assembly inside as shown in the figure, pry slightly, and remove the timing chain case assembly

and install the timing chain to the sprockets.
24. Install the timing chain tension side guide.
25. Install the timing chain tensioner.
   a. Check that the sprocket timing chain mating marks align with the link plates (orange or blue) of the timing chain, and install the timing chain tensioner to the cylinder block.
   b. Remove the hard wire or L-shaped hexagon wrench fixing the plunger of the timing chain tensioner to apply tension to the timing chain.
26. Install the timing chain upper guide.
27. Install the crankshaft front oil seal.
   a. Apply a small amount of engine oil to the entire inner diameter of the crankshaft front oil seal lip.

> ✳✳ **WARNING**
>
> **When installing the crankshaft front oil seal, be careful to avoid damage to the crankshaft front oil seal.**

   b. Using a seal installer, press in the crankshaft front oil seal up to

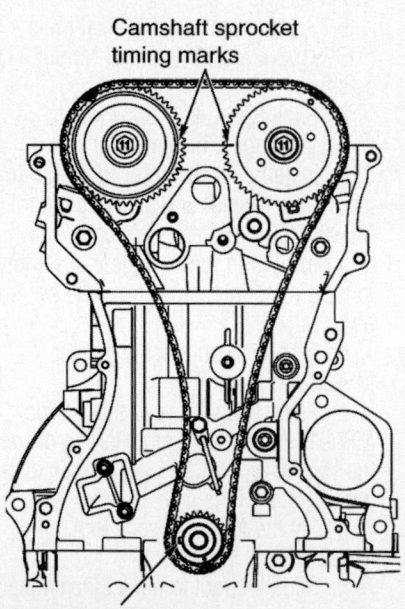

Fig. 158 Turn the crankshaft clockwise to align the sprocket timing marks as shown in the figure and set the cylinder No. 1 to the top dead center of compression stroke

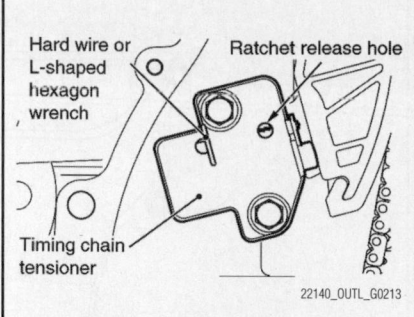

Fig. 159 Using a flat-tipped precision screwdriver, release the ratchet of timing chain tensioner. Compress the plunger of timing chain tensioner and insert hard wire (such as piano wire) or an L-shaped hexagon wrench to hold the plunger of the timing chain tensioner

the chamfered surface of timing chain case.
28. Install the timing chain case.

> ✳✳ **WARNING**
>
> **Be sure to remove the sealant inside the mounting holes and the O-ring grooves. After degreasing with white gasoline or the like, check that there is no oil on the surface where the sealant is applied. After degreasing, never touch the degreased area with fingers.**

   a. Remove sealant from the timing chain case assembly and the timing chain case assembly mounting surface of the cylinder block and the cylinder head, and degrease the surface where the sealant is applied by white gasoline or the like.
   b. Remove all the sealant adhering to the gasket between the cylinder head and cylinder block (three-surface aligned part). Then, degrease the surfaces with the white gasoline.
   c. As for the three-surface aligned part that is indicated in Step 2 above, the engine oil oozes from the cylinder head gasket. Thus, quickly apply the sealant to it after degreasing.
   d. Apply a bead of Three bond 1217G or equivalent sealant to the timing chain case assembly mounting surface as shown. The bead diameter should be 0.1 ± 0.02 inch (2.5 ± 0.5 mm). Overlap the part "A" with the diameter of 0.18 ± 0.02 inch (4.5 ± 0.5 mm) or 0.1 ± 0.02 inch (2.5 ± 0.5 mm) as shown, and apply the sealant.

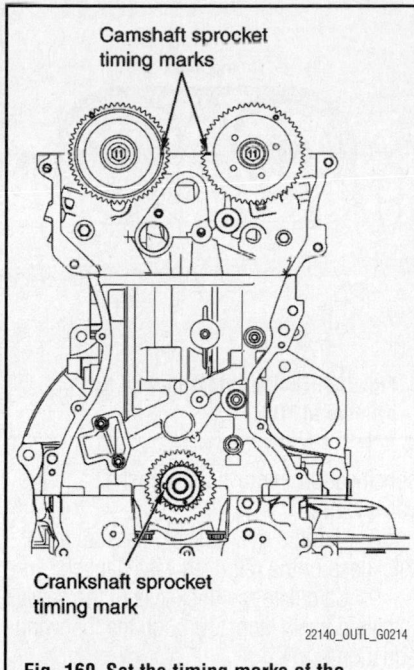

Fig. 160 Set the timing marks of the camshaft sprockets and the crankshaft sprocket as shown

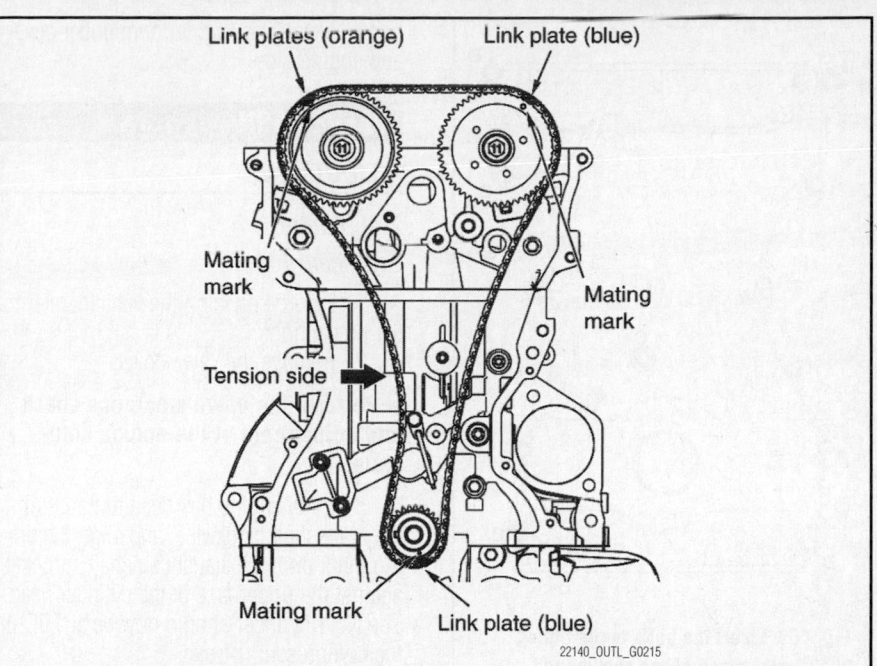

Fig. 161 Align each sprocket timing chain mating mark with the link plate (orange or blue) of timing chain to avoid slack of the timing chain tension side, and install the timing chain to the sprockets

✳✳ WARNING

If the sealant contacts any other part during installation of the timing chain case assembly, apply sealant again before installing the timing chain case assembly. Do not apply oil or water to the sealant-applied area or start up the engine within 2 hours after the installation of the timing chain case assembly.

   e. Install the timing chain case assembly to the cylinder block and cylinder head so that the sealant does not contact other parts.

➡**Install the timing chain case assembly within 3 minutes after the application of sealant.**

   f. Insert the bolts to the timing chain case assembly as shown, and tighten them as follows:
- Flange bolt (A)–M6 _ 25–Tighten to 89 inch lbs. (10 Nm)
- Flange bolt (B)–M8 _ 28–Tighten to 18 ft. lbs. (24 Nm)
- Bolt (C)–M6 _ 25–Tighten to 89 inch lbs. (10 Nm)

29. Install the gasket.
30. Install the cylinder block engine front mounting bracket
31. Install the engine mounting insulator.
32. Install the engine and transaxle assembly holding fixture.

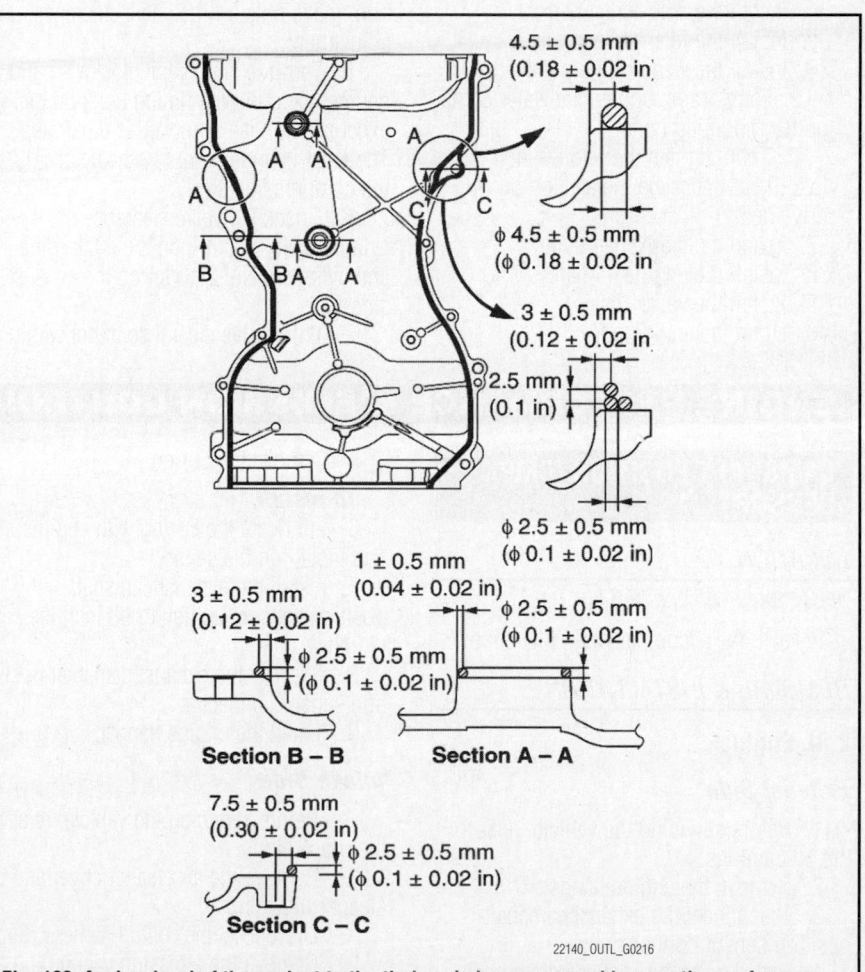

Fig. 162 Apply a bead of the sealant to the timing chain case assembly mounting surface as shown

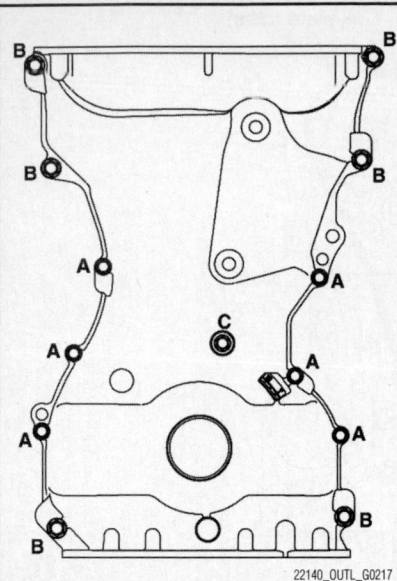

Fig. 163 Insert the bolts to the timing chain case assembly as shown, and tighten them to the specified torque

33. Install the power steering oil pump.
34. Install the auto-tensioner.
35. Install the idler pulley.
36. Install the water pump pulley.
    a. Temporarily tighten the water pump pulley mounting bolts.
    b. Then, tighten them to 80 inch lbs. (9 Nm) after the installation of drive belt.
37. Install the crankshaft pulley.
38. Install the engine oil pan.
39. Install the valve cover.
40. Drain the engine oil.

41. Install the engine room under cover and engine side.

VALVE LASH

ADJUSTMENT

2.4L Engine

See Figure 164.

1. Before servicing the vehicle, refer to the precautions.
2. Remove the valve cover.

➡**Perform the valve clearance check and adjustment at the engine cold state.**

3. If the engine has been turned, turn the crankshaft clockwise, and align the timing mark on the exhaust camshaft sprocket against the upper face of the cylinder head. The No.1 cylinder should now be at TDC on the compression stroke.
4. Using a thickness gauge, measure the valve clearance of cylinders No. 1 and 2 on the intake side and cylinders No. 1 and 3 on the exhaust side. If not within specification, make note for the valve clearance adjustment.
5. Turn the crankshaft clockwise 360 degrees, and put the timing mark on the exhaust camshaft sprocket as illustrated. The No.1 cylinder should now be at TDC on the compression stroke.
6. Check the valve clearance of cylinders No. 3 and 4 on the intake side and cylinders No. 2 and 4 on the exhaust side.
7. If the valve clearance is not within

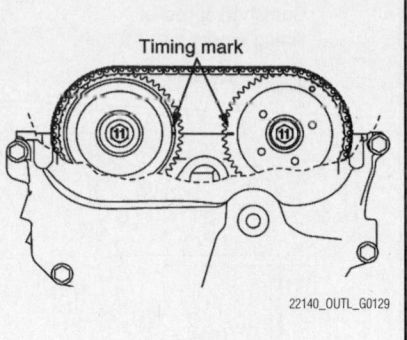

Fig. 164 Timing sprockets with No. 1 cylinder at TDC

specification, remove the camshaft and the valve tappet.
8. Using a micrometer, measure the thickness of the removed valve tappet.
9. Calculate the thickness of the newly installed valve tappet through the following equation.
    • A: thickness of newly installed valve tappet
    • B: thickness of removed valve tappet
    • C: measured valve clearance
10. Calculate the thickness of the valve tappet as follows:
    • Intake valve: $A = B + [C - 0.20\ mm\ (0.008\ inch)]$
    • Exhaust valve: $A = B + [C - 0.30\ mm\ (0.012\ inch)]$
11. Install the valve tappet selected and put the camshaft in position.
12. After installing the timing chain, measure the valve clearance again.
13. Install the valve cover.

ENGINE PERFORMANCE & EMISSION CONTROLS

CAMSHAFT POSITION (CMP) SENSOR

LOCATION

See Figures 165 and 166.

Refer to the accompanying illustrations.

REMOVAL & INSTALLATION

2.4L Engine

Exhaust Side

1. Before servicing the vehicle, refer to the precautions.
2. Remove the engine hanger.
3. Disconnect the exhaust camshaft position sensor connector.
4. Remove the exhaust camshaft position sensor.

5. Remove the O-ring.

To install:
6. Lubricate the O-ring with engine oil and install on the sensor.
7. Install the exhaust camshaft position sensor. Tighten to 98 inch lbs. (11 Nm).
8. Connect the exhaust camshaft position sensor connector.
9. Install the engine hanger.

Intake Side

1. Before servicing the vehicle, refer to the precautions.
2. Remove the air cleaner cover and air cleaner intake hose.
3. Disconnect the control harness clamp.
4. Remove the emission vacuum hose and pipe assembly mounting bolt.

5. Disconnect the intake camshaft position sensor connector.
6. Remove the intake camshaft position sensor.
7. Remove the O-ring.

To install:
8. Lubricate the O-ring with engine oil and install on the sensor.
9. Install the exhaust camshaft position sensor. Tighten to 98 inch lbs. (11 Nm).
10. Connect the exhaust camshaft position sensor connector.
11. Install the emission vacuum hose and pipe assembly mounting bolt.
12. Connect the control harness clamp.
13. Install the air cleaner cover and air cleaner intake hose.

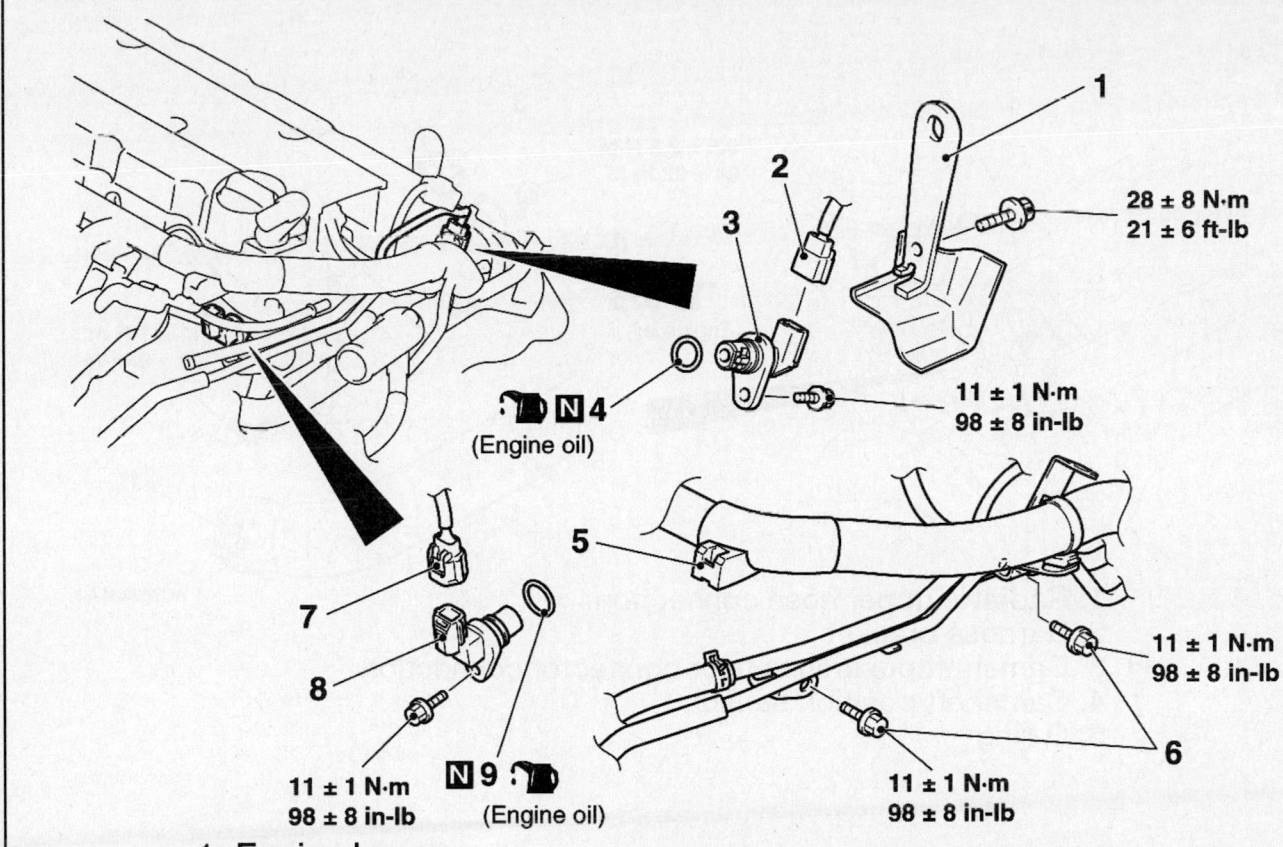

28 ± 8 N·m
21 ± 6 ft-lb

11 ± 1 N·m
98 ± 8 in-lb

N 4
(Engine oil)

11 ± 1 N·m
98 ± 8 in-lb

N 9
(Engine oil)

11 ± 1 N·m
98 ± 8 in-lb

1. Engine hanger
2. Exhaust camshaft position sensor connector connection
3. Exhaust camshaft position sensor
4. O-ring
5. Control harness clamp
6. Emission vacuum hose and pipe assembly mounting bolt
7. Intake camshaft position sensor connector connection
8. Intake camshaft position sensor
9. O-ring

22140_OUTL_G0231

Fig. 165 CMP sensor location—2.4L engine

3.0L Engine

1. Before servicing the vehicle, refer to the precautions.
2. Drain the engine coolant.
3. Remove the battery and battery tray.
4. Disconnect the radiator upper house.
5. Remove the harness bracket.
6. Disconnect the camshaft position sensor connector.
7. Remove the camshaft position sensor.
8. Remove the O-ring.

To install:

9. Lubricate the O-ring with engine oil and install on the sensor.

10. Install the exhaust camshaft position sensor. Tighten to 84 inch lbs. (10 Nm).
11. Connect the exhaust camshaft position sensor connector.
12. Install the harness bracket.
13. Connect the radiator upper house.
14. Install the battery and battery tray.
15. Fill the engine with coolant.

CRANKSHAFT POSITION (CKP) SENSOR

LOCATION

See Figures 167 and 168.

Refer to the accompanying illustrations.

REMOVAL & INSTALLATION

2.4L Engine

1. Before servicing the vehicle, refer to the precautions.
2. Remove the air cleaner assembly.
3. Disconnect the crankshaft position sensor connector.
4. Remove the crankshaft position sensor.
5. Remove the O-ring.

To install:

6. Lubricate the O-ring with engine oil and install on the sensor.
7. Install the exhaust crankshaft position sensor. Tighten to 98 inch lbs. (11 Nm)

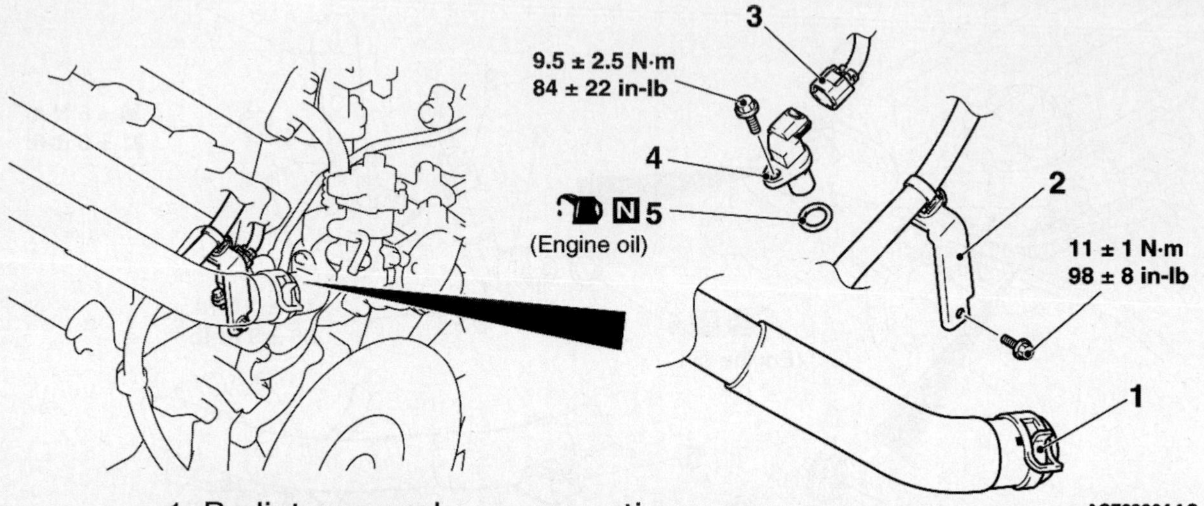

9.5 ± 2.5 N·m
84 ± 22 in-lb

N 5
(Engine oil)

11 ± 1 N·m
98 ± 8 in-lb

1. Radiator upper hose connection
2. Harness bracket
3. Camshaft position sensor connector connection
4. Camshaft position sensor
5. O-ring

AC703804 AC

22140_OUTL_G0232

Fig. 166 CMP sensor location—3.0L engine

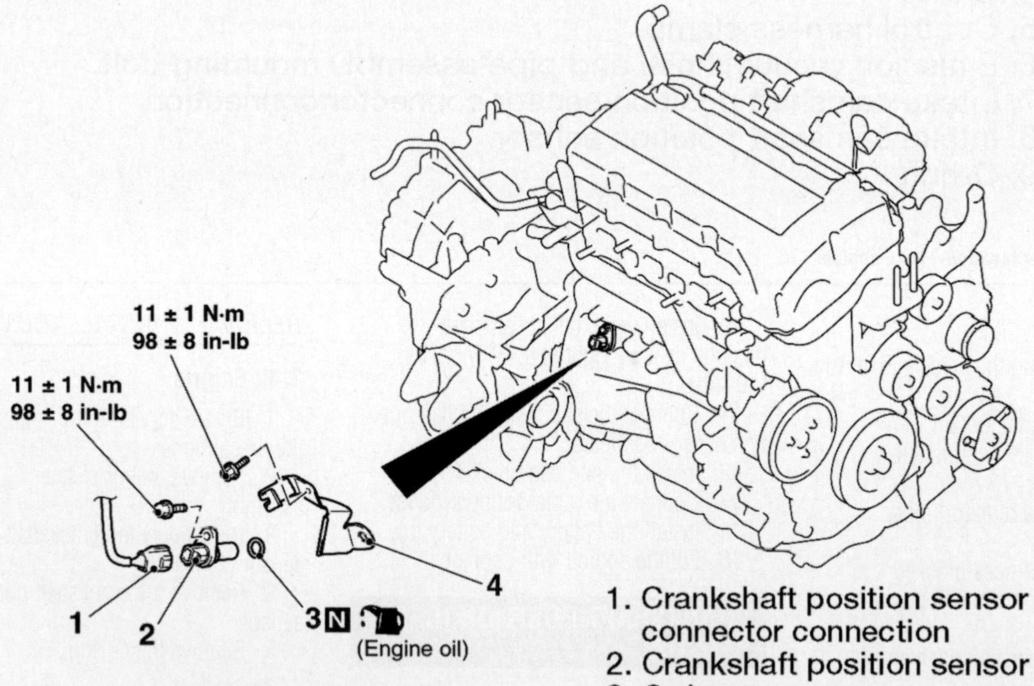

11 ± 1 N·m
98 ± 8 in-lb

11 ± 1 N·m
98 ± 8 in-lb

N
(Engine oil)

1. Crankshaft position sensor
   connector connection
2. Crankshaft position sensor
3. O-ring
4. Crankshaft position sensor cover

22140_OUTL_G0233

Fig. 167 CKP sensor location—2.4L engine

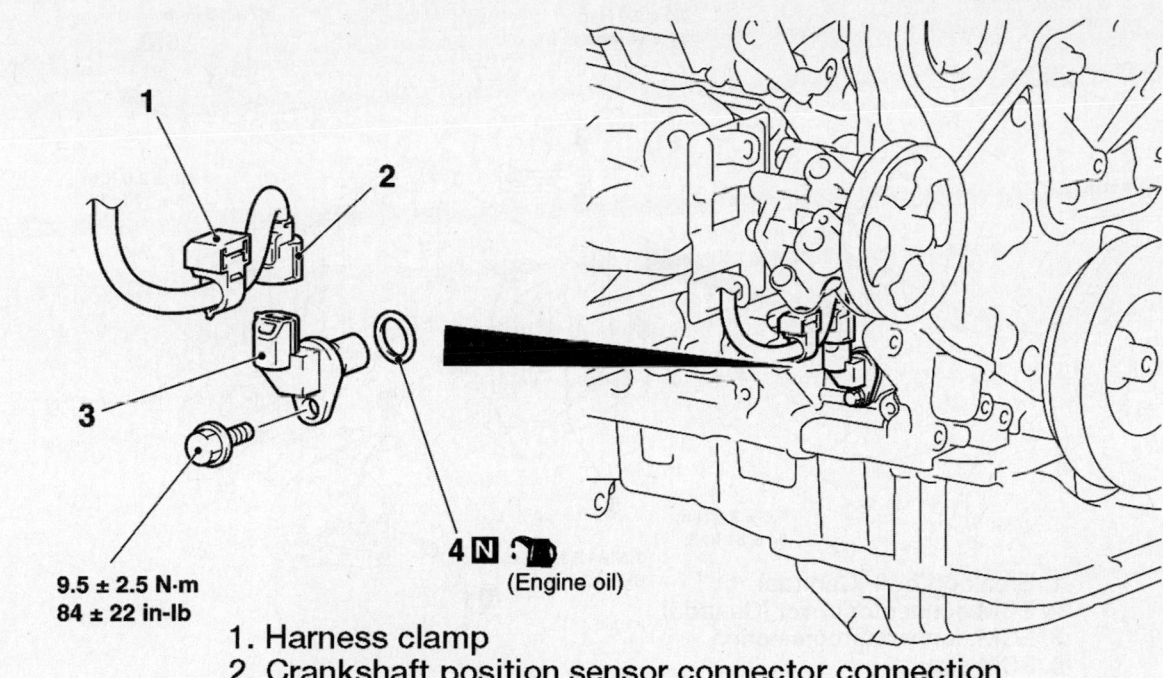

9.5 ± 2.5 N·m
84 ± 22 in-lb

(Engine oil)

1. Harness clamp
2. Crankshaft position sensor connector connection
3. Crankshaft position sensor
4. O-ring

22140_OUTL_G0234

Fig. 168 CKP sensor location—3.0L engine

8. Connect the crankshaft position sensor connector.

9. Install the air cleaner assembly.

3.0L Engine

1. Before servicing the vehicle, refer to the precautions.

2. Remove the harness clamp.

3. Disconnect the crankshaft position sensor connector connection

4. Remove the crankshaft position sensor.

5. Remove the O-ring.

To install:

6. Lubricate the O-ring with engine oil and install on the sensor.

7. Install the exhaust crankshaft position sensor. Tighten to 84 inch lbs. (10 Nm).

8. Connect the crankshaft position sensor connector.

9. Install the harness clamp.

ELECTRONIC CONTROL MODULE (ECM)

LOCATION

See Figure 169.

Refer to the accompanying illustration.

REMOVAL & INSTALLATION

> ※※ **WARNING**
>
> **When the ECM is replaced, do not replace the immobilizer ECU (WCM) or KOS-ECU simultaneously. When multiple ECUs are to be replaced, always replace only one ECU at a time and complete necessary ID registration using a scan tool.**

> ※※ **WARNING**
>
> **After the ECM replacement, idling speed may be unstable because the MFI engine learning is not completed. To make it stable, let the system learn the idling.**

> ※※ **WARNING**
>
> **After the ECM replacement, register a key code using a scan tool.**

> ※※ **WARNING**
>
> **To replace the ECM, use an ECM for which the vehicle identification num-**

ber entry and coding are completed. If DTC P0630 vehicle identification number-related or DTC P1676 variant coding-related is set in the ECM, replace it with an ECM to which the same vehicle identification number has been entered and the same coding has been made as for the relevant vehicle.

1. Remove the air cleaner assembly.

2. Disconnect the ECM connector.

3. Remove the ECM bracket.

4. Remove the ECM.

To install:

5. Installation is the reverse of removal.

6. Perform the initialization procedure. Turn the ignition switch to the **ON** position and then to **OFF** position and hold it for at least 10 seconds.

RESET

1. Perform the initialization procedure. Turn the ignition switch to the **ON** position and then to **OFF** position and hold it for at least 10 seconds.

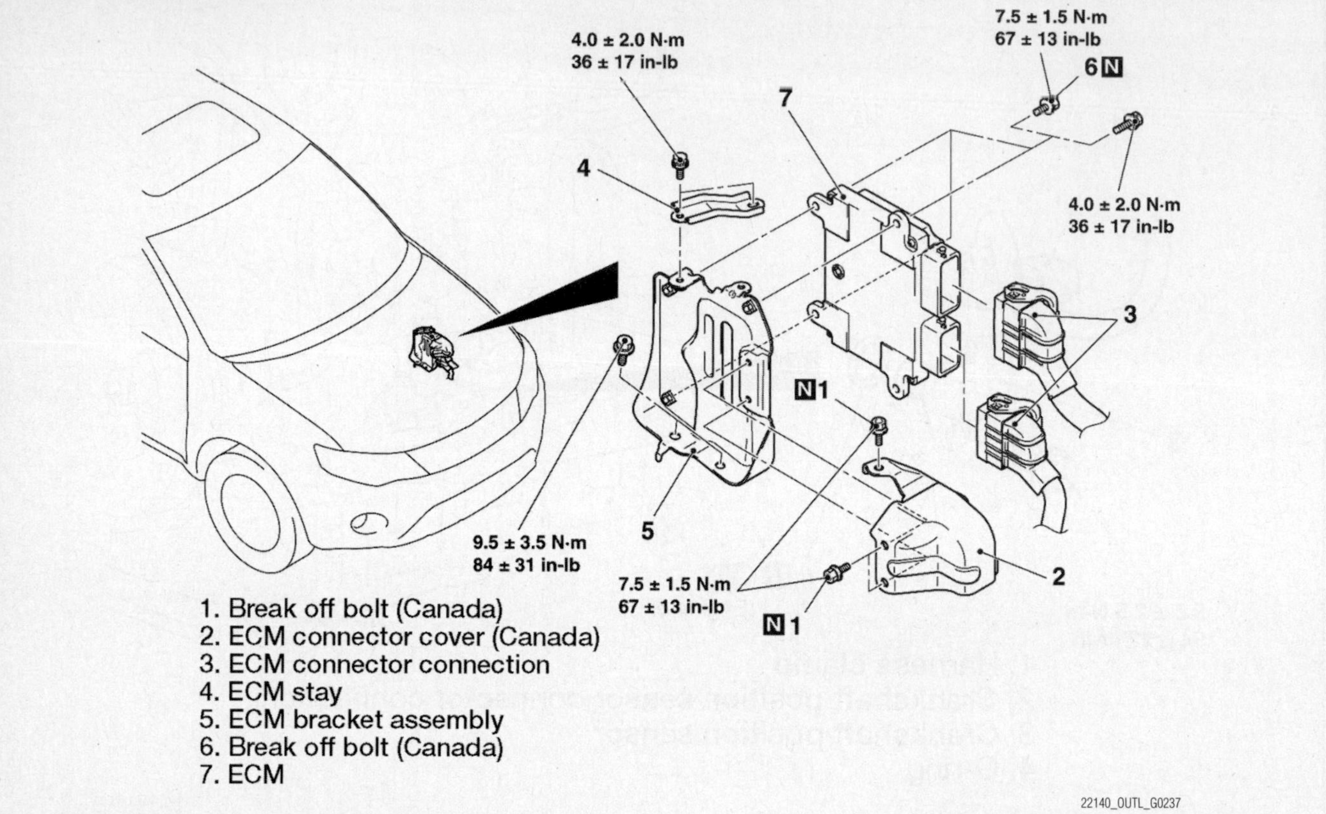

4.0 ± 2.0 N·m
36 ± 17 in-lb

7.5 ± 1.5 N·m
67 ± 13 in-lb

6 N

4

7

4.0 ± 2.0 N·m
36 ± 17 in-lb

3

N 1

5

9.5 ± 3.5 N·m
84 ± 31 in-lb

7.5 ± 1.5 N·m
67 ± 13 in-lb

N 1

2

1. Break off bolt (Canada)
2. ECM connector cover (Canada)
3. ECM connector connection
4. ECM stay
5. ECM bracket assembly
6. Break off bolt (Canada)
7. ECM

22140_OUTL_G0237

Fig. 169 ECM location

ENGINE COOLANT TEMPERATURE (ECT) SENSOR

LOCATION

See Figures 170 and 171.

Refer to the accompanying illustrations.

REMOVAL & INSTALLATION

1. Before servicing the vehicle, refer to the precautions.
2. Drain the engine coolant to a level below the intake manifold.
3. Unplug the sensor wiring harness.
4. Unplug the ECT sensor electrical connector

5. Unthread and remove the sensor from the engine.

➡**Use a deep socket and an extension to reach the ECT sensor, remove the ECT sensor from the thermostat housing**

To install:

6. Coat the threads of the sensor with a suitable sealant and thread into the housing. Tighten to 22 ft. lbs. (30 Nm).
7. Refill the cooling system to the proper level.
8. Connect the electrical connector to the sensor.

HEATED OXYGEN (HO2S) SENSOR

REMOVAL & INSTALLATION

✳✳ CAUTION

The temperature of the exhaust system is extremely high after the engine has been run. To prevent personal injury, allow the exhaust system to cool completely before removing sensor from the exhaust system.

1. Detach the negative battery cable.
2. Raise and safely support the vehicle.
3. Detach the electrical connector from the oxygen sensor.
4. Using socket MD998770, or equivalent oxygen sensor socket, remove the oxygen sensor.

To install:

5. If installing old oxygen sensor, coat the threads with anti-seize compound. New sensors are already coated. Take care not to contaminate the oxygen sensor probe with the anti-seize compound.
6. Install the oxygen sensor into the exhaust manifold. Tighten the sensor, using the correct tool, to 37 ft. lbs. (50 Nm)
7. Attach the wiring to the sensor.

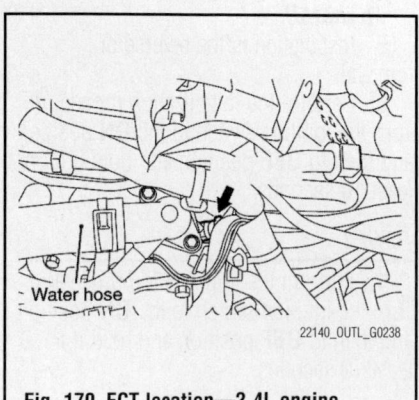

Water hose

22140_OUTL_G0238

Fig. 170 ECT location—2.4L engine

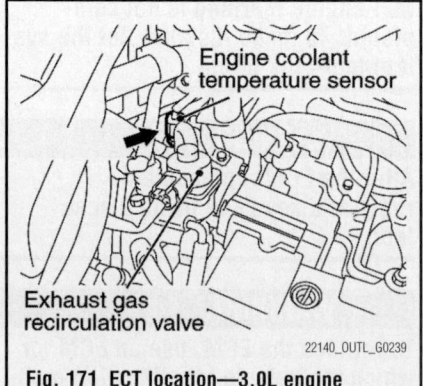

Engine coolant
temperature sensor

Exhaust gas
recirculation valve

22140_OUTL_G0239

Fig. 171 ECT location—3.0L engine

8. Carefully lower the vehicle.

9. Connect the negative battery cable.

INTAKE AIR TEMPERATURE (IAT) SENSOR

LOCATION

See Figure 172.

Refer to the accompanying illustration.

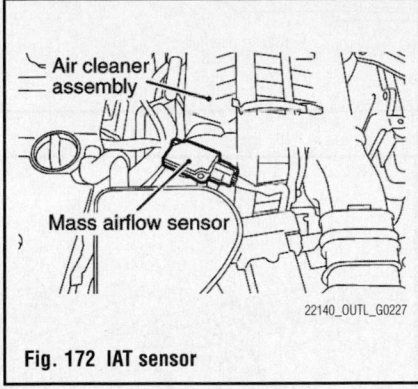

Fig. 172 IAT sensor

REMOVAL & INSTALLATION

1. Before servicing the vehicle, refer to the precautions.

2. Replacing the sensor requires disconnecting the electrical connector, then carefully removing the lid of the air filter housing.

➡**Handle the sensor assembly carefully, protecting it from impact, extremes of temperature and/or exposure to shop chemicals.**

To install:

3. Installation is the reverse of the removal procedure.

KNOCK SENSOR (KS)

LOCATION

See Figures 173 and 174.

Refer to the accompanying illustrations.

REMOVAL & INSTALLATION

1. Before servicing the vehicle, refer to the precautions.

✴ WARNING

Do not drop or hit the knock sensor against other components. Internal damage may result, and the knock sensor will need to be replaced.

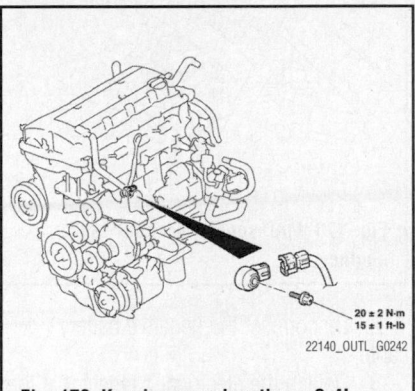

Fig. 173 Knock sensor location—2.4L engine

2. Remove the intake manifold assembly.

3. Disconnect the knock sensor connector.

4. Remove the knock sensor.

To install:

5. Install the knock sensor. Tighten to 15 ft. lbs. (20 Nm).

6. Connect the knock sensor connector.

7. Install the intake manifold assembly.

8. When the knock sensor replacement is performed, use a scan tool to initialize the learning value.

MALFUNCTION INDICATOR LIGHT (MIL)

RESET PROCEDURE

1. Connect an OBD II scan tool to the diagnostic connector.

2. Clear DTCs.

3. The MIL should turn **OFF**.

MASS AIR FLOW (MAF) SENSOR

LOCATION

See Figures 175 and 176.

Refer to the accompanying illustrations.

REMOVAL & INSTALLATION

1. Before servicing the vehicle, refer to the precautions.

2. Disconnect the negative battery cable.

3. Replacing the sensor requires disconnecting the electrical connector, then

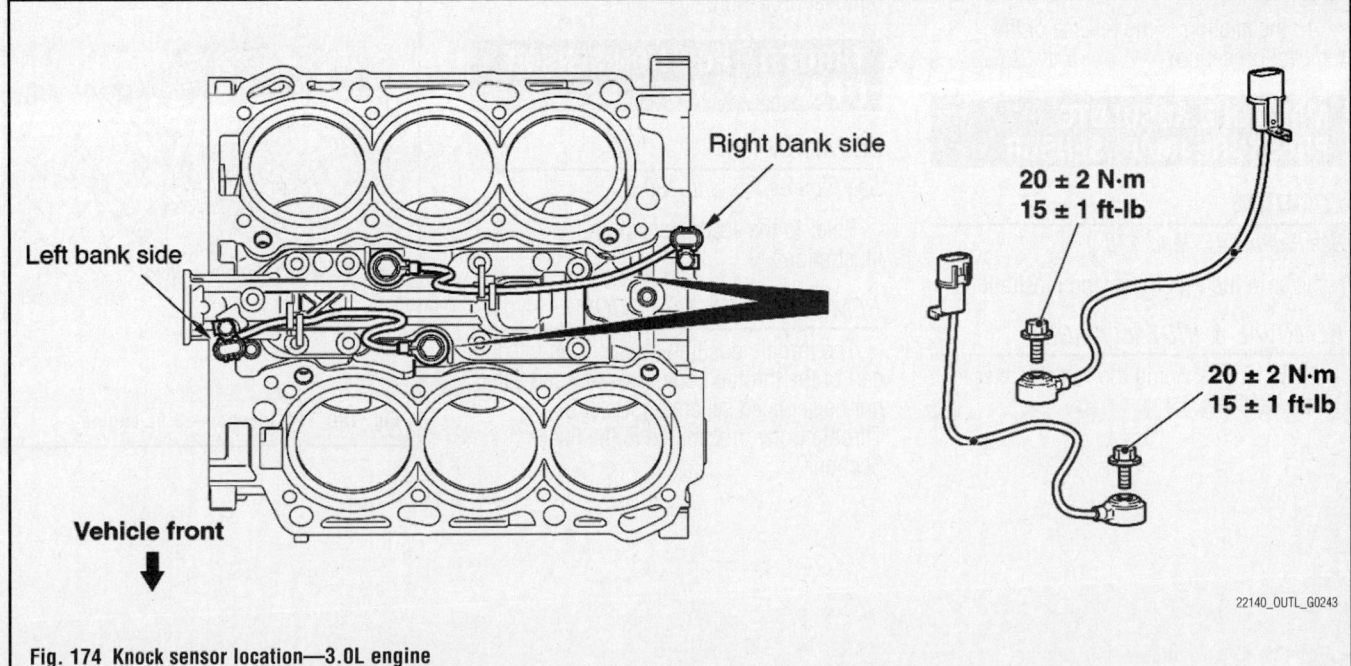

Fig. 174 Knock sensor location—3.0L engine

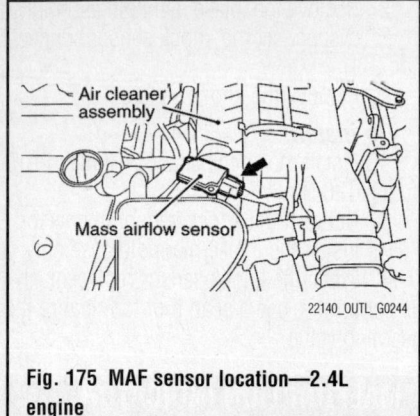

Fig. 175 MAF sensor location—2.4L engine

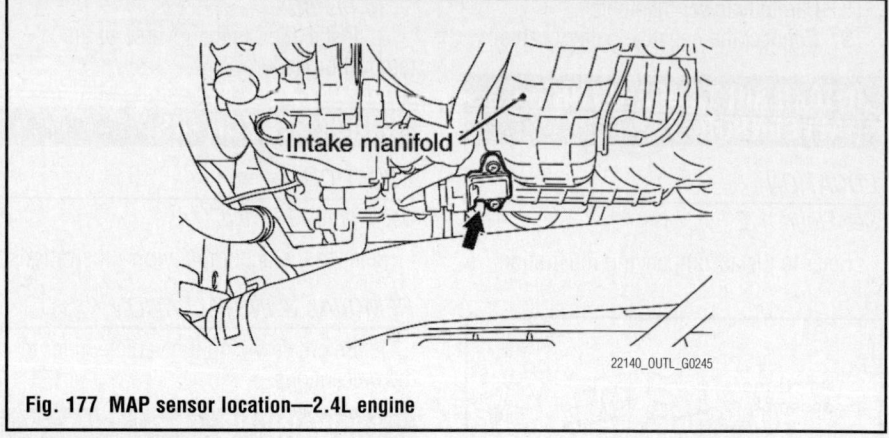

Fig. 177 MAP sensor location—2.4L engine

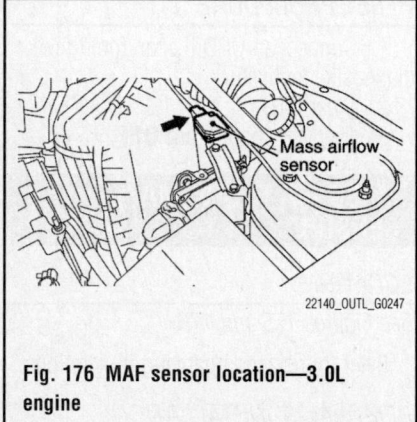

Fig. 176 MAF sensor location—3.0L engine

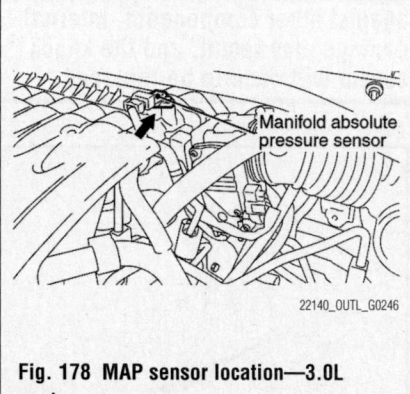

Fig. 178 MAP sensor location—3.0L engine

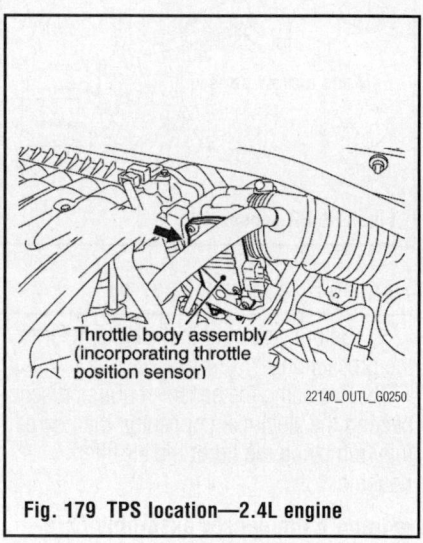

Fig. 179 TPS location—2.4L engine

carefully removing the lid of the air filter housing.

➡**Handle the sensor assembly carefully, protecting it from impact, extremes of temperature and/or exposure to shop chemicals.**

4. Installation is the reverse of the removal procedure.

MANIFOLD ABSOLUTE PRESSURE (MAP) SENSOR

LOCATION

See Figures 177 and 178.

Refer to the accompanying illustrations.

REMOVAL & INSTALLATION

1. Before servicing the vehicle, refer to the precautions.

2. Disconnect the negative battery cable.
3. Detach the electrical connector at the sensor.
4. Remove the sensor.
5. Installation is the reverse of the removal procedure.

THROTTLE POSITION SENSOR (TPS)

LOCATION

See Figures 179 and 180.

Refer to the accompanying illustrations.

REMOVAL & INSTALLATION

The throttle position sensor is an integral part of the throttle body assembly and cannot be serviced separately. Refer to the Throttle Body procedures in the Fuel Section.

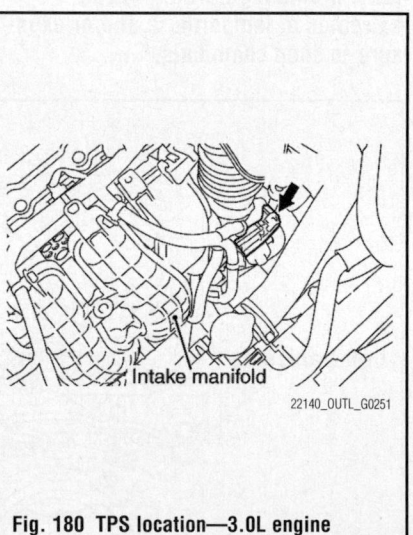

Fig. 180 TPS location—3.0L engine

FUEL   **GASOLINE FUEL INJECTION SYSTEM**

FUEL SYSTEM SERVICE PRECAUTIONS

Safety is the most important factor when performing not only fuel system maintenance, but any type of maintenance. Failure to conduct maintenance and repairs in a safe manner may result in serious personal injury or death. Work on a vehicle's fuel system components can be accomplished safely and effectively by adhering to the following rules and guidelines.

• To avoid the possibility of fire and personal injury, always disconnect the negative battery cable unless the repair or test procedure requires that battery voltage be applied.

• Always relieve the fuel system pressure prior to disconnecting any fuel system component (injector, fuel rail, pressure regulator, etc.) fitting or fuel line connection. Exercise extreme caution whenever relieving fuel system pressure to avoid exposing skin, face and eyes to fuel spray. Please be advised that fuel under pressure may penetrate the skin or any part of the body that it contacts.

• Always place a shop towel or cloth around the fitting or connection prior to loosening to absorb any excess fuel due to spillage. Ensure that all fuel spillage is quickly removed from engine surfaces. Ensure that all fuel-soaked cloths or towels are deposited into a flame-proof waste container with a lid.

• Always keep a dry chemical (Class B) fire extinguisher near the work area.

• Do not allow fuel spray or fuel vapors to come into contact with a spark or open flame.

• Always use a second wrench when loosening or tightening fuel line connection fittings. This will prevent unnecessary stress and torsion on fuel piping. Always follow the proper torque specifications.

• Always replace worn fuel fitting O-rings with new ones. Do not substitute fuel hose where rigid pipe is installed.

FUEL SYSTEM PRESSURE

RELIEVING

See Figure 181.

1. Before servicing the vehicle, refer to the precautions.
2. Turn the ignition to the **OFF** position.
3. Loosen the fuel filler cap to release fuel tank pressure.
4. Remove the rear seat assembly, then

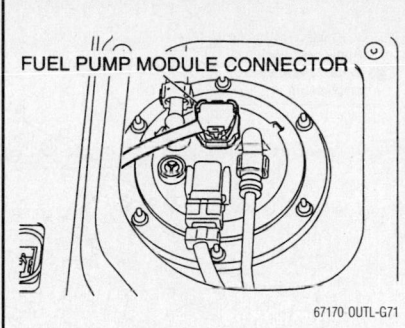

Fig. 181 Fuel pump module connector location

remove the protector and disconnect the fuel pump module connector.

5. Start the vehicle and allow it to run until it stalls from lack of fuel. Turn the key to the **OFF** position.
6. Disconnect the negative battery cable, then reconnect the fuel pump connector. Install the rear seat assembly.

FUEL FILTER

REMOVAL & INSTALLATION

The fuel filter is an integral part of the fuel pump module and is not normally serviced.

FUEL LEVEL SENDING UNIT

LOCATION

See Figure 182.

Refer to the accompanying illustration.

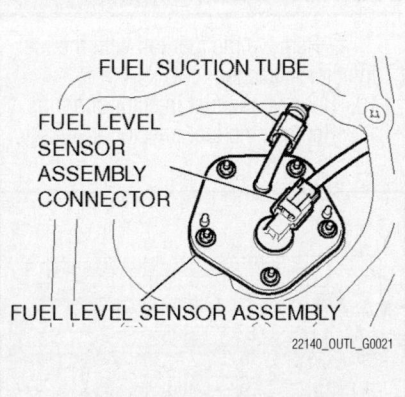

Fig. 182 Fuel level sending unit location

REMOVAL & INSTALLATION

See Figure 183.

1. Before servicing the vehicle, refer to the precautions.

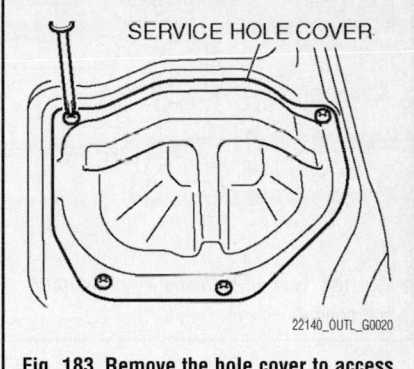

Fig. 183 Remove the hole cover to access the fuel level sending unit.

2. Disconnect the negative battery cable.
3. Remove the rear seats.
4. Remove hole cover and remove the fuel pump module and fuel level sensor.

To install:

5. Installation is the reverse of the removal procedure.
6. Connect the negative battery cable.

FUEL PUMP MODULE

REMOVAL & INSTALLATION

See Figure 184.

1. Before servicing the vehicle, refer to the precautions.
2. Remove the harness electrical connector(s).
3. Properly relieve the fuel system pressure.
4. Remove the rear seat cushion.
5. Remove the retainer screws and service hole cover.
6. Remove the fuel lines.
7. Remove the fuel pump module mounting nuts and plate.

✳✳ WARNING

Be careful not to damage the module unit and float, when removing the fuel pump module from the fuel tank.

8. Remove the fuel pump module (FWD) or Fuel pump and fuel level sensor (AWD) from the service hole.
9. Disassemble the fuel pump module as necessary.

To install:

10. Installation is the reverse of the removal procedure.
11. Replace the fuel pump module gasket with a new one.
12. Torque the fuel pump module place nuts to 22 inch lbs. (2.5 Nm).

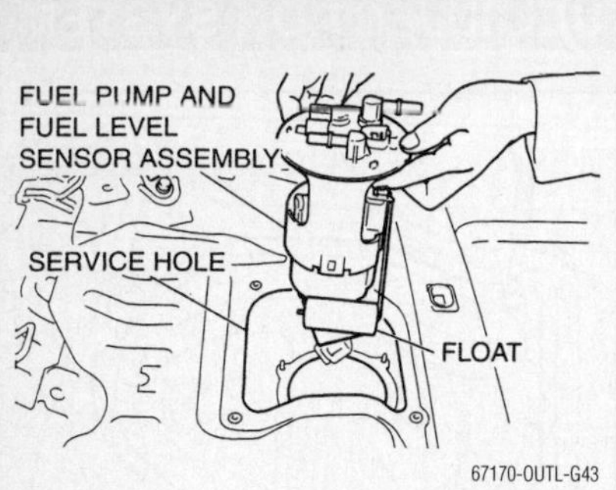
FUEL PUMP AND
FUEL LEVEL
SENSOR ASSEMBLY
SERVICE HOLE
FLOAT
67170-OUTL-G43

Fig. 184 Fuel pump module and related components

FUEL RAIL AND INJECTOR

REMOVAL & INSTALLATION

2.4L Engine

See Figures 185 through 189.

1. Disconnect the fuel pump connector.
2. Remove the engine upper cover.
3. Remove the air cleaner intake hose.
4. Disconnect the control wiring harness.
5. Disconnect the fuel high-pressure hose.
   a. Follow the steps below to unlock the fuel high-pressure hose connector.
   b. Insert a flat-tipped screwdriver [6 mm (0.24 inch) wide and 1 mm (0.04inch) thick] into the retainer of the fuel high-pressure hose connector.

WARNING

When pushing up the retainer of the fuel high-pressure hose connector, pay attention to avoid damage to the retainer.

   c. Turn the flat-tipped screwdriver inserted into the retainer by 90° to push up the retainer and unlock the fuel high-pressure hose connector.

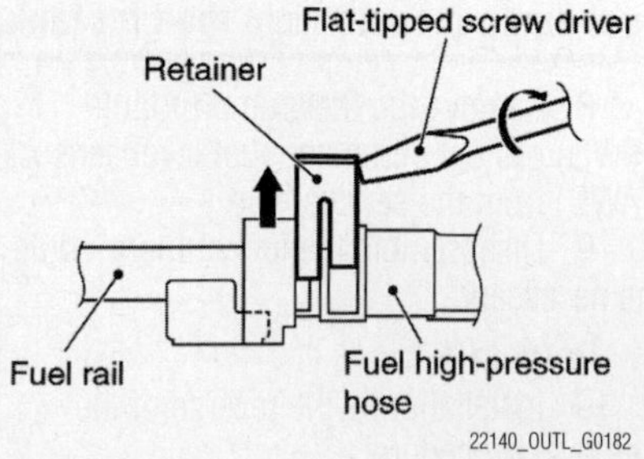
Flat-tipped screw driver
Retainer
Fuel rail
Fuel high-pressure hose
22140_OUTL_G0182

Fig. 185 Turn the flat-tipped screwdriver approximately 90° to the right, and lift the retainer to unlock

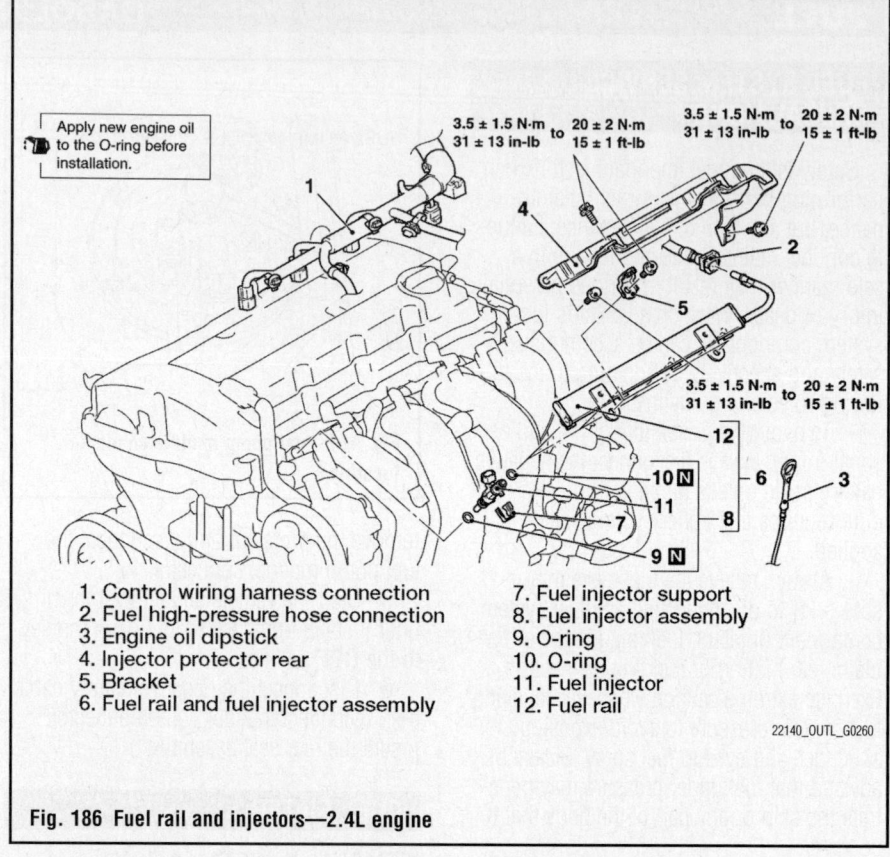

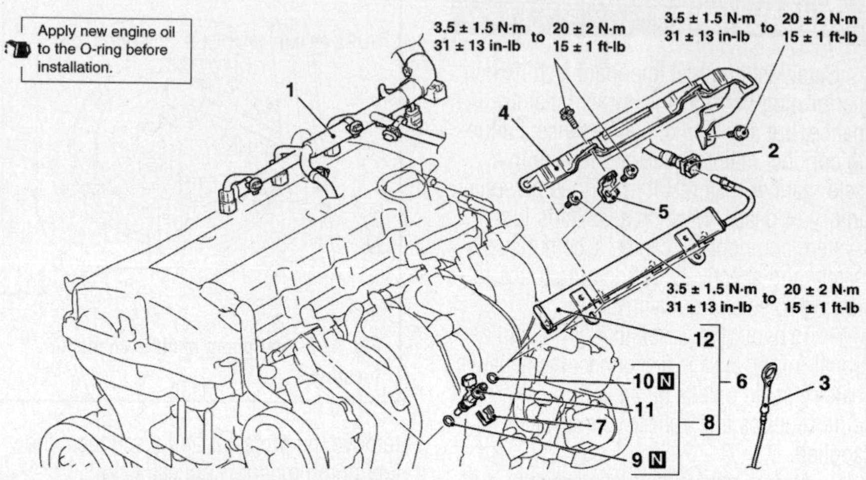
Apply new engine oil to the O-ring before installation.
3.5 ± 1.5 N·m
31 ± 13 in-lb
to
20 ± 2 N·m
15 ± 1 ft-lb
1. Control wiring harness connection
2. Fuel high-pressure hose connection
3. Engine oil dipstick
4. Injector protector rear
5. Bracket
6. Fuel rail and fuel injector assembly
7. Fuel injector support
8. Fuel injector assembly
9. O-ring
10. O-ring
11. Fuel injector
12. Fuel rail
22140_OUTL_G0260

Fig. 186 Fuel rail and injectors—2.4L engine

   d. Disconnect the fuel high-pressure hose.
6. Remove the engine oil dipstick.
7. Remove the rear injector protector.
8. Remove the bracket
9. Remove the fuel rail and fuel injector assembly

WARNING

Do not drop the fuel injector.

   a. Remove the fuel rail with the fuel injectors attached to it.
10. Remove the fuel injector support.
11. Remove the fuel injector assembly.

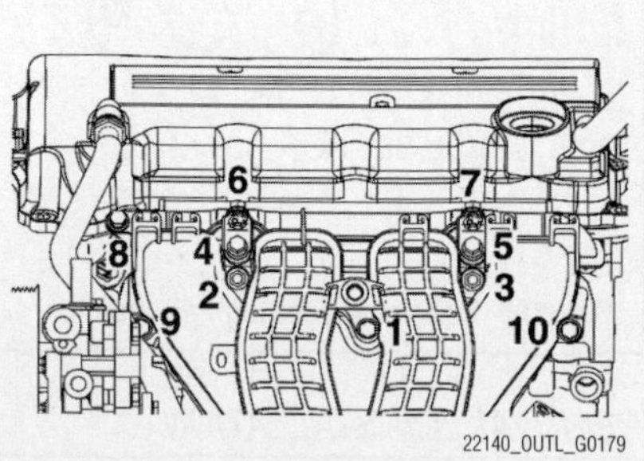
22140_OUTL_G0179

Fig. 187 Exploded view of the intake manifold assembly—2.4L engine

12. Remove the O-ring.

To install:

13. Install the top O-ring
   a. Apply a small amount of new engine oil to the O-ring.
   b. While turning the fuel injector to right and left, install the O-ring to the fuel injector with care to avoid damage to the O-ring.
14. Install the bottom O-ring
   a. Apply a small amount of new engine oil to the O-ring.
   b. Using special tool MB992106, install the O-ring onto the fuel injector paying attention to avoid damage to the O-ring.
15. Install the fuel injector assembly

WARNING

When applying the engine oil, make sure not to allow the engine oil to enter the fuel rail inside.

   a. Apply a small amount of new engine oil to the O-ring.
   b. Turning the fuel injector assembly to right and left, install it to the fuel rail with care not to damage the O-ring. After the installation, check for its smooth rotation. At this time, check that the pro-

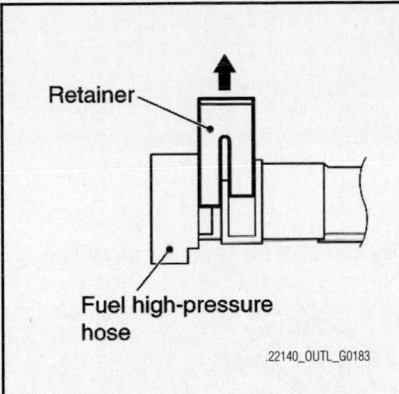

Fig. 188 Pull up the lock of fuel high-pressure hose to unlock before installing

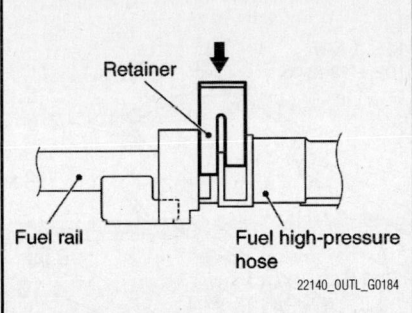

Fig. 189 Install the fuel high-pressure hose to the fuel rail securely and push the lock of fuel high-pressure hose downward and lock thoroughly

jection of the fuel injector assembly is in the centre.

c. If the rotation is not smooth, the O-ring may be caught. Remove the fuel injector assembly and check the O-ring for damage. After this, re-insert it to the fuel rail and check for its smooth rotation.

16. Install the fuel injector support.

a. Install the fuel injector support to the fuel injector groove and fuel rail flange, and fix the fuel injector assembly and fuel rail.

17. Install the fuel rail and fuel injector assembly.

a. Apply a small amount of new engine oil to the O-ring at the end of fuel injector assembly.

✻✻ WARNING

When installing the fuel rail and fuel injector assembly to the intake manifold, pay attention to avoid damage to the O-ring at the end of the fuel injector assembly.

b. Install the fuel rail and fuel injector assembly to the intake manifold.

c. Install the bracket and injector protector rear.

d. Loosen the intake manifold mounting bolts and nuts (Bolts and nuts 1, 2, 3, and 9 shown in the figure).

e. Remove the EGR valve.

f. Loosen the EGR valve support and intake manifold coupling bolts.

g. Loosen the intake manifold stay mounting bolts.

h. Temporarily tighten the mounting bolts and nuts of the inlet manifold, bracket, fuel rail and injector protector rear 31 inch lbs. (4 Nm) in the order of number shown in the figure.

i. Tighten the mounting bolts and nuts of the inlet manifold, bracket, fuel rail and injector protector rear to 15 ft. lbs. (20 Nm) in the order of number shown in the figure again.

j. Tighten the inlet manifold stay mounting bolts to 15 ft. lbs. (20 Nm).

k. Tighten the EGR valve support and intake manifold coupling bolts to 15 ft. lbs. (20 Nm).

l. Install the EGR valve.

18. Install the bracket

19. Install the rear injector protector.

20. Install the engine oil dipstick.

21. Connect the fuel high-pressure hose.

a. Pull up the retainer of fuel high-pressure hose to unlock before installing.

b. Securely insert the fuel rail stopper into the fuel high-pressure hose connector groove to install the fuel high-pressure hose to the fuel rail.

✻✻ WARNING

When pushing in the retainer of the fuel high-pressure hose connector, pay attention to avoid damage to the retainer.

c. After the installation of the fuel high-pressure hose, slightly pull the fuel high-pressure hose to check that it is connected securely. At this time, also check that there is approximately 0.04inch (1 mm) play.

d. Push in the retainer of the fuel high-pressure hose connector to lock the fuel high-pressure hose and fuel rail.

22. Connect the control wiring harness.

23. Install the air cleaner intake hose.

24. Install the engine upper cover.

3.0L Engine

See Figures 189 through 190.

1. Disconnect the fuel pump connector.

2. Remove the intake manifold plenum.

3. Disconnect the fuel injector connectors.

4. Disconnect the fuel high pressure hose.

a. Follow the steps below to unlock the fuel high-pressure hose connector.

b. Insert a flat-tipped screwdriver [6 mm (0.24 inch) wide and 1 mm (0.04inch) thick] into the retainer of the fuel high-pressure hose connector.

✻✻ WARNING

When pushing up the retainer of the fuel high-pressure hose connector, pay attention to avoid damage to the retainer.

c. Turn the flat-tipped screwdriver inserted into the retainer by 90 degrees to push up the retainer and unlock the fuel high-pressure hose connector.

d. Disconnect the fuel high-pressure hose.

5. Remove the insulator.

6. Remove the fuel injector support.

7. Remove the fuel injector assembly.

✻✻ WARNING

Do not drop the fuel injector.

a. Remove the fuel rail with the fuel injectors attached to it.

8. Remove the O-rings.

9. Repeat the process for the other bank.

To install:

10. Install the top O-ring

a. Apply a small amount of new engine oil to the O-ring.

b. While turning the fuel injector to right and left, install the O-ring to the fuel injector with care to avoid damage to the O-ring.

11. Install the bottom O-ring

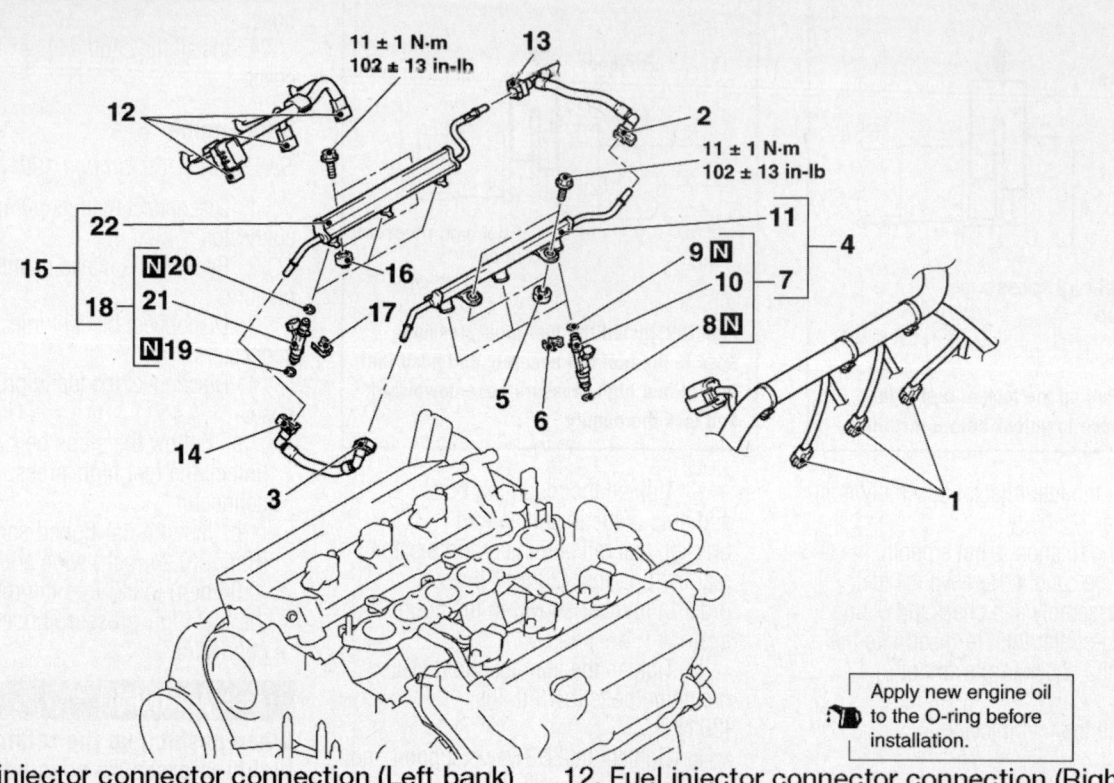

1. Fuel injector connector connection (Left bank)
2. Fuel high pressure hose connection
3. Fuel high pressure hose connection
4. Fuel rail and fuel injector assembly (Left bank)
5. Insulator
6. Fuel injector support
7. Fuel injector assembly
8. O-ring
9. O-ring
10. Fuel injector
11. Fuel rail (Left bank)

12. Fuel injector connector connection (Right bank)
13. Fuel high pressure hose connection
14. Fuel high pressure hose connection
15. Fuel rail and injector assembly (Right bank)
16. Insulator
17. Fuel injector support
18. Fuel injector assembly
19. O-ring
20. O-ring
21. Fuel injector
22. Fuel rail (Right bank)

Apply new engine oil to the O-ring before installation.

22140_OUTL_G0261

Fig. 190 Fuel rail and injectors—3.0L engine

a. Apply a small amount of new engine oil to the O-ring.

b. Using special tool MB992106, install the O-ring onto the fuel injector paying attention to avoid damage to the O-ring.

c. Install the fuel rail with the fuel injectors attached to it.

12. Install the fuel injector assembly.

✳✳ WARNING

When applying the engine oil, make sure not to allow the engine oil to enter the fuel rail inside.

a. Apply a small amount of new engine oil to the O-ring.

b. Turning the fuel injector assembly to

right and left, install it to the fuel rail with care not to damage the O-ring. After the installation, check for its smooth rotation. At this time, check that the projection of the fuel injector assembly is in the centre.

c. If the rotation is not smooth, the O-ring may be caught. Remove the fuel injector assembly and check the O-ring for damage. After this, re-insert it to the fuel rail and check for its smooth rotation.

13. Install the fuel injector support.

a. Install the fuel injector support to the fuel injector groove and fuel rail flange, and fix the fuel injector assembly and fuel rail.

14. Install the insulator.

15. Install the fuel rail and injector assembly.

✳✳ WARNING

When applying the engine oil, make sure not to allow the engine oil to enter the intake manifold inside.

a. Apply a small amount of new engine oil to the O-ring at the end of fuel injector assembly.

✳✳ WARNING

When installing the fuel rail and fuel injector assembly to the intake manifold, pay attention to avoid damage to the O-ring at the end of the fuel injector assembly.

b. Install the fuel rail and fuel injector assembly to the intake manifold.

c. Tighten the fuel rail mounting bolts to 102 inch lbs. (11 Nm).

16. Connect the fuel high pressure hose.

   a. Pull up the retainer of fuel high-pressure hose to unlock before installing.

   b. Securely insert the fuel rail stopper into the fuel high-pressure hose connector groove to install the fuel high-pressure hose to the fuel rail.

❄ WARNING

When pushing in the retainer of the fuel high-pressure hose connector, pay attention to avoid damage to the retainer.

   c. After the installation of the fuel high-pressure hose, slightly pull the fuel high-pressure hose to check that it is connected securely. At this time, also check that there is approximately 0.04inch (1 mm) play.

   d. Push in the retainer of the fuel high-pressure hose connector to lock the fuel high-pressure hose and fuel rail.

17. Connect the fuel injector connectors.
18. Install the intake manifold plenum.
19. Connect the fuel pump connector.

FUEL TANK

REMOVAL & INSTALLATION

1. Disconnect the negative battery cable.
2. Raise and safely support the vehicle securely on jackstands.
3. Drain the fuel tank.
4. On AWD, remove the driveshaft.
5. Remove the center exhaust pipe.
6. Remove the engine under cover.
7. Remove the rear seat assembly.
8. Disconnect the fuel pump module connector.
9. Disconnect the fuel tank differential pressure sensor connector.
10. Disconnect the fuel main pipe.
11. Disconnect the fuel filler hose.
12. Disconnect the fuel leveling hose.
13. Disconnect the fuel tank vapor hoses.
14. Disconnect the parking brake rear cable clamp.
15. Hold the fuel tank assembly with a transaxle jack, and remove the connecting bolts of fuel tank band and the connecting nuts of fuel tank assembly.
16. Remove the fuel tank band.
17. On AWD, remove the rear deferential support member mounting bolt.

   a. Remove the mounting bolts of rear differential support member and tilt the differential carrier assembly.

   b. Hold the fuel tank assembly with the transaxle jack, and remove the connecting bolts of fuel tank band and the connecting nuts of fuel tank assembly.

   c. Remove the fuel tank assembly in the tilt direction, paying attention not to bump it against the rear differential carrier.

18. Slowly lower the fuel tank assembly from the vehicle.

To install:

19. Slowly raise the fuel tank assembly into the vehicle.
20. Install the fuel tank band and tighten to 31 ft. lbs. (41 Nm).
21. On AWD, install the rear deferential support member mounting bolt.
22. Connect the parking brake rear cable clamp.
23. Connect the fuel tank vapor hoses.
24. Connect the fuel leveling hose.
25. Connect the fuel filler hose.
26. Connect the fuel main pipe.
27. Connect the fuel tank differential pressure sensor connector.
28. Connect the fuel pump module connector.
29. Install the rear seat assembly.
30. Install the engine under cover.
31. Install the center exhaust pipe.
32. On AWD, install the driveshaft.
33. Fill the fuel tank.
34. Lower the vehicle.
35. Connect the negative battery cable.
36. Check for leaks.

IDLE SPEED

ADJUSTMENT

Idle speed is maintained by the Powertrain Control Module (PCM). No adjustment is necessary or possible.

THROTTLE BODY

REMOVAL & INSTALLATION

2.4L Engine
See Figure 191.

The throttle position sensor is an integral part of the throttle body assembly and cannot be serviced separately.

1. Before servicing the vehicle, refer to the precautions section.
2. Remove the upper engine cover.
3. Drain the engine coolant.
4. Remove the air cleaner intake hose.
5. Disconnect the throttle body connector.

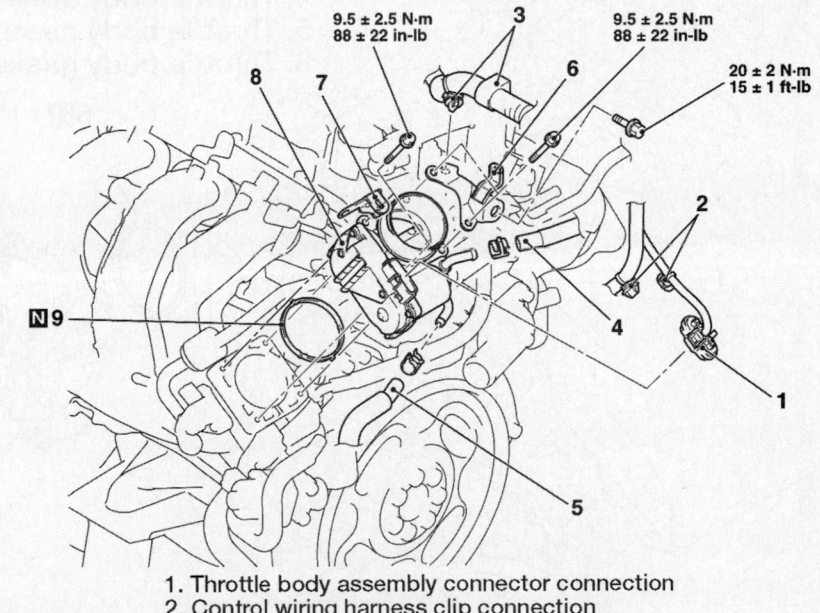

1. Throttle body assembly connector connection
2. Control wiring harness clip connection
3. Battery wiring harness clamp connection
4. Throttle body water feed hose connection
5. Throttle body water return hose connection
6. Throttle body stay
7. Throttle body wiring harness connector bracket
8. Throttle body assembly
9. Throttle body gasket

22140_OUTL_G0252

Fig. 191 Exploded view of the throttle body components—2.4L engine

6. Remove the control wiring harness clip.

7. Remove the battery wiring harness clamp.

8. Disconnect the throttle body water feed and return hoses.

9. Remove the throttle body bracket.

10. Remove the throttle body wiring harness connector bracket

11. Remove the throttle body and gasket.

To install:

12. Install the throttle body and gasket. Tighten to 88 inch lbs. (10 Nm).

    a. Fit the throttle body gasket in the intake manifold groove securely without twisting or damage.

13. Install the throttle body wiring harness connector bracket

14. Install the throttle body bracket.

15. Disconnect the throttle body water feed and return hoses.

16. Install the battery wiring harness clamp.

17. Install the control wiring harness clip.

18. Disconnect the throttle body connector.

19. Install the air cleaner intake hose.

20. Fill the engine with coolant.

21. Install the upper engine cover.

22. Start the engine and check for leaks.

23. Perform the initialization procedure. Turn the ignition switch to the **ON** position and then to **OFF** position and hold it for at least 10 seconds.

3.0L Engine

See Figure 192.

1. Before servicing the vehicle, refer to the precautions section.

2. Drain the engine coolant.

3. Remove the air cleaner intake hose.

4. Disconnect the throttle body connector.

5. Disconnect the throttle body water feed and return hoses.

6. Remove the throttle body bracket.

7. Remove the throttle body wiring harness connector bracket

8. Remove the throttle body and gasket.

To install:

9. Install the throttle body and gasket. Tighten to 15 ft. lbs. (20 Nm).

10. Install the throttle body wiring harness connector bracket

11. Install the throttle body bracket.

12. Disconnect the throttle body water feed and return hoses.

13. Disconnect the throttle body connector.

14. Install the air cleaner intake hose.

15. Fill the engine with coolant.

16. Start the engine and check for leaks.

17. Perform the initialization procedure. Turn the ignition switch to the **ON** position and then to **OFF** position and hold it for at least 10 seconds.

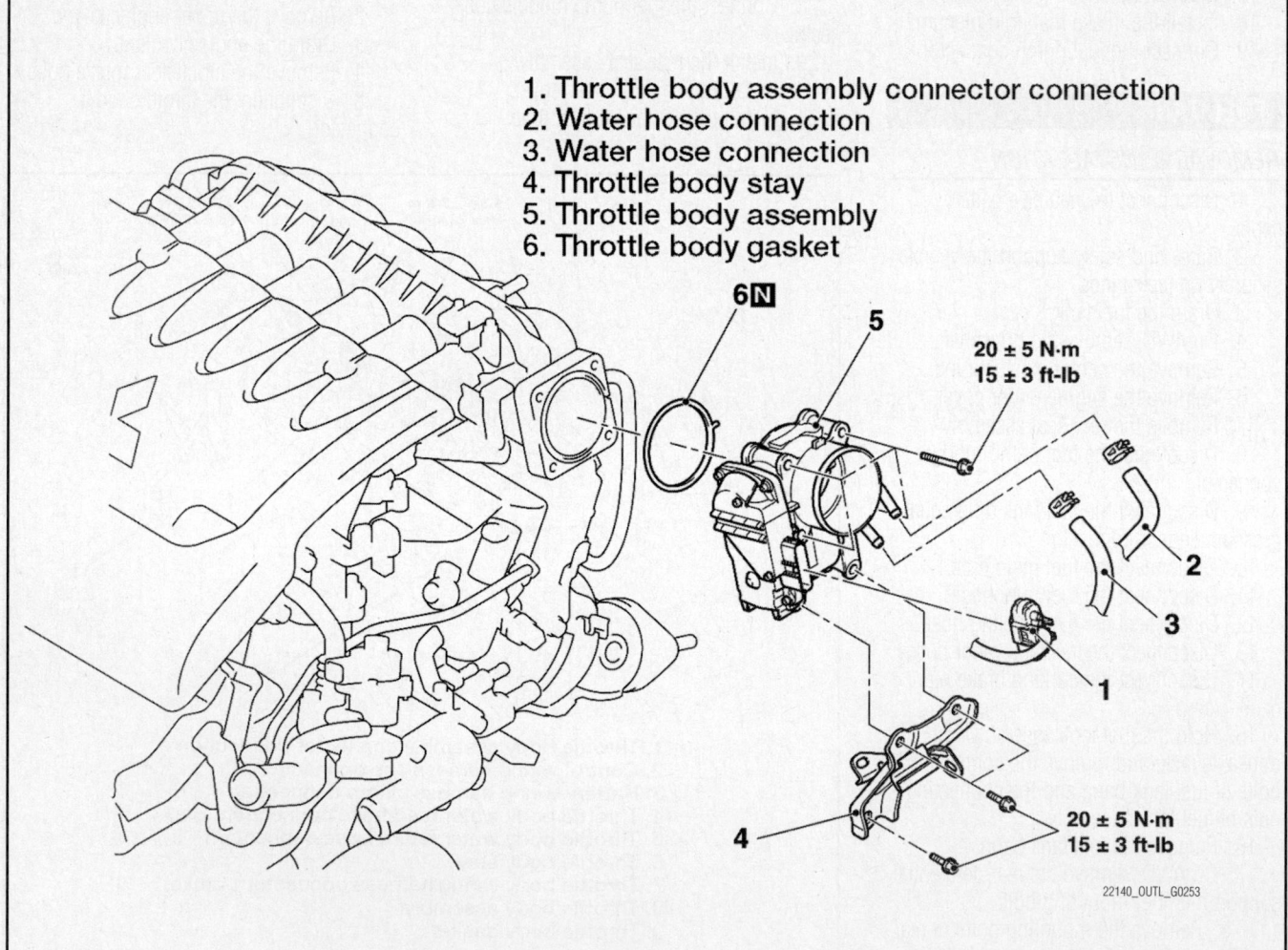

1. Throttle body assembly connector connection
2. Water hose connection
3. Water hose connection
4. Throttle body stay
5. Throttle body assembly
6. Throttle body gasket

20 ± 5 N·m
15 ± 3 ft-lb

20 ± 5 N·m
15 ± 3 ft-lb

22140_OUTL_G0253

Fig. 192 Exploded view of the throttle body components—2.4L engine

HEATING & AIR CONDITIONING SYSTEM

BLOWER MOTOR

REMOVAL & INSTALLATION

See Figure 193.

1. Before servicing the vehicle, refer to the precautions.
2. Remove the bottom cover (passenger side).
3. Disconnect the blower hose.
4. Remove the blower motor

To install:

5. Installation is the reverse of the removal procedure

HEATER CORE

REMOVAL & INSTALLATION

See Figures 194 through 198.

1. Remove the heater unit assembly.
2. Remove the heater core from the heater unit.

To install:

3. Install the heater core into the heater unit.
4. Install the heater unit assembly. Tighten fasteners to 98 inch lbs. (11 Nm).

HEATER UNIT

REMOVAL & INSTALLATION

See Figures 194 through 197 and 199.

1. Drain the engine coolant.
2. Discharge and recycle the refrigerant.
3. Remove the front seat assembly.
4. Remove the rear heater duct
   a. Remove the inner front scuff plate and cowl side trim.
   b. Remove the selector lever, selector lever panel, glove box, lower panel, console side cover and center console.

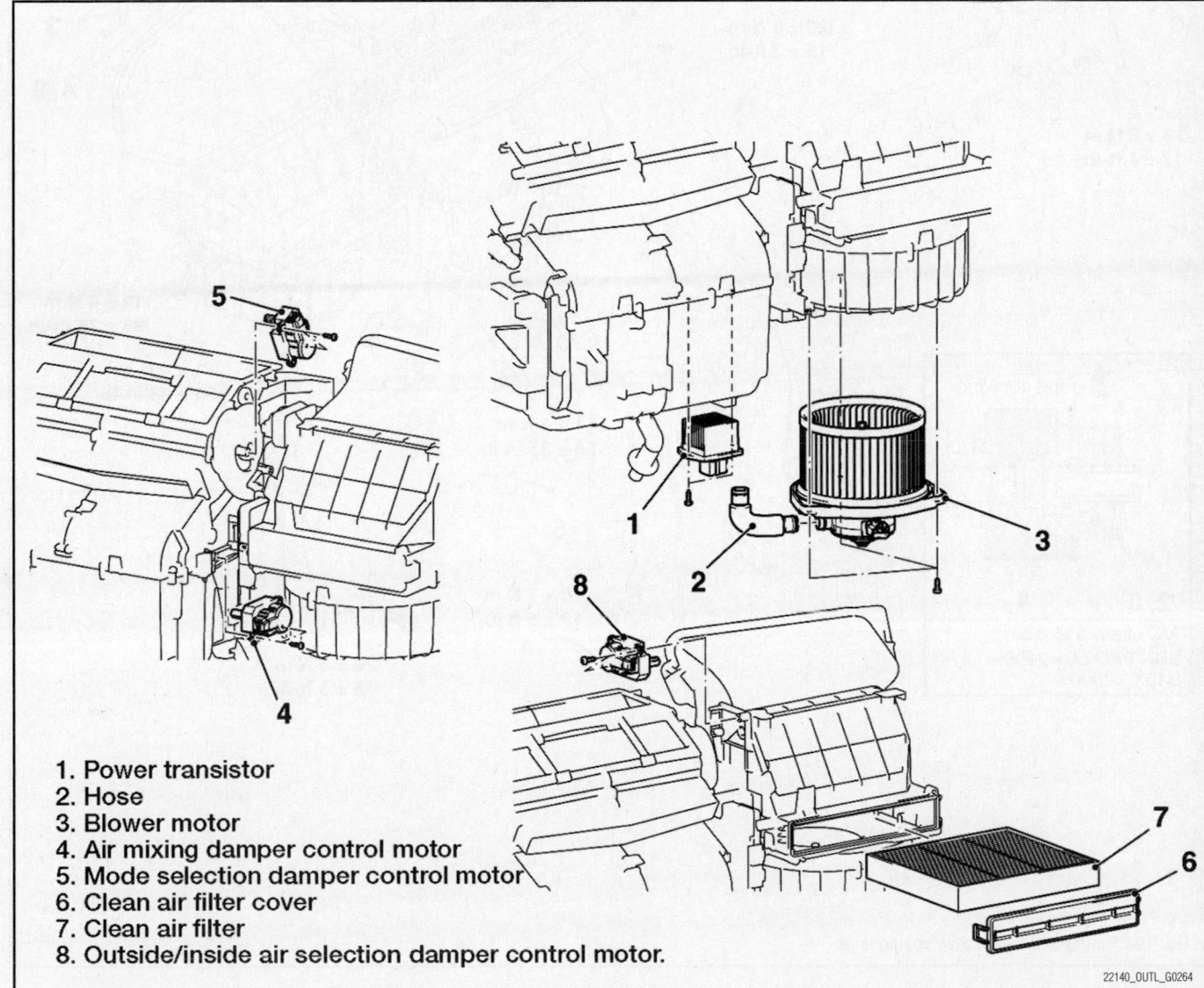

1. Power transistor
2. Hose
3. Blower motor
4. Air mixing damper control motor
5. Mode selection damper control motor
6. Clean air filter cover
7. Clean air filter
8. Outside/inside air selection damper control motor.

22140_OUTL_G0264

Fig. 193 Exploded view of the blower motor and related components

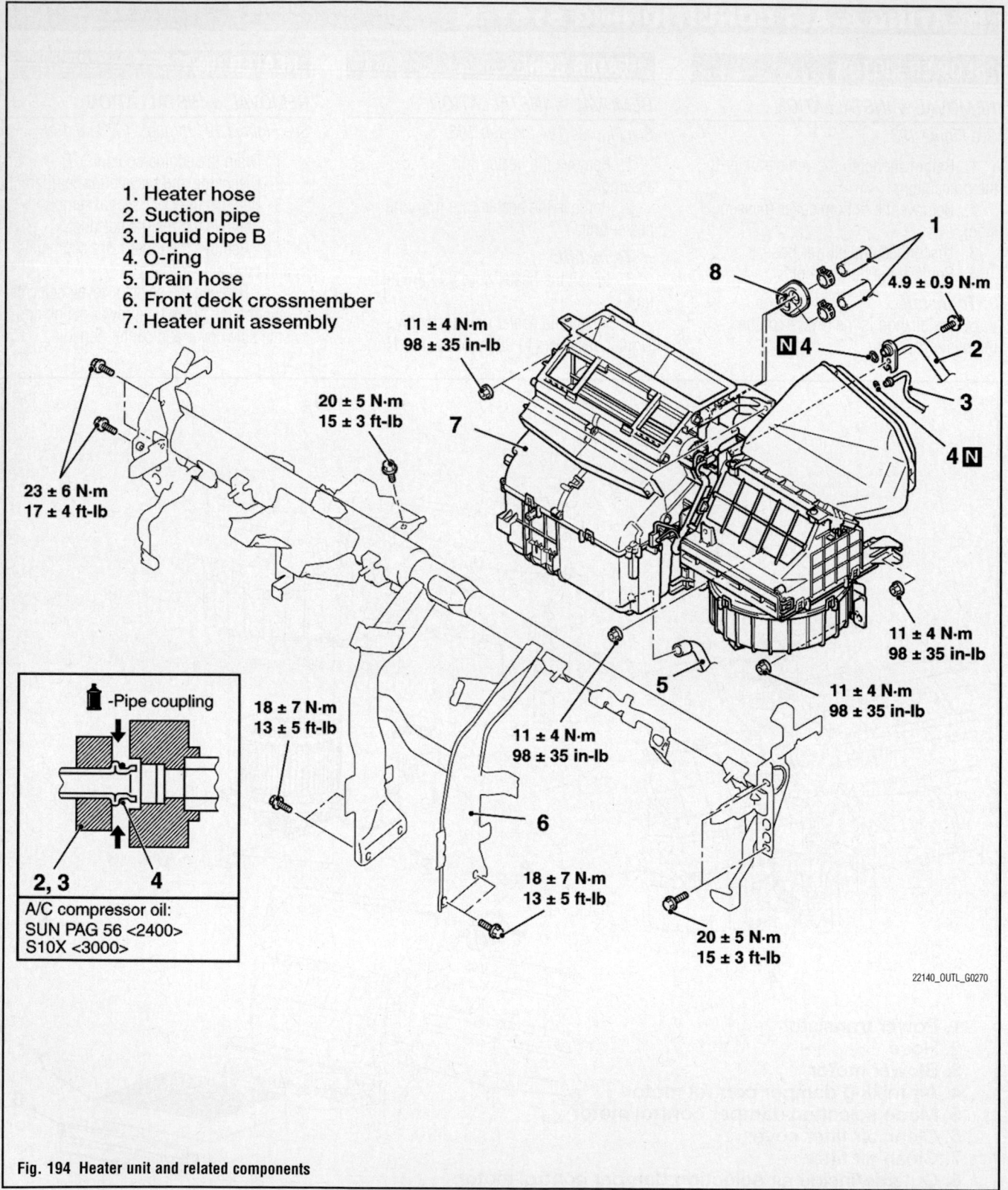

1. Heater hose
2. Suction pipe
3. Liquid pipe B
4. O-ring
5. Drain hose
6. Front deck crossmember
7. Heater unit assembly

4.9 ± 0.9 N·m

11 ± 4 N·m
98 ± 35 in-lb

20 ± 5 N·m
15 ± 3 ft-lb

23 ± 6 N·m
17 ± 4 ft-lb

11 ± 4 N·m
98 ± 35 in-lb

11 ± 4 N·m
98 ± 35 in-lb

18 ± 7 N·m
13 ± 5 ft-lb

11 ± 4 N·m
98 ± 35 in-lb

18 ± 7 N·m
13 ± 5 ft-lb

20 ± 5 N·m
15 ± 3 ft-lb

-Pipe coupling

2, 3   4

A/C compressor oil:
SUN PAG 56 <2400>
S10X <3000>

22140_OUTL_G0270

Fig. 194 Heater unit and related components

c. Remove the rear heater ducts.
5. Remove the instrument panel.
6. Disconnect the wiring harness and clamps.
7. Disconnect the heater hose.
8. Remove the heat protector.
9. Disconnect the suction pipe.

❊❊ WARNING

Use the plug which is not breathable because A/C compressor oil or receiver are highly hygroscopic.

10. Disconnect the liquid pipe.
11. Remove the O-ring.
12. Disconnect the drain hose.
13. Remove the front deck crossmember.
14. Remove the heater unit assembly.

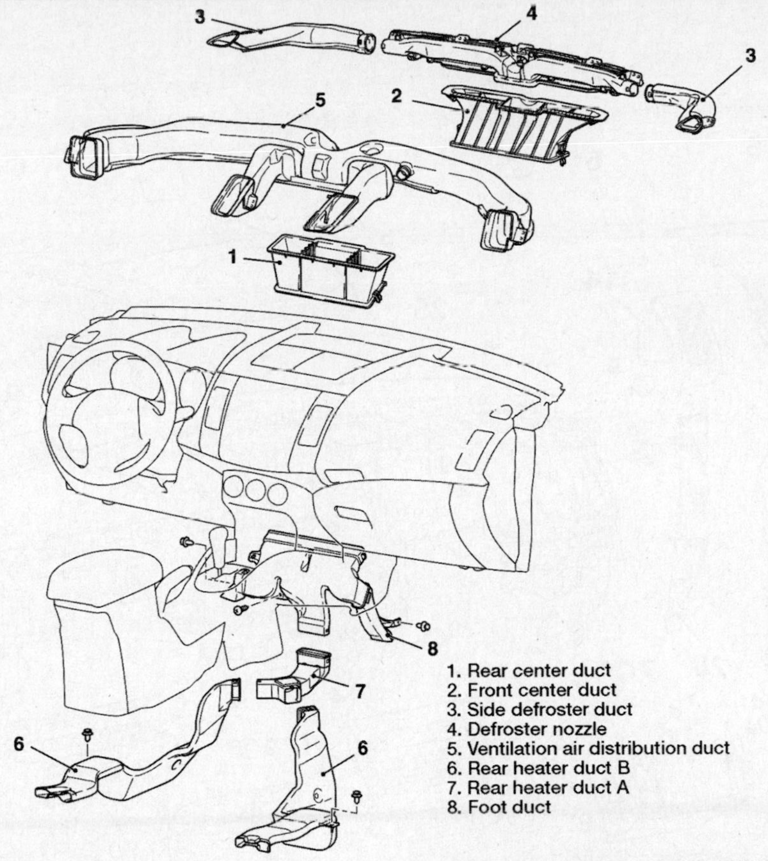

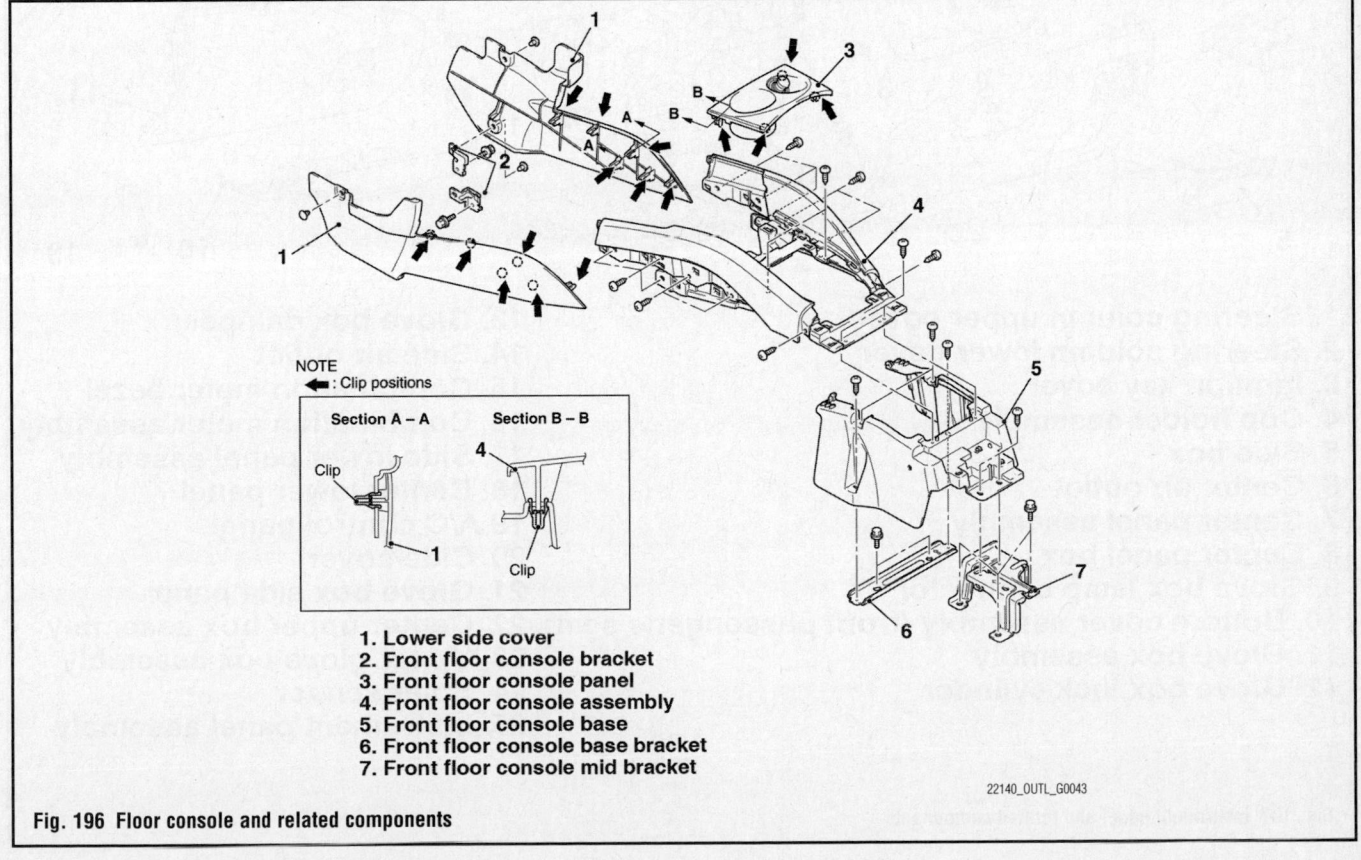

3
4
3
2
5
1
8
7
6
6
1. Rear center duct
2. Front center duct
3. Side defroster duct
4. Defroster nozzle
5. Ventilation air distribution duct
6. Rear heater duct B
7. Rear heater duct A
8. Foot duct
22140_OUTL_G0272

Fig. 195 Ventilation and air distribution ducts

1
3
B
B
A
A
2
4
1
5
7
6
NOTE
: Clip positions
Section A – A
Section B – B
Clip
4
1
Clip
1. Lower side cover
2. Front floor console bracket
3. Front floor console panel
4. Front floor console assembly
5. Front floor console base
6. Front floor console base bracket
7. Front floor console mid bracket
22140_OUTL_G0043

Fig. 196 Floor console and related components

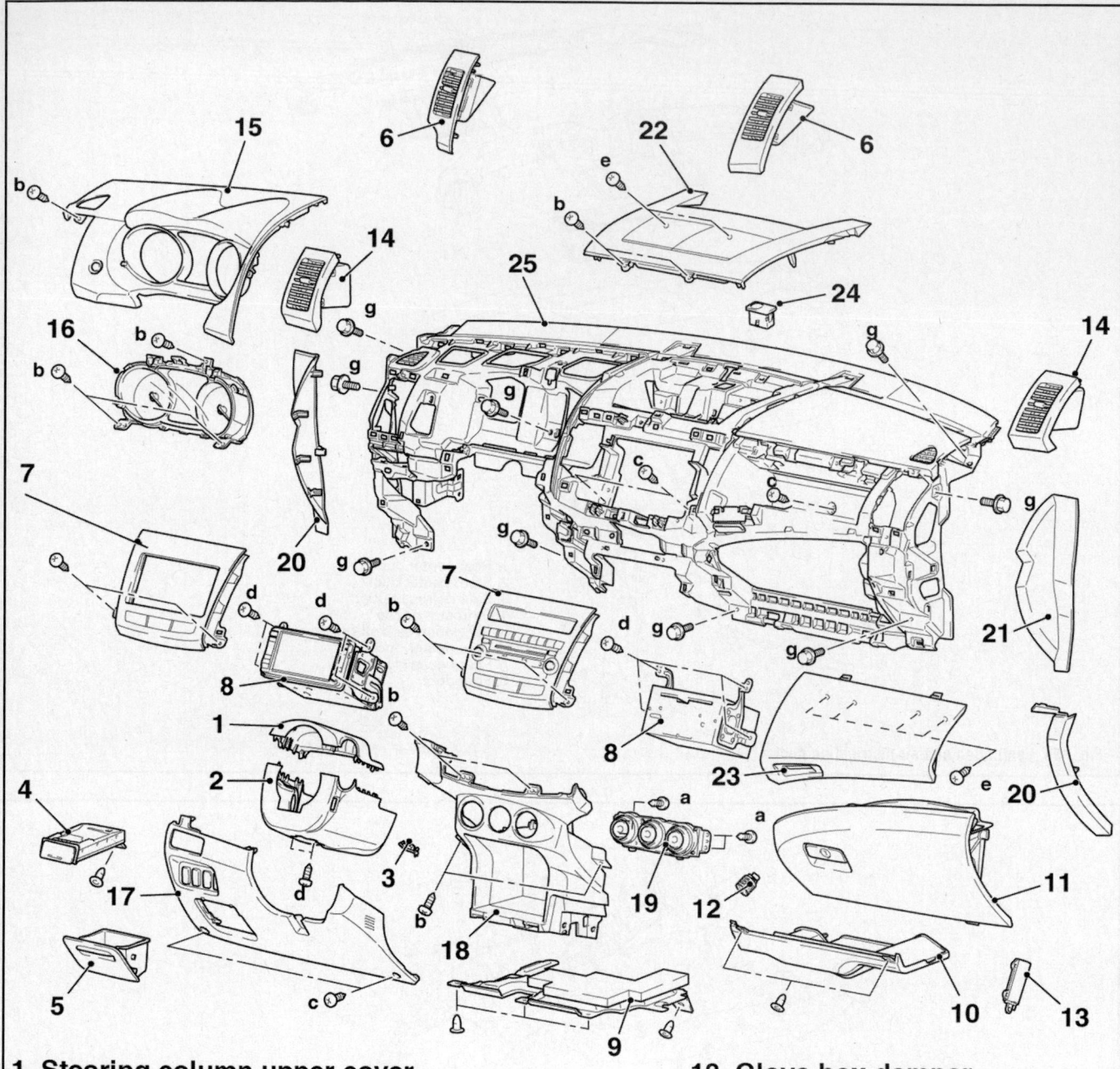

1. **Steering column upper cover**
2. **Steering column lower cover**
3. **Ignition key cover**
4. **Cup holder assembly**
5. **Side box**
6. **Center air outlet**
7. **Center panel assembly**
8. **Center panel box**
9. **Glove box lamp connector**
10. **Bottom cover assembly (front passenger's seat)**
11. **Glove box assembly**
12. **Glove box lock cylinder**
13. **Glove box damper**
14. **Side air outlet**
15. **Combination meter bezel**
16. **Combination meter assembly**
17. **Side lower panel assembly**
18. **Center lower panel**
19. **A/C control panel**
20. **Side cover**
21. **Glove box side panel**
22. **Center upper box assembly**
23. **Upper glove box assembly**
24. **Solar sensor**
25. **Instrument panel assembly**

22140_OUTL_G0044

Fig. 197 Instrument panel and related components

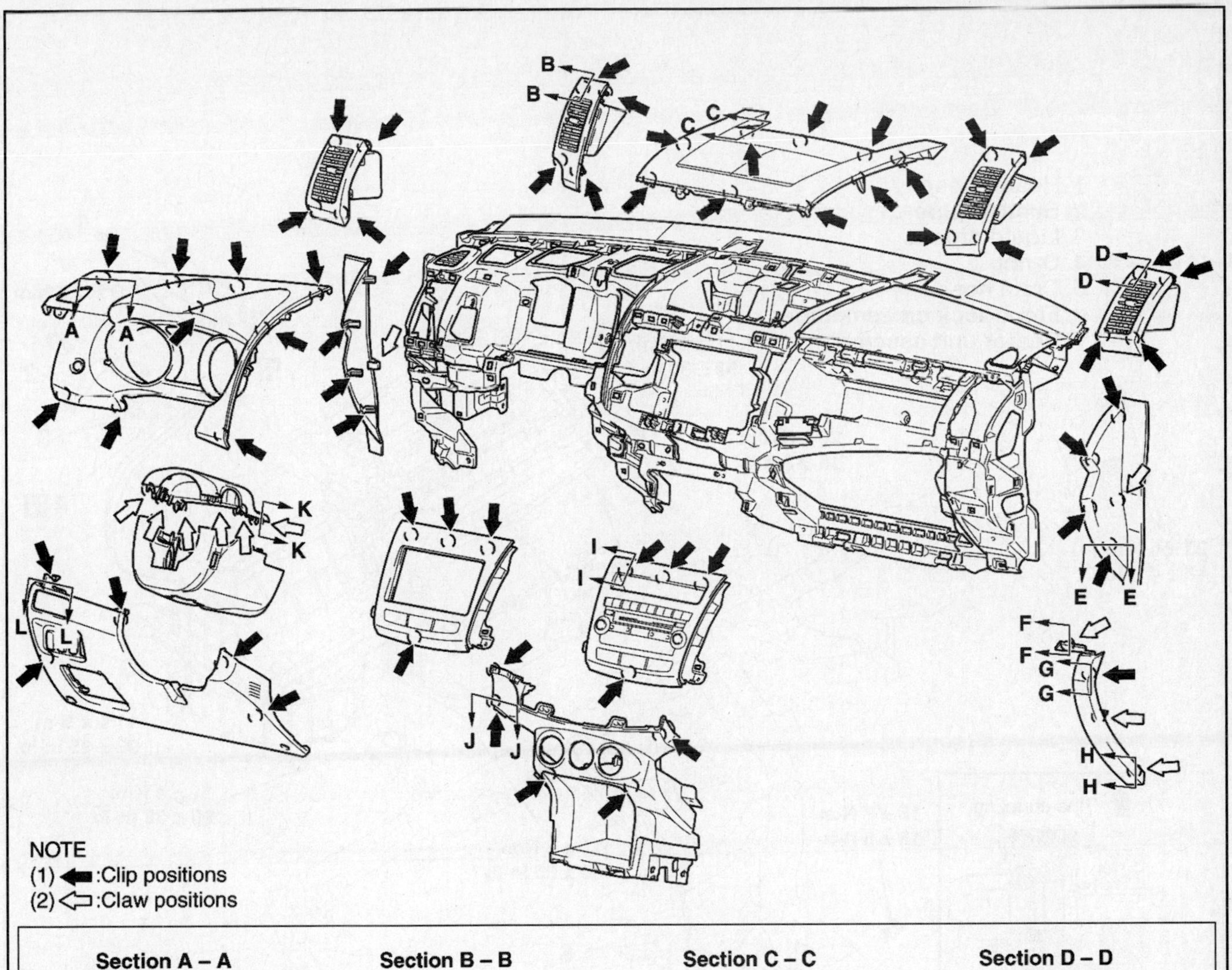

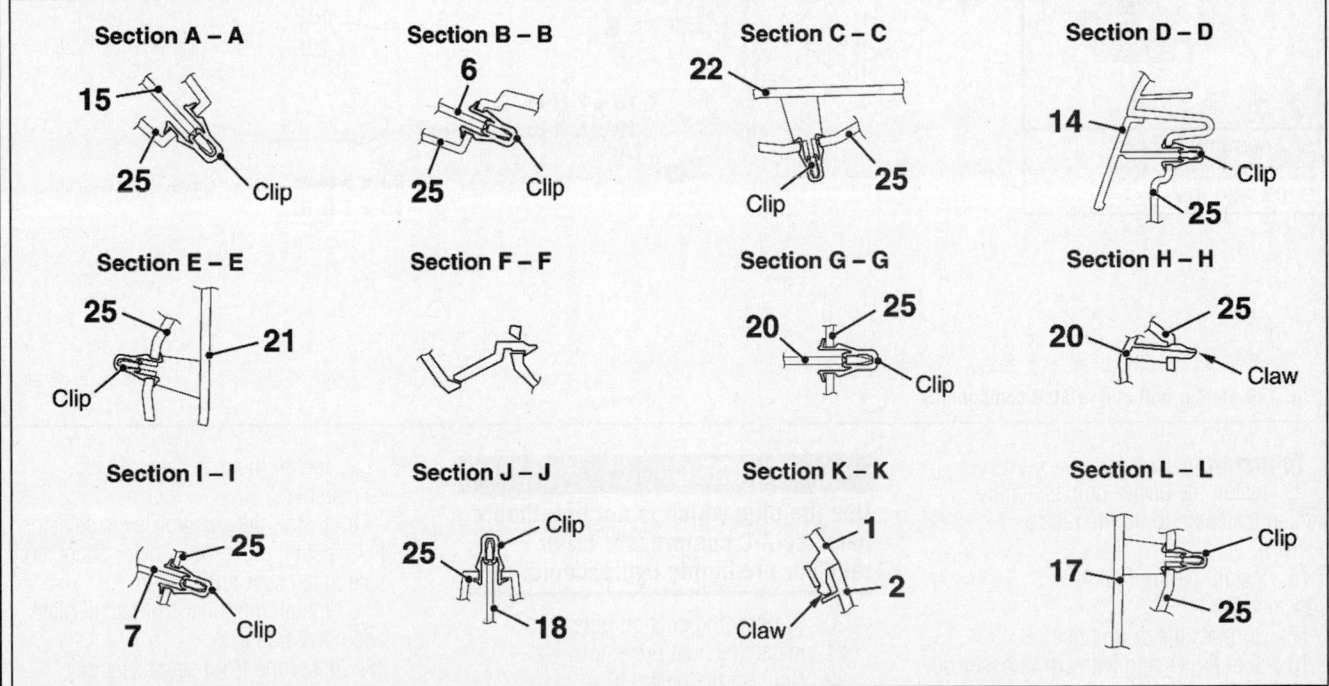

NOTE
(1) ◄■ :Clip positions
(2) ⇦ :Claw positions

Section A – A

15
25 — Clip

Section B – B

6
25 — Cllp

Section C – C

22
Clip   25

Section D – D

14
Clip
25

Section E – E

25
21
Clip

Section F – F

Section G – G

20   25
Clip

Section H – H

20   25
Claw

Section I – I

25
7   Clip

Section J – J

25   Clip
18

Section K – K

1
2
Claw

Section L – L

17
Clip
25

22140_OUTL_G0045

Fig. 198 Instrument panel retention clip locations

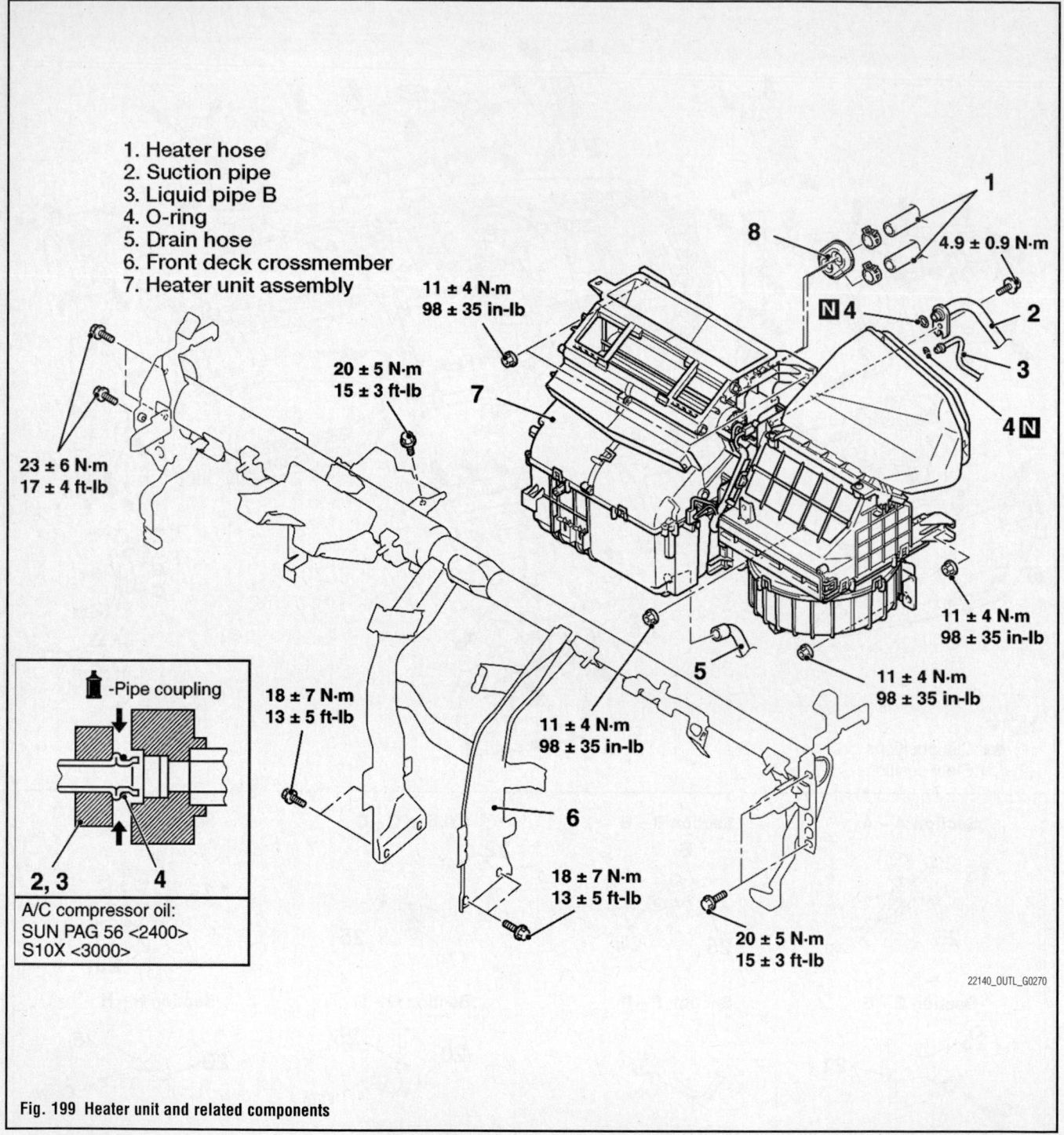

1. Heater hose
2. Suction pipe
3. Liquid pipe B
4. O-ring
5. Drain hose
6. Front deck crossmember
7. Heater unit assembly

8

1

4.9 ± 0.9 N·m

N 4

2

3

4 N

11 ± 4 N·m
98 ± 35 in-lb

7

20 ± 5 N·m
15 ± 3 ft-lb

23 ± 6 N·m
17 ± 4 ft-lb

11 ± 4 N·m
98 ± 35 in-lb

11 ± 4 N·m
98 ± 35 in-lb

5

11 ± 4 N·m
98 ± 35 in-lb

18 ± 7 N·m
13 ± 5 ft-lb

-Pipe coupling

6

2, 3   4

18 ± 7 N·m
13 ± 5 ft-lb

A/C compressor oil:
SUN PAG 56 <2400>
S10X <3000>

20 ± 5 N·m
15 ± 3 ft-lb

22140_OUTL_G0270

Fig. 199 Heater unit and related components

To install:

15. Install the heater unit assembly. Tighten fasteners to 98 inch lbs. (11 Nm).

16. Install the front deck crossmember.

17. Connect the drain hose.

18. Coat the O-ring with compressor oil and install.

19. Connect the liquid pipe. Tighten to 48 inch lbs. (5 Nm).

✷✷ WARNING

Use the plug which is not breathable because A/C compressor oil or receiver are highly hygroscopic.

20. Connect the suction pipe.
21. Install the heat protector.
22. Connect the heater hose.
23. Connect the wiring harness and clamps.
24. Install the instrument panel.

a. Install the rear heater ducts. console.

b. Install the selector lever, selector lever panel, glove box, lower panel, console side cover and center

c. Install the inner front scuff plate and cowl side trim.

25. Install the front seat assembly.

26. Evacuate and charge the refrigerant system.

27. Fill the engine with coolant.

28. Check for leaks.

STEERING

POWER RACK & PINION STEERING GEAR

REMOVAL & INSTALLATION

See Figures 200 and 201.

➡**Prior to removal of the steering gear box, center the front wheels and remove the ignition key. Failure to do so may damage the Supplemental Restraint System (SRS) clock spring and render SRS system inoperative.**

1. Before servicing the vehicle, refer to the precautions.
2. Raise and safely support the vehicle securely on jackstands.
3. Drain the power steering system.
4. Disconnect the negative battery cable.
5. Insulate the cable end with high-quality electrical tape or similar non-conductive wrapping and Wait at least 60 seconds before proceeding.
6. Remove the steering wheel.
7. Remove the steering shaft cover.
8. Remove the engine room under covers.
9. Remove the headlight front height sensor.
10. Remove the front exhaust pipe.
11. Remove the front roll stopper and center member.
12. Remove the steering gear and joint connecting bolt.
13. Remove the self-lock nut connecting the tie-rod end and knuckle.

✳✳ WARNING

Loosen the self-lock nut from the ball joint. Do not remove.

a. Install the ball joint remover (Special tool MB991897 or MB992011).
b. Turn the bolt and knob to make the special tool jaws parallel, then hand-tighten the bolt. After tightening, check that the jaws are still parallel.
c. Unscrew the bolt to disconnect the ball joint.
14. Remove the eye bolt.
15. Remove the pressure and return hoses with their gaskets. connection
16. Remove the flanged nut.
17. Remove the engine rear roll stopper bracket connecting bolt.
18. Remove the front axle crossmember.
a. Jack up and support the crossmember, and remove it.
b. Check the hoses and harnesses for roughness, and then remove the front axle No.1 crossmember assembly with the rear roll stopper and the steering gear and linkage installed.
19. Remove the engine rear roll stopper.
20. Remove the heat protector.
21. Remove the joint cover grommet
22. Remove the flanged bolt.
23. Remove the steering gear and linkage

To install:

24. Install the steering gear and linkage. Tighten the flanged bolt to 52 ft. lbs. (70 Nm).
25. Install the joint cover grommet.
a. Install the joint cover grommet to the steering gear and linkage by aligning the mating marks.
26. Install the heat protector. Tighten to 44 inch lbs. (5 Nm).
27. Install the engine rear roll stopper. Tighten to 37 ft. lbs. (50 Nm).
28. Install the front axle crossmember. Tighten to 82 ft. lbs. (110 Nm).
29. Install the engine rear roll stopper bracket connecting bolt and flanged nut. Tighten to 39 ft. lbs. (53 Nm).
30. Install the pressure and return hoses with their gaskets.
31. Install the eye bolt and tighten to 42 ft. lbs. (57 Nm).
32. Install the self-lock nut connecting the tie-rod end and knuckle. Tighten to 19 ft. lbs. (25 Nm).
33. Install the steering gear and joint connecting bolt. Tighten to 15 ft. lbs. (20 Nm).
34. Install the front roll stopper and center member.
35. Install the front exhaust pipe.
36. Install the headlight front height sensor.
37. Install the engine room under covers.
38. Install the steering shaft cover.
39. Install the steering wheel.
40. Install the wheels.
41. Connect the negative battery cable.
42. Refill the power steering reservoir and bleed the system.
43. Perform a front end alignment.
44. Perform the headlight aiming procedure.

POWER STEERING PUMP

REMOVAL & INSTALLATION

See Figures 202 and 203.

1. Before servicing the vehicle, refer to the precautions.
2. Drain the power steering fluid.
3. Remove the right engine room side cover.
4. Remove the accessory drive belt.
5. Remove the eye bolt and gasket.
6. Disconnect the suction and pressure hoses.

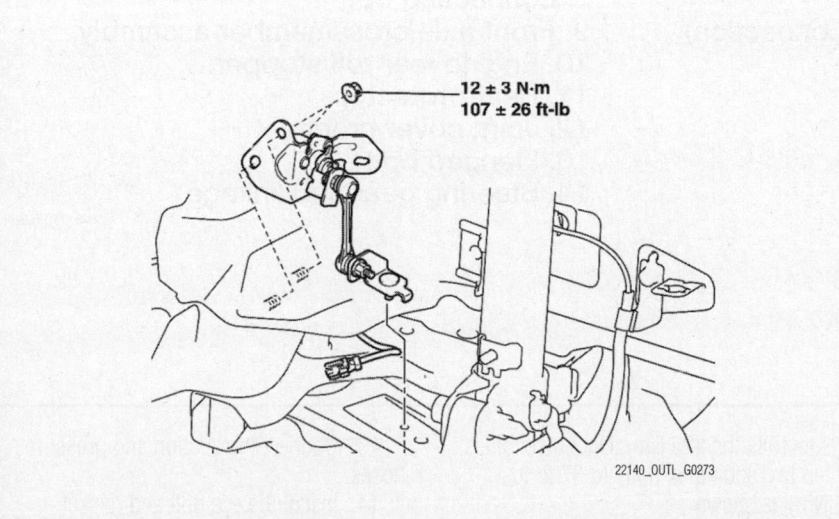

12 ± 3 N·m
107 ± 26 ft-lb

22140_OUTL_G0273

Fig. 200 Front headlight height sensor

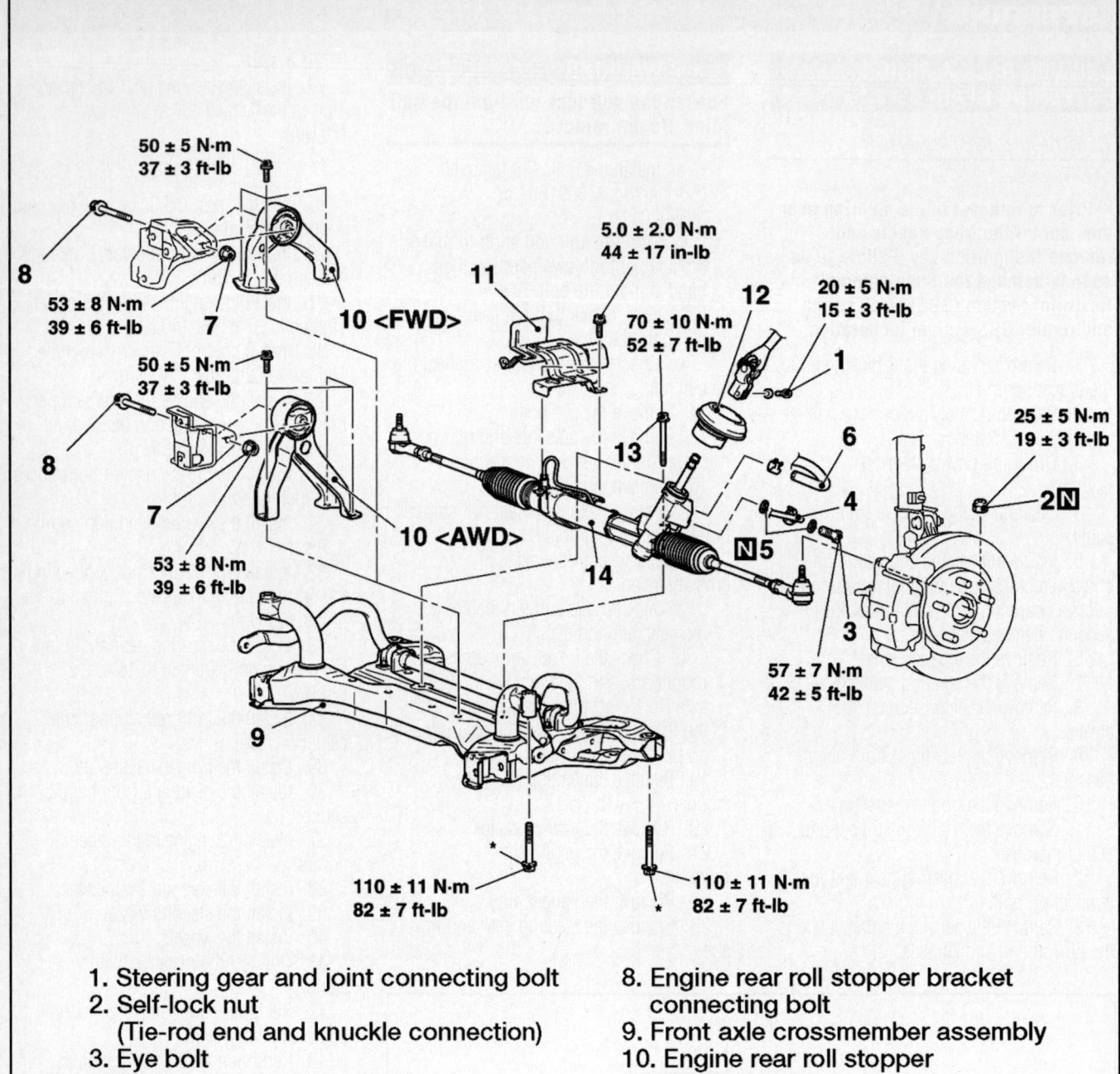

50 ± 5 N·m
37 ± 3 ft-lb

53 ± 8 N·m
39 ± 6 ft-lb

50 ± 5 N·m
37 ± 3 ft-lb

53 ± 8 N·m
39 ± 6 ft-lb

5.0 ± 2.0 N·m
44 ± 17 in-lb

70 ± 10 N·m
52 ± 7 ft-lb

20 ± 5 N·m
15 ± 3 ft-lb

25 ± 5 N·m
19 ± 3 ft-lb

57 ± 7 N·m
42 ± 5 ft-lb

110 ± 11 N·m
82 ± 7 ft-lb

110 ± 11 N·m
82 ± 7 ft-lb

10 <FWD>
10 <AWD>

1. Steering gear and joint connecting bolt
2. Self-lock nut
   (Tie-rod end and knuckle connection)
3. Eye bolt
4. Pressure hose connection
5. Gasket
6. Return hose connection
7. Flanged nut
8. Engine rear roll stopper bracket
   connecting bolt
9. Front axle crossmember assembly
10. Engine rear roll stopper
11. Heat protector
12. Joint cover grommet
13. Flanged bolt
14. Steering gear and linkage

22140_OUTL_G0274

Fig. 201 Rack and pinion steering components

7. Remove the pressure switch.
8. Remove the harness bracket
9. Remove the oil pump assembly.

To install:
10. Install the oil pump assembly. Tighten
bolts to 19 ft. lbs. (25 Nm). On 3.0L engines,
tighten two additional bolts to 37 ft. lbs.
(50 Nm) as shown.
11. Install the harness bracket
12. Install the pressure switch.

13. Connect the suction and pressure
hoses.
14. Install the eye bolt and gasket.
Tighten to 42 ft. lbs. (57 Nm).
15. Install the accessory drive belt.

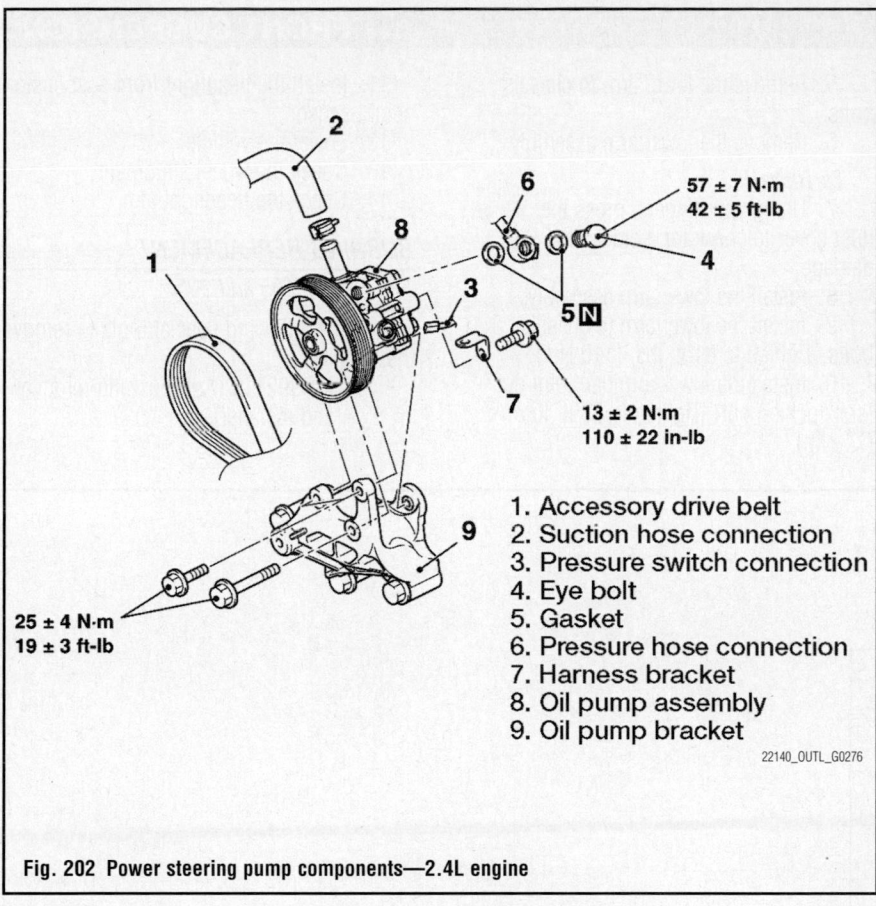

1. Accessory drive belt
2. Suction hose connection
3. Pressure switch connection
4. Eye bolt
5. Gasket
6. Pressure hose connection
7. Harness bracket
8. Oil pump assembly
9. Oil pump bracket

57 ± 7 N·m
42 ± 5 ft-lb

13 ± 2 N·m
110 ± 22 in-lb

25 ± 4 N·m
19 ± 3 ft-lb

22140_OUTL_G0276

Fig. 202 Power steering pump components—2.4L engine

1. Accessory drive belt
2. Pressure switch connection
3. Suction hose connection
4. Eye bolt
5. Gasket
6. Pressure hose connection
7. Oil pump assembly
8. Heat protector
9. Power steering pump bracket

57 ± 7 N·m
42 ± 5 ft-lb

13 ± 2 N·m
110 ± 22 in-lb

25 ± 4 N·m
19 ± 3 ft-lb

50 ± 7 N·m
37 ± 5 ft-lb

22140_OUTL_G0277

Fig. 203 Power steering pump components—3.0L engine

16. Install the right engine room side cover.

17. Fill and bleed the power steering system.

BLEEDING

1. Raise and support the front wheels.

2. Disconnect the all of the injector connectors.

❋❋ WARNING

Perform air bleeding only while cranking the engine. If air bleeding is performed while the engine is running, air could enter the fluid. During air bleeding, refill the steering fluid supply so that the level never falls below the lower mark on the dipstick.

3. Turn the steering wheel all the way to the left and right five or six times while using the starter motor to crank the engine intermittently several times (for 15 to 20 seconds).

4. Connect the all of the injector connectors.

5. Start the engine (idling).

6. Turn the steering wheel to the left and right until there are no air bubbles in the oil reservoir.

7. Confirm that the fluid is not milky, and that the level is between the high and low dipstick marks.

8. Confirm that there is very little change in the fluid level when the steering wheel is turned left and right.

❋❋ WARNING

If the fluid level rises suddenly after the engine is stopped, the air has not been completely bled. If air bleeding is not complete, there will be abnormal noises from the pump and the flow-control valve, and this condition could cause reduce the life of the power steering components.

9. Confirm that the change in the fluid level is no more than 5 mm (0.2 inch) when the engine is stopped and when it is running.

10. If the change of the fluid level is 5 mm (0.2 inch) or more, the air has not been completely bled from the system. The air bleeding procedure must be repeated.

11. Use a scan tool to check if a DTC is set. If the diagnosis code is set, erase it.

LOWER CONTROL ARM

REMOVAL & INSTALLATION

See Figure 204.

1. Before servicing the vehicle, refer to the precautions.
2. Raise and safely support the vehicle securely on jackstands.
3. Remove the headlight front suspension height sensor.
4. Remove the lower arm ball joint nut (self-locking nut).
5. Remove the lower arm to chassis bolts.
6. Remove the lower arm assembly

To install:

7. Using your fingers, press the dust cover to check for a crack or damage.
8. Install the lower arm assembly.
9. Install the lower arm to chassis bolts. Tighten to 81 ft. lbs. (110 Nm).
10. Install the lower arm ball joint nut (self-locking nut). Tighten to 53 ft. lbs. (71 Nm).
11. Install the headlight front suspension height sensor.
12. Lower the vehicle.
13. Check the wheel alignment.
14. Check the headlight aim.

BUSHING REPLACEMENT

See Figures 205 and 206.

Use the following special tools to remove the bushing:

- MB992119 (Arm bushing remover and installer)

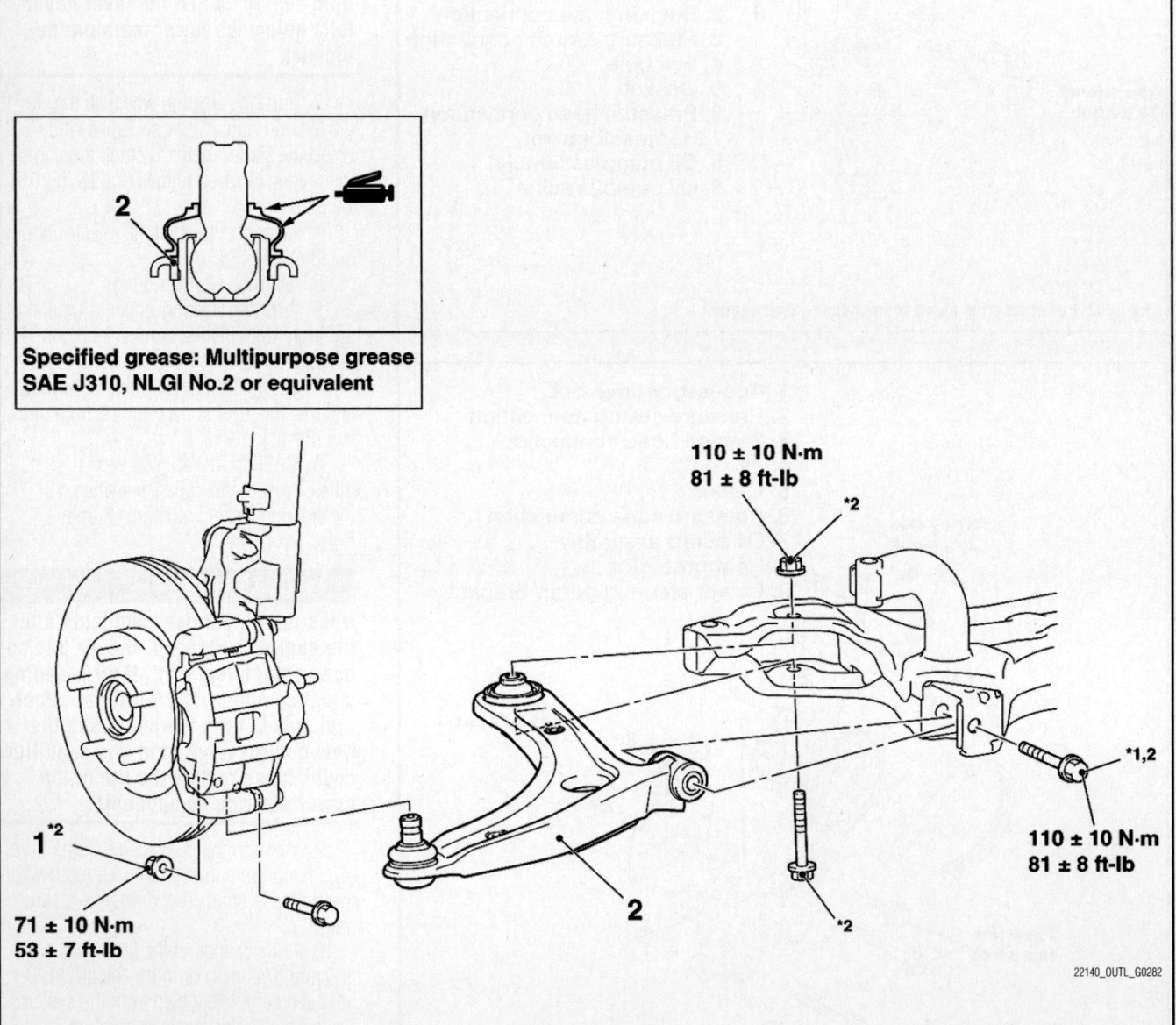

Specified grease: Multipurpose grease SAE J310, NLGI No.2 or equivalent

110 ± 10 N·m
81 ± 8 ft-lb

71 ± 10 N·m
53 ± 7 ft-lb

110 ± 10 N·m
81 ± 8 ft-lb

22140_OUTL_G0282

Fig. 204 Lower control arm assembly

MB992119
Lower arm assembly
MB990979
MB990890
22140_OUTL_G0283

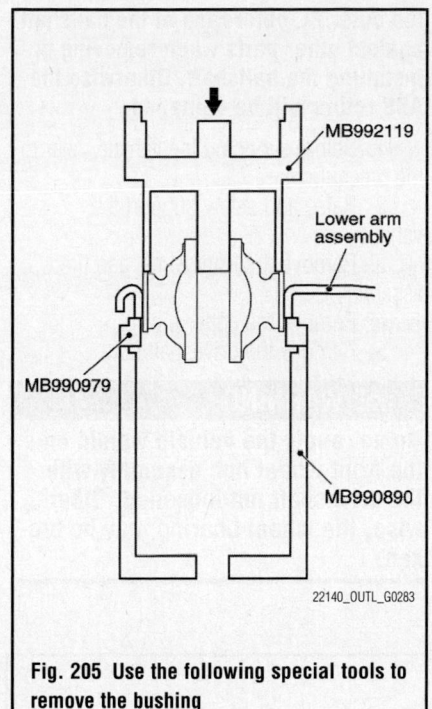

Fig. 205 Use the following special tools to remove the bushing

MB992119
Lower arm assembly
MB990979
MB990890
30° ± 3°
Projection
Lower arm assembly
22140_OUTL_G0284

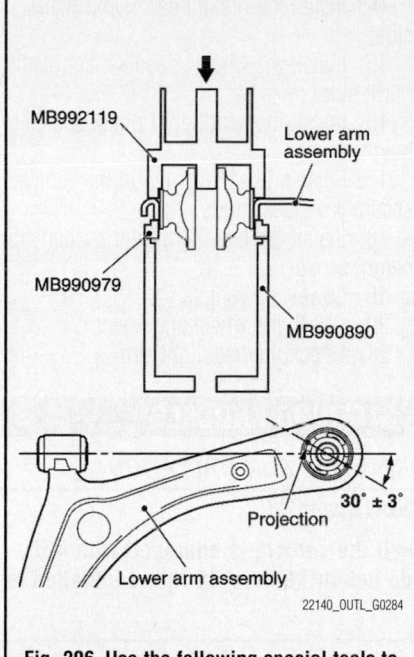

Fig. 206 Use the following special tools to press-fit the bushing

- MB990979 (Ring)
- MB990890 (Rear suspension bushing base)

Use the following special tools to press-fit the bushing.

- MB992119 (Arm bushing remover and installer)
- MB990979 (Ring)
- MB990890 (Rear suspension bushing base)

1. Press-fit the bushing so that the bushing protrusion is in the direction shown in the figure.
2. Press-fit the bushing until the special tool contacts with the lower arm assembly.

MACPHERSON STRUT

REMOVAL & INSTALLATION

See Figure 207.

1. Before servicing the vehicle, refer to the precautions.
2. Raise and safely support the vehicle securely on jackstands.

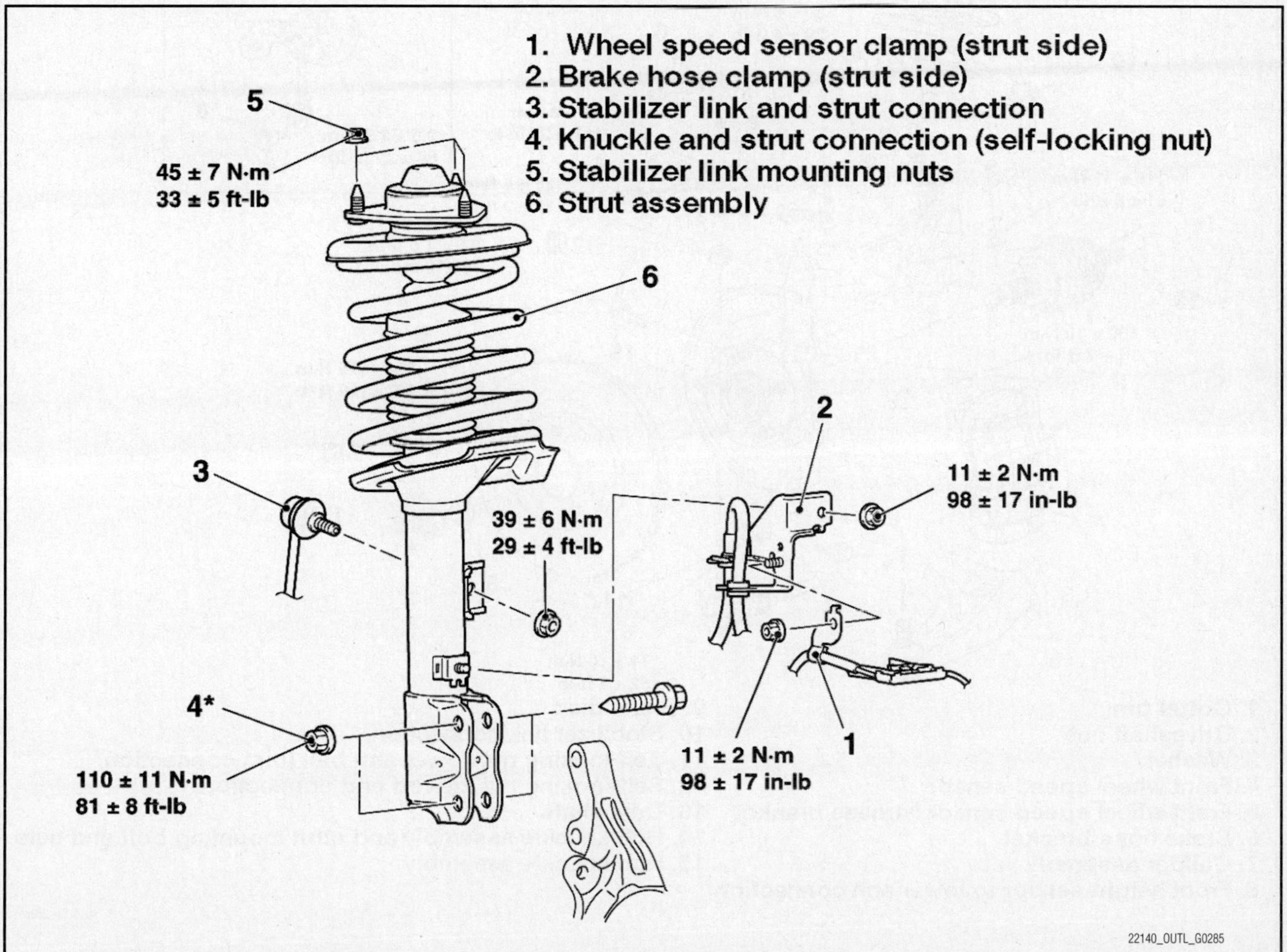

1. Wheel speed sensor clamp (strut side)
2. Brake hose clamp (strut side)
3. Stabilizer link and strut connection
4. Knuckle and strut connection (self-locking nut)
5. Stabilizer link mounting nuts
6. Strut assembly
5
45 ± 7 N·m
33 ± 5 ft-lb
6
2
3
11 ± 2 N·m
98 ± 17 in-lb
39 ± 6 N·m
29 ± 4 ft-lb
4*
110 ± 11 N·m
81 ± 8 ft-lb
11 ± 2 N·m
98 ± 17 in-lb
1
22140_OUTL_G0285

Fig. 207 Front MacPherson strut assembly

3. Remove the headlight front suspension height sensor.

4. Remove the wheel speed sensor clamp (strut side).

5. Remove the brake hose clamp (strut side).

6. Disconnect the stabilizer link and strut.

7. Disconnect the knuckle and strut connection (self-locking nut).

8. Remove the stabilizer link mounting nuts.

9. Remove the strut assembly.

To install:

10. Install the strut assembly.

11. Install the stabilizer link mounting nuts.

12. Connect the knuckle and strut connection (self-locking nut).

13. Connect the stabilizer link and strut.

14. Install the brake hose clamp (strut side).

15. Install the wheel speed sensor clamp (strut side).

16. Install the headlight front suspension height sensor.

17. Raise and safely support the vehicle securely on jackstands.

18. Install the headlight front suspension height sensor.

19. Lower the vehicle.

20. Check the wheel alignment.

21. Check the headlight aim.

STEERING KNUCKLE

REMOVAL & INSTALLATION

See Figure 208.

➡ If the vehicle is equipped with ABS, do not strike the ABS rotors installed to the outer BL outer race of the halfshaft against other parts when removing or installing the halfshaft. Otherwise the ABS rotors will be damaged.

1. Before servicing the vehicle, refer to the precautions.

2. Raise and safely support the vehicle.

3. Remove the front wheel and tire assembly.

4. Remove the cotter pin.

5. Remove the driveshaft nut.

✶✶ WARNING

Do not apply the vehicle weight on the front wheel hub assembly with the driveshaft nut loosened. Otherwise, the wheel bearing may be broken.

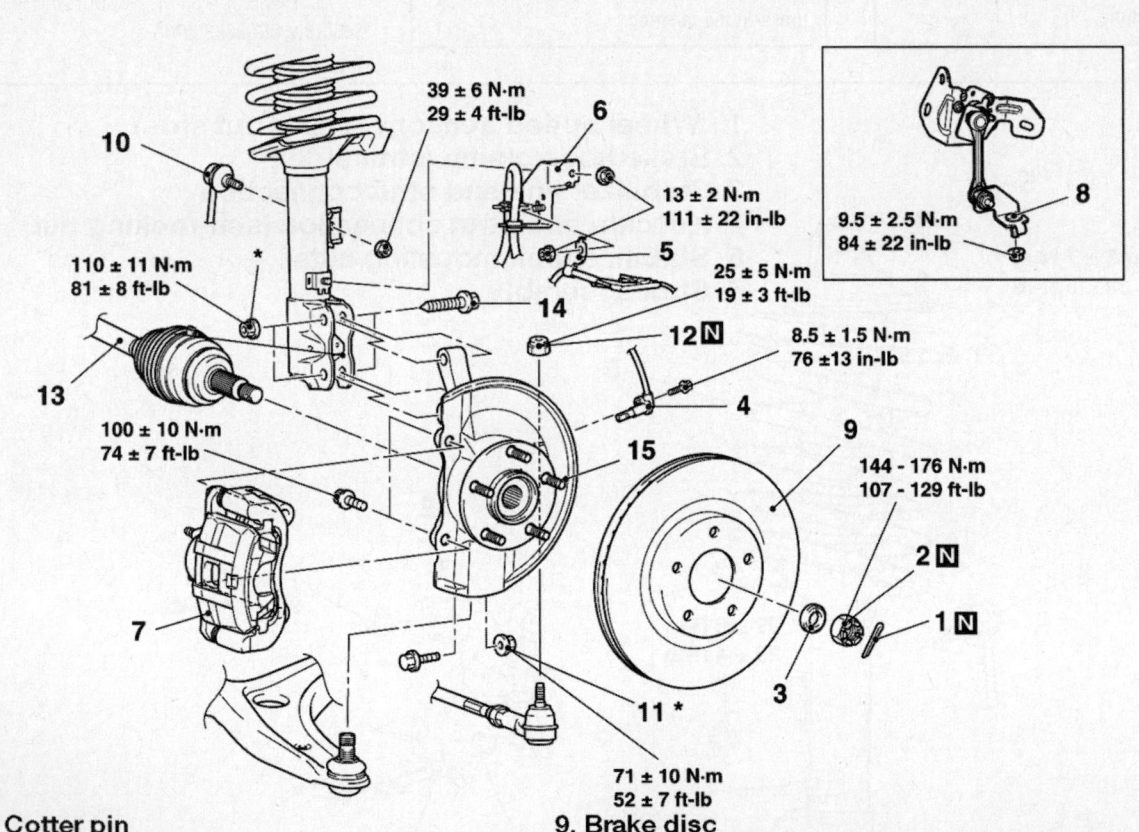

1. Cotter pin
2. Driveshaft nut
3. Washer
4. Front wheel speed sensor
5. Front wheel speed sensor harness bracket
6. Brake hose bracket
7. Caliper assembly
8. Front height sensor to lower arm connection
9. Brake disc
10. Stabilizer link connection
11. Self-locking nut (lower arm ball joint connection)
12. Self-locking nut (tie-rod end connection)
13. Driveshaft
14. Hub knuckle assembly and strut mounting bolt and nuts
15. Hub knuckle assembly

Fig. 208 Steering knuckle assembly

22140_OUTL_G0286

a. Use special tool MD990767 to hold the hub when removing the driveshaft nut.

6. Remove the washer.

7. Remove the front wheel speed sensor.

8. Remove the front wheel speed sensor harness bracket.

9. Remove the brake hose bracket.

10. Remove the caliper assembly with brake hose.

11. Secure the removed caliper assembly with a wire or other similar material at a position where it will not interfere with the removal and installation of the hub knuckle assembly.

12. Remove the headlight front height sensor to lower arm connection.

13. Remove the brake disc.

14. Disconnect the stabilizer link.

15. Remove the self-locking nut for the lower arm ball joint.

16. Remove the self-locking nut for the tie-rod end.

✳✳ WARNING

Loosen the self-locking nut (tie-rod end connection) from the ball joint, but do not remove here. Use the special tool. To prevent the special tool from dropping off, suspend it with a cord.

a. Install special tool MB991897 or MB992011 as shown in the figure.

b. Turn the bolt and knob to make the special tool jaws parallel, then hand-tighten the bolt. After tightening, check that the jaws are still parallel.

c. Unscrew the bolt to disconnect the ball joint.

17. Remove the driveshaft.

a. If the driveshaft is seized, use special tools MB990242 and MB990244, MB991354 and MB990767 to push the driveshaft out from the hub.

18. Remove the hub knuckle assembly and strut mounting bolt and nuts.

19. Remove the hub knuckle assembly.

To install:

20. Install the hub knuckle assembly.

21. Install the hub knuckle assembly and strut mounting bolt and nuts. Tighten to 81 ft. lbs. (110 Nm).

22. Install the driveshaft.

23. Install the self-locking nut for the tie-rod end. Tighten to 19 ft. lbs. (25 Nm).

24. Install the self-locking nut for the lower arm ball joint. Tighten to 53 ft. lbs. (71 Nm).

25. Connect the stabilizer link. Tighten to 29 ft. lbs. (39 Nm).

26. Install the brake disc.

27. Install the headlight front height sensor to lower arm connection.

28. Install the caliper assembly with brake hose.

29. Install the brake hose bracket.

30. Install the front wheel speed sensor harness bracket.

31. Install the front wheel speed sensor.

32. Install the washer.

a. Be sure to install the driveshaft washer in the illustrated direction.

b. Use special tool MB990767 to tighten the driveshaft nuts. At this time, tighten the nuts within the specified torque range considering the final tightening. Tighten to 107–129 ft. lbs. (144–176 Nm).

c. If the pin holes do not align with the pins, tighten the driveshaft nut less than 147 ft. lbs. (200 Nm)] and find the nearest hole then bend the cotter pin to fit it.

33. Install the cotter pin.

34. Install the front wheel and tire assembly.

35. Lower the vehicle.

STABILIZER BAR

REMOVAL & INSTALLATION

See Figure 209.

1. Remove the stabilizer nuts.

2. Remove the stabilizer link.

3. Remove the front axle crossmember.

4. Remove the stabilizer bar bracket.

5. Remove the stabilizer bushing.

6. Remove the stabilizer bar.

To install:

7. Install the stabilizer bar.

8. Install the stabilizer bushing.

9. Install the stabilizer bar bracket. Tighten bushing bracket bolts to 23 ft. lbs. (31 Nm).

10. Install the front axle crossmember.

11. Install the stabilizer link.

12. Install the stabilizer link nuts. Tighten to 29 ft. lbs. (39 Nm).

WHEEL HUB & BEARING

REMOVAL & INSTALLATION

See Figures 210 through 219.

1. Raise and safely support the vehicle securely on jackstands.

2. Remove the front wheel.

3. Remove the hub.

a. Replace special tool MB992250 with a guide of special MB991355 as shown in the figure.

b. Insert special tool MB992250 in

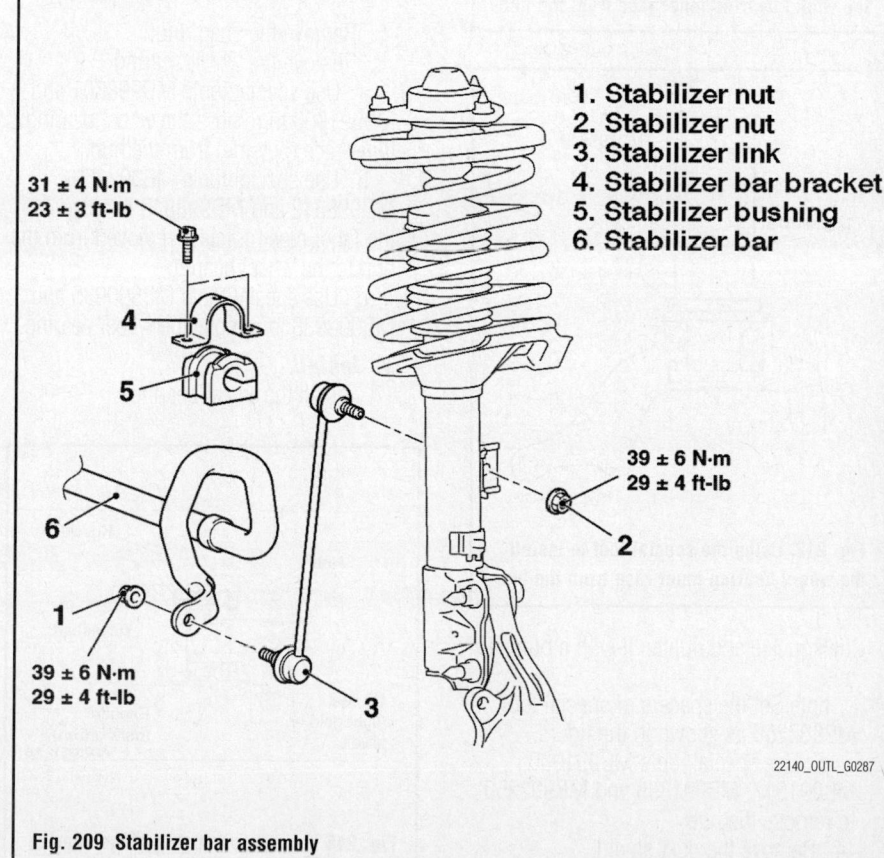

1. **Stabilizer nut**
2. **Stabilizer nut**
3. **Stabilizer link**
4. **Stabilizer bar bracket**
5. **Stabilizer bushing**
6. **Stabilizer bar**

31 ± 4 N·m
23 ± 3 ft-lb

39 ± 6 N·m
29 ± 4 ft-lb

39 ± 6 N·m
29 ± 4 ft-lb

22140_OUTL_G0287

Fig. 209 Stabilizer bar assembly

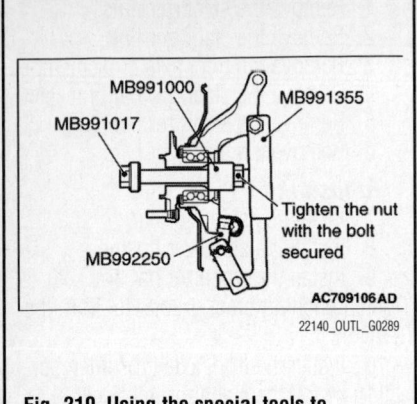

Fig. 210 Using the special tools to remove the hub

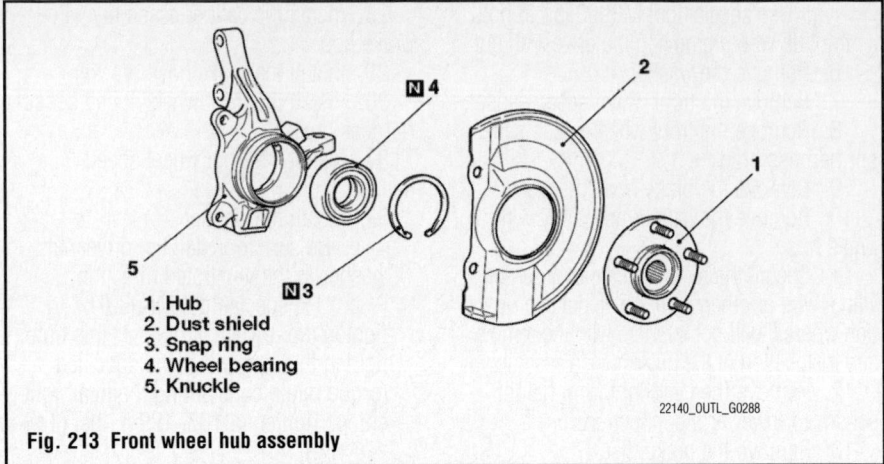

1. Hub
2. Dust shield
3. Snap ring
4. Wheel bearing
5. Knuckle

Fig. 213 Front wheel hub assembly

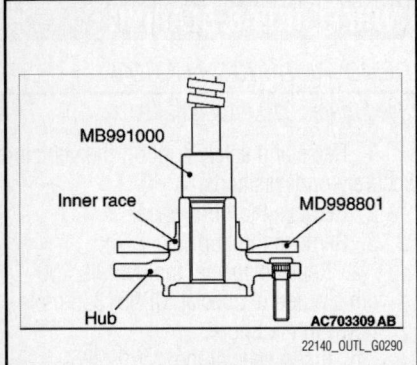

Fig. 211 Using the special tool to remove the wheel bearing inner race from the hub

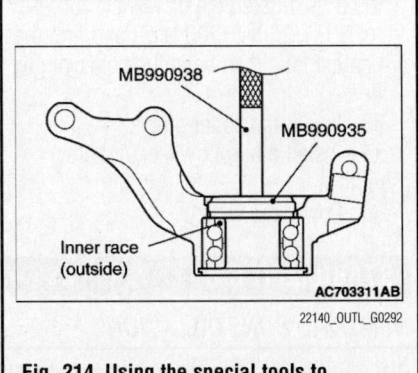

Fig. 214 Using the special tools to remove the wheel bearing

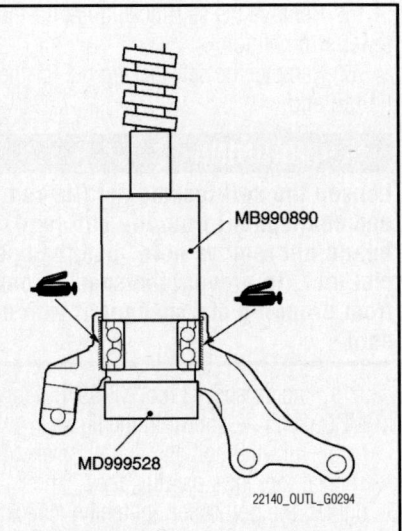

Fig. 216 Using the special tools to press-fit the wheel bearing

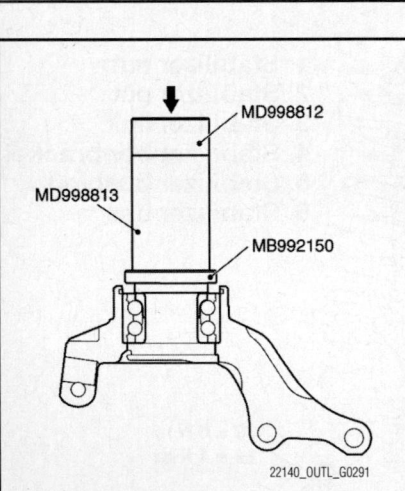

Fig. 212 Using the special tool to install the wheel bearing inner race from the hub

the knuckle and tighten it with a bolt and nut.

note Set the spacers of special tool MB992250 as shown in the figure.

c. Use special tools MB991000, MB991017, MB991355 and MB992250 to remove the hub.
4. Remove the dust shield.

5. Remove the snap ring.
6. Remove the wheel bearing.

a. Use special tools MD998801 and MB991000 to remove the wheel bearing inner race (outside) from the hub.

b. Use special tools MB992150, MD998812 and MD998813 to assemble the inner race (outside) removed from the hub to the wheel bearing.

c. Use special tools MB990935 and MB990938 to remove the wheel bearing.

To install:
7. Install the wheel bearing.

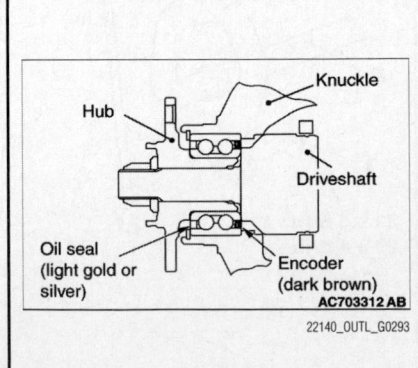

Fig. 215 Proper positioning of the bearing

☀☀ WARNING

The magnetic encoder for wheel speed sensor is installed in the wheel bearing.

a. Install the wheel bearing so that the encoder is positioned in the direction shown in the figure.

b. When press-fit the wheel bearing, push the outer race. After press-fit the wheel bearing, wipe off the extra grease in order not to remain on the magnetic encoder.

c. Remove grease and foreign material cleanly from the inside of knuckle bore.

d. Apply Molykote BR2 Plus thinly and evenly to the inside of knuckle as shown in the figure.

e. Use special tools MB990883 and MB 990890 to press-fit the wheel bearing.

f. Remove excessive grease seeped

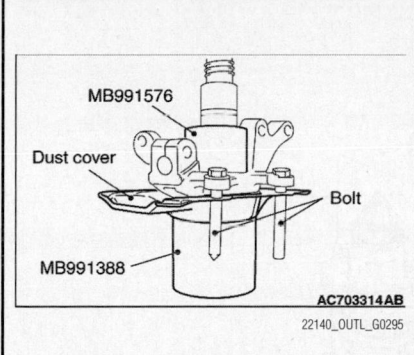

Fig. 217 Using the special tools to press-fit the knuckle into the dust shield

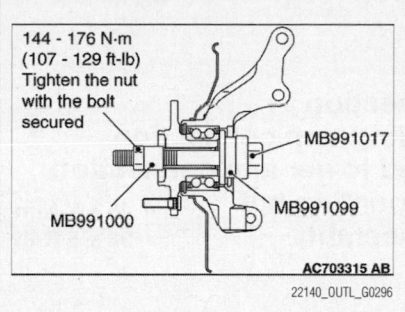

Fig. 218 Using the special tools as shown in the figure, tighten the nut to the specified torque, and press-fit the hub into the knuckle

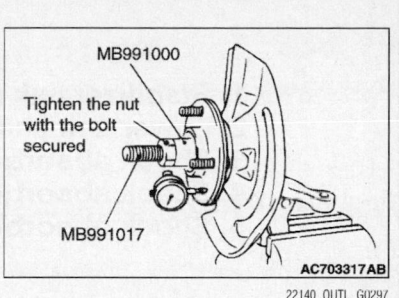

Fig. 219 Using the special tools to measure to determine whether the wheel bearing end play is within the specified limit

out between knuckle and wheel bearing outer race after press-fitting the wheel bearing.

8. Install the dust shield.

a. Use special tools MB991388 and MB991576 to press-fit the knuckle into the dust shield.

➡**Use the bolts (M12) to align the caliper mounting holes.**

9. Install the hub.

10. Check the hub rotation starting torque.

a. Set special tools MB991000 and MB991017 as shown in the figure,

tighten the nut to 107–129 ft. lbs. (144–176 Nm), and press-fit the hub into the knuckle.

b. Rotate the hub to make the bearing well-greased.

c. Use a beam type torque wrench to measure the hub rotation starting torque. Starting torque should be 13 inch lbs. (1.5 Nm).

d. Hub rotation starting torque should be within the limit value, and there should be no roughness and gritty feeling in rotation.

11. Check the wheel bearing end play.

a. Use special tools MB991000 and

MB991017 to measure to determine whether the wheel bearing end play is 0.002 inch (0.05 mm).

b. If the end play is not within the limit range while the nut is tightened to specified torque, the bearing, hub and/or knuckle have probably not been installed correctly. Replace the bearing and re-install.

ADJUSTMENT

The front wheel bearings are not adjustable. If the bearings are noisy or become loose, they must be replaced.

SUSPENSION

COIL SPRING

REMOVAL & INSTALLATION

2.4L Engine

See Figures 220 and 221.

1. Remove the coil over shock assembly from the vehicle.

2. Remove the self-locking nut.

✳✳ WARNING

To hold the coil spring securely, install the spring compressor evenly, and so that the space between both arms of the special tool will be maximum within the installation range.

a. The locking nut for the piston rod inside the shock absorber may be loose. Do not use an impact wrench to loosen the self-locking nut.

b. While holding the piston rod, remove the self-locking nut.

3. Remove the washer.

4. Remove the bushing "B".

5. Remove the collar.

6. Remove the spring upper bracket assembly.

7. Remove the spring upper pad.

8. Remove the bushing "A".

9. Remove the plate.

10. Remove the bump rubber.

11. Remove the coil spring.

To install:

12. Install a spring compressor to compress the coil spring, and install it to the lower spring pad.

13. Align the end of the coil spring with the shock absorber as shown in the figure.

14. Assemble the coil over shock in the reverse of the removal procedure.

15. Counterhold the piston rod of the shock absorber, and tighten the self-locking nut to 19ft. lbs. (25 Nm).

16. Install the coil over shock assembly on the vehicle.

3.0L Engine

See Figure 222.

1. Before servicing the vehicle, refer to the precautions.

REAR SUSPENSION

2. Raise and safely support the vehicle.

3. Disconnect the rear height sensor to control link connection.

4. Disconnect the stabilizer link.

5. Disconnect the lower arm from the trailing arm.

a. While jacking-up the lower arm with the garage jack, remove the mounting bolts.

6. Disconnect the shock absorber from the lower arm.

7. Remove the coil spring.

8. Remove the coil spring upper and lower pads.

To install:

➡**The suspension components should not be fully tightened until the vehicle's weight is resting on its wheels.**

9. Install the coil spring upper and lower pads.

10. Install the coil spring.

11. Connect the shock absorber to the lower arm.

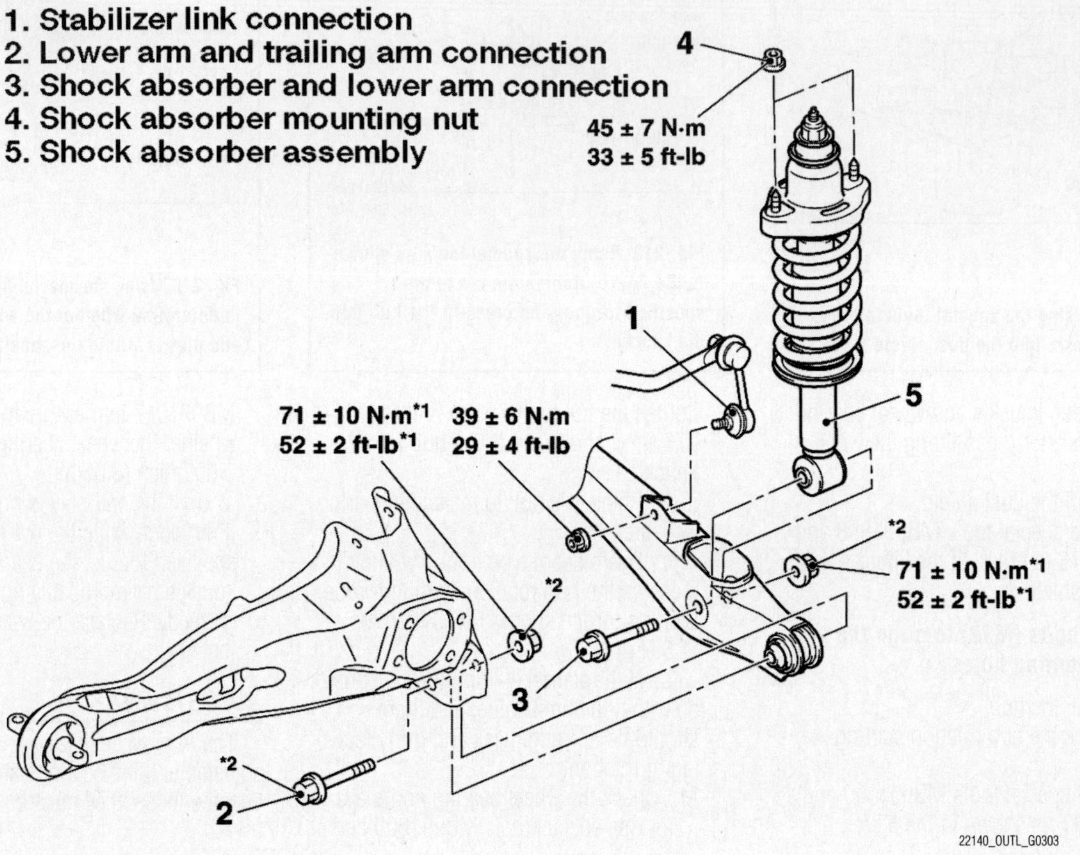

1. Stabilizer link connection
2. Lower arm and trailing arm connection
3. Shock absorber and lower arm connection
4. Shock absorber mounting nut
5. Shock absorber assembly

4
45 ± 7 N·m
33 ± 5 ft-lb

1

71 ± 10 N·m*1   39 ± 6 N·m
52 ± 2 ft-lb*1   29 ± 4 ft-lb

5

*2

*2
71 ± 10 N·m*1
52 ± 2 ft-lb*1

3

*2

2

22140_OUTL_G0303

Fig. 220 Rear shock absorber assembly—2.4L engine

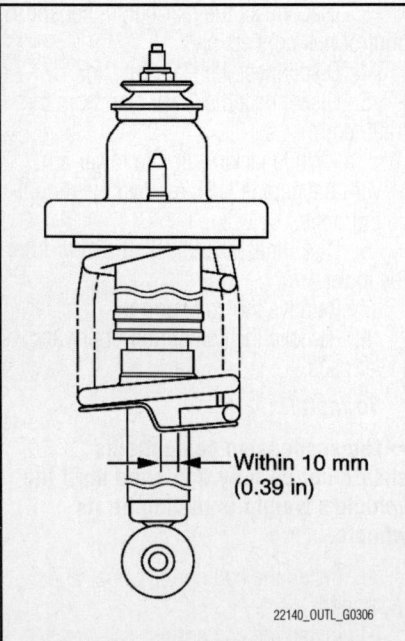

Within 10 mm
(0.39 in)

22140_OUTL_G0306

Fig. 221 Align the end of the coil spring with the shock absorber as shown

a. While jacking-up the lower arm with the garage jack, Install the mounting bolts.

12. Connect the lower arm and trailing arm.

13. Connect the stabilizer link.

14. Temporarily tighten the suspension bolts.

15. Connect the rear height sensor to control link connection.

16. Lower the vehicle.

17. Finish tightening the bolts as specified in the illustration.

18. Check the wheel alignment.

19. Check the headlight aiming.

CONTROL ARMS/LINKS

REMOVAL & INSTALLATION

2.4L Engine
See Figure 223.

Lower Control Arm

1. Before servicing the vehicle, refer to the precautions.

2. Raise and safely support the vehicle.

3. Disconnect the stabilizer link.

4. Disconnect the lower arm from the trailing arm.

5. Disconnect the shock absorber.

6. Remove the rear suspension cross-member bracket.

7. Remove the lower arm.

To install:

➡ The suspension components should not be fully tightened until the vehicle's weight is resting on its wheels.

8. Install the lower arm.

9. Install the rear suspension cross-member bracket.

10. Connect the shock absorber.

11. Connect the lower arm from the trailing arm.

12. Connect the stabilizer link.

13. Temporarily tighten the suspension bolts.

14. Lower the vehicle.

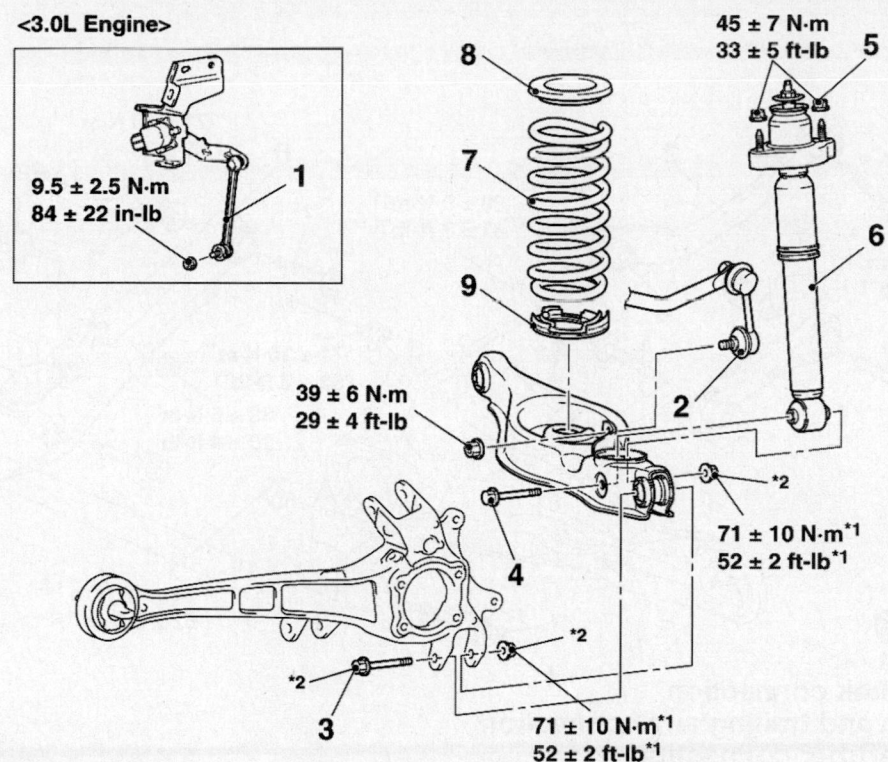

<3.0L Engine>

9.5 ± 2.5 N·m
84 ± 22 in-lb

45 ± 7 N·m
33 ± 5 ft-lb

39 ± 6 N·m
29 ± 4 ft-lb

71 ± 10 N·m*1
52 ± 2 ft-lb*1

71 ± 10 N·m*1
52 ± 2 ft-lb*1

1. **Rear height sensor to control link connection**
2. **Stabilizer link connection**
3. **Lower arm and trailing arm connection**
4. **Shock absorber and lower arm connection**
5. **Shock absorber mounting nut**
6. **Shock absorber assembly**
7. **Coil spring**
8. **Coil spring upper pad**
9. **Coil spring lower pad**

22140_OUTL_G0304

Fig. 222 Rear shock absorber assembly—3.0L engine

15. Finish tightening the bolts as specified in the illustration.
16. Check the wheel alignment.
17. Check the headlight aiming.

Upper Control Arm

1. Before servicing the vehicle, refer to the precautions.
2. Raise and safely support the vehicle.
3. Remove the control link.
  a. Make a mating mark on the toe adjusting bolt, and remove the control link.
4. Disconnect the fuel tank vapor hose.
5. Remove the upper arm.

To install:

➡The suspension components should not be fully tightened until the vehicle's weight is resting on its wheels.

6. Install the upper arm.
  a. Install the upper arm so that the hole faces the body side.
7. Connect the fuel tank vapor hose.
8. Install the control link.
  a. Align the mating mark on the toe adjusting bolt.
9. Temporarily tighten the suspension bolts.
10. Lower the vehicle.
11. Finish tightening the bolts as specified in the illustration.
12. Check the wheel alignment.
13. Check the headlight aiming.

3.0L Engine

See Figure 224.

Lower Control Arm

1. Before servicing the vehicle, refer to the precautions.

2. Raise and safely support the vehicle.
3. Disconnect the rear height sensor to control link.
4. Disconnect the stabilizer link.
5. Disconnect the lower arm and trailing arm connection.
6. Disconnect the shock absorber.
7. Remove the coil spring.
8. Remove the coil spring upper and lower pads.
9. Remove the lower arm.

To install:

➡The suspension components should not be fully tightened until the vehicle's weight is resting on its wheels.

10. Install the lower arm.
11. Install the coil spring upper and lower pads.
12. Install the coil spring.
13. Connect the shock absorber.

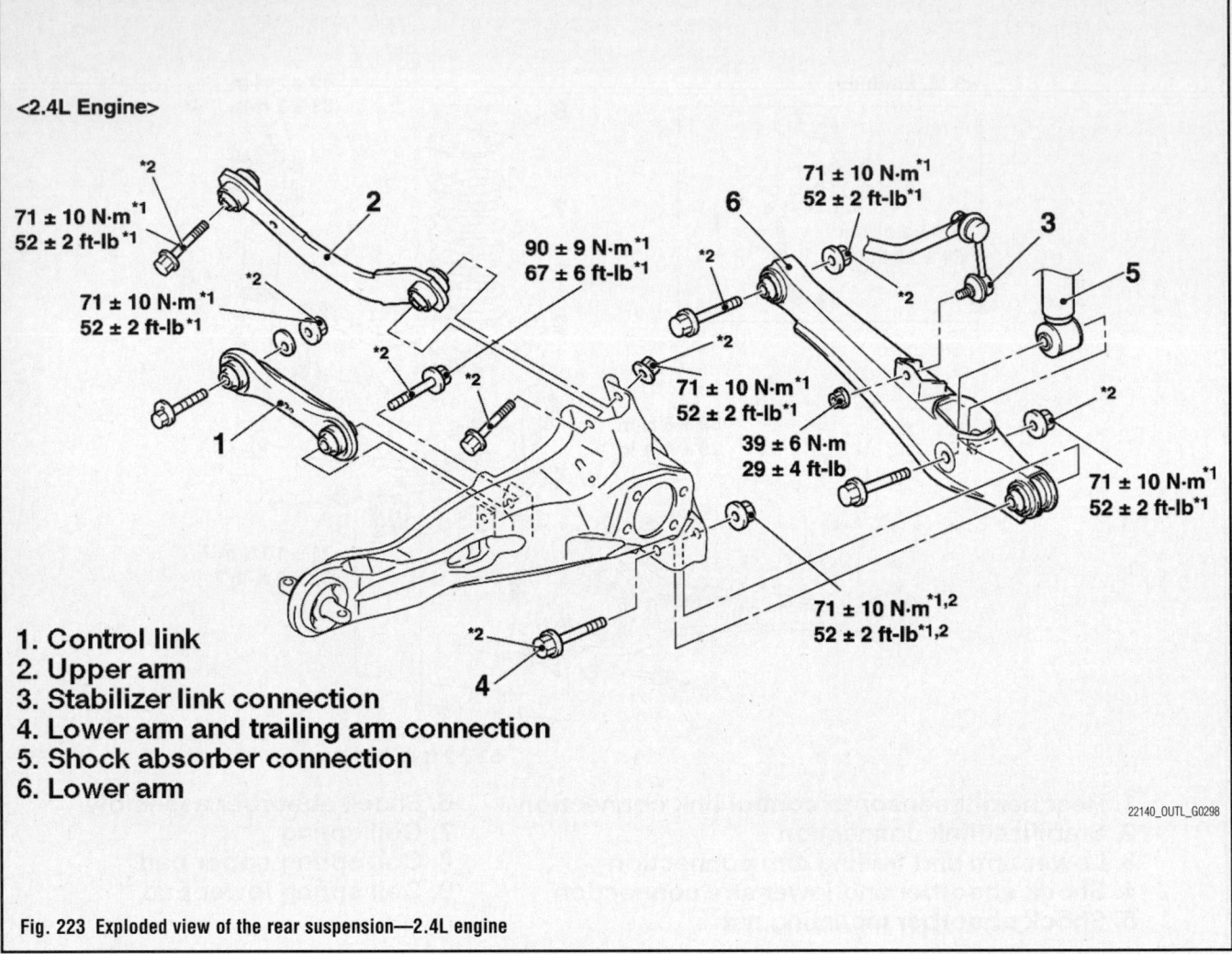

<2.4L Engine>

71 ± 10 N·m[1]
52 ± 2 ft-lb[1]

71 ± 10 N·m[1]
52 ± 2 ft-lb[1]

71 ± 10 N·m[1]
52 ± 2 ft-lb[1]

90 ± 9 N·m[1]
67 ± 6 ft-lb[1]

71 ± 10 N·m[1]
52 ± 2 ft-lb[1]

39 ± 6 N·m
29 ± 4 ft-lb

71 ± 10 N·m[1]
52 ± 2 ft-lb[1]

71 ± 10 N·m[1,2]
52 ± 2 ft-lb[1,2]

1. Control link
2. Upper arm
3. Stabilizer link connection
4. Lower arm and trailing arm connection
5. Shock absorber connection
6. Lower arm

22140_OUTL_G0298

Fig. 223 Exploded view of the rear suspension—2.4L engine

14. Connect the lower arm and trailing arm connection.

15. Connect the stabilizer link.

16. Connect the rear height sensor to control link.

17. Temporarily tighten the suspension bolts.

18. Lower the vehicle.

19. Finish tightening the bolts as specified in the illustration.

20. Check the wheel alignment.

21. Check the headlight aiming.

Upper Control Arm

1. Before servicing the vehicle, refer to the precautions.

2. Raise and safely support the vehicle.

3. Disconnect the rear height sensor to control link.

4. Remove the control link.

a. Make a mating mark on the toe adjusting bolt, and remove the control link.

5. Remove the upper arm.

To install:

➡The suspension components should not be fully tightened until the vehicle's weight is resting on its wheels.

6. Install the upper arm.

a. Install the upper arm so that the hole faces the body side.

7. Install the control link.

a. Align the mating mark on the toe adjusting bolt.

8. Connect the rear height sensor to control link.

9. Temporarily tighten the suspension bolts.

10. Lower the vehicle.

11. Finish tightening the bolts as specified in the illustration.

12. Check the wheel alignment.

13. Check the headlight aiming.

BUSHING REPLACEMENT

See Figures 225 and 226.

Use the following special tools to remove and install the bushing:

• MB992123: Arm Bushing Remover and Installer
• MB991448: Bushing Remover and Installer Base
• MB991449: Bushing Remover and Installer Supporter

※※ WARNING

As the bushing has different outer diameters at both ends, be careful not to confuse the removal direction with the press-fit direction.

Use the special tools MB992123, MB991448 and MB991449 to remove and press-fit the lower arm bushing.

SHOCK ABSORBER

REMOVAL & INSTALLATION

2.4L Engine

See Figures 227 and 228.

1. Before servicing the vehicle, refer to the precautions.

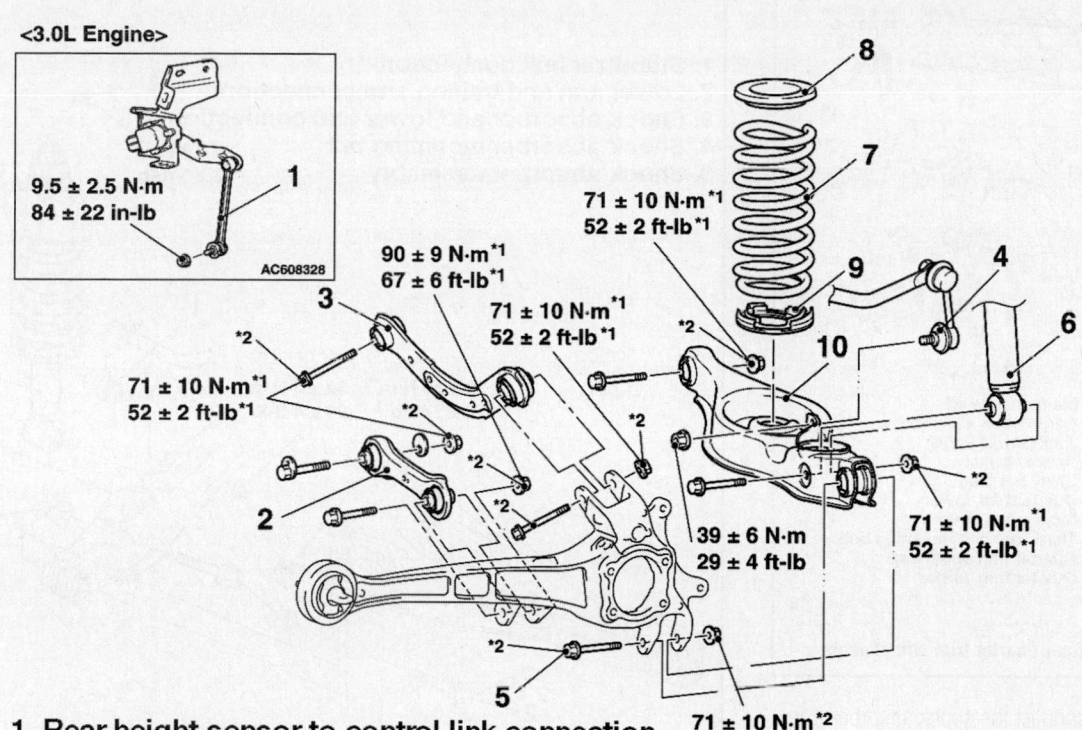

<3.0L Engine>

9.5 ± 2.5 N·m
84 ± 22 in-lb

AC608328

71 ± 10 N·m*1
52 ± 2 ft-lb*1

90 ± 9 N·m*1
67 ± 6 ft-lb*1

71 ± 10 N·m*1
52 ± 2 ft-lb*1

71 ± 10 N·m*1
52 ± 2 ft-lb*1

71 ± 10 N·m*1
52 ± 2 ft-lb*1

39 ± 6 N·m
29 ± 4 ft-lb

71 ± 10 N·m*2
52 ± 2 ft-lb*2

1. **Rear height sensor to control link connection**
2. **Control link**
3. **Upper arm**
4. **Stabilizer link connection**
5. **Lower arm and trailing arm connection**
6. **Shock absorber connection**
7. **Coil spring**
8. **Coil spring upper pad**
9. **Coil spring lower pad**
10. **Lower arm**

22140_OUTL_G0299

Fig. 224 Exploded view of the rear suspension—3.0L engine

2. Raise and safely support the vehicle.
3. Remove the quarter trim.
  a. Remove the maintenance lid.
  b. Remove the cargo hook.
  c. Remove the jack lid.
  d. Remove the tumble switch.
  e. Remove the utility bar plug.
  f. Remove the lower quarter trim.
  g. Remove the sash guide cover.
  h. Remove the third seat belt mounting bolt.
  i. Remove the quarter trim upper cap.
  j. Remove the upper quarter trim.
4. Disconnect the stabilizer link.
5. Disconnect the lower arm from the trailing arm.
  a. While jacking-up the lower arm with the garage jack, remove the mounting bolts.

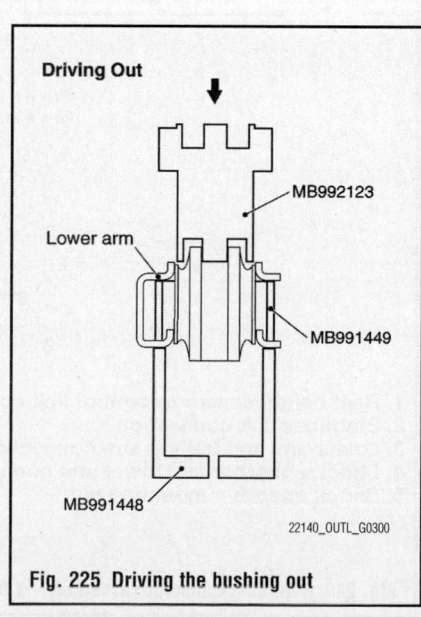

Driving Out

MB992123

Lower arm

MB991449

MB991448

22140_OUTL_G0300

Fig. 225 Driving the bushing out

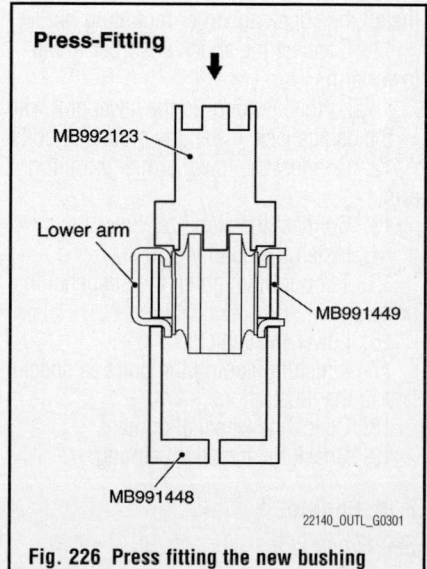

Press-Fitting

MB992123

Lower arm

MB991449

MB991448

22140_OUTL_G0301

Fig. 226 Press fitting the new bushing

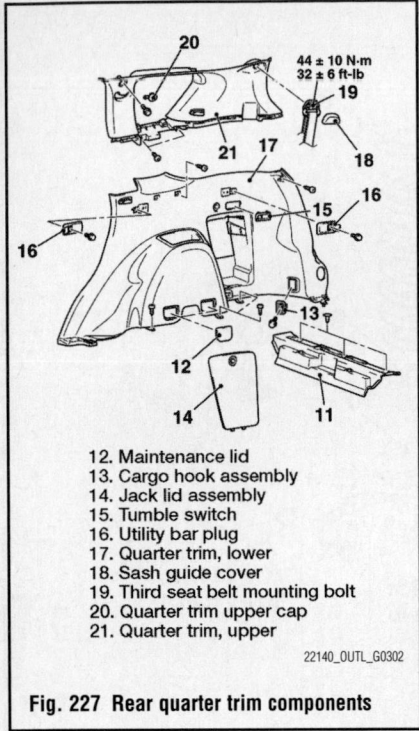

Fig. 227 Rear quarter trim components

12. Maintenance lid
13. Cargo hook assembly
14. Jack lid assembly
15. Tumble switch
16. Utility bar plug
17. Quarter trim, lower
18. Sash guide cover
19. Third seat belt mounting bolt
20. Quarter trim upper cap
21. Quarter trim, upper

22140_OUTL_G0302

6. Disconnect the shock absorber from the lower arm.

7. Remove the maintenance lid, and then remove the shock absorber mounting nut.

8. Remove the shock absorber assembly.

To install:

➡ The suspension components should not be fully tightened until the vehicle's weight is resting on its wheels.

9. Install the shock absorber assembly.

a. Install the shock absorber assembly so that the coil spring end faces the rear of the vehicle.

10. Install the maintenance lid, and then Install the shock absorber mounting nut.

11. Connect the shock absorber to the lower arm.

a. While jacking-up the lower arm with the garage jack, Install the mounting bolts.

12. Connect the lower arm and trailing arm.

13. Connect the stabilizer link.

14. Install the quarter trim.

15. Temporarily tighten the suspension bolts.

16. Lower the vehicle.

17. Finish tightening the bolts as specified in the illustration.

18. Check the wheel alignment.

19. Check the headlight aiming.

3.0L Engine

See Figure 229.

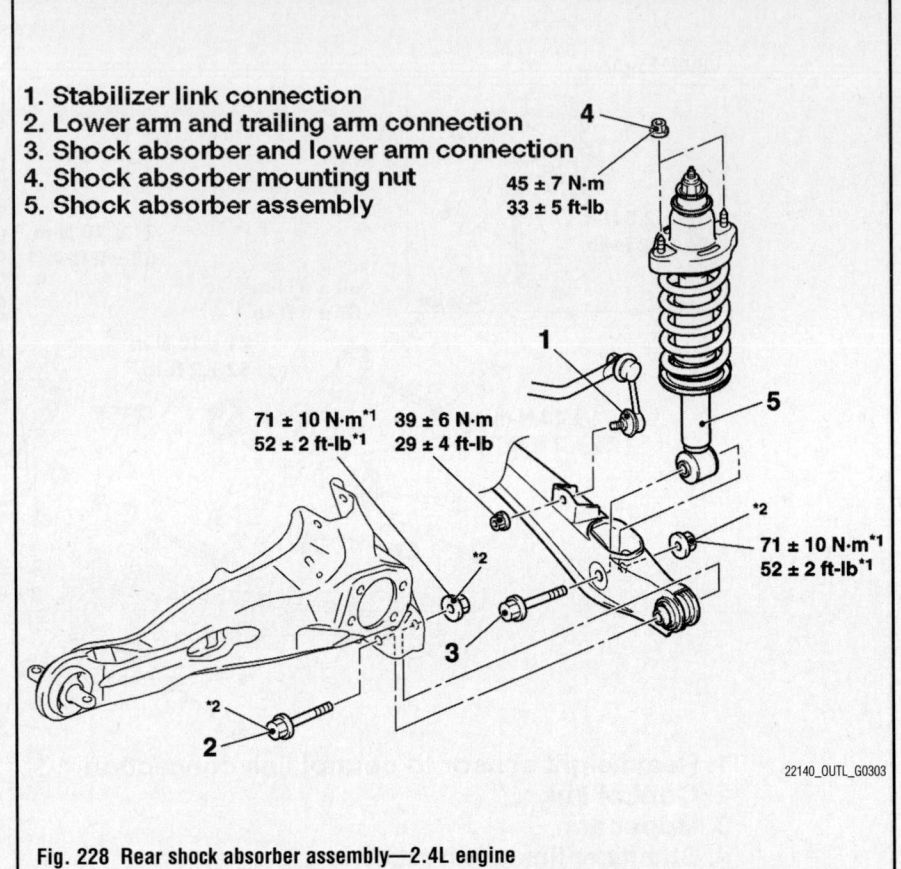

1. Stabilizer link connection
2. Lower arm and trailing arm connection
3. Shock absorber and lower arm connection
4. Shock absorber mounting nut
5. Shock absorber assembly

22140_OUTL_G0303

Fig. 228 Rear shock absorber assembly—2.4L engine

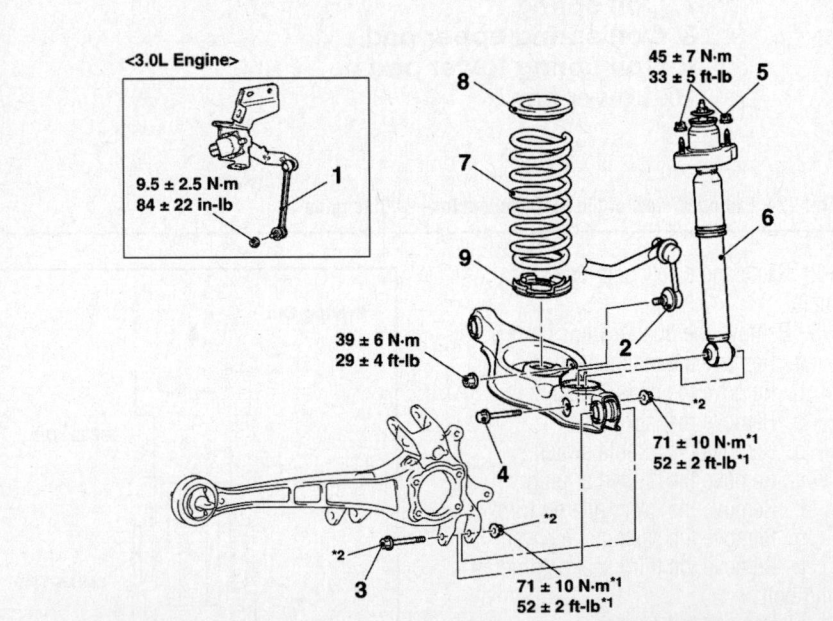

1. Rear height sensor to control link connection
2. Stabilizer link connection
3. Lower arm and trailing arm connection
4. Shock absorber and lower arm connection
5. Shock absorber mounting nut
6. Shock absorber assembly
7. Coil spring
8. Coil spring upper pad
9. Coil spring lower pad

22140_OUTL_G0304

Fig. 229 Rear shock absorber assembly—3.0L engine

1. Before servicing the vehicle, refer to the precautions.

2. Raise and safely support the vehicle.

3. Disconnect the rear height sensor to control link connection.

4. Disconnect the stabilizer link.

5. Disconnect the lower arm from the trailing arm.

   a. While jacking-up the lower arm with the garage jack, remove the mounting bolts.

6. Disconnect the shock absorber from the lower arm.

7. Remove the shock absorber mounting nut.

8. Remove the shock absorber assembly.

To install:

➡**The suspension components should not be fully tightened until the vehicle's weight is resting on its wheels.**

9. Install the shock absorber assembly.

   a. Install the shock absorber assembly so that the coil spring end faces the rear of the vehicle.

10. Install the shock absorber mounting nut.

11. Connect the shock absorber to the lower arm.

   a. While jacking-up the lower arm with the garage jack, Install the mounting bolts.

12. Connect the lower arm and trailing arm.

13. Connect the stabilizer link.

14. Temporarily tighten the suspension bolts.

15. Connect the rear height sensor to control link connection.

16. Lower the vehicle.

17. Finish tightening the bolts as specified in the illustration.

18. Check the wheel alignment.

19. Check the headlight aiming.

TESTING

1. Check the rubber parts for cracks and wear.

2. Check the shock absorber for malfunctions, oil leakage, or abnormal noise.

3. If shock absorber does not perform properly, replace the shock absorber.

STABILIZER BAR

REMOVAL & INSTALLATION

See Figures 230 through 233.

1. Before servicing the vehicle, refer to the precautions.

2. Raise and support the vehicle safely.

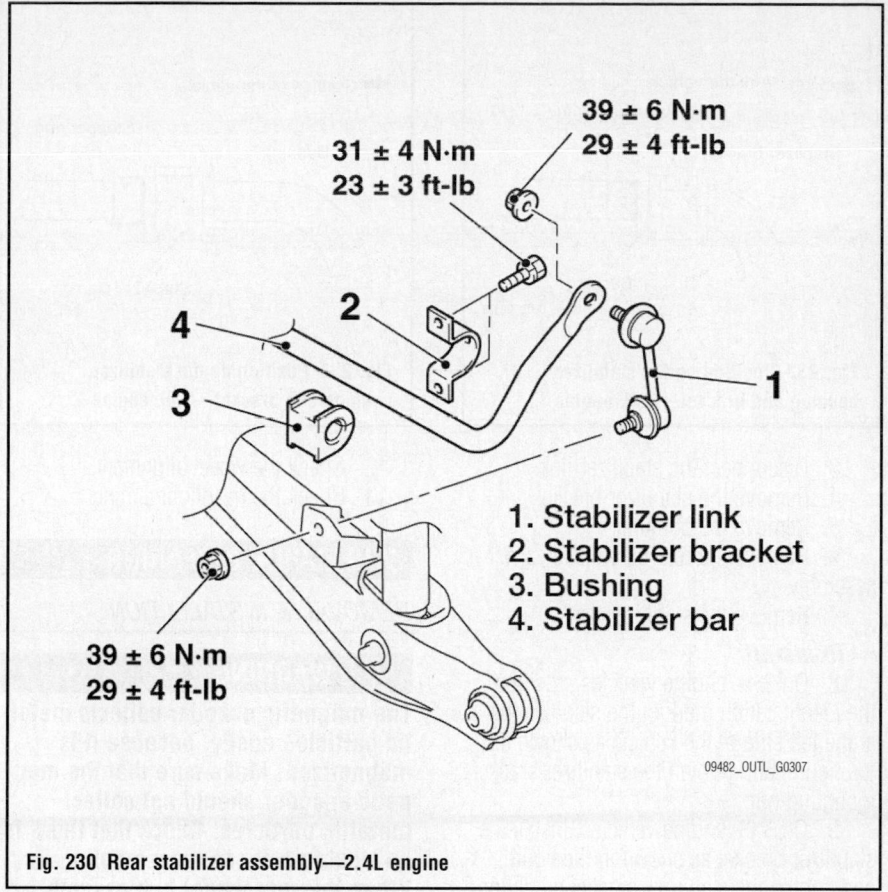

1. Stabilizer link
2. Stabilizer bracket
3. Bushing
4. Stabilizer bar

31 ± 4 N·m
23 ± 3 ft-lb

39 ± 6 N·m
29 ± 4 ft-lb

39 ± 6 N·m
29 ± 4 ft-lb

09482_OUTL_G0307

Fig. 230 Rear stabilizer assembly—2.4L engine

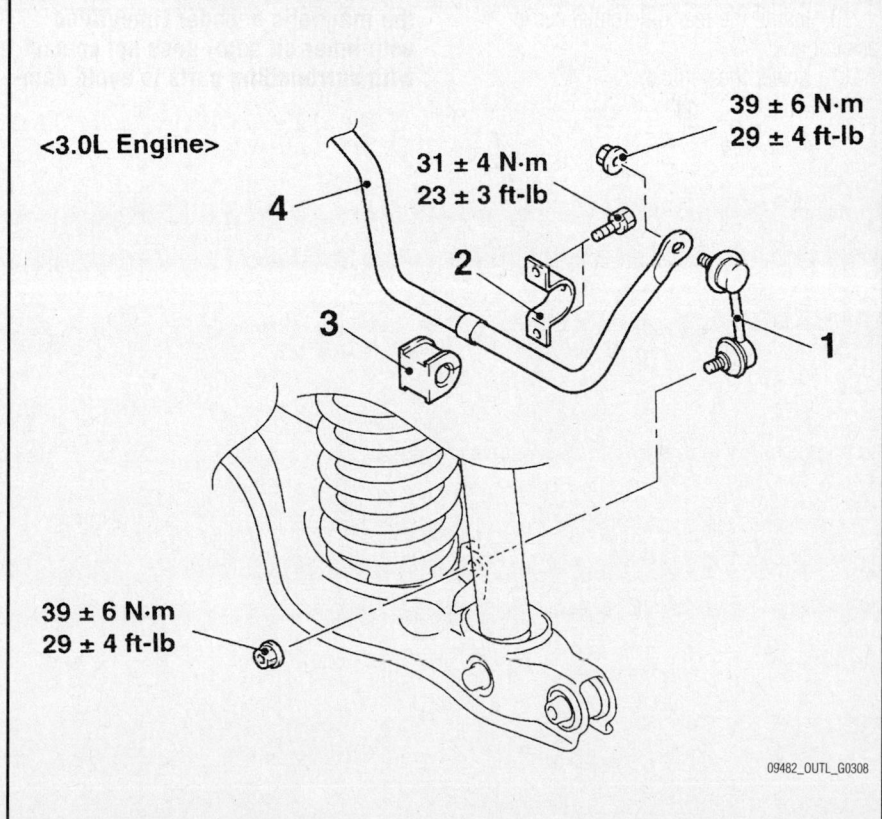

<3.0L Engine>

31 ± 4 N·m
23 ± 3 ft-lb

39 ± 6 N·m
29 ± 4 ft-lb

39 ± 6 N·m
29 ± 4 ft-lb

09482_OUTL_G0308

Fig. 231 Rear stabilizer assembly—3.0L engine

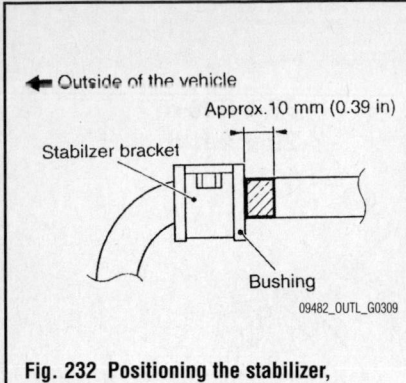

Fig. 232 Positioning the stabilizer, bushing and bracket—2.4L engine

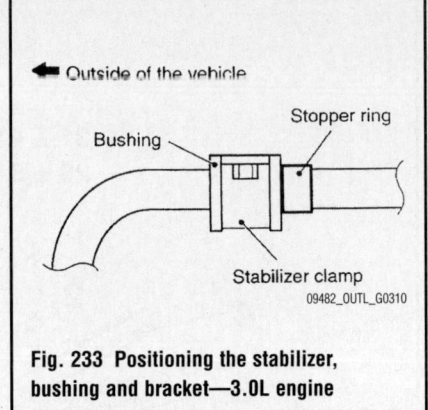

Fig. 233 Positioning the stabilizer, bushing and bracket—3.0L engine

3. Disconnect the stabilizer link.
4. Remove the stabilizer bracket.
5. Remove the bushing.
6. Remove the rear differential carrier assembly.
7. Remove the stabilizer bar.

To install:
8. On 2.4L engine vehicles, position the identification mark of the stabilizer bar at the left side of the vehicle as shown in the figure, and tighten the stabilizer bracket mounting nut.
9. On 3.0L engine vehicles, install the stabilizer bracket as shown in the figure, and tighten the stabilizer bracket mounting nut.
10. Install the rear differential carrier assembly.
11. Lower the vehicle.

12. Check the wheel alignment.
13. Check the headlight aiming.

WHEEL HUB & BEARING

REMOVAL & INSTALLATION

❊❊ WARNING

The magnetic encoder collects metallic particles easily, because it is magnetized. Make sure that the magnetic encoder should not collect metallic particles. Check that there is no trouble prior to reassembling it. When the rear wheel hub assembly is removed/installed, make sure that the magnetic encoder (integrated with inner oil seal) does not contact with surrounding parts to avoid damage. When removing and installing the rear wheel speed sensor, make sure that its pole piece at the end does not contact with surrounding parts to avoid damage.

1. Before servicing the vehicle, refer to the precautions.
2. Raise and safely support the vehicle.
3. Remove the tire and wheel assembly.
4. Remove the rear wheel speed sensor.
5. Remove the brake caliper and wire it to the side. Do not disconnect the brake line.
6. Remove the brake rotor.
7. Remove the rear hub assembly.

To install:
8. Install the rear hub assembly. Tighten the bolts to 70 ft. lbs. (95 Nm).
9. Install the brake rotor.
10. Install the brake caliper.
11. Install the rear wheel speed sensor.
12. Install the wheel and lower the vehicle.

ADJUSTMENT

The rear wheel bearings are not adjustable. If the bearings are noisy or become loose, they must be replaced.

SPECIFICATIONS AND MAINTENANCE CHARTS

ENGINE AND VEHICLE IDENTIFICATION

Engine							Model Year	
Code ①	Liters (cc)	Cu. In.	Cyl.	Fuel Sys.	Engine Type	Eng. Mfg.	Code ②	Year
K	3.7 (3701)	226	6	SEFI	SOHC	Chrysler	9	2009

① 8th position of VIN

② 10th position of VIN

37671_RAID_C0001

GENERAL ENGINE SPECIFICATIONS

Year	Engine Displacement Liters	Engine VIN ①	Net Horsepower @ rpm	Net Torque @ rpm (ft. lbs.)	Bore x Stroke (in.)	Com- pression Ratio	Oil Pressure @ rpm
2009	3.7	K	211@5200	236@4000	3.66x3.40	9.6:1	25-110@3000

① Eighth digit of VIN

37671_RAID_C0002

GASOLINE ENGINE TUNE-UP SPECIFICATIONS

Year	Engine Displacement Liters	Engine VIN	Spark Plug Gap (in.)	Ignition Timing (deg.)	Fuel Pump (psi)	Idle Speed (rpm)	Valve Clearance	
							Intake	Exhaust
2009	3.7	K	0.043	①	56-60	②	HYD	HYD

NOTE: The Vehicle Emission Control Information (VECI) label often reflects specification changes made during production.

The label figures must be used if they differ from those in this chart.

HYD: Hydraulic

① Ignition timing is controlled by the PCM and is not adjustable.

② Idle speed is controlled by the PCM and is not adjustable

37671_RAID_C0003

CAPACITIES

Year	Engine Displ. Liters	Engine VIN	Oil with Filter (qts.)	Transmission (pts.) Manual	Transmission (pts.) Auto.	Transfer Case (pts.)	Drive Axle Front (pts.)	Drive Axle Rear (pts.)	Fuel Tank (gal.)	Cooling System (qts.)
2009	3.7	K	5.0	4.65	①	②	3.5	③	22.0	13.3

NOTE: All capacities are approximate. Add fluid gradually and check to be sure a proper fluid level is obtained.

① 42RLE: 8.0 pts.

　545RFE 2wd: 11.0 pts.

　545RFE 4wd: 13.0 pts.

② NV233: 2.5 pts.

　NV244: 2.85 pts.

③ The following values include 0.25 pt. of friction

　modifier for LSD axles:

　8.25 in. axle: 4.4 pts.

　9.25 in. axle: 4.9 pts.

37671_RAID_C0004

FLUID SPECIFICATIONS

Year	Model	Engine Displacement Liters	Engine VIN	Engine Oil	Auto. Trans.	Manual Trans.	Drive Axle	Transfer Case	Power Steering Fluid	Brake Master Cylinder
2009	Raider	3.7	K	5W-30	ATF+4	ATF+4	①	ATF+4	ATF+4	DOT 3 or 4

DOT: Department Of Transpotation

① Front: 75W-90

　Rear: 75W-140 Synthetic (LSD add MS-10111 limited slip additive)

37671_RAID_C0014

VALVE SPECIFICATIONS

Year	Engine Displ. Liters	Engine VIN	Seat Angle (deg.)	Face Angle (deg.)	Spring Test Pressure (lbs. @ in.)	Spring Installed Height (in.)	Stem-to-Guide Clearance (in.) Intake	Stem-to-Guide Clearance (in.) Exhaust	Stem Diameter (in.) Intake	Stem Diameter (in.) Exhaust
2009	3.7	K	44.5-45	45-45.5	①	1.579	0.0008-0.0028	0.0019-0.0039	0.2729-0.2739	0.2717-0.2728

① Intake: 213-234 lbs @1.107 in.

　Exhaust: 197-215 lbs @1.067 in.

37671_RAID_C0005

CAMSHAFT AND BEARING SPECIFICATIONS CHART

All measurements are given in inches.

Year	Engine Displ. Liters	Engine ID/VIN	Journal Dia.	Brg. Oil Clearance	Shaft End-play	Runout	Journal Bore	Lobe Height Intake	Lobe Height Exhaust
2009	3.7	K	1.0227-1.0235	0.0010-0.0026	0.0030-0.0079	0.0008	1.0245-1.0252	NA	NA

NA: Not Available

37671_RAID_C0006

CRANKSHAFT AND CONNECTING ROD SPECIFICATIONS

All measurements are given in inches.

Year	Engine Displ. Liters	Engine VIN	Crankshaft Main Brg. Journal Dia.	Crankshaft Main Brg. Oil Clearance	Crankshaft Shaft End-play	Crankshaft Thrust on No.	Connecting Rod Journal Diameter	Connecting Rod Oil Clearance	Connecting Rod Side Clearance
2009	3.7	K	2.4996-2.5005	0.0008-0.0018	0.0021-0.0112	2	2.2792-2.2798	0.0002-0.0017	0.0040-0.0138

37671_RAID_C0007

PISTON AND RING SPECIFICATIONS

All measurements are given in inches.

Year	Engine Displ. Liters	Engine VIN	Piston Clearance	Ring Gap Top Comp.	Ring Gap Bottom Comp.	Ring Gap Oil Control	Ring Side Clearance Top Comp.	Ring Side Clearance Bottom Comp.	Ring Side Clearance Oil Control
2009	3.7	K	0.0014	0.0079-0.0142	0.0146-0.0249	0.0099-0.0300	0.0020-0.0037	0.0016-0.0031	0.0007-0.0091

37671_RAID_C0008

TORQUE SPECIFICATIONS

All readings in ft. lbs.

Year	Engine Displ. Liters	Engine VIN	Cylinder Head Bolts	Main Bearing Bolts	Rod Bearing Bolts	Crankshaft Damper Bolts	Flywheel Bolts	Manifold Intake	Manifold Exhaust	Spark Plugs	Oil Pan Drain Plug
2009	3.7	K	①	①	②	130	70	9	18	20	25

① See text

② 20 ft. lbs. plus 90 degrees

37671_RAID_C0009

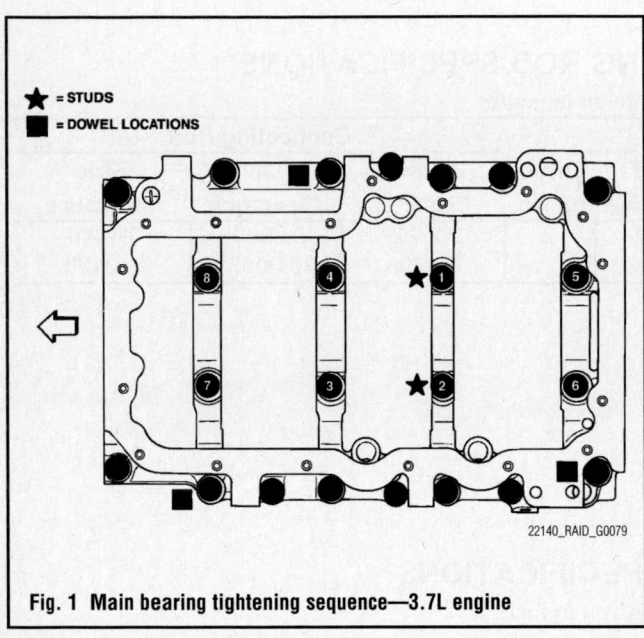

★ = STUDS

■ = DOWEL LOCATIONS

22140_RAID_G0079

Fig. 1 Main bearing tightening sequence—3.7L engine

WHEEL ALIGNMENT

Year	Model			Caster Range (+/-Deg.)	Caster Preferred Setting (Deg.)	Camber Range (+/-Deg.)	Camber Preferred Setting (Deg.)	Toe-in (Deg.)
2009	Raider	F	Left	0.05	+3.50	0.50	-0.25	0.20+/-0.10
			Right	0.05	+3.70	0.50	-0.25	0.20+/-0.10
		R		—	—	0.35	-0.10	0.30+/-0.35

37671_RAID_C0011

TIRE, WHEEL AND BALL JOINT SPECIFICATIONS

Year	Model	OEM Tires Standard	OEM Tires Optional	Tire Pressures (psi) Front	Tire Pressures (psi) Rear	Wheel Size	Ball Joint Inspection	Lug Nut Torque (ft. lbs.)
2009	Raider	P245/70R16	P255/65R16	①	①	NA	0.020 in.	135
			P265/70R16	①	①	NA		

OEM: Original Equipment Manufacturer

PSI: Pounds Per Square Inch

① See the tire placard on the vehicle

37671_RAID_C0010

BRAKE SPECIFICATIONS

All measurements in inches unless noted

Year			Brake Disc Original Thickness	Brake Disc Minimum Thickness	Brake Disc Maximum Run-out	Brake Drum Original Inside Diameter	Brake Drum Max. Wear Limit	Brake Drum Maximum Machine Diameter	Minimum Lining Thickness Front	Minimum Lining Thickness Rear	Brake Caliper Bracket Bolts (ft. lbs.)	Brake Caliper Mounting Bolts (ft. lbs.)
2009	Raider	F	1.100	1.039	0.0010	—	—	—	①	—	130	26
		R	—	—	—	11.50	②	11.693	—	③	—	—

F: Front

R: Rear

NA: Not Available

① Riveted brake pads: 0.0625 in.

Bonded brake pads: 0.1875 in.

② Maximum allowable drum diameter, either from wear or machining, is stamped on the drum.

③ Riveted brake shoes: 0.031 in.

Bonded brake shoes: 0.0625 in.

37671_RAID_C0012

SCHEDULED MAINTENANCE INTERVALS
Mistubishi Raider

TO BE SERVICED	SERVICE	VEHICLE MILEAGE INTERVAL (x1000)													
		6	12	18	24	30	36	42	48	54	60	66	72	78	84
Engine oil & filter	R	✓	✓	✓	✓	✓	✓	✓	✓	✓	✓	✓	✓	✓	✓
Tires	Rotate	✓	✓	✓	✓	✓	✓	✓	✓	✓	✓	✓	✓	✓	✓
Manual trans level	I	✓	✓	✓	✓	✓	✓	✓	✓	✓	✓	✓	✓	✓	✓
Brake Hoses and lines	I	✓	✓	✓	✓	✓	✓	✓	✓	✓	✓	✓	✓	✓	✓
Brake linings	S/I		✓		✓		✓		✓		✓		✓		✓
CV-Joints	S/I		✓		✓				✓				✓		
Exhaust system	S/I		✓		✓				✓				✓		
Front and rear axle fluid	S/I			✓			✓			✓			✓		
Front suspension	S/I				✓				✓				✓		
Air cleaner element	R					✓					✓				
Transfer case fluid level	I					✓					✓				
Spark plugs	R					✓					✓				
Spark plug wires	R										✓				
Engine coolant	R	Replace every 60 months or 102,000 miles, whichever comes first													
Automatic trans fluid	R	Replace every 120 months or 120,000 miles, whichever comes first													
Accessory drive belt	R	Replace every 120 months or 120,000 miles, whichever comes first													

R: Replace   S/I: Service or Inspect   L: Lubricate   Adj: Adjust

FREQUENT OPERATION MAINTENANCE (SEVERE SERVICE)

If a vehicle is operated under any of the following conditions it is considered severe service:

- Extremely dusty areas.
- 50%or more of the vehicle operation is in 32°C (90°F) or higher temperatures, or constant operation in temperatures
  below 0°C (32°F).
- Prolonged idling (vehicle operation in stop and go traffic.
- Frequent short running periods (engine does not warm to normal operating temperatures).
- Police, taxi, delivery usage or trailer towing usage.

Air filter (dusty or off road conditions): inspect every 12,000 miles.

Front and rear axle fluid: Change every 18,000 miles.

Automatic transmission fluid & filter: change every 60,000 miles.

Transfer case fluid: change every 60,000 miles.

Manual transmission fluid: change every 60,000 miles.

Oil Change Indicator System

On Electronic Vehicle Information Center (EVIC) equipped vehicles, "Oil Change Required" is displayed in the EVIC and a single chime sounds, indicating that an oil change is necessary. On non-EVIC equipped vehicles, "Change Oil" flashes in the instrument cluster and a single chime sounds indicating that an oil change is necessary. Illumination of the oil change message is based on the operating conditions of the vehicle. When the message is illuminated, the vehicle must be serviced within 500 miles.

The oil change indicator will not monitor the time since the last oil change. Change the oil if it has been more than 6 months since the last oil change, even if the oil change indicator message is not illuminated.

Under no circumstances should oil change intervals exceed 6,000 miles or 6 months, whichever comes first.

To reset the oil change indicator, perform the following procedure:

1. Turn the ignition switch to the ON position. Do not start the engine.

2. Fully press the accelerator pedal 3 times within 10 seconds.

3. Turn the ignition switch to the LOCK position.

If the indicator message illuminates when the vehicle is started, repeat the procedure.

BRAKES / INFORMATION AND PRECAUTIONS

ANTI-LOCK SYSTEMS

- Certain components within the ABS system are not intended to be serviced or repaired individually.
- Do not use rubber hoses or other parts not specifically specified for and ABS system. When using repair kits, replace all parts included in the kit. Partial or incorrect repair may lead to functional problems and require the replacement of components.
- Lubricate rubber parts with clean, fresh brake fluid to ease assembly. Do not use shop air to clean parts; damage to rubber components may result.
- Use only DOT 3 brake fluid from an unopened container.
- If any hydraulic component or line is removed or replaced, it may be necessary to bleed the entire system.
- A clean repair area is essential. Always clean the reservoir and cap thoroughly before removing the cap. The slightest amount of dirt in the fluid may plug an orifice and impair the system function. Perform repairs after components have been thoroughly cleaned; use only denatured alcohol to clean components. Do not allow ABS components to come into contact with any substance containing mineral oil; this includes used shop rags.

- The Anti-Lock control unit is a microprocessor similar to other computer units in the vehicle. Ensure that the ignition switch is **OFF** before removing or installing controller harnesses. Avoid static electricity discharge at or near the controller.
- If any arc welding is to be done on the vehicle, the control unit should be unplugged before welding operations begin.

DISC AND DRUM SYSTEMS

> **✳✳ CAUTION**
>
> **Dust and dirt accumulating on brake parts during normal use may contain asbestos fibers from production or aftermarket brake linings. Breathing excessive concentrations of asbestos fibers can cause serious bodily harm. Exercise care when servicing brake parts. Do not sand or grind brake lining unless equipment used is designed to contain the dust residue. Do not clean brake parts with compressed air or by dry brushing. Cleaning should be done by dampening the brake components with a fine mist of water, then wiping the brake components clean with a dampened cloth. Dispose of cloth and all residue containing asbestos fibers in an impermeable container with the appropriate label. Follow practices prescribed by the Occupational Safety and Health Administration (OSHA) and the Environmental Protection Agency (EPA) for the handling, processing, and disposing of dust or debris that may contain asbestos fibers.**

BRAKES / BLEEDING THE BRAKE SYSTEM

BLEEDING PROCEDURE

BLEEDING PROCEDURE

Use Mitsubishi brake fluid, or an equivalent quality fluid meeting SAE J1703-F and DOT 3 standards only. Use fresh, clean fluid from a sealed container at all times.

1. Before servicing the vehicle, refer to the precautions.
2. Remove reservoir filler caps and fill reservoir.
3. If calipers were overhauled, open all caliper bleed screws. Then close each bleed screw as fluid starts to drip from it. Top off master cylinder reservoir once more before proceeding.
4. Attach one end of bleed hose to bleed screw and insert opposite end in glass container (2) partially filled with brake fluid. Be sure end of bleed hose is immersed in fluid.

➡ **Bleed procedure should be in this order (1) Right rear (2) Left rear (3) Right front (4) Left front.**

5. Open up bleeder, then have a helper press down the brake pedal. Once the pedal is down close the bleeder. Repeat bleeding until fluid stream is clear and free of bubbles. Then move to the next wheel.
6. Before moving the vehicle verify the pedal is firm and not mushy.
7. Top off the brake fluid and install the reservoir cap.

BLEEDING THE ABS SYSTEM

ABS system bleeding requires conventional bleeding methods plus use of a scan tool. The procedure involves performing a base brake bleeding, followed by use of the scan tool to cycle and bleed the Hydraulic Control Unit (HCU) pump and solenoids. A second base brake bleeding procedure is then required to remove any air remaining in the system.

1. Perform base brake bleeding.
2. Connect the scan tool to the data link connector.
3. Select "Anti-lock Brakes" followed by "Miscellaneous", then "Bleed Brakes". Follow the instructions displayed until the unit displays "Test Complete", then disconnect the scan tool and proceed.
4. Perform a base brake bleeding a second time.
5. Top up the master cylinder.

BRAKES   ANTI-LOCK BRAKE SYSTEM (ABS)

WHEEL SPEED SENSORS

REMOVAL & INSTALLATION

Front

See Figure 2.

1. Before servicing the vehicle, refer to the precautions.
2. Raise and safely support the vehicle securely on jackstands.
3. Remove the wheel.
4. Remove the brake caliper.
5. Remove the rotor.
6. Remove the sensor attaching bolts.
7. Disconnect the wire and remove the sensor from the vehicle.

To install:

8. Install the sensor from the vehicle. Tighten the bolts to 190 inch. lbs. (21 Nm).
9. Connect the sensor wire.
10. Install the sensor attaching bolts.
11. Install the rotor.
12. Install the brake caliper.
13. Install the wheel.
14. Lower the vehicle.

Rear

See Figure 3.

1. Before servicing the vehicle, refer to the precautions.
2. Raise and safely support the vehicle securely on jackstands.

3. Remove brake line mounting nut and remove the brake line from the sensor stud.
4. Remove the park brake cable and bracket from the sensor stud.
5. Disconnect the electrical connector.
6. Remove mounting stud from the sensor.
7. Remove sensor from differential housing.

To install:

8. Install O-ring on sensor if removed.
9. Install sensor on the differential housing. Tighten to 18 ft. lbs. (24 Nm).
10. Connect harness to sensor. Be sure the seal is securely in place between the sensor and wiring connector.
11. Install the park brake cable and bracket from the sensor bolt.
12. Install brake line mounting nut and Install the brake line from the sensor bolt.
13. Lower the vehicle.

WHEEL SPEED SENSOR RINGS (TOOTHED RINGS)

REMOVAL & INSTALLATION

The ABS wheel speed sensor ring is an integral part of either the wheel hub (front) or the differential (rear) and cannot be disassembled.

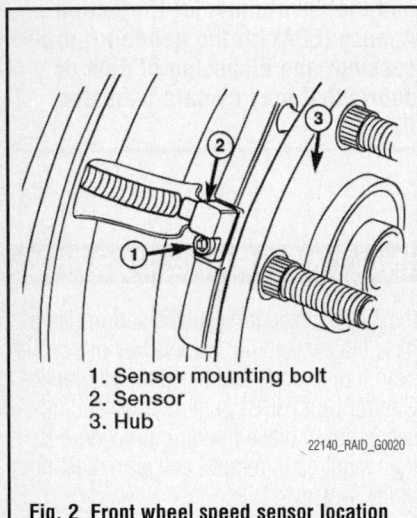

1. Sensor mounting bolt
2. Sensor
3. Hub

22140_RAID_G0020

Fig. 2 Front wheel speed sensor location

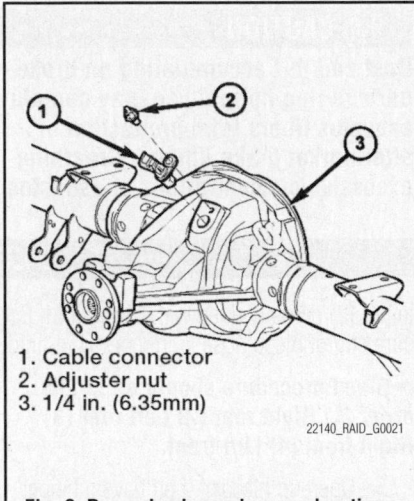

1. Cable connector
2. Adjuster nut
3. 1/4 in. (6.35mm)

22140_RAID_G0021

Fig. 3 Rear wheel speed sensor location

BRAKES   FRONT DISC BRAKES

BRAKE CALIPER

REMOVAL & INSTALLATION

See Figure 4.

1. Before servicing the vehicle, refer to the precautions.

✳✳ CAUTION

Never allow the disc brake caliper to hang from the brake hose. Damage to the brake hose with result. Provide a suitable support to hang the caliper securely.

2. Remove the wheel.
3. Compress the disc brake caliper.
4. Remove the banjo bolt and discard the copper washers.

5. Remove the caliper slide pin bolts.
6. Remove the disc brake caliper from the caliper adapter.
7. Remove the caliper slide pins from the adapter.

To install:

✳✳ WARNING

Petroleum based grease should not be used on any of the rubber components of the caliper, Use only Non-Petroleum based grease.

→Clean slide pin bores thoroughly to remove any old grease.

→Use grease packets included with the kit or Dow Corning-807T grease.

8. Thoroughly coat the new slide pins on all working surfaces.
9. Install the boot onto the slide pin and then insert into the adapter.
10. Push the pin all the way into the adapter and carefully expel the trapped air by gently pushing on the boot near the slide pin head.

→Install a new copper washers on the banjo bolt when installing

11. Install the disc brake caliper to the brake caliper adapter.
12. Install the banjo bolt with new copper washers to the caliper. Tighten to 21 ft. lbs. (28 Nm).
13. Install the caliper slide pin bolts. Tighten to 24 ft. lbs. (32 Nm).

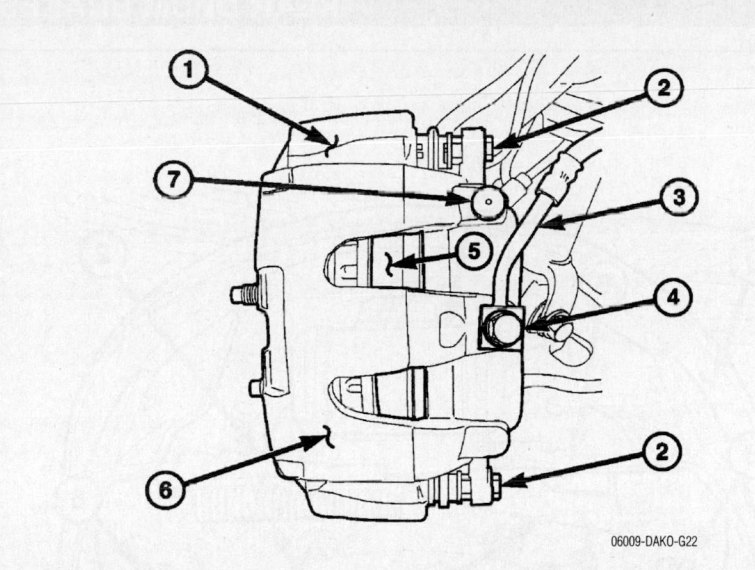

Fig. 4 (1) Caliper mounting adaptor, (2) Caliper mounting bolts, (3) Brake line, (4) Banjo bolt, (5) Pad, (6) Caliper

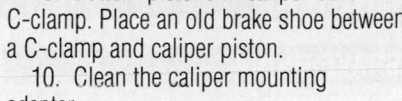

9. Bottom pistons in caliper bore with C-clamp. Place an old brake shoe between a C-clamp and caliper piston.

10. Clean the caliper mounting adapter.

11. Install new anti-rattle clips to the brake pads.

12. Install the inboard brake pad in the adapter.

13. Install the outboard brake pad in the adapter.

14. Install the caliper over rotor. Then, push the caliper onto the adapter.

15. Install the caliper slide pin bolts.

16. Install the wheel.

17. Apply brakes several times to seat caliper pistons and brake shoes and obtain firm pedal.

18. Top off master cylinder fluid level if required.

14. Bleed the brake system.
15. Install the wheel.

DISC BRAKE PADS

REMOVAL & INSTALLATION

See Figures 5 and 6.

1. Before servicing the vehicle, refer to the precautions.
2. Remove the wheel.
3. Compress the caliper.
4. Remove the caliper slide pin bolts.
5. Remove the caliper from the caliper adapter.

➡**Do not allow brake hose to support caliper assembly.**

6. Remove the inboard brake pad from the caliper adapter.
7. Remove the outboard brake pad from the caliper adapter.
8. Remove the anti-rattle clips from the pad.

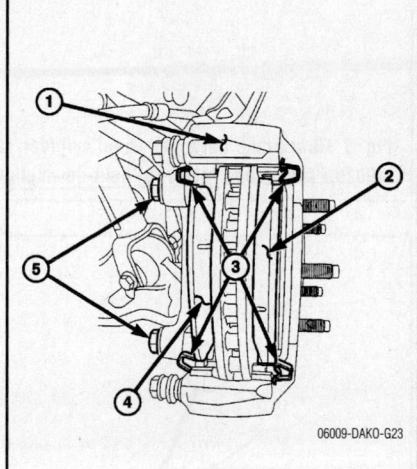

Fig. 5 (1) Caliper adapter, (2) Outboard pad, (3) Anti-rattle clips, (4) Inboard pad, (5) Caliper mounting bolts

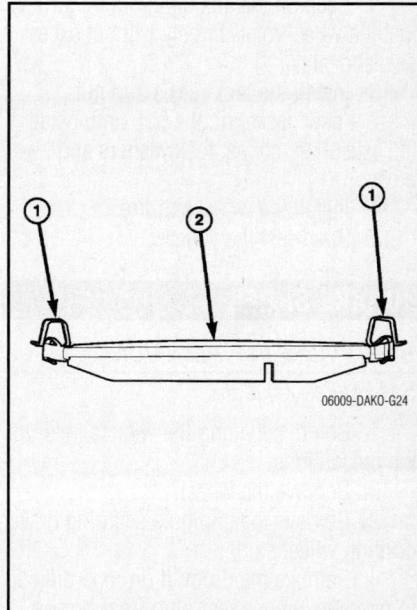

Fig. 6 Anti-rattle clips (1) installed on the pad (2)

BRAKE DRUM

REMOVAL & INSTALLATION

1. Before servicing the vehicle, refer to the precautions.

2. Remove the axle shaft nuts, washers and cones. If the cones do not readily release, rap the axle shaft sharply in the center.

3. Remove the axle shaft.

4. Remove the outer hub nut.

5. Straighten the lockwasher tab and remove it along with the inner nut and bearing.

6. Carefully remove the drum.

To install:

7. Position the drum on the axle housing.

8. Install the bearing and inner nut. While rotating the wheel and tire, tighten the adjusting nut until a slight drag is felt.

9. Back off the adjusting nut ⅙ turn so that the wheel rotates freely without excessive end-play.

10. Install the lock ring s and nut. Place a new gasket on the hub and install the axle shaft, cones, lockwashers and nuts.

11. Install the wheel and tire.

12. Road-test the vehicle.

BRAKE SHOES

REMOVAL & INSTALLATION

See Figures 7 and 8.

1. Before servicing the vehicle, refer to the precautions.

2. Remove the wheel.

3. Remove the clip nuts securing brake drum to wheel studs.

4. Remove the drum. If drum is difficult to remove, remove rear plug from access hole in support plate. Back-off self adjusting by inserting a thin screwdriver into access hole and push lever away from adjuster star wheel. Then insert an adjuster tool into brake adjusting hole rotate adjuster star wheel to retract brake shoes.

5. Vacuum brake components to remove brake lining dust.

6. Remove shoe return spring with brake spring pliers tool.

7. Remove adjuster spring and lever. Disengage lever from spring by sliding lever forward to clear pivot and work lever out from under spring.

8. Disengage and remove shoe return spring from brake shoes.

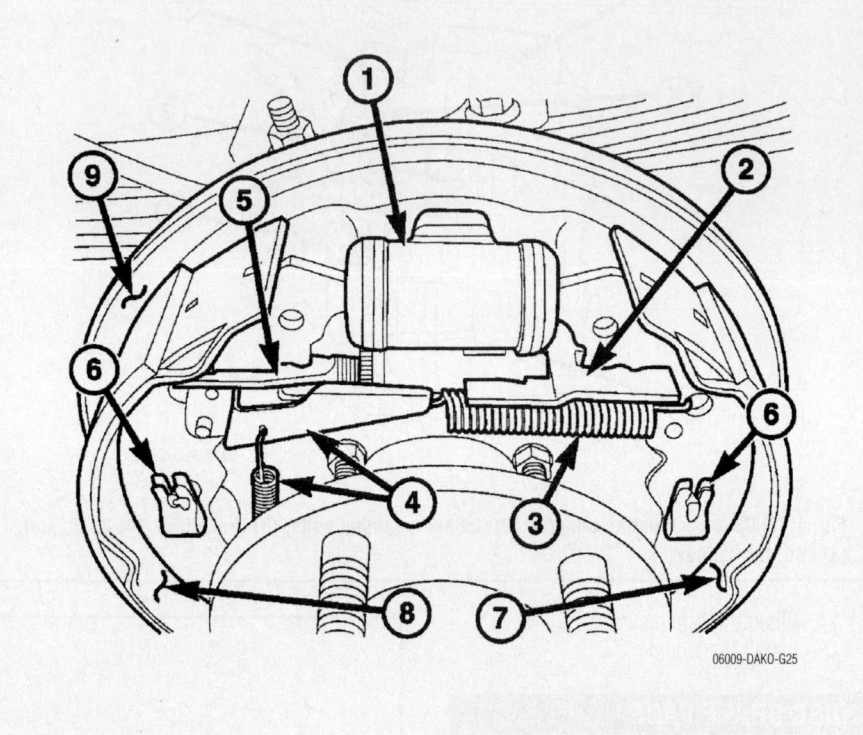

Fig. 7 Rear brake parts (1) wheel cylinder, (2) parking brake lever, (3) return spring, (4 & 5) adjuster spring and lever, (6) hold-down clips, (7 & 8) brake shoes, (9) backing plate

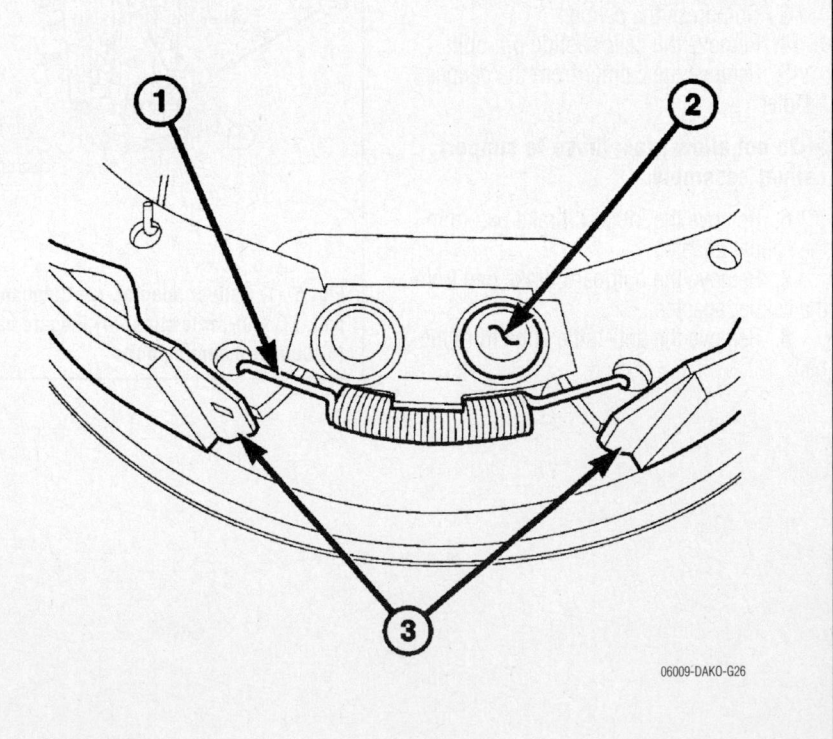

Fig. 8 Lower return spring (1), cam (2) and brake shoes, (3)

9. Remove the brake shoe hold down cllps.

10. Remove the rear brake shoe from support plate.

11. Remove the front brake shoe from support plate.

12. Remove the park brake lever from the brake shoe.

To install:

13. Clean and inspect individual brake components.

14. Lubricate where the brake shoe contacts the support plate with high temperature grease or Lubriplate®.

15. Lubricate adjuster screw socket, nut, button and screw thread surfaces with grease or Lubriplate®.

16. Install parking brake lever to the rear shoe and install the hold down clip.

17. Install the adjuster strut onto the shoes and park brake lever.

18. Install the front shoe on support plate, and install the hold down clip.

19. Install the adjuster spring and lever in the slot in the adjuster strut.

20. Install the lower return spring to the shoes.

21. Verify adjuster operation. Pull both shoes outward to move the adjuster lever to rotate the star wheel. Be sure adjuster lever properly engages star wheel teeth.

22. Adjust brake shoes to drum with a brake gauge.

23. Install wheel and tire assembly.

ADJUSTMENT

The rear drum brakes are equipped with a self-adjusting mechanism. Under normal circumstances, the only time adjustment is required is when the shoes are replaced, removed for access to other parts, or when one or both drums are replaced. Adjustment can be made with a standard brake gauge or with adjusting tool. Adjustment is performed with the complete brake assembly installed on the backing plate.

Adjustment with a Brake Gauge

See Figures 9 and 10.

1. Before servicing the vehicle, refer to the precautions.

2. Be sure parking brakes are fully released.

3. Raise rear of vehicle and remove wheels and brake drums.

4. Verify that left and right automatic adjuster levers and cables are properly connected.

5. Insert brake gauge in drum. Expand gauge until gauge inner legs contact drum braking surface. Then lock gauge in position.

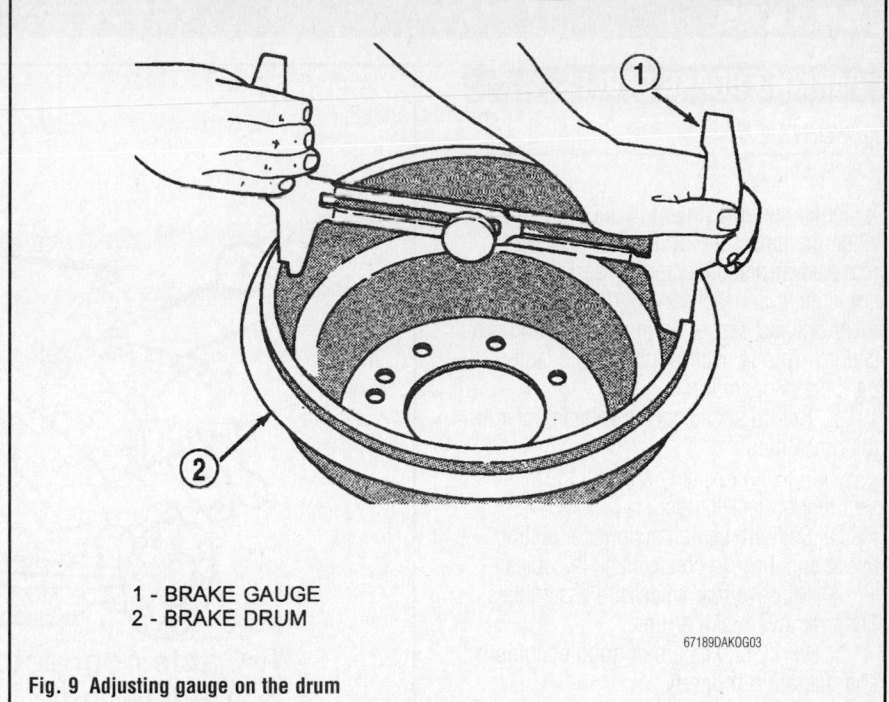

1 - BRAKE GAUGE
2 - BRAKE DRUM

67189DAKOG03

Fig. 9 Adjusting gauge on the drum

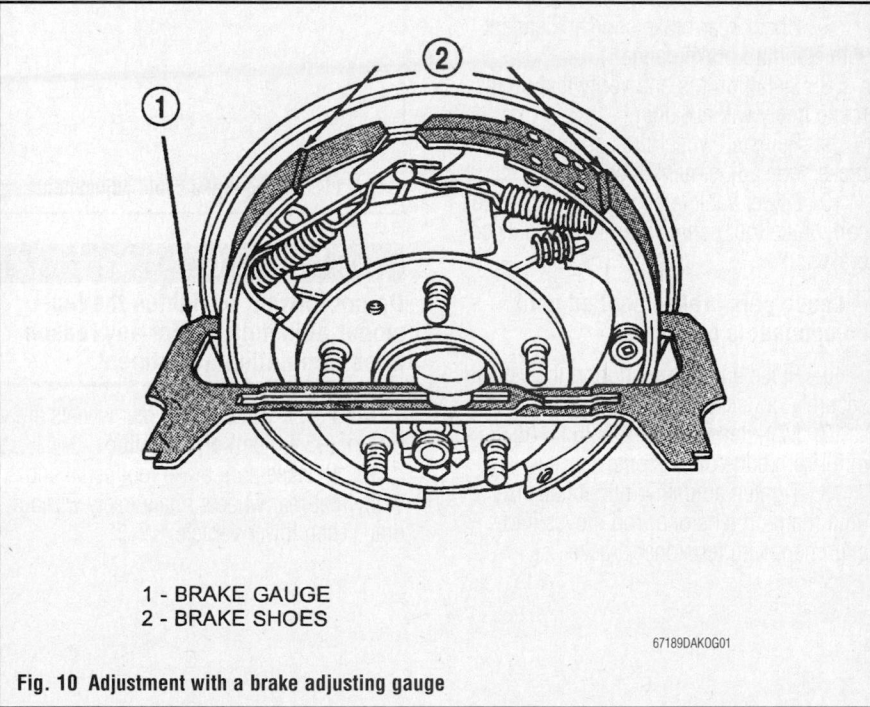

1 - BRAKE GAUGE
2 - BRAKE SHOES

67189DAKOG01

Fig. 10 Adjustment with a brake adjusting gauge

6. Reverse gauge and install it on brake shoes. Position gauge legs at shoe centers as shown. If gauge does not fit (too loose/too tight), adjust shoes.

7. Pull shoe adjuster lever away from adjuster screw star wheel.

8. Turn adjuster screw star wheel (by hand) to expand or retract brake shoes. Continue adjustment until gauge outside legs are light drag-fit on shoes.

9. Install brake drums and wheels and lower vehicle.

10. Drive vehicle and make one forward stop followed by one reverse stop. Repeat procedure 8-10 times to operate automatic adjusters and equalize adjustment.

➡**Bring vehicle to complete standstill at each stop. Incomplete, rolling stops will not activate automatic adjusters.**

PARKING BRAKE CABLES

ADJUSTMENT

See Figure 11.

Tensioner adjustment is only necessary when the tensioner, or a cable has been replaced or disconnected for service. When adjustment is necessary, perform adjustment only as described in the following procedure. This is necessary to avoid faulty park brake operation.

1. Before servicing the vehicle, refer to the precautions.

2. Raise and safely support the vehicle securely on jackstands.

3. Back off cable tensioner adjusting nut at equalizer to create slack in cables.

4. Remove rear wheel/tire assemblies. Then remove brake drums.

5. Verify brakes are in good condition and operating properly.

6. Verify park brake cables operate freely and are not binding, or seized.

7. Check rear brake shoe adjustment with standard brake gauge.

8. Install drums and verify that drums rotate freely without drag.

9. Reinstall wheel/tire assemblies after brake shoe adjustment is complete.

10. Lower vehicle enough for access to park brake foot pedal. Then fully apply park brakes.

➡**Leave park brakes applied until adjustment is complete.**

11. Raise and safely support the vehicle securely on jackstands.

12. Mark tensioner rod 0.25 in. (6.35 mm) from edge of tensioner bracket.

13. Tighten adjusting nut at equalizer until mark on tensioner rod moves into alignment with tensioner bracket.

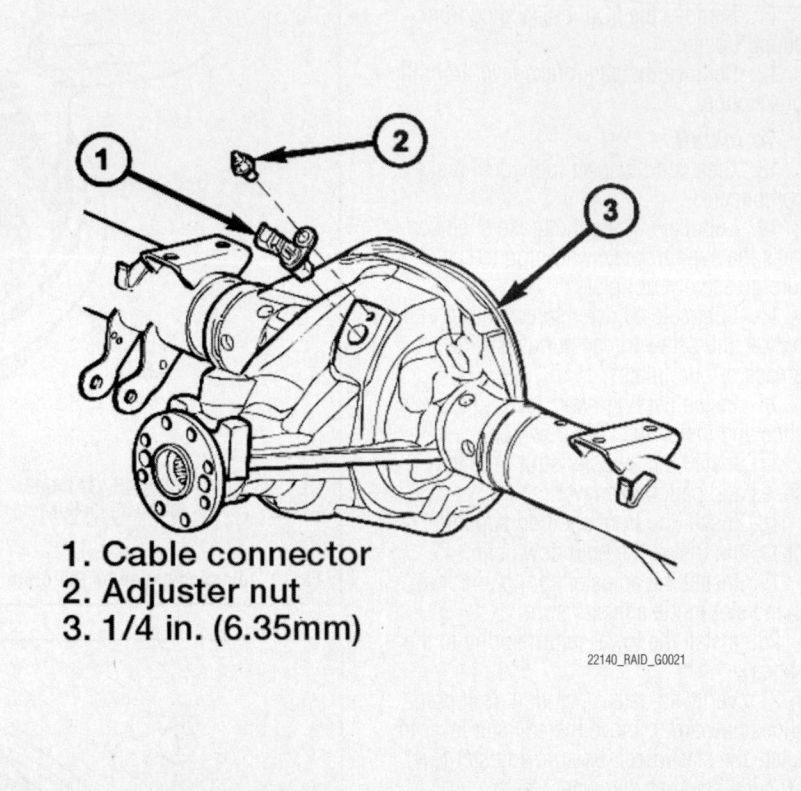

1. Cable connector
2. Adjuster nut
3. 1/4 in. (6.35mm)

22140_RAID_G0021

Fig. 11 Parking brake cable adjustment

❋❋ WARNING

Do not loosen, or tighten the tensioner adjusting nut for any reason after completing adjustment.

14. Lower vehicle until rear wheels are 6-8 in. (15-20 cm) off shop floor.

15. Release park brake foot pedal and verify that rear wheels rotate freely without drag. Then lower vehicle.

PARKING BRAKE SHOES

REMOVAL & INSTALLATION

The rear brake shoes serve as the parking brakes. Refer to Rear Drum Brakes, Brake Shoes, Removal & Installation.

ADJUSTMENT

The rear brake shoes serve as the parking brakes. Refer to Rear Drum Brakes, Brake Shoes, Adjustment.

CHASSIS ELECTRICAL   AIR BAG (SUPPLEMENTAL RESTRAINT SYSTEM)

GENERAL INFORMATION

✳✳ CAUTION

Some vehicles are equipped with an air bag system. The system must be disarmed before performing service on, or around, system components, the steering column, instrument panel components, wiring and sensors. Failure to follow the safety precautions and the disarming procedure could result in accidental air bag deployment, possible injury and unnecessary system repairs.

PRECAUTIONS

Disconnect and isolate the battery negative cable before beginning any airbag system component diagnosis, testing, removal, or installation procedures. Allow system capacitor to discharge for two minutes before beginning any component service. This will disable the airbag system. Failure to disable the airbag system may result in accidental airbag deployment, personal injury, or death.

DISARMING THE SYSTEM

Disconnect and isolate the negative battery cable. Wait 2 minutes for the system capacitor to discharge before performing any service.

ARMING THE SYSTEM

When repairs are completed, connect the negative battery cable.

CLOCKSPRING CENTERING

See Figure 12.

The clockspring is mounted on the steering column behind the steering wheel. Its purpose is to maintain a continuous electrical circuit between the wiring harness and the driver's side air bag module. This assembly consists of a flat, ribbon-like electrically conductive tape that winds and unwinds with the steering wheel rotation.

Service replacement clocksprings are shipped pre-centered and with a molded plastic locking pin that snaps into a receptacle on the rotor and is engaged between two tabs on the upper surface of the rotor case. The locking pin secures the centered

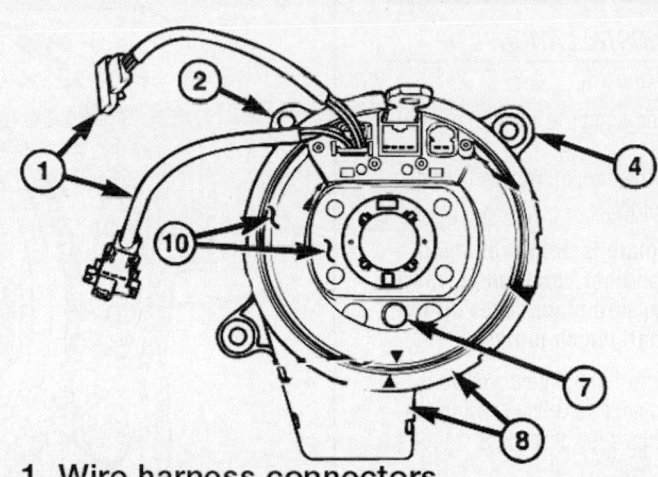

1. Wire harness connectors
2. Locating tab
4. Screws
7. Engagement dowel and yellow rubber boot
8. Alignment arrows on the clockspring case
10. Clockspring rotor

22140_RAID_G0024

Fig. 12 Clockspring alignment marks

clockspring rotor to the clockspring case during shipment, but the locking pin must be removed from the clockspring after it is installed on the steering column. This locking pin should not be removed until the clockspring has been installed on the steering column. If the locking pin is removed before the clockspring is installed on a steering column, the clockspring centering procedure must be performed.

➡**The clockspring cannot be repaired. If the clockspring is faulty, damaged, or if the driver airbag has been deployed, the clockspring must be replaced.**

1. Before servicing the vehicle, refer to the precautions.
2. Disconnect the negative battery cable.

➡**Wait at least 2 minutes after disconnecting the cable from the negative (-) battery terminal to prevent airbag activation.**

➡**Before starting this procedure, be certain to turn the steering wheel until the front wheels are in the straight-ahead position.**

3. Place the front wheels in the straight-ahead position.
4. Remove the clockspring from the steering column.
5. Rotate the clockspring rotor clockwise to the end of its travel. Do not apply excessive torque.
6. From the end of the clockwise travel, rotate the rotor about two and one-half turns counterclockwise.
7. The engagement dowel and yellow rubber boot should end up at the bottom, and the arrows on the clockspring rotor and case should be in alignment. The clockspring is now centered.
8. The front wheels should still be in the straight-ahead position. Reinstall the clockspring onto the steering column.

DRIVE TRAIN

CLUTCH DRIVEN DISC & PRESSURE PLATE

REMOVAL & INSTALLATION

See Figures 13 and 14.

1. Before servicing the vehicle, refer to the precautions.

2. Remove the transmission and clutch housing as assembly.

➡**If pressure plate is being removed for access to another component, mark position of pressure plate cover on flywheel with small punch marks.**

3. Loosen pressure plate cover bolts evenly and in rotation to relieve spring tension. Loosen bolts a few threads at a time to avoid warping cover.

4. Remove cover bolts, pressure plate and clutch disc.

To install:

➡**Clean flywheel surface with solvent. Scuff sand the surface with 120/180 grit emery cloth to remove minor scratches and glazing.**

5. Check new clutch disc for runout and free operation on input shaft splines.

6. Lubricate crankshaft pilot bearing with a NLGI—2 rated grease.

7. Position clutch disc with pressure plate on the flywheel.

8. Insert alignment tool or spare input shaft through clutch disc and into pilot bearing.

9. Verify that the disc hub is positioned correctly. The raised portion of the hub faces away from the flywheel.

10. Install the cover bolts finger tight.

11. Tighten cover bolts evenly (and in rotation) a few threads at a time. Cover bolts

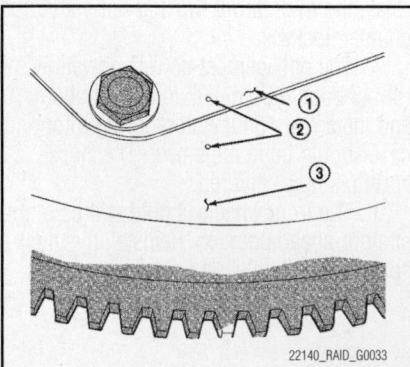

Fig. 13 Mark position of pressure plate cover on flywheel with small punch marks

22140_RAID_G0033

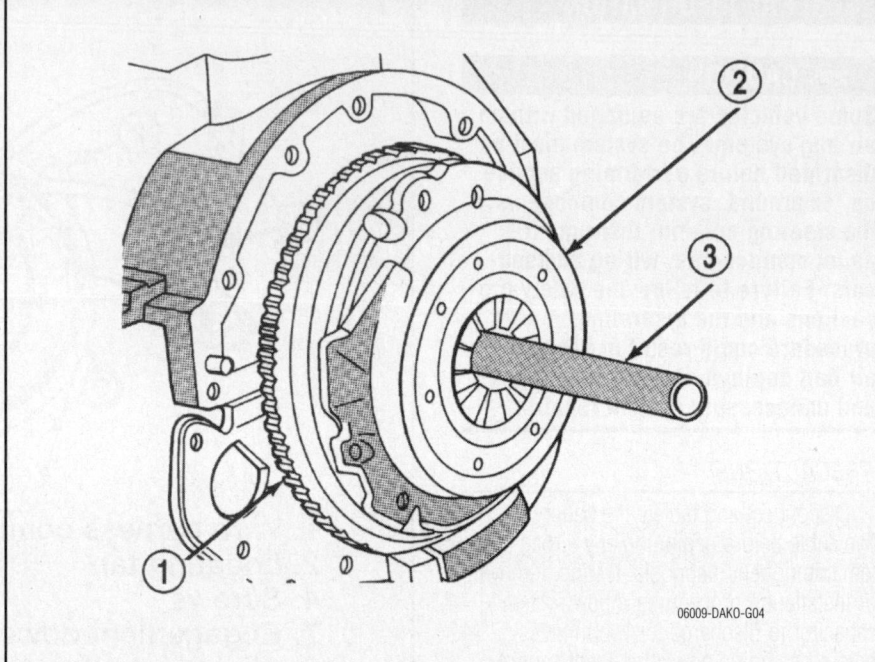

Fig. 14 Clutch installation. (1) flywheel, (2) pressure plate, (3) alignment tool

06009-DAKO-G04

must be tightened evenly and to specified tighten to avoid distorting cover.

12. Tighten the cover bolts as follows:
   a. $5/16$ in. bolts to 17 ft. lbs. (23 Nm).
   b. $3/8$ in. bolts to 30 ft. lbs. (41 Nm).

13. Apply light coat of high temperature bearing grease to splines of transmission input shaft and to release bearing slide surface of front bearing retainer.

✳✳ WARNING

Do not over-lubricate shaft splines. This could result in grease contamination of disc.

14. Install the transmission assembly.

HYDRAULIC SYSTEM BLEEDING

BLEEDING PROCEDURE

The system is self-bleeding. Press the clutch pedal repeatedly to release air from the fluid. The air will be vented from the reservoir.

FRONT AXLE SHAFT, BEARING & SEAL

REMOVAL & INSTALLATION

Front Differential

See Figures 15 through 17.

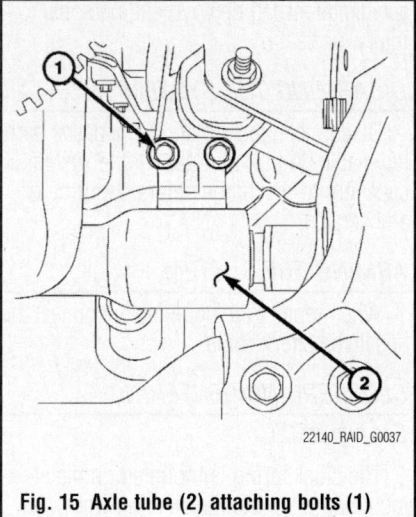

22140_RAID_G0037

Fig. 15 Axle tube (2) attaching bolts (1)

1. Before servicing the vehicle, refer to the precautions.

2. With vehicle in neutral, position vehicle on hoist.

3. Remove the skid plate, if equipped.

4. Remove skid plate support crossmember, if necessary.

5. Remove hub/bearing wheel speed sensor .and dust shield from knuckles .

6. Remove both halfshafts.

7. Mark the propeller shaft, transfer case, and pinion companion flange for installation reference.

8. Remove the front propeller shaft.

9. Remove the axle vent tube.

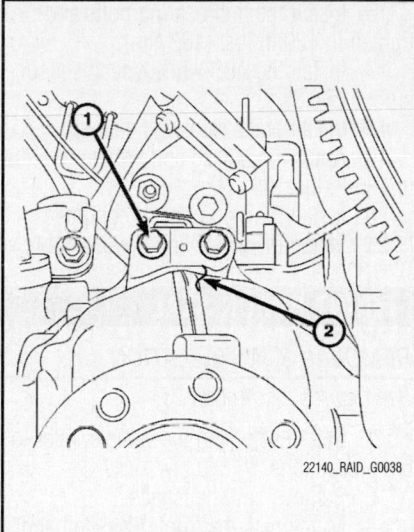

Fig. 16 Axle differential housing (2) attaching bolts (1)

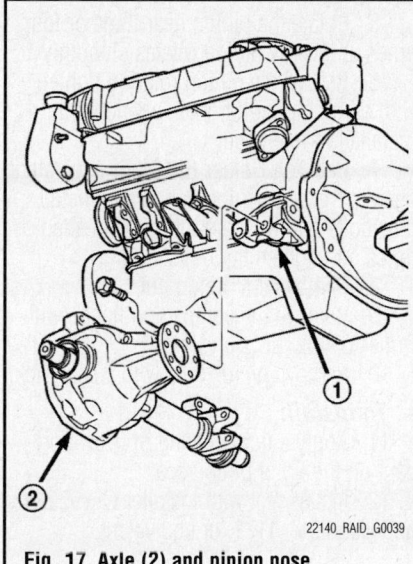

Fig. 17 Axle (2) and pinion nose bracket (1)

19. Tighten all bolts to 85 ft. lbs. (115 Nm).

20. Install the axle vent tube.

21. Align the reference marks on the propeller shaft, transfer case, and pinion companion flange.

22. Install propeller shaft.

23. Install halfshafts.

24. Install hub/bearing wheel speed sensor and dust shield on knuckles .

25. Install skid plate support crossmember, if necessary.

26. Install skid plate, if necessary.

27. Check differential lubricant level and add lubricant, if necessary.

Axle Bearings

See Figure 18.

1. Before servicing the vehicle, refer to the precautions.

9. Install a new axle shaft seal with installer 8402 and handle C-4171, or equivalent.

10. Install the axle shaft and halfshaft.

Fig. 19 Install a new axle shaft seal with Installer 8402 or equivalent

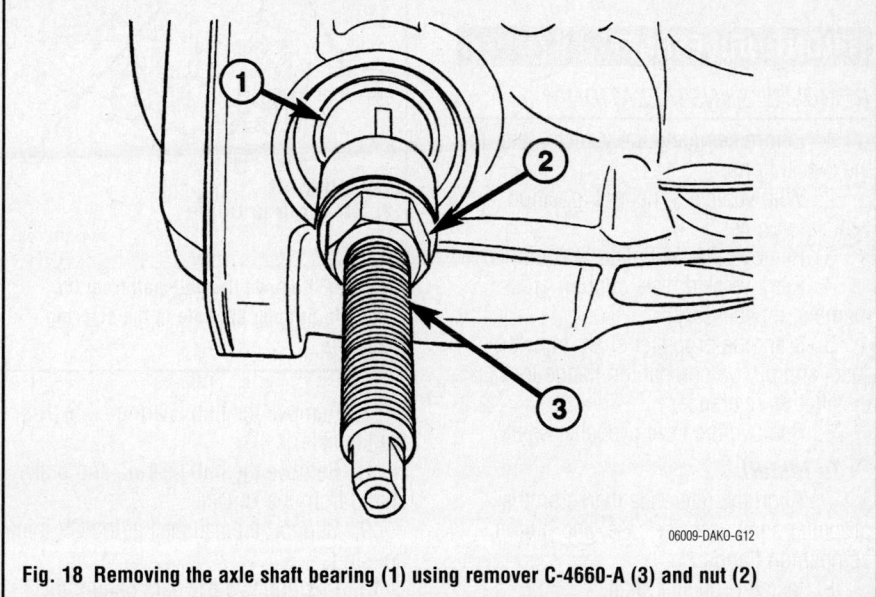

Fig. 18 Removing the axle shaft bearing (1) using remover C-4660-A (3) and nut (2)

10. Position and secure a jack under differential housing.

11. Remove bolts from the axle tube .

12. Remove bolts from axle differential housing .

13. Remove bolts from axle pinion nose bracket .

14. Lower jack and remove axle from vehicle.

To install:

15. Raise the axle into position.

16. Loosely install the bolts and nuts to hold the axle pinion nose bracket .

17. Loosely install bolts from axle differential housing .

18. Loosely install bolts from the axle tube.

2. Remove the axle shaft.

3. Remove the axle shaft seal.

4. Install the axle shaft bearing remover C-4660-A, or equivalent, in the bearing. Then tighten the nut to spread the remover in the bearing.

5. Install the bearing remove cup, bearing and nut. Then tighten the nut to draw the bearing out.

6. Inspect the axle shaft tube bore for roughness and burrs. Remove as necessary.

To install:

7. Wipe the axle shaft tube bore clean.

8. Install the axle shaft bearing with installer 5063 and handle C-4171, or equivalent.

Axle Shaft Seal

See Figure 19.

1. Before servicing the vehicle, refer to the precautions.

2. Remove the axle shaft.

3. Remove the axle shaft seal with a small prybar.

To install:

4. Wipe the axle shaft tube bore clean.

5. Install a new axle shaft seal with installer 8402 and handle C-4171,.

6. Install the axle shaft and halfshaft.

FRONT AXLE SHAFT

REMOVAL & INSTALLATION

1. Before servicing the vehicle, refer to the precautions.
2. With vehicle in neutral, position vehicle on hoist.
3. Remove the skid plate, if equipped.
4. Remove skid plate support crossmember, if necessary.
5. Remove hub/bearing wheel speed sensor .and dust shield from knuckles .
6. Remove both halfshafts.

To install:

7. Install halfshafts.
8. Install hub/bearing wheel speed sensor and dust shield on knuckles .
9. Install skid plate support crossmember, if necessary.
10. Install skid plate, if necessary.
11. Check differential lubricant level and add lubricant, if necessary.

FRONT DRIVESHAFT

REMOVAL & INSTALLATION

1. Before servicing the vehicle, refer to the precautions.
2. With vehicle in neutral, position vehicle on hoist.
3. Remove the skid plate, if equipped.
4. Remove skid plate support crossmember, if necessary.
5. Mark the propeller shaft, transfer case, and pinion companion flange for installation reference.
6. Remove the front propeller shaft.

To install:

7. Align the reference marks on the propeller shaft, transfer case, and pinion companion flange.
8. Install propeller shaft.
9. Install skid plate support crossmember, if necessary.
10. Install skid plate, if necessary.
11. Check differential lubricant level and add lubricant, if necessary.

FRONT HALFSHAFTS

REMOVAL & INSTALLATION

See Figures 20 and 21.

1. Before servicing the vehicle, refer to the precautions.
2. Remove the front wheel.
3. Remove the skid plate, if equipped.
4. Remove the hub nut and washer.
5. Remove the brake caliper and rotor.
6. Remove the ABS wheel speed sensor if equipped.

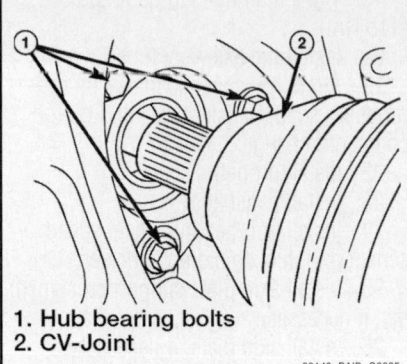

1. Hub bearing bolts
2. CV-Joint

22140_RAID_G0035

Fig. 20 Hub bearing bolts on the steering knuckle

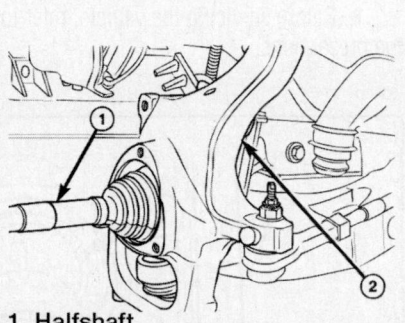

1. Halfshaft
2. Steering knuckle

22140_RAID_G0036

Fig. 21 Remove the halfshaft from the vehicle through the hole in the steering knuckle

7. Remove the hub bearing bolts from the knuckle.
8. Remove the hub bearing and brake shield from the knuckle.
9. Support the halfshaft at the CV joint housings.
10. Position two pry bars behind the inner CV housing and disengage the CV joint from the axle.
11. Remove the halfshaft from the vehicle.

To install:

12. Apply a light coating of wheel bearing grease on the axle splines.
13. Insert the halfshaft stub through the steering knuckle and onto the axle. Verify the shaft snapring engages with the groove on the inside of the joint housing.
14. Clean the hub bearing bore and hub bearing mating surface of all foreign materials. Apply a light coating of grease to all mating surfaces.
15. Install the hub bearing onto the axle halfshaft and steering knuckle.

16. Install the hub bearing bolts and tighten to 120 ft. lbs. (163 Nm).
17. Install the ABS wheel speed sensor, If equipped.
18. Install brake rotor and caliper.
19. Apply the brakes and tighten hub nut to 185 ft. lbs. (251 Nm).
20. Install the skid plate, if equipped.
21. Install the wheel and tire assembly.

FRONT PINION SEAL

REMOVAL & INSTALLATION

See Figures 22 through 25.

1. Before servicing the vehicle, refer to the precautions.
2. Remove both halfshafts.
3. Matchmark the front driveshaft and pinion companion flange for installation reference, if equipped.
4. Remove the front driveshaft, if equipped
5. Rotate the pinion gear three or four times and verify pinion rotates smoothly.
6. Record the pinion rotating tighten with an inch pound tighten wrench, for installation reference.
7. Position Holder 6719A or equivalent against the companion flange and install four bolts and washers into the threaded holes and tighten the bolts.
8. Remove the pinion nut.
9. Remove the companion flange with Puller C-452, or equivalent.
10. Remove pinion seal with a seal pick.

To install:

11. Apply a light coating of gear lubricant on the lip of pinion seal.
12. Install seal with Installer C-3972-A and Handle C-4171, or equivalent.

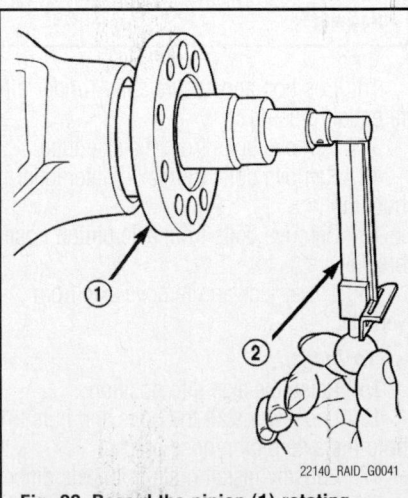

22140_RAID_G0041

Fig. 22 Record the pinion (1) rotating torque with an inch pound torque wrench (2)

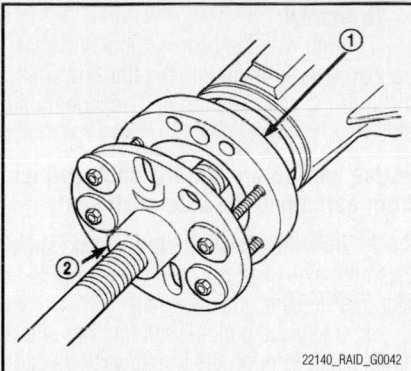

Fig. 23 Remove the companion flange (1) with Puller C-452, or equivalent (2)

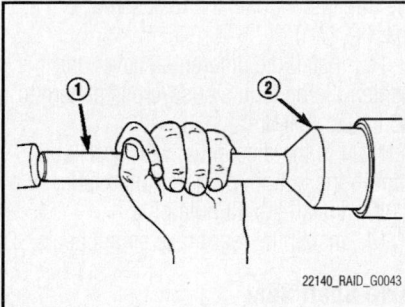

Fig. 24 Install seal with Installer C-3972-A (2) and Handle C-4171 (1), or equivalent

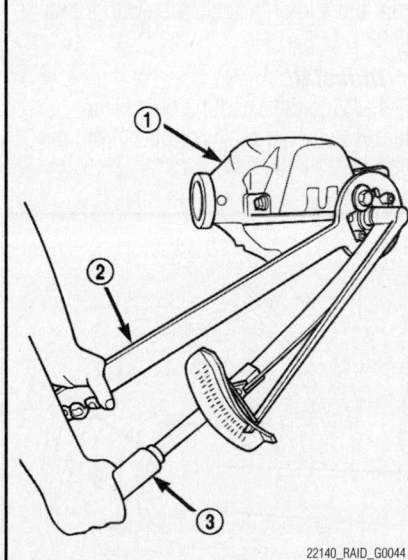

Fig. 25 Position holder (2) against the companion flange. Tighten the differential (1) bolt and washer using a torque wrench (3) to obtain the rotating torque referenced during removal

13. Install the companion flange onto the pinion with Installer C-3718 and Holder 6719A, or equivalent.

14. Position holder against the companion flange and install four bolts and washers into the threaded holes. Tighten the bolt and washer so that the holder is held to the flange.

15. Install a new pinion nut onto the pinion shaft and tighten the pinion nut until there is zero bearing end-play.

✳✳ WARNING

Do not exceed the minimum tightening tighten when installing the companion flange at this point. Damage to the collapsible spacer or bearings may result.

16. Tighten the nut to 200 ft. lbs. (271 Nm).

➡**Never loosen pinion nut to decrease pinion bearing rotating tighten and never exceed specified preload tighten. If preload tighten or rotating tighten is exceeded a new collapsible spacer must be installed.**

17. Record the pinion rotating tighten using a tighten wrench. The rotating tighten should be equal to the reading recorded during removal plus 5 inch lbs. (0.56 Nm).

18. If the rotating tighten is low, tighten the pinion nut in 5 ft. lbs. (6.8 Nm) increments until the proper rotating tighten is achieved.

➡**If the maximum tightening tighten is reached prior to reaching the required rotating tighten, the collapsible spacer may have been damaged. Replace the collapsible spacer.**

19. Install the driveshaft with the reference marks aligned.

20. Install halfshafts.

REAR AXLE HOUSING

REMOVAL & INSTALLATION

See Figure 26.

1. Before servicing the vehicle, refer to the precautions.

2. Place the transmission in neutral.

3. Raise and safely support the vehicle securely on jackstands.

4. Position a lift under axle and secure axle to lift.

5. Remove the wheels and rear brake components.

6. Remove ABS sensor from the differential housing.

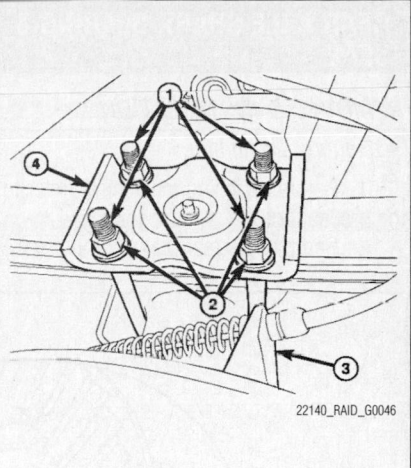

Fig. 26 Spring clamp (1), nuts (2), emergency brake cable (3) and spring plate (4)

7. Disconnect the brake hose at the axle junction block.

8. Disconnect the vent hose from the axle shaft tube.

9. Mark the propeller shaft and companion flange for installation alignment reference.

10. Remove propeller shaft.

11. Remove shock absorbers axle mounting bolts.

12. Remove stabilizer bar retainer clamp bolts and retainer clamps from the axle.

13. Remove spring clamps nuts and spring plates.

14. Remove the axle from the vehicle.

To install:

15. Raise axle with lifting device and align to the leaf spring centering bolts.

16. Install spring clamps and spring plates. Tighten the U-bolt nuts to 110 ft. lbs. (149 Nm).

17. Install shock absorbers and tighten nuts to 75 ft. lbs. (102 Nm).

18. Install ABS sensor into the differential housing.

19. Install rear brake and park brake components.

20. Connect brake hose to axle junction block.

21. Install stabilizer bar and center it with equal spacing on both sides. Tighten retainer clamp bolts to 45 ft. lbs. (61 Nm).

22. Install axle vent hose.

23. Install propeller shaft with reference marks aligned. Tighten bolts to 80 ft. lbs. (108 Nm).

24. Add gear lubricant, if necessary.

REAR AXLE SHAFT, BEARING & SEAL

REMOVAL & INSTALLATION

See Figures 27 and 28.

1. Before servicing the vehicle, refer to the precautions.
2. Remove the rear brake components.

3. Remove the differential cover and drain the fluid.
4. Rotate the differential case to access the pinion mate shaft lock screw. Remove the lock screw and pinion mate shaft from the differential case.
5. Push the axle shaft inward then remove the C-clip from the axle shaft.
6. Remove the axle shaft.

To install:

7. Lubricate the bearing bore and seal lip with gear lubricant. Insert the axle shaft through the seal, bearing and engage it into side gear splines.

➡**Use care to prevent the shaft splines from damaging the axle shaft seal.**

8. Insert the C-clip in end of axle shaft. Push the axle shaft outward to seat the C-clip in side gear.
9. Insert the pinion shaft into the differential case, through the thrust washers and differential pinions.
10. Align the hole in the pinion mate shaft with the hole in the differential case and install the lock screw with Loctite®on the threads. Tighten the lock screw to 8 ft. lbs. (11 Nm).
11. Install the differential cover and tighten the bolts in a criss-cross pattern to 30 ft. lbs. (41 Nm).
12. Fill the differential with gear lubricant to the bottom of the fill plug hole.
13. Install the fill hole plug.
14. Install the rear brake components.

Axle Shaft Seal

See Figure 29.

1. Before servicing the vehicle, refer to the precautions.
2. Remove the axle shaft.
3. Remove the axle shaft seal from the end of the axle tube with a seal pick.

To install:

4. Wipe the axle tube bore clean. Remove any old sealer or burrs from the tube.

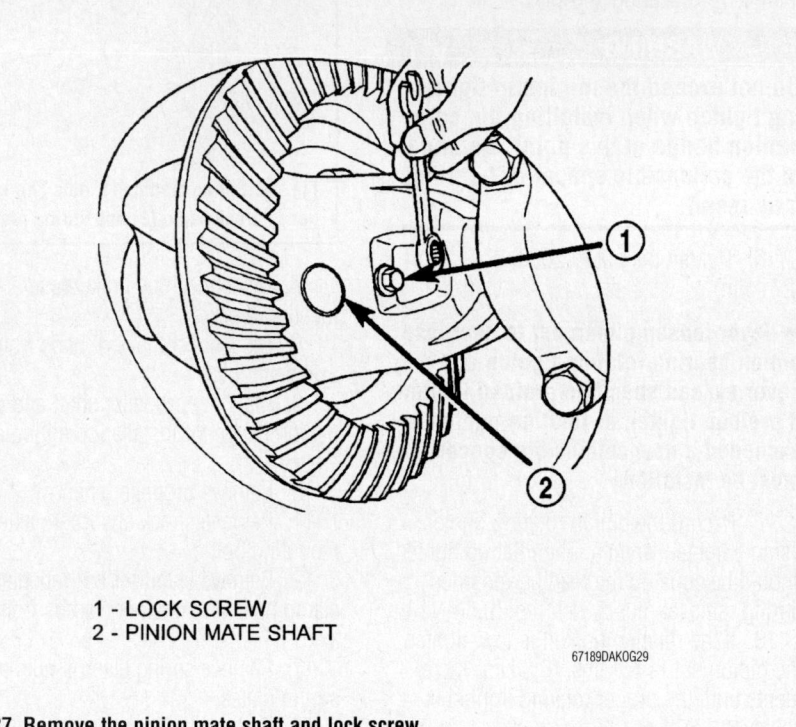

1 - LOCK SCREW
2 - PINION MATE SHAFT

67189DAK0G29

Fig. 27 Remove the pinion mate shaft and lock screw

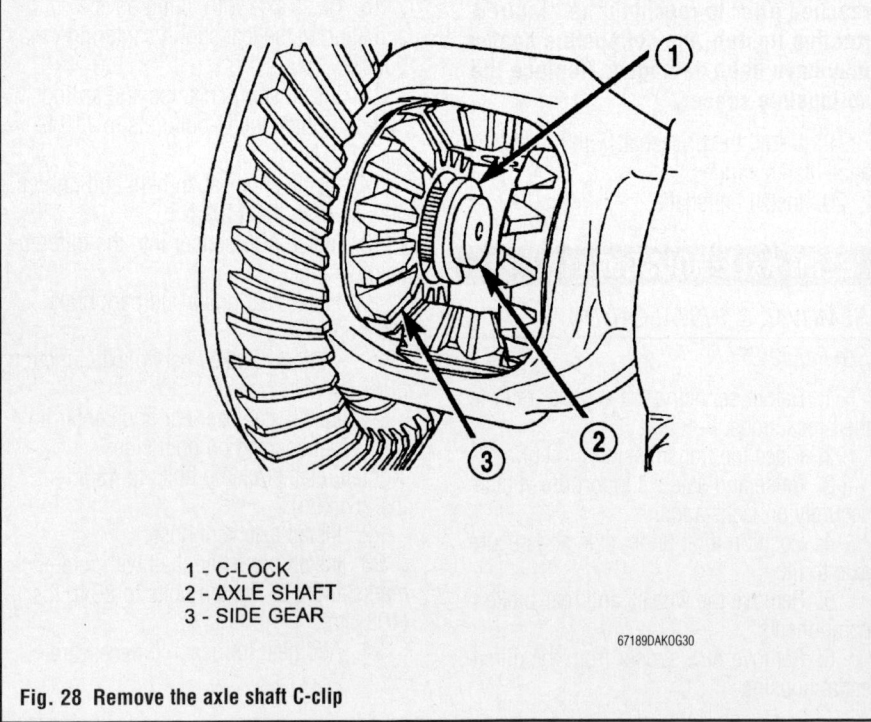

1 - C-LOCK
2 - AXLE SHAFT
3 - SIDE GEAR

67189DAK0G30

Fig. 28 Remove the axle shaft C-clip

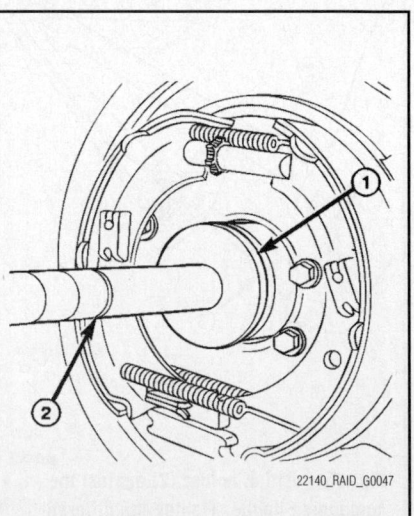

22140_RAID_G0047

Fig. 29 Install the new axle seal with Installer C-4198 (1) and Handle C-4171 (2), or equivalent

5. Coat the lip of the new seal with axle lubricant for protection prior to installing the axle shaft.

6. Install the new axle seal with Installer C-4198 and Handle C-4171, or equivalent. When the tool contacts the axle tube, the seal is installed to the correct depth.

7. Install the axle shaft.

Axle Bearings

See Figure 30.

1. Before servicing the vehicle, refer to the precautions.

2. Remove the axle shaft.

3. Remove the axle shaft seal from the axle tube with a seal pick.

➡**The seal and bearing can be removed at the same time with the bearing removal tool.**

4. Remove the axle shaft bearing with Bearing Removal Tool Set 6310 and Adapter Foot 6310-9, or equivalent.

To install:

5. Wipe the axle tube bore clean. Remove any old sealer or burrs from the tube.

6. Install the axle shaft bearing with Installer C-4198 and Handle C-4171, or equivalent.

➡**Install the bearing with part number against the installer.**

7. Install a new axle seal.
8. Install the axle shaft

REAR DRIVESHAFT

REMOVAL & INSTALLATION

1. Before servicing the vehicle, refer to the precautions.

2. Matchmark the universal joint, companion flange and pinion shaft for installation reference.

3. Disconnect the rear driveshaft.

4. Remove companion flange bolts and secure the shaft in an upright position to prevent damage to the rear universal joint.

To install:

5. Install driveshaft with the installation reference marks aligned.

6. Tighten companion flange bolts to 80 ft. lbs. (108 Nm).

REAR PINION SEAL

REMOVAL & INSTALLATION

See Figures 31 through 34.

1. Before servicing the vehicle, refer to the precautions.

2. Matchmark the universal joint, companion flange and pinion shaft for installation reference.

3. Disconnect the rear driveshaft.

4. Remove companion flange bolts and secure the shaft in an upright position to prevent damage to the rear universal joint.

5. Remove brake drums to prevent any drag.

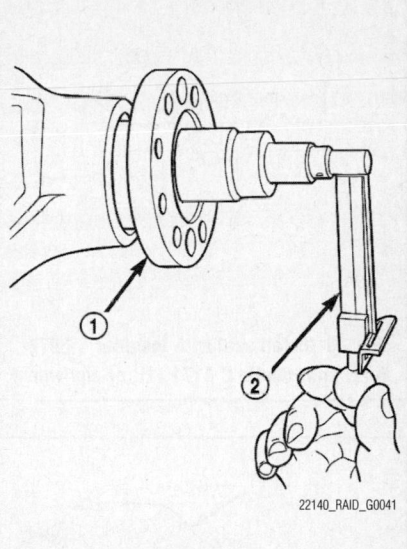

Fig. 31 Record the pinion (1) rotating torque with an inch pound torque wrench (2)

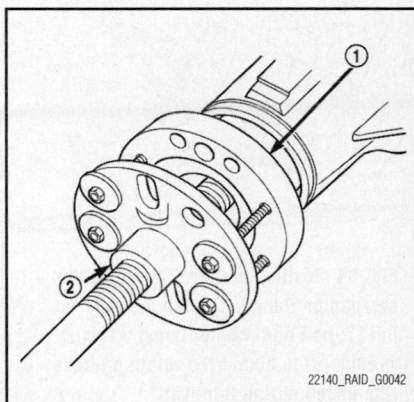

Fig. 32 Remove the companion flange (1) with Puller C-452, or equivalent (2)

6. Rotate companion flange three or four times and verify flange rotates smoothly.

7. Measure rotating torque of the pinion with an inch pound torque wrench and record the reading for installation reference.

8. Install bolts into two of the threaded holes in the companion flange 180° apart.

9. Position Holder 6719 or equivalent against the companion flange and install a bolt and washer into one of the remaining threaded holes. Tighten the bolts so the Holder 6719 or equivalent is held to the flange.

10. Remove the pinion nut and washer.

11. Remove companion flange with Remover C-452, or equivalent.

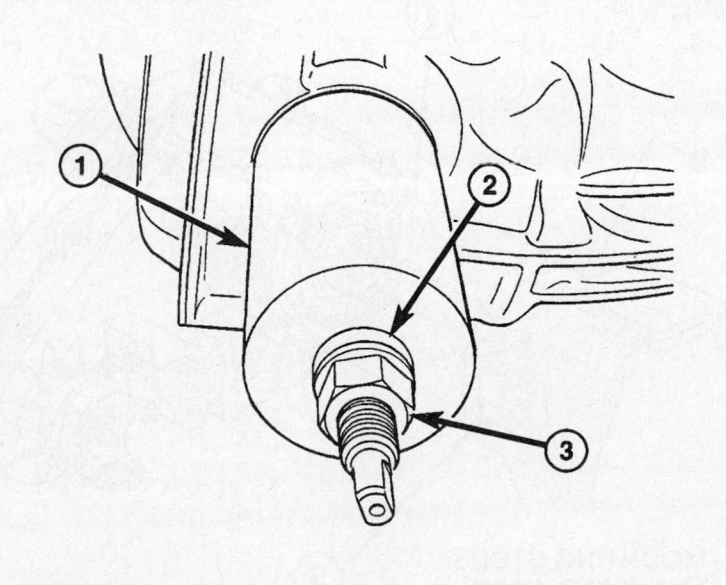

1 - REMOVER CUP
2 - BEARING
3 - NUT

67189DAKOG32

Fig. 30 Bearing Removal Tool Set 6310

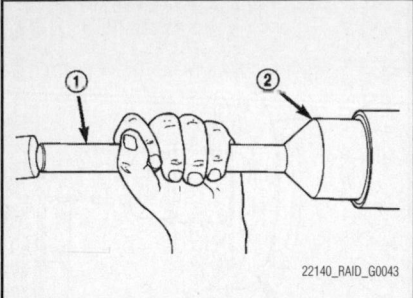

Fig. 33 Install seal with Installer C-3972-A (2) and Handle C-4171 (1), or equivalent

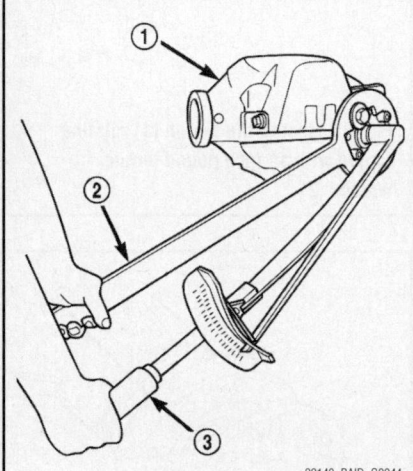

Fig. 34 Position holder (2) against the companion flange. Tighten the differential (1) bolt and washer using a torque wrench (3) to obtain the rotating torque referenced during removal

12. Remove pinion seal with a pry tool or slide hammer mounted screw.

To install:

13. Apply a light coating of gear lubricant on the lip of pinion seal.

14. Install a new pinion seal with Installer C-4076-B and Handle C-4735, or equivalent.

15. Install companion flange on the end of the shaft with the reference marks aligned.

16. Install bolts into two of the threaded holes in the companion flange 180°apart.

17. Position Holder 6719, or equivalent, against the companion flange and install a bolt and washer into one of the remaining threaded holes. Tighten the bolts so Holder 6719 is held to the flange.

18. Install companion flange on pinion shaft with Installer C-3718 and Holder 6719, or equivalent.

19. Install the pinion washer and a new pinion nut. The convex side of the washer must face outward.

➡**Do not exceed the minimum tightening tighten when installing the companion flange retaining nut at this point. Damage to collapsible spacer or bearings may result.**

20. Hold companion flange with Holder 6719 and tighten the pinion nut to 210 ft. lbs. (285 Nm). Rotate pinion several revolutions to ensure the bearing rollers are seated.

21. Rotate the pinion flange with an inch pound tighten wrench. Rotating tighten should be equal to the reading recorded during removal plus 5 inch lbs. (0.56 Nm).

➡**Never loosen pinion nut to decrease pinion bearing rotating tighten and never exceed specified preload tighten. If rotating tighten is exceeded, a new collapsible spacer must be installed.**

22. If rotating tighten is too low, use Holder 6719 to hold the companion flange and tighten pinion nut in 5 ft. lbs. (6.8 Nm) increments until proper rotating tighten is achieved.

➡**The seal replacement is unacceptable if final pinion nut tighten is less than 210 ft. lbs. (285 Nm).**

➡**The bearing rotating tighten should be constant during a complete revolution of the pinion. If the rotating tighten varies, this indicates a binding condition.**

23. Install driveshaft with the installation reference marks aligned.

24. Tighten companion flange bolts to 80 ft. lbs. (108 Nm).

25. Install the rear brake components.

26. Check the differential housing lubricant level.

TRANSFER CASE ASSEMBLY

REMOVAL & INSTALLATION

See Figure 35.

1. Before servicing the vehicle, refer to the precautions.

2. Drain the transfer case fluid.

3. Shift the transfer case into 2WD.

4. Mark the front and rear propeller shafts for alignment reference.

5. Support the transmission with a suitable jack stand.

6. Remove the rear crossmember and skid plate, if equipped.

7. Disconnect the front and rear driveshafts at the transfer case.

8. Disconnect the transfer case shift motor and mode sensor wire connectors.

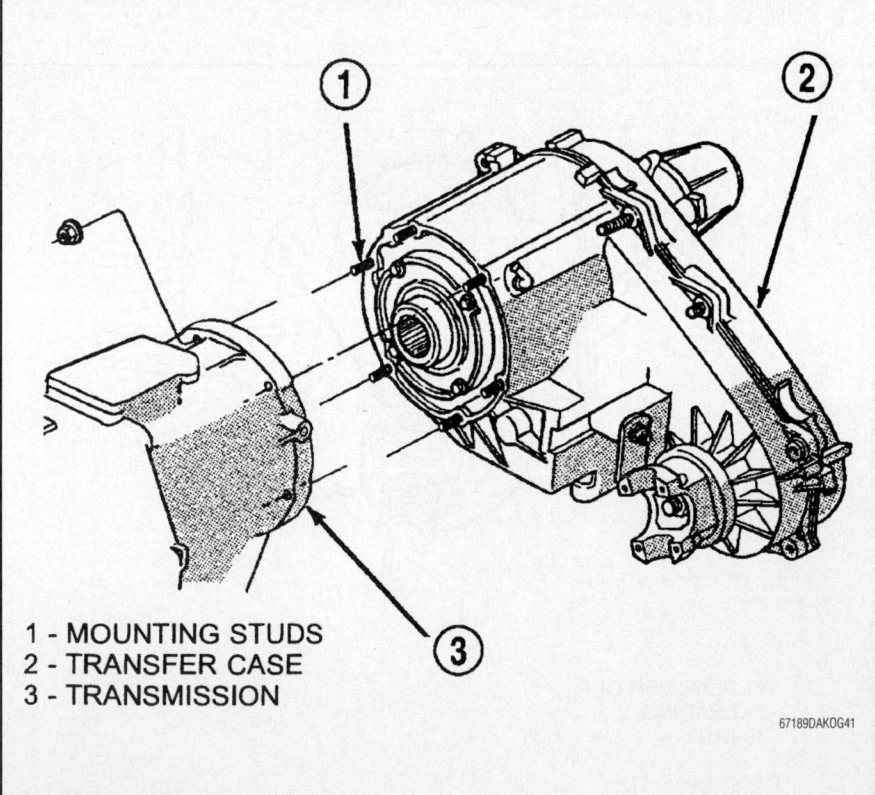

1 - MOUNTING STUDS
2 - TRANSFER CASE
3 - TRANSMISSION

Fig. 35 Typical transfer case mounting

9. Disconnect the transfer case vent hose.

10. Support the transfer case with a suitable transmission jack and secure the transfer case to the jack with chains.

11. Remove the nuts attaching transfer case to the transmission.

12. Pull the transfer case and jack rearward to disengage the transfer case.

13. Remove the transfer case assembly.

To install:

14. Align the transfer case and transmission shafts and install the transfer case onto the transmission.

15. Install and the transfer case attaching nuts. Tighten to 20–25 ft. lbs. (27–34 Nm).

16. Connect the vent hose.

17. Connect the shift motor and mode sensor wiring connectors. Secure the wire harness to clips on the transfer case.

18. Align and connect the driveshafts.

19. Install the rear crossmember and skid plate, if equipped. Tighten the crossmember bolts to 30 ft. lbs. (41 Nm).

20. Refill the transfer case with fluid to the correct level.

21. Verify proper transfer case shift operation.

ENGINE COOLING

ENGINE FAN

REMOVAL & INSTALLATION

See Figure 36.

1. Before servicing the vehicle, refer to the precautions.

2. Partially drain the cooling system.

3. Remove the upper radiator hose.

4. Remove the air filter housing assembly.

5. Using special tool 6958 and adapter 8346 or similar, remove the fan/viscous fan drive assembly from the water pump. Do not remove the fan/viscous fan drive assembly from the vehicle at this time.

6. Position the fan/fan drive assembly in the radiator shroud.

7. Remove the two shroud mounting screws.

8. Remove the radiator shroud and fan drive assembly.

9. Place the viscous fan drive in a horizontal position.

10. Remove the four bolts securing the fan blade assembly to viscous fan drive.

To install:

11. Install fan blade assembly to viscous fan drive. Tighten bolts to 17 ft. lbs. (23 Nm).

12. Position fan blade/viscous fan drive assembly into the radiator shroud.

13. Install the radiator shroud and fan drive assembly into the vehicle.

14. Install fan shroud retaining screws. Tighten screws to 50 inch lbs. (6 Nm)

15. Install fan blade/viscous fan drive assembly to water pump shaft .

16. Install the upper radiator hose.

17. Connect battery negative cable.

18. Fill and bleed cooling system.

RADIATOR

REMOVAL & INSTALLATION

See Figure 37.

1. Before servicing the vehicle, refer to the precautions.

2. Drain the cooling system.

3. Disconnect the negative battery cable.

4. Remove the pushpins and upper condenser/radiator seal.

5. Disconnect the upper radiator hose.

6. Disconnect the power steering hoses from power steering fluid/transmission cooler.

7. Disconnect the overflow tube.

8. Remove the radiator fan shroud from the radiator and position over the radiator fan.

9. Raise and safely support the vehicle securely on jackstands.

10. Disconnect the power steering cooler lines.

11. Remove the lower radiator hose.

12. Lower the vehicle.

13. Remove the upper radiator mounting bolts.

14. Remove the radiator.

To install:

15. Install the radiator.

16. Install the upper radiator mounting bolts. Tighten radiator support bolts to 200 inch lbs. (23 Nm).

17. Lower the vehicle.

18. Install the lower radiator hose.

19. Connect the power steering cooler lines.

20. Raise and safely support the vehicle securely on jackstands.

21. Install the radiator fan shroud to the radiator and position over the radiator fan.

22. Connect the overflow tube.

23. Connect the power steering hoses to power steering fluid/transmission cooler.

24. Connect the upper radiator hose.

25. Install the pushpins and upper condenser/radiator seal.

26. Connect the negative battery cable.

27. Fill and bleed the cooling system.

THERMOSTAT

REMOVAL & INSTALLATION

See Figure 38.

1. Before servicing the vehicle, refer to the precautions.

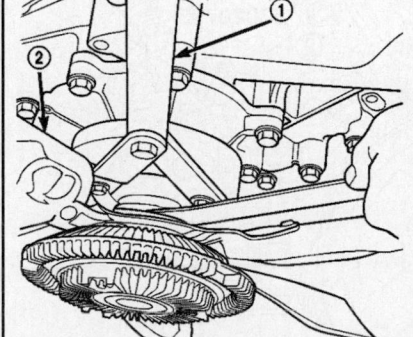

22140_RAID_G0052

Fig. 36 Using special tool 6958 and adapter 8346 (1) or similar, remove the fan/viscous fan drive assembly (2)

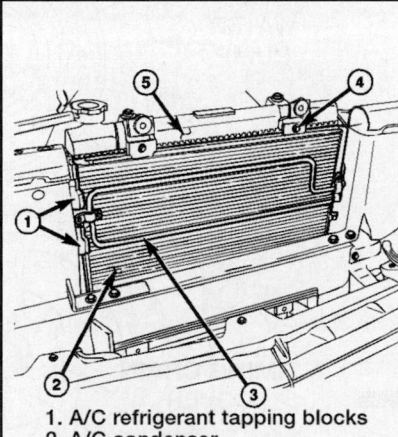

1. A/C refrigerant tapping blocks
2. A/C condenser
3. Power steering cooler
4. Mounting bolt
5. Radiator

22140_RAID_G0051

Fig. 37 Radiator assembly

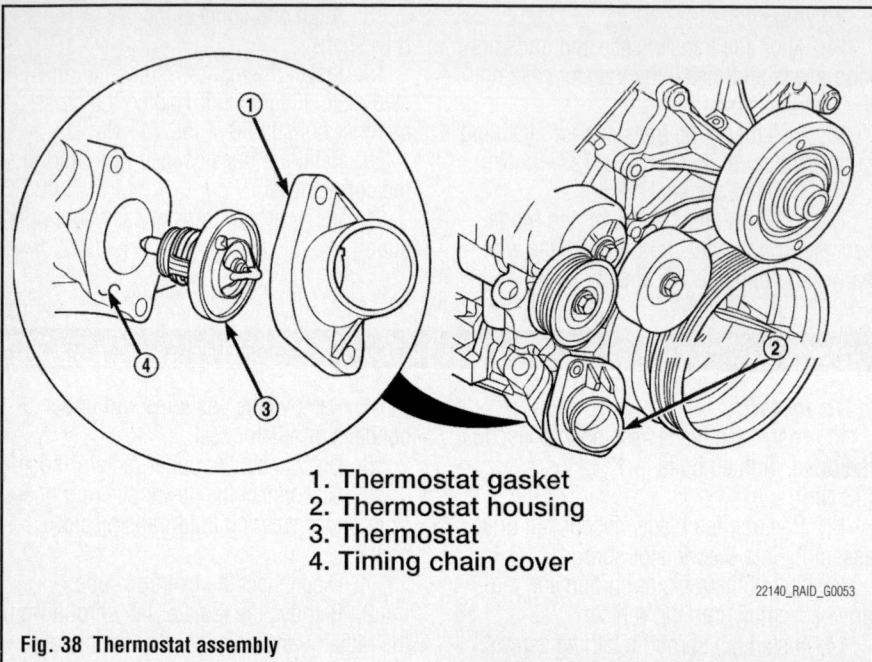

1. Thermostat gasket
2. Thermostat housing
3. Thermostat
4. Timing chain cover

22140_RAID_G0053

Fig. 38 Thermostat assembly

2. Drain the cooling system.
3. Raise the vehicle.
4. Remove the splash shield.
5. Remove the lower radiator hose at the thermostat housing.
6. Remove the thermostat housing mounting bolts and remove housing and thermostat.

To install:

7. Clean mating areas of timing chain cover and thermostat housing.
8. Install thermostat (spring side down) into recessed machined groove on timing chain cover .
9. Position thermostat housing on timing chain cover.
10. Install two housing-to-timing chain cover bolts. Tighten bolts to 112 inch lbs. (13 Nm).

❄❄ WARNING

Housing must be tightened evenly and thermostat must be centered into recessed groove in timing chain cover. If not, it may result in a cracked housing, damaged timing chain cover threads or coolant leaks.

11. Install lower radiator hose on thermostat housing.
12. Install splash shield.
13. Lower vehicle.
14. Fill cooling system.
15. Connect negative battery cable to battery.
16. Start and warm the engine. Check for leaks.

WATER PUMP

REMOVAL & INSTALLATION

See Figure 39.

1. Before servicing the vehicle, refer to the precautions.
2. Disconnect negative battery cable.
3. Drain cooling system.

4. Remove fan/viscous fan drive assembly from water pump. Do not attempt to remove fan/viscous fan drive assembly from vehicle at this time.
5. If water pump is being replaced, do not unbolt fan blade assembly from thermal viscous fan drive.
6. Remove two fan shroud-to-radiator screws.
7. Remove fan shroud and fan blade/viscous fan drive assembly from vehicle.
8. After removing fan blade/viscous fan drive assembly, do not place thermal viscous fan drive in horizontal position. If stored horizontally, silicone fluid in viscous fan drive could drain into its bearing assembly and contaminate lubricant.
9. Remove accessory drive belt from the water pump pulley.
10. Remove upper radiator hose clamp and remove upper hose at water pump.
11. Remove seven water pump mounting bolts and one stud bolt.

❄❄ WARNING

Do not pry water pump at timing chain case/cover. The machined surfaces may be damaged resulting in leaks.

12. Remove water pump and gasket. Discard gasket.

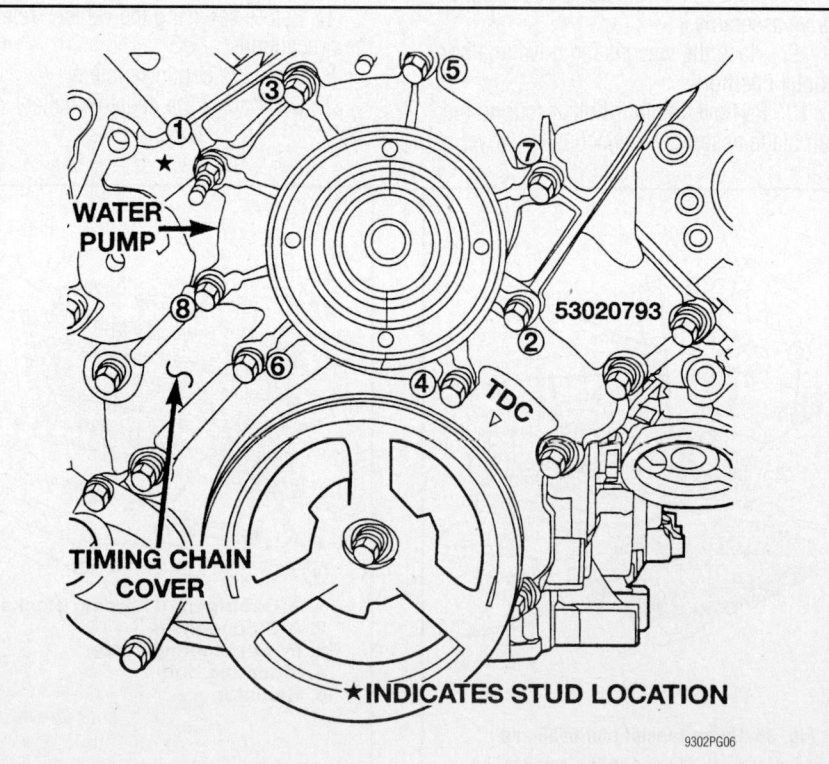

9302PG06

Fig. 39 Water pump tightening sequence

To install:

13. Clean gasket mating surfaces.

14. Using a new gasket, position water pump and install mounting bolts as shown. Tighten water pump mounting bolts to 43 ft. lbs. (58 Nm).

15. Spin water pump to be sure that pump impeller does not rub against timing chain case/cover.

16. Connect radiator upper hose to water pump.

17. Install accessory drive belt.

18. Position fan shroud and fan blade/viscous fan drive assembly.

19. Install two fan shroud-to-radiator screws.

20. Be sure of at least 1.0 inches (25 mm) between tips of fan blades and fan shroud.

21. Install fan blade/viscous fan drive assembly to water pump shaft.

22. Fill cooling system.

23. Connect negative battery cable.

24. Start and warm the engine. Check for leaks.

ENGINE ELECTRICAL

CHARGING SYSTEM

ALTERNATOR

REMOVAL & INSTALLATION

See Figure 40.

> **✳✳ WARNING**
>
> **Disconnect negative battery cable before removing battery output wire (B+ wire) from alternator. Failure to do so can result in injury or damage to electrical system.**

1. Disconnect the negative battery cable.

2. Remove generator drive belt.

3. Unsnap plastic insulator cap from B+ output terminal.

4. Remove B+ terminal mounting nut at rear of generator. Disconnect terminal from generator.

5. Disconnect field wire connector at rear of generator by pushing on connector tab.

6. Remove 1 rear vertical generator mounting bolt.

7. Remove 2 front horizontal generator mounting bolts.

8. Remove generator from vehicle.

To install:

9. Position generator to engine and install 2 horizontal bolts and 1 vertical bolt.

10. Tighten the short horizontal bolt to 55 ft. lbs. (74 Nm) and the vertical bolt and long horizontal bolt to 40 ft. lbs. (55 Nm).

11. Snap field wire connector into rear of generator.

12. Install B+ terminal eyelet to generator output stud.

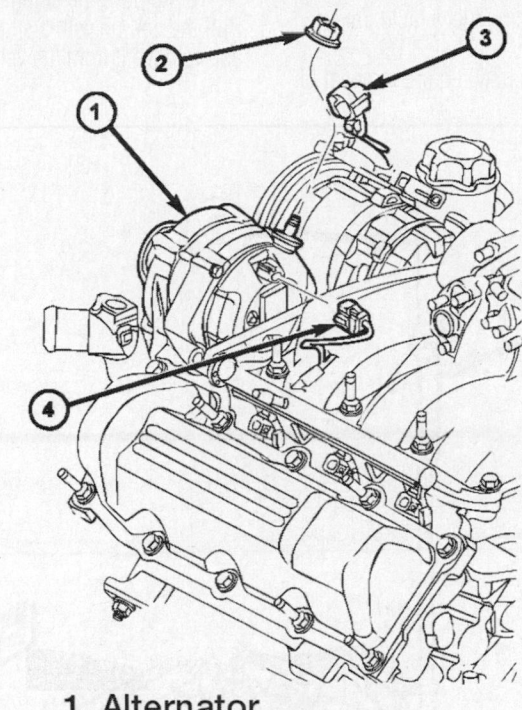

1. Alternator
2. B+ terminal mounting nut
3. Plastic insulator cap
4. Field wire connector

22140_RAID_G0054

Fig. 40 Alternator mounting

> **✳✳ WARNING**
>
> **Never force a belt over a pulley rim using a screwdriver. The synthetic fiber of the belt can be damaged.**

> **✳✳ WARNING**
>
> **When installing a serpentine accessory drive belt, the belt MUST be routed correctly. The water pump may be rotating in the wrong direction if the belt is installed incorrectly, causing the engine to overheat.**

13. Install generator drive belt.

14. Connect the negative battery cable.

ENGINE ELECTRICAL　　　　　　　　**IGNITION SYSTEM**

FIRING ORDER

Engine firing order is 1–6–5–4–3–2. The No.1 cylinder is located on the left bank.

IGNITION COIL

REMOVAL & INSTALLATION

See Figure 41.

1. Before servicing the vehicle, refer to the precautions.
2. Certain coils may require removal of the throttle body air intake tube or intake box for access.
3. Disconnect the negative battery cable.
4. Detach the electrical connector from the coil by pushing downward on the release lock on top of the connector and pulling the connector from the coil.
5. Clean the area at the base of each coil with compressed air.
6. Remove the coil mounting nut(s). Pull the coil up with a slight twisting action and remove it from the vehicle.

To install:

7. Smear the coil O–ring with silicone grease.
8. Install the coil by pushing down with a slight twisting action.
9. Install the coil mounting nut(s). Tighten the mounting nut to 70 inch lbs. (8 Nm).
10. Connect the electrical connector.
11. Connect the negative battery cable.
12. Install the throttle body air intake tube or intake box if removed.

IGNITION TIMING

ADJUSTMENT

The ignition timing is controlled by the Engine Control Module and is not adjustable.

SPARK PLUGS

REMOVAL & INSTALLATION

1. Before servicing the vehicle, refer to the precautions.
2. Remove air filter tubing at throttle body.
3. Use compressed air around coil base at cylinder head and into cylinder head opening. This will help prevent foreign material from entering combustion chamber.
4. Remove ignition coil.
5. Remove spark plug from cylinder head.
6. Inspect spark plug condition.

To install:

✳✳ WARNING

Do not attempt to clean any of the spark plugs. Replace only.

7. Start the spark plug into the cylinder head by hand to avoid cross threading.
8. Tighten spark plugs. Tighten to 20 ft. lbs. (30 Nm).
9. Before installing ignition coil(s), check condition of coil o-ring and replace as necessary. To aid in coil installation, apply silicone to coil o-ring.
10. Install ignition coil(s).
11. Install air filter tubing at throttle body.

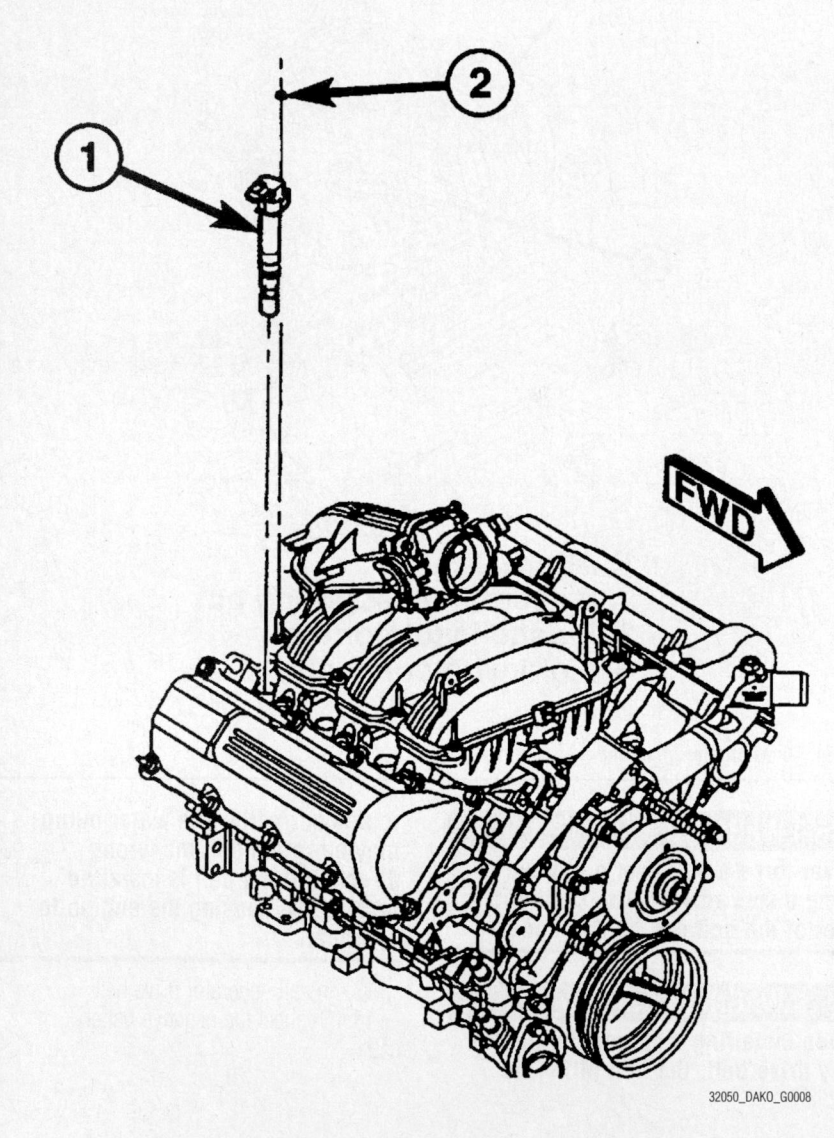

32050_DAKO_G0008

Fig. 41 3.7L Engine V6 Coil Location (1) Ignition Coil (2) Coil Mounting Nut

STARTER

REMOVAL & INSTALLATION

Manual Transmission

See Figure 42.

1. Before servicing the vehicle, refer to the precautions.

2. Disconnect and isolate negative battery cable.

3. If equipped with 4WD:

  a. Remove the bracket bolts for the support bracket between the front axle and side of the transmission.

  b. Pry the support bracket slightly to gain access to the starter lower mounting bolt.

4. Remove the starter mounting bolts.

5. Move the starter motor towards front of vehicle far enough for nose of starter pinion housing to clear the housing.

6. Tilt the nose downwards and lower starter motor far enough to access and remove the nut that secures the battery positive cable wire harness connector eyelet to solenoid battery terminal stud.

7. Remove the positive cable wire harness connector eyelet from solenoid battery terminal stud.

8. Disconnect the battery positive cable wire harness connector from the solenoid terminal connector receptacle.

9. Remove the starter motor.

To install:

10. Position starter motor to transmission housing.

11. Connect the battery cable solenoid terminal wire harness connector to connector receptacle on starter solenoid.

❊❊ CAUTION

Always support the starter motor during this process. Do not let the starter motor hang from the wire harness.

12. Install the battery cable eyelet terminal onto solenoid B (+) terminal stud.

13. Install the nut securing battery cable eyelet terminal to starter solenoid B (+) terminal stud. Tighten nut to 120 inch lbs. (13.6 Nm).

14. Position the starter motor over stud on transmission housing.

15. Loosely install the washers, bolt, and nut to starter. Tighten bolt and nut to 50 ft. lbs. (67.8 Nm).

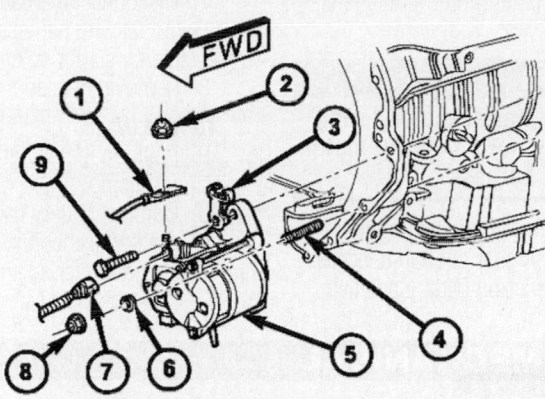

1. Wire harness connector eyelet
2. Nut
3. Wire harness clamp
4. Stud
5. Starter motor
6. Washer
7. Positive cable wire harness
8. Nut
9. Bolt

22140_RAID_G0056

Fig. 42 Starter motor assembly—manual transmission

16. Connect the negative battery cable.

Automatic Transmission

See Figure 43.

1. Before servicing the vehicle, refer to the precautions.

2. Disconnect and isolate negative battery cable.

3. If equipped with 4WD:

  a. Remove the bracket bolts for the support bracket between the front axle and side of the transmission.

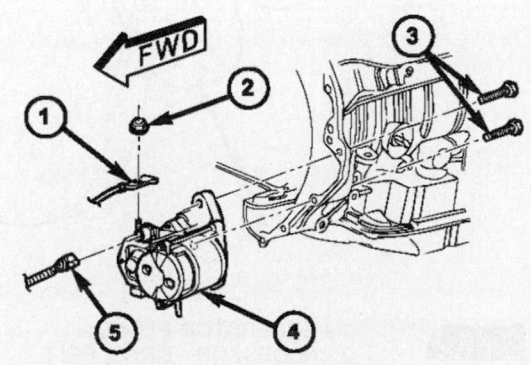

1. Wire harness connector eyelet
2. Nut
3. Mounting bolts
4. Starter motor
5. Positive cable wire harness

22140_RAID_G0057

Fig. 43 Starter motor assembly—automatic transmission

b. Pry the support bracket slightly to gain access to the starter lower mounting bolt.

4. Remove mounting bolts (rearward facing) securing starter motor to the transmission housing.

> ✳✳ **CAUTION**
>
> **Always support starter motor during this process. Do not let starter motor hang from wire harness.**

5. Lower starter motor from front of transmission housing far enough to access and remove nut securing battery positive cable eyelet terminal to the starter solenoid B (+) terminal stud.

6. Remove the battery cable eyelet terminal from solenoid B (+) terminal stud.

7. Disconnect the battery cable solenoid terminal wire harness connector from receptacle on starter solenoid.

8. Remove the starter motor.

To install:

9. Position the starter motor to transmission housing.

10. Connect battery cable solenoid terminal wire harness connector to connector receptacle on starter solenoid.

11. Install the battery cable eyelet terminal onto solenoid B (+) terminal stud.

12. Install and tighten nut securing battery cable eyelet terminal to starter solenoid B (+) terminal stud. Tighten nut to 120 inch lbs. (13.6 Nm).

13. Position starter motor to transmission housing and loosely install two bolts/washers.

14. Tighten bolts to 50 ft. lbs. (67.8 Nm).

15. Connect negative battery cable.

ENGINE MECHANICAL

ACCESSORY DRIVE BELTS

ACCESSORY BELT ROUTING

See Figure 44.

Refer to the accompanying illustration.

INSPECTION

See Figure 45.

Inspect the drive belt for signs of glazing or cracking. A glazed belt will be perfectly smooth from slippage, while a good belt will have a slight texture of fabric visible. Cracks will usually start at the inner edge of the belt and run outward. All worn or damaged drive belts should be replaced immediately.

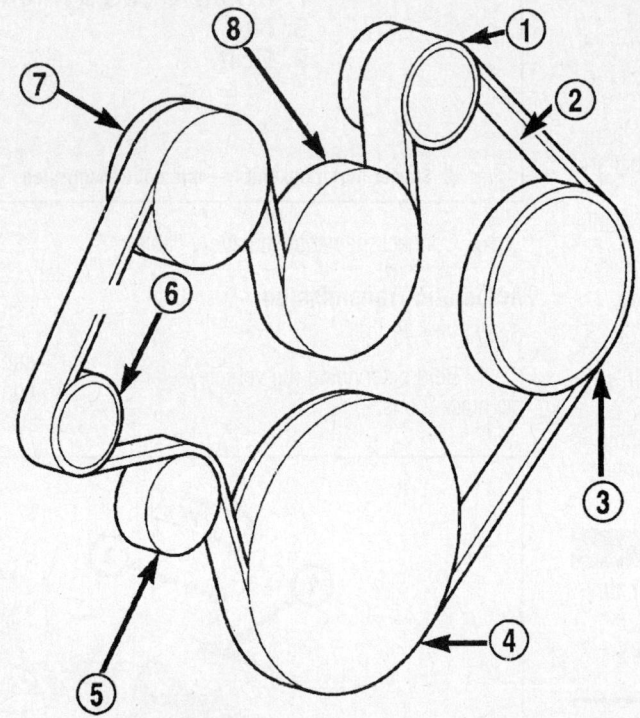

1 - GENERATOR PULLEY
2 - ACCESSORY DRIVE BELT
3 - POWER STEERING PUMP PULLEY
4 - CRANKSHAFT PULLEY
5 - IDLER PULLEY
6 - TENSIONER
7 - A/C COMPRESSOR PULLEY
8 - WATER PUMP PULLEY

67189DAK0G51

Fig. 44 Accessory drive belt routing 3.7L Engine and 4.7L engine

ACCESSORY DRIVE BELT DIAGNOSIS CHART

CONDITION	POSSIBLE CAUSES	CORRECTION
RIB CHUNKING (One or more ribs has separated from belt body)	1. Foreign objects imbedded in pulley grooves.	1. Remove foreign objects from pulley grooves. Replace belt.
	2. Installation damage	2. Replace belt
RIB OR BELT WEAR	1. Pulley misaligned	1. Align pulley(s)
	2. Abrasive environment	2. Clean pulley(s). Replace belt if necessary
	3. Rusted pulley(s)	3. Clean rust from pulley(s)
	4. Sharp or jagged pulley groove tips	4. Replace pulley. Inspect belt.
	5. Belt rubber deteriorated	5. Replace belt
BELT SLIPS	1. Belt slipping because of insufficient tension	1. Inspect/Replace tensioner if necessary
	2. Belt or pulley exposed to substance that has reduced friction (belt dressing, oil, ethylene glycol)	2. Replace belt and clean pulleys
	3. Driven component bearing failure (seizure)	3. Replace faulty component or bearing
	4. Belt glazed or hardened from heat and excessive slippage	4. Replace belt.
LONGITUDE BELT CRACKING	1. Belt has mistracked from pulley groove	1. Replace belt
	2. Pulley groove tip has worn away rubber to tensile member	2. Replace belt
"GROOVE JUMPING" (Belt does not maintain correct position on pulley)	1. Incorrect belt tension	1. Inspect/Replace tensioner if necessary
	2. Pulley(s) not within design tolerance	2. Replace pulley(s)
	3. Foreign object(s) in grooves	3. Remove foreign objects from grooves
	4. Pulley misalignment	4. Align component
	5. Belt cordline is broken	5. Replace belt
BELT BROKEN (Note: Identify and correct problem before new belt is installed)	1. Incorrect belt tension	1. Replace Inspect/Replace tensioner if necessary
	2. Tensile member damaged during belt installation	2. Replace belt
	3. Severe misalignment	3. Align pulley(s)
	4. Bracket, pulley, or bearing failure	4. Replace defective component and belt
NOISE (Objectionable squeal, squeak, or rumble is heard or felt while drive belt is in operation)	1. Incorrect belt tension	1. Inspect/Replace tensioner if necessary
	2. Bearing noise	2. Locate and repair
	3. Belt misalignment	3. Align belt/pulley(s)
	4. Belt to pulley mismatch	4. Install correct belt
	5. Driven component induced vibration	5. Locate defective driven component and repair
TENSION SHEETING FABRIC FAILURE (Woven fabric on outside, circumference of belt has cracked or separated from body of belt)	1. Tension sheeting contacting stationary object	1. Correct rubbing condition
	2. Excessive heat causing woven fabric to age	2. Replace belt
	3. Tension sheeting splice has fractured	3. Replace belt
CORD EDGE FAILURE (Tensile member exposed at edges of belt or separated from belt body)	1. Incorrect belt tension	1. Inspect/Replace tensioner if necessary
	2. Belt contacting stationary object	2. Replace belt
	3. Pulley(s) out of tolerance	3. Replace pulley
	4. Insufficient adhesion between tensile member and rubber matrix	4. Replace belt

22140_RAID_G0018A

Fig. 45 Accessory drive belt diagnosis chart

ADJUSTMENT

See Figure 46.

Belt tension is not adjustable. Belt adjustment is maintained by an automatic (spring load) belt tensioner.

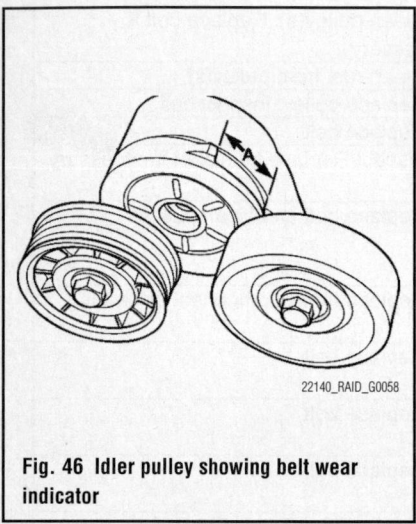

22140_RAID_G0058

Fig. 46 Idler pulley showing belt wear indicator

REMOVAL & INSTALLATION

See Figure 46.

1. Disconnect negative battery cable.
2. Rotate the belt tensioner until it contacts its stops.
3. Remove the belt, then slowly release the tensioner

To install:

4. Check condition of all pulleys.

❊ WARNING

When installing the serpentine accessory drive belt, the belt MUST be routed correctly. If not, the engine may overheat due to the water pump rotating in the wrong direction.

5. Install new belt . Route the belt around all pulleys except the idler pulley
6. Rotate the tensioner arm until it contacts it's stop position.
7. Route the belt around the idler and slowly let the tensioner rotate into the belt. Make sure the belt is seated onto all pulleys.
8. With the drive belt installed, inspect the belt wear indicator.

BALANCE SHAFT

REMOVAL & INSTALLATION

See Figures 47 through 53.

1. Before servicing the vehicle, refer to the precautions.
2. Drain the cooling system.

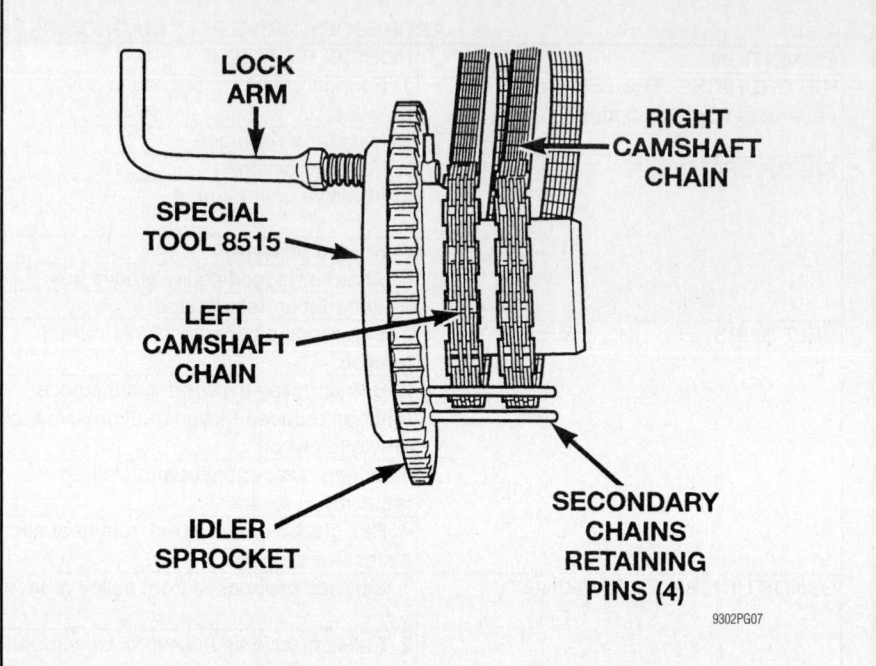

Fig. 47 Use the Timing Chain Locking tool to lock the timing chains on the idler gear—3.7L engine

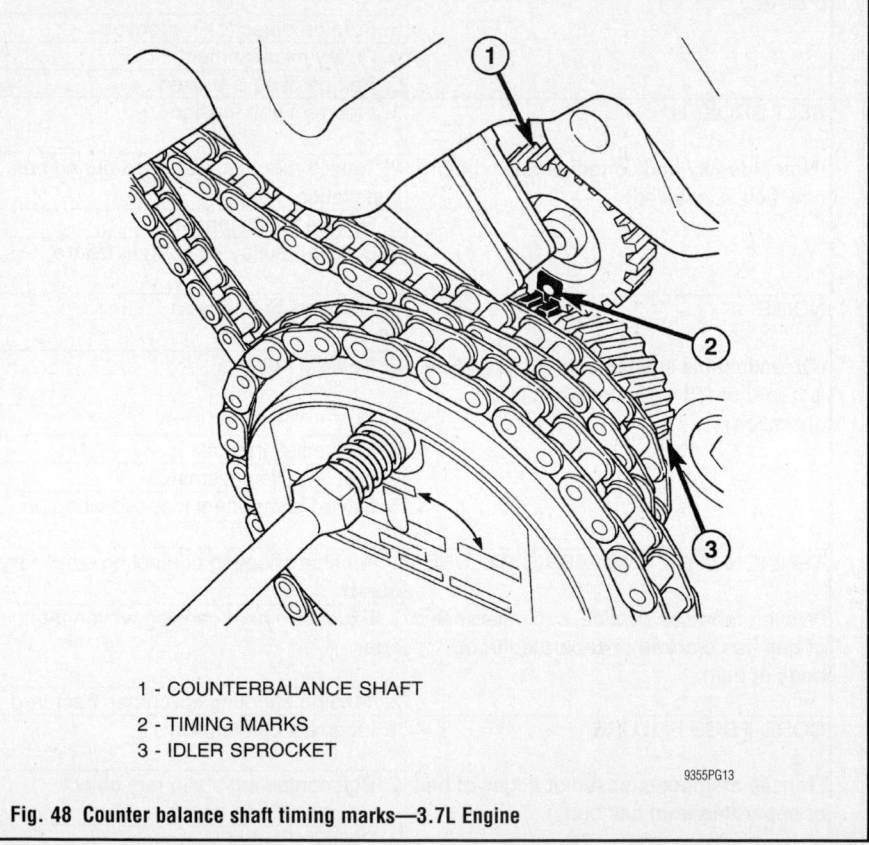

1 - COUNTERBALANCE SHAFT
2 - TIMING MARKS
3 - IDLER SPROCKET

9355PG13

Fig. 48 Counter balance shaft timing marks—3.7L Engine

3. Disconnect the negative battery cable.
4. Remove the cylinder head covers.
5. Remove the radiator fan.
6. Rotate the crankshaft so that the crankshaft timing mark aligns with the Top Dead Center (TDC) mark on the front cover, and the **V6** marks on the camshaft sprockets are at 12 O'clock.
7. Remove the power steering pump.

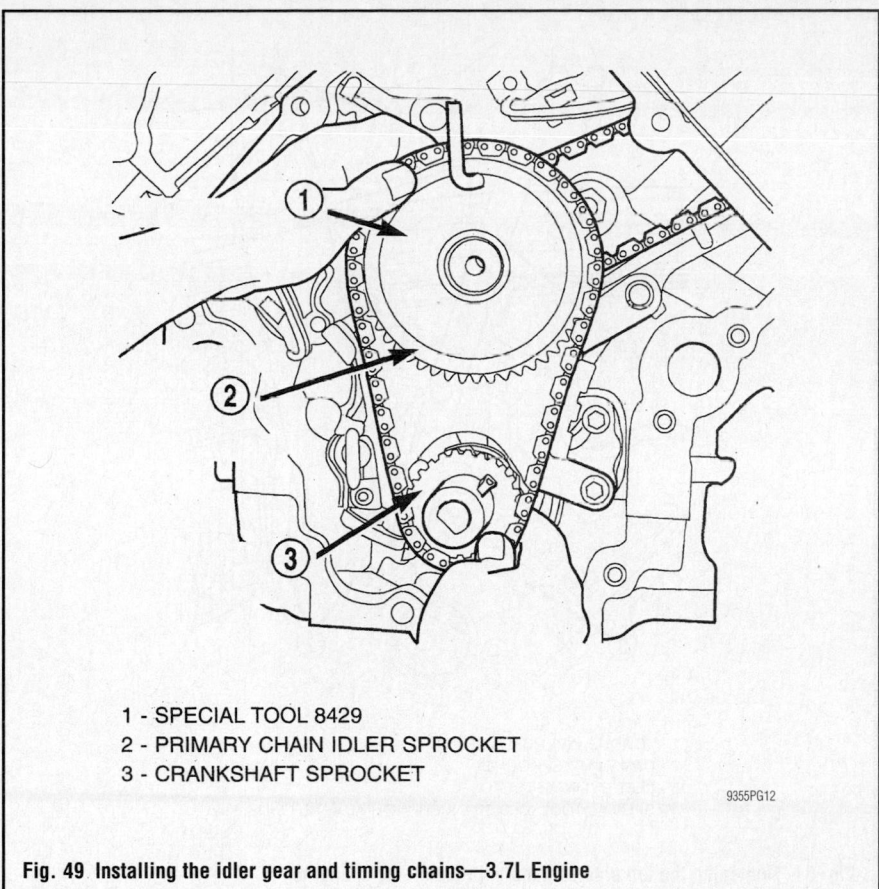

1 - SPECIAL TOOL 8429
2 - PRIMARY CHAIN IDLER SPROCKET
3 - CRANKSHAFT SPROCKET

9355PG12

Fig. 49 Installing the idler gear and timing chains—3.7L Engine

8. Remove the access plugs from the cylinder heads.

9. Remove the oil fill housing.

10. Remove the crankshaft damper.

11. Compress the primary timing chain tensioner and install a lock pin.

12. Remove the secondary timing chain tensioners.

13. Hold the left camshaft with adjustable pliers and remove the sprocket and chain. Rotate thc **left** camshaft 5 degrees **clockwise** to the neutral position.

14. Hold the right camshaft with adjustable pliers and remove the camshaft sprocket. Rotate the **right** camshaft 45 degrees **counterclockwise** to the neutral position.

15. Remove the primary and secondary timing chain and sprockets.

16. Remove the balance shaft from engine using special tool 8641 (balance shaft remover/installer tool) or equivalent.

To install:

17. Coat balance shaft bearing journals with clean engine oil.

18. Install balance shaft into engine using special tool 8641 (balance shaft remover/installer tool).

RIGHT
CAMSHAFT
SPROCKET
AND
SECONDARY
CHAIN

SECONDARY
TIMING CHAIN
TENSIONER

SECONDARY
TENSIONER ARM

LEFT
CAMSHAFT
SPROCKET
AND
SECONDARY
CHAIN

CHAIN GUIDE

SECONDARY
TENSIONER ARM

TWO PLATED
LINKS ON RIGHT
CAMSHAFT CHAIN

TWO PLATED
LINKS ON LEFT
CAMSHAFT CHAIN

PRIMARY CHAIN

IDLER SPROCKET

PRIMARY CHAIN
TENSIONER

CRANKSHAFT
SPROCKET

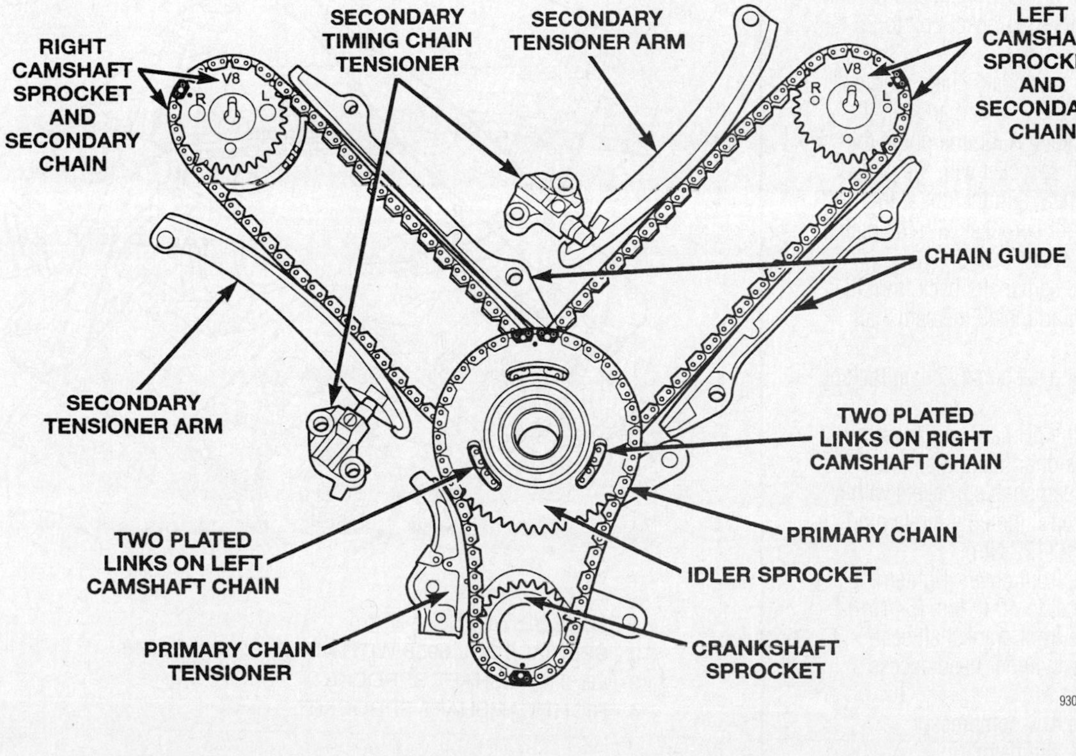

9302PG24

Fig. 50 Timing chain system and alignment marks—3.7L engine

19. Install balance shaft thrust plate retaining bolt finger tight. Do not tighten at this time.

20. Position the right side of the thrust plate with the right chain guide bolt, install bolt finger tight.

21. Tighten the thrust plate retaining bolt to 250 inch lbs.

22. Remove the chain guide bolt so that the guide can be installed.

23. Use a small prytool to hold the ratchet pawl and compress the secondary timing chain tensioners in a vise and install locking pins.

➡ **The black bolts fasten the guide to the engine block and the silver bolts fasten the guide to the cylinder head.**

24. Install the secondary timing chain guides. Tighten the bolts to 21 ft. lbs. (28 Nm).

25. Install the secondary timing chains to the idler sprocket so that the double plated links on each chain are visible through the slots in the primary idler sprocket

26. Lock the secondary timing chains to the idler sprocket with Timing Chain Locking tool as shown.

27. Align the primary chain double plated links with the idler sprocket timing mark and the single plated link with the crankshaft sprocket timing mark.

28. Install the primary chain and sprockets. Tighten the idler sprocket bolt to 25 ft. lbs. (34 Nm).

29. Align the secondary chain single plated links with the timing marks on the secondary sprockets. Align the dot at the **L**mark on the left sprocket with the plated link on the left chain and the dot at the **R**mark on the right sprocket with the plated link on the right chain.

30. Rotate the camshafts back from the neutral position and install the camshaft sprockets.

31. Remove the secondary chain locking tool.

32. Remove the primary and secondary timing chain tensioner locking pins.

33. Hold the camshaft sprockets with a spanner wrench and tighten the retaining bolts to 90 ft. lbs. (122 Nm).

34. Install the front cover. Tighten the bolts, in sequence, to 40 ft. lbs. (54 Nm).

35. Install the front crankshaft seal.

36. Install the cylinder head access plugs.

37. Install the A/C compressor.

38. Install the alternator.

39. Install the accessory drive belt tensioner. tighten the bolt to 40 ft. lbs. (54 nm).

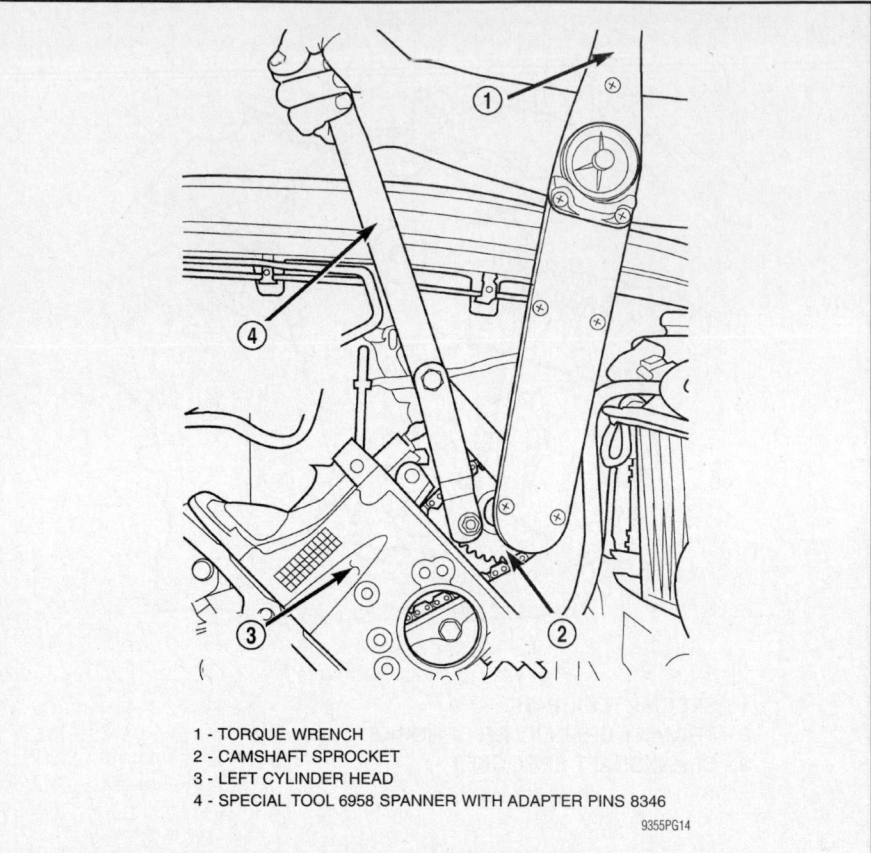

1 - TORQUE WRENCH
2 - CAMSHAFT SPROCKET
3 - LEFT CYLINDER HEAD
4 - SPECIAL TOOL 6958 SPANNER WITH ADAPTER PINS 8346

9355PG14

Fig. 51 Tightening the left side camshaft sprocket—3.7L Engine

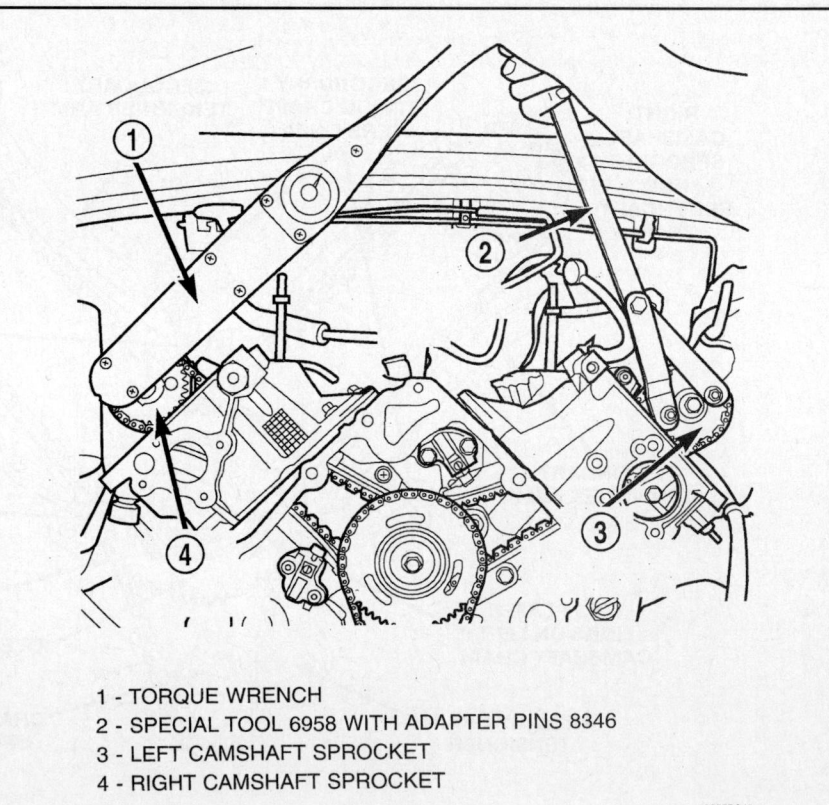

1 - TORQUE WRENCH
2 - SPECIAL TOOL 6958 WITH ADAPTER PINS 8346
3 - LEFT CAMSHAFT SPROCKET
4 - RIGHT CAMSHAFT SPROCKET

9355PG15

Fig. 52 Tightening the right side camshaft sprocket—3.7L Engine

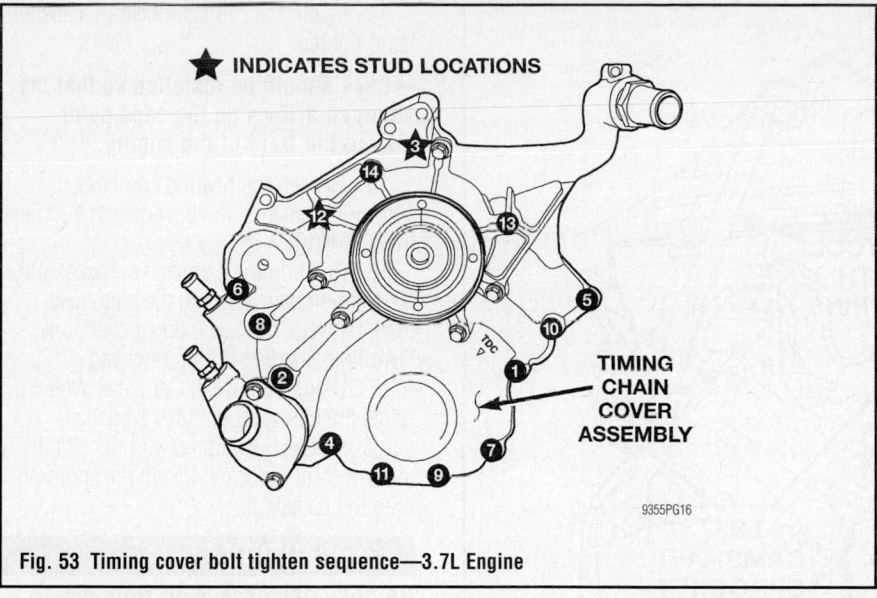

★ INDICATES STUD LOCATIONS

TIMING
CHAIN
COVER
ASSEMBLY

TDC

9355PG16

Fig. 53 Timing cover bolt tighten sequence—3.7L Engine

40. Install the oil fill housing.
41. Install the crankshaft damper. tighten the bolt to 130 ft. lbs. (175 nm).
42. Install the power steering pump.
43. Install the lower radiator hose.
44. Install the heater hoses.
45. Install the accessory drive belt.
46. Install the engine cooling fan and shroud.
47. Install the camshaft position (CMP) sensor.
48. Install the valve covers.
49. Connect the negative battery cable.
50. Fill and bleed the cooling system.
51. Start the engine, check for leaks and repair if necessary.

CAMSHAFT AND VALVE LIFTERS

REMOVAL & INSTALLATION

See Figures 54 through 59.

1. Before servicing the vehicle, refer to the precautions.
2. Disconnect the negative battery cable.
3. Remove the cylinder head cover.
4. Set the No. 1 cylinder to Top Dead Center (TDC). The camshaft sprocket alignment mark should be at the 12 o'clock position.

➡Keep all valve train components in order for assembly.

5. Loosen, but DO NOT remove the camshaft sprocket retaining bolt. Leave the bolt snug against the sprocket.
6. Position Special Tool 8379 timing chain wedge between the timing chain strands, tap the tool to securely wedge the timing chain against the tensioner arm and guide.

7. Remove the camshaft position sensor, right camshaft only.
8. Hold the camshaft with Special Tool 8428 Camshaft Wrench, while removing the camshaft sprocket bolt and sprocket.
9. Using Special Tool 8428 Camshaft Wrench, gently allow the camshaft to rotate 5°clockwise until the camshaft is in the neutral position (no valve load).
10. Starting at the outside working inward, loosen the camshaft bearing cap retaining bolts ½turn at a time. Repeat until the load is off the bearing caps.
11. Remove the camshaft bearing caps and camshaft.

➡When the camshaft is removed the rocker arms may slide downward. Mark the rocker arms before removing camshaft.

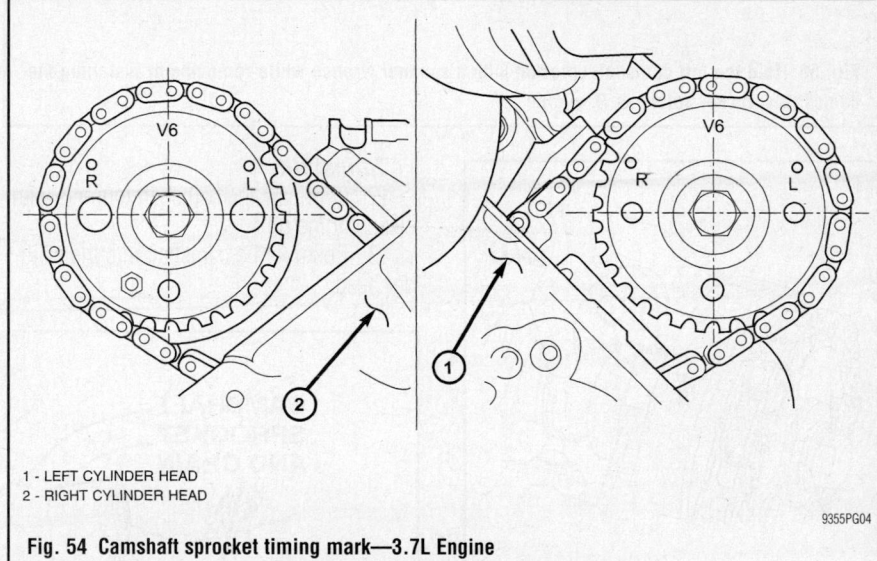

1 - LEFT CYLINDER HEAD
2 - RIGHT CYLINDER HEAD

9355PG04

Fig. 54 Camshaft sprocket timing mark—3.7L Engine

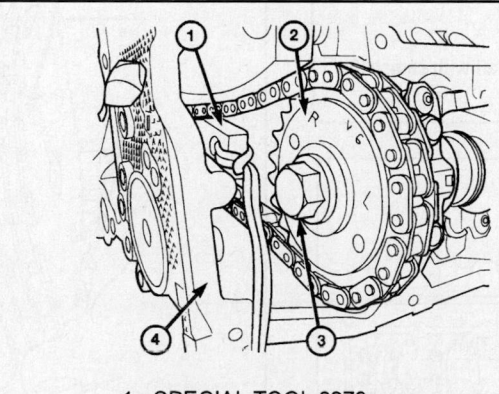

1 - SPECIAL TOOL 8379
2 - CAMSHAFT SPROCKET
3 - CAMSHAFT SPROCKET BOLT
4 - CYLINDER HEAD

09482_RAID_G0002

Fig. 55 Chain tensioner retaining wedge installation

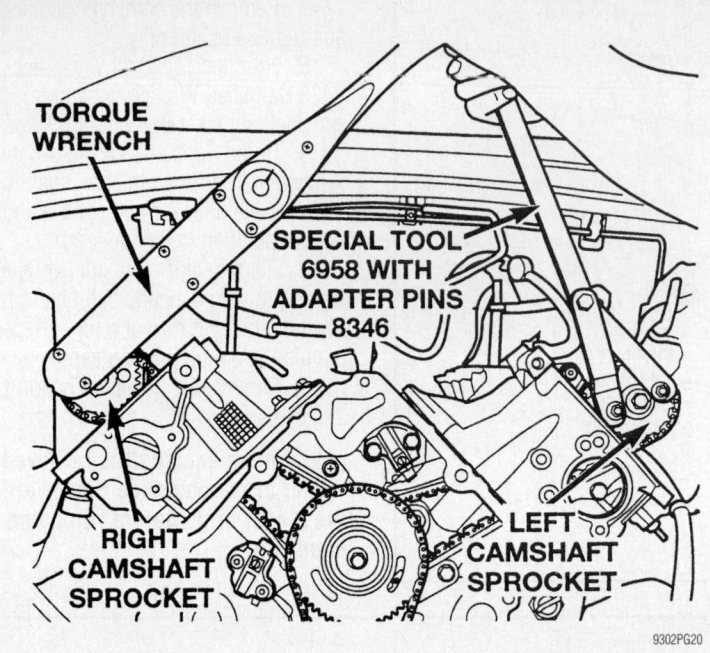

Fig. 56 Hold the left camshaft sprocket with a spanner wrench while removing or installing the camshaft sprocket bolts—3.7L engine

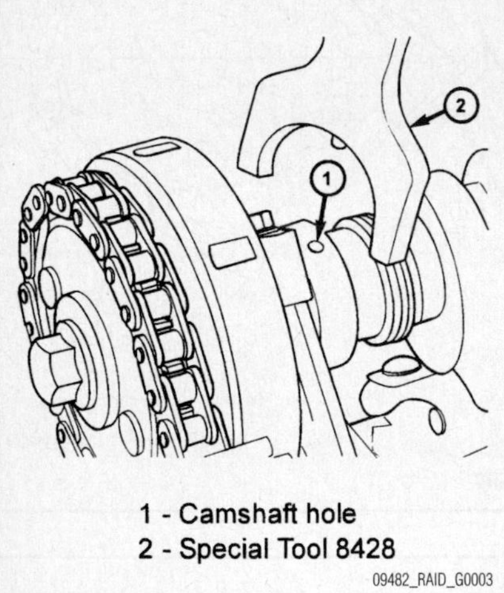

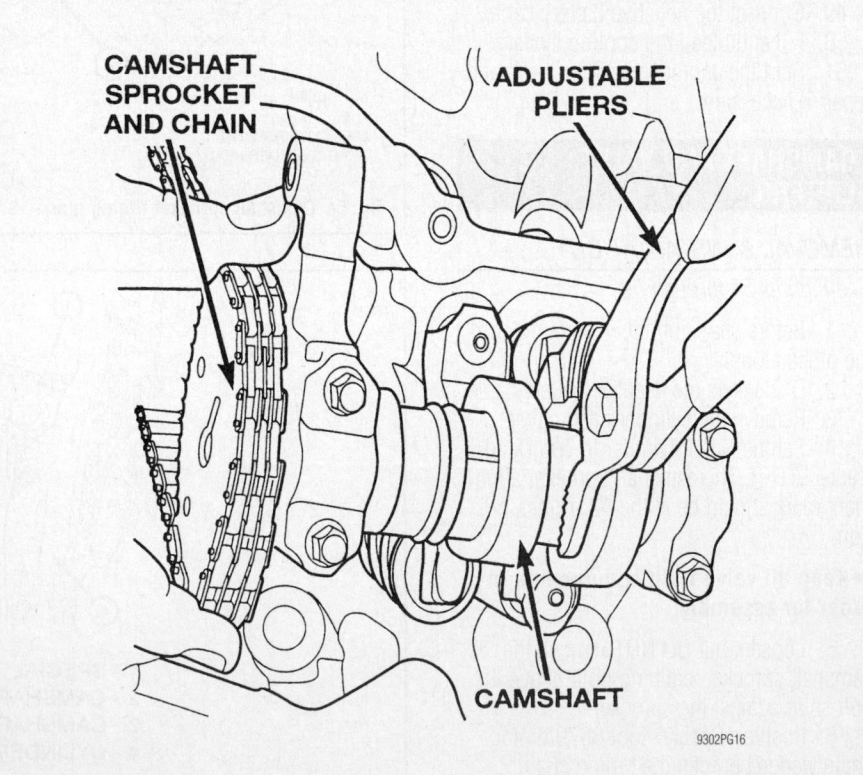

Fig. 57 Hold the camshaft with camshaft wrench to remove

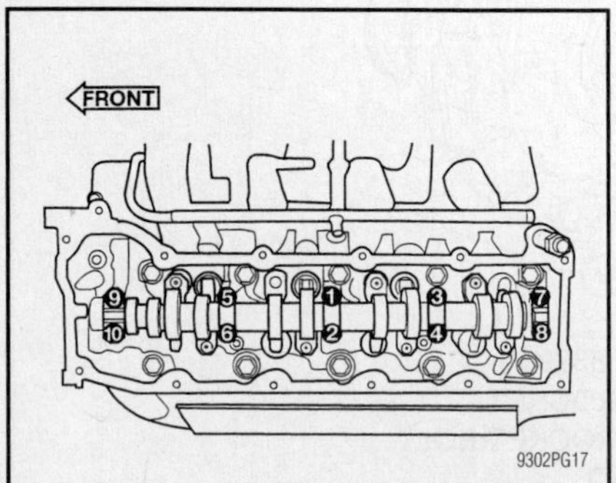

1 - Camshaft hole
2 - Special Tool 8428

09482_RAID_G0003

Fig. 58 Camshaft bearing cap bolt tightening sequence—3.7L engine

9302PG17

Fig. 59 Turn the camshaft with pliers, if needed, to align the dowel in the sprocket—3.7L engine

9302PG16

14. Install the camshaft bearing caps and hand tighten.

➡ **Caps should be installed so that the stamped arrows on the caps point toward the front of the engine.**

15. Tighten the bearing cap bolts in ½ turn increments, in sequence, to 100 inch lbs. (11 Nm).

16. Position the camshaft sprocket into the timing chain aligning the alignment mark between the two marked chain links (Two links marked during removal).

17. Using Tool 8428 Camshaft Wrench, rotate the camshaft until the camshaft sprocket dowel is aligned with the slot in the camshaft sprocket. Install the sprocket onto the camshaft.

⁕⁕ CAUTION

Remove any excess oil from the camshaft sprocket bolt. Failure to do so can cause bolt over-tighten.

18. Remove excess oil from bolt, then install the camshaft sprocket retaining bolt and hand tighten.

19. Remove Special Tool 8379 timing chain wedge.

20. Using Special Tool 6958 spanner wrench with adapter pins 8346, tighten the

To install:

12. Lubricate the camshaft journals with clean engine oil

13. Position the camshaft into the cylinder head.

camshaft sprocket retaining bolt to 90 ft. lbs. (122 Nm).

21. Install or connect the following:
   * Camshaft position sensor, right side only
   * Cylinder head cover
22. Start the engine and check for leaks.

CATALYTIC CONVERTER

REMOVAL & INSTALLATION

See Figure 60.

✳✳ CAUTION

When servicing or replacing exhaust system components, disconnect the oxygen sensor connector(s). Allowing the exhaust to hang by the oxygen sensor wires will damage the harness and/or sensor.

1. Raise and support the vehicle.
2. Saturate all exhaust bolts and nuts with Rust Penetrant. Allow 5 minutes for penetration.
3. Disconnect the oxygen sensor electrical connectors.
4. Remove the catalytic converter-to-manifold bolts.
5. Remove the catalytic converter to the exhaust pipe clamp.
6. If present, grind tack weld.
7. Remove the catalytic converter. You may have to loosen up other sections of the exhaust system.

To install:

➡**The band clamps are not reusable. After removal, they must be replaced with new one.**

8. Make sure the catalytic converter pipe is free of burrs.
9. Insert the catalytic converter into the exhaust pipe.
10. Make sure the alignment tang is fully seated in the alignment slot.
11. If other sections of the exhaust system where loosened in removal, tighten them.
12. At the catalytic converter-to-extension pipe connection, install new clamp and nuts. Tighten the clamp nuts to 45 ft. lbs. (61 Nm).
13. Lower the vehicle.
14. Start the engine, inspect for exhaust leaks. Repair exhaust leaks as necessary.
15. Check the exhaust system for contact with the body panels. Make the necessary adjustments, if necessary.

CRANKSHAFT DAMPER

REMOVAL & INSTALLATION

See Figures 61 and 62.

1. Disconnect negative cable from battery.
2. Remove radiator fan.
3. Remove accessory drive belt,
4. Remove crankshaft damper bolt.

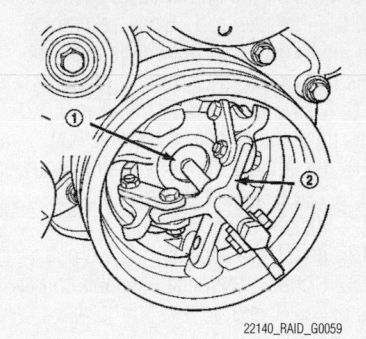

Fig. 61 Remove damper using the Crankshaft Insert 8513 (1) and Three Jaw Puller 1026 (2)

5. Remove damper using Mitsubishi special tools 8513 insert and 1026 three jaw puller.

To install:

6. To prevent severe damage to the Crankshaft, Damper or Damper Installer 8512-A, thoroughly clean the damper bore and the crankshaft nose before installing Damper.

7. Align vibration damper slot with key in crankshaft. Slide damper onto crankshaft slightly. caution Damper Installer 8512-A, is assembled in a specific sequence. Failure to assemble this tool in this sequence can result in tool failure and severe damage to either the tool or the crankshaft.

8. Assemble Damper Installer 8512-A as follows: The nut is threaded onto the threaded rod first. Then the roller bearing is placed onto the threaded rod (The hardened bearing surface of the bearing MUST face

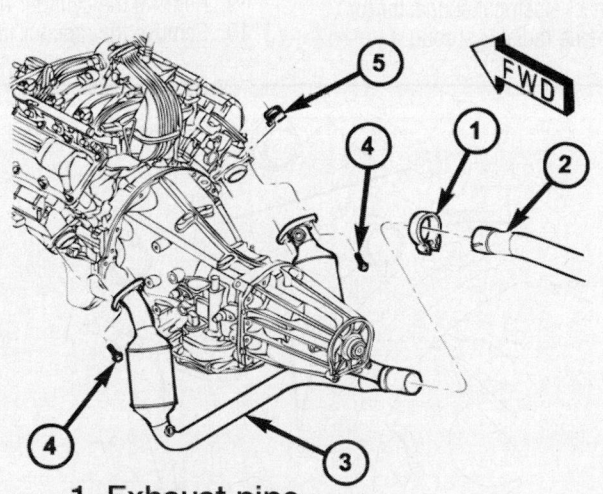

1. Exhaust pipe
2. Catalytic converter
3. Clamp
4. Extension pipe to muffler

Fig. 61 Catalytic converter

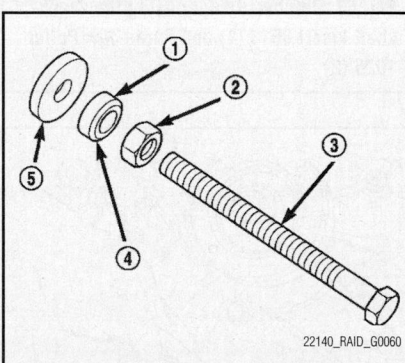

Fig. 62 Assemble the Damper Installer as follows: The nut (2) is threaded onto the threaded rod (3) first. Then the roller bearing (1) is placed onto the threaded rod. The hardened bearing surface of the bearing MUST face the nut (4). Then the hardened washer (5) slides onto the threaded rod

the nut). Then the hardened washer slides onto the threaded rod . Once assembled coat the threaded rod's threads with Mitsubishi Nickel Anti-Seize or equivalent.

9. Using the Damper Installer 8512A, press the damper onto the crankshaft.

10. Install and tighten vibration damper bolt to 130 ft. lbs. (175 Nm).

11. Install the radiator fan.

12. Install the accessory drive belt.

13. Connect the negative cable to battery.

CRANKSHAFT FRONT SEAL

REMOVAL & INSTALLATION

See Figures 63 through 66.

1. Disconnect negative cable from battery.

2. Remove accessory drive belt.

3. Remove A/C compressor mounting fasteners and set aside.

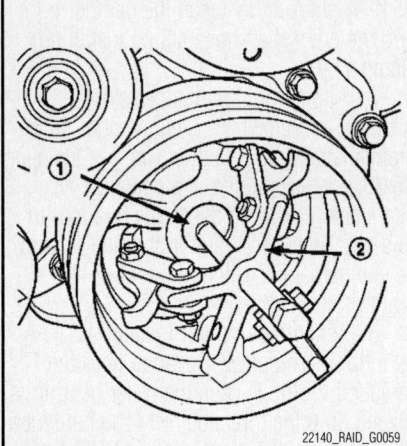

Fig. 63 Remove damper using the Crankshaft Insert 8513 (1) and Three Jaw Puller 1026 (2)

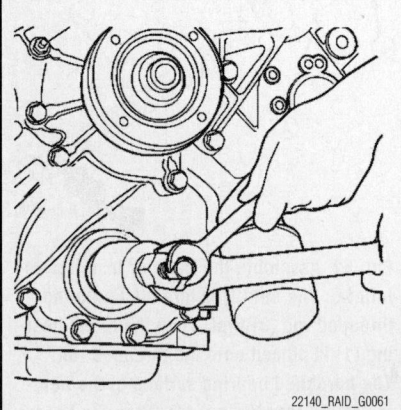

Fig. 64 Using the Seal Remover 8511, remove crankshaft front seal

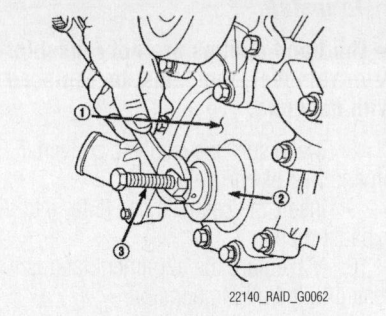

22140_RAID_G0062

Fig. 65 Using the Seal Installer 8348 (2) and Damper Installer 8512A (3), install the crankshaft front seal

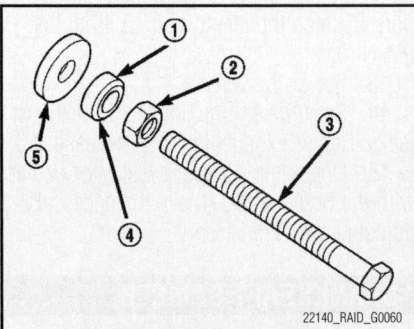

22140_RAID_G0060

Fig. 66 Assemble the Damper Installer as follows: The nut (2) is threaded onto the threaded rod (3) first. Then the roller bearing (1) is placed onto the threaded rod. The hardened bearing surface of the bearing MUST face the nut (4). Then the hardened washer (5) slides onto the threaded rod

4. Drain cooling system.

5. Remove upper radiator hose.

6. Disconnect electrical connector for fan mounted inside radiator shroud.

7. Remove radiator cooling fan.

8. Remove crankshaft damper bolt.

9. Remove damper using Mitsubishi special tools 8513 and tool 1026 three jaw puller.

10. Using the Seal Remover 8511, remove crankshaft front seal.

To install:

11. Using Mitsubishi special tool 8348 and 8512, install crankshaft front seal.

12. Install vibration damper.

13. Install radiator cooling fan and shroud.

14. Install upper radiator hose.

15. Install A/C compressor and tighten fasteners to 40 ft. lbs. (54 Nm).

16. Install accessory drive belt.

17. Refill cooling system.

18. Connect negative cable to battery.

CYLINDER HEAD

REMOVAL & INSTALLATION

See Figures 67 through 72.

1. Before servicing the vehicle, refer to the precautions.

2. Disconnect the negative battery cable.

3. Raise and safely support the vehicle securely on jackstands.

4. Disconnect the exhaust pipe at the exhaust manifold.

5. Drain the engine coolant.

6. Lower the vehicle.

7. Remove the intake manifold.

8. On the left side, remove the master cylinder and booster assembly.

9. Remove the cylinder head cover.

10. Remove the radiator fan.

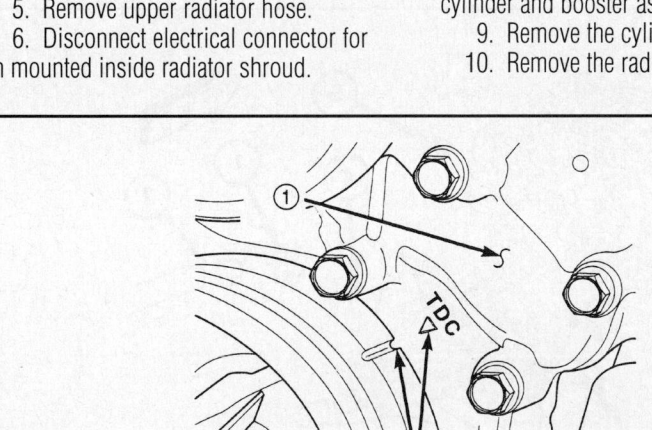

1 – TIMING CHAIN COVER
2 – CRANKSHAFT TIMING MARKS

9308PG04

Fig. 67 Rotate the crankshaft until the damper timing mark is aligned with TDC indicator mark

11. Remove oil fill housing from cylinder head.

12. Remove accessory drive belt.

13. On the left side, remove the power steering pump and set it aside.

14. Rotate the crankshaft until the damper timing mark is aligned with TDC indicator mark.

15. Verify the V6 mark on the camshaft sprocket is at the 12 o'clock position. Rotate the crankshaft one turn if necessary.

16. Remove the crankshaft damper.

17. Remove the timing chain cover.

18. Lock the secondary timing chains to the idler sprocket using Special Tool 8429 Timing Chain Holding Fixture.

➡Mark the secondary timing chain prior to removal to aid in installation.

19. Mark the secondary timing chain, one link on each side of the V6 mark on the camshaft drive gear.

20. Remove the secondary chain tensioner.

21. Remove the cylinder head access plug.

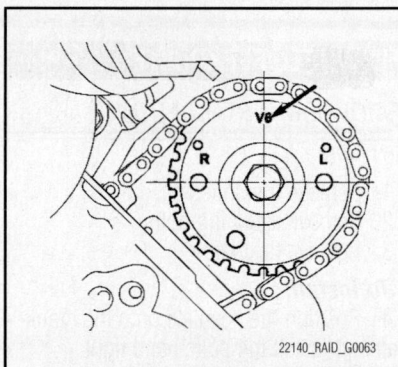

Fig. 68 Verify the V6 mark on the camshaft sprocket is at the 12 o'clock position. Rotate the crankshaft one turn if necessary

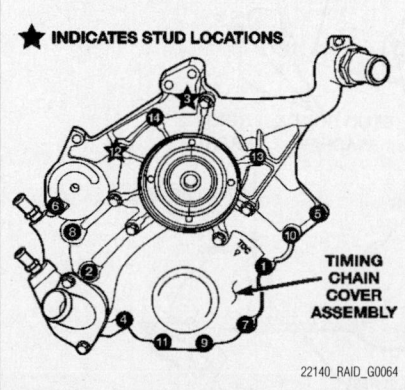

★ INDICATES STUD LOCATIONS

TIMING CHAIN COVER ASSEMBLY

22140_RAID_G0064

Fig. 69 Timing cover bolt locations—3.7L engine

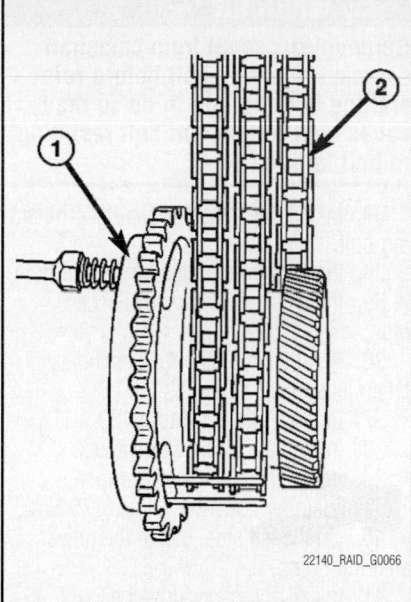

22140_RAID_G0066

Fig. 70 Lock the secondary timing chains (2) to the idler sprocket using Secondary Camshaft Chain Holder 8429 (1)—3.7L engine

22. Remove the secondary chain guide.

✳✳ WARNING

The nut on the camshaft sprocket should not be removed for any reason, as the sprocket and camshaft sensor target wheel is serviced as an assembly. If the nut was removed retorque nut to 44 inch lbs. (5 Nm).

23. Remove the retaining bolt and the camshaft drive gear.

✳✳ WARNING

Do not allow the engine to rotate. Severe damage to the valve train can occur.

➡Do not overlook the four smaller bolts at the front of the cylinder head. Do not attempt to remove the cylinder head without removing these four bolts.

✳✳ WARNING

Do not hold or pry on the camshaft target wheel for any reason. A damaged target wheel can result in a vehicle no start condition.

24. Remove the 12 cylinder head retaining bolts.

25. Remove the cylinder head and gasket. Discard the gasket.

✳✳ WARNING

Do not lay the cylinder head on its gasket sealing surface, due to the design of the cylinder head gasket any distortion to the cylinder head sealing surface may prevent the gasket from properly sealing resulting in leaks.

To install:

✳✳ WARNING

The cylinder head bolts are tightened using a torque plus angle procedure. The bolts must be examined BEFORE reuse. If the threads are necked down the bolts should be replaced. Necking can be checked by holding a straight edge against the threads. If all the threads do not contact the scale, the bolt should be replaced.

✳✳ WARNING

When cleaning cylinder head and cylinder block surfaces, DO NOT use a metal scraper because the surfaces could be cut or ground. Use only a wooden or plastic scraper.

26. Clean the cylinder head and cylinder block mating surfaces.

27. Position the new cylinder head gasket on the locating dowels.

✳✳ WARNING

When installing cylinder head, use care not damage the tensioner arm or the guide arm.

28. Position the cylinder head onto the cylinder block. Make sure the cylinder head seats fully over the locating dowels.

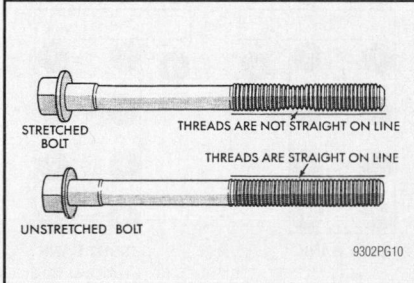

STRETCHED BOLT

THREADS ARE NOT STRAIGHT ON LINE

THREADS ARE STRAIGHT ON LINE

UNSTRETCHED BOLT

9302PG10

Fig. 71 Examine the head bolts for signs of stretching—3.7L engine

➡ The four M8 cylinder head mounting bolts require sealant to be added to them before installing. Failure to do so may cause leaks.

29. Lubricate the cylinder head bolt threads with clean engine oil and install the eight M10 bolts.

30. Coat the four M8 cylinder head bolts with Loctite 242 Lock and Seal Adhesive (or equivalent) then install the bolts.

➡ The cylinder head bolts are tightened using an angle torque procedure, however, the bolts are not a torque-to-yield design.

31. Tighten the cylinder head bolts, in sequence, as follows:
   - Step 1: Bolts 1–8 to 20 ft. lbs. (27 Nm)
   - Step 2: Bolts 1–8 verify tighten without loosening
   - Step 3: Bolts 9–12 to 10 ft. lbs. (14 Nm)
   - Step 4: Bolts 1–8 plus 90° rotation
   - Step 5: Bolts 1–8 plus 90° rotation
   - Step 6: Bolts 9–12 to 19 ft. lbs. (26 Nm)

✳✳ WARNING

The nut on the camshaft sprocket should not be removed for any reason, as the sprocket and camshaft sensor target wheel is serviced as an assembly. If the nut was removed retorque nut to 60 inch lbs. (5 Nm).

32. Position the secondary chain onto the camshaft drive gear, making sure one marked chain link is on either side of the V6 mark on the gear then using Tool 8428 Camshaft wrench, position the gear onto the camshaft.

✳✳ WARNING

Remove excess oil from camshaft sprocket retaining bolt before reinstalling bolt. Failure to do so may cause over-torquing of bolt resulting in bolt failure.

33. Install the camshaft drive gear retaining bolt.
34. Install the secondary chain guide.
35. Install the cylinder head access plug.
36. Re-set and install the secondary chain tensioner.
37. Remove Special Tool 8429.
38. Install the timing chain cover.
39. Install the crankshaft damper. Tighten damper bolt 130 ft. lbs. (175 Nm).
40. On the left side, install the power steering pump.
41. Install accessory drive belt.
42. Install the radiator fan and shroud.
43. Install the cylinder head cover.
44. On the left side, install the master cylinder and booster assembly.
45. Install the intake manifold.
46. Install oil fill housing onto cylinder head.
47. Refill the cooling system.
48. Raise the vehicle.
49. Install the exhaust pipe onto the right exhaust manifold.
50. Lower the vehicle.
51. Reconnect battery negative cable.
52. Start the engine and check for leaks.

EXHAUST MANIFOLD

REMOVAL & INSTALLATION

See Figure 73.

1. Before servicing the vehicle, refer to the precautions.
2. Disconnect the negative battery cable.
3. Remove the exhaust manifold heat shields.
4. Remove the exhaust gas recirculation (EGR) tube.
5. Remove the exhaust y-pipe.
6. Remove the exhaust manifolds.

To install:

➡ If the exhaust manifold studs came out with the nuts when removing the exhaust manifolds, replace them with new studs.

7. Install the exhaust manifolds. Tighten the fasteners to 20 ft. lbs. (27 Nm), starting with the center nuts and work out to the ends.
8. Install the exhaust y-pipe.
9. Install the EGR tube.
10. Install the exhaust manifold heat shields.
11. Connect the negative battery cable.
12. Start the engine, check for leaks and repair if necessary.

FLYWHEEL/FLEXPLATE

REMOVAL & INSTALLATION

See Figure 74.

1. Remove transmission.
2. Remove attaching bolts.
3. Remove flexplate

To install:

4. Position the flexplate onto the crankshaft and install the bolts hand tight.
5. Tighten the flexplate retaining bolts to 70 ft lbs (95 Nm) in a criss-cross sequence.
6. Install transmission.

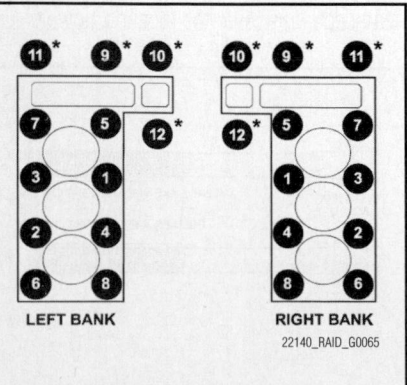

Fig. 72 Cylinder head bolt tighten sequence—3.7L Engine

22140_RAID_G0065

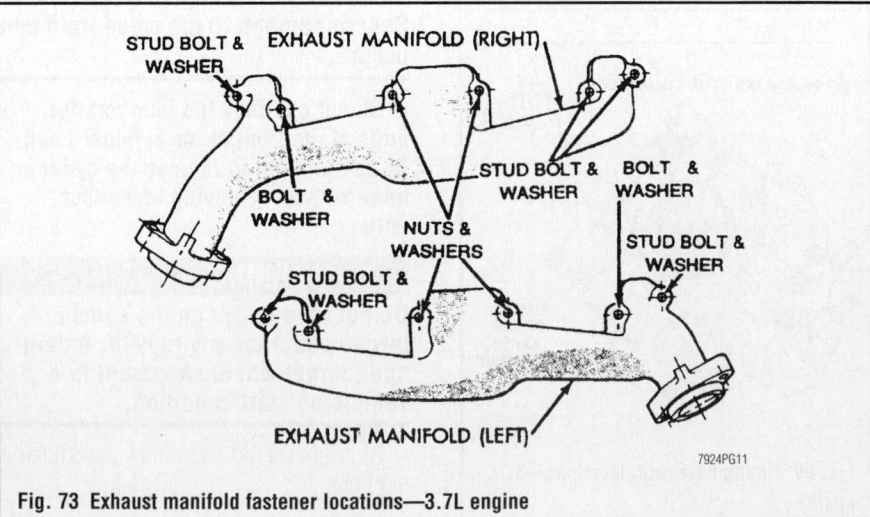

Fig. 73 Exhaust manifold fastener locations—3.7L engine

7924PG11

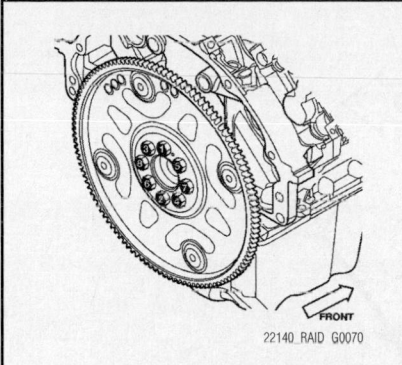

Fig. 74 Tighten the flexplate retaining bolts in a criss-cross sequence

INTAKE MANIFOLD

REMOVAL & INSTALLATION

See Figure 75.

1. Before servicing the vehicle, refer to the precautions.
2. Drain the cooling system.
3. Relieve the fuel system pressure.
4. Disconnect the negative battery cable.
5. Remove the air cleaner assembly.
6. Remove the accelerator cable.
7. Remove the cruise control cable.
8. Disconnect the following electrical connectors:
   a. Manifold Absolute Pressure (MAP) sensor
   b. Intake Air Temperature (IAT) sensor
   c. Throttle Position (TP) sensor
   d. Idle Air Control (IAC) valve
9. Remove the Positive Crankcase Ventilation (PCV) valve and hose.
10. Remove the canister purge vacuum line.
11. Remove the brake booster vacuum line.
12. Remove the cruise control servo hose.
13. Remove the accessory drive belt.
14. Disconnect the alternator electrical connections.
15. Remove the engine ground straps.
16. Remove the ignition coil towers.
17. Remove the oil dipstick tube.
18. Remove the fuel line.
19. Remove the fuel rail.
20. Remove the throttle body assembly.
21. Remove the heater hoses.
22. Disconnect the coolant temperature sensor connector.
23. Remove the intake manifold.

➡**Remove the fasteners in reverse of the tightening sequence.**

To install:

24. Install the intake manifold using new gaskets. Tighten the bolts, in sequence, to 105 inch lbs. (12 Nm).

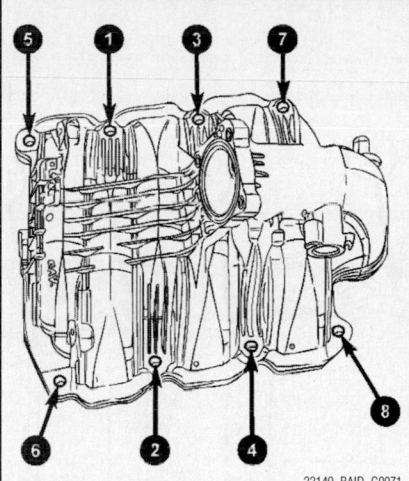

Fig. 75 Intake manifold tighten sequence—3.7L engine

25. Install the coolant temperature sensor connector.
26. Install the heater hoses.
27. Install the throttle body assembly.
28. Install the fuel rail.
29. Install the fuel line.
30. Install the oil dipstick tube.
31. Install the ignition coil towers.
32. Install the engine ground straps.
33. Connect the alternator electrical connections.
34. Install the accessory drive belt.
35. Install the cruise control servo hose.
36. Install the brake booster vacuum line.
37. Install the canister purge vacuum line.
38. Install the positive crankcase ventilation (PCV) valve and hose.
39. Connect the following electrical connectors:
   a. Idle Air Control (IAC) valve
   b. Throttle Position (TP) sensor
   c. Intake Air Temperature (IAT) sensor
   d. Manifold Absolute Pressure (MAP) sensor
40. Install the cruise control cable.
41. Install the accelerator cable.
42. Install the air cleaner assembly.
43. Connect the negative battery cable.
44. Fill the cooling system.
45. Start the engine and check for leaks.

OIL PAN

REMOVAL & INSTALLATION

See Figures 76 and 77.

1. Before servicing the vehicle, refer to the precautions.

2. Drain the engine oil.
3. Disconnect the negative battery cable.
4. Install engine support fixture 8354. Do not raise engine at this time.
5. Loosen both left and right side engine mount through bolts. Do not remove bolts.
6. Remove the structural dust cover, if equipped.
7. Remove the front crossmember.
8. Raise engine to provide clearance to remove oil pan.

➡**Raise the engine just enough to provide clearance for oil pan removal. Check for proper clearance at fan shroud to fan and cowl to intake manifold.**

9. If equipped with 4WD, remove or disconnect the following:
   a. Pinion bracket
   b. Front driveshaft at the front axle
   c. Front axle mounting bolts
   d. Lower the front axle using a suitable jack.
10. Remove the oil pan mounting bolts and oil pan.

➡**Do not pry on oil pan or oil pan gasket. Gasket is integral to engine windage tray and does not come out with oil pan.**

11. Unbolt oil pump pickup tube and remove tube.

To install:

12. Inspect the integral windage tray and gasket and replace as needed.
13. Clean the oil pan gasket mating surface of the bedplate and oil pan.
14. Position the integrated oil pan gasket/windage tray assembly.
15. Install the oil pickup tube
16. If removed, install stud at position No. 9.
17. Install the mounting bolt and nuts. Tighten nuts to 20 ft. lbs. (28 Nm).
18. Position the oil pan and install the mounting bolts. Tighten the mounting bolts to 11 ft. lbs. (15 Nm) in the sequence shown.
19. Lower the engine into mounts.
20. Install both the left and right side engine mount through bolts. Tighten the nuts to 50 ft. lbs. (68 Nm).
21. Remove the lifting device.
22. If equipped with 4WD, install or connect the following:
   a. Front axle mounting bolts
   b. Pinion bracket
   c. Front driveshaft to front axle
23. Install structural dust cover, if equipped.
24. Install the front crossmember.
25. Refill the engine oil to the correct level.

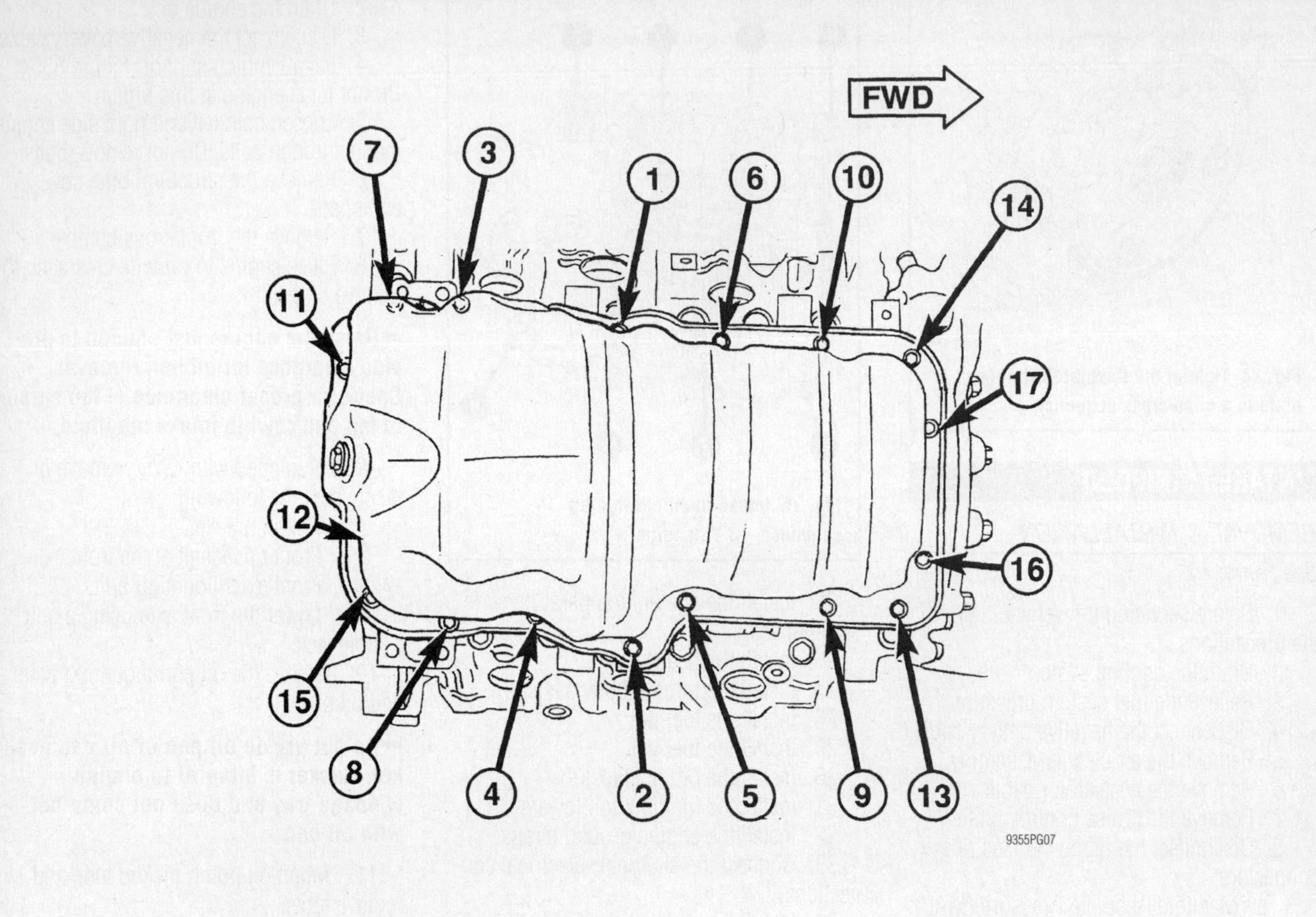

FWD →

Fig. 76 Oil pan mounting bolt tighten sequence—3.7L Engine

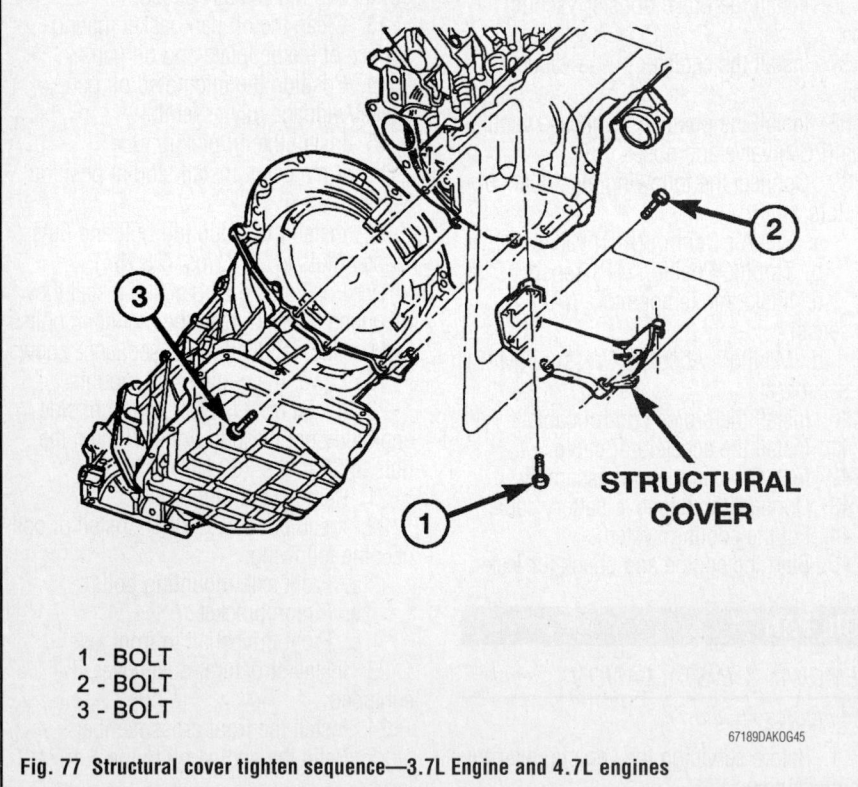

1 - BOLT
2 - BOLT
3 - BOLT

STRUCTURAL COVER

Fig. 77 Structural cover tighten sequence—3.7L Engine and 4.7L engines

26. Reconnect the negative battery cable.

27. Start engine and check for leaks.

OIL PUMP

REMOVAL & INSTALLATION

See Figure 78.

1. Before servicing the vehicle, refer to the precautions.

2. Drain the engine oil.

3. Remove the oil pan.

4. Remove the timing chain cover.

5. Remove the timing chains and tensioners.

6. Remove the oil pump.

To install:

7. Remove the oil pump. Tighten the pump bolts, in sequence, to 21 ft. lbs. (28 Nm).

8. Remove the timing chains and tensioners.

9. Remove the timing chain cover.

10. Remove the oil pan.

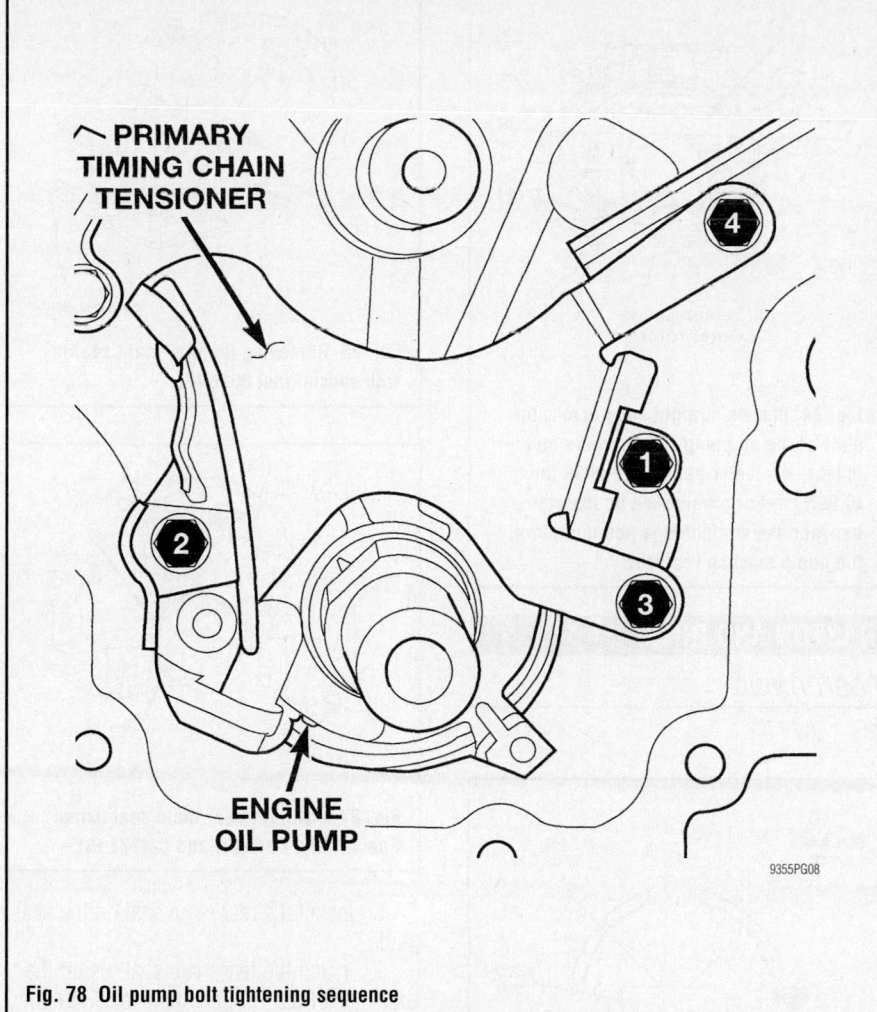

**PRIMARY
TIMING CHAIN
TENSIONER**

**ENGINE
OIL PUMP**

9355PG08

Fig. 78 Oil pump bolt tightening sequence

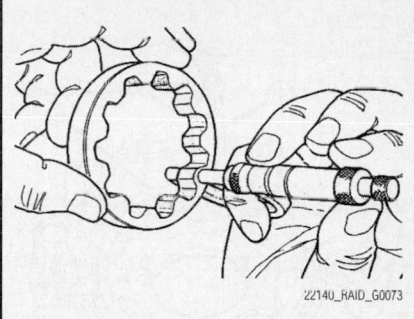

22140_RAID_G0073

Fig. 80 Measure the thickness of the outer rotor. If the outer rotor thickness measures at 0.472 in. (12.005 mm) or less the oil pump assembly must be replaced

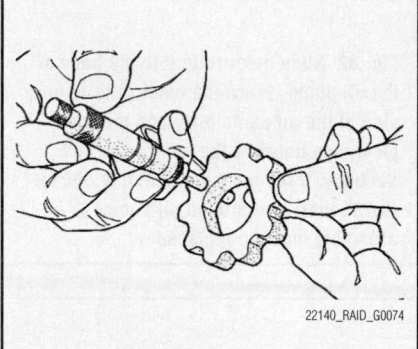

22140_RAID_G0074

Fig. 81 Measure the thickness of the inner rotor . If the inner rotor thickness measures at 0.472 in. (12.005 mm) or less then the oil pump assembly must be replaced

11. Refill the engine oil to the correct level.

INSPECTION

See Figures 79 through 84.

❊❊ WARNING

The oil pump pressure relief valve and spring should not be removed from the oil pump. If these components are disassembled and or removed from the pump the entire oil pump assembly must be replaced.

1. Clean all parts thoroughly. Mating surface of the oil pump housing should be smooth. If the pump cover is scratched or grooved, the oil pump assembly should be replaced.

2. Lay a straight edge across the pump cover surface. If a 0.001 in. (0.025 mm) feeler gauge can be inserted between the cover and the straight edge, the oil pump assembly should be replaced.

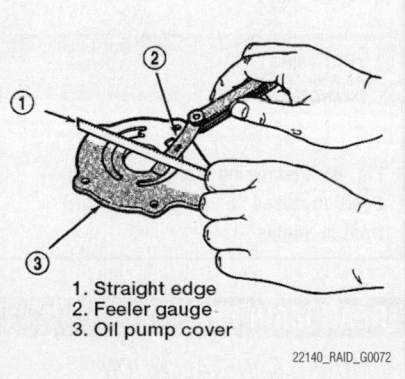

1. Straight edge
2. Feeler gauge
3. Oil pump cover

22140_RAID_G0072

Fig. 79 Lay a straight edge across the pump cover surface. If a 0.001 in. (0.025 mm) feeler gauge can be inserted between the cover and the straight edge, the oil pump assembly should be replaced

3. Measure the thickness of the outer rotor. If the outer rotor thickness measures at 0.472 in. (12.005 mm) or less the oil pump assembly must be replaced.

4. Measure the diameter of the outer rotor. If the outer rotor diameter measures at 3.382 in. (85.925 mm) or less the oil pump assembly must be replaced.

5. Measure the thickness of the inner rotor . If the inner rotor thickness measures at 0.472 in. (12.005 mm) or less then the oil pump assembly must be replaced.

6. Slide outer rotor into the body of the oil pump. Press the outer rotor to one side of the oil pump body and measure clearance between the outer rotor and the body. If the measurement is 0.009 in. (0.235 mm) or more the oil pump assembly must be replaced.

7. Install the inner rotor in the into the oil pump body. Measure the clearance between the inner and outer rotors. If the clearance between the rotors is 0.006 in. (0.150 mm) or more the oil pump assembly must be replaced.

8. Place a straight edge across the body of the oil pump (between the bolt holes), if a feeler gauge of 0.0038 in. (0.095 mm) or greater can be inserted between the

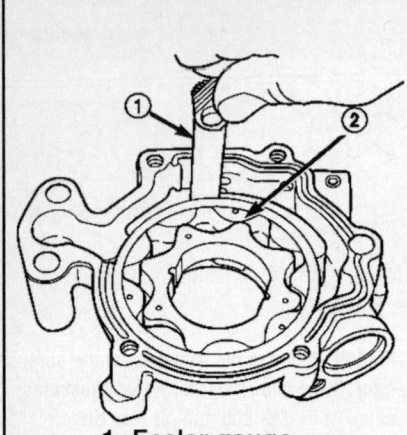

1. Feeler gauge
2. Outer rotor

22140_RAID_G0075

Fig. 82 Slide outer rotor into the body of the oil pump. Press the outer rotor to one side of the oil pump body and measure clearance between the outer rotor and the body. If the measurement is 0.009 in. (0.235 mm) or more the oil pump assembly must be replaced

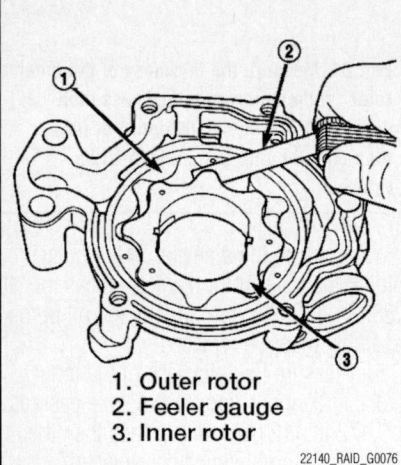

1. Outer rotor
2. Feeler gauge
3. Inner rotor

22140_RAID_G0076

Fig. 83 Install the inner rotor in the into the oil pump body. Measure the clearance between the inner and outer rotors. If the clearance between the rotors is 0.006 in. (0.150 mm) or more the oil pump assembly must be replaced

straightedge and the rotors, the pump must be replaced.

➡ The oil pump is sold as an assembly. There are no Mitsubishi part numbers for sub-assembly components. In the event the oil pump is functioning or out of specification it must be replaced as an assembly.

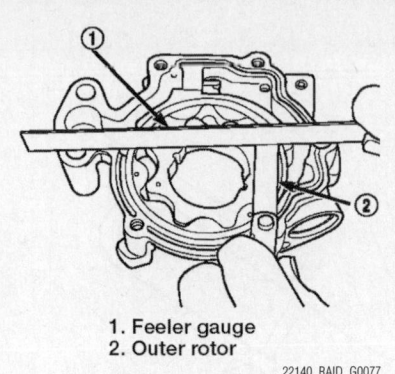

1. Feeler gauge
2. Outer rotor

22140_RAID_G0077

Fig. 84 Place a straight edge across the body of the oil pump (between the bolt holes), if a feeler gauge of 0.0038 in. (0.095 mm) or greater can be inserted between the straightedge and the rotors, the pump must be replaced

PISTON AND RING

POSITIONING

See Figure 85.

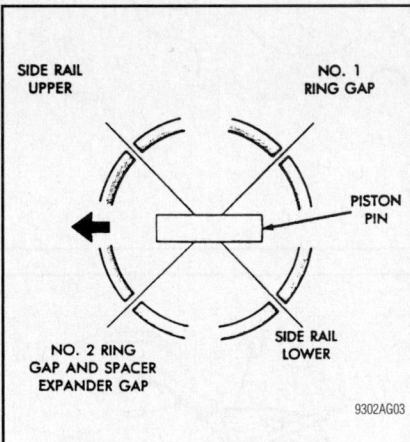

SIDE RAIL UPPER

NO. 1 RING GAP

PISTON PIN

NO. 2 RING GAP AND SPACER EXPANDER GAP

SIDE RAIL LOWER

9302AG03

Fig. 85 Piston ring end-gap spacing. Position raised "F" on piston toward front of engine

REAR MAIN SEAL

REMOVAL & INSTALLATION

See Figures 86 and 87.

1. Before servicing the vehicle, refer to the precautions.
2. Remove the transmission.
3. Remove the flexplate.
4. Thread Oil Seal Remover 8506 into the rear main seal as far as possible and remove the rear main seal.

To install:
5. Install the Seal Guide 8349-2 onto the crankshaft

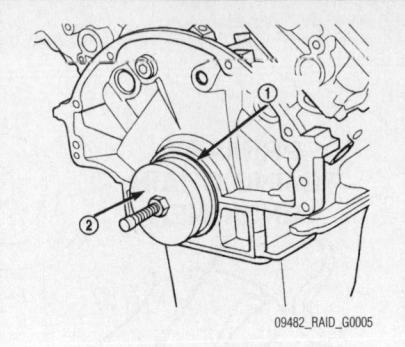

09482_RAID_G0005

Fig. 86 Removing the rear main seal (1) with special tool 8506 (2)

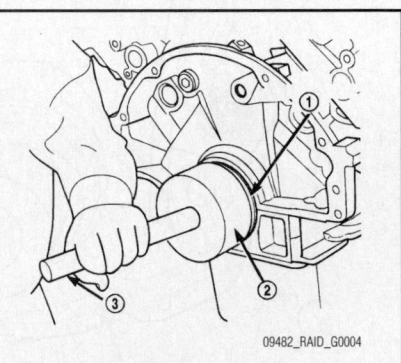

09482_RAID_G0004

Fig. 87 Installing rear main seal using Special Tool 8349 (2) and C-4171 (3)

6. Install the rear main seal on the seal guide
7. Install the rear main seal, using the Crankshaft Rear Oil Seal Installer 8349 and Driver Handle C-4171; tap it into place until the installer is flush with the cylinder block
8. Install the flexplate. Tighten the bolts to 45 ft. lbs. (60 Nm).
9. Install the transmission.
10. Start the engine and check for leaks.

ROCKER ARMS

REMOVAL & INSTALLATION

See Figure 88.

1. Before servicing the vehicle, refer to the precautions.
2. Disconnect the negative battery cable.
3. Remove the cylinder head cover.
4. Rotate the crankshaft so that the piston of the cylinder to be serviced is at Bottom Dead Center (BDC) and both valves are closed.
5. Use special tool 8516 to depress the valve and remove the rocker arm.
6. Repeat for each rocker arm to be serviced.

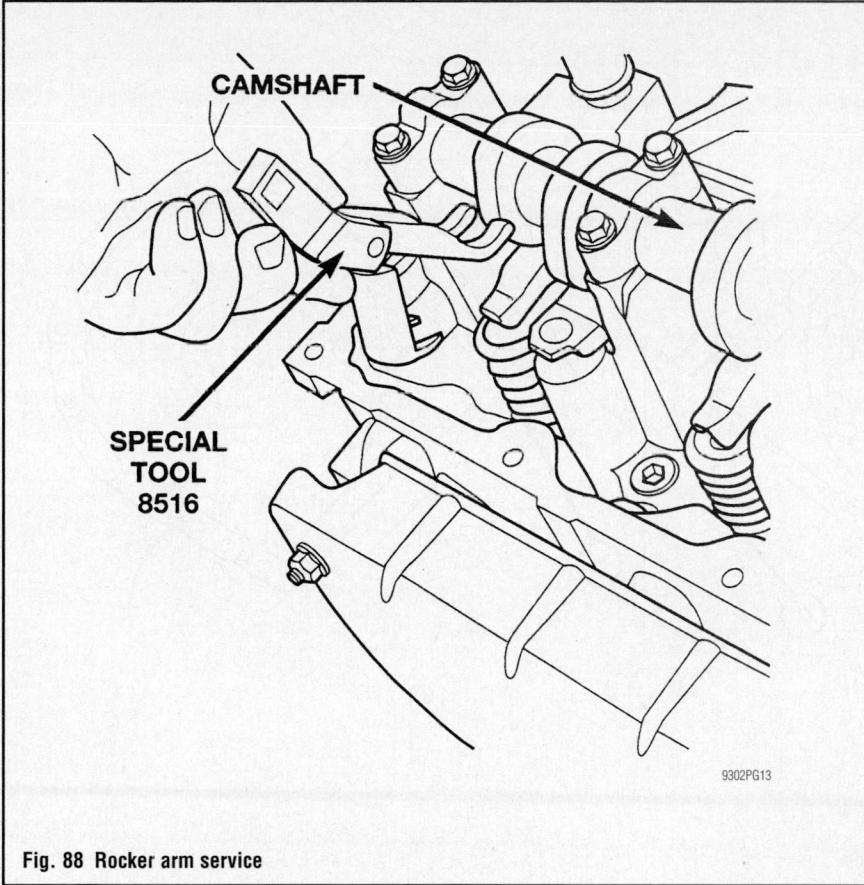

Fig. 88 Rocker arm service

➡ **Keep valvetrain components in order for reassembly.**

To install:

7. Rotate the crankshaft so that the piston of the cylinder to be serviced is at BDC.

8. Compress the valve spring and install each rocker arm in its original position.

9. Repeat for each rocker arm to be installed.

10. Install the cylinder head cover.

11. Connect the negative battery cable.

TIMING CHAIN FRONT COVER

REMOVAL & INSTALLATION

See Figures 89 through 92.

1. Disconnect the negative battery cable.

2. Drain cooling system.

3. Remove right and left cylinder head covers.

4. Remove radiator fan.

5. Rotate engine until timing mark on crankshaft damper aligns with TDC mark on timing chain cover.

6. Make sure the camshaft sprocket

"V6" marks (1) are at the 12 o'clock position (No. 1 TDC exhaust stroke).

7. Remove power steering pump.

8. Remove access plugs (1 and 2) from left and right cylinder heads for access to chain guide fasteners.

9. Remove the oil fill housing to gain access to the right side tensioner arm fastener.

10. Remove crankshaft damper and timing chain cover.

To install:

11. Install timing chain cover and crankshaft damper.

12. Install cylinder head covers.

➡ **Before installing threaded plug in right cylinder head, the plug must be coated with sealant to prevent leaks.**

13. Coat the large threaded access plug with Mitsubishi Thread Sealant with Teflon, then install into the right cylinder head and tighten to 60 ft. lbs. (81 Nm).

14. Install the oil fill housing.

15. Install access plug in left cylinder head.

16. Install power steering pump.

17. Fill cooling system.

18. Connect the negative battery cable.

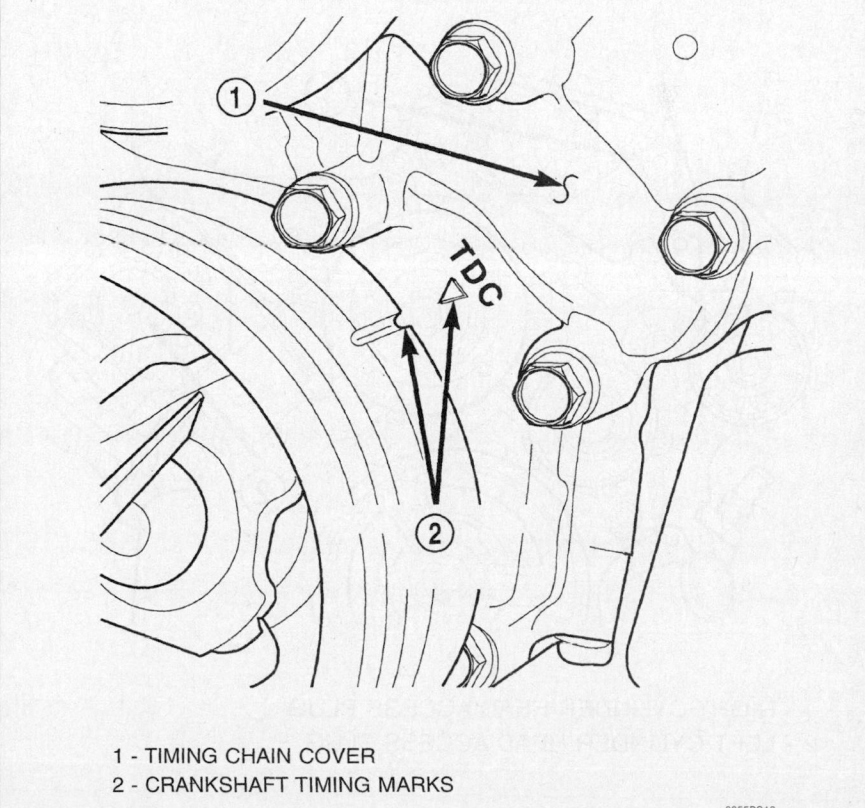

1 - TIMING CHAIN COVER
2 - CRANKSHAFT TIMING MARKS

Fig. 89 Crankshaft timing marks—3.7L engine

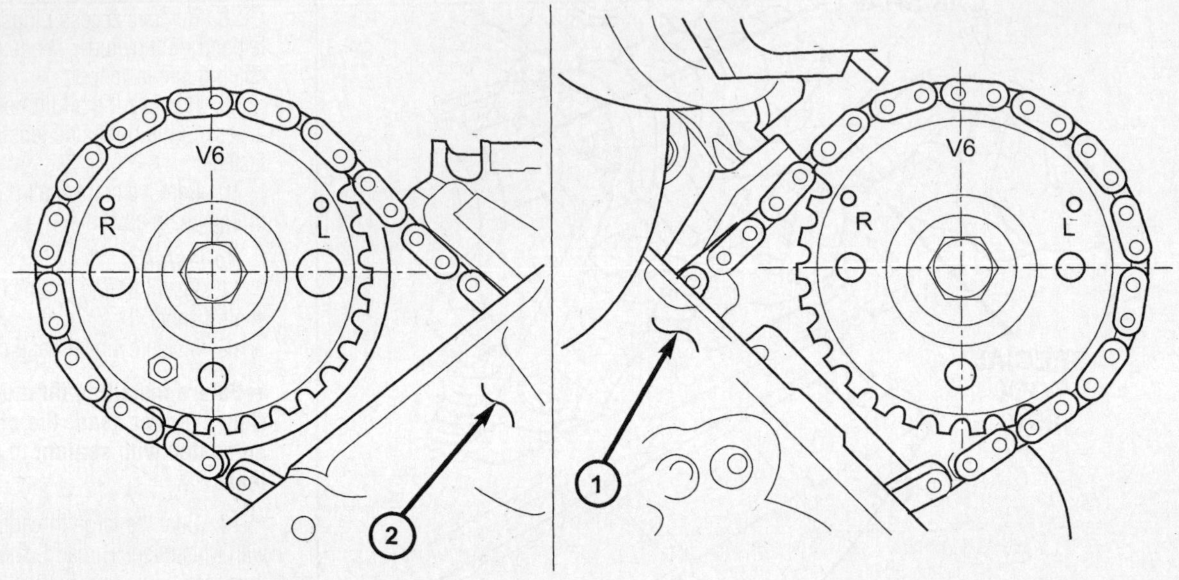

1 - LEFT CYLINDER HEAD
2 - RIGHT CYLINDER HEAD

9355PG09

Fig. 90 Camshaft sprocket timing marks—3.7L engine

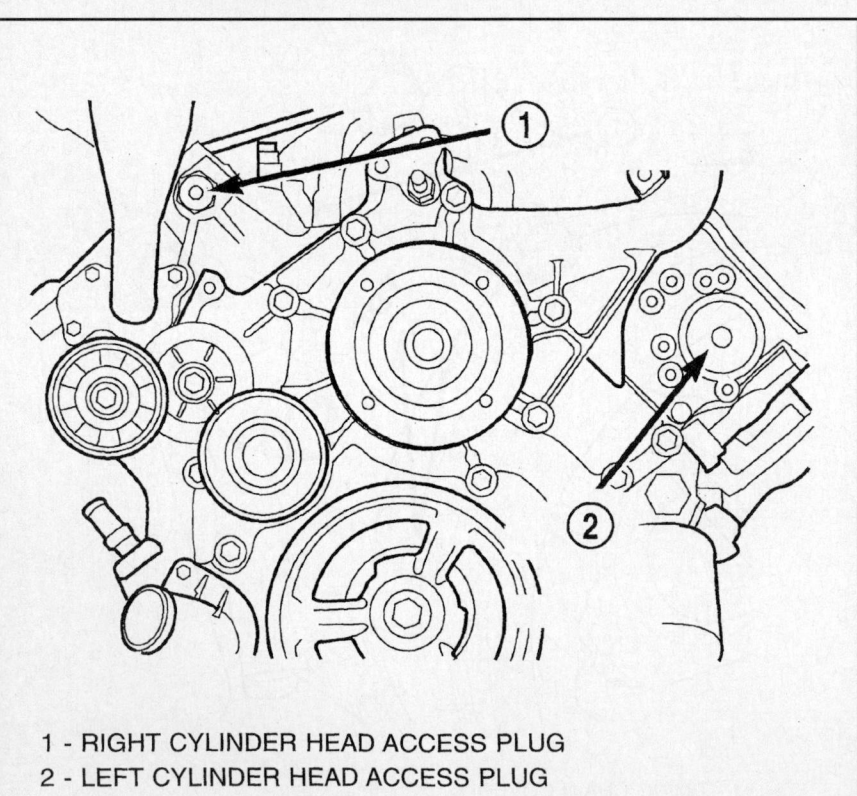

1 - RIGHT CYLINDER HEAD ACCESS PLUG
2 - LEFT CYLINDER HEAD ACCESS PLUG

9355PG11

Fig. 91 Cylinder head access plugs—3.7L engine

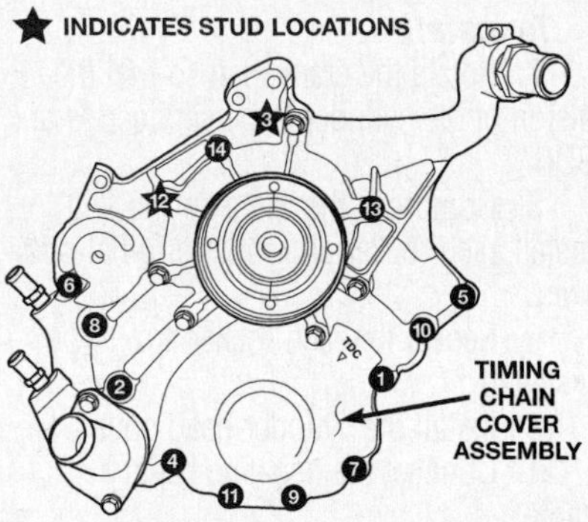

★ INDICATES STUD LOCATIONS

TIMING CHAIN COVER ASSEMBLY

9355PG16

Fig. 92 Timing cover bolt tighten sequence—3.7L engine

TIMING CHAIN & SPROCKETS

REMOVAL & INSTALLATION

See Figures 90, 91, 93 through 101.

1. Disconnect the negative battery cable.
2. Drain cooling system.
3. Remove right and left cylinder head covers.
4. Remove radiator fan.

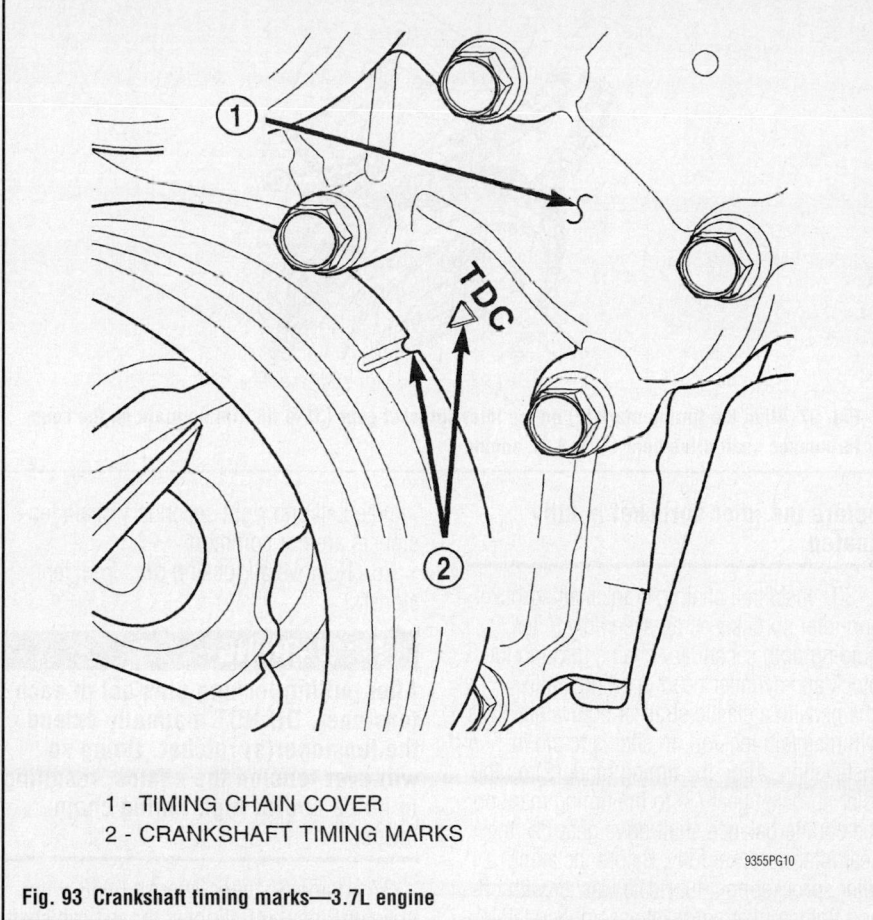

1 - TIMING CHAIN COVER
2 - CRANKSHAFT TIMING MARKS

9355PG10

Fig. 93 Crankshaft timing marks—3.7L engine

22140_RAID_G0080

Fig. 94 While holding the camshaft steel tube (1) with Camshaft Holder 8428 (2), remove the camshaft sprocket—3.7L engine

5. Rotate engine until timing mark on crankshaft damper aligns with TDC mark on timing chain cover.

6. Make sure the camshaft sprocket "V6" marks (1) are at the 12 o'clock position (No. 1 TDC exhaust stroke).

7. Remove power steering pump.

8. Remove access plugs (1 and 2) from left and right cylinder heads for access to chain guide fasteners.

9. Remove the oil fill housing to gain access to the right side tensioner arm fastener.

10. Remove crankshaft damper and timing chain cover.

11. Collapse and pin primary chain tensioner.

⁂ WARNING

Plate behind left secondary chain tensioner could fall into oil pan. Therefore, cover pan opening.

12. Remove secondary chain tensioners.
13. Remove camshaft position sensor.

⁂ WARNING

Care should be taken not to damage camshaft target wheel. Do not hold

target wheel while loosening or tightening camshaft sprocket. Do not place the target wheel near a magnetic source of any kind. A damaged or magnetized target wheel could cause a vehicle no start condition.

⁂ WARNING

Do not forcefully rotate the camshafts or crankshaft independently of each other. Damaging intake valve to piston contact will occur. Ensure negative battery cable is disconnected to guard against accidental starter engagement.

14. Remove left and right camshaft sprocket bolts.

15. While holding the left camshaft steel tube with Camshaft Holder 8428, remove the left camshaft sprocket. Slowly rotate the camshaft approximately 5 degrees clockwise to a neutral position.

16. While holding the right camshaft steel tube with Camshaft Holder 8428, remove the right camshaft sprocket.

17. Remove idler sprocket assembly bolt.

18. Slide the idler sprocket assembly and crank sprocket forward simultaneously to remove the primary and secondary chains.

19. Remove both pivoting tensioner arms and chain guides.

20. Remove primary chain tensioner.

To install:

21. Using a vise, lightly compress the secondary chain tensioner piston until the piston step (5) is flush with the tensioner body. Using a pin or suitable tool, release ratchet pawl (4) by pulling pawl back against spring force through access hole on side of tensioner. While continuing to hold pawl back, Push ratchet device to approximately 2 mm from the tensioner body.

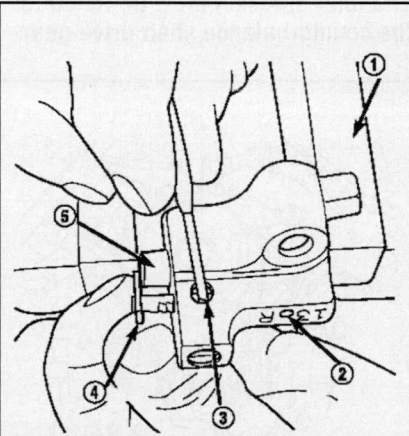

1. Secondary chain tensioner
2. Tensioner pins
3. Pin
4. Ratchet pawl
5. Piston step

22140_RAID_G0081

Fig. 95 Servicing the secondary chain tensioner—3.7L engine

Install Tensioner Pins 8514 (2) into hole on front of tensioner. Slowly open vise to transfer piston spring force to lock pin.

22. Position primary chain tensioner over oil pump and insert bolts into lower two holes on tensioner bracket. Tighten bolts to 21 ft. lbs. (28 Nm).

23. Install right side chain tensioner arm. Install Torx® bolt. Tighten Torx® bolt to 21 ft. lbs. (28 Nm). caution The silver bolts retain the guides to the cylinder heads and the black bolts retain the guides to the engine block.

24. Install the left side chain guide. Tighten the bolts to 21 ft. lbs. (28 Nm).

25. Install left side chain tensioner arm, and Torx® bolt. Tighten Torx® bolt to 21 ft. lbs. (28 Nm).

26. Install the right side chain guide. Tighten the bolts to 21 ft. lbs. (28 Nm).

27. Install both secondary chains onto the idler sprocket. Align two plated links on the secondary chains to be visible through the two lower openings on the idler sprocket (4 o'clock and 8 o'clock). Once the secondary timing chains are installed, install the Secondary Camshaft Chain Holder 8429 to hold chains in place for installation.

28. Align primary chain double plated links with the timing mark at 12 o'clock on the idler sprocket. Align the primary chain single plated link with the timing mark at 6 o'clock on the crankshaft sprocket.

29. Lubricate idler shaft and bushings with clean engine oil.

✳✳ WARNING

The idler sprocket must be timed to the counterbalance shaft drive gear

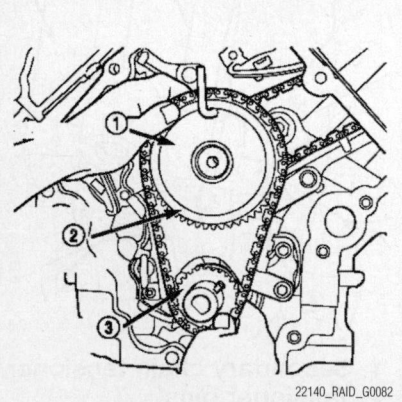

22140_RAID_G0082

Fig. 96 Install both secondary chains onto the idler sprocket (2) and crankshaft sprocket (3). Then install the Secondary Camshaft Chain Holder 8429 (1) to hold chains in place for installation—3.7L engine

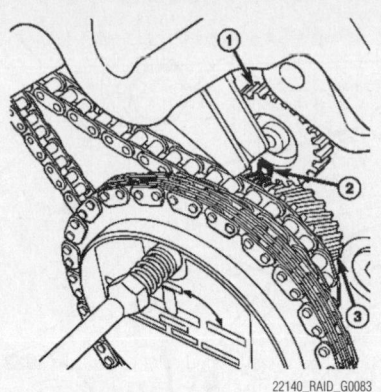

22140_RAID_G0083

Fig. 97 Align the timing mark (2) on the idler sprocket gear (3) to the timing mark on the counterbalance shaft drive gear (1) —3.7L engine

before the idler sprocket is fully seated.

30. Install all chains, crankshaft sprocket, and idler sprocket as an assembly. After guiding both secondary chains through the block and cylinder head openings, affix chains with a elastic strap or equivalent. This will maintain tension on chains to aid in installation. Align the timing mark (2) on the idler sprocket gear (3) to the timing mark on the counterbalance shaft drive gear (1), then seat idler sprocket fully. Before installing idler sprocket bolt, lubricate washer with oil, and tighten idler sprocket assembly retaining bolt to 25 ft. lbs. (34 Nm).

➡It will be necessary to slightly rotate camshafts for sprocket installation.

31. Align left camshaft sprocket **L**dot to plated link on chain.

32. Align right camshaft sprocket **R**dot to plated link on chain.

✳✳ WARNING

Remove excess oil from the camshaft sprocket bolt. Failure to do so can result in over-torque of bolt resulting in bolt failure.

33. Remove Secondary Camshaft Chain Holder 8429, then attach both sprockets to camshafts. Remove excess oil from bolts, then Install sprocket bolts, but do not tighten at this time.

34. Verify that all plated links are aligned with the marks on all sprockets and the "V6" marks on camshaft sprockets are at the 12 o'clock position. caution Ensure the plate between the left secondary chain tensioner and block is correctly installed.

35. Install both secondary chain tensioners. Tighten bolts to 250 inch lbs. (28 Nm).

note Left and right secondary chain tensioners are not common.

36. Remove all locking pins from tensioners.

✳✳ WARNING

After pulling locking pins out of each tensioner, DO NOT manually extend the tensioner(s) ratchet. Doing so will over tension the chains, resulting in noise and/or high timing chain loads.

37. Using Spanner Wrench 6958, with Adaptor Pins 8346, tighten the left camshaft sprocket bolts to 90 ft. lbs. (122 Nm).

18.Using Spanner Wrench 6958, with Adaptor Pins 8346, tighten the right camshaft sprocket bolts to 90 ft. lbs. (122 Nm).

38. Rotate engine two full revolutions. Verify timing marks are at the follow locations:
- Primary chain idler sprocket dot is at 12 o'clock
- Primary chain crankshaft sprocket dot is at 6 o'clock
- Secondary chain camshaft sprockets "V6" marks are at 12 o'clock
- Balance shaft drive gear dot is aligned to the idler sprocket gear dot

39. Lubricate all three chains with engine oil.

40. After installing all chains, it is recommended that the idler gear end play be checked. The end play must be within 0.004– 0.010 in. (0.10–0.25 mm). If not within specification, the idler gear must be replaced.

41. Install timing chain cover and crankshaft damper.

42. Install cylinder head covers.

➡Before installing threaded plug in right cylinder head, the plug must be coated with sealant to prevent leaks.

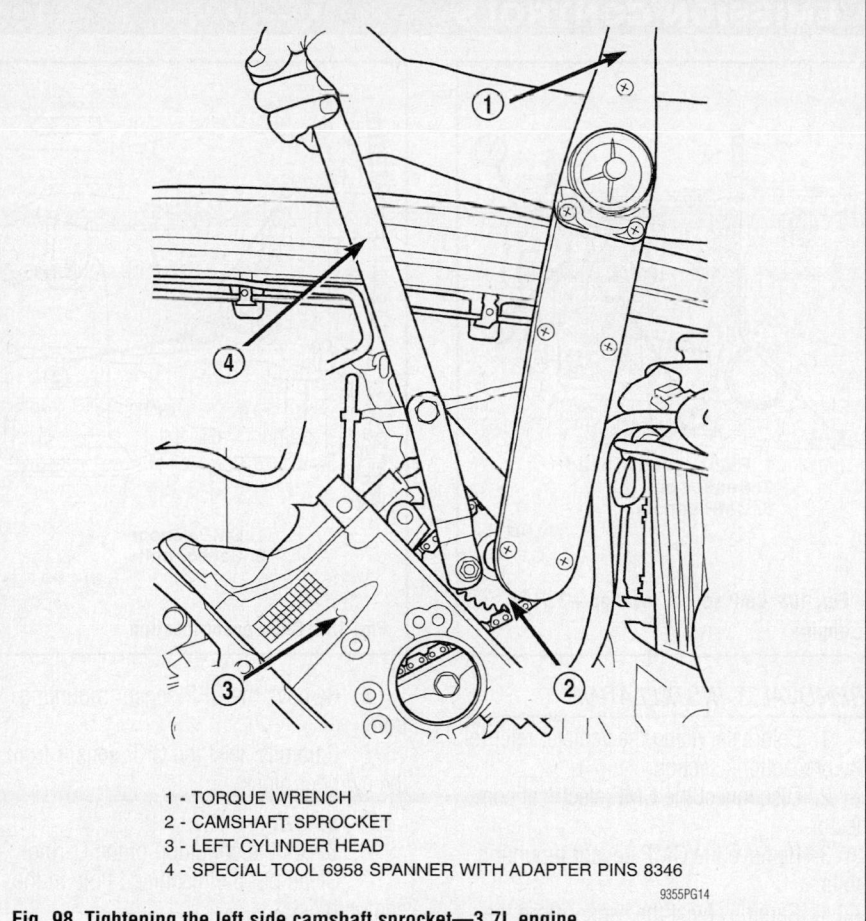

1 - TORQUE WRENCH
2 - CAMSHAFT SPROCKET
3 - LEFT CYLINDER HEAD
4 - SPECIAL TOOL 6958 SPANNER WITH ADAPTER PINS 8346

9355PG14

Fig. 98 Tightening the left side camshaft sprocket—3.7L engine

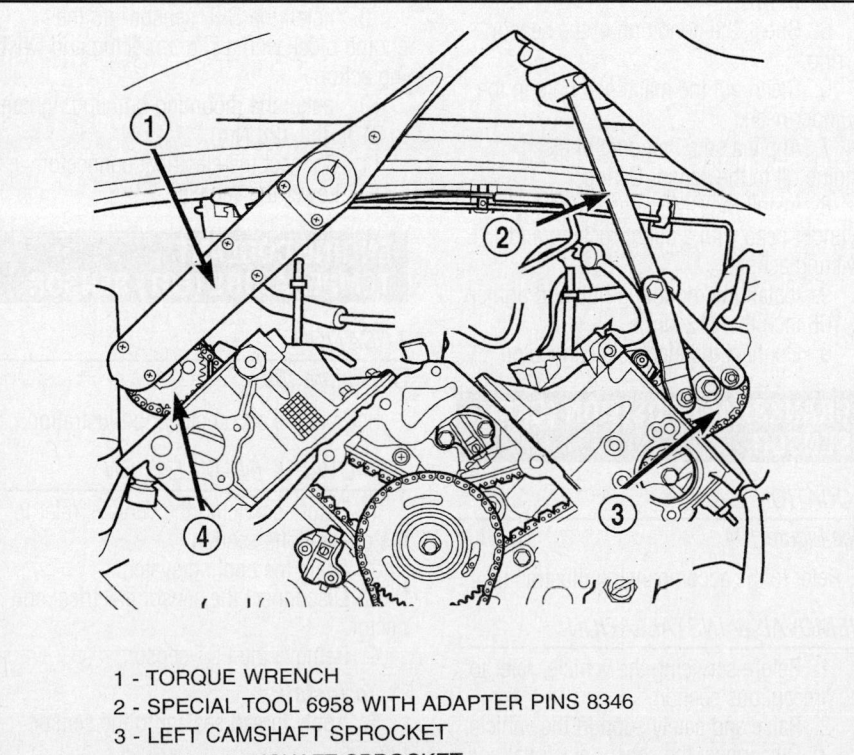

1 - TORQUE WRENCH
2 - SPECIAL TOOL 6958 WITH ADAPTER PINS 8346
3 - LEFT CAMSHAFT SPROCKET
4 - RIGHT CAMSHAFT SPROCKET

9355PG15

Fig. 99 Tightening the right side camshaft sprocket—3.7L engine

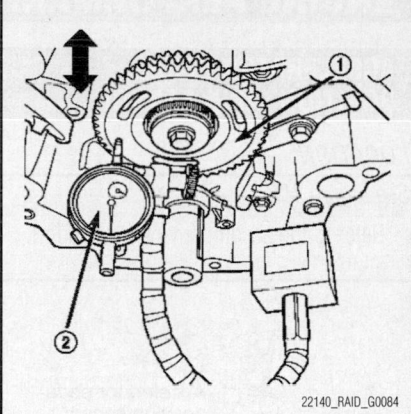

22140_RAID_G0084

Fig. 100 Check idler gear end play using a dial indicator (2). If not within specification, the idler gear (1) must be replaced—3.7L engine

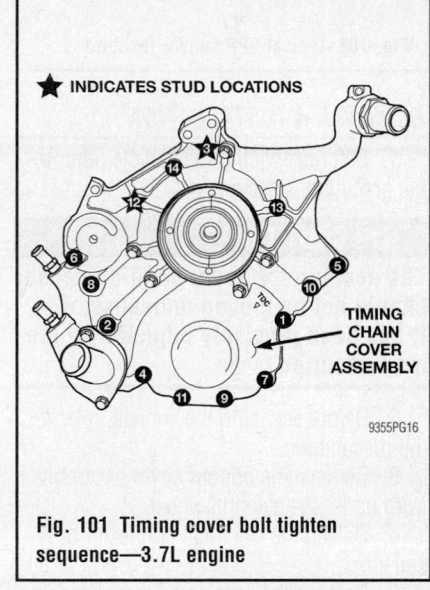

★ INDICATES STUD LOCATIONS

TIMING CHAIN COVER ASSEMBLY

9355PG16

Fig. 101 Timing cover bolt tighten sequence—3.7L engine

43. Coat the large threaded access plug with Mitsubishi Thread Sealant with Teflon, then install into the right cylinder head and tighten to 60 ft. lbs. (81 Nm).

44. Install the oil fill housing.

45. Install access plug in left cylinder head.

46. Install power steering pump.

47. Fill cooling system.

48. Connect the negative battery cable.

VALVE LASH

ADJUSTMENT

All gasoline engines use hydraulic lifters. No maintenance or periodic adjustment is required.

ENGINE PERFORMANCE & EMISSION CONTROLS

ACCELERATOR PEDAL POSITION (APP) SENSOR

LOCATION

See Figure 102.

Refer to the accompanying illustration.

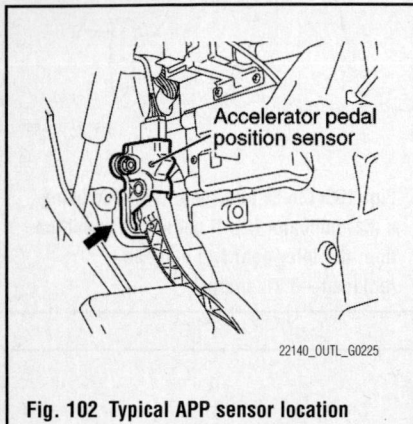

Fig. 102 Typical APP sensor location

REMOVAL & INSTALLATION

1. Before servicing the vehicle, refer to the precautions section.

❋❋ CAUTION

The accelerator pedal position sensor should not be moved unnecessarily; it has been precisely adjusted by the manufacturer.

2. Before servicing the vehicle, refer to the precautions.
3. Remove the bottom cover assembly from under the steering wheel.
4. Detach the electrical connector at the sensor.
5. Remove the accelerator pedal assembly from the vehicle.

To install:

6. Install the accelerator pedal and tighten nuts to 111 inch lbs. (13 Nm)
7. Install the accelerator pedal assembly from the vehicle.
8. Connect the electrical connector at the sensor.
9. Install the bottom cover assembly from under the steering wheel.

CAMSHAFT POSITION (CMP) SENSOR

LOCATION

See Figure 103.

Refer to the accompanying illustration.

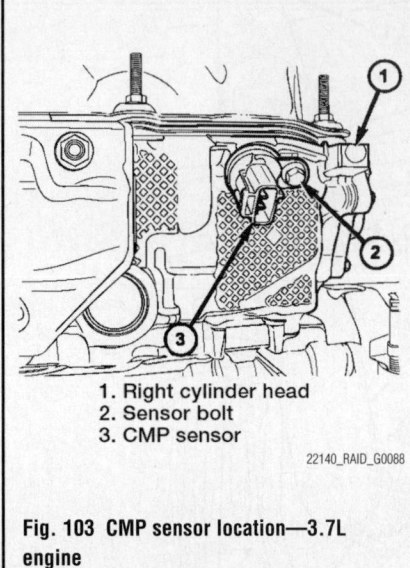

1. Right cylinder head
2. Sensor bolt
3. CMP sensor

22140_RAID_G0088

Fig. 103 CMP sensor location—3.7L engine

REMOVAL & INSTALLATION

1. Before servicing the vehicle, refer to the precautions section.
2. Disconnect the CMP electrical connector.
3. Remove the CMP sensor mounting bolts.
4. Carefully twist the sensor from the cylinder.

To install:

5. Check the condition of the sensor O-ring.
6. Clean out the machined hole in the cylinder head.
7. Apply a small amount of clean engine oil to the sensor O-ring.
8. Install the CMP sensor into the cylinder head with a slight rocking and twisting action.
9. Install the mounting bolt and tighten to 106 inch lbs. (12 Nm).
10. Connect the electrical connector.

CRANKSHAFT POSITION (CKP) SENSOR

LOCATION

See Figure 104.

Refer to the accompanying illustration.

REMOVAL & INSTALLATION

1. Before servicing the vehicle, refer to the precautions section.
2. Raise and safely support the vehicle.
3. Disconnect the sensor electrical connector.

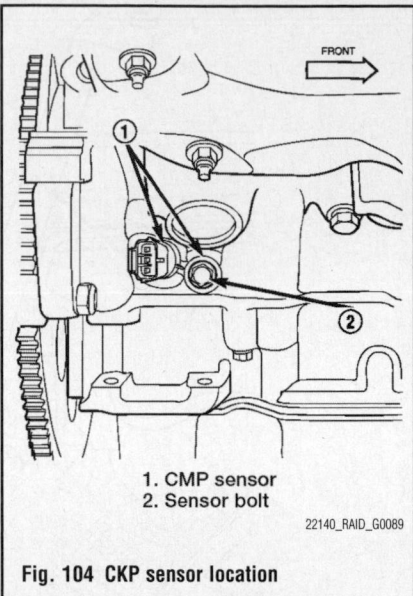

1. CMP sensor
2. Sensor bolt

22140_RAID_G0089

Fig. 104 CKP sensor location

4. Remove the CKP sensor mounting bolt.
5. Carefully twist the CKP sensor from the cylinder block.

To install:

6. Check the condition of the O-ring.
7. Clean out the machined hole in the engine block.
8. Apply a small amount of clean engine oil to the sensor O-ring.
9. Install the CKP sensor into the engine block with a slight rocking and twisting action.
10. Install the mounting bolt and tighten to 21 ft. lbs. (28 Nm).
11. Connect the electrical connector.
12. Lower the vehicle.

ENGINE COOLANT TEMPERATURE (ECT) SENSOR

LOCATION

See Figure 105.

Refer to the accompanying illustration.

REMOVAL & INSTALLATION

1. Before servicing the vehicle, refer to the precautions section.
2. Drain the cooling system.
3. Disconnect the sensor electrical connector.
4. Remove the ECT sensor.

To install:

5. Apply thread sealant to the sensor threads.
6. Install the ECT sensor into the engine

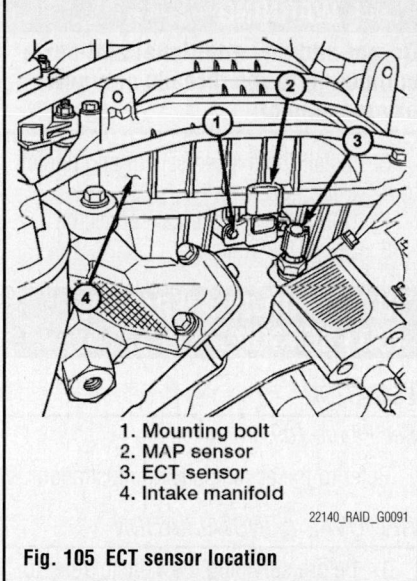

1. Mounting bolt
2. MAP sensor
3. ECT sensor
4. Intake manifold

22140_RAID_G0091

Fig. 105 ECT sensor location

block and tighten the mounting bolt to 8 ft. lbs. (11 Nm).

7. Connect the electrical connector.

8. Refill the cooling system to the correct level.

EVAPORATIVE EMISSIONS (EVAP) CANISTER

LOCATION

See Figure 106.

Refer to the accompanying illustration.

REMOVAL & INSTALLATION

4 Door Cab

➡**On 4 door Double Cab models, the fuel tank must be lowered for canister removal.**

1. Lower the fuel tank.

➡**The EVAP canister is mounted to the top of the fuel tank.**

2. Remove vapor hose at ESIM switch

3. Remove vapor line from EVAP canister by pressing on both tabs simultaneously.

4. Using a screwdriver, pry up and remove plastic mounting clip and remove canister from top of fuel tank. The opposite end of the canister is supported by two plastic legs.

To install:

5. Guide the two plastic canister locating legs into the top of fuel tank.

6. Install pin and press down.

7. Raise fuel tank.

8. Install electrical connector ESIM switch.

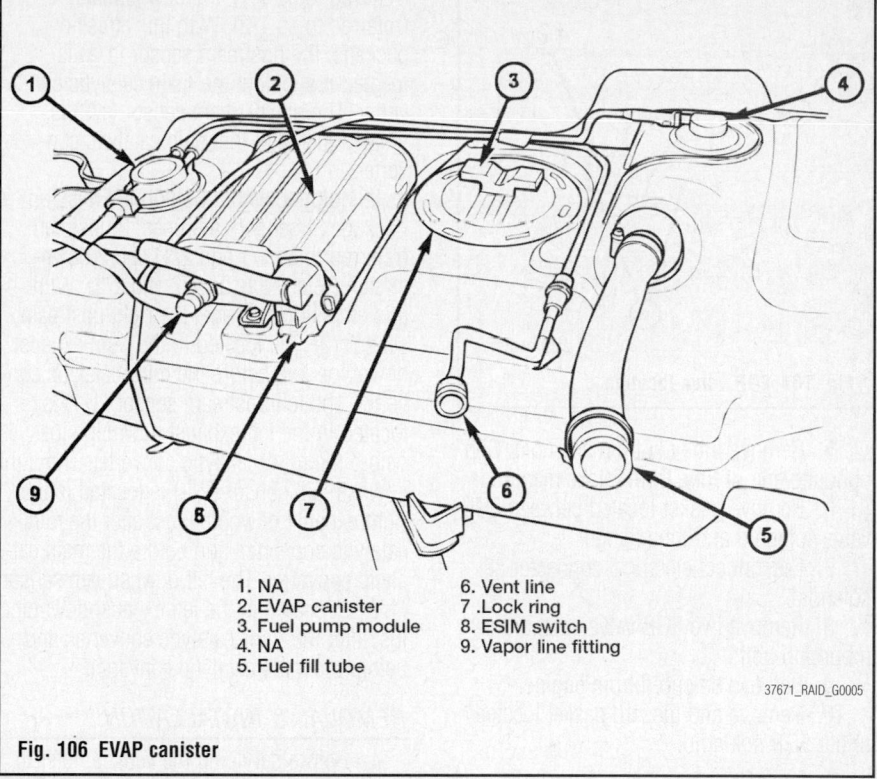

1. NA
2. EVAP canister
3. Fuel pump module
4. NA
5. Fuel fill tube
6. Vent line
7. .Lock ring
8. ESIM switch
9. Vapor line fitting

37671_RAID_G0005

Fig. 106 EVAP canister

9. Install all vapor hoses and lines to both the EVAP canister and ESIM switch.

10. The vapor/vacuum lines and hoses must be firmly connected. Also check the vapor/vacuum lines at the ESIM switch, filter and EVAP canister purge solenoid for damage or leaks. If a leak is present, a Diagnostic Trouble Code (DTC) may be set.

11. Install fuel tank.

Extended Cab

➡**The EVAP canister is mounted to the top of the fuel tank. Both the EVAP canister and ESIM switch can be removed without removing the fuel tank.**

1. Raise and support vehicle.

2. Remove left-rear tire.

3. Remove plastic shield in front of left-rear tire. Access to both the EVAP canister and ESIM switch can be achieved from the area in front of the removed tire and plastic shield.

4. Remove vapor hose at ESIM switch.

5. Remove vapor line from EVAP canister by pressing on both tabs simultaneously.

6. Using a screwdriver, pry up and remove plastic mounting clip and remove canister from top of fuel tank. The opposite end of the canister is supported by two plastic legs.

To install:

7. Guide the two plastic canister locating legs into the top of fuel tank.

8. Install pin and press down.

9. Install electrical connector ESIM switch.

10. Install all vapor hoses and lines to both the EVAP canister and ESIM switch.

11. The vapor/vacuum lines and hoses must be firmly connected. Also check the vapor/vacuum lines at the ESIM switch, filter and EVAP canister purge solenoid for damage or leaks. If a leak is present, a Diagnostic Trouble Code (DTC) may be set.

12. Install plastic shield in front of left-rear tire.

13. Install left-rear tire.

EXHAUST GAS RECIRCULATION (EGR) VALVE

LOCATION

See Figure 107.

Refer to the accompanying illustration.

REMOVAL & INSTALLATION

1. Use a diagnostic scan tool to record any DTC's (Diagnostic Trouble Codes).

2. Disconnect and isolate the negative battery cable.

3. Remove EVAP purge solenoid. Solenoid lifts from tongue-type bracket.

➡**An exhaust gas routing tube connects the EGR valve to the intake manifold.**

4. Remove two tube mounting bolts.

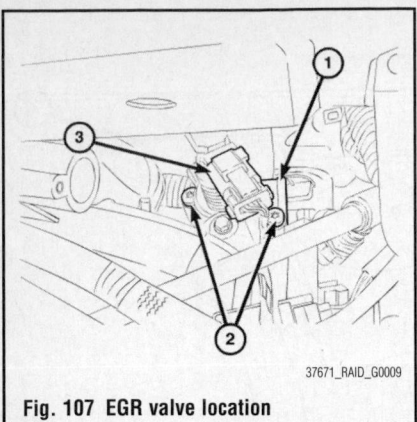

Fig. 107 EGR valve location

5. Remove tube (1) from solenoid. Slip opposite end of tube from intake manifold.

6. Remove gasket located between EGR valve solenoid and tube flange.

7. Disconnect electrical connector at solenoid.

8. Remove two EGR valve solenoid mounting bolts.

9. Remove solenoid from engine.

10. Remove and discard gasket located under EGR solenoid.

To install:

11. Clean gasket area at rear of left cylinder head where it joins base of EGR valve.

12. Clean EGR tube where it joins EGR valve.

13. Position new gasket between EGR valve and cylinder head.

14. Position EGR valve to cylinder head. Install and tighten two bolts to 80 inch lbs. (9 Nm).

15. Position new gasket between EGR tube flange and EGR valve assembly.

16. Position EGR tube to side of EGR valve. Position end of tube into intake manifold. Install two bolts and tighten to 8.5 ft. lbs. (11 Nm).

17. Connect electrical connector to top of EGR valve solenoid.

18. Install EVAP canister purge solenoid by slipping assembly to tongue-type bracket.

19. Connect negative battery cable.

20. Using a diagnostic scan tool, erase any previously recorded DTC's (Diagnostic Trouble Codes).

HEATED OXYGEN SENSOR (HO2S)

LOCATION

See Figure 108.

If equipped with a Federal Emission Package, two sensors are used: upstream (referred to as 1/1) and downstream (referred to as 1/2). With this emission package, the upstream sensor (1/1) is located just before the main catalytic converter. The downstream sensor (1/2) is located just after the main catalytic converter.

If equipped with a California Emission Package, 4 sensors are used: 2 upstream (referred to as 1/1 and 2/1) and 2 downstream (referred to as 1/2 and 2/2). With this emission package, the right upstream sensor (2/1) is located in the right exhaust downpipe just before the mini-catalytic converter. The left upstream sensor (1/1) is located in the left exhaust downpipe just before the mini-catalytic converter. The right downstream sensor (2/2) is located in the right exhaust downpipe just after the mini-catalytic converter, and before the main catalytic converter. The left downstream sensor (1/2) is located in the left exhaust downpipe just after the mini-catalytic converter, and before the main catalytic converter.

REMOVAL & INSTALLATION

1. Before servicing the vehicle, refer to the precautions section.

2. Raise and safely support the vehicle.

3. Disconnect the wire connector from oxygen sensor.

✲✲ WARNING

When disconnecting sensor electrical connector, do not pull directly on wire going into sensor.

4. Remove the sensor with an oxygen sensor removal and installation tool.

5. Clean threads in exhaust pipe using appropriate tap.

To install:

➡**Threads of new oxygen sensors are factory coated with anti-seize compound.**

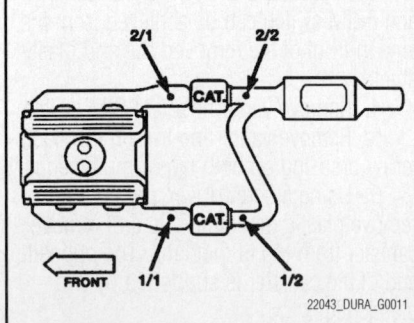

Fig. 108 Common oxygen sensor mounting points

✲✲ WARNING

Do not add any additional anti-seize compound to the threads of a new oxygen sensor.

6. Install the oxygen sensor and tighten to 22 ft. lbs. (30 Nm).

7. Connect the electrical connector.

8. Lower the vehicle.

INTAKE AIR TEMPERATURE (IAT) SENSOR

LOCATION

See Figure 109.

Refer to the accompanying illustration.

REMOVAL & INSTALLATION

1. Before servicing the vehicle, refer to the precautions section.

2. Disconnect the electrical connector form the Intake Air Temperature (IAT) sensor.

3. Clean any dirt from the air inlet tube at the sensor base.

4. Gently lift the small plastic release tab and rotate the sensor about ¼ turn counterclockwise to remove.

To install:

5. Check the condition of the sensor O-ring.

6. Clean the sensor mounting hole.

7. Position the sensor into the intake air tube and rotate clockwise until the release tab clicks into place.

8. Install the electrical connector.

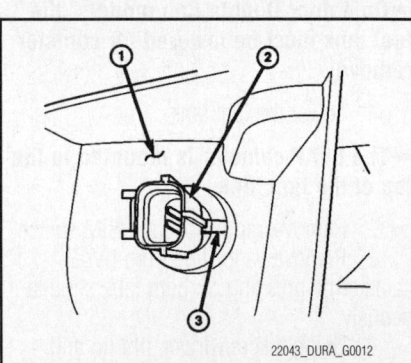

Fig. 109 The IAT sensor (2) is located in the air inlet (1). Lift the tab (3) to remove it

KNOCK SENSOR (KS)

LOCATION

See Figure 110.

Refer to the accompanying illustration.

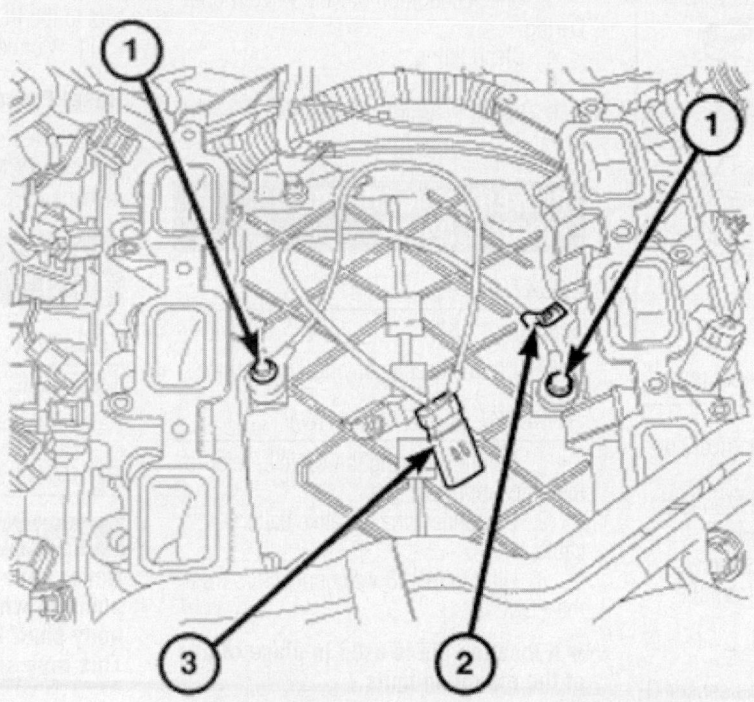

1. Knock sensor mounting bolts
2. Left sensor ID tag
3. Electrical connector

22043_DAKO_G0020

Fig. 110 Knock sensor mounting

REMOVAL & INSTALLATION

1. Before servicing the vehicle, refer to the precautions section.

2. Disconnect the knock sensor dual pigtail harness from engine wiring harness. This connection is made near rear of engine.

3. Remove the intake manifold

4. Remove the Knock Sensor (KS) mounting bolts.

5. Remove the sensors from engine.

To install:

6. Thoroughly clean the KS mounting holes.

7. Install the sensors into the engine block. Tighten the mounting bolts to 15 ft. lbs. (20 Nm).

8. Install the intake manifold.

9. Connect the KS wiring harness to the engine wiring harness at the rear of the engine.

MALFUNCTION INDICATOR LIGHT (MIL)

RESET PROCEDURE

1. Connect an OBD II scan tool to the diagnostic connector.
2. Clear DTCs.
3. The MIL should turn **OFF**.

MANIFOLD ABSOLUTE PRESSURE (MAP) SENSOR

LOCATION

See Figure 111.

Refer to the accompanying illustration.

REMOVAL & INSTALLATION

See Figure 112.

1. Before servicing the vehicle, refer to the precautions section.

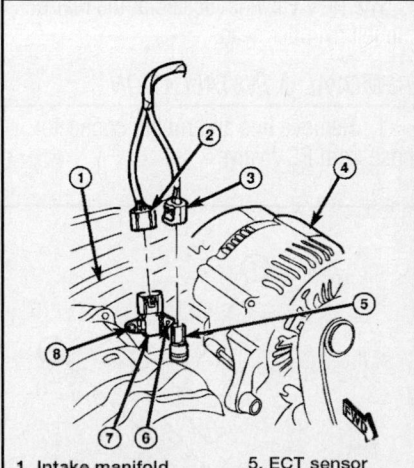

1. Intake manifold
2. Electrical connector
3. Electrical connector
4. Alternator
5. ECT sensor
6. Locating pin
7. MAP sensor
8. Mounting screw

22140_RAID_G0093

Fig. 111 MAP sensor location

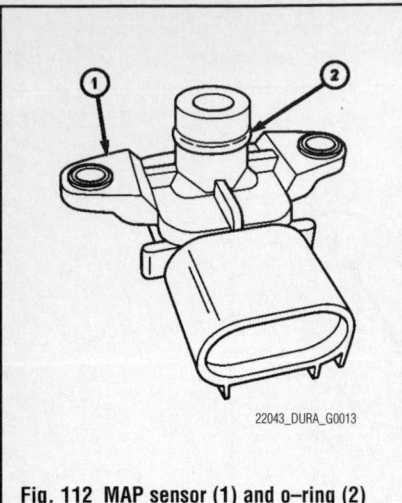

Fig. 112 MAP sensor (1) and o–ring (2)

2. Disconnect the sensor electrical connector.

3. Clean the area around the Manifold Absolute Pressure (MAP) sensor.

4. Remove the two mounting screws.

5. Remove the MAP sensor from the intake manifold.

To install:

6. Inspect the condition of the sensor O-ring and replace if necessary.

7. Position the MAP sensor into the manifold and install the two mounting screws.

8. Connect the electrical connector.

POSITIVE CRANKCASE VENTILATION (PCV) VALVE

LOCATION

See Figure 113.

The PCV valve is located at the rear of the left cylinder head.

REMOVAL & INSTALLATION

1. Remove line and rubber connector hose from PCV valve.

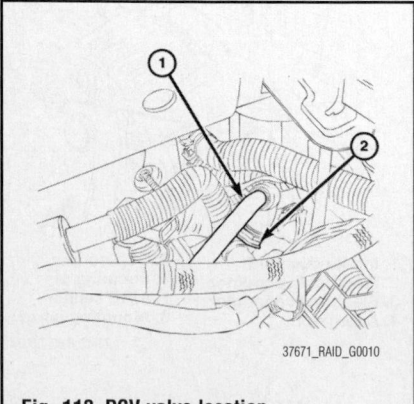

Fig. 113 PCV valve location

2. Unthread PCV valve from metal fitting.

To install:

3. Check condition of PCV valve rubber O-ring.

4. Clean fitting.

5. Install PCV valve into fitting.

6. Install PCV line and rubber connector to valve.

POWERTRAIN CONTROL MODULE (PCM)

LOCATION

See Figure 114.

Refer to the accompanying illustration.

REMOVAL & INSTALLATION

1. Before servicing the vehicle, refer to the precautions section.

2. Disconnect the negative battery cable.

3. Unplug the 38-way connectors from the PCM.

➡A locating pin is used in place of one of the mounting bolts.

4. Pry the clip from the locating pin.

5. Remove the two remaining mounting bolts.

6. Remove the PCM from the vehicle.

To install:

7. Position the PCM to the body and install the two mounting bolts.

➡Position the ground strap in place before tightening the mounting bolts.

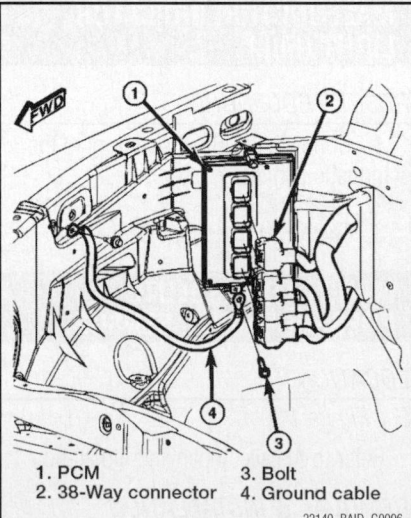

| 1. PCM | 3. Bolt |
| 2. 38-Way connector | 4. Ground cable |

Fig. 114 Exploded view of the PCM mounting

8. Install the clip to the locating pin.

9. Tighten the mounting bolts to 35 inch lbs. (4 Nm).

10. Carefully plug in the 38-way connectors to the PCM.

11. Connect the negative battery cable.

Reset Procedure

1. Use a diagnostic scan tool to reprogram the PCM with the VIN and original mileage if PCM has been replaced.

THROTTLE POSITION SENSOR (TPS)

LOCATION

See Figure 115.

Refer to the accompanying illustration.

REMOVAL & INSTALLATION

✳✳ WARNING

Never have the ignition key in the ON position when/if checking the throttle body shaft for a binding condition. This may set DTC's.

➡A (factory adjusted) set screw is used to mechanically limit the position of the throttle body throttle plate. Never attempt to adjust the engine idle speed using this screw. All idle speed functions are controlled by the Powertrain Control Module (PCM).

1. Before servicing the vehicle, refer to the precautions.

2. Disconnect the negative battery cable.

3. Remove air intake tube at throttle body flange.

4. Disconnect throttle body electrical connector.

5. Disconnect necessary vacuum lines at throttle body.

6. Remove four throttle body mounting bolts.

7. Remove throttle body from intake manifold.

8. Check condition of old throttle body-to-intake manifold o-ring.

To install:

9. Check condition of throttle body-to-intake manifold O-ring. Replace as necessary.

10. Clean mating surfaces of throttle body and intake manifold.

11. Install O-ring between throttle body and intake manifold.

12. Position throttle body (1) to intake manifold.

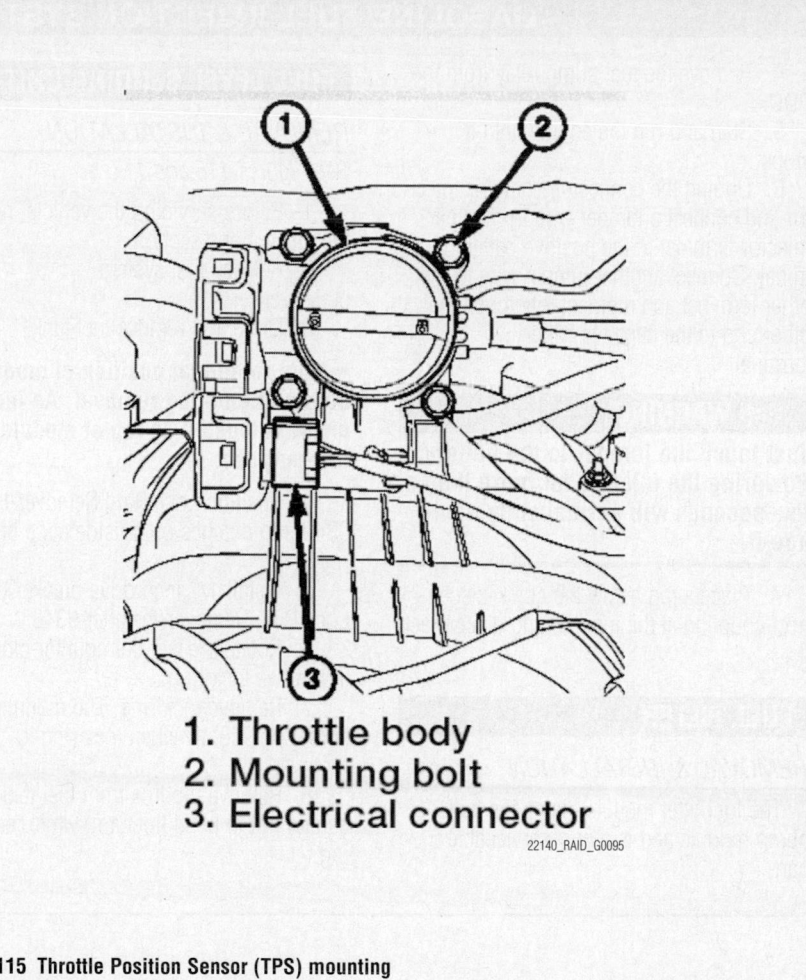

1. **Throttle body**
2. **Mounting bolt**
3. **Electrical connector**

22140_RAID_G0095

Fig. 115 Throttle Position Sensor (TPS) mounting

13. Install all throttle body mounting bolts (2) finger tight.

❋❋ WARNING

The throttle body mounting bolts MUST be torqued to specifications. DO NOT OVER TIGHTEN MOUNTING BOLTS. Over tightening can cause damage to the throttle body, throttle plate, gaskets, bolts and/or the intake manifold. Proper torque of the mounting bolts is critical to normal operation.

14. Tighten mounting bolts in a criss-cross pattern sequence to 65 inch lbs. (7.5 Nm).

15. Install electrical connector.

16. Install necessary vacuum lines.

17. Install air cleaner duct at throttle body.

18. Connect the negative battery cable.

19. Using the diagnostic scan tool, erase all previous DTC's and perform the ETC Relearn function.

VEHICLE SPEED SENSOR (VSS)

LOCATION

The Vehicle Speed Sensor (VSS) is located on the left side of the transmission case.

REMOVAL & INSTALLATION

1. Before servicing the vehicle, refer to the precautions section.

2. Raise and safely support the vehicle.

3. Place a suitable catch pan under the transmission for any fluid.

4. Remove the wiring connector from the output speed sensor.

5. Remove the mounting bolt and remove the speed sensor from the transmission case.

To install:

6. Install the speed sensor into the transmission case and tighten the bolt to 105 inch lbs. (12 Nm).

7. Install the wiring connector to the speed sensor.

8. Verify the proper transmission fluid level and refill as necessary.

9. Lower the vehicle.

FUEL SYSTEM SERVICE PRECAUTIONS

Safety is the most important factor when performing not only fuel system maintenance, but any type of maintenance. Failure to conduct maintenance and repairs in a safe manner may result in serious personal injury or death. Work on a vehicle's fuel system components can be accomplished safely and effectively by adhering to the following rules and guidelines.

• To avoid the possibility of fire and personal injury, always disconnect the negative battery cable unless the repair or test procedure requires that battery voltage be applied.

• Always relieve the fuel system pressure prior to disconnecting any fuel system component (injector, fuel rail, pressure regulator, etc.) fitting or fuel line connection. Exercise extreme caution whenever relieving fuel system pressure to avoid exposing skin, face and eyes to fuel spray. Please be advised that fuel under pressure may penetrate the skin or any part of the body that it contacts.

• Always place a shop towel or cloth around the fitting or connection prior to loosening to absorb any excess fuel due to spillage. Ensure that all fuel spillage is quickly removed from engine surfaces. Ensure that all fuel-soaked cloths or towels are deposited into a flame-proof waste container with a lid.

• Always keep a dry chemical (Class B) fire extinguisher near the work area.

• Do not allow fuel spray or fuel vapors to come into contact with a spark or open flame.

• Always use a second wrench when loosening or tightening fuel line connection fittings. This will prevent unnecessary stress and torsion on fuel piping. Always follow the proper torque specifications.

• Always replace worn fuel fitting O-rings with new ones. Do not substitute fuel hose where rigid pipe is installed.

FUEL SYSTEM PRESSURE

RELIEVING

1. Before servicing the vehicle, refer to the precautions.
2. Disconnect the negative battery cable.
3. Remove the fuel tank filler cap to release any fuel tank pressure.

4. Remove the fuel pump relay from the PDC.
5. Start and run the engine until it stops.
6. Unplug the connector from any injector and connect a jumper wire from either injector terminal to the positive battery terminal. Connect another jumper wire to the other terminal and momentarily touch the other end to the negative battery terminal.

❊❊ WARNING

Just touch the jumper to the battery. Powering the injector for more than a few seconds will permanently damage it.

7. Place a rag below the quick-disconnect coupling at the fuel rail and disconnect it.

FUEL FILTER

REMOVAL & INSTALLATION

The fuel filter mounts inside the fuel pump module and is a non-serviceable part.

FUEL LEVEL SENDING UNIT

REMOVAL & INSTALLATION

See Figures 116 and 117.

1. Before servicing the vehicle, refer to the precautions section.
2. Release fuel system pressure.
3. Drain and remove fuel tank.

➡Note rotational position of module before attempting removal. An indexing arrow is located on top of module for this purpose.

4. Position Lock Ring Remover/Installer 9340 into notches on outside edge of lock ring.
5. Install 1/2 inch drive breaker bar to Lock Ring Remover/Installer 9340.
6. Rotate breaker bar counter-clockwise to remove lock ring.
7. Remove lock ring. The module will spring up slightly when lock ring is removed.
8. Remove module from fuel tank. Be careful not to bend float arm while removing.

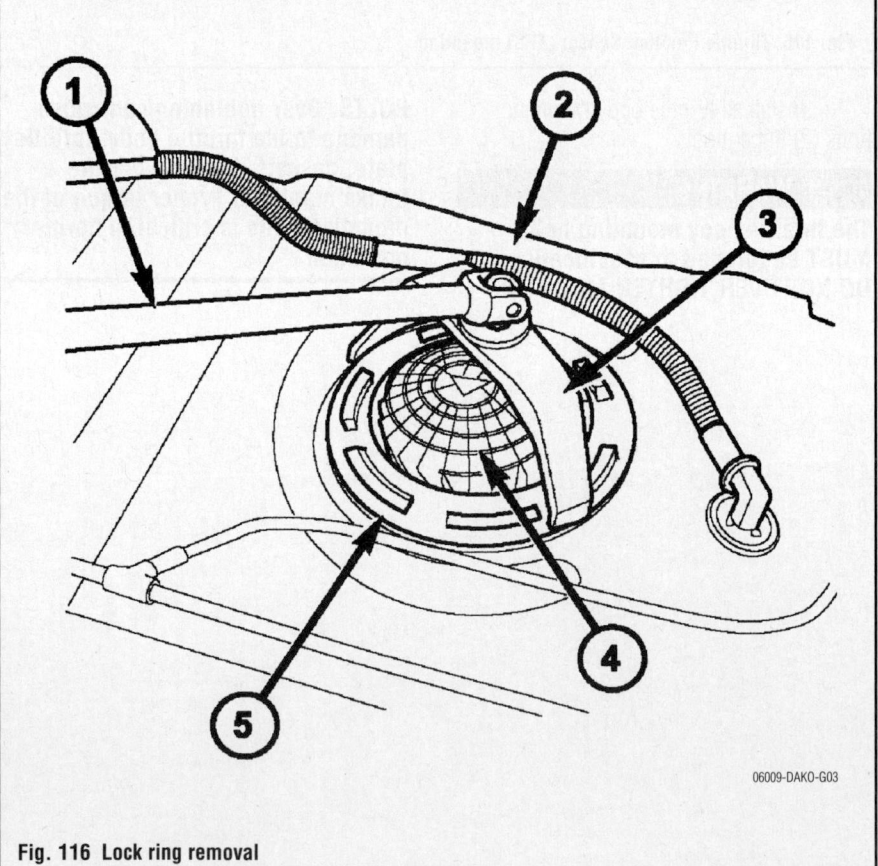

06009-DAKO-G03

Fig. 116 Lock ring removal

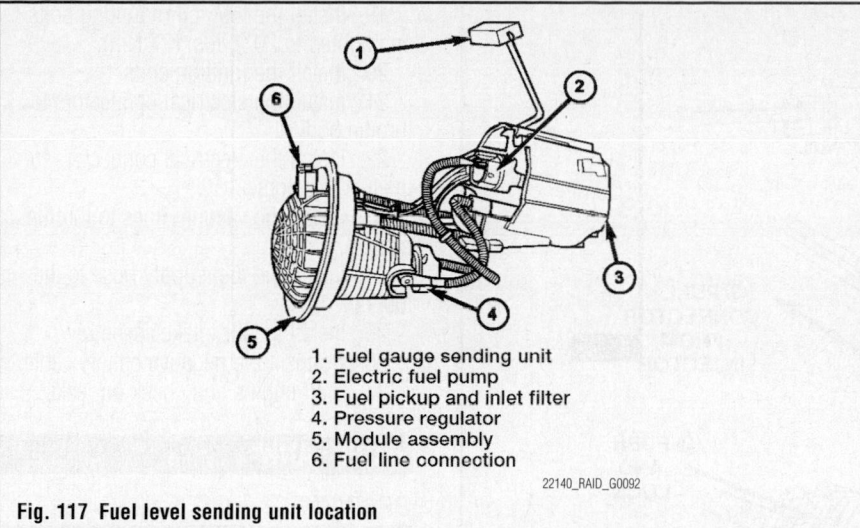

1. Fuel gauge sending unit
2. Electric fuel pump
3. Fuel pickup and inlet filter
4. Pressure regulator
5. Module assembly
6. Fuel line connection

22140_RAID_G0092

Fig. 117 Fuel level sending unit location

To install:

➡**Whenever the fuel pump module is serviced, the rubber seal (gasket) must be replaced.**

9. Using a new seal (gasket), position fuel pump module into opening in fuel tank.

10. Position lock ring over top of fuel pump module.

11. Rotate module until embossed alignment arrow points to center alignment mark. This step must be performed to prevent float from contacting side of fuel tank. Also be sure fuel fitting on top of pump module is pointed to drivers side of vehicle.

12. Install Lock Ring Remover/Installer 9340 to lock ring.

13. Install 1/2 inch drive breaker into Lock Ring Remover/Installer 9340.

14. Tighten lock ring (clockwise) until all seven notches have engaged.

15. Install fuel tank.

16. Fill fuel tank with fuel.

17. Start engine and check for fuel leaks near top of module.

FUEL PUMP MODULE

REMOVAL & INSTALLATION

See Figure 116.

1. Before servicing the vehicle, refer to the precautions section.

2. Release fuel system pressure.

3. Drain and remove fuel tank.

➡**Note rotational position of module before attempting removal. An indexing arrow is located on top of module for this purpose.**

4. Position Lock Ring Remover/Installer 9340 into notches on outside edge of lock ring.

5. Install 1/2 inch drive breaker bar to Lock Ring Remover/Installer 9340.

6. Rotate breaker bar counter-clockwise to remove lock ring.

7. Remove lock ring. The module will spring up slightly when lock ring is removed.

8. Remove module from fuel tank. Be careful not to bend float arm while removing.

To install:

➡**Whenever the fuel pump module is serviced, the rubber seal (gasket) must be replaced.**

9. Using a new seal (gasket), position fuel pump module into opening in fuel tank.

10. Position lock ring over top of fuel pump module.

11. Rotate module until embossed alignment arrow points to center alignment mark. This step must be performed to prevent float from contacting side of fuel tank. Also be sure fuel fitting on top of pump module is pointed to drivers side of vehicle.

12. Install Lock Ring Remover/Installer 9340 to lock ring.

13. Install 1/2 inch drive breaker into Lock Ring Remover/Installer 9340.

14. Tighten lock ring (clockwise) until all seven notches have engaged.

15. Install fuel tank.

16. Fill fuel tank with fuel.

17. Start engine and check for fuel leaks near top of module.

FUEL RAIL AND INJECTOR

REMOVAL & INSTALLATION

See Figures 118 and 119.

1. Before servicing the vehicle, refer to the precautions.

2. Relieve the fuel system pressure.

> ❋❋ **CAUTION**
>
> **The fuel system is under constant pressure even with engine off. Before servicing fuel rail, fuel system pressure must be released.**

> ❋❋ **CAUTION**
>
> **The left and right fuel rails are replaced as an assembly. Do not attempt to separate rail halves at connector tube. Due to design of tube, it does not use any clamps. Never attempt to install a clamping device of any kind to tube. When removing fuel rail assembly for any reason, be careful not to bend or kink tube.**

3. Remove the fuel tank filler tube cap.

4. Remove the negative battery cable.

5. Remove the air intake assembly.

6. Remove the fuel line latch clip and fuel line at fuel rail.

7. Remove the vacuum lines at throttle body.

8. Remove the fuel injector electrical connectors.

9. Remove the electrical connectors at throttle body sensors.

10. Remove the ignition coils.

11. Remove the fuel rail mounting bolts.

12. Remove the fuel rail with injectors attached.

13. Disconnect the clip that retains the fuel injector to fuel rail to remove the injector.

To install:

➡**Apply a small amount of clean engine oil to each fuel injector o-ring. This will help in fuel rail installation.**

14. Install the injectors and injector clips to fuel rail.

15. Position fuel rail/fuel injector assembly to machined injector openings in cylinder head.

16. Guide each injector into cylinder head.

> ❋❋ **CAUTION**
>
> **Be careful not to tear injector o-rings.**

17. Push right side of fuel rail down until fuel injectors have bottomed on cylinder head shoulder.

18. Push left fuel rail down until injectors have bottomed on cylinder head shoulder.

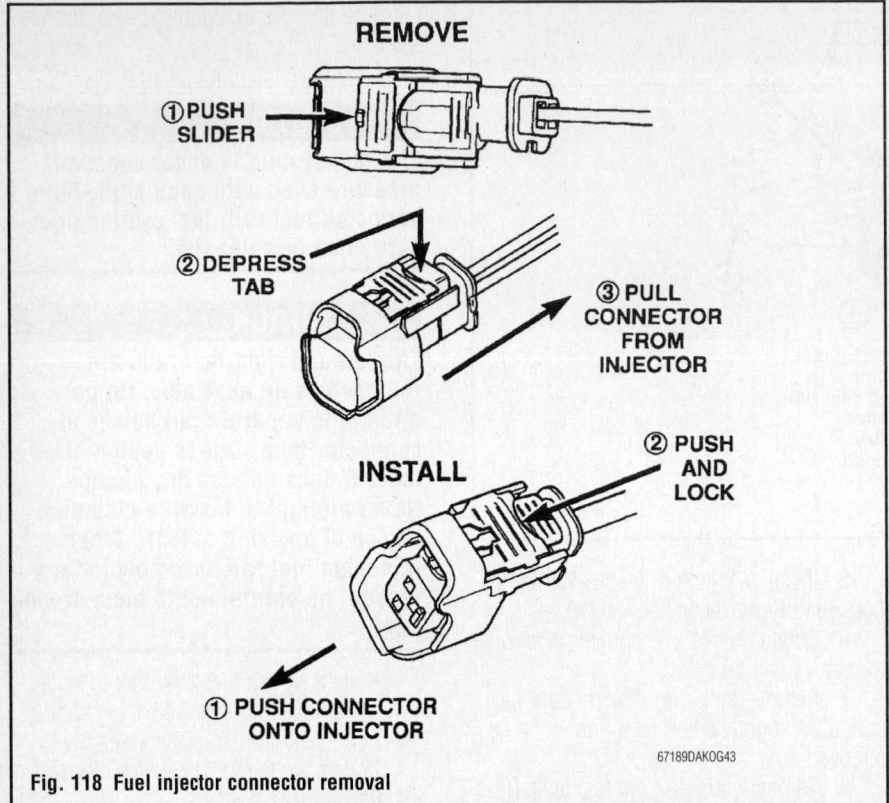

Fig. 118 Fuel injector connector removal

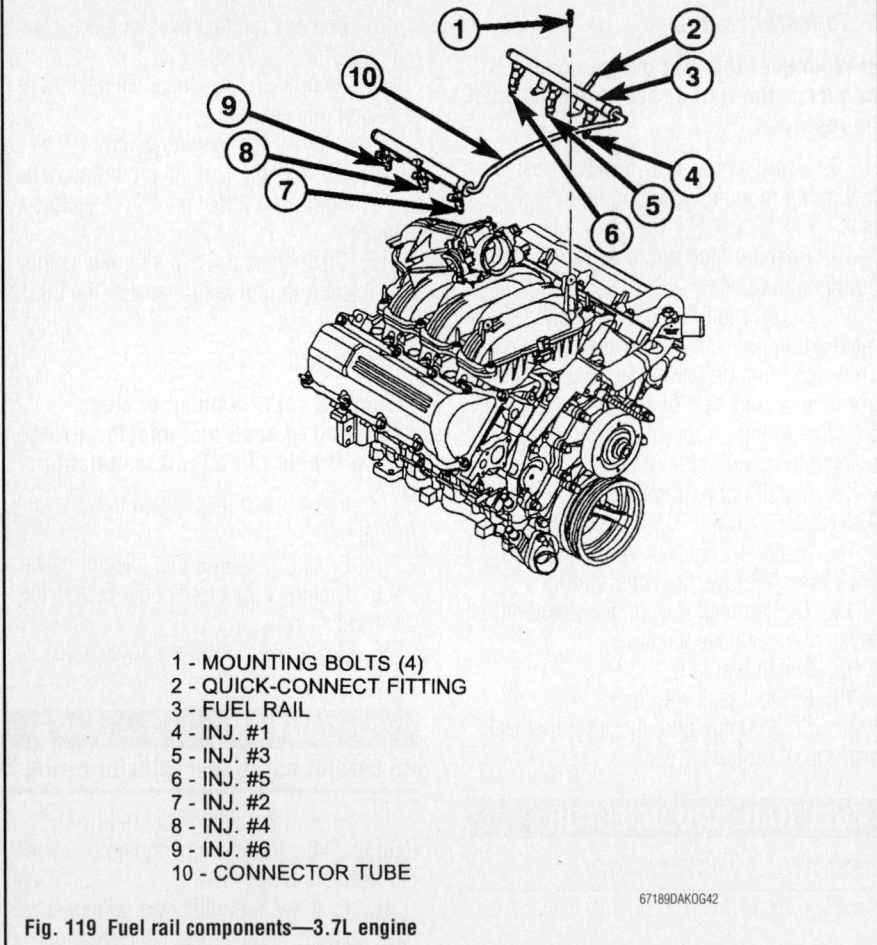

1 - MOUNTING BOLTS (4)
2 - QUICK-CONNECT FITTING
3 - FUEL RAIL
4 - INJ. #1
5 - INJ. #3
6 - INJ. #5
7 - INJ. #2
8 - INJ. #4
9 - INJ. #6
10 - CONNECTOR TUBE

Fig. 119 Fuel rail components—3.7L engine

19. Install the fuel rail mounting bolts and tighten to 20 ft. lbs. (27 Nm).

20. Install the ignition coils.

21. Install the electrical connectors to throttle body.

22. Install the electrical connectors to the fuel injectors.

23. Install the vacuum lines to throttle body.

24. Install the fuel supply hose to the fuel rail.

25. Install the air intake assembly.

26. Connect the negative battery cable.

27. Start engine and check for leaks.

FUEL TANK

DRAINING

✷✷ WARNING

The fuel system may be under constant fuel pressure even with the engine off. This pressure must be released before servicing fuel tank.

Two different procedures may be used to drain fuel tank: through the fuel fill fitting on tank, or using a diagnostic scan tool to activate the fuel pump relay.

The quickest draining procedure involves removing the rubber fuel fill hose at the fuel tank fill fitting.

As an alternative procedure, the electric fuel pump may be activated allowing tank to be drained at fuel rail connection. Refer to scan tool for fuel pump activation procedures.

Before disconnecting fuel line at fuel rail, release fuel pressure. Attach end of Fuel Pressure Test Adapter 6631 or 6539 at fuel rail disconnection (Disconnect Tool 6631 is used on 5/16" fuel lines while 6539 is used on 3/8" fuel lines). Position opposite end of this hose tool to an approved gasoline draining station. Activate fuel pump and drain tank until empty.

If electric fuel pump is not operating, fuel must be drained through fuel fill fitting at tank.

1. Release fuel system pressure.

2. Raise vehicle.

3. Remove left-rear tire/wheel.

4. Remove plastic fender liner in front of left-rear tire/wheel.

5. Thoroughly clean area around fuel fill fitting and rubber fuel fill hose at fuel tank.

6. Loosen clamps and disconnect rubber fuel fill hose at fuel tank fitting. Using an approved gas holding tank, drain fuel tank through this fitting.

REMOVAL & INSTALLATION

See Figure 120.

1. Before servicing the vehicle, refer to the precautions section.
2. Release fuel system pressure.
3. Drain fuel tank.
4. Disconnect vent line from tank.
5. Remove clamp and disconnect fill hose at fuel fill tube.
6. If equipped, remove fuel tank skid plate.
7. Disconnect electrical connector at ESIM switch.
8. Disconnect ESIM, ORVR and EVAP lines at front of tank.
9. Support tank with a hydraulic jack.
10. Remove two fuel tank strap nuts and remove both tank support straps.
11. Carefully lower tank a few inches and disconnect fuel pump module electrical connector at top of tank. To disconnect electrical connector: Push upward on red colored tab to unlock. Push on black colored tab while removing connector.
12. Disconnect fuel line at fuel pump module fitting by pressing on tabs at side of quick-connect fitting.
13. Continue to lower tank for removal.
14. If fuel tank is to be replaced, remove fuel pump module from tank.

To install:

15. If fuel tank is to be replaced, install fuel pump module into tank.
16. Position fuel tank to hydraulic jack.
17. Raise tank until positioned near body.

18. Connect fuel pump module electrical connector at top of tank
19. Connect fuel line quick-connect fitting to pump module.
20. Continue raising tank until positioned snug to body.
21. Install and position both tank support straps. Install two fuel tank strap nuts and tighten.
22. Connect EVAP, ORVR and ESIM lines at front of tank.
23. Connect electrical connector to ESIM switch.
24. Connect vent line.
25. Connect rubber fill hose to fuel tank fitting and tighten hose clamps.
26. The vapor/vacuum lines and hoses must be firmly connected. Also check the vapor/vacuum lines at the ESIM switch, filter and EVAP canister purge solenoid for damage or leaks. If a leak is present, a Diagnostic Trouble Code (DTC) may be set.
27. If equipped, install fuel tank skid plate.
28. Install plastic liner in front of left-rear tire/wheel.
29. Install left-rear tire/wheel.
30. Lower vehicle.
31. Fill fuel tank with fuel.
32. Start engine and check for fuel leaks near top of module.

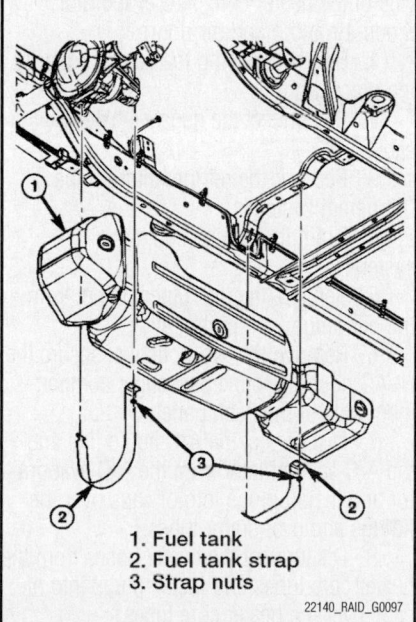

1. Fuel tank
2. Fuel tank strap
3. Strap nuts

22140_RAID_G0097

Fig. 120 Exploded view of the fuel tank assembly

IDLE SPEED

ADJUSTMENT

Idle speed is maintained by the Powertrain Control Module (PCM). No adjustment is necessary or possible.

THROTTLE BODY

REMOVAL & INSTALLATION

See Figure 121.

✳✳ WARNING

Never have the ignition key in the ON position when/if checking the throttle body shaft for a binding condition. This may set DTC's.

➡**A (factory adjusted) set screw is used to mechanically limit the position of the throttle body throttle plate. Never attempt to adjust the engine idle speed using this screw. All idle speed functions are controlled by the Powertrain Control Module (PCM).**

1. Before servicing the vehicle, refer to the precautions.
2. Disconnect the negative battery cable.

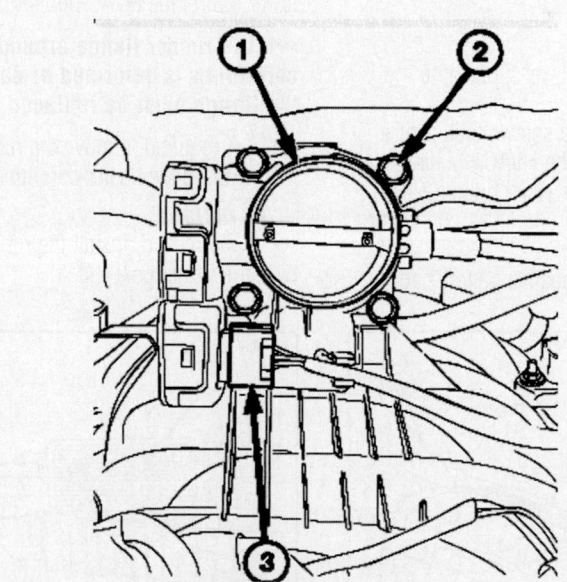

1. Throttle body
2. Mounting bolt
3. Electrical connector

22140_RAID_G0095

Fig. 121 Throttle Position Sensor (TPS) mounting

3. Remove air intake tube at throttle body flange.

4. Disconnect throttle body electrical connector.

5. Disconnect necessary vacuum lines at throttle body.

6. Remove four throttle body mounting bolts.

7. Remove throttle body from intake manifold.

8. Check condition of old throttle body-to-intake manifold o-ring.

To install:

9. Check condition of throttle body-to-intake manifold O-ring. Replace as necessary.

10. Clean mating surfaces of throttle body and intake manifold.

11. Install O-ring between throttle body and intake manifold.

12. Position throttle body (1) to intake manifold.

13. Install all throttle body mounting bolts (2) finger tight.

❋❋ WARNING

The throttle body mounting bolts MUST be torqued to specifications. DO NOT OVER TIGHTEN MOUNTING BOLTS. Over tightening can cause damage to the throttle body, throttle plate, gaskets, bolts and/or the intake manifold. Proper torque of the mounting bolts is critical to normal operation.

14. Tighten mounting bolts in a criss-cross pattern sequence to 65 inch lbs. (7.5 Nm).

15. Install electrical connector.

16. Install necessary vacuum lines.

17. Install air cleaner duct at throttle body.

18. Connect the negative battery cable.

19. Using the diagnostic scan tool, erase all previous DTC's and perform the ETC Relearn function.

HEATING & AIR CONDITIONING SYSTEM

BLOWER MOTOR

REMOVAL & INSTALLATION

See Figure 122.

1. Before servicing the vehicle, refer to the precautions.

2. Disconnect the negative battery cable.

3. Disconnect the wire harness connector (1) from the blower motor (2).

4. Remove the three screws (3) that secure the blower motor to the HVAC housing (4).

5. Remove the blower motor from the HVAC housing.

To install:

6. Position the blower motor into the HVAC housing.

7. Install the three screws that secure the blower motor to the HVAC housing. Tighten the screws to 17 inch lbs. (2 Nm).

8. Connect the wire harness connector to the blower motor.

9. Connect the negative battery cable.

HEATER CORE

REMOVAL & INSTALLATION

See Figure 123.

1. Before servicing the vehicle, refer to the precautions.

2. Remove the HVAC housing and place it on a workbench.

➡**If the foam seal around the heater core is deformed or damaged, the seal must be replaced.**

3. Carefully lift the heater core out of the lower half of the HVAC housing.

➡**If the rubber flange around the heater core tubes is deformed or damaged, the flange must be replaced.**

4. If required, remove the rubber flange from around the heater core tubes.

To install:

5. If removed, install the rubber flange over the heater core tubes.

➡**Make sure that the foam seal and the rubber flange are properly positioned in the HVAC housing.**

6. Carefully install the heater core into the lower half of the HVAC housing.

7. Assemble the HVAC housing.

➡**If the heater core is being replaced, flush the cooling system.**

8. Install the HVAC housing.

HVAC HOUSING

REMOVAL & INSTALLATION

See Figures 124 and 125.

The HVAC housing must be removed from the vehicle and disassembled for service of the heater core, A/C evaporator, mode-air and blend-air doors.

1. Before servicing the vehicle, refer to the precautions.

2. Disconnect the negative battery cable.

3. Recover the refrigerant from the refrigerant system.

4. Drain the engine cooling system.

5. Remove the instrument panel from the passenger compartment.

6. Remove the two bolts that secure the HVAC housing to the passenger compartment side of the dash panel.

7. Disconnect the A/C liquid line and the A/C accumulator from the A/C evaporator and install plugs into or caps over the fittings and evaporator tubes.

8. Disconnect the heater hoses from the heater core tubes and install plugs into or caps over the heater core tubes.

9. Remove the three nuts that secure the HVAC housing and the fresh air inlet

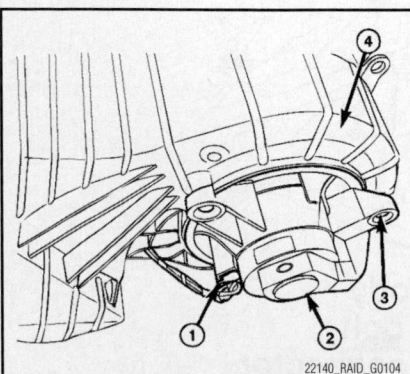

Fig. 122 Disconnect the wire harness connector (1) from the blower motor (2). Remove the three screws (3) that secure the blower motor to the HVAC housing (4)

22140_RAID_G0104

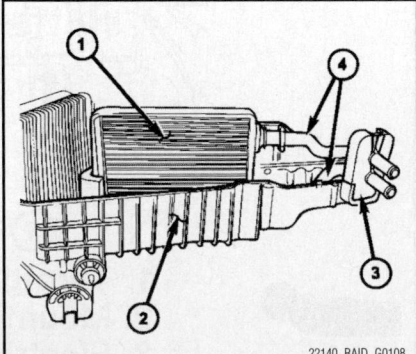

Fig. 123 Heater core assembly (1) inside the HVAC housing (2). If required, remove the rubber flange (3) from around the heater core tubes (4)

22140_RAID_G0108

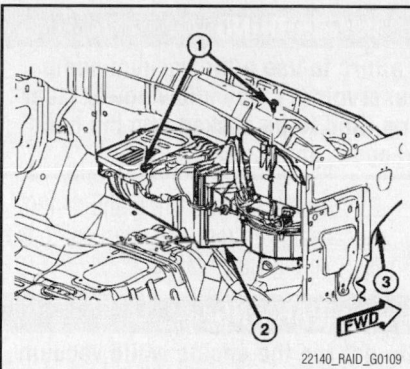

Fig. 124 Remove the two bolts (1) that secure the HVAC housing (2) to the passenger compartment side of the dash panel (3)

screen to the engine compartment side of the dash panel.

10. Pull the HVAC housing assembly rearward so that the mounting studs and condensate drain tube clear the dash panel and remove the HVAC housing from the passenger compartment.

11. If required, remove the fresh air inlet screen from the dash panel.

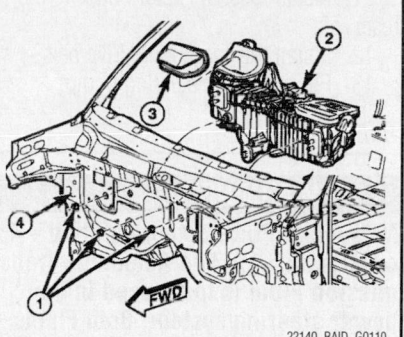

Fig. 125 Remove the three nuts (1) that secure the HVAC housing (2) and the fresh air inlet screen (3) to the engine compartment side of the dash panel (4)

To install:

12. If removed, install the fresh air screen onto the dash panel.

13. Position the HVAC housing into the passenger compartment with the mounting studs and the condensate drain tube in their proper locations in the dash panel.

14. Loosely install the three nuts that

secure the HVAC housing to the engine compartment side of the dash panel.

15. Loosely install the two bolts that secure the HVAC housing to the passenger compartment side of the dash panel.

16. Tighten the three nuts that secure the HVAC housing to the engine compartment side of the dash panel to 62 inch lbs. (7 Nm).

17. Tighten the two bolts that secure the HVAC housing to the passenger compartment side of the dash panel to 26 inch lbs. (3 Nm).

18. Install the instrument panel.

19. Remove the previously installed plugs or caps and connect the heater hoses to the heater core tubes.

20. Remove the previously installed plugs or caps and connect the A/C liquid line and the A/C accumulator to the A/C evaporator.

21. Connect the negative battery cable.

22. If the heater core is being replaced, flush the cooling system.

23. Refill the engine cooling system.

24. Evacuate and charge the refrigerant system.

25. Initiate the Actuator Calibration function using a scan tool.

STEERING

POWER RACK & PINION STEERING GEAR

REMOVAL & INSTALLATION

See Figure 126.

1. Before servicing the vehicle, refer to the precautions.

2. Siphon out as much power steering fluid as possible from the pump.

3. Lock the steering wheel.

4. Remove the front wheels.

5. Remove the nuts from the tie rod ends.

6. Disconnect the tie rod ends from the knuckles.

7. Remove the steering gear pinch bolt.

8. Remove the lower steering coupling from the steering gear.

9. Turn the steering gear to the full right position.

➡**Protect the end of hoses to prevent contamination to the system and damage to the O-rings.**

10. Remove the power steering lines from the gear.

11. Remove the steering gear mounting bolts and nuts.

12. Tip the steering gear assembly

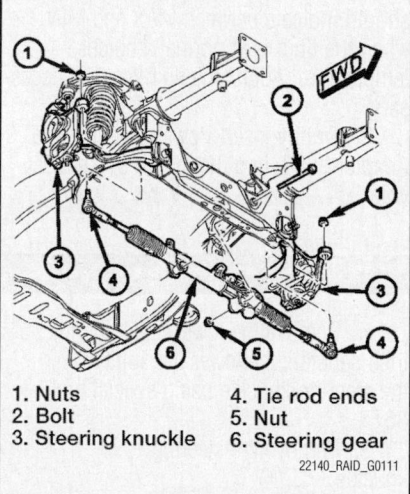

1. Nuts	4. Tie rod ends
2. Bolt	5. Nut
3. Steering knuckle	6. Steering gear

Fig. 126 Power steering gear assembly

forward to allow clearance and move to the right then tip the gear downward on the left side to remove from the vehicle.

To install:

➡**Before installing gear inspect bushings and replace if worn or damaged.**

13. Install the steering gear assembly to the vehicle and tighten mounting nuts and bolts to 190 ft. lbs. (258 Nm).

14. Install the power steering lines to the steering gear. Tighten the pressure hose to 23 ft. lbs. (31 Nm) and tighten the return hose to 27 ft. lbs. (37 Nm).

15. Slide the shaft coupler onto the steering gear. Install a new bolt and tighten to 42 ft. lbs. (57 Nm).

16. Clean the tie rod end studs and knuckle tapers.

17. Install the tie rod ends into the steering knuckles and tighten the nuts to 60 ft. lbs. (81 Nm).

18. Install the front wheels.

19. Refill the power steering system with fluid to the correct level.

20. Check the wheel alignment and adjust, as necessary.

POWER STEERING PUMP

REMOVAL & INSTALLATION

See Figure 127.

1. Before servicing the vehicle, refer to the precautions.

2. Drain and siphon the power steering fluid from the pump.

3. Remove the serpentine drive belt.

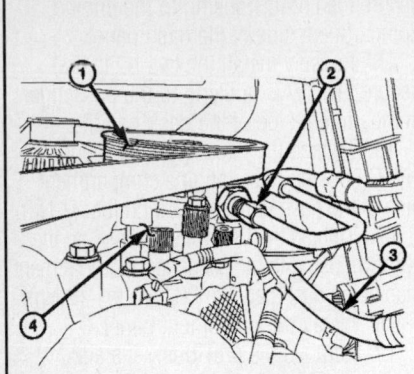

1. Access hole in pulley
2. Pressure hose connection
3. Suction hose connection
4. Power steering pump

22140_RAID_G0112

Fig. 127 Power steering pump connections

4. Remove the reservoir return hose at the reservoir.

5. Remove the pressure hose from the pump.

6. Remove 3 pump mounting bolts through pulley access holes.

7. Remove the pump from the engine.

To install:

8. Align the pump with the mounting holes on the engine.

9. Install 3 pump mounting bolts through the pulley access holes. Tighten the bolts to 21 ft. lbs. (28 Nm).

10. Install the pressure hose to the pump. Tighten the tube nut to 23 ft. lbs. (31 Nm).

11. Install reservoir return hose to the reservoir.

12. Install the serpentine drive belt.

13. Fill the power steering pump.

BLEEDING

❊❊ WARNING

Mitsubishi Power Steering Fluid + 4 or Mitsubishi ATF+4 Automatic Transmission Fluid is to be used in the power steering system. Both Fluids have the same material standard specifications (MS-9602). No other power steering or automatic transmission fluid is to be used in the system. Damage may result to the power steering pump and system if another fluid is used. Do not overfill the system.

❊❊ WARNING

If the air is not purged from the power steering system correctly, pump failure could result.

➡**Be sure the vacuum tool used in the following procedure is clean and free of any fluids.**

1. Check the fluid level. As measured on the side of the reservoir, the level should indicate between MAX and MIN when the fluid is at normal ambient temperature. Adjust the fluid level as necessary.

2. Tightly insert Power Steering Cap Adapter 9688, into the mouth of the reservoir.

❊❊ WARNING

Failure to use a the vacuum pump reservoir (1) may allow power steering fluid to be sucked into the hand vacuum pump.

3. Attach Hand Vacuum Pump C-4207 or equivalent, with reservoir attached, to the Power Steering Cap Adapter.

❊❊ WARNING

Do not run the engine while vacuum is applied to the power steering system. Damage to the power steering pump can occur.

➡**When performing the following step make sure the vacuum level is maintained during the entire time period.**

4. Using Hand Vacuum Pump, apply 20-25 in. Hg (68-85 kPa) of vacuum to the system for a minimum of three minutes.

5. Slowly release the vacuum and remove the special tools.

6. Adjust the fluid level as necessary.

7. Repeat steps until the fluid no longer drops when vacuum is applied.

8. Start the engine and cycle the steering wheel lock-to-lock three times.

❊❊ WARNING

Do not hold the steering wheel at the stops.

9. Stop the engine and check for leaks at all connections.

10. Check for any signs of air in the reservoir and check the fluid level. If air is present, repeat the procedure as necessary.

SUSPENSION

LOWER BALL JOINT

REMOVAL & INSTALLATION

See Figure 128.

1. Before servicing the vehicle, refer to the precautions.

2. Remove the front wheel.

3. Remove the brake caliper and rotor.

4. Remove the outer tie rod retaining nut from the knuckle.

5. Separate the tie rod from the steering knuckle using Special Tool C-3894-A.

6. Remove the upper ball joint nut, then separate the upper ball joint from the knuckle using special tool 8677, or equivalent.

7. Remove the lower ball joint nut, then separate the lower ball joint from the steering knuckle using special tool 8677.

8. Remove the steering knuckle.

9. If equipped with 4WD, move the halfshaft to the side and support the halfshaft out of the way.

10. Remove the snapring from the ball joint flange.

➡**Extreme pressure lubrication must be used on the threaded portions of the tool. This will increase the longevity of the tool and insure proper operation during the removal and installation process.**

FRONT SUSPENSION

11. Press the ball joint (3) from the lower control arm (4) using special tools C-4212-F (Press) (1), 8445-3 (Driver) (2) and 9604 (Receiver) (5).

To install:

➡**Extreme pressure lubrication must be used on the threaded portions of the tool. This will increase the longevity of the tool and insure proper operation during the removal and installation process.**

12. Install the ball joint (3) into the control arm and press in using special tools C-4212-F (press) (1), 8441-4 (Receiver) (2) and 9654-1 (Driver) (4).

13. Install the snapring around the ball joint flange.

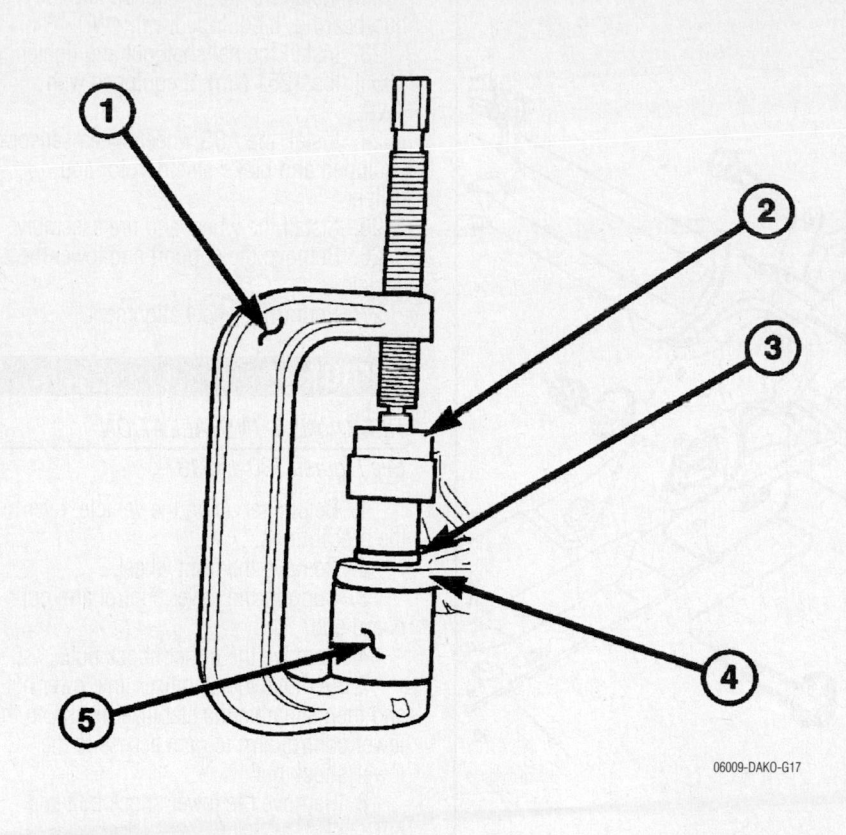

Fig. 128 Install the ball joint (3) into the control arm and press in using special tools C-4212-F (1), 8441-4 (2) and 9654-1 (4).

06009-DAKO-G17

14. Remove the support for the halfshaft and install into position, 4WD models only.

15. Install the steering knuckle.

16. Install the tie rod end into the steering knuckle, then install the retaining nut and tighten to 55 ft. lbs. (75 Nm).

17. Install and tighten the halfshaft nut (if equipped) to 185 ft. lbs. (251 Nm).

18. Install the brake caliper and rotor.

19. Install the front wheel.

20. Check the vehicle ride height.

21. Check the wheel alignment and adjust, as necessary.

LOWER CONTROL ARM

REMOVAL & INSTALLATION

See Figure 129.

1. Before servicing the vehicle, refer to the precautions.

2. Remove the front wheel.

3. Remove the brake caliper assembly and rotor.

4. Disconnect the wheel speed sensor at the wheel well.

5. Remove tie rod end jam nut.

6. Disconnect the tie rod from the knuckle using special tool C-3894-A, or equivalent.

7. Remove the front halfshaft nut, if equipped with 4WD.

8. Remove the upper ball joint nut. Separate the upper ball joint from the steering knuckle with a remover.

9. Remove the lower ball joint nut. Separate the lower ball joint from the steering knuckle with a remover.

10. Remove or disconnect the following:
   - Steering knuckle
   - Stabilizer bar link
   - Shock absorber lower bolt and nut
   - Lower control arm bolts, nuts and washers
   - Lower control arm

   To install:

➡**All suspension components should be tightened with the weight of the vehicle on them (curb height).**

11. Position the lower control arm at the frame rail brackets. Install the pivot bolts washers and nuts. Tighten the nuts finger-tight.

The ball joint stud taper must be CLEAN and DRY before installing the knuckle. Clean the stud taper with mineral spirits to remove dirt and grease.

12. Install the steering knuckle.

13. Insert the lower ball joint into the steering knuckle. Install and tighten the retaining nut to 95 ft. lbs. (129 Nm).

14. Install shock absorber lower bolt and nut. Tighten to 155 ft. lbs. (210 Nm).

15. Install the front halfshaft nut, if equipped with 4WD.

16. Insert the upper ball joint into the steering knuckle. Install and tighten the retaining nut to 70 ft. lbs. (95 Nm).

17. Install the stabilizer bar link and tighten to 75 ft. lbs. (102 Nm).

18. Tighten the lower control arm pivot nut and bolts to 105 ft. lbs. (142 Nm).

19. Insert the outer tie rod end into the steering knuckle. Install and tighten the retaining nut to 55 ft. lbs. (75 Nm).

20. Install the brake caliper and rotor.

21. Install the front wheel.

22. Check the wheel alignment and adjust, as necessary.

TESTING

1. Check the rubber parts for cracks and wear.

2. Check the shock absorber for malfunctions, oil leakage, or abnormal noise.

3. If shock absorber does not perform properly, replace the shock absorber.

STEERING KNUCKLE

REMOVAL & INSTALLATION

See Figure 129.

1. Before servicing the vehicle, refer to the precautions.

2. Raise and support the vehicle.

3. Remove the wheel and tire assembly.

4. Remove the brake caliper, rotor, shield and ABS wheel speed sensor.

5. Remove the front halfshaft nut, if equipped with 4WD.

6. Remove the tie rod end nut.

7. Separate the tie rod from the knuckle with puller C-3894-A.

When installing puller 8677 to separate the ball joint, be careful not to damage the ball joint seal.

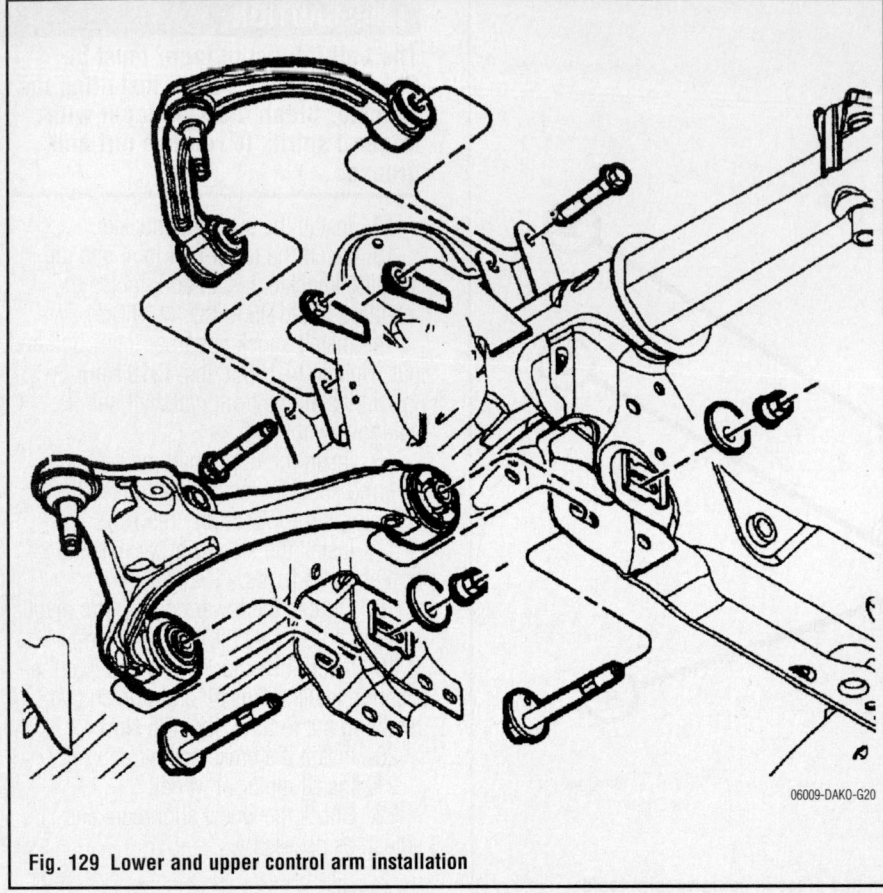

Fig. 129 Lower and upper control arm installation

8. Remove the upper ball joint nut.

9. Separate the ball joint from the knuckle with puller 867.

10. Install an hydraulic jack to support the lower control arm.

11. Remove the lower ball joint nut.

12. Separate the ball joint from the knuckle with puller 8677 and remove the knuckle.

13. Remove the hub/bearing bolts from the knuckle.

14. Remove the hub/bearing from the steering knuckle.

15. Remove the steering knuckle.

To install:

➡The ball joint stud tapers must be clean and dry before installing the knuckle. Clean the stud tapers with mineral spirits to remove dirt and grease.

16. .Install the hub/bearing to the steering knuckle and tighten the bolts to 120 ft. lbs. (163 Nm).

17. Install the knuckle onto the upper and lower ball joints.

18. Install the upper ball joint nut. Tighten the nut to 70 ft. lbs. (95 Nm).

19. Install the lower ball joint nut. Tighten the nut to 95 ft. lbs. (129 Nm)

20. Remove the hydraulic jack from the lower control arm.

21. Install the tie rod end and tighten the nut to 80 ft. lbs. (108 Nm).

22. Install the front halfshaft into the hub/bearing, if equipped with 4WD.

23. Install the halfshaft nut and tighten to 185 ft. lbs (251 Nm), if equipped with 4WD.

24. Install the ABS wheel speed sensor if equipped and brake shield, rotor and caliper.

25. Install the wheel and tire assembly.

26. Remove the support and lower the vehicle.

27. Perform a wheel alignment.

STRUT

REMOVAL & INSTALLATION

See Figures 130 and 131.

1. Before servicing the vehicle, refer to the precautions.

2. Remove the front wheel.

3. Support the lower control arm outboard end.

4. Remove the upper shock nuts.

5. Remove the stabilizer link lower nut and then separate the stabilizer link from the lower control arm to gain access to the lower shock nut.

6. Remove the lower shock bolt and nut.

7. Remove the shock.

To install:

➡All suspension components should be tightened with the weight of the vehicle on them (curb height).

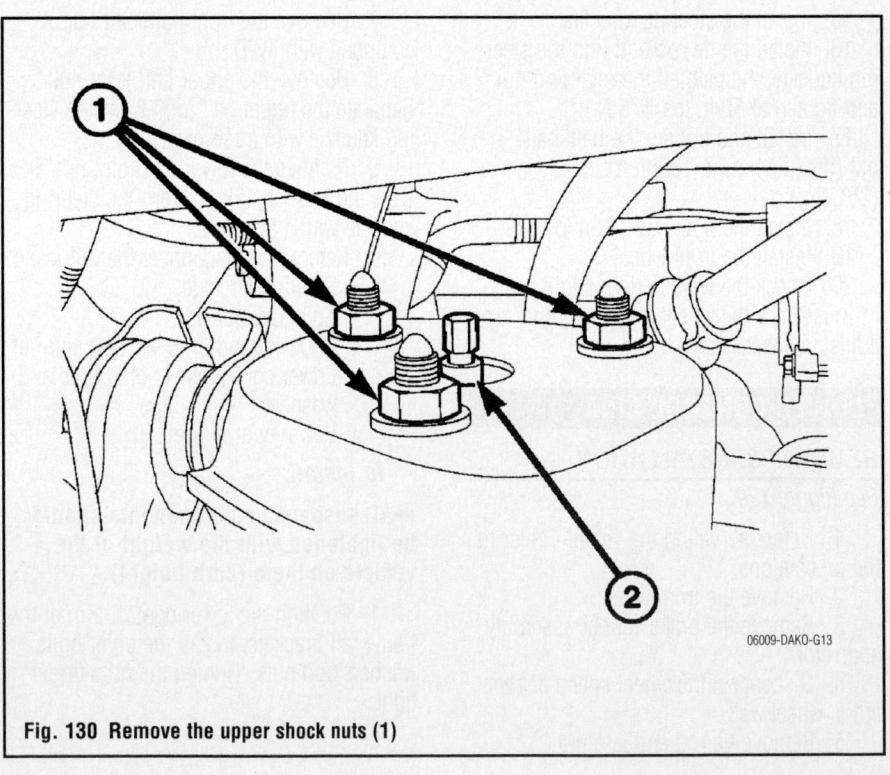

Fig. 130 Remove the upper shock nuts (1)

on the shock shaft retaining nut. Insert an 8 mm socket though Wrench onto hex located on end of shock shaft. While holding shock shaft from turning, remove nut from shock shaft using Wrench.

6. Remove the upper shock nut.

7. Remove the shock.

8. Remove the shock upper mounting plate.

9. Remove and inspect the upper and lower spring isolators.

To install:

10. Install the lower isolator.

11. Install the upper isolator.

12. Position the shock into the coil spring.

13. Install the upper shock mounting plate.

14. Install Wrench (on end of a tighten wrench), Special Tool 9362, on shock shaft retaining nut. Next, insert 8 mm socket though Wrench onto hex located on end of shock shaft. While holding shock shaft from turning, tighten nut using Wrench to 33 ft. lbs. (45 Nm).

15. Install the shock upper mounting nut.

16. Decompress the spring.

17. Remove the shock assembly from the spring compressor tool.

18. Install the shock assembly.

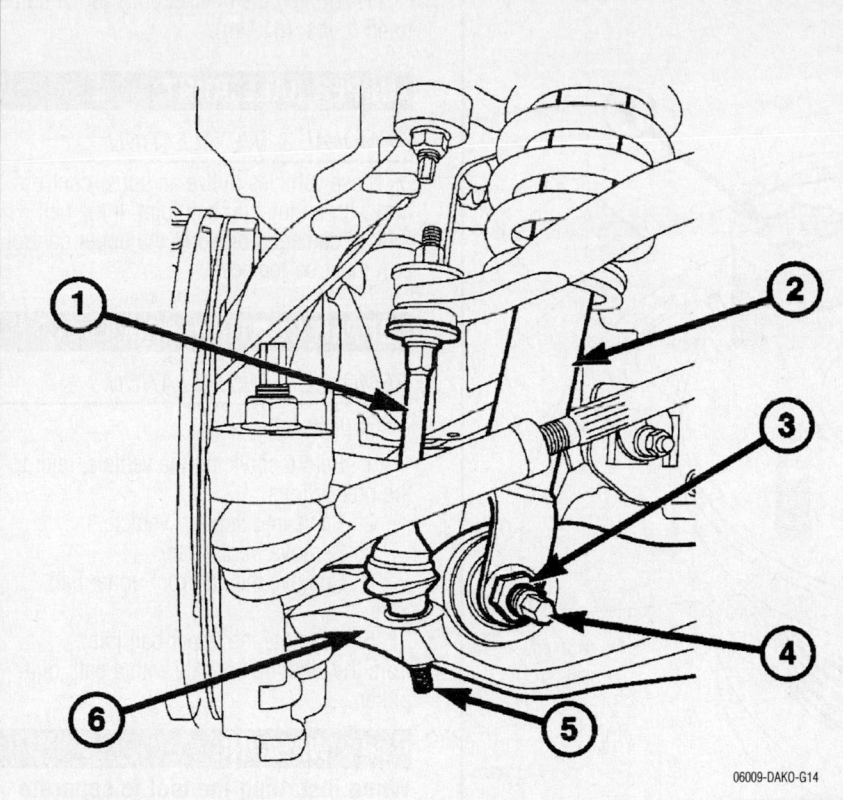

Fig. 131 Stabilizer link (1), strut (2), nut (3), lower strut bolt (4), link lower nut (5), lower control arm (6)

8. Install the upper part of the shock into the frame bracket.

9. Install the upper nuts. Tighten to 45 ft. lbs. (61 Nm).

10. Install the lower part of the shock into the lower control arm shock bushing.

11. Position shock module clevis to lower control arm. Install bolt so head of bolt is facing rear of vehicle and hand start nut. Tighten the bolt and nut to 155 ft. lbs. (210 Nm).

12. Install the stabilizer link lower nut to the lower control arm. Tighten to 75 ft. lbs. (102 Nm).

13. Remove the support from the lower control arm outboard end.

14. Install the front wheel.

OVERHAUL

See Figures 132 and 133.

1. Before servicing the vehicle, refer to the precautions.

2. Remove the shock assembly.

3. Install the shock assembly in the Branick 7200T spring removal/installation tool or equivalent press.

4. Compress the spring.

5. Position Special Tool 9362 Wrench

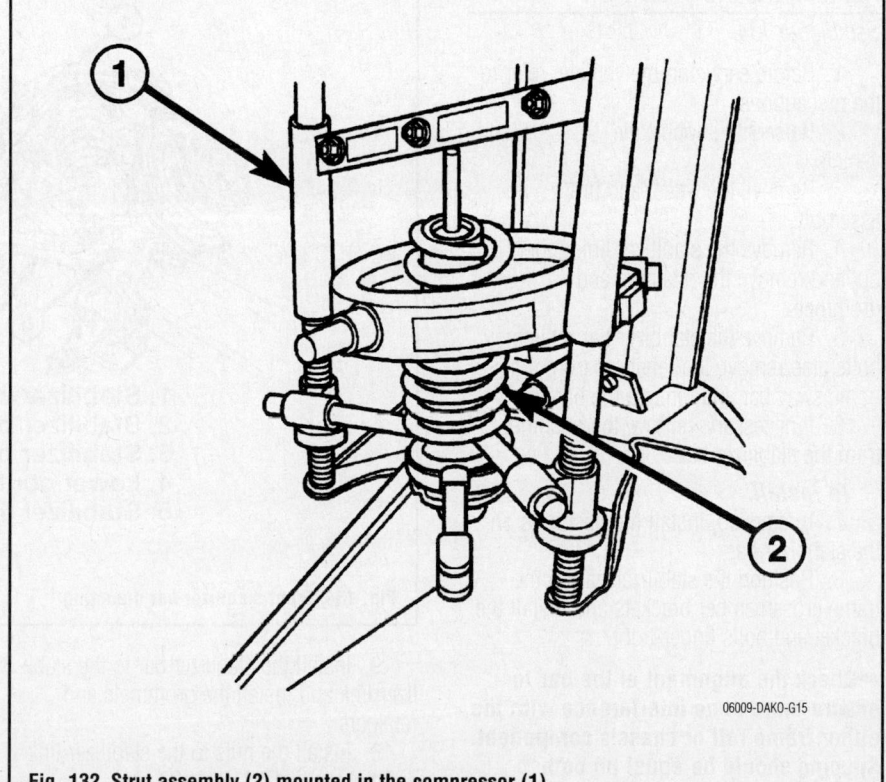

Fig. 132 Strut assembly (2) mounted in the compressor (1)

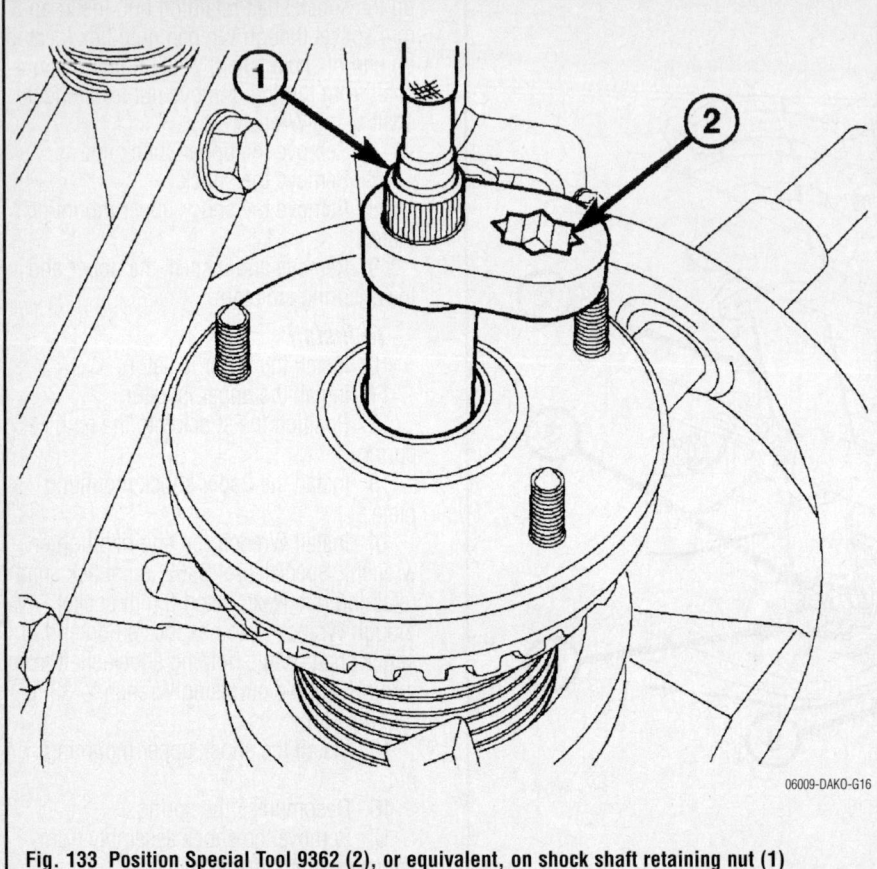

Fig. 133 Position Special Tool 9362 (2), or equivalent, on shock shaft retaining nut (1)

STABILIZER BAR

REMOVAL & INSTALLATION

See Figure 134.

1. Before servicing the vehicle, refer to the precautions.
2. Raise and support the vehicle.
3. Remove the wheel and tire assembly.
4. Remove the stabilizer link upper nut and remove the retainers and grommets.
5. Remove the stabilizer bar retainer bolts also remove the retainers from the frame sway bar and remove the bar.
6. If necessary, remove the bushings from the stabilizer bar.

To install:

7. If removed, install the bushings on the stabilizer bar.
8. Position the stabilizer bar on the frame crossmember brackets and install the bracket and bolts finger-tight.

➡ **Check the alignment of the bar to ensure there is no interference with the either frame rail or chassis component. Spacing should be equal on both sides.**

11. Tighten the bracket bolts to the frame to 45 ft. lbs. (61 Nm).

UPPER BALL JOINT

REMOVAL & INSTALLATION

These vehicles utilize an upper control arm with an integral ball joint. If the ball joint is damaged or worn, the upper control arm must be replaced.

UPPER CONTROL ARM

REMOVAL & INSTALLATION

See Figure 135.

1. Before servicing the vehicle, refer to the precautions.
2. Raise and support vehicle.
3. Remove front wheel.
4. Remove the nut from upper ball joint.
5. Separate the upper ball joint from the steering knuckle with a ball joint puller.

❊❊ WARNING

When installing the tool to separate the ball joint, be careful not to damage the ball joint seal.

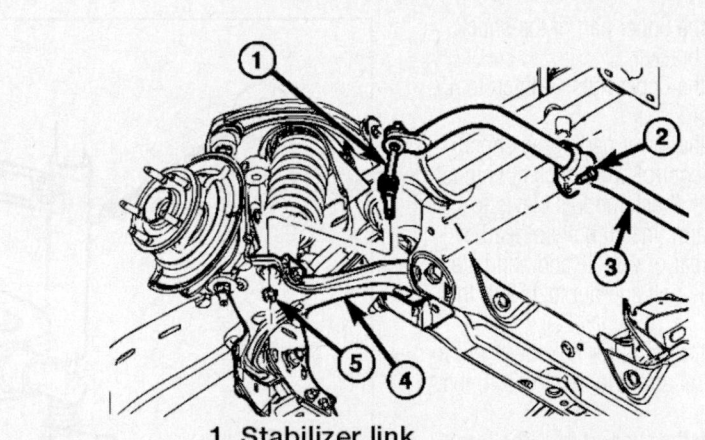

1. Stabilizer link
2. Stabilizer bar retaining bolts
3. Stabilizer bar
4. Lower control arm
5. Stabilizer link nut

Fig. 134 Front stabilizer bar mounting

9. Install the stabilizer bar to the stabilizer link and install the grommets and retainers.
10. Install the nuts to the stabilizer link and tighten to 17 ft. lbs. (23 Nm).

6. Remove the wheel speed sensor wire from the retaining brackets to the upper control arm.
7. Remove the control arm pivot bolts and nuts and remove control arm.

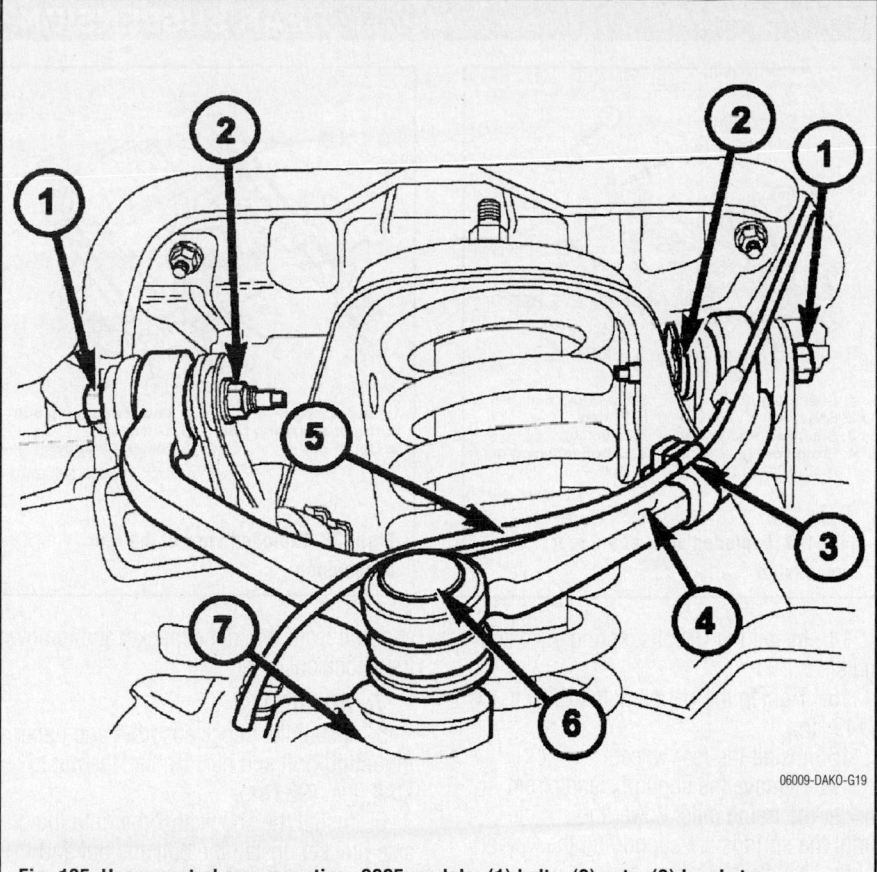

Fig. 135 Upper control arm mounting, 2005 models. (1) bolts, (2) nuts, (3) brackets, (4) control arm, (5) ABS wheel speed wire, (6) ball joint, (7) knuckle

To install:

➡All suspension components should be tightened with the weight of the vehicle on them (curb height).

8. Position the control arm into the frame brackets. Install bolts and nuts. Tighten to 75 ft. lbs. (102 Nm).

9. Reposition the wheel speed wire into the retaining brackets.

10. Insert the ball joint in steering knuckle and tighten ball joint nut to 70 ft. lbs. (95 Nm).

11. Install the front wheel.

12. Check the wheel alignment and adjust, as necessary.

WHEEL HUB & BEARING

REMOVAL & INSTALLATION

See Figure 136.

1. Before servicing the vehicle, refer to the precautions.

2. Remove the wheel.

3. Remove the brake caliper and rotor.

4. Remove the ABS wheel speed sensor if equipped.

5. Remove the halfshaft nut, if equipped with 4WD.

❊❊ WARNING

Do not strike the knuckle with a hammer to remove the tie rod end or the ball joint. Damage to the steering knuckle will occur.

6. Pull down on the steering knuckle to separate the halfshaft from the hub/bearing (4WD).

7. Remove the three hub/bearing mounting bolts from the steering knuckle.

8. Slide the hub/bearing out of the steering knuckle.

9. Remove the brake dust shield.

To install:

10. Install the brake dust shield.

11. Install the hub/bearing into the steering knuckle and tighten the bolts to 120 ft. lbs. (163 Nm).

12. Install the brake rotor and caliper.

13. Install the ABS wheel speed sensor if equipped.

14. Install the halfshaft nut, if equipped with 4WD. Tighten to 185 ft. lbs. (251 Nm).

15. Install the wheel.

ADJUSTMENT

The hub/bearing assembly is not adjustable.

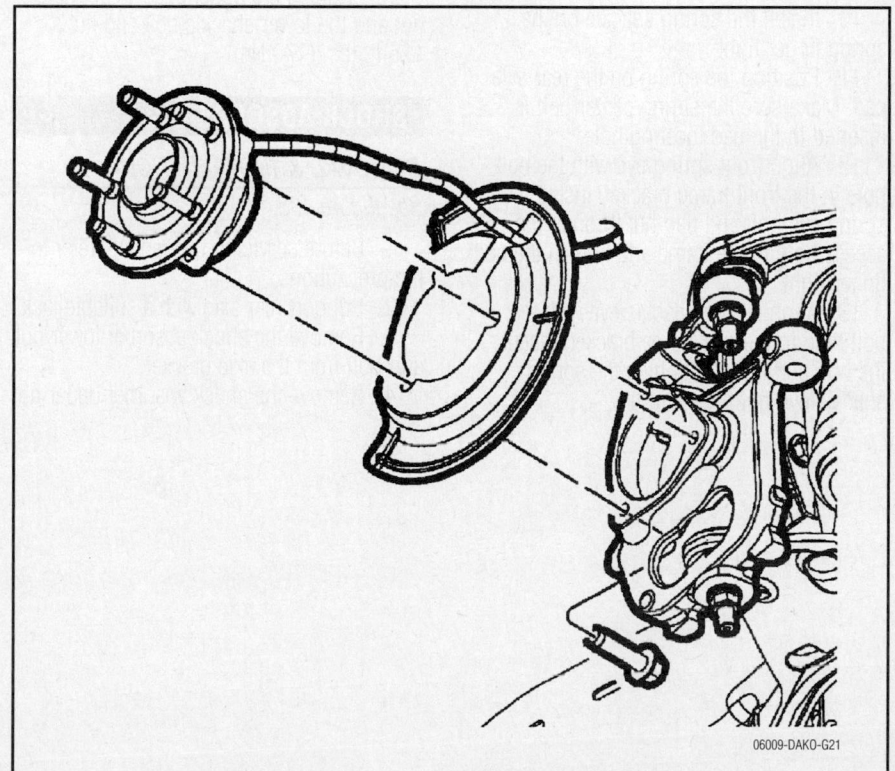

Fig. 136 Front hub assembly

SUSPENSION

LEAF SPRING

REMOVAL & INSTALLATION

See Figure 137.

1. Before servicing the vehicle, refer to the precautions.

✳✳ CAUTION

The rear of the vehicle must be lifted only with a jack or hoist. The lift must be placed under the frame rail crossmember located aft of the rear axle. Use care to avoid bending the side rail flange.

2. Raise the vehicle at the frame.
3. Use a hydraulic jack to relieve the axle weight.
4. Remove the rear wheel.
5. Remove the nuts, the U-bolts and spring plate from the axle.
6. Loosen and remove the bolt and then remove the flag nut through the access hole in the bracket from the spring front eye.
7. Remove the nut and bolt that attaches the spring shackle to the rear frame bracket.
8. Remove the spring from the vehicle.
9. Remove the shackle from the spring.

To install:

10. Install the spring shackle on the spring finger tight.
11. Position the spring on the rear axle pad. Make sure the spring center bolt is inserted in the pad locating hole.
12. Align front spring eye with the bolt hole in the front frame bracket. Install the spring eye bolt and flag nut through the access hole in the frame and tighten the bolt finger-tight.
13. Align spring shackle eye with the bolt hole in the rear frame bracket. Install the bolt and nut and tighten the spring shackle eye nut finger-tight.

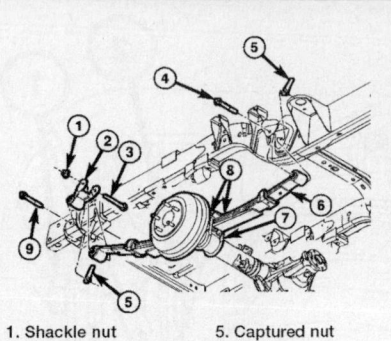

1. Shackle nut
2. Shackle
3. Shackle through bolt
4. Front spring mount through bolt
5. Captured nut
6. Spring
7. U-bolt
8. U-bolt retaining nut
9. Bolt

22140_RAID_G0119

Fig. 137 Exploded view of the rear suspension

14. Install the U-bolts, spring plate and nuts.
15. Tighten the U-bolt nuts to 110 ft. lbs. (149 Nm).
16. Install the rear wheel.
17. Remove the support stands from under the frame rails. Lower the vehicle until the springs are supporting the weight of the vehicle.
18. Tighten the spring eye pivot bolt and flag nut to 125 ft. lbs. (170 Nm).
19. Tighten the upper shackle bolt and nut and the lower shackle bolt and nut to 125 ft. lbs. (170 Nm).

SHOCK ABSORBER

REMOVAL & INSTALLATION

See Figure 138.

1. Before servicing the vehicle, refer to the precautions.
2. Support rear axle with a suitable jack.
3. Remove the shock absorber lower nut and bolt from the axle bracket.
4. Remove the shock absorber upper nut

REAR SUSPENSION

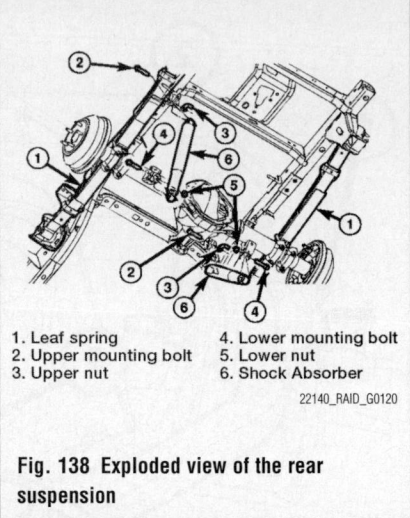

1. Leaf spring
2. Upper mounting bolt
3. Upper nut
4. Lower mounting bolt
5. Lower nut
6. Shock Absorber

22140_RAID_G0120

Fig. 138 Exploded view of the rear suspension

and bolt from the frame bracket and remove the shock absorber.

To install:

5. Install the shock absorber and upper mounting bolt and nut. Tighten the nut to 70 ft. lbs. (95 Nm).
6. Install the shock absorber into the axle bracket. Install the bolt and nut and tighten the nut to 70 ft. lbs. (95 Nm).

TESTING

1. Check the rubber parts for cracks and wear.
2. Check the shock absorber for malfunctions, oil leakage, or abnormal noise.
3. If shock absorber does not perform properly, replace the shock absorber.

WHEEL BEARINGS

REMOVAL & INSTALLATION

Wheel bearing service is covered under Rear Drive Axle, Axle Shaft, Bearing & Seal, Removal & Installation.

MITSUBISHI

Diagnostic Trouble Codes

17

DIAGNOSTIC TROUBLE CODES

OBD II VEHICLE APPLICATIONS

MITSUBISHI

Eclipse
2009-2010
- 2.4L I4 . . . . VIN F
- 3.8L V6. VIN T

Eclipse Spyder
2009-2010
- 2.4L I4 . . . . VIN F
- 3.8L V6. VIN T

Endeavor
2009-2010
- 3.8L V6. VIN S

Evolution
2009-2010
- 2.0L I4 . . . . VIN V

Galant
2009-2010
- 2.4L I4 . . . . VIN F
- 3.8L V6. VIN S
- 3.8L V6. VIN T

Lancer
2009-2010
- 2.0L I4 . . . . VIN U
- 2.4L I4 . . . . VIN W

Outlander
2009
- 2.4L I4 . . . . VIN W
- 3.0L V6. VIN X

Raider
2009-2010
- 3.7L V6. VIN K

OBD II Trouble Code List (P0XXX Codes)

DTC	Trouble Code Title, Conditions & Possible Causes
DTC: P0010 **1T ECM, MIL: Yes** **Year:** 2009, 2010 **Model:** Lancer, Outlander **Engine:** 2.0L L4 VIN U, 2.0L L4 VIN V, 2.4L L4 VIN W **Transmission:** All	**Intake Engine Oil Control Valve Circuit:** Ignition switch is "ON" position. Battery positive voltage is between 10 and 16.5 volts. ON duty cycle of the intake engine oil control valve circuit is higher than 20 percent. The ECM terminal current of intake engine oil control valve circuit is less than 0.1 ampere for 2 seconds. **Possible Causes:** • Intake engine oil control valve failed. • Open or shorted intake engine oil control valve circuit, or harness damage, or connector damage. • ECM failed.
DTC: P0011 **1T ECM, MIL: Yes** **Year:** 2009, 2010 **Model:** Lancer, Outlander **Engine:** 2.0L L4 VIN U, 2.0L L4 VIN V, 2.4L L4 VIN W **Transmission:** All	**Intake Variable Valve Timing System Target Error:** More than 20 seconds have passed since the engine starting sequence was completed. Engine speed is more than 1,188 r/min. Engine coolant temperature is more than 76°C (169°F). The difference between the actual intake valve opening timing and the intake valve target opening timing is more than 5 degrees for 5 seconds. **Possible Causes:** • Intake engine oil control valve failed. • Oil passage of variable valve timing control system clogged. • Intake variable valve timing sprocket operation mechanism stuck. • ECM failed.
DTC: P0012 **1T ECM, MIL: Yes** **Year:** 2009, 2010 **Model:** Lancer **Engine:** 2.0L L4 VIN U, 2.0L L4 VIN V **Transmission:** All	**Camshaft Position - Timing Over-Retarded:** Ignition switch is "ON" position. The learning value for the current phase angle of V.V.T. at the intake (retard angle) side and the exhaust (advanced angle) side is different from the initial phase angle of V.V.T. by more than the specified value. **NOTE: If the vehicle equipped with 2.0L ENGINE Turbo engine continues the rough driving like competitive running (the running that constantly repeats the cycle of the full opened position of the accelerator pedal and the full closed position of the accelerator pedal.), the amount of carbon mixed into the engine oil tends to increase. This can possibly cause the timing chain to gradually elongate.** **Possible Causes:** • Timing chain elongated.
DTC: P0013 **1T ECM, MIL: Yes** **Year:** 2009, 2010 **Model:** Lancer, Outlander **Engine:** 2.0L L4 VIN U, 2.0L L4 VIN V, 2.4L L4 VIN W **Transmission:** All	**Exhaust Engine Oil Control Valve Circuit:** Ignition switch is "ON" position. Battery positive voltage is between 10 and 16.5 volts. The ECM terminal current of exhaust engine oil control valve circuit is less than 0.1 ampere for 2 seconds. Or. The ECM terminal current of exhaust engine oil control valve circuit is higher than 2.9 amperes for 0.1 second. **Possible Causes:** • Exhaust engine oil control valve failed. • Open or shorted exhaust engine oil control valve circuit, or harness damage, or connector damage. • ECM failed.
DTC: P0014 **1T ECM, MIL: Yes** **Year:** 2009, 2010 **Model:** Lancer, Outlander **Engine:** 2.0L L4 VIN U, 2.0L L4 VIN V, 2.4L L4 VIN W **Transmission:** All	**Exhaust Variable Valve Timing System Target Error:** More than 20 seconds have passed since the engine starting sequence was completed. Engine speed is 1,188 r/min or more. Engine coolant temperature is higher than 76°C (169°F). The difference between the actual exhaust valve closing timing and the exhaust valve target closing timing is more than 5 degrees for 5 seconds. **Possible Causes:** • Exhaust engine oil control valve failed. • Oil passage of variable valve timing control system clogged. • Exhaust variable valve timing sprocket operation mechanism stuck. • ECM failed.
DTC: P0016 **1T ECM, MIL: Yes** **Year:** 2009, 2010 **Model:** Lancer **Engine:** 2.0L L4 VIN V **Transmission:** All	**Crankshaft/camshaft (Intake) Position Sensor Phase Problem (Turbo Charged):** Engine speed is between 594 r/min and 1,500 r/min. Engine coolant temperature is between 20°C (68°F) and 88°C (190°F). Intake engine oil control valve is "OFF". After 1 second has passed after the above mentions have been met. The open timing of the intake valve is less than -17.0 degrees (ATDC) for 10 seconds. Or, the open timing of the intake valve is more than -3.9 degrees (ATDC) for 10 seconds. **Possible Causes:** • Timing chain in out of place. • Loose timing chain. • Intake variable valve timing sprocket tooth coming off. • ECM failed.

DTC	Trouble Code Title, Conditions & Possible Causes
DTC: P0016 **1T ECM, MIL:** Yes **Year:** 2009, 2010 **Model:** Lancer, Outlander **Engine:** 2.4L L4 VIN W **Transmission:** All	**Crankshaft/camshaft (Intake) Position Sensor Phase Problem:** Engine speed is between 594 r/min and 1,500 r/min. Engine coolant temperature is between 20°C (68°F) and 88°C (190°F). Intake engine oil control valve is "OFF". After 1 second has elapsed after the above mentions have been met. The open timing of the intake valve is faster than -7.0 degrees (ATDC) for 10 seconds. Or. The open timing of the intake valve is slower than 6.1 degrees (ATDC) for 10 seconds. **Possible Causes:** • Timing chain in out of place. • Loose timing chain. • Intake variable valve timing sprocket tooth coming off. • ECM failed.
DTC: P0016 **1T ECM, MIL:** Yes **Year:** 2009, 2010 **Model:** Lancer **Engine:** 2.0L L4 VIN U **Transmission:** All	**Crankshaft/camshaft (Intake) Position Sensor Phase Problem (Non-Turbo):** Engine speed is between 594 r/min and 1,500 r/min. Engine coolant temperature is between 20°C (68°F) and 88°C (190°F). Intake engine oil control valve is "OFF". After 1 second has elapsed after the above mentions have been met. The open timing of the intake valve is faster than -7.0 degrees (ATDC) <Except for California> or -10.0 degrees (ATDC) <California> for 10 seconds. Or, the open timing of the intake valve is slower than 6.1 degrees (ATDC) <Except for California> or 3.1 degrees (ATDC) <California> for 10 seconds. **Possible Causes:** • Timing chain in out of place. • Loose timing chain. • Intake variable valve timing sprocket tooth coming off. • ECM failed.
DTC: P0016 **1T PCM, MIL:** Yes **Year:** 2009 **Model:** Raider **Engine:** 3.7L V6 VIN K **Transmission:** All	**Crankshaft/ Camshaft Timing Misalignment:** With the engine speed between 480 and 6816 RPM and no CMP or CKP sensor DTCs detected. Powertrain Control Module detects an error when the camshaft position is out of phase with the crankshaft position **Possible Causes:** • Erratic camshaft position sensor signal • Camshaft position sensor tone wheel/pulse ring • Erratic crankshaft position sensor signal • Crankshaft position sensor tone wheel/pulse ring • Timing gear alignment • Camshaft position sensor • Crankshaft position sensor
DTC: P0017 **1T ECM, MIL:** Yes **Year:** 2009, 2010 **Model:** Lancer, Outlander **Engine:** 2.4L L4 VIN W **Transmission:** All	**Crankshaft/camshaft (Exhaust) Position Sensor Phase Problem:** Engine speed is between 594 r/min and 1,500 r/min. Engine coolant temperature is between 20°C (68°F) and 88°C (190°F). Exhaust engine oil control valve is "OFF". After 1 second has elapsed after the above mentions have been met. The close timing of the exhaust valve is faster than -7.7 degrees (ATDC) for 10 seconds. Or, the close timing of the exhaust valve is slower than 5.4 degrees (ATDC) for 10 seconds. **Possible Causes:** • Timing chain in out of place. • Loose timing chain. • Exhaust variable valve timing sprocket tooth coming off. • ECM failed.
DTC: P0017 **1T ECM, MIL:** Yes **Year:** 2009, 2010 **Model:** Lancer **Engine:** 2.0L L4 VIN U, 2.0L L4 VIN V **Transmission:** All	**Crankshaft/camshaft (Exhaust) Position Sensor Phase Problem:** Engine speed is between 594 r/min and 1,500 r/min. Engine coolant temperature is between 20°C (68°F) and 88°C (190°F). Exhaust engine oil control valve is "OFF". After 1 second has elapsed after the above mentions have been met. The close timing of the exhaust valve is faster than -7.7 degrees (ATDC) <Except for California> or -4.7 degrees (ATDC) <California> for 10 seconds. Or, the close timing of the exhaust valve is slower than 5.4 degrees (ATDC) <Except for California> or 8.4 degrees (ATDC) <California> for 10 seconds. **Possible Causes:** • Timing chain in out of place. • Loose timing chain. • Exhaust variable valve timing sprocket tooth coming off. • ECM failed.

DTC	Trouble Code Title, Conditions & Possible Causes
DTC: P001A **1T PCM, MIL: Yes** **Year:** 2009, 2010 **Model:** Eclipse, Galant, Outlander **Engine:** 2.4L L4 VIN F, 3.0L V6 VIN X, 3.8L V6 VIN S, 3.8L V6 VIN T **Transmission:** All	**Camshaft Profile Control (engine oil control valve) Circuit:** With the ignition switch ON, engine oil control valve is OFF, battery voltage is between 10-16.5 volts, 0.1 second has elapsed after the above mentions have been met. The ECM <M/T> or the PCM <A/T> terminal voltage of engine control valve circuit is less than 1.5 volts for 2 seconds. **Possible Causes:** • Engine oil control valve failed • Open or shorted engine oil control valve circuit, or harness damage or connector damage • ECM failed <M/T> • PCM failed <A/T>
DTC: P0031 **2T PCM, MIL: Yes** **Year:** 2009, 2010 **Model:** Eclipse, Galant **Engine:** 2.4L L4 VIN F **Transmission:** All	**Cylinder 1, 4 Heated Oxygen Sensor Heater (front) Control Circuit Low:** More than 2 seconds have passed since the engine starting sequence was completed. While the cylinder 1, 4 heated oxygen sensor (front) heater is on. Battery positive voltage is between 11 and 16.5 volts. The cylinder 1, 4 heated oxygen sensor (front) heater current has continued to be lower than 0.17 ampere for 2 seconds, heater voltage has continued to be lower than 2.0 volts for 2 seconds. **Possible Causes:** • Open or shorted cylinder 1, 4 heated oxygen sensor (front) heater circuit, harness damage or connector damage. • Cylinder 1, 4 heated oxygen sensor (front) heater failed. • ECM failed. <M/T> • PCM failed. <A/T>
DTC: P0031 **2T PCM, MIL: Yes** **Year:** 2009, 2010 **Model:** Eclipse, Endeavor, Galant, Outlander **Engine:** 3.0L V6 VIN X, 3.8L V6 VIN S, 3.8L V6 VIN T **Transmission:** All	**Heated Oxygen Sensor Heater Control Circuit Low (Bank 1, Sensor 1):** More than 2 seconds have passed since the engine starting sequence was completed. While the right bank heated oxygen sensor (front) heater is on. Battery positive voltage is between 11 and 16.5 volts. The right bank heated oxygen sensor (front) heater current has continued to be lower than 0.17 ampere for 2 seconds. **Possible Causes:** • Open or shorted right bank heated oxygen sensor (front) heater circuit, harness damage, or connector damage • Right bank heated oxygen sensor (front) heater • ECM failed <M/T> • PCM failed <A/T>
DTC: P0031 **1T ECM, MIL: Yes** **Year:** 2009, 2010 **Model:** Lancer, Outlander **Engine:** 2.0L L4 VIN U, 2.0L L4 VIN V, 2.4L L4 VIN W **Transmission:** All	**Heated Oxygen Sensor (Front) Heater Control Circuit Low :** More than 2 seconds have passed since the engine starting sequence was completed. While the heated oxygen sensor (Front) heater is on. Battery positive voltage is between 11 and 16.5 volts. The heated oxygen sensor (Front) heater current has continued to be lower than 0.17 ampere for 2 seconds. While the heated oxygen sensor (Front) heater is off. The heated oxygen sensor (front) heater voltage has continued to be lower than 2.0 volts for 2 seconds. **Possible Causes:** • Open or shorted heated oxygen sensor (Front) heater circuit, harness damage or connector damage. • Heated oxygen sensor (Front) heater failed. • ECM failed.
DTC: P0031 **1T PCM, MIL: Yes** **Year:** 2009 **Model:** Raider **Engine:** 3.7L V6 VIN K **Transmission:** All	**O2 Sensor 1/1 Heater Circuit Low:** Continuously during O2 heater operation with battery voltage between 10.4 and 15.75 volts. The PCM detects that the O2 sensor heater element input is below the minimum acceptable voltage. **Possible Causes:** • O2 sensor 1/1 heater control circuit shorted to ground • Excessive resistance in the O2 sensor 1/1 heater control circuit • O2 sensor • PCM
DTC: P0032 **1T PCM, MIL: Yes** **Year:** 2009 **Model:** Raider **Engine:** 3.7L V6 VIN K **Transmission:** All	**O2 Sensor 1/1 Heater Circuit High:** Continuously during O2 heater operation with battery voltage between 10.4 and 15.75 volts. The PCM detects that the O2 sensor heater element input is above the maximum acceptable voltage. **Possible Causes:** • O2 1/1 heater control circuit open • O2 1/1 heater ground circuit open • O2 1/1 heater control circuit shorted to voltage • O2 Sensor • PCM

DTC	Trouble Code Title, Conditions & Possible Causes
DTC: P0032 **1T ECM, MIL: Yes** **Year:** 2009, 2010 **Model:** Outlander **Engine:** 2.4L L4 VIN W **Transmission:** All	**Linear Air-Fuel Ratio Sensor Heater Control Circuit High (California):** More than 2 seconds have passed since the engine starting sequence was completed. While the linear air-fuel ratio sensor heater is on. Battery positive voltage is between 11 and 16.5 volts. The linear air-fuel ratio sensor heater current has continued to be higher than average 10 amperes for 2 seconds. **Possible Causes:** • Shorted linear air-fuel ratio sensor heater circuit or connector damage. • Linear air-fuel ratio sensor heater failed. • ECM failed.
DTC: P0032 **1T ECM, MIL: Yes** **Year:** 2009, 2010 **Model:** Outlander **Engine:** 2.4L L4 VIN W **Transmission:** All	**Heated Oxygen Sensor (Front) Heater Control Circuit High:** More than 2 seconds have passed since the engine starting sequence was completed. While the heated oxygen sensor (front) heater is on. Battery positive voltage is between 11 and 16.5 volts. The heated oxygen sensor (front) heater current has continued to be higher than 10.5 amperes for 2 seconds. **Possible Causes:** • Heated oxygen sensor (front) heater failed. • ECM failed.
DTC: P0032 **2T PCM, MIL: Yes** **Year:** 2009, 2010 **Model:** Eclipse, Galant **Engine:** 2.4L L4 VIN F **Transmission:** All	**Cylinder 1, 4 Heated Oxygen Sensor Heater (front) Control Circuit High:** More than 2 seconds have passed since the engine starting sequence was completed. While the cylinder 1, 4 heated oxygen sensor (front) heater is on. Battery positive voltage is between 11 and 16.5 volts. The cylinder 1, 4 heated oxygen sensor (front) heater current has continued to be higher than 10.5 ampere for 2 seconds. **Possible Causes:** • Cylinder 1, 4 heated oxygen sensor (front) heater failed. • ECM failed. <M/T> • PCM failed. <A/T>
DTC: P0032 **2T PCM, MIL: Yes** **Year:** 2009, 2010 **Model:** Eclipse, Endeavor, Galant, Outlander **Engine:** 3.0L V6 VIN X, 3.8L V6 VIN S, 3.8L V6 VIN T **Transmission:** All	**Heated Oxygen Sensor Heater Control Circuit High (Bank 1, Sensor 1):** More than 2 seconds have passed since the engine starting sequence was completed. While the right bank heated oxygen sensor (front) heater is on. Battery positive voltage is between 11 and 16.5 volts. The right bank heated oxygen sensor (front) heater current has continued to be higher than 10.5 ampere for 2 seconds. **Possible Causes:** • Right bank heated oxygen sensor (front) heater • ECM failed <M/T> • PCM failed <A/T>
DTC: P0037 **2T PCM, MIL: Yes** **Year:** 2009, 2010 **Model:** Eclipse, Endeavor, Galant, Outlander **Engine:** 3.0L V6 VIN X, 3.8L V6 VIN S, 3.8L V6 VIN T **Transmission:** All	**Heated Oxygen Sensor Heater Control Circuit Low (Bank 1, Sensor 2):** More than 2 seconds have passed since the engine starting sequence was completed. While the right bank heated oxygen sensor (rear) heater is on. Battery positive voltage is at between 11 and 16.5 volts. The right bank heated oxygen sensor (rear) heater voltage has continued to be lower than 2.0 voltage for 2 seconds. **Possible Causes:** • Open or shorted right bank heated oxygen sensor (rear) heater circuit, or harness damage • Right bank heated oxygen sensor (rear) heater failed • Connector damage • ECM failed <M/T> • PCM failed <A/T>
DTC: P0037 **2T PCM, MIL: Yes** **Year:** 2009, 2010 **Model:** Eclipse, Galant **Engine:** 2.4L L4 VIN F **Transmission:** All	**Cylinder 1, 4 Heated Oxygen Sensor Heater (rear) Control Circuit Low:** More than 2 seconds have passed since the engine starting sequence was completed. While the cylinder 1, 4 heated oxygen sensor (rear) heater is on. Battery positive voltage is between 11 and 16.5 volts. The cylinder 1, 4 heated oxygen sensor (rear) heater current has continued to be lower than 0.17 ampere for 2 seconds. While the cylinder 1, 4 heated oxygen sensor (rear) heater is off. The cylinder 1, 4 heated oxygen sensor (rear) heater voltage has continued to be lower than 2.0 volts for 2 seconds. **Possible Causes:** • Open or shorted cylinder 1, 4 heated oxygen sensor (rear) heater circuit, or harness damage. • Cylinder 1, 4 heated oxygen sensor (rear) heater failed. • Connector damage. • ECM failed. <M/T> • PCM failed. <A/T>

DTC	Trouble Code Title, Conditions & Possible Causes
DTC: P0037 **1T PCM, MIL: Yes** **Year:** 2009 **Model:** Raider **Engine:** 3.7L V6 VIN K **Transmission:** All	**O2 1/2 Heater Circuit Low:** Continuously during O2 heater operation with battery voltage between 10.4 and 15.75 volts. The PCM detects that the O2 sensor heater element input is below the minimum acceptable voltage. **Possible Causes:** • O2 1/2 heater control circuit shorted to ground • Excessive resistance in the O2 heater control circuit • O2 Sensor • PCM
DTC: P0037 **1T ECM, MIL: Yes** **Year:** 2009, 2010 **Model:** Lancer, Outlander **Engine:** 2.0L L4 VIN U, 2.0L L4 VIN V, 2.4L L4 VIN W **Transmission:** All	**Heated Oxygen Sensor (Rear) Heater Control Circuit Low :** More than 2 seconds have passed since the engine starting sequence was completed. While the heated oxygen sensor (rear) heater is on. Battery positive voltage is between 11 and 16.5 volts. The heated oxygen sensor (rear) heater current has continued to be lower than 0.17 ampere for 2 seconds. While the heated oxygen sensor (rear) heater is off. The heated oxygen sensor (rear) heater voltage has continued to be lower than 2.0 volts for 2 seconds. **Possible Causes:** • Open or shorted heated oxygen sensor (rear) heater circuit, or harness damage. • Heated oxygen sensor (rear) heater failed. • Connector damage. • ECM failed.
DTC: P0038 **2T PCM, MIL: Yes** **Year:** 2009, 2010 **Model:** Eclipse, Galant **Engine:** 2.4L L4 VIN F **Transmission:** All	**Cylinder 1, 4 Heated Oxygen Sensor Heater (rear) Control Circuit High:** More than 2 seconds have passed since the engine starting sequence was completed. While the cylinder 1, 4 heated oxygen sensor (rear) heater is on. Battery positive voltage is between 11 and 16.5 volts. The cylinder 1, 4 heated oxygen sensor (rear) heater current has continued to be higher than 10.5 ampere for 2 seconds. **Possible Causes:** • Cylinder 1, 4 heated oxygen sensor (rear) failed. • Connector damage. • ECM failed. <M/T> • PCM failed. <A/T>
DTC: P0038 **1T PCM, MIL: Yes** **Year:** 2009 **Model:** Raider **Engine:** 3.7L V6 VIN K **Transmission:** All	**O2 Sensor 1/2 Heater Circuit High:** Continuously during O2 heater operation with battery voltage between 10.4 and 15.75 volts. The PCM detects that the O2 sensor heater element input is above the maximum acceptable voltage. **Possible Causes:** • O2 1/2 heater control circuit open • O2 1/2 heater ground circuit open • O2 SENSOR 1/2 heater control circuit shorted to voltage • O2 Sensor • PCM
DTC: P0038 **1T ECM, MIL: Yes** **Year:** 2009, 2010 **Model:** Lancer, Outlander **Engine:** 2.0L L4 VIN U, 2.0L L4 VIN V, 2.4L L4 VIN W **Transmission:** All	**Heated Oxygen Sensor (Rear) Heater Control Circuit High:** More than 2 seconds have passed since the engine starting sequence was completed. While the heated oxygen sensor (rear) heater is on. Battery positive voltage is between 11 and 16.5 volts. The heated oxygen sensor (rear) heater current has continued to be higher than 10.5 amperes for 2 seconds **Possible Causes:** • Heated oxygen sensor (rear) heater failed. • ECM failed.
DTC: P0038 **2T PCM, MIL: Yes** **Year:** 2009, 2010 **Model:** Eclipse, Endeavor, Galant, Outlander **Engine:** 3.0L V6 VIN X, 3.8L V6 VIN S, 3.8L V6 VIN T **Transmission:** All	**Heated Oxygen Sensor Heater Control Circuit High (Bank 1, Sensor 2):** More than 2 seconds have passed since the engine starting sequence was completed. While the right bank heated oxygen sensor (rear) heater is on. Battery positive voltage is at between 11 and 16.5 volts. The right bank heated oxygen sensor (rear) heater current has continued to be higher than 10.5 ampere for 2 seconds. **Possible Causes:** • Right bank heated oxygen sensor (rear) failed • Connector damage • ECM failed <M/T> • PCM failed <A/T>
DTC: P003C **1T PCM** **Year:** 2009, 2010 **Model:** Eclipse, Galant, Outlander **Engine:** 2.4L L4 VIN F, 3.0L V6 VIN X, 3.8L V6 VIN S, 3.8L V6 VIN T **Transmission:** All	**Mitsubishi Innovative Valve Timing Electronic Control System (MIVEC) Performance Problem:** Engine speed is less than 3,300 r/min, engine coolant temperature is 77°C (171°F) or more, battery positive voltage is between 10 and 16.5 volts. 30 seconds or more have passed since the engine starting sequence was completed. Engine oil pressure switch (for MIVEC) has been OFF/ON for 5 seconds. **Possible Causes:** • Engine oil pressure switch (for MIVEC) failed • Engine oil control valve failed • Open or shorted engine oil pressure switch (for MIVEC) circuit, harness damage or connector damage

DTC	Trouble Code Title, Conditions & Possible Causes
DTC: P0043 **1T ECM, MIL: Yes** **Year:** 2009, 2010 **Model:** Lancer, Outlander **Engine:** 2.0L L4 VIN U, 2.0L L4 VIN V, 2.4L L4 VIN W **Transmission:** All	**Heated Oxygen Sensor (3rd) Heater Control Circuit Low :** More than 2 seconds have passed since the engine starting sequence was completed. While the heated oxygen sensor (3rd) heater is on. Battery positive voltage is between 11 and 16.5 volts. The heated oxygen sensor (3rd) heater current has continued to be lower than 0.17 ampere for 2 seconds. While the heated oxygen sensor (3rd) heater is off. The heated oxygen sensor (3rd) heater voltage has continued to be lower than 2.0 volts for 2 seconds. **Possible Causes:** • Open or shorted heated oxygen sensor (3rd) heater circuit, or harness damage. • Heated oxygen sensor (3rd) heater failed. • Connector damage. • ECM failed.
DTC: P0043 **1T ECM, MIL: Yes** **Year:** 2009, 2010 **Model:** Outlander **Engine:** 3.0L V6 VIN X **Transmission:** All	**Heated Oxygen Sensor Heater Circuit Low (Bank 1, Sensor 3):** More than 2 seconds have passed since the engine starting sequence was completed. While the center exhaust pipe heater was OFF. Battery voltage is between 11 and 16.5 volts. The center exhaust pipe heated oxygen sensor heater voltage has continued to be lower than 2.0 volts for 2 seconds. **Possible Causes:** • Connector damage. • Center exhaust pipe heated oxygen sensor heater failed. • Harness damage. • ECM failed.
DTC: P0044 **1T ECM, MIL: Yes** **Year:** 2009, 2010 **Model:** Outlander **Engine:** 3.0L V6 VIN X **Transmission:** All	**Heated Oxygen Sensor Heater Circuit High (Bank 1, Sensor 3):** More than 2 seconds have passed since the engine starting sequence was completed. While the center exhaust pipe heater was ON. Battery voltage is between 11 and 16.5 volts. The center exhaust pipe heated oxygen sensor heater current has continued to be higher than 10.5 amps for 2 seconds. **Possible Causes:** • Center exhaust pipe heated oxygen sensor heater. • ECM failed.
DTC: P0044 **1T PCM, MIL: Yes** **Year:** 2009, 2010 **Model:** Lancer, Outlander **Engine:** 2.0L L4 VIN U, 2.0L L4 VIN V, 2.4L L4 VIN W **Transmission:** All	**Heated Oxygen Sensor (3rd) Heater Control Circuit High:** More than 2 seconds have passed since the engine starting sequence was completed. While the heated oxygen sensor (3rd) heater is on. Battery positive voltage is between 11 and 16.5 volts. The heated oxygen sensor (3rd) heater current has continued to be higher than 10.5 amperes for 2 seconds. **Possible Causes:** • Heated oxygen sensor (3rd) heater failed. • ECM failed.
DTC: P0051 **2T PCM, MIL: Yes** **Year:** 2009, 2010 **Model:** Eclipse, Galant **Engine:** 2.4L L4 VIN F **Transmission:** All	**Cylinder 2, 3 Heated Oxygen Sensor Heater (front) Control Circuit Low:** More than 2 seconds have passed since the engine starting sequence was completed. While the cylinder 2, 3 heated oxygen sensor (front) heater is on. Battery positive voltage is between 11 and 16.5 volts. The cylinder 2, 3 heated oxygen sensor (front) heater current has continued to be lower than 0.17 ampere for 2 seconds. While the cylinder 2, 3 heated oxygen sensor (front) heater is off, the (front) heater voltage has continued to be lower than 2.0 volts for 2 seconds. **Possible Causes:** • Open or shorted cylinder 2, 3 heated oxygen sensor (front) heater circuit, or harness damage. • Cylinder 2, 3 heated oxygen sensor (front) heater failed. • Connector damage. • ECM failed. <M/T> • PCM failed. <A/T>
DTC: P0051 **1T PCM, MIL: Yes** **Year:** 2009 **Model:** Raider **Engine:** 3.7L V6 VIN K **Transmission:** All	**O2 Sensor 2/1 Heater Circuit Low:** Continuously during O2 heater operation with battery voltage between 10.4 and 15.75 volts. The PCM detects that the O2 sensor heater element input is below the minimum acceptable voltage. **Possible Causes:** • O2 2/1 Heater control circuit shorted to ground • Excessive resistance in the O2 2/1 heater control circuit • O2 Sensor • PCM

DTC	Trouble Code Title, Conditions & Possible Causes
DTC: P0051 **2T PCM, MIL: Yes** **Year:** 2009, 2010 **Model:** Eclipse, Endeavor, Galant, Outlander **Engine:** 3.0L V6 VIN X, 3.8L V6 VIN S, 3.8L V6 VIN T **Transmission:** All	**Heated Oxygen Sensor Heater Control Circuit Low (Bank 2, Sensor 1):** More than 2 seconds have passed since the engine starting sequence was completed. While the left bank heated oxygen sensor (front) heater is ON. Battery positive voltage is at between 11 and 16.5 volts. The left bank heated oxygen sensor (front) heater current has continued to be lower than 0.17 ampere for 2 seconds. **Possible Causes:** • Open or shorted left bank heated oxygen sensor (front) heater circuit, or harness damage • Left bank heated oxygen sensor (front) heater • Connector damage • ECM failed <M/T> • PCM failed <A/T>
DTC: P0052 **2T PCM, MIL: Yes** **Year:** 2009, 2010 **Model:** Eclipse, Endeavor, Galant, Outlander **Engine:** 3.0L V6 VIN X, 3.8L V6 VIN S, 3.8L V6 VIN T **Transmission:** All	**Heated Oxygen Sensor Heater Control Circuit High (Bank 2, Sensor 1):** More than 2 seconds have passed since the engine starting sequence was completed. While the left bank heated oxygen sensor (front) heater is on. Battery positive voltage is at between 11 and 16.5 volts. The left bank heated oxygen sensor (front) heater current has continued to be higher than 10.5 ampere for 2 seconds. **Possible Causes:** • Left bank heated oxygen sensor (front) heater failed • Connector damage. • ECM failed <M/T> • PCM failed <A/T>
DTC: P0052 **1T PCM, MIL: Yes** **Year:** 2009 **Model:** Raider **Engine:** 3.7L V6 VIN K **Transmission:** All	**O2 Sensor 2/1 heater Circuit High:** Continuously during O2 heater operation with battery voltage between 10.4 and 15.75 volts. The PCM detects that the O2 sensor heater element input is above the maximum acceptable voltage. **Possible Causes:** • O2 2/1 Heater control circuit open • O2 2/1 Heater ground circuit open • O2 2/1 Heater control circuit shorted to voltage • O2 Sensor • PCM
DTC: P0052 **2T PCM, MIL: Yes** **Year:** 2009, 2010 **Model:** Eclipse, Galant **Engine:** 2.4L L4 VIN F **Transmission:** All	**Cylinder 2, 3 Heated Oxygen Sensor Heater (front) Control Circuit High:** More than 2 seconds have passed since the engine starting sequence was completed. While the cylinder 2, 3 heated oxygen sensor (front) heater is on. Battery positive voltage is between 11 and 16.5 volts. The cylinder 2, 3 heated oxygen sensor (front) heater current has continued to be higher than 10.5 ampere for 2 seconds. **Possible Causes:** • Cylinder 2, 3 heated oxygen sensor (front) heater failed. • ECM failed. <M/T> • PCM failed. <A/T>
DTC: P0053 **2T PCM, MIL: Yes** **Year:** 2009, 2010 **Model:** Eclipse **Engine:** 3.8L V6 VIN T **Transmission:** All	**Heated Oxygen Sensor Heater Resistance (Bank 1, Sensor 1):** More than 60 seconds have passed since the engine starting sequence was completed. While the right bank heated oxygen sensor (front) heater is on. Battery positive voltage is between 11 and 16.5 volts. Intake air temperature is more than -10°C (14°F). On duty cycle of the right bank heated oxygen sensor (front) heater is between 3 and 97 percent. The right bank heated oxygen sensor (front) heater current has continued to be lower than average 0.6 ampere for 10 seconds. **Possible Causes:** • Harness damage in right bank heated oxygen sensor (front) heater circuit or connector damage • Right bank heated oxygen sensor (front) heater • ECM failed <M/T> • PCM failed <A/T>
DTC: P0053 **1T ECM, MIL: Yes** **Year:** 2009, 2010 **Model:** Outlander **Engine:** 2.4L L4 VIN W **Transmission:** All	**Linear Air-Fuel Ratio Sensor Heater Resistance (California):** More than 220 seconds have passed since the engine starting sequence was completed. While the linear air-fuel ratio sensor heater is on. Battery positive voltage is between 11 and 16.5 volts. Intake air temperature is more than -10°C (14°F). On duty cycle of the linear air-fuel ratio sensor heater is between 3 and 97 percent. The linear air-fuel ratio sensor heater current has continued to be lower than average 1.25 ampere for 10 seconds. **Possible Causes:** • Linear air-fuel ratio sensor circuit harness damage or connector damage. • Linear air-fuel ratio sensor heater failed. • ECM failed.

DTC	Trouble Code Title, Conditions & Possible Causes
DTC: P0053 **2T PCM, MIL: Yes** **Year:** 2009, 2010 **Model:** Eclipse **Engine:** 2.4L L4 VIN F **Transmission:** All	**Cylinder 1, 4 Heated Oxygen Sensor Heater (front) Resistance:** More than 60 seconds have passed since the engine starting sequence was completed. While the cylinder 1, 4 heated oxygen sensor heater is on. Battery positive voltage is between 11 and 16.5 volts. Intake air temperature is more than -10°C (14°F). On duty cycle of the cylinder 1, 4 heated oxygen sensor heater is between 3 and 97 percent. The cylinder 1, 4 heated oxygen sensor heater current has continued to be lower than average 0.6 ampere for 10 seconds. **Possible Causes:** • Harness damage in cylinder 1, 4 heated oxygen sensor (front) heater circuit or connector damage. • Cylinder 1, 4 heated oxygen sensor (front) heater failed. • ECM failed. <M/T> • PCM failed. <A/T>
DTC: P0057 **2T PCM, MIL: Yes** **Year:** 2009, 2010 **Model:** Eclipse, Galant **Engine:** 2.4L L4 VIN F **Transmission:** All	**Cylinder 2, 3 Heated Oxygen Sensor Heater (rear) Control Circuit Low:** More than 2 seconds have passed since the engine starting sequence was completed. While the cylinder 2, 3 heated oxygen sensor (rear) heater is on. Battery positive voltage is between 11 and 16.5 volts. The cylinder 2, 3 heated oxygen sensor (rear) heater current has continued to be lower than 0.17 ampere for 2 seconds. While the cylinder 2, 3 heated oxygen sensor (rear) heater is off. The cylinder 2, 3 heated oxygen sensor (rear) heater voltage has continued to be lower than 2.0 volts for 2 seconds. **Possible Causes:** • Open or shorted cylinder 2, 3 heated oxygen sensor (rear) heater circuit, or harness damage. • Cylinder 2, 3 heated oxygen sensor (rear) heater failed. • Connector damage failed. • ECM failed. <M/T> • PCM failed. <A/T>
DTC: P0057 **1T PCM, MIL: Yes** **Year:** 2009 **Model:** Raider **Engine:** 3.7L V6 VIN K **Transmission:** All	**O2 Sensor 2/2 Heater Circuit Low:** Continuously during O2 heater operation with battery voltage between 10.4 and 15.75 volts. The PCM detects that the O2 sensor heater element input is above the maximum acceptable voltage. **Possible Causes:** • O2 SENSOR 2/2 heater control circuit shorted to ground • Excessive resistance in the O2 sensor 2/2 heater control circuit • O2 Sensor • PCM
DTC: P0057 **2T PCM, MIL: Yes** **Year:** 2009, 2010 **Model:** Eclipse, Endeavor, Galant, Outlander **Engine:** 3.0L V6 VIN X, 3.8L V6 VIN S, 3.8L V6 VIN T **Transmission:** All	**Heated Oxygen Sensor Heater Control Circuit Low (Bank 2, Sensor 2):** More than 2 seconds have passed since the engine starting sequence was completed. While the left bank heated oxygen sensor (rear) heater is on. Battery positive voltage is between 11 and 16.5 volts. The left bank heated oxygen sensor (rear) heater current has continued to be lower than 0.17 ampere for 2 seconds. **Possible Causes:** • Open or shorted left bank heated oxygen sensor (rear) heater circuit, or harness damage • Left bank heated oxygen sensor (rear) heater failed • Connector damage • Left bank heated oxygen sensor (rear) failed • ECM failed <M/T> • PCM failed <A/T>
DTC: P0058 **2T PCM, MIL: Yes** **Year:** 2009, 2010 **Model:** Eclipse, Galant **Engine:** 2.4L L4 VIN F **Transmission:** All	**Cylinder 2, 3 Heated Oxygen Sensor Heater (rear) Control Circuit High:** More than 2 seconds have passed since the engine starting sequence was completed. While the cylinder 2, 3 heated oxygen sensor (rear) heater is on. Battery positive voltage is between 11 and 16.5 volts. The cylinder 2, 3 heated oxygen sensor (rear) heater current has continued to be higher than 10.5 ampere for 2 seconds. **Possible Causes:** • Connector damage. • Cylinder 2, 3 heated oxygen sensor (rear) failed. • ECM failed. <M/T> • PCM failed. <A/T>
DTC: P0058 **2T PCM, MIL: Yes** **Year:** 2009, 2010 **Model:** Eclipse, Endeavor, Galant, Outlander **Engine:** 3.0L V6 VIN X, 3.8L V6 VIN S, 3.8L V6 VIN T **Transmission:** All	**Heated Oxygen Sensor Heater Control Circuit High (Bank 2, Sensor 2):** More than 2 seconds have passed since the engine starting sequence was completed. While the left bank heated oxygen sensor (rear) heater is on. Battery positive voltage is between 11 and 16.5 volts. The left bank heated oxygen sensor (rear) heater current has continued to be higher than 10.5 ampere for 2 seconds. **Possible Causes:** • Left bank heated oxygen sensor (rear) heater failed • Connector damage • Left bank heated oxygen sensor (rear) failed • ECM failed <M/T> • PCM failed <A/T>

DTC	Trouble Code Title, Conditions & Possible Causes
DTC: P0058 **1T PCM, MIL:** Yes **Year:** 2009 **Model:** Raider **Engine:** 3.7L V6 VIN K **Transmission:** All	**O2 Sensor 2/2 Heater Circuit High:** Continuously during O2 heater operation with battery voltage between 10.4 and 15.75 volts. The PCM detects that the O2 sensor heater element input is above the maximum acceptable voltage. **Possible Causes:** • O2 2/2 Heater control circuit open • O2 2/2 Heater ground circuit open • O2 2/2 Heater control circuit shorted to voltage • 2/2 O2 Sensor • PCM
DTC: P0059 **2T PCM, MIL:** Yes **Year:** 2009, 2010 **Model:** Eclipse **Engine:** 2.4L L4 VIN F **Transmission:** All	**Cylinder 2, 3 Heated Oxygen Sensor Heater (front) Resistance:** More than 60 seconds have passed since the engine starting sequence was completed. While the cylinder 2, 3 heated oxygen sensor heater is on. Battery positive voltage is between 11 and 16.5 volts. Intake air temperature is more than -10°C (14°F). On duty cycle of the cylinder 2, 3 heated oxygen sensor heater is between 3 and 97 percent. The cylinder 2, 3 heated oxygen sensor heater current has continued to be lower than average 0.6 ampere for 10 seconds. **Possible Causes:** • Harness damage in cylinder 2, 3 heated oxygen sensor (front) heater circuit or connector damage. • Cylinder 2, 3 heated oxygen sensor (front) heater failed. • ECM failed. <M/T> • PCM failed. <A/T>
DTC: P0059 **2T PCM, MIL:** Yes **Year:** 2009, 2010 **Model:** Eclipse **Engine:** 3.8L V6 VIN T **Transmission:** All	**Heated Oxygen Sensor Heater Resistance (Bank 2, Sensor 1):** More than 60 seconds have passed since the engine starting sequence was completed. While the left bank heated oxygen sensor (front) heater is on. Battery positive voltage is between 11 and 16.5 volts. Intake air temperature is more than -10°C (14°F). On duty cycle of the left bank heated oxygen sensor (front) heater is between 3 and 97 percent. The left bank heated oxygen sensor (front) heater current has continued to be lower than average 0.6 ampere for 10 seconds. **Possible Causes:** • Harness damage in left bank heated oxygen sensor (front) heater circuit or connector damage • Left bank heated oxygen sensor (front) heater • ECM failed <M/T> • PCM failed <A/T>
DTC: P0069 **1T ECM, MIL:** Yes **Year:** 2009, 2010 **Model:** Lancer **Engine:** 2.0L L4 VIN U, 2.0L L4 VIN V, 2.4L L4 VIN W **Transmission:** All	**Abnormal Correlation Between Manifold Absolute Pressure Sensor And Barometric Pressure Sensor:** Ignition switch is in "LOCK" (OFF) position. After 2 seconds pass from the time when the engine is stopped. Engine coolant temperature is more than 0°C (32°F). The difference between barometric pressure sensor output and manifold absolute pressure sensor output is more than 9 kPa (2.7 in. Hg) for 2 seconds. **Possible Causes:** • Manifold absolute pressure sensor failed. • Barometric pressure sensor failed. • ECM failed.
DTC: P0069 **1T PCM, MIL:** Yes **Year:** 2009, 2010 **Model:** Eclipse, Galant, Outlander **Engine:** 2.4L L4 VIN F, 2.4L L4 VIN W, 3.0L V6 VIN X, 3.8L V6 VIN S, 3.8L V6 VIN T **Transmission:** All	**Abnormal Correlation Between Manifold Absolute Pressure Sensor And Barometric Pressure Sensor:** Ignition switch is in "LOCK" (OFF) position. After two seconds pass from the time when the engine is stopped. Engine coolant temperature is higher than 0°C (32°F). The difference between manifold absolute pressure sensor output and barometric pressure sensor output is more than 9 kPa (2.7 in. Hg) for 2 seconds. **Possible Causes:** • Manifold absolute pressure sensor failed • Barometric pressure sensor failed • ECM failed <M/T> • PCM failed <A/T>

DTC	Trouble Code Title, Conditions & Possible Causes
DTC: P0071 **2T PCM, MIL: Yes** **Year:** 2009 **Model:** Raider **Engine:** 3.7L V6 VIN K **Transmission:** All	**Ambient Air Temperature Sensor Performance:** Engine off time is greater than 480 minutes and the vehicle has been driven for one minute over 35 mph. Ambient temperature is greater than -64°C (-83°F). The PCM compares the ambient, engine coolant, and intake air temperature sensor values. If engine coolant and intake air temperature sensors agree with each other but ambient air temperature does not agree with them, the ambient air temperature sensor is declared as irrational. **Possible Causes:** • AAT signal circuit high resistance • AAT signal circuit shorted to ground • AAT signal circuit shorted to the sensor ground circuit • AAT signal circuit shorted to voltage • Sensor ground high resistance • ATT sensor • Can C-Bus Circuits • Front Control Module • PCM
DTC: P0072 **1T PCM, MIL: Yes** **Year:** 2009 **Model:** Raider **Engine:** 3.7L V6 VIN K **Transmission:** All	**Ambient Air Temperature Sensor Circuit Low:** The ignition key on. Ambient Temperature Sensor is less than the minimum acceptable voltage. **Possible Causes:** • AAT signal circuit shorted to ground • AAT signal circuit shorted to the sensor ground circuit* Excessive Resistance in the * AAT signal circuit • ATT Sensor • Front control module
DTC: P0073 **1T PCM, MIL: Yes** **Year:** 2009 **Model:** Raider **Engine:** 3.7L V6 VIN K **Transmission:** All	**Ambient Air Temperature Sensor Circuit High:** With the ignition on. Battery voltage greater than 10 volts. The Ambient Temperature Sensor voltage is greater than maximum allowable voltage. **Possible Causes:** • AAT Signal circuit shorted to voltage • AAT Signal circuit open • ATT Sensor ground circuit open • ATT Sensor • Front control module
DTC: P0096 **1T ECM, MIL: Yes** **Year:** 2009, 2010 **Model:** Lancer **Engine:** 2.0L L4 VIN V **Transmission:** All	**Intake Air Temperature Circuit Range/Performance Problem (sensor 2):** Engine coolant temperature is more than 76°C (169°F). Repeat 2 times or more: drive*1, stop*2. Drive*1: Vehicle speed is more than 50 km/h (31 mph) lasting a total of 60 seconds or more. Stop*2: Vehicle speed is less than 1.5 km/h (1.0 mph) lasting 30 seconds or more. Changes in the intake air temperature is less than 1°C (1.8°F). **Possible Causes:** • Intake air temperature sensor 2 failed. • Harness damage or connector damage. • ECM failed.
DTC: P0097 **1T ECM, MIL: Yes** **Year:** 2009, 2010 **Model:** Lancer **Engine:** 2.0L L4 VIN V **Transmission:** All	**Intake Air Temperature Circuit Low Input (sensor 2):** More than 2 seconds have passed since the engine starting sequence was completed. The intake air temperature sensor 2 output voltage is less than 0.2 volt [corresponding to an intake air temperature of 115°C (239°F) or more] for 2 seconds. **Possible Causes:** • Intake air temperature sensor 2 failed. • Shorted intake air temperature sensor 2 circuit, or connector damage. • ECM failed.
DTC: P0098 **1T ECM, MIL: Yes** **Year:** 2009, 2010 **Model:** Lancer **Engine:** 2.0L L4 VIN V **Transmission:** All	**Intake Air Temperature Circuit High Input (sensor 2):** More than 2 seconds have passed since the engine starting sequence was completed. The intake air temperature sensor 2 output voltage is more than 4.6 volts [corresponding to an intake air temperature of -40°C (-40°F) or less] for 2 seconds. **Possible Causes:** • Intake air temperature sensor 2 failed. • Open intake air temperature sensor 2 circuit, or connector damage. • ECM failed.

DTC	Trouble Code Title, Conditions & Possible Causes
DTC: P0101 **1T PCM, MIL: Yes** **Year:** 2009, 2010 **Model:** Eclipse, Endeavor, Galant, Outlander **Engine:** 2.4L L4 VIN F, 2.4L L4 VIN W, 3.0L V6 VIN X, 3.8L V6 VIN S, 3.8L V6 VIN T **Transmission:** All	**Mass Airflow Circuit Range/Performance Problem:** Engine speed is higher than 1,500 r/min. Throttle position sensor output voltage is 1.5 volt or higher. Mass airflow sensor output voltage is 0.2 volt or higher. Mass airflow sensor output voltage has continued to be 1.8 volts or lower for 2 seconds. Or mass airflow sensor output voltage has continued to be 2.5 volts or higher for 2 seconds. **Possible Causes:** • Mass airflow sensor failed • Mass airflow sensor circuit harness damage, or connector damage • ECM failed <M/T> • PCM failed <A/T>
DTC: P0101 **1T ECM, MIL: Yes** **Year:** 2009, 2010 **Model:** Lancer **Engine:** 2.0L L4 VIN U, 2.0L L4 VIN V, 2.4L L4 VIN W **Transmission:** All	**Mass Airflow Circuit Range/Performance Problem:** The throttle position sensor output voltage is 0.8 volt or lower. Mass airflow sensor output voltage is 4.9 volts (corresponding to an air flow rate of 295 g/sec) or lower. The mass airflow sensor output voltage has continued to be 2.5 volts (corresponding to an air flow rate of 28 g/sec) or higher for 2 seconds. Or, The throttle position sensor output voltage is 1.5 volts or higher. Mass airflow sensor output voltage is 0.2 volt (corresponding to an air flow rate of 0 g/sec) or higher. Engine speed is 1,500 r/min or more. The mass airflow sensor output voltage has continued to be 1.8 volts (corresponding to an air flow rate of 8 g/sec) or lower for 2 seconds. **Possible Causes:** • Mass airflow sensor failed. • Mass airflow sensor circuit harness damage, or connector damage. • ECM failed.
DTC: P0102 **1T ECM, MIL: Yes** **Year:** 2009, 2010 **Model:** Lancer **Engine:** 2.0L L4 VIN U, 2.0L L4 VIN V, 2.4L L4 VIN W **Transmission:** All	**Mass Airflow Circuit Low Input:** After 3 seconds or more have passed since the ignition switch was turned to "ON" position. The mass airflow sensor output voltage has continued to be lower than 0.2 volt (corresponding to an air flow rate of 0 g/sec) for 2 seconds. **Possible Causes:** • Mass airflow sensor failed. • Open or shorted mass airflow sensor circuit, or connector damage. • ECM failed.
DTC: P0102 **1T PCM, MIL: Yes** **Year:** 2009, 2010 **Model:** Eclipse, Endeavor, Galant, Outlander **Engine:** 2.4L L4 VIN F, 2.4L L4 VIN W, 3.0L V6 VIN X, 3.8L V6 VIN S, 3.8L V6 VIN T **Transmission:** All	**Mass Airflow Circuit Low Input:** After 3 seconds or more have passed since the ignition switch was turned to "ON" position. The mass airflow sensor output voltage has continued to be lower than 0.2 volt for 2 seconds. **Possible Causes:** • Mass airflow sensor failed • Open or shorted mass airflow sensor circuit, or connector damage • ECM failed <M/T> • PCM failed <A/T>
DTC: P0103 **1T ECM, MIL: Yes** **Year:** 2009, 2010 **Model:** Lancer **Engine:** 2.0L L4 VIN V **Transmission:** All	**Mass Airflow Circuit High Input:** More than 3 seconds have passed since the ignition switch was turned to "ON" position. The mass airflow sensor output voltage is more than 4.9 volts (corresponding to an air flow rate of 387 g/sec) for 2 seconds. **Possible Causes:** • Mass airflow sensor failed. • Open mass airflow sensor circuit, or connector damage. • ECM failed.
DTC: P0103 **1T PCM, MIL: Yes** **Year:** 2009, 2010 **Model:** Eclipse, Endeavor, Galant, Outlander **Engine:** 2.4L L4 VIN F, 2.4L L4 VIN W, 3.0L V6 VIN X, 3.8L V6 VIN S, 3.8L V6 VIN T **Transmission:** All	**Mass Airflow Circuit High Input:** After 3 seconds or more have passed since the ignition switch was turned to "ON" position. The mass airflow sensor output voltage has continued to be higher than 4.9 volts for 2 seconds. **Possible Causes:** • Mass airflow sensor failed. • Open mass airflow sensor circuit, or connector damage • ECM failed <M/T> • PCM failed <A/T>
DTC: P0103 **1T ECM, MIL: Yes** **Year:** 2009, 2010 **Model:** Lancer **Engine:** 2.4L L4 VIN W **Transmission:** All	**Mass Airflow Circuit High Input:** After 3 seconds or more have passed since the ignition switch was turned to "ON" position. Mass airflow sensor output voltage has continued to be higher than 4.9 volts (corresponding to an air flow rate of 295 g/sec) for 2 seconds. **Possible Causes:** • Mass airflow sensor failed. • Open mass airflow sensor circuit, or connector damage. • ECM failed.

DTC	Trouble Code Title, Conditions & Possible Causes
DTC: P0103 **1T ECM, MIL: Yes** **Year:** 2009, 2010 **Model:** Lancer **Engine:** 2.0L L4 VIN U **Transmission:** All	**Mass Airflow Circuit High Input (Non Turbo):** After More than 3 seconds have passed since the ignition switch was turned to "ON" position. Mass airflow sensor output voltage has continued to be higher than 4.9 volts (corresponding to an air flow rate of 263 g/sec) for 2 seconds. **Possible Causes:** • Mass airflow sensor failed. • Open mass airflow sensor circuit, or connector damage. • ECM failed.
DTC: P0106 **1T ECM, MIL: Yes** **Year:** 2009, 2010 **Model:** Lancer **Engine:** 2.0L L4 VIN U, 2.0L L4 VIN V, 2.4L L4 VIN W **Transmission:** All	**Manifold Absolute Pressure Circuit Range/Performance Problem:** After 8 minutes or more have passed since the engine starting sequence was completed, when the engine coolant temperature at engine start is 0°C (32°F) or lower. Engine speed is between 500 r/min and 1,500 r/min. Throttle position sensor output voltage is 0.8 volt or lower. The manifold absolute pressure sensor output voltage has continued to be 3.5 volts [corresponding to a manifold absolute pressure of 89 kPa (26.3 in. Hg)] or higher for 2 seconds. Or, The engine speed is 1,500 r/min or higher. Throttle position sensor output voltage is 3.5 volts or higher. Manifold absolute pressure sensor output voltage has continued to be 1.8 volts [corresponding to a manifold absolute pressure of 45 kPa (13.3 in. Hg)] or lower for 2 seconds. **Possible Causes:** • Manifold absolute pressure sensor failed. • Harness damage. • Connector damage. • ECM failed.
DTC: P0106 **1T PCM, MIL: Yes** **Year:** 2009, 2010 **Model:** Eclipse, Endeavor, Galant, Outlander **Engine:** 2.4L L4 VIN F, 2.4L L4 VIN W, 3.0L V6 VIN X, 3.8L V6 VIN S, 3.8L V6 VIN T **Transmission:** All	**Manifold Absolute Pressure Circuit Range/Performance Problem:** After 8 minutes or more have passed since the engine starting sequence was completed, when the engine coolant temperature at engine start is 0°C (32°F) or lower. Engine speed is higher than 1,500 r/min. Throttle position sensor output voltage is higher than 3.5 volts. Or Throttle position sensor output voltage is lower than 0.9 volt. The manifold absolute pressure is 89 kPa (26.3 in. Hg) or higher for 2 seconds. **Possible Causes:** • Manifold absolute pressure sensor failed • Harness damage • Connector damage • ECM failed <M/T> • PCM failed <A/T>
DTC: P0107 **1T PCM, MIL: Yes** **Year:** 2009 **Model:** Raider **Engine:** 3.7L V6 VIN K **Transmission:** All	**Manifold Absolute Pressure Circuit Low Input:** Engine speed between 600 to 3500 RPM. Battery voltage greater than 10 volts. The PCM detects that the MAP Sensor input voltage is below the minimum acceptable value. **Possible Causes:** • 5-Volt supply circuit open • 5-Volt supply circuit shorted to ground • MAP signal circuit shorted to ground • MAP signal circuit shorted to the sensor ground circuit • Excessive resistance in the MAP signal circuit • MAP sensor • PCM
DTC: P0107 **1T ECM, MIL: Yes** **Year:** 2009, 2010 **Model:** Lancer **Engine:** 2.0L L4 VIN V **Transmission:** All	**Manifold Absolute Pressure Circuit Low Input (Turbo Charged):** More than 8 minutes have passed since the engine starting sequence was completed, when the engine coolant temperature at engine start is less than 0°C (32°F). Volumetric efficiency is more than 20 percent. Manifold absolute pressure sensor output voltage is less than 0.2 volt [corresponding to a manifold absolute pressure of 13 kPa (3.8 in. Hg)] for 2 seconds. **Possible Causes:** • Manifold absolute pressure sensor failed. • Open or shorted manifold absolute pressure sensor circuit, harness damage or connector damage. • ECM failed.
DTC: P0107 **1T ECM, MIL: Yes** **Year:** 2009, 2010 **Model:** Lancer **Engine:** 2.0L L4 VIN U, 2.4L L4 VIN W **Transmission:** All	**Manifold Absolute Pressure Circuit Low Input:** After 8 minutes or more have passed since the engine starting sequence was completed, when the engine coolant temperature at engine start is 0°C (32°F) or lower. Volumetric efficiency is higher than 20 percent. Manifold absolute pressure sensor output voltage has continued to be 0.2 volt [corresponding to a manifold absolute pressure of 5 kPa (1.5 in. Hg)] or lower for 2 seconds. **Possible Causes:** • Manifold absolute pressure sensor failed. • Open or shorted manifold absolute pressure sensor circuit, harness damage or connector damage. • ECM failed.

DTC	Trouble Code Title, Conditions & Possible Causes
DTC: P0107 **1T PCM, MIL: Yes** **Year:** 2009, 2010 **Model:** Eclipse, Endeavor, Galant, Outlander **Engine:** 2.4L L4 VIN F, 2.4L L4 VIN W, 3.0L V6 VIN X, 3.8L V6 VIN S, 3.8L V6 VIN T **Transmission:** All	**Manifold Absolute Pressure Circuit Low Input:** After 8 minutes or more have passed since the engine starting sequence was completed, when the engine coolant temperature at engine start is 0°C (32°F) or lower. Volumetric efficiency is higher than 20 percent. The manifold absolute pressure is 5 kPa (1.5 in. Hg) or lower for 2 seconds. **Possible Causes:** • Manifold absolute pressure sensor failed • Open or shorted manifold absolute pressure sensor circuit, harness damage or connector damage • ECM failed <M/T> • PCM failed <A/T>
DTC: P0108 **1T ECM, MIL: Yes** **Year:** 2009, 2010 **Model:** Lancer **Engine:** 2.0L L4 VIN U, 2.4L L4 VIN W **Transmission:** All	**Manifold Absolute Pressure Circuit High Input:** After 8 minutes or more have passed since the engine starting sequence was completed, when the engine coolant temperature at engine start is 0°C (32°F) or lower. The manifold absolute pressure sensor output voltage has continued to be 4.6 volts [corresponding to a manifold absolute pressure of 117 kPa (34.6 in. Hg)] or higher for 2 seconds. **Possible Causes:** • Manifold absolute pressure sensor failed. • Open manifold absolute pressure sensor circuit, or connector damage. • ECM failed.
DTC: P0108 **1T PCM, MIL: Yes** **Year:** 2009, 2010 **Model:** Eclipse, Endeavor, Galant, Outlander **Engine:** 2.4L L4 VIN F, 2.4L L4 VIN W, 3.0L V6 VIN X, 3.8L V6 VIN S, 3.8L V6 VIN T **Transmission:** All	**Manifold Absolute Pressure Circuit High Input:** After 8 minutes or more have passed since the engine starting sequence was completed, when the engine coolant temperature at engine start is 0°C (32°F) or lower. The manifold absolute pressure is 117 kPa (34.6 in. Hg) or higher for 2 seconds. **Possible Causes:** • Manifold absolute pressure sensor failed • Open manifold absolute pressure sensor circuit, or connector damage • ECM failed <M/T> • PCM failed <A/T>
DTC: P0108 **1T ECM, MIL: Yes** **Year:** 2009, 2010 **Model:** Lancer **Engine:** 2.0L L4 VIN V **Transmission:** All	**Manifold Absolute Pressure Circuit High Input (Turbo Charged):** More than 8 minutes have passed since the engine starting sequence was completed, when the engine coolant temperature at engine start is less than 0°C (32°F). The manifold absolute pressure sensor output voltage is more than 4.6 volts [corresponding to a manifold absolute pressure of 314 kPa (92.7 in. Hg)] for 2 seconds. **Possible Causes:** • Manifold absolute pressure sensor failed. • Open manifold absolute pressure sensor circuit, or connector damage. • ECM failed.
DTC: P0108 **1T PCM, MIL: Yes** **Year:** 2009 **Model:** Raider **Engine:** 3.7L V6 VIN K **Transmission:** All	**Manifold Absolute Pressure Circuit High Input:** Engine speed between 600 to 3500 RPM. Battery voltage greater than 10.37 volts. The MAP sensor signal voltage is greater than the maximum allowable voltage. **Possible Causes:** • MAP signal circuit shorted to voltage • MAP signal circuit open • MAP signal circuit shorted to 5-volt supply circuit • MAP sensor ground circuit open • MAP sensor • PCM
DTC: P0111 **1T ECM, MIL: Yes** **Year:** 2009, 2010 **Model:** Lancer **Engine:** 2.0L L4 VIN U, 2.0L L4 VIN V, 2.4L L4 VIN W **Transmission:** All	**Intake Air Temperature Circuit Range/Performance Problem:** Engine coolant temperature is higher than 76°C (169°F). Repeat 2 or more times: drive*1, stop*2. Drive*1: vehicle speed higher than 50 km/h (31 mph) lasting a total of more than 60 seconds. Stop*2: vehicle speed lower than 1.5 km/h (1.0 mph) lasting more than 30 seconds. Changes in the intake air temperature is lower than 1°C (1.8°F). **Possible Causes:** • Intake air temperature sensor failed. • Harness damage or connector damage. • ECM failed.

DTC	Trouble Code Title, Conditions & Possible Causes
DTC: P0111 **2T PCM, MIL: Yes** **Year:** 2009 **Model:** Raider **Engine:** 3.7L V6 VIN K **Transmission:** All	**Intake Air Temperature Sensor Rationality:** The engine off time is greater than 480 minutes. Ambient Temperature if greater than -64°C (-83°F). Once the vehicle is soaked for a calibrated engine off time and then driven over calibrated speed and load conditions for some calibrated time, the PCM compares the ambient, engine coolant, and intake air temperature sensor values. If engine coolant and ambient air temperature sensors agree with each other but intake air temperature does not agree with them, the intake air temperature sensor is declared as irrational. **Possible Causes:** • Excessive resistance in the IAT signal circuit • IAT signal circuit shorted to voltage • IAT signal circuit shorted to ground • IAT signal circuit shorted to the sensor ground circuit • Excessive resistance in the IAT in the sensor ground circuit • IAT Sensor • POWERTRAIN CONTROL MODULE (PCM)
DTC: P0111 **1T PCM, MIL: Yes** **Year:** 2009, 2010 **Model:** Eclipse, Endeavor, Galant, Outlander **Engine:** 2.4L L4 VIN F, 2.4L L4 VIN W, 3.0L V6 VIN X, 3.8L V6 VIN S, 3.8L V6 VIN T **Transmission:** All	**Intake Air Temperature Circuit Range/Performance Problem:** Engine coolant temperature is higher than 76°C (169°F). Repeat 2 or more times: drive the vehicle with speed higher than 50 km/h (31 mph) lasting a total of more than 60 seconds and vehicle speed lower than 1.5 km/h (1.0 mph) lasting more than 30 seconds. The change in the intake air temperature is lower than 1°C (1.8°F). **Possible Causes:** • Intake air temperature sensor failed • Open or intake air temperature sensor circuit, harness damage, or connector damage • ECM failed <M/T> • PCM failed <A/T>
DTC: P0112 **1T PCM, MIL: Yes** **Year:** 2009 **Model:** Raider **Engine:** 3.7L V6 VIN K **Transmission:** All	**Intake Air Temperature Sensor Circuit Low:** With the ignition on and battery voltage greater than 10.4 volts. When the Inlet Air Temp Sensor Signal circuit voltage is less than the minimum acceptable voltage. **Possible Causes:** • IAT Signal circuit shorted to ground • IAT Signal circuit shorted to the sensor ground circuit • Excessive resistance in the IAT signal circuit • IAT Sensor • PCM
DTC: P0112 **1T PCM, MIL: Yes** **Year:** 2009, 2010 **Model:** Eclipse, Endeavor, Galant, Outlander **Engine:** 2.4L L4 VIN F, 2.4L L4 VIN W, 3.0L V6 VIN X, 3.8L V6 VIN S, 3.8L V6 VIN T **Transmission:** All	**Intake Air Temperature Circuit Low Input:** After 2 seconds or more have passed since the engine starting sequence was completed. The intake air temperature sensor output voltage has continued to be 0.2 volt or lower [corresponding to an air intake temperature of 115°C (239°F) or higher] for 2 seconds. **Possible Causes:** • Intake air temperature sensor failed • Shorted intake air temperature sensor circuit, or connector damage • ECM failed <M/T> • PCM failed <A/T>
DTC: P0112 **1T PCM, MIL: Yes** **Year:** 2009, 2010 **Model:** Lancer **Engine:** 2.0L L4 VIN V **Transmission:** All	**Intake Air Temperature Circuit Low Input (Sensor 1) (Turbo Charged):** After 2 seconds or more have passed since the engine starting sequence was completed. Intake air temperature sensor output voltage has continued to be 0.2 volt or lower [corresponding to an intake air temperature of 115°C (239°F) or higher] for 2 seconds. **Possible Causes:** • Intake air temperature sensor failed. • Shorted intake air temperature sensor circuit, or connector damage. • ECM failed.
DTC: P0112 **1T ECM, MIL: Yes** **Year:** 2009, 2010 **Model:** Lancer **Engine:** 2.0L L4 VIN U, 2.4L L4 VIN W **Transmission:** All	**Intake Air Temperature Circuit Low Input:** After 2 seconds or more have passed since the engine starting sequence was completed. Intake air temperature sensor output voltage has continued to be 0.2 volt or lower [corresponding to an intake air temperature of 115°C (239°F) or higher] for 2 seconds. **Possible Causes:** • Intake air temperature sensor failed. • Shorted intake air temperature sensor circuit, or connector damage. • ECM failed.

DTC	Trouble Code Title, Conditions & Possible Causes
DTC: P0113 **1T ECM, MIL: Yes** **Year:** 2009, 2010 **Model:** Lancer **Engine:** 2.0L L4 VIN V **Transmission:** All	**Intake Air Temperature Circuit High Input (Sensor 1) (Turbo Charged):** More than 2 seconds have passed since the engine starting sequence was completed. Intake air temperature sensor 1 output voltage is more than 4.6 volts [corresponding to an intake air temperature of -40°C (-40°F) or less] for 2 seconds. **Possible Causes:** • Intake air temperature sensor 1 failed. • Open intake air temperature sensor 1 circuit, or connector damage. • ECM failed.
DTC: P0113 **1T PCM, MIL: Yes** **Year:** 2009 **Model:** Raider **Engine:** 3.7L V6 VIN K **Transmission:** All	**Intake Air Temperature Sensor Circuit High:** With the ignition on. Battery voltage greater than 10.4 volts. The PCM detects that the Intake Air Temperature Sensor input voltage is above the maximum acceptable value. **Possible Causes:** • IAT Signal circuit shorted to voltage • IAT Signal circuit open • Sensor ground circuit open • IAT Sensor • PCM
DTC: P0113 **1T ECM, MIL: Yes** **Year:** 2009, 2010 **Model:** Lancer **Engine:** 2.0L L4 VIN U, 2.4L L4 VIN W **Transmission:** All	**Intake Air Temperature Circuit High Input:** After 2 seconds or more have passed since the engine starting sequence was completed. The intake air temperature sensor output voltage has continued to be 4.6 volts or higher [corresponding to an intake air temperature of -40°C (-40°F) or lower] for 2 seconds. **Possible Causes:** • Intake air temperature sensor failed. • Open intake air temperature sensor circuit, or connector damage. • ECM failed.
DTC: P0113 **1T PCM, MIL: Yes** **Year:** 2009, 2010 **Model:** Eclipse, Endeavor, Galant, Outlander **Engine:** 2.4L L4 VIN F, 2.4L L4 VIN W, 3.0L V6 VIN X, 3.8L V6 VIN S, 3.8L V6 VIN T **Transmission:** All	**Intake Air Temperature Circuit High Input:** After 2 seconds or more have passed since the engine starting sequence was completed. The intake air temperature sensor output voltage has continued to be 4.6 volts or higher [corresponding to an air intake temperature of -40°C (-40°F) or lower] for 2 seconds. **Possible Causes:** • Intake air temperature sensor failed • Open intake air temperature sensor circuit, or connector damage • ECM failed <M/T> • PCM failed <A/T>
DTC: P0116 **2T PCM, MIL: Yes** **Year:** 2009 **Model:** Raider **Engine:** 3.7L V6 VIN K **Transmission:** All	**Engine Coolant Temperature Sensor Circuit Performance:** Engine off time is greater than 480 minutes and the vehicle has been driven for one minute over 35 mph. Ambient temperature is greater than -64°C (-83°F). Once the vehicle is soaked for a calibrated engine off time and then driven over calibrated speed and load conditions for some calibrated time, the PCM compares the ambient, engine coolant, and intake air temperature sensor values. If ambient air and intake air temperature sensors agree with each other but engine coolant temperature does not agree with them, the engine coolant temperature sensor is declared as irrational **Possible Causes:** • Excessive resistance in the ECT signal circuit • ECT signal circuit shorted to ground • ECT signal circuit shorted to the sensor ground circuit • ECT signal circuit shorted to voltage • Excessive resistance in the sensor ground circuit • ECT Sensor • PCM
DTC: P0116 **1T ECM, MIL: Yes** **Year:** 2009, 2010 **Model:** Lancer **Engine:** 2.0L L4 VIN U, 2.0L L4 VIN V, 2.4L L4 VIN W **Transmission:** All	**Engine Coolant Temperature Circuit Range/Performance Problem:** Engine coolant temperature was more than 0°C (32°F) when the engine started. The accumulation is more than 300 seconds during the acceleration having the mass airflow rate of 12 g/sec or more. The accumulation is more than 30 seconds during the deceleration having the mass airflow rate of 9 g/sec or less. Engine coolant temperature fluctuates within 1°C (1.8°F) after 330 seconds have passed since the engine was started. However, time is not counted if any of the following conditions are met. * Intake air temperature is more than 60°C (140°F). * During fuel shut-off operation. **Possible Causes:** • Engine coolant temperature sensor failed. • Harness damage or connector damage. • ECM failed.

DTC	Trouble Code Title, Conditions & Possible Causes
DTC: P0116 **1T PCM, MIL: Yes** **Year:** 2009, 2010 **Model:** Eclipse, Endeavor, Galant, Outlander **Engine:** 2.4L L4 VIN F, 2.4L L4 VIN W, 3.0L V6 VIN X, 3.8L V6 VIN S, 3.8L V6 VIN T **Transmission:** All	**Engine Coolant Temperature Circuit Range/Performance Problem:** Engine coolant temperature was 7°C (45°F) or more when the engine started. Engine coolant temperature fluctuates within 1°C (1.8°F) after 330 seconds have passed since the engine was started. The accumulated time in the acceleration is 300 sec or more. Acceleration*: The air flow rate is 12 g/sec or more. Deceleration: The accumulated time in the deceleration is 30 sec or more. The air flow rate is 9 g/sec or more. **NOTE: However, time is not counted if any of the following conditions are met.** * Intake air temperature is 60°C (140°F) or more. * Mass airflow sensor output is 15 g/sec or less. * During fuel shut-off operation. **Possible Causes:** • Engine coolant temperature sensor failed • Harness damage or connector damage • ECM failed <M/T> • PCM failed <A/T>
DTC: P0117 **1T ECM, MIL: Yes** **Year:** 2009, 2010 **Model:** Lancer **Engine:** 2.0L L4 VIN U, 2.0L L4 VIN V, 2.4L L4 VIN W **Transmission:** All	**Engine Coolant Temperature Circuit Low Input:** After 2 seconds or more have passed since the engine starting sequence was completed. The engine coolant temperature sensor output voltage has continued to be 0.1 volt or lower for 2 seconds. **Possible Causes:** • Engine coolant temperature sensor failed. • Shorted engine coolant temperature sensor circuit, or connector damage. • ECM failed.
DTC: P0117 **3T PCM, MIL: Yes** **Year:** 2009 **Model:** Raider **Engine:** 3.7L V6 VIN K **Transmission:** All	**Transmission Temperature Sensor Low :** Continuously with the ignition on and engine running. The DTC will set when the monitored Temperature Sensor voltage drops below 0.078 volts for the period of 1.45 seconds. When the fault is set, calculated temperature is substituted for measured temperature, however the fault code is stored only after three consecutive occurrences of the fault. **Possible Causes:** • Related DTCs are present • Temperature temperature sensor signal circuit is shorted to ground • Temperature temperature sensor is damaged or has failed (it may be shorted) • Intermittent wiring or connector problems • PCM has failed
DTC: P0117 **1T PCM, MIL: Yes** **Year:** 2009, 2010 **Model:** Eclipse, Endeavor, Galant, Outlander **Engine:** 2.4L L4 VIN F, 2.4L L4 VIN W, 3.0L V6 VIN X, 3.8L V6 VIN S, 3.8L V6 VIN T **Transmission:** All	**Engine Coolant Temperature Circuit Low Input:** After 2 seconds or more have passed since the engine starting sequence was completed. The engine coolant temperature sensor output voltage has continued to be 0.1 volt or lower [corresponding to coolant temperature of 140°C (284°F) or higher] for 2 seconds. **Possible Causes:** • Engine coolant temperature sensor failed • Shorted engine coolant temperature sensor circuit, or connector damage • ECM failed <M/T> • PCM failed <A/T>
DTC: P0117 **1T PCM, MIL: Yes** **Year:** 2009 **Model:** Raider **Engine:** 3.7L V6 VIN K **Transmission:** All	**Engine Coolant Temperature Circuit Low :** With the ignition on. Battery voltage greater than 10.4 volts. The PCM detects that the Engine Coolant Temperature Sensor input voltage is below the minimum acceptable value. **Possible Causes:** • ECT Signal circuit shorted to voltage • ECT Signal circuit shorted to the sensor ground circuit • Excessive resistance in the ECT signal circuit • ECT Sensor • PCM
DTC: P0118 **1T PCM, MIL: Yes** **Year:** 2009, 2010 **Model:** Eclipse, Endeavor, Galant **Engine:** 3.8L V6 VIN S, 3.8L V6 VIN T **Transmission:** All	**Engine Coolant Temperature Circuit High Input:** After 2 seconds or more have passed since the engine starting sequence was completed. The engine coolant temperature sensor output voltage has continued to be 0.1 volt or lower [corresponding to coolant temperature of 140°C (284°F) or higher] for 2 seconds. **Possible Causes:** • Engine coolant temperature sensor failed • Open engine coolant temperature sensor circuit, or connector damage • ECM failed <M/T> • PCM failed <A/T>

DTC	Trouble Code Title, Conditions & Possible Causes
DTC: P0118 **1T PCM, MIL: Yes** **Year:** 2009 **Model:** Raider **Engine:** 3.7L V6 VIN K **Transmission:** All	**Engine Coolant Temperature Circuit High Input:** With the ignition on. Battery voltage greater than 10.4 volts. The PCM detects that the Engine Coolant Temperature Sensor input voltage is above the maximum acceptable value. **Possible Causes:** • ECT signal circuit shorted to voltage • ECT signal circuit open • Sensor ground circuit open • ECT Sensor • PCM
DTC: P0118 **1T PCM, MIL: Yes** **Year:** 2009, 2010 **Model:** Eclipse, Galant, Outlander **Engine:** 2.4L L4 VIN F, 2.4L L4 VIN W, 3.0L V6 VIN X **Transmission:** All	**Engine Coolant Temperature Circuit High Input:** After 2 seconds or more have passed since the engine starting sequence was completed. The engine coolant temperature sensor output voltage has continued to be 4.6 volts or higher [corresponding to coolant temperature of -45°C (-49°F) or lower] for 2 seconds. **Possible Causes:** • Engine coolant temperature sensor failed. • Open engine coolant temperature sensor circuit, or connector damage. • ECM failed. <M/T> • PCM failed. <A/T>
DTC: P0118 **1T ECM, MIL: Yes** **Year:** 2009, 2010 **Model:** Lancer **Engine:** 2.0L L4 VIN U, 2.0L L4 VIN V, 2.4L L4 VIN W **Transmission:** All	**Engine Coolant Temperature Circuit High Input:** After 2 seconds or more have passed since the engine starting sequence was completed. The engine coolant temperature sensor output voltage has continued to be 4.6 volts or higher for 2 seconds. **Possible Causes:** • Engine coolant temperature sensor failed. • Open engine coolant temperature sensor circuit, or connector damage. • ECM failed.
DTC: P0121 **2T PCM, MIL: Yes** **Year:** 2009 **Model:** Raider **Engine:** 3.7L V6 VIN K **Transmission:** All	**Throttle Position Sensor 1 Performance:** Ignition on and No MAP Sensor DTCs set. The PCM detects that the sensor input voltage does not fall within a valid range based on engine speed and load. **NOTE: P2135 should set with this code.** **Possible Causes:** • Excessive resistance in the TP sensor NO.1 signal circuit • Excessive resistance in the TP sensor NO.2 signal circuit • Excessive resistance in the 5-volt supply circuit • Excessive resistance in the TP sensor return circuit • TP Sensor NO.1 signal circuit shorted to the TP sensor NO.2 signal circuit • TP Sensor/Throttle body • PCM
DTC: P0122 **1T ECM, MIL: Yes** **Year:** 2009, 2010 **Model:** Lancer, Outlander **Engine:** 2.0L L4 VIN U, 2.0L L4 VIN V, 2.4L L4 VIN W, 3.0L V6 VIN X **Transmission:** All	**Throttle Position Sensor (Main) Circuit Low Input:** With the ignition switch is "ON" position. The throttle position sensor (main) output voltage should be 0.2 volt or less for 0.3 second. **Possible Causes:** • Throttle position sensor failed. • Shorted throttle position sensor (main) circuit, harness damage, or connector damage. • ECM failed.
DTC: P0122 **1T PCM, MIL: Yes** **Year:** 2009 **Model:** Raider **Engine:** 3.7L V6 VIN K **Transmission:** All	**Throttle Position Sensor 1 Circuit Low:** With the ignition on. Battery voltage greater than 10 volts. The PCM detects the TP Sensor 1 voltage below the minimum acceptable voltage. **Possible Causes:** • 5-Volt supply circuit open • 5-Volt supply circuit shorted to ground • TP Sensor NO.1 signal circuit shorted to ground • TP Sensor NO.1 signal circuit shorted to the TP sensor return circuit • Excessive resistance in the TP sensor NO.1 signal circuit • TP Sensor/Throttle body • PCM

DTC	Trouble Code Title, Conditions & Possible Causes
DTC: P0122 **1T PCM, MIL: Yes** **Year:** 2009 **Model:** Raider **Engine:** 3.7L V6 VIN K **Transmission:** All	**TPS/APP Circuit Low :** Continuously with the ignition on and engine running. This DTC will set if the monitored TPS voltage drops below .078 volts for the period of 0.48 seconds. **Possible Causes:** • Related engine TPS/APPS DTC's present • PCM
DTC: P0122 **1T PCM, MIL: Yes** **Year:** 2009, 2010 **Model:** Eclipse, Endeavor, Galant **Engine:** 2.4L L4 VIN F, 3.8L V6 VIN S, 3.8L V6 VIN T **Transmission:** All	**Throttle Position Sensor (Main) Circuit Low Input:** When the ignition switch is in the "ON" position. Throttle position sensor (main) output voltage should be 0.35 volt or less for 0.5 second. Throttle position sensor (main) output voltage is out of specified range. **Possible Causes:** • Throttle position sensor failed • Open or shorted throttle position sensor (main) circuit, harness damage, or connector damage • ECM failed <M/T> • PCM failed <A/T>
DTC: P0123 **1T PCM, MIL: Yes** **Year:** 2009, 2010 **Model:** Eclipse, Endeavor, Galant, Outlander **Engine:** 2.4L L4 VIN F, 2.4L L4 VIN W, 3.0L V6 VIN X, 3.8L V6 VIN S, 3.8L V6 VIN T **Transmission:** All	**Throttle Position Sensor (main) Circuit High Input:** With the ignition switch ON, Throttle position sensor (main) output voltage should be 4.8 volts or more for 0.5 second. Throttle position sensor (main) output voltage is out of specified range. **Possible Causes:** • Throttle position sensor failed • Open throttle position sensor (main) circuit, or harness damage or connector damage • ECM failed <M/T> • PCM failed <A/T>
DTC: P0123 **1T PCM, MIL: Yes** **Year:** 2009 **Model:** Raider **Engine:** 3.7L V6 VIN K **Transmission:** All	**Throttle Position Sensor 1 Circuit High:** With the ignition on. Battery voltage greater than 10 volts. The PCM detects TP Sensor 1 voltage above the maximum acceptable voltage. **Possible Causes:** • 5-Volt supply circuit open • 5-Volt supply circuit shorted to ground • TP Sensor No.1 signal circuit shorted to ground • TP Sensor No.1 signal circuit shorted to the TP sensor return circuit • Excessive resistance in the TP sensor NO.1 signal circuit • TP Sensor/Throttle body • PCM
DTC: P0123 **1T PCM, MIL: Yes** **Year:** 2009 **Model:** Raider **Engine:** 3.7L V6 VIN K **Transmission:** All	**TPS/APP Circuit High:** Continuously with the ignition on and engine running. This DTC will set if the monitored TPS voltage rises above 4.94 volts for the period of 0.48 seconds. **Possible Causes:** • Related engine TPS/APPS DTC's present • PCM
DTC: P0123 **1T ECM, MIL: Yes** **Year:** 2009, 2010 **Model:** Lancer **Engine:** 2.0L L4 VIN U, 2.0L L4 VIN V, 2.4L L4 VIN W **Transmission:** All	**Throttle Position Sensor (Main) Circuit High Input:** With the ignition switch is "ON". The throttle position sensor (main) output voltage should be 4.8 volts or more for 0.3 second. **Possible Causes:** • Throttle position sensor failed. • Open or shorted throttle position sensor (main) circuit, harness damage or connector damage. • ECM failed.
DTC: P0124 **1T PCM, MIL: Yes** **Year:** 2009 **Model:** Raider **Engine:** 3.7L V6 VIN K **Transmission:** All	**TPS/APP Sensor Intermittent :** Continuously with the ignition on and engine running. This DTC will set if the monitored TPS throttle angle between the angles of 6° and 120° and the degree change is greater than 5° within a period of less than 7.0 msec. **Possible Causes:** • Related engine TPS DTC's present • TPS Sensor • PCM

DTC	Trouble Code Title, Conditions & Possible Causes
DTC: P0125 **2T PCM, MIL: Yes** **Year:** 2009 **Model:** Raider **Engine:** 3.7L V6 VIN K **Transmission:** All	**Insufficient Coolant Temperature for Closed Loop Fuel Control:** With battery voltage greater than 10.4 volts and after engine is started. The engine temperature does not go above -10°C (15°F). Failure time depends on start-up coolant temperature and ambient temperature. (i.e. 2 minutes for a start temp of -10°C (15° F) or up to 10 minutes for a vehicle with a start-up temp of -28°C (5°F). **Possible Causes:** • Low coolant level • Thermostat • ECT Sensor
DTC: P0125 **1T ECM, MIL: Yes** **Year:** 2009, 2010 **Model:** Lancer, Outlander **Engine:** 2.0L L4 VIN U, 2.0L L4 VIN V, 2.4L L4 VIN W, 3.0L V6 VIN X **Transmission:** All	**Insufficient Coolant Temperature for Closed Loop Fuel Control:** About 90 - 300 seconds have passed for the engine coolant temperature to rise to about 0°C (32°F) after starting sequence was completed. However, time is not counted when fuel is shut off. Control as if the engine coolant temperature is 80 °C (176°F). **Possible Causes:** • Engine coolant temperature sensor failed. • Harness damage in engine coolant temperature sensor circuit or connector damage. • ECM failed.
DTC: P0125 **1T PCM, MIL: Yes** **Year:** 2009, 2010 **Model:** Eclipse, Endeavor, Galant **Engine:** 2.4L L4 VIN F, 3.8L V6 VIN S, 3.8L V6 VIN T **Transmission:** All	**Insufficient Coolant Temperature for Closed Loop Fuel Control:** Engine coolant temperature decreases from higher than 40°C (104°F) to lower than 40°C (104°F). Then the engine coolant temperature has continued to be 40°C (104°F) or lower for 5 minutes. However, time is not counted when fuel is shut off. About 90 - 300 seconds have passed for the engine coolant temperature to rise to about 7°C (45°F) after starting sequence was completed. **Possible Causes:** • Engine coolant temperature sensor failed • Harness damage in engine coolant temperature sensor circuit or connector damage • ECM failed.<M/T> • PCM failed <A/T>
DTC: P0128 **1T PCM, MIL: Yes** **Year:** 2009, 2010 **Model:** Eclipse, Endeavor, Galant, Outlander **Engine:** 2.4L L4 VIN F, 2.4L L4 VIN W, 3.0L V6 VIN X, 3.8L V6 VIN S, 3.8L V6 VIN T **Transmission:** All	**Coolant Thermostat (Coolant Temperature Below Thermostat Regulating Temperature):** Engine coolant temperature is between -10°C (14°F) and 60°C (140°F) when the engine is started. Intake air temperature is -10°C (14°F) or higher. The intake air temperature when the engine is started - intake air temperature is 5°C (9°F) or less. The total time of the mass airflow sensor output whose state is below 10 g/sec within 300 seconds. The time for the engine coolant temperature to rise to 77°C (171°F) takes longer than approximately 9.5 to 33.5 minutes. Engine coolant temperature does not reach 77°C (171°F) within specified period after cold start. **Possible Causes:** • The thermostat is faulty • ECM failed <M/T> • PCM failed <A/T>
DTC: P0128 **2T PCM, MIL: Yes** **Year:** 2009 **Model:** Raider **Engine:** 3.7L V6 VIN K **Transmission:** All	**Thermostat Rationality:** Engine running. This diagnostic will fail when the model temperature reaches 91°C (196°F) before the ECT. reaches 80°C (176°F. This monitor only runs after a cold start condition ECT, ACT, AMB are all within 10°C (18°F) of one another. **Possible Causes:** • Low coolant Level • Thermostat operation • Signal circuit shorted to battery voltage • Temperature sensor • Signal circuit open • Sensor ground circuit open • Signal circuit shorted to ground • Signal circuit shorted to sensor ground • PCM
DTC: P0128 **1T ECM, MIL: Yes** **Year:** 2009, 2010 **Model:** Lancer **Engine:** 2.0L L4 VIN U, 2.0L L4 VIN V, 2.4L L4 VIN W **Transmission:** All	**Coolant Thermostat (Coolant Temperature Below Thermostat Regulating Temperature):** Engine coolant temperature is between -10°C (14°F) and 60°C (140°F) when the engine is started. Intake air temperature is -10°C (14°F) or higher. The intake air temperature subtracted from the intake air temperature when the engine is started is 10°C (18°F) or less. The total time when the amount of intake air is small is less than the specified time After the estimated engine coolant temperature rises above 77°C (171°F), the actual engine coolant temperature is less than 77°C (171°F) even though the specified time has passed **Possible Causes:** • The engine cooling system is faulty. • ECM failed.

DTC	Trouble Code Title, Conditions & Possible Causes
DTC: P0129 **1T PCM, MIL: Yes** **Year:** 2009 **Model:** Raider **Engine:** 3.7L V6 VIN K **Transmission:** All	**Barometric Pressure Out-Of-Range Low:** With the ignition key on. No Cam or Crank signal within 75 ms. Engine speed less than 250 RPM. The PCM senses the voltage from the MAP sensor to be less than 2.2 volts but above 0.04 of a volt for 300 milliseconds. **Possible Causes:** • 5-volt supply circuit shorted to battery • 5-volt supply circuit open • 5-volt supply circuit shorted to ground • MAP signal circuit open • MAP signal circuit shorted to ground • MAP signal circuit shorted to the sensor ground circuit • MAP Sensor • PCM
DTC: P0130 **1T ECM, MIL: Yes** **Year:** 2009, 2010 **Model:** Outlander **Engine:** 2.4L L4 VIN W **Transmission:** All	**Linear Air-Fuel Ratio Sensor Circuit (California):** Battery positive voltage is between 10 and 16.5 volts. More than 220 seconds have passed since the engine starting sequence was completed. The negative circuit voltage subtracted from the pumping circuit voltage is be lower than -2.1 volts or higher than 2.3 volts for 12 seconds. The pumping current is be lower than 0 milliampere or higher than 0 milliampere for 12 seconds. **Possible Causes:** • Linear air-fuel ratio sensor failed. • Connector damage. • ECM failed
DTC: P0131 **2T PCM, MIL: Yes** **Year:** 2009 **Model:** Raider **Engine:** 3.7L V6 VIN K **Transmission:** All	**O2 1/1 Circuit Low:** Engine running for less than 30 seconds and the O2 Sensor Heater Temperature is less than 251°C (484°F) with battery voltage greater 10.4 volts. The oxygen sensor signal voltage is below 2.5196 volts for 6 seconds after starting engine. **Possible Causes:** • O2 Return upstream circuit shorted to ground • O2 1/1 Signal circuit shorted to ground • O2 1/1 Signal circuit shorted to the O2 return upstream circuit • O2 1/1 Signal circuit shorted to the O2 1/1 heater ground circuit • O2 Sensor • PCM
DTC: P0131 **2T PCM, MIL: Yes** **Year:** 2009, 2010 **Model:** Eclipse, Galant **Engine:** 2.4L L4 VIN F **Transmission:** All	**Cylinder 1, 4 Heated Oxygen Sensor (front) Circuit Low Voltage:** Heated oxygen sensor offset voltage is between 0.4 and 0.6 volt. Battery positive voltage is between 11 and 16.5 volts. After 180 seconds have passed since the engine starting sequence was completed. Cylinder 1, 4 heated oxygen sensor (front) output voltage is lower than 0.2 volt. **Possible Causes:** • Cylinder 1, 4 heated oxygen sensor (front) failed. • Open or shorted circuit in cylinder1, 4 heated oxygen sensor (front) output line, or harness damage. • Open circuit in cylinder 1, 4 heated oxygen sensor (front) ground line, or harness damage. • Connector damage. • ECM failed. <M/T> • PCM failed. <A/T>
DTC: P0131 **2T PCM, MIL: Yes** **Year:** 2009, 2010 **Model:** Eclipse, Endeavor, Galant, Outlander **Engine:** 3.0L V6 VIN X, 3.8L V6 VIN S, 3.8L V6 VIN T **Transmission:** All	**Heated Oxygen Sensor Circuit Low Voltage (Bank 1, Sensor 1):** Heated oxygen sensor offset voltage is between 0.4 and 0.6 volts. Battery positive voltage is between 11 and 16.5 volts. 180 seconds have passed since the engine starting sequence was completed. Right bank heated oxygen sensor (front) output voltage is lower than 0.2 volt. **Possible Causes:** • Right bank heated oxygen sensor (front) failed • Open or shorted circuit in right bank heated oxygen sensor (front) output line, or harness damage • Open circuit in right bank heated oxygen sensor (front) ground line, or harness damage • Connector damage • ECM failed <M/T> • PCM failed <A/T>

DTC	Trouble Code Title, Conditions & Possible Causes
DTC: P0131 **1T ECM, MIL: Yes** **Year:** 2009, 2010 **Model:** Lancer, Outlander **Engine:** 2.0L L4 VIN U, 2.0L L4 VIN V, 2.4L L4 VIN W **Transmission:** All	**Heated Oxygen Sensor (Front) Circuit Low Voltage:** Heated oxygen sensor offset voltage is between 0.4 and 0.6 volt. Battery positive voltage is between 11 and 16.5 volts. More than 180 seconds have passed since the engine starting sequence was completed. The heated oxygen sensor (front) output voltage is lower than 0.2 volt for 2 seconds. **Possible Causes:** • Heated oxygen sensor (front) failed. • Open or shorted circuit in heated oxygen sensor (front) output line, or harness damage. • Open circuit in heated oxygen sensor (front) ground line, or harness damage. • Connector damage. • ECM failed
DTC: P0132 **2T PCM, MIL: Yes** **Year:** 2009 **Model:** Raider **Engine:** 3.7L V6 VIN K **Transmission:** All	**O2 1/1 Circuit High:** Continuously with the engine running, no O2 sensor heater DTCs present, 1/1 Oxygen Sensor heater temperature within a specific range, and battery voltage greater than 10.4 volts. The Oxygen Sensor voltage is above the maximum acceptable value. **Possible Causes:** • O2 Return upstream circuit shorted to voltage • O2 1/1 Signal circuit shorted to voltage • O2 1/1 Signal circuit open • O2 Return upstream circuit open • O2 Sensor • PCM
DTC: P0132 **1T ECM, MIL: Yes** **Year:** 2009, 2010 **Model:** Lancer, Outlander **Engine:** 2.0L L4 VIN U, 2.0L L4 VIN V, 2.4L L4 VIN W **Transmission:** All	**Heated Oxygen Sensor (Front) Circuit High Voltage :** After 2 seconds or more have passed since the engine starting sequence was completed. Heated oxygen sensor offset voltage is between 0.4 and 0.6 volt. The heated oxygen sensor (front) output voltage has continued to be 1.8 volts or higher for 2 seconds. **Possible Causes:** • Short circuit in heated oxygen sensor (front) output line. • Connector damage. • ECM failed.
DTC: P0132 **2T PCM, MIL: Yes** **Year:** 2009, 2010 **Model:** Eclipse, Galant **Engine:** 2.4L L4 VIN F **Transmission:** All	**Cylinder 1, 4 Heated Oxygen Sensor (front) Circuit High Voltage:** After 2 seconds or more have passed since the engine starting sequence was completed. Heated oxygen sensor offset voltage is between 0.4 and 0.6 volt. Cylinder 1, 4 heated oxygen sensor (front) output voltage has continued to be 1.8 volts or higher for 2 seconds. **Possible Causes:** • Short circuit in cylinder 1, 4 heated oxygen sensor (front) output line. • Connector damage. • ECM failed. <M/T> • PCM failed. <A/T>
DTC: P0132 **2T PCM, MIL: Yes** **Year:** 2009, 2010 **Model:** Eclipse, Endeavor, Galant, Outlander **Engine:** 3.0L V6 VIN X, 3.8L V6 VIN S, 3.8L V6 VIN T **Transmission:** All	**Heated Oxygen Sensor Circuit High Voltage (Bank 1, Sensor 1):** After 2 seconds or more have passed since the engine starting sequence was completed. Heated oxygen sensor off set voltage is between 0.4 and 0.6 volt. The right bank heated oxygen sensor (front) output voltage has continued to be 1.8 volts or higher for 2 seconds. **Possible Causes:** • Short circuit in right bank heated oxygen sensor (front) output line • Connector damage • ECM failed <M/T> • PCM failed <A/T>
DTC: P0133 **2T PCM** **Year:** 2009, 2010 **Model:** Eclipse, Galant **Engine:** 2.4L L4 VIN F **Transmission:** All	**Cylinder 1, 4 Heated Oxygen Sensor (front) Circuit Slow Response:** Engine coolant temperature is higher than 60°C (140°F). Engine speed is between 1,500 and 3,500 r/min <M/T> or 1,200 and 3,500 r/min <A/T>. Volumetric efficiency is between 18 and 70 percent <M/T> or 25 and 70 percent <A/T>. Under the closed loop air/fuel control. The accelerator pedal is depressed. Short-term fuel trim is between -25 and +25 percent. More than 2 seconds have elapsed after the above mentioned conditions have been met. During the drive cycle, the ECM <M/T> or the PCM <A/T> performs monitoring with the accumulated total time of 10 seconds, 5 times. The average of the cylinder 1, 4 heated oxygen sensor (front) rich/lean switching frequency is less than 12 times for the accumulated time of 10 seconds. The cylinder 1, 4 heated oxygen sensor (front) rich/lean switching frequency is less than 25 times for the accumulated time of 10 seconds. **Possible Causes:** • Cylinder 1, 4 heated oxygen sensor (front) deteriorated. • Connector damage. • ECM failed. <M/T> • PCM failed. <A/T>

DTC	Trouble Code Title, Conditions & Possible Causes
DTC: P0133 **2T PCM, MIL: Yes** **Year:** 2009 **Model:** Raider **Engine:** 3.7L V6 VIN K **Transmission:** All	**O2 1/1 Slow Response:** With the ECT above 70 C (158 F), engine RPM between 1400 and 2300, vehicle speed between 64 and 96 kph (40 and 60 mph), and engine run time greater than 3 minutes. The PCM detects that the oxygen sensor signal does not switch adequately during monitoring. **Possible Causes:** • Exhaust leak • O2 1/1 Signal circuit • O2 Return upstream circuit • O2 Sensor
DTC: P0133 **2T PCM, MIL: Yes** **Year:** 2009, 2010 **Model:** Eclipse, Endeavor, Galant, Outlander **Engine:** 3.0L V6 VIN X, 3.8L V6 VIN S, 3.8L V6 VIN T **Transmission:** All	**Heated Oxygen Sensor Circuit Slow Response (Bank 1, Sensor 1):** The engine coolant temperature is higher than 50°C (122°F). Engine speed is between 1,500 and 3,000 r/min <M/T> or 1,300 and 3,000 r/min <A/T>. Volumetric efficiency is between 12 and 60 <M/T> or 20 and 65 <A/T> percent. Under the closed loop air/fuel control. The accelerator pedal is dcprcsscd. Short-term fuel trim is between -30 and +25 percent. More than 2 seconds <A/T> or 3 seconds <M/T> have elapsed after the above mentioned conditions have been met. During the drive cycle, the ECM <M/T> or the PCM <A/T> performs monitoring with the accumulated total time of 12 seconds, 7 times <M/T> or 5 times <A/T>. The average of the right bank heated oxygen sensor (front) rich/lean switching frequency is less than 10 times for the accumulated time of 12 seconds. Or the right bank heated oxygen sensor (front) rich/lean switching frequency is less than 20 times for the accumulated time of 12 seconds. **Possible Causes:** • Right bank heated oxygen sensor (front) deteriorated • Connector damage • ECM failed <M/T> • PCM failed <A/T>
DTC: P0133 **1T ECM, MIL: Yes** **Year:** 2009, 2010 **Model:** Lancer, Outlander **Engine:** 2.0L L4 VIN U, 2.0L L4 VIN V, 2.4L L4 VIN W **Transmission:** All	**Heated Oxygen Sensor (Front) Circuit Slow Response :** Engine coolant temperature is higher than 60°C (140°F). Engine speed is between 1,375 and 3,500 r/min <M/T> or 1,219 and 3,000 r/min <CVT>. Volumetric efficiency is between 18 and 65 percent <M/T> or 27.5 and 65 percent <CVT>. Under the closed loop air/fuel control. The accelerator pedal is depressed. Short-term fuel trim is between -25 and +25 percent. More than 2 seconds have elapsed after the above mentioned conditions have been met. During the drive cycle, the ECM performs monitoring with the accumulated total time of 10 seconds, 5 times. The average of the heated oxygen sensor (front) rich/lean switching frequency is less than 8 times for the accumulated time of 10 seconds. The heated oxygen sensor (front) rich/lean switching frequency is less than 16 times for the accumulated time of 10 seconds. **Possible Causes:** • Heated oxygen sensor (front) deteriorated. • Connector damage. • ECM failed.
DTC: P0134 **1T ECM, MIL: Yes** **Year:** 2009, 2010 **Model:** Lancer **Engine:** 2.0L L4 VIN U, 2.0L L4 VIN V, 2.4L L4 VIN W **Transmission:** All	**Heated Oxygen Sensor (Front) No Activity Detected:** More than 120 seconds have passed since the engine starting sequence was completed. Engine coolant temperature is higher than 0°C (32°F). Engine speed is higher than 1,188 r/min. Volumetric efficiency is higher than 30 percent. Throttle position sensor output voltage is lower than 3.0 volts <M/T> or 3.8 volts <CVT>. Except while fuel is being shut off. The heated oxygen sensor (front) output voltage does not get across lean/rich criteria (about 0.5 volt) within about 30 seconds. **NOTE: When the heated oxygen sensor (front) begins to deteriorate, the heated oxygen sensor output voltage will deviate from the voltage when the sensor was new (normally 0.5 volt at stoichiometric ratio). This deviation will be corrected by the heated oxygen sensor (rear).** **If the heated oxygen sensor (rear) responds poorly because it has deteriorated, it will improperly correct the heated oxygen sensor (front). Thus, even when closed loop control is being effected, the fluctuation of the heated oxygen sensor (front) output voltage decreases, without intersecting with 0.5 volt.** **Possible Causes:** • Heated oxygen sensor (front) deteriorated. • Harness damage in heated oxygen sensor (front) output line. • Heated oxygen sensor (rear) deteriorated. • Open circuit in injector. • Harness damage in injector circuit. • Connector damage. • ECM failed. • Exhaust leak. • Air drawn in from gaps in gasket, seals, etc. • Incorrect fuel pressure

DTC	Trouble Code Title, Conditions & Possible Causes
DTC: P0134 **2T PCM, MIL: Yes** **Year:** 2009, 2010 **Model:** Eclipse, Endeavor, Galant **Engine:** 3.8L V6 VIN S, 3.8L V6 VIN T **Transmission:** All	**Heated Oxygen Sensor Circuit No Activity Detected (Bank 1, Sensor 1):** The engine coolant temperature is higher than 7°C (45°F). 128 seconds or more have passed since the engine starting sequence was completed. Engine speed is higher than 1,200 r/min. Volumetric efficiency is higher than 30 percent. Throttle position sensor output voltage is lower than 3 volts. Except while fuel is being shut off. Monitoring time: 30 seconds. Right bank heated oxygen sensor (front) output voltage does not get across lean/rich criteria (about 0.5 volt) within about 30 seconds. **NOTE: When the right bank heated oxygen sensor (front) begins to deteriorate, the heated oxygen sensor output voltage will deviate from the voltage when the sensor was new (normally 0.5 volt at stoichiometric ratio). This deviation will be corrected by the right bank heated oxygen sensor (rear).** **If the right bank heated oxygen sensor (rear) responds poorly because it has deteriorated, it will improperly correct the right bank heated oxygen sensor (front). Thus, even when closed loop control is being effected, the fluctuation of the right bank heated oxygen sensor (front) output voltage decreases, without intersecting with 0.5 volt. As a result, there is a possibility of DTC P0134 becoming registered.** **Possible Causes:** • Right bank heated oxygen sensor (front) deteriorated • Harness damage in right bank heated oxygen sensor (front) output line • Right bank heated oxygen sensor (rear) deteriorated • Open circuit in right bank injector • Harness damage in right bank injector circuit • Connector damage • Exhaust leak • Air drawn in from gaps in gasket, seals, etc • Incorrect fuel pressure • ECM failed <M/T> • PCM failed <A/T>
DTC: P0134 **2T PCM, MIL: Yes** **Year:** 2009, 2010 **Model:** Eclipse, Galant **Engine:** 2.4L L4 VIN F **Transmission:** All	**Cylinder 1, 4 Heated Oxygen Sensor (front) No Activity Detected:** After 128 seconds or more have passed since the engine starting sequence was completed. Engine coolant temperature is higher than 7°C (45°F). Engine speed is higher than 1,200 r/min. Volumetric efficiency is higher than 30 percent. Throttle position sensor output voltage is lower than 3 volts. Except while fuel is being shut off. Monitoring time: 30 seconds. Cylinder 1, 4 heated oxygen sensor (front) output voltage does not get across lean/rich criteria (about 0.5 volt) within about 30 seconds. **NOTE: When the cylinder 1, 4 heated oxygen sensor (front) begins to deteriorate, the heated oxygen sensor output voltage will deviate from the voltage when the sensor was new (normally 0.5 volt at stoichiometric ratio). This deviation will be corrected by the cylinder 1, 4 heated oxygen sensor (rear).** **If the cylinder 1, 4 heated oxygen sensor (rear) responds poorly because it has deteriorated, it will improperly correct the cylinder 1, 4 heated oxygen sensor (front). Thus, even when closed loop control is being effected, the fluctuation of the cylinder 1, 4 heated oxygen sensor (front) output voltage decreases, without intersecting with 0.5 volt. As a result, there is a possibility of DTC P0134 becoming registered.** **Possible Causes:** • Cylinder 1, 4 heated oxygen sensor (front) deteriorated. • Harness damage in cylinder 1, 4 heated oxygen sensor (front) output line. • Cylinder 1, 4 heated oxygen sensor (rear) deteriorated. • Open circuit in cylinder 1, 4 injector. • Harness damage in cylinder 1, 4 injector circuit. • Connector damage. • ECM failed. <M/T> • PCM failed. <A/T> • Exhaust leak. • Air drawn in from gaps in gasket, seals, etc. • Incorrect fuel pressure.
DTC: P0135 **2T PCM, MIL: Yes** **Year:** 2009 **Model:** Raider **Engine:** 3.7L V6 VIN K **Transmission:** All	**O2 Sensor 1/1 Heater Performance :** Continuously during O2 sensor heater operation with battery voltage between 10.4 and 15.75 volts and no O2 sensor circuit DTCs present. The PCM detects no temperature change in the O2 sensor heater element when the heater circuit is active. The heater temperature is obtained by measuring the heater resistance and calculating the heater temperature. **Possible Causes:** • O2 Heater control circuit open • O2 1/1 Heater ground circuit open • O2 Sensor • PCM

DTC	Trouble Code Title, Conditions & Possible Causes
DTC: P0137 **2T PCM, MIL: Yes** **Year:** 2009, 2010 **Model:** Eclipse, Endeavor, Galant, Outlander **Engine:** 3.0L V6 VIN X, 3.8L V6 VIN S, 3.8L V6 VIN T **Transmission:** All	**Heated Oxygen Sensor Circuit Low Voltage (Bank 1, Sensor 2):** The heated oxygen sensor offset voltage is between 0.4 and 0.6 volt. Battery positive voltage is between 11 and 16.5 volts. 3 minutes or more have passed since the engine starting sequence was completed. The right bank heated oxygen sensor output voltage is lower than 0.2 volt for 2 seconds. **Possible Causes:** • Right bank heated oxygen sensor (rear) failed • Open or shorted circuit in right bank heated oxygen sensor (rear) output line or harness damage • Open circuit in right bank heated oxygen sensor (rear) ground line or harness damage • Connector damage • ECM failed <M/T> • PCM failed <A/T>
DTC: P0137 **1T ECM, MIL: Yes** **Year:** 2009, 2010 **Model:** Lancer, Outlander **Engine:** 2.0L L4 VIN U, 2.0L L4 VIN V, 2.4L L4 VIN W **Transmission:** All	**Heated Oxygen Sensor (Rear) Circuit Low Voltage :** Heated oxygen sensor offset voltage is between 0.4 and 0.6 volt. Battery positive voltage is between 11 and 16.5 volts. After 3 minutes or more have passed since the engine starting sequence was completed. The heated oxygen sensor (rear) output voltage is lower than 0.2 volt for 2 seconds. **Possible Causes:** • Heated oxygen sensor (rear) failed. • Open or shorted circuit in heated oxygen sensor (rear) output line or harness damage. • Open circuit in heated oxygen sensor (rear) ground line or harness damage. • Connector damage. • ECM failed.
DTC: P0137 **2T PCM, MIL: Yes** **Year:** 2009, 2010 **Model:** Eclipse, Galant **Engine:** 2.4L L4 VIN F **Transmission:** All	**Cylinder 1, 4 Heated Oxygen Sensor (rear) Circuit Low Voltage:** Heated oxygen sensor offset voltage is between 0.4 and 0.6 volt. Battery positive voltage is between 11 and 16.5 volts. After 3 minutes or more have passed since the engine starting sequence was completed. Cylinder 1, 4 heated oxygen sensor output voltage is lower than 0.2 volt for 2 seconds. **Possible Causes:** • Cylinder 1, 4 heated oxygen sensor (rear) failed. • Open or shorted circuit in cylinder 1, 4 heated oxygen sensor (rear) output line or harness damage. • Open circuit in cylinder 1, 4 heated oxygen sensor (rear) ground line or harness damage. • Connector damage. • ECM failed. <M/T> • PCM failed. <A/T>
DTC: P0137 **2T PCM, MIL: Yes** **Year:** 2009 **Model:** Raider **Engine:** 3.7L V6 VIN K **Transmission:** All	**O2 Sensor 1/2 Circuit Low:** Engine running for less than 30 seconds and the O2 Sensor Heater Temperature is less than 251°C (484°F) with battery voltage greater 10.99 volts. The PCM detects that the 1/2 Oxygen Sensor signal voltage is below approximately 1.5 volts for 2.8 seconds after engine startup. **Possible Causes:** • O2 Return downstream circuit shorted to ground • O2 1/2 Signal circuit shorted to ground • O2 1/2 Signal circuit shorted to the O2 return downstream circuit • O2 1/2 Signal circuit shorted to the O2 1/2 heater ground circuit • O2 Sensor • PCM
DTC: P0138 **1T ECM, MIL: Yes** **Year:** 2009, 2010 **Model:** Lancer, Outlander **Engine:** 2.0L L4 VIN U, 2.0L L4 VIN V, 2.4L L4 VIN W **Transmission:** All	**Heated Oxygen Sensor (Rear) Circuit High Voltage:** After 2 seconds or more have passed since the engine starting sequence was completed. Heated oxygen sensor offset voltage is between 0.4 and 0.6 volt. The heated oxygen sensor (rear) output voltage has continued to be 1.8 volts or higher for 2 seconds. **Possible Causes:** • Short circuit in heated oxygen sensor (rear) output line. • Connector damage. • ECM failed.
DTC: P0138 **2T PCM, MIL: Yes** **Year:** 2009, 2010 **Model:** Eclipse, Galant **Engine:** 2.4L L4 VIN F **Transmission:** All	**Cylinder 1, 4 Heated Oxygen Sensor (rear) Circuit High Voltage:** After 2 seconds or more have passed since the engine starting sequence was completed. Heated oxygen sensor offset voltage is between 0.4 and 0.6 volt. Cylinder 1, 4 heated oxygen sensor (rear) output voltage has continued to be 1.8 volts or higher for 2 seconds. **Possible Causes:** • Short circuit in cylinder 1, 4 heated oxygen sensor (rear) output line. • Connector damage. • ECM failed. <M/T> • PCM failed. <A/T>

DTC	Trouble Code Title, Conditions & Possible Causes
DTC: P0138 **2T PCM, MIL:** Yes **Year:** 2009, 2010 **Model:** Eclipse, Endeavor, Galant, Outlander **Engine:** 3.0L V6 VIN X, 3.8L V6 VIN S, 3.8L V6 VIN T **Transmission:** All	**Heated Oxygen Sensor Circuit High Voltage (Bank 1, Sensor 2):** After 2 seconds or more have passed since the engine starting sequence was completed. Heated oxygen sensor off set voltage is between 0.4 and 0.6 volt. The right bank heated oxygen sensor (rear) output voltage has continued to be 1.8 volts or higher for 2 seconds. **Possible Causes:** • Short circuit in right bank heated oxygen sensor (rear) output line • Connector damage • ECM failed <M/T> • PCM failed <A/T>
DTC: P0138 **2T PCM, MIL:** Yes **Year:** 2009 **Model:** Raider **Engine:** 3.7L V6 VIN K **Transmission:** All	**O2 1/2 Circuit High:** Continuously with the engine running, no O2 sensor heater DTCs present, 1/2 Oxygen Sensor heater temperature within a specific range, and battery voltage greater than 10.4 volts. The PCM detects that the 1/2 Oxygen Sensor voltage is greater than the maximum acceptable value for a specific amount of time, based on O2 sensor heater temperature. **Possible Causes:** • O2 1/2 Signal circuit shorted to voltage • O2 1/2 Return downstream signal circuit shorted to voltage • O2 1/2 Signal circuit open • O2 Return downstream signal circuit open • O2 Sensor • PCM
DTC: P0139 **2T PCM, MIL:** Yes **Year:** 2009, 2010 **Model:** Eclipse, Endeavor, Galant **Engine:** 3.8L V6 VIN S, 3.8L V6 VIN T **Transmission:** All	**Heated Oxygen Sensor Circuit Slow Response (Bank 1, Sensor 2):** Engine coolant temperature is higher than 76°C (169°F). The right bank heated oxygen sensor (front) is active. The cumulative mass airflow sensor output is higher than 1,638 g. Fuel is being shut off. The right bank heated oxygen sensor (rear) output voltage is higher than 0.5 volt when fuel cut is started. It has been taking more than 180 seconds since the drive signal of the right bank heated oxygen sensor (rear) heater was turned on. Barometric pressure is higher than 76 kPa (22.4 in. Hg). The right bank heated oxygen sensor (rear) output voltage does not reach 0.2 volt for 0.5 second from 0.4 volt while fuel is being shut off. During the drive cycle, the ECM <M/T> or the PCM <A/T> performs monitoring with the accumulated total time of 12 seconds, 7 times <M/T> or 5 times <A/T>. **Possible Causes:** • Right bank heated oxygen sensor (rear) failed • Connector damage • ECM failed <M/T> • PCM failed <A/T>
DTC: P0139 **2T PCM, MIL:** Yes **Year:** 2009 **Model:** Raider **Engine:** 3.7L V6 VIN K **Transmission:** All	**O2 Sensor 1/2 Slow Response:** Vehicle is started and driven between 32 and 88.5 km/h (20 and 55 mph) with the Throttle open for a minimum of 120 seconds. Coolant greater than 70°C (158°F). Catalytic Converter Temp greater than 600°C (1112°F) and EVAP Purge is active. The oxygen sensor signal voltage switches less than 16 times from lean to rich within 20 seconds during monitoring. **Possible Causes:** • Exhaust leak • O2 1/2 Signal circuit • O2 Return downstream circuit • O2 Sensor
DTC: P0139 **2T PCM, MIL:** Yes **Year:** 2009, 2010 **Model:** Eclipse, Galant **Engine:** 2.4L L4 VIN F **Transmission:** All	**Cylinder 1, 4 Heated Oxygen Sensor (rear) Circuit Slow Response:** Engine coolant temperature is higher than 76°C (169°F). The cylinder 1, 4 heated oxygen sensor (front) is active. The cumulative mass airflow sensor output is higher than 1,638 g. Fuel is being shut off. Cylinder 1, 4 heated oxygen sensor (rear) output voltage is higher than 0.5 volt when fuel cut is started. It has been taking more than 180 seconds since the drive signal of the cylinder 1,4 heated oxygen sensor (rear) heater was turned on. Barometric pressure is higher than 76 kPa (22.4 in. Hg). Cylinder 1, 4 heated oxygen sensor (rear) output voltage does not reach 0.2 volt for 6 seconds from fuel cut start. **Possible Causes:** • Cylinder 1, 4 heated oxygen sensor (rear) deteriorated. • Connector damage. • ECM failed. <M/T> • PCM failed. <A/T>

DTC	Trouble Code Title, Conditions & Possible Causes
DTC: P0139 **1T ECM, MIL:** Yes **Year:** 2009, 2010 **Model:** Lancer **Engine:** 2.0L L4 VIN U, 2.0L L4 VIN V, 2.4L L4 VIN W **Transmission:** All	**Heated Oxygen Sensor (Rear) Circuit Slow Response :** Engine coolant temperature is higher than 76°C (169°F). The heated oxygen sensor (front) is active. The cumulative mass airflow sensor output is higher than 1,638 g <M/T> or 1,741 g <CVT>. Fuel is being shut off. Heated oxygen sensor (rear) output voltage is higher than 0.5 volt when fuel cut is started. It has been taking more than 180 seconds since the drive signal of the heated oxygen sensor (rear) heater was turned on. Barometric pressure is higher than 76 kPa (22.4 in. Hg). The heated oxygen sensor (rear) output voltage does not reach 0.2 volt for 6 seconds from fuel cut start. **Possible Causes:** • Heated oxygen sensor (rear) deteriorated. • Connector damage. • ECM failed.
DTC: P013A **1T PCM, MIL:** Yes **Year:** 2009 **Model:** Raider **Engine:** 3.7L V6 VIN K **Transmission:** All	**O2 Sensor 1/2 Slow Response (Rich To Lean):** With the engine running, vehicle speed above 96 kph (60 mph), throttle open for a minimum of 120 seconds, ECT greater than 70 C (158 F), catalytic converter temperature greater than 600 C (1112 F), and downstream oxygen sensor in a rich state. During a decel fuel shutoff event, the downstream oxygen sensor should switch from rich to lean within a specific time. The PCM monitors the downstream O2 sensor. If the PCM does not detect a rich to lean switch within a specific time during a decel fuel shutoff event, the monitor will fail. **Possible Causes:** • Exhaust leak • O2 1/2 Signal circuit • O2 Return downstream circuit • O2 Sensor
DTC: P013C **1T PCM, MIL:** Yes **Year:** 2009 **Model:** Raider **Engine:** 3.7L V6 VIN K **Transmission:** All	**O2 Sensor 2/2 Slow Response (Rich To Lean):** With the engine running, vehicle speed above 96 kph (60 mph), throttle open for a minimum of 120 seconds, ECT greater than 70 C (158 F), catalytic converter temperature greater than 600 C (1112 F), and downstream oxygen sensor in a rich state. During a decel fuel shutoff event, the downstream oxygen sensor should switch from rich to lean within a specific time. The PCM monitors the downstream O2 sensor. If the PCM does not detect a rich to lean switch within a specific time during a decel fuel shutoff event, the monitor will fail. **Possible Causes:** • Exhaust leak • O2 2/2 Signal circuit • O2 Return downstream circuit • O2 Sensor
DTC: P0140 **2T PCM, MIL:** Yes **Year:** 2009, 2010 **Model:** Eclipse, Galant **Engine:** 2.4L L4 VIN F **Transmission:** All	**Cylinder 1, 4 Heated Oxygen Sensor (rear) Circuit No Activity Detected:** Engine coolant temperature is higher than 76°C (169°F). The cylinder 1, 4 heated oxygen sensor (front) is active. The cumulative mass airflow sensor output is higher than 1,638 g. Repeat 1 or more times: drive*1, stop*2. Drive*1: Engine speed is higher than 1,500 r/min. Volumetric efficiency is higher than 40 percent. Vehicle speed is higher than 30 km/h (19 mph). A total of more than 60 seconds have elapsed with the above mentioned conditions, and more than 3 seconds have elapsed with the fuel shut off. Stop*2: Vehicle speed is lower than 1.5 km/h (1 mph). Change in the output voltage of the cylinder 1, 4 heated oxygen sensor (rear) is lower than 0.313 volt. Or. The maximum output voltage of the cylinder 1, 4 heated oxygen sensor is 0.508 volt or less. **Possible Causes:** • Cylinder 1, 4 heated oxygen sensor (rear) deteriorated. • Connector damage. • ECM failed. <M/T> • PCM failed. <A/T>

DTC	Trouble Code Title, Conditions & Possible Causes
DTC: P0140 **1T ECM, MIL: Yes** **Year:** 2009, 2010 **Model:** Lancer, Outlander **Engine:** 2.0L L4 VIN U, 2.0L L4 VIN V, 2.4L L4 VIN W **Transmission:** All	**Heated Oxygen Sensor (Rear) Circuit No Activity Detected :** Engine coolant temperature is higher than 76°C (169°F). The heated oxygen sensor (front) is active. The cumulative mass airflow sensor output is higher than 1,638 g <M/T> or 1,741 g <CVT>. Repeat 1 or more times: drive*1, stop*2 Drive*1: Engine speed is higher than 1,500 r/min. Volumetric efficiency is higher than 40 percent. Vehicle speed is higher than 30 km/h (19 mph). A total of more than 60 seconds have elapsed with the above mentioned conditions, and more than 3 seconds have elapsed with the fuel shut off. Stop*2: Vehicle speed is lower than 1.5 km/h (1 mph). Change in the output voltage of the heated oxygen sensor (rear) is lower than 0.313 volt. Ort, the maximum output voltage of the heated oxygen sensor (rear) is less than 0.508 volt. **Possible Causes:** • Heated oxygen sensor (rear) deteriorated. • Connector damage. • ECM failed.
DTC: P0140 **2T PCM, MIL: Yes** **Year:** 2009, 2010 **Model:** Eclipse, Endeavor, Galant, Outlander **Engine:** 3.0L V6 VIN X, 3.8L V6 VIN S, 3.8L V6 VIN T **Transmission:** All	**Heated Oxygen Sensor Circuit No Activity Detected (Bank 1, Sensor 2):** Engine coolant temperature is higher than 76°C (169°F). The right bank heated oxygen sensor (front) is active. The cumulative mass airflow sensor output is higher than 1,638 g. Engine speed is higher than 1,500 r/min. Volumetric efficiency is higher than 40 percent. Vehicle speed is higher than 30 km/h (19 mph). A total of more than 60 seconds have elapsed with the above mentioned conditions, and more than 3 seconds have elapsed with the fuel shut off. Change in the output voltage of the right bank heated oxygen sensor (rear) is lower than 0.313 volt. Or. The maximum output voltage of the right bank heated oxygen sensor (rear) is lower than 0.508 volt. **Possible Causes:** • Right bank heated oxygen sensor (rear) failed • Connector damage • ECM failed <M/T> • PCM failed <A/T>
DTC: P0141 **2T PCM, MIL: Yes** **Year:** 2009 **Model:** Raider **Engine:** 3.7L V6 VIN K **Transmission:** All	**O2 Sensor 1/2 Heater Performance:** Continuously during O2 sensor heater operation with battery voltage between 10.4 and 15.75 volts and no O2 sensor circuit DTCs present. The PCM detects no temperature change in the O2 sensor heater element when the heater circuit is active. The heater temperature is obtained by measuring the heater resistance and calculating the heater temperature. **Possible Causes:** • O2 1/2 Heater control circuit open • O2 Heater ground circuit open • O2 Sensor • PCM
DTC: P0143 **1T ECM, MIL: Yes** **Year:** 2009, 2010 **Model:** Outlander **Engine:** 2.4L L4 VIN W **Transmission:** All	**Heated Oxygen Sensor (3rd) Circuit Low Voltage (California):** Heated oxygen sensor offset voltage is between 0.4 and 0.6 volt. Battery positive voltage is between 11 and 16.5 volts. After 3 minutes or more have passed since the engine starting sequence was completed. The heated oxygen sensor (3rd) output voltage is lower than 0.2 volt for 2 seconds. **Possible Causes:** • Heated oxygen sensor (3rd) failed. • Open or shorted circuit in heated oxygen sensor (3rd) output line or harness damage. • Open circuit in heated oxygen sensor (3rd) ground line or harness damage. • Connector damage. • ECM failed.
DTC: P0144 **1T ECM, MIL: Yes** **Year:** 2009, 2010 **Model:** Outlander **Engine:** 2.4L L4 VIN W **Transmission:** All	**Heated Oxygen Sensor (3rd) Circuit High Voltage (California):** After 2 seconds or more have passed since the engine starting sequence was completed. Heated oxygen sensor offset voltage is between 0.4 and 0.6 volt. The heated oxygen sensor (3rd) output voltage has continued to be 1.8 volts or higher for 2 seconds. **Possible Causes:** • Short circuit in heated oxygen sensor (3rd) output line. • Connector damage. • ECM failed.

DTC	Trouble Code Title, Conditions & Possible Causes
DTC: P0145 **1T ECM, MIL: Yes** **Year:** 2009, 2010 **Model:** Outlander **Engine:** 2.4L L4 VIN W **Transmission:** All	**Heated Oxygen Sensor (3rd) Circuit Slow Response (California):** Engine coolant temperature is higher than 76°C (169°F). The linear air-fuel ratio sensor is active. The cumulative mass airflow sensor output is higher than 2,900 g. Fuel is being shut off. Heated oxygen sensor (3rd) output voltage is higher than 0.5 volt when fuel cut is started. The heated oxygen sensor (3rd) output voltage does not reach 0.2 volt for 7 seconds from fuel cut start. Or. The heated oxygen sensor (3rd) output voltage does not reach 0.2 volt for 1 second from 0.4 volt while fuel is being shut off **Possible Causes:** • Heated oxygen sensor (3rd) deteriorated. • Connector damage. • ECM failed.
DTC: P0146 **1T ECM, MIL: Yes** **Year:** 2009, 2010 **Model:** Outlander **Engine:** 2.4L L4 VIN W **Transmission:** All	**Heated Oxygen Sensor (3rd) Circuit Slow Response (California):** Engine coolant temperature is higher than 76°C (169°F). The linear air-fuel ratio sensor is active. The cumulative mass airflow sensor output is higher than 2,900 g. Repeat 3 times or more: drive*1, stop*2. Drive*1: Engine speed is higher than 1,500 r/min. Volumetric efficiency is higher than 40 percent. Vehicle speed is higher than 30 km/h (19 mph). A total of more than 10 seconds have elapsed with the above mentioned conditions, and more than 3 seconds have elapsed with the fuel shut off. Stop*2: Vehicle speed is lower than 1.5 km/h (1 mph). Change in the output voltage of the heated oxygen sensor (3rd) is lower than 0.098 volt. **Possible Causes:** • Heated oxygen sensor (3rd) deteriorated. • Connector damage. • ECM failed.
DTC: P0151 **2T PCM, MIL: Yes** **Year:** 2009, 2010 **Model:** Eclipse, Endeavor, Galant, Outlander **Engine:** 3.0L V6 VIN X, 3.8L V6 VIN S, 3.8L V6 VIN T **Transmission:** All	**Heated Oxygen Sensor Circuit Low Voltage (Bank 2, Sensor 1):** Heated oxygen sensor offset voltage is between 0.4 and 0.6 volt. Battery positive voltage is between 11 and 16.5 volts. 180 seconds have passed since the engine starting sequence was completed. The left bank heated oxygen sensor (front) output voltage is lower than 0.2 volts. **Possible Causes:** • Left bank heated oxygen sensor (front) failed • Open or shorted circuit in left bank heated oxygen sensor (front) output line, or harness damage • Open circuit in left bank heated oxygen sensor (front) ground line, or harness damage • Connector damage • ECM failed <M/T> • PCM failed <A/T>
DTC: P0151 **2T PCM, MIL: Yes** **Year:** 2009 **Model:** Raider **Engine:** 3.7L V6 VIN K **Transmission:** All	**O2 Sensor 2/1 Circuit Low:** With the engine running, battery voltage greater than 10.4 volts, and no O2 sensor heater DTCs present. The PCM detects that the 2/1 Oxygen Sensor signal voltage is below approximately 1.5 volts for 2.8 seconds after engine startup. The **Possible Causes:** • O2 Return upstream circuit shorted to ground • O2 2/1 Signal circuit shorted to ground • O2 2/1 Signal circuit shorted to the O2 return upstream circuit • O2 2/1 Signal circuit shorted to the O2 heater ground circuit • O2 Sensor • PCM
DTC: P0151 **2T PCM, MIL: Yes** **Year:** 2009, 2010 **Model:** Eclipse, Galant **Engine:** 2.4L L4 VIN F **Transmission:** All	**Cylinder 2, 3 Heated Oxygen Sensor (front) Circuit Low Voltage:** Heated oxygen sensor offset voltage is between 0.4 and 0.6 volt. Battery positive voltage is between 11 and 16.5 volts. After 180 seconds have passed since the engine starting sequence was completed, Cylinder 1, 4 heated oxygen sensor (front) output voltage is lower than 0.2 volt. **Possible Causes:** • Cylinder 2, 3 heated oxygen sensor (front) failed. • Open or shorted circuit in cylinder 2, 3 heated oxygen sensor (front) output line, or harness damage. • Open circuit in cylinder 2, 3 heated oxygen sensor (front) ground line, or harness damage. • Connector damage. • ECM failed. <M/T> • PCM failed. <A/T>

DTC	Trouble Code Title, Conditions & Possible Causes
DTC: P0152 **2T PCM, MIL:** Yes **Year:** 2009, 2010 **Model:** Eclipse, Endeavor, Galant, Outlander **Engine:** 3.0L V6 VIN X, 3.8L V6 VIN S, 3.8L V6 VIN T **Transmission:** All	**Heated Oxygen Sensor Circuit High Voltage (Bank 2, Sensor 1):** After 2 seconds or more have passed since the engine starting sequence was completed. Heated oxygen sensor offset voltage is between 0.4 and 0.6 volt. The left bank heated oxygen sensor (front) output voltage has continued to be 1.8 volts or higher for 2 seconds. Output voltage is over specified range. **Possible Causes:** • Short circuit in left bank heated oxygen sensor (front) output line • Connector damage • ECM failed <M/T> • PCM failed <A/T>
DTC: P0152 **2T PCM, MIL:** Yes **Year:** 2009, 2010 **Model:** Eclipse, Galant **Engine:** 2.4L L4 VIN F **Transmission:** All	**Cylinder 2, 3 Heated Oxygen Sensor (front) Circuit High Voltage:** After 2 seconds or more have passed since the engine starting sequence was completed. Heated oxygen sensor offset voltage is between 0.4 and 0.6 volt. Cylinder 2, 3 heated oxygen sensor (front) output voltage has continued to be 1.8 volts or higher for 2 seconds. **Possible Causes:** • Short circuit in cylinder 2, 3 heated oxygen sensor (front) output line. • Connector damage. • ECM failed. <M/T> • PCM failed. <A/T>
DTC: P0152 **2T PCM, MIL:** Yes **Year:** 2009 **Model:** Raider **Engine:** 3.7L V6 VIN K **Transmission:** All	**O2 Sensor 2/1 Circuit High:** Continuously with the engine running, no O2 sensor heater DTCs present, 2/1 Oxygen Sensor heater temperature within a specific range, and battery voltage greater than 10.4 volts. The PCM detects that the 2/1 Oxygen Sensor voltage is greater than the maximum acceptable value for a specific amount of time, based on O2 sensor heater temperature. **Possible Causes:** • O2 2/1 Signal circuit shorted to voltage • O2 Return upstream circuit shorted to voltage • O2 2/1 Signal circuit open • O2 Return upstream circuit open • O2 Sensor • PCM
DTC: P0153 **2T PCM, MIL:** Yes **Year:** 2009, 2010 **Model:** Eclipse, Galant **Engine:** 2.4L L4 VIN F **Transmission:** All	**Cylinder 2, 3 Heated Oxygen Sensor (front) Circuit Slow Response:** Engine coolant temperature is higher than 60°C (140°F). Engine speed is between 1,500 and 3,500 r/min <M/T> or 1,200 and 3,500 r/min <A/T>. Volumetric efficiency is between 18 and 70 percent <M/T> or 25 and 70 percent <A/T>. Under the closed loop air/fuel control. The accelerator pedal is depressed. Short-term fuel trim is at between -25 and +25 percent. More than 2 seconds have elapsed after the above mentioned conditions have been met. During the drive cycle, the ECM <M/T> or the PCM <A/T> performs monitoring with the accumulated total time of 10 seconds, 5 times. The average of the cylinder 2, 3 heated oxygen sensor (front) rich/lean switching frequency is less than 12 times for the accumulated time of 10 seconds. The cylinder 2, 3 heated oxygen sensor (front) rich/lean switching frequency is less than 25 times for the accumulated time of 10 seconds. **NOTE: If the sensor switching frequency is lower than the Judgment Criteria due to the M.U.T.-III OBD-II test Mode - HO2S Test Results, it is assumed that the heated oxygen sensor has deteriorated. If it is higher, it is assumed that the harness is damaged or has a short circuit.** **If the heated oxygen sensor signal voltage has not changed even once (lean/rich) after the DTC was erased, the sensor switch time will display as 0 seconds.** **Possible Causes:** • Cylinder 2, 3 heated oxygen sensor (front) deteriorated. • Connector damage. • ECM failed. <M/T> • PCM failed. <A/T>
DTC: P0153 **2T PCM, MIL:** Yes **Year:** 2009 **Model:** Raider **Engine:** 3.7L V6 VIN K **Transmission:** All	**O2 Sensor 2/1 Slow Response:** Vehicle is started and driven between 32 and 88.5 km/h (20 and 55 mph) with the Throttle open for a minimum of 120 seconds. Coolant greater than 70°C (158°F). Catalytic Converter Temp greater than 600°C (1112°F) and EVAP Purge is active. The PCM detects that the oxygen sensor signal switches from lean to rich less than 16 times within 20 seconds during monitoring. **Possible Causes:** • Exhaust leak • O2 2/1 Signal circuit • O2 Return upstream circuit • O2 Sensor

DTC	Trouble Code Title, Conditions & Possible Causes
DTC: P0153 **2T PCM, MIL: Yes** **Year:** 2009, 2010 **Model:** Eclipse, Endeavor, Galant, Outlander **Engine:** 3.0L V6 VIN X, 3.8L V6 VIN S, 3.8L V6 VIN T **Transmission:** All	**Heated Oxygen Sensor Circuit Slow Response (Bank 2, Sensor 1):** Engine coolant temperature is higher than 50°C (122°F). Engine speed is between 1,500 and 3,000 r/min <M/T> or 1,300 and 3,000 r/min <A/T>. Volumetric efficiency is between 12 and 60 <M/T> or 20 and 60 <A/T> percent. Under the closed loop air/fuel control. The accelerator pedal depressed. Short-term fuel trim is at between -30 and +25 percent. More than 2 seconds have elapsed after the above mentioned conditions have been met. During the drive cycle, the ECM <M/T> or the PCM <A/T> performs monitoring with the accumulated total time of 12 seconds, 7 times <M/T> or 5 times <A/T>. The average of the left bank heated oxygen sensor (front) rich/lean switching frequency is less than 10 times for the accumulated time of 12 seconds. Or. The left bank heated oxygen sensor (front) rich/lean switching frequency is less than 20 times for the accumulated time of 12 seconds. **Possible Causes:** • Left bank heated oxygen sensor (front) deteriorated • Connector damage • ECM failed <M/T> • PCM failed <A/T>
DTC: P0154 **2T PCM, MIL: Yes** **Year:** 2009, 2010 **Model:** Eclipse, Endeavor, Galant, Outlander **Engine:** 3.0L V6 VIN X, 3.8L V6 VIN S, 3.8L V6 VIN T **Transmission:** All	**Heated Oxygen Sensor Circuit No Activity Detected (Bank 2, Sensor 1):** After 128 seconds or more have passed since the engine starting sequence was completed. Engine coolant temperature is higher than 7°C (45° F). Engine speed is higher than 1,200 r/min. Volumetric efficiency is higher than 30 percent. Throttle position sensor output voltage is lower than 4 volts. Except while fuel is being shut off. Monitoring time: 30 seconds. The left bank heated oxygen sensor (front) output voltage does not get across lean/rich criteria (about 0.5 volt) within about 30 seconds. **Possible Causes:** • Left bank heated oxygen sensor (front) deteriorated • Harness damage in left bank heated oxygen sensor (front) output line • Left bank heated oxygen sensor (rear) deteriorated • Open circuit in left bank injector • Harness damage in left bank injector circuit • Connector damage. • Exhaust leak • Air drawn in from gaps in gasket, seals, etc • Incorrect fuel pressure • ECM failed <M/T> • PCM failed <A/T>
DTC: P0154 **2T PCM, MIL: Yes** **Year:** 2009, 2010 **Model:** Eclipse, Galant **Engine:** 2.4L L4 VIN F **Transmission:** All	**Cylinder 2, 3 Heated Oxygen Sensor (front) No Activity Detected:** After 128 seconds or more have passed since the engine starting sequence was completed. Engine coolant temperature is higher than 7°C (45° F). Engine speed is higher than 1,200 r/min. Volumetric efficiency is higher than 30 percent. Throttle position sensor output voltage is lower than 4 volts. Except while fuel is being shut off. Monitoring time: 30 seconds. Cylinder 2, 3 heated oxygen sensor (front) output voltage does not get across lean/rich criteria (about 0.5 volt) within about 30 seconds. **NOTE: When the cylinder 2, 3 heated oxygen sensor (front) begins to deteriorate, the heated oxygen sensor output voltage will deviate from the voltage when the sensor was new (normally 0.5 volt at stoichiometric ratio). This deviation will be corrected by the cylinder 2, 3 heated oxygen sensor (rear).** **If the cylinder 2, 3 heated oxygen sensor (rear) responds poorly because it has deteriorated, it will improperly correct the cylinder 2, 3 heated oxygen sensor (front). Thus, even when closed loop control is being effected, the fluctuation of the cylinder 2, 3 heated oxygen sensor (front) output voltage decreases, without intersecting with 0.5 volt. As a result, there is a possibility of DTC P0154 becoming registered.** **Possible Causes:** • Cylinder 2, 3 heated oxygen sensor (front) deteriorated. • Harness damage in cylinder 2, 3 heated oxygen sensor (front) output line. • Cylinder 2, 3 heated oxygen sensor (rear) deteriorated. • Open circuit in cylinder 2, 3 injector. • Harness damage in cylinder 2, 3 injector circuit. • Connector damage. • ECM failed. <M/T> • PCM failed. <A/T> • Exhaust leak. • Air drawn in from gaps in gasket, seals, etc. • Incorrect fuel pressure.

DTC	Trouble Code Title, Conditions & Possible Causes
DTC: P0155 **2T PCM, MIL:** Yes **Year:** 2009 **Model:** Raider **Engine:** 3.7L V6 VIN K **Transmission:** All	**O2 Sensor 2/1 Heater Performance:** Continuously during O2 sensor heater operation with battery voltage between 10.4 and 15.75 volts and no O2 sensor circuit DTCs present. The PCM detects no temperature change in the O2 sensor heater element when the heater circuit is active. The heater temperature is obtained by measuring the heater resistance and calculating the heater temperature. **Possible Causes:** • O2 2/1 Heater control circuit open • O2 Heater ground circuit open • O2 Sensor • PCM
DTC: P0157 **2T PCM, MIL:** Yes **Year:** 2009, 2010 **Model:** Eclipse, Endeavor, Galant, Outlander **Engine:** 3.0L V6 VIN X, 3.8L V6 VIN S, 3.8L V6 VIN T **Transmission:** All	**Heated Oxygen Sensor Circuit Low Voltage (Bank 2, Sensor 2):** The heated oxygen sensor offset voltage is between 0.4 and 0.6 volt. Battery positive voltage is between 11 and 16.5 volts. After 3 minutes or more have passed since the engine starting sequence was completed. The left bank heated oxygen sensor output voltage is lower than 0.2 volt for 2 seconds. **Possible Causes:** • Left bank heated oxygen sensor (rear) failed • Open or shorted circuit in left bank heated oxygen sensor (rear) output line, or harness damage • Open circuit in left bank heated oxygen sensor (rear) ground line, or harness damage • Connector damage • ECM failed <M/T> • PCM failed <A/T>
DTC: P0157 **2T PCM, MIL:** Yes **Year:** 2009, 2010 **Model:** Eclipse, Galant **Engine:** 2.4L L4 VIN F **Transmission:** All	**Cylinder 2, 3 Heated Oxygen Sensor (rear) Circuit Low Voltage:** Heated oxygen sensor offset voltage is between 0.4 and 0.6 volt. Battery positive voltage is between 11 and 16.5 volts. After 3 minutes or more have passed since the engine starting sequence was completed. Cylinder 1, 4 heated oxygen sensor output voltage is lower than 0.2 volt for 2 seconds. **Possible Causes:** • Cylinder 2, 3 heated oxygen sensor (rear) failed. • Open or shorted circuit in cylinder 2, 3 heated oxygen sensor (rear) output line, or harness damage. • Open circuit in cylinder 2, 3 heated oxygen sensor (rear) ground line, or harness damage. • Connector damage. • ECM failed. <M/T> • PCM failed. <A/T>
DTC: P0157 **2T PCM, MIL:** Yes **Year:** 2009 **Model:** Raider **Engine:** 3.7L V6 VIN K **Transmission:** All	**O2 Sensor 2/2 Circuit Low:** With the engine running, battery voltage greater than 10.4 volts, and no O2 sensor heater DTCs present. The PCM detects that the 2/2 Oxygen Sensor signal voltage is below approximately 1.5 volts for 2.8 seconds after engine startup. **Possible Causes:** • O2 Return upstream circuit shorted to ground • O2 2/2 Signal circuit shorted to ground • O2 2/2 Signal circuit shorted to the O2 return upstream circuit • O2 2/2 Signal circuit shorted to the O2 2/2 heater ground circuit • O2 Sensor • PCM
DTC: P0158 **2T PCM, MIL:** Yes **Year:** 2009, 2010 **Model:** Eclipse, Endeavor, Galant, Outlander **Engine:** 3.0L V6 VIN X, 3.8L V6 VIN S, 3.8L V6 VIN T **Transmission:** All	**Heated Oxygen Sensor Circuit High Voltage (bank 2, sensor 2):** After 2 seconds or more have passed since the engine starting sequence was completed. Heated oxygen sensor off set voltage is between 0.4 and 0.6 volt. The left bank heated oxygen sensor (rear) output voltage has continued to be 1.8 volts or higher for 2 seconds. **Possible Causes:** • Short circuit in left bank heated oxygen sensor (rear) output line • Connector damage • ECM failed.<M/T> • PCM failed <A/T>
DTC: P0158 **2T PCM, MIL:** Yes **Year:** 2009, 2010 **Model:** Eclipse, Galant **Engine:** 2.4L L4 VIN F **Transmission:** All	**Cylinder 2, 3 Heated Oxygen Sensor (rear) Circuit High Voltage:** After 2 seconds or more have passed since the engine starting sequence was completed. Heated oxygen sensor offset voltage is between 0.4 and 0.6 volt. Cylinder 2, 3 heated oxygen sensor (rear) output voltage has continued to be 1.8 volts or higher for 2 seconds. **Possible Causes:** • Short circuit in cylinder 2, 3 heated oxygen sensor (rear) output line. • Connector damage. • ECM failed. <M/T> • PCM failed. <A/T>

DTC	Trouble Code Title, Conditions & Possible Causes
DTC: P0158 **2T PCM, MIL: Yes** **Year:** 2009 **Model:** Raider **Engine:** 3.7L V6 VIN K **Transmission:** All	**O2 Sensor 2/2 Circuit High:** Continuously with the engine running, no O2 sensor heater DTCs present, 2/2 Oxygen Sensor heater temperature within a specific range, and battery voltage greater than 10.4 volts. The PCM detects that the 2/2 Oxygen Sensor voltage is greater than the maximum acceptable value for a specific amount of time, based on O2 sensor heater temperature **Possible Causes:** • O2 2/2 Signal circuit shorted to voltage • O2 Return downstream circuit shorted to voltage • O2 2/2 Signal circuit open • O2 Return downstream circuit open • O2 Sensor • PCM
DTC: P0159 **2T PCM, MIL: Yes** **Year:** 2009, 2010 **Model:** Eclipse, Galant **Engine:** 2.4L L4 VIN F **Transmission:** All	**Cylinder 2, 3 Heated Oxygen Sensor (rear) Circuit Slow Response:** Engine coolant temperature is higher than 76°C (169°F). The cylinder 2, 3 heated oxygen sensor (front) is active. The cumulative mass airflow sensor output is higher than 1,638 g. Cylinder 2, 3 heated oxygen sensor (rear) output voltage is higher than 0.5 volt when fuel cut is started. It has been taking more than 180 seconds since the drive signal of the cylinder 2, 3 heated oxygen sensor (rear) heater was turned on. Barometric pressure is higher than 76 kPa (22.4 in. Hg). The ECM <M/T> or the PCM <A/T> monitors for this condition for 3 cycles of 0.5 second each during drive cycles. Cylinder 2, 3 heated oxygen sensor (rear) output voltage does not reach 0.2 volt for 0.5 second from 0.4 volt while fuel is being shut off. The ECM <M/T> or the PCM <A/T> monitors for this condition once during the drive cycle. **Possible Causes:** • Cylinder 2, 3 heated oxygen sensor (rear) deteriorated. • Connector damage. • ECM failed. <M/T> • PCM failed. <A/T>
DTC: P0159 **2T PCM, MIL: Yes** **Year:** 2009, 2010 **Model:** Eclipse, Endeavor, Galant, Outlander **Engine:** 3.0L V6 VIN X, 3.8L V6 VIN S, 3.8L V6 VIN T **Transmission:** All	**Heated Oxygen Sensor Circuit Slow Response (bank 2, sensor 2):** Engine coolant temperature is higher than 76°C (169°F). The left bank heated oxygen sensor (front) is active. The cumulative mass airflow sensor output is higher than 1,638 g. Fuel is being shut off. The left bank heated oxygen sensor (rear) output voltage is higher than 0.5 volt when fuel cut is started. It has been taking more than 180 seconds since the drive signal of the left bank heated oxygen sensor (rear) heater was turned on. Barometric pressure is higher than 76 kPa (22.4 in. Hg). The left bank heated oxygen sensor (rear) output voltage does not reach 0.2 volt for 6 seconds from fuel cut start. **Possible Causes:** • Left bank heated oxygen sensor (rear) deteriorated • Connector damage • ECM failed <M/T> • PCM failed <A/T>
DTC: P0159 **2T PCM, MIL: Yes** **Year:** 2009 **Model:** Raider **Engine:** 3.7L V6 VIN K **Transmission:** All	**O2 Sensor 2/2 Slow Response:** Vehicle is started and driven between 32 and 88.5 km/h (20 and 55 mph) with the Throttle open for a minimum of 120 seconds. Coolant greater than 70°C (158°F). Catalytic Converter Temp greater than 600°C (1112°F) and Evap Purge is active. The oxygen sensor signal voltage switches less than 16 times from lean to rich within 20 seconds during monitoring. **Possible Causes:** • Exhaust leak • O2 2/2 Signal circuit • O2 Return downstream circuit • O2 Sensor
DTC: P01590 **1T ECM, MIL: Yes** **Year:** 2009, 2010 **Model:** Lancer **Engine:** 2.0L L4 VIN U, 2.0L L4 VIN V, 2.4L L4 VIN W **Transmission:** All	**TC-SST-ECU to ECM Communication Error in Torque Reduction Request:** Ignition switch is "ON" position. ECM detects an error in communication between ECM and TC-SST-ECU. **NOTE: Engine output is restricted.** **Possible Causes:** • CAN line harness damage or connector damage. • ECM failed.
DTC: P0160 **2T PCM, MIL: Yes** **Year:** 2009, 2010 **Model:** Eclipse, Endeavor, Galant, Outlander **Engine:** 3.0L V6 VIN X, 3.8L V6 VIN S, 3.8L V6 VIN T **Transmission:** All	**Heated Oxygen Sensor Circuit No Activity Detected (bank 2, sensor 2):** Engine coolant temperature is higher than 76°C (169°F). The left bank heated oxygen sensor (front) is active. The cumulative mass airflow sensor output is higher than 1,638 g. Change in the output voltage of the left bank heated oxygen sensor (rear) is lower than 0.313 volt. Or. The maximum output voltage of the left bank heated oxygen sensor (rear) is lower than 0.508 volt. **Possible Causes:** • Left bank heated oxygen sensor (rear) deteriorated • Connector damage • ECM failed <M/T> • PCM failed <A/T>

DTC	Trouble Code Title, Conditions & Possible Causes
DTC: P0160 **2T PCM, MIL: Yes** **Year:** 2009, 2010 **Model:** Eclipse, Galant **Engine:** 2.4L L4 VIN F **Transmission:** All	**Cylinder 2, 3 Heated Oxygen Sensor (rear) Circuit No Activity Detected:** Engine coolant temperature is higher than 76°C (169°F). The cylinder 2, 3 heated oxygen sensor (front) is active. The cumulative mass airflow sensor output is higher than 1,638 g. Repeat 1 or more times: drive*1, stop*2. Drive*1: Engine speed is higher than 1,500 r/min. Volumetric efficiency is higher than 40 percent. Vehicle speed is higher than 30 km/h (19 mph). A total of more than 60 seconds have elapsed with the above mentioned conditions, and more than 3 seconds have elapsed with the fuel shut off. Stop*2: Vehicle speed is lower than 1.5 km/h (1 mph). Change in the output voltage of the cylinder 2, 3 heated oxygen sensor (rear) is lower than 0.313 volt. Or. The maximum output voltage of the cylinder 2, 3 heated oxygen sensor is 0.508 volt or less. **Possible Causes:** • Cylinder 2, 3 heated oxygen sensor (rear) deteriorated. • Connector damage. • ECM failed. <M/T> • PCM failed. <A/T>
DTC: P0161 **2T PCM, MIL: Yes** **Year:** 2009 **Model:** Raider **Engine:** 3.7L V6 VIN K **Transmission:** All	**O2 Sensor 2/2 Heater Performance:** Continuously during O2 sensor heater operation with battery voltage between 10.4 and 15.75 volts and no O2 sensor circuit DTCs present. The PCM detects no temperature change in the O2 sensor heater element when the heater circuit is active. The heater temperature is obtained by measuring the heater resistance and calculating the heater temperature. **Possible Causes:** • O2 2/2 Heater control circuit open • O2 2/2 Heater ground circuit open • O2 Sensor • PCM
DTC: P0171 **2T PCM, MIL: Yes** **Year:** 2009 **Model:** Raider **Engine:** 3.7L V6 VIN K **Transmission:** All	**Fuel System 1/1 Lean:** With the engine running in closed loop mode, the ambient/battery temperature above -6.7°C (20°F) and altitude below 2590.8 m (8500 ft). If the PCM multiplies short term compensation by long term adaptive and a certain percentage is exceeded for two trips, a freeze frame is stored, the MIL illuminates and a trouble code is stored. **Possible Causes:** • Restricted fuel supply line • Fuel pump inlet strainer plugged • Fuel pump module • O2 signal circuit • O2 return circuit • O2 Sensor heater operation • O2 Sensor • Map Sensor operation • ECT Sensor operation • Engine mechanical • Fuel filter/Pressure regulator (HIGH) • PCM
DTC: P0171 **1T ECM, MIL: Yes** **Year:** 2009, 2010 **Model:** Lancer, Outlander **Engine:** 2.0L L4 VIN U, 2.0L L4 VIN V, 2.4L L4 VIN W **Transmission:** All	**System Too Lean:** Engine coolant temperature is lower than 100°C (212°F) when the engine is started. Intake air temperature is lower than 60°C (140°F) when the engine is started. Under the closed loop air/fuel ratio control. Engine coolant temperature is higher than 76°C (169°F). Mass airflow sensor output is 9 g/sec or more. The long-term fuel trim has continued to be higher than +12.5 percent for 5 seconds. Or, the short-term fuel trim has continued to be higher than +7.0 percent <Except for California> or +5.5 percent <California> for 5 seconds. **Possible Causes:** • Mass airflow sensor failed. • Injector failed. • Incorrect fuel pressure. • Air drawn in from gaps in gasket, seals, etc. • Engine coolant temperature sensor failed. • Intake air temperature sensor failed. • Barometric pressure sensor failed. • Manifold absolute pressure sensor failed. • Exhaust leak. • Use of incorrect or contaminated fuel. • Harness damage in injector circuit or connector damage. • ECM failed.

DTC	Trouble Code Title, Conditions & Possible Causes
DTC: P0171 **1T PCM, MIL: Yes** **Year:** 2009, 2010 **Model:** Eclipse, Endeavor, Galant, Outlander **Engine:** 3.0L V6 VIN X, 3.8L V6 VIN S, 3.8L V6 VIN T **Transmission:** All	**System Too Lean (Bank 1):** Engine coolant temperature is lower than 100°C (212°F) when the engine is started. Intake air temperature is lower than 60°C (140°F) when the engine is started. Under the closed loop air/fuel ratio control. Engine coolant temperature is higher than 76°C (169°F). Mass airflow sensor output is 15 g/sec or more. The long-term fuel trim has continued to be higher than +12.5 percent for 5 seconds. Or. The short-term fuel trim has continued to be higher than +7.4 percent for 5 seconds. **Possible Causes:** • Mass airflow sensor failed. • Injector (Number 1, 3, 5) failed. • Incorrect fuel pressure. • Air drawn in from gaps in gasket, seals, etc. • Engine coolant temperature sensor failed. • Intake air temperature sensor failed. • Barometric pressure sensor failed. • Manifold absolute pressure sensor failed. • Exhaust leak. • Use of incorrect or contaminated fuel. • Harness damage in right bank injector circuit, or connector damage. • ECM failed. <M/T> • PCM failed. <A/T>
DTC: P0171 **1T PCM, MIL: Yes** **Year:** 2009, 2010 **Model:** Eclipse, Galant **Engine:** 2.4L L4 VIN F **Transmission:** All	**System Too Lean (Cylinder 1, 4):** Engine coolant temperature is lower than 100°C (212°F) when the engine is started. Intake air temperature is lower than 60°C (140°F) when the engine is started. Under the closed loop air/fuel ratio control. Engine coolant temperature is higher than 76°C (169°F). Mass airflow sensor output is 6 g/sec or less. Long-term fuel trim has continued to be higher than +12.5 percent for 5 seconds. Or. Short-term fuel trim has continued to be higher than +17.4 percent for 5 seconds. **Possible Causes:** • Mass airflow sensor failed. • Injector (Number 1, 4) failed. • Incorrect fuel pressure. • Air drawn in from gaps in gasket, seals, etc. • Engine coolant temperature sensor failed. • Intake air temperature sensor failed. • Barometric pressure sensor failed. • Manifold absolute pressure sensor failed. • Exhaust leak. • Use of incorrect or contaminated fuel. • Harness damage in cylinder 1, 4 injector circuit or connector damage. • ECM failed. <M/T> • PCM failed. <A/T>
DTC: P0172 **1T PCM, MIL: Yes** **Year:** 2009, 2010 **Model:** Eclipse, Endeavor, Galant, Outlander **Engine:** 3.0L V6 VIN X, 3.8L V6 VIN S, 3.8L V6 VIN T **Transmission:** All	**System Too Rich (Bank 1):** Under the closed loop air/fuel ratio control. Engine coolant temperature is higher than 76°C (169°F). Mass airflow sensor output is 10 g/sec or more. The long-term fuel trim has continued to be lower than -12.5 percent for 5 seconds. The short-term fuel trim has continued to be lower than -5.5 percent for 5 seconds. **Possible Causes:** • Mass airflow sensor failed. • Injector (Number 1, 3, 5) failed. • Incorrect fuel pressure. • Engine coolant temperature sensor failed. • Intake air temperature sensor failed. • Barometric pressure sensor failed. • Manifold absolute pressure sensor failed. • Harness damage in right bank injector circuit, or connector damage. • ECM failed. <M/T> • PCM failed. <A/T>

DTC	Trouble Code Title, Conditions & Possible Causes
DTC: P0172 **2T PCM, MIL: Yes** **Year:** 2009 **Model:** Raider **Engine:** 3.7L V6 VIN K **Transmission:** All	**Fuel System 1/1 Rich:** With the engine running in closed loop mode, the ambient/battery temperature above -6.7°C (20°F) and altitude below 2590.8 m (8500 ft). If the PCM multiplies short term compensation by long term adaptive and a purge fuel multiplier and the result is below a certain value for 30 seconds over two trips, a freeze frame is stored, the MIL illuminates and a trouble code is stored. **Possible Causes:** • O2 Sensor heater operation • EVAP purge solenoid operation • O2 Signal circuit • O2 Return circuit • O2 Sensor • MAP Sensor • ECT Sensor • Engine mechanical • Fuel filter/Pressure regulator • PCM
DTC: P0172 **1T PCM, MIL: Yes** **Year:** 2009, 2010 **Model:** Eclipse, Galant **Engine:** 2.4L L4 VIN F **Transmission:** All	**System Too Rich (cylinder 1, 4):** Under the closed loop air/fuel ratio control. Engine coolant temperature is higher than 76°C (169°F). Mass airflow sensor output is 6 g/sec or more. Long-term fuel trim has continued to be lower than -12.5 percent for 5 seconds. Or. Short-term fuel trim has continued to be lower than -10.0 percent for 5 seconds. **Possible Causes:** • Mass airflow sensor failed. • Injector (Number 1, 4) failed. • Incorrect fuel pressure. • Engine coolant temperature sensor failed. • Intake air temperature sensor failed. • Barometric pressure sensor failed. • Manifold absolute pressure sensor failed. • Harness damage in cylinder 1, 4 injector circuit or connector damage. • ECM failed. \<M/T\> • PCM failed. \<A/T\>
DTC: P0172 **1T ECM, MIL: Yes** **Year:** 2009, 2010 **Model:** Lancer, Outlander **Engine:** 2.0L L4 VIN U, 2.0L L4 VIN V, 2.4L L4 VIN W **Transmission:** All	**System Too Rich:** Under the closed loop air/fuel ratio control. Engine coolant temperature is higher than 76°C (169°F). Mass airflow sensor output is 9 g/sec or more. Long-term fuel trim has continued to be lower than -12.5 percent for 5 seconds. Or, short-term fuel trim has continued to be lower than -10.2 percent \<Except for California\> or -5.5 percent \<California\> for 5 seconds. **Possible Causes:** • Mass airflow sensor failed. • Injector failed. • Incorrect fuel pressure. • Engine coolant temperature sensor failed. • Intake air temperature sensor failed. • Barometric pressure sensor failed. • Manifold absolute pressure sensor failed. • ECM failed.
DTC: P0174 **1T PCM, MIL: Yes** **Year:** 2009, 2010 **Model:** Eclipse, Galant **Engine:** 2.4L L4 VIN F **Transmission:** All	**System Too Lean (Cylinder 2, 3):** Engine coolant temperature is lower than 100°C (212°F) when the engine is started. Intake air temperature is lower than 60°C (140°F) when the engine is started. Under the closed loop air/fuel ratio control. Engine coolant temperature is higher than 76°C (169°F). Mass airflow sensor output is 6 g/sec or more. Long-term fuel trim has continued to be higher than +12.5 percent for 5 seconds. Or. Short-term fuel trim has continued to be higher than +7.4 percent for 5 seconds. **Possible Causes:** • Mass airflow sensor failed. • Injector (Number 2, 3) failed. • Incorrect fuel pressure. • Air drawn in from gaps in gasket, seals, etc. • Engine coolant temperature sensor failed. • Intake air temperature sensor failed. • Barometric pressure sensor failed. • Manifold absolute pressure sensor failed. • Exhaust leak. • Use of incorrect or contaminated fuel. • Harness damage in cylinder 2, 3 injector circuit or connector damage. • ECM failed. \<M/T\> • PCM failed. \<A/T\>

DTC	Trouble Code Title, Conditions & Possible Causes
DTC: P0174 **2T PCM, MIL: Yes** **Year:** 2009 **Model:** Raider **Engine:** 3.7L V6 VIN K **Transmission:** All	**Fuel System 2/1 Lean:** With the engine running in closed loop mode, the ambient/battery temperature above -6.7°C (20°F) and altitude below 2590.8 m (8500 ft). If the PCM multiplics short term compensation by long term adaptive and a certain percentage is exceeded for two trips, a freeze frame is stored, the MIL illuminates and a trouble code is stored. **Possible Causes:** • Restricted fuel supply line • Fuel pump inlet strainer plugged • Fuel pump module • O2 signal circuit • O2 return circuit • O2 Sensor heater operation • O2 Sensor • Map Sensor operation • ECT Sensor operation • Engine mechanical • Fuel filter/Pressure regulator • PCM
DTC: P0174 **1T PCM, MIL: Yes** **Year:** 2009, 2010 **Model:** Eclipse, Endeavor, Galant, Outlander **Engine:** 3.0L V6 VIN X, 3.8L V6 VIN S, 3.8L V6 VIN T **Transmission:** All	**System Too Lean (Bank 2):** Engine coolant temperature is lower than approximately 100°C (212°F) when the engine is started. Intake air temperature is lower than 60°C (140°F) when the engine is started. Under the closed loop air/fuel ratio control. Engine coolant temperature is higher than 76°C (169°F). Mass airflow sensor output is 10 g/sec or more. Long-term fuel trim has continued to be higher than +12.5 percent for 5 seconds. Or, short-term fuel trim has continued to be higher than +7.4 percent for 5 seconds. **Possible Causes:** • Mass airflow sensor failed. • Injector (Number 2, 4, 6) failed. • Incorrect fuel pressure. • Air drawn in from gaps in gasket, seals, etc. • Engine coolant temperature sensor failed. • Intake air temperature sensor failed. • Barometric pressure sensor failed. • Manifold absolute pressure sensor failed • Exhaust leak. • Use of incorrect or contaminated fuel. • Harness damage in left bank injector circuit or connector damage. • PCM failed.
DTC: P0175 **2T PCM, MIL: Yes** **Year:** 2009, 2010 **Model:** Eclipse, Endeavor, Galant, Outlander **Engine:** 3.0L V6 VIN X, 3.8L V6 VIN S, 3.8L V6 VIN T **Transmission:** All	**System Too Rich (Bank 2):** Under the closed loop air/fuel ratio control. Engine coolant temperature is higher than 76°C (169°F). Mass airflow sensor output is 10 g/sec or more. Long-term fuel trim has continued to be lower than -12.5 percent for 5 seconds. Or short-term fuel trim has continued to be lower than -5.5 percent for 5 seconds. **Possible Causes:** • Mass airflow sensor failed • Injector (Number 2, 4, 6) failed • Incorrect fuel pressure • Engine coolant temperature sensor failed • Intake air temperature sensor failed • Barometric pressure sensor failed • Manifold absolute pressure sensor failed • Harness damage in left bank injector circuit or connector damage • ECM failed <M/T> • PCM failed <A/T>

DTC	Trouble Code Title, Conditions & Possible Causes
DTC: P0175 **1T PCM, MIL: Yes** **Year:** 2009, 2010 **Model:** Eclipse, Galant **Engine:** 2.4L L4 VIN F **Transmission:** All	**System Too Rich (Cylinder 2, 3):** Under the closed loop air/fuel ratio control. Engine coolant temperature is higher than 76°C (169°F). Mass airflow sensor output is 6 g/sec or more. Long-term fuel trim has continued to be lower than -12.5 percent for 5 seconds. Or. Short-term fuel trim has continued to be lower than -10.0 percent for 5 seconds. **Possible Causes:** • Mass airflow sensor failed. • Injector (Number 2, 3) failed. • Incorrect fuel pressure. • Engine coolant temperature sensor failed. • Intake air temperature sensor failed. • Barometric pressure sensor failed. • Manifold absolute pressure sensor failed. • Harness damage in cylinder 2, 3 injector circuit or connector damage. • ECM failed. <M/T> • PCM failed. <A/T>
DTC: P0175 **2T PCM, MIL: Yes** **Year:** 2009 **Model:** Raider **Engine:** 3.7L V6 VIN K **Transmission:** All	**Fuel System 2/1 Rich:** With the engine running in closed loop mode, the ambient/battery temperature above -6.7°C (20°F) and altitude below 2590.8 m (8500 ft). If the PCM multiplies short term compensation by long term adaptive and a purge fuel multiplier and the result is below a certain value for 30 seconds over two trips, a freeze frame is stored, the MIL illuminates and a trouble code is stored. **Possible Causes:** • O2 Sensor heater operation • EVAP purge solenoid operation • O2 Signal circuit • O2 Return circuit • O2 Sensor • MAP Sensor • ECT Sensor • Engine mechanical • Fuel filter/Pressure regulator (High) • PCM
DTC: P0181 **1T PCM, MIL: Yes** **Year:** 2009, 2010 **Model:** Eclipse, Endeavor, Galant, Outlander **Engine:** 2.4L L4 VIN F, 2.4L L4 VIN W, 3.0L V6 VIN X, 3.8L V6 VIN S, 3.8L V6 VIN T **Transmission:** All	**Fuel Tank Temperature Sensor Circuit Range/Performance:** The engine coolant temperature - intake air temperature is 5°C (9°F) or less when the engine is started. The engine coolant temperature is between -10°C (14°F) and 36°C (97°F) when the engine is started. The engine coolant temperature is higher than 60°C (140°F). Maximum vehicle speed is higher than 30 km/h (19 mph) after the engine starting sequence has been completed. The fuel tank temperature - engine coolant temperature is 15°C (27°F) or more when the engine is started. **Possible Causes:** • Fuel tank temperature sensor failed. • Open or shorted fuel tank temperature sensor circuit, harness damage, or connector damage. • ECM failed. <M/T> • PCM failed. <A/T>
DTC: P0181 **1T ECM, MIL: Yes** **Year:** 2009, 2010 **Model:** Lancer **Engine:** 2.0L L4 VIN U, 2.0L L4 VIN V, 2.4L L4 VIN W **Transmission:** All	**Fuel Tank Temperature Sensor Circuit Range/Performance:** The engine coolant temperature - intake air temperature is 5°C (9°F) or less when the engine is started. The engine coolant temperature is between -10°C (14°F) and 36°C (97°F) when the engine is started. The engine coolant temperature is higher than 60°C (140°F). Maximum vehicle speed is higher than 30 km/h (19 mph) after the engine starting sequence has been completed. The fuel tank temperature - engine coolant temperature is 15°C (27°F) or more when the engine is started. **NOTE: A diagnostic trouble code (DTC) could be output if the engine and the radiator have been flushed repeatedly when the engine coolant temperature was high (or the fuel tank temperature was high). Because this is not a failure, the DTC must be erased. Make sure to test drive the vehicle in accordance with the OBD-II drive cycle pattern in order to verify that a DTC will not be output.** **Possible Causes:** • Fuel tank temperature sensor failed. • Open or shorted fuel tank temperature sensor circuit, harness damage, or connector damage. • ECM failed.
DTC: P0182 **1T PCM, MIL: Yes** **Year:** 2009, 2010 **Model:** Eclipse, Endeavor, Galant, Outlander **Engine:** 2.4L L4 VIN F, 2.4L L4 VIN W, 3.0L V6 VIN X, 3.8L V6 VIN S, 3.8L V6 VIN T **Transmission:** All	**Fuel Tank Temperature Sensor Circuit Low Input:** After 2 seconds or more have passed since the engine starting sequence was completed. Fuel tank temperature sensor output voltage is out of specified range. Sensor output voltage has continued to be 0.1 volt or lower for 2 seconds. **Possible Causes:** • Fuel tank temperature sensor failed. • Shorted fuel tank temperature sensor circuit or connector damage. • ECM failed. <M/T> • PCM failed. <A/T>

DTC	Trouble Code Title, Conditions & Possible Causes
DTC: P0182 **1T ECM, MIL: Yes** **Year:** 2009, 2010 **Model:** Lancer **Engine:** 2.0L L4 VIN U, 2.0L L4 VIN V, 2.4L L4 VIN W **Transmission:** All	**Fuel Tank Temperature Sensor Circuit Low Input:** After 2 seconds or more have passed since the engine starting sequence was completed. Sensor output voltage has continued to be 0.1 volt or lower for 2 seconds. **Possible Causes:** • Fuel tank temperature sensor failed. • Shorted fuel tank temperature sensor circuit or connector damage. • ECM failed.
DTC: P0183 **1T PCM, MIL: Yes** **Year:** 2009, 2010 **Model:** Eclipse, Endeavor, Galant, Outlander **Engine:** 2.4L L4 VIN F, 2.4L L4 VIN W, 3.0L V6 VIN X, 3.8L V6 VIN S, 3.8L V6 VIN T **Transmission:** All	**Fuel Tank Temperature Sensor Circuit High Input:** After 2 seconds or more have passed since the engine starting sequence was completed. Fuel tank temperature sensor output voltage is out of specified range. Sensor output voltage has continued to be 4.6 volts or higher for 2 seconds. **Possible Causes:** • Fuel tank temperature sensor failed. • Open fuel tank temperature sensor circuit, harness damage, or connector damage. • ECM failed. <M/T> • PCM failed. <A/T>
DTC: P0183 **1T ECM, MIL: Yes** **Year:** 2009, 2010 **Model:** Lancer **Engine:** 2.0L L4 VIN U, 2.0L L4 VIN V, 2.4L L4 VIN W **Transmission:** All	**Fuel Tank Temperature Sensor Circuit High Input:** After 2 seconds or more have passed since the engine starting sequence was completed. The sensor output voltage has continued to be 4.6 volts or higher for 2 seconds. **Possible Causes:** • Fuel tank temperature sensor failed. • Open fuel tank temperature sensor circuit, harness damage, or connector damage. • ECM failed.
DTC: P0201 **1T PCM, MIL: Yes** **Year:** 2009, 2010 **Model:** Eclipse, Endeavor, Galant, Outlander **Engine:** 2.4L L4 VIN F, 2.4L L4 VIN W, 3.0L V6 VIN X, 3.8L V6 VIN S, 3.8L V6 VIN T **Transmission:** All	**Injector Circuit Malfunction - Cylinder 1:** Engine is running. The ECM <M/T> or the PCM <A/T> detects open circuit and short malfunction. The supply voltage is 3 volts or less without the injector driving. The coil current is 4.5 ampere or more with the injector driving. **Possible Causes:** • No. 1 cylinder injector failed. • Open or shorted No.1cylinder injector circuit, harness damage, or connector damage. • ECM failed. <M/T> • PCM failed. <A/T>
DTC: P0201 **1T PCM, MIL: Yes** **Year:** 2009 **Model:** Raider **Engine:** 3.7L V6 VIN K **Transmission:** All	**Fuel Injector 1 Circuit:** With battery voltage greater than 10 volts. Auto Shutdown Relay energized. Engine speed less than 3000 rpm. The PCM monitors the continuity of the injector circuits as well as the voltage spike created by the collapse of the magnetic field in the injector coil. Any condition that reduces the maximum current flow or the magnitude of the voltage spike can cause this DTC to set. **Possible Causes:** • ASD Relay output circuit • Injector control No.1 circuit open • Injector control No.1 circuit shorted to ground • Fuel injector • PCM
DTC: P0201 **1T ECM, MIL: Yes** **Year:** 2009, 2010 **Model:** Lancer **Engine:** 2.0L L4 VIN U, 2.0L L4 VIN V, 2.4L L4 VIN W **Transmission:** All	**Injector Circuit-Cylinder 1:** Engine is running. The supply voltage is 3 volts or less without the injector driving for 2 seconds. The coil current is 4.5 amperes or more with the injector driving for 2 seconds. **Possible Causes:** • No. 1 cylinder injector failed. • Open or shorted No. 1 cylinder injector circuit, harness damage or connector damage. • ECM failed.
DTC: P0202 **1T ECM, MIL: Yes** **Year:** 2009, 2010 **Model:** Lancer **Engine:** 2.0L L4 VIN U, 2.0L L4 VIN V, 2.4L L4 VIN W **Transmission:** All	**Injector Circuit-Cylinder 2:** Engine is running. The supply voltage is 3 volts or less without the injector driving for 2 seconds. The coil current is 4.5 amperes or more with the injector driving for 2 seconds. **Possible Causes:** • No. 2 cylinder injector failed. • Open or shorted No. 2 cylinder injector circuit, harness damage or connector damage. • ECM failed.

DTC	Trouble Code Title, Conditions & Possible Causes
DTC: P0202 **1T PCM, MIL: Yes** **Year:** 2009 **Model:** Raider **Engine:** 3.7L V6 VIN K **Transmission:** All	**Fuel Injector 2 Circuit:** With battery voltage greater than 10 volts. Auto Shutdown Relay energized. Engine speed less than 3000 rpm. The PCM monitors the continuity of the injector circuits as well as the voltage spike created by the collapse of the magnetic field in the injector coil. Any condition that reduces the maximum current flow or the magnitude of the voltage spike can cause this DTC to set. **Possible Causes:** • ASD Relay output circuit • Injector control No.2 circuit open • Injector control No.2 circuit shorted to ground • Fuel injector • PCM
DTC: P0202 **1T PCM, MIL: Yes** **Year:** 2009, 2010 **Model:** Eclipse, Endeavor, Galant, Outlander **Engine:** 2.4L L4 VIN F, 2.4L L4 VIN W, 3.0L V6 VIN X, 3.8L V6 VIN S, 3.8L V6 VIN T **Transmission:** All	**Injector Circuit Malfunction - Cylinder 2:** Engine is running. The ECM <M/T> or the PCM <A/T> detects open circuit and short malfunction. The supply voltage is 3 volts or less without the injector driving. The coil current is 4.5 ampere or more with the injector driving. **Possible Causes:** • No. 2 cylinder injector failed. • Open or shorted No.1cylinder injector circuit, harness damage, or connector damage. • ECM failed. <M/T> • PCM failed. <A/T>
DTC: P0203 **1T ECM, MIL: Yes** **Year:** 2009, 2010 **Model:** Lancer **Engine:** 2.0L L4 VIN U, 2.0L L4 VIN V, 2.4L L4 VIN W **Transmission:** All	**Injector Circuit-Cylinder 3:** Engine is running. The supply voltage is 3 volts or less without the injector driving for 2 seconds. The coil current is 4.5 amperes or more with the injector driving for 2 seconds. **Possible Causes:** • No. 3 cylinder injector failed. • Open or shorted No. 3 cylinder injector circuit, harness damage or connector damage. • ECM failed.
DTC: P0203 **1T PCM, MIL: Yes** **Year:** 2009, 2010 **Model:** Eclipse, Endeavor, Galant, Outlander **Engine:** 2.4L L4 VIN F, 2.4L L4 VIN W, 3.0L V6 VIN X, 3.8L V6 VIN S, 3.8L V6 VIN T **Transmission:** All	**Injector Circuit Malfunction - Cylinder 3:** Engine is running. The ECM <M/T> or the PCM <A/T> detects open circuit and short malfunction. The supply voltage is 3 volts or less without the injector driving. The coil current is 4.5 ampere or more with the injector driving. **Possible Causes:** • No. 3 cylinder injector failed. • Open or shorted No.1cylinder injector circuit, harness damage, or connector damage. • ECM failed. <M/T> • PCM failed. <A/T>
DTC: P0203 **1T PCM, MIL: Yes** **Year:** 2009 **Model:** Raider **Engine:** 3.7L V6 VIN K **Transmission:** All	**Fuel Injector 3 Circuit:** With battery voltage greater than 10 volts. Auto Shutdown Relay energized. Engine speed less than 3000 rpm. The PCM monitors the continuity of the injector circuits as well as the voltage spike created by the collapse of the magnetic field in the injector coil. Any condition that reduces the maximum current flow or the magnitude of the voltage spike can cause this DTC to set **Possible Causes:** • ASD Relay output circuit • Injector control No.3 circuit open • Injector control No.3 circuit shorted to ground • Fuel injector • PCM
DTC: P0204 **1T PCM, MIL: Yes** **Year:** 2009, 2010 **Model:** Eclipse, Endeavor, Galant, Outlander **Engine:** 2.4L L4 VIN F, 2.4L L4 VIN W, 3.0L V6 VIN X, 3.8L V6 VIN S, 3.8L V6 VIN T **Transmission:** All	**Injector Circuit Malfunction - Cylinder 4:** Engine is running. The ECM <M/T> or the PCM <A/T> detects open circuit and short malfunction. The supply voltage is 3 volts or less without the injector driving. The coil current is 4.5 ampere or more with the injector driving. **Possible Causes:** • No. 4 cylinder injector failed. • Open or shorted No.1cylinder injector circuit, harness damage, or connector damage. • ECM failed. <M/T> • PCM failed. <A/T>

DTC	Trouble Code Title, Conditions & Possible Causes
DTC: P0204 **1T PCM, MIL: Yes** **Year:** 2009 **Model:** Raider **Engine:** 3.7L V6 VIN K **Transmission:** All	**Fuel Injector 4 Circuit:** With battery voltage greater than 10 volts. Auto Shutdown Relay energized. Engine speed less than 3000 rpm. The PCM monitors the continuity of the injector circuits as well as the voltage spike created by the collapse of the magnetic field in the injector coil. Any condition that reduces the maximum current flow or the magnitude of the voltage spike can cause this DTC to set. **Possible Causes:** • ASD Relay output circuit • Injector control No.4 circuit open • Injector control No.4 circuit shorted to ground • Fuel injector • PCM
DTC: P0204 **1T ECM, MIL: Yes** **Year:** 2009, 2010 **Model:** Lancer **Engine:** 2.0L L4 VIN U, 2.0L L4 VIN V, 2.4L L4 VIN W **Transmission:** All	**Injector Circuit-Cylinder 4:** Engine is running. The supply voltage is 3 volts or less without the injector driving for 2 seconds. The coil current is 4.5 amperes or more with the injector driving for 2 seconds. **Possible Causes:** • No. 4 cylinder injector failed. • Open or shorted No. 4 cylinder injector circuit, harness damage or connector damage. • ECM failed.
DTC: P0205 **1T PCM, MIL: Yes** **Year:** 2009 **Model:** Raider **Engine:** 3.7L V6 VIN K **Transmission:** All	**Fuel Injector 5 Circuit:** With battery voltage greater than 10 volts. Auto Shutdown Relay energized. Engine speed less than 3000 rpm. The PCM monitors the continuity of the injector circuits as well as the voltage spike created by the collapse of the magnetic field in the injector coil. Any condition that reduces the maximum current flow or the magnitude of the voltage spike can cause this DTC to set. **Possible Causes:** • ASD Relay output circuit • Injector control No.5 circuit open • Injector control No.5 circuit shorted to ground • Fuel injector • PCM
DTC: P0205 **1T PCM, MIL: Yes** **Year:** 2009, 2010 **Model:** Eclipse, Endeavor, Galant, Outlander **Engine:** 3.0L V6 VIN X, 3.8L V6 VIN S, 3.8L V6 VIN T **Transmission:** All	**Injector Circuit Malfunction - Cylinder 5:** Engine is running. The ECM <M/T> or the PCM <A/T> detects open circuit and short malfunction. The supply voltage is 3 volts or less without the injector driving. The coil current is 4.5 ampere or more with the injector driving. **Possible Causes:** • No. 5 cylinder injector failed. • Open or shorted No.1 cylinder injector circuit, harness damage, or connector damage. • ECM failed. <M/T> • PCM failed. <A/T>
DTC: P0206 **1T PCM, MIL: Yes** **Year:** 2009 **Model:** Raider **Engine:** 3.7L V6 VIN K **Transmission:** All	**Fuel Injector 6 Circuit:** With battery voltage greater than 10 volts. Auto Shutdown Relay energized. Engine speed less than 3000 rpm. The PCM monitors the continuity of the injector circuits as well as the voltage spike created by the collapse of the magnetic field in the injector coil. Any condition that reduces the maximum current flow or the magnitude of the voltage spike can cause this DTC to set. **Possible Causes:** • ASD Relay output circuit • Injector control No.6 circuit open • Injector control No.6 circuit shorted to ground • Fuel injector • PCM
DTC: P0206 **1T PCM, MIL: Yes** **Year:** 2009, 2010 **Model:** Eclipse, Endeavor, Galant, Outlander **Engine:** 3.0L V6 VIN X, 3.8L V6 VIN S, 3.8L V6 VIN T **Transmission:** All	**Injector Circuit Malfunction - Cylinder 6:** Engine is running. The ECM <M/T> or the PCM <A/T> detects open circuit and short malfunction. The supply voltage is 3 volts or less without the injector driving. The coil current is 4.5 ampere or more with the injector driving. **Possible Causes:** • No. 6 cylinder injector failed. • Open or shorted No.1 cylinder injector circuit, harness damage, or connector damage. • ECM failed. <M/T> • PCM failed. <A/T>

DTC	Trouble Code Title, Conditions & Possible Causes
DTC: P0218 **1T PCM, MIL: Yes** **Year:** 2009 **Model:** Raider **Engine:** 3.7L V6 VIN K **Transmission:** All	**High Temperature Operation Activated (45RFE/545RFE Transmission):** Whenever the engine is running. Immediately after a Overheat shift schedule is activated when the Transmission temperature exceeds 127° C or (260° F). **Possible Causes:** • High temperature operation activated • Torque converter clutch Slipping/Not activating • Operating the vehicle with the transmission in limp-in mode for a extended period of time • Excessive idling time
DTC: P0218 **1T PCM, MIL: Yes** **Year:** 2009 **Model:** Raider **Engine:** 3.7L V6 VIN K **Transmission:** All	**High Temperature Operation Activated (42RLE Transmission):** Whenever the engine is running. Immediately when a Overheat shift schedule is activated when the Transmission Oil Temperature reaches 155° C or (240° F). **Possible Causes:** • Engine cooling system operation • Transmission oil cooler plugged • High temperature operation activated
DTC: P0219 **1T ECM, MIL: Yes** **Year:** 2009, 2010 **Model:** Lancer **Engine:** 2.0L L4 VIN V **Transmission:** All	**Engine Over Speed Condition (Turbo Charged):** Ignition switch is "ON" position. The maximum engine rotational speed stored in the past was more than 8,398 r/min. **NOTE: Except for an engine malfunction, there are possible causes attributed to the excessive engine rotational speed. Check whether there is another malfunction except the engine failure.** **Possible Causes:** • The failure is supposed to be attributed to the excessive engine rotational speed.
DTC: P0221 **2T PCM, MIL: Yes** **Year:** 2009 **Model:** Raider **Engine:** 3.7L V6 VIN K **Transmission:** All	**Throttle Position Sensor 2 Performance:** Ignition on and No MAP Sensor DTCs set. The PCM detects that the sensor input voltage does not fall within a valid range based on engine speed and load. **NOTE: P2135 should set with this code.** **Possible Causes:** • Excessive resistance in the TP Sensor No.1 or No.2 Signal circuit • Excessive resistance in the 5-volt Supply circuit • Excessive resistance in the TP sensor return circuit • Short between the TP Sensor No.2 Signal circuit and the TP Sensor No.1 Signal circuit • TP Sensor/Throttle body • PCM
DTC: P0222 **1T ECM, MIL: Yes** **Year:** 2009, 2010 **Model:** Lancer, Outlander **Engine:** 2.0L L4 VIN U, 2.0L L4 VIN V, 2.4L L4 VIN W, 3.0L V6 VIN X **Transmission:** All	**Throttle Position Sensor (Sub) Circuit Low Input:** Ignition switch is "ON" position. Throttle position sensor (sub) output voltage should be 0.2 volt or less for 0.3 second. **Possible Causes:** • Throttle position sensor failed. • Open or shorted throttle position sensor (sub) circuit, harness damage, or connector damage. • ECM failed.
DTC: P0222 **1T PCM, MIL: Yes** **Year:** 2009, 2010 **Model:** Eclipse, Endeavor, Galant **Engine:** 2.4L L4 VIN F, 3.8L V6 VIN S, 3.8L V6 VIN T **Transmission:** All	**Throttle Position Sensor (Sub) Circuit Low Input:** Ignition switch is "ON" position. Throttle position sensor (sub) output voltage is out of specified range. The throttle position sensor (sub) output voltage should be 2.25 volts or less for 0.5 second. **Possible Causes:** • Throttle position sensor failed. • Open or shorted throttle position sensor (sub) circuit, harness damage, or connector damage. • ECM failed. <M/T> • PCM failed. <A/T>
DTC: P0222 **1T PCM, MIL: Yes** **Year:** 2009 **Model:** Raider **Engine:** 3.7L V6 VIN K **Transmission:** All	**Throttle Position Sensor 2 Circuit Low:** With the ignition on. Battery voltage greater than 10 volts. The PCM detects that the TP Sensor 2 voltage is lower than the acceptable value. **Possible Causes:** • 5-volt supply circuit open • 5-volt supply circuit shorted to ground • TP Sensor No.2 signal circuit shorted to ground • TP Sensor No.2 signal circuit shorted to the TP sensor return circuit • TP Sensor No.2 signal circuit open • TP Sensor/Throttle body • Throttle plate jammed against the maximum stop • PCM

DTC	Trouble Code Title, Conditions & Possible Causes
DTC: P0223 **1T ECM, MIL: Yes** **Year:** 2009, 2010 **Model:** Lancer **Engine:** 2.0L L4 VIN U, 2.0L L4 VIN V, 2.4L L4 VIN W **Transmission:** All	**Throttle Position Sensor (Sub) Circuit High Input:** Ignition switch is "ON" position. Throttle position sensor (sub) output voltage should be 4.8 volts or more for 0.3 second. **Possible Causes:** • Throttle position sensor failed. • Open throttle position sensor (sub) circuit, harness damage, or connector damage. • ECM failed.
DTC: P0223 **1T PCM, MIL: Yes** **Year:** 2009, 2010 **Model:** Eclipse, Endeavor, Galant, Outlander **Engine:** 2.4L L4 VIN F, 2.4L L4 VIN W, 3.0L V6 VIN X, 3.8L V6 VIN S, 3.8L V6 VIN T **Transmission:** All	**Throttle Position Sensor (Sub) Circuit High Input:** Ignition switch is "ON" position. Throttle position sensor (sub) output voltage should be 4.8 volts or more for 0.5 second. Throttle position sensor (sub) output voltage is out of specified range. **Possible Causes:** • Throttle position sensor failed. • Open throttle position sensor (sub) circuit, harness damage, or connector damage. • ECM failed. <M/T> • PCM failed. <A/T>
DTC: P0223 **1T PCM, MIL: Yes** **Year:** 2009 **Model:** Raider **Engine:** 3.7L V6 VIN K **Transmission:** All	**Throttle Position Sensor 2 Circuit High:** With the ignition on. Throttle Position Sensor No.2 Signal circuit voltage is greater than the maximum acceptable value. **Possible Causes:** • TP Sensor No.2 signal circuit shorted to voltage • TP Sensor No.2 signal circuit shorted to the 5-volt supply circuit • TP sensor return circuit open • TP Sensor/Throttle body • PCM
DTC: P0300 **2T PCM, MIL: Yes** **Year:** 2009 **Model:** Raider **Engine:** 3.7L V6 VIN K **Transmission:** All	**Multiple Cylinder Misfire:** Any time the engine is running, and the adaptive numerator has been successfully updated. The threshold to set the fault is application specific; it is tied to the level of misfire that will cause emissions to increase to 1.5 times the standard or in some cases 1%. It is always a two trip fault above the calibrated RPM. It takes 1 soft fail to set a malfunction and two trips to set the MIL. **Possible Causes:** • ASD Relay output circuit • Injector control circuit • Coil control circuit • Spark plug • Ignition wire • Ignition coil • Fuel pump/Strainer • Restricted fuel supply line • Fuel injector • Engine mechanical problem • PCM
DTC: P0300 **2T PCM, MIL: Yes** **Year:** 2009, 2010 **Model:** Eclipse, Endeavor, Galant, Lancer, Outlander **Engine:** 2.0L L4 VIN U, 2.0L L4 VIN V, 2.4L L4 VIN F, 2.4L L4 VIN W, 3.0L V6 VIN X, 3.8L V6 VIN S, 3.8L V6 VIN T **Transmission:** All	**Random/Multiple Cylinder Misfire Detected :** Engine speed is between 440 and 6,500 r/min. Engine coolant temperature is higher than -10°C (14°F). Barometric pressure is higher than 76 kPa (22.4 in. Hg). Volumetric efficiency is between 30 and 60 percent. Adaptive learning is complete for the vane which generates a crankshaft position signal. While the engine is running, excluding gear shifting, deceleration and sudden acceleration/deceleration. The throttle deviation is -0.06 V/25 ms to +0.06 V/10 ms. Misfire has occurred more frequently than allowed during the last 200 revolutions [when the catalyst temperature is higher than 1,000°C (1,832°F)]. Or. Misfire has occurred in 15 or more of the last 1,000 revolutions (corresponding to 1.5 times the limit of emission standard). **Possible Causes:** • Ignition system related part(s) failed. • Poor crankshaft position sensor. • Incorrect air/fuel ratio. • Low compression pressure. • Skipping of timing belt teeth. • EGR system and EGR valve failed. • ECM failed. <M/T> • PCM failed. <A/T>

DTC	Trouble Code Title, Conditions & Possible Causes
DTC: P0301 **2T PCM, MIL: Yes** **Year:** 2009, 2010 **Model:** Eclipse, Endeavor, Galant, Lancer, Outlander **Engine:** 2.0L L4 VIN U, 2.0L L4 VIN V, 2.4L L4 VIN F, 2.4L L4 VIN W, 3.0L V6 VIN X, 3.8L V6 VIN S, 3.8L V6 VIN T **Transmission:** All	**Cylinder 1 Misfire Detected:** Engine speed is between 440 and 6,500 r/min. Engine coolant temperature is higher than -10°C (14°F). Barometric pressure is higher than 76 kPa (22.4 in. Hg). Volumetric efficiency is at between 30 and 60 percent. Adaptive learning is complete for the vane which generates a crankshaft position signal, While the engine is running, excluding gear shifting, deceleration and sudden acceleration/deceleration. The throttle deviation is -0.06 V/25 ms to +0.06 V/10 ms. Misfire has occurred more frequently than allowed during the last 200 revolutions [when the catalyst temperature is higher than 1,000°C (1,832°F)]. Or. Misfire has occurred in 15 or more of the last 1,000 revolutions (corresponding to 1.5 times the limit of emission standard). **Possible Causes:** • Ignition system related part(s) failed. • Low compression pressure. • ECM failed. <M/T> • PCM failed. <A/T>
DTC: P0301 **2T PCM, MIL: Yes** **Year:** 2009 **Model:** Raider **Engine:** 3.7L V6 VIN K **Transmission:** All	**Cylinder 1 Misfire:** Any time the engine is running, and the adaptive numerator has been successfully updated. The threshold to set the fault is application specific; it is tied to the level of misfire that will cause emissions to increase to 1.5 times the standard or in some cases 1%. It is always a two trip fault above the calibrated RPM. It takes 1 soft fail to set a malfunction and two trips to set the MIL. **Possible Causes:** • ASD Relay output circuit • Injector control circuit • Coil control circuit • Spark plug • Ignition wire • Ignition coil • Fuel pump/Strainer • Restricted fuel supply line • Fuel injector • Engine mechanical problem • PCM
DTC: P0302 **2T PCM, MIL: Yes** **Year:** 2009, 2010 **Model:** Eclipse, Endeavor, Galant, Lancer, Outlander **Engine:** 2.0L L4 VIN U, 2.0L L4 VIN V, 2.4L L4 VIN F, 2.4L L4 VIN W, 3.0L V6 VIN X, 3.8L V6 VIN S, 3.8L V6 VIN T **Transmission:** All	**Cylinder 2 Misfire Detected:** Engine speed is between 440 and 6,500 r/min. Engine coolant temperature is higher than -10°C (14°F). Barometric pressure is higher than 76 kPa (22.4 in. Hg). Volumetric efficiency is at between 30 and 60 percent. Adaptive learning is complete for the vane which generates a crankshaft position signal, while the engine is running, excluding gear shifting, deceleration and sudden acceleration/deceleration. The throttle deviation is -0.06 V/25 ms to +0.06 V/10 ms. Misfire has occurred more frequently than allowed during the last 200 revolutions [when the catalyst temperature is higher than 1,000°C (1,832°F)]. Or. Misfire has occurred in 15 or more of the last 1,000 revolutions (corresponding to 1.5 times the limit of emission standard). **Possible Causes:** • Ignition system related part(s) failed. • Low compression pressure. • ECM failed. <M/T> • PCM failed. <A/T>
DTC: P0302 **2T PCM, MIL: Yes** **Year:** 2009 **Model:** Raider **Engine:** 3.7L V6 VIN K **Transmission:** All	**Cylinder 2 Misfire:** Any time the engine is running, and the adaptive numerator has been successfully updated. The threshold to set the fault is application specific; it is tied to the level of misfire that will cause emissions to increase to 1.5 times the standard or in some cases 1%. It is always a two trip fault above the calibrated RPM. It takes 1 soft fail to set a malfunction and two trips to set the MIL. **Possible Causes:** • ASD Relay output circuit • Injector control circuit • Coil control circuit • Spark plug • Ignition wire • Ignition coil • Fuel pump/Strainer • Restricted fuel supply line • Fuel injector • Engine mechanical problem • PCM

DTC	Trouble Code Title, Conditions & Possible Causes
DTC: P0303 **2T PCM, MIL: Yes** **Year:** 2009 **Model:** Raider **Engine:** 3.7L V6 VIN K **Transmission:** All	**Cylinder 3 Misfire:** Any time the engine is running, and the adaptive numerator has been successfully updated. The threshold to set the fault is application specific; it is tied to the level of misfire that will cause emissions to increase to 1.5 times the standard or in some cases 1%. It is always a two trip fault above the calibrated RPM. It takes 1 soft fail to set a malfunction and two trips to set the MIL. **Possible Causes:** • ASD Relay output circuit • Injector control circuit • Coil control circuit • Spark plug • Ignition wire • Ignition coil • Fuel pump/Strainer • Restricted fuel supply line • Fuel injector • Engine mechanical problem • PCM
DTC: P0303 **2T PCM, MIL: Yes** **Year:** 2009, 2010 **Model:** Eclipse, Endeavor, Galant, Lancer, Outlander **Engine:** 2.0L L4 VIN U, 2.0L L4 VIN V, 2.4L L4 VIN F, 2.4L L4 VIN W, 3.0L V6 VIN X, 3.8L V6 VIN S, 3.8L V6 VIN T **Transmission:** All	**Cylinder 3 Misfire Detected:** Engine speed is between 440 and 6,500 r/min. Engine coolant temperature is higher than -10°C (14°F). Barometric pressure is higher than 76 kPa (22.4 in. Hg). Volumetric efficiency is at between 30 and 60 percent. Adaptive learning is complete for the vane which generates a crankshaft position signal, while the engine is running, excluding gear shifting, deceleration and sudden acceleration/deceleration. The throttle deviation is -0.06 V/25 ms to +0.06 V/10 ms. Misfire has occurred more frequently than allowed during the last 200 revolutions [when the catalyst temperature is higher than 1,000°C (1,832°F)]. Or. Misfire has occurred in 15 or more of the last 1,000 revolutions (corresponding to 1.5 times the limit of emission standard). **Possible Causes:** • Ignition system related part(s) failed. • Low compression pressure. • ECM failed. <M/T> • PCM failed. <A/T>
DTC: P0304 **2T PCM, MIL: Yes** **Year:** 2009, 2010 **Model:** Eclipse, Endeavor, Galant, Lancer, Outlander **Engine:** 2.0L L4 VIN U, 2.0L L4 VIN V, 2.4L L4 VIN F, 2.4L L4 VIN W, 3.0L V6 VIN X, 3.8L V6 VIN S, 3.8L V6 VIN T **Transmission:** All	**Cylinder 4 Misfire Detected:** Engine speed is between 440 and 6,500 r/min. Engine coolant temperature is higher than -10°C (14°F). Barometric pressure is higher than 76 kPa (22.4 in. Hg). Volumetric efficiency is at between 30 and 60 percent. Adaptive learning is complete for the vane which generates a crankshaft position signal, while the engine is running, excluding gear shifting, deceleration and sudden acceleration/deceleration. The throttle deviation is -0.06 V/25 ms to +0.06 V/10 ms. Misfire has occurred more frequently than allowed during the last 200 revolutions [when the catalyst temperature is higher than 1,000°C (1,832°F)]. Or. Misfire has occurred in 15 or more of the last 1,000 revolutions (corresponding to 1.5 times the limit of emission standard). **Possible Causes:** • Ignition system related part(s) failed. • Low compression pressure. • ECM failed. <M/T> • PCM failed. <A/T>
DTC: P0304 **2T PCM, MIL: Yes** **Year:** 2009 **Model:** Raider **Engine:** 3.7L V6 VIN K **Transmission:** All	**Cylinder 4 Misfire:** Any time the engine is running, and the adaptive numerator has been successfully updated. The threshold to set the fault is application specific; it is tied to the level of misfire that will cause emissions to increase to 1.5 times the standard or in some cases 1%. It is always a two trip fault above the calibrated RPM. It takes 1 soft fail to set a malfunction and two trips to set the MIL. **Possible Causes:** • ASD Relay output circuit • Injector control circuit • Coil control circuit • Spark plug • Ignition wire • Ignition coil • Fuel pump/Strainer • Restricted fuel supply line • Fuel injector • Engine mechanical problem • PCM

DTC	Trouble Code Title, Conditions & Possible Causes
DTC: P0305 **2T PCM, MIL: Yes** **Year:** 2009 **Model:** Raider **Engine:** 3.7L V6 VIN K **Transmission:** All	**Cylinder 5 Misfire:** Any time the engine is running, and the adaptive numerator has been successfully updated. The threshold to set the fault is application specific; it is tied to the level of misfire that will cause emissions to increase to 1.5 times the standard or in some cases 1%. It is always a two trip fault above the calibrated RPM. It takes 1 soft fail to set a malfunction and two trips to set the MIL. **Possible Causes:** • ASD Relay output circuit • Injector control circuit • Coil control circuit • Spark plug • Ignition wire • Ignition coil • Fuel pump/Strainer • Restricted fuel supply line • Fuel injector • Engine mechanical problem • PCM
DTC: P0305 **2T PCM, MIL: Yes** **Year:** 2009, 2010 **Model:** Eclipse, Endeavor, Galant, Outlander **Engine:** 3.0L V6 VIN X, 3.8L V6 VIN S, 3.8L V6 VIN T **Transmission:** All	**Cylinder 5 Misfire Detected:** Engine speed is between 440 and 6,500 r/min. Engine coolant temperature is higher than -10°C (14°F). Barometric pressure is higher than 76 kPa (22.4 in. Hg). Volumetric efficiency is at between 30 and 60 percent. Adaptive learning is complete for the vane which generates a crankshaft position signal, while the engine is running, excluding gear shifting, deceleration and sudden acceleration/deceleration. The throttle deviation is -0.06 V/25 ms to +0.06 V/10 ms. Misfire has occurred more frequently than allowed during the last 200 revolutions [when the catalyst temperature is higher than 1,000°C (1,832°F)]. Or. Misfire has occurred in 15 or more of the last 1,000 revolutions (corresponding to 1.5 times the limit of emission standard). **Possible Causes:** • Ignition system related part(s) failed. • Low compression pressure. • ECM failed. <M/T> • PCM failed. <A/T>
DTC: P0306 **2T PCM, MIL: Yes** **Year:** 2009, 2010 **Model:** Eclipse, Endeavor, Galant, Outlander **Engine:** 3.0L V6 VIN X, 3.8L V6 VIN S, 3.8L V6 VIN T **Transmission:** All	**Cylinder 6 Misfire Detected:** Engine speed is between 440 and 6,500 r/min. Engine coolant temperature is higher than -10°C (14°F). Barometric pressure is higher than 76 kPa (22.4 in. Hg). Volumetric efficiency is at between 30 and 60 percent. Adaptive learning is complete for the vane which generates a crankshaft position signal, while the engine is running, excluding gear shifting, deceleration and sudden acceleration/deceleration. The throttle deviation is -0.06 V/25 ms to +0.06 V/10 ms. Misfire has occurred more frequently than allowed during the last 200 revolutions [when the catalyst temperature is higher than 1,000°C (1,832°F)]. Or. Misfire has occurred in 15 or more of the last 1,000 revolutions (corresponding to 1.5 times the limit of emission standard). **Possible Causes:** • Ignition system related part(s) failed. • Low compression pressure. • ECM failed. <M/T> • PCM failed. <A/T>
DTC: P0306 **2T PCM, MIL: Yes** **Year:** 2009 **Model:** Raider **Engine:** 3.7L V6 VIN K **Transmission:** All	**Cylinder 6 Misfire:** Any time the engine is running, and the adaptive numerator has been successfully updated. The threshold to set the fault is application specific; it is tied to the level of misfire that will cause emissions to increase to 1.5 times the standard or in some cases 1%. It is always a two trip fault above the calibrated RPM. It takes 1 soft fail to set a malfunction and two trips to set the MIL. **Possible Causes:** • ASD Relay output circuit • Injector control circuit • Coil control circuit • Spark plug • Ignition wire • Ignition coil • Fuel pump/Strainer • Restricted fuel supply line • Fuel injector • Engine mechanical problem • PCM

DTC	Trouble Code Title, Conditions & Possible Causes
DTC: P0315 **1T PCM, MIL: Yes** **Year:** 2009 **Model:** Raider **Engine:** 3.7L V6 VIN K **Transmission:** All	**No Crankshaft Position Sensor Learned:** Under closed throttle decel and A/C off. ECT above 75° C (167° F). Engine start time is greater than 50 seconds. One of the CKP sensor target windows has more than 2.86% variance from the reference. **Possible Causes:** • Tone wheel/ Pulse ring • Wire harness • CKP sensor
DTC: P0325 **2T PCM, MIL: Yes** **Year:** 2009 **Model:** Raider **Engine:** 3.7L V6 VIN K **Transmission:** All	**Knock Sensor No. 1 Circuit:** This monitor runs above 2000 rpm, under open throttle conditions. The Knock diagnostic does not run at idle or during decelerations. The high voltage test runs all the times the engine is running. The Knock Sensor voltage is not within specification. **Possible Causes:** • Short to voltage in the Knock Sensor No.1 Signal circuit • Open in the Knock Sensor No.1 Signal circuit • Open in the Knock Sensor No.1 Return circuit • Short between the Knock Sensor No.1 Signal circuit and the Knock Sensor No.1 Return circuit • Short to ground in the Knock Sensor No.1 Signal circuit • Knock Sensor • PCM
DTC: P0325 **1T PCM, MIL: Yes** **Year:** 2009, 2010 **Model:** Eclipse, Galant **Engine:** 2.4L L4 VIN F **Transmission:** All	**Knock Sensor Circuit:** After 2 seconds or more have passed since the starting sequence was completed. Engine speed is higher than 2,500 r/min. Volumetric efficiency is 40% or more. Knock sensor output voltage (knock sensor peak voltage in each 1/2 turn of the crankshaft) has not changed for 2 seconds. **Possible Causes:** • Knock sensor failed. • Open or shorted knock sensor circuit, harness damage or connector damage. • ECM failed. <M/T> • PCM failed. <A/T>
DTC: P0325 **1T PCM, MIL: Yes** **Year:** 2009, 2010 **Model:** Eclipse, Endeavor, Galant **Engine:** 3.8L V6 VIN S, 3.8L V6 VIN T **Transmission:** All	**Knock Sensor Circuit:** Engine speed is higher than 2,500 r/min. Volumetric efficiency is 40 percent or higher. After 2 seconds or more have passed since the engine starting sequence was completed. The knock sensor output voltage (knock sensor peak voltage in each 1/3 turn of the crankshaft) has not changed for 2 seconds. **Possible Causes:** • Knock sensor failed. • Open or shorted knock sensor circuit, or harness damage, or connector damage. • ECM failed. <M/T> • PCM failed. <A/T>
DTC: P0326 **1T ECM, MIL: Yes** **Year:** 2009, 2010 **Model:** Lancer **Engine:** 2.0L L4 VIN U, 2.0L L4 VIN V, 2.4L L4 VIN W **Transmission:** All	**Knock Sensor Circuit Performance:** After 2 seconds or more have passed since the engine starting sequence was completed. Engine speed is higher than 3,000 rpm. Volumetric efficiency is 40% or more. The knock sensor output voltage (Knock sensor peak voltage in each 1/2 turn of the crankshaft) has not changed more than 0.06 volts in the last consecutive 200 periods. **Possible Causes:** • Knock sensor has failed. • Open or shorted knock sensor circuit, harness or connector damage. • ECM has failed.
DTC: P0327 **1T ECM, MIL: Yes** **Year:** 2009, 2010 **Model:** Outlander **Engine:** 3.0L V6 VIN X **Transmission:** All	**Knock Sensor Circuit Low (Bank 1):** After 2 seconds or more have passed since the engine starting sequence was completed. Knock sensor 1 output voltage has continued to be lower than 0.5 volt for 2 seconds. **Possible Causes:** • Knock sensor 1 failed. • Open or shorted knock sensor 1 circuit or connector damage. • ECM failed.
DTC: P0327 **1T ECM, MIL: Yes** **Year:** 2009, 2010 **Model:** Lancer, Outlander **Engine:** 2.0L L4 VIN U, 2.0L L4 VIN V, 2.4L L4 VIN W **Transmission:** All	**Knock Sensor Circuit Low:** After 2 seconds or more have passed since the engine starting sequence was completed. The knock sensor output voltage has continued to be lower than 0.5 volt for 2 seconds. **Possible Causes:** • Knock sensor failed. • Open or shorted knock sensor circuit or connector damage. • ECM failed.

DTC	Trouble Code Title, Conditions & Possible Causes
DTC: P0328 **1T ECM, MIL: Yes** **Year:** 2009, 2010 **Model:** Outlander **Engine:** 3.0L V6 VIN X **Transmission:** All	**Knock Sensor Circuit High (Bank 1):** After 2 seconds or more have passed since the engine starting sequence was completed. The knock sensor 1 output voltage has continued to be higher than 2.3 volts for 2 seconds. **Possible Causes:** • Shorted knock sensor 1 circuit or connector damage. • ECM failed.
DTC: P0328 **1T ECM, MIL: Yes** **Year:** 2009, 2010 **Model:** Lancer, Outlander **Engine:** 2.0L L4 VIN U, 2.0L L4 VIN V, 2.4L L4 VIN W **Transmission:** All	**Knock Sensor Circuit High:** After 2 seconds or more have passed since the engine starting sequence was completed. The knock sensor output voltage has continued to be higher than 2.3 volts for 2 seconds. **Possible Causes:** • Shorted knock sensor circuit or connector damage. • ECM failed.
DTC: P0330 **2T PCM, MIL: Yes** **Year:** 2009 **Model:** Raider **Engine:** 3.7L V6 VIN K **Transmission:** All	**Knock Sensor No. 2 Circuit:** This monitor runs above 2000 rpm, under open throttle conditions. The Knock diagnostic does not run at idle or during decelerations. The high voltage test runs all the times the engine is running. The Knock Sensor voltage is not within specification. **Possible Causes:** • Short to voltage in the Knock Sensor No.2 Signal circuit • Open in the Knock Sensor No.2 Signal circuit • Open in the Knock Sensor No.2 Return circuit • Short between the Knock Sensor No.2 Signal circuit and the Knock Sensor No.2 Return circuit • Short to ground in the Knock Sensor No.2 Signal circuit • Knock Sensor • PCM
DTC: P0333 **1T ECM, MIL: Yes** **Year:** 2009, 2010 **Model:** Outlander **Engine:** 3.0L V6 VIN X **Transmission:** All	**Knock Sensor Circuit High (Bank 2):** After 2 seconds or more have passed since the engine starting sequence was completed. The knock sensor output voltage has continued to be higher than 2.3 volts for 2 seconds. **Possible Causes:** • Shorted knock sensor 2 circuit or connector damage. • ECM failed.
DTC: P0335 **1T PCM, MIL: Yes** **Year:** 2009, 2010 **Model:** Eclipse, Endeavor, Galant, Lancer, Outlander **Engine:** 2.0L L4 VIN U, 2.0L L4 VIN V, 2.4L L4 VIN F, 2.4L L4 VIN W, 3.0L V6 VIN X, 3.8L V6 VIN S, 3.8L V6 VIN T **Transmission:** All	**Crankshaft Position Sensor Circuit:** Engine is being cranked, or engine speed is higher than 500 r/min excluding during cranking. Normal signal pattern has not been input for cylinder identification from the crankshaft position sensor signal and camshaft position sensor signal for 2 seconds. Crankshaft position sensor signal does not change, or crankshaft position sensor signal is not normal pattern. **Possible Causes:** • Crankshaft position sensor failed. • Open or shorted crankshaft position sensor circuit, or harness damage, or connector damage. • ECM failed. <M/T> • PCM failed. <A/T>
DTC: P0335 **1T PCM, MIL: Yes** **Year:** 2009 **Model:** Raider **Engine:** 3.7L V6 VIN K **Transmission:** All	**Crankshaft Position Sensor Circuit:** Engine cranking. No CKP signal is present during engine cranking, and at least 8 camshaft position sensor signals have occurred. **Possible Causes:** • 5-Volt supply circuit shorted to voltage • 5-Volt supply circuit open • 5-Volt supply circuit shorted to ground • CKP Signal circuit open • CKP Signal circuit shorted to voltage • CKP Signal circuit shorted to ground • CKP Signal circuit shorted to the 5-Volt supply circuit • Sensor ground circuit open • CKP Sensor • CMP Sensor • PCM

DTC	Trouble Code Title, Conditions & Possible Causes
DTC: P0339 **1T PCM, MIL: Yes** **Year:** 2009 **Model:** Raider **Engine:** 3.7L V6 VIN K **Transmission:** All	**Crankshaft Position Sensor Circuit Intermittent :** While cranking the engine and with the engine running. When the CKP Sensor failure counter reaches 20 **Possible Causes:** • 5-Volt supply circuit open • 5-Volt supply circuit shorted to ground • CKP Signal circuit open • CKP Signal circuit shorted to voltage • CKP Signal circuit shorted to ground • CKP Signal circuit shorted to the 5-Volt supply circuit • CKP Sensor • Tone wheel/Pulse ring • PCM
DTC: P0340 **1T PCM, MIL: Yes** **Year:** 2009, 2010 **Model:** Eclipse, Endeavor, Galant, Outlander **Engine:** 2.4L L4 VIN F, 2.4L L4 VIN W, 3.0L V6 VIN X, 3.8L V6 VIN S, 3.8L V6 VIN T **Transmission:** All	**Camshaft Position Sensor Circuit:** Engine speed is being cranked, or engine speed is higher than 500 r/min excluding during cranking. Camshaft position sensor output voltage has not changed (no pulse signal is input) for 2 seconds. Camshaft position sensor signal does not change. Camshaft position sensor signal is not normal pattern. **Possible Causes:** • Camshaft position sensor failed. • Open or shorted camshaft position sensor circuit, or harness damage, or connector damage. • ECM failed. <M/T> • PCM failed. <A/T>
DTC: P0340 **1T ECM, MIL: Yes** **Year:** 2009, 2010 **Model:** Lancer **Engine:** 2.0L L4 VIN U, 2.0L L4 VIN V, 2.4L L4 VIN W **Transmission:** All	**Intake Camshaft Position Sensor Circuit:** Engine is being cranked, or the engine speed is higher than 500 r/min excluding during cranking. The intake camshaft position sensor output voltage has not changed (no pulse signal is input) for 2 seconds. **Possible Causes:** • Intake camshaft position sensor failed. • Open or shorted intake camshaft position sensor circuit, or harness damage, or connector damage. • Intake camshaft failed. • ECM failed.
DTC: P0340 **1T PCM, MIL: Yes** **Year:** 2009 **Model:** Raider **Engine:** 3.7L V6 VIN K **Transmission:** All	**Camshaft Position Sensor Circuit:** During engine cranking and with the engine running. Battery voltage greater than 10 volts. At least 5 seconds or 2.5 engine revolutions have elapsed with crankshaft position sensor signals present but no camshaft position sensor signal **Possible Causes:** • 5-Volt supply circuit shorted to voltage • 5-Volt supply circuit shorted open • 5-Volt supply circuit shorted to ground • CMP Signal circuit shorted to voltage • CMP Signal circuit open • CMP Signal circuit shorted to ground • CMP Signal circuit shorted to the 5-Volt supply circuit • CMP Sensor • CKP Sensor • PCM
DTC: P0344 **1T PCM, MIL: Yes** **Year:** 2009 **Model:** Raider **Engine:** 3.7L V6 VIN K **Transmission:** All	**Camshaft Position Sensor Circuit Intermittent :** While cranking the engine and engine running. When the failure counter reaches 20. **Possible Causes:** • 5-Volt supply circuit open • 5 Volt supply circuit shorted to ground • CMP Signal circuit open • CMP Signal circuit shorted to voltage • CMP Signal circuit shorted to ground • CMP Signal circuit shorted to the 5-Volt supply circuit • CMP Sensor • Tone wheel/Pulse ring • PCM

DTC	Trouble Code Title, Conditions & Possible Causes
DTC: P0365 **1T ECM, MIL: Yes** **Year:** 2009, 2010 **Model:** Lancer **Engine:** 2.0L L4 VIN U, 2.0L L4 VIN V, 2.4L L4 VIN W **Transmission:** All	**Exhaust Camshaft Position Sensor Circuit:** Engine is being cranked, or the engine speed is higher than 500 r/min excluding during cranking. The exhaust camshaft position sensor output voltage has not changed (no pulse signal is input) for 2 seconds. **Possible Causes:** • Exhaust camshaft position sensor failed. • Open or shorted exhaust camshaft position sensor circuit, or harness damage, or connector damage. • Exhaust camshaft failed. • ECM failed.
DTC: P0401 **1T PCM, MIL: Yes** **Year:** 2009, 2010 **Model:** Eclipse, Galant, Lancer **Engine:** 2.0L L4 VIN U, 2.0L L4 VIN V, 2.4L L4 VIN F, 2.4L L4 VIN W **Transmission:** All	**Exhaust Gas Recirculation Flow Insufficient Detected:** After at least 20 seconds have passed since the last monitor was complete. Engine coolant temperature is higher than 76°C (169°F). Engine speed is at between 1,200 and 2,500 r/min. <M/T> Engine speed is at between 1,100 and 1,750 r/min. <A/T> Intake air temperature is higher than -10°C (14°F). Barometric pressure is higher than 76 kPa (22.4 in. Hg). Fuel is being shut off. <M/T> Vehicle speed is 1.5 km/h (1 mph) or more. <M/T> Vehicle speed is 20 km/h (32 mph) or more. <A/T> At least 90 seconds have passed since manifold absolute pressure sensor output voltage fluctuated 1.5 volts or more. Battery positive voltage is higher than 10.3 volts. Accelerator pedal is closed. Volumetric efficiency is lower than 15 percent. <M/T> Volumetric efficiency is lower than 24 percent. <A/T> The ECM <M/T> or the PCM <A/T> monitors for this condition for 3 cycles of 2.0 seconds each during the drive cycle. The EGR valve opens to the prescribed opening, when intake manifold pressure fluctuation width is lower than 1.5 kPa (0.44 in. Hg). **Possible Causes:** • Contaminated EGR valve and EGR passage. • EGR valve (stopper motor) failed. • Open or shorted EGR valve (stopper motor) circuit, or connector damage. • Manifold absolute pressure sensor failed. • ECM failed. <M/T> • PCM failed. <A/T>
DTC: P0401 **1T PCM, MIL: Yes** **Year:** 2009, 2010 **Model:** Eclipse, Endeavor, Galant, Outlander **Engine:** 2.4L L4 VIN W, 3.0L V6 VIN X, 3.8L V6 VIN S, 3.8L V6 VIN T **Transmission:** All	**Exhaust Gas Recirculation Flow Insufficient Detected:** At least 10 seconds have passed since the last monitor was complete. <M/T> At least 20 seconds have passed since the last monitor was complete. <A/T> Engine coolant temperature is higher than 76°C (169°F). Engine speed is at between 940 and 2,500 r/min. <A/T> Engine speed is at between 1,000 and 2,500 r/min. <M/T> Intake air temperature is higher than -10°C (14°F). Barometric pressure is higher than 76 kPa (22.4 in. Hg). Fuel is being shut off. <M/T> Vehicle speed is 30 km/h (19 mph) or more. <A/T> Vehicle speed is 1.5 km/h (1 mph) or more. <M/T> At least 90 seconds have passed since manifold absolute pressure sensor output voltage fluctuated 1.5 volts or more. Battery positive voltage is higher than 10.3 volts. The throttle valve is closed. Volume efficiency is lower than 29 percent. <A/T> Volumetric efficiency is lower than 22 percent. <M/T> The ECM <M/T> or the PCM <A/T> monitors for this condition for 3 cycles of 1.4 seconds each during the drive cycle. When the EGR valve opens to the prescribed opening, when intake manifold pressure fluctuation width is lower than 2.0 kPa (0.59 in. Hg). <A/T> When the EGR valve opens to the prescribed opening, when intake manifold pressure fluctuation width is lower than 3.0 kPa (0.88 in. Hg). <M/T> **Possible Causes:** • Contaminated EGR valve and EGR passage. • EGR valve (stopper motor) failed. • Open or shorted EGR valve (stopper motor) circuit, or connector damage. • Manifold absolute pressure sensor failed. • ECM failed. <M/T> • PCM failed. <A/T>
DTC: P0401 **2T PCM, MIL: Yes** **Year:** 2009 **Model:** Raider **Engine:** 3.7L V6 VIN K **Transmission:** All	**Exhaust Gas Recirculation (EGR) System Performance:** Engine running for greater than two minutes with the Engine Coolant Temp greater than 70°C (158°F). EGR active. Less than 8500 feet. Ambient temperature greater than -6°C (20°F). The PCM closes the EGR valve while monitoring the O2 Sensor signal. Once a closed EGR fueling sample has been established the PCM then ramps in EGR and additional fueling while monitoring the O2 sensor signal in the open state. A fueling sample is again established. The PCM then compares the different O2 Sensor signal readings (fueling samples). If a larger than expected variation is detected, a soft failure is recorded. **Possible Causes:** • EGR Solenoid ground circuit open • EGR Solenoid control circuit shorted to voltage • EGR Solenoid control circuit open • EGR Solenoid assembly • PCM

DTC	Trouble Code Title, Conditions & Possible Causes
DTC: P0403 **1T PCM, MIL: Yes** **Year:** 2009, 2010 **Model:** Eclipse, Endeavor, Galant **Engine:** 2.4L L4 VIN F, 3.8L V6 VIN S, 3.8L V6 VIN T **Transmission:** All	**Exhaust Gas Regulation Control Circuit:** Battery positive voltage is higher than 10.3 volts. In a few seconds, just after ignition switch is turned to the "ON" position from the "LOCK"(OFF) position. (While EGR valve is initialized.) The EGR valve motor coil surge voltage (battery positive voltage +2 volts) is not detected for 3-30 seconds. Off-surge does not occur after stepper motor is operated on to off. **Possible Causes:** • EGR valve (stepper motor) failed. • Open or shorted EGR valve (stepper motor) circuit, harness damage or connector damage. • ECM failed. <M/T> • PCM failed. <A/T>
DTC: P0403 **1T PCM, MIL: Yes** **Year:** 2009 **Model:** Raider **Engine:** 3.7L V6 VIN K **Transmission:** All	**Exhaust Gas Recirculation (EGR) Solenoid Circuit:** Engine running. Battery voltage greater than 10 volts. The EGR solenoid control circuit is not in the expected state when requested to operate by the PCM. **Possible Causes:** • EGR Solenoid ground circuit open • EGR Solenoid control circuit shorted to voltage • EGR Solenoid control circuit shorted to ground • EGR Solenoid control circuit open • EGR Solenoid assembly • PCM
DTC: P0404 **2T PCM, MIL: Yes** **Year:** 2009 **Model:** Raider **Engine:** 3.7L V6 VIN K **Transmission:** All	**Exhaust Gas Recirculation (EGR) Sensor Position Rationality Open:** Engine running. The EGR flow or valve movement is not what is expected. A rationality error has been detected from the EGR Close Position Performance **Possible Causes:** • Excessive resistance in the EGR sensor signal circuit • Excessive resistance in the 5-volt supply circuit • Excessive resistance in the EGR solenoid control circuit • Excessive resistance in the sensor ground circuit • EGR solenoid ground circuit open • EGR Solenoid assembly • PCM
DTC: P0405 **1T PCM, MIL: Yes** **Year:** 2009 **Model:** Raider **Engine:** 3.7L V6 VIN K **Transmission:** All	**Exhaust Gas Recirculation (EGR) Position Sensor Circuit Low Voltage:** With the ignition on. Battery voltage above 10.0 volts. EGR Position Sensor Signal is less than the minimum acceptable value. **Possible Causes:** • 5-volt supply circuit shorted to ground • 5-volt supply circuit open • EGR Position sensor signal circuit shorted to ground • EGR Position sensor signal circuit shorted to sensor ground circuit • Excessive resistance in the EGR signal circuit • EGR Solenoid assembly • PCM
DTC: P0406 **1T PCM, MIL: Yes** **Year:** 2009 **Model:** Raider **Engine:** 3.7L V6 VIN K **Transmission:** All	**Exhaust Gas Recirculation (EGR) Position Sensor Circuit High :** With the ignition on. Battery voltage greater than 10 volts. EGR position sensor signal is greater than the maximum acceptable value. **Possible Causes:** • EGR Position sensor signal circuit shorted to the 5-volt supply circuit • EGR Position sensor signal circuit shorted to voltage • EGR Position sensor signal circuit open • Sensor ground circuit open • EGR Solenoid assembly • PCM

DTC	Trouble Code Title, Conditions & Possible Causes
DTC: P0420 **1T ECM, MIL: Yes** **Year:** 2009, 2010 **Model:** Outlander **Engine:** 2.4L L4 VIN W **Transmission:** All	**Warm Up Catalyst Efficiency Below Threshold :** Engine speed is lower than 3,000 r/min. Accelerator pedal is depressed. Mass airflow is between 10 and 45 g/sec. More than 3 seconds have elapsed after the above-mentioned three conditions have been met. Intake air temperature is higher than -10°C (14°F). Barometric pressure is higher than 76 kPa (22.4 in. Hg). Under the closed loop air/fuel ratio control. Vehicle speed is 1.5 km/h (1.0 mph) or more. The ECM monitors the maximum 3 times per drive cycle under these conditions. Short-term fuel trim is higher than -25 percent and lower than +25 percent. The cumulative mass airflow is higher than 1,741 g. When the monitoring for 10 seconds is carried out 5 times, the frequency ratio of rear and front signals is the specified value or more. **Possible Causes:** • Catalytic converter deteriorated. • ECM failed.
DTC: P0420 **2T PCM, MIL: Yes** **Year:** 2009 **Model:** Raider **Engine:** 3.7L V6 VIN K **Transmission:** All	**Catalyst Efficiency Below Threshold (Bank 1):** The monitor will run at between 1400 and 2300 RPM. It also runs between 40 and 70 kPa. If the final State of Change index is within the calibrated fail threshold. **Possible Causes:** • O2 Return upstream circuit shorted to ground • O2 Return downstream circuit shorted to ground • O2 Return upstream circuit open • O2 Return downstream circuit open • Exhaust leak • Engine mechanical problem • Aging O2 Sensor • Catalytic Converter
DTC: P0420 **1T ECM, MIL: Yes** **Year:** 2009, 2010 **Model:** Lancer **Engine:** 2.0L L4 VIN U, 2.0L L4 VIN V, 2.4L L4 VIN W **Transmission:** All	**Warm Up Catalyst Efficiency Below Threshold :** Engine speed is lower than 3,500 r/min <M/T> or 3,000 r/min <CVT>. Accelerator pedal is depressed. Mass airflow is between 10 and 45 g/sec. More than 3 seconds have elapsed after the above-mentioned three conditions have been met. Intake air temperature is higher than -10°C (14°F). Barometric pressure is higher than 76 kPa (22.4 in. Hg). Under the closed loop air/fuel ratio control. Vehicle speed is 1.5 km/h (1.0 mph) or more. The ECM monitors the maximum 4 times <M/T> or 3 times <CVT> per drive cycle under these conditions. When the monitoring for 10 seconds is carried out 5 times, the frequency ratio of rear and front signals is the specified value or more. **NOTE: The specified value varies depending on the average air flow rate.** **Short-term fuel trim is higher than -25 percent and lower than +25 percent.** **The cumulative mass airflow is higher than 1,638 g <M/T> or 1,741 g <CVT>.** **The ECM monitors the maximum 4 times <M/T> or 3 times <CVT> per drive cycle under these conditions.** **Short-term fuel trim is higher than -25 percent and lower than +25 percent.** **The cumulative mass airflow is higher than 1,638 g <M/T> or 1,741 g <CVT>.** **Possible Causes:** • Catalytic converter deteriorated. • Heated oxygen sensor failed. • Exhaust leak. • ECM failed.

DTC	Trouble Code Title, Conditions & Possible Causes
DTC: P0421 **2T PCM, MIL: Yes** **Year:** 2009, 2010 **Model:** Eclipse, Galant **Engine:** 2.4L L4 VIN F **Transmission:** All	**Warm Up Catalyst Efficiency Below Threshold (Cylinder 1, 4):** * Engine speed is lower than 3,500 r/min. Accelerator pedal is open. Mass airflow is between 9 and 45 g/sec. <M/T> Mass airflow is between 14 and 45 g/sec. </VT> More than 3 seconds have elapsed after the above-mentioned three conditions have been met. Intake air temperature is higher than -10°C (14°F). Barometric pressure is higher than 76 kPa (22.4 in. Hg). Under the closed loop air/fuel ratio control. Vehicle speed is 1.5 km/h (1.0 mph) or more. The ECM <M/T> or the PCM <A/T> monitors for this condition for 5 cycles of 10 seconds each during the drive cycle. When the cylinder 1, 4 heated oxygen sensor (rear) signal frequency is divided by the cylinder 1, 4 heated oxygen sensor (front) signal frequency, the average is more than 0.7. The cylinder 1, 4 heated oxygen sensor (rear) signal frequency divided by the cylinder 1, 4 heated oxygen sensor (front) signal frequency is more than 0.4. Short-term fuel trim is higher than -25 percent and lower than +25 percent. The cumulative mass airflow is higher than 1,638 g. **Possible Causes:** • Cylinder 1, 4 side catalytic converter deteriorated. • Cylinder 1, 4 heated oxygen sensor failed. • Exhaust leak. • ECM failed. <M/T> • PCM failed. <A/T>
DTC: P0421 **1T ECM, MIL: Yes** **Year:** 2009, 2010 **Model:** Outlander **Engine:** 2.4L L4 VIN W **Transmission:** All	**Warm Up Catalyst Efficiency Below Threshold:** Engine speed is lower than 3,000 r/min. Accelerator pedal is depressed. Mass airflow is between 10 and 45 g/sec. More than 3 seconds have elapsed after the above-mentioned three conditions have been met. Intake air temperature is higher than -10°C (14°F). Barometric pressure is higher than 76 kPa (22.4 in. Hg). Under the closed loop air/fuel ratio control. Vehicle speed is 1.5 km/h (1.0 mph) or more. The ECM monitors the maximum 3 times per drive cycle under these conditions. Short-term fuel trim is higher than -25 percent and lower than +25 percent. The cumulative mass airflow is higher than 1,741 g. When the monitoring for 10 seconds is carried out 5 times, the frequency ratio of heated oxygen sensor (rear) and linear air-fuel ratio sensor signals is the specified value or more. **Possible Causes:** • Catalytic converter deteriorated. • ECM failed.
DTC: P0421 **2T PCM, MIL: Yes** **Year:** 2009, 2010 **Model:** Eclipse, Endeavor, Galant, Outlander **Engine:** 3.0L V6 VIN X, 3.8L V6 VIN S, 3.8L V6 VIN T **Transmission:** All	**Warm Up Catalyst Efficiency Below Threshold (Bank 1):** Engine speed is lower than 3,000 r/min. Accelerator pedal is open. Mass airflow is between 10 and 35 g/sec. <A/T> Mass airflow is between 9 and 37 g/sec. <M/T> More than 3 seconds have elapsed after the above-mentioned three conditions have been met. Intake air temperature is higher than -10°C (14°F). Barometric pressure is higher than 76 kPa (22.4 in. Hg). Under the closed loop air/fuel ratio control. Vehicle speed is 1.5 km/h (1.0 mph) or more. The PCM monitors for this condition for 5 cycles of 12 seconds each during the drive cycle. <A/T> The ECM monitors for this condition for 7 cycles of 12 seconds each during the drive cycle. <M/T> Short-term fuel trim is higher than -30 percent and lower than +25 percent. The cumulative mass airflow is higher than 1,638 g. When the right bank heated oxygen sensor (rear) signal frequency is divided by the right bank heated oxygen sensor (front) signal frequency, the average is more than 0.7. The right bank heated oxygen sensor (rear) signal frequency divided by the right bank heated oxygen sensor (front) signal frequency is more than 0.4. **Possible Causes:** • Right bank side catalytic converter deteriorated. • Right bank heated oxygen sensor failed. • Exhaust leak. • ECM failed. <M/T> • PCM failed. <A/T>
DTC: P0430 **2T PCM, MIL: Yes** **Year:** 2009 **Model:** Raider **Engine:** 3.7L V6 VIN K **Transmission:** All	**Catalyst Efficiency Below Threshold (Bank 2):** The monitor will run at between 1400 and 2300 RPM. It also runs between 40 and 70 kPa. If the final State of Change index is within the calibrated fail threshold. **Possible Causes:** • O2 Return upstream circuit shorted to ground • O2 Return downstream circuit shorted to ground • O2 Return upstream circuit open • O2 Return downstream circuit open • Exhaust leak • Engine mechanical problem • Aging O2 Sensor • Catalytic Converter

DTC	Trouble Code Title, Conditions & Possible Causes
DTC: P0430 **2T PCM, MIL:** Yes **Year:** 2009 **Model:** Raider **Engine:** 3.7L V6 VIN K **Transmission:** All	**General EVAP System Failure:** Engine running after a cold start with the difference between ECT and AAT is less than 10° C (19° F). Fuel Level between 12% and 88% full. Manifold vacuum greater than a calculated minimum value. Ambient Temperature between 4° C and 32° C (39° F and 89° F). When the monitor conditions are met, the PCM will ramp in purge flow. If the PCM does not sense an ESM switch closure after a calculated amount of purge flow accumulation, an error is detected. **Possible Causes:** • EVAP Purge solenoid vacuum supply • EVAP Purge hose, tube and/or fresh air filter obstruction • Ground circuit open • ESM Signal circuit open • EVAP leak • EVAP System monitor switch • EVAP Purge solenoid • PCM
DTC: P0431 **2T PCM, MIL:** Yes **Year:** 2009, 2010 **Model:** Eclipse, Galant **Engine:** 2.4L L4 VIN F **Transmission:** All	**Warm Up Catalyst Efficiency Below Threshold (cylinder 2, 3):** Engine speed is lower than 3,500 r/min. Accelerator pedal is open. Mass airflow is between 9 and 45 g/sec. <M/T> Mass airflow is between 14 and 45 g/sec. <A/T> More than 3 seconds have elapsed after the above-mentioned three conditions have been met. Intake air temperature is higher than -10°C (14°F). Barometric pressure is higher than 76 kPa (22.4 in. Hg). Under the closed loop air/fuel ratio control. Vehicle speed is 1.5 km/h (1.0 mph) or more. The ECM <M/T> or the PCM <A/T> monitors for this condition for 5 cycles of 10 seconds each during the drive cycle. Short-term fuel trim is higher than -25 percent and lower than +25 percent. The cumulative mass airflow is higher than 1,638 g. When the cylinder 2, 3 heated oxygen sensor (rear) signal frequency is divided by the cylinder 2, 3 heated oxygen sensor (front) signal frequency, the average is more than 0.7. The cylinder 2, 3 heated oxygen sensor (rear) signal frequency divided by the cylinder 2, 3 heated oxygen sensor (front) signal frequency is more than 0.4. **Possible Causes:** • Cylinder 2, 3 side catalytic converter deteriorated. • Cylinder 2, 3 heated oxygen sensor failed. • Exhaust leak. • ECM failed. <M/T> • PCM failed. <A/T>
DTC: P0431 **2T PCM, MIL:** Yes **Year:** 2009, 2010 **Model:** Eclipse, Endeavor, Galant, Outlander **Engine:** 3.0L V6 VIN X, 3.8L V6 VIN S, 3.8L V6 VIN T **Transmission:** All	**Warm Up Catalyst Efficiency Below Threshold (Bank 2):** Engine speed is lower than 3,000 r/min. Accelerator pedal is open. Mass airflow is between 10 and 35 g/sec. <A/T> Mass airflow is between 9 and 37 g/sec. <M/T> More than 3 seconds have elapsed after the above-mentioned three conditions have been met. Intake air temperature is higher than -10°C (14°F). Barometric pressure is higher than 76 kPa (22.4 in. Hg). Under the closed loop air/fuel ratio control. Vehicle speed is 1.5 km/h (1.0 mph) or more. The PCM monitors for this condition for 5 cycles of 12 seconds each during the drive cycle. <A/T> The ECM monitors for this condition for 7 cycles of 12 seconds each during the drive cycle. <M/T> When the left bank heated oxygen sensor (rear) signal frequency is divided by the left bank heated oxygen sensor (front) signal frequency, the average is more than 0.7. The left bank heated oxygen sensor (rear) signal frequency divided by the left bank heated oxygen sensor (front) signal frequency is more than 0.4. Short-term fuel trim is higher than -30 percent and lower than +25 percent. The cumulative mass airflow is higher than 1,638 g. **Possible Causes:** • Left bank side catalytic converter deteriorated. • Left bank heated oxygen sensor failed. • Exhaust leak. • ECM failed. <M/T> • PCM failed. <A/T>

DTC	Trouble Code Title, Conditions & Possible Causes
DTC: P0440 **2T PCM, MIL:** Yes **Year:** 2009 **Model:** Raider **Engine:** 3.7L V6 VIN K **Transmission:** All	**General EVAP System Failure:** Engine running after a cold start with the difference between ECT and AAT is less than 10° C (19° F). Fuel Level between 12% and 88% full. Manifold vacuum greater than a calculated minimum value. Ambient Temperature between 4° C and 32° C (39° F and 89° F). When the monitor conditions are met, the PCM will ramp in purge flow. If the PCM does not sense an ESM switch closure after a calculated amount of purge flow accumulation, an error is detected. **Possible Causes:** • EVAP Purge solenoid vacuum supply • EVAP Purge hose, tube and/or fresh air filter obstruction • Ground circuit open • ESM signal circuit open • EVAP leak • EVAP System monitor switch • EVAP Purge solenoid • PCM
DTC: P0441 **2T PCM, MIL:** Yes **Year:** 2009 **Model:** Raider **Engine:** 3.7L V6 VIN K **Transmission:** All	**EVAP Purge System Performance:** After the Evap System small leak test has passed, with the engine running, ambient temperature between 4° C (39° F) and 35° (95° F), with the engine at idle after a calibrated amount of drive time has accumulated. If the PCM detects that the purge vapor ratio and the ESM switch closed ratio are below a calculated value, the PCM commands the purge solenoid to flow at a specified rate to update the purge vapor ratio. If the ratio remains below a specified value, a one trip failure is recorded. **Possible Causes:** • EVAP Purge solenoid vacuum supply • EVAP Purge hose, tube and/or fresh air filter obstruction • EVAP Purge solenoid • PCM
DTC: P0441 **2T PCM, MIL:** Yes **Year:** 2009, 2010 **Model:** Eclipse, Endeavor, Galant, Lancer, Outlander **Engine:** 2.0L L4 VIN U, 2.0L L4 VIN V, 2.4L L4 VIN F, 2.4L L4 VIN W, 3.0L V6 VIN X, 3.8L V6 VIN S, 3.8L V6 VIN T **Transmission:** All	**Evaporative Emission Control System Incorrect Purge Flow:** ON duty cycle of the evaporative emission purge solenoid is 0 percent. Engine is running. After 20 seconds have elapsed since the duty cycle of the evaporative emission purge solenoid has turned to 0 percent. The pressure in the fuel tank is -1,961 Pa (-0.58 in. Hg) or less for 0.1 second. Fuel tank pressure decreases largely during purge-cut. **Possible Causes:** • Evaporative emission purge solenoid failed. • Evaporative emission ventilation solenoid failed. • Fuel tank differential pressure sensor circuit related part(s) failed.
DTC: P0442 **2T PCM, MIL:** Yes **Year:** 2009, 2010 **Model:** Eclipse, Endeavor, Galant, Lancer, Outlander **Engine:** 2.0L L4 VIN U, 2.0L L4 VIN V, 2.4L L4 VIN F, 2.4L L4 VIN W, 3.0L V6 VIN X, 3.8L V6 VIN S, 3.8L V6 VIN T **Transmission:** All	**Evaporative Emission Control System Leak Detected (Small Leak):** Intake air temperature is greater than -10 °C (14°F). Engine coolant temperature is greater than 20 °C (68 °F). When the evaporative emission purge solenoid and evaporative emission ventilation solenoid are closed, the pressure in the fuel tank is less than 451 Pa (0.065 psi). After 10 seconds have elapsed since the start of the previous monitoring. Monitoring time: 10 - 14 minutes. Internal pressure of the fuel tank has changed more than 1,922 Pa (0.29 psi) in 128 seconds after the tank and vapor line were closed. Measure reverting pressure after depressurizing by intake manifold negative pressure and detect malfunction if reverting pressure rises largely. **Possible Causes:** • Loose fuel cap. • Fuel cap relief pressure is incorrect. • Evaporative emission canister seal is leaking. • Fuel tank, purge line or vapor line seal is leaking. • Evaporative emission ventilation solenoid does not seal.
DTC: P0443 **1T PCM, MIL:** Yes **Year:** 2009 **Model:** Raider **Engine:** 3.7L V6 VIN K **Transmission:** All	**EVAP Purge Solenoid Control Circuit :** The ignition on or engine running. Battery voltage greater than 10 volts. The PCM will set a trouble code if the actual state of the solenoid does not match the intended state. **Possible Causes:** • EVAP Purge solenoid control circuit shorted to voltage • EVAP Purge solenoid signal circuit shorted to voltage • EVAP Purge solenoid control circuit shorted to ground • Excessive resistance in the EVAP Purge control circuit • Excessive resistance in the EVAP Purge solenoid signal circuit • EVAP Purge solenoid • PCM

DTC	Trouble Code Title, Conditions & Possible Causes
DTC: P0443 **1T PCM, MIL: Yes** **Year:** 2009, 2010 **Model:** Eclipse, Endeavor, Galant, Lancer, Outlander **Engine:** 2.0L L4 VIN U, 2.0L L4 VIN V, 2.4L L4 VIN F, 2.4L L4 VIN W, 3.0L V6 VIN X, 3.8L V6 VIN S, 3.8L V6 VIN T **Transmission:** All	**Evaporative Emission Control System Purge Control Valve Circuit:** Engine is being cranked. Battery positive voltage is between 10 and 16.5 volts. The evaporative emission purge solenoid coil surge voltage (battery positive voltage + 2 volts) is not detected for 0.2 second. The ECM <M/T> or the PCM <A/T> monitors for this condition once during the drive cycle. Off-surge does not occur after solenoid is operated from on to off. **Possible Causes:** • Evaporative emission purge solenoid failed. • Open or shorted evaporative emission purge solenoid circuit, harness damage, or connector damage. • ECM failed. <M/T> • PCM failed. <A/T>
DTC: P0446 **2T PCM, MIL: Yes** **Year:** 2009, 2010 **Model:** Eclipse, Galant, Lancer, Outlander **Engine:** 2.0L L4 VIN U, 2.0L L4 VIN V, 2.4L L4 VIN F, 2.4L L4 VIN W, 3.0L V6 VIN X, 3.8L V6 VIN S, 3.8L V6 VIN T **Transmission:** All	**Evaporative Emission Control System Vent Control Circuit:** Battery positive voltage is between 10 and 16.5 volts. ON duty cycle of the evaporative emission purge solenoid is 0 percent. Evaporative emission ventilation solenoid is ON. More than 1 second has elapsed after the above mentioned conditions have been met. The evaporative emission ventilation solenoid coil surge voltage (battery positive voltage + 2 volts) is not detected for 1 second after the evaporative emission ventilation solenoid is turned OFF. **Possible Causes:** • Evaporative emission ventilation solenoid failed. • Open or shorted evaporative emission ventilation solenoid circuit, harness damage, or connector damage. • ECM failed. <M/T> • PCM failed. <A/T>
DTC: P0450 **1T PCM, MIL: Yes** **Year:** 2009, 2010 **Model:** Eclipse, Endeavor, Galant, Lancer, Outlander **Engine:** 2.0L L4 VIN U, 2.0L L4 VIN V, 2.4L L4 VIN F, 2.4L L4 VIN W, 3.0L V6 VIN X, 3.8L V6 VIN S, 3.8L V6 VIN T **Transmission:** All	**Evaporative Emission Control System Pressure Sensor Malfunction:** Intake air temperature is between 5°C (41°F) and 45°C (113°F) or greater. Engine speed is 1,594 r/min or greater. Volumetric efficiency is between 20 and 70 percent. When the evaporative emission purge solenoid valve is fully operational (100 percent ratio), the fuel differential pressure sensor output voltage remains at 4.0 volts or greater for ten seconds. **Possible Causes:** • Malfunction of the fuel tank differential pressure sensor. • A damaged harness in the fuel tank differential pressure sensor circuit. • Malfunction of the ECM <M/T> or PCM <A/T>.
DTC: P0451 **2T PCM, MIL: Yes** **Year:** 2009, 2010 **Model:** Eclipse, Endeavor, Galant, Lancer, Outlander **Engine:** 2.0L L4 VIN U, 2.0L L4 VIN V, 2.4L L4 VIN F, 2.4L L4 VIN W, 3.0L V6 VIN X, 3.8L V6 VIN S, 3.8L V6 VIN T **Transmission:** All	**Evaporative Emission Control System Pressure Sensor Range/Performance:** Throttle valve is closed. Vehicle speed is 1.5 km/h (0.93 mph) or less. If the voltage signal from the fuel tank differential pressure sensor changes by 0.2 volt or more, DTC P0451 will set. The code may also set if a sudden pressure fluctuation occurs twenty times while the engine is idling, and then four consecutive times during normal driving. Detect malfunction if change of fuel tank differential pressure sensor output voltage during idling stays large during specified go/stop operations. **NOTE: If the number of sudden pressure fluctuations does not reach twenty during any one period of engine idling, or if the ignition switch is turned OFF, the counter will reset to zero.** **Possible Causes:** • Malfunction of the fuel tank differential pressure sensor. • A damaged harness in the fuel tank differential pressure sensor circuit. • Malfunction of the ECM <M/T> or PCM <A/T>.
DTC: P0452 **1T PCM, MIL: Yes** **Year:** 2009, 2010 **Model:** Eclipse, Endeavor, Galant, Lancer, Outlander **Engine:** 2.0L L4 VIN U, 2.0L L4 VIN V, 2.4L L4 VIN F, 2.4L L4 VIN W, 3.0L V6 VIN X, 3.8L V6 VIN S, 3.8L V6 VIN T **Transmission:** All	**Evaporative Emission Control System Pressure Sensor Low Input:** After 2 seconds or more have passed since the starting sequence was completed. The fuel tank differential pressure sensor output voltage remains 0.2 volt or less for 5 seconds. **Possible Causes:** • Malfunction of the fuel tank differential pressure sensor. • A damaged harness in the fuel tank differential pressure sensor circuit. • Malfunction of the ECM <M/T> or PCM <A/T>.

DTC	Trouble Code Title, Conditions & Possible Causes
DTC: P0452 **2T PCM, MIL: Yes** **Year:** 2009 **Model:** Raider **Engine:** 3.7L V6 VIN K **Transmission:** All	**EVAP Pressure Switch Stuck Closed:** Immediately after the ignition has been turned off. At key off, the PCM energizes the Purge Solenoid for a calibrated amount of time (30 seconds maximum) and stores the state of the ESIM switch. The state is evaluated again at the next key on. If the PCM does not detect that the ESIM switch is open, an error is detected. **Possible Causes:** • EVAP Purge hose, tube and/or fresh air filter obstruction • EVAP Purge solenoid control circuit shorted to ground • ESIM Switch signal circuit shorted to ground • ESIM Switch • EVAP Purge solenoid • PCM
DTC: P0453 **1T PCM, MIL: Yes** **Year:** 2009, 2010 **Model:** Eclipse, Endeavor, Galant, Lancer, Outlander **Engine:** 2.0L L4 VIN U, 2.0L L4 VIN V, 2.4L L4 VIN F, 2.4L L4 VIN W, 3.0L V6 VIN X, 3.8L V6 VIN S, 3.8L V6 VIN T **Transmission:** All	**Evaporative Emission Control System Pressure Sensor High Input:** After 2 seconds or more have passed since the starting sequence was completed. Remaining fuel level is 80% or less when the engine is started. The fuel tank differential pressure sensor output voltage remains 4.8 volt or greater for 5 seconds. **Possible Causes:** • Malfunction of the fuel tank differential pressure sensor. • Open or shorted fuel tank differential pressure sensor circuit. • Malfunction of the ECM <M/T> or PCM <A/T>.
DTC: P0455 **2T PCM, MIL: Yes** **Year:** 2009 **Model:** Raider **Engine:** 3.7L V6 VIN K **Transmission:** All	**EVAP Purge System Large Leak Detected:** With the engine running, during a cold start test with the fuel level above 12%, ambient temperature between 4° C and 32° C (39° F and 89° F) and the fuel system in closed loop. The test runs when the small leak test is maturing. The PCM activates the Evap Purge Solenoid to pull the Evap system into a vacuum to close the ESM switch. Once the ESM switch is closed, the PCM turns the Evap Purge solenoid off to seal the Evap system. If the ESM switch reopens before the calibrated amount of time, a large leak error is detected. **Possible Causes:** • EVAP Purge system leak • EVAP Purge solenoid • EVAP System monitor switch
DTC: P0455 **2T PCM, MIL: Yes** **Year:** 2009, 2010 **Model:** Eclipse, Endeavor, Galant, Lancer, Outlander **Engine:** 2.0L L4 VIN U, 2.0L L4 VIN V, 2.4L L4 VIN F, 2.4L L4 VIN W, 3.0L V6 VIN X, 3.8L V6 VIN S, 3.8L V6 VIN T **Transmission:** All	**Evaporative Emission Control System Leak Detected (Gross Leak):** The intake air temperature is greater than 5°C (41°F). When the evaporative emission purge solenoid and evaporative emission ventilation solenoid are closed, the pressure in the fuel tank rises to 451 Pa (0.065 psi) or less and the amount of remaining fuel is 15 - 40 percent of capacity upon engine start-up. When the evaporative emission purge solenoid and evaporative emission ventilation solenoid are closed, the pressure in the fuel tank rises to 324 Pa (0.047 psi) or less and the amount of remaining fuel is 40 - 85 percent of capacity upon engine start-up. After 10 seconds have elapsed from the start of the previous monitoring. Monitoring time: 150 seconds. The fuel tank internal pressure is 2 kPa (0.29 psi) or more after the evaporative emission purge solenoid valve has been driven when the fuel tank and vapor line were closed. **Possible Causes:** • Loose fuel cap. • Fuel cap relief pressure is incorrect. • Fuel overflow limiter valve failed. • Purge line or vapor line is clogged. • Fuel tank, purge line or vapor line seal failed. • Evaporative emission purge solenoid valve failed. • Evaporative emission ventilation solenoid valve failed. • Fuel tank differential pressure sensor failed. • Evaporative emission canister seal is faulty. • Evaporative emission canister is clogged.

DTC	Trouble Code Title, Conditions & Possible Causes
DTC: P0456 **1T ECM, MIL:** Yes **Year:** 2009, 2010 **Model:** Outlander **Engine:** 2.4L L4 VIN W, 3.0L V6 VIN X **Transmission:** All	**Evaporative Emission Control System Leak Detected (very small leak):** Engine coolant temperature is 36°C (97°F) or less when the engine is started. Intake air temperature is 36°C (97°F) or less when the engine is started. Engine coolant temperature is higher than 20°C (67°F). When the evaporative emission purge solenoid and evaporative emission ventilation solenoid are closed, the pressure in the fuel tank is less than 324 Pa (0.09 in. Hg). Amount of remaining fuel is 40 - 85 percent of capacity. Barometric pressure is higher than 76 kPa (22.4 in. Hg). Intake air temperature is higher than -10°C (14°F) Fuel tank temperature is less than 33°C (91°F). Fuel tank differential pressure sensor output voltage is 1.0 - 4.0 volts. At least 10 seconds have passed since the last monitor was complete. The internal pressure of the fuel tank has changed more than 1,177 - 1,497 Pa (0.177 - 0.217 psi) in 88 seconds after the tank and vapor line were closed. **Possible Causes:** • Loose fuel cap. • Fuel cap relief pressure is incorrect. • Malfunction of the evaporative emission canister seal. • Malfunction of the fuel tank, purge line or vapor line seal. • Malfunction of the evaporative emission ventilation solenoid.
DTC: P0456 **1T PCM, MIL:** Yes **Year:** 2009 **Model:** Raider **Engine:** 3.7L V6 VIN K **Transmission:** All	**EVAP Purge System Small Leak Detected:** With the ignition off, fuel level less than 88%, ambient temperature between 4° C and 43° C (39° F and 109° F), and the fuel system in closed loop. As temperatures change, a vacuum is created in the fuel tank and Evap system. With the Evap system sealed, the PCM monitors the ESM switch. If the ESM switch does not close within a calibrated time, an error is detected by the PCM. **Possible Causes:** • EVAP Purge system leak • EVAP Purge solenoid • EVAP System monitor switch
DTC: P0456 **2T PCM, MIL:** Yes **Year:** 2009, 2010 **Model:** Eclipse, Endeavor, Galant, Lancer **Engine:** 2.0L L4 VIN U, 2.0L L4 VIN V, 2.4L L4 VIN F, 2.4L L4 VIN W, 3.8L V6 VIN S, 3.8L V6 VIN T **Transmission:** All	**Evaporative Emission Control System Leak Detected (Very Small Leak):** Engine coolant temperature is greater than 20 °C (68 °F). Intake air temperature is greater than -10 °C (14 °F). When the evaporative emission purge solenoid and evaporative emission ventilation solenoid are closed, the pressure rises in the fuel tank is less than 324 Pa (0.047 psi). After 10 seconds have elapsed from the start of the previous monitoring. Monitoring time: 10 - 14 minutes. Internal pressure of the fuel tank has changed more than 1,177 - 1,373 Pa (0.177 - 0.199 psi) in 128 seconds after the tank and vapor line were closed. **Possible Causes:** • Loose fuel cap. • Fuel cap relief pressure is incorrect. • Malfunction of the evaporative emission canister seal. • Malfunction of the fuel tank, purge line or vapor line seal. • Malfunction of the evaporative emission ventilation solenoid.
DTC: P0457 **2T PCM, MIL:** Yes **Year:** 2009 **Model:** Raider **Engine:** 3.7L V6 VIN K **Transmission:** All	**Loose Fuel Cap:** Ignition on. Ambient Temperature between 4°C and 32°C (39°F and 89°F). Close Loop fuel system. Test runs after the medium leak test is inconclusive and the PCM has senses a fuel increase. The PCM activates the Evap Purge Solenoid to pull the Evap system into a vacuum to close the ESIM switch. Once the ESIM switch is closed, the PCM turns the Evap Purge Solenoid off to seal the Evap system. If the ESIM switch reopens before the calibrated amount of time after a fuel tank fill, an error is detected. **Possible Causes:** • Loose or missing fuel cap
DTC: P0461 **1T ECM, MIL:** Yes **Year:** 2009, 2010 **Model:** Outlander **Engine:** 2.4L L4 VIN W, 3.0L V6 VIN X **Transmission:** All	**Fuel Level Sensor <FWD> or Fuel Level Sensor (main) <AWD> Circuit Range/performance:** When the fuel consumption calculated from the operation time of the injector amounts to 20 liters (5.3 gal) <FWD> or 30 liters (7.9 gal) <AWD>, the diversity of the amount of fuel in tank calculated from the fuel level sensor is 2 liters (0.5 gal) or less. **Possible Causes:** • Combination meters assembly failed. • Fuel level sensor failed. • ECM failed.
DTC: P0461 **1T PCM, MIL:** Yes **Year:** 2009, 2010 **Model:** Eclipse, Endeavor, Galant **Engine:** 2.4L L4 VIN F, 3.8L V6 VIN S, 3.8L V6 VIN T **Transmission:** All	**Fuel Level Sensor (Main) Circuit Range/Performance:** Detect malfunction if change of fuel level sensor output voltage is small when sum of fuel injection is large. When the fuel consumption calculated from the operation time of the injector amounts to 30 liters (7.8 gal), the diversity of the amount of fuel in tank calculated from the fuel level sensor is 2 liters (0.5 gal) or less. **Possible Causes:** • Fuel pump module or fuel level sensor failed. • Harness damage or connector damage in fuel level sensor circuit. • ECM failed. <M/T> • PCM failed. <A/T>

DTC	Trouble Code Title, Conditions & Possible Causes
DTC: P0461 **T PCM, MIL:** Yes **Year:** 2009 **Model:** Raider **Engine:** 3.7L V6 VIN K **Transmission:** All	**Fuel Level Sensor 1 Performance:** TEST No.1: WIth the ignition on, the fuel level is compared to the previous key down after a 20 second delay. TFST No.2: The PCM monitors the fuel level at ignition on. TEST No.1: If the PCM does not see a difference in fuel level of greater than 0.1 volt the test will fail. TEST No.2: If the PCM does not see a change in the fuel level of .1765 over a set amount of miles the test will fail. **Possible Causes:** • Fuel level signal circuit open • Fuel level signal circuit shorted to ground • Fuel level sensor return circuit open • Fuel tank • Fuel level sensor
DTC: P0461 **1T ECM, MIL:** Yes **Year:** 2009, 2010 **Model:** Lancer **Engine:** 2.0L L4 VIN U, 2.0L L4 VIN V, 2.4L L4 VIN W **Transmission:** All	**Fuel Level Sensor Circuit Range/Performance:** When the fuel consumption calculated from the operation time of the injector amounts to 20 liters (5.2 gal), the diversity of the amount of fuel in tank calculated from the fuel level sensor is 2 liters (0.5 gal) or less. Detect malfunction if change of fuel level sensor output voltage is small when sum of fuel injection is large. **Possible Causes:** • Combination meters assembly failed. • Fuel level sensor failed. • ECM failed.
DTC: P0462 **1T ECM, MIL:** Yes **Year:** 2009, 2010 **Model:** Lancer **Engine:** 2.0L L4 VIN U, 2.0L L4 VIN V, 2.4L L4 VIN W **Transmission:** All	**Fuel Level Sensor Circuit Low Input:** Battery positive voltage is between 11 and 16.5 volts. After 2 seconds or more have passed since the engine staring sequence was completed. Fuel level sensor resistance has continued to be lower than 2 ohms for 2 seconds. **Possible Causes:** • Fuel level sensor failed. • Combination meters assembly failed. • Connector damage • Harness damage • ECM failed.
DTC: P0462 **1T PCM, MIL:** Yes **Year:** 2009, 2010 **Model:** Eclipse, Endeavor, Galant **Engine:** 2.4L L4 VIN F, 3.8L V6 VIN S, 3.8L V6 VIN T **Transmission:** All	**Fuel Level Sensor Circuit Low Input:** Battery positive voltage is between 11 and 16.5 volts. After 2 seconds or more have passed since the engine staring sequence was completed. Fuel level sensor resistance has continued to be lower than 2 ohms for 2 seconds. **Possible Causes:** • Fuel level sensor failed. • Shorted fuel level sensor circuit or connector damage. • ECM failed. <M/T> • PCM failed. <A/T>
DTC: P0462 **1T PCM, MIL:** Yes **Year:** 2009 **Model:** Raider **Engine:** 3.7L V6 VIN K **Transmission:** All	**Fuel Level Sensor 1 Circuit Low:** Ignition on and battery voltage above 10.4 volts. The fuel level sensor signal voltage goes below the acceptable value. **Possible Causes:** • Fuel level signal circuit shorted to ground • Fuel level signal circuit shorted to the fuel level return circuit • Excessive resistance in the fuel level signal circuit • Fuel level sensor • Instrument cluster
DTC: P0462 **1T ECM, MIL:** Yes **Year:** 2009, 2010 **Model:** Outlander **Engine:** 2.4L L4 VIN W, 3.0L V6 VIN X **Transmission:** All	**Fuel Level Sensor <FWD> or Fuel Level Sensor (main) <AWD> Circuit Low Input:** Battery positive voltage is between 11 and 16.5 volts. After 2 seconds or more have passed since the engine staring sequence was completed. The fuel level sensor resistance has continued to be lower than 2 ohms for 2 seconds. **Possible Causes:** • Fuel level sensor failed. • Combination meters assembly failed. • ECM failed.
DTC: P0463 **1T ECM, MIL:** Yes **Year:** 2009, 2010 **Model:** Outlander **Engine:** 2.4L L4 VIN W, 3.0L V6 VIN X **Transmission:** All	**Fuel Level Sensor <FWD> or Fuel Level Sensor (main) <AWD> Circuit High Input:** Battery positive voltage is between 11 and 16.5 volts. After 2 seconds or more have passed since the engine staring sequence was completed. The fuel level sensor resistance has continued to be higher than 186 ohms for 2 seconds **Possible Causes:** • Fuel level sensor failed. • Combination meters assembly failed. • Connector damage • Harness damage • ECM failed.

DTC	Trouble Code Title, Conditions & Possible Causes
DTC: P0463 **1T PCM, MIL: Yes** **Year:** 2009 **Model:** Raider **Engine:** 3.7L V6 VIN K **Transmission:** All	**Fuel Level Sensor 1 Circuit High:** Ignition on and battery voltage above 10.4 volts. The fuel level sensor signal voltage at the PCM goes above the maximum acceptable value. **Possible Causes:** • Fuel level signal circuit shorted to voltage • Fuel level signal circuit open • Fuel level sensor return circuit open • Fuel level sensor • Instrument cluster
DTC: P0463 **1T ECM, MIL: Yes** **Year:** 2009, 2010 **Model:** Lancer **Engine:** 2.0L L4 VIN U, 2.0L L4 VIN V, 2.4L L4 VIN W **Transmission:** All	**Fuel Level Sensor Circuit High Input:** Battery positive voltage is between 11 and 16.5 volts. After 2 seconds or more have passed since the engine staring sequence was completed. Fuel level sensor resistance has continued to be higher than 186 ohms for 2 seconds. **Possible Causes:** • Fuel level sensor failed. • Combination meters assembly failed. • Connector damage • Harness damage • ECM failed.
DTC: P0463 **1T PCM, MIL: Yes** **Year:** 2009, 2010 **Model:** Eclipse, Endeavor, Galant **Engine:** 2.4L L4 VIN F, 3.8L V6 VIN S, 3.8L V6 VIN T **Transmission:** All	**Fuel Level Sensor Circuit High Input:** Battery positive voltage is between 11 and 16.5 volts. After 2 seconds or more have passed since the engine staring sequence was completed. Fuel level sensor resistance has continued to be higher than 186 ohms for 2 seconds. **Possible Causes:** • Fuel level sensor failed. • Open or shorted fuel level sensor circuit, or connector damage. • ECM failed. <M/T> • PCM failed. <A/T>
DTC: P0489 **1T ECM, MIL: Yes** **Year:** 2009, 2010 **Model:** Lancer, Outlander **Engine:** 2.0L L4 VIN U, 2.4L L4 VIN W, 3.0L V6 VIN X **Transmission:** All	**EGR Valve (Stepper Motor) Circuit Malfunction (Ground Short):** Ignition switch is "ON" position. Battery positive voltage is between 10 and 16.5 volts. When EGR valve is de-energized, the EGR voltage should be 1.5 volts or less for 1.4 seconds **Possible Causes:** • EGR valve (stepper motor) failed. • Open or shorted EGR valve (stepper motor) circuit, harness damage or connector damage. • ECM failed.
DTC: P0490 **1T ECM, MIL: Yes** **Year:** 2009, 2010 **Model:** Lancer, Outlander **Engine:** 2.0L L4 VIN U, 2.4L L4 VIN W, 3.0L V6 VIN X **Transmission:** All	**EGR Valve (Stepper Motor) Circuit Malfunction (Battery Short):** Ignition switch is "ON" position. Battery positive voltage is between 10 and 16.5 volts. When EGR valve is energized, the ECM voltage should be 6.7 volts or more for 1.4 seconds. **Possible Causes:** • EGR valve (stepper motor) failed. • Shorted EGR valve (stepper motor) circuit, harness damage or connector damage. • ECM failed.
DTC: P0500 **1T ECM, MIL: Yes** **Year:** 2009, 2010 **Model:** Lancer **Engine:** 2.0L L4 VIN V **Transmission:** All	**Vehicle Speed Signal Malfunction:** More than 2 seconds have passed since the engine starting sequence was completed. Battery positive voltage is between 10 and 16.5 volts. Vehicle speed information from ASC-ECU is fail for 2 seconds. **Possible Causes:** • Wheel speed sensor of ASC system failed. • ASC-ECU failed. • ECM failed.
DTC: P0500 **1T ECM, MIL: Yes** **Year:** 2009, 2010 **Model:** Eclipse, Lancer **Engine:** 2.0L L4 VIN U, 2.4L L4 VIN F, 2.4L L4 VIN W, 3.8L V6 VIN T **Transmission:** All	**Vehicle Speed Sensor Malfunction <M/T>:** After 2 seconds or more have passed the engine starting sequence was completed. Engine speed is at between 2,000 and 4,000 r/min. Volumetric efficiency is at between 40 and 75 percent. The vehicle speed sensor output voltage has not changed (no pulse signal is input) for 2 seconds. **Possible Causes:** • Vehicle speed sensor failed. • Open or shorted vehicle speed sensor circuit, or harness damage, or connector damage. • ECM failed.

DTC	Trouble Code Title, Conditions & Possible Causes
DTC: P0501 **2T PCM, MIL: Yes** **Year:** 2009 **Model:** Raider **Engine:** 3.7L V6 VIN K **Transmission:** All	**Vehicle Speed Sensor 1 Performance:** With the engine running, transmission not in park or neutral, brakes not applied, and engine rpm greater than 1500. This code will set if no vehicle speed signal is received from the ABS Module up to 120 seconds for 2 consecutive trips. **Possible Causes:** • Active BUS or communications DTC • Tire Circumference • PCM
DTC: P0503 **1T PCM, MIL: Yes** **Year:** 2009 **Model:** Raider **Engine:** 3.7L V6 VIN K **Transmission:** All	**Vehicle Speed Sensor 1 Erratic:** Ignition on and battery voltage greater than 10 volts. Transmission in Drive or Reverse. This code will set if no vehicle speed signal is received from the ABS Module up to 120 seconds for 2 consecutive trips. **Possible Causes:** • Active BUS or communications DTC • Tire Circumference • PCM
DTC: P0506 **1T PCM, MIL: Yes** **Year:** 2009, 2010 **Model:** Eclipse, Endeavor, Galant, Lancer, Outlander **Engine:** 2.0L L4 VIN U, 2.0L L4 VIN V, 2.4L L4 VIN F, 2.4L L4 VIN W, 3.0L V6 VIN X, 3.8L V6 VIN S, 3.8L V6 VIN T **Transmission:** All	**Idle Control System RPM Lower Than Expected:** Under the closed loop idle speed control. The engine coolant temperature is more than 41°C (105°F). Battery positive voltage is higher than 10 volts. Barometric pressure is higher than 76 kPa (22.4 in. Hg). Intake air temperature is higher than -10°C (14°F). After 3 seconds have elapsed from the start of the previous monitoring. Target throttle actuator control motor position is more than 512 steps. The actual idle speed is more than 100 r/min lower than the target idle speed for 10 seconds. **Possible Causes:** • Throttle valve area is dirty. • Throttle body assembly failed. • ECM failed. <M/T> • PCM failed. <A/T>
DTC: P0506 **2T PCM, MIL: Yes** **Year:** 2009 **Model:** Raider **Engine:** 3.7L V6 VIN K **Transmission:** All	**Idle Speed Performance Lower Than Expected:** With the engine idling in drive, the brake applied, engine run time above a minimum calibrated value, and no VSS, MAF/MAP, ECT, TPS, ETC, CKP Sensor, fuel system, or injector DTCs present. Engine speed is 100 RPM or more below target idle speed for 30 seconds. **Possible Causes:** • Air induction system restrictions • PCV system restrictions • Throttle body • PCM
DTC: P0507 **2T PCM, MIL: Yes** **Year:** 2009 **Model:** Raider **Engine:** 3.7L V6 VIN K **Transmission:** All	**Idle Speed Performance Higher Than Expected:** With the engine idling in drive, the brake applied, engine run time above a minimum calibrated value, and no VSS, MAF/MAP, ECT, TPS, ETC, CKP Sensor, fuel system, or injector DTCs present. Engine speed is 200 RPM or more above idle speed for 7 seconds. **Possible Causes:** • Air induction system leaks • PCV system leaks • Engine vacuum leaks • Throttle body • PCM
DTC: P0507 **1T PCM, MIL: Yes** **Year:** 2009, 2010 **Model:** Eclipse, Endeavor, Galant, Lancer, Outlander **Engine:** 2.0L L4 VIN U, 2.0L L4 VIN V, 2.4L L4 VIN F, 2.4L L4 VIN W, 3.0L V6 VIN X, 3.8L V6 VIN S, 3.8L V6 VIN T **Transmission:** All	**Idle Control System RPM Higher Than Expected:** Under the closed loop idle speed control. Engine coolant temperature is higher than 41°C (105°F). Battery positive voltage is higher than 10 volts. Barometric pressure is higher than 76 kPa (22.4 in. Hg). Intake air temperature is higher than -10°C (14°F). After 3 seconds have elapsed from the start of the previous monitoring. Target throttle actuator control motor position is 0 step. Actual idle speed has continued to be higher than the target idle speed by 200 r/min (300 r/min*) or more for 10 seconds. **Possible Causes:** • Intake system vacuum leak. • Throttle body assy failed. • ECM failed. <M/T> • PCM failed. <A/T>

DTC	Trouble Code Title, Conditions & Possible Causes
DTC: P050B **1T PCM, MIL: Yes** **Year:** 2009, 2010 **Model:** Eclipse, Lancer, Outlander **Engine:** 2.0L L4 VIN U, 2.0L L4 VIN V, 2.4L L4 VIN F, 2.4L L4 VIN W, 3.0L V6 VIN X **Transmission:** All	**Ignition Timing Retard Insufficient:** Under the ignition timing retard control. Engine coolant temperature is between 7°C (45°F) and 41°C (106°F). For 10 seconds, the difference between the basic ignition timing and the target/specified ignition timing is 3°CA or less on average during the retard control. After starting the engine in the cold state, the retard ignition timing control is abnormal, which is caused by a decrease in idle speed and so on. **Possible Causes:** • The thermostat is faulty. • ECM failed. <M/T> • PCM failed. <A/T>
DTC: P050B **2T PCM, MIL: Yes** **Year:** 2009 **Model:** Raider **Engine:** 3.7L V6 VIN K **Transmission:** All	**Cold Start Ignition Timing Performance:** Cold start condition. Ambient Air temperature between -7°C and 50°C (19.4°F and 122°F). Engine Coolant temperature between -7°C and 50°C (19.4°F and 122°F). The difference between the Ambient Air temp and ECT temp at Start is equal to or less than 10°C (50°F). Engine running at idle only. Engine RPM is 50 RPM or more (depending on vehicle specifications), below idle speed for at least 3 seconds and the average spark advance is above the threshold, too much spark advance, for a specified time limit. **Possible Causes:** • Restricted intake air system • Low battery voltage • Fuel contamination • Excessive resistance in the ETC positive circuit • Excessive resistance in the ETC negative circuit • ETC Motor/Throttle body • PCM
DTC: P050D **2T PCM, MIL: Yes** **Year:** 2009 **Model:** Raider **Engine:** 3.7L V6 VIN K **Transmission:** All	**Cold Start Rough Idle:** Cold start condition. Ambient Air temperature between -7° C and 50° C (19.4°F and 122° F). Engine Coolant temperature between -7°C and 50° C (19.4° F and 122° F). The difference between the Ambient Air temp and ECT temp at Start is equal to or less than 10°C (50°F). Engine running at idle only. If a rough idle is detected and the Dynamic Crankshaft Fuel Control remains or returns to the high limit window for a calibrated time **Possible Causes:** • Engine Mechanical problem • 1/1 O2 Sensor • MAP Sensor • ECT Sensor • Ignition coil • Spark plug • Fuel injector • Internal fuel leak • Fuel pump module/strainer • Restricted fuel supply • PCM
DTC: P0513 **1T PCM, MIL: Yes** **Year:** 2009 **Model:** Raider **Engine:** 3.7L V6 VIN K **Transmission:** All	**Invalid Skim Key:** Ignition on. The PCM detects an invalid SKREEM key. **Possible Causes:** • Incorrect VIN programmed in the PCM • No communication with the SKREEM • Ignition key • PCM
DTC: P0513 **T ECM, MIL: Yes** **Year:** 2009, 2010 **Model:** Lancer **Engine:** 2.0L L4 VIN U, 2.0L L4 VIN V, 2.4L L4 VIN W **Transmission:** All	**Immobilizer Malfunction:** Ignition switch is "ON" position. When it was different in the encrypted code sent from immobilizer-ECU and operation result by ECM. **Possible Causes:** • CAN system failed. • Immobilizer system failed. • ECM failed.

DTC	Trouble Code Title, Conditions & Possible Causes
DTC: P0513 **1T PCM, MIL: Yes** **Year:** 2009, 2010 **Model:** Eclipse, Endeavor, Galant, Outlander **Engine:** 2.4L L4 VIN F, 2.4L L4 VIN W, 3.0L V6 VIN X, 3.8L V6 VIN S, 3.8L V6 VIN T **Transmission:** All	**Immobilizer Malfunction:** With the ignition switch ON, A communication error between ECM <M/T> or PCM <A/T> and the immobilizer-ECU continues for 2 seconds or more. Or. When it was different in the encrypted code sent from immobilizer-ECU and operation result by ECM <M/T> or PCM <A/T>. **Possible Causes:** • Malfunction of harness or connector. • Malfunction of immobilizer-ECU. • ECM failed. <M/T> • PCM failed. <A/T>
DTC: P0522 **1T PCM, MIL: Yes** **Year:** 2009 **Model:** Raider **Engine:** 3.7L V6 VIN K **Transmission:** All	**Engine Oil Pressure Low:** With the ignition key on and battery voltage above 10.4 volts. The oil pressure sensor voltage at PCM is not within the calibrated specification. **Possible Causes:** • Oil pressure signal circuit open • Oil pressure signal circuit shorted to voltage • Oil pressure signal circuit shorted to ground • Oil pressure switch • PCM
DTC: P0532 **1T PCM, MIL: Yes** **Year:** 2009 **Model:** Raider **Engine:** 3.7L V6 VIN K **Transmission:** All	**A/C Pressure Sensor Circuit Low :** Engine running, AC is learned, and AC Clutch Relay energized. The PCM detects that the A/C Pressure Transducer input voltage is below the minimum acceptable value. **Possible Causes:** • 5-volt supply circuit open • 5-volt supply circuit shorted to ground • A/C Pressure transducer • A/C Pressure signal circuit shorted to ground • A/C Pressure signal circuit shorted to the sensor return signal • Excessive resistance in the A/C pressure signal circuit • Front Control Module (FCM)
DTC: P0533 **1T PCM, MIL: Yes** **Year:** 2009 **Model:** Raider **Engine:** 3.7L V6 VIN K **Transmission:** All	**A/C Pressure Sensor Circuit High :** Engine running and the AC Clutch Relay energized. The A/C pressure transducer signal at the PCM goes above 4.92 volts. **Possible Causes:** • A/C Pressure signal circuit shorted to voltage • A/C Pressure signal circuit open • A/C Pressure signal circuit shorted to the 5-volt supply circuit • Sensor return circuit open • A/C Pressure transducer • Front Control Module (FCM)
DTC: P0551 **1T PCM, MIL: Yes** **Year:** 2009, 2010 **Model:** Eclipse, Endeavor, Galant, Lancer, Outlander **Engine:** 2.0L L4 VIN U, 2.0L L4 VIN V, 2.4L L4 VIN F, 2.4L L4 VIN W, 3.0L V6 VIN X, 3.8L V6 VIN S, 3.8L V6 VIN T **Transmission:** All	**Power Steering Pressure Switch Circuit Range/Performance:** Engine coolant temperature is higher than 15°C (86°F). Drive for 4 seconds or more with the vehicle speed is 50 km/h (31 mph) or more. Stop the vehicle [vehicle speed is 1.5 km/h (1.0 mph) or less]. Repeat 10 times or more. The power steering pressure switch continues to be "ON". **Possible Causes:** • Power steering pressure switch failed. • Open or shorted power steering pressure switch circuit, harness damage, or connector damage. • PCM failed. <M/T> • PCM failed. <A/T>
DTC: P0554 **1T PCM, MIL: Yes** **Year:** 2009, 2010 **Model:** Eclipse, Endeavor, Galant, Lancer, Outlander **Engine:** 2.0L L4 VIN U, 2.0L L4 VIN V, 2.4L L4 VIN F, 2.4L L4 VIN W, 3.0L V6 VIN X, 3.8L V6 VIN S, 3.8L V6 VIN T **Transmission:** All	**Power Steering Pressure Switch Circuit Intermittent:** Engine coolant temperature is higher than 15°C (59°F). Vehicle speed is higher than 50 km/h (31 mph). The ON/OFF frequency of a power steering pressure switch is 10 Hz or more for 20 seconds. The power steering pressure switch changes from off to on more than 10 times for 1 second. **Possible Causes:** • Power steering pressure switch failed. • Incorrect power steering fluid level. • Incorrect oil pump pressure. • Harness damage in power steering pressure switch circuit, or connector damage. • ECM failed. <M/T> • PCM failed. <A/T>

DTC	Trouble Code Title, Conditions & Possible Causes
DTC: P0562 **1T PCM, MIL: Yes** **Year:** 2009 **Model:** Raider **Engine:** 3.7L V6 VIN K **Transmission:** All	**Battery Voltage Low :** With the engine running and the PCM has closed the Transmission Control Relay. If the battery voltage of the Transmission Control Relay Output Sense circuit(s) to the PCM is less than 10.0 volts for the period of 15 seconds. **NOTE: P0562 generally indicates a gradually falling battery voltage or a resistive connection(s) to the PCM. The DTC will also set if the battery voltage sensed at the PCM is less than 6.5-volts for 200 msec or where Transmission Control Relay Output circuits are less than 7.2-volts for 200 msec.** **Possible Causes:** • Related charging system DTC • Ground circuit open or high resistance • FUSED B (+) circuit to PCM open or high resistance • Transmission control relay output circuit open or high resistance • Transmission control relay • PCM
DTC: P0563 **1T ECM, MIL: Yes** **Year:** 2009, 2010 **Model:** Lancer **Engine:** 2.0L L4 VIN U, 2.0L L4 VIN V, 2.4L L4 VIN W **Transmission:** All	**Power Supply System:** The shift lever-ECU checks that the power supply circuit is normal. The voltage from the battery is determined to be overvoltage. **Possible Causes:** • Malfunction of the shift lever-ECU • Generator malfunction
DTC: P0563 **1T PCM, MIL: Yes** **Year:** 2009 **Model:** Raider **Engine:** 3.7L V6 VIN K **Transmission:** All	**Battery Voltage High:** With the ignition on. Engine RPM greater than 1000 RPM. With no other charging system codes set. Battery voltage is 1 volt greater than desired voltage. Battery voltage greater than 15.75 volts. **Possible Causes:** • Generator field control circuit shorted to voltage • Generator • PCM
DTC: P0564 **1T PCM, MIL: Yes** **Year:** 2009 **Model:** Raider **Engine:** 3.7L V6 VIN K **Transmission:** All	**Primary 5-Volt Supply Circuit High :** Ignition on. When the PCM recognizes the Primary 5-volt Supply circuit voltage is too high. ETC light is flashing. **Possible Causes:** • Primary 5-Volt Supply shorted to voltage • PCM
DTC: P0571 **1T PCM, MIL: Yes** **Year:** 2009 **Model:** Raider **Engine:** 3.7L V6 VIN K **Transmission:** All	**Brake Switch No. 1 Performance:** Ignition on. If the output of Brake Switch No.1 to the PCM looks like it is not applied, while Brake Lamp Switch Output circuit is applied the fault will mature in 60ms. **Possible Causes:** • Brake switch No.1 signal shorted to ground • Brake switch No.2 signal circuit open • Brake switch No.2 signal shorted to ground • Ground circuit open • Stop lamp switch • PCM
DTC: P0572 **2T PCM, MIL: Yes** **Year:** 2009 **Model:** Raider **Engine:** 3.7L V6 VIN K **Transmission:** All	**Brake Switch No. 1 Stuck On:** Ignition on. When the PCM recognizes Brake Switch No.1 is mechanically stuck in the low/on position. **Possible Causes:** • Brake switch No.1 signal circuit shorted to ground • Brake switch No.2 signal circuit open • Stop lamp switch • PCM
DTC: P0573 **2T PCM, MIL: Yes** **Year:** 2009 **Model:** Raider **Engine:** 3.7L V6 VIN K **Transmission:** All	**Brake Switch No.1 Stuck Off:** Ignition on. When the PCM recognizes Brake Switch No.1 is stuck in the high/off position. **Possible Causes:** • Brake switch No.2 signal shorted to ground • Brake switch No.1 signal circuit open • Ground circuit open • Stop lamp switch • PCM

DTC	Trouble Code Title, Conditions & Possible Causes
DTC: P0579 **T PCM** **Year:** 2009 **Model:** Raider **Engine:** 3.7L V6 VIN K **Transmission:** All	**Speed Control Switch No. 1 Performance:** With the ignition key on. Cruise switch voltage output is not out of range but it does not equal any of the values for any of the button positions. **Possible Causes:** • S/C Switch No.1 signal circuit shorted to voltage • S/C Switch No.1 signal circuit open • Switch return circuit open • S/C Switch No.1 signal circuit shorted to ground • S/C Switch No.1 signal circuit shorted to the switch return circuit • Clockspring • Speed control switch • PCM
DTC: P0580 **1T PCM, MIL: Yes** **Year:** 2009 **Model:** Raider **Engine:** 3.7L V6 VIN K **Transmission:** All	**Speed Control Switch No. 1 Circuit Low :** With the ignition key on. Speed control switch input No.1 is below the minimum acceptable voltage at the PCM. **Possible Causes:** • S/C Switch No.1 signal circuit shorted to the S/C switch return circuit • S/C Switch No.1 signal circuit shorted to ground • Clockspring • Speed control On/Off switch • Speed control Resume/Accel switch • PCM
DTC: P0581 **1T PCM, MIL: Yes** **Year:** 2009 **Model:** Raider **Engine:** 3.7L V6 VIN K **Transmission:** All	**Speed Control Switch No. 1 Circuit High:** With the ignition key on. Speed control switch input above the maximum acceptable voltage at the PCM. **Possible Causes:** • S/C Switch No.1 signal circuit shorted to voltage • S/C Switch No.1 signal circuit open • Switch return circuit open • S/C Switch return circuit shorted open • Clockspring • Speed control switch • PCM
DTC: P0585 **1T PCM, MIL: Yes** **Year:** 2009 **Model:** Raider **Engine:** 3.7L V6 VIN K **Transmission:** All	**Speed Control Switch 1/2 Correlation:** Ignition on. Cruise Switch inputs are not coherent with each other. Example: PCM is reading Switch No.1 as Accel and Switch No.2 as Coast at the same time. **Possible Causes:** • S/C Switch signal circuits shorted to voltage • Resistance in the S/C switch signal circuits • Resistance in the S/C switch return circuit • S/C Switch signal circuits shorted to the switch return circuit • S/C Switch No.1 signal circuit shorted to the S/C No.2 switch signal circuit • S/C Switch signal circuits shorted to ground • Clockspring • Speed control ON/OFF switch • Speed control RESUME/ACCEL switch • PCM
DTC: P0591 **1T PCM, MIL: Yes** **Year:** 2009 **Model:** Raider **Engine:** 3.7L V6 VIN K **Transmission:** All	**Speed Control Switch No. 2 Performance:** With the ignition key on. Cruise switch voltage output is not out of range but it does not equal any of the values for any of the button positions. **Possible Causes:** • S/C Switch No.2 signal circuit shorted to voltage • S/C Switch No.2 signal circuit open • S/C switch return circuit open • S/C Switch No.2 signal circuit shorted to ground • S/C Switch No.2 signal circuit shorted to the switch return circuit • Clockspring • Speed control switch • PCM

DTC	Trouble Code Title, Conditions & Possible Causes
DTC: P0592 **1T PCM, MIL: Yes** **Year:** 2009 **Model:** Raider **Engine:** 3.7L V6 VIN K **Transmission:** All	**Speed Control Switch No. 2 Circuit Low:** With the ignition key on. Speed control switch input No.2 is below the minimum acceptable voltage at the PCM. **Possible Causes:** • S/C Switch No.2 signal circuit ground • S/C Switch No.2 signal circuit shorted to the S/C switch return circuit • S/C Switch No.1 signal circuit open • Clockspring • Speed control ON/OFF switch • Speed control RESUME/ACCEL switch • PCM
DTC: P0593 **1T PCM, MIL: Yes** **Year:** 2009 **Model:** Raider **Engine:** 3.7L V6 VIN K **Transmission:** All	**Speed Control Switch No. 2 Circuit High:** With the ignition key on. Speed Control Switch No.2 input above the maximum acceptable voltage at the PCM. **Possible Causes:** • S/C Switch No.2 signal circuit shorted to voltage • S/C Switch No.2 signal circuit open • Switch return circuit open between the PCM and clockspring • Switch No.2 return circuit open between the PCM and clockspring • S/C Switch No.2 signal circuit open between the clockspring S/C switch • Clockspring • Speed control switch • PCM
DTC: P0600 **1T PCM, MIL: Yes** **Year:** 2009 **Model:** Raider **Engine:** 3.7L V6 VIN K **Transmission:** All	**Serial Communication Link:** With the ignition on. Internal Bus communication failure between processors. **Possible Causes:** • PCM
DTC: P0601 **1T PCM, MIL: Yes** **Year:** 2009 **Model:** Raider **Engine:** 3.7L V6 VIN K **Transmission:** All	**Internal Memory Check Sum Invalid:** With the ignition on. Internal checksum for software failed, does not match calculated value. **Possible Causes:** • PCM
DTC: P0602 **1T PCM, MIL: Yes** **Year:** 2009 **Model:** Raider **Engine:** 3.7L V6 VIN K **Transmission:** All	**Control Module Programming Error:** Check for generic software is made at power-up. If generic software is found, the MIL will light immediately. This DTC is designed to inform the technician that the controller still has generic software installed. **Possible Causes:** • PCM programming error
DTC: P0603 **1T PCM, MIL: Yes** **Year:** 2009, 2010 **Model:** Eclipse, Endeavor, Galant, Lancer, Outlander **Engine:** 2.0L L4 VIN U, 2.0L L4 VIN V, 2.4L L4 VIN F, 2.4L L4 VIN W, 3.0L V6 VIN X, 3.8L V6 VIN S, 3.8L V6 VIN T **Transmission:** All	**EEPROM Malfunction:** Ignition switch is in "ON" position. The latest data that was flashed while the ignition switch was in "LOCK" (OFF) position are not stored correctly. **Possible Causes:** • ECM failed. <M/T> • PCM failed. <A/T>
DTC: P0604 **1T PCM** **Year:** 2009 **Model:** Raider **Engine:** 3.7L V6 VIN K **Transmission:** All	**Internal Control Module RAM Error:** One time after the controller is reset (ignition turned to the RUN position). Whenever the Powertrain Control Module (PCM) detects an internal controller problem. **Possible Causes:** • Power or ground circuit • PCM

DTC	Trouble Code Title, Conditions & Possible Causes
DTC: P0606 **1T PCM, MIL: Yes** **Year:** 2009 **Model:** Raider **Engine:** 3.7L V6 VIN K **Transmission:** All	**Internal ECM Processor:** Engine running. When the PCM recognizes an internal failure to communicate with the ECM or the CMP and CKP Sensor count periods are too short. **Possible Causes:** • PCM
DTC: P0606 **1T PCM, MIL: Yes** **Year:** 2009, 2010 **Model:** Eclipse, Endeavor, Galant, Lancer, Outlander **Engine:** 2.0L L4 VIN U, 2.0L L4 VIN V, 2.4L L4 VIN F, 2.4l I 4 VIN W, 3.0L V6 VIN X, 3.8L V6 VIN S, 3.8L V6 VIN T **Transmission:** All	**Engine Control Module <M/T> or Powertrain Control Module <A/T> Main Processor Malfunction:** Ignition switch is "ON" position. No surveillance pulse signals should be input for 0.5 second. **Possible Causes:** • MFI relay failed. • Shorted MFI relay circuit or connector damage. • Open or shorted ignition switch-IG circuit, harness damage or connector damage. • ECM failed. <M/T> • PCM failed. <A/T>
DTC: P060B **1T PCM, MIL: Yes** **Year:** 2009 **Model:** Raider **Engine:** 3.7L V6 VIN K **Transmission:** All	**ETC A/D Ground Performance:** When the Throttle Motor is powered. When ASD reading does not return to ground within a set period of time of test activation, this fault sets. The test typically runs a couple of times per second, and is the reason why APP2 signal spikes to ground a couple of times per second in normal running. Reprogramming the module may not always fix this fault. **Possible Causes:** • PCM needs to be programmed • PCM
DTC: P060D **1T PCM, MIL: Yes** **Year:** 2009 **Model:** Raider **Engine:** 3.7L V6 VIN K **Transmission:** All	**ETC Level 2 APP Performance:** Throttle motor is powered and no matured faults related to APP Sensors. When secondary software determines that APPS 1 and APPS 2 signals do not match for a period of time. One trip fault. ETC lamp will flash. **Possible Causes:** • PCM needs to be programmed • PCM
DTC: P060E **1T PCM, MIL: Yes** **Year:** 2009 **Model:** Raider **Engine:** 3.7L V6 VIN K **Transmission:** All	**ETC Level 2 TPS Performance:** Throttle motor is powered and no matured faults related to TP Sensors. When secondary software determines that TPS 1 and TPS 2 signals do not match for a period of time. One trip fault. ETC lamp will flash. **Possible Causes:** • PCM needs to be programmed • PCM
DTC: P060F **1T , MIL: Yes** **Year:** 2009 **Model:** Raider **Engine:** 3.7L V6 VIN K **Transmission:** All	**ETC Level 2 ECT Performance:** Throttle motor is powered and no matured faults related to the Engine Coolant Temp Sensor. When secondary software determines that the Coolant Temperature is implausible for a period of time. One trip fault. ETC lamp will flash. **Possible Causes:** • PCM needs to be programmed • PCM
DTC: P0613 **T PCM** **Year:** 2009 **Model:** Raider **Engine:** 3.7L V6 VIN K **Transmission:** All	**Internal TCM Error (45RFE/545RFE Transmission):** One time after the controller is reset (ignition turned to the RUN position) and every 60 seconds thereafter. The Delay Test is executed after a reset only. Or, 2 seconds after an invalid test. If either of the following conditions occur 3 times: 1. The watchdog fault line remains high after the period has elapsed for the too early - too late watchdog test. 2. The Transmission Control Relay remains on after the watchdog delay expired. **Possible Causes:** • Power or ground circuit • PCM
DTC: P0613 **T PCM** **Year:** 2009 **Model:** Raider **Engine:** 3.7L V6 VIN K **Transmission:** All	**Internal Transmission Processor (42RLE Transmission):** After the ignition key is turned to the run position and 60 seconds thereafter. Either of the following conditions occur 3 times in less than 590 milliseconds: The watchdog line remains high after the watchdog test or the transmission relay coil is energized and remains on after the watchdog delay expires. **Possible Causes:** • PCM Internal error

DTC	Trouble Code Title, Conditions & Possible Causes
DTC: P061C **1T PCM, MIL:** Yes **Year:** 2009 **Model:** Raider **Engine:** 3.7L V6 VIN K **Transmission:** All	**ETC Level 2 RPM Performance:** Throttle motor is powered and no camshaft or crankshaft electrical signal related DTCs are set. When secondary software determines that the engine speed is implausible for a period of time. ETC lamp will flash. **Possible Causes:** • PCM needs to be programmed • PCM
DTC: P0622 **1T PCM, MIL:** Yes **Year:** 2009, 2010 **Model:** Eclipse, Endeavor, Galant, Lancer, Outlander **Engine:** 2.0L L4 VIN U, 2.0L L4 VIN V, 2.4L L4 VIN F, 2.4L L4 VIN W, 3.0L V6 VIN X, 3.8L V6 VIN S, 3.8L V6 VIN T **Transmission:** All	**Generator FR Terminal Circuit Malfunction:** Engine speed is higher than 50 r/min. Input voltage from the generator FR terminal has continued to be approximately battery positive voltage for 20 seconds. **Possible Causes:** • Generator failed. • Open or shorted circuit in generator FR terminal circuit, harness damage or connector damage. • ECM failed. <M/T> • PCM failed. <A/T>
DTC: P0622 **1T PCM, MIL:** Yes **Year:** 2009 **Model:** Raider **Engine:** 3.7L V6 VIN K **Transmission:** All	**Generator Field Control Circuit :** With the ignition on. Engine running. When the PCM tries to regulate the generator field with no result during monitoring. **Possible Causes:** • Generator Field Control Circuit shorted to voltage • Generator Field Control Circuit open • Generator Field Control Circuit shorted to ground • Generator • PCM
DTC: P0627 **1T PCM, MIL:** Yes **Year:** 2009 **Model:** Raider **Engine:** 3.7L V6 VIN K **Transmission:** All	**Fuel Pump Relay Control Circuit :** With the ignition on. Battery voltage greater than 10.4 volts. An open or shorted condition is detected in the fuel pump relay control circuit. **Possible Causes:** • Internal fused B+ circuit • Fused ignition switch circuit • Fuel pump relay control circuit open • Fuel pump relay control circuit shorted to ground • Fuel pump relay • PCM
DTC: P062C **1T PCM, MIL:** Yes **Year:** 2009 **Model:** Raider **Engine:** 3.7L V6 VIN K **Transmission:** All	**ETC Level 2 MPH Performance:** Throttle motor is powered and no vehicle speed related DTCs have matured. When secondary software determines that the vehicle speed is implausible for a period of time. ETC lamp will flash. **Possible Causes:** • PCM needs to be programmed • PCM
DTC: P0630 **T PCM, MIL:** Yes **Year:** 2009, 2010 **Model:** Eclipse, Endeavor, Galant, Lancer, Outlander **Engine:** 2.0L L4 VIN U, 2.0L L4 VIN V, 2.4L L4 VIN F, 2.4L L4 VIN W, 3.0L V6 VIN X, 3.8L V6 VIN S, 3.8L V6 VIN T **Transmission:** All	**Vehicle Identification Number (VIN) Malfunction:** Ignition switch is in "ON" position. EEPROM is normal. The VIN (current) has not been written. **Possible Causes:** • ECM failed. <M/T> • PCM failed. <A/T>
DTC: P0630 **1T ECM, MIL:** Yes **Year:** 2009, 2010 **Model:** Lancer **Engine:** 2.0L L4 VIN U, 2.0L L4 VIN V, 2.4L L4 VIN W **Transmission:** All	**VIN Not Recorded:** The chassis number is determined to be written abnormally. **Possible Causes:** • The CAN bus line is defective. • Malfunction of engine control module • Malfunction of TC-SST-ECU

DTC	Trouble Code Title, Conditions & Possible Causes

DTC: P0630
1T PCM, MIL: Yes
Year: 2009
Model: Raider
Engine: 3.7L V6 VIN K
Transmission: All
	VIN Not Programmed In PCM:
	At initialization. The VIN has not been programmed into the PCM.
	Possible Causes:
	  • Programming VIN into PCM
	  • PCM

DTC: P0632
1T PCM, MIL: Yes
Year: 2009
Model: Raider
Engine: 3.7L V6 VIN K
Transmission: All
	Odometer Not Programmed Into The PCM:
	Ignition on. Odometer is not programmed into the PCM.
	Possible Causes:
	  • Programming mileage into the PCM
	  • PCM

DTC: P0633
1T PCM, MIL: Yes
Year: 2009
Model: Raider
Engine: 3.7L V6 VIN K
Transmission: All
	SKIM Secret Key Not Stored In The PCM:
	Ignition on. The SKIM Key information has not been programmed into the PCM.
	Possible Causes:
	  • Programming SKIM key into the PCM
	  • PCM

DTC: P0638
1T PCM, MIL: Yes
Year: 2009, 2010
Model: Eclipse, Endeavor, Galant, Lancer, Outlander
Engine: 2.0L L4 VIN U, 2.0L L4 VIN V, 2.4L L4 VIN F, 2.4L L4 VIN W, 3.0L V6 VIN X, 3.8L V6 VIN S, 3.8L V6 VIN T
Transmission: All
	Throttle Actuator Control Motor Circuit Range/Performance Problem:
	Battery positive voltage is higher than 8.3 volts. Throttle position sensor (main) output voltage is between 0.35 and 4.8 volts. The difference between throttle position sensor (main) output voltage and target throttle position sensor (main) voltage is 1 volt or higher for 1 seconds.
	Possible Causes:
	  • Throttle valve return spring failed.
	  • Throttle valve operation failed.
	  • Throttle actuator control motor failed.
	  • Harness damage in throttle actuator control motor circuit, or connector damage.
	  • ECM failed. <M/T>
	  • PCM failed. <A/T>

DTC: P063A
1T PCM, MIL: Yes
Year: 2009
Model: Raider
Engine: 3.7L V6 VIN K
Transmission: All
	Generator Voltage Sense Circuit:
	The engine running. The engine speed greater than 1157 RPM.
	The PCM recognizes the alternator output voltage is less than the Battery feed circuit voltage. The Generator light will illuminate. The fault will be checked again on the next key cycle.
	Possible Causes:
	  • Excessive resistance in the generator sense circuit
	  • Generator sense circuit shorted to ground
	  • Generator
	  • PCM

DTC: P0642
1T PCM, MIL: Yes
Year: 2009
Model: Raider
Engine: 3.7L V6 VIN K
Transmission: All
	Sensor Reference Voltage 1 Circuit Low:
	Ignition on. When the PCM recognizes the Primary 5-volt Supply circuit voltage is too low. ETC light is flashing.
	Possible Causes:
	  • Primary 5-volt supply shorted to ground
	  • Sensor shorted to ground
	  • 5-volt sensor
	  • PCM

DTC: P0642
1T PCM, MIL: Yes
Year: 2009, 2010
Model: Eclipse, Endeavor, Galant, Lancer, Outlander
Engine: 2.0L L4 VIN U, 2.0L L4 VIN V, 2.4L L4 VIN F, 2.4L L4 VIN W, 3.0L V6 VIN X, 3.8L V6 VIN S, 3.8L V6 VIN T
Transmission: All
	Throttle Position Sensor Power Supply:
	Battery positive voltage is higher than 6.3 volts. Throttle position sensor power voltage should be 4.1 volts or less for 0.3 second. Throttle position sensor source voltage is smaller than the specified value.
	Possible Causes:
	  • ECM failed. <M/T>
	  • PCM failed. <A/T>

DTC	Trouble Code Title, Conditions & Possible Causes
DTC: P0645 **1T PCM, MIL: Yes** **Year:** 2009 **Model:** Raider **Engine:** 3.7L V6 VIN K **Transmission:** All	**A/C Clutch Relay Control Circuit:** With the ignition on. Battery voltage greater than 10 volts. A/C Switch on. An open or shorted condition is detected in the A/C clutch relay control circuit. **Possible Causes:** • Internal fused B+ circuits • A/C clutch relay control Circuit open • A/C clutch relay control Circuit shorted to ground • A/C clutch relay • PCM
DTC: P0652 **1T PCM, MIL: Yes** **Year:** 2009 **Model:** Raider **Engine:** 3.7L V6 VIN K **Transmission:** All	**Sensor Reference Voltage 2 Circuit Low:** Ignition on. When the PCM recognizes the Auxiliary 5-volt Supply circuit voltage is too low. ETC light is flashing. **Possible Causes:** • Auxiliary 5-volt supply circuit shorted to ground • Sensor shorted to ground • CMP Sensor • PCM
DTC: P0653 **1T PCM, MIL: Yes** **Year:** 2009 **Model:** Raider **Engine:** 3.7L V6 VIN K **Transmission:** All	**Sensor Reference Voltage 2 Circuit High:** Ignition on. When the PCM recognizes the Auxiliary 5-volt Supply circuit voltage is too high. ETC light is flashing. **Possible Causes:** • Auxiliary 5-volt supply circuit shorted to ground • PCM
DTC: P0657 **1T PCM, MIL: Yes** **Year:** 2009, 2010 **Model:** Eclipse, Endeavor, Galant, Lancer, Outlander **Engine:** 2.0L L4 VIN U, 2.0L L4 VIN V, 2.4L L4 VIN F, 2.4L L4 VIN W, 3.0L V6 VIN X, 3.8L V6 VIN S, 3.8L V6 VIN T **Transmission:** All	**Throttle Actuator Control Motor Relay Circuit Malfunction:** Battery positive voltage is higher than 8.3 volts. The power line voltage of the electronic controlled throttle system should be 6.0 volts or less for 0.35 seconds. Throttle actuator control motor relay circuit voltage is smaller than the specified value. **Possible Causes:** • Throttle actuator control motor relay failed or maladjusted. • Open or shorted throttle actuator control motor relay circuit, or loose connector. • ECM failed. <M/T> • PCM failed. <A/T>
DTC: P0685 **1T PCM, MIL: Yes** **Year:** 2009 **Model:** Raider **Engine:** 3.7L V6 VIN K **Transmission:** All	**ASD Relay Control Circuit :** With ignition on. Battery voltage above 10 volts. An open or shorted condition is detected in the ASD relay control circuit. **Possible Causes:** • Internal fused B+ circuits • ASD relay control circuit open • ASD relay control circuit shorted to ground • ASD Relay • PCM
DTC: P0688 **1T PCM, MIL: Yes** **Year:** 2009 **Model:** Raider **Engine:** 3.7L V6 VIN K **Transmission:** All	**ASD Relay Sense Circuit Low:** With ignition key on. Battery voltage greater than 10 volts. No voltage sensed at the PCM when the ASD relay is energized. **Possible Causes:** • Internal fused B+ circuits • ASD relay output circuit open • ASD Relay • PCM
DTC: P0689 **T PCM** **Year:** 2009 **Model:** Raider **Engine:** 3.7L V6 VIN K **Transmission:** All	**ASD Relay Sense Circuit Low:** Ignition on. Battery voltage between 9 and 16 volts and the ASD Relay is powered on. The ASD Output circuit voltage drops below an acceptable value at the FCM. The circuit is continuously monitored. **Possible Causes:** • ASD power supply (fused B+) problem • ASD relay output circuit is open or is shorted to ground • ASD relay has failed • PCM has failed

DTC	Trouble Code Title, Conditions & Possible Causes
DTC: P0690 **T PCM** **Year:** 2009 **Model:** Raider **Engine:** 3.7L V6 VIN K **Transmission:** All	**ASD Relay Sense Circuit High:** With the ignition on and battery voltage greater than 10 volts. If the FCM detects high voltage on the ASD Relay Sense circuit for more than 3.5 seconds. **Possible Causes:** • ASD relay output circuit is shorted to voltage • ASD relay is damaged or it has failed • PCM has failed
DTC: P0700 **1T TCM, MIL: Yes** **Year:** 2009 **Model:** Raider **Engine:** 3.7L V6 VIN K **Transmission:** All	**Automatic Transmission Control System Malfunction (MIL Request):** Ignition on and battery voltage greater than 10 volts. An active DTC is stored in the TCM. **Possible Causes:** • This code is for information only - check for other TCM codes
DTC: P0701 **1T ECM, MIL: Yes** **Year:** 2009, 2010 **Model:** Lancer **Engine:** 2.0L L4 VIN U, 2.0L L4 VIN V, 2.4L L4 VIN W **Transmission:** All	**EEPROM System (Malfunction):** TC-SST-ECU checks that the EEPROM and RAM in the TC-SST-ECU is normal. The EEPROM writing data is determined to be abnormal. WARNING: If there is any problem in the CAN bus lines, an incorrect diagnostic trouble code may be set. Prior to this diagnosis, diagnose the CAN bus lines. Whenever the ECU is replaced, ensure that the CAN bus lines are normal. **Possible Causes:** • Malfunction of TC-SST-ECU
DTC: P0702 **1T ECM, MIL: Yes** **Year:** 2009, 2010 **Model:** Lancer **Engine:** 2.0L L4 VIN U, 2.0L L4 VIN V, 2.4L L4 VIN W **Transmission:** All	**Internal control module, monitoring processor system (Malfunction):** TC-SST-ECU checks that the internal module and monitoring processor are normal. The internal module and monitoring processor are determined to be abnormal. WARNING: If there is any problem in the CAN bus lines, an incorrect diagnostic trouble code may be set. Prior to this diagnosis, diagnose the CAN bus lines. Whenever the ECU is replaced, ensure that the CAN bus lines are normal. **Possible Causes:** • Malfunction of TC-SST-ECU
DTC: P0703 **1T ECM, MIL: Yes** **Year:** 2009, 2010 **Model:** Lancer, Outlander **Engine:** 2.0L L4 VIN U, 2.4L L4 VIN W, 3.0L V6 VIN X **Transmission:** All	**Malfunction of Stoplight Switch:** Vehicle speed: 30 km/h (19 mph) or more. Voltage of battery: 9 volts or more. Voltage of battery: 16 volts or less. The change of stoplight switch signal during driving cycle: no occurrence (10 seconds × 2 times). **Possible Causes:** • Malfunction of the CAN bus • Malfunction of the stoplight switch • Damaged wiring harness and connectors • Malfunction of TCM • Malfunction of ETACS-ECU
DTC: P0705 **1T PCM, MIL: Yes** **Year:** 2009, 2010 **Model:** Eclipse **Engine:** 2.4L L4 VIN F, 3.8L V6 VIN T **Transmission:** All	**Transmission Range Switch System (Open Circuit):** Battery positive voltage is applied to the transmission range switch when the ignition switch is turned "ON." Battery positive voltage is applied to the PCM when the transmission range is in the "P" range. The PCM judges that the transmission range is in the "P" range when the battery positive voltage is applied. Battery positive voltage is applied to the PCM when the selector lever is in the "R" range ("N," "D" range). The PCM judges that the selector lever is in the "R" range ("N," "D" range) when the battery positive voltage is applied. The transmission range switch has no signal detected. (30 seconds) **Possible Causes:** • Malfunction of the transmission range switch • Malfunction of the ignition switch • Damaged harness or connector • Malfunction of the PCM
DTC: P0705 **1T ECM, MIL: Yes** **Year:** 2009, 2010 **Model:** Lancer, Outlander **Engine:** 2.0L L4 VIN U, 2.4L L4 VIN W, 3.0L V6 VIN X **Transmission:** All	**Malfunction of Transmission Range Switch:** Vehicle speed over 1 km/h (0.6 mph): 10 seconds or more. Throttle position sensor voltage: 1.37 volts or more. Engine speed: 450 r/min or more. Voltage of battery: 9 volts or more. Voltage of battery: 16 volts or less. Transmission range switch: multiple signal. (2 seconds) Transmission range switch: no signal. (5 seconds) **Possible Causes:** • Malfunction of the transmission range switch • Improper adjustment of transaxle control cable • Damaged wiring harness and connectors • Malfunction of TCM

DTC	Trouble Code Title, Conditions & Possible Causes
DTC: P0706 **1T PCM, MIL:** Yes **Year:** 2009 **Model:** Raider **Engine:** 3.7L V6 VIN K **Transmission:** All	**Transaxle Range Sensor Rationality:** Continuously with the ignition on. The DTC will set if the controller detects an invalid PRNDL code which lasts for more than 0.042 seconds. **Possible Causes:** • TRS Sense circuit open • TRS Sense circuit short to ground • TRS Sense circuit short to voltage • Metal debris in oil pan • Transmission Range Sensor (TRS) • PCM
DTC: P0710 **1T ECM, MIL:** Yes **Year:** 2009, 2010 **Model:** Lancer **Engine:** 2.0L L4 VIN U, 2.4L L4 VIN W **Transmission:** All	**Malfunction of Transmission Fluid Temperature Sensor:** After vehicle speed: 10 km/h (6.2 mph) or more. Voltage of battery: 9 volts or more. Voltage of battery: 16 volts or less. Value of temperature of transmission fluid: -40°C (-40°F) or less. (5 seconds) **Possible Causes:** • Malfunction of the valve body assembly (Faulty transmission fluid temperature sensor) • Damaged wiring harness and connectors • Malfunction of TCM
DTC: P0711 **1T ECM, MIL:** Yes **Year:** 2009, 2010 **Model:** Lancer, Outlander **Engine:** 2.0L L4 VIN U, 2.4L L4 VIN W **Transmission:** All	**Malfunction of Transmission Fluid Temperature Sensor:** Transmission range switch position: D. Vehicle speed: 10 km/h (6.2 mph) or more. Throttle position sensor voltage: 1.37 V or more. Engine speed: 450 r/min or more. Transmission fluid temperature is less than -20°C (-4°F) for 600 seconds or more Transmission fluid temperature is -20°C (-4°F) or more and less than 0°C (32°F) for 600 seconds or more Transmission fluid temperature is 0°C (32°F) or more and less than 20°C (68°F) for 600 seconds or more **Possible Causes:** • Malfunction of transmission fluid temperature sensor • Malfunction of transmission fluid cooler piping and oil pump • Malfunction of TCM
DTC: P0711 **1T PCM, MIL:** Yes **Year:** 2009 **Model:** Raider **Engine:** 3.7L V6 VIN K **Transmission:** All	**Transmission Temperature Sensor Performance:** Continuously with the ignition on and engine running. This DTC will set when the desired transmission temperature does not reach a normal operating temperature within a given time frame. Time is variable due to ambient temperature. Approximate times are starting temperature to warm up time: (-40° F / -40° C - 35 min) (-20° F / -28° C - 25 min) (20° F / -6.6° C - 20 min) (60° F / 15.5 ° C - 10 min) **Possible Causes:** • Related DTCs will be present • Transmission temperature sensor has failed • Intermittent wiring or connector problems • PCM has failed
DTC: P0712 **3T PCM, MIL:** Yes **Year:** 2009 **Model:** Raider **Engine:** 3.7L V6 VIN K **Transmission:** All	**Transmission Temperature Sensor Low :** Continuously with the ignition on and engine running. The DTC will set when the monitored Temperature Sensor voltage drops below 0.078 volts for the period of 1.45 seconds. When the fault is set, calculated temperature is substituted for measured temperature, however the fault code is stored only after three consecutive occurrences of the fault. **Possible Causes:** • Related DTCs are present • Transmission temperature sensor signal circuit is shorted to ground • Transmission temperature sensor has failed • Intermittent wiring or connector problems • PCM has failed
DTC: P0712 **1T ECM, MIL:** Yes **Year:** 2009, 2010 **Model:** Lancer, Outlander **Engine:** 2.0L L4 VIN U, 2.4L L4 VIN W, 3.0L V6 VIN X **Transmission:** All	**Malfunction of Transmission Fluid :** TCM detects the fluid temperature 180°C (356°F) or more for 5 seconds. **Possible Causes:** • Malfunction of the valve body assembly (Faulty transmission fluid temperature sensor) • Damaged wiring harness and connectors • Malfunction of TCM

DTC	Trouble Code Title, Conditions & Possible Causes
DTC: P0713 **1T ECM, MIL: Yes** **Year:** 2009, 2010 **Model:** Lancer, Outlander **Engine:** 2.0L L4 VIN U, 2.4L L4 VIN W, 3.0L V6 VIN X **Transmission:** All	**Malfunction of Transmission Fluid Temperature Sensor (Open):** After vehicle speed: 10 km/h (6.2 mph) or more. Voltage of battery: 9 volts or more. Voltage of battery: 16 volts or less. Value of temperature of transmission fluid: -40°C (-40°F) or less. (5 seconds) **Possible Causes:** • Malfunction of the valve body assembly (Faulty transmission fluid temperature sensor) • Damaged wiring harness and connectors • Malfunction of TCM
DTC: P0713 **T PCM** **Year:** 2009 **Model:** Raider **Engine:** 3.7L V6 VIN K **Transmission:** All	**Transmission Temperature Sensor High :** Continuously with the ignition on and engine running. The DTC will set when the monitored Temperature Sensor voltage rises above 4.94 volts for the period of 0.45 seconds. **Possible Causes:** • Related DTCs are present • Transmission temperature sensor signal circuit is shorted to other circuits • Transmission temperature sensor is damaged or has failed • Intermittent wiring or connector problems • Transmission temperature sensor • PCM has failed
DTC: P0714 **3T PCM, MIL: Yes** **Year:** 2009 **Model:** Raider **Engine:** 3.7L V6 VIN K **Transmission:** All	**Transmission Temperature Sensor Intermittent:** Continuously with the ignition on and engine running. The DTC will set when the monitored Temperature Sensor voltage fluctuates or changes abruptly within a predetermined period of time. **Possible Causes:** • Related DTCs are present • Transmission temperature sensor • PCM
DTC: P0715 **1T ECM, MIL: Yes** **Year:** 2009, 2010 **Model:** Lancer **Engine:** 2.0L L4 VIN U, 2.4L L4 VIN W **Transmission:** All	**Input shaft 1 (Odd number gear axle) speed sensor system (Output high range out) (TCC-SST):** Voltage of battery: 8 V or more. Voltage of battery: 16.5 V or less. Time after engine start: 1.5 seconds or more. Sensor current: 17 mA or more. (500 millisecond) **Possible Causes:** • Malfunction of TC-SST-ECU. • Malfunction of input shaft 1 speed sensor.
DTC: P0715 **1T CCM, MIL: Yes** **Year:** 2009, 2010 **Model:** Eclipse **Engine:** 2.4L L4 VIN F, 3.8L V6 VIN T **Transmission:** All	**A/T Input/Turbine Speed Sensor Circuit Malfunction:** Engine started, vehicle driven at cruise speed for 3-5 minutes, and the TCM detected too large a change in the gear/speed ratio from the Input Speed sensor signal during the CCM Rationality test. **Possible Causes:** • Input speed sensor signal circuit is open or shorted to ground • Input speed sensor signal circuit is shorted to VREF or power • Input speed sensor is damaged or has failed • TCM has failed
DTC: P0715 **1T ECM, MIL: Yes** **Year:** 2009, 2010 **Model:** Lancer, Outlander **Engine:** 2.0L L4 VIN U, 2.4L L4 VIN W **Transmission:** All	**Malfunction of Primary Pulley Speed Sensor (CVT):** Transmission range switch position: D. Secondary pulley speed: 500 r/min or more. Primary pulley speed at 10 millisecond before: 1,000 r/min or more. Throttle position sensor voltage: 1.37 volts or more. Engine speed: 450 r/min or more. Voltage of battery: 9 volts or more. Voltage of battery: 16 volts or less. Primary pulley speed [secondary pulley speed: 500 r/min or more]: 300 r/min or less. (5 seconds) Primary pulley speed [Primary pulley speed at 10 millisecond before: 1,000 r/min or more]: 300 r/min or less. (0.5 second) **Possible Causes:** • Malfunction of primary pulley speed sensor • Damaged wiring harness and connectors • Malfunction of TCM
DTC: P0715 **1T ECM, MIL: Yes** **Year:** 2009, 2010 **Model:** Outlander **Engine:** 3.0L V6 VIN X **Transmission:** All	**Input Shaft Speed Sensor System:** Idle switch: OFF. Vehicle speed: 40 km/h (24.9 mph) or more. Engine speed: 1,600 r/min or more. The input shaft speed sensor signal: 600 r/min. or less (5 seconds). **Possible Causes:** • Malfunction of the Input shaft speed sensor system circuit • Damaged harness or connector • Malfunction of the Input shaft speed sensor • Malfunction of the TCM • Malfunction of the input shaft speed sensor rotor (A/T assembly)

DTC	Trouble Code Title, Conditions & Possible Causes
DTC: P0716 **1T ECM, MIL: Yes** **Year:** 2009, 2010 **Model:** Lancer **Engine:** 2.0L L4 VIN U, 2.4L L4 VIN W **Transmission:** All	**Input shaft 1 (Odd number gear axle) speed sensor system (Poor performance):** Voltage of battery: 8 V or more. Voltage of battery: 16.5 V or less. Input shaft [odd] gear: engaged. Input shaft [even] gear: engaged. The rotation speed of the input shaft 1 (odd number gear axle) is determined to be abnormal. **Possible Causes:** • Malfunction of TC-SST-ECU • Malfunction of input shaft 1 speed sensor
DTC: P0716 **1T PCM, MIL: Yes** **Year:** 2009 **Model:** Raider **Engine:** 3.7L V6 VIN K **Transmission:** All	**Input Speed Sensor 1 Circuit Performance:** The transmission gear ratio is monitored continuously while the transmission is in gear. If there is an excessive change in the Input rpm in any valid gear (R, 1st, 2nd, 3rd, or 4th). **Possible Causes:** • Input speed signal circuit open • Speed sensor ground circuit open • Input speed signal circuit shorted to ground • Input speed signal circuit shorted to voltage • Speed sensor ground circuit shorted to voltage • Input speed sensor • PCM
DTC: P0717 **1T ECM, MIL: Yes** **Year:** 2009, 2010 **Model:** Lancer **Engine:** 2.0L L4 VIN U, 2.4L L4 VIN W **Transmission:** All	**Input shaft 1 (Odd number gear axle) speed sensor system (Output current low range out):** Voltage of battery: 8 V or more. Voltage of battery: 16.5 V or less. Time after engine start: 1.5 seconds or more. Sensor current: 4 mA or less. (500 millisecond) **Possible Causes:** • Malfunction of TC-SST-ECU • Malfunction of input shaft 1 speed sensor
DTC: P0720 **1T CCM, MIL: Yes** **Year:** 2009, 2010 **Model:** Eclipse **Engine:** 2.4L L4 VIN F, 3.8L V6 VIN T **Transmission:** All	**A/T Output Speed Sensor Circuit Malfunction:** Engine started, vehicle driven at cruise speed for 3-5 minutes, and the TCM detected too large a change in the gear/speed ratio from the Output Speed sensor signal during the CCM Rationality test. **Possible Causes:** • Output speed sensor signal circuit is open or shorted to ground • Output speed sensor signal circuit is shorted to VREF or power • Output speed sensor is damaged or has failed • TCM has failed
DTC: P0720 **1T ECM, MIL: Yes** **Year:** 2009, 2010 **Model:** Lancer, Outlander **Engine:** 2.0L L4 VIN U, 2.4L L4 VIN W **Transmission:** All	**Malfunction of Secondary Pulley Speed Sensor:** Transmission range switch position: D. Throttle position sensor voltage: 1.37 volts or more. Primary pulley speed: 1,000 r/min or more. Secondary pulley speed at 10 millisecond before: 500 r/min or more. Voltage of battery: 9 volts or more. Voltage of battery: 16 volts or less. Secondary pulley speed [primary pulley speed: 1,000 r/min or more]: 150 r/min or less. (5 seconds) Secondary pulley speed [secondary speed at 10 millisecond before: 500 r/min or more]: 260 r/min or less. (0.5 second) **Possible Causes:** • Malfunction of secondary pulley speed sensor • Damaged wiring harness and connectors • Malfunction of TCM
DTC: P0721 **1T PCM, MIL: Yes** **Year:** 2009 **Model:** Raider **Engine:** 3.7L V6 VIN K **Transmission:** All	**Output Speed Sensor Circuit Performance:** The transmission gear ratio is monitored continuously while the transmission is in gear. If there is an excessive change in output rpm in any gear. This DTC can take up to five minutes of problem identification before illuminating the MIL. **Possible Causes:** • Output speed signal circuit open • Speed sensor ground circuit open • Output speed signal circuit shorted to ground • Output speed signal circuit shorted to voltage • Speed sensor ground circuit shorted to voltage • Output speed sensor • PCM

DTC	Trouble Code Title, Conditions & Possible Causes
DTC: P0725 **1T ECM, MIL: Yes** **Year:** 2009, 2010 **Model:** Lancer **Engine:** 2.0L L4 VIN U, 2.4L L4 VIN W **Transmission:** All	**Engine speed signal abnormality (TC-SST):** TC-SST-ECU receives the periodic communication data from the engine control module via the CAN bus lines, and checks the data for abnormality. The engine speed signal from the engine control module is determined to be abnormal. **Possible Causes:** • The CAN bus line is defective. • Malfunction of crankshaft position sensor • Malfunction of engine control module • Malfunction of TC-SST-ECU
DTC: P0725 **1T ECM, MIL: Yes** **Year:** 2009, 2010 **Model:** Lancer, Outlander **Engine:** 2.0L L4 VIN U, 2.4L L4 VIN W **Transmission:** All	**Malfunction of Engine Speed (CVT):** Transmission range switch position: D. Throttle position sensor voltage: 1.37 volts or more. Primary pulley speed: 1,000 r/min or more. The absolute value of the Primary pulley speed - secondary pulley speed: 1,000 r/min or less. Voltage of battery: 9 volts or more. Voltage of battery: 16 volts or less. Engine revolution [primary pulley speed: 1,000 r/min or more]: less than 450 r/min. The difference of engine speed and primary speed: more than 1,000 r/min. (5 seconds) The difference of engine speed and secondary speed: more than 1,000 r/min. (5 seconds) **Possible Causes:** • Malfunction of the CAN bus • Malfunction of the engine system • Malfunction of TCM
DTC: P0726 **1T PCM, MIL: Yes** **Year:** 2009 **Model:** Raider **Engine:** 3.7L V6 VIN K **Transmission:** All	**Engine Speed Input Circuit Range/Performance:** Whenever the engine is running. The Engine RPM is less than 390 or greater than 8000 for more than 2 seconds while the engine is running. The MIL illuminates after 10 seconds of vehicle operation and the transmission system defaults to Limp-in mode. **Possible Causes:** • Engine DTCs are present • PCM
DTC: P0729 **1T ECM, MIL: Yes** **Year:** 2009, 2010 **Model:** Outlander **Engine:** 3.0L V6 VIN X **Transmission:** All	**Incorrect 6th Gear Ratio:** Shift stage: 6th gear. Time after shift changing finish: 2 seconds or more. Gear ratio: 0.753 or more. (0.25 second) **Possible Causes:** • Transaxle assembly powertrain parts failure • Malfunction of the P0715: input shaft speed sensor system • Malfunction of the P0720: output shaft speed sensor system • Malfunction of the P0748: line pressure linear solenoid valve system circuit • Malfunction of the P0753: low clutch linear solenoid valve system circuit • Malfunction of the P0758: 2-6 brake linear solenoid valve system circuit • Malfunction of the P0763: 3-5 reverse clutch linear solenoid valve system circuit • Malfunction of the P0768: high clutch linear solenoid valve system circuit • Malfunction of the P1753: low clutch shift solenoid valve system circuit • Malfunction of the low clutch, 2-6 brake, 3-5 reverse clutch, high clutch • Malfunction of the valve body assembly
DTC: P0731 **1T PCM, MIL: Yes** **Year:** 2009 **Model:** Raider **Engine:** 3.7L V6 VIN K **Transmission:** All	**Gear Ratio Error In 1st:** Continuously with the ignition on, engine running, and after the transmission has achieved the proper gear ratio. If the ratio of the Input rpm to the Output rpm does not match the current gear ratio. This DTC can take up to five minutes of problem identification before illuminating the MIL. **Possible Causes:** • Related DTCs will be present • Low fluid level • Internal transmission mechanical problems may exist • Intermittent gear ratio errors are present • PCM

DTC	Trouble Code Title, Conditions & Possible Causes
DTC: P0731 **1T PCM, MIL: Yes, TCIL: Yes** **Year:** 2009, 2010 **Model:** Eclipse **Engine:** 2.4L L4 VIN F, 3.8L V6 VIN T **Transmission:** All	**1st Gear Incorrect Ratio:** Engine speed: 450 r/min or more. Output speed: 350 r/min or more. Shift stage: 1st gear. Input speed: more than 0 r/min. Transmission fluid temperature sensor voltage: 4.5 volts or less. Voltage of battery: 10 volts or more. Transmission range switch rationality: only one signal. Time after shift changing finish: 2 seconds or more. Output speed: [(input speed - 200 r/min) / 1st gear ratio] or less. (4 seconds) If DTC P1779 (P0731) is set consecutively four times, the transaxle is locked into 3rd gear as a fail-safe measure, and the selector lever position indicator light (D, 1 through 4 or 5) flashes once per second. **Possible Causes:** • Malfunction of the input shaft speed sensor • Malfunction of the output shaft speed sensor • Malfunction of the PCM • Malfunction of the underdrive clutch retainer • Malfunction of the transfer drive gear or driven gear <4A/T> • Malfunction of the direct planetary carriers <5A/T> • Malfunction of clutch system and / or brake system • Malfunction of the valve body • Malfunction of the accumulator • Electrical noise generated
DTC: P0731 **1T ECM, MIL: Yes** **Year:** 2009, 2010 **Model:** Outlander **Engine:** 3.0L V6 VIN X **Transmission:** All	**1st Gear Incorrect Ratio:** Shift stage: 1st gear. Time after shift changing finish: 2 seconds or more. Gear ratio: 4.978 or more. (0.25 second) Gear ratio: 2.114 or more and 2.584 or less. (2 seconds) Gear ratio: 1.392 or more and 1.701 or less. (2 seconds) Gear ratio: 1.021 or more and 1.247 or less. (2 seconds) **Possible Causes:** • Transaxle assembly powertrain parts failure • Malfunction of the P0715: input shaft speed sensor system • Malfunction of the P0720: output shaft speed sensor system • Malfunction of the P0748: line pressure linear solenoid valve system circuit • Malfunction of the P0753: low clutch linear solenoid valve system circuit • Malfunction of the P1758: low-reverse brake shift solenoid valve system circuit • Malfunction of the low clutch • Malfunction of the low one-way clutch • Malfunction of the valve body assembly
DTC: P0732 **1T ECM, MIL: Yes** **Year:** 2009, 2010 **Model:** Outlander **Engine:** 3.0L V6 VIN X **Transmission:** All	**Incorrect 2nd Gear Ratio:** Shift stage: 2nd gear. Time after shift changing finish: 2 seconds or more. Gear ratio: 3.225 or more. (0.25 second) **Possible Causes:** • Transaxle assembly powertrain parts failure • Malfunction of the P0715: input shaft speed sensor system • Malfunction of the P0720: output shaft speed sensor system • Malfunction of the P0748: line pressure linear solenoid valve system circuit • Malfunction of the P0753: low clutch linear solenoid valve system circuit • Malfunction of the P0758: 2-6 brake linear solenoid valve system circuit • Malfunction of the P0763: 3-5 reverse clutch linear solenoid valve system circuit • Malfunction of the P0768: high clutch linear solenoid valve system circuit • Malfunction of the P1753: low clutch shift solenoid valve system circuit • Malfunction of the low clutch, 2-6 brake, 3-5 reverse clutch, high clutch • Malfunction of the valve body assembly
DTC: P0732 **1T PCM, MIL: Yes** **Year:** 2009 **Model:** Raider **Engine:** 3.7L V6 VIN K **Transmission:** All	**Gear Ratio Error In 2nd:** Continuously with the ignition on, engine running, and after the transmission has achieved the proper gear ratio. If the ratio of the Input rpm to the Output rpm does not match the current gear ratio. This DTC can take up to five minutes of problem identification before illuminating the MIL **Possible Causes:** • Related DTCs will be present • Low fluid level • Internal transmission mechanical problems may exist • Intermittent gear ratio errors are present • PCM

DTC	Trouble Code Title, Conditions & Possible Causes
DTC: P0732 **1T PCM, MIL: Yes, TCIL: Yes** **Year:** 2009, 2010 **Model:** Eclipse **Engine:** 2.4L L4 VIN F, 3.8L V6 VIN T **Transmission:** All	**2nd Gear Incorrect Ratio :** Engine speed: 450 r/min or more. Output speed: 500 r/min or more. Shift stage: 2nd gear. Input speed: more than 0 r/min. Transmission fluid temperature sensor voltage: 4.5 volts or less. Voltage of battery: 10 volts or more. Transmission range switch rationality: only one signal. Time after shift changing finish: 2 seconds or more. Output speed: [(input speed + 200 r/min) / 2nd gear ratio] or more. (4 seconds) Output speed: [(input speed - 200 r/min) / 2nd gear ratio] or less. (4 seconds) If DTC P1780 (P0732) is set consecutively four times, the transaxle is locked into 3rd gear as a fail-safe measure, and the selector lever position indicator light (D, 1 through 4 or 5) flashes once per second. **Possible Causes:** • Malfunction of the input shaft speed sensor • Malfunction of the output shaft speed sensor • Malfunction of the PCM • Malfunction of the underdrive clutch retainer • Malfunction of the transfer drive gear or driven gear <4A/T> • Malfunction of the direct planetary carriers <5A/T> • Malfunction of clutch system and / or brake system • Malfunction of the valve body • Malfunction of the accumulator • Electrical noise generated
DTC: P0733 **1T ECM, MIL: Yes** **Year:** 2009, 2010 **Model:** Outlander **Engine:** 3.0L V6 VIN X **Transmission:** All	**Incorrect 3rd Gear Ratio:** Shift stage: 3rd gear. Time after shift changing finish: 2 seconds or more. Gear ratio: 1.947 or more. (0.25 second) **Possible Causes:** • Transaxle assembly powertrain parts failure • Malfunction of the P0715: input shaft speed sensor system • Malfunction of the P0720: output shaft speed sensor system • Malfunction of the P0748: line pressure linear solenoid valve system circuit • Malfunction of the P0753: low clutch linear solenoid valve system circuit • Malfunction of the P0758: 2-6 brake linear solenoid valve system circuit • Malfunction of the P0763: 3-5 reverse clutch linear solenoid valve system circuit • Malfunction of the P0768: high clutch linear solenoid valve system circuit • Malfunction of the P1753: low clutch shift solenoid valve system circuit • Malfunction of the low clutch, 2-6 brake, 3-5 reverse clutch, high clutch • Malfunction of the valve body assembly
DTC: P0733 **1T PCM, MIL: Yes** **Year:** 2009 **Model:** Raider **Engine:** 3.7L V6 VIN K **Transmission:** All	**Gear Ratio Error In 3rd:** Continuously with the ignition on, engine running, and after the transmission has achieved the proper gear ratio. If the ratio of the Input rpm to the Output rpm does not match the current gear ratio. This DTC can take up to five minutes of problem identification before illuminating the MIL **Possible Causes:** • Related DTCs will be present • Low fluid level • Internal transmission mechanical problems may exist • Intermittent gear ratio errors are present • PCM
DTC: P0733 **1T PCM, MIL: Yes, TCIL: Yes** **Year:** 2009, 2010 **Model:** Eclipse **Engine:** 2.4L L4 VIN F, 3.8L V6 VIN T **Transmission:** All	**3rd Gear Incorrect Ratio:** Engine speed: 450 r/min or more. Output speed: 900 r/min or more. Shift stage: 3rd gear. Input speed: more than 0 r/min. Transmission fluid temperature sensor voltage: 4.5 volts or less. Voltage of battery: 10 volts or more. Transmission range switch rationality: only one signal. Time after shift changing finish: 2 seconds or more. Output speed: [(input speed + 200 r/min) / 3rd gear ratio] or more. (4 seconds) Output speed: [(input speed - 200 r/min) / 3rd gear ratio] or less. (4 seconds) If DTC P1781 (P0733) is set consecutively four times, the transaxle is locked into 3rd gear as a fail-safe measure, and the selector lever position indicator light (D, 1 through 4 or 5) flashes once per second. **Possible Causes:** • Malfunction of the input shaft speed sensor • Malfunction of the output shaft speed sensor • Malfunction of the PCM • Malfunction of the underdrive clutch retainer • Malfunction of the transfer drive gear or driven gear <4A/T> • Malfunction of the direct planetary carriers <5A/T> • Malfunction of clutch system and / or brake system • Malfunction of the valve body • Malfunction of the accumulator • Electrical noise generated

DTC	Trouble Code Title, Conditions & Possible Causes
DTC: P0734 **1T PCM, MIL: Yes** **Year:** 2009 **Model:** Raider **Engine:** 3.7L V6 VIN K **Transmission:** All	**Gear Ratio Error In 4th:** Continuously with the ignition on, engine running, and after the transmission has achieved the proper gear ratio. If the ratio of the Input rpm to the Output rpm does not match the current gear ratio. This DTC can take up to five minutes of problem identification before illuminating the MIL **Possible Causes:** • Related DTCs will be present • Low fluid level • Internal transmission mechanical problems may exist • Intermittent gear ratio errors are present • PCM
DTC: P0734 **1T ECM, MIL: Yes** **Year:** 2009, 2010 **Model:** Outlander **Engine:** 3.0L V6 VIN X **Transmission:** All	**Incorrect 4th Gear Ratio:** Shift stage: 4th gear. Time after shift changing finish: 2 seconds or more. Gear ratio: 1.340 or more. (0.25 second) **Possible Causes:** • Transaxle assembly powertrain parts failure • Malfunction of the P0715: input shaft speed sensor system • Malfunction of the P0720: output shaft speed sensor system • Malfunction of the P0748: line pressure linear solenoid valve system circuit • Malfunction of the P0753: low clutch linear solenoid valve system circuit • Malfunction of the P0758: 2-6 brake linear solenoid valve system circuit • Malfunction of the P0763: 3-5 reverse clutch linear solenoid valve system circuit • Malfunction of the P0768: high clutch linear solenoid valve system circuit • Malfunction of the P1753: low clutch shift solenoid valve system circuit • Malfunction of the low clutch, 2-6 brake, 3-5 reverse clutch, high clutch • Malfunction of the valve body assembly
DTC: P0734 **1T PCM, MIL: Yes, TCIL: Yes** **Year:** 2009, 2010 **Model:** Eclipse **Engine:** 2.4L L4 VIN F, 3.8L V6 VIN T **Transmission:** All	**4th Gear Incorrect Ratio:** Engine speed: 450 r/min or more. Output speed: 900 r/min or more. Shift stage: 4th gear. Input speed: more than 0 r/min. Transmission fluid temperature sensor voltage: 4.5 volts or less. Voltage of battery: 10 volts or more. Transmission range switch rationality: only one signal. Time after shift changing finish: 2 seconds or more. Output speed: [(input speed + 200 r/min) / 4th gear ratio] or more. (4 seconds) Output speed: [(input speed - 200 r/min) / 4th gear ratio] or less. (4 seconds) If DTC P1782 (P0734) is set consecutively four times, the transaxle is locked into 3rd gear as a fail-safe measure, and the selector lever position indicator light (D, 1 through 4 or 5) flashes once per second. **Possible Causes:** • Malfunction of the input shaft speed sensor • Malfunction of the output shaft speed sensor • Malfunction of the PCM • Malfunction of the underdrive clutch retainer • Malfunction of the transfer drive gear or driven gear <4A/T> • Malfunction of the direct planetary carriers <5A/T> • Malfunction of clutch system and / or brake system • Malfunction of the valve body • Malfunction of the accumulator • Electrical noise generated
DTC: P0735 **1T ECM, MIL: Yes** **Year:** 2009, 2010 **Model:** Outlander **Engine:** 3.0L V6 VIN X **Transmission:** All	**Incorrect 5th Gear Ratio:** Shift stage: 5th gear. Time after shift changing finish: 2 seconds or more. Gear ratio: 0.984 or more. (0.25 second) **Possible Causes:** • Transaxle assembly powertrain parts failure • Malfunction of the P0715: input shaft speed sensor system • Malfunction of the P0720: output shaft speed sensor system • Malfunction of the P0748: line pressure linear solenoid valve system circuit • Malfunction of the P0753: low clutch linear solenoid valve system circuit • Malfunction of the P0758: 2-6 brake linear solenoid valve system circuit • Malfunction of the P0763: 3-5 reverse clutch linear solenoid valve system circuit • Malfunction of the P0768: high clutch linear solenoid valve system circuit • Malfunction of the P1753: low clutch shift solenoid valve system circuit • Malfunction of the low clutch, 2-6 brake, 3-5 reverse clutch, high clutch • Malfunction of the valve body assembly

DTC	Trouble Code Title, Conditions & Possible Causes
DTC: P0735 **1T PCM, MIL: Yes, TCIL: Yes** **Year:** 2009, 2010 **Model:** Eclipse **Engine:** 2.4L L4 VIN F, 3.8L V6 VIN T **Transmission:** All	**5th Gear Incorrect Ratio:** Engine speed: 450 r/min or more. Output speed: 900 r/min or more. Shift stage: 5th gear. Input speed: more than 0 r/min. Transmission fluid temperature sensor voltage: 4.5 volts or less. Voltage of battery: 10 volts or more. Transmission range switch rationality: only one signal. Time after shift changing finish: 2 seconds or more. Output speed: [(input speed + 200 r/min) / 5th gear ratio] or more. (4 seconds) Output speed: [(input speed - 200 r/min) / 5th gear ratio] or less. (4 seconds) If DTC P1782 (P0734) is set consecutively four times, the transaxle is locked into 3rd gear as a fail-safe measure, and the selector lever position indicator light (D, 1 through 4 or 5) flashes once per second. **Possible Causes:** • Malfunction of the input shaft speed sensor • Malfunction of the output shaft speed sensor • Malfunction of the PCM • Malfunction of the underdrive clutch retainer • Malfunction of the transfer drive gear or driven gear <4A/T> • Malfunction of the direct planetary carriers <5A/T> • Malfunction of clutch system and / or brake system • Malfunction of the valve body • Malfunction of the accumulator • Electrical noise generated
DTC: P0735 **1T PCM, MIL: Yes** **Year:** 2009 **Model:** Raider **Engine:** 3.7L V6 VIN K **Transmission:** All	**Gear Ratio Error In 5th:** Continuously with the ignition on, engine running, and after the transmission has achieved the proper gear ratio. If the ratio of the Input rpm to the Output rpm does not match the current gear ratio. This DTC can take up to five minutes of problem identification before illuminating the MIL. **Possible Causes:** • Related DTCs will be present • Low fluid level • Internal transmission mechanical problems may exist • Intermittent gear ratio errors are present • PCM
DTC: P0735 **1T PCM, MIL: Yes** **Year:** 2009 **Model:** Raider **Engine:** 3.7L V6 VIN K **Transmission:** All	**Gear Ratio Error In Reverse:** Continuously with the ignition on, engine running, and after the transmission has achieved the proper gear ratio. If the ratio of the Input rpm to the Output rpm does not match the current gear ratio. This DTC can take up to five minutes of problem identification before illuminating the MIL. **Possible Causes:** • Related DTCs will be present • Low fluid level • Internal transmission mechanical problems may exist • Intermittent gear ratio errors are present • PCM
DTC: P0736 **1T PCM, MIL: Yes, TCIL: Yes** **Year:** 2009, 2010 **Model:** Eclipse **Engine:** 2.4L L4 VIN F, 3.8L V6 VIN T **Transmission:** All	**Reverse Gear Incorrect Ratio:** Engine speed: 450 r/min or more. Output speed: 100 r/min or more. Shift stage: reverse gear. Input speed: more than 0 r/min. Transmission fluid temperature sensor voltage: 4.5 volts or less. Voltage of battery: 10 volts or more. Transmission range switch rationality: only one signal. Time after shift changing finish: 0.5 second or more. Output speed: [(input speed + 200 r/min) / reverse gear ratio] or more. (1 second) Output speed: [(input speed - 200 r/min) / reverse gear ratio] or less. (1 second) If DTC P1784 (P0736) is set consecutively four times, the transaxle is locked into 3rd gear as a fail-safe measure, and the selector lever position indicator light (D, 1 through 4 or 5) flashes once per second. **Possible Causes:** • Malfunction of the input shaft speed sensor • Malfunction of the output shaft speed sensor • Malfunction of the PCM • Malfunction of the underdrive clutch retainer • Malfunction of the transfer drive gear or driven gear <4A/T> • Malfunction of the direct planetary carriers <5A/T> • Malfunction of clutch system and / or brake system • Malfunction of the valve body • Malfunction of the accumulator • Electrical noise generated

DTC	Trouble Code Title, Conditions & Possible Causes
DTC: P0736 **1T ECM, MIL:** Yes **Year:** 2009, 2010 **Model:** Outlander **Engine:** 3.0L V6 VIN X **Transmission:** All	**Reverse Gear Incorrect Ratio:** Shift stage: reverse gear. Gear ratio: more than -3.457. (0.25 second) **Possible Causes:** • Transaxle assembly powertrain parts failure • Malfunction of the P0715: input shaft speed sensor system • Malfunction of the P0720: output shaft speed sensor system • Malfunction of the P0748: line pressure linear solenoid valve system circuit • Malfunction of the P0763: 3-5 reverse clutch linear solenoid valve system circuit • Malfunction of the P0743: lock-up and low-reverse brake linear solenoid valve system circuit • Malfunction of the P1758: low-reverse brake shift solenoid valve system circuit • Malfunction of the 3-5 reverse clutch • Malfunction of the low-reverse brake • Malfunction of the valve body assembly
DTC: P0740 **1T ECM, MIL:** Yes **Year:** 2009, 2010 **Model:** Lancer, Outlander **Engine:** 2.0L L4 VIN U, 2.4L L4 VIN W **Transmission:** All	**Malfunction of Lockup Solenoid Valve:** TCM determines malfunction by detecting the abnormality in the lockup solenoid valve. Specified amount of current is not flown to the lockup solenoid valve because of an open or short circuit. Monitored current [target current: 750 mA or more]: 400 mA or less. (5 seconds). Monitored current: 1.34 A or more (1 second). **Possible Causes:** • Malfunction of valve body assembly (Faulty lockup solenoid valve) • Damaged wiring harness and connectors • Malfunction of TCM
DTC: P0740 **1T PCM, MIL:** Yes **Year:** 2009 **Model:** Raider **Engine:** 3.7L V6 VIN K **Transmission:** All	**TCC Out Of Range:** During Electronically Modulated Converter Clutch (EMCC) Operation. Transmission must be in EMCC, with input speed greater than 1750 rpm. L/R-TCC Solenoid achieves the maximum duty cycle and can not pull engine speed within 60 rpm of input speed. Also when the transmission is in FEMCC and the engine slips TCC more than 100 rpm for 10 seconds. This DTC can take up to five minutes of problem identification before illuminating the MIL. **Possible Causes:** • Related DTCs will be present • Internal transmission mechanical problems may exist • Intermittent gear ratio errors are present • PCM
DTC: P0741 **1T ECM, MIL:** Yes **Year:** 2009, 2010 **Model:** Outlander **Engine:** 3.0L V6 VIN X **Transmission:** All	**Torque Converter Clutch System (Stuck Off):** Lock-up status: engaging. Transmission range switch: D Calculated slip (engine speed - input shaft speed): (40 + 0.035 × output shaft speed>)/2 or more. (30 seconds) **Possible Causes:** • Malfunction of the P0743: lock-up and low-reverse brake linear solenoid valve system circuit • Malfunction of the P0712: transmission fluid temperature sensor system (short circuit) • Malfunction of the P0713: transmission fluid temperature sensor system (open circuit) • Poor installation of the engine and A/T (deviation to the axial direction) • Malfunction of the torque converter • Malfunction of the valve body assembly
DTC: P0741 **1T ECM, MIL:** Yes **Year:** 2009, 2010 **Model:** Lancer, Outlander **Engine:** 2.0L L4 VIN U, 2.4L L4 VIN W **Transmission:** All	**Abnormality in Lock-up Function (CVT):** Transmission range switch position: D. Throttle position sensor voltage: 1.37 volts or more. Engine speed: 450 r/min or more. Transmission fluid temperature: 20°C (68°F) or more. Transmission fluid temperature: 180°C (356°F) or less. Voltage of battery: 9 volts or more. Voltage of battery: 16 volts or less. Lock-up pressure: 0.2 MPa (29 psi) or more. (30 seconds) Torque converter slip revolution: 40+(vehicle speed/2) or more. (30 seconds) **Possible Causes:** • Abnormal line pressure • Malfunction of TCM
DTC: P0741 **1T PCM, MIL:** Yes **Year:** 2009, 2010 **Model:** Eclipse **Engine:** 2.4L L4 VIN F, 3.8L V6 VIN T **Transmission:** All	**Torque Converter Clutch System (Stuck Off):** Solenoid status: plunging into connecting condition. Transmission range switch position: D. Time during 100% duty: 4 seconds or more. At start of lock-up operation, if lock-up clutch cannot be engaged even when duty ratio of torque converter clutch solenoid remains 100% for more than specified time, PCM judges that torque converter clutch is stuck OFF. **Possible Causes:** • Malfunction of the torque converter clutch solenoid valve • Malfunction of the input shaft speed sensor • Malfunction of the valve body • Damaged harness or connector • Malfunction of the PCM • Malfunction of the torque converter

DTC	Trouble Code Title, Conditions & Possible Causes
DTC: P0742 **1T ECM, MIL: Yes** **Year:** 2009, 2010 **Model:** Outlander **Engine:** 3.0L V6 VIN X **Transmission:** All	**Torque Converter Clutch System (Stuck ON):** Output speed: more than 1,000 r/min. Lock-up status: disengaging. Transmission range switch position: D. Calculated slip (engine speed - input shaft speed): 5 r/min or less. (5 seconds **Possible Causes:** • Malfunction of the P0743: lock-up and low-reverse brake linear solenoid valve system circuit • Malfunction of the P0712: transmission fluid temperature sensor system (short circuit) • Malfunction of the P0713: transmission fluid temperature sensor system (open circuit) • Poor installation of the engine and A/T (deviation to the axial direction) • Malfunction of the torque converter • Malfunction of the valve body assembly
DTC: P0742 **1T PCM, MIL: Yes** **Year:** 2009, 2010 **Model:** Eclipse **Engine:** 2.4L L4 VIN F, 3.8L V6 VIN T **Transmission:** All	**Torque Converter Clutch System (Stuck On):** Throttle position sensor voltage: 1.5 volts or more. Output speed: 1,000 r/min or more. Solenoid status: OFF. Transmission range switch position: D. Time after lock up clutch release: 5 seconds or more. Calculated slip (engine speed - input speed): 5 r/min or less. (5 seconds) Or calculated slip (engine speed - input speed): -5 r/min or more. (5 seconds) **Possible Causes:** • Malfunction of the torque converter clutch solenoid valve • Malfunction of the valve body • Damaged harness or connector • Malfunction of the PCM
DTC: P0743 **1T PCM, MIL: Yes, TCIL: Yes** **Year:** 2009, 2010 **Model:** Eclipse **Engine:** 2.4L L4 VIN F, 3.8L V6 VIN T **Transmission:** All	**Torque Converter Clutch Solenoid Valve System:** Solenoid status: either solid ON or OFF. Voltage of battery: 10 volts or more. Solenoid voltage: 3 volts or less. (0.3 second) If DTC P1778 (P0743) is set consecutively four times, the transaxle is locked into 3rd gear as a fail-safe measure, and the selector lever position indicator light (D, 1 through 4 or 5) flashes once per second. **Possible Causes:** • Malfunction of the torque converter clutch solenoid valve • Damaged harness or connector • Malfunction of the PCM
DTC: P0743 **1T ECM, MIL: Yes** **Year:** 2009, 2010 **Model:** Outlander **Engine:** 3.0L V6 VIN X **Transmission:** All	**Lock-up and Low-reverse Brake Linear Solenoid Valve System:** Sport mode: 1st or Lock-up status: engaging (2nd to 6th). Lock-up and low-reverse brake linear solenoid valve actual command: more than 612 mA. Lock-up and low-reverse brake linear solenoid valve actual current: less than 190 mA. (5 seconds) **Possible Causes:** • Malfunction of the lock-up and low-reverse brake linear solenoid valve system circuit • Damaged harness or connector • Malfunction of the TCM • Malfunction of the lock-up and low-reverse brake linear solenoid valve (valve body assembly)
DTC: P0745 **1T ECM, MIL: Yes** **Year:** 2009, 2010 **Model:** Lancer, Outlander **Engine:** 2.0L L4 VIN U, 2.4L L4 VIN W **Transmission:** All	**Malfunction of Line Pressure Solenoid Valve (CVT):** The TCM determines malfunction by detecting the abnormality in the line pressure solenoid valve. Monitored current [target current: 750 mA or more]: 400 mA or less. (5 seconds) Monitored current: 1.34 A or more. (1 second) **Possible Causes:** • Malfunction of valve body assembly (Faulty line pressure solenoid valve) • Damaged wiring harness and connectors • Malfunction of TCM
DTC: P0746 **1T ECM, MIL: Yes** **Year:** 2009, 2010 **Model:** Lancer, Outlander **Engine:** 2.0L L4 VIN U, 2.4L L4 VIN W **Transmission:** All	**Line Pressure Solenoid System (Drive current range out) (CVT):** Transmission range switch position: D. Throttle position sensor voltage: 1.37 volts or more. Engine speed: 450 r/min or more. Transmission fluid temperature: 20°C (68°F) or more. Transmission fluid temperature: 180°C (356°F) or less. Voltage of battery: 9 volts or more. Voltage of battery: 16 volts or less. Pulley ratio (primary pulley ratio/secondary pulley ratio): more than 3.5. (0.1 second) Pulley ratio (primary pulley ratio/secondary pulley ratio): more than 2.7. (0.2 second) **Possible Causes:** • Abnormal line pressure • Malfunction of TCM
DTC: P0746 **1T ECM, MIL: Yes** **Year:** 2009, 2010 **Model:** Lancer **Engine:** 2.0L L4 VIN U, 2.4L L4 VIN W **Transmission:** All	**Line Pressure Solenoid System (Drive current range out) (TC-SST):** Voltage of battery: 8 V or more. Voltage of battery: 16.5 V or less. Calculated current (actual current - target current): 500 mA or more. (1 second) **Possible Causes:** • Malfunction of TC-SST-ECU • Malfunction of line pressure solenoid

DTC	Trouble Code Title, Conditions & Possible Causes
DTC: P0748 **1T ECM, MIL: Yes** **Year:** 2009, 2010 **Model:** Outlander **Engine:** 3.0L V6 VIN X **Transmission:** All	**Line Pressure Linear Solenoid Valve System:** Voltage of battery: 2 volts or more. Line pressure linear solenoid valve activation command: more than 625 mA. Line pressure linear solenoid valve actual current: less than 190 mA. (5 seconds) **Possible Causes:** • Malfunction of the line pressure linear solenoid valve system circuit • Damaged harness or connector • Malfunction of the TCM • Malfunction of the line pressure linear solenoid valve (valve body assembly)
DTC: P0750 **1T PCM, MIL: Yes** **Year:** 2009 **Model:** Raider **Engine:** 3.7L V6 VIN K **Transmission:** All	**LR Solenoid Circuit:** Initially at ignition on, then every 10 seconds thereafter. The solenoids will also be tested immediately after a gear ratio error or pressure switch error is detected. Three consecutive solenoid continuity test failures, or one failure if test is run in response to a gear ratio or pressure switch error. **Possible Causes:** • Related TCM power input or transmission control output DTC's • L/R control circuit open • L/R control circuit shorted to ground • L/R control circuit short to voltage • Transmission Solenoid/Pressure switch assembly • PCM
DTC: P0753 **1T ECM, MIL: Yes** **Year:** 2009, 2010 **Model:** Outlander **Engine:** 3.0L V6 VIN X **Transmission:** All	**Low Clutch Linear Solenoid Valve System:** Gear position: 1st to 4th. Low clutch linear solenoid valve activation command: more than 937 mA. Low clutch linear solenoid valve actual current: less than 500 mA. (5 seconds) **Possible Causes:** • Malfunction of the low clutch linear solenoid valve system circuit • Damaged harness or connector • Malfunction of the TCM • Malfunction of the low clutch linear solenoid valve (valve body assembly)
DTC: P0753 **1T PCM, MIL: Yes** **Year:** 2009, 2010 **Model:** Eclipse **Engine:** 2.4L L4 VIN F, 3.8L V6 VIN T **Transmission:** All	**Low-Reverse Solenoid Valve :** Solenoid status: either solid ON or OFF. Shift status: in-gear. Voltage of battery: 10 volts or more. Solenoid voltage: 3 volts or less. (0.3 second) If DTC P1773 (P0753) is set consecutively four times, the transaxle is locked into 3rd gear as a fail-safe measure, and the selector lever position indicator light (D, 1 through 4 or 5) flashes once per second. **Possible Causes:** • Malfunction of the low-reverse solenoid valve • Damaged harness or connector • Malfunction of the PCM
DTC: P0753 **1T ECM, MIL: Yes** **Year:** 2009, 2010 **Model:** Lancer **Engine:** 2.0L L4 VIN U, 2.4L L4 VIN W **Transmission:** All	**Shift Select Solenoid 1 System (Open circuit):** Voltage of battery: 8 V or more. Voltage of battery: 16.5 V or less. FET (Field Effect Transistor) output: 1 V or more. (1 second) The shift select solenoid 1 circuit is determined to be open. **Possible Causes:** • Malfunction of TC-SST-ECU. • Malfunction of shift select solenoid 1.
DTC: P0755 **1T PCM, MIL: Yes** **Year:** 2009 **Model:** Raider **Engine:** 3.7L V6 VIN K **Transmission:** All	**2C Solenoid Circuit (45RFE/545RFE Transmission):** Initially at power-up, then every 10 seconds thereafter. The solenoid circuits will also be tested immediately after a gear ratio or pressure switch error is detected. After three consecutive solenoid continuity test failures, or one failure if test is run in response to a gear ratio or pressure switch error. **Possible Causes:** • Related DTCs present • Transmission control relay output circuit open • 2C Solenoid control circuit is open or shorted to ground • 2C Solenoid control circuit is shorted to system power • Transmission Solenoid/TRS Assembly • Intermittent wiring or connector problems • PCM has failed

DTC	Trouble Code Title, Conditions & Possible Causes
DTC: P0755 **1T PCM, MIL: Yes** **Year:** 2009 **Model:** Raider **Engine:** 3.7L V6 VIN K **Transmission:** All	**2/4 Solenoid Circuit Failure (42RLE Transmission):** Initially at ignition on, then every 10 seconds thereafter. The solenoids will also be tested immediately after a gear ratio error or pressure switch error is detected. Three consecutive solenoid continuity test failures, or one failure if test is run in response to a gear ratio or pressure switch error. **Possible Causes:** • Related DTCs present • Transmission control relay output circuit open • 2/4 Solenoid control circuit is open or shorted to ground • 2/4 Solenoid control circuit is shorted to system power • 2/4 Solenoid is damaged or has failed • Transmission Solenoid/Pressure switch assembly • Intermittent wiring or connector problems • PCM has failed
DTC: P0758 **1T ECM, MIL: Yes** **Year:** 2009, 2010 **Model:** Lancer **Engine:** 2.0L L4 VIN U, 2.4L L4 VIN W **Transmission:** All	**Shift Select Solenoid 2 System (Open circuit):** Voltage of battery: 8 V or more. Voltage of battery: 16.5 V or less. FET (Field Effect Transistor) output: 1 V or more. (1 second) The shift select solenoid 2 circuit is determined to be open. **Possible Causes:** • Malfunction of TC-SST-ECU. • Malfunction of shift select solenoid 2.
DTC: P0758 **1T PCM, MIL: Yes** **Year:** 2009, 2010 **Model:** Eclipse **Engine:** 2.4L L4 VIN F, 3.8L V6 VIN T **Transmission:** All	**Underdrive Solenoid Valve System:** Solenoid status: either solid ON or OFF. Shift status: in-gear. Voltage of battery: 10 volts or more. Solenoid voltage: 3 volts or less. (0.3 second) If DTC P1774 (P0758) is set consecutively four times, the transaxle is locked into 3rd gear as a fail-safe measure, and the selector lever position indicator light (D, 1 through 4 or 5) flashes once per second. **Possible Causes:** • Malfunction of the underdrive solenoid valve • Damaged harness or connector • Malfunction of the PCM
DTC: P0758 **1T ECM, MIL: Yes** **Year:** 2009, 2010 **Model:** Outlander **Engine:** 3.0L V6 VIN X **Transmission:** All	**2-6 Brake Linear Solenoid Valve System:** Gear position: 2nd or 6th. 2-6 brake linear solenoid valve activation command: more than 937 mA. 2-6 brake linear solenoid valve actual current: less than 500 mA. (5 seconds) **Possible Causes:** • Malfunction of the 2-6 brake linear solenoid valve system circuit • Damaged harness or connector • Malfunction of the TCM • Malfunction of the 2-6 brake linear solenoid valve (valve body assembly)
DTC: P0760 **1T PCM, MIL: Yes** **Year:** 2009 **Model:** Raider **Engine:** 3.7L V6 VIN K **Transmission:** All	**OD Solenoid Circuit:** Initially at power-up, then every 10 seconds thereafter. The solenoid circuits will also be tested immediately after a gear ratio or pressure switch error is detected. After three consecutive solenoid continuity test failures, or one failure if test is run in response to a gear ratio or pressure switch error. **Possible Causes:** • Related TCM power input DTCs present • OD Solenoid control circuit is shorted to other circuits • OD Solenoid control circuit is open or shorted to ground • Transmission Solenoid/TRS Assembly • PCM
DTC: P0763 **1T PCM, MIL: Yes, TCIL: Yes** **Year:** 2009, 2010 **Model:** Eclipse **Engine:** 2.4L L4 VIN F, 3.8L V6 VIN T **Transmission:** All	**Second Solenoid Valve System:** Solenoid status: either solid ON or OFF. Shift status: in-gear. Voltage of battery: 10 volts or more. Solenoid voltage: 3 volts or less. (0.3 second) If DTC P1775 (P0763) is set consecutively four times, the transaxle is locked into 3rd gear as a fail-safe measure, and the selector lever position indicator light (D, 1 through 4 or 5) flashes once per second. **Possible Causes:** • Malfunction of the second solenoid valve • Damaged harness or connector • Malfunction of the PCM

DTC	Trouble Code Title, Conditions & Possible Causes
DTC: P0763 **1T ECM, MIL: Yes** **Year:** 2009, 2010 **Model:** Outlander **Engine:** 3.0L V6 VIN X **Transmission:** All	**3-5 Reverse Clutch Linear Solenoid Valve System:** Gear position: 1st or 2nd or 4th or 6th. 3-5 reverse clutch linear solenoid valve activation command: more than 937 mA. 3-5 reverse clutch linear solenoid valve actual current: less than 500 mA. (5 seconds) **Possible Causes:** • Malfunction of the 3-5 reverse clutch linear solenoid valve system circuit • Damaged harness or connector • Malfunction of the TCM • Malfunction of the 3-5 reverse clutch linear solenoid valve (valve body assembly)
DTC: P0765 **1T PCM, MIL: Yes** **Year:** 2009 **Model:** Raider **Engine:** 3.7L V6 VIN K **Transmission:** All	**UD Solenoid Circuit:** Initially at power-up, then every 10 seconds thereafter. The solenoid circuits will also be tested immediately after a gear ratio or pressure switch error is detected. After three consecutive solenoid continuity test failures, or one failure if test is run in response to a gear ratio or pressure switch error. **Possible Causes:** • Related TCM power input DTCs present • UD Solenoid control circuit is shorted to other circuits • UD Solenoid control circuit is open or shorted to ground • Transmission Solenoid/TRS Assembly • PCM
DTC: P0768 **1T PCM, MIL: Yes, TCIL: Yes** **Year:** 2009, 2010 **Model:** Eclipse **Engine:** 2.4L L4 VIN F, 3.8L V6 VIN T **Transmission:** All	**Overdrive Solenoid Valve System:** Solenoid status: either solid ON or OFF. Shift status: in-gear. Voltage of battery: 10 volts or more. Solenoid voltage: 3 volts or less. (0.3 second) If DTC P1776 (P0768) is set consecutively four times, the transaxle is locked into 3rd gear as a fail-safe measure, and the selector lever position indicator light (D, 1 through 4 or 5) flashes once per second. **Possible Causes:** • Malfunction of the overdrive solenoid valve • Damaged harness or connector • Malfunction of the PCM
DTC: P0770 **1T PCM, MIL: Yes** **Year:** 2009 **Model:** Raider **Engine:** 3.7L V6 VIN K **Transmission:** All	**4C Solenoid Circuit (45RFE/545RFE Transmission):** Initially at power-up, then every 10 seconds thereafter. The solenoid circuits will also be tested immediately after a gear ratio or pressure switch error is detected. After three consecutive solenoid continuity test failures, or one failure if test is run in response to a gear ratio or pressure switch error. **Possible Causes:** • Related TCM power input DTCs present • 4C Solenoid control circuit is shorted to other circuits • 4C Solenoid control circuit is open or shorted to ground • Transmission Solenoid/TRS Assembly • PCM
DTC: P0773 **1T PCM, MIL: Yes, TCIL: Yes** **Year:** 2009, 2010 **Model:** Eclipse **Engine:** 2.4L L4 VIN F, 3.8L V6 VIN T **Transmission:** All	**Reduction solenoid valve system <5A/T>:** Solenoid status: either solid ON or OFF. Shift status: in-gear. Voltage of battery: 10 volts or more. Solenoid voltage: 3 volts or less. (0.3 second) If DTC P1777 (P0773) is set consecutively four times, the transaxle is locked into 3rd gear as a fail-safe measure, and the selector lever position indicator light (D, 1 through 4 or 5) flashes once per second. **Possible Causes:** • Malfunction of the reduction solenoid valve • Damaged harness or connector • Malfunction of the PCM
DTC: P0776 **1T ECM, MIL: Yes** **Year:** 2009, 2010 **Model:** Lancer, Outlander **Engine:** 2.0L L4 VIN U, 2.4L L4 VIN W **Transmission:** All	**Abnormality in Secondary Pressure Solenoid Valve Function (CVT):** The following three conditions are met for three seconds. * Engine speed is 450 r/min or more. * Power supply voltage is 10 volts or more. *-Difference between target secondary pressure and actual secondary pressure is 1.20 MPa (174 psi) or more. TCM conducts fault detection by measuring the difference between the target value and the actual value for the secondary pressure. **Possible Causes:** • Trouble in CAN bus system • Abnormal line pressure • Malfunction of TCM

DTC	Trouble Code Title, Conditions & Possible Causes
DTC: P0776 **1T ECM, MIL: Yes** **Year:** 2009, 2010 **Model:** Lancer **Engine:** 2.0L L4 VIN U, 2.4L L4 VIN W **Transmission:** All	**Clutch Cooling Flow Solenoid System (Drive current range out) (TC-SST):** Voltage of battery: 8 V or more. Voltage of battery: 16.5 V or less. Engine speed: 650 r/min or more. Time since above engine condition: 1.5 seconds or more. Calculated current (actual current - target current): 1,000 mA or more. (1 second) Or. Calculated current (target current - actual current): 1,000 mA or more. (1 second) The difference between the actual current of the clutch cooling flow solenoid and target current is large. **Possible Causes:** • Malfunction of TC-SST-ECU. • Malfunction of clutch cooling flow solenoid.
DTC: P0777 **1T ECM, MIL: Yes** **Year:** 2009, 2010 **Model:** Lancer **Engine:** 2.0L L4 VIN U, 2.4L L4 VIN W **Transmission:** All	**Clutch Cooling Flow Solenoid System (Stuck):** Voltage of battery: 8 V or more. Voltage of battery: 16.5 V or less. Engine speed: 650 r/min or more. Time since above engine condition: 1.5 seconds or more. Clutch 1 target pressure: 380 hPa or more. Clutch 2 target pressure: 380 hPa or more. Clutch 1 pressure: 100 hPa or less, and clutch 2 pressure: 100 hPa or less. (8 seconds) The clutch cooling flow solenoid is determined to be seized. **Possible Causes:** • Malfunction of TC-SST-ECU. • Malfunction of clutch cooling flow solenoid. • Insufficient fluid level. • Improper installation of mechatronic assembly.
DTC: P0778 **1T ECM, MIL: Yes** **Year:** 2009, 2010 **Model:** Outlander **Engine:** 2.4L L4 VIN W **Transmission:** All	**Malfunction of Secondary Pressure Solenoid Valve:** Ignition switch : ON (start the engine and keep it for 10 seconds or more). Monitored current [target current: 750 mA or more]: 400 mA or less. (5 seconds) Monitored current: 1.34 A or more. (1 second) **Possible Causes:** • Malfunction of valve body assembly (Faulty secondary pressure solenoid valve) • Damaged wiring harness and connectors • Malfunction of TCM
DTC: P0815 **1T ECM, MIL: Yes** **Year:** 2009, 2010 **Model:** Lancer, Outlander **Engine:** 2.0L L4 VIN U, 2.4L L4 VIN W, 3.0L V6 VIN X **Transmission:** All	**Malfunction of Paddle Shift Up Switch:** When the paddle shift UP switch being ON consecutively for 60 seconds is detected. TCM detects that the paddle shift UP switch is stuck to ON. **Possible Causes:** • Malfunction of paddle shift switch. • Damaged wiring harness and connectors. • Malfunction of TCM.
DTC: P0816 **1T ECM, MIL: Yes** **Year:** 2009, 2010 **Model:** Lancer, Outlander **Engine:** 2.0L L4 VIN U, 2.4L L4 VIN W, 3.0L V6 VIN X **Transmission:** All	**Malfunction of Paddle Shift Down Switch:** When the paddle shift DOWN switch being ON consecutively for 60 seconds is detected. TCM detects that the paddle shift DOWN switch is stuck to ON. **Possible Causes:** • Malfunction of paddle shift switch. • Damaged wiring harness and connectors. • Malfunction of TCM.
DTC: P0826 **1T ECM, MIL: Yes** **Year:** 2009, 2010 **Model:** Lancer, Outlander **Engine:** 2.0L L4 VIN U, 2.4L L4 VIN W, 3.0L V6 VIN X **Transmission:** All	**Malfunction of Shift Switch Assembly:** TCM detects the UP/DOWN operation of the shift switch assembly. TCM determines that the malfunction is present if an abnormal value is input. Input to the shift switch assembly remains abnormal for 1 second. **Possible Causes:** • Malfunction of selector lever assembly (Faulty shift switch assembly) • Damaged wiring harness and connectors • Malfunction of TCM
DTC: P0840 **1T ECM, MIL: Yes** **Year:** 2009, 2010 **Model:** Lancer, Outlander **Engine:** 2.0L L4 VIN U, 2.4L L4 VIN W **Transmission:** All	**Malfunction of Secondary Pressure Sensor:** Secondary pressure sensor voltage [transmission fluid temperature: more than -20°C (-4°F)]: 4.69 volts or more. (5 seconds) Or. Secondary pressure sensor voltage [transmission fluid temperature: more than -20°C (-4°F)]: 0.09 volt or less. (5 seconds) **Possible Causes:** • Malfunction of valve body assembly (Faulty secondary pressure sensor) • Damaged wiring harness and connectors • Malfunction of TCM

DTC	Trouble Code Title, Conditions & Possible Causes
DTC: P0841 **1T PCM, MIL:** Yes **Year:** 2009 **Model:** Raider **Engine:** 3.7L V6 VIN K **Transmission:** All	**LR Pressure Switch Rationality:** Continuously with the ignition on and engine running. The DTC will set if the L/R Pressure Switch reads open or closed at the wrong time in a given gear. **Possible Causes:** • Related TCM power input DTCs present • Low fluid level • Internal transmission mechanical problems may exist • OD Solenoid control circuit is shorted to other circuits • OD Solenoid control circuit is open or shorted to ground • Transmission Solenoid/TRS Assembly • PCM
DTC: P0841 **1T ECM, MIL:** Yes **Year:** 2009, 2010 **Model:** Lancer, Outlander **Engine:** 2.0L L4 VIN U, 2.4L L4 VIN W **Transmission:** All	**Abnormality in Line Pressure Sensor Function (CVT):** Transmission range switch position: D. Throttle position sensor voltage: 1.37 volts or more. Engine speed: 450 r/min or more. Transmission fluid temperature: 20°C (68°F) or more. Transmission fluid temperature: 180°C (356°F) or less. Voltage of battery: 9 volts or more. Voltage of battery: 16 volts or less. The actual secondary pressure reading is 0.675 MPa (98 psi) or more, and the status over the target line pressure remains for 5 seconds. **Possible Causes:** • Abnormal line pressure • Abnormality in secondary pressure sensor system • Malfunction of TCM
DTC: P0841 **1T ECM, MIL:** Yes **Year:** 2009, 2010 **Model:** Lancer **Engine:** 2.0L L4 VIN U, 2.4L L4 VIN W **Transmission:** All	**Clutch 1 Pressure Sensor System (Poor performance) (TC-SST):** Voltage of battery: 8 V or more. Voltage of battery: 16.5 V or less. Engine speed: 650 r/min or more. Time since above engine condition: 1.5 seconds or more. Engine speed: 6,800 r/min or less. Clutch 1 (odd) slip state: Slip or engaged. Clutch 1 (odd) slip speed: 20 r/min or more. Clutch 2 (even) state: Disengaged Calculated torque (Clutch 1 (odd) permit torque - engine torque): 200 Nm (148 ft. lbs.) or more. (4 seconds) **Possible Causes:** • Malfunction of TC-SST-ECU • Malfunction of clutch 1 pressure sensor • Malfunction of clutch assembly • Malfunction of engine system • Insufficient fluid
DTC: P0842 **1T ECM, MIL:** Yes **Year:** 2009, 2010 **Model:** Lancer **Engine:** 2.0L L4 VIN U, 2.4L L4 VIN W **Transmission:** All	**Clutch 1 Pressure Sensor System (Output low range out):** Voltage of battery: 8 V or more. Voltage of battery: 16.5 V or less. Sensor voltage: 1.16 V or less. (160 millisecond) The output of the clutch 1 pressure sensor is too low. **Possible Causes:** • Malfunction of TC-SST-ECU. • Malfunction of clutch 1 pressure sensor.
DTC: P0843 **1T ECM, MIL:** Yes **Year:** 2009, 2010 **Model:** Lancer **Engine:** 2.0L L4 VIN U, 2.4L L4 VIN W **Transmission:** All	**Clutch 1 Pressure Sensor System (Output high range out):** Voltage of battery: 8 V or more. Voltage of battery: 16.5 V or less. Sensor voltage: 2.48 V or more. (160 millisecond) The output of the clutch 1 pressure sensor is too high. **Possible Causes:** • Malfunction of TC-SST-ECU • Malfunction of clutch 1 pressure sensor

DTC	Trouble Code Title, Conditions & Possible Causes
DTC: P0845 **1T PCM, MIL: Yes** **Year:** 2009 **Model:** Raider **Engine:** 3.7L V6 VIN K **Transmission:** All	**2C Hydraulic Pressure Test (45RFE/545RFE Transmission):** In any forward gear with engine speed above 1000 rpm shortly after a shift and every minute thereafter. After a shift into a forward gear, with engine speed above 1000 rpm, the PCM momentarily turns on element pressure to the Clutch circuits that don't have pressure to identify the correct Pressure Switch closes. If the Pressure Switch does not close 2 times, the DTC sets. **Possible Causes:** • Line pressure sensor or connector • Transmission fluid contamination • Related DTCs present • Transmission Solenoid/TRS Assembly • Excessive debris in oil pan • TCM power input DTCs present • 5-volt supply circuit open • 2C Pressure switch is shorted to other circuits • 5-volt supply circuit short to ground • Internal transmission mechanical problems may exist • PCM
DTC: P0845 **1T PCM, MIL: Yes** **Year:** 2009 **Model:** Raider **Engine:** 3.7L V6 VIN K **Transmission:** All	**2/4 Hydraulic Pressure Test Malfunction (42RLE Transmission):** In any forward gear with engine speed above 1000 rpm, shortly after a shift and every minute thereafter. After a shift into a forward gear, with engine speed greater than 1000 rpm, the PCM momentarily turns on element pressure to the clutch circuits that don't have pressure to verify that the correct pressure switch closes. If the pressure switch does not close 2 times the DTC sets. **Possible Causes:** • Loss of prime P0944 present • Transmission control unit output circuit open • 2/4 Pressure control circuit open • 2/4 Pressure control circuit short to ground • 2/4 Pressure control circuit short to voltage • Transmission Solenoid/Pressure switch assembly • Internal transmission mechanical problems may exist • PCM
DTC: P0846 **1T ECM, MIL: Yes** **Year:** 2009, 2010 **Model:** Outlander **Engine:** 3.0L V6 VIN X **Transmission:** All	**2-6 Brake Pressure Switch System Malfunction:** Transmission range switch position: D or R. Gear position: 1st, 3rd, 4th, 5th, reverse Time after shift changing finish: 2 seconds or more. 2-6 brake pressure switch status: ON (2 seconds × 2times **Possible Causes:** • Malfunction of the 2-6 brake pressure switch system circuit • Damaged harness or connector • Malfunction of the TCM • Malfunction of the 2-6 brake pressure switch (valve body assembly) • Malfunction of the valve body assembly, abnormal hydraulic circuit valve.
DTC: P0846 **1T PCM, MIL: Yes** **Year:** 2009 **Model:** Raider **Engine:** 3.7L V6 VIN K **Transmission:** All	**2C Pressure Switch Rationality (45RFE/545RFE Transmission):** Continuously with the ignition on, engine running, with the transmission in gear. The DTC is set if the 2C Pressure Switch reads open or closed at the wrong time in a given gear. **Possible Causes:** • Related TCM power input DTCs present • Low line pressure • 2/C Pressure switch circuit is open or shorted to ground • 2/C Pressure switch circuit is shorted to system power • Transmission Solenoid/TRS Assembly • Internal transmission mechanical problems may exist • PCM
DTC: P0846 **1T ECM, MIL: Yes** **Year:** 2009, 2010 **Model:** Lancer **Engine:** 2.0L L4 VIN U, 2.4L L4 VIN W **Transmission:** All	**Clutch 2 Pressure Sensor System (Poor performance):** Voltage of battery: 8 V or more. Voltage of battery: 16.5 V or less. Engine speed: 650 r/min or more. Time since above engine condition: 1.5 seconds or more. Engine speed: 6,800 r/min or less. Clutch 2 (even) slip state: Slip or engaged. Clutch 2 (even) slip speed: 20 r/min or more. Clutch 1 (odd) state: Disengaged. Calculated torque (Clutch 2 (even) permit torque - engine torque): 200 Nm (148 ft-lb) or more. (4 seconds) **Possible Causes:** • Malfunction of TC-SST-ECU • Malfunction of clutch 2 pressure sensor • Malfunction of clutch assembly • Malfunction of engine system • Insufficient fluid

DTC	Trouble Code Title, Conditions & Possible Causes
DTC: P0846 **3T PCM, MIL: Yes** **Year:** 2009 **Model:** Raider **Engine:** 3.7L V6 VIN K **Transmission:** All	**2/4 Pressure Switch Rationality (42RLE Transmission):** Whenever the engine is running. The DTC is set if one of the pressure switches are open or closed at the wrong time in a given gear. If the problem is identified for 3 successive key starts, the transmission will go into Limp-in mode and the MIL will turn on after 10 seconds of vehicle operation. **Possible Causes:** • Related TCM power input DTCs present • Loss of prime DTC present • 2/4 Pressure signal circuit is open or shorted to ground • 2/4 Pressure signal circuit is shorted to system power • Transmission Solenoid/Pressure switch assembly • PCM has failed
DTC: P0847 **1T ECM, MIL: Yes** **Year:** 2009, 2010 **Model:** Lancer **Engine:** 2.0L L4 VIN U, 2.4L L4 VIN W **Transmission:** All	**Clutch 2 Pressure Sensor System (Output low range out):** Voltage of battery: 8 V or more. Voltage of battery: 16.5 V or less. The sensor voltage: 0.69 V or less. (160 millisecond) The output of the clutch 2 pressure sensor is too low. **Possible Causes:** • Malfunction of TC-SST-ECU • Malfunction of clutch 2 pressure sensor
DTC: P0848 **1T ECM, MIL: Yes** **Year:** 2009, 2010 **Model:** Lancer **Engine:** 2.0L L4 VIN U, 2.4L L4 VIN W **Transmission:** All	**Clutch 2 Pressure Sensor System (Output high range out):** Voltage of battery: 8 V or more. Voltage of battery: 16.5 V or less. The sensor voltage: 2.66 V or more. (160 millisecond) The output of the clutch 2 pressure sensor is too high. **Possible Causes:** • Malfunction of TC-SST-ECU • Malfunction of clutch 2 pressure sensor
DTC: P0850 **1T PCM, MIL: Yes** **Year:** 2009 **Model:** Raider **Engine:** 3.7L V6 VIN K **Transmission:** All	**Park/Neutral Switch Performance:** Continuously with the transmission in Park, Neutral, or Drive and NOT in Limp-in mode. This code will set if the PCM detects an incorrect Park/Neutral switch state for a given mode of vehicle operation. **Possible Causes:** • TRS SENSE (P/N SENSE) circuit open • TRS SENSE (P/N SENSE) circuit shorted to ground • Faulty TRS • PCM
DTC: P0868 **1T ECM, MIL: Yes** **Year:** 2009, 2010 **Model:** Lancer, Outlander **Engine:** 2.0L L4 VIN U, 2.4L L4 VIN W **Transmission:** All	**Secondary Pressure Drop:** Engine speed: 450 r/min or more. Voltage of battery: 9 volts or more. Voltage of battery: 16 volts or less. Secondary linear solenoid performance fail [target secondary pressure - measured secondary pressure: 0.25 MPa (36 psi) or more. (1.52 seconds)]: 2 count or more. **Possible Causes:** • Damaged wiring harness and connectors • Malfunction of CVT assembly • Malfunction of TCM
DTC: P0868 **1T PCM, MIL: Yes** **Year:** 2009 **Model:** Raider **Engine:** 3.7L V6 VIN K **Transmission:** All	**Line Pressure Low:** Continuously while driving in a forward gear. The PCM continuously monitors Actual Line Pressure and compares it to Desired Line Pressure. If the Actual Line Pressure is more than 15 psi below Desired Line Pressure, for 3.5 seconds continuously this DTC will set. **Possible Causes:** • Fluid level low • Related Relay DTCs may be present • 5v supply circuit is open or is shorted to ground or to voltage • Poor Line Pressure Sensor connection • Pressure Control Solenoid control circuit is shorted to voltage • Internal transmission problems • Line Pressure Sensor has failed • Plugged filter • PCM

DTC	Trouble Code Title, Conditions & Possible Causes
DTC: P0869 **1T PCM, MIL: Yes** **Year:** 2009 **Model:** Raider **Engine:** 3.7L V6 VIN K **Transmission:** All	**Line Pressure High:** Continuously while driving in a forward gear. The PCM continuously monitors Actual Line Pressure. If the Actual Line Pressure reading is greater than 827 kPa (120 psi) for 3.5 seconds continuously, while the Pressure Control Solenoid duty cycle is at or near its maximum value (which should result in minimum line pressure), the DTC will set. **Possible Causes:** • Related Relay DTCs may be present • 5V supply circuit is open or is shorted to ground • Poor Line Pressure Sensor connection • Pressure Control Solenoid control circuit is open or is shorted to ground • Internal transmission problems (line pressure high) • Line Pressure Sensor has failed • PCM has failed • Intermittent wiring and connector problems
DTC: P0870 **1T PCM** **Year:** 2009 **Model:** Raider **Engine:** 3.7L V6 VIN K **Transmission:** All	**OD Hydraulic Pressure Test:** In any forward gear with engine speed above 1000 rpm shortly after a shift and every minute thereafter. After a shift into a forward gear, with engine speed above 1000 rpm, the PCM momentarily turns on element pressure to the Clutch circuits that don't have pressure to identify the correct Pressure Switch closes. If the Pressure Switch does not close 2 times, the DTC sets. **Possible Causes:** • Check for related line pressure trouble codes • Check for related speed ratio and pressure switch codes • 5V supply circuit is open or is shorted to ground • Transmission Control Relay output circuit is open • OD Pressure Switch sense circuit is shorted to ground or to voltage • Excessive debris in the oil pan • Internal transmission problems • Line pressure Sensor connector is loose or damaged • Oil pressure switch is damaged or has failed • Intermittent wiring or connector problems exist • PCM has failed
DTC: P0871 **1T PCM, MIL: Yes** **Year:** 2009 **Model:** Raider **Engine:** 3.7L V6 VIN K **Transmission:** All	**O/D Pressure Switch Rationality:** Whenever the engine is running. The DTC is set if one of the pressure switches are open or closed at the wrong time in a given gear. If the problem is identified for 3 successive key starts, the transmission will go into Limp-in mode and the MIL will turn on after 10 seconds of vehicle operation. **Possible Causes:** • Related TCM power input DTCs present • Loss of prime DTC present • OD Pressure signal circuit open • OD Pressure signal circuit short to ground • OD Pressure signal circuit short to voltage • Transmission Solenoid/Pressure switch assembly • Internal transmission mechanical problems may exist • PCM
DTC: P0875 **1T PCM, MIL: Yes** **Year:** 2009 **Model:** Raider **Engine:** 3.7L V6 VIN K **Transmission:** All	**U/D Hydraulic Pressure Test :** In any forward gear with engine speed above 1000 rpm shortly after a shift and every minute thereafter. After a shift into a forward gear, with engine speed above 1000 rpm, the PCM momentarily turns on element pressure to the Clutch circuits that don't have pressure to identify the correct Pressure Switch closes. If the Pressure Switch does not close 2 times, the DTC sets. **Possible Causes:** • Fluid low or contamination • Related Relay DTCs may be present • 5v supply circuit is open or is shorted to ground or to voltage • Poor Line Pressure Sensor connection • Pressure Control Solenoid control circuit is shorted to voltage • Internal transmission problems • Transmission solenoids/TRS assembly is damaged or has failed • Line Pressure Sensor has failed • Plugged filter • PCM

DTC	Trouble Code Title, Conditions & Possible Causes
DTC: P0876 **1T ECM, MIL: Yes** **Year:** 2009, 2010 **Model:** Outlander **Engine:** 3.0L V6 VIN X **Transmission:** All	**High Clutch Pressure Switch System:** Check Conditions <Stuck-off>:Transmission range switch position: D. Gear position: 4th, 5th, 6th. Time after shift changing finish: 2 seconds or more. Check Conditions <Stuck-on>: Transmission range switch position: D or R. Gear position: 1st, 2nd, 3rd, reverse. Time after shift changing finish: 2 seconds or more. High clutch pressure switch status: ON/Off. (2 seconds × 2times) **Possible Causes:** • Malfunction of the high clutch pressure switch system circuit • Damaged harness or connector • Malfunction of the TCM • Malfunction of the high clutch pressure switch system (valve body assembly) • Malfunction of the valve body assembly, abnormal hydraulic circuit valve.
DTC: P0876 **1T PCM, MIL: Yes** **Year:** 2009 **Model:** Raider **Engine:** 3.7L V6 VIN K **Transmission:** All	**U/D Pressure Switch Rationality:** Continuously with the ignition on and engine running. This DTC is set if the UD pressure switch is in the wrong state for the current gear. For example, this code would be set if the UD pressure switch remained off while the transmission was in second gear. **Possible Causes:** • Fluid level low • Related Relay DTCs may be present • 5v supply circuit is open or is shorted to ground or to voltage • Poor Line Pressure Sensor connection • Pressure Control Solenoid control circuit is shorted to voltage • Internal transmission problems • Transmission solenoids/TRS assembly is damaged or has failed • Line Pressure Sensor has failed • Plugged filter • PCM
DTC: P0882 **1T PCM** **Year:** 2009 **Model:** Raider **Engine:** 3.7L V6 VIN K **Transmission:** All	**TCM Power Input Low (42RLE Transmission):** This DTC is set when the Transmission Control Relay output circuit voltage at the Powertrain Control Module is less than 3 volts when the PCM is energizing the relay. **NOTE: Due to the integration of the Powertrain and Transmission Control Modules, the transmission part of the PCM has its own specific power and ground circuits.** **Possible Causes:** • Fused B (+) circuit open • Transmission control output circuit open • Transmission control circuit open • Ground circuit open • Transmission control circuit shorted to ground • Transmission control output circuit shorted to ground • Transmission control relay • Transmission Solenoids/Pressure switch assembly • PCM
DTC: P0882 **1T ECM, MIL: Yes** **Year:** 2009, 2010 **Model:** Lancer, Outlander **Engine:** 2.0L L4 VIN U, 2.4L L4 VIN W **Transmission:** All	**Malfunction of power supply system (Low):** When the engine speed is 450 r/min or more and secondary pressure is above 0.3 MPa (44 psi), the power supply voltage is 9.0 volts or less (for 5 seconds) The TCM monitors if power supply system is normal or not and judges the trouble. **Possible Causes:** • Damaged wiring harness and connectors • Malfunction of TCM
DTC: P0882 **1T PCM, MIL: Yes** **Year:** 2009 **Model:** Raider **Engine:** 3.7L V6 VIN K **Transmission:** All	**TCM Power Input Low (45RFE/545RFE Transmission):** When the ignition is turned from "OFF" position to "RUN" position and/or the ignition is turned from "START" position to "RUN" position. This DTC is set when less than 3.0 volts are present at the transmission control relay output circuits at the Powertrain Control Module when the PCM is energizing the relay. **Possible Causes:** • FUSED B(+) circuit open • Transmission control relay output open or shorted to ground • Transmission control relay control circuit open or shorted to ground • Transmission control relay ground circuit open • Transmission control relay stuck open • PCM

DTC	Trouble Code Title, Conditions & Possible Causes
DTC: P0883 **1T PCM, MIL: Yes** **Year:** 2009 **Model:** Raider **Engine:** 3.7L V6 VIN K **Transmission:** All	**TCM Power Input High (42RLE Transmission):** After a reset (ignition key turned to the RUN position) and after a power down. Relay output (Switched Battery) is higher than 3 volts when relay is not energized by the controller. Fault Sct Time: Less than 100 msec. **Possible Causes:** • Transmission control relay stuck closed • Transmission control circuit shorted to voltage • Transmission control output circuit shorted to voltage • PCM
DTC: P0883 **1T ECM, MIL: Yes** **Year:** 2009, 2010 **Model:** Outlander **Engine:** 2.4L L4 VIN W **Transmission:** All	**Malfunction of power supply system (High):** When the engine speed is 450 r/min or more, the secondary pressure is above 0.3 MPa (44 psi) and the vehicle speed is 1 km/h (0.6 mph) or more, the power supply voltage is 16.0 volts or more. (for 5 seconds) **Possible Causes:** • Damaged wiring harness and connectors • Malfunction of TCM
DTC: P0883 **1T PCM, MIL: Yes** **Year:** 2009 **Model:** Raider **Engine:** 3.7L V6 VIN K **Transmission:** All	**TCM Power Input High (45RFE/545RFE Transmission):** When the ignition is turned from "OFF" position to "RUN" position and/or the ignition is turned from "START" position to "RUN" position. This DTC is set if the Powertrain Control Module senses greater than 3.0 volts on the Transmission Control Relay Output circuits prior to a request from the PCM to energize the Transmission Control Relay. **Possible Causes:** • Transmission control relay control circuit shorted to voltage • Transmission control relay output circuit shorted to voltage • Transmission control relay • Transmission solenoids/TRS assembly • PCM
DTC: P0884 **1T PCM, MIL: Yes** **Year:** 2009 **Model:** Raider **Engine:** 3.7L V6 VIN K **Transmission:** All	**Power-Up At Speed (42RLE Transmission):** One time after each controller reset. **NOTE: the Transmission Control Module is integrated with Powertrain Control Module. The Transmission Control Module has separate powers and grounds specifically to its portion of the PCM. This DTC will set if the PCM powers up and senses the vehicle in a valid forward gear (no PRNDL DTCs) with a output speed above 800 rpm, approximately 32 Km/h or 20 mph.** **Possible Causes:** • Intermittent power and ground circuit
DTC: P0884 **1T PCM, MIL: Yes** **Year:** 2009 **Model:** Raider **Engine:** 3.7L V6 VIN K **Transmission:** All	**Power-Up At Speed (45RFE/545RFE Transmission):** When Powertrain Control Module initially powers up. Due to the integration of the Powertrain and Transmission Control Modules, the transmission part of the PCM has its own specific power and ground circuits. This DTC will set if the PCM powers up and senses the vehicle in a valid forward gear, with no PRNDL DTCs, and an output speed above 800 rpm, approximately 32Km/h or 20 mph. **Possible Causes:** • Power-Up at speed
DTC: P0890 **1T PCM, MIL: Yes** **Year:** 2009 **Model:** Raider **Engine:** 3.7L V6 VIN K **Transmission:** All	**Switched Battery (42RLE Transmission):** One time after a reset (ignition key turned to the RUN position or after cranking engine). A fault is set if voltage greater than 4.5 volts is detected for 7 msec on any of the pressure switch circuits before the relay is energized. The transmission is placed in Limp-In. The MIL is on after 10 seconds. of vehicle operation. **Possible Causes:** • 2/4 Pressure signal circuit short to voltage • L/R Pressure signal circuit short to voltage • OD Pressure signal circuit short to voltage • PCM
DTC: P0890 **1T PCM, MIL: Yes** **Year:** 2009 **Model:** Raider **Engine:** 3.7L V6 VIN K **Transmission:** All	**Switched Battery (45RFE/545RFE Transmission):** When the ignition is turned from "OFF" position to "RUN" position and/or the ignition is turned from "START" position to "RUN" position. This DTC is set if the Powertrain Control Module senses voltage on any of the pressure switch inputs prior to the Transmission Control Relay being energized. **Possible Causes:** • Pressure switch sense circuits Short to voltage • Transmission solenoids/TRS assembly • PCM

DTC	Trouble Code Title, Conditions & Possible Causes
DTC: P0893 **1T ECM, MIL: Yes** **Year:** 2009, 2010 **Model:** Outlander **Engine:** 3.0L V6 VIN X **Transmission:** All	**Interlock Detection Malfunction:** Fluid pressure switch status combination: not designed combination. Stoplight switch signal: OFF. Deceleration: more than threshold value. (1.9 second) **Possible Causes:** • Malfunction of the P0720: output shaft speed sensor system circuit • Damaged harness or connector • Malfunction of the P0743: lock-up and low-reverse brake linear solenoid valve system circuit • Malfunction of the P0753: low clutch linear solenoid valve system circuit • Malfunction of the P0758: 2-6 brake linear solenoid valve system circuit • Malfunction of the P0763: 3-5 reverse clutch linear solenoid valve system circuit • Malfunction of the P0768: high clutch linear solenoid valve system circuit • Malfunction of the P1753: low clutch shift solenoid valve system circuit • Malfunction of the P1758: low-reverse brake shift solenoid valve system circuit • Malfunction of the P0846: 2-6 brake pressure switch system circuit • Malfunction of the P0876: high clutch pressure switch system circuit • Malfunction of the P0988: low-reverse brake pressure switch system circuit
DTC: P0897 **1T PCM, MIL: Yes** **Year:** 2009 **Model:** Raider **Engine:** 3.7L V6 VIN K **Transmission:** All	**Transmission Fluid Deterioration:** Each transition from full EMCC to partial EMCC for A/C bump prevention. DTC set if 20 occurrences of a turbine acceleration sum. Fault Set Time: 20 transitions from full EMCC to partial EMCC. Transmission will not use partial EMCC. Established for A/C bump prevention. **Possible Causes:** • Transmission fluid burnt or worn out
DTC: P0930 **T PCM** **Year:** 2009 **Model:** Raider **Engine:** 3.7L V6 VIN K **Transmission:** All	**BTSI Control Circuit Low:** With the ignition in the Unlock or Run position and the driver's foot applied to the brake pedal. The Instrument Cluster detects a low value on the BTSI Solenoid Control circuit. **Possible Causes:** • BTSI Solenoid control circuit shorted to the BTSI solenoid ground circuit • BTSI Solenoid control circuit shorted to ground • Transmission unlock solenoid (BTSI) • Instrument cluster
DTC: P0931 **T PCM** **Year:** 2009 **Model:** Raider **Engine:** 3.7L V6 VIN K **Transmission:** All	**BTSI Control Circuit High:** With the ignition in the Unlock or Run position and the driver's foot applied to the brake pedal. The Instrument Cluster detects a high value on the BTSI Solenoid Control circuit. **Possible Causes:** • BTSI Solenoid ground circuit open • BTSI Solenoid control circuit open • Transmission unlock solenoid (BTSI) • Instrument cluster
DTC: P0933 **T PCM** **Year:** 2009 **Model:** Raider **Engine:** 3.7L V6 VIN K **Transmission:** All	**Hydraulic Pressure Sensor Range/Performance:** Continuously with the ignition on, engine running, with the transmission in gear. The PCM continuously monitors Actual Line Pressure and compares it to Desired Line Pressure. If the Actual Line Pressure reading is more than 172.4 kPa (25 psi) higher than the Desired Line Pressure, but is less than the highest Line Pressure ever used in the current gear, the DTC sets. **Possible Causes:** • Related DTC present • Line pressure connector and wiring • Line pressure sensor • Internal transmission problems • PCM
DTC: P0934 **T PCM, MIL: Yes** **Year:** 2009 **Model:** Raider **Engine:** 3.7L V6 VIN K **Transmission:** All	**Line Pressure Sensor Low:** Continuously with the ignition on and engine running. This DTC will set when the monitored Line Pressure Sensor voltage is less than or equal to 0.35 volts for 0.18 seconds. **Possible Causes:** • 5-Volt supply circuit open • 5-Volt supply circuit shorted to ground • Line pressure sensor signal circuit shorted to ground • Line pressure sensor • PCM

DTC	Trouble Code Title, Conditions & Possible Causes
DTC: P0935 **T PCM, MIL: Yes** **Year:** 2009 **Model:** Raider **Engine:** 3.7L V6 VIN K **Transmission:** All	**Line Pressure Sensor High:** Continuously with ignition on and engine running. This DTC will set if the monitored Line Pressure Sensor voltage is greater than or equal to 4.75 volts for the period of 0.18 seconds **Possible Causes:** • Line pressure sensor signal circuit open • Sensor ground circuit open • Line pressure sensor signal circuit shorted to voltage • Line pressure sensor • PCM
DTC: P0944 **T PCM, MIL: Yes** **Year:** 2009 **Model:** Raider **Engine:** 3.7L V6 VIN K **Transmission:** All	**Loss Of Hydraulic Pump Prime (42RLE Transmission):** Set every 350 msec. If the transmission begins to slip in any forward gear, and the pressure switch or switches that should be closed for a given gear are open, a loss of prime test begins. All available elements (in 1st gear L/R, 2/4 and OD, in 2nd, 3rd, and 4th gear 2/4 and OD) are turned on by the PCM to see if pump prime exists. The code is set if none of the pressure switches respond. The PCM will continue to run the loss of prime test until pump pressure returns. The vehicle will not move or the transmission will slip. Normal operation will continue if pump prime returns. **Possible Causes:** • Low transmission fluid level • Shift lever position • Plugged transmission filter • Transmission oil pump
DTC: P0944 **T PCM, MIL: Yes** **Year:** 2009 **Model:** Raider **Engine:** 3.7L V6 VIN K **Transmission:** All	**Loss Of Hydraulic Pump Prime (45RFE/545RFE Transmission):** If the transmission is slipping in any forward gear and all the pressure switches are not indicating pressure, a loss of prime test is run. If the transmission begins to slip in a forward gear and all the pressure switch(s) that should be closed are open a loss of prime test begins. Available elements are turned on by the PCM to see if pump prime exists. The DTC sets if no pressure switch(s) respond. **Possible Causes:** • Shift lever out of adjustment • Improper fluid level • Cracked or improperly installed primary oil filter or seal • Loose cooler return filter • Stuck or sticking main regulator valve • Transmission oil pump
DTC: P0960 **1T ECM, MIL: Yes** **Year:** 2009, 2010 **Model:** Lancer **Engine:** 2.0L L4 VIN U, 2.4L L4 VIN W **Transmission:** All	**Line Pressure Solenoid System (Open circuit):** Voltage of battery: 8 V or more. Voltage of battery: 16.5 V or less. FET (Field Effect Transistor) output: 1 V or more. (500 millisecond) The line pressure solenoid circuit is determined to be open. **Possible Causes:** • Malfunction of TC-SST-ECU • Malfunction of line pressure solenoid
DTC: P0961 **1T ECM, MIL: Yes** **Year:** 2009, 2010 **Model:** Lancer **Engine:** 2.0L L4 VIN U, 2.4L L4 VIN W **Transmission:** All	**Line Pressure Solenoid System (Overcurrent):** Voltage of battery: 8 V or more. Voltage of battery: 16.5 V or less. FET (Field Effect Transistor) current: 3.5 A or more, and FET (Field Effect Transistor) output: 100 mV or less (300 millisecond). **Possible Causes:** • Malfunction of TC-SST-ECU • Malfunction of line pressure solenoid
DTC: P0962 **1T ECM, MIL: Yes** **Year:** 2009, 2010 **Model:** Lancer **Engine:** 2.0L L4 VIN U, 2.4L L4 VIN W **Transmission:** All	**Line Pressure Solenoid System (Short to ground):** Voltage of battery: 8 V or more. Voltage of battery: 16.5 V or less. FET (Field Effect Transistor) output: 100 mV or less. (12 seconds) **Possible Causes:** • Malfunction of TC-SST-ECU • Malfunction of line pressure solenoid
DTC: P0963 **1T ECM, MIL: Yes** **Year:** 2009, 2010 **Model:** Lancer **Engine:** 2.0L L4 VIN U, 2.4L L4 VIN W **Transmission:** All	**Line Pressure Solenoid System (Short to power supply):** Voltage of battery: 8 V or more. Voltage of battery: 16.5 V or less. Line pressure solenoid: OFF. FET (Field Effect Transistor) output: (Battery voltage - 2 V) or more. (1 second) **Possible Causes:** • Malfunction of TC-SST-ECU • Malfunction of line pressure solenoid

DTC	Trouble Code Title, Conditions & Possible Causes
DTC: P0964 **1T ECM, MIL: Yes** **Year:** 2009, 2010 **Model:** Lancer **Engine:** 2.0L L4 VIN U, 2.4L L4 VIN W **Transmission:** All	**Clutch Cooling Flow Solenoid System (Open circuit):** Voltage of battery: 8 V or more. Voltage of battery: 16.5 V or less. FET (Field Effect Transistor) output: 1 V or more. (180 millisecond) The clutch cooling flow solenoid circuit is determined to be open. **Possible Causes:** • Malfunction of TC-SST-ECU • Malfunction of clutch cooling flow solenoid
DTC: P0965 **1T ECM, MIL: Yes** **Year:** 2009, 2010 **Model:** Lancer **Engine:** 2.0L L4 VIN U, 2.4L L4 VIN W **Transmission:** All	**Clutch Cooling Flow Solenoid System (Overcurrent):** Voltage of battery: 8 V or more. Voltage of battery: 16.5 V or less. FET (Field Effect Transistor) current: 3.5 A or more, and FET (Field Effect Transistor) output: 100 mV or less (300 millisecond). **Possible Causes:** • Malfunction of TC-SST-ECU • Malfunction of clutch cooling flow solenoid
DTC: P0966 **1T ECM, MIL: Yes** **Year:** 2009, 2010 **Model:** Lancer **Engine:** 2.0L L4 VIN U, 2.4L L4 VIN W **Transmission:** All	**Clutch Cooling Flow Solenoid System (Short to ground):** Voltage of battery: 8 V or more. Voltage of battery: 16.5 V or less. FET (Field Effect Transistor) output: 100 mV or less. (180 millisecond) The clutch cooling flow solenoid circuit is determined to be short to ground **Possible Causes:** • Malfunction of TC-SST-ECU • Malfunction of clutch cooling flow solenoid
DTC: P0967 **1T ECM, MIL: Yes** **Year:** 2009, 2010 **Model:** Lancer **Engine:** 2.0L L4 VIN U, 2.4L L4 VIN W **Transmission:** All	**Clutch Cooling Flow Solenoid System (Short to power supply):** Voltage of battery: 8 V or more. Voltage of battery: 16.5 V or less. Clutch cooling flow solenoid: OFF. FET (Field Effect Transistor) output: (Battery voltage - 2 V) or more. (5 seconds) **Possible Causes:** • Malfunction of TC-SST-ECU • Malfunction of clutch cooling flow solenoid
DTC: P0968 **1T ECM, MIL: Yes** **Year:** 2009, 2010 **Model:** Lancer **Engine:** 2.0L L4 VIN U, 2.4L L4 VIN W **Transmission:** All	**Shift/Cooling Switching Solenoid System (Open circuit):** Voltage of battery: 8 V or more. Voltage of battery: 16.5 V or less. FET (Field Effect Transistor) output: 1 V or more. (220 millisecond) **Possible Causes:** • Malfunction of TC-SST-ECU • Malfunction of shift/cooling switching solenoid
DTC: P0970 **1T ECM, MIL: Yes** **Year:** 2009, 2010 **Model:** Lancer **Engine:** 2.0L L4 VIN U, 2.4L L4 VIN W **Transmission:** All	**Shift/Cooling Switching Solenoid System (Short to ground):** Voltage of battery: 8 V or more. Voltage of battery: 16.5 V or less. FET (Field Effect Transistor) output: 100 mV or less. (220 millisecond) The shift/cooling switching solenoid circuit is determined to be short to ground. **Possible Causes:** • Malfunction of TC-SST-ECU • Malfunction of shift/cooling switching solenoid
DTC: P0971 **1T ECM, MIL: Yes** **Year:** 2009, 2010 **Model:** Lancer **Engine:** 2.0L L4 VIN U, 2.4L L4 VIN W **Transmission:** All	**Shift/Cooling Switching Solenoid System (Short to power supply):** Voltage of battery: 8 V or more. Voltage of battery: 16.5 V or less. Shift/cooling switching solenoid: OFF. FET (Field Effect Transistor) output: (Battery voltage - 2 V) or more. (220 millisecond) The shift/cooling switching solenoid circuit is determined to be short to power supply. **Possible Causes:** • Malfunction of TC-SST-ECU • Malfunction of shift/cooling switching solenoid
DTC: P0973 **1T ECM, MIL: Yes** **Year:** 2009, 2010 **Model:** Lancer **Engine:** 2.0L L4 VIN U, 2.4L L4 VIN W **Transmission:** All	**Shift Select Solenoid 1 System (Short to ground):** Voltage of battery: 8 V or more. Voltage of battery: 16.5 V or less. FET (Field Effect Transistor) output: 100 mV or less. (200 millisecond) The shift select solenoid 1 circuit is determined to be short to ground. **Possible Causes:** • Malfunction of TC-SST-ECU • Malfunction of shift select solenoid 1

DTC	Trouble Code Title, Conditions & Possible Causes
DTC: P0974 **1T ECM, MIL: Yes** **Year:** 2009, 2010 **Model:** Lancer **Engine:** 2.0L L4 VIN U, 2.4L L4 VIN W **Transmission:** All	**Shift Select Solenoid 1 System (Short to power supply):** Voltage of battery: 8 V or more. Voltage of battery: 16.5 V or less. Shift select solenoid 1: OFF. FET (Field Effect Transistor) output: (Battery voltage - 2 V) or more. (160 millisecond) **Possible Causes:** • Malfunction of TC-SST-ECU • Malfunction of shift select solenoid 1
DTC: P0976 **1T ECM, MIL: Yes** **Year:** 2009, 2010 **Model:** Lancer **Engine:** 2.0L L4 VIN U, 2.4L L4 VIN W **Transmission:** All	**Shift Select Solenoid 2 System (Short to ground):** Voltage of battery: 8 V or more. Voltage of battery: 16.5 V or less. FET (Field Effect Transistor) output: 100 mV or less. (200 millisecond) The shift select solenoid 2 circuit is determined to be short to ground. **Possible Causes:** • Malfunction of TC-SST-ECU • Malfunction of shift select solenoid 2
DTC: P0977 **1T ECM, MIL: Yes** **Year:** 2009, 2010 **Model:** Lancer **Engine:** 2.0L L4 VIN U, 2.4L L4 VIN W **Transmission:** All	**Shift Select Solenoid 2 System (Short to power supply):** The shift select solenoid 2 circuit is determined to be short to power supply. **NOTE: If there is any problem in the CAN bus lines, an incorrect diagnostic trouble code may be set. Prior to this diagnosis, diagnose the CAN bus lines.** **Whenever the ECU is replaced, ensure that the CAN bus lines are normal.** **Possible Causes:** • Malfunction of TC-SST-ECU • Malfunction of shift select solenoid 2
DTC: P0987 **T PCM, MIL: Yes** **Year:** 2009 **Model:** Raider **Engine:** 3.7L V6 VIN K **Transmission:** All	**4C Hydraulic Pressure Test :** In any forward gear with engine speed above 1000 rpm shortly after a shift and every minute thereafter. After a shift into a forward gear, with engine speed above 1000 rpm, the PCM momentarily turns on element pressure to the Clutch circuits that don't have pressure to identify the correct Pressure Switch closes. If the Pressure Switch does not close 2 times, the DTC sets. **Possible Causes:** • Check for related speed ratio and pressure switch codes • Excessive debris in the oil pan • Contaminated transmission fluid • 5-Volt supply circuit is open or is shorted to ground • Transmission Control Relay output circuit is open • 4C Pressure Switch sense circuit is shorted to ground or to voltage • Line Pressure Sensor connector is loose or damaged • 4C Line Pressure Sensor has failed • Transmission Solenoid/TRS Assembly has failed • Internal transmission problems • PCM has failed • Intermittent wiring and connector problems
DTC: P0988 **T PCM, MIL: Yes** **Year:** 2009 **Model:** Raider **Engine:** 3.7L V6 VIN K **Transmission:** All	**4C Pressure Switch Rationality:** Continuously with the ignition on, engine running, with the transmission in gear. The DTC is set if the 4C Pressure Switch reads open or closed at the wrong time in a given gear. **Possible Causes:** • Related DTCs may be present • DTC P0871 also present • Loss or 12 volt feed • 4C Pressure Switch sense circuit is open, shorted to ground or power • Excessive fluid leakage with 4C clutch circuit • Check ball cut or damaged • Low line pressure • Transmission Solenoid/TRS Assembly has failed • PCM has failed • Intermittent wiring and connector problems

DTC	Trouble Code Title, Conditions & Possible Causes
DTC: P0988 **1T ECM, MIL: Yes** **Year:** 2009, 2010 **Model:** Outlander **Engine:** 3.0L V6 VIN X **Transmission:** All	**Low-Reverse Brake Pressure Switch System:** Check Conditions <Stuck off>: Transmission range switch position: R. Gear position: reverse. Time after shift changing finish: 2 seconds or more. Check Conditions <Stuck-on>: Transmission range switch position: D. Gear position: 1st, 2nd, 3rd, 4th, 5th, 6th. Time after shift changing finish: 2 seconds or more. Low-reverse brake pressure switch status: ON/OFF (2 seconds × 2times). **Possible Causes:** • Malfunction of the low-reverse brake pressure switch system circuit • Damaged harness or connector • Malfunction of the TCM • Malfunction of the low-reverse brake pressure switch system (valve body assembly) • Malfunction of the valve body assembly, abnormal hydraulic circuit valve.
DTC: P0992 **T PCM, MIL: Yes** **Year:** 2009 **Model:** Raider **Engine:** 3.7L V6 VIN K **Transmission:** All	**2/4 & O/D Hydraulic Pressure Test :** In any forward gear with engine speed above 1000 rpm, shortly after a shift and every minute thereafter. After a shift into a forward gear, with engine speed greater than 1000 rpm, the PCM momentarily turns on element pressure to the clutch circuits that do not have pressure to identify that the correct pressure switch closes. If the pressure switch does not close 2 times the DTC sets. **Possible Causes:** • 2/4 pressure switch circuit is open, shorted to ground or power • 2/4 pressure switch is damaged or it has failed • O/D pressure switch circuit is open, shorted to ground or power • O/D pressure switch is damaged or it has failed • Internal transmission faults exist • TCM relay power circuit to 2/4 or O/D switch open (loss of B+) • PCM has failed

OBD II Trouble Code List (P1XXX Codes)

DTC	Trouble Code Title, Conditions & Possible Causes
DTC: P1115 **2T PCM, MIL: Yes** **Year:** 2009 **Model:** Raider **Engine:** 3.7L V6 VIN K **Transmission:** All	**General Temperature Sensor Rationality:** Engine off time is greater than 480 minutes and the vehicle has been driven for one minute over 35 mph. Ambient temperature is greater than -64°C (-83°F). Once the vehicle is soaked for a calibrated engine off time and then driven over calibrated speed and load conditions for some calibrated time, the PCM compares the ambient air, engine coolant, and intake air temperature sensor values. If the values of all the three sensors disagree with one another, a general temperature sensor irrationality is declared. **Possible Causes:** • Excessive resistance in the sensor signal circuit • Excessive resistance in the sensor ground circuit • Sensor signal circuit shorted to ground • Sensor signal circuit shorted to voltage • Sensor signal circuit shorted to the sensor ground circuit • Temperature sensor • PCM
DTC: P1128 **2T PCM, MIL: Yes** **Year:** 2009 **Model:** Raider **Engine:** 3.7L V6 VIN K **Transmission:** All	**Closed Loop Fuelling Not Achieved (BANK 1):** Engine running in closed loop mode. Enable conditions are met and the 02 sensor has not been in closed loop control at least once on each of the two consecutive trips, the MIL illuminates and the DTC is set. **Possible Causes:** • Fuel delivery system • ECT sensor, wiring or connectors • MAP sensor, wiring or connectors • 02 sensor, wiring or connectors • Engine mechanical • PCM
DTC: P1129 **2T PCM, MIL: Yes** **Year:** 2009 **Model:** Raider **Engine:** 3.7L V6 VIN K **Transmission:** All	**Closed Loop Fuelling Not Achieved (BANK 2):** Engine running in closed loop mode. Enable conditions are met and the 02 sensor has not been in closed loop control at least once on each of the two consecutive trips, the MIL illuminates and the DTC is set. **Possible Causes:** • Fuel delivery system • ECT sensor, wiring or connectors • MAP sensor, wiring or connectors • 02 sensor, wiring or connectors • Engine mechanical • PCM

DTC	Trouble Code Title, Conditions & Possible Causes
DTC: P1231 **2T ECM, MIL: Yes** **Year:** 2009, 2010 **Model:** Lancer, Outlander **Engine:** 2.0L L4 VIN U, 2.0L L4 VIN V, 2.4L L4 VIN W, 3.0L V6 VIN X **Transmission:** All	**Active Stability Control Plausibility:** Checks for an abnormal signal of active stability control (ASC) via the CAN communication. Ignition switch is "ON" position. A torque demand signal from the active stability control is not normal. **Possible Causes:** • Active stability control system failed. • ECM failed.
DTC: P1232 **1T ECM, MIL: Yes** **Year:** 2009, 2010 **Model:** Lancer, Outlander **Engine:** 2.0L L4 VIN U, 2.0L L4 VIN V, 2.4L L4 VIN W, 3.0L V6 VIN X **Transmission:** All	**Fail Safe System:** Ignition switch is "ON" position. Power supply to the throttle actuator control motor cannot be shut down (though power supply is stopped). **Possible Causes:** • Throttle actuator control motor relay circuit failed. • ECM failed.
DTC: P1233 **1T ECM, MIL: Yes** **Year:** 2009, 2010 **Model:** Lancer, Outlander **Engine:** 2.0L L4 VIN U, 2.4L L4 VIN W, 3.0L V6 VIN X **Transmission:** All	**Throttle Position Sensor (main) Plausibility:** The difference between the actual volumetric efficiency and the volumetric efficiency estimated by the throttle position sensor (main) is 0 percent or more. Or, the volumetric efficiency is 60 percent or less. The engine speed is between 750 and 3,000 r/min. Or, the throttle position sensor (main) output voltage is 3 volts or less. For 0.4 second, the difference between the actual volumetric efficiency and the volumetric efficiency estimated by the throttle position sensor (main) is 33 percent or more. **Possible Causes:** • Throttle position sensor (main) system failed. • Intake system vacuum leak. • There is some trash around mass airflow sensor. • ECM failed.
DTC: P1233 **1T ECM, MIL: Yes** **Year:** 2009, 2010 **Model:** Lancer **Engine:** 2.0L L4 VIN V **Transmission:** All	**Throttle Position Sensor (main) Plausibility:** The engine speed is between 750 and 3,000 r/min. Or, the throttle position sensor (main) output voltage is less than 3 volts. Except while fuel is being shut off. For 0.8 second, the difference between the actual volumetric efficiency and the volumetric efficiency estimated by the throttle position sensor (main) is more than 43 percent. **Possible Causes:** • Throttle opening degree is restricted. • Throttle opening degree position is in default position if throttle position sensor (sub) fails.
DTC: P1234 **1T ECM, MIL: Yes** **Year:** 2009, 2010 **Model:** Lancer **Engine:** 2.0L L4 VIN V **Transmission:** All	**Throttle Position Sensor (Sub) Plausibility:** The engine speed is between 750 and 3,000 r/min. Or, the throttle position sensor (main) output voltage is less than 3 volts. Except while fuel is being shut off. For 0.8 second, the difference between the actual volumetric efficiency and the volumetric efficiency estimated by the throttle position sensor (sub) is more than 43 percent **Possible Causes:** • Throttle opening degree is restricted. • Throttle opening degree position is in default position if throttle position sensor (main) fails.
DTC: P1234 **1T ECM, MIL: Yes** **Year:** 2009, 2010 **Model:** Lancer, Outlander **Engine:** 2.0L L4 VIN U, 2.4L L4 VIN W, 3.0L V6 VIN X **Transmission:** All	**Throttle Position Sensor (Sub) Plausibility:** The difference between the actual volumetric efficiency and the volumetric efficiency estimated by the throttle position sensor (sub) is 0 percent or more. Or, the volumetric efficiency is 60 percent or less. The engine speed is between 750 and 3,000 r/min. Or, the throttle position sensor (main) output voltage is 3 volts or less. For 0.4 second, the difference between the actual volumetric efficiency and the volumetric efficiency estimated by the throttle position sensor (sub) is 33 percent or more. **Possible Causes:** • Throttle position sensor (sub) system failed. • Intake system vacuum leak. • There is some trash around mass airflow sensor. • ECM failed.

DTC	Trouble Code Title, Conditions & Possible Causes
DTC: P1235 **1T ECM, MIL: Yes** **Year:** 2009, 2010 **Model:** Lancer, Outlander **Engine:** 2.0L L4 VIN U, 2.0L L4 VIN V, 2.4L L4 VIN W, 3.0L V6 VIN X **Transmission:** All	**Mass Airflow Sensor Plausibility:** The plausibility error of the throttle position sensor (main) is detected. The plausibility error of the throttle position sensor (sub) is detected. For 0.36 second, the difference between the volumetric efficiency estimated by the throttle position sensor (main) and the volumetric efficiency estimated by the throttle position sensor (sub) is 8.8 percent or less. **Possible Causes:** • Mass airflow sensor system failed. • Intake system vacuum leak. • There is some trash around mass airflow sensor. • ECM failed.
DTC: P1236 **1T ECM, MIL: Yes** **Year:** 2009, 2010 **Model:** Lancer, Outlander **Engine:** 2.0L L4 VIN U, 2.0L L4 VIN V, 2.4L L4 VIN W, 3.0L V6 VIN X **Transmission:** All	**A/D Converter:** Ignition switch is "ON" position. When the input voltage from the accelerator pedal position sensor (sub) is periodically 0 V for 0.45 second, the digital value of the input voltage indicates 0.2 V or more. **Possible Causes:** • ECM failed.
DTC: P1237 **1T ECM, MIL: Yes** **Year:** 2009, 2010 **Model:** Lancer, Outlander **Engine:** 2.0L L4 VIN U, 2.0L L4 VIN V, 2.4L L4 VIN W, 3.0L V6 VIN X **Transmission:** All	**Accelerator Pedal Position Sensor Plausibility:** Change of accelerator pedal position sensor (sub) output voltage per 40 milliseconds is lower than 0.06 volt. Range/performance error of accelerator pedal position sensor (main and sub) circuit is not detected. Voltage obtained with the formula given below is 0.4 volt or higher for 0.5 second: accelerator pedal position sensor (main) output voltage - accelerator pedal position sensor (sub) output voltage. **NOTE: The accelerator pedal position sensor voltage used for the judgment is converted into the accelerator pedal position sensor voltage for the internal processing by the ECM.** **Possible Causes:** • Accelerator pedal position sensor failed. • ECM failed.
DTC: P1238 **1T ECM, MIL: Yes** **Year:** 2009, 2010 **Model:** Lancer **Engine:** 2.0L L4 VIN V **Transmission:** All	**Mass Airflow Sensor Plausibility (torque monitor):** Engine speed is more than 750 r/min. Except while fuel is being shut off. For 0.9 second, the difference between the actual volumetric efficiency and the volumetric efficiency estimated by the throttle position sensor (main) and (sub) is more than 45 percent. **NOTE: Throttle opening degree position is in default position.** **Possible Causes:** • Mass airflow sensor system failed. • ECM failed.
DTC: P1238 **1T ECM, MIL: Yes** **Year:** 2009, 2010 **Model:** Lancer, Outlander **Engine:** 2.0L L4 VIN U, 2.4L L4 VIN W, 3.0L V6 VIN X **Transmission:** All	**Mass Airflow Sensor Plausibility (Torque Monitor):** The difference between the actual volumetric efficiency and the volumetric efficiency estimated by the throttle position sensor (main) is 0 percent or more. Or, the volumetric efficiency is 60 percent or less. Engine speed is 750 r/min or higher. For 0.5 second, the difference between the actual volumetric efficiency and the volumetric efficiency estimated by the throttle position sensor (main) and (sub) is 35 percent or more. **Possible Causes:** • Mass airflow sensor system failed. • ECM failed
DTC: P1239 **1T ECM, MIL: Yes** **Year:** 2009, 2010 **Model:** Lancer, Outlander **Engine:** 2.0L L4 VIN U, 2.0L L4 VIN V, 2.4L L4 VIN W, 3.0L V6 VIN X **Transmission:** All	**Engine RPM Plausibility:** The engine speed monitored with a 180 degree- cycle pulse is 500 r/min or more. The difference between the engine speed monitored with a 180 degree-cycle pulse and the engine speed monitored with a 10 degree-cycle pulse is 500 r/min or more for 0.5 second. Or, the engine speed monitored with a 10 degree- cycle pulse is 1,000 r/min or more for 0.5 second. **Possible Causes:** • Crankshaft position sensor system failed. • ECM failed.
DTC: P1240 **1T ECM, MIL: Yes** **Year:** 2009, 2010 **Model:** Lancer, Outlander **Engine:** 2.0L L4 VIN U, 2.0L L4 VIN V, 2.4L L4 VIN W, 3.0L V6 VIN X **Transmission:** All	**Ignition Angle:** Ignition switch is "ON" position. The ignition timing retard angle demand signal is not normal. Checks for an abnormal ignition timing retard angle demand signal from the Active Stability Control (ASC). **Possible Causes:** • ECM failed.

DTC	Trouble Code Title, Conditions & Possible Causes
DTC: P1241 **2T ECM, MIL: Yes** **Year:** 2009, 2010 **Model:** Lancer **Engine:** 2.0L L4 VIN V **Transmission:** All	**Torque Monitor:** Engine speed is more than 500 r/min. Volumetric efficiency is more than 16 percent. The requested torque signal subtracted from the actual torque signal is more than 130 Nm (96 ft-lb) for 1 second. **NOTE: Throttle opening degree position is in default position.** **Possible Causes:** • Throttle actuator control motor failed. • Connector damage. • Harness damage. • Intake system vacuum leak. • There is some trash around mass airflow sensor. • Intake charge pressure failed. • Intake charge pressure control system failed. • ECM failed.
DTC: P1241 **2T ECM, MIL: Yes** **Year:** 2009, 2010 **Model:** Lancer, Outlander **Engine:** 2.0L L4 VIN U, 2.4L L4 VIN W, 3.0L V6 VIN X **Transmission:** All	**Torque Monitoring:** Engine speed is 500 r/min or higher. Volumetric efficiency is 16 percent or higher. For 1 second, the difference between the actual torque signal and the requested torque signal is 50 Nm (37 ft-lb) or more. **Possible Causes:** • Throttle actuator control motor failed. • Connector damage. • Harness damage. • Intake system vacuum leak. • There is some trash around mass airflow sensor. • ECM failed
DTC: P1242 **1T ECM, MIL: Yes** **Year:** 2009, 2010 **Model:** Lancer, Outlander **Engine:** 2.0L L4 VIN U, 2.0L L4 VIN V, 2.4L L4 VIN W, 3.0L V6 VIN X **Transmission:** All	**Fail Safe Control Monitor:** Monitors the engine speed during fail-safe control. The engine speed is higher than assumed. **Possible Causes:** • ECM failed.
DTC: P1243 **1T ECM, MIL: Yes** **Year:** 2009, 2010 **Model:** Lancer, Outlander **Engine:** 2.0L L4 VIN U, 2.0L L4 VIN V, 2.4L L4 VIN W, 3.0L V6 VIN X **Transmission:** All	**Inquiry/Response Error:** Ignition switch is "ON" position. ECM cannot calculate input data **Possible Causes:** • ECM failed.
DTC: P1244 **1T ECM, MIL: Yes** **Year:** 2009, 2010 **Model:** Outlander **Engine:** 2.4L L4 VIN W, 3.0L V6 VIN X **Transmission:** All	**RAM Test For All Area:** Ignition switch is "ON" position. All RAM data of ECM are defect. **Possible Causes:** • ECM failed.
DTC: P1245 **1T ECM, MIL: Yes** **Year:** 2009, 2010 **Model:** Lancer, Outlander **Engine:** 2.0L L4 VIN U, 2.0L L4 VIN V, 2.4L L4 VIN W, 3.0L V6 VIN X **Transmission:** All	**Cycle RAM Test (Engine):** Ignition switch is "ON" position. RAM data (engine) is defect. **Possible Causes:** • ECM failed.
DTC: P1247 **2T ECM, MIL: Yes** **Year:** 2009, 2010 **Model:** Lancer **Engine:** 2.0L L4 VIN U, 2.0L L4 VIN V, 2.4L L4 VIN W **Transmission:** All	**TC-SST Plausibility <TC-SST>:** Ignition switch is "ON" position. A torque demand signal from TC-SST-ECU is not normal. **Possible Causes:** • TC-SST system failed. • ECM failed.

DTC	Trouble Code Title, Conditions & Possible Causes
DTC: P1247 **1T ECM, MIL: Yes** **Year:** 2009, 2010 **Model:** Outlander **Engine:** 3.0L V6 VIN X **Transmission:** All	**A/T Plausibility:** Checks for an abnormal signal of transmission control module. Ignition switch is "ON" position. A torque demand signal from TCM is not normal. **Possible Causes:** • Automatic transaxle system failed. • ECM failed.
DTC: P1247 **2T ECM, MIL: Yes** **Year:** 2009, 2010 **Model:** Lancer, Outlander **Engine:** 2.0L L4 VIN U, 2.4L L4 VIN W **Transmission:** All	**CVT Plausibility <CVT>:** Ignition switch is "ON" position. A torque demand signal from TCM is not normal. **Possible Causes:** • CVT system failed. • ECM failed.
DTC: P1248 **1T ECM, MIL: Yes** **Year:** 2009, 2010 **Model:** Outlander **Engine:** 2.4L L4 VIN W, 3.0L V6 VIN X **Transmission:** All	**AWD Plausibility :** Checks for an abnormal signal of AWD-ECU. Ignition switch is "ON" position. The torque reduction requested signal from AWD-ECU is not normal. **Possible Causes:** • AWD system failed. • ECM failed.
DTC: P1404 **2T PCM, MIL: Yes** **Year:** 2009 **Model:** Raider **Engine:** 3.7L V6 VIN K **Transmission:** All	**EGR Valve Position Rationality Closed:** With the engine running. The EGR flow or valve movement is not what is expected. A rationality error has been detected for the EGR Open Position Performance. **Possible Causes:** • Excessive resistance in the 5-volt supply circuit • EGR SOLENOID control circuit open • EGR signal circuit shorted to ground • Ground circuit open • EGR solenoid assembly • PCM
DTC: P1501 **1T PCM, MIL: Yes** **Year:** 2009 **Model:** Raider **Engine:** 3.7L V6 VIN K **Transmission:** All	**Vehicle Speed Sensor 1/2 Correlation Drive Wheels:** Ignition on and vehicle moving. Cruise is learned and customer is trying to use the Cruise. The PCM recognizes rear wheel speed is greater than front wheel speed. **Possible Causes:** • Active BUS or Communications DTC's • Tire circumference • PCM
DTC: P1502 **1T PCM, MIL: Yes** **Year:** 2009 **Model:** Raider **Engine:** 3.7L V6 VIN K **Transmission:** All	**Vehicle Speed Sensor 1/2 Correlation Non Drive Wheels:** Ignition on. In plant mode only. PCM recognizes the Brake Pedal could not electrically indicate the applied (On) position with both switch inputs. **Possible Causes:** • Active BUS or Communications DTC's • Tire circumference • PCM
DTC: P1506 **1T ECM, MIL: Yes** **Year:** 2009, 2010 **Model:** Lancer **Engine:** 2.0L L4 VIN U, 2.0L L4 VIN V, 2.4L L4 VIN W **Transmission:** All	**Idle Control System RPM Lower Than Expected at Low Temperature:** Under the closed loop idle speed control. The engine coolant temperature is between 7°C (45°F) and 41°C (106°F). Battery positive voltage is higher than 10 volts. Barometric pressure is higher than 76 kPa (22.4 in. Hg). Intake air temperature is higher than -10°C (14°F). More than 3 seconds have elapsed from the previous monitoring. Target air flow rate is more than 30 L/sec. The actual idle speed is more than 100 r/min lower than the target idle speed for 10 seconds. **Possible Causes:** • Throttle valve area is dirty. • Throttle body assembly failed. • ECM failed.

DTC	Trouble Code Title, Conditions & Possible Causes
DTC: P1506 **1T PCM, MIL: Yes** **Year:** 2009, 2010 **Model:** Eclipse, Endeavor, Galant, Outlander **Engine:** 2.4L L4 VIN F, 2.4L L4 VIN W, 3.0L V6 VIN X, 3.8L V6 VIN S, 3.8L V6 VIN T **Transmission:** All	**Idle Control System RPM Lower Than Expected at Low Temperature:** Under the closed loop idle speed control. The engine coolant temperature is between 7°C (45°F) and 41°C (106°F). Battery positive voltage is higher than 10 volts. Barometric pressure is higher than 76 kPa (22.4 in. Hg). Intake air temperature is higher than -10°C (14°F). More than 3 seconds have elapsed from the previous monitoring. Target throttle actuator control motor position is more than 512 steps. The actual idle speed is more than 100 r/min lower than the target idle speed for 10 seconds. **Possible Causes:** • Throttle valve area is dirty. • Throttle body assembly failed. • ECM failed. <M/T> • PCM failed. <A/T>
DTC: P1507 **1T PCM, MIL: Yes** **Year:** 2009, 2010 **Model:** Eclipse, Endeavor, Galant, Lancer, Outlander **Engine:** 2.0L L4 VIN U, 2.0L L4 VIN V, 2.4L L4 VIN F, 2.4L L4 VIN W, 3.0L V6 VIN X, 3.8L V6 VIN S, 3.8L V6 VIN T **Transmission:** All	**Idle Control System RPM Higher Than Expected at Low Temperature:** Under the closed loop idle speed control. Engine coolant temperature is between 7°C (45°F) and 41°C (106°F). Battery positive voltage is higher than 10 volts. Barometric pressure is higher than 76 kPa (22.4 in. Hg). Intake air temperature is higher than -10°C (14°F). More than 3 seconds have elapsed from the previous monitoring. Target throttle actuator control motor position is 0 step. The actual idle speed has continued to be higher than the target idle speed by 200 r/min (300 r/min*) or more for 10 seconds. **Possible Causes:** • Intake system vacuum leak. • Throttle body assembly failed. • ECM failed. <M/T> • PCM failed. <A/T>
DTC: P1540 **2T PCM, MIL: Yes** **Year:** 2009, 2010 **Model:** Galant **Engine:** 2.4L L4 VIN F **Transmission:** All	**LIN Communication Check by PCM (Sending Message to Radiator Sensor):** The battery voltage is applied from the ignition switch to the radiator sensor, and it is grounded to the vehicle body from the radiator sensor. The radiator sensor sends the sensor signal to the powertrain control module. The PCM judges the tampering of the DOR radiator according to the signal from the radiator sensor. The radiator sensor cannot receive the data from the PCM <When P1540 is set>. **Possible Causes:** • Tampering of the DOR radiator • A poor connection, open circuit, short circuit or damaged circuit is present in the radiator sensor circuit. • PCM has failed.
DTC: P1541 **1T PCM, MIL: Yes** **Year:** 2009, 2010 **Model:** Galant **Engine:** 2.4L L4 VIN F **Transmission:** All	**LIN Communication Check by PCM (Receiving Message from Radiator Sensor):** Ignition switch: "ON" Battery positive voltage is higher than 10 volts Engine coolant temp -30°C (-22°F) to 105°C (221°F) The PCM judges the tampering of the DOR radiator according to the signal from the radiator sensor. The PCM cannot receive the data from the radiator sensor <When P1541 is set>. When the PCM is defective **Possible Causes:** • Tampering of the DOR radiator • Malfunction of the PCM • A poor connection, open circuit, short circuit or damaged circuit is present in the radiator sensor circuit.
DTC: P1543 **1T PCM, MIL: Yes** **Year:** 2009, 2010 **Model:** Galant **Engine:** 2.4L L4 VIN F **Transmission:** All	**DOR Radiator Tampering Monitor (Temperature Check After Thermostat Open):** Battery positive voltage is higher than 11 volts. Intake air temp from 0 °C(32°F) to 40 °C(104°F). Engine coolant temperature is less than 40 °C(104°F). The engine coolant temperature - intake air temperature is 5°C (41°F) or less at engine is start. When the engine coolant temperature is 77°C (171°F) or higher under the following conditions. * Drive the vehicle at the engine speed of 1,700 r/min or more for 10 minutes. * The engine load is 30 percent or more. When the output temperature of the radiator sensor after the thermostat valve is opened is low. **Possible Causes:** • Tampering of the DOR radiator • Engine coolant is insufficient or has deteriorated. • Malfunction of the thermostat • Malfunction of the water pump • Malfunction of the PCM

DTC	Trouble Code Title, Conditions & Possible Causes
DTC: P1544 **1T PCM, MIL: Yes** **Year:** 2009, 2010 **Model:** Galant **Engine:** 2.4L L4 VIN F **Transmission:** All	**DOR Radiator Tampering Monitor (Temperature Check Before Thermostat Open):** Battery positive voltage is higher than 11 volts. Intake air temp from 0 °C (32°F) to 40 °C (104°F). Engine coolant temperature is less than 40 °C (104°F) when at engine start. The engine coolant temperature - intake air temperature is 5°C (41°F) or less at engine start. Vehicle speed is lower than 1.5km/h (1mph). When the output temperature of the radiator sensor before the thermostat valve is opened is high, the sensor is defective. **Possible Causes:** • Tampering of the DOR radiator • Malfunction of the thermostat • Engine coolant is insufficient or has deteriorated. • Malfunction of the water pump • Malfunction of the PCM
DTC: P1545 **1T PCM, MIL: Yes** **Year:** 2009, 2010 **Model:** Galant **Engine:** 2.4L L4 VIN F **Transmission:** All	**DOR Radiator Tampering Monitor (Temperature Change After Thermostat Open):** Battery positive voltage is higher than 11 volts. Intake air temp from 0 °C (32°F) to 40 °C (104°F). Engine coolant temperature is less than 40 °C (104°F) at engine start. The engine coolant temperature - intake air temperature is 5°C (41°F) or less when at engine start. Engine coolant temperature is higher than 77°C (171°F). When the output temperature of the radiator sensor after the thermostat valve is opened is rising by 6°C (43°F) or more in 1 second. **Possible Causes:** • Tampering of the DOR radiator
DTC: P1546 **1T PCM, MIL: Yes** **Year:** 2009, 2010 **Model:** Galant **Engine:** 2.4L L4 VIN F **Transmission:** All	**DOR Radiator Tampering Monitor (Temperature Check at Cold Start):** Battery positive voltage is higher than 11 volts. Intake air temp from 0 °C (32°F) to 40 °C (104°F). Engine coolant temperature is less than 40 °C (104°F) at engine start. The engine coolant temperature - intake air temperature is 5°C (41°F) or less at engine start. When the difference between the radiator sensor output temperature after the engine is started and the coolant temperature is 15°C (59°F) or more. **Possible Causes:** • Tampering of the DOR radiator
DTC: P1547 **1T PCM, MIL: Yes** **Year:** 2009, 2010 **Model:** Galant **Engine:** 2.4L L4 VIN F **Transmission:** All	**DOR Radiator Tampering Monitor (Rationality Check):** Battery positive voltage is higher than 11 volts. Intake air temp from 0 °C (32°F) to 40 °C (104°F). Engine coolant temperature is less than 40 °C (104°F) when the engine started. The engine coolant temperature - intake air temperature is 5°C (41°F) or less at engine start. When the engine coolant temperature is 77°C (171°F) or higher under the following conditions. * Drive the vehicle at the engine speed of 1,700 r/min or more for 10 minutes. * The engine load is 30 percent or more. The temperature difference between thermistor 1 and thermistor 2 is 6°C (43°F) or more. **Possible Causes:** • Tampering of the DOR radiator
DTC: P1572 **1T PCM, MIL: Yes** **Year:** 2009 **Model:** Raider **Engine:** 3.7L V6 VIN K **Transmission:** All	**Brake Pedal Stuck ON:** Ignition on. In plant mode only. PCM recognizes the Brake Pedal could not electrically indicate the applied (On) position with both switch inputs. **Possible Causes:** • Brake switch 1 signal is open • Brake switch 2 signal is open • Stop lamp switch has failed • PCM has failed
DTC: P1573 **1T PCM, MIL: Yes** **Year:** 2009 **Model:** Raider **Engine:** 3.7L V6 VIN K **Transmission:** All	**Brake Switch Stuck Off:** Ignition on, In plant mode passed the Applied test. PCM recognizes the Brake Pedal could not electronically indicate the released (Off) position with both switches. If P1572 sets, P1573 will also set. **Possible Causes:** • Brake switch 2 signal circuit is shorted voltage • Brake switch 1 signal circuit is shorted to ground • Stop lamp switch has failed • PCM has failed
DTC: P1580 **1T PCM, MIL: Yes** **Year:** 2009, 2010 **Model:** Galant **Engine:** 2.4L L4 VIN F **Transmission:** All	**DOR Radiator Tampering Monitor (Encrypted Message):** Battery positive voltage is higher than 11 volts. Engine coolant between -30°C (-22°F) and 110°C (230°F). Communication error. **Possible Causes:** • Malfunction of the radiator sensor • Malfunction of the PCM

DTC	Trouble Code Title, Conditions & Possible Causes
DTC: P1590 **1T ECM, MIL: Yes** **Year:** 2009, 2010 **Model:** Lancer, Outlander **Engine:** 2.0L L4 VIN U, 2.4L L4 VIN W, 3.0L V6 VIN X **Transmission:** All	**TCM to ECM Communication Error in Torque Reduction Request:** Ignition switch is "ON" position. ECM detects an error in communication between ECM and TCM. **Possible Causes:** • CAN line harness damage or connector damage. • ECM failed.
DTC: P1590 **1T ECM, MIL: Yes** **Year:** 2009, 2010 **Model:** Lancer **Engine:** 2.0L L4 VIN U, 2.4L L4 VIN W **Transmission:** All	**TC-SST-ECU to ECM Communication Error in Torque Reduction Request:** Ignition switch is "ON" position. ECM detects an error in communication between ECM and TC-SST-ECU. **NOTE: Engine output is restricted.** **Possible Causes:** • CAN line harness damage or connector damage. • ECM failed.
DTC: P1593 **1T PCM, MIL: Yes** **Year:** 2009 **Model:** Raider **Engine:** 3.7L V6 VIN K **Transmission:** All	**Speed Control Switch 1/2 Stuck :** With the ignition on. One of the S/C Switches is mechanically stuck in the On/Off, Resume/Accel, or Set position for too long. **Possible Causes:** • S/C Switch NO.1 signal circuit open • S/C Switch NO.2 signal circuit open • Switch return circuit open • S/C Switch NO.1 signal circuit shorted to voltage • S/C Switch NO.2 signal circuit shorted to voltage • S/C Switch NO.1 signal circuit shorted to ground • S/C Switch NO.2 signal circuit shorted to ground • S/C Switch NO.1 signal circuit shorted to the S/C Switch return circuit • S/C Switch NO.2 signal circuit shorted to the S/C Switch return circuit • Speed control switch • PCM
DTC: P1602 **T PCM, MIL: Yes** **Year:** 2009 **Model:** Raider **Engine:** 3.7L V6 VIN K **Transmission:** All	**PCM Not Programmed:** Ignition on and battery voltage greater than 10 volts. The PCM has not been programmed. **Possible Causes:** • Program the PCM and then retest for this same trouble code • PCM has failed
DTC: P1602 **1T PCM, MIL: Yes** **Year:** 2009, 2010 **Model:** Eclipse, Endeavor, Galant **Engine:** 2.4L L4 VIN F, 3.8L V6 VIN S, 3.8L V6 VIN T **Transmission:** All	**Communication Malfunction (between ECM <M/T> or PCM <A/T> Main Processor and System LSI):** Ignition switch is "ON" position. The ECM <M/T> or PCM <A/T> detects an error in communication between ECM <M/T> or PCM <A/T> main processor and system LSI for 0.07 second. **Possible Causes:** • ECM failed. <M/T> • PCM failed. <A/T>
DTC: P1603 **1T PCM, MIL: Yes** **Year:** 2009, 2010 **Model:** Eclipse, Endeavor, Galant, Lancer, Outlander **Engine:** 2.0L L4 VIN U, 2.0L L4 VIN V, 2.4L L4 VIN F, 2.4L L4 VIN W, 3.0L V6 VIN X, 3.8L V6 VIN S, 3.8L V6 VIN T **Transmission:** All	**Battery Backup Line Malfunction:** Engine starting sequence was completed. Battery positive voltage is higher than 10 volts. The battery backup line voltage has continued to be 6 volts or lower for 2 seconds. Battery backup line voltage is under specified value. **Possible Causes:** • Open or shorted battery backup line, harness damage or connector damage. • ECM failed. <M/T> • PCM failed. <A/T>
DTC: P1606 **1T PCM, MIL: Yes** **Year:** 2009, 2010 **Model:** Eclipse, Endeavor, Galant **Engine:** 2.4L L4 VIN F, 3.8L V6 VIN S, 3.8L V6 VIN T **Transmission:** All	**EEPROM Malfunction:** DTC P1606 will be set when abnormal conditions are encountered in the A/T side area of EEPROM. **Possible Causes:** • Malfunction of the PCM

DTC	Trouble Code Title, Conditions & Possible Causes
DTC: P1607 **2T PCM, MIL: Yes** **Year:** 2009 **Model:** Raider **Engine:** 3.7L V6 VIN K **Transmission:** All	**PCM Internal Shutdown Timer Rationality:** With the engine running after a cycle when a complete engine warm up was achieved, the difference between engine coolant temperature and ambient air temperature less than or equal to 10°C (50°F), and battery voltage greater than 10 volts. This DTC sets if the engine coolant temp does not drop enough or drops too much during engine off time. This DTC may also set if the controller timer is inaccurate. **Possible Causes:** • Engine cooling conditions may set this code • Engine coolant level • PCM internal timer
DTC: P1618 **1T PCM, MIL: Yes** **Year:** 2009 **Model:** Raider **Engine:** 3.7L V6 VIN K **Transmission:** All	**Sensor Reference Voltage 1 Circuit Erratic:** With the ignition on. When the PCM recognizes the Primary 5-volt Supply circuit voltage is varying too much too quickly. ETC light is flashing. **Possible Causes:** • 5-volt supply circuit ground • 5-volt supply circuit shorted to voltage • 5-volt supply circuit open • 5-volt sensor • PCM
DTC: P1628 **1T PCM, MIL: Yes** **Year:** 2009 **Model:** Raider **Engine:** 3.7L V6 VIN K **Transmission:** All	**Sensor Reference Voltage Circuit Erratic:** With the ignition on. When the PCM recognizes the Auxiliary 5-volt Supply circuit voltage is varying too much too quickly. ETC light is flashing. **Possible Causes:** • 5-volt supply circuit ground • 5-volt supply circuit shorted to voltage • 5-volt supply circuit open • 5-volt sensor • PCM
DTC: P1637 **1T ECM, MIL: Yes** **Year:** 2009, 2010 **Model:** Lancer, Outlander **Engine:** 2.0L L4 VIN U, 2.4L L4 VIN W **Transmission:** All	**Malfunction of Memory Backup (CVT):** Ignition switch : ON (start the engine and keep it for 10 seconds or more) Calculated checksum is not same as memorized checksum. **Possible Causes:** • Damaged wiring harness and connectors • Malfunction of TCM (Faulty EEPROM)
DTC: P1637 **1T ECM, MIL: Yes** **Year:** 2009, 2010 **Model:** Lancer **Engine:** 2.0L L4 VIN U, 2.4L L4 VIN W **Transmission:** All	**EEPROM System (DTC storing malfunction) (TC-SST):** Voltage of battery: 8 V or more. Voltage of battery: 16.5 V or less. Calculated checksum: Not equal the memorized checksum. **Possible Causes:** • Malfunction of TC-SST-ECU
DTC: P1676 **1T ECM, MIL: Yes** **Year:** 2009, 2010 **Model:** Lancer **Engine:** 2.0L L4 VIN U, 2.4L L4 VIN W **Transmission:** All	**Coding incomplete (TC-SST):** Ignition switch: ON, Coding state: Not programmed. (Immediately) **Possible Causes:** • Malfunction of TC-SST-ECU.
DTC: P1676 **1T ECM, MIL: Yes** **Year:** 2009, 2010 **Model:** Lancer, Outlander **Engine:** 2.0L L4 VIN U, 2.0L L4 VIN V, 2.4L L4 VIN W, 3.0L V6 VIN X **Transmission:** All	**Variant Coding:** Ignition switch is "ON" position. Vehicle information is not entered into the ECM. **Possible Causes:** • ECM failed.

DTC	Trouble Code Title, Conditions & Possible Causes
DTC: P1684 **T PCM, MIL: Yes** **Year:** 2009 **Model:** Raider **Engine:** 3.7L V6 VIN K **Transmission:** All	**Battery Was Disconnected:** Whenever the ignition is in the Run/Start position. This DTC will set whenever Powertrain Control Module is disconnected from Fused B(+) or ground. It will also be set using the scan tool to perform a Battery Disconnect and/or Quick Learn procedure. **Possible Causes:** • Open B(+) or ground circuit to PCM • Battery was disconnected • Scan tool battery disconnect performed • PCM was disconnected or replaced • Quick learn was performed
DTC: P1696 **1T PCM, MIL: Yes** **Year:** 2009 **Model:** Raider **Engine:** 3.7L V6 VIN K **Transmission:** All	**EEPROM Memory Write Denied/Invalid:** Continuously with the ignition on. An attempt to program/write to the internal EEPROM failed, Also checks at power down. **Possible Causes:** • PCM software update • Excessive resistance in the PCM power and ground circuits • PCM
DTC: P1705 **1T ECM, MIL: Yes** **Year:** 2009, 2010 **Model:** Outlander **Engine:** 3.0L V6 VIN X **Transmission:** All	**Throttle Position Sensor Information (Engine):** TCM is started: more than 1.05 seconds. Vehicle speed: 5 km/h(3.1 mph) or more. Throttle position sensor signal: fail. (5.5 seconds) **Possible Causes:** • Malfunction of the CAN • Malfunction of the ECM • Damaged harness or connector • Malfunction of the throttle position sensor • Malfunction of the TCM
DTC: P1706 **1T ECM, MIL: Yes** **Year:** 2009, 2010 **Model:** Outlander **Engine:** 3.0L V6 VIN X **Transmission:** All	**Accelerator Pedal Position Information:** TCM is started: more than 1.05 seconds. Accelerator pedal position sensor signal: fail. (3 seconds) **Possible Causes:** • Malfunction of the CAN • Malfunction of the ECM • Damaged harness or connector • Malfunction of the throttle position sensor
DTC: P1706 **1T ECM, MIL: Yes** **Year:** 2009, 2010 **Model:** Lancer, Outlander **Engine:** 2.0L L4 VIN U, 2.4L L4 VIN W **Transmission:** All	**Malfunction of Throttle Signal (CVT):** Transmission range switch position: D. Voltage of battery: 9 volts or more. Voltage of battery: 16 volts or less. Throttle position toggle signal on CAN: fail. (1 second) **Possible Causes:** • Malfunction of the CAN bus • Malfunction of engine system • Malfunction of TCM
DTC: P1710 **1T ECM, MIL: Yes** **Year:** 2009, 2010 **Model:** Lancer, Outlander **Engine:** 2.0L L4 VIN U, 2.4L L4 VIN W **Transmission:** All	**Malfunction of Vehicle Speed Signal (CVT):** Difference between the estimated vehicle speed and the one measured before 0.1 second is 29 km/h (18 mph) or more. The status with the vehicle speed of 10 km/h (6.2 mph) or more and with the estimated vehicle speed of 2 km/h (1.2 mph) or less continues for 20 seconds. **Possible Causes:** • Malfunction of the CAN bus • Malfunction of TCM
DTC: P1713 **T PCM** **Year:** 2009 **Model:** Raider **Engine:** 3.7L V6 VIN K **Transmission:** All	**Restricted Manual Valve In T2 Range:** Ignition on, engine running with the gear shift selector in a valid forward gear. This DTC sets whenever Transmission control system detects the manual valve is in the T2 range when it should be in OD. This is mainly an informational DTC. **Possible Causes:** • Related transmission DTC's present • Customer driving habits • Mis-adjusted shifter

DTC	Trouble Code Title, Conditions & Possible Causes
DTC: P1715 **T PCM** **Year:** 2009 **Model:** Raider **Engine:** 3.7L V6 VIN K **Transmission:** All	**Restricted Manual Valve In T3 Range:** Whenever the PRNDL code indicates Temp 3. This DTC sets when conditions for the DTC P1776 are satisfied or 3 unsuccessful attempts to engage 1st gear while the shifter is in the temp 3 zone. This indicates a restricted port at the manual valve because the shifter is not fully engaged in the drive position. **Possible Causes:** • Related transmission DTC's present • Customer driving habits • Mis-adjusted shifter
DTC: P1723 **1T ECM, MIL: Yes** **Year:** 2009, 2010 **Model:** Lancer, Outlander **Engine:** 2.0L L4 VIN U, 2.4L L4 VIN W **Transmission:** All	**Abnormality in Speed Sensor System Function:** Voltage of battery: 9 volts or more. Voltage of battery: 16 volts or less. Noise of primary pulley or primary pulley speed sensor signal: more than 30 r/min. (1 second) Noise of primary pulley or secondary pulley speed sensor signal: more than 30 r/min. (1 second) **Possible Causes:** • Malfunction of primary pulley speed sensor • Malfunction of secondary pulley speed sensor • Malfunction of TCM
DTC: P1731 **1T ECM, MIL: Yes** **Year:** 2009, 2010 **Model:** Outlander **Engine:** 3.0L V6 VIN X **Transmission:** All	**1st Engine Brake Detection:** With the mode other than SPORT MODE and with the accelerator angle is smaller than 6/8, the correlation between the lock-up and low-reverse brake linear solenoid valve control current and the low-reverse brake pressure switch ON/OFF is abnormal. **Possible Causes:** • Transaxle assembly powertrain parts failure • Malfunction of the valve assembly (Malfunction of hydraulic valve and hydraulic switch) • Malfunction of the TCM • Malfunction of the lock-up and low-reverse brake linear solenoid valve (valve assembly) • Malfunction of the low-reverse brake pressure switch (valve assembly) • Malfunction of the low clutch pressure switch (valve assembly) • Malfunction of the low clutch linear solenoid valve (valve assembly) • Malfunction of the low-reverse brake shift solenoid valve (valve assembly) • Malfunction of the low clutch, low-reverse brake
DTC: P1736 **1T PCM, MIL: Yes** **Year:** 2009 **Model:** Raider **Engine:** 3.7L V6 VIN K **Transmission:** All	**Gear Ratio Error In 2nd Prime:** Continuously with the ignition on, engine running, and after the transmission has achieved the proper gear ratio. If the ratio of the Input rpm to the Output rpm does not match the current gear ratio. This DTC can take up to five minutes of problem identification before illuminating the MIL. **Possible Causes:** • Related DTC's present • Input sensor or wiring • Low fluid level • Cracked or mis-installed primary oil filter or seal • Worn solenoid switch valve or plugs • Stuck or sticking main regulator valve • Burnt UD or 4C clutch • Cut 4C or UD piston seal • Broken 4C piston casting • Broken or missing 4C bleed orifice • Broken or missing UD bleed orifice • Cut 4C or UD accumulator piston seal • Cracked 4C or UD accumulator piston • Transmission Solenoid/TRS Assembly • Broken weld-reaction carrier to reverse sun gear • Missing teeth on input clutch hub tone wheel
DTC: P1740 **1T ECM, MIL: Yes** **Year:** 2009, 2010 **Model:** Lancer, Outlander **Engine:** 2.0L L4 VIN U, 2.4L L4 VIN W **Transmission:** All	**Malfunction of Lockup/Select Switching Solenoid Valve (CVT):** Lockup/select switching solenoid valve activation command (ON voltage): 1.42 volts or less (0.2 second) Lockup/select switching solenoid valve activation command (OFF voltage): more than 1.42 volts (0.2 second) **Possible Causes:** • Malfunction of valve body assembly (Faulty lockup/select switching solenoid valve) • Damaged wiring harness and connectors • Malfunction of TCM

DTC	Trouble Code Title, Conditions & Possible Causes
DTC: P1745 **1T ECM, MIL: Yes** **Year:** 2009, 2010 **Model:** Lancer, Outlander **Engine:** 2.0L L4 VIN U, 2.4L L4 VIN W **Transmission:** All	**Monitoring of Percentage Change in Pulley Ratio (CVT):** The TCM conducts fault detection by monitoring the internal calculated value. Percentage change in pulley ratio is larger than the standard value. **Possible Causes:** • Malfunction of TCM.
DTC: P1745 **1T PCM, MIL: Yes** **Year:** 2009 **Model:** Raider **Engine:** 3.7L V6 VIN K **Transmission:** All	**Transmission Line Pressure Too High For Too Long:** Continuously with ignition on. If the transmission has been operating in an open-loop line pressure control for 3220 kilometers (2000 milcs) or 1000 2-3 upshifts. **Possible Causes:** • Line pressure DTC present
DTC: P1751 **1T PCM, MIL: Yes, TCIL: Yes** **Year:** 2009, 2010 **Model:** Eclipse **Engine:** 2.4L L4 VIN F, 3.8L V6 VIN T **Transmission:** All	**A/T Control Relay System:** Voltage of battery: 9 volts or more. Time after PCM turns on A/T control relay: 0.5 second or more. A/T control relay output voltage: 7 volts or less. (0.1 second) **NOTE: If DTC P1788 (P1751) is set consecutively four times, the transaxle is locked into 3rd gear as a fail-safe measure, and the selector lever position indicator light (D, 1 through 4 or 5) flashes once per second.** **Possible Causes:** • Malfunction of the A/T control relay • Damaged harness or connector • Malfunction of the PCM
DTC: P1753 **1T ECM, MIL: Yes** **Year:** 2009, 2010 **Model:** Outlander **Engine:** 3.0L V6 VIN X **Transmission:** All	**Low Clutch Shift Solenoid Valve System:** Check Conditions <Circuit continuity ground>: Lock-up status: engaged. Gear position: 5th or 6th. Judgment Criteria: <Circuit continuity ground> Low clutch shift solenoid valve actual current: more than 1.7 A. Check Conditions <Circuit continuity open>: Lock-up status: engaged. Gear position: 1st to 4th. Low clutch shift solenoid valve activation command: more than 937 mA Judgment Criteria: <Circuit continuity open> Low clutch shift solenoid valve actual current: less than 500 mA. (0.2 seconds) **Possible Causes:** • Malfunction of the low clutch shift solenoid valve system circuit • Damaged harness or connector • Malfunction of the TCM • Malfunction of the low clutch shift solenoid valve (valve body Assembly)
DTC: P1753 **1T ECM, MIL: Yes** **Year:** 2009, 2010 **Model:** Outlander **Engine:** 3.0L V6 VIN X **Transmission:** All	**Low Clutch Shift Solenoid Valve System:** Check Conditions: <Circuit continuity ground> Lock-up status: engaged. Gear position: 5th or 6th. Judgment Criteria: <Circuit continuity ground> Low clutch shift solenoid valve actual current: more than 1.7 A Check Conditions: <Circuit continuity open> Lock-up status: engaged. Gear position: 1st to 4th.Low clutch shift solenoid valve activation command: more than 937 mA Judgment Criteria: <Circuit continuity open> Low clutch shift solenoid valve actual current: less than 500 mA. (0.2 seconds) **Possible Causes:** • Malfunction of the low clutch shift solenoid valve system circuit • Damaged harness or connector • Malfunction of the TCM • Malfunction of the low clutch shift solenoid valve (valve body Assembly)
DTC: P1758 **1T ECM, MIL: Yes** **Year:** 2009, 2010 **Model:** Outlander **Engine:** 3.0L V6 VIN X **Transmission:** All	**Low-Reverse Brake Shift Solenoid Valve System:** Check Conditions: <Circuit continuity ground> Voltage of battery: 2 volts or more. Judgment Criteria: <Circuit continuity ground> Low-reverse brake shift solenoid valve actual current: more than 1.7 A (0.2 second) Check Conditions: <Circuit continuity open> Voltage of battery: 2 volts or more. Low-reverse brake shift solenoid valve activation command: more than 937 mA Judgment Criteria: <Circuit continuity open> Low-reverse brake shift solenoid valve actual current: less than 500 mA. (0.2 seconds) **Possible Causes:** • Malfunction of the low-reverse brake shift solenoid valve system circuit • Damaged harness or connector • Malfunction of the TCM • Malfunction of the low-reverse brake shift solenoid valve (valve body assembly)

DTC	Trouble Code Title, Conditions & Possible Causes
DTC: P1763 **1T PCM, MIL: Yes** **Year:** 2009, 2010 **Model:** Eclipse, Endeavor, Galant **Engine:** 2.4L L4 VIN F, 3.8L V6 VIN S, 3.8L V6 VIN T **Transmission:** All	**Transmission Fluid Temperature Sensor System (Open Circuit):** Engine speed: 1,000 r/min or more. Output speed: 1,000 r/min or more. Accumulated time in above condition: 10 minutes. Transmission fluid temperature sensor voltage: 4.5 volts or more. (1 second) **NOTE: If transmission fluid temperature is below specified value even after test driving for more than the specified period, the PCM judges that the transmission fluid temperature sensor has a failure.** **Possible Causes:** • Malfunction of the transmission fluid temperature sensor circuit • Damaged harness or connector • Malfunction of the PCM
DTC: P1764 **1T PCM, MIL: Yes** **Year:** 2009, 2010 **Model:** Eclipse, Endeavor, Galant **Engine:** 2.4L L4 VIN F, 3.8L V6 VIN S, 3.8L V6 VIN T **Transmission:** All	**Transmission Fluid Temperature Sensor System (Short Circuit):** With the ignition "ON". The transmission fluid temperature sensor voltage: 0.2 volt or less. (1 second) **NOTE: If transmission fluid temperature equals or exceeds specified value, PCM judges that transmission fluid temperature sensor has a failure.** **Possible Causes:** • Malfunction of the transmission fluid temperature sensor circuit • Damaged harness or connector • Malfunction of the PCM
DTC: P1766 **1T PCM, MIL: Yes, TCIL: Yes** **Year:** 2009, 2010 **Model:** Eclipse, Endeavor, Galant **Engine:** 2.4L L4 VIN F, 3.8L V6 VIN S, 3.8L V6 VIN T **Transmission:** All	**Input Shaft Speed Sensor System (Short Circuit/Open Circuit):** Transmission range switch position: D. Output speed: 1,000 r/min or more. <4A/T> Vehicle speed: 30 km/h (19 mph). <5A/T> Transmission fluid temperature sensor voltage: 4.5 volts or less. Input shaft speed sensor signal: no signal change. (4 seconds) If DTC P1766 (P0715) is set consecutively four times, the transaxle is locked into 3rd gear or 2nd gear as a fail-safe measure, and the selector lever position indicator light (D, 1 through 4 or 5) flashes once per second. **Possible Causes:** • Malfunction of the input shaft speed sensor • Malfunction of the underdrive clutch retainer • Damaged harness or connector • Malfunction of the PCM
DTC: P1767 **1T PCM, MIL: Yes** **Year:** 2009, 2010 **Model:** Eclipse, Endeavor, Galant **Engine:** 2.4L L4 VIN F, 3.8L V6 VIN S, 3.8L V6 VIN T **Transmission:** All	**Output Shaft Speed Sensor System (Short Circuit/Open Circuit):** Transmission range switch position: D. Input speed: 1,000 r/min or more. Engine speed: 1,000 r/min or more. Transmission fluid temperature sensor voltage: 4.5 volts or less. Calculated slip (engine speed - input speed): 100 r/min or more. Output speed: no signal change. (4 seconds) If DTC P1767 (P0720) is set consecutively four times, the transaxle is locked into 3rd gear or 2nd gear as a fail-safe measure, and the selector lever position indicator light (D, 1 through 4 or 5) flashes once per second. **Possible Causes:** • Malfunction of the output shaft speed sensor • Malfunction of the transfer drive gear or driven gear <4A/T> • Malfunction of the direct planetary carrier <5A/T> • Damaged harness or connector • Malfunction of the PCM
DTC: P1769 **1T PCM, MIL: Yes** **Year:** 2009, 2010 **Model:** Eclipse, Endeavor, Galant **Engine:** 2.4L L4 VIN F, 3.8L V6 VIN S, 3.8L V6 VIN T **Transmission:** All	**Stoplight Switch System:** If the stoplight switch is on for five minutes or more while driving above 50 km/h (31 mph), or all of the stop light bulbs are blown, it is judged there is a short circuit or open circuit in the stoplight switch. This causes DTC P1769 to be set. **Possible Causes:** • Malfunction of the stoplight switch • Malfunction of stoplight valve • Damaged harness or connector • Malfunction of the PCM
DTC: P1770 **1T PCM, MIL: Yes** **Year:** 2009, 2010 **Model:** Eclipse, Endeavor, Galant **Engine:** 2.4L L4 VIN F, 3.8L V6 VIN S, 3.8L V6 VIN T **Transmission:** All	**Transmission Range Switch System (Open Circuit):** Battery positive voltage is applied to the transmission range switch (terminal 8) when the ignition switch is turned "ON." Battery positive voltage is applied to the PCM (terminal 67) when the transmission range is in the "P" range. The PCM judges that the transmission range is in the "P" range when the battery positive voltage is applied. Battery positive voltage is applied to the PCM terminal 61 (77, 73) when the selector lever is in the "R" range ("N," "D" range). The PCM judges that the selector lever is in the "R" range ("N," "D" range) when the battery positive voltage is applied. The transmission range switch has no signal detected. (30 seconds) **Possible Causes:** • Malfunction of the transmission range switch • Malfunction of the ignition switch • Damaged harness or connector • Malfunction of the PCM

DTC	Trouble Code Title, Conditions & Possible Causes
DTC: P1771 **1T PCM, MIL: Yes** **Year:** 2009, 2010 **Model:** Eclipse, Endeavor, Galant **Engine:** 2.4L L4 VIN F, 3.8L V6 VIN S, 3.8L V6 VIN T **Transmission:** All	**Transmission Range Switch System (Short Circuit):** Transmission range switch: multiple signal. (30 seconds) If two types or more of signals are input from transmission range switch for more than 30 seconds, PCM judges that transmission range switch has a failure. **Possible Causes:** • Malfunction of the transmission range switch circuit • Damaged harness or connector • Malfunction of the PCM
DTC: P1773 **1T PCM, MIL: Yes** **Year:** 2009, 2010 **Model:** Eclipse, Endeavor, Galant **Engine:** 2.4L L4 VIN F, 3.8L V6 VIN S, 3.8L V6 VIN T **Transmission:** All	**Low-Reverse Solenoid Valve :** Solenoid status: either solid ON or OFF. Shift status: in-gear. Voltage of battery: 10 volts or more. Solenoid voltage: 3 volts or less. (0.3 second) **NOTE: If DTC P1773 (P0753) is set consecutively four times, the transaxle is locked into 3rd gear as a fail-safe measure, and the selector lever position indicator light (D, 1 through 4 or 5) flashes once per second.** **Possible Causes:** • Malfunction of the low-reverse solenoid valve • Damaged harness or connector • Malfunction of the PCM
DTC: P1773 **2T ECM, MIL: Yes** **Year:** 2009, 2010 **Model:** Lancer, Outlander **Engine:** 2.0L L4 VIN U, 2.4L L4 VIN W **Transmission:** All	**Malfunction of ABS (CVT):** Time after TCM start: more than 1.05 seconds. ABS status: malfunction **Possible Causes:** • Malfunction of the ABS system. • Malfunction of TCM.
DTC: P1773 **1T ECM, MIL: Yes** **Year:** 2009, 2010 **Model:** Outlander **Engine:** 3.0L V6 VIN X **Transmission:** All	**ABS Information (ABS/ASC) :** TCM is started: more than 1.05 seconds. Vehicle speed: more than 55 km/h (34.4 mph). ABS status signal: fail. (0.5 second) **Possible Causes:** • Malfunction of the CAN • Malfunction of the ABS/ASC-ECU
DTC: P1774 **1T PCM, MIL: Yes** **Year:** 2009, 2010 **Model:** Eclipse, Endeavor, Galant **Engine:** 2.4L L4 VIN F, 3.8L V6 VIN S, 3.8L V6 VIN T **Transmission:** All	**Underdrive Solenoid Valve System:** Solenoid status: either solid ON or OFF. Shift status: in-gear. Voltage of battery: 10 volts or more. Solenoid voltage: 3 volts or less. (0.3 second) If DTC P1774 (P0758) is set consecutively four times, the transaxle is locked into 3rd gear as a fail-safe measure, and the selector lever position indicator light (D, 1 through 4 or 5) flashes once per second. **Possible Causes:** • Malfunction of the underdrive solenoid valve • Damaged harness or connector • Malfunction of the PCM
DTC: P1775 **1T PCM, MIL: Yes** **Year:** 2009 **Model:** Raider **Engine:** 3.7L V6 VIN K **Transmission:** All	**Solenoid Switch Valve Latched In TCC Position (42RLE Transmission):** During an attempted shift into 1st gear. This DTC is set if three unsuccessful attempts are made to shift the Solenoid Switch Valve (SSV) into the downshifted position in one given ignition start. This DTC can take up to five minutes to mature before illuminating the MIL. **Possible Causes:** • Related DTC P0841 may be present. • Transmission control output circuit is open • L/R Pressure control circuit is open • L/R Pressure control circuit is shorted to ground • L/R Pressure control circuit is shorted to voltage • Internal Transmission problem • PCM has failed
DTC: P1775 **1T PCM, MIL: Yes, TCIL: Yes** **Year:** 2009, 2010 **Model:** Eclipse, Endeavor, Galant **Engine:** 2.4L L4 VIN F, 3.8L V6 VIN S, 3.8L V6 VIN T **Transmission:** All	**Second Solenoid Valve System:** Solenoid status: either solid ON or OFF. Shift status: in-gear. Voltage of battery: 10 volts or more. Solenoid voltage: 3 volts or less. (0.3 second) If DTC P1775 (P0763) is set consecutively four times, the transaxle is locked into 3rd gear as a fail-safe measure, and the selector lever position indicator light (D, 1 through 4 or 5) flashes once per second. **Possible Causes:** • Malfunction of the second solenoid valve • Damaged harness or connector • Malfunction of the PCM

DTC	Trouble Code Title, Conditions & Possible Causes
DTC: P1775 **1T PCM, MIL: Yes** **Year:** 2009 **Model:** Raider **Engine:** 3.7L V6 VIN K **Transmission:** All	**Solenoid Switch Valve Latched In TCC Position (45RFE/545RFE Transmission):** During an attempted shift into 1st gear. This DTC is set if three unsuccessful attempts are made to shift the Solenoid Switch Valve (SSV) into the downshifted position in one given ignition start. This DTC can take up to five minutes to mature before illuminating the MIL. **Possible Causes:** • Related DTC P0841 may be present. • Intermittent wiring or connector problems • L/R Solenoid pressure switch circuit is shorted to other circuits • Transmission Solenoid/TRS Assembly • PCM
DTC: P1776 **1T PCM, MIL: Yes, TCIL: Yes** **Year:** 2009, 2010 **Model:** Eclipse, Endeavor, Galant **Engine:** 2.4L L4 VIN F, 3.8L V6 VIN S, 3.8L V6 VIN T **Transmission:** All	**Overdrive Solenoid Valve System:** Solenoid status: either solid ON or OFF. Shift status: in-gear. Voltage of battery: 10 volts or more. Solenoid voltage: 3 volts or less. (0.3 second) If DTC P1776 (P0768) is set consecutively four times, the transaxle is locked into 3rd gear as a fail-safe measure, and the selector lever position indicator light (D, 1 through 4 or 5) flashes once per second. **Possible Causes:** • Malfunction of the overdrive solenoid valve • Damaged harness or connector • Malfunction of the PCM
DTC: P1776 **1T PCM, MIL: Yes** **Year:** 2009 **Model:** Raider **Engine:** 3.7L V6 VIN K **Transmission:** All	**Solenoid Switch Valve Latched In LR Position (42RLE Transmission):** Every 7 ms when doing PEMCC or FEMCC. Must be in partial or full EMCC. The DTC is set if L/R pressure is detected high for the fourth time. **Possible Causes:** • Related DTC P0841 may be present. • Event date shows TRS Code TR2 - shift lever or manual control valve in a invalid position • Shift cable out of adjustment • Solenoid switch valve sticking in its bore • Transmission control output circuit open • L/R Pressure signal circuit open • L/R Pressure signal circuit shorted to ground • L/R Pressure signal circuit shorted to voltage • Valve body • PCM
DTC: P1776 **1T PCM, MIL: Yes** **Year:** 2009 **Model:** Raider **Engine:** 3.7L V6 VIN K **Transmission:** All	**Solenoid Switch Valve Latched In LR Position (45RFE/545RFE Transmission):** Continuously when performing partial or full EMCC - PEMCC or FEMCC. If the transmission senses the L/R pressure switch closing while performing PEMCC or FEMCC. This DTC will set after two unsuccessful attempts to perform PEMCC or FEMCC. This DTC can take up to five minutes of problem identification before illuminating the MIL. **Possible Causes:** • Related DTC P0841 may be present. • L/R pressure switch sense circuit open • L/R pressure switch sense circuit shorted to ground • Solenoid switch valve • PCM
DTC: P1777 **1T PCM, MIL: Yes, TCIL: Yes** **Year:** 2009, 2010 **Model:** Eclipse, Galant **Engine:** 2.4L L4 VIN F, 3.8L V6 VIN S, 3.8L V6 VIN T **Transmission:** All	**Reduction solenoid valve system <5A/T>:** Solenoid status: either solid ON or OFF. Shift status: in-gear. Voltage of battery: 10 volts or more. Solenoid voltage: 3 volts or less. (0.3 second) If DTC P1777 (P0773) is set consecutively four times, the transaxle is locked into 3rd gear as a fail-safe measure, and the selector lever position indicator light (D, 1 through 4 or 5) flashes once per second. **Possible Causes:** • Malfunction of the reduction solenoid valve • Damaged harness or connector • Malfunction of the PCM
DTC: P1777 **1T ECM, MIL: Yes** **Year:** 2009, 2010 **Model:** Lancer, Outlander **Engine:** 2.0L L4 VIN U, 2.4L L4 VIN W **Transmission:** All	**Malfunction of Stepper Motor:** Stepping motor activation command (ON voltage): 2.8 volts or less. (0.2 second) Stepping motor activation command (OFF voltage): More than 2.8 volts. (0.2 second) **Possible Causes:** • Malfunction of valve body assembly (Faulty stepper motor) • Damaged wiring harness and connectors • Malfunction of TCM

DTC	Trouble Code Title, Conditions & Possible Causes
DTC: P1778 **1T PCM, MIL: Yes, TCIL: Yes** **Year:** 2009, 2010 **Model:** Eclipse, Endeavor, Galant **Engine:** 2.4L L4 VIN F, 3.8L V6 VIN S, 3.8L V6 VIN T **Transmission:** All	**Torque Converter Clutch Solenoid Valve System:** Solenoid status: either solid ON or OFF. Voltage of battery: 10 volts or more. Solenoid voltage: 3 volts or less. (0.3 second) If DTC P1778 (P0743) is set consecutively four times, the transaxle is locked into 3rd gear as a fail-safe measure, and the selector lever position indicator light (D, 1 through 4 or 5) flashes once per second. **Possible Causes:** • Malfunction of the torque converter clutch solenoid valve • Damaged harness or connector • Malfunction of the PCM
DTC: P1778 **1T ECM, MIL: Yes** **Year:** 2009, 2010 **Model:** Lancer **Engine:** 2.0L L4 VIN U, 2.4L L4 VIN W **Transmission:** All	**Malfunction of Stepper Motor (CVT):** Stepping motor activation command (ON voltage): 2.8 volts or less. (0.2 second) Stepping motor activation command (OFF voltage): More than 2.8 volts. (0.2 second) **Possible Causes:** • Malfunction of valve body assembly (Faulty stepper motor) • Damaged wiring harness and connectors • Malfunction of TCM
DTC: P1778 **1T ECM, MIL: Yes** **Year:** 2009, 2010 **Model:** Lancer, Outlander **Engine:** 2.0L L4 VIN U, 2.4L L4 VIN W **Transmission:** All	**Malfunction of Stepper Motor (CVT):** The difference between the target value of the primary pulley speed and the actual measurement value is greater than the standard value, the difference between the target pulley ratio and the actual pulley ratio is 0.3 or greater, and this status continues for 5 seconds. **Possible Causes:** • Malfunction of TCM • Malfunction of valve body assembly (Faulty stepper motor)
DTC: P1779 **1T PCM, MIL: Yes, TCIL: Yes** **Year:** 2009, 2010 **Model:** Eclipse, Endeavor, Galant **Engine:** 2.4L L4 VIN F, 3.8L V6 VIN S, 3.8L V6 VIN T **Transmission:** All	**1st Gear Incorrect Ratio :** Engine speed: 450 r/min or more. Output speed: 350 r/min or more. Shift stage: 1st gear. Input speed: more than 0 r/min. Transmission fluid temperature sensor voltage: 4.5 volts or less. Voltage of battery: 10 volts or more. Transmission range switch rationality: only one signal. Time after shift changing finish: 2 seconds or more. Output speed: [(input speed - 200 r/min) / 1st gear ratio] or less. (4 seconds) If DTC P1779 (P0731) is set consecutively four times, the transaxle is locked into 3rd gear as a fail-safe measure, and the selector lever position indicator light (D, 1 through 4 or 5) flashes once per second. **Possible Causes:** • Malfunction of the input shaft speed sensor • Malfunction of the output shaft speed sensor • Malfunction of the PCM • Malfunction of the underdrive clutch retainer • Malfunction of the transfer drive gear or driven gear <4A/T> • Malfunction of the direct planetary carriers <5A/T> • Malfunction of clutch system and / or brake system • Malfunction of the valve body • Malfunction of the accumulator • Electrical noise generated
DTC: P1780 **1T PCM, MIL: Yes, TCIL: Yes** **Year:** 2009, 2010 **Model:** Eclipse, Endeavor, Galant **Engine:** 2.4L L4 VIN F, 3.8L V6 VIN S, 3.8L V6 VIN T **Transmission:** All	**2nd Gear Incorrect Ratio :** Engine speed: 450 r/min or more. Output speed: 500 r/min or more. Shift stage: 2nd gear. Input speed: more than 0 r/min. Transmission fluid temperature sensor voltage: 4.5 volts or less. Voltage of battery: 10 volts or more. Transmission range switch rationality: only one signal. Time after shift changing finish: 2 seconds or more. Output speed: [(input speed + 200 r/min) / 2nd gear ratio] or more. (4 seconds) Output speed: [(input speed - 200 r/min) / 2nd gear ratio] or less. (4 seconds) If DTC P1780 (P0732) is set consecutively four times, the transaxle is locked into 3rd gear as a fail-safe measure, and the selector lever position indicator light (D, 1 through 4 or 5) flashes once per second. **Possible Causes:** • Malfunction of the input shaft speed sensor • Malfunction of the output shaft speed sensor • Malfunction of the PCM • Malfunction of the underdrive clutch retainer • Malfunction of the transfer drive gear or driven gear <4A/T> • Malfunction of the direct planetary carriers <5A/T> • Malfunction of clutch system and / or brake system • Malfunction of the valve body • Malfunction of the accumulator • Electrical noise generated

DTC	Trouble Code Title, Conditions & Possible Causes
DTC: P1781 **1T PCM, MIL: Yes, TCIL: Yes** **Year:** 2009, 2010 **Model:** Eclipse, Endeavor, Galant **Engine:** 2.4L L4 VIN F, 3.8L V6 VIN S, 3.8L V6 VIN T **Transmission:** All	**3rd Gear Incorrect Ratio :** Engine speed: 450 r/min or more. Output speed: 900 r/min or more. Shift stage: 3rd gear. Input speed: more than 0 r/min. Transmission fluid temperature sensor voltage: 4.5 volts or less. Voltage of battery: 10 volts or more. Transmission range switch rationality: only one signal. Time after shift changing finish: 2 seconds or more. Output speed: [(input speed + 200 r/min) / 3rd gear ratio] or more. (4 seconds) Output speed: [(input speed - 200 r/min) / 3rd gear ratio] or less. (4 seconds) If DTC P1781 (P0733) is set consecutively four times, the transaxle is locked into 3rd gear as a fail-safe measure, and the selector lever position indicator light (D, 1 through 4 or 5) flashes once per second. **Possible Causes:** • Malfunction of the input shaft speed sensor • Malfunction of the output shaft speed sensor • Malfunction of the PCM • Malfunction of the underdrive clutch retainer • Malfunction of the transfer drive gear or driven gear <4A/T> • Malfunction of the direct planetary carriers <5A/T> • Malfunction of clutch system and / or brake system • Malfunction of the valve body • Malfunction of the accumulator • Electrical noise generated
DTC: P1782 **1T PCM, MIL: Yes, TCIL: Yes** **Year:** 2009, 2010 **Model:** Eclipse, Endeavor, Galant **Engine:** 2.4L L4 VIN F, 3.8L V6 VIN S, 3.8L V6 VIN T **Transmission:** All	**4th Gear Incorrect Ratio :** Engine speed: 450 r/min or more. Output speed: 900 r/min or more. Shift stage: 4th gear. Input speed: more than 0 r/min. Transmission fluid temperature sensor voltage: 4.5 volts or less. Voltage of battery: 10 volts or more. Transmission range switch rationality: only one signal. Time after shift changing finish: 2 seconds or more. Output speed: [(input speed + 200 r/min) / 4th gear ratio] or more. (4 seconds) Output speed: [(input speed - 200 r/min) / 4th gear ratio] or less. (4 seconds) If DTC P1782 (P0734) is set consecutively four times, the transaxle is locked into 3rd gear as a fail-safe measure, and the selector lever position indicator light (D, 1 through 4 or 5) flashes once per second. **Possible Causes:** • Malfunction of the input shaft speed sensor • Malfunction of the output shaft speed sensor • Malfunction of the PCM • Malfunction of the underdrive clutch retainer • Malfunction of the transfer drive gear or driven gear <4A/T> • Malfunction of the direct planetary carriers <5A/T> • Malfunction of clutch system and / or brake system • Malfunction of the valve body • Malfunction of the accumulator • Electrical noise generated
DTC: P1783 **1T PCM, MIL: Yes, TCIL: Yes** **Year:** 2009, 2010 **Model:** Eclipse, Endeavor, Galant **Engine:** 2.4L L4 VIN F, 3.8L V6 VIN S, 3.8L V6 VIN T **Transmission:** All	**5th Gear Incorrect Ratio <5A/T> :** Engine speed: 450 r/min or more. Output speed: 900 r/min or more. Shift stage: 5th gear. Input speed: more than 0 r/min. Transmission fluid temperature sensor voltage: 4.5 volts or less. Voltage of battery: 10 volts or more. Transmission range switch rationality: only one signal. Time after shift changing finish: 2 seconds or more. Output speed: [(input speed + 200 r/min) / 5th gear ratio] or more. (4 seconds) Output speed: [(input speed - 200 r/min) / 5th gear ratio] or less. (4 seconds) If DTC P1782 (P0734) is set consecutively four times, the transaxle is locked into 3rd gear as a fail-safe measure, and the selector lever position indicator light (D, 1 through 4 or 5) flashes once per second. **Possible Causes:** • Malfunction of the input shaft speed sensor • Malfunction of the output shaft speed sensor • Malfunction of the PCM • Malfunction of the underdrive clutch retainer • Malfunction of the transfer drive gear or driven gear <4A/T> • Malfunction of the direct planetary carriers <5A/T> • Malfunction of clutch system and / or brake system • Malfunction of the valve body • Malfunction of the accumulator • Electrical noise generated

DTC	Trouble Code Title, Conditions & Possible Causes
DTC: P1784 **1T PCM, MIL: Yes, TCIL: Yes** **Year:** 2009, 2010 **Model:** Eclipse, Endeavor, Galant **Engine:** 2.4L L4 VIN F, 3.8L V6 VIN S, 3.8L V6 VIN T **Transmission:** All	**Reverse Gear Incorrect Ratio:** Engine speed: 450 r/min or more. Output speed: 100 r/min or more. Shift stage: reverse gear. Input speed: more than 0 r/min. Transmission fluid temperature sensor voltage: 4.5 volts or less. Voltage of battery: 10 volts or more. Transmission range switch rationality: only one signal. Time after shift changing finish: 0.5 second or more. Output speed: [(input speed + 200 r/min) / reverse gear ratio] or more. (1 second) Output speed: [(input speed - 200 r/min) / reverse gear ratio] or less. (1 second) If DTC P1784 (P0736) is set consecutively four times, the transaxle is locked into 3rd gear as a fail-safe measure, and the selector lever position indicator light (D, 1 through 4 or 5) flashes once per second. **Possible Causes:** • Malfunction of the input shaft speed sensor • Malfunction of the output shaft speed sensor • Malfunction of the PCM • Malfunction of the underdrive clutch retainer • Malfunction of the transfer drive gear or driven gear <4A/T> • Malfunction of the direct planetary carriers <5A/T> • Malfunction of clutch system and / or brake system • Malfunction of the valve body • Malfunction of the accumulator • Electrical noise generated
DTC: P1786 **1T PCM, MIL: Yes, TCIL: Yes** **Year:** 2009, 2010 **Model:** Eclipse, Endeavor, Galant **Engine:** 2.4L L4 VIN F, 3.8L V6 VIN S, 3.8L V6 VIN T **Transmission:** All	**Torque Converter Clutch System (Stuck Off):** Solenoid status: plunging into connecting condition. Transmission range switch position: D. Time during 100% duty: 4 seconds or more. At start of lock-up operation, if lock-up clutch cannot be engaged even when duty ratio of torque converter clutch solenoid remains 100% for more than specified time, PCM judges that torque converter clutch is stuck OFF. **Possible Causes:** • Malfunction of the torque converter clutch solenoid valve • Malfunction of the input shaft speed sensor • Malfunction of the valve body • Damaged harness or connector • Malfunction of the PCM • Malfunction of the torque converter
DTC: P1787 **1T PCM, MIL: Yes** **Year:** 2009, 2010 **Model:** Eclipse, Endeavor, Galant **Engine:** 2.4L L4 VIN F, 3.8L V6 VIN S, 3.8L V6 VIN T **Transmission:** All	**Torque Converter Clutch System (Stuck On):** Throttle position sensor voltage: 1.5 volts or more. Output speed: 1,000 r/min or more. Solenoid status: OFF. Transmission range switch position: D. Time after lock up clutch release: 5 seconds or more. Calculated slip (engine speed - input speed): 5 r/min or less. (5 seconds) Or calculated slip (engine speed - input speed): -5 r/min or more. (5 seconds) **Possible Causes:** • Malfunction of the torque converter clutch solenoid valve • Malfunction of the valve body • Damaged harness or connector • Malfunction of the PCM
DTC: P1788 **1T PCM, MIL: Yes, TCIL: Yes** **Year:** 2009, 2010 **Model:** Eclipse, Endeavor, Galant **Engine:** 2.4L L4 VIN F, 3.8L V6 VIN S, 3.8L V6 VIN T **Transmission:** All	**A/T Control Relay System:** Voltage of battery: 9 volts or more. Time after PCM turns on A/T control relay: 0.5 second or more. A/T control relay output voltage: 7 volts or less. (0.1 second) **NOTE: If DTC P1788 (P1751) is set consecutively four times, the transaxle is locked into 3rd gear as a fail-safe measure, and the selector lever position indicator light (D, 1 through 4 or 5) flashes once per second.** **Possible Causes:** • Malfunction of the A/T control relay • Damaged harness or connector • Malfunction of the PCM
DTC: P1790 **1T PCM, MIL: Yes** **Year:** 2009 **Model:** Raider **Engine:** 3.7L V6 VIN K **Transmission:** All	**Fault Immediately After Shift:** After a Gear Ratio Error code is stored. After a Gear Ratio Error DTC has already been set. The DTC is set if the fault happened within 1.3 seconds of a shift. The DTC set time will vary from 1.214 seconds to 15 seconds. **Possible Causes:** • Fault after shift

DTC	Trouble Code Title, Conditions & Possible Causes
DTC: P1794 **1T PCM, MIL: Yes** **Year:** 2009 **Model:** Raider **Engine:** 3.7L V6 VIN K **Transmission:** All	**Speed Sensor Ground Error (45RFE/545RFE Transmission):** The gear ratio is monitored continuously while the Transmission is in gear. After a controller reset in neutral and a ratio of input to output, of 1 to 2. This DTC can take up to five minutes of problem identification before illuminating the MIL **Possible Causes:** • Speed sensor ground circuit open • Speed sensor ground circuit shorted to ground • Speed sensor ground circuit shorted to voltage • PCM
DTC: P1794 **1T PCM, MIL: Yes** **Year:** 2009 **Model:** Raider **Engine:** 3.7L V6 VIN K **Transmission:** All	**Speed Sensor Ground Error (42RLE Transmission):** Every 7ms after a controller reset with transmission in neutral. After a PCM reset in neutral and Input and Output sensor ratio equals 2.50 to 1.0 ± 50.0 rpm. **Possible Causes:** • Speed sensor ground circuit open • PCM
DTC: P1796 **1T ECM, MIL: Yes** **Year:** 2009, 2010 **Model:** Outlander **Engine:** 3.0L V6 VIN X **Transmission:** All	**Idle neutral control malfunction:** With the idle neutral control, the status that the "actual slip speed - target slip speed" is faster than 300 r/min continues for 10 seconds. After 1.05 seconds after the TCM activation, the failure signal of longitudinal G-sensor is received. **Possible Causes:** • Malfunction of the CAN circuit • Malfunction of the P0753: low clutch linear solenoid valve system • Malfunction of the P0731: 1st gear incorrect ratio • Malfunction of the TCM
DTC: P1797 **T PCM, MIL: Yes** **Year:** 2009 **Model:** Raider **Engine:** 3.7L V6 VIN K **Transmission:** All	**Manual Shift Overheat:** Continuously with engine running. f the Engine Temperature exceeds 123° C (255° F) or the Transmission Temperature exceeds 135° C (275° F). **NOTE: Aggressive driving or driving in low for extended periods of time will set this DTC.** **Possible Causes:** • Manual shift overheat
DTC: P1802 **1T ECM, MIL: Yes** **Year:** 2009, 2010 **Model:** Lancer **Engine:** 2.0L L4 VIN U, 2.4L L4 VIN W **Transmission:** All	**Shift Lever System (LIN communication malfunction):** The CAN back-up communication is determined to be abnormal. **Possible Causes:** • Malfunction of the shift lever-ECU. • Malfunction of the LIN bus. • Malfunction of TC-SST-ECU.
DTC: P1803 **1T ECM, MIL: Yes** **Year:** 2009, 2010 **Model:** Lancer **Engine:** 2.0L L4 VIN U, 2.4L L4 VIN W **Transmission:** All	**Shift Lever System (CAN, LIN Time-out Error):** The CAN and LIN communication with the shift lever-ECU is determined to be abnormal. **Possible Causes:** • Malfunction of the shift lever-ECU. • Malfunction of the LIN bus. • The CAN bus line is defective. • Malfunction of TC-SST-ECU.
DTC: P1804 **1T ECM, MIL: Yes** **Year:** 2009, 2010 **Model:** Lancer **Engine:** 2.0L L4 VIN U, 2.4L L4 VIN W **Transmission:** All	**Shift Fork Position Sensor 1 and 2 System (Power supply voltage low range out):** Voltage of battery: 8 V or more. Voltage of battery: 16.5 V or less. Supply voltage: 3.07 V or less. (140 millisecond) **Possible Causes:** • Malfunction of TC-SST-ECU • Malfunction of shift fork position sensor 1 and 2
DTC: P1805 **1T ECM, MIL: Yes** **Year:** 2009, 2010 **Model:** Lancer **Engine:** 2.0L L4 VIN U, 2.4L L4 VIN W **Transmission:** All	**Shift Fork Position Sensor 1 and 2 System (Power supply voltage high range out):** Voltage of battery: 8 V or more. Voltage of battery: 16.5 V or less. The supply voltage remains 3.25 V or less for 140 milliseconds. **Possible Causes:** • Malfunction of TC-SST-ECU • Malfunction of shift fork position sensor 1 and 2

DTC	Trouble Code Title, Conditions & Possible Causes
DTC: P1806 **1T ECM, MIL: Yes** **Year:** 2009, 2010 **Model:** Lancer **Engine:** 2.0L L4 VIN U, 2.4L L4 VIN W **Transmission:** All	**Shift Fork Position Sensor 3 and 4 System (Power supply voltage low range out):** Voltage of battery: 8 V or more. Voltage of battery: 16.5 V or less. The supply voltage remains 3.07 V or more for 140 milliseconds. **Possible Causes:** • Malfunction of TC-SST-ECU. • Malfunction of shift fork position sensor 3 and 4.
DTC: P1807 **1T ECM, MIL: Yes** **Year:** 2009, 2010 **Model:** Lancer **Engine:** 2.0L L4 VIN U, 2.4L L4 VIN W **Transmission:** All	**Shift Fork Position Sensor 3 and 4 System (Power supply voltage high range out):** Voltage of battery: 8 V or more. Voltage of battery: 16.5 V or less. The supply voltage remains 3.25 V or less for 140 milliseconds. **Possible Causes:** • Malfunction of TC-SST-ECU • Malfunction of shift fork position sensor 3 and 4
DTC: P1808 **1T ECM, MIL: Yes** **Year:** 2009, 2010 **Model:** Lancer **Engine:** 2.0L L4 VIN U, 2.4L L4 VIN W **Transmission:** All	**TC-SST-ECU temperature, fluid temperature sensor system (Correlation error):** Voltage of battery: 8 V or more. Voltage of battery: 16.5 V or less. Calculated temperature ("Temperature B: Fluid temperature sensor output" - "Temperature A: TC-SST-ECU temperature sensor output"): 20°C (36°F) or more. (12 seconds) **Possible Causes:** • Malfunction of TC-SST-ECU.
DTC: P180C **1T ECM, MIL: Yes** **Year:** 2009, 2010 **Model:** Lancer **Engine:** 2.0L L4 VIN U, 2.4L L4 VIN W **Transmission:** All	**Clutch pressure cut spool sticking:** The clutch pressure cut spool is determined to be seized. **Possible Causes:** • Malfunction of TC-SST-ECU.
DTC: P181B **1T ECM, MIL: Yes** **Year:** 2009, 2010 **Model:** Lancer **Engine:** 2.0L L4 VIN U, 2.4L L4 VIN W **Transmission:** All	**Clutch 1 (Pressure low range out):** Voltage of battery: 8 V or more. Voltage of battery: 16.5 V or less. Engine speed: 650 r/min or more. Time since above engine condition: 1.5 seconds or more. Clutch 1 (odd) state: Slip or engaged. Clutch/shift pressure solenoid 1: Not OFF or not in valve cleaning mode. The calculated pressure remains 2,000 hPa or less for 5 seconds. **Possible Causes:** • Malfunction of TC-SST-ECU. • Insufficient fluid level. • Improper installation of mechatronic assembly.
DTC: P181C **1T ECM, MIL: Yes** **Year:** 2009, 2010 **Model:** Lancer **Engine:** 2.0L L4 VIN U, 2.4L L4 VIN W **Transmission:** All	**Clutch 1 (Pressure high range out):** The pressure of the clutch 1 is too high. **Possible Causes:** • Malfunction of TC-SST-ECU.
DTC: P181E **1T ECM, MIL: Yes** **Year:** 2009, 2010 **Model:** Lancer **Engine:** 2.0L L4 VIN U, 2.4L L4 VIN W **Transmission:** All	**Clutch 2 (Pressure low range out):** Voltage of battery: 8 V or more. Voltage of battery: 16.5 V or less. Engine speed: 650 r/min or more. Time since above engine condition: 1.5 seconds or more. Clutch 2 (even) state: Slip or engaged. Clutch/shift pressure solenoid 2: Not OFF or not in valve cleaning mode. The calculated pressure remains 2,000 hPa or less for 5 seconds. **Possible Causes:** • Malfunction of TC-SST-ECU • Insufficient fluid level • Improper installation of mechatronic assembly
DTC: P181F **1T ECM, MIL: Yes** **Year:** 2009, 2010 **Model:** Lancer **Engine:** 2.0L L4 VIN U, 2.4L L4 VIN W **Transmission:** All	**Clutch 2 (Pressure high range out):** Engine speed: 650 r/min or more. Time since above engine condition: 1.5 seconds or more. Clutch 2 (even) state: Slip or engaged. Target clutch 2 (even) pressure: 3,000 hPa or more. Calculated pressure: -2,000 hPa or less. **Possible Causes:** • Malfunction of TC-SST-ECU.

DTC	Trouble Code Title, Conditions & Possible Causes
DTC: P1820 **1T ECM, MIL: Yes** **Year:** 2009, 2010 **Model:** Lancer **Engine:** 2.0L L4 VIN U, 2.4L L4 VIN W **Transmission:** All	**Shift Fork Position Sensor 1 System (Voltage low range out):** Voltage of battery: 8 V or more. Voltage of battery: 16.5 V or less. The position sensor voltage remains 0.128 V or more for 160 milliseconds. **Possible Causes:** • Malfunction of TC-SST-ECU. • Malfunction of shift fork position sensor 1.
DTC: P1821 **1T ECM, MIL: Yes** **Year:** 2009, 2010 **Model:** Lancer **Engine:** 2.0L L4 VIN U, 2.4L L4 VIN W **Transmission:** All	**Shift Fork Position Sensor 1 System (Voltage high range out):** Voltage of battery: 8 V or more. Voltage of battery: 16.5 V or less. The position sensor voltage remains 3.008 V or less for 160 milliseconds. **Possible Causes:** • Malfunction of TC-SST-ECU. • Malfunction of shift fork position sensor 1.
DTC: P1822 **1T ECM, MIL: Yes** **Year:** 2009, 2010 **Model:** Lancer **Engine:** 2.0L L4 VIN U, 2.4L L4 VIN W **Transmission:** All	**Shift Fork Position Sensor 1 System (Output range out):** Voltage of battery: 8 V or more. Voltage of battery: 16.5 V or less. Engine speed: 650 r/min or more. Time since above engine condition: 1.5 seconds or more. Shift system status: Gear change mode (during shift fork moving). The status with the shift fork 1 position counter (digitized sensor value) -500 counts or more, or with the shift fork 1 position counter (digitized sensor value) 500 counts or less continues for 180 milliseconds. **Possible Causes:** • Malfunction of TC-SST-ECU. • Malfunction of shift fork position sensor 1.
DTC: P1823 **1T ECM, MIL: Yes** **Year:** 2009, 2010 **Model:** Lancer **Engine:** 2.0L L4 VIN U, 2.4L L4 VIN W **Transmission:** All	**Shift Fork Position Sensor 1 System (Neutral):** Voltage of battery: 8 V or more. Voltage of battery: 16.5 V or less. Engine speed: 650 r/min or more. Time since above engine condition: 1.5 seconds or more. Input shaft 1 (odd) speed: 500 r/min or more. Input shaft 2 (even) speed: 500 r/min or more. Shift fork 1 position: Neutral. Input shaft 2 (even) gear: Engaged. Engine speed - input shaft 2 speed: 50 r/min or less. The calculated speed remains 40 r/min or more for 500 millisecond. **Possible Causes:** • Malfunction of TC-SST-ECU. • Malfunction of shift fork position sensor 1.
DTC: P1824 **1T ECM, MIL: Yes** **Year:** 2009, 2010 **Model:** Lancer **Engine:** 2.0L L4 VIN U, 2.4L L4 VIN W **Transmission:** All	**Shift Fork Position Sensor 1 System (Poor performance):** Voltage of battery: 8 V or more. Voltage of battery: 16.5 V or less. Engine speed: 650 r/min or more. Time since above engine condition: 1.5 seconds or more. Input shaft 1 (odd) speed: 500 r/min or more. Shift fork 1 current gear: 1st gear. Clutch 1 (odd) transmit torque: 40 Nm (30 ft-lb) or more. Requested shift fork: Not shift fork 1. The travel distance of the shift fork 1 remains 6 mm (0.24 inch) or more for 250 millisecond. **Possible Causes:** • Malfunction of TC-SST-ECU. • Malfunction of shift fork position sensor 1.
DTC: P1825 **1T ECM, MIL: Yes** **Year:** 2009, 2010 **Model:** Lancer **Engine:** 2.0L L4 VIN U, 2.4L L4 VIN W **Transmission:** All	**Shift Fork Position Sensor 2 System (Voltage low range out):** Voltage of battery: 8 V or more. Voltage of battery: 16.5 V or less. The position sensor voltage remains 0.128 V or more for 160 millisecond. **Possible Causes:** • Malfunction of TC-SST-ECU • Malfunction of shift fork position sensor 2
DTC: P1826 **1T ECM, MIL: Yes** **Year:** 2009, 2010 **Model:** Lancer **Engine:** 2.0L L4 VIN U, 2.4L L4 VIN W **Transmission:** All	**Shift Fork Position Sensor 2 System (Voltage high range out):** Voltage of battery: 8 V or more. Voltage of battery: 16.5 V or less. The position sensor voltage remains 3.008 V or less for 160 millisecond. **Possible Causes:** • Malfunction of TC-SST-ECU • Malfunction of shift fork position sensor 2

DTC	Trouble Code Title, Conditions & Possible Causes
DTC: p1828 **1T ECM, MIL: Yes** **Year:** 2009, 2010 **Model:** Lancer **Engine:** 2.0L L4 VIN U, 2.4L L4 VIN W **Transmission:** All	**Shift Fork Position Sensor 2 System (Neutral):** Voltage of battery: 8 V or more. Voltage of battery: 16.5 V or less. Engine speed: 650 r/min or more. Time since above engine condition: 1.5 seconds or more. Input shaft 1 (odd) speed: 500 r/min or more. Input shaft 2 (even) speed: 500 r/min or more. Shift fork 2 position: Neutral. Input shaft 2 (even) gear: Engaged. Engine speed - input shaft 2 speed: 50 r/min or less. The calculated speed remains 40 r/min or more for 500 millisecond. **Possible Causes:** • Malfunction of TC-SST-ECU • Malfunction of shift fork position sensor 2
DTC: P1829 **1T ECM, MIL: Yes** **Year:** 2009, 2010 **Model:** Lancer **Engine:** 2.0L L4 VIN U, 2.4L L4 VIN W **Transmission:** All	**Shift Fork Position Sensor 2 System (Poor performance):** Voltage of battery: 8 V or more. Voltage of battery: 16.5 V or less. Engine speed: 650 r/min or more. Time since above engine condition: 1.5 seconds or more. Input shaft 1 (odd) speed: 500 r/min or more. Shift fork 2 current gear: 5th gear. Clutch 1 (odd) transmit torque: 40 Nm (30 ft-lb) or more. Requested shift fork: Not shift fork 2. The travel distance of the shift fork 2 remains 6 mm (0.24 inch) or more for 250 millisecond. **Possible Causes:** • Malfunction of TC-SST-ECU • Malfunction of shift fork position sensor 2
DTC: P182A **1T ECM, MIL: Yes** **Year:** 2009, 2010 **Model:** Lancer **Engine:** 2.0L L4 VIN U, 2.4L L4 VIN W **Transmission:** All	**Shift Fork Position Sensor 3 System (Voltage low range out:** Voltage of battery: 8 V or more. Voltage of battery: 16.5 V or less. The position sensor voltage remains 0.128 V or more for 160 milliseconds. **Possible Causes:** • Malfunction of TC-SST-ECU • Malfunction of shift fork position sensor 3
DTC: P182B **1T ECM, MIL: Yes** **Year:** 2009, 2010 **Model:** Lancer **Engine:** 2.0L L4 VIN U, 2.4L L4 VIN W **Transmission:** All	**Shift Fork Position Sensor 3 System (Voltage high range out):** Voltage of battery: 8 V or more. Voltage of battery: 16.5 V or less. The position sensor voltage remains 3.008 V or less for 160 millisecond. **Possible Causes:** • Malfunction of TC-SST-ECU • Malfunction of shift fork position sensor 3
DTC: P182C **1T ECM, MIL: Yes** **Year:** 2009, 2010 **Model:** Lancer **Engine:** 2.0L L4 VIN U, 2.4L L4 VIN W **Transmission:** All	**Shift Fork Position Sensor 3 System (Output range out):** Voltage of battery: 8 V or more. Voltage of battery: 16.5 V or less. Engine speed: 650 r/min or more. Time since above engine condition: 1.5 seconds or more. Shift system status: Gear change mode (during shift fork moving). The status with the shift fork 3 position counter (digitized sensor value) -500 counts or more, or with the shift fork 3 position counter (digitized sensor value) 500 counts or less continues for 180 millisecond. **Possible Causes:** • Malfunction of TC-SST-ECU • Malfunction of shift fork position sensor 3
DTC: P182D **1T ECM, MIL: Yes** **Year:** 2009, 2010 **Model:** Lancer **Engine:** 2.0L L4 VIN U, 2.4L L4 VIN W **Transmission:** All	**Shift Fork Position Sensor 3 System (Neutral):** Voltage of battery: 8 V or more. Voltage of battery: 16.5 V or less. Engine speed: 650 r/min or more. Time since above engine condition: 1.5 seconds or more. Input shaft 1 (odd) speed: 500 r/min or more. Input shaft 2 (even) speed: 500 r/min or more. Shift fork 3 position: Neutral. Input shaft 1 (odd) gear: Engaged. Engine speed - input shaft 1 speed: 50 r/min or less. The calculated speed remains 40 r/min or more for 500 millisecond. **Possible Causes:** • Malfunction of TC-SST-ECU • Malfunction of shift fork position sensor 3
DTC: P182E **1T ECM, MIL: Yes** **Year:** 2009, 2010 **Model:** Lancer **Engine:** 2.0L L4 VIN U, 2.4L L4 VIN W **Transmission:** All	**Shift Fork Position Sensor 3 System (Poor performance):** Voltage of battery: 8 V or more. Voltage of battery: 16.5 V or less. Engine speed: 650 r/min or more. Time since above engine condition: 1.5 seconds or more. Input shaft 2 (even) speed: 500 r/min or more. Shift fork 3 current gear: 6th gear. Clutch 2 (even) transmit torque: 40 Nm (30 ft-lb) or more. Requested shift fork: Not shift fork 3. The travel distance of the shift fork 3 remains -6 mm (-0.24 inch) or less for 250 millisecond. **Possible Causes:** • Malfunction of TC-SST-ECU • Malfunction of shift fork position sensor 3

DTC	Trouble Code Title, Conditions & Possible Causes

DTC: P1831
1T ECM, MIL: Yes
Year: 2009, 2010
Model: Lancer
Engine: 2.0L L4 VIN U, 2.4L L4 VIN W
Transmission: All
	Shift Fork Position Sensor 4 System (Voltage low range out):
	Voltage of battery: 8 V or more. Voltage of battery: 16.5 V or less. The position sensor voltage remains 0.128 V or more for 160 milliseconds.
	Possible Causes:
	   • Malfunction of TC-SST-ECU
	   • Malfunction of shift fork position sensor 4

DTC: P1832
1T ECM, MIL: Yes
Year: 2009, 2010
Model: Lancer
Engine: 2.0L L4 VIN U, 2.4L L4 VIN W
Transmission: All
	Shift Fork Position Sensor 4 System (Voltage high range out):
	Voltage of battery: 8 V or more. Voltage of battery: 16.5 V or less. The position sensor voltage remains 3.008 V or less for 160 milliseconds.
	Possible Causes:
	   • Malfunction of TC-SST-ECU
	   • Malfunction of shift fork position sensor 4

DTC: P1833
1T ECM, MIL: Yes
Year: 2009, 2010
Model: Lancer
Engine: 2.0L L4 VIN U, 2.4L L4 VIN W
Transmission: All
	Shift Fork Position Sensor 4 System (Output range out):
	Voltage of battery: 8 V or more. Voltage of battery: 16.5 V or less. Engine speed: 650 r/min or more. Time since above engine condition: 1.5 seconds or more. Shift system status: Gear change mode (during shift fork moving). The status with the shift fork 4 position counter (digitized sensor value) -500 counts or more, or with the shift fork 4 position counter (digitized sensor value) 500 counts or less continues for 180 millisecond.
	Possible Causes:
	   • Malfunction of TC-SST-ECU
	   • Malfunction of shift fork position sensor 4

DTC: P1834
1T ECM, MIL: Yes
Year: 2009, 2010
Model: Lancer
Engine: 2.0L L4 VIN U, 2.4L L4 VIN W
Transmission: All
	Shift Fork Position Sensor 4 System (Neutral):
	Voltage of battery: 8 V or more. Voltage of battery: 16.5 V or less. Engine speed: 650 r/min or more. Time since above engine condition: 1.5 seconds or more. Input shaft 1 (odd) speed: 500 r/min or more. Input shaft 2 (even) speed: 500 r/min or more. Shift fork 4 position: Neutral. Input shaft 1 (odd) gear: Engaged. Engine speed - input shaft 1 speed: 50 r/min or less. The calculated speed remains 40 r/min or more for 500 millisecond.
	Possible Causes:
	   • Malfunction of TC-SST-ECU
	   • Malfunction of shift fork position sensor 4

DTC: P1835
1T ECM, MIL: Yes
Year: 2009, 2010
Model: Lancer
Engine: 2.0L L4 VIN U, 2.4L L4 VIN W
Transmission: All
	Shift Fork Position Sensor 4 System (Poor performance):
	Voltage of battery: 8 V or more. Voltage of battery: 16.5 V or less. Engine speed: 650 r/min or more. Time since above engine condition: 1.5 seconds or more. Input shaft 2 (even) speed: 500 r/min or more. Shift fork 4 current gear: 4th gear. Clutch 2 (even) transmit torque: 40 Nm (30 ft-lb) or more. Requested shift fork: Not shift fork 4. The travel distance of the shift fork 4 remains 6 mm (0.24 inch) or more for 250 millisecond.
	Possible Causes:
	   • Malfunction of TC-SST-ECU
	   • Malfunction of shift fork position sensor 4

DTC: P1836
1T ECM, MIL: Yes
Year: 2009, 2010
Model: Lancer
Engine: 2.0L L4 VIN U, 2.4L L4 VIN W
Transmission: All
	Shift Fork 1 Malfunction:
	Voltage of battery: 8 V or more. Voltage of battery: 16.5 V or less. Engine speed: 650 r/min or more. Time since above engine condition: 1.5 seconds or more.
	Shift fork 1 current gear: Neutral. The shift fork position remains 1.8 mm (0.071 inch) or less for 250 millisecond.
	Possible Causes:
	   • Malfunction of TC-SST-ECU
	   • Malfunction of TC-SST shift fork

DTC: P183D
1T ECM, MIL: Yes
Year: 2009, 2010
Model: Lancer
Engine: 2.0L L4 VIN U, 2.4L L4 VIN W
Transmission: All
	Shift Fork 2 Malfunction:
	Voltage of battery: 8 V or more. Voltage of battery: 16.5 V or less. Engine speed: 650 r/min or more. Time since above engine condition: 1.5 seconds or more.
	Requested shift fork: Shift fork 2. The status with the shift fork position 6 mm (0.24 inch) [1st gear side] or more, or with -6 mm (-0.24 inch) [reverse gear side] or less, or with 1.8 mm (0.071 inch) [neutral position] or less and with the initialized fork position - shift fork 2 position 2.5 mm (0.098 inch) or more continues for 200 millisecond.
	Possible Causes:
	   • Malfunction of TC-SST-ECU
	   • Malfunction of TC-SST shift fork

DTC	Trouble Code Title, Conditions & Possible Causes
DTC: P1844 **1T ECM, MIL: Yes** **Year:** 2009, 2010 **Model:** Lancer **Engine:** 2.0L L4 VIN U, 2.4L L4 VIN W **Transmission:** All	**Shift Fork 3 Malfunction:** Voltage of battery: 8 V or more. Voltage of battery: 16.5 V or less. Engine speed: 650 r/min or more. Time since above engine condition: 1.5 seconds or more. Requested shift fork: Shift fork 3. The status with the shift fork position 6 mm (0.24 inch) [4th gear side] or more, or with -6 mm (-0.24 inch) [2nd gear side] or less, or with 1.8 mm (0.071 inch) [neutral position] or less and with the initialized fork position - shift fork 3 position 2.5 mm (0.098 inch) or more continues for 200 millisecond. **Possible Causes:** • Malfunction of TC-SST-ECU • Malfunction of TC-SST shift fork
DTC: P184B **1T ECM, MIL: Yes** **Year:** 2009, 2010 **Model:** Lancer **Engine:** 2.0L L4 VIN U, 2.4L L4 VIN W **Transmission:** All	**Shift Fork 4 Malfunction:** Voltage of battery: 8 V or more. Voltage of battery: 16.5 V or less. Engine speed: 650 r/min or more. Time since above engine condition: 1.5 seconds or more. Requested shift fork: Shift fork 4. The status of the shift fork position -6 mm (-0.24 inch) [6th gear side] or less, or with 1.8 mm (0.071 inch) [neutral position] or less and with the initialized fork position - shift fork 3 position 2.5 mm (0.098 inch) or more continues for 200 millisecond. **Possible Causes:** • Malfunction of TC-SST-ECU • Malfunction of TC-SST shift fork
DTC: P1852 **1T ECM, MIL: Yes** **Year:** 2009, 2010 **Model:** Lancer **Engine:** 2.0L L4 VIN U, 2.4L L4 VIN W **Transmission:** All	**Shift Fork 1 or 2 opposite direction movement:** Voltage of battery: 8 V or more. Voltage of battery: 16.5 V or less. Engine speed: 650 r/min or more. Time since above engine condition: 1.5 seconds or more. Shift fork 1 and 2 current gear: Neutral. Shift fork 1 and 2 target direction: 0 mm (0 inch) or more. The shift fork 1 and 2 positions remain -1.8 mm (-0.071 inch) or more for 250 millisecond. Shift fork 1 and 2 target direction: 0 mm (0 inch) or more **Possible Causes:** • Malfunction of TC-SST-ECU • Malfunction of valve body
DTC: P1855 **1T ECM, MIL: Yes** **Year:** 2009, 2010 **Model:** Lancer **Engine:** 2.0L L4 VIN U, 2.4L L4 VIN W **Transmission:** All	**Shift Fork 3 or 4 opposite direction movement:** Voltage of battery: 8 V or more. Voltage of battery: 16.5 V or less. Engine speed: 650 r/min or more. Time since above engine condition: 1.5 seconds or more. Shift fork 3 and 4 current gear: Neutral. Shift fork 3 and 4 target direction: 0 mm (0 inch) or more. The shift fork 3 and 4 positions remain -1.8 mm (-0.071 inch) or more for 250 millisecond. **Possible Causes:** • Malfunction of TC-SST-ECU • Malfunction of valve body
DTC: P1857 **1T ECM, MIL: Yes** **Year:** 2009, 2010 **Model:** Lancer **Engine:** 2.0L L4 VIN U, 2.4L L4 VIN W **Transmission:** All	**Odd number gear axle interlock:** Voltage of battery: 8 V or more. Voltage of battery: 16.5 V or less. Engine speed: 650 r/min or more. Time since above engine condition: 1.5 seconds or more. The shift fork 1 and 2 positions remain 3.8 mm (0.15 inch) or less for 500 millisecond. **Possible Causes:** • Malfunction of TC-SST-ECU • Malfunction of TC-SST gear
DTC: P1858 **1T ECM, MIL: Yes** **Year:** 2009, 2010 **Model:** Lancer **Engine:** 2.0L L4 VIN U, 2.4L L4 VIN W **Transmission:** All	**Even number gear axle interlock:** Voltage of battery: 8 V or more. Voltage of battery: 16.5 V or less. Engine speed: 650 r/min or more. Time since above engine condition: 1.5 seconds or more. The shift fork 3 and 4 positions remain 3.8 mm (0.15 inch) or less for 500 millisecond. **Possible Causes:** • Malfunction of TC-SST-ECU • Malfunction of TC-SST gear
DTC: P185D **1T ECM, MIL: Yes** **Year:** 2009, 2010 **Model:** Lancer **Engine:** 2.0L L4 VIN U, 2.4L L4 VIN W **Transmission:** All	**Clutch open not possible:** The disengagement of the clutch 1 and 2 are determined to be impossible. **Possible Causes:** • Malfunction of clutch assembly.

DTC	Trouble Code Title, Conditions & Possible Causes
DTC: P1862 **1T ECM, MIL: Yes** **Year:** 2009, 2010 **Model:** Lancer **Engine:** 2.0L L4 VIN U, 2.4L L4 VIN W **Transmission:** All	**High side 1 system (Overcurrent):** Voltage of battery: 8 V or more. Voltage of battery: 16.5 V or less. High side 1 switch: ON. The status with the current of the FET channel shunt 8.1 A or less and with the FET channel output 100 mV or more continues for 1 second. **Possible Causes:** • Malfunction of TC-SST-ECU.
DTC: P1863 **1T ECM, MIL: Yes** **Year:** 2009, 2010 **Model:** Lancer **Engine:** 2.0L L4 VIN U, 2.4L L4 VIN W **Transmission:** All	**High side 1 system (Open circuit):** Voltage of battery: 8 V or more. Voltage of battery: 16.5 V or less. High side 1 switch: ON. The status with the FET switch ON and with the FET channel output 100 mV or more continues for 1 second. **Possible Causes:** • Malfunction of TC-SST-ECU.
DTC: P1864 **1T ECM, MIL: Yes** **Year:** 2009, 2010 **Model:** Lancer **Engine:** 2.0L L4 VIN U, 2.4L L4 VIN W **Transmission:** All	**High side 1 system (Short to power supply):** Voltage of battery: 8 V or more. Voltage of battery: 16.5 V or less. High side 1 switch: OFF. The FET switch remains OFF for 12 seconds. **Possible Causes:** • Malfunction of TC-SST-ECU.
DTC: P1866 **1T ECM, MIL: Yes** **Year:** 2009, 2010 **Model:** Lancer **Engine:** 2.0L L4 VIN U, 2.4L L4 VIN W **Transmission:** All	**High side 2 system (Overcurrent):** Voltage of battery: 8 V or more. Voltage of battery: 16.5 V or less. High side 2 switch: ON. The status with the current of the FET channel shunt 8.1 A or less and with the FET channel output 100 mV or more continues for 1 second. **Possible Causes:** • Malfunction of TC-SST-ECU.
DTC: P1867 **1T ECM, MIL: Yes** **Year:** 2009, 2010 **Model:** Lancer **Engine:** 2.0L L4 VIN U, 2.4L L4 VIN W **Transmission:** All	**High side 2 system (Open circuit):** Voltage of battery: 8 V or more. Voltage of battery: 16.5 V or less. High side 2 switch: ON. The status with the FET switch ON and with the FET channel output 100 mV or more continues for 1 second. **Possible Causes:** • Malfunction of TC-SST-ECU.
DTC: P1868 **1T ECM, MIL: Yes** **Year:** 2009, 2010 **Model:** Lancer **Engine:** 2.0L L4 VIN U, 2.4L L4 VIN W **Transmission:** All	**High side 2 system (Short to power supply):** Voltage of battery: 8 V or more. Voltage of battery: 16.5 V or less. High side 2 switch: OFF. The FET switch remains OFF for 12 seconds. **Possible Causes:** • Malfunction of TC-SST-ECU.
DTC: P186A **1T ECM, MIL: Yes** **Year:** 2009, 2010 **Model:** Lancer **Engine:** 2.0L L4 VIN U, 2.4L L4 VIN W **Transmission:** All	**High side 3 system (Overcurrent):** Voltage of battery: 8 V or more. Voltage of battery: 16.5 V or less. High side 3 switch: ON. The status with the current of the FET channel shunt 8.1 A or less and with the FET channel output 100 mV or more continues for 1 second. **Possible Causes:** • Malfunction of TC-SST-ECU.
DTC: P186B **1T ECM, MIL: Yes** **Year:** 2009, 2010 **Model:** Lancer **Engine:** 2.0L L4 VIN U, 2.4L L4 VIN W **Transmission:** All	**High side 3 system (Open circuit):** Voltage of battery: 8 V or more. Voltage of battery: 16.5 V or less. High side 3 switch: ON. The status with the FET switch ON and with the FET channel output 100 mV or more continues for 1 second. **Possible Causes:** • Malfunction of TC-SST-ECU.

DTC	Trouble Code Title, Conditions & Possible Causes
DTC: P186C **1T ECM, MIL: Yes** **Year:** 2009, 2010 **Model:** Lancer **Engine:** 2.0L L4 VIN U, 2.4L L4 VIN W **Transmission:** All	**High side 3 system (Short to power supply):** Voltage of battery: 8 V or more. Voltage of battery: 16.5 V or less. High side 3 switch: OFF. The FET (Field Effect Transistor): Can't be switched off. (12 seconds) **Possible Causes:** • Malfunction of TC-SST-ECU.
DTC: P186D **1T ECM, MIL: Yes** **Year:** 2009, 2010 **Model:** Lancer **Engine:** 2.0L L4 VIN U, 2.4L L4 VIN W **Transmission:** All	**High side 1 system (Voltage low range out):** Voltage of battery: 8 V or more. Voltage of battery: 16.5 V or less. Engine speed: 650 r/min or more. Time since above engine condition: 1.5 seconds or more. FET (Field Effect Transistor) of high side 1: Switched off. The FET channel output remains 100 mV or more for 5 seconds. **Possible Causes:** • Malfunction of TC-SST-ECU • Malfunction of power supply circuit (open circuit)
DTC: P186E **1T ECM, MIL: Yes** **Year:** 2009, 2010 **Model:** Lancer **Engine:** 2.0L L4 VIN U, 2.4L L4 VIN W **Transmission:** All	**High side 2 system (Voltage low range out):** Voltage of battery: 8 V or more. Voltage of battery: 16.5 V or less. Engine speed: 650 r/min or more. Time since above engine condition: 1.5 seconds or more. FET (Field Effect Transistor) of high side 2: Switched off. The FET channel output remains 100 mV or more for 5 seconds. **Possible Causes:** • Malfunction of TC-SST-ECU • Malfunction of power supply circuit (open circuit)
DTC: P186F **1T ECM, MIL: Yes** **Year:** 2009, 2010 **Model:** Lancer **Engine:** 2.0L L4 VIN U, 2.4L L4 VIN W **Transmission:** All	**High side 3 system (Voltage low range out):** Voltage of battery: 8 V or more. Voltage of battery: 16.5 V or less. Engine speed: 650 r/min or more. Time since above engine condition: 1.5 seconds or more. FET (Field Effect Transistor) of high side 3: Switched off. The FET channel output remains 100 mV or more for 5 seconds. **Possible Causes:** • Malfunction of TC-SST-ECU • Malfunction of power supply circuit (open circuit)
DTC: P1870 **1T ECM, MIL: Yes** **Year:** 2009, 2010 **Model:** Lancer **Engine:** 2.0L L4 VIN U, 2.4L L4 VIN W **Transmission:** All	**Engine torque signal abnormality:** Voltage of battery: 8 V or more. Voltage of battery: 16.5 V or less. Time after TC-SST-ECU start: 5 seconds or more. Engine torque signal: SNA, or parity/toggle error. (immediately) **Possible Causes:** • The CAN bus line is defective. • Malfunction of engine control module • Malfunction of TC-SST-ECU
DTC: P1871 **1T ECM, MIL: Yes** **Year:** 2009, 2010 **Model:** Lancer **Engine:** 2.0L L4 VIN U, 2.4L L4 VIN W **Transmission:** All	**APS system (Signal abnormality):** Voltage of battery: 8 V or more. Voltage of battery: 16.5 V or less. Time after TC-SST-ECU start: 5 seconds or more. APS signal: SNA, or parity/toggle error. **Possible Causes:** • The CAN bus line is defective. • APS malfunction • Malfunction of engine control module • Malfunction of TC-SST-ECU
DTC: P1872 **1T ECM, MIL: Yes** **Year:** 2009, 2010 **Model:** Lancer **Engine:** 2.0L L4 VIN U, 2.4L L4 VIN W **Transmission:** All	**Between shift lever and TC-SST system (Q-A function abnormality):** The shift lever-ECU is determined to be abnormal. **Possible Causes:** • Malfunction of the shift lever-ECU • Malfunction of TC-SST-ECU
DTC: P1873 **1T ECM, MIL: Yes** **Year:** 2009, 2010 **Model:** Lancer **Engine:** 2.0L L4 VIN U, 2.4L L4 VIN W **Transmission:** All	**Clutch 1 System (Pressure abnormality):** The clutch 1 pressure is determined to be abnormal. **Possible Causes:** • Malfunction of TC-SST-ECU • Malfunction of clutch assembly • Malfunction of engine system

DTC	Trouble Code Title, Conditions & Possible Causes
DTC: P1874 **1T ECM, MIL:** Yes **Year:** 2009, 2010 **Model:** Lancer **Engine:** 2.0L L4 VIN U, 2.4L L4 VIN W **Transmission:** All	**Clutch 2 System (Pressure abnormality):** The clutch 2 pressure is determined to be abnormal. **Possible Causes:** • Malfunction of TC-SST-ECU • Malfunction of clutch assembly • Malfunction of engine system
DTC: P1875 **1T ECM, MIL:** Yes **Year:** 2009, 2010 **Model:** Lancer **Engine:** 2.0L L4 VIN U, 2.4L L4 VIN W **Transmission:** All	**Damper Speed Sensor System (Poor performance):** Voltage of battery: 8 V or more. Voltage of battery: 16.5 V or less. Engine speed: 650 r/min or more. Time since above engine condition: 1.5 seconds. The output of the sensor remains 16,000 r/min or less for 400 millisecond. The absolute value of the "Engine speed via CAN" - "Sensor output speed" remains 500 r/min or less for 500 millisecond. **Possible Causes:** • Malfunction of damper speed sensor • Malfunction of TC-SST-ECU
DTC: P1876 **1T ECM, MIL:** Yes **Year:** 2009, 2010 **Model:** Lancer **Engine:** 2.0L L4 VIN U, 2.4L L4 VIN W **Transmission:** All	**Gear Block 1st:** The engagement of the 1st gear is determined to be impossible. **Possible Causes:** • Malfunction of TC-SST-ECU • Malfunction of TC-SST gear • Malfunction of clutch assembly
DTC: P1877 **1T ECM, MIL:** Yes **Year:** 2009, 2010 **Model:** Lancer **Engine:** 2.0L L4 VIN U, 2.4L L4 VIN W **Transmission:** All	**Gear Block 2nd:** Engine speed: 650 r/min or more. Time since above engine condition: 1.5 seconds or more. Target gear: 2nd gear. Gear (synchro) engagement: 3 times blocked. (immediately) **Possible Causes:** • Malfunction of TC-SST-ECU • Malfunction of TC-SST gear • Malfunction of clutch assembly
DTC: P1878 **1T ECM, MIL:** Yes **Year:** 2009, 2010 **Model:** Lancer **Engine:** 2.0L L4 VIN U, 2.4L L4 VIN W **Transmission:** All	**Gear Block 3rd:** Engine speed: 650 r/min or more. Time since above engine condition: 1.5 seconds or more. Target gear: 3rd gear. Gear (synchro) engagement: 3 times blocked. (immediately) **Possible Causes:** • Malfunction of TC-SST-ECU • Malfunction of TC-SST gear • Malfunction of clutch assembly
DTC: P1879 **1T ECM, MIL:** Yes **Year:** 2009, 2010 **Model:** Lancer **Engine:** 2.0L L4 VIN U, 2.4L L4 VIN W **Transmission:** All	**Gear Block 4th:** Engine speed: 650 r/min or more. Time since above engine condition: 1.5 seconds or more. Target gear: 4th gear. Gear (synchro) engagement: 3 times blocked. (immediately) **Possible Causes:** • Malfunction of TC-SST-ECU • Malfunction of TC-SST gear • Malfunction of clutch assembly
DTC: P187A **1T ECM, MIL:** Yes **Year:** 2009, 2010 **Model:** Lancer **Engine:** 2.0L L4 VIN U, 2.4L L4 VIN W **Transmission:** All	**Gear Block 5th:** Engine speed: 650 r/min or more. Time since above engine condition: 1.5 seconds or more. Target gear: 5th gear. Gear (synchro) engagement: 3 times blocked. (immediately) **Possible Causes:** • Malfunction of TC-SST-ECU • Malfunction of TC-SST gear • Malfunction of clutch assembly
DTC: P187B **1T ECM, MIL:** Yes **Year:** 2009, 2010 **Model:** Lancer **Engine:** 2.0L L4 VIN U, 2.4L L4 VIN W **Transmission:** All	**Gear Block 6th:** Engine speed: 650 r/min or more. Time since above engine condition: 1.5 seconds or more. Target gear: 6th gear. Gear (synchro) engagement: 3 times blocked. (immediately) **Possible Causes:** • Malfunction of TC-SST-ECU • Malfunction of TC-SST gear • Malfunction of clutch assembly

DTC	Trouble Code Title, Conditions & Possible Causes
DTC: P187C **1T ECM, MIL: Yes** **Year:** 2009, 2010 **Model:** Lancer **Engine:** 2.0L L4 VIN U, 2.4L L4 VIN W **Transmission:** All	**Gear Block Reverse:** The engagement of the reverse gear is determined to be impossible. **Possible Causes:** • Malfunction of TC-SST-ECU • Malfunction of TC-SST gear • Malfunction of clutch assembly
DTC: P1880 **1T ECM, MIL: Yes** **Year:** 2009, 2010 **Model:** Lancer **Engine:** 2.0L L4 VIN U, 2.4L L4 VIN W **Transmission:** All	**EOL Mode Active:** The TC-SST setting mode is determined to be EOL (end of line) mode. **Possible Causes:** • The setting mode changeover mistake when TC-SST is shipped. • Malfunction of TC-SST-ECU.
DTC: P1881 **1T ECM, MIL: Yes** **Year:** 2009, 2010 **Model:** Lancer **Engine:** 2.0L L4 VIN U, 2.4L L4 VIN W **Transmission:** All	**Twin clutch SST control mode switch system (Malfunction):** The "+" and "-" signals of the twin clutch SST control mode switch is determined to be stuck on. **Possible Causes:** • Twin clutch SST control mode switch malfunction • Malfunction of TC-SST-ECU
DTC: P1885 **1T ECM, MIL: Yes** **Year:** 2009, 2010 **Model:** Lancer **Engine:** 2.0L L4 VIN U, 2.4L L4 VIN W **Transmission:** All	**SHIFT FORK 1 JUMP OUT:** The movement of the shift fork 1 is determined to be abnormal. **Possible Causes:** • Malfunction of TC-SST-ECU • Malfunction of TC-SST shift fork • Malfunction of valve body
DTC: P1886 **1T ECM, MIL: Yes** **Year:** 2009, 2010 **Model:** Lancer **Engine:** 2.0L L4 VIN U, 2.4L L4 VIN W **Transmission:** All	**SHIFT FORK 2 JUMP OUT:** The movement of the shift fork 2 is determined to be abnormal **Possible Causes:** • Malfunction of TC-SST-ECU • Malfunction of TC-SST shift fork • Malfunction of valve body
DTC: P1887 **1T ECM, MIL: Yes** **Year:** 2009, 2010 **Model:** Lancer **Engine:** 2.0L L4 VIN U, 2.4L L4 VIN W **Transmission:** All	**SHIFT FORK 3 JUMP OUT:** The movement of the shift fork 3 is determined to be abnormal. **Possible Causes:** • Malfunction of TC-SST-ECU • Malfunction of TC-SST shift fork • Malfunction of valve body
DTC: P1888 **1T ECM, MIL: Yes** **Year:** 2009, 2010 **Model:** Lancer **Engine:** 2.0L L4 VIN U, 2.4L L4 VIN W **Transmission:** All	**SHIFT FORK 4 JUMP OUT:** The movement of the shift fork 4 is determined to be abnormal. **Possible Causes:** • Malfunction of TC-SST-ECU • Malfunction of TC-SST shift fork • Malfunction of valve body
DTC: P1890 **1T ECM, MIL: Yes** **Year:** 2009, 2010 **Model:** Lancer **Engine:** 2.0L L4 VIN U, 2.4L L4 VIN W **Transmission:** All	**Teach-In Not Completed:** It is judged that Teach-In is not completed normally. **Possible Causes:** • Teach-In not completed • Malfunction of TC-SST-ECU • Malfunction of clutch assembly

DTC	Trouble Code Title, Conditions & Possible Causes
DTC: P1897 **1T PCM, MIL:** Yes **Year:** 2009 **Model:** Raider **Engine:** 3.7L V6 VIN K **Transmission:** All	**Level 1 RPM BUS Unlock:** With the engine running. When the PCM recognizes an internal failure to communicate with the FCM or the CMP and CKP Sensor count periods are too short. ETC light is flashing. **Possible Causes:** • PCM
DTC: P1902 **1T ECM, MIL:** Yes **Year:** 2009, 2010 **Model:** Lancer, Outlander **Engine:** 2.0L L4 VIN U, 2.4L L4 VIN W **Transmission:** All	**Malfunction of Engine System:** The TCM receives the limp-home signal from ECM via CAN. **Possible Causes:** • Malfunction of the engine system (ETV)
DTC: P198D **1T ECM, MIL:** Yes **Year:** 2009, 2010 **Model:** Lancer **Engine:** 2.0L L4 VIN U, 2.4L L4 VIN W **Transmission:** All	**EEPROM System:** The EEPROM writing data is determined to be abnormal. **Possible Causes:** • Malfunction of the shift lever-ECU
DTC: P198E **1T ECM, MIL:** Yes **Year:** 2009, 2010 **Model:** Lancer **Engine:** 2.0L L4 VIN U, 2.4L L4 VIN W **Transmission:** All	**Lever Position Sensor System:** When one position signal is determined not to be. **Possible Causes:** • Malfunction of the shift lever-ECU • Malfunction of the lever position sensor
DTC: P198F **1T ECM, MIL:** Yes **Year:** 2009, 2010 **Model:** Lancer **Engine:** 2.0L L4 VIN U, 2.4L L4 VIN W **Transmission:** All	**Lever Position Sensor System:** When two position signals are determined not to be. **Possible Causes:** • Malfunction of the shift lever-ECU • Malfunction of the lever position sensor

OBD II Trouble Code List (P2XXX Codes)

DTC	Trouble Code Title, Conditions & Possible Causes
DTC: P2066 **1T PCM, MIL:** Yes **Year:** 2009, 2010 **Model:** Eclipse, Endeavor, Galant, Lancer, Outlander **Engine:** 2.0L L4 VIN U, 2.0L L4 VIN V, 2.4L L4 VIN F, 2.4L L4 VIN W, 3.0L V6 VIN X, 3.8L V6 VIN S, 3.8L V6 VIN T **Transmission:** All	**Fuel Level Sensor (sub) Circuit Range/Performance :** When the fuel consumption calculated from the operation time of the injector amounts to 30 litters (7.8 gal), the diversity of the amount of fuel in tank calculated from the fuel level sensor is 2 litters (0.5 gal) or less. **Possible Causes:** • Fuel pump module or fuel level sensor (sub) failed. • ECM failed. <M/T> • PCM failed. <A/T>
DTC: P2072 **1T PCM** **Year:** 2009 **Model:** Raider **Engine:** 3.7L V6 VIN K **Transmission:** All	**Electronic Throttle Control System (Ice Blockage):** Ignition on. The PCM recognizes the Throttle plate is stuck during extremely cold Ambient Temperature operation. The throttle plate goes through a de-icing procedure. If the throttle blade still doesn't move this fault sets. The MIL will not illuminate. ETC light will illuminate. The vehicle will be in Limp home condition, limiting rpm and vehicle speed. **Possible Causes:** • Throttle plate frozen

DTC	Trouble Code Title, Conditions & Possible Causes
DTC: P2096 **2T PCM, MIL: Yes** **Year:** 2009 **Model:** Raider **Engine:** 3.7L V6 VIN K **Transmission:** All	**Downstream Fuel System 1 Lean:** With the engine running in closed loop mode, the ambient/battery temperature above -6.7°C (20°F) and altitude below 2590.8 m (8500 ft) and fuel level greater than 15%. The conditions that cause this diagnostic to fail is when the upstream O2 sensor becomes biased from an exhaust leak, O2 sensor contamination, or some other extreme operating condition. The downstream O2 sensor is considered to be protected from extreme environments by the catalyst. The PCM monitors the downstream O2 sensor feedback control, called downstream fuel trim, to detect any shift in the upstream O2 sensor target voltage from nominal target voltage. The value of the downstream fuel trim is compared with the lean thresholds. Every time the value exceeds the calibrated threshold, a fail timer is incremented and mass flow through the exhaust is accumulated. If the fail timer and accumulated mass flow exceed the fail thresholds, the test fails and the diagnostic stops running for that trip. If the test fails on consecutive trips, a DTC is set. **Possible Causes:** • Exhaust leak • Engine mechanical problem • O2 Sensor has failed • O2 Sensor signal circuit or return circuit problem • Fuel delivery • PCM
DTC: P2096 **2T ECM, MIL: Yes** **Year:** 2009, 2010 **Model:** Lancer **Engine:** 2.0L L4 VIN U, 2.0L L4 VIN V, 2.4L L4 VIN W **Transmission:** All	**Post Catalyst Fuel Trim System too Lean:** Engine coolant temperature is more than 76°C (169°F). Under the closed loop air/fuel control. More than 90 seconds have passed since the engine starting sequence was completed. Mass airflow sensor output is more than 5 g/sec. The amount of compensated rich air/fuel ratio is output by the heated oxygen sensor (rear). The status that the amount of compensated rich air/fuel ratio is greater than the specified amount continues for 5 seconds. **Possible Causes:** • Heated oxygen sensor (front) failed. • Heated oxygen sensor (rear) failed. • Injector failed. • ECM failed.
DTC: P2097 **2T PCM, MIL: Yes** **Year:** 2009 **Model:** Raider **Engine:** 3.7L V6 VIN K **Transmission:** All	**Downstream Fuel System 1 Rich:** With the engine running in closed loop mode, the ambient/battery temperature above -6.7°C (20°F) and altitude below 2590.8 m (8500 ft). The conditions that cause this diagnostic to fail is when the upstream O2 sensor becomes biased from an exhaust leak, O2 sensor contamination, or some other extreme operating condition. The downstream O2 sensor is considered to be protected from extreme environments by the catalyst. The PCM monitors the downstream O2 sensor feedback control, called downstream fuel trim, to detect any shift in the upstream O2 sensor target voltage from nominal target voltage. The value of the downstream fuel trim is compared with the rich thresholds. Every time the value exceeds the calibrated threshold, a fail timer is incremented and mass flow through the exhaust is accumulated. If the fail timer and accumulated mass flow exceed the fail thresholds, the test fails and the diagnostic stops running for that trip. If the test fails on consecutive trips, a DTC is set. **Possible Causes:** • Exhaust leak • Engine mechanical problem • O2 Sensor, wiring or connectors • Fuel delivery system • PCM
DTC: P2097 **2T ECM, MIL: Yes** **Year:** 2009, 2010 **Model:** Lancer **Engine:** 2.0L L4 VIN U, 2.0L L4 VIN V, 2.4L L4 VIN W **Transmission:** All	**Post Catalyst Fuel Trim System too Rich:** Engine coolant temperature is more than 76°C (169°F). Under the closed loop air/fuel control. More than 90 seconds have passed since the engine starting sequence was completed. Mass airflow sensor output is more than 5 g/sec. The amount of compensated lean air/fuel ratio is output by the heated oxygen sensor (rear). The status that the amount of compensated lean air/fuel ratio is greater than the specified amount continues for 5 seconds. **Possible Causes:** • Heated oxygen sensor (front) failed. • Heated oxygen sensor (rear) failed. • Injector failed. • ECM failed.

DTC	Trouble Code Title, Conditions & Possible Causes
DTC: P2098 **2T PCM, MIL: Yes** **Year:** 2009 **Model:** Raider **Engine:** 3.7L V6 VIN K **Transmission:** All	**Downstream Fuel System 2 Lean:** With the engine running in closed loop mode, the ambient/battery temperature above -6.7°C (20°F) and altitude below 2590.8 m (8500 ft) and fuel level greater than 15%. The conditions that cause this diagnostic to fail is when the upstream O2 sensor becomes biased from an exhaust leak, O2 sensor contamination, or some other extreme operating condition. The downstream O2 sensor is considered to be protected from extreme environments by the catalyst. The PCM monitors the downstream O2 sensor feedback control, called downstream fuel trim, to detect any shift in the upstream O2 sensor target voltage from nominal target voltage. The value of the downstream fuel trim is compared with the lean thresholds. Every time the value exceeds the calibrated threshold, a fail timer is incremented and mass flow through the exhaust is accumulated. If the fail timer and accumulated mass flow exceed the fail thresholds, the test fails and the diagnostic stops running for that trip. If the test fails on consecutive trips, a DTC is set **Possible Causes:** • Exhaust leak • Engine mechanical problem • O2 Sensor has failed • O2 Sensor signal circuit or return circuit problem • Fuel delivery • PCM
DTC: P2099 **2T PCM, MIL: Yes** **Year:** 2009 **Model:** Raider **Engine:** 3.7L V6 VIN K **Transmission:** All	**Downstream Fuel System 2 Rich:** With the engine running in closed loop mode, the ambient/battery temperature above -6.7°C (20°F) and altitude below 2590.8 m (8500 ft). The conditions that cause this diagnostic to fail is when the upstream O2 sensor becomes biased from an exhaust leak, O2 sensor contamination, or some other extreme operating condition. The downstream O2 sensor is considered to be protected from extreme environments by the catalyst. The PCM monitors the downstream O2 sensor feedback control, called downstream fuel trim, to detect any shift in the upstream O2 sensor target voltage from nominal target voltage. The value of the downstream fuel trim is compared with the rich thresholds. Every time the value exceeds the calibrated threshold, a fail timer is incremented and mass flow through the exhaust is accumulated. If the fail timer and accumulated mass flow exceed the fail thresholds, the test fails and the diagnostic stops running for that trip. If the test fails on consecutive trips, a DTC is set. **Possible Causes:** • Exhaust leak • Engine mechanical problem • O2 Sensor, wiring or connectors • Fuel delivery system • PCM
DTC: P2100 **1T PCM, MIL: Yes** **Year:** 2009, 2010 **Model:** Eclipse, Endeavor, Galant, Lancer, Outlander **Engine:** 2.0L L4 VIN U, 2.0L L4 VIN V, 2.4L L4 VIN F, 2.4L L4 VIN W, 3.0L V6 VIN X, 3.8L V6 VIN S, 3.8L V6 VIN T **Transmission:** All	**Throttle Actuator Control Motor Circuit (Open):** Battery positive voltage is higher than 8.3 volts. The output voltage of the (main) throttle position sensor minus the proposed output voltage becomes 0.1 volt or more. The output voltage of the (sub) throttle position sensor minus the proposed output voltage becomes 0.1 volt or more. The drive duty of the throttle actuator control motor is 100 percent or more. Or. Battery positive voltage is higher than 8.3 volts. The proposed output voltage minus the output voltage of the (main) throttle position sensor becomes 0.1 volt or more. The proposed output voltage minus the output voltage of the (sub) throttle position sensor becomes 0.1 volt or more. The drive duty of the throttle actuator control motor is 100 percent or more. **Possible Causes:** • Throttle actuator control motor failed. • Open throttle actuator control motor circuit, harness damage, or connector damage. • ECM failed. <M/T> • PCM failed. <A/T>
DTC: P2100 **1T PCM, MIL: Yes** **Year:** 2009 **Model:** Raider **Engine:** 3.7L V6 VIN K **Transmission:** All	**Electronic Throttle Control Motor Circuit Malfunction:** With the ignition on and the ETC Motor is not in Limp Home mode. When the PCM detects an internal error or a short between the ETC Motor- and ETC Motor + circuits in the ETC Motor Driver. ETC light is flashing. **Possible Causes:** • Intermittent condition • Throttle plate or bore may have foreign object blockage • ETC positive circuit is open or is shorted to battery voltage, to ground, or to ETC negative circuit • ETC negative circuit is open or is shorted to battery voltage or to ground • Low battery voltage • ETC Motor or Throttle Body has failed • PCM has failed

DTC	Trouble Code Title, Conditions & Possible Causes
DTC: P2101 **1T PCM, MIL: Yes** **Year:** 2009 **Model:** Raider **Engine:** 3.7L V6 VIN K **Transmission:** All	**Electronic Throttle Control Motor Performance:** With vehicle running and ETC motor is not is Limp Home mode, and the TPS adaptation is complete. The PCM recognizes too large of an error between the actual position of the throttle plate and the set point position. This DTC will set within 5 seconds. The ETC light will be flashing. **Possible Causes:** • Throttle body assembly may have failed • Low battery voltage • PCM has failed
DTC: P2101 **1T PCM, MIL: Yes** **Year:** 2009, 2010 **Model:** Eclipse, Endeavor, Galant, Lancer, Outlander **Engine:** 2.0L L4 VIN U, 2.0L L4 VIN V, 2.4L L4 VIN F, 2.4L L4 VIN W, 3.0L V6 VIN X, 3.8L V6 VIN S, 3.8L V6 VIN T **Transmission:** All	**Throttle Actuator Control Motor Magneto Malfunction:** Battery positive voltage is higher than 8.3 volts. The coil current of the throttle actuator control motor is 8 ampere or more for 0.3 second. The throttle actuator control motor intelligent power device detects shorted-high/low and overheat of itself. **Possible Causes:** • Throttle actuator control motor failed. • Shorted throttle actuator control motor circuit, harness damage or connector damage. • ECM failed. <M/T> • PCM failed. <A/T>
DTC: P2106 **1T PCM, MIL: Yes** **Year:** 2009 **Model:** Raider **Engine:** 3.7L V6 VIN K **Transmission:** All	**Electronic Throttle Control System (Forced Limited Power):** Ignition on. This DTC sets for OBDII MIL illumination purposes. This DTC will always have associated DTCs indicating a system failure. Engine speed is being limited and/or throttle motor is power free. ETC light is flashing. **Possible Causes:** • Other DTC's related to ETC have set
DTC: P2107 **1T PCM, MIL: Yes** **Year:** 2009 **Model:** Raider **Engine:** 3.7L V6 VIN K **Transmission:** All	**Electronic Throttle Control Module Processor :** Ignition is on. This condition is caused by an internal PCM failure. The module will attempt to reset, so you will be able to hear the throttle relearning. If the condition is continuous, the vehicle may not be drivable. The ETC light will be flashing. **Possible Causes:** • PCM requires reprogramming
DTC: P2108 **1T PCM, MIL: Yes** **Year:** 2009 **Model:** Raider **Engine:** 3.7L V6 VIN K **Transmission:** All	**Electronic Throttle Control Module Performance:** Ignition is on. This condition is caused by an internal PCM failure. The module will attempt to reset, so you will be able to hear the throttle relearning. If the condition is continuous, the vehicle may not be drivable. The ETC light will be flashing. **Possible Causes:** • PCM requires reprogramming
DTC: P2110 **1T PCM, MIL: Yes** **Year:** 2009 **Model:** Raider **Engine:** 3.7L V6 VIN K **Transmission:** All	**Electronic Throttle Control Forced Limited RPM:** Ignition is on and ETC motor is working. When the PCM requests to limit engine speed, if the PWM is too high for 20.5 seconds and before P2118 sets. This one-trip fault will set within 5 seconds. The ETC light will be illuminated. **Possible Causes:** • Throttle plate stuck • ETC positive circuit is open or is shorted to ground • ETC negative circuit is open or is shorted to ground • ETC motor has failed • PCM has failed
DTC: P2111 **1T PCM, MIL: Yes** **Year:** 2009 **Model:** Raider **Engine:** 3.7L V6 VIN K **Transmission:** All	**Electronic Throttle Control Unable To Close:** Ignition on and battery voltage greater than 10 volts. Just after key on, the throttle is opened and closed to test the system. If the TP Sensor does not return to Limp Home Position at the end of this test, this DTC will set. ETC light is flashing. **Possible Causes:** • Throttle plate stuck above Limp Home position • TP Sensors 1 & 2 both read 2.5V • ETC positive circuit is open or is shorted to ground or to battery voltage • ETC negative circuit is open or is shorted to ground • PCM has failed

DTC	Trouble Code Title, Conditions & Possible Causes
DTC: P2112 **1T PCM, MIL: Yes** **Year:** 2009 **Model:** Raider **Engine:** 3.7L V6 VIN K **Transmission:** All	**Electronic Throttle Control Unable To Open:** Ignition on and battery voltage greater than 10 volts. Just after key on, the throttle is opened and closed to test the system. If the TP Sensor does not return to Limp Home Position at the end of this test, this DTC will set. ETC light is flashing. **Possible Causes:** • Throttle plate stuck • ETC positive circuit is open or is shorted to ground • ETC negative circuit is open or is shorted to ground • ETC motor has failed • PCM has failed
DTC: P2115 **1T PCM, MIL: Yes** **Year:** 2009 **Model:** Raider **Engine:** 3.7L V6 VIN K **Transmission:** All	**Accelerator Pedal Position Sensor 1 Minimum Stop Performance:** Ignition is on. During in-plant mode the APP Sensors need to be checked to make sure that idle and full pedal travel can be reached on both Sensors. The test for this DTC is enabled once the test for DTC P2166 has passed. This DTC will set if the APP Sensor No. 1 has failed to achieve the required minimum value during in-plant testing. The engine will only idle. **Possible Causes:** • In plant test failure • APP Sensors must be reprogrammed to relearn
DTC: P2116 **1T PCM, MIL: Yes** **Year:** 2009 **Model:** Raider **Engine:** 3.7L V6 VIN K **Transmission:** All	**Accelerator Pedal Position Sensor 2 Minimum Stop Performance:** Ignition is on. During in-plant mode the APP Sensors need to be checked to make sure that idle and full pedal travel can be reached on both Sensors. The test for this DTC is enabled once the test for DTC P2167 has passed. This DTC will set if the APP Sensor No. 2 has failed to achieve the required minimum value during in-plant testing. This one-trip fault will set within 5 seconds. The engine will only idle. **Possible Causes:** • In plant test failure • APP Sensors must be reprogrammed to relearn
DTC: P2118 **1T PCM, MIL: Yes** **Year:** 2009 **Model:** Raider **Engine:** 3.7L V6 VIN K **Transmission:** All	**Electronic Throttle Control Motor Circuit :** Ignition is on and ETC motor is not in limp-home mode. When the PCM detects an internal error or short between the ETC motor and ETC motor positive circuits in the ETC motor driver. The ETC light will be flashing **Possible Causes:** • Throttle plate or bore malfunctions • ETC positive circuit is open or is shorted to ground, battery voltage or ETC negative circuit • ETC negative circuit is open or is shorted to ground or to battery voltage • ETC motor has malfunctioned • PCM has failed
DTC: P2122 **1T PCM, MIL: Yes** **Year:** 2009, 2010 **Model:** Eclipse, Endeavor, Galant, Lancer, Outlander **Engine:** 2.0L L4 VIN U, 2.0L L4 VIN V, 2.4L L4 VIN F, 2.4L L4 VIN W, 3.0L V6 VIN X, 3.8L V6 VIN S, 3.8L V6 VIN T **Transmission:** All	**Accelerator Pedal Position Sensor (main) Circuit Low Input:** With the ignition switch in the "ON" position. The accelerator pedal position sensor (main) output voltage is 0.6 volt or less for 0.3 second. Accelerator pedal position sensor (main) output voltage is out of specified range. **Possible Causes:** • Accelerator pedal position sensor failed. • Open or shorted accelerator pedal position sensor (main) circuit, harness damage or connector damage. • ECM failed. <M/T> • PCM failed. <A/T>
DTC: P2123 **1T PCM, MIL: Yes** **Year:** 2009, 2010 **Model:** Eclipse, Endeavor, Galant, Lancer, Outlander **Engine:** 2.0L L4 VIN U, 2.0L L4 VIN V, 2.4L L4 VIN F, 2.4L L4 VIN W, 3.0L V6 VIN X, 3.8L V6 VIN S, 3.8L V6 VIN T **Transmission:** All	**Accelerator Pedal Position Sensor (main) Circuit High Input:** Ignition switch is "ON" position. Accelerator pedal position sensor (main) output voltage should be 4.8 volts or higher for 0.3 second. The accelerator pedal position sensor (main) output voltage is out of specified range. **Possible Causes:** • Accelerator pedal position sensor failed. • Open accelerator pedal position sensor (main) circuit, harness damage or connector damage. • ECM failed. <M/T> • PCM failed. <A/T>

DTC	Trouble Code Title, Conditions & Possible Causes
DTC: P2123 **T PCM, MIL: Yes** **Year:** 2009 **Model:** Raider **Engine:** 3.7L V6 VIN K **Transmission:** All	**Accelerator Pedal Position Sensor 1 Circuit High:** Ignition is on and no other APP Sensor No. 1 DTCs are present. When APP Sensor No. 1 voltage is too high, the engine will additionally idle, if the brake pedal is pressed or has failed. Acceleration rate and engine output are limited. This one-trip fault will set within 5 seconds. The ETC light will be flashing. **Possible Causes:** • APP Sensor No. 1 return circuit is open • APP Sensor No. 1 signal circuit is shorted to either 5V supply circuit • APP Sensor No. 1 has failed • PCM has failed
DTC: P2127 **1T PCM, MIL: Yes** **Year:** 2009 **Model:** Raider **Engine:** 3.7L V6 VIN K **Transmission:** All	**Accelerator Pedal Position Sensor 2 Circuit Low :** With the ignition on and no other APPS No.2 DTCs. When the APP Sensor No.2 voltage is too low. Engine will only idle if the Brake pedal is Pressed or has failed. Acceleration rate and Engine output are limited. ETC light is flashing. **Possible Causes:** • 5V supply circuit is open or shorted to ground • APP Sensor No. 2 signal circuit is open, shorted to ground or to sensor return circuit • APP Sensor No. 2 has failed • PCM has failed
DTC: P2127 **1T PCM, MIL: Yes** **Year:** 2009, 2010 **Model:** Eclipse, Endeavor, Galant, Lancer, Outlander **Engine:** 2.0L L4 VIN U, 2.0L L4 VIN V, 2.4L L4 VIN F, 2.4L L4 VIN W, 3.0L V6 VIN X, 3.8L V6 VIN S, 3.8L V6 VIN T **Transmission:** All	**Accelerator Pedal Position Sensor (sub) Circuit Low Input:** Ignition switch is "ON" position. Accelerator pedal position sensor (sub) output voltage is 0.2 volt or less for 0.3 second. The accelerator pedal position sensor (sub) output voltage is out of specified range. **Possible Causes:** • Accelerator pedal position sensor failed. • Open or shorted accelerator pedal position sensor (sub) circuit, harness damage or connector damage. • ECM failed. <M/T> • PCM failed. <A/T>
DTC: P2128 **1T PCM, MIL: Yes** **Year:** 2009 **Model:** Raider **Engine:** 3.7L V6 VIN K **Transmission:** All	**Accelerator Pedal Position Sensor 2 Circuit High:** With the ignition on and no other APPS No.2 DTCs present. When APP Sensor No.2 voltage is too high. Idle is additionally forced any time the brake is applied or failed. Acceleration rate and Engine output are limited. ETC light is flashing. **Possible Causes:** • APP Sensor No. 2 return circuit is open • APP Sensor No. 2 signal circuit is shorted to 5-volt supply circuit • APP Sensor No. 2 has failed • PCM has failed
DTC: P2128 **1T PCM, MIL: Yes** **Year:** 2009, 2010 **Model:** Eclipse, Endeavor, Galant, Lancer, Outlander **Engine:** 2.0L L4 VIN U, 2.0L L4 VIN V, 2.4L L4 VIN F, 2.4L L4 VIN W, 3.0L V6 VIN X, 3.8L V6 VIN S, 3.8L V6 VIN T **Transmission:** All	**Accelerator Pedal Position Sensor (Sub) Circuit High Input:** Ignition switch is "ON" position. Accelerator pedal position sensor (sub) output voltage is 4.8 volts or higher for 0.3 second. The accelerator pedal position sensor (sub) output voltage is out of specified range. **Possible Causes:** • Accelerator pedal position sensor failed. • Open accelerator pedal position sensor (sub) circuit, harness damage or connector damage. • ECM failed. <M/T> • PCM failed. <A/T>
DTC: P2135 **1T PCM, MIL: Yes** **Year:** 2009, 2010 **Model:** Eclipse, Endeavor, Galant, Lancer, Outlander **Engine:** 2.0L L4 VIN U, 2.0L L4 VIN V, 2.4L L4 VIN F, 2.4L L4 VIN W, 3.0L V6 VIN X, 3.8L V6 VIN S, 3.8L V6 VIN T **Transmission:** All	**Throttle Position Sensor (Main and Sub) Range/Performance Problem:** Ignition switch is "ON" position. Throttle position sensor (main) output voltage is between 0.35 and 2.5 volts. Throttle position sensor (sub) output voltage is between 2.25 and 4.8 volts. Voltage obtained with the formula given below is 0.3 volt or higher for 0.5 second: throttle position sensor (main) output voltage - [throttle position sensor (sub) output voltage - 2 volts] Detect malfunction if the relation between throttle position sensor (main) and throttle position sensor (sub) is wrong. **Possible Causes:** • Throttle position sensor failed. • Shorted throttle position sensor circuit or connector damage. • ECM failed. <M/T> • PCM failed. <A/T>

DTC	Trouble Code Title, Conditions & Possible Causes
DTC: P2135 **1T PCM, MIL: Yes** **Year:** 2009 **Model:** Raider **Engine:** 3.7L V6 VIN K **Transmission:** All	**Throttle Position Sensors 1/2 Correlation:** With the ignition on and no other DTCs present for TP Sensor No.1 or No.2. PCM recognizes TP Sensors No.1 and No.2 are not coherent. ETC light is illuminated. **Possible Causes:** • TP Sensor No. 1 or 2 signal circuit is shorted to ground or to battery voltage • TP Sensor No. 1 or 2 signal circuit has high resistance • 5V supply circuit has high resistance • 5V supply circuit shorted to ground • TP Sensor ground circuit has high resistance • TP Sensor No. 1 signal circuit is shorted to Sensor No. 2 signal circuit • TP Sensor has failed • PCM has failed
DTC: P2138 **1T PCM, MIL: Yes** **Year:** 2009, 2010 **Model:** Eclipse, Endeavor, Galant, Lancer, Outlander **Engine:** 2.0L L4 VIN U, 2.0L L4 VIN V, 2.4L L4 VIN F, 2.4L L4 VIN W, 3.0L V6 VIN X, 3.8L V6 VIN S, 3.8L V6 VIN T **Transmission:** All	**Accelerator Pedal Position Sensor (main and sub) Circuit Range/Performance Problem:** Ignition switch is "ON" position. Accelerator pedal position sensor (main) output voltage is between 0.6 and 4.8 volts. Accelerator pedal position sensor (sub) output voltage is between 0.2 and 4.8 volts. Change of accelerator pedal position sensor (sub) output voltage per 25 milliseconds is lower than 0.06 volt. Battery voltage is higher than 6.3 volts. Voltage obtained with the formula given below is 1.0 volt or higher for 1 second: [accelerator pedal position sensor (sub) output voltage +0.3 volt] - accelerator pedal position sensor (main) output voltage. **Possible Causes:** • Accelerator pedal position sensor failed. • Harness damage in accelerator pedal position sensor circuit or connector damage. • ECM failed. <M/T> • PCM failed. <A/T>
DTC: P2138 **1T PCM, MIL: Yes** **Year:** 2009 **Model:** Raider **Engine:** 3.7L V6 VIN K **Transmission:** All	**Throttle Position Sensors 1/2 Correlation:** With the ignition on and no APPS No.1 and APPS No.2. APPS values No.1 and No.2 are not coherent. Idle is additionally forced when the brake pedal is pressed or failed. Acceleration rate and Engine output are limited. ETC light is flashing. **Possible Causes:** • APP Sensor No. 1 or 2 signal circuit has high resistance • APP Sensor No. 1 or 2 return circuit has high resistance • 5- volt supply circuit has high resistance • APP Sensor has failed • PCM has failed
DTC: P2161 **T PCM** **Year:** 2009 **Model:** Raider **Engine:** 3.7L V6 VIN K **Transmission:** All	**Vehicle Speed Sensor 2 Erratic:** Ignition is on and battery voltage is greater than 10v. Transmission is in Drive or Reverse. The PCM recognizes the VSS 2 speed signal is erratic or high. No MIL and no ETC light. The cruise control is disabled. **Possible Causes:** • Active Bus or Communications DTCs • Incorrect tire circumference • PCM has failed
DTC: P2166 **1T PCM, MIL: Yes** **Year:** 2009 **Model:** Raider **Engine:** 3.7L V6 VIN K **Transmission:** All	**Accelerator Pedal Position Sensor 1 Maximum Stop Performance:** Ignition is on. During in-plant mode the APP Sensors need to be checked to make sure that idle and full pedal travel can be reached on both Sensors. This DTC will set if the APP Sensor No. 1 has failed to achieve the required maximum value during in-plant testing. This one-trip fault will set within 5 seconds. The engine will only idle. **Possible Causes:** • In-Plant test failure • APP Sensors must be reprogrammed to relearn
DTC: P2167 **1T PCM, MIL: Yes** **Year:** 2009 **Model:** Raider **Engine:** 3.7L V6 VIN K **Transmission:** All	**Accelerator Pedal Position Sensor 2 Maximum Stop Performance:** Ignition is on. During in-plant mode the APP Sensors need to be checked to make sure that idle and full pedal travel can be reached on both Sensors. This DTC will set if the APP Sensor No. 2 has failed to achieve the required maximum value during in-plant testing. This one-trip fault will set within 5 seconds. The engine will only idle. **Possible Causes:** • In-Plant test failure • APP Sensors must be reprogrammed to relearn

DTC	Trouble Code Title, Conditions & Possible Causes
DTC: P2172 **1T PCM, MIL: Yes** **Year:** 2009 **Model:** Raider **Engine:** 3.7L V6 VIN K **Transmission:** All	**High Airflow/Vacuum Leak Detected (Instantaneous Accumulation):** Ignition is on and engine running with no MAP Sensor DTCs present. A large vacuum leak has been detected or both of the TP Sensors have failed, based on their position being 2.5V and the calculated MAP value is less than the actual MAP, minus an Offset value. This one-trip fault will set within 5 seconds. The ETC light will flash. **Possible Causes:** • Vacuum leak • 5V supply circuit has high resistance or is shorted to ground • MAP signal circuit has high resistance or is shorted to ground • TP Sensor ground circuit has high resistance • TP Sensor signal circuit is shorted to ground • TP Sensor return circuit has high resistance • MAP Sensor has failed • TP Sensor has failed • PCM has failed
DTC: P2173 **1T PCM, MIL: Yes** **Year:** 2009 **Model:** Raider **Engine:** 3.7L V6 VIN K **Transmission:** All	**High Airflow/Vacuum Leak Detected (Slow Accumulation):** Ignition is on and engine running with no MAP Sensor DTCs present. A large vacuum leak has been detected or both of the TP Sensors have failed, based on their position being 2.5V and the calculated MAP value is less than the Gas Flow Adaptation value. This one-trip fault will set within 5 seconds. The ETC light will flash. **Possible Causes:** • Vacuum leak • 5V supply circuit has high resistance or is shorted to ground • MAP signal circuit has high resistance or is shorted to ground • TP Sensor ground circuit has high resistance • TP Sensor signal circuit is shorted to ground • TP Sensor return circuit has high resistance • MAP Sensor has failed • TP Sensor has failed • PCM has failed
DTC: P2174 **1T PCM, MIL: Yes** **Year:** 2009 **Model:** Raider **Engine:** 3.7L V6 VIN K **Transmission:** All	**Low Airflow/Vacuum Leak Detected (Instantaneous Accumulation):** Ignition is on and engine running with no MAP Sensor DTCs present. The PCM calculated the MAP value is greater than actual MAP value, plus an Offset value. 3 good trips required to turn off MIL. The ETC light will flash. **Possible Causes:** • Restricted air inlet system • 5V supply circuit has high resistance or is shorted to ground • MAP signal circuit has high resistance or is shorted to ground • TP Sensor ground circuit has high resistance • TP Sensor signal circuit is shorted to ground • TP Sensor return circuit has high resistance • MAP Sensor has failed • TP Sensor has failed • PCM has failed
DTC: P2175 **1T PCM, MIL: Yes** **Year:** 2009 **Model:** Raider **Engine:** 3.7L V6 VIN K **Transmission:** All	**Low Airflow/Vacuum Leak Detected (Slow Accumulation):** Ignition is on and engine running with no MAP Sensor DTCs present. The PCM calculated the MAP value is greater than actual MAP value, plus an Offset value. This DTC will set in 5 seconds after occurrence. 3 good trips required to turn off MIL. The ETC light will flash. **Possible Causes:** • Restricted air inlet system • 5V supply circuit has high resistance or is shorted to ground • MAP signal circuit has high resistance or is shorted to ground • TP Sensor ground circuit has high resistance • TP Sensor signal circuit is shorted to ground • TP Sensor return circuit has high resistance • MAP Sensor has failed • TP Sensor has failed • PCM has failed

DTC	Trouble Code Title, Conditions & Possible Causes

DTC: P2181
1T PCM, MIL: Yes
Year: 2009
Model: Raider
Engine: 3.7L V6 VIN K
Transmission: All
	Cooling System Performance:
	Ignition is on and engine running with no ECT Sensor DTCs present. The PCM recognizes that the ECT has failed its self-coherence test. The coolant temperature should only change at a certain rate. If this rate is too slow or too fast, this DTC will set. 3 good trips required to turn off MIL. The ETC light will illuminate on first trip failure.
	Possible Causes:
	   • Low coolant level
	   • ECT signal circuit is open or shorted to ground, Sensor ground, or battery voltage
	   • ECT Sensor ground circuit is open
	   • Thermostat has failed
	   • ECT Sensor has failed
	   • PCM has failed

DTC: P2195
2T PCM, MIL: Yes
Year: 2009, 2010
Model: Eclipse, Galant
Engine: 2.4L L4 VIN F
Transmission: All
	Cylinder 1, 4 Heated Oxygen Sensor (front) Inactive:
	More than 128 seconds have passed since the engine starting sequence was completed. Engine coolant temperature is higher than 7°C (45°F) or more.
	Under the closed loop air/fuel ratio control. Intake air temperature is more than -10°C (14° F). Cylinder 1, 4 heated oxygen sensor (front) output voltage is less than lean/rich criteria (about 0.5 volt) for 128 seconds.
	Possible Causes:
	   • Cylinder 1, 4 heated oxygen sensor (front) deteriorated.
	   • PCM failed.

DTC: P2195
2T PCM, MIL: Yes
Year: 2009, 2010
Model: Eclipse, Endeavor, Galant, Outlander
Engine: 3.0L V6 VIN X, 3.8L V6 VIN S, 3.8L V6 VIN T
Transmission: All
	Heated Oxygen Sensor Inactive (Bank 1, Sensor 1):
	After 128 seconds or more have passed since the engine starting sequence was completed. Engine coolant temperature is 7°C (45°F) or more. Under the closed loop air/fuel ratio control. Intake air temperature is more than -10°C (14°F). The right bank heated oxygen sensor (front) output voltage is less than lean/rich criteria (about 0.5 volt) for 128 seconds.
	Possible Causes:
	   • Right bank heated oxygen sensor (front) deteriorated.
	   • ECM failed. <M/T>
	   • PCM failed. <A/T>

DTC: P2195
2T ECM, MIL: Yes
Year: 2009, 2010
Model: Lancer, Outlander
Engine: 2.0L L4 VIN U, 2.0L L4 VIN V, 2.4L L4 VIN W
Transmission: All
	Heated Oxygen Sensor (front) Inactive:
	More than 300 seconds have passed since the engine starting sequence was completed. Engine coolant temperature is more than 7°C (45°F). Under the closed loop air/fuel ratio control. Intake air temperature is more than -10°C (14° F).
	The heated oxygen sensor (front) output voltage is less than lean/rich criteria (about 0.5 volt) for 128 seconds.
	Possible Causes:
	   • Heated oxygen sensor (front) deteriorated.
	   • ECM failed.

DTC: P2197
2T PCM, MIL: Yes
Year: 2009, 2010
Model: Eclipse, Endeavor, Galant, Outlander
Engine: 3.0L V6 VIN X, 3.8L V6 VIN S, 3.8L V6 VIN T
Transmission: All
	Heated Oxygen Sensor Inactive (Bank 2, Sensor 1):
	After 128 seconds or more have passed since the engine starting sequence was completed. Engine coolant temperature is 7°C (45°F) or more. Under the closed loop air/fuel ratio control. Intake air temperature is more than -10°C (14°F). The left bank heated oxygen sensor (front) output voltage is less than lean/rich criteria (about 0.5 volt) for 128 seconds.
	Possible Causes:
	   • Left bank heated oxygen sensor (front) deteriorated.
	   • ECM failed. <M/T>
	   • PCM failed. <A/T>

DTC: P2197
2T PCM, MIL: Yes
Year: 2009, 2010
Model: Eclipse, Galant
Engine: 2.4L L4 VIN F
Transmission: All
	Cylinder 2, 3 Heated Oxygen Sensor (front) Inactive:
	More than 128 seconds have passed since the engine starting sequence was completed. Engine coolant temperature is 7°C (45° F) or more. Under the closed loop air/fuel ratio control. Intake air temperature is more than -10°C (14° F). Cylinder 2, 3 heated oxygen sensor (front) output voltage is less than lean/rich criteria (about 0.5 volt) for 128 seconds.
	Possible Causes:
	   • Cylinder 2, 3 heated oxygen sensor (front) deteriorated.
	   • PCM failed.

DTC: P2228
1T PCM, MIL: Yes
Year: 2009, 2010
Model: Eclipse, Endeavor, Galant, Lancer, Outlander
Engine: 2.0L L4 VIN U, 2.0L L4 VIN V, 2.4L L4 VIN F, 2.4L L4 VIN W, 3.0L V6 VIN X, 3.8L V6 VIN S, 3.8L V6 VIN T
Transmission: All
	Barometric Pressure Circuit Low Input:
	After 2 seconds or more have passed since the engine starting sequence was completed. Battery positive voltage is higher than 8 volts. Barometric pressure sensor output signal has continued to be 49 kPa (14.5 in. Hg) or lower (approximately 15,000 feet above sea level) for 10 seconds.
	Possible Causes:
	   • ECM failed. <M/T>
	   • PCM failed. <A/T>

DTC	Trouble Code Title, Conditions & Possible Causes
DTC: P2229 **1T PCM, MIL: Yes** **Year:** 2009, 2010 **Model:** Endeavor, Galant, Lancer, Outlander **Engine:** 2.0L L4 VIN U, 2.0L L4 VIN V, 2.4L L4 VIN F, 2.4L L4 VIN W, 3.0L V6 VIN X, 3.8L V6 VIN S, 3.8L V6 VIN T **Transmission:** All	**Barometric Pressure Circuit High Input:** After 2 seconds or more have passed since the engine starting sequence was completed. Battery positive voltage is higher than 8 volts. The barometric pressure sensor output signal has continued to be approximately 113 kPa (33.4 in. Hg) or higher (approximately 4,000 feet below sea level) for 10 seconds. **Possible Causes:** • PCM failed
DTC: P2229 **2T CCM, MIL: Yes** **Year:** 2009, 2010 **Model:** Eclipse, Galant, Lancer, Outlander **Engine:** 2.0L L4 VIN U, 2.0L L4 VIN V, 2.4L L4 VIN F, 2.4L L4 VIN W, 3.0L V6 VIN X, 3.8L V6 VIN T **Transmission:** All	**Barometric Pressure Circuit High Input:** Engine running for 2 seconds or more and battery voltage is greater than 8 volts, barometric pressure sensor output has continued to be 33.3 in. Hg or less (about 4,000 ft above sea level) for 10 seconds. **Possible Causes:** • PCM has failed
DTC: P2237 **1T ECM, MIL: Yes** **Year:** 2009, 2010 **Model:** Outlander **Engine:** 2.4L L4 VIN W **Transmission:** All	**Linear Air-Fuel Ratio Sensor Positive Current Control Circuit/Open <California>:** Battery positive voltage is between 10 and 16.5 volts. More than 220 seconds have passed since the engine starting sequence was completed. The negative circuit voltage subtracted from the pumping circuit voltage is be lower than -2.1 volts or higher than 2.3 volts for 2 seconds. The pumping current is 0 milli ampere for 2 seconds. **Possible Causes:** • Linear air-fuel ratio sensor failed. • Open linear air-fuel ratio sensor circuit, or connector damage. • ECM failed.
DTC: P2243 **1T ECM, MIL: Yes** **Year:** 2009, 2010 **Model:** Outlander **Engine:** 2.4L L4 VIN W **Transmission:** All	**Linear Air-Fuel Ratio Sensor Reference Voltage Circuit/Open <California>:** Battery positive voltage is between 10 and 16.5 volts. More than 220 seconds have passed since the engine starting sequence was completed. The sensor impedance is be higher than 450 ohms for 2 seconds. The reference circuit voltage is more than 5.3 volts for 2 seconds. **Possible Causes:** • Linear air-fuel ratio sensor failed. • Open linear air-fuel ratio sensor circuit, or connector damage. • ECM failed.
DTC: P2251 **1T ECM, MIL: Yes** **Year:** 2009, 2010 **Model:** Outlander **Engine:** 2.4L L4 VIN W **Transmission:** All	**Linear Air-Fuel Ratio Sensor Negative Current Control Circuit/Open <California>:** Battery positive voltage is between 10 and 16.5 volts. More than 220 seconds have passed since the engine starting sequence was completed. The sensor impedance is be higher than 450 ohms for 2 seconds. The reference circuit voltage is be lower than 5.3 volts for 2 seconds. **Possible Causes:** • Linear air-fuel ratio sensor failed. • Open linear air-fuel ratio sensor circuit, or connector damage. • ECM failed.
DTC: P2252 **2T PCM, MIL: Yes** **Year:** 2009, 2010 **Model:** Eclipse, Galant **Engine:** 2.4L L4 VIN F **Transmission:** All	**Heated Oxygen Sensor Offset Circuit Low Voltage :** After 2 seconds or more have passed since the engine starting sequence was completed. Heated oxygen sensor offset voltage has continued to be 0.4 volt or lower for 2 seconds. **Possible Causes:** • Cylinder 1, 4 heated oxygen sensor (front) failed. • Cylinder 1, 4 heated oxygen sensor (rear) failed. • Cylinder 2, 3 heated oxygen sensor (front) failed. • Cylinder 2, 3 heated oxygen sensor (rear) failed. • Shorted cylinder 1, 4 heated oxygen sensor (front) circuit or connector damage. • Shorted cylinder 1, 4 heated oxygen sensor (rear) circuit or connector damage. • Shorted cylinder 2, 3 heated oxygen sensor (front) circuit or connector damage. • Shorted cylinder 2, 3 heated oxygen sensor (rear) circuit or connector damage. • PCM failed. <AT> • ECM failed. <MT>

DTC	Trouble Code Title, Conditions & Possible Causes
DTC: P2252 **1T ECM, MIL: Yes** **Year:** 2009, 2010 **Model:** Lancer, Outlander **Engine:** 2.0L L4 VIN U, 2.0L L4 VIN V, 2.4L L4 VIN W **Transmission:** All	**Heated Oxygen Sensor Offset Circuit Low Voltage :** More than 2 seconds have passed since the engine starting sequence was completed. The heated oxygen sensor offset voltage is less than 0.4 volt for 2 seconds. **Possible Causes:** • Heated oxygen sensor (front) failed. • Heated oxygen sensor (rear) failed. • Shorted heated oxygen sensor (front) circuit or connector damage. • Shorted heated oxygen sensor (rear) circuit or connector damage. • ECM failed.
DTC: P2252 **2T PCM, MIL: Yes** **Year:** 2009, 2010 **Model:** Eclipse, Endeavor, Galant, Outlander **Engine:** 3.0L V6 VIN X, 3.8L V6 VIN S, 3.8L V6 VIN T **Transmission:** All	**Heated Oxygen Sensor Offset Circuit Low Voltage:** After 2 seconds or more have passed since the engine starting sequence was completed. The heated oxygen sensor offset voltage is lower than 0.4 volt for 2 seconds. **Possible Causes:** • Right bank heated oxygen sensor (front) failed. • Right bank heated oxygen sensor (rear) failed. • Left bank heated oxygen sensor (front) failed. • Left bank heated oxygen sensor (rear) failed. • Shorted right bank heated oxygen sensor (front) circuit or connector damage. • Shorted right bank heated oxygen sensor (rear) circuit or connector damage. • Shorted left bank heated oxygen sensor (front) circuit or connector damage. • Shorted left bank heated oxygen sensor (rear) circuit or connector damage. • ECM failed. <M/T> • PCM failed. <A/T>
DTC: P2253 **2T PCM, MIL: Yes** **Year:** 2009, 2010 **Model:** Eclipse, Galant **Engine:** 2.4L L4 VIN F **Transmission:** All	**Heated Oxygen Sensor Offset Circuit High Voltage:** After 2 seconds or more have passed since the engine starting sequence was completed. Heated oxygen sensor offset voltage has continued to be 0.6 volt or higher for 2 seconds. **Possible Causes:** • Cylinder 1, 4 heated oxygen sensor (front) failed. • Cylinder 1, 4 heated oxygen sensor (rear) failed. • Cylinder 2, 3 heated oxygen sensor (front) failed. • Cylinder 2, 3 heated oxygen sensor (rear) failed. • Shorted cylinder 1, 4 heated oxygen sensor (front) circuit or connector damage. • Shorted cylinder 1, 4 heated oxygen sensor (rear) circuit or connector damage. • Shorted cylinder 2, 3 heated oxygen sensor (front) circuit or connector damage. • Shorted cylinder 2, 3 heated oxygen sensor (rear) circuit or connector damage. • PCM failed. <AT> • ECM failed. <MT>
DTC: P2253 **2T PCM, MIL: Yes** **Year:** 2009, 2010 **Model:** Eclipse, Endeavor, Galant, Outlander **Engine:** 3.0L V6 VIN X, 3.8L V6 VIN S, 3.8L V6 VIN T **Transmission:** All	**Heated Oxygen Sensor Offset Circuit High Voltage:** After 2 seconds or more have passed since the engine starting sequence was completed. The heated oxygen sensor offset voltage is higher than 0.6 volt for 2 seconds. **Possible Causes:** • Right bank heated oxygen sensor (front) failed. • Right bank heated oxygen sensor (rear) failed. • Left bank heated oxygen sensor (front) failed. • Left bank heated oxygen sensor (rear) failed. • Shorted right bank heated oxygen sensor (front) circuit or connector damage. • Shorted right bank heated oxygen sensor (rear) circuit or connector damage. • Shorted left bank heated oxygen sensor (front) circuit or connector damage. • Shorted left bank heated oxygen sensor (rear) circuit or connector damage. • ECM failed. <M/T> • PCM failed. <A/T>

DTC	Trouble Code Title, Conditions & Possible Causes
DTC: P2253 **1T ECM, MIL: Yes** **Year:** 2009, 2010 **Model:** Lancer **Engine:** 2.0L L4 VIN U, 2.0L L4 VIN V, 2.4L L4 VIN W **Transmission:** All	**Heated Oxygen Sensor Offset Circuit High Voltage:** More than 2 seconds have passed since the engine starting sequence was completed. Heated oxygen sensor offset voltage is more than 0.6 volt for 2 seconds. **Possible Causes:** • Heated oxygen sensor (front) failed. • Heated oxygen sensor (rear) failed. • Heated oxygen sensor (3rd) failed. • Shorted heated oxygen sensor (front) circuit or connector damage. • Shorted heated oxygen sensor (rear) circuit or connector damage. • Shorted heated oxygen sensor (3rd) circuit or connector damage. • ECM failed.
DTC: P2253 **1T ECM, MIL: Yes** **Year:** 2009, 2010 **Model:** Outlander **Engine:** 2.4L L4 VIN W **Transmission:** All	**Heated Oxygen Sensor Offset Circuit High Voltage:** More than 2 seconds have passed since the engine starting sequence was completed. The heated oxygen sensor offset voltage is more than 0.6 volt for 2 seconds. **Possible Causes:** • Heated oxygen sensor (front) failed. • Heated oxygen sensor (rear) failed. • Shorted heated oxygen sensor (front) circuit or connector damage. • Shorted heated oxygen sensor (rear) circuit or connector damage. • ECM failed.
DTC: P2253 **1T ECM, MIL: Yes** **Year:** 2009, 2010 **Model:** Outlander **Engine:** 2.4L L4 VIN W **Transmission:** All	**Heated Oxygen Sensor Offset Circuit Low Voltage <California>:** More than 2 seconds have passed since the engine starting sequence was completed. The heated oxygen sensor offset voltage is less than 0.4 volt for 2 seconds. **Possible Causes:** • Heated oxygen sensor (rear) failed. • Heated oxygen sensor (3rd) failed. • Shorted heated oxygen sensor (rear) circuit or connector damage. • Shorted heated oxygen sensor (3rd) circuit or connector damage. • ECM failed.
DTC: P2263 **1T ECM, MIL: Yes** **Year:** 2009, 2010 **Model:** Lancer **Engine:** 2.0L L4 VIN V **Transmission:** All	**Intake Charge System Malfunction:** Engine speed is between 3,000 and 5,000 r/min. Throttle position sensor (main) output voltage is more than 2 volts. Volumetric efficiency is less than 100 percent for 2 seconds. **NOTE: Fuel is cut in engine overboost condition.** **Possible Causes:** • Turbocharger wastegate actuator failed. • Turbocharger wastegate regulating valve failed. • Charging pressure control system failed. • Intake charge pressure leak. • ECM failed.
DTC: P2271 **2T PCM, MIL: Yes** **Year:** 2009 **Model:** Raider **Engine:** 3.7L V6 VIN K **Transmission:** All	**O2 Sensor 1/2 Signal Stuck Rich:** With the engine running, vehicle speed above 96 kph (60 mph), throttle open for a minimum of 120 seconds, ECT greater than 70 C (158 F), catalytic converter temperature greater than 600 C (1112 F), and downstream oxygen sensor in a rich state. During a decel fuel shutoff event, the downstream oxygen sensor should switch from rich to lean within a specific time. The PCM monitors the downstream O2 sensor. If the PCM does not detect a rich to lean switch within a specific time during a decel fuel shutoff event, the monitor will fail. Two trip fault. Three good trips to turn off the MIL. **Possible Causes:** • Exhaust leak • O2 sensor wiring or connections • O2 sensor signal circuit • O2 sensor return circuit • O2 Sensor • PCM has failed

DTC	Trouble Code Title, Conditions & Possible Causes
DTC: P2273 **2T PCM, MIL: Yes** **Year:** 2009 **Model:** Raider **Engine:** 3.7L V6 VIN K **Transmission:** All	**O2 Sensor 2/2 Signal Stuck Rich:** With the engine running, vehicle speed above 96 kph (60 mph), throttle open for a minimum of 120 seconds, ECT greater than 70 C (158 F), catalytic converter temperature greater than 600 C (1112 F), and downstream oxygen sensor in a rich state. During a decel fuel shutoff event, the downstream oxygen sensor should switch from rich to lean within a specific time. The PCM monitors the downstream O2 sensor. If the PCM does not detect a rich to lean switch within a specific time during a decel fuel shutoff event, the monitor will fail. Two trip fault. Three good trips to turn off the MIL. **Possible Causes:** • O2 sensor wiring or connections • O2 sensor signal circuit • O2 sensor return circuit • O2 Sensor • PCM has failed
DTC: P2299 **1T PCM, MIL: Yes** **Year:** 2009 **Model:** Raider **Engine:** 3.7L V6 VIN K **Transmission:** All	**Brake Pedal Position/Accelerator Pedal Position Incompatible:** Ignition is on and no Brake or APPS DTCs present. The PCM recognizes that a brake application following the APPS showing a fixed pedal opening. Temporary or permanent in nature. Internally, the PCM will reduce throttle opening below driver demand. This one-trip fault code will set in 5 seconds. The ETC light will illuminate and will only stay on while the DTC is active. **Possible Causes:** • Customer pressing accelerator pedal, then pressing brake pedal and holds both down at the same time • Stop lamp switch has failed • APP Sensor has failed
DTC: P2302 **1T PCM, MIL: Yes** **Year:** 2009 **Model:** Raider **Engine:** 3.7L V6 VIN K **Transmission:** All	**Ignition Coil No. 1 Secondary Circuit Insufficient Ionization:** Engine running and battery voltage greater than 10 volts. If PCM detects that the secondary ignition burn time is incorrect, too short, or not present, an error is detected. **Possible Causes:** • Intermittent condition • Cylinder No. 1 spark plug or wire is damaged or it has failed • Ignition Coil No. 1 is damaged or it has failed • Ignition coil control circuit is open or shorted to ground • ASD relay output circuit problems • PCM has failed
DTC: P2305 **1T PCM, MIL: Yes** **Year:** 2009 **Model:** Raider **Engine:** 3.7L V6 VIN K **Transmission:** All	**Ignition Coil No. 2 Secondary Circuit Insufficient Ionization:** Engine running and battery voltage greater than 10 volts. If PCM detects that the secondary ignition burn time is incorrect, too short, or not present, an error is detected. **Possible Causes:** • Intermittent condition • Cylinder No. 2 spark plug or wire is damaged or it has failed • Ignition Coil No. 2 is damaged or it has failed • Ignition coil control circuit is open or shorted to ground • ASD relay output circuit problems • PCM has failed
DTC: P2308 **1T PCM, MIL: Yes** **Year:** 2009 **Model:** Raider **Engine:** 3.7L V6 VIN K **Transmission:** All	**Ignition Coil No. 3 Secondary Circuit Insufficient Ionization:** Engine running and battery voltage greater than 10 volts. If PCM detects that the secondary ignition burn time is incorrect, too short, or not present, an error is detected. **Possible Causes:** • Intermittent condition • Cylinder No. 3 spark plug or wire is damaged or it has failed • Ignition Coil No. 3 is damaged or it has failed • Ignition coil control circuit is open or shorted to ground • ASD relay output circuit problems • PCM has failed
DTC: P2311 **1T PCM, MIL: Yes** **Year:** 2009 **Model:** Raider **Engine:** 3.7L V6 VIN K **Transmission:** All	**Ignition Coil No. 4 Secondary Circuit Insufficient Ionization:** Engine running and battery voltage greater than 10 volts. If PCM detects that the secondary ignition burn time is incorrect, too short, or not present, an error is detected. **Possible Causes:** • Intermittent condition • Cylinder No. 4 spark plug or wire is damaged or it has failed • Ignition Coil No. 4 is damaged or it has failed • Ignition coil control circuit is open or shorted to ground • ASD relay output circuit problems • PCM has failed

DTC	Trouble Code Title, Conditions & Possible Causes
DTC: P2314 **1T PCM, MIL: Yes** **Year:** 2009 **Model:** Raider **Engine:** 3.7L V6 VIN K **Transmission:** All	**Ignition Coil No. 5 Secondary Circuit Insufficient Ionization:** Engine running and battery voltage greater than 10 volts. If PCM detects that the secondary ignition burn time is incorrect, too short, or not present, an error is detected. **Possible Causes:** • Intermittent condition • Cylinder No. 4 spark plug or wire is damaged or it has failed • Ignition Coil No. 4 is damaged or it has failed • Ignition coil control circuit is open or shorted to ground • ASD relay output circuit problems • PCM has failed
DTC: P2317 **1T PCM, MIL: Yes** **Year:** 2009 **Model:** Raider **Engine:** 3.7L V6 VIN K **Transmission:** All	**Ignition Coil No. 6 Secondary Circuit Insufficient Ionization:** Engine running and battery voltage greater than 10 volts. If PCM detects that the secondary ignition burn time is incorrect, too short, or not present, an error Is detected. **Possible Causes:** • Intermittent condition • Cylinder No. 6 spark plug or wire is damaged or it has failed • Ignition Coil No. 6 is damaged or it has failed • Ignition coil control circuit is open or shorted to ground • ASD relay output circuit problems • PCM has failed
DTC: P2320 **1T PCM, MIL: Yes** **Year:** 2009 **Model:** Raider **Engine:** 3.7L V6 VIN K **Transmission:** All	**Ignition Coil No. 7 Secondary Circuit Insufficient Ionization:** Engine running and battery voltage greater than 10 volts. If PCM detects that the secondary ignition burn time is incorrect, too short, or not present, an error is detected. **Possible Causes:** • Intermittent condition • Cylinder No. 7 spark plug or wire is damaged or it has failed • Ignition Coil No. 7 is damaged or it has failed • Ignition coil control circuit is open or shorted to ground • ASD relay output circuit problems • PCM has failed
DTC: P2323 **1T PCM, MIL: Yes** **Year:** 2009 **Model:** Raider **Engine:** 3.7L V6 VIN K **Transmission:** All	**Ignition Coil No. 8 Secondary Circuit Insufficient Ionization:** Engine running and battery voltage greater than 10 volts. If PCM detects that the secondary ignition burn time is incorrect, too short, or not present, an error is detected. **Possible Causes:** • Intermittent condition • Cylinder No. 8 spark plug or wire is damaged or it has failed • Ignition Coil No. 8 is damaged or it has failed • Ignition coil control circuit is open or shorted to ground • ASD relay output circuit problems • PCM has failed
DTC: P2423 **1T ECM, MIL: Yes** **Year:** 2009, 2010 **Model:** Lancer **Engine:** 2.0L L4 VIN U, 2.4L L4 VIN W **Transmission:** All	**HC absorber (HC trap catalyst) efficiency below threshold:** Engine speed is lower than 1,656 r/min <M/T> or 1,313 r/min <CVT>. Volumetric efficiency is lower than 25 percent <M/T> or 35 percent <CVT>. The accelerator pedal is not depressed. Barometric pressure is higher than 76 kPa (22.4 in. Hg). Intake air temperature is higher than -10°C (14°F). The accumulative mass airflow is higher than 2,900 g. The fuel shut-off mode continues for 3 seconds or more. After the fuel shut-off mode is terminated, the output voltage of the heated oxygen sensor (3rd) is 0.2 volt or less for 0.5 seconds. After the output voltage of the heated oxygen sensor (rear) reaches 0.5 volt, the output voltage of the heated oxygen sensor (3rd) reaches 0.5 volt within 1.5 seconds. **Possible Causes:** • HC trap catalyst deterioration within center exhaust pipe. • ECM failed.
DTC: P2503 **1T PCM, MIL: Yes** **Year:** 2009 **Model:** Raider **Engine:** 3.7L V6 VIN K **Transmission:** All	**Charging System Output Low:** The engine running. The engine RPM is high enough to assure sufficient generator current output to satisfy the electrical loads. The battery sensed voltage is less than the target charging voltage, during engine operation, for a calibrated amount of time. One Trip Fault. Generator light will illuminate. **Possible Causes:** • Loose drive belt • Excessive resistance in the battery positive circuit • Generator • PCM

DTC	Trouble Code Title, Conditions & Possible Causes
DTC: P2504 **1T PCM, MIL: Yes** **Year:** 2009 **Model:** Raider **Engine:** 3.7L V6 VIN K **Transmission:** All	**Charging System Output High:** The engine running. The engine speed greater than 1157 RPM. The alternator B+ voltage sense circuit voltage reading exceeds the direct Battery B+ sense circuit. The Generator Output terminal in not connected to the Battery B+ post. **Possible Causes:** • Excessive resistance in the battery positive circuit • PCM
DTC: P2567 **1T PCM, MIL: Yes** **Year:** 2009, 2010 **Model:** Galant **Engine:** 2.4L L4 VIN F **Transmission:** All	**LIN Communication Check by Radiator Sensor (Thermistor Error):** Ignition switch: "ON". Battery positive voltage is higher than 11 volts. Engine coolant temp -30°C (-22°F) to 110°C (230°F). When the radiator sensor is defective. When the PCM is defective. **Possible Causes:** • Malfunction of the radiator sensor • Malfunction of the PCM
DTC: P2610 **2T PCM, MIL: Yes** **Year:** 2009 **Model:** Raider **Engine:** 3.7L V6 VIN K **Transmission:** All	**PCM Internal Shutdown Timer Rationality Too Fast:** With the engine running after a cycle when a complete engine warm up was achieved, the difference between engine coolant temperature and ambient air temperature greater than 10°C (50°F), after a minimum temperature drop of 10°C (50°F) during ignition off, and battery voltage greater than 10 volts. The PCM detects that the engine coolant temperature drops a specified amount during the measured engine off time. Two trip fault. Three good trips to turn off the MIL. **Possible Causes:** • PCM software • PCM power or ground circuit resistance • PCM
DTC: P2700 **1T PCM, MIL: Yes** **Year:** 2009 **Model:** Raider **Engine:** 3.7L V6 VIN K **Transmission:** All	**L/R Inadequate Element Volume Detected:** Engine started; transmission fluid temperature more than 110°F, vehicle driven, and the PCM updated the L/R volume (during a 3-1 or 2-1 Manual downshift) with the throttle angle less than 5 degrees, and it detected that the L/R volume fell below 16 during the test. **Possible Causes:** • Hydraulic leak in valve body • Broken L/R clutch piston return spring or spring retainers • Broken L/R accumulator springs • Transmission Solenoid/TRS Assembly • PCM
DTC: P2701 **1T PCM, MIL: Yes** **Year:** 2009 **Model:** Raider **Engine:** 3.7L V6 VIN K **Transmission:** All	**Inadequate Element Volume2C:** Whenever the engine is running. The 2C clutch volume index (CVI) is updated during a 3-2 kickdown with throttle angle between 10 and 54 degrees. Transmission temperature must be at least 43° C (110° F). When the 2C CVI falls below 5. **Possible Causes:** • Hydraulic leak in valve body • Broken 2C clutch piston return spring, dislodged snap ring • Broken accumulator springs • Transmission Solenoid/TRS Assembly • PCM
DTC: P2702 **1T PCM, MIL: Yes** **Year:** 2009 **Model:** Raider **Engine:** 3.7L V6 VIN K **Transmission:** All	**Inadequate Element Volume OD:** Whenever the engine is running. The OD clutch volume index (CVI) is updated during a 2-3 upshift with throttle angle between 10 and 54 degrees. Transmission temperature must be at least 43° C (110° F). When the OD CVI falls below 5. **Possible Causes:** • Hydraulic leak in valve body • Broken OD/Reverse piston belleville spring or dislodged snap ring • Broken OD accumulator springs • Internal leakage in pump assembly • Transmission Solenoid/TRS Assembly • PCM
DTC: P2703 **1T PCM, MIL: Yes** **Year:** 2009 **Model:** Raider **Engine:** 3.7L V6 VIN K **Transmission:** All	**Inadequate Element Volume UD:** Whenever the engine is running. The UD clutch volume index (CVI) is updated during a 4-3 kickdown with throttle angle between 10 and 54 degrees. Transmission temperature must be at least 43° C (110° F). When the UD CVI falls below 11. **Possible Causes:** • Broken UD clutch piston spring or dislodged snap ring • Broken UD accumulator springs • Internal leakage in pump or valve body assembly • Transmission Solenoid/TRS Assembly • PCM

DTC	Trouble Code Title, Conditions & Possible Causes
DTC: P2704 **1T PCM, MIL: Yes** **Year:** 2009 **Model:** Raider **Engine:** 3.7L V6 VIN K **Transmission:** All	**Inadequate Element Volume 4C:** Whenever the engine is running. The 4C clutch volume index (CVI) is updated during a 3-4 upshift with throttle angle between 10 and 54 degrees. Transmission temperature must be at least 43° C (110° F). When the 4C CVI falls below 5. **Possible Causes:** • Broken 4C return spring or dislodged snap ring • Broken $c accumulator springs • Internal leakage in valve body assembly • Transmission Solenoid/TRS Assembly • PCM
DTC: P2706 **1T PCM, MIL: Yes** **Year:** 2009 **Model:** Raider **Engine:** 3.7L V6 VIN K **Transmission:** All	**MS Solenoid Circuit Malfunction:** Initially at power-up, then every 10 seconds thereafter. The solenoid circuits will also be tested immediately after a gear ratio or pressure switch error is detected. After three consecutive solenoid continuity test failures, or one failure if test is run in response to a gear ratio or pressure switch error. **Possible Causes:** • Related DTC's present • MS Solenoid control circuit shorted to another circuit • MS Solenoid control circuit open • MS Solenoid control circuit shorted to ground • Transmission Solenoid/TRS Assembly • PCM
DTC: P2718 **1T ECM, MIL: Yes** **Year:** 2009, 2010 **Model:** Lancer **Engine:** 2.0L L4 VIN U, 2.4L L4 VIN W **Transmission:** All	**Clutch/Shift Pressure Solenoid 1 System (Open circuit):** Voltage of battery: 8 V or more. Voltage of battery: 16.5 V or less. The FET channel output remains 1 V or less for 400 milliseconds **Possible Causes:** • Malfunction of TC-SST-ECU • Malfunction of clutch/shift pressure solenoid 1
DTC: P2719 **1T ECM, MIL: Yes** **Year:** 2009, 2010 **Model:** Lancer **Engine:** 2.0L L4 VIN U, 2.4L L4 VIN W **Transmission:** All	**Clutch/Shift Pressure Solenoid 1 System (Overcurrent):** Voltage of battery: 8 V or more. Voltage of battery: 16.5 V or less. The status with the current of the FET channel shunt 3.5 A or less and with the FET channel output 100 mV or more continues for 300 milliseconds. **Possible Causes:** • Malfunction of TC-SST-ECU • Malfunction of clutch/shift pressure solenoid 1
DTC: P2720 **1T ECM, MIL: Yes** **Year:** 2009, 2010 **Model:** Lancer **Engine:** 2.0L L4 VIN U, 2.4L L4 VIN W **Transmission:** All	**Clutch/Shift Pressure Solenoid 1 System (Short to ground):** Voltage of battery: 8 V or more. Voltage of battery: 16.5 V or less. The FET channel output remains 100 mV or more for 400 milliseconds. **Possible Causes:** • Malfunction of TC-SST-ECU • Malfunction of clutch/shift pressure solenoid 1
DTC: P2721 **1T ECM, MIL: Yes** **Year:** 2009, 2010 **Model:** Lancer **Engine:** 2.0L L4 VIN U, 2.4L L4 VIN W **Transmission:** All	**Clutch/Shift Pressure Solenoid 1 System (Short to power supply):** Voltage of battery: 8 V or more. Voltage of battery: 16.5 V or less. Clutch/shift pressure solenoid 1: OFF. The FET channel output remains the battery voltage - 2 V or less for 160 milliseconds. **Possible Causes:** • Malfunction of TC-SST-ECU • Malfunction of clutch/shift pressure solenoid 1
DTC: P2727 **1T ECM, MIL: Yes** **Year:** 2009, 2010 **Model:** Lancer **Engine:** 2.0L L4 VIN U, 2.4L L4 VIN W **Transmission:** All	**Clutch/Shift Pressure Solenoid 2 System (Open circuit):** Voltage of battery: 8 V or more. Voltage of battery: 16.5 V or less. The FET channel output remains 1 V or less for 400 milliseconds **Possible Causes:** • Malfunction of TC-SST-ECU • Malfunction of clutch/shift pressure solenoid 2

DTC	Trouble Code Title, Conditions & Possible Causes
DTC: P2728 **1T ECM, MIL: Yes** **Year:** 2009, 2010 **Model:** Lancer **Engine:** 2.0L L4 VIN U, 2.4L L4 VIN W **Transmission:** All	**Clutch/Shift Pressure Solenoid 2 System (Overcurrent):** Voltage of battery: 8 V or more. Voltage of battery: 16.5 V or less. The status with the current of the FET channel shunt 3.5 A or less and with the FET channel output 100 mV or more continues for 300 milliseconds. **Possible Causes:** • Malfunction of TC-SST-ECU • Malfunction of clutch/shift pressure solenoid 2
DTC: P2729 **1T ECM, MIL: Yes** **Year:** 2009, 2010 **Model:** Lancer **Engine:** 2.0L L4 VIN U, 2.4L L4 VIN W **Transmission:** All	**Clutch/Shift Pressure Solenoid 2 System (Short to ground):** Voltage of battery: 8 V or more. Voltage of battery: 16.5 V or less. The FET channel output remains 100 mV or more for 400 milliseconds. **Possible Causes:** • Malfunction of TC-SST-ECU • Malfunction of clutch/shift pressure solenoid 2
DTC: P2730 **1T ECM, MIL: Yes** **Year:** 2009, 2010 **Model:** Lancer **Engine:** 2.0L L4 VIN U, 2.4L L4 VIN W **Transmission:** All	**Clutch/Shift Pressure Solenoid 2 System (Short to power supply):** Voltage of battery: 8 V or more. Voltage of battery: 16.5 V or less. Clutch/shift pressure solenoid 2: OFF. The FET channel output remains (Battery voltage - 2 V) or less for 160 milliseconds. **Possible Causes:** • Malfunction of TC-SST-ECU • Malfunction of clutch/shift pressure solenoid 2
DTC: P2733 **1T ECM, MIL: Yes** **Year:** 2009, 2010 **Model:** Lancer **Engine:** 2.0L L4 VIN U, 2.4L L4 VIN W **Transmission:** All	**Clutch/Shift Switching Solenoid 1, spool stuck:** Voltage of battery: 8 V or more. Voltage of battery: 16.5 V or less. Engine speed: 650 r/min or more. Time since above engine condition: 1.5 seconds or more. Clutch/shift switching solenoid 1: ON. Clutch/shift switching solenoid 2: OFF. Clutch cooling flow solenoid: OFF. Clutch/shift pressure solenoid 1: ON. The status with the clutch 1 pressure 1,400 hPa or less, or with the clutch 1 pressure sensor (odd) and shift fork position sensor 1 and 2 changed continues for 200 millisecond. **Possible Causes:** • Malfunction of TC-SST-ECU • Insufficient fluid level • Malfunction of clutch/shift switching solenoid 1
DTC: P2736 **1T ECM, MIL: Yes** **Year:** 2009, 2010 **Model:** Lancer **Engine:** 2.0L L4 VIN U, 2.4L L4 VIN W **Transmission:** All	**Clutch/Shift Switching Solenoid 1 System (Open circuit):** Voltage of battery: 8 V or more. Voltage of battery: 16.5 V or less. The FET channel output remains 1 V or less for 400 milliseconds. **Possible Causes:** • Malfunction of TC-SST-ECU • Malfunction of clutch/shift switching solenoid 1
DTC: P2738 **1T ECM, MIL: Yes** **Year:** 2009, 2010 **Model:** Lancer **Engine:** 2.0L L4 VIN U, 2.4L L4 VIN W **Transmission:** All	**Clutch/Shift Switching Solenoid 1 System (Short to ground):** Voltage of battery: 8 V or more. Voltage of battery: 16.5 V or less. The FET channel output remains 100 mV or more for 200 milliseconds. **Possible Causes:** • Malfunction of TC-SST-ECU • Malfunction of clutch/shift switching solenoid 1
DTC: P2739 **1T ECM, MIL: Yes** **Year:** 2009, 2010 **Model:** Lancer **Engine:** 2.0L L4 VIN U, 2.4L L4 VIN W **Transmission:** All	**Clutch/Shift Switching Solenoid 1 System (Short to power supply):** Voltage of battery: 8 V or more. Voltage of battery: 16.5 V or less. Clutch / shift switching solenoid 1: OFF. The FET channel output remains (Battery voltage - 2 V) or less for 160 milliseconds. **Possible Causes:** • Malfunction of TC-SST-ECU • Malfunction of clutch/shift switching solenoid
DTC: P2742 **1T ECM, MIL: Yes** **Year:** 2009, 2010 **Model:** Lancer **Engine:** 2.0L L4 VIN U, 2.4L L4 VIN W **Transmission:** All	**Fluid Temperature Sensor System (Output low range out):** Voltage of battery: 8 V or more. Voltage of battery: 16.5 V or less. The fluid temperature remains -39°C (-38.2°F) or more for 400 milliseconds **Possible Causes:** • Malfunction of TC-SST-ECU

DTC	Trouble Code Title, Conditions & Possible Causes
DTC: P2743 **1T ECM, MIL: Yes** **Year:** 2009, 2010 **Model:** Lancer **Engine:** 2.0L L4 VIN U, 2.4L L4 VIN W **Transmission:** All	**Fluid Temperature Sensor System (Output high range out):** Voltage of battery: 8 V or more. Voltage of battery: 16.5 V or less. The fluid temperature remains 149°C (300.2°F) or more for 400 milliseconds. **Possible Causes:** • Malfunction of TC-SST-ECU
DTC: P2766 **1T ECM, MIL: Yes** **Year:** 2009, 2010 **Model:** Lancer **Engine:** 2.0L L4 VIN U, 2.4L L4 VIN W **Transmission:** All	**Input Shaft 2 (Even number gear axle) Speed Sensor System (Poor performance):** Voltage of battery: 8 V or more. Voltage of battery: 16.5 V or less. Input shaft 2 (even) speed: Refer to Logic Flow Charts (Monitor Sequence) <Rationality plausibility failure>. (250 millisecond) **Possible Causes:** • Malfunction of TC-SST-ECU
DTC: P2809 **1T ECM, MIL: Yes** **Year:** 2009, 2010 **Model:** Lancer **Engine:** 2.0L L4 VIN U, 2.4L L4 VIN W **Transmission:** All	**Clutch/Shift Switching Solenoid 2, spool stuck:** Voltage of battery: 8 V or more. Voltage of battery: 16.5 V or less. Engine speed: 650 r/min or more. Time since above engine condition: 1.5 seconds or more. Clutch/shift switching solenoid 2: ON. Clutch cooling flow solenoid: OFF. Clutch/shift pressure solenoid 2: ON. The status with the clutch 2 pressure 1,400 hPa or less, or with the clutch 2 pressure sensor (even) and shift fork position sensor 3 and 4 changed continues for 200 millisecond. **Possible Causes:** • Malfunction of TC-SST-ECU • Insufficient fluid level • Malfunction of clutch/shift switching solenoid 2
DTC: P2812 **1T ECM, MIL: Yes** **Year:** 2009, 2010 **Model:** Lancer **Engine:** 2.0L L4 VIN U, 2.4L L4 VIN W **Transmission:** All	**Clutch/Shift Switching Solenoid 2 System (Open circuit):** Voltage of battery: 8 V or more. Voltage of battery: 16.5 V or less. The FET channel output remains 1 V or less for 400 milliseconds. **Possible Causes:** • Malfunction of TC-SST-ECU • Malfunction of clutch/shift switching solenoid 2
DTC: P2814 **1T ECM, MIL: Yes** **Year:** 2009, 2010 **Model:** Lancer **Engine:** 2.0L L4 VIN U, 2.4L L4 VIN W **Transmission:** All	**Clutch/Shift Switching Solenoid 2 System (Short to ground):** Voltage of battery: 8 V or more. Voltage of battery: 16.5 V or less. The FET channel output remains 100 mV or more for 200 milliseconds. **Possible Causes:** • Malfunction of TC-SST-ECU • Malfunction of clutch/shift switching solenoid 2
DTC: P2815 **1T ECM, MIL: Yes** **Year:** 2009, 2010 **Model:** Lancer **Engine:** 2.0L L4 VIN U, 2.4L L4 VIN W **Transmission:** All	**Clutch/Shift Switching Solenoid 2 System (Short to power supply):** Voltage of battery: 8 V or more. Voltage of battery: 16.5 V or less. Clutch / shift switching solenoid 2: OFF. The FET channel output remains (Battery voltage - 2 V) or less for 160 milliseconds. **Possible Causes:** • Malfunction of TC-SST-ECU • Malfunction of clutch/shift switching solenoid 2

OBD II Trouble Code List (U0XXX Codes)

DTC	Trouble Code Title, Conditions & Possible Causes
DTC: U0001 **T ECM, MIL: Yes** **Year:** 2009, 2010 **Model:** Lancer, Outlander **Engine:** 2.0L L4 VIN U, 2.0L L4 VIN V, 2.4L L4 VIN W, 3.0L V6 VIN X **Transmission:** All	**Bus-Off (MFI):** Always, bus-off error detected. **NOTE: If the ECM output the DTC U0001, make sure to diagnose the CAN bus line.** **Before replacing the ECU, make sure that the communication circuit is operating normally.** **Possible Causes:** • CAN line harness damage or connector damage. • ECM failed.

DTC	Trouble Code Title, Conditions & Possible Causes
DTC: U0001 **T ECM** **Year:** 2009, 2010 **Model:** Lancer, Outlander **Engine:** 2.0L L4 VIN U, 2.0L L4 VIN V, 2.4L L4 VIN W, 3.0L V6 VIN X **Transmission:** All	**Bus-Off (ASC):** This DTC is set when ASC-ECU has ceased the CAN communication (bus off). Malfunction of wiring harness, connector (s), or ASC-ECU may be present. **Possible Causes:** • Wiring harness or connector failure of CAN bus line • ASC-ECU malfunction • Other ECU malfunction
DTC: U0001 **T ECM** **Year:** 2009, 2010 **Model:** Lancer **Engine:** 2.0L L4 VIN U, 2.4L L4 VIN W **Transmission:** All	**Bus-Off (AWC-ECU):** The code is set when AWC-ECU ceases (bus-off). **Possible Causes:** • The CAN bus line is defective. • Malfunction of AWC-ECU • ECU malfunction of other system
DTC: U0001 **T ECM** **Year:** 2009, 2010 **Model:** Lancer, Outlander **Engine:** 2.0L L4 VIN U, 2.4L L4 VIN W **Transmission:** All	**Malfunction of CAN Communication Circuit (CVT):** Time after TCM start: more than 1.05 seconds. Voltage of battery: 9 volts or more Voltage of battery: 16 volts or less. CAN communication: fail **Possible Causes:** • Malfunction of the CAN bus
DTC: U0001 **T ECM** **Year:** 2009, 2010 **Model:** Lancer **Engine:** 2.0L L4 VIN U, 2.4L L4 VIN W **Transmission:** All	**Bus-Off (TC-SST):** Voltage of battery: 8 V or more. Voltage of battery: 16.5 V or less. Time after TC-SST-ECU start: 5 seconds or more. The CAN communication: Error. (60 millisecond) **Possible Causes:** • The CAN bus line is defective. • Malfunction of TC-SST-ECU
DTC: U0001 **T ECM** **Year:** 2009, 2010 **Model:** Outlander **Engine:** 3.0L V6 VIN X **Transmission:** All	**Can Bus Off (A/T):** TCM is started: more than 1.05 seconds. A/T control relay voltage: 10 volts or more. A/T control relay voltage: 16 volts or less. CAN communication: fail. **Possible Causes:** • Malfunction of the CAN circuit • Malfunction of the TCM • Damaged harness or connector • Malfunction of the CAN system module
DTC: U0001 **1T PCM, MIL: Yes** **Year:** 2009 **Model:** Raider **Engine:** 3.7L V6 VIN K **Transmission:** All	**CAN C BUS Circuit:** When Monitored: Ignition run time is greater than 1 second. Battery voltage between 9 and 16 volts. Engine run time greater than 3 seconds. Set Condition: The PCM loses communication over the CAN C Bus circuit. The circuit is continuously monitored. One trip fault. **Possible Causes:** • CAN C BUS Failure open or short • PCM
DTC: U0019 **T ECM** **Year:** 2009, 2010 **Model:** Lancer **Engine:** 2.0L L4 VIN U, 2.0L L4 VIN V, 2.4L L4 VIN W **Transmission:** All	**Bus off (CAN-B) (Satellite Radio) :** With the ignition switch at the ON position and the system voltage at 10 - 16 volts (data from ETACS-ECU), if the satellite radio tuner becomes unable to transmit data normally due to the CAN-B bus circuit malfunction, the satellite radio tuner determines that a problem has occurred. **Possible Causes:** • The satellite radio tuner, power supply for the satellite radio tuner, earth circuit, or CAN bus line may have a problem.
DTC: U0019 **T ECM** **Year:** 2009, 2010 **Model:** Lancer **Engine:** 2.0L L4 VIN U, 2.0L L4 VIN V, 2.4L L4 VIN W **Transmission:** All	**Bus Off (CAN-B) (SRS):** Because of the CAN-B bus circuit malfunction, if SRS-ECU becomes unable to perform the normal data transmission, SRS-ECU determines that an abnormality is present. **Possible Causes:** • The CAN bus line may be defective

DTC	Trouble Code Title, Conditions & Possible Causes
DTC: U0019 **T ECM** **Year:** 2009, 2010 **Model:** Lancer **Engine:** 2.0L L4 VIN U, 2.0L L4 VIN V, 2.4L L4 VIN W **Transmission:** All	**Bus off (CAN-B):** With the ignition switch at the ON position and the system voltage at 10 - 16 volts (data from ETACS-ECU), if the CAN box unit becomes unable to transmit data normally due to the CAN-B bus circuit malfunction, the CAN box unit determines that a problem has occurred. **Possible Causes:** • The CAN bus line may be defective
DTC: U0019 **T ECM** **Year:** 2009, 2010 **Model:** Lancer **Engine:** 2.0L L4 VIN U, 2.0L L4 VIN V, 2.4L L4 VIN W **Transmission:** All	**Bus off (CAN-B) Hands Free Module:** When the hands free module is returned from the bus off state, or when the bus error is indicated to the hands free module state, the DTC U0019 (CAN-B) is set. **Possible Causes:** • The hands free module may be defective. • The CAN bus line may be defective.
DTC: U0020 **T ECM** **Year:** 2009, 2010 **Model:** Lancer **Engine:** 2.0L L4 VIN U, 2.0L L4 VIN V, 2.4L L4 VIN W **Transmission:** All	**CAN-B Bus Off Performance:** Because of the CAN-B bus circuit malfunction, if occupant classification-ECU becomes unable to perform the normal data transmission, occupant classification-ECU determines that an abnormality is present. **NOTE: If the CAN-B circuit malfunction is present, the occupant classification-ECU sets the DTC U0020, U0021, U0022, U0023, U0024, U0025, U0026.** **Possible Causes:** • The CAN bus line may be defective
DTC: U0100 **T ECM** **Year:** 2009, 2010 **Model:** Lancer **Engine:** 2.0L L4 VIN U, 2.0L L4 VIN V, 2.4L L4 VIN W **Transmission:** All	**Engine Time-out Error (TC-SST):** The periodic communication data from the engine control module cannot be received. **Possible Causes:** • The CAN bus line is defective. • ECM malfunction • Malfunction of TC-SST-ECU
DTC: U0100 **T ECM** **Year:** 2009, 2010 **Model:** Outlander **Engine:** 3.0L V6 VIN X **Transmission:** All	**Engine Time-out Error (A/T):** The code is set when the signal sent from the ECM cannot be received for a certain period. **Possible Causes:** • Malfunction of the CAN bus line • Malfunction of power supply circuit system (ECM) • Malfunction of ECU (ECM) • Malfunction of AWC-ECU
DTC: U0100 **T ECM** **Year:** 2009, 2010 **Model:** Lancer, Outlander **Engine:** 2.0L L4 VIN U, 2.4L L4 VIN W **Transmission:** All	**CAN Time-out Error (Engine) (CVT):** Time after TCM start: more than 1.05 seconds. Voltage of battery: 9 volts or more Voltage of battery: 16 volts or less. The CAN communication with ECM: fail. (500 millisecond) The TCM cannot receive the periodic communication data from the ECM. **Possible Causes:** • Malfunction of the CAN bus • Engine ECU malfunction • Malfunction of TCM
DTC: U0101 **1T TCM** **Year:** 2009 **Model:** Raider **Engine:** 3.7L V6 VIN K **Transmission:** All	**No CAN Bus Message:** The PCM does not receive a Bus message. The circuit is continuously monitored. **Possible Causes:** • CAN C Bus failure open or shorted • PCM has failed
DTC: U0114 **T ECM** **Year:** 2009, 2010 **Model:** Outlander **Engine:** 3.0L V6 VIN X **Transmission:** All	**AWD-ECU Time-Out <AWD> :** Some of the possible causes are a harness or connector damage between the ECM and the AWD-ECU on the CAN bus line, a failure in the AWD-ECU power supply system, a failure in the AWD-ECU, or a failure in the ECM. **Possible Causes:** • CAN line harness damage or connector damage.

DTC	Trouble Code Title, Conditions & Possible Causes
DTC: U0121 **T ECM** **Year:** 2009, 2010 **Model:** Outlander **Engine:** 3.0L V6 VIN X **Transmission:** All	**ASC Time-out Error (A/T):** The code is set when the signal sent from the ASC-ECU cannot be received for a certain period. **Possible Causes:** • Malfunction of the CAN bus line • Malfunction of power supply circuit system (ASC-ECU) • Malfunction of ECU (ASC-ECU) • Malfunction of AWC-ECU
DTC: U0121 **1T PCM** **Year:** 2009 **Model:** Raider **Engine:** 3.7L V6 VIN K **Transmission:** All	**Lost Communication With ABS Module:** Ignition is on and battery voltage is 9-16v. Engine is running for more than 3 seconds. The PCM does not receive an ABS message over the CAN C circuit for 7 consecutive seconds. The circuit is continuously monitored. **Possible Causes:** • CAN C Bus failure open or shorted • ABS module has failed • PCM has failed
DTC: U0121 **T ECM** **Year:** 2009, 2010 **Model:** Outlander **Engine:** 2.4L L4 VIN W **Transmission:** All	**CAN Time-out Error (ABS):** TCM determines that malfunction is present if the periodic communication data sent from the ABS-ECU via CAN bus lines is abnormal. TCM cannot receive the periodic communication data from ABS-ECU. **Possible Causes:** • Malfunction of the CAN bus • ABS-ECU malfunction • Malfunction of TCM
DTC: U0141 **1T TCM** **Year:** 2009 **Model:** Raider **Engine:** 3.7L V6 VIN K **Transmission:** All	**Lost Communication With Front Control Module:** Ignition is on and battery voltage is 9-16v. Engine is running for more than 3 seconds. The PCM does not receive an FCM message over the CAN C circuit for 7 consecutive seconds. The circuit is continuously monitored. **Possible Causes:** • CAN C Bus failure open or shorted • Front control module has failed • PCM has failed
DTC: U0141 **T ECM** **Year:** 2009, 2010 **Model:** Outlander **Engine:** 3.0L V6 VIN X **Transmission:** All	**ETACS-ECU Time Out:** Open circuit or short circuit occurs in the CAN communication line (CAN_H, CAN_L). Reception from ETACS-ECU becomes impossible for 0.5 seconds or more. **Possible Causes:** • Damaged harness or connector • Malfunction of the CAN communication • Malfunction of the ETACS-ECU • Malfunction of the TCM
DTC: U0155 **1T PCM** **Year:** 2009 **Model:** Raider **Engine:** 3.7L V6 VIN K **Transmission:** All	**No Cluster Bus Message:** Engine running. Battery voltage more than 10v. No Bus messages are received from the MIC (instrument cluster) for 20 seconds. **Possible Causes:** • Communication link with instrument cluster has failed • Instrument cluster operation improper or has failed • PCM has failed
DTC: U0167 **T ECM** **Year:** 2009, 2010 **Model:** Outlander **Engine:** 2.4L L4 VIN W, 3.0L V6 VIN X **Transmission:** All	**Immobilizer Communication Error:** Ignition switch is in "ON" position. Battery positive voltage is 8 volts or higher. Unable to receive KOS-ECU/WCM (immobilizer-ECU) signals through the CAN bus line. **Possible Causes:** • CAN line harness damage or connector damage. • KOS-ECU/WCM (immobilizer-ECU) failed. • ECM failed.
DTC: U0401 **T ECM** **Year:** 2009, 2010 **Model:** Lancer, Outlander **Engine:** 2.0L L4 VIN U, 2.0L L4 VIN V, 2.4L L4 VIN W, 3.0L V6 VIN X **Transmission:** All	**Engine signal malfunction detected:** This DTC is set when the engine control module (ECM) malfunction has been detected. **Possible Causes:** • Malfunction of engine system • Malfunction of engine control module (ECM) • Malfunction of ASC-ECU • External noise interference

DTC	Trouble Code Title, Conditions & Possible Causes
DTC: U0431 T ECM Year: 2009, 2010 Model: Lancer Engine: 2.0L L4 VIN U, 2.4L L4 VIN W Transmission: All	**ETACS Data Error:** The code is set when abnormality is detected in the signal received from the ETACS-ECU. **Possible Causes:** • The CAN bus line is defective. • Malfunction of the ETACS-ECU • Malfunction of AWC-ECU

OBD II Trouble Code List (U1XXX Codes)

DTC	Trouble Code Title, Conditions & Possible Causes
DTC: U1073 T PCM Year: 2010 Model: Endeavor Engine: 3.8L V6 VIN S Transmission: All	**Bus Off (Multi-Center Display Unit):** The diagnostic trouble code will be stored when the center display unit ceases communicating once (i.e. bus off) and returns to communication from "Penalty mode". **NOTE: If DTC U1073 is set in the multi-center display unit (middle grade type), always diagnose the CAN bus lines.** **WARNING: Whenever the multi-center display unit (middle grade type) is replaced, ensure that the power supply circuit and the ground circuit are normal.** **Possible Causes:** • The wiring harness or connectors may have loose, corroded, or damaged terminals, or terminals pushed back in the connector • Malfunction of multi-center display unit (middle grade type)
DTC: U1073 1T BCM, MIL: Yes Year: 2009, 2010 Model: Eclipse, Endeavor, Galant Engine: 2.4L L4 VIN F, 3.8L V6 VIN S, 3.8L V6 VIN T Transmission: All	**Bus Off (SRS System):** Diagnosis code No.U1073 will be stored when the SRS-ECU ceases CAN communication (bus off) and then resumes the communication by turning the ignition switch to the "LOCK" (OFF) position. **NOTE: The wiring harness wire or connectors may have loose, corroded, or damage terminals, or terminals pushed back in the connector, or the SRS-ECU may be defective.** **Possible Causes:** • Defective connector(s) or wiring harness • Malfunction of the SRS-ECU
DTC: U1073 2T BCM, MIL: Yes Year: 2009, 2010 Model: Eclipse, Endeavor, Galant Engine: 2.4L L4 VIN F, 3.8L V6 VIN S, 3.8L V6 VIN T Transmission: All	**Bus Off (Traction Control):** This code is stored when the TCL/ASC-ECU has ceased the CAN communication error (bus off). Then, if a penalty mode is entered after approximately five minutes, regular data transmission will be cancelled. **NOTE: If DTC U1073 is set in the TCL/ASC-ECU, always diagnose the CAN main bus line. If there is any fault in the CAN bus lines, an incorrect DTC may be set.** **Whenever the ECU is replaced, ensure that the communication circuit is normal.** **Possible Causes:** • Damaged wiring harness or connector • Malfunction of the hydraulic unit (integrated with TCL/ASC-ECU)
DTC: U1073 2T CCM, MIL: Yes Year: 2009, 2010 Model: Galant Engine: 2.4L L4 VIN F, 3.8L V6 VIN S, 3.8L V6 VIN T Transmission: All	**Bus Off:** Bus off error detected **Possible Causes:** • CAN line harness or connectors damage • PCM failed
DTC: U1073 1T BCM, MIL: Yes Year: 2009, 2010 Model: Eclipse, Endeavor Engine: 2.4L L4 VIN F, 3.8L V6 VIN S, 3.8L V6 VIN T Transmission: All	**Bus Off (ECM PCM):** Bus off error detected. **NOTE: If the ECM <M/T> or PCM <A/T> output the DTC U1073, make sure to diagnose the CAN bus line.** **Before replacing the ECU, make sure that the communication circuit is operating normally.** **Possible Causes:** • CAN line harness damage or connector damage. • ECM failed. <M/T> • PCM failed. <A/T>
DTC: U1073 2T BCM, MIL: Yes Year: 2009, 2010 Model: Eclipse, Galant Engine: 2.4L L4 VIN F, 3.8L V6 VIN S, 3.8L V6 VIN T Transmission: All	**Bus Off (Body):** If the CAN circuit malfunction occurs, the convertible top control module sets DTC U1073. **Possible Causes:** • Malfunction of CAN bus line

DTC	Trouble Code Title, Conditions & Possible Causes
DTC: U1100 **1T PCM, MIL: Yes** **Year:** 2009, 2010 **Model:** Eclipse, Endeavor, Galant **Engine:** 2.4L L4 VIN F, 3.8L V6 VIN S, 3.8L V6 VIN T **Transmission:** All	**CAN Communications System Time Out Error Engine Related Data:** The TCL/ASC-ECU receives engine system-related signals from the ECM <M/T> or PCM <A/T> via CAN bus lines. **NOTE: If DTC U1100 is set in the TCL/ASC-ECU, always diagnose the CAN main bus line. If there is any fault in the CAN bus lines, an incorrect DTC may be set.** **Whenever the TCL/ASC-ECU is replaced, ensure that the communication circuit is normal** **Possible Causes:** • Damaged harness or connector. • Malfunction of the ECM <M/T> or PCM <A/T>. • Malfunction of the TCL/ASC-ECU.
DTC: U1101 **1T PCM, MIL: Yes** **Year:** 2009, 2010 **Model:** Eclipse **Engine:** 2.4L L4 VIN F, 3.8L V6 VIN T **Transmission:** All	**CAN Communications System Time Out Error A/T Related Data:** The TCL/ASC-ECU receives A/T system-related signals from the PCM via CAN bus lines. **NOTE: If DTC U1101 is set in the TCL/ASC-ECU, always diagnose the CAN main bus line. If there is any fault in the CAN bus lines, an incorrect DTC may be set.** **Whenever the TCL/ASC-ECU is replaced, ensure that the communication circuit is normal.** **Possible Causes:** • Damaged harness or connector. • Malfunction of the PCM. • Malfunction of the TCL/ASC-ECU.
DTC: U1102 **1T PCM, MIL: Yes** **Year:** 2009, 2010 **Model:** Eclipse, Endeavor, Galant **Engine:** 2.4L L4 VIN F, 3.8L V6 VIN S, 3.8L V6 VIN T **Transmission:** All	**ABS/TCL-ECU OR TCL/ASC-ECU TIME-OUT:** Battery positive voltage is 10 volts or higher. Engine is not cranked, or at least 3 seconds have passed since engine was cranked. Unable to receive ABS/TCL-ECU or the TCL/ASC-ECU signals through the CAN bus line. **NOTE: If the ECM <M/T> or PCM <A/T> output the DTC U1102, make sure to diagnose the CAN bus line.** **Before replacing the ECU, make sure that the communication circuit is operating normally.** **Possible Causes:** • CAN line harness damage or connector damage. • ABS/TCL-ECU failed <vehicles without ASC> • TCL/ASC-ECU failed <vehicles with ASC> • ECM failed. <M/T> • PCM failed. <A/T> • .
DTC: U1104 **1T PCM, MIL: Yes** **Year:** 2009, 2010 **Model:** Eclipse, Endeavor **Engine:** 2.4L L4 VIN F, 3.8L V6 VIN S, 3.8L V6 VIN T **Transmission:** All	**SAS CAN Timeout/Not equipped:** Connector(s) or wiring harness in the CAN bus lines between the TCL/ASC-ECU and the steering wheel sensor, the power supply system to the steering wheel sensor, the steering wheel sensor itself, or the TCL/ASC-ECU may be defective. **NOTE: If DTC U1104 is set in the TCL/ASC-ECU, always diagnose the CAN bus lines.** **Whenever the ECU is replaced, ensure that the CAN bus lines are normal.** **Possible Causes:** • CAN bus line is defective. • Malfunction of the hydraulic unit (integrated with TCL/ASC-ECU) • Malfunction of the steering wheel sensor
DTC: U1105 **1T PCM, MIL: Yes** **Year:** 2010 **Model:** Endeavor **Engine:** 3.8L V6 VIN S **Transmission:** All	**YR/G CAN Timeout/Not equipped:** The TCL/ASC-ECU and the G and yaw rate sensor are connected each other via CAN bus line, and the sensor sends longitudinal and lateral acceleration information to the ECU. **Possible Causes:** • The CAN bus line is defective. • Malfunction of the hydraulic unit (integrated with TCL/ASC-ECU) • Malfunction of the G and yaw rate sensor
DTC: U1108 **2T CCM, MIL: Yes** **Year:** 2009, 2010 **Model:** Eclipse, Endeavor, Galant **Engine:** 2.4L L4 VIN F, 3.8L V6 VIN S, 3.8L V6 VIN T **Transmission:** All	**Combination Meter Timeout:** Engine not cranking or 3 seconds have passed since the engine was cranked and battery voltage is 10 volts or greater, the combination meter was unable to receive a signal through the CAN bus line. **Possible Causes:** • CAN line harness or connectors damage • Combination meter power supply system failed • Combination meter failed • PCM failed

DTC	Trouble Code Title, Conditions & Possible Causes
DTC: U1109 **2T CCM, MIL: Yes** **Year:** 2009, 2010 **Model:** Eclipse, Endeavor **Engine:** 2.4L L4 VIN F, 3.8L V6 VIN S, 3.8L V6 VIN T **Transmission:** All	**ETACS-ECU Timeout:** Engine not cranking or 3 seconds have passed since the engine was cranked and battery voltage is 10 volts or greater, the ETACS-ECU was unable to receive a signal through the CAN bus line. **Possible Causes:** • CAN line harness or connectors damage • ETACS-ECU power supply system failed • ETACS-ECU failed • PCM failed
DTC: U110C **1T TCM** **Year:** 2009 **Model:** Raider **Engine:** 3.7L V6 VIN K **Transmission:** All	**No Fuel Level Bus Message:** Ignition is on. PCM does not receive a fuel level signal from the FCM over the CAN C circuit. The circuit is constantly monitored. **Possible Causes:** • CAN C Bus failure open or shorted • Front Control Module (FCM) has failed • PCM has failed
DTC: U1110 **2T CCM, MIL: Yes** **Year:** 2009, 2010 **Model:** Eclipse, Endeavor **Engine:** 2.4L L4 VIN F, 3.8L V6 VIN S, 3.8L V6 VIN T **Transmission:** All	**A/C–ECU Timeout:** Engine not cranking or 3 seconds have passed since the engine was cranked and battery voltage is 10 volts or greater, the A/C-ECU was unable to receive a signal through the CAN bus line. **Possible Causes:** • CAN line harness or connectors damage • A/C–ECU power supply system failed • A/C–ECU failed • PCM failed
DTC: U1110 **1T TCM** **Year:** 2009 **Model:** Raider **Engine:** 3.7L V6 VIN K **Transmission:** All	**No Vehicle Speed Message:** Ignition is on. PCM does not receive a vehicle speed signal from the ABS module or FCM (non-ABS) over the CAN C circuit. **Possible Causes:** • CAN C Bus failure open or shorted • Front Control Module (FCM) has failed • ABS Module has failed • PCM has failed
DTC: U1110 **2T CCM, MIL: Yes** **Year:** 2009, 2010 **Model:** Galant **Engine:** 2.4L L4 VIN F, 3.8L V6 VIN S, 3.8L V6 VIN T **Transmission:** All	**A/C-ECU Timeout:** Engine not cranking or 3 seconds have passed since the engine was cranked and battery voltage is 10 volts or greater, the A/C-ECU was unable to receive a signal through the CAN bus line. **Possible Causes:** • CAN line harness or connectors damage • A/C-ECU power supply system failed • A/C-ECU failed • PCM failed
DTC: U1113 **1T PCM** **Year:** 2009 **Model:** Raider **Engine:** 3.7L V6 VIN K **Transmission:** All	**Lost A/C Pressure Message:** When Monitored: Ignition on. Set Condition: The PCM doesn't receive the A/C Pressure signal over the CAN C bus from the FCM. The circuit is continuously monitored. One trip fault. **Possible Causes:** • CAN C BUS circuit open or shorted • Front Control Module (FCM) • PCM
DTC: U1117 **2T CCM, MIL: Yes** **Year:** 2009, 2010 **Model:** Eclipse, Endeavor, Galant **Engine:** 2.4L L4 VIN F, 3.8L V6 VIN S, 3.8L V6 VIN T **Transmission:** All	**Immobilizer-ECU Timeout:** Ignition switch is ON, the ETACS-ECU (immobilizer-ECU) was unable to receive a signal through the CAN bus line. **Possible Causes:** • CAN line harness or connectors damage • ETACS-ECU (immobilizer-ECU) power supply system failed • ETACS-ECU (immobilizer-ECU) failed • ECM failed <M/T> • PCM failed <A/T>

DTC	Trouble Code Title, Conditions & Possible Causes
DTC: U1120 **T PCM** **Year:** 2010 **Model:** Endeavor **Engine:** 3.8L V6 VIN S **Transmission:** All	**Failure Information on Powertrain Control Module (Related to Engine):** The multi-center display unit (middle grade type) communicates with the powertrain control module via CAN bus line. If failure information is sent to the powertrain control module, DTC U1120 will be set. **Possible Causes:** • The wiring harness or connectors may have loose, corroded, or damaged terminals, or terminals pushed back in the connector • Malfunction of powertrain control module • Malfunction of multi-center display unit (middle grade type)
DTC: U1120 **1T TCM** **Year:** 2009 **Model:** Raider **Engine:** 3.7L V6 VIN K **Transmission:** All	**No Wheel Distance Message:** Ignition is on. PCM does not receive a wheel distance signal from the ABS module or FCM (non-ABS) over the CAN C circuit. **Possible Causes:** • CAN C Bus failure open or shorted • Front Control Module (FCM) has failed • ABS Module has failed • PCM has failed
DTC: U1120 **1T PCM, MIL: Yes** **Year:** 2009, 2010 **Model:** Eclipse, Galant **Engine:** 2.4L L4 VIN F, 3.8L V6 VIN S, 3.8L V6 VIN T **Transmission:** All	**CAN Communications System TCL Uncontrollable by Engine Malfunction:** The TCL/ASC-ECU receives engine system-related signals from the ECM <M/T> or PCM <A/T> via CAN bus lines. If a fail-safe related data is contained in the signal from the ECM <M/T> or PCM <A/T>, DTC U1120 will be stored. **NOTE: If DTC U1120 is set in the TCL/ASC-ECU, always diagnose the CAN main bus line. If there is any fault in the CAN bus lines, an incorrect DTC may be set.** **Whenever the TCL/ASC-ECU is replaced, ensure that the communication circuit is normal. The engine control system-related DTC may be set when DTC U1120 is set. Diagnose the engine control system first when the engine control system-related DTC is set.** **Possible Causes:** • Damaged harness or connector. • Malfunction of the ECM <M/T> or PCM <A/T>. • Malfunction of the TCL/ASC-ECU.
DTC: U1128 **1T PCM** **Year:** 2010 **Model:** Endeavor **Engine:** 3.8L V6 VIN S **Transmission:** All	**Failure information on combination meter:** The multi-center display unit (middle grade type) receives combination meter-related signal from the combination meter via the CAN bus lines. If there is any trouble in the communication data of the combination meter, DTC U1128 will be set. **Possible Causes:** • The multi-center display unit (middle grade type) may be defective • The combination meter may be defective
DTC: U1130 **T PCM** **Year:** 2009, 2010 **Model:** Endeavor, Galant **Engine:** 2.4L L4 VIN F, 3.8L V6 VIN S, 3.8L V6 VIN T **Transmission:** All	**Failure information on A/C-ECU:** The multi-center display unit (middle grade type) receives A/C-ECU-related signal from the A/C-ECU via the CAN bus lines. If there is any trouble in the communication data of the A/C-ECU, DTC U1130 will be set. **NOTE: If DTC U1130 are stored as a past trouble, carry out diagnosis with particular emphasis on wiring and connector(s) in the CAN bus line between the A/C-ECU and the multi-center display unit (middle grade type), and the power supply system to the A/C-ECU.** **Possible Causes:** • The multi-center display unit (middle grade type) may be defective • The A/C-ECU may be defective
DTC: U1180 **T ECM, MIL: Yes** **Year:** 2009, 2010 **Model:** Lancer **Engine:** 2.0L L4 VIN U, 2.0L L4 VIN V, 2.4L L4 VIN W **Transmission:** All	**Combination Meter Time-out:** Some of the possible causes are a harness or connector damage between the ECM and the combination meter on the CAN bus line, a failure in the combination meter power supply system, a failure in the combination meter, or a failure in the ECM. **Possible Causes:** • CAN line harness damage or connector damage. • Combination meter failed. • ECM failed.
DTC: U1403 **1T TCM** **Year:** 2009 **Model:** Raider **Engine:** 3.7L V6 VIN K **Transmission:** All	**Implausible Fuel Level Signal:** Ignition is on. The fuel level message that the PCM is receiving is implausible. The circuit is continuously monitored. **Possible Causes:** • CAN B Bus failure open or shorted • Instrument Cluster Module has failed • Front Control Module has failed • PCM has failed

DTC	Trouble Code Title, Conditions & Possible Causes
DTC: U1411 **1T PCM, MIL: Yes** **Year:** 2009 **Model:** Raider **Engine:** 3.7L V6 VIN K **Transmission:** All	**Implausible Fuel Volume Signal Received:** Ignition on. The fuel volume message the PCM is receiving is implausible. The circuit is continuously monitored. **Possible Causes:** • CAN B Bus failure open or shorted • Instrument Cluster Module has failed • Front Control Module has failed • PCM has failed
DTC: U1415 **T ECM** **Year:** 2009, 2010 **Model:** Lancer **Engine:** 2.0L L4 VIN U, 2.0L L4 VIN V, 2.4L L4 VIN W **Transmission:** All	**Coding incomplete/fail:** The diagnostic trouble code is set when the AWC-ECU coding has not been performed. **Possible Causes:** • Variant coding for ETACS-ECU has not been implemented. • Malfunction of AWC-ECU
DTC: U1417 **T ECM** **Year:** 2009, 2010 **Model:** Outlander **Engine:** 2.4L L4 VIN W, 3.0L V6 VIN X **Transmission:** All	**Coding Data Malfunction:** AWC-ECU receives the vehicle information stored in ETACS-ECU via CAN bus lines. The code is set when the vehicle information received from ETACS-ECU is not correct. **Possible Causes:** • Malfunction of the CAN bus line • ETACS-ECU malfunction • Malfunction of AWC-ECU • ETACS-ECU has been interchanged between two vehicles
DTC: U1417 **1T TCM** **Year:** 2009 **Model:** Raider **Engine:** 3.7L V6 VIN K **Transmission:** All	**Implausible Left Wheel Distance Signal:** Ignition is on. The left wheel distance message that the PCM is receiving over the CAN C circuit from the ABS module or FCM (non-ABS) is implausible. The circuit is continuously monitored. **Possible Causes:** • Vehicle speed Sensor fault active in ABS module • CAN C Bus failure open or shorted • ABS Module has failed • Front Control Module has failed • PCM has failed
DTC: U1418 **1T TCM** **Year:** 2009 **Model:** Raider **Engine:** 3.7L V6 VIN K **Transmission:** All	**Implausible Right Wheel Distance Signal:** Ignition is on. The left wheel distance message that the PCM is receiving over the CAN C circuit from the ABS module or FCM (non-ABS) is implausible. The circuit is continuously monitored. **Possible Causes:** • Vehicle speed Sensor fault active in ABS module • CAN C Bus failure open or shorted • ABS Module has failed • Front Control Module has failed • PCM has failed
DTC: U1425 **T ECM** **Year:** 2009, 2010 **Model:** Lancer **Engine:** 2.0L L4 VIN U, 2.4L L4 VIN W **Transmission:** All	**Data Error (TC-SST):** The code is set when abnormality is detected in the signal received from the TC-SST-ECU. **Possible Causes:** • The CAN bus line is defective. • Malfunction of TC-SST-ECU • Malfunction of AWC-ECU
DTC: U1433 **1T PCM, MIL: Yes** **Year:** 2010 **Model:** Endeavor **Engine:** 3.8L V6 VIN S **Transmission:** All	**Dynamic range error TPS:** This code is set when the MFI system sets either of DTC code Nos. P0222, P0223 or P2135. **Possible Causes:** • Malfunction of the MFI system

DTC	Trouble Code Title, Conditions & Possible Causes
DTC: U1606 **T PCM** **Year:** 2009, 2010 **Model:** Endeavor, Galant **Engine:** 2.4L L4 VIN F, 3.8L V6 VIN S, 3.8L V6 VIN T **Transmission:** All	**M-BUS error:** With the ignition switch in the "ACC" or "ON" position, if the communication between the radio, CD player and CD changer and the multi-center display unit (middle grade type) cannot be made for 30 seconds or more, or if there is a problem in the communication data, DTC U1606 will be set. **NOTE: Carry out diagnosis with particular emphasis on connector(s) or wiring harness between the radio, CD player and CD changer and the multi-center display unit (middle grade type).** **Possible Causes:** • The wiring harness or connectors may have loose, corroded, or damaged terminals, or terminals pushed back in the connector. • Malfunction of radio, CD player and CD changer • Malfunction of multi-center display unit (middle grade type)

SPECIFICATIONS AND MAINTENANCE CHARTS

ENGINE AND VEHICLE IDENTIFICATION

Engine Code							Model Year	
Code ①	Liters	Cu. In.	Cyl.	Fuel Sys.	Type	Eng. Mfg.	Code ②	Year
6	2.5L	150	4	MFI	DOHC	Subaru	9	2009
							A	2010

MFI: Multiport Fuel Injection

DOHC: Double Overhead Camshafts

SOHC: Single Overhead Camshaft

① 6th digit of the VIN

② 10th digit of the VIN

③ Turbocharged

37671_FORE_C0001

GENERAL ENGINE SPECIFICATIONS

Year	Model	Engine Displacement Liters (VIN)	Net Horsepower @ rpm	Net Torque @ rpm (ft. lbs.)	Bore x Stroke (in.)	Com-pression Ratio	Oil Pressure psi @ rpm
2009	Forester	2.5 (6) ①	173@5600	166@4400	3.92x3.11	10.0:1	14@600
		2.5 (6) ②	230@5600	235@3600	3.92x3.11	8.2:1	14@600
2010	Forester	2.5 (6) ①	173@5600	166@4400	3.92x3.11	10.0:1	14@600
		2.5 (6) ②	230@5600	235@3600	3.92x3.11	8.2:1	14@600

① SOHC

② DOHC

37671_FORE_C0002

ENGINE TUNE-UP SPECIFICATIONS

Year	Engine Displacement Liters (VIN)	Spark Plugs Gap (in.)	Ignition Timing (deg.) ① ② MT	AT	Fuel Pump (psi)	Idle Speed (rpm) MT	AT	Valve Clearance ③ In.	Ex.
2009	2.5 (6) ④	0.039-0.043	⑤	⑥	⑦	550-750	600-800	0.0063-0.0095	0.0082-0.0114
	2.5 (6) ⑧	0.028-0.031	⑨	⑨	⑩	600-800	600-800	0.0055-0.0095	0.0118-0.0158
2010	2.5 (6) ④	0.039-0.043	⑤	⑥	⑦	550-750	600-800	0.0063-0.0095	0.0082-0.0114
	2.5 (6) ⑧	0.028-0.031	⑨	⑨	⑩	600-800	600-800	0.0055-0.0095	0.0118-0.0158

① At idle

② BTDC: before top dead center

③ With engine cold

④ SOHC

⑤ 10 degrees +/-8 at 650rpm

⑥ 15 degrees +/-8 at 700rpm

⑦ 41-46 while disconnecting pressure regulator vacuum hose from intake manifold

33-38 after connecting pressure regulator vacuum hose

⑧ DOHC

⑨ 17 degrees +/-10 at 700rpm

⑩ 48-53 while disconnecting pressure regulator vacuum hose from intake manifold

40-45 after connecting pressure regulator vacuum hose

37671_FORE_C0005

CAPACITIES

Year	Model	Engine Displacement Liters (VIN)	Engine Oil with Filter (qts.)	Transmission (pts.) 4-Spd	Transmission (pts.) 5-Spd	Transmission (pts.) Auto. ②	Transfer Case (pts.)	Drive Axle Front (pts.)	Drive Axle Rear (pts.)	Fuel Tank (gal.)	Cooling System (qts.)
2009	Forester	2.5 (6)	4.2	—	7.4	19.6	—	2.6 ①	1.6	15.9	③
2010	Forester	2.5 (6)	4.2	—	7.4	19.6	—	2.6 ①	1.6	15.9	③

① Automatic transmission differential

② Overhaul specification

③ SOHC: manual transmission: 7.3. Automatic transmission: 7.2

   DOHC: manual transmission: 7.8. Automatic transmission: 7.7

37671_FORE_C0004

FLUID SPECIFICATIONS

Year	Model	Engine Displacement Liters	Engine ID/VIN	Engine Oil	Auto. Trans.	Drive Axle	Power Steering Fluid	Brake Master Cylinder	Engine Coolant
2009	Forester	2.5L ①	6	5W-30	Subaru ATF HP	75W-90	Dexron III	DOT 3	Subaru Coolant
		2.5L	6	5W-30	Subaru ATF HP	75W-90	Dexron III	DOT 3	Subaru Coolant
2010	Forester	2.5L ①	6	5W-30	Subaru ATF HP	75W-90	Dexron III	DOT 3	Subaru Coolant
		2.5L	6	5W-30	Subaru ATF HP	75W-90	Dexron III	DOT 3	Subaru Coolant

DOT: Department Of Transpotation

① DOHC

37671_FORE_C0003

VALVE SPECIFICATIONS
All measurements are given in inches.

Year	Engine Displacement Liters (VIN)	Seat Angle (deg.)	Face Angle (deg.)	Spring Test Pressure (lbs. @ in.)	Spring Installed Height (in.)	Stem-to-Guide Clearance (in.) Intake	Stem-to-Guide Clearance (in.) Exhaust	Stem Diameter (in.) Intake	Stem Diameter (in.) Exhaust
2009	2.5 (6) ①	90	NA	102 - 118@ 1.315	②	0.0014-0.0024	0.0016-0.0026	0.2343-0.2348	0.2341-0.2346
	2.5 (6) ③	90	NA	102 - 118@ 1.315	⑤	0.0012-0.0022	0.0016-0.0026	0.2344-0.2350	0.2341-0.2346
2010	2.5 (6) ①	90	NA	102 - 118@ 1.315	②	0.0014-0.0024	0.0016-0.0026	0.2343-0.2348	0.2341-0.2346
	2.5 (6) ③	90	NA	102 - 118@ 1.315	⑤	0.0012-0.0022	0.0016-0.0026	0.2344-0.2350	0.2341-0.2346

NA: Not Available

① SOHC

② Free length: 2.173 in.

③ DOHC

④ Free length: 1.863 in.
   Lift: 95.8-110.0@1.043

37671_FORE_C0006

CAMSHAFT SPECIFICATIONS

All measurements are given in inches.

Year	Engine Displacement Liters (VIN)	Journal Diameter	Brg. Oil Clearance	Shaft End-play ①	Runout	Lobe Height Intake	Lobe Height Exhaust
2009	2.5 (6) ②	1.2570-1.2577	0.0022-0.0035	0.0012-0.0035	NA	1.5545-1.5585	③
	2.5 (6) ④	⑤	0.0015-0.0028	0.0027-0.0047	NA	1.8330-1.8370	1.8410-1.8440
2010	2.5 (6) ②	1.2570-1.2577	0.0022-0.0035	0.0012-0.0035	NA	1.5545-1.5585	③
	2.5 (6) ④	⑤	0.0015-0.0028	0.0027-0.0047	NA	1.8330-1.8370	1.8410-1.8440

NA: Not Available

① Side clearance

② SOHC

③ California: 1.5686-1.5726
Except California: 1.5638-1.5677

④ DOHC

⑤ Front: 1.4939-1.4946
Except front: 1.1790-1.1796

37671_FORE_C0007

CRANKSHAFT AND CONNECTING ROD SPECIFICATIONS

All measurements are given in inches.

Year	Engine Displacement Liters (VIN)	Crankshaft Main Brg. Journal Dia.	Crankshaft Main Brg. Oil Clearance	Crankshaft Shaft End-play	Crankshaft Thrust on No.	Connecting Rod Journal Diameter	Connecting Rod Oil Clearance	Connecting Rod Side Clearance
2009	2.5 (6) ①	2.3619-2.3625	0.0004-0.0012	0.0012-0.0045	NA	NA	0.0006-0.0017	0.0028-0.0130
	2.5 (6) ③	2.3619-2.3625	0.0004-0.0012	0.0012-0.0045	NA	NA	0.0007-0.0018	0.0028-0.0130
2010	2.5 (6) ①	2.3619-2.3625	0.0004-0.0012	0.0012-0.0045	NA	NA	0.0006-0.0017	0.0028-0.0130
	2.5 (6) ③	2.3619-2.3625	0.0004-0.0012	0.0012-0.0045	NA	NA	0.0007-0.0018	0.0028-0.0130

NA: Not Available

① SOHC

② DOHC

37671_FORE_C0008

PISTON AND RING SPECIFICATIONS
All measurements are given in inches.

Year	Engine Displacement Liters (VIN)	Piston Clearance	Ring Gap			Ring Side Clearance		
			Top Compression	Bottom Compression	Oil Control	Top Compression	Bottom Compression	Oil Control
2009	2.5 (6) ①	0.0004-0.0012	0.0079-0.0138	0.0144-0.0203	0.0079-0.0197	0.0016-0.0031	0.0012-0.0028	NA
	2.5 (6) ②	0.0004-0.0012	0.0079-0.0098	0.0150-0.0203	0.0079-0.0197	0.0016-0.0031	0.0012-0.0028	NA
2010	2.5 (6) ①	0.0004-0.0012	0.0079-0.0138	0.0144-0.0203	0.0079-0.0197	0.0016-0.0031	0.0012-0.0028	NA
	2.5 (6) ②	0.0004-0.0012	0.0079-0.0098	0.0150-0.0203	0.0079-0.0197	0.0016-0.0031	0.0012-0.0028	NA

NA: Not Available

① SOHC

② DOHC

37671_FORE_C0009

TORQUE SPECIFICATIONS
All readings in ft. lbs.

Year	Engine Displacement Liters (VIN)	Cylinder Head Bolts	Main ① Bearing Bolts	Rod Bearing Bolts	Crankshaft Damper Bolts	Flywheel Bolts	Manifold		Spark Plugs	Oil Pan Drain Plug
							Intake	Exhaust		
2009	2.5 (6) ②	③	④	33.3	132.7	52.8	18	19-26	13-17	33
	2.5 (6) ⑤	⑥	④	38.4	132.7	⑦	18	⑧	13-17	33
2010	2.5 (6) ②	③	④	33.3	132.7	52.8	18	19-26	13-17	33
	2.5 (6) ⑤	⑥	④	38.4	132.7	⑦	18	⑧	13-17	33

① Engine block connecting bolts

② SOHC

③ Step 1: Tighten all bolts to 22 ft. lbs.
Step 2: Tighten all bolts to 51 ft. lbs.
Step 3: Loosen all bolts 180 degrees, and again 180 degrees
Step 4: Tighten all bolts to 36 ft. lbs.
Step 5: + 80-90 degrees
Step 6: + 40-45 degrees

④ Step 1: Left side 10mm bolts (A-D) 7.2 ft. lbs.
Step 2: Right side 10mm bolts (E-J) 7.2 ft. lbs.
Step 3: Left side 10mm bolts (A-D) 13 ft. lbs.
Step 4: Right side 10mm bolts (E-J) 13 ft. lbs.
Step 5: Left side bolts (A and C) +90 degrees, bolts (B and D) 29.5 ft. lbs.
Step 6: Right side 10mm bolts (E-J) + 90 degrees
Step 7: Left side cylinder block connecting bolts (A-G) 18.1 ft. lbs., bolt (H) 4.7 ft. lbs.

⑤ DOHC

⑥ Step 1: Tighten all bolts to 21.4 ft. lbs.
Step 2: Tighten all bolts to 51 ft. lbs.
Step 3: Loosen all bolts 180 degrees, and again 180 degrees
Step 4: Tighten all bolts to 36 ft. lbs.
Step 5: + 80-90 degrees
Step 6: + 40-45 degrees

⑦ Automatic: 53.1 ft. lbs.
Manual: 38.4 ft. lbs.

⑧ Exhaust pipe to manifold 26 ft. lbs.
Right upper cover 5.5 ft. lbs.
Front exhaust pipe 13.7 ft. lbs.
Right manifold to turbocharger joint pipe 26-28 ft. lbs.
Lower covers 13.7 ft. lbs.

37671_FORE_C0010

WHEEL ALIGNMENT

Year	Model		Caster Range (+/-Deg.)	Caster Preferred Setting (Deg.)	Camber Range (+/-Deg.)	Camber Preferred Setting (Deg.)	Toe-in (in.)
2009	Forester	F	—	①	0.50	-0.25	0+/-0.12
		R	—	—	0.75	②	③
2010	Forester	F	—	①	0.50	-0.25	0+/-0.12
		R	—	—	0.75	②	③

① Reference: 3 degrees 03'

② SOHC: -0 degrees 50'. DOHC: -0 degrees 55'

③ 0.079 +/-0.12 inch. Toe angle (sum of both wheels) 0 degrees 10' +/-0 degrees 15'

37671_FORE_C0011

TIRE, WHEEL AND BALL JOINT SPECIFICATIONS

Year	Model	OEM Tires Standard	OEM Tires Optional	Tire Pressures (psi) Front	Tire Pressures (psi) Rear	Wheel Size	Ball Joint Inspection	Lug Nut (ft. lbs.)
2009	Forester	P215/60R16 94H	—	①	①	16X6.5 JJ	0.012 in. ②	74
		P215/55R17 93H	—	①	①	17X7 JJ	0.012 in. ②	74
2010	Forester	P215/60R16 94H	—	①	①	16X6.5 JJ	0.012 in. ②	74
		P215/55R17 93H	—	①	①	17X7 JJ	0.012 in. ②	74

OEM: Original Equipment Manufacturer

PSI: Pounds Per Square Inch

STD: Standard

OPT: Optional

① See drivers door placard

② Apply 154 lbs. vertical force

37671_FORE_C0012

BRAKE SPECIFICATIONS

All measurements in inches unless noted

Year	Model		Brake Disc Original Thickness	Brake Disc Minimum Thickness	Brake Disc Maximum Runout	Brake Drum Diameter Original Inside Diameter	Brake Drum Diameter Max. Wear Limit	Brake Drum Diameter Maximum Machine Diameter	Minimum Lining Thickness Front	Minimum Lining Thickness Rear	Brake Caliper Bracket Bolts (ft. lbs.)	Brake Caliper Mounting Bolts (ft. lbs.)
2009	Forester	F	0.940	0.870	0.0030	—	—	—	0.059	0.059	59	19
		R	0.390	0.335	0.0028	9.00 ①	9.08 ②	NA	—	0.059	38	28
2010	Forester	F	0.940	0.870	0.0030	—	—	—	0.059	0.059	59	19
		R	0.390	0.335	0.0028	9.00 ①	9.08 ②	NA	—	0.059	38	28

NA: Not Available

① Parking brake drum on vehicles with rear disc brakes: 6.69

② Parking brake drum on vehicles with rear disc brakes: 6.73

37671_FORE_C0013

SCHEDULED MAINTENANCE INTERVALS
SUBARU FORESTER

TO BE SERVICED	SERVICE	3	7.5	15	22.5	30	37.5	45	52.5	60	67.5	75	82.5	90	97.5	105	112.5	120
Accessory drive belts	R									✓								✓
Accessory drive belts	S/I					✓				✓						✓		
Air condition filter	S/I		✓	✓	✓	✓	✓	✓	✓	✓	✓	✓	✓	✓	✓	✓	✓	✓
Air cleaner filter	R					✓				✓				✓				✓
Auto transmission fluid	S/I					✓				✓				✓				✓
Axle shaft joints	S/I			✓		✓		✓		✓		✓		✓		✓		✓
Brake fluid	R					✓				✓				✓				✓
Brake system lines	S/I			✓		✓		✓		✓		✓		✓		✓		✓
Clutch operation	S/I			✓		✓		✓		✓		✓		✓		✓		✓
Disc brake pads & rotors	S/I			✓		✓		✓		✓		✓		✓		✓		✓
Drums brake linings & drums	S/I					✓				✓				✓				✓
Engine coolant	R					✓				✓				✓				✓
Engine cooling system, hoses & connections	S/I					✓				✓				✓				✓
Engine oil & filter	R	✓	✓	✓	✓	✓	✓	✓	✓	✓	✓	✓	✓	✓	✓	✓	✓	✓
Front & rear axle boots	S/I			✓		✓		✓		✓		✓		✓		✓		✓
Fuel Filter	S/I					✓				✓				✓				
Front & rear wheel bearings	S/I			✓		✓		✓		✓		✓		✓		✓		✓
Parking & service brake systems' operation	S/I			✓		✓		✓		✓		✓		✓		✓		✓
Spark plugs (non turbo)	R					✓				✓				✓				✓
Spark plugs (turbo)	R									✓								✓
Steering & suspension	S/I			✓		✓		✓		✓		✓		✓		✓		✓
Timing belt	R															✓		
Timing belt	S/I					✓				✓				✓				
Tire rotation	S/I	✓	✓	✓	✓	✓	✓	✓	✓	✓	✓	✓	✓	✓	✓	✓	✓	✓
Transmission & differential fluid levels	S/I					✓				✓				✓				✓
Valve clearance	S/I															✓		

R: Replace   S/I: Inspect and service, if needed   L: Lubricate

FREQUENT OPERATION MAINTENANCE (SEVERE SERVICE)

If a vehicle is operated under any of the following conditions it is considered severe service

- Towing a trailer or using a camper or car-top carrier.

- Repeated short trips of less than 5 miles in temperatures below freezing, or trips of less than 10 miles in any temperature.

- Extensive idling or low-speed driving for long distances as in heavy commercial use, such as delivery, taxi or police cars.

- Operating on rough, muddy or salt-covered roads, or extensive mountain driving.

- Operating on unpaved or dusty roads.

- Driving in extremely hot (over 90°) conditions.

Engine oil and filter: replace every 3000 miles or 3 months, whichever occurs first.

Fuel system, hoses & connections: inspect every 7500 miles or 7.5 months, whichever occurs first.

Transmission & differential fluid: replace every 15,000 miles.

Automatic transmission fluid: replace every 15,000 miles.

Brake fluid: replace every 15,000 miles.

Disc brake pads & rotors: inspect every 7500 miles or 7.5 months, whichever occurs first.

Front & rear axle boots: inspect every 7500 miles or 7.5 months, whichever occurs first.

Axle shaft boots: inspect every 7500 miles or 7.5 months, whichever occurs first.

Drum brake linings & drums: inspect every 7500 miles or 7.5 months, whichever occurs first.

Brake lines: inspect every 7500 miles or 7.5 months, whichever occurs first.

Parking & service brake system operation: inspect every 7500 miles or 7.5 months, whichever occurs first.

Clutch operation: inspect every 7500 miles or 7.5 months, whichever occurs first.

BRAKES — INFORMATION AND PRECAUTIONS

ANTI-LOCK SYSTEMS

- Certain components within the ABS system are not intended to be serviced or repaired individually.
- Do not use rubber hoses or other parts not specifically specified for and ABS system. When using repair kits, replace all parts included in the kit. Partial or incorrect repair may lead to functional problems and require the replacement of components.
- Lubricate rubber parts with clean, fresh brake fluid to ease assembly. Do not use shop air to clean parts; damage to rubber components may result.
- Use only DOT 3 brake fluid from an unopened container.
- If any hydraulic component or line is removed or replaced, it may be necessary to bleed the entire system.
- A clean repair area is essential. Always clean the reservoir and cap thoroughly before removing the cap. The slightest amount of dirt in the fluid may plug an orifice and impair the system function. Perform repairs after components have been thoroughly cleaned; use only denatured alcohol to clean components. Do not allow ABS components to come into contact with any substance containing mineral oil; this includes used shop rags.
- The Anti-Lock control unit is a microprocessor similar to other computer units in the vehicle. Ensure that the ignition switch is **OFF** before removing or installing controller harnesses. Avoid static electricity discharge at or near the controller.
- If any arc welding is to be done on the vehicle, the control unit should be unplugged before welding operations begin.

DISC AND DRUM SYSTEMS

> ❊❊ **CAUTION**
>
> Dust and dirt accumulating on brake parts during normal use may contain asbestos fibers from production or aftermarket brake linings. Breathing excessive concentrations of asbestos fibers can cause serious bodily harm. Exercise care when servicing brake parts. Do not sand or grind brake lining unless equipment used is designed to contain the dust residue. Do not clean brake parts with compressed air or by dry brushing. Cleaning should be done by dampening the brake components with a fine mist of water, then wiping the brake components clean with a dampened cloth. Dispose of cloth and all residue containing asbestos fibers in an impermeable container with the appropriate label. Follow practices prescribed by the Occupational Safety and Health Administration (OSHA) and the Environmental Protection Agency (EPA) for the handling, processing, and disposing of dust or debris that may contain asbestos fibers.

BRAKES — BLEEDING THE BRAKE SYSTEM

BLEEDING PROCEDURE

BLEEDING PROCEDURE

> ❊❊ **WARNING**
>
> Do not let brake fluid come into contact with the painted surface of the vehicle body. Wash away with water immediately and wipe off if it is spilled by accident.

> ❊❊ **CAUTION**
>
> Avoid mixing brake fluid of different brands to prevent fluid performance from degrading.

> ❊ **CAUTION**
>
> Be careful not to allow dirt or dust to enter the reservoir tank.

Master Cylinder

See Figures 1 and 2.

➡ Note the following:

- When the master cylinder is disassembled or the reservoir tank is empty, bleed the master cylinder.
- If bleeding of the master cylinder is not necessary, omit the following procedures, and perform bleeding of the brake line. Refer to Brake Line in this section.

1. Fill the reservoir tank of the master cylinder with brake fluid.

➡ While bleeding air, keep the reservoir tank filled with brake fluid to prevent entry of air.

2. Disconnect the brake line at primary and secondary sides.
3. Wrap the master cylinder with a plastic bag.
4. Depress the brake pedal slowly and hold it.

➡ If the vehicle is equipped with a brake assist mechanism, the following may result when the brake pedal is depressed. They are normal conditions that occur when the mechanism is performing properly:

- Brake feel is soft when brake pedal is depressed hard or quicker than usual.
- A light crack is heard when brake pedal is depressed hard or quicker than usual.

5. Plug the outlet plug with your finger, and then release the brake pedal.
6. Repeat the step 4) and 5) several times.
7. Remove the plastic bag.
8. Install the brake pipe to the master cylinder. Tighten to 14 ft. lbs. (19 Nm)

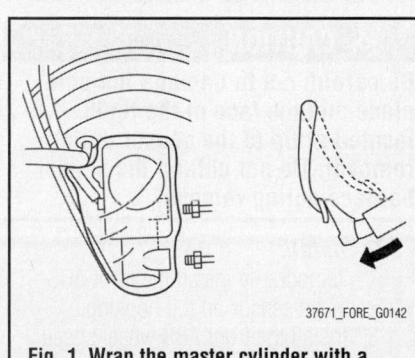

37671_FORE_G0142

Fig. 1 Wrap the master cylinder with a plastic bag and depress the brake pedal slowly

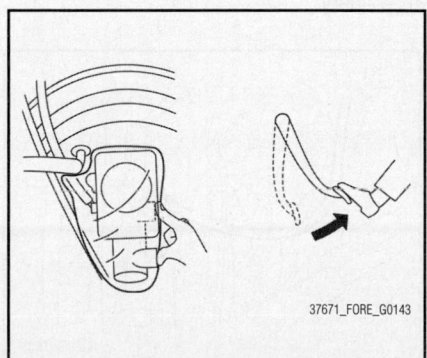

37671_FORE_G0143

Fig. 2 Plug the outlet plug with your finger, and then release the brake pedal

9. Bleed air from the brake line. Refer to Brake Line in this section.

Brake Line

See Figures 3 and 4.

1. When the master cylinder is disassembled or the reservoir tank is empty, bleed the master cylinder before bleeding the brake line. Refer to MASTER CYLINDER in this section.

2. Fill the reservoir tank of the master cylinder with brake fluid.

➡ **While bleeding air, keep the reservoir tank filled with brake fluid to prevent entry of air.**

3. Attach one end of the vinyl tube to the air bleeder and the other end to the brake fluid container.

4. Depress the brake pedal several times, and hold it.

5. Loosen the air bleeder screw to drain brake fluid. Tighten the air bleeder quickly, and release the brake pedal.

6. Repeat the steps 4) to 5) until there are no more air bubbles in the vinyl tube.

7. Repeat the steps from 2) to 6) above to bleed air from each wheel.

➡ **Perform air bleed starting in the order from the farthest wheel cylinder from the master cylinder.**

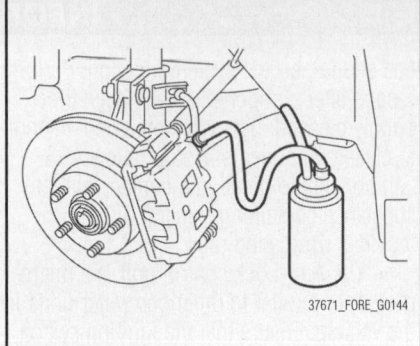

Fig. 3 Attach one end of the vinyl tube to the air bleeder, and the other end to the brake fluid container

37671_FORE_G0144

8. Securely tighten the air bleeder screws. Tighten screws to 61 inch lbs. (8 Nm)

9. Check that there are no brake fluid leaks in the entire system.

10. Check the pedal stroke:

a. Run the engine at idle after warming up the engine, and depress the brake pedal with a force of 500 N (51 kgf, 112 lbf). Measure the distance between the brake pedal and steering wheel. Release the pedal, and measure the distance between pedal and steering wheel again.

b. Specification of pedal stroke: When depressing the pedal with a force of 112 lbf. (500 N). 3.74 inch (95 mm or less).

c. If the distance is more than specifi-

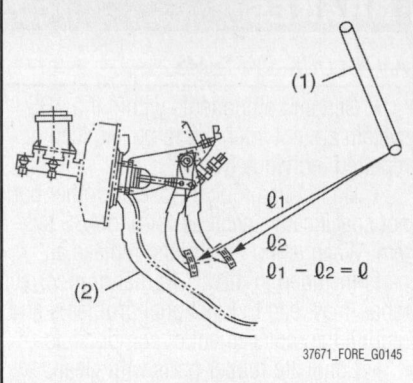

Fig. 4 Measure the distance between the brake pedal, and steering wheel

$\ell_1 - \ell_2 = \ell$

37671_FORE_G0145

cation, there is a possibility of air being caught in the brake line. Bleed the brake line of all air until the pedal stroke meets the specification.

11. Operate the hydraulic control unit in the sequence control mode. Refer to ABS Sequence Control in this section.

12. Check the pedal stroke again.

13. If the distance is more than specification, there is a possibility of air being caught in the hydraulic unit. Repeat above steps 2) to 9) until the pedal stroke meets the specification.

14. Fill the reservoir tank with brake fluid up to the "MAX" level.

15. Test run the vehicle and ensure that the brakes operate normally.

BRAKES

ANTI-LOCK BRAKE SYSTEM (ABS)

WHEEL SPEED SENSORS

REMOVAL & INSTALLATION

Front

See Figure 5.

1. Before servicing the vehicle, refer to the Precautions Section.

2. Disconnect the ground cable from the battery.

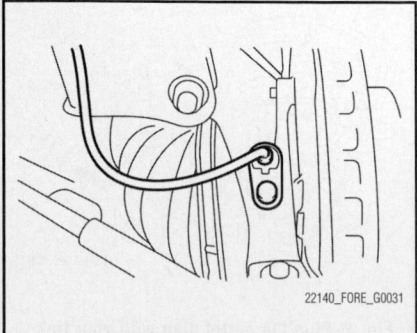

Fig. 5 Front ABS wheel speed sensor view

22140_FORE_G0031

3. Disconnect the ABS wheel speed sensor connector located next to the front strut mounting house in the engine compartment.

4. Remove the bolts which secure the sensor harness to the strut.

5. Remove the bolts which secure the sensor harness to the body.

6. Remove the bolts which secure front ABS wheel speed sensor to the housing, and remove the front ABS wheel speed sensor.

☀ CAUTION

Be careful not to damage the pole piece and the face of the teeth located at tip of the sensor during removal. Do not pull on the sensor harness during removal.

To install:

7. Temporarily install the front ABS wheel speed sensor on the housing.

8. Install the front ABS wheel speed sensor on the strut and the wheel apron bracket. Tighten mounting bolt for bracket to 25 ft. lbs. (33 Nm).

9. Check the clearance of the sensor.

10. ABS wheel speed sensor gap standard value is as follows:

a. 0.012–0.031 inch (0.3–0.8mm)

11. If clearance is outside of the standard value, readjust by using the spacer (Part No. 26755AA000).

12. After confirmation of the ABS wheel speed sensor clearance, connect the connector to the ABS wheel speed sensor.

13. Connect the ground cable to the battery.

Rear

See Figures 6 through 9.

1. Disconnect the ground cable from the battery.

2. Disconnect the connector from the rear ABS wheel speed sensor.

3. Remove the sensor harness clamp of the rear sub frame.

4. Remove the sensor harness bracket from the upper arm.

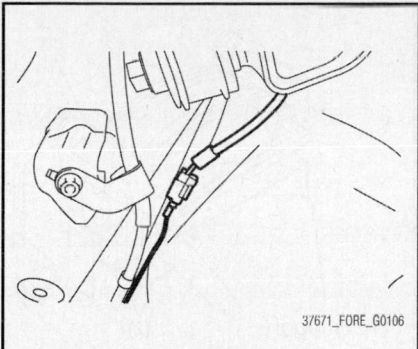

Fig. 6 Disconnect the connector from the rear ABS wheel speed sensor

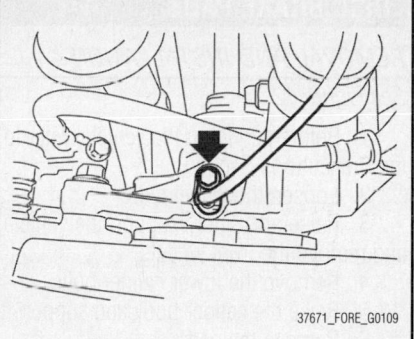

Fig. 9 Rear ABS wheel speed sensor from the rear axle

5. Remove the rear ABS wheel speed sensor from the rear axle.

✴✴ WARNING

Be careful not to damage the sensor. Do not apply excessive force to the sensor harness.

To install:

✴✴ CAUTION

Be careful not to damage the sensor.

6. Install the rear ABS wheel speed sensor from the rear axle.

7. Install the sensor harness bracket from the upper arm.

8. Install the sensor harness clamp of the rear sub frame.

9. Connect the connector from the rear ABS wheel speed sensor.

10. Connect the ground cable from the battery.

11. Tighten the Sensor and Bracket to 61 inch lbs. (8 Nm).

➡**Check the identification (mark) on the harness to make sure there is no warpage. (RH: P3 (light blue), LH: P4 (brown))**

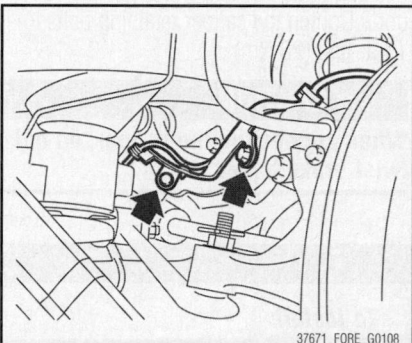

Fig. 8 Sensor harness bracket from the upper arm

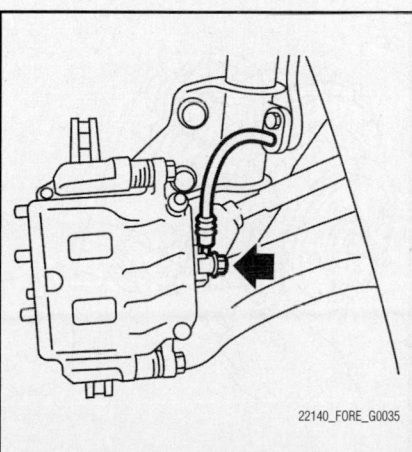

Fig. 7 Sensor harness clamp of the rear sub frame

BRAKES

FRONT DISC BRAKES

BRAKE CALIPER

REMOVAL & INSTALLATION

See Figures 10 and 11.

1. Before servicing the vehicle, refer to the Precautions Section.
2. Loosen the wheel nuts.
3. Raise and support the vehicle safely.
4. Remove the tire and wheel.

5. Remove the brake hose mounting bolt. Disconnect the brake line from the brake caliper. Be sure to properly catch the fluid to avoid damage to painted surfaces and improper disposal.

6. Remove the bolt securing the lock pin (yellow) to caliper body assembly.

7. Raise the caliper body assembly and move it toward the center of the vehicle to separate it from the support.

8. Remove the caliper from its mounting.

To install:

9. Installation is the reverse of the removal procedure.

10. Apply a thin coat of Molykote M7439 (Part No. 003602001) to the support.

11. Apply a thin coat of Molykote AS880N (Part No. K0777YA010) to the pad clip and both sides of inner shim.

12. Tighten the caliper retaining bolts to 19 ft. lbs. (26 Nm).

✴✴ CAUTION

When connecting the brake hose, do not twist it.

13. Be sure to use new brake hose gaskets. Tighten the banjo bolt to 13 ft. lbs. (18 Nm).

14. Bleed the hydraulic system, as required.

15. Check the brake fluid level, correct as required.

16. Test drive and burnish brake system.

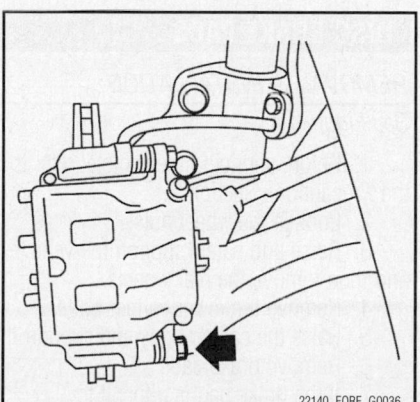

Fig. 10 Brake hose mounting bolt

Fig. 11 Brake caliper mounting bolt securing the lock pin (yellow)

DISC BRAKE PADS

REMOVAL AND INSTALLATION

See Figure 12.

1. Before servicing the vehicle, refer to the Precautions Section.
2. Loosen the wheel nuts.
3. Raise and safely support the vehicle, and remove the front wheels.
4. Remove the lower caliper bolts.
5. Raise the caliper body and support it.
6. Remove the pad.
7. If the brake pad is difficult to remove, proceed as follows:
   - Remove the caliper body from support.
   - Remove the support.
   - Place the support between wooden blocks in the vise.
   - Apply a rod with less than 0.47 inch (12mm) diameter to the shaded area of brake pad, and strike the rod with a hammer to remove brake pad.

To install:

8. Compress the caliper pistons.

➡**If it is difficult to push the pistons during pad replacement, loosen the air bleeder to facilitate work.**

9. Apply a thin coat of Molykote M7439 (Part No. 003602001) to the support.
10. Apply a thin coat of Molykote AS880N (Part No. K0777YA010) to the pad clip and both sides of inner shim.
11. Install the pad to support.
12. Install the caliper body to the support. Tighten the caliper retaining bolts to 19 ft. lbs. (26 Nm).

✳✳ CAUTION

When connecting the caliper, do not twist brake hose.

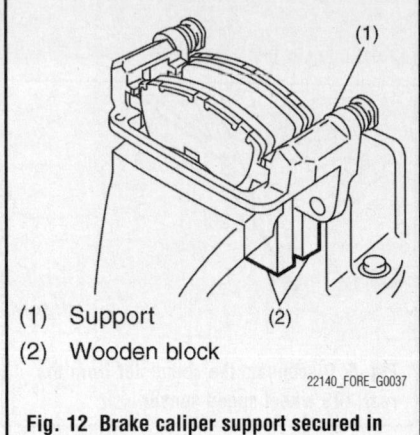

(1) Support
(2) Wooden block

22140_FORE_G0037

Fig. 12 Brake caliper support secured in vise

13. Bleed the hydraulic system, as required.
14. Check the brake fluid level, correct as required.
15. Test drive and burnish brake system.

BRAKES

BRAKE CALIPER

REMOVAL & INSTALLATION

See Figure 13.

1. Before servicing the vehicle, refer to the Precautions Section.
2. Loosen the wheel nuts.
3. Raise and support the vehicle safely.
4. Remove the tire and wheel.
5. Disconnect the brake hose from the brake caliper. Be sure to properly catch the fluid to avoid damage to painted surfaces and improper disposal.
6. Remove the bolt securing the lock pin (yellow) to caliper body assembly.
7. Raise the caliper body assembly and move it toward the center of the vehicle to separate it from the support.
8. Remove the caliper from its mounting.

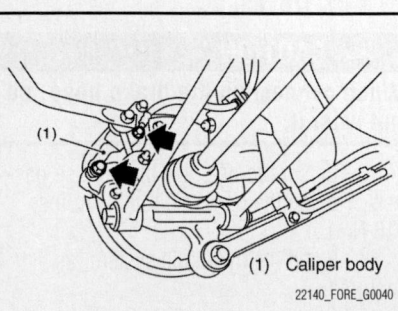

(1) Caliper body

22140_FORE_G0040

Fig. 13 Caliper body (1), bolt securing the lock pin (yellow) and brake hose mounting bolt

To install:

9. Installation is the reverse of the removal procedure.
10. Tighten the caliper retaining bolts to 28 ft. lbs. (37 Nm).

✳✳ CAUTION

When connecting the brake hose, do not twist it.

11. Be sure to use new brake hose gaskets. Tighten banjo bolt to 13 ft. lbs. (18 Nm).
12. Bleed the hydraulic system, as required.
13. Check the brake fluid level, correct as required.
14. Test drive and burnish brake system.

DISC BRAKE PADS

REMOVAL & INSTALLATION

See Figure 14.

1. Before servicing the vehicle, refer to the Precautions Section.
2. Loosen the wheel nuts.
3. Raise and safely support the vehicle, and then remove the rear wheels.
4. Remove the lower caliper bolts.
5. Raise the caliper body and support it.
6. Remove brake pads.
7. If the brake pad is difficult to remove, proceed as follows:
   - Remove the caliper body from support.

REAR DISC BRAKES

   - Remove the support.
   - Place the support between wooden blocks in the vise.
   - Apply a rod with less than 0.47 inch (12mm) diameter to the shaded area of brake pad, and strike the rod with a hammer to remove brake pad.

To install:

8. Compress the caliper piston.

➡**If it is difficult to push the pistons during pad replacement, loosen the air bleeder to facilitate work.**

9. Apply a thin coat of Molykote M7439 (Part No. 003602001) to the support.

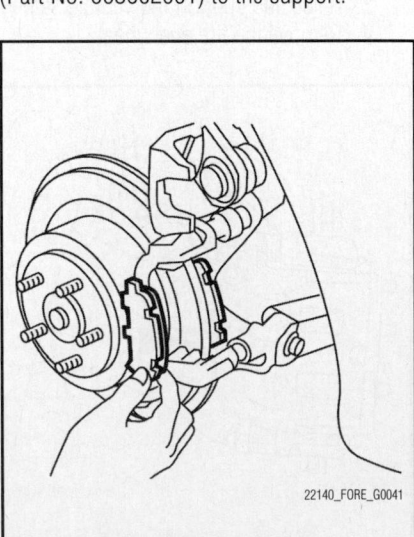

22140_FORE_G0041

Fig. 14 Brake pad removal shown

10. Apply a thin coat of Molykote AS880N (Part No. K0777YA010) to the pad clip and both sides of inner shim.

11. Install the pad to support.

12. Install the caliper body to the sup-port. Tighten the caliper retaining bolts to 28 ft. lbs. (37 Nm).

> **✻✻ CAUTION**
>
> **When connecting the caliper, do not twist brake hose.**

13. Bleed the hydraulic system, as required.

14. Check the brake fluid level, correct as required.

15. Test drive and burnish brake system.

BRAKES

BRAKE DRUM

REMOVAL & INSTALLATION

See Figure 15.

1. Before servicing the vehicle, refer to the Precautions Section.

2. Loosen the wheel nuts.

3. Raise and support the vehicle safely.

4. Remove the rear wheels.

5. Release the parking brake.

6. If necessary, remove the adjusting hole cover from the backing plate and using a suitable tool, back off the shoe adjuster.

7. If the drum is difficult to remove, insert an 8mm bolt into the hole on the drum to push it off.

8. Remove the drum.

To install:

9. Adjust the brake shoes. Set the outside diameter of brake shoes 0.020–0.031 inch (0.5–0.8mm) smaller in comparison with the inside diameter of the brake drum.

10. Install the drum.

11. Install the rear wheels.

12. Bleed the hydraulic system, as required.

13. Check the brake fluid level, correct as required.

14. Test drive and burnish brake system.

BRAKE SHOES

REMOVAL & INSTALLATION

See Figure 16.

1. Before servicing the vehicle, refer to the Precautions Section.

2. Loosen the wheel nuts.

3. Raise and safely support the vehicle.

4. Remove the rear wheel.

5. Release the parking brake.

6. Remove the brake drum from the brake assembly.

7. Hold the hold-down pin by securing rear of back plate with your hand.

8. Disconnect the hold-down cup from hold-down pin by rotating hold-down cup.

9. Disconnect the lower shoe return spring from shoes.

REAR DRUM BRAKES

10. Remove the shoes one by one from back plate with the adjuster.

11. Disconnect the parking brake cable from parking lever.

12. Remove the upper shoe return spring and adjusting spring from the brake shoe.

To install:

13. Apply brake grease to the backing plate where the brake shoes contact it.

14. Apply brake grease to the adjusting screw and both ends of the adjuster.

15. Connect the upper shoe return spring to the shoes.

16. Connect the parking brake cable to the parking lever.

17. While positioning the shoes one at a time in the groove on the wheel cylinder, secure the shoes.

18. Secure the shoes by connecting the hold-down cup to the hold-down pin.

19. Connect the lower shoe return spring to both shoes.

20. Set the outside diameter of the shoes less than 0.020–0.031 in. (0.5–0.8mm) compared to the inside diameter of the drum.

21. Install the drum and adjust the brake shoes.

22. Install the wheels and lower the vehicle.

23. Bleed the hydraulic system, as required.

24. Check the brake fluid level, correct as required.

25. Test drive and burnish brake system.

ADJUSTMENT

See Figure 15.

1. Remove the adjusting hole cover from back plate.

2. Turn adjusting screw using a slot-type screwdriver until brake shoe is in close contact with the drum.

3. Turn back (downward) the adjusting screw 3 or 4 notches.

4. Install the adjusting hole cover to back plate.

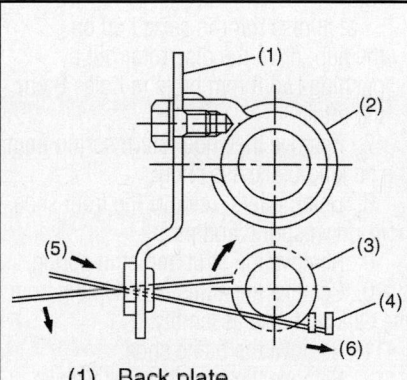

(1) Back plate
(2) Wheel cylinder ASSY
(3) Adjuster ASSY claws
(4) Adjusting lever
(5) Tightening direction

22140_FORE_G0042

Fig. 15 Inside adjusting lever and related parts

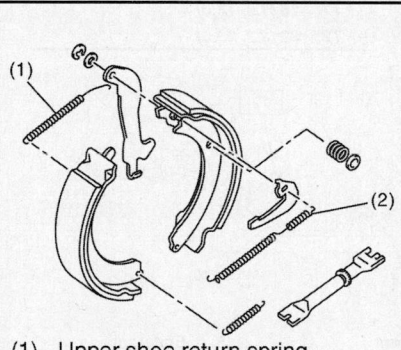

(1) Upper shoe return spring
(2) Adjusting spring

22140_FORE_G0043

Fig. 16 Rear brake shoes and related parts

PARKING BRAKE CABLES

ADJUSTMENT

See Figure 17.

1. Before servicing the vehicle, refer to the Precautions Section.
2. Remove the console cover.
3. Forcibly pull the parking brake lever 3 to 5 times.
4. Adjust the parking brake lever by turning adjuster until parking brake lever stroke is set at 7 to 8 notches with operating force of 44 lbs. (196 N).
5. Tighten the lock nut.
6. Install the console cover.

PARKING BRAKE SHOES

REMOVAL & INSTALLATION

See Figures 18 and 19.

1. Before servicing the vehicle, refer to the Precautions Section.
2. Release the parking brake.
3. Raise and safely support the vehicle.
4. Remove the two mounting bolts and remove the brake caliper assembly.

Suspend the brake caliper assembly so that the hose is not stretched.

5. Remove the brake rotor.
6. If the disc rotor is difficult to remove, try the following two methods in order:

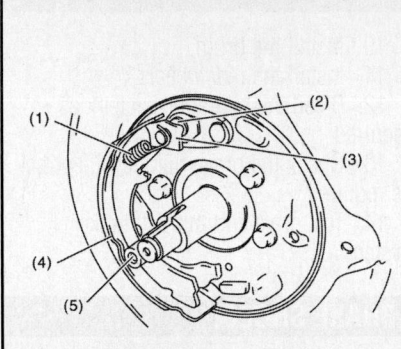

(1) Primary return spring
(2) Anchor pin
(3) Plate
(4) Primary shoe
(5) Shoe hold-down pin & cup

42050_FORE_G0023

Fig. 18 Assembled primary components of the parking brake

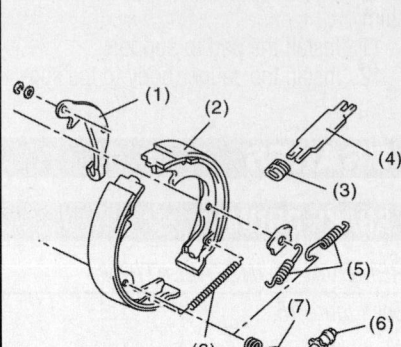

(1) Lever
(2) Secondary brake shoe
(3) Strut spring
(4) Strut
(5) Secondary return spring
(6) Adjuster
(7) Hold-down cup
(8) Adjusting spring

42050_FORE_G0024

Fig. 19 Exploded view of the parking brake secondary components

a. Turn the adjusting screw using a flat tip screwdriver until the brake shoe moves adequately away from the disc rotor.

b. If disc rotor is seized up on the hub, drive the disc rotor out by pushing two 8 mm bolts in holes B on the rotor.

7. Remove the shoe return spring from the parking brake assembly.
8. Using pliers, remove the front shoe hold-down spring and pin.
9. Remove the strut and strut spring.
10. Remove the adjuster assembly from the parking brake assembly.
11. Remove the brake shoe.
12. Remove the rear shoe hold-down spring and pin with pliers.
13. Remove the parking brake cable from the parking brake lever.
14. Using a standard screwdriver, raise the retainer. Remove the parking brake lever and washer from the brake shoe.

To install:

15. Apply brake grease to the backing plate where the brake shoes contact it.
16. Insert the primary side brake shoe into the anchor pin groove.
17. Secure the brake shoe with the shoe-hold down pin and cup.

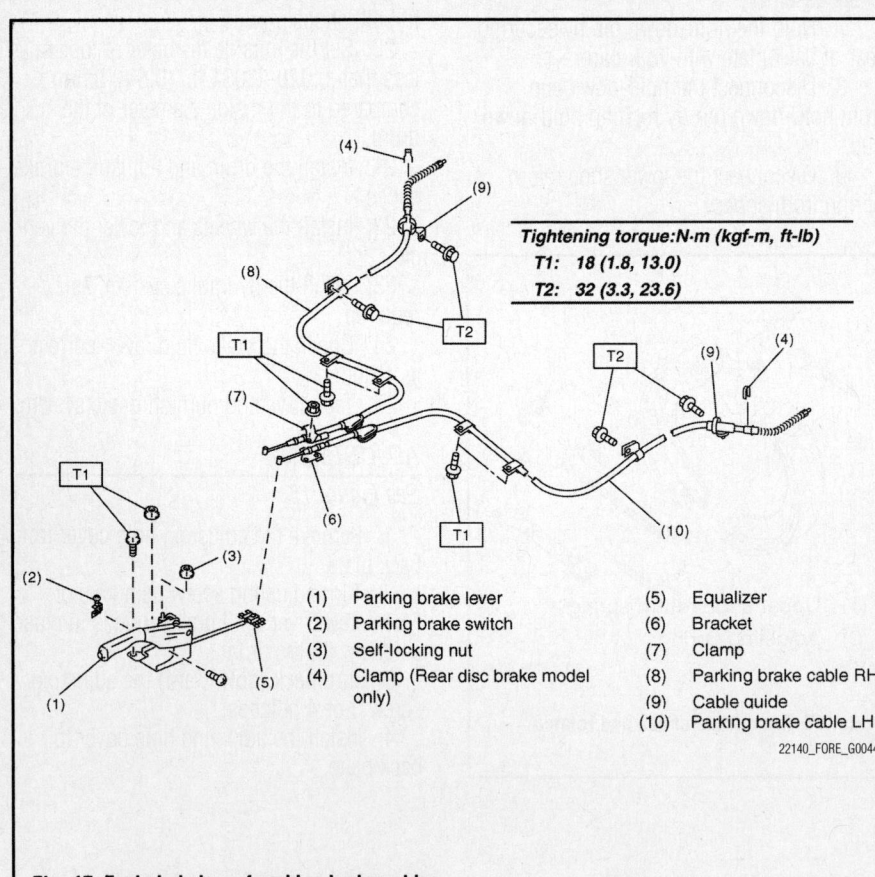

Tightening torque:N·m (kgf-m, ft-lb)
T1: 18 (1.8, 13.0)
T2: 32 (3.3, 23.6)

(1) Parking brake lever
(2) Parking brake switch
(3) Self-locking nut
(4) Clamp (Rear disc brake model only)
(5) Equalizer
(6) Bracket
(7) Clamp
(8) Parking brake cable RH
(9) Cable guide
(10) Parking brake cable LH

22140_FORE_G0044

Fig. 17 Exploded view of parking brake cables

18. Install the plate to the anchor pin, and then assemble the primary return spring to the anchor pin.

19. Install the parking brake cable to the parking brake lever.

20. Assemble the strut and adjuster, and then secure the secondary side brake shoe with the shoe hold-down pin and cup.

➡**Install the strut spring of both right and left wheel facing vehicle front.**

21. Install the adjuster assembly with screw on left side.

22. Install the secondary return spring and adjuster spring.

23. Adjust the parking brakes.

24. Install the rotor and brake assembly.

25. Lower the vehicle.

ADJUSTMENT

See Figure 20.

1. Remove the adjusting hole cover from the back plate.

2. Turn the adjusting screw using a flat tip screwdriver until the brake shoe is in close contact with the disc rotor.

3. Turn back (downward) the adjusting screw 3 to 4 notches.

4. Install the adjusting hole cover to the back plate.

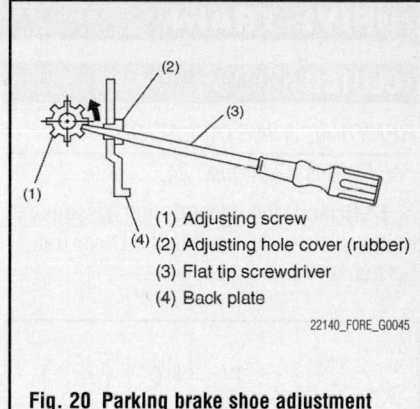

(1) Adjusting screw
(2) Adjusting hole cover (rubber)
(3) Flat tip screwdriver
(4) Back plate

22140_FORE_G0045

Fig. 20 Parking brake shoe adjustment shown

CHASSIS ELECTRICAL

GENERAL INFORMATION

> ❊❊ **CAUTION**
>
> **These vehicles are equipped with an air bag system. The system must be disarmed before performing service on, or around, system components, the steering column, instrument panel components, wiring and sensors. Failure to follow the safety precautions and the disarming procedure could result in accidental air bag deployment, possible injury and unnecessary system repairs.**

SERVICE PRECAUTIONS

> ❊❊ **CAUTION**
>
> **Disconnect and isolate the battery negative cable before beginning any airbag system component diagnosis, testing, removal, or installation procedures. Wait at least 90 seconds**

AIR BAG (SUPPLEMENTAL RESTRAINT SYSTEM)

after the ignition switch is turned off and the negative (-) terminal cable is disconnected from the battery before starting the operation. The SRS is equipped with a backup power source, so if work is started within 90 seconds after disconnecting the negative (-) terminal cable from the battery, the SRS may be deployed. Failure to disable the airbag system may result in accidental airbag deployment, personal injury, or death.

DISARMING THE SYSTEM

1. Be sure to position the front wheels in the straight ahead position.

2. Disconnect the negative battery cable. Tape the battery cable for added protection.

3. Wait more than 20 seconds before starting work.

ARMING THE SYSTEM

Connect the negative battery cable.

CLOCKSPRING CENTERING

See Figure 21.

1. Check that front wheels are positioned in straight ahead direction.

2. Turn the clockspring pin (A) clockwise until it stops.

3. Turn the clockspring pins (A) approx. 3.25 turns until marks are aligned.

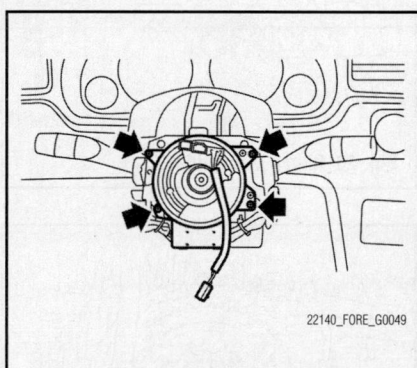

22140_FORE_G0049

Fig. 21 Clockspring alignment marks

DRIVE TRAIN

CLUTCH

REMOVAL & INSTALLATION

See Figures 22 through 24.

1. Remove the transmission assembly from vehicle body. Refer to the Drive Train Section.

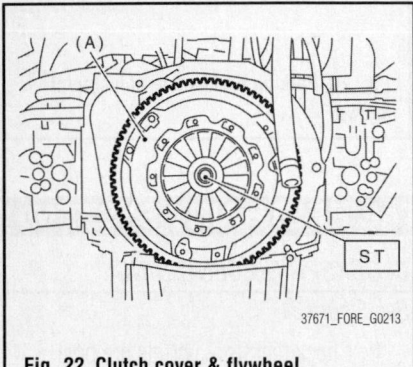

Fig. 22 Clutch cover & flywheel

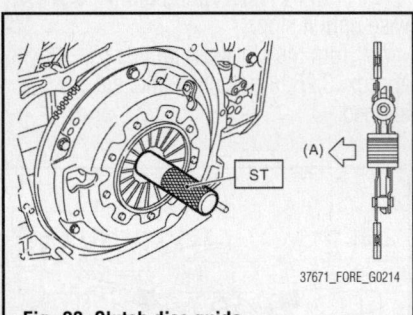

Fig. 23 Clutch disc guide

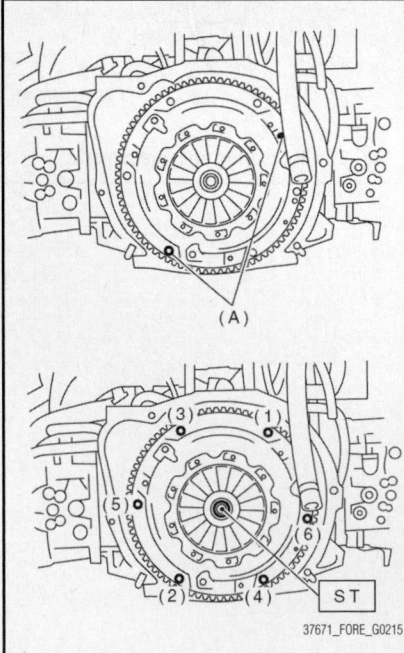

Fig. 24 Clutch unbalance marks

2. Attach the ST on the flywheel. ST 499747100 CLUTCH DISC GUIDE
3. Remove the clutch cover and clutch disc.

➡**Take care not to allow oil to touch the clutch disc face.**

❊ WARNING

Do not disassemble the clutch cover or clutch disc.

To install:

4. Insert the ST into the clutch disc and attach to the flywheel by inserting the ST end into pilot bearing.

➡**When installing the clutch disc, be careful to attach in the correct direction. ST 499747100 CLUTCH DISC GUIDE**

5. Install the clutch cover to the flywheel and tighten the bolts to the specified torque:

   a. When installing the clutch cover to the fly wheel, position the clutch cover so that the distance between unbalance marks (paint marks) is at least 120°. (The unbalance marks indicate the directions of residual unbalance.)

   b. Note the front and rear of the clutch disc when installing.

   c. Temporarily tighten the bolts by hand. Each bolt should be tightened to the specified torque in a crisscross order. Tighten bolts to 142 inch lbs. (16 Nm)

6. Remove the ST. ST 499747100 CLUTCH DISC GUIDE
7. Install the transmission assembly. Refer to the Drive Train Section.

FRONT HUB UNIT BEARING

REMOVAL & INSTALLATION

See Figures 25 through 30.

1. Disconnect the ground cable from the battery.
2. Lift up the vehicle, and remove the front wheels.
3. Lift the crimped section of axle nut.
4. Remove the axle nut using a socket wrench while depressing the brake pedal.

❊ CAUTION

Remove the wheel before loosening the axle nut. Failure to follow this rule may damage the wheel bearings.

5. Remove the disc brake caliper from the housing, and suspend it from strut using a wire.
6. Remove the disc rotor from the hub.

➡**If it is difficult to remove the disc rotor from the hub, drive the 8 mm bolt into the threaded end of rotor, and then remove the rotor.**

7. Remove the four bolts from the housing.

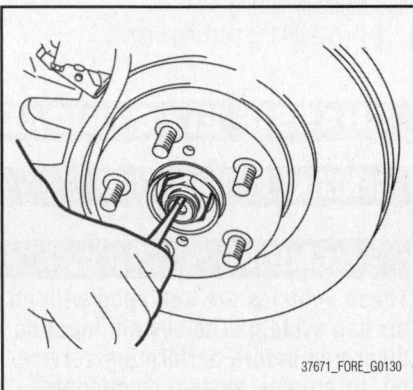

Fig. 25 Lift the crimped section of axle nut

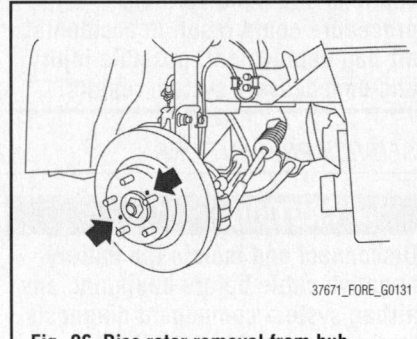

Fig. 26 Disc rotor removal from hub

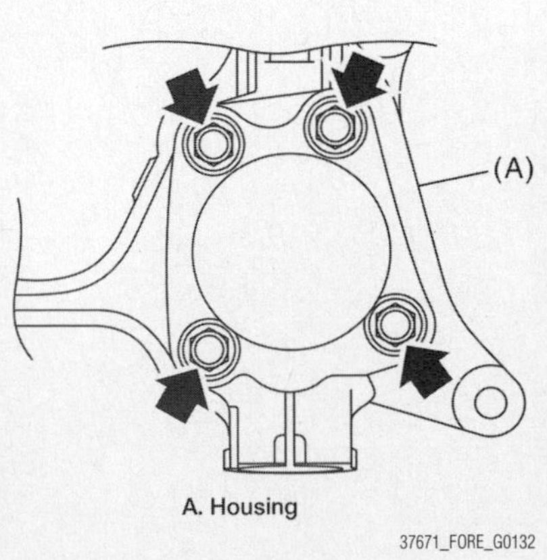

Fig. 27 Housing bolt removal

A. Housing

8. Remove the front hub unit bearing. If it is hard to remove, use the ST.

To install:

9. Place the disc cover between housing and front hub unit, and tighten the four bolts. Tighten bolts to 48 ft. lbs. (65 Nm)

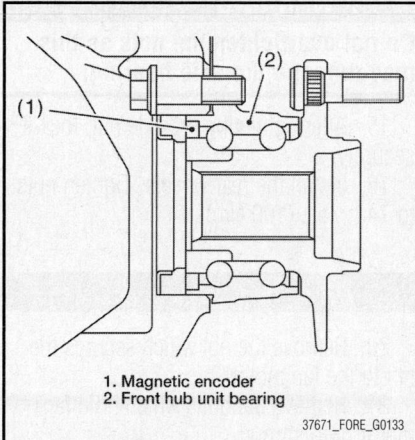

1. Magnetic encoder
2. Front hub unit bearing

37671_FORE_G0133

Fig. 28 Magnetic encoder and front hub bearing

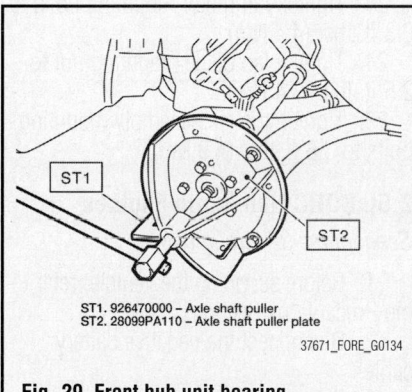

ST1. 926470000 – Axle shaft puller
ST2. 28099PA110 – Axle shaft puller plate

37671_FORE_G0134

Fig. 29 Front hub unit bearing

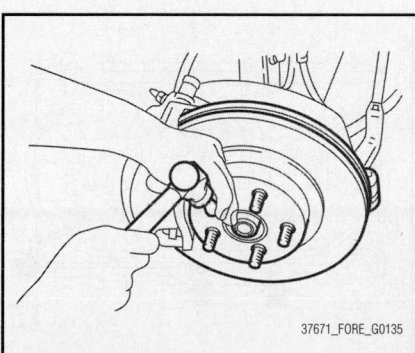

37671_FORE_G0135

Fig. 30 Securely lock axle nut

10. Tighten the axle nut temporarily.
11. Install the disc rotor to hub.
12. Install the disc brake caliper on the housing. Tighten to 59 ft. lbs. (80 Nm)
13. While depressing the brake pedal, tighten a new axle nut to the specified torque and lock it securely. Tighten nut to 162 ft. lbs. (220 Nm)

14. After tightening the axle nut, lock it securely.
15. Install the wheel. Tighten to 74 ft. lbs. (100 Nm)

REAR HUB UNIT BEARING

REMOVAL & INSTALLATION

See Figures 31 through 36.

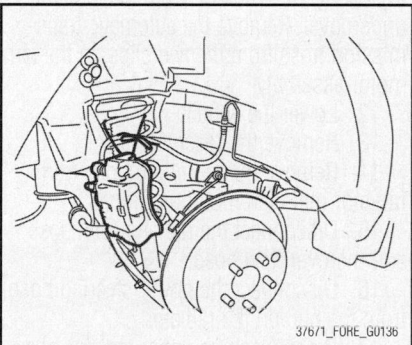

37671_FORE_G0136

Fig. 31 Remove disc brake caliper from rear housing

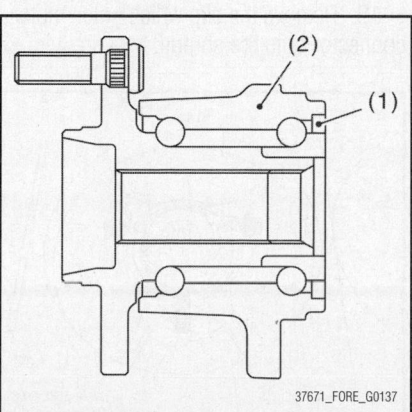

37671_FORE_G0137

Fig. 32 Rear hub unit bearing & magnetic encoder

1. Disconnect the ground cable from the battery.
2. Lift up the vehicle, and then remove the rear wheels.
3. Lift the crimped section of axle nut.
4. Remove the axle nut using a socket wrench while depressing the brake pedal.

5. Remove the disc brake caliper from the rear housing, and suspend it from the vehicle using a string.
6. Remove the rear disc rotor.
7. Remove the four bolts from the rear housing.

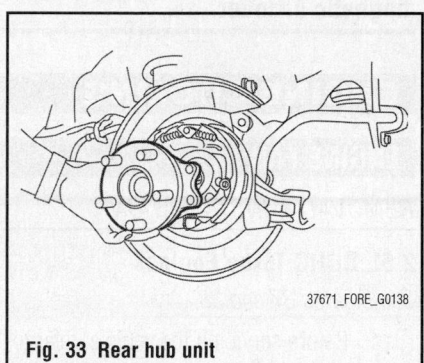

37671_FORE_G0138

Fig. 33 Rear hub unit

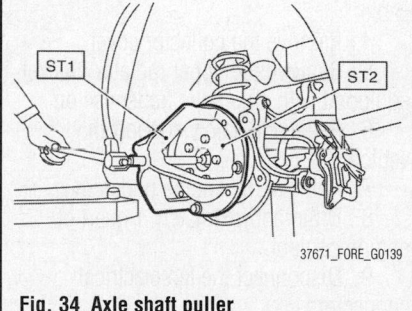

37671_FORE_G0139

Fig. 34 Axle shaft puller

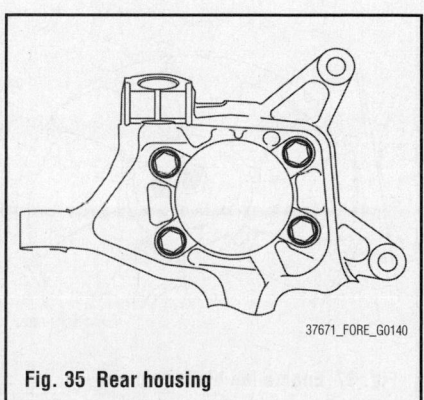

37671_FORE_G0140

Fig. 35 Rear housing

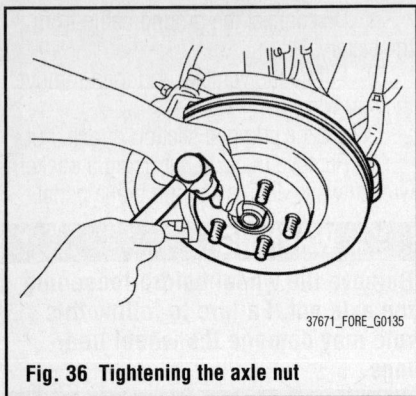

Fig. 36 Tightening the axle nut

8. Remove the rear hub unit bearing.

Be careful not to damage the magnetic encoder. Do not get closer the tool which charged magnetism to magnetic encoder.

→ **If it is hard to remove, use the ST.**

To install:

9. Aligning with the mounting hole of the rear back plate, temporarily tighten the rear hub unit bearing to the rear housing.

※ WARNING

Be careful not to damage the magnetic encoder. Do not get closer the tool which charged magnetism to magnetic encoder.

10. Tighten the four bolts of the rear housing. Tighten bolts to 48 ft. lbs. (65 Nm)

11. Tighten the new axle nut temporarily.

※ CAUTION

Use new axle nuts.

12. Install the rear disc rotor.

13. Install the disc brake caliper to the rear housing. Tighten to 49 ft. lbs. (66 Nm)

14. While pressing the brake pedal, tighten the new axle nuts to the specified torque. Tighten nuts to 140 ft. lbs. (190 Nm)

※ WARNING

Do not install wheel and let it touch the ground before tightening the axle nut. Failure to follow this rule may damage the axle bearing.

※ WARNING

Do not overtighten the nuts as this may damage the axle bearing.

15. After tightening the axle nut, lock it securely.

16. Install the rear wheels. Tighten nuts to 74 ft. lbs. (100 Nm)

ENGINE COOLING

ENGINE FAN

REMOVAL & INSTALLATION

2.5L DOHC Turbo Engines

See Figures 37 and 38.

1. Before servicing the vehicle, refer to the Precautions Section.

2. Raise and safely support the vehicle.

3. Disconnect the negative battery cable.

4. Remove the collector cover.

5. Remove the upper radiator bracket, if equipped with automatic transmission.

6. Raise and safely support the vehicle.

7. Remove the engine undercover.

8. Drain approximately 1 quart of engine coolant.

9. Disconnect the fan electrical connectors.

10. Remove the radiator undercover mounting bolt, if equipped with automatic transmission.

11. Pull the lower radiator hose upward to separate the radiator from the radiator undercover. Remove the automatic transmission hose from the two clips on the fan motor assembly.

12. Lower the vehicle.

13. Remove the reservoir tank.

14. Remove the mounting bolts from radiator main fan motor assembly.

15. Disconnect the radiator inlet hose and the over flow hose.

16. Disconnect the power steering hose from the clip on the radiator.

17. Disconnect the upper radiator hose from the radiator.

18. Lift the engine fan assembly up and out of the engine compartment.

19. Remove the clip which holds motor connector onto the shroud.

20. Remove the nut which secures the fan to the fan motor.

21. Remove the bolts which hold fan motor onto shroud.

To install:

22. Installation is the reverse order of assembly.

23. Tighten fan motor mounting bolts to 3.3 ft. lbs. (4.4 Nm).

24. Tighten fan blade mounting nut to 2.5 ft. lbs. (3.4 Nm).

25. Tighten the fan assembly mounting bolts to 3.6 ft. lbs. (4.9 Nm).

2.5L SOHC Non-Turbo Engines

See Figures 37 through 40.

1. Before servicing the vehicle, refer to the Precautions Section.

2. Disconnect the negative battery cable.

3. Raise and safely support the vehicle.

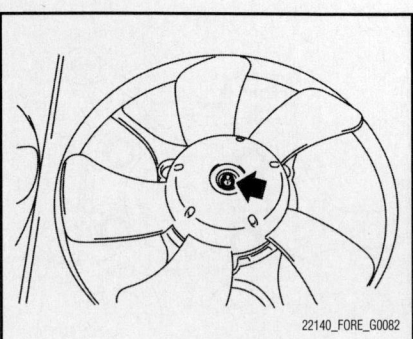

Fig. 37 Engine fan blade removal

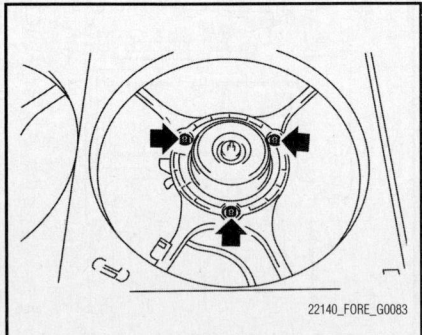

Fig. 38 Engine main fan motor removal

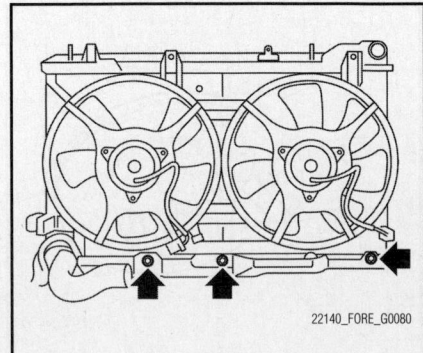

Fig. 39 Attaching bolts

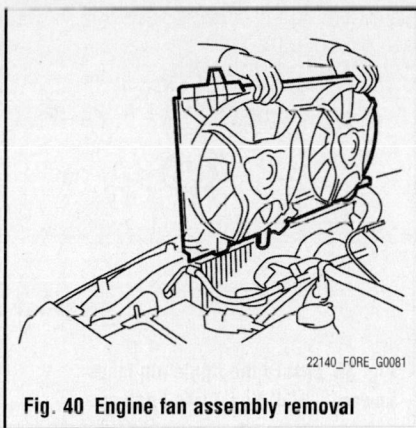

Fig. 40 Engine fan assembly removal

4. Remove the engine undercover attaching bolts.

5. Drain approximately 1 quart of engine coolant.

6. Disconnect the fan electrical connectors.

7. Remove the transmission cooler hose from the clip of the radiator fan motor assembly, if equipped.

8. Lower the vehicle.

9. Remove the accessory drive belt cover.

10. Disconnect the overflow hose.

11. Remove the reservoir tank.

12. Remove the three upper fan assembly mounting bolts.

13. Disconnect the upper radiator hose from the radiator.

14. Detach the power steering hose from the clip on the radiator.

15. Lift the engine fan assembly up and out of the engine compartment.

16. Remove the clip which holds motor connector onto the shroud.

17. Remove the nut which secures the fan to the fan motor.

18. Remove the bolts which hold fan motor onto shroud.

To install:

19. Installation is the reverse order of assembly.

20. Tighten fan motor mounting bolts to 3.3 ft. lbs. (4.4 Nm).

21. Tighten fan blade mounting nut to 2.5 ft. lbs. (3.4 Nm).

22. Tighten the fan assembly mounting bolts to 3.6 ft. lbs. (4.9 Nm).

RADIATOR

REMOVAL & INSTALLATION

2.5L DOHC Turbo Engine

See Figures 41 through 43.

1. Raise and safely support the vehicle.

2. Remove the collector cover.

3. Disconnect the ground cable from the battery.

4. Lift up the vehicle.

5. Remove the undercover.

6. Drain engine coolant completely.

7. Disconnect the connectors of radiator main fan and sub fan motor.

8. Disconnect the radiator outlet hose from thermostat cover.

9. Disconnect the ATF hoses from ATF pipes, A/T models. Apply the cap to the ATF pipe to prevent ATF leaks.

10. Lower the vehicle.

11. Remove the reservoir tank.

12. Disconnect the radiator inlet hose, air breather hose and the overflow hose from radiator.

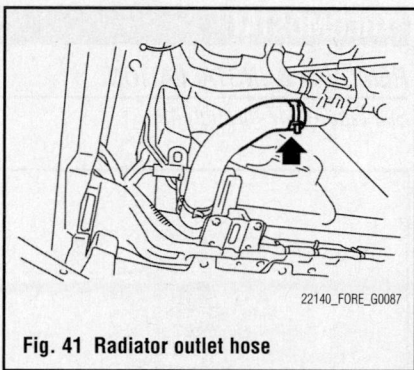

Fig. 41 Radiator outlet hose

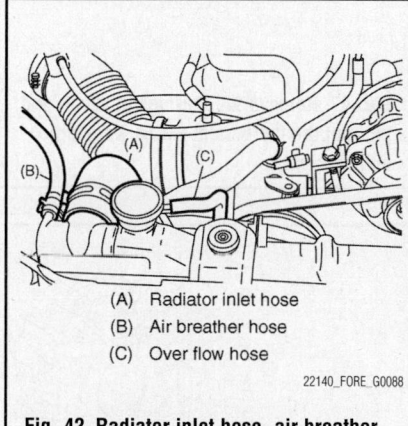

(A) Radiator inlet hose
(B) Air breather hose
(C) Over flow hose

Fig. 42 Radiator inlet hose, air breather hose and the overflow hose

13. Remove the radiator upper brackets.

14. Disconnect the power steering hose from the clip on the radiator.

15. Lift the radiator up and away from vehicle.

16. Remove the radiator undercover, A/T models.

To install:

17. Attach the radiator undercover to the radiator, A/T models.

18. Attach the radiator lower cushion to the hole of the radiator lower bracket.

19. Install the radiator to vehicle.

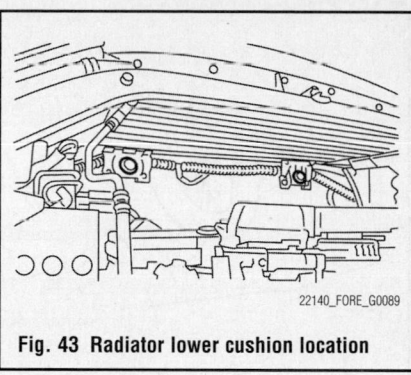

Fig. 43 Radiator lower cushion location

➡ **Make pins on the lower side of radiator be fitted into the radiator lower cushions on body side.**

20. Install the radiator upper brackets and tighten the bolts to 13.3 ft. lbs. (18 Nm).

21. Attach the power steering hose to the radiator.

22. Connect the radiator inlet hose, air breather hose and the overflow hose (B).

23. Install the reservoir tank.

24. Lift up the vehicle.

25. Connect the ATF hoses, A/T models.

26. Connect the radiator outlet hose.

27. Connect the connectors to the radiator main fan motor (A) and sub fan motor (B).

28. Install the undercover.

29. Lower the vehicle.

30. Connect the ground cable to the battery.

31. Fill engine coolant.

32. Check the ATF level.

33. Install the collector cover.

2.5L SOHC Non-Turbo Engine

See Figures 44 through 46.

1. Before servicing the vehicle, refer to the Precautions Section.

2. Raise and safely support the vehicle.

3. Disconnect the negative battery cable.

4. Lift up the vehicle.

Fig. 44 Connector locations of radiator main fan (A) and sub fan (B)

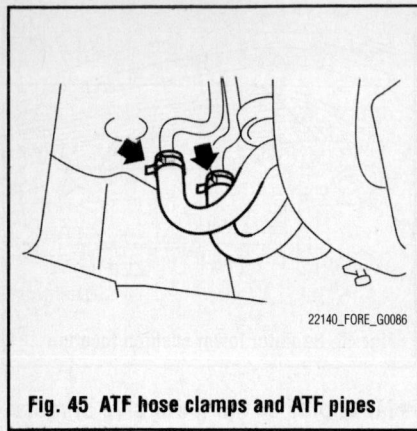

Fig. 45 ATF hose clamps and ATF pipes

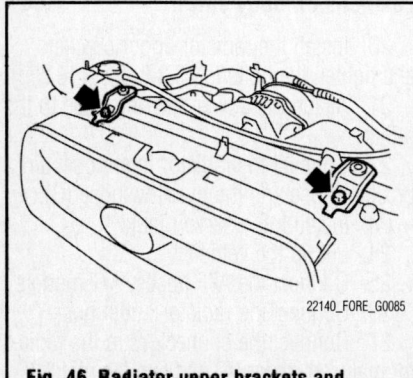

Fig. 46 Radiator upper brackets and mounting bolts

5. Remove the undercover.

6. Drain engine coolant completely.

7. Disconnect the connectors of radiator main fan (A) and sub fan (B) motor.

8. Disconnect the radiator outlet hose from thermostat cover.

9. Disconnect the ATF hoses from ATF pipes, A/T models. Apply the cap to the ATF pipe to prevent ATF leaks.

10. Lower the vehicle.

11. Disconnect the over flow hose.

12. Remove the reservoir tank.

13. Disconnect the radiator inlet hose from radiator.

14. Remove the radiator upper brackets.

15. Detach the power steering hose from the clip on the radiator.

16. Lift the radiator up and away from vehicle.

17. Remove the radiator undercover, A/T models.

To install:

18. Attach the radiator undercover to the radiator, A/T models.

19. Attach the radiator lower cushion to the hole of the radiator lower bracket.

20. Install the radiator to vehicle.

21. Install the radiator upper brackets and tighten the bolts to 13.3 ft. lbs. (18 Nm).

22. Attach the power steering hose to the radiator.

23. Connect the radiator inlet hose.

24. Install the reservoir tank.

25. Connect the over flow hose.

26. Lift up the vehicle.

27. Connect the ATF hoses, A/T models.

28. Connect the radiator outlet hose.

29. Connect the connectors to the radiator main fan motor (A) and sub fan motor (B).

30. Install the undercover.

31. Lower the vehicle.

32. Connect the ground cable to the battery. Fill engine coolant.

33. Check the ATF level, A/T models.

THERMOSTAT

REMOVAL & INSTALLATION

See Figures 47 through 49.

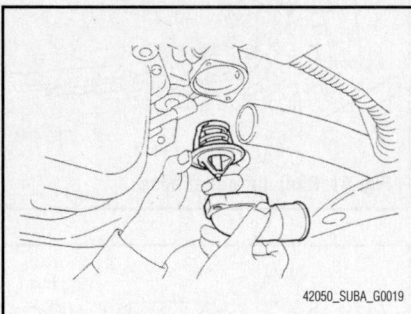

Fig. 47 Remove the thermostat cover, and pull out the thermostat—2.5L Non-Turbo engine shown

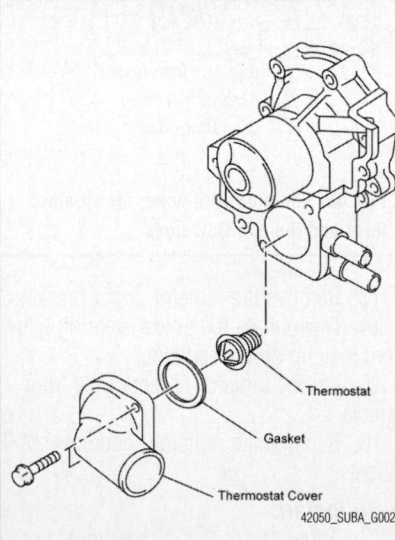

Fig. 48 Exploded view of the thermostat and related components—2.5L Turbo engine shown

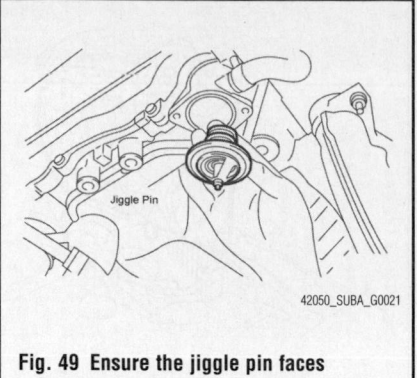

Fig. 49 Ensure the jiggle pin faces upward—2.5L Non-Turbo Engines

1. Raise and safely support the vehicle.

2. Remove the engine undercover.

3. Drain the engine coolant.

4. Loosen the hose clamp and disconnect the lower radiator hose from the thermostat cover.

5. Remove the thermostat cover and gasket, and pull out the thermostat.

To install:

6. Install the thermostat in the intake manifold.

➡On non-Turbo engines, the thermostat must be installed with the jiggle pin upward. On turbo engines, the thermostat must be install with the jiggle pin facing front.

7. Install the thermostat cover together with a new gasket.

8. Install the lower radiator hose to the thermostat cover and tighten the hose clamp.

9. Install the engine undercover.

10. Lower the vehicle.

11. Refill the engine cooling system to the correct level.

12. Start the engine and check for leaks.

WATER PUMP

REMOVAL & INSTALLATION

Non-Turbo Engines

See Figures 50 and 51.

1. Before servicing the vehicle, refer to the Precautions Section.

2. Disconnect the negative battery cable.

3. Drain the cooling system. Remove the radiator.

4. To remove the front side V-belt, remove the belt covers. Loosen the lock bolt. Loosen the slider bolt. Remove the front side belt.

5. To remove the rear side V-belt, remove the belt covers. Loosen the lock bolt. Loosen the slider bolt. Remove the rear side belt. Remove the belt tensioner.

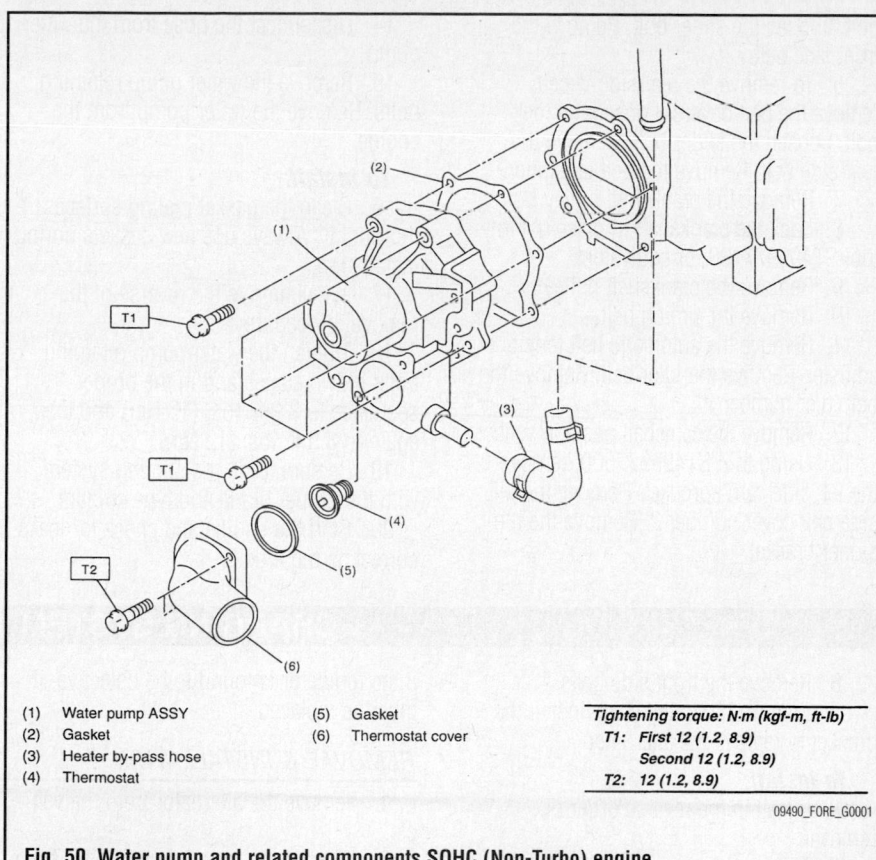

(1) Water pump ASSY
(2) Gasket
(3) Heater by-pass hose
(4) Thermostat
(5) Gasket
(6) Thermostat cover

Tightening torque: N·m (kgf-m, ft-lb)		
T1:	First 12 (1.2, 8.9)	
	Second 12 (1.2, 8.9)	
T2:	12 (1.2, 8.9)	

09490_FORE_G0001

Fig. 50 Water pump and related components SOHC (Non-Turbo) engine

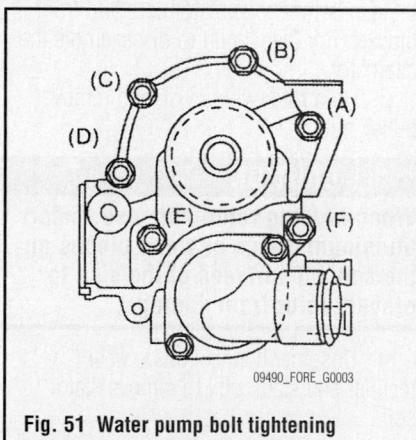

09490_FORE_G0003

Fig. 51 Water pump bolt tightening sequence SOHC (Non-Turbo) engine

6. Remove the crankshaft pulley bolt.

7. Lock the crankshaft in place using tool ST499977100, or equivalent.

8. Remove the crankshaft pulley.

9. Remove the timing belt.

10. Remove the automatic belt tensioner adjuster. Remove the belt idler number 2.

11. Using tool ST18231AA010, remove the left side cam sprocket. Remove the left side belt cover number 2. Remove the tensioner bracket.

12. Disconnect the hose from the water pump.

13. Remove the water pump retaining bolts. Remove the water pump from the engine.

To install:

14. Clean the gasket mating surfaces thoroughly. Always use new gaskets during installation.

15. Installation is the reverse of the removal procedure.

16. Tighten the water pump retaining bolts in two stages and in the proper sequence to 8.9 ft. lbs. (12 Nm) and then again to 8.9 ft. lbs. (12 Nm).

17. Be sure to fill the cooling system with the proper grade and type coolant.

18. Start the engine and check for leaks, correct as required.

Turbo Engines

See Figures 52 and 53.

1. Before servicing the vehicle, refer to the Precautions Section.

2. Disconnect the negative battery cable.

3. Drain the cooling system. Remove the radiator.

4. Remove the collector cover.

5. To remove the front side V-belt, remove the belt covers. Loosen the lock

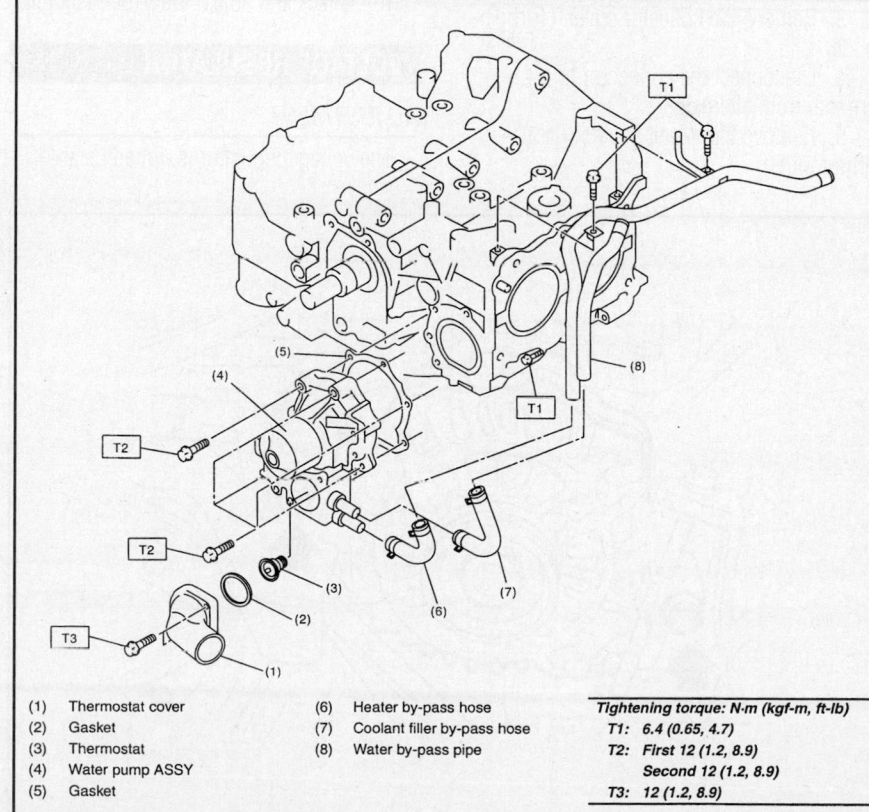

(1) Thermostat cover
(2) Gasket
(3) Thermostat
(4) Water pump ASSY
(5) Gasket
(6) Heater by-pass hose
(7) Coolant filler by-pass hose
(8) Water by-pass pipe

Tightening torque: N·m (kgf-m, ft-lb)		
T1:	6.4 (0.65, 4.7)	
T2:	First 12 (1.2, 8.9)	
	Second 12 (1.2, 8.9)	
T3:	12 (1.2, 8.9)	

09490_FORE_G0002

Fig. 52 Water pump and related components DOHC (Turbo) engine

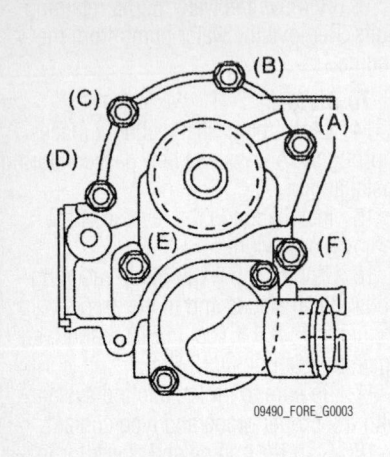

Fig. 53 Water pump bolt tightening sequence DOHC (Turbo) engine

bolt. Loosen the slider bolt. Remove the front side belt.

6. To remove the rear side V-belt, remove the belt covers. Loosen the lock bolt. Loosen the slider bolt. Remove the rear side belt. Remove the belt tensioner.

7. Remove the crankshaft pulley bolt.

8. Lock the crankshaft in place using tool ST499977100, or equivalent.

9. Remove the crankshaft pulley.

10. Remove the timing belt.

11. Remove the automatic belt tensioner adjuster. Remove the idler belt. Remove the belt idler number 2.

12. Remove the camshaft position sensor.

13. Using tool ST499977500, remove the left side cam sprocket. Remove the left side belt cover number 2. Remove the tensioner bracket.

14. Disconnect the hose from the water pump.

15. Remove the water pump retaining bolts. Remove the water pump from the engine.

To install:

16. Clean the gasket mating surfaces thoroughly. Always use new gaskets during installation.

17. Installation is the reverse of the removal procedure.

18. Tighten the water pump retaining bolts in two stages and in the proper sequence to 8.9 ft. lbs. (12 Nm) and then again to 8.9 ft. lbs. (12 Nm).

19. Be sure to fill the cooling system with the proper grade and type coolant.

20. Start the engine and check for leaks, correct as required.

ENGINE ELECTRICAL

ALTERNATOR

REMOVAL & INSTALLATION

See Figure 54.

1. Before servicing the vehicle, refer to the Precautions Section.

2. Disconnect the ground cable from the battery.

3. Remove the collector cover. (Turbo model)

4. Disconnect the connector and terminal from alternator.

5. Remove the V-belt covers. (Non-turbo model)

6. Remove the front side belts.

7. Remove the bolts which install the bracket to remove the alternator.

To install:

8. Install in the reverse order of removal.

9. Tighten mounting bolts to 18 ft. lbs. (25 Nm).

10. Check and adjust the V-belt tension.

VOLTAGE REGULATOR

ADJUSTMENT

The voltage regulator is not adjustable.

CHARGING SYSTEM

If the regulator is found to be defective, it must be replaced.

REMOVAL & INSTALLATION

1. Remove the alternator from the vehicle.

2. Remove the four through bolts. Then insert the tip of a flat-head screwdriver into the gap between the stator core and front bracket. Pry them apart to disassemble the alternator.

3. Hold rotor with a vise and remove pulley nut.

✳✳ CAUTION

When holding rotor with vise, insert aluminum plates or wood pieces on the contact surfaces of the vise to prevent rotor from damage.

4. Unsolder connection between rectifier and stator coil to remove stator coil.

✳✳ CAUTION

Finish the work rapidly (less than three seconds) because the rectifier cannot withstand heat very well.

5. Remove screws which secure voltage regulator to rear cover, and unsolder connection between voltage regulator and rectifier to remove the regulator.

6. Installation is the reverse order of assembly. After assembly, turn the pulley by hand to check that the rotor turns smoothly

Fig. 54 Alternator mounting bolts

FIRING ORDER

See Figure 56.

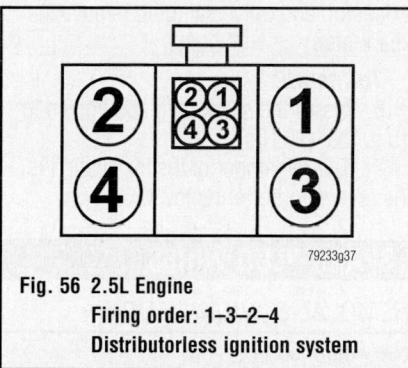

Fig. 56 2.5L Engine
Firing order: 1–3–2–4
Distributorless ignition system

IGNITION COIL

REMOVAL & INSTALLATION

See Figure 57.

1. Disconnect the ground cable from the battery.
2. Remove the collector cover.
3. Remove the air cleaner case (RH side).
4. Remove the secondary air pump for LH side
5. Disconnect the connector from ignition coil.
6. Remove the ignition coil.

To install:

7. Install in the reverse order of removal.
8. Tighten the ignition coil mounting bolt to 12 ft. lbs. (16 Nm).

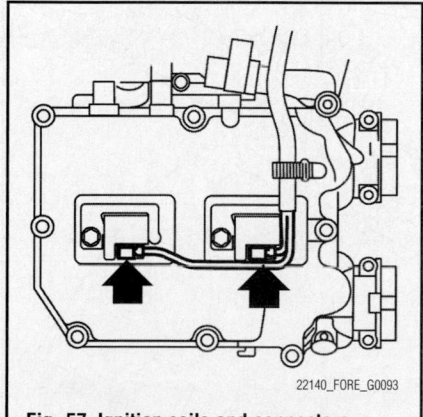

Fig. 57 Ignition coils and connectors shown

IGNITION COIL PACK

REMOVAL & INSTALLATION

See Figure 58.

1. Disconnect the negative battery cable.

2. Disconnect the spark plug wires from the ignition coil.
3. Disconnect the electrical connector from the ignition coil.
4. Unbolt and remove the ignition coil and igniter assembly.

To install:

5. Installation is the reverse order of assembly.
6. Tighten the ignition coil mounting bolts to 5 ft. lbs. (7 Nm).

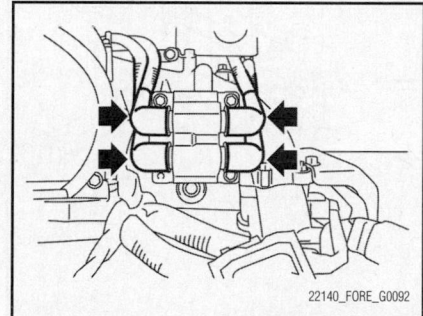

Fig. 58 Spark plug wires at ignition coil assembly

IGNITION TIMING

INSPECTION

2.5L DOHC Turbo Engine

See Figure 59.

1. Before checking the ignition timing, ensure the air cleaner element is free from debris, the spark plugs are in good condition, and all hoses are connected properly.
2. Set the transmission in Park for automatic transmissions or Neutral for manual transmissions.
3. Set the parking brake, start and run the engine until normal operating temperature is obtained. Then turn the engine **OFF**.
4. Insert the cartridge to SUBARU SELECT MONITOR.
5. Connect SUBARU SELECT MONITOR to the data link connector.

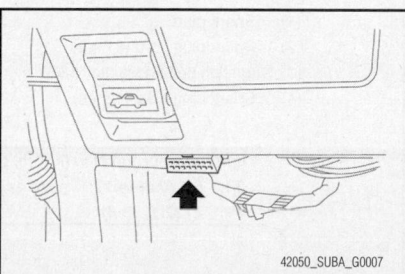

Fig. 59 Location of the data link connector under the driver's side instrument panel.

6. Turn ignition switch to ON, and SUBARU SELECT MONITOR switch to ON.
7. Select "2. Each System Check" in Main Menu.
8. Select "Engine Control System" in Selection Menu.
9. Select "1. Current Data Display & Save" in Engine Control System Diagnosis.
10. Select "1.12 Data Display" in Data Display Menu.
11. Start engine at idle speed and check the ignition timing.

2.5L SOHC Non-Turbo Engine

See Figure 60.

1. Before checking the ignition timing, ensure the air cleaner element is free from debris, the spark plugs are in good condition, and all hoses are connected properly.
2. Set the transmission in Park for automatic transmissions or Neutral for manual transmissions.
3. Set the parking brake, start and run the engine until normal operating temperature is obtained. Then turn the engine **OFF**.
4. Connect a conventional power timing light to the No. 1 cylinder spark plug wire. Start the engine and run at idle.
5. Aim the timing light at the timing mark located near the crankshaft pulley.

ADJUSTMENT

The ignition timing is controlled by the Electronic Control Module (ECM). No adjustment is necessary or possible.

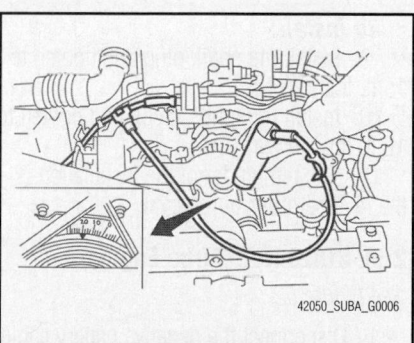

Fig. 60 Using a conventional timing light to check the ignition timing

SPARK PLUGS

REMOVAL & INSTALLATION

2.5L DOHC Turbo Engine

See Figure 61.

1. Disconnect the negative battery cable.

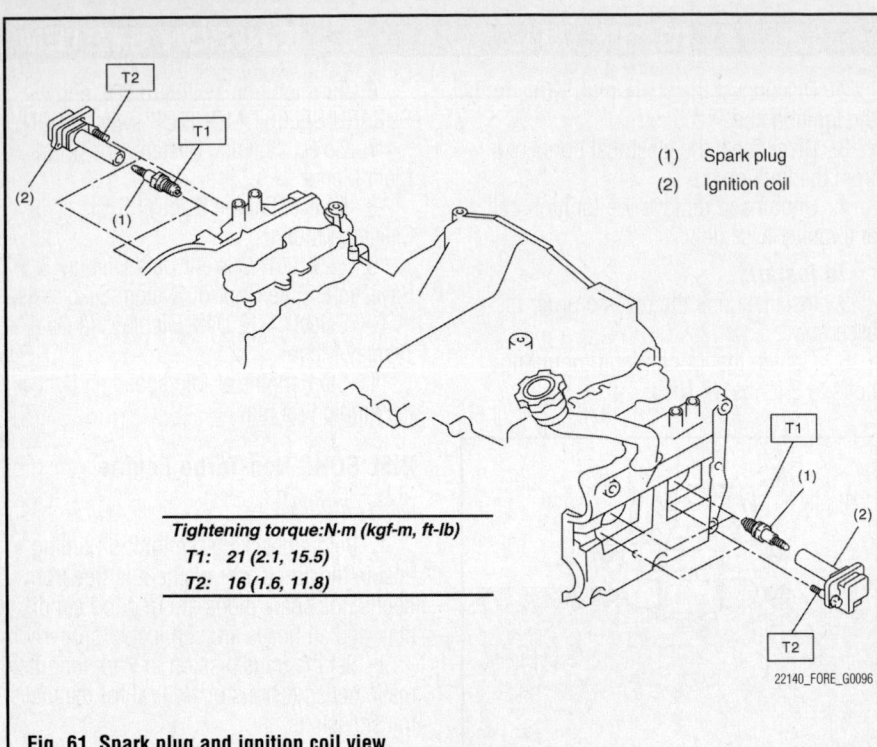

Tightening torque:N·m (kgf-m, ft-lb)
T1: 21 (2.1, 15.5)
T2: 16 (1.6, 11.8)

(1) Spark plug
(2) Ignition coil

22140_FORE_G0096

Fig. 61 Spark plug and ignition coil view

2. Remove the collector cover.

3. Remove the secondary air pump (LH side).

4. Remove the air cleaner case (RH side).

5. Disconnect the electrical connector form the ignition coil.

6. Unbolt and remove the ignition coil.

7. Remove the spark plug with a spark plug socket.

8. Using a spark plug socket with an extension and universal joint, remove the spark plug.

To install:

9. Install the spark plug and tighten to 15 ft. lbs. (20 Nm).

10. Install the ignition coil and tighten to 12 ft. lbs. (16 Nm).

11. The remainder of the installation is the reverse order of removal.

2.5L SOHC Non-Turbo Engine

See Figure 62.

1. Disconnect the negative battery cable.

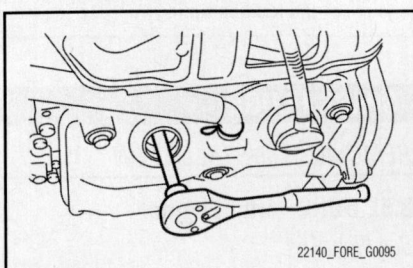

22140_FORE_G0095

Fig. 62 Spark plug removal

2. Disconnect the mass air flow sensor connector.

3. Remove the air intake assembly.

4. Remove the spark plug cable by the grasping the boot firmly.

5. Using a spark plug socket with an extension and universal joint, remove the spark plug.

To install:

6. Install the spark plug and tighten to 15 ft. lbs. (20 Nm).

7. The remainder of the installation is the reverse order of removal.

SPARK PLUG WIRES

REMOVAL & INSTALLATION

See Figure 63.

1. Disconnect the negative battery cable.

2. Disconnect the spark plug wires from the ignition coil and igniter assembly.

3. Grasp the boot firmly and remove the spark plug wire from the spark plug.

4. Installation is the reverse order of removal.

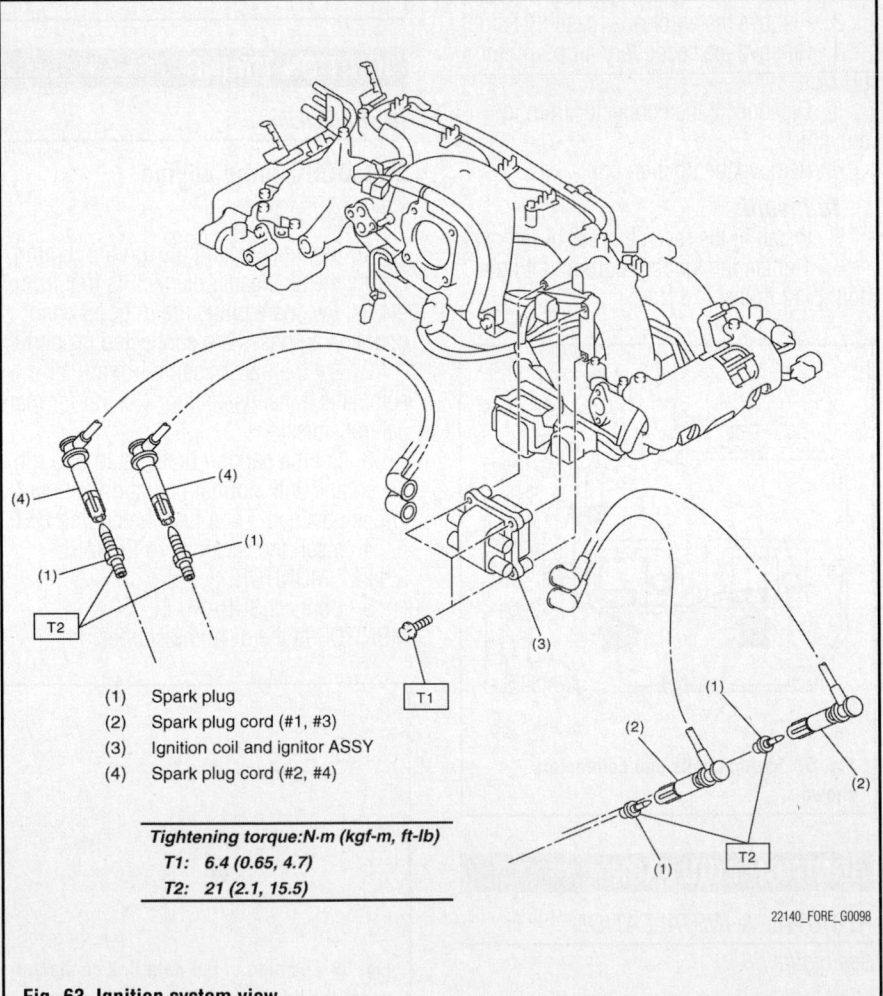

(1) Spark plug
(2) Spark plug cord (#1, #3)
(3) Ignition coil and ignitor ASSY
(4) Spark plug cord (#2, #4)

Tightening torque:N·m (kgf-m, ft-lb)
T1: 6.4 (0.65, 4.7)
T2: 21 (2.1, 15.5)

22140_FORE_G0098

Fig. 63 Ignition system view

STARTER

REMOVAL & INSTALLATION

See Figure 64.

1. Disconnect the negative battery cable.
2. Remove the air intake chamber, if equipped with a non-turbocharged engine.
3. Remove the intercooler, if equipped with a turbocharged engine.
4. Remove the air intake chamber stay, if equipped with a non turbocharged engine.
5. Disconnect the electrical connectors from the starter.
6. Remove the starter retaining bolts.
7. Remove the starter from the vehicle.

To install:
8. Installation is the reverse of the removal procedure.
9. Torque the starter retaining bolts to 37 ft. lbs. (50 Nm).

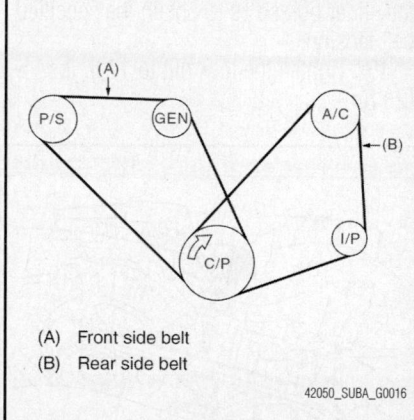

(A) Terminal S
(B) Terminal M

22140_FORE_G0100

Fig. 64 Starter motor and terminals

ENGINE MECHANICAL

ACCESSORY DRIVE BELTS

ACCESSORY BELT ROUTING

See Figure 65.

Refer to the accompanying illustration.

INSPECTION

Inspect the drive belt for signs of glazing or cracking. A glazed belt will be perfectly smooth from slippage, while a good belt will have a slight texture of fabric visible. Cracks will usually start at the inner edge of the belt and run outward. All worn or damaged drive belts should be replaced immediately.

ADJUSTMENT

See Figures 66 and 67.

1. Loosen the lock bolt of the belt you are adjusting. Turn the slider bolt until the correct tension is achieved. Tighten the lock bolt.

The belt tension may be check with or without a belt tension gauge.

2. If using a belt tension gauge, the tension should be as follows:
    a. Used front belt: 110.2–143.9 lbs. (490–640 N)
    b. New front belt: 143–175 lbs. (640–780 N)
    c. Used rear belt: 78.7–101.2 lbs. (350–450 N)
    d. New rear belt: 166–198 lbs. (740–880 N)
3. If you are not using a belt tension gauge, using 22 lbs. (98 Nm) of force, push on the belt as shown in the illustration. The total distance the belt travels up and down should be:
    a. Used front belt: 0.354–0.433 in. (9–11mm)
    b. New front belt: 0.276–0.354 in. (7–9mm)
    c. Used rear belt: 0.354–0.394 in. (9.0–10.0mm)
    d. New rear belt: 0.295–0.335 in. (7.5–8.5mm)

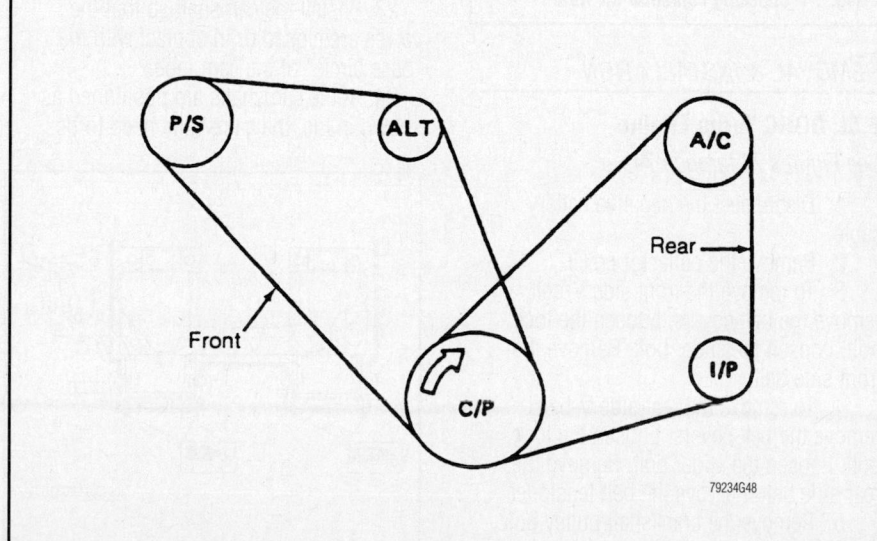

Fig. 65 Accessory drive belt routing—2.5L engines

79234G48

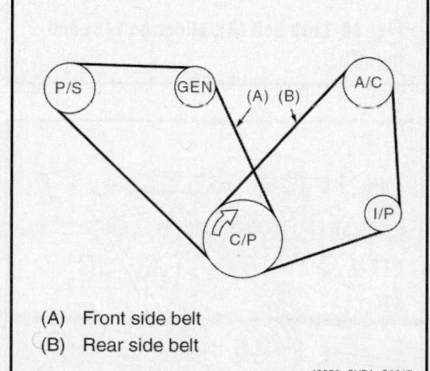

(A) Front side belt
(B) Rear side belt

42050_SUBA_G0016

Fig. 66 Location to check the belt tension if you are using a belt tension gauge

(A) Front side belt
(B) Rear side belt

42050_SUBA_G0017

Fig. 67 Location to check to the belt tension if you are not using a belt tension gauge

REMOVAL & INSTALLATION

See Figures 68 and 69.

FRONT SIDE BELT

1. Remove the accessory drive belt cover, if equipped.
2. Loosen the lock bolt.
3. Loosen the slider bolt.
4. Remove the front side belt.

To install:

5. Install the front side belt, and tighten the slider bolt so as to obtain the specified belt tension.
6. Tighten the lock bolt to 18 ft. lbs. (25 Nm).
7. Tighten the slider bolt 6 ft. lbs. (8 Nm).

REAR SIDE BELT

8. Remove the front side belts.
9. Loosen the lock nut.
10. Loosen the slider bolt.
11. Remove the rear side belt.
12. Remove the belt tensioner.

To install:

13. Install the belt tensioner.
14. Install a rear side belt, and tighten the slider bolt so as to obtain the specified belt tension.
15. Tighten the lock nut to 17 ft. lbs. (23 Nm).

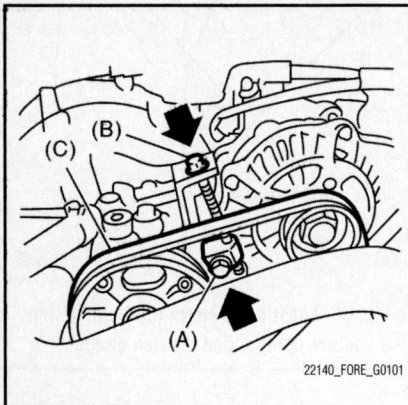

Fig. 68 Lock bolt (A), slider bolt (B) and belt (C)

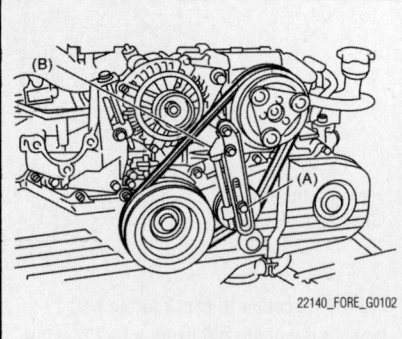

Fig. 69 Lock nut, slider bolt.

CAMSHAFT AND VALVE LIFTERS

INSPECTION

See Figure 70.

1. Remove the camshaft from the engine.
2. Check the camshaft bearing journals for damage and binding.
3. If the journals are binding, check the cylinder head for damage.
4. Check the cylinder head for clogged oil holes.
5. Check the camshaft surface for abnormal wear and damage. Replace the camshaft, as required.
6. Measure the camshaft lobe surface and replace the camshaft if not within specification.
7. Measure the camshaft journal diameter and replace the camshaft if not within specification.
8. Measure the camshaft run out and replace the camshaft if not within specification.

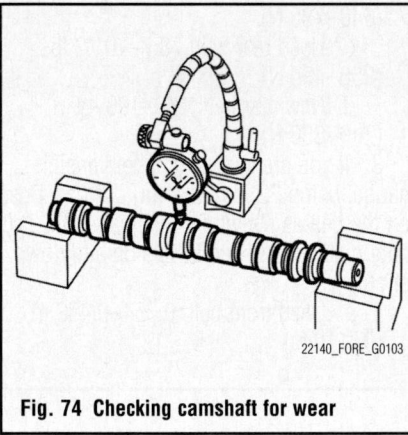

Fig. 74 Checking camshaft for wear

REMOVAL & INSTALLATION

2.5L DOHC Turbo Engine

See Figures 71 through 74.

1. Disconnect the negative battery cable.
2. Remove the collector cover.
3. To remove the front side V-belt, remove the belt covers. Loosen the lock bolt. Loosen the slider bolt. Remove the front side belt.
4. To remove the rear side V-belt, remove the belt covers. Loosen the lock bolt. Loosen the slider bolt. Remove the rear side belt. Remove the belt tensioner.
5. Remove the crankshaft pulley bolt.
6. Lock the crankshaft in place using tool ST499977100, or equivalent.

7. Remove the crankshaft pulley.
8. Remove the left side timing belt cover.
9. Remove the right side timing belt cover.
10. Remove the front timing belt cover.
11. Remove the timing belt.
12. Remove the camshaft position sensor.
13. Remove the camshaft sprockets.

➡ **Be sure to lock the camshaft in place using tool ST499207400, or equivalent.**

14. Lock the crankshaft in place using tool ST499977100, or equivalent.
15. Remove the crankshaft pulley.
16. Remove the tensioner bracket. Remove the right and left timing belt No.2 covers.
17. Remove the spark plug wires. Remove the oil level gauge, on the left side.
18. Remove the rocker cover and gasket. Remove the oil pipe.
19. Loosen the oil flow control solenoid valve assembly and the intake camshaft cap bolts equally and in the proper sequence.
20. Loosen the exhaust camshaft cap bolts equally and in the proper sequence.
21. Remove the oil flow control solenoid valve assembly, intake camshaft cap and camshaft.
22. Remove the exhaust camshaft caps and camshaft.

➡ **Arrange the camshafts caps so that they can be installed in their original positions.**

To install:

➡ **Lubricate the camshaft journals with clean engine oil prior to installation.**

23. Install the camshaft so that the valves are closed or in contact with the "base circle" of the cam lobe.
24. If the camshafts are positioned as shown in the, the camshafts need to be

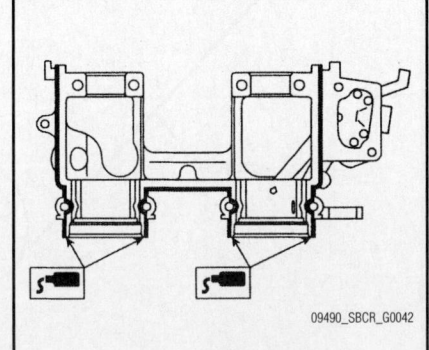

Fig. 71 Camshaft cap liquid gasket application points DOHC engine

rotated at a minimum to align the timing belt during installation.

25. The right hand camshaft need not be rotated when set at the position illustrated, the left hand intake camshaft should be rotated 80 degrees clockwise. The left hand exhaust camshaft should be rotated 45 degrees counterclockwise.

26. To install the camshaft cap and oil flow control solenoid valve, apply a small amount of liquid gasket to the mating surface of the cap. Do not apply an excessive amount of sealant, as it will squish out and flow toward the seal resulting in an oil leak.

27. Apply a thin coat of engine oil to the cap bearing surface and install the cap according to the cap identification mark. Gradually tighten the cap bolts in two stages, first to 7.2 ft. lbs. (9.75 Nm). and then to 14.8 ft. lbs. (20 Nm) in the proper sequence.

➡ **After tightening the camshaft cap, ensure that the camshaft rotates slightly while holding it at base circle.**

28. Using tools ST49587600 and ST499597200, install a new seal on the camshaft. Be sure to coat the seal with clean engine oil before installation. Use tool ST499587700 and install the plug.

29. Install a new gasket on the rocker cover. Apply liquid gasket to the cylinder head (see illustration).

➡ **Apply an extra amount of liquid gasket around the semicircular plugs, 5mm or more.**

30. Temporarily tighten the rocker cover retaining bolts, in the proper sequence, and then tighten to 4.7 ft. lbs. Install the oil pipe.

31. Continue the installation in the reverse order of the removal procedure.

2.5L SOHC Non-Turbo Engine

See Figures 75 through 82.

1. Disconnect the negative battery cable.
2. To remove the front side V-belt, remove the belt covers. Loosen the lock bolt. Loosen the slider bolt. Remove the front side belt.

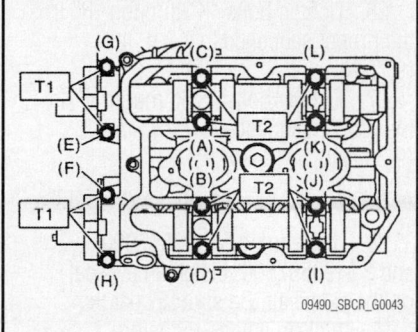

Fig. 72 Camshaft bolt tightening sequence DOHC engine

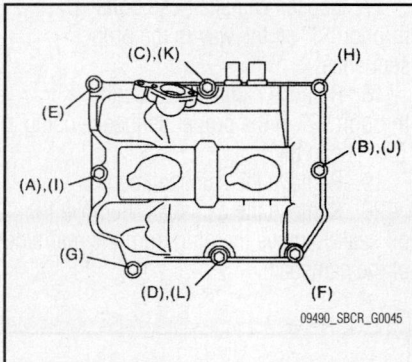

Fig. 74 Rocker cover tightening sequence DOHC engine

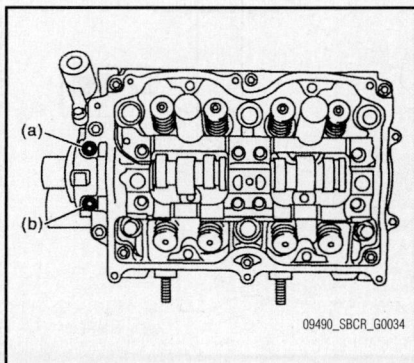

Fig. 75 Camshaft bolt loosening sequence SOHC engine

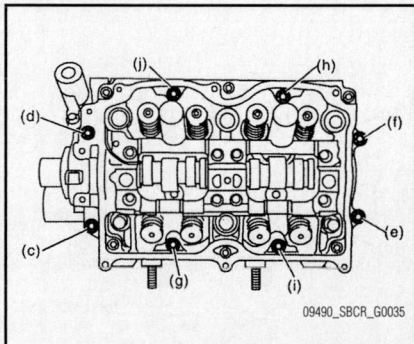

Fig. 76 Camshaft bolt loosening sequence SOHC engine

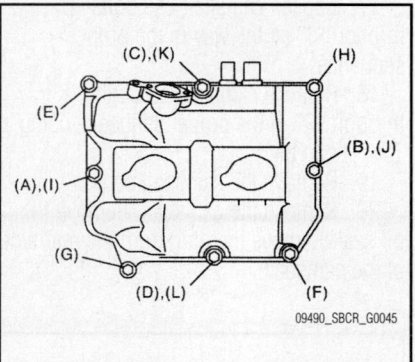

Fig. 73 Rocker cover liquid gasket application points DOHC engine

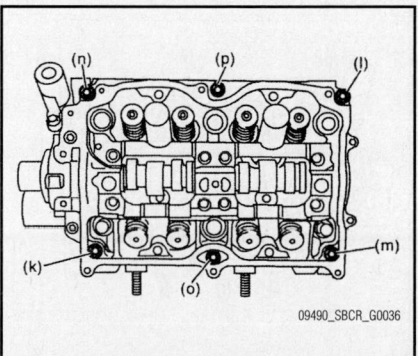

Fig. 77 Camshaft bolt loosening sequence SOHC engine

3. To remove the rear side V-belt, remove the belt covers. Loosen the lock bolt. Loosen the slider bolt. Remove the rear side belt. Remove the belt tensioner.

4. Remove the crankshaft pulley bolt.

5. Lock the crankshaft in place using tool ST499977100, or equivalent.

6. Remove the crankshaft pulley.

7. Remove the left side timing belt cover.

8. Remove the front timing belt cover.

9. Remove the timing belt.

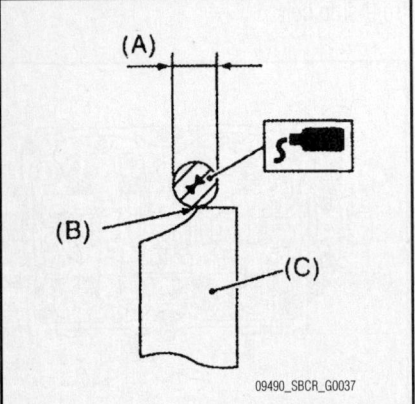

Fig. 78 Camshaft cap sealant application SOHC engine

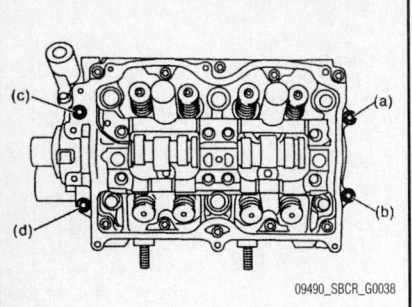

Fig. 79 Camshaft bolt tightening sequence SOHC engine

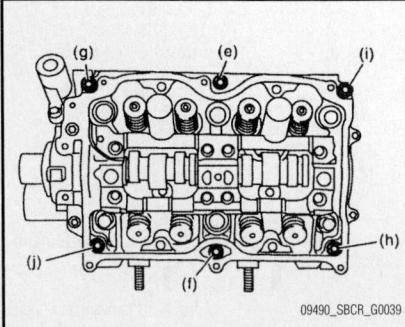

Fig. 80 Camshaft bolt tightening sequence SOHC engine

10. Remove the camshaft position sensor.

11. Remove the camshaft sprockets.

➡**Be sure to lock the camshaft in place using tool ST18231AA010, or equivalent.**

12. Remove the left and right timing belt N02 covers.

➡**Do not damage or lose the rubber seal when removing the covers.**

13. Remove the tensioner bracket. Remove the camshaft position sensor support, on the left side.

14. Remove the oil level gauge guide, on the left side.

15. Remove the valve rocker arm assembly.

16. Remove camshaft cap retaining bolts "A" and "B" in the proper sequence.

17. Loosen camshaft cap bolts "C" through "J" all the way in the proper sequence.

18. Remove camshaft cap bolts "K" through "P" in the proper sequence using a Torx® head bit.

19. Remove the camshaft caps.

20. Remove the camshaft. Remove the oil seal. Remove the plug from the rear side of the camshaft.

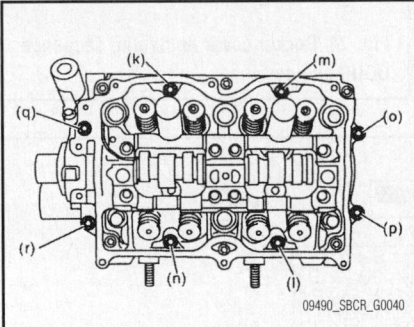

Fig. 81 Camshaft bolt tightening sequence SOHC engine

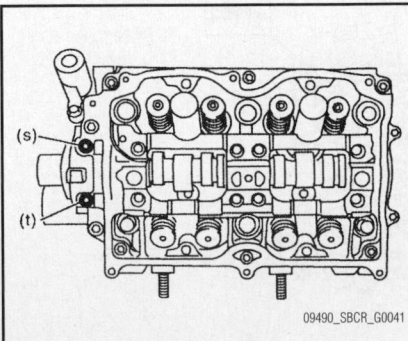

Fig. 82 Camshaft bolt tightening sequence SOHC engine

➡**Do not remove the oil seal unless necessary.**

To install:

➡**Lubricate the camshaft journals with clean engine oil prior to installation.**

21. Install the camshaft into the cylinder head.

22. Apply liquid gasket to the mating surfaces of the camshaft cap.

23. Apply a bead of sealant (0.12 inch in diameter) along the edge of the camshaft cap mating surface. Install with 20 minutes after applying the sealant.

24. Temporarily tighten the bolts "A" through "D" in the proper sequence.

25. Install the valve rocker arm assembly. Tighten Torx® head bolts "E" through "J" in the proper sequence to 13 ft. lbs. (18 Nm).

26. Tighten bolts "K" through "R" in the proper sequence to 7.2 ft. lbs. (9.75 Nm).

27. Tighten bolts "S" through "T" in the proper sequence to 7.2 ft. lbs. (9.75 Nm).

➡**Be sure to use a new seal washer.**

28. Using tools ST499597000 and ST499587500, install a new seal on the camshaft. Be sure to coat the seal with clean engine oil before installation. Use tool ST499587700 and install the plug.

29. Adjust the valve clearance.

30. Continue the installation in the reverse order of the removal procedure.

CATALYTIC CONVERTER

REMOVAL & INSTALLATION

Front

1. Remove the center exhaust pipe.

2. Disconnect the front catalytic converter from the rear catalytic converter.

To install:

3. Connect the front catalytic converter from the rear catalytic converter.

4. Install the center exhaust pipe.

Rear

1. Remove the center exhaust pipe.

2. Disconnect the rear catalytic converter from the front catalytic converter.

To install:

3. Connect the rear catalytic converter from the front catalytic converter.

4. Install the center exhaust pipe.

CRANKSHAFT DAMPER

REMOVAL & INSTALLATION

See Figure 84.

1. Remove the accessory drive belt. Refer to "Accessory Drive Belts, Removal & Installation".

2. Using Special Tool 499977100 to lock the crankshaft in place, remove the crankshaft pulley bolt.

3. Remove the crankshaft pulley.

To install:

4. Clean the crankshaft pulley bolt threads with an air gun. Apply clean engine oil to the crankshaft pulley bolt seat and threads.

5. Install the crankshaft pulley and pulley bolt, and temporarily tighten the pulley bolt.

6. Lock the crankshaft in place with Special Tool 499977100 and tighten the crankshaft pulley bolt to 132 ft. lbs. (180 Nm).

7. The remainder of the installation is the reverse order of removal.

Fig. 84 Use Special Tool 49997700 to lock the crankshaft pulley into place.

CYLINDER HEAD

REMOVAL & INSTALLATION

2.5L DOHC Turbo Engine

See Figures 85 through 87.

1. Before servicing the vehicle, refer to the Precautions section.

2. Properly relieve the fuel system pressure.

3. Remove or disconnect the following:

4. Disconnect the negative battery cable.

5. Remove the drive belts.

6. Remove the crankshaft pulley.

7. Remove the timing belt cover, timing belt assembly and camshaft sprockets.

8. Remove the intake manifold.

9. Remove the bolt attaching the A/C compressor bracket to the head.

10. Remove the camshaft.

11. Remove the bolts in the sequence illustrated. Leave bolts **A** and **D** installed loosely to prevent the cylinder head from falling.

12. Separate the cylinder head from the block. Use a plastic-faced hammer, if needed.

13. Remove bolts A and D. Remove the cylinder head and gasket.

14. Clean all gasket material from both mating surfaces.

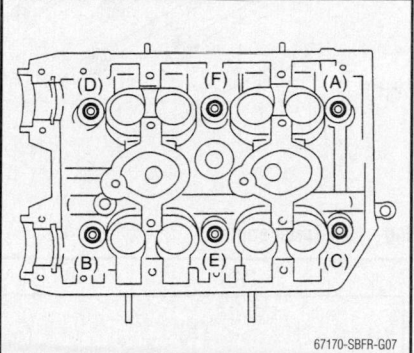

Fig. 85 Cylinder head bolt loosening sequence–Turbo engines

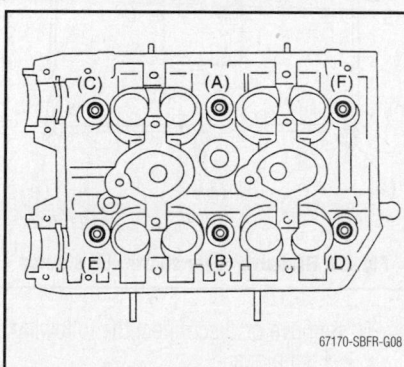

Fig. 86 Cylinder head torque sequence— Turbo engines

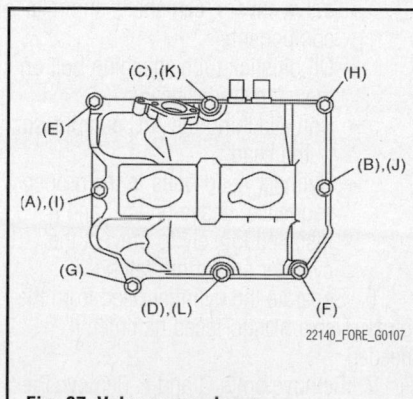

Fig. 87 Valve cover shown

To install:

15. Inspect the cylinder head for warpage. Warpage should not exceed 0.0020 in. (0.05mm).

16. Install the cylinder head(s) on the block using new gaskets. Secure in place with the mounting bolts. Coat each bolts with clean engine oil, and hand-tighten.

17. Tighten the cylinder head bolts as follows:

   a. Step 1: Torque all the bolts in sequence to 22 ft. lbs. (29 Nm).

   b. Step 2: Torque all the bolts in sequence to 51 ft. lbs. (69 Nm).

   c. Step 3: Loosen all bolts in sequence by 180 degrees, then loosen an additional 180 degrees.

   d. Step 4: Tighten all bolts in sequence to 36 ft. lbs. (49 Nm).

   e. Step 5: Tighten all bolts in sequence an additional 80–90 degrees.

   f. Step 6: All bolts: turn an additional 40–45 degrees.

   g. Step 7: Bolts **A** and **B** an additional 40–45 degrees.

** WARNING

Do not exceed 90 degrees total tightening in the previous two steps.

18. Temporarily tighten the valve cover tightening bolts in alphabetical sequence shown in the figure, and then tighten to 4.7 ft. lbs. (6.4 Nm) in alphabetical sequence.

19. Install the remaining components in the reverse order of removal.

20. Start the engine and allow it to reach operating temperature.

21. Check for leaks.

2.5L SOHC Non-Turbo Engine

See Figures 88 through 91.

1. Before servicing the vehicle, refer to the Precautions section.

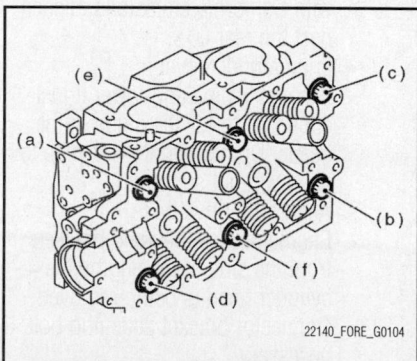

Fig. 88 Cylinder head bolt loosening sequence—Non-Turbo engine

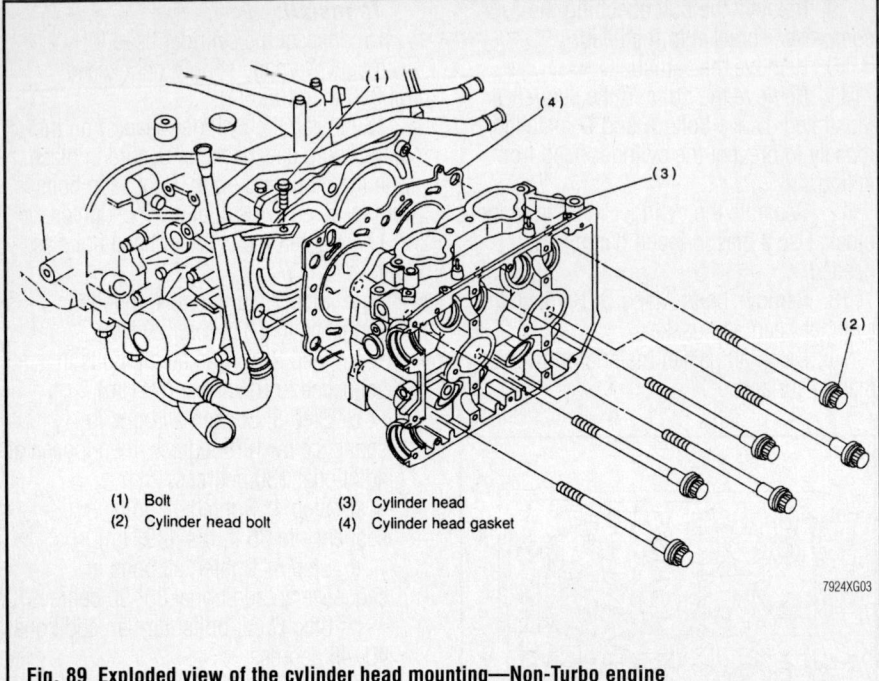

(1) Bolt
(2) Cylinder head bolt
(3) Cylinder head
(4) Cylinder head gasket

7924XG03

Fig. 89 Exploded view of the cylinder head mounting—Non-Turbo engine

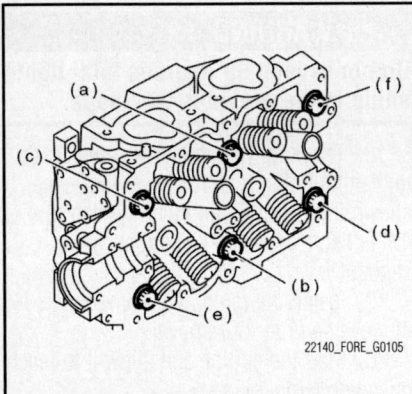

22140_FORE_G0105

Fig. 90 Cylinder head torque sequence—Non-Turbo engine

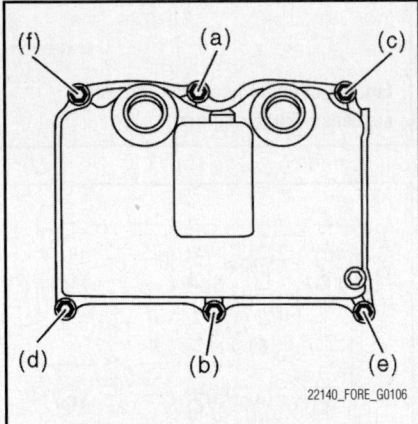

22140_FORE_G0106

Fig. 91 RH valve cover shown, LH similar

2. Properly relieve the fuel system pressure.

3. Remove or disconnect the following:
- Negative battery cable
- Oxygen Sensor (O₂S). If equipped with California emissions, disconnect the rear O₂S.
- Engine undercover
- Exhaust Y-pipe and lower it just enough to clear the studs in the heads. Do not allow the Y-pipe to hang without support.
- Accessory drive belts
- Engine accessories and brackets from the side of the engine the cylinder head is being removed
- Connector bracket attaching bolt, if necessary

4. On the left cylinder head, remove the CMP sensor.

5. Remove or disconnect the following:
- Fuel pipes
- Intake manifold and gasket
- Timing belt, camshaft sprockets, and related components
- Valve covers, camshafts and related components
- Oil dipstick tube attaching bolt on the left cylinder head
- Bolt attaching the A/C compressor to the head
- Cylinder head bolts in the proper sequence. Leave bolts C and F installed loosely to prevent the cylinder head from falling.

6. Separate the cylinder head from the block. Use a plastic-faced hammer, if needed.

7. Remove bolts C and F. Remove the cylinder head and gasket.

8. Clean all gasket material from both mating surfaces.

To install:

9. Inspect the cylinder head for warpage. Warpage should not exceed 0.0020 in. (0.05mm).

10. Install the cylinder head(s) on the block using new gaskets. Secure in place with the mounting bolts. Coat each bolts with clean engine oil, and hand-tighten.

11. Tighten the cylinder head bolts as follows:

   a. Step 1: Torque all the bolts in sequence to 22 ft. lbs. (29 Nm).

   b. Step 2: Torque all the bolts in sequence to 51 ft. lbs. (69 Nm).

   c. Step 3: Loosen all bolts in sequence by 180 degrees, then loosen an additional 180 degrees.

   d. Step 4: Tighten all bolts in sequence: 40–45 degrees.

✳✳ **CAUTION**

Do not tighten the bolts more that 45 degrees in step 4.

   e. Step 5: Tighten bolts A and B in sequence: 40–45 degrees.

✳✳ **WARNING**

Do not exceed 90 degrees total tightening.

12. Install or connect the following:
- Oil dipstick tube attaching bolt on the left cylinder head
- Camshafts and related components
- Camshaft sprocket, timing belt, and related components
- Intake manifold
- Fuel delivery pipes

13. On the left cylinder head, install the CMP sensor.

14. Tighten the bolts to the valve covers in two stages in alphabetical sequence as shown in figure:
- First stage: 4.7 ft. lbs. (6.4 Nm).
- Second stage: 4.7 ft. lbs. (6.4 Nm).

15. Install or connect the following:
- Connector bracket attaching bolt
- Spark plug wires
- Engine accessories and brackets
- Accessory drive belts
- Exhaust Y-pipe. Torque the fasteners to 19–26 ft. lbs. (25–35 Nm).
- Engine undercover
- Front O₂S and rear O₂S, if removed
- Negative battery cable

16. Start the engine and allow it to reach operating temperature.

17. Check for leaks.

EXHAUST MANIFOLD

REMOVAL & INSTALLATION

2.5L DOHC Turbo Engine

See Figures 92 and 93.

Due to the unique design of the Subaru engine an exhaust manifold is not used. The exhaust enters directly into the front Y-pipe.

✳✳ CAUTION

The exhaust pipe may be hot; DO NOT perform any work until the system has completely cooled.

1. Before servicing the vehicle, refer to the Precautions section.
2. Raise and safely support the vehicle.

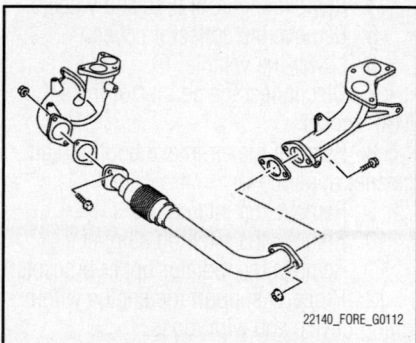

Fig. 92 Front exhaust pipe and the exhaust manifold

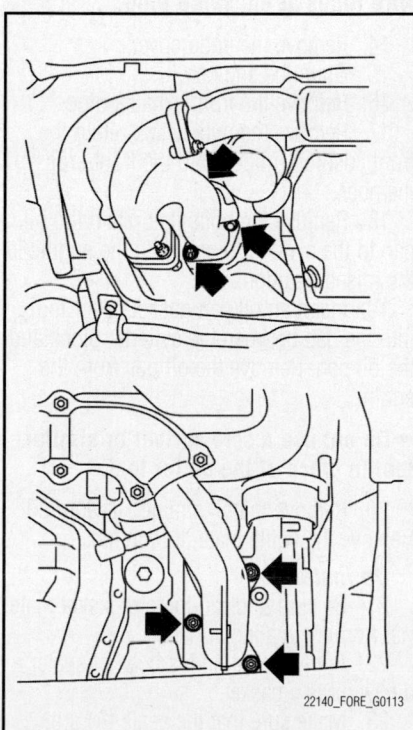

Fig. 93 Front exhaust pipe assembly

3. Disconnect the ground cable from battery.
4. Remove the collector cover.
5. Remove the front oxygen (A/F) sensor.
6. Remove the undercover.
7. Remove the exhaust manifold lower cover (RH) and the exhaust manifold cover (LH).
8. Remove the bolts and nuts which hold front exhaust pipe assembly onto turbocharger joint pipe.
9. While holding the front exhaust pipe assembly with one hand, remove the nuts which hold the front exhaust pipe assembly to cylinder head exhaust port.
10. Remove the front exhaust pipe assembly.
11. Remove the covers from exhaust manifold and front exhaust pipe.
12. Separate the front exhaust pipe from exhaust manifolds.

To install:
13. Assemble the front exhaust pipe and the exhaust manifold. Use a new gasket and tighten to 31.3 ft. lbs. (42.5 Nm).
14. Install the front exhaust pipe cover. Tighten to 5.5 ft. lbs. (7.5 Nm).
15. Install the exhaust manifold upper cover (RH). Tighten to 14 ft. lbs. (19 Nm).
16. Install the front exhaust pipe assembly. Use a new gasket and tighten to 28.9 ft. lbs. (40 Nm).
17. Connect the exhaust manifold (RH) to turbocharger joint pipe. Use a new gasket and tighten to 31.3 ft. lbs. (42.5 Nm).
18. Install the exhaust manifold lower cover (RH) and the exhaust manifold cover (LH).
19. Install the front oxygen (A/F) sensor.
20. Install the undercover.
21. Lower the vehicle.
22. Connect the ground cable to battery.
23. Install the collector cover.

2.5L SOHC Non-Turbo Engine

See Figure 94.

Due to the unique design of the Subaru engine an exhaust manifold is not used. The exhaust enters directly into the front Y-pipe.

✳✳ CAUTION

The exhaust pipe may be hot; DO NOT perform any work until the system has completely cooled.

1. Before servicing the vehicle, refer to the Precautions section.
2. Disconnect the ground cable from battery.
3. Unclip the clip fastening the harness and disconnect the following connectors.

a. Front oxygen (A/F) sensor connector.
b. Rear oxygen sensor connector.
4. Lift-up the vehicle.
5. Separate the center exhaust pipe from rear exhaust pipe.
6. Remove the undercover.
7. Remove the nuts which hold front exhaust pipe onto cylinder heads.

✳✳ CAUTION

Be careful not to pull down the front and center exhaust pipe assembly.

8. Remove the bolt which holds center exhaust pipe to hanger bracket.
9. Remove the front and center exhaust pipe assembly from the vehicle.
10. Separate the front exhaust pipe from center exhaust pipe.
11. Remove the front oxygen (A/F) sensor and rear oxygen sensor.

To install:
12. Install the front oxygen (A/F) sensor and rear oxygen sensor to the front exhaust pipe
13. Install the front exhaust pipe to center exhaust pipe. Use a new gasket and tighten to 30 ft. lbs. (40 Nm).
14. Install the front and center exhaust pipe assembly to the vehicle.
15. Temporarily tighten the nuts which hold front exhaust pipe to cylinder heads.
16. Tighten the bolt which holds the center exhaust pipe to exhaust pipe. Use a new gasket and tighten to 13 ft. lbs. (18 Nm).
17. Install the bolt which holds center exhaust pipe to hanger bracket. Tighten to 26 ft. lbs. (35 Nm).
18. Install the nuts which hold front exhaust pipe to cylinder heads.
19. Use a new gasket and tighten to 22.4 ft. lbs. (30 Nm).
20. Install the undercover.
21. Lower the vehicle.

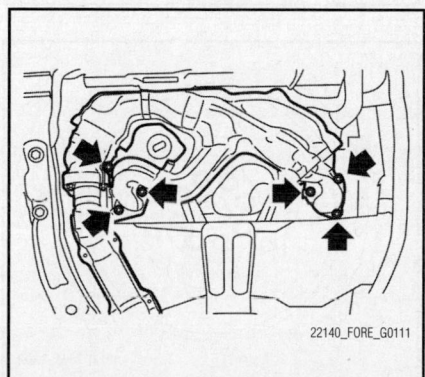

Fig. 94 Exhaust pipe to cylinder heads

22. Connect the following connectors.
  a. Front oxygen (A/F) sensor connector
  b. Rear oxygen sensor connector
23. Connect the ground cable to battery.

FLYWHEEL

REMOVAL & INSTALLATION

See Figures 95 and 96.

1. Remove the transmission assembly.
2. If equipped with a manual transmission, remove the clutch housing cover.
3. Using Special Tool 498497100 to lock the crankshaft into place, remove the flywheel (manual transmission) or driveplate (automatic transmission).

To install:

4. Install the flywheel (M/T) or drive plate (A/T) and use Special Tool 498497100 to lock the crankshaft into place.

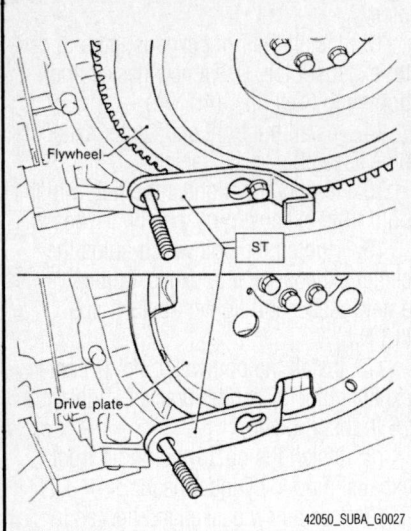

42050_SUBA_G0027

Fig. 95 Use Special Tool 498497100 to lock the crankshaft into place and remove the flywheel

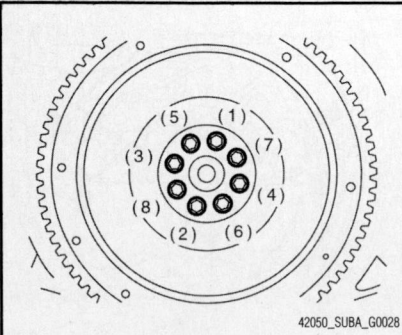

42050_SUBA_G0028

Fig. 96 Flywheel torque sequence

5. Tighten the bolts in the sequence shown to 52.8 ft. lbs. (72 Nm).
6. Install the clutch housing cover, if equipped with manual transmission.
7. Install the transmission assembly.

INTAKE MANIFOLD

REMOVAL & INSTALLATION

See Figures 97 and 98.

1. Before servicing the vehicle, refer to the Precautions Section.
2. Properly relieve the fuel system pressure. Remove the fuel cap.
3. Disconnect the negative battery cable. Drain the engine coolant.
4. If equipped, remove the undercover.
5. Remove the air intake duct, air cleaner case and air intake chamber.
6. If equipped, remove the intercooler.
7. Remove the alternator.
8. If equipped with DOHC engine, remove the coolant filler tank.
9. If equipped with SOHC engine, remove the spark plug wires.
10. Disconnect the engine coolant hoses from the throttle body. Disconnect the brake booster hose.
11. Disconnect the PCV hose from the intake manifold. Disconnect the engine harness electrical connectors from the bulkhead harness connectors.
12. Disconnect the engine coolant temperature sensor electrical connector, knock sensor electrical connector and crankshaft position sensor connector.
13. Disconnect the power steering pump switch electrical connector, oil pressure switch connector and camshaft sensor connector.
14. If equipped with DOHC engine, disconnect the oil flow solenoid valve electrical connector and the ignition coil connector.
15. If equipped with SOHC engine, remove the EGR pipe from the intake manifold and disconnect the fuel lines from the fuel pipe.
16. If equipped with DOHC engine, disconnect the fuel delivery hose, return hose and evaporation hose.
17. Remove the intake manifold retaining bolts. Remove the intake manifold from the engine.

To install:

18. Installation is the reverse of the removal procedure.
19. Be sure to use new intake manifold gaskets. Torque the manifold retaining

bolts to specification and in alternating sequence.
20. Be sure to fill the cooling system with the proper grade and type engine coolant.
21. Start the engine and check for leaks, correct as required.

OIL PAN

REMOVAL & INSTALLATION

2.5L DOHC Turbo Engine

1. Before servicing the vehicle, refer to the Precautions Section.
2. Disconnect the negative battery cable.
3. Raise and support the vehicle safely.
4. Remove the front tires and wheels.
5. Remove the collector cover.
6. Lower the vehicle.
7. Disconnect the connector from the MAF sensor.
8. Remove the air intake boot and air cleaner upper cover.
9. Remove the intercooler.
10. Remove the pitching stopper.
11. Remove the radiator upper brackets.
12. Properly support the engine with a lifting device and wire ropes.
13. Lift the vehicle and support it safely.

➡**When lifting the vehicle, raise the wire ropes at the same time.**

14. Remove the undercover.
15. Drain the engine oil.
16. Remove the front exhaust pipe.
17. Remove the nuts which retain the front cushion rubber onto the front crossmember.
18. Remove the bolts that retain the oil pan to the cylinder block, with the engine in the raised position.
19. Insert an oil pan gasket cutter tool into the gap between the cylinder block and the oil pan. Remove the oil pan from the engine.

➡**Do not use a screwdriver or similar tool in place of the cutter tool.**

20. Remove the oil strainer, if required. Remove the baffle plate, if required.

To install:

21. Be sure to clean the old gasket material from the mating surfaces.
22. Apply a continuous bead of sealer to a new oil pan gasket.
23. Make sure that the seals (A) are installed securely on the baffle plate and in

the direction shown in the illustration. Install the baffle plate; tighten the retaining bolts to 4.7 ft. lbs. (6.4 Nm).

24. Replace the O-ring and install the oil strainer. Tighten the bolt to 7 ft. lbs. (10 Nm)

25. Apply liquid gasket, part number 004403012 or equivalent, to the oil pan mating surface. Install the oil pan. Torque the retaining bolts to specification.

26. Continue the installation in the reverse order of the removal procedure.

27. Torque the front cushion mounting bolts to 63 ft. lbs. (85 Nm).

28. Be sure to fill the engine with the correct grade and type engine oil.

29. Start the engine and check for leaks. Correct as required.

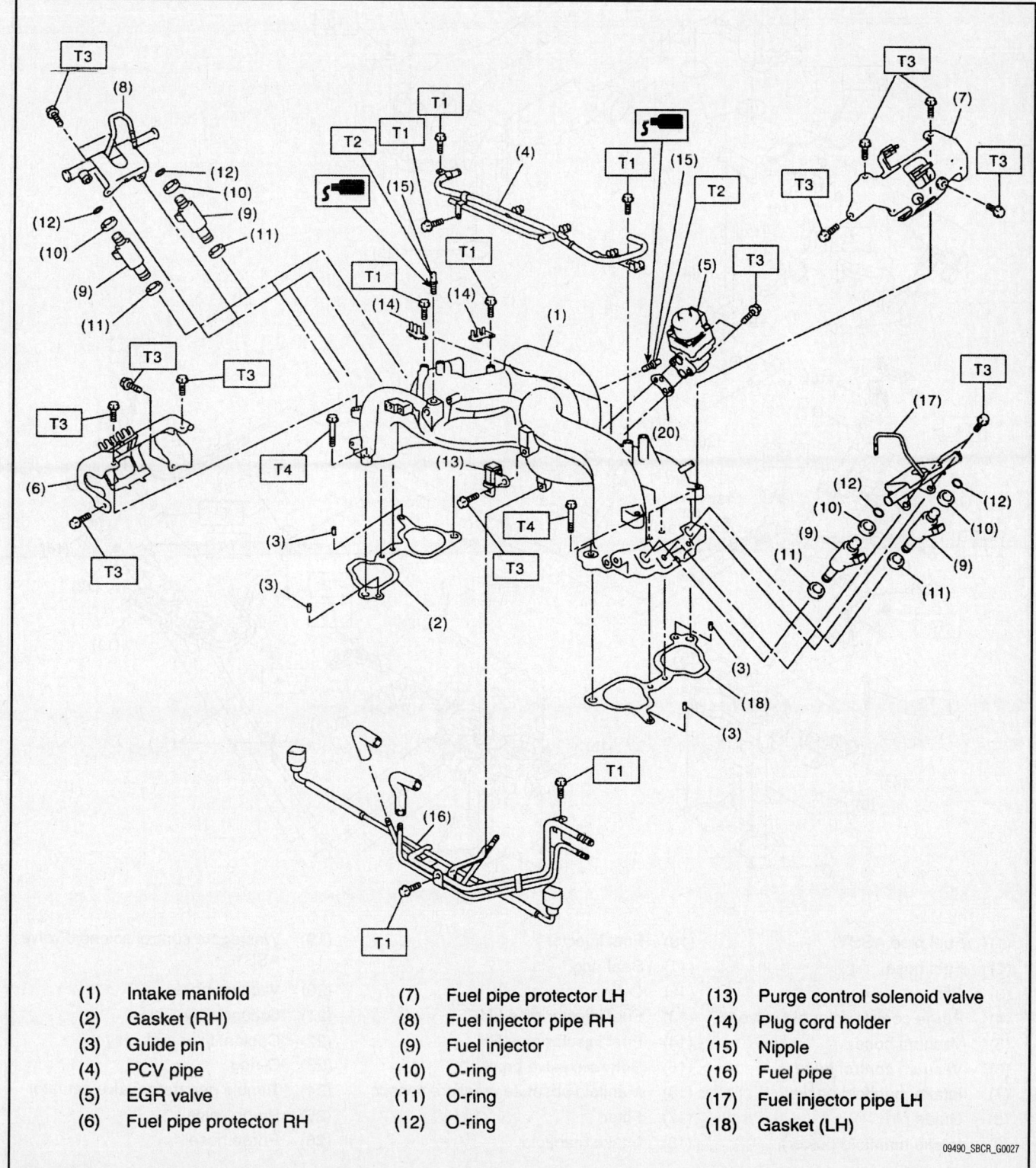

(1)	Intake manifold	(7)	Fuel pipe protector LH	(13)	Purge control solenoid valve
(2)	Gasket (RH)	(8)	Fuel injector pipe RH	(14)	Plug cord holder
(3)	Guide pin	(9)	Fuel injector	(15)	Nipple
(4)	PCV pipe	(10)	O-ring	(16)	Fuel pipe
(5)	EGR valve	(11)	O-ring	(17)	Fuel injector pipe LH
(6)	Fuel pipe protector RH	(12)	O-ring	(18)	Gasket (LH)

09490_SBCR_G0027

Fig. 97 Intake manifold and related components SOHC (Non-Turbo) engine

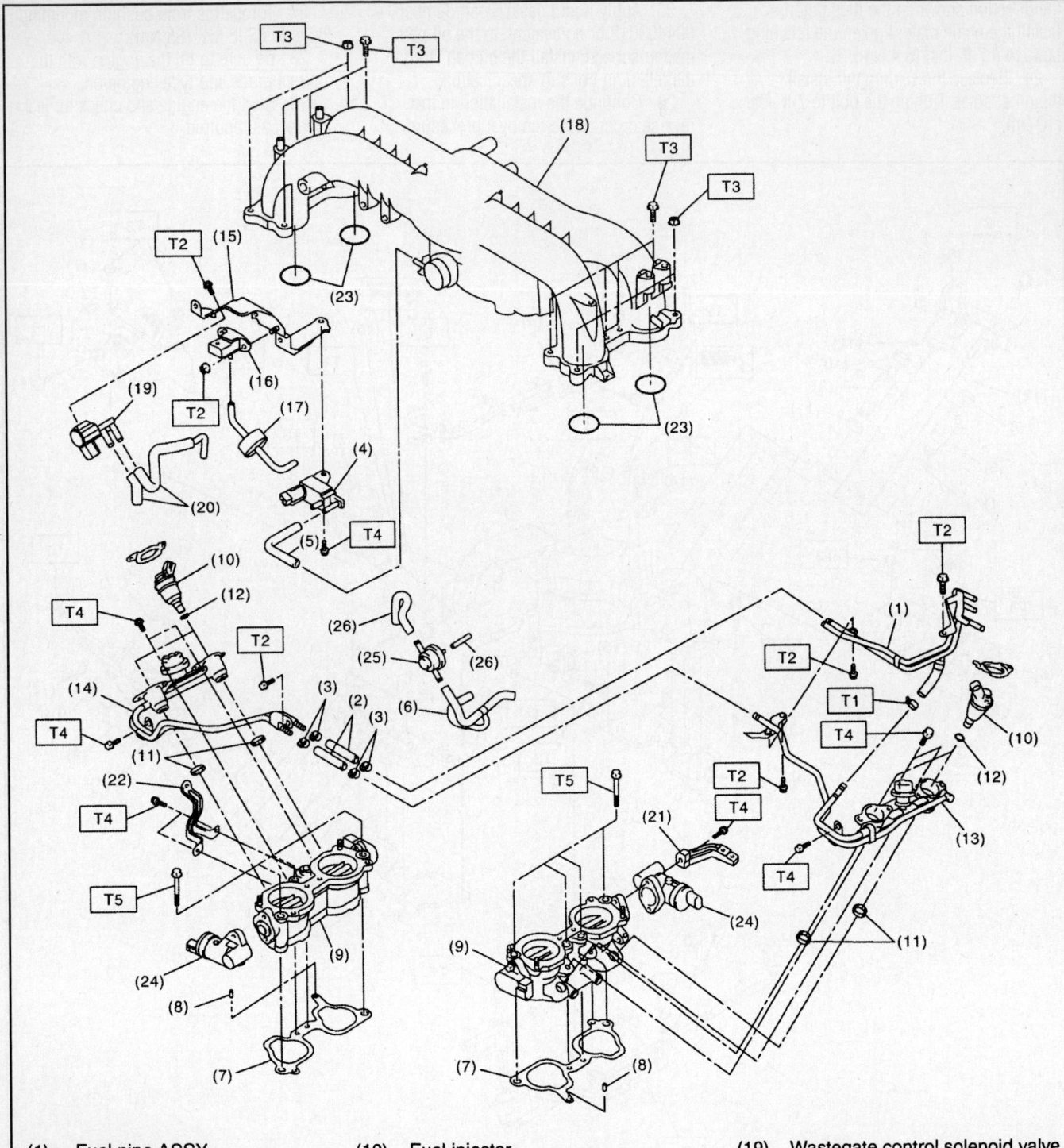

(1) Fuel pipe ASSY	(10) Fuel injector	(19) Wastegate control solenoid valve ASSY
(2) Fuel hose	(11) Seal ring	
(3) Clip	(12) O-ring	(20) Vacuum hose
(4) Purge control solenoid valve	(13) Fuel injector pipe LH	(21) Ground stay
(5) Vacuum hose	(14) Fuel injector pipe RH	(22) Coolant filler tank stay
(6) Vacuum control hose	(15) Solenoid valve bracket	(23) O-ring
(7) Intake manifold gasket	(16) Manifold absolute pressure sensor	(24) Tumble generator valve actuator
(8) Guide pin	(17) Filter	(25) Purge valve
(9) Intake manifold (lower)	(18) Intake manifold	(26) Purge hose

09490_SBCR_G0028

Fig. 98 Intake manifold and related components DOHC (Turbo) engine

2.5L SOHC Non-Turbo Engine

See Figure 99.

1. Before servicing the vehicle, refer to the Precautions Section.
2. Disconnect the negative battery cable.
3. Raise and support the vehicle safely.
4. Remove the front tires and wheels.
5. Lower the vehicle.
6. Remove the air intake duct and the air cleaner case. Remove the air intake chamber.
7. Remove the pitching stopper.
8. Remove the hood stay holder and the radiator upper brackets.
9. Properly support the engine with a lifting device and wire ropes.
10. Lift the vehicle and support it safely.

➡ **When lifting the vehicle, raise the wire ropes at the same time.**

11. Remove the undercover.
12. Drain the engine oil.
13. Remove the front and center exhaust pipes.
14. Remove the nuts which retain the front cushion rubber onto the front crossmember.
15. Remove the bolts that retain the oil pan to the cylinder block, with the engine in the raised position.
16. Insert an oil pan gasket cutter tool into the gap between the cylinder block and the oil pan. Remove the oil pan from the engine.

➡ **Do not use a screwdriver or similar tool in place of the cutter tool.**

17. Remove the oil strainer, if required. Remove the baffle plate, if required.

To install:

18. Be sure to clean the old gasket material from the mating surfaces.
19. Apply a continuous bead of sealer to a new oil pan gasket.
20. Make sure that the seals (A) are installed securely on the baffle plate and in the direction shown in the illustration. Install the baffle plate; tighten the retaining bolts to 4.7 ft. lbs. (6.4 Nm).

21. Replace the O-ring and install the oil strainer. Tighten the bolt to 7 ft. lbs. (10 Nm).
22. Apply liquid gasket, part number 004403012 or equivalent, to the oil pan mating surface. Install the oil pan. Torque the retaining bolts to specification.
23. Continue the installation in the reverse order of the removal procedure.
24. Torque the front cushion mounting bolts to 63 ft. lbs.
25. Be sure to fill the engine with the correct grade and type engine oil.
26. Start the engine and check for leaks. Correct as required.

OIL PUMP

REMOVAL & INSTALLATION

See Figures 100 and 101.

1. Before servicing the vehicle, refer to the Precautions Section.
2. Disconnect the negative battery cable. Drain the cooling system.
3. On the DOHC (Turbo) engine, remove the collector cover.
4. Raise and support the vehicle safely.
5. Remove the undercover.
6. On the DOHC engine, remove the bolts which retain the water pipe of the oil cooler to the oil pump. Remove the water pipe and hoses between the oil cooler and the water pump.
7. Lower the vehicle.
8. Remove the radiator.
9. To remove the front side V-belt, remove the belt covers. Loosen the lock bolt. Loosen the slider bolt. Remove the front side belt.
10. To remove the rear side V-belt, remove the belt covers. Loosen the lock bolt. Loosen the slider bolt. Remove the rear side belt.
11. On the SOHC engine remove the belt tensioner.
12. On the DOHC engine remove the rear side V-belt tensioner
13. Remove the crankshaft position sensor.

14. Remove the crankshaft pulley bolt.
15. Lock the crankshaft in place using tool ST499977100 for the SOHC engine and tool ST499207400 for the DOHC engine, or equivalent.
16. Remove the crankshaft pulley.
17. Remove the water pump.
18. If equipped, remove the timing belt guide. Remove the crankshaft sprocket.
19. Remove the oil pump retaining bolts.

➡ **When disassembling and checking the oil pump, loosen the relief valve plug before removing the oil pump from its mounting.**

20. Using a flat tip tool remove the oil pump from the engine.

To install:

21. Be sure all mating surfaces are clean and free of dirt.
22. Apply liquid gasket part number 004403007, or equivalent to the mating surfaces of the oil pump.
23. Be sure to replace the O-ring with a new one.
24. Using the ST-499587100, install the front oil seal.
25. Apply a thin coat of clean engine oil to the inside of the oil seal.
26. Position the oil pump to its mounting, aligning the notched area with the crankshaft and push the pump straight.

➡ **Be sure that the oil seal lip is not folded.**

27. Install the oil pump. Apply liquid gasket part number 004403042, or equivalent to the three retaining bolt threads.
28. Install the bolts and tighten to 4.7 ft. lbs. (6.4 Nm).
29. Continue the installation in the reverse order of the removal procedure.
30. Be sure to fill the cooling system with the proper grade and type coolant.
31. Start the engine and check for leaks. Correct, as required.

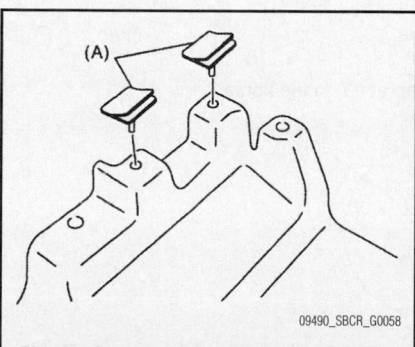

09490_SBCR_G0058

Fig. 99 Oil pan baffle plate seal location and positioning

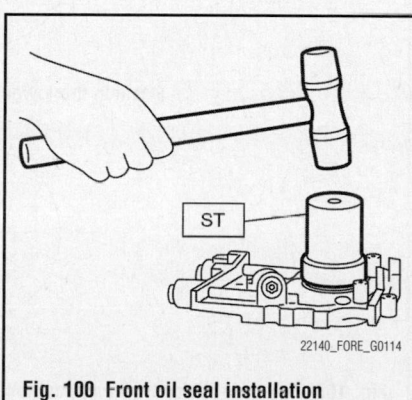

22140_FORE_G0114

Fig. 100 Front oil seal installation

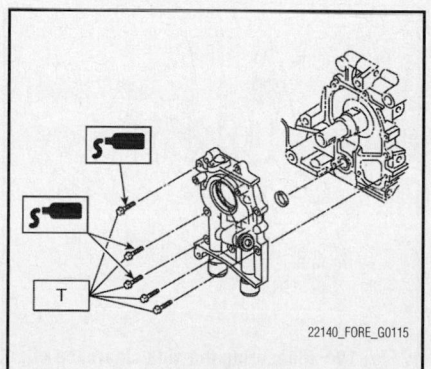

22140_FORE_G0115

Fig. 101 Oil pump and mounting bolts (T)

INSPECTION

See Figures 102 through 104.

1. Check the oil pump case for worn shaft hole, clogged oil passage, worn rotor chamber, cracks, and other faults.

2. Check the oil seal lips for deformation, hardening, wear, etc. and replace if defective.

3. Measure the tip clearance of rotors. If clearance exceeds the standard, replace the rotors as a matched set.

    a. 0.0016–0.0055 in. (0.04–0.14mm)

4. Measure the clearance between the outer rotor and oil pump rotor housing. If clearance exceeds the standard, replace the rotor.

    a. 0.0039–0.0069 in. (0.10–0.18mm)

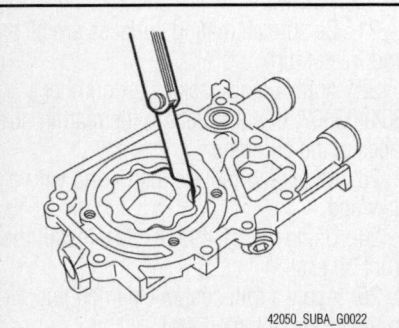

Fig. 102 Measuring for tip clearance of the rotors of the oil pump.

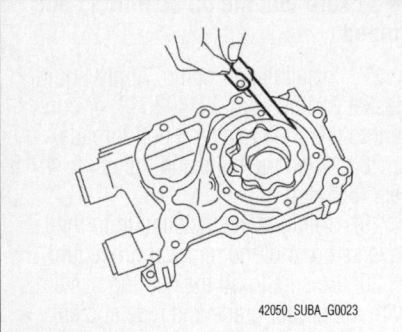

Fig. 103 Measuring the case clearance of the oil pump.

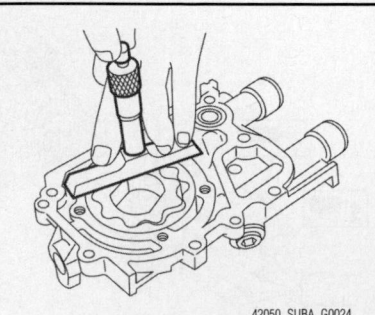

Fig. 104 Measuring the side clearance of the oil pump.

5. Measure the clearance between the oil pump inner rotor and pump cover. If clearance exceeds the standard, replace the rotor or pump body.

    a. 0.0008–0.0028 in. (0.02–0.07mm)

6. Check the valve for fitting condition and damage, and the relief valve spring for damage and deterioration. Replace the parts if defective.

PISTON AND RING

POSITIONING

See Figures 105 and 106.

Position the top ring gap at (A) or (B) in the figure.

Position the second ring gap at 180° on the reverse side the top ring gap.

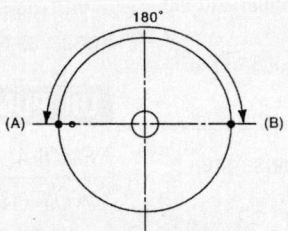

Position the upper rail gap at (C) in the figure.

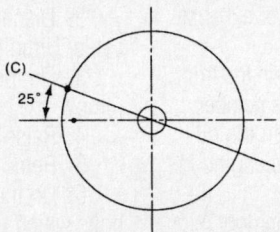

Align the upper rail spin stopper (D) to the side hole (E) on the piston.

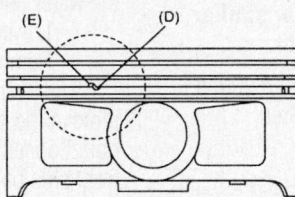

Position the expander gap at (F) in the figure on the 180° opposite direction of (C).

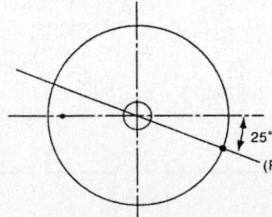

Position the lower rail gap at (G) in the figure.

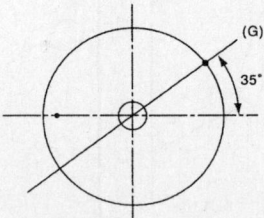

Fig. 105 Piston ring alignment and positioning SOHC engine

Position the top ring gap at (A) or (B) in the figure.

NOTE:
Assemble so that the piston ring mark "R" faces the upper side of the piston.

Position the second ring gap at 180° on the reverse side the top ring gap.

NOTE:
Assemble so that the piston ring mark "R" faces the upper side of the piston.

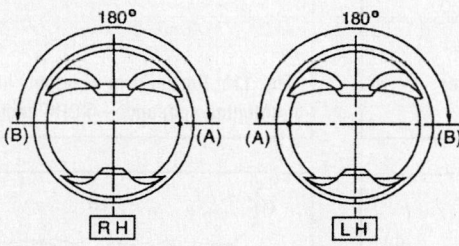

Position the upper rail gap at (C) in the figure.

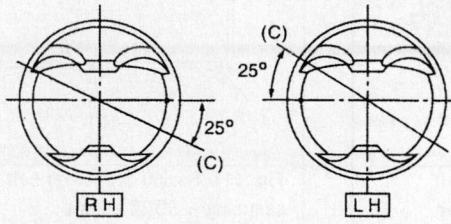

Align the upper rail spin stopper (E) to the side hole (D) on the piston.

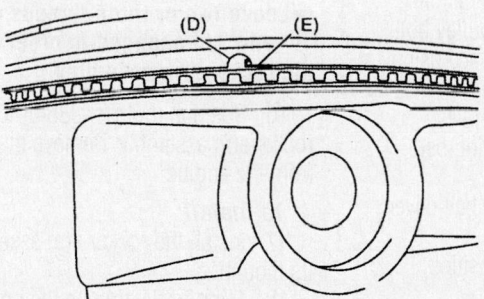

Position the expander gap at (F) in the figure.

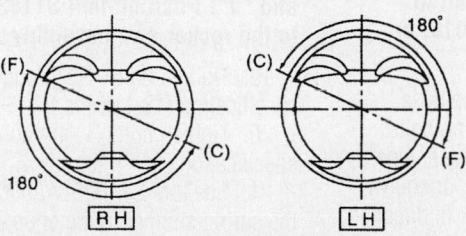

Position the lower rail gap at (G) in the figure.

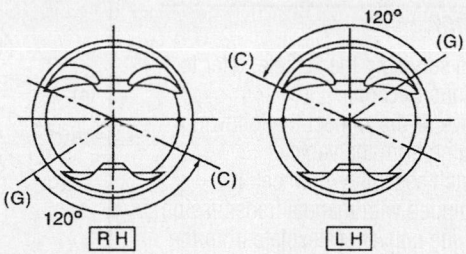

CAUTION:
• Make sure ring gaps do not face the same direction.
• Make sure ring gaps are not within the piston skirt area.

Install the snap ring.
Install the snap rings in the piston holes located opposite to the service holes in cylinder block when positioning all pistons in corresponding cylinders.

NOTE:
Use new snap rings.

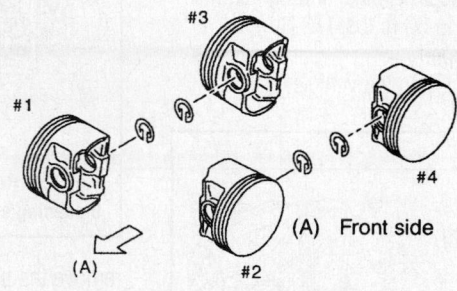

CAUTION:
Piston front mark faces towards the front of engine.

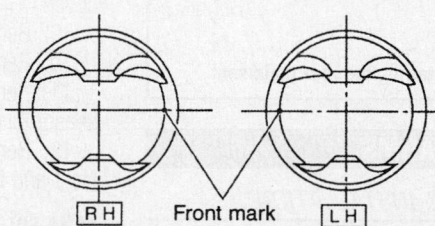

Fig. 106 Piston ring alignment and positioning DOHC engine

09490_SBCR_G0080

REAR MAIN SEAL

REMOVAL & INSTALLATION

See Figure 107.

1. Before servicing the vehicle, refer to the Precautions Section.
2. Remove or disconnect the following:
   - Engine from the vehicle
   - Clutch assembly/flywheel, if equipped with manual transmission
   - Torque converter flexplate from the crankshaft, if equipped with an automatic transmission
3. Using a seal removal tool, pry the oil seal from the housing.

To install:

4. Utilizing the appropriate seal installer, install or connect the following:
   - New oil seal and press it into the housing using oil seal guide ST 499597100 and installer ST 499587200.
   - Clutch assembly/flywheel, if equipped
   - Flywheel/flexplate and tighten the bolts to 53 ft. lbs. (72 Nm), if equipped
   - Engine into the vehicle

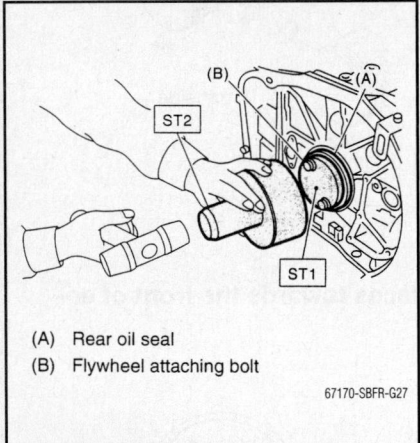

(A) Rear oil seal
(B) Flywheel attaching bolt

67170-SBFR-G27

Fig. 107 Installing the rear main seal

ROCKER ARMS

REMOVAL & INSTALLATION

2.5L SOHC Non-Turbo Engine

See Figures 108 through 111.

1. Before servicing the vehicle, refer to the Precautions Section.
2. Disconnect the negative battery cable.
3. To remove the front side V-belt, remove the belt covers. Loosen the lock bolt. Loosen the slider bolt. Remove the front side belt.
4. To remove the rear side V-belt,

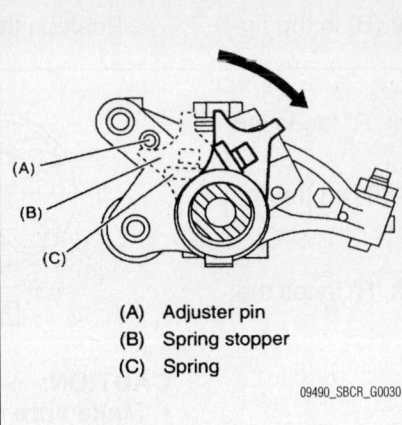

(A) Adjuster pin
(B) Spring stopper
(C) Spring

09490_SBCR_G0030

Fig. 108 Rocker arm spring stopper removal—SOHC engine

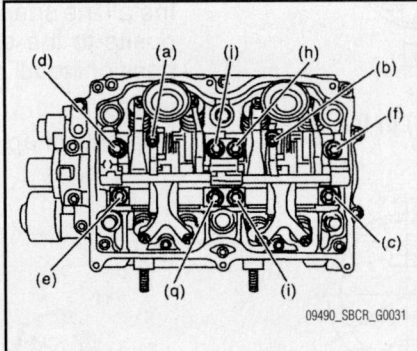

09490_SBCR_G0031

Fig. 109 Rocker arm assembly bolt loosening sequence—SOHC engine

remove the belt covers. Loosen the lock bolt. Loosen the slider bolt. Remove the rear side belt. Remove the belt tensioner.

5. Remove the crankshaft pulley bolt.
6. Lock the crankshaft in place using tool ST499977100, or equivalent.
7. Remove the crankshaft pulley.
8. Remove the left side timing belt cover.
9. Remove the front timing belt cover.
10. Remove the timing belt.
11. Remove the camshaft position sensor.
12. Remove the camshaft sprockets, No.1 and No.2.

➡**Be sure to lock the camshaft in place using tool ST18231AA010, or equivalent.**

13. Disconnect the PCV valve hose. Remove the rocker cover retaining bolts. Remove the rocker cover from the engine.
14. Use tool St18258AA000, or equivalent, to rotate the spring stopper in direction of the arrow, see illustration, and remove the spring stopper from the adjuster spring.
15. Remove bolts "A" through "J" in

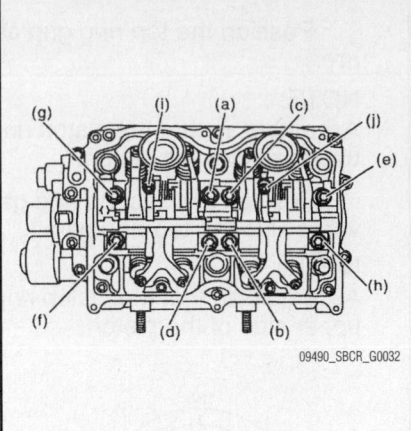

09490_SBCR_G0032

Fig. 110 Rocker arm assembly bolt tightening sequence—SOHC engine

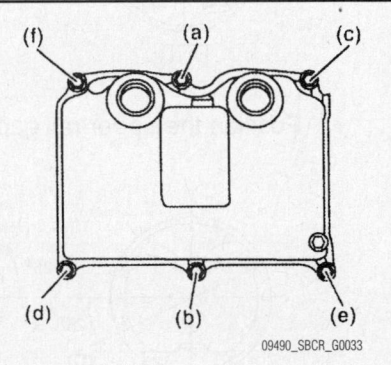

09490_SBCR_G0033

Fig. 111 Rocker arm cover bolt tightening sequence—SOHC engine

sequence. Remove the assembly from the engine.

➡**Leave two or three threads on bolts "I" and "J" engaged in order to retain the rocker arm assembly.**

16. Position tool ST18354AA000 on the rocker arm assembly. Remove the assembly from the engine.

To install:

17. Install the rocker arm assembly on its mounting.
18. Temporarily tighten the bolts equally in the proper sequence.

➡**Do not temporarily tighten bolts "I" and "J". Position tool ST18354AA000 to the rocker arm assembly.**

19. Tighten bolts "A" through "H" to specification (18.1 ft. lbs.).
20. Tighten bolts "I" through "J" to specification (5.9 ft. lbs.).
21. Use tool ST18258AA000 to rotate the spring stopper in the opposite direction of the arrow, see illustration, to fasten the adjuster pin.
22. Adjust the valve clearance.

23. Use a new gasket and install the rocker arm cover. Torque the retaining bolts to 4.7 ft. lbs and in the proper sequence. Recheck and torque to 4.7 ft. lbs in the proper sequence.

24. Continue the installation in the reverse order of the removal procedure.

TIMING BELT FRONT COVER

REMOVAL & INSTALLATION

Refer to "Timing Belt and Sprockets, Removal & Installation" procedure in this section to remove the timing belt front cover.

TIMING BELT & SPROCKETS

REMOVAL & INSTALLATION

2.5L DOHC Turbo Engine

See Figures 112 through 123.

1. Before servicing the vehicle, refer to the Precautions Section.

2. Disconnect the negative battery cable.

3. Remove the collector cover.

4. To remove the front side V-belt, remove the belt covers. Loosen the lock bolt. Loosen the slider bolt. Remove the front side belt.

5. To remove the rear side V-belt, remove the belt covers. Loosen the lock bolt. Loosen the slider bolt. Remove the rear side belt. Remove the belt tensioner.

6. Remove the crankshaft pulley bolt.

7. Lock the crankshaft in place using tool ST499977100, or equivalent.

8. Remove the crankshaft pulley.

9. Remove the left side timing belt cover.

10. Remove the right side timing belt cover.

11. Remove the front timing belt cover.

12. Remove the timing belt guide, if equipped.

➡ **If the belt is going to be reused and the alignment mark on the belt is not readable, put a new mark on the belt to indicate the direction of rotation. Using tool ST499987500, turn the crankshaft to align the mark on the crankshaft sprocket, intake camshaft sprocket (left), exhaust camshaft sprocket (left), intake camshaft sprocket (right), exhaust camshaft sprocket (right) with the notches of the timing belt cover and cylinder block. Paint an alignment mark on the belts in relation to the camshaft sprockets. Z1 measurement is 54.4 teeth. Z2 measurement is 51.0 teeth and Z3 measurement is 28.0 teeth.**

13. Remove the belt idler belt "A". Remove the timing belt.

14. Remove the belt idlers belt "B" and "C".

15. Remove the belt idler number two. Remove the automatic belt tension adjuster assembly.

To install:

16. Attach the automatic belt tension adjuster assembly to a vertical pressing tool.

➡ **Always use a vertical type pressing tool to move the adjuster rod downward. Do not use a lateral type vise. Push the adjuster rod vertically. Press in the push adjuster rod gradually, which should take three minutes or more. Do not allow pressure to exceed 2,205 lb. force.**

17. Slowly move the adjuster rod down until the adjuster rod is aligned with the stopper pin hole in the cylinder.

➡ **Press the adjuster rod as far as the end surface of the cylinder. Do not press the adjuster rod into the cylinder. Doing so may damage the cylinder.**

18. Using a 0.08 inch stopper pin, insert it into the stopper pin hole in the cylinder. Secure the adjuster rod.

➡ **Do not release the press pressure until the stopper pin is completely inserted in the hole.**

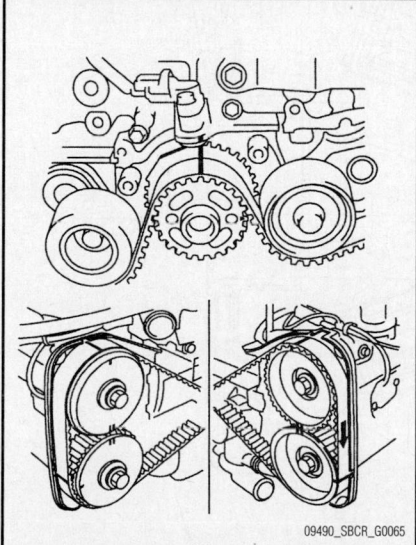

Fig. 112 Timing belt alignment—DOHC (Turbo) engine

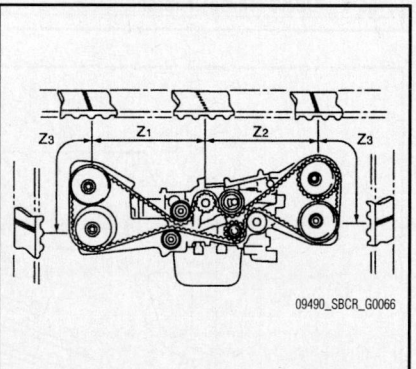

Fig. 113 Timing belt Z1, Z2 and Z3 teeth measurement— DOHC (Turbo) engine

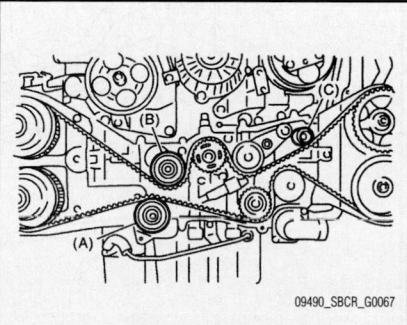

Fig. 114 Belt idler identification and location— DOHC (Turbo) engine

Fig. 115 Crankshaft sprocket mark A to oil pump cover alignment— DOHC (Turbo) engine

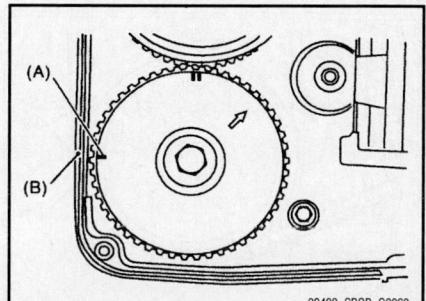

Fig. 116 Right exhaust camshaft sprocket alignment mark A with timing belt cover B alignment mark— DOHC (Turbo) engine

19. Install the automatic belt tensioner assembly. Tighten the retaining bolt to 28.9 ft. lbs.

20. Install the belt idler number two. Tighten the retaining bolt to 28.9 ft. lbs.

21. Install the belt idlers. Tighten the retaining bolts to 28.9 ft. lbs.

22. Align the mark "A" on the crankshaft sprocket with the mark on the oil pump at the cylinder block.

23. Align the single line mark "A" on the right exhaust camshaft sprocket with the notch "B" on the timing belt cover.

24. Align single line mark "A" on the right intake camshaft sprocket with the notch "B" on the timing cover. Ensure that the double lines "C" on the intake and exhaust camshaft sprockets are aligned.

25. Align single line mark "A" on the left exhaust camshaft sprocket with the notch "B" on the timing cover by turning the sprocket counterclockwise as viewed from the front of the engine.

26. Align single line mark "A" on the left intake camshaft sprocket with the notch "B" on the timing cover, by turning the sprocket clockwise as viewed from the front of the

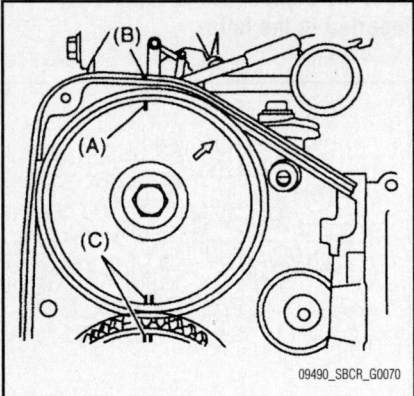

Fig. 117 Right intake camshaft sprocket alignment mark A with timing belt cover B alignment mark an double line C alignment mark— DOHC (Turbo) engine

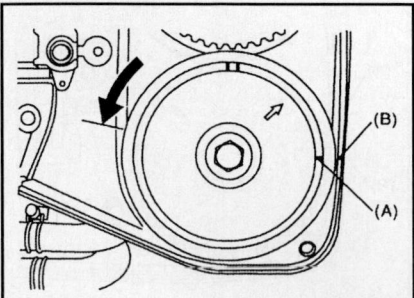

Fig. 118 Left exhaust camshaft sprocket alignment mark A with timing belt cover B alignment mark—DOHC (Turbo) engine

engine. Ensure that the double lines "C" on the intake and exhaust camshaft sprockets are aligned.

27. Make sure that the camshaft and crankshaft sprockets are positioned properly.

➡The intake and exhaust camshafts on this engine can be rotated independently with the timing belt removed. By looking at the illustration it will show you that if the intake and exhaust valve are lift together the heads will hit each other and bend.

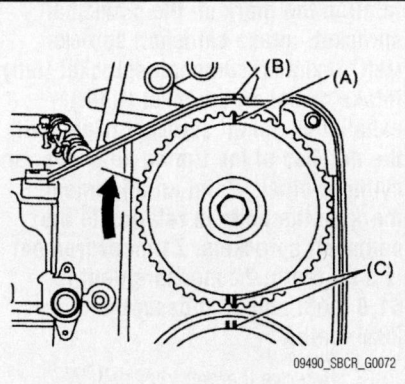

Fig. 119 Left intake camshaft sprocket alignment mark A with timing belt cover B alignment mark an double line C alignment mark—DOHC (Turbo) engine

➡When the timing belts are not installed, 4 camshafts are held at "zero lift" position, where all cams on the camshafts do not push the intake and exhaust valves down (under this condition all valves remain unlifted). When the camshafts are rotated to install the timing belts, No. 2 intake and No. 4 exhaust cam of the left hand camshafts are held to push their corresponding valves down. Under this condition these valves are held lifted. The right side camshafts are held in so that their cams do not push the valves down. The left hand camshafts must be rotated from the "zero lift" position to the position where the timing belt is to be installed at as small an angle as possible, in order to prevent mutual interference of intake and exhaust valve heads. Do not allow the camshafts to rotate in the direction illustrated as this causes both the intake and exhaust valves to lift off at the same time with will cause valve damage.

28. When installing the belt, make sure to align the marks made during removal or if using a new belt, align the in alphabetical order as shown in the illustration.

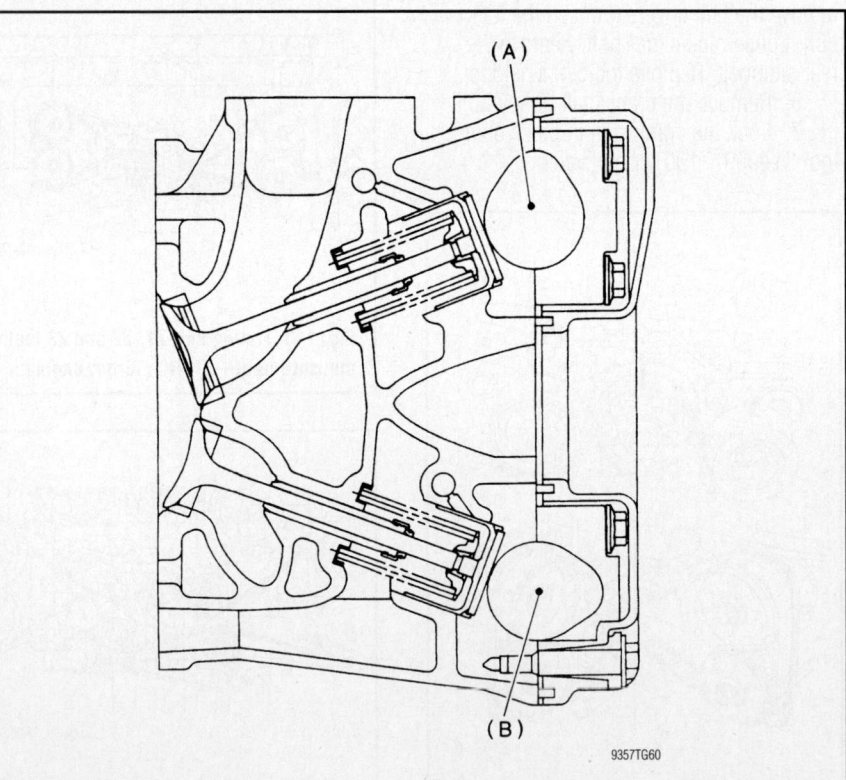

Fig. 120 If the intake and exhaust valve are lifted together the heads will hit each other and bend — DOHC (Turbo) engine

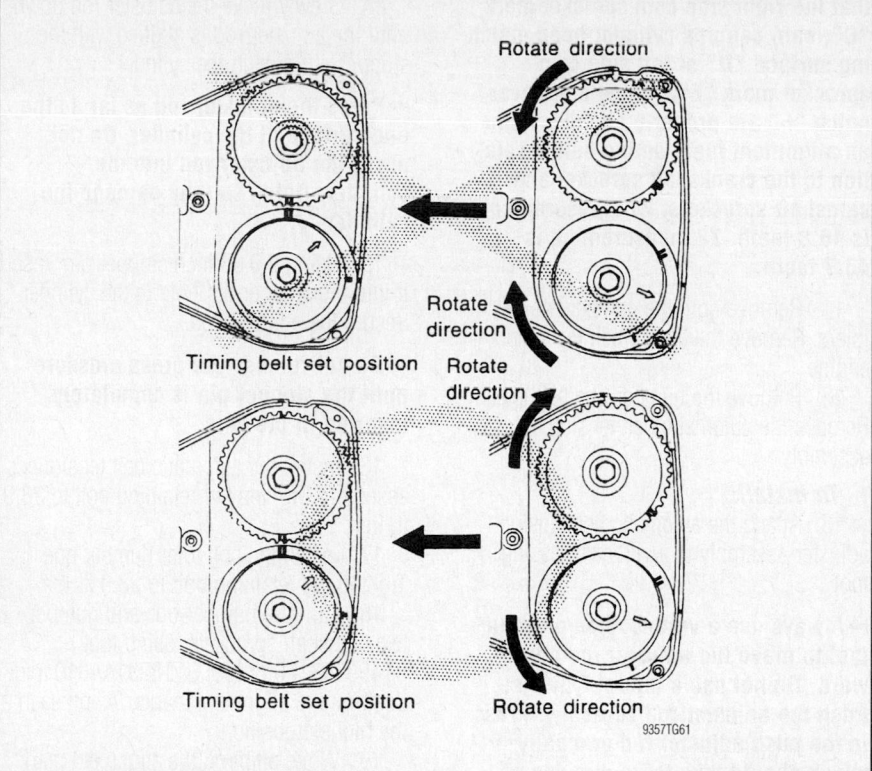

Fig. 121 Do not allow the camshafts to rotate in the direction shown as this causes both the intake and exhaust valves to lift off at the same time with will cause valve damage— DOHC (Turbo) engine

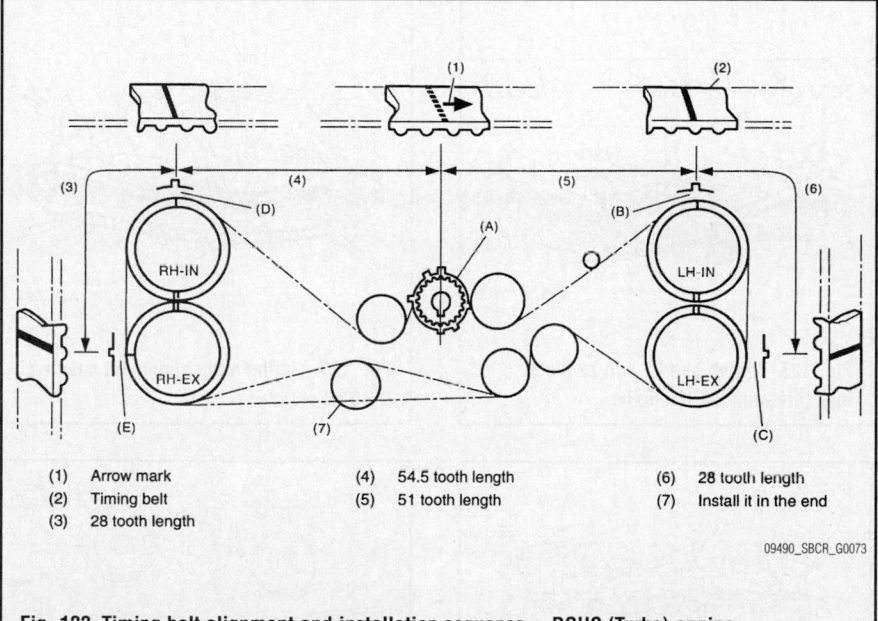

(1)	Arrow mark	(4)	54.5 tooth length	(6)	28 tooth length
(2)	Timing belt	(5)	51 tooth length	(7)	Install it in the end
(3)	28 tooth length				

Fig. 122 Timing belt alignment and installation sequence— DOHC (Turbo) engine

---

❋❋ WARNING

Disengagement of more than 3 timing belt teeth may result in contact between the valve and piston. Always make sure the belts rotation is correct.

29. Install the timing belt.

➡**Align the alignment mark on the timing belt with marks on the sprocket in the order shown in the illustration. While aligning the timing marks, position the timing belt properly.**

Fig. 123 Timing belt guide bolt location and tightening specification— DOHC (Turbo) engine

30. Install the belt idlers. Tighten the retaining bolts to 28.9 ft. lbs.

➡**Make sure that the marks on the timing belt and sprockets are aligned.**

31. After checking to be sure the marks on the timing belt and the camshaft sprockets are aligned remove the stopper pin from the belt tension adjuster.

32. Install the timing belt guide, if equipped with manual transmission. Temporarily tighten the bolts. Check and adjust the clearance between the belt and the guide. It should be 0.039 +/- 0.020 inch. Tighten the guide bolts to specification, see illustration.

33. Continue the installation in the reverse order of the removal procedure.

2.5L SOHC Non-Turbo Engine

See Figures 124 through 128.

1. Before servicing the vehicle, refer to the Precautions Section.

2. Disconnect the negative battery cable.

3. To remove the front side V-belt, remove the belt covers. Loosen the lock bolt. Loosen the slider bolt. Remove the front side belt.

4. To remove the rear side V-belt, remove the belt covers. Loosen the lock bolt. Loosen the slider bolt. Remove the rear side belt. Remove the belt tensioner.

5. Remove the crankshaft pulley bolt.

6. Lock the crankshaft in place using tool ST499977100, or equivalent.

7. Remove the crankshaft pulley.

8. Remove the left side timing belt cover.

9. Remove the front timing belt cover.

10. If equipped with manual transmission, remove the timing belt guide.

➡ **If the belt is going to be reused and the alignment mark on the belt is not readable, put a new mark on the belt to indicate the direction of rotation. Using tool ST499987500, turn the crankshaft to align the mark of the sprocket "A" to the cylinder mark notch "B". Ensure**

that the right side cam sprocket mark "C", cam cap and cylinder head matching surface "D" or left side cam sprocket mark "E", timing belt cover notch "F" are properly aligned. Paint an alignment mark on the belt in relation to the crankshaft sprocket and camshaft sprockets. Z1 measurement is 46.8 teeth. Z2 measurement is 43.7 teeth.

11. Remove both the number two belt idlers. Remove the timing belt from the engine.

12. Remove the number one belt idler. Remove the automatic belt tension adjuster assembly.

To install:

13. Attach the automatic belt tension adjuster assembly to a vertical pressing tool.

➡ **Always use a vertical type pressing tool to move the adjuster rod downward. Do not use a lateral type vise. Push the adjuster rod vertically. Press in the push adjuster rod gradually, which should take three minutes or more. Do not allow pressure to exceed 2,205 lb. force.**

14. Slowly move the adjuster rod down until the adjuster rod is aligned with the stopper pin hole in the cylinder.

➡ **Press the adjuster rod as far as the end surface of the cylinder. Do not press the adjuster rod into the cylinder. Doing so may damage the cylinder.**

15. Using a 0.08 inch stopper pin, insert it into the stopper pin hole in the cylinder. Secure the adjuster rod.

➡ **Do not release the press pressure until the stopper pin is completely inserted in the hole.**

16. Install the automatic belt tensioner assembly. Tighten the retaining bolt to 28.9 ft. lbs.

17. Install the belt idler number one. Tighten the retaining bolt to 28.9 ft. lbs.

18. Turn the number one and number two camshaft sprockets, using tool ST499207100 or tool ST18231AA010 and position the alignment marks "A" on each at the highest position.

19. While aligning the alignment mark "B" on the timing belt with mark "A" on the sprockets, position the timing belt properly.

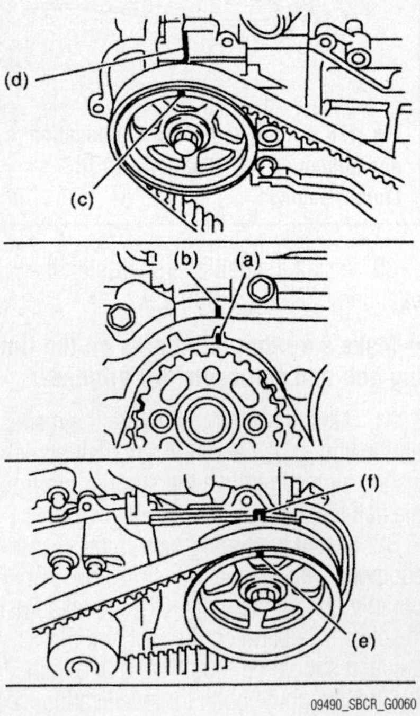

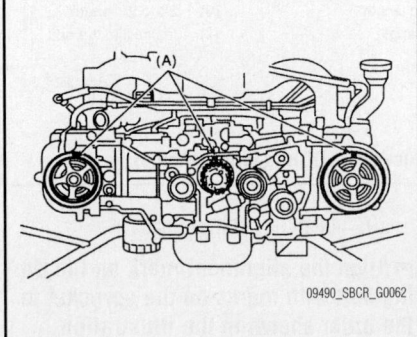

Fig. 124 Timing belt alignment SOHC engine

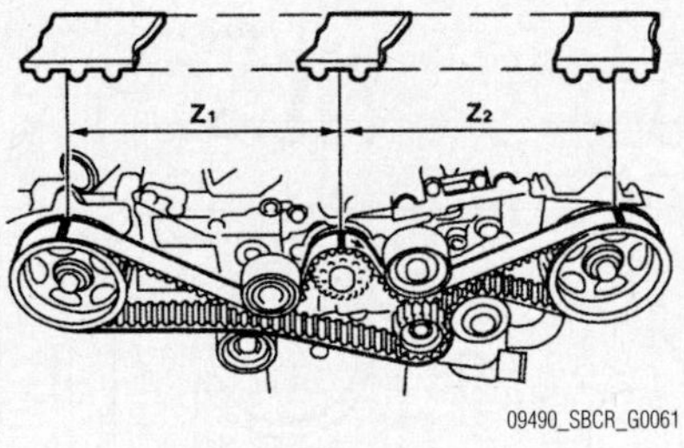

Fig. 125 Timing belt Z1 and Z2 teeth measurement SOHC engine

09490_SBCR_G0061

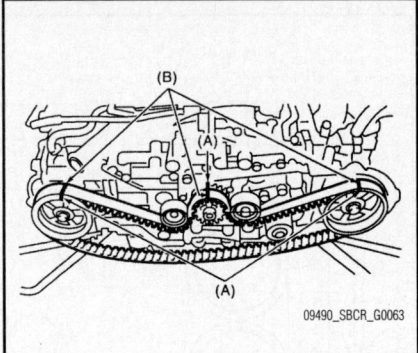

Fig. 127 Timing mark alignment position B SOHC engine

09490_SBCR_G0063

Fig. 126 Timing mark alignment position A SOHC engine

09490_SBCR_G0062

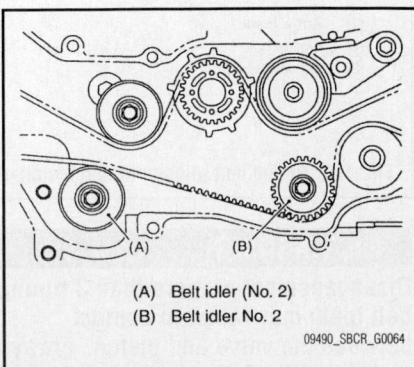

(A) Belt idler (No. 2)
(B) Belt idler No. 2

09490_SBCR_G0064

Fig. 128 Belt idler number two locations SOHC engine

20. Install both belt idler number two's. Tighten the retaining bolt to 28.9 ft. lbs.

21. After checking to be sure the marks on the timing belt and the camshaft sprockets are aligned remove the stopper pin from the belt tension adjuster.

22. Install the timing belt guide, if equipped with manual transmission. Temporarily tighten the bolts. Check and adjust the clearance between the belt and the guide. It should be 0.039 +/- 0.020 inch. Tighten the bolts to 7.2 ft. lbs.

23. Continue the installation in the reverse order of the removal procedure.

TURBOCHARGER

REMOVAL & INSTALLATION

See Figure 129.

1. Before servicing the vehicle, refer to the Precautions Section.

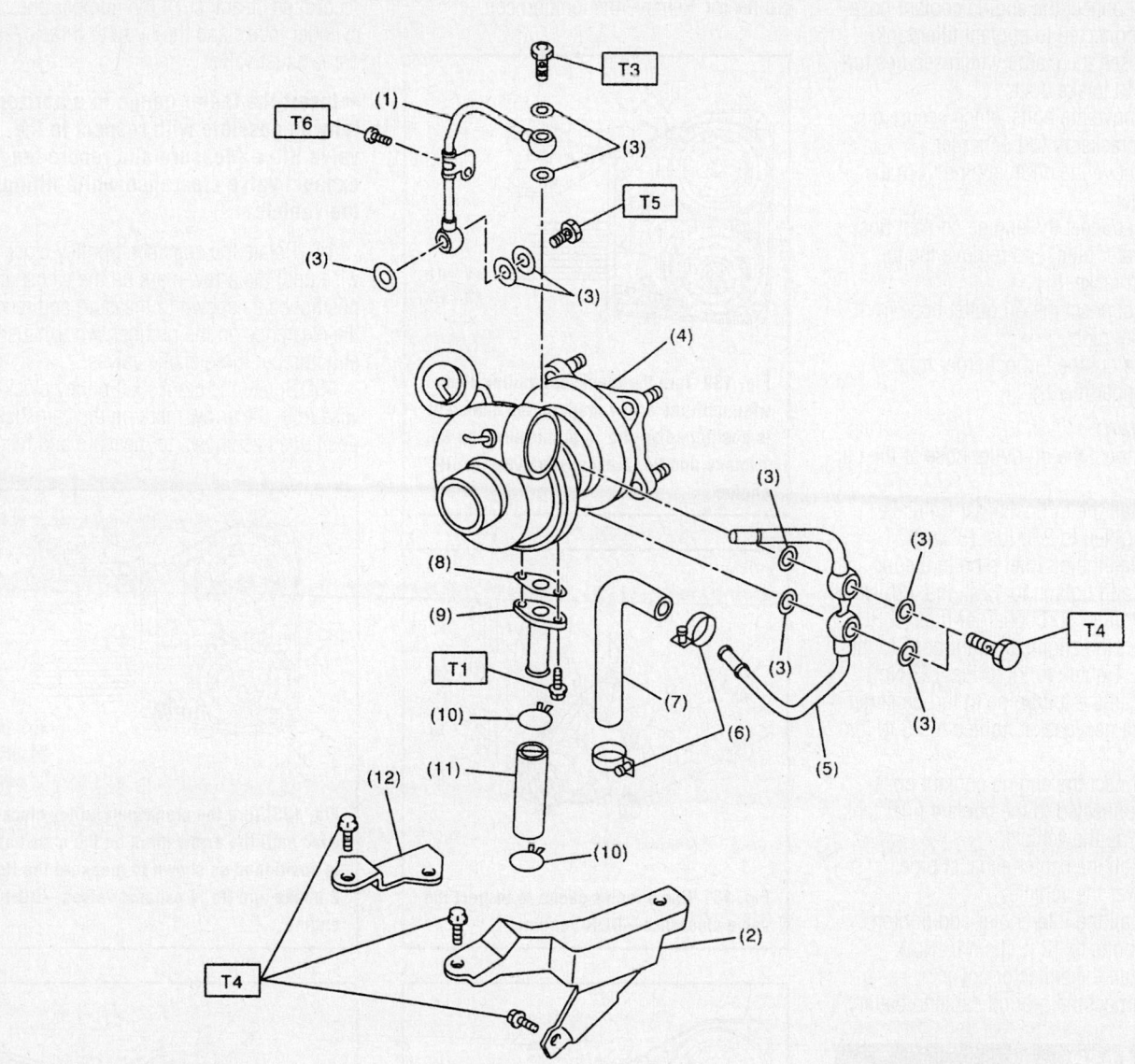

(1)	Oil inlet pipe A	(8)	Gasket
(2)	Turbocharger bracket LH	(9)	Oil outlet pipe
(3)	Metal gasket	(10)	Clip
(4)	Turbocharger	(11)	Oil outlet hose
(5)	Water pipe	(12)	Turbocharger bracket RH
(6)	Clamp		
(7)	Engine coolant hose		

Tightening torque: N·m (kgf-m, ft-lb)
T1: 4.4 (0.45, 3.3)
T2: 20 (2.0, 14.8)
T3: 16 (1.6, 11.6)
T4: 33 (3.4, 24.6)
T5: 29 (3.0, 21.7)
T6: 4.9 (0.50, 3.6)

09490_FORE_G0008

Fig. 129 Turbocharger and related components

2. Raise and safely support the vehicle.

3. Remove the collector cover.

4. Disconnect the ground cable from the battery.

5. Remove the intercooler.

6. Remove the center exhaust pipe.

7. Lower the vehicle.

8. Separate the turbocharger joint pipe from turbocharger.

9. Disconnect the engine coolant hose which is connected to coolant filler tank.

10. Loosen the clamp which secures turbocharger to intake duct.

11. Remove the bolts which secure oil inlet pipe bracket to turbocharger.

12. Remove the oil inlet pipe from the turbocharger.

13. Disconnect the engine coolant hose from the water pipe, and remove the turbocharger bracket RH.

14. Disconnect the oil outlet hose from the oil outlet pipe.

15. Take out the turbocharger from engine compartment.

To install:

16. Connect the oil outlet hose to the oil outlet pipe.

17. Install the turbocharger to intake duct and tighten to 2 ft. lbs. (3 Nm).

18. Install the oil inlet pipe to the turbocharger and tighten to 12 ft. lbs. (16 Nm).

19. Install the turbocharger bracket RH, and connect the engine coolant hose to the water pipe. Tighten to 24 ft. lbs. (33 Nm).

20. Install the joint pipe to turbocharger and using a new gasket tighten to 25 ft. lbs. (35 Nm).

21. Connect the engine coolant hose which is connected to the coolant filler tank.

22. Lift up the vehicle.

23. Install the center exhaust pipe.

24. Lower the vehicle.

25. Install the intercooler and tighten mounting bolts to 12 ft. lbs. (16 Nm).

26. Install the collector cover.

27. Connect the ground cable to battery.

VALVE LASH

ADJUSTMENT

2.5L DOHC Turbo Engine

See Figures 130 through 136.

➡The valve adjustment should be performed while the engine is cold.

1. Before servicing the vehicle, refer to the Precautions Section.

2. Raise and support the vehicle safely.

3. Remove the undercover.

4. Lower the vehicle.

5. Remove the collector cover.

6. Disconnect the negative battery cable.

7. Remove the air intake duct.

8. Remove the bolt that retains the right side timing belt cover. Remove the remaining bolts and remove the right side timing belt cover.

9. Disconnect the ignition coil electrical connector. Remove the ignition coil.

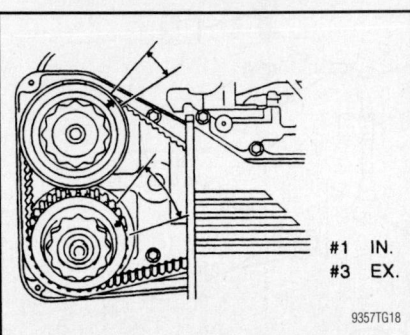

Fig. 130 Turn the crankshaft pulley clockwise until the arrow mark on the camshaft is positioned as shown to measure the No. 1 intake and No. 3 exhaust valves—DOHC engine

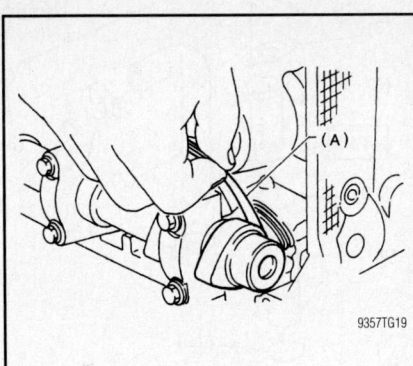

Fig. 131 Use a feeler gauge to inspect the valve clearance—DOHC engine

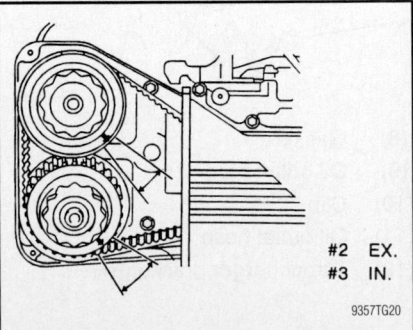

Fig. 132 Turn the crankshaft pulley clockwise until the arrow mark on the camshaft is positioned as shown to measure the No. 2 exhaust and No. 3 intake valves—DOHC engine

10. Position a suitable container under the vehicle.

11. Disconnect the PCV hose from the rocker cover. Remove the rocker cover retaining bolts. Remove the rocker cover from the vehicle.

12. Position the number one piston at TDC of the compression stroke.

13. Using a feeler gauge, measure and record the clearance of the number one cylinder intake and the number three cylinder exhaust valves.

➡Insert the feeler gauge in a horizontally as possible with respect to the valve lifter. Measure and record the exhaust valve clearance while lifting up the vehicle.

14. Rotate the crankshaft pulley clockwise until the arrow mark on the camshaft is positioned as shown to measure and record the clearance on the number two exhaust and number three intake valves.

15. Rotate the crankshaft pulley clockwise until the arrow mark on the camshaft is positioned as shown to measure and record

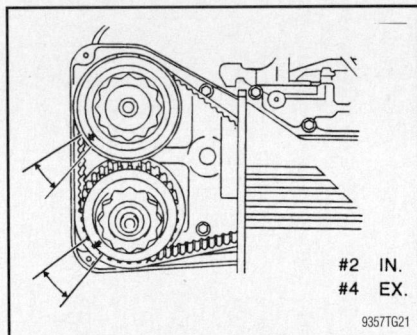

Fig. 133 Turn the crankshaft pulley clockwise until the arrow mark on the camshaft is positioned as shown to measure the No. 2 intake and No. 4 exhaust valves—DOHC engine

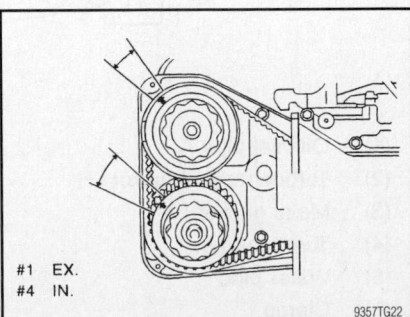

Fig. 134 Turn the crankshaft pulley clockwise until the arrow mark on the camshaft is positioned as shown to measure the No. 1 exhaust and No. 4 intake valves—DOHC engine

the number two intake and number four exhaust valves.

16. Rotate the crankshaft pulley clockwise until the arrow mark on the camshaft is positioned as shown to measure and record the number one exhaust and number four intake valves.

17. If adjustment is required, remove the camshafts.

	Unit: (mm)
Intake valve: $S = (V + T) - 0.20$	
Exhaust valve: $S = (V + T) - 0.35$	
S: Valve lifter thickness required	
V: Measured valve clearance	
T: Valve lifter thickness to be used	

09490 SBCR G0051

Fig. 135 Use this table to help you select a suitable shim—DOHC engine

Part No.	Thickness mm (in)	Part No.	Thickness mm (in)
13228 AB102	4.68 (0.1843)	13228 AB622	5.20 (0.2047)
13228 AB112	4.69 (0.1846)	13228 AB632	5.21 (0.2051)
13228 AB122	4.70 (0.1850)	13228 AB642	5.22 (0.2055)
13228 AB132	4.71 (0.1854)	13228 AB652	5.23 (0.2059)
13228 AB142	4.72 (0.1858)	13228 AB662	5.24 (0.2063)
13228 AB152	4.73 (0.1862)	13228 AB672	5.25 (0.2067)
13228 AB162	4.74 (0.1866)	13228 AB682	5.26 (0.2071)
13228 AB172	4.75 (0.1870)	13228 AB692	5.27 (0.2075)
13228 AB182	4.76 (0.1874)	13228 AB702	4.38 (0.1724)
13228 AB192	4.77 (0.1878)	13228 AB712	4.40 (0.1732)
13228 AB202	4.78 (0.1882)	13228 AB722	4.42 (0.1740)
13228 AB212	4.79 (0.1886)	13228 AB732	4.44 (0.1748)
13228 AB222	4.80 (0.1890)	13228 AB742	4.46 (0.1756)
13228 AB232	4.81 (0.1894)	13228 AB752	4.48 (0.1764)
13228 AB242	4.82 (0.1898)	13228 AB762	4.50 (0.1771)
13228 AB252	4.83 (0.1902)	13228 AB772	4.52 (0.1780)
13228 AB262	4.84 (0.1906)	13228 AB782	4.54 (0.1787)
13228 AB272	4.85 (0.1909)	13228 AB792	4.56 (0.1795)
13228 AB282	4.86 (0.1913)	13228 AB802	4.58 (0.1803)
13228 AB292	4.87 (0.1917)	13228 AB812	4.60 (0.1811)
13228 AB302	4.88 (0.1921)	13228 AB822	4.62 (0.1819)
13228 AB312	4.89 (0.1925)	13228 AB832	4.64 (0.1827)
13228 AB322	4.90 (0.1929)	13228 AB842	4.66 (0.1835)
13228 AB332	4.91 (0.1933)	13228 AB852	5.29 (0.2083)
13228 AB342	4.92 (0.1937)	13228 AB862	5.31 (0.2091)
13228 AB352	4.93 (0.1941)	13228 AB872	5.33 (0.2098)
13228 AB362	4.94 (0.1945)	13228 AB882	5.35 (0.2106)
13228 AB372	4.95 (0.1949)	13228 AB892	5.37 (0.2114)
13228 AB382	4.96 (0.1953)	13228 AB902	5.39 (0.2122)
13228 AB392	4.97 (0.1957)	13228 AB912	5.41 (0.2123)
13228 AB402	4.98 (0.1961)	13228 AB922	5.43 (0.2138)
13228 AB412	4.99 (0.1965)	13228 AB932	5.45 (0.2146)
13228 AB422	5.00 (0.1969)	13228 AB942	5.47 (0.2154)
13228 AB432	5.01 (0.1972)	13228 AB952	5.49 (0.2161)
13228 AB442	5.02 (0.1976)	13228 AB962	5.51 (0.2169)
13228 AB452	5.03 (0.1980)	13228 AB972	5.53 (0.2177)
13228 AB462	5.04 (0.1984)	13228 AB982	5.55 (0.2185)
13228 AB472	5.05 (0.1988)	13228 AB992	5.57 (0.2193)
13228 AB482	5.06 (0.1992)	13228 AC002	5.59 (0.2201)
13228 AB492	5.07 (0.1996)	13228 AC012	5.61 (0.2209)
13228 AB502	5.08 (0.2000)	13228 AC022	5.63 (0.2217)
13228 AB512	5.09 (0.2004)	13228 AC032	5.65 (0.2224)
13228 AB522	5.10 (0.2008)		
13228 AB532	5.11 (0.2012)		
13228 AB542	5.12 (0.2016)		
13228 AB552	5.13 (0.2020)		
13228 AB562	5.14 (0.2024)		
13228 AB572	5.15 (0.2028)		
13228 AB582	5.16 (0.2031)		
13228 AB592	5.17 (0.2035)		
13228 AB602	5.18 (0.2039)		
13228 AB612	5.19 (0.2043)		

09490_SBCR_G0052

Fig. 136 Valve adjusting shim chart—DOHC engine

18. Remove and measure the thickness of the valve lifter. Select a suitable shim, using the shim selection chart.

19. Install the replacement shim to the lifter.

20. After all shims have been adjusted, inspect the valve clearances again.

21. After completion, install all removed components.

2.5L SOHC Non-Turbo Engine

See Figure 137.

➡The valve adjustment should be performed while the engine is cold.

1. Before servicing the vehicle, refer to the Precautions Section.

2. Before servicing the vehicle, refer to the Precautions section.

3. Raise and support the vehicle safely.

4. Remove the undercover.

5. Lower the vehicle.

6. Disconnect the negative battery cable.

7. To remove the front side V-belt, remove the belt covers. Loosen the lock bolt. Loosen the slider bolt. Remove the front side belt.

8. To remove the rear side V-belt, remove the belt covers. Loosen the lock bolt. Loosen the slider bolt. Remove the rear side belt. Remove the belt tensioner.

9. Remove the crankshaft pulley bolt.

10. Lock the crankshaft in place using tool ST499977100, or equivalent.

11. Remove the crankshaft pulley.

12. Remove the left side timing belt cover.

13. Remove the fuel injector.

14. Remove the rocker cover.

15. Position the number one piston at TDC of the compression stroke.

➡When the arrow (see illustration) on the camshaft sprocket (left side) comes exactly to the top, number one cylinder piston is at TDC of the compression stroke.

16. Measure the valve clearance, using a feeler gauge.

17. If adjustment is needed, loosen the valve rocker nut and screw. Position the feeler gauge.

➡Insert the feeler gauge in a horizontally as possible with respect to the valve stem end face. Adjust the exhaust valve clearance while lifting up the vehicle.

18. While noting the valve clearance, tighten the rocker adjusting screw.

19. When the proper valve clearance is obtained, tighten the valve rocker nut to 7.2 ft. lbs.

20. Adjust the valve clearance on the remaining cylinders, following the above procedure.

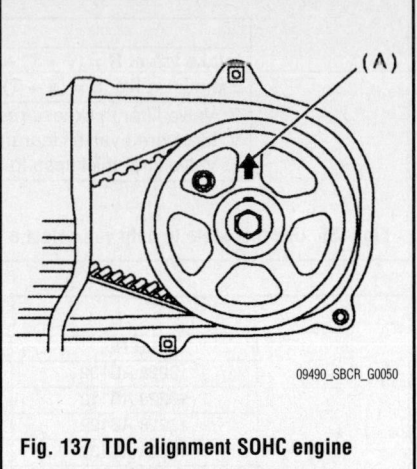

09490_SBCR_G0050

Fig. 137 TDC alignment SOHC engine

➡Be sure to position the pistons to their respective TDC positions on the compression stroke, before checking and adjusting the valves. By rotating the crankshaft pulley clockwise every 180 degrees from the state that number one piston is on TDC of the compression stroke, the remaining pistons come to TDC of the compression stroke in the following order, No. 3, No. 2, and No. 4.

21. After adjustment, replace any removed components.

22. Be sure to use new gaskets and seals, as required.

ENGINE PERFORMANCE & EMISSION CONTROLS

CAMSHAFT POSITION (CMP) SENSOR

REMOVAL & INSTALLATION

2.5L SOHC Non-Turbo Engine

1. Disconnect the negative battery cable.

2. Disconnect the connector from the sensor.

3. Remove the bolt that retains the Camshaft Position (CMP) sensor to the sensor support.

4. Remove the bolt that retains the sensor support to the camshaft cap.

5. Remove the CMP sensor and the sensor support as a unit.

6. Separate the CMP sensor from the support.

To install:

7. Installation is the reverse of the removal procedure.

8. Tighten the CMP sensor support to 4.7 ft. lbs. (6.4 Nm).

9. Tighten the CMP sensor to 4.7 ft. lbs. (6.4 Nm).

2.5L DOHC Turbo Engine

1. Disconnect the negative battery cable.

2. Remove the collector cover.

3. Disconnect the connector from the Camshaft Position (CMP) sensor RH.

4. Remove the CMP sensor RH from the rear side of the cylinder head.

5. Remove the CMP sensor LH in the same way as RH.

To install:

6. Installation is the reverse of the removal procedure.

7. Tighten the sensor to 4.7 ft. lbs. (6.4 Nm).

CRANKSHAFT POSITION (CKP) SENSOR

REMOVAL & INSTALLATION

1. Disconnect the negative battery cable.

2. Remove the collector cover, as required.

3. Remove the bolt that retains the Crankshaft Position (CKP) sensor in place at the cylinder block.

4. Remove the sensor from its mounting.

5. Disconnect the connector from the sensor.

To install:

6. Installation is the reverse of the removal procedure.

7. Tighten the sensor to 4.7 ft. lbs. (6.4 Nm).

ELECTRONIC CONTROL MODULE (ECM)

REMOVAL & INSTALLATION

1. Disconnect the negative battery cable.

2. Remove the lower inner trim on the passenger's side of the vehicle.

3. Detach the floor mat. Remove the protective cover.

4. Remove the Electronic Control Module (ECM) bracket retaining nuts. Remove the clip from the bracket.

5. Disconnect the connectors.

6. Remove the ECM from the vehicle.

To install:

7. Installation is the reverse of the removal procedure.

8. Tighten the retaining screws to 5.5 ft. lbs. (7.5 Nm).

➡**When replacing the ECM, be careful not to use the wrong part number, as damage to the injection system could occur.**

ENGINE COOLANT TEMPERATURE (ECT) SENSOR

REMOVAL & INSTALLATION

See Figure 138.

1. Remove the collector cover (Turbo only).

2. Disconnect the ground cable from the battery.

3. Remove the alternator.

4. Drain engine coolant.

5. Disconnect the connectors from the engine coolant temperature sensor.

6. Remove the engine coolant temperature sensor.

To install:

7. Install the engine coolant temperature sensor.

8. Connect the connectors from the engine coolant temperature sensor.

9. Fill engine coolant.

10. Install the alternator.

11. Connect the ground cable from the battery.

12. Install the collector cover (Turbo only).

➡**Use a new gasket. Tighten 13 ft. lbs. (18 Nm).**

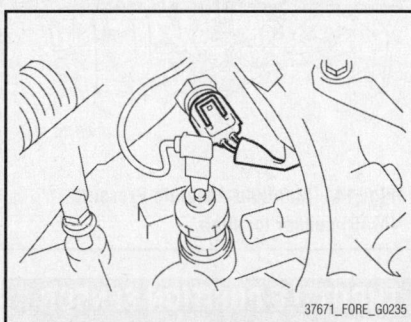

Fig. 138 Engine coolant temperature sensor connectors

EXHAUST GAS RECIRCULATION (EGR) VALVE

LOCATION

On the SOHC engine the Exhaust Gas Recirculation (EGR) valve is mounted on the intake manifold. The DOHC engine does not use a conventional EGR valve.

REMOVAL & INSTALLATION

See Figure 139.

1. Disconnect the negative battery cable.

2. Disconnect the connector.

3. Remove the Exhaust Gas Recirculation (EGR) valve from the intake manifold.

To install:

4. Installation is the reverse of the removal procedure.

5. Tighten the retaining bolts to 14.0 ft. lbs. (19 Nm).

Fig. 139 EGR valve and connector shown

HEATED OXYGEN (HO2S) SENSOR

REMOVAL & INSTALLATION

2.5L DOHC Turbo Engine

Front Sensor

1. Disconnect the negative battery cable.

2. Disconnect the connector. Disconnect the engine harness clip.

3. Raise and support the vehicle safely.

4. Remove the tire and wheel.

5. Remove the service hole cover.

6. Remove the oxygen sensor.

To install:

7. Installation is the reverse of the removal procedure.

➡**Apply anti-seize compound to the threaded portion of the sensor, prior to installation. Never apply anti-seize compound to the protector of the sensor.**

8. Tighten the sensor to 22 ft. lbs. (30 Nm).

Rear Sensor

1. Disconnect the negative battery cable.

2. Raise and support the vehicle safely.

3. Disconnect the connector.

4. Remove the clip by pulling it out from the upper side of the crossmember.

5. Remove the oxygen sensor.

To install:

6. Installation is the reverse of the removal procedure.

➡**Apply anti-seize compound to the threaded portion of the sensor, prior to installation. Never apply anti-seize compound to the protector of the sensor.**

7. Tighten the sensor to 22 ft. lbs. (30 Nm).

2.5L SOHC Non-Turbo Engine

Front Sensor

1. Disconnect the negative battery cable.

2. Disconnect the clip fastening the harness. Disconnect the connector.

3. Raise and support the vehicle safely.

4. Remove the engine undercover, as required.

5. Remove the oxygen sensor.

To install:

6. Installation is the reverse of the removal procedure.

➡**Apply anti-seize compound to the threaded portion of the sensor, prior to installation. Never apply anti-seize compound to the protector of the sensor.**

7. Tighten the sensor to 16 ft. lbs. (21 Nm).

Rear Sensor

1. Disconnect the negative battery cable.

2. Disconnect the clip fastening the harness. Disconnect the connector.

3. Raise and support the vehicle safely.

4. Remove the engine undercover, as required.

5. Remove the oxygen sensor.

To install:

6. Installation is the reverse of the removal procedure.

➡**Apply anti-seize compound to the threaded portion of the sensor, prior to installation. Never apply anti-seize compound to the protector of the sensor.**

7. Tighten the sensor to 16 ft. lbs. (21 Nm).

INTAKE AIR TEMPERATURE (IAT) SENSOR

REMOVAL & INSTALLATION

1. Disconnect the negative battery cable.
2. Disconnect the connector from the Mass Air Flow (MAF) and Intake Air Temperature (IAT) sensor.
3. Remove the MAF and IAT sensor.

To install:

4. Installation is the reverse of the removal procedure.
5. Tighten the sensor mounting screws to 0.74 ft. lbs. (1.0 Nm).

KNOCK SENSOR (KS)

OPERATION

The Knock Sensor (KS) is a piezo-electric (pressure-sensitive) electronic device that enables the Powertrain Control Module (PSM) to control the ignition timing for optimum performance, and to protect the engine from potentially damaging levels of detonation. The PCM uses the KS system to test for abnormal engine noise that may indicate detonation, which is also known as spark knock.

REMOVAL & INSTALLATION

See Figure 140.

1. Disconnect the negative battery cable.
2. Remove the air intake chamber.
3. Remove the collector cover, as required.
4. On DOHC engine, remove the intercooler.
5. Disconnect the Knock Sensor (KS) connector.
6. Remove the KS from its mounting.

To install:

7. Installation is the reverse of the removal procedure.

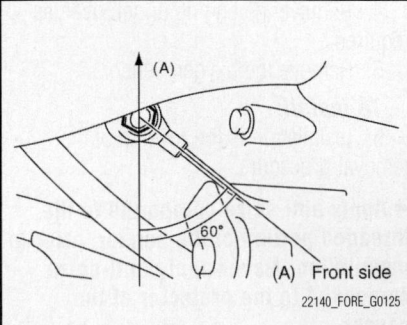

(A)

(A) Front side

22140_FORE_G0125

Fig. 140 Knock Sensor (KS) positioned at a 60 degree angle

8. Tighten the sensor to 17.7 ft. lbs. (24 Nm).

➡ **The extraction area of the knock sensor wire must be positioned at a 60 degree angle relative to the engine rear.**

MALFUNCTION INDICATOR LIGHT (MIL)

RESET PROCEDURE

Subaru Select Monitor (OBD MODE):

1. On the (Main Menu) display screen, select the (Each System Check).
2. On the (System Selection Menu) display screen, select the (Engine Control System).
3. Select (OK) after the information of engine type has been displayed.
4. On the (Engine Diagnosis) display screen, select the (OBD System).
5. On the «OBD Menu» display screen, select the (Clear Diagnostic Code).
6. When the (Clear Diagnostic Code) is shown on the screen, click the (Yes) button.
7. When (Done) and (Turn ignition switch to OFF) are shown on the display screen, turn the ignition switch to OFF.

General Scan Tool:

8. For procedures clearing memory using the general scan tool, refer to the general scan tool operation manual.

➡ **Initial diagnosis of electronic throttle control is performed after memory clearance. For this reason, start the engine after 10 seconds or more have elapsed since the ignition switch was turned to ON.**

➡ **The Malfunction Indicator Light (MIL) must be reset with a scan tool.**

MASS AIR FLOW (MAF) SENSOR

LOCATION

The Mass Air Flow (MAF) sensor is mounted in the intake air hose of the air cleaner assembly. This sensor is combined with the Intake Air Temperature (IAT) sensor.

REMOVAL & INSTALLATION

1. Disconnect the negative battery cable.
2. Disconnect the connector from the Mass Air Flow (MAF) and Intake Air Temperature (IAT) sensor.

3. Remove the MAF and IAT sensor.

To install:

4. Installation is the reverse of the removal procedure.
5. Tighten the sensor mounting screws to 0.74 ft. lbs. (1.0 Nm).

MANIFOLD ABSOLUTE PRESSURE (MAP) SENSOR

LOCATION

See Figure 141.

The sensor is located on the throttle body unit.

REMOVAL & INSTALLATION

1. Disconnect the negative battery cable.
2. Disconnect the connector from the sensor.
3. Remove the Manifold Absolute Pressure (MAP) sensor from its mounting.

To install:

4. Installation is the reverse of the removal procedure, noting the following:
   a. Be sure to use new O-rings.
   b. Tighten the retaining screw to 1.5 ft. lbs. (2.0 Nm).

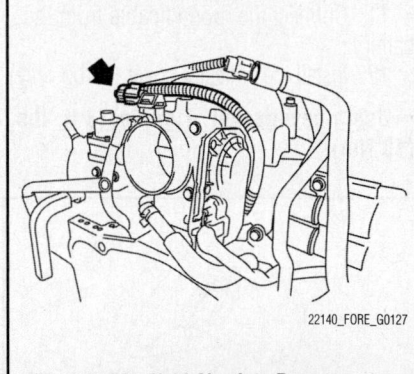

22140_FORE_G0127

Fig. 141 Manifold Absolute Pressure (MAP) sensor location

THROTTLE POSITION SENSOR (TPS)

REMOVAL & INSTALLATION

1. Disconnect the ground cable from the battery.
2. Remove the air intake chamber.
3. Disconnect the connectors from throttle position sensor and manifold absolute pressure sensor.
4. Disconnect the engine coolant hoses from throttle body.

5. Remove the bolts which install throttle body to the intake manifold.

To install:

6. Install in the reverse order of removal.

7. Install a new gasket.

8. Tighten throttle body mounting bolts to 6 ft. lbs. (8 Nm).

VEHICLE SPEED SENSOR (VSS)

REMOVAL & INSTALLATION

1. Raise and support the vehicle safely.

2. Place a drip pan below the speed sensor to catch any spilled fluid.

3. Disconnect the connector from the Vehicle Speed Sensor (VSS).

4. Remove the VSS from its mounting.

To install:

5. Installation is the reverse of the removal procedure.

6. Replace any lost fluid.

FUEL   GASOLINE FUEL INJECTION SYSTEM

FUEL SYSTEM SERVICE PRECAUTIONS

Safety is the most important factor when performing not only fuel system maintenance, but any type of maintenance. Failure to conduct maintenance and repairs in a safe manner may result in serious personal injury or death. Work on a vehicle's fuel system components can be accomplished safely and effectively by adhering to the following rules and guidelines.

• To avoid the possibility of fire and personal injury, always disconnect the negative battery cable unless the repair or test procedure requires that battery voltage be applied.

• Always relieve the fuel system pressure prior to disconnecting any fuel system component (injector, fuel rail, pressure regulator, etc.) fitting or fuel line connection. Exercise extreme caution whenever relieving fuel system pressure to avoid exposing skin, face and eyes to fuel spray. Please be advised that fuel under pressure may penetrate the skin or any part of the body that it contacts.

• Always place a shop towel or cloth around the fitting or connection prior to loosening to absorb any excess fuel due to spillage. Ensure that all fuel spillage is quickly removed from engine surfaces. Ensure that all fuel-soaked cloths or towels are deposited into a flame-proof waste container with a lid.

• Always keep a dry chemical (Class B) fire extinguisher near the work area.

• Do not allow fuel spray or fuel vapors to come into contact with a spark or open flame.

• Always use a second wrench when loosening or tightening fuel line connection fittings. This will prevent unnecessary stress and torsion on fuel piping. Always follow the proper torque specifications.

• Always replace worn fuel fitting O-rings with new ones. Do not substitute fuel hose where rigid pipe is installed.

FUEL SYSTEM PRESSURE

RELIEVING

See Figure 142.

1. Before servicing the vehicle, refer to the Precautions Section.

➡**This procedure must be performed prior to servicing any component of the fuel injection system.**

2. Disconnect the fuel pump connector from the fuel pump relay.

3. Start the engine and let it stall.

4. Crank the engine for 5 seconds or more to ensure the fuel pressure is properly relieved. If the engine starts during this time, allow it to run until it stalls.

5. After performing the required service, connect the fuel pump harness.

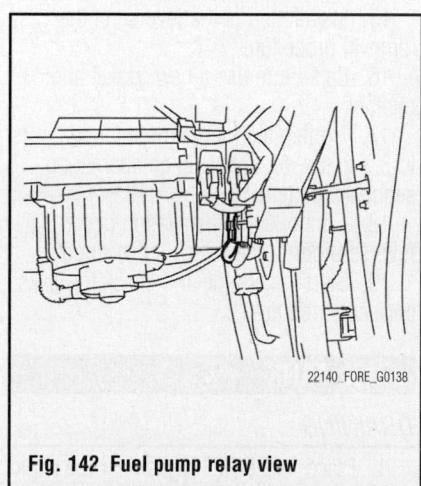

Fig. 142 Fuel pump relay view

FUEL FILTER

REMOVAL & INSTALLATION

See Figures 143 and 144.

➡**The fuel filter is an integral part of the fuel pump assembly.**

➡**This procedure must be performed prior to servicing any component of the fuel injection system.**

1. Before servicing the vehicle, refer to the Precautions Section.

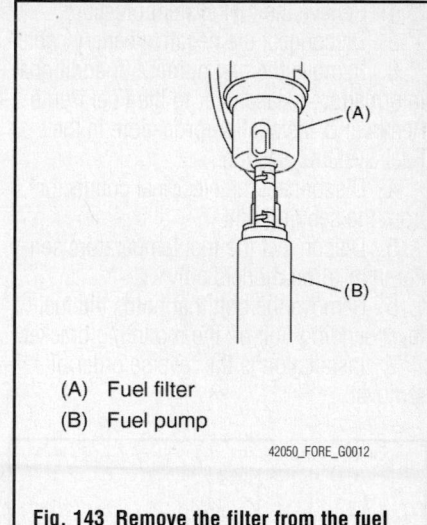

(A) Fuel filter
(B) Fuel pump

Fig. 143 Remove the filter from the fuel pump

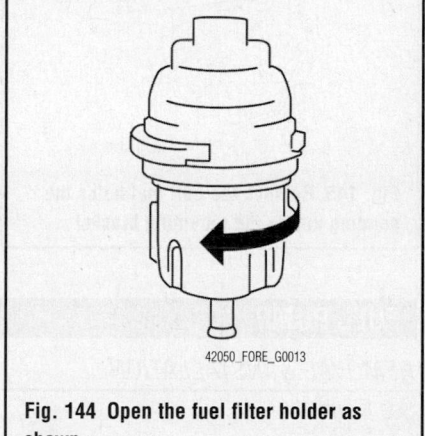

Fig. 144 Open the fuel filter holder as shown

2. Disconnect the fuel pump connector from the fuel pump relay.

3. Properly relieve the fuel system pressure.

4. Disconnect the negative battery cable.

5. Remove the fuel pump assembly.

6. Separate the fuel filter from the fuel pump.

7. Turn the filter holder in the direction shown and then remove the filter.

8. Installation is the reverse of the removal procedure.

FUEL LEVEL SENDING UNIT

LOCATION

The fuel level sending unit is located in the fuel tank.

REMOVAL & INSTALLATION

See Figure 145.

➡ **The fuel level sending unit is built into the fuel pump assembly.**

1. Relieve the fuel system pressure.
2. Disconnect the negative battery cable.
3. Remove the fuel pump. For additional information, please refer to the Fuel Pump Removal & Installation procedure in the Fuel Systems Section.
4. Disconnect the electrical connector from the sending unit.
5. Disconnect the fuel temperature sensor, non-turbo models only.
6. Remove the bolt that holds the fuel level sending unit on the mounting bracket.
7. Installation is the reverse order of removal.

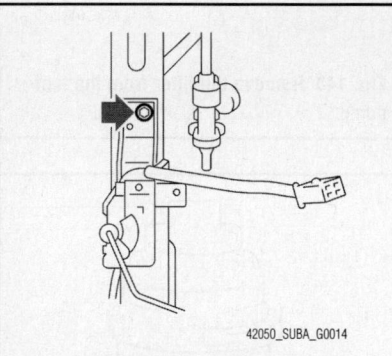

42050_SUBA_G0014

Fig. 145 Remove the bolt that holds the sending unit to the mounting bracket

FUEL PUMP

REMOVAL & INSTALLATION

See Figure 146.

➡ **This procedure must be performed prior to servicing any component of the fuel injection system.**

1. Before servicing the vehicle, refer to the Precautions Section.
2. Disconnect the fuel pump connector from the fuel pump relay.
3. Properly relieve the fuel system pressure.
4. Disconnect the negative battery cable.
5. Remove the fuel filler cap.
6. Raise and support the vehicle safely.
7. Drain the fuel tank.

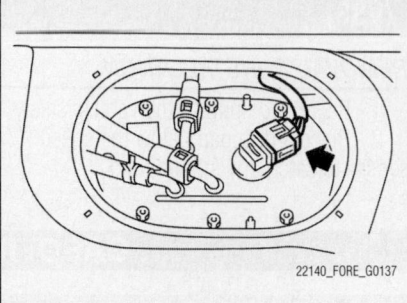

22140_FORE_G0137

Fig. 146 Fuel pump electrical connector

8. Remove the luggage floor mat.
9. Remove the fuel pump access cover retaining screws. Remove the access cover.
10. Disconnect the electrical connector from the fuel pump.
11. Disconnect and plug the fuel line hoses at the fuel pump.
12. Remove the clips and disconnect the jet pump hose.
13. Remove the fuel pump assembly retaining nuts. Remove the fuel pump from the vehicle.

To install:

14. Before servicing the vehicle, refer to the Precautions Section.
15. Installation is the reverse of the removal procedure.
16. Be sure to use a new gasket and retainer.
17. Tighten the fuel pump retaining nuts to 3.3 ft. lbs. (4.4 Nm). in an alternating sequence pattern.
18. Continue the installation in the reverse order of the removal procedure.
19. Start the engine and check for leaks, correct as required.

FUEL TANK

DRAINING

1. Place "NO FIRE" signs near the working area.
2. Before servicing the vehicle, refer to the Precautions Section.

➡ **This procedure must be performed prior to servicing any component of the fuel injection system.**

3. Disconnect the fuel pump connector from the fuel pump relay.
4. Release the fuel pressure.
5. Disconnect the ground cable from the battery.
6. Remove the fuel pump.

7. Drain fuel from the fuel pump attachment section using pump, etc.

✳✳ CAUTION

Be sure to use an anti-gasoline pump.

8. Install the fuel pump.

REMOVAL & INSTALLATION

See Figures 147 through 149.

✳✳ CAUTION

Place "NO FIRE" signs near the working area. Be careful not to spill fuel.

➡ **This procedure must be performed prior to servicing any component of the fuel injection system.**

1. Before servicing the vehicle, refer to the Precautions Section.
2. Disconnect the fuel pump connector from the fuel pump relay.
3. Properly relieve the fuel system pressure.
4. Remove the fuel cap.
5. Raise and safely support the vehicle.
6. Drain fuel from fuel tank.
7. Remove the rear seat.
8. Disconnect the connector between the fuel tank cord and the rear harness.
9. Push the grommet which holds the fuel tank cord on the floor panel into under the body.
10. Remove the rear crossmember.
11. Remove the canisters.
12. Disconnect the connector from the pressure control solenoid valve.
13. Loosen the clamp and disconnect the fuel filler hose (A) and evaporation hose (B) from the fuel filler pipe.
14. Move the retainer for fuel delivery tube, and disconnect the quick connector.
15. Support the fuel tank with transmission jack, remove the bolts from bands and dismount the fuel tank from the vehicle.

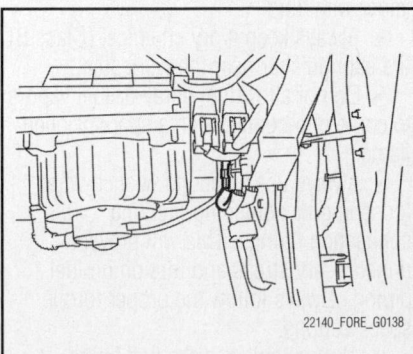

22140_FORE_G0138

Fig. 147 Fuel pump relay view next to blower motor

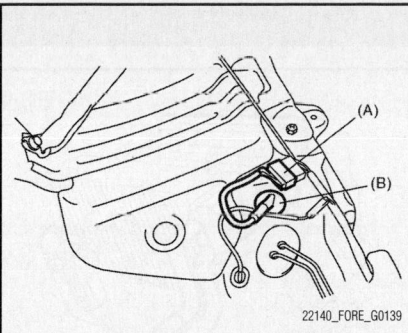

Fig. 148 Connector (A), and grommet (B) shown

A helper is required to perform this work. Fuel may remain in the fuel tank. Be careful not to lose balance and drop the fuel tank when removing it, as this remaining fuel may offset the weight balance of the tank.

To install:

16. Support the fuel tank with transmission jack and push the fuel tank harness into access hole along with the grommet.

17. Set the fuel tank and temporarily tighten the bolts of fuel tank bands.

18. Insert the fuel filler hose (A) approx. 35 to 40 mm (1.38 to 1.57 in) over the lower end of fuel filler pipe and tighten the clamp.

19. Insert the evaporation hose (B) into

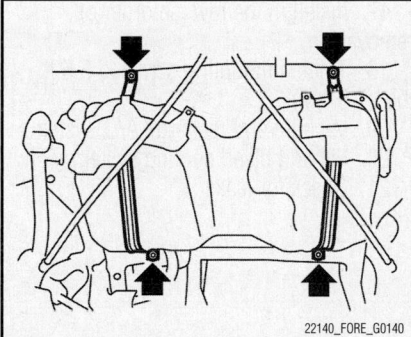

Fig. 149 Mounting bolts, and fuel tank bands

the lower end of evaporation pipe, and secure the clamps and clips.

20. Connect the fuel hoses, and hold them with clips and quick connector.

21. Connect the connector to pressure control solenoid valve.

22. Install the canister.

23. Tighten the band mounting bolts to 24.3 ft. lbs. (33 Nm).

24. Install the rear crossmember.

25. Connect the connector to the fuel tank cord and plug the service hole with grommet.

26. Set the rear seat and floor mats.

27. Connect the connector to fuel pump relay.

IDLE SPEED

ADJUSTMENT

Idle speed is maintained by the Electronic Control Module (ECM). No adjustment is necessary or possible.

THROTTLE BODY

REMOVAL & INSTALLATION

2.5L DOHC Turbo Engine

See Figure 150.

1. Disconnect the negative battery cable.

2. Remove the intercooler.

3. Disconnect the throttle position sensor and manifold absolute pressure (MAP) sensor.

4. Disconnect the coolant hoses from the throttle body.

5. Remove the throttle body mounting bolts.

6. Remove the throttle body and gasket.

To install:

7. Install the throttle body with a new gasket.

8. Tighten the mounting bolts to 6 ft. lbs. (8 Nm).

9. The remainder of the installation is the reverse order of removal. What is head

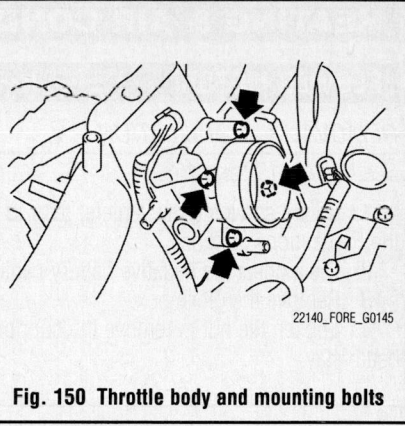

Fig. 150 Throttle body and mounting bolts

2.5L SOHC Non-Turbo Engine

See Figure 151.

1. Before servicing the vehicle, refer to the Precautions Section.

2. Disconnect the ground cable from battery.

3. Remove the air intake chamber.

4. Disconnect the connectors from throttle position sensor and manifold absolute pressure sensor.

5. Disconnect the engine coolant hoses from throttle body.

6. Remove the bolt which installs throttle body on intake manifold.

To install:

7. Tighten throttle body mounting bolts to 6 ft. lbs. (8 Nm).

8. The remainder of the installation is the reverse order of removal.

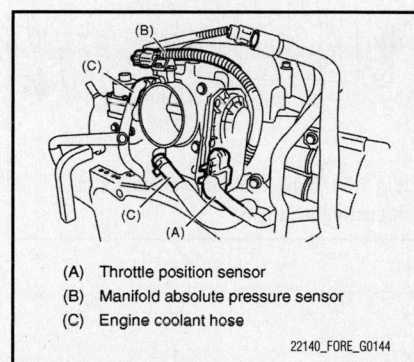

(A) Throttle position sensor
(B) Manifold absolute pressure sensor
(C) Engine coolant hose

Fig. 151 2.5L SOHC Non-Turbo Engine throttle body

HEATING & AIR CONDITIONING SYSTEM

BLOWER MOTOR

REMOVAL & INSTALLATION

See Figures 152 and 153.

1. Before servicing the vehicle, refer to the Precautions Section.
2. Disconnect the negative battery cable.
3. Remove the glove box.
4. Loosen the nut to remove the support beam stay.

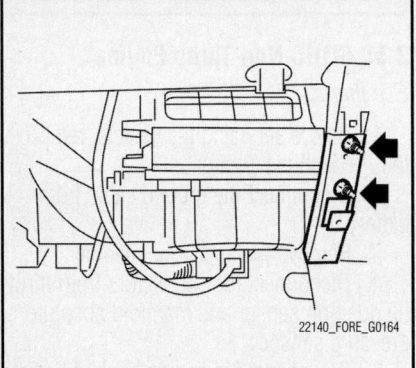

Fig. 152 Support beam stay and mounting nuts

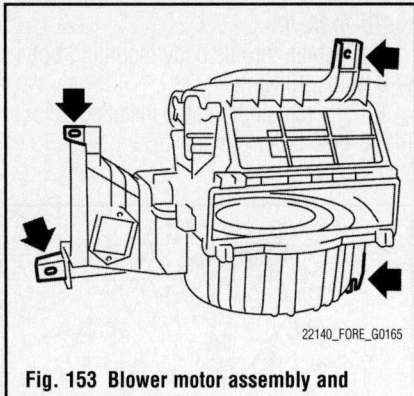

Fig. 153 Blower motor assembly and mounting locations

5. Disconnect the blower motor wiring harness.
6. Disconnect the blower resistor connector.
7. Loosen the bolt and nut to remove the blower motor unit assembly.
8. Installation is the reverse order of removal.

HEATER CORE

REMOVAL & INSTALLATION

See Figures 154 and 155.

1. Before servicing the vehicle, refer to the Precautions Section.
2. Disconnect the ground cable from the battery.
3. Using the refrigerant recovery system, discharge refrigerant.
4. Drain coolant from the radiator.
5. Remove the bolts securing expansion valve and pipe in engine compartment. Release the heater hose clamps in engine compartment to disconnect the hoses.

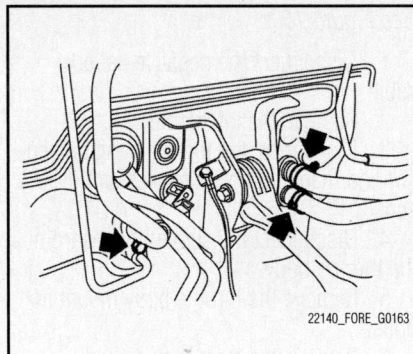

Fig. 154 Remove expansion valve, pipe and disconnect the heater hoses

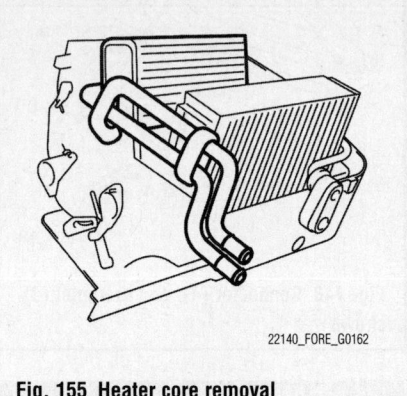

Fig. 155 Heater core removal

6. Remove the instrument panel.
7. Remove the support beam.
8. Remove the blower motor unit assembly.
9. Disconnect the servo motor connector.
10. Loosen the bolt and nuts and remove the heater and cooling unit.
11. Open the heater core pipe cover.
12. Loosen the screws to remove the mode actuator.
13. Loosen the screws to remove the foot duct.
14. Loosen the screws to remove evaporator cover.
15. Loosen the screws to remove lower case.
16. Remove the heater core.

To install:

17. Install in the reverse order of removal.
18. Tighten mounting screw to 5.5 ft. lbs. (7.5 Nm).
19. Vacuum and recharge A/C system.
20. Fill and bleed cooling system.
21. Check for leaks

STEERING

POWER RACK & PINION STEERING GEAR

REMOVAL & INSTALLATION

See Figures 156 and 157.

1. Before servicing the vehicle, refer to the Precautions Section.
2. Disconnect the negative battery cable.
3. Loosen the front wheel nut.
4. Raise and support the vehicle safely.
5. Remove the tires and wheels.
6. Remove the undercover.
7. Remove the sub frame. Leave bolt (1) connected by a few threads and remove the bolts in the sequence illustrated. Once the other bolts are removed, remove bolt (1) and the sub frame.
8. Remove the front exhaust pipe.
9. Remove the cotter pin and castle nut. Using a puller, remove the tie rod end from the knuckle arm.
10. Remove the jack up plate. Remove the front stabilizer.
11. Disconnect the power steering fluid pipe at the center of the gearbox and attach a vinyl hose. Discharge the fluid into a suitable container by turning the steering wheel fully clockwise and counterclockwise. Disconnect the other fluid line, and repeat the discharge procedure.
12. Remove the steering wheel. Make a matchmark on the universal joint. Remove the universal joint bolts and remove the joint from the vehicle.
13. Disconnect the fluid lines from the steering gear, pressure hose first.
14. Remove the steering gear retaining bolts and clamps securing the steering gear to the crossmember. Remove the steering gear from the vehicle.

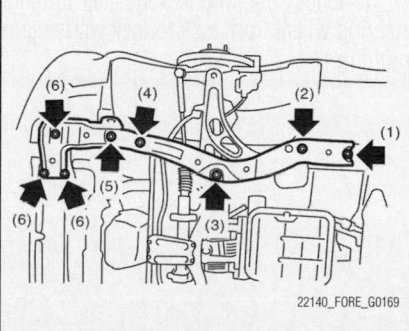

22140_FORE_G0169

Fig. 156 Sub frame and mounting bolt view

To install:

15. Insert the steering gear into the crossmember.

➡**Be careful not to damage the gearbox boot.**

16. Tighten the steering gear to the crossmember bracket to 44.1 ft. lbs. (55 Nm).
17. Connect the fluid lines.
18. Align the cutout at the serrated section of the column shaft and yoke. Insert the universal joint into the column shaft.
19. Align the mating marks and insert the universal joint to serrated section of the steering gear assembly. Tighten the bolt to 17.4 ft. lbs.
20. Continue the installation in the reverse order of the removal procedure.
21. When installing the sub frame be sure to torque the bolts to specification and in the proper sequence. Tighten bolts marked (T1) to 41 ft. lbs. (55 Nm) and the bolts marked (T2) to 52 ft. lbs. (71 Nm).
22. Fill the power steering system with the proper grade and type fluid.
23. Start the engine and check for leaks. Correct as required.

POWER STEERING PUMP

REMOVAL & INSTALLATION

See Figures 158 through 161.

1. Before servicing the vehicle, refer to the Precautions Section.
2. Disconnect the negative battery cable.
3. Drain about ⅓ of the power steering fluid from the power steering reservoir.
4. Remove the accessory drive belt cover.

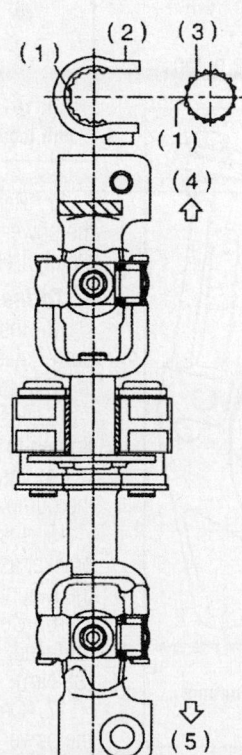

(1) Cutout portion
(2) Yoke
(3) Column shaft
(4) Column shaft side
(5) Gearbox side

09490_SBCR_G0013

Fig. 157 Steering column joint alignment

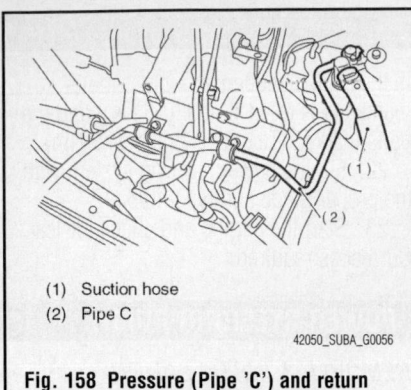

(1) Suction hose
(2) Pipe C

42050_SUBA_G0056

Fig. 158 Pressure (Pipe 'C') and return (suction) hose component locations

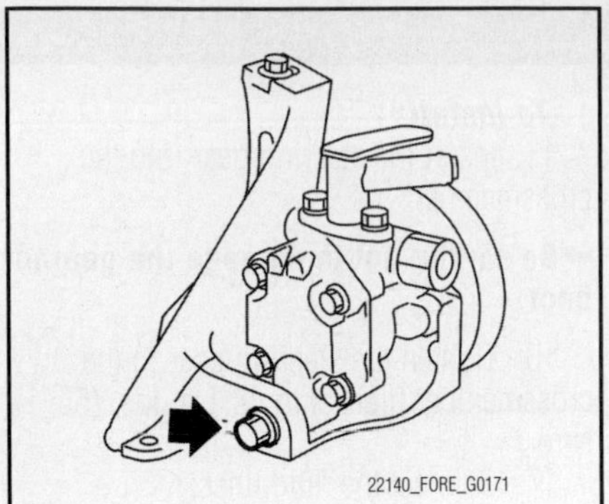

5. Loosen the lock bolt and slider bolt, and remove the power steering pump belt.

6. Disconnect the electrical connector form the power steering pump switch.

7. Disconnect pressure and return hoses from the power steering pump.

8. Remove power steering pump bracket mounting bolts.

9. Place the power steering pump in a vise and remove the bolts from the front side of the pump.

10. Remove the socket from the pump.

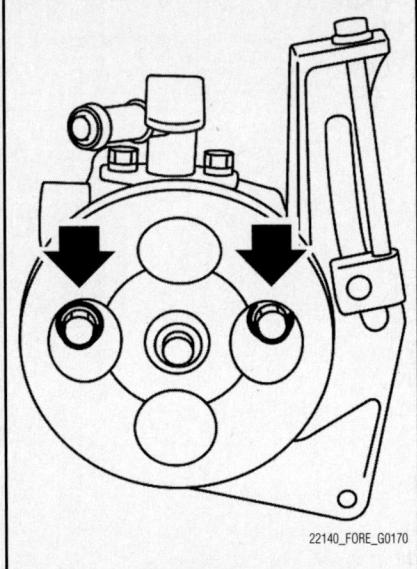

22140_FORE_G0170

Fig. 159 Front power steering mounting bolts

22140_FORE_G0171

Fig. 160 Rear power steering mounting bolt

42050_SUBA_G0057

Fig. 161 Power steering pump mounting bolt locations—2.5L engines

11. Remove the bolt from the rear side of the power steering pump to remove it completely from the mounting bracket.

To install:

12. Install the mounting bracket to the power steering pump. Tighten the rear bolt to 28 ft. lbs. (37 Nm)

13. Install the socket to the pump and tighten to 5 ft. lbs. (7 Nm).

14. Install the bolts on the front side of the pump. Tighten to 12 ft. lbs. (16 Nm).

15. Install the power steering pump bracket assembly into the vehicle. Tighten the mounting bolts to 16 ft. lbs. (22 Nm).

16. Connect the pipes that were removed. Tighten the joint nuts to 29 ft. lbs. (39 Nm).

17. Connect the electrical connector for the power steering oil pressure switch.

18. The remainder of the installation is the reverse order of removal.

19. Bleed the power steering system.

20. Start the engine and check for leaks.

BLEEDING

1. Before servicing the vehicle, refer to the Precautions Section.

2. Fill the power steering fluid reservoir about half way with the specified fluid.

3. Continue to turn the steering wheel slowly from lock to lock until bubbles stop appearing on oil surface while keeping the fluid at that level.

4. If turning the steering wheel in low fluid level condition, air will be sucked in pipe. In this case, leave it about half an hour and then repeat the previous step.

5. Lift up the vehicle, start the engine and let it idle.

6. Continue to turn the steering wheel slowly from lock to lock again until bubbles stop appearing on oil surface while keeping the fluid at that level. It is normal that bubbles stop appearing after three times turning of steering wheel from lock to lock.

7. In case the bubbles do not stop appearing in the tank, leave it about half an hour and then begin the process again.

8. Lower the vehicle, and then idle the engine.

9. Continue to turn the steering wheel from lock to lock until bubbles stop appearing and change of the fluid level is within 3 mm (0.12 in).

10. In case the following happens, leave it about half an hour and then do step 5–8 again.

 a. The fluid level changes over 3 mm (0.12 in).

 b. Bubbles remain on the upper surface of the fluid.

 c. Grinding noise is generated from oil pump.

11. Check the fluid leakage after turning steering wheel from lock to lock with engine running.

SUSPENSION   **FRONT SUSPENSION**

COIL SPRING

REMOVAL & INSTALLATION

See Figure 162.

1. Before servicing the vehicle, refer to the Precautions Section.
2. Remove the strut from the vehicle.
3. Using a coil spring compressor tool, carefully compress the spring. Remove the self locking nut.
4. Remove the strut mount, upper spring and rubber seat from the strut.
5. Gradually decrease the compression force of the spring compressor tool. Remove the coil spring.
6. Remove the dust cover and helper spring.
7. Check for the presence of air in the damping force generating mechanism.
8. Using the spring compression tool, compress the coil spring.
9. Remove coil spring.

To install:

➡**Be sure to properly install the coil spring.**

10. Position the coil spring so that its end face fits good into the spring seat.
11. Install the helper spring and dust cover to the piston rod.
12. Pull the piston rod fully upward, and install the rubber seat and spring seat.
13. Install the strut mount to the piston rod, and then tighten the self locking nut, temporarily. Be sure to use a new self locking nut.
14. Use a hexagon wrench to prevent the strut rod from turning.
15. Tighten the self locking nut to 41 ft. lbs. (55 Nm).
16. Carefully loosen the coil spring.

LOWER BALL JOINT

REMOVAL & INSTALLATION

See Figure 163.

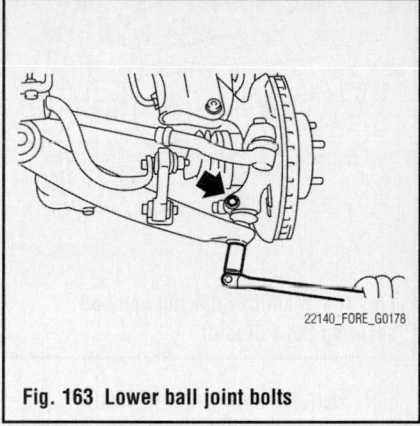

22140_FORE_G0178

Fig. 163 Lower ball joint bolts

1. Before servicing the vehicle, refer to the Precautions Section.
2. Remove the wheels.
3. Pull out the cotter pin from the ball stud, remove the castle nut, and extract the ball stud from the transverse link.
4. Remove the bolts which secure the ball joint to the housing.
5. Extract the ball joint from housing.

To install:

6. Insert the ball joint into housing and tighten to 37 ft. lbs. (50 Nm).
7. Connect the ball joint to transverse link and tighten the castle nut to 30 ft. lbs. (40 Nm).
8. Tighten the castle nut further but within 60° until the hole in ball stud is aligned with a slot in castle nut. Then, insert a new cotter pin and bend it around the castle nut
9. Install the front wheels.

LOWER CONTROL ARM

REMOVAL & INSTALLATION

See Figures 164 and 165.

1. Before servicing the vehicle, refer to the Precautions Section.
2. Disconnect the negative battery cable.
3. Raise and support the vehicle safely.
4. Remove the tire and wheel.
5. Remove the sub frame.
6. Disconnect the stabilizer link from the lower control arm.
7. Remove the bolt securing the ball joint of the lower control arm.
8. Remove the nut (do not remove the bolt) securing the lower control arm to the crossmember.

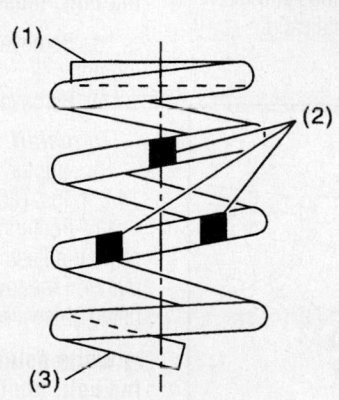

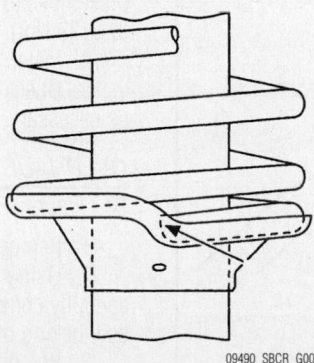

(1) Flat (top side)

(2) Identification paint

(3) Inclined (bottom side)

09490_SBCR_G0099

Fig. 162 Strut spring alignment

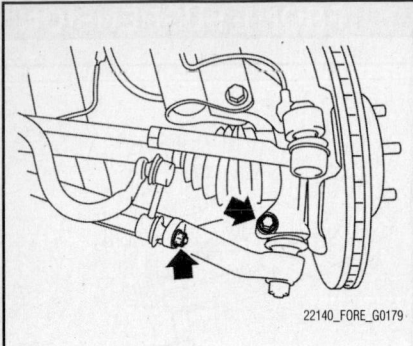

Fig. 164 Stabilizer link nut and bolt securing the ball joint

9. Remove the two bolts securing the bushing bracket of the lower control arm to the vehicle body at the rear bushing.

10. Remove the ball joint from the housing.

11. Remove the bolt securing the lower control arm to the crossmember. Remove the lower control arm from the crossmember.

To install:

12. Install the lower control arm to its mounting.

13. Temporarily tighten the two bolts used to secure the rear bushing of the lower control arm

➡These bolts should be tightened so that they can still move back and forth in the oblong shaped hole in the bracket, which holds the bushing.

14. Continue the installation in the reverse order of the removal procedure.

15. Always fully tighten the rubber bushings when the wheels are in full contact with the ground and the vehicle is at curb height.

16. Tighten stabilizer link nut to 33 ft. lbs. (45 Nm).

17. Tighten lower control arm rear bushing bracket to 92.3 ft. lbs. (125 Nm).

18. Tighten the lower control arm rear bushing to body to 184 ft. lbs. (250 Nm)

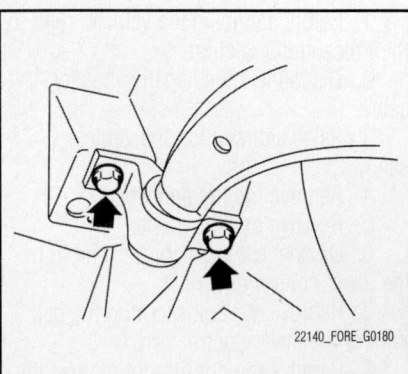

Fig. 165 Stabilizer link nut and bolt securing the ball joint

19. Check and adjust alignment, as required.

STABILIZER BAR

REMOVAL & INSTALLATION

See Figures 166 and 167.

1. Before servicing the vehicle, refer to the Precautions Section.

2. Raise and support the vehicle safely.

3. Remove the jack up plate from the lower part of the crossmember.

4. Remove the sub frame.

5. Remove the nut which secures the stabilizer link to the front transverse link.

6. Remove the bolts that secure the stabilizer bar to the crossmember.

7. Remove the stabilizer bar from the vehicle.

To install:

8. Installation is the reverse of the removal procedure.

9. Install the rubber bushing, on the front crossmember side, while aligning it with the paint mark on the stabilizer bar.

10. Be sure that the bushings and the stabilizer have the same identification colors.

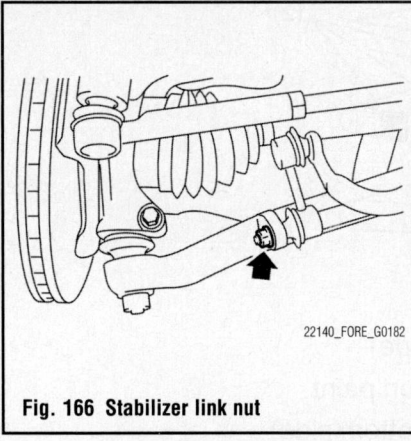

Fig. 166 Stabilizer link nut

11. Always fully tighten the rubber bushings when the wheels are in full contact with the ground and the vehicle is at curb height.

12. Tighten stabilizer link nuts to 33 ft. lbs. (45 Nm).

13. Tighten stabilizer to crossmember to 18 ft. lbs. (25 Nm)

STRUT

REMOVAL & INSTALLATION

See Figure 168.

1. Before servicing the vehicle, refer to the Precautions Section.

2. Raise and support the vehicle safely.

3. Remove the tire and wheel.

4. Remove the bolt retaining the brake hose to the strut.

5. Make and alignment mark on the camber adjusting bolt which secures the strut to the housing.

6. Remove the bolt retaining the ABS wheel speed sensor harness.

7. Remove the two bolts retaining the strut to the housing.

➡While holding the head of the adjusting bolt, loosen the self locking nut.

8. Remove the three upper strut retaining nuts.

9. Remove the strut from the vehicle.

To install:

10. Tighten the upper retaining nuts to 14.8 ft. lbs. (20 Nm).

11. Position the alignment mark on the camber adjusting bolt with the alignment mark on the lower side of the strut. Install using a new self locking nut.

➡While holding the head of the adjusting bolt, tighten the self locking nut to 129 ft. lbs. (175 Nm).

12. Secure the ABS wheel speed sensor harness to the strut and tighten to 24.3 ft. lbs. (33 Nm).

13. Install the bolts which secure the brake hose to the strut. Tighten to 24.3 ft. lbs. (33 Nm).

14. Install the wheel.

15. Check and adjust wheel alignment, as necessary.

OVERHAUL

See Figure 169.

1. Remove the strut from the vehicle.

2. Using a coil spring compressor tool, carefully compress the spring. Remove the self locking nut.

3. Remove the strut mount, upper spring and rubber seat from the strut.

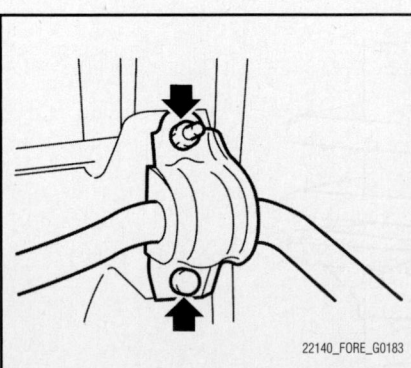

Fig. 167 Bolts that secure the stabilizer bar

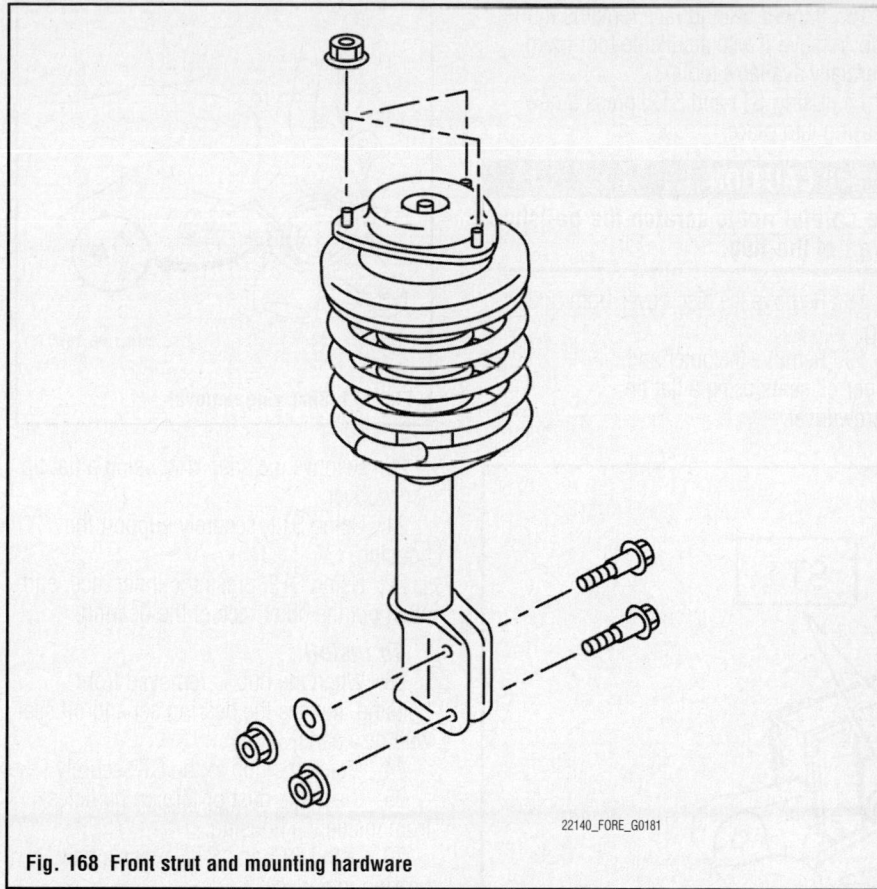

Fig. 168 Front strut and mounting hardware

(1) Flat (top side)
(2) Identification paint
(3) Inclined (bottom side)

Fig. 169 Strut spring alignment

4. Gradually decrease the compression force of the spring compressor tool. Remove the coil spring.

5. Remove the dust cover and helper spring.

6. Check for the presence of air in the damping force generating mechanism.

7. Using the spring compression tool, compress the coil spring.

8. Remove coil spring.

To install:

➡**Be sure to properly install the coil spring.**

9. Position the coil spring so that its end face fits good into the spring seat.

10. Install the helper spring and dust cover to the piston rod.

11. Pull the piston rod fully upward, and install the rubber seat and spring seat.

12. Install the strut mount to the piston rod, and then tighten the self locking nut, temporarily. Be sure to use a new self locking nut.

13. Use a hexagon wrench to prevent the strut rod from turning.

14. Tighten the self locking nut to 41 ft. lbs. (55 Nm).

15. Carefully loosen the coil spring.

WHEEL BEARINGS

REMOVAL & INSTALLATION
See Figures 170 and 171.

1. Before servicing the vehicle, refer to the Precautions Section.

2. Raise and safely support the vehicle.

3. Remove the front wheel.

4. While depressing the brake pedal, un-stake and remove axle nut.

✲✲ CAUTION

Be sure to loosen and retighten axle nut after removing the wheel from vehicle or damage to the wheel bearings may occur.

5. Remove the stabilizer link.

6. Remove the brake caliper without disconnect the brake housing. Securely hang the caliper from the strut assembly.

7. Remove the brake rotor.

8. Remove the cotter pin and castle nut that secure the tie-rod end to the knuckle.

9. Using a suitable puller, remove the tie rod joint from the knuckle.

10. Remove the ABS sensor assembly, if equipped.

11. Remove the bolt that holds the sensor wiring harness to the strut.

12. Disconnect the lower ball joint from the steering knuckle.

13. Remove the front axle shaft assembly from the hub.

➡**If the axle shaft is difficult to remove, use Special Tool 926470000 Puller and 927140000 Plate to assist with the removal.**

14. Using ST1, securely support the housing and hub.

15. Attach ST2 to housing and drive hub out.

16. If inner bearing race remains in the hub, remove it with a suitable tool (commercially available tools).

17. Using ST1 and ST2, press a new bearing into place.

✳✳ CAUTION

Be careful not to scratch the polished area of the hub.

18. Remove the disc cover from housing.

19. Remove the outer and inner oil seals using a flat tip screwdriver.

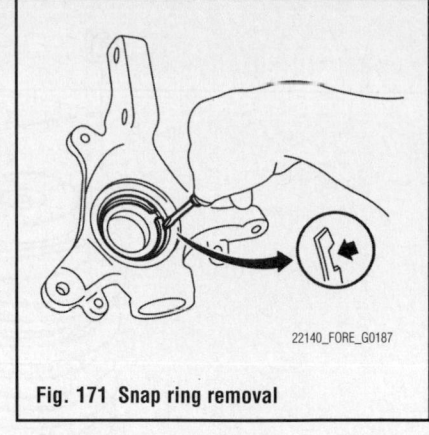

22140_FORE_G0187

Fig. 171 Snap ring removal

20. Remove the snap ring using a flat tip screwdriver.

21. Using ST1, securely support the housing.

22. Using ST2, press the inner race, and push out the outer race of the bearing.

To install:

23. When the hub is removed from housing, replace the bearing set and oil seal with new parts.

24. Attach the hub to the ST securely.

25. Clean the dust or foreign particles from inside the housing.

26. Using ST1 and ST2, press a new bearing into place.

✳✳ CAUTION

Always press the outer race when installing bearings. Be careful not to remove the plastic lock from the inner race when installing the bearings.

27. Using a pliers, securely install the snap ring.

28. Using the ST1 and ST2, press the outer oil seal until it contacts the bottom of housing.

29. Using the ST1 and ST2, press the inner oil seal until it contacts the circlip.

30. Invert the ST and housing (up and down).

31. Apply sufficient grease to the oil seal lip.

32. Install the disc cover to housing with three bolts. Tighten to 13 ft. lbs. (18 Nm).

33. Attach the hub to ST1 securely

34. Clean dust and foreign particles from the polished surface of hub.

35. Place ST2 against the bearing inner race to press the bearing into the hub.

36. The remainder of the installation is the reverse of the removal procedure.

ST1

ST2

ST1

ST2

22140_FORE_G0186

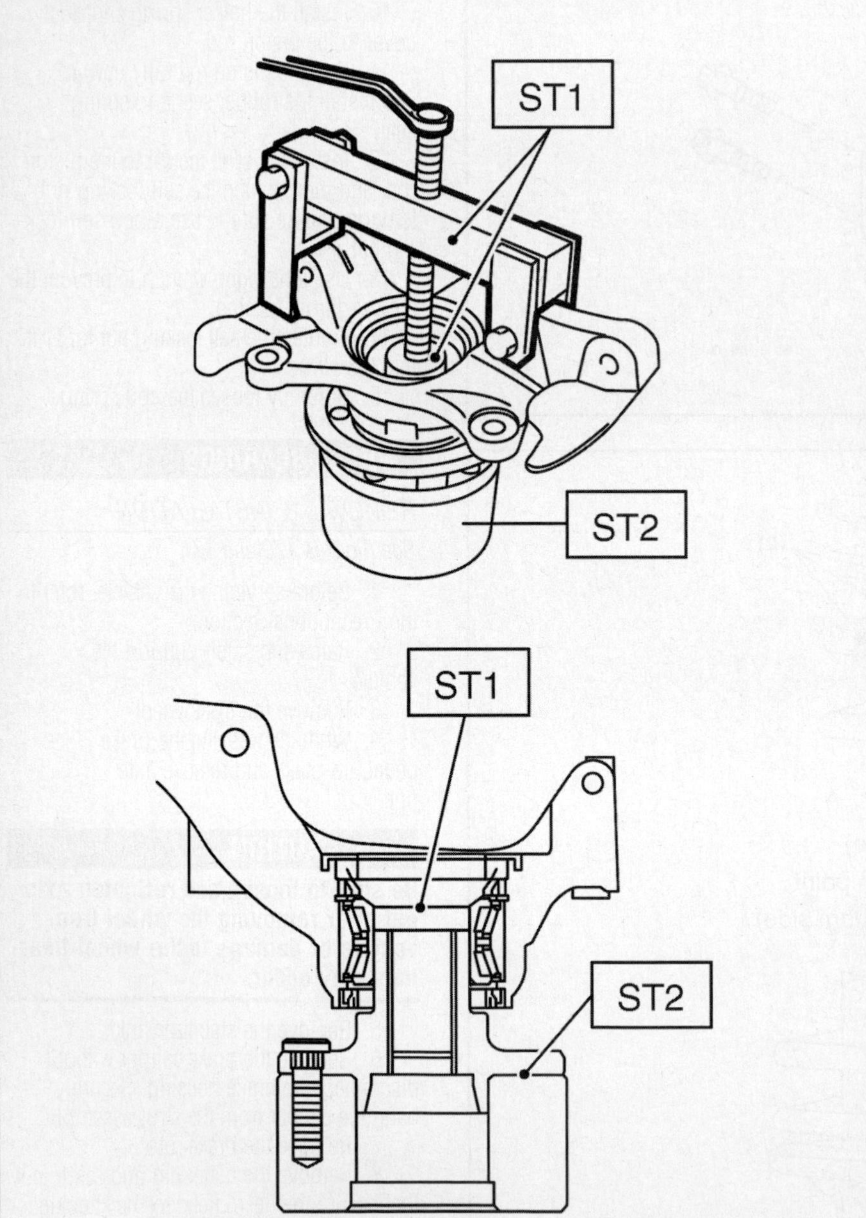

Fig. 170 Special tools ST1 and ST2

SUSPENSION   **REAR SUSPENSION**

COIL SPRING

REMOVAL & INSTALLATION

See Figures 172 through 174.

1. Before servicing the vehicle, refer to the Precautions Section.

2. Remove the strut from the vehicle.

3. Using a coil spring compressor tool, carefully compress the spring. Remove the self locking nut.

4. Remove the strut mount, upper spring and rubber seat from the strut.

5. Gradually decrease the compression force of the spring compressor tool.

6. Remove the coil spring.

To install:

7. Before installing the coil spring, strut mount, etc. on the strut, check the condition of air inside the strut damper mechanism to make sure that excessive air is not inhibiting the creation of appropriate damping force.

8. Using a coil spring compressor, compress the coil spring.

9. Set the coil spring correctly so that its end face seats well in the spring seat as shown in the figure.

10. Install the helper and dust cover to the piston rod.

11. Pull the piston rod fully upward, and install the rubber seat and spring seat.

12. Position the upper spring seat.

13. Install the strut mount to piston rod, and temporarily attach and tighten a new self locking nut.

14. Using a hex wrench to prevent strut rod from turning, tighten the self-locking nut with the ST-92776000 strut mount socket. Tighten to 41 ft. lbs. (55 Nm).

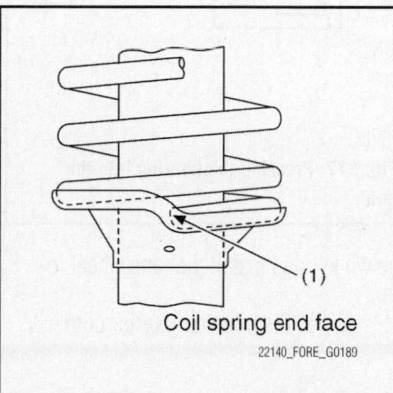

(1)

Coil spring end face

22140_FORE_G0189

Fig. 173 Coil spring seat alignment

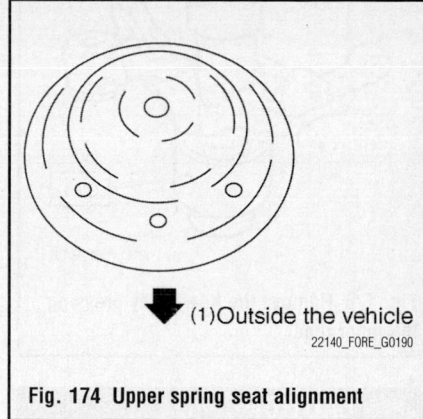

⬇ (1)Outside the vehicle

22140_FORE_G0190

Fig. 174 Upper spring seat alignment

15. Carefully loosen the coil spring tension.

WHEEL BEARINGS

REMOVAL & INSTALLATION

See Figures 175 through 177.

1. Before servicing the vehicle, refer to the Precautions Section.

2. Before servicing the vehicle, refer to the Precautions Section.

3. Remove rear hub and bearing housing assembly.

4. Remove the back plate from the rear housing

5. Remove the outer and inner oil seals using a flat tip screwdriver.

6. Remove the snap ring using a flat tip screwdriver.

➡**Be careful not to damage housing during removal.**

7. Using ST1 and ST2, remove the bearing by pressing the inner race.

8. Remove the tone wheel bolts and remove the tone wheel from the hub.

9. Using ST, press the hub bolt out.

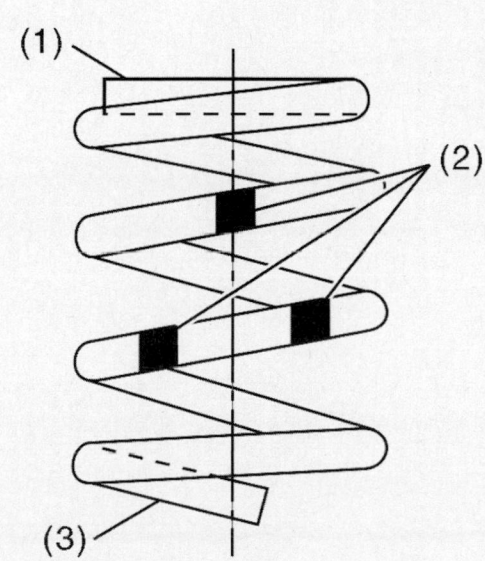

(1) ──

(2)

(2)

(3) ──

(1) Flat (top side)
(2) Identification paint
(3) Inclined (bottom side)

22140_FORE_G0188

Fig. 172 Coil spring alignment

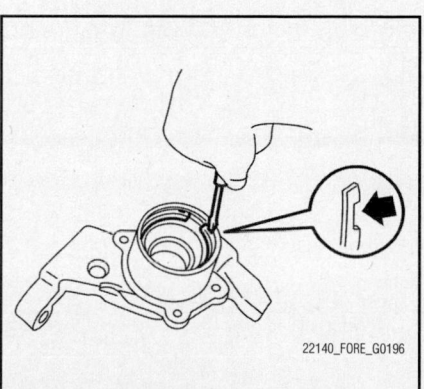

22140_FORE_G0196

Fig. 175 Remove the snap ring using a flat tip screwdriver

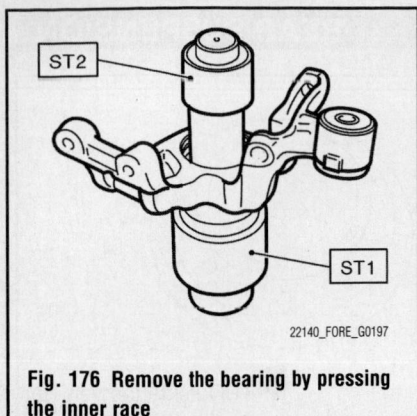

Fig. 176 Remove the bearing by pressing the inner race

✳✳ CAUTION

Be careful not to hammer the hub bolts. Doing so may deform the hub.

To install:

➡ **When the hub is removed from the housing, replace the bearing set and oil seal.**

10. Remove the foreign particles (dust, rust, etc.) from the mating surfaces of hub

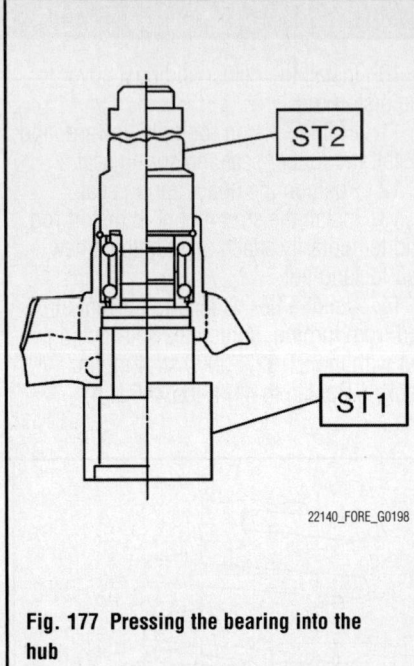

Fig. 177 Pressing the bearing into the hub

tone wheel, and install the tone wheel to hub.

11. Clean the housing interior com-pletely. Using ST1 and ST2, press the bearing into the housing.

12. Be careful not to remove the plastic lock from the inner race when installing the bearings.

13. Using snap ring pliers, securely install the snap ring.

14. Using the ST1 and ST2, press the outer oil seal unit until it comes in contact with snap ring.

15. Invert both ST1 and housing (upside down).

16. Using ST2, press the inner oil seal into the housing until it touches the bottom.

17. Using ST1 and ST2, press the sub seal into place.

18. Apply sufficient grease to the oil seal lip.

19. Install back plate to the rear housing.

20. Using ST1 and ST2, press the bearing into the hub.

21. Install the rear hub and bearing housing assembly.

ADJUSTMENT

The wheel bearings are not adjustable.

SPECIFICATIONS AND MAINTENANCE CHARTS

ENGINE AND VEHICLE IDENTIFICATION CHART

Code ①	Liters (cc)	Cu. In.	Cyl.	Fuel Sys.	Type	Eng. Mfg.	Code ②	Year
6	2.5 (2457)	150	4	MFI	DOHC	Subaru	9	2009
7	2.5 (2457)	150	4	MFI	SOHC	Subaru	A	2010

MFI: Multiport Fuel Injection

SOHC: Single Overhead Camshaft

DOHC: Double Overhead Camshaft

① 6th digit of the VIN

② 10th digit of the VIN

37671_IMPZ_C0001

GENERAL ENGINE SPECIFICATIONS

Year	Model	Engine Displacement Liters (cc)	Engine ID/VIN	Net Horsepower @ rpm	Net Torque @ rpm (ft. lbs.)	Bore x Stroke (in.)	Compression Ratio	Oil Pressure @ rpm
2009	Impreza	2.5 (2457)	6	165@5600	166@4000	3.92x3.11	10.0:1	14@600
	Impreza	2.5 (2457)	7	165@5600	166@4000	3.92x3.11	10.0:1	14@600
2010	Impreza	2.5 (2457)	6	165@5600	166@4000	3.92x3.11	10.0:1	14@600
	Impreza	2.5 (2457)	7	165@5600	166@4000	3.92x3.11	10.0:1	14@600

ENGINE TUNE-UP SPECIFICATIONS

Year	Model	Engine Displacement Liters	Spark Plug Gap (in.)	Ignition Timing (deg.) ① MT	AT	Fuel Pump (psi)	Idle Speed (rpm) MT	AT	Valve Clearance ② In.	Ex.
2009	Impreza	2.5	0.028-0.031	12 ③	17 ③	34-38	650-750	650-750	0.0071-0.0087	0.0090-0.0106
2010	Impreza	2.5	0.028-0.031	12 ③	17 ③	34-38	650-750	650-750	0.0071-0.0087	0.0090-0.0106

Note: The Vehicle Emission Control Information label often reflects specification changes made during production. The lable figures mudst be used if they differ from those in this chart.

① Before Top Dead Center

② Measured with engine cold

③ Plus or minus (10 deg.) at 750 rpm

④ Plus or minus (8 deg.) at 650 rpm

37671_IMPZ_C0003

CAPACITIES

Year	Model	Engine Displacement Liters	Engine Oil with Filter (qts.)	Transmission (pts.) 5-Spd	Transmission (pts.) Auto. ①	Transfer Case (pts.)	Drive Axle Front (pts.)	Drive Axle Rear (pts.)	Fuel Tank (gal.)	Cooling System (qts.)
2009	Impreza	2.5	4.2	7.4	20.0	—	2.6	1.6	16.9	6.9
2010	Impreza	2.5	4.2	7.4	20.0	—	2.6	1.6	16.9	6.9

Note: All capacities are approximate. Add fluid gradually and check to be sure a proper fluid level is obtained.

① Ovehaul specification

37671_IMPZ_C0004

FLUID SPECIFICATIONS

Year	Model	Engine Displacement Liters	Engine ID/VIN	Engine Oil	Auto. Trans.	Drive Axle	Power Steering Fluid	Brake Master Cylinder	Engine Coolant
2009	Impreza	2.5	6/7	5W-30	Subaru ATF HP	75W-90	Dexron III	DOT 3	Subaru Coolant
2010	Impreza	2.5	6/7	5W-30	Subaru ATF HP	75W-90	Dexron III	DOT 3	Subaru Coolant

DOT: Department Of Transpotation

37671_IMPZ_C0005

VALVE SPECIFICATIONS

Year	Engine Displacement Liters	Seat Angle (deg.)	Face Angle (deg.)	Spring Test Pressure (lbs. @ in.)	Spring Installed Height (in.)	Stem-to-Guide Clearance (in.) Intake	Stem-to-Guide Clearance (in.) Exhaust	Stem Diameter (in.) Intake	Stem Diameter (in.) Exhaust
2009	2.5	45	45	102 - 118@ 1.315	①	0.0014- 0.0024	0.0016- 0.0026	0.2343- 0.2348	0.2341- 0.2346
2010	2.5	45	45	102 - 118@ 1.315	①	0.0014- 0.0024	0.0016- 0.0026	0.2343- 0.2348	0.2341- 0.2346

① Free length: 1.8913 in.
② Free length: 1.8421 in.

37671_IMPZ_C0007

CAMSHAFT AND BEARING SPECIFICATIONS CHART

All measurements are given in inches.

Year	Engine Displacement Liters	Engine ID/VIN	Journal Dia.	Brg. Oil Clearance	Shaft End-play	Runout	Journal Bore	Lobe Height Intake	Lobe Height Exhaust
2009	2.5 SOHC	6	1.2570-1.2577	0.0022-0.0035	0.0012-0.0035	0.0024	NA	1.8330-1.8370	1.8410-1.8440
	2.5 DOHC	7	1.0614-1.0620	0.0022-0.0035	0.0027-0.0047	0.0024	NA	1.5778-1.5817	1.5468-1.5507
2010	2.5 SOHC	6	1.2570-1.2577	0.0022-0.0035	0.0012-0.0035	0.0024	NA	1.8330-1.8370	1.8410-1.8440
	2.5 DOHC	7	1.0614-1.0620	0.0022-0.0035	0.0027-0.0047	0.0024	NA	1.5778-1.5817	1.5468-1.5507

NA: Not Available

① Intake Journals 4 and 5: 0.0010 - 0.0022 in.
   All Others: 0.0010 - 0.0024 in.

37671_IMPZ_C0006

CRANKSHAFT AND CONNECTING ROD SPECIFICATIONS

All measurements are given in inches.

Year	Engine Displacement Liters	Crankshaft Main Brg. Journal Dia.	Crankshaft Main Brg. Oil Clearance	Crankshaft Shaft End-play	Crankshaft Thrust on No.	Connecting Rod Journal Diameter	Connecting Rod Oil Clearance	Connecting Rod Side Clearance
2009	2.5	2.3619-2.3625	①	0.0012-0.0098	3	1.8891-1.8898	0.0005-0.0015	0.0028-0.0130
2010	2.5	2.3619-2.3625	①	0.0012-0.0098	3	1.8891-1.8898	0.0005-0.0015	0.0028-0.0130

① Journals 1 and 5: 0.0001 - 0.0016 in.
   Journals 2 and 4: 0.0004 - 0.0014 in.
   Journal 3: 0.0004 - 0.0014 in.
② Journals 1 and 5: 0.0001 - 0.0016 in.
   Journals 2 and 4: 0.0004 - 0.0018 in.
   Journal 3: 0.0004 - 0.0016 in.

37671_IMPZ_C0008

PISTON AND RING SPECIFICATIONS

All measurements are given in inches.

Year	Engine Displacement Liters	Piston Clearance	Ring Gap Top Compression	Ring Gap Bottom Compression	Ring Gap Oil Control	Ring Side Clearance Top Compression	Ring Side Clearance Bottom Compression	Ring Side Clearance Oil Control
2009	2.5	0.0004-0.0012	0.0079-0.0138	0.0146-0.0250	0.0079-0.0197	0.0016-0.0031	0.0012-0.0028	NA
2010	2.5	0.0004-0.0012	0.0079-0.0138	0.0146-0.0250	0.0079-0.0197	0.0016-0.0031	0.0012-0.0028	NA

NA: Not Available

37671_IMPZ_C0009

TORQUE SPECIFICATIONS
All readings in ft. lbs.

Year	Engine Displacement Liters	Cylinder Head Bolts	Main ① Bearing Bolts	Rod Bearing Bolts	Crankshaft Damper Bolts	Flywheel Bolts	Manifold Intake	Manifold Exhaust	Spark Plugs	Oil Pan Drain Plug
2009	2.5	②	③	38	133	51 - 55	18	26	15	33
2010	2.5	②	③	38	133	51 - 55	18	26	15	33

① Engine block connecting bolts

② Step 1: Tighten all bolts to 22 ft. lbs.

Step 2: Tighten all bolts to 51 ft. lbs.

Step 3: Loosen all botls 180 degrees.

Step 4: Repeat Step 3.

Step 5: Tighten bolts 1 and 2 to 25 ft. lbs.

Step 6: Tighten bolts 3, 4, 5 and 6 to 11 ft. lbs.

Step 7: Tighten all bolts 80 to 90 degrees.

Step 8: Repeat Step 7. Do not exceed 180 degrees total tightening.

③ Split engine case connecting bolts:

Short bolts: 17-20 ft. lbs.

Long bolts: 33-37 ft. lbs.

Smaller short bolts (if used) 5 ft. lbs.

④ Step 1: Tighten all bolts to 14 ft. lbs.

Step 2: Tighten all bolts to 37 ft. lbs.

Step 3: Loosen all bolts 180 degrees the an additional 180 degrees in two steps in the reverse order of tightening sequence.

Step 4: Tighten all bolts to 18 ft. lbs.

Step 5: Tighten all bolts to 18 ft. lbs.

Step 6: Tighten bolts 90 degrees

Step 7: Tighten bolts 1, 2, 3 and 4 90 degrees.

Step 8: Tighten bolts 5, 6, 7 and 8 45 degrees.

37671_IMPZ_C0010

NOTE:
Remove oil on the mating surface of bearing and cylinder block before installation. Apply engine oil to crankshaft pins.
1) Position the crankshaft and O-rings on the #1 and #3 cylinder block.
2) Apply liquid gasket to the mating surface of #1 and #3 cylinder block, and position #2 and #4 cylinder block.

Liquid gasket:
Three bond 1215 (Part No. 004403007) or equivalent

NOTE:
Do not allow liquid gasket to jut into O-ring grooves, oil passages, bearing grooves, etc.

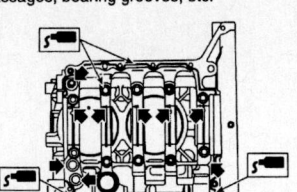

3) Apply a coat of engine oil to the washer and bolt thread.

4) Tighten the 10 mm cylinder block connecting bolts on LH side (A — D) in alphabetical sequence.

Tightening torque:
9.75 N·m (1.0 kgf-m, 7.2 ft-lb)

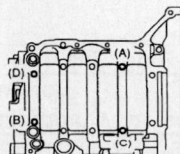

5) Tighten the 10 mm cylinder block connecting bolts on RH side (E — J) in alphabetical sequence.

Tightening torque:
9.75 N·m (1.0 kgf-m, 7.2 ft-lb)

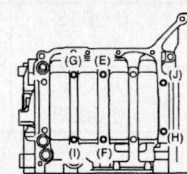

6) Further tighten the LH side bolts (A — D) in alphabetical sequence.

Tightening torque:
18 N·m (1.8 kgf-m, 13 ft-lb)

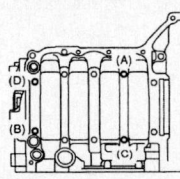

7) Further tighten the RH side bolts (E — J) in alphabetical sequence.

Tightening torque:
18 N·m (1.8 kgf-m, 13 ft-lb)

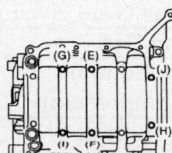

8) Further tighten the LH side bolts (A — D) in alphabetical sequence.
• (A), (C): Angle tightening

Tightening angle:
90°
• (B), (D): Torque tightening

Tightening torque:
40 N·m (4.1 kgf-m, 29.6 ft-lb)

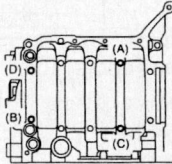

9) Tighten the RH side bolts (E — J) 90° further in alphabetical sequence.

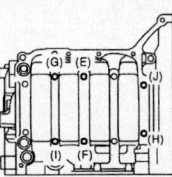

10) Tighten the 8 mm and 6 mm cylinder block connecting bolts on LH side (A — H) in alphabetical sequence.

Tightening torque:
(A) — (G): 25 N·m (2.5 kgf-m, 18.1 ft-lb)
(H): 6.4 N·m (0.65 kgf-m, 4.7 ft-lb)

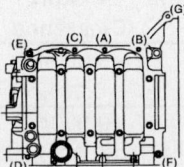

09490_SBCR_G0004

Fig. 1 Main bearing torque sequence—2.5L SOHC engine

1) Remove oil on the mating surface of bearing and cylinder block before installation. Apply engine oil to crankshaft pins.
2) Position the crankshaft and O-rings on the #1 and #3 cylinder block.
3) Apply liquid gasket to the mating surface of #1 and #3 cylinder blocks, and position cylinder block #2 and #4.

NOTE:
Do not allow liquid gasket to run over to O-ring grooves, oil passages, bearing grooves, etc.

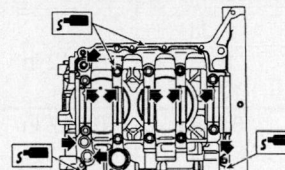

4) Apply a coat of engine oil to the washer and bolt thread.
5) Tighten the 10 mm cylinder block connecting bolts on LH side (A — D) in alphabetical sequence.

Tightening torque:
10 N·m (1.0 kgf-m, 7.2 ft-lb)

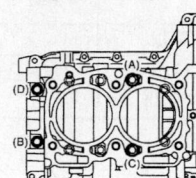

6) Tighten the 10 mm cylinder block connecting bolts on RH side (E — J) in alphabetical sequence.

Tightening torque:
10 N·m (1.0 kgf-m, 7.2 ft-lb)

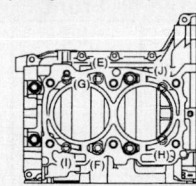

7) Further tighten the LH side bolts (A — D) in alphabetical sequence.

Tightening torque:
18 N·m (1.8 kgf-m, 13.0 ft-lb)

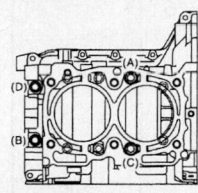

8) Further tighten the RH side bolts (E — J) in alphabetical sequence.

Tightening torque:
18 N·m (1.8 kgf-m, 13.0 ft-lb)

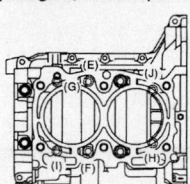

9) Further tighten the LH side bolts (A — D) in alphabetical sequence.
(A), (C): 90°
(B), (D): 40 N·m (4.1 kgf-m, 29.5 ft-lb)

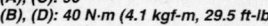
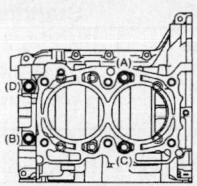

10) Tighten the RH side bolts (E — J) 90° further in alphabetical sequence.

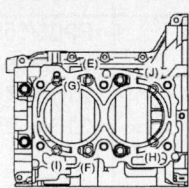

11) Tighten the 8 mm and 6 mm cylinder block connecting bolts on LH side (A — H) in alphabetical sequence.

Tightening torque:
(A) — (G): 25 N·m (2.5 kgf-m, 18.1 ft-lb)
(H): 6.4 N·m (0.65 kgf-m, 4.7 ft-lb)

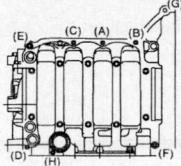

09490_SBCR_G0007

Fig. 2 Main bearing torque sequence—2.5L DOHC engine

WHEEL ALIGNMENT

Year	Model		Caster Range (+/-Deg.)	Caster Preferred Setting (Deg.)	Camber Range (+/-Deg.)	Camber Preferred Setting (Deg.)	Toe-in (in.)
2009	Impreza Sedan	F	1.00	+3.25	0.45	-0.15	0+/-0.12
		R	—	—	0.45	-1.25	0+/-0.12
	Impreza Sedan Turbo	F	1.00	+3.30	0.45	-0.25	0+/-0.12
		R	—	—	0.45	-1.25	0+/-0.12
	Impreza Sedan STI	F	1.00	+4.50	0.45	-0.30	0+/-0.12
		R	—	—	0.45	-1.15	0+/-0.12
	Impreza Wagon	F	1.00	+3.25	0.45	-0.10	0+/-0.12
		R	—	—	0.45	-1.15	0+/-0.12
	Impreza Wagon Turbo	F	1.00	+3.30	0.45	-0.20	0+/-0.12
		R	—	—	0.45	-1.20	0+/-0.12
2010	Impreza Sedan	F	1.00	+3.25	0.45	-0.15	0+/-0.12
		R	—	—	0.45	-1.25	0+/-0.12
	Impreza Sedan Turbo	F	1.00	+3.30	0.45	-0.25	0+/-0.12
		R	—	—	0.45	-1.25	0+/-0.12
	Impreza Sedan STI	F	1.00	+4.50	0.45	-0.30	0+/-0.12
		R	—	—	0.45	-1.15	0+/-0.12
	Impreza Wagon	F	1.00	+3.25	0.45	-0.10	0+/-0.12
		R	—	—	0.45	-1.15	0+/-0.12
	Impreza Wagon Turbo	F	1.00	+3.30	0.45	-0.20	0+/-0.12
		R	—	—	0.45	-1.20	0+/-0.12

TIRE, WHEEL AND BALL JOINT SPECIFICATIONS

Year	Model	OEM Tires Standard	OEM Tires Optional	Tire Pressures (psi) Front	Tire Pressures (psi) Rear	Wheel Size	Ball Joint Inspection	Lug Nut (ft. lbs.)
2009	Impreza	P205/55R16 P205/50R17 215/45ZR18 P225/55R17	T135/70 D17 T135/80R16	35	35	6.5-JJ 7-JJ	0.012 in.	74
	Impreza WRX	P215/45R17	None	35	35	7-JJ		74
	Impreza STI	P225/45R17	P205/55R16 T135/70 D17	35	35	8-JJ 17x4T	0.012 in.	74
2010	Impreza	P205/55R16 P205/50R17 215/45ZR17 215/45ZR18	T135/70 D17 T135/80R16	35	35	6.5-JJ 7-JJ	0.012 in.	74
	Impreza WRX	P215/45R17	None	35	35	7-JJ		74
	Impreza STI	P225/45R17	P205/55R16 T135/70 D17	35	35	8-JJ 17x4T	0.012 in.	74

OEM: Original Equipment Manufacturer

PSI: Pounds Per Square Inch

37671_IMPZ_C0012

BRAKE SPECIFICATIONS

All measurements in inches unless noted

Year	Model		Brake Disc Original Thickness	Brake Disc Minimum Thickness	Brake Disc Maximum Runout	Brake Drum Diameter Original Inside Diameter	Brake Drum Diameter Max. Wear Limit	Brake Drum Diameter Maximum Machine Diameter	Minimum Lining Thickness Front	Minimum Lining Thickness Rear	Brake Caliper Bracket Bolts (ft. lbs.)	Brake Caliper Mounting Bolts (ft. lbs.)
2009	Impreza	F	0.940 ①	0.870 ②	0.003	NA	NA	NA	0.059	—	59	20
		R	0.390 ③	0.335 ④	0.003	NA	NA	NA	—	0.059	39	28
	WRX	F	0.940 ①	0.870 ②	0.003	NA	NA	NA	0.059	—	59	20
		R	0.390 ③	0.335 ④	0.003	NA	NA	NA	—	0.059	39	28
	STI	F	0.940 ①	0.870 ②	0.003	NA	NA	NA	0.059	—	59	20
		R	0.390 ③	0.335 ④	0.003	NA	NA	NA	—	0.059	39	28
2010	Impreza	F	0.940 ①	0.870 ②	0.003	NA	NA	NA	0.059	—	59	20
		R	0.390 ③	0.335 ④	0.003	NA	NA	NA	—	0.059	39	28
	WRX	F	0.940 ①	0.870 ②	0.003	—	NA	—	—	—	59	20
		R	0.390 ③	0.335 ④	0.003	10.000	10.080	10.080	0.177	0.177	39	28
	STI	F	0.940 ①	0.870 ②	0.003	NA	NA	NA	0.059	—	59	20
		R	0.390 ③	0.335 ④	0.003	NA	NA	NA	—	0.059	39	28

NA: Not Available

① 17 inch type rotors original thickness: 1.18

② 17 inch type rotors minimum thickness: 1.10

③ Vented type rotors original thickness: 0.71

④ Vented type rotors minimum thickness: 0.63

37671_IMPZ_C0013

SCHEDULED MAINTENANCE INTERVALS
SUBARU—IMPREZA

TO BE ERVICED	TYPE OF SERVICE	VEHICLE MILEAGE INTERVAL (x1000)												
		7.5	15	22.5	30	37.5	45	52.5	60	67.5	75	82.5	90	97.5
Engine oil & filter ①	R	✓	✓	✓	✓	✓	✓	✓	✓	✓	✓	✓	✓	✓
Brake lines	S/I		✓		✓		✓		✓		✓		✓	
Clutch & hill holder system	S/I		✓		✓		✓		✓		✓		✓	
Disc brake pads & discs, front & rear axle boots & axle shaft joint portions	S/I		✓		✓		✓		✓		✓		✓	
Parking brake	S/I		✓		✓		✓		✓		✓		✓	
Steering & suspension	S/I		✓		✓		✓		✓		✓		✓	
Air filter element	R				✓				✓				✓	
Engine coolant	R				✓				✓				✓	
Fuel filter	R								✓				✓	
Spark plugs	R				✓				✓					
Automatic transmission fluid & filter	S/I				✓				✓				✓	
Brake fluid	S/I				✓				✓				✓	
Camshaft drive belt ②	S/I				✓				✓				✓	
Coolant level, hoses & clamps	S/I				✓				✓				✓	
Drive belts	S/I				✓				✓				✓	
Fuel system, hoses & connections	S/I				✓				✓				✓	
Transmission and/or differential gear fluid	S/I				✓								✓	
Front & rear wheel bearing	S/I								✓					

R: Replace   S/I: Service or Inspect

① 2.5L turbo models change every 3875 miles or 3.75 months.

② Replace camshaft drive belt at 105,000 miles or 105 months

FREQUENT OPERATION MAINTENANCE (SEVERE SERVICE)

If a vehicle is operated under any of the following conditions it is considered severe service:

- **Extremely dusty areas.**
- **50% or more of the vehicle operation is in 32°C (90°F) or higher temperatures, or constant operation in temperatures below 0°C (32°F).**
- **Prolonged idling (vehicle operation in stop and go traffic).**
- **Frequent short running periods (engine does not warm to normal operating temperatures).**
- **Police, taxi, delivery usage or trailer towing usage.**

Oil & oil filter change: change every 3750 miles.

Clutch & hill holder system: service or inspect every 7500 miles.

Disc brake pads & discs, front & rear axle boots & axle shaft joint portions: service or inspect every 7500 miles.

Steering & suspension: service or inspect every 7500 miles.

Air filter element: service or inspect every 15,000 miles.

Automatic transmission fluid: service or inspect every 15,000 miles.

Brake linings & drums: service or inspect every 15,000 miles.

Coolant level, hoses & clamps: service or inspect every 15,000 miles.

Drive belts: service or inspect every 15,000 miles.

Transmission/differential gear oil service or inspect every 15,000 miles.

Front & rear wheel bearing repack: service or inspect every 30,000 miles.

37671_IMPZ_C0014

BRAKES   INFORMATION AND PRECAUTIONS

ANTI-LOCK SYSTEMS

• Certain components within the ABS system are not intended to be serviced or repaired individually.

• Do not use rubber hoses or other parts not specifically specified for and ABS system. When using repair kits, replace all parts included in the kit. Partial or incorrect repair may lead to functional problems and require the replacement of components.

• Lubricate rubber parts with clean, fresh brake fluid to ease assembly. Do not use shop air to clean parts; damage to rubber components may result.

• Use only DOT 3 brake fluid from an unopened container.

• If any hydraulic component or line is removed or replaced, it may be necessary to bleed the entire system.

• A clean repair area is essential. Always clean the reservoir and cap thoroughly before removing the cap. The slightest amount of dirt in the fluid may plug an

orifice and impair the system function. Perform repairs after components have been thoroughly cleaned; use only denatured alcohol to clean components. Do not allow ABS components to come into contact with any substance containing mineral oil; this includes used shop rags.

• The Anti-Lock control unit is a microprocessor similar to other computer units in the vehicle. Ensure that the ignition switch is **OFF** before removing or installing controller harnesses. Avoid static electricity discharge at or near the controller.

• If any arc welding is to be done on the vehicle, the control unit should be unplugged before welding operations begin.

DISC AND DRUM SYSTEMS

✳✳ CAUTION

Dust and dirt accumulating on brake parts during normal use may contain asbestos fibers from production or

aftermarket brake linings. Breathing excessive concentrations of asbestos fibers can cause serious bodily harm. Exercise care when servicing brake parts. Do not sand or grind brake lining unless equipment used is designed to contain the dust residue. Do not clean brake parts with compressed air or by dry brushing. Cleaning should be done by dampening the brake components with a fine mist of water, then wiping the brake components clean with a dampened cloth. Dispose of cloth and all residue containing asbestos fibers in an impermeable container with the appropriate label. Follow practices prescribed by the Occupational Safety and Health Administration (OSHA) and the Environmental Protection Agency (EPA) for the handling, processing, and disposing of dust or debris that may contain asbestos fibers.

BRAKES   BLEEDING THE BRAKE SYSTEM

BLEEDING PROCEDURE

BLEEDING PROCEDURE

See Figures 3 through 5.

1. Raise the vehicle and safely support it.
2. Remove both front and rear wheels.
3. Draw out the brake fluid from master cylinder with syringe.
4. Refill reservoir tank with recommended brake fluid.
5. Install one end of a vinyl tube onto the air bleeder and insert the other end of the tube into a container to collect the brake fluid.
6. Instruct your co-worker to depress the brake pedal slowly two or three times and then hold it depressed.
7. Loosen bleeder screw approximately 1/4 turn until a small amount of brake fluid drains into container, and then quickly tighten screw.
8. Repeat again from the two former procedures above until there are no air bubbles in drained brake fluid and new fluid flows through vinyl tube.

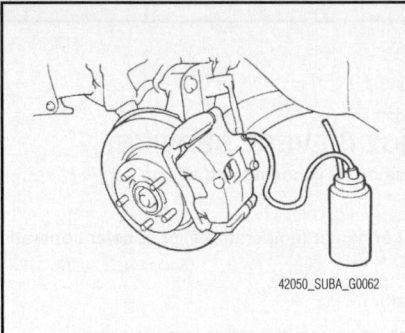

42050_SUBA_G0062

Fig. 3 Connect one end of the tube onto the air bleeder.

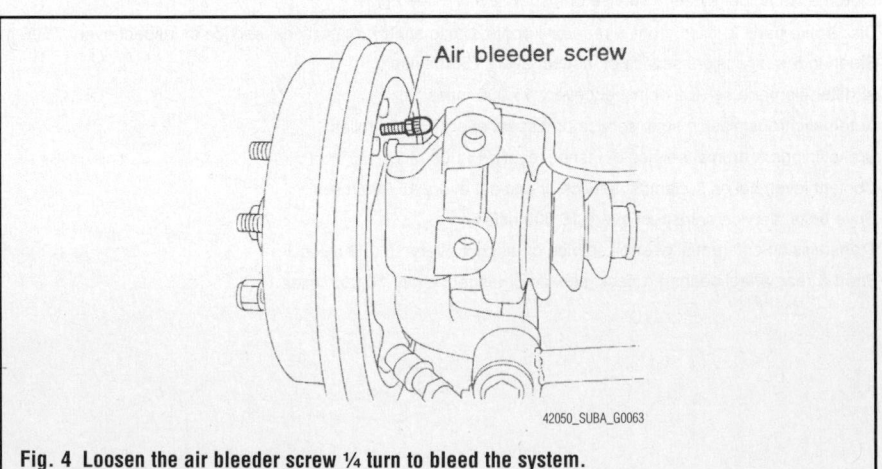

Air bleeder screw

42050_SUBA_G0063

Fig. 4 Loosen the air bleeder screw ¼ turn to bleed the system.

➡Add brake fluid as necessary while performing the air bleed operation, in order to prevent the tank from running short of brake fluid.

9. After completing the bleeding operation, hold brake pedal depressed and tighten screw and install bleeder cap.
10. Bleed air from each wheel cylinder using the same procedures as described above.
11. Depress brake pedal and

hold it there for approximately 20 seconds. At this time check pedal to see if it shows any unusual movement. Visually inspect bleeder screws and brake pipe joints to make sure that there is no fluid leakage.

12. Install the wheels, and drive car for a short distance (between 1–2 miles to make sure that brakes are operating properly.

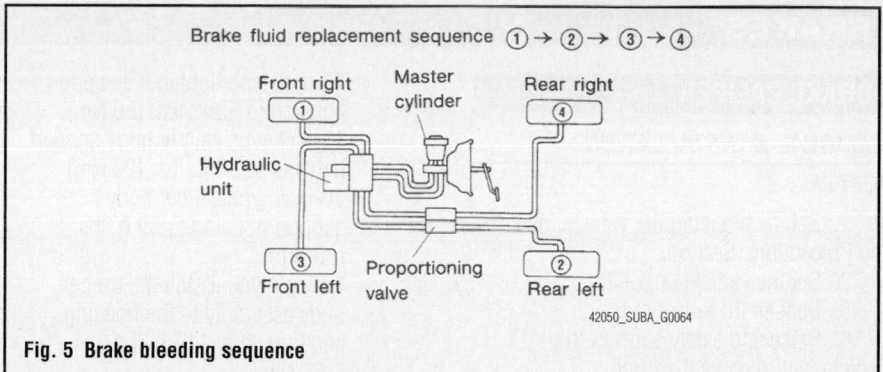

Brake fluid replacement sequence ① → ② → ③ → ④

Fig. 5 Brake bleeding sequence

BRAKES

ANTI-LOCK BRAKE SYSTEM (ABS)

WHEEL SPEED SENSORS

REMOVAL & INSTALLATION

Front

See Figure 6.

1. Before servicing the vehicle, refer to the Precautions Section.
2. Disconnect the battery ground cable from the battery.
3. Disconnect the ABS wheel speed sensor connector located next to the front strut mounting house in the engine compartment.
4. Remove the sensor harness bracket.
5. Remove the bolts which secure the sensor harness to the front strut.
6. Remove the front ABS wheel speed sensor from housing.

> ✳✳ **CAUTION**
>
> **Be careful not to damage the sensor. Do not apply excessive force to the sensor harness.**

To install:

7. Installation is the reverse order of assembly.
8. Tighten the sensor mounting bolt to 5.5 ft. lbs. (7.5 Nm).

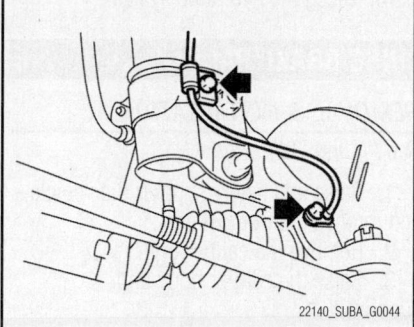

22140_SUBA_G0044

Fig. 6 Front wheel speed sensor and mounting bolt location

9. Tighten the sensor harness bolts to 24 ft. lbs. (33 Nm).

➡**Check if the harness is not pulled and does not come in contact with the suspension or body during steering wheel effort.**

Rear

1. Before servicing the vehicle, refer to the Precautions Section.
2. Disconnect the negative battery cable.
3. Disconnect the connector from the rear ABS wheel speed sensor.
4. Remove the sensor harness clamp of the rear sub frame.

5. Remove the sensor harness bracket from the upper arm.
6. Remove the rear ABS wheel speed sensor from the rear axle.

> ✳✳ **CAUTION**
>
> **Be careful not to damage the sensor. Do not apply excessive force to the sensor harness.**

To install:

7. Installation is the reverse order of assembly.
8. Tighten the sensor mounting bolt to 5.5 ft. lbs. (7.5 Nm).
9. Tighten the harness mounting bolts to 5.5 ft. lbs. (7.5 Nm).

WHEEL SPEED SENSOR RINGS (TOOTHED RINGS)

REMOVAL & INSTALLATION

Front Wheel

Refer to Front Hub Bearing for removal, because the front magnetic encoder is integrated with front hub bearing.

Rear Wheel

Refer to Front Hub Bearing for removal, because the front magnetic encoder is integrated with front hub bearing.

BRAKE CALIPER

REMOVAL & INSTALLATION

See Figure 7.

1. Before servicing the vehicle, refer to the Precautions Section.
2. Set the vehicle on a lift.
3. Loosen the wheel nuts.
4. Raise and safely support the vehicle, and remove the front wheels.
5. Remove the union bolt, and disconnect the brake hose from the caliper body assembly.
6. Remove the bolt securing the lock pin to caliper body.
7. Raise the caliper body, and then move it toward vehicle center to separate it from the support.
8. Remove the support from housing.

➡**Remove the support only when replacing the rotor or support. It need not be removed when servicing the caliper body assembly.**

9. Remove mud and foreign matter from the caliper body assembly and the support.

To install:

10. Apply a thin coat of Molykote M7439 (Part No. 003602001) to the support.
11. Apply a thin coat of Molykote M7439 (Part No. 003602001) to the contact surface between the pad and pad clip.
12. Apply a thin coat of Molykote AS-880N (Part No. K0777YA010) to both surfaces of the inner shim.
13. Install the pad to support.
14. Tightening specifications are as follows:

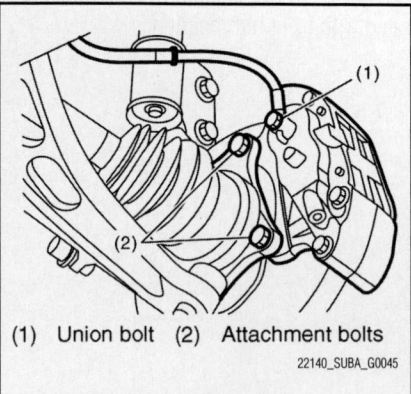

(1) Union bolt   (2) Attachment bolts

22140_SUBA_G0045

Fig. 7 Rear brake caliper view

- 15-inch type tighten the support housing to 59 ft. lbs. (80 Nm)
- 15-inch type caliper body support housing to 20 ft. lbs. (26 Nm)
- 16-inch type caliper body support housing to 59 ft. lbs. (80 Nm)
- 17-inch type install the caliper body assembly to the housing and tighten to 114 ft. lbs. (155 Nm).

15. Connect the brake hose using a new brake hose gasket and tighten to 13 ft. lbs. (18 Nm).
16. Bleed the hydraulic system.

DISC BRAKE PADS

REMOVAL & INSTALLATION

See Figures 8 through 11.

1. Raise and safely support the vehicle, and remove the front wheels.
2. Remove the caliper bolt.
3. Raise the caliper body and support it.

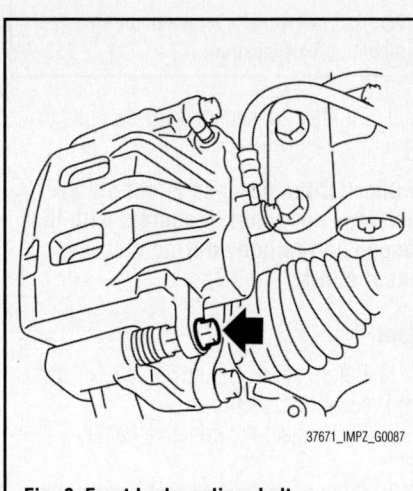

37671_IMPZ_G0087

Fig. 8 Front brake caliper bolt

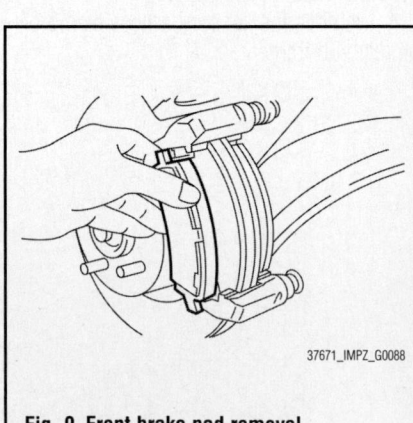

37671_IMPZ_G0088

Fig. 9 Front brake pad removal

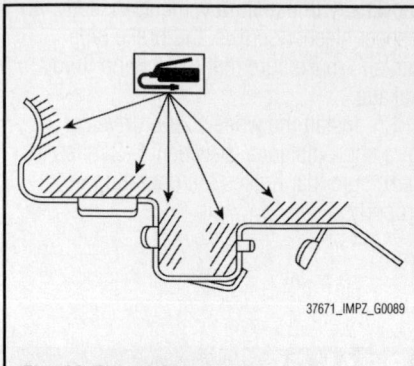

37671_IMPZ_G0089

Fig. 10 Front lubricant or grease application to pad inner shim (1 of 2)

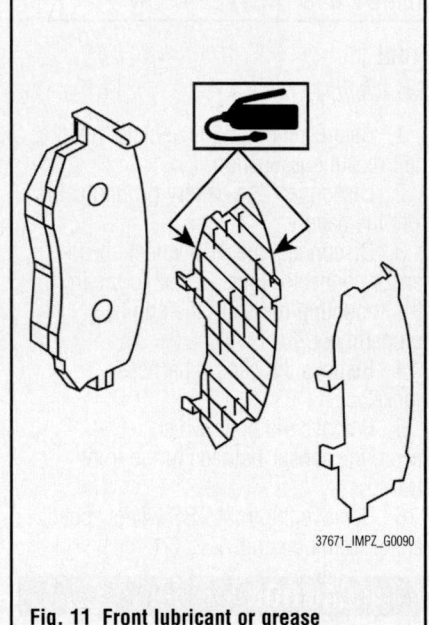

37671_IMPZ_G0090

Fig. 11 Front lubricant or grease application to pad inner shim (2 of 2)

➡**Do not disconnect the brake hose from the caliper body.**

4. Remove the pad.

To install:

5. Apply a thin coat of Molykote M7439 (Part No. K0770YA000) or grease contained in the pad kit to the pad clip.
6. Apply a thin coat of Molykote AS880N (Part No. K0777YA010) or grease contained in the pad kit to both surfaces of the pad inner shim.
7. Install the pad to support.
8. Install the caliper body to the support. Tighten to 20 ft. lbs. (27 Nm).
9. Install the wheels, and lower the vehicle.

BRAKE CALIPER

REMOVAL & INSTALLATION

See Figure 12.

1. Before servicing the vehicle, refer to the Precautions Section.
2. Lift up the vehicle, then remove the rear wheels.
3. Remove the caliper body.
4. Remove mud and foreign matter from the caliper body assembly.

❊❊ CAUTION

Be careful not to allow foreign matter to enter the brake hose connector.

To install:

5. Install the caliper body to the housing and tighten to 48 ft. lbs. (65 Nm).
6. Apply a thin coat of Molykote AS880N (Part No. K0777YA010) or the pad kit grease to both surfaces of the pad side and pad inner shim.
7. Install the pads to the caliper body.
8. Install the cross spring and pad pins.
9. Install the clips.
10. Replace the brake hose gaskets with new ones, and then connect the brake hose and tighten to 13 ft. lbs. (18 Nm).
11. Bleed air from the brake system.

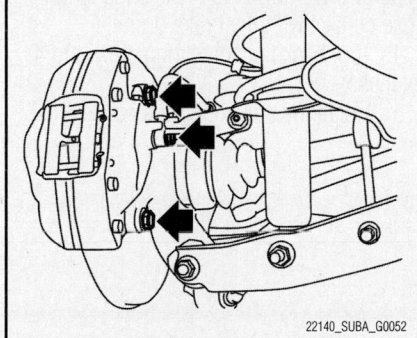

Fig. 12 Rear brake caliper mounting bolts and brake hose banjo bolt

DISC BRAKE PADS

REMOVAL & INSTALLATION

See Figures 13 through 18.

1. Lift up the vehicle, and then remove the rear wheels.

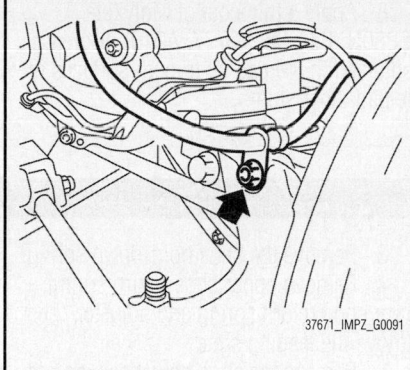

Fig. 13 Rear brake hose bracket

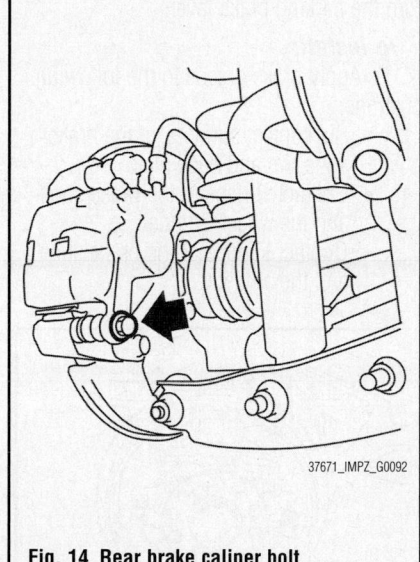

Fig. 14 Rear brake caliper bolt

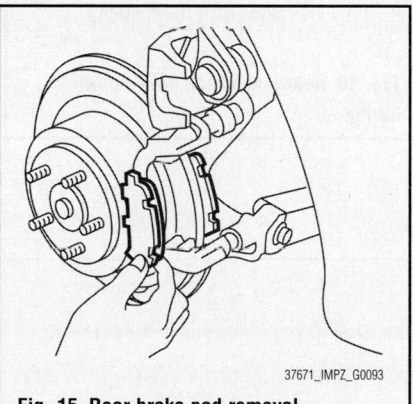

Fig. 15 Rear brake pad removal

2. Remove the brake hose bracket.
3. Remove the caliper bolt.
4. Raise the caliper body and support it.

➡**Do not disconnect the brake hose from the caliper body.**

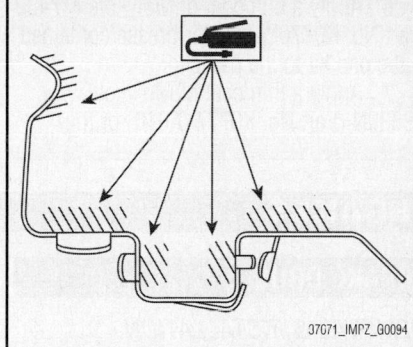

Fig. 16 Rear brake lubricant or grease application to pad inner shim (1 of 3)

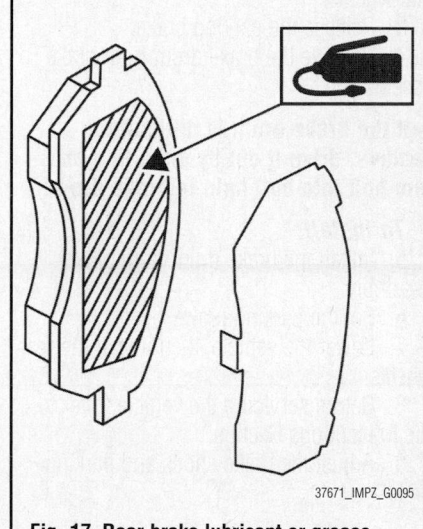

Fig. 17 Rear brake lubricant or grease application to pad inner shim (2 of 3)

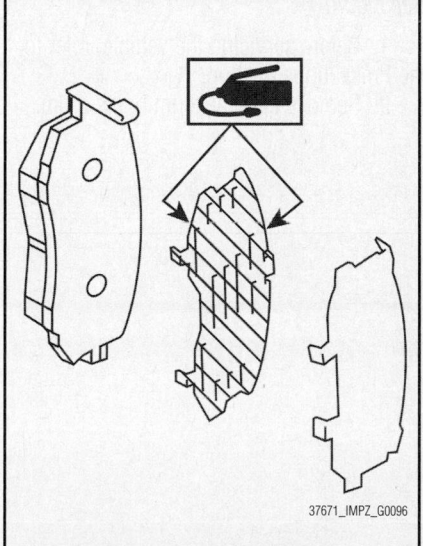

Fig. 18 Rear brake lubricant or grease application to pad inner shim (3 of 3)

5. Remove the pad.

To install:

6. Apply a thin coat of Molykote M7439 (Part No. K0770YA000) or grease contained in the pad kit to the pad clip.

7. Apply a thin coat of Molykote AS880N (Part No. K0777YA010) or the grease contained in the pad kit to the contact surface between the pad and shim.

8. Apply a thin coat of Molykote AS880N (Part No. K0777YA010) or grease contained in the pad kit to both surfaces of the pad inner shim.

9. Install the pad to support.

10. Install the caliper body to the support. Tighten to 20 ft. lbs. (27 Nm)

11. Install the brake hose bracket. Tighten bracket to 24 ft. lbs. (33 Nm)

BRAKES

BRAKE DRUM

REMOVAL & INSTALLATION

1. Before servicing the vehicle, refer to the Precautions Section.

2. Lift up the vehicle, then remove the rear wheels.

3. Release the parking brake.

4. Remove the brake drum from brake assembly.

➡**If the brake drum is difficult to remove, drive it out by installing an 8-mm bolt into bolt hole in brake drum.**

To install:

5. Install the brake drum from brake assembly.

6. Set the parking brake.

7. Lower the vehicle, then install the rear wheels.

8. Before servicing the vehicle, refer to the Precautions Section.

9. Adjust the brake shoes and parking brake.

BRAKE SHOES

REMOVAL & INSTALLATION

See Figure 19.

1. Before servicing the vehicle, refer to the Precautions Section.

2. Remove the rear drum brake drum.

3. Remove the shoe hold-down spring.

4. Remove upper shoe return spring, lower shoe return spring and adjuster, then remove the leading shoe

5. Remove the shoe hold-down spring and trailing shoe.

6. Disconnect the parking brake cable from the parking brake lever.

To install:

7. Apply brake grease to the following locations:

- Six contact surfaces of the brake shoe rim and back plate
- Contact surface of the brake shoe and the wheel cylinder
- Contact surface of the brake shoe and the adjuster

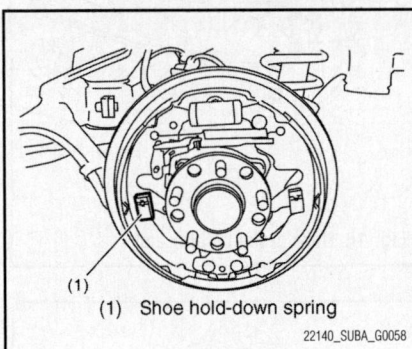

(1) Shoe hold-down spring

22140_SUBA_G0058

Fig. 19 Rear brake shoe hold—down spring

REAR DRUM BRAKES

- Contact surface of lever and brake shoe

8. Connect the parking brake cable to the parking brake lever.

9. Check that the parking brake cable is not coming off from the cable guide.

10. Install the trailing shoe and hold it down with the shoe hold-down spring.

11. Install the adjuster and adjusting lever to the leading shoe.

➡**Install the adjuster with the adjusting wheel facing toward the front side of vehicle.**

12. Install the leading shoe and hold it down with the shoe hold-down spring.

13. Install the adjusting spring and return spring.

14. Install the rear brake drum.

15. Adjust the brake shoes and parking brake.

16. Road test and burnish brake shoes.

17. Re-adjust brake shoes if needed.

ADJUSTMENT

1. Return the parking brake lever fully.

2. Remove the adjusting hole cover.

3. Turn the adjusting screw using a flat tip screwdriver until the brake shoe is in close contact with the brake drum.

4. Turn back the adjusting screw 3 to 4 notches.

5. Install the adjusting hole cover.

BRAKES | **PARKING BRAKE**

PARKING BRAKE CABLES

ADJUSTMENT

See Figure 20.

1. Remove the console cover.
2. Forcefully pull the parking brake lever 3 to 5 times.

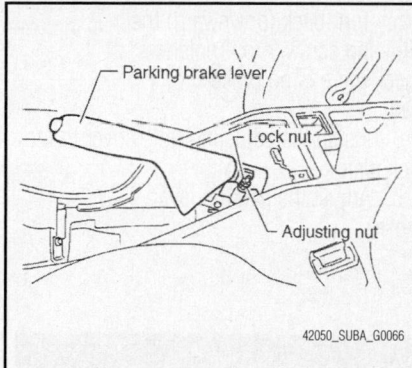

Fig. 20 Turn the adjusting nut to adjust the parking brake cable.

3. Adjust parking brake cable by turning the adjusting nut until the parking brake lever stroke is set at 7 to 8 notches with operating force of 44 lbs. (196 N).
4. Tighten lock nut.
5. Install the console cover.

PARKING BRAKE SHOES

REMOVAL & INSTALLATION

See Figure 21.

1. Before servicing the vehicle, refer to the Precautions Section.
2. Release the parking brake.
3. Remove the two mounting bolts and remove the brake caliper assembly.
4. Suspend the brake caliper assembly so that the hose is not stretched.
5. Remove the disc rotor.
6. If disc rotor is seized on the hub, drive the disc rotor out by pushing two 8mm bolts in holes B on the rotor.

7. Remove the shoe return spring from the parking brake assembly.
8. Remove the front shoe hold down spring and pin.
9. Remove the strut and strut spring.
10. Remove the adjuster assembly from the parking brake assembly.
11. Remove the brake shoe.
12. Remove the rear shoe hold down spring and pin with pliers.
13. Remove the parking brake cable from lever.
14. Using a flat tip screwdriver, raise the retainer. Remove the parking lever and washer from brake shoe.

To install:

15. Apply brake grease to the following locations:

- Six contact surfaces of the shoe rim and back plate gasket
- Contact surface of the shoe wave and the anchor pin
- Contact surface of the lever and strut

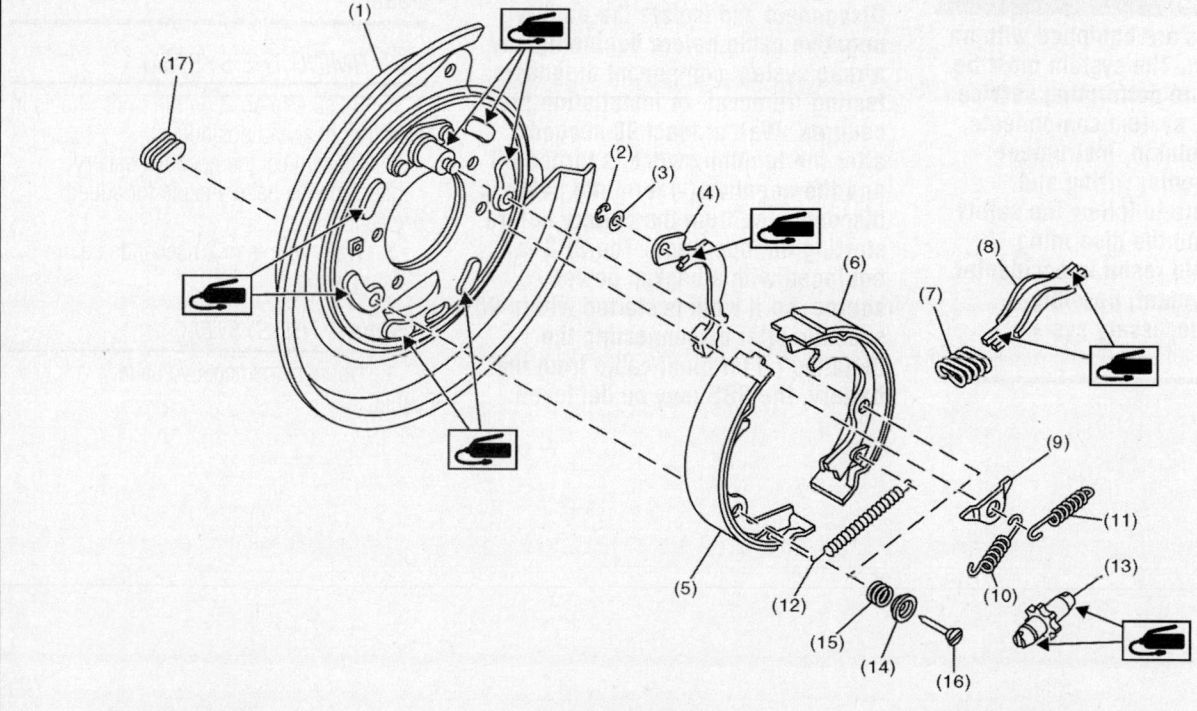

(1)	Back plate	(7)	Strut spring	(13)	Adjuster
(2)	Retainer	(8)	Strut	(14)	Shoe hold-down cup
(3)	Spring washer	(9)	Shoe guide plate	(15)	Shoe hold-down spring
(4)	Lever	(10)	Primary return spring	(16)	Shoe hold-down pin
(5)	Parking brake shoe (Primary)	(11)	Secondary return spring	(17)	Adjusting hole cover
(6)	Parking brake shoe (Secondary)	(12)	Adjusting spring		

22140_SUBA_G0059

Fig. 21 Exploded view of parking brake components

- Contact surface of the shoe wave and the adjuster assembly
- Contact surface of the shoe wave and the strut
- Contact surface of the lever and the shoe wave

16. Insert the primary side brake shoe into the anchor pin groove.

17. Secure the brake shoe with the shoe hold-down pin and cup.

18. Install the plate to the anchor pin, then install the primary return spring.

19. Install the parking brake cable to the lever.

20. Install the strut and adjuster, then secure the secondary side brake shoe with the shoe hold-down pin and cup.

➡**Install the strut spring of both right and left wheel facing vehicle front.**

Install the adjuster assembly with screw section on the left side.

21. Install the secondary return spring and the adjusting spring.

22. Adjust the parking brake.

23. Drive the vehicle to break-in the parking brake lining.

   a. Drive the vehicle at about 22 mph (35 km/h).

   b. With the parking brake release button pushed in, pull the parking brake lever gently.

   c. Drive the vehicle for about 0.12 miles (200 m) in this condition.

   d. Wait 5 to 10 minutes for the parking brake to cool down. Repeat again from step (A).

   e. After breaking-in, Re-adjust the parking brakes.

ADJUSTMENT

1. Before servicing the vehicle, refer to the Precautions Section.

2. Return the parking brake lever fully.

3. Remove the adjusting hole cover from the back plate.

4. Turn the adjusting screw using a flat tip screwdriver until the brake shoe is in close contact with the disc rotor.

5. Turn back (downward) the adjusting screw 3 to 4 notches. Check there is no brake drag.

6. Install the adjusting hole cover to the back plate.

7. Adjust the parking lever stroke.

CHASSIS ELECTRICAL

GENERAL INFORMATION

❊❊ CAUTION

These vehicles are equipped with an air bag system. The system must be disarmed before performing service on, or around, system components, the steering column, instrument panel components, wiring and sensors. Failure to follow the safety precautions and the disarming procedure could result in accidental air bag deployment, possible injury and unnecessary system repairs.

SERVICE PRECAUTIONS

❊❊ CAUTION

Disconnect and isolate the battery negative cable before beginning any airbag system component diagnosis, testing, removal, or installation procedures. Wait at least 90 seconds after the ignition switch is turned off and the negative (-) terminal cable is disconnected from the battery before starting the operation. The SRS is equipped with a backup power source, so if work is started within 90 seconds after disconnecting the negative (-) terminal cable from the battery, the SRS may be deployed.

Failure to disable the airbag system may result in accidental airbag deployment, personal injury, or death.

DISARMING THE SYSTEM

1. Be sure to position the front wheels in the straight ahead position.

2. Disconnect the negative battery cable. Tape the battery cable for added protection.

3. Wait more than 20 seconds before starting work.

ARMING THE SYSTEM

1. Connect the negative battery cable.

DRIVE TRAIN

CLUTCH

REMOVAL & INSTALLATION

See Figures 22 through 24.

1. Remove the transmission assembly.
2. Attach the ST on the flywheel.
3. Remove the clutch cover and clutch disc.

✳✳ WARNING

Take care not to allow oil to touch the clutch disc face. Do not disassemble the clutch cover or clutch disc.

To install:

4. Insert the ST into the clutch disc and attach to the flywheel by inserting the ST end into pilot bearing.

➡**When installing the clutch disc, be careful to attach in the correct direction.**

5. Install the clutch cover to the flywheel and tighten the bolts to the specified torque.

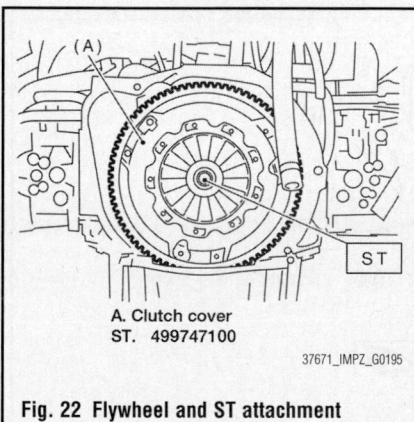

A. Clutch cover
ST. 499747100

37671_IMPZ_G0195

Fig. 22 Flywheel and ST attachment

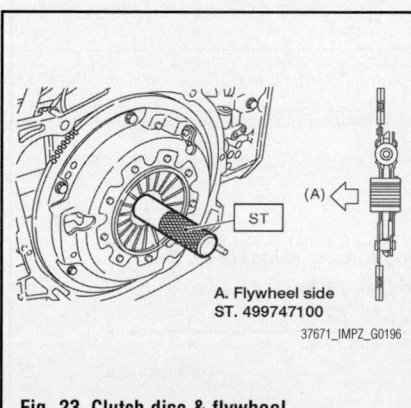

A. Flywheel side
ST. 499747100

37671_IMPZ_G0196

Fig. 23 Clutch disc & flywheel

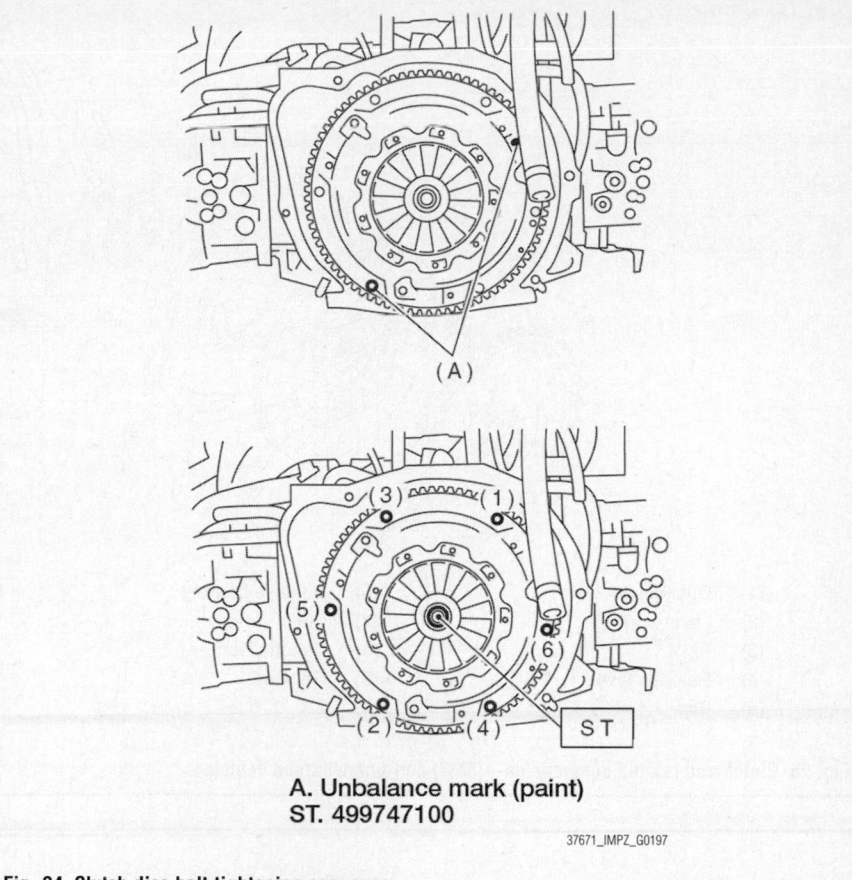

A. Unbalance mark (paint)
ST. 499747100

37671_IMPZ_G0197

Fig. 24 Clutch disc bolt tightening sequence

➡**When installing the clutch cover to the flywheel, position the clutch cover so that the distance between unbalance marks (paint marks) is at least 120°. (The unbalance mark indicates the direction of residual unbalance.)**

a. Note the front and rear of the clutch disc when installing.
b. Temporarily tighten the bolts by hand. Each bolt should be tightened to the specified torque in a crisscross order.
c. Tighten bolts to 142 inch lbs. (16 Nm).
6. Remove the CLUTCH DISC GUIDE.
7. Install the transmission assembly as outlined in the Drive Train Section.

CLUTCH DRIVEN DISC & PRESSURE PLATE

REMOVAL & INSTALLATION

See Figures 25 and 26.

1. Before servicing the vehicle, refer to the Precautions Section.

2. Disconnect the negative battery cable.
3. Remove the transmission.
4. Install tool ST499747100, or equivalent on the flywheel.
5. Remove the clutch cover and clutch disc. Do not disassemble

➡**Be sure to put alignment marks on the flywheel and clutch cover before removing the clutch cover.**

To install:

6. Remove the clutch cover and clutch disc. Do not disassemble.
7. Install tool ST499747100, or equivalent on the flywheel.
8. Remove the transmission.
9. Disconnect the negative battery cable.
10. Before servicing the vehicle, refer to the Precautions Section.

ADJUSTMENTS

Models are equipped with a hydraulic system that is not adjustable.

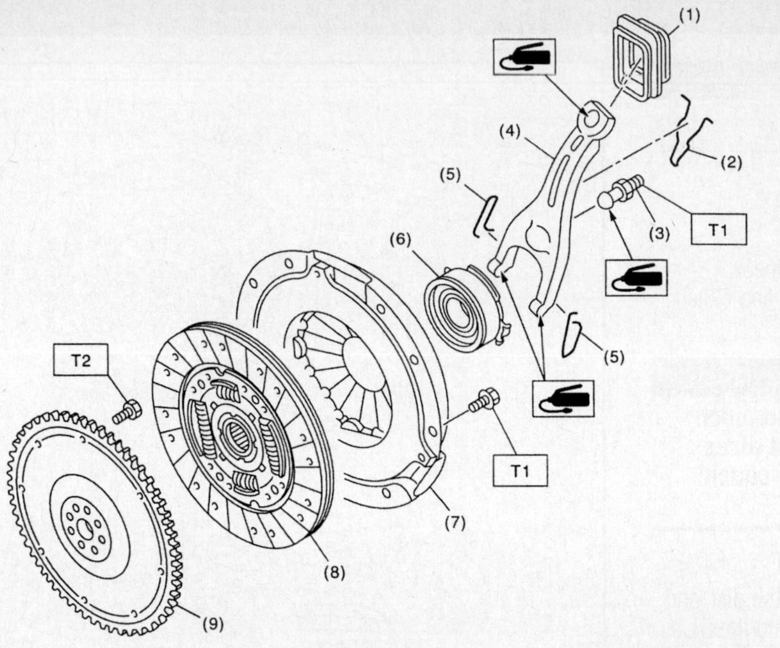

(1)	Dust cover	(6)	Release bearing
(2)	Lever spring	(7)	Clutch cover
(3)	Pivot	(8)	Clutch disc
(4)	Release lever	(9)	Flywheel
(5)	Clip		

Tightening torque: N·m (kgf-m, ft-lb)
T1: 16 (1.6, 11.8)
T2: 72 (7.3, 52.8)

09490_SBCR_G0084

Fig. 25 Clutch and related components—(5MT) non-turbocharged vehicles

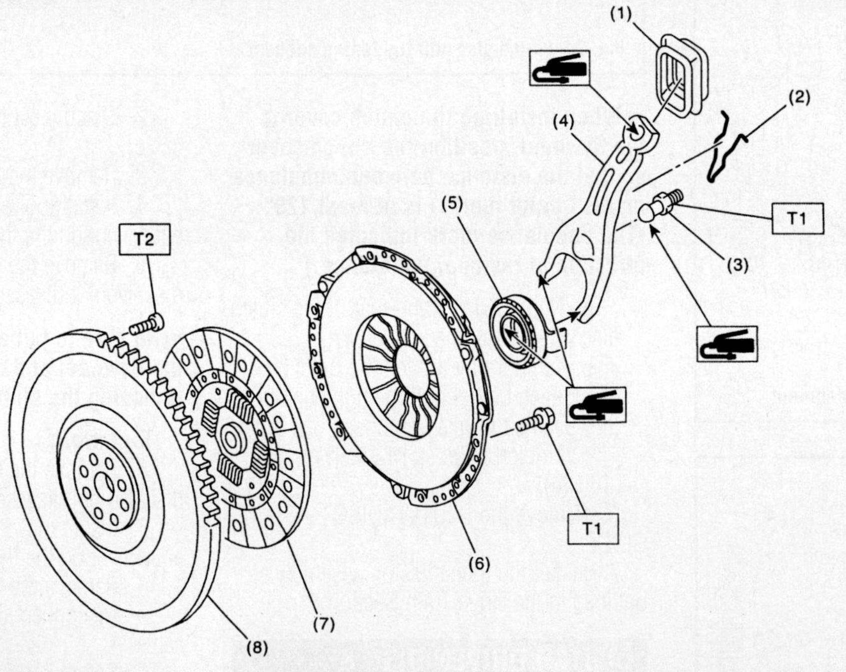

(1)	Dust cover	(5)	Release bearing
(2)	Lever spring	(6)	Clutch cover
(3)	Pivot	(7)	Clutch disc
(4)	Release lever	(8)	Flywheel

Tightening torque: N·m (kgf-m, ft-lb)
T1: 16 (1.6, 11.8)
T2: 72 (7.3, 52.8)

09490_SBCR_G0085

Fig. 26 Clutch and related components—(5MT) turbocharged vehicles

HYDRAULIC SYSTEM BLEEDING

BLEEDING PROCEDURE

See Figure 27.

1. On non turbocharged engine, remove the air intake chamber.
2. On turbocharged engine, remove the intercooler.
3. Connect a vinyl tube to the air bleeder on the master cylinder. Put the other end in a jar with clean clutch fluid.
4. Slowly depress the clutch pedal and keep it depressed. Open the air bleeder to discharge air and fluid.
5. Release the air bleeder for one or two seconds. With the bleeder closed, slowly release the clutch pedal.
6. Repeat the procedure until there are no more air bubbles in the vinyl tube.
7. Tighten the air bleeder.
8. Connect a vinyl tube to the air bleeder on the clutch operating (slave) cylinder. Put the other end in a jar with clean clutch fluid.
9. Slowly depress the clutch pedal and keep it depressed. Open the air bleeder to discharge air and fluid.
10. Release the air bleeder for one or two seconds. With the bleeder closed, slowly release the clutch pedal.
11. Repeat the procedure until there are no more air bubbles in the vinyl tube.
12. Tighten the air bleeder.
13. After depressing the clutch pedal, make sure that there are no leaks in the entire system
14. Recheck to ensure that the clutch is operating correctly.

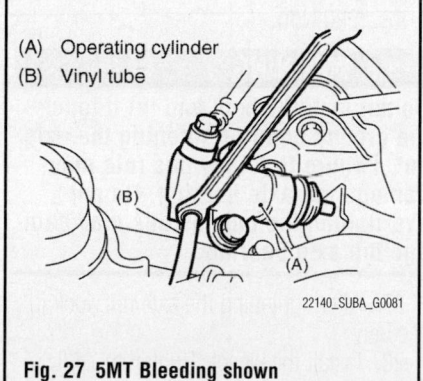

(A) Operating cylinder
(B) Vinyl tube

22140_SUBA_G0081

Fig. 27 5MT Bleeding shown

FRONT AXLE SHAFT SEALS

REMOVAL & INSTALLATION

See Figure 28.

1. Before servicing the vehicle, refer to the Precautions Section.
2. Remove the rear differential.
3. Remove the rear differential side oil seal using a screwdriver wrapped with vinyl tape to prevent the side retainer from scratches.

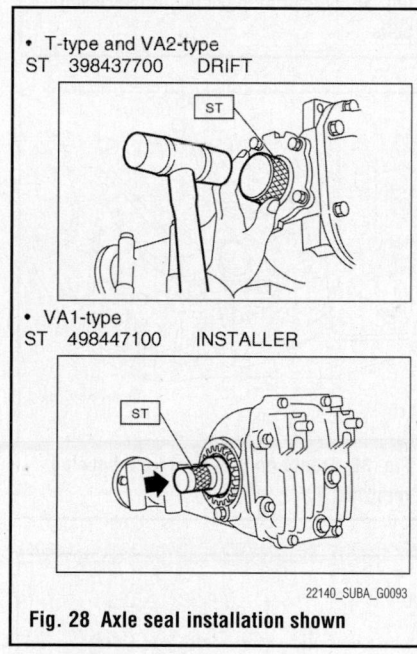

• T-type and VA2-type
ST   398437700   DRIFT

• VA1-type
ST   498447100   INSTALLER

22140_SUBA_G0093

Fig. 28 Axle seal installation shown

To install:

4. Using the ST, install the oil seal to the side retainer.
5. Install the rear differential.

AXLE SHAFT

REMOVAL & INSTALLATION

See Figures 29 through 39.

1. Disconnect the ground cable from the battery.
2. Raise and safely support the vehicle, and remove the front wheels.
3. Lift the crimped section of axle nut.
4. Remove the axle nut using a socket wrench while depressing the brake pedal.

✷✷ CAUTION

Remove the wheel before loosening the axle nut. Failure to follow this rule may damage the wheel bearings.

5. Remove the stabilizer link.
6. Remove the disc brake caliper from the housing, and suspend it from strut using a wire.
7. Remove the disc rotor from the hub.

➡ **If it is difficult to remove the disc rotor from the hub, drive the 8mm bolt into the threaded end of rotor, and then remove the rotor.**

8. Remove the cotter pin and castle nut securing the tie-rod end to the housing knuckle arm.
9. Using a puller, remove the tie-rod ball joint from knuckle arm.
10. Remove the ABS wheel speed sensor assembly and harness.
11. Remove the bolts which secure the sensor harness to the strut.
12. Remove the front arm ball joint from the housing.
13. Remove the front drive shaft from the transmission.
14. Remove the front drive shaft assembly from the hub. If it is hard to remove, use the ST.
15. After scribing an alignment mark on camber adjusting bolt head, remove the bolts which connect the housing and strut, and disconnect the housing from strut.

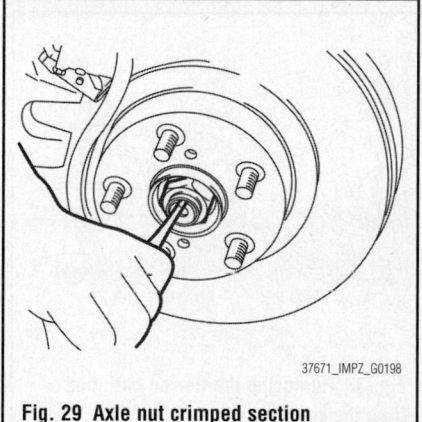

37671_IMPZ_G0198

Fig. 29 Axle nut crimped section

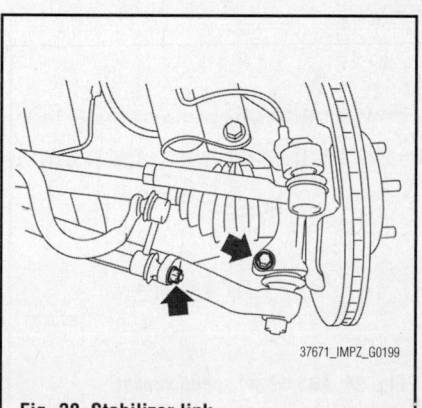

37671_IMPZ_G0199

Fig. 30 Stabilizer link

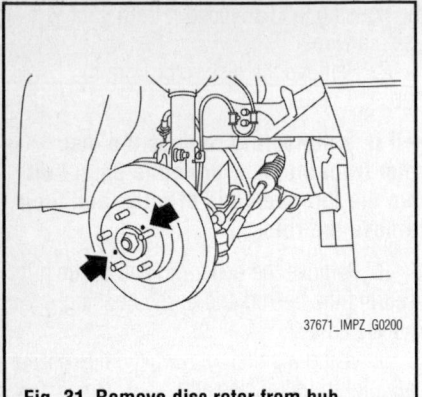

Fig. 31 Remove disc rotor from hub

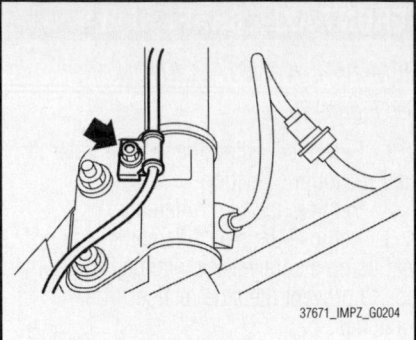

Fig. 35 Sensor harness to strut securing bolts

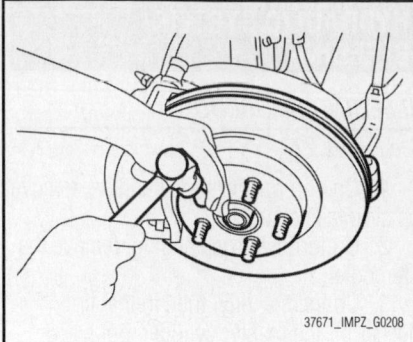

Fig. 39 After tightening the axle nut, lock it securely

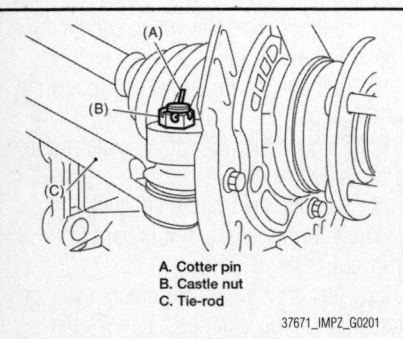

(A)
(B)
(C)

A. Cotter pin
B. Castle nut
C. Tie-rod

Fig. 32 Remove the cotter pin and castle nut securing the tie-rod end to the housing knuckle arm

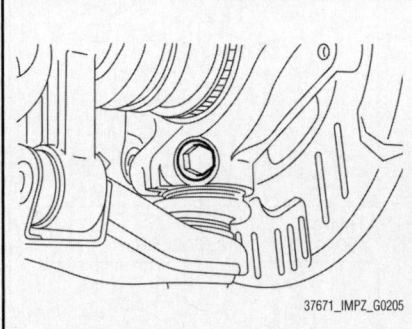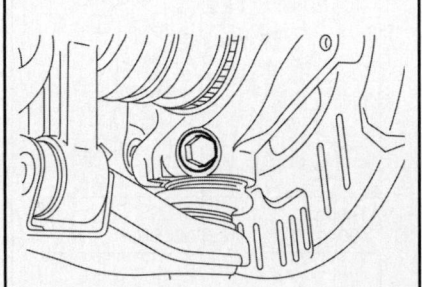

Fig. 36 Front arm ball joint from housing removal

To install:

16. Align the alignment mark on the camber adjusting bolt head, and tighten the housing and strut using a new flange nut. Tighten nut 114 ft. lbs. (155 Nm).

17. Install the front drive shaft.

18. Install the front arm ball joint to the housing. Tighten to 37 ft. lbs. (50 Nm)

19. Install the ABS sensor harness to the strut.

20. Install the ABS wheel speed sensor on the housing. Tighten to 61 inch lbs. (7.5 Nm).

21. Install the disc rotor to hub.

22. Install the disc brake caliper on the housing. Tighten to 59 ft. lbs. (80 Nm).

23. Install the stabilizer link.

24. Connect the tie-rod end ball joint to the knuckle arm with a castle nut. Tighten to 20 ft. lbs. (27 Nm).

✳✳ CAUTION

When connecting the tie-rod, do not hit the cap at bottom of tie-rod end with a hammer.

25. Tighten the castle nut to specified torque and tighten further within 60° until the pin hole is aligned with the slot in the nut. Bend the cotter pin to lock.

26. While depressing the brake pedal, tighten a new axle nut to the specified torque and lock it securely. Tighten to 162 ft. lbs. (220 Nm).

✳✳ WARNING

Do not install wheel and let it touch the ground before tightening the axle nut. Failure to follow this rule may damage the axle bearing. Do not overtighten the nuts as this may damage the axle bearing.

27. After tightening the axle nut, lock it securely.

28. Install the wheel. Tighten to 74 ft. lbs. (100 Nm).

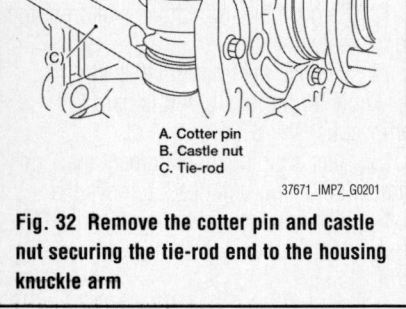

Fig. 33 Removing the tie-rod ball joint from the knuckle arm

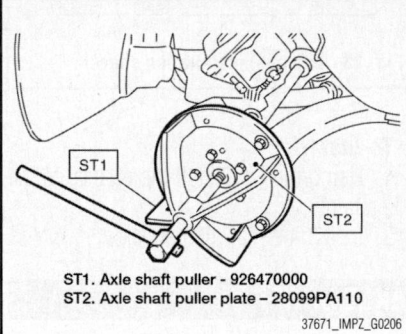

ST1
ST2

ST1. Axle shaft puller – 926470000
ST2. Axle shaft puller plate – 28099PA110

Fig. 37 Front drive shaft assembly from hub removal

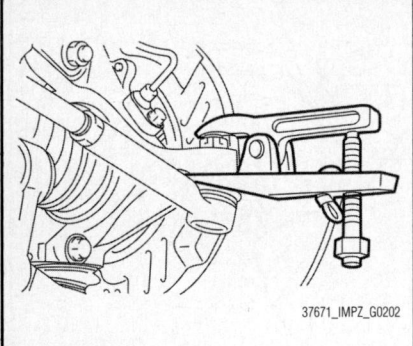

Fig. 34 ABS wheel speed sensor assembly and harness removal

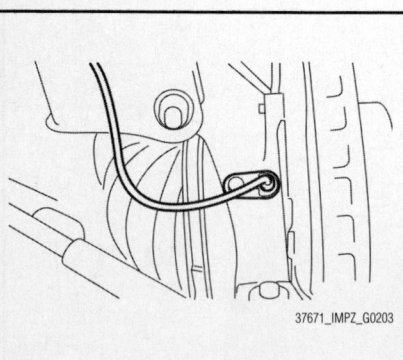

Fig. 38 After scribing an alignment mark on camber adjusting bolt head, remove the bolts that connect the housing and strut

29. Connect the ground cable to battery.
30. Inspect the wheel alignment and adjust if necessary.

CV-JOINT OVERHAUL

See Figures 40 through 45.

1. Place alignment marks on the shaft and outer race.
2. Remove the inner boot band and boot.
3. Remove the circlip from the inner joint outer race using a suitable pry tool.
4. Outer race from the shaft assembly and wipe off the grease.

5. Place alignment marks on the free ring and trunnion as shown in the illustration.
6. Remove the free ring from the trunnion.
7. Place alignment marks on the trunnion and shaft as shown in the illustration.
8. Remove the snapring and trunnion.
9. Place the shaft in a vise between wooden blocks.
10. Using a suitable pry tool, raise the outer boot band claws.
11. Cut and remove the boot.
12. Only the boot can be replaced, the

joint is not serviceable and must be replaced if damaged.

To install:

13. Place the half shaft in a vise.
14. Place the outer boot and small band on the shaft.
15. Apply 2.12–2.47 oz. (60–70g) of supplied grease to the joint.
16. Apply 0.71–1.06 oz. (20–30g) of supplied grease to the whole inner surface of the boot, and apply some grease to the shaft.
17. Install the boot to the joint groove, and attach the large boot band as shown.

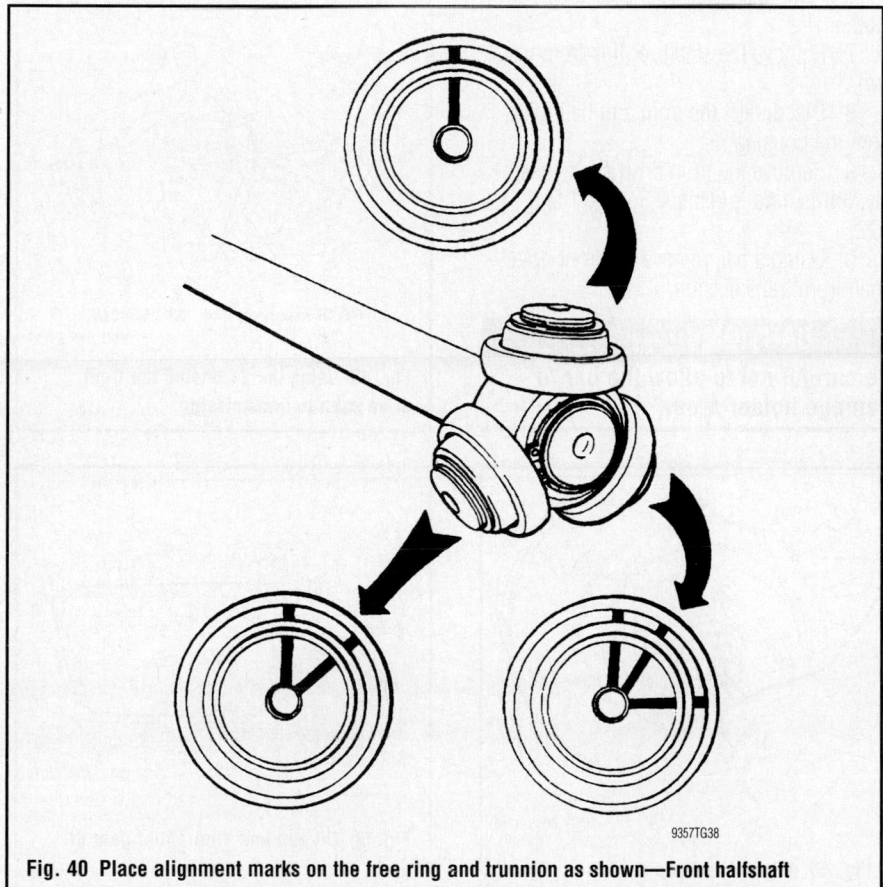

Fig. 40 Place alignment marks on the free ring and trunnion as shown—Front halfshaft

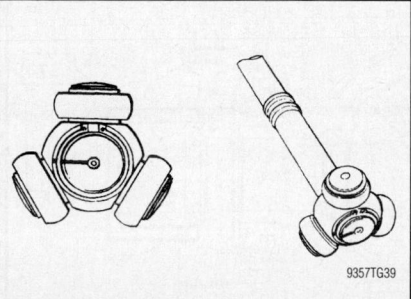

Fig. 41 Place alignment marks on the trunnion and shaft as shown—Front halfshaft

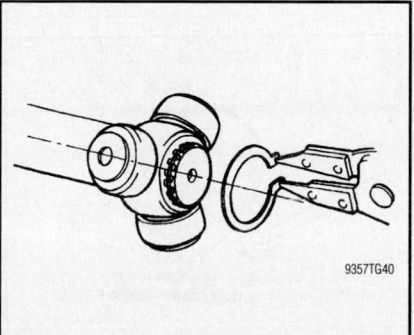

Fig. 42 Remove the snapring and trunnion—Front halfshaft

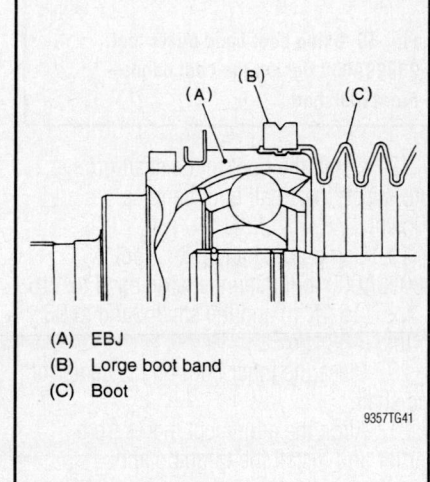

(A) EBJ
(B) Lorge boot band
(C) Boot

Fig. 43 Position the boot to the joint groove, and attach the large boot band as shown—Front halfshaft

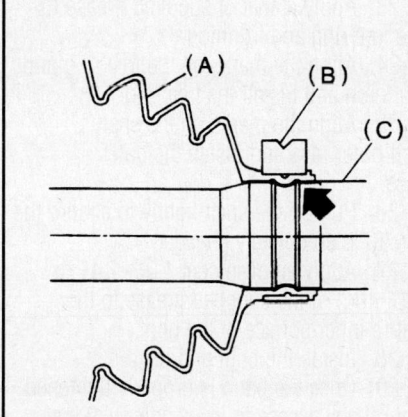

(A) Boot
(B) Small boot band
(C) Shaft

Fig. 44 Position the boot to the shaft groove, and attach the small boot band as shown—Front halfshaft

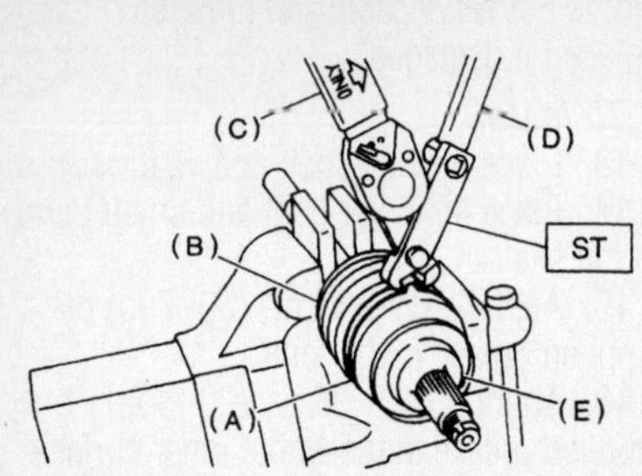

(A) Large boot band
(B) Boot
(C) Torque wrench
(D) Socket flex handle
(E) BJ

9357TG43

Fig. 45 Using boot band pliers tool 28099A000 tighten the boot bands—Front halfshaft

18. Install the boot to the shaft groove, and attach the small boot band as shown.
19. Using boot band pliers tool 28099A000 to tighten the large band to 116 ft. lbs. (157 Nm) and the small band to 98 ft. lbs. (133 Nm).
20. Place the inner boot on the center of the shaft.
21. Align the alignment marks from earlier and install the trunnion and snapring. Make sure the snapring is fully engaged.
22. Apply 3.53–3.88 oz. (100–110g) of supplied grease to the joint outer race.
23. Apply a coat of supplied grease to the free ring and trunnion.
24. Align the marks on the free ring and trunnion and install the free ring.
25. Align the marks on the shaft and outer race and install the outer race.
26. Pull on the shaft lightly to ensure the circlip is completely engaged.
27. Apply an even coat 1.06–1.41 oz. (30–40g) of the supplied grease to the entire inner surface of the boot.
28. Install the boot and band.
29. Once the band is properly tightened, cut off any excess to leave only 0.39 inch (10mm) and bend it over.
30. Install the shaft.

FRONT DRIVESHAFT

REMOVAL & INSTALLATION

See Figures 46 through 53.

1. Disconnect the ground cable from the battery.
2. Raise and safely support the vehicle, and remove the front wheels.
3. Lift the crimped section of axle nut.
4. Remove the axle nut using a socket wrench while depressing the brake pedal.

⁂ CAUTION

Remove the wheel before loosening the axle nut. Failure to follow this rule may damage the wheel bearings.

5. Drain the transmission gear oil—MT model.
6. Drain the differential gear oil—AT model.
7. Remove the stabilizer link from front arm.
8. Disconnect the front arm ball joint from the housing.
9. Remove the front drive shaft assembly. If it is hard to remove, use ST1 and ST2.
10. Using a bar, remove the front drive shaft from transmission.

⁂ CAUTION

Be careful not to allow the bar to damage holder area.

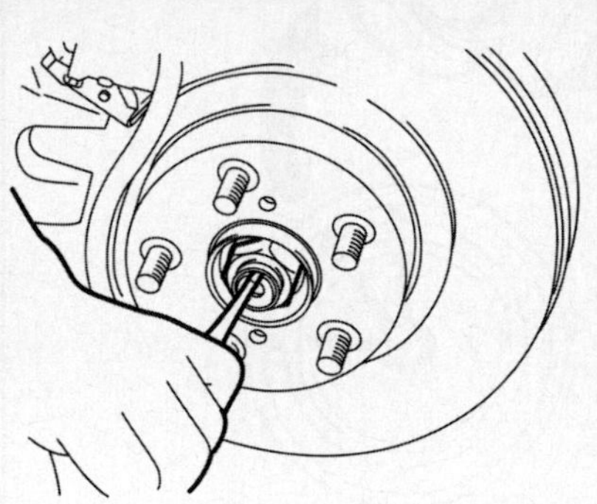

37671_IMPZ_G0198

Fig. 46 Axle nut crimped section

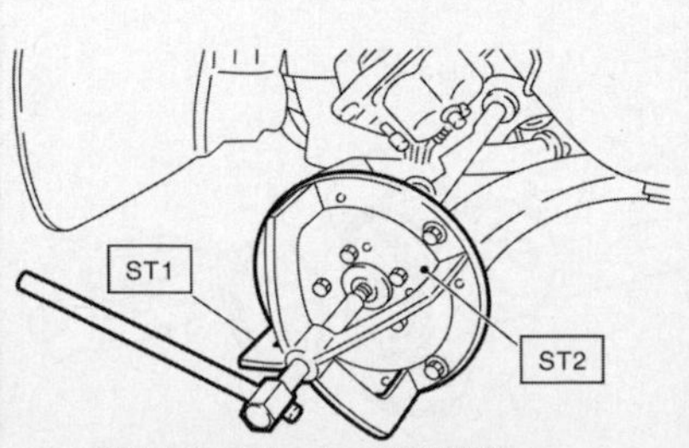

ST1. Axle shaft puller - 926470000
ST2. Axle shaft puller plate - 28099PA110

37671_IMPZ_G0206

Fig. 47 Front drive shaft assembly from hub removal

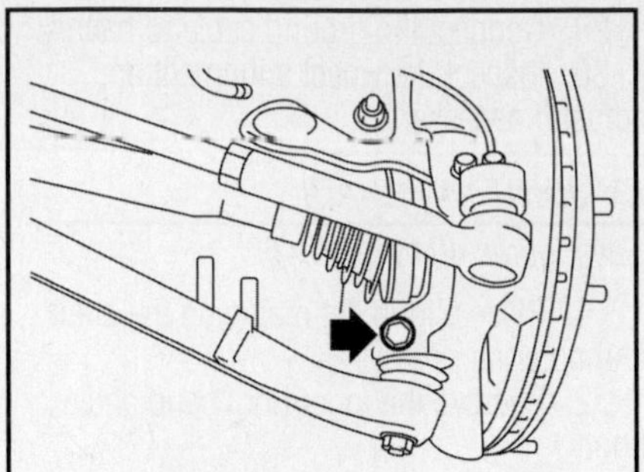

37671_IMPZ_G0209

Fig. 48 Front arm ball joint removal from the housing

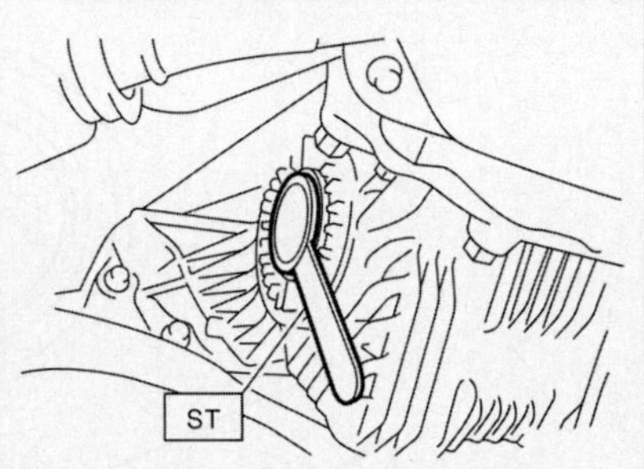

ST. Oil seal protector - 28399SA010

37671_IMPZ_G0210

Fig. 49 Using the ST, install the front drive shaft to transmission

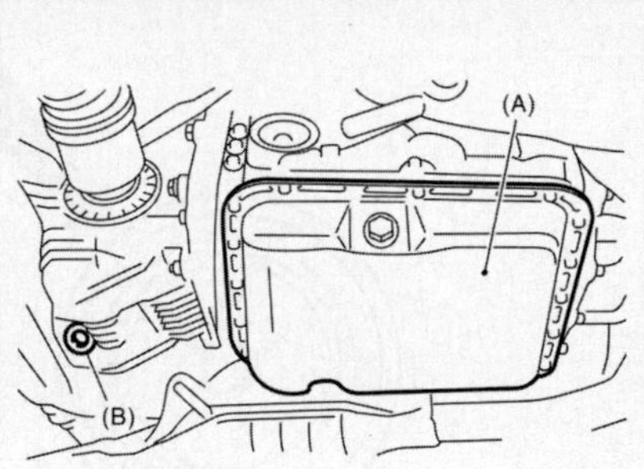

37671_IMPZ_G0211

Fig. 50 Oil pan and differential gear oil drain plug

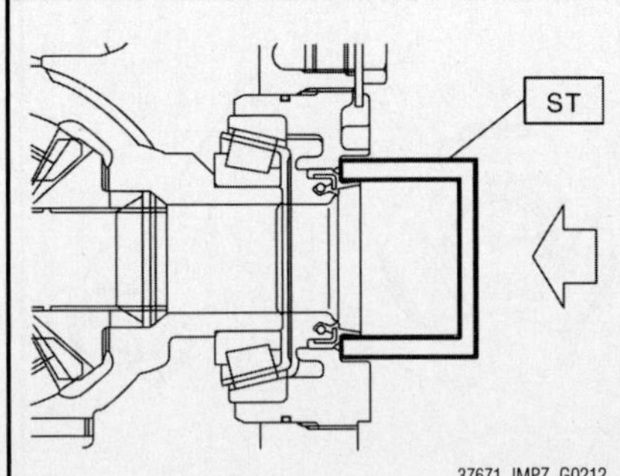

37671_IMPZ_G0212

Fig. 51 Differential side retainer oil seal installation—A/T

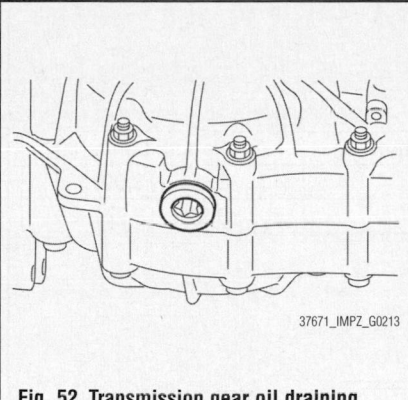

37671_IMPZ_G0213

Fig. 52 Transmission gear oil draining

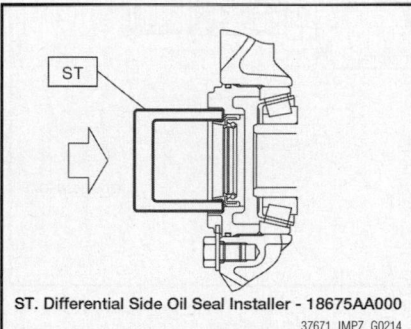

ST. Differential Side Oil Seal Installer - 18675AA000

37671_IMPZ_G0214

Fig. 53 Differential side retainer oil seal installation—M/T

To install:

11. Replace the differential side retainer oil seal with
a new part.

➡**After pulling out the drive shaft, be sure to replace with a new oil seal as outlined in this section.**

12. Insert the AC into hub splines.
13. Draw the drive shaft into specified position.

✳ CAUTION

Do not hammer drive shaft when installing it.

14. Tighten the axle nut temporarily.
15. Using the ST, install the front drive shaft to transmission.
16. Connect the front arm ball joint to the housing. Tighten to 36.9 ft. lbs. (50 Nm).
17. Install the stabilizer link.

✳ CAUTION

Be sure to use a new self-locking nut. Tighten nut to 33.2 ft. lbs. (45 Nm).

18. While pressing the brake pedal, tighten the new axle nuts to the specified

torque. Tighten nuts to 162.3 ft. lbs. (220 Nm)

✳ WARNING

Do not install wheel and let it touch the ground before tightening the axle nut. Failure to follow this rule may damage the axle bearing. Do not overtighten the nuts as this may damage the axle bearing.

19. After tightening axle nut, lock it securely.
20. Fill the transmission gear oil. (MT model)
21. Fill the differential gear oil. (AT model)
22. Install the front wheels. Tighten to 74 ft. lbs. (100 Nm).
23. Connect the ground cable to battery.
24. Inspect the wheel alignment and adjust if necessary.

FRONT HALFSHAFT

REMOVAL & INSTALLATION

See Figure 54.

1. Before servicing the vehicle, refer to the Precautions Section.
2. Disconnect the negative battery cable.
3. Jack-up vehicle, support it with safety stands, and remove front wheels.
4. Unlock axle nut.
5. Remove axle nut while depressing brake pedal to prevent front driveshaft from turning.

✳ CAUTION

Be sure to loose and retighten axle nut after removing wheel from vehicle. Failure to follow this rule may damage wheel bearings.

6. Remove stabilizer link.
7. Remove disc brake caliper from housing, and suspend it from strut using a wire.
8. Remove disc rotor from hub.
9. If disc rotor seizes up within hub, drive disc rotor out by installing an 8-mm bolt in screw hole on the rotor.
10. Remove cotter pin and castle nut which secure tie-rod end to housing knuckle arm.
11. Using a puller, remove tie-rod ball joint from knuckle arm.
12. Remove ABS sensor assembly and harness.

13. Remove bolt which secures sensor harness to strut.
14. Remove transverse link ball joint from housing.
15. Remove inner joint from transmission spindle.
16. Remove front driveshaft assembly from hub. If it is hard to remove, use a ST-926470000 puller.

✳ CAUTION

Be careful not to damage oil seal lip and tone wheel when removing front driveshaft. If front driveshaft is removed, replace inner oil seal with new one.

17. After scribing an alignment mark on camber adjusting bolt head, remove bolts which connect housing and strut, and disconnect housing from strut.

To install:

18. While aligning alignment mark on camber adjusting bolt head, connect housing and strut. Tighten to 130 ft. lbs. (177 Nm).
19. Install front driveshaft.
20. Install transverse link ball joint to housing and tighten to 36 ft. lbs. (49 Nm).
21. Install ABS sensor harness on strut.
22. Install ABS sensor on housing. Tighten the mounting bolt to 24 ft. lbs. (32 Nm).
23. Install disc rotor on hub.
24. Install disc brake caliper on housing and tighten to 58 ft. lbs. (78 Nm).
25. Connect stabilizer link.
26. Install tie-rod end and tighten castle nut to 20 ft. lbs. (27 Nm).
27. After tightening castle nut to specified torque, retighten it further within 60° until a slot in castle nut is aligned with the ball joint hole. Insert cotter pin, and then bend the cotter pin around castle nut to secure it.
28. While depressing brake pedal to prevent front driveshaft from turning, tighten axle nut to 162 ft. lbs. (220 Nm).

✳ CAUTION

When axle nut is removed, replace it with new one. Be sure to tighten axle nut to specified torque. Do not over tighten it as this may damage wheel bearing.

29. After tightening axle nut, lock it securely.
30. Install wheel and tighten wheel nuts.

2. FRONT DRIVE SHAFT ASSEMBLY

Model	Type of drive shaft	Axle diameterΦ D mm (in)	Axle length L mm (in)
Turbo 5MT, 6MT	EBJ + PTJ	26 (1.0)	332.5 (13.09)
Other than above	EBJ + PTJ	26 (1.0)	349.6 (13.76)

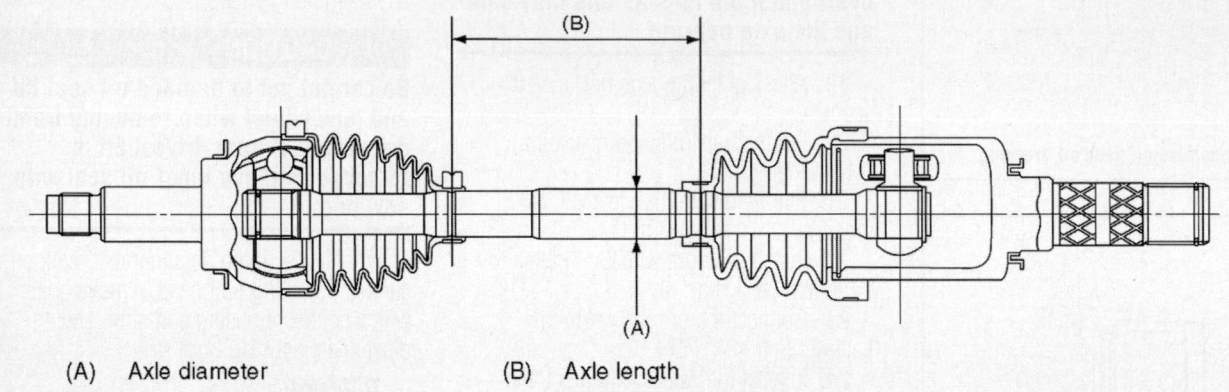

(A)   Axle diameter   (B)   Axle length

22140_SUBA_G0075

Fig. 54 Front driveshaft assembly

REAR AXLE HOUSING

REMOVAL & INSTALLATION

T-Type Rear Differential

See Figures 55 through 57.

1. Before servicing the vehicle, refer to the Precautions Section.
2. Disconnect the ground cable from the battery.
3. Shift the select lever or gear shift lever to neutral.
4. Release the parking brake.
5. Lift up the vehicle.
6. Remove the rear exhaust pipe and muffler.
7. Remove the heat shield cover.
8. Remove the propeller shaft.
9. Prepare the transmission jack and band.
10. Loosen the self-lock nuts which hold the rear differential to rear crossmember.
11. Remove the DOJ of rear driveshaft from rear differential using ST-28099PA100.
12. Remove the rear differential front member.
13. Support the rear differential with the transmission jack.
14. Secure the rear differential using band.
15. Remove the self-lock nuts which hold the rear differential to the crossmember.

16. Remove the rear differential stud bolt from rear crossmember bushing.

➡ **When removing the stud bolt, carefully adjust the angle and location of transmission jack and jack stand, if necessary.**

17. Lower the transmission jack stand after removing the rear differential stud bolt from the rear crossmember. Rear driveshaft should not come into contact with the lateral link bolt.
18. Pull out the axle shaft from rear differential.
19. Lower the transmission jack.
20. Secure the rear driveshaft to lateral link using wire.
21. Remove the rear differential member plate from rear differential.

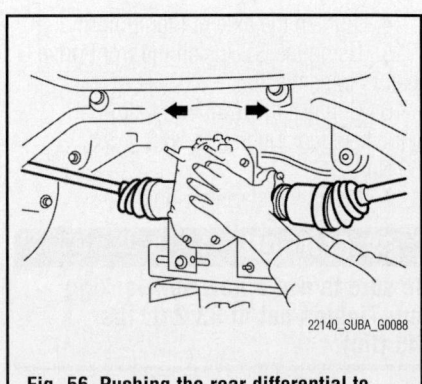

22140_SUBA_G0087

Fig. 55 Lowering the rear differential

To install:

22. Install the rear differential member plate to the rear differential.
23. Set the rear differential to transmission jack.

➡ **Secure the rear differential to transmission jack using band.**

24. Insert the spline shaft until the spline portion comes inside the side oil seal. Use seal protector ST-28099PA090.
25. Remove ST from rear differential.
26. Push the rear differential to insert the axle shaft into rear differential.
27. Adjust the transmission jack, if necessary, and insert the rear differential stud bolt into rear crossmember bushing properly.
28. After inserting the rear differential stud bolt into the rear crossmember bush-

22140_SUBA_G0088

Fig. 56 Pushing the rear differential to insert the axle shaft

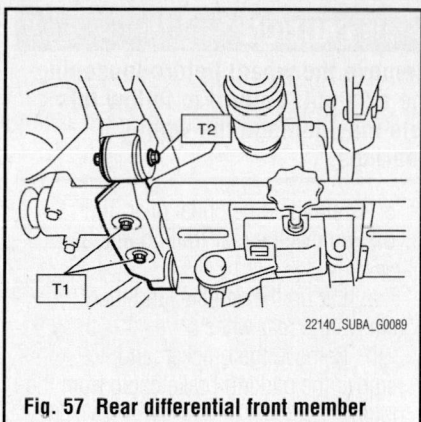

Fig. 57 Rear differential front member

ing, lift up the transmission jack and align the rear differential to its attachment position.

29. Tighten a new self-locking nut temporarily to rear crossmember.

30. Remove the band from rear differential. Lift up the rear differential until the rear differential is separated from the transmission jack.

31. Install the rear differential front member with a new self-locking nut. Tighten to specifications.

32. Tighten the self-locking nut. Tighten to 51 ft. lbs. (70 Nm).

33. Lower the transmission jack.

34. Install the propeller shaft.

35. Install the heat shield cover.

36. Install the rear exhaust pipe and muffler.

37. After installing the rear differential carrier to the vehicle, remove the filler plug, and refill with gear oil up to the lower lip of the plug hole.

38. Oil capacity specifications:
   a. Except for GT specifications B 6MT model: 0.8 qt. (0.8 L)
   b. GT specifications B 6MT model: 1.1 qt. (1.0 L)

39. Tighten the filler plug.

40. Road test and check for noise and leaks.

VA-Type Rear Differential

See Figures 58 through 60.

1. Before servicing the vehicle, refer to the Precautions Section.

2. Disconnect the ground cable from the battery.

3. Shift the select lever or gear shift lever to neutral.

4. Release the parking brake.

5. Lift up the vehicle.

6. Remove the rear exhaust pipe and muffler.

7. Remove the heat shield cover.

8. Remove the propeller shaft.

9. Prepare the transmission jack and band.

10. Loosen the self-lock nuts which hold the rear differential to rear crossmember.

11. Remove the DOJ of rear driveshaft from rear differential.

12. Remove the nuts which hold the rear differential front member.

13. Support the rear differential with the transmission jack.

14. Remove the rear differential front member.

15. Secure the rear differential using band.

16. Remove the self-lock nuts which hold the rear differential to rear crossmember.

17. Remove the rear differential stud bolt from rear crossmember bushing.

➡**When removing the stud bolt, carefully adjust the angle and location of transmission jack and jack stand, if necessary**

18. Lower the transmission jack stand after removing the rear differential stud bolt from the rear crossmember. Rear driveshaft should not come into contact with the lateral link bolt.

19. Pull out the axle shafts from the rear differential.

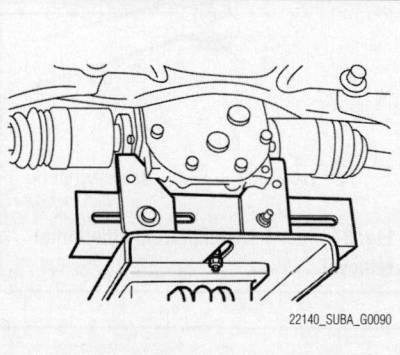

Fig. 58 Rear differential supported with transmission jack

20. Lower the transmission jack.

21. Secure the rear driveshaft to lateral link using wire.

22. Remove the rear differential member plate from rear differential.

To install:

23. Insert the rear differential member plate into rear differential.

24. Set the rear differential to transmission jack.

25. Attach the ST to rear differential.

26. Insert the spline shaft until the spline portion comes inside the side oil seal.

27. Remove ST from rear differential.

28. Push the rear differential to insert the axle shaft into rear differential.

29. Adjust the transmission jack, if necessary, and insert the rear differential stud bolt into rear crossmember bushing properly.

30. After inserting the rear differential stud bolt into the rear crossmember bushing, lift up the transmission jack and align the rear differential to its attachment position.

31. Tighten a new self-locking nut temporarily to rear crossmember.

32. Remove the band from rear differential. Lift up the rear differential until the rear differential is separated from the transmission jack.

33. Install the rear differential front member with a new self-locking nut. Tighten to specifications.

34. Tighten the self-locking nut and tighten to 51 ft. lbs. (70 Nm).

35. Lower the transmission jack.

36. Install the propeller shaft.

37. Install the heat shield cover.

38. Install the rear exhaust pipe and muffler.

39. Install correct differential oil and tighten the filler plug.

40. Road test and check for noise and leaks.

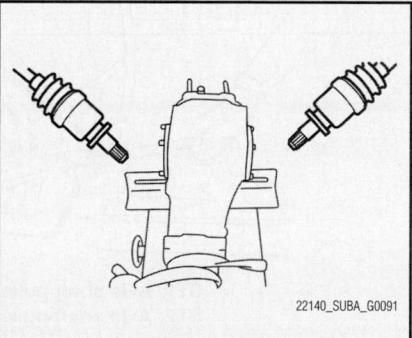

Fig. 59 Axle shaft removal from rear differential shown

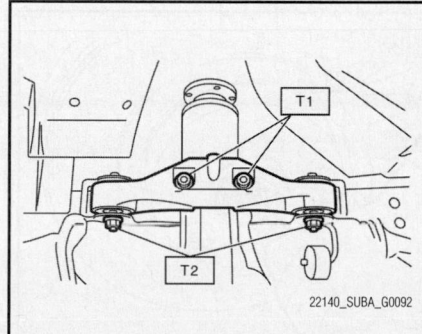

Fig. 60 Rear differential front member and locking nuts shown

REAR AXLE SHAFT SEALS

REMOVAL & INSTALLATION

See Figure 61.

1. Before servicing the vehicle, refer to the Precautions Section.
2. Remove the rear differential.
3. Remove the rear differential side oil seal using a screwdriver wrapped with vinyl tape to prevent the side retainer from scratches.

To install:

4. Using the ST, install the oil seal to the side retainer.
5. Install the rear differential.

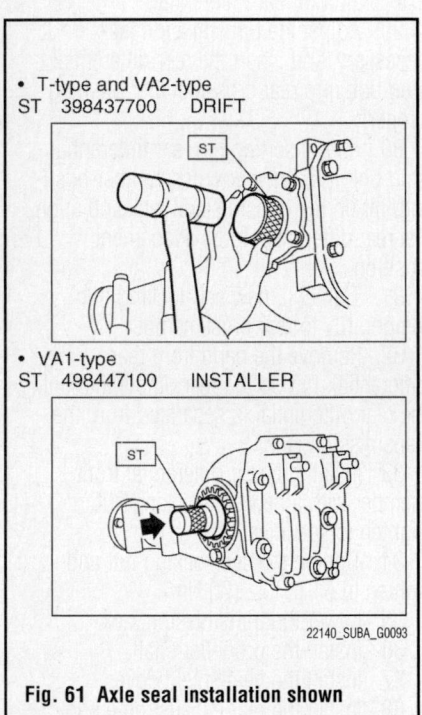

Fig. 61 Axle seal installation shown

REAR DRIVESHAFT

REMOVAL & INSTALLATION

See Figures 62 through 70.

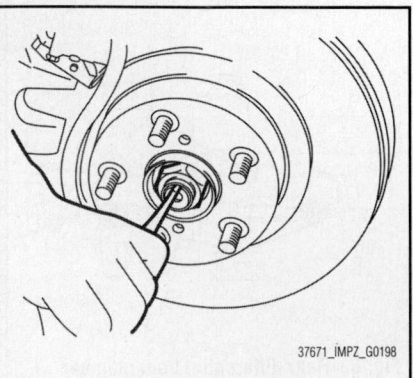

Fig. 62 Axle nut crimped section

1. Disconnect the ground cable from battery.
2. Lift up the vehicle, and remove the rear wheels.
3. Lift the crimped section of axle nut.
4. Remove the axle nut using a socket wrench while depressing the brake pedal.

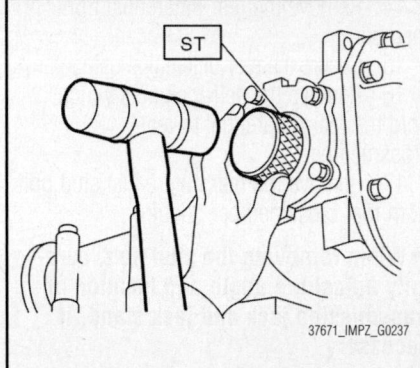

Fig. 63 Installing the oil seal to the side retainer

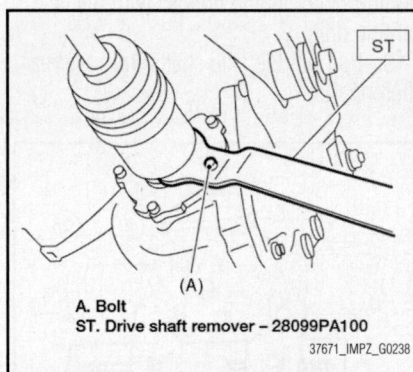

A. Bolt
ST. Drive shaft remover – 28099PA100

Fig. 64 Rear drive shaft from differential removal

✳✳ CAUTION

Remove the wheel before loosening the axle nut. Failure to follow this rule may damage the wheel bearings.

5. Drain the differential gear oil.
6. Remove the rear trailing link as follows:
   a. Lift up the vehicle, and then remove the rear wheels.
   b. Remove the bracket, and remove the parking brake cable from the guide.
   c. Remove the trailing link.
7. Remove the rear lateral link as follows:
   a. Lift up the vehicle, and then remove the rear wheels.
   b. Remove the nut and disconnect the stabilizer link.
   c. Remove the shock absorber lower bolt.
   d. Remove the bolt, and remove rear lateral link.
8. Remove the rear drive shaft from the rear differential by using the ST.

➡ **Fit the ST to the bolts as shown in the figure to prevent damage of the side bearing retainer.**

9. Remove the rear drive shaft from the rear axle. If it is hard to remove, use ST1 and ST2.

To install:

10. Replace the rear differential side oil seal as follows:
    a. Remove the rear differential as shown in this section.

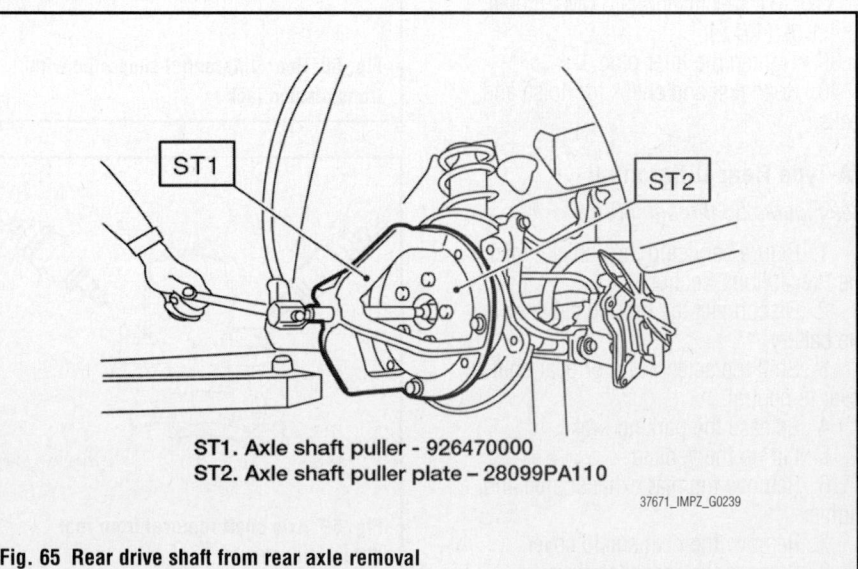

ST1. Axle shaft puller - 926470000
ST2. Axle shaft puller plate - 28099PA110

Fig. 65 Rear drive shaft from rear axle removal

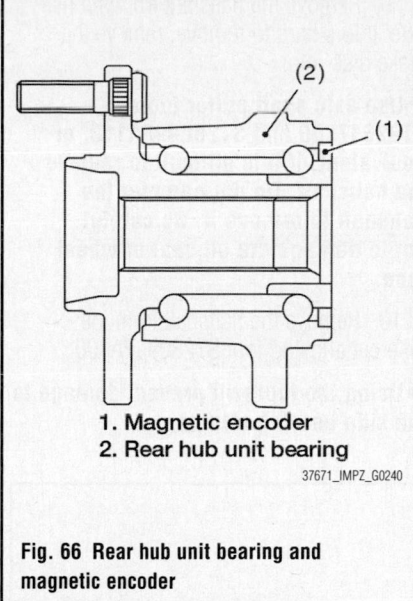

1. Magnetic encoder
2. Rear hub unit bearing

37671_IMPZ_G0240

Fig. 66 Rear hub unit bearing and magnetic encoder

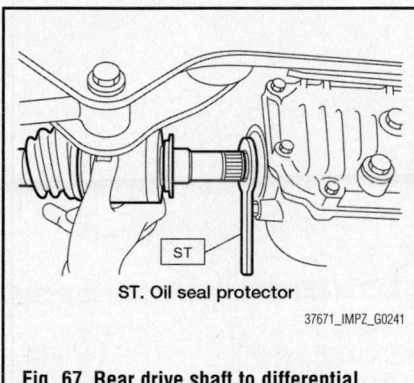

ST. Oil seal protector

37671_IMPZ_G0241

Fig. 67 Rear drive shaft to differential installation

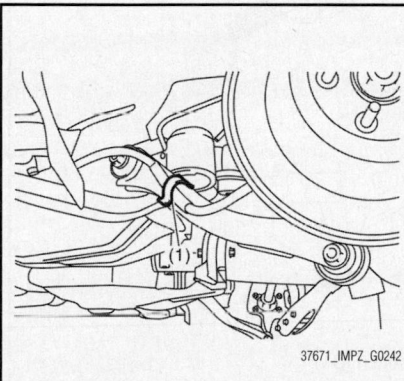

37671_IMPZ_G0242

Fig. 68 Rear trailing link—parking brake cable removal from guide

b. Remove the rear differential side oil seal using

c. Using the ST, install the oil seal to the side retainer.

➡**Apply differential gear oil to the oil seal lips.**

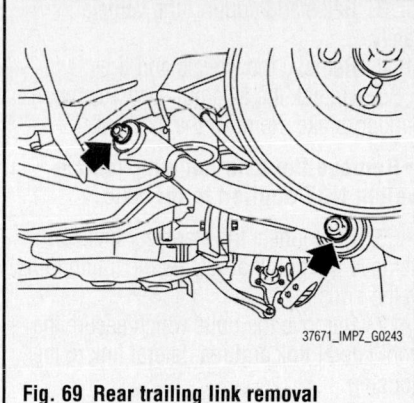

37671_IMPZ_G0243

Fig. 69 Rear trailing link removal

37671_IMPZ_G0244

Fig. 70 Rear lateral link removal

d. Install the rear differential as shown in this section.

➡**After pulling out the drive shaft, be sure to replace with a new oil seal.**

11. Insert the EBJ into rear hub splines.

❋❋ WARNING

Be careful not to damage the magnetic encoder. Do not get closer the tool which charged magnetism to magnetic encoder.

12. Draw the rear drive shaft into specified position.

❋❋ WARNING

Do not hammer drive shaft when installing it.

13. Tighten the axle nut temporarily.
14. Using the ST, install the rear drive shaft to the rear differential.
15. Install the rear lateral link as follows:

a. Remove the bolt, and remove rear lateral link.

b. Remove the shock absorber lower bolt.

c. Remove the nut and disconnect the stabilizer link.

d. Lift up the vehicle, and then remove the rear wheels.

❋❋ WARNING

Be sure to use a new self-locking nut. Always tighten the bushing in the state where the vehicle is at curb weight and the wheels are in full contact with the ground.

e. Tightening torque:
• Rear lateral link to 88.5 ft. lbs. (120 Nm).
• Shock absorber to 88.5 ft. lbs. (120 Nm).
• Stabilizer link to 33.2 ft. lbs. (45 Nm).

f. Inspect the wheel alignment and adjust if necessary.

16. Install the rear trailing link as follows:

a. Install the trailing link.

b. Install the bracket, and remove the parking brake cable from the guide.

c. Lower the vehicle, and then install the rear wheels.

❋❋ WARNING

Be sure to use a new self-locking nut. Always tighten the bushing in the state where the vehicle is at curb weight and the wheels are in full contact with the ground. Tighten nut to 66.4 ft. lbs. (90 Nm).

17. Attach the links to the rear housing and tighten them to the specified torque:
• Stabilizer link to 33.2 ft. lbs. (45 Nm).
• Shock absorber to 88.5 ft. lbs. (120 Nm).
• Rear lateral link to 88.5 ft. lbs. (120 Nm).
• Trailing link to 66.4 ft. lbs. (90 Nm).

18. While pressing the brake pedal, tighten the new axle nuts to the specified torque. Tighten nuts to 140 ft. lbs. (190 Nm)

❋❋ CAUTION

Note the following:

• Do not install wheel and let it touch the ground before tightening the axle nut. Failure to follow this rule may damage the axle bearing.
• Do not overtighten the nuts as this may damage the axle bearing.

19. Lock the axle nut securely.

20. Fill the differential gear oil.

21. Install the rear wheels. Tighten to 74 ft. lbs. (100 Nm).

22. Connect the ground cable to battery.

23. Inspect the wheel alignment and adjust if necessary.

REAR HALFSHAFTS

REMOVAL & INSTALLATION

See Figure 71.

1. Before servicing the vehicle, refer to the Precautions Section.

2. Disconnect the negative battery cable.

3. Raise and support the vehicle safely.

4. Remove the wheels and tires.

5. Unlock the axle nut. Depress the parking brake. Remove the axle nut.

➡ **Remove the axle nut with vehicle weight NOT applied to the axle.**

6. Disconnect the stabilizer link. Remove the bolt that retains the trailing link to the housing.

7. Remove the bolts which secure the front lateral link and rear lateral link to the housing.

8. Remove the ABS wheel speed sensor from the back plate.

9. Remove the halfshaft from the rear axle. If it is hard to remove, remove the brake disk rotor.

➡ **Use axle shaft puller tools ST92647000 and ST28099PA110, or equivalent, if it is difficult to remove the halfshaft. Do not hammer the halfshaft to remove it. Be careful not to damage the oil seal or wheel tone.**

10. Remove the halfshaft from the differential using tool ST28099PA100.

➡ **Using the tool will prevent damage to the side bearing retainer.**

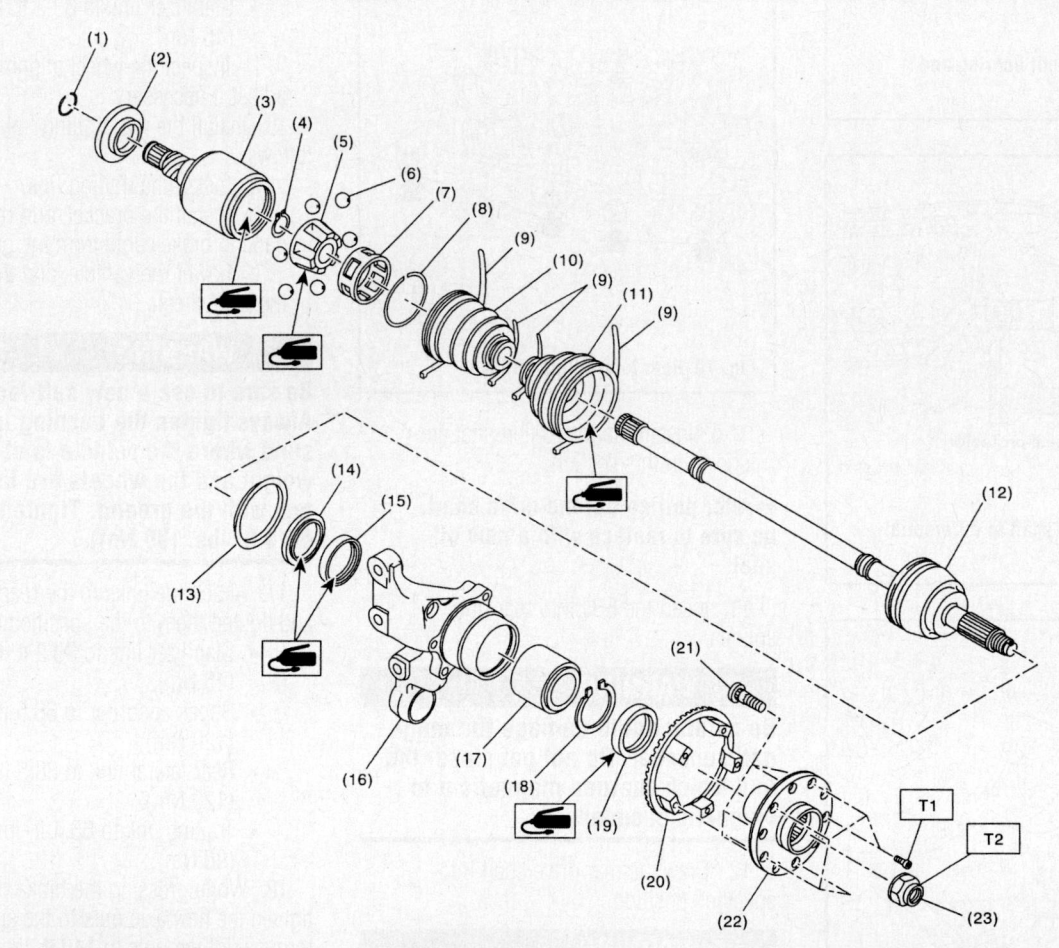

(1)	Circlip	(9)	Boot band	(19)	Oil seal (OUT)
(2)	Baffle plate (DOJ)	(10)	Boot (DOJ)	(20)	Tone wheel
(3)	Outer race DOJ: Except STI model	(11)	Boot	(21)	Hub bolt
	Outer race EDJ: STI model	(12)	EBJ ASSY	(22)	Hub
(4)	Snap ring	(13)	Baffle plate	(23)	Axle nut
(5)	Inner race	(14)	Oil seal (IN. No. 2)		
(6)	Ball	(15)	Oil seal (IN.)		
(7)	Cage	(16)	Housing		
(8)	Snap ring	(17)	Bearing		
		(18)	Snap ring		

Tightening torque: N·m (kgf-m, ft-lb)
T1: 13 (1.3, 9.4)
T2: 190 (19.4, 140)

09490_SBCR_G0092

Fig. 71 Rear halfshaft and related components

To install:

11. Installation is the reverse of the removal procedure.

12. Tighten the axle nut to 140 ft. lbs. with the parking brake applied.

REAR PINION SEAL

REMOVAL & INSTALLATION

See Figures 72 through 74.

1. Before servicing the vehicle, refer to the Precautions Section.

2. Move the gear shift level to "N".

3. Secure the vehicle with wheel chocks and release the parking brake.

4. Remove the rear differential protector, if equipped.

5. Remove the drain plug and drain the gear oil. Install the drain plug.

6. Jack up the rear wheels and safely support the vehicle with suitable jack stands.

7. Remove the rear exhaust pipe and muffler.

8. Remove the driveshaft.

9. Remove the self-locking nut while holding the companion flange with Special Tool 498427200.

10. Remove the protector mounting nut.

11. Remove the tank cover.

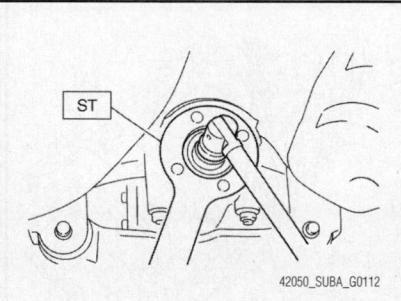

Fig. 72 Hold the companion flange with Special Tool 498427200 and remove the pinion nut.

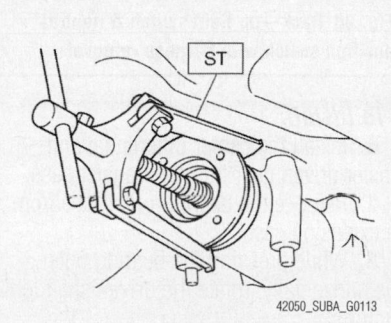

Fig. 73 Extract the companion flange using Special Tool 399703600

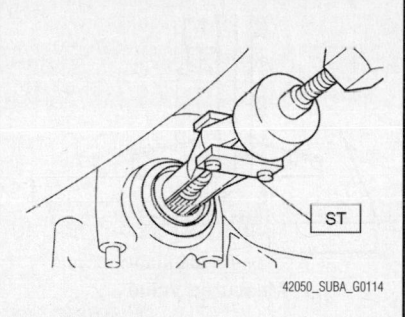

Fig. 74 Remove the oil seal using Special Tool 398527700

12. Extract the companion flange using Special Tool 399703602.

13. Remove the oil seal using a screw driver or Special Tool 398527700.

To install:

14. Drive in a new oil seal using a seal driver Special Tool 498447120.

15. Install the companion flange. Use a plastic hammer to install the flange.

16. Tighten a new self-locking nut for T—type differentials to 134 ft. lbs. (181 Nm). For VA—type tighten to 141 ft. lbs. (191 Nm). Use Special Tool 498427200 Flange Wrench to hold the companion flange.

17. The remainder of the installation is the reverse order of removal.

18. Refill the differential with fluid to the correct level.

TRANSFER CASE AND EXTENSION CASE ASSEMBLY

REMOVAL & INSTALLATION

See Figures 75 through 90.

1. Remove the manual transmission assembly from the vehicle.

2. Remove the back-up light switch & neutral position switch as follows:

   a. Disconnect the ground cable from battery.

   b. Remove the air intake chamber and intake boot—non-turbo.

   c. Remove the intercooler—turbo.

   d. Disconnect the connector back-up light switch & neutral position switch.

   e. Lift up the vehicle.

   f. Remove the back-up light switch & neutral position switch with the harness.

3. Remove the transfer case together with the extension case assembly.

4. Remove the shifter arm.

5. Remove the extension case assembly.

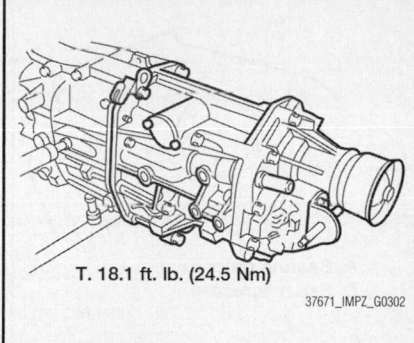

T. 18.1 ft. lb. (24.5 Nm)

Fig. 75 Transfer case & extension case assembly (1 of 2)

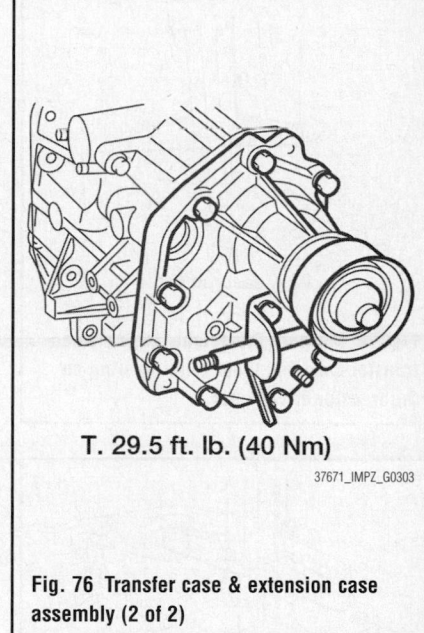

T. 29.5 ft. lb. (40 Nm)

Fig. 76 Transfer case & extension case assembly (2 of 2)

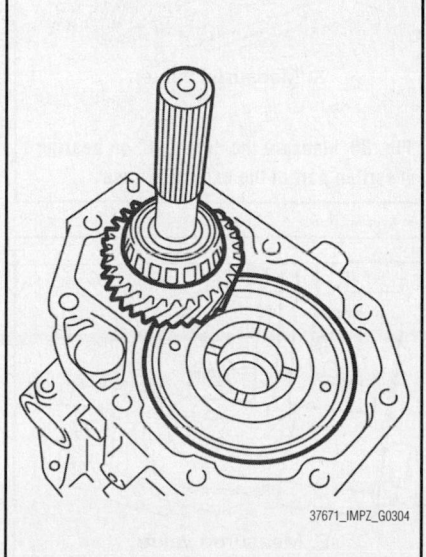

Fig. 77 Center differential and transfer driven gear installation

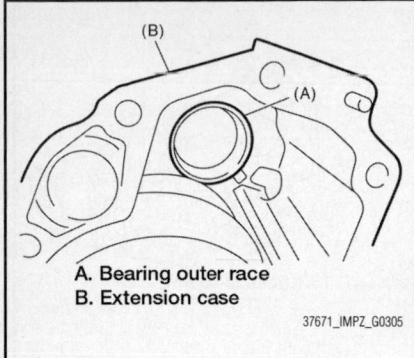

A. Bearing outer race
B. Extension case

37671_IMPZ_G0305

Fig. 78 Bearing outer race from the extension case removal

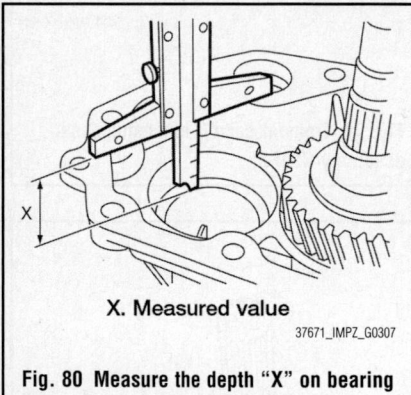

W. Measured value

37671_IMPZ_G0306

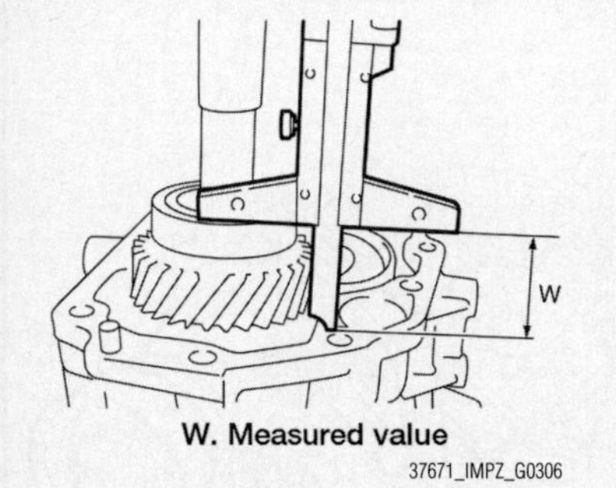

Fig. 79 Measure the height "W" between transfer case and taper roller bearing on the transfer driven gear.

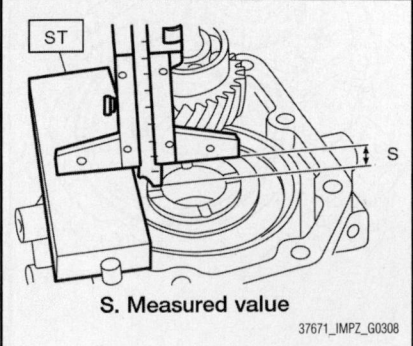

X. Measured value

37671_IMPZ_G0307

Fig. 80 Measure the depth "X" on bearing insertion part of the extension case

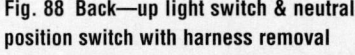

S. Measured value

37671_IMPZ_G0308

Fig. 81 Measure the depth "S" between transfer case and center differential

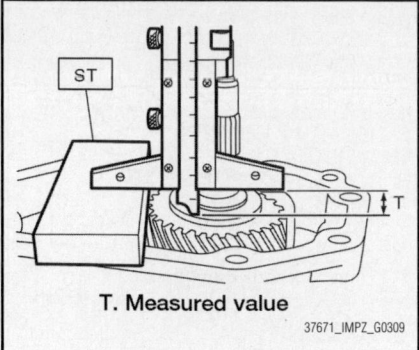

T. Measured value

37671_IMPZ_G0309

Fig. 82 Measure the height "T" between the extension case and transfer drive gear

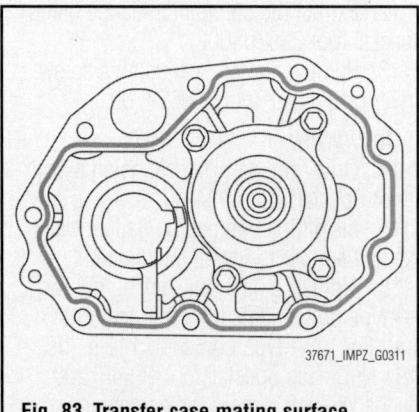

37671_IMPZ_G0311

Fig. 83 Transfer case mating surface

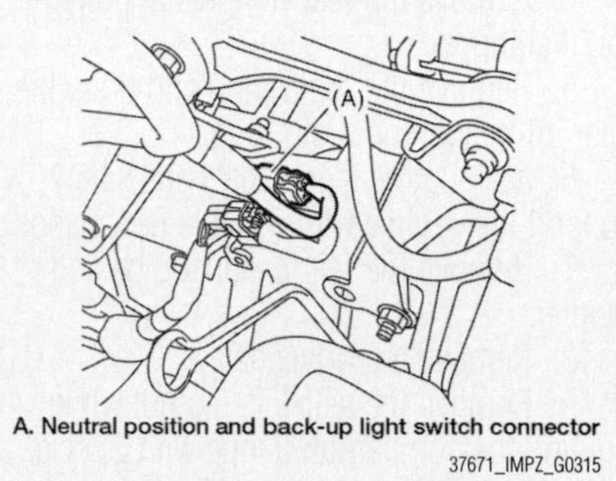

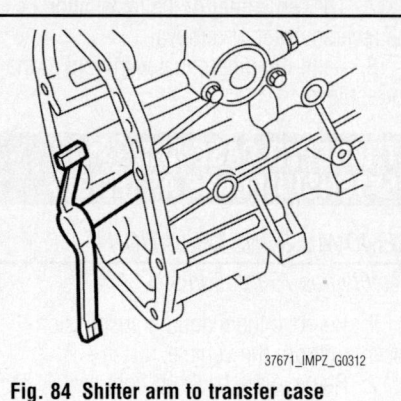

37671_IMPZ_G0312

Fig. 84 Shifter arm to transfer case installation

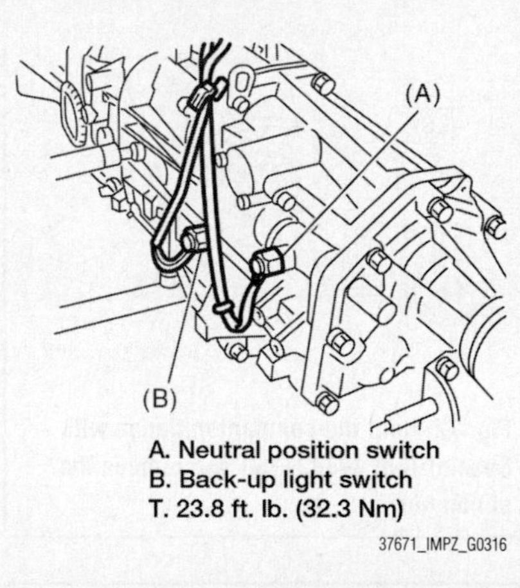

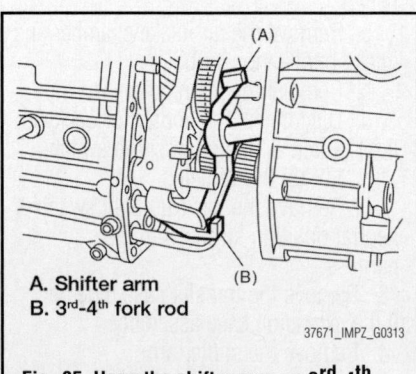

A. Shifter arm
B. 3rd-4th fork rod

37671_IMPZ_G0313

Fig. 85 Hang the shifter arm on 3rd-4th fork rod

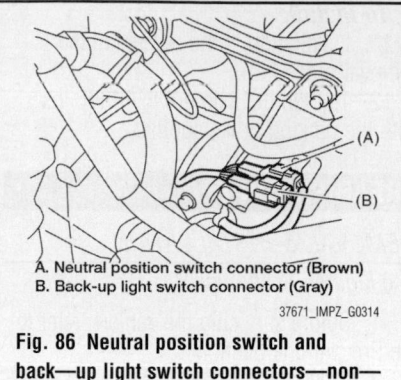

A. Neutral position switch connector (Brown)
B. Back-up light switch connector (Gray)

37671_IMPZ_G0314

Fig. 86 Neutral position switch and back—up light switch connectors—non—turbo

A. Neutral position and back-up light switch connector

37671_IMPZ_G0315

Fig. 87 Neutral position switch and back—up light switch connectors—turbo

A. Neutral position switch
B. Back-up light switch
T. 23.8 ft. lb. (32.3 Nm)

37671_IMPZ_G0316

Fig. 88 Back—up light switch & neutral position switch with harness removal

To install:

6. Install the center differential and transfer driven gear into the transfer case.

7. Remove the bearing outer race from the extension case.

8. While pressing the bearing outer race horizontally, rotate the driven shaft for ten turns.

9. Measure the height "W" between transfer case and taper roller bearing on the transfer driven gear.

10. Measure depth "X" on bearing insertion part of the extension case.

➡**Measure with bearing outer race and thrust washer removed.**

11. Calculate the thrust washer thickness "t" using the following calculation:
   - $t = X - W + (0.15 - 0.20 \text{ mm} (0.006 - 0.008 \text{ in}))$

12. Select the washer with the nearest value in the following table.

Preload (standard protrusion) of the taper roller bearing: 0.15 — 0.20 mm (0.006 — 0.008 in)

➡**Be sure that it is always within the preload.**

13. Fit the thrust washers on the transfer drive shaft.

14. Install the bearing outer race into the extension case.

15. Measure the depth "S" between transfer case and center differential. ST 398643600 GAUGE

16. Measure the height "T" between the extension case and transfer drive gear. ST 398643600 GAUGE

➡**Thickness of ST: 0.59 in (15 mm).**

17. Calculate the thrust washer thickness "U" using the following calculation:
   a. $U = S + T - 30 \text{ mm} (1.18 \text{ in})$ [Thickness of ST]

18. Select a suitable washer in the following table.
   a. Standard clearance:
   - 0.15 — 0.35 mm (0.0059 — 0.0138 in)

19. Fit the thrust washer onto the center differential.

20. Apply a proper amount of liquid gasket to the transfer case mating surface.

Thrust washer	
Part No.	Thickness mm (in)
803036050	0.9 (0.035)
803036054	1.0 (0.039)
803036051	1.1 (0.043)
803036055	1.2 (0.047)
803036052	1.3 (0.051)
803036056	1.4 (0.055)
803036053	1.5 (0.059)
803036057	1.6 (0.063)
803036058	1.7 (0.067)
803036080	1.8 (0.071)
803036081	1.9 (0.075)

37671_IMPZ_G0310

Fig. 90 Thrust washer thickness value table using the following calculations— U=S + T — 1.18 in (30 mm) [Thickness of ST]

Liquid gasket: THREE BOND 1215 (Part No. 004403007) or equivalent

21. Install the extension assembly into the transfer case. Tighten to 29 ft. lbs. (40 Nm).

22. Attach the shifter arm to transfer case.

23. Hang the shifter arm on 3rd-4th fork rod.

24. Install the extension case assembly along with the transfer case to the transmission case. Tighten to 18.1 ft. lbs. (24.5 Nm)

25. Install the back-up light switch & neutral position switch as follows:
   a. Install the back-up light switch & neutral position switch with the harness. Tighten to 23.8 ft. lbs. (32.3 Nm).
   b. Connect the connectors of back-up light switch & neutral position switch.
   c. Install the air intake chamber and intake boot—non—turbo.
   d. Install the intercooler—turbo.
   e. Connect the battery ground cable to the battery.

26. Install the manual transmission assembly to the vehicle.

Thrust washer (50 × 61 × t)	
Part No.	Thickness mm (in)
803050060	0.50 (0.0197)
803050061	0.55 (0.0217)
803050062	0.60 (0.0236)
803050063	0.65 (0.0256)
803050064	0.70 (0.0276)
803050065	0.75 (0.0295)
803050066	0.80 (0.0315)
803050067	0.85 (0.0335)
803050068	0.90 (0.0354)
803050069	0.95 (0.0374)
803050070	1.00 (0.0394)
803050071	1.05 (0.0413)
803050072	1.10 (0.0433)
803050073	1.15 (0.0453)
803050074	1.20 (0.0472)
803050075	1.25 (0.0492)
803050076	1.30 (0.0512)
803050077	1.35 (0.0531)
803050078	1.40 (0.0551)
803050079	1.45 (0.0571)

37671_IMPZ_G0301

Fig. 89 Thrust washer thickness value table using the following calculations—t = X – W + (0.008 in (0.15 – 0.02 mm)

ENGINE COOLING

ENGINE FAN

REMOVAL & INSTALLATION
See Figures 91 through 95.

1. Set the vehicle on a lift.
2. Disconnect the ground cable from battery.
3. Lift up the vehicle.
4. Remove the undercover.
5. Disconnect the connector of the main fan motor assembly.
6. Remove the heat shield cover—A/T model
7. Remove the ATF hose from the clip of the radiator main fan shroud—A/T model
8. Lower the vehicle.
9. Drain the coolant.
10. Disconnect the radiator inlet hose from engine side.
11. Disconnect the over flow hose.
12. Remove the reservoir tank as follows:
    a. Disconnect the over flow hose connected to the radiator filler neck.
    b. Push in the hook, and pull the reserve tank in the direction of the arrow to remove.

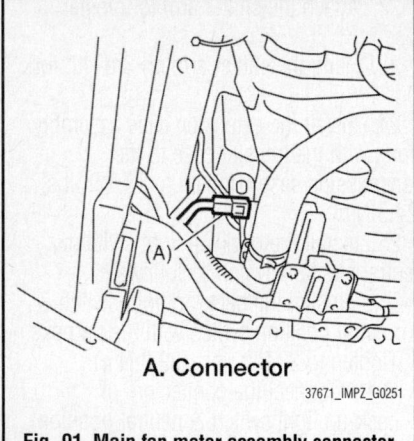

A. Connector

37671_IMPZ_G0251

Fig. 91 Main fan motor assembly connector

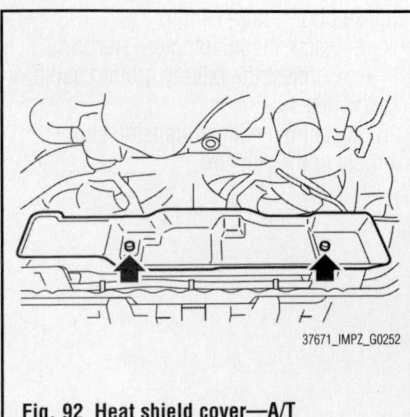

37671_IMPZ_G0252

Fig. 92 Heat shield cover—A/T

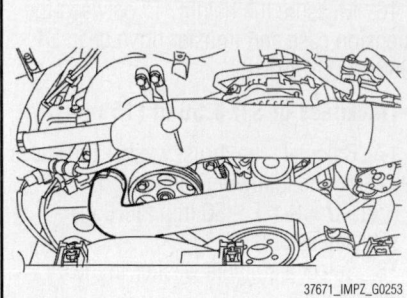

37671_IMPZ_G0253

Fig. 93 Radiator inlet hose to engine side removal

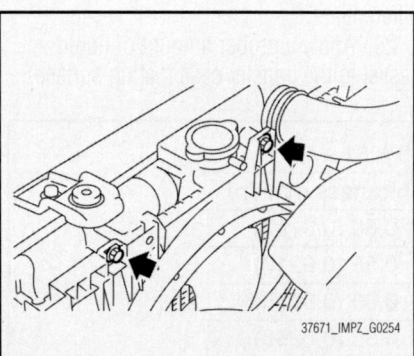

37671_IMPZ_G0254

Fig. 94 Radiator main fan shroud bolt removal

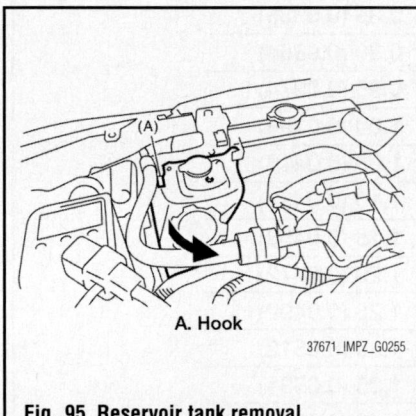

A. Hook

37671_IMPZ_G0255

Fig. 95 Reservoir tank removal

13. Remove the bolts which hold the radiator main fan shroud onto the radiator.
14. Remove the radiator main fan and fan motor.

To install:
15. Check if the radiator hose and the over flow hose are properly connected.

➡ **If the installation of the radiator main fan and fan motor is difficult, attempt installation after loosening the bolts which hold the radiator sub fan shroud to the radiator.**

16. Install the radiator main fan and fan motor.
17. Install the bolts which hold the radiator main fan shroud onto the radiator.
18. Install the reservoir tank as follows:
    a. Push in the hook, and push the reserve tank in the opposite direction of the arrow to install.
    b. Connect the over flow hose connected to the radiator filler neck.
19. Connect the over flow hose.
20. Connect the radiator inlet hose from engine side.
21. Lower the vehicle.
22. Fill the engine cooling system with the proper type and amount of coolant.
23. Install the ATF hose from the clip of the radiator main fan shroud—A/T model
24. Install the heat shield cover—A/T model
25. Connect the connector of the main fan motor assembly.
26. Install the undercover.
27. Lift up the vehicle.
28. Connect the ground cable to the battery.
29. Remove the vehicle on a lift.

RADIATOR

REMOVAL & INSTALLATION
See Figures 91, 92, 96 through 103.

✳✳ CAUTION

The radiator is pressurized when the engine and radiator are hot. Wait until engine and radiator cool down before working on the radiator.

1. Set the vehicle on a lift.
2. Disconnect the ground cable from battery.
3. Lift up the vehicle.
4. Remove the undercover.
5. Remove the heat shield cover from radiator—A/T.
6. Drain the engine coolant as outlined in this section.
7. Disconnect the main fan motor assembly connector and sub fan motor assembly connector.
8. Disconnect the radiator outlet hose from thermostat cover.
9. Disconnect the ATF cooler hoses from ATF pipes—A/T.

➡ **Plug the ATF pipe to prevent ATF from leaking.**

10. Lower the vehicle.
11. Disconnect the over flow hose.
12. Remove the reservoir tank.

13. Remove the reservoir tank as follows:

   a. Disconnect the over flow hose connected to the radiator filler neck.

   b. Push in the hook, and pull the reserve tank in the direction of the arrow to remove.

14. Remove the air intake duct.

15. Disconnect the radiator inlet hose from engine.

16. Remove the radiator upper brackets.

17. Lift the radiator up and remove the radiator from vehicle.

To install:

18. Attach the radiator lower cushion to the hole on the radiator lower bracket.

19. Install the radiator to vehicle.

➡**Make pins on the lower side of radiator be fitted into the radiator lower cushions.**

20. Install the radiator upper brackets and tighten the bolts to 106.8 inch lbs. (12 Nm).

21. Connect the radiator inlet hose.

22. Install the air intake duct.

23. Install the reservoir tank as follows:

   a. Push in the hook, and pull the reserve tank in the direction of the arrow to remove.

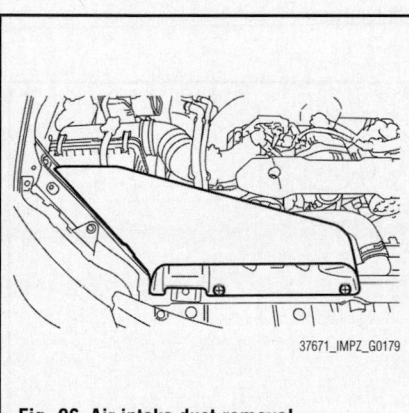

Fig. 96 Air intake duct removal

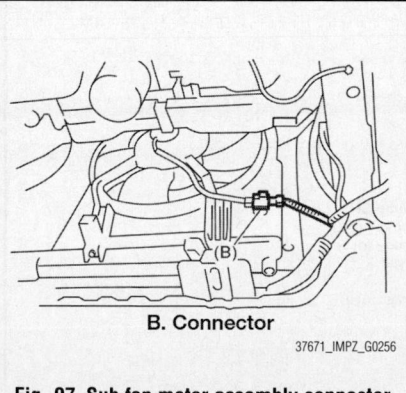

B. Connector

Fig. 97 Sub fan motor assembly connector

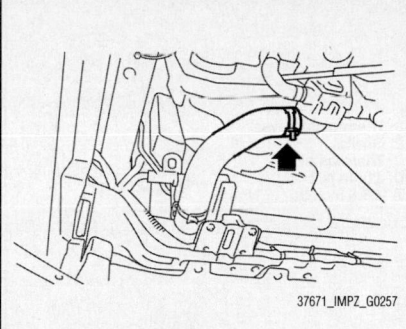

Fig. 98 Radiator outlet hose from thermostat cover

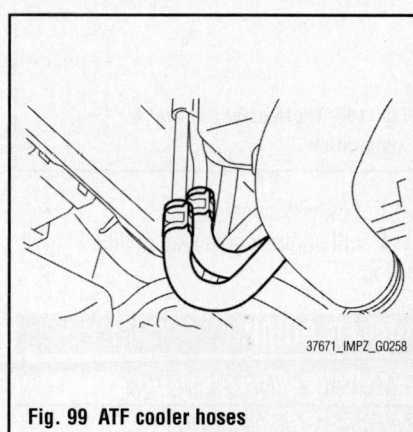

Fig. 99 ATF cooler hoses

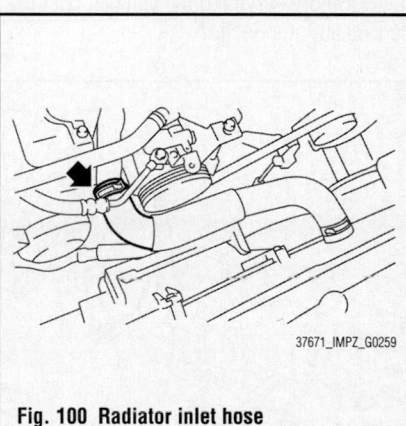

Fig. 100 Radiator inlet hose

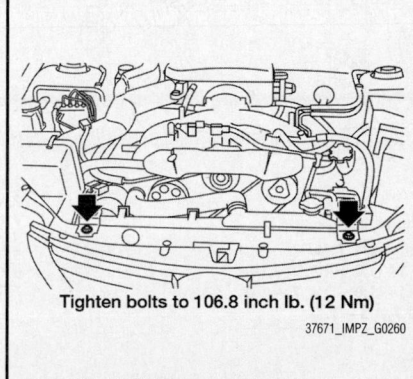

Tighten bolts to 106.8 inch lb. (12 Nm)

Fig. 101 Radiator upper brackets

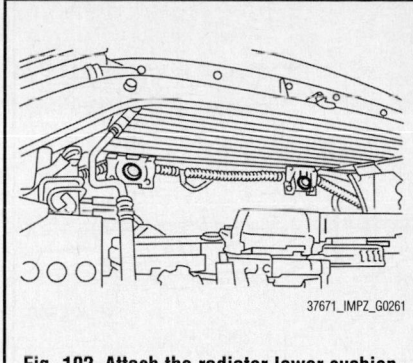

Fig. 102 Attach the radiator lower cushion to the hole on the radiator lower bracket

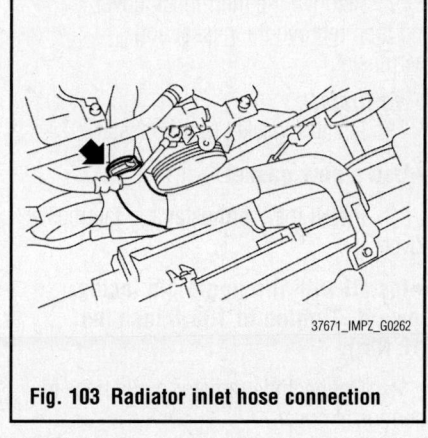

Fig. 103 Radiator inlet hose connection

   b. Disconnect the over flow hose connected to the radiator filler neck.

24. Connect the over flow hose.

25. Lift up the vehicle.

26. Connect the ATF cooler hoses—A/T.

27. Connect the radiator outlet hose.

28. Connect the main fan motor assembly connector and sub fan motor assembly connector.

29. Install the heat shield cover—AT model. Tighten bolts to 26.4 inch lbs. (3 Nm).

30. Install the undercover.

31. Lower the vehicle.

32. Connect the ground cable to battery.

33. Fill engine coolant as outlined in this section.

34. Check the ATF level.

THERMOSTAT

REMOVAL & INSTALLATION

See Figures 104 and 105.

1. Set the vehicle on a lift.

2. Lift up the vehicle.

3. Remove the undercover.

4. Drain the engine coolant as outlined in this section.

5. Remove the front exhaust pipe.

6. Disconnect the radiator outlet hose from thermostat cover.

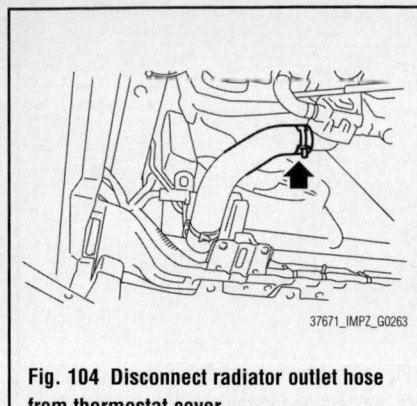

Fig. 104 Disconnect radiator outlet hose from thermostat cover

7. Remove the thermostat cover, and then remove the gasket and thermostat.

To install:

8. Install a gasket to thermostat.

➡**Use a new gasket.**

9. Install the thermostat and thermostat cover.

➡**Install with the jiggle pin facing upward. Tighten to 106.8 inch lbs. (12 Nm).**

10. Connect the radiator outlet hose to thermostat cover.
11. Install the front exhaust pipe.
12. Install the undercover.

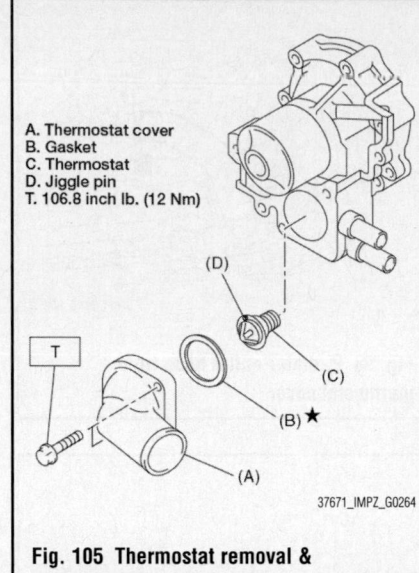

A. Thermostat cover
B. Gasket
C. Thermostat
D. Jiggle pin
T. 106.8 inch lb. (12 Nm)

Fig. 105 Thermostat removal & installation

13. Lower the vehicle.
14. Fill engine coolant as outlined in this section.

WATER PUMP

REMOVAL & INSTALLATION

See Figures 106 and 107.

1. Before servicing the vehicle, refer to the Precautions Section.

2. Remove or disconnect the following:
   • Negative battery cable
   • Engine undercover, if equipped
3. Drain the coolant into a suitable container.
4. Remove or disconnect the following:
   • Radiator fan connectors
   • Radiator outlet and heater hoses
   • Heater bypass hose or overflow hose, if equipped
   • Radiator fan motor assembly
   • Accessory drive belts
   • Timing belt
   • Belt tension adjuster
   • Belt idler No. 2
   • Camshaft Position (CMP) sensor
   • Left side camshaft pulley
   • Left side rear timing belt cover
   • Tensioner bracket
   • Radiator and heater hoses from water pump
   • Water pump retainer bolts
   • Water pump
5. Inspect the radiator hoses for deterioration and replace as necessary.

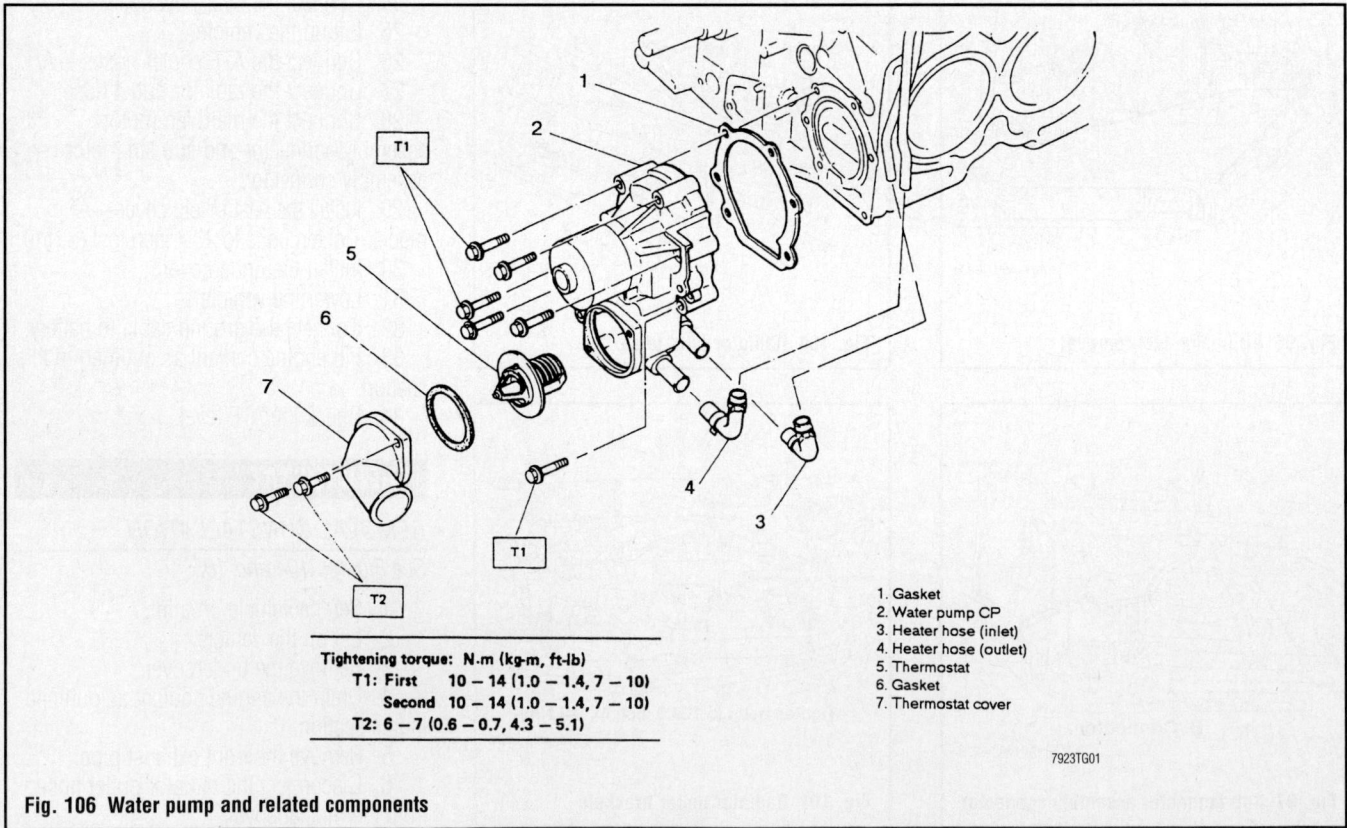

Tightening torque: N·m (kg-m, ft-lb)
T1: First   10 – 14 (1.0 – 1.4, 7 – 10)
     Second 10 – 14 (1.0 – 1.4, 7 – 10)
T2: 6 – 7 (0.6 – 0.7, 4.3 – 5.1)

1. Gasket
2. Water pump CP
3. Heater hose (inlet)
4. Heater hose (outlet)
5. Thermostat
6. Gasket
7. Thermostat cover

Fig. 106 Water pump and related components

To install:

6. Clean the gasket mating surfaces thoroughly. Always use new gaskets during installation.

7. Install or connect the following:

8. Tighten the pump bolts in sequence to 8.9 ft. lbs. (12 Nm). After tightening the bolts once, retighten to the same specification again.

- Radiator heater hoses to water pump
- Tensioner bracket and tighten to 18 ft. lbs. (25 Nm)
- Left side rear timing belt cover
- Left side camshaft pulley(s). Tighten to 58 ft. lbs. (78 Nm) on non turbocharged engine and 72 ft. lbs. (98 Nm) on turbocharged engine.

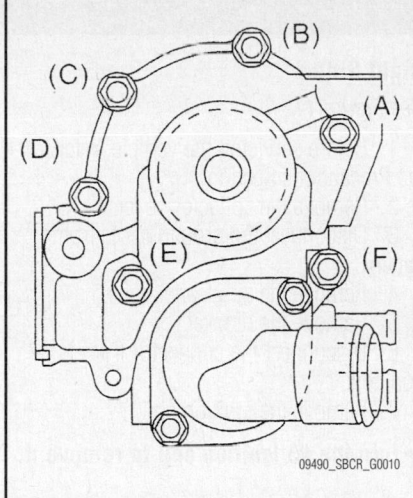

09490_SBCR_G0010

Fig. 107 Tighten the water pump bolts in two steps using the following sequence

- CMP sensor
- Belt idler No. 2 and tighten to 29 ft. lbs. (39 Nm)
- Belt tension adjuster
- Timing belt
- Accessory drive belts
- Radiator fan assembly
- Reservoir tank, if removed
- Heater bypass hose or overflow hose, if equipped
- Air intake duct
- Radiator outlet and heater hoses
- Radiator fan connectors
- Engine undercover, if removed

9. Fill the system with coolant and connect the negative battery cable.

10. Start the engine and allow it to reach operating temperature.

11. Check for leaks.

ENGINE ELECTRICAL

ALTERNATOR

REMOVAL & INSTALLATION

See Figure 108.

1. Before servicing the vehicle, refer to the Precautions Section.

2. Remove or disconnect the following:
- Negative battery cable
- Connector and terminal from the alternator
- V-belt cover, if equipped
- Front side V-belt
- Alternator to bracket bolts
- Alternator from the vehicle

To install:
3. Install or connect the following:
- Alternator into the vehicle
- Alternator to bracket bolts

- Front side V-belt
- V-belt cover, if equipped
- Connector and terminal to the alternator
- Negative battery cable

4. Check and adjust the belt tension.

VOLTAGE REGULATOR

ADJUSTMENT

The voltage regulator is not adjustable. If the regulator is found to be defective, it must be replaced.

REMOVAL & INSTALLATION

See Figure 109.

1. Before servicing the vehicle, refer to the Precautions Section.

2. Remove the alternator from the vehicle.

3. Remove the four through bolts. Then insert the tip of a flat-head screwdriver into the gap between the stator core and front bracket. Pry them apart to disassemble the alternator.

4. Hold rotor with a vise and remove pulley nut.

✳✳ CAUTION

When holding rotor with vise, insert aluminum plates or wood pieces on the contact surfaces of the vise to prevent rotor from damage.

CHARGING SYSTEM

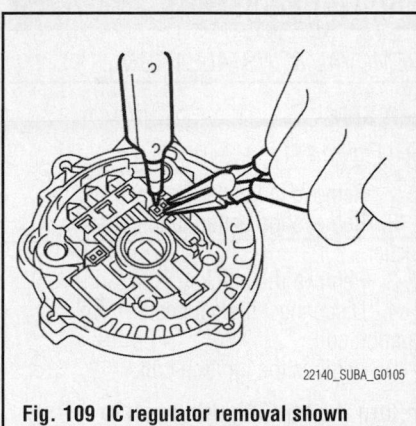

22140_SUBA_G0105

Fig. 109 IC regulator removal shown

5. Unsolder connection between rectifier and stator coil to remove stator coil.

✳✳ CAUTION

Finish the work rapidly (less than three seconds) because the rectifier cannot withstand heat very well.

6. Remove screws which secure voltage regulator to rear cover, and unsolder connection between voltage regulator and rectifier to remove the regulator.

To install:
7. Installation is the reverse order of assembly.

8. After assembly, turn the pulley by hand to check that the rotor turns smoothly.

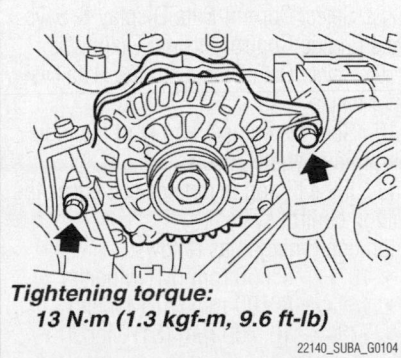

Tightening torque:
13 N·m (1.3 kgf-m, 9.6 ft-lb)

22140_SUBA_G0104

Fig. 108 Alternator and mounting bolts

FIRING ORDER

See Figure 110.

Fig. 110 Firing order: 1–3–2–4
Distributorless ignition system

IGNITION COIL

REMOVAL & INSTALLATION

Left Side

See Figure 111.

1. Remove the collector cover.
2. Remove the battery and battery carrier.
3. Remove the bracket.
4. Disconnect the connector from ignition coil.
5. Remove the ignition coil.

➡**Turn the no. 6 ignition coil to remove it.**

To install:

6. To install reverse the removal procedures:
   • Tighten the ignition coil to 12 ft. lbs. (16 Nm)

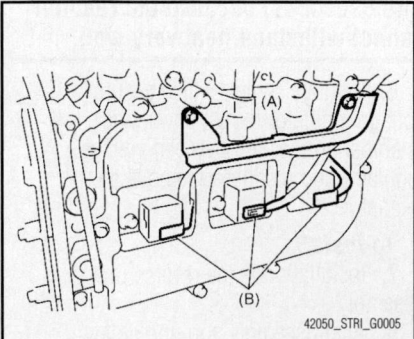

Fig. 111 Removing bracket (A) and connector (B) from the ignition coil—left side

Right Side

See Figure 112.

1. Before servicing the vehicle, refer to the Precautions Section.
2. Remove the collector cover.
3. Disconnect the ground cable from battery.
4. Remove the air cleaner case.
5. Remove the bracket.
6. Disconnect the connector from ignition coil.
7. Remove the ignition coil.

➡**Turn the #5 ignition coil to remove it.**

To install:

To install reverse the removal procedures:
   • Tighten the ignition coil to 12 ft. lbs. (16 Nm)

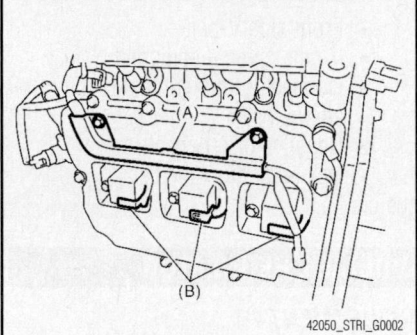

Fig. 112 Removing bracket (A) and connector (B) from ignition coil—right side

IGNITION COIL MODULE

REMOVAL & INSTALLATION

See Figure 113.

1. Before servicing the vehicle, refer to the Precautions Section.
2. Disconnect the negative battery cable.
3. Disconnect the spark plug cables from the ignition coil pack.
4. Disconnect the electrical connector from the ignition coil pack.
5. Remove the ignition coil pack assembly.

To install:

6. Installation is the reverse order of removal.
7. Tighten the ignition coil pack mounting bolts to 5 ft. lbs. (6 Nm).

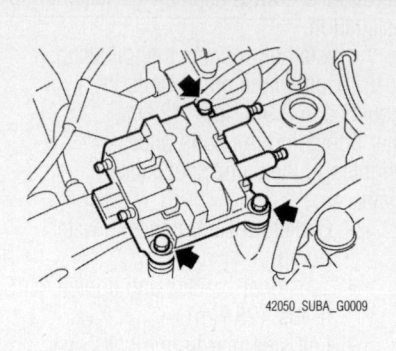

Fig. 113 Location of the ignition coil pack mounting bolts

IGNITION TIMING

INSPECTION

METHOD WITH SUBARU SELECT MONITOR

See Figure 114.

Before checking the ignition timing, check the following:
   • Ensure the air cleaner element is free from clogging, spark plugs are in good condition, and that hoses are connected properly.
   • Ensure the malfunction indicator light does not illuminate.

1. Idle the engine.
2. Stop the engine, and turn the ignition switch to OFF.
3. Insert the cartridge to the Subaru Select Monitor.
4. Connect the Subaru Select Monitor to data link connector.
5. Turn the ignition switch to ON, and Subaru Select Monitor power switch to ON.
6. Select Each System Check in the Main Menu.
7. Select Engine in the Selection Menu.
8. Select Current Data Display & Save in the Engine Control System Diagnosis.
9. Select Data Display in the Data Display Menu.
10. Start the engine, and read the ignition timing by idle speed.
11. If the timing is not correct, check the ignition control system.
   Ignition timing is as follows:
   • 10°+/−8°650 rpm (MT Models)
   • 15°+/−8°700 rpm (AT Models)
   • 17°+/−10°700 rpm (STI Model)

METHOD WITH TIMING LIGHT

12. Before checking the ignition timing, check the following item:

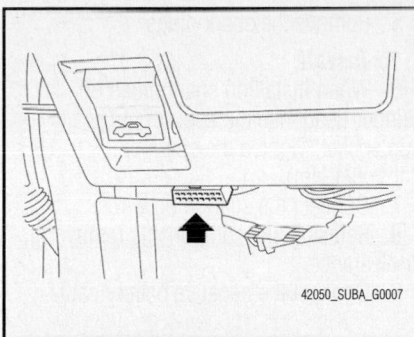

Fig. 114 Location of the data link connector under the driver's side instrument panel.

13. Check the air cleaner element is free from clogging, spark plugs are in good condition, and hoses are connected properly.

14. Check the malfunction indicator light does not illuminate.

15. Warm-up the engine.

16. Stop the engine, and turn the ignition switch to OFF.

17. Remove the air intake duct.

18. Disconnect the connectors of the mass air flow and intake air temperature sensor.

19. Remove the air cleaner case and element.

20. Connect the timing light to the power wire of #1 ignition coil.

21. Attach the air cleaner case, element and connector of mass air flow and intake air temperature sensor.

22. Start the engine, turn the timing light to the crank pulley, and check the ignition timing by means of crank pulley indicator.

23. If the timing is not correct, check the ignition control system.

Ignition timing is as follows:
- 10°+/−8°650 rpm (MT Models)
- 15°+/−8°700 rpm (AT Models)
- 17°+/−10°700 rpm (STI Model)

ADJUSTMENT

The ignition timing is controlled by the Powertrain Control Module (PCM). No adjustment is necessary or possible.

SPARK PLUGS

REMOVAL & INSTALLATION

2.5L DOHC Turbo Engine

Left Side

See Figures 115 and 116.

1. Before servicing the vehicle, refer to the Precautions Section.

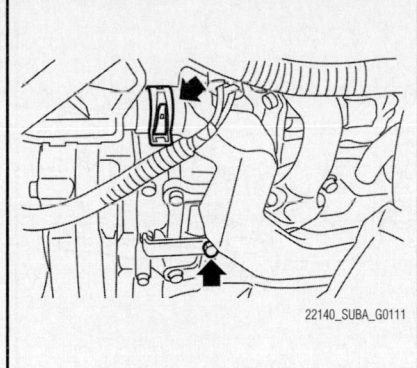

Fig. 115 Secondary air pump duct view

2. Remove the collector cover.

3. Remove the battery and battery carrier.

4. Disconnect the secondary air pump duct from the secondary air pump.

5. Remove the bolts that attach the secondary air pump duct to the rocker cover (LH), and raise the secondary air pump duct.

6. Disconnect the connector from ignition coil.

7. Remove the ignition coil.

➡**Turn number 4 ignition coil by 180° to remove it.**

8. Remove the spark plug with a spark plug socket.

To install:

9. When installing spark plugs on cylinder head, use the appropriate spark plug socket.

10. Tighten the spark plugs to 15 ft. lbs. (21 Nm).

11. Tighten ignition coil mounting bolts to 12 ft. lbs. (16 Nm).

12. Reinstall any components removed for clearance.

13. Connect the negative battery cable.

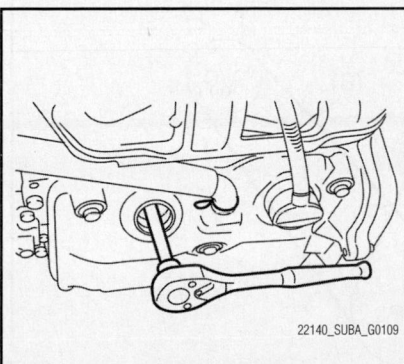

Fig. 116 Spark plug removal shown

Right Side

See Figures 116 and 117.

1. Before servicing the vehicle, refer to the Precautions Section.

2. Remove the collector cover.

3. Disconnect the ground cable from the battery.

4. Remove the air cleaner case.

5. Disconnect the connector from ignition coil.

6. Remove the ignition coil.

➡**Turn number 3 ignition coil by 180° to remove it.**

7. Remove the spark plug with a spark plug socket.

To install:

8. When installing spark plugs on cylinder head, use the appropriate spark plug socket.

9. Tighten the spark plugs to 15 ft. lbs. (21 Nm).

10. Tighten ignition coil mounting bolts to 12 ft. lbs. (16 Nm).

11. Reinstall any components removed for clearance.

12. Connect the negative battery cable.

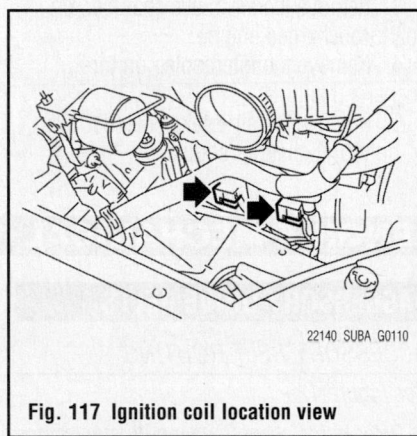

Fig. 117 Ignition coil location view

2.5L SOHC Engine

See Figures 118 and 119.

1. Before servicing the vehicle, refer to the Precautions Section.

2. Disconnect the negative battery cable.

3. Remove the following components if necessary for access to remove the spark plugs:
- Battery
- Air intake assembly

✳✳ CAUTION

Do not disconnect the washer tank hoses.

4. Gripping the spark plug wire by the boot, remove the spark plug wires.

Fig. 118 Spark plug wires and boots shown

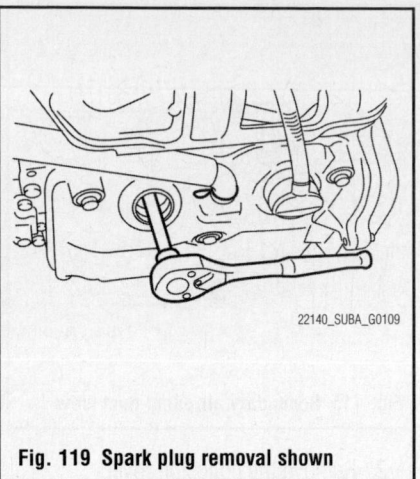

Fig. 119 Spark plug removal shown

5. Remove the spark plugs.

To install:

6. When installing spark plugs on cylinder head, use the appropriate spark plug socket. Tighten the spark plugs to 15 ft. lbs. (21 Nm).

7. Connect the spark plug wires.

8. Reinstall any components removed for clearance.

9. Connect the negative battery cable.

SPARK PLUG WIRES

REMOVAL & INSTALLATION

Refer to Spark plug "Removal & Installation".

ENGINE ELECTRICAL

STARTER

REMOVAL & INSTALLATION

1. Before servicing the vehicle, refer to the Precautions Section.

2. Disconnect the negative battery cable.

3. Remove the air intake chamber, on non turbocharged engine.

4. Remove the intercooler, on turbocharged engine.

5. Remove the air intake chamber stay, on non turbocharged engine.

6. Disconnect the electrical connectors from the starter.

7. Remove the starter retaining bolts. Remove the starter from the vehicle.

To install:

8. Installation is the reverse of the removal procedure.

9. Torque the starter retaining bolts to 37 ft. lbs. (50 Nm).

SOLENOID OR RELAY REPLACEMENT

1. Before servicing the vehicle, refer to the Precautions Section.

STARTING SYSTEM

2. Remove the starter from the vehicle.

3. Disconnect the lead wire from the solenoid switch.

4. Remove the through-bolts from the end frame.

5. Remove the yoke from the solenoid switch.

6. Remove the mounting screws that hold the housing to the solenoid switch.

7. Separate the solenoid switch from the housing.

8. Installation is the reverse order of removal.

ENGINE MECHANICAL

ACCESSORY DRIVE BELTS

ACCESSORY BELT ROUTING

See Figure 120.

Refer to the accompanying illustration for belt routing.

INSPECTION

Front Side Belt

> ☀ **CAUTION**

Check and adjust the front side belt tension so that it is within the

specified range. Using the belt with a tension out of the specified range may result in a fault such as the following:

- If the front side belt tension is higher, unexpected force is generated at the power steering oil pump, alternator and crankshaft bearing, causing abnormal noise due to abnormal wear of the bearing.
- If the front side belt tension is lower, the front side belt and crank pulley slip, causing abnormally high temperature on the crank pulley due to frictional heat. If this condition repeatedly occurs, the front side belt may abnormally wear, causing abnormal noise, front side belt damage or crank pulley damage.

1. Replace the front side belt, if crack, fraying or wear is found.

2. Check the front side belt tension and

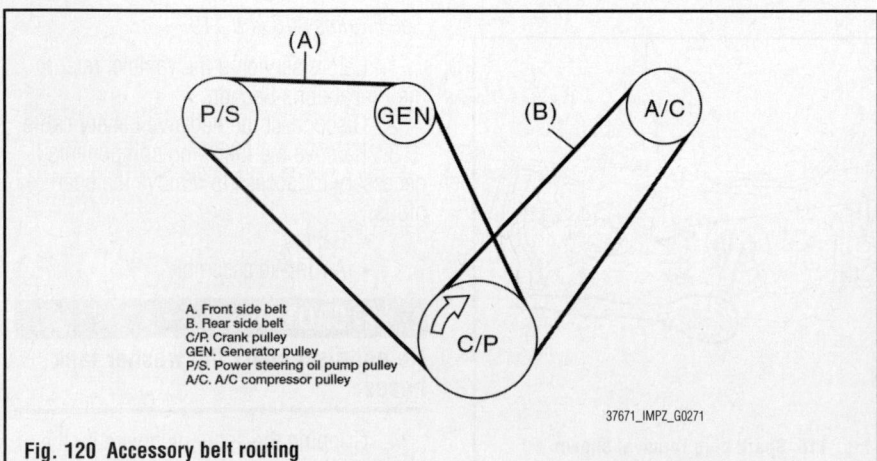

A. Front side belt
B. Rear side belt
C/P. Crank pulley
GEN. Generator pulley
P/S. Power steering oil pump pulley
A/C. A/C compressor pulley

Fig. 120 Accessory belt routing

adjust it if necessary by changing the alternator installing position.

Rear Side Belt

If cracks, fraying or wear is found, and when abnormal noise is produced, replace the rear side belt.

➡**Because the rear side belt is a stretch type belt, it is not necessary to check deflection and tension.**

ADJUSTMENT

➡**The belt tension may be checked with or without a belt tension gauge.**

1. If using a belt tension gauge, the tension should be as follows:
   a. Used front belt: 110.2–143.9 lbs. (490–640 N)
   b. New front belt: 143–175 lbs. (640–780 N)
   c. Used rear belt: 78.7–101.2 lbs. (350–450 N)
   d. New rear belt: 166–198 lbs. (740–880 N)

2. If you are not using a belt tension gauge, using 22 lbs. (98 Nm) of force, push on the belt as shown in the illustration. The total distance the belt travels up and down should be:
   a. Used front belt: 0.354–0.433 in. (9–11 mm)
   b. New front belt: 0.276–0.354 in. (7–9 mm)
   c. Used rear belt: 0.354–0.394 in. (9.0–10.0 mm)
   d. New rear belt: 0.295–0.335 in. (7.5–8.5 mm)

REMOVAL & INSTALLATION

See Figure 120.

1. Before servicing the vehicle, refer to the Precautions Section.
2. Remove the accessory drive belt cover, if equipped.
3. Loosen the lock bolt (bottom) of the front belt tensioner.
4. Loosen the slider bolt (top) of the front belt tensioner.
5. Remove the front side belt.
6. Loosen the lock bolt (bottom) of the rear belt tensioner.
7. Loosen the slider bolt (top) of the rear belt tensioner.
8. Remove the rear side belt.

To install:

➡**Wipe off any oil or water on the belt and pulley.**

9. Install the rear side belt, then tighten the slider bolt until the belt reaches the specified tension. Please refer to the following topic: "Accessory Drive Belts", "Adjustment."

10. Tighten the lock nut of the rear belt tensioner to 18 ft. lbs. (25 Nm).

11. Install the front side belt, then tighten the slider bolt until the belt reaches the specified tension. Please refer to the following topic: "Accessory Drive Belts", "Adjustment."

12. Tighten the lock nut of the front belt tensioner to 17 ft. lbs. (23 Nm).

CAMSHAFT AND VALVE LIFTERS

INSPECTION

1. Remove the camshaft from the engine.
2. Check the camshaft bearing journals for damage and binding.
3. If the journals are binding, check the cylinder head for damage.
4. Check the cylinder head for clogged oil holes.
5. Check the camshaft surface for abnormal wear and damage. Replace the camshaft, as required.
6. Measure the camshaft lobe surface and replace the camshaft if not within specification.
7. Measure the camshaft journal diameter and replace the camshaft if not within specification.
8. Measure the camshaft run out and replace the camshaft if not within specification.

REMOVAL & INSTALLATION

2.5L DOHC Engine

See Figures 121 through 124.

1. Before servicing the vehicle, refer to the Precautions Section.
2. Disconnect the negative battery cable.
3. Remove the collector cover.
4. To remove the front side V belt, remove the belt covers. Loosen the lock bolt. Loosen the slider bolt. Remove the front side belt.
5. To remove the rear side V belt, remove the belt covers. Loosen the lock bolt. Loosen the slider bolt. Remove the rear side belt. Remove the belt tensioner.
6. Remove the crankshaft pulley bolt.
7. Lock the crankshaft in place using tool ST499977100, or equivalent.
8. Remove the crankshaft pulley.

9. Remove the left side timing belt cover.
10. Remove the right side timing belt cover.
11. Remove the front timing belt cover.
12. Remove the timing belt.
13. Remove the camshaft position sensor.
14. Remove the camshaft sprockets.

➡**Be sure to lock the camshaft in place using tool ST499207400, or equivalent.**

15. Lock the crankshaft in place using tool ST499977100, or equivalent.
16. Remove the crankshaft pulley.
17. Remove the tensioner bracket. Remove the right and left timing belt No.2 covers.
18. Remove the spark plug wires. Remove the oil level gauge, on the left side.
19. Remove the rocker cover and gasket. Remove the oil pipe.
20. Loosen the oil flow control solenoid valve assembly and the intake camshaft cap bolts equally and in the proper sequence.
21. Loosen the exhaust camshaft cap bolts equally and in the proper sequence.
22. Remove the oil flow control solenoid valve assembly, intake camshaft cap and camshaft.
23. Remove the exhaust camshaft caps and camshaft.

➡**Arrange the camshafts caps so that they can be installed in their original positions.**

To install:

➡**Lubricate the camshaft journals with clean engine oil prior to installation.**

24. Install the camshaft so that the valves are closed or in contact with the "base circle" of the cam lobe.
25. If the camshafts are positioned as shown in the, the camshafts need to be rotated at a minimum to align the timing belt during installation.
26. The right hand camshaft need not be rotated when set at the position illustrated, the left hand intake camshaft should be rotated 80 degrees clockwise. The left hand exhaust camshaft should be rotated 45 degrees counterclockwise.
27. To install the camshaft cap and oil flow control solenoid valve, apply a small amount of liquid gasket to the mating surface of the cap.

➡**Do not apply an excessive amount of sealant, as it will squish out and flow toward the seal resulting in an oil leak.**

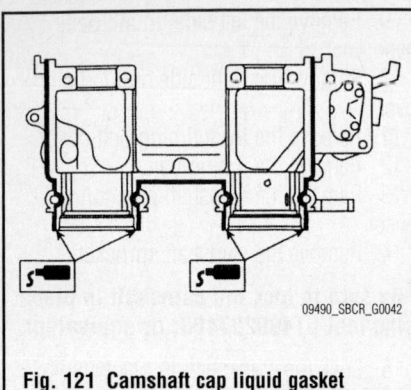

Fig. 121 Camshaft cap liquid gasket application points—2.5L DOHC engine

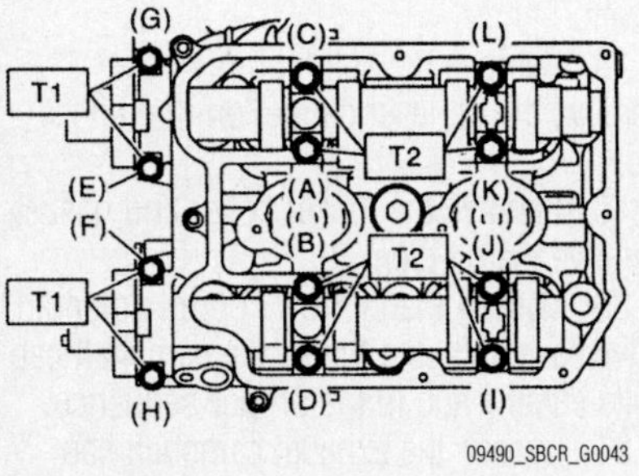

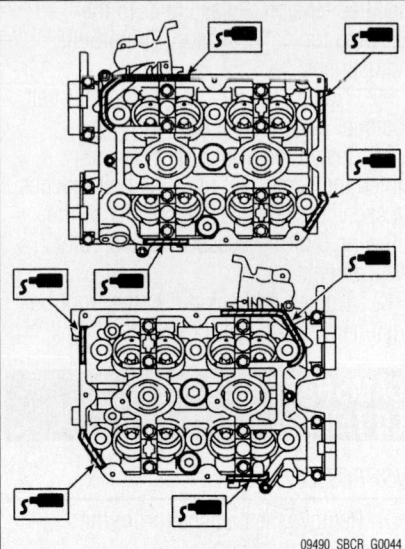

Fig. 122 Camshaft bolt tightening sequence—2.5L DOHC engine

28. Apply a thin coat of engine oil to the cap bearing surface and install the cap according to the cap identification mark. Gradually tighten the cap bolts in two stages, first to 7.2 ft. lbs. and then to 14.5 ft. lbs. in the proper sequence.

➡**After tightening the camshaft cap, ensure that the camshaft rotates slightly while holding it at base circle.**

29. Using tools ST49587600 and ST499597200, install a new seal on the camshaft. Be sure to coat the seal with clean engine oil before installation. Use tool ST499587700 and install the plug.

30. Install a new gasket on the rocker cover. Apply liquid gasket to the cylinder head (see illustration).

➡**Apply an extra amount of liquid gasket around the semicircular plugs, 5mm or more.**

31. Temporarily tighten the rocker cover retaining bolts, in the proper sequence, and then tighten to 4.7 ft. lbs. Install the oil pipe.

32. Continue the installation in the reverse order of the removal procedure.

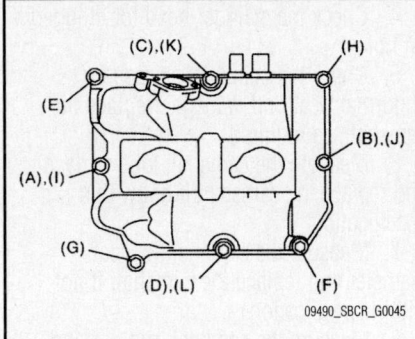

Fig. 123 Rocker cover liquid gasket application points—2.5L DOHC engine

Fig. 124 Rocker cover tightening sequence—2.5L DOHC engine

2.5L SOHC Engine

See Figures 125 through 132.

1. Before servicing the vehicle, refer to the Precautions Section.
2. Disconnect the negative battery cable.
3. To remove the front side V belt, remove the belt covers. Loosen the lock bolt. Loosen the slider bolt. Remove the front side belt.
4. To remove the rear side V belt, remove the belt covers. Loosen the lock bolt. Loosen the slider bolt. Remove the rear side belt. Remove the belt tensioner.
5. Remove the crankshaft pulley bolt.
6. Lock the crankshaft in place using tool ST499977100, or equivalent.
7. Remove the crankshaft pulley.
8. Remove the left side timing belt cover.
9. Remove the front timing belt cover.
10. Remove the timing belt.

11. Remove the camshaft position sensor.
12. Remove the camshaft sprockets.

➡**Be sure to lock the camshaft in place using tool ST18231AA010, or equivalent.**

13. Remove the left and right timing belt N02 covers.

➡**Do not damage or lose the rubber seal when removing the covers.**

14. Remove the tensioner bracket. Remove the camshaft position sensor support, on the left side.
15. Remove the oil level gauge guide, on the left side.
16. Remove the valve rocker arm assembly.
17. Remove camshaft cap retaining bolts "A" and "B" in the proper sequence.
18. Loosen camshaft cap bolts "C" through "J" all the way in the proper sequence.
19. Remove camshaft cap bolts "K" through "P" in the proper sequence using a Torx® head bit.
20. Remove the camshaft caps.
21. Remove the camshaft. Remove the oil seal. Remove the plug from the rear side of the camshaft.

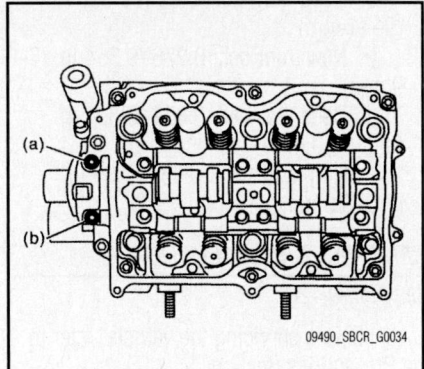

Fig. 125 Camshaft bolt loosening sequence—2.5L SOHC engine

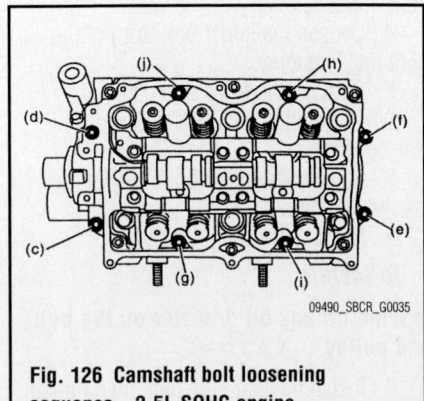

Fig. 126 Camshaft bolt loosening sequence—2.5L SOHC engine

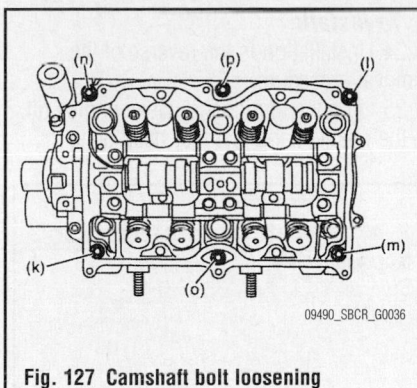

Fig. 127 Camshaft bolt loosening sequence—2.5L SOHC engine

➡ **Do not remove the oil seal unless necessary.**

To install:

➡ **Lubricate the camshaft journals with clean engine oil prior to installation.**

22. Install the camshaft into the cylinder head.

23. Apply liquid gasket to the mating surfaces of the camshaft cap.

24. Apply a bead of sealant (0.12 inch in diameter) along the edge of the camshaft cap mating surface. Install with 20 minutes after applying the sealant.

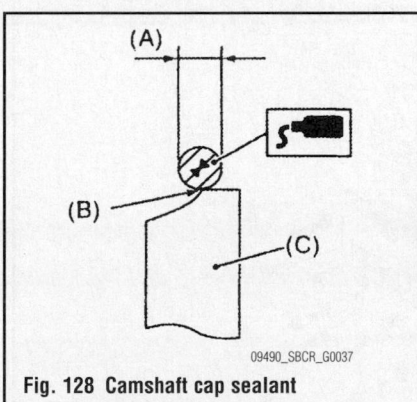

Fig. 128 Camshaft cap sealant application—2.5L SOHC engine

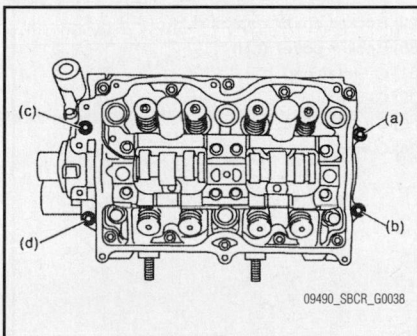

Fig. 129 Camshaft bolt tightening sequence—2.5L SOHC engine

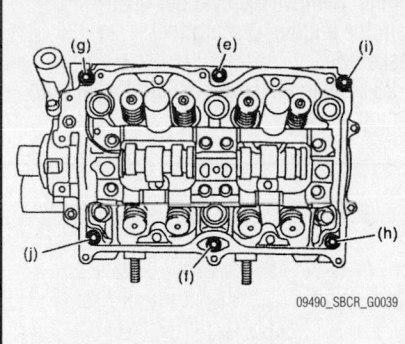

Fig. 130 Camshaft bolt tightening sequence—2.5L SOHC engine

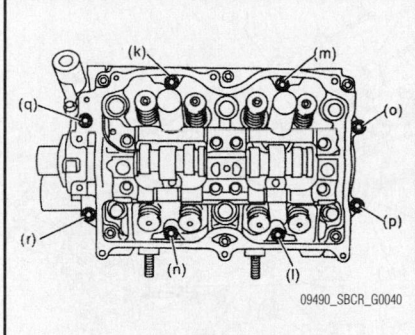

Fig. 131 Camshaft bolt tightening sequence—2.5L SOHC engine

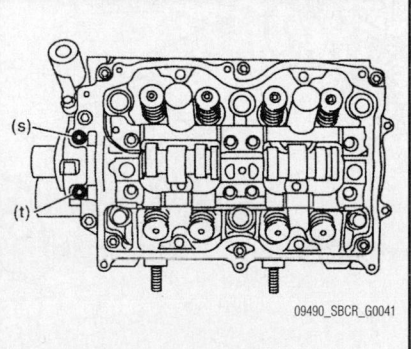

Fig. 132 Camshaft bolt tightening sequence—2.5L SOHC engine

25. Temporarily tighten the bolts "A" through "D" in the proper sequence.

26. Install the valve rocker arm assembly. Tighten Torx® head bolts "E" through "J" in the proper sequence to 13 ft. lbs.

27. Tighten bolts "K" through "R" in the proper sequence to 7.2 ft. lbs.

28. Tighten bolts "S" through "T" in the proper sequence to 7.2 ft. lbs.

➡ **Be sure to use a new seal washer.**

29. Using tools ST499597000 and ST499587500, install a new seal on the camshaft. Be sure to coat the seal with clean

engine oil before installation. Use tool ST499587700 and install the plug.

30. Adjust the valve clearance.

31. Continue the installation in the reverse order of the removal procedure.

CYLINDER HEAD

REMOVAL & INSTALLATION

See Figures 133 through 138.

1. Before servicing the vehicle, refer to the Precautions Section.

2. Disconnect the negative battery cable.

3. If equipped with DOHC engine, remove the collector cover.

4. To remove the front side V belt, remove the belt covers. Loosen the lock bolt. Loosen the slider bolt. Remove the front side belt.

5. To remove the rear side V belt, remove the belt covers. Loosen the lock bolt. Loosen the slider bolt. Remove the rear side belt. Remove the belt tensioner.

6. Remove the crankshaft pulley bolt.

7. Lock the crankshaft in place using tool ST499977100, or equivalent.

8. Remove the crankshaft pulley.

9. Remove the left side timing belt cover.

10. If equipped with DOHC engine, remove the right side timing belt cover.

11. Remove the front timing belt cover.

12. Remove the timing belt.

13. Remove the camshaft position sensor.

14. Remove the camshaft sprockets.

➡ **Be sure to lock the camshaft in place using tool ST18231AA010 (SOHC engine) and ST499207400 (DOHC engine), or equivalent.**

15. Remove the intake manifold.

16. If equipped with SOHC engine, remove the bolt that retains the air conditioning compressor bracket to the cylinder head.

17. Remove the rocker cover retaining bolts. Remove the rocker cover.

18. If equipped with SOHC engine, remove the rocker arm assembly.

19. If equipped with SOHC engine remove the camshaft. If equipped with DOHC engine, remove the camshafts.

20. Remove the cylinder head, in the proper sequence. On the DOHC engine leave bolts A and D installed loosely to prevent the cylinder head from falling. On the SOHC engine leave bolts A and C installed loosely to prevent the cylinder head from falling.

21. Loosen the cylinder head from the block using a plastic-faced hammer, if needed.

22. Remove bolts A and D on the DOHC engine and bolts A and C on the SOHC engine. Remove the cylinder head from the engine. Discard the gasket

23. Clean all gasket material from both mating surfaces.

To install:

24. Installation is the reverse of the removal procedure.

25. Apply a thin coat of clean engine oil to the washers and cylinder head bolts.

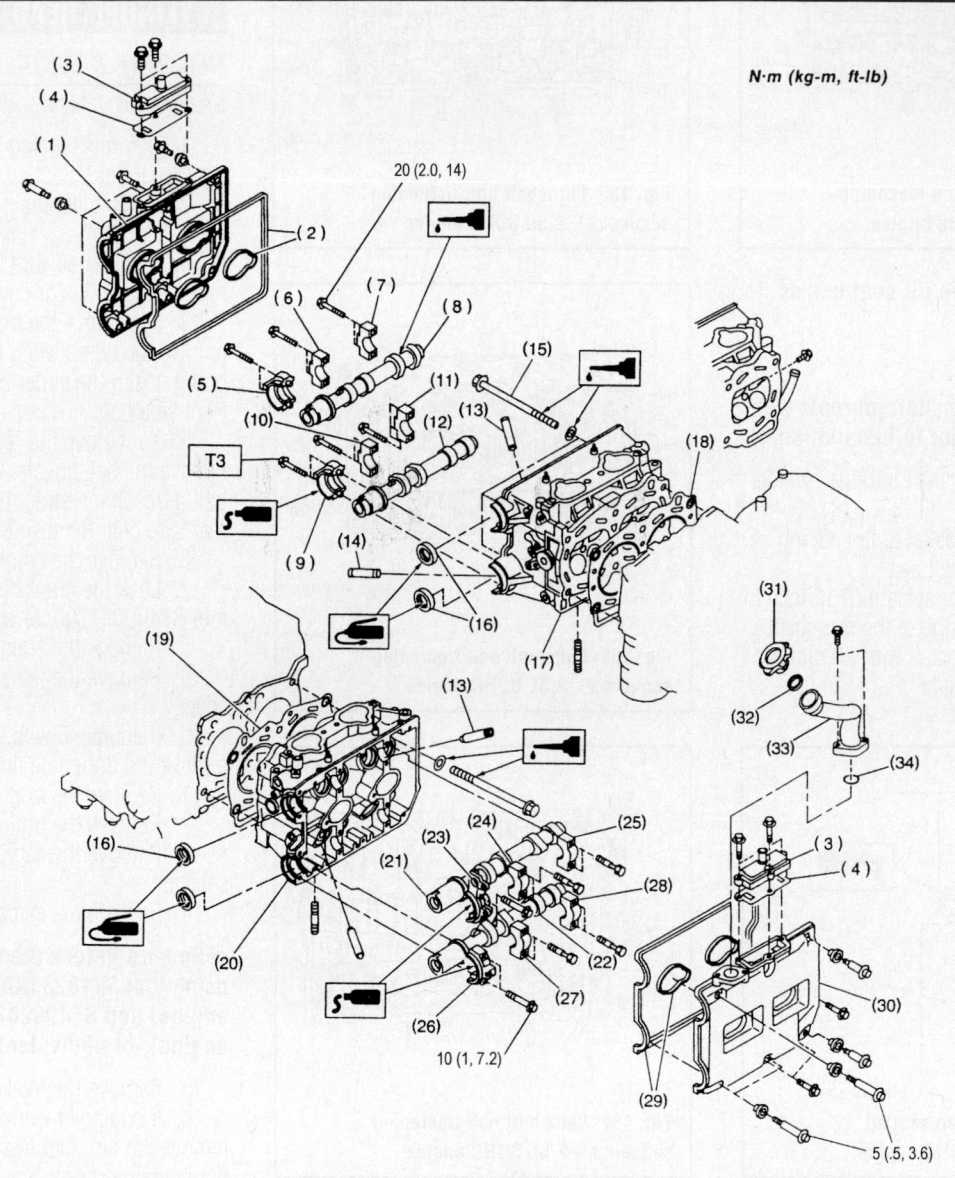

N·m (kg-m, ft-lb)

(1) Rocker cover (RH)
(2) Rocker cover gasket (RH)
(3) Oil separator cover
(4) Gasket
(5) Intake camshaft cap (Front RH)
(6) Intake camshaft cap (Center RH)
(7) Intake camshaft cap (Rear RH)
(8) Intake camshaft (RH)
(9) Exhaust camshaft cap (Front RH)
(10) Exhaust camshaft cap (Center RH)
(11) Exhaust camshaft cap (Rear RH)
(12) Exhaust camshaft (RH)
(13) Intake valve guide
(14) Exhaust valve guide

(15) Cylinder head bolt
(16) Oil seal
(17) Cylinder head (RH)
(18) Cylinder head gasket (RH)
(19) Cylinder head gasket (LH)
(20) Cylinder head (LH)
(21) Intake camshaft (LH)
(22) Exhaust camshaft (LH)
(23) Intake camshaft cap (Front LH)
(24) Intake camshaft cap (Center LH)
(25) Intake camshaft cap (Rear LH)
(26) Exhaust camshaft (Front LH)
(27) Exhaust camshaft cap (Center LH)
(28) Exhaust camshaft cap (Rear LH)

(29) Rocker cover gasket (LH)
(30) Rocker cover (LH)
(31) Oil filler cap
(32) Gasket
(33) Oil filler duct
(34) O-ring

7923TG07

Fig. 133 Cylinder head and related components—2.5L DOHC engine

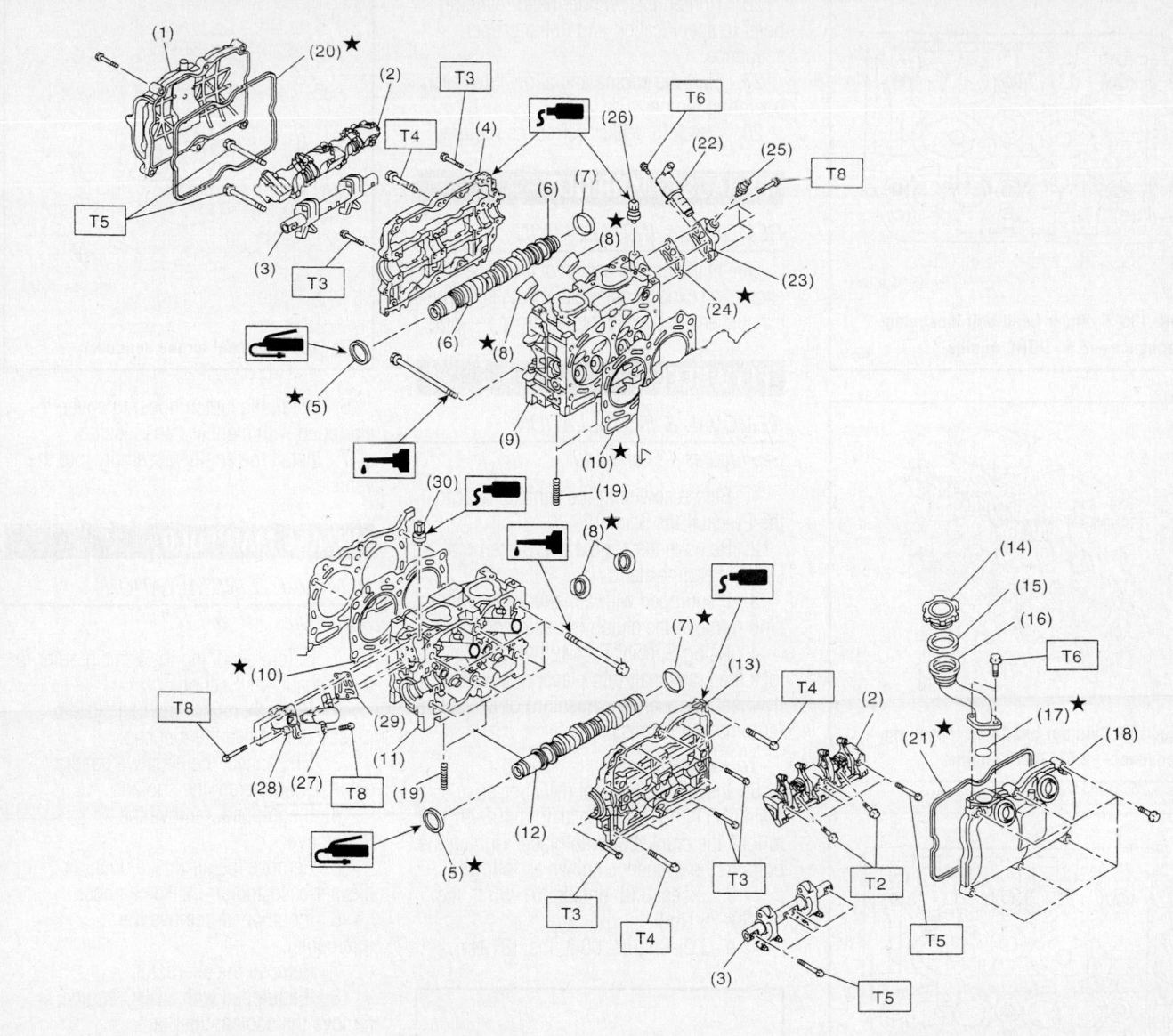

(1) Rocker cover (RH)
(2) Intake valve rocker assembly
(3) Exhaust valve rocker assembly
(4) Camshaft cap (RH)
(5) Oil seal
(6) Camshaft (RH)
(7) Plug
(8) Spark plug pipe gasket
(9) Cylinder head (RH)
(10) Cylinder head gasket
(11) Cylinder head (LH)
(12) Camshaft (LH)
(13) Camshaft cap (LH)
(14) Oil filler cap
(15) Gasket
(16) Oil filler duct

(17) O-ring
(18) Rocker cover (LH)
(19) Stud bolt
(20) Rocker cover gasket (RH)
(21) Rocker cover gasket (LH)
(22) Oil switching solenoid valve (RH)
(23) Oil switching solenoid valve holder (RH)
(24) Gasket
(25) Oil temperature sensor
(26) Variable valve lift diagnosis oil pressure switch (RH)
(27) Oil switching solenoid valve (LH)
(28) Oil switching solenoid valve holder (LH)
(29) Gasket

(30) Variable valve lift diagnosis oil pressure switch (LH)

Tightening torque: N·m (kgf-m, ft-lb)

T3:	9.75 (1.0, 7.2)
T4:	18 (1.8, 13.0)
T5:	25 (2.5, 18.1)
T6:	6.4 (0.65, 4.7)
T7:	8 (0.8, 5.9)
T8:	10 (1.0, 7.4)

09490_SBCR_G0016

Fig. 134 Cylinder head and related components—2.5L SOHC engine

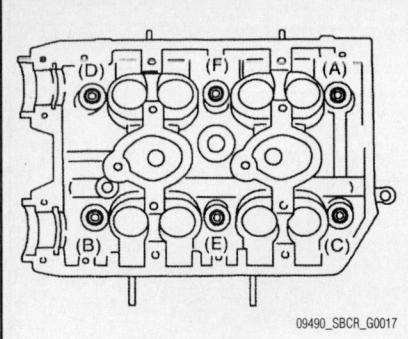

Fig. 135 Cylinder head bolt loosening sequence—2.5L DOHC engine

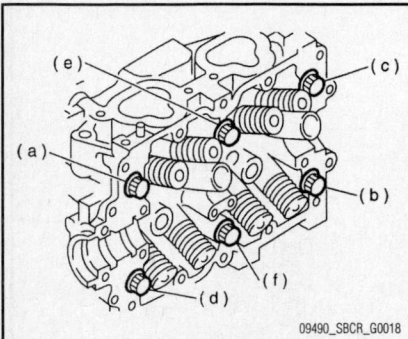

Fig. 136 Cylinder head bolt loosening sequence—2.5L SOHC engine

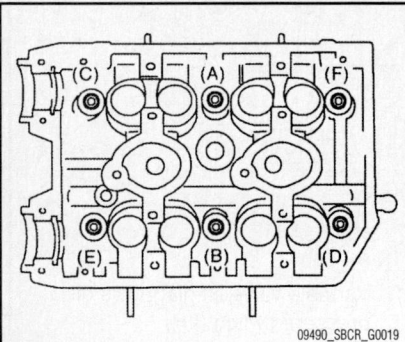

Fig. 137 Cylinder head bolt tightening sequence—2.5L DOHC engine

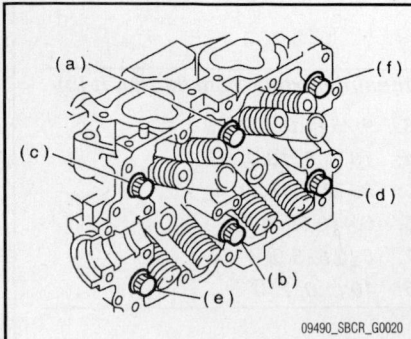

Fig. 138 Cylinder head bolt tightening sequence—2.5L SOHC engine

26. Tighten the cylinder head retaining bolts to specification and in the proper sequence.

27. Start the engine and allow it to reach operating temperature.

28. Check for leaks, correct as required.

EXHAUST MANIFOLD

REMOVAL & INSTALLATION

Due to the unique design of the Subaru engine, an exhaust manifold is not used. The exhaust enters directly into the front Y—pipe.

FLYWHEEL

REMOVAL & INSTALLATION

See Figures 139 and 140.

1. Before servicing the vehicle, refer to the Precautions Section.

2. Remove the engine and place on a suitable engine stand.

3. If equipped with a manual transmission, remove the clutch housing cover.

4. Using Special Tool 498497100 to lock the crankshaft into place, remove the flywheel (manual transmission) or driveplate (automatic transmission).

To install:

5. Install the flywheel (M/T) or drive plate (A/T) and use Special Tool 498497100 to lock the crankshaft into place. Tighten the bolts in the sequence shown as follows:

   a. Except 3.0L engine: 51–55 ft. lbs. (69–75 Nm).

   b. 3.0L engine: 60 ft. lbs. (81 Nm).

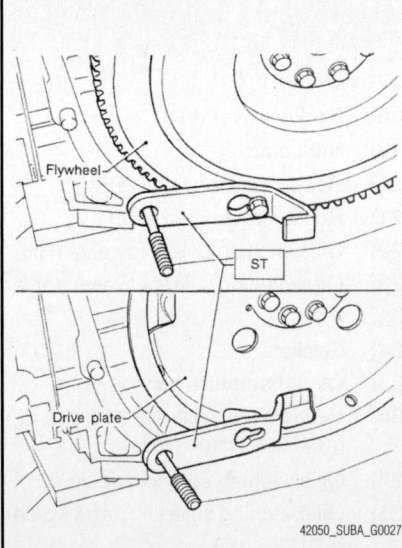

Fig. 139 Use Special Tool 498497100 to lock the crankshaft into place and remove the flywheel

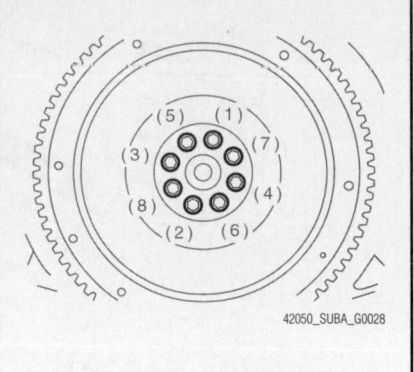

Fig. 140 Flywheel torque sequence

6. Install the clutch housing cover, if equipped with manual transmission.

7. Install the engine assembly into the vehicle.

INTAKE MANIFOLD

REMOVAL & INSTALLATION

See Figures 141 and 142.

1. Before servicing the vehicle, refer to the Precautions Section.

2. Properly relieve the fuel system pressure. Remove the fuel cap.

3. Disconnect the negative battery cable. Drain the engine coolant.

4. If equipped, remove the undercover.

5. Remove the air intake duct, air cleaner case and air intake chamber.

6. If equipped, remove the intercooler.

7. Remove the alternator.

8. If equipped with DOHC engine, remove the coolant filler tank.

9. If equipped with SOHC engine, remove the spark plug wires.

10. Disconnect the engine coolant hoses from the throttle body. Disconnect the brake booster hose.

11. Disconnect the PCV hose from the intake manifold. Disconnect the engine harness electrical connectors from the bulkhead harness connectors.

12. Disconnect the engine coolant temperature sensor electrical connector, knock sensor electrical connector and crankshaft position sensor connector.

13. Disconnect the power steering pump switch electrical connector, oil pressure switch connector and camshaft sensor connector.

14. If equipped with DOHC engine, disconnect the oil flow solenoid valve electrical connector and the ignition coil connector.

15. If equipped with SOHC engine, remove the EGR pipe from the intake manifold and disconnect the fuel lines from the fuel pipe.

16. If equipped with DOHC engine, disconnect the fuel delivery hose, return hose and evaporation hose.

17. Remove the intake manifold retaining bolts. Remove the intake manifold from the engine.

To install:

18. Installation is the reverse of the removal procedure.

19. Be sure to use new intake manifold gaskets. Tighten the manifold retaining bolts

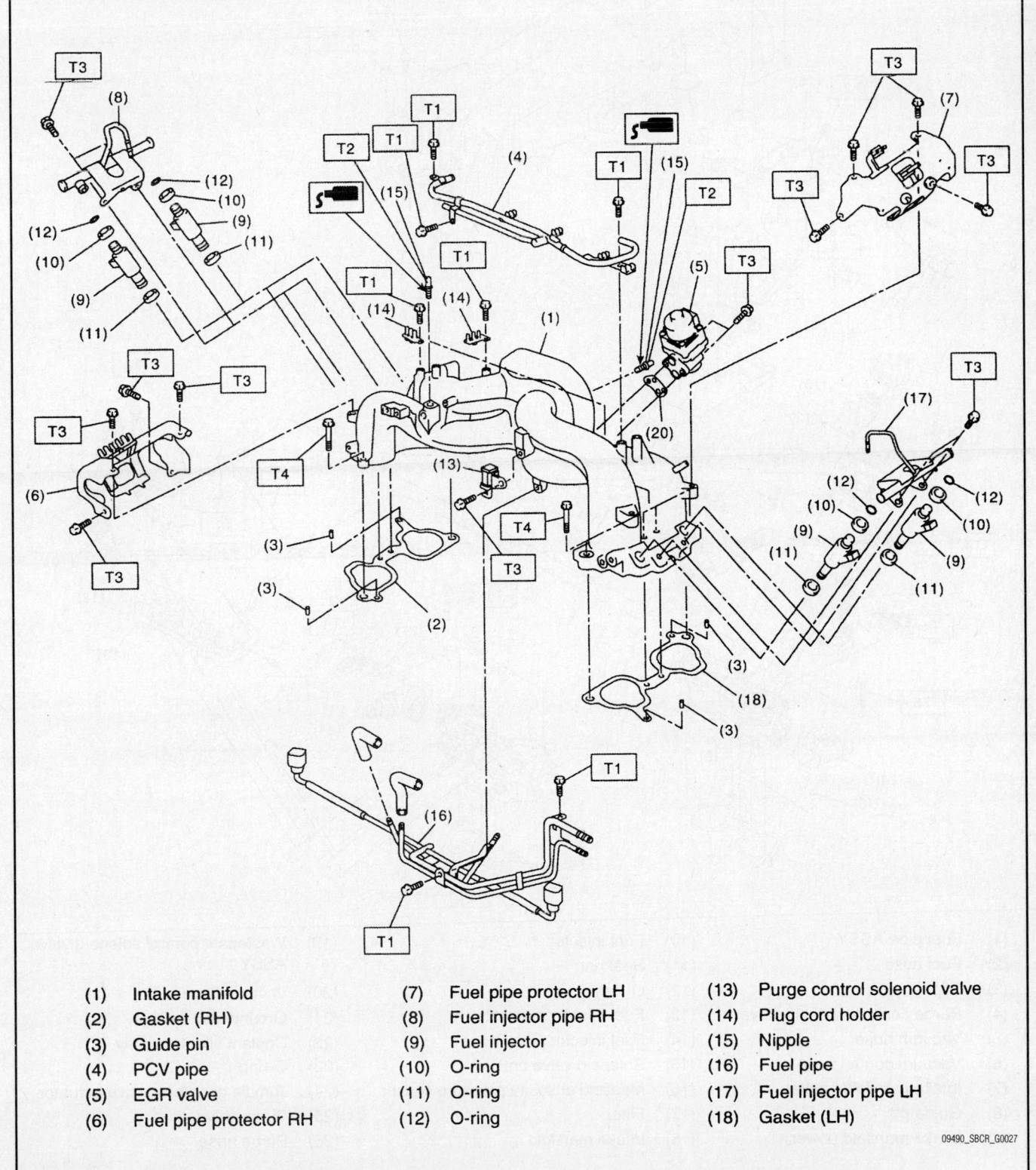

(1)	Intake manifold	(7)	Fuel pipe protector LH	(13)	Purge control solenoid valve
(2)	Gasket (RH)	(8)	Fuel injector pipe RH	(14)	Plug cord holder
(3)	Guide pin	(9)	Fuel injector	(15)	Nipple
(4)	PCV pipe	(10)	O-ring	(16)	Fuel pipe
(5)	EGR valve	(11)	O-ring	(17)	Fuel injector pipe LH
(6)	Fuel pipe protector RH	(12)	O-ring	(18)	Gasket (LH)

09490_SBCR_G0027

Fig. 141 Intake manifold and related components—2.5L SOHC engine

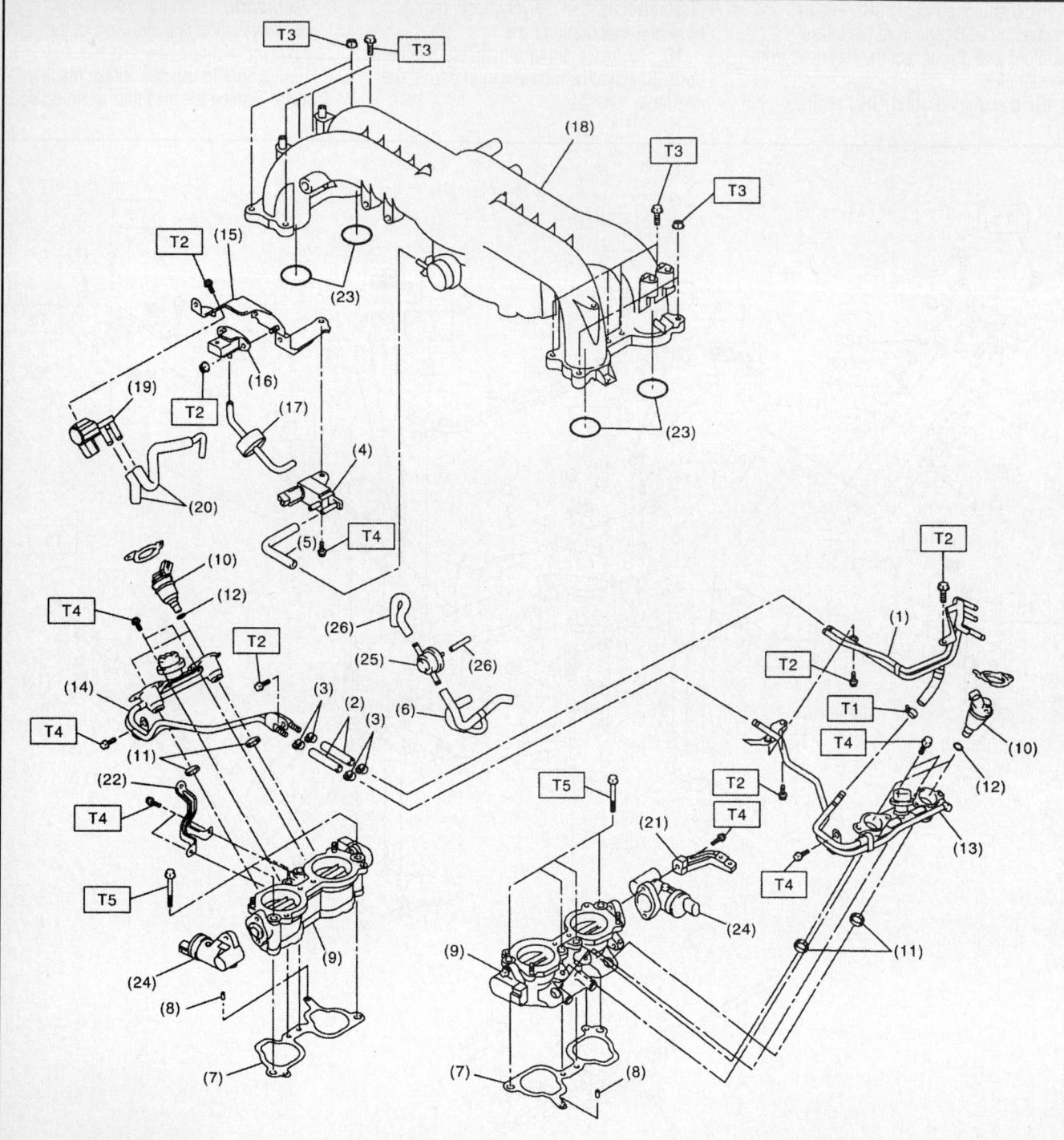

(1)	Fuel pipe ASSY	(10)	Fuel injector	(19)	Wastegate control solenoid valve ASSY
(2)	Fuel hose	(11)	Seal ring	(20)	Vacuum hose
(3)	Clip	(12)	O-ring	(21)	Ground stay
(4)	Purge control solenoid valve	(13)	Fuel injector pipe LH	(22)	Coolant filler tank stay
(5)	Vacuum hose	(14)	Fuel injector pipe RH	(23)	O-ring
(6)	Vacuum control hose	(15)	Solenoid valve bracket	(24)	Tumble generator valve actuator
(7)	Intake manifold gasket	(16)	Manifold absolute pressure sensor	(25)	Purge valve
(8)	Guide pin	(17)	Filter	(26)	Purge hose
(9)	Intake manifold (lower)	(18)	Intake manifold		

09490_SBCR_G0028

Fig. 142 Intake manifold and related components—2.5L DOHC engine

to 18 ft. lbs. (25 Nm). and in alternating sequence.

20. Be sure to fill and bleed the cooling system with the proper grade and type engine coolant.

21. Start the engine and check for leaks, correct as required.

OIL PAN

REMOVAL & INSTALLATION

See Figures 143 through 145.

1. Before servicing the vehicle, refer to the Precautions Section.

2. Disconnect the negative battery cable.

3. Raise and support the vehicle safely.

4. Remove the front tires and wheels.

5. Lower the vehicle.

6. Remove the air intake duct and the air cleaner case. Remove the air intake chamber.

7. Remove the pitching stopper.

8. Remove the hood stay holder and the radiator upper brackets.

9. Properly support the engine with a lifting device and wire ropes.

10. Lift the vehicle and support it safely.

➡**When lifting the vehicle, raise the wire ropes at the same time.**

11. Remove the undercover.

12. Drain the engine oil.

13. Remove the front and center exhaust pipes.

14. Remove the nuts which retain the front cushion rubber onto the front crossmember.

15. Remove the bolts that retain the oil pan to the cylinder block, with the engine in the raised position.

16. Insert an oil pan gasket cutter tool into the gap between the cylinder block and the oil pan. Remove the oil pan from the engine.

Fig. 143 Engine supported with lifting device

Tightening torque:
T1: 5 N·m (0.5 kgf-m, 3.6 ft-lb)
T2: 6.4 N·m (0.65 kgf-m, 4.7 ft-lb)
T3: 10 N·m (1.0 kgf-m, 7.2 ft-lb)

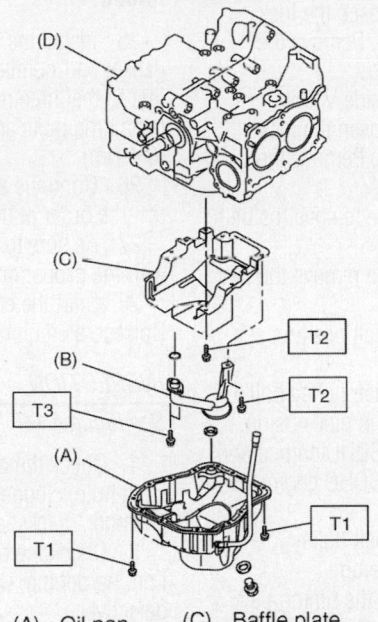

(A) Oil pan   (C) Baffle plate
(B) Oil strainer (D) Cylinder block

22140_SUBA_G0125

Fig. 144 Exploded view of oil pan, strainer, baffle plate and cylinder block

➡**Do not use a screwdriver or similar tool in place of the cutter tool.**

17. Remove the oil strainer, if required. Remove the baffle plate, if required.

To install:

18. Be sure to clean the old gasket material from the mating surfaces.

19. Apply a continuous bead of sealer to a new oil pan gasket.

20. Make sure that the seals (A) are installed securely on the baffle plate and in the direction shown in the illustration. Install the baffle plate; tighten the retaining bolts to 4.7 ft. lbs.

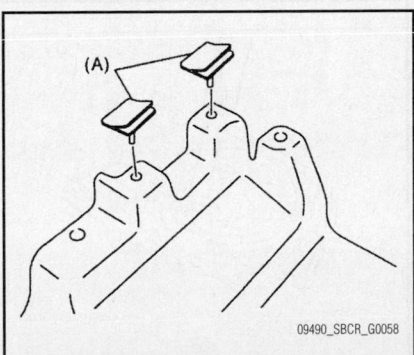

09490_SBCR_G0058

Fig. 145 Oil pan baffle plate seal location and positioning—2.5L engine

21. Replace the O-ring and install the oil strainer. Tighten the bolt to 7.2 ft. lbs.

22. Apply liquid gasket, part number 004403012 or equivalent, to the oil pan mating surface. Install the oil pan. Torque the retaining bolts to specification.

23. Continue the installation in the reverse order of the removal procedure.

24. Tighten the front cushion mounting bolts to 63 ft. lbs.

25. Be sure to fill the engine with the correct grade and type engine oil.

26. Start the engine and check for leaks. Correct as required.

OIL PUMP

REMOVAL & INSTALLATION

See Figure 146.

1. Before servicing the vehicle, refer to the Precautions Section.

2. Disconnect the negative battery cable. Drain the cooling system.

3. On the DOHC engine, remove the collector cover.

4. Raise and support the vehicle safely.

5. Remove the undercover.

6. On the DOHC engine, remove the bolts which retain the water pipe of the oil cooler to the oil pump. Remove the water

pipe and hoses between the oil cooler and the water pump.

7. Lower the vehicle. Remove the radiator.

8. To remove the front side V belt, remove the belt covers. Loosen the lock bolt. Loosen the slider bolt. Remove the front side belt.

9. To remove the rear side V belt, remove the belt covers. Loosen the lock bolt. Loosen the slider bolt. Remove the rear side belt.

10. On the SOHC engine remove the belt tensioner.

11. On the DOHC engine remove the rear side V belt tensioner

12. Remove the crankshaft position sensor.

13. Remove the crankshaft pulley bolt.

14. Lock the crankshaft in place using tool ST499977100 for the SOHC engine and tool ST499207400 for the DOHC engine, or equivalent.

15. Remove the crankshaft pulley.

16. Remove the water pump.

17. If equipped, remove the timing belt guide. Remove the crankshaft sprocket.

18. Remove the oil pump retaining bolts.

➡**When disassembling and checking the oil pump, loosen the relief valve plug before removing the oil pump from its mounting.**

19. Using a flat tip tool remove the oil pump from the engine.

To install:

20. Be sure all mating surfaces are clean and free of dirt.

21. Apply liquid gasket part number 004403007, or equivalent to the mating surfaces of the oil pump.

22. Be sure to replace the O-ring with a new one.

23. Apply a thin coat of clean engine oil to the inside of the oil seal.

24. Position the oil pump to its mounting, aligning the notched area with the crankshaft and push the pump straight

➡**Be sure that the oil seal lip is not folded.**

25. Install the oil pump. Apply liquid gasket part number 004403042, or equivalent to the three retaining bolt threads. Install the bolts and tighten to 5 ft. lbs (6.4 Nm)

26. Continue the installation in the reverse order of the removal procedure.

27. Be sure to fill the cooling system with the proper grade and type coolant.

28. Start the engine and check for leaks. Correct, as required.

INSPECTION

See Figures 147 through 149.

1. Check the oil pump case for worn shaft hole, clogged oil passage, worn rotor chamber, cracks, and other faults.

2. Check the oil seal lips for deformation, hardening, wear, etc. and replace if defective.

3. Measure the tip clearance of rotors. If clearance exceeds the standard, replace the rotors as a matched set.
   a. 0.0016–0.0055 in. (0.04–0.14 mm)

4. Measure the clearance between the outer rotor and oil pump rotor housing. If clearance exceeds the standard, replace the rotor.
   a. 0.0039–0.0069 in. (0.10–0.18 mm)

5. Measure the clearance between the oil pump inner rotor and pump cover. If clearance exceeds the standard, replace the rotor or pump body.
   a. 0.0008–0.0028 in. (0.02–0.07 mm)

6. Check the valve for fitting condition and damage, and the relief valve spring for damage and deterioration. Replace the parts if defective.

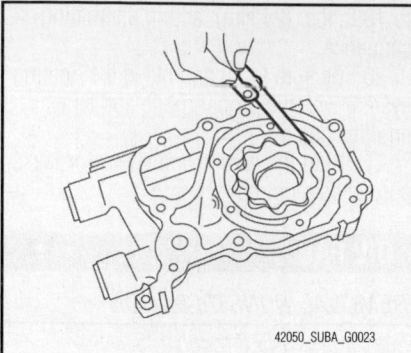

Fig. 148 Measuring the case clearance of the oil pump.

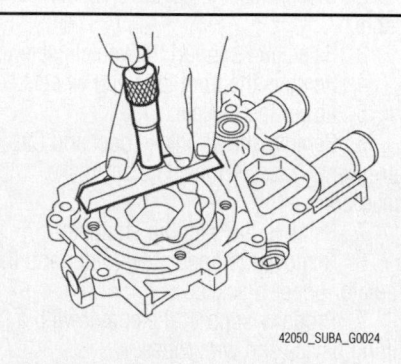

Fig. 149 Measuring the side clearance of the oil pump.

TIMING BELT FRONT COVER

REMOVAL & INSTALLATION

Refer to the Timing Belt and Sprockets, Removal & Installation procedure in this section to remove the timing belt front cover.

TIMING BELT & SPROCKETS

REMOVAL & INSTALLATION

2.5L DOHC Engine

See Figures 150 through 161.

1. Disconnect the negative battery cable.

2. Remove the collector cover.

3. To remove the front side V belt, remove the belt covers. Loosen the lock bolt. Loosen the slider bolt. Remove the front side belt.

4. To remove the rear side V belt, remove the belt covers. Loosen the lock bolt. Loosen the slider bolt. Remove the rear side belt. Remove the belt tensioner.

5. Remove the crankshaft pulley bolt.

6. Lock the crankshaft in place using tool ST499977100, or equivalent.

7. Remove the crankshaft pulley.

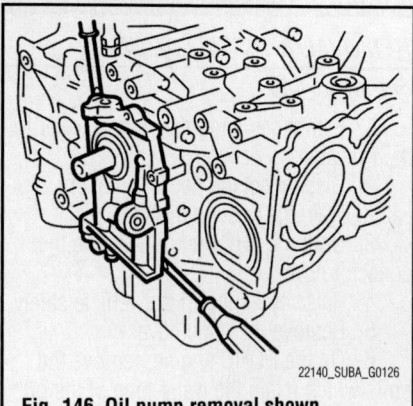

Fig. 146 Oil pump removal shown

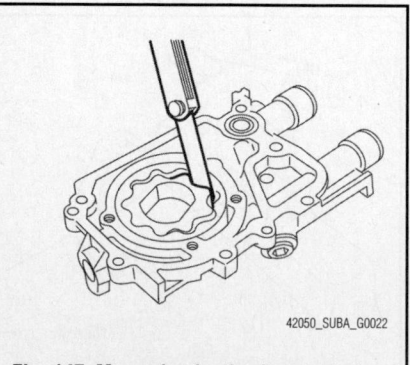

Fig. 147 Measuring for tip clearance of the rotors of the oil pump.

8. Remove the left side timing belt cover.

9. Remove the right side timing belt cover.

10. Remove the front timing belt cover.

11. Remove the timing belt guide, if equipped.

➡If the belt is going to be reused and the alignment mark on the belt is not readable, put a new mark on the belt to indicate the direction of rotation. Using tool ST499987500, turn the crankshaft to align the mark on the crankshaft sprocket, intake camshaft sprocket (left), exhaust camshaft sprocket (left), intake camshaft sprocket (right), exhaust camshaft sprocket (right) with the notches of the timing belt cover and cylinder block. Paint an alignment mark on the belts in relation to the camshaft sprockets. Z1 measurement is 54.4 teeth. Z2 measurement is 51.0 teeth and Z3 measurement is 28.0 teeth.

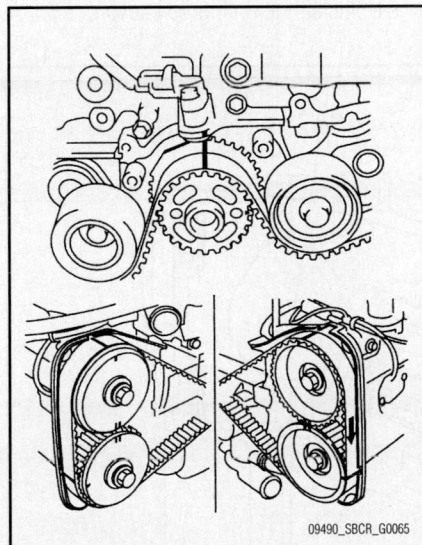

Fig. 150 Timing belt alignment—2.5L DOHC engine

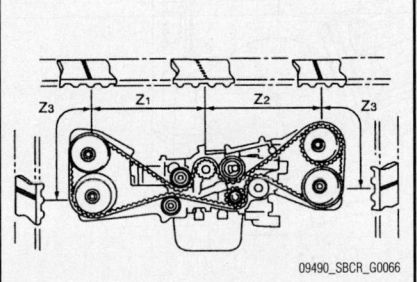

Fig. 151 Timing belt Z1, Z2 and Z3 teeth measurement—2.5L DOHC engine

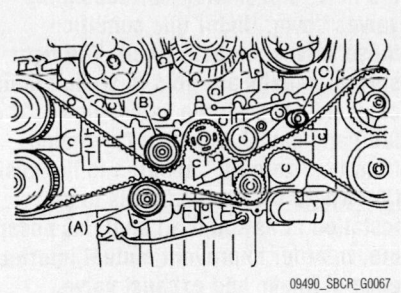

Fig. 152 Belt idler identification and location—2.5L DOHC engine

12. Remove the belt idler belt "A". Remove the timing belt.

13. Remove the belt idlers belt "B" and "C".

14. Remove the belt idler number two. Remove the automatic belt tension adjuster assembly.

To install:

15. Attach the automatic belt tension adjuster assembly to a vertical pressing tool.

➡Always use a vertical type pressing tool to move the adjuster rod downward. Do not use a lateral type vise. Push the adjuster rod vertically. Press in the push adjuster rod gradually, which should take three minutes or more. Do not allow pressure to exceed 2,205 lb. force.

16. Slowly move the adjuster rod down until the adjuster rod is aligned with the stopper pin hole in the cylinder.

➡Press the adjuster rod as far as the end surface of the cylinder. Do not press the adjuster rod into the cylinder. Doing so may damage the cylinder.

17. Using a 0.08 inch stopper pin, insert it into the stopper pin hole in the cylinder. Secure the adjuster rod.

➡Do not release the press pressure until the stopper pin is completely inserted in the hole.

18. Install the automatic belt tensioner assembly. Tighten the retaining bolt to 28.9 ft. lbs.

19. Install the belt idler number two. Tighten the retaining bolt to 28.9 ft. lbs.

20. Install the belt idlers. Tighten the retaining bolts to 28.9 ft. lbs.

21. Align the mark "A" on the crankshaft sprocket with the mark on the oil pump at the cylinder block. Align the single line mark "A" on the right exhaust camshaft sprocket with the notch "B" on the timing belt cover.

22. Align single line mark "A" on the right intake camshaft sprocket with the notch "B" on the timing cover. Ensure that the double lines "C" on the intake and exhaust camshaft sprockets are aligned.

23. Align single line mark "A" on the left exhaust camshaft sprocket with the notch "B" on the timing cover by turning the sprocket counterclockwise as viewed from the front of the engine.

24. Align single line mark "A" on the left intake camshaft sprocket with the notch "B" on the timing cover, by turning the sprocket clockwise as viewed from the front of the engine. Ensure that the double lines "C" on the intake and exhaust camshaft sprockets are aligned.

25. Make sure that the camshaft and crankshaft sprockets are positioned properly.

➡The intake and exhaust camshafts on this engine can be rotated independently with the timing belt removed. By looking at the illustration it will show you that if the intake and exhaust valve are lift together the heads will hit each other and bend.

Fig. 153 Crankshaft sprocket mark A to oil pump cover alignment—2.5L DOHC engine

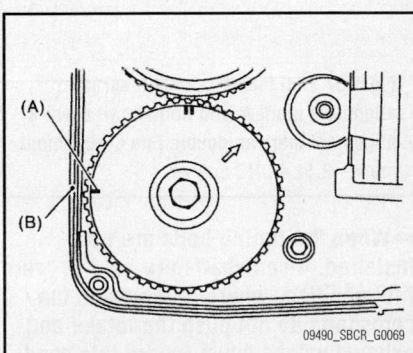

Fig. 154 Right exhaust camshaft sprocket alignment mark A with timing belt cover B alignment mark—2.5L DOHC engine

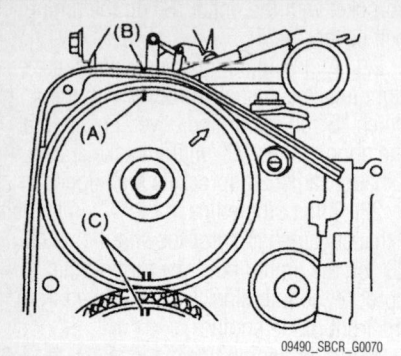

Fig. 155 Right intake camshaft sprocket alignment mark A with timing belt cover B alignment mark an double line C alignment mark—2.5L DOHC engine

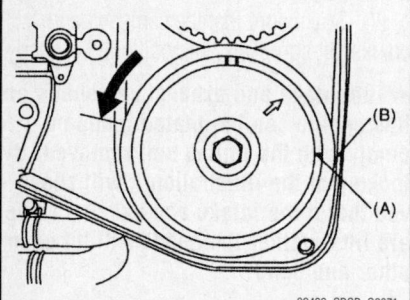

Fig. 156 Left exhaust camshaft sprocket alignment mark A with timing belt cover B alignment mark—2.5L DOHC engine

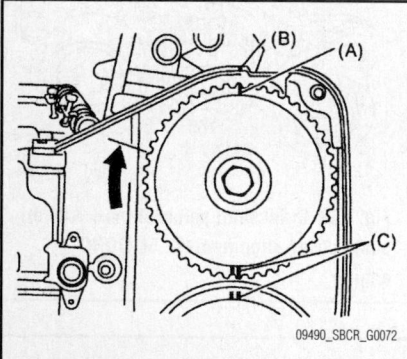

Fig. 157 Left intake camshaft sprocket alignment mark A with timing belt cover B alignment mark an double line C alignment mark—2.5L DOHC engine

➡When the timing belts are not installed, 4 camshafts are held at "zero lift" position, where all cams on the camshafts do not push the intake and exhaust valves down (under this condition all valves remain unlifted). When the camshafts are rotated to install the timing belts, No. 2 intake and No. 4 exhaust cam of the left hand camshafts

are held to push their corresponding valves down. Under this condition these valves are held lifted. The right side camshafts are held in so that their cams do not push the valves down. The left hand camshafts must be rotated from the "zero lift" position to the position where the timing belt is to be installed at as small an angle as possible, in order to prevent mutual interference of intake and exhaust valve heads. Do not allow the camshafts to rotate in the direction illustrated as this causes both the intake and exhaust valves to lift off at the same time with will cause valve damage.

26. When installing the belt, make sure to align the marks made during removal or if using a new belt, align the in alphabetical order as shown in the illustration.

✳✳ WARNING

Disengagement of more than 3 timing belt teeth may result in contact between the valve and piston.

Always make sure the belts rotation is correct.

27. Install the timing belt.

➡Align the alignment mark on the timing belt with marks on the sprocket in the order shown in the illustration. While aligning the timing marks, position the timing belt properly.

28. Install the belt idlers. Tighten the retaining bolts to 28.9 ft. lbs. (39 Nm).

➡Make sure that the marks on the timing belt and sprockets are aligned.

29. After checking to be sure the marks on the timing belt and the camshaft sprockets are aligned remove the stopper pin from the belt tension adjuster.

30. Install the timing belt guide, if equipped with manual transmission. Temporarily tighten the bolts. Check and adjust the clearance between the belt and the guide. It should be 0.039 +/− 0.020 inch.

31. Install the timing belt cover.

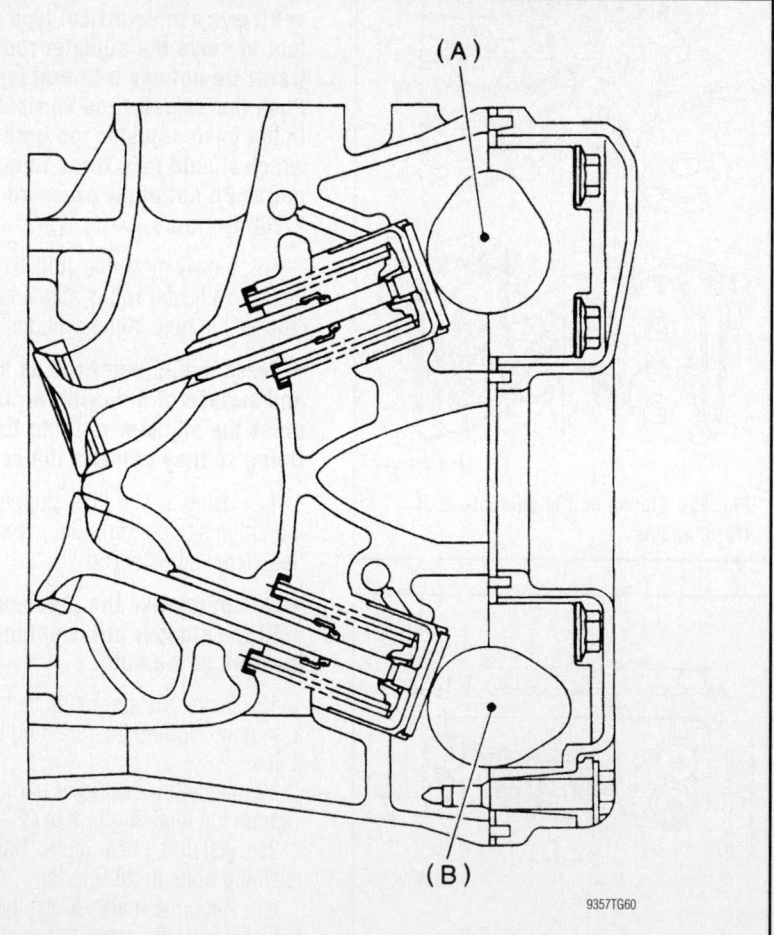

Fig. 158 If the intake and exhaust valve are lift together the heads will hit each other and bend –2.5L DOHC engine

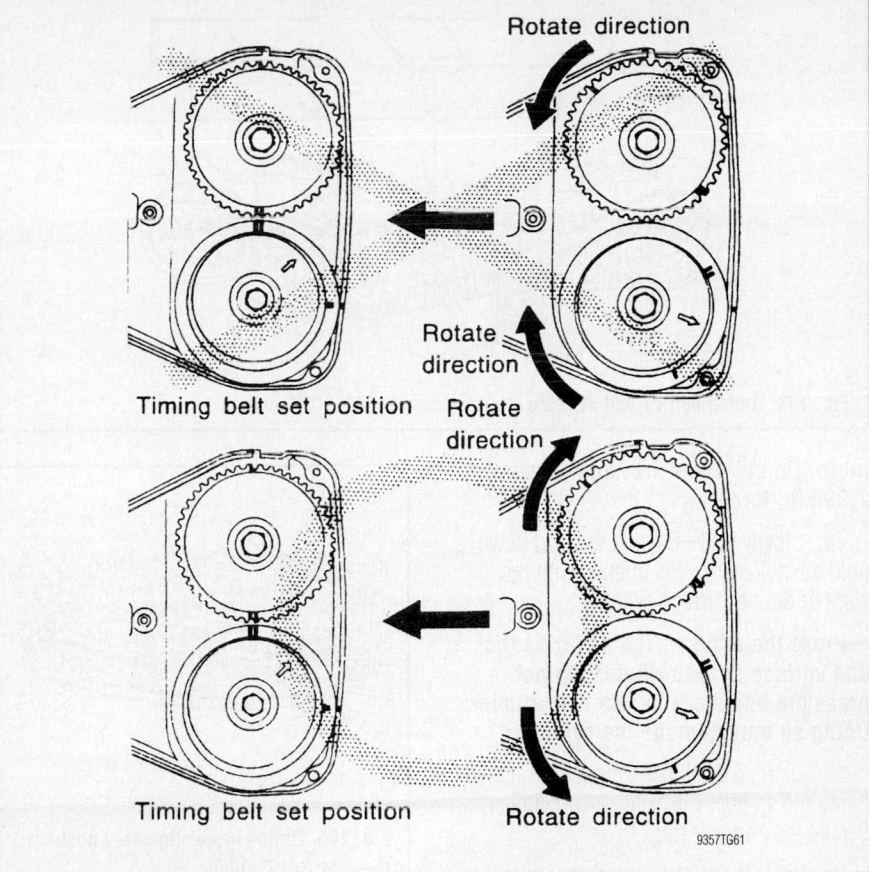

Fig. 159 Do not allow the camshafts to rotate in the direction shown as this causes both the intake and exhaust valves to lift off at the same time with will cause valve damage on DOHC engines

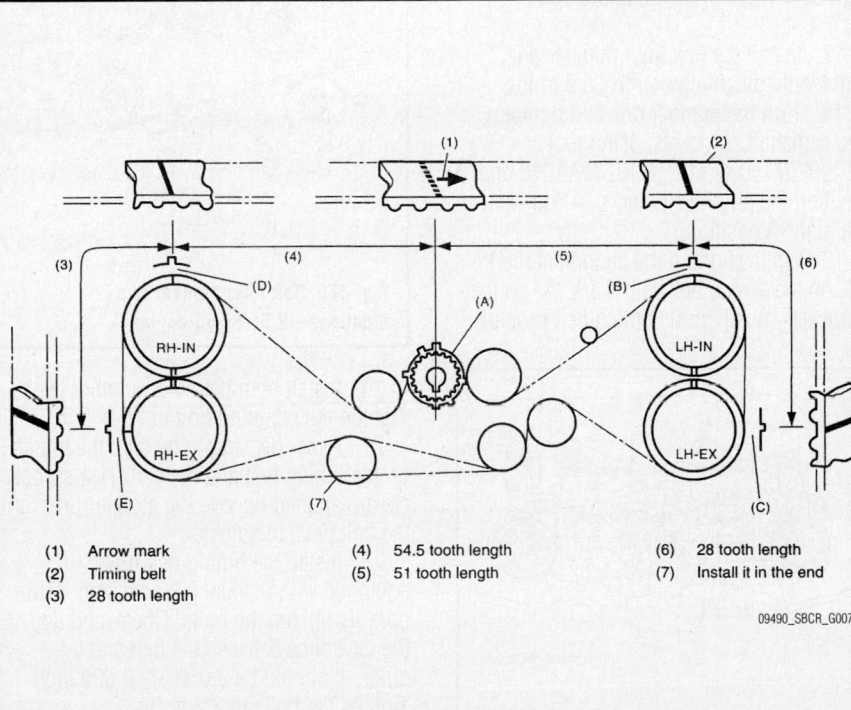

(1)	Arrow mark	(4)	54.5 tooth length	(6)	28 tooth length
(2)	Timing belt	(5)	51 tooth length	(7)	Install it in the end
(3)	28 tooth length				

Fig. 160 Timing belt alignment and installation sequence—2.5L DOHC engine

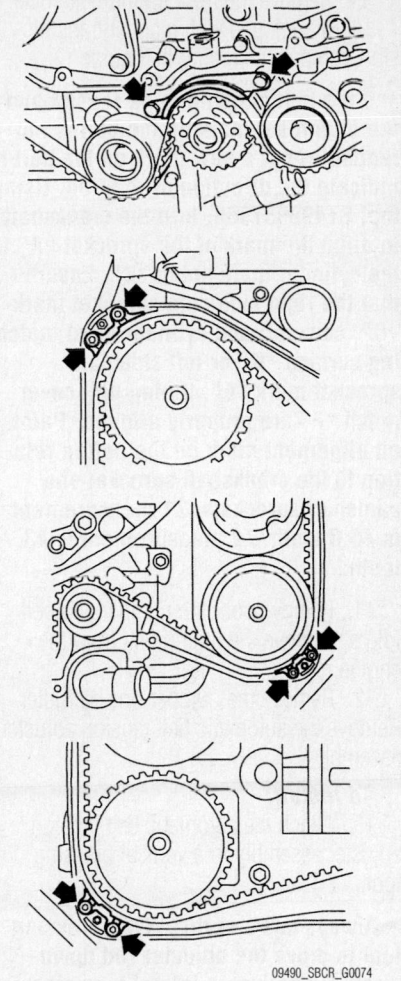

Fig. 161 Timing belt guide bolt location—2.5L DOHC engine

32. Install the crank pulley.
33. Install the V-belts.

2.5L SOHC Engine

See Figures 162 through 166.

1. Before servicing the vehicle, refer to the Precautions Section.

2. Disconnect the negative battery cable.

3. To remove the front side V belt, remove the belt covers. Loosen the lock bolt. Loosen the slider bolt. Remove the front side belt.

4. To remove the rear side V belt, remove the belt covers. Loosen the lock bolt. Loosen the slider bolt. Remove the rear side belt. Remove the belt tensioner.

5. Remove the crankshaft pulley bolt.

6. Lock the crankshaft in place using tool ST499977100, or equivalent.

7. Remove the crankshaft pulley.

8. Remove the left side timing belt cover.

9. Remove the front timing belt cover.

10. If equipped with manual transmission, remove the timing belt guide.

→If the belt is going to be reused and the alignment mark on the belt is not readable, put a new mark on the belt to indicate the direction of rotation. Using tool ST499987500, turn the crankshaft to align the mark of the sprocket "A" to the cylinder mark notch "B". Ensure that the right side cam sprocket mark "C", cam cap and cylinder head matching surface "D" or left side cam sprocket mark "E", timing belt cover notch "F" are properly aligned. Paint an alignment mark on the belt in relation to the crankshaft sprocket and camshaft sprockets. Z1 measurement is 46.8 teeth. Z2 measurement is 43.7 teeth.

11. Remove both the number two belt idlers. Remove the timing belt from the engine.

12. Remove the number one belt idler. Remove the automatic belt tension adjuster assembly.

To install:

13. Attach the automatic belt tension adjuster assembly to a vertical pressing tool.

→Always use a vertical type pressing tool to move the adjuster rod downward. Do not use a lateral type vise. Push the adjuster rod vertically. Press in the push adjuster rod gradually, which should take three minutes or

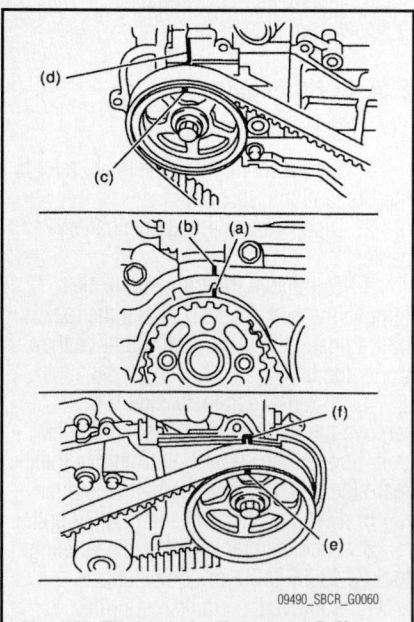

Fig. 162 Timing belt alignment—2.5L SOHC engine

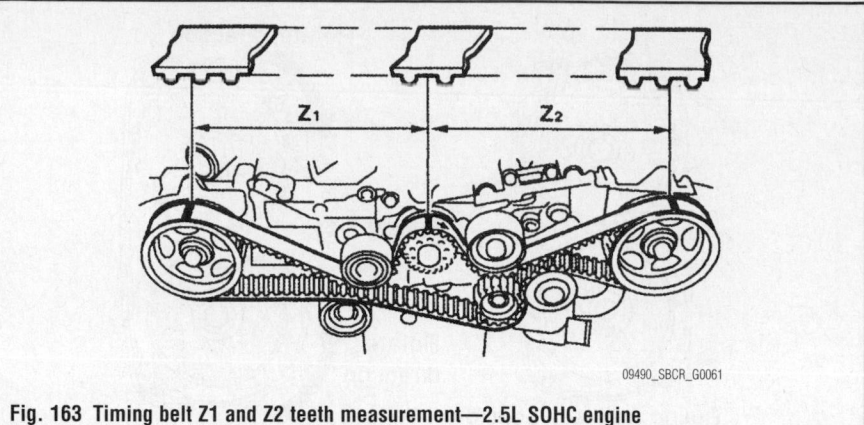

Fig. 163 Timing belt Z1 and Z2 teeth measurement—2.5L SOHC engine

more. Do not allow pressure to exceed 2,205 lb. force.

14. Slowly move the adjuster rod down until the adjuster rod is aligned with the stopper pin hole in the cylinder.

→Press the adjuster rod as far as the end surface of the cylinder. Do not press the adjuster rod into the cylinder. Doing so may damage the cylinder.

15. Using a 0.08 inch stopper pin, insert it into the stopper pin hole in the cylinder. Secure the adjuster rod.

→Do not release the press pressure until the stopper pin is completely inserted in the hole.

16. Install the automatic belt tensioner assembly. Tighten the retaining bolt to 28.9 ft. lbs.

17. Install the belt idler number one. Tighten the retaining bolt to 28.9 ft. lbs.

18. Turn the number one and number two camshaft sprockets, using tool ST499207100 or tool ST18231AA010 and position the alignment marks "A" on each at the highest position.

19. While aligning the alignment mark "B" on the timing belt with mark "A" on the sprockets, position the timing belt properly.

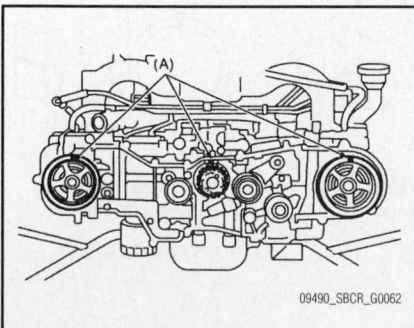

Fig. 164 Timing mark alignment position A—2.5L SOHC engine

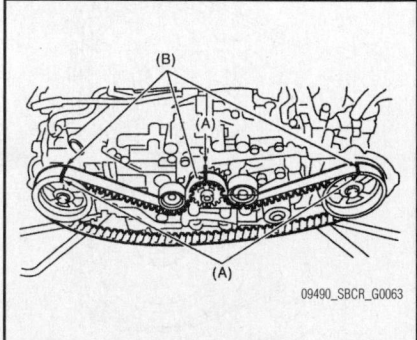

Fig. 165 Timing mark alignment position B—2.5L SOHC engine

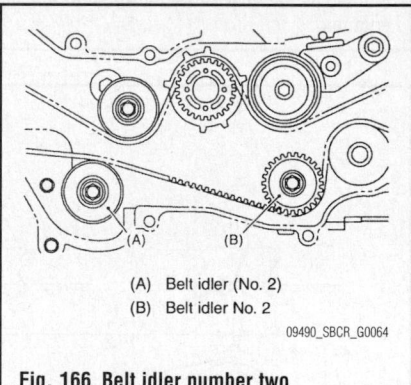

(A) Belt idler (No. 2)
(B) Belt idler No. 2

Fig. 166 Belt idler number two locations—2.5L SOHC engine

20. Install both belt idler number two's. Tighten the retaining bolt to 28.9 ft. lbs.

21. After checking to be sure the marks on the timing belt and the camshaft sprockets are aligned remove the stopper pin from the belt tension adjuster.

22. Install the timing belt guide, if equipped with manual transmission. Temporarily tighten the bolts. Check and adjust the clearance between the belt and the guide. It should be 0.039 +/- 0.020 inch. Tighten the bolts to 7.2 ft. lbs.

23. Continue the installation in the reverse order of the removal procedure.

ENGINE PERFORMANCE & EMISSION CONTROLS

ACCELERATOR PEDAL POSITION (APP) SENSOR

LOCATION

The Accelerator Pedal Position (APP) Sensor is located inside the vehicle at the accelerator pedal.

REMOVAL & INSTALLATION

See Figure 167.

1. Disconnect the ground cable from the battery.
2. Disconnect the connector.
3. Remove the nut securing accelerator pedal assembly.

To install:

4. Install the nut securing accelerator pedal assembly.
5. Connect the connector.
6. Connect the ground cable from the battery.
7. Tighten to 13 ft. lbs. (18 Nm).

CAMSHAFT POSITION (CMP) SENSOR

REMOVAL & INSTALLATION

2.5L DOHC Engine

1. Disconnect the negative battery cable.
2. Remove the collector cover.
3. Disconnect the connector from camshaft position sensor RH.
4. Remove the camshaft position sensor RH from the rear side of the cylinder head.
5. Remove the cam shaft position sensor LH in the same way as RH.

To install:

6. Remove the cam shaft position sensor LH in the same way as RH.
7. Remove the camshaft position sensor RH from the rear side of the cylinder head.
8. Disconnect the connector from camshaft position sensor RH.
9. Remove the collector cover.
10. Disconnect the negative battery cable.
11. Tighten the sensor to 4.7 ft. lbs. (6.4 Nm).

2.5L SOHC Engine

1. Disconnect the negative battery cable.
2. Disconnect the connector from the Camshaft Position (CMP) sensor.
3. Remove the bolt that retains the CMP sensor to the sensor support.
4. Remove the bolt that retains the CMP sensor support to the camshaft cap.
5. Remove the sensor and the CMP sensor support as a unit.
6. Separate the sensor from the support.

To install:

7. Separate the sensor from the support.
8. Remove the sensor and the CMP sensor support as a unit.
9. Remove the bolt that retains the CMP sensor support to the camshaft cap.
10. Remove the bolt that retains the CMP sensor to the sensor support.
11. Disconnect the connector from the Camshaft Position (CMP) sensor.
12. Disconnect the negative battery cable.
13. Tighten the CMP sensor support to 4.7 ft. lbs. (6.4 Nm).
14. Tighten the sensor to 4.7 ft. lbs. (6.4 Nm).

CRANKSHAFT POSITION (CKP) SENSOR

LOCATION

See Figures 168 and 169.

Refer to the accompanying illustration.

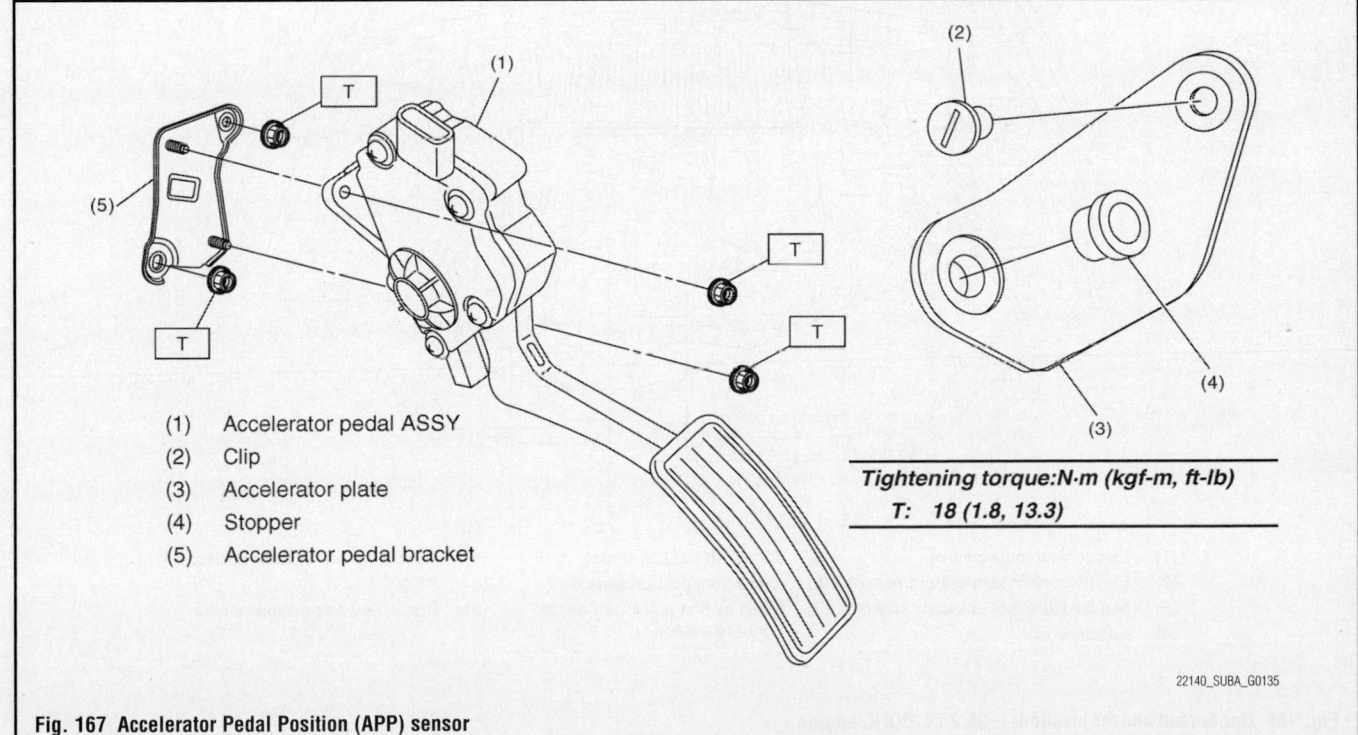

(1) Accelerator pedal ASSY
(2) Clip
(3) Accelerator plate
(4) Stopper
(5) Accelerator pedal bracket

Tightening torque:N·m (kgf-m, ft-lb)
T: 18 (1.8, 13.3)

22140_SUBA_G0135

Fig. 167 Accelerator Pedal Position (APP) sensor

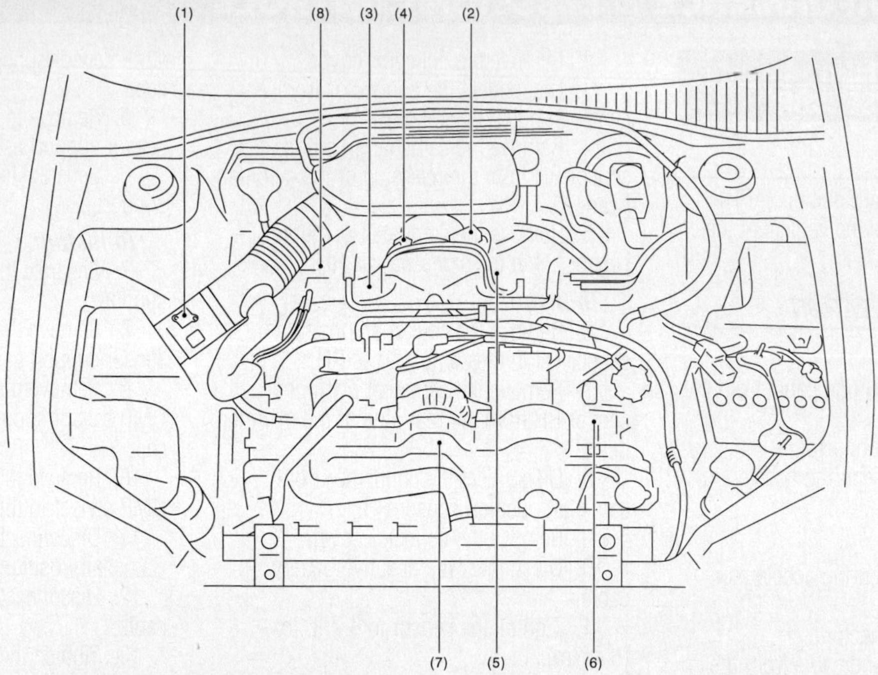

(1) Mass air flow and intake air temperature sensor
(2) Manifold absolute pressure sensor
(3) Engine coolant temperature sensor
(4) Throttle position sensor
(5) Knock sensor
(6) Camshaft position sensor
(7) Crankshaft position sensor
(8) Oil temperature sensor

29157_SUBA_G0066

Fig. 168 Underhood sensor locations—2.5L SOHC engine

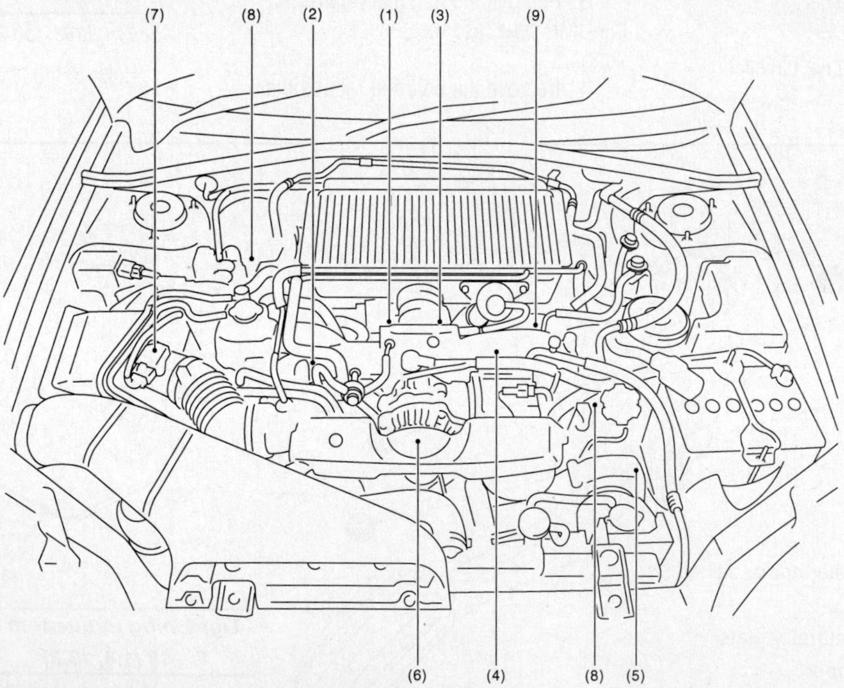

(1) Electronic throttle control
(2) Engine coolant temperature sensor
(3) Manifold absolute pressure sensor
(4) Knock sensor
(5) Camshaft position sensor
(6) Crankshaft position sensor
(7) Mass air flow and intake air temperature sensor
(8) Tumble generator valve position sensor
(9) Secondary air pressure sensor

29157_SUBA_G0058

Fig. 169 Underhood sensor locations—08 2.5L DOHC engine

REMOVAL & INSTALLATION

See Figures 170 and 171.

1. Disconnect the ground cable from battery.
2. Remove the bolt which installs crankshaft position sensor to cylinder block.
3. Remove the crankshaft position sensor, and then disconnect the connector from it.

To install:

4. Install the crankshaft position sensor, and then disconnect the connector from it.
5. Install the bolt which installs crankshaft position sensor to cylinder block.
6. Connect the ground cable to the battery.
7. Tighten the sensor to 4.7 ft. lbs. (6.4 Nm).

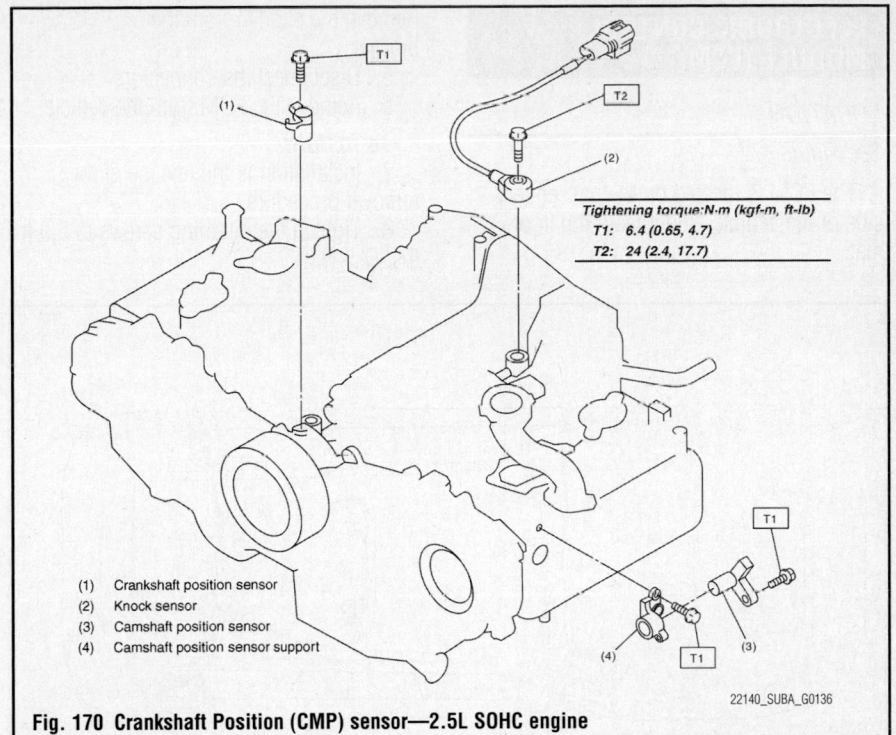

T1
T2
T1
T1
(1)
(2)
(3)
(4)
Tightening torque:N·m (kgf-m, ft-lb)
T1: 6.4 (0.65, 4.7)
T2: 24 (2.4, 17.7)
(1) Crankshaft position sensor
(2) Knock sensor
(3) Camshaft position sensor
(4) Camshaft position sensor support
22140_SUBA_G0136

Fig. 170 Crankshaft Position (CMP) sensor—2.5L SOHC engine

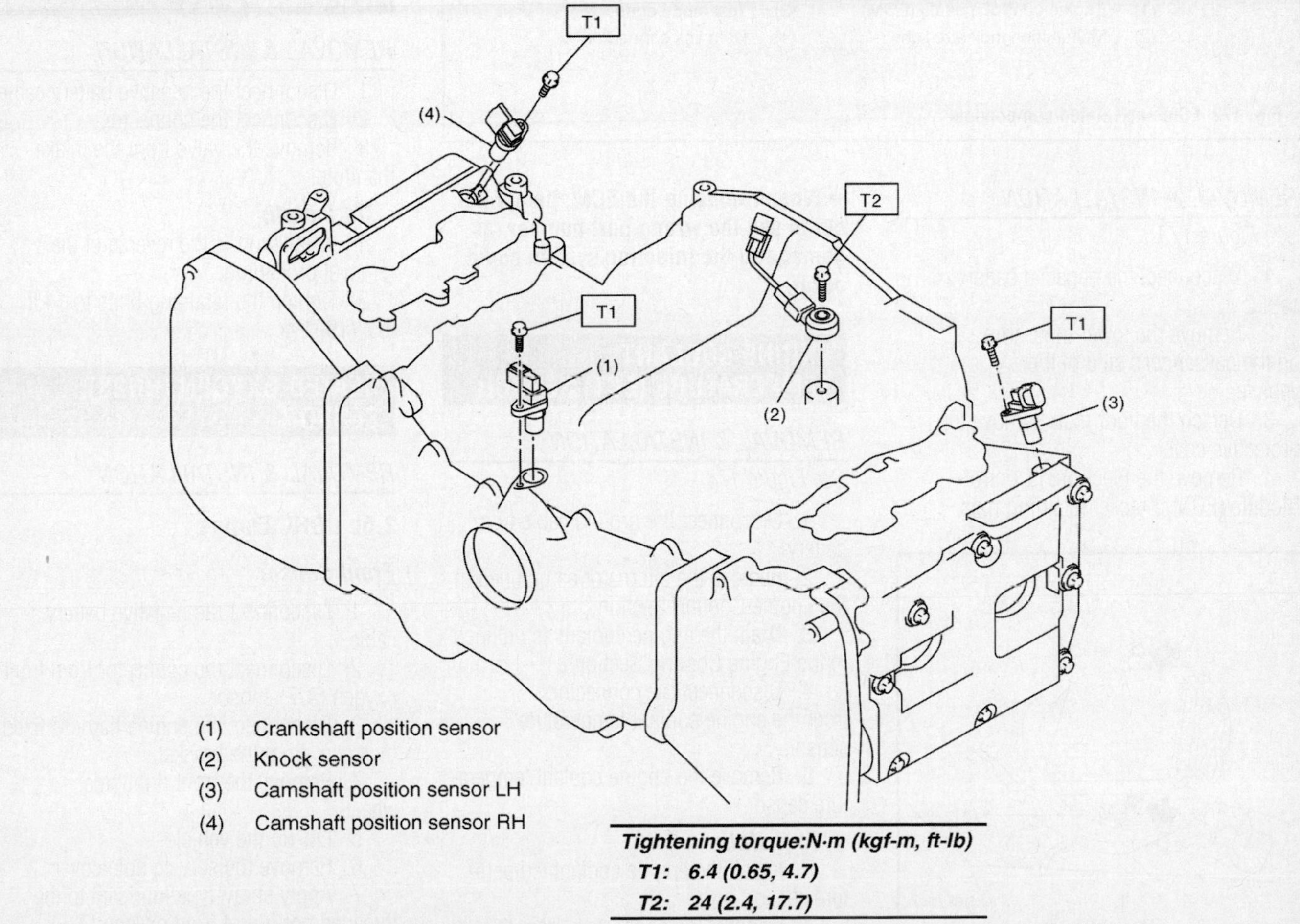

T1
T2
T1
T1
(1)
(2)
(3)
(4)
(1) Crankshaft position sensor
(2) Knock sensor
(3) Camshaft position sensor LH
(4) Camshaft position sensor RH
Tightening torque:N·m (kgf-m, ft-lb)
T1: 6.4 (0.65, 4.7)
T2: 24 (2.4, 17.7)
22140_SUBA_G0138

Fig. 171 Crankshaft Position (CMP) sensor—2.5L DOHC engine

ELECTRONIC CONTROL MODULE (ECM)

LOCATION

See Figure 172.

The ECM is located on the passenger's side of the vehicle, underneath the floor mat.

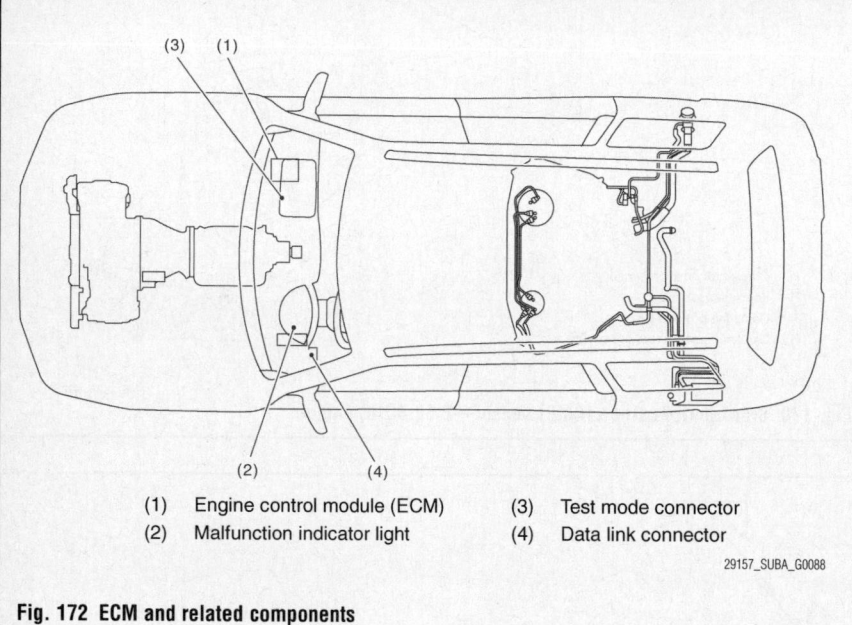

(1) Engine control module (ECM)
(2) Malfunction indicator light
(3) Test mode connector
(4) Data link connector

29157_SUBA_G0088

Fig. 172 ECM and related components

REMOVAL & INSTALLATION

See Figure 173.

1. Disconnect the negative battery cable.
2. Remove the lower inner trim on the passenger's side of the vehicle.
3. Detach the floor mat. Remove the protective cover.
4. Remove the Electronic Control Module (ECM) bracket retaining nuts.

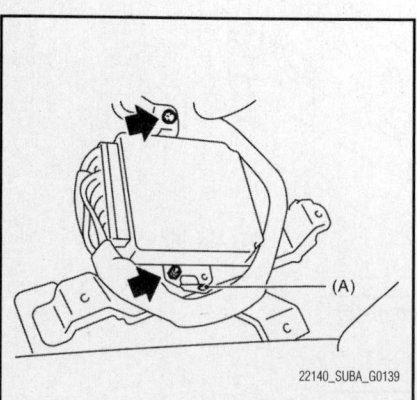

22140_SUBA_G0139

Fig. 173 Electronic Control Module (ECM)

Remove the clip (A) from the bracket.

5. Disconnect the connectors.
6. Remove the ECM from the vehicle.

To install:

7. Installation is the reverse of the removal procedure.
8. Tighten the retaining screws to 5.5 ft. lbs. (7.5 Nm).

➡When replacing the ECM, be careful not to use the wrong part number, as damage to the injection system could occur.

ENGINE COOLANT TEMPERATURE (ECT) SENSOR

REMOVAL & INSTALLATION

See Figure 174.

1. Disconnect the ground cable from battery.
2. Remove the alternator as outlined in the Engine Cooling Section.
3. Drain the engine coolant as outlined in the Engine Cooling Section.
4. Disconnect the connectors from the engine coolant temperature sensor.
5. Remove the engine coolant temperature sensor.

To install:

6. Install the engine coolant temperature sensor.
7. Connect the connectors from the engine coolant temperature sensor.

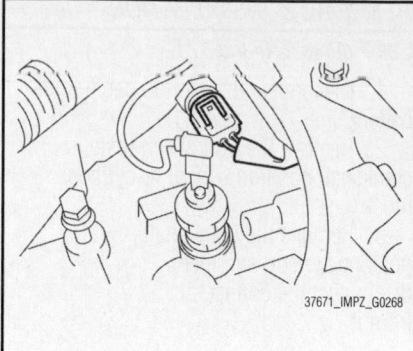

37671_IMPZ_G0268

Fig. 174 Engine Coolant Temperature (ECT) Sensor & connectors

8. Fill the engine with the proper type and amount of coolant.
9. Install the alternator as outlined in the Engine Electrical Section.
10. Connect the ground cable to the battery.

➡Use a new gasket. Tighten to 13 ft. lbs. (18 Nm).

EXHAUST GAS RECIRCULATION (EGR) VALVE

REMOVAL & INSTALLATION

1. Disconnect the negative battery cable.
2. Disconnect the connector.
3. Remove the valve from the intake manifold.

To install:

4. Installation is the reverse of the removal procedure.
5. Tighten the retaining bolts to 14 ft. lbs. (19 Nm).

HEATED OXYGEN (HO2S) SENSOR

REMOVAL & INSTALLATION

2.5L DOHC Engine

Front Sensor

1. Disconnect the negative battery cable.
2. Disconnect the connector from front oxygen (A/F) sensor.
3. Disconnect the engine harness fixed by a clip, from the bracket.
4. Remove the front right side wheel.
5. Lift-up the vehicle.
6. Remove the service hole cover.
7. Apply spray-type lubricant to the threaded portion of front oxygen (A/F) sensor, and leave it for one minute or more.

8. Remove the front oxygen (A/F) sensor.

✳✳ CAUTION

When removing the oxygen (A/F) sensor, wait until exhaust pipe cools, otherwise it will damage the exhaust pipe.

To install:

9. Before installing front oxygen (A/F) sensor, apply the anti-seize compound only to the threaded portion of front oxygen (A/F) sensor. This facilitates the next removal.

10. Install the front oxygen (A/F) sensor and tighten to 22 ft. lbs. (30 Nm).

11. Install the service hole cover.

12. Lower the vehicle.

13. Install the front right side wheel.

14. Connect the engine harness to the bracket using a clip.

15. Connect the connector of front oxygen (A/F) sensor.

16. Connect the battery ground cable to the battery.

Rear Sensor

1. Disconnect the negative battery cable.

2. Lift-up the vehicle.

3. Disconnect the connector from the rear oxygen sensor.

4. Apply spray-type lubricant to the threaded portion of rear oxygen sensor, and leave it for one minute or more.

5. Remove the rear oxygen sensor.

✳✳ CAUTION

When removing the rear oxygen sensor, wait until exhaust pipe cools, otherwise it may damage the exhaust pipe.

To install:

6. Before installing rear oxygen sensor, apply the anti-seize compound only to the threaded portion of rear oxygen sensor to make the next removal easier.

7. Install the rear oxygen sensor and tighten to 15 ft. lbs. (21 Nm).

8. Connect the connector to rear oxygen sensor.

9. Lower the vehicle.

10. Connect the battery ground cable to the battery.

2.5L SOHC Engine

Front Sensor

1. Disconnect the negative battery cable.

2. Remove the clip fastening the har-

ness and disconnect the front oxygen (A/F) sensor connector.

3. Lift-up the vehicle.

4. Remove the undercover.

5. Apply spray-type lubricant to the threaded portion of front oxygen (A/F) sensor, and leave it for one minute or more.

6. Remove the front oxygen (A/F) sensor.

✳✳ CAUTION

When removing the front oxygen (A/F) sensor, wait until exhaust pipe cools, because it can damage the exhaust pipe.

To install:

7. Before installing front oxygen (A/F) sensor, apply anti-seize compound only to the threaded portion of front oxygen (A/F) sensor to make the next removal easier.

8. Install the front oxygen (A/F) sensor and tighten to 15 ft. lbs. (21 Nm).

9. Install the undercover.

10. Lower the vehicle.

11. Connect the connector of front oxygen (A/F) sensor connector and fasten the harness with clips.

12. Connect the battery ground cable to the battery.

Rear Sensor

1. Disconnect the negative battery cable.

2. Remove the clip fastening the harness and disconnect the rear oxygen sensor connector.

3. Lift-up the vehicle.

4. Remove the undercover.

5. Apply spray-type lubricant to the threaded portion of rear oxygen sensor, and leave it for one minute or more.

6. Remove the rear oxygen sensor.

To install:

7. Before installing rear oxygen sensor, apply the anti-seize compound only to the threaded portion of rear oxygen sensor to make the next removal easier.

8. Install the rear oxygen sensor and tighten to 15 ft. lbs. (21 Nm).

9. Install the undercover.

10. Lower the vehicle.

11. Connect the connector to rear oxygen sensor and mount the harness clips to the bracket for fastening.

12. Connect the battery ground cable to the battery.

KNOCK SENSOR (KS)

REMOVAL & INSTALLATION

1. Disconnect the negative battery cable.

2. Remove the air cleaner case.

3. On DOHC engine, remove the intercooler.

4. Disconnect the sensor connector.

5. Remove the sensor from its mounting.

To install:

6. Installation is the reverse of the removal procedure.

7. Tighten the sensor to 18 ft. lbs. (24 Nm).

➡ **The extraction area of the knock sensor wire must be positioned at a 60 degree angle relative to the engine rear.**

MASS AIR FLOW AND INTAKE AIR TEMPERATURE (IAT) SENSOR

REMOVAL & INSTALLATION

See Figure 175.

1. Disconnect the ground cable from battery.

2. Disconnect the connector from the mass air flow and intake air temperature sensor, and remove the mass air flow and intake air temperature sensor.

To install:

3. Disconnect the connector from the mass air flow and intake air temperature sensor, and remove the mass air flow and intake air temperature sensor.

4. Tighten sensor to 8.4 inch lbs. (1 Nm).

5. Disconnect the ground cable from battery.

37671_IMPZ_G0319

Fig. 175 Mass air flow and intake air temperature sensor removal

MANIFOLD ABSOLUTE PRESSURE (MAP) SENSOR

REMOVAL & INSTALLATION

1. Disconnect the ground cable from the battery.
2. Disconnect the connector from Manifold Absolute Pressure (MAP) sensor.
3. Remove the MAP sensor from throttle body.

To install:

4. Install in the reverse order of removal, noting the following:
   a. Use new O-rings.
   b. Tighten the MAP sensor to 1.5 ft. lbs. (2 Nm).

VEHICLE SPEED SENSOR (VSS)

REMOVAL & INSTALLATION

1. Raise and support the vehicle safely.

2. Place a drip pan below the speed sensor to catch any spilled fluid.
3. Disconnect the connector.
4. Remove the sensor from its mounting.

To install:

5. Installation is the reverse of the removal procedure.
6. Replace any lost fluid.

FUEL   GASOLINE FUEL INJECTION SYSTEM

FUEL SYSTEM SERVICE PRECAUTIONS

Safety is the most important factor when performing not only fuel system maintenance, but any type of maintenance. Failure to conduct maintenance and repairs in a safe manner may result in serious personal injury or death. Work on a vehicle's fuel system components can be accomplished safely and effectively by adhering to the following rules and guidelines.

• To avoid the possibility of fire and personal injury, always disconnect the negative battery cable unless the repair or test procedure requires that battery voltage be applied.

• Always relieve the fuel system pressure prior to disconnecting any fuel system component (injector, fuel rail, pressure regulator, etc.) fitting or fuel line connection. Exercise extreme caution whenever relieving fuel system pressure to avoid exposing skin, face and eyes to fuel spray. Please be advised that fuel under pressure may penetrate the skin or any part of the body that it contacts.

• Always place a shop towel or cloth around the fitting or connection prior to loosening to absorb any excess fuel due to spillage. Ensure that all fuel spillage is quickly removed from engine surfaces. Ensure that all fuel-soaked cloths or towels are deposited into a flame-proof waste container with a lid.

• Always keep a dry chemical (Class B) fire extinguisher near the work area.

• Do not allow fuel spray or fuel vapors to come into contact with a spark or open flame.

• Always use a second wrench when loosening or tightening fuel line connection fittings. This will prevent unnecessary stress and torsion on fuel piping. Always follow the proper torque specifications.

• Always replace worn fuel fitting O-rings with new ones. Do not substitute fuel hose where rigid pipe is installed.

FUEL SYSTEM PRESSURE

RELIEVING
See Figure 176.

✳✳ WARNING
Place "NO OPEN FLAMES" signs near the working area.

✳✳ CAUTION
Be careful not to spill fuel.

1. Remove the fuse of fuel pump from main fuse box.
2. Start the engine and run it until it stalls.
3. After the engine stalls, crank it for five more seconds.
4. Turn the ignition switch to OFF.
5. Install the fuse of fuel pump to the main fuse box.

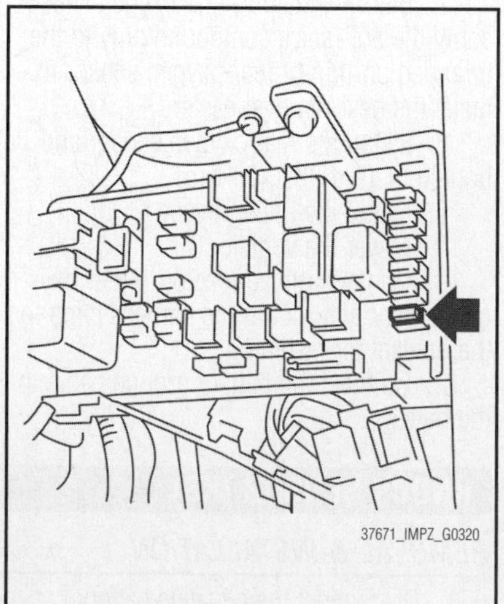

37671_IMPZ_G0320

Fig. 176 Fuel pump fuse location

FUEL FILTER

REMOVAL & INSTALLATION
See Figures 177 through 187.

✳✳ WARNING
Place "NO OPEN FLAMES" signs near the working area.

✳✳ CAUTION
Note the following:

• Be careful not to spill fuel.
• If the fuel gauge indicates that two thirds or more of the fuel is remaining, be sure to drain fuel before starting work to avoid the fuel to spill.
• Be careful not to drop or apply any impact to the fuel pump during work. This may deteriorate its performance.

➡**The fuel filter is built in fuel pump assembly.**

1. Remove the fuel pump assembly as outlined in this section.
2. Remove the fuel level sensor and fuel temperature sensor as outlined in this section.
3. Disconnect the pump assembly connector from the sub tank bracket assembly.
4. Cut off the tab holders connecting the sub tank bracket assembly and the sub tank in four locations, and separate the two.

✳✳ CAUTION
Be careful not to damage the sub tank.

➡**If the O-ring is remaining on the sub tank, remove.**

5. Disconnect the fuel piping connector from the fuel filter assembly in two locations.

6. Push to compress the fuel filter assembly in the direction of the arrow, remove clip, and separate the sub tank bracket assembly and the fuel filter assembly.

✳✳ CAUTION

When separating the sub tank bracket assembly and the fuel filter assembly, be careful not to damage the ground wire.

7. Disconnect the connector from the pump assembly.

8. Lift the two tab holders connecting the pump assembly to the fuel filter using a flat tip screwdriver (with a shaft diameter of approx. (0.12 in (3 mm)), etc., and separate the fuel filter and pump assembly.

✳✳ WARNING

To prevent damaging the tabs of the pump assembly, wrap the tip of flat tip screwdriver, etc. with tape. Be careful not to drop or apply any impact to the pump assembly.

➡️ **If the spacer and O-ring is remaining on the pump assembly, remove these.**

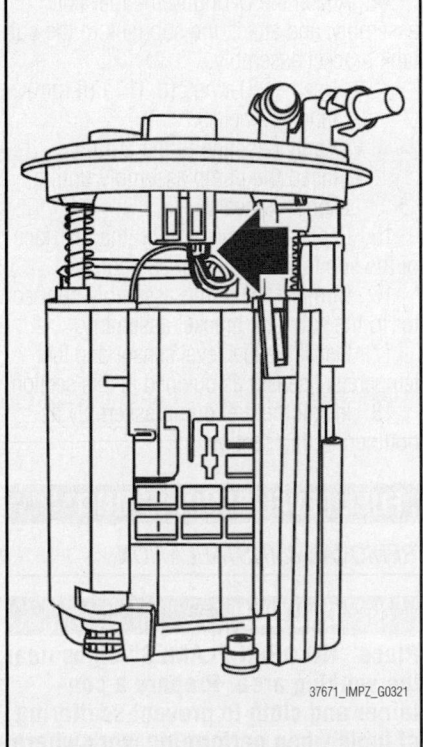

Fig. 177 Fuel pump assembly connector from sub tank bracket assembly removal

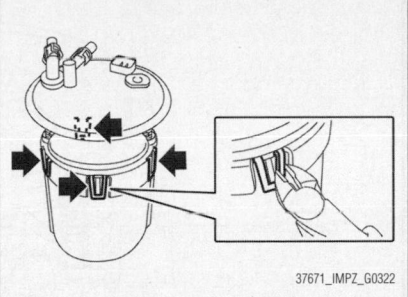

37671_IMPZ_G0322

Fig. 178 Cut off the tab holders connecting the sub tank bracket assembly and sub tank

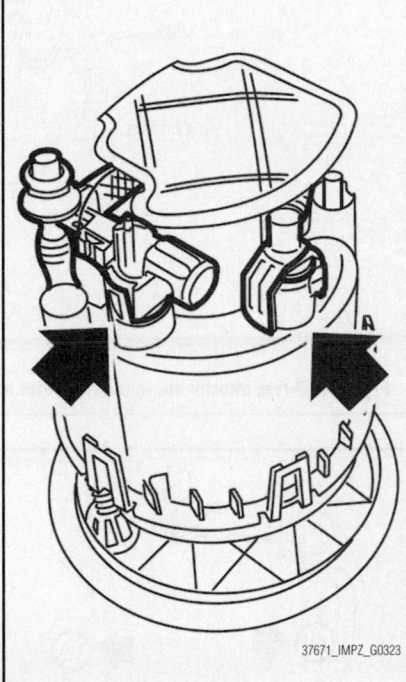

37671_IMPZ_G0323

Fig. 179 Disconnect fuel piping connector from fuel filter assembly

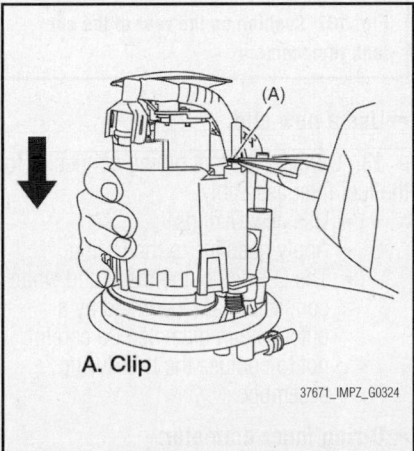

(A)

A. Clip

37671_IMPZ_G0324

Fig. 180 Separate the sub tank bracket assembly and fuel filter assembly

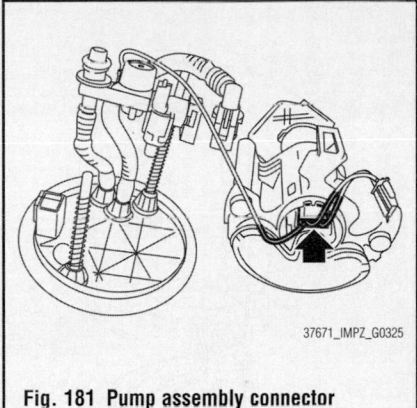

37671_IMPZ_G0325

Fig. 181 Pump assembly connector

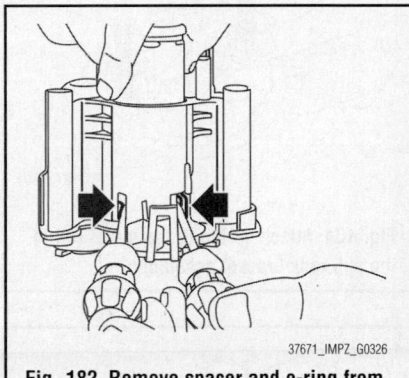

37671_IMPZ_G0326

Fig. 182 Remove spacer and o-ring from pump assembly

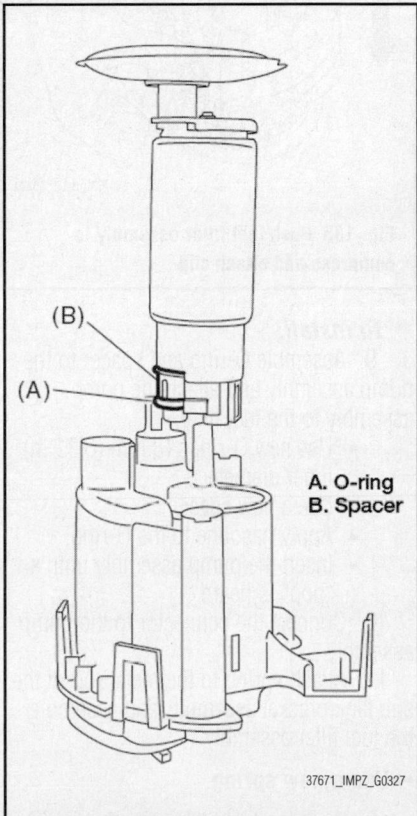

(B)

(A)

A. O-ring
B. Spacer

37671_IMPZ_G0327

Fig. 183 Assemble o-ring and spacer to the pump assembly and attach to fuel filter

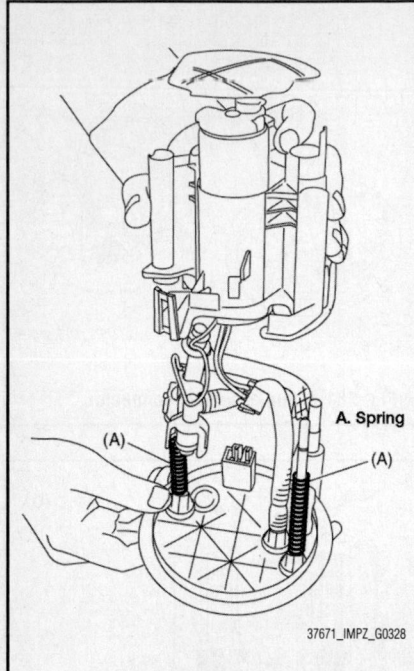

Fig. 184 Attach spring to the metal rod of the sub tank bracket assembly

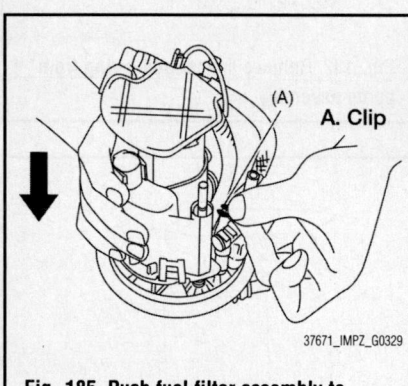

Fig. 185 Push fuel filter assembly to compress and attach clip

To install:

9. Assemble O-ring and spacer to the pump assembly and attach the pump assembly to the fuel filter:
- Use new O-rings (8 mm (0.31 in) inner diameter).
- Use a new spacer.
- Apply gasoline to the O-ring.
- Insert the pump assembly until a "pop" is heard.

10. Connect the connector to the pump assembly.

11. Attach spring to the metal rod of the sub tank bracket assembly, and assemble the fuel filter assembly.

➡**Use a new spring.**

12. Push the fuel filter assembly in the direction of the arrow to compress, and attach clip.

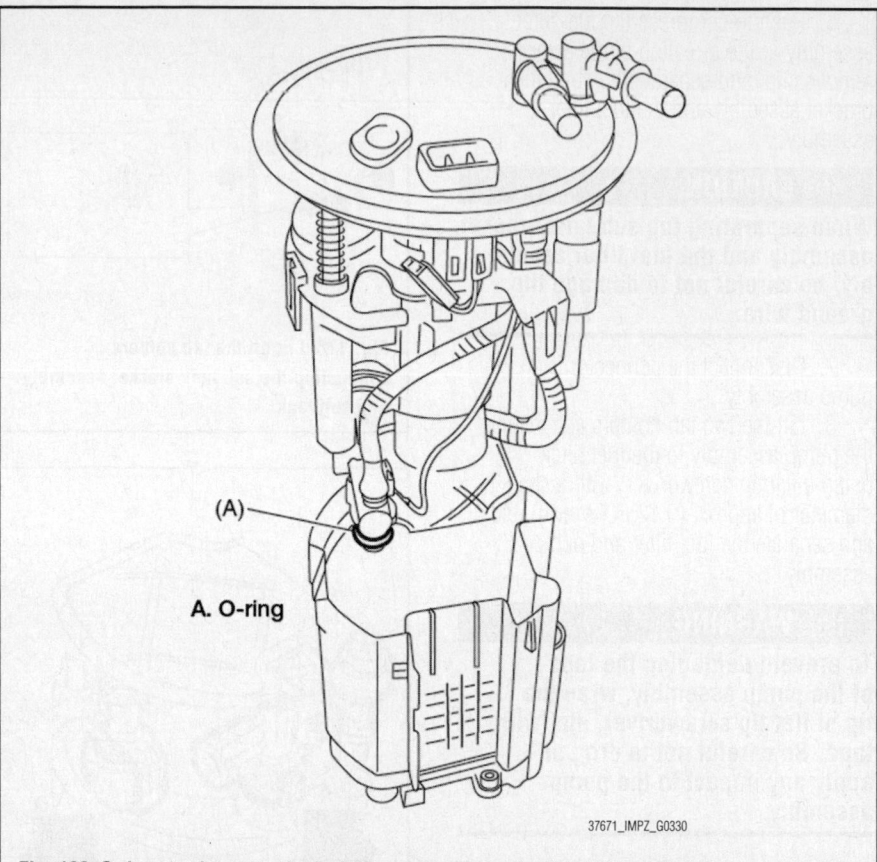

Fig. 186 O-ring attachment to the fuel filter assembly, and sub tank to sub tank bracket assembly

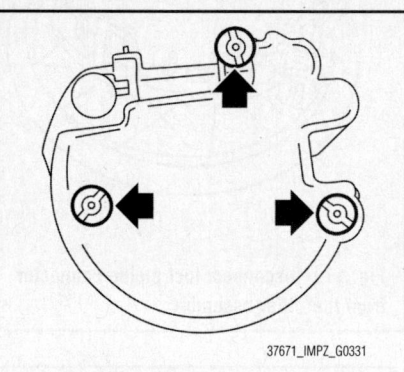

Fig. 187 Cushion on the rear of the sub tank replacement

➡**Use a new clip.**

13. Connect the fuel piping connector to the fuel filter assembly:
- Use new O-rings.
- Apply gasoline to the O-ring.
- The O-rings of the black and white connectors are identified by a difference in diameter. Be careful not to confuse the two during assembly.

➡**O-ring inner diameter:**

- Black connector O-ring [Approx. 0.28 in (7 mm)]

- White connector O-ring [Approx. 0.31 in (8 mm)]

14. Attach the O-ring to the fuel filter assembly, and attach the sub tank to the sub tank bracket assembly:
- Use new O-rings (0.31 in (8 mm) inner diameter).
- Apply gasoline to the O-ring.
- Insert the pump assembly until a "pop" is heard.

15. Replace the cushion on the rear face of the sub tank with a new cushion.

16. Connect the pump assembly connector to the sub tank bracket assembly.

17. Install the fuel level sensor and fuel temperature sensor as outlined in this section.

18. Install the fuel pump assembly as outlined in this section.

FUEL LEVEL SENDING UNIT

REMOVAL & INSTALLATION

❊❊ CAUTION

Place "NO OPEN FLAMES" signs near the working area. Prepare a container and cloth to prevent scattering of fuels when performing work where fuels can be spilled. If the fuel spills, wipe it off immediately to prevent

from penetrating into floor or flowing out for environmental protection. Follow all government and local regulations concerning disposal of refuse when disposing fuel.

1. Before servicing the vehicle, refer to the Precautions Section.
2. Relieve the fuel system pressure.
3. Disconnect the negative battery cable.
4. Remove the fuel pump assembly.
5. Disconnect the connector from fuel pump bracket.
6. Remove the fuel level sensor and fuel temperature sensor.

To install:
7. Install in the reverse order of removal.
8. Tighten sensor to 3 ft. lbs. (4 Nm).

FUEL PUMP

REMOVAL & INSTALLATION
See Figure 188.

❊❊ CAUTION

When the fuel gauge pointer is at two third or more, the fuel may spill out. Be sure to drain the fuel before the operation. Be careful not to spill fuel.

1. Before servicing the vehicle, refer to the Precautions Section.
2. Release the fuel pressure.
3. Drain fuel.
4. Disconnect the ground cable from the battery.
5. Remove the rear seat.
6. Remove the service hole cover.
7. Disconnect the connector from fuel pump.
8. Disconnect the quick connector, then disconnect the fuel delivery tube and jet pump tube.

9. Remove the nuts which install fuel pump assembly onto fuel tank.
10. Remove the fuel pump assembly from the fuel tank.

To install:
11. Install in the reverse order of removal while being careful of the following:
- Make sure the sealing portion is free from fuel or foreign matter before installation.
- When assembling, point the protrusion (A) of the gasket towards the front of the vehicle.
- Insert the protrusion (B) of the gasket into the upper plate.
- Align the protrusion (C) of the fuel pump assembly to the cut out in the upper plate.
12. Tighten the nuts to 3.2 ft. lbs. (4.4 Nm) in the order as shown in the figure below.

FUEL TANK

REMOVAL & INSTALLATION
See Figures 189 through 192.

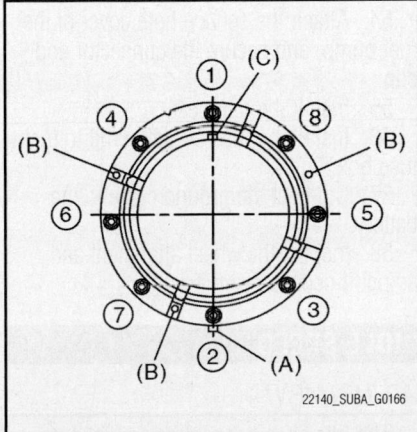

Fig. 188 Fuel pump assembly tightening procedure

❊❊ CAUTION

Place "NO OPEN FLAMES" signs near the working area.

1. Before servicing the vehicle, refer to the Precautions Section.
2. Set vehicle on the lift.
3. Release fuel pressure.
4. Drain fuel from fuel tank.
5. Disconnect the ground cable from the battery.
6. Remove the rear seat.
7. Disconnect fuel pump connector, and remove clip.
8. Remove the bolts.
9. Push the grommet down and remove the service hole cover.
10. Remove the service hole cover of fuel sub level sensor.
11. Disconnect the quick connector on the fuel delivery tube.
12. Remove the rear wheels.
13. Lift up the vehicle.
14. Remove the rear ABS wheel speed sensor from the rear housing.
15. Remove the bolt holding the rear brake hose bracket.

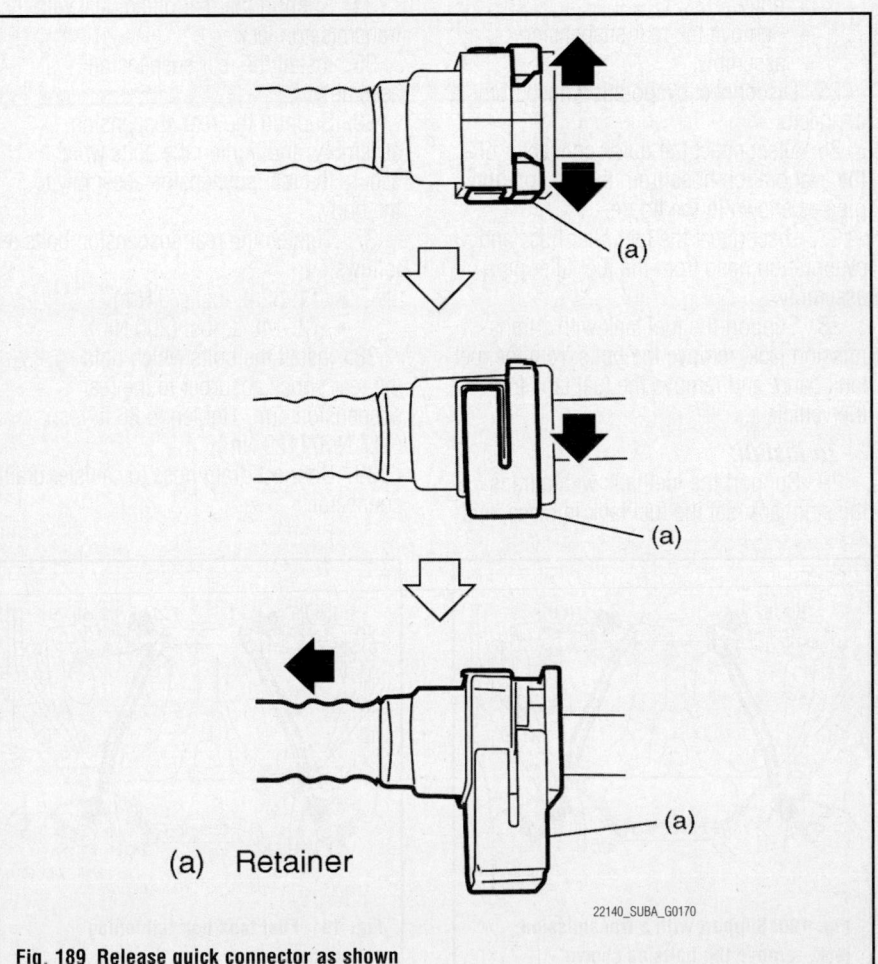

(a) Retainer

Fig. 189 Release quick connector as shown

16. Remove the rear brake caliper and tie it to the body side of the vehicle.

17. Remove the parking brake cable from the parking brake assembly.

18. Remove the rear exhaust pipe.

19. Remove the propeller shaft.

20. Remove the heat shield cover and fuel tank protector.

21. Disconnect the connector from the rear ABS wheel speed sensor.

22. Remove the bolts securing the parking brake cable clamp.

23. Disconnect drain hose from the canister drain connector.

✳✳ CAUTION

A helper is required to perform this work.

24. Remove the rear suspension assembly as follows:
- Support the rear differential with the transmission jack.
- Remove the bolts which hold the rear shock absorber to the rear suspension arm.
- Remove the bolts which secure the rear suspension assembly to the body.
- Remove the rear suspension assembly.

25. Disconnect evaporation hose from connector.

26. Disconnect the quick connector of the evaporation hose from the evaporation pipe as shown in the figure.

27. Disconnect the fuel filler hose and evaporation hose from the fuel filler pipe assembly.

28. Support the fuel tank with a transmission jack, remove the bolts from the fuel tank band, and remove the fuel tank from the vehicle.

To install:

29. Support the fuel tank with a transmission jack, set the fuel tank in place, and temporarily tighten the bolts of the fuel tank band.

30. Securely insert the fuel filler hose and evaporation hose into the specified position, then attach the clamp and clip as shown in the figure.

31. Connect the quick connector of the evaporation hose to the evaporation pipe.

✳✳ CAUTION

Check that there is no damage or dust on the quick connector. If necessary, clean seal surface of pipe. When connecting the quick connector, insert the pipe all the way in securely, then operate the push lock. If it is not possible to perform the push lock operation of the retainer, recheck whether the pipe is securely inserted. Confirm that the quick connector is securely connected.

32. Connect evaporation hose to connector.

33. Tighten the fuel tank band bolts in the order shown in the figure to 25 ft. lbs. (33 Nm).

34. Support the rear differential with the transmission jack.

35. Install the rear suspension assembly.

36. Support the rear suspension assembly, and tighten the bolts which secure the rear suspension assembly to the body.

37. Tighten the rear suspension bolts as follows:
- T1: 52 ft. lbs. (70 Nm)
- T2: 148 ft. lbs. (200 Nm)

38. Install the bolts which hold the rear shock absorber to the rear suspension arm. Tighten to 86 ft. lbs. (117 Nm).(120 Nm).

39. Connect drain hose to canister drain connector.

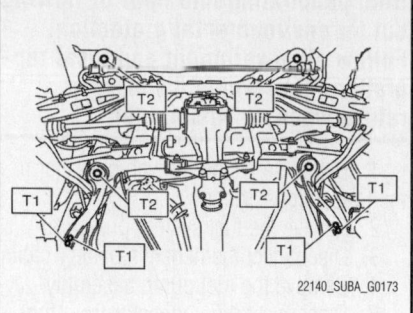

Fig. 192 Tighten the rear suspension bolts as shown

40. Tighten the bolts holding the parking brake cable clamp to 13 ft. lbs. (18 Nm)

41. Connect the connector to the rear ABS wheel speed sensor.

42. Install the heat shield cover.

43. Install the fuel tank protector. Tighten the nuts to 7 ft. lbs. (9 Nm). Tighten the bolts to 13 ft. lbs. (18 Nm).

44. Install the propeller shaft.

45. Install the rear exhaust pipe.

46. Lower the vehicle.

47. Connect the parking brake cable to the parking brake assembly.

48. Install the rear brake caliper.

49. Tighten the bolts which hold the rear brake hose bracket to 25 ft. lbs. (33 Nm).

50. Attach the rear ABS vehicle speed sensor to the rear housing. Tighten the mounting bolt to 6 ft. lbs. (8 Nm).

51. Install the rear wheels.

52. Connect quick connector of the fuel delivery tube.

53. Install the service hole cover of fuel sub level sensor.

54. Attach the service hole cover of the fuel pump, and secure the connector and clip.

55. Install the rear seat.

56. Install the fuse of fuel pump to main fuse box.

57. Connect the ground cable to the battery.

58. Inspect the wheel alignment and adjust if necessary.

IDLE SPEED

ADJUSTMENT

The idle speed cannot be adjusted manually, because the idle speed is automatically adjusted.

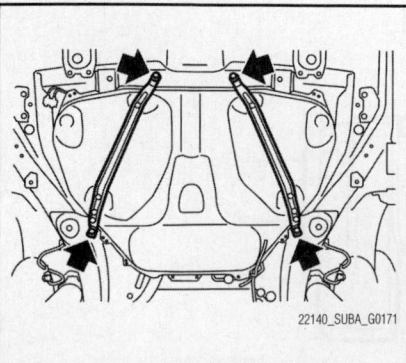

Fig. 190 Support with a transmission jack, remove the bolts as shown

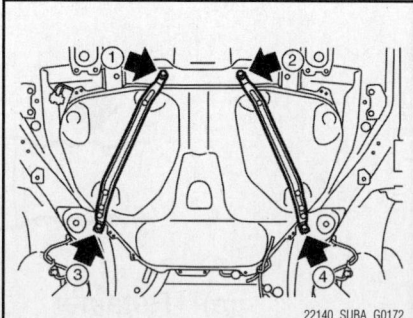

Fig. 191 Fuel tank bolt tightening sequence

THROTTLE BODY

REMOVAL & INSTALLATION

See Figures 193 and 194.

1. Disconnect the ground cable from battery.
2. Raise and safely support the vehicle.
3. Remove the undercover.
4. Drain approximately 3.0 2 (3.2 US qt, 2.6 Imp qt) of coolant.
5. Remove the air intake chamber.
6. Disconnect the throttle position sensor connector and manifold absolute pressure sensor connector.
7. Disconnect the engine coolant hoses from throttle body.
8. Remove the bolts which secure the throttle body to the intake manifold, and remove the throttle body.

To install:

9. Installation is the reverse of the removal procedure, noting the following:

   a. Use a new gasket when installing the throttle body and tighten the mounting bolts to 71 inch lbs. (8 Nm).

   b. Refill the cooling system with the proper type and amount of coolant.

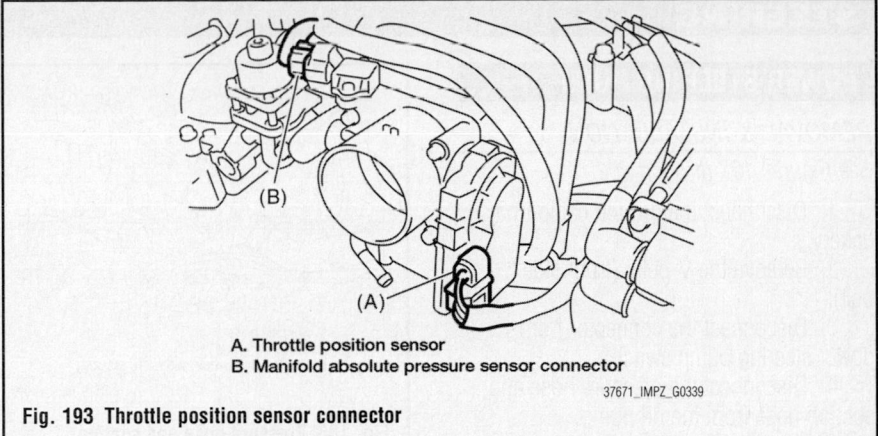

A. Throttle position sensor
B. Manifold absolute pressure sensor connector

37671_IMPZ_G0339

Fig. 193 Throttle position sensor connector

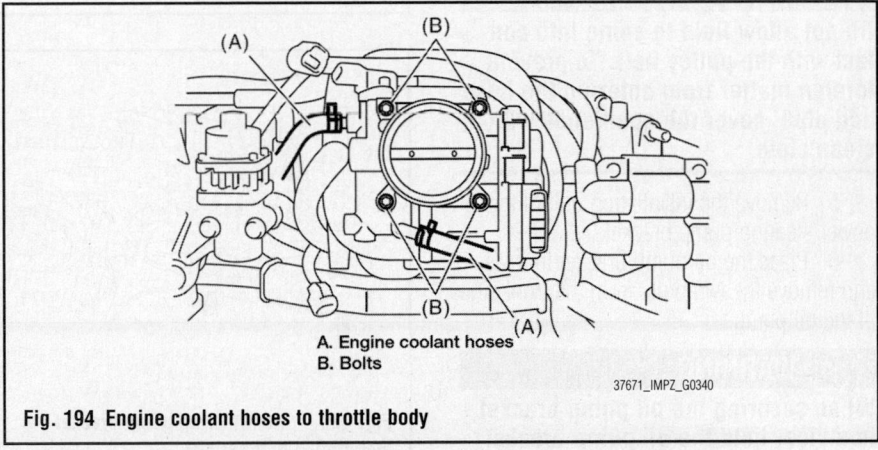

A. Engine coolant hoses
B. Bolts

37671_IMPZ_G0340

Fig. 194 Engine coolant hoses to throttle body

HEATING & AIR CONDITIONING SYSTEM

BLOWER MOTOR

REMOVAL & INSTALLATION

See Figure 195.

1. Before servicing the vehicle, refer to the Precautions Section.
2. Disconnect the negative battery cable.
3. Remove the glove box.
4. Loosen the nut to remove the support beam stay.
5. Disconnect the blower motor wiring harness.
6. Disconnect the blower resistor connector.
7. Loosen the bolt and nut to remove the blower motor unit assembly.
8. Installation is the reverse order of removal.

HEATER CORE

REMOVAL & INSTALLATION

See Figure 196.

1. Before servicing the vehicle, refer to the Precautions Section.
2. Disconnect the negative battery cable.
3. Discharge the air conditioning system.
4. Drain the cooling system.
5. Remove the bolts securing the expansion valve and pipe in the engine compartment. Release the heater hose clamps in the engine compartment to remove the heater hoses.
6. Remove the instrument panel.
7. Remove the support beam.
8. Remove the blower motor assembly.
9. Loosen the bolts and nuts to remove the heater and cooling unit.
10. Loosen the screws to remove the heater core cover.
11. Remove the heater core.
12. Installation is the reverse order of removal.

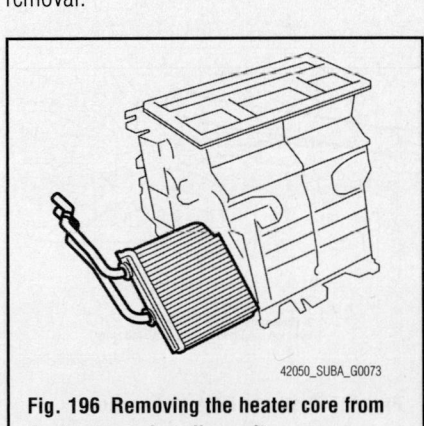

42050_SUBA_G0073

Fig. 196 Removing the heater core from the heater and cooling unit

22140_SUBA_G0193

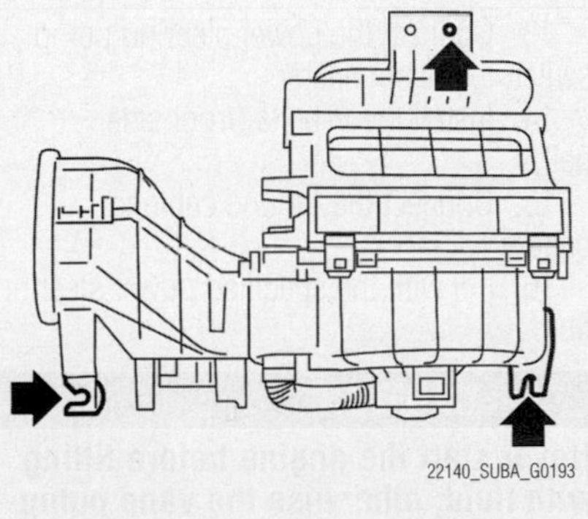

Fig. 195 Blower motor unit assembly

STEERING

POWER STEERING PUMP

REMOVAL & INSTALLATION

See Figures 197 through 202.

1. Disconnect the ground cable from battery.
2. Remove the V-belts (front side belt).
3. Disconnect the connector from power steering pump switch.
4. Disconnect the pressure hose and suction hose from the oil pump.

※ WARNING

Do not allow fluid to come into contact with the pulley belt. To prevent foreign matter from entering the hose and pipe, cover the open ends with clean cloth.

5. Remove the installation bolt of the power steering pump bracket.
6. Place the oil pump bracket in a vise, and remove the two bolts from the front side of the oil pump.

※ WARNING

When securing the oil pump bracket in a vice, hold the oil pump bracket with the least possible force between two pieces of wood.

7. Remove the bolt from the rear side of oil pump.
8. Disassemble the oil pump and bracket by inserting a flat tip screwdriver as shown in the figure.

To install:

9. Install the oil pump to bracket.
10. Attach the installation bolts of the power steering pump bracket, and observe the tightening specifications in the accompanying illustration.

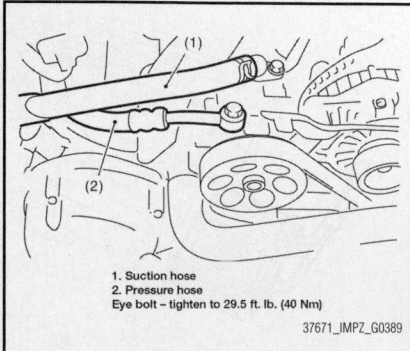

1. Suction hose
2. Pressure hose
Eye bolt – tighten to 29.5 ft. lb. (40 Nm)
37671_IMPZ_G0389

Fig. 197 Pressure hose and suction hose—non—turbo

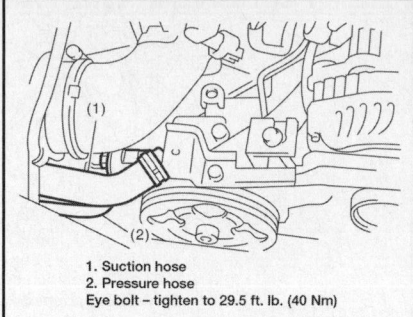

1. Suction hose
2. Pressure hose
Eye bolt – tighten to 29.5 ft. lb. (40 Nm)
37671_IMPZ_G0390

Fig. 198 Pressure hose and suction hose—turbo

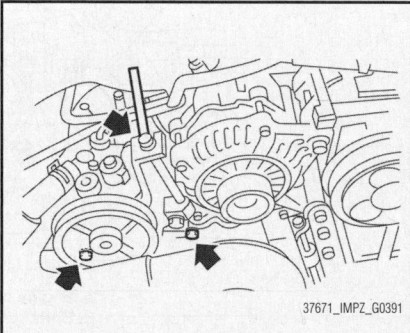

37671_IMPZ_G0391

Fig. 199 Power steering pump bracket installation bolt removal

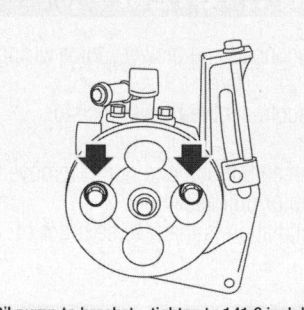

Oil pump to bracket – tighten to 141.6 inch lb. (16 Nm)
Oil pump–non-turbo – tighten to 26.6 ft. lb. (36 Nm)
Oil pump–turbo – tighten to 35.4 ft. lb. (48 Nm)
37671_IMPZ_G0392

Fig. 200 Remove the two bolts from the front side of the oil pump

11. After installing the oil pump, fill the oil pump with fluid while rotating the pulley by hand and bleed the air from the oil pump.

※ CAUTION

Always fill the oil pump with the fluid to prevent abnormal noise and seizure of the oil pump.

12. Connect the pressure hose and suction hose.

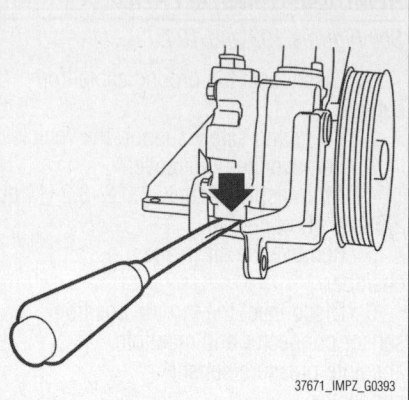

37671_IMPZ_G0393

Fig. 201 Remove the rear bolt from the oil pump. Disassemble the pump and bracket using a flat tip screwdriver.

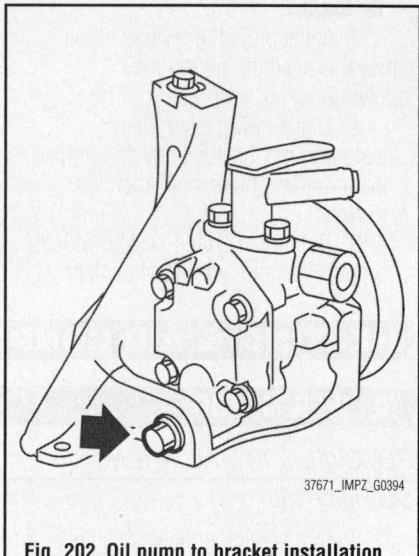

37671_IMPZ_G0394

Fig. 202 Oil pump to bracket installation

※ CAUTION

Be careful when installing; If the hose is twisted it may come into contact with other parts.

13. Connect the power steering pump switch to the connector.
14. Install the V-belts (front side belt).
15. Connect the ground cable to battery.
16. Fill with the specified power steering fluid.

※ CAUTION

Never start the engine before filling with fluid; otherwise the vane pump may become seized.

BLEEDING & FLUID FILL PROCEDURE

See Figure 203.

1. Lift up the vehicle.
2. Remove the crossmember support.
3. Remove the pipe joint in the center of gearbox, and connect the vinyl hose to the pipe and joint. Wipe fluid off while turning the steering wheel.
4. Add the specified fluid to reservoir tank at "MAX" level.
5. Continue to turn the steering wheel slowly from lock to lock until the bubbles stop appearing on oil surface while keeping the fluid at the level in the Step 4).
6. If the steering wheel is turned in a low fluid level condition, air will be sucked into the pipe. If air has entered, leave it for about half an hour and then repeat step 5) again.
7. Start the engine and let it idle.
8. Continue to turn the steering wheel slowly from lock to lock again until the bubbles stop appearing on oil surface, while keeping the fluid at the level in Step 4). Normally bubbles will stop appearing after turning the steering wheel from lock to lock three times.
9. In case bubbles do not stop appearing in the tank, leave it for about half an hour and then repeat from step 4) again.
10. Lower the vehicle, and then idle the engine.
11. Continue to turn the steering wheel from lock to lock until the bubbles stop appearing and change of the fluid level is within 3 mm (0.12 in).
12. In case the following happens, leave it about half an hour and then perform step 8) to 11) again.
   a. The fluid level changes by 3 mm (0.12 in) or more.
   b. Bubbles remain on the upper surface of the fluid.
   c. Grinding noise is generated from oil pump.
13. Check the fluid leakage after turning steering wheel from lock to lock with engine running.

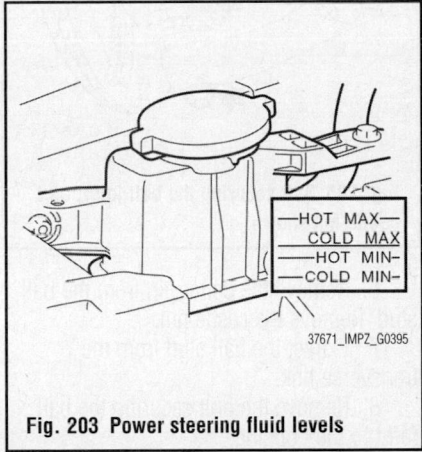

Fig. 203 Power steering fluid levels

SUSPENSION

COIL SPRING

REMOVAL & INSTALLATION

See Figure 204.

1. Before servicing the vehicle, refer to the Precautions Section.
2. Disconnect the negative battery cable.
3. Raise and support the vehicle safely.
4. Remove front strut, Refer to front MacPherson Strut.
5. Using a coil spring compressor, compress the coil spring.
6. Using a strut mount socket remove the self-locking nut.
7. Remove the strut mount, spacer and upper spring seat from strut.
8. Gradually decrease the compression force of compressor, and remove the coil spring.

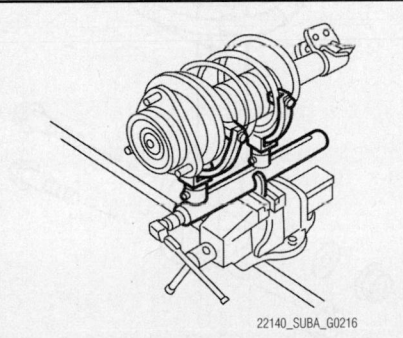

Fig. 204 Coil spring being compressed by spring compressor tool

9. Remove the dust cover and helper spring.

To install:

10. Using a coil spring compressor, compress the coil spring.
11. Set the coil spring correctly so that its end face seats well in the spring seat as shown in the figure.
12. Install the helper and dust cover to the piston rod.
13. Pull the piston rod fully upward, and install the spring seat.
14. Install spacer and the strut mount to piston rod, and tighten a new self-locking nut temporarily.
15. Using a hexagon wrench to prevent strut rod from turning, tighten the new self-locking nut to 41 ft. lbs. (55 Nm), with a strut mount socket.
16. Loosen the coil spring compressor carefully.
17. Install front strut. "Refer to front MacPherson Strut".
18. Connect the negative battery cable

CONTROL LINKS

REMOVAL & INSTALLATION

See Figure 205.

1. Before servicing the vehicle, refer to the Precautions Section.
2. Lift up the vehicle, and then remove the front wheels.
3. Remove the front undercover.
4. Remove the stabilizer control link upper and lower mounting nuts.

FRONT SUSPENSION

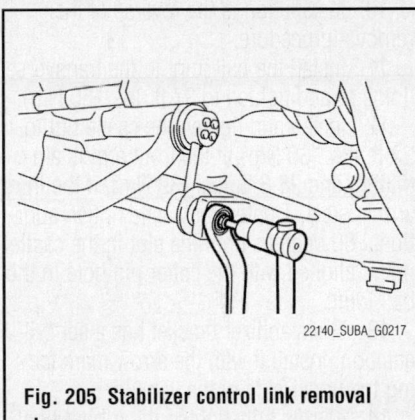

Fig. 205 Stabilizer control link removal shown

To install:

5. Install the stabilizer control link.
6. Tighten the stabilizer link upper and lower mounting nuts to 44 ft. lbs. (60 Nm).
7. Install front wheels, and lower the vehicle.

LOWER BALL JOINT

REMOVAL & INSTALLATION

See Figure 206.

1. Before servicing the vehicle, refer to the Precautions Section.
2. Disconnect the negative battery cable.
3. Raise and support the vehicle safely.
4. Remove the tire and wheel.
5. Remove the stabilizer bar brackets and bushings (both sides).

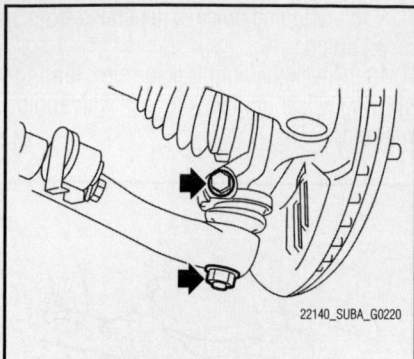

Fig. 206 Bolt securing the ball joint, and castle nut shown

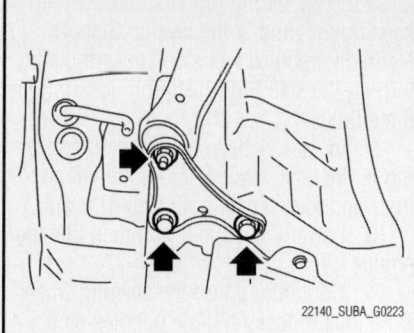

Fig. 207 Front arm support plate mounting points

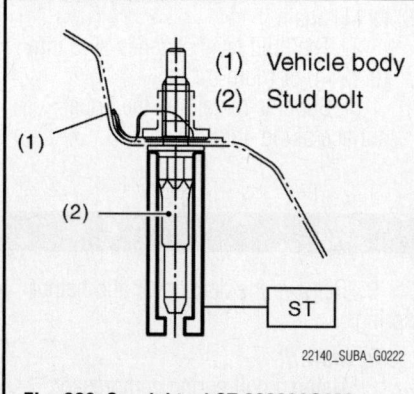

Fig. 208 Special tool ST-20299AG020

6. Remove the cotter pin from the ball stud. Remove the castle nut.

7. Extract the ball stud from the transverse link.

8. Remove the bolt securing the ball joint to the housing.

9. Extract the ball joint from the housing.

To install:

10. Installation is the reverse of the removal procedure.

11. Install the ball joint to the transverse link arm and tighten to 37 ft. lbs. (50 Nm).

12. Install and tighten the castle nut to 22 ft. lbs. (30 Nm), if the front arm is aluminum and 28.8 ft. lbs. (39 Nm), if the front arm is steel. Tighten the castle nut an additional 60 degrees until the slot in the castle nut is aligned with the cotter pin hole in the ball joint.

13. The stabilizer bracket has a set orientation. Install it with the arrow mark facing the upper side of the vehicle.

14. Always fully tighten the rubber bushings when the wheels are in full contact with the ground and the vehicle is at curb height.

15. Tighten the stabilizer bracket bolts to 18 ft. lbs. (25 Nm).

16. Check and adjust alignment, as required.

LOWER CONTROL ARM

REMOVAL & INSTALLATION

See Figures 207 and 208.

1. Before servicing the vehicle, refer to the Precautions Section.

2. Lift up the vehicle, and then remove the front wheels.

3. Remove the front crossmember support plate.

4. Remove the front stabilizer.

5. Remove the ball joint of front arm.

6. Remove the nut securing the front arm to crossmember. (Do not remove the bolt.)

7. Remove the front arm support plate.

8. Remove the bolt securing front arm to crossmember and pull the front arm out of the crossmember.

9. To remove the stud bolt, use the ST-20299AG020

❋❋ CAUTION

Do not remove the stud bolt without necessity. Always replace the parts with new parts when removed.

To install:

10. Using the ST-20299AG020, install the stud bolt. Tighten to 81 ft. lbs. (119 Nm).

11. Using new bolts and self-locking nuts, temporarily tighten the front arm to crossmember.

12. Secure the front arm to body, and then install the support plate with new bolts and self-locking nuts.

13. Tighten as follows:
- Support plate to Front arm: 65 ft. lbs. (88 Nm)
- Support plate to Body: 111 ft. lbs. (150 Nm)

14. Install the ball joint into housing, and tighten retaining bolt to 37 ft. lbs. (50 Nm).

15. Install the stabilizer.

16. Lower the vehicle from lift, and tighten the bolt which secures the front arm to crossmember with wheels in full contact with the ground and the vehicle at curb weight.

17. Tighten the stabilizer as follows:
- Stabilizer link: 33 ft. lbs. (45 Nm).
- Stabilizer bracket: 18 ft. lbs. (25 Nm).

18. Check and adjust alignment, as required.

MACPHERSON STRUT

REMOVAL & INSTALLATION

See Figure 209.

1. Disconnect the negative battery cable.

2. Raise and support the vehicle safely.

3. Remove the tire and wheel.

4. Remove the bolt retaining the brake hose to the strut.

5. Make and alignment mark on the camber adjusting bolt which secures the strut to the housing.

6. Remove the bolt retaining the ABS wheel speed sensor harness.

7. Remove the two bolts retaining the strut to the housing.

➡**While holding the head of the adjusting bolt, loosen the self locking nut.**

8. Remove the three upper strut retaining nuts.

9. Remove the strut from the vehicle.

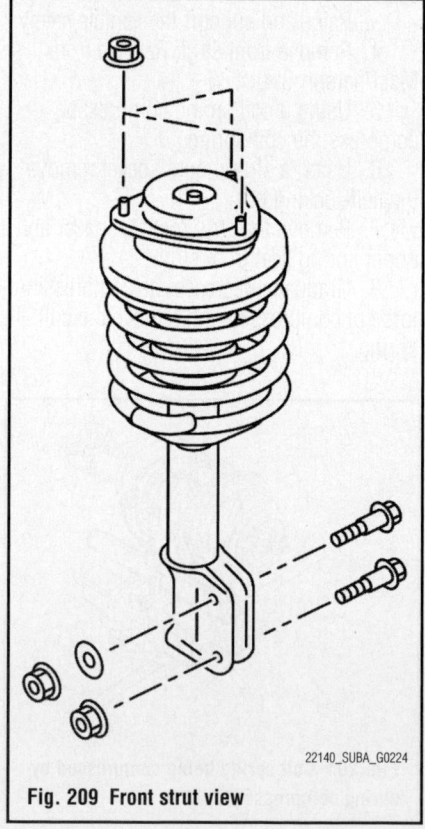

Fig. 209 Front strut view

To install:

10. Installation is the reverse of the removal procedure.

11. Tighten the upper retaining nuts to 14.5 ft. lbs. (20 Nm). Tighten the lower retaining bolts to 129 ft. lbs. (175 Nm).

12. Position the alignment mark on the camber adjusting bolt with the alignment mark on the lower side of the strut. Install using a new self locking nut.

13. Install the ABS wheel speed sensor harness, and brake hose mounting bolts. Tighten both to 24 ft. lbs. (33 Nm).

➡**While holding the head of the adjusting bolt, tighten the self locking nut.**

14. Check and adjust wheel alignment, as necessary.

OVERHAUL

See Figure 210.

1. Remove the strut from the vehicle.

2. Using a coil spring compressor tool, carefully compress the spring. Remove the self locking nut.

3. Remove the strut mount, upper spring and rubber seat from the strut.

4. Gradually decrease the compression force of the spring compressor tool. Remove the coil spring.

5. Remove the dust cover and helper spring.

6. Check for the presence of air in the damping force generating mechanism.

7. Using the spring compression tool, compress the coil spring.

➡**Be sure to properly install the coil spring.**

8. Position the coil spring so that its end face fits good into the spring seat.

9. Install the helper spring and dust cover to the piston rod.

10. Pull the piston rod fully upward, and install the rubber seat and spring seat.

11. Install the strut mount to the piston rod, and then tighten the self locking nut, temporarily. Be sure to use a new self locking nut.

12. Use a hexagon wrench to prevent the strut rod from turning. Tighten the self locking nut to 41 ft. lbs.

13. Carefully loosen the coil spring.

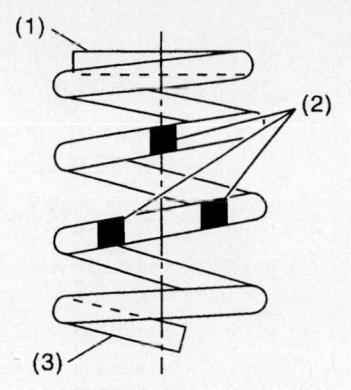

(1) Flat (top side)
(2) Identification paint
(3) Inclined (bottom side)

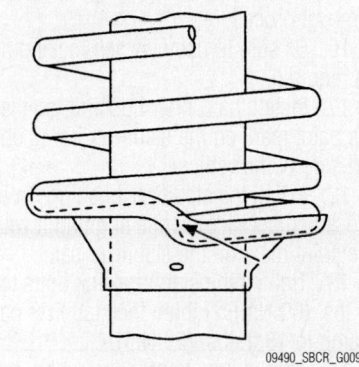

09490_SBCR_G0099

Fig. 210 Front strut spring alignment

KNUCKLES

REMOVAL & INSTALLATION

See Figures 211 through 214.

1. Before servicing the vehicle, refer to the Precautions Section.

2. Raise and support the vehicle safely.

3. Remove the tire and wheels.

4. Remove the front undercover.

5. Remove the stabilizer link.

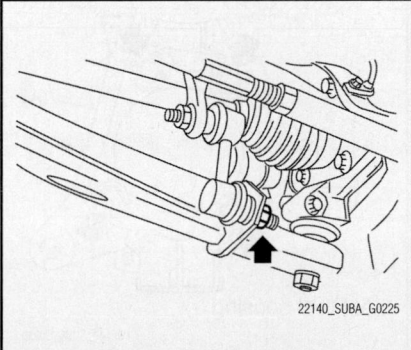

22140_SUBA_G0225

Fig. 211 Stabilizer control link and mounting nut shown

6. Remove the disc brake caliper from the front housing, and suspend it from strut using a piece of wire.

7. Remove the disc rotor from the hub.

8. Remove the cotter pin and castle nut securing the tie-rod end to the front housing knuckle arm.

9. Using a puller, remove the tie rod ball joint from knuckle arm.

10. Remove the ABS wheel speed sensor assembly and harness.

11. Remove the lower strut mounting bolts.

12. Remove the front arm ball joint from the front housing.

13. Remove the front driveshaft assembly from the steering knuckle hub.

14. After scribing an alignment mark on the camber adjusting bolt head, remove the bolts which connect the front housing and strut, and disconnect the steering knuckle from the strut.

15. Remove the front steering knuckle hub.

To install:

16. Install the front steering knuckle hub.

17. Align the alignment mark on the camber adjusting bolt head, and affix the

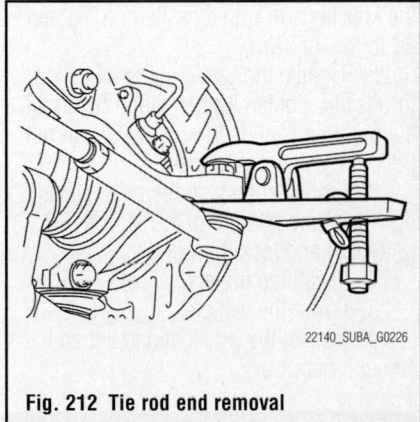

22140_SUBA_G0226

Fig. 212 Tie rod end removal

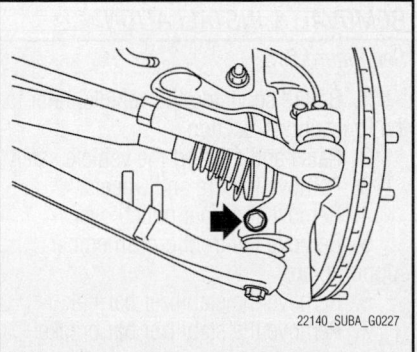

22140_SUBA_G0227

Fig. 213 Ball joint knuckle retaining bolt shown

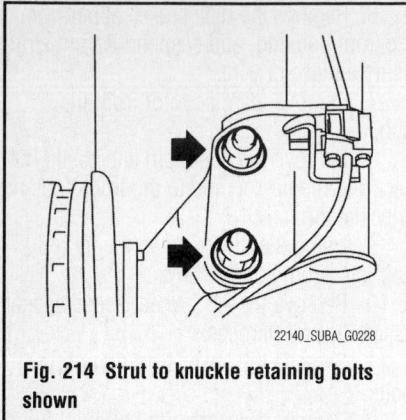

Fig. 214 Strut to knuckle retaining bolts shown

front housing and strut together using a new self—locking nut. Tighten to 129 ft. lbs. (175 Nm).

18. Install the front driveshaft.

19. Install the front arm ball joint to the front housing. Tighten mounting bolt to 37 ft. lbs. (50 Nm).

20. Install the ABS wheel speed sensor on the front housing. Tighten to 5.5 ft. lbs. (7.5 Nm).

21. Install the disc rotor to hub.

22. Install the disc brake caliper to the front housing. Tighten to 88.5 ft. lbs. (120 Nm).

23. Install the stabilizer link.

24. Connect the tie-rod end ball joint to the knuckle arm with a castle nut. Tighten to 20 ft. lbs. (27 Nm).

25. Tighten the castle nut to specified torque and tighten further within 60° until the pin hole is aligned with the slot in nut. Bend the cotter pin to lock.

26. While depressing the brake pedal, tighten a new axle nut to 177 ft. lbs. (240 Nm) and lock it securely.

27. Install the tire and wheels.

28. Lower the vehicle.

29. Inspect the wheel alignment and adjust if necessary.

STABILIZER BAR

REMOVAL & INSTALLATION

See Figure 215.

1. Before servicing the vehicle, refer to the Precautions Section.

2. Raise and support the vehicle safely.

3. Remove the tire and wheels.

4. Remove the front undercover.

5. Remove the front crossmember support plate.

6. Remove the stabilizer bar link.

7. Remove the stabilizer bar bracket bolts and bushings.

8. Remove the stabilizer bar from the vehicle.

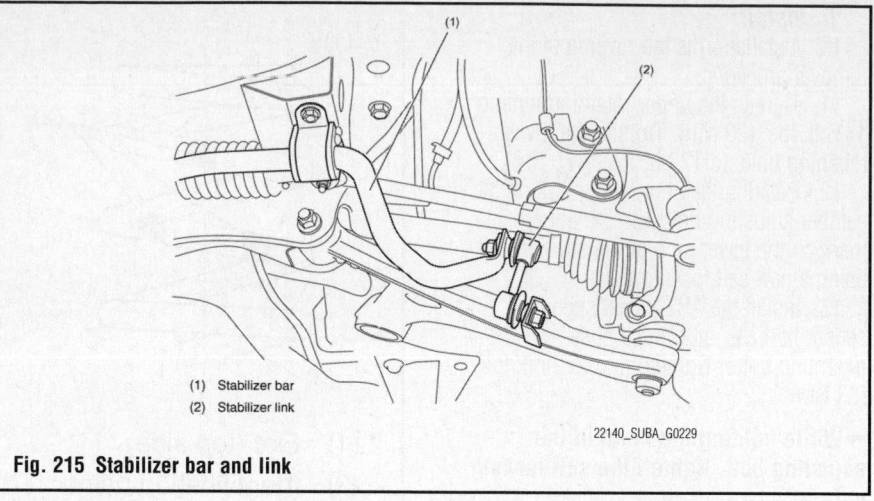

(1) Stabilizer bar
(2) Stabilizer link

Fig. 215 Stabilizer bar and link

To install:

9. Installation is the reverse of the removal procedure.

10. Be sure to use new self locking nuts, as required.

11. Install the rubber bushing, so that the paint mark on the stabilizer bar is on the left side of the vehicle.

12. Install the stabilizer bushing (front crossmember side), while aligning it with the paint mark on the stabilizer bar.

13. Tighten the stabilizer link bolts to 44 ft. lbs. (60 Nm). Tighten the stabilizer bar clamp to 18 ft. lbs. (25 Nm).

14. Always fully tighten the rubber bushings when the wheels are in full contact with the ground and the vehicle is at curb height.

WHEEL BEARINGS

REMOVAL & INSTALLATION

See Figure 216.

➡**It may be necessary to remove the front halfshaft from the vehicle.**

1. Disconnect the negative battery cable.
2. Raise and support the vehicle safely.
3. Remove the tire and wheel.

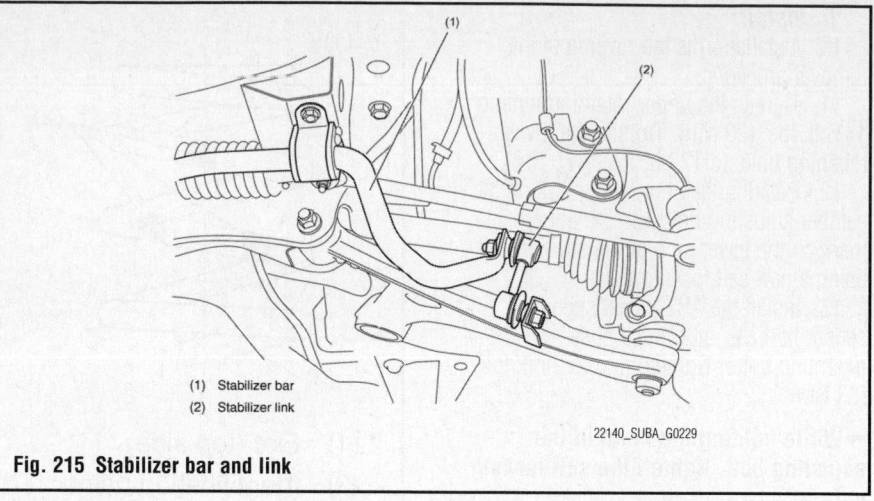

(A) Front housing

Fig. 216 Front axle housing bearing assembly retaining bolts

4. Remove the crimped section of the axle nut. Depress the brake pedal and remove the axle nut.

➡**Be sure to loosen the axle nut after removing the tire and wheel from the vehicle. Failure to do this may damage the wheel bearings.**

5. Remove the disc brake caliper from its mounting as suspend it to the side, with wire. Do not disconnect the brake line.

6. Remove the rotor.

➡**If the rotor is seized within the hub, remove the rotor by installing an 8mm bolt in the screw hole on the rotor.**

7. Remove the ABS wheel speed sensor assembly and harness.

8. Remove the four bolts from the housing.

9. Remove the front hub unit bearing. Use tools ST92647000 and ST28099PA100 if necessary.

To install:

10. Tighten the front hub unit bearing to housing bolts to 47.9 ft. lbs. (65 Nm).

11. Be sure to use a new axle nut. Tighten the axle nut, temporarily.

12. Install the rotor. Install the caliper.

13. While depressing the brake pedal, tighten the axle nut (olive color) to 162 ft. lbs. (220 Nm). Lock it securely in place.

➡**Install the tire and wheel after installation of the axle nut. Failure to do this may result in wheel bearing damage. Do not over tighten, as this too could cause wheel bearing damage.**

14. Continue the installation in the reverse order of the removal procedure.

ADJUSTMENT

The wheel bearings are of a sealed unit and are not adjustable.

SUSPENSION

REAR SUSPENSION

LATERAL LINK

REMOVAL & INSTALLATION

Front

See Figures 217 and 218.

1. Before servicing the vehicle, refer to the Precautions Section.
2. Lift up the vehicle, and then remove the rear wheels.
3. Remove the rear trailing link.
4. Remove the snap pin and nut.
5. Using a puller, remove the ball joint.
6. Scribe an alignment mark on the front lateral link adjusting bolt and rear sub frame.
7. Remove the adjusting bolt, and remove the front lateral link.

To install:

8. Install the front lateral link, and adjusting bolt. (Match alignment marks)
9. Install the remove the ball joint.

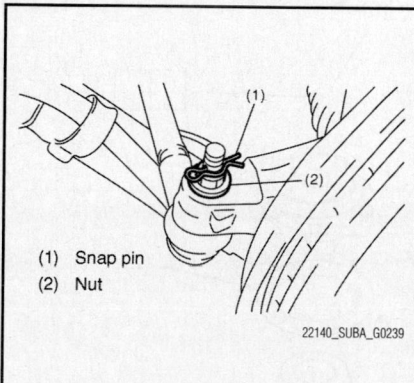

(1) Snap pin
(2) Nut

22140_SUBA_G0239

Fig. 217 Remove the snap pin and nut

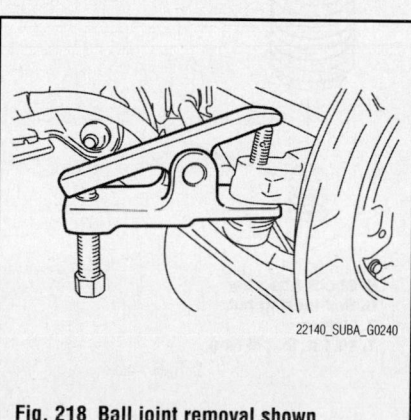

22140_SUBA_G0240

Fig. 218 Ball joint removal shown

10. Install the rear trailing link.
11. Observe the following tightening specifications:
    - Front lateral link to sub-frame 74 ft. lbs. (100 Nm).
    - Front lateral link to rear axle housing 44 ft. lbs. (60 Nm).
12. Inspect the wheel alignment and adjust if necessary.

Rear

See Figure 219.

1. Before servicing the vehicle, refer to the Precautions Section.
2. Lift up the vehicle, and then remove the rear wheels.
3. Remove the nut and disconnect the stabilizer link.
4. Remove the shock absorber lower bolt.
5. Remove the bolt, and remove rear lateral link.

To install:

6. Install rear lateral link, and bolt.
7. Install shock absorber lower bolt.
8. Connect the stabilizer link.
9. Observe the following tightening specifications:
    - Rear lateral link 89 ft. lbs. (120 Nm).
    - Shock absorber lower bolt 89 ft. lbs. (120 Nm).
    - Stabilizer link 33.2 ft. lbs. (45 Nm).
10. Inspect the wheel alignment and adjust if necessary.

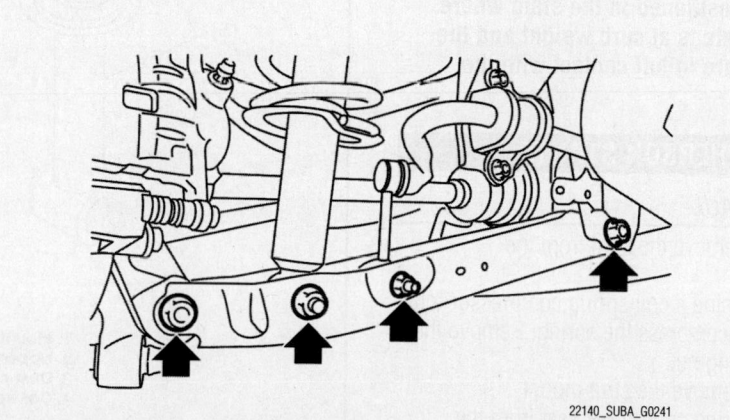

22140_SUBA_G0241

Fig. 219 Rear lateral link, and bolt locations

TRAILING LINK

REMOVAL & INSTALLATION

1. Before servicing the vehicle, refer to the Precautions Section.
2. Remove the bracket, and remove the parking brake cable from the guide.
3. Remove the trailing link.

To install:

4. Install the trailing link, and mounting nuts.
5. Tighten the trailing link nuts to 66.4 ft. lbs. (90 Nm). Be sure to use a new self-locking nut.

➡**Always tighten the bushing in the state where the vehicle is at curb weight and the wheels are in full contact with the ground.**

UPPER CONTROL ARMS

REMOVAL & INSTALLATION

See Figure 220.

1. Before servicing the vehicle, refer to the Precautions Section.
2. Separate the front exhaust pipe and rear exhaust pipe
3. Remove the rear exhaust pipe and muffler.
4. Remove the propeller shaft.
5. Remove the brake hose bracket.
6. Remove the rear disc brake caliper and suspend it from the shock absorber. (disc brake model)
7. Remove the rear parking brake cable from the parking brake assembly

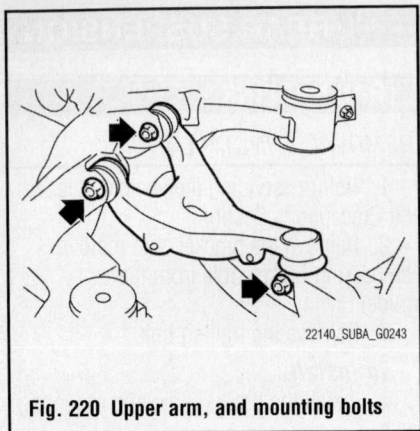

Fig. 220 Upper arm, and mounting bolts

8. Disconnect the ABS wheel speed sensor connector.

9. Disconnect the drain hose from the canister drain connector.

10. Remove the shock absorber lower bolt.

11. Support the sub frame using a jack.

12. Remove the support plate.

13. Remove the rear sub frame.

14. Remove the rear ABS wheel speed sensor bracket.

15. Remove the bolts, then remove the upper arm.

To install:

16. Install in the reverse order of removal.

17. Tighten the upper arm to rear sub frame to 111 ft. lbs. (150 Nm).

18. Tighten the upper arm to rear housing to 66.4 ft. lbs. (90 Nm).

19. Tighten the rear ABS wheel speed sensor bracket to 5.5 ft. lbs. (7.5 Nm).

20. Inspect the wheel alignment and adjust if necessary.

➡**Always tighten the bushing when the arm is positioned in the state where the vehicle is at curb weight and the wheels are in full contact with the ground.**

MACPHERSON STRUTS

OVERHAUL

1. Remove the strut from the vehicle.

2. Using a coil spring compressor tool, carefully compress the spring. Remove the self locking nut.

3. Remove the strut mount, upper spring and rubber seat from the strut.

4. Gradually decrease the compression

force of the spring compressor tool. Remove the coil spring.

5. Remove the dust cover and helper spring.

6. Check for the presence of air in the damping force generating mechanism.

7. Using the spring compression tool, compress the coil spring.

➡**Be sure to properly install the coil spring.**

8. Position the coil spring so that its end face fits good into the spring seat.

9. Install the helper spring and dust cover to the piston rod.

10. Pull the piston rod fully upward, and install the rubber seat and spring seat.

11. Install the strut mount to the piston rod, and then tighten the self locking nut, temporarily. Be sure to use a new self locking nut.

12. Use a hexagon wrench to prevent the strut rod from turning. Tighten the self locking nut to 41 ft. lbs.

13. Carefully loosen the coil spring.

SHOCK ABSORBER

REMOVAL & INSTALLATION

See Figure 221.

1. Remove the luggage floor mat—5 door.

2. Remove the trunk room mat—4 door.

3. Remove the strut cap of the quarter trim.

4. Lift up the vehicle, and then remove the rear wheels.

5. Remove the nut and disconnect the rear stabilizer link.

6. Remove the shock absorber lower bolt.

7. Disconnect the rear lateral link.

8. Remove the shock absorber mount nut.

9. Remove the shock absorber.

To install:

10. Install in the reverse order of removal, noting the following:

• Be sure to use a new self-locking nut.

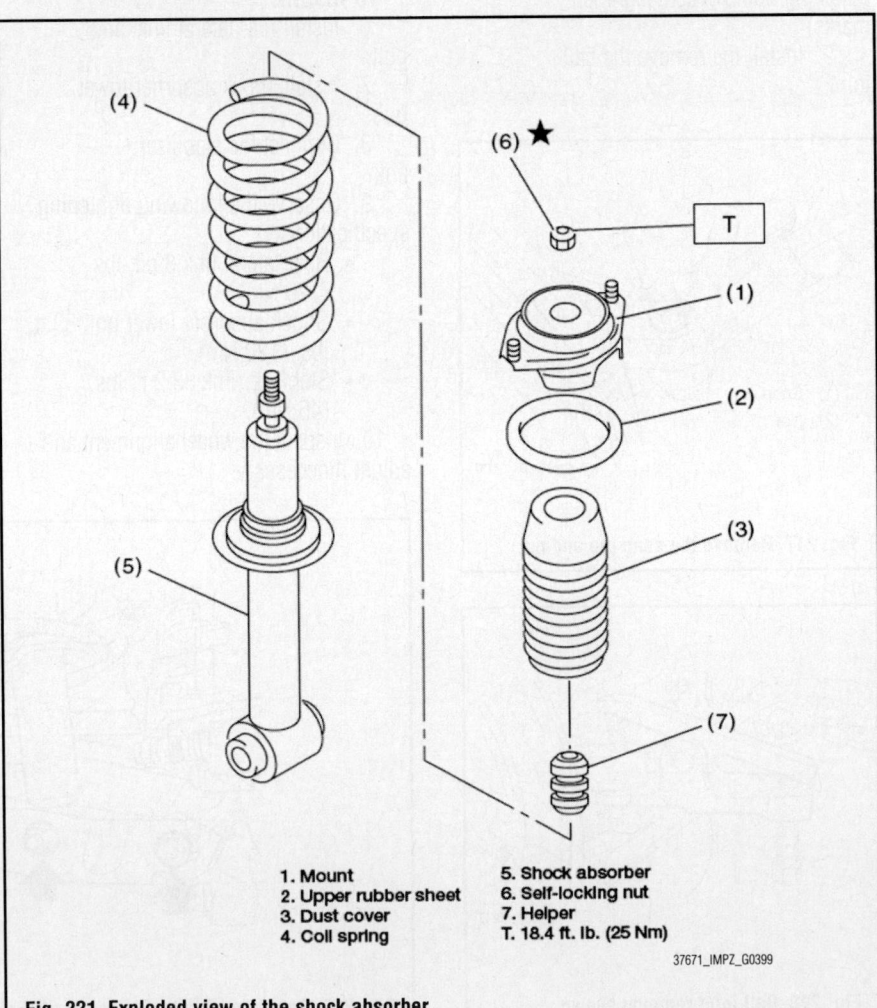

1. Mount	5. Shock absorber
2. Upper rubber sheet	6. Self-locking nut
3. Dust cover	7. Helper
4. Coil spring	T. 18.4 ft. lb. (25 Nm)

Fig. 221 Exploded view of the shock absorber

- Always tighten the bushing in the state where the vehicle is at curb weight and the wheels are in full contact with the ground.

11. Check the wheel alignment and adjust it if necessary.

STABILIZER BAR

REMOVAL & INSTALLATION

See Figure 222.

1. Loosen the rear wheel lug nuts.
2. Raise and support the vehicle safely.
3. Remove the tire and wheel.
4. Remove the bolts that secure the stabilizer link to the rear arm.
5. Remove the bolts which secure the stabilizer bar to the sub frame.
6. Remove the stabilizer bar from the vehicle.

To install:

7. Installation is the reverse of the removal procedure.
8. Be sure that the stabilizer bar and the bushings have the same identification markings and/or colors.
9. Be sure to use new bolts and nuts, as required.
10. Always fully tighten the rubber bushings when the wheels are in full contact with the ground and the vehicle is at curb height.
11. Check and adjust the wheel alignment, as necessary.

WHEEL BEARINGS

REMOVAL & INSTALLATION

See Figures 223 and 224.

1. Before servicing the vehicle, refer to the Precautions Section.
2. Disconnect the ground cable from the battery.
3. Lift up the vehicle, and then remove the rear wheels.
4. Lift the crimped section of axle nut.
5. Remove the axle nut using a socket wrench while depressing the brake pedal.

✳✳ CAUTION

Remove the wheel before loosening the axle nut. Failure to follow this rule may damage the wheel bearings.

6. Remove the disc brake caliper from the rear housing, and suspend it from the vehicle using a string.

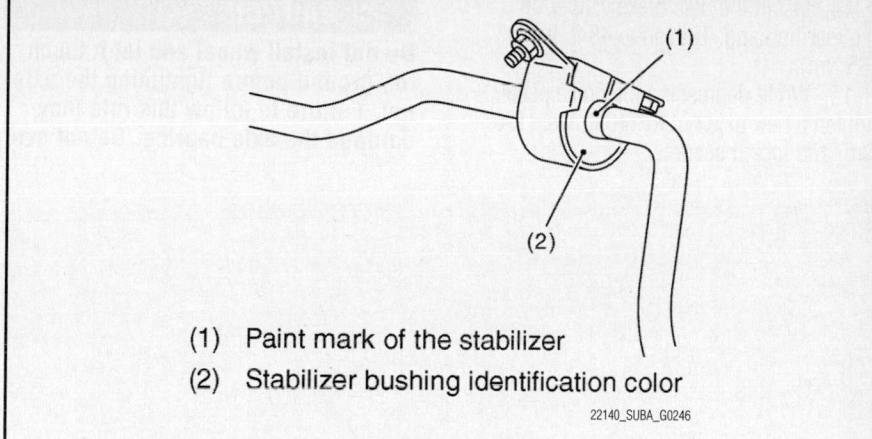

(1) Paint mark of the stabilizer
(2) Stabilizer bushing identification color

22140_SUBA_G0246

Fig. 222 Stabilizer bar and bushing identification markings

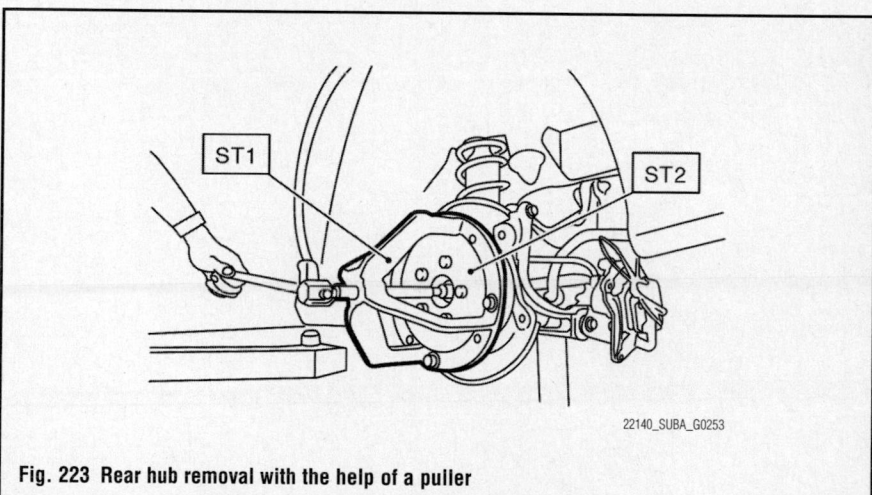

22140_SUBA_G0253

Fig. 223 Rear hub removal with the help of a puller

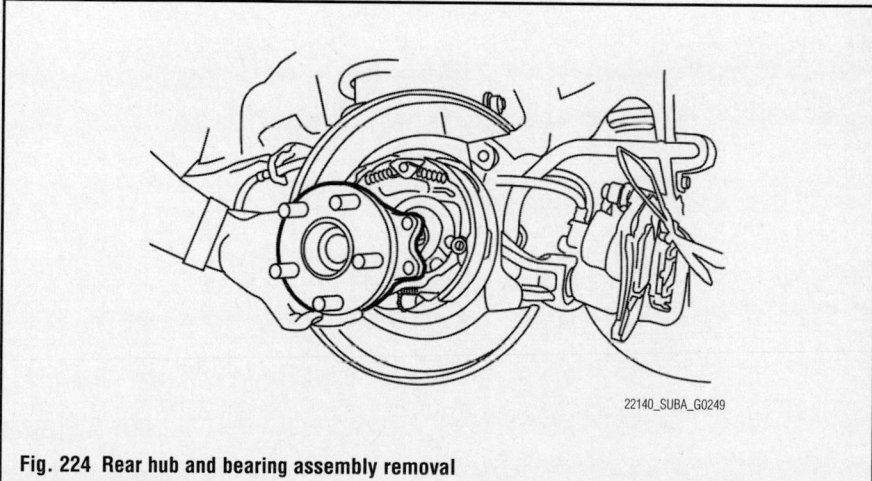

22140_SUBA_G0249

Fig. 224 Rear hub and bearing assembly removal

7. Remove the rear disc rotor.
8. Remove the four bolts from the rear housing.
9. Remove the rear hub unit bearing.

➥**If it is hard to remove, use a puller.**

To install:
10. Aligning with the mounting hole of the rear brake back plate, temporarily tighten the rear hub unit bearing to the rear housing.
11. Tighten the four bolts of the rear housing to 48 ft. lbs. (65 Nm).
12. Tighten the new axle nut temporarily.
13. Install the rear disc rotor.

14. Install the disc brake caliper on the rear housing. Tighten to 48 ft. lbs. (65 Nm).

15. While depressing the brake pedal, tighten a new axle nut to 140 ft. lbs. (190 Nm) and lock it securely.

❋❋ CAUTION

Do not install wheel and let it touch the ground before tightening the axle nut. Failure to follow this rule may damage the axle bearing. Do not over

tighten the nuts as this may damage the axle bearing.

ADJUSTMENT

The wheel bearings are not adjustable.

SPECIFICATIONS AND MAINTENANCE CHARTS

ENGINE AND VEHICLE IDENTIFICATION CHART

		Engine Code							Model Year	
Code ①	Liters (cc)	Cu. In.	Cyl.	Fuel Sys.	Type	Eng. Mfg.			Code ②	Year
6	2.5 (2457)	150	4	MFI	SOHC	Subaru			9	2009
7	2.5 (2457)	150	4	MFI	DOHC	Subaru			A	2010
8	3.0 (3000)	185	6	MFI	DOHC	Subaru				
9	3.6 (3630)	221.4	6	MFI	DOHC	Subaru				

MFI: Multiport Fuel Injection

SOHC: Single Overhead Camshaft

DOHC: Double Overhead Camshaft

① 6th digit of the VIN

② 10th digit of the VIN

37671_LEGC_C0001

GENERAL ENGINE SPECIFICATIONS

Year	Model	Engine Disp. Liters	Engine SOHC/DOHC Turbo/Non-Turbo	Engine ID/VIN	Net Horsepower @ rpm	Net Torque @ rpm (ft. lbs.)	Bore x Stroke (in.)	Compression Ratio	Oil Pressure @ rpm
2009	Legacy	2.5	SOHC Non-turbo	6	170@6000	170@4400	3.92x3.11	10.0:1	14@600
	Legacy	2.5	DOHC Turbo	6	243@6000	241@3600	3.92x3.11	8.4:1	14@600
	Legacy	3.0	DOHC Non-turbo	8	245@6600	215@4200	3.51x3.15	10.7:1	20@600
2010	Legacy	2.5	SOHC Non-turbo	6	170@5600	170@4000	3.92x3.11	10.0:1	14@600
	Legacy	2.5	DOHC Turbo	7	265@5600	258@2000-5200	3.92x3.11	10.0:1	14@600
	Legacy	3.6	DOHC Non-turbo	9	256@6000	247@4400	3.62x3.58	10.5:1	15@700

37671_LEGC_C0002

ENGINE TUNE-UP SPECIFICATIONS

Year	Model	Engine Displacement Liters	Spark Plug Gap (in.)	Ignition Timing (deg.) ①		Fuel Pump (psi)	Idle Speed (rpm)		Valve Clearance ②	
				MT	AT		MT	AT	In.	Ex.
2009	Legacy	2.5	0.028-0.031	10 ③	15 ⑤	49-50	650-750	650-750	0.0079-0.0095	0.0098-0.0114
	Legacy	2.5	0.028-0.031	12 ③	17 ③	33-38	650-750	650-750	0.0078-0.0087	0.0138-0.0146
	Legacy	3.0	0.028-0.031	15 ④	15 ④	49-50.5	650-750	600-800	0.0079-0.0087 ⑧	0.0138-0.0158
2010	Legacy	2.5	0.028-0.031	10 ④	15 ⑦	49-50	650-750	675-775	0.0079-0.0095	0.0098-0.0114
	Legacy	2.5	0.028-0.031	15 ⑥	15 ⑥	40-44	N/A	700-800	0.0079-0.0083 ⑨	0.0138-0.0146
	Legacy	3.6	0.028-0.031	15 ⑤	15 ⑤	49-50	N/A	700-800	0.0079-0.0087 ⑧	0.0138-0.0158

Note: The Vehicle Emission Control Information label often reflects specification changes made during production. The lable figures mudst be used if they differ from those in this chart.

① Before Top Dead Center
② Measured with engine cold
③ Plus or minus (10 deg.) at 750 rpm
④ Plus or minus (8 deg.) at 650 rpm
⑤ Plus or minus (8 deg) at 700 rpm
⑥ Plus or minus (10 deg.) at 700 rpm
⑦ Plus or minus (10 deg) at 675 rpm
⑧ Minus 0.0024
⑨ Minus 0.0012

37671_LEGC_C0003

CAPACITIES

Year	Model	Engine Displacement Liters	Engine Oil with Filter (qts.)	Transmission (pts.)		Drive Axle		Fuel Tank (gal.)	Cooling System (qts.)
				5-Spd	Auto.	Front ① (pts.)	Rear (pts.)		
2009	Legacy	2.5	4.4	NA	20.0	2.6	1.6	16.9	6.7
	Legacy	2.5	4.3	NA	20.0	3.0	1.6	16.9	7.6
	Legacy	3.0	6	NA	20.0	3.0	1.6	16.9	7.6
2010	Legacy	2.5	4.4	NA	26.0	3.0	1.6	18.5	6.8
	Legacy	2.5	4.5	NA	26.0	—	1.6	18.5	6.9
	Legacy	3.6	6.9	NA	26.0	3.0	1.6	18.5	6.9

Note: All capacities are approximate. Add fluid gradually and check to be sure a proper fluid level is obtained.

NA: Not Applicable
① Overhaul specification

37671_LEGC_C0004

FLUID SPECIFICATIONS

Year	Model	Engine Displacement Liters	Engine ID/VIN	Engine Oil	Auto. Trans. ①	Drive Axle	Power Steering Fluid	Brake Master Cylinder	Engine Coolant
2009	Legacy	2.5	6	5W-30	Subaru ATF HP	75W-90	Dexron III	DOT 3	Subaru Super Coolant
		3.0	8	5W-30	Subaru ATF HP	75W-90	Dexron III	DOT 3	Subaru SuperCoolant
2010	Legacy	2.5	6	5W-30	Subaru ATF HP	75W-90	Dexron III	DOT 3	Subaru Super Coolant
		2.5	7	5W-30	Subaru ATF HP	75W-90	Dexron III	DOT 3	Subaru Super Coolant
		3.6	9	5W-30	Subaru ATF HP	75W-90	Dexron III	DOT 3	Subaru Super Coolant

DOT: Department Of Transpotation

① Subaru CVT for Lineartronic

37671_LEGC_C0005

VALVE SPECIFICATIONS

Year	Engine Displacement Liters	Seat Angle (deg.)	Face Angle (deg.)	Spring Test Pressure (lbs. @ in.)	Spring Installed Height (in.)	Stem-to-Guide Clearance (in.) Intake	Exhaust	Stem Diameter (in.) Intake	Exhaust
2009	2.5	90	NA	NA	①	0.0012-0.0022	0.0016-0.0026	0.2344-0.2350	0.2341-0.2346
	3.0	90	NA	NA	②	0.0012-0.0022	0.0016-0.0026	002148-0.2154	0.2144-0.2150
2010	2.5	90	NA	NA	③	0.0012-0.0022	0.0016-0.0026	0.2344-0.2350	0.2341-0.2346
	3.6	90	NA	NA	④	0.0012-0.0022	0.0016-0.0026	002148-0.2154	0.2148-0.2150

NA: Not Available

① Free length: 1.8913 in.

② Free length: 1.5571 in. (Inner), 1.6213 in. (Outer)

③ Free length: 1.8630 in.

④ Free length: 1.9315

37671_LEGC_C0007

CAMSHAFT AND BEARING SPECIFICATIONS CHART

All measurements are given in inches.

Year	Engine Displacement Liters	Engine ID/VIN	Journal Dia.	Brg. Oil Clearance	Shaft End-play	Runout	Journal Bore	Lobe Height Intake	Exhaust
2009	2.5 SOHC	6	1.2570-1.2577	0.0022-0.0035	0.0012-0.0035	0.0024	1.2598-1.2605	1.5778-1.5817	1.5468-1.5507
	2.5 DOHC	6	1.1790-1.1796	0.0015-0.0028	0.0027-0.0047	0.0024	NA	1.8330-1.8370	1.8410-1.8440
	3.0 DOHC	8	1.0215-1.0222	0.0015-0.0028-①	NA	0.0024	NA	1.6571-1.6610	1.5925-1.5965
2010	2.5 SOHC	6	1.2570-1.2577	0.0022-0.0035	0.0012-0.0035	0.0024	1.2598-1.2605	1.5778-1.5817	1.5468-1.5507
	2.5 DOHC	7	1.1790-1.1796	0.0015-0.0028	0.0027-0.0047	0.0024	NA	1.833-1.8370	1.805-1.8090
	3.6 DOHC	9	1.0215-1.0222	0.0015-0.0028	NA	0.0024	NA	1.8071-1.8110	1.7579-1.7618

NA: Not Available

① Intake Journals 4 and 5: 0.0010 - 0.0022 in.

All Others: 0.0010 - 0.0024 in.

37671_LEGC_C0006

CRANKSHAFT AND CONNECTING ROD SPECIFICATIONS

All measurements are given in inches.

Year	Engine Displacement Liters	Crankshaft Main Brg. Journal Dia.	Main Brg. Oil Clearance	Shaft End-play	Thrust on No.	Connecting Rod Journal Diameter	Oil Clearance	Side Clearance
2009	2.5	2.3619-2.3625	①	0.0012-0.0098	3	1.8891-1.8898	0.0004-0.0012	0.0028-0.0130
	3.0	2.5194-2.5200	②	0.0012-0.0098	3	1.8891-1.8898	0.0006-0.0017	0.0028-0.0130
2010	2.5	2.3619-2.3625	①	0.0012-0.0098	3	1.8891-1.8898	0.0006-0.0017	0.0028-0.0130
	3.6	2.3619-2.3625	②	0.0012-0.0098	3	1.8891-1.8898	0.0009-0.0020	0.0028-0.0130

① Journals 1 and 5: 0.0001 - 0.0016 in.

Journals 2 and 4: 0.0004 - 0.0014 in.

Journal 3: 0.0004 - 0.0014 in.

② Journals 1 and 5: 0.0001 - 0.0016 in.

Journals 2 and 4: 0.0004 - 0.0018 in.

Journal 3: 0.0004 - 0.0016 in.

37671_LEGC_C0008

PISTON AND RING SPECIFICATIONS
All measurements are given in inches.

Year	Engine Displacement Liters	Piston Clearance	Ring Gap			Ring Side Clearance		
			Top Compression	Bottom Compression	Oil Control	Top Compression	Bottom Compression	Oil Control
2009	2.5	0.0004-0.0012	0.0079-0.0138	0.0146-0.0250	0.0079-0.0197	0.0016-0.0031	0.0012-0.0028	NA
	3.0	0.0004-0.0012	0.0079-0.0138	0.0138-0.0197	0.0079-0.0236	0.0016-0.0031	0.0012-0.0028	NA
2010	2.5	0.0004-0.0012	0.0079-0.0138	0.0146-0.0250	0.0079-0.0197	0.0016-0.0031	0.0012-0.0028	NA
	3.6	0.0004-0.0012	0.0079-0.0138	0.0138-0.0197	0.0079-0.0236	0.0016-0.0031	0.0012-0.0028	NA

NA: Not Available

37671_LEGC_C0009

TORQUE SPECIFICATIONS
All readings in ft. lbs.

Year	Engine Displacement Liters	Cylinder Head Bolts	Main ① Bearing Bolts	Rod Bearing Bolts	Crankshaft Damper Bolts	Flywheel Bolts	Manifold		Spark Plugs	Drain Plug
							Intake	Exhaust		
2009	2.5	②	③	38	133	51 - 55	18	26	15	33
	3.0	④	14	39	131	60	18	26	15	33
2010	2.5	②	③	38	133	51 - 55	18	26	15	33
	3.6	④	14	39	131	60	18	26	15	33

① Engine block connecting bolts

② Step 1: Tighten all bolts to 22 ft. lbs.

Step 2: Tighten all bolts to 51 ft. lbs.

Step 3: Loosen all botls 180 degrees.

Step 4: Repeat Step 3.

Step 5: Tighten bolts 1 and 2 to 25 ft. lbs.

Step 6: Tighten bolts 3, 4, 5 and 6 to 11 ft. lbs.

Step 7: Tighten all bolts 80 to 90 degrees.

Step 8: Repeat Step 7. Do not exceed 180 degrees total tightening.

③ Split engine case connecting bolts:

Short bolts: 17-20 ft. lbs.

Long bolts: 33-37 ft. lbs.

Smaller short bolts (if used) 5 ft. lbs.

④ Step 1: Tighten all bolts to 14 ft. lbs.

Step 2: Tighten all bolts to 37 ft. lbs.

Step 3: Loosen all bolts 180 degrees the an additional 180 degrees in two steps in the reverse order of tightening sequence.

Step 4: Tighten all bolts to 18 ft. lbs.

Step 5: Tighten all bolts to 18 ft. lbs.

Step 6: Tighten bolts 90 degrees

Step 7: Tighten bolts 1, 2, 3 and 4 90 degrees.

Step 8: Tighten bolts 5, 6, 7 and 8 45 degrees.

37671_LEGC_C0010

NOTE:
Remove oil on the mating surface of bearing and cylinder block before installation. Apply engine oil to crankshaft pins.

1) Position the crankshaft and O-rings on the #1 and #3 cylinder block.

2) Apply liquid gasket to the mating surface of #1 and #3 cylinder block, and position #2 and #4 cylinder block.

Liquid gasket:
Three bond 1215 (Part No. 004403007) or equivalent

NOTE:
Do not allow liquid gasket to jut into O-ring grooves, oil passages, bearing grooves, etc.

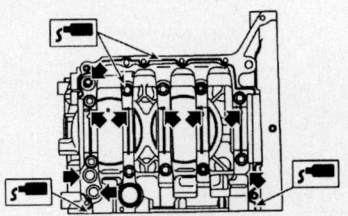

3) Apply a coat of engine oil to the washer and bolt thread.

4) Tighten the 10 mm cylinder block connecting bolts on LH side (A — D) in alphabetical sequence.

Tightening torque:
9.75 N·m (1.0 kgf-m, 7.2 ft-lb)

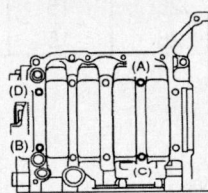

5) Tighten the 10 mm cylinder block connecting bolts on RH side (E — J) in alphabetical sequence.

Tightening torque:
9.75 N·m (1.0 kgf-m, 7.2 ft-lb)

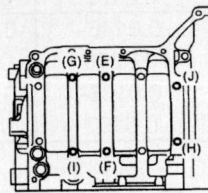

6) Further tighten the LH side bolts (A — D) in alphabetical sequence.

Tightening torque:
18 N·m (1.8 kgf-m, 13 ft-lb)

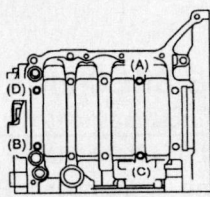

7) Further tighten the RH side bolts (E — J) in alphabetical sequence.

Tightening torque:
18 N·m (1.8 kgf-m, 13 ft-lb)

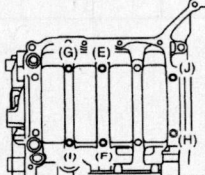

8) Further tighten the LH side bolts (A — D) in alphabetical sequence.
• (A), (C): Angle tightening

Tightening angle:
90°
• (B), (D): Torque tightening

Tightening torque:
40 N·m (4.1 kgf-m, 29.6 ft-lb)

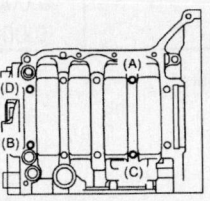

9) Tighten the RH side bolts (E — J) 90° further in alphabetical sequence.

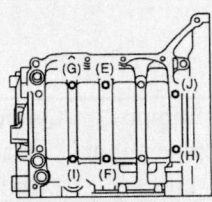

10) Tighten the 8 mm and 6 mm cylinder block connecting bolts on LH side (A — H) in alphabetical sequence.

Tightening torque:
(A) — (G): 25 N·m (2.5 kgf-m, 18.1 ft-lb)
(H): 6.4 N·m (0.65 kgf-m, 4.7 ft-lb)

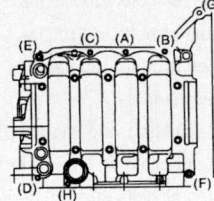

09490_SBCR_G0004

Fig. 1 Main bearing torque sequence—2.5L SOHC engine

1) Remove oil on the mating surface of bearing and cylinder block before installation. Apply engine oil to crankshaft pins.

2) Position the crankshaft and O-rings on the #1 and #3 cylinder block.

3) Apply liquid gasket to the mating surface of #1 and #3 cylinder blocks, and position cylinder block #2 and #4.

NOTE:

Do not allow liquid gasket to run over to O-ring grooves, oil passages, bearing grooves, etc.

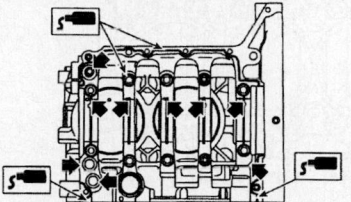

4) Apply a coat of engine oil to the washer and bolt thread.

5) Tighten the 10 mm cylinder block connecting bolts on LH side (A — D) in alphabetical sequence.

Tightening torque:
10 N·m (1.0 kgf-m, 7.2 ft-lb)

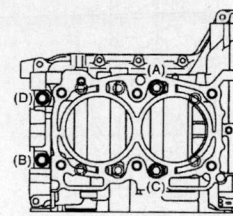

6) Tighten the 10 mm cylinder block connecting bolts on RH side (E — J) in alphabetical sequence.

Tightening torque:
10 N·m (1.0 kgf-m, 7.2 ft-lb)

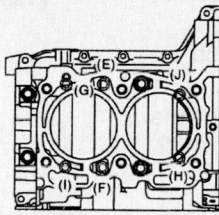

7) Further tighten the LH side bolts (A — D) in alphabetical sequence.

Tightening torque:
18 N·m (1.8 kgf-m, 13.0 ft-lb)

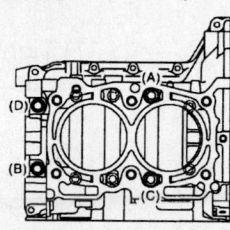

8) Further tighten the RH side bolts (E — J) in alphabetical sequence.

Tightening torque:
18 N·m (1.8 kgf-m, 13.0 ft-lb)

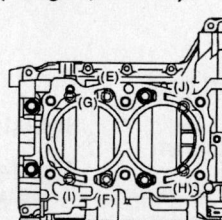

9) Further tighten the LH side bolts (A — D) in alphabetical sequence.

(A), (C): 90°
(B), (D): 40 N·m (4.1 kgf-m, 29.5 ft-lb)

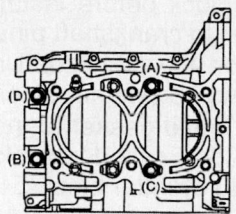

10) Tighten the RH side bolts (E — J) 90° further in alphabetical sequence.

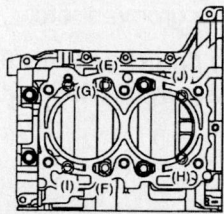

11) Tighten the 8 mm and 6 mm cylinder block connecting bolts on LH side (A — H) in alphabetical sequence.

Tightening torque:
(A) — (G): 25 N·m (2.5 kgf-m, 18.1 ft-lb)
(H): 6.4 N·m (0.65 kgf-m, 4.7 ft-lb)

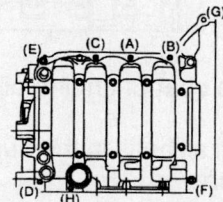

09490_SBCR_G0007

Fig. 2 Main bearing torque sequence—2.5L DOHC engine

1) After setting the cylinder block to ST, install the crankshaft bearing.

NOTE:
Remove oil on the mating surface of bearing and cylinder block before installation. Apply a coat of engine oil to crankshaft pins.

2) Position the crankshaft and connecting rod on #2, #4 and #6 cylinder block.

3) Apply liquid gasket to the mating surface of the #1, #3 and #5 cylinder blocks, and position it on #2, #4 and #6 cylinder blocks.

NOTE:
Do not allow liquid gasket to run over to oil passages, bearing grooves, etc.

Applying liquid gasket diameter:
1.0±0.2 mm (0.039±0.008 in)

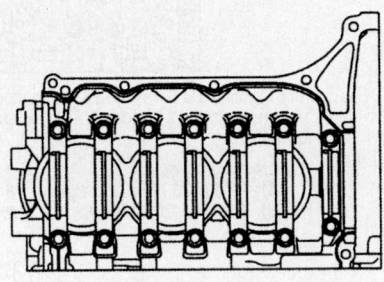

4) Apply a coat of engine oil to the washer and bolt thread.

5) Tighten all bolts in the numerical order as shown in the figure.

Tightening torque:
(1) — (11), (13): 25 N·m (2.5 kgf-m, 18 ft-lb)
(12), (14): 20 N·m (2.0 kgf-m, 14 ft-lb)

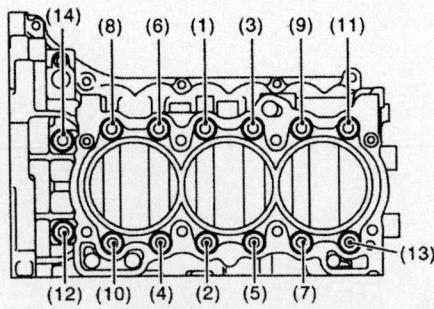

6) Retighten all bolts in the numerical order as shown in the figure.

Tightening torque:
(1) — (11), (13): 25 N·m (2.5 kgf-m, 18.4 ft-lb)
(12), (14): 20 N·m (2.0 kgf-m, 14 ft-lb)

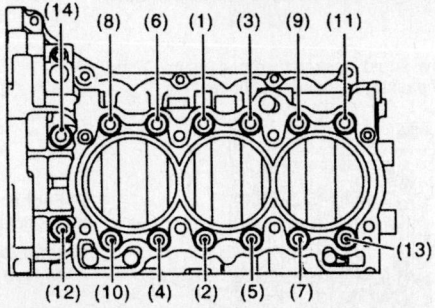

7) Tighten all bolts 90° — 110° in the numerical order as shown in the figure.

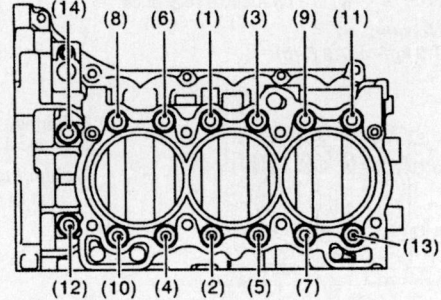

8) Install the upper bolt to cylinder block.

Tightening torque:
25 N·m (2.5 kgf-m, 18 ft-lb)

NOTE:
Remove the liquid gasket which is running over to sealing surface between cylinder block and rear chain cover, cylinder block and oil pan upper, after tightening the bolts which combine the cylinder block.

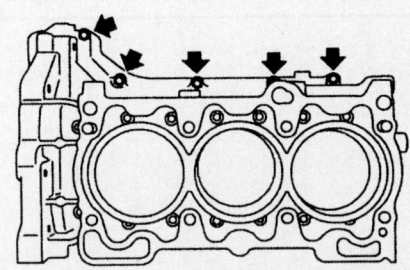

09490_SBCR_G0009

Fig. 3 Main bearing torque sequence—3.0L & 3.6L engine

WHEEL ALIGNMENT

Year	Model		Caster Range (+/-Deg.)	Caster Preferred Setting (Deg.)	Camber Range (+/-Deg.)	Camber Preferred Setting (Deg.)	Toe-in (in.)
2009	Legacy Sedan	F	0.75	+3.05	0.50	-0.05	0+/-0.12
		R	—		0.75	-0.50	0+/-0.12
	Legacy Wagon	F	0.75	+2.05	0.50	-0.05	0+/-0.12
		R	—		0.75	-0.20	0+/-0.12
2010	Legacy Sedan	F	0.75	+3.05	0.50	-0.05	0+/-0.12
		R	—		0.75	-0.50	0+/-0.12
	Legacy Wagon	F	0.75	+2.05	0.50	-0.05	0+/-0.12
		R	—		0.75	-0.20	0+/-0.12

37671_LEGC_C0011

TIRE, WHEEL AND BALL JOINT SPECIFICATIONS

Year	Model	OEM Tires Standard	OEM Tires Optional	Tire Pressures (psi) Front	Tire Pressures (psi) Rear	Wheel Size	Ball Joint Inspection	Lug Nut (ft. lbs.)
2009	Legacy	P205/55R16 P205/50R17 215/45R17 P215/45R18	T135/70 D17 T135/80R16	35	35	6.5-JJ 7-JJ	0.012 in.	86
2010	Legacy	P205/60R16 P215/50R17 P225/50R17 P225/45R18 P225/60R17 P215/70R16	T135/70 D17 T135/80R16	35	35	6.5-JJ 7-JJ 7.5-JJ	0.012 in.	86

OEM: Original Equipment Manufacturer

PSI: Pounds Per Square Inch

37671_LEGC_C0012

BRAKE SPECIFICATIONS

All measurements in inches unless noted

Year	Model		Brake Disc Original Thickness	Brake Disc Minimum Thickness	Brake Disc Maximum Runout	Brake Drum Diameter Original Inside Diameter	Brake Drum Diameter Max. Wear Limit	Brake Drum Diameter Maximum Machine Diameter	Minimum Lining Thickness Front	Minimum Lining Thickness Rear	Brake Caliper Bracket Bolts (ft. lbs.)	Brake Caliper Mounting Bolts (ft. lbs.)
2009	Legacy	F	0.940 ①	0.870 ②	0.003	NA	NA	NA	0.059	—	89	20
		R	0.390 ③	0.335 ④	0.003	NA	NA	NA	—	0.059	47	27 ⑤
2010	Legacy	F	0.940 ①	0.870 ②	0.002	NA	NA	NA	0.059	—	89	20
		R	0.390 ③	0.335 ④	0.002	NA	NA	NA	—	0.059	47	27 ⑤

NA: Not Available

① 17 inch type rotors original thickness: 1.18

② 17 inch type rotors minimum thickness: 1.10

③ Vented type rotors original thickness: 0.71

④ Vented type rotors minimum thickness: 0.63

⑤ Solid type rotors tighten to 20 ft. lbs.

37671_LEGC_C0013

SCHEDULED MAINTENANCE INTERVALS
SUBARU—LEGACY

TO BE SERVICED	TYPE OF SERVICE	VEHICLE MILEAGE INTERVAL (x1000)													
		3	7.5	15	22.5	30	37.5	45	52.5	60	67.5	75	82.5	90	97.5
Engine oil & filter ① ②	R	✓	✓	✓	✓	✓	✓	✓	✓	✓	✓	✓	✓	✓	✓
Brake lines	S/I			✓		✓		✓		✓		✓		✓	
Disc brake pads & discs, front & rear axle boots & axle shaft joint portions	S/I			✓		✓		✓		✓		✓		✓	
Parking brake	S/I			✓		✓		✓		✓		✓		✓	
Steering & suspension	S/I			✓		✓		✓		✓		✓		✓	
Air filter element	R					✓				✓				✓	
Engine coolant	R					✓				✓				✓	
Fuel filter	R					✓				✓				✓	
Spark plugs	R					✓				✓					
Automatic transmission fluid & filter	S/I					✓				✓				✓	
Brake fluid	R					✓				✓				✓	
Camshaft drive belt ③	S/I					✓				✓				✓	
Coolant level, hoses & clamps	S/I					✓				✓				✓	
Drive belts	S/I					✓				✓				✓	
Fuel system, hoses & connections	S/I					✓				✓				✓	
Transmission and/or differential gear fluid	S/I					✓								✓	
Front & rear wheel bearing	S/I									✓					

R: Replace   S/I: Service or Inspect

① 3.0L & 3.6L engine change every 3000 miles.

② 2.5L turbo models change every 3875 miles or 3.75 months.

③ Replace camshaft drive belt at 105,000 miles or 105 months

FREQUENT OPERATION MAINTENANCE (SEVERE SERVICE)

If a vehicle is operated under any of the following conditions it is considered severe service:

- Extremely dusty areas.

- 50% or more of the vehicle operation is in 32°C (90°F) or higher temperatures, or constant operation in temperatures below 0°C (32°F).

- Prolonged idling (vehicle operation in stop and go traffic).

- Frequent short running periods (engine does not warm to normal operating temperatures).

- Police, taxi, delivery usage or trailer towing usage.

Oil & oil filter change: change every 3750 miles.

Clutch & hill holder system: service or inspect every 7500 miles.

Disc brake pads & discs, front & rear axle boots & axle shaft joint portions: service or inspect every 7500 miles.

Steering & suspension: service or inspect every 7500 miles.

Air filter element: service or inspect every 15,000 miles.

Automatic transmission fluid: service or inspect every 15,000 miles.

Brake linings & drums: service or inspect every 15,000 miles.

Coolant level, hoses & clamps: service or inspect every 15,000 miles.

Drive belts: service or inspect every 15,000 miles.

Transmission/differential gear oil service or inspect every 15,000 miles.

Front & rear wheel bearing repack: service or inspect every 30,000 miles.

37671_LEGC_C0014

BRAKES · INFORMATION AND PRECAUTIONS

ANTI-LOCK SYSTEMS

• Certain components within the ABS system are not intended to be serviced or repaired individually.

• Do not use rubber hoses or other parts not specifically specified for and ABS system. When using repair kits, replace all parts included in the kit. Partial or incorrect repair may lead to functional problems and require the replacement of components.

• Lubricate rubber parts with clean, fresh brake fluid to ease assembly. Do not use shop air to clean parts; damage to rubber components may result.

• Use only DOT 3 brake fluid from an unopened container.

• If any hydraulic component or line is removed or replaced, it may be necessary to bleed the entire system.

• A clean repair area is essential. Always clean the reservoir and cap thoroughly before removing the cap. The slightest amount of dirt in the fluid may plug an orifice and impair the system function. Perform repairs after components have been thoroughly cleaned; use only denatured alcohol to clean components. Do not allow ABS components to come into contact with any substance containing mineral oil; this includes used shop rags.

• The Anti-Lock control unit is a microprocessor similar to other computer units in the vehicle. Ensure that the ignition switch is **OFF** before removing or installing controller harnesses. Avoid static electricity discharge at or near the controller.

• If any arc welding is to be done on the vehicle, the control unit should be unplugged before welding operations begin.

DISC AND DRUM SYSTEMS

> ✳✳ **CAUTION**
>
> **Dust and dirt accumulating on brake parts during normal use may contain asbestos fibers from production or aftermarket brake linings. Breathing excessive concentrations of asbestos fibers can cause serious bodily harm. Exercise care when servicing brake parts. Do not sand or grind brake lining unless equipment used is designed to contain the dust residue. Do not clean brake parts with compressed air or by dry brushing. Cleaning should be done by dampening the brake components with a fine mist of water, then wiping the brake components clean with a dampened cloth. Dispose of cloth and all residue containing asbestos fibers in an impermeable container with the appropriate label. Follow practices prescribed by the Occupational Safety and Health Administration (OSHA) and the Environmental Protection Agency (EPA) for the handling, processing, and disposing of dust or debris that may contain asbestos fibers.**

BRAKES · BLEEDING THE BRAKE SYSTEM

BLEEDING PROCEDURE

BLEEDING PROCEDURE

See Figures 4 through 6.

1. Raise the vehicle and safely support it.

2. Remove both front and rear wheels.

3. Draw out the brake fluid from master cylinder with syringe.

4. Refill reservoir tank with recommended brake fluid.

5. Install one end of a vinyl tube onto the air bleeder and insert the other end of the tube into a container to collect the brake fluid.

6. Instruct your assistant to depress the brake pedal slowly two or three times and then hold it depressed.

7. Loosen bleeder screw approximately 1/4 turn until a small amount of brake fluid drains into container, and then quickly tighten screw.

8. Repeat again from the two former procedures above until there are no air bubbles in drained brake fluid and new fluid flows through vinyl tube.

➡**Add brake fluid as necessary while performing the air bleed operation, in order to prevent the tank from running short of brake fluid.**

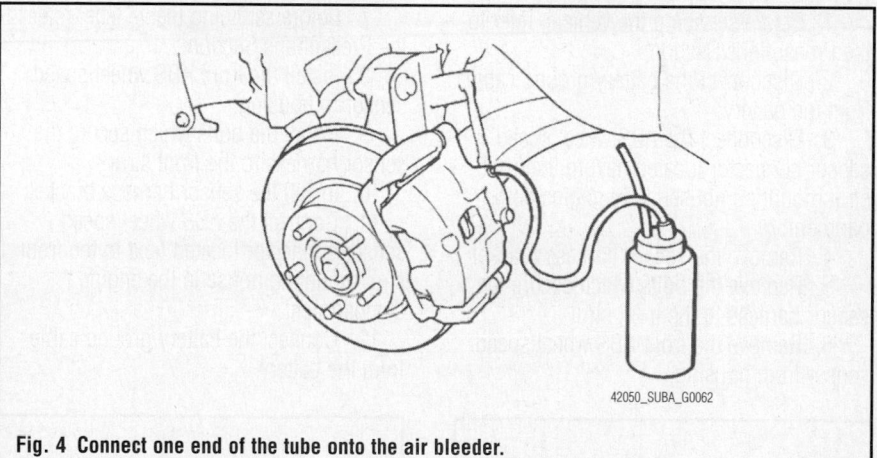

42050_SUBA_G0062

Fig. 4 Connect one end of the tube onto the air bleeder.

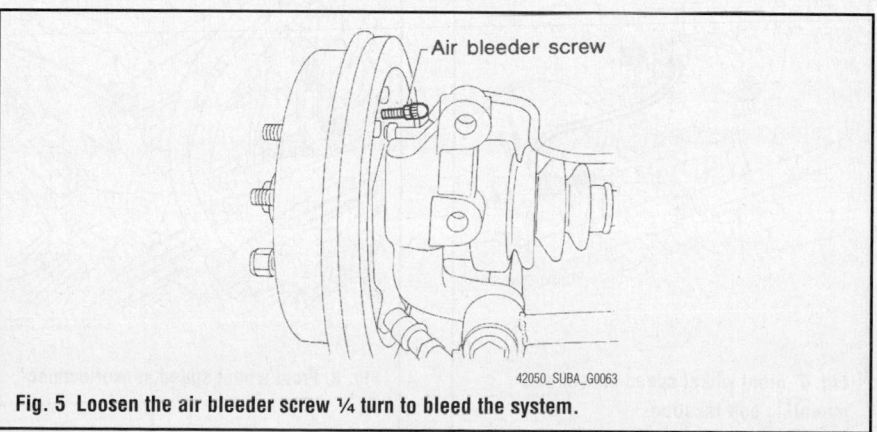

Air bleeder screw

42050_SUBA_G0063

Fig. 5 Loosen the air bleeder screw ¼ turn to bleed the system.

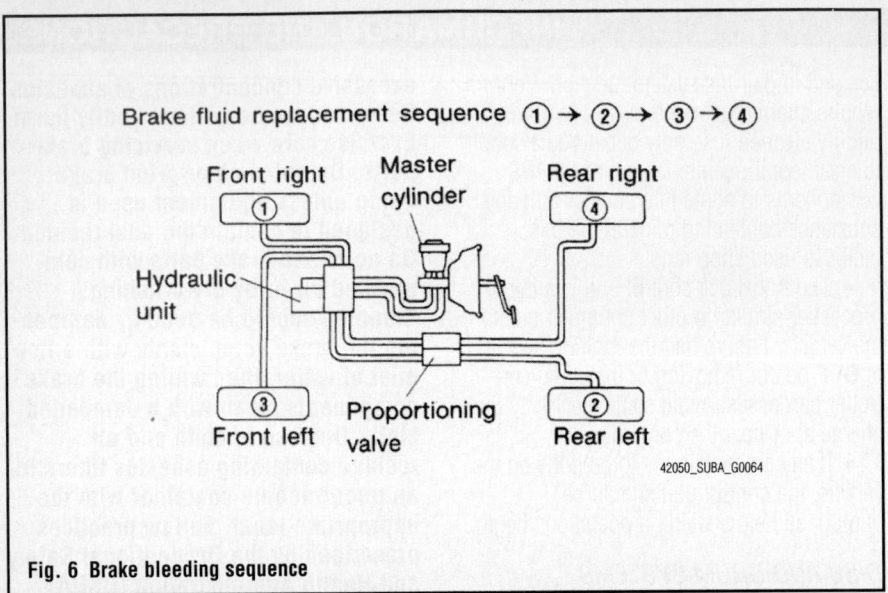

Brake fluid replacement sequence ① → ② → ③ → ④

Front right — Master cylinder — Rear right
① ④

Hydraulic unit

③ Front left — Proportioning valve — Rear left ②

42050_SUBA_G0064

Fig. 6 Brake bleeding sequence

9. After completing the bleeding operation, hold brake pedal depressed and tighten screw and install bleeder cap.

10. Bleed air from each wheel cylinder using the same procedures as described above.

11. Depress brake pedal and hold it there for approximately 20 seconds. At this time check pedal to see if it shows any unusual movement. Visually inspect bleeder screws and brake pipe joints to make sure that there is no fluid leakage.

12. Install the wheels, and drive car for a short distance (between 1–2 miles to make sure that brakes are operating properly.

BLEEDING THE ABS SYSTEM

There is no separate bleeding procedure for ABS. Refer to Bleeding the Brake System.

BRAKES   **ANTI-LOCK BRAKE SYSTEM (ABS)**

WHEEL SPEED SENSORS

REMOVAL & INSTALLATION

Front

See Figures 7 through 10.

1. Before servicing the vehicle, refer to the Precautions Section.

2. Disconnect the battery ground cable from the battery.

3. Disconnect the ABS wheel speed sensor connector located next to the front strut mounting house in the engine compartment.

4. Remove the sensor harness bracket.

5. Remove the bolts which secure the sensor harness to the front strut.

6. Remove the front ABS wheel speed sensor from housing.

✳ CAUTION

Be careful not to damage the sensor. Do not apply excessive force to the sensor harness.

To install:

7. Before servicing the vehicle, refer to the Precautions Section.

8. Install the front ABS wheel speed sensor to housing.

9. Install the bolts which secure the sensor harness to the front strut.

10. Install the sensor harness bracket.

11. Connect the ABS wheel speed sensor connector located next to the front strut mounting house in the engine compartment.

12. Connect the battery ground cable from the battery.

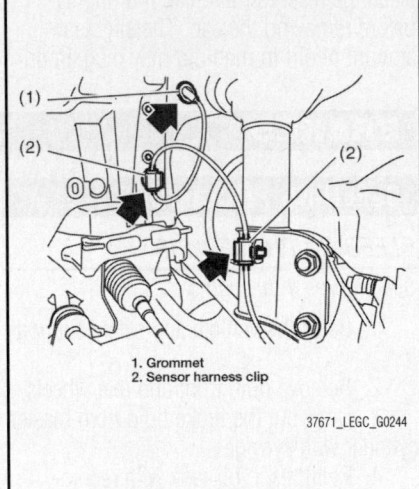

1. Grommet
2. Sensor harness clip

37671_LEGC_G0244

Fig. 9 Front wheel speed sensor harness clip and grommet

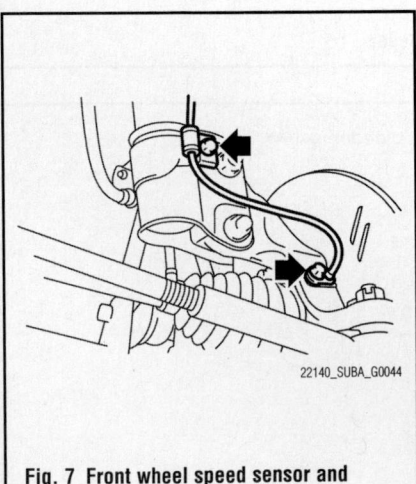

22140_SUBA_G0044

Fig. 7 Front wheel speed sensor and mounting bolt location

37671_LEGC_G0243

Fig. 8 Front wheel speed sensor connector

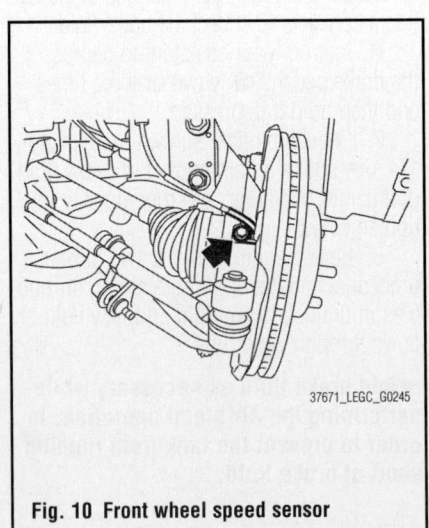

37671_LEGC_G0245

Fig. 10 Front wheel speed sensor

13. Tighten the sensor mounting bolt to 5.5 ft. lbs. (7.5 Nm).

14. Tighten the sensor harness bolts to 24 ft. lbs. (33 Nm).

➡**Check that the harness is not pulled and does not come in contact with the suspension or body during steering wheel effort.**

Rear

See Figures 11 through 14.

1. Disconnect the ground cable from battery.

2. Remove the clips, turn over the mud guard, and disconnect the rear ABS wheel speed sensor connector.

3. Remove the rear ABS wheel speed sensor from the rear housing.

✳✳ WARNING

Be careful not to damage the sensor. Do not apply excessive force to the sensor harness.

---

4. Remove the sensor harness clip.

5. Remove the sensor harness bracket from the upper arm.

To install:

✳✳ WARNING

Be careful not to damage the sensor.

---

6. Install the sensor harness bracket from the upper arm.

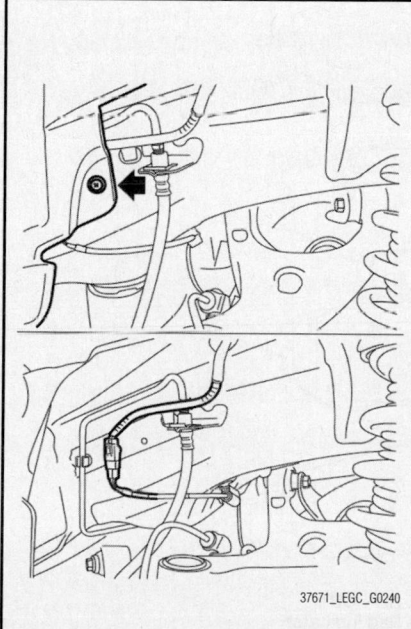

Fig. 11 Rear wheel speed sensor connector

7. Install the sensor harness clip.

8. Install the rear ABS wheel speed sensor to the rear housing.

9. Install the clips, turn over the mud guard, and
connect the rear ABS wheel speed sensor connector.

10. Connect the ground cable from battery.

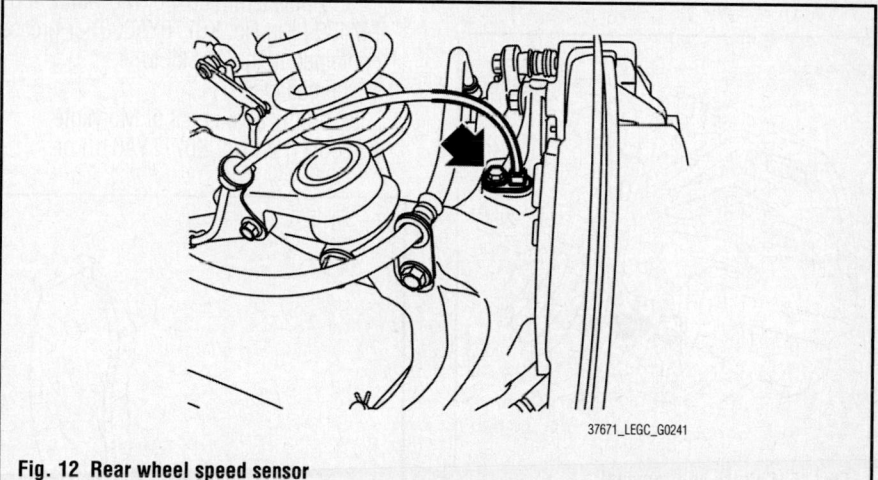
37671_LEGC_G0241

Fig. 12 Rear wheel speed sensor

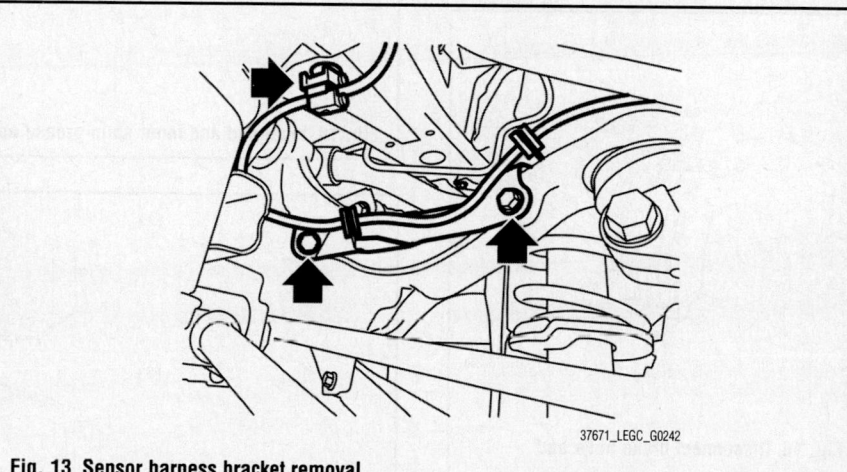
37671_LEGC_G0242

Fig. 13 Sensor harness bracket removal

11. Tighten the sensor mounting bolt to 66 inch lbs. (7.5 Nm), and the sensor bracket bolt to 66 inch lbs. (7.5 Nm).

➡**Check the identification (mark) on the harness to make sure there is no warpage. Refer to the accompanying illustration.**

Item			Specification or identification
ABS wheel speed sensor	ABS wheel speed sensor gap (for reference)	Front	0.77 — 1.43 mm (0.030 — 0.056 in)
		Rear	0.64 — 1.56 mm (0.025 — 0.061 in)
	Identifications of harness (marks, color)	Front (Except for OBK model)	S1 (Pink)
		Front OBK model	S2 (Blue)
		Rear RH (Except for OBK model)	S3 (Yellow)
		Rear LH (Except for OBK model)	S4 (White)
		Rear RH (OBK model)	S5 (Red)
		Rear RH (OBK model)	S6 (Light green)
VDCCM&H/U Identification			V1

37671_LEGC_G0249

Fig. 14 Vehicle dynamics control specifications

DISC BRAKE PADS

REMOVAL & INSTALLATION

See Figures 15 through 21.

1. Lift up the vehicle, and remove the front wheels.
2. Remove the caliper bolt.
3. Raise the caliper body and support it.

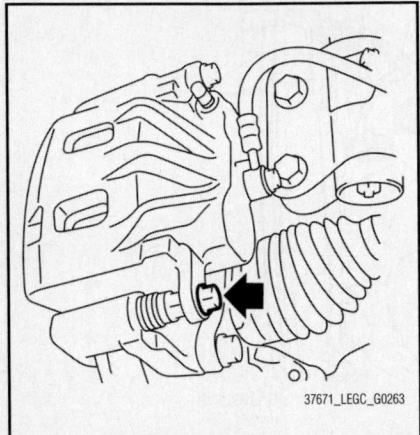

Fig. 15 Caliper bolt removal

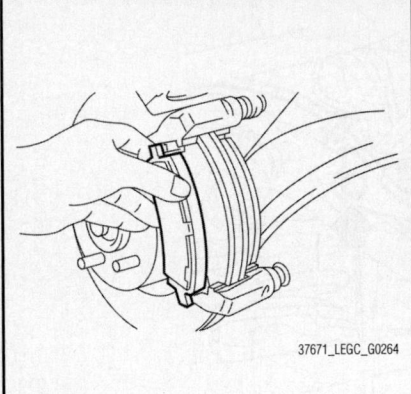

Fig. 16 Disconnect brake hose and remove pad

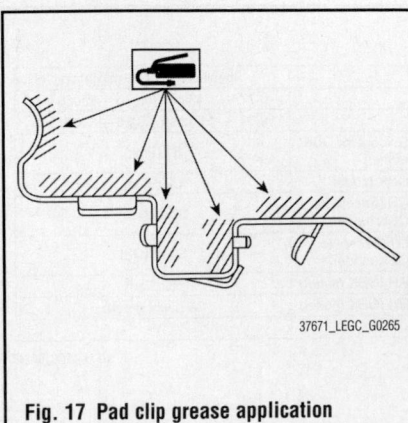

Fig. 17 Pad clip grease application

➡**Do not disconnect the brake hose from the caliper body.**

4. Remove the pad.

➡**If the pad return spring is deformed or damaged, replace the brake pad.**

To install:

5. **16-INCH TYPE:**

a. Apply a thin coat of Molykote M7439 (Part No. K0770YA000) or grease contained in the pad kit to the pad clip.

b. Apply a thin coat of Molykote AS880N (Part No. K0777YA010) or

grease contained in the pad kit to both surfaces of the pad and inner shim.

c. Install the pad to support.

d. Install the caliper body to the support. Tighten to 20 ft. lbs. (27 Nm).

6. **17-INCH TYPE:**

a. Apply a thin coat of Molykote M7439 (Part No. K0770YA000) or grease contained in the pad kit to the pad clip.

b. Apply a thin coat of Molykote AS880N (Part No. K0777YA010) or

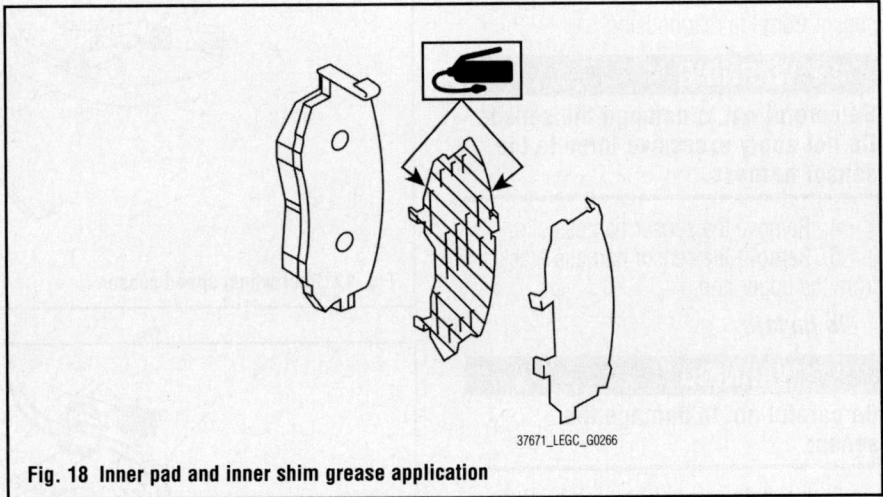

Fig. 18 Inner pad and inner shim grease application

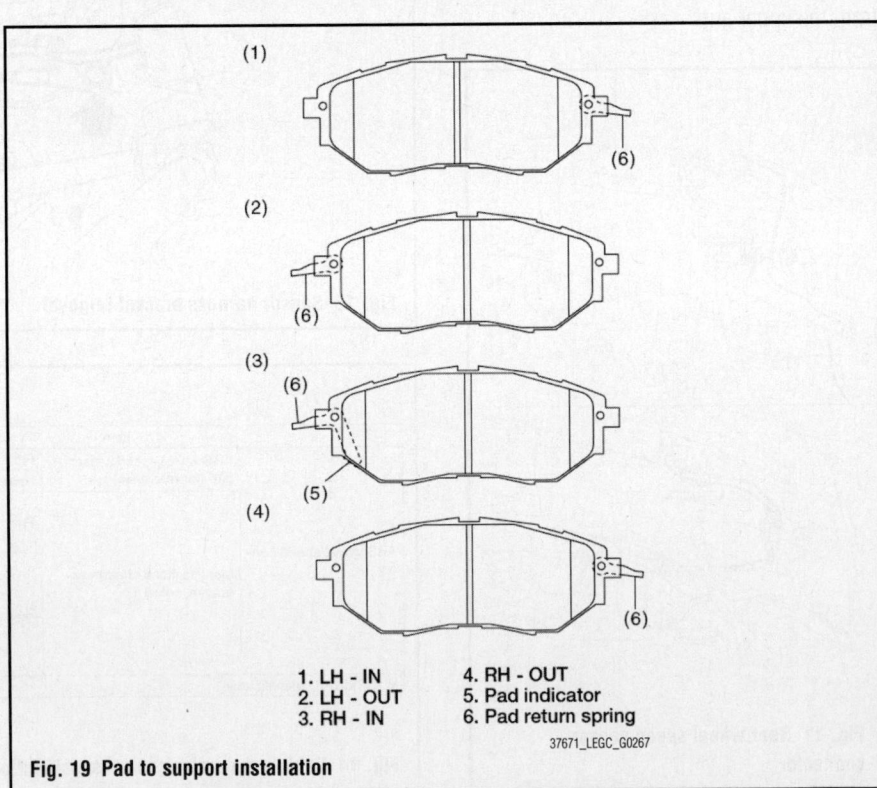

1. LH - IN	4. RH - OUT
2. LH - OUT	5. Pad indicator
3. RH - IN	6. Pad return spring

Fig. 19 Pad to support installation

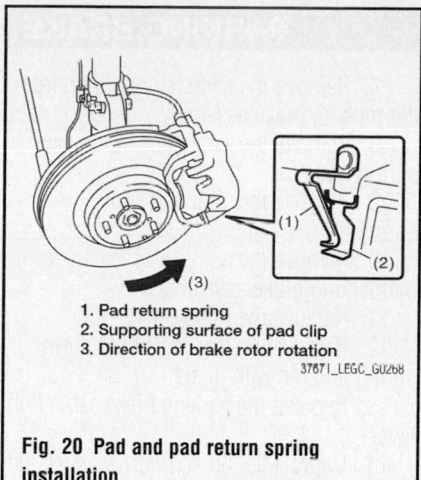

1. Pad return spring
2. Supporting surface of pad clip
3. Direction of brake rotor rotation

37671_LEGC_G0268

Fig. 20 Pad and pad return spring installation

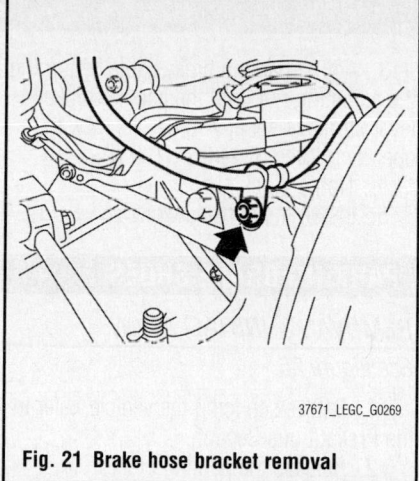

37671_LEGC_G0269

Fig. 21 Brake hose bracket removal

grease contained in the pad kit to both surfaces of the pad and inner shim.

7. Install the pad to support.

➡**Install the pad indicator in proper direction.**

➡**Be sure to install so that the pad return spring faces the input side of the direction of brake rotor rotation, as shown in the figure. Correctly install the pad return spring to the supporting surface of the pad clip as shown in the figure.**

8. Install the caliper body to the support. Tighten to 20 ft. lbs. (27 Nm).

BRAKES

REAR DISC BRAKES

DISC BRAKE PADS

REMOVAL & INSTALLATION

See Figures 17, 18, 21 through 24.

1. Lift up the vehicle, and then remove the rear wheels.
2. Remove the brake hose bracket.
3. Remove the caliper bolt.
4. Raise the caliper body and support it.

➡**Do not disconnect the brake hose from the caliper body.**

5. Remove the pad.

To install:

6. Apply a thin coat of Molykote M7439 (Part No. K0770YA000) or grease contained in the pad kit to the pad clip.

7. Apply a thin coat of Molykote AS880N (Part No. K0777YA010) or grease contained in the pad kit between the outer pad and shim.

8. Apply a thin coat of Molykote AS880N (Part No. K0777YA010) or grease contained in the pad kit to both surfaces of the inner pad inner shim.

9. Install the pad to support.

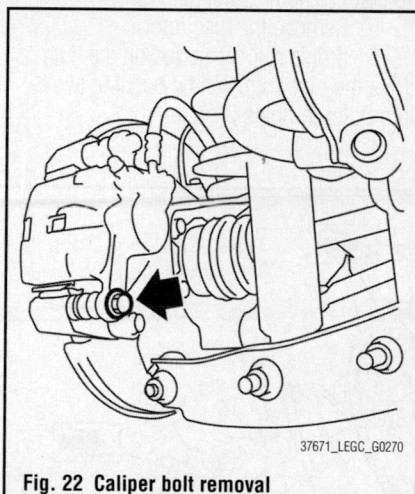

37671_LEGC_G0270

Fig. 22 Caliper bolt removal

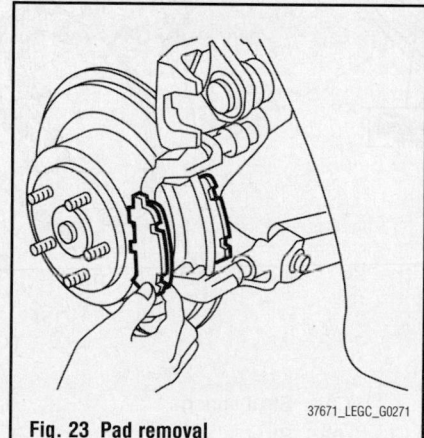

37671_LEGC_G0271

Fig. 23 Pad removal

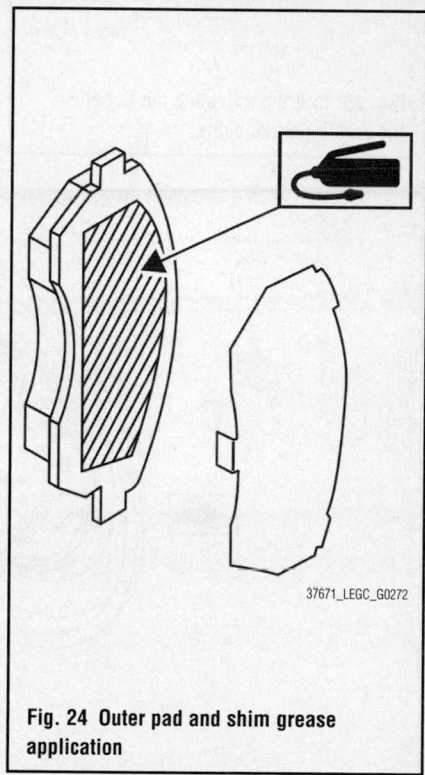

37671_LEGC_G0272

Fig. 24 Outer pad and shim grease application

10. Install the caliper body to the support. Tighten to 20 ft. lbs. (27 Nm).

11. Install the brake hose bracket. Tighten 24.3 ft. lbs. (33 Nm).

PARKING BRAKE CABLES

ADJUSTMENT

See Figure 25.

1. Remove the console cover.
2. Forcefully pull the parking brake lever 3 to 5 times.

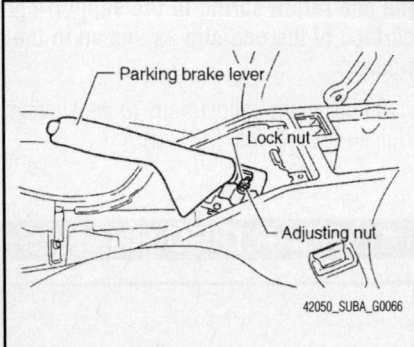

Fig. 25 Turn the adjusting nut to adjust the parking brake cable.

3. Adjust parking brake cable by turning the adjusting nut until the parking brake lever stroke is set at 7 to 8 notches with operating force of 44 lbs. (196 N).
4. Tighten lock nut.
5. Install the console cover.

PARKING BRAKE SHOES

REMOVAL & INSTALLATION

See Figure 26.

1. Before servicing the vehicle, refer to the Precautions Section.
2. Release the parking brake.
3. Remove the two mounting bolts and remove the brake caliper assembly.
4. Suspend the brake caliper assembly so that the hose is not stretched.
5. Remove the disc rotor.
6. If disc rotor is seized on the hub, drive the disc rotor out by pushing two 8 mm bolts in holes B on the rotor.

7. Remove the shoe return spring from the parking brake assembly.
8. Remove the front shoe hold down spring and pin.
9. Remove the strut and strut spring.
10. Remove the adjuster assembly from the parking brake assembly.
11. Remove the brake shoe.
12. Remove the rear shoe hold down spring and pin with pliers.
13. Remove the parking brake cable from lever.
14. Using a flat tip screwdriver, raise the retainer. Remove the parking lever and washer from brake shoe.

To install:

15. Apply brake grease to the following locations:
- Six contact surfaces of the shoe rim and back plate gasket
- Contact surface of the shoe wave and the anchor pin

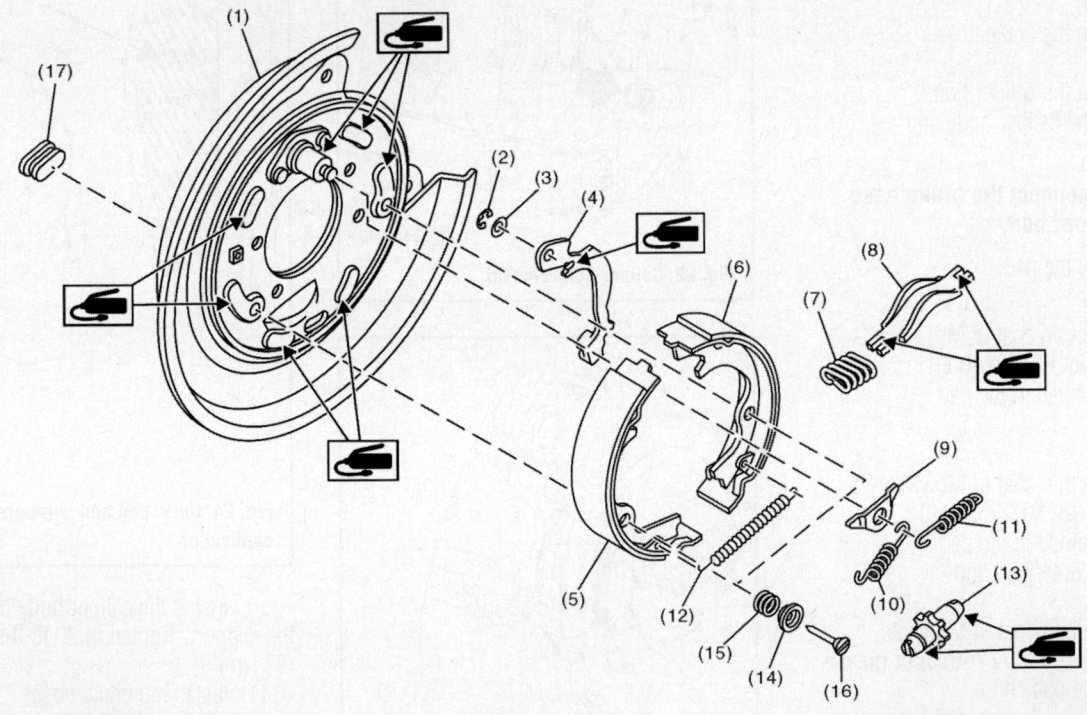

(1)	Back plate	(7)	Strut spring	(13)	Adjuster
(2)	Retainer	(8)	Strut	(14)	Shoe hold-down cup
(3)	Spring washer	(9)	Shoe guide plate	(15)	Shoe hold-down spring
(4)	Lever	(10)	Primary return spring	(16)	Shoe hold-down pin
(5)	Parking brake shoe (Primary)	(11)	Secondary return spring	(17)	Adjusting hole cover
(6)	Parking brake shoe (Secondary)	(12)	Adjusting spring		

22140_SUBA_G0059

Fig. 26 Exploded view of parking brake components

- Contact surface of the lever and strut
- Contact surface of the shoe wave and the adjuster assembly
- Contact surface of the shoe wave and the strut
- Contact surface of the lever and the shoe wave

16. Insert the primary side brake shoe into the anchor pin groove.

17. Secure the brake shoe with the shoe hold-down pin and cup.

18. Install the plate to the anchor pin, then install the primary return spring.

19. Install the parking brake cable to the lever.

20. Install the strut and adjuster, then secure the secondary side brake shoe with the shoe hold-down pin and cup.

➡Install the strut spring of both right and left wheel facing vehicle front. Install the adjuster assembly with screw section on the left side.

21. Install the secondary return spring and the adjusting spring.

22. Adjust the parking brake.

23. Drive the vehicle to break-in the parking brake lining.

   a. Drive the vehicle at about 22 mph (35 km/h).

   b. With the parking brake release button pushed in, pull the parking brake lever gently.

   c. Drive the vehicle for about 0.12 miles (200 m) in this condition.

   d. Wait 5 to 10 minutes for the parking brake to cool down. Repeat again from step (A).

   e. After breaking-in, re—adjust the parking brakes.

ADJUSTMENT

1. Before servicing the vehicle, refer to the Precautions Section.

2. Return the parking brake lever fully.

3. Remove the adjusting hole cover from the back plate.

4. Turn the adjusting screw using a flat tip screwdriver until the brake shoe is in close contact with the disc rotor.

5. Turn back (downward) the adjusting screw 3 to 4 notches. Check there is no brake drag.

6. Install the adjusting hole cover to the back plate.

7. Adjust the parking lever stroke.

CHASSIS ELECTRICAL   AIR BAG (SUPPLEMENTAL RESTRAINT SYSTEM)

GENERAL INFORMATION

✵✵ CAUTION

These vehicles are equipped with an air bag system. The system must be disarmed before performing service on, or around, system components, the steering column, instrument panel components, wiring and sensors. Failure to follow the safety precautions and the disarming procedure could result in accidental air bag deployment, possible injury and unnecessary system repairs.

SERVICE PRECAUTIONS

✵ CAUTION

Disconnect and isolate the battery negative cable before beginning any airbag system component diagnosis, testing, removal, or installation procedures. Wait at least 90 seconds after the ignition switch is turned off and the negative (-) terminal cable is disconnected from the battery before starting the operation. The SRS is equipped with a backup power source, so if work is started within 90 seconds after disconnecting the

negative (-) terminal cable from the battery, the SRS may be deployed. Failure to disable the airbag system may result in accidental airbag deployment, personal injury, or death.

DISARMING THE SYSTEM

1. Be sure to position the front wheels in the straight ahead position.

2. Disconnect the negative battery cable. Tape the battery cable for added protection.

3. Wait more than 20 seconds before starting work.

ARMING THE SYSTEM

1. Connect the negative battery cable.

CLOCKSPRING CENTERING

See Figure 27.

✵✵ CAUTION

When servicing a vehicle, be sure to turn the ignition switch to OFF, disconnect the ground cable from battery, and wait for more than 20 seconds before starting work. The airbag system is fitted with a backup power source. After disconnecting the

battery ground cable, the airbag may deploy if you do not wait for more than 20 seconds before starting the service of airbag system.

1. Turn the ignition switch to OFF.

2. Disconnect the ground cable from battery and wait for at least 20 seconds before starting work.

3. Check that front wheels are positioned in straight ahead direction.

4. Turn the clock spring pin clockwise until it stops.

5. Turn the clock spring connector pins approx. 3.25 turns until the marks are aligned.

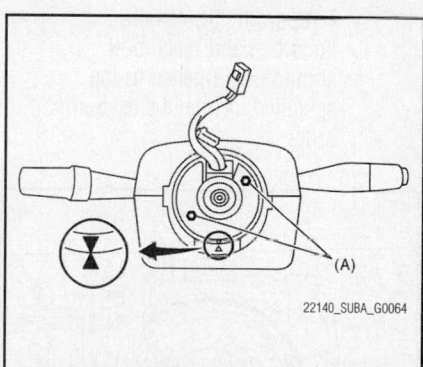

22140_SUBA_G0064

Fig. 27 Clock spring pins (A) and alignment marks shown

DRIVE TRAIN

CLUTCH

REMOVAL & INSTALLATION

See Figures 28 through 30.

1. Remove the transmission assembly.
2. Attach the ST on the flywheel.
3. Remove the clutch cover and clutch disc.

➡**Take care not to allow oil to touch the clutch disc face. Do not disassemble the clutch cover or clutch disc.**

To install:

4. Insert the ST into the clutch disc and attach to the flywheel by inserting the ST end into pilot bearing.

➡**When installing the clutch disc, be careful to attach in the correct direction.**

5. Install the clutch cover to the flywheel and tighten the bolts to the specified torque.

➡**When installing the clutch cover to the flywheel, position the clutch cover so that the distance between unbalance marks (paint marks) is at least 120°. (The unbalance mark indicates the direction of residual unbalance.)**

- Note the front and rear of the clutch disc when installing.
- Temporarily tighten the bolts by hand. Each bolt should be tightened to the specified torque in a crisscross order.

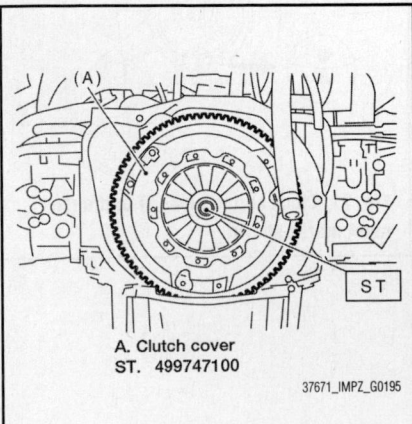

A. Clutch cover
ST. 499747100

37671_IMPZ_G0195

Fig. 28 Flywheel and ST attachment

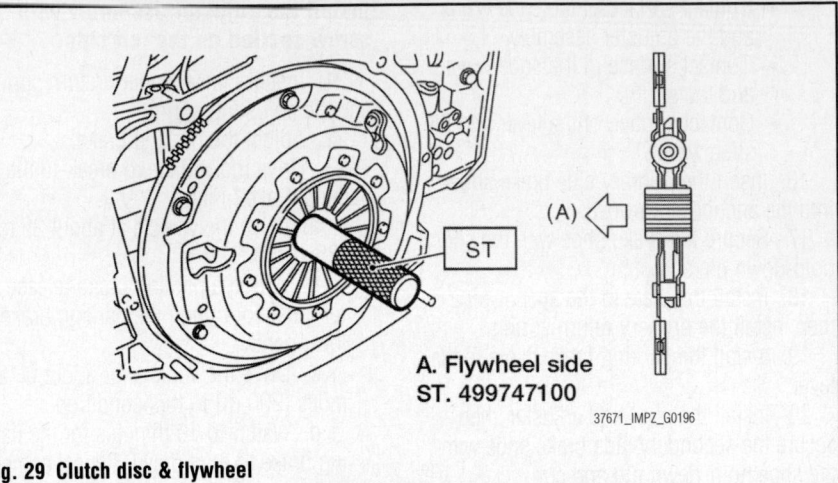

A. Flywheel side
ST. 499747100

37671_IMPZ_G0196

Fig. 29 Clutch disc & flywheel

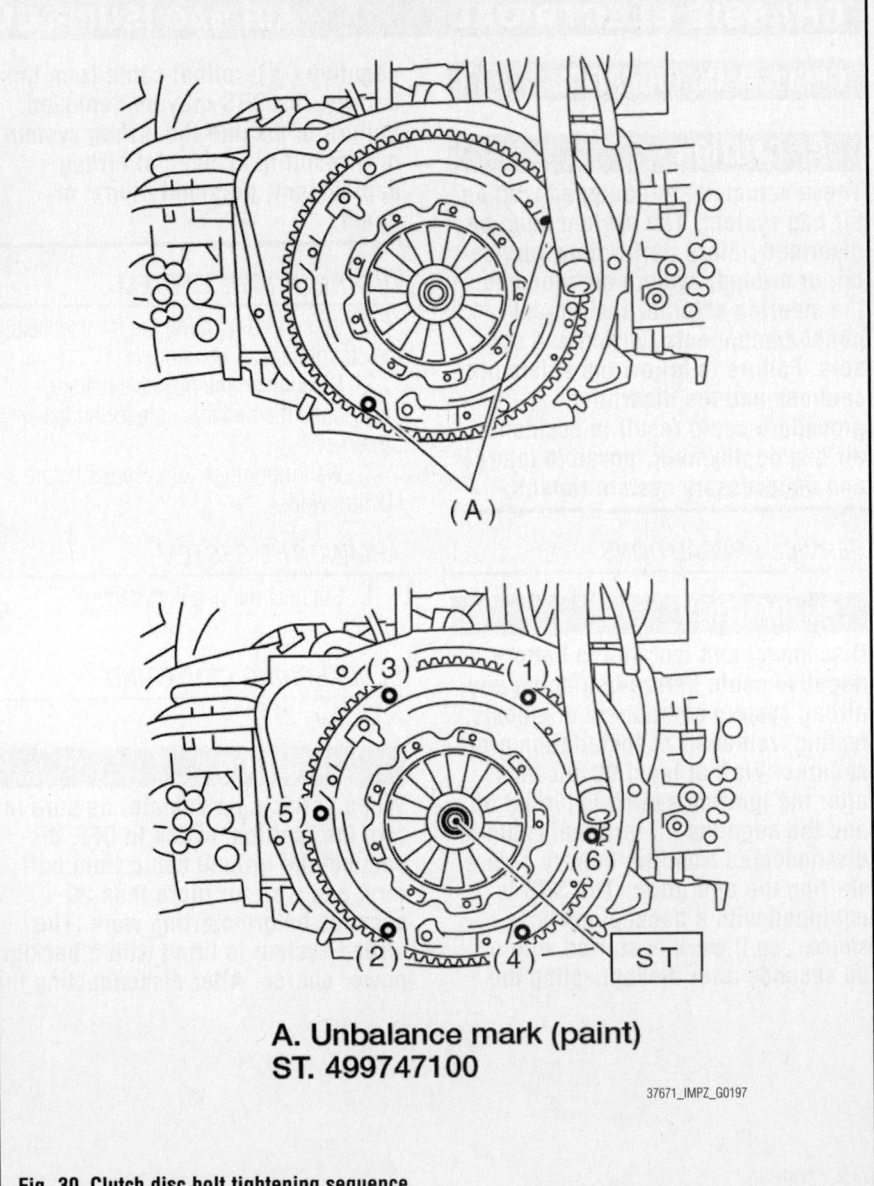

A. Unbalance mark (paint)
ST. 499747100

37671_IMPZ_G0197

Fig. 30 Clutch disc bolt tightening sequence

6. Tighten bolts to 142 inch lbs. (16 Nm).

7. Remove the CLUTCH DISC GUIDE.

8. Install the transmission assembly.

CLUTCH DRIVEN DISC & PRESSURE PLATE

REMOVAL & INSTALLATION

See Figures 31 through 33.

1. Before servicing the vehicle, refer to the Precautions Section.

2. Disconnect the negative battery cable.

3. Remove the transmission.

4. Install tool ST499747100, or equivalent on the flywheel.

5. Remove the clutch cover and clutch disc. Do not disassemble

➡**Be sure to put alignment marks on the flywheel and clutch cover before removing the clutch cover.**

To install:

6. Before servicing the vehicle, refer to the Precautions Section.

➡**Be sure to use alignment marks on the flywheel and clutch cover before installing the clutch cover.**

7. Install the clutch cover and clutch disc. Do not disassemble

8. Install tool ST499747100, or equivalent on the flywheel.

9. Install the transmission.

10. Connect the negative battery cable.

ADJUSTMENTS

Models are equipped with a hydraulic system that is not adjustable.

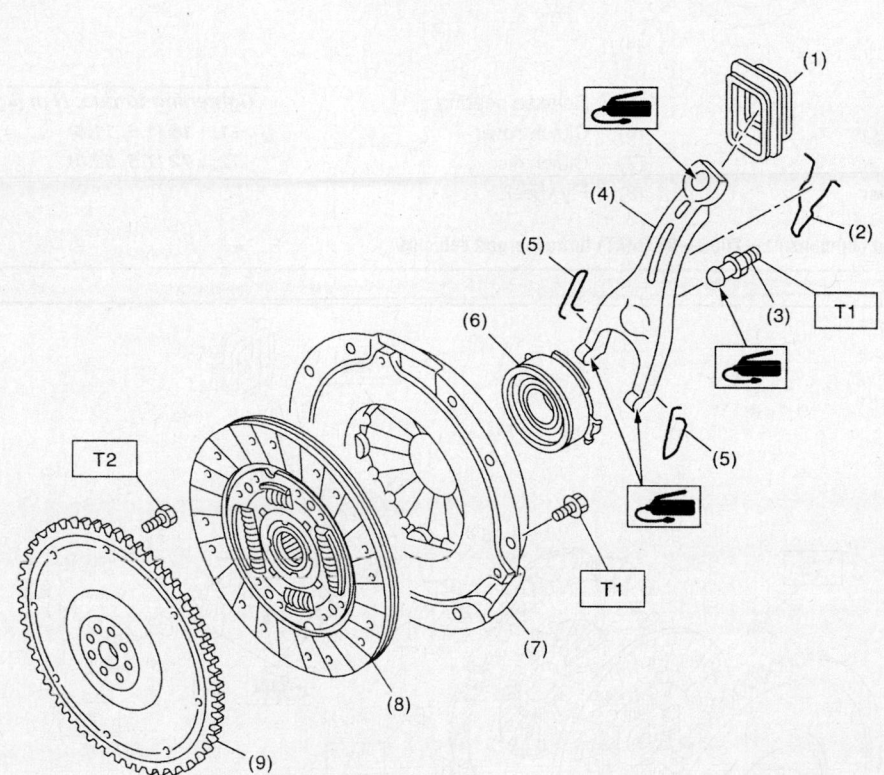

(1)	Dust cover	(6)	Release bearing
(2)	Lever spring	(7)	Clutch cover
(3)	Pivot	(8)	Clutch disc
(4)	Release lever	(9)	Flywheel
(5)	Clip		

Tightening torque: N·m (kgf-m, ft-lb)
T1: 16 (1.6, 11.8)
T2: 72 (7.3, 52.8)

09490_SBCR_G0084

Fig. 31 Clutch and related components—2006—08 (5MT) non-turbocharged vehicles

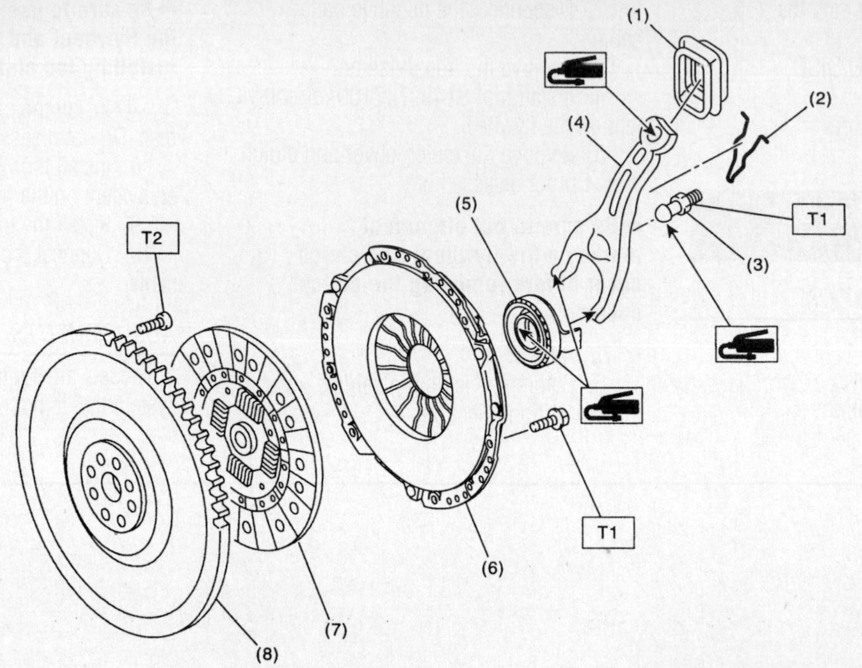

(1)	Dust cover	(5)	Release bearing		
(2)	Lever spring	(6)	Clutch cover		
(3)	Pivot	(7)	Clutch disc		
(4)	Release lever	(8)	Flywheel		

Tightening torque: N·m (kgf-m, ft-lb)
T1: 16 (1.6, 11.8)
T2: 72 (7.3, 52.8)

09490_SBCR_G0085

Fig. 32 Clutch and related components—2006—08 (5MT) turbocharged vehicles

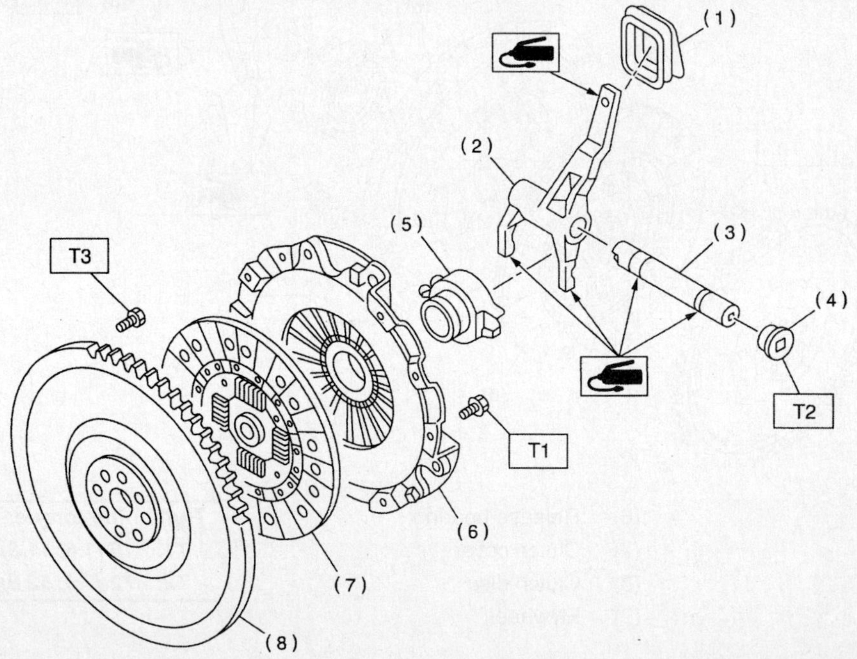

(1)	Dust cover	(5)	Release bearing		
(2)	Release lever	(6)	Clutch cover		
(3)	Clutch release lever shaft	(7)	Clutch disc		
(4)	Plug	(8)	Flywheel		

Tightening torque: N·m (kgf-m, ft-lb)
T1: 16 (1.6, 11.8)
T2: 44 (4.5, 32.5)
T3: 75 (7.6, 55.3)

09490_SBCR_G0086

Fig. 33 Clutch and related components—2006—08 (6MT) vehicles

HYDRAULIC SYSTEM BLEEDING

BLEEDING PROCEDURE

5MT Transmission

See Figure 34.

1. On non turbocharged engine, remove the air intake chamber.
2. On turbocharged engine, remove the intercooler.
3. Connect a vinyl tube to the air bleeder on the master cylinder. Put the other end in a jar with clean clutch fluid.
4. Slowly depress the clutch pedal and keep it depressed. Open the air bleeder to discharge air and fluid.
5. Release the air bleeder for one or two seconds. With the bleeder closed, slowly release the clutch pedal.
6. Repeat the procedure until there are no more air bubbles in the vinyl tube.
7. Tighten the air bleeder.
8. Connect a vinyl tube to the air bleeder on the clutch operating (slave) cylinder. Put the other end in a jar with clean clutch fluid.
9. Slowly depress the clutch pedal and keep it depressed. Open the air bleeder to discharge air and fluid.
10. Release the air bleeder for one or two seconds. With the bleeder closed, slowly release the clutch pedal.
11. Repeat the procedure until there are no more air bubbles in the vinyl tube.
12. Tighten the air bleeder.
13. After depressing the clutch pedal, make sure that there are no leaks in the entire system
14. Recheck to ensure that the clutch is operating correctly.

6MT Transmission

See Figure 35.

1. Remove the intercooler.
2. Remove the clutch operating cylinder. Do not remove the clutch hose.
3. Using a service clamp, fix the piston to avoid the piston from jumping out.
4. Connect a vinyl tube to the air bleeder on the clutch operating (slave) cylinder. Put the other end in a jar with clean clutch fluid.
5. Slowly depress the clutch pedal and keep it depressed. Open the air bleeder to discharge air and fluid.
6. Release the air bleeder for one or two seconds. With the bleeder closed, slowly release the clutch pedal.

➡ **Set the air breather screw position higher than the tip of the operating**

cylinder when performing this procedure.

7. Repeat the procedure until there are no more air bubbles in the vinyl tube.
8. Tighten the air bleeder.
9. After depressing the clutch pedal, make sure that there are no leaks in the entire system
10. Recheck to ensure that the clutch is operating correctly.

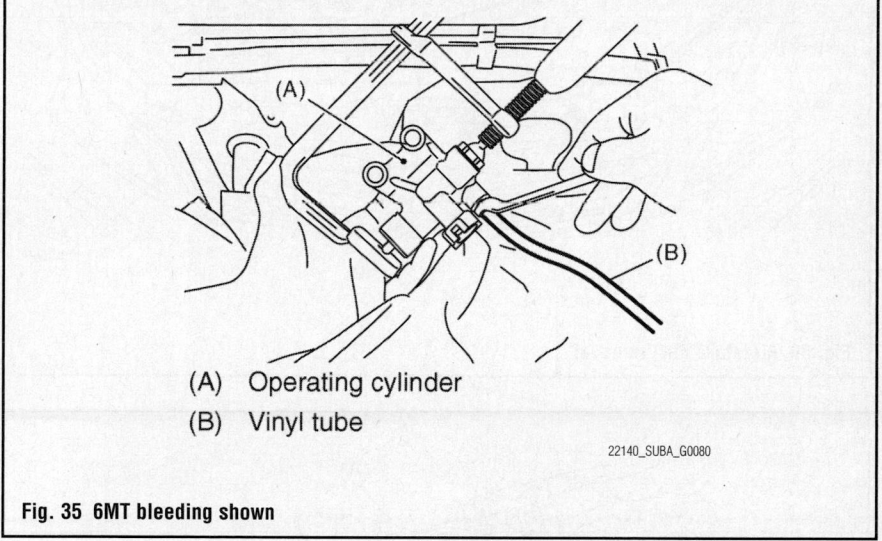

(A) Operating cylinder
(B) Vinyl tube

22140_SUBA_G0080

Fig. 35 6MT bleeding shown

TRANSFER CASE ASSEMBLY

REMOVAL & INSTALLATION

See Figures 36 through 58.

1. Remove the manual transmission assembly from the vehicle.
2. Remove the back-up light switch neutral position switch.
3. Remove the transfer case together with the extension case assembly.
   a. Remove the transmission cover.
   b. Set and hold the selector lever COMPL to the 1st-2nd side, and remove the transfer case and extension case assembly as a unit.
4. Remove the transfer cover.
5. Use the ST to push out the spring pin and remove the shift lever COMPL from shifter arm No. 2. ST 398791700 SPRING PIN REMOVER 2
6. Remove the extension case assembly.

To install:

7. Install the center differential and transfer driven gear into the transfer case.

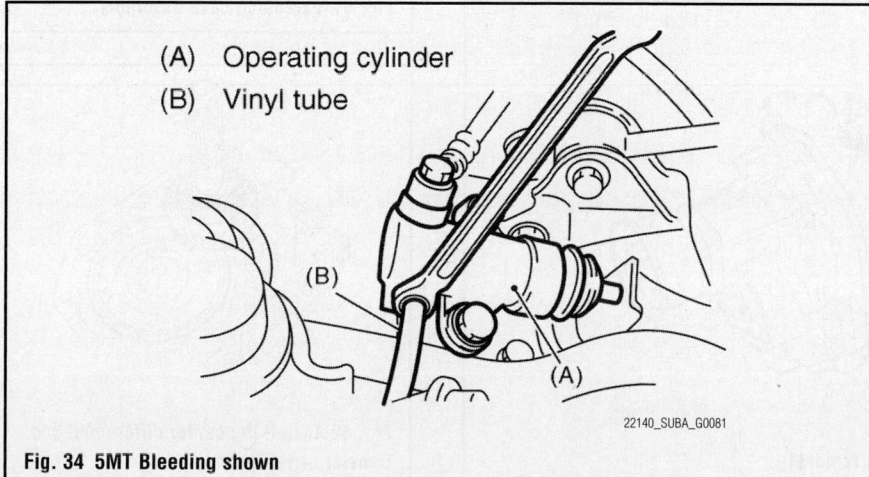

(A) Operating cylinder
(B) Vinyl tube

22140_SUBA_G0081

Fig. 34 5MT Bleeding shown

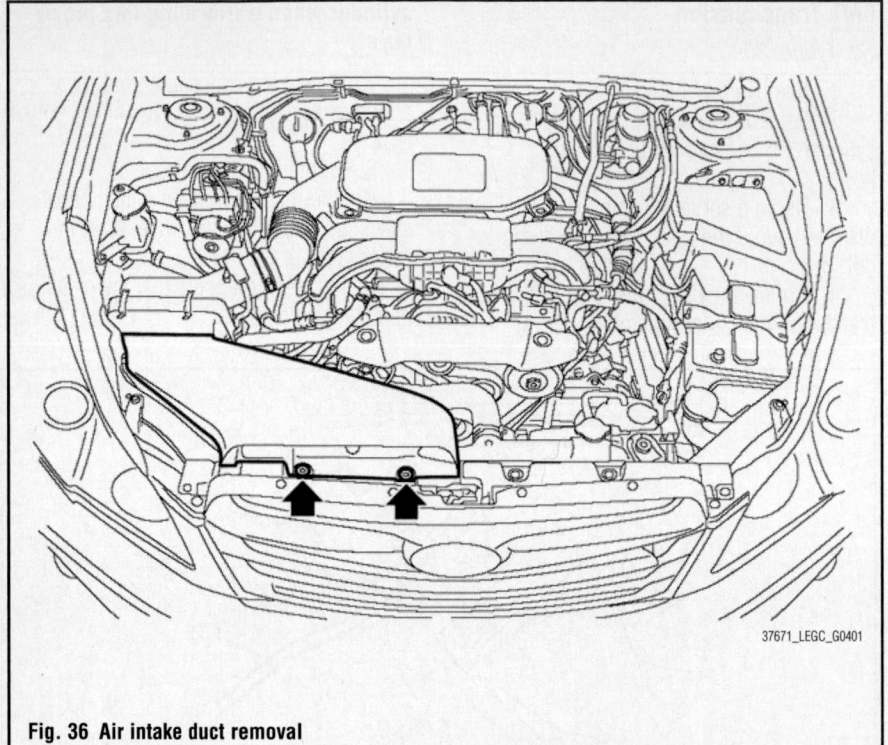

37671_LEGC_G0401

Fig. 36 Air intake duct removal

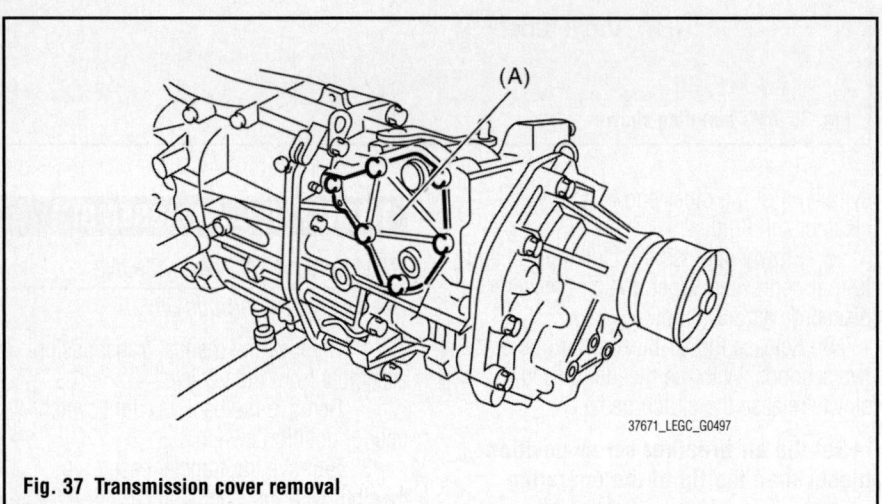

37671_LEGC_G0497

Fig. 37 Transmission cover removal

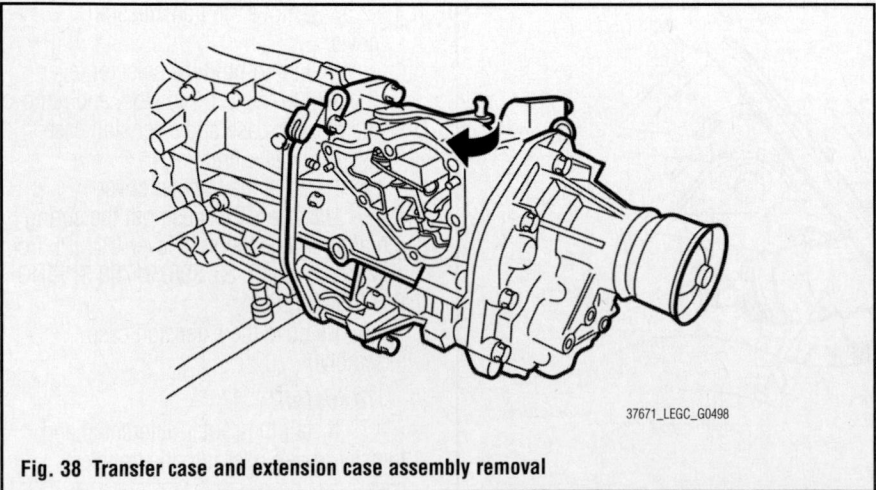

37671_LEGC_G0498

Fig. 38 Transfer case and extension case assembly removal

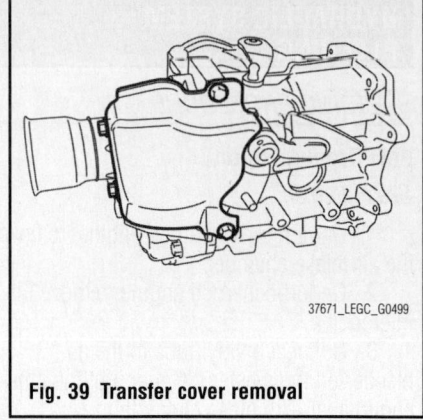

37671_LEGC_G0499

Fig. 39 Transfer cover removal

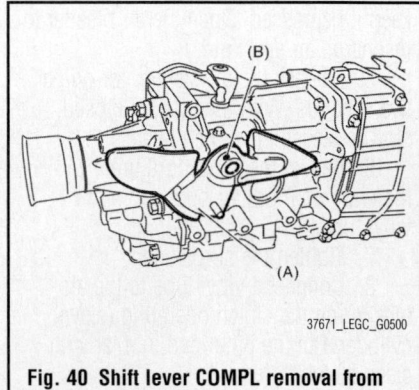

37671_LEGC_G0500

Fig. 40 Shift lever COMPL removal from shifter arm No. 2

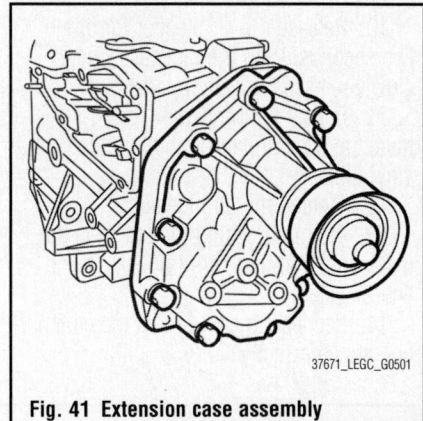

37671_LEGC_G0501

Fig. 41 Extension case assembly

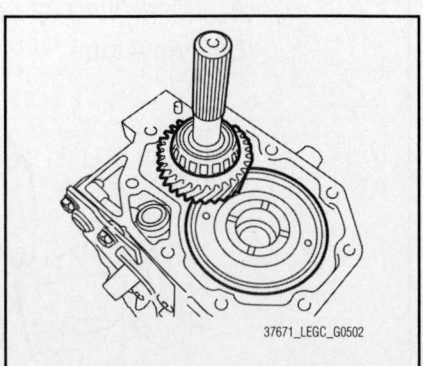

37671_LEGC_G0502

Fig. 42 Install the center differential and transfer driven gear

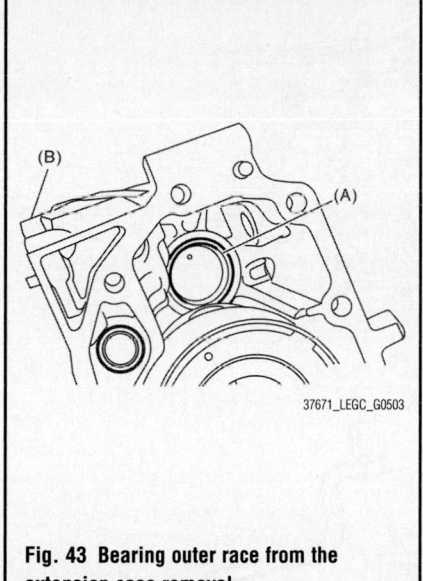

Fig. 43 Bearing outer race from the extension case removal

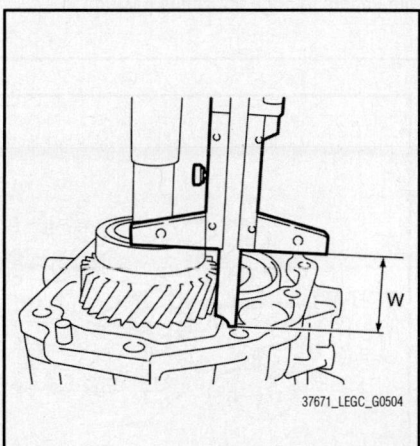

Fig. 44 Measure the height "W" between transfer case and taper roller bearing on the transfer driven gear

Thrust washer (50 × 61 × t)	
Part number	Thickness mm (in)
803050060	0.50 (0.0197)
803050061	0.55 (0.0217)
803050062	0.60 (0.0236)
803050063	0.65 (0.0256)
803050064	0.70 (0.0276)
803050065	0.75 (0.0295)
803050066	0.80 (0.0315)
803050067	0.85 (0.0335)
803050068	0.90 (0.0354)
803050069	0.95 (0.0374)
803050070	1.00 (0.0394)
803050071	1.05 (0.0413)
803050072	1.10 (0.0433)
803050073	1.15 (0.0453)
803050074	1.20 (0.0472)
803050075	1.25 (0.0492)
803050076	1.30 (0.0512)
803050077	1.35 (0.0531)
803050078	1.40 (0.0551)
803050079	1.45 (0.0571)

Fig. 46 Thrush washer part number and thickness table—thickness "t"

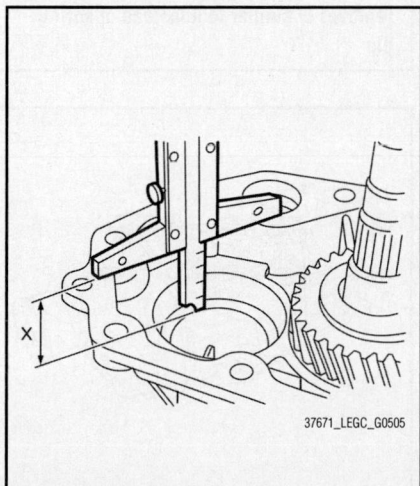

Fig. 45 Measure depth "X" on bearing insertion part of the extension

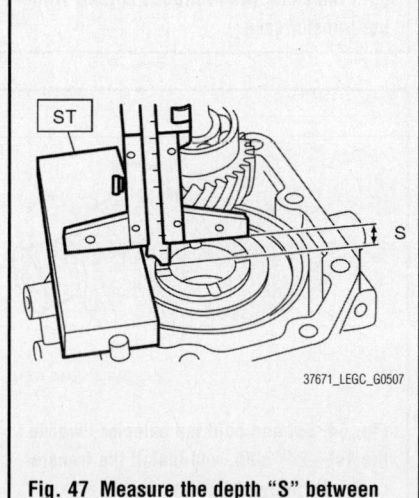

Fig. 47 Measure the depth "S" between transfer case and center differential

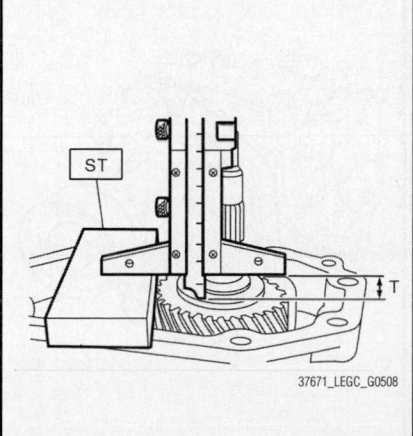

Fig. 48 Measure the height "T" between the extension case and transfer drive gear

Thrust washer	
Part number	Thickness mm (in)
803036050	0.9 (0.035)
803036054	1.0 (0.039)
803036051	1.1 (0.043)
803036055	1.2 (0.047)
803036052	1.3 (0.051)
803036056	1.4 (0.055)
803036053	1.5 (0.059)
803036057	1.6 (0.063)
803036058	1.7 (0.067)
803036080	1.8 (0.071)
803036081	1.9 (0.075)

37671_LEGC_G0509

Fig. 49 Thrush washer part number and thickness table—thickness "U"

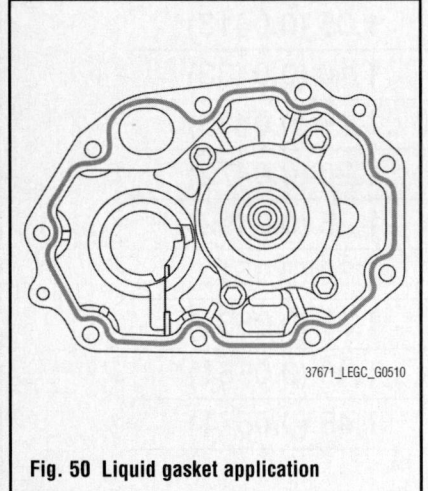

37671_LEGC_G0510

Fig. 50 Liquid gasket application

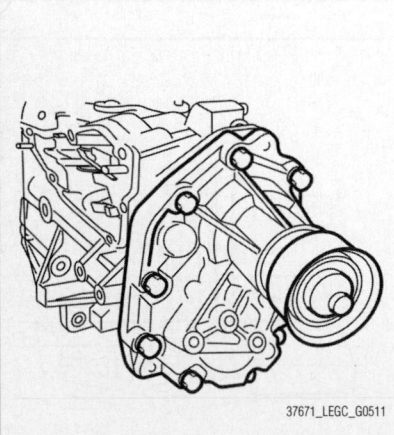

37671_LEGC_G0511

Fig. 51 Installing the extension case assembly to the transfer case

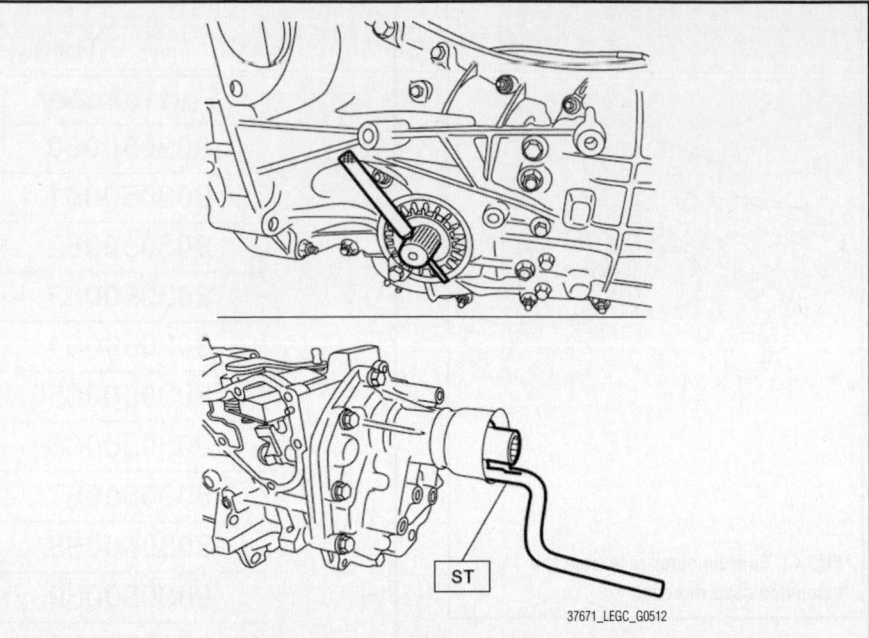

37671_LEGC_G0512

Fig. 52 Insert the axle shaft and hold, then move the handle to align the spline position to install

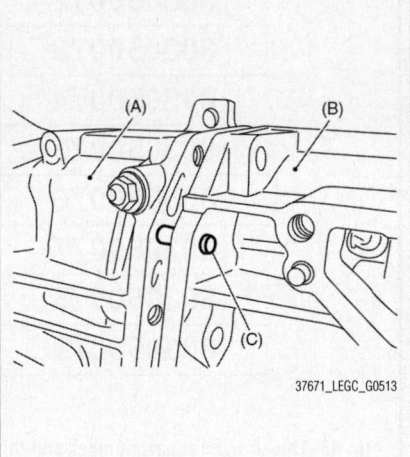

37671_LEGC_G0513

Fig. 53 Push the transfer case assembly until the knock pin protrudes slightly from the transfer case

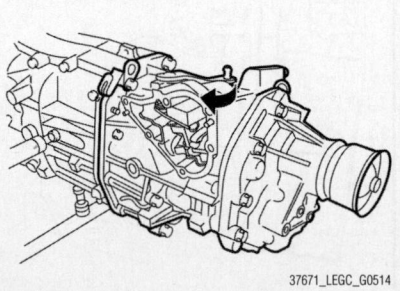

37671_LEGC_G0514

Fig. 54 Set and hold the selector lever to the 1st—2nd side, and install the transfer case assembly so that the shifter arm edge and fork rod does not contact

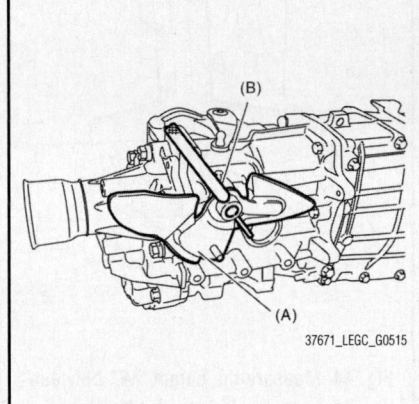

37671_LEGC_G0515

Fig. 55 Install the shift lever COMPL to the shifter arm No. 2, then insert the remover or similar tool instead of spring pin

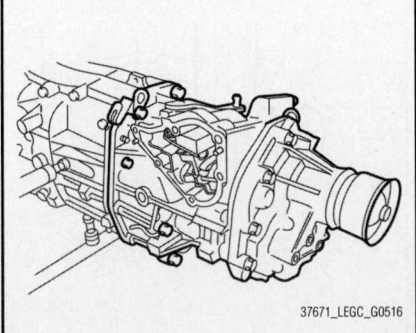

37671_LEGC_G0516

Fig. 56 Tightening of the transfer case assembly

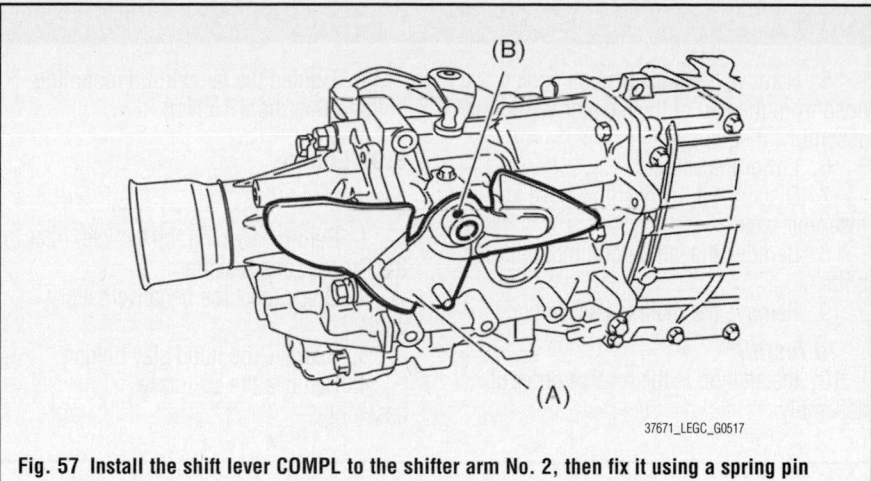

37671_LEGC_G0517

Fig. 57 Install the shift lever COMPL to the shifter arm No. 2, then fix it using a spring pin

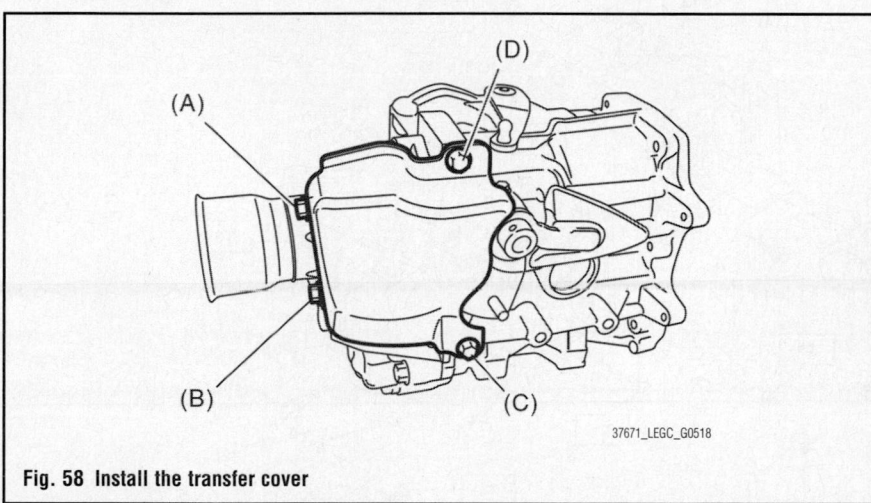

37671_LEGC_G0518

Fig. 58 Install the transfer cover

8. Remove the bearing outer race from the extension case.

9. While pressing the bearing outer race horizontally, rotate the driven shaft for ten turns.

10. Measure the height "W" between transfer case and taper roller bearing on the transfer driven gear.

11. Measure depth "X" on bearing insertion part of the extension case.

➡**Measure with bearing outer race and thrust washer removed.**

12. Calculate the thrust washer thickness "t" using the following calculation:

a. t = X - W + (0.15–0.20 mm (0.006–0.008 in))

13. Select the washer with the nearest value.

14. Preload of the taper roller bearing (amount of standard protrusion): 0.15–0.20 mm (0.006–0.008 in)

➡**Be sure that it is always within the preload.**

15. Fit the thrust washers on the transfer drive shaft.

16. Install the bearing outer race into the extension case.

17. Measure the depth "S" between transfer case and center differential.

a. ST 398643600 GAUGE

18. Measure the height "T" between the extension case and transfer drive gear.

a. ST 398643600 GAUGE

19. Calculate the thrust washer thickness "U" using the following calculation.

a. U = S + T - 30 mm (1.18 in) [Thickness of ST]

➡**Thickness of ST: 15 mm (0.59 in)**

20. Select a suitable washer.

a. Standard clearance: 0.15–0.35 mm (0.0059–0.0138 in)

21. Fit the thrust washer onto the center differential.

22. Apply a proper amount of liquid gasket to the transfer case mating surface.

a. Liquid gasket: THREE BOND 1215 (Part No. 004403007) or equivalent

23. Install the extension case assembly to the transfer case. Tighten to 29.5 ft. lbs. (40 Nm).

24. Install the extension case assembly along with the transfer case to the transmission case.

a. Insert the axle shaft and hold, then move the HANDLE to align the spline position to install.

• ST 18631AA000 HANDLE

• Part No. 38415AA100 Axle shaft

➡**Push the transfer case assembly until the knock pin protrudes slightly from the transfer case.**

b. Set and hold the selector lever to the 1st- 2nd side, and install the transfer case assembly

so that the shifter arm edge and fork rod does not contact.

25. Temporarily tighten the transfer case assembly.

26. Install the shift lever COMPL to the shifter arm No. 2, then insert the remover or similar tool instead of spring pin.

27. Operate the selector arm and remover to confirm the correct shifting into each gear is possible.

28. Tighten the transfer case assembly to 18 ft. lbs. (25 Nm).

29. Install the shift lever COMPL to the shifter arm No. 2, then fix it using a spring pin.

➡**Use new spring pin.**

30. Install the transfer cover.

a. Temporarily tighten the transfer cover

mounting bolts.

b. Tighten bolts in order to the specified torque. Tightening torque: 133 inch lbs. (15 Nm).

31. Install the transmission cover.

ENGINE COOLING

ENGINE FAN

REMOVAL & INSTALLATION

2.5L Engine

1. Before servicing the vehicle, refer to the Precautions Section.
2. Disconnect the negative battery cable.
3. Raise and safely support the vehicle.
4. Disconnect the fan electrical connector.

5. Remove the transmission cooler hose from the clip of the radiator fan motor assembly, if equipped.
6. Lower the vehicle.
7. Disconnect the overflow hose and reservoir tank.
8. Remove the fan shroud mounting bolts.
9. Remove the main fan assembly.

To install:

10. Installation is the reverse order of assembly.

11. Tighten the fan shroud mounting bolts to 6 ft. lbs. (7.5 Nm).

3.0L Engine

See Figure 59.

1. Before servicing the vehicle, refer to the Precautions Section.
2. Disconnect the negative battery cable.
3. Remove the hood stay holder.
4. Remove the air intake assembly.

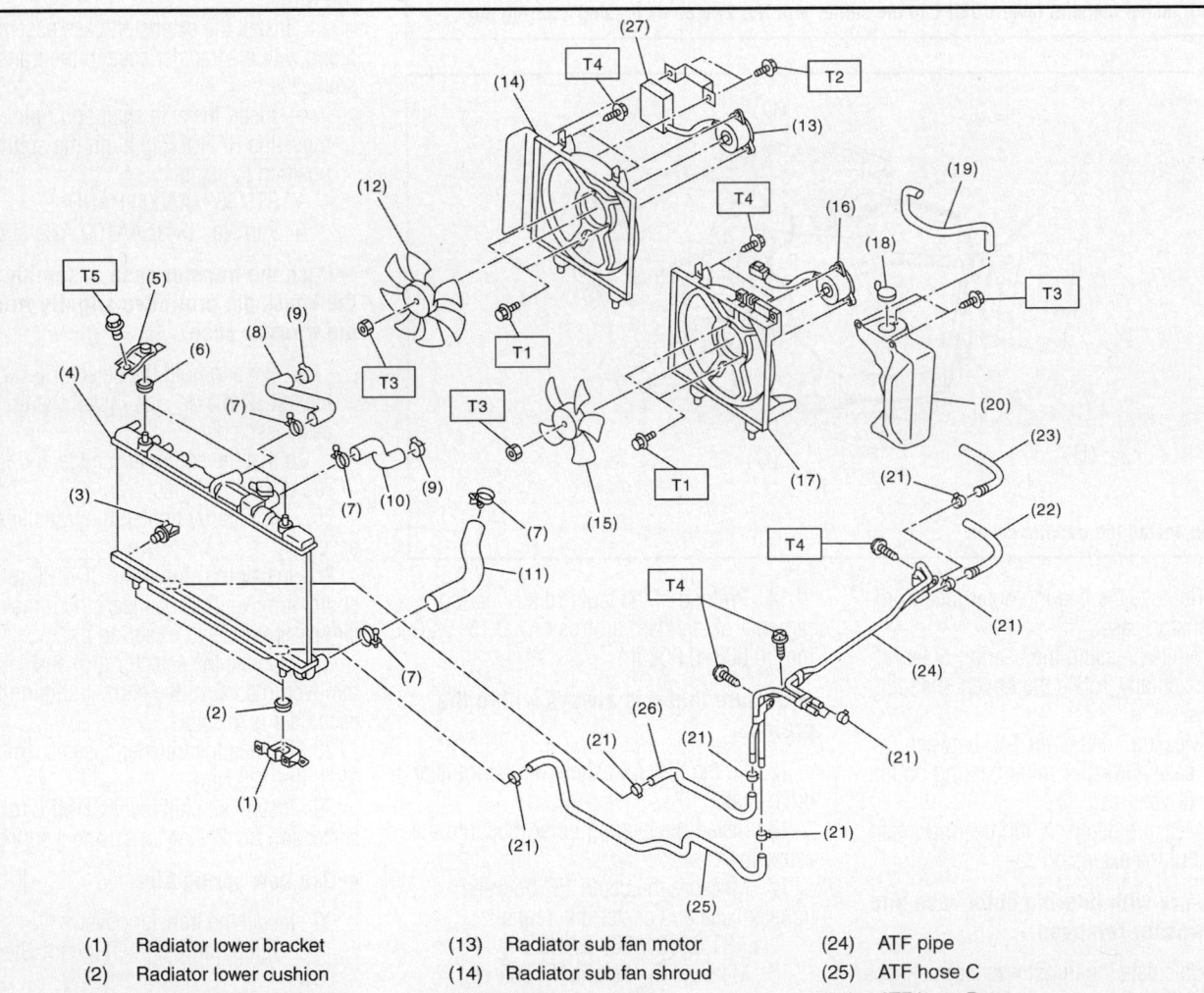

(1)	Radiator lower bracket	(13)	Radiator sub fan motor	(24)	ATF pipe	
(2)	Radiator lower cushion	(14)	Radiator sub fan shroud	(25)	ATF hose C	
(3)	Engine coolant drain plug	(15)	Radiator main fan	(26)	ATF hose D	
(4)	Radiator	(16)	Radiator main fan motor	(27)	Radiator fan control unit	
(5)	Radiator upper bracket	(17)	Radiator main fan shroud			
(6)	Radiator upper cushion	(18)	Engine coolant reservoir tank cap			
(7)	Clamp	(19)	Over flow hose			
(8)	Radiator hose A	(20)	Engine coolant reservoir tank			
(9)	Clamp	(21)	ATF hose clamp			
(10)	Radiator hose B	(22)	ATF hose A			
(11)	Radiator hose C	(23)	ATF hose B			
(12)	Radiator sub fan					

Tightening torque:N·m (kgf-m, ft-lb)

T1:	3.8 (0.39, 2.8)
T2:	5.4 (0.55, 4.0)
T3:	6.2 (0.63, 4.6)
T4:	7.5 (0.76, 5.5)
T5:	12 (1.2, 8.9)

22140_SUBA_G0100

Fig. 59 Exploded view of radiator and fan system

5. Disconnect the connector from the radiator fan control unit.

6. Raise and safely support the vehicle.

7. Remove the engine undercover.

8. Drain the engine coolant.

9. Remove the transmission cooler hose from the clip of the radiator fan motor assembly, if equipped.

10. Remove the radiator fan motor harness from the clip.

11. Lower the vehicle.

12. Remove the reservoir tank.

13. Remove the radiator sub-fan assembly.

14. Remove the radiator fan assembly.

➡**When removing the main fan assembly, lifting it straight up will cause the main fan shroud contacts to inlet part of engine coolant. To avoid contacting it, move the main fan assembly to sub fan assembly side before removal.**

To install:

15. Installation is the reverse order of assembly.

16. Tighten the fan shroud mounting bolts to 6 ft. lbs. (7.5 Nm).

17. Refill the engine cooling system to the correct level.

RADIATOR

REMOVAL & INSTALLATION

See Figures 60 and 61.

1. Before servicing the vehicle, refer to the Precautions Section.

2. Set the vehicle on a lift.

3. Disconnect the ground cable from battery.

4. Lift-up the vehicle.

5. Remove the undercover.

6. Drain engine coolant completely.

7. Remove the radiator undercover.

8. Disconnect the radiator main and sub fan motor connectors.

9. Disconnect the radiator outlet hose from thermostat cover.

10. Disconnect the ATF cooler hoses from ATF pipes. (AT models)

Plug the ATF pipe to prevent ATF from leaking.

11. Lower the vehicle.

12. Disconnect the over flow hose.

13. Remove the V-belt covers for (2.5L DOHC) engines.

14. Remove the reservoir tank.

15. Remove the hood stay holder.

16. Remove the air intake duct.

17. Disconnect the connector from radiator fan control unit for (3.0L) engines.

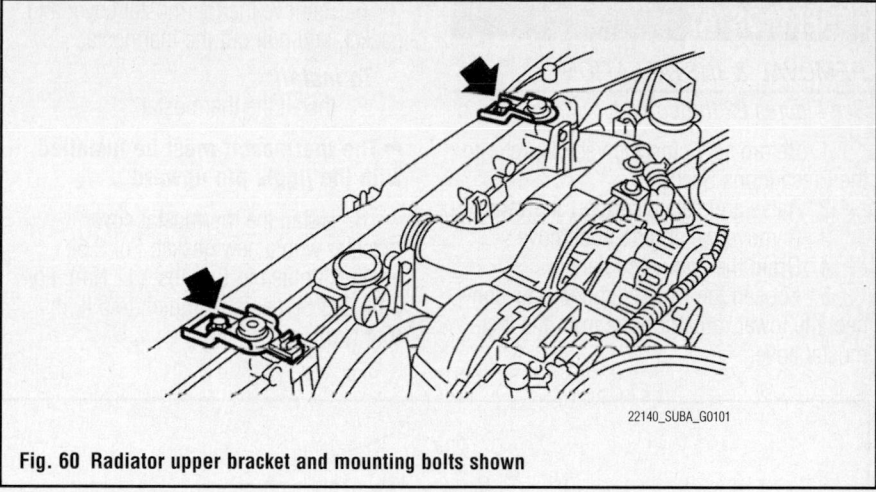

Fig. 60 Radiator upper bracket and mounting bolts shown

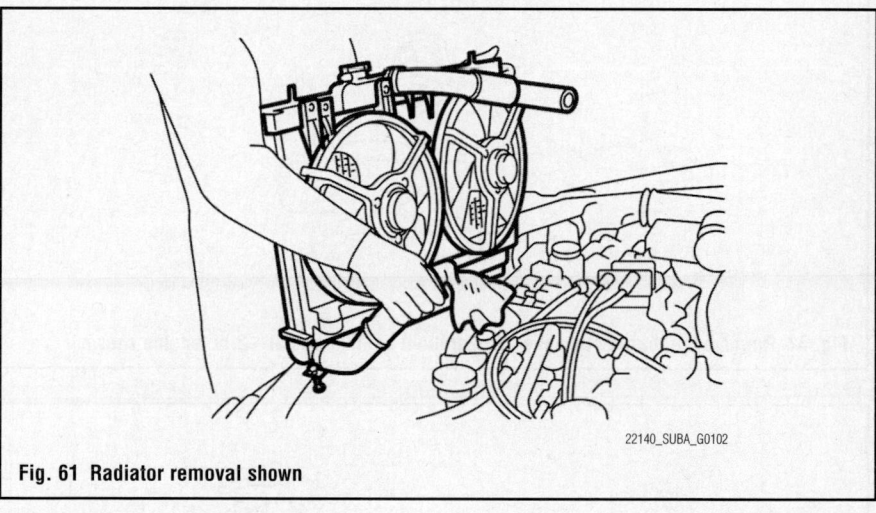

Fig. 61 Radiator removal shown

18. Disconnect the radiator inlet hoses from the engine.

19. Remove the radiator upper brackets.

20. Detach the power steering hose from the clip on the radiator for (2.5L DOHC) engines.

21. Move the radiator to the left while lifting it upward.

22. Lift the radiator up and away from vehicle.

To install:

23. Attach the radiator lower cushion to the hole on the radiator lower bracket.

24. Install the radiator to vehicle.

➡**Make pins on the lower side of radiator be fitted into the radiator lower cushions on body side.**

25. Install the radiator upper brackets and tighten the bolts 9 ft. lbs. (12 Nm).

26. Attach the power steering hose to the radiator for (2.5L DOHC) engines.

27. Connect the radiator inlet hose.

28. Install the air intake duct.

29. Connect the connector from radiator fan control unit for (3.0L) engines.

30. Install the hood stay holder.

31. Install the reservoir tank.

32. Connect the over flow hose.

33. Install the V-belt covers for (2.5L DOHC) engines.

34. Lift-up the vehicle.

35. Connect the ATF cooler hoses. (AT models)

36. Connect the radiator outlet hose.

37. Connect the radiator main and sub fan motor connectors.

38. Install the radiator undercover.

39. Install the undercover.

40. Lower the vehicle.

41. Connect the battery ground cable to the battery.

42. Fill engine coolant and bleed the cooling system.

43. Check the ATF level.

THERMOSTAT

REMOVAL & INSTALLATION

See Figures 62 through 64.

1. Before servicing the vehicle, refer to the Precautions Section.
2. Raise and safely support the vehicle.
3. Remove the engine undercover.
4. Drain the engine coolant.
5. Loosen the hose clamp and disconnect the lower radiator hose from the thermostat cover.

6. Remove the thermostat cover and gasket, and pull out the thermostat.

To install:

7. Install the thermostat.

➡ **The thermostat must be installed with the jiggle pin upward.**

8. Install the thermostat cover together with a new gasket. For 2.5L engines tighten to 9 ft. lbs. (12 Nm). For 3.0L & 3.6L engines tighten to 5 ft. lbs. (6.4 Nm).

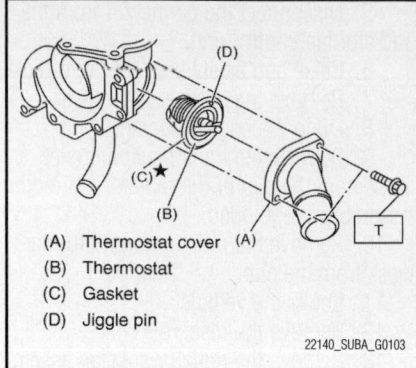

(A) Thermostat cover   (A)
(B) Thermostat
(C) Gasket
(D) Jiggle pin

22140_SUBA_G0103

Fig. 64 Exploded view of the thermostat and related components—3.0L engine shown 3.6L similar

9. Install the lower radiator hose to the thermostat cover and tighten the hose clamp.
10. Install the engine undercover.
11. Lower the vehicle.
12. Refill the engine cooling system to the correct level.
13. Start the engine and check for leaks.

WATER PUMP

REMOVAL & INSTALLATION

2.5L Engine

See Figures 65 and 66.

1. Before servicing the vehicle, refer to the Precautions Section.
2. Remove or disconnect the following:
   - Negative battery cable
   - Engine undercover, if equipped
3. Drain the coolant into a suitable container.
4. Remove or disconnect the following:
   - Radiator fan connectors
   - Radiator outlet and heater hoses
   - Heater bypass hose or overflow hose, if equipped
   - Reservoir tank, on Legacy
   - Radiator fan motor assembly
   - Accessory drive belts
   - Timing belt
   - Belt tension adjuster
   - Belt idler No. 2
   - Camshaft Position (CMP) sensor
   - Left side camshaft pulley
   - Left side rear timing belt cover
   - Tensioner bracket
   - Radiator and heater hoses from water pump
   - Water pump retainer bolts
   - Water pump
5. Inspect the radiator hoses for deterioration and replace as necessary.

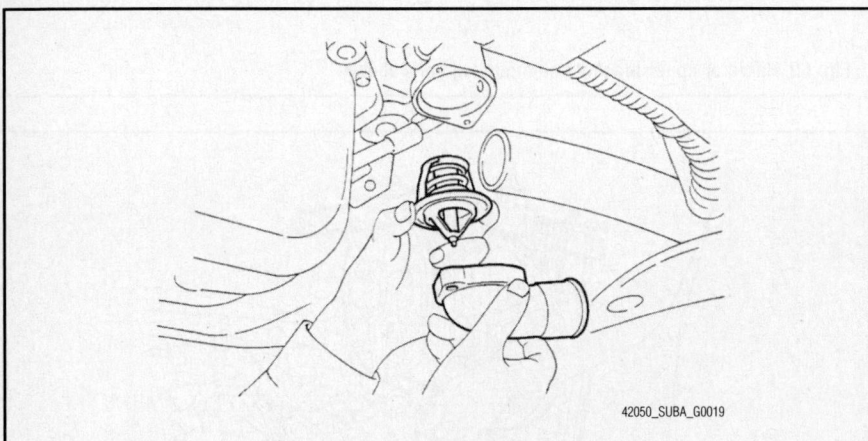

42050_SUBA_G0019

Fig. 62 Remove the thermostat cover, and pull out the thermostat—2.5L engine shown

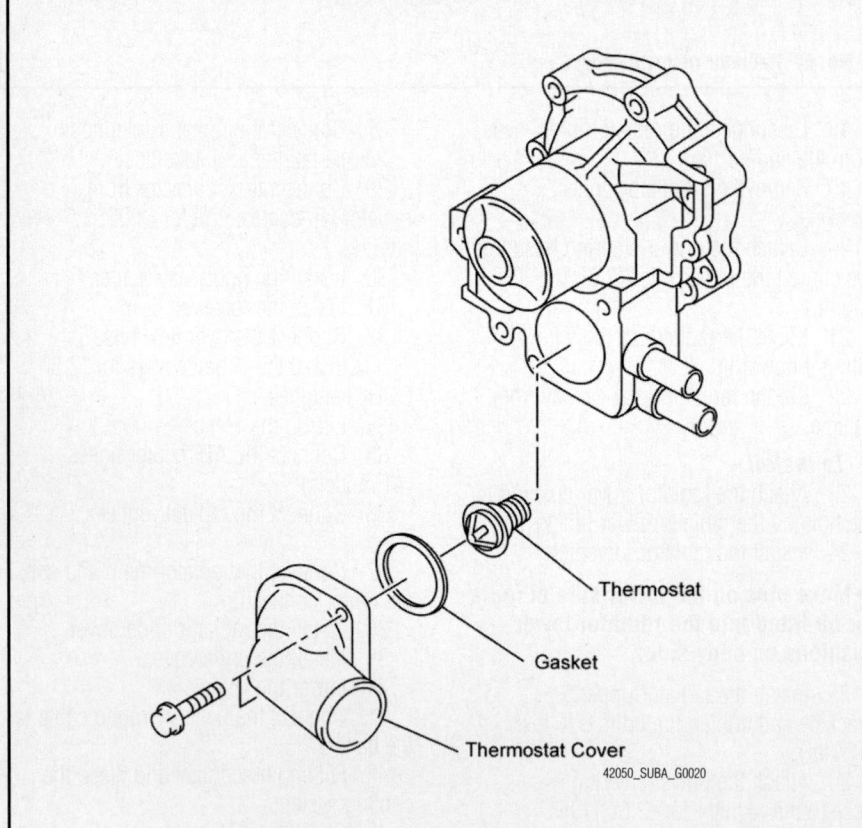

Thermostat

Gasket

Thermostat Cover

42050_SUBA_G0020

Fig. 63 Exploded view of the thermostat and related components—2.5L Turbo engine shown

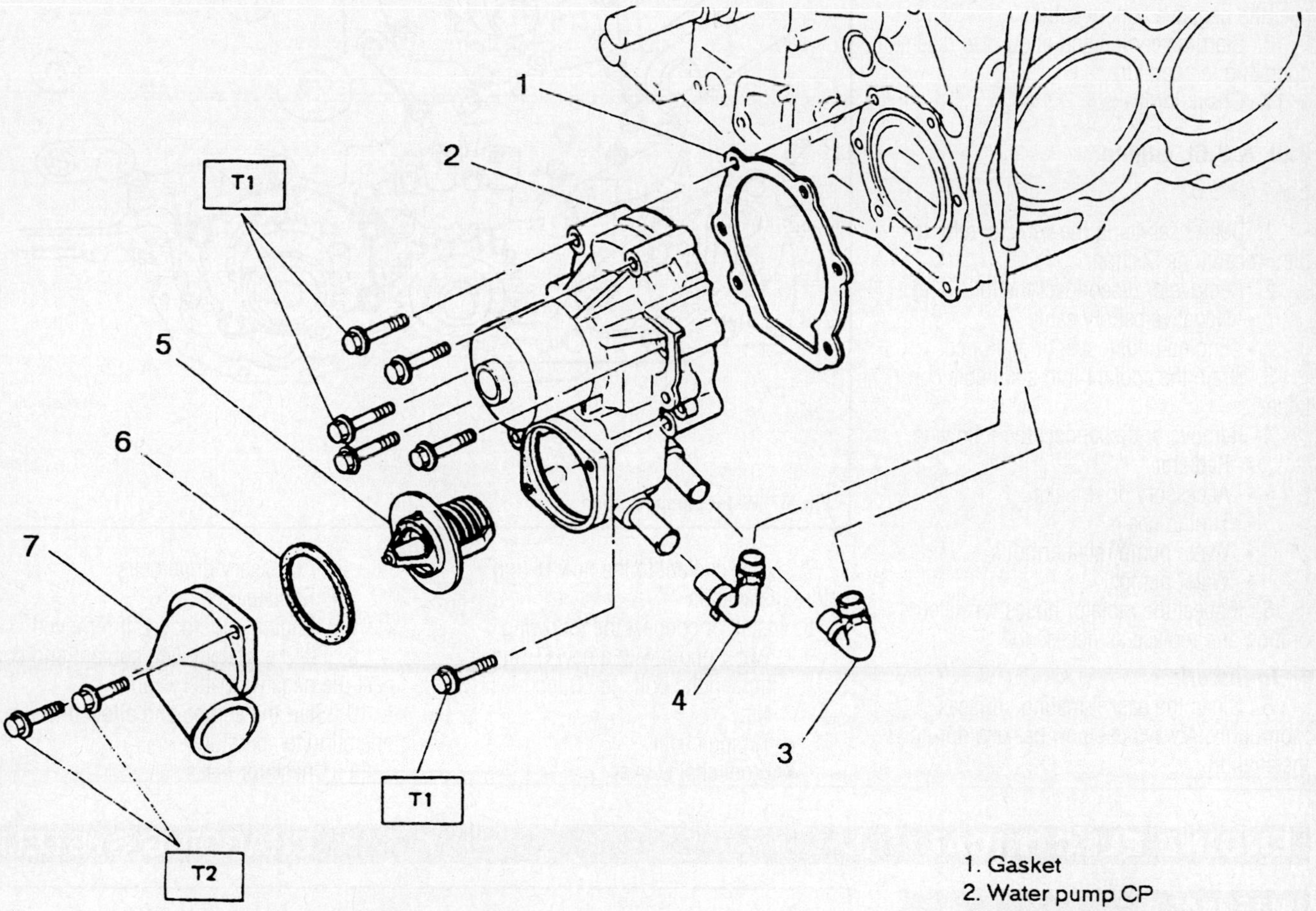

Tightening torque: N.m (kg-m, ft-lb)
T1: First   10 − 14 (1.0 − 1.4, 7 − 10)
      Second 10 − 14 (1.0 − 1.4, 7 − 10)
T2: 6 − 7 (0.6 − 0.7, 4.3 − 5.1)

1. Gasket
2. Water pump CP
3. Heater hose (inlet)
4. Heater hose (outlet)
5. Thermostat
6. Gasket
7. Thermostat cover

7923TG01

Fig. 65 Water pump and related components

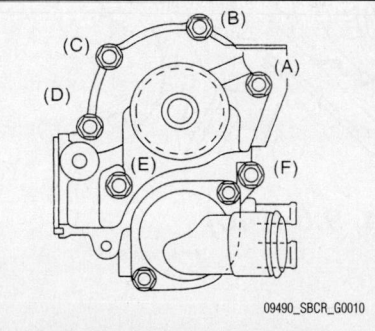

09490_SBCR_G0010

Fig. 66 Tighten the water pump bolts in two steps using the following sequence— 2.5L engine

To install:

6. Clean the gasket mating surfaces thoroughly. Always use new gaskets during installation.

7. Install or connect the following:

8. Tighten the pump bolts in sequence to 8.9 ft. lbs. (12 Nm). After tightening the bolts once, retighten to the same specification again.

- Radiator heater hoses to water pump
- Tensioner bracket and tighten to 18 ft. lbs. (25 Nm)
- Left side rear timing belt cover
- Left side camshaft pulley(s).

Tighten to 58 ft. lbs. (78 Nm) on non turbocharged engine and 72 ft. lbs. (98 Nm) on turbocharged engine.

- CMP sensor
- Belt idler No. 2 and tighten to 29 ft. lbs. (39 Nm)
- Belt tension adjuster
- Timing belt
- Accessory drive belts
- Radiator fan assembly
- Reservoir tank, if removed
- Heater bypass hose or overflow hose, if equipped
- Air intake duct

- Radiator outlet and heater hoses
- Radiator fan connectors
- Engine undercover, if removed

9. Fill the system with coolant and connect the negative battery cable.

10. Start the engine and allow it to reach operating temperature.

11. Check for leaks.

3.0L & 3.6L Engine

See Figure 67.

1. Before servicing the vehicle, refer to the Precautions Section.

2. Remove or disconnect the following:
- Negative battery cable
- Engine undercover, if equipped

3. Drain the coolant into a suitable container.

4. Remove or disconnect the following:
- Radiator
- Accessory drive belts
- Timing chain
- Water pump retainer bolts
- Water pump

5. Inspect the radiator hoses for deterioration and replace as necessary.

To install:

6. Clean the gasket mating surfaces thoroughly. Always use new gaskets during installation.

Fig. 67 Water pump location

7. Apply coolant to the new O-ring before installation

8. Install or connect the following:
- Water pump with a new O-ring, tighten the bolts to 60 inch lbs. (7 Nm).
- Timing chain
- Front chain cover

- Accessory drive belts
- Radiator
- Engine undercover, if removed

9. Fill the system with coolant and connect the negative battery cable.

10. Start the engine and allow it to reach operating temperature.

11. Check for leaks.

ENGINE ELECTRICAL   CHARGING SYSTEM

ALTERNATOR

REMOVAL & INSTALLATION

See Figure 68.

1. Before servicing the vehicle, refer to the Precautions Section.

2. Remove or disconnect the following:
- Negative battery cable
- Connector and terminal from the alternator
- V-belt cover, if equipped
- Front side V-belt
- Alternator to bracket bolts
- Alternator from the vehicle

To install:

3. Install or connect the following:
- Alternator into the vehicle
- Alternator to bracket bolts
- Front side V-belt
- V-belt cover, if equipped
- Connector and terminal to the alternator
- Negative battery cable

4. Check and adjust the belt tension.

Tightening torque:
13 N·m (1.3 kgf-m, 9.6 ft-lb)

Fig. 68 Alternator and mounting bolts

VOLTAGE REGULATOR

ADJUSTMENT

The voltage regulator is not adjustable. If the regulator is found to be defective, it must be replaced.

REMOVAL & INSTALLATION

See Figure 69.

1. Before servicing the vehicle, refer to the Precautions Section.
2. Remove the alternator from the vehicle.
3. Remove the four through bolts. Then insert the tip of a flat-head screwdriver into the gap between the stator core and front bracket. Pry them apart to disassemble the alternator.
4. Hold rotor with a vise and remove pulley nut.

✳✳ CAUTION

When holding rotor with vise, insert aluminum plates or wood pieces on

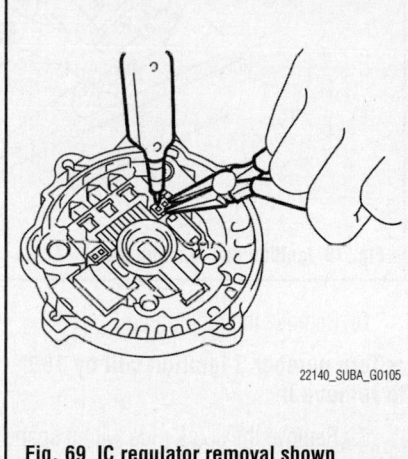

22140_SUBA_G0105

Fig. 69 IC regulator removal shown

the contact surfaces of the vise to prevent rotor from damage.

5. Unsolder connection between rectifier and stator coil to remove stator coil.

✳✳ CAUTION

Finish the work rapidly (less than three seconds) because the rectifier cannot withstand heat very well.

6. Remove screws which secure voltage regulator to rear cover, and unsolder connection between voltage regulator and rectifier to remove the regulator.

To install:

7. Installation is the reverse order of assembly.
8. After assembly, turn the pulley by hand to check that the rotor turns smoothly.

ENGINE ELECTRICAL

IGNITION TIMING

INSPECTION

See Figure 70.

METHOD WITH SUBARU SELECT MONITOR

Before checking the ignition timing, check the following:

• Ensure the air cleaner element is free from clogging, spark plugs are in good condition, and that hoses are connected properly.

• Ensure the malfunction indicator light does not illuminate.

1. Idle the engine.
2. Stop the engine, and turn the ignition switch to OFF.
3. Insert the cartridge to the Subaru Select Monitor.
4. Connect the Subaru Select Monitor to data link connector.
5. Turn the ignition switch to ON, and Subaru Select Monitor power switch to ON.
6. Select Each System Check in the Main Menu.
7. Select Engine in the Selection Menu.
8. Select Current Data Display & Save in the Engine Control System Diagnosis.
9. Select Data Display in the Data Display Menu.
10. Start the engine, and read the ignition timing by idle speed.
11. If the timing is not correct, check the ignition control system.

Ignition timing is as follows:

• 10° +/− 8°650 rpm (MT Models)

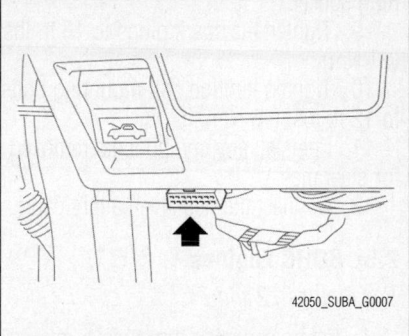

42050_SUBA_G0007

Fig. 70 Location of the data link connector under the driver's side instrument panel.

• 15° +/− 8°700 rpm (AT Models)
• 17° +/− 10°700 rpm (STI Model)

METHOD WITH TIMING LIGHT

12. Before checking the ignition timing, check the following item:
13. Check the air cleaner element is free from clogging, spark plugs are in good condition, and hoses are connected properly.
14. Check the malfunction indicator light does not illuminate.
15. Warm-up the engine.
16. Stop the engine, and turn the ignition switch to OFF.
17. Remove the air intake duct.
18. Disconnect the connectors of the mass air flow and intake air temperature sensor.
19. Remove the air cleaner case and element.
20. Connect the timing light to the power wire of #1 ignition coil.

21. Attach the air cleaner case, element and connector of mass air flow and intake air temperature sensor.
22. Start the engine, turn the timing light to the crank pulley, and check the ignition timing by means of crank pulley indicator.
23. If the timing is not correct, check the ignition control system.

Ignition timing is as follows:

• 10° +/− 8°650 rpm (MT Models)
• 15° +/− 8°700 rpm (AT Models)
• 17° +/− 10°700 rpm (STI Model)

IGNITION SYSTEM

ADJUSTMENT

The ignition timing is controlled by the Powertrain Control Module (PCM). No adjustment is necessary or possible.

SPARK PLUGS

REMOVAL & INSTALLATION

2.5L DOHC Turbo Engine

Left Side

See Figures 71 and 72.

1. Before servicing the vehicle, refer to the Precautions Section.
2. Remove the collector cover.
3. Remove the battery and battery carrier.
4. Disconnect the secondary air pump duct from the secondary air pump.
5. Remove the bolts that attach the secondary air pump duct to the rocker cover (LH), and raise the secondary air pump duct.

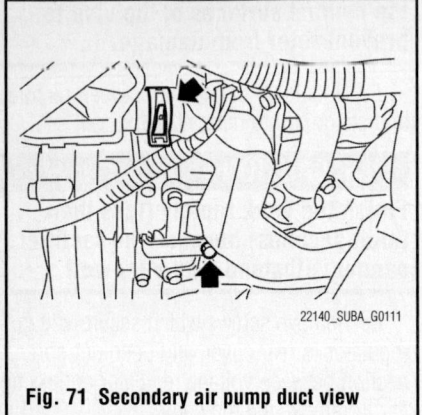

Fig. 71 Secondary air pump duct view

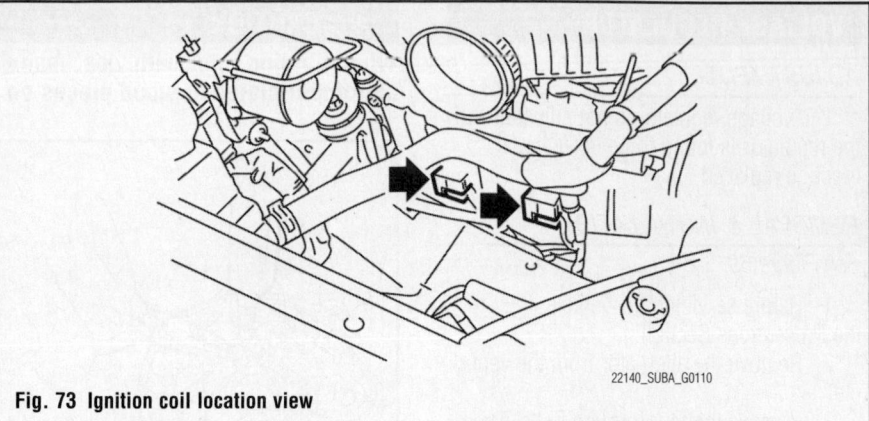

Fig. 73 Ignition coil location view

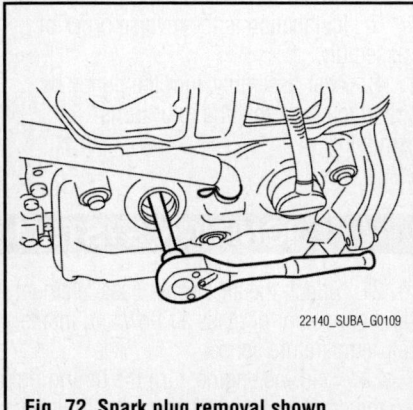

Fig. 72 Spark plug removal shown

6. Disconnect the connector from ignition coil.

7. Remove the ignition coil.

→**Turn number 4 ignition coil by 180° to remove it.**

8. Remove the spark plug with a spark plug socket.

To install:

9. When installing spark plugs on cylinder head, use the appropriate spark plug socket.

10. Tighten the spark plugs to 15 ft. lbs. (21 Nm).

11. Tighten ignition coil mounting bolts to 12 ft. lbs. (16 Nm).

12. Reinstall any components removed for clearance.

13. Connect the negative battery cable.

Right Side

See Figures 72 and 73.

1. Before servicing the vehicle, refer to the Precautions Section.

2. Remove the collector cover.

3. Disconnect the ground cable from the battery.

4. Remove the air cleaner case.

5. Disconnect the connector from ignition coil.

6. Remove the ignition coil.

→**Turn number 3 ignition coil by 180° to remove it.**

7. Remove the spark plug with a spark plug socket.

To install:

8. When installing spark plugs on cylinder head, use the appropriate spark plug socket.

9. Tighten the spark plugs to 15 ft. lbs. (21 Nm).

10. Tighten ignition coil mounting bolts to 12 ft. lbs. (16 Nm).

11. Reinstall any components removed for clearance.

12. Connect the negative battery cable.

2.5L SOHC Engines

See Figures 72 and 74.

1. Before servicing the vehicle, refer to the Precautions Section.

2. Disconnect the negative battery cable.

3. Remove the following components if necessary for access to remove the spark plugs:

- Battery
- Air intake assembly

Fig. 74 Spark plug wires and boots shown

✶✶ CAUTION

Do not disconnect the washer tank hoses.

4. Gripping the spark plug wire by the boot, remove the spark plug wires.

5. Remove the spark plugs.

To install:

6. When installing spark plugs on cylinder head, use the appropriate spark plug socket. Tighten the spark plugs to 15 ft. lbs. (21 Nm).

7. Connect the spark plug wires.

8. Reinstall any components removed for clearance.

9. Connect the negative battery cable.

3.0L & 3.6L Engines

Left Side

See Figures 75 and 76.

1. Before servicing the vehicle, refer to the Precautions Section.

2. Remove the collector cover.

3. Remove the battery and battery carrier.

4. Remove the bracket.

5. Disconnect the connector from ignition coil.

6. Remove the ignition coil.

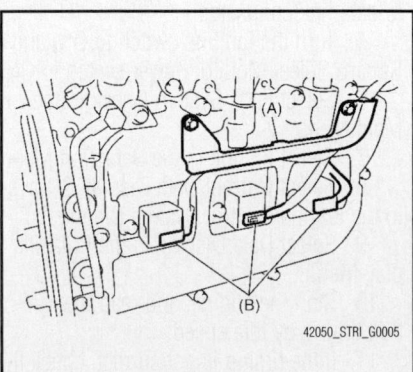

Fig. 75 Removing bracket and connector from the ignition coil—left side

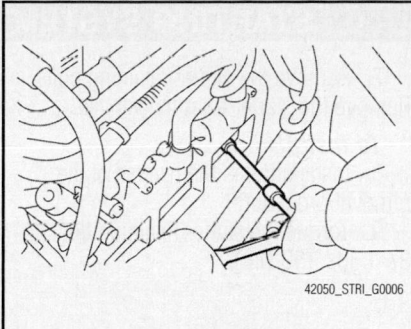

Fig. 76 Removing the spark plug with a spark plug socket—left side

➡️**Turn the no. 6 ignition coil to remove it.**

7. Remove the spark plug with a spark plug socket.

To Install:

8. To install reverse the removal procedures, and note the following:
- Tighten the spark plug to 15 ft. lbs. (21 Nm)
- Tighten the ignition coil to 12 ft. lbs. (16 Nm)

➡️**The tightening torque described above should be applied to only new spark plugs without oil on their threads. In case their threads are lubricated, the torque should be reduced by approximately ⅓ of the specified torque in order to avoid over—stressing.**

Right Side

See Figures 77 through 79.

1. Before servicing the vehicle, refer to the Precautions Section.
2. Remove the collector cover.
3. Disconnect the ground cable from battery.
4. Remove the air cleaner case.
5. Remove the bracket.
6. Disconnect the connector from ignition coil.
7. Remove the ignition coil.

➡️**Turn the #5 ignition coil to remove it.**

8. Remove the spark plug with a spark plug socket.

To install:

9. To install reverse the removal procedures, and note the following:
- Tighten the spark plug to 15 ft. lbs. (21 Nm)
- Tighten the ignition coil to 12 ft. lbs. (16 Nm)

➡️**The tightening torque described above should be applied to only new spark plugs without oil on their threads. In case their threads are**

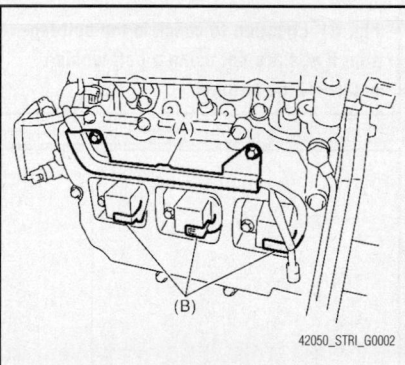

Fig. 77 Removing bracket and connector from ignition coil—right side

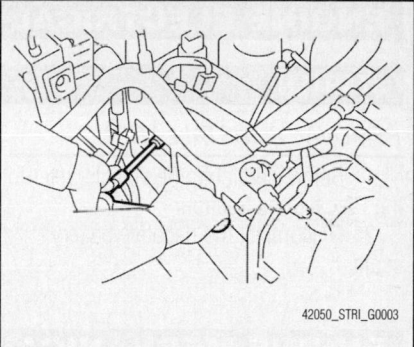

Fig. 78 Remove the spark plug with a spark plug socket—right side

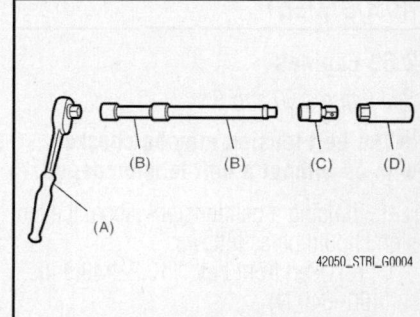

Fig. 79 Identifying (A) ratchet handle, (B) extension bar, (C) universal joint, (D) spark plug socket

lubricated, the torque should be reduced by approximately⅓of the specified torque in order to avoid over—stressing.

SPARK PLUG WIRES

REMOVAL & INSTALLATION

Refer to Spark plug "Removal & Installation".

STARTER

REMOVAL & INSTALLATION

1. Before servicing the vehicle, refer to the Precautions Section.

2. Disconnect the negative battery cable.

3. Remove the air intake chamber, on non turbocharged engine.

4. Remove the intercooler, on turbocharged engine.

5. Remove the air intake chamber stay, on non turbocharged engine.

6. Disconnect the electrical connectors from the starter.

7. Remove the starter retaining bolts. Remove the starter from the vehicle.

To install:

8. Installation is the reverse of the removal procedure.

9. Torque the starter retaining bolts to 37 ft. lbs. (50 Nm).

ENGINE MECHANICAL

ACCESSORY DRIVE BELTS

ADJUSTMENT

2.5L Engines

See Figures 80 and 81.

➡The belt tension may be checked with or without a belt tension gauge.

1. If using a belt tension gauge, the tension should be as follows:

   a. Used front belt: 110.2–143.9 lbs. (490–640 N)

   b. New front belt: 143–175 lbs. (640–780 N)

   c. Used rear belt: 78.7–101.2 lbs. (350–450 N)

   d. New rear belt: 166–198 lbs. (740–880 N)

2. If you are not using a belt tension gauge, using 22 lbs. (98 Nm) of force, push on the belt as shown in the illustration. The total distance the belt travels up and down should be:

   a. Used front belt: 0.354–0.433 in. (9–11 mm)

   b. New front belt: 0.276–0.354 in. (7–9 mm)

   c. Used rear belt: 0.354–0.394 in. (9.0–10.0 mm)

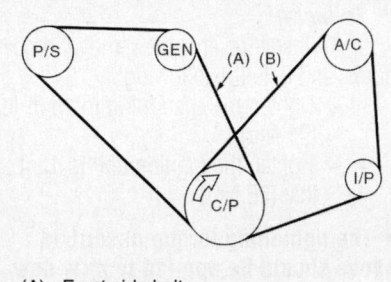

(A) Front side belt
(B) Rear side belt

42050_SUBA_G0017

Fig. 81 Location to check to the belt tension if you are not using a belt tension gauge—2.5L engine

   d. New rear belt: 0.295–0.335 in. (7.5–8.5 mm)

3.0L & 3.6L Engines

See Figure 82.

1. With the belt installed, check that the automatic belt tension indicator is within the range.

2. Replace the belt if the indicator is outside of the service limit.

REMOVAL & INSTALLATION

2.5L Engine

1. Before servicing the vehicle, refer to the Precautions Section.

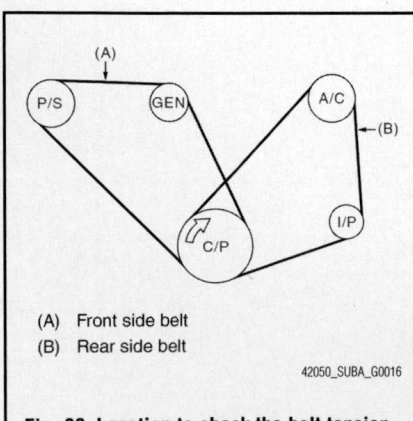

(A) Front side belt
(B) Rear side belt

42050_SUBA_G0016

Fig. 80 Location to check the belt tension if you are using a belt tension gauge—2.5L engine

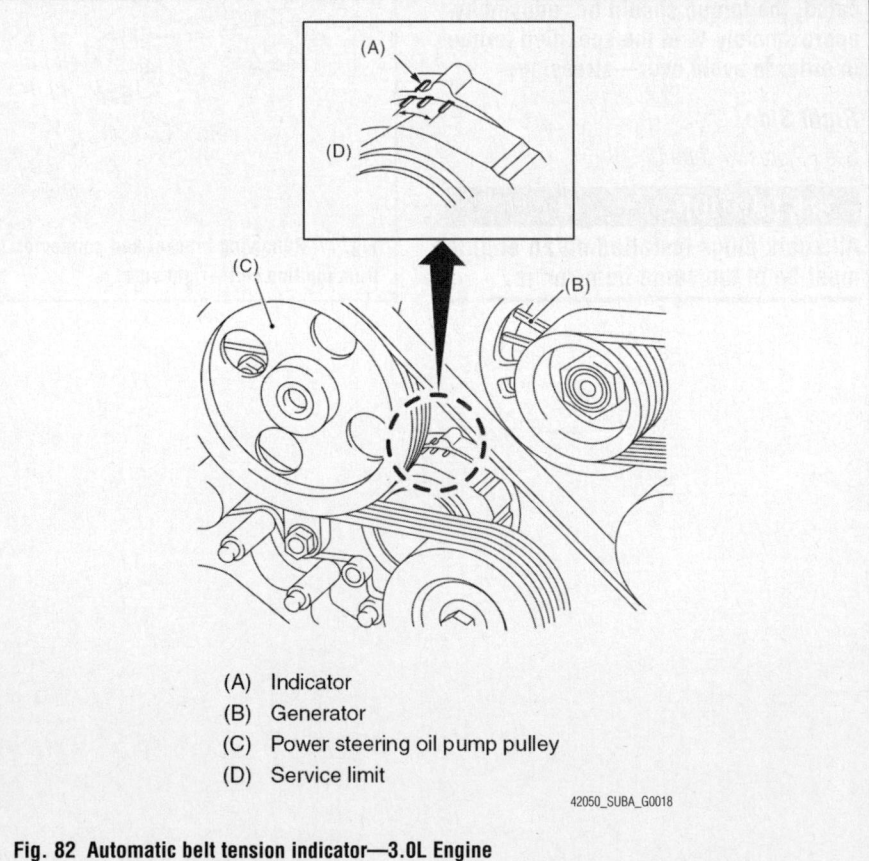

(A) Indicator
(B) Generator
(C) Power steering oil pump pulley
(D) Service limit

42050_SUBA_G0018

Fig. 82 Automatic belt tension indicator—3.0L Engine

2. Remove the accessory drive belt cover, if equipped.

3. Loosen the lock bolt (bottom) of the front belt tensioner.

4. Loosen the slider bolt (top) of the front belt tensioner.

5. Remove the front side belt.

6. Loosen the lock bolt (bottom) of the rear belt tensioner.

7. Loosen the slider bolt (top) of the rear belt tensioner.

8. Remove the rear side belt.

To install:

➡**Wipe off any oil or water on the belt and pulley.**

9. Install the rear side belt, then tighten the slider bolt until the belt reaches the specified tension. Please refer to the following topic: "Accessory Drive Belts", "Adjustment."

10. Tighten the lock nut of the rear belt tensioner to 18 ft. lbs. (25 Nm).

11. Install the front side belt, then tighten the slider bolt until the belt reaches the specified tension. Please refer to the following topic: "Accessory Drive Belts", "Adjustment."

12. Tighten the lock nut of the front belt tensioner to 17 ft. lbs. (23 Nm).

3.0L & 3.6L DOHC Engines

See Figure 83.

1. Before servicing the vehicle, refer to the Precautions Section.

2. Remove the collector cover.

3. Install the tool to belt tension adjuster assembly installation bolt.

4. Rotate the tool clockwise and loosen the V—belt to remove.

5. Remove the accessory drive belt.

To install:

6. Installation is the reverse order of removal.

22140_SUBA_G0117

Fig. 83 Rotate the tool clockwise

CAMSHAFT AND VALVE LIFTERS

INSPECTION

1. Remove the camshaft from the engine.

2. Check the camshaft bearing journals for damage and binding.

3. If the journals are binding, check the cylinder head for damage.

4. Check the cylinder head for clogged oil holes.

5. Check the camshaft surface for abnormal wear and damage. Replace the camshaft, as required.

6. Measure the camshaft lobe surface and replace the camshaft if not within specification.

7. Measure the camshaft journal diameter and replace the camshaft if not within specification.

8. Measure the camshaft run out and replace the camshaft if not within specification.

REMOVAL & INSTALLATION

2.5L DOHC Engine

See Figures 84 through 87.

1. Before servicing the vehicle, refer to the Precautions Section.

2. Disconnect the negative battery cable.

3. Remove the collector cover.

4. To remove the front side V belt, remove the belt covers. Loosen the lock bolt. Loosen the slider bolt. Remove the front side belt.

5. To remove the rear side V belt, remove the belt covers. Loosen the lock bolt. Loosen the slider bolt. Remove the rear side belt. Remove the belt tensioner.

6. Remove the crankshaft pulley bolt.

7. Lock the crankshaft in place using tool ST499977100, or equivalent.

8. Remove the crankshaft pulley.

9. Remove the left side timing belt cover.

10. Remove the right side timing belt cover.

11. Remove the front timing belt cover.

12. Remove the timing belt.

13. Remove the camshaft position sensor.

14. Remove the camshaft sprockets.

➡**Be sure to lock the camshaft in place using tool ST499207400, or equivalent.**

15. Lock the crankshaft in place using tool ST499977100, or equivalent.

16. Remove the crankshaft pulley.

17. Remove the tensioner bracket. Remove the right and left timing belt No.2 covers.

18. Remove the spark plug wires. Remove the oil level gauge, on the left side.

19. Remove the rocker cover and gasket. Remove the oil pipe.

20. Loosen the oil flow control solenoid valve assembly and the intake camshaft cap bolts equally and in the proper sequence.

21. Loosen the exhaust camshaft cap bolts equally and in the proper sequence.

22. Remove the oil flow control solenoid valve assembly, intake camshaft cap and camshaft.

23. Remove the exhaust camshaft caps and camshaft.

➡**Arrange the camshafts caps so that they can be installed in their original positions.**

To install:

➡**Lubricate the camshaft journals with clean engine oil prior to installation.**

24. Install the camshaft so that the valves are closed or in contact with the "base circle" of the cam lobe.

25. If the camshafts are positioned as shown in the, the camshafts need to be rotated at a minimum to align the timing belt during installation.

26. The right hand camshaft need not be rotated when set at the position illustrated, the left hand intake camshaft should be rotated 80 degrees clockwise. The left hand exhaust camshaft should be rotated 45 degrees counterclockwise.

27. To install the camshaft cap and oil flow control solenoid valve, apply a small amount of liquid gasket to the mating surface of the cap.

➡**Do not apply an excessive amount of sealant, as it will squish out and flow toward the seal resulting in an oil leak.**

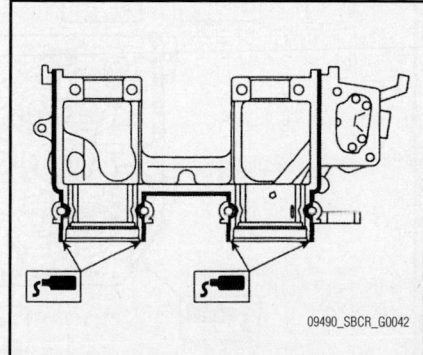

09490_SBCR_G0042

Fig. 84 Camshaft cap liquid gasket application points—2.5L DOHC engine

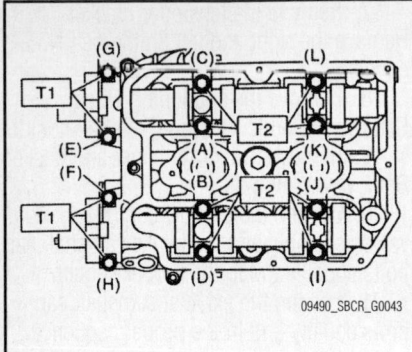

Fig. 85 Camshaft bolt tightening sequence—2.5L DOHC engine

28. Apply a thin coat of engine oil to the cap bearing surface and install the cap according to the cap identification mark. Gradually tighten the cap bolts in two stages, first to 7.2 ft. lbs. and then to 14.5 ft. lbs. in the proper sequence.

➡**After tightening the camshaft cap, ensure that the camshaft rotates slightly while holding it at base circle.**

29. Using tools ST49587600 and ST499597200, install a new seal on the camshaft. Be sure to coat the seal with clean engine oil before installation. Use tool ST499587700 and install the plug.

30. Install a new gasket on the rocker cover. Apply liquid gasket to the cylinder head (see illustration).

➡**Apply an extra amount of liquid gasket around the semicircular plugs, 5mm or more.**

31. Temporarily tighten the rocker cover retaining bolts, in the proper sequence, and then tighten to 4.7 ft. lbs. Install the oil pipe.

32. Continue the installation in the reverse order of the removal procedure.

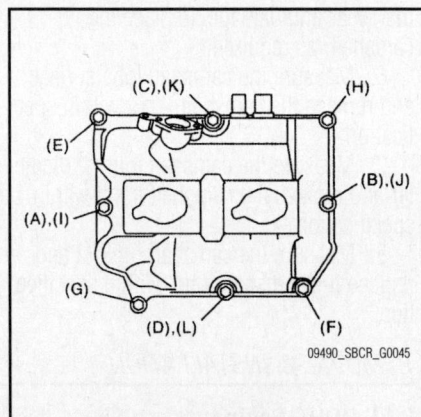

Fig. 87 Rocker cover tightening sequence—2.5L DOHC engine

2.5L SOHC Engine

See Figures 88 through 95.

1. Before servicing the vehicle, refer to the Precautions Section.

2. Disconnect the negative battery cable.

3. To remove the front side V belt, remove the belt covers. Loosen the lock bolt. Loosen the slider bolt. Remove the front side belt.

4. To remove the rear side V belt, remove the belt covers. Loosen the lock bolt. Loosen the slider bolt. Remove the rear side belt. Remove the belt tensioner.

5. Remove the crankshaft pulley bolt.

6. Lock the crankshaft in place using tool ST499977100, or equivalent.

7. Remove the crankshaft pulley.

8. Remove the left side timing belt cover.

9. Remove the front timing belt cover.

10. Remove the timing belt.

11. Remove the camshaft position sensor.

12. Remove the camshaft sprockets.

➡**Be sure to lock the camshaft in place using tool ST18231AA010, or equivalent.**

13. Remove the left and right timing belt N02 covers.

➡**Do not damage or lose the rubber seal when removing the covers.**

14. Remove the tensioner bracket. Remove the camshaft position sensor support, on the left side.

15. Remove the oil level gauge guide, on the left side.

16. Remove the valve rocker arm assembly.

17. Remove camshaft cap retaining bolts "A" and "B" in the proper sequence.

18. Loosen camshaft cap bolts "C" through "J" all the way in the proper sequence.

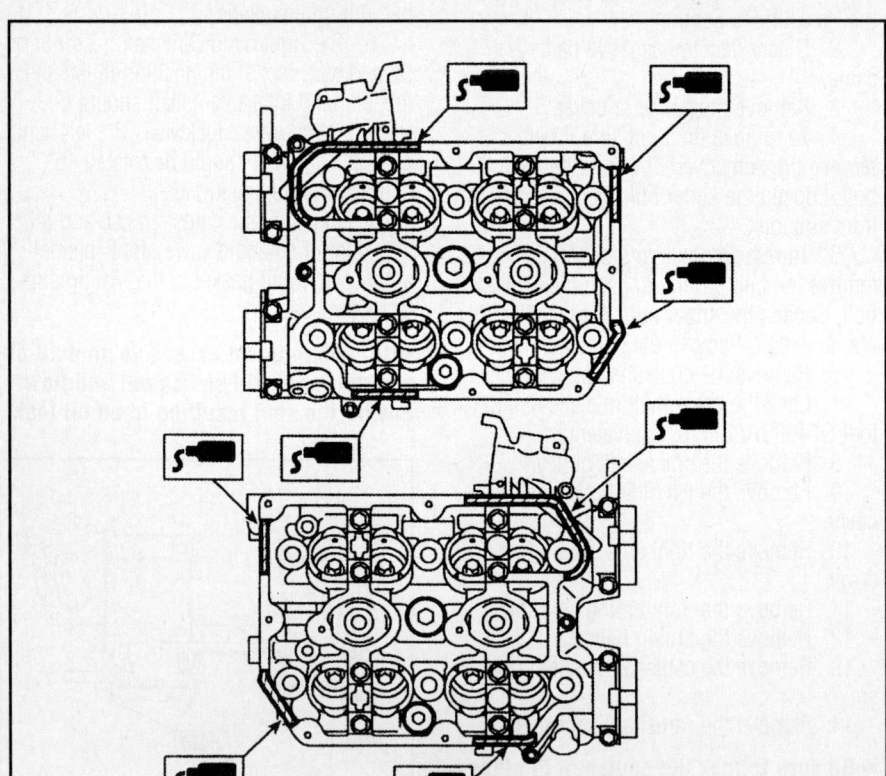

Fig. 86 Rocker cover liquid gasket application points—2.5L DOHC engine

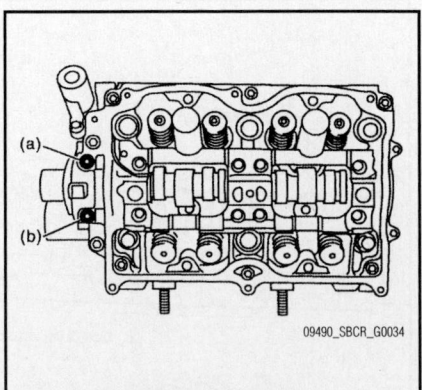

Fig. 88 Camshaft bolt loosening sequence—2.5L SOHC engine

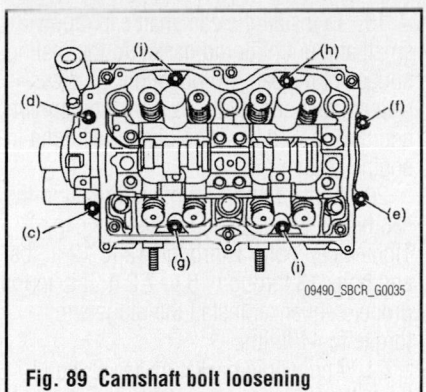

Fig. 89 Camshaft bolt loosening sequence—2.5L SOHC engine

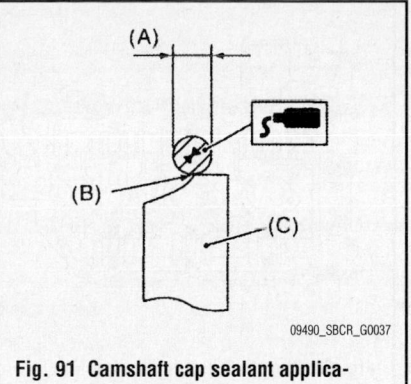

Fig. 91 Camshaft cap sealant application—2005 2.5L SOHC engine

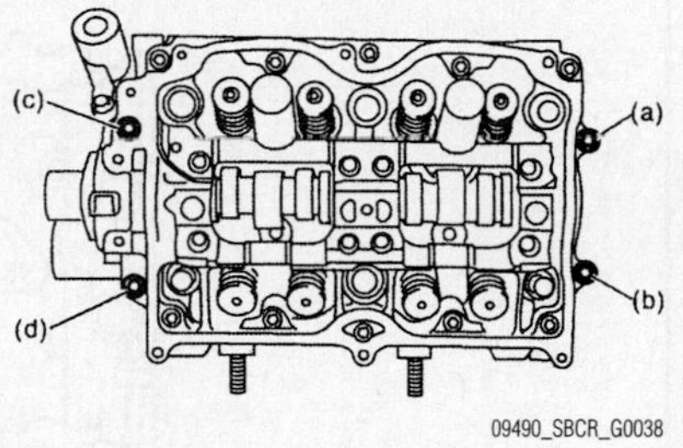

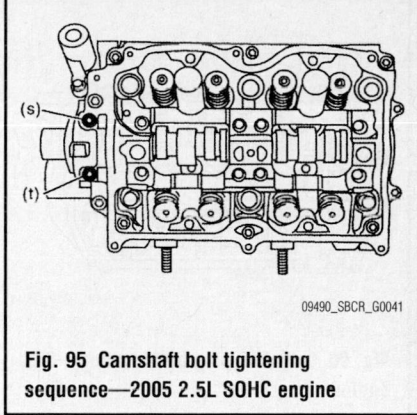

Fig. 95 Camshaft bolt tightening sequence—2005 2.5L SOHC engine

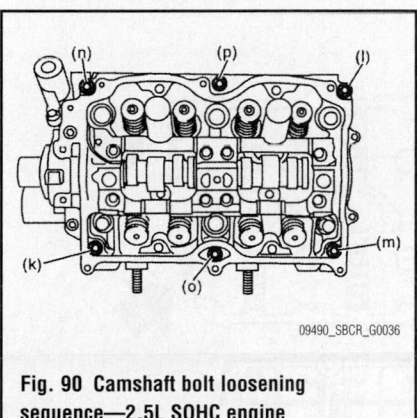

Fig. 90 Camshaft bolt loosening sequence—2.5L SOHC engine

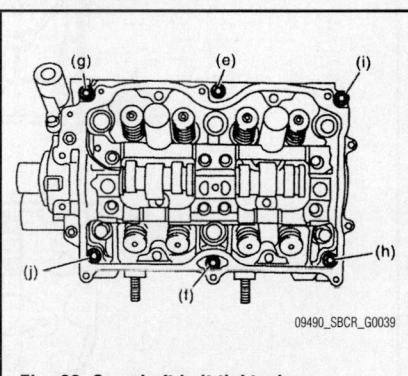

Fig. 92 Camshaft bolt tightening sequence—2005 2.5L SOHC engine

Fig. 93 Camshaft bolt tightening sequence—2005 2.5L SOHC engine

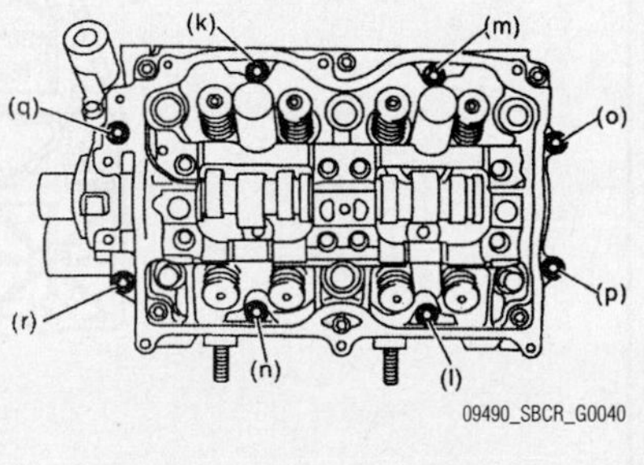

Fig. 94 Camshaft bolt tightening sequence—2005 2.5L SOHC engine

19. Remove camshaft cap bolts "K" through "P" in the proper sequence using a Torx® head bit.

20. Remove the camshaft caps.

21. Remove the camshaft. Remove the oil seal. Remove the plug from the rear side of the camshaft.

➡**Do not remove the oil seal unless necessary.**

To install:

➡**Lubricate the camshaft journals with clean engine oil prior to installation.**

22. Install the camshaft into the cylinder head.

23. Apply liquid gasket to the mating surfaces of the camshaft cap.

24. Apply a bead of sealant (0.12 inch in diameter) along the edge of the camshaft cap mating surface. Install with 20 minutes after applying the sealant.

25. Temporarily tighten the bolts "A" through "D" in the proper sequence.

26. Install the valve rocker arm assembly. Tighten Torx® head bolts "E" through "J" in the proper sequence to 13 ft. lbs.

27. Tighten bolts "K" through "R" in the proper sequence to 7.2 ft. lbs.

28. Tighten bolts "S" through "I" in the proper sequence to 7.2 ft. lbs.

➡**Be sure to use a new seal washer.**

29. Using tools ST499597000 and ST499587500, install a new seal on the camshaft. Be sure to coat the seal with clean engine oil before installation. Use tool ST499587700 and install the plug.

30. Adjust the valve clearance.

31. Continue the installation in the reverse order of the removal procedure.

3.0L & 3.6L Engines

See Figures 96 through 105.

1. Before servicing the vehicle, refer to the Precautions Section.

2. Disconnect the negative battery cable.

3. Remove the crankshaft pulley cover.

4. Remove the crankshaft pulley bolt.

5. Lock the crankshaft in place using tool ST499977100, or equivalent.

6. Remove the crankshaft pulley.

7. Remove the front timing chain cover.

➡**Bolts are three different sizes. Be careful to install the correct bolt in the correct hole.**

8. Remove the timing chain.

9. Remove the camshaft sprocket.

➡**Be sure to lock the camshaft in place using tool ST499977500, or equivalent.**

10. Remove the crankshaft sprocket.

11. Remove the oil pump.

12. Remove the water pump.

13. Remove the rear timing chain cover retaining bolts. Remove the rear timing chain cover.

➡**There are seven different size bolts. Be sure not to confuse them on installation.**

14. Disconnect the oil pipe.

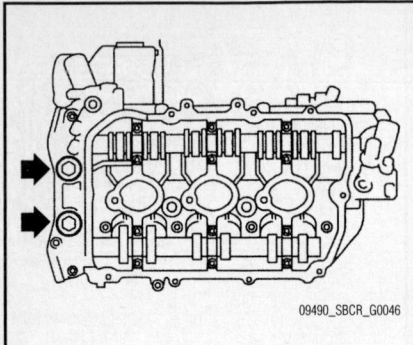

Fig. 96 Camshaft plug bolt location—3.0L engine

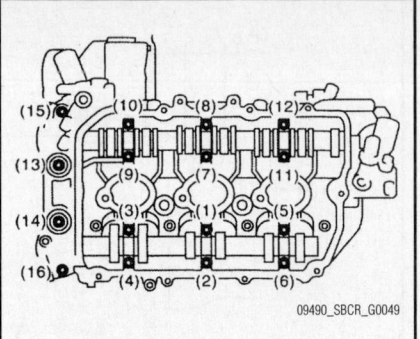

Fig. 99 Camshaft bolt tightening sequence—3.0L engine

19. To install the camshaft cap, apply a small amount of liquid gasket to the mating surface of the cap. Do not apply an excessive amount of sealant, as it will squish out and flow toward the cam journal resulting in engine seizure.

20. Apply a thin coat of engine oil to the cap bearing surface and install the cap. Tighten cap bolts 1 through 12 to 12 ft. lbs. and bolts 13 through 16 to 7.2 ft. lbs. in the proper sequence. Install the plugs and torque to 44 ft. lbs.

21. Apply fluid gasket maker of the of the cylinder heads and valve covers.

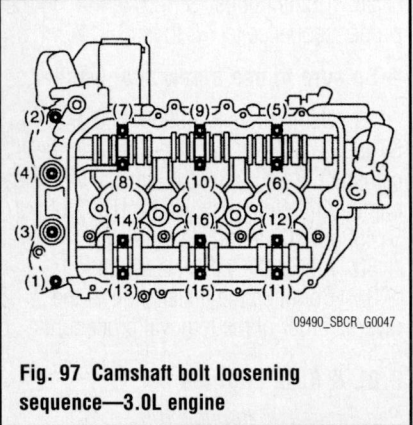

Fig. 97 Camshaft bolt loosening sequence—3.0L engine

15. Remove the rocker cover retaining bolts. Remove the rocker cover. Discard the gasket.

16. Remove the plugs, see illustration for location.

17. Loosen the camshaft cap bolts in the proper sequence. Remove the camshaft caps and remove the camshaft.

To install:

18. Apply a coat of engine oil to the journals on the camshafts and place the camshafts into position.

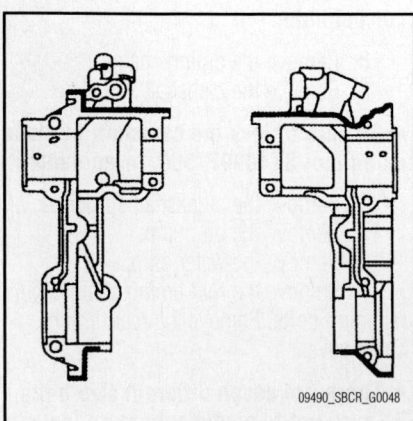

Fig. 98 Camshaft cap liquid gasket application points—3.0L engine

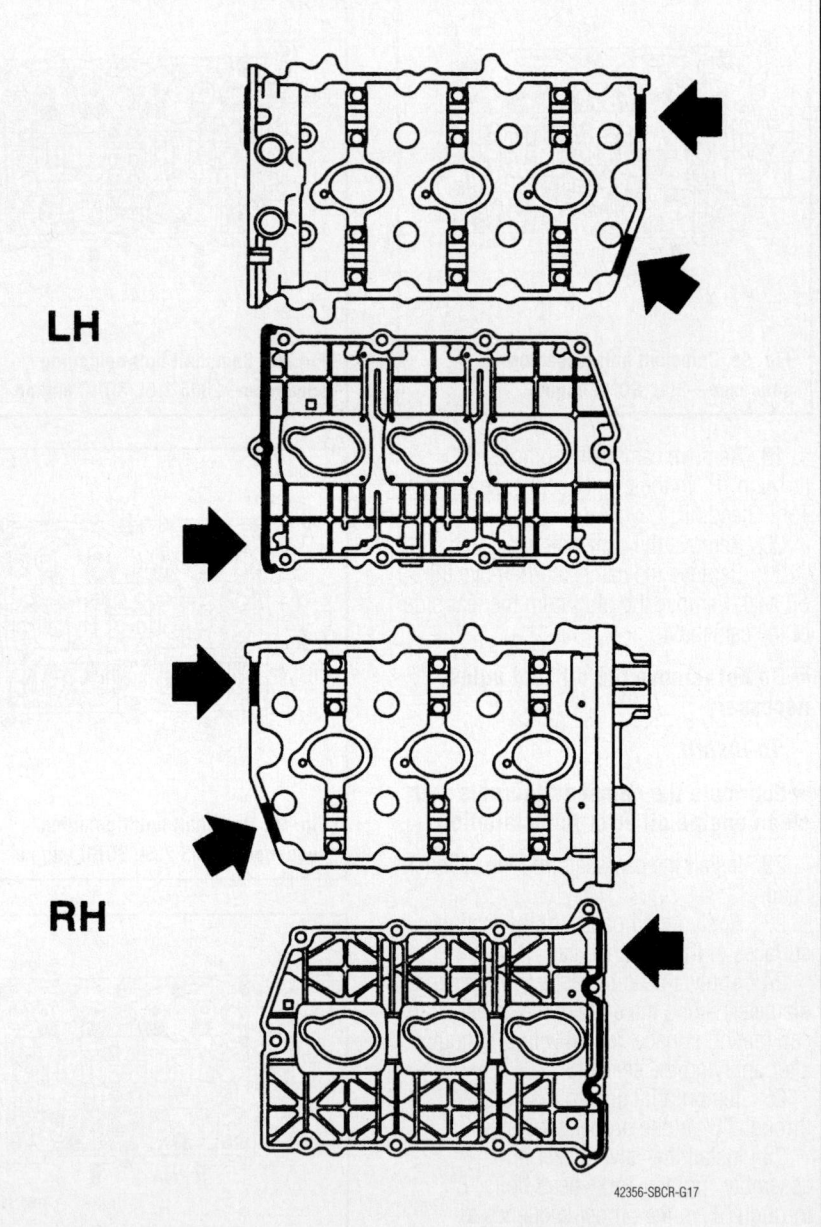

LH

RH

Fig. 100 Apply fluid gasket maker on the cylinder heads and valve covers as shown—3.0L engine

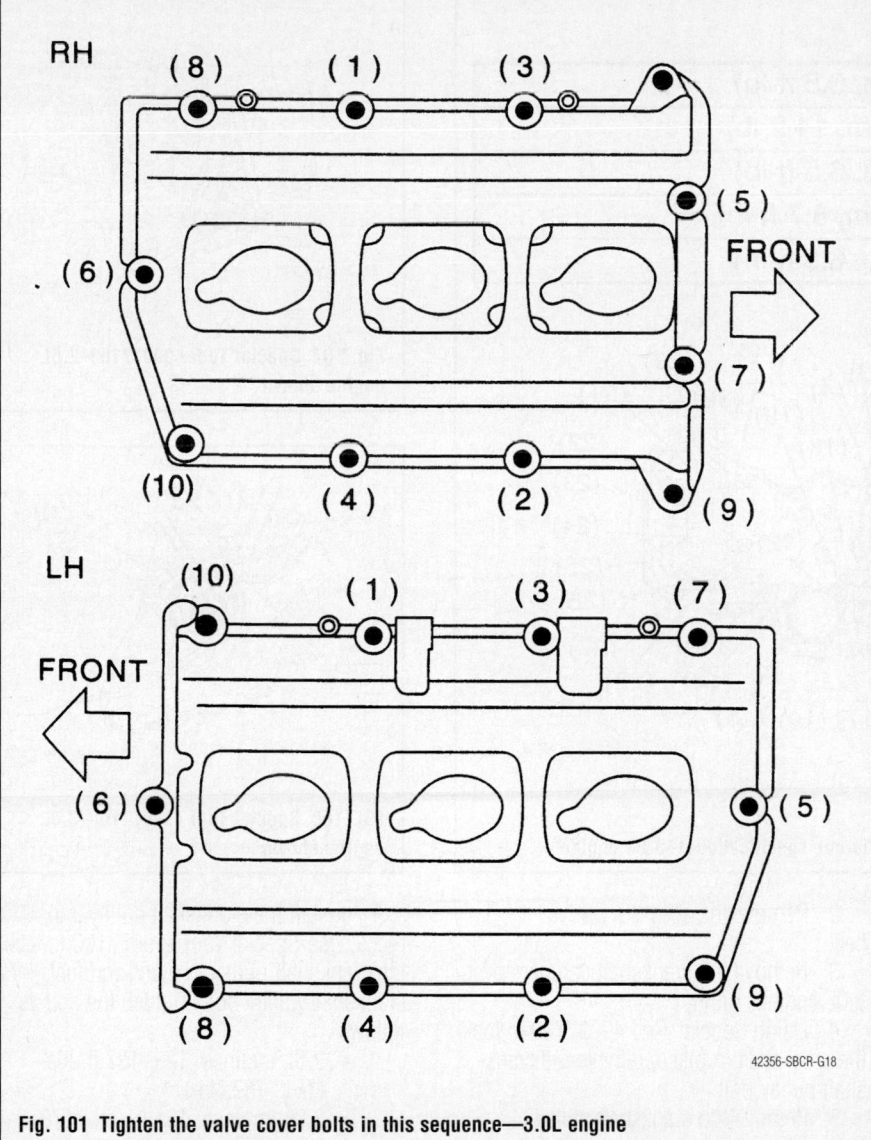

RH

FRONT

LH

FRONT

42356-SBCR-G18

Fig. 101 Tighten the valve cover bolts in this sequence—3.0L engine

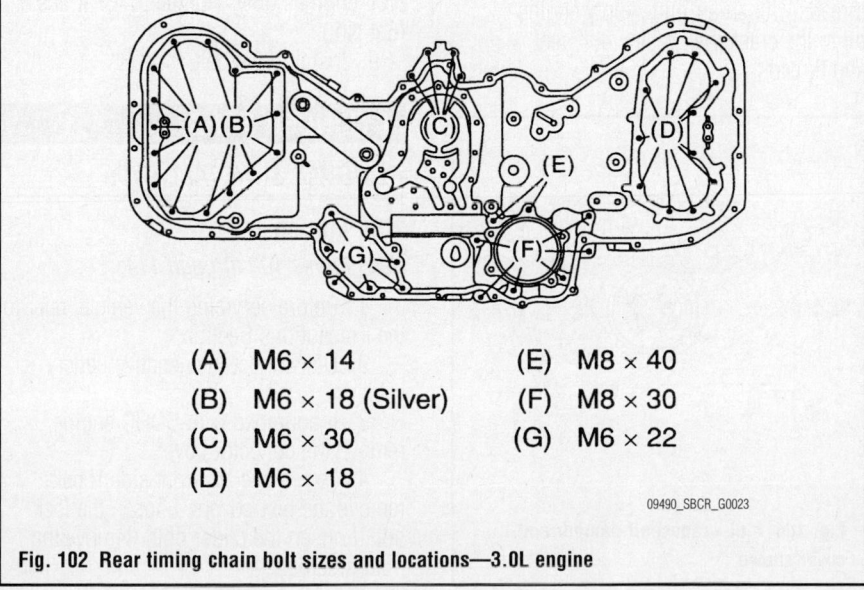

(A)	M6 × 14	(E)	M8 × 40
(B)	M6 × 18 (Silver)	(F)	M8 × 30
(C)	M6 × 30	(G)	M6 × 22
(D)	M6 × 18		

09490_SBCR_G0023

Fig. 102 Rear timing chain bolt sizes and locations—3.0L engine

Fluid gasket application diameter:
(A) 1.0±0.5 mm (0.039±0.020 in)
(B) 3.0±1.0 mm (0.118±0.039 in)

42356-SBCR-G06

Fig. 103 Apply liquid gasket maker to the mating surfaces of the cover as shown—3.0L engine

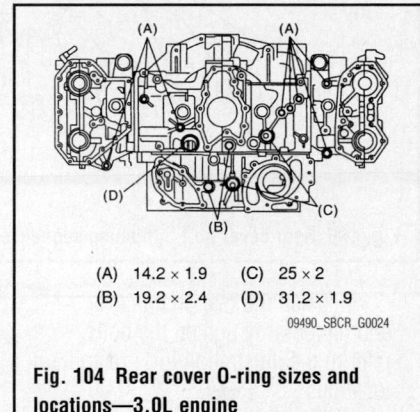

(A)	14.2 × 1.9	(C)	25 × 2
(B)	19.2 × 2.4	(D)	31.2 × 1.9

09490_SBCR_G0024

Fig. 104 Rear cover O-ring sizes and locations—3.0L engine

✳✳ CAUTION

Do not apply too much gasket maker. This may cause excess fluid gasket maker to come out and flow towards the camshaft journal resulting in engine damage.

22. Tighten the valve cover bolts to 4.7 ft. lbs. and in the proper sequence.

23. Install the rear chain cover as follows:

➡**There are several size bolts used, refer to the illustration for size and locations.**

a. Rear chain cover gasket and clean the mating surfaces.

b. Apply liquid gasket maker to the mating surfaces of the cover. Refer to the illustration for gasket maker application and diameter.

c. Install new O-rings. Refer to the illustration for O-ring location and size

(1) — (11)	9 N·m (0.9 kgf-m, 6.5 ft-lb)
(12) — (19)	20 N·m (2.0 kgf-m, 14 ft-lb)
(20) — (30)	9 N·m (0.9 kgf-m, 6.5 ft-lb)
(31) — (38)	12 N·m (1.2 kgf-m, 8.7 ft-lb)
(39) — (45)	9 N·m (0.9 kgf-m, 6.5 ft-lb)

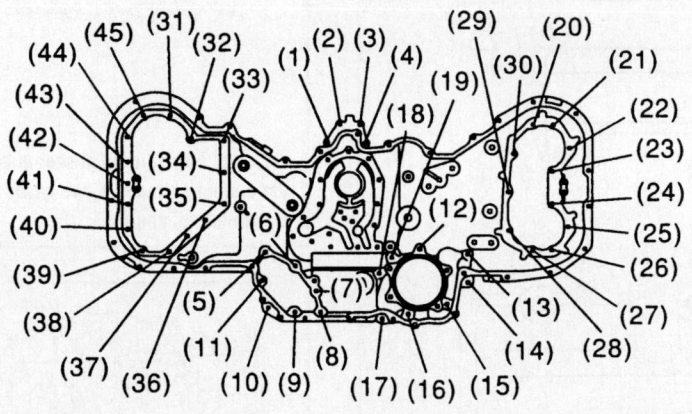

Fig. 105 Rear cover bolt tightening sequence and torque specifications—3.0L engine

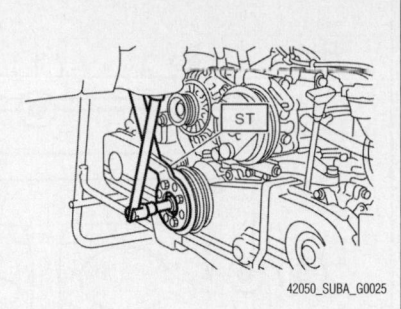

Fig. 107 Special Tool 499977100–2.5L engine shown

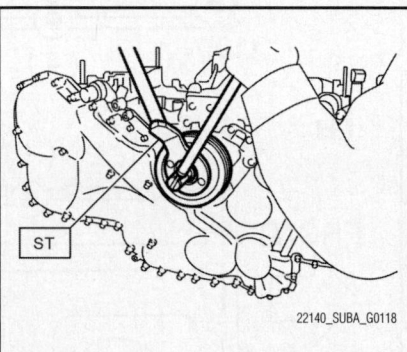

Fig. 108 Special Tool 499977100–3.0L engine shown

d. Install the rear chain cover and temporarily tighten the bolts, refer to the illustration for size and locations.

e. Tighten the cover bolts in the sequence illustrated to the specifications shown in the illustration.

24. Continue the installation in the reverse order of the removal procedure.

CATALYTIC CONVERTER

REMOVAL & INSTALLATION

Front

The front catalytic converter is integrated into the front exhaust pipe.

Rear

The rear catalytic converter is integrated into the center exhaust pipe.

CRANKSHAFT DAMPER

REMOVAL & INSTALLATION

See Figures 106 through 108.

1. Before servicing the vehicle, refer to the Precautions Section.

2. Remove the accessory drive belt.

3. Remove the crankshaft damper cover, 3.0L engines only.

4. Using Special Tool 49997700 to lock the crankshaft in place, remove the crankshaft pulley bolt.

5. Remove the crankshaft pulley.

To install:

6. Clean the crankshaft pulley threads with an air gun. Apply engine oil to the crankshaft pulley bolt seat and threads.

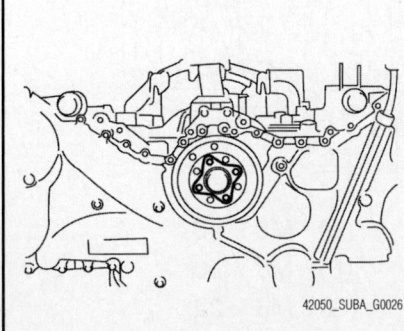

Fig. 106 3.0L crankshaft damper and cover shown

7. Install the crankshaft pulley into place. Use Special Tool 499977100 to lock the crankshaft pulley in place and install the crankshaft pulley bolt. Tighten the bolt as follows

- 2.5L engines: 123–137 ft. lbs. (167–187 Nm).
- 3.0L engines: 131 ft. lbs. (178 Nm).

8. Install the crankshaft damper cover, 3.0L engines only. Tighten to 4.7 ft. lbs. (6.4 Nm).

9. Install drive belt.

CYLINDER HEAD

REMOVAL & INSTALLATION

2.5L Engine

See Figures 109 through 114.

1. Before servicing the vehicle, refer to the Precautions Section.

2. Disconnect the negative battery cable.

3. If equipped with DOHC engine, remove the collector cover.

4. To remove the front side V belt, remove the belt covers. Loosen the lock bolt. Loosen the slider bolt. Remove the front side belt.

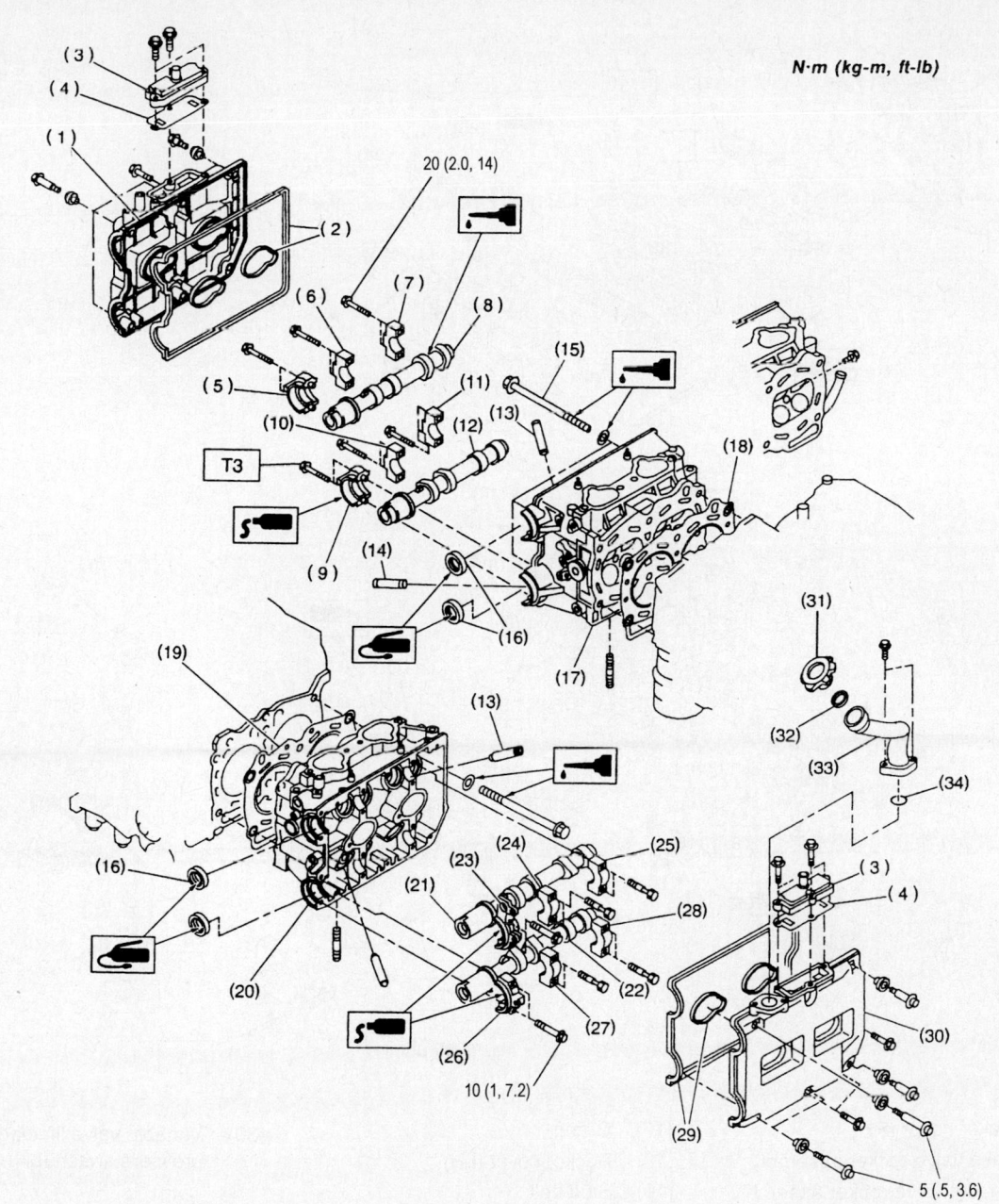

N·m (kg-m, ft-lb)

Fig. 109 Cylinder head and related components—2.5L DOHC engine

(1) Rocker cover (RH)
(2) Rocker cover gasket (RH)
(3) Oil separator cover
(4) Gasket
(5) Intake camshaft cap (Front RH)
(6) Intake camshaft cap (Center RH)
(7) Intake camshaft cap (Rear RH)
(8) Intake camshaft (RH)
(9) Exhaust camshaft cap (Front RH)
(10) Exhaust camshaft cap (Center RH)
(11) Exhaust camshaft cap (Rear RH)
(12) Exhaust camshaft (RH)
(13) Intake valve guide
(14) Exhaust valve guide

(15) Cylinder head bolt
(16) Oil seal
(17) Cylinder head (RH)
(18) Cylinder head gasket (RH)
(19) Cylinder head gasket (LH)
(20) Cylinder head (LH)
(21) Intake camshaft (LH)
(22) Exhaust camshaft (LH)
(23) Intake camshaft cap (Front LH)
(24) Intake camshaft cap (Center LH)
(25) Intake camshaft cap (Rear LH)
(26) Exhaust camshaft (Front LH)
(27) Exhaust camshaft cap (Center LH)
(28) Exhaust camshaft cap (Rear LH)

(29) Rocker cover gasket (LH)
(30) Rocker cover (LH)
(31) Oil filler cap
(32) Gasket
(33) Oil filler duct
(34) O-ring

7923TG07

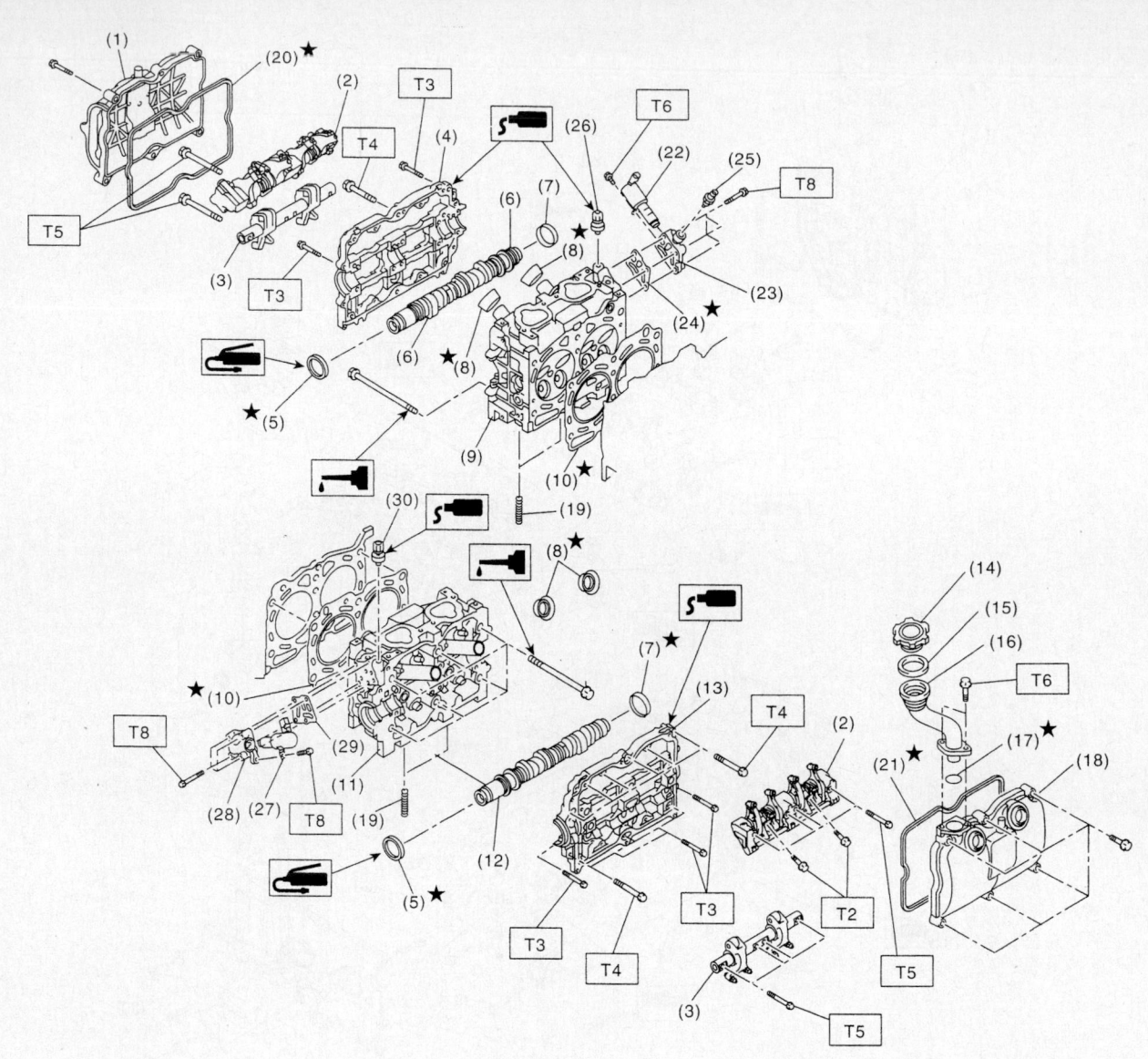

(1)	Rocker cover (RH)	(17)	O-ring	(30)	Variable valve lift diagnosis oil pressure switch (LH)
(2)	Intake valve rocker assembly	(18)	Rocker cover (LH)		
(3)	Exhaust valve rocker assembly	(19)	Stud bolt		
(4)	Camshaft cap (RH)	(20)	Rocker cover gasket (RH)		
(5)	Oil seal	(21)	Rocker cover gasket (LH)		
(6)	Camshaft (RH)	(22)	Oil switching solenoid valve (RH)		
(7)	Plug	(23)	Oil switching solenoid valve holder (RH)		
(8)	Spark plug pipe gasket				
(9)	Cylinder head (RH)	(24)	Gasket		
(10)	Cylinder head gasket	(25)	Oil temperature sensor		
(11)	Cylinder head (LH)	(26)	Variable valve lift diagnosis oil pressure switch (RH)		
(12)	Camshaft (LH)				
(13)	Camshaft cap (LH)	(27)	Oil switching solenoid valve (LH)		
(14)	Oil filler cap	(28)	Oil switching solenoid valve holder (LH)		
(15)	Gasket				
(16)	Oil filler duct	(29)	Gasket		

Tightening torque: N·m (kgf-m, ft-lb)

T3:	9.75 (1.0, 7.2)
T4:	18 (1.8, 13.0)
T5:	25 (2.5, 18.1)
T6:	6.4 (0.65, 4.7)
T7:	8 (0.8, 5.9)
T8:	10 (1.0, 7.4)

09490_SBCR_G0016

Fig. 110 Cylinder head and related components—2.5L SOHC engine

5. To remove the rear side V belt, remove the belt covers. Loosen the lock bolt. Loosen the slider bolt. Remove the rear side belt. Remove the belt tensioner.

6. Remove the crankshaft pulley bolt.

7. Lock the crankshaft in place using tool ST499977100, or equivalent.

8. Remove the crankshaft pulley.

9. Remove the left side timing belt cover.

10. If equipped with DOHC engine, remove the right side timing belt cover.

11. Remove the front timing belt cover.

12. Remove the timing belt.

13. Remove the camshaft position sensor.

14. Remove the camshaft sprockets.

➡ **Be sure to lock the camshaft in place using tool ST18231AA010 (SOHC engine) and ST499207400 (DOHC engine), or equivalent.**

15. Remove the intake manifold.

16. If equipped with SOHC engine, remove the bolt that retains the air conditioning compressor bracket to the cylinder head.

17. Remove the rocker cover retaining bolts. Remove the rocker cover.

18. If equipped with SOHC engine, remove the rocker arm assembly.

19. If equipped with SOHC engine remove the camshaft. If equipped with DOHC engine, remove the camshafts.

20. Remove the cylinder head, in the proper sequence. On the DOHC engine leave bolts A and D installed loosely to prevent the cylinder head from falling. On the SOHC engine leave bolts A and C installed loosely to prevent the cylinder head from falling.

21. Loosen the cylinder head from the block using a plastic-faced hammer, if needed.

22. Remove bolts A and D on the DOHC engine and bolts A and C on the SOHC engine. Remove the cylinder head from the engine. Discard the gasket

23. Clean all gasket material from both mating surfaces.

To install:

24. Installation is the reverse of the removal procedure.

25. Apply a thin coat of clean engine oil to the washers and cylinder head bolts.

26. Tighten the cylinder head retaining bolts to specification and in the proper sequence.

27. Start the engine and allow it to reach operating temperature.

28. Check for leaks, correct as required.

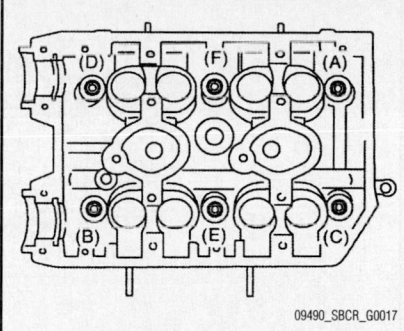

Fig. 111 Cylinder head bolt loosening sequence—2.5L DOHC engine

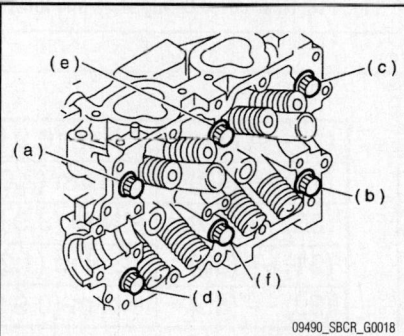

Fig. 112 Cylinder head bolt loosening sequence—2.5L SOHC engine

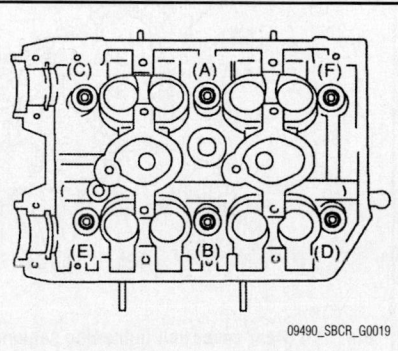

Fig. 113 Cylinder head bolt tightening sequence—2.5L DOHC engine

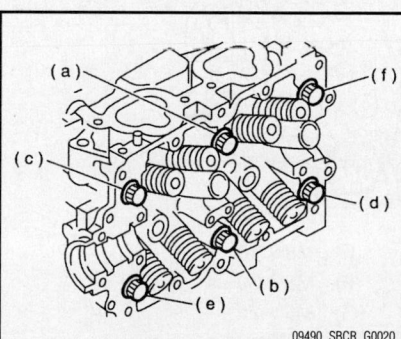

Fig. 114 Cylinder head bolt tightening sequence—2.5L SOHC engine

3.0L & 3.6L Engines

See Figures 115 through 119.

1. Before servicing the vehicle, refer to the Precautions Section.

2. Remove the crankshaft pulley cover.

3. Remove the crankshaft pulley bolt.

4. Lock the crankshaft in place using tool ST499977100, or equivalent.

5. Remove the crankshaft pulley.

6. Remove the front timing chain cover.

➡ **Bolts are three different sizes. Be careful to install the correct bolt in the correct hole.**

7. Remove the timing chain.

8. Remove the camshaft sprocket.

➡ **Be sure to lock the camshaft in place using tool ST499977500, or equivalent.**

9. Remove the crankshaft sprocket.

10. Remove the oil pump.

11. Remove the water pump.

12. Remove the rear timing chain cover retaining bolts. Remove the rear timing chain cover.

➡ **There are seven different size bolts. Be sure not to confuse them on installation.**

13. Remove the camshafts.

14. Remove the cylinder head bolts in the proper sequence. Leave bolts 2 and 4 connected by a few threads to prevent the head from falling. Tap the head with a plastic mallet to separate it from the block.

15. Remove bolts 2 and 4 from the cylinder head. Remove the cylinder head from the engine. Discard the gasket.

16. Clean all gasket material from both mating surfaces.

To install:

17. Installation is the reverse of the removal procedure.

18. Apply a thin coat of clean engine oil to the washers and cylinder head bolts.

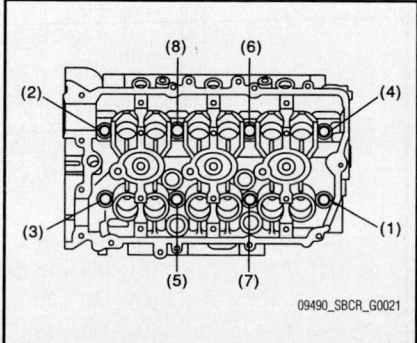

Fig. 115 Cylinder head bolt loosening sequence—3.0L engine

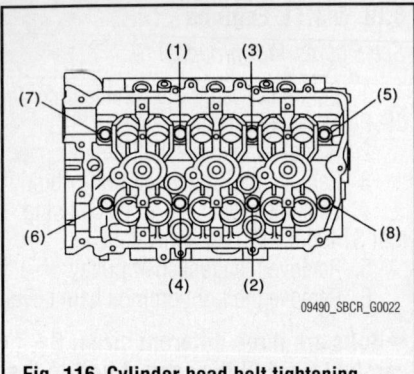

Fig. 116 Cylinder head bolt tightening sequence—3.0L engine

19. Tighten the cylinder head retaining bolts to specification and in the proper sequence.

20. Install the rear chain cover as follows:

➡ **There are several size bolts used, refer to the illustration for size and locations.**

   a. Rear chain cover gasket and clean the mating surfaces.

   b. Apply liquid gasket maker to the mating surfaces of the cover. Refer to the illustration for gasket maker application and diameter.

   c. Install new O-rings. Refer to the illustration for O-ring location and size

   d. Install the rear chain cover and temporarily tighten the bolts, refer to the illustration for size and locations.

   e. Tighten the cover bolts in the sequence illustrated to the specifications shown in the illustration.

21. Start the engine and allow it to reach operating temperature.

22. Check for leaks, correct as required.

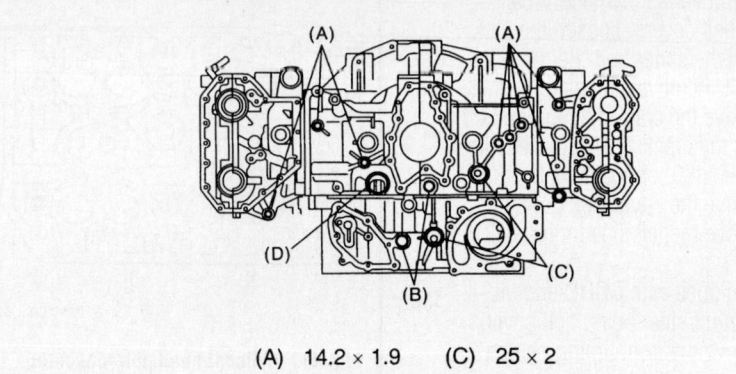

| (A) | 14.2 × 1.9 | (C) | 25 × 2 |
| (B) | 19.2 × 2.4 | (D) | 31.2 × 1.9 |

Fig. 118 Rear cover O-ring sizes and locations—3.0L engine

(1) — (11)	9 N·m (0.9 kgf-m, 6.5 ft-lb)
(12) — (19)	20 N·m (2.0 kgf-m, 14 ft-lb)
(20) — (30)	9 N·m (0.9 kgf-m, 6.5 ft-lb)
(31) — (38)	12 N·m (1.2 kgf-m, 8.7 ft-lb)
(39) — (45)	9 N·m (0.9 kgf-m, 6.5 ft-lb)

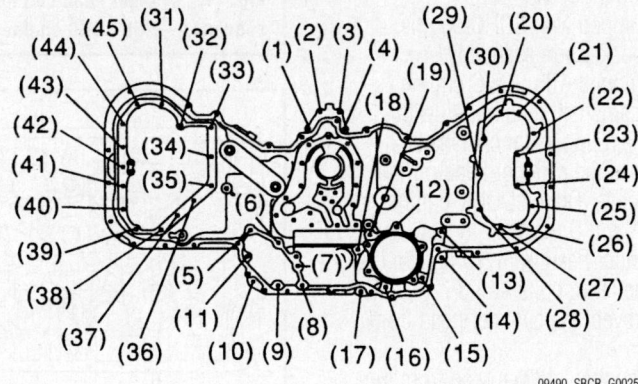

Fig. 119 Rear cover bolt tightening sequence and torque specifications—3.0L engine

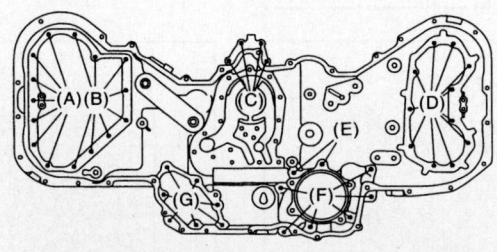

(A)	M6 × 14	(E)	M8 × 40
(B)	M6 × 18 (Silver)	(F)	M8 × 30
(C)	M6 × 30	(G)	M6 × 22
(D)	M6 × 18		

Fig. 117 Rear timing chain bolt sizes and locations—3.0L engine

EXHAUST MANIFOLD

REMOVAL & INSTALLATION

Due to the unique design of the Subaru engine, an exhaust manifold is not used. The exhaust enters directly into the front Y—pipe.

FLYWHEEL

REMOVAL & INSTALLATION

See Figures 120 and 121.

1. Before servicing the vehicle, refer to the Precautions Section.

2. Remove the engine and place on a suitable engine stand.

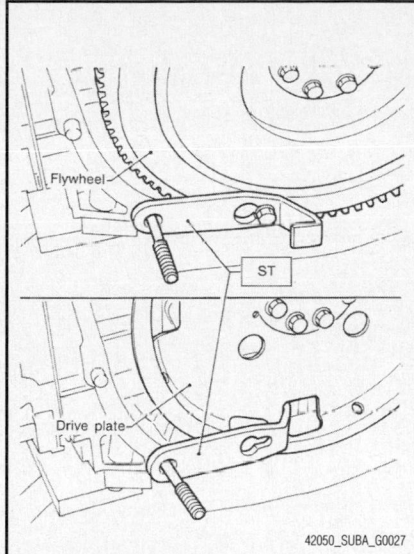

Fig. 120 Use Special Tool 498497100 to lock the crankshaft into place and remove the flywheel

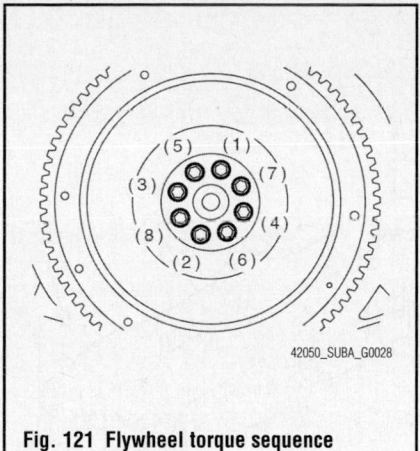

Fig. 121 Flywheel torque sequence

3. If equipped with a manual transmission, remove the clutch housing cover.

4. Using Special Tool 498497100 to lock the crankshaft into place, remove the flywheel (manual transmission) or driveplate (automatic transmission).

To install:

5. Install the flywheel (M/T) or drive plate (A/T) and use Special Tool 498497100 to lock the crankshaft into place. Tighten the bolts in the sequence shown as follows:

    a. Except 3.0L engine: 51–55 ft. lbs. (69–75 Nm).

    b. 3.0L engine: 60 ft. lbs. (81 Nm).

6. Install the clutch housing cover, if equipped with manual transmission.

7. Install the engine assembly into the vehicle.

INTAKE MANIFOLD

REMOVAL & INSTALLATION

2.5L Engines

See Figures 122 and 123.

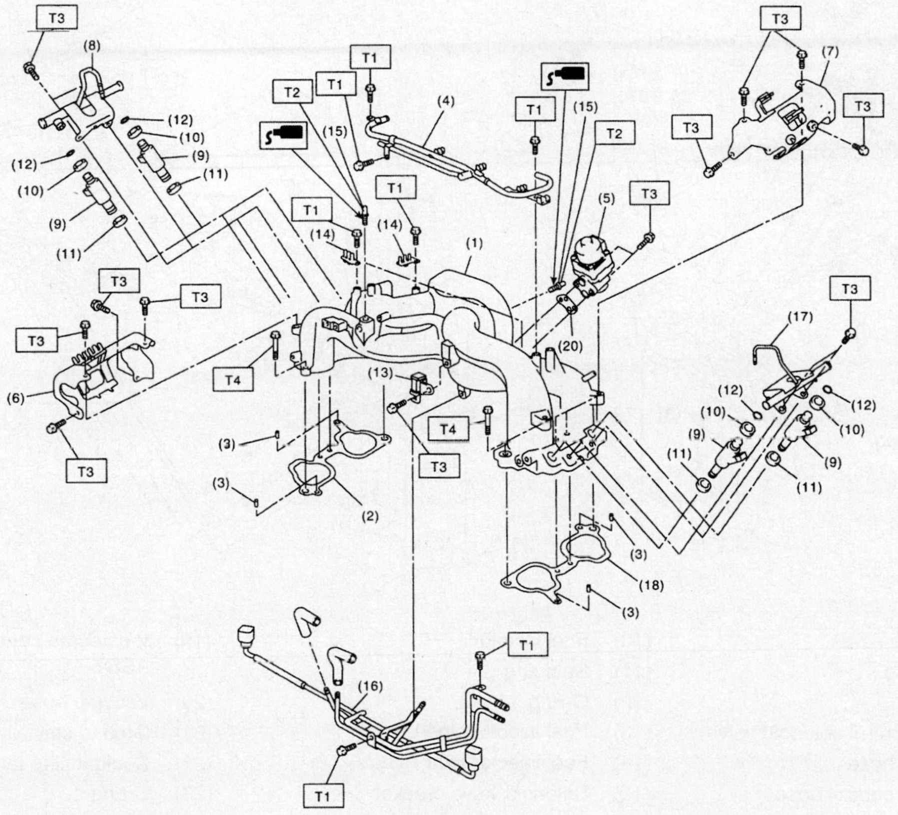

(1)	Intake manifold	(7)	Fuel pipe protector LH	(13)	Purge control solenoid valve
(2)	Gasket (RH)	(8)	Fuel injector pipe RH	(14)	Plug cord holder
(3)	Guide pin	(9)	Fuel injector	(15)	Nipple
(4)	PCV pipe	(10)	O-ring	(16)	Fuel pipe
(5)	EGR valve	(11)	O-ring	(17)	Fuel injector pipe LH
(6)	Fuel pipe protector RH	(12)	O-ring	(18)	Gasket (LH)

Fig. 122 Intake manifold and related components—2.5L SOHC engine

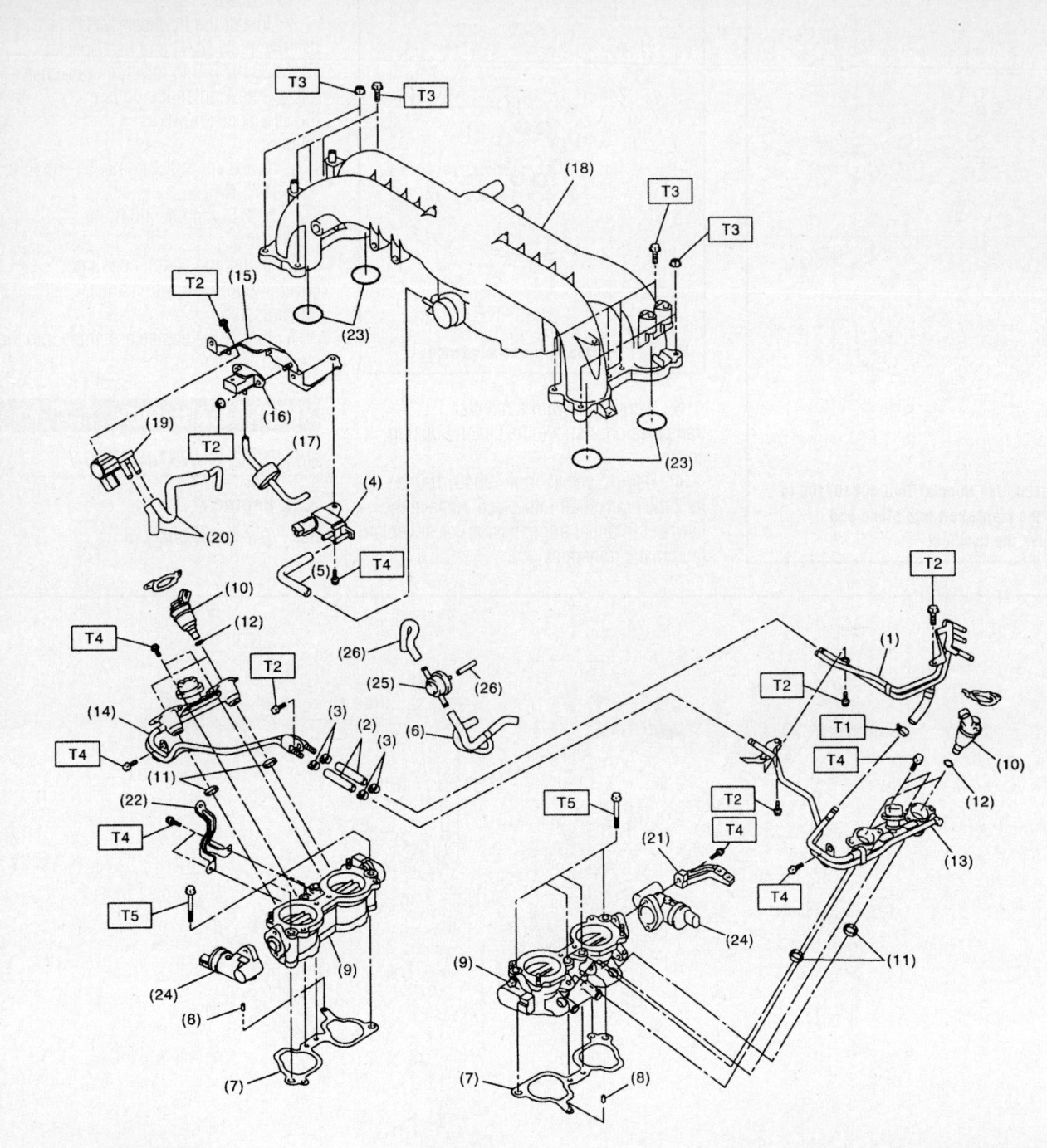

(1) Fuel pipe ASSY
(2) Fuel hose
(3) Clip
(4) Purge control solenoid valve
(5) Vacuum hose
(6) Vacuum control hose
(7) Intake manifold gasket
(8) Guide pin
(9) Intake manifold (lower)

(10) Fuel injector
(11) Seal ring
(12) O-ring
(13) Fuel injector pipe LH
(14) Fuel injector pipe RH
(15) Solenoid valve bracket
(16) Manifold absolute pressure sensor
(17) Filter
(18) Intake manifold

(19) Wastegate control solenoid valve ASSY
(20) Vacuum hose
(21) Ground stay
(22) Coolant filler tank stay
(23) O-ring
(24) Tumble generator valve actuator
(25) Purge valve
(26) Purge hose

09490_SBCR_G0028

Fig. 123 Intake manifold and related components—2.5L DOHC engine

1. Before servicing the vehicle, refer to the Precautions Section.

2. Properly relieve the fuel system pressure. Remove the fuel cap.

3. Disconnect the negative battery cable. Drain the engine coolant.

4. If equipped, remove the undercover.

5. Remove the air intake duct, air cleaner case and air intake chamber.

6. If equipped, remove the intercooler.

7. Remove the alternator.

8. If equipped with DOHC engine, remove the coolant filler tank.

9. If equipped with SOHC engine, remove the spark plug wires.

10. Disconnect the engine coolant hoses from the throttle body. Disconnect the brake booster hose.

11. Disconnect the PCV hose from the intake manifold. Disconnect the engine har-ness electrical connectors from the bulk-head harness connectors.

12. Disconnect the engine coolant temperature sensor electrical connector, knock sensor electrical connector and crankshaft position sensor connector.

13. Disconnect the power steering pump switch electrical connector, oil pressure switch connector and camshaft sensor connector.

14. If equipped with DOHC engine, disconnect the oil flow solenoid valve electrical connector and the ignition coil connector.

15. If equipped with SOHC engine, remove the EGR pipe from the intake manifold and disconnect the fuel lines from the fuel pipe.

16. If equipped with DOHC engine, disconnect the fuel delivery hose, return hose and evaporation hose.

17. Remove the intake manifold retaining bolts. Remove the intake manifold from the engine.

To install:

18. Installation is the reverse of the removal procedure.

19. Be sure to use new intake manifold gaskets. Tighten the manifold retaining bolts to 18 ft. lbs. (25 Nm). and in alternating sequence.

20. Be sure to fill and bleed the cooling system with the proper grade and type engine coolant.

21. Start the engine and check for leaks, correct as required.

3.0L & 3.6L Engines

See Figure 124.

1. Before servicing the vehicle, refer to the Precautions Section.

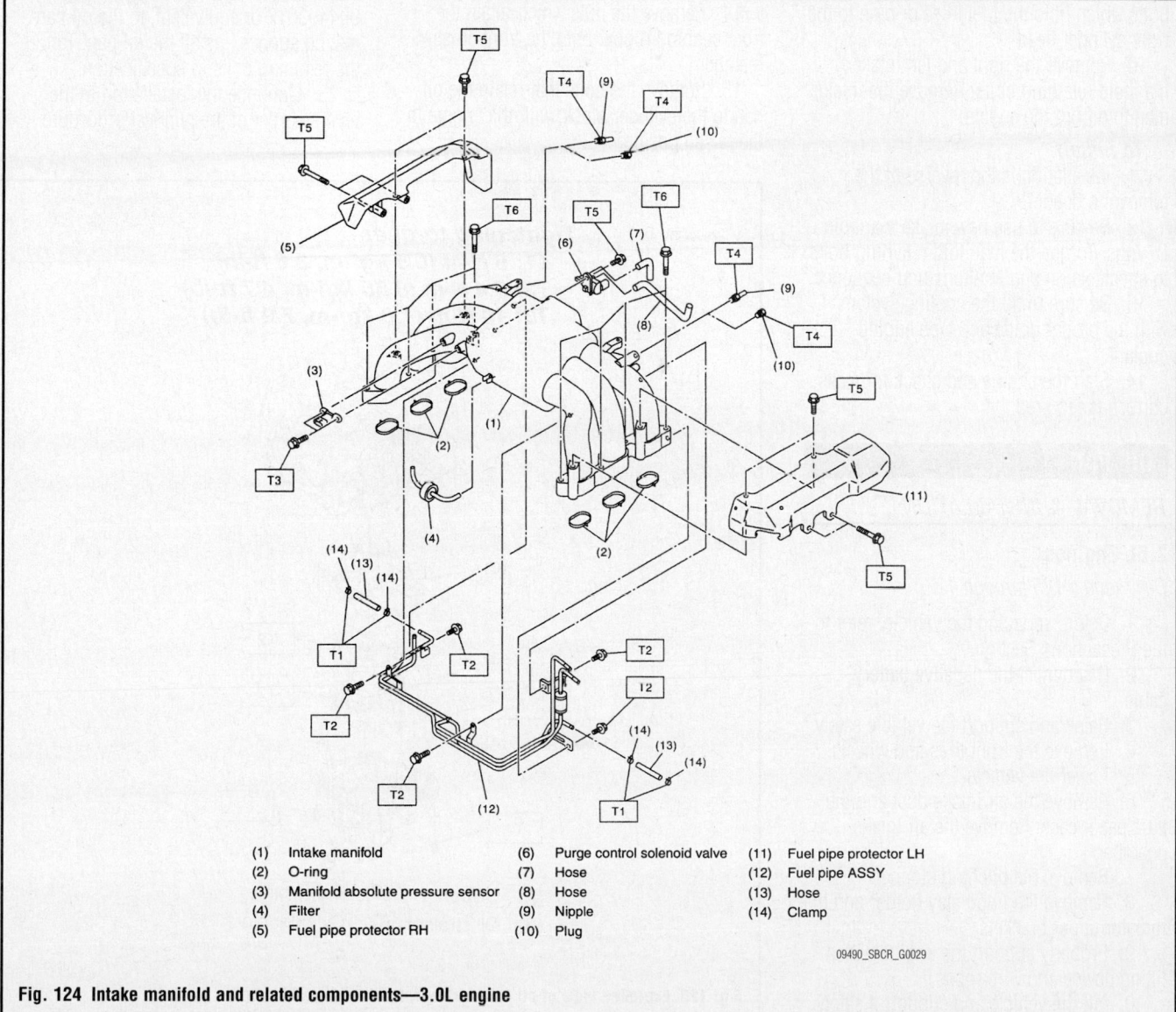

(1)	Intake manifold	(6)	Purge control solenoid valve	(11)	Fuel pipe protector LH
(2)	O-ring	(7)	Hose	(12)	Fuel pipe ASSY
(3)	Manifold absolute pressure sensor	(8)	Hose	(13)	Hose
(4)	Filter	(9)	Nipple	(14)	Clamp
(5)	Fuel pipe protector RH	(10)	Plug		

09490_SBCR_G0029

Fig. 124 Intake manifold and related components—3.0L engine

2. Properly relieve the fuel system pressure. Remove the fuel cap.

3. Disconnect the negative battery cable. Drain the engine coolant.

4. Remove the air cleaner case and air intake chamber.

5. Remove the alternator.

6. Disconnect the electrical connector from the throttle body. Disconnect the engine coolant hoses from the throttle body.

7. Disconnect the engine harness connector. Disconnect the PCV hose. Disconnect the brake booster hose. Disconnect the fuel hoses from the fuel pipe.

8. Remove the left side fuel line protector. Remove the engine harness from the left side fuel injector pipe. Remove the bolts which hold the fuel injector pipe to the left cylinder head.

9. Remove the right side fuel line protector. Remove the engine harness from the right side fuel injector pipe. Remove the bolts which hold the fuel injector pipe to the right cylinder head.

10. Remove the right and left intake manifold retaining bolts. Remove the intake manifold from the engine.

To install:

11. Installation is the reverse of the removal procedure.

12. Be sure to use new intake manifold O-rings. Torque the manifold retaining bolts to specification and in alternating sequence.

13. Be sure to fill the cooling system with the proper grade and type engine coolant.

14. Start the engine and check for leaks, correct as required.

OIL PAN

REMOVAL & INSTALLATION

2.5L Engines

See Figures 125 through 127.

1. Before servicing the vehicle, refer to the Precautions Section.

2. Disconnect the negative battery cable.

3. Raise and support the vehicle safely.

4. Remove the front tires and wheels.

5. Lower the vehicle.

6. Remove the air intake duct and the air cleaner case. Remove the air intake chamber.

7. Remove the pitching stopper.

8. Remove the hood stay holder and the radiator upper brackets.

9. Properly support the engine with a lifting device and wire ropes.

10. Lift the vehicle and support it safely.

22140_SUBA_G0121

Fig. 125 Engine supported with lifting device

➡**When lifting the vehicle, raise the wire ropes at the same time.**

11. Remove the undercover.

12. Drain the engine oil.

13. Remove the front and center exhaust pipes.

14. Remove the nuts which retain the front cushion rubber onto the front crossmember.

15. Remove the bolts that retain the oil pan to the cylinder block, with the engine in the raised position.

16. Insert an oil pan gasket cutter tool into the gap between the cylinder block and the oil pan. Remove the oil pan from the engine.

➡**Do not use a screwdriver or similar tool in place of the cutter tool.**

17. Remove the oil strainer, if required. Remove the baffle plate, if required.

To install:

18. Be sure to clean the old gasket material from the mating surfaces.

19. Apply a continuous bead of sealer to a new oil pan gasket.

20. Make sure that the seals (A) are installed securely on the baffle plate and in the direction shown in the illustration. Install the baffle plate; tighten the retaining bolts to 4.7 ft. lbs.

21. Replace the O-ring and install the oil strainer. Tighten the bolt to 7.2 ft. lbs.

22. Apply liquid gasket, part number 004403012 or equivalent, to the oil pan mating surface. Install the oil pan. Torque the retaining bolts to specification.

23. Continue the installation in the reverse order of the removal procedure.

Tightening torque:
T1: 5 N·m (0.5 kgf-m, 3.6 ft-lb)
T2: 6.4 N·m (0.65 kgf-m, 4.7 ft-lb)
T3: 10 N·m (1.0 kgf-m, 7.2 ft-lb)

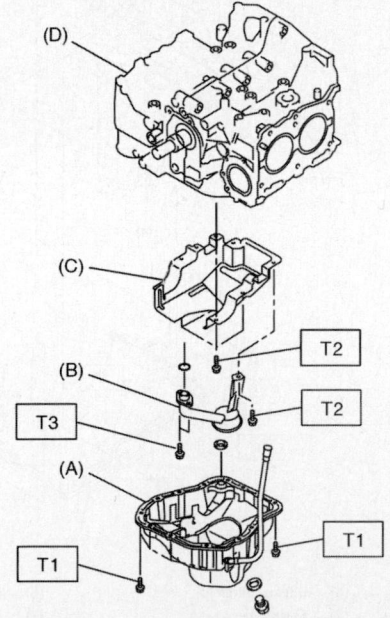

(A) Oil pan   (C) Baffle plate
(B) Oil strainer   (D) Cylinder block

22140_SUBA_G0125

Fig. 126 Exploded view of oil pan, strainer, baffle plate and cylinder block

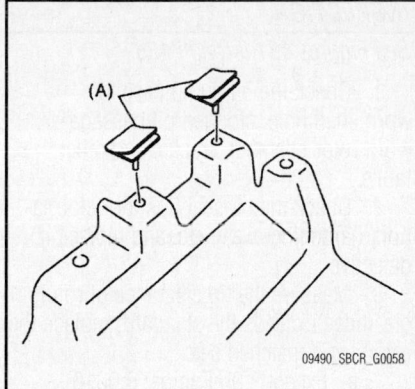

Fig. 127 Oil pan baffle plate seal location and positioning—2.5L engine

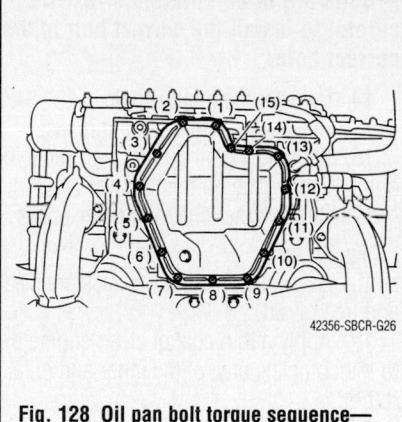

Fig. 128 Oil pan bolt torque sequence— 3.0L engine

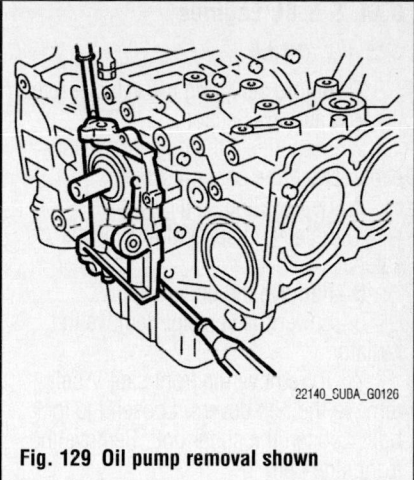

Fig. 129 Oil pump removal shown

24. Tighten the front cushion mounting bolts to 63 ft. lbs.

25. Be sure to fill the engine with the correct grade and type engine oil.

26. Start the engine and check for leaks. Correct as required.

3.0L & 3.6L Engines

See Figure 128.

➡ **If removing the upper oil pan, the engine must first be removed from the vehicle.**

1. Before servicing the vehicle, refer to the Precautions Section.

2. Disconnect the negative battery cable.

3. Raise and support the vehicle safely.

4. Remove the undercover. Drain the engine oil.

5. Remove the lower oil pan retaining bolts.

6. Insert an oil pan gasket cutter tool into the gap between the upper oil pan and the lower oil pan. Remove the lower oil pan from the engine.

➡ **Do not use a screwdriver or similar tool in place of the cutter tool.**

7. Remove the oil strainer, if required.

To install:

8. Be sure to clean the old gasket material from the mating surfaces.

9. Replace the O-ring and install the oil strainer. Tighten the bolt to 5 ft. lbs (6.4 Nm)

10. Apply a continuous bead (0.039 inch thick) of liquid gasket, part number K0877YA018 or equivalent, to the mating surfaces. Install the oil pan. Tighten the retaining bolts to 5 ft. lbs (6.4 Nm) and in the proper sequence.

11. Continue the installation in the reverse order of the removal procedure.

12. Be sure to fill the engine with the correct grade and type engine oil.

13. Start the engine and check for leaks. Correct as required.

OIL PUMP

REMOVAL & INSTALLATION

2.5L Engines

See Figure 129.

1. Before servicing the vehicle, refer to the Precautions Section.

2. Disconnect the negative battery cable. Drain the cooling system.

3. On the DOHC engine, remove the collector cover.

4. Raise and support the vehicle safely.

5. Remove the undercover.

6. On the DOHC engine, remove the bolts which retain the water pipe of the oil cooler to the oil pump. Remove the water pipe and hoses between the oil cooler and the water pump.

7. Lower the vehicle. Remove the radiator.

8. To remove the front side V belt, remove the belt covers. Loosen the lock bolt. Loosen the slider bolt. Remove the front side belt.

9. To remove the rear side V belt, remove the belt covers. Loosen the lock bolt. Loosen the slider bolt. Remove the rear side belt.

10. On the SOHC engine remove the belt tensioner.

11. On the DOHC engine remove the rear side V belt tensioner

12. Remove the crankshaft position sensor.

13. Remove the crankshaft pulley bolt.

14. Lock the crankshaft in place using tool ST499977100 for the SOHC engine and tool ST499207400 for the DOHC engine, or equivalent.

15. Remove the crankshaft pulley.

16. Remove the water pump.

17. If equipped, remove the timing belt guide. Remove the crankshaft sprocket.

18. Remove the oil pump retaining bolts.

➡ **When disassembling and checking the oil pump, loosen the relief valve plug before removing the oil pump from its mounting.**

19. Using a flat tip tool remove the oil pump from the engine.

To install:

20. Be sure all mating surfaces are clean and free of dirt.

21. Apply liquid gasket part number 004403007, or equivalent to the mating surfaces of the oil pump.

22. Be sure to replace the O-ring with a new one.

23. Apply a thin coat of clean engine oil to the inside of the oil seal.

24. Position the oil pump to its mounting, aligning the notched area with the crankshaft and push the pump straight.

➡ **Be sure that the oil seal lip is not folded.**

25. Install the oil pump. Apply liquid gasket part number 004403042, or equivalent to the three retaining bolt threads. Install the bolts and tighten to 5 ft. lbs (6.4 Nm)

26. Continue the installation in the reverse order of the removal procedure.

27. Be sure to fill the cooling system with the proper grade and type coolant.

28. Start the engine and check for leaks. Correct, as required.

3.0L & 3.6L Engines

See Figure 130.

1. Before servicing the vehicle, refer to the Precautions Section.
2. Disconnect the negative battery cable. Drain the cooling system.
3. Remove the collector cover.
4. Raise and support the vehicle safely.
5. Remove the undercover.
6. Lower the vehicle. Remove the radiator.
7. To remove the front side V belt, remove the belt covers. Loosen the lock bolt. Loosen the slider bolt. Remove the front side belt.
8. To remove the rear side V belt, remove the belt covers. Loosen the lock bolt. Loosen the slider bolt. Remove the rear side belt.
9. Remove the crankshaft pulley cover.
10. Remove the crankshaft pulley bolt.
11. Lock the crankshaft in place using tool ST499977100, or equivalent.
12. Remove the crankshaft pulley.
13. Remove the front timing chain cover.

➡Bolts are three different sizes. Be careful to install the correct bolt in the correct hole.

14. Remove the timing chain.
15. Remove the crankshaft sprocket.
16. Remove the oil pump cover retaining bolts. Remove the oil pump cover.
17. Remove the inner and outer rotors.

To install:

18. Be sure all mating surfaces are clean and free of dirt.
19. Apply a thin coat of clean engine oil to the complete area of the inner and outer rotors.
20. Position the inner rotor in place. Position the outer rotor in place.
21. Install the pump cover. Tighten the retaining bolts to 5 ft. lbs (6.4 Nm). and in the proper sequence.

➡Make sure that the bolts are installed in the correct positions.

22. Continue the installation in the reverse order of the removal procedure.
23. Be sure to fill the cooling system with the proper grade and type coolant.
24. Start the engine and check for leaks. Correct, as required.

INSPECTION

See Figures 131 through 133.

1. Check the oil pump case for worn shaft hole, clogged oil passage, worn rotor chamber, cracks, and other faults.
2. Check the oil seal lips for deformation, hardening, wear, etc. and replace if defective.
3. Measure the tip clearance of rotors. If clearance exceeds the standard, replace the rotors as a matched set.
   a. Except 3.0L Engine: 0.0016–0.0055 in. (0.04–0.14 mm)
   b. 3.0L Engine: 0.0016–0.0055 in. (0.04–0.14 mm)
4. Measure the clearance between the outer rotor and oil pump rotor housing. If clearance exceeds the standard, replace the rotor.
   a. Except 3.0L Engine: 0.0039–0.0069 in. (0.10–0.18 mm)
   b. 3.0L Engine: 0.0043–0.0069 in. (0.11–0.18 mm)
5. Measure the clearance between the oil pump inner rotor and pump cover. If clearance exceeds the standard, replace the rotor or pump body.

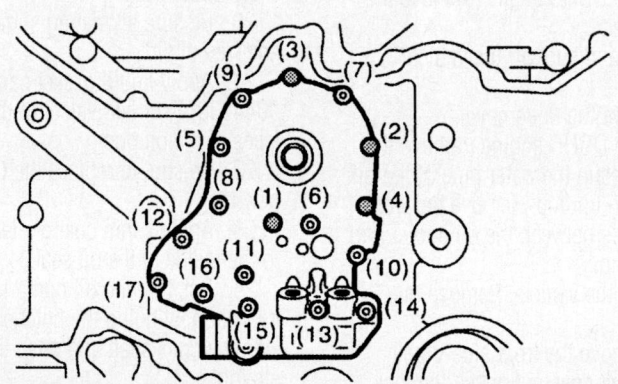

Bolt installing position	Bolt dimension
(1) and (3)	6 × 14 × 14
(2) and (4)	6 × 35 × 18
(5), (6), (7), (8), (9), (10) and (11)	6 × 35 × 15
(12), (15), (16) and (17)	6 × 16 × 16
(13) and (14)	6 × 26 × 15

09490_SBCR_G0059

Fig. 130 Oil pump tightening sequence and bolt location—3.0L engine

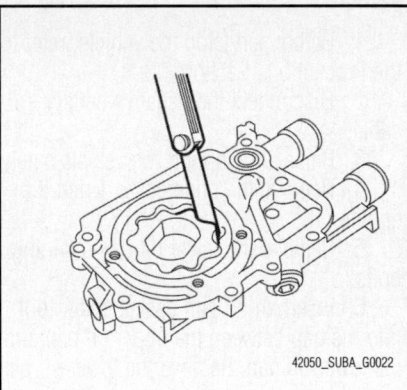

42050_SUBA_G0022

Fig. 131 Measuring for tip clearance of the rotors of the oil pump.

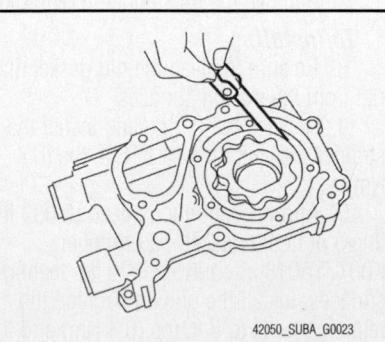

42050_SUBA_G0023

Fig. 132 Measuring the case clearance of the oil pump.

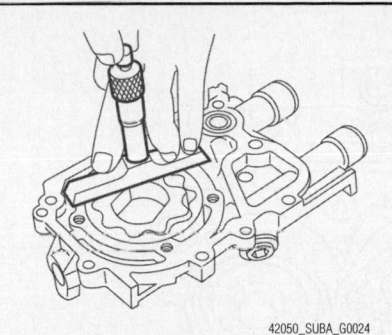

Fig. 133 Measuring the side clearance of the oil pump.

a. Except 3.0L Engine: 0.0008–0.0028 in. (0.02–0.07 mm)

b. 3.0L Engine: 0.0008–0.0018 in. (0.02–0.05 mm)

6. Check the valve for fitting condition and damage, and the relief valve spring for damage and deterioration. Replace the parts if defective.

REAR MAIN SEAL

REMOVAL & INSTALLATION

See Figure 134.

1. Before servicing the vehicle, refer to the Precautions Section.
2. Remove or disconnect the following:
   • Engine from the vehicle
   • Clutch assembly/flywheel using the Clutch Disc Guide tool 499747000, if equipped with a manual transmission
   • Torque converter flexplate from the crankshaft, if equipped with an automatic transmission
   • Oil seal from the cylinder block using a small pry bar

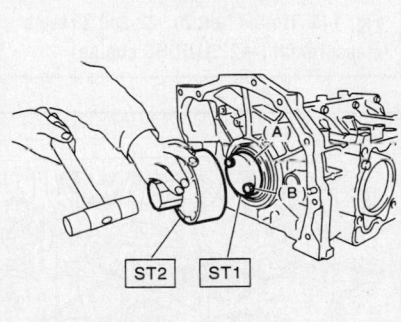

(A) Rear oil seal
(B) Drive plate attaching bolt

42356-SBCR-G28

Fig. 134 Installing the rear main seal using oil seal guide ST1 499597100 and ST2 499598200

To install:

3. Install or connect the following:
   • New oil seal by pressing it into the cylinder block using the appropriate driver and hammer
   • Flywheel housing using new gaskets and sealant where necessary.
   • Flywheel and tighten the bolts to specification.
   • Engine

ROCKER ARMS

REMOVAL & INSTALLATION

2.5L SOHC Engine

See Figures 135 through 138.

1. Before servicing the vehicle, refer to the Precautions Section.
2. Disconnect the negative battery cable.
3. To remove the front side V belt, remove the belt covers. Loosen the lock bolt. Loosen the slider bolt. Remove the front side belt.
4. To remove the rear side V belt, remove the belt covers. Loosen the lock bolt. Loosen the slider bolt. Remove the rear side belt. Remove the belt tensioner.
5. Remove the crankshaft pulley bolt.
6. Lock the crankshaft in place using tool ST499977100, or equivalent.
7. Remove the crankshaft pulley.
8. Remove the left side timing belt cover.
9. Remove the front timing belt cover.
10. Remove the timing belt.
11. Remove the camshaft position sensor.
12. Remove the camshaft sprockets, No.1 and No.2.

➡**Be sure to lock the camshaft in place using tool ST18231AA010, or equivalent.**

13. Disconnect the PCV valve hose. Remove the rocker cover retaining bolts. Remove the rocker cover from the engine.
14. Use tool St18258AA000, or equivalent, to rotate the spring stopper in direction of the arrow, see illustration, and remove the spring stopper from the adjuster spring.
15. Remove bolts "A" through "J" in sequence. Remove the assembly from the engine.

➡**Leave two or three threads on bolts "I" and "J" engaged in order to retain the rocker arm assembly.**

16. Position tool ST18354AA000 on the rocker arm assembly. Remove the assembly from the engine.

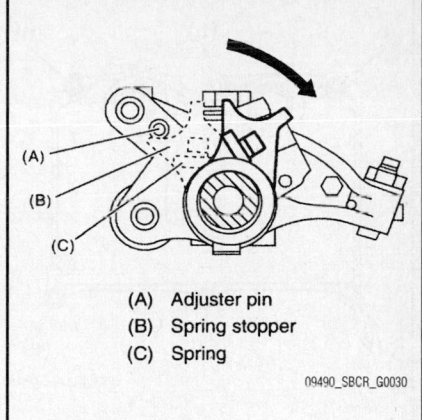

(A) Adjuster pin
(B) Spring stopper
(C) Spring

09490_SBCR_G0030

Fig. 135 Rocker arm spring stopper removal—2.5L SOHC engine

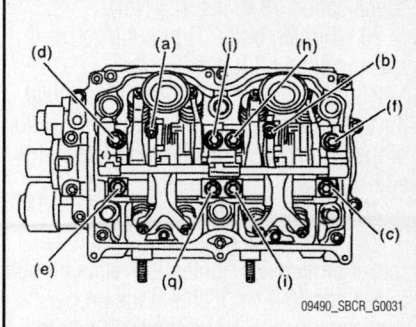

09490_SBCR_G0031

Fig. 136 Rocker arm assembly bolt loosening sequence—2.5L SOHC engine

To install:

17. Install the rocker arm assembly on its mounting.
18. Temporarily tighten the bolts equally in the proper sequence.

➡**Do not temporarily tighten bolts "I" and "J". Position tool ST18354AA000 to the rocker arm assembly.**

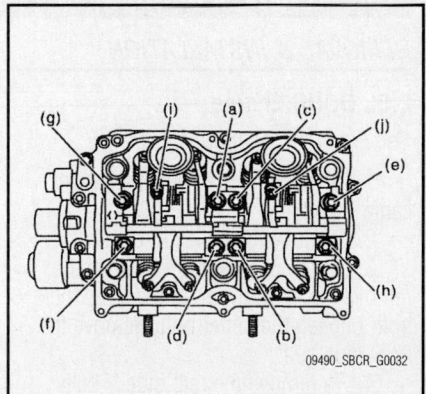

09490_SBCR_G0032

Fig. 137 Rocker arm assembly bolt tightening sequence—2.5L SOHC engine

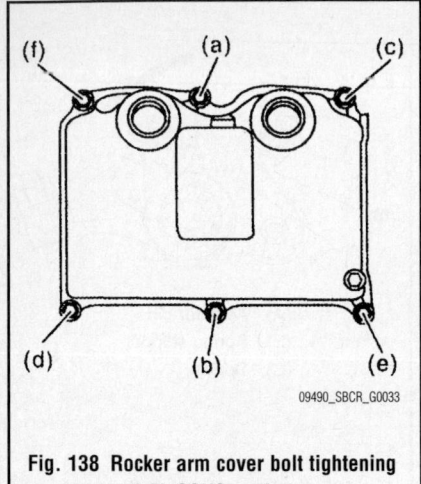

Fig. 138 Rocker arm cover bolt tightening sequence—2.5L SOHC engine

19. Tighten bolts "A" through "H" to specification 18 ft. lbs. (25 Nm).

20. Tighten bolts "I" through "J" to specification 4.3 ft. lbs. (5 Nm).

21. Use tool ST18258AA000 to rotate the spring stopper in the opposite direction of the arrow, see illustration, to fasten the adjuster pin.

22. Adjust the valve clearance.

23. Use a new gasket and install the rocker arm cover. Tighten the retaining bolts to 5 ft. lbs (6.4 Nm). and in the proper sequence. Recheck and tighten again to 5 ft. lbs (6.4 Nm) in the proper sequence.

24. Continue the installation in the reverse order of the removal procedure.

TIMING BELT FRONT COVER

REMOVAL & INSTALLATION

Refer to the Timing Belt and Sprockets, Removal & Installation procedure in this section to remove the timing belt front cover.

TIMING BELT & SPROCKETS

REMOVAL & INSTALLATION

2.5L DOHC Engine

See Figures 139 through 150.

1. Disconnect the negative battery cable.

2. Remove the collector cover.

3. To remove the front side V belt, remove the belt covers. Loosen the lock bolt. Loosen the slider bolt. Remove the front side belt.

4. To remove the rear side V belt, remove the belt covers. Loosen the lock bolt. Loosen the slider bolt. Remove the rear side belt. Remove the belt tensioner.

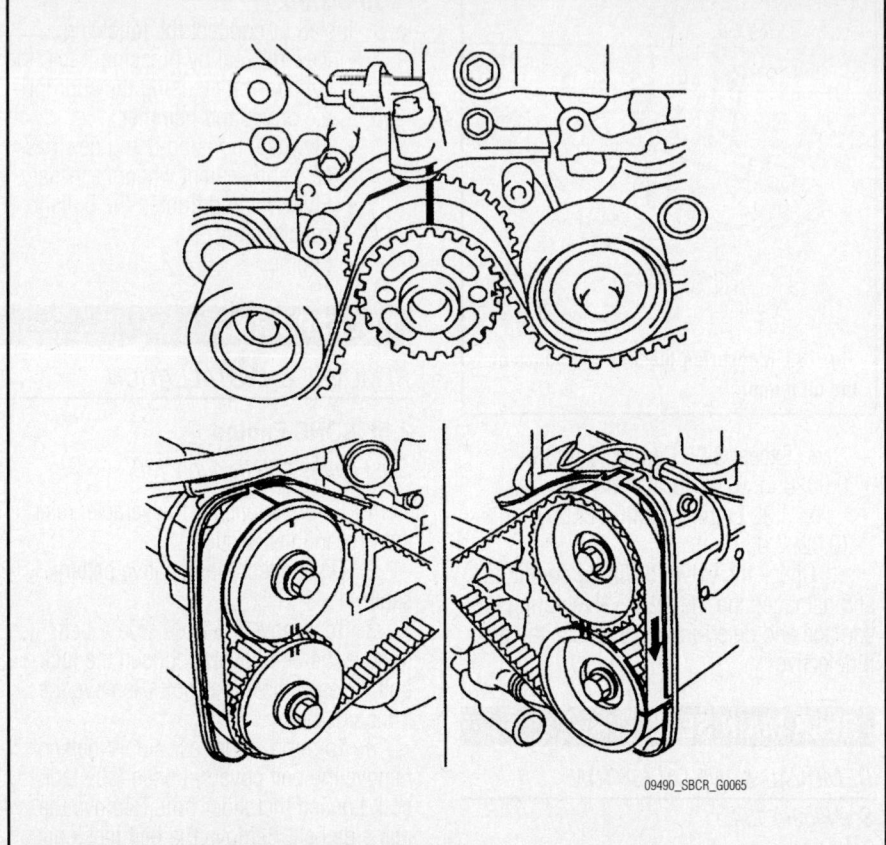

Fig. 139 Timing belt alignment—2.5L DOHC engine

5. Remove the crankshaft pulley bolt.

6. Lock the crankshaft in place using tool ST499977100, or equivalent.

7. Remove the crankshaft pulley.

8. Remove the left side timing belt cover.

9. Remove the right side timing belt cover.

10. Remove the front timing belt cover.

11. Remove the timing belt guide, if equipped.

➡If the belt is going to be reused and the alignment mark on the belt is not readable, put a new mark on the belt to indicate the direction of rotation. Using tool ST499987500, turn the crankshaft to align the mark on the crankshaft sprocket, intake camshaft sprocket (left), exhaust camshaft sprocket (left), intake camshaft sprocket (right), exhaust camshaft sprocket (right) with the notches of the timing belt cover and cylinder block. Paint an alignment mark on the belts in relation to the camshaft sprockets. Z1 measurement is 54.4 teeth. Z2 measurement is 51.0 teeth and Z3 measurement is 28.0 teeth.

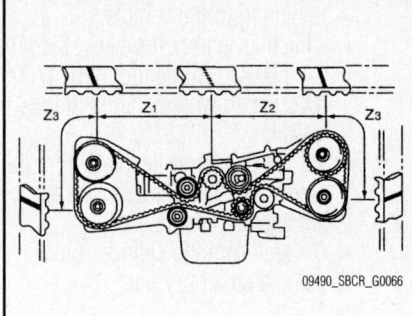

Fig. 140 Timing belt Z1, Z2 and Z3 teeth measurement—2.5L DOHC engine

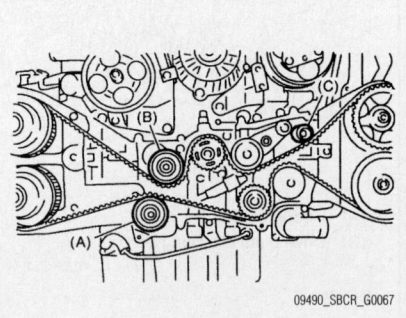

Fig. 141 Belt idler identification and location—2.5L DOHC engine

12. Remove the belt idler belt "A". Remove the timing belt.

13. Remove the belt idlers belt "B" and "C".

14. Remove the belt idler number two. Remove the automatic belt tension adjuster assembly.

To install:

15. Attach the automatic belt tension adjuster assembly to a vertical pressing tool.

➡ **Always use a vertical type pressing tool to move the adjuster rod downward. Do not use a lateral type vise. Push the adjuster rod vertically. Press in the push adjuster rod gradually, which should take three minutes or more. Do not allow pressure to exceed 2,205 lb. force.**

16. Slowly move the adjuster rod down until the adjuster rod is aligned with the stopper pin hole in the cylinder.

➡ **Press the adjuster rod as far as the end surface of the cylinder. Do not press the adjuster rod into the cylinder. Doing so may damage the cylinder.**

17. Using a 0.08 inch stopper pin, insert it into the stopper pin hole in the cylinder. Secure the adjuster rod.

➡ **Do not release the press pressure until the stopper pin is completely inserted in the hole.**

18. Install the automatic belt tensioner assembly. Tighten the retaining bolt to 28.9 ft. lbs.

19. Install the belt idler number two. Tighten the retaining bolt to 28.9 ft. lbs.

20. Install the belt idlers. Tighten the retaining bolts to 28.9 ft. lbs.

21. Align the mark "A" on the crankshaft sprocket with the mark on the oil pump at the cylinder block. Align the single line mark "A" on the right exhaust camshaft sprocket with the notch "B" on the timing belt cover.

22. Align single line mark "A" on the right intake camshaft sprocket with the notch "B" on the timing cover. Ensure that the double lines "C" on the intake and exhaust camshaft sprockets are aligned.

23. Align single line mark "A" on the left exhaust camshaft sprocket with the notch "B" on the timing cover by turning the sprocket counterclockwise as viewed from the front of the engine.

24. Align single line mark "A" on the left

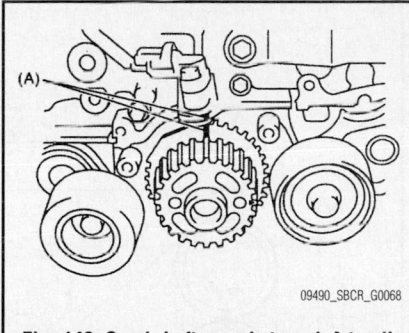

Fig. 142 Crankshaft sprocket mark A to oil pump cover alignment—2.5L DOHC engine

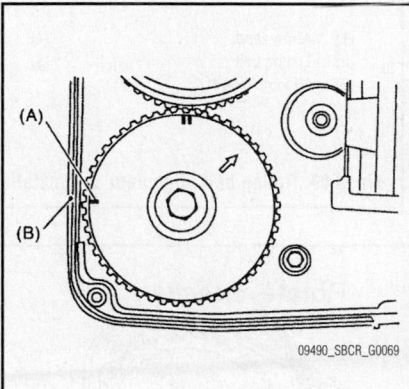

Fig. 143 Right exhaust camshaft sprocket alignment mark A with timing belt cover B alignment mark—2.5L DOHC engine

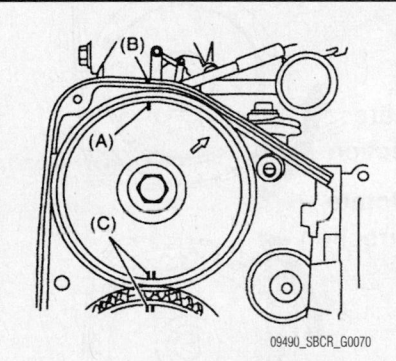

Fig. 144 Right intake camshaft sprocket alignment mark A with timing belt cover B alignment mark an double line C alignment mark—2.5L DOHC engine

intake camshaft sprocket with the notch "B" on the timing cover, by turning the sprocket clockwise as viewed from the front of the engine. Ensure that the double lines "C" on the intake and exhaust camshaft sprockets are aligned.

25. Make sure that the camshaft and crankshaft sprockets are positioned properly.

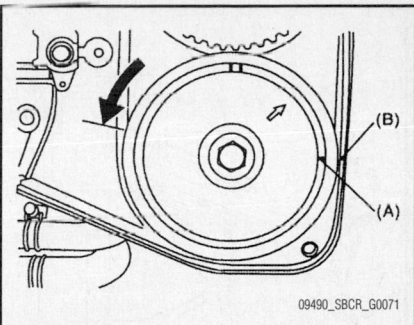

Fig. 145 Left exhaust camshaft sprocket alignment mark A with timing belt cover B alignment mark—2.5L DOHC engine

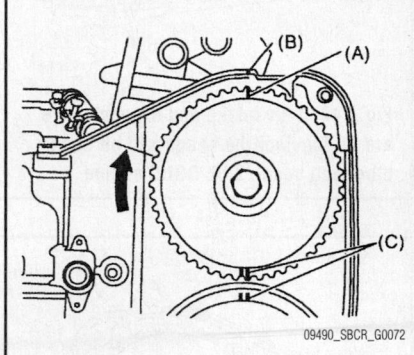

Fig. 146 Left intake camshaft sprocket alignment mark A with timing belt cover B alignment mark an double line C alignment mark—2.5L DOHC engine

➡ The intake and exhaust camshafts on this engine can be rotated independently with the timing belt removed. By looking at the illustration it will show you that if the intake and exhaust valve are lift together the heads will hit each other and bend.

➡ When the timing belts are not installed, 4 camshafts are held at "zero lift" position, where all cams on the camshafts do not push the intake and exhaust valves down (under this condition all valves remain unlifted). When the camshafts are rotated to install the timing belts, No. 2 intake and No. 4 exhaust cam of the left hand camshafts are held to push their corresponding valves down. Under this condition these valves are held lifted. The right side camshafts are held in so that their cams do not push the valves down. The left hand camshafts must be rotated from the "zero lift" position to the position where the timing belt is to be installed at as small an angle as possible, in order to prevent mutual interference of intake and exhaust valve heads. Do not allow the camshafts to

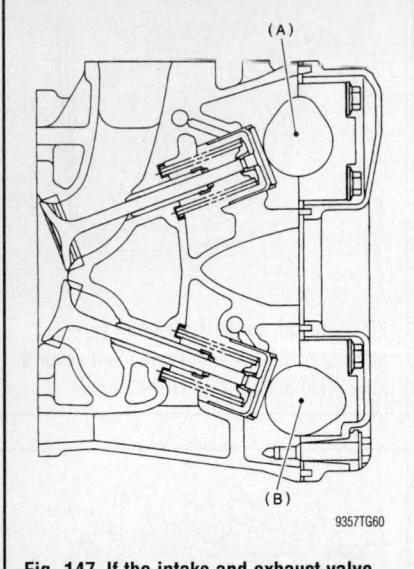

Fig. 147 If the intake and exhaust valve are lift together the heads will hit each other and bend —2.5L DOHC engine

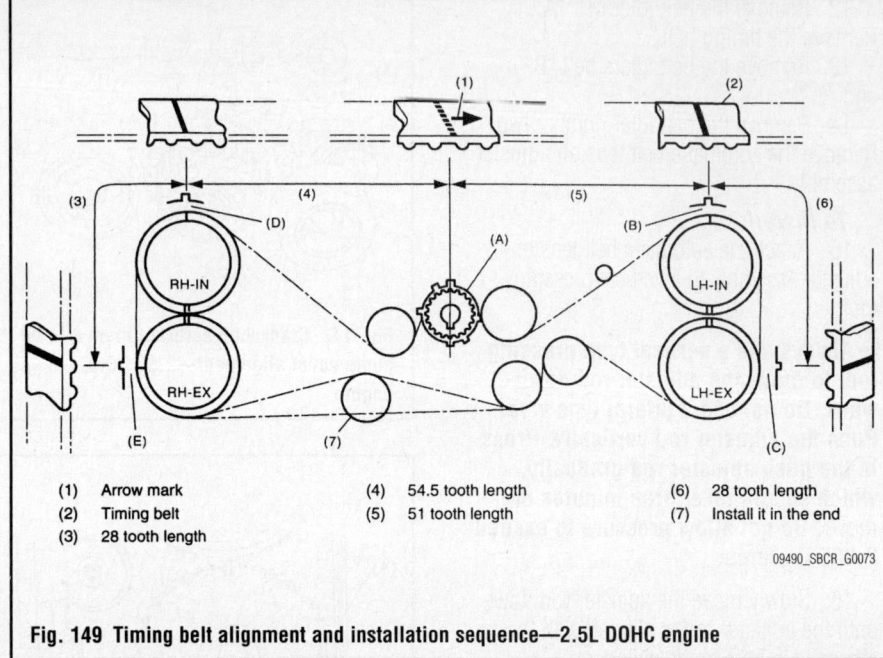

(1)	Arrow mark	(4)	54.5 tooth length	(6)	28 tooth length
(2)	Timing belt	(5)	51 tooth length	(7)	Install it in the end
(3)	28 tooth length				

Fig. 149 Timing belt alignment and installation sequence—2.5L DOHC engine

Rotate direction

Rotate direction

Rotate direction

Timing belt set position

Timing belt set position

Rotate direction

Fig. 148 Do not allow the camshafts to rotate in the direction shown as this causes both the intake and exhaust valves to lift off at the same time with will cause valve damage on DOHC engines

rotate in the direction illustrated as this causes both the intake and exhaust valves to lift off at the same time with will cause valve damage.

26. When installing the belt, make sure to align the marks made during removal or if using a new belt, align the in alphabetical order as shown in the illustration.

☀ WARNING

Disengagement of more than 3 timing belt teeth may result in contact between the valve and piston. Always make sure the belts rotation is correct.

27. Install the timing belt.

➡ Align the alignment mark on the timing belt with marks on the sprocket in the order shown in the illustration. While aligning the timing marks, position the timing belt properly.

28. Install the belt idlers. Tighten the retaining bolts to 28.9 ft. lbs. (39 Nm).

➡ Make sure that the marks on the timing belt and sprockets are aligned.

29. After checking to be sure the marks on the timing belt and the camshaft sprockets are aligned remove the stopper pin from the belt tension adjuster.

30. Install the timing belt guide, if equipped with manual transmission. Temporarily tighten the bolts. Check and adjust the clearance between the belt and the guide. It should be 0.039/ 0.020 inch.

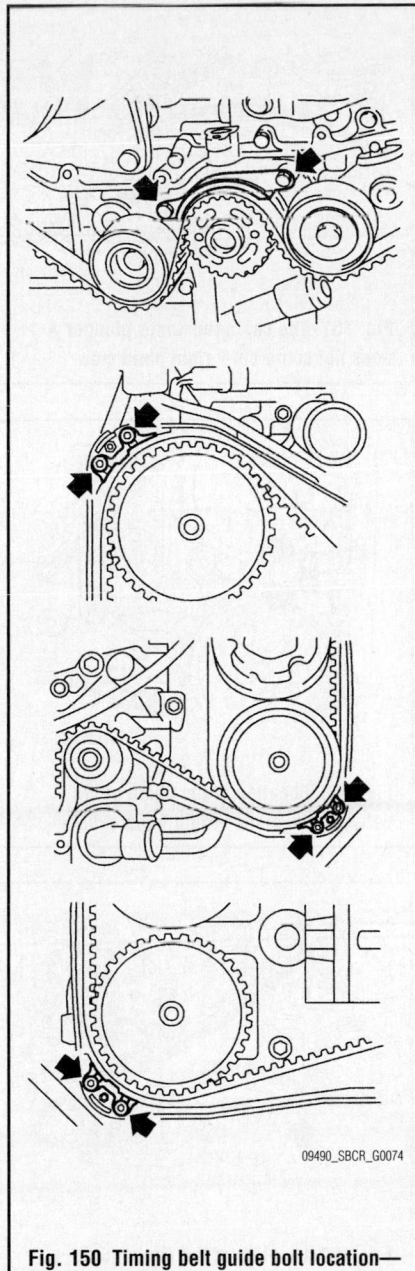

Fig. 150 Timing belt guide bolt location—
2.5L DOHC engine

31. Install the timing belt cover.
32. Install the crank pulley.
33. Install the V-belts.

2.5L SOHC Engine

See Figures 151 through 155.

1. Before servicing the vehicle, refer to the Precautions Section.
2. Disconnect the negative battery cable.
3. To remove the front side V belt, remove the belt covers. Loosen the lock bolt. Loosen the slider bolt. Remove the front side belt.
4. To remove the rear side V belt, remove the belt covers. Loosen the lock bolt. Loosen the slider bolt. Remove the rear side belt. Remove the belt tensioner.
5. Remove the crankshaft pulley bolt.
6. Lock the crankshaft in place using tool ST499977100, or equivalent.
7. Remove the crankshaft pulley.
8. Remove the left side timing belt cover.
9. Remove the front timing belt cover.
10. If equipped with manual transmission, remove the timing belt guide.

➡ If the belt is going to be reused and the alignment mark on the belt is not readable, put a new mark on the belt to indicate the direction of rotation. Using tool ST499987500, turn the crankshaft to align the mark of the sprocket "A" to the cylinder mark notch "B". Ensure that the right side cam sprocket mark "C", cam cap and cylinder head matching surface "D" or left side cam sprocket mark "E", timing belt cover notch "F" are properly aligned. Paint an alignment mark on the belt in relation to the crankshaft sprocket and camshaft sprockets. Z1 measurement is 46.8 teeth. Z2 measurement is 43.7 teeth.

11. Remove both the number two belt idlers. Remove the timing belt from the engine.

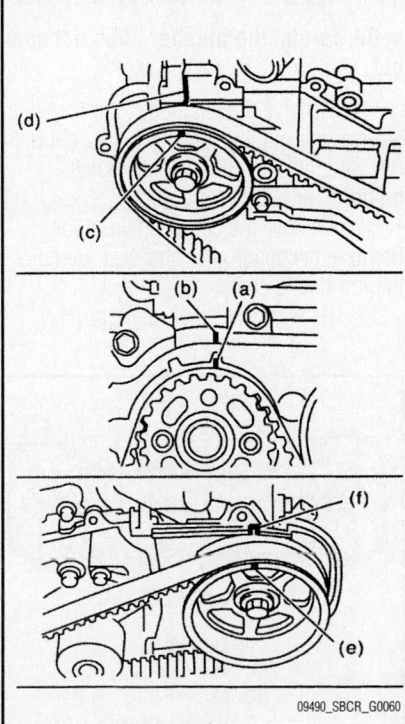

Fig. 151 Timing belt alignment—2.5L SOHC engine

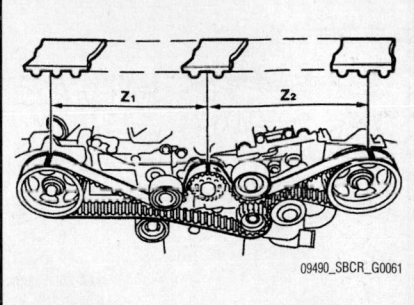

Fig. 152 Timing belt Z1 and Z2 teeth measurement—2.5L SOHC engine

12. Remove the number one belt idler. Remove the automatic belt tension adjuster assembly.

To install:

13. Attach the automatic belt tension adjuster assembly to a vertical pressing tool.

➡ Always use a vertical type pressing tool to move the adjuster rod downward. Do not use a lateral type vise. Push the adjuster rod vertically. Press in the push adjuster rod gradually, which should take three minutes or more. Do not allow pressure to exceed 2,205 lb. force.

14. Slowly move the adjuster rod down until the adjuster rod is aligned with the stopper pin hole in the cylinder.

➡ Press the adjuster rod as far as the end surface of the cylinder. Do not press the adjuster rod into the cylinder. Doing so may damage the cylinder.

15. Using a 0.08 inch stopper pin, insert it into the stopper pin hole in the cylinder. Secure the adjuster rod.

➡ Do not release the press pressure until the stopper pin is completely inserted in the hole.

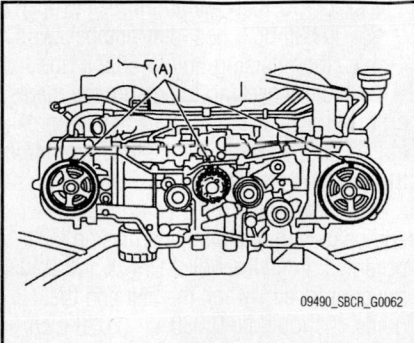

Fig. 153 Timing mark alignment position A—2.5L SOHC engine

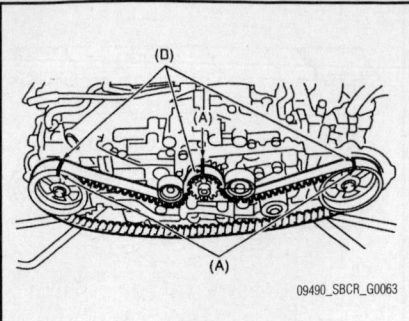

Fig. 154 Timing mark alignment position B—2.5L SOHC engine

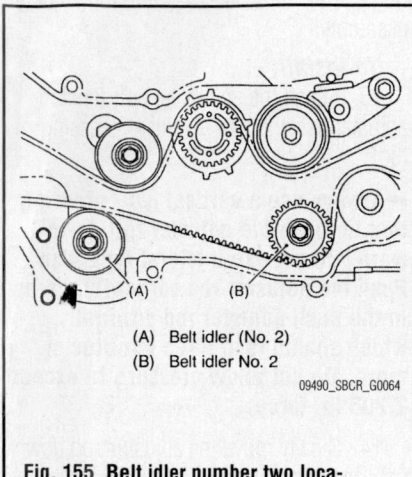

(A) Belt idler (No. 2)
(B) Belt idler No. 2

Fig. 155 Belt idler number two locations—2.5L SOHC engine

16. Install the automatic belt tensioner assembly. Tighten the retaining bolt to 28.9 ft. lbs.

17. Install the belt idler number one. Tighten the retaining bolt to 28.9 ft. lbs.

18. Turn the number one and number two camshaft sprockets, using tool ST499207100 or tool ST18231AA010 and position the alignment marks "A" on each at the highest position.

19. While aligning the alignment mark "B" on the timing belt with mark "A" on the sprockets, position the timing belt properly.

20. Install both belt idler number two's. Tighten the retaining bolt to 28.9 ft. lbs.

21. After checking to be sure the marks on the timing belt and the camshaft sprockets are aligned remove the stopper pin from the belt tension adjuster.

22. Install the timing belt guide, if equipped with manual transmission. Temporarily tighten the bolts. Check and adjust the clearance between the belt and the guide. It should be 0.039 +/- 0.020 inch. Tighten the bolts to 7.2 ft. lbs.

23. Continue the installation in the reverse order of the removal procedure.

TIMING CHAIN & SPROCKETS

REMOVAL & INSTALLATION

3.0L & 3.6L Engines

See Figures 156 through 168.

1. Disconnect the negative battery cable.

2. Remove the crankshaft pulley cover.

3. Remove the crankshaft pulley bolt.

4. Lock the crankshaft in place using tool ST499977100, or equivalent.

5. Remove the crankshaft pulley.

6. Remove the front timing chain cover.

➡**Bolts are three different sizes. Be careful to install the correct bolt in the correct hole.**

7. Remove the right side chain tensioner.

➡**Be careful the plunger does not come out.**

8. Remove the right chain guide, between the right side cams. Remove the right side chain guide. Remove the right side chain tensioner lever. Remove the right timing chain.

9. Remove the left side chain tensioner.

➡**Be careful the plunger does not come out.**

10. Remove the left side chain tensioner lever. Remove the left chain guide, between the left side cams. Remove the chain guide.

11. Remove the center chain guide. Remove the upper idler sprocket. Remove the left timing chain.

12. Remove the lower idler sprocket.

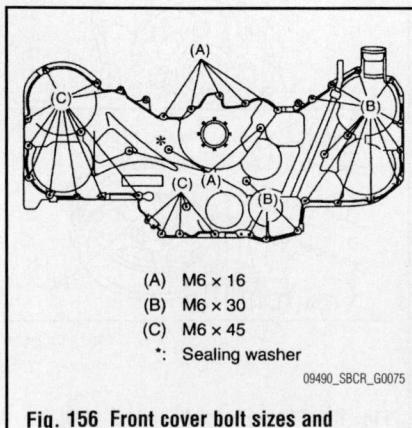

(A) M6 × 16
(B) M6 × 30
(C) M6 × 45
*: Sealing washer

Fig. 156 Front cover bolt sizes and locations

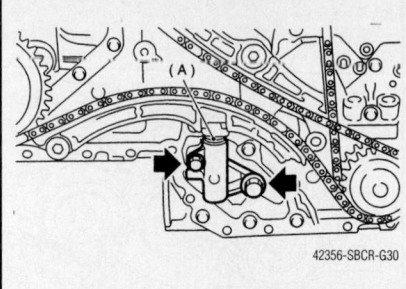

Fig. 157 The chain tensioner plunger A does not come out—right hand side

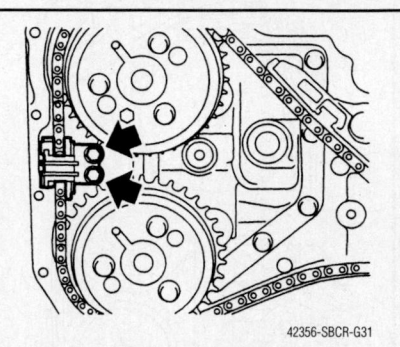

Fig. 158 Location of the chain guide between the cams—right hand side

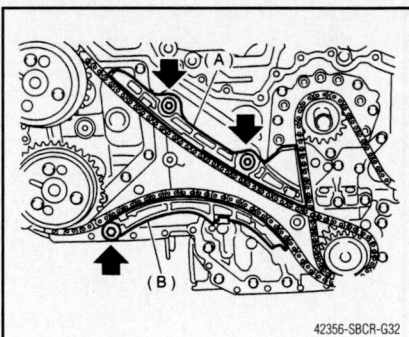

Fig. 159 Location of the chain guide—right hand side

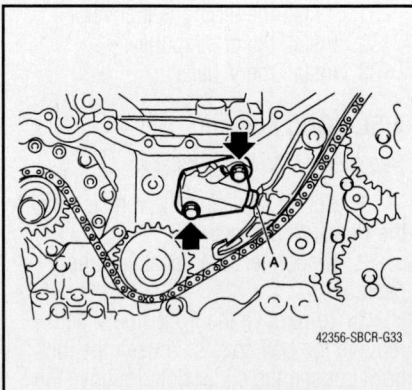

Fig. 160 The chain tensioner plunger A does not come out—left hand side

Fig. 161 Location of the chain guide between the cams—left hand side

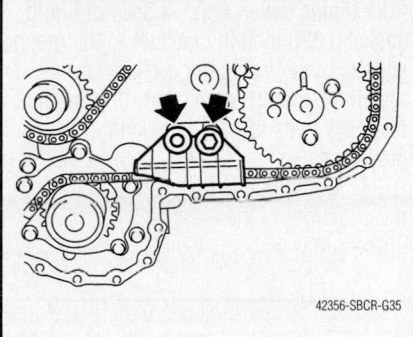

Fig. 162 Location of the chain guide—left hand side

To install:

13. Make sure all components are clean. Apply oil to the chain guide, tensioner lever and idler sprockets.

14. Place the screw, spring, pin and tension rod into the tensioner body.

15. While pressing the tensioner onto a rubber mat, twist it to the left and right to

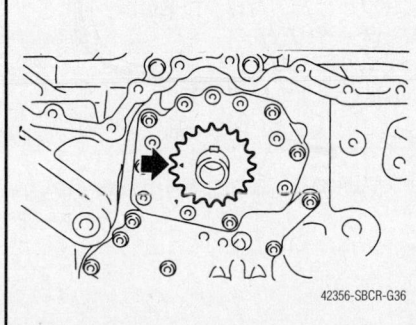

Fig. 163 Align the TOP MARK on the crankshaft sprocket to the 9 O'clock position

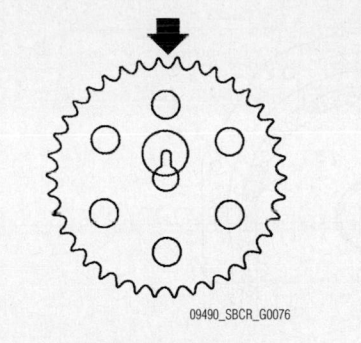

Fig. 164 Align the TOP MARK on the camshaft sprocket to the 12 O'clock position

shorten the rod. Place a thin pin into the holes between the rod and body to hold it in place. Always perform this task on a rubber mat.

16. Using the crankshaft socket tool, align the **TOP MARK** on the crankshaft sprocket to the 9 O'clock position, as shown in the illustration.

17. Align the key groove on the exhaust camshaft sprocket to the 12 O'clock position, as shown in the illustration.

18. Align the intake camshaft sprocket, as shown in the illustration.

19. Turn the crankshaft sprocket clockwise; align the **TOP MARK** to the 12 O'clock position. Piston number one is now at Top Dead Center (TDC).

✳✳ CAUTION

Do not rotate the camshaft or crankshaft sprockets until the chain is completely routed or damage will occur.

20. Install the lower idler sprocket and tighten the bolt to 50.6 ft. lbs. (69 Nm).

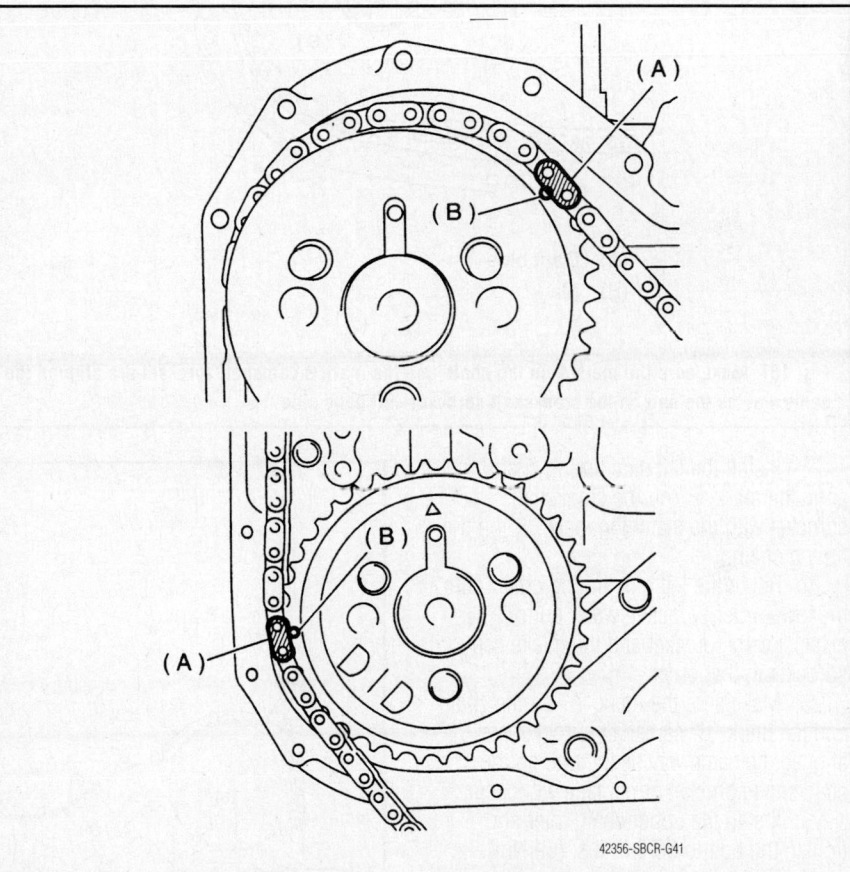

Fig. 165 Make sure the mark A on the chain and the mark B camshaft sprocket are aligned the same way as the one on the crankshaft sprocket–right hand side

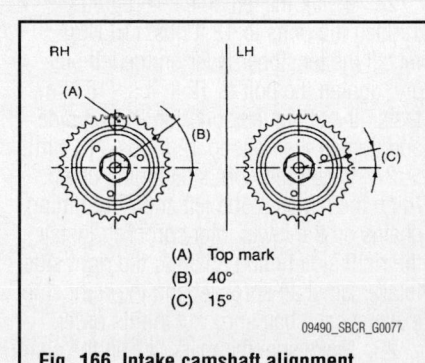

(A) Top mark
(B) 40°
(C) 15°

Fig. 166 Intake camshaft alignment

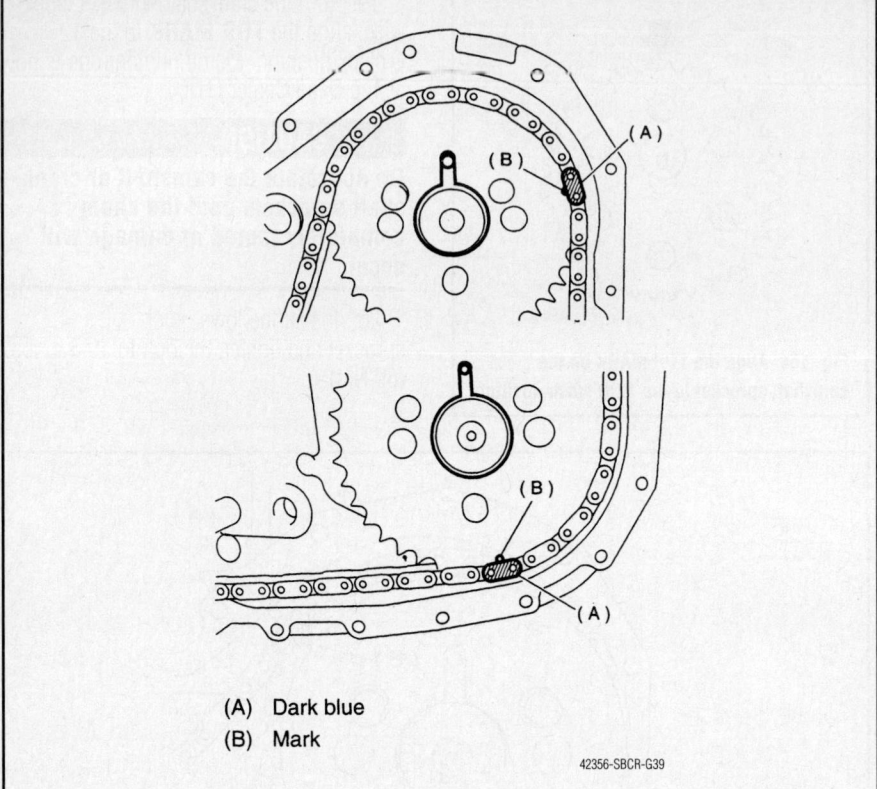

(A) Dark blue
(B) Mark

42356-SBCR-G39

Fig. 167 Make sure the mark A on the chain and the mark B camshaft sprocket are aligned the same way as the one on the crankshaft sprocket–left hand side

aligned the same way as the one on the crankshaft sprocket or damage will occur.

29. Install the right side chain guide. Install the right side chain tensioner lever and tighten the bolts to 12 ft. lbs. (16 Nm). Install the right side chain guide and tighten the NEW bolt to 4.7 ft. lbs. (6 Nm). Install the right side chain tensioner and tighten the bolts to 12 ft. lbs. (16 Nm).

30. Adjust the clearance between the chain guide on the right side and the center chain guide so that there is range between 0.331–0.339 inch (8.4–8.6mm).

31. Install the center chain guide and tighten the NEW bolt to 5.8 ft. lbs. (8 Nm).

32. Check the match marks on each sprocket and corresponding timing chain are correct, remove the stopper from the tensioner.

33. Clean the mating surfaces on the front timing cover. Apply a bead of liquid gasket 0.020 inch in diameter to the mating surface of the front timing chain cover. Install the timing chain cover. Torque the retaining bolts to 4.8 ft. lbs. and in the proper sequence.

34. Continue the installation in the reverse order of the removal procedure.

21. Install the left side timing chain, align the mark "B" on the crankshaft sprocket with the matching mark "A" on the timing chain.

22. Route the left side timing chain onto the lower idler sprocket, water pump, exhaust cam sprocket and the intake cam sprocket in that order.

23. Make sure the mark "A" on the chain and the mark "B" camshaft sprocket are aligned the same way as the one on the crankshaft sprocket or damage will occur.

24. Install the upper chain idler and tighten the bolt to 50.6 ft. lbs. (69 Nm).

25. Install the left side chain guide, between the camshafts. Tighten the bolt to 4.7 ft. lbs. (6 Nm) using a NEW bolt.

26. Install the left side chain guide. Tighten the bolts to 12 ft. lbs. (16 Nm). Install the tensioner lever on the left side and tighten the bolt to 12 ft. lbs. (16 Nm). Install the chain tensioner on the left side and tighten the bolts to 12 ft. lbs. (16 Nm).

27. Install the right side timing chain. Align the marks of the left and right timing chains on the lower idler sprocket. Install the right side timing chain to the right side intake camshaft sprocket and the right side exhaust camshaft sprocket in this order.

28. Make sure the mark "A" on the chain and the mark "B" camshaft sprocket are

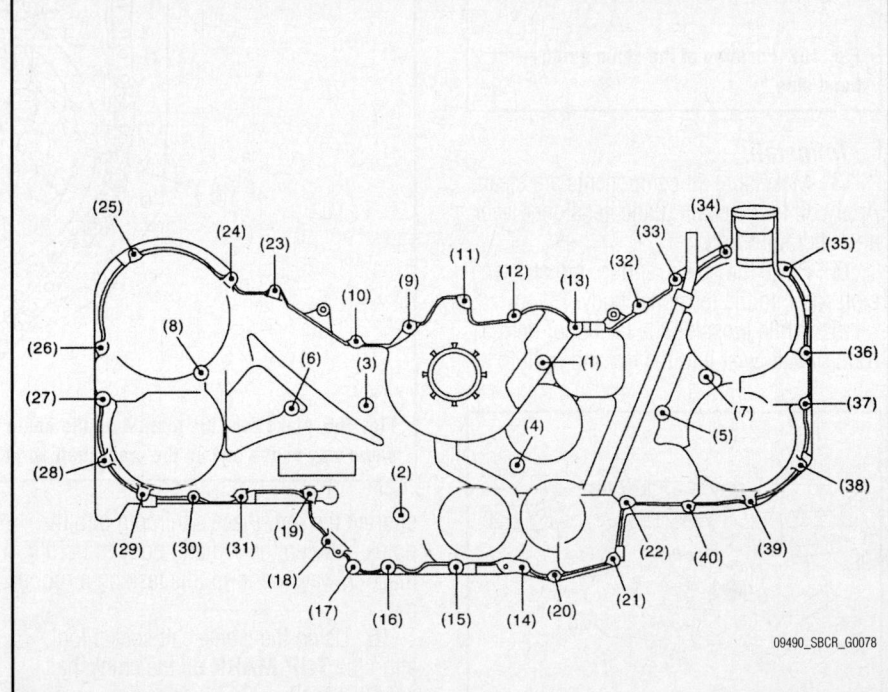

09490_SBCR_G0078

Fig. 168 Front timing cover bolt tightening sequence

VALVE LASH

ADJUSTMENT

2.5L DOHC Engine

See Figures 169 through 176.

➡️The valve adjustment should be performed while the engine is cold.

1. Raise and support the vehicle safely.
2. Remove the undercover.
3. Lower the vehicle.
4. Remove the collector cover.
5. Disconnect the negative battery cable.
6. Remove the air intake duct.
7. Remove the bolt that retains the right side timing belt cover. Remove the remaining bolts and remove the right side timing belt cover.
8. Disconnect the ignition coil electrical connector. Remove the ignition coil.
9. Position a suitable container under the vehicle.
10. Disconnect the PCV hose from the rocker cover. Remove the rocker cover retaining bolts. Remove the rocker cover from the vehicle.
11. Position the number one piston at TDC of the compression stroke.
12. Using a feeler gauge, measure and record the clearance of the number one cylinder intake and the number three cylinder exhaust valves.

➡️Insert the feeler gauge in a horizontally as possible with respect to the valve lifter. Measure and record the exhaust valve clearance while lifting up the vehicle.

13. Rotate the crankshaft pulley clockwise until the arrow mark on the camshaft is positioned as shown to measure and record the clearance on the number two exhaust and number three intake valves.
14. Rotate the crankshaft pulley clockwise until the arrow mark on the camshaft is positioned as shown to measure and record the number two intake and number four exhaust valves.
15. Rotate the crankshaft pulley clockwise until the arrow mark on the camshaft is positioned as shown to measure and record the number one exhaust and number four intake valves.
16. If adjustment is required, remove the camshafts.

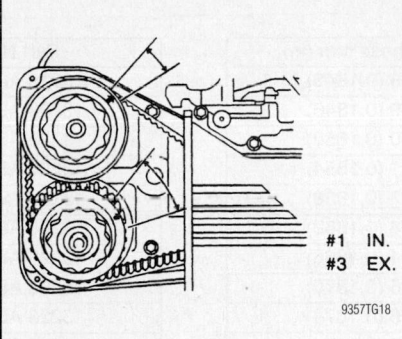

Fig. 169 Turn the crankshaft pulley clockwise until the arrow mark on the camshaft is positioned as shown to measure the No. 1 intake and No. 3 exhaust valves—2.5L DOHC engine

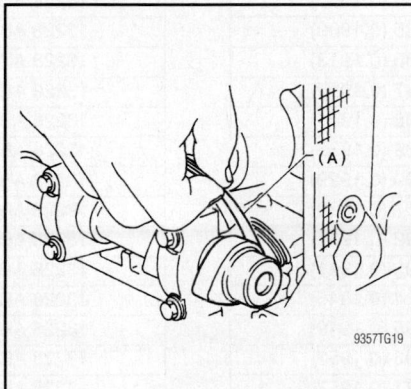

Fig. 170 Use a feeler gauge to inspect the valve clearance—2.5L DOHC engine

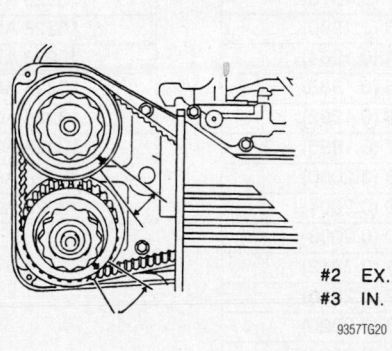

Fig. 171 Turn the crankshaft pulley clockwise until the arrow mark on the camshaft is positioned as shown to measure the No. 2 exhaust and No. 3 intake valves—2.5L DOHC engine

17. Remove and measure the thickness of the valve lifter. Select a suitable shim, using the shim selection chart.

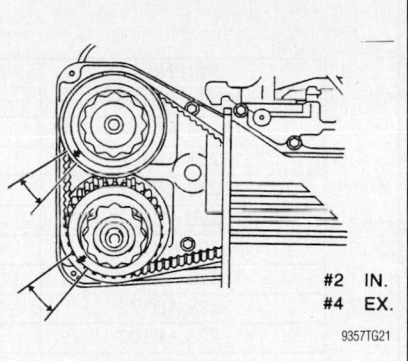

Fig. 172 Turn the crankshaft pulley clockwise until the arrow mark on the camshaft is positioned as shown to measure the No. 2 intake and No. 4 exhaust valves—2.5L DOHC engine

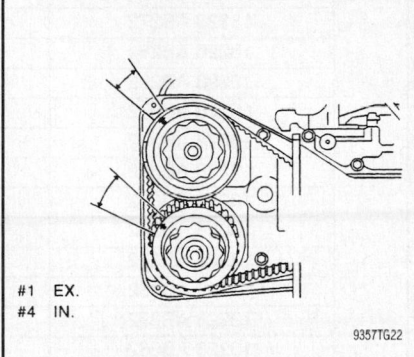

Fig. 173 Turn the crankshaft pulley clockwise until the arrow mark on the camshaft is positioned as shown to measure the No. 1 exhaust and No. 4 intake valves—2.5L DOHC engine

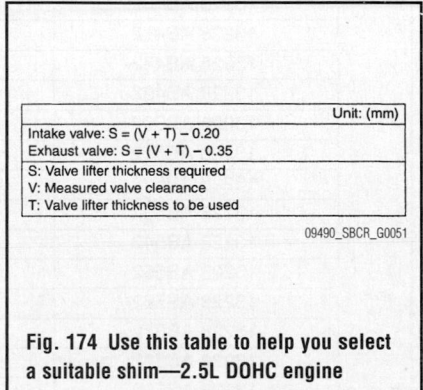

	Unit: (mm)
Intake valve: S = (V + T) − 0.20	
Exhaust valve: S = (V + T) − 0.35	
S: Valve lifter thickness required	
V: Measured valve clearance	
T: Valve lifter thickness to be used	

Fig. 174 Use this table to help you select a suitable shim—2.5L DOHC engine

18. Install the replacement shim to the lifter.
19. After all shims have been adjusted, inspect the valve clearances again.
20. After completion, install all removed components.

Part No.	Thickness mm (in)
13228 AB102	4.68 (0.1843)
13228 AB112	4.69 (0.1846)
13228 AB122	4.70 (0.1850)
13228 AB132	4.71 (0.1854)
13228 AB142	4.72 (0.1858)
13228 AB152	4.73 (0.1862)
13228 AB162	4.74 (0.1866)
13228 AB172	4.75 (0.1870)
13228 AB182	4.76 (0.1874)
13228 AB192	4.77 (0.1878)
13228 AB202	4.78 (0.1882)
13228 AB212	4.79 (0.1886)
13228 AB222	4.80 (0.1890)
13228 AB232	4.81 (0.1894)
13228 AB242	4.82 (0.1898)
13228 AB252	4.83 (0.1902)
13228 AB262	4.84 (0.1906)
13228 AB272	4.85 (0.1909)
13228 AB282	4.86 (0.1913)
13228 AB292	4.87 (0.1917)
13228 AB302	4.88 (0.1921)
13228 AB312	4.89 (0.1925)
13228 AB322	4.90 (0.1929)
13228 AB332	4.91 (0.1933)
13228 AB342	4.92 (0.1937)
13228 AB352	4.93 (0.1941)
13228 AB362	4.94 (0.1945)
13228 AB372	4.95 (0.1949)
13228 AB382	4.96 (0.1953)
13228 AB392	4.97 (0.1957)
13228 AB402	4.98 (0.1961)
13228 AB412	4.99 (0.1965)
13228 AB422	5.00 (0.1969)
13228 AB432	5.01 (0.1972)
13228 AB442	5.02 (0.1976)
13228 AB452	5.03 (0.1980)
13228 AB462	5.04 (0.1984)
13228 AB472	5.05 (0.1988)
13228 AB482	5.06 (0.1992)
13228 AB492	5.07 (0.1996)
13228 AB502	5.08 (0.2000)
13228 AB512	5.09 (0.2004)
13228 AB522	5.10 (0.2008)
13228 AB532	5.11 (0.2012)
13228 AB542	5.12 (0.2016)
13228 AB552	5.13 (0.2020)
13228 AB562	5.14 (0.2024)
13228 AB572	5.15 (0.2028)
13228 AB582	5.16 (0.2031)
13228 AB592	5.17 (0.2035)
13228 AB602	5.18 (0.2039)
13228 AB612	5.19 (0.2043)

Part No.	Thickness mm (in)
13228 AB622	5.20 (0.2047)
13228 AB632	5.21 (0.2051)
13228 AB642	5.22 (0.2055)
13228 AB652	5.23 (0.2059)
13228 AB662	5.24 (0.2063)
13228 AB672	5.25 (0.2067)
13228 AB682	5.26 (0.2071)
13228 AB692	5.27 (0.2075)
13228 AB702	4.38 (0.1724)
13228 AB712	4.40 (0.1732)
13228 AB722	4.42 (0.1740)
13228 AB732	4.44 (0.1748)
13228 AB742	4.46 (0.1756)
13228 AB752	4.48 (0.1764)
13228 AB762	4.50 (0.1771)
13228 AB772	4.52 (0.1780)
13228 AB782	4.54 (0.1787)
13228 AB792	4.56 (0.1795)
13228 AB802	4.58 (0.1803)
13228 AB812	4.60 (0.1811)
13228 AB822	4.62 (0.1819)
13228 AB832	4.64 (0.1827)
13228 AB842	4.66 (0.1835)
13228 AB852	5.29 (0.2083)
13228 AB862	5.31 (0.2091)
13228 AB872	5.33 (0.2098)
13228 AB882	5.35 (0.2106)
13228 AB892	5.37 (0.2114)
13228 AB902	5.39 (0.2122)
13228 AB912	5.41 (0.2123)
13228 AB922	5.43 (0.2138)
13228 AB932	5.45 (0.2146)
13228 AB942	5.47 (0.2154)
13228 AB952	5.49 (0.2161)
13228 AB962	5.51 (0.2169)
13228 AB972	5.53 (0.2177)
13228 AB982	5.55 (0.2185)
13228 AB992	5.57 (0.2193)
13228 AC002	5.59 (0.2201)
13228 AC012	5.61 (0.2209)
13228 AC022	5.63 (0.2217)
13228 AC032	5.65 (0.2224)

09490_SBCR_G0052

Fig. 175 Valve adjusting shim chart—2.5L DOHC engine

2.5L SOHC Engine

See Figure 176.

➡ **The valve adjustment should be performed while the engine is cold.**

1. Raise and support the vehicle safely.
2. Remove the undercover.
3. Lower the vehicle.
4. Disconnect the negative battery cable.
5. To remove the front side V belt, remove the belt covers. Loosen the lock bolt. Loosen the slider bolt. Remove the front side belt.
6. To remove the rear side V belt, remove the belt covers. Loosen the lock bolt. Loosen the slider bolt. Remove the rear side belt. Remove the belt tensioner.
7. Remove the crankshaft pulley bolt.
8. Lock the crankshaft in place using tool ST499977100, or equivalent.
9. Remove the crankshaft pulley.
10. Remove the left side timing belt cover.
11. Remove the fuel injector.
12. Remove the rocker cover.
13. Position the number one piston at TDC of the compression stroke.

➡ **When the arrow (see illustration) on the camshaft sprocket (left side) comes exactly to the top, number one cylinder piston is at TDC of the compression stroke.**

14. Measure the valve clearance, using a feeler gauge.
15. If adjustment is needed, loosen the valve rocker nut and screw. Position the feeler gauge.

➡ **Insert the feeler gauge in a horizontally as possible with respect to the valve stem end face. Adjust the exhaust valve clearance while lifting up the vehicle.**

16. While noting the valve clearance, tighten the rocker adjusting screw.
17. When the proper valve clearance is obtained, tighten the valve rocker nut to 7.2 ft. lbs.
18. Adjust the valve clearance on the remaining cylinders, following the above procedure.

➡ **Be sure to position the pistons to their respective TDC positions on the compression stroke, before checking and adjusting the valves. By rotating the crankshaft pulley clockwise every 180 degrees from the state that number one piston is on TDC of the compression stroke, the remaining pistons come to TDC of the compression stroke in the following order, #3, #2, and #4.**

19. After adjustment, replace any removed components.
20. Be sure to use new gaskets and seals, as required.

3.0L Engine

See Figures 177 through 181.

➡ **The valve adjustment should be performed while the engine is cold.**

1. Raise and support the vehicle safely.
2. Remove the undercover.
3. Lower the vehicle.
4. Remove the collector cover.
5. Disconnect the negative battery cable.
6. On the right side, remove the air intake duct and air cleaner case. Remove the fuel tank protector. Disconnect the oil pressure switch electrical connector. Remove the ignition coil.
7. On the left side, remove the battery and battery carrier. Disconnect the PCV hose from the rocker cover. Remove the ignition coil.

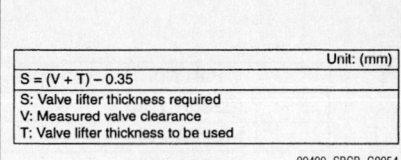

	Unit: (mm)
$S = (V + T) - 0.35$	
S: Valve lifter thickness required	
V: Measured valve clearance	
T: Valve lifter thickness to be used	

09490_SBCR_G0054

Fig. 178 Use this table to help you select a suitable exhaust valve shim—3.0L engine

	Unit: (mm)
$S = (V + T) - 0.20$	
S: Required shim thickness	
V: Measured valve clearance	
T: Shim thickness to be used	

09490_SBCR_G0055

Fig. 179 Use this table to help you select a suitable intake valve shim—3.0L engine

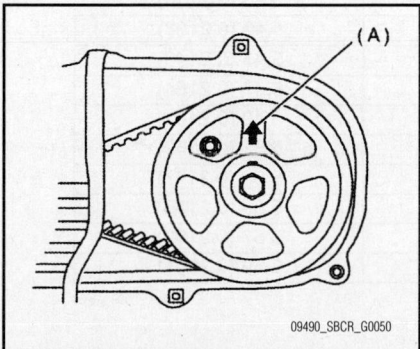

09490_SBCR_G0050

Fig. 176 TDC alignment—2.5L SOHC engine

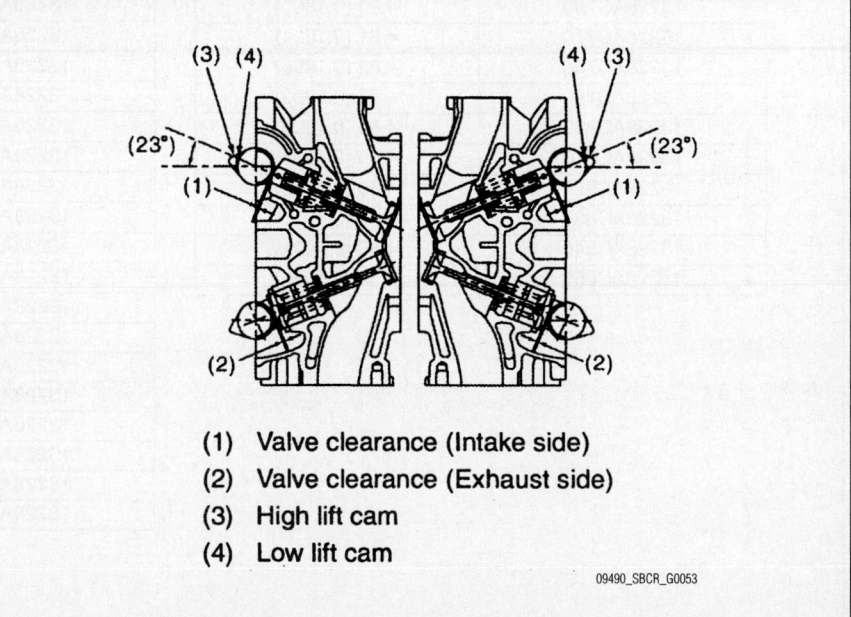

(1) Valve clearance (Intake side)
(2) Valve clearance (Exhaust side)
(3) High lift cam
(4) Low lift cam

09490_SBCR_G0053

Fig. 177 Valve adjustment crankshaft positioning—3.0L engine

Part No.	Thickness mm (in)	Part No.	Thickness mm (in)
13228AD180	4.32 (0.1701)	13228AC860	4.90 (0.1929)
13228AD190	4.34 (0.1709)	13228AC870	4.91 (0.1933)
13228AD200	4.36 (0.1717)	13228AC880	4.92 (0.1937)
13228AD210	4.38 (0.1724)	13228AC890	4.93 (0.1941)
13228AD220	4.40 (0.1732)	13228AC900	4.94 (0.1945)
13228AD230	4.42 (0.1740)	13228AC910	4.95 (0.1949)
13228AD240	4.44 (0.1748)	13228AC920	4.96 (0.1953)
13228AD250	4.46 (0.1756)	13228AC930	4.97 (0.1957)
13228AD260	4.48 (0.1764)	13228AC940	4.98 (0.1961)
13228AD270	4.50 (0.1772)	13228AC950	4.99 (0.1965)
13228AD280	4.52 (0.1780)	13228AC960	5.00 (0.1969)
13228AD290	4.54 (0.1787)	13228AC970	5.01 (0.1972)
13228AD300	4.56 (0.1795)	13228AC980	5.02 (0.1976)
13228AD310	4.58 (0.1803)	13228AC990	5.03 (0.1980)
13228AD320	4.60 (0.1811)	13228AD000	5.04 (0.1984)
13228AC580	4.62 (0.1819)	13228AD010	5.05 (0.1988)
13228AC590	4.63 (0.1823)	13228AD020	5.06 (0.1992)
13228AC600	4.64 (0.1827)	13228AD030	5.07 (0.1996)
13228AC610	4.65 (0.1831)	13228AD040	5.08 (0.2000)
13228AC620	4.66 (0.1835)	13228AD050	5.09 (0.2004)
13228AC630	4.67 (0.1839)	13228AD060	5.10 (0.2008)
13228AC640	4.68 (0.1843)	13228AD070	5.11 (0.2012)
13228AC650	4.69 (0.1846)	13228AD080	5.12 (0.2016)
13228AC660	4.70 (0.1850)	13228AD090	5.13 (0.2020)
13228AC670	4.71 (0.1854)	13228AD100	5.14 (0.2024)
13228AC680	4.72 (0.1858)	13228AD110	5.15 (0.2028)
13228AC690	4.73 (0.1862)	13228AD120	5.16 (0.2032)
13228AC700	4.74 (0.1866)	13228AD130	5.17 (0.2035)
13228AC710	4.75 (0.1870)	13228AD140	5.18 (0.2039)
13228AC720	4.76 (0.1874)	13228AD150	5.19 (0.2043)
13228AC730	4.77 (0.1878)	13228AD160	5.20 (0.2047)
13228AC740	4.78 (0.1882)	13228AD170	5.21 (0.2051)
13228AC750	4.79 (0.1886)	13228AD330	5.23 (0.2059)
13228AC760	4.80 (0.1890)	13228AD340	5.25 (0.2067)
13228AC770	4.81 (0.1894)	13228AD350	5.27 (0.2075)
13228AC780	4.82 (0.1898)	13228AD360	5.29 (0.2083)
13228AC790	4.83 (0.1902)	13228AD370	5.31 (0.2091)
13228AC800	4.84 (0.1906)	13228AD380	5.33 (0.2098)
13228AC810	4.85 (0.1909)	13228AD390	5.35 (0.2106)
13228AC820	4.86 (0.1913)	13228AD400	5.37 (0.2114)
13228AC830	4.87 (0.1917)	13228AD410	5.39 (0.2122)
13228AC840	4.88 (0.1921)	13228AD420	5.41 (0.2130)
13228AC850	4.89 (0.1925)	13228AD430	5.43 (0.2138)
		13228AD440	5.45 (0.2146)
		13228AD450	5.47 (0.2154)
		13228AD460	5.49 (0.2161)
		13228AD470	5.51 (0.2169)
		13228AD480	5.53 (0.2177)
		13228AD490	5.55 (0.2185)
		13228AD500	5.57 (0.2193)
		13228AD510	5.59 (0.2201)

09490_SBCR_G0056

Fig. 180 Exhaust valve adjusting shim chart—3.0L engine

Part No.	Thickness mm (in)		Part No.	Thickness mm (in)
13218AK890	1.92 (0.0756)		13218AL320	2.42 (0.0953)
13218AK900	1.94 (0.0764)		13218AL330	2.43 (0.0957)
13218AK910	1.96 (0.0772)		13218AL340	2.44 (0.0961)
13218AK920	1.98 (0.0780)		13218AL350	2.45 (0.0965)
13218AK930	2.00 (0.0787)		13218AL360	2.46 (0.0969)
13218AK940	2.02 (0.0795)		13218AL370	2.47 (0.0972)
13218AK950	2.04 (0.0803)		13218AL380	2.48 (0.0976)
13218AK960	2.06 (0.0811)		13218AL390	2.49 (0.0980)
13218AK970	2.07 (0.0815)		13218AL400	2.50 (0.0984)
13218AK980	2.08 (0.0819)		13218AL410	2.51 (0.0988)
13218AK990	2.09 (0.0823)		13218AL420	2.52 (0.0992)
13218AL000	2.10 (0.0827)		13218AL430	2.53 (0.0996)
13218AL010	2.11 (0.0831)		13218AL440	2.54 (0.1000)
13218AL020	2.12 (0.0835)		13218AL450	2.55 (0.1004)
13218AL030	2.13 (0.0839)		13218AL460	2.56 (0.1008)
13218AL040	2.14 (0.0843)		13218AL470	2.57 (0.1012)
13218AL050	2.15 (0.0846)		13218AL480	2.58 (0.1016)
13218AL060	2.16 (0.0850)		13218AL490	2.59 (0.1020)
13218AL070	2.17 (0.0854)		13218AL500	2.60 (0.1024)
13218AL080	2.18 (0.0858)		13218AL510	2.61 (0.1028)
13218AL090	2.19 (0.0862)		13218AL520	2.62 (0.1032)
13218AL100	2.20 (0.0866)		13218AL530	2.64 (0.1039)
13218AL110	2.21 (0.0870)		13218AL540	2.66 (0.1047)
13218AL120	2.22 (0.0874)		13218AL550	2.68 (0.1055)
13218AL130	2.23 (0.0878)		13218AL560	2.70 (0.1063)
13218AL140	2.24 (0.0882)		13218AL570	2.72 (0.1071)
13218AL150	2.25 (0.0886)		13218AL580	2.74 (0.1079)
13218AL160	2.26 (0.0890)		13218AL590	2.76 (0.1087)
13218AL170	2.27 (0.0894)			
13218AL180	2.28 (0.0898)			
13218AL190	2.29 (0.0902)			
13218AL200	2.30 (0.0906)			
13218AL210	2.31 (0.0909)			
13218AL220	2.32 (0.0913)			
13218AL230	2.33 (0.0917)			
13218AL240	2.34 (0.0921)			
13218AL250	2.35 (0.0925)			
13218AL260	2.36 (0.0929)			
13218AL270	2.37 (0.0933)			
13218AL280	2.38 (0.0937)			
13218AL290	2.39 (0.0941)			
13218AL300	2.40 (0.0945)			
13218AL310	2.41 (0.0949)			

09490_SBCR_G0057

Fig. 181 Intake valve adjusting shim chart—3.0L engine

8. Remove the rocker cover retaining bolts. Remove the rocker covers from the engine.

9. Rotate the crankshaft clockwise until the cam is set in position, see illustration.

10. Using a feeler gauge measure and record the clearance of the intake and exhaust valve.

➡️**Measure it within the range of +/− 30 degrees from the specified position,** shown in the illustration. Measure it in the low cam for the intake side. Insert the feeler gauge In a horizontally as possible with respect to the valve lifter.

11. Further turn the crankshaft pulley clockwise and then measure and record the valve clearance again.

12. If adjustment is required, remove the camshafts.

13. Remove and measure the thickness of the valve lifter. Select a suitable shim, using the shim selection chart.

14. Install the replacement shim to the lifter.

15. After all shims have been adjusted, inspect the valve clearances again.

16. After completion, install all removed components.

ENGINE PERFORMANCE & EMISSION CONTROLS

CAMSHAFT POSITION (CMP) SENSOR

REMOVAL & INSTALLATION

2.5L DOHC Engine

1. Disconnect the negative battery cable.
2. Remove the collector cover.
3. Disconnect the connector from camshaft position sensor RH.
4. Remove the camshaft position sensor RH from the rear side of the cylinder head.
5. Remove the cam shaft position sensor LH in the same way as RH.

To install:

6. Installation is the reverse of the removal procedure.
7. Tighten the sensor to 4.7 ft. lbs. (6.4 Nm).

2.5L SOHC Engine

1. Disconnect the negative battery cable.
2. Disconnect the connector from the Camshaft Position (CMP) sensor.
3. Remove the bolt that retains the CMP sensor to the sensor support.
4. Remove the bolt that retains the CMP sensor support to the camshaft cap.
5. Remove the sensor and the CMP sensor support as a unit.
6. Separate the sensor from the support.

To install:

7. Installation is the reverse of the removal procedure.
8. Tighten the CMP sensor support to 4.7 ft. lbs. (6.4 Nm).
9. Tighten the sensor to 4.7 ft. lbs. (6.4 Nm).

3.0L Engine

1. Disconnect the negative battery cable.
2. Remove the collector cover, as required.
3. Remove the alternator harness from the fuel pipe. Remove the fuel pipe protector.
4. Disconnect the connector from the sensor.
5. Remove the bolt that retains the sensor.
6. Remove the sensor.

To install:

7. Installation is the reverse of the removal procedure.
8. Tighten the sensor to 4.7 ft. lbs. (6.4 Nm).

3.6L Engine

Intake Side

See Figures 182 through 187.

1. Remove the collector cover.

➡**Follow these procedures for removal of the collector cover.**

   a. Lift up the rear side holding two positions.
   b. Lift up the front side holding two

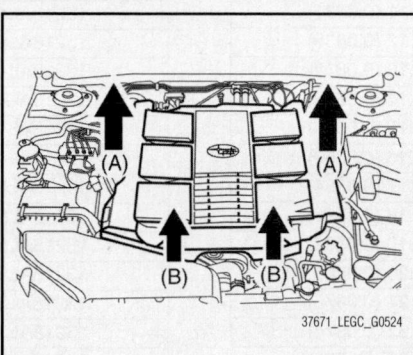

Fig. 182 Collector cover removal—intake side

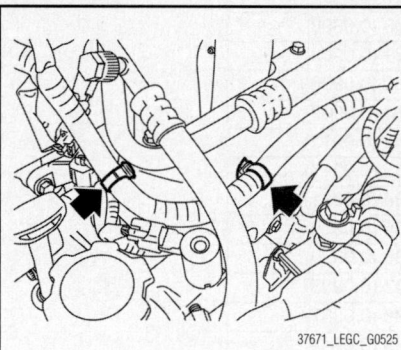

Fig. 183 Securing clips on the fuel pipe protector LH—intake side

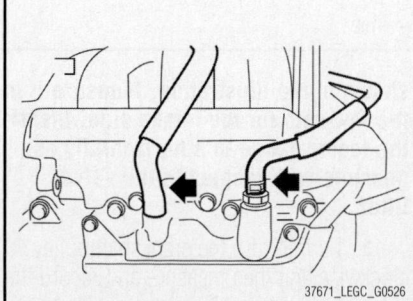

Fig. 184 PCV hose and blow—by—hose disconnection—intake side

Fig. 185 Fuel pipe protector LH—intake side

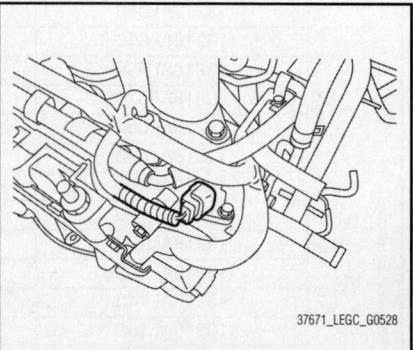

Fig. 186 Intake camshaft position sensor LH connector—intake side

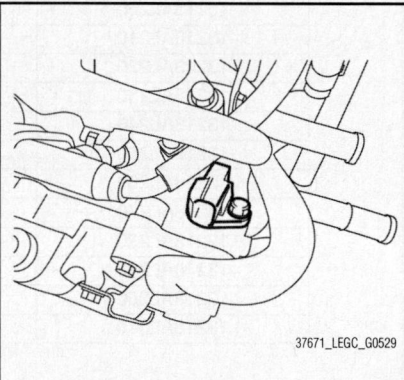

Fig. 187 Intake camshaft position sensor LH—intake side

positions while moving it in the forward direction of the vehicle.

2. Disconnect the ground cable from battery.
3. Remove the securing clips from the two locations on the fuel pipe protector LH.
4. Disconnect the PCV hose and blow-by hose.
5. Remove the fuel pipe protector LH.
6. Disconnect the connector from intake camshaft position sensor LH.

7. Remove the intake camshaft position sensor LH.

8. Remove the intake camshaft position sensor RH in the same procedure as LH side.

To install:

9. Install in the reverse order of removal.

10. Tighten the camshaft position sensor to 56 inch lbs. (6 Nm) and the fuel pipe protector to 14 ft. lbs. (19 Nm).

Exhaust Side

See Figures 188 and 189.

1. Disconnect the ground cable from battery.

2. Lift up the vehicle.

3. Remove the under cover.

4. Disconnect the connector from exhaust camshaft position sensor LH.

5. Remove the exhaust camshaft position sensor LH.

6. Remove the exhaust camshaft position sensor RH in the same procedure as LH side.

To install:

7. Install in the reverse order of removal. Tighten the exhaust camshaft position sensor to 56 inch lbs. (6 Nm).

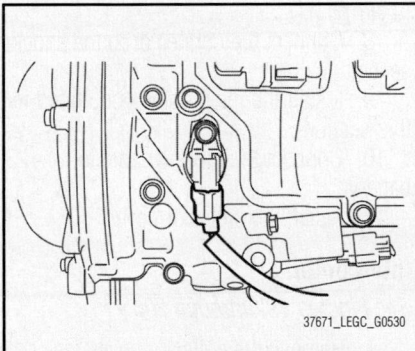

Fig. 188 Camshaft position sensor LH connector—exhaust side

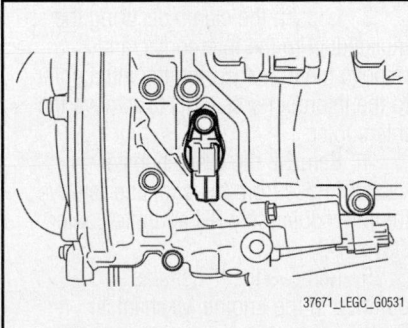

Fig. 189 Camshaft position sensor LH—exhaust

CRANKSHAFT POSITION (CKP) SENSOR

REMOVAL & INSTALLATION

2.5L Engine

1. Remove the collector cover.

2. Disconnect the ground cable from battery.

3. Remove the alternator.

4. Remove the bolt which installs crankshaft position sensor to cylinder block.

5. Remove the crankshaft position sensor, and then disconnect the connector from it.

To install:

6. Installation is the reverse of the removal procedure.

7. Tighten the sensor to 4.7 ft. lbs. (6.4 Nm).

3.0L Engine

1. Before servicing the vehicle, refer to the precautions section.

2. Remove the collector cover.

3. Disconnect the ground cable from battery.

4. Remove the air intake chamber.

5. Remove the service hole cover.

6. Remove the crankshaft position sensor.

7. Disconnect the connector from crankshaft position sensor.

To install:

8. Install in the reverse order of removal.

9. Tighten the crankshaft sensor mounting bolt to 5 ft. lbs. (6.8 Nm).

3.6L Engine

See Figure 190.

1. Remove the collector cover.

➡**Follow these procedures for removal of the collector cover.**

a. Lift up the rear side holding two positions.

b. Lift up the front side holding two positions while moving it in the forward direction of the vehicle.

2. Disconnect the ground cable from battery.

3. Remove the air intake boot assembly.

4. Remove the throttle body from the intake manifold.

➡**Do not disconnect the inlet and outlet hoses of engine coolant.**

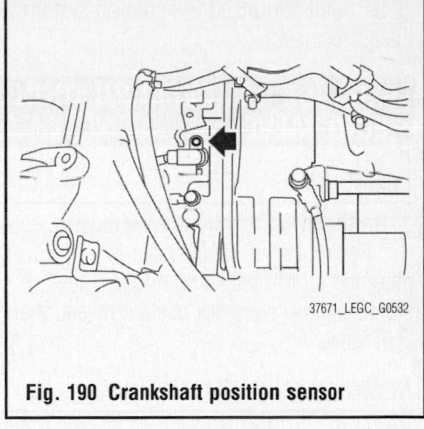

Fig. 190 Crankshaft position sensor

5. Remove the service hole plug.

6. Remove the crankshaft position sensor.

7. Disconnect the connector from crankshaft position sensor.

To install:

8. Install in the reverse order of removal.

9. Tighten the crankshaft position sensor to 56.4 inch lbs. (6.4 Nm).

ENGINE COOLANT TEMPERATURE (ECT) SENSOR

REMOVAL & INSTALLATION

See Figure 191.

1. Before servicing the vehicle, refer to the Precautions Section.

2. Disconnect the negative battery cable.

3. Drain the cooling system.

4. Remove the alternator.

5. Disconnect the electrical connector from the temperature sensor.

6. Remove the coolant temperature sensor.

To install:

7. Installation is the reverse order of removal. Tighten the sensor to 13 ft. lbs. (18 Nm)

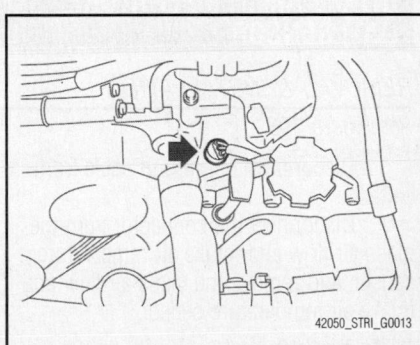

Fig. 191 Removing and installing coolant temperature sensor

8. Refill and bleed the cooling system to the correct level.

EXHAUST GAS RECIRCULATION (EGR) VALVE

LOCATION

On the SOHC engine the Exhaust Gas Recirculation (EGR) valve is mounted on the intake manifold. The DOHC engine does not use a conventional EGR valve.

REMOVAL & INSTALLATION

See Figure 192.

1. Disconnect the negative battery cable.
2. Disconnect the connector.
3. Remove the Exhaust Gas Recirculation (EGR) valve from the intake manifold.

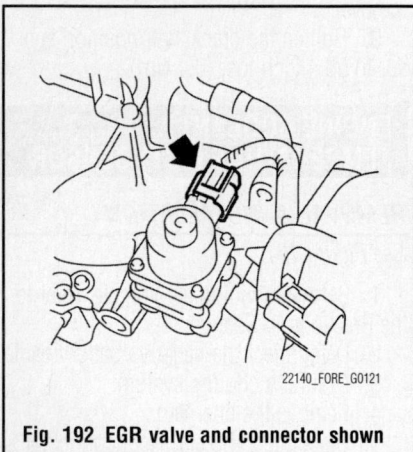

Fig. 192 EGR valve and connector shown

To install:
4. Installation is the reverse of the removal procedure.
5. Tighten the retaining bolts to 14.0 ft. lbs. (19 Nm).

INTAKE AIR TEMPERATURE (IAT) SENSOR

REMOVAL & INSTALLATION

See Figure 193.

1. Disconnect the ground cable from battery.
2. Disconnect the connector from the mass air flow and intake air temperature sensor, and remove the mass air flow and intake air temperature sensor.

To install:
3. Connect the connector from the mass air flow and intake air temperature sensor,

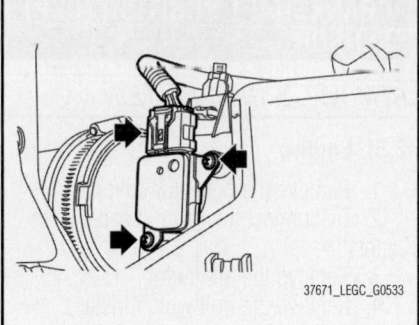

Fig. 193 Mass air flow and intake air temperature sensors

and install the mass air flow and intake air temperature sensor.
4. Connect the ground cable from battery.
5. Tighten sensor to 8.4 inch lbs. (1 Nm).

KNOCK SENSOR (KS)

REMOVAL & INSTALLATION

See Figures 194 through 196.

1. Remove the collector cover.
2. Disconnect the ground cable from battery.

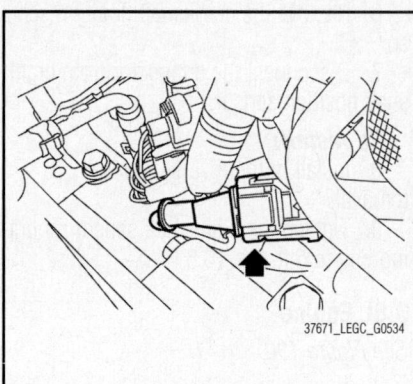

Fig. 194 Knock sensor connector

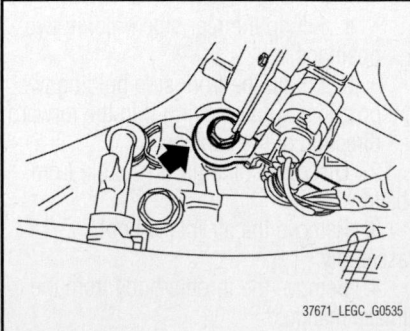

Fig. 195 Remove the knock sensor from cylinder block

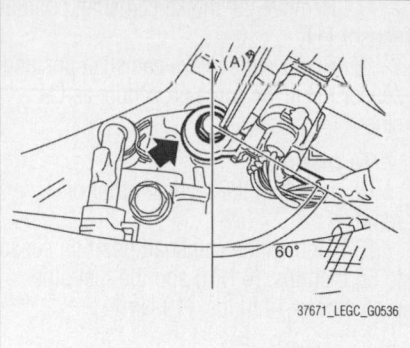

Fig. 196 Installing the knock sensor to the cylinder block

3. Remove the intercooler as outlined in this section.
4. Disconnect the connector from the knock sensor.
5. Remove the knock sensor from cylinder block.

To install:
6. Install the knock sensor to the cylinder block.

➡**The portion of the knock sensor cord that is pulled out must be positioned at a 60° angle relative to the engine rear.**

7. Tighten sensor to 18 inch lbs. (4 Nm).
8. Connect the connector to the knock sensor.
9. Install the intercooler as outlined in this section.
10. Connect the ground cable to battery.
11. Install the collector cover.

Intercooler

See Figures 197 through 206.

1. Remove the collector cover.
2. Remove the clamp which holds the intake duct to the intercooler, and remove bolts which secure the intercooler to the intercooler stay LH.
3. Loosen the clamp securing the intake duct to the intercooler, and remove the bolt securing the intercooler to the intercooler stay RH to remove the intercooler.
4. Remove the brake booster vacuum hose from the clip, and remove the intercooler stay RH and intercooler stay LH.
5. Remove the V-belts as outlined in the Engine Mechanical Section.
6. Remove the alternator as outlined in the Engine Electrical Section.

7. Remove the power steering pump.

a. Remove the suction hose from intake duct.

b. Disconnect the connector (A) from power steering pump, and remove the power steering pump together with the power steering pump bracket from engine and stopper rod.

c. Mount the power steering pump together with the power steering

pump bracket on RH side wheel apron.

8. Remove the radiator sub fan motor assembly as outlined in the Engine Cooling Section.

9. Remove the engine coolant hose from the clip.

10. Remove the bolts which secure the intake duct to the turbocharger, and remove the air by-pass valve from the intake duct.

11. Remove the intake duct from intake manifold.

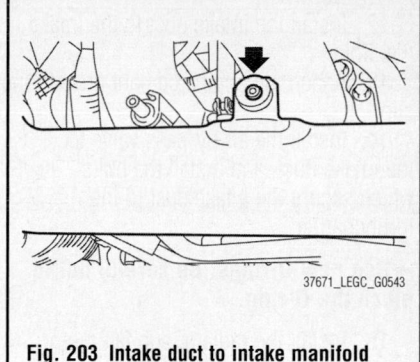

Fig. 203 Intake duct to intake manifold removal & installation

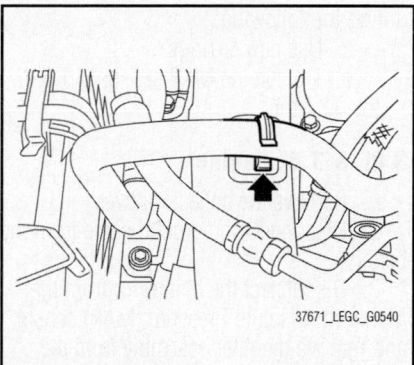

Fig. 197 Collector cover removal

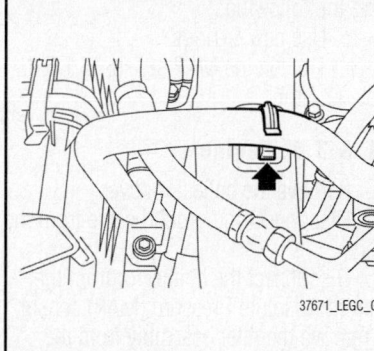

Fig. 200 Suction hose from intake duct removal

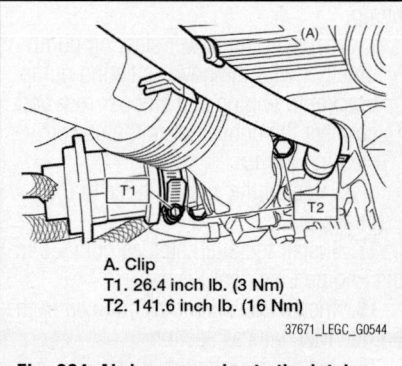

Fig. 204 Air by-pass valve to the intake duct installation

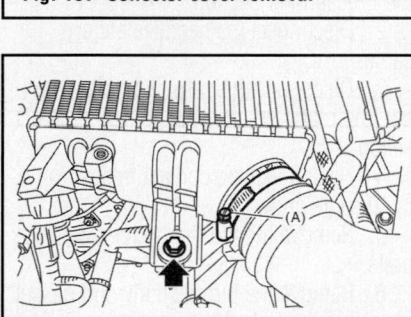

Fig. 198 Removal of the bolt securing the intercooler to the intercooler stay RH to remove the intercooler

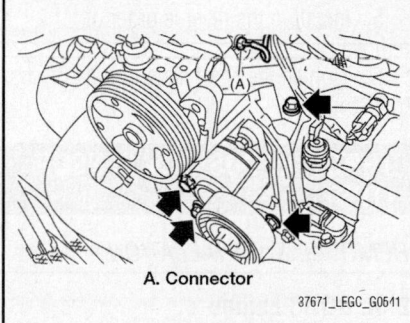

Fig. 201 Removal & installation of the power steering pump together with power steering pump bracket from engine and stopper rod

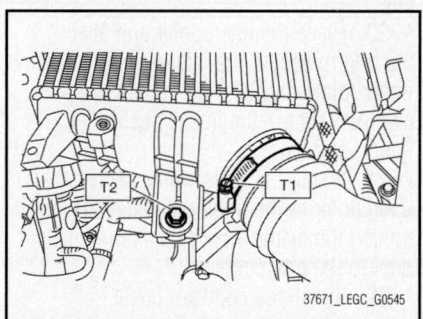

Fig. 205 Intercooler to the intercooler stay RH, and intake duct to the intercooler installation

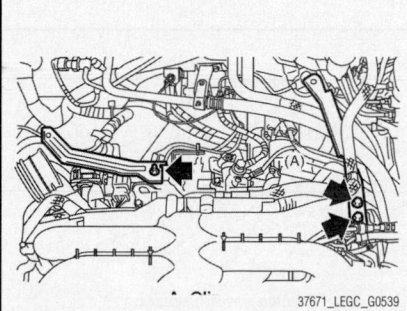

Fig. 199 Remove the brake booster vacuum hose from the clip, and remove the inter cooler stay RH, and intercooler stay LH

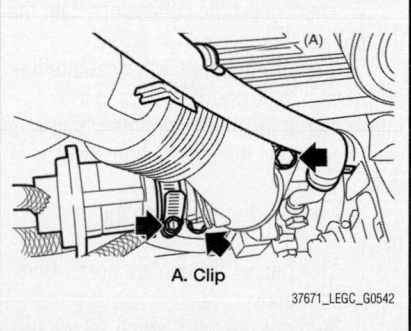

Fig. 202 Air by-pass valve from intake duct removal

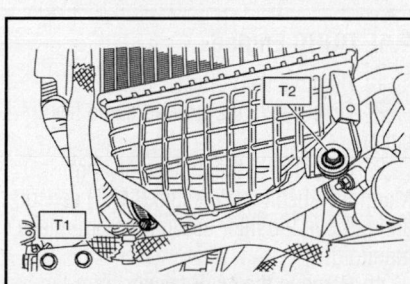

Fig. 206 Intercooler to the intercooler stay installation

To install:

12. Install the intake duct to the intake manifold.

13. Attach the engine coolant hose to the clip.

14. Install the air by-pass valve to the intake duct, and install the bolts which secure the intake duct to the turbocharger.

➡**Use new O-rings. Be careful not to pinch the O-ring.**

15. Install the radiator sub fan motor assembly as outlined in the Engine Cooling Section.

16. Install the power steering pump.

   a. Install the power steering pump together with the power steering pump bracket to engine and stopper rod, and connect the connector to the power steering pump.

   b. Install the suction hose to the intake duct.

17. Install the alternator, as outlined in the Engine Electrical Section.

18. Install the V-belts as outlined in the Engine Mechanical Section.

19. Install the intercooler stay RH and intercooler stay LH, and install the brake booster vacuum hose to the clip.

20. Place the intercooler and attach the bolt to secure the intercooler to the intercooler stay RH. And tighten the clamp to secure the intake duct to the intercooler.

21. Install the bolts which secure the intercooler to the intercooler stay LH, and tighten the clamp which holds the intake duct to the intercooler.

22. Install the collector cover.

MANIFOLD ABSOLUTE PRESSURE (MAP) SENSOR

REMOVAL & INSTALLATION

2.5L DOHC Engine

1. Remove the collector cover.
2. Disconnect the ground cable from battery.
3. Disconnect the connector from Manifold Absolute Pressure (MAP) sensor, and remove the filter assembly from intake manifold.
4. Remove the MAP sensor from the solenoid valve bracket.

To install:

5. Install in the reverse order of removal.

6. Tighten the MAP sensor to 4.7 ft. lbs. (6.4 Nm).

2.5L SOHC Engine

1. Disconnect the ground cable from the battery.
2. Disconnect the connector from Manifold Absolute Pressure (MAP) sensor.
3. Remove the MAP sensor from throttle body.

To install:

4. Install in the reverse order of removal, noting the following:

   a. Use new O-rings.

   b. Tighten the MAP sensor to 1.5 ft. lbs. (2 Nm).

3.0L & 3.6L Engine

1. Remove the collector cover.
2. Disconnect the ground cable from the battery.
3. Disconnect the connector from the Manifold Absolute Pressure (MAP) sensor, and remove the filter assembly from the intake manifold.
4. Remove the MAP sensor from intake manifold.

To install:

5. Install in the reverse order of removal.
6. Tighten the MAP sensor to 4.7 ft. lbs. (6.4 Nm).

THROTTLE POSITION SENSOR (TPS)

REMOVAL & INSTALLATION

2.5L DOHC Engine

See Figure 207.

1. Before servicing the vehicle, refer to the Precautions Section.
2. Disconnect the negative battery cable.
3. Remove the intercooler.
4. Disconnect the connectors from the throttle position sensor.
5. Remove the bolts which secure the air by-pass pipe and PCV pipe to the intake manifold, and loosen the clamp which connects the throttle body and duct.
6. Remove the duct from the throttle body.
7. Disconnect the coolant hoses from the throttle body.
8. Remove the bolts which secure the throttle body to the intake manifold, and remove the throttle body.

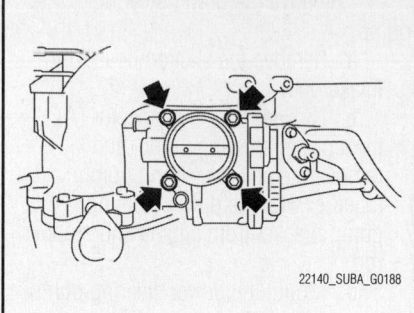

22140_SUBA_G0188

Fig. 207 Remove the bolts which secure the throttle body

To install:

9. The remainder of the installation is the reverse order of removal.
10. Install the throttle body with a new gasket.
11. Tighten the mounting bolts to 6 ft. lbs. (8 Nm).

2.5L SOHC Engines

See Figure 208.

1. Before servicing the vehicle, refer to the Precautions Section.
2. Disconnect the negative battery cable.
3. Disconnect the throttle position sensor and manifold absolute pressure (MAP) sensor.
4. Disconnect the coolant hoses from the throttle body.
5. Remove the throttle body mounting bolts.
6. Remove the throttle body and gasket.

To install:

7. Install the throttle body with a new gasket.
8. Tighten the mounting bolts to 6 ft. lbs. (8 Nm).
9. The remainder of the installation is the reverse order of removal.

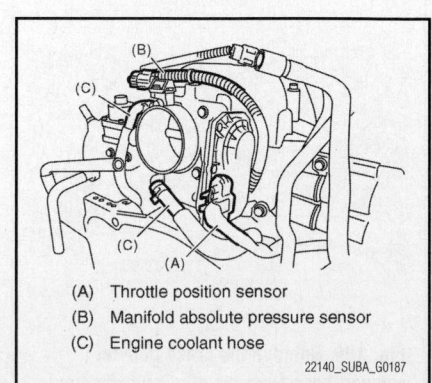

(A) Throttle position sensor
(B) Manifold absolute pressure sensor
(C) Engine coolant hose

22140_SUBA_G0187

Fig. 208 Throttle body view SOHC—2.5L engine

3.0L & 3.6L Engines

See Figure 209.

1. Before servicing the vehicle, refer to the Precautions Section.
2. Disconnect the negative battery cable.
3. Remove the collector cover.
4. Remove the air intake chamber.
5. Disconnect the connector from throttle body.
6. Disconnect the engine coolant hoses from the throttle body.
7. Remove the bolts which secure throttle body to intake manifold.

To install:

8. The remainder of the installation is the reverse order of removal.

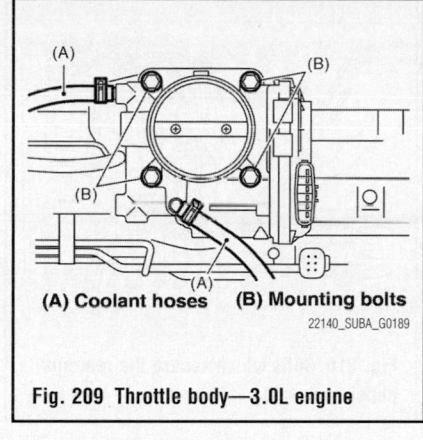

(A) Coolant hoses   (B) Mounting bolts

22140_SUBA_G0189

Fig. 209 Throttle body—3.0L engine

9. Install the throttle body with a new gasket.

10. Tighten the mounting bolts to 6 ft. lbs. (8 Nm).

VEHICLE SPEED SENSOR (VSS)

REMOVAL & INSTALLATION

1. Raise and support the vehicle safely.
2. Place a drip pan below the speed sensor to catch any spilled fluid.
3. Disconnect the connector.
4. Remove the sensor from its mounting.

To install:

5. Installation is the reverse of the removal procedure.
6. Replace any lost fluid.

FUEL

FUEL SYSTEM SERVICE PRECAUTIONS

Safety is the most important factor when performing not only fuel system maintenance, but any type of maintenance. Failure to conduct maintenance and repairs in a safe manner may result in serious personal injury or death. Work on a vehicle's fuel system components can be accomplished safely and effectively by adhering to the following rules and guidelines.

• To avoid the possibility of fire and personal injury, always disconnect the negative battery cable unless the repair or test procedure requires that battery voltage be applied.

• Always relieve the fuel system pressure prior to disconnecting any fuel system component (injector, fuel rail, pressure regulator, etc.) fitting or fuel line connection. Exercise extreme caution whenever relieving fuel system pressure to avoid exposing skin, face and eyes to fuel spray. Please be advised that fuel under pressure may penetrate the skin or any part of the body that it contacts.

• Always place a shop towel or cloth around the fitting or connection prior to loosening to absorb any excess fuel due to spillage. Ensure that all fuel spillage is quickly removed from engine surfaces. Ensure that all fuel-soaked cloths or towels are deposited into a flame-proof waste container with a lid.

• Always keep a dry chemical (Class B) fire extinguisher near the work area.

• Do not allow fuel spray or fuel vapors to come into contact with a spark or open flame.

• Always use a second wrench when loosening or tightening fuel line connection fittings. This will prevent unnecessary stress and torsion on fuel piping. Always follow the proper torque specifications.

• Always replace worn fuel fitting O-rings with new ones. Do not substitute fuel hose where rigid pipe is installed.

FUEL SYSTEM PRESSURE

RELIEVING

➡This procedure must be performed prior to servicing any component of the fuel injection system.

1. Remove the fuel pump fuse from the main fuse box.
2. Start the engine and run until it stalls.
3. Crank the engine for 5 seconds or more to ensure the fuel pressure is properly relieved. If the engine starts during this time, allow it to run until it stalls.
4. Turn the ignition switch to the **OFF** position. Remove the key.
5. Disconnect the negative battery cable.
6. Remove the fuel cap.

FUEL FILTER

REMOVAL & INSTALLATION

The fuel filter is an integral part of the fuel pump assembly.

FUEL LEVEL SENDING UNIT

REMOVAL & INSTALLATION

❋❋ CAUTION

Place "NO OPEN FLAMES" signs near the working area. Prepare a

GASOLINE FUEL INJECTION SYSTEM

container and cloth to prevent scattering of fuels when performing work where fuels can be spilled. If the fuel spills, wipe it off immediately to prevent from penetrating into floor or flowing out for environmental protection. Follow all government and local regulations concerning disposal of refuse when disposing fuel.

1. Before servicing the vehicle, refer to the Precautions Section.
2. Relieve the fuel system pressure.
3. Disconnect the negative battery cable.
4. Remove the fuel pump assembly.
5. Disconnect the connector from fuel pump bracket.
6. Remove the fuel level sensor and fuel temperature sensor.

To install:

7. Install in the reverse order of removal.
8. Tighten sensor to 3 ft. lbs. (4 Nm).

FUEL TANK

REMOVAL & INSTALLATION

See Figures 210 through 213.

❋❋ CAUTION

Place "NO OPEN FLAMES" signs near the working area.

1. Before servicing the vehicle, refer to the Precautions Section.
2. Set vehicle on the lift.
3. Release fuel pressure.

4. Drain fuel from fuel tank.

5. Remove the rear seat.

6. Remove the service hole cover of fuel pump.

7. Disconnect the connector from fuel pump.

8. Remove the connector and clip.

9. Remove the bolts.

10. Push the grommet down and remove the service hole cover.

11. Disconnect connector from fuel sub level sensor.

12. Disconnect the quick connector on the fuel delivery hose.

13. Remove the trunk room trim. (Sedan models)

14. Remove the rear quarter trim. (Wagon models)

15. Remove the pipe protector.

16. Remove the grommet and disconnect the quick connector of the evaporation pipe.

17. Remove the rear wheels.

18. Lift up the vehicle.

19. Remove the bolts which secure the rear brake hose mounting bracket.

20. Remove the rear brake caliper and tie it to the body side of the vehicle.

21. Remove the parking brake cable from the parking brake assembly.

22. Remove the rear exhaust pipe.

23. Remove the propeller shaft.

24. Remove the heat shield cover.

25. Disconnect the connector from the rear ABS wheel speed sensor.

26. Remove the bolts securing the parking brake cable clamp.

✳✳ CAUTION

A helper is required to perform this work.

27. Remove the rear suspension assembly as follows:
- Support the rear differential with the transmission jack.
- Remove the bolts which hold the rear shock absorber to the rear suspension arm.
- Remove the bolts which secure the rear suspension assembly to the body.
- Remove the rear suspension assembly.

28. Disconnect the connector.

29. Disconnect the evaporation hose.

30. Disconnect the fuel filler hose and evaporation hose.

✳✳ CAUTION

A helper is required to perform this work. Fuel may remain in the fuel

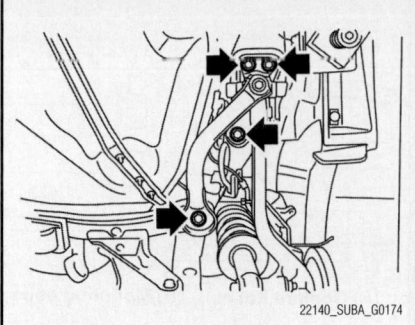

Fig. 210 Bolts which secure the rear suspension view (1)

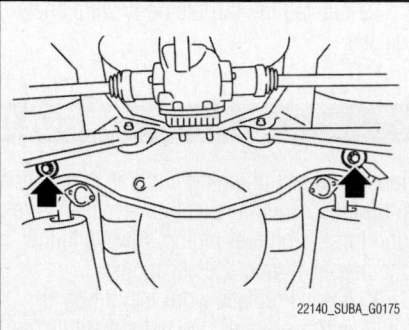

Fig. 211 Bolts which secure the rear suspension view (2)

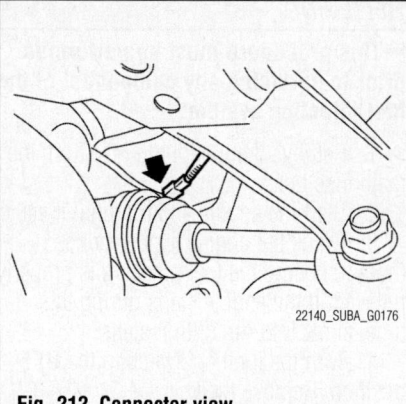

Fig. 212 Connector view

tank. Be careful not to let the fuel tank fall off when removing as it is bad balance on either side.

31. Support the fuel tank with a transmission jack, remove the bolts from the fuel tank band, and remove the fuel tank from the vehicle.

To install:

32. Support the fuel tank with a transmission jack, set the fuel tank in place, and temporarily tighten the bolts of the fuel tank band.

33. Securely insert the fuel filler hose

and evaporation hose to the specified position, then tighten the clamp.

34. Tighten the fuel tank band bolts to 25 ft. lbs. (33 Nm).

35. Install the rear suspension assembly as follows:
- Support the rear differential with the transmission jack.
- Support the rear suspension assembly, and tighten the bolts which secure the rear suspension assembly to the body.
- Tighten the bolts which hold the rear shock absorber to the rear suspension arm.

36. Tighten the suspension bolts.

37. Install the bolts which hold the rear shock absorber to the rear suspension arm. Tighten to 46 ft. lbs. (62 Nm).

38. Tighten the bolts holding the parking brake cable clamp to 13 ft. lbs. (18 Nm).

39. Connect the connector to the rear ABS wheel speed sensor.

40. Install the heat shield cover.

41. Install the propeller shaft.

42. Install the rear exhaust pipe.

43. Lower the vehicle.

44. Connect the parking brake cable to the parking brake assembly.

45. Install the rear brake caliper.

46. Tighten the bolts which secure the

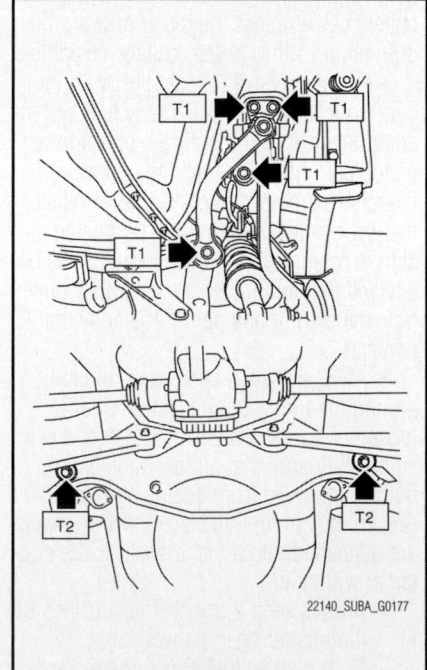

Fig. 213 Tighten the suspension bolts T1: 92 ft. lbs. (125 Nm) and T1: 129 ft. lbs. (175 Nm) as shown

rear brake hose mounting bracket to 25 ft. lbs. (33 Nm).

47. Install the rear wheels.

48. Connect the quick connector of the evaporation pipe.

49. Install the pipe protector.

50. Install the trunk room trim. (Sedan model)

51. Install the rear quarter trim. (Wagon model)

52. Connect connector to the fuel sub level sensor.

53. Connect the quick connector of the fuel delivery hose.

54. Install the service hole cover of fuel sub level sensor.

55. Connect connector, and install clip.

56. Connect the connector to the fuel pump.

57. Install the service hole cover of fuel pump.

58. Install the rear seat.

59. Install the fuse of fuel pump to the main fuse box.

60. Connect the battery ground cable to battery.

61. Inspect the wheel alignment and adjust if necessary.

IDLE SPEED

ADJUSTMENT

The idle speed cannot be adjusted manually, because the idle speed is automatically adjusted.

THROTTLE BODY

REMOVAL & INSTALLATION

2.5L DOHC Engines

See Figure 214.

1. Before servicing the vehicle, refer to the Precautions Section.

2. Disconnect the negative battery cable.

3. Remove the intercooler.

4. Disconnect the connectors from the throttle position sensor.

5. Remove the bolts which secure the air by-pass pipe and PCV pipe to the intake manifold, and loosen the clamp which connects the throttle body and duct.

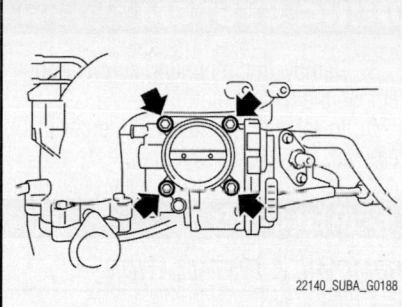

22140_SUBA_G0188

Fig. 214 Remove the bolts which secure the throttle body

6. Remove the duct from the throttle body.

7. Disconnect the coolant hoses from the throttle body.

8. Remove the bolts which secure the throttle body to the intake manifold, and remove the throttle body.

To install:

9. The remainder of the installation is the reverse order of removal.

10. Install the throttle body with a new gasket.

11. Tighten the mounting bolts to 6 ft. lbs. (8 Nm).

2.5L SOHC Engines

See Figure 215.

1. Before servicing the vehicle, refer to the Precautions Section.

2. Disconnect the negative battery cable.

(A) Throttle position sensor
(B) Manifold absolute pressure sensor
(C) Engine coolant hose

22140_SUBA_G0187

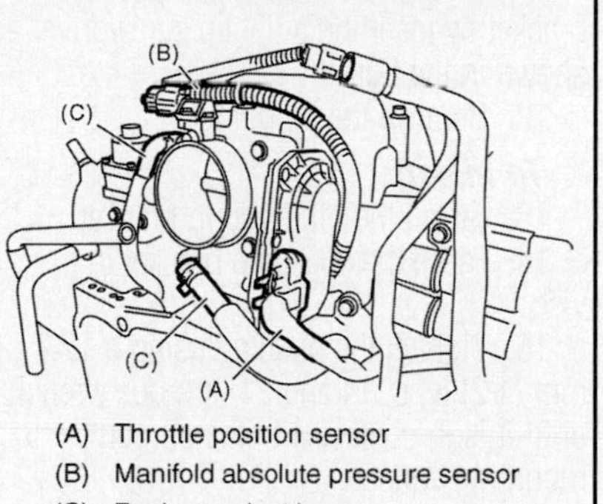

Fig. 215 Throttle body view SOHC—2.5L engine

3. Disconnect the throttle position sensor and manifold absolute pressure (MAP) sensor.

4. Disconnect the coolant hoses from the throttle body.

5. Remove the throttle body mounting bolts.

6. Remove the throttle body and gasket.

To install:

7. Install the throttle body with a new gasket.

8. Tighten the mounting bolts to 6 ft. lbs. (8 Nm).

9. The remainder of the installation is the reverse order of removal.

3.0L & 3.6L Engines

See Figure 216.

1. Before servicing the vehicle, refer to the Precautions Section.

2. Disconnect the negative battery cable.

3. Remove the collector cover.

4. Remove the air intake chamber.

5. Disconnect the connector from throttle body.

6. Disconnect the engine coolant hoses (A) from the throttle body.

7. Remove the bolts which secure throttle body to intake manifold.

To install:

8. The remainder of the installation is the reverse order of removal.

9. Install the throttle body with a new gasket.

10. Tighten the mounting bolts to 6 ft. lbs. (8 Nm).

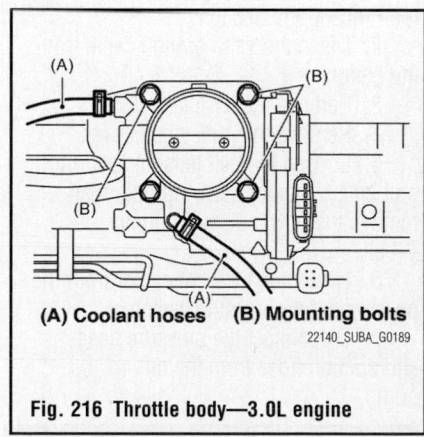

(A) Coolant hoses **(B) Mounting bolts**

22140_SUBA_G0189

Fig. 216 Throttle body—3.0L engine

HEATING & AIR CONDITIONING SYSTEM

BLOWER MOTOR

REMOVAL & INSTALLATION

See Figure 217.

1. Before servicing the vehicle, refer to the Precautions Section.
2. Disconnect the negative battery cable.
3. Remove the glove box lower cover.
4. Disconnect the blower motor wiring harness.

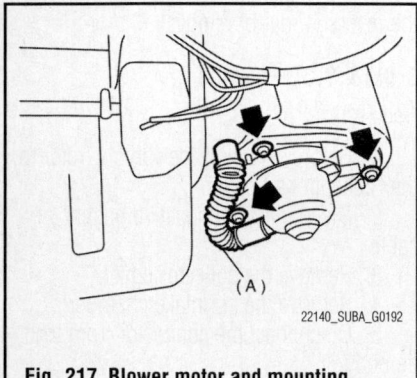

Fig. 217 Blower motor and mounting screws

5. Remove the mounting screws and remove the blower motor.
6. Installation is the reverse order of removal.

HEATER CORE

REMOVAL & INSTALLATION

See Figure 218.

1. Before servicing the vehicle, refer to the Precautions Section.
2. Disconnect the negative battery cable.
3. Discharge the air conditioning system.
4. Drain the cooling system.
5. Remove the bolts securing the expansion valve and pipe in the engine compartment. Release the heater hose clamps in the engine compartment to remove the heater hoses.
6. Remove the instrument panel.
7. Remove the support beam.
8. Remove the blower motor assembly.

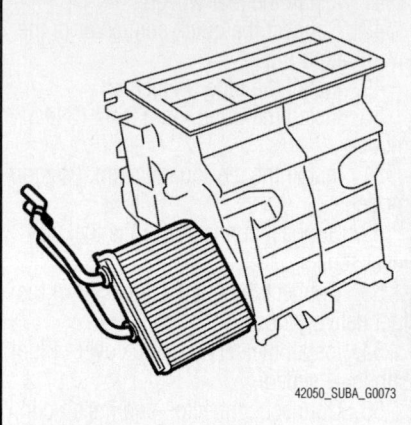

Fig. 218 Removing the heater core from the heater and cooling unit

9. Loosen the bolts and nuts to remove the heater and cooling unit.
10. Loosen the screws to remove the heater core cover.
11. Remove the heater core.

To install:
12. Installation is the reverse order of removal.

STEERING

POWER STEERING PUMP

REMOVAL & INSTALLATION

2.5L Engine

See Figures 219 and 220.

1. Before servicing the vehicle, refer to the Precautions Section.
2. Disconnect the ground cable from the battery.
3. Remove the air intake duct.
4. Remove the pulley belt cover.
5. Loosen the belt tension securing bolt and generator securing bolt, then remove the power steering pump V-belt.
6. Disconnect the connector from the power steering pressure switch.
7. Disconnect the pressure hose and suction hose from the oil pump.

✳✳ CAUTION

Prevent foreign matter from entering the hose and pipe, cover the open ends with clean cloth.

8. Remove the installation bolt of the power steering pump bracket.

9. Place the oil pump bracket in a vise, and remove the two bolts from the front side of the oil pump.
10. Remove the bolt from the rear side of oil pump.
11. Disassemble the oil pump and bracket by inserting a flat tip screwdriver as shown in the figure.
12. Remove the oil pump.

To install:
13. Install the oil pump to bracket.
14. Place the oil pump bracket in a vise.
15. Tighten the bushing using a 12.7 mm (1/2) type, 14 and 21 mm box wrench until it is in contact with the oil pump mounting surface.

✳✳ CAUTION

When securing the oil pump bracket in a vice, hold the oil pump bracket with the least possible force between two pieces of wood.

16. Tighten the two front bolts which hold the oil pump to the bracket to 12 ft. lbs. (16 Nm).
17. Tighten the rear pump to bracket bolt to 27.5 ft. lbs. (37 Nm).

18. Attach the installation bolts of the power steering pump bracket.
19. Connect the pressure hose and suction hose. Tighten the pressure hose eye bolt to 29.5 ft. lbs. (40 Nm).
20. Connect the power steering pressure switch connector.
21. Install the V-belts to the oil pump.
22. Check the tension of the V-belt.
23. Tighten the belt tension bolt to 18.4 ft. lbs. (25 Nm).
24. Install the pulley belt cover.
25. Install the air intake duct.
26. Connect the battery ground cable to the battery.
27. Fill with the specified power steering fluid. (ATF DEXRON III®)

✳✳ CAUTION

Never start the engine before feeding the fluid otherwise the vane pump might be seized.

28. Bleed the power steering system.
29. Start the engine and check for leaks.

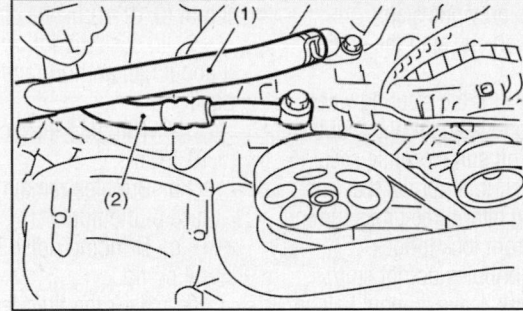

(1) Suction hose
(2) Pressure hose

Turbo model

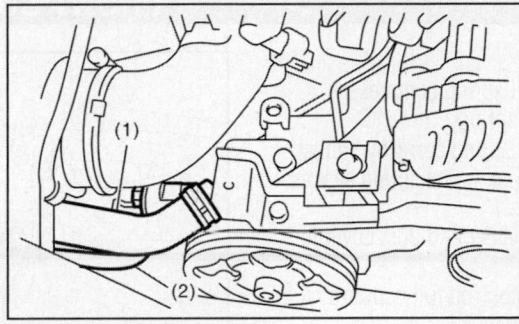

(1) Suction hose
(2) Pressure hose

22140_SUBA_G0210

Fig. 219 Pressure hose and suction hose—2.5L—turbo and non-turbo engines

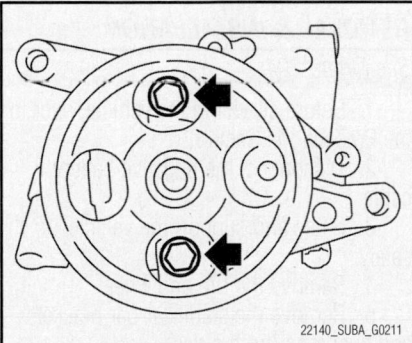

22140_SUBA_G0211

Fig. 220 Front side oil pump to bracket bolts

3.0L & 3.6L Engines

See Figures 220 and 221.

1. Disconnect the ground cable from the battery.
2. Remove the cover of the pulley belt.
3. Remove the drive belts.
4. Remove the power steering pressure switch connector.

5. Remove the tensioner adjuster.
6. Disconnect the pressure hose and suction hose from the oil pump.

✳✳ CAUTION

Prevent foreign matter from entering the hose and pipe, cover the open ends with clean cloth.

7. Remove the bolts which holds the power steering pump bracket.
8. Place the oil pump bracket in a vise, and remove the two bolts from the front side of the oil pump.

✳✳ CAUTION

When securing the oil pump bracket in a vice, hold the oil pump bracket with the least possible force between two pieces of wood.

9. Remove the bolt from the back side of the oil pump.
10. Remove the oil pump from the bracket.

To install:

11. Install the oil pump to bracket.
12. Tighten the bushing using a 12.7 mm (1⁄2) type, 14 and 21 mm box wrench until it is in contact with the oil pump mounting surface.

✳✳ CAUTION

When securing the oil pump bracket in a vice, hold the oil pump bracket with the least possible force between two pieces of wood.

13. Tighten the two front bolts which hold the oil pump to the bracket to 12 ft. lbs. (16 Nm).
14. Tighten the rear pump to bracket bolt to 27.5 ft. lbs. (37 Nm).
15. Attach the installation bolts of the power steering pump bracket.
16. Connect the pressure hose and suction hose. Tighten the pressure hose eye bolt to 29.5 ft. lbs. (40 Nm).
17. Connect the power steering pressure switch connector.
18. Install the tensioner adjuster.
19. Install the drive belts.
20. Install the cover of the pulley belt.
21. Connect the battery ground cable to the battery.
22. Fill with the specified power steering fluid. (ATF DEXRON III®)

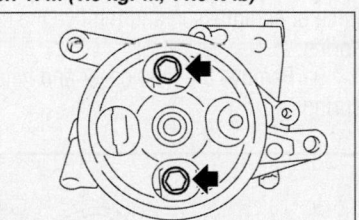

Tightening torque:
15.7 N·m (1.6 kgf-m, 11.6 ft-lb)

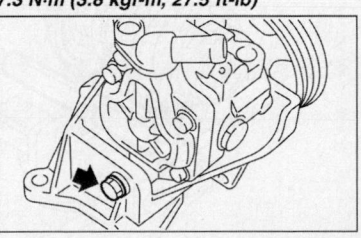

Tightening torque:
37.3 N·m (3.8 kgf-m, 27.5 ft-lb)

22140_SUBA_G0212

Fig. 221 Pump to bracket bolt location, front and rear

❄ CAUTION

Never start the engine before feeding the fluid otherwise the vane pump might be seized.

23. Bleed the power steering system.
24. Start the engine and check for leaks.

BLEEDING

1. Fill the power steering fluid reservoir about half way with the specified fluid.
2. Continue to turn the steering wheel slowly from lock to lock until bubbles stop appearing on oil surface while keeping the fluid at that level.

3. If turning the steering wheel in low fluid level condition, air will be sucked in pipe. In this case, leave it about half an hour and then repeat the previous step.
4. Lift up the vehicle, start the engine and let it idle.
5. Continue to turn the steering wheel slowly from lock to lock again until bubbles stop appearing on oil surface while keeping the fluid at that level. It is normal that bubbles stop appearing after three times turning of steering wheel from lock to lock.
6. In case the bubbles do not stop appearing in the tank, leave it about half an hour and then begin the process again.
7. Lower the vehicle, and then idle the engine.

8. Continue to turn the steering wheel from lock to lock until bubbles stop appearing and change of the fluid level is within 3 mm (0.12 in).
9. In case the following happens, leave it about half an hour and then do step 5–8 again.
   a. The fluid level changes over 3 mm (0.12 in).
   b. Bubbles remain on the upper surface of the fluid.
   c. Grinding noise is generated from oil pump.
10. Check the fluid leakage after turning steering wheel from lock to lock with engine running.

SUSPENSION

COIL SPRING

REMOVAL & INSTALLATION

See Figure 222.

1. Before servicing the vehicle, refer to the Precautions Section.
2. Disconnect the negative battery cable.
3. Raise and support the vehicle safely.
4. Remove front strut.
5. Using a coil spring compressor, compress the coil spring.
6. Using a strut mount socket remove the self-locking nut.
7. Remove the strut mount, spacer and upper spring seat from strut.
8. Gradually decrease the compression force of compressor, and remove the coil spring.
9. Remove the dust cover and helper spring.

To install:
10. Using a coil spring compressor, compress the coil spring.
11. Set the coil spring correctly so that its end face seats well in the spring seat as shown in the figure.
12. Install the helper and dust cover to the piston rod.
13. Pull the piston rod fully upward, and install the spring seat.
14. Install spacer and the strut mount to piston rod, and tighten a new self-locking nut temporarily.
15. Using a hexagon wrench to prevent strut rod from turning, tighten the new self-locking nut to 41 ft. lbs. (55 Nm), with a strut mount socket.
16. Loosen the coil spring compressor carefully.
17. Install front strut.
18. Connect the negative battery cable

CONTROL LINKS

REMOVAL & INSTALLATION

See Figure 223.

1. Before servicing the vehicle, refer to the Precautions Section.
2. Lift up the vehicle, and then remove the front wheels.
3. Remove the front undercover.
4. Remove the stabilizer control link upper and lower mounting nuts.

To install:
5. Install the stabilizer control link.
6. Tighten the stabilizer link upper and lower mounting nuts to 44 ft. lbs. (60 Nm).
7. Install front wheels, and lower the vehicle.

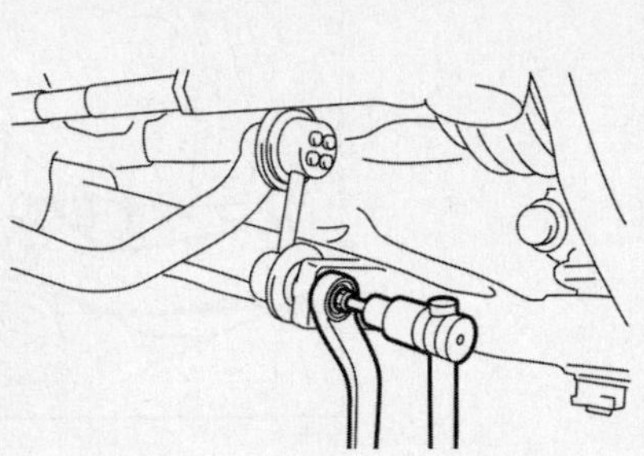

FRONT SUSPENSION

22140_SUBA_G0217

Fig. 223 Stabilizer control link removal shown

LOWER BALL JOINT

REMOVAL & INSTALLATION

See Figure 224.

1. Before servicing the vehicle, refer to the Precautions Section.
2. Disconnect the negative battery cable.
3. Raise and support the vehicle safely.
4. Remove the tire and wheel.
5. Remove the stabilizer bar brackets and bushings (both sides).
6. Remove the cotter pin from the ball stud. Remove the castle nut.
7. Extract the ball stud from the transverse link.
8. Remove the bolt securing the ball joint to the housing.
9. Extract the ball joint from the housing.

To install:
10. Installation is the reverse of the removal procedure.

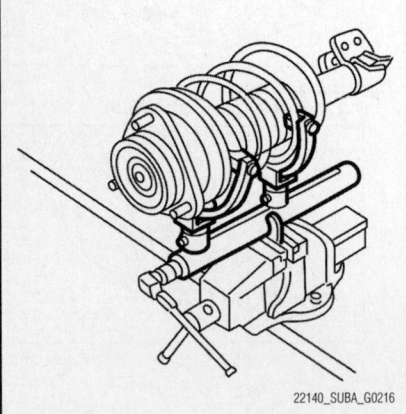

22140_SUBA_G0216

Fig. 222 Coil spring being compressed by spring compressor tool

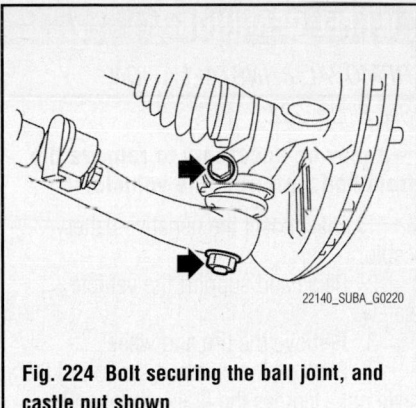

Fig. 224 Bolt securing the ball joint, and castle nut shown

11. Install the ball joint to the transverse link arm and tighten to 37 ft. lbs. (50 Nm).

12. Install and tighten the castle nut to 22 ft. lbs. (30 Nm), if the front arm is aluminum and 28.8 ft. lbs. (39 Nm), if the front arm is steel. Tighten the castle nut an additional 60 degrees until the slot in the castle nut is aligned with the cotter pin hole in the ball joint.

13. The stabilizer bracket has a set orientation. Install it with the arrow mark facing the upper side of the vehicle.

14. Always fully tighten the rubber bushings when the wheels are in full contact with the ground and the vehicle is at curb height.

15. Tighten the stabilizer bracket bolts to 18 ft. lbs. (25 Nm).

16. Check and adjust alignment, as required.

LOWER CONTROL ARM

REMOVAL & INSTALLATION

See Figures 225 and 226.

1. Before servicing the vehicle, refer to the Precautions Section.

2. Lift up the vehicle, and then remove the front wheels.

3. Remove the front crossmember support plate.

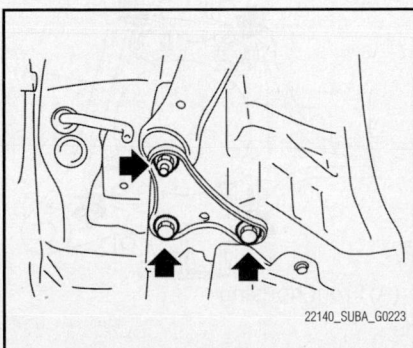

Fig. 225 Front arm support plate mounting points

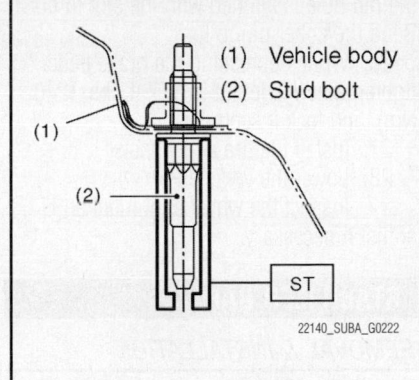

Fig. 226 Special tool ST-20299AG020

4. Remove the front stabilizer.

5. Remove the ball joint of front arm.

6. Remove the nut securing the front arm to crossmember. (Do not remove the bolt.)

7. Remove the front arm support plate.

8. Remove the bolt securing front arm to crossmember and pull the front arm out of the crossmember.

9. To remove the stud bolt, use the ST-20299AG020

❊❊ CAUTION

Do not remove the stud bolt without necessity. Always replace the parts with new parts when removed.

To install:

10. Using the ST-20299AG020, install the stud bolt. Tighten to 81 ft. lbs. (119 Nm).

11. Using new bolts and self-locking nuts, temporarily tighten the front arm to crossmember.

12. Secure the front arm to body, and then install the support plate with new bolts and self-locking nuts.

13. Tighten as follows:
- Support plate to Front arm: 65 ft. lbs. (88 Nm)
- Support plate to Body: 111 ft. lbs. (150 Nm)

14. Install the ball joint into housing, and tighten retaining bolt to 37 ft. lbs. (50 Nm).

15. Install the stabilizer.

16. Lower the vehicle from lift, and tighten the bolt which secures the front arm to crossmember with wheels in full contact with the ground and the vehicle at curb weight.

17. Tighten the stabilizer as follows:
- Stabilizer link: 33 ft. lbs. (45 Nm).
- Stabilizer bracket: 18 ft. lbs. (25 Nm).

18. Check and adjust alignment, as required.

KNUCKLE

REMOVAL & INSTALLATION

See Figures 227 through 230.

1. Before servicing the vehicle, refer to the Precautions Section.

2. Raise and support the vehicle safely.

3. Remove the tire and wheels.

4. Remove the front undercover.

5. Remove the stabilizer link.

6. Remove the disc brake caliper from the front housing, and suspend it from strut using a piece of wire.

7. Remove the disc rotor from the hub.

8. Remove the cotter pin and castle nut securing the tie—rod end to the front housing knuckle arm.

9. Using a puller, remove the tie rod ball joint from knuckle arm.

10. Remove the ABS wheel speed sensor assembly and harness.

11. Remove the lower strut mounting bolts.

12. Remove the front arm ball joint from the front housing.

13. Remove the front driveshaft assembly from the steering knuckle hub.

14. After scribing an alignment mark on the camber adjusting bolt head, remove the

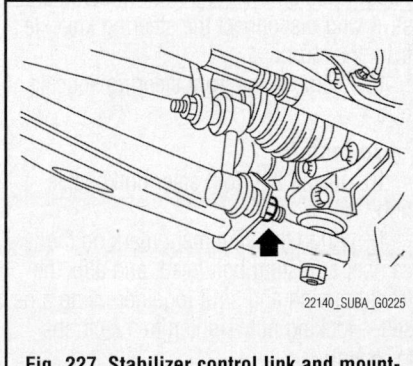

Fig. 227 Stabilizer control link and mounting nut shown

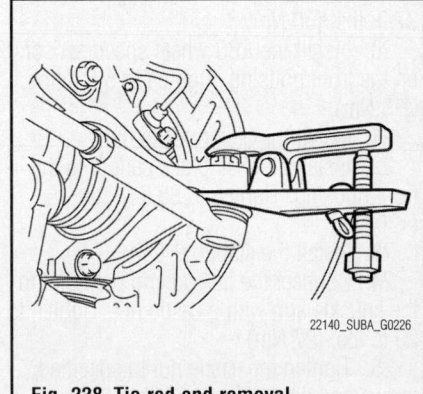

Fig. 228 Tie rod end removal

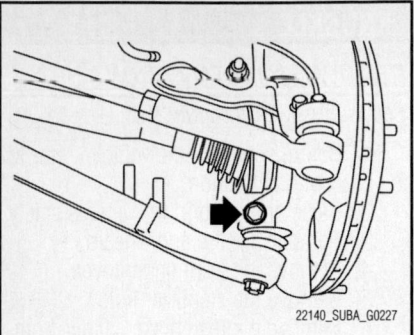

Fig. 229 Ball joint knuckle retaining bolt shown

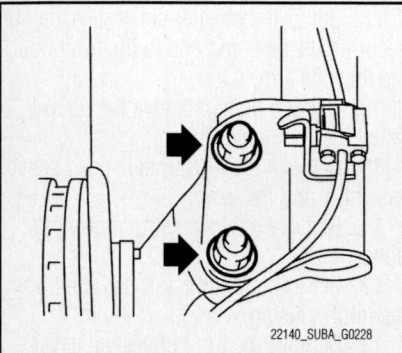

Fig. 230 Strut to knuckle retaining bolts shown

bolts which connect the front housing and strut, and disconnect the steering knuckle from the strut.

15. Remove the front steering knuckle hub.

To install:

16. Install the front steering knuckle hub.

17. Align the alignment mark on the camber adjusting bolt head, and affix the front housing and strut together using a new self—locking nut. Tighten to 129 ft. lbs. (175 Nm).

18. Install the front driveshaft.

19. Install the front arm ball joint to the front housing. Tighten mounting bolt to 37 ft. lbs. (50 Nm).

20. Install the ABS wheel speed sensor on the front housing. Tighten to 5.5 ft. lbs. (7.5 Nm).

21. Install the disc rotor to hub.

22. Install the disc brake caliper to the front housing. Tighten to 88.5 ft. lbs. (120 Nm).

23. Install the stabilizer link.

24. Connect the tie-rod end ball joint to the knuckle arm with a castle nut. Tighten to 20 ft. lbs. (27 Nm).

25. Tighten the castle nut to specified torque and tighten further within 60° until

the pin hole is aligned with the slot in nut. Bend the cotter pin to lock.

26. While depressing the brake pedal, tighten a new axle nut to 177 ft. lbs. (240 Nm). and lock it securely.

27. Install the tire and wheels.

28. Lower the vehicle.

29. Inspect the wheel alignment and adjust if necessary.

STABILIZER BAR

REMOVAL & INSTALLATION

See Figure 231.

1. Before servicing the vehicle, refer to the Precautions Section.

2. Raise and support the vehicle safely.

3. Remove the tire and wheels.

4. Remove the front undercover.

5. Remove the front crossmember support plate.

6. Remove the stabilizer bar link.

7. Remove the stabilizer bar bracket bolts and bushings.

8. Remove the stabilizer bar from the vehicle.

To install:

9. Installation is the reverse of the removal procedure.

10. Be sure to use new self locking nuts, as required.

11. Install the rubber bushing, so that the paint mark on the stabilizer bar is on the left side of the vehicle.

12. Install the stabilizer bushing (front crossmember side), while aligning it with the paint mark on the stabilizer bar.

13. Tighten the stabilizer link bolts to 44 ft. lbs. (60 Nm). Tighten the stabilizer bar clamp to 18 ft. lbs. (25 Nm).

14. Always fully tighten the rubber bushings when the wheels are in full contact with the ground and the vehicle is at curb height.

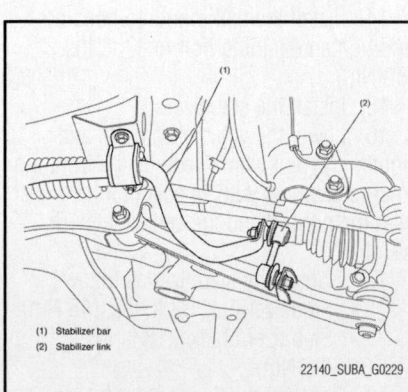

(1) Stabilizer bar
(2) Stabilizer link

Fig. 231 Stabilizer bar and link

WHEEL BEARINGS

REMOVAL & INSTALLATION

See Figure 232.

➡️**It may be necessary to remove the front halfshaft from the vehicle.**

1. Disconnect the negative battery cable.

2. Raise and support the vehicle safely.

3. Remove the tire and wheel.

4. Remove the crimped section of the axle nut. Depress the brake pedal and remove the axle nut.

➡️**Be sure to loosen the axle nut after removing the tire and wheel from the vehicle. Failure to do this may damage the wheel bearings.**

5. Remove the disc brake caliper from its mounting as suspend it to the side, with wire. Do not disconnect the brake line.

6. Remove the rotor.

➡️**If the rotor is seized within the hub, remove the rotor by installing an 8mm bolt in the screw hole on the rotor.**

7. Remove the ABS wheel speed sensor assembly and harness.

8. Remove the four bolts from the housing.

9. Remove the front hub unit bearing. Use tools ST92647000 and ST28099PA100 if necessary.

To install:

10. Tighten the front hub unit bearing to housing bolts to 47.9 ft. lbs. (65 Nm).

11. Be sure to use a new axle nut. Tighten the axle nut, temporarily.

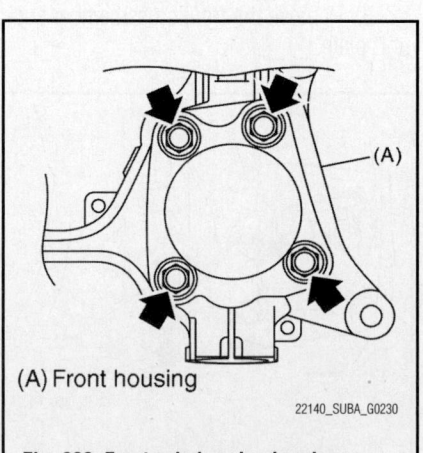

(A) Front housing

Fig. 232 Front axle housing bearing assembly retaining bolts

12. Install the rotor. Install the caliper.

13. While depressing the brake pedal, tighten the axle nut (olive color) to 162 ft. lbs. (220 Nm). Lock it securely in place.

➡ **Install the tire and wheel after**

SUSPENSION

FRONT LINK

REMOVAL & INSTALLATION

See Figures 233 and 234.

1. Before servicing the vehicle, refer to the Precautions Section.

2. Loosen wheel nuts. Lift-up vehicle and remove wheel.

3. Use transmission jack to support rear arm horizontally.

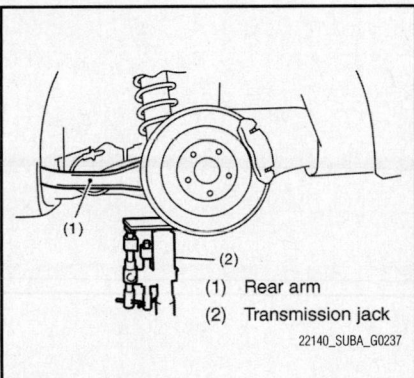

(1) Rear arm
(2) Transmission jack

22140_SUBA_G0237

Fig. 233 Use transmission jack to support rear arm

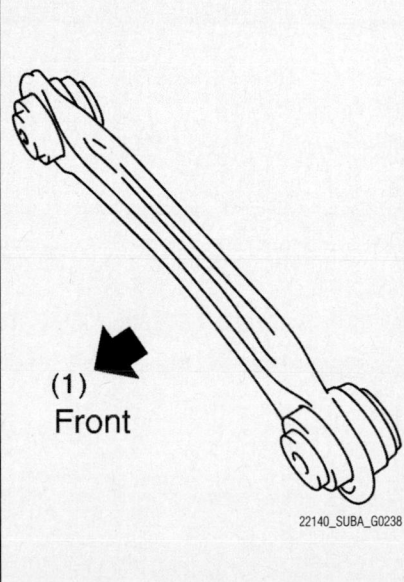

(1) Front

22140_SUBA_G0238

Fig. 234 Install front link with protruding side facing the front side of the vehicle

Installation of the axle nut. Failure to do this may result in wheel bearing damage. Do not over tighten, as this too could cause wheel bearing damage.

14. Continue the installation in

4. Remove bolt securing link front to sub frame.

5. Remove bolts which secure link front to rear arm and detach link front.

➡ **Link front bushing cannot be replaced alone. Always replace link front and bushing as a single unit.**

To install:

6. Installation is the reverse of the removal procedure.

7. Tighten bolts which secure link to 89 ft. lbs. (120 Nm).

8. Check and adjust the wheel alignment, as necessary.

REAR LINK

REMOVAL & INSTALLATION

See Figures 235 through 237.

1. Lift up the vehicle, and then remove the rear wheels.

2. Remove the rear trailing link as outlined in this section.

3. Remove the snap pin and nut.

4. Using a puller, remove the ball joint.

5. Place an alignment mark on the front lateral link adjusting bolt and rear sub frame.

6. Remove the adjusting bolt, and remove the front lateral link.

❋❋ CAUTION

When removing the adjusting bolt, make sure to fix the bolt head in place, and loosen the nut.

To install:

❋❋ CAUTION

Note the following:

* Be sure to use a new self-locking nut.
* Always tighten the bushing in the state where the vehicle is at curb weight and the wheels are in full contact with the ground.

the reverse order of the removal procedure.

ADJUSTMENT

The wheel bearings are of a sealed unit and are not adjustable.

REAR SUSPENSION

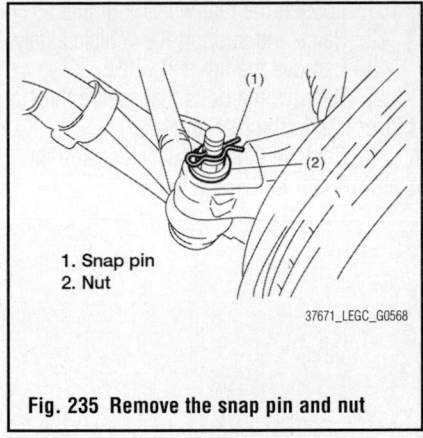

1. Snap pin
2. Nut

37671_LEGC_G0568

Fig. 235 Remove the snap pin and nut

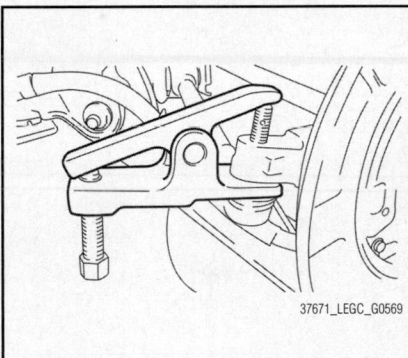

37671_LEGC_G0569

Fig. 236 Using a puller, remove the ball joint

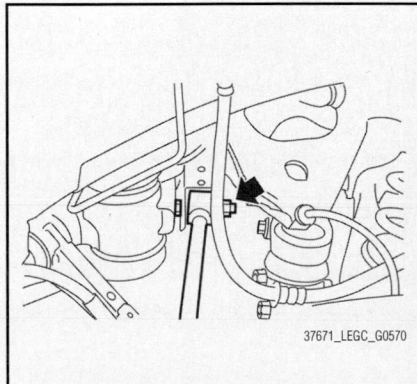

37671_LEGC_G0570

Fig. 237 Remove the adjusting bolt, and remove the front lateral link

7. Install in the reverse order of removal.

8. Tighten Front lateral link to Sub frame to 88.5 ft. lbs. (120 Nm), Front lateral

link to Rear axle housing 44.3 ft. lbs. (60 Nm)

9. Inspect the wheel alignment and adjust if necessary.

STABILIZER BAR

REMOVAL & INSTALLATION

See Figure 238.

1. Loosen the rear wheel lug nuts.
2. Raise and support the vehicle safely.
3. Remove the tire and wheel.
4. Remove the bolts that secure the stabilizer link to the rear arm.
5. Remove the bolts which secure the stabilizer bar to the sub frame.

6. Remove the stabilizer bar from the vehicle.

To install:

7. Installation is the reverse of the removal procedure.
8. Be sure that the stabilizer bar and the bushings have the same identification markings and/or colors.
9. Be sure to use new bolts and nuts, as required.
10. Always fully tighten the rubber bushings when the wheels are in full contact with the ground and the vehicle is at curb height.
11. Check and adjust the wheel alignment, as necessary.

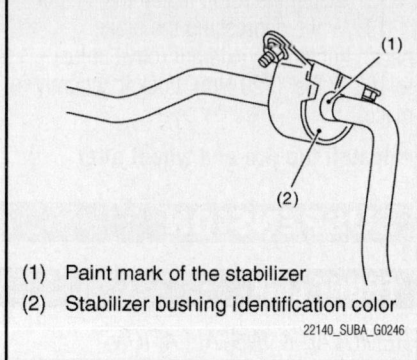

(1) Paint mark of the stabilizer
(2) Stabilizer bushing identification color

22140_SUBA_G0246

Fig. 238 Stabilizer bar and bushing identification markings

SUBARU

Outback

21

SPECIFICATIONS AND MAINTENANCE CHARTS

ENGINE AND VEHICLE IDENTIFICATION CHART

		Engine Code						Model Year	
Code ①	Liters (cc)	Cu. In.	Cyl.	Fuel Sys.	Type	Eng. Mfg.		Code ②	Year
6	2.5 (2457)	150	4	MFI	SOHC	Subaru		9	2009
7	2.5 (2457)	150	4	MFI	DOHC	Subaru		A	2010
8	3.0 (3000)	183.06	6	MFI	DOHC	Subaru			
9	3.6 (3630)	221.4	6	MFI	DOHC	Subaru			

MFI: Multiport Fuel Injection

SOHC: Single Overhead Camshaft

DOHC: Double Overhead Camshaft

① 6th digit of the VIN

② 10th digit of the VIN

37671_OUTB_C0001

GENERAL ENGINE SPECIFICATIONS

Year	Model	Engine Disp. Liters	Engine SOHC/DOHC Turbo/Non-Turbo	Engine ID/VIN	Net Horsepower @ rpm	Net Torque @ rpm (ft. lbs.)	Bore x Stroke (in.)	Com- pression Ratio	Oil Pressure @ rpm
2009	Outback	2.5	SOHC Non-Turbo	6	170@6000	170@4400	3.92x3.11	10.0:1	14@600
	Outback	2.5	DOHC Turbo	7	243@6000	241@3600	3.92x3.11	8.4:1	14@600
	Outback	3.0	DOHC Non-turbo	8	245@6600	215@4200	3.51x3.15	10.7:1	14@600
2010	Outback	2.5	SOHC Non-Turbo	6	170@5600	170@4000	3.92x3.11	10.0:1	14@600
	Outback	3.6	DOHC Non-turbo	9	256@6000	247@4400	3.62x3.61	10.5:1	15@700

37671_OUTB_C0002

ENGINE TUNE-UP SPECIFICATIONS

Year	Model	Engine Displacement Liters	Spark Plug Gap (in.)	Ignition Timing (deg.) ① MT	AT	Fuel Pump (psi)	Idle Speed (rpm) MT	AT	Valve Clearance ② In.	Ex.
2009	Outback	2.5	0.039-0.043	10 ③	15 ⑤	49-50	650-750	700-750	0.0079-0.0095	0.0098-0.0114
	Outback	2.5 Turbo	0.028-0.031	12 ③	17 ③	33-38	650-750	650-750	0.0078-0.0087	0.0138-0.0146
	Outback	3.0	0.028-0.031	15 ④	15 ④	49-50.5	650-750	600-800	0.0079-0.0087 ⑧	0.0138-0.0158
2010	Outback	2.5	0.039-0.043	10 ④	15 ⑦	49-50	650-750	675-775	0.0079-0.0095	0.0098-0.0114
	Outback	3.6	0.039-0.043	15 ⑤	15 ⑤	49-50.5	N/A	700-800	0.0079-0.0087 ⑧	0.0138-0.0158

Note: The Vehicle Emission Control Information label often reflects specification changes made during production. The lable figures mudst be used if they differ from those in this chart.

① Before Top Dead Center
② Measured with engine cold
③ Plus or minus (10 deg.) at 750 rpm
④ Plus or minus (8 deg.) at 650 rpm
⑤ Plus or minus (8 deg) at 700 rpm
⑥ Plus or minus (10 deg) at 700 rpm
⑦ Plus or minus (10 deg) at 675 rpm
⑧ Minus 0.0024
⑨ Minus 0.0012

37671_OUTB_C0003

CAPACITIES

Year	Model	Engine Displacement Liters	Engine Oil with Filter (qts.)	Transmission (pts.) 5-Spd	Auto.	Drive Axle Front (pts.)	Rear (pts.)	Fuel Tank (gal.)	Cooling System (qts.)
2009	Outback	2.5	4.4	7.4	19.6	2.6	1.6	16.9	6.7-6.8
	Outback	2.5 ①	4.5	7.4	20.8	3.0	1.6	16.9	7.6-7.7
	Outback	3.0	6.0	NA	20.2	2.6	1.6	16.9	7.6
2010	Outback	2.5	4.4	7.8	25.4-26.4	3.0	1.6	16.9	6.9
	Outback	3.5	6.9	NA	20.8	3.0	1.6	16.9	6.9

Note: All capacities are approximate. Add fluid gradually and check to be sure a proper fluid level is obtained.
NA: Not Applicable
① Turbo

37671_OUTB_C0004

FLUID SPECIFICATIONS

Year	Model	Engine Displacement Liters	Engine ID/VIN	Engine Oil	Auto. Trans.	Drive Axle	Power Steering Fluid	Brake Master Cylinder	Engine Coolant
2009	Outback	2.5	6	5W-30	Subaru ATF HP	75W-90	Dexron III	DOT 3	Subaru Coolant
		2.5	7	5W-30	Subaru ATF HP	75W-90	Dexron III	DOT 3	Subaru Coolant
		3.0	8	5W-30	Subaru ATF HP	75W-90	Dexron III	DOT 3	Subaru Coolant
2010	Outback	2.5	6	5W-30	Subaru ATF HP	75W-90	Dexron III	DOT 3	Subaru Coolant
		3.6	9	5W-30	Subaru ATF HP	75W-90	Dexron III	DOT 3	Subaru Coolant

DOT: Department Of Transpotation

37671_OUTB_C0005

VALVE SPECIFICATIONS

Year	Engine Displacement Liters	Seat Angle (deg.)	Face Angle (deg.)	Spring Test Pressure (lbs. @ in.)	Spring Installed Height (in.)	Stem-to-Guide Clearance (in.) Intake	Stem-to-Guide Clearance (in.) Exhaust	Stem Diameter (in.) Intake	Stem Diameter (in.) Exhaust
2009	2.5	90	NA	NA	①	0.0012-0.0022	0.0016-0.0026	0.2344-0.2350	0.2341-0.2346
	3.0	90	NA	NA	②	0.0012-0.0022	0.0016-0.0026	002148-0.2154	0.2144-0.2150
2010	2.5	90	NA	NA	③	0.0012-0.0022	0.0016-0.0026	0.2344-0.2350	0.2341-0.2346
	3.6	90	NA	NA	④	0.0012-0.0022	0.0016-0.0026	002148-0.2154	0.2148-0.2150

NA: Not Available

① Free length: 1.8913 in.
② Free length: 1.5571 in. (Inner), 1.6213 in. (Outer)
③ Free length: 1.8630 in.
④ Free length: 1.9315

37671_OUTB_C0007

CAMSHAFT AND BEARING SPECIFICATIONS CHART
All measurements are given in inches.

Year	Engine Displacement Liters	Engine ID/VIN	Journal Dia.	Brg. Oil Clearance	Shaft End-play	Runout	Journal Bore	Lobe Height Intake	Lobe Height Exhaust
2009	2.5 SOHC	6	1.2570-1.2577	0.0022-0.0035	0.0012-0.0035	0.0024	1.2598-1.2605	1.5778-1.5817	1.5468-1.5507
	2.5 DOHC	7	1.1790-1.1796	0.0015-0.0028	0.0027-0.0047	0.0024	NA	1.8330-1.8370	1.8410-1.8440
	3.0 DOHC	8	1.0215-1.0222	0.0015-0.0028-	NA	0.0024	NA	1.6571-1.6610	1.5925-1.5965
2010	2.5 SOHC	6	1.2570-1.2577	0.0022-0.0035	0.0012-0.0035	0.0024	1.2598-1.2605	1.5778-1.5817	1.5468-1.5507
	3.6 DOHC	9	1.0215-1.0222	0.0015-0.0028-	NA	0.0024	NA	1.6571-1.6610	1.5925-1.5965

NA: Not Available

① Intake Journals 4 and 5: 0.0010 - 0.0022 in.

All Others: 0.0010 - 0.0024 in.

37671_OUTB_C0006

CRANKSHAFT AND CONNECTING ROD SPECIFICATIONS

All measurements are given in inches.

Year	Engine Displacement Liters	Crankshaft				Connecting Rod		
		Main Brg. Journal Dia.	Main Brg. Oil Clearance	Shaft End-play	Thrust on No.	Journal Diameter	Oil Clearance	Side Clearance
2009	2.5	2.3619-2.3625	①	0.0012-0.0098	3	1.8891-1.8898	0.0004-0.0012	0.0028-0.0130
	3.0	2.5194-2.5200	②	0.0012-0.0098	3	1.8891-1.8898	0.0006-0.0017	0.0028-0.0130
2010	2.5	2.3619-2.3625	①	0.0012-0.0098	3	1.8891-1.8898	0.0006-0.0017	0.0028-0.0130
	3.6	2.3619-2.3625	②	0.0012-0.0098	3	1.8891-1.8898	0.0009-0.0020	0.0028-0.0130

① Journals 1 and 5: 0.0001 - 0.0016 in.

  Journals 2 and 4: 0.0004 - 0.0014 in.

  Journal 3: 0.0004 - 0.0014 in.

② Journals 1 and 5: 0.0001 - 0.0016 in.

  Journals 2 and 4: 0.0004 - 0.0018 in.

  Journal 3: 0.0004 - 0.0016 in.

37671_OUTB_C0008

PISTON AND RING SPECIFICATIONS

All measurements are given in inches.

Year	Engine Displacement Liters	Piston Clearance	Ring Gap			Ring Side Clearance		
			Top Compression	Bottom Compression	Oil Control	Top Compression	Bottom Compression	Oil Control
2009	2.5	0.0004-0.0012	0.0079-0.0138	0.0146-0.0250	0.0079-0.0197	0.0016-0.0031	0.0012-0.0028	NA
	3.0	0.0004-0.0012	0.0079-0.0138	0.0138-0.0197	0.0079-0.0236	0.0016-0.0031	0.0012-0.0028	NA
2010	2.5	0.0004-0.0012	0.0079-0.0138	0.0146-0.0250	0.0079-0.0197	0.0016-0.0031	0.0012-0.0028	NA
	3.6	0.0004-0.0012	0.0079-0.0138	0.0138-0.0197	0.0079-0.0236	0.0016-0.0031	0.0012-0.0028	NA

NA: Not Available

37671_OUTB_C009

TORQUE SPECIFICATIONS
All readings in ft. lbs.

Year	Engine Displacement Liters	Cylinder Head Bolts	Main ① Bearing Bolts	Rod Bearing Bolts	Crankshaft Damper Bolts	Flywheel Bolts	Manifold		Spark Plugs	Drain Plug
							Intake	Exhaust		
2009	2.5	②	③	38	133	51 - 55	18	26	15	33
	3.0	④	14	39	131	60	18	26	15	33
2010	2.5	②	③	38	133	51 - 55	18	26	15	33
	3.6	④	14	39	131	60	18	26	15	33

① Engine block connecting bolts

② Step 1: Tighten all bolts to 22 ft. lbs.

Step 2: Tighten all bolts to 51 ft. lbs.

Step 3: Loosen all botls 180 degrees.

Step 4: Repeat Step 3.

Step 5: Tighten bolts 1 and 2 to 25 ft. lbs.

Step 6: Tighten bolts 3, 4, 5 and 6 to 11 ft. lbs.

Step 7: Tighten all bolts 80 to 90 degrees.

Step 8: Repeat Step 7. Do not exceed 180 degrees total tightening.

③ Split engine case connecting bolts:

Short bolts: 17-20 ft. lbs.

Long bolts: 33-37 ft. lbs.

Smaller short bolts (if used) 5 ft. lbs.

④ Step 1: Tighten all bolts to 14 ft. lbs.

Step 2: Tighten all bolts to 37 ft. lbs.

Step 3: Loosen all bolts 180 degrees the an additional 180 degrees

in two steps in the reverse order of tightening sequence.

Step 4: Tighten all bolts to 18 ft. lbs.

Step 5: Tighten all bolts to 18 ft. lbs.

Step 6: Tighten bolts 90 degrees

Step 7: Tighten bolts 1, 2, 3 and 4 90 degrees.

Step 8: Tighten bolts 5, 6, 7 and 8 45 degrees.

37671_OUTB_C0010

NOTE:
Remove oil on the mating surface of bearing and cylinder block before installation. Apply engine oil to crankshaft pins.

1) Position the crankshaft and O-rings on the #1 and #3 cylinder block.

2) Apply liquid gasket to the mating surface of #1 and #3 cylinder block, and position #2 and #4 cylinder block.

Liquid gasket:
   Three bond 1215 (Part No. 004403007) or equivalent

NOTE:
Do not allow liquid gasket to jut into O-ring grooves, oil passages, bearing grooves, etc.

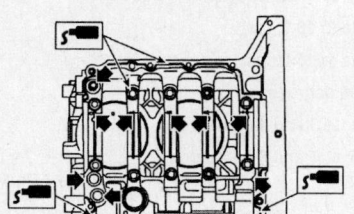

3) Apply a coat of engine oil to the washer and bolt thread.

4) Tighten the 10 mm cylinder block connecting bolts on LH side (A — D) in alphabetical sequence.

Tightening torque:
   9.75 N·m (1.0 kgf-m, 7.2 ft-lb)

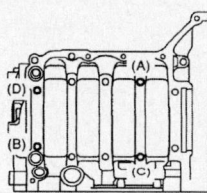

5) Tighten the 10 mm cylinder block connecting bolts on RH side (E — J) in alphabetical sequence.

Tightening torque:
   9.75 N·m (1.0 kgf-m, 7.2 ft-lb)

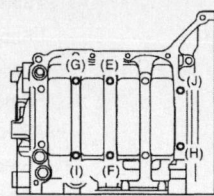

6) Further tighten the LH side bolts (A — D) in alphabetical sequence.

Tightening torque:
   18 N·m (1.8 kgf-m, 13 ft-lb)

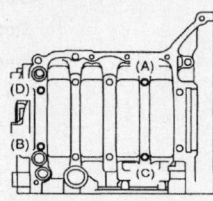

7) Further tighten the RH side bolts (E — J) in alphabetical sequence.

Tightening torque:
   18 N·m (1.8 kgf-m, 13 ft-lb)

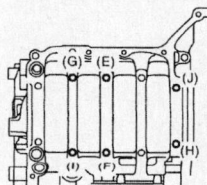

8) Further tighten the LH side bolts (A — D) in alphabetical sequence.
   • (A), (C): Angle tightening

Tightening angle:
   90°
   • (B), (D): Torque tightening

Tightening torque:
   40 N·m (4.1 kgf-m, 29.6 ft-lb)

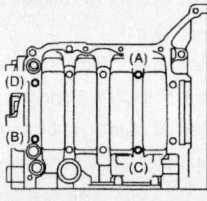

9) Tighten the RH side bolts (E — J) 90° further in alphabetical sequence.

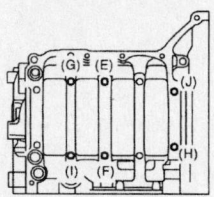

10) Tighten the 8 mm and 6 mm cylinder block connecting bolts on LH side (A — H) in alphabetical sequence.

Tightening torque:
   (A) — (G): 25 N·m (2.5 kgf-m, 18.1 ft-lb)
   (H): 6.4 N·m (0.65 kgf-m, 4.7 ft-lb)

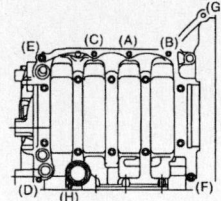

09490_SBCR_G0004

Fig. 1 Main bearing torque sequence—2.5L SOHC engine

1) Remove oil on the mating surface of bearing and cylinder block before installation. Apply engine oil to crankshaft pins.

2) Position the crankshaft and O-rings on the #1 and #3 cylinder block.

3) Apply liquid gasket to the mating surface of #1 and #3 cylinder blocks, and position cylinder block #2 and #4.

NOTE:
Do not allow liquid gasket to run over to O-ring grooves, oil passages, bearing grooves, etc.

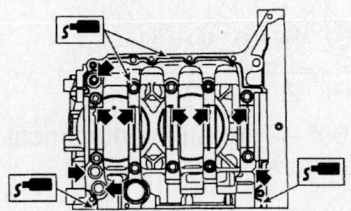

4) Apply a coat of engine oil to the washer and bolt thread.

5) Tighten the 10 mm cylinder block connecting bolts on LH side (A — D) in alphabetical sequence.

Tightening torque:
10 N·m (1.0 kgf-m, 7.2 ft-lb)

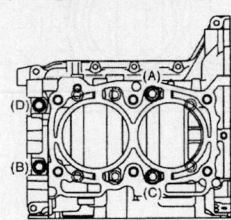

6) Tighten the 10 mm cylinder block connecting bolts on RH side (E — J) in alphabetical sequence.

Tightening torque:
10 N·m (1.0 kgf-m, 7.2 ft-lb)

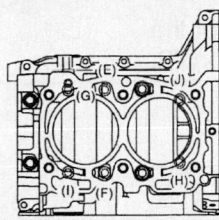

7) Further tighten the LH side bolts (A — D) in alphabetical sequence.

Tightening torque:
18 N·m (1.8 kgf-m, 13.0 ft-lb)

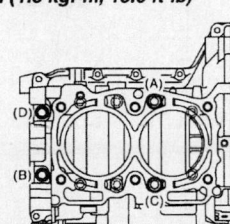

8) Further tighten the RH side bolts (E — J) in alphabetical sequence.

Tightening torque:
18 N·m (1.8 kgf-m, 13.0 ft-lb)

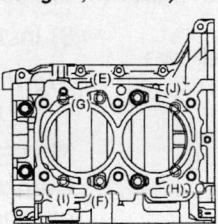

9) Further tighten the LH side bolts (A — D) in alphabetical sequence.

(A), (C): 90°
(B), (D): 40 N·m (4.1 kgf-m, 29.5 ft-lb)

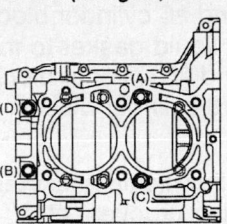

10) Tighten the RH side bolts (E — J) 90° further in alphabetical sequence.

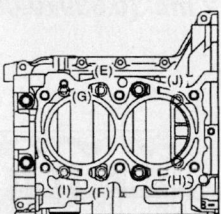

11) Tighten the 8 mm and 6 mm cylinder block connecting bolts on LH side (A — H) in alphabetical sequence.

Tightening torque:
(A) — (G): 25 N·m (2.5 kgf-m, 18.1 ft-lb)
(H): 6.4 N·m (0.65 kgf-m, 4.7 ft-lb)

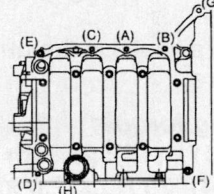

09490_SBCR_G0007

Fig. 2 Main bearing torque sequence—2.5L DOHC engine

1) After setting the cylinder block to ST, install the crankshaft bearing.

NOTE:
Remove oil on the mating surface of bearing and cylinder block before installation. Apply a coat of engine oil to crankshaft pins.

2) Position the crankshaft and connecting rod on #2, #4 and #6 cylinder block.

3) Apply liquid gasket to the mating surface of the #1, #3 and #5 cylinder blocks, and position it on #2, #4 and #6 cylinder blocks.

NOTE:
Do not allow liquid gasket to run over to oil passages, bearing grooves, etc.

Applying liquid gasket diameter:
1.0±0.2 mm (0.039±0.008 in)

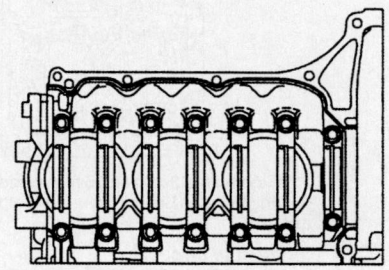

4) Apply a coat of engine oil to the washer and bolt thread.

5) Tighten all bolts in the numerical order as shown in the figure.

Tightening torque:
(1) — (11), (13): 25 N·m (2.5 kgf-m, 18 ft-lb)
(12), (14): 20 N·m (2.0 kgf-m, 14 ft-lb)

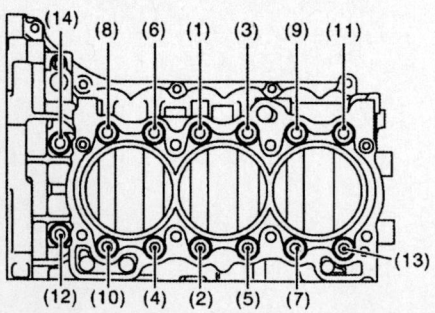

6) Retighten all bolts in the numerical order as shown in the figure.

Tightening torque:
(1) — (11), (13): 25 N·m (2.5 kgf-m, 18.4 ft-lb)
(12), (14): 20 N·m (2.0 kgf-m, 14 ft-lb)

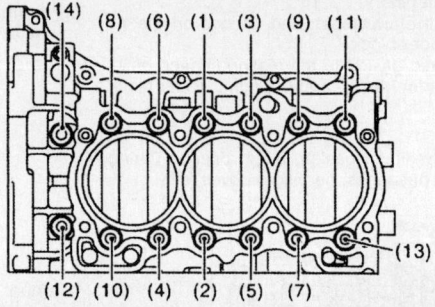

7) Tighten all bolts 90° — 110° in the numerical order as shown in the figure.

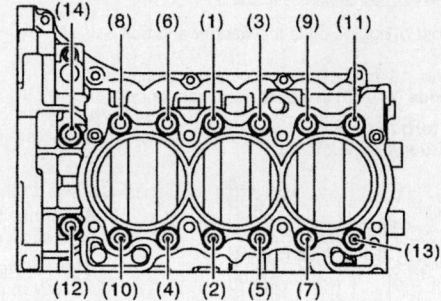

8) Install the upper bolt to cylinder block.

Tightening torque:
25 N·m (2.5 kgf-m, 18 ft-lb)

NOTE:
Remove the liquid gasket which is running over to sealing surface between cylinder block and rear chain cover, cylinder block and oil pan upper, after tightening the bolts which combine the cylinder block.

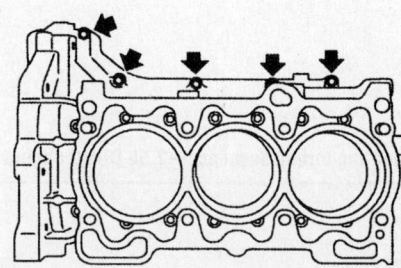

09490_SBCR_G0009

Fig. 3 Main bearing torque sequence—3.0L & 3.6L engine

WHEEL ALIGNMENT

Year	Model		Caster Range (+/-Deg.)	Caster Preferred Setting (Deg.)	Camber Range (+/-Deg.)	Camber Preferred Setting (Deg.)	Toe-in (in.)
2009	Outback	F	0.75	+2.45	0.50	0.05	0+/-0.12
		R	—		0.75	-0.17	0+/-0.12
2010	Outback	F	0.75	+2.45	0.50	0.05	0+/-0.12
		R	—		0.75	-0.17	0+/-0.12

37671_OUTB_C0011

TIRE, WHEEL AND BALL JOINT SPECIFICATIONS

Year	Model	OEM Tires Standard	OEM Tires Optional	Tire Pressures (psi) Front	Tire Pressures (psi) Rear	Wheel Size	Ball Joint Inspection	Lug Nut (ft. lbs.)
2009	Outback	P225/60R16	T135/80R16	32	30	6.5-JJ	0.012 in.	88.5
		P225/55R17	T155/70D17	32	30	7-JJ		
			T135/70D17	32	30			
2010	Outback	P205/60R16	T145/80R17	33	35	6.5-JJ/6.5-J	0.012 in.	88.5
		P215/50R17	T135/80R16	33	32	7.5-J		
		P225/50R17	T135/70R17	33	32	7.5-J		
		P225/45R18		33	32	7.5-J		
		P225/60R17		32	31	7-JJ		
		P215/70R16		32	31	6.5-JJ		

OEM: Original Equipment Manufacturer

PSI: Pounds Per Square Inch

37671_OUTB_C0012

BRAKE SPECIFICATIONS

All measurements in inches unless noted

Year	Model		Brake Disc Original Thickness	Brake Disc Minimum Thickness	Brake Disc Maximum Runout	Brake Drum Diameter Original Inside Diameter	Brake Drum Diameter Max. Wear Limit	Brake Drum Diameter Maximum Machine Diameter	Minimum Lining Thickness Front	Minimum Lining Thickness Rear	Brake Caliper Bracket Bolts (ft. lbs.)	Brake Caliper Mounting Bolts (ft. lbs.)
2009	Outback	F	0.940 ①	0.870 ②	0.003	NA	NA	NA	0.059	—	89	20
		R	0.390 ③	0.335 ④	0.003	NA	NA	NA	—	0.059	47	27 ⑤
2010	Outback	F	0.940 ①	0.870 ②	0.003	NA	NA	NA	0.059	—	89	20
		R	0.390 ③	0.335 ④	0.003	NA	NA	NA	—	0.059	47	27 ⑤

NA: Not Available

① 17 inch type rotors original thickness: 1.18

② 17 inch type rotors minimum thickness: 1.10

③ Vented type rotors original thickness: 0.71

④ Vented type rotors minimum thickness: 0.63

⑤ Solid type rotors tighten to 20 ft. lbs.

37671_OUTB_C0013

SCHEDULED MAINTENANCE INTERVALS
SUBARU—OUTBACK

TO BE SERVICED	TYPE OF SERVICE	VEHICLE MILEAGE INTERVAL (x1000)													
		3	7.5	15	22.5	30	37.5	45	52.5	60	67.5	75	82.5	90	97.5
Engine oil & filter ① ②	R	✓	✓	✓	✓	✓	✓	✓	✓	✓	✓	✓	✓	✓	✓
Brake lines	S/I			✓		✓		✓		✓		✓		✓	
Clutch & hill holder system	S/I			✓		✓		✓		✓		✓		✓	
Disc brake pads & discs, front & rear axle boots & axle shaft joint portions	S/I			✓		✓		✓		✓		✓		✓	
Parking brake	S/I			✓		✓		✓		✓		✓		✓	
Steering & suspension	S/I			✓		✓		✓		✓		✓		✓	
Air filter element	R					✓				✓				✓	
Engine coolant	R					✓				✓				✓	
Fuel filter	R									✓				✓	
Spark plugs	R					✓				✓					
Automatic transmission fluid & filter	S/I					✓				✓				✓	
Brake fluid	R					✓				✓				✓	
Camshaft drive belt ③	S/I					✓				✓				✓	
Coolant level, hoses & clamps	S/I					✓				✓				✓	
Drive belts	S/I					✓				✓				✓	
Fuel system, hoses & connections	S/I					✓				✓				✓	
Transmission and/or differential gear fluid	S/I					✓								✓	
Front & rear wheel bearing	S/I									✓					

R: Replace   S/I: Service or Inspect

① 3.0L & 3.6L engine change every 3000 miles.

② 2.5L turbo models change every 3875 miles or 3.75 months.

③ Replace camshaft drive belt at 105,000 miles or 105 months

FREQUENT OPERATION MAINTENANCE (SEVERE SERVICE)

If a vehicle is operated under any of the following conditions it is considered severe service:

- Extremely dusty areas.
- 50% or more of the vehicle operation is in 32°C (90°F) or higher temperatures, or constant operation in temperatures below 0°C (32°F).
- Prolonged idling (vehicle operation in stop and go traffic).
- Frequent short running periods (engine does not warm to normal operating temperatures).
- Police, taxi, delivery usage or trailer towing usage.

Oil & oil filter change: change every 3750 miles.

Clutch & hill holder system: service or inspect every 7500 miles.

Disc brake pads & discs, front & rear axle boots & axle shaft joint portions: service or inspect every 7500 miles.

Steering & suspension: service or inspect every 7500 miles.

Air filter element: service or inspect every 15,000 miles.

Automatic transmission fluid: service or inspect every 15,000 miles.

Brake linings & drums: service or inspect every 15,000 miles.

Coolant level, hoses & clamps: service or inspect every 15,000 miles.

Drive belts: service or inspect every 15,000 miles.

Transmission/differential gear oil service or inspect every 15,000 miles.

Front & rear wheel bearing repack: service or inspect every 30,000 miles.

BRAKES — INFORMATION AND PRECAUTIONS

ANTI-LOCK SYSTEMS

- Certain components within the ABS system are not intended to be serviced or repaired individually.
- Do not use rubber hoses or other parts not specifically specified for and ABS system. When using repair kits, replace all parts included in the kit. Partial or incorrect repair may lead to functional problems and require the replacement of components.
- Lubricate rubber parts with clean, fresh brake fluid to ease assembly. Do not use shop air to clean parts; damage to rubber components may result.
- Use only DOT 3 brake fluid from an unopened container.
- If any hydraulic component or line is removed or replaced, it may be necessary to bleed the entire system.
- A clean repair area is essential. Always clean the reservoir and cap thoroughly before removing the cap. The slightest amount of dirt in the fluid may plug an orifice and impair the system function. Perform repairs after components have been thoroughly cleaned; use only denatured alcohol to clean components. Do not allow ABS components to come into contact with any substance containing mineral oil; this includes used shop rags.
- The Anti-Lock control unit is a microprocessor similar to other computer units in the vehicle. Ensure that the ignition switch is **OFF** before removing or installing controller harnesses. Avoid static electricity discharge at or near the controller.
- If any arc welding is to be done on the vehicle, the control unit should be unplugged before welding operations begin.

DISC AND DRUM SYSTEMS

> **✳✳ CAUTION**
>
> Dust and dirt accumulating on brake parts during normal use may contain asbestos fibers from production or aftermarket brake linings. Breathing excessive concentrations of asbestos fibers can cause serious bodily harm. Exercise care when servicing brake parts. Do not sand or grind brake lining unless equipment used is designed to contain the dust residue. Do not clean brake parts with compressed air or by dry brushing. Cleaning should be done by dampening the brake components with a fine mist of water, then wiping the brake components clean with a dampened cloth. Dispose of cloth and all residue containing asbestos fibers in an impermeable container with the appropriate label. Follow practices prescribed by the Occupational Safety and Health Administration (OSHA) and the Environmental Protection Agency (EPA) for the handling, processing, and disposing of dust or debris that may contain asbestos fibers.

BRAKES — BLEEDING THE BRAKE SYSTEM

BLEEDING PROCEDURE

BLEEDING PROCEDURE

See Figures 4 through 6.

1. Raise the vehicle and safely support it.
2. Remove both front and rear wheels.
3. Draw out the brake fluid from master cylinder with syringe.
4. Refill reservoir tank with recommended brake fluid.
5. Install one end of a vinyl tube onto the air bleeder and insert the other end of the tube into a container to collect the brake fluid.
6. Instruct your assistant to depress the brake pedal slowly two or three times and then hold it depressed.
7. Loosen bleeder screw approximately 1/4 turn until a small amount of brake fluid drains into container, and then quickly tighten screw.
8. Repeat again from the two former procedures above until there are no air bubbles in drained brake fluid and new fluid flows through vinyl tube.

➡ **Add brake fluid as necessary while performing the air bleed operation, in order to prevent the tank from running short of brake fluid.**

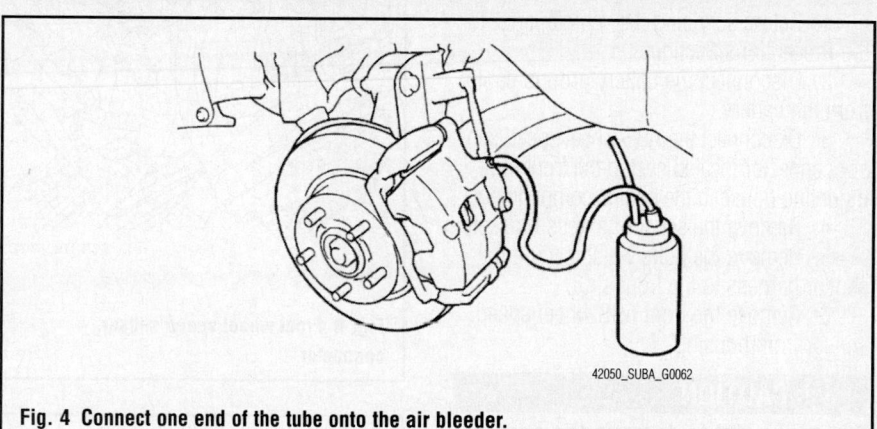

Fig. 4 Connect one end of the tube onto the air bleeder.

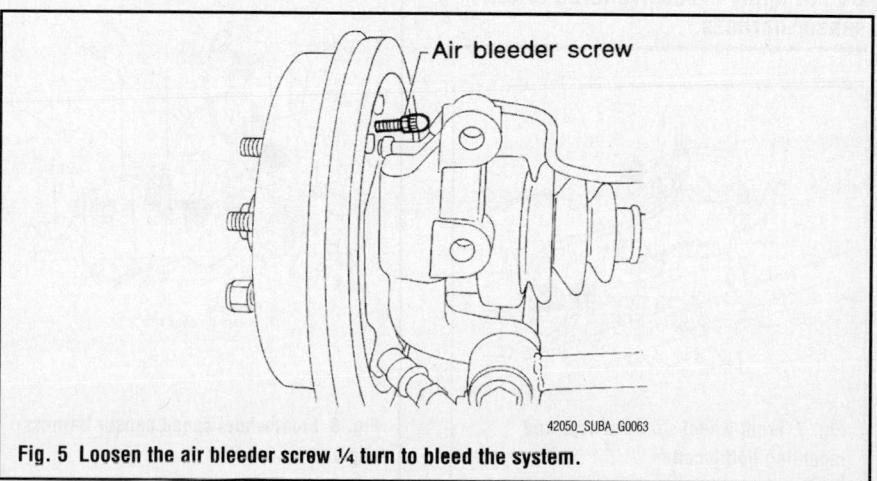

Fig. 5 Loosen the air bleeder screw 1/4 turn to bleed the system.

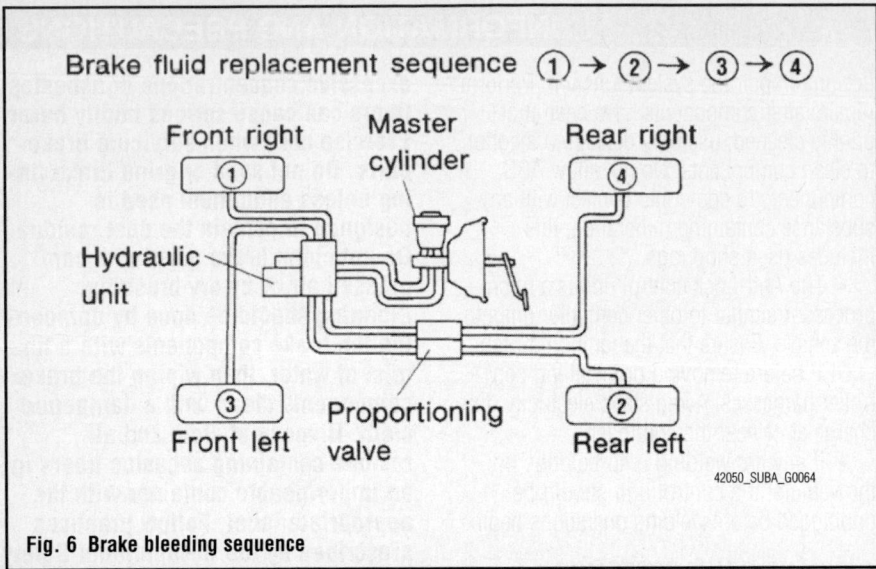

Fig. 6 Brake bleeding sequence

9. After completing the bleeding operation, hold brake pedal depressed and tighten screw and install bleeder cap.

10. Bleed air from each wheel cylinder using the same procedures as described above.

11. Depress brake pedal and hold it there for approximately 20 seconds. At this time check pedal to see if it shows any unusual movement. Visually inspect bleeder screws and brake pipe joints to make sure that there is no fluid leakage.

12. Install the wheels, and drive car for a short distance (between 1–2 miles to make sure that brakes are operating properly.

BLEEDING THE ABS SYSTEM

There is no separate bleeding procedure for ABS. Refer to Bleeding the Brake System.

BRAKES   ANTI-LOCK BRAKE SYSTEM (ABS)

WHEEL SPEED SENSORS

REMOVAL & INSTALLATION

Front

See Figures 7 through 10.

1. Before servicing the vehicle, refer to the Precautions Section.

2. Disconnect the battery ground cable from the battery.

3. Disconnect the ABS wheel speed sensor connector located next to the front strut mounting house in the engine compartment.

4. Remove the sensor harness bracket.

5. Remove the bolts which secure the sensor harness to the front strut.

6. Remove the front ABS wheel speed sensor from housing.

❈ CAUTION

Be careful not to damage the sensor. Do not apply excessive force to the sensor harness.

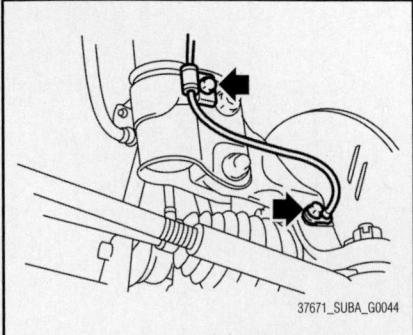

Fig. 7 Front wheel speed sensor and mounting bolt location

1. Grommet
2. Sensor harness clip

Fig. 9 Front wheel speed sensor harness clip and grommet

Fig. 8 Front wheel speed sensor connector

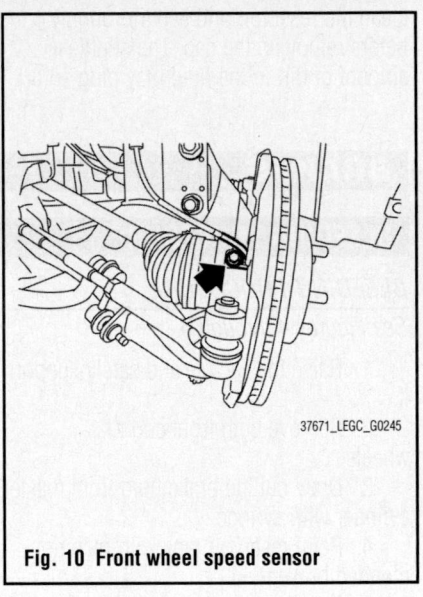

Fig. 10 Front wheel speed sensor

To install:

7. Before servicing the vehicle, refer to the Precautions Section.

8. Install the front ABS wheel speed sensor to housing.

9. Install the bolts which secure the sensor harness to the front strut.

10. Install the sensor harness bracket.

11. Connect the ABS wheel speed sensor connector located next to the front strut mounting house in the engine compartment.

12. Connect the battery ground cable from the battery.

13. Tighten the sensor mounting bolt to 5.5 ft. lbs. (7.5 Nm).

14. Tighten the sensor harness bolts to 24 ft. lbs. (33 Nm).

➥**Check that the harness is not pulled and does not come in contact with the suspension or body during steering wheel effort.**

Rear

See Figures 11 through 14.

1. Disconnect the ground cable from battery.
2. Remove the clips, turn over the mud guard, and disconnect the rear ABS wheel speed sensor connector.
3. Remove the rear ABS wheel speed sensor from the rear housing.

✳✳ WARNING

Be careful not to damage the sensor. Do not apply excessive force to the sensor harness.

Fig. 11 Rear wheel speed sensor connector

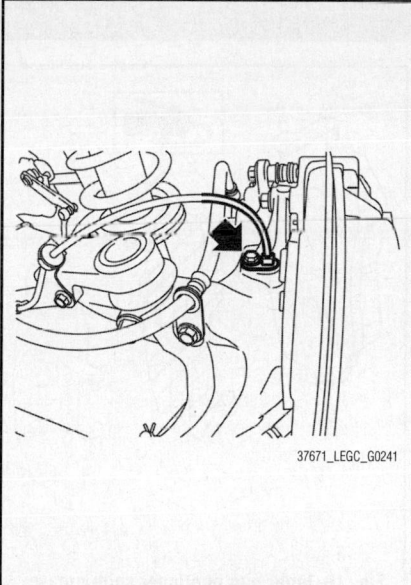

Fig. 12 Rear wheel speed sensor

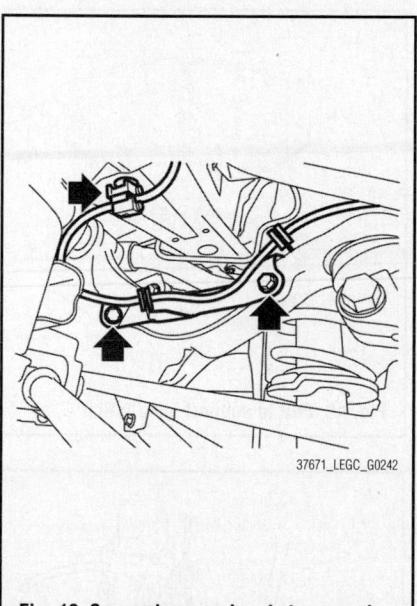

Fig. 13 Sensor harness bracket removal

4. Remove the sensor harness clip.
5. Remove the sensor harness bracket from the upper arm.

To install:

✳✳ WARNING

Be careful not to damage the sensor.

6. Install the sensor harness bracket from the upper arm.
7. Install the sensor harness clip.
8. Install the rear ABS wheel speed sensor to the rear housing.
9. Install the clips, turn over the mud guard, and connect the rear ABS wheel speed sensor connector.
10. Connect the ground cable from battery.
11. Tighten the sensor mounting bolt to 66 inch lbs. (7.5 Nm), and the sensor bracket bolt to 66 inch lbs. (7.5 Nm).

➥**Check the identification (mark) on the harness to make sure there is no warpage. Refer to the accompanying illustration.**

WHEEL SPEED SENSOR RINGS (TOOTHED RINGS)

REMOVAL & INSTALLATION

Front Wheel

Refer to Front Hub Bearing for removal, because the front magnetic encoder is integrated with front hub bearing.

Rear Wheel

Refer to Front Hub Bearing for removal, because the front magnetic encoder is integrated with front hub bearing.

Item				Specification or identification
ABS wheel speed sensor	ABS wheel speed sensor gap (for reference)	Front		0.77 — 1.43 mm (0.030 — 0.056 in)
		Rear		0.64 — 1.56 mm (0.025 — 0.061 in)
	Identifications of harness (marks, color)	Front	(Except for OBK model)	S1 (Pink)
			OBK model	S2 (Blue)
		Rear	RH (Except for OBK model)	S3 (Yellow)
			LH (Except for OBK model)	S4 (White)
			RH (OBK model)	S5 (Red)
			RH (OBK model)	S6 (Light green)
VDCCM&H/U Identification				V1

Fig. 14 Vehicle dynamics control specifications

BRAKES

FRONT DISC BRAKES

DISC BRAKE PADS

REMOVAL & INSTALLATION

See Figures 15 through 21.

1. Lift up the vehicle, and remove the front wheels.
2. Remove the caliper bolt.
3. Raise the caliper body and support it.

➡**Do not disconnect the brake hose from the caliper body.**

4. Remove the pad.

➡**If the pad return spring is deformed or damaged, replace the brake pad.**

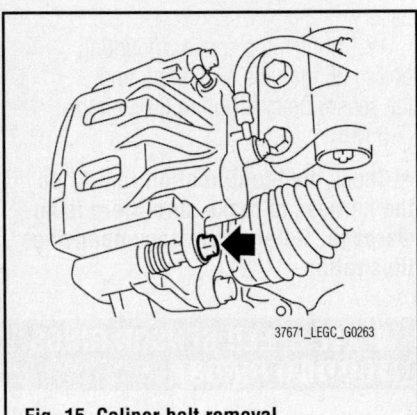

Fig. 15 Caliper bolt removal

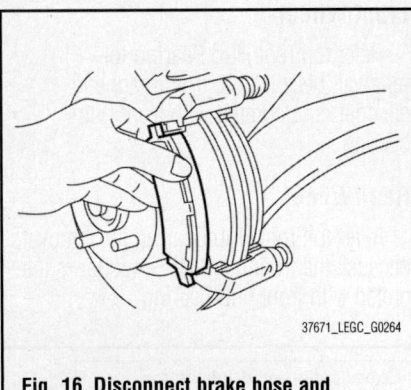

Fig. 16 Disconnect brake hose and remove pad

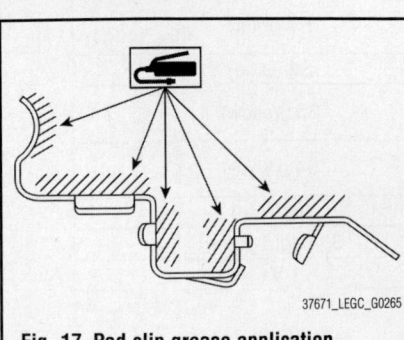

Fig. 17 Pad clip grease application

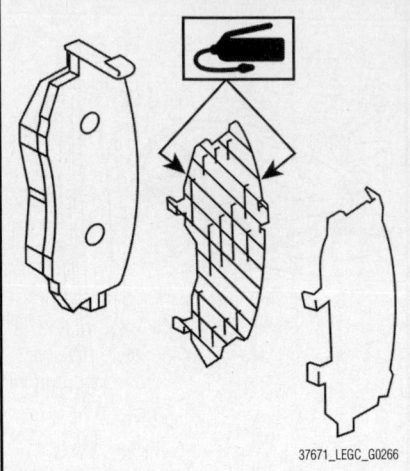

Fig. 18 Inner pad and inner shim grease application

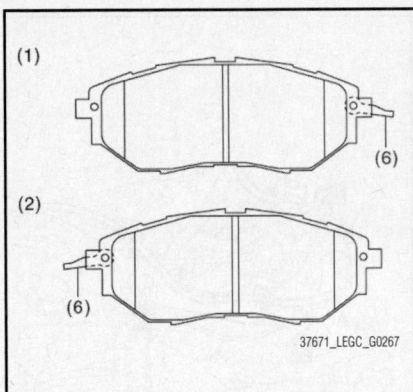

Fig. 19 Pad to support installation

To install:

5. **16-INCH TYPE:**
   a. Apply a thin coat of Molykote M7439 (Part No. K0770YA000) or grease contained in the pad kit to the pad clip.
   b. Apply a thin coat of Molykote AS880N (Part No. K0777YA010) or grease contained in the pad kit to both surfaces of the pad and inner shim.
   c. Install the pad to support.
   d. Install the caliper body to the support. Tighten to 20 ft. lbs. (27 Nm).
6. **17-INCH TYPE:**
   a. Apply a thin coat of Molykote M7439 (Part No. K0770YA000) or grease contained in the pad kit to the pad clip.
   b. Apply a thin coat of Molykote AS880N (Part No. K0777YA010) or grease contained in the pad kit to both surfaces of the pad and inner shim.
7. Install the pad to support.

➡**Install the pad indicator in proper direction.**

➡**Be sure to install so that the pad return spring faces the input side of the direction of brake rotor rotation, as shown in the figure. Correctly install the pad return spring to the supporting surface of the pad clip as shown in the figure.**

8. Install the caliper body to the support. Tighten to 20 ft. lbs. (27 Nm).

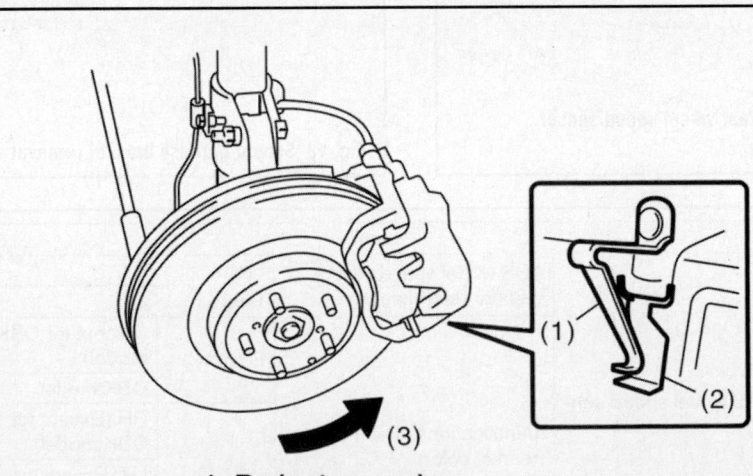

1. Pad return spring
2. Supporting surface of pad clip
3. Direction of brake rotor rotation

Fig. 20 Pad and pad return spring installation

BRAKES

DISC BRAKE PADS

REMOVAL & INSTALLATION

See Figures 17, 18, 21 through 24.

1. Lift up the vehicle, and then remove the rear wheels.
2. Remove the brake hose bracket.
3. Remove the caliper bolt.
4. Raise the caliper body and support it.

➡ **Do not disconnect the brake hose from the caliper body.**

5. Remove the pad.

To install:

6. Apply a thin coat of Molykote M7439 (Part No. K0770YA000) or grease contained in the pad kit to the pad clip.
7. Apply a thin coat of Molykote

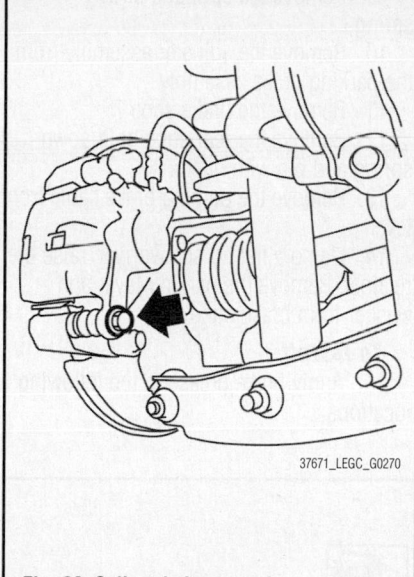

Fig. 22 Caliper bolt removal

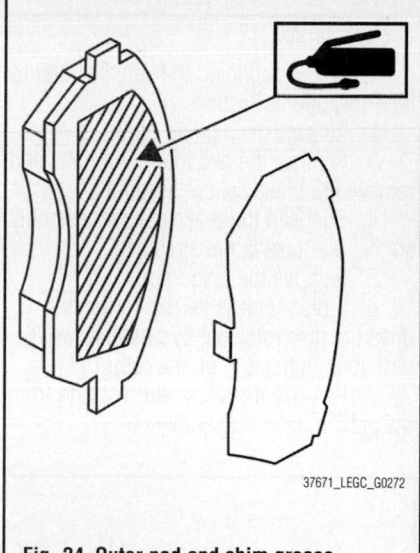

Fig. 24 Outer pad and shim grease application

AS880N (Part No. K0777YA010) or grease contained in the pad kit between the outer pad and shim.

8. Apply a thin coat of Molykote AS880N (Part No. K0777YA010) or grease contained in the pad kit to both surfaces of the inner pad inner shim.
9. Install the pad to support.
10. Install the caliper body to the support. Tighten to 20 ft. lbs. (27 Nm).
11. Install the brake hose bracket. Tighten 24.3 ft. lbs. (33 Nm).

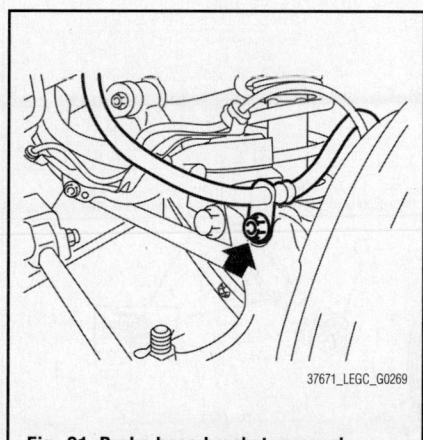

Fig. 21 Brake hose bracket removal

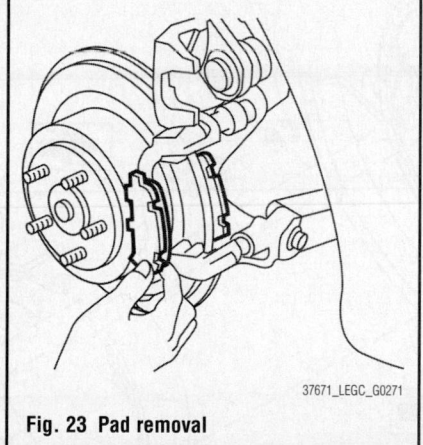

Fig. 23 Pad removal

BRAKES

PARKING BRAKE CABLES

ADJUSTMENT

See Figure 25.

1. Remove the console cover.
2. Forcefully pull the parking brake lever 3 to 5 times.
3. Adjust parking brake cable by turning the adjusting nut until the parking brake lever stroke is set at 7 to 8 notches with operating force of 44 lbs. (196 N).
4. Tighten lock nut.
5. Install the console cover.

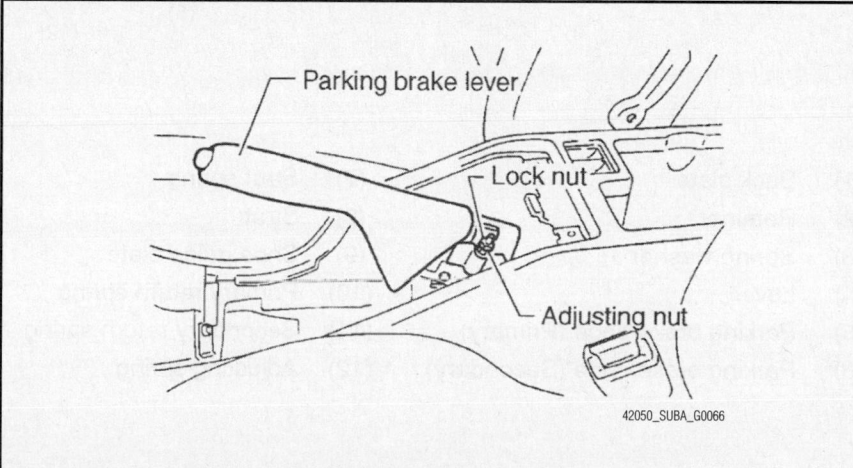

Fig. 25 Turn the adjusting nut to adjust the parking brake cable.

PARKING BRAKE SHOES

REMOVAL & INSTALLATION

See Figure 26.

1. Before servicing the vehicle, refer to the Precautions Section.

2. Release the parking brake.

3. Remove the two mounting bolts and remove the brake caliper assembly.

4. Suspend the brake caliper assembly so that the hose is not stretched.

5. Remove the disc rotor.

6. If disc rotor is seized on the hub, drive the disc rotor out by pushing two 8 mm bolts in holes B on the rotor.

7. Remove the shoe return spring from the parking brake assembly.

8. Remove the front shoe hold down spring and pin.

9. Remove the strut and strut spring.

10. Remove the adjuster assembly from the parking brake assembly.

11. Remove the brake shoe.7

12. Remove the rear shoe hold down spring and pin with pliers.

13. Remove the parking brake cable from lever.

14. Using a flat tip screwdriver, raise the retainer. Remove the parking lever and washer from brake shoe.

To install:

15. Apply brake grease to the following locations:

- Six contact surfaces of the shoe rim and back plate gasket
- Contact surface of the shoe wave and the anchor pin
- Contact surface of the lever and strut
- Contact surface of the shoe wave and the adjuster assembly
- Contact surface of the shoe wave and the strut
- Contact surface of the lever and the shoe wave

16. Insert the primary side brake shoe into the anchor pin groove.

17. Secure the brake shoe with the shoe hold-down pin and cup.

18. Install the plate to the anchor pin, then install the primary return spring.

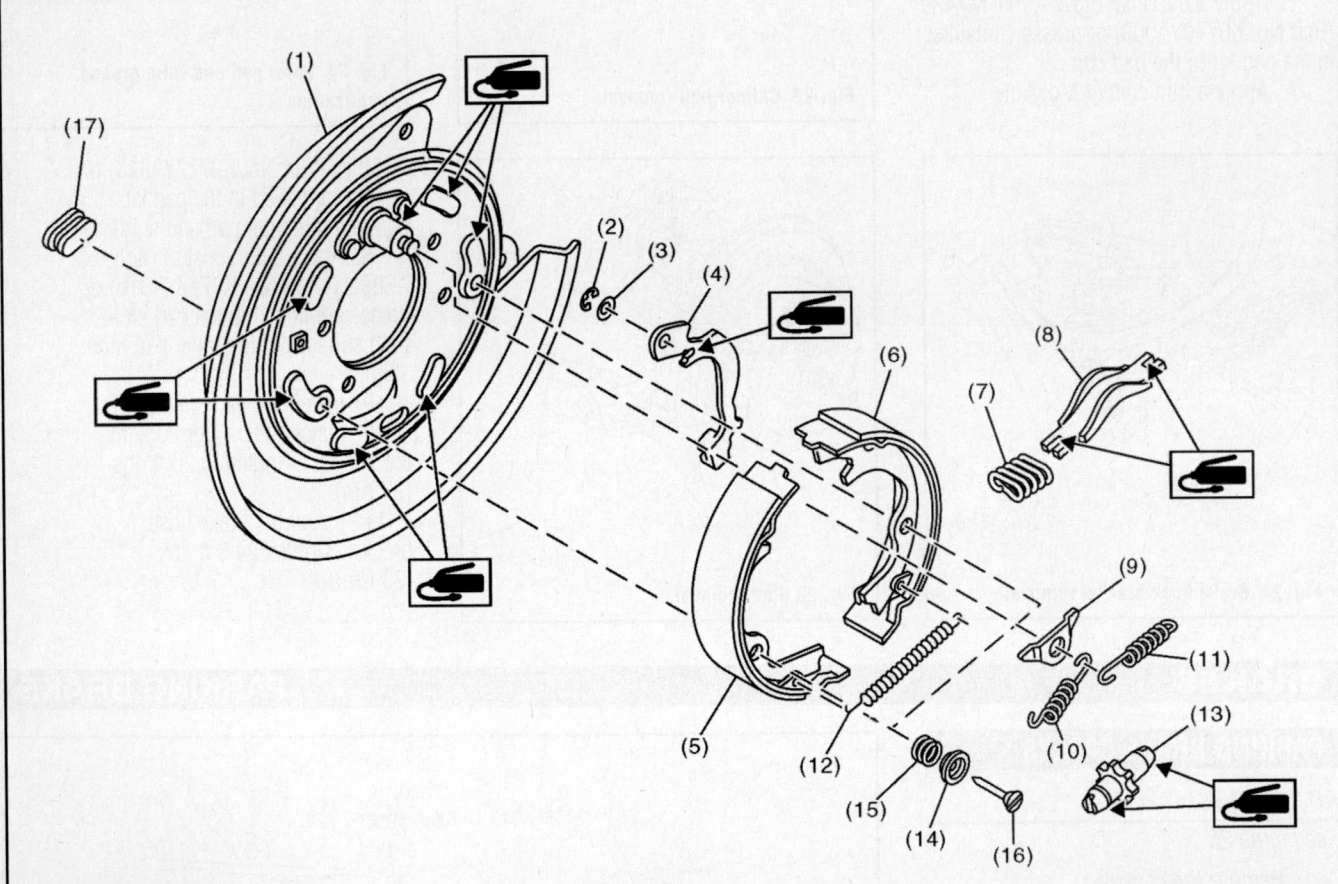

(1)	Back plate	(7)	Strut spring	(13)	Adjuster
(2)	Retainer	(8)	Strut	(14)	Shoe hold-down cup
(3)	Spring washer	(9)	Shoe guide plate	(15)	Shoe hold-down spring
(4)	Lever	(10)	Primary return spring	(16)	Shoe hold-down pin
(5)	Parking brake shoe (Primary)	(11)	Secondary return spring	(17)	Adjusting hole cover
(6)	Parking brake shoe (Secondary)	(12)	Adjusting spring		

22140_SUBA_G0059

Fig. 26 Exploded view of parking brake components

19. Install the parking brake cable to the lever.

20. Install the strut and adjuster, then secure the secondary side brake shoe with the shoe hold-down pin and cup.

➡**Install the strut spring of both right and left wheel facing vehicle front. Install the adjuster assembly with screw section on the left side.**

21. Install the secondary return spring and the adjusting spring.

22. Adjust the parking brake.

23. Drive the vehicle to break-in the parking brake lining.

a. Drive the vehicle at about 22 mph (35 km/h).

b. With the parking brake release button pushed in, pull the parking brake lever gently.

c. Drive the vehicle for about 0.12 miles (200 m) in this condition.

d. Wait 5 to 10 minutes for the parking brake to cool down. Repeat again from step (A).

e. After breaking-in, re-adjust the parking brakes.

ADJUSTMENT

1. Before servicing the vehicle, refer to the Precautions Section.

2. Return the parking brake lever fully.

3. Remove the adjusting hole cover from the back plate.

4. Turn the adjusting screw using a flat tip screwdriver until the brake shoe is in close contact with the disc rotor.

5. Turn back (downward) the adjusting screw 3 to 4 notches. Check there is no brake drag.

6. Install the adjusting hole cover to the back plate.

7. Adjust the parking lever stroke.

CHASSIS ELECTRICAL

GENERAL INFORMATION

❊❊ CAUTION

These vehicles are equipped with an air bag system. The system must be disarmed before performing service on, or around, system components, the steering column, instrument panel components, wiring and sensors. Failure to follow the safety precautions and the disarming procedure could result in accidental air bag deployment, possible injury and unnecessary system repairs.

SERVICE PRECAUTIONS

❊ CAUTION

Disconnect and isolate the battery negative cable before beginning any airbag system component diagnosis, testing, removal, or installation procedures. Wait at least 90 seconds after the ignition switch is turned off and the negative (-) terminal cable is disconnected from the battery before starting the operation. The SRS is equipped with a backup power source, so if work is started within 90

AIR BAG (SUPPLEMENTAL RESTRAINT SYSTEM)

seconds after disconnecting the negative (-) terminal cable from the battery, the SRS may be deployed. Failure to disable the airbag system may result in accidental airbag deployment, personal injury, or death.

DISARMING THE SYSTEM

1. Be sure to position the front wheels in the straight ahead position.

2. Disconnect the negative battery cable. Tape the battery cable for added protection.

3. Wait more than 20 seconds before starting work.

ARMING THE SYSTEM

1. Connect the negative battery cable.

CLOCKSPRING CENTERING

See Figure 27.

❊❊ CAUTION

When servicing a vehicle, be sure to turn the ignition switch to OFF, disconnect the ground cable from battery, and wait for more than 20 seconds before starting work. The airbag system is fitted with a backup power source. After disconnecting the

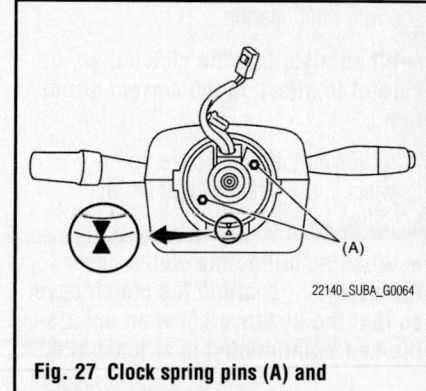

Fig. 27 Clock spring pins (A) and alignment marks shown

battery ground cable, the airbag may deploy if you do not wait for more than 20 seconds before starting the service of airbag system.

1. Turn the ignition switch to OFF.

2. Disconnect the ground cable from battery and wait for at least 20 seconds before starting work.

3. Check that front wheels are positioned in straight ahead direction.

4. Turn the clock spring pin clockwise until it stops.

5. Turn the clock spring connector pins approx. 3.25 turns until the marks are aligned.

DRIVE TRAIN

CLUTCH

REMOVAL & INSTALLATION

See Figures 28 through 30.

1. Remove the transmission assembly.

2. Attach the ST on the flywheel.

3. Remove the clutch cover and clutch disc.

➡**Take care not to allow oil to touch the clutch disc face. Do not disassemble the clutch cover or clutch disc.**

To install:

4. Insert the ST into the clutch disc and attach to the flywheel by inserting the ST end into pilot bearing.

➡**When installing the clutch disc, be careful to attach in the correct direction.**

5. Install the clutch cover to the flywheel and tighten the bolts to the specified torque.

➡**When installing the clutch cover to the flywheel, position the clutch cover so that the distance between unbalance marks (paint marks) is at least 120°.**

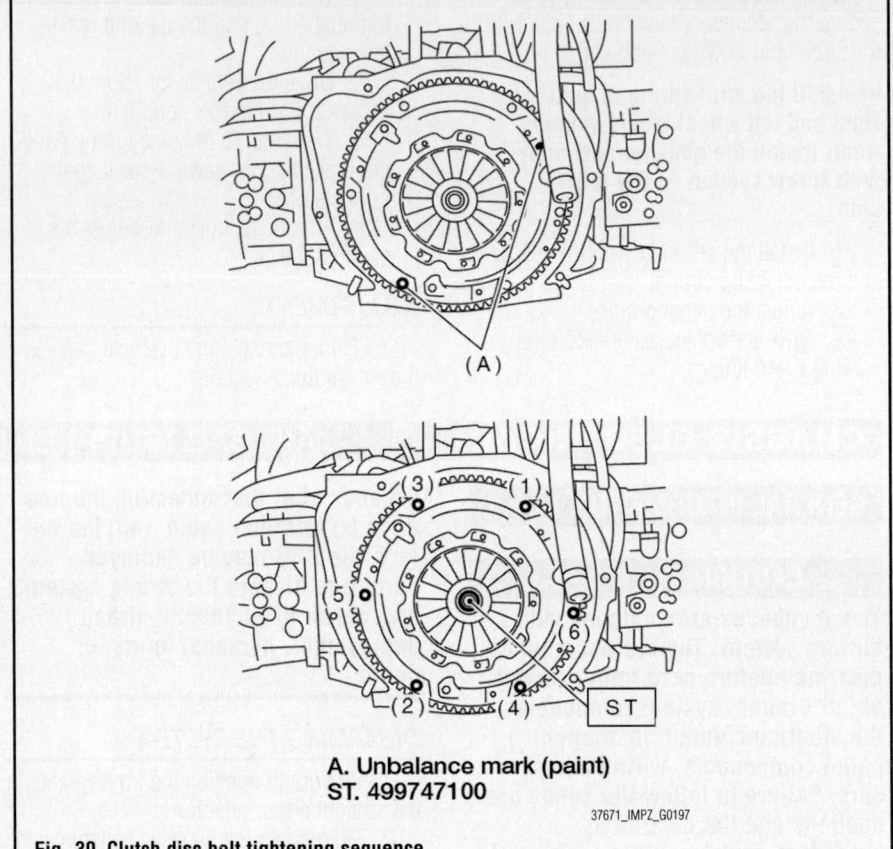

A. Unbalance mark (paint)
ST. 499747100

37671_IMPZ_G0197

Fig. 30 Clutch disc bolt tightening sequence

(The unbalance mark indicates the direction of residual unbalance.)

- Note the front and rear of the clutch disc when installing.
- Temporarily tighten the bolts by hand. Each bolt should be tightened to the specified torque in a crisscross order.

6. Tighten bolts to 142 inch lbs. (16 Nm).

7. Remove the CLUTCH DISC GUIDE.

8. Install the transmission assembly.

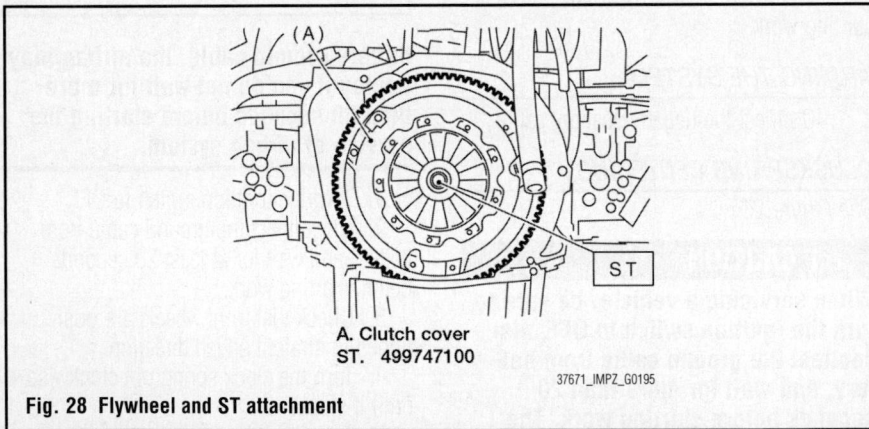

A. Clutch cover
ST. 499747100

37671_IMPZ_G0195

Fig. 28 Flywheel and ST attachment

DRIVEN DISC & PRESSURE PLATE

REMOVAL & INSTALLATION

See Figures 31 through 33.

1. Before servicing the vehicle, refer to the Precautions Section.

2. Disconnect the negative battery cable.

3. Remove the transmission.

4. Install tool ST499747100, or equivalent on the flywheel.

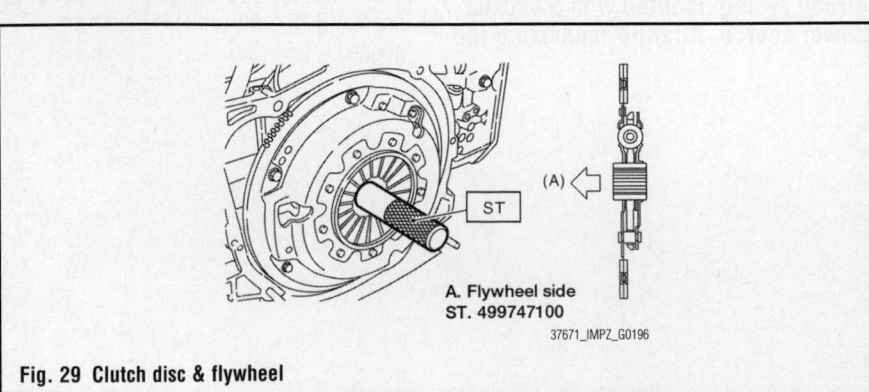

A. Flywheel side
ST. 499747100

37671_IMPZ_G0196

Fig. 29 Clutch disc & flywheel

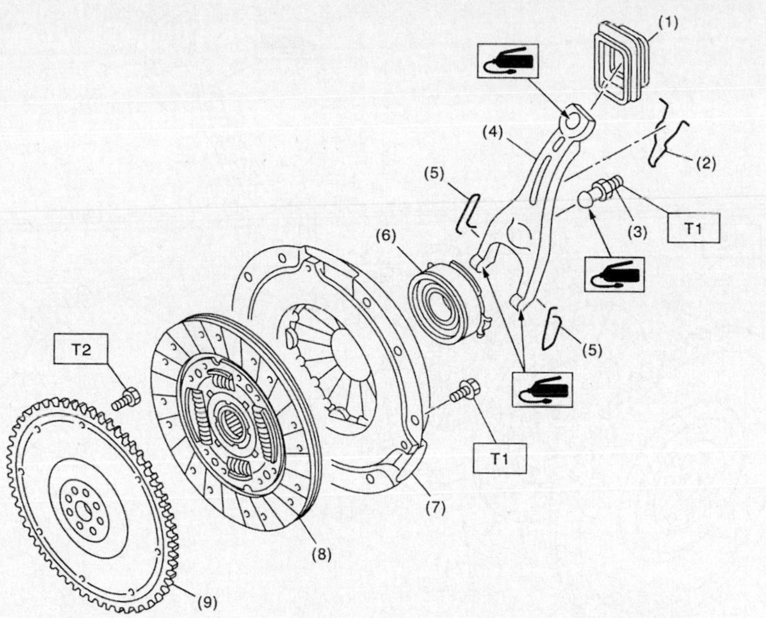

(1)	Dust cover	(6)	Release bearing
(2)	Lever spring	(7)	Clutch cover
(3)	Pivot	(8)	Clutch disc
(4)	Release lever	(9)	Flywheel
(5)	Clip		

Tightening torque: N·m (kgf-m, ft-lb)
T1: 16 (1.6, 11.8)
T2: 72 (7.3, 52.8)

09490_SBCR_G0084

Fig. 31 Clutch and related components—2006—08 (5MT) non-turbocharged vehicles

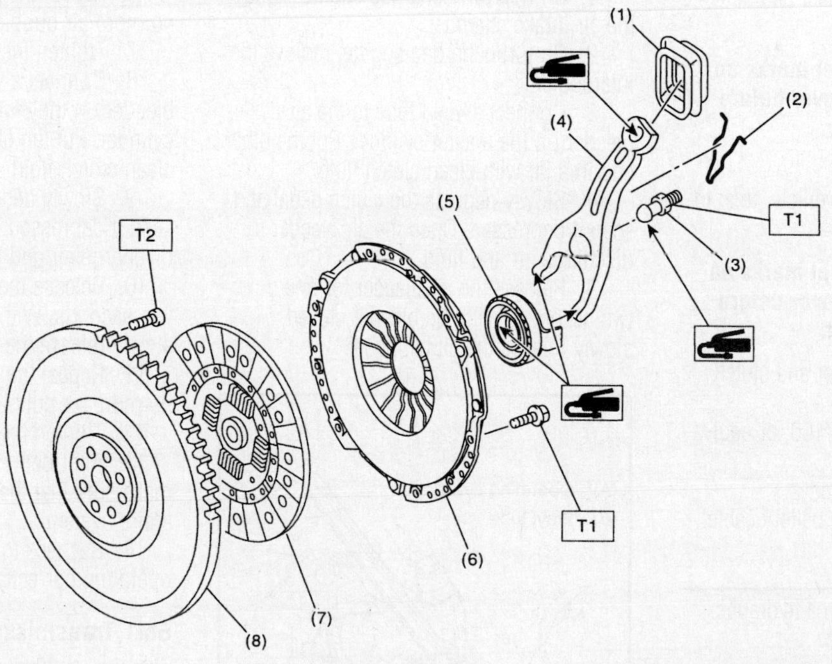

(1)	Dust cover	(5)	Release bearing
(2)	Lever spring	(6)	Clutch cover
(3)	Pivot	(7)	Clutch disc
(4)	Release lever	(8)	Flywheel

Tightening torque: N·m (kgf-m, ft-lb)
T1: 16 (1.6, 11.8)
T2: 72 (7.3, 52.8)

09490_SBCR_G0085

Fig. 32 Clutch and related components—2006—08 (5MT) turbocharged vehicles

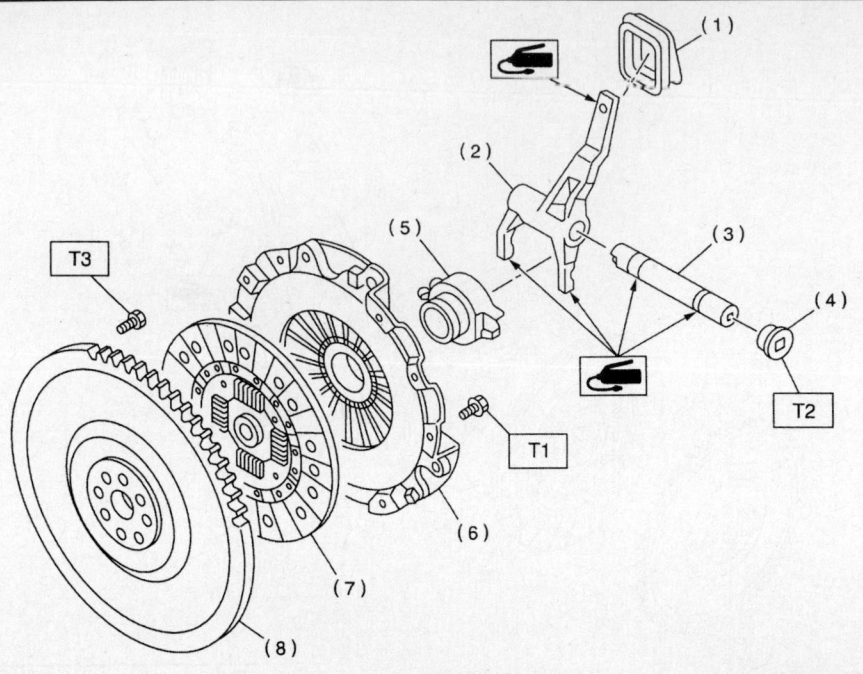

(1)	Dust cover	(5)	Release bearing
(2)	Release lever	(6)	Clutch cover
(3)	Clutch release lever shaft	(7)	Clutch disc
(4)	Plug	(8)	Flywheel

Tightening torque: N·m (kgf-m, ft-lb)
T1: 16 (1.6, 11.8)
T2: 44 (4.5, 32.5)
T3: 75 (7.6, 55.3)

09490_SBCR_G0086

Fig. 33 Clutch and related components—2006—08 (6MT) vehicles

5. Remove the clutch cover and clutch disc. Do not disassemble

➡**Be sure to put alignment marks on the flywheel and clutch cover before removing the clutch cover.**

To install:

6. Before servicing the vehicle, refer to the Precautions Section.

➡**Be sure to use alignment marks on the flywheel and clutch cover before installing the clutch cover.**

7. Install the clutch cover and clutch disc. Do not disassemble

8. Install tool ST499747100, or equivalent on the flywheel.

9. Install the transmission.

10. Connect the negative battery cable.

ADJUSTMENTS

Models are equipped with a hydraulic system that is not adjustable.

HYDRAULIC SYSTEM BLEEDING

BLEEDING PROCEDURE

5MT Transmission

See Figure 34.

1. On non turbocharged engine, remove the air intake chamber.

2. On turbocharged engine, remove the intercooler.

3. Connect a vinyl tube to the air bleeder on the master cylinder. Put the other end in a jar with clean clutch fluid.

4. Slowly depress the clutch pedal and keep it depressed. Open the air bleeder to discharge air and fluid.

5. Release the air bleeder for one or two seconds. With the bleeder closed, slowly release the clutch pedal.

(A) Operating cylinder
(B) Vinyl tube

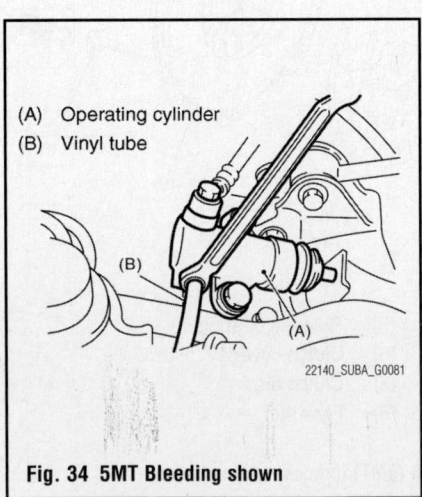

22140_SUBA_G0081

Fig. 34 5MT Bleeding shown

6. Repeat the procedure until there are no more air bubbles in the vinyl tube.

7. Tighten the air bleeder.

8. Connect a vinyl tube to the air bleeder on the clutch operating (slave) cylinder. Put the other end in a jar with clean clutch fluid.

9. Slowly depress the clutch pedal and keep it depressed. Open the air bleeder to discharge air and fluid.

10. Release the air bleeder for one or two seconds. With the bleeder closed, slowly release the clutch pedal.

11. Repeat the procedure until there are no more air bubbles in the vinyl tube.

12. Tighten the air bleeder.

13. After depressing the clutch pedal, make sure that there are no leaks in the entire system

14. Recheck to ensure that the clutch is operating correctly.

6MT Transmission

See Figure 35.

1. Remove the intercooler.

2. Remove the clutch operating cylinder. Do not remove the clutch hose.

3. Using a service clamp, fix the piston to avoid the piston from jumping out.

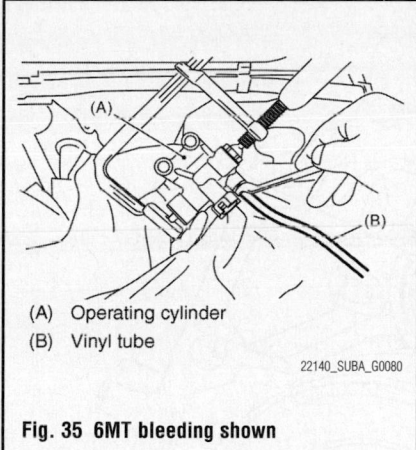

(A) Operating cylinder
(B) Vinyl tube

22140_SUBA_G0080

Fig. 35 6MT bleeding shown

4. Connect a vinyl tube to the air bleeder on the clutch operating (slave) cylinder. Put the other end in a jar with clean clutch fluid.

5. Slowly depress the clutch pedal and keep it depressed. Open the air bleeder to discharge air and fluid.

6. Release the air bleeder for one or two seconds. With the bleeder closed, slowly release the clutch pedal.

➡**Set the air breather screw position higher than the tip of the operating cylinder when performing this procedure.**

7. Repeat the procedure until there are no more air bubbles in the vinyl tube.

8. Tighten the air bleeder.

9. After depressing the clutch pedal, make sure that there are no leaks in the entire system

10. Recheck to ensure that the clutch is operating correctly.

TRANSFER CASE ASSEMBLY

REMOVAL & INSTALLATION

See Figures 36 through 58.

1. Remove the manual transmission assembly from the vehicle.

2. Remove the back-up light switch neutral position switch.

3. Remove the transfer case together with the extension case assembly.

  a. Remove the transmission cover.

  b. Set and hold the selector lever COMPL to the 1st-2nd side, and remove the transfer case and extension case assembly as a unit.

4. Remove the transfer cover.

5. Use the ST to push out the spring pin and remove the shift lever COMPL from shifter arm No. 2. ST 398791700 SPRING PIN REMOVER 2

6. Remove the extension case assembly.

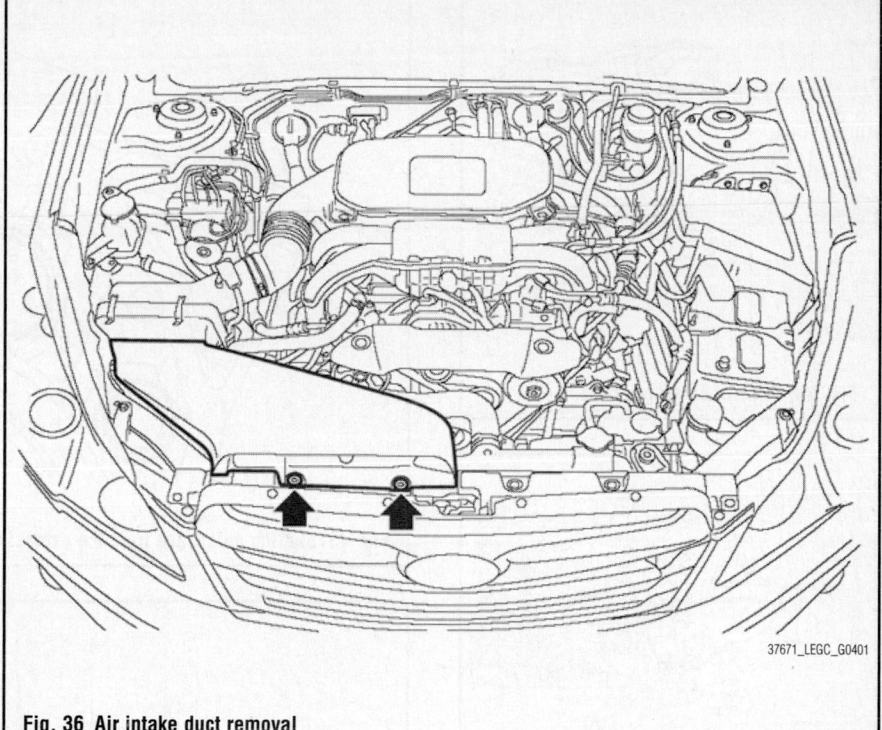

Fig. 36 Air intake duct removal

37671_LEGC_G0401

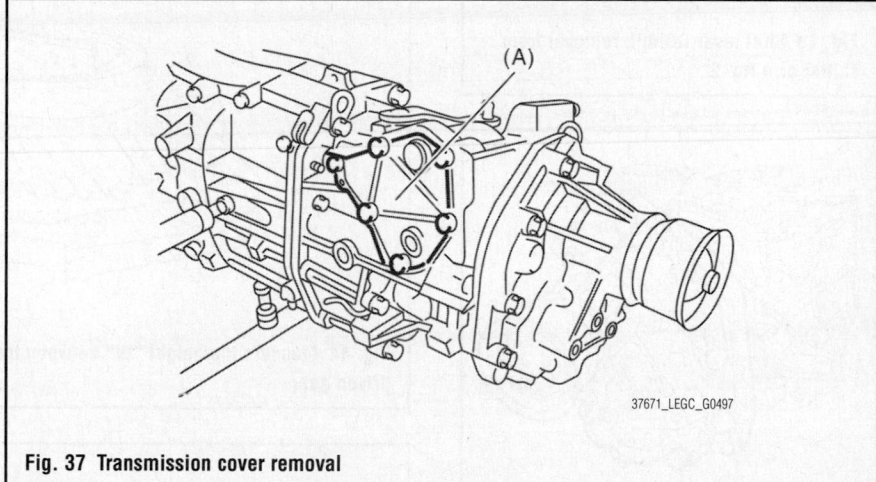

(A)

Fig. 37 Transmission cover removal

37671_LEGC_G0497

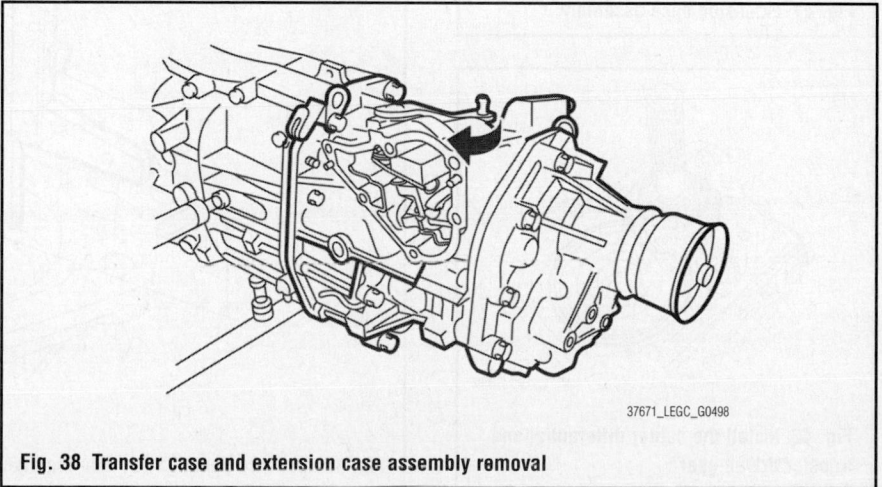

Fig. 38 Transfer case and extension case assembly removal

37671_LEGC_G0498

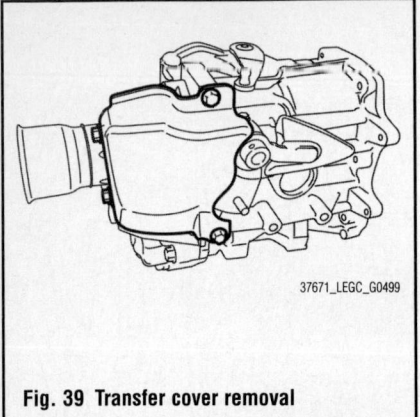

Fig. 39 Transfer cover removal

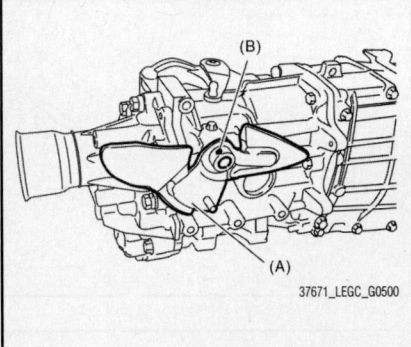

Fig. 40 Shift lever COMPL removal from shifter arm No. 2

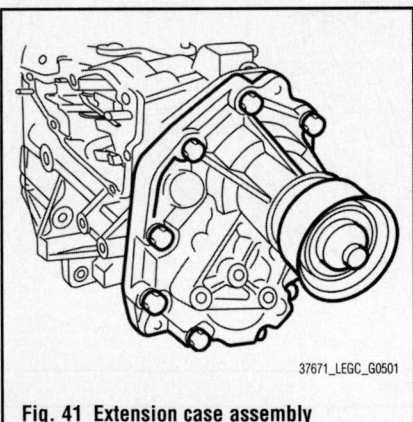

Fig. 41 Extension case assembly

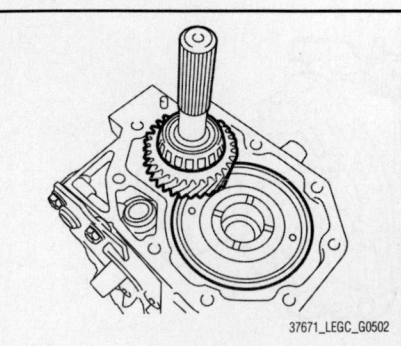

Fig. 42 Install the center differential and transfer driven gear

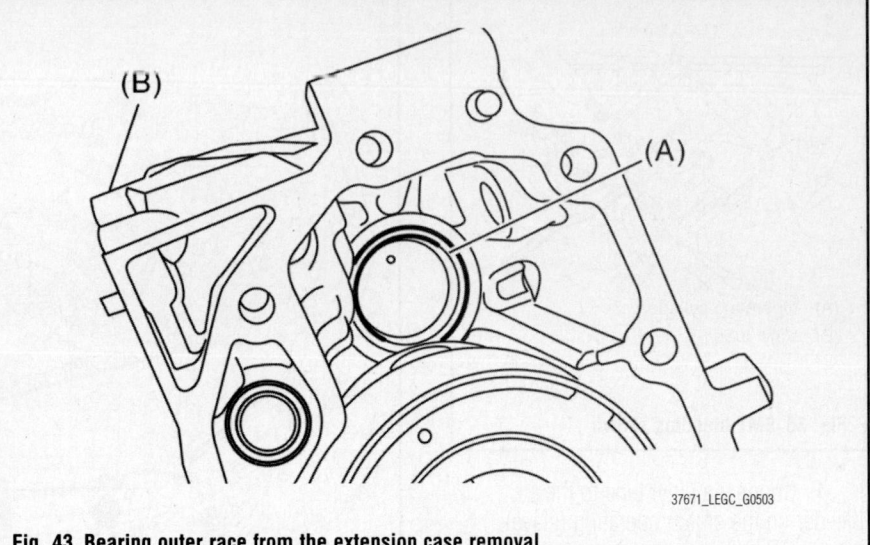

Fig. 43 Bearing outer race from the extension case removal

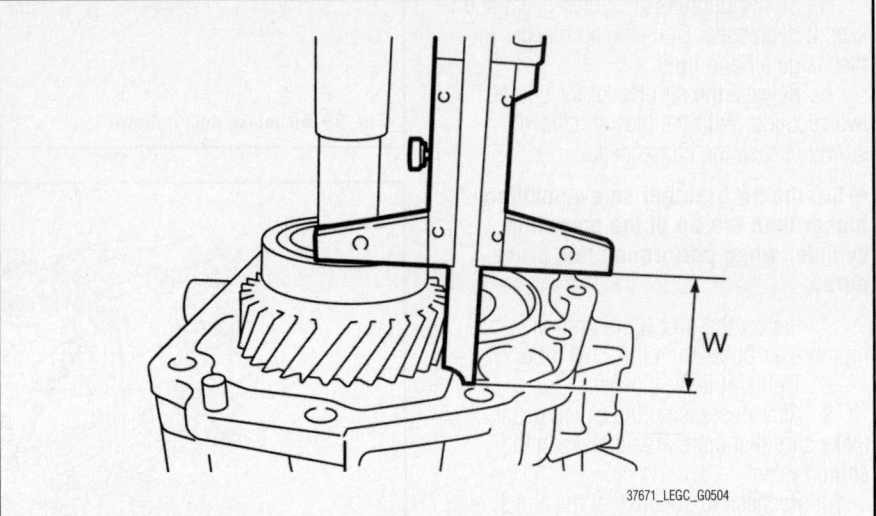

Fig. 44 Measure the height "W" between transfer case and taper roller bearing on the transfer driven gear

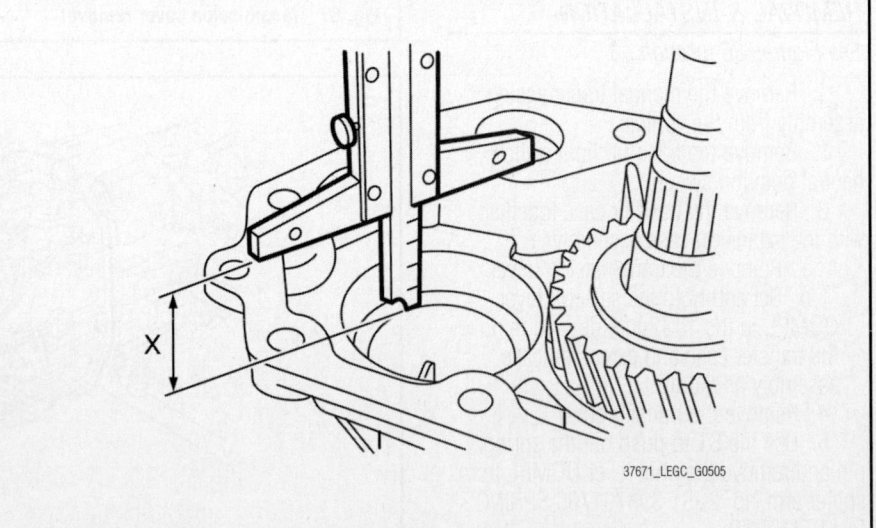

Fig. 45 Measure depth "X" on bearing insertion part of the extension

Thrust washer (50 × 61 × t)	
Part number	Thickness mm (in)
803050060	0.50 (0.0197)
803050061	0.55 (0.0217)
803050062	0.60 (0.0236)
803050063	0.65 (0.0256)
803050064	0.70 (0.0276)
803050065	0.75 (0.0295)
803050066	0.80 (0.0315)
803050067	0.85 (0.0335)
803050068	0.90 (0.0354)
803050069	0.95 (0.0374)
803050070	1.00 (0.0394)
803050071	1.05 (0.0413)
803050072	1.10 (0.0433)
803050073	1.15 (0.0453)
803050074	1.20 (0.0472)
803050075	1.25 (0.0492)
803050076	1.30 (0.0512)
803050077	1.35 (0.0531)
803050078	1.40 (0.0551)
803050079	1.45 (0.0571)

Fig. 46 Thrush washer part number and thickness table—thickness "t"

Thrust washer	
Part number	Thickness mm (in)
803036050	0.9 (0.035)
803036054	1.0 (0.039)
803036051	1.1 (0.043)
803036055	1.2 (0.047)
803036052	1.3 (0.051)
803036056	1.4 (0.055)
803036053	1.5 (0.059)
803036057	1.6 (0.063)
803036058	1.7 (0.067)
803036080	1.8 (0.071)
803036081	1.9 (0.075)

Fig. 49 Thrush washer part number and thickness table—thickness "U"

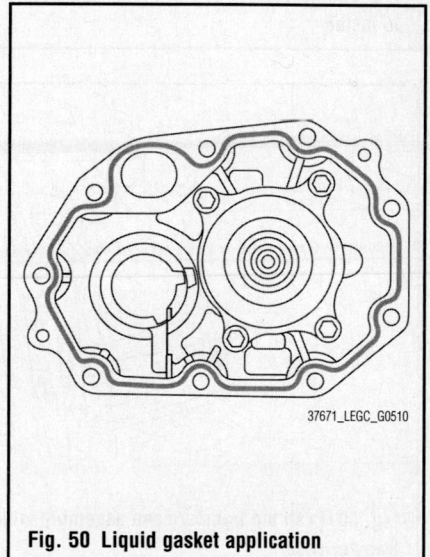

Fig. 50 Liquid gasket application

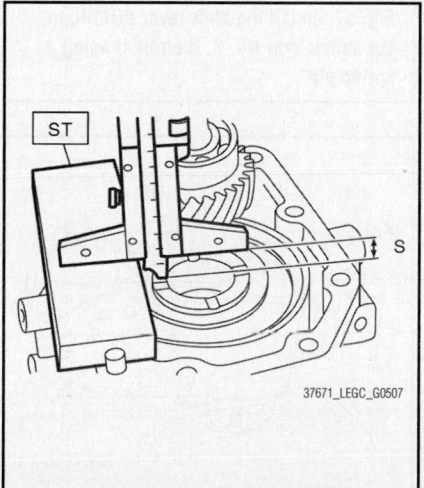

Fig. 47 Measure the depth "S" between transfer case and center differential

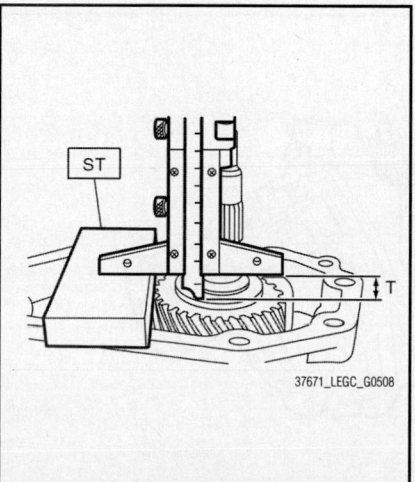

Fig. 48 Measure the height "T" between the extension case and transfer drive gear

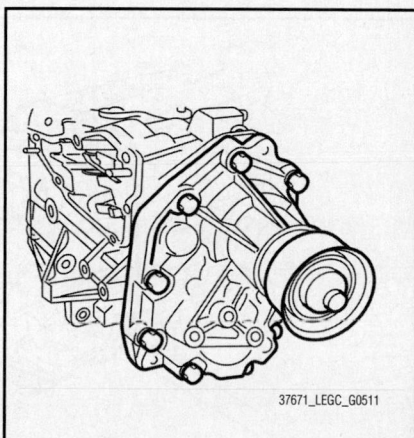

Fig. 51 Installing the extension case assembly to the transfer case

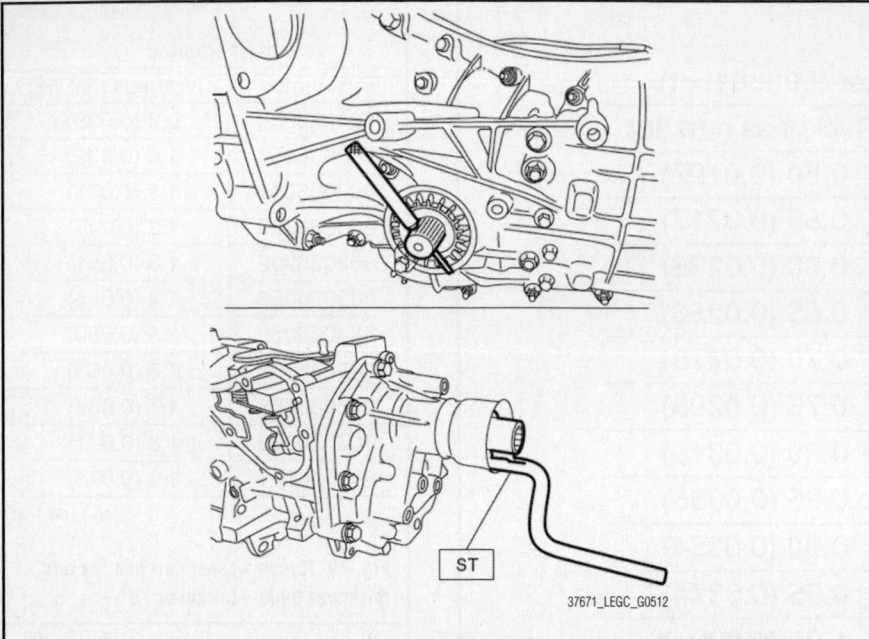

Fig. 52 Insert the axle shaft and hold, then move the handle to align the spline position to install

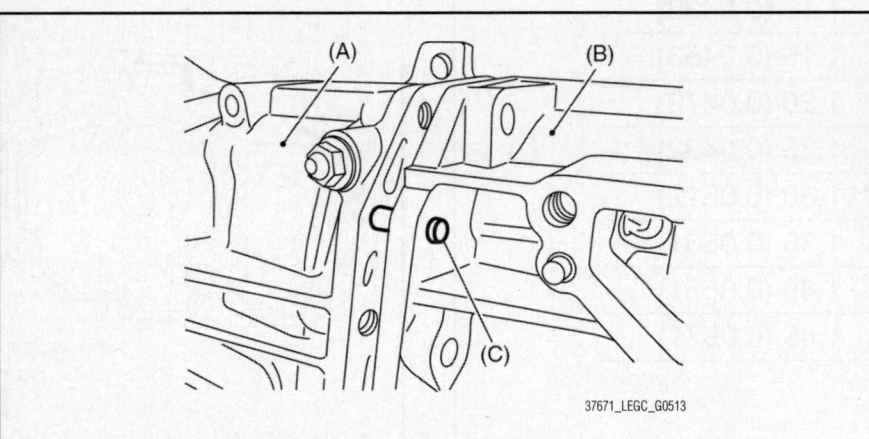

Fig. 53 Push the transfer case assembly until the knock pin protrudes slightly from the transfer case

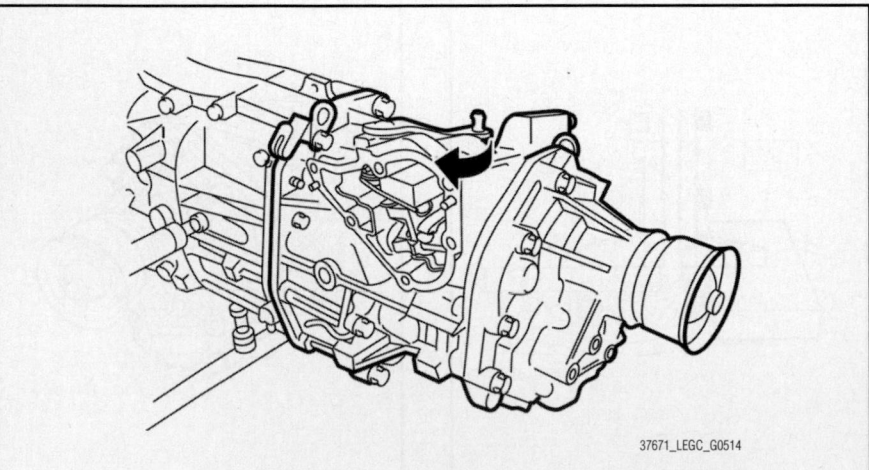

Fig. 54 Set and hold the selector lever to the 1st—2nd side, and install the transfer case assembly so that the shifter arm edge and fork rod does not contact

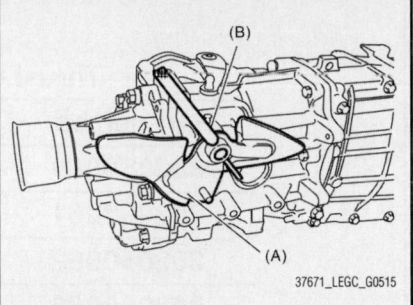

Fig. 55 Install the shift lever COMPL to the shifter arm No. 2, then insert the remover or similar tool instead of spring pin

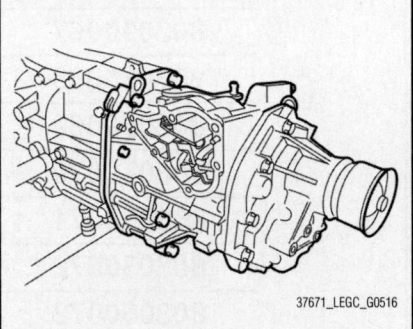

Fig. 56 Tightening of the transfer case assembly

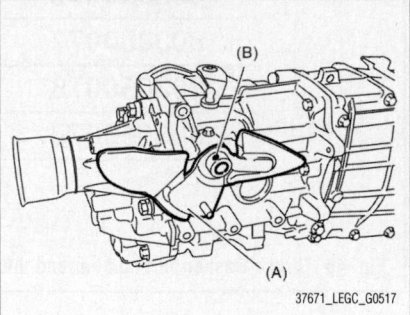

Fig. 57 Install the shift lever COMPL to the shifter arm No. 2, then fix it using a spring pin

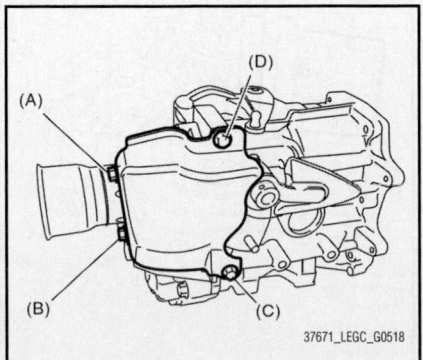

Fig. 58 Install the transfer cover

To install:

7. Install the center differential and transfer driven gear into the transfer case.

8. Remove the bearing outer race from the extension case.

9. While pressing the bearing outer race horizontally, rotate the driven shaft for ten turns.

10. Measure the height "W" between transfer case and taper roller bearing on the transfer driven gear.

11. Measure depth "X" on bearing insertion part of the extension case.

➡**Measure with bearing outer race and thrust washer removed.**

12. Calculate the thrust washer thickness "t" using the following calculation:

  a. t = X - W + (0.15–0.20 mm (0.006–0.008 in))

13. Select the washer with the nearest value.

14. Preload of the taper roller bearing (amount of standard protrusion): 0.15–0.20 mm (0.006–0.008 in)

➡**Be sure that it is always within the preload.**

15. Fit the thrust washers on the transfer drive shaft.

16. Install the bearing outer race into the extension case.

17. Measure the depth "S" between transfer case and center differential.

  a. ST 398643600 GAUGE

18. Measure the height "T" between extension case and transfer drive gear.

  a. ST 398643600 GAUGE

19. Calculate the thrust washer thickness "U" using the following calculation.

  a. U = S + T - 30 mm (1.18 in) [Thickness of ST]

➡**Thickness of ST: 15 mm (0.59 in)**

20. Select a suitable washer.

  a. Standard clearance: 0.15–0.35 mm (0.0059–0.0138 in)

21. Fit the thrust washer onto the center differential.

22. Apply a proper amount of liquid gasket to the transfer case mating surface.

  a. Liquid gasket: THREE BOND 1215 (Part No. 004403007) or equivalent

23. Install the extension case assembly to the transfer case. Tighten to 29.5 ft. lbs. (40 Nm).

24. Install the extension case assembly along with the transfer case to the transmission case.

  a. Insert the axle shaft and hold, then move the HANDLE to align the spline position to install.

  • ST 18631AA000 HANDLE

  • Part No. 38415AA100 Axle shaft

➡**Push the transfer case assembly until the knock pin protrudes slightly from the transfer case.**

  b. Set and hold the selector lever to the 1st- 2nd side, and install the transfer case assembly so that the shifter arm edge and fork rod does not contact.

25. Temporarily tighten the transfer case assembly.

26. Install the shift lever COMPL to the shifter arm No. 2, then insert the remover or similar tool instead of spring pin.

27. Operate the selector arm and remover to confirm the correct shifting into each gear is possible.

28. Tighten the transfer case assembly to 18 ft. lbs. (25 Nm).

29. Install the shift lever COMPL to the shifter arm No. 2, then fix it using a spring pin.

➡**Use new spring pin.**

30. Install the transfer cover.

  a. Temporarily tighten the transfer cover mounting bolts.

  b. Tighten bolts in order to the specified torque. Tightening torque: 133 inch lbs. (15 Nm).

31. Install the transmission cover.

ENGINE COOLING

ENGINE FAN

REMOVAL & INSTALLATION

2.5L Engine

1. Before servicing the vehicle, refer to the Precautions Section.

2. Disconnect the negative battery cable.

3. Raise and safely support the vehicle.

4. Disconnect the fan electrical connector.

5. Remove the transmission cooler hose from the clip of the radiator fan motor assembly, if equipped.

6. Lower the vehicle.

7. Disconnect the overflow hose and reservoir tank.

8. Remove the fan shroud mounting bolts.

9. Remove the main fan assembly.

To install:

10. Installation is the reverse order of assembly.

11. Tighten the fan shroud mounting bolts to 6 ft. lbs. (7.5 Nm).

3.0L Engine

See Figure 59.

1. Before servicing the vehicle, refer to the Precautions Section.

2. Disconnect the negative battery cable.

3. Remove the hood stay holder.

4. Remove the air intake assembly.

5. Disconnect the connector from the radiator fan control unit.

6. Raise and safely support the vehicle.

7. Remove the engine undercover.

8. Drain the engine coolant.

9. Remove the transmission cooler hose from the clip of the radiator fan motor assembly, if equipped.

10. Remove the radiator fan motor harness from the clip.

11. Lower the vehicle.

12. Remove the reservoir tank.

13. Remove the radiator sub-fan assembly.

14. Remove the radiator fan assembly.

➡**When removing the main fan assembly, lifting it straight up will cause the** main fan shroud contacts to inlet part of engine coolant. To avoid contacting it, move the main fan assembly to sub fan assembly side before removal.

To install:

15. Installation is the reverse order of assembly.

16. Tighten the fan shroud mounting bolts to 6 ft. lbs. (7.5 Nm).

17. Refill the engine cooling system to the correct level.

RADIATOR

REMOVAL & INSTALLATION

See Figures 60 and 61.

1. Before servicing the vehicle, refer to the Precautions Section.

2. Set the vehicle on a lift.

3. Disconnect the ground cable from battery.

4. Lift-up the vehicle.

5. Remove the undercover.

6. Drain engine coolant completely.

7. Remove the radiator undercover.

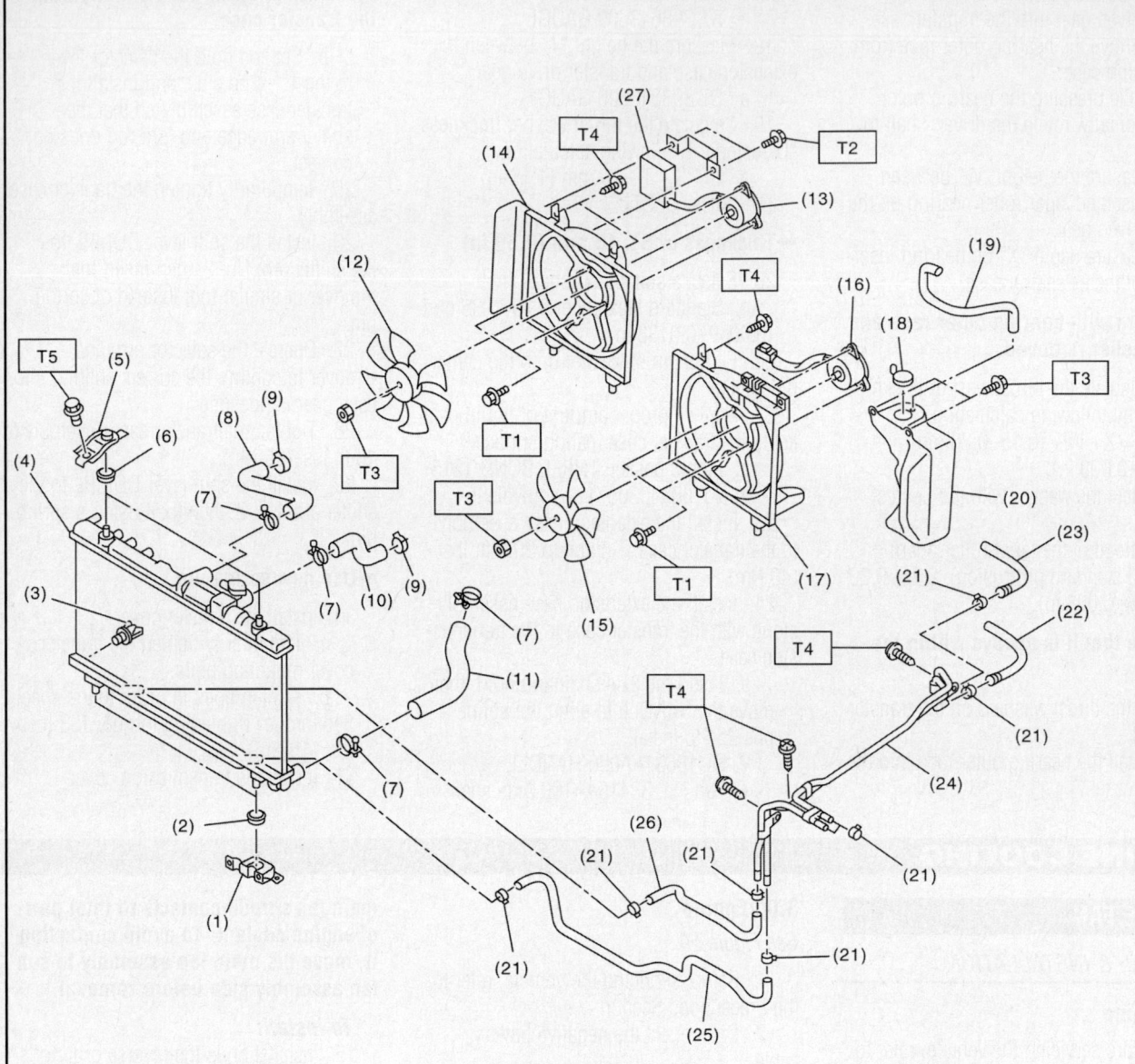

(1)	Radiator lower bracket	(13)	Radiator sub fan motor	(24)	ATF pipe
(2)	Radiator lower cushion	(14)	Radiator sub fan shroud	(25)	ATF hose C
(3)	Engine coolant drain plug	(15)	Radiator main fan	(26)	ATF hose D
(4)	Radiator	(16)	Radiator main fan motor	(27)	Radiator fan control unit
(5)	Radiator upper bracket	(17)	Radiator main fan shroud		
(6)	Radiator upper cushion	(18)	Engine coolant reservoir tank cap		
(7)	Clamp	(19)	Over flow hose		
(8)	Radiator hose A	(20)	Engine coolant reservoir tank		
(9)	Clamp	(21)	ATF hose clamp		
(10)	Radiator hose B	(22)	ATF hose A		
(11)	Radiator hose C	(23)	ATF hose B		
(12)	Radiator sub fan				

Tightening torque:N·m (kgf-m, ft-lb)
T1: 3.8 (0.39, 2.8)
T2: 5.4 (0.55, 4.0)
T3: 6.2 (0.63, 4.6)
T4: 7.5 (0.76, 5.5)
T5: 12 (1.2, 8.9)

22140_SUBA_G0100

Fig. 59 Exploded view of radiator and fan system

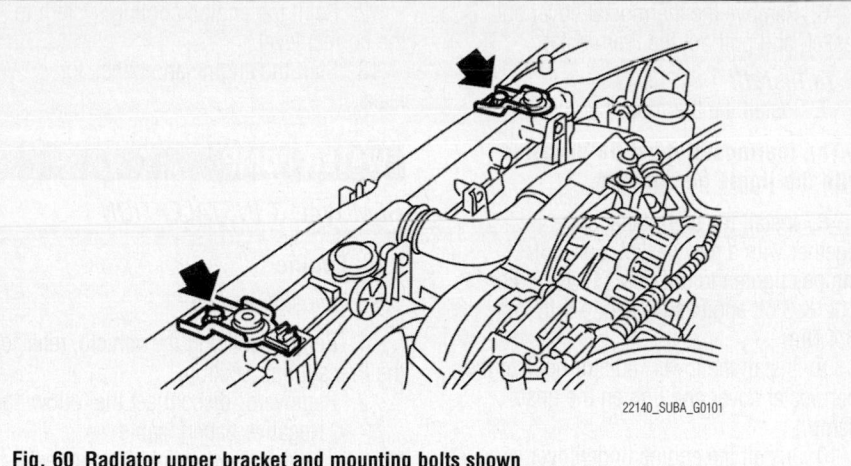

Fig. 60 Radiator upper bracket and mounting bolts shown

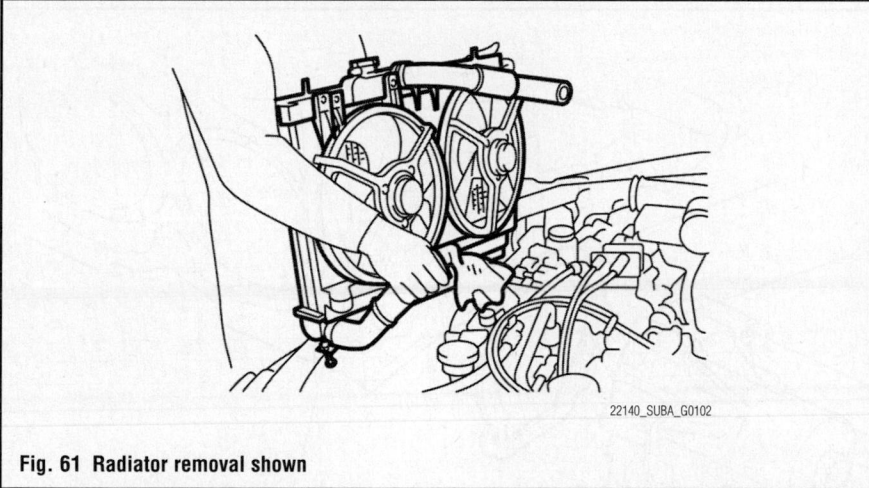

Fig. 61 Radiator removal shown

8. Disconnect the radiator main and sub fan motor connectors.

9. Disconnect the radiator outlet hose from thermostat cover.

10. Disconnect the ATF cooler hoses from ATF pipes. (AT models) Plug the ATF pipe to prevent ATF from leaking.

11. Lower the vehicle.

12. Disconnect the over flow hose.

13. Remove the V—belt covers for (2.5L DOHC) engines.

14. Remove the reservoir tank.

15. Remove the hood stay holder.

16. Remove the air intake duct.

17. Disconnect the connector from radiator fan control unit for (3.0L) engines.

18. Disconnect the radiator inlet hoses from the engine.

19. Remove the radiator upper brackets.

20. Detach the power steering hose from the clip on the radiator for (2.5L DOHC) engines.

21. Move the radiator to the left while lifting it upward.

22. Lift the radiator up and away from vehicle.

To install:

23. Attach the radiator lower cushion to the hole on the radiator lower bracket.

24. Install the radiator to vehicle.

➡**Make pins on the lower side of radiator be fitted into the radiator lower cushions on body side.**

25. Install the radiator upper brackets and tighten the bolts 9 ft. lbs. (12 Nm).

26. Attach the power steering hose to the radiator for (2.5L DOHC) engines.

27. Connect the radiator inlet hose.

28. Install the air intake duct.

29. Connect the connector from radiator fan control unit for (3.0L) engines.

30. Install the hood stay holder.

31. Install the reservoir tank.

32. Connect the over flow hose.

33. Install the V—belt covers for (2.5L DOHC) engines.

34. Lift-up the vehicle.

35. Connect the ATF cooler hoses. (AT models)

36. Connect the radiator outlet hose.

37. Connect the radiator main and sub fan motor connectors.

38. Install the radiator undercover.

39. Install the undercover.

40. Lower the vehicle.

41. Connect the battery ground cable to the battery.

42. Fill engine coolant and bleed the cooling system.

43. Check the ATF level.

THERMOSTAT

REMOVAL & INSTALLATION

See Figures 62 through 64.

1. Before servicing the vehicle, refer to the Precautions Section.

2. Raise and safely support the vehicle.

3. Remove the engine undercover.

4. Drain the engine coolant.

5. Loosen the hose clamp and disconnect the lower radiator hose from the thermostat cover.

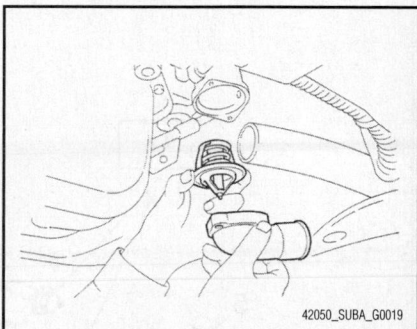

Fig. 62 Remove the thermostat cover, and pull out the thermostat—2.5L engine shown

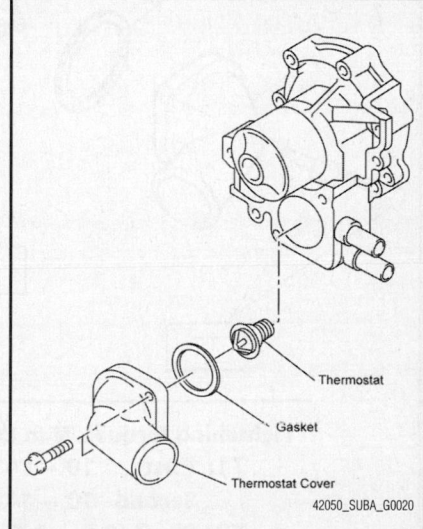

Fig. 63 Exploded view of the thermostat and related components—2.5L Turbo engine shown

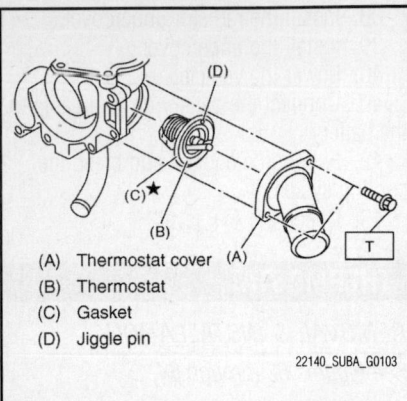

(A) Thermostat cover
(B) Thermostat
(C) Gasket
(D) Jiggle pin

22140_SUBA_G0103

Fig. 64 Exploded view of the thermostat and related components—3.0L engine shown 3.6L similar

6. Remove the thermostat cover and gasket, and pull out the thermostat.

To install:

7. Install the thermostat.

→**The thermostat must be installed with the jiggle pin upward.**

8. Install the thermostat cover together with a new gasket. For 2.5L engines tighten to 9 ft. lbs. (12 Nm). For 3.0L & 3.6L engines tighten to 5 ft. lbs. (6.4 Nm).

9. Install the lower radiator hose to the thermostat cover and tighten the hose clamp.

10. Install the engine undercover.

11. Lower the vehicle.

12. Refill the engine cooling system to the correct level.

13. Start the engine and check for leaks.

WATER PUMP

REMOVAL & INSTALLATION

2.5L Engine

See Figures 65 and 66.

1. Before servicing the vehicle, refer to the Precautions Section.

2. Remove or disconnect the following:
- Negative battery cable
- Engine undercover, if equipped

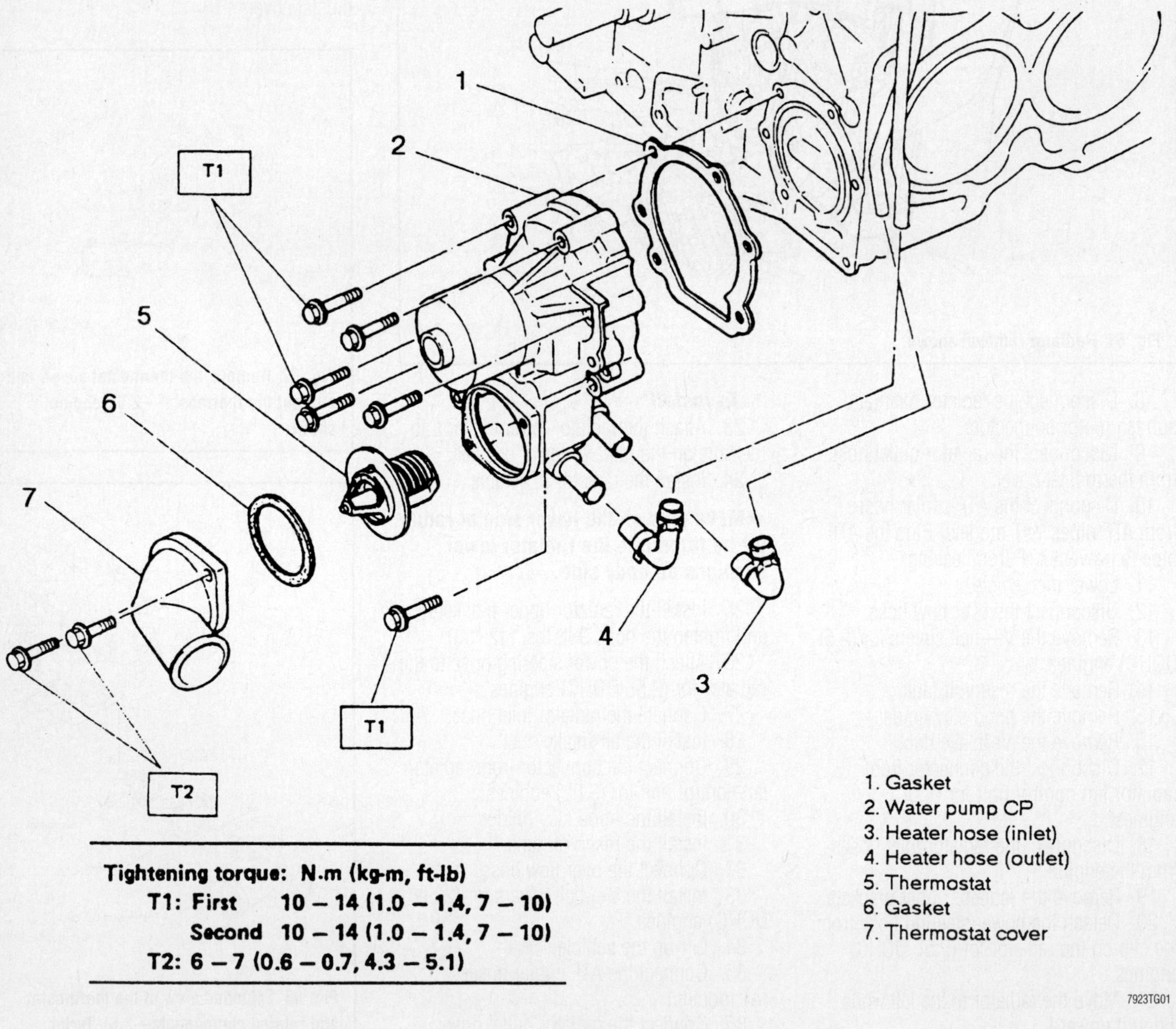

T1

5

6

7

T1

T2

1. Gasket
2. Water pump CP
3. Heater hose (inlet)
4. Heater hose (outlet)
5. Thermostat
6. Gasket
7. Thermostat cover

Tightening torque: N·m (kg-m, ft-lb)
T1: First   10 — 14 (1.0 — 1.4, 7 — 10)
Second 10 — 14 (1.0 — 1.4, 7 — 10)
T2: 6 — 7 (0.6 — 0.7, 4.3 — 5.1)

7923TG01

Fig. 65 Water pump and related components

3. Drain the coolant into a suitable container.

4. Remove or disconnect the following:
- Radiator fan connectors
- Radiator outlet and heater hoses
- Heater bypass hose or overflow hose, if equipped
- Reservoir tank, on Legacy
- Radiator fan motor assembly
- Accessory drive belts
- Timing belt
- Belt tension adjuster
- Belt idler No. 2
- Camshaft Position (CMP) sensor
- Left side camshaft pulley
- Left side rear timing belt cover
- Tensioner bracket
- Radiator and heater hoses from water pump
- Water pump retainer bolts
- Water pump

5. Inspect the radiator hoses for deterioration and replace as necessary.

To install:

6. Clean the gasket mating surfaces thoroughly. Always use new gaskets during installation.

7. Install or connect the following:

8. Tighten the pump bolts in sequence to 8.9 ft. lbs. (12 Nm). After tightening the bolts once, retighten to the same specification again.
- Radiator heater hoses to water pump
- Tensioner bracket and tighten to 18 ft. lbs. (25 Nm)
- Left side rear timing belt cover
- Left side camshaft pulley(s). Tighten to 58 ft. lbs. (78 Nm) on non turbocharged engine and 72 ft. lbs. (98 Nm) on turbocharged engine.
- CMP sensor
- Belt idler No. 2 and tighten to 29 ft. lbs. (39 Nm)
- Belt tension adjuster
- Timing belt

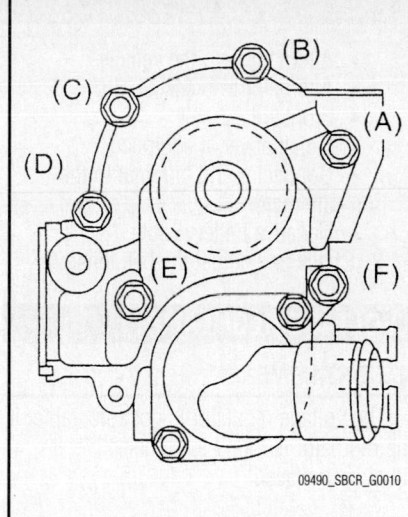

09490_SBCR_G0010

Fig. 66 Tighten the water pump bolts in two steps using the following sequence— 2.5L engine

- Accessory drive belts
- Radiator fan assembly
- Reservoir tank, if removed
- Heater bypass hose or overflow hose, if equipped
- Air intake duct
- Radiator outlet and heater hoses
- Radiator fan connectors
- Engine undercover, if removed

9. Fill the system with coolant and connect the negative battery cable.

10. Start the engine and allow it to reach operating temperature.

11. Check for leaks.

3.0L & 3.6L Engine

See Figure 67.

1. Before servicing the vehicle, refer to the Precautions Section.

2. Remove or disconnect the following:
- Negative battery cable
- Engine undercover, if equipped

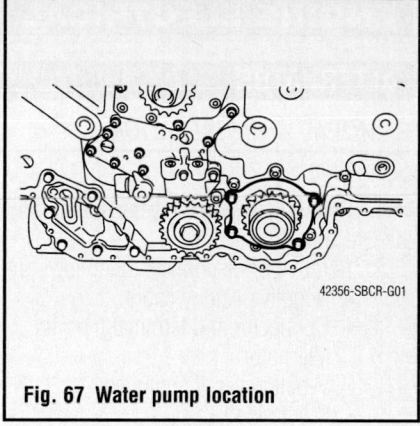

42356-SBCR-G01

Fig. 67 Water pump location

3. Drain the coolant into a suitable container.

4. Remove or disconnect the following:
- Radiator
- Accessory drive belts
- Timing chain
- Water pump retainer bolts
- Water pump

5. Inspect the radiator hoses for deterioration and replace as necessary.

To install:

6. Clean the gasket mating surfaces thoroughly. Always use new gaskets during installation.

7. Apply coolant to the new O-ring before installation

8. Install or connect the following:
- Water pump with a new O-ring, tighten the bolts to 60 inch lbs. (7 Nm).
- Timing chain
- Front chain cover
- Accessory drive belts
- Radiator
- Engine undercover, if removed

9. Fill the system with coolant and connect the negative battery cable.

10. Start the engine and allow it to reach operating temperature.

11. Check for leaks.

ENGINE ELECTRICAL — CHARGING SYSTEM

ALTERNATOR

REMOVAL & INSTALLATION

See Figure 68.

1. Before servicing the vehicle, refer to the Precautions Section.
2. Remove or disconnect the following:
   - Negative battery cable
   - Connector and terminal from the alternator
   - V-belt cover, if equipped
   - Front side V-belt
   - Alternator to bracket bolts
   - Alternator from the vehicle.

To install:

3. Install or connect the following:

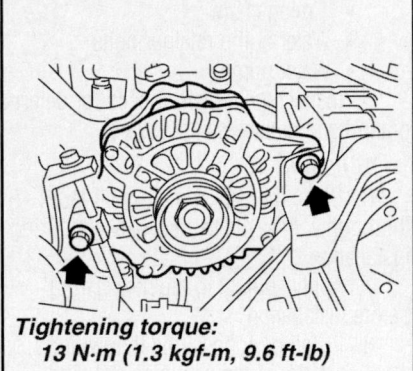

Tightening torque:
13 N·m (1.3 kgf-m, 9.6 ft-lb)

22140_SUBA_G0104

Fig. 68 Alternator and mounting bolts

- Alternator into the vehicle
- Alternator to bracket bolts
- Front side V-belt
- V-belt cover, if equipped
- Connector and terminal to the alternator
- Negative battery cable

4. Check and adjust the belt tension.

VOLTAGE REGULATOR

ADJUSTMENT

The voltage regulator is not adjustable. If the regulator is found to be defective, it must be replaced.

REMOVAL & INSTALLATION

See Figure 69.

1. Before servicing the vehicle, refer to the Precautions Section.
2. Remove the alternator from the vehicle.
3. Remove the four through bolts. Then insert the tip of a flat-head screwdriver into the gap between the stator core and front bracket. Pry them apart to disassemble the alternator.
4. Hold rotor with a vise and remove pulley nut.

✳✳ CAUTION

When holding rotor with vise, insert aluminum plates or wood pieces on the contact surfaces of the vise to prevent rotor from damage.

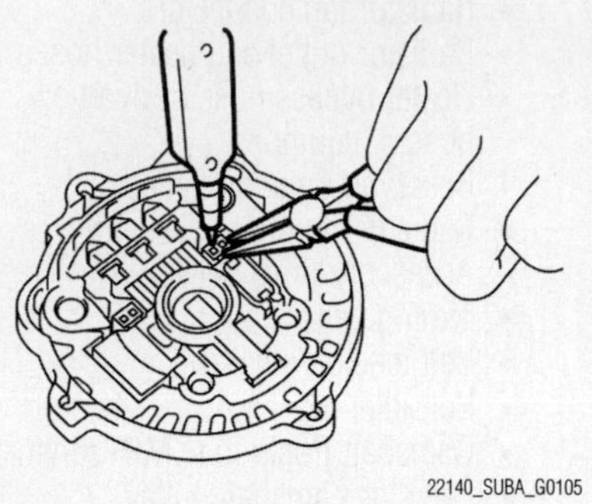

22140_SUBA_G0105

Fig. 69 IC regulator removal shown

5. Unsolder connection between rectifier and stator coil to remove stator coil.

✳✳ CAUTION

Finish the work rapidly (less than three seconds) because the rectifier cannot withstand heat very well.

6. Remove screws which secure voltage regulator to rear cover, and unsolder connection between voltage regulator and rectifier to remove the regulator.

To install:

7. Installation is the reverse order of assembly.
8. After assembly, turn the pulley by hand to check that the rotor turns smoothly.

ENGINE ELECTRICAL — IGNITION SYSTEM

IGNITION TIMING

INSPECTION

See Figure 70.

METHOD WITH SUBARU SELECT MONITOR

Before checking the ignition timing, check the following:

- Ensure the air cleaner element is free from clogging, spark plugs are in good condition, and that hoses are connected properly.
- Ensure the malfunction indicator light does not illuminate.

1. Idle the engine.
2. Stop the engine, and turn the ignition switch to OFF.
3. Insert the cartridge to the Subaru Select Monitor.

4. Connect the Subaru Select Monitor to data link connector.
5. Turn the ignition switch to ON, and Subaru Select Monitor power switch to ON.
6. Select Each System Check in the Main Menu.
7. Select Engine in the Selection Menu.
8. Select Current Data Display & Save in the Engine Control System Diagnosis.
9. Select Data Display in the Data Display Menu.
10. Start the engine, and read the ignition timing by idle speed.
11. If the timing is not correct, check the ignition control system.

Ignition timing is as follows:
- 10°+/-8°650 rpm (MT Models)
- 15°+/-8°700 rpm (AT Models)
- 17°+/-10°700 rpm (STI Model)

METHOD WITH TIMING LIGHT

12. Before checking the ignition timing, check the following item:

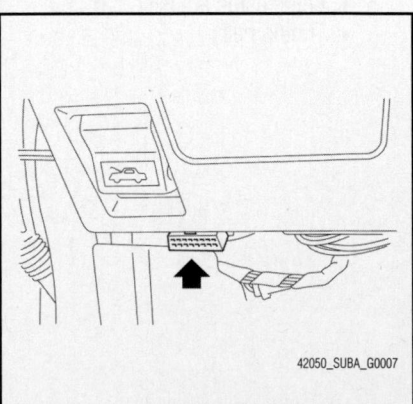

42050_SUBA_G0007

Fig. 70 Location of the data link connector under the driver's side instrument panel.

13. Check the air cleaner element is free from clogging, spark plugs are in good condition, and hoses are connected properly.

14. Check the malfunction indicator light does not illuminate.

15. Warm-up the engine.

16. Stop the engine, and turn the ignition switch to OFF.

17. Remove the air intake duct.

18. Disconnect the connectors of the mass air flow and intake air temperature sensor.

19. Remove the air cleaner case and element.

20. Connect the timing light to the power wire of #1 ignition coil.

21. Attach the air cleaner case, element and connector of mass air flow and intake air temperature sensor.

22. Start the engine, turn the timing light to the crank pulley, and check the ignition timing by means of crank pulley indicator.

23. If the timing is not correct, check the ignition control system.

Ignition timing is as follows:

• 10°+/-8°650 rpm (MT Models)
• 15°+/-8°700 rpm (AT Models)
• 17°+/-10°700 rpm (STI Model)

ADJUSTMENT

The ignition timing is controlled by the Powertrain Control Module (PCM). No adjustment is necessary or possible.

SPARK PLUGS

REMOVAL & INSTALLATION

2.5L DOHC Turbo Engine

Left Side

See Figures 71 and 72.

1. Before servicing the vehicle, refer to the Precautions Section.

2. Remove the collector cover.

3. Remove the battery and battery carrier.

4. Disconnect the secondary air pump duct from the secondary air pump.

5. Remove the bolts that attach the secondary air pump duct to the rocker cover (LH), and raise the secondary air pump duct.

6. Disconnect the connector from ignition coil.

7. Remove the ignition coil.

➡**Turn number 4 ignition coil by 180° to remove it.**

8. Remove the spark plug with a spark plug socket.

Fig. 71 Secondary air pump duct view

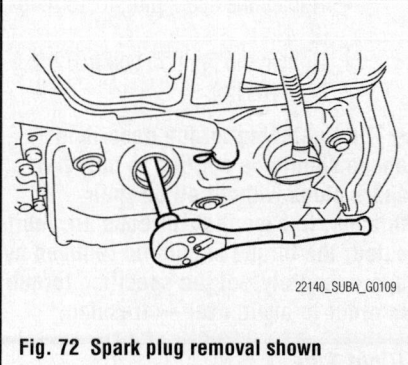

Fig. 72 Spark plug removal shown

To install:

9. When installing spark plugs on cylinder head, use the appropriate spark plug socket.

10. Tighten the spark plugs to 15 ft. lbs. (21 Nm).

11. Tighten ignition coil mounting bolts to 12 ft. lbs. (16 Nm).

12. Reinstall any components removed for clearance.

13. Connect the negative battery cable.

Right Side

See Figures 72 and 73.

1. Before servicing the vehicle, refer to the Precautions Section.

2. Remove the collector cover.

3. Disconnect the ground cable from the battery.

4. Remove the air cleaner case.

5. Disconnect the connector from ignition coil.

6. Remove the ignition coil.

➡**Turn number 3 ignition coil by 180° to remove it.**

7. Remove the spark plug with a spark plug socket.

To install:

8. When installing spark plugs on cylinder head, use the appropriate spark plug socket.

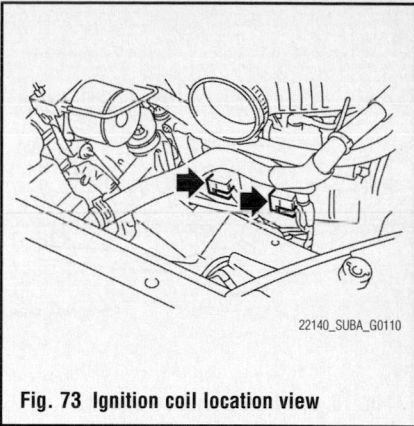

Fig. 73 Ignition coil location view

9. Tighten the spark plugs to 15 ft. lbs. (21 Nm).

10. Tighten ignition coil mounting bolts to 12 ft. lbs. (16 Nm).

11. Reinstall any components removed for clearance.

12. Connect the negative battery cable.

2.5L SOHC Engines

See Figures 74 and 75.

1. Before servicing the vehicle, refer to the Precautions Section.

2. Disconnect the negative battery cable.

3. Remove the following components if necessary for access to remove the spark plugs:

• Battery
• Air intake assembly

✳✳ CAUTION

Do not disconnect the washer tank hoses.

4. Gripping the spark plug wire by the boot, remove the spark plug wires.

5. Remove the spark plugs.

To install:

6. When installing spark plugs on cylinder head, use the appropriate spark plug

Fig. 74 Spark plug wires and boots shown

Fig. 75 Spark plug removal shown

socket. Tighten the spark plugs to 15 ft. lbs. (21 Nm).

7. Connect the spark plug wires.

8. Reinstall any components removed for clearance.

9. Connect the negative battery cable.

3.0L & 3.6L Engines

Left Side

See Figures 76 and 77.

1. Before servicing the vehicle, refer to the Precautions Section.

2. Remove the collector cover.

3. Remove the battery and battery carrier.

4. Remove the bracket.

5. Disconnect the connector from ignition coil.

6. Remove the ignition coil.

➡**Turn the no. 6 ignition coil to remove it.**

7. Remove the spark plug with a spark plug socket.

To install:

8. To install reverse the removal procedures, and note the following:

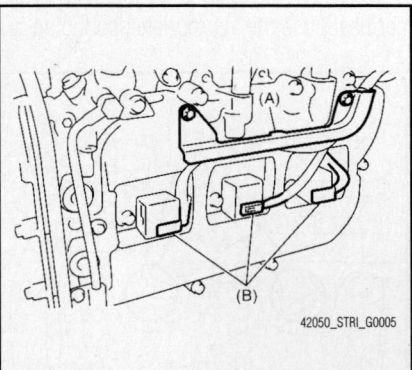

Fig. 76 Removing bracket and connector from the ignition coil—left side

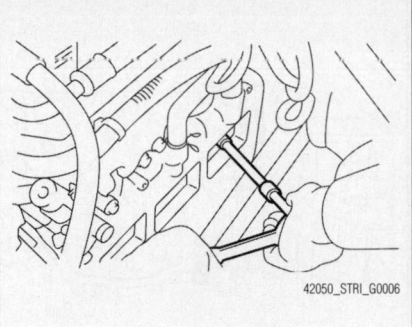

Fig. 77 Removing the spark plug with a spark plug socket—left side

- Tighten the spark plug to 15 ft. lbs. (21 Nm)

- Tighten the ignition coil to 12 ft. lbs. (16 Nm)

➡**The tightening torque described above should be applied to only new spark plugs without oil on their threads. In case their threads are lubricated, the torque should be reduced by approximately⅓of the specified torque in order to avoid over—stressing.**

Right Side

See Figures 78 through 80.

✳✳ CAUTION

All spark plugs installed on an engine must be of the same heat range.

1. Before servicing the vehicle, refer to the Precautions Section.

2. Remove the collector cover.

3. Disconnect the ground cable from battery.

4. Remove the air cleaner case.

5. Remove the bracket.

6. Disconnect the connector from ignition coil.

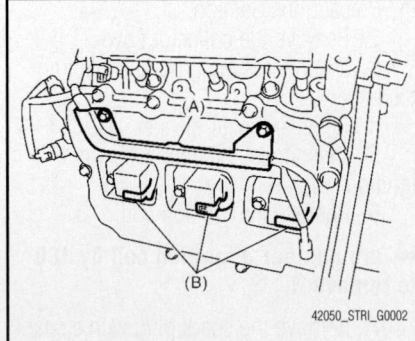

Fig. 78 Removing bracket and connector from ignition coil—right side

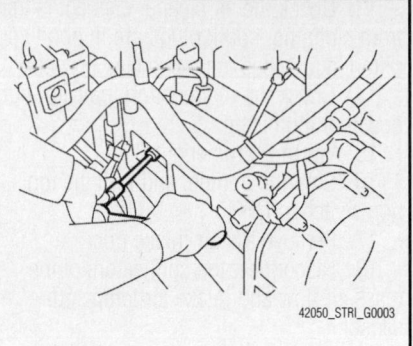

Fig. 79 Remove the spark plug with a spark plug socket—right side

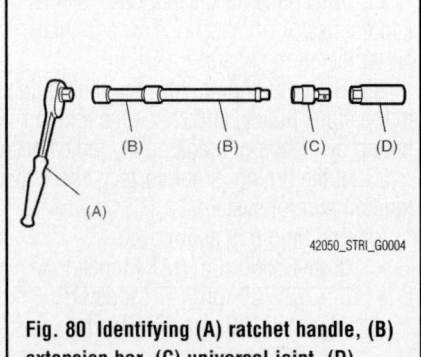

Fig. 80 Identifying (A) ratchet handle, (B) extension bar, (C) universal joint, (D) spark plug socket

7. Remove the ignition coil.

➡**Turn the #5 ignition coil to remove it.**

8. Remove the spark plug with a spark plug socket.

To install:

9. To install reverse the removal procedures, and note the following:

- Tighten the spark plug to 15 ft. lbs. (21 Nm)

- Tighten the ignition coil to 12 ft. lbs. (16 Nm)

➡**The tightening torque described above should be applied to only new spark plugs without oil on their threads. In case their threads are lubricated, the torque should be reduced by approximately⅓of the specified torque in order to avoid over—stressing.**

SPARK PLUG WIRES

REMOVAL & INSTALLATION

Refer to Spark plug "Removal & Installation".

ENGINE ELECTRICAL

STARTER

REMOVAL & INSTALLATION

1. Before servicing the vehicle, refer to the Precautions Section.
2. Disconnect the negative battery cable.
3. Remove the air intake chamber, on non turbocharged engine.

4. Remove the intercooler, on turbocharged engine.
5. Remove the air intake chamber stay, on non turbocharged engine.
6. Disconnect the electrical connectors from the starter.
7. Remove the starter retaining bolts. Remove the starter from the vehicle.

STARTING SYSTEM

To install:
8. Installation is the reverse of the removal procedure.
9. Torque the starter retaining bolts to 37 ft. lbs. (50 Nm).

ENGINE MECHANICAL

ACCESSORY DRIVE BELTS

ADJUSTMENT

2.5L Engines
See Figures 81 and 82.

➡**The belt tension may be checked with or without a belt tension gauge.**

1. If using a belt tension gauge, the tension should be as follows:
   a. Used front belt: 110.2–143.9 lbs. (490–640 N)
   b. New front belt: 143–175 lbs. (640–780 N)
   c. Used rear belt: 78.7–101.2 lbs. (350–450 N)
   d. New rear belt: 166–198 lbs. (740–880 N)
2. If you are not using a belt tension gauge, using 22 lbs. (98 Nm) of force, push on the belt as shown in the illustration. The total distance the belt travels up and down should be:

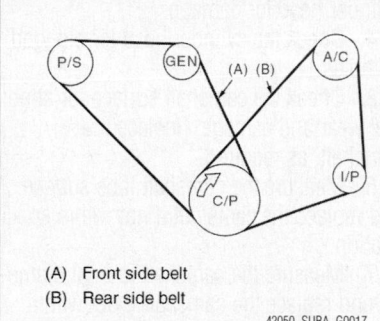

(A) Front side belt
(B) Rear side belt

42050_SUBA_G0017

Fig. 82 Location to check to the belt tension if you are not using a belt tension gauge—2.5L engine

a. Used front belt: 0.354–0.433 in. (9–11 mm)
b. New front belt: 0.276–0.354 in. (7–9 mm)
c. Used rear belt: 0.354–0.394 in. (9.0–10.0 mm)
d. New rear belt: 0.295–0.335 in. (7.5–8.5 mm)

3.0L & 3.6L Engines
See Figure 83.

1. With the belt installed, check that the automatic belt tension indicator is within the range.
2. Replace the belt if the indicator is outside of the service limit.

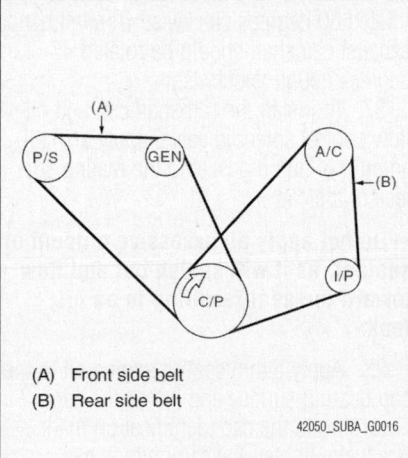

(A) Front side belt
(B) Rear side belt

42050_SUBA_G0016

Fig. 81 Location to check the belt tension if you are using a belt tension gauge—2.5L engine

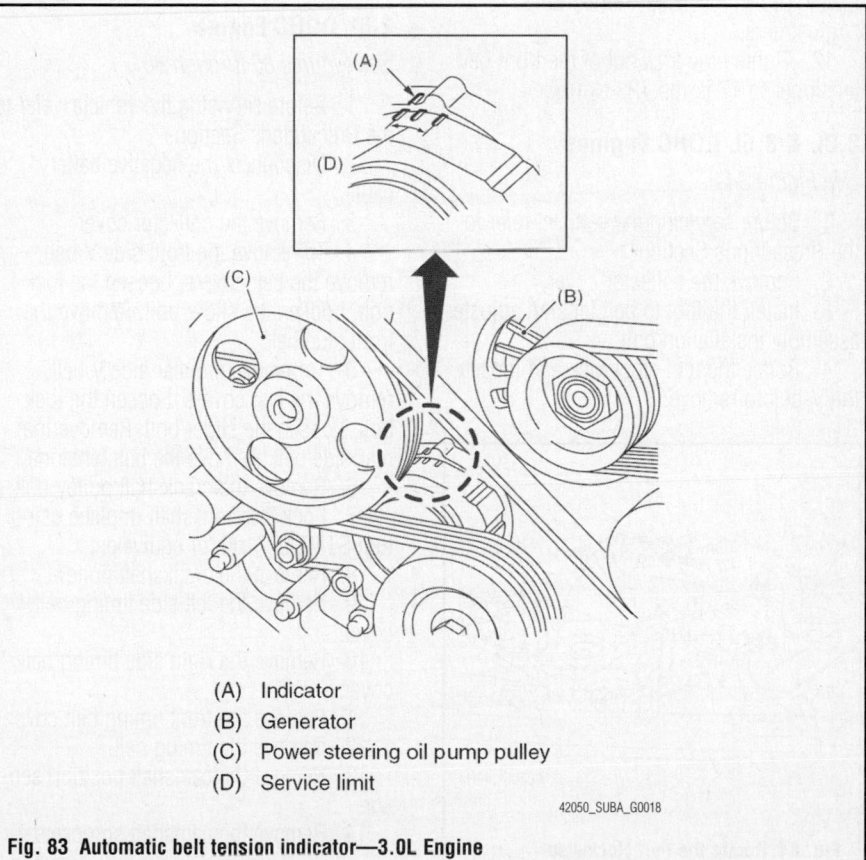

(A) Indicator
(B) Generator
(C) Power steering oil pump pulley
(D) Service limit

42050_SUBA_G0018

Fig. 83 Automatic belt tension indicator—3.0L Engine

REMOVAL & INSTALLATION

2.5L Engine

1. Before servicing the vehicle, refer to the Precautions Section.
2. Remove the accessory drive belt cover, if equipped.
3. Loosen the lock bolt (bottom) of the front belt tensioner.
4. Loosen the slider bolt (top) of the front belt tensioner.
5. Remove the front side belt.
6. Loosen the lock bolt (bottom) of the rear belt tensioner.
7. Loosen the slider bolt (top) of the rear belt tensioner.
8. Remove the rear side belt.

To install:

➡ **Wipe off any oil or water on the belt and pulley.**

9. Install the rear side belt, then tighten the slider bolt until the belt reaches the specified tension. Please refer to the following topic: "Accessory Drive Belts", "Adjustment."
10. Tighten the lock nut of the rear belt tensioner to 18 ft. lbs. (25 Nm).
11. Install the front side belt, then tighten the slider bolt until the belt reaches the specified tension. Please refer to the following topic: "Accessory Drive Belts", "Adjustment."
12. Tighten the lock nut of the front belt tensioner to 17 ft. lbs. (23 Nm).

3.0L & 3.6L DOHC Engines

See Figure 84.

1. Before servicing the vehicle, refer to the Precautions Section.
2. Remove the collector cover.
3. Install the tool to belt tension adjuster assembly installation bolt.
4. Rotate the tool clockwise and loosen the V-belt to remove.

22140_SUBA_G0117

Fig. 84 Rotate the tool clockwise

5. Remove the accessory drive belt.

To install:

6. Installation is the reverse order of removal.

CAMSHAFT AND VALVE LIFTERS

INSPECTION

1. Remove the camshaft from the engine.
2. Check the camshaft bearing journals for damage and binding.
3. If the journals are binding, check the cylinder head for damage.
4. Check the cylinder head for clogged oil holes.
5. Check the camshaft surface for abnormal wear and damage. Replace the camshaft, as required.
6. Measure the camshaft lobe surface and replace the camshaft if not within specification.
7. Measure the camshaft journal diameter and replace the camshaft if not within specification.
8. Measure the camshaft run out and replace the camshaft if not within specification.

REMOVAL & INSTALLATION

2.5L DOHC Engine

See Figures 85 through 88.

1. Before servicing the vehicle, refer to the Precautions Section.
2. Disconnect the negative battery cable.
3. Remove the collector cover.
4. To remove the front side V belt, remove the belt covers. Loosen the lock bolt. Loosen the slider bolt. Remove the front side belt.
5. To remove the rear side V belt, remove the belt covers. Loosen the lock bolt. Loosen the slider bolt. Remove the rear side belt. Remove the belt tensioner.
6. Remove the crankshaft pulley bolt.
7. Lock the crankshaft in place using tool ST499977100, or equivalent.
8. Remove the crankshaft pulley.
9. Remove the left side timing belt cover.
10. Remove the right side timing belt cover.
11. Remove the front timing belt cover.
12. Remove the timing belt.
13. Remove the camshaft position sensor.
14. Remove the camshaft sprockets.

➡ **Be sure to lock the camshaft in place using tool ST499207400, or equivalent.**

15. Lock the crankshaft in place using tool ST499977100, or equivalent.
16. Remove the crankshaft pulley.
17. Remove the tensioner bracket. Remove the right and left timing belt No.2 covers.
18. Remove the spark plug wires. Remove the oil level gauge, on the left side.
19. Remove the rocker cover and gasket. Remove the oil pipe.
20. Loosen the oil flow control solenoid valve assembly and the intake camshaft cap bolts equally and in the proper sequence.
21. Loosen the exhaust camshaft cap bolts equally and in the proper sequence.
22. Remove the oil flow control solenoid valve assembly, intake camshaft cap and camshaft.
23. Remove the exhaust camshaft caps and camshaft.

➡ **Arrange the camshafts caps so that they can be installed in their original positions.**

To install:

➡ **Lubricate the camshaft journals with clean engine oil prior to installation.**

24. Install the camshaft so that the valves are closed or in contact with the "base circle" of the cam lobe.
25. If the camshafts are positioned as shown in the, the camshafts need to be rotated at a minimum to align the timing belt during installation.
26. The right hand camshaft need not be rotated when set at the position illustrated, the left hand intake camshaft should be rotated 80 degrees clockwise. The left hand exhaust camshaft should be rotated 45 degrees counterclockwise.
27. To install the camshaft cap and oil flow control solenoid valve, apply a small amount of liquid gasket to the mating surface of the cap.

➡ **Do not apply an excessive amount of sealant, as it will squish out and flow toward the seal resulting in an oil leak.**

28. Apply a thin coat of engine oil to the cap bearing surface and install the cap according to the cap identification mark. Gradually tighten the cap bolts in two stages, first to 7.2 ft. lbs. and then to 14.5 ft. lbs. in the proper sequence.

➡ **After tightening the camshaft cap, ensure that the camshaft rotates slightly while holding it at base circle.**

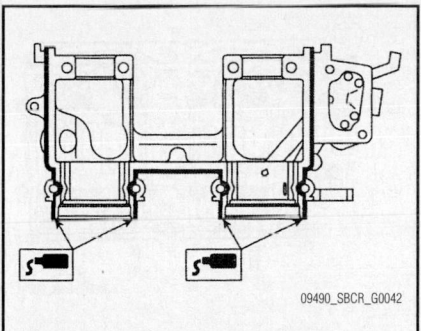

Fig. 85 Camshaft cap liquid gasket application points—2.5L DOHC engine

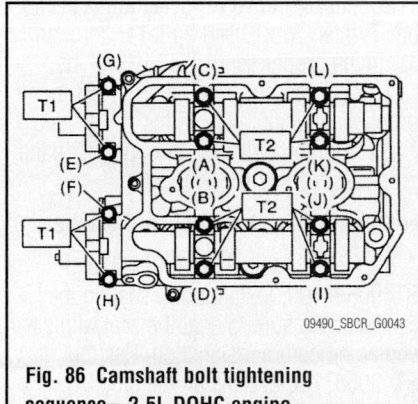

Fig. 86 Camshaft bolt tightening sequence—2.5L DOHC engine

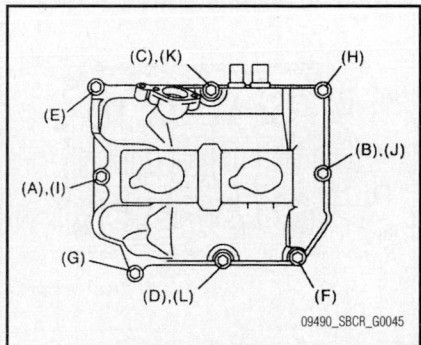

Fig. 88 Rocker cover tightening sequence—2.5L DOHC engine

29. Using tools ST49587600 and ST499597200, install a new seal on the camshaft. Be sure to coat the seal with clean engine oil before installation. Use tool ST499587700 and install the plug.

30. Install a new gasket on the rocker cover. Apply liquid gasket to the cylinder head (see illustration).

➡ **Apply an extra amount of liquid gasket around the semicircular plugs, 5mm or more.**

31. Temporarily tighten the rocker cover retaining bolts, in the proper sequence, and then tighten to 4.7 ft. lbs. Install the oil pipe.

32. Continue the installation in the reverse order of the removal procedure.

2.5L SOHC Engine

See Figures 89 through 96.

1. Before servicing the vehicle, refer to the Precautions Section.

2. Disconnect the negative battery cable.

3. To remove the front side V belt, remove the belt covers. Loosen the lock bolt. Loosen the slider bolt. Remove the front side belt.

4. To remove the rear side V belt, remove the belt covers. Loosen the lock bolt. Loosen the slider bolt. Remove the rear side belt. Remove the belt tensioner.

5. Remove the crankshaft pulley bolt.

6. Lock the crankshaft in place using tool ST499977100, or equivalent.

7. Remove the crankshaft pulley.

8. Remove the left side timing belt cover.

9. Remove the front timing belt cover.

10. Remove the timing belt.

11. Remove the camshaft position sensor.

12. Remove the camshaft sprockets.

➡ **Be sure to lock the camshaft in place using tool ST18231AA010, or equivalent.**

13. Remove the left and right timing belt N02 covers.

➡ **Do not damage or lose the rubber seal when removing the covers.**

14. Remove the tensioner bracket. Remove the camshaft position sensor support, on the left side.

15. Remove the oil level gauge guide, on the left side.

16. Remove the valve rocker arm assembly.

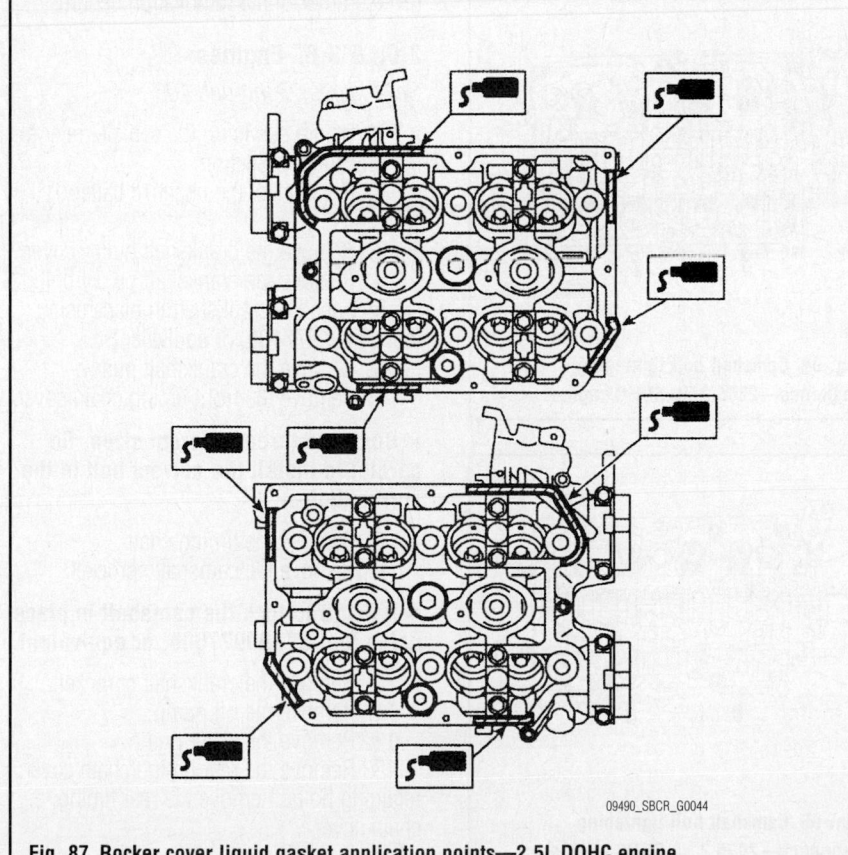

Fig. 87 Rocker cover liquid gasket application points—2.5L DOHC engine

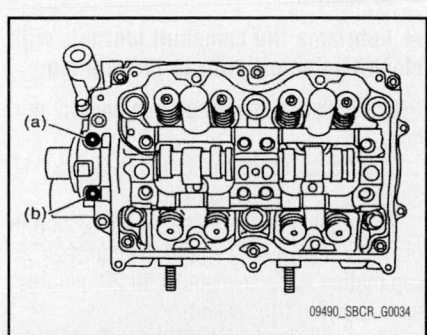

Fig. 89 Camshaft bolt loosening sequence—2.5L SOHC engine

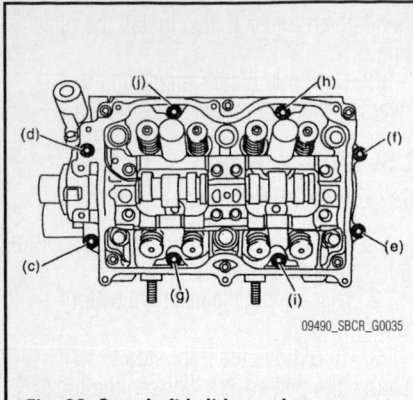

Fig. 90 Camshaft bolt loosening sequence—2.5L SOHC engine

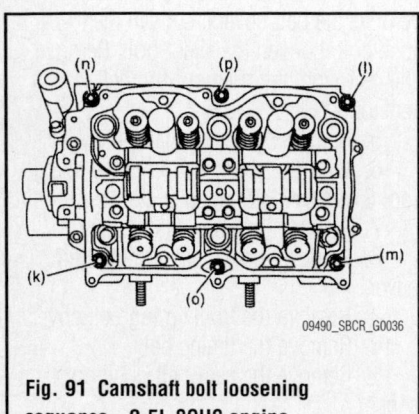

Fig. 91 Camshaft bolt loosening sequence—2.5L SOHC engine

17. Remove camshaft cap retaining bolts "A" and "B" in the proper sequence.

18. Loosen camshaft cap bolts "C" through "J" all the way in the proper sequence.

19. Remove camshaft cap bolts "K" through "P" in the proper sequence using a Torx® head bit.

20. Remove the camshaft caps.

21. Remove the camshaft. Remove the oil seal. Remove the plug from the rear side of the camshaft.

➡**Do not remove the oil seal unless necessary.**

To install:

➡ **Lubricate the camshaft journals with clean engine oil prior to installation.**

22. Install the camshaft into the cylinder head.

23. Apply liquid gasket to the mating surfaces of the camshaft cap.

24. Apply a bead of sealant (0.12 inch in diameter) along the edge of the camshaft cap mating surface. Install with 20 minutes after applying the sealant.

25. Temporarily tighten the bolts "A" through "D" in the proper sequence.

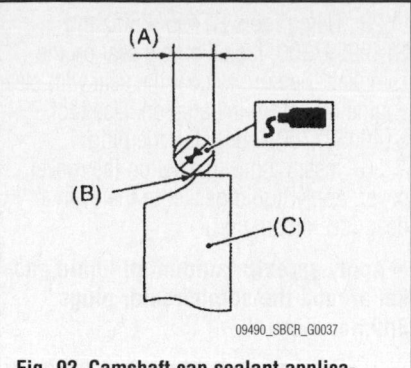

Fig. 92 Camshaft cap sealant application—2005 2.5L SOHC engine

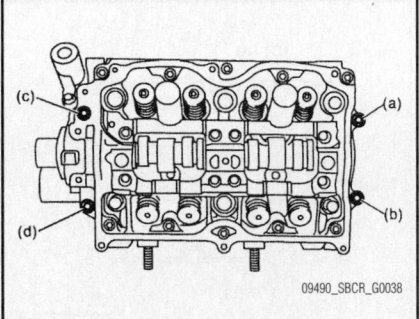

Fig. 93 Camshaft bolt tightening sequence—2005 2.5L SOHC engine

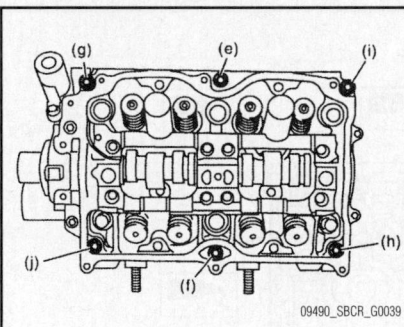

Fig. 94 Camshaft bolt tightening sequence—2005 2.5L SOHC engine

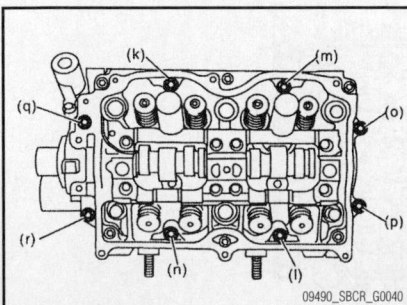

Fig. 95 Camshaft bolt tightening sequence—2005 2.5L SOHC engine

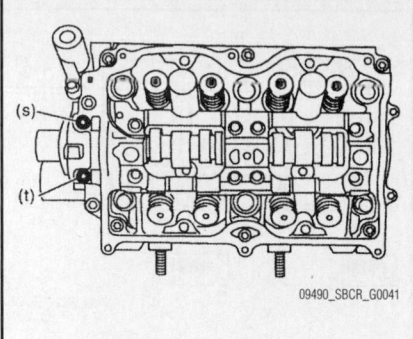

Fig. 96 Camshaft bolt tightening sequence—2005 2.5L SOHC engine

26. Install the valve rocker arm assembly. Tighten Torx® head bolts "E" through "J" in the proper sequence to 13 ft. lbs.

27. Tighten bolts "K" through "R" in the proper sequence to 7.2 ft. lbs.

28. Tighten bolts "S" through "T" in the proper sequence to 7.2 ft. lbs.

➡**Be sure to use a new seal washer.**

29. Using tools ST499597000 and ST499587500, install a new seal on the camshaft. Be sure to coat the seal with clean engine oil before installation. Use tool ST499587700 and install the plug.

30. Adjust the valve clearance.

31. Continue the installation in the reverse order of the removal procedure.

3.0L & 3.6L Engines

See Figures 97 through 106.

1. Before servicing the vehicle, refer to the Precautions Section.

2. Disconnect the negative battery cable.

3. Remove the crankshaft pulley cover.

4. Remove the crankshaft pulley bolt.

5. Lock the crankshaft in place using tool ST499977100, or equivalent.

6. Remove the crankshaft pulley.

7. Remove the front timing chain cover.

➡**Bolts are three different sizes. Be careful to install the correct bolt in the correct hole.**

8. Remove the timing chain.

9. Remove the camshaft sprocket.

➡**Be sure to lock the camshaft in place using tool ST499977500, or equivalent.**

10. Remove the crankshaft sprocket.

11. Remove the oil pump.

12. Remove the water pump.

13. Remove the rear timing chain cover retaining bolts. Remove the rear timing chain cover.

➡ **There are seven different size bolts. Be sure not to confuse them on installation.**

14. Disconnect the oil pipe.

15. Remove the rocker cover retaining bolts. Remove the rocker cover. Discard the gasket.

16. Remove the plugs, see illustration for location.

17. Loosen the camshaft cap bolts in the proper sequence. Remove the camshaft caps and remove the camshaft.

To install:

18. Apply a coat of engine oil to the journals on the camshafts and place the camshafts into position.

19. To install the camshaft cap, apply a small amount of liquid gasket to the mating surface of the cap. Do not apply an excessive amount of sealant, as it will squish out and flow toward the cam journal resulting in engine seizure.

20. Apply a thin coat of engine oil to the cap bearing surface and install the cap. Tighten cap bolts 1 through 12 to 12 ft. lbs. and bolts 13 through 16 to 7.2 ft. lbs. in the proper sequence. Install the plugs and torque to 44 ft. lbs.

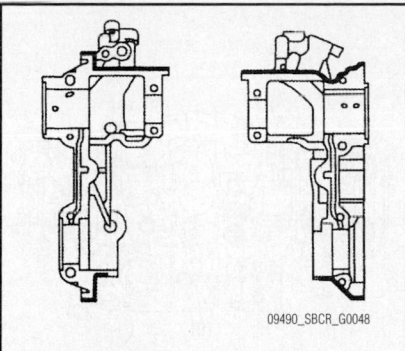

Fig. 99 Camshaft cap liquid gasket application points—3.0L engine

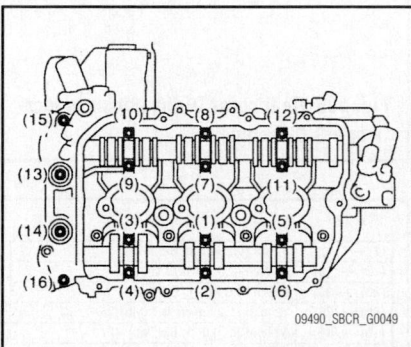

Fig. 100 Camshaft bolt tightening sequence—3.0L engine

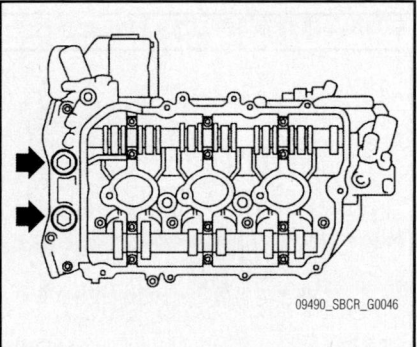

Fig. 97 Camshaft plug bolt location—3.0L engine

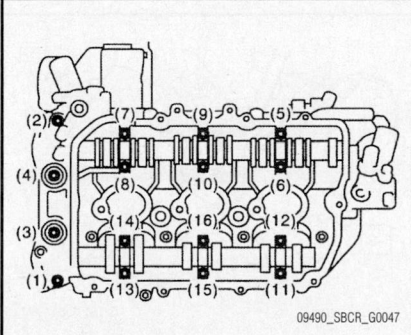

Fig. 98 Camshaft bolt loosening sequence—3.0L engine

21. Apply fluid gasket maker of the of the cylinder heads and valve covers.

❋❋ **CAUTION**

Do not apply too much gasket maker. This may cause excess fluid gasket maker to come out and flow towards the camshaft journal resulting in engine damage.

22. Tighten the valve cover bolts to 4.7 ft. lbs. and in the proper sequence.

23. Install the rear chain cover as follows:

➡ **There are several size bolts used, refer to the illustration for size and locations.**

a. Rear chain cover gasket and clean the mating surfaces.

b. Apply liquid gasket maker to the mating surfaces of the cover. Refer to the illustration for gasket maker application and diameter.

c. Install new O-rings. Refer to the illustration for O-ring location and size

d. Install the rear chain cover and temporarily tighten the bolts, refer to the illustration for size and locations.

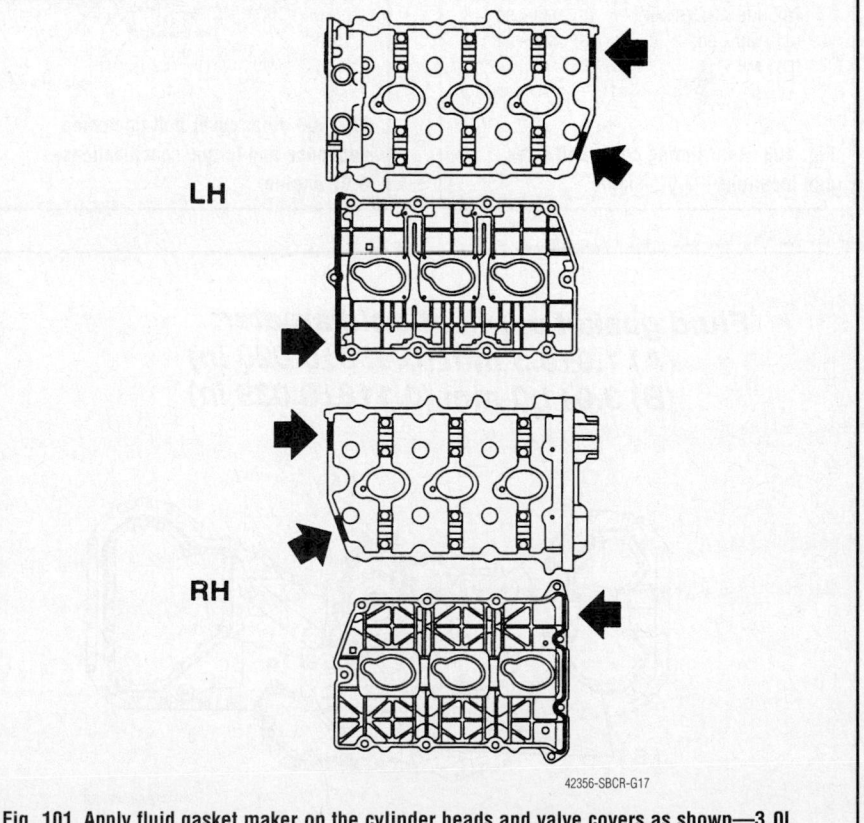

Fig. 101 Apply fluid gasket maker on the cylinder heads and valve covers as shown—3.0L engine

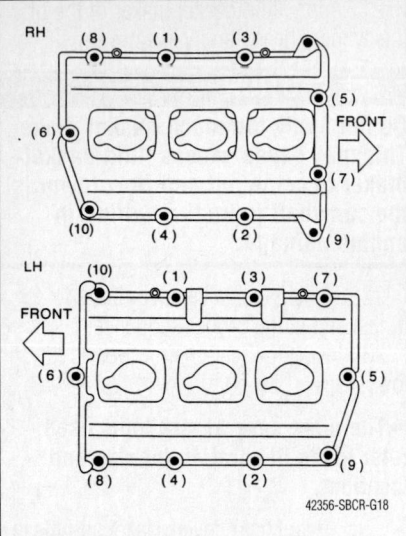

Fig. 102 Tighten the valve cover bolts in this sequence—3.0L engine

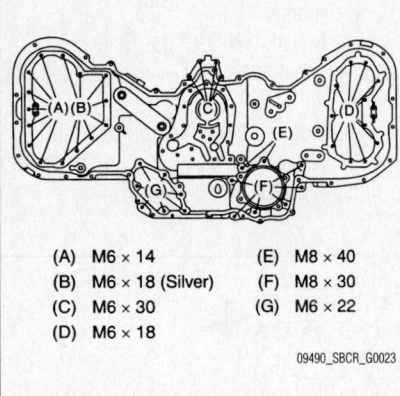

(A) M6 × 14   (E) M8 × 40
(B) M6 × 18 (Silver) (F) M8 × 30
(C) M6 × 30   (G) M6 × 22
(D) M6 × 18

09490_SBCR_G0023

Fig. 103 Rear timing chain bolt sizes and locations—3.0L engine

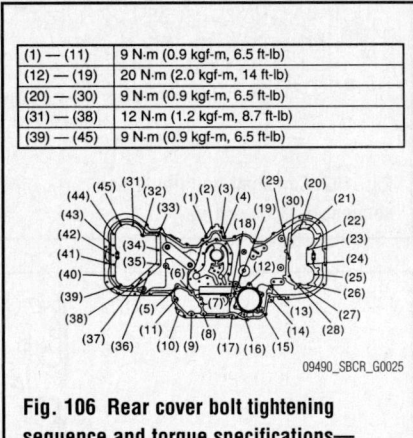

(A) 14.2 × 1.9   (C) 25 × 2
(B) 19.2 × 2.4   (D) 31.2 × 1.9

09490_SBCR_G0024

Fig. 105 Rear cover O-ring sizes and locations—3.0L engine

(1) — (11)	9 N·m (0.9 kgf-m, 6.5 ft-lb)
(12) — (19)	20 N·m (2.0 kgf-m, 14 ft-lb)
(20) — (30)	9 N·m (0.9 kgf-m, 6.5 ft-lb)
(31) — (38)	12 N·m (1.2 kgf-m, 8.7 ft-lb)
(39) — (45)	9 N·m (0.9 kgf-m, 6.5 ft-lb)

09490_SBCR_G0025

Fig. 106 Rear cover bolt tightening sequence and torque specifications—3.0L engine

e. Tighten the cover bolts in the sequence illustrated to the specifications shown in the illustration.

24. Continue the installation in the reverse order of the removal procedure.

CRANKSHAFT DAMPER

REMOVAL & INSTALLATION

See Figures 107 through 109.

1. Before servicing the vehicle, refer to the Precautions Section.

2. Remove the accessory drive belt.

3. Remove the crankshaft damper cover, 3.0L engines only.

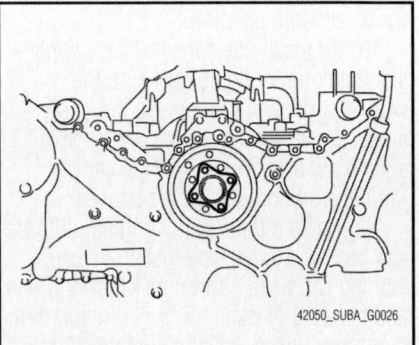

42050_SUBA_G0026

Fig. 107 3.0L crankshaft damper and cover shown

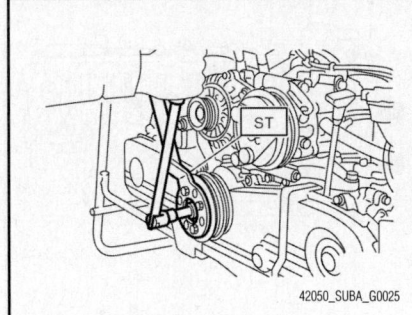

42050_SUBA_G0025

Fig. 108 Special Tool 499977100–2.5L engine shown

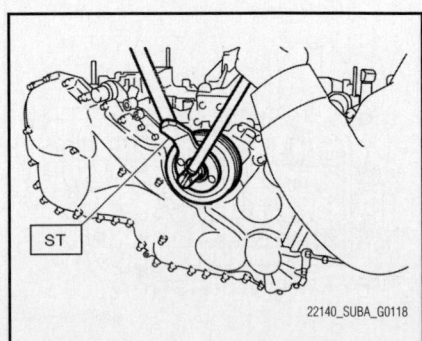

22140_SUBA_G0118

Fig. 109 Special Tool 499977100–3.0L engine shown

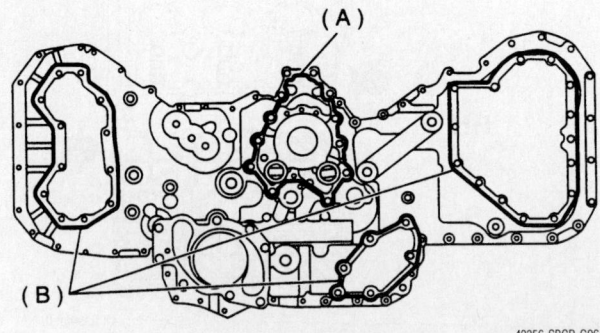

Fluid gasket application diameter:
(A) 1.0±0.5 mm (0.039±0.020 in)
(B) 3.0±1.0 mm (0.118±0.039 in)

42356-SBCR-G06

Fig. 104 Apply liquid gasket maker to the mating surfaces of the cover as shown—3.0L engine

4. Using Special Tool 49997700 to lock the crankshaft in place, remove the crankshaft pulley bolt.

5. Remove the crankshaft pulley.

To install:

6. Clean the crankshaft pulley threads with an air gun. Apply engine oil to the crankshaft pulley bolt seat and threads.

7. Install the crankshaft pulley into place. Use Special Tool 499977100 to lock the crankshaft pulley in place and install the crankshaft pulley bolt. Tighten the bolt as follows

- 2.5L engines: 123–137 ft. lbs. (167–187 Nm).
- 3.0L engines: 131 ft. lbs. (178 Nm).

8. Install the crankshaft damper cover, 3.0L engines only. Tighten to 4.7 ft. lbs. (6.4 Nm).

9. Install drive belt.

CYLINDER HEAD

REMOVAL & INSTALLATION

2.5L Engine

See Figures 110 through 115.

1. Before servicing the vehicle, refer to the Precautions Section.

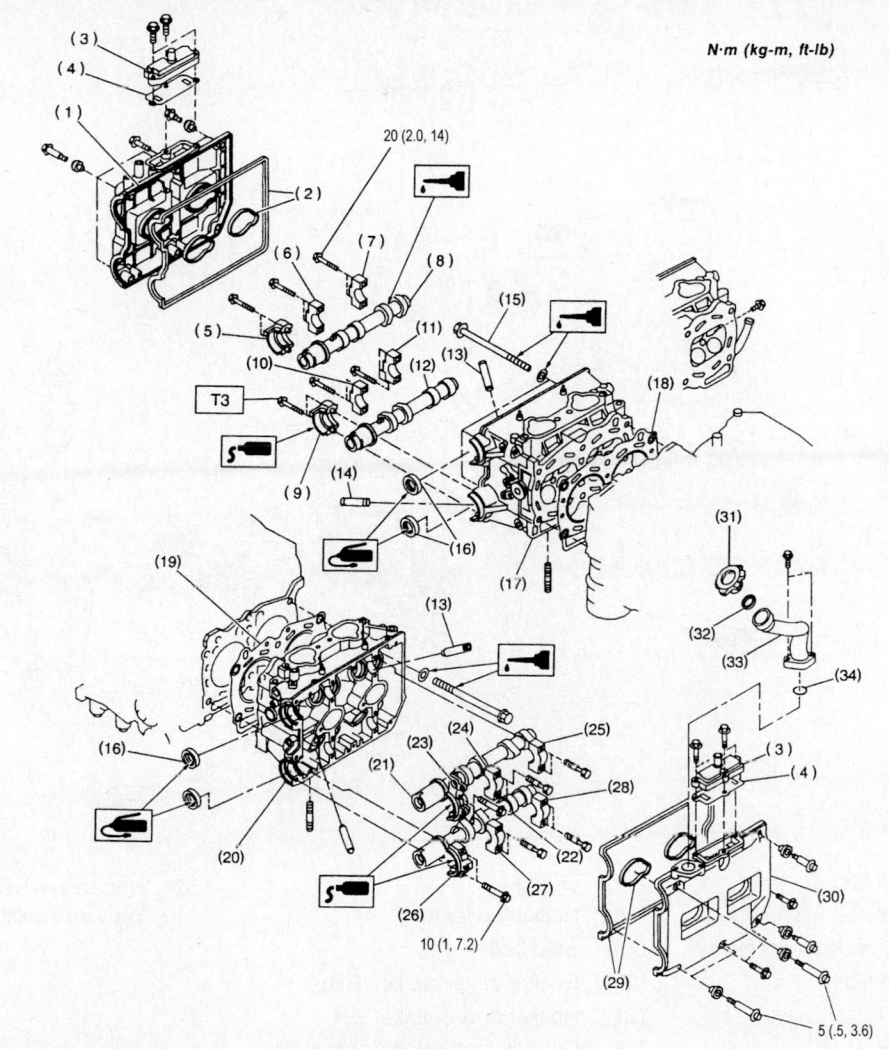

N·m (kg-m, ft-lb)

(1) Rocker cover (RH)	(15) Cylinder head bolt
(2) Rocker cover gasket (RH)	(16) Oil seal
(3) Oil separator cover	(17) Cylinder head (RH)
(4) Gasket	(18) Cylinder head gasket (RH)
(5) Intake camshaft cap (Front RH)	(19) Cylinder head gasket (LH)
(6) Intake camshaft cap (Center RH)	(20) Cylinder head (LH)
(7) Intake camshaft cap (Rear RH)	(21) Intake camshaft (LH)
(8) Intake camshaft (RH)	(22) Exhaust camshaft (LH)
(9) Exhaust camshaft cap (Front RH)	(23) Intake camshaft cap (Front LH)
(10) Exhaust camshaft cap (Center RH)	(24) Intake camshaft cap (Center LH)
(11) Exhaust camshaft cap (Rear RH)	(25) Intake camshaft cap (Rear LH)
(12) Exhaust camshaft (RH)	(26) Exhaust camshaft (Front LH)
(13) Intake valve guide	(27) Exhaust camshaft cap (Center LH)
(14) Exhaust valve guide	(28) Exhaust camshaft cap (Rear LH)

(29) Rocker cover gasket (LH)
(30) Rocker cover (LH)
(31) Oil filler cap
(32) Gasket
(33) Oil filler duct
(34) O-ring

7923TG07

Fig. 110 Cylinder head and related components—2.5L DOHC engine

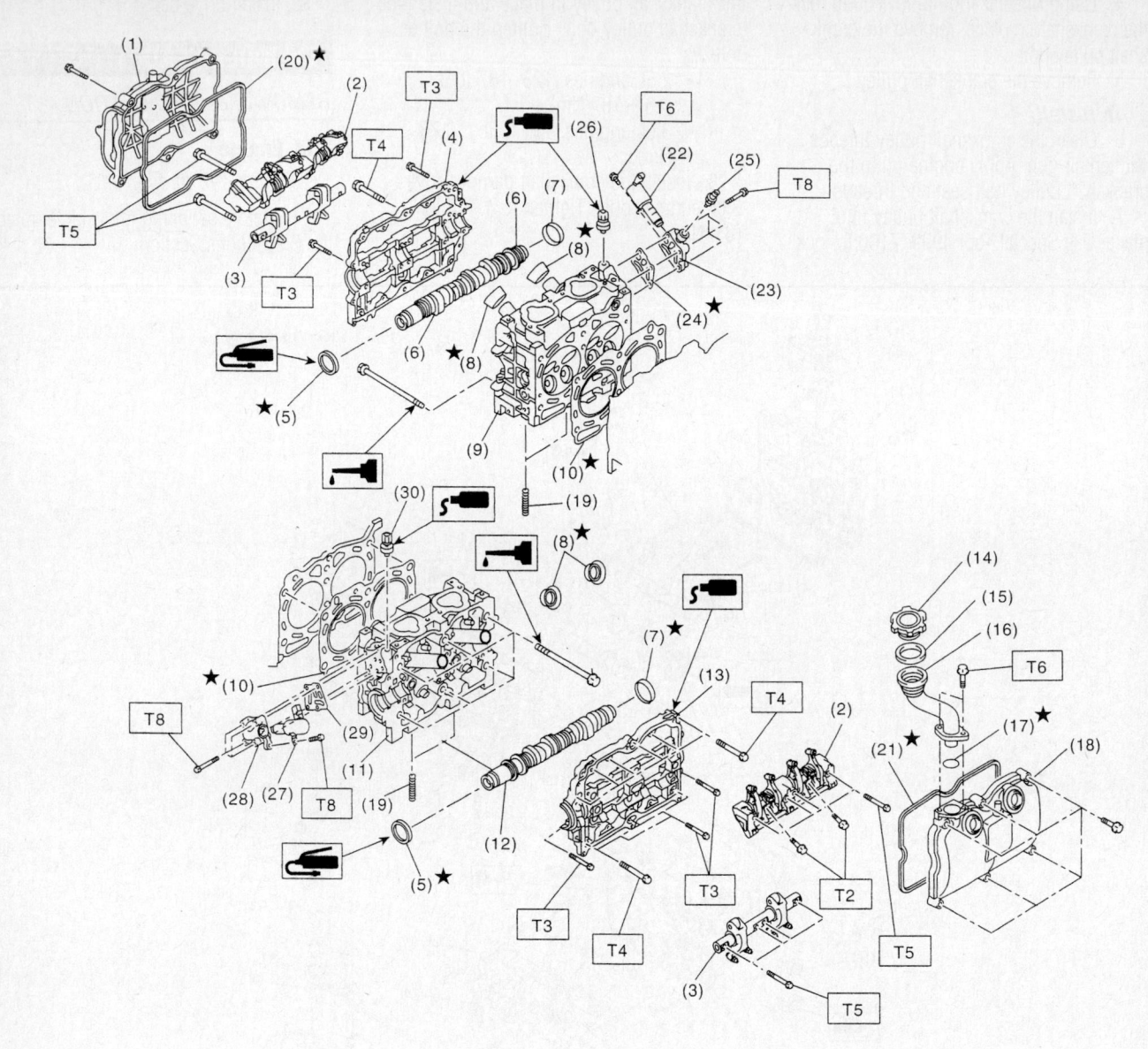

(1) Rocker cover (RH)

(2) Intake valve rocker assembly

(3) Exhaust valve rocker assembly

(4) Camshaft cap (RH)

(5) Oil seal

(6) Camshaft (RH)

(7) Plug

(8) Spark plug pipe gasket

(9) Cylinder head (RH)

(10) Cylinder head gasket

(11) Cylinder head (LH)

(12) Camshaft (LH)

(13) Camshaft cap (LH)

(14) Oil filler cap

(15) Gasket

(16) Oil filler duct

(17) O-ring

(18) Rocker cover (LH)

(19) Stud bolt

(20) Rocker cover gasket (RH)

(21) Rocker cover gasket (LH)

(22) Oil switching solenoid valve (RH)

(23) Oil switching solenoid valve holder (RH)

(24) Gasket

(25) Oil temperature sensor

(26) Variable valve lift diagnosis oil pressure switch (RH)

(27) Oil switching solenoid valve (LH)

(28) Oil switching solenoid valve holder (LH)

(29) Gasket

(30) Variable valve lift diagnosis oil pressure switch (LH)

Tightening torque: N·m (kgf-m, ft-lb)

T3: 9.75 (1.0, 7.2)

T4: 18 (1.8, 13.0)

T5: 25 (2.5, 18.1)

T6: 6.4 (0.65, 4.7)

T7: 8 (0.8, 5.9)

T8: 10 (1.0, 7.4)

Fig. 111 Cylinder head and related components—2.5L SOHC engine

09490_SBCR_G0016

2. Disconnect the negative battery cable.

3. If equipped with DOHC engine, remove the collector cover.

4. To remove the front side V belt, remove the belt covers. Loosen the lock bolt. Loosen the slider bolt. Remove the front side belt.

5. To remove the rear side V belt, remove the belt covers. Loosen the lock bolt. Loosen the slider bolt. Remove the rear side belt. Remove the belt tensioner.

6. Remove the crankshaft pulley bolt.

7. Lock the crankshaft in place using tool ST499977100, or equivalent.

8. Remove the crankshaft pulley.

9. Remove the left side timing belt cover.

10. If equipped with DOHC engine, remove the right side timing belt cover.

11. Remove the front timing belt cover.

12. Remove the timing belt.

13. Remove the camshaft position sensor.

14. Remove the camshaft sprockets.

➡**Be sure to lock the camshaft in place using tool ST18231AA010 (SOHC engine) and ST499207400 (DOHC engine), or equivalent.**

15. Remove the intake manifold.

16. If equipped with SOHC engine, remove the bolt that retains the air conditioning compressor bracket to the cylinder head.

17. Remove the rocker cover retaining bolts. Remove the rocker cover.

18. If equipped with SOHC engine, remove the rocker arm assembly.

19. If equipped with SOHC engine remove the camshaft. If equipped with DOHC engine, remove the camshafts.

20. Remove the cylinder head, in the proper sequence. On the DOHC engine leave bolts A and D installed loosely to prevent the cylinder head from falling. On the SOHC engine leave bolts A and C installed loosely to prevent the cylinder head from falling.

21. Loosen the cylinder head from the block using a plastic-faced hammer, if needed.

22. Remove bolts A and D on the DOHC engine and bolts A and C on the SOHC engine. Remove the cylinder head from the engine. Discard the gasket

23. Clean all gasket material from both mating surfaces.

To install:

24. Installation is the reverse of the removal procedure.

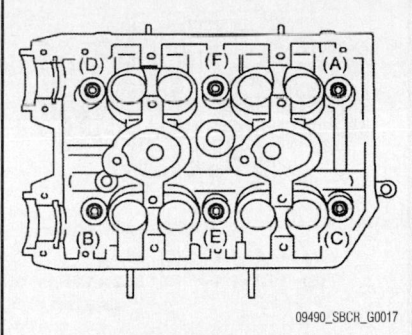

Fig. 112 Cylinder head bolt loosening sequence—2.5L DOHC engine

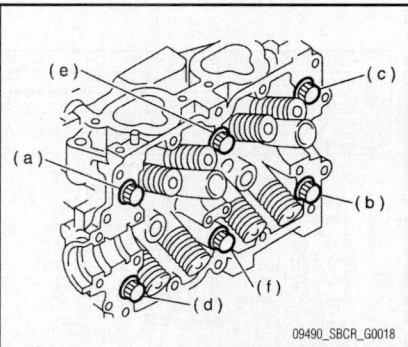

Fig. 113 Cylinder head bolt loosening sequence—2.5L SOHC engine

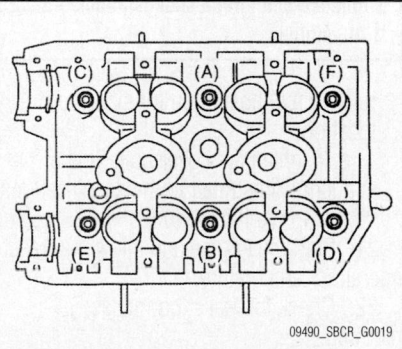

Fig. 114 Cylinder head bolt tightening sequence—2.5L DOHC engine

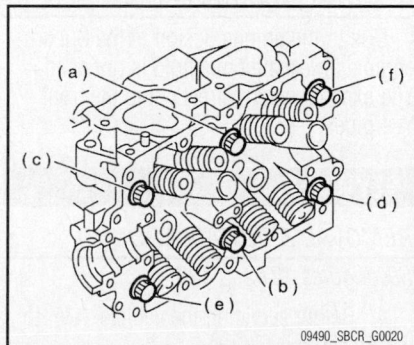

Fig. 115 Cylinder head bolt tightening sequence—2.5L SOHC engine

25. Apply a thin coat of clean engine oil to the washers and cylinder head bolts.

26. Tighten the cylinder head retaining bolts to specification and in the proper sequence.

27. Start the engine and allow it to reach operating temperature.

28. Check for leaks, correct as required.

3.0L & 3.6L Engines

See Figures 116 through 120.

1. Before servicing the vehicle, refer to the Precautions Section.

2. Remove the crankshaft pulley cover.

3. Remove the crankshaft pulley bolt.

4. Lock the crankshaft in place using tool ST499977100, or equivalent.

5. Remove the crankshaft pulley.

6. Remove the front timing chain cover.

➡**Bolts are three different sizes. Be careful to install the correct bolt in the correct hole.**

7. Remove the timing chain.

8. Remove the camshaft sprocket.

➡**Be sure to lock the camshaft in place using tool ST499977500, or equivalent.**

9. Remove the crankshaft sprocket.

10. Remove the oil pump.

11. Remove the water pump.

12. Remove the rear timing chain cover retaining bolts. Remove the rear timing chain cover.

➡**There are seven different size bolts. Be sure not to confuse them on installation.**

13. Remove the camshafts.

14. Remove the cylinder head bolts in the proper sequence. Leave bolts 2 and 4 connected by a few threads to prevent the head from falling. Tap the head with a plastic mallet to separate it from the block.

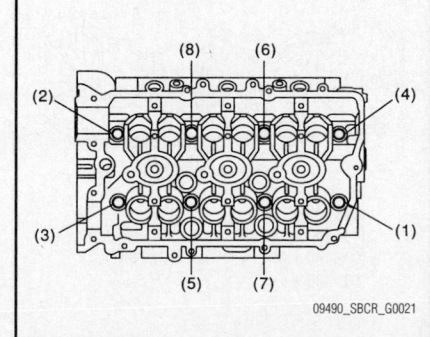

Fig. 116 Cylinder head bolt loosening sequence—3.0L engine

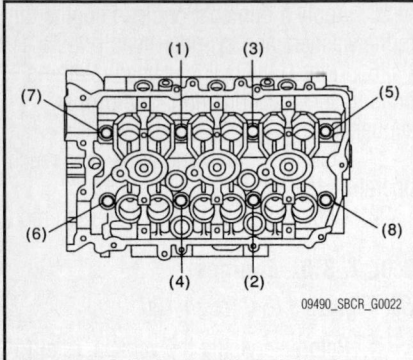

Fig. 117 Cylinder head bolt tightening sequence—3.0L engine

15. Remove bolts 2 and 4 from the cylinder head. Remove the cylinder head from the engine. Discard the gasket.

16. Clean all gasket material from both mating surfaces.

To install:

17. Installation is the reverse of the removal procedure.

18. Apply a thin coat of clean engine oil to the washers and cylinder head bolts.

19. Tighten the cylinder head retaining bolts to specification and in the proper sequence.

20. Install the rear chain cover as follows:

➡**There are several size bolts used, refer to the illustration for size and locations.**

a. Rear chain cover gasket and clean the mating surfaces.

b. Apply liquid gasket maker to the mating surfaces of the cover. Refer to the illustration for gasket maker application and diameter.

c. Install new O-rings. Refer to the illustration for O-ring location and size

d. Install the rear chain cover and temporarily tighten the bolts,

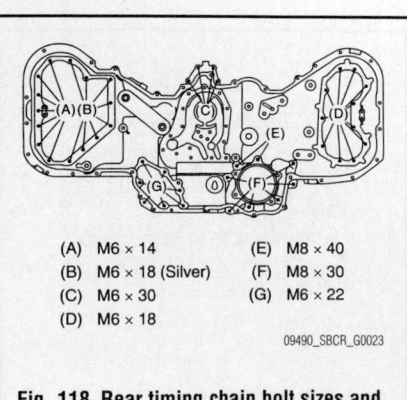

(A)	M6 × 14	(E)	M8 × 40
(B)	M6 × 18 (Silver)	(F)	M8 × 30
(C)	M6 × 30	(G)	M6 × 22
(D)	M6 × 18		

Fig. 118 Rear timing chain bolt sizes and locations—3.0L engine

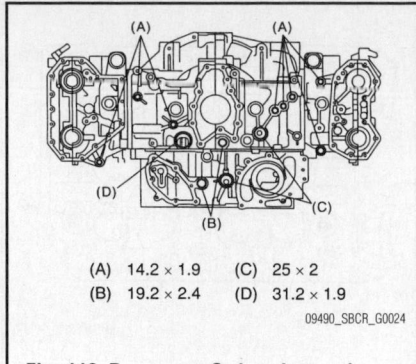

| (A) | 14.2 × 1.9 | (C) | 25 × 2 |
| (B) | 19.2 × 2.4 | (D) | 31.2 × 1.9 |

Fig. 119 Rear cover O-ring sizes and locations—3.0L engine

(1) — (11)	9 N·m (0.9 kgf-m, 6.5 ft-lb)
(12) — (19)	20 N·m (2.0 kgf-m, 14 ft-lb)
(20) — (30)	9 N·m (0.9 kgf-m, 6.5 ft-lb)
(31) — (38)	12 N·m (1.2 kgf-m, 8.7 ft-lb)
(39) — (45)	9 N·m (0.9 kgf-m, 6.5 ft-lb)

Fig. 120 Rear cover bolt tightening sequence and torque specifications—3.0L engine

refer to the illustration for size and locations.

e. Tighten the cover bolts in the sequence illustrated to the specifications shown in the illustration.

21. Start the engine and allow it to reach operating temperature.

22. Check for leaks, correct as required.

EXHAUST MANIFOLD

REMOVAL & INSTALLATION

Due to the unique design of the Subaru engine, an exhaust manifold is not used. The exhaust enters directly into the front Y—pipe.

FLYWHEEL

REMOVAL & INSTALLATION

See Figures 121 and 122.

1. Before servicing the vehicle, refer to the Precautions Section.

2. Remove the engine and place on a suitable engine stand.

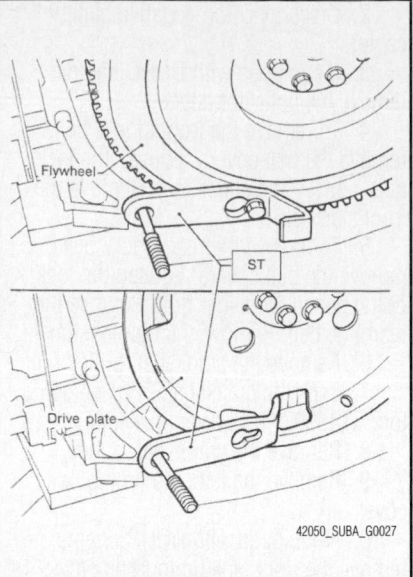

Fig. 121 Use Special Tool 498497100 to lock the crankshaft into place and remove the flywheel

3. If equipped with a manual transmission, remove the clutch housing cover.

4. Using Special Tool 498497100 to lock the crankshaft into place, remove the flywheel (manual transmission) or driveplate (automatic transmission).

To install:

5. Install the flywheel (M/T) or drive plate (A/T) and use Special Tool 498497100 to lock the crankshaft into place. Tighten the bolts in the sequence shown as follows:

a. Except 3.0L engine: 51–55 ft. lbs. (69–75 Nm).

b. 3.0L engine: 60 ft. lbs. (81 Nm).

6. Install the clutch housing cover, if equipped with manual transmission.

7. Install the engine assembly into the vehicle.

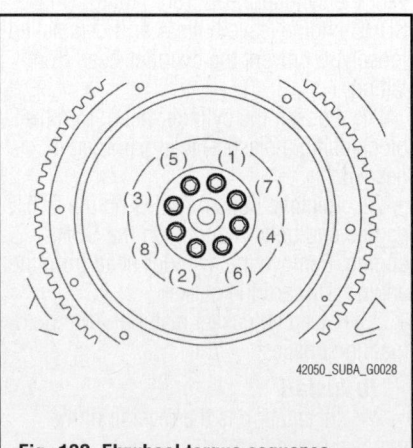

Fig. 122 Flywheel torque sequence

INTAKE MANIFOLD

REMOVAL & INSTALLATION

2.5L Engines

See Figures 123 and 124.

1. Before servicing the vehicle, refer to the Precautions Section.

2. Properly relieve the fuel system pressure. Remove the fuel cap.

3. Disconnect the negative battery cable. Drain the engine coolant.

4. If equipped, remove the undercover.

5. Remove the air intake duct, air cleaner case and air intake chamber.

6. If equipped, remove the intercooler.

7. Remove the alternator.

8. If equipped with DOHC engine, remove the coolant filler tank.

9. If equipped with SOHC engine, remove the spark plug wires.

10. Disconnect the engine coolant hoses from the throttle body. Disconnect the brake booster hose.

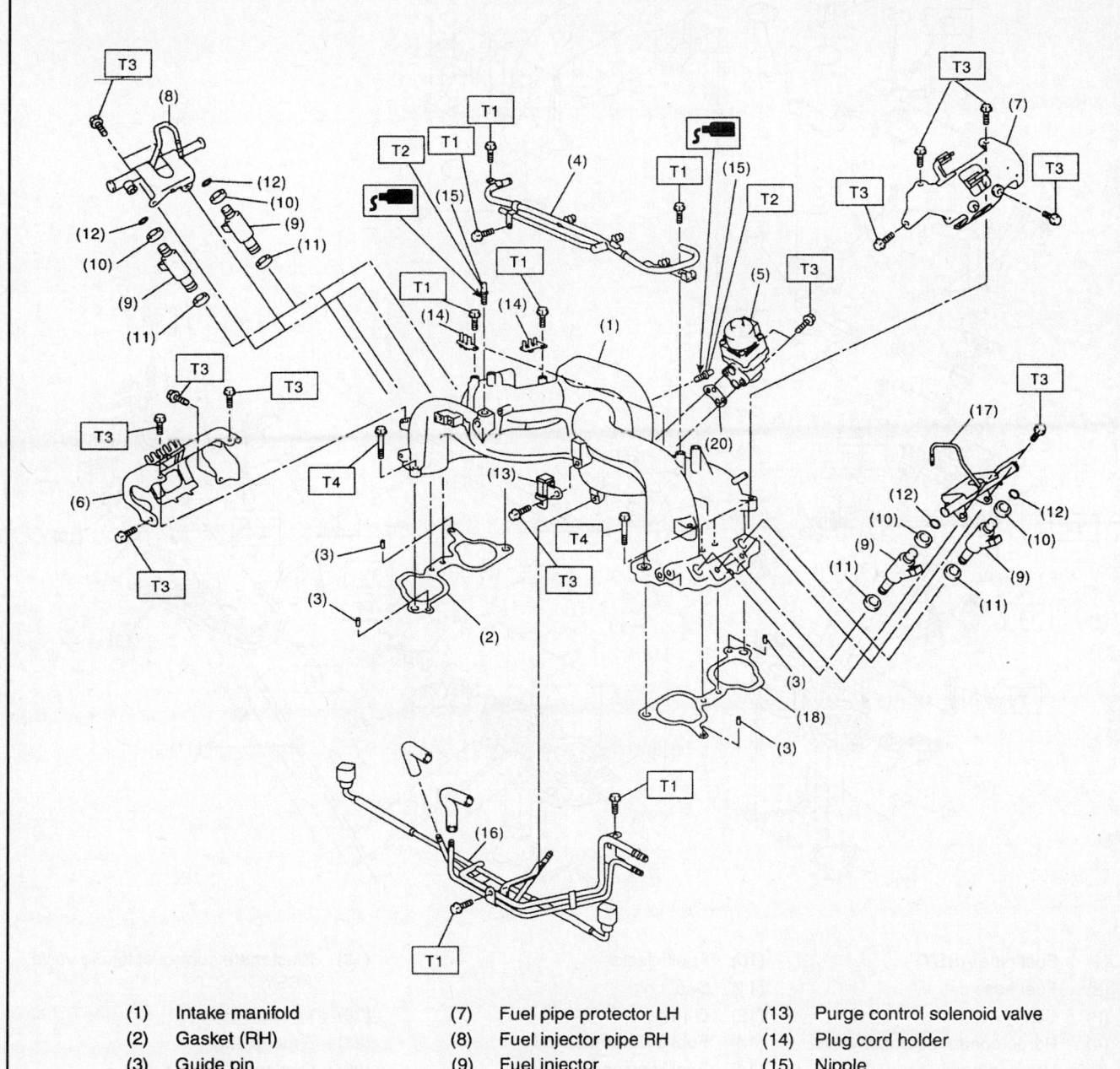

(1)	Intake manifold	(7)	Fuel pipe protector LH	(13)	Purge control solenoid valve
(2)	Gasket (RH)	(8)	Fuel injector pipe RH	(14)	Plug cord holder
(3)	Guide pin	(9)	Fuel injector	(15)	Nipple
(4)	PCV pipe	(10)	O-ring	(16)	Fuel pipe
(5)	EGR valve	(11)	O-ring	(17)	Fuel injector pipe LH
(6)	Fuel pipe protector RH	(12)	O-ring	(18)	Gasket (LH)

09490_SBCR_G0027

Fig. 123 Intake manifold and related components—2.5L SOHC engine

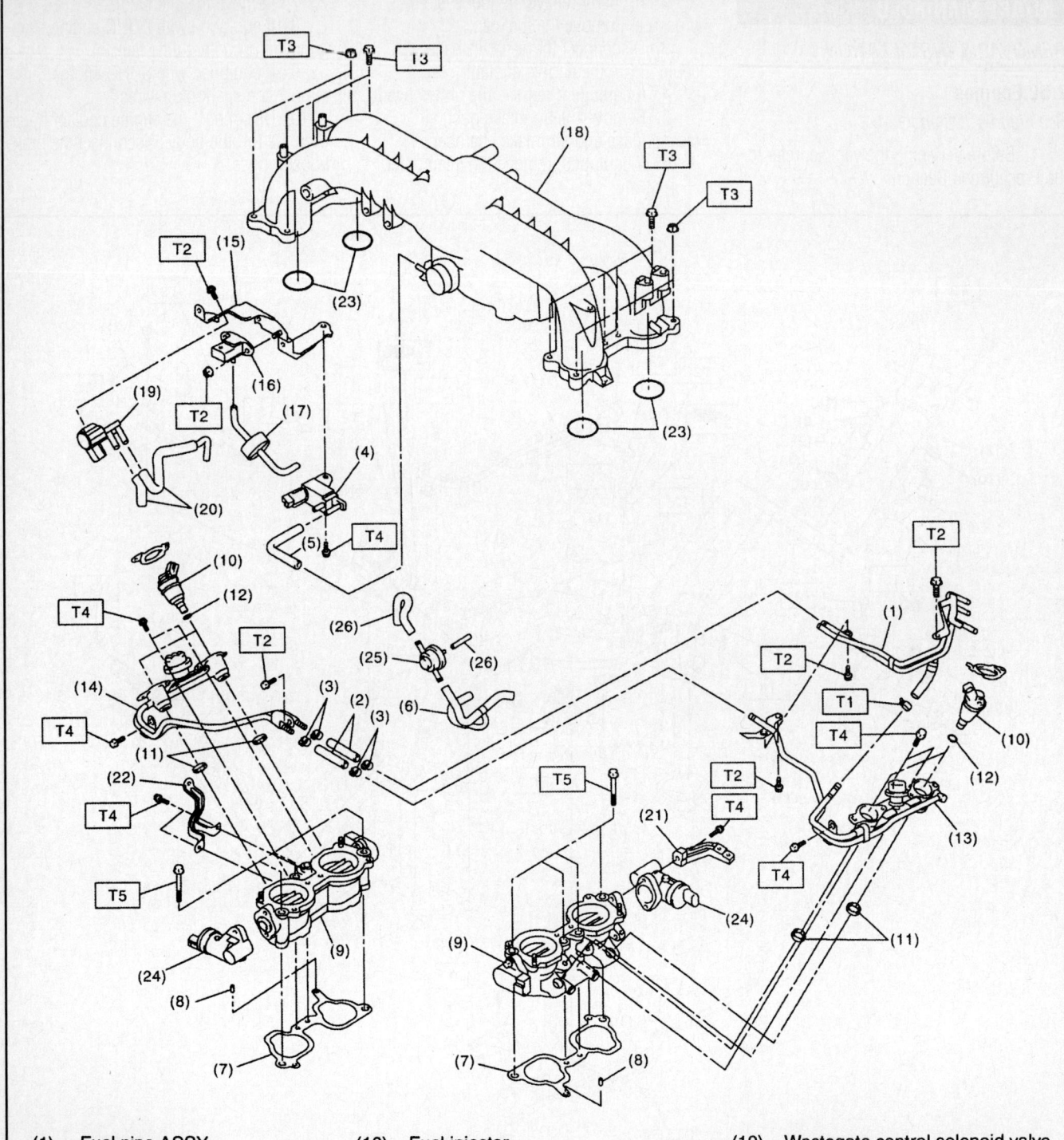

(1)	Fuel pipe ASSY	(10)	Fuel injector	(19)	Wastegate control solenoid valve ASSY
(2)	Fuel hose	(11)	Seal ring	(20)	Vacuum hose
(3)	Clip	(12)	O-ring	(21)	Ground stay
(4)	Purge control solenoid valve	(13)	Fuel injector pipe LH	(22)	Coolant filler tank stay
(5)	Vacuum hose	(14)	Fuel injector pipe RH	(23)	O-ring
(6)	Vacuum control hose	(15)	Solenoid valve bracket	(24)	Tumble generator valve actuator
(7)	Intake manifold gasket	(16)	Manifold absolute pressure sensor	(25)	Purge valve
(8)	Guide pin	(17)	Filter	(26)	Purge hose
(9)	Intake manifold (lower)	(18)	Intake manifold		

09490_SBCR_G0028

Fig. 124 Intake manifold and related components—2.5L DOHC engine

11. Disconnect the PCV hose from the intake manifold. Disconnect the engine harness electrical connectors from the bulkhead harness connectors.

12. Disconnect the engine coolant temperature sensor electrical connector, knock sensor electrical connector and crankshaft position sensor connector.

13. Disconnect the power steering pump switch electrical connector, oil pressure switch connector and camshaft sensor connector.

14. If equipped with DOHC engine, disconnect the oil flow solenoid valve electrical connector and the ignition coil connector.

15. If equipped with SOHC engine, remove the EGR pipe from the intake manifold and disconnect the fuel lines from the fuel pipe.

16. If equipped with DOHC engine, disconnect the fuel delivery hose, return hose and evaporation hose.

17. Remove the intake manifold retaining bolts. Remove the intake manifold from the engine.

To install:

18. Installation is the reverse of the removal procedure.

19. Be sure to use new intake manifold gaskets. Tighten the manifold retaining bolts to 18 ft. lbs. (25 Nm). and in alternating sequence.

20. Be sure to fill and bleed the cooling system with the proper grade and type engine coolant.

21. Start the engine and check for leaks, correct as required.

3.0L & 3.6L Engines

See Figure 125.

1. Before servicing the vehicle, refer to the Precautions Section.

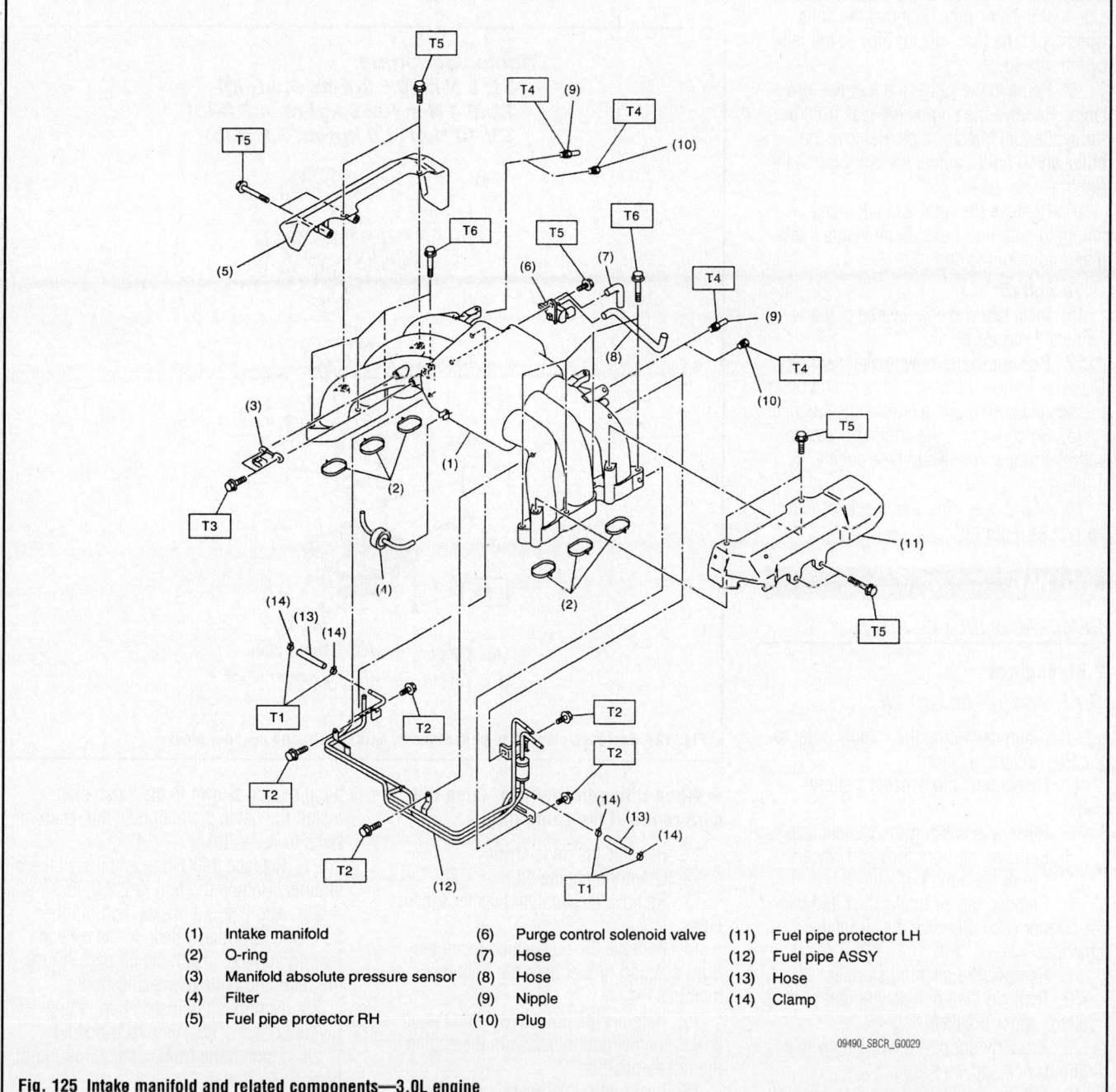

(1)	Intake manifold	(6)	Purge control solenoid valve	(11) Fuel pipe protector LH
(2)	O-ring	(7)	Hose	(12) Fuel pipe ASSY
(3)	Manifold absolute pressure sensor	(8)	Hose	(13) Hose
(4)	Filter	(9)	Nipple	(14) Clamp
(5)	Fuel pipe protector RH	(10)	Plug	

09490_SBCR_G0029

Fig. 125 Intake manifold and related components—3.0L engine

2. Properly relieve the fuel system pressure. Remove the fuel cap.

3. Disconnect the negative battery cable. Drain the engine coolant.

4. Remove the air cleaner case and air intake chamber.

5. Remove the alternator.

6. Disconnect the electrical connector from the throttle body. Disconnect the engine coolant hoses from the throttle body.

7. Disconnect the engine harness connector. Disconnect the PCV hose. Disconnect the brake booster hose. Disconnect the fuel hoses from the fuel pipe.

8. Remove the left side fuel line protector. Remove the engine harness from the left side fuel injector pipe. Remove the bolts which hold the fuel injector pipe to the left cylinder head.

9. Remove the right side fuel line protector. Remove the engine harness from the right side fuel injector pipe. Remove the bolts which hold the fuel injector pipe to the right cylinder head.

10. Remove the right and left intake manifold retaining bolts. Remove the intake manifold from the engine.

To install:

11. Installation is the reverse of the removal procedure.

12. Be sure to use new intake manifold O-rings. Torque the manifold retaining bolts to specification and in alternating sequence.

13. Be sure to fill the cooling system with the proper grade and type engine coolant.

14. Start the engine and check for leaks, correct as required.

OIL PAN

REMOVAL & INSTALLATION

2.5L Engines

See Figures 126 through 128.

1. Before servicing the vehicle, refer to the Precautions Section.

2. Disconnect the negative battery cable.

3. Raise and support the vehicle safely.

4. Remove the front tires and wheels.

5. Lower the vehicle.

6. Remove the air intake duct and the air cleaner case. Remove the air intake chamber.

7. Remove the pitching stopper.

8. Remove the hood stay holder and the radiator upper brackets.

9. Properly support the engine with a lifting device and wire ropes.

10. Lift the vehicle and support it safely.

Fig. 126 Engine supported with lifting device

➡ **When lifting the vehicle, raise the wire ropes at the same time.**

11. Remove the undercover.

12. Drain the engine oil.

13. Remove the front and center exhaust pipes.

14. Remove the nuts which retain the front cushion rubber onto the front cross-member.

15. Remove the bolts that retain the oil pan to the cylinder block, with the engine in the raised position.

16. Insert an oil pan gasket cutter tool into the gap between the cylinder block and the oil pan. Remove the oil pan from the engine.

➡ **Do not use a screwdriver or similar tool in place of the cutter tool.**

17. Remove the oil strainer, if required. Remove the baffle plate, if required.

To install:

18. Be sure to clean the old gasket material from the mating surfaces.

19. Apply a continuous bead of sealer to a new oil pan gasket.

20. Make sure that the seals (A) are installed securely on the baffle plate and in

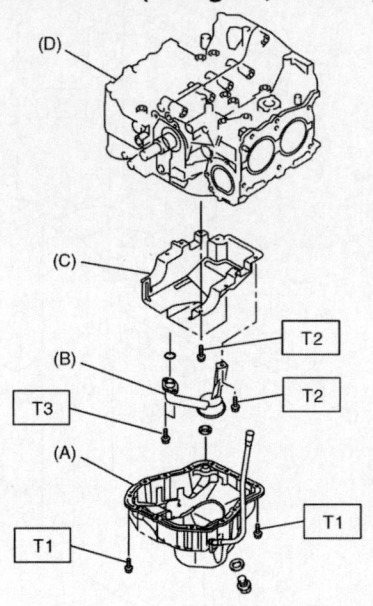

Tightening torque:
T1: 5 N·m (0.5 kgf-m, 3.6 ft-lb)
T2: 6.4 N·m (0.65 kgf-m, 4.7 ft-lb)
T3: 10 N·m (1.0 kgf-m, 7.2 ft-lb)

(A) Oil pan   (C) Baffle plate
(B) Oil strainer   (D) Cylinder block

Fig. 127 Exploded view of oil pan, strainer, baffle plate and cylinder block

the direction shown in the illustration. Install the baffle plate; tighten the retaining bolts to 4.7 ft. lbs.

21. Replace the O-ring and install the oil strainer. Tighten the bolt to 7.2 ft. lbs.

22. Apply liquid gasket, part number 004403012 or equivalent, to the oil pan mating surface. Install the oil pan. Torque the retaining bolts to specification.

23. Continue the installation in the reverse order of the removal procedure.

24. Tighten the front cushion mounting bolts to 63 ft. lbs.

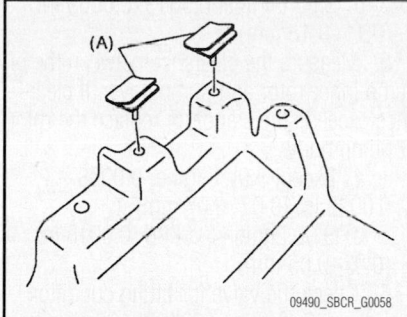

Fig. 128 Oil pan baffle plate seal location and positioning—2.5L engine

09490_SBCR_G0058

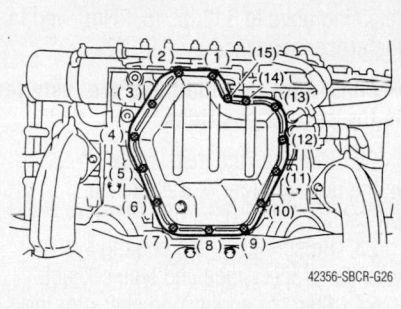

Fig. 129 Oil pan bolt torque sequence— 3.0L engine

42356-SBCR-G26

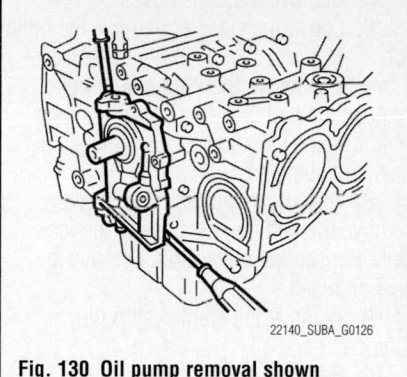

Fig. 130 Oil pump removal shown

22140_SUBA_G0126

25. Be sure to fill the engine with the correct grade and type engine oil.

26. Start the engine and check for leaks. Correct as required.

3.0L & 3.6L Engines

See Figure 129.

➡ **If removing the upper oil pan, the engine must first be removed from the vehicle.**

1. Before servicing the vehicle, refer to the Precautions Section.

2. Disconnect the negative battery cable.

3. Raise and support the vehicle safely.

4. Remove the undercover. Drain the engine oil.

5. Remove the lower oil pan retaining bolts.

6. Insert an oil pan gasket cutter tool into the gap between the upper oil pan and the lower oil pan. Remove the lower oil pan from the engine.

➡ **Do not use a screwdriver or similar tool in place of the cutter tool.**

7. Remove the oil strainer, if required.

To install:

8. Be sure to clean the old gasket material from the mating surfaces.

9. Replace the O-ring and install the oil strainer. Tighten the bolt to 5 ft. lbs (6.4 Nm)

10. Apply a continuous bead (0.039 inch thick) of liquid gasket, part number K0877YA018 or equivalent, to the mating surfaces. Install the oil pan. Tighten the retaining bolts to 5 ft. lbs (6.4 Nm) and in the proper sequence.

11. Continue the installation in the reverse order of the removal procedure.

12. Be sure to fill the engine with the correct grade and type engine oil.

13. Start the engine and check for leaks. Correct as required.

OIL PUMP

REMOVAL & INSTALLATION

2.5L Engines

See Figure 130.

1. Before servicing the vehicle, refer to the Precautions Section.

2. Disconnect the negative battery cable. Drain the cooling system.

3. On the DOHC engine, remove the collector cover.

4. Raise and support the vehicle safely.

5. Remove the undercover.

6. On the DOHC engine, remove the bolts which retain the water pipe of the oil cooler to the oil pump. Remove the water pipe and hoses between the oil cooler and the water pump.

7. Lower the vehicle. Remove the radiator.

8. To remove the front side V belt, remove the belt covers. Loosen the lock bolt. Loosen the slider bolt. Remove the front side belt.

9. To remove the rear side V belt, remove the belt covers. Loosen the lock bolt. Loosen the slider bolt. Remove the rear side belt.

10. On the SOHC engine remove the belt tensioner.

11. On the DOHC engine remove the rear side V belt tensioner

12. Remove the crankshaft position sensor.

13. Remove the crankshaft pulley bolt.

14. Lock the crankshaft in place using tool ST499977100 for the SOHC engine and tool ST499207400 for the DOHC engine, or equivalent.

15. Remove the crankshaft pulley.

16. Remove the water pump.

17. If equipped, remove the timing belt guide. Remove the crankshaft sprocket.

18. Remove the oil pump retaining bolts.

➡ **When disassembling and checking the oil pump, loosen the relief valve plug before removing the oil pump from its mounting.**

19. Using a flat tip tool remove the oil pump from the engine.

To install:

20. Be sure all mating surfaces are clean and free of dirt.

21. Apply liquid gasket part number 004403007, or equivalent to the mating surfaces of the oil pump.

22. Be sure to replace the O-ring with a new one.

23. Apply a thin coat of clean engine oil to the inside of the oil seal.

24. Position the oil pump to its mounting, aligning the notched area with the crankshaft and push the pump straight.

➡ **Be sure that the oil seal lip is not folded.**

25. Install the oil pump. Apply liquid gasket part number 004403042, or equivalent to the three retaining bolt threads. Install the bolts and tighten to 5 ft. lbs (6.4 Nm)

26. Continue the installation in the reverse order of the removal procedure.

27. Be sure to fill the cooling system with the proper grade and type coolant.

28. Start the engine and check for leaks. Correct, as required.

3.0L & 3.6L Engines

See Figure 131.

1. Before servicing the vehicle, refer to the Precautions Section.

2. Disconnect the negative battery cable. Drain the cooling system.

3. Remove the collector cover.

4. Raise and support the vehicle safely.

5. Remove the undercover.

6. Lower the vehicle. Remove the radiator.

7. To remove the front side V belt, remove the belt covers. Loosen the lock bolt. Loosen the slider bolt. Remove the front side belt.

8. To remove the rear side V belt, remove the belt covers. Loosen the lock bolt. Loosen the slider bolt. Remove the rear side belt.

9. Remove the crankshaft pulley cover.

10. Remove the crankshaft pulley bolt.

11. Lock the crankshaft in place using tool ST499977100, or equivalent.

12. Remove the crankshaft pulley.

13. Remove the front timing chain cover.

➡**Bolts are three different sizes. Be careful to install the correct bolt in the correct hole.**

14. Remove the timing chain.

15. Remove the crankshaft sprocket.

16. Remove the oil pump cover retaining bolts. Remove the oil pump cover.

17. Remove the inner and outer rotors.

To install:

18. Be sure all mating surfaces are clean and free of dirt.

19. Apply a thin coat of clean engine oil to the complete area of the inner and outer rotors.

20. Position the inner rotor in place. Position the outer rotor in place.

21. Install the pump cover. Tighten the retaining bolts to 5 ft. lbs (6.4 Nm). and in the proper sequence.

➡**Make sure that the bolts are installed in the correct positions.**

22. Continue the installation in the reverse order of the removal procedure.

23. Be sure to fill the cooling system with the proper grade and type coolant.

24. Start the engine and check for leaks. Correct, as required.

INSPECTION

See Figures 132 through 134.

1. Check the oil pump case for worn shaft hole, clogged oil passage, worn rotor chamber, cracks, and other faults.

2. Check the oil seal lips for deformation, hardening, wear, etc. and replace if defective.

3. Measure the tip clearance of rotors. If clearance exceeds the standard, replace the rotors as a matched set.

   a. Except 3.0L Engine: 0.0016–0.0055 in. (0.04–0.14 mm)

   b. 3.0L Engine: 0.0016–0.0055 in. (0.04–0.14 mm)

4. Measure the clearance between the outer rotor and oil pump rotor housing. If clearance exceeds the standard, replace the rotor.

   a. Except 3.0L Engine: 0.0039–0.0069 in. (0.10–0.18 mm)

   b. 3.0L Engine: 0.0043–0.0069 in. (0.11–0.18 mm)

5. Measure the clearance between the oil pump inner rotor and pump cover. If clearance exceeds the standard, replace the rotor or pump body.

   a. Except 3.0L Engine: 0.0008–0.0028 in. (0.02–0.07 mm)

   b. 3.0L Engine: 0.0008–0.0018 in. (0.02–0.05 mm)

6. Check the valve for fitting condition and damage, and the relief valve spring for damage and deterioration. Replace the parts if defective.

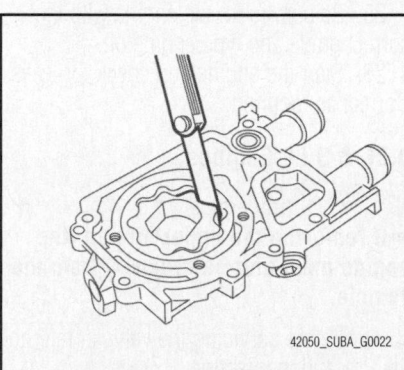

42050_SUBA_G0022

Fig. 132 Measuring for tip clearance of the rotors of the oil pump.

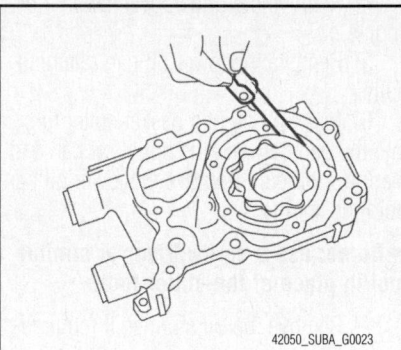

42050_SUBA_G0023

Fig. 133 Measuring the case clearance of the oil pump.

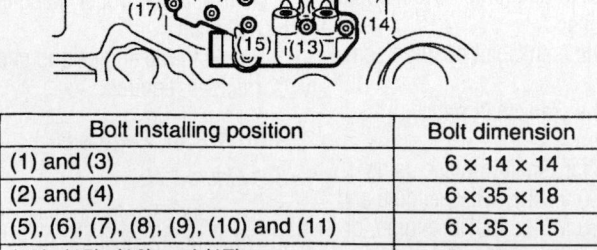

Bolt installing position	Bolt dimension
(1) and (3)	6 × 14 × 14
(2) and (4)	6 × 35 × 18
(5), (6), (7), (8), (9), (10) and (11)	6 × 35 × 15
(12), (15), (16) and (17)	6 × 16 × 16
(13) and (14)	6 × 26 × 15

09490_SBCR_G0059

Fig. 131 Oil pump tightening sequence and bolt location—3.0L engine

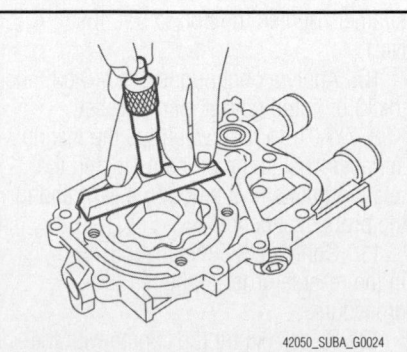

42050_SUBA_G0024

Fig. 134 Measuring the side clearance of the oil pump.

REAR MAIN SEAL

REMOVAL & INSTALLATION

See Figure 135.

1. Before servicing the vehicle, refer to the Precautions Section.

2. Remove or disconnect the following:
   - Engine from the vehicle
   - Clutch assembly/flywheel using the Clutch Disc Guide tool 499747000, if equipped with a manual transmission
   - Torque converter flexplate from the crankshaft, if equipped with an automatic transmission
   - Oil seal from the cylinder block using a small pry bar

To install:

3. Install or connect the following:
   - New oil seal by pressing it into the cylinder block using the appropriate driver and hammer
   - Flywheel housing using new gaskets and sealant where necessary.
   - Flywheel and tighten the bolts to specification.
   - Engine

3. To remove the front side V belt, remove the belt covers. Loosen the lock bolt. Loosen the slider bolt. Remove the front side belt.

4. To remove the rear side V belt, remove the belt covers. Loosen the lock bolt. Loosen the slider bolt. Remove the rear side belt. Remove the belt tensioner.

5. Remove the crankshaft pulley bolt.

6. Lock the crankshaft in place using tool ST499977100, or equivalent.

7. Remove the crankshaft pulley.

8. Remove the left side timing belt cover.

9. Remove the front timing belt cover.

10. Remove the timing belt.

11. Remove the camshaft position sensor.

12. Remove the camshaft sprockets, No.1 and No.2.

➡**Be sure to lock the camshaft in place using tool ST18231AA010, or equivalent.**

13. Disconnect the PCV valve hose. Remove the rocker cover retaining bolts. Remove the rocker cover from the engine.

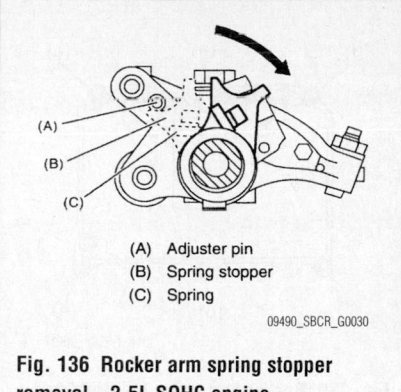

(A) Adjuster pin
(B) Spring stopper
(C) Spring

09490_SBCR_G0030

Fig. 136 Rocker arm spring stopper removal—2.5L SOHC engine

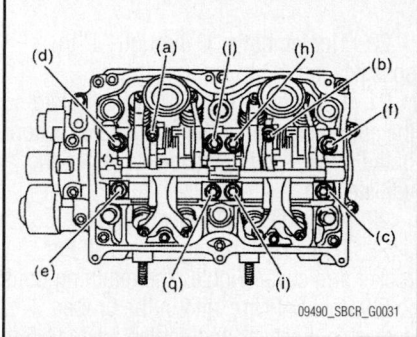

09490_SBCR_G0031

Fig. 137 Rocker arm assembly bolt loosening sequence—2.5L SOHC engine

16. Position tool ST18354AA000 on the rocker arm assembly. Remove the assembly from the engine.

To install:

17. Install the rocker arm assembly on its mounting.

18. Temporarily tighten the bolts equally in the proper sequence.

➡**Do not temporarily tighten bolts "I" and "J". Position tool ST18354AA000 to the rocker arm assembly.**

19. Tighten bolts "A" through "H" to specification 18 ft. lbs. (25 Nm).

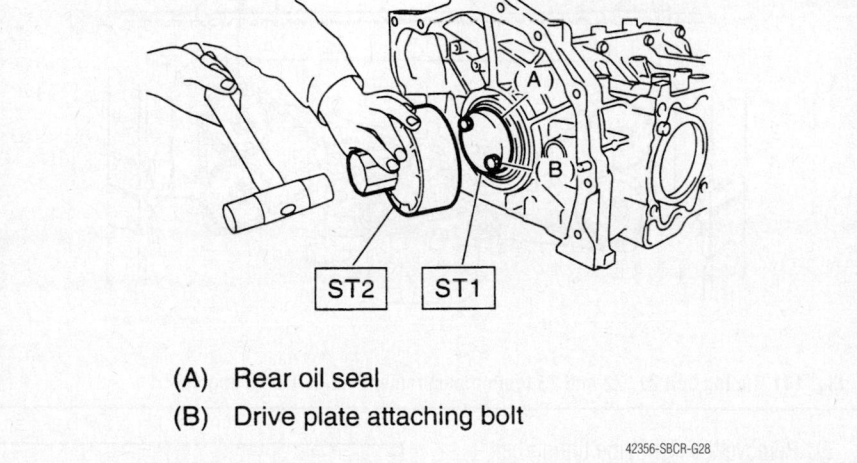

(A) Rear oil seal
(B) Drive plate attaching bolt

42356-SBCR-G28

Fig. 135 Installing the rear main seal using oil seal guide ST1 499597100 and ST2 499598200

ROCKER ARMS

REMOVAL & INSTALLATION

2.5L SOHC Engine

See Figures 136 through 139.

1. Before servicing the vehicle, refer to the Precautions Section.

2. Disconnect the negative battery cable.

14. Use tool St18258AA000, or equivalent, to rotate the spring stopper in direction of the arrow, see illustration, and remove the spring stopper from the adjuster spring.

15. Remove bolts "A" through "J" in sequence. Remove the assembly from the engine.

➡**Leave two or three threads on bolts "I" and "J" engaged in order to retain the rocker arm assembly.**

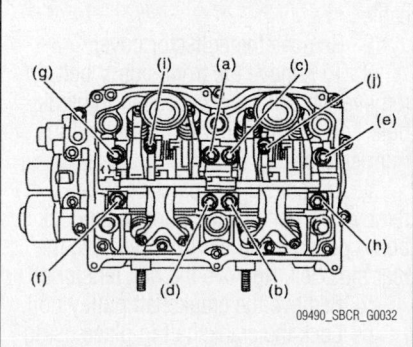

09490_SBCR_G0032

Fig. 138 Rocker arm assembly bolt tightening sequence—2.5L SOHC engine

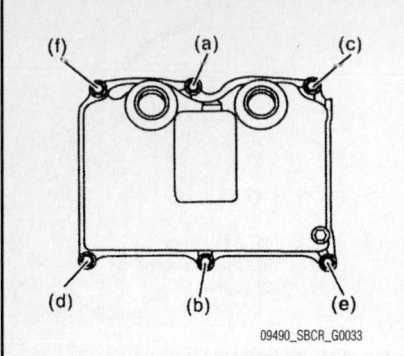

Fig. 139 Rocker arm cover bolt tightening sequence—2.5L SOHC engine

20. Tighten bolts "I" through "J" to specification 4.3 ft. lbs. (5 Nm).

21. Use tool ST18258AA000 to rotate the spring stopper in the opposite direction of the arrow, see illustration, to fasten the adjuster pin.

22. Adjust the valve clearance.

23. Use a new gasket and install the rocker arm cover. Tighten the retaining bolts to 5 ft. lbs (6.4 Nm). and in the proper sequence. Recheck and tighten again to 5 ft. lbs (6.4 Nm) in the proper sequence.

24. Continue the installation in the reverse order of the removal procedure.

TIMING BELT FRONT COVER

REMOVAL & INSTALLATION

Refer to the Timing Belt and Sprockets, Removal & Installation procedure in this section to remove the timing belt front cover.

TIMING BELT & SPROCKETS

REMOVAL & INSTALLATION

2.5L DOHC Engine

See Figures 140 through 151.

1. Disconnect the negative battery cable.

2. Remove the collector cover.

3. To remove the front side V belt, remove the belt covers. Loosen the lock bolt. Loosen the slider bolt. Remove the front side belt.

4. To remove the rear side V belt, remove the belt covers. Loosen the lock bolt. Loosen the slider bolt. Remove the rear side belt. Remove the belt tensioner.

5. Remove the crankshaft pulley bolt.

6. Lock the crankshaft in place using tool ST499977100, or equivalent.

7. Remove the crankshaft pulley.

8. Remove the left side timing belt cover.

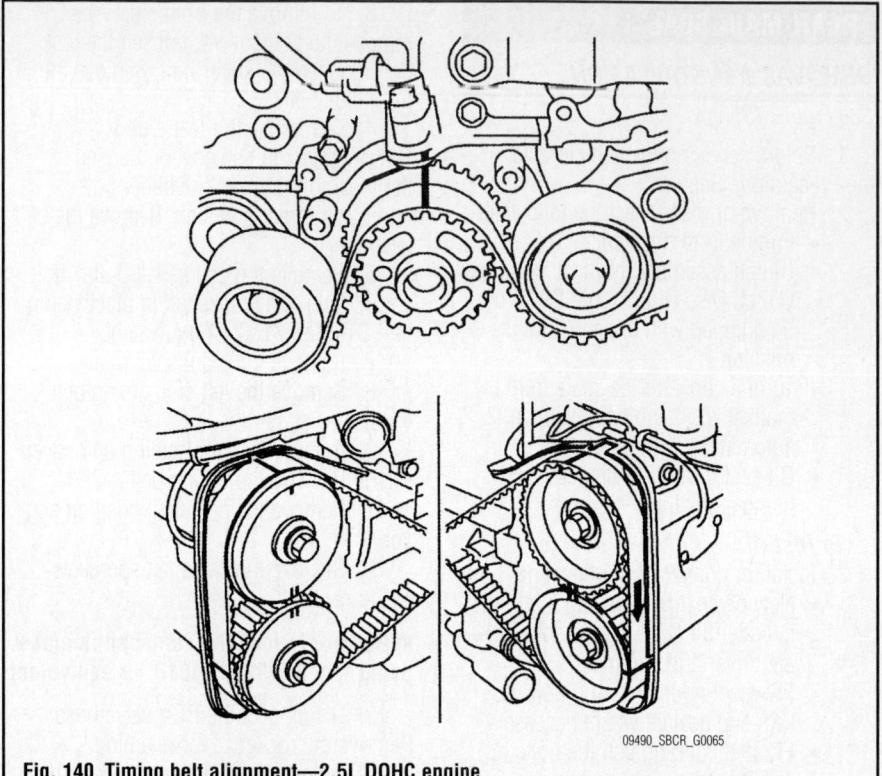

Fig. 140 Timing belt alignment—2.5L DOHC engine

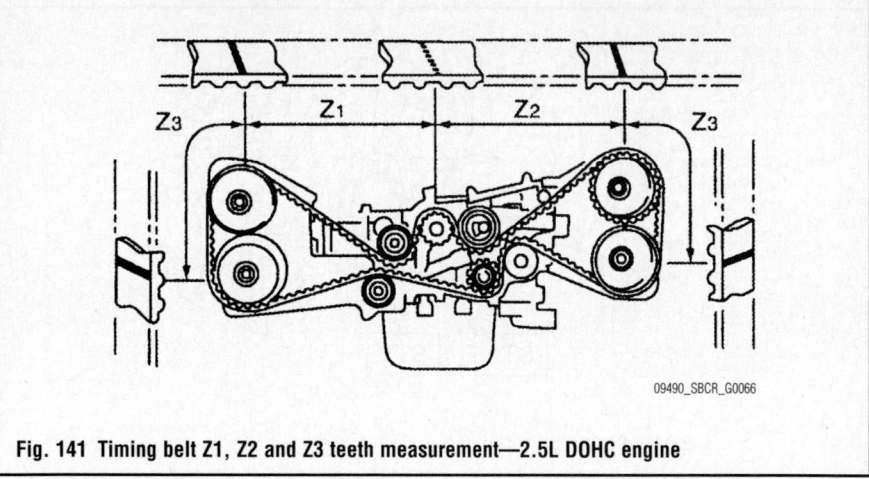

Fig. 141 Timing belt Z1, Z2 and Z3 teeth measurement—2.5L DOHC engine

9. Remove the right side timing belt cover.

10. Remove the front timing belt cover.

11. Remove the timing belt guide, if equipped.

➡️**If the belt is going to be reused and the alignment mark on the belt is not readable, put a new mark on the belt to indicate the direction of rotation. Using tool ST499987500, turn the crankshaft to align the mark on the crankshaft sprocket, intake camshaft sprocket (left), exhaust camshaft sprocket (left), intake camshaft sprocket (right), exhaust camshaft sprocket (right) with**

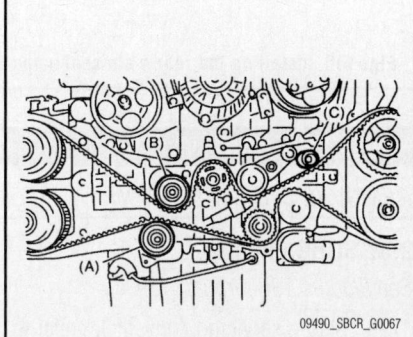

Fig. 142 Belt idler identification and location—2.5L DOHC engine

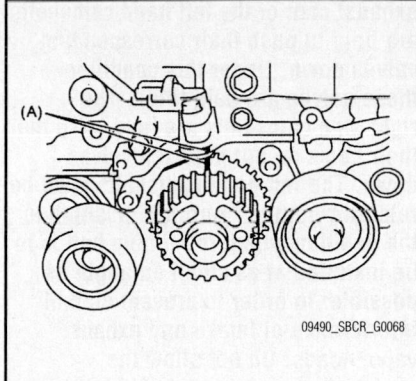

Fig. 143 Crankshaft sprocket mark A to oil pump cover alignment—2.5L DOHC engine

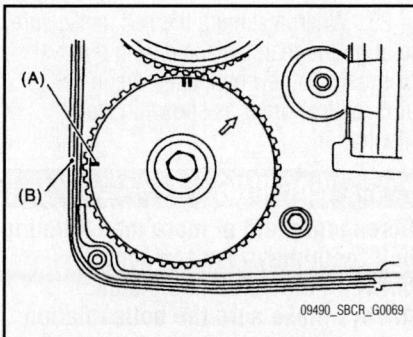

Fig. 144 Right exhaust camshaft sprocket alignment mark A with timing belt cover B alignment mark—2.5L DOHC engine

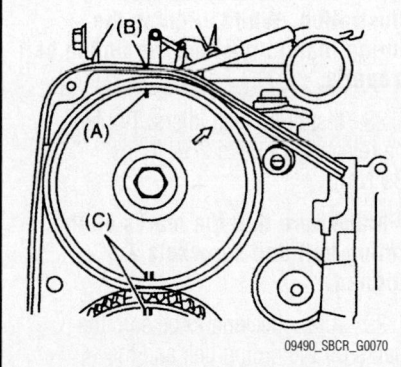

Fig. 145 Right intake camshaft sprocket alignment mark A with timing belt cover B alignment mark an double line C alignment mark—2.5L DOHC engine

the notches of the timing belt cover and cylinder block. Paint an alignment mark on the belts in relation to the camshaft sprockets. Z1 measurement is 54.4 teeth. Z2 measurement is 51.0 teeth and Z3 measurement is 28.0 teeth.

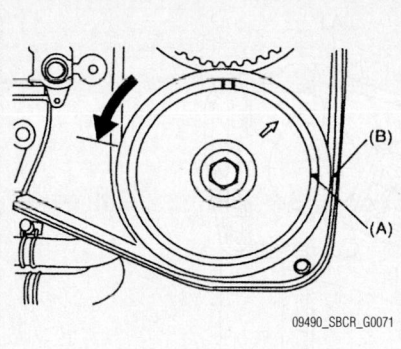

Fig. 146 Left exhaust camshaft sprocket alignment mark A with timing belt cover B alignment mark—2.5L DOHC engine

12. Remove the belt idler belt "A". Remove the timing belt.

13. Remove the belt idlers belt "B" and "C".

14. Remove the belt idler number two. Remove the automatic belt tension adjuster assembly.

To install:

15. Attach the automatic belt tension adjuster assembly to a vertical pressing tool.

➡**Always use a vertical type pressing tool to move the adjuster rod downward. Do not use a lateral type vise. Push the adjuster rod vertically. Press in the push adjuster rod gradually, which should take three minutes or more. Do not allow pressure to exceed 2,205 lb. force.**

16. Slowly move the adjuster rod down until the adjuster rod is aligned with the stopper pin hole in the cylinder.

➡**Press the adjuster rod as far as the end surface of the cylinder. Do not press the adjuster rod into the cylinder. Doing so may damage the cylinder.**

17. Using a 0.08 inch stopper pin, insert it into the stopper pin hole in the cylinder. Secure the adjuster rod.

➡**Do not release the press pressure until the stopper pin is completely inserted in the hole.**

18. Install the automatic belt tensioner assembly. Tighten the retaining bolt to 28.9 ft. lbs.

19. Install the belt idler number two. Tighten the retaining bolt to 28.9 ft. lbs.

20. Install the belt idlers. Tighten the retaining bolts to 28.9 ft. lbs.

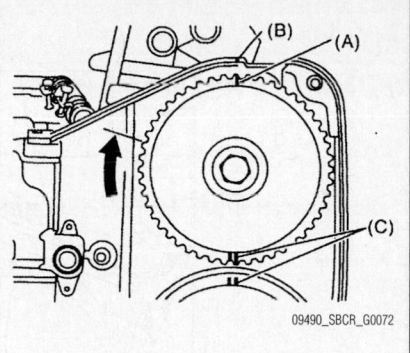

Fig. 147 Left intake camshaft sprocket alignment mark A with timing belt cover B alignment mark an double line C alignment mark—2.5L DOHC engine

21. Align the mark "A" on the crankshaft sprocket with the mark on the oil pump at the cylinder block. Align the single line mark "A" on the right exhaust camshaft sprocket with the notch "B" on the timing belt cover.

22. Align single line mark "A" on the right intake camshaft sprocket with the notch "B" on the timing cover. Ensure that the double lines "C" on the intake and exhaust camshaft sprockets are aligned.

23. Align single line mark "A" on the left exhaust camshaft sprocket with the notch "B" on the timing cover by turning the sprocket counterclockwise as viewed from the front of the engine.

24. Align single line mark "A" on the left intake camshaft sprocket with the notch "B" on the timing cover, by turning the sprocket clockwise as viewed from front of the engine. Ensure that the double lines "C" on the intake and exhaust camshaft sprockets are aligned.

25. Make sure that the camshaft and crankshaft sprockets are positioned properly.

➡**The intake and exhaust camshafts on this engine can be rotated independently with the timing belt removed. By looking at the illustration it will show you that if the intake and exhaust valve are lift together the heads will hit each other and bend.**

➡**When the timing belts are not installed, 4 camshafts are held at "zero lift" position, where all cams on the camshafts do not push the intake and exhaust valves down (under this condition all valves remain unlifted). When the camshafts are rotated to install the timing belts, No. 2 intake and No. 4**

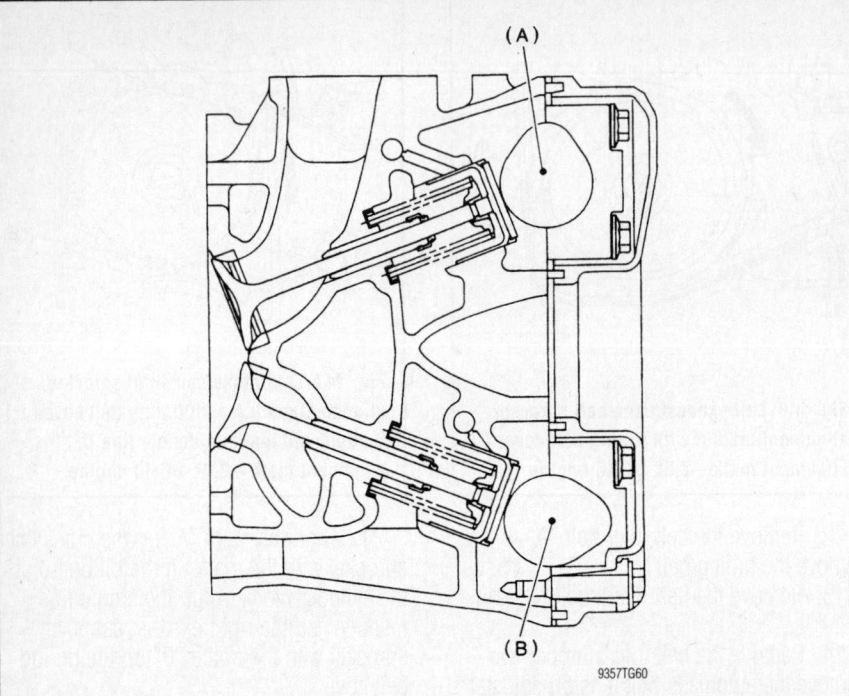

Fig. 148 If the intake and exhaust valve are lift together the heads will hit each other and bend –2.5L DOHC engine

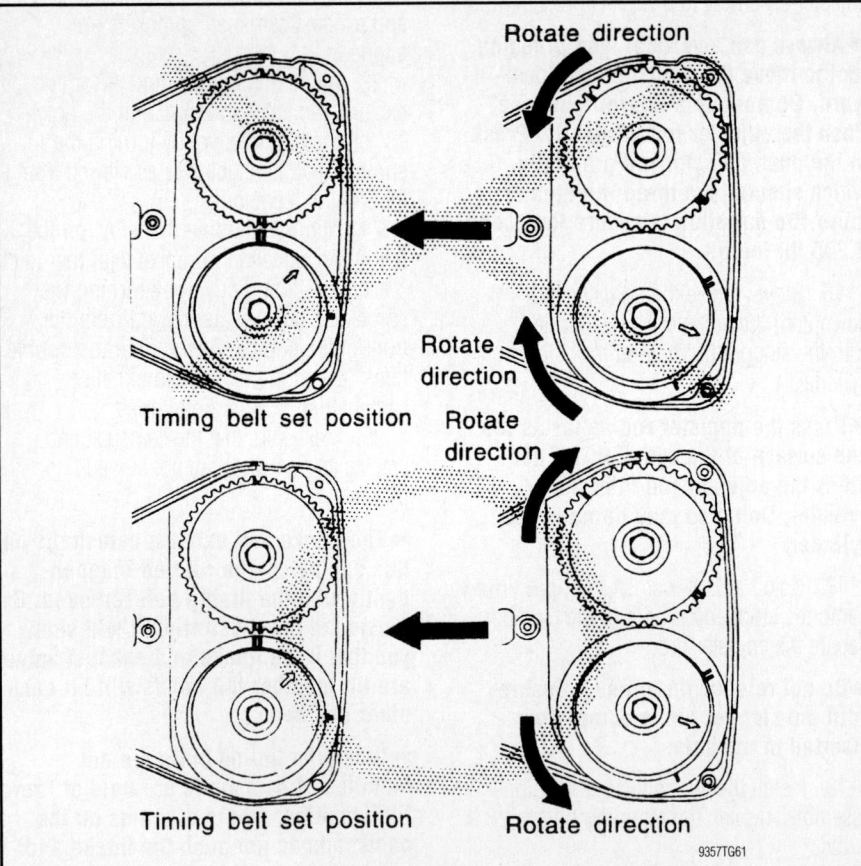

Fig. 149 Do not allow the camshafts to rotate in the direction shown as this causes both the intake and exhaust valves to lift off at the same time with will cause valve damage on DOHC engines

exhaust cam of the left hand camshafts are held to push their corresponding valves down. Under this condition these valves are held lifted. The right side camshafts are held in so that their cams do not push the valves down. The left hand camshafts must be rotated from the "zero lift" position to the position where the timing belt is to be installed at as small an angle as possible, in order to prevent mutual interference of intake and exhaust valve heads. Do not allow the camshafts to rotate in the direction illustrated as this causes both the intake and exhaust valves to lift off at the same time with will cause valve damage.

26. When installing the belt, make sure to align the marks made during removal or if using a new belt, align the in alphabetical order as shown in the illustration.

✳✳ WARNING

Disengagement of more than 3 timing belt teeth may result in contact between the valve and piston. Always make sure the belts rotation is correct.

27. Install the timing belt.

➥Align the alignment mark on the timing belt with marks on the sprocket in the order shown in the illustration. While aligning the timing marks, position the timing belt properly.

28. Install the belt idlers. Tighten the retaining bolts to 28.9 ft. lbs. (39 Nm).

➥Make sure that the marks on the timing belt and sprockets are aligned.

29. After checking to be sure the marks on the timing belt and the camshaft sprockets are aligned remove the stopper pin from the belt tension adjuster.

30. Install the timing belt guide, if equipped with manual transmission. Temporarily tighten the bolts. Check and adjust the clearance between the belt and the guide. It should be 0.039/ 0.020 inch.

31. Install the timing belt cover.
32. Install the crank pulley.
33. Install the V-belts.

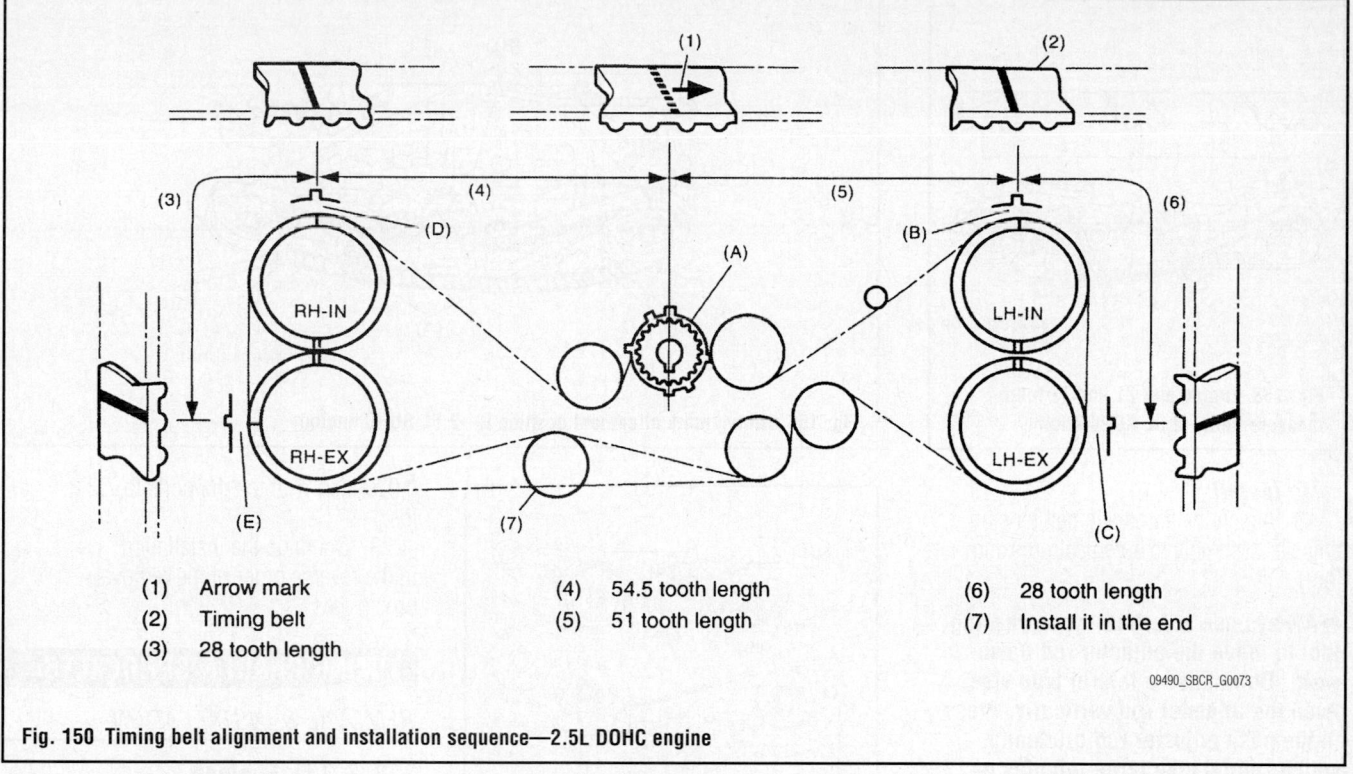

(1) Arrow mark
(2) Timing belt
(3) 28 tooth length
(4) 54.5 tooth length
(5) 51 tooth length
(6) 28 tooth length
(7) Install it in the end

09490_SBCR_G0073

Fig. 150 Timing belt alignment and installation sequence—2.5L DOHC engine

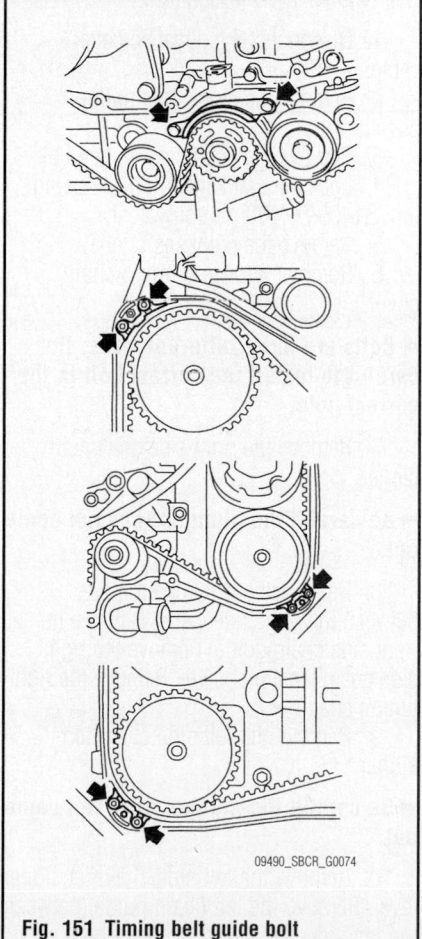

09490_SBCR_G0074

Fig. 151 Timing belt guide bolt location—2.5L DOHC engine

2.5L SOHC Engine

See Figures 152 through 156.

1. Before servicing the vehicle, refer to the Precautions Section.

2. Disconnect the negative battery cable.

3. To remove the front side V belt, remove the belt covers. Loosen the lock bolt. Loosen the slider bolt. Remove the front side belt.

4. To remove the rear side V belt, remove the belt covers. Loosen the lock bolt. Loosen the slider bolt. Remove the rear side belt. Remove the belt tensioner.

5. Remove the crankshaft pulley bolt.

6. Lock the crankshaft in place using tool ST499977100, or equivalent.

7. Remove the crankshaft pulley.

8. Remove the left side timing belt cover.

9. Remove the front timing belt cover.

10. If equipped with manual transmission, remove the timing belt guide.

➡If the belt is going to be reused and the alignment mark on the belt is not readable, put a new mark on the belt to indicate the direction of rotation. Using tool ST499987500, turn the crankshaft to align the mark of the sprocket "A" to the cylinder mark notch "B". Ensure that the right side cam sprocket mark "C", cam cap and cylinder head matching surface "D" or left side cam sprocket mark "E", timing belt cover notch "F"

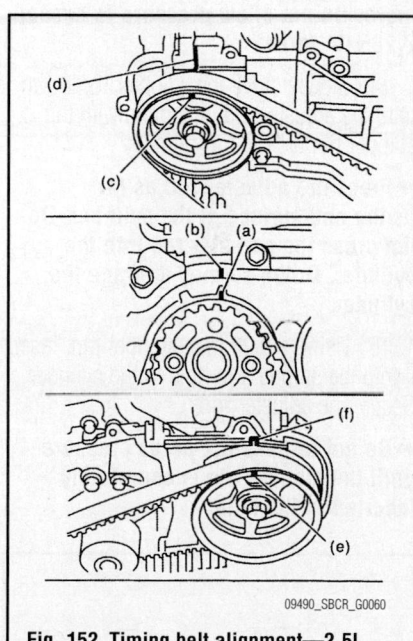

09490_SBCR_G0060

Fig. 152 Timing belt alignment—2.5L SOHC engine

are properly aligned. Paint an alignment mark on the belt in relation to the crankshaft sprocket and camshaft sprockets. Z1 measurement is 46.8 teeth. Z2 measurement is 43.7 teeth.

11. Remove both the number two belt idlers. Remove the timing belt from the engine.

12. Remove the number one belt idler. Remove the automatic belt tension adjuster assembly.

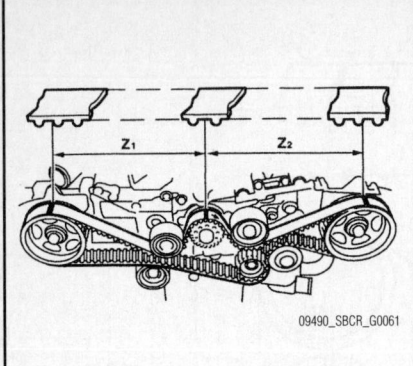

Fig. 153 Timing belt Z1 and Z2 teeth measurement—2.5L SOHC engine

To install:

13. Attach the automatic belt tension adjuster assembly to a vertical pressing tool.

➡ **Always use a vertical type pressing tool to move the adjuster rod downward. Do not use a lateral type vise. Push the adjuster rod vertically. Press in the push adjuster rod gradually, which should take three minutes or more. Do not allow pressure to exceed 2,205 lb. force.**

14. Slowly move the adjuster rod down until the adjuster rod is aligned with the stopper pin hole in the cylinder.

➡ **Press the adjuster rod as far as the end surface of the cylinder. Do not press the adjuster rod into the cylinder. Doing so may damage the cylinder.**

15. Using a 0.08 inch stopper pin, insert it into the stopper pin hole in the cylinder. Secure the adjuster rod.

➡ **Do not release the press pressure until the stopper pin is completely inserted in the hole.**

Fig. 154 Timing mark alignment position A—2.5L SOHC engine

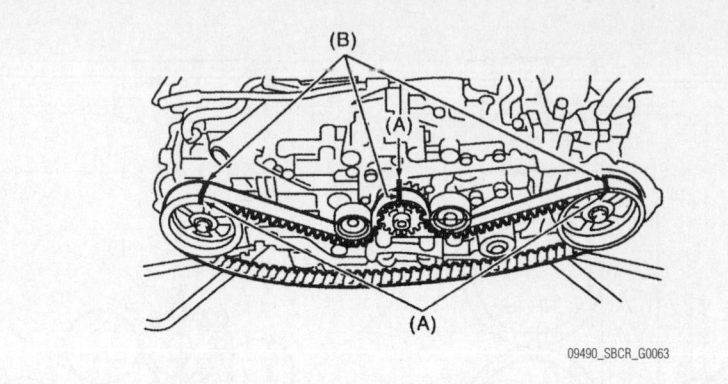

Fig. 155 Timing mark alignment position B—2.5L SOHC engine

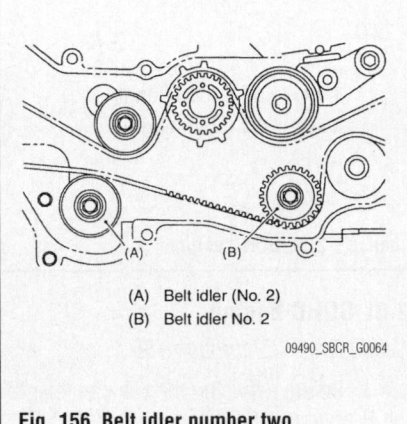

(A) Belt idler (No. 2)
(B) Belt idler No. 2

Fig. 156 Belt idler number two locations—2.5L SOHC engine

16. Install the automatic belt tensioner assembly. Tighten the retaining bolt to 28.9 ft. lbs.

17. Install the belt idler number one. Tighten the retaining bolt to 28.9 ft. lbs.

18. Turn the number one and number two camshaft sprockets, using tool ST499207100 or tool ST18231AA010 and position the alignment marks "A" on each at the highest position.

19. While aligning the alignment mark "B" on the timing belt with mark "A" on the sprockets, position the timing belt properly.

20. Install both belt idler number two's. Tighten the retaining bolt to 28.9 ft. lbs.

21. After checking to be sure the marks on the timing belt and the camshaft sprockets are aligned remove the stopper pin from the belt tension adjuster.

22. Install the timing belt guide, if equipped with manual transmission. Temporarily tighten the bolts. Check and adjust the clearance between the belt and the guide. It should be 0.039 +/-

0.020 inch. Tighten the bolts to 7.2 ft. lbs.

23. Continue the installation in the reverse order of the removal procedure.

TIMING CHAIN & SPROCKETS

REMOVAL & INSTALLATION

3.0L & 3.6L Engines

See Figures 157 through 169.

1. Disconnect the negative battery cable.

2. Remove the crankshaft pulley cover.

3. Remove the crankshaft pulley bolt.

4. Lock the crankshaft in place using tool ST499977100, or equivalent.

5. Remove the crankshaft pulley.

6. Remove the front timing chain cover.

➡ **Bolts are three different sizes. Be careful to install the correct bolt in the correct hole.**

7. Remove the right side chain tensioner.

➡ **Be careful the plunger does not come out.**

8. Remove the right chain guide, between the right side cams. Remove the right side chain guide. Remove the right side chain tensioner lever. Remove the right timing chain.

9. Remove the left side chain tensioner.

➡ **Be careful the plunger does not come out.**

10. Remove the left side chain tensioner lever. Remove the left chain guide, between the left side cams. Remove the chain guide.

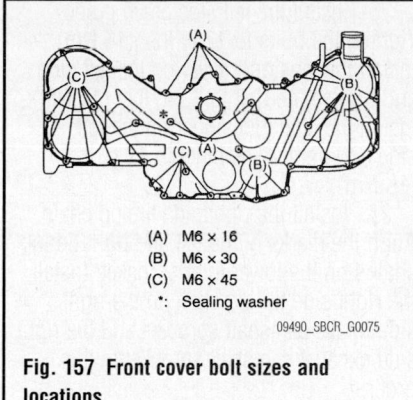

(A) M6 × 16
(B) M6 × 30
(C) M6 × 45
*: Sealing washer

09490_SBCR_G0075

Fig. 157 Front cover bolt sizes and locations

11. Remove the center chain guide. Remove the upper idler sprocket. Remove the left timing chain.

12. Remove the lower idler sprocket.

To install:

13. Make sure all components are clean. Apply oil to the chain guide, tensioner lever and idler sprockets.

14. Place the screw, spring, pin and tension rod into the tensioner body.

15. While pressing the tensioner onto a rubber mat, twist it to the left and right to shorten the rod. Place a thin pin into the holes between the rod and body to hold it in

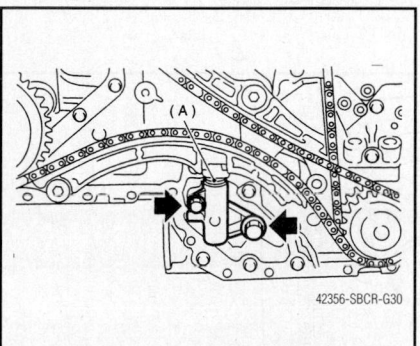

42356-SBCR-G30

Fig. 158 The chain tensioner plunger A does not come out—right hand side

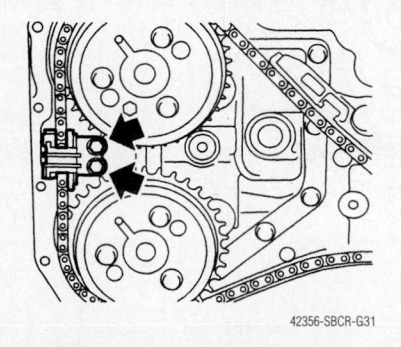

42356-SBCR-G31

Fig. 159 Location of the chain guide between the cams—right hand side

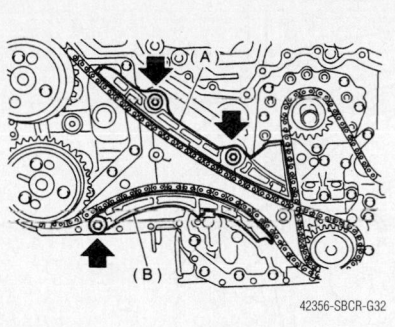

42356-SBCR-G32

Fig. 160 Location of the chain guide—right hand side

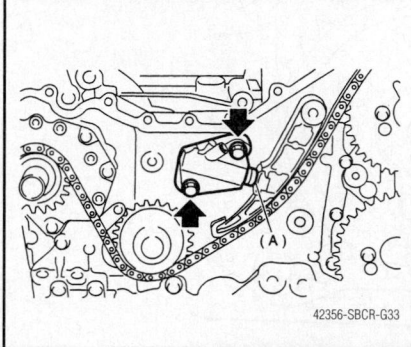

42356-SBCR-G33

Fig. 161 The chain tensioner plunger A does not come out—left hand side

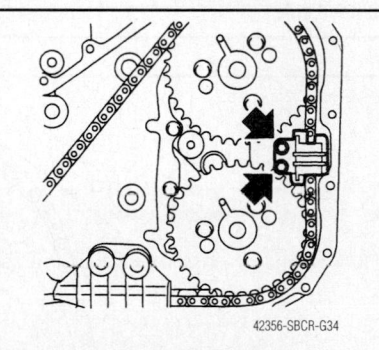

42356-SBCR-G34

Fig. 162 Location of the chain guide between the cams—left hand side

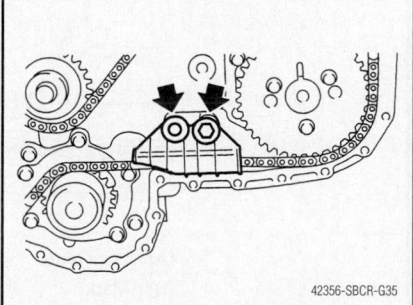

42356-SBCR-G35

Fig. 163 Location of the chain guide—left hand side

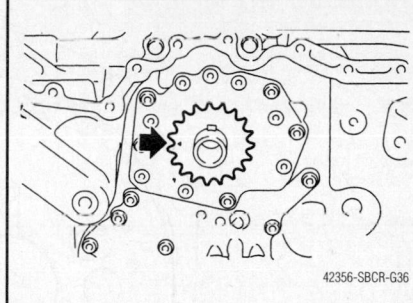

42356-SBCR-G36

Fig. 164 Align the TOP MARK on the crankshaft sprocket to the 9 O'clock position

09490_SBCR_G0076

Fig. 165 Align the TOP MARK on the camshaft sprocket to the 12 O'clock position

place. Always perform this task on a rubber mat.

16. Using the crankshaft socket tool, align the **TOP MARK** on the crankshaft sprocket to the 9 O'clock position, as shown in the illustration.

17. Align the key groove on the exhaust camshaft sprocket to the 12 O'clock position, as shown in the illustration.

18. Align the intake camshaft sprocket, as shown in the illustration.

19. Turn the crankshaft sprocket clockwise; align the **TOP MARK** to the 12 O'clock position. Piston number one is now at Top Dead Center (TDC).

※※ CAUTION

Do not rotate the camshaft or crankshaft sprockets until the chain is completely routed or damage will occur.

20. Install the lower idler sprocket and tighten the bolt to 50.6 ft. lbs. (69 Nm).

21. Install the left side timing chain, align the mark "B" on the crankshaft sprocket with the matching mark "A" on the timing chain.

22. Route the left side timing chain onto the lower idler sprocket, water pump,

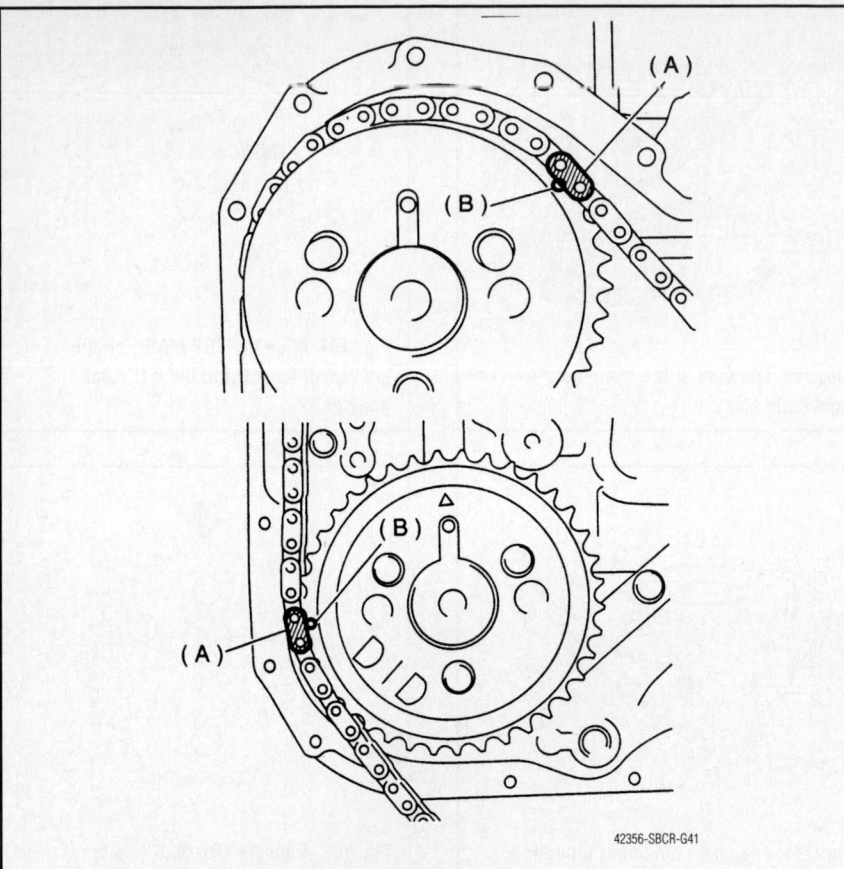

Fig. 166 Make sure the mark A on the chain and the mark B camshaft sprocket are aligned the same way as the one on the crankshaft sprocket—right hand side

26. Install the left side chain guide. Tighten the bolts to 12 ft. lbs. (16 Nm). Install the tensioner lever on the left side and tighten the bolt to 12 ft. lbs. (16 Nm). Install the chain tensioner on the left side and tighten the bolts to 12 ft. lbs. (16 Nm).

27. Install the right side timing chain. Align the marks of the left and right timing chains on the lower idler sprocket. Install the right side timing chain to the right side intake camshaft sprocket and the right side exhaust camshaft sprocket in this order.

28. Make sure the mark "A" on the chain and the mark "B" camshaft sprocket are aligned the same way as the one on the crankshaft sprocket or damage will occur.

29. Install the right side chain guide. Install the right side chain tensioner lever and tighten the bolts to 12 ft. lbs. (16 Nm). Install the right side chain guide and tighten the NEW bolt to 4.7 ft. lbs. (6 Nm). Install the right side chain tensioner and tighten the bolts to 12 ft. lbs. (16 Nm).

30. Adjust the clearance between the chain guide on the right side and the center chain guide so that there is range between 0.331–0.339 inch (8.4–8.6mm).

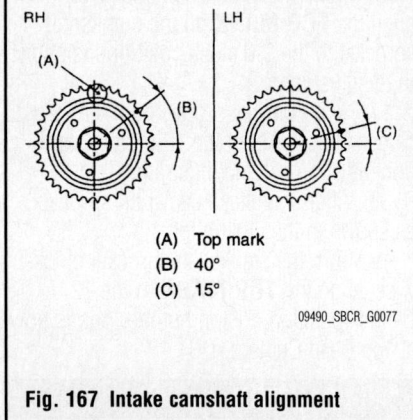

(A) Top mark
(B) 40°
(C) 15°

Fig. 167 Intake camshaft alignment

exhaust cam sprocket and the intake cam sprocket in that order.

23. Make sure the mark "A" on the chain and the mark "B" camshaft sprocket are aligned the same way as the one on the crankshaft sprocket or damage will occur.

24. Install the upper chain idler and tighten the bolt to 50.6 ft. lbs. (69 Nm).

25. Install the left side chain guide, between the camshafts. Tighten the bolt to 4.7 ft. lbs. (6 Nm) using a NEW bolt.

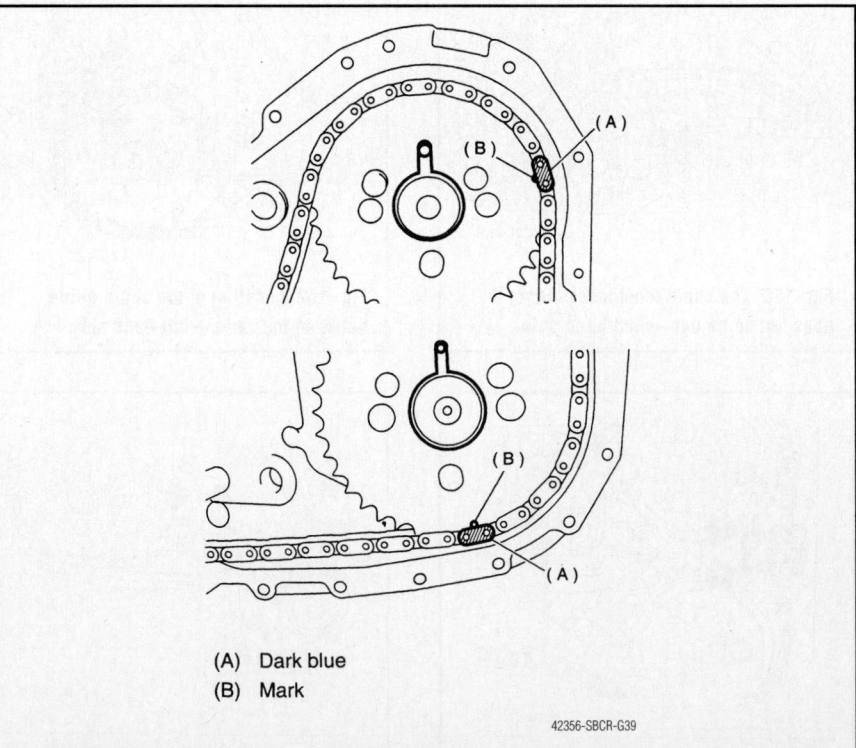

(A) Dark blue
(B) Mark

Fig. 168 Make sure the mark A on the chain and the mark B camshaft sprocket are aligned the same way as the one on the crankshaft sprocket—left hand side

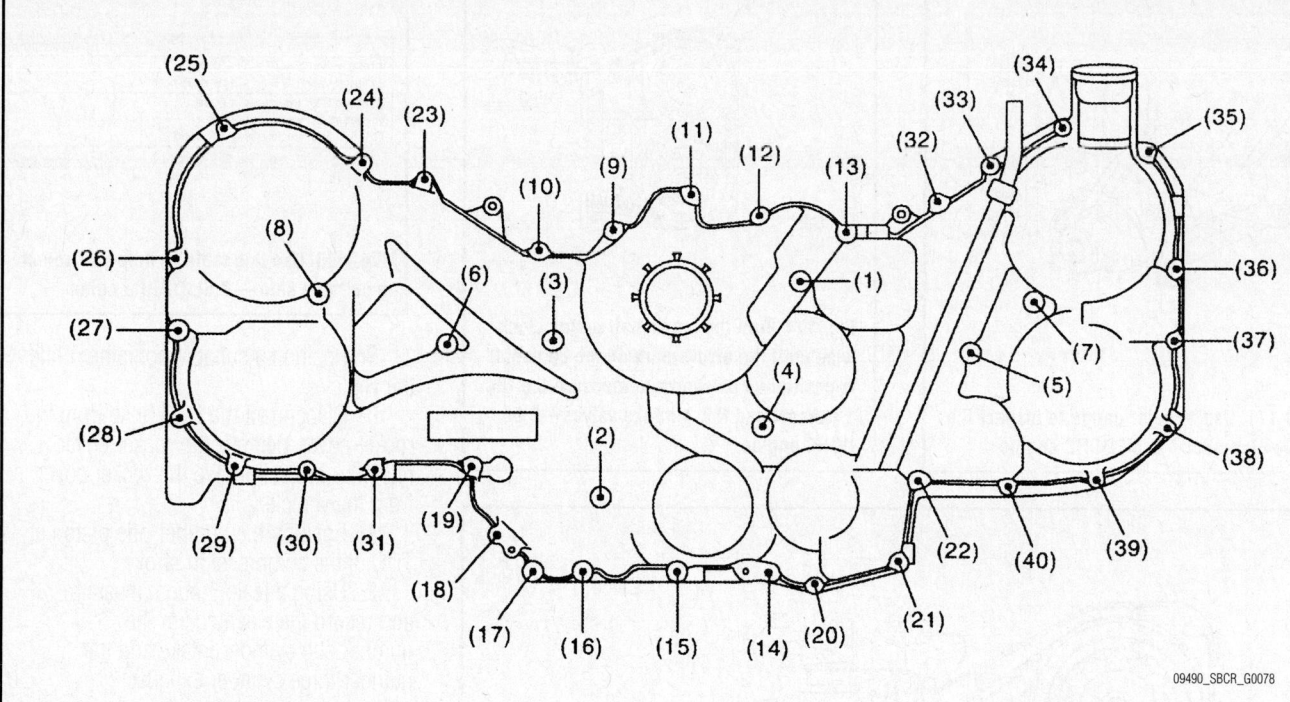

Fig. 169 Front timing cover bolt tightening sequence

31. Install the center chain guide and tighten the NEW bolt to 5.8 ft. lbs. (8 Nm).

32. Check the match marks on each sprocket and corresponding timing chain are correct, remove the stopper from the tensioner.

33. Clean the mating surfaces on the front timing cover. Apply a bead of liquid gasket 0.020 inch in diameter to the mating surface of the front timing chain cover. Install the timing chain cover. Torque the retaining bolts to 4.8 ft. lbs. and in the proper sequence.

34. Continue the installation in the reverse order of the removal procedure.

VALVE LASH

ADJUSTMENT

2.5L DOHC Engine

See Figures 170 through 177.

➡**The valve adjustment should be performed while the engine is cold.**

1. Raise and support the vehicle safely.

2. Remove the undercover.
3. Lower the vehicle.
4. Remove the collector cover.
5. Disconnect the negative battery cable.
6. Remove the air intake duct.
7. Remove the bolt that retains the

right side timing belt cover. Remove the remaining bolts and remove the right side timing belt cover.

8. Disconnect the ignition coil electrical connector. Remove the ignition coil.

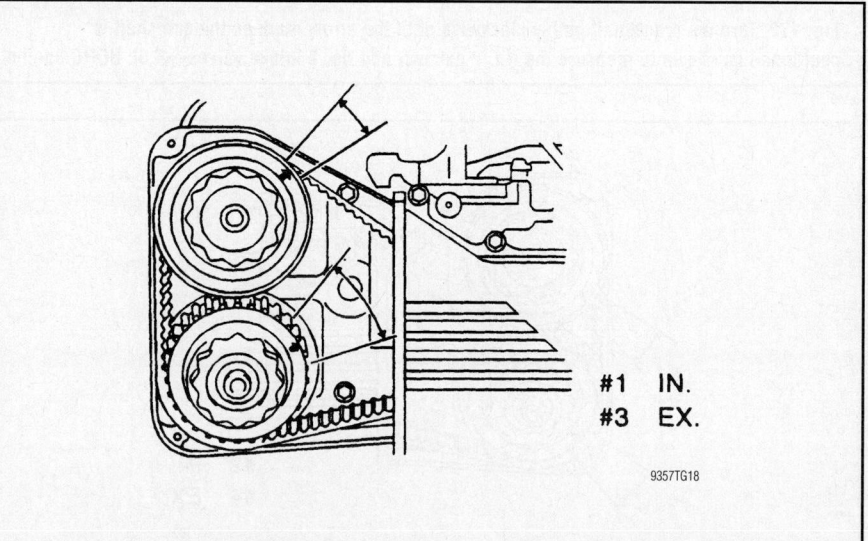

#1 IN.
#3 EX.

Fig. 170 Turn the crankshaft pulley clockwise until the arrow mark on the camshaft is positioned as shown to measure the No. 1 intake and No. 3 exhaust valves—2.5L DOHC engine

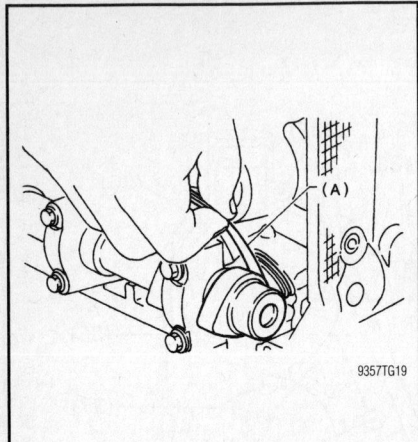

Fig. 171 Use a feeler gauge to inspect the valve clearance—2.5L DOHC engine

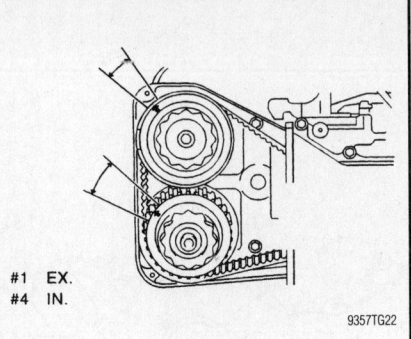

#1 EX.
#4 IN.

Fig. 174 Turn the crankshaft pulley clockwise until the arrow mark on the camshaft is positioned as shown to measure the No. 1 exhaust and No. 4 intake valves—2.5L DOHC engine

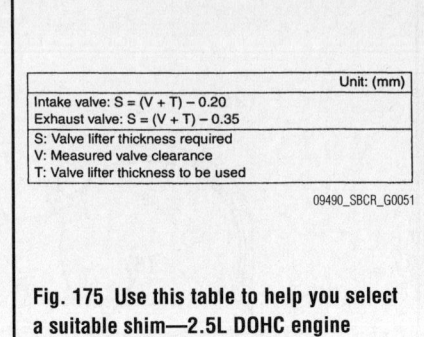

	Unit: (mm)
Intake valve: S = (V + T) − 0.20	
Exhaust valve: S = (V + T) − 0.35	
S: Valve lifter thickness required	
V: Measured valve clearance	
T: Valve lifter thickness to be used	

09490_SBCR_G0051

Fig. 175 Use this table to help you select a suitable shim—2.5L DOHC engine

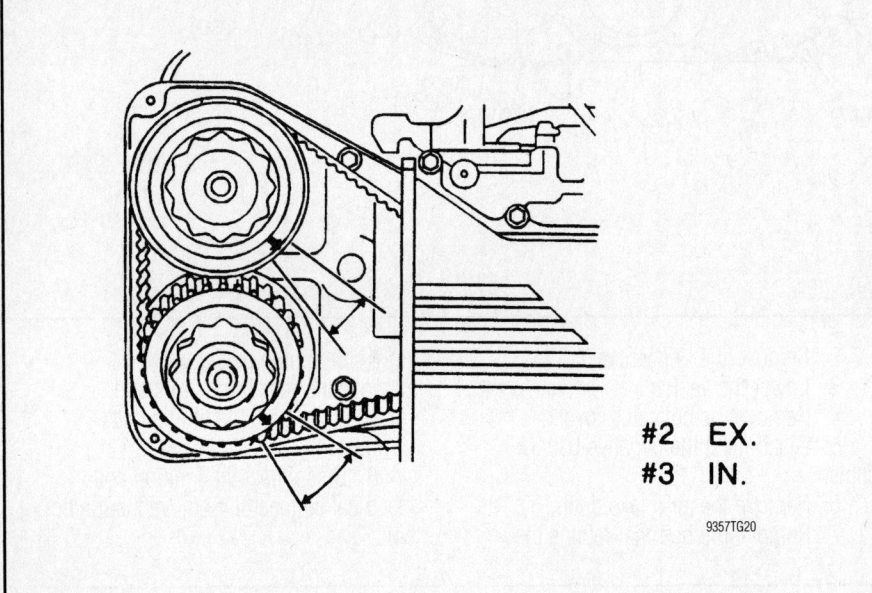

#2 EX.
#3 IN.

9357TG20

Fig. 172 Turn the crankshaft pulley clockwise until the arrow mark on the camshaft is positioned as shown to measure the No. 2 exhaust and No. 3 intake valves—2.5L DOHC engine

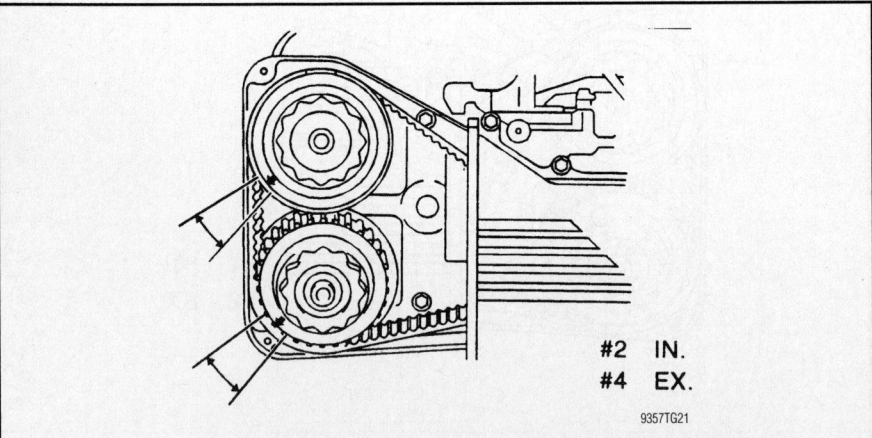

#2 IN.
#4 EX.

9357TG21

Fig. 173 Turn the crankshaft pulley clockwise until the arrow mark on the camshaft is positioned as shown to measure the No. 2 intake and No. 4 exhaust valves—2.5L DOHC engine

9. Position a suitable container under the vehicle.

10. Disconnect the PCV hose from the rocker cover. Remove the rocker cover retaining bolts. Remove the rocker cover from the vehicle.

11. Position the number one piston at TDC of the compression stroke.

12. Using a feeler gauge, measure and record the clearance of the number one cylinder intake and the number three cylinder exhaust valves.

➡ **Insert the feeler gauge in a horizontally as possible with respect to the valve lifter. Measure and record the exhaust valve clearance while lifting up the vehicle.**

13. Rotate the crankshaft pulley clockwise until the arrow mark on the camshaft is positioned as shown to measure and record the clearance on the number two exhaust and number three intake valves.

14. Rotate the crankshaft pulley clockwise until the arrow mark on the camshaft is positioned as shown to measure and record the number two intake and number four exhaust valves.

15. Rotate the crankshaft pulley clockwise until the arrow mark on the camshaft is positioned as shown to measure and record the number one exhaust and number four intake valves.

16. If adjustment is required, remove the camshafts.

17. Remove and measure the thickness of the valve lifter. Select a suitable shim, using the shim selection chart.

18. Install the replacement shim to the lifter.

19. After all shims have been adjusted, inspect the valve clearances again.

20. After completion, install all removed components.

Part No.	Thickness mm (in)	Part No.	Thickness mm (in)
13228 AB102	4.68 (0.1843)	13228 AB622	5.20 (0.2047)
13228 AB112	4.69 (0.1846)	13228 AB632	5.21 (0.2051)
13228 AB122	4.70 (0.1850)	13228 AB642	5.22 (0.2055)
13228 AB132	4.71 (0.1854)	13228 AB652	5.23 (0.2059)
13228 AB142	4.72 (0.1858)	13228 AB662	5.24 (0.2063)
13228 AB152	4.73 (0.1862)	13228 AB672	5.25 (0.2067)
13228 AB162	4.74 (0.1866)	13228 AB682	5.26 (0.2071)
13228 AB172	4.75 (0.1870)	13228 AB692	5.27 (0.2075)
13228 AB182	4.76 (0.1874)	13228 AB702	4.38 (0.1724)
13228 AB192	4.77 (0.1878)	13228 AB712	4.40 (0.1732)
13228 AB202	4.78 (0.1882)	13228 AB722	4.42 (0.1740)
13228 AB212	4.79 (0.1886)	13228 AB732	4.44 (0.1748)
13228 AB222	4.80 (0.1890)	13228 AB742	4.46 (0.1756)
13228 AB232	4.81 (0.1894)	13228 AB752	4.48 (0.1764)
13228 AB242	4.82 (0.1898)	13228 AB762	4.50 (0.1771)
13228 AB252	4.83 (0.1902)	13228 AB772	4.52 (0.1780)
13228 AB262	4.84 (0.1906)	13228 AB782	4.54 (0.1787)
13228 AB272	4.85 (0.1909)	13228 AB792	4.56 (0.1795)
13228 AB282	4.86 (0.1913)	13228 AB802	4.58 (0.1803)
13228 AB292	4.87 (0.1917)	13228 AB812	4.60 (0.1811)
13228 AB302	4.88 (0.1921)	13228 AB822	4.62 (0.1819)
13228 AB312	4.89 (0.1925)	13228 AB832	4.64 (0.1827)
13228 AB322	4.90 (0.1929)	13228 AB842	4.66 (0.1835)
13228 AB332	4.91 (0.1933)	13228 AB852	5.29 (0.2083)
13228 AB342	4.92 (0.1937)	13228 AB862	5.31 (0.2091)
13228 AB352	4.93 (0.1941)	13228 AB872	5.33 (0.2098)
13228 AB362	4.94 (0.1945)	13228 AB882	5.35 (0.2106)
13228 AB372	4.95 (0.1949)	13228 AB892	5.37 (0.2114)
13228 AB382	4.96 (0.1953)	13228 AB902	5.39 (0.2122)
13228 AB392	4.97 (0.1957)	13228 AB912	5.41 (0.2123)
13228 AB402	4.98 (0.1961)	13228 AB922	5.43 (0.2138)
13228 AB412	4.99 (0.1965)	13228 AB932	5.45 (0.2146)
13228 AB422	5.00 (0.1969)	13228 AB942	5.47 (0.2154)
13228 AB432	5.01 (0.1972)	13228 AB952	5.49 (0.2161)
13228 AB442	5.02 (0.1976)	13228 AB962	5.51 (0.2169)
13228 AB452	5.03 (0.1980)	13228 AB972	5.53 (0.2177)
13228 AB462	5.04 (0.1984)	13228 AB982	5.55 (0.2185)
13228 AB472	5.05 (0.1988)	13228 AB992	5.57 (0.2193)
13228 AB482	5.06 (0.1992)	13228 AC002	5.59 (0.2201)
13228 AB492	5.07 (0.1996)	13228 AC012	5.61 (0.2209)
13228 AB502	5.08 (0.2000)	13228 AC022	5.63 (0.2217)
13228 AB512	5.09 (0.2004)	13228 AC032	5.65 (0.2224)
13228 AB522	5.10 (0.2008)		
13228 AB532	5.11 (0.2012)		
13228 AB542	5.12 (0.2016)		
13228 AB552	5.13 (0.2020)		
13228 AB562	5.14 (0.2024)		
13228 AB572	5.15 (0.2028)		
13228 AB582	5.16 (0.2031)		
13228 AB592	5.17 (0.2035)		
13228 AB602	5.18 (0.2039)		
13228 AB612	5.19 (0.2043)		

09490_SBCR_G0052

Fig. 176 Valve adjusting shim chart—2.5L DOHC engine

2.5L SOHC Engine

See Figure 177.

➡**The valve adjustment should be performed while the engine is cold.**

1. Raise and support the vehicle safely.
2. Remove the undercover.
3. Lower the vehicle.
4. Disconnect the negative battery cable.
5. To remove the front side V belt, remove the belt covers. Loosen the lock bolt. Loosen the slider bolt. Remove the front side belt.
6. To remove the rear side V belt, remove the belt covers. Loosen the lock bolt. Loosen the slider bolt. Remove the rear side belt. Remove the belt tensioner.
7. Remove the crankshaft pulley bolt.
8. Lock the crankshaft in place using tool ST499977100, or equivalent.
9. Remove the crankshaft pulley.
10. Remove the left side timing belt cover.
11. Remove the fuel injector.
12. Remove the rocker cover.
13. Position the number one piston at TDC of the compression stroke.

➡**When the arrow (see illustration) on the camshaft sprocket (left side) comes exactly to the top, number one cylinder piston is at TDC of the compression stroke.**

14. Measure the valve clearance, using a feeler gauge.
15. If adjustment is needed, loosen the valve rocker nut and screw. Position the feeler gauge.

➡**Insert the feeler gauge in a horizontally as possible with respect to the valve stem end face. Adjust the exhaust valve clearance while lifting up the vehicle.**

16. While noting the valve clearance, tighten the rocker adjusting screw.

17. When the proper valve clearance is obtained, tighten the valve rocker nut to 7.2 ft. lbs.
18. Adjust the valve clearance on the remaining cylinders, following the above procedure.

➡**Be sure to position the pistons to their respective TDC positions on the compression stroke, before checking and adjusting the valves. By rotating the crankshaft pulley clockwise every 180 degrees from the state that number one piston is on TDC of the compression stroke, the remaining pistons come to TDC of the compression stroke in the following order, #3, #2, and #4.**

19. After adjustment, replace any removed components.
20. Be sure to use new gaskets and seals, as required.

3.0L Engine

See Figures 178 through 182.

➡**The valve adjustment should be performed while the engine is cold.**

1. Raise and support the vehicle safely.
2. Remove the undercover.
3. Lower the vehicle.
4. Remove the collector cover.
5. Disconnect the negative battery cable.
6. On the right side, remove the air intake duct and air cleaner case. Remove the fuel tank protector. Disconnect the oil pressure switch electrical connector. Remove the ignition coil.
7. On the left side, remove the battery and battery carrier. Disconnect the PCV hose from the rocker cover. Remove the ignition coil.
8. Remove the rocker cover retaining bolts. Remove the rocker covers from the engine.
9. Rotate the crankshaft clockwise until the cam is set in position, see illustration.

Unit: (mm)
$S = (V + T) - 0.35$
S: Valve lifter thickness required
V: Measured valve clearance
T: Valve lifter thickness to be used

09490_SBCR_G0054

Fig. 179 Use this table to help you select a suitable exhaust valve shim—3.0L engine

Unit: (mm)
$S = (V + T) - 0.20$
S: Required shim thickness
V: Measured valve clearance
T: Shim thickness to be used

09490_SBCR_G0055

Fig. 180 Use this table to help you select a suitable intake valve shim—3.0L engine

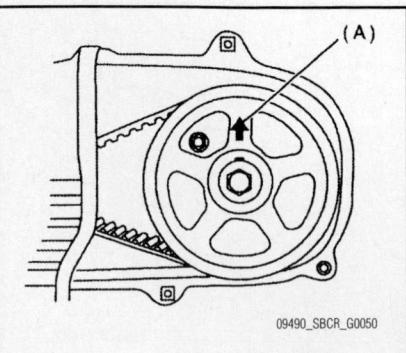

09490_SBCR_G0050

Fig. 177 TDC alignment—2.5L SOHC engine

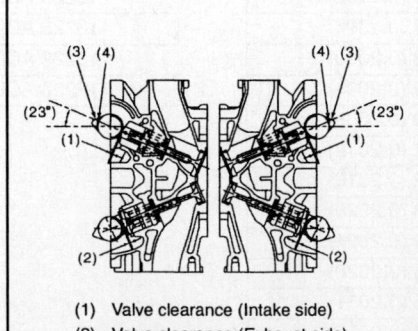

(1) Valve clearance (Intake side)
(2) Valve clearance (Exhaust side)
(3) High lift cam
(4) Low lift cam

09490_SBCR_G0053

Fig. 178 Valve adjustment crankshaft positioning—3.0L engine

Part No.	Thickness mm (in)	Part No.	Thickness mm (in)
13228AD180	4.32 (0.1701)	13228AC860	4.90 (0.1929)
13228AD190	4.34 (0.1709)	13228AC870	4.91 (0.1933)
13228AD200	4.36 (0.1717)	13228AC880	4.92 (0.1937)
13228AD210	4.38 (0.1724)	13228AC890	4.93 (0.1941)
13228AD220	4.40 (0.1732)	13228AC900	4.94 (0.1945)
13228AD230	4.42 (0.1740)	13228AC910	4.95 (0.1949)
13228AD240	4.44 (0.1748)	13228AC920	4.96 (0.1953)
13228AD250	4.46 (0.1756)	13228AC930	4.97 (0.1957)
13228AD260	4.48 (0.1764)	13228AC940	4.98 (0.1961)
13228AD270	4.50 (0.1772)	13228AC950	4.99 (0.1965)
13228AD280	4.52 (0.1780)	13228AC960	5.00 (0.1969)
13228AD290	4.54 (0.1787)	13228AC970	5.01 (0.1972)
13228AD300	4.56 (0.1795)	13228AC980	5.02 (0.1976)
13228AD310	4.58 (0.1803)	13228AC990	5.03 (0.1980)
13228AD320	4.60 (0.1811)	13228AD000	5.04 (0.1984)
13228AC580	4.62 (0.1819)	13228AD010	5.05 (0.1988)
13228AC590	4.63 (0.1823)	13228AD020	5.06 (0.1992)
13228AC600	4.64 (0.1827)	13228AD030	5.07 (0.1996)
13228AC610	4.65 (0.1831)	13228AD040	5.08 (0.2000)
13228AC620	4.66 (0.1835)	13228AD050	5.09 (0.2004)
13228AC630	4.67 (0.1839)	13228AD060	5.10 (0.2008)
13228AC640	4.68 (0.1843)	13228AD070	5.11 (0.2012)
13228AC650	4.69 (0.1846)	13228AD080	5.12 (0.2016)
13228AC660	4.70 (0.1850)	13228AD090	5.13 (0.2020)
13228AC670	4.71 (0.1854)	13228AD100	5.14 (0.2024)
13228AC680	4.72 (0.1858)	13228AD110	5.15 (0.2028)
13228AC690	4.73 (0.1862)	13228AD120	5.16 (0.2032)
13228AC700	4.74 (0.1866)	13228AD130	5.17 (0.2035)
13228AC710	4.75 (0.1870)	13228AD140	5.18 (0.2039)
13228AC720	4.76 (0.1874)	13228AD150	5.19 (0.2043)
13228AC730	4.77 (0.1878)	13228AD160	5.20 (0.2047)
13228AC740	4.78 (0.1882)	13228AD170	5.21 (0.2051)
13228AC750	4.79 (0.1886)	13228AD330	5.23 (0.2059)
13228AC760	4.80 (0.1890)	13228AD340	5.25 (0.2067)
13228AC770	4.81 (0.1894)	13228AD350	5.27 (0.2075)
13228AC780	4.82 (0.1898)	13228AD360	5.29 (0.2083)
13228AC790	4.83 (0.1902)	13228AD370	5.31 (0.2091)
13228AC800	4.84 (0.1906)	13228AD380	5.33 (0.2098)
13228AC810	4.85 (0.1909)	13228AD390	5.35 (0.2106)
13228AC820	4.86 (0.1913)	13228AD400	5.37 (0.2114)
13228AC830	4.87 (0.1917)	13228AD410	5.39 (0.2122)
13228AC840	4.88 (0.1921)	13228AD420	5.41 (0.2130)
13228AC850	4.89 (0.1925)	13228AD430	5.43 (0.2138)
		13228AD440	5.45 (0.2146)
		13228AD450	5.47 (0.2154)
		13228AD460	5.49 (0.2161)
		13228AD470	5.51 (0.2169)
		13228AD480	5.53 (0.2177)
		13228AD490	5.55 (0.2185)
		13228AD500	5.57 (0.2193)
		13228AD510	5.59 (0.2201)

09490_SBCR_G0056

Fig. 181 Exhaust valve adjusting shim chart—3.0L engine

Part No.	Thickness mm (in)
13218AK890	1.92 (0.0756)
13218AK900	1.94 (0.0764)
13218AK910	1.96 (0.0772)
13218AK920	1.98 (0.0780)
13218AK930	2.00 (0.0787)
13218AK940	2.02 (0.0795)
13218AK950	2.04 (0.0803)
13218AK960	2.06 (0.0811)
13218AK970	2.07 (0.0815)
13218AK980	2.08 (0.0819)
13218AK990	2.09 (0.0823)
13218AL000	2.10 (0.0827)
13218AL010	2.11 (0.0831)
13218AL020	2.12 (0.0835)
13218AL030	2.13 (0.0839)
13218AL040	2.14 (0.0843)
13218AL050	2.15 (0.0846)
13218AL060	2.16 (0.0850)
13218AL070	2.17 (0.0854)
13218AL080	2.18 (0.0858)
13218AL090	2.19 (0.0862)
13218AL100	2.20 (0.0866)
13218AL110	2.21 (0.0870)
13218AL120	2.22 (0.0874)
13218AL130	2.23 (0.0878)
13218AL140	2.24 (0.0882)
13218AL150	2.25 (0.0886)
13218AL160	2.26 (0.0890)
13218AL170	2.27 (0.0894)
13218AL180	2.28 (0.0898)
13218AL190	2.29 (0.0902)
13218AL200	2.30 (0.0906)
13218AL210	2.31 (0.0909)
13218AL220	2.32 (0.0913)
13218AL230	2.33 (0.0917)
13218AL240	2.34 (0.0921)
13218AL250	2.35 (0.0925)
13218AL260	2.36 (0.0929)
13218AL270	2.37 (0.0933)
13218AL280	2.38 (0.0937)
13218AL290	2.39 (0.0941)
13218AL300	2.40 (0.0945)
13218AL310	2.41 (0.0949)

Part No.	Thickness mm (in)
13218AL320	2.42 (0.0953)
13218AL330	2.43 (0.0957)
13218AL340	2.44 (0.0961)
13218AL350	2.45 (0.0965)
13218AL360	2.46 (0.0969)
13218AL370	2.47 (0.0972)
13218AL380	2.48 (0.0976)
13218AL390	2.49 (0.0980)
13218AL400	2.50 (0.0984)
13218AL410	2.51 (0.0988)
13218AL420	2.52 (0.0992)
13218AL430	2.53 (0.0996)
13218AL440	2.54 (0.1000)
13218AL450	2.55 (0.1004)
13218AL460	2.56 (0.1008)
13218AL470	2.57 (0.1012)
13218AL480	2.58 (0.1016)
13218AL490	2.59 (0.1020)
13218AL500	2.60 (0.1024)
13218AL510	2.61 (0.1028)
13218AL520	2.62 (0.1032)
13218AL530	2.64 (0.1039)
13218AL540	2.66 (0.1047)
13218AL550	2.68 (0.1055)
13218AL560	2.70 (0.1063)
13218AL570	2.72 (0.1071)
13218AL580	2.74 (0.1079)
13218AL590	2.76 (0.1087)

09490_SBCR_G0057

Fig. 182 Intake valve adjusting shim chart—3.0L engine

10. Using a feeler gauge measure and record the clearance of the intake and exhaust valve.

➡**Measure it within the range of +/-30 degrees from the specified position, shown in the illustration. Measure it in the low cam for the intake side. Insert the feeler gauge in a** horizontally as possible with respect to the valve lifter.

11. Further turn the crankshaft pulley clockwise and then measure and record the valve clearance again.

12. If adjustment is required, remove the camshafts.

13. Remove and measure the thickness of the valve lifter. Select a suitable shim, using the shim selection chart.

14. Install the replacement shim to the lifter.

15. After all shims have been adjusted, inspect the valve clearances again.

16. After completion, install all removed components.

ENGINE PERFORMANCE & EMISSION CONTROLS

CAMSHAFT POSITION (CMP) SENSOR

REMOVAL & INSTALLATION

2.5L DOHC Engine

1. Disconnect the negative battery cable.
2. Remove the collector cover.
3. Disconnect the connector from camshaft position sensor RH.
4. Remove the camshaft position sensor RH from the rear side of the cylinder head.
5. Remove the cam shaft position sensor LH in the same way as RH.

To install:
6. Installation is the reverse of the removal procedure.
7. Tighten the sensor to 4.7 ft. lbs. (6.4 Nm).

2.5L SOHC Engine

1. Disconnect the negative battery cable.
2. Disconnect the connector from the Camshaft Position (CMP) sensor.
3. Remove the bolt that retains the CMP sensor to the sensor support.
4. Remove the bolt that retains the CMP sensor support to the camshaft cap.
5. Remove the sensor and the CMP sensor support as a unit.
6. Separate the sensor from the support.

To install:
7. Installation is the reverse of the removal procedure.
8. Tighten the CMP sensor support to 4.7 ft. lbs. (6.4 Nm).
9. Tighten the sensor to 4.7 ft. lbs. (6.4 Nm).

3.0L Engine

1. Disconnect the negative battery cable.
2. Remove the collector cover, as required.
3. Remove the alternator harness from the fuel pipe. Remove the fuel pipe protector.
4. Disconnect the connector from the sensor.

5. Remove the bolt that retains the sensor.
6. Remove the sensor.

To install:
7. Installation is the reverse of the removal procedure.
8. Tighten the sensor to 4.7 ft. lbs. (6.4 Nm).

3.6L Engine

Intake Side

See Figures 183 through 188.

1. Remove the collector cover.

➡**Follow these procedures for removal of the collector cover.**

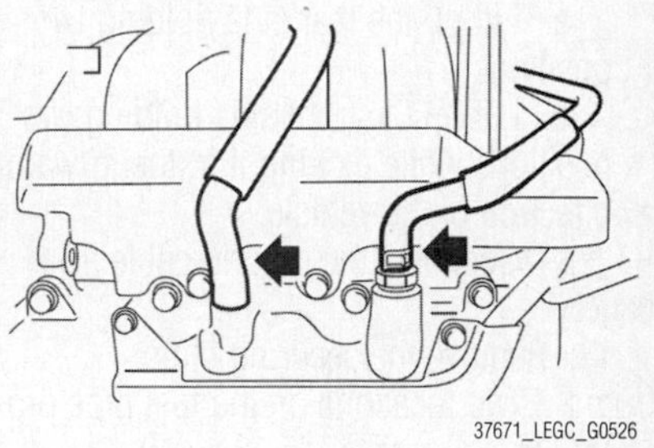

Fig. 185 PCV hose and blow—by—hose disconnection—intake side

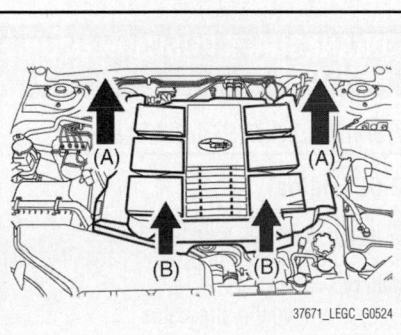

Fig. 183 Collector cover removal—intake side

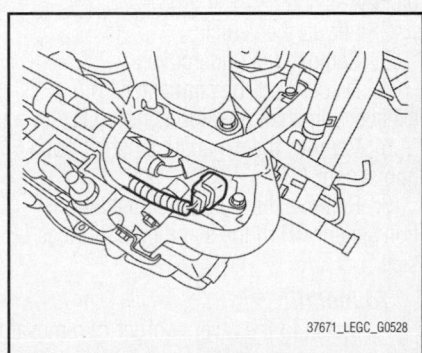

Fig. 186 Fuel pipe protector LH—intake side

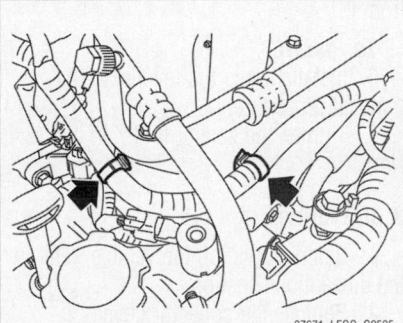

Fig. 184 Securing clips on the fuel pipe protector LH—intake side

Fig. 187 Intake camshaft position sensor LH connector—intake side

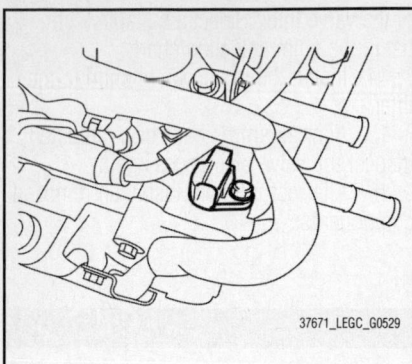

Fig. 188 Intake camshaft position sensor LH—intake side

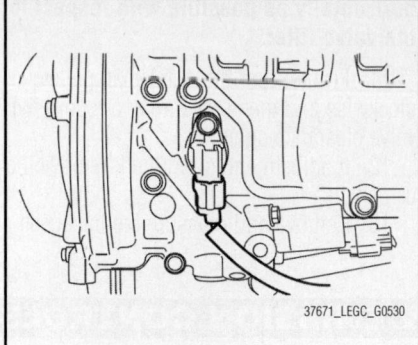

Fig. 189 Camshaft position sensor LH connector—exhaust side

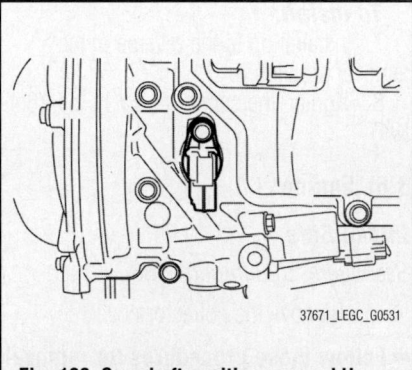

Fig. 190 Camshaft position sensor LH—exhaust

a. Lift up the rear side holding two positions.

b. Lift up the front side holding two positions while moving it in the forward direction of the vehicle.

2. Disconnect the ground cable from battery.

3. Remove the securing clips from the two locations on the fuel pipe protector LH.

4. Disconnect the PCV hose and blow-by hose.

5. Remove the fuel pipe protector LH.

6. Disconnect the connector from intake camshaft position sensor LH.

7. Remove the intake camshaft position sensor LH.

8. Remove the intake camshaft position sensor RH in the same procedure as LH side.

To install:

9. Install in the reverse order of removal.

10. Tighten the camshaft position sensor to 56 inch lbs. (6 Nm) and the fuel pipe protector to 14 ft. lbs. (19 Nm)

Exhaust Side

See Figures 189 and 190.

1. Disconnect the ground cable from battery.

2. Lift up the vehicle.

3. Remove the undercover.

4. Disconnect the connector from exhaust camshaft position sensor LH.

5. Remove the exhaust camshaft position sensor LH.

6. Remove the exhaust camshaft position sensor RH in the same procedure as LH side.

To install:

7. Install in the reverse order of removal. Tighten the exhaust camshaft position sensor to 56 inch lbs. (6 Nm).

CRANKSHAFT POSITION (CKP) SENSOR

REMOVAL & INSTALLATION

2.5L Engine

1. Remove the collector cover.

2. Disconnect the ground cable from battery.

3. Remove the alternator.

4. Remove the bolt which installs crankshaft position sensor to cylinder block.

5. Remove the crankshaft position sensor, and then disconnect the connector from it.

To install:

6. Installation is the reverse of the removal procedure.

7. Tighten the sensor to 4.7 ft. lbs. (6.4 Nm).

3.0L Engine

1. Before servicing the vehicle, refer to the precautions section.

2. Remove the collector cover.

3. Disconnect the ground cable from battery.

4. Remove the air intake chamber.

5. Remove the service hole cover.

6. Remove the crankshaft position sensor.

7. Disconnect the connector from crankshaft position sensor.

To install:

8. Install in the reverse order of removal.

9. Tighten the crankshaft sensor mounting bolt to 5 ft. lbs. (6.8 Nm).

3.6L Engine

See Figure 191.

1. Remove the collector cover.

➡**Follow these procedures for removal of the collector cover.**

a. Lift up the rear side holding two positions.

b. Lift up the front side holding two positions while moving it in the forward direction of the vehicle.

2. Disconnect the ground cable from battery.

3. Remove the air intake boot assembly.

4. Remove the throttle body from the intake manifold.

➡**Do not disconnect the inlet and outlet hoses of engine coolant.**

5. Remove the service hole plug.

6. Remove the crankshaft position sensor.

7. Disconnect the connector from crankshaft position sensor.

To install:

8. Install in the reverse order of removal.

9. Tighten the crankshaft position sensor to 56.4 inch lbs. (6.4 Nm).

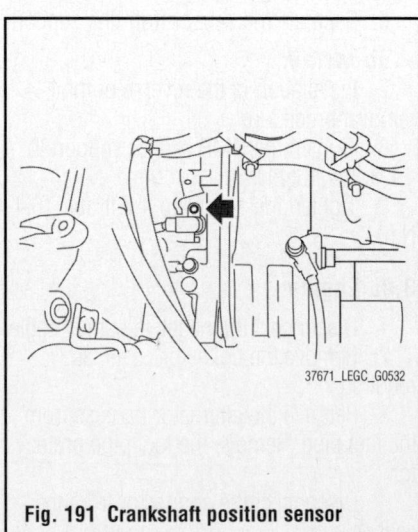

Fig. 191 Crankshaft position sensor

ENGINE COOLANT TEMPERATURE (ECT) SENSOR

REMOVAL & INSTALLATION

See Figure 192.

1. Before servicing the vehicle, refer to the Precautions Section.
2. Disconnect the negative battery cable.
3. Drain the cooling system.
4. Remove the alternator.
5. Disconnect the electrical connector from the temperature sensor.
6. Remove the coolant temperature sensor.

To install:

7. Installation is the reverse order of removal. Tighten the sensor to 13 ft. lbs. (18 Nm)
8. Refill and bleed the cooling system to the correct level.

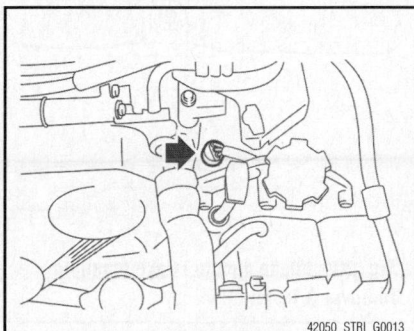

Fig. 192 Removing and installing coolant temperature sensor

EXHAUST GAS RECIRCULATION (EGR) VALVE

REMOVAL & INSTALLATION

See Figure 193.

1. Disconnect the negative battery cable.

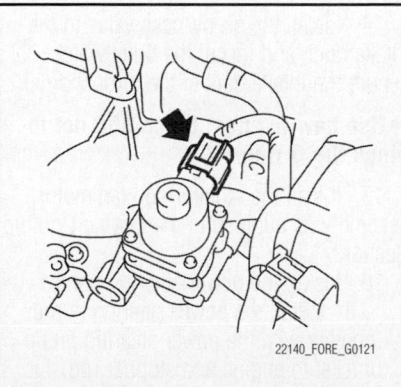

Fig. 193 EGR valve and connector shown

2. Disconnect the connector.
3. Remove the Exhaust Gas Recirculation (EGR) valve from the intake manifold.

To install:

4. Installation is the reverse of the removal procedure.
5. Tighten the retaining bolts to 14 ft. lbs. (19 Nm).

INTAKE AIR TEMPERATURE (IAT) SENSOR

REMOVAL & INSTALLATION

See Figure 194.

1. Disconnect the ground cable from battery.
2. Disconnect the connector from the mass air flow and intake air temperature sensor, and remove the mass air flow and intake air temperature sensor.

To install:

3. Connect the connector from the mass air flow and intake air temperature sensor, and install the mass air flow and intake air temperature sensor.
4. Connect the ground cable from battery.
5. Tighten sensor to 8.4 inch lbs. (1 Nm).

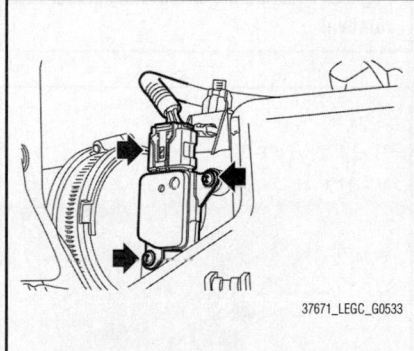

Fig. 194 Mass air flow and intake air temperature sensors

KNOCK SENSOR (KS)

REMOVAL & INSTALLATION

See Figures 195 through 197.

1. Remove the collector cover.
2. Disconnect the ground cable from battery.
3. Remove the intercooler as outlined in this section.
4. Disconnect the connector from the knock sensor.
5. Remove the knock sensor from cylinder block.

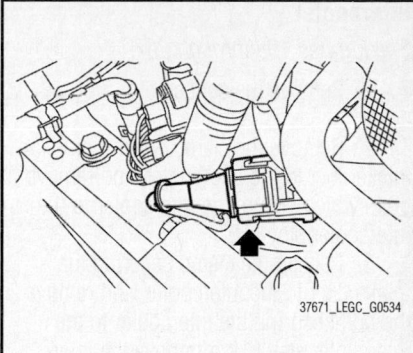

Fig. 195 Knock sensor connector

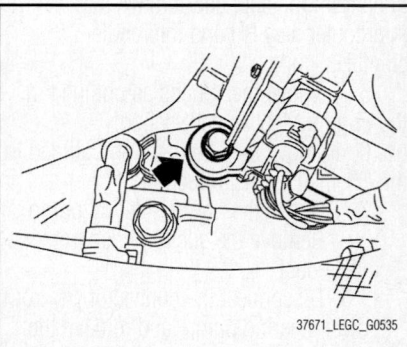

Fig. 196 Remove the knock sensor from cylinder block

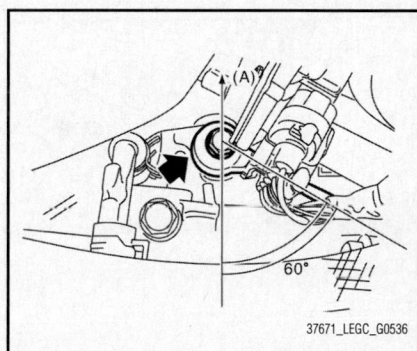

Fig. 197 Installing the knock sensor to the cylinder block

To install:

6. Install the knock sensor to the cylinder block.

➡**The portion of the knock sensor cord that is pulle d out must be positioned at a 60° angle relative to the engine rear.**

7. Tighten sensor to 18 inch lbs. (4 Nm).
8. Connect the connector to the knock sensor.
9. Install the intercooler as outlined in this section.
10. Connect the ground cable to battery.
11. Install the collector cover.

Intercooler

See Figures 198 through 207.

1. Remove the collector cover.

2. Remove the clamp which holds the intake duct to the intercooler, and remove bolts which secure the intercooler to the intercooler stay LH.

3. Loosen the clamp securing the intake duct to the intercooler, and remove the bolt securing the intercooler to the intercooler stay RH to remove the intercooler.

4. Remove the brake booster vacuum hose from the clip, and remove the intercooler stay RH and intercooler stay LH.

5. Remove the V-belts as outlined in the Engine Mechanical Section.

6. Remove the alternator as outlined in the Engine Electrical Section.

7. Remove the power steering pump.

  a. Remove the suction hose from intake duct.

  b. Disconnect the connector (A) from power steering pump, and remove the

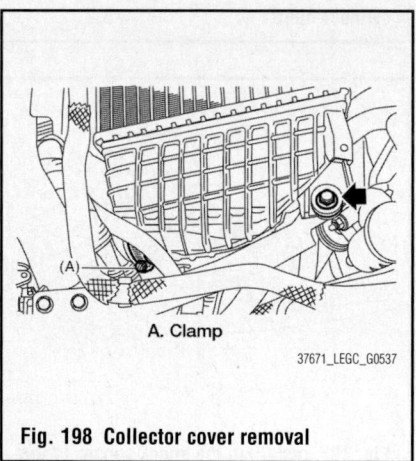

A. Clamp

37671_LEGC_G0537

Fig. 198 Collector cover removal

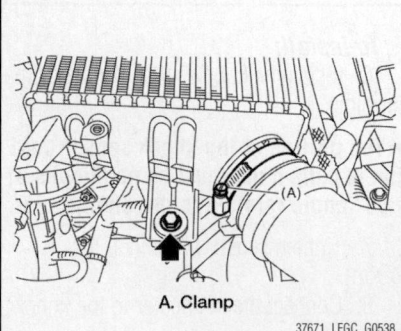

A. Clamp

37671_LEGC_G0538

Fig. 199 Removal of the bolt securing the intercooler to the intercooler stay RH to remove the intercooler

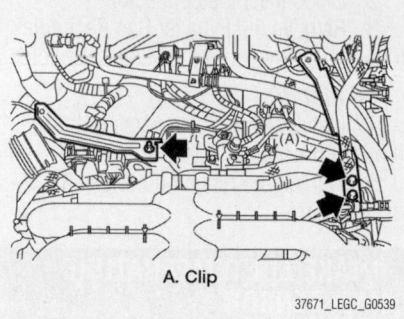

A. Clip

37671_LEGC_G0539

Fig. 200 Remove the brake booster vacuum hose from the clip, and remove the inter cooler stay RH, and intercooler stay LH

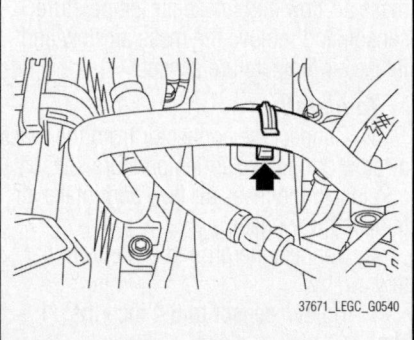

37671_LEGC_G0540

Fig. 201 Suction hose from intake duct removal

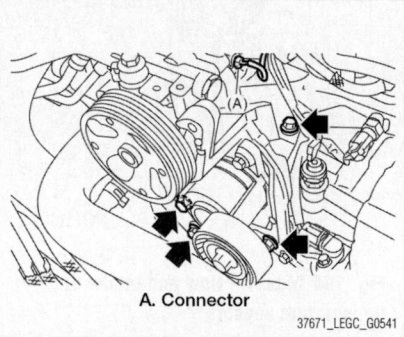

A. Connector

37671_LEGC_G0541

Fig. 202 Removal & installation of the power steering pump together with the power steering pump bracket from engine and stopper rod

power steering pump together with the power steering pump bracket from engine and stopper rod.

  c. Mount the power steering pump together with the power steering pump bracket on RH side wheel apron.

8. Remove the radiator sub fan motor assembly as outlined in the Engine Cooling Section.

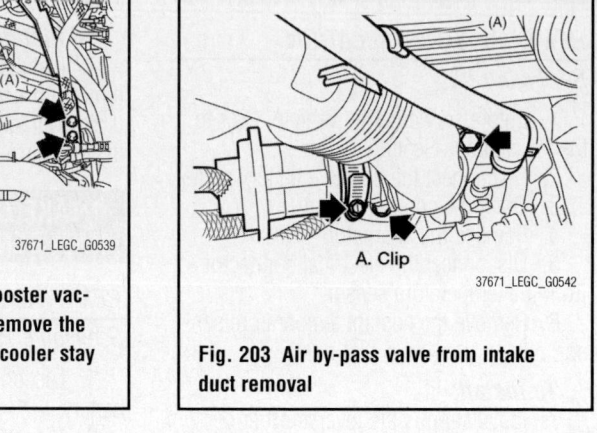

A. Clip

37671_LEGC_G0542

Fig. 203 Air by-pass valve from intake duct removal

37671_LEGC_G0543

Fig. 204 Intake duct to intake manifold removal & installation

9. Remove the engine coolant hose from the clip.

10. Remove the bolts which secure the intake duct to the turbocharger, and remove the air by-pass valve from the intake duct.

11. Remove the intake duct from intake manifold.

To install:

12. Install the intake duct to the intake manifold.

13. Attach the engine coolant hose to the clip.

14. Install the air by-pass valve to the intake duct, and install the bolts which secure the intake duct to the turbocharger.

➡ **Use new O-rings. Be careful not to pinch the O-ring.**

15. Install the radiator sub fan motor assembly as outlined in the Engine Cooling Section.

16. Install the power steering pump.

  a. Install the power steering pump together with the power steering pump bracket to engine and stopper rod, and connect the connector to the power steering pump.

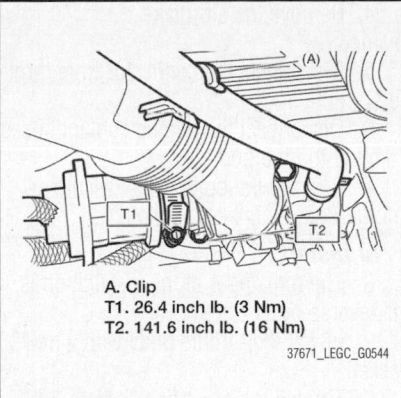

A. Clip
T1. 26.4 inch lb. (3 Nm)
T2. 141.6 inch lb. (16 Nm)

37671_LEGC_G0544

Fig. 205 Air by-pass valve to the intake duct installation

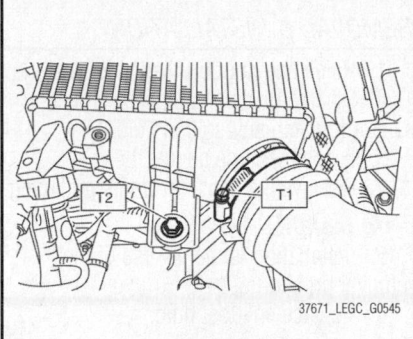

37671_LEGC_G0545

Fig. 206 Intercooler to the intercooler stay RH, and intake duct to the intercooler installation

T1. 26.4 inch lb. (3 Nm)
T2. 141.6 inch lb. (16 Nm)

37671_LEGC_G0546

Fig. 207 Intercooler to the intercooler stay installation

b. Install the suction hose to the intake duct.

17. Install the alternator, as outlined in the Engine Electrical Section.

18. Install the V-belts as outlined in the Engine Mechanical Section.

19. Install the intercooler stay RH and intercooler stay LH, and install

the brake booster vacuum hose to the clip.

20. Place the intercooler and attach the bolt to secure the intercooler to the intercooler stay RH. And tighten the clamp to secure the intake duct to the intercooler.

21. Install the bolts which secure the intercooler to the intercooler stay LH, and tighten the clamp which holds the intake duct to the intercooler.

22. Install the collector cover.

MANIFOLD ABSOLUTE PRESSURE (MAP) SENSOR

REMOVAL & INSTALLATION

2.5L DOHC Engine

1. Remove the collector cover.
2. Disconnect the ground cable from battery.
3. Disconnect the connector from Manifold Absolute Pressure (MAP) sensor, and remove the filter assembly from intake manifold.
4. Remove the MAP sensor from the solenoid valve bracket.

To install:

5. Install in the reverse order of removal.
6. Tighten the MAP sensor to 4.7 ft. lbs. (6.4 Nm).

2.5L SOHC Engine

1. Disconnect the ground cable from the battery.
2. Disconnect the connector from Manifold Absolute Pressure (MAP) sensor.
3. Remove the MAP sensor from throttle body.

To install:

4. Install in the reverse order of removal, noting the following:
   a. Use new O-rings.
   b. Tighten the MAP sensor to 1.5 ft. lbs. (2 Nm).

3.0L & 3.6L Engine

1. Remove the collector cover.
2. Disconnect the ground cable from the battery.
3. Disconnect the connector from the Manifold Absolute Pressure (MAP) sensor, and remove the filter assembly from the intake manifold.
4. Remove the MAP sensor from intake manifold.

To install:

5. Install in the reverse order of removal.

6. Tighten the MAP sensor to 4.7 ft. lbs. (6.4 Nm).

THROTTLE POSITION SENSOR (TPS)

REMOVAL & INSTALLATION

2.5L DOHC Engine

See Figure 208.

1. Before servicing the vehicle, refer to the Precautions Section.
2. Disconnect the negative battery cable.
3. Remove the intercooler.
4. Disconnect the connectors from the throttle position sensor.
5. Remove the bolts which secure the air by-pass pipe and PCV pipe to the intake manifold, and loosen the clamp which connects the throttle body and duct.
6. Remove the duct from the throttle body.
7. Disconnect the coolant hoses from the throttle body.
8. Remove the bolts which secure the throttle body to the intake manifold, and remove the throttle body.

To install:

9. The remainder of the installation is the reverse order of removal.
10. Install the throttle body with a new gasket.
11. Tighten the mounting bolts to 6 ft. lbs. (8 Nm).

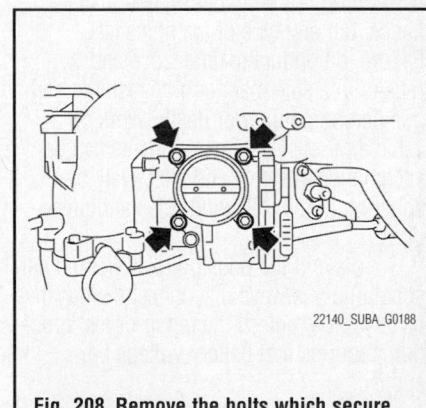

22140_SUBA_G0188

Fig. 208 Remove the bolts which secure the throttle body

2.5L SOHC Engines

See Figure 209.

1. Before servicing the vehicle, refer to the Precautions Section.
2. Disconnect the negative battery cable.

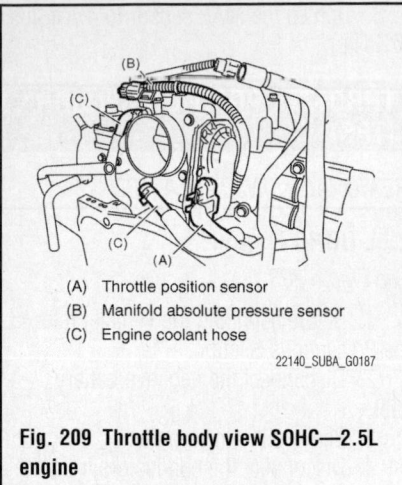

(A) Throttle position sensor
(B) Manifold absolute pressure sensor
(C) Engine coolant hose

22140_SUBA_G0187

Fig. 209 Throttle body view SOHC—2.5L engine

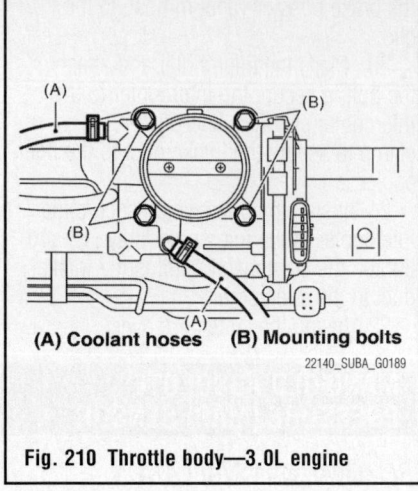

(A) Coolant hoses   (B) Mounting bolts

22140_SUBA_G0189

Fig. 210 Throttle body—3.0L engine

3. Disconnect the throttle position sensor and manifold absolute pressure (MAP) sensor.
4. Disconnect the coolant hoses from the throttle body.
5. Remove the throttle body mounting bolts.
6. Remove the throttle body and gasket.

To install:
7. Install the throttle body with a new gasket.

8. Tighten the mounting bolts to 6 ft. lbs. (8 Nm).
9. The remainder of the installation is the reverse order of removal.

3.0L & 3.6L Engines
See Figure 210.

1. Before servicing the vehicle, refer to the Precautions Section.
2. Disconnect the negative battery cable.
3. Remove the collector cover.

4. Remove the air intake chamber.
5. Disconnect the connector from throttle body.
6. Disconnect the engine coolant hoses from the throttle body.
7. Remove the bolts which secure throttle body to intake manifold.

To install:
8. The remainder of the installation is the reverse order of removal.
9. Install the throttle body with a new gasket.
10. Tighten the mounting bolts to 6 ft. lbs. (8 Nm).

VEHICLE SPEED SENSOR (VSS)

REMOVAL & INSTALLATION

1. Raise and support the vehicle safely.
2. Place a drip pan below the speed sensor to catch any spilled fluid.
3. Disconnect the connector.
4. Remove the sensor from its mounting.

To install:
5. Installation is the reverse of the removal procedure.
6. Replace any lost fluid.

FUEL

GASOLINE FUEL INJECTION SYSTEM

FUEL SYSTEM SERVICE PRECAUTIONS

Safety is the most important factor when performing not only fuel system maintenance, but any type of maintenance. Failure to conduct maintenance and repairs in a safe manner may result in serious personal injury or death. Work on a vehicle's fuel system components can be accomplished safely and effectively by adhering to the following rules and guidelines.

• To avoid the possibility of fire and personal injury, always disconnect the negative battery cable unless the repair or test procedure requires that battery voltage be applied.

• Always relieve the fuel system pressure prior to disconnecting any fuel system component (injector, fuel rail, pressure regulator, etc.) fitting or fuel line connection. Exercise extreme caution whenever relieving fuel system pressure to avoid exposing skin, face and eyes to fuel spray. Please be advised that fuel under pressure may penetrate the skin or any part of the body that it contacts.

• Always place a shop towel or cloth around the fitting or connection prior to loosening to absorb any excess fuel due to spillage. Ensure that all fuel spillage is quickly removed from engine surfaces. Ensure that all fuel-soaked cloths or towels are deposited into a flame-proof waste container with a lid.

• Always keep a dry chemical (Class B) fire extinguisher near the work area.

• Do not allow fuel spray or fuel vapors to come into contact with a spark or open flame.

• Always use a second wrench when loosening or tightening fuel line connection fittings. This will prevent unnecessary stress and torsion on fuel piping. Always follow the proper torque specifications.

• Always replace worn fuel fitting O-rings with new ones. Do not substitute fuel hose where rigid pipe is installed.

FUEL SYSTEM PRESSURE

RELIEVING

➡**This procedure must be performed prior to servicing any component of the fuel injection system.**

1. Remove the fuel pump fuse from the main fuse box.
2. Start the engine and run until it stalls.
3. Crank the engine for 5 seconds or more to ensure the fuel pressure is properly relieved. If the engine starts during this time, allow it to run until it stalls.
4. Turn the ignition switch to the **OFF** position. Remove the key.
5. Disconnect the negative battery cable.
6. Remove the fuel cap.

FUEL FILTER

REMOVAL & INSTALLATION

The fuel filter is an integral part of the fuel pump assembly.

FUEL LEVEL SENDING UNIT

REMOVAL & INSTALLATION

✳✳ CAUTION

Place "NO OPEN FLAMES" signs near the working area. Prepare a container and cloth to prevent scattering of fuels when performing work where fuels can be spilled. If the fuel spills, wipe it off immediately to prevent from penetrating into floor or flowing out for environmental protection. Follow all government and local regulations concerning disposal of refuse when disposing fuel.

1. Before servicing the vehicle, refer to the Precautions Section.
2. Relieve the fuel system pressure.
3. Disconnect the negative battery cable.
4. Remove the fuel pump assembly.
5. Disconnect the connector from fuel pump bracket.
6. Remove the fuel level sensor and fuel temperature sensor.

To install:

7. Install in the reverse order of removal.
8. Tighten sensor to 3 ft. lbs. (4 Nm).

FUEL TANK

REMOVAL & INSTALLATION

See Figures 211 through 214.

✳✳ CAUTION

Place "NO OPEN FLAMES" signs near the working area.

1. Before servicing the vehicle, refer to the Precautions Section.
2. Set vehicle on the lift.
3. Release fuel pressure.
4. Drain fuel from fuel tank.
5. Remove the rear seat.
6. Remove the service hole cover of fuel pump.
7. Disconnect the connector from fuel pump.
8. Remove the connector and clip.
9. Remove the bolts.
10. Push the grommet down and remove the service hole cover.
11. Disconnect connector from fuel sub level sensor.
12. Disconnect the quick connector on the fuel delivery hose.
13. Remove the trunk room trim. (Sedan models)
14. Remove the rear quarter trim. (Wagon models)

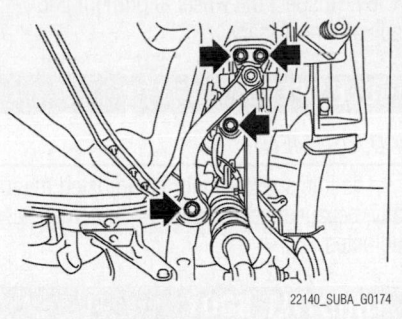

Fig. 211 Bolts which secure the rear suspension view (1)

15. Remove the pipe protector.
16. Remove the grommet and disconnect the quick connector of the evaporation pipe.
17. Remove the rear wheels.
18. Lift up the vehicle.
19. Remove the bolts which secure the rear brake hose mounting bracket.
20. Remove the rear brake caliper and tie it to the body side of the vehicle.
21. Remove the parking brake cable from the parking brake assembly.
22. Remove the rear exhaust pipe.
23. Remove the propeller shaft.
24. Remove the heat shield cover.
25. Disconnect the connector from the rear ABS wheel speed sensor.
26. Remove the bolts securing the parking brake cable clamp.

✳✳ CAUTION

A helper is required to perform this work.

27. Remove the rear suspension assembly as follows:
   - Support the rear differential with the transmission jack.
   - Remove the bolts which hold the rear shock absorber to the rear suspension arm.
   - Remove the bolts which secure the rear suspension assembly to the body.
   - Remove the rear suspension assembly.
28. Disconnect the connector.
29. Disconnect the evaporation hose.
30. Disconnect the fuel filler hose and evaporation hose.

✳✳ CAUTION

A helper is required to perform this work. Fuel may remain in the fuel tank. Be careful not to let the fuel tank fall off when removing as it is bad balance on either side.

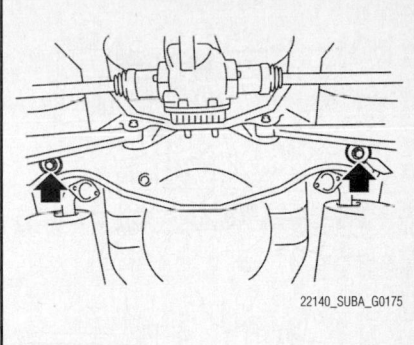

Fig. 212 Bolts which secure the rear suspension view (2)

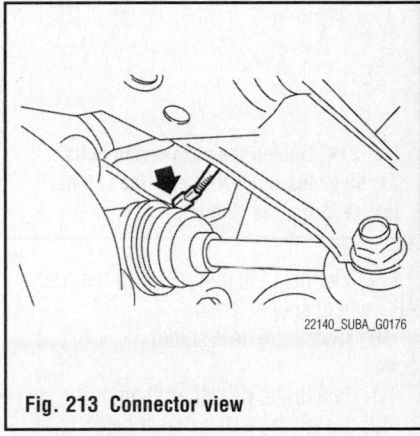

Fig. 213 Connector view

31. Support the fuel tank with a transmission jack, remove the bolts from the fuel tank band, and remove the fuel tank from the vehicle.

To install:

32. Support the fuel tank with a transmission jack, set the fuel tank in place, and temporarily tighten the bolts of the fuel tank band.
33. Securely insert the fuel filler hose and evaporation hose to the specified position, then tighten the clamp.
34. Tighten the fuel tank band bolts to 25 ft. lbs. (33 Nm).
35. Install the rear suspension assembly as follows:
   - Support the rear differential with the transmission jack.
   - Support the rear suspension assembly, and tighten the bolts which secure the rear suspension assembly to the body.
   - Tighten the bolts which hold the rear shock absorber to the rear suspension arm.
36. Tighten the suspension bolts.
37. Install the bolts which hold the rear shock absorber to the rear suspension arm. Tighten to 46 ft. lbs. (62 Nm).
38. Tighten the bolts holding the parking brake cable clamp to 13 ft. lbs. (18 Nm).

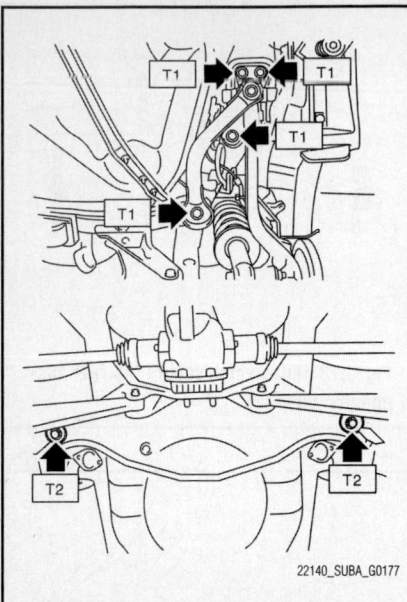

Fig. 214 Tighten the suspension bolts T1: 92 ft. lbs. (125 Nm) and T1: 129 ft. lbs. (175 Nm) as shown

39. Connect the connector to the rear ABS wheel speed sensor.
40. Install the heat shield cover.
41. Install the propeller shaft.
42. Install the rear exhaust pipe.
43. Lower the vehicle.
44. Connect the parking brake cable to the parking brake assembly.
45. Install the rear brake caliper.
46. Tighten the bolts which secure the rear brake hose mounting bracket to 25 ft. lbs. (33 Nm).
47. Install the rear wheels.
48. Connect the quick connector of the evaporation pipe.
49. Install the pipe protector.
50. Install the trunk room trim. (Sedan model)
51. Install the rear quarter trim. (Wagon model)
52. Connect connector to the fuel sub level sensor.
53. Connect the quick connector of the fuel delivery hose.
54. Install the service hole cover of fuel sub level sensor.
55. Connect connector, and install clip.
56. Connect the connector to the fuel pump.
57. Install the service hole cover of fuel pump.
58. Install the rear seat.
59. Install the fuse of fuel pump to the main fuse box.
60. Connect the battery ground cable to battery.

61. Inspect the wheel alignment and adjust if necessary.

IDLE SPEED

ADJUSTMENT

The idle speed cannot be adjusted manually, because the idle speed is automatically adjusted.

THROTTLE BODY

REMOVAL & INSTALLATION

2.5L DOHC Engines

See Figure 215.

1. Before servicing the vehicle, refer to the Precautions Section.
2. Disconnect the negative battery cable.
3. Remove the intercooler.
4. Disconnect the connectors from the throttle position sensor.
5. Remove the bolts which secure the air by-pass pipe and PCV pipe to the intake manifold, and loosen the clamp which connects the throttle body and duct.
6. Remove the duct from the throttle body.
7. Disconnect the coolant hoses from the throttle body.
8. Remove the bolts which secure the throttle body to the intake manifold, and remove the throttle body.

To install:

9. The remainder of the installation is the reverse order of removal.
10. Install the throttle body with a new gasket.
11. Tighten the mounting bolts to 6 ft. lbs. (8 Nm).

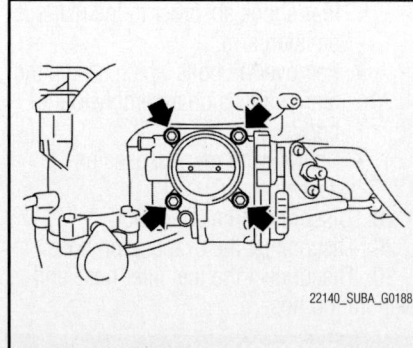

Fig. 215 Remove the bolts which secure the throttle body

2.5L SOHC Engines

See Figure 216.

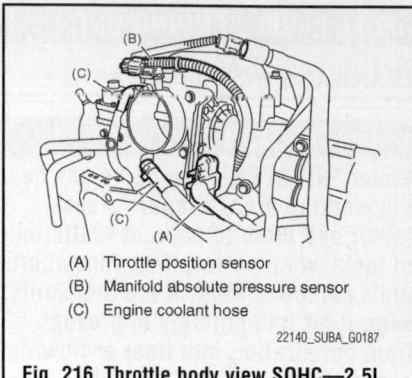

(A) Throttle position sensor
(B) Manifold absolute pressure sensor
(C) Engine coolant hose

Fig. 216 Throttle body view SOHC—2.5L engine

1. Before servicing the vehicle, refer to the Precautions Section.
2. Disconnect the negative battery cable.
3. Disconnect the throttle position sensor and manifold absolute pressure (MAP) sensor.
4. Disconnect the coolant hoses from the throttle body.
5. Remove the throttle body mounting bolts.
6. Remove the throttle body and gasket.

To install:

7. Install the throttle body with a new gasket.
8. Tighten the mounting bolts to 6 ft. lbs. (8 Nm).
9. The remainder of the installation is the reverse order of removal.

3.0L & 3.6L Engines

See Figure 217.

1. Before servicing the vehicle, refer to the Precautions Section.
2. Disconnect the negative battery cable.
3. Remove the collector cover.
4. Remove the air intake chamber.
5. Disconnect the connector from throttle body.

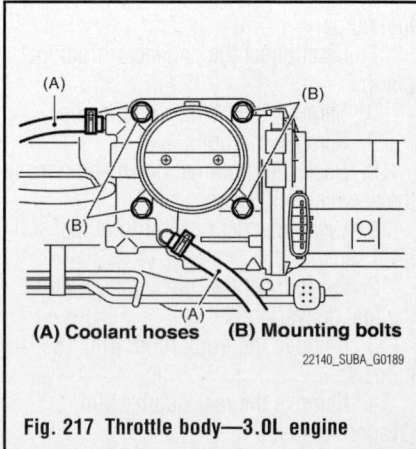

(A) Coolant hoses   (B) Mounting bolts

Fig. 217 Throttle body—3.0L engine

6. Disconnect the engine coolant hoses (A) from the throttle body.

7. Remove the bolts which secure throttle body to intake manifold.

To install:

8. The remainder of the installation is the reverse order of removal.

9. Install the throttle body with a new gasket.

10. Tighten the mounting bolts to 6 ft. lbs. (8 Nm).

HEATING & AIR CONDITIONING SYSTEM

BLOWER MOTOR

REMOVAL & INSTALLATION

See Figure 218.

1. Before servicing the vehicle, refer to the Precautions Section.

2. Disconnect the negative battery cable.

3. Remove the glove box lower cover.

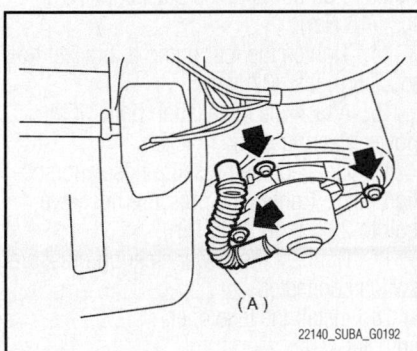

Fig. 218 Blower motor and mounting screws

4. Disconnect the blower motor wiring harness.

5. Remove the mounting screws and remove the blower motor.

6. Installation is the reverse order of removal.

HEATER CORE

REMOVAL & INSTALLATION

See Figure 219.

1. Before servicing the vehicle, refer to the Precautions Section.

2. Disconnect the negative battery cable.

3. Discharge the air conditioning system.

4. Drain the cooling system.

5. Remove the bolts securing the expansion valve and pipe in the engine compartment. Release the heater hose clamps in the engine compartment to remove the heater hoses.

6. Remove the instrument panel.

7. Remove the support beam.

8. Remove the blower motor assembly.

9. Loosen the bolts and nuts to remove the heater and cooling unit.

10. Loosen the screws to remove the heater core cover.

11. Remove the heater core.

To install:

12. Installation is the reverse order of removal.

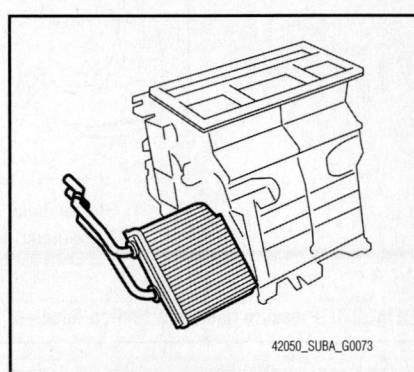

Fig. 219 Removing the heater core from the heater and cooling unit

STEERING

POWER STEERING PUMP

REMOVAL & INSTALLATION

2.5L Engine

See Figures 220 and 221.

1. Before servicing the vehicle, refer to the Precautions Section.

2. Disconnect the ground cable from the battery.

3. Remove the air intake duct.

4. Remove the pulley belt cover.

5. Loosen the belt tension securing bolt and generator securing bolt, then remove the power steering pump V-belt.

6. Disconnect the connector from the power steering pressure switch.

7. Disconnect the pressure hose and suction hose from the oil pump.

✳✳ CAUTION

Prevent foreign matter from entering the hose and pipe, cover the open ends with clean cloth.

8. Remove the installation bolt of the power steering pump bracket.

9. Place the oil pump bracket in a vise, and remove the two bolts from the front side of the oil pump.

10. Remove the bolt from the rear side of oil pump.

11. Disassemble the oil pump and bracket by inserting a flat tip screwdriver as shown in the figure.

12. Remove the oil pump.

To install:

13. Install the oil pump to bracket.

14. Place the oil pump bracket in a vise.

15. Tighten the bushing using a 12.7 mm (1/2) type, 14 and 21 mm box wrench until it is in contact with the oil pump mounting surface.

✳✳ CAUTION

When securing the oil pump bracket in a vice, hold the oil pump bracket with the least possible force between two pieces of wood.

16. Tighten the two front bolts which hold the oil pump to the bracket to 12 ft. lbs. (16 Nm).

17. Tighten the rear pump to bracket bolt to 27.5 ft. lbs. (37 Nm).

18. Attach the installation bolts of the power steering pump bracket.

19. Connect the pressure hose and suction hose. Tighten the pressure hose eye bolt to 29.5 ft. lbs. (40 Nm).

20. Connect the power steering pressure switch connector.

21. Install the V-belts to the oil pump.

22. Check the tension of the V-belt.

23. Tighten the belt tension bolt to 18.4 ft. lbs. (25 Nm).

24. Install the pulley belt cover.

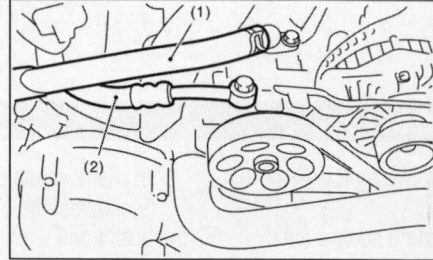

(1) Suction hose
(2) Pressure hose

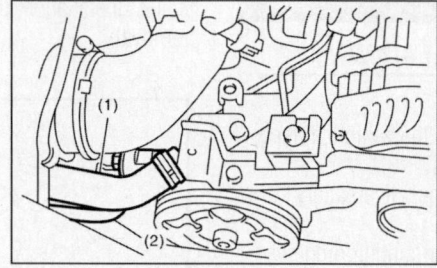

(1) Suction hose
(2) Pressure hose

22140_SUBA_G0210

Fig. 220 Pressure hose and suction hose—2.5L—turbo and non-turbo engines

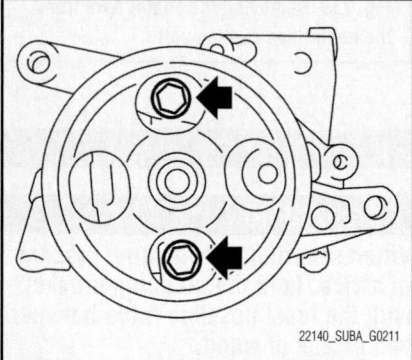

22140_SUBA_G0211

Fig. 221 Front side oil pump to bracket bolts

25. Install the air intake duct.
26. Connect the battery ground cable to the battery.
27. Fill with the specified power steering fluid. (ATF DEXRON III®)

> ❖❖ **CAUTION**
>
> **Never start the engine before feeding the fluid otherwise the vane pump might be seized.**

28. Bleed the power steering system.
29. Start the engine and check for leaks.

3.0L & 3.6L Engines

See Figures 221 and 222.

1. Disconnect the ground cable from the battery.
2. Remove the cover of the pulley belt.
3. Remove the drive belts.
4. Remove the power steering pressure switch connector.
5. Remove the tensioner adjuster.
6. Disconnect the pressure hose and suction hose from the oil pump.

> ❖❖ **CAUTION**
>
> **Prevent foreign matter from entering the hose and pipe, cover the open ends with clean cloth.**

7. Remove the bolts which holds the power steering pump bracket.
8. Place the oil pump bracket in a vise, and remove the two bolts from the front side of the oil pump.

> ❖❖ **CAUTION**
>
> **When securing the oil pump bracket in a vice, hold the oil pump bracket with the least possible force between two pieces of wood.**

9. Remove the bolt from the back side of the oil pump.
10. Remove the oil pump from the bracket.

To install:

11. Install the oil pump to bracket.
12. Tighten the bushing using a 12.7 mm (1⁄2) type, 14 and 21 mm box wrench until it is in contact with the oil pump mounting surface.

> ❖❖ **CAUTION**
>
> **When securing the oil pump bracket in a vice, hold the oil pump bracket with the least possible force between two pieces of wood.**

13. Tighten the two front bolts which hold the oil pump to the bracket to 12 ft. lbs. (16 Nm).
14. Tighten the rear pump to bracket bolt to 27.5 ft. lbs. (37 Nm).
15. Attach the installation bolts of the power steering pump bracket.
16. Connect the pressure hose and suction hose. Tighten the pressure hose eye bolt to 29.5 ft. lbs. (40 Nm).
17. Connect the power steering pressure switch connector.
18. Install the tensioner adjuster.
19. Install the drive belts.
20. Install the cover of the pulley belt.
21. Connect the battery ground cable to the battery.
22. Fill with the specified power steering fluid. (ATF DEXRON III®)

Tightening torque:
15.7 N·m (1.6 kgf-m, 11.6 ft-lb)

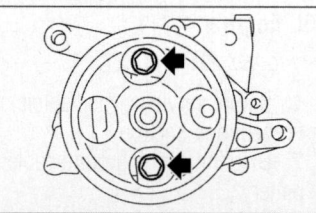

Tightening torque:
37.3 N·m (3.8 kgf-m, 27.5 ft-lb)

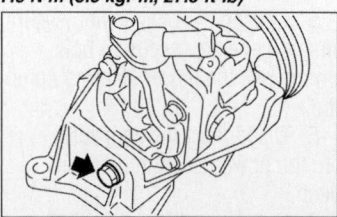

22140_SUBA_G0212

Fig. 222 Pump to bracket bolt location, front and rear

✳✳ CAUTION

Never start the engine before feeding the fluid otherwise the vane pump might be seized.

23. Bleed the power steering system.
24. Start the engine and check for leaks.

BLEEDING

1. Fill the power steering fluid reservoir about half way with the specified fluid.
2. Continue to turn the steering wheel slowly from lock to lock until bubbles stop appearing on oil surface while keeping the fluid at that level.
3. If turning the steering wheel in low fluid level condition, air will be sucked in pipe. In this case, leave it about half an hour and then repeat the previous step.
4. Lift up the vehicle, start the engine and let it idle.
5. Continue to turn the steering wheel slowly from lock to lock again until bubbles stop appearing on oil surface while keeping the fluid at that level. It is normal that bubbles stop appearing after three times turning of steering wheel from lock to lock.
6. In case the bubbles do not stop appearing in the tank, leave it about half an hour and then begin the process again.
7. Lower the vehicle, and then idle the engine.
8. Continue to turn the steering wheel from lock to lock until bubbles stop appearing and change of the fluid level is within 3 mm (0.12 in).
9. In case the following happens, leave it about half an hour and then do step 5–8 again.
   a. The fluid level changes over 3 mm (0.12 in).
   b. Bubbles remain on the upper surface of the fluid.
   c. Grinding noise is generated from oil pump.
10. Check the fluid leakage after turning steering wheel from lock to lock with engine running.

SUSPENSION

COIL SPRING

REMOVAL & INSTALLATION

See Figure 223.

1. Before servicing the vehicle, refer to the Precautions Section.
2. Disconnect the negative battery cable.
3. Raise and support the vehicle safely.
4. Remove front strut.
5. Using a coil spring compressor, compress the coil spring.
6. Using a strut mount socket remove the self-locking nut.
7. Remove the strut mount, spacer and upper spring seat from strut.
8. Gradually decrease the compression force of compressor, and remove the coil spring.
9. Remove the dust cover and helper spring.

To install:

10. Using a coil spring compressor, compress the coil spring.
11. Set the coil spring correctly so that its end face seats well in the spring seat as shown in the figure.
12. Install the helper and dust cover to the piston rod.
13. Pull the piston rod fully upward, and install the spring seat.
14. Install spacer and the strut mount to piston rod, and tighten a new self-locking nut temporarily.
15. Using a hexagon wrench to prevent strut rod from turning, tighten the new self-locking nut to 41 ft. lbs. (55 Nm), with a strut mount socket.
16. Loosen the coil spring compressor carefully.
17. Install front strut.
18. Connect the negative battery cable

CONTROL LINKS

REMOVAL & INSTALLATION

See Figure 224.

1. Before servicing the vehicle, refer to the Precautions Section.

FRONT SUSPENSION

2. Lift up the vehicle, and then remove the front wheels.
3. Remove the front undercover.
4. Remove the stabilizer control link upper and lower mounting nuts.

To install:

5. Install the stabilizer control link.
6. Tighten the stabilizer link upper and lower mounting nuts to 44 ft. lbs. (60 Nm).
7. Install front wheels, and lower the vehicle.

LOWER BALL JOINT

REMOVAL & INSTALLATION

See Figure 225.

1. Before servicing the vehicle, refer to the Precautions Section.
2. Disconnect the negative battery cable.
3. Raise and support the vehicle safely.
4. Remove the tire and wheel.
5. Remove the stabilizer bar brackets and bushings (both sides).

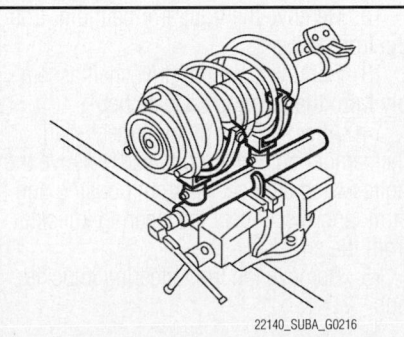

22140_SUBA_G0216

Fig. 223 Coil spring being compressed by spring compressor tool

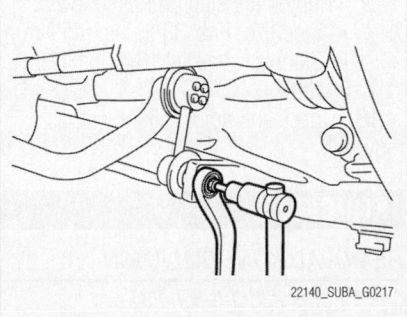

22140_SUBA_G0217

Fig. 224 Stabilizer control link removal shown

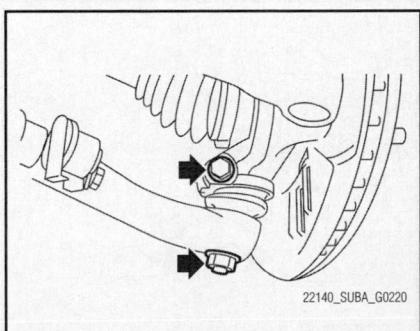

22140_SUBA_G0220

Fig. 225 Bolt securing the ball joint, and castle nut shown

6. Remove the cotter pin from the ball stud. Remove the castle nut.

7. Extract the ball stud from the transverse link.

8. Remove the bolt securing the ball joint to the housing.

9. Extract the ball joint from the housing.

To install:

10. Installation is the reverse of the removal procedure.

11. Install the ball joint to the transverse link arm and tighten to 37 ft. lbs. (50 Nm).

12. Install and tighten the castle nut to 22 ft. lbs. (30 Nm), if the front arm is aluminum and 28.8 ft. lbs. (39 Nm), if the front arm is steel. Tighten the castle nut an additional 60 degrees until the slot in the castle nut is aligned with the cotter pin hole in the ball joint.

13. The stabilizer bracket has a set orientation. Install it with the arrow mark facing the upper side of the vehicle.

14. Always fully tighten the rubber bushings when the wheels are in full contact with the ground and the vehicle is at curb height.

15. Tighten the stabilizer bracket bolts to 18 ft. lbs. (25 Nm).

16. Check and adjust alignment, as required.

LOWER CONTROL ARM

REMOVAL & INSTALLATION

See Figures 226 and 227.

1. Before servicing the vehicle, refer to the Precautions Section.

2. Lift up the vehicle, and then remove the front wheels.

3. Remove the front crossmember support plate.

4. Remove the front stabilizer.

5. Remove the ball joint of front arm.

6. Remove the nut securing the front arm to crossmember. (Do not remove the bolt.)

7. Remove the front arm support plate.

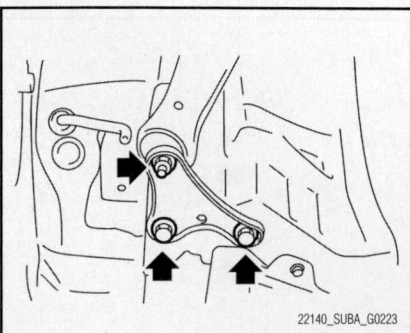

Fig. 226 Front arm support plate mounting points

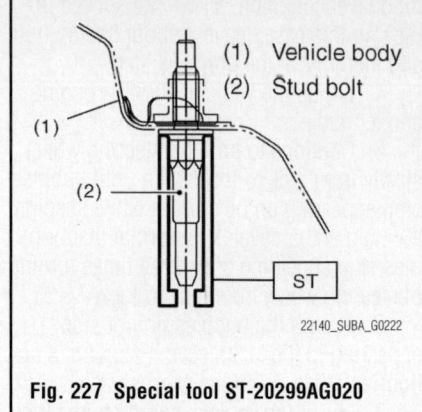

(1) Vehicle body
(2) Stud bolt

Fig. 227 Special tool ST-20299AG020

8. Remove the bolt securing front arm to crossmember and pull the front arm out of the crossmember.

9. To remove the stud bolt, use the ST-20299AG020

※※ CAUTION

Do not remove the stud bolt without necessity. Always replace the parts with new parts when removed.

To install:

10. Using the ST-20299AG020, install the stud bolt. Tighten to 81 ft. lbs. (119 Nm).

11. Using new bolts and self-locking nuts, temporarily tighten the front arm to crossmember.

12. Secure the front arm to body, and then install the support plate with new bolts and self-locking nuts.

13. Tighten as follows:
- Support plate to Front arm: 65 ft. lbs. (88 Nm)
- Support plate to Body: 111 ft. lbs. (150 Nm)

14. Install the ball joint into housing, and tighten retaining bolt to 37 ft. lbs. (50 Nm).

15. Install the stabilizer.

16. Lower the vehicle from lift, and tighten the bolt which secures the front arm to crossmember with wheels in full contact with the ground and the vehicle at curb weight.

17. Tighten the stabilizer as follows:
- Stabilizer link: 33 ft. lbs. (45 Nm).
- Stabilizer bracket: 18 ft. lbs. (25 Nm).

18. Check and adjust alignment, as required.

KNUCKLES

REMOVAL & INSTALLATION

See Figures 228 through 231.

1. Before servicing the vehicle, refer to the Precautions Section.

2. Raise and support the vehicle safely.

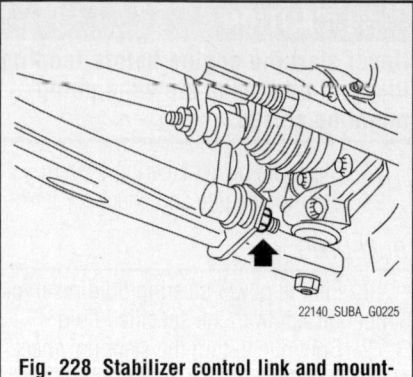

Fig. 228 Stabilizer control link and mounting nut shown

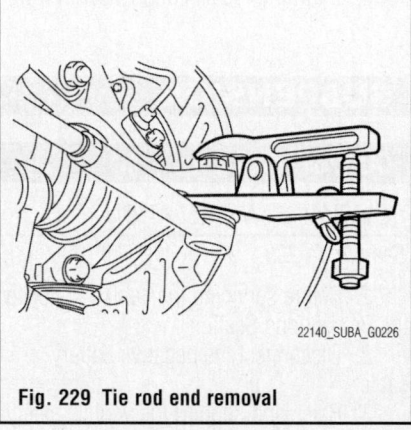

Fig. 229 Tie rod end removal

3. Remove the tire and wheels.

4. Remove the front undercover.

5. Remove the stabilizer link.

6. Remove the disc brake caliper from the front housing, and suspend it from strut using a piece of wire.

7. Remove the disc rotor from the hub.

8. Remove the cotter pin and castle nut securing the tie-rod end to the front housing knuckle arm.

9. Using a puller, remove the tie rod ball joint from knuckle arm.

10. Remove the ABS wheel speed sensor assembly and harness.

11. Remove the lower strut mounting bolts.

12. Remove the front arm ball joint from the front housing.

13. Remove the front driveshaft assembly from the steering knuckle hub.

14. After scribing an alignment mark on the camber adjusting bolt head, remove the bolts which connect the front housing and strut, and disconnect the steering knuckle from the strut.

15. Remove the front steering knuckle hub.

To install:

16. Install the front steering knuckle hub.

17. Align the alignment mark on the

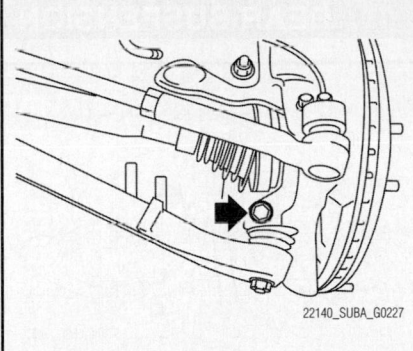

Fig. 230 Ball joint knuckle retaining bolt shown

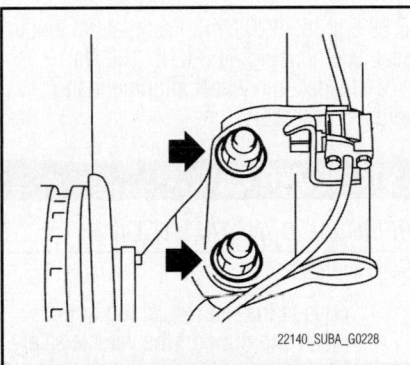

Fig. 231 Strut to knuckle retaining bolts shown

camber adjusting bolt head, and affix the front housing and strut together using a new self—locking nut. Tighten to 129 ft. lbs. (175 Nm).

18. Install the front driveshaft.

19. Install the front arm ball joint to the front housing. Tighten mounting bolt to 37 ft. lbs. (50 Nm).

20. Install the ABS wheel speed sensor on the front housing. Tighten to 5.5 ft. lbs. (7.5 Nm).

21. Install the disc rotor to hub.

22. Install the disc brake caliper to the front housing. Tighten to 88.5 ft. lbs. (120 Nm).

23. Install the stabilizer link.

24. Connect the tie-rod end ball joint to the knuckle arm with a castle nut. Tighten to 20 ft. lbs. (27 Nm).

25. Tighten the castle nut to specified torque and tighten further within 60° until the pin hole is aligned with the slot in nut. Bend the cotter pin to lock.

26. While depressing the brake pedal, tighten a new axle nut to 177 ft. lbs. (240 Nm). and lock it securely.

27. Install the tire and wheels.

28. Lower the vehicle.

29. Inspect the wheel alignment and adjust if necessary.

STABILIZER BAR

REMOVAL & INSTALLATION

See Figure 232.

1. Before servicing the vehicle, refer to the Precautions Section.

2. Raise and support the vehicle safely.

3. Remove the tire and wheels.

4. Remove the front undercover.

5. Remove the front crossmember support plate.

6. Remove the stabilizer bar link.

7. Remove the stabilizer bar bracket bolts and bushings.

8. Remove the stabilizer bar from the vehicle.

To install:

9. Installation is the reverse of the removal procedure.

10. Be sure to use new self locking nuts, as required.

11. Install the rubber bushing, so that the paint mark on the stabilizer bar is on the left side of the vehicle.

12. Install the stabilizer bushing (front crossmember side), while aligning it with the paint mark on the stabilizer bar.

13. Tighten the stabilizer link bolts to 44 ft. lbs. (60 Nm). Tighten the stabilizer bar clamp to 18 ft. lbs. (25 Nm).

14. Always fully tighten the rubber bushings when the wheels are in full contact with the ground and the vehicle is at curb height.

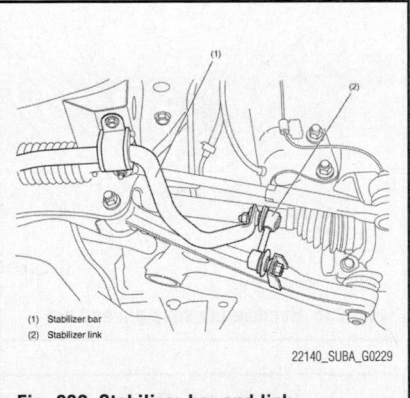

(1) Stabilizer bar
(2) Stabilizer link

Fig. 232 Stabilizer bar and link

WHEEL BEARINGS

REMOVAL & INSTALLATION

See Figure 233.

➡**It may be necessary to remove the front halfshaft from the vehicle.**

1. Disconnect the negative battery cable.

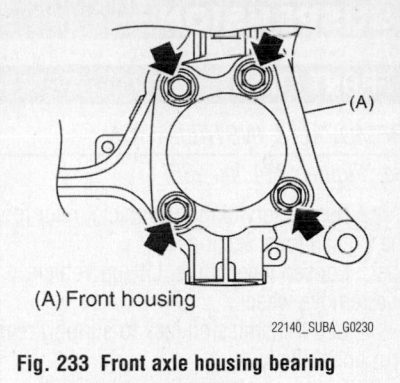

(A) Front housing

Fig. 233 Front axle housing bearing assembly retaining bolts

2. Raise and support the vehicle safely.

3. Remove the tire and wheel.

4. Remove the crimped section of the axle nut. Depress the brake pedal and remove the axle nut.

➡**Be sure to loosen the axle nut after removing the tire and wheel from the vehicle. Failure to do this may damage the wheel bearings.**

5. Remove the disc brake caliper from its mounting as suspend it to the side, with wire. Do not disconnect the brake line.

6. Remove the rotor.

➡**If the rotor is seized within the hub, remove the rotor by installing an 8mm bolt in the screw hole on the rotor.**

7. Remove the ABS wheel speed sensor assembly and harness.

8. Remove the four bolts from the housing.

9. Remove the front hub unit bearing. Use tools ST92647000 and ST28099PA100 if necessary.

To install:

10. Tighten the front hub unit bearing to housing bolts to 47.9 ft. lbs. (65 Nm).

11. Be sure to use a new axle nut. Tighten the axle nut, temporarily.

12. Install the rotor. Install the caliper.

13. While depressing the brake pedal, tighten the axle nut (olive color) to 162 ft. lbs. (220 Nm). Lock it securely in place.

➡**Install the tire and wheel after installation of the axle nut. Failure to do this may result in wheel bearing damage. Do not over tighten, as this too could cause wheel bearing damage.**

14. Continue the installation in the reverse order of the removal procedure.

ADJUSTMENT

The wheel bearings are of a sealed unit and are not adjustable.

SUSPENSION

REAR SUSPENSION

FRONT LINK

REMOVAL & INSTALLATION

See Figures 234 and 235.

1. Before servicing the vehicle, refer to the Precautions Section.
2. Loosen wheel nuts. Lift-up vehicle and remove wheel.
3. Use transmission jack to support rear arm horizontally.
4. Remove bolt securing link front to sub frame.
5. Remove bolts which secure link front to rear arm and detach link front.

➡**Link front bushing cannot be replaced alone. Always replace link front and bushing as a single unit.**

To install:

6. Installation is the reverse of the removal procedure.
7. Tighten bolts which secure link to 89 ft. lbs. (120 Nm).
8. Check and adjust the wheel alignment, as necessary.

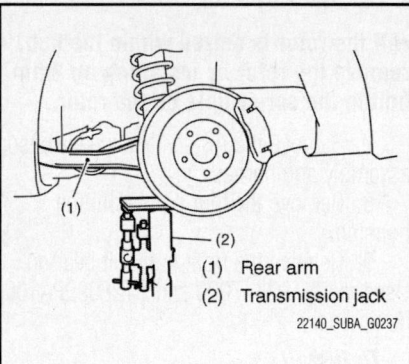

(1) Rear arm
(2) Transmission jack

22140_SUBA_G0237

Fig. 234 Use transmission jack to support rear arm

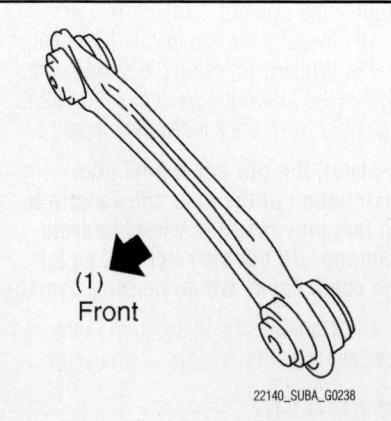

(1) Front

22140_SUBA_G0238

Fig. 235 Install front link with protruding side facing the front side of the vehicle

REAR LINK

REMOVAL & INSTALLATION

See Figures 236 through 238.

1. Lift up the vehicle, and then remove the rear wheels.
2. Remove the rear trailing link as outlined in this section.
3. Remove the snap pin and nut.
4. Using a puller, remove the ball joint.
5. Place an alignment mark on the front lateral link adjusting bolt and rear sub frame.
6. Remove the adjusting bolt, and remove the front lateral link.

✳✳ CAUTION

When removing the adjusting bolt, make sure to fix the bolt head in place, and loosen the nut.

To install:

✳✳ WARNING

Be sure to use a new self-locking nut. Always tighten the bushing in the state where the vehicle is at curb weight and the wheels are in full contact with the ground.

7. Install in the reverse order of removal.
8. Tighten Front lateral link to Sub frame

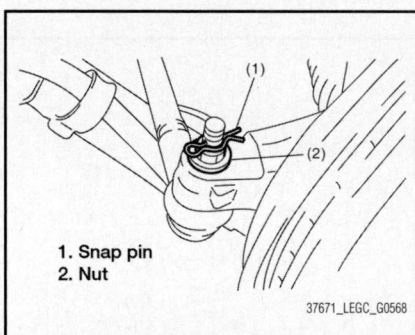

1. Snap pin
2. Nut

37671_LEGC_G0568

Fig. 236 Remove the snap pin and nut

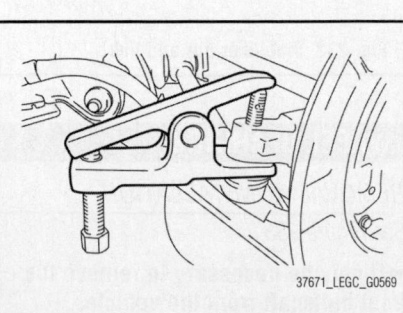

37671_LEGC_G0569

Fig. 237 Using a puller, remove the ball joint

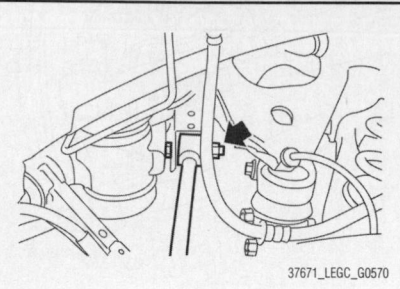

37671_LEGC_G0570

Fig. 238 Remove the adjusting bolt, and remove the front lateral link

to 88.5 ft. lb. (120 Nm), Front lateral link to Rear axle housing 44.3 ft. lb. (60 Nm)

9. Inspect the wheel alignment and adjust if necessary.

STABILIZER BAR

REMOVAL & INSTALLATION

See Figure 239.

1. Loosen the rear wheel lug nuts.
2. Raise and support the vehicle safely.
3. Remove the tire and wheel.
4. Remove the bolts that secure the stabilizer link to the rear arm.
5. Remove the bolts which secure the stabilizer bar to the sub frame.
6. Remove the stabilizer bar from the vehicle.

To install:

7. Installation is the reverse of the removal procedure.
8. Be sure that the stabilizer bar and the bushings have the same identification markings and/or colors.
9. Be sure to use new bolts and nuts, as required.
10. Always fully tighten the rubber bushings when the wheels are in full contact with the ground and the vehicle is at curb height.
11. Check and adjust the wheel alignment, as necessary.

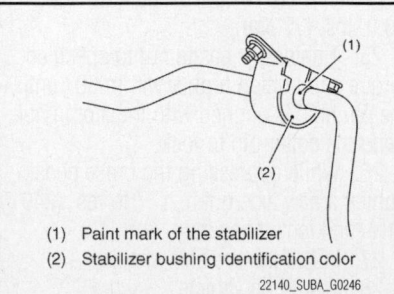

(1) Paint mark of the stabilizer
(2) Stabilizer bushing identification color

22140_SUBA_G0246

Fig. 239 Stabilizer bar and bushing identification markings

SPECIFICATIONS AND MAINTENANCE CHARTS

ENGINE AND VEHICLE IDENTIFICATION CHART

	Engine Code						Model Year	
Code ①	Liters (cc)	Cu. In.	Cyl.	Fuel Sys.	Type	Eng. Mfg.	Code ②	Year
9	3.6 (3630)	221	6	MFI	DOHC	Subaru	9	2009
							A	2010

MFI: Multiport Fuel Injection

DOHC: Double Overhead Camshaft

① 6th digit of the VIN

② 10th digit of the VIN

37671_STRI_C0001

GENERAL ENGINE SPECIFICATIONS

Year	Model	Engine Displacement Liters (VIN)	Net Horsepower @ rpm	Net Torque @ rpm (ft. lbs.)	Bore x Stroke (in.)	Com-pression Ratio	Oil Pressure @ rpm
2009	Tribeca	3.6 (9)	256@6000	247@4400	3.62x3.58	10.5:1	15@700
2010	Tribeca	3.6 (9)	256@6000	247@4400	3.62x3.58	10.5:1	15@700

37671_STRI_C0002

ENGINE TUNE-UP SPECIFICATIONS

Year	Model	Engine Displacement Liters (VIN)	Spark Plug Gap (in.)	Ignition Timing (deg.) ① MT	AT	Fuel Pump (psi)	Idle Speed (rpm) MT	AT	Valve Clearance ② In.	Ex.
2009	Tribeca	3.6 (9)	0.028-0.031	NA	15	③	NA	④	⑤	⑥
2010	Tribeca	3.6 (9)	0.028-0.031	NA	15	③	NA	④	⑤	⑥

NOTE: The Vehicle Emission Control Information label often reflects specification changes made during production.

The lable figures must be used if they differ from those in this chart.

NA: Not Applicable

① Before Top Dead Center. At idle. +/- 8 degrees.

② Measured with engine cold

③ 49-50.5 at operating temperature

④ 600-800 in N. 705-905 in N with AC on.

⑤ 0.0079 +0.0016/-0.0024

⑥ 0.0138 +/-0.0020

37671_STRI_C0003

CAPACITIES

Year	Model	Engine Displacement Liters (VIN)	Engine Oil with Filter (qts.)	Transmission (pts.) 5-Spd	Transmission (pts.) Auto.	Transfer Case (pts.)	Drive Axle Front ① (pts.)	Drive Axle Rear (pts.)	Fuel Tank (gal.)	Cooling System (qts.)
2009	Tribeca	3.6 (9)	6.9	NA	20.8	NA	2.4	1.4	16.9	8.0
2010	Tribeca	3.6 (9)	6.9	NA	20.8	NA	2.4	1.4	16.9	8.0

Note: All capacities are approximate. Add fluid gradually and check to be sure a proper fluid level is obtained.

NA: Not Applicable

① A/T differential

37671_STRI_C0005

FLUID SPECIFICATIONS

Year	Model	Engine Displacement Liters	Engine ID/VIN	Engine Oil	Auto. Trans.	Drive Axle	Power Steering Fluid	Brake Master Cylinder	Engine Coolant
2009	Tribeca	3.0	9	5W-30	Subaru ATF	75W-90	Dexron III	DOT 3	Subaru Coolant
2010	Tribeca	3.0	9	5W-30	Subaru ATF	75W-90	Dexron III	DOT 3	Subaru Coolant

DOT: Department Of Transpotation

37671_STRI_C0004

VALVE SPECIFICATIONS

All measurements are given in inches.

Year	Engine Displacement Liters (VIN)	Seat Angle (deg.)	Face Angle (deg.)	Spring Test Pressure (lbs. @ in.)	Spring Installed Height (in.)	Stem-to-Guide Clearance (in.) Intake	Stem-to-Guide Clearance (in.) Exhaust	Stem Diameter (in.) Intake	Stem Diameter (in.) Exhaust
2009	3.6 (9)	90	NA	NA	②	0.0012-0.0022	0.0016-0.0026	002148-0.2154	0.2144-0.2150
2010	3.6 (9)	90	NA	NA	②	0.0012-0.0022	0.0016-0.0026	002148-0.2154	0.2144-0.2150

NA: Not Available

① Free length: Intake inner 1.557 in., outer 1.621 in. exhaust 1.824 in.

② Free length: Intake 1.6342 in., exhaust 1.6342 in.

37671_STRI_C0007

CAMSHAFT SPECIFICATIONS

All measurements arc given in inches.

Year	Engine Displacement Liters (VIN)	Journal Dia.	Brg. Oil Clearance	Shaft End-play	Runout	Lobe Height Intake	Lobe Height Exhaust
2009	3.6 (9)	①	0.0015-0.0028	②	NA	1.8071-1.8110	1.7579-1.7618
2010	3.6 (9)	①	0.0015-0.0028	②	NA	1.8071-1.8110	1.7579-1.7618

NA: Not Available

① Front: 1.4939-1.4946

   Except front: 1.0215-1.0222

② Intake: 0.0030-0.0053

   Exhaust: 0.0012-0.0035

③ High: 1.6571-1.6610

   Low 1: 1.5016-1.5055

   Low 2: 1.3756-1.3795

37671_STRI_C0006

CRANKSHAFT AND CONNECTING ROD SPECIFICATIONS

All measurements are given in inches.

Year	Engine Displacement Liters (VIN)	Crankshaft Main Brg. Journal Dia.	Crankshaft Main Brg. Oil Clearance	Crankshaft Shaft End-play	Crankshaft Thrust on No.	Connecting Rod Journal Diameter	Connecting Rod Oil Clearance	Connecting Rod Side Clearance
2009	3.6 (9)	2.5194-2.5200	0.0004-0.0012	0.0012-0.0045	NA	NA	0.0006-0.0017	0.0028-0.0130
2010	3.6 (9)	2.5194-2.5200	0.0004-0.0012	0.0012-0.0045	NA	NA	0.0006-0.0017	0.0028-0.0130

NA: Not Available

37671_STRI_C0008

PISTON AND RING SPECIFICATIONS

All measurements are given in inches.

Year	Engine Displacement Liters (VIN)	Piston Clearance	Ring Gap Top Compression	Ring Gap Bottom Compression	Ring Gap Oil Control	Ring Side Clearance Top Compression	Ring Side Clearance Bottom Compression	Ring Side Clearance Oil Control
2009	3.6 (9)	0.0004	0.0079-0.0138	0.0138-0.0197	0.0079-0.0236	0.0016-0.0031	0.0012-0.0028	0.0018-0.0049
2010	3.6 (9)	0.0004	0.0079-0.0138	0.0138-0.0197	0.0079-0.0236	0.0016-0.0031	0.0012-0.0028	0.0018-0.0049

37671_STRI_C0009

TORQUE SPECIFICATIONS
All readings in ft. lbs.

Year	Engine Displacement Liters (VIN)	Cylinder Head Bolts	Main Bearing Bolts	Rod Bearing Bolts	Crankshaft Damper Bolts	Flywheel Bolts	Manifold Intake	Manifold Exhaust	Spark Plugs	Oil Pan Drain Plug
2009	3.6 (9)	①	②	39	144	60	18	52	15	33
2010	3.6 (9)	①	②	39	144	60	18	52	15	33

① Step 1: Tighten all bolts to 14 ft. lbs.

Step 2: Tighten all bolts to 37 ft. lbs.

Step 3: Loosen all bolts 180 degrees, and again 180 degrees

Step 4: Tighten all bolts to 14 ft. lbs.

Step 5: Bolts (1-4) 35.4 ft. lbs.

Step 6: Bolts (5-8) 33 ft. lbs.

Step 7: + 90 degrees

Step 8: Bolts (1-4) 45 degrees

② Step 1: Bolts (1-11 and 13) 18 ft. lbs.

Bolts (12 and 14) 14. ft. lbs.

Step 2: Retighten bolts (1-11 and 13) 18. ft. lbs.

Retighten bolts (12 and 14) 14. ft. lbs.

Step 3: + 90 degrees

Step 4: Upper bolt to cylinder block 18 ft. lbs.

Step 7: + 90 degrees

Step 8: Bolts (1-4) 45 degrees

37671_STRI_C0010

WHEEL ALIGNMENT

Year	Model		Caster Range (+/-Deg.)	Caster Preferred Setting (Deg.)	Camber Range (+/-Deg.)	Camber Preferred Setting (Deg.)	Toe-in (in.)
2009	Tribeca	F	—	①	—	②	③
		R	—	—	—	④	⑤
2010	Tribeca	F	—	①	—	②	③
		R	—	—	—	④	⑤

① Referential value: 4 degrees 04'

② Difference between right and left is 45' or less: 0 degrees 00'

③ 0 +/-0.12: toe angle (sum of both wheels) 0 degrees +/-0 degrees 14'

④ Difference between right and left is 45' or less: -0 degrees 31'

⑤ 0.08 +/-0.08: toe angle (sum of both wheels) 0 degrees +/-0 degrees 14'

37671_STRI_C0011

TIRE, WHEEL AND BALL JOINT SPECIFICATIONS

Year	Model	OEM Tires Standard	OEM Tires Optional	Tire Pressures (psi) Front	Tire Pressures (psi) Rear	Wheel Size	Ball Joint Inspection	Lug Nut
2009	Tribeca	P255/55 R18	T165/80 R17	33	32	18 x 8JJ ①	0.012	②
2010	Tribeca	P255/55 R18	T165/80 R17	33	32	18 x 8JJ ①	0.012	②

OEM: Original Equipment Manufacturer

PSI: Pounds Per Square Inch

① 17 inch wheel size 17 x 4T and tire (psi) is 60

② Chromed wheel 111 ft. lbs. Other than chrome 88 ft. lbs.

37671_STRI_C0012

BRAKE SPECIFICATIONS

All measurements in inches unless noted

Year	Model		Brake Disc Original Thickness	Brake Disc Minimum Thickness	Brake Disc Maximum Runout	Brake Drum Diameter Original Inside Diameter	Brake Drum Diameter Max. Wear Limit	Brake Drum Diameter Maximum Machine Diameter	Minimum Lining Thickness Front	Minimum Lining Thickness Rear	Brake Caliper Bracket Bolts (ft. lbs.)	Brake Caliper Mounting Bolts (ft. lbs.)
2009	Tribeca	F	1.180	1.100	0.0020	NA	NA	NA	0.059	—	89	20
		R	0.710	0.630	0.0020	NA	NA	NA	—	0.059	49	20
2010	Tribeca	F	1.180	1.100	0.0020	NA	NA	NA	0.059	—	89	20
		R	0.710	0.630	0.0020	NA	NA	NA	—	0.059	49	20

NA: Not Available

37671_STRI_C0013

SCHEDULED MAINTENANCE INTERVALS
SUBARU TRIBECA

TO BE SERVICED	TYPE OF SERVICE	VEHICLE MILEAGE INTERVAL (x1000)												
		7.5	15	22.5	30	37.5	45	52.5	60	67.5	75	82.5	90	97.5
Engine oil & filter ①	R	✓	✓	✓	✓	✓	✓	✓	✓	✓	✓	✓	✓	✓
Brake lines	S/I		✓		✓		✓		✓		✓		✓	
Disc brake pads & discs, front & rear axle boots & axle shaft joint portions	S/I		✓		✓		✓		✓		✓		✓	
Parking brake	S/I		✓		✓		✓		✓		✓		✓	
Steering & suspension	S/I		✓		✓		✓		✓		✓		✓	
A/C cabin filter	R		✓		✓		✓		✓		✓		✓	
Air filter element	R				✓				✓				✓	
Engine coolant	R				✓				✓				✓	
Spark plugs	R				✓				✓				✓	
Automatic transmission fluid & filter	S/I				✓				✓				✓	
Brake fluid	R				✓				✓				✓	
Brake linings & drums	S/I				✓				✓				✓	
Camshaft drive belt	S/I				✓				✓				✓	
Coolant level, hoses & clamps	S/I				✓				✓				✓	
Drive belts	S/I				✓				✓				✓	
Fuel system, hoses & connections	S/I				✓				✓				✓	
Transmission and/or differential gear fluid	S/I				✓								✓	
Tires (rotate)	S/I	✓	✓	✓	✓	✓	✓	✓	✓	✓	✓	✓	✓	✓
Front & rear wheel bearing	I								✓					

R: Replace   S/I: Service or Inspect

① First oil change 3,000

FREQUENT OPERATION MAINTENANCE (SEVERE SERVICE)

If a vehicle is operated under any of the following conditions it is considered severe service:

- Extremely dusty areas.

- 50% or more of the vehicle operation is in 32°C (90°F) or higher temperatures, or constant operation in temperatures below 0°C (32°F).

- Prolonged idling (vehicle operation in stop and go traffic).

- Frequent short running periods (engine does not warm to normal operating temperatures).

- Police, taxi, delivery usage or trailer towing usage.

Oil & oil filter change: change every 3750 miles.

Air filter element: service or inspect every 15,000 miles.

Automatic transmission fluid: service or inspect every 15,000 miles.

Brake linings & drums: service or inspect every 15,000 miles.

Coolant level, hoses & clamps: service or inspect every 15,000 miles.

Drive belts: service or inspect every 15,000 miles.

Transmission/differential gear oil: service or inspect every 15,000 miles.

Front & rear wheel bearing: service or inspect every 30,000 miles.

Suspension & Steering: inspect every 7,500 miles.

Axle boots & Joints: inspect every 7,500 miles.

Note: Inspect SRS system every 10 years.

37671_STRI_C0014

BRAKES · INFORMATION AND PRECAUTIONS

ANTI-LOCK SYSTEMS

- Certain components within the ABS system are not intended to be serviced or repaired individually.
- Do not use rubber hoses or other parts not specifically specified for and ABS system. When using repair kits, replace all parts included in the kit. Partial or incorrect repair may lead to functional problems and require the replacement of components.
- Lubricate rubber parts with clean, fresh brake fluid to ease assembly. Do not use shop air to clean parts; damage to rubber components may result.
- Use only DOT 3 brake fluid from an unopened container.
- If any hydraulic component or line is removed or replaced, it may be necessary to bleed the entire system.
- A clean repair area is essential. Always clean the reservoir and cap thoroughly before removing the cap. The slightest amount of dirt in the fluid may plug an orifice and impair the system function. Perform repairs after components have been thoroughly cleaned; use only denatured alcohol to clean components. Do not allow ABS components to come into contact with any substance containing mineral oil; this includes used shop rags.
- The Anti-Lock control unit is a microprocessor similar to other computer units in the vehicle. Ensure that the ignition switch is **OFF** before removing or installing controller harnesses. Avoid static electricity discharge at or near the controller.
- If any arc welding is to be done on the vehicle, the control unit should be unplugged before welding operations begin.

DISC AND DRUM SYSTEMS

> ※※ **CAUTION**
>
> **Dust and dirt accumulating on brake parts during normal use may contain asbestos fibers from production or aftermarket brake linings. Breathing excessive concentrations of asbestos fibers can cause serious bodily harm. Exercise care when servicing brake parts. Do not sand or grind brake lining unless equipment used is designed to contain the dust residue. Do not clean brake parts with compressed air or by dry brushing. Cleaning should be done by dampening the brake components with a fine mist of water, then wiping the brake components clean with a dampened cloth. Dispose of cloth and all residue containing asbestos fibers in an impermeable container with the appropriate label. Follow practices prescribed by the Occupational Safety and Health Administration (OSHA) and the Environmental Protection Agency (EPA) for the handling, processing, and disposing of dust or debris that may contain asbestos fibers.**

BRAKES · BLEEDING THE BRAKE SYSTEM

BLEEDING PROCEDURE

BLEEDING PROCEDURE

> ※※ **CAUTION**
>
> **Do not let brake fluid come into contact with the painted surface of the vehicle body. Wash away with water immediately and wipe off if it is spilled by accident.**

> ※※ **CAUTION**
>
> **Avoid mixing brake fluids of different brands to prevent fluid performance from degrading.**

> ※※ **CAUTION**
>
> **Be careful not to allow dirt or dust to enter the reservoir tank.**

Brake Lines

See Figures 1 and 2.

1. When the master cylinder is disassembled or the reservoir tank is empty, bleed the master cylinder before bleeding the brake line.
2. Fill the reservoir tank of the master cylinder with brake fluid.

➡ While bleeding air, keep the reservoir tank filled with brake fluid to prevent entry of air.

3. Attach one end of the vinyl tube to the air bleeder and the other end to the brake fluid container.
4. Depress the brake pedal several times, and keep it pressed.
5. Loosen the air bleeder screw to drain brake fluid. Tighten the air bleeder quickly, and release the brake pedal.
6. Repeat the steps 4) to 5) until there are no more air bubbles in the vinyl tube.
7. Repeat the steps from 2) to 6) above to bleed air from each wheel.

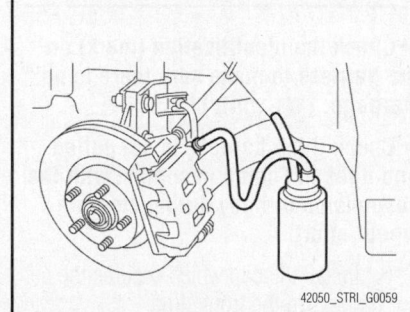

42050_STRI_G0059

Fig. 1 Attach one end of the vinyl tube to the air bleeder and the other end to the brake fluid container

➡ Perform the operation in the order from closest wheel cylinder to the master cylinder.

8. Securely tighten the air bleeder screws. Tighten to 5.8 ft. lbs. (8 Nm)
9. Check that there are no brake fluid leaks in the entire system.
10. Check the pedal stroke.
- Run the engine at idle after warming up the engine, and depress the brake pedal with a force of 112 lbs. (500 N). Measure the distance between the brake pedal and (1) steering wheel. Release the pedal, and measure the distance between pedal and steering wheel again.
- When depressing the pedal with a force of 112 lbs. (500 N). Less than 4.53 inch (115 mm)
11. If the distance is more than specification, there is a possibility of air being caught in the brake line. Bleed the brake line of all air until the pedal stroke meets the specification.
12. Operate the hydraulic control unit in the sequence control mode.
13. Check the pedal stroke again.
14. If the distance is more than specification, there is a possibility of air being caught in the hydraulic unit. Repeat above

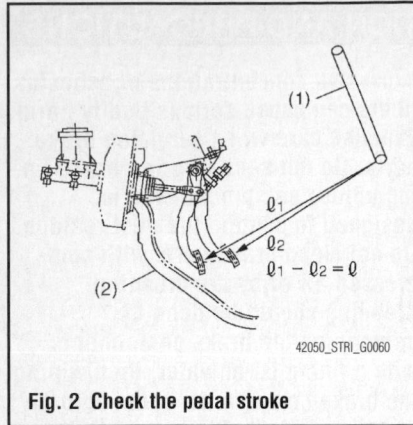

42050_STRI_G0060

Fig. 2 Check the pedal stroke

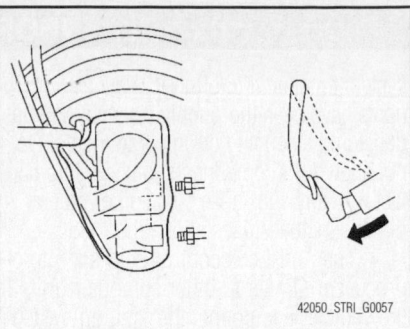

42050_STRI_G0057

Fig. 3 Disconnect the brake line at primary and second sides and depress the brake pedal

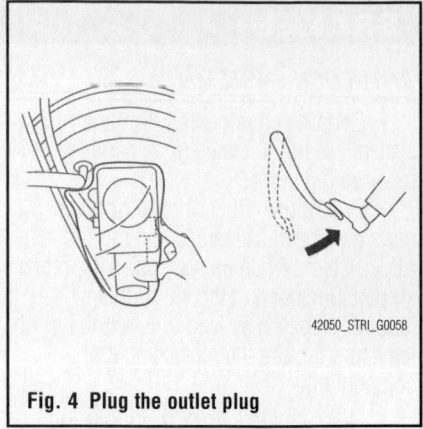

42050_STRI_G0058

Fig. 4 Plug the outlet plug

steps 2) to 9) until the pedal stroke meets the specification.

15. Fill brake fluid up to the "MAX" level of the reservoir tank.

16. Test run the vehicle and ensure that the brakes operate normally.

Master Cylinder

See Figures 3 and 4.

➡When the master cylinder is disassembled or the reservoir tank is empty, bleed the master cylinder. If bleeding the master cylinder is not necessary, omit the following procedures, and perform bleeding of the brake line.

1. Fill the reservoir tank of the master cylinder with brake fluid.

➡While bleeding air, keep the reservoir tank filled with brake fluid to prevent entry of air.

2. Disconnect the brake line at primary and secondary sides.

3. Wrap the master cylinder with a plastic bag.

4. Depress the brake pedal slowly and hold it.

5. Plug the outlet plug with your finger, and then release the brake pedal.

6. Repeat the step 4) and 5) several times.

7. Remove the plastic bag.

8. Install the brake pipe to the master cylinder. Tighten to 14 ft. lbs. (19 Nm).

9. Bleed air from the brake line.

BLEEDING THE ABS SYSTEM

Refer to the procedures under Bleeding the Brake System.

BRAKES   ANTI-LOCK BRAKE SYSTEM (ABS)

WHEEL SPEED SENSORS

REMOVAL & INSTALLATION

Front Wheel

See Figures 5 and 6.

1. Disconnect the ground cable from the battery.

2. Disconnect the ABS wheel speed sensor connector in the engine compartment.

3. Remove the sensor harness from the clip.

4. Remove the clip which secures the sensor harness to the front strut.

> ❄ **CAUTION**
>
> Be careful not to damage the sensor.

> ❄ **CAUTION**
>
> Do not apply excessive force to the sensor harness.

To install:

5. Install the sensor and tighten to 60 inch lbs. (7.5 Nm).

> ❄❄ **CAUTION**
>
> Be careful not to damage the sensor.

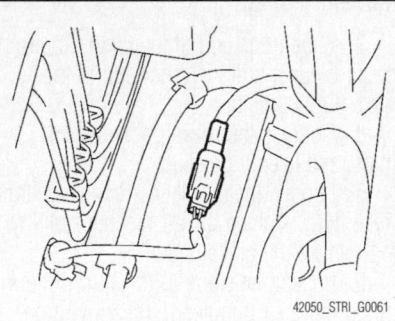

42050_STRI_G0061

Fig. 5 Disconnect the ABS front wheel speed sensor connector in the engine compartment

➡Check the identification (mark) on the harness to make sure there is no warpage. (W1 white)

➡Check if the harness is not pulled and does not come in contact with the suspension or body during steering wheel effort.

6. Install the clip which secures the sensor harness to the front strut.

7. Install the sensor harness from the clip.

8. Connect the ABS wheel speed sensor connector in the engine compartment.

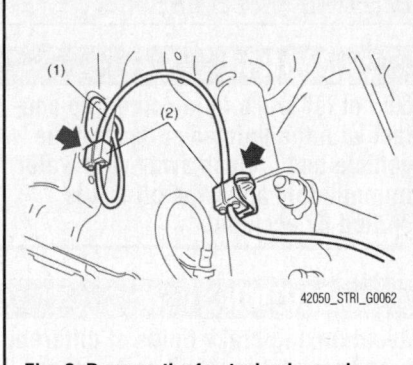

42050_STRI_G0062

Fig. 6 Remove the front wheel speed sensor harness from the clip

9. Connect the ground cable from the battery.

Rear Wheel

See Figures 7 and 8.

1. Disconnect the ground cable from the battery.

2. Disconnect the connector from the rear ABS wheel speed sensor.

3. Remove the sensor harness from the rear arm clip.

4. Remove the rear ABS wheel speed sensor from the rear housing.

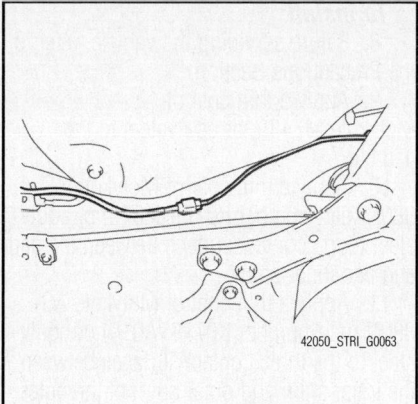

42050_STRI_G0063

Fig. 7 Remove the rear wheel speed sensor harness from the rear arm clip

---

※ **CAUTION**

Be careful not to damage the sensor.

BRAKES

BRAKE CALIPER

REMOVAL & INSTALLATION

See Figures 9 and 10.

1. Before servicing the vehicle, refer to the Precautions Section.
2. Loosen the wheel nuts.
3. Raise and support the vehicle safely.
4. Remove the tire and wheel.
5. Remove the union bolt. Disconnect the brake line from the brake caliper. Be sure to properly catch the fluid to avoid damage to painted surfaces and improper disposal.
6. Remove the bolt retaining the lock pin to the caliper body.
7. Remove the caliper from its mounting.

To install:

8. Before servicing the vehicle, refer to the Precautions Section.
9. Install the caliper from its mounting.
10. Install the caliper from its mounting.
11. Install the bolt retaining the lock pin to the caliper body.
12. Install the union bolt. Disconnect the brake line from the brake caliper. Be sure to properly catch the fluid to avoid damage to painted surfaces and improper disposal.
13. Install the tire and wheel.

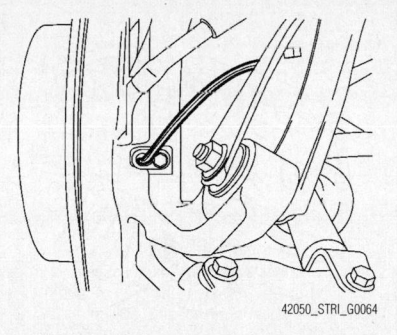

42050_STRI_G0064

Fig. 8 Remove the rear ABS wheel speed sensor from the rear housing

---

※ **CAUTION**

Do not apply excessive force to the sensor harness.

To install:

5. Install in the reverse order of removal. Tighten to 60 inch lbs. (7.5 Nm).

---

14. Lower the vehicle safely.
15. Tighten the wheel nuts.
16. Tighten the support bolts to 88 ft. lbs. (120 Nm).

※ **CAUTION**

Be careful not to damage the sensor.

→**Check the identification (mark) on the harness to make sure there is no warpage. [W3 (white)]**

6. Install the rear ABS wheel speed sensor from the rear housing.
7. Install the sensor harness from the rear arm clip.
8. Connect the connector from the rear ABS wheel speed sensor.
9. Connect the ground cable from the battery.

WHEEL SPEED SENSOR RINGS (TOOTHED RINGS)

REMOVAL & INSTALLATION

The wheel speed sensor ring is part of the hub and bearing assembly.

FRONT DISC BRAKES

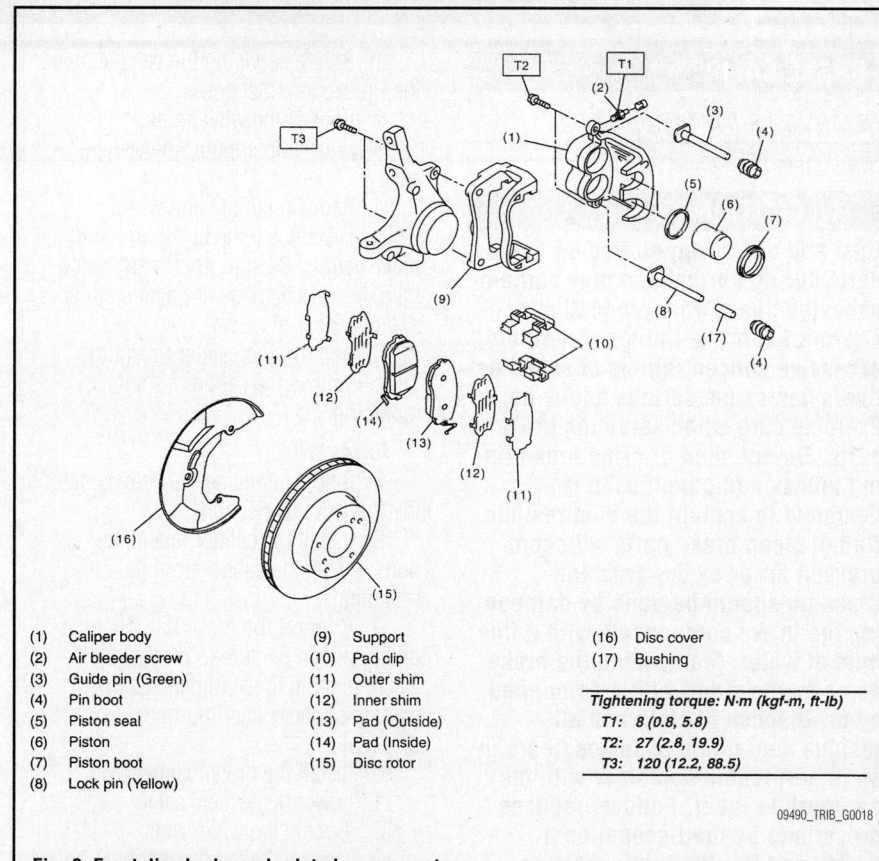

(1)	Caliper body	(9)	Support	(16)	Disc cover
(2)	Air bleeder screw	(10)	Pad clip	(17)	Bushing
(3)	Guide pin (Green)	(11)	Outer shim		
(4)	Pin boot	(12)	Inner shim		
(5)	Piston seal	(13)	Pad (Outside)		
(6)	Piston	(14)	Pad (Inside)		
(7)	Piston boot	(15)	Disc rotor		
(8)	Lock pin (Yellow)				

Tightening torque: N·m (kgf-m, ft-lb)
T1: 8 (0.8, 5.8)
T2: 27 (2.8, 19.9)
T3: 120 (12.2, 88.5)

09490_TRIB_G0018

Fig. 9 Front disc brake and related components

---

17. Check the brake fluid level, correct as required.
18. Bleed the hydraulic system, as required.

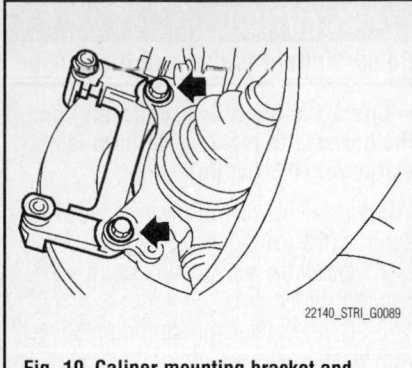

22140_STRI_G0089

Fig. 10 Caliper mounting bracket and mounting bolt

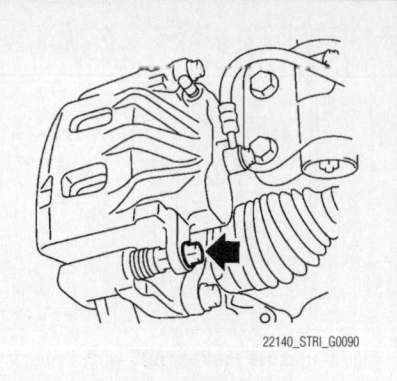

22140_STRI_G0090

Fig. 11 Caliper mounting bolt view

DISC BRAKE PADS

REMOVAL & INSTALLATION

See Figure 11.

1. Before servicing the vehicle, refer to the Precautions Section.
2. Loosen the wheel nuts.
3. Raise and support the vehicle safely.

4. Remove the tire and wheel.
5. Remove the caliper bolt.
6. Raise the caliper body and properly support it.

➡ **Do not disconnect the brake fluid line from the caliper.**

7. Remove the pads.

To install:

8. Before servicing the vehicle, refer to the Precautions Section.
9. Apply a thin coat of Molykote M7439, or equivalent to the pad clip.
10. Apply a thin coat of Molykote AS-880N (part number K0779YA010) or equivalent to the contact surface between the pad and pad inner shim.
11. Apply a thin coat of Molykote AS-880N (part number K0779YA010) or equivalent to the three contact surfaces between the inner shim and outer shim of the outer pads.
12. Check the brake fluid level, correct as required.
13. Bleed the hydraulic system, as required.
14. Install the caliper bolt.
15. Install the tire and wheel.
16. Lower the vehicle safely.
17. Tighten the wheel nuts.

BRAKES

BRAKE CALIPER

REMOVAL & INSTALLATION

See Figures 12 and 13.

✷✷ CAUTION

Dust and dirt accumulating on brake parts during normal use may contain asbestos fibers from production or aftermarket brake linings. Breathing excessive concentrations of asbestos fibers can cause serious bodily harm. Exercise care when servicing brake parts. Do not sand or grind brake lining unless equipment used is designed to contain the dust residue. Do not clean brake parts with compressed air or by dry brushing. Cleaning should be done by dampening the brake components with a fine mist of water, then wiping the brake components clean with a dampened cloth. Dispose of cloth and all residue containing asbestos fibers in an impermeable container with the appropriate label. Follow practices prescribed by the Occupational Safety and Health Administration (OSHA) and the Environmental Protection Agency (EPA) for the handling, processing, and disposing of dust or debris that may contain asbestos fibers.

1. Before servicing the vehicle, refer to the Precautions Section.
2. Loosen the wheel nuts.
3. Raise and support the vehicle safely.
4. Remove the tire and wheel.
5. Disconnect the brake line from the brake caliper. Be sure to properly catch the fluid to avoid damage to painted surfaces and improper disposal.
6. Remove the caliper retaining bolts. Remove the caliper from its mounting.

To install:

7. Before servicing the vehicle, refer to the Precautions Section.
8. Install the caliper retaining bolts. Install the caliper from its mounting.
9. Connect the brake line from the brake caliper. Be sure to properly catch the fluid to avoid damage to painted surfaces and improper disposal.
10. Install the tire and wheel.
11. Lower the vehicle safely.
12. Tighten the wheel nuts.
13. Tighten the caliper retaining bolts to 20 ft. lbs. (36 Nm).
14. Check the brake fluid level, correct as required.
15. Bleed the hydraulic system, as required.

REAR DISC BRAKES

DISC BRAKE PADS

REMOVAL & INSTALLATION

See Figure 14.

✷✷ CAUTION

Dust and dirt accumulating on brake parts during normal use may contain asbestos fibers from production or aftermarket brake linings. Breathing excessive concentrations of asbestos fibers can cause serious bodily harm. Exercise care when servicing brake parts. Do not sand or grind brake lining unless equipment used is designed to contain the dust residue. Do not clean brake parts with compressed air or by dry brushing. Cleaning should be done by dampening the brake components with a fine mist of water, then wiping the brake components clean with a dampened cloth. Dispose of cloth and all residue containing asbestos fibers in an impermeable container with the appropriate label. Follow practices prescribed by the Occupational Safety and Health Administration (OSHA) and the Environmental Protection Agency (EPA) for the handling, processing, and disposing of dust or debris that may contain asbestos fibers.

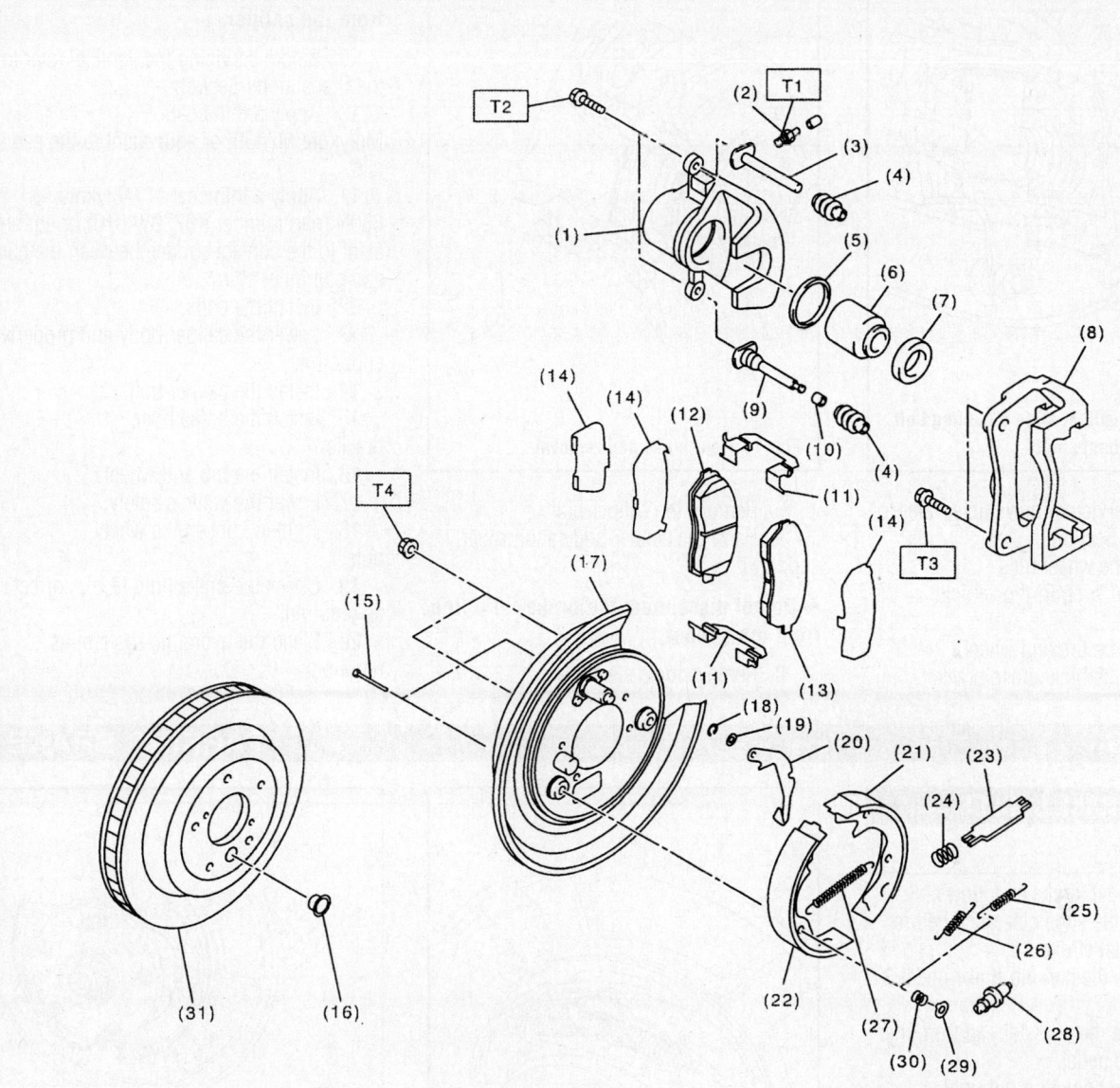

(1)	Caliper body	(13)	Outer pad	(25)	Secondary shoe return spring
(2)	Air bleeder screw	(14)	Shim	(26)	Primary shoe return spring
(3)	Guide pin (Green)	(15)	Shoe hold pin	(27)	Adjusting spring
(4)	Pin boot	(16)	Cover	(28)	Adjuster
(5)	Piston seal	(17)	Back plate	(29)	Brake shoe cup
(6)	Piston	(18)	Retainer	(30)	Brake shoe spring
(7)	Piston boot	(19)	Spring washer		
(8)	Support	(20)	Parking brake lever		
(9)	Lock pin (Yellow)	(21)	Parking brake shoe (Secondary)		
(10)	Bushing	(22)	Parking brake shoe (Primary)		
(11)	Pad clip	(23)	Strut		
(12)	Inner pad	(24)	Strut shoe spring		

Tightening torque: N·m (kgf-m, ft-lb)

T1:	**8 (0.8, 5.8)**
T2:	**27 (2.8, 19.9)**
T3:	**37 (3.7, 27.2)**
T4:	**66 (6.7, 48.7)**

09490_TRIB_G0019

Fig. 12 Rear disc brake and related components

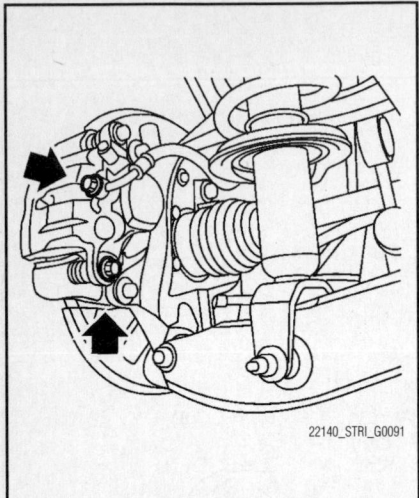

Fig. 13 Lower caliper slide mounting bolt and brake line banjo bolt

1. Before servicing the vehicle, refer to the Precautions Section.
2. Loosen the wheel nuts.
3. Raise and support the vehicle safely.
4. Remove the tire and wheel.
5. Remove the brake hose bracket.

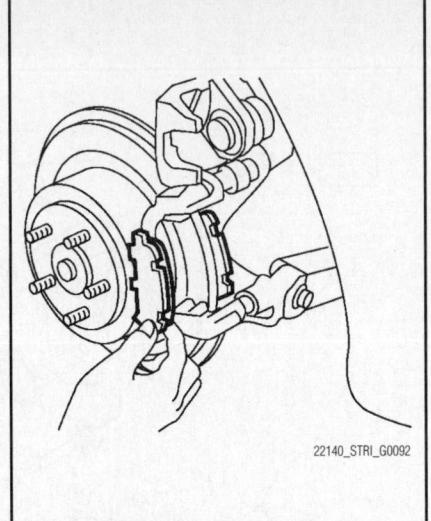

Fig. 14 Rear brake pad removal

6. Remove the caliper bolt.
7. Raise the caliper body and properly support it.

➡ **Do not disconnect the brake fluid line from the caliper.**

8. Remove the pads.

To install:

➡ **Do not disconnect the brake fluid line from the caliper.**

9. Before servicing the vehicle, refer to the Precautions Section.
10. Apply a thin coat of Molykote M7439, or equivalent to the pad clip.
11. Apply a thin coat of Molykote AS-880N (part number K0779YA010) or equivalent to the contact surface between the pad and pad inner shim.
12. Install the pads.
13. Lower the caliper body and properly support it.
14. Install the caliper bolt.
15. Install the brake hose bracket.
16. Install the tire and wheel.
17. Lower the vehicle safely.
18. Tighten Loosen the wheel nuts.
19. Check the brake fluid level, correct as required.
20. Bleed the hydraulic system, as required.

BRAKES

PARKING BRAKE CABLES

ADJUSTMENT

Adjust the pedal stroke as follows:
   a. Adjust the shoe clearance before adjusting pedal stroke.
   b. Operate the parking brake pedal 3 to 4 times.
   c. Remove the adjuster cap from the parking brake pedal.
   d. Turn the adjusting nut until the pedal stroke is at the specified value.

➡ **The pedal stroke when stepped on with a force of 67 lbs. (300 N) should be about 5–6 notches.**

1. Check there is no brake drag.
2. Install the adjuster cap.

PARKING BRAKE SHOES

REMOVAL & INSTALLATION

See Figures 15 through 17.

1. Release the parking brake.
2. Lift up the vehicle, and then remove the rear wheels.
3. Remove the two mounting bolts and remove the disc brake caliper assembly.

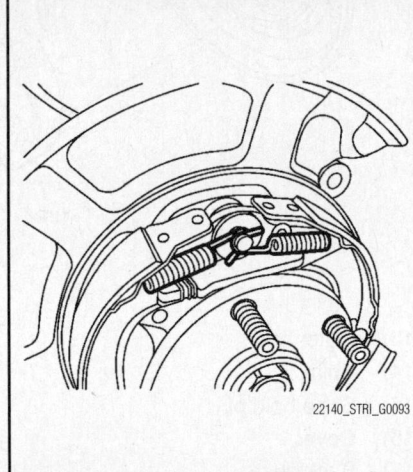

Fig. 15 Parking brake return springs shown

4. Suspend the rear disc brake caliper assembly so that the brake hose is not stretched.
5. Remove the rear disc brake rotor.

➡ **If the disc rotor is difficult to remove loosen the parking brake shoes. If it is difficult to remove the disc rotor from the hub, drive an 8 mm bolt into the threads of the rotor, then remove the rotor.**

PARKING BRAKE

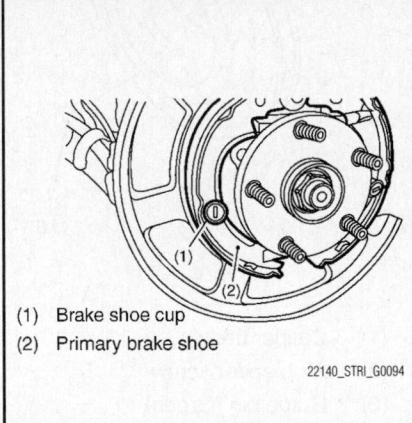

(1) Brake shoe cup
(2) Primary brake shoe

Fig. 16 Brake shoe cup and primary brake shoe

6. Remove the shoe return spring.
7. Remove the brake shoe cup and brake shoe spring, then remove the primary brake shoe.
8. Remove the strut and strut spring.
9. Remove the adjuster assembly.
10. Remove the brake shoe cup and brake shoe spring, then remove the secondary brake shoe.
11. Remove the parking brake cable from lever.

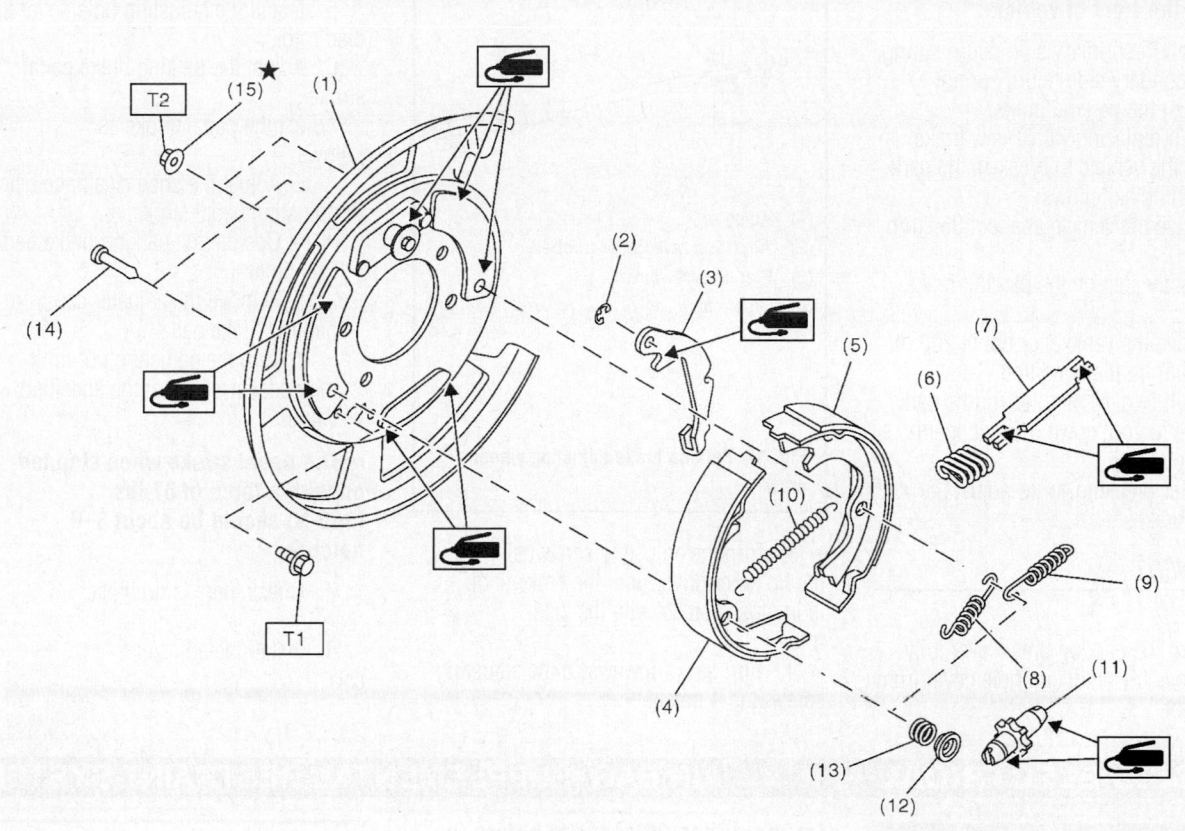

Fig. 17 Exploded view of parking brake shoe components

(1)	Back plate	(8)	Primary return spring	(14)	Shoe hold pin
(2)	Retainer	(9)	Secondary return spring	(15)	Self locking nut (with WAX)
(3)	Lever	(10)	Adjusting spring		
(4)	Parking brake shoe (Primary)	(11)	Adjuster		
(5)	Parking brake shoe (Secondary)	(12)	Brake shoe cup		
(6)	Strut spring	(13)	Brake shoe spring		
(7)	Strut				

Tightening torque:N·m (kgf-m, ft-lb)

T1: 8 (0.8, 5.9)

T2: 75 (7.6, 55.33)

22140_STRI_G0095

12. Remove the retainer from the secondary side brake shoe. Remove the parking lever and washer from brake shoe.

To install:

✳✳ CAUTION

Be sure the lining surface is free from oil and grease.

13. Apply Brake Grease (Part No. 003602002) to the following locations:
- Six contact surfaces of brake shoe rim and back plate gasket
- Contact surface of the brake shoe and the anchor pin
- Contact surface of the lever and strut
- Contact surface of the brake shoe and the adjuster assembly
- Contact surface of the brake shoe and the strut
- Contact surface of the lever and brake shoe

14. Install the parking lever to the secondary side brake shoe, then clamp and lock the retainer securely.

15. Attach the parking brake cable to the parking brake lever.

16. Attach the adjuster assembly and adjusting spring to the brake shoe.

➡Install the adjuster assembly so that the screw section will be towards the rear of the vehicle.

17. Check the parking brake cable is not fallen from the cable guide.

18. Install the brake shoe to the back plate using shoe hold pins, brake shoe spring and brake shoe cup.

19. Install the strut and strut spring to the brake shoe.

➡**Install the strut spring so that it is placed on the front of vehicle.**

20. Install the primary side return spring, then the secondary side return spring.

21. Adjust the parking brake.

22. When replacing with a new brake shoe, drive the vehicle to break-in the parking brake lining as follows:

    a. Drive the vehicle at about 35 km/h (22 MPH).

    b. lightly step on the parking brake pedal.

    c. Drive the vehicle for about 200 m (0.12 mile) in this condition.

    d. Wait 5 to 10 minutes for the parking brake to cool down. Repeat again from step (1).

    e. After breaking-in, re-adjust the parking brakes.

ADJUSTMENT

See Figure 18.

1. Return the parking brake lever fully.

2. Remove the adjusting hole cover from the disc rotor.

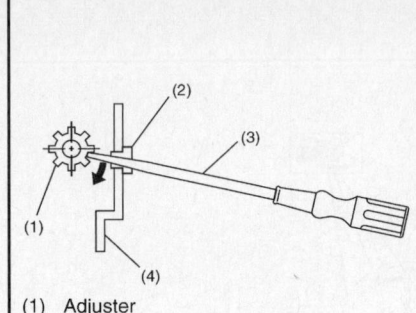

(1) Adjuster
(2) Adjusting hole cover (rubber)
(3) Flat tip screwdriver
(4) Disc rotor

22140_STRI_G0096

Fig. 18 Parking brake adjusting wheel view

3. Turn the adjusting screw using a flat tip screwdriver until the brake shoe is in close contact with the disc rotor.

4. Turn back (downward) the adjusting screw 3 to 4 notches.

✳✳ CAUTION
Check that there is no brake drag.

5. Install the adjusting hole cover to the disc rotor.

6. Adjust the parking brake pedal stroke.

Adjust the pedal stroke as follows:

    a. Adjust the shoe clearance before adjusting pedal stroke.

    b. Operate the parking brake pedal 3 to 4 times.

    c. Remove the adjuster cap from the parking brake pedal.

    d. Turn the adjusting nut until the pedal stroke is at the specified value.

➡**The pedal stroke when stepped on with a force of 67 lbs. (300 N) should be about 5–6 notches.**

7. Check there is no brake drag.

8. Install the adjuster cap.

CHASSIS ELECTRICAL

GENERAL INFORMATION

PRECAUTIONS

Disconnect and isolate the battery negative cable before beginning any airbag system component diagnosis, testing, removal, or installation procedures. Allow system capacitor to discharge for two minutes before beginning any component service. This will disable the airbag system. Failure to disable the airbag system may result in accidental airbag deployment, personal injury, or death.

DISARMING THE SYSTEM

1. Turn the ignition switch to OFF.

2. Disconnect the ground cable from the battery and wait for at least 20 seconds before starting work.

✳✳ WARNING
The airbag system is fitted with a backup power source. After disconnecting the battery ground cable, the airbag may deploy if you do not wait

AIR BAG (SUPPLEMENTAL RESTRAINT SYSTEM)

for more than 20 seconds before starting the service of airbag system.

ARMING THE SYSTEM

1. Install negative battery cable.

CLOCKSPRING CENTERING

See Figure 19.

✳✳ CAUTION
When servicing a vehicle, be sure to turn the ignition switch to OFF, disconnect the ground cable from battery, and wait for more than 20 seconds before starting work. The airbag system is fitted with a backup power source. After disconnecting the battery ground cable, the airbag may deploy if you do not wait for more than 20 seconds before starting the service of airbag system.

1. Turn the ignition switch to OFF.

2. Disconnect the ground cable from

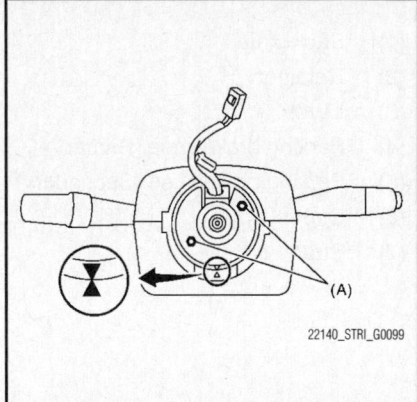

22140_STRI_G0099

Fig. 19 Clock spring pins (A) and alignment marks shown

battery and wait for at least 20 seconds before starting work.

3. Check that front wheels are positioned in straight ahead direction.

4. Turn the clock spring pin (A) clockwise until it stops.

5. Turn the clock spring connector pins approx. 3.25 turns until the marks are aligned.

DRIVE TRAIN

FRONT DRIVESHAFT

REMOVAL & INSTALLATION

See Figures 20 through 24.

1. Disconnect the ground cable from the battery.
2. Lift up the vehicle, and remove the front wheels.
3. Drain the differential gear oil.
4. Lift the crimped section of axle nut.
5. Remove the axle nut using a socket wrench while depressing the brake pedal.

※※ CAUTION

Remove the wheel before loosening the axle nut. Failure to follow this rule may damage the wheel bearings.

6. Remove the stabilizer link from front arm.
7. Disconnect the front arm ball joint from the front housing.
8. Remove the front drive shaft assembly. If it is hard to remove, use ST1 and ST2 as shown in the illustration.

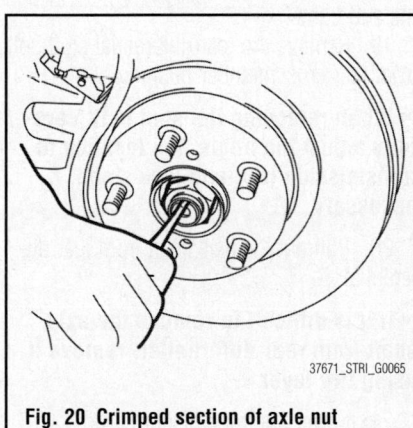

37671_STRI_G0065

Fig. 20 Crimped section of axle nut

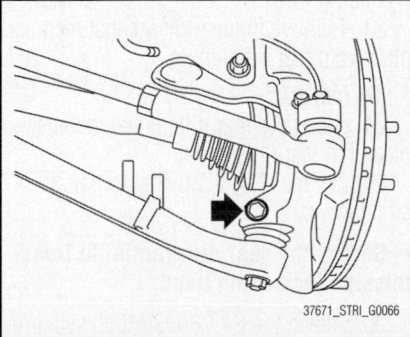

37671_STRI_G0066

Fig. 21 Disconnect the front arm ball joint from the front housing

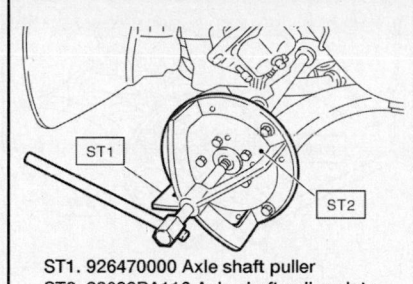

ST1. 926470000 Axle shaft puller
ST2. 28099PA110 Axle shaft puller plate

37671_STRI_G0067

Fig. 22 Front driveshaft assembly removal, showing special tools

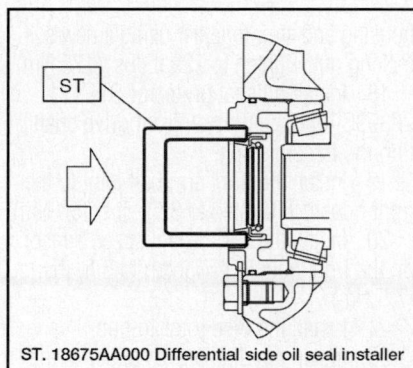

ST. 18675AA000 Differential side oil seal installer

37671_STRI_G0068

Fig. 23 Replace the differential side retainer oil seal with a new seal

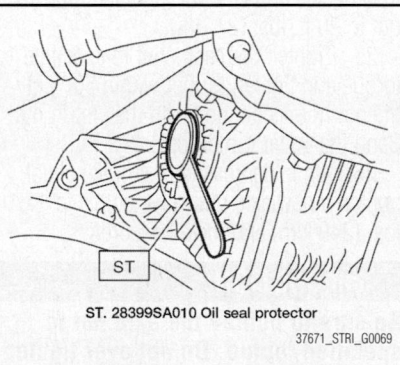

ST. 28399SA010 Oil seal protector

37671_STRI_G0069

Fig. 24 Install the front driveshaft to transmission

9. Using a bar, remove the front drive shaft from transmission.

※※ CAUTION

Be careful not to allow the bar to damage holder area.

To install:

10. Using the ST, replace the differential side retainer oil seal with a new seal as shown in the illustration.

→**After pulling out the drive shaft, be sure to replace with a new oil seal.**

11. Insert the EBJ into hub splines.
12. Draw the drive shaft into specified position.

※※ WARNING

Do not hammer drive shaft when installing it.

13. Tighten the axle nut temporarily.
14. Using the ST, install the front drive shaft to transmission as shown in the illustration.
15. Connect the front arm ball joint to the front housing. Tighten to 37 ft. lbs. (50 Nm).
16. Install the stabilizer link. Tighten to 44 ft. lbs. (60 Nm).

→**Be sure to use a new self-locking nut.**

17. While depressing the brake pedal, tighten a new axle nut to the specified torque and lock it securely. Tighten to 177 ft. lbs. (240 Nm).

※※ WARNING

Install the wheel after installation of axle nut. Failure to follow this rule may damage the wheel bearing.

※※ WARNING

Be sure to tighten axle nut to specified torque. Do not overtighten it as this may damage the wheel bearing.

18. After tightening axle nut, lock it securely.
19. Fill the differential gear oil.
20. Install the front wheel and tighten the wheel nuts, as follows:
   - Chromed wheel: 110 ft. lbs. (150 Nm)
   - Other than Chromed wheel: 88 ft. lbs. (120 Nm)

FRONT HALFSHAFT

REMOVAL & INSTALLATION

See Figures 25, 26 and 38.

1. Before servicing the vehicle, refer to the Precautions Section.
2. Disconnect the ground cable from the battery.
3. Lift-up the vehicle, and remove the front wheels.

4. Lift the crimped section of axle nut.

5. Remove the axle nut using a socket wrench while depressing the brake pedal.

✳ CAUTION

Remove the wheel before loosening the axle nut. Failure to follow this rule may damage the wheel bearings.

6. Remove the stabilizer link.

7. Remove the disc brake caliper from the front housing, and suspend it from strut using a piece of wire.

8. Remove the disc rotor from the hub.

➡**If it is difficult to remove the disc rotor from the hub, drive the 8 mm of bolt into the threaded end of rotor, then remove the rotor.**

9. Remove the cotter pin and castle nut securing the tie rod end to the front housing knuckle arm.

10. Using a puller, remove the tie–rod ball joint from knuckle arm.

✳ CAUTION

When removing tie-rod, do not hit the tie-rod end with hammer.

11. Remove the ABS wheel speed sensor assembly and harness.

12. Remove the front arm ball joint from the front housing.

13. Remove the PTJ from transmission.

14. Remove the front drive shaft assembly from the hub. If it is hard to remove, use the ST1-92647000 and ST2-28099PA110.

15. After scribing an alignment mark on the camber adjusting bolt head, remove the bolts which connect the front housing and strut, and disconnect the front housing from the strut.

16. Remove the front axle.

To install:

17. Align the alignment mark on the camber adjusting bolt head, and affix the front

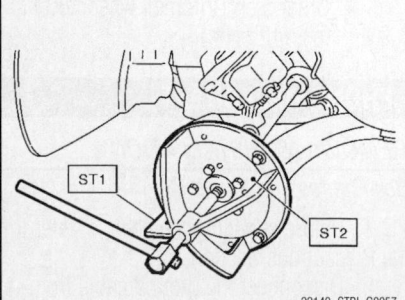

Fig. 25 Front arm ball joint removal location shown

22140_STRI_G0057

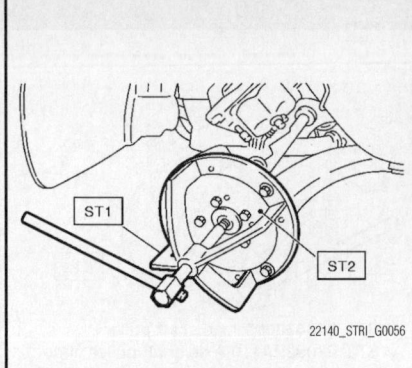

Fig. 26 Special tools shown for removal of axle shaft

22140_STRI_G0056

housing and strut together using a new self locking nut. Tighten to 129 ft. lbs. (175 Nm).

18. Install oil seal protector ST-28399SA000 then install front drive shaft into the transmission.

19. Install the front arm ball joint to the front housing. Tighten to 37 ft. lbs. (50 Nm).

20. Install the ABS wheel speed sensor on the front housing. Tighten to 6 ft. lbs. (7.5 Nm).

21. Install the disc rotor to hub.

22. Install the disc brake caliper to the front housing. Tighten mounting bolts to 88 ft. lbs. (120 Nm).

23. Install the stabilizer link.

24. Connect the tie-rod end ball joint to the knuckle arm with a castle nut. Tighten nut to 20 ft. lbs. (27 Nm).

25. Tighten the castle nut to specified torque and tighten further within 60° until the pin hole is aligned with the slot in nut. Bend the cotter pin to lock.

26. While depressing the brake pedal, tighten a new axle nut (olive color) to 177 ft. lbs. (240 Nm) and lock it securely.

✳✳ CAUTION

Be sure to tighten the axle nut to specified torque. Do not over tighten as this may damage the wheel bearing.

27. After tightening the axle nut, lock it securely.

28. Install the wheel and tighten the wheel nuts to 81 ft. lbs. (110 Nm).

REAR AXLE HOUSING

REMOVAL & INSTALLATION

See Figures 27 through 29.

1. Before servicing the vehicle, refer to the Precautions Section.

2. Disconnect the ground cable from the battery.

3. Position the select lever to neutral.

4. Release the parking brake.

5. Lift up the vehicle.

6. Remove the rear exhaust pipe and muffler.

7. Remove the heat shield cover.

8. Remove the propeller shaft.

9. Remove the connector from the rear differential oil temperature switch.

10. Prepare the transmission jack and band.

11. Remove the DOJ of rear drive shaft from rear differential.

12. Remove the nuts which hold the rear differential front member.

13. Support the rear differential with the transmission jack.

14. Loosen the self-lock nuts which hold the rear differential to rear crossmember.

15. Remove the rear differential front member.

16. Secure the rear differential using band.

17. Remove the self-lock nuts which hold the rear differential to rear crossmember.

18. Remove the air breather hose from the sub frame.

19. Remove the rear differential stud bolt from rear crossmember bushing.

➡**When removing the stud bolt, carefully adjust the angle and location of transmission jack and jack stand, if necessary.**

20. Pull out the axle shaft from rear differential.

➡**If it is difficult to remove the axle shaft from rear differential, remove it using tire lever.**

21. Lower the transmission jack.

22. Secure the rear drive shaft to lateral link using wire.

23. Remove the rear differential member plate from rear differential.

To install:

24. Insert the rear differential member plate into rear differential.

25. Set the rear differential to transmission jack.

➡**Secure the rear differential to transmission jack using band.**

26. Attach the ST-28099PA090 seal protector to rear differential.

27. Insert the shaft until the spline portion comes inside the side oil seal.

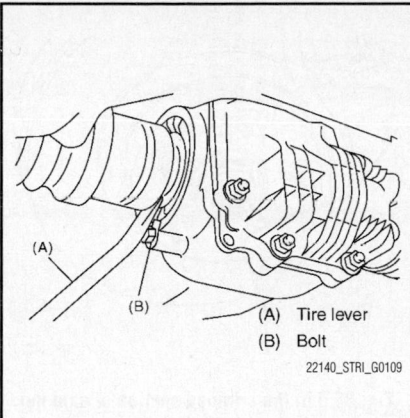

Fig. 27 Drive axle removal, tire lever (A) bolt (B)

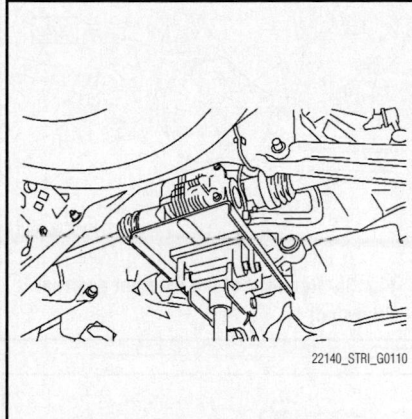

Fig. 28 Rear differential supported with transmission jack

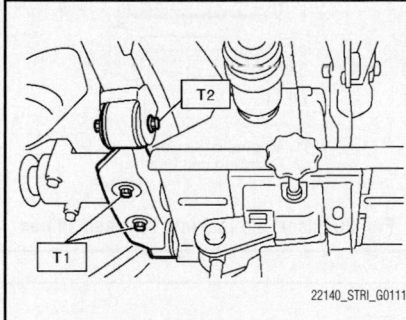

Fig. 29 Front member and (T1) and (T2) locations shown

28. Remove ST from rear differential.
29. Push the rear differential to insert the axle shaft into rear differential.
30. Adjust the transmission jack, if necessary, and insert the rear differential stud bolt into rear crossmember bushing properly.
31. After inserting the rear differential stud bolt into the rear crossmember bush-

ing, lift up the transmission jack and align the rear differential to the its attachment position.
32. Tighten a new self-locking nut temporarily to rear crossmember.
33. Remove the band from rear differential. Lift up the rear differential until the rear differential is separated from the transmission jack.
34. Install the rear differential front member with new self-locking nuts. Tighten (T1) to 37 ft. lbs. (50 Nm). Tighten (T2) to 81 ft. lbs. (110 Nm).
35. Tighten the self-locking nut to 51 ft. lbs. (70 Nm).
36. Lower the transmission jack.
37. Install the air breather hose to the sub frame.
38. Install the propeller shaft.
39. Install the heat shield cover.
40. Install the rear exhaust pipe and muffler.
41. Fill rear differential with approved fluid.

REAR AXLE SHAFT SEAL

REMOVAL & INSTALLATION
See Figure 30.

1. Before servicing the vehicle, refer to the Precautions Section.
2. Remove the rear differential.
3. Remove the rear differential side oil seal using a screwdriver wrapped with vinyl tape to prevent the side retainer from scratches.

To install:
4. Using the ST-398437700, install the oil seal to the side retainer.
5. Install the rear differential

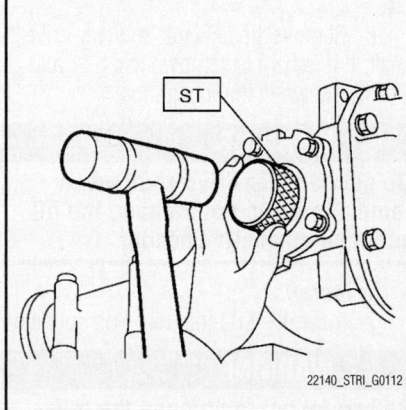

Fig. 30 Rear differential axle side oil seal installation

REAR AXLE UNIT

REMOVAL & INSTALLATION
See Figures 31 through 33.

1. Before servicing the vehicle, refer to the Precautions Section.
2. Disconnect the ground cable from the battery.
3. Position the select lever to neutral.
4. Release the parking brake.
5. Lift up the vehicle.
6. Remove the rear exhaust pipe and muffler.
7. Remove the heat shield cover.
8. Remove the propeller shaft.
9. Remove the connector from the rear differential oil temperature switch.
10. Prepare the transmission jack and band.
11. Remove the DOJ of rear drive shaft from rear differential.
12. Remove the nuts which hold the rear differential front member.
13. Support the rear differential with the transmission jack.
14. Loosen the self-lock nuts which hold the rear differential to rear crossmember.
15. Remove the rear differential front member.
16. Secure the rear differential using band.
17. Remove the self-lock nuts which hold the rear differential to rear crossmember.
18. Remove the air breather hose from the sub frame.
19. Remove the rear differential stud bolt from rear crossmember bushing.

➡**When removing the stud bolt, carefully adjust the angle and location of transmission jack and jack stand, if necessary.**

20. Pull out the axle shaft from rear differential.

➡**If it is difficult to remove the axle shaft from rear differential, remove it using tire lever.**

21. Lower the transmission jack.
22. Secure the rear drive shaft to lateral link using wire.
23. Remove the rear differential member plate from rear differential.

To install:
24. Insert the rear differential member plate into rear differential.
25. Set the rear differential to transmission jack.

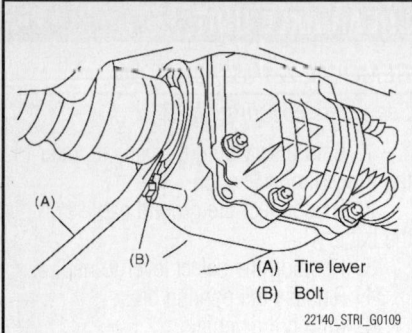

Fig. 31 Drive axle removal, tire lever (A) bolt (B)

(A) Tire lever
(B) Bolt

22140_STRI_G0109

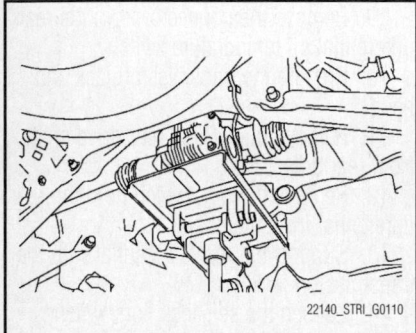

Fig. 32 Rear differential supported with transmission jack

22140_STRI_G0110

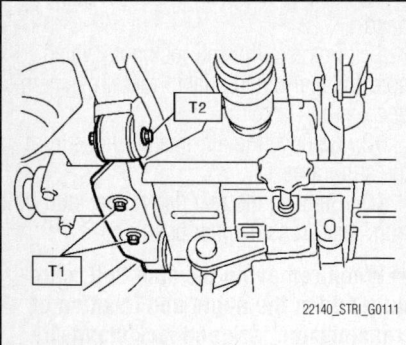

Fig. 33 Front member and (T1) and (T2) locations shown

22140_STRI_G0111

➡️Secure the rear differential to transmission jack using band.

26. Attach the ST-28099PA090 seal protector to rear differential.

27. Insert the shaft until the spline portion comes inside the side oil seal.

28. Remove ST from rear differential.

29. Push the rear differential to insert the axle shaft into rear differential.

30. Adjust the transmission jack, if necessary, and insert the rear differential stud bolt into rear crossmember bushing properly.

31. After inserting the rear differential stud bolt into the rear crossmember bushing, lift up the transmission jack and align the rear differential to the Its attachment position.

32. Tighten a new self-locking nut temporarily to rear crossmember.

33. Remove the band from rear differential. Lift up the rear differential until the rear differential is separated from the transmission jack.

34. Install the rear differential front member with new self-locking nuts.

35. Tighten the self-locking nut to 51 ft. lbs. (70 Nm).

36. Lower the transmission jack.

37. Install the air breather hose to the sub frame.

38. Install the propeller shaft.

39. Install the heat shield cover.

40. Install the rear exhaust pipe and muffler.

41. Fill rear differential with approved fluid.

REAR DRIVESHAFT

REMOVAL & INSTALLATION

See Figures 34 through 37.

1. Disconnect the ground cable from battery.

2. Lift up the vehicle, and then remove the rear wheels.

3. Lift the crimped section of axle nut.

4. While applying the parking brake, remove the axle nut using a socket wrench.

✳️ CAUTION

Remove the wheel before loosening the axle nut. Failure to follow this rule may damage the wheel bearings.

5. Remove the rear differential assembly.

6. Remove the axle nut and rear drive shaft. If it is hard to remove, use ST1 and ST2 as shown in the illustration.

✳️ WARNING

Do not hammer drive shaft when removing it. Do not damage the oil seal and magnetic encoder.

To install:

7. Insert the EBJ into rear hub splines.

✳️ WARNING

Be careful not to damage the magnetic encoder. Do not get closer the tool which charged magnetism to magnetic encoder.

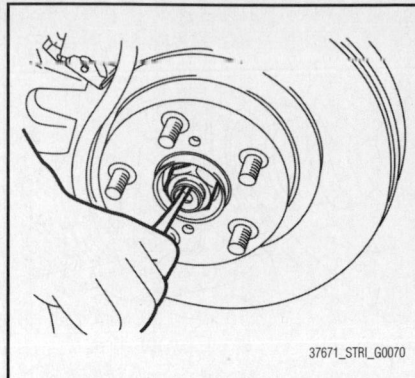

37671_STRI_G0070

Fig. 34 Lift the crimped section of axle nut

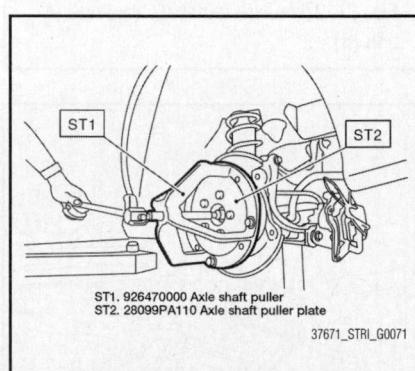

ST1. 926470000 Axle shaft puller
ST2. 28099PA110 Axle shaft puller plate

37671_STRI_G0071

Fig. 35 Removal of the axle nut and rear driveshaft

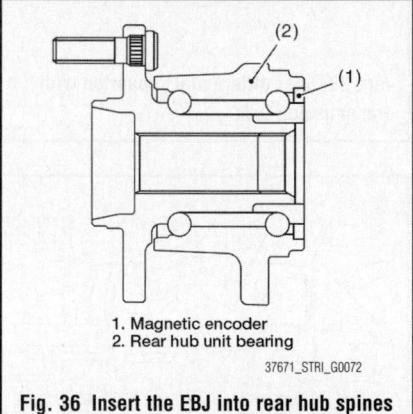

1. Magnetic encoder
2. Rear hub unit bearing

37671_STRI_G0072

Fig. 36 Insert the EBJ into rear hub spines

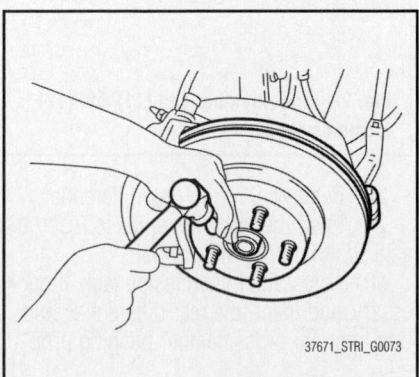

37671_STRI_G0073

Fig. 37 Lock the axle nut securely

8. Draw the rear drive shaft into specified position.

✳✳ CAUTION

Do not hammer drive shaft when installing it.

9. Tighten the axle nut temporarily.
10. Install the rear differential assembly.
11. While applying the parking brake and depressing the brake pedal, tighten a new axle nut to 177 ft. lbs. (240 Nm) and lock it securely.

✳ WARNING

Install the wheel after installation of axle nut. Failure to follow this rule may damage the wheel bearing.

- Be sure to tighten the axle nut to specified
torque. Do not overtighten it as this may damage
the wheel bearing.
12. Lock the axle nut securely.
13. Install the rear wheel and tighten, as follows:
- Chromed wheel: 110 ft. lbs. (150 Nm).
- Except chromed wheel: 88 ft. lbs. (120 Nm)

REAR HALFSHAFT

REMOVAL & INSTALLATION

See Figure 38.

1. Before servicing the vehicle, refer to the Precautions Section.
2. Disconnect the negative battery cable.
3. Raise and support the vehicle safely.
4. Remove the wheels and tires.
5. Unlock the axle nut. Depress the parking brake. Remove the axle nut.

➡**Be sure to loosen the axle nut after removing the tire and wheel from the vehicle. Failure to do this may damage the wheel bearings.**

6. Remove the rear differential assembly, VA-type.
7. Remove the axle nut and rear halfshaft.

➡**Use axle shaft puller tools ST-92647000 and ST-28099PA110, or equivalent, if it is difficult to remove the halfshaft. Do not hammer the halfshaft to remove it. Be careful not to damage the oil seal or magnetic encoder.**

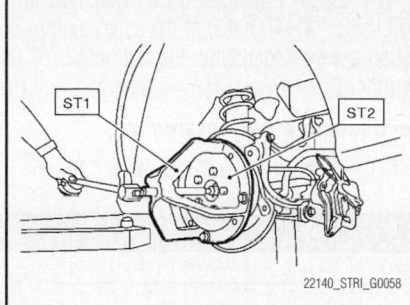

Fig. 38 Axle shaft puller tools ST-92647000 and ST-28099PA110

To install:

8. Insert the EBJ into hub splines.

✳✳ CAUTION

Be careful not to damage the magnetic encoder. Do not get closer the tool which charged magnetism to magnetic encoder.

9. Draw the rear drive shaft into specified position.
10. Tighten the axle nut temporarily.
11. Install the rear differential assembly, VA-type.
12. While applying the parking brake and depressing the brake pedal, tighten a new axle nut to (240 Nm) and lock it securely.

✳✳ CAUTION

Be sure to tighten the axle nut to specified torque. Do not over tighten as this may damage the wheel bearing.

13. Lock the axle nut securely.
14. Install the rear wheel.
15. For chrome wheels tighten to 111 ft. lbs. (150 Nm).
16. All other wheels tighten to 89 ft. lbs. (120 Nm).

REAR PINION SEAL

REMOVAL & INSTALLATION

See Figures 39 through 41.

1. Before servicing the vehicle, refer to the Precautions Section.
2. Release the parking brake.
3. Remove the oil drain plug, and drain gear oil.
4. Install the oil drain plug. Use a new metal gasket and tighten to 22 ft. lbs. (29 Nm).
5. Jack-up the rear wheel and support the body with rigid racks.

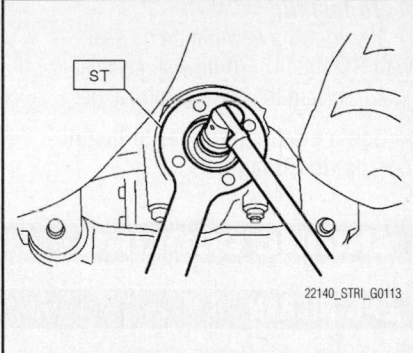

Fig. 39 Removing the self locking nut using ST-498427200 Flange wrench.

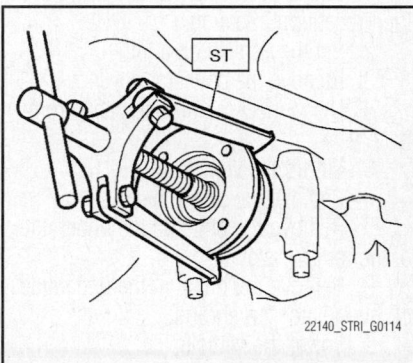

Fig. 40 Removal of companion flange with ST-399703600 Puller Assembly

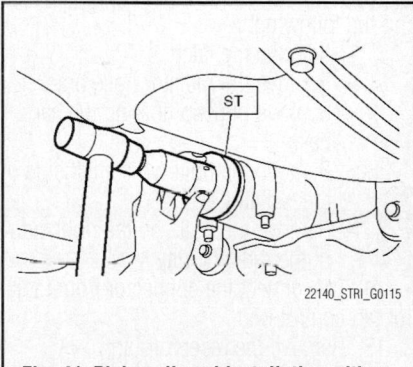

Fig. 41 Pinion oil seal installation with ST-498447120 oil seal installer

6. Remove the rear exhaust pipe and muffler.
7. Remove the heat shield cover.
8. Remove the propeller shaft.
9. Remove the self-locking nut while holding the companion flange with ST-498427200 Flange wrench.
10. Extract the companion flange using ST-399703600 Puller Assembly.
11. Remove the oil seal using ST-398527700 puller or screwdriver.

To install:

12. Install a new pinion oil seal with ST-498447120 oil seal installer.
13. Install the companion flange.

➡**Use a plastic hammer to install companion flange.**

14. Apply Loctite®THREE BOND 1324 (Part No. 004403042) to the drive pinion shaft screw threads and the nut seat surface.

➡**Use a new self-locking nut.**

15. Tighten the self-locking nut to 141 ft. lbs. (191 Nm) so that the rotating resistance of companion flange becomes the same as that of before oil seal replacement.

16. Hereafter, reassemble in the reverse order of disassembly.

ENGINE COOLING

ENGINE FAN

REMOVAL & INSTALLATION

See Figures 42 through 48.

1. Before servicing the vehicle, refer to the Precautions Section.
2. Set the vehicle on a lift.
3. Remove the collector cover.
4. Disconnect the ground cable from battery.
5. Lift up the vehicle.
6. Remove the undercover.
7. Remove the bolts on the underside of the radiator stay.
8. Remove the bolts on the underside of the radiator fan shroud.
9. Lower the vehicle.
10. Remove the air intake duct.
11. Remove the front upper cover.
12. Remove the radiator upper brackets.
13. Remove the radiator stay by performing the following:
   • Remove the latch
   • Remove the radiator hose bracket
   • Remove the clip holding the harness
   • Remove the bolts on the left side of the radiator stay
   • Remove the bolts on the right side of the radiator stay
14. Disconnect the connector from radiator fan control unit.
15. Remove the reservoir tank.
16. Remove the bolts on the upper side of the radiator fan shroud.

➡**When pulling the radiator fan shroud up, be careful not to hit and damage the radiator and ATF hoses.**

17. Remove the radiator fan shroud.

To install:

18. Before servicing the vehicle, refer to the Precautions Section.

➡**When pushing the radiator fan shroud down, be careful not to hit and damage the radiator and ATF hoses.**

19. Install the radiator fan shroud. Tighten the bolts on the top side of the radiator fan shroud to 60 inch lbs. (7 Nm).
20. Install the bolts on the upper side of the radiator fan shroud.

Fig. 42 Remove the bolts from the underside of the radiator stay

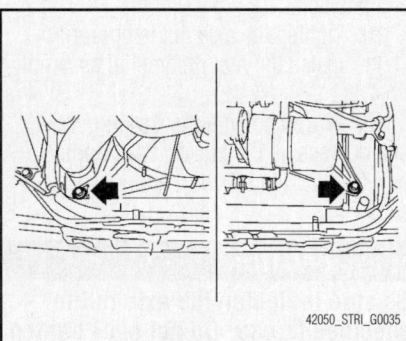

Fig. 43 Remove the bolts from the underside of the radiator fan shroud

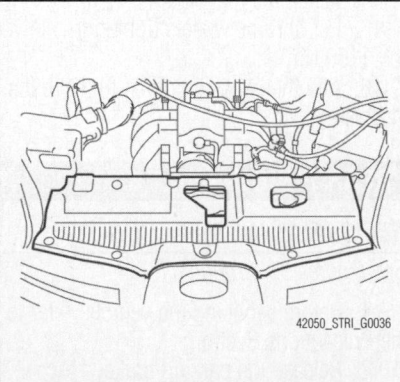

Fig. 44 Remove the front upper cover

21. Install the reservoir tank.
22. Connect the connector from radiator fan control unit.
23. Install the radiator stay by performing the following:

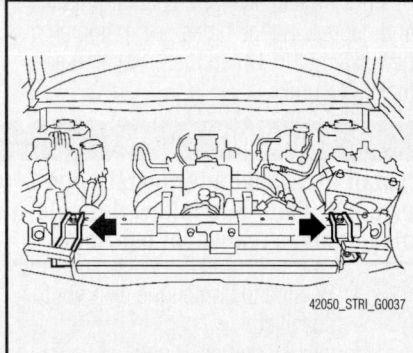

Fig. 45 Remove the radiator upper brackets

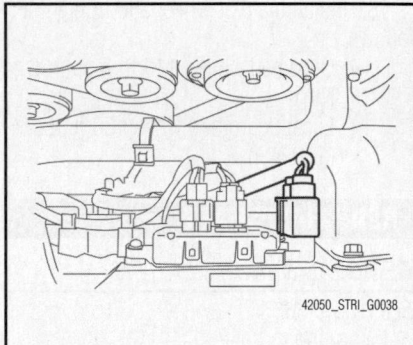

Fig. 46 Disconnect the connector from the radiator fan control unit

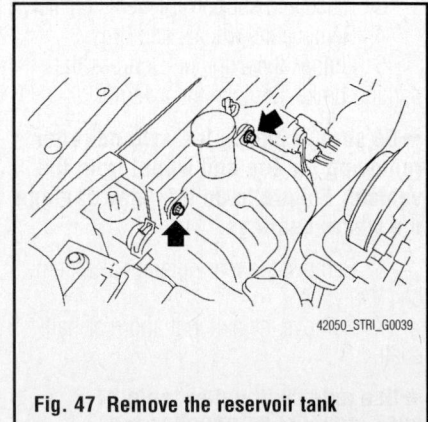

Fig. 47 Remove the reservoir tank

   • Install the bolts on the right side of the radiator stay
   • Install the bolts on the left side of the radiator stay
   • Install the clip holding the harness

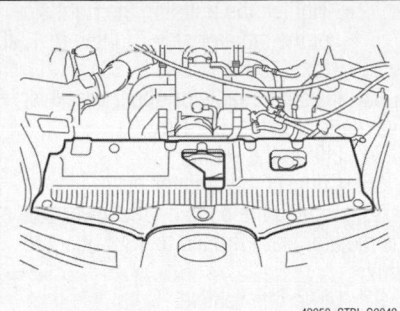

Fig. 48 Remove the bolts from the upper side of the radiator fan shroud

• Install the radiator hose bracket
• Install the latch
24. Install the radiator upper brackets.
25. Install the front upper cover.
26. Install the air intake duct.
27. Raise the vehicle.
28. Install the bolts on the underside of the radiator fan shroud.
29. Install the bolts on the underside of the radiator stay.
30. Install the undercover.
31. Lower the vehicle.
32. Connect the ground cable from battery.
33. Install the collector cover.
34. Remove the vehicle from the lift.

RADIATOR

REMOVAL & INSTALLATION

See Figures 49 through 62.

❉❉ WARNING

The radiator is pressurized. Wait until engine cools down before working on the radiator.

1. Before servicing the vehicle, refer to the Precautions Section.
2. Set the vehicle on a lift.
3. Remove the collector cover.
4. Disconnect the ground cable from battery.
5. Lift-up the vehicle.
6. Remove the undercover.
7. Drain engine coolant completely.
8. Disconnect the radiator hose from radiator.
9. Remove the bolts on the underside of the radiator stay.
10. Lower the vehicle.
11. Remove the air intake duct.
12. Remove the front upper cover.
13. Remove the radiator upper brackets.
14. Remove the radiator stay by performing the following:

• Remove the latch
• Remove the radiator hose bracket
• Remove the clip holding the harness
• Remove the bolts on the left side of the radiator stay
• Remove the bolts on the right side of the radiator stay
15. Disconnect the connector from radiator fan control unit.
16. Remove the reservoir tank.
17. Disconnect the radiator hose from radiator.
18. Disconnect the ATF cooler hose from the radiator.

➡**Plug the ATF pipe to prevent ATF leaks.**

19. Remove the front bumper.
20. Remove the bolts which hold the radiator and condenser.
21. Remove the clip holding the ambient temperature sensor harness.
22. Lift the radiator up and away from vehicle.

To install:

23. Install the radiator lower cushion to the vehicle.
24. Install the radiator to vehicle.

➡**Insert the pins on the lower side of radiator into the radiator lower cushions on the vehicle side.**

25. Hold the harness of the ambient temperature sensor with a clip.
26. Hold the radiator and condenser with bolts. Tighten to 60 inch lbs. (7 Nm).
27. Install the front bumper. Tighten the front bumper beam to 24 ft. lbs. (33 Nm).
28. Connect the ATF cooler hoses.
29. Connect the radiator hose.
30. Install the reservoir tank.
31. Connect the connector of the radiator fan control unit.
32. Install the radiator stay by performing the following:

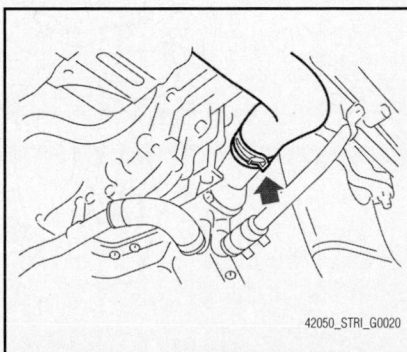

Fig. 49 Disconnect the radiator hose from the radiator

Fig. 50 Remove the bolts on the underside of the radiator stay

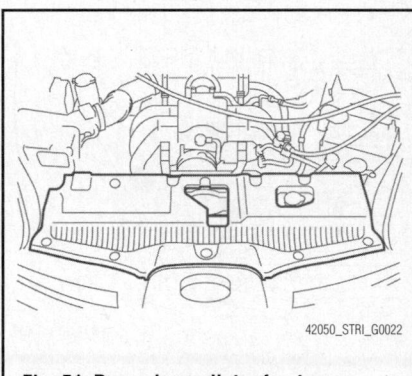

Fig. 51 Removing radiator front upper cover

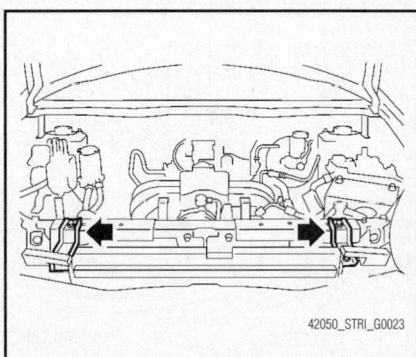

Fig. 52 Remove the radiator upper brackets

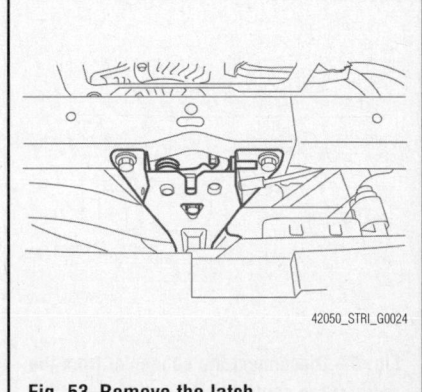

Fig. 53 Remove the latch

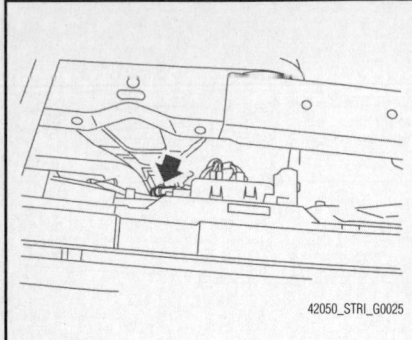

Fig. 54 Remove the clip holding the harness

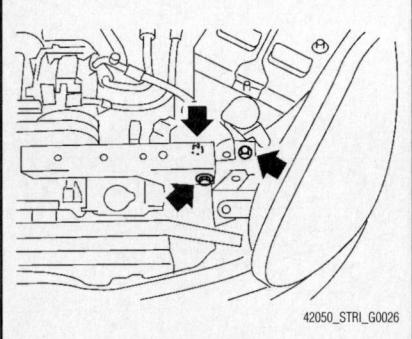

Fig. 55 Remove the bolts from the left side of the radiator stay

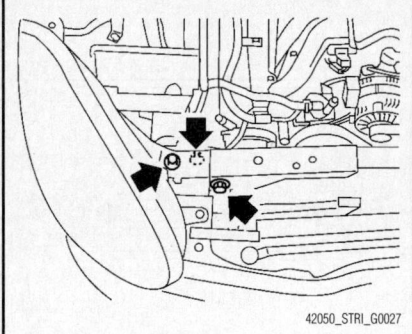

Fig. 56 Remove the bolts from the right side of the radiator stay

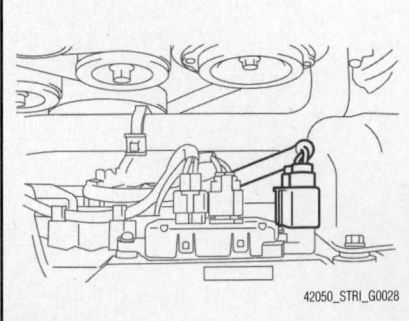

Fig. 57 Disconnect the connector from the radiator fan control unit

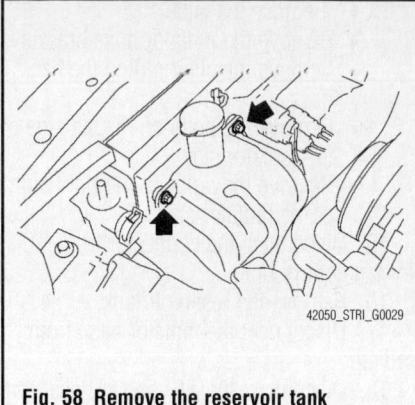

Fig. 58 Remove the reservoir tank

- Install the latch. Torque to 25 ft. lbs.(33 Nm)
- Install the radiator hose bracket
- Hold the harness to the radiator stay with a clip
- Tighten the bolts on the left side of the radiator stay. Tighten to 13 ft. lbs. (18 Nm)
- Tighten the bolts on the right side of the radiator stay. Tighten to 13 ft. lbs. (18 Nm)

33. Install the radiator upper brackets. Tighten to 9 ft. lbs. (12 Nm)

34. Lift up the vehicle.

35. Connect the radiator hose.

36. Tighten the bolts on the lower side of the radiator stay. Tighten to 13 ft. lbs. (18 Nm).

37. Lower the vehicle.

38. Connect the ground cable to battery.

39. Fill engine coolant.

40. Check the ATF level and replenish it if necessary.

41. Lift up the vehicle.

42. Install the undercover.

43. Lower the vehicle.

44. Install the front upper cover.

45. Install the air intake duct and fasten clip holing the air intake duct.

46. Install the collector cover.

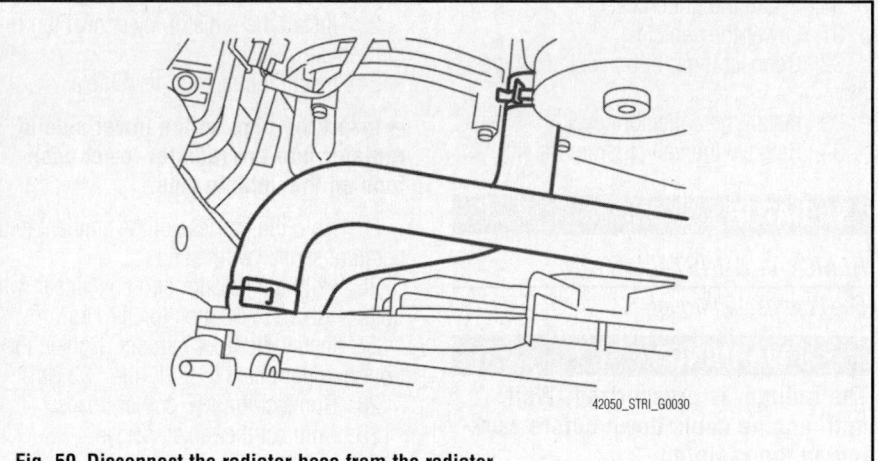

Fig. 59 Disconnect the radiator hose from the radiator

Fig. 60 Disconnect the ATF cooler hose from the radiator

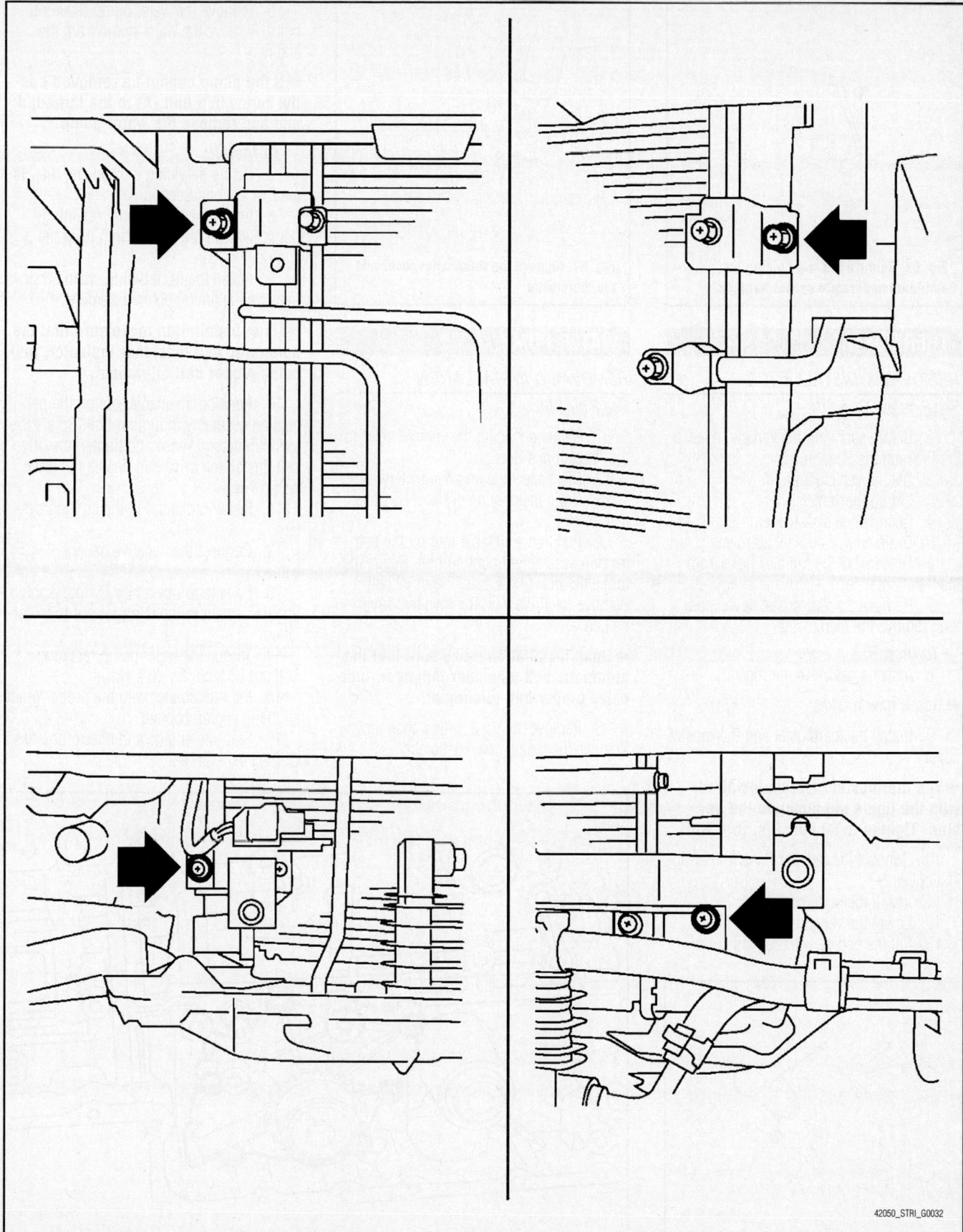

Fig. 61 Remove the bolts holding the radiator and condenser

42050_STRI_G0032

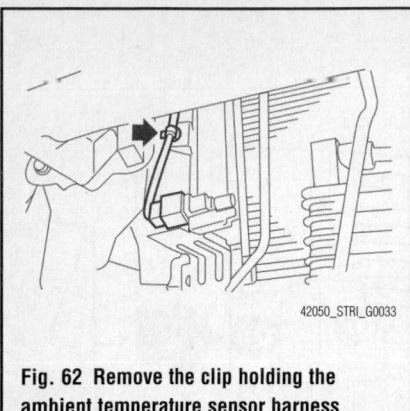

Fig. 62 Remove the clip holding the ambient temperature sensor harness

THERMOSTAT

REMOVAL & INSTALLATION

See Figures 63 and 64.

1. Before servicing the vehicle, refer to the Precautions Section.
2. Set the vehicle on a lift.
3. Lift up the vehicle.
4. Remove the undercover.
5. Drain engine coolant completely.
6. Disconnect the radiator hose from the thermostat cover.
7. Remove the thermostat cover, and then remove the thermostat.

To install:

8. Install a gasket to thermostat.

➡**Use a new gasket.**

9. Install the thermostat and thermostat cover.

➡**The thermostat must be installed with the jiggle pin facing to the up side. Tighten to 60 inch lbs. (6.4 Nm).**

10. Connect the radiator hose to thermostat cover.
11. Install the undercover.
12. Lower the vehicle.
13. Fill the engine with coolant.

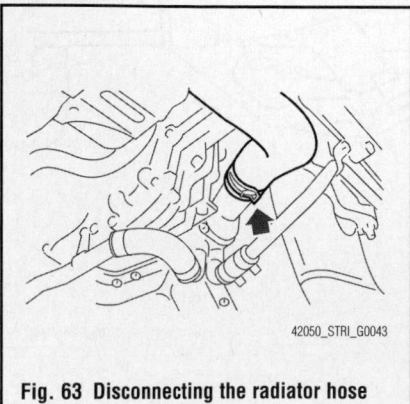

Fig. 63 Disconnecting the radiator hose from the thermostat cover

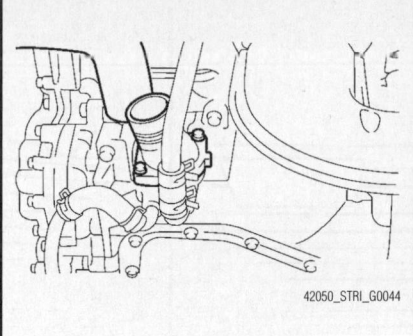

Fig. 64 Remove the thermostat cover and the thermostat

WATER PUMP

REMOVAL & INSTALLATION

See Figure 65.

1. Before servicing the vehicle, refer to the Precautions Section.
2. Disconnect the negative battery cable.
3. Drain the cooling system. Remove the radiator.
4. Position a suitable tool on the belt tension assembly mounting bolt. Rotate the tool clockwise and loosen the drive belt. Remove the drive belt. Remove the drive belt cover.

➡**Upon installation make sure that the automatic belt tensioner indicator, indicates proper belt alignment.**

5. Remove the front timing chain cover. Remove the timing chain assembly.

6. Remove the water pump retaining bolts. Remove the water pump from the engine.

➡**If the pump cannot be removed easily; screw in a bolt (A) to the threaded end and remove the water pump.**

To install:

7. Before servicing the vehicle, refer to the Precautions Section.
8. Install the water pump retaining bolts. Remove the water pump from the engine.
9. Install the front timing chain cover. Remove the timing chain assembly.

➡**Upon installation make sure that the automatic belt tensioner indicator, indicates proper belt alignment.**

10. Position a suitable tool on the belt tension assembly mounting bolt. Rotate the tool counterclockwise and tighten the drive belt. Install the drive belt. Install the drive belt cover.
11. Install radiator. Fill the cooling system.
12. Connect the negative battery cable.
13. Be sure to use a new O ring. Apply engine coolant to the O ring, prior to installation.
14. Torque the water pump retaining bolts to 60 inch lbs. (6.4 Nm).
15. Fill the radiator with the proper grade and type engine coolant.
16. Start the engine and check for leaks. Correct as required.

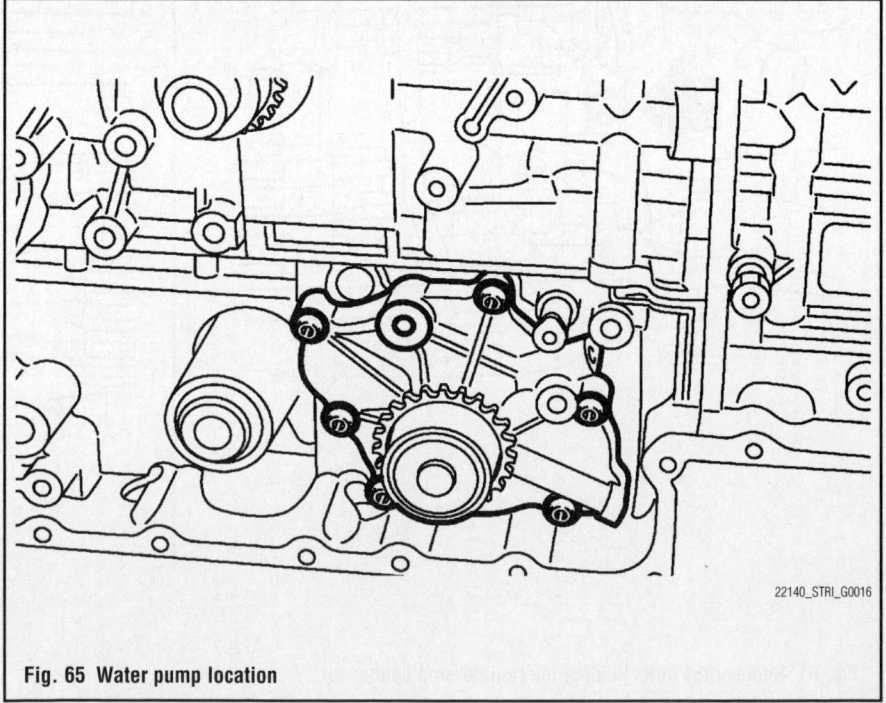

Fig. 65 Water pump location

ENGINE ELECTRICAL   **CHARGING SYSTEM**

ALTERNATOR

REMOVAL & INSTALLATION

See Figure 66.

1. Before servicing the vehicle, refer to the Precautions Section.
2. Disconnect the negative battery cable.
3. Remove the collector cover.
4. Disconnect the electrical connector and the terminal from the alternator.
5. Position a suitable tool on the belt tension assembly mounting bolt. Rotate the tool clockwise and loosen the drive belt. Remove the drive belt. Remove the drive belt cover.

➡**Upon installation make sure that the automatic belt tensioner indicator indicates proper belt alignment.**

6. Remove the alternator retaining bolts. Remove the alternator from the vehicle.

To install:

7. Before servicing the vehicle, refer to the Precautions Section.

22140_STRI_G0002

Fig. 66 Alternator and retaining bolt locations shown

➡**Upon installation make sure that the automatic belt tensioner indicator indicates proper belt alignment.**

8. Install the alternator retaining bolts. Install the alternator from the vehicle.
9. Position a suitable tool on the belt tension assembly mounting bolt. Rotate the tool counterclockwise and loosen the drive belt. Install the drive belt. Install the drive belt cover.
10. Connect the electrical connector and the terminal to the alternator.
11. Install the collector cover.
12. Connect the negative battery cable.
13. Tighten the retaining bolts to 18 ft. lbs. (25 Nm).

ENGINE ELECTRICAL   **IGNITION SYSTEM**

FIRING ORDER

The firing order is: 1–6–3–2–5–4.

IGNITION COIL

REMOVAL & INSTALLATION

Right Side

See Figure 67.

1. Remove the collector cover.

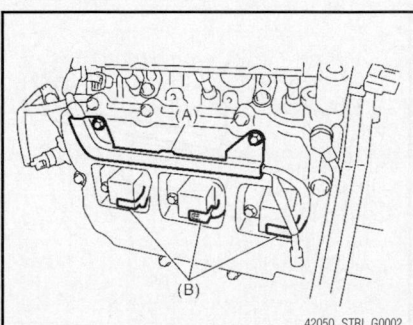

42050_STRI_G0002

Fig. 67 Removing bracket and connector from ignition coil—right side

2. Disconnect the ground cable from battery.
3. Remove the air cleaner case.
4. Remove the bracket.
5. Disconnect the connector from ignition coil.
6. Remove the ignition coil.

➡**Turn the #5 ignition coil to remove it.**

To install:

7. Install the ignition coil.

➡**Turn the #5 ignition coil to install it.**

8. Connect the connector from ignition coil.
9. Install the bracket.
10. Install the air cleaner case.
11. Connect the ground cable from battery.
12. Install the collector cover. Tighten the ignition coil to 12 ft. lbs. (16 Nm)

Left Side

See Figure 68.

1. Remove the collector cover.
2. Remove the battery and battery carrier.

3. Remove the bracket.
4. Disconnect the connector from ignition coil.
5. Remove the ignition coil.

➡**Turn the no. 6 ignition coil to install it.**

To install:

6. Install the Ignition coil.

➡**Turn the no. 6 ignition coil to install it.**

7. Connect the connector from ignition coil.
8. Install the bracket.

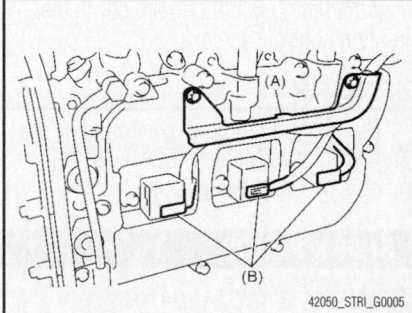

42050_STRI_G0005

Fig. 68 Removing bracket and connector from the ignition coil—left side

9. Install the battery and battery carrier.

10. Install the collector cover. Tighten the ignition coil to 12 ft. lbs. (16 Nm)

IGNITION TIMING

INSPECTION

※※ CAUTION

After warming—up, engine becomes very hot. Be careful not to burn yourself at measurement.

1. Before checking the ignition timing, check the following item:
   - Check the air cleaner element is free from clogging, spark plugs are in good condition, and hoses are connected properly
   - Check the malfunction indicator light does not illuminate
2. Idle the engine.
3. Stop the engine, and turn the ignition switch to OFF.
4. Insert the cartridge to Subaru Select Monitor.
5. Connect the Subaru Select Monitor to data link connector.
6. Turn the ignition switch to ON and Subaru Select Monitor switch to ON.
7. Select **Each System Check** in Main Menu.
8. Select **Engine** in Selection Menu.
9. Select **Current Data Display & Save** in Engine Control System Diagnosis.
10. Select **Data Display** in Data Display Menu.
11. Start the engine and check the ignition timing at idle speed. Ignition timing (BTDC/RPM): 15° / 8°/650 to 700

If the timing is not correct, check the ignition control system.

ADJUSTMENT

The ignition timing is controlled by the Engine Control Module (ECM). No adjustment is necessary or possible.

SPARK PLUGS

REMOVAL & INSTALLATION

Right Side
See Figures 69 through 71.

※※ CAUTION

All spark plugs installed on an engine must be of the same heat range.

1. Remove the collector cover.
2. Disconnect the ground cable from battery.
3. Remove the air cleaner case.
4. Remove the bracket.
5. Disconnect the connector from ignition coil.
6. Remove the ignition coil.

➡**Turn the #5 ignition coil to remove it.**

7. Remove the spark plug with a spark plug socket.

To install:

8. Install the spark plug with a spark plug socket.

➡**Turn the #5 ignition coil to remove it.**

9. Install the ignition coil.
10. Connect the connector from ignition coil.
11. Install the bracket.
12. Install the air cleaner case.
13. Connect the ground cable from battery.
14. Install the collector cover.

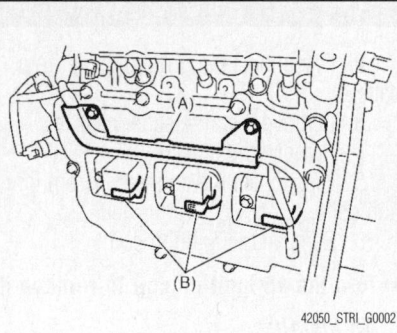

Fig. 69 Removing bracket and connector from ignition coil—right side

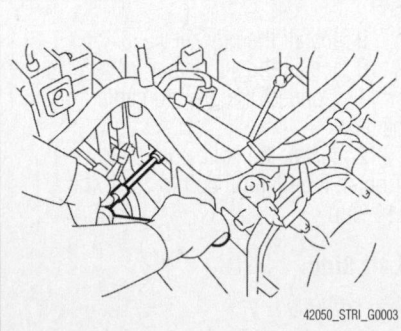

Fig. 70 Remove the spark plug with a spark plug socket—right side

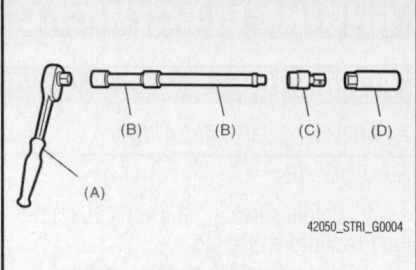

Fig. 71 Identifying (A) ratchet handle, (B) extension bar, (C) universal joint, (D) spark plug socket

※※ CAUTION

All spark plugs installed on an engine must be of the same heat range.

- Tighten the spark plug to 15 ft. lbs. (21 Nm)
- Tighten the ignition coil to 12 ft. lbs. (16 Nm)

➡**The tightening torque described above should be applied to only new spark plugs without oil on their threads. In case their threads are lubricated, the torque should be reduced by approximately⅓of the specified torque in order to avoid over—stressing.**

Left Side
See Figures 72 and 73.

1. Remove the collector cover.
2. Remove the battery and battery carrier.
3. Remove the bracket.
4. Disconnect the connector from ignition coil.
5. Remove the ignition coil.

➡**Turn the no. 6 ignition coil to remove it.**

6. Remove the spark plug with a spark plug socket.

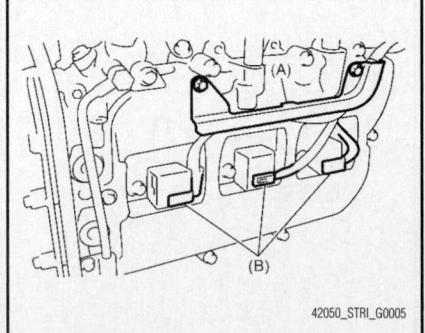

Fig. 72 Removing bracket and connector from the ignition coil—left side

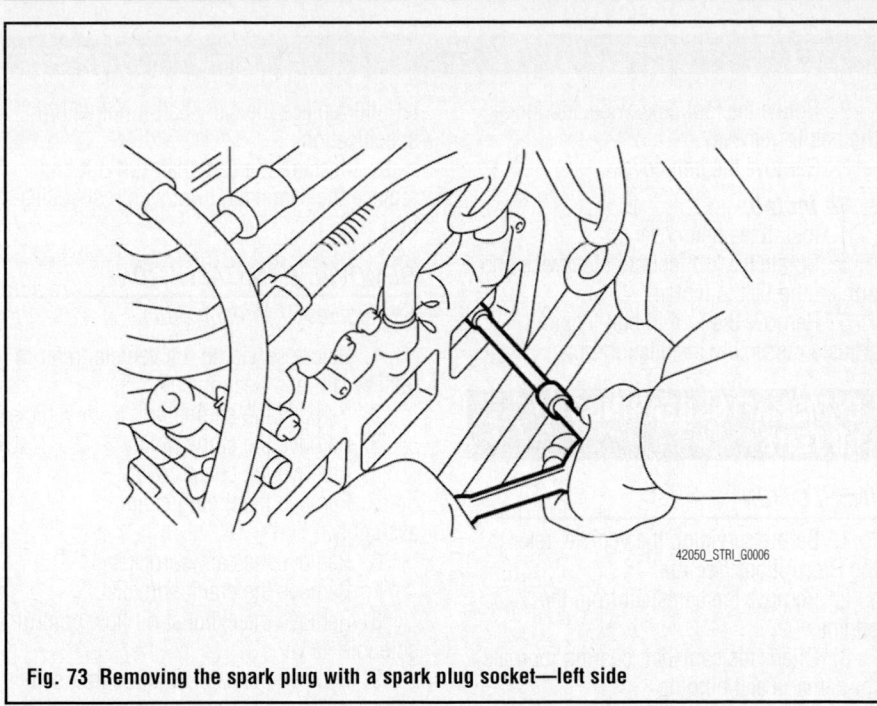

Fig. 73 Removing the spark plug with a spark plug socket—left side

To install:

7. Install the spark plug with a spark plug socket.

➡**Turn the no. 6 ignition coil to install it.**

8. Install the ignition coil.
9. Connect the connector from ignition coil.
10. Install the bracket.
11. Install the battery and battery carrier.
12. Install the collector cover.
   • Tighten the spark plug to 15 ft. lbs. (21 Nm)
   • Tighten the ignition coil to 12 ft. lbs. (16 Nm)

➡**The tightening torque described above should be applied to only new spark plugs without oil on their threads. In case their threads are lubricated, the torque should be reduced by approximately⅓of the specified torque in order to avoid over—stressing.**

ENGINE ELECTRICAL   STARTING SYSTEM

STARTER

REMOVAL & INSTALLATION

See Figure 74.

1. Before servicing the vehicle, refer to the Precautions Section.
2. Disconnect the negative battery cable.
3. Remove the collector cover.
4. Remove the air intake chamber.
5. Disconnect the electrical connectors from the starter.
6. Remove the starter retaining bolts. Remove the starter from the vehicle.

To install:

7. Before servicing the vehicle, refer to the Precautions Section.
8. Install the starter retaining bolts. Install the starter from the vehicle.
9. Connect the electrical connectors from the starter.

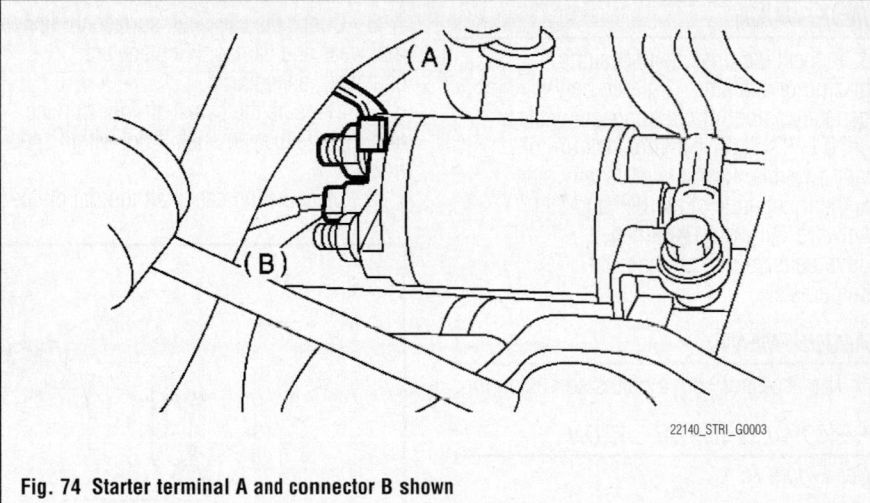

Fig. 74 Starter terminal A and connector B shown

10. Install the air intake chamber.
11. Install the collector cover.
12. Connect the negative battery cable.

13. Before servicing the vehicle, refer to the Precautions Section.
14. Torque the starter retaining bolts to 37 ft. lbs. (50 Nm).

ENGINE MECHANICAL

ACCESSORY DRIVE BELTS

ACCESSORY BELT ROUTING

See Figure 75.

Refer to the accompanying illustration for belt routing.

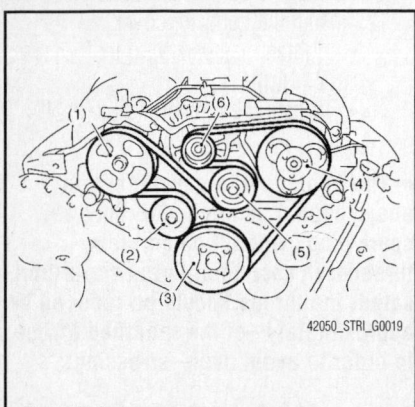

Fig. 75 Accessory belt routing

INSPECTION

Inspect the drive belt for signs of glazing or cracking. A glazed belt will be perfectly smooth from slippage, while a good belt will have a slight texture of fabric visible. Cracks will usually start at the inner edge of the belt and run outward. All worn or damaged drive belts should be replaced immediately.

ADJUSTMENT

The accessory drive belt is self adjusting.

REMOVAL & INSTALLATION

See Figure 76.

1. Install the tool to belt tension adjuster assembly installation bolt.

Fig. 76 Removing accessory drive belt

2. Rotate the tool clockwise and loosen the belt to remove.
3. Remove the belt cover.

To install:

4. Install the belt cover.
5. Rotate the tool counterclockwise and tighten the belt to install.
6. Remove the tool to belt tension adjuster assembly installation bolt.

CAMSHAFTS, BEARINGS & LIFTERS

INSPECTION

1. Before servicing the vehicle, refer to the Precautions Section.
2. Remove the camshaft from the engine.
3. Check the camshaft bearing journals for damage and binding.
4. If the journals are binding, check the cylinder head for damage.
5. Check the cylinder head for clogged oil holes.
6. Check the camshaft surface for abnormal wear and damage. Replace the camshaft, as required.
7. Measure the camshaft lobe surface and replace the camshaft if not within specification.
8. Measure the camshaft journal diameter and replace the camshaft if not within specification.
9. Measure the camshaft run out and replace the camshaft if not within specification.

REMOVAL & INSTALLATION

See Figures 77 through 84.

1. Before servicing the vehicle, refer to the Precautions Section.
2. Remove the engine unit from vehicle.
3. Remove the crank pulley.
4. Remove the chain cover.
5. Remove the timing chain assembly.
6. Remove the cam sprocket.
7. Remove the crank sprocket.
8. Remove the exhaust oil flow control solenoid valve.
9. Remove the camshaft position sensor.
10. Loosen the rocker cover bolt in the numerical order as shown in the figure, and then remove the rocker cover.
11. Loosen the camshaft cap bolts equally, a little at a time in numerical sequence shown in the figure.
12. Remove the camshaft caps and camshaft RH.
13. Similarly, remove the camshafts LH and related parts.

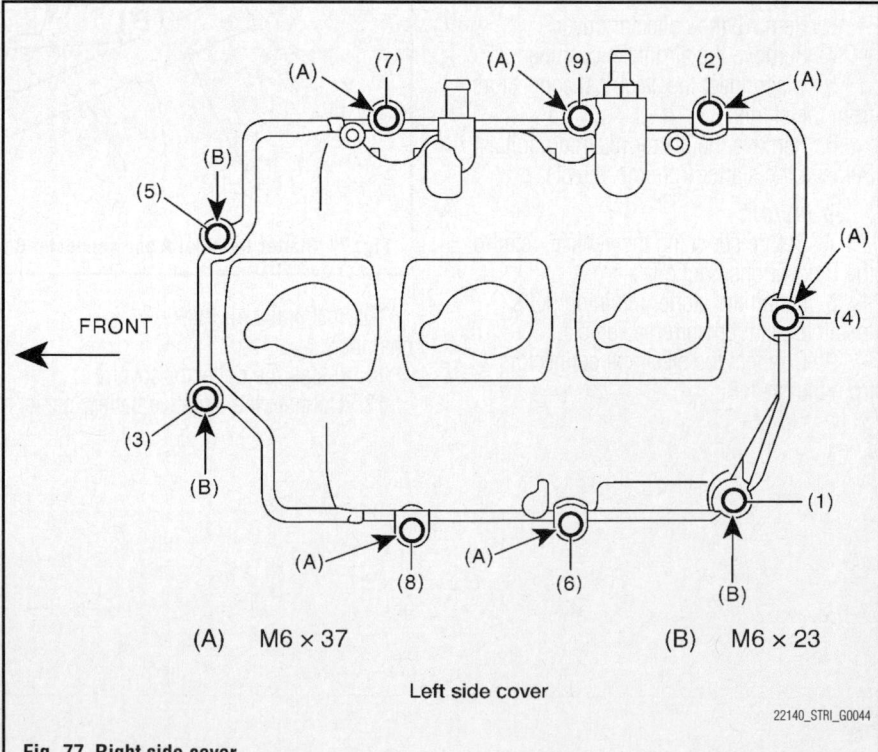

(A) M6 × 37   (B) M6 × 23

Left side cover

Fig. 77 Right side cover

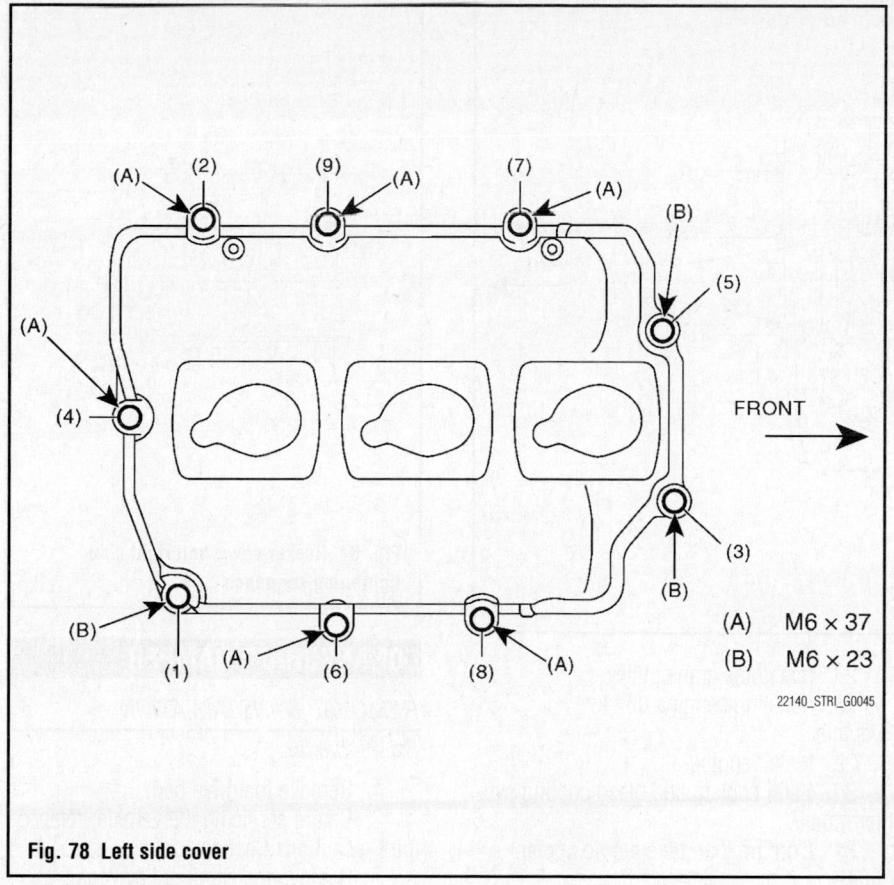

Fig. 78 Left side cover

(A) M6 × 37
(B) M6 × 23

22140_STRI_G0045

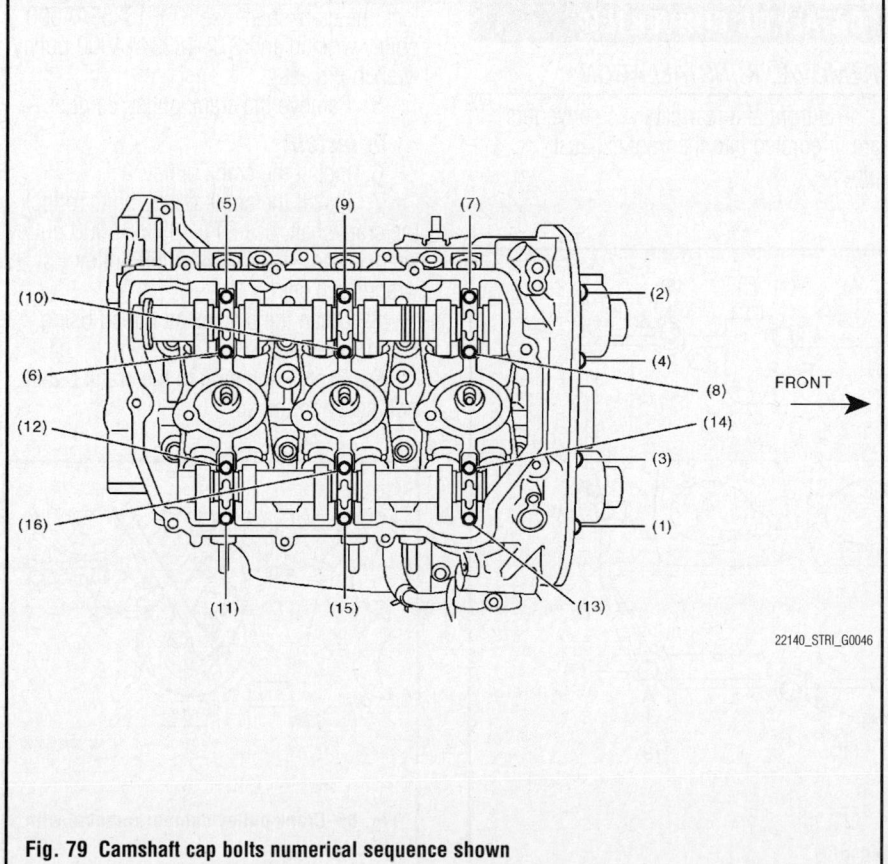

Fig. 79 Camshaft cap bolts numerical sequence shown

22140_STRI_G0046

To install:

14. Apply engine oil to camshaft journals, and install the camshaft.

15. Install the camshaft cap.

16. Apply liquid gasket sparingly to back side of front camshaft cap as shown in the figure.

✳✳ CAUTION

Do not apply liquid gasket excessively. Applying excessively may cause excess gasket to flow toward camshaft journal, resulting in engine seizure. Applying liquid gasket diameter: 0.079+/-0.020 inch. (2.0+/-0.5 mm).

17. Apply engine oil to cap bearing surface, and install the cap to camshaft.

18. Tighten the camshaft cap bolts in the numerical order as shown in the figure.

19. Install the rocker cover gasket to the rocker cover.

20. Apply liquid gasket sparingly to the mating surface of cylinder head and rocker cover as shown in the figure.

21. Apply liquid gasket THREE BOND 1217G (Part No. K0877Y0100) or equivalent. Applying liquid gasket diameter: 0.138+/-0.020 inch. (3.5+/-0.5 mm).

22. Tighten the rocker cover bolts in the numerical order as shown in the figure.

23. Install the crank sprocket.

24. Install the cam sprocket.

25. Install the timing chain assembly.

26. Install the chain cover.

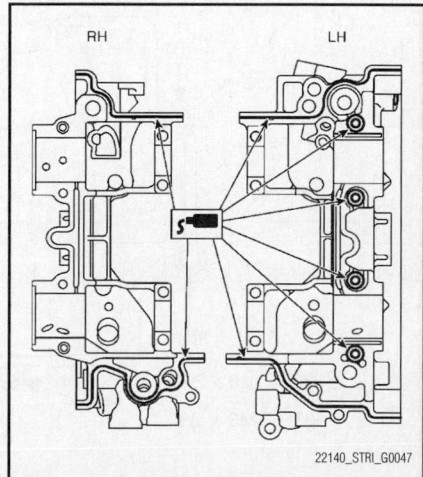

22140_STRI_G0047

Fig. 80 Liquid gasket installed to back side of front camshaft cap shown

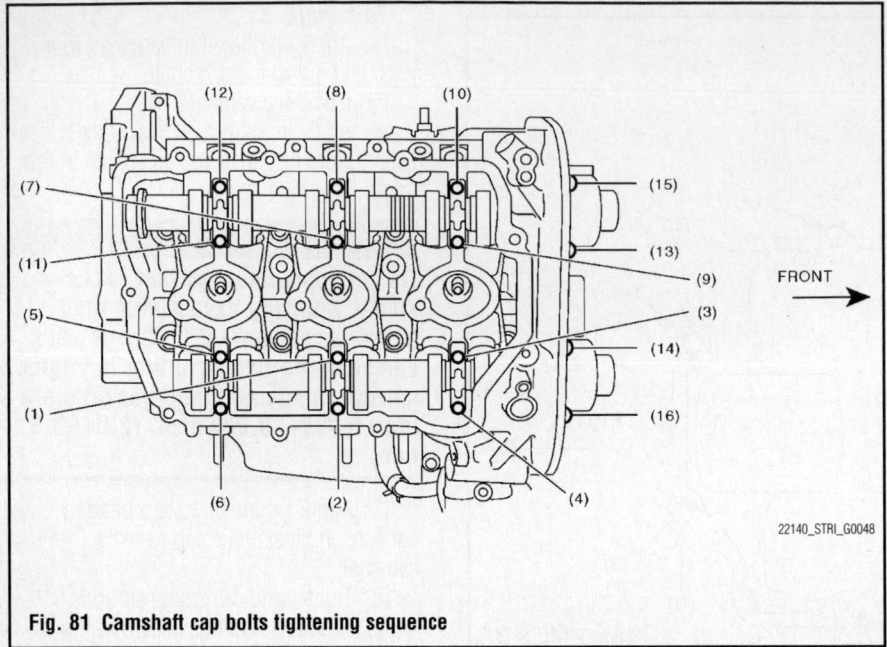

Fig. 81 Camshaft cap bolts tightening sequence

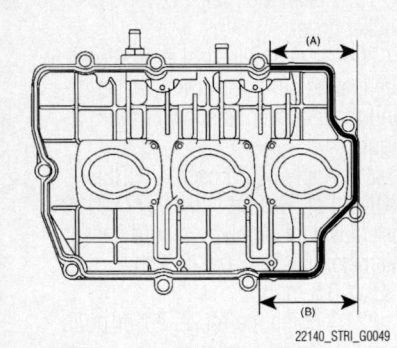

Fig. 82 Left side mating surface of cylinder head and rocker cover as shown right side similar

27. Install the crank pulley.
28. Install the engine unit to vehicle.
29. Install engine oil.
30. Refill coolant and bleed cooling system of air.
31. Confirm that there are no coolant leaks.

CATALYTIC CONVERTER

REMOVAL & INSTALLATION

The front and rear catalytic converters are integrated into the front exhaust pipe.

Fig. 84 Rocker cover bolt right side tightening sequence

CRANKSHAFT DAMPER

REMOVAL & INSTALLATION

See Figure 85.

1. Remove the drive belt.
2. Before servicing the vehicle, refer to the Precautions Section.
3. Remove the crank pulley cover.
4. Remove the crank pulley bolt. To lock the crankshaft, use ST1-18355AA000 pulley wrench and ST2-18334AA000 pulley wrench pin set.
5. Remove the crank pulley damper.

To install:
6. Install the crank pulley.
7. Install the crank pulley bolt. To lock the crankshaft, use ST1-18355AA000 pulley wrench and ST2-18334AA000 pulley wrench pin set.
8. Clean the crankshaft thread using compressed air.
9. Apply engine oil to the crank pulley bolt seat and thread.

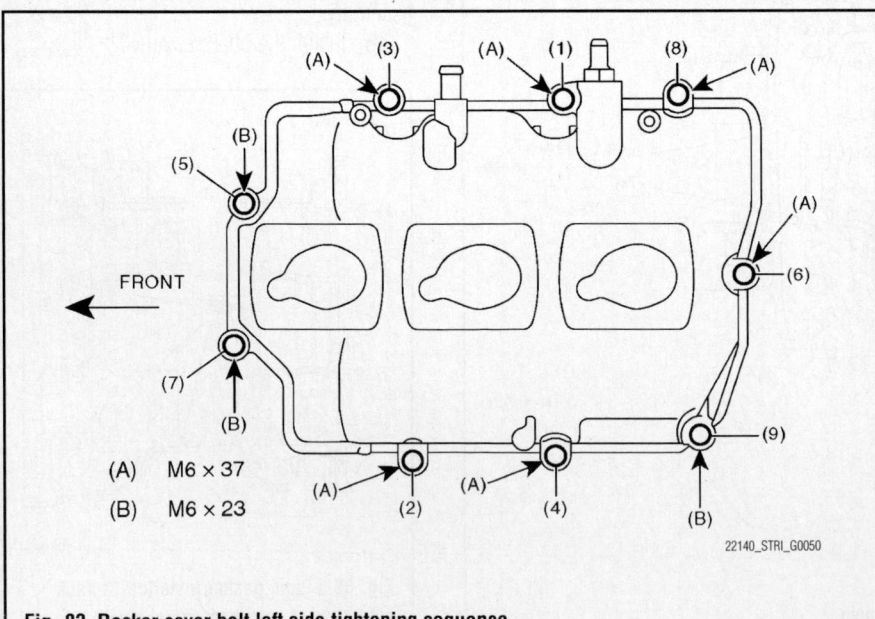

(A) M6 × 37
(B) M6 × 23

Fig. 83 Rocker cover bolt left side tightening sequence

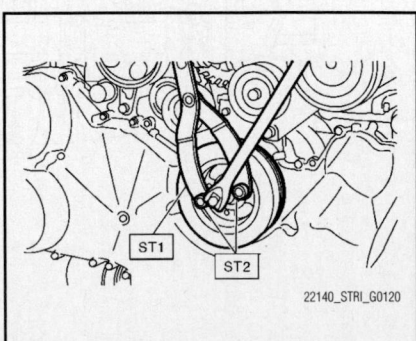

Fig. 85 Crank pulley damper removal with special tools

10. Tighten the crank pulley bolts to 144ft. lbs. (195 Nm).

11. Install the crank pulley cover. Assemble the O-ring to crank pulley cover and tighten to 5 ft lbs. (6.4 Nm).

12. Install the drive belt.

CRANKSHAFT FRONT SEAL

REMOVAL & INSTALLATION

See Figure 86.

1. Before servicing the vehicle, refer to the Precautions Section.

2. Remove or disconnect the following:
- Negative battery cable
- Drive belt

3. Secure the crankshaft pulley with tool No. 499977100.

4. Remove or disconnect the following:
- Crankshaft pulley bolt and pulley
- Front chain cover
- Timing chain
- Crankshaft seal

To install:

5. Using a suitable seal driver, install a new crankshaft seal.

6. Install or connect the following:
- Timing chain
- Front chain cover. Refer to the timing chain procedure in this section.
- Crankshaft pulley and tighten the bolt to 131 ft. lbs. (178 Nm). for 3.0L engine. For 3.6L engines tighten bolt to 144 ft. lbs. (195 Nm)
- Drive belt
- Negative battery cable

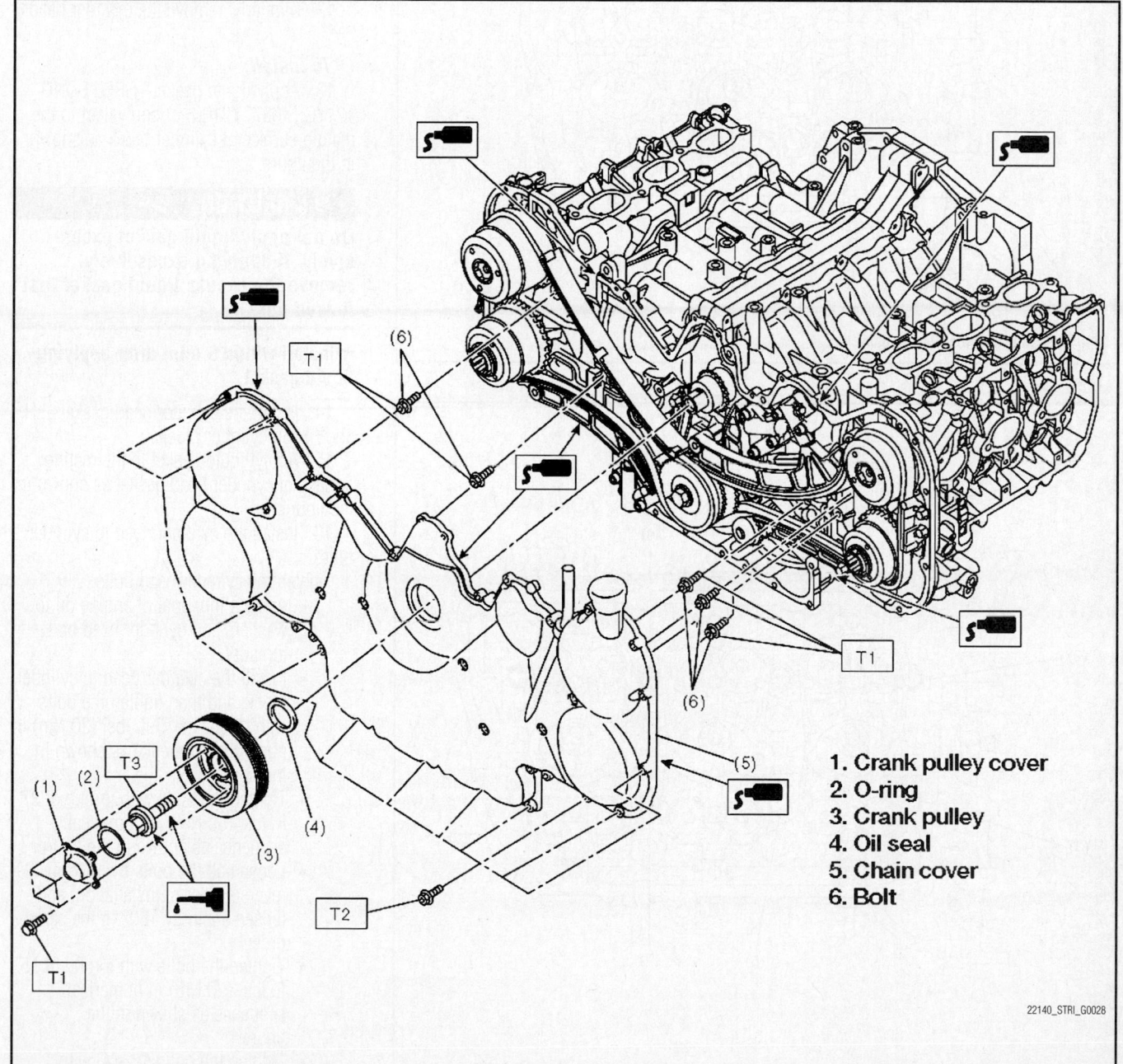

1. Crank pulley cover
2. O-ring
3. Crank pulley
4. Oil seal
5. Chain cover
6. Bolt

22140_STRI_G0028

Fig. 86 Front engine view 3.6L engine shown

CYLINDER HEAD

REMOVAL & INSTALLATION

See Figures 87 through 95.

1. Before servicing the vehicle, refer to the Precautions Section.
2. Remove the crank pulley.
3. Remove the chain cover.
4. Remove the timing chain assembly.
5. Remove the cam sprocket. To lock the camshaft, use the ST499977100, or equivalent.
6. Remove the crank sprocket.
7. Remove the seal bolts as shown in the figure.
8. Remove the cylinder head bolts in the numerical order as shown in the figure.

Leave bolts (2) and (4) engaged by three or four threads to prevent the cylinder head from falling.

9. While tapping the cylinder head with a plastic hammer, separate it from cylinder block.
10. Remove the bolts (2) and (4) to remove cylinder head.
11. Remove the cylinder head gasket.

✳✳ CAUTION

Be careful not to scratch the mating surface of cylinder head and cylinder block.

12. Similarly, remove the cylinder head RH.

To install:

13. Apply liquid gasket THREE BOND 1217G (K0877Y0100) or equivalent to the mating surface of cylinder block as shown in the figure.

✳✳ CAUTION

Do not apply liquid gasket excessively. If applying excessively, remove the excess liquid gasket that flowed out.

➡ **Install within 5 min. after applying liquid gasket.**

14. Install the cylinder head gaskets LH and RH on cylinder block.
15. Apply liquid gasket to the mating surface of cylinder head gasket as shown in the figure.
16. Install the cylinder head to cylinder block.

Tighten the cylinder head bolts.

- Apply a thin coat of engine oil to washers and cylinder head bolt threads.
- Install the cylinder head to cylinder block, and then tighten the bolts with torque of 15 ft. lbs. (20 Nm) in numerical sequence as shown in the figure.
- Tighten the bolts with torque of 37 ft. lbs. (50 Nm) in numerical sequence as shown in the figure.
- Loosen all the bolts by 180° in reverse order of installation, and loosen again by 180° in the same order.
- Tighten the bolts with torque of 15 ft. lbs. (20 Nm) in numerical sequence as shown in the figure.
- Tighten the bolts (1)–(4) with torque of 35 ft. lbs. (48 Nm) in numerical sequence.

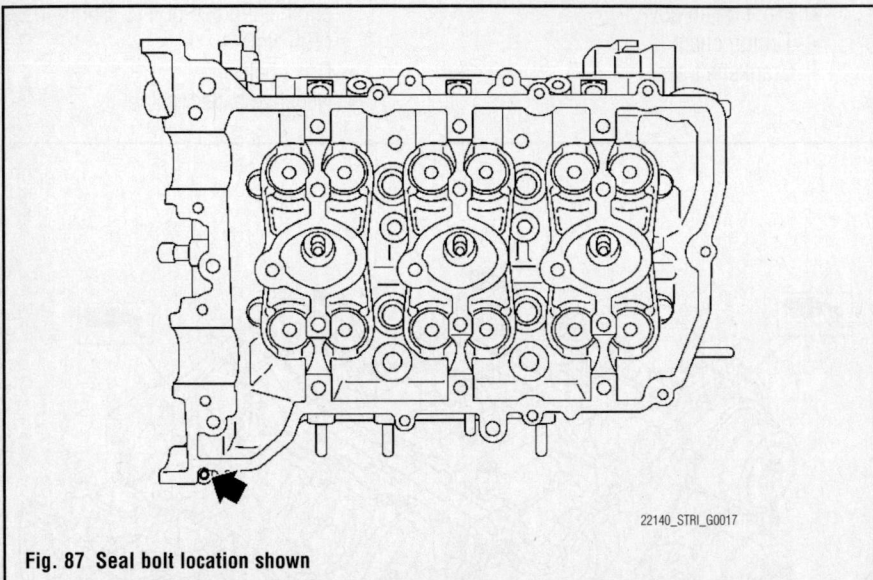

22140_STRI_G0017

Fig. 87 Seal bolt location shown

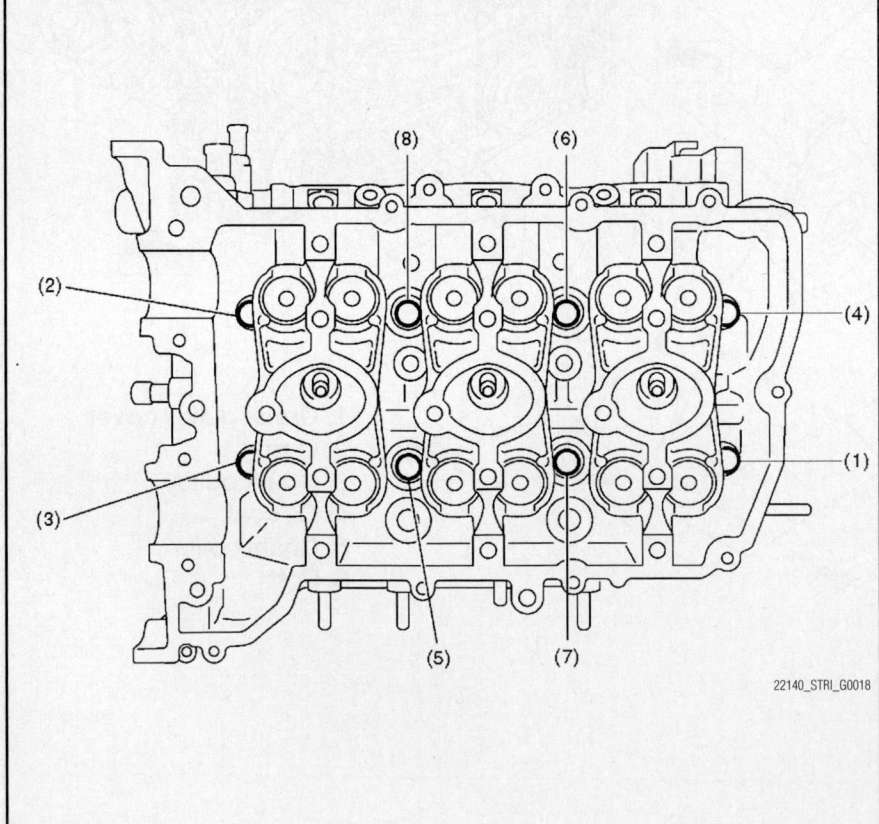

22140_STRI_G0018

Fig. 88 Cylinder head bolt loosening sequence

(A)
(C)
(B)
(D)
(E)
(F)
22140_STRI_G0019

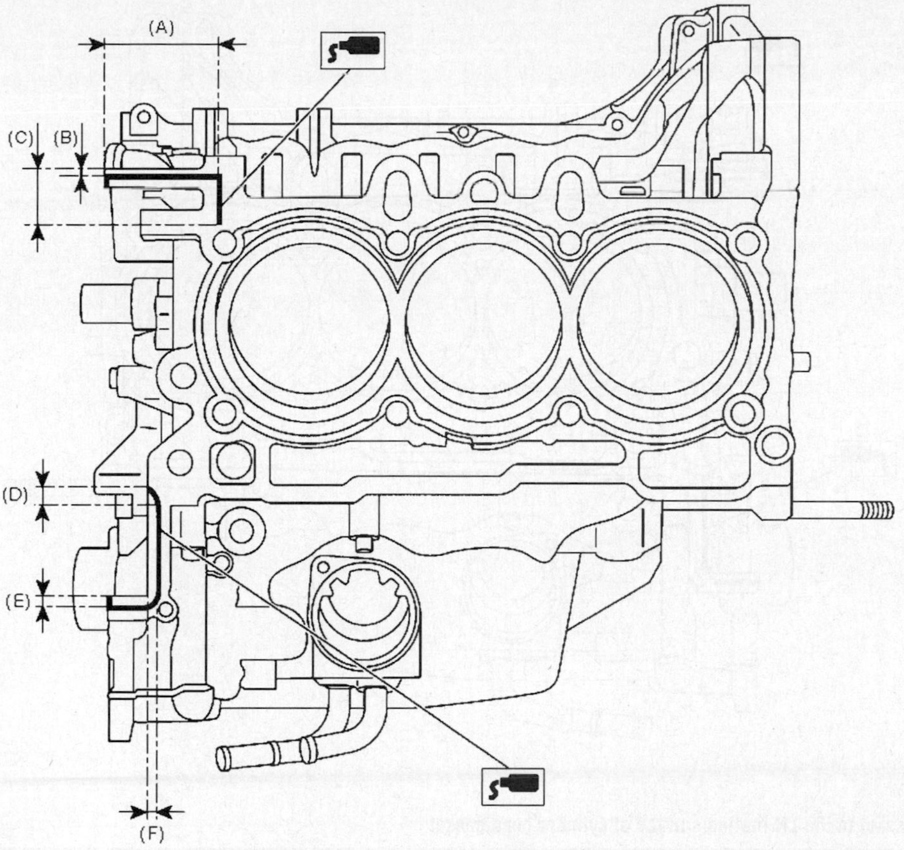

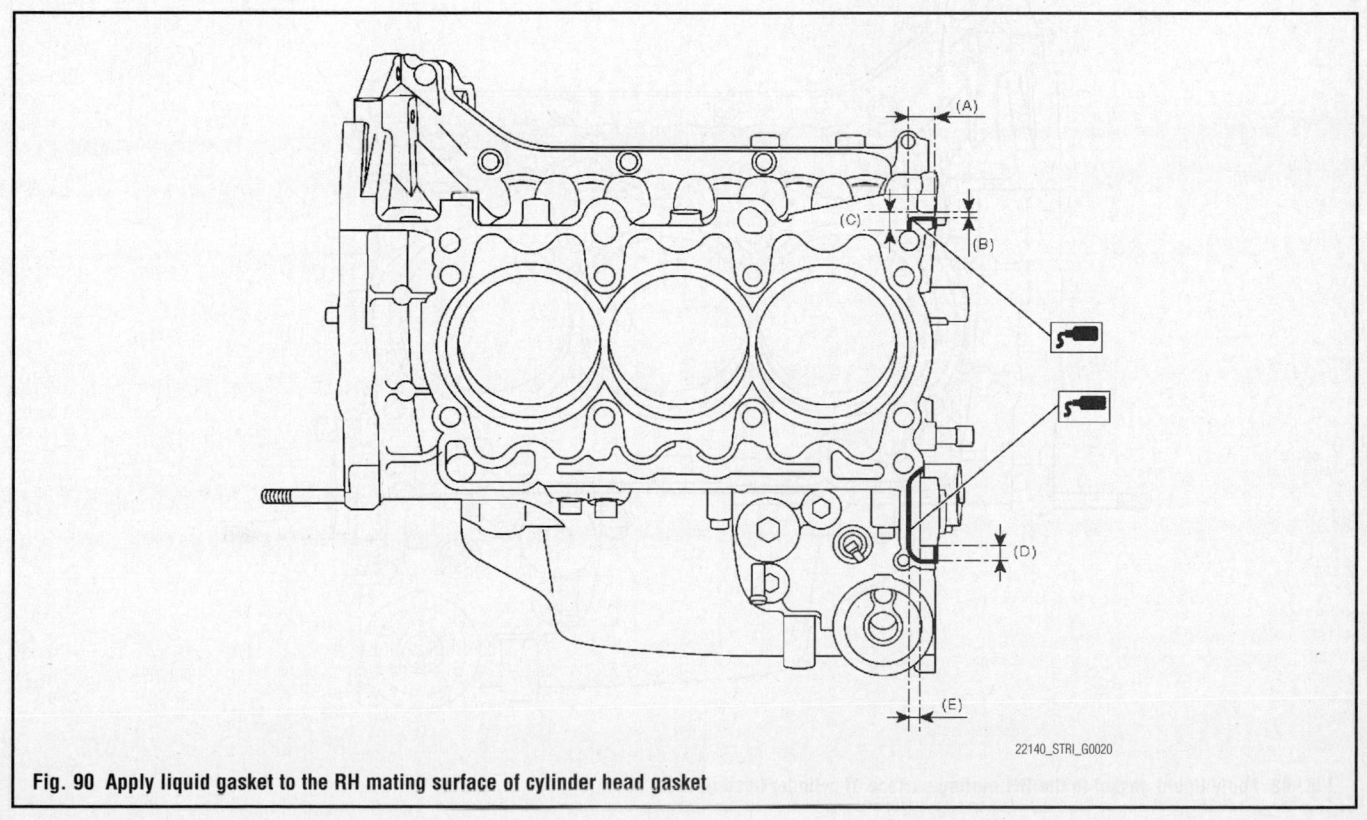

Fig. 89 Apply liquid gasket to the LH mating surface of cylinder head gasket

(A)
(C)
(B)
(D)
(E)
22140_STRI_G0020

Fig. 90 Apply liquid gasket to the RH mating surface of cylinder head gasket

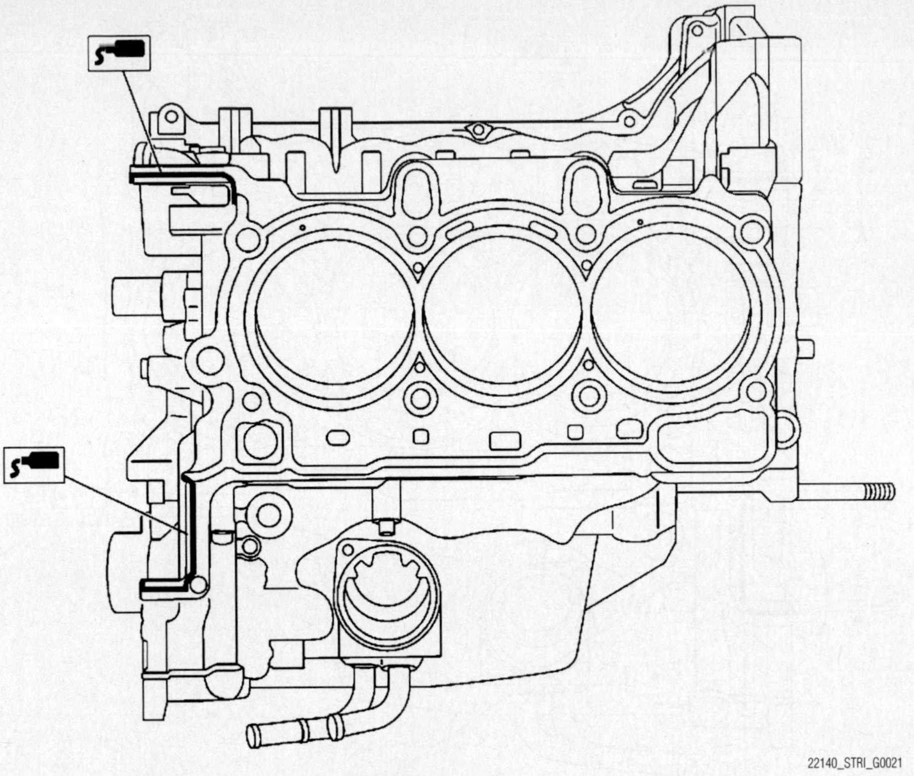

22140_STRI_G0021

Fig. 91 Apply liquid gasket to the LH mating surface of cylinder head gasket

22140_STRI_G0022

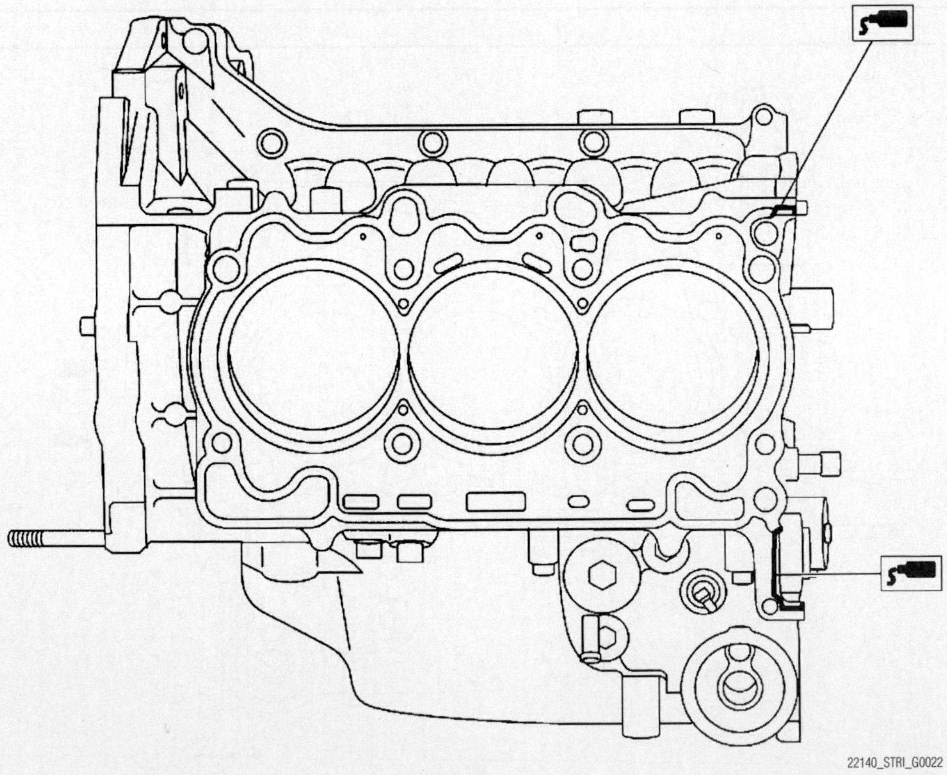

Fig. 92 Apply liquid gasket to the RH mating surface of cylinder head gasket

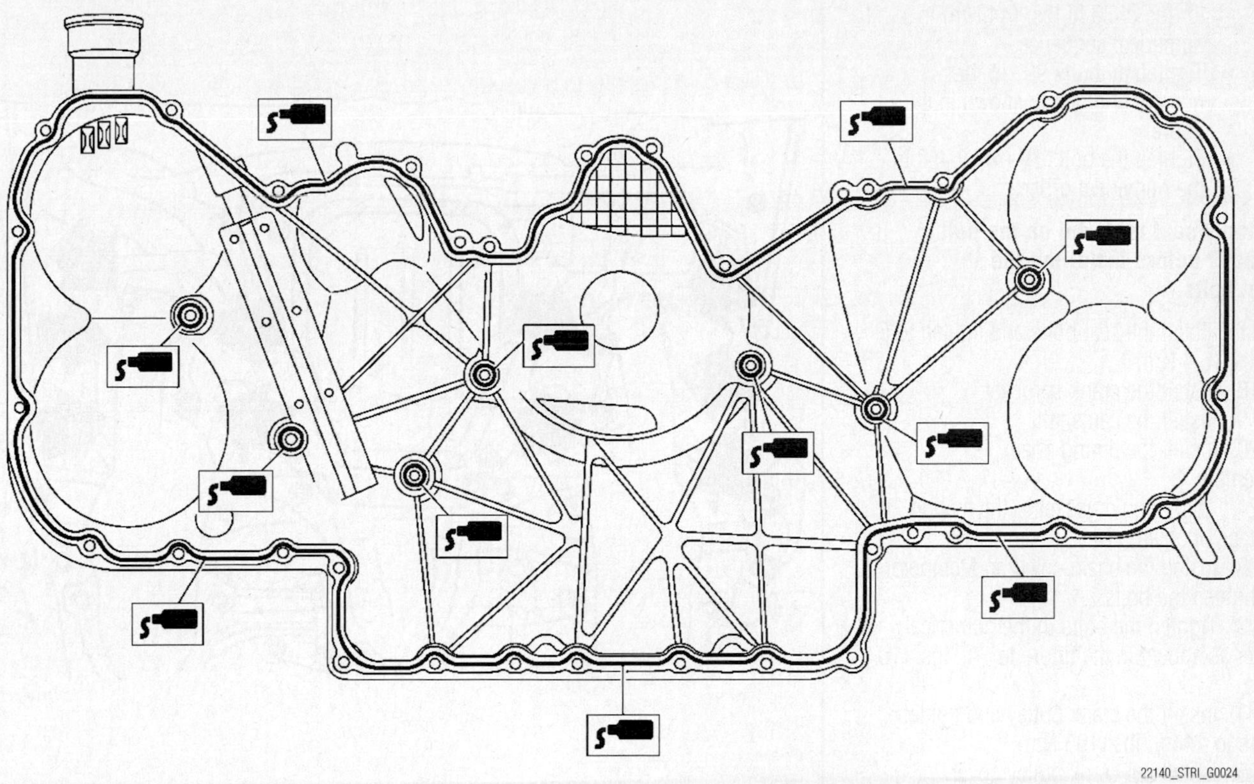

Fig. 94 Apply liquid gasket to the mating surface of chain cover

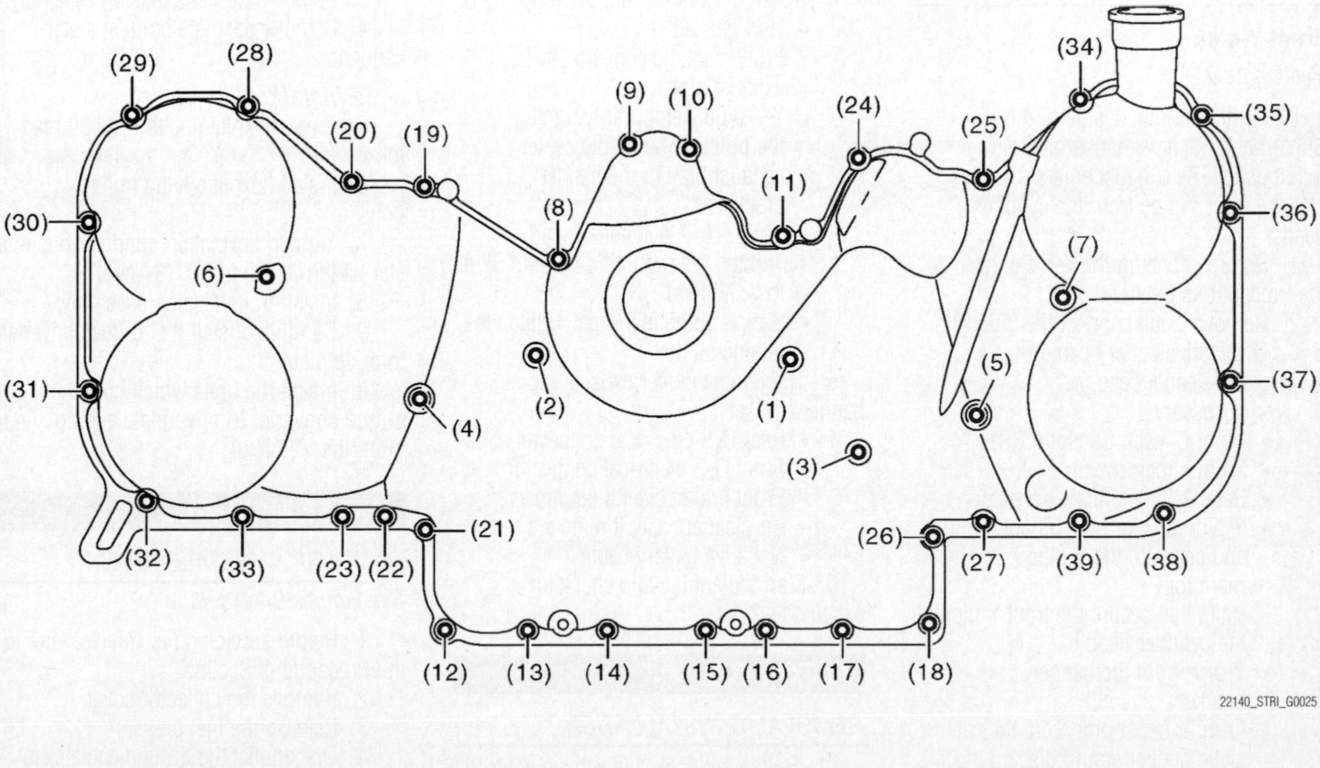

Fig. 95 Timing chain cover bolt tightening sequence

- Tighten the bolts (5)–(8) with torque of 33 ft. lbs. (44 Nm) in numerical sequence.
- Tighten all bolts 90° in the numerical order as shown in the figure.
- Tighten the bolt (1)–(4) by 45° in the numerical order.

➡**Apply seal material on the bolt threads before installing the seal bolts.**

17. Install the seal bolts and tighten to 5 ft. lbs. (6.4 Nm).
18. Install the crank sprocket.
19. Install the camshaft.
20. Install the timing chain assembly.
21. Apply liquid gasket to the mating surface of chain cover.
22. Install the chain cover and temporarily tighten the bolts.
23. Tighten the bolts in the numerical order as shown in the figure to 7 ft. lbs. (10 Nm).
24. Install the crank pulley and tighten bolts to 144 ft. lbs.(195 Nm).
25. Install the crank pulley cover.
26. Install the engine unit to vehicle.

EXHAUST MANIFOLD

REMOVAL & INSTALLATION

Front Y-pipe

See Figure 96.

Due to the unique design of the Subaru engine, an exhaust manifold is not used. The exhaust enters directly into the front Y-pipe from the cylinder heads.

1. Before servicing the vehicle, refer to the Precautions Section.
2. Remove or disconnect the following:
   - Negative battery cable
   - Air cleaner case, if necessary
   - Front Oxygen Sensor (O$_2$S)
   - Front undercover
   - Rear O$_2$S electrical connector
   - Y-pipe-to-rear pipe mounting nuts and separate the Y-pipe from the rear pipe
   - Bolts that secure the front Y-pipe to the cylinder head
   - Y-pipe from the hanger bracket
   - Front exhaust pipe from the catalytic converter and discard the gaskets

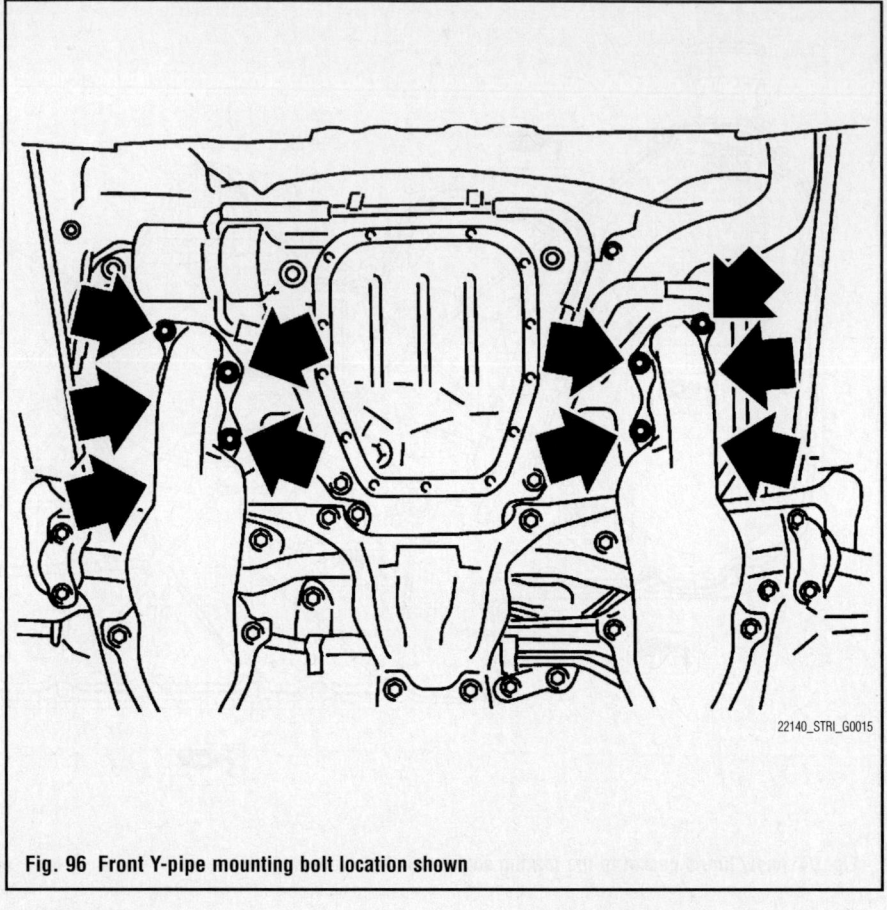

Fig. 96 Front Y-pipe mounting bolt location shown

22140_STRI_G0015

To install:
3. Clean all gasket surfaces completely.
4. Install or connect the following:
   - New gaskets
   - Front catalytic converter to front exhaust pipe
   - Y—pipe. Temporarily tighten the bolt that holds the center exhaust pipe to the hanger bracket.
   - Y-pipe, to the cylinder head. Tighten the mounting bolts to 52 ft. lbs. (70 Nm).
   - Y-pipe to the rear pipe. Tighten the retainers.
5. Tighten the center exhaust pipe to hanger bracket.
   - Rear O$_2$S electrical connector
   - Front O$_2$S electrical connector
   - Front undercover, if equipped
   - Air cleaner case, if removed
   - Negative battery cable
6. Start the engine and check for exhaust leaks.

FLEXPLATE

REMOVAL & INSTALLATION

1. Before servicing the vehicle, refer to the Precautions Section.

2. Separate the torque converter from drive plate.
3. Remove the transmission assembly.
4. Remove flexplate bolts in a star sequence.

To install:
5. Clean flexplate bolts install thread locker.
6. Install flexplate hand tighten bolts.
7. Tighten flexplate mounting bolts in a star pattern to 60 ft. lbs. (81 Nm).
8. Install transmission assembly.
9. Carefully install the torque converter to drive plate.
10. Install the bolts which hold the torque converter to drive plate and tighten to 18 ft. lbs. (25 Nm).

INTAKE MANIFOLD

REMOVAL & INSTALLATION

See Figures 97 and 98.

1. Before servicing the vehicle, refer to the Precautions Section.
2. Remove the collector cover.
3. Release the fuel pressure.
4. Disconnect the ground cable from the battery.

Right side of Intake Manifold

22140_STRI_G0013

Fig. 97 Right side of intake manifold mounting bolts—3.6L engine

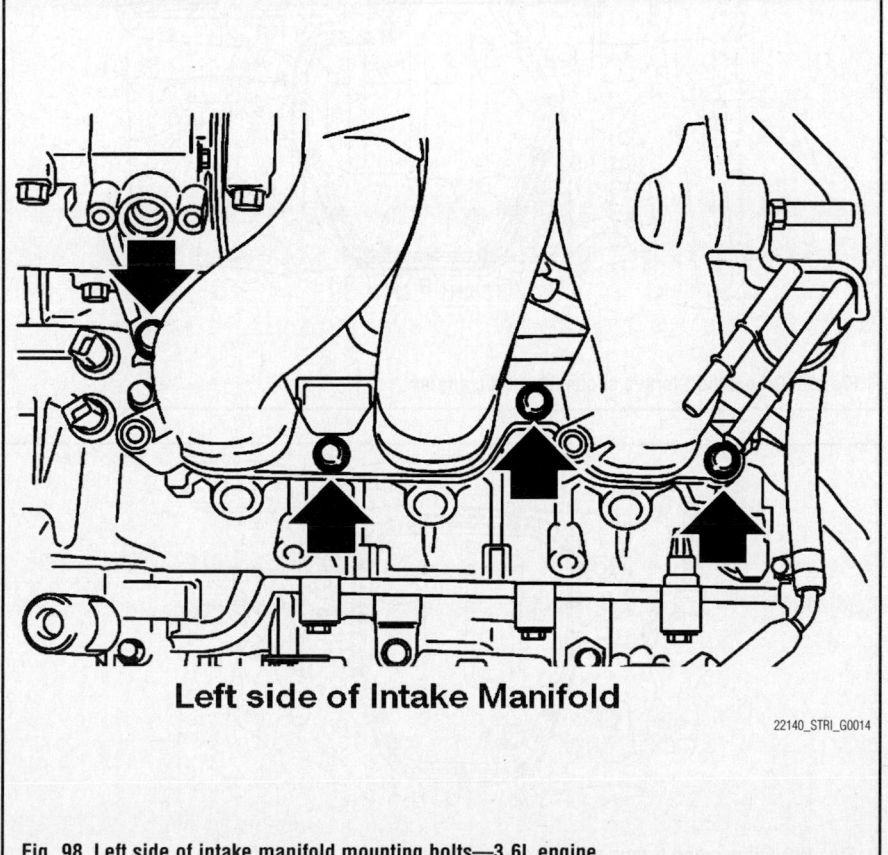

Left side of Intake Manifold

22140_STRI_G0014

Fig. 98 Left side of intake manifold mounting bolts—3.6L engine

5. Open the fuel filler flap lid, and remove the fuel filler cap.

6. Remove the air cleaner case and air intake chamber.

7. Remove the generator.

8. Disconnect the connector from the throttle body.

9. Disconnect the engine harness connector.

10. Disconnect the engine coolant hoses from the throttle body.

11. Disconnect the PCV hose.

12. Disconnect the brake booster hose.

13. Disconnect the fuel hoses from the fuel pipe.

   a. Disconnect the connector of fuel pipe by pushing the ST in the direction of arrow.

   b. Remove the clip, and disconnect the evaporation hose from the pipe.

✳✳ CAUTION

Be careful not to spill fuel and catch the fuel from hoses using a container or cloth.

14. Remove the fuel pipe protector LH.

15. Remove the engine harness from the fuel injector pipe LH.

16. Disconnect the connectors of the oil temperature sensor, engine coolant temperature sensor and intake oil flow control solenoid valve LH.

17. Remove the bolts which hold fuel injector pipe LH to cylinder head.

18. Disconnect the connector of intake oil flow control solenoid valve RH.

19. Remove the fuel pipe protector RH.

20. Remove the engine harness from the fuel injector pipe RH.

21. Remove the bolts which hold the fuel injector pipe RH to the cylinder head.

22. Remove the bolts which hold the intake manifold to the cylinder head.

23. Remove the EGR pipe fixing bolt.

24. Remove the intake manifold.

To install:

25. Use new O-rings and install the intake manifold onto cylinder heads.

26. Install and tighten the intake mounting bolts to 18 ft. lbs. (25 Nm).

27. Use new gasket and install the EGR pipe fixing bolt. Tighten the bolt to 4.7 ft. lbs. (6.4 Nm).

28. Use new O-rings and install the bolts which hold fuel injector pipe RH to cylinder head. Tighten the bolts to 14 ft. lbs. (19 Nm).

29. Install the engine harness to fuel injector pipe RH.

30. Install the fuel pipe protector RH. Tighten the bolts to 14 ft. lbs. (19 Nm).

31. Connect the connector of intake oil flow control solenoid valve RH.

32. Install the bolts which hold fuel injector pipe LH to cylinder head.

33. Use new O-rings and install the bolts which hold fuel injector pipe LH to cylinder head. Tighten the bolts to 14 ft. lbs. (19 Nm).

34. Connect the connectors of the oil temperature sensor, engine coolant temperature sensor and intake oil flow control solenoid valve.

35. Install the engine harness to fuel injector pipe LH.

36. Install the fuel pipe protector LH. Tighten the bolts to 14 ft. lbs. (19 Nm).

37. Connect the fuel hoses to fuel pipe.

38. Connect the brake booster hose.

39. Connect the PCV hose.

40. Connect the engine coolant hoses to throttle body.

41. Connect the connector to throttle body.

42. Connect the engine harness connector.

43. Install the generator.

44. Install the air cleaner case and air intake chamber.

45. Install the fuse of fuel pump to the main fuse box.

46. Connect the battery ground cable to the battery.

47. Install the collector cover.

OIL PAN

REMOVAL & INSTALLATION

See Figures 99 and 100.

➡**If removing the upper oil pan, the engine must first be removed from the vehicle.**

1. Before servicing the vehicle, refer to the Precautions Section.

2. Disconnect the negative battery cable.

3. Raise and support the vehicle safely.

4. Remove the undercover. Drain the engine oil.

5. Remove the lower oil pan retaining bolts.

6. Insert an oil pan gasket cutter tool into the gap between the upper oil pan and the lower oil pan. Remove the lower oil pan from the engine.

➡**Do not use a screwdriver or similar tool in place of the cutter tool.**

7. Remove the oil strainer, if required.

To install:

8. Be sure to clean the old gasket material from the mating surfaces.

9. Replace the O-ring and install the oil strainer. Tighten the bolt to 4.7 ft. lbs. (6.4 Nm).

10. Apply a continuous bead (0.039 inch thick) of liquid gasket, part number K0877YA018 or equivalent, to the mating surfaces. Install the oil pan.

11. Tighten the oil pan lower installing bolts in the numerical order as shown in the figure to 5 ft. lbs. (6.4 Nm).

12. Continue the installation in the reverse order of the removal procedure.

13. Be sure to fill the engine with the correct grade and type engine oil.

14. Start the engine and check for leaks. Correct as required.

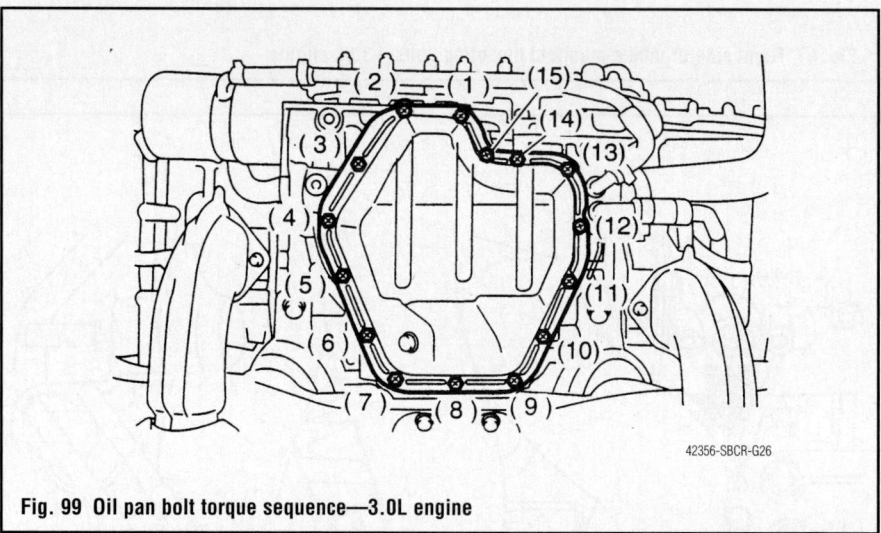

42356-SBCR-G26

Fig. 99 Oil pan bolt torque sequence—3.0L engine

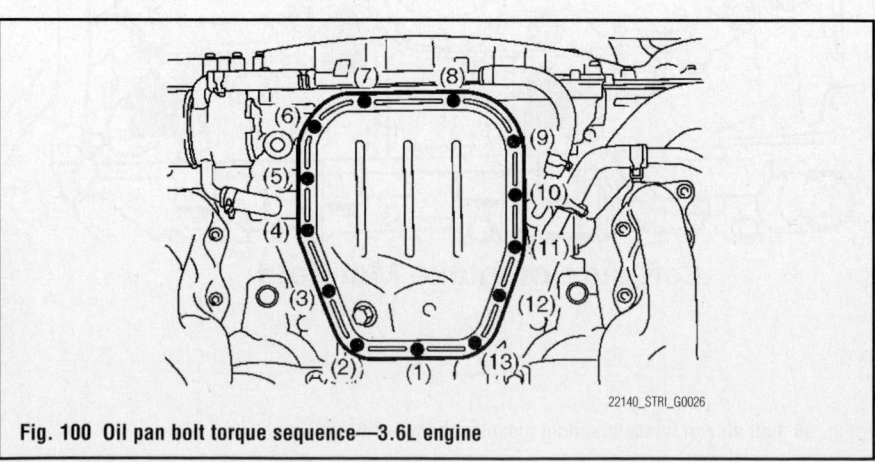

22140_STRI_G0026

Fig. 100 Oil pan bolt torque sequence—3.6L engine

OIL PUMP

REMOVAL & INSTALLATION

See Figures 100 and 101.

1. Before servicing the vehicle, refer to the Precautions Section.
2. Set the vehicle on a lift.
3. Remove the collector cover.
4. Disconnect the ground cable from the battery.
5. Lift up the vehicle.

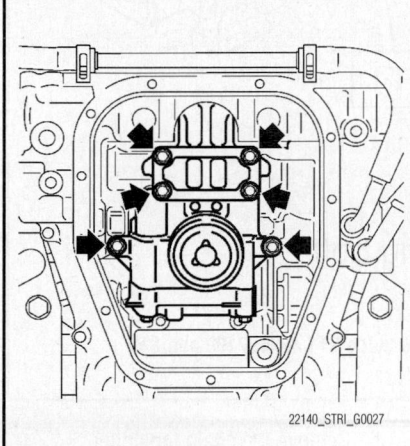

Fig. 101 Oil pump mounting bolt location—3.6L engine

6. Remove the undercover.
7. Drain engine coolant.
8. Lower the vehicle.
9. Remove the radiator. Refer to radiator removal.
10. Remove the V-belt.
11. Remove the crank pulley cover.
12. Remove the crank pulley bolt. To lock the crankshaft, use ST 18355AA000 wrench and ST 1834AA000 wrench pin set.
13. Remove the chain cover.
14. Remove the timing chain. Refer to timing chain removal.
15. Remove the oil pan.
16. Remove the oil pump.
17. Remove the O-ring from the oil pan upper.

To install:

18. Apply the engine oil to the O-ring and install it to the oil pan upper.
19. Install the oil pump and tighten mounting bolts to 10 ft. lbs. (13 Nm).
20. Install the oil pan lower.
21. Tighten the oil pan lower installing bolts in the numerical order as shown in the figure to 5 ft. lbs. (6.4 Nm).
22. Install the timing chain.
23. Install the chain cover.
24. Install the crank pulley and tighten to 144 ft. lbs. (195 Nm).

25. Install the crank pulley cover and tighten to 5 ft. lbs. (6.4 Nm).
26. Install the V-belt.
27. Install the radiator.
28. Lift up the vehicle.
29. Install the undercover.
30. Lower the vehicle.
31. Be sure to fill the cooling system and engine oil with the proper grade oil and type coolant.
32. Start the engine and check for leaks. Correct, as required.

INSPECTION

Inspect the oil pump exterior for the following:

- Cracks, damage or oil leakage
- Play of pulley shaft

Inspect the oil pump interior for the following:

- Faulty or seized pump vane
- Bends in the shaft or damage to bearing
- Visually check the strainer part for clogging

PISTON AND RING

POSITIONING

See Figures 102 through 104.

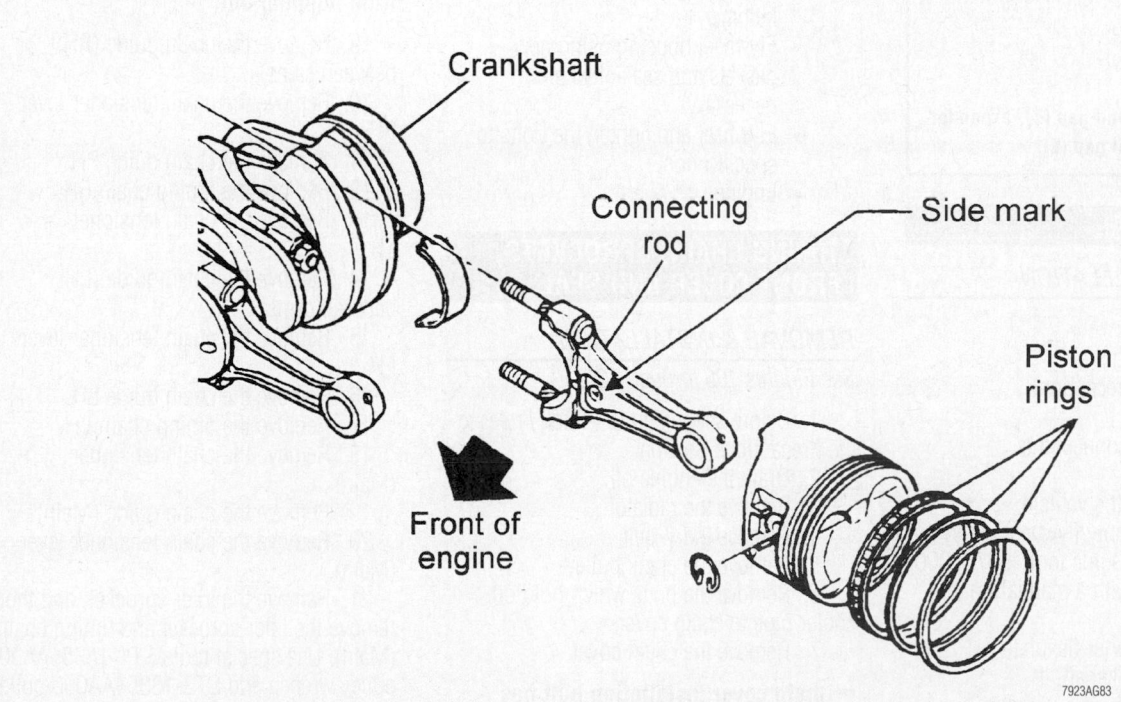

Fig. 102 Piston and connecting rod assembly positioning

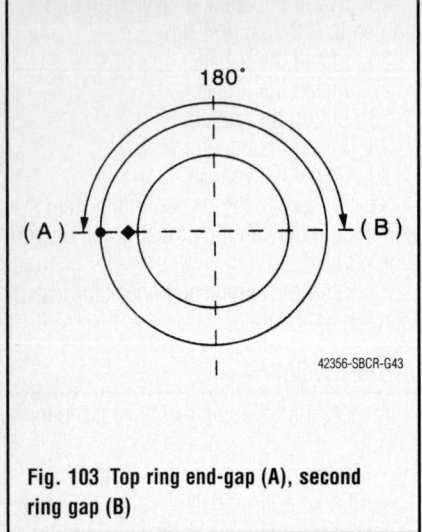

Fig. 103 Top ring end-gap (A), second ring gap (B)

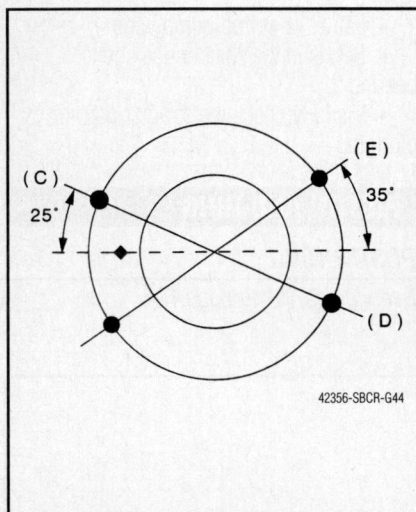

Fig. 104 Upper rail end-gap (C), expander gap (D) and lower rail gap (E)

REAR MAIN SEAL

REMOVAL & INSTALLATION

See Figure 105.

1. Before servicing the vehicle, refer to the Precautions Section.

2. Remove or disconnect the following:
   - Engine from the vehicle
   - Clutch assembly/flywheel using the Clutch Disc Guide tool 499747000, if equipped with a manual transmission
   - Torque converter flexplate from the crankshaft, if equipped with an automatic transmission
   - Oil seal from the cylinder block using a small prybar

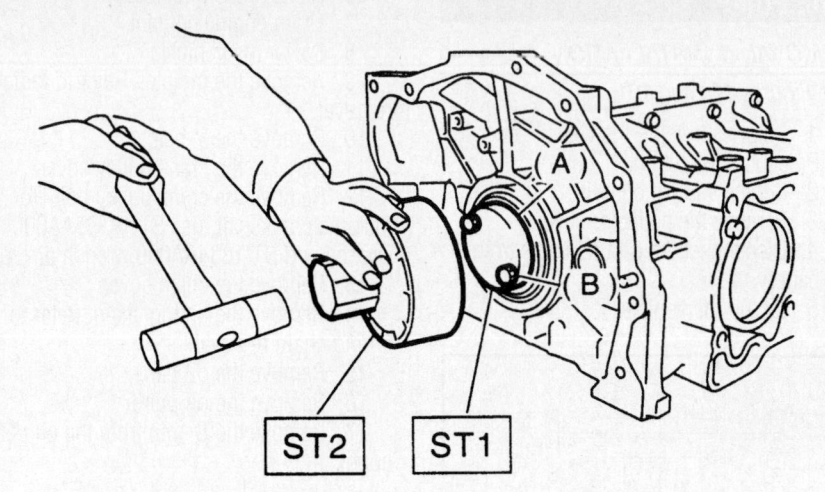

(A) Rear oil seal

(B) Drive plate attaching bolt

Fig. 105 Installing the rear main seal using oil seal guide ST1 499597100 and ST2 499598200

To install:

3. Install or connect the following:
   - New oil seal by pressing it into the cylinder block using the appropriate driver and hammer
   - Flywheel housing using new gaskets and sealant where necessary.
   - Flywheel and tighten the bolts to specification.
   - Engine

TIMING CHAIN, SPROCKETS, FRONT COVER AND SEAL

REMOVAL & INSTALLATION

See Figures 106 through 120.

1. Before servicing the vehicle, refer to the Precautions Section.

2. Drain the engine oil.

3. Remove the radiator.

4. Remove the V-belt.

5. Remove the crank pulley.

6. Remove the bolts which hold oil cooler pipe to chain cover.

7. Remove the chain cover.

➡Chain cover installation bolt has three different sizes. To prevent the confusion in installation, keep these bolts on container individually.

8. Remove the chain tensioner RH.

➡During removal of chain tensioner RH from chain tensioner lever RH, press the plunger by hand to prevent it from popping out.

9. Remove the chain guide (RH: between cams).

10. Remove the chain tensioner lever RH.

11. Remove the chain guide RH.

12. Remove the timing chain RH.

13. Remove the chain tensioner LH.

14. Remove the chain guide (LH: between cams).

15. Remove the chain tensioner lever LH.

16. Remove the chain guide LH.

17. Remove the timing chain LH.

18. Remove the chain tensioner (Main).

19. Remove the chain guide (Main).

20. Remove the chain tensioner lever (Main).

21. Remove the idler sprocket, and then remove the idler sprocket and timing chain (Main). Use special tools ST1-18355AA000 pulley wrench and ST2-18354AA000 pulley wrench pin set.

22. Remove cam sprocket and lock the cam shaft using ST-499977500

23. Remove crank sprocket.

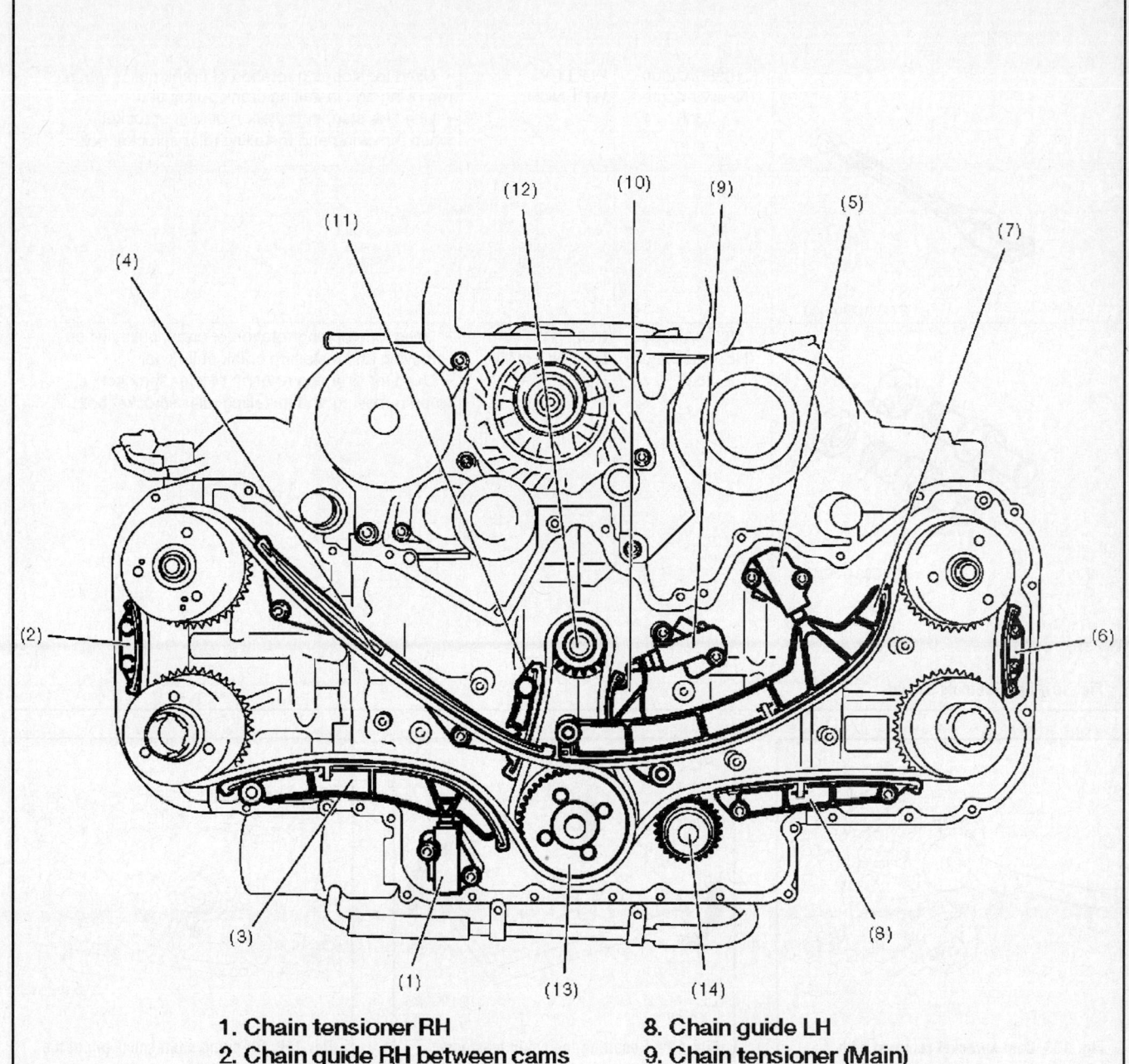

1. Chain tensioner RH
2. Chain guide RH between cams
3. Chain tensioner lever RH
4. Chain guide RH
5. Chain tensioner LH
6. Chain guide LH between cams
7. Chain tensioner lever LH
8. Chain guide LH
9. Chain tensioner (Main)
10. Chain tensioner lever (Main)
11. Chain guide (MAIN)
12. Crank sprocket
13. Idler sprocket
14. Water pump sprocket

22140_STRI_G0031

Fig. 106 Timing chain assembly parts location—3.6L engine

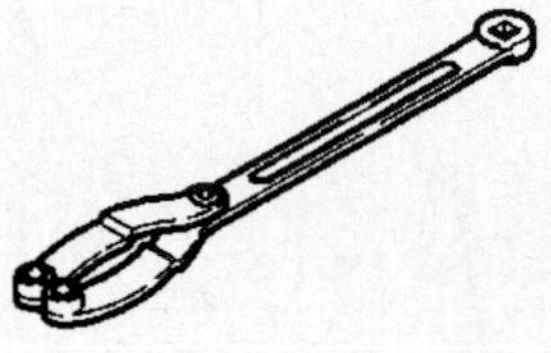 ST18355AA000	18355AA000 (Newly adopted tool)	PULLEY WRENCH	• Used for stopping rotation of crank pulley when removing and installing crank pulley bolt. • Used for stopping rotation of idler sprocket when removing and installing idler sprocket bolt.
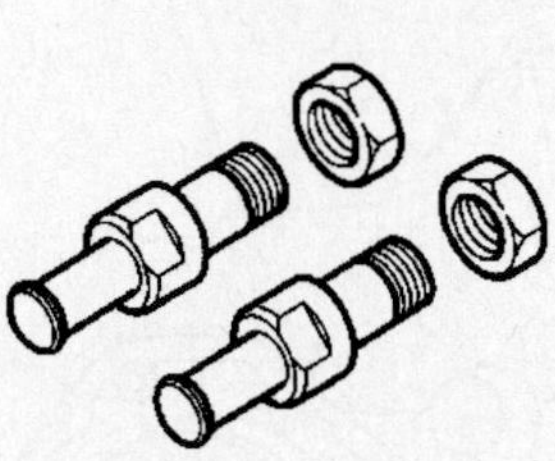 ST18334AA000	18334AA000 (Newly adopted tool)	PULLEY WRENCH PIN SET	• Used for stopping rotation of crank pulley when removing and installing crank pulley bolt. • Used for stopping rotation of idler sprocket when removing and installing idler sprocket bolt.

22140_STRI_G0032

Fig. 107 Special tools shown

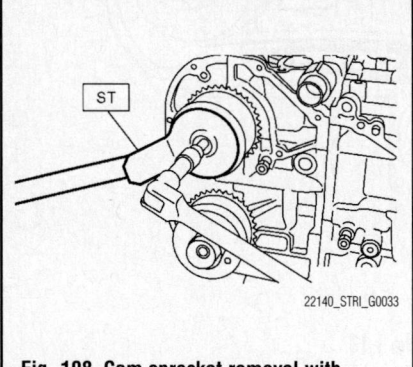

22140_STRI_G0033

Fig. 108 Cam sprocket removal with special tool ST-499977500 shown

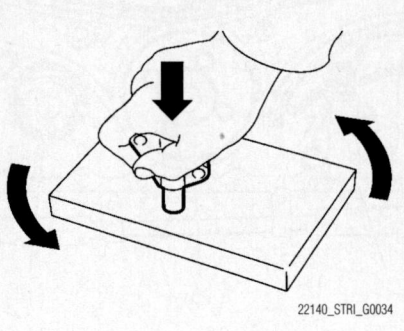

22140_STRI_G0034

Fig. 109 Resetting the chain tensioner shown

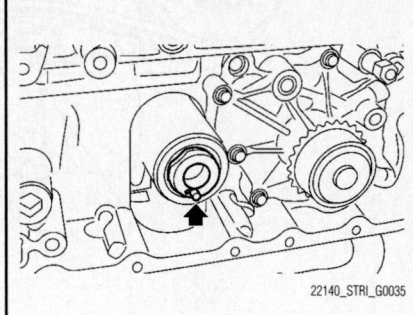

22140_STRI_G0035

Fig. 110 Oil pump shaft knock pin at the six o'clock position shown

To install:

24. Install crank sprocket.

25. Install cam sprocket holding cam with special tool ST-499977500. Tighten the bolt to 22 ft. lbs. (29 Nm).

➥**Be careful that the foreign matter is not into or onto assembled component during installation. Apply engine oil to the all timing chain components.**

26. The preparation for chain tensioner installation is as follows:

• Insert the screw, spring pin and plunger into tensioner body.

• Rotate the rubber mat counterclockwise while pressing the chain tensioner from upper side by hand as shown in the figure.

• Insert the stopper pin into the chain tensioner body hole.

27. Set the oil pump shaft knock pin at the six o'clock position as shown in the figure.

28. Using the ST, align the "Top mark" on crank sprocket to nine o'clock position as shown in the figure.

29. Align the intake cam sprocket to 12 o'clock position as shown in the figure.

30. Align the exhaust cam sprocket to

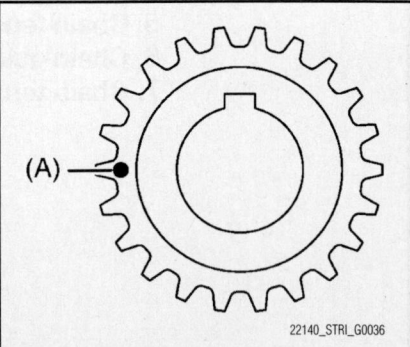

22140_STRI_G0036

Fig. 111 Crank sprocket at nine o'clock position shown

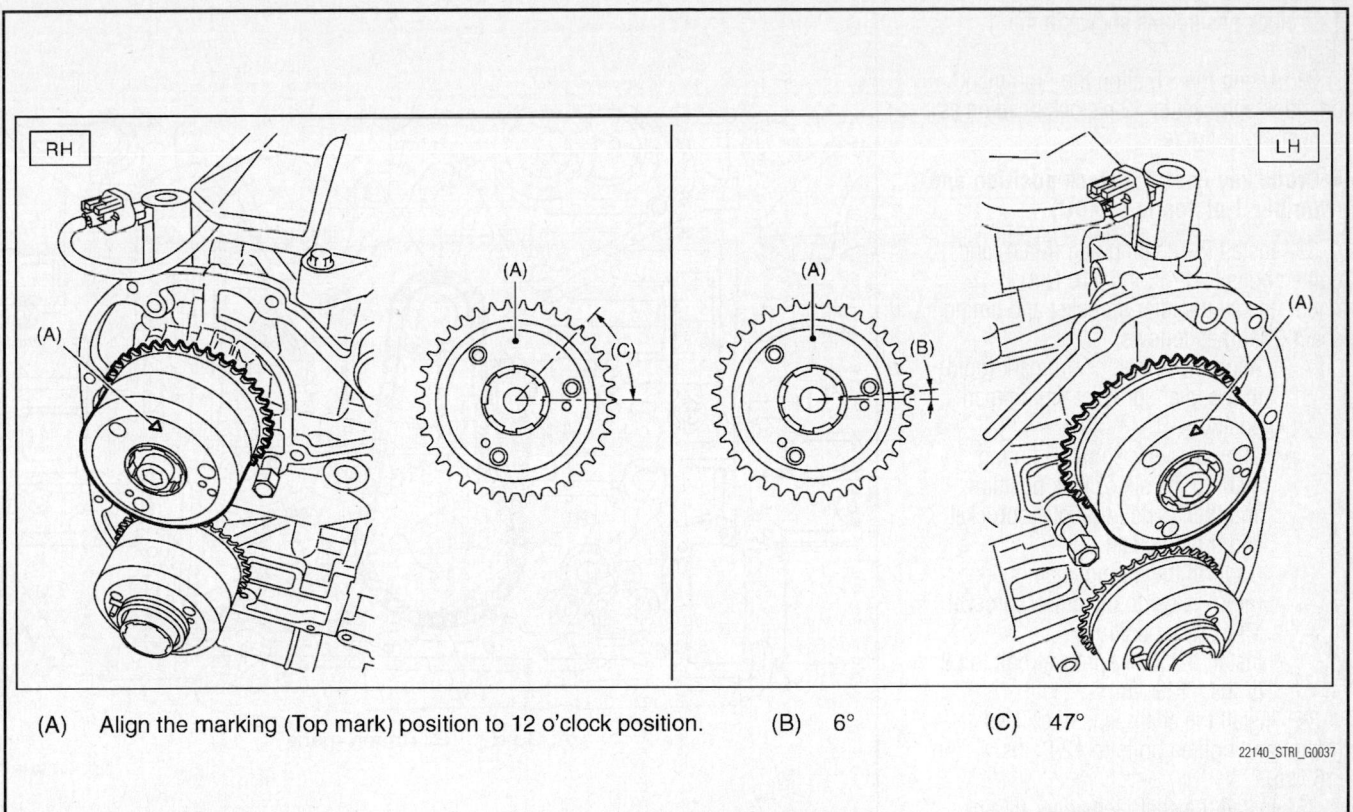

(A)   Align the marking (Top mark) position to 12 o'clock position.   (B)   6°   (C)   47°

22140_STRI_G0037

Fig. 112 Intake cam sprocket at 12 o'clock position as shown

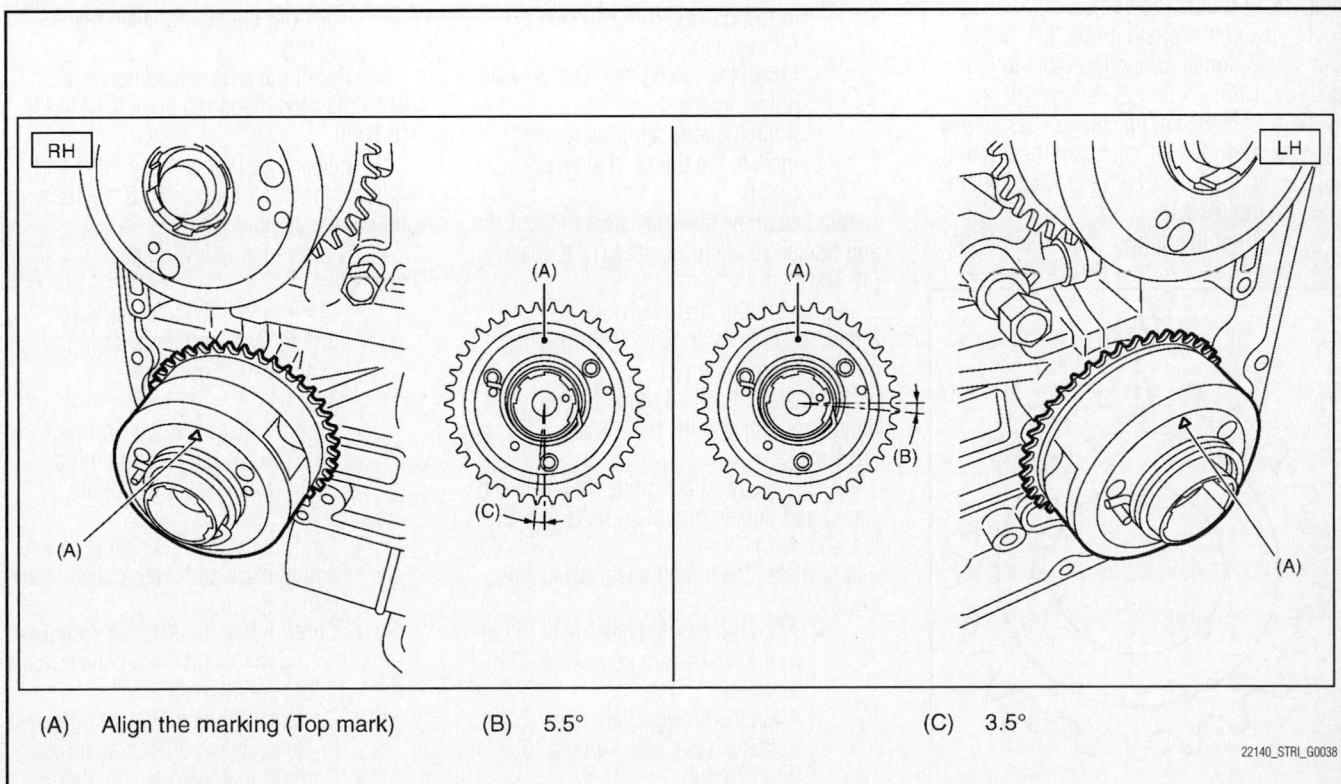

(A)   Align the marking (Top mark)   (B)   5.5°   (C)   3.5°

22140_STRI_G0038

Fig. 113 Exhaust cam sprocket at 12 o'clock position as shown

12 o'clock position as shown in the figure.

31. Using the ST, align the "Top mark" on crank sprocket to 12 o'clock position as shown in the figure.

➡**Crank key is at 3 o'clock position and (Number 1 piston is at TDC).**

32. Install the chain guide (Main) and tighten bolts to 12 ft. lbs. (16 Nm).

33. Install the idler sprocket and timing chain (Main) as follows:

- Align the timing chain mark (gold) to the idler sprocket timing mark position.
- Align the idler sprocket timing mark to the six o'clock position, and then install the idler sprocket and timing chain.
- Confirm the timing chain mark (gold) is set to the crank sprocket 12 o'clock position.
- Install the idler sprocket bolt to 88 ft. lbs. (120 Nm).

34. Install the chain tensioner lever (Main) and tighten bolts to 12 ft. lbs. (16 Nm).

35. Install the chain tensioner (Main) tighten bolts to 12 ft. lbs. (16 Nm). Then pull out the stopper pin.

36. Install the chain guide LH tighten bolts to 12 ft. lbs. (16 Nm).

37. Install the chain guide (LH: between cams) and tighten mounting bolts to 5 ft. lbs. (6.4 Nm).

38. Install the timing chain LH as follows:

- Align the timing chain mark (blue) to the intake cam sprocket LH timing mark.
- Align the timing chain mark (blue)

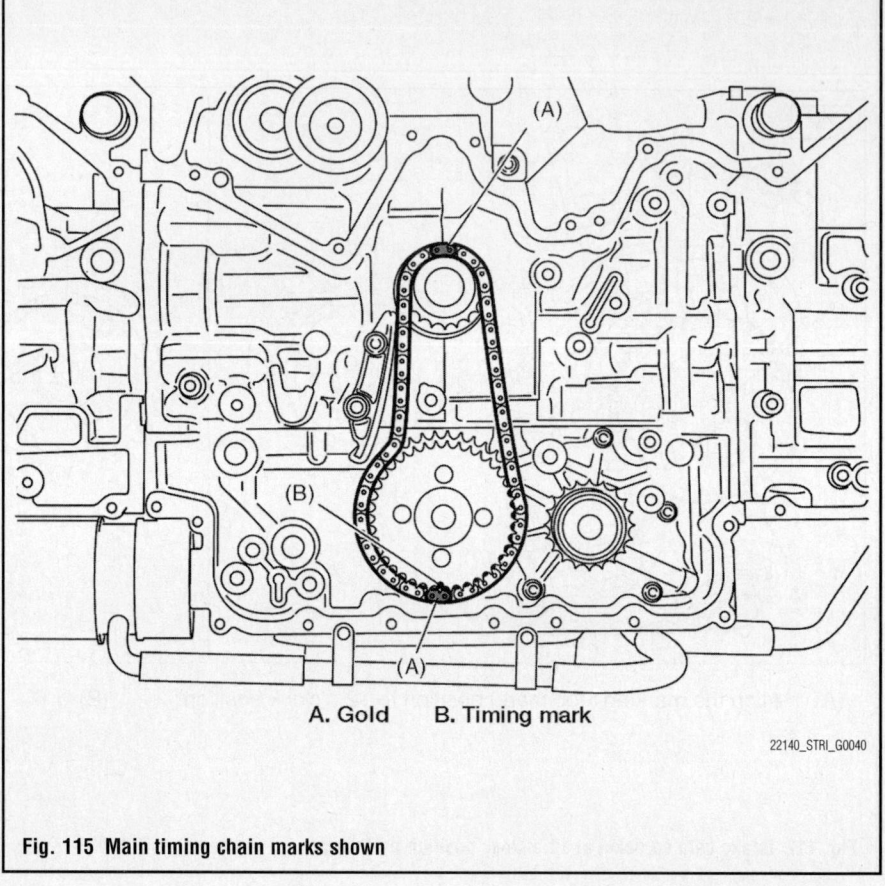

A. Gold   B. Timing mark

22140_STRI_G0040

Fig. 115 Main timing chain marks shown

to the exhaust cam sprocket LH timing mark.

- Install the timing chain to the water pump sprocket.
- Align the idler sprocket timing mark to the timing chain mark (gold).

39. Install the chain tensioner lever LH and tighten mounting bolts to 12 ft. lbs. (16 Nm).

40. Install the chain tensioner LH and tighten bolts to 12 ft. lbs. (16 Nm). Then pull out the stopper pin.

41. Install the chain guide RH and tighten mounting bolts to 12 ft. lbs. (16 Nm).

42. Install the chain guide (RH: between cams) and tighten mounting bolts to 5 ft. lbs. (6.4 Nm).

43. Install the timing chain RH as follows:

- Align the timing chain mark (blue) to the intake cam sprocket RH timing mark.
- Align the timing chain mark (blue) to the exhaust cam sprocket RH timing mark.
- Install the timing chain to the water pump sprocket.
- Align the idler sprocket timing

mark to the timing chain mark (gold).

44. Install the chain tensioner lever RH and tighten mounting bolts to 12 ft. lbs. (16 Nm).

45. Install the chain tensioner RH and tighten bolts to 12 ft. lbs. (16 Nm). Then pull out the stopper pin.

46. Confirm the following after installation:

- Confirm that the idler sprocket timing mark and three timing chain marks (gold) are aligned.
- Confirm that the crank sprocket 12 o'clock position and timing chain (Main) mark (gold) are aligned.
- Confirm that the LH cam sprocket timing mark and timing chain mark (blue) are aligned.
- Confirm that the RH cam sprocket timing mark and timing chain mark (blue) are aligned.
- Confirm that the bolts are tightened to the specified tightening torque.
- Confirm that there are no abnormalities, while turning the crankshaft in the engine rotating direction using ST.

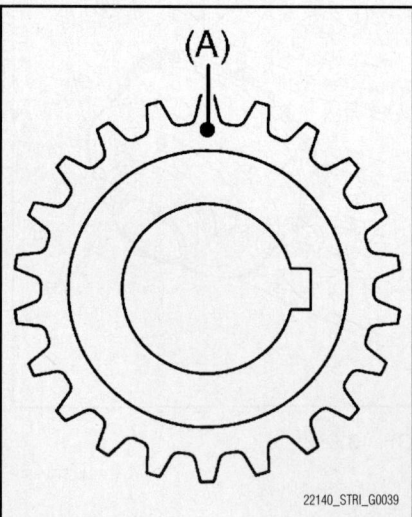

(A)

22140_STRI_G0039

Fig. 114 Top mark (A) on crank sprocket at 12 o'clock position shown

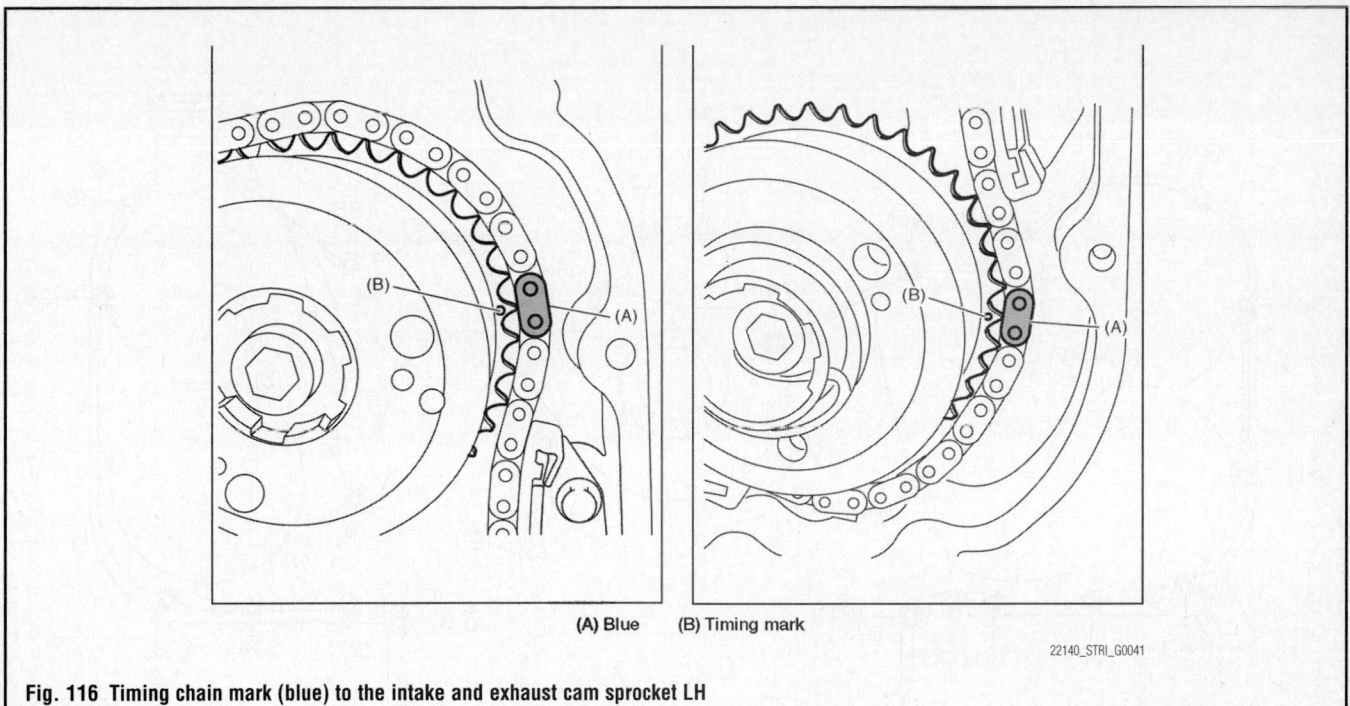

(A) Blue **(B) Timing mark**

22140_STRI_G0041

Fig. 116 Timing chain mark (blue) to the intake and exhaust cam sprocket LH

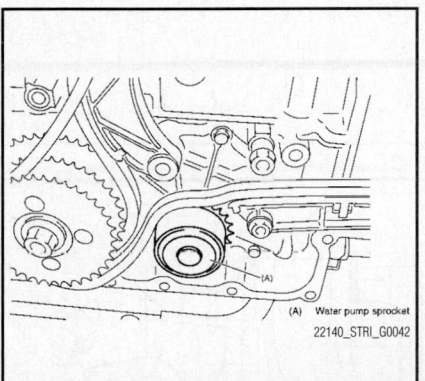

(A) Water pump sprocket

22140_STRI_G0042

Fig. 117 Timing chain to the water pump sprocket shown

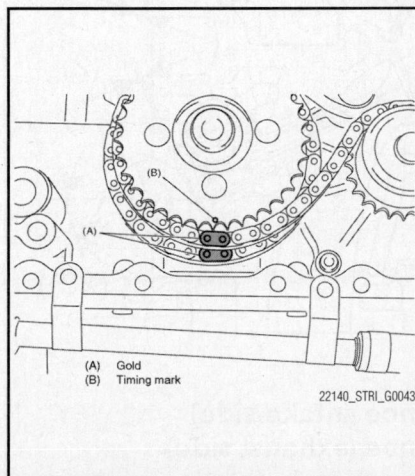

(A) Gold
(B) Timing mark

22140_STRI_G0043

Fig. 118 Sprocket timing mark to the timing chain mark shown

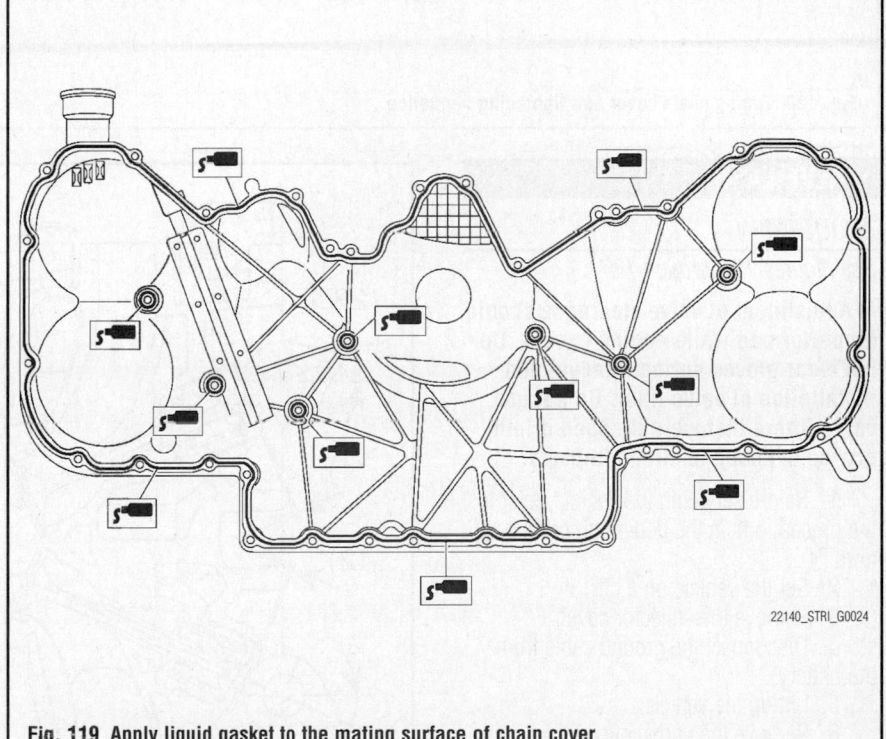

22140_STRI_G0024

Fig. 119 Apply liquid gasket to the mating surface of chain cover

47. Apply liquid gasket to the mating surface of chain cover.

48. Install the chain cover and temporarily tighten the bolts.

49. Tighten the bolts in the numerical order as shown in the figure to 7 ft. lbs. (10 Nm).

50. Install the crank pulley.

51. Install the V-belt.

52. Fill engine oil.

53. Install the radiator.

54. Refill coolant and bleed cooling system of air.

55. Confirm that there are no oil leakage at the chain cover mating surface.

56. Confirm that there are no coolant leaks.

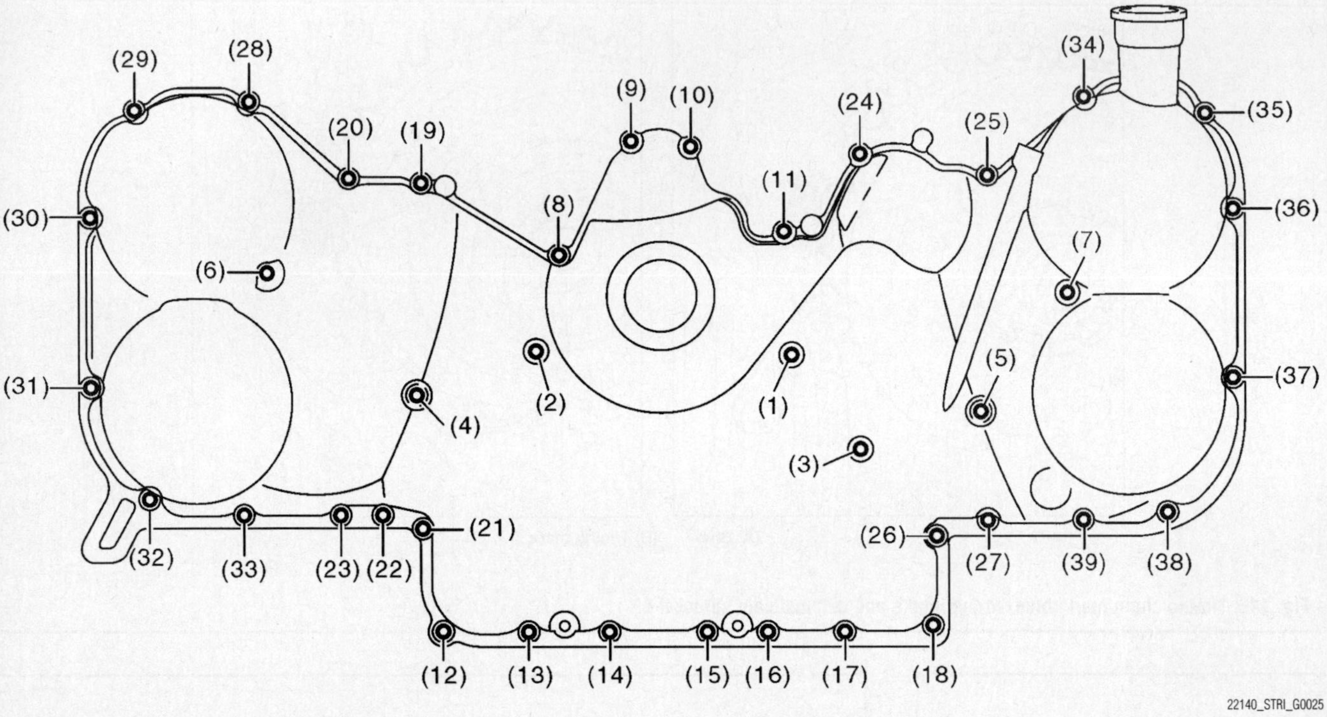

Fig. 120 Timing chain cover bolt tightening sequence

VALVE LASH

ADJUSTMENT

See Figures 121 through 130.

➡️**Adjustment of valve clearance should be performed while engine is cold. Do not wear gloves during removal and installation of valve lifter. Do not use valve lifters that were dropped or otherwise exposed to strong impacts.**

   1. Before servicing the vehicle, refer to the precautions in the beginning of this section.

   2. Set the vehicle on a lift.

   3. Remove the collector cover.

   4. Disconnect the ground cable from the battery.

   5. Lift up the vehicle.

   6. Remove the undercover.

   7. Lower the vehicle.

   8. When inspecting RH side cylinders:

     a. Remove the air intake duct and air cleaner case.

     b. Remove the fuel pipe protector RH.

     c. Disconnect the connector of oil pressure switch.

     d. Remove the ignition coil.

     e. Remove the rocker cover RH.

   9. When inspecting LH side cylinders:

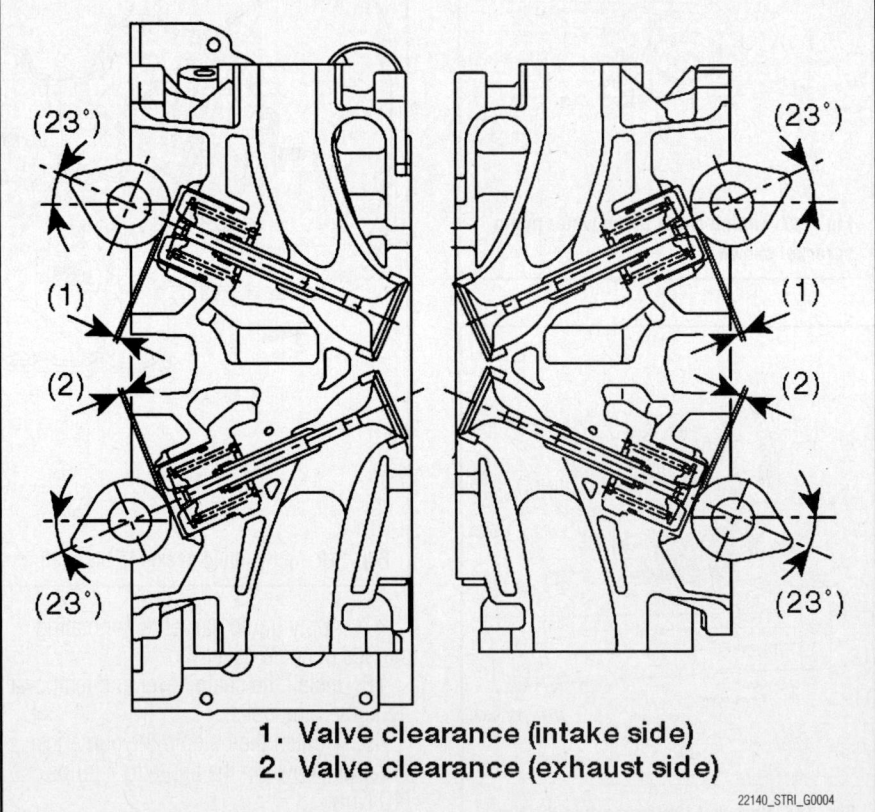

1. Valve clearance (intake side)
2. Valve clearance (exhaust side)

Fig. 121 Crankshaft clockwise until the cam is set to position

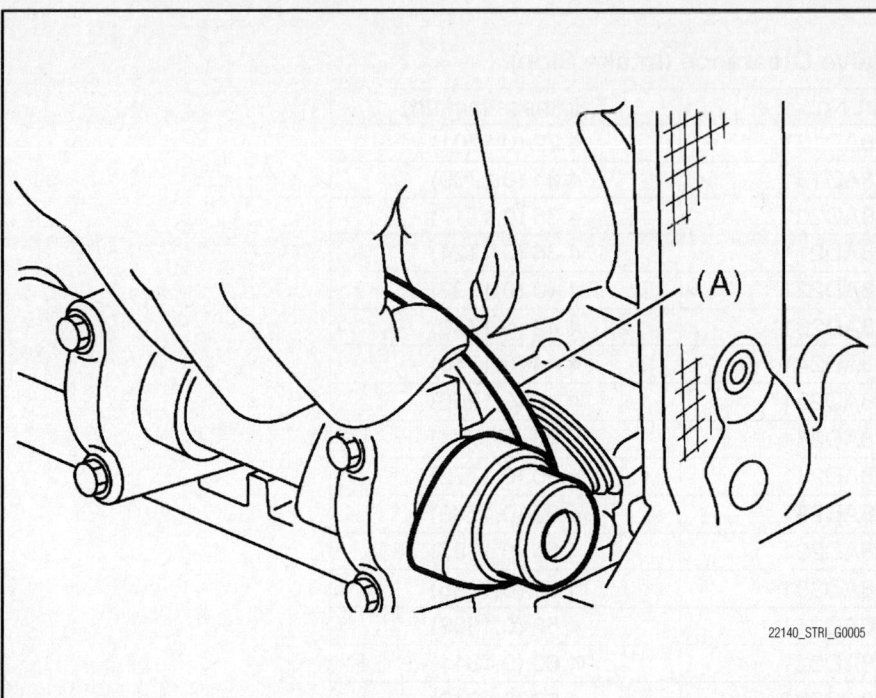

Fig. 122 Measure the clearance of intake valve and exhaust valve using thickness gauge (A).

a. Disconnect the battery cable, and then remove the battery and battery carrier.

b. Disconnect the PCV hose and blow-by hose from the rocker cover LH.

c. Remove the fuel pipe protector LH.

d. Remove the ignition coil.

e. Remove the rocker cover LH.

10. Turn the crankshaft clockwise until the cam is set to position shown in the figure.

11. Measure the clearance of intake valve and exhaust valve using thickness gauge (A).

➡ **Measure it within the range of +/-30°from specified position shown in the figure. Insert a thickness gauge in a direction as horizontal as possible with respect to the valve lifter. If the measured value is not within specification, take notes of the value in order to adjust the valve clearance later on.**

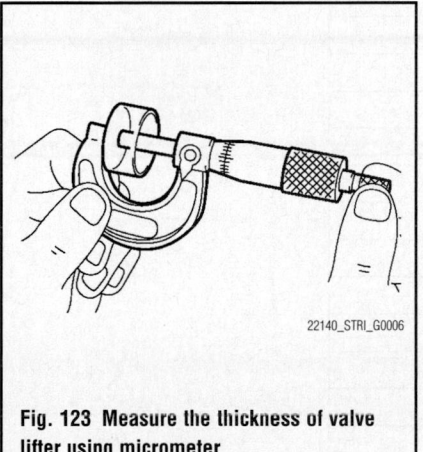

Fig. 123 Measure the thickness of valve lifter using micrometer

Use this table to help you select a suitable Exhaust valve shim

	Unit: mm (in)
$S = (V + T) - 0.35 \ (0.0138)$	
S: Valve lifter thickness required V: Measured valve clearance T: Valve lifter thickness to be used	

22140_STRI_G0010

Fig. 125 Use this table to help you select a suitable exhaust valve shim

Use this table to help you select a suitable Intake valve shim

	Unit: mm (in)
$S = (V + T) - 0.20 \ (0.0079)$	
S: Valve lifter thickness required V: Measured valve clearance T: Valve lifter thickness to be used	

22140_STRI_G0007

Fig. 124 Use this table to help you select a suitable intake valve shim

Valve Clearance (Intake Side)

Part No.	Thickness mm (in)
13228AD181	4.32 (0.1701)
13228AD191	4.34 (0.1709)
13228AD201	4.36 (0.1717)
13228AD211	4.38 (0.1724)
13228AD221	4.40 (0.1732)
13228AD231	4.42 (0.1740)
13228AD241	4.44 (0.1748)
13228AD251	4.46 (0.1756)
13228AD261	4.48 (0.1764)
13228AD271	4.50 (0.1772)
13228AD281	4.52 (0.1780)
13228AD291	4.54 (0.1787)
13228AD301	4.56 (0.1795)
13228AD311	4.58 (0.1803)
13228AD321	4.60 (0.1811)
13228AC581	4.62 (0.1819)
13228AC591	4.63 (0.1823)
13228AC601	4.64 (0.1827)
13228AC611	4.65 (0.1831)
13228AC621	4.66 (0.1835)
13228AC631	4.67 (0.1839)
13228AC641	4.68 (0.1843)
13228AC651	4.69 (0.1846)
13228AC661	4.70 (0.1850)
13228AC671	4.71 (0.1854)
13228AC681	4.72 (0.1858)
13228AC691	4.73 (0.1862)
13228AC701	4.74 (0.1866)
13228AC711	4.75 (0.1870)
13228AC721	4.76 (0.1874)
13228AC731	4.77 (0.1878)
13228AC741	4.78 (0.1882)
13228AC751	4.79 (0.1886)
13228AC761	4.80 (0.1890)
13228AC771	4.81 (0.1894)
13228AC781	4.82 (0.1898)
13228AC791	4.83 (0.1902)
13228AC801	4.84 (0.1906)
13228AC811	4.85 (0.1909)
13228AC821	4.86 (0.1913)
13228AC831	4.87 (0.1917)
13228AC841	4.88 (0.1921)
13228AC851	4.89 (0.1925)

22140_STRI_G0008

Fig. 126 Intake valve adjusting shim chart

Valve Clearance (Intake side)

Part No.	Thickness mm (in)
13228AC861	4.90 (0.1929)
13228AC871	4.91 (0.1933)
13228AC881	4.92 (0.1937)
13228AC891	4.93 (0.1941)
13228AC901	4.94 (0.1945)
13228AC911	4.95 (0.1949)
13228AC921	4.96 (0.1953)
13228AC931	4.97 (0.1957)
13228AC941	4.98 (0.1961)
13228AC951	4.99 (0.1965)
13228AC961	5.00 (0.1969)
13228AC971	5.01 (0.1972)
13228AC981	5.02 (0.1976)
13228AC991	5.03 (0.1980)
13228AD001	5.04 (0.1984)
13228AD011	5.05 (0.1988)
13228AD021	5.06 (0.1992)
13228AD031	5.07 (0.1996)
13228AD041	5.08 (0.2000)
13228AD051	5.09 (0.2004)
13228AD061	5.10 (0.2008)
13228AD071	5.11 (0.2012)
13228AD081	5.12 (0.2016)
13228AD091	5.13 (0.2020)
13228AD101	5.14 (0.2024)
13228AD111	5.15 (0.2028)
13228AD121	5.16 (0.2032)
13228AD131	5.17 (0.2035)
13228AD141	5.18 (0.2039)
13228AD151	5.19 (0.2043)
13228AD161	5.20 (0.2047)
13228AD171	5.21 (0.2051)
13228AD331	5.23 (0.2059)
13228AD341	5.25 (0.2067)
13228AD351	5.27 (0.2075)
13228AD361	5.29 (0.2083)
13228AD371	5.31 (0.2091)
13228AD381	5.33 (0.2098)
13228AD391	5.35 (0.2106)
13228AD401	5.37 (0.2114)
13228AD411	5.39 (0.2122)
13228AD421	5.41 (0.2130)
13228AD431	5.43 (0.2138)
13228AD441	5.45 (0.2146)
13228AD451	5.47 (0.2154)
13228AD461	5.49 (0.2161)
13228AD471	5.51 (0.2169)
13228AD481	5.53 (0.2177)
13228AD491	5.55 (0.2185)
13228AD501	5.57 (0.2193)
13228AD511	5.59 (0.2201)

22140_STRI_G0009

Fig. 127 Intake valve adjusting shim chart

Valve Clearance (Exhaust Side)

Part No.	Thickness mm (in)
13228AD081	5.12 (0.2016)
13228AD091	5.13 (0.2020)
13228AD101	5.14 (0.2024)
13228AD111	5.15 (0.2028)
13228AD121	5.16 (0.2032)
13228AD131	5.17 (0.2035)
13228AD141	5.18 (0.2039)
13228AD151	5.19 (0.2043)
13228AD161	5.20 (0.2047)
13228AD171	5.21 (0.2051)
13228AD331	5.23 (0.2059)
13228AD341	5.25 (0.2067)
13228AD351	5.27 (0.2075)
13228AD361	5.29 (0.2083)
13228AD371	5.31 (0.2091)
13228AD381	5.33 (0.2098)
13228AD391	5.35 (0.2106)
13228AD401	5.37 (0.2114)
13228AD411	5.39 (0.2122)
13228AD421	5.41 (0.2130)
13228AD431	5.43 (0.2138)
13228AD441	5.45 (0.2146)
13228AD451	5.47 (0.2154)
13228AD461	5.49 (0.2161)
13228AD471	5.51 (0.2169)
13228AD481	5.53 (0.2177)
13228AD491	5.55 (0.2185)
13228AD501	5.57 (0.2193)
13228AD511	5.59 (0.2201)

22140_STRI_G0011

Fig. 128 Exhaust valve adjusting shim chart

Valve Clearance (Exhaust Side)	
Part No.	Thickness mm (in)
13228AD181	4.32 (0.1701)
13228AD191	4.34 (0.1709)
13228AD201	4.36 (0.1717)
13228AD211	4.38 (0.1724)
13228AD221	4.40 (0.1732)
13228AD231	4.42 (0.1740)
13228AD241	4.44 (0.1748)
13228AD251	4.46 (0.1756)
13228AD261	4.48 (0.1764)
13228AD271	4.50 (0.1772)
13228AD281	4.52 (0.1780)
13228AD291	4.54 (0.1787)
13228AD301	4.56 (0.1795)
13228AD311	4.58 (0.1803)
13228AD321	4.60 (0.1811)

22140_STRI_G0012

Fig. 129 Exhaust valve adjusting shim chart

Right side of Intake Manifold

22140_STRI_G0013

Fig. 130 Exhaust valve adjusting shim chart

12. If adjustment is required, remove the camshafts.

13. Remove and measure the thickness of the valve lifter. Select a suitable shim, using the shim selection chart.

To install:

14. Install the replacement shim to the lifter.

15. After all shims have been adjusted, inspect the valve clearances again.

16. After completion, install all removed components.

ENGINE PERFORMANCE & EMISSION CONTROLS

ACCELERATOR PEDAL POSITION (APP) SENSOR

LOCATION

The Accelerator Pedal Position (APP) Sensor is located inside the vehicle. It is part of the pedal assembly.

REMOVAL & INSTALLATION

1. Before servicing the vehicle, refer to the precautions section.

2. Disconnect the ground cable from battery.

3. Disconnect the connector.

4. Remove the nut securing accelerator pedal assembly.

5. Install in the reverse order of removal. Tighten the nut securing accelerator pedal assembly to 6 ft. lbs. (7.5 Nm).

CAMSHAFT POSITION (CMP) SENSOR

LOCATION

See Figures 131 and 132.

Refer to the accompanying illustrations

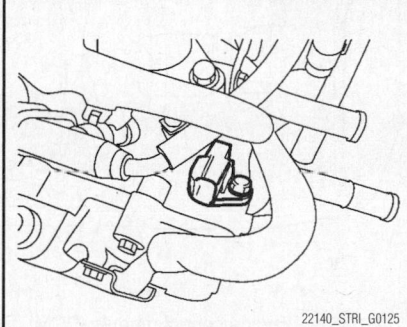

22140_STRI_G0125

Fig. 131 Camshaft Position (CMP) Sensor LH intake shown

for Camshaft Position (CMP) sensor location.

REMOVAL & INSTALLATION

1. Before servicing the vehicle, refer to the precautions section.

2. Disconnect the negative battery cable.

3. Remove the collector cover, as required.

4. Remove the alternator harness from the fuel pipe. Remove the fuel pipe protector.

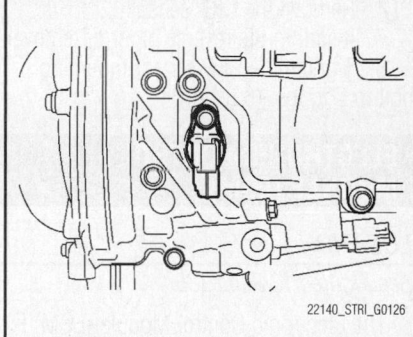

22140_STRI_G0126

Fig. 132 Camshaft Position (CMP) Sensor LH exhaust shown

5. Disconnect the connector from the Crankshaft Position (CKP) sensor.

6. Remove the bolt that retains the CMP sensor, then remove the sensor.

7. Remove the RH CMP sensor using the same procedure as the LH sensor.

8. Installation is the reverse of the removal procedure. Tighten the camshaft sensor mounting bolt to 5 ft. lbs. (6.8 Nm).

CRANKSHAFT POSITION (CKP) SENSOR

LOCATION

See Figure 133.

The Crankshaft Position (CKP) sensor is located in the front of the engine near the crankshaft pulley.

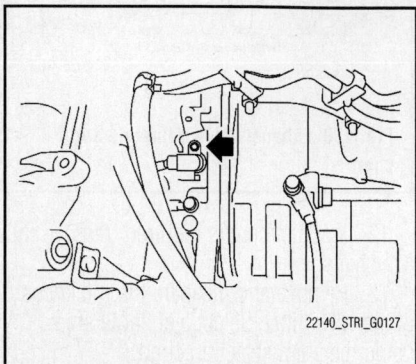

22140_STRI_G0127

Fig. 133 Crankshaft Position (CKP) Sensor location view

REMOVAL & INSTALLATION

1. Before servicing the vehicle, refer to the precautions section.
2. Remove the collector cover.
3. Disconnect the ground cable from battery.
4. Remove the air intake chamber.
5. Remove the service hole cover.
6. Remove the CKP sensor.
7. Install in the reverse order of removal.
8. Tighten the CKP sensor mounting bolt to 5 ft. lbs. (6.8 Nm).

ELECTRONIC CONTROL MODULE (ECM)

LOCATION

See Figures 134 and 135.

The Electronic Control Module (ECM) is located on the passenger's side of the vehicle, underneath the floor mat.

REMOVAL & INSTALLATION

1. Before servicing the vehicle, refer to the precautions section.
2. Disconnect the negative battery cable.
3. Remove the lower inner trim on the passenger's side of the vehicle.
4. Detach the floor mat. Remove the protective cover.
5. Remove the Electronic Control Module (ECM) bracket retaining nuts. Remove the clip from the bracket.

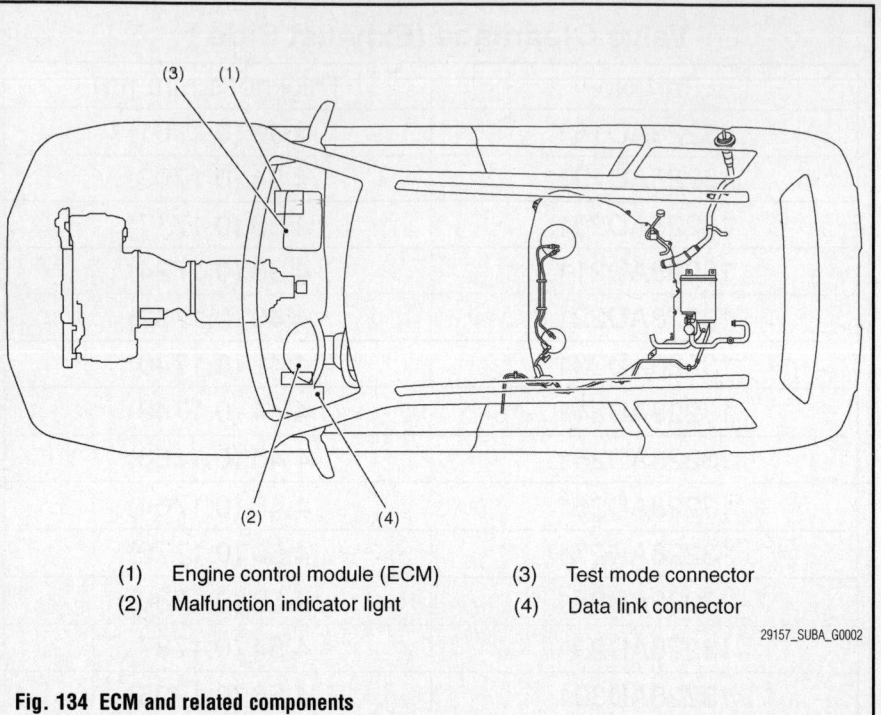

| (1) | Engine control module (ECM) | (3) | Test mode connector |
| (2) | Malfunction indicator light | (4) | Data link connector |

29157_SUBA_G0002

Fig. 134 ECM and related components

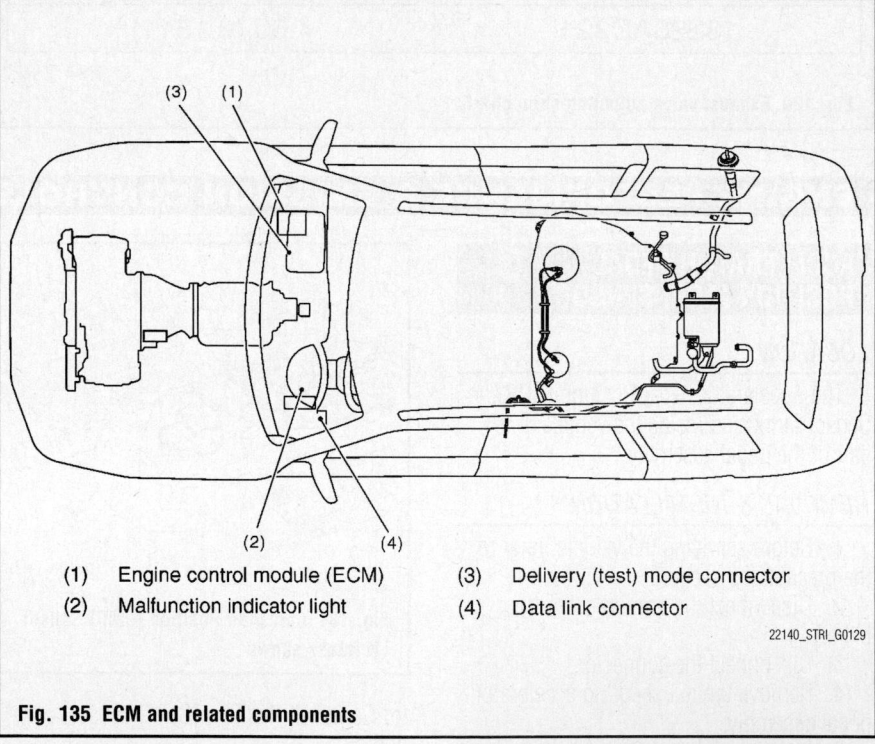

| (1) | Engine control module (ECM) | (3) | Delivery (test) mode connector |
| (2) | Malfunction indicator light | (4) | Data link connector |

22140_STRI_G0129

Fig. 135 ECM and related components

6. Disconnect the connectors.
7. Remove the ECM from the vehicle.
8. Installation is the reverse of the removal procedure. Tighten the retaining screws to 3.7 ft. lbs.

➡**When replacing the ECM, be careful not to use the wrong part number, as damage to the injection system could occur.**

ENGINE COOLANT TEMPERATURE (ECT) SENSOR

LOCATION

See Figure 136.

The Engine Coolant Temperature (ECT) Sensor is located to the right of the A/C compressor.

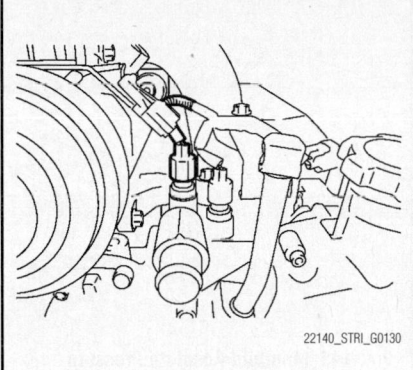

Fig. 136 Engine Coolant Temperature (ECT) Sensor location view

REMOVAL & INSTALLATION

1. Before servicing the vehicle, refer to the precautions section.
2. Remove the collector cover.
3. Disconnect the negative battery cable.
4. Disconnect the connector from the ECT sensor.
5. Remove the sensor from its mounting.
6. Installation is the reverse of the removal procedure. Tighten the sensor to 13 ft. lbs. (18 Nm).

HEATED OXYGEN SENSOR (HO2S)

LOCATION

See Figures 137 and 138.

The front Heated Oxygen Sensors (HO2S) are located in the front section of the front catalytic converter. The rear oxygen sensors are located in the rear section of the front catalytic converter. There are two front catalytic converters, but only one rear catalytic converter.

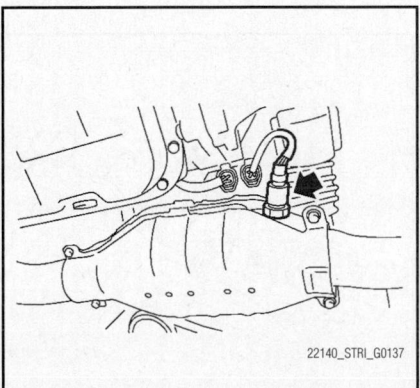

Fig. 137 Front Heated Oxygen Sensor (HO2S) location

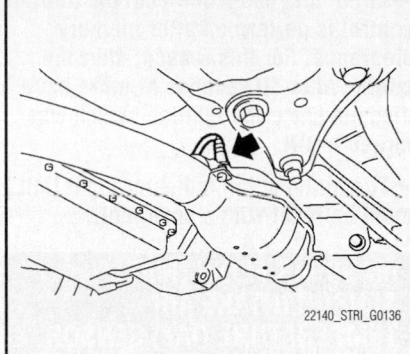

Fig. 138 Rear Heated Oxygen Sensor (HO2S) location

REMOVAL & INSTALLATION

Front Sensor

1. Before servicing the vehicle, refer to the precautions section.
2. Disconnect the negative battery cable.
3. Raise and support the vehicle safely.
4. Disconnect the connector.
5. Remove the oxygen sensor.

To install:

6. Installation is the reverse of the removal procedure.

➡ Apply anti-seize compound to the threaded portion of the sensor, prior to installation. Never apply anti-seize compound to the protector of the sensor.

7. Tighten the sensor to 16 ft. lbs. (21 Nm).

Rear Sensor

1. Before servicing the vehicle, refer to the precautions section.
2. Disconnect the negative battery cable.
3. Raise and support the vehicle safely.
4. Disconnect the connector. Remove the clip by pulling it out from the upper side of the crossmember.
5. Remove the oxygen sensor.

To install:

6. Installation is the reverse of the removal procedure.

➡ Apply anti-seize compound to the threaded portion of the sensor, prior to installation. Never apply anti-seize compound to the protector of the sensor.

7. Tighten the sensor to 16 ft. lbs. (21 Nm).

INTAKE AIR TEMPERATURE (IAT) SENSOR

LOCATION

See Figure 139.

The IAT is mounted in the intake air hose of the air cleaner assembly. This sensor is combined with the mass air flow sensor.

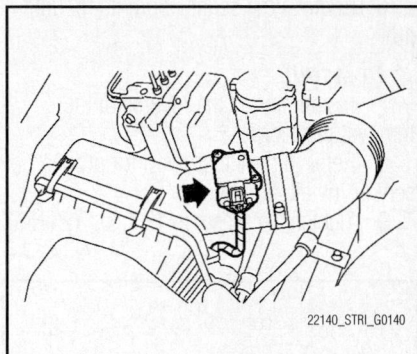

Fig. 139 Intake Air Temperature (IAT) and Mass Air Flow (MAF) sensor location view

REMOVAL & INSTALLATION

1. Before servicing the vehicle, refer to the precautions section.
2. Disconnect the negative battery cable.
3. Remove the mounting screws from the sensor.
4. Remove the sensor from its mounting.
5. Installation is the reverse of the removal procedure.
6. Tighten the mounting screws to 0.7 ft. lbs. (1 Nm).

KNOCK SENSOR (KS)

LOCATION

See Figure 140.

The Knock Sensor (KS) is located under the intake manifold.

Fig. 140 Knock Sensor (KS) location view

REMOVAL & INSTALLATION

See Figure 141.

1. Before servicing the vehicle, refer to the precautions section.
2. Disconnect the negative battery cable.
3. Remove the collector cover, as required.
4. Remove the intake manifold.
5. Disconnect the sensor connector.
6. Remove the sensor from its mounting.

To install:

7. Installation is the reverse of the removal procedure.
8. Refer to the illustration for proper sensor installation angle.
9. Tighten the sensor to 18 ft. lbs. (25 Nm).

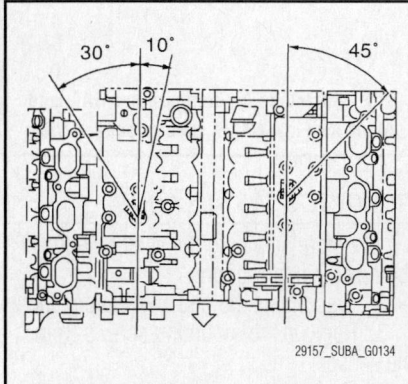

Fig. 141 Knock sensor installation angle

MALFUNCTION INDICATOR LIGHT (MIL)

RESET PROCEDURE

Subaru Select Monitor (OBD MODE)

1. On the (Main Menu) display screen, select the (Each System Check).
2. On the (System Selection Menu) display screen, select the (Engine Control System).
3. Select (OK) after the information of engine type has been displayed.
4. On the (Engine Diagnosis) display screen, select the (OBD System).
5. On the «OBD Menu» display screen, select the (Clear Diagnostic Code).
6. When the (Clear Diagnostic Code) is shown on the screen, click the (Yes) button.
7. When (Done) and (Turn ignition switch to OFF) are shown on the display screen, turn the ignition switch to OFF.

General Scan Tool

8. For procedures clearing memory using the general scan tool, refer to the general scan tool operation manual.

➡Initial diagnosis of electronic throttle control is performed after memory clearance. For this reason, start the engine after 10 seconds or more have elapsed since the ignition switch was turned to ON.

➡The Malfunction Indicator Light (MIL) must be reset with a scan tool.

MASS AIR FLOW (MAF) SENSOR/INTAKE AIR TEMPERATURE (IAT) SENSOR

LOCATION

See Figure 142.

The Mass Air Flow (MAF) and Intake Air Temperature (IAT) Sensor is mounted in the intake air hose of the air cleaner assembly.

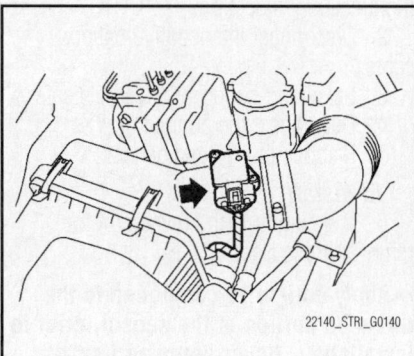

Fig. 142 Mass Air Flow and Intake Air Temperature location view

REMOVAL & INSTALLATION

1. Before servicing the vehicle, refer to the precautions section.
2. Disconnect the negative battery cable.
3. Remove the mounting screws from the sensor.
4. Remove the sensor from its mounting.

To install:

5. Installation is the reverse of the removal procedure.
6. Tighten the mounting screws to 0.7 ft. lbs. (1 Nm).

MANIFOLD ABSOLUTE PRESSURE (MAP) SENSOR

LOCATION

See Figure 143.

The Manifold Absolute Pressure (MAP) sensor is located on the intake manifold.

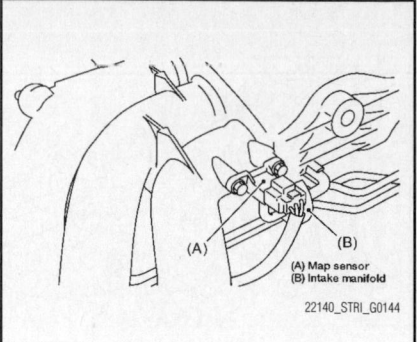

(A) Map sensor
(B) Intake manifold

Fig. 143 Manifold Absolute Pressure (MAP) Sensor location view

REMOVAL & INSTALLATION

1. Before servicing the vehicle, refer to the precautions section.
2. Disconnect the negative battery cable.
3. Remove the collector cover, if equipped.
4. Disconnect the connector from the sensor.
5. Remove the filter assembly from the intake manifold, if equipped.
6. Remove the MAP sensor from its mounting.

To install:

7. Installation is the reverse of the removal procedure.
8. Be sure to use new O-rings.
9. Torque the retaining bolts to 5 ft. lbs. (6.4 Nm).

THROTTLE POSITION SENSOR (TPS)

LOCATION

See Figure 144.

The Throttle Position Sensor (TPS) is located near the center of the engine by the air cleaner assembly.

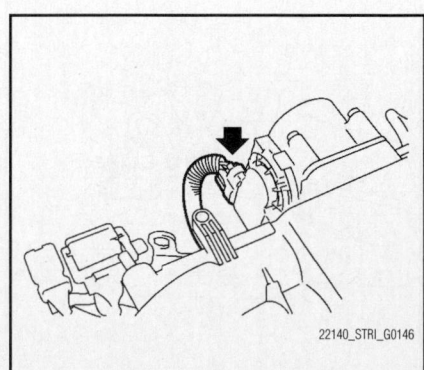

Fig. 144 Throttle Position Sensor (TPS) location view

➡The throttle body is a non-disassembled part, so do not remove the throttle position sensor from throttle body. Replace as an assembly.

REMOVAL & INSTALLATION

➡The throttle body is a non-disassembled part, so do not remove the throttle position sensor from throttle body. Replace as an assembly.

1. Before servicing the vehicle, refer to the precautions section.
2. Remove the collector cover.
3. Disconnect the negative battery cable.
4. Remove the air intake chamber.
5. Disconnect the sensor connector.
6. Disconnect the engine coolant hoses from the throttle body.
7. Remove the bolts which secure throttle body to intake manifold.
8. Remove throttle body assembly.

To install:

9. Installation is the reverse of the removal procedure.
10. Install new gaskets and tighten throttle body mounting bolts to 6 ft. lbs. (8 Nm).

VEHICLE SPEED SENSOR (VSS)

LOCATION

See Figures 145 and 146.

This vehicle uses a front and rear Vehicle Speed Sensor (VSS). Both sensors are mounted on the transaxle.

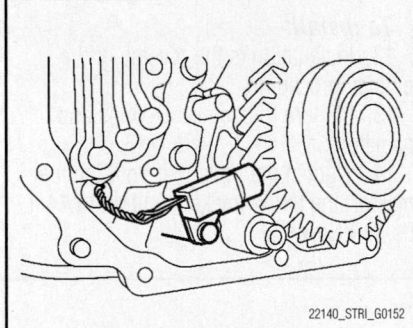

22140_STRI_G0152

Fig. 145 Front Vehicle Speed Sensor (VSS) location view

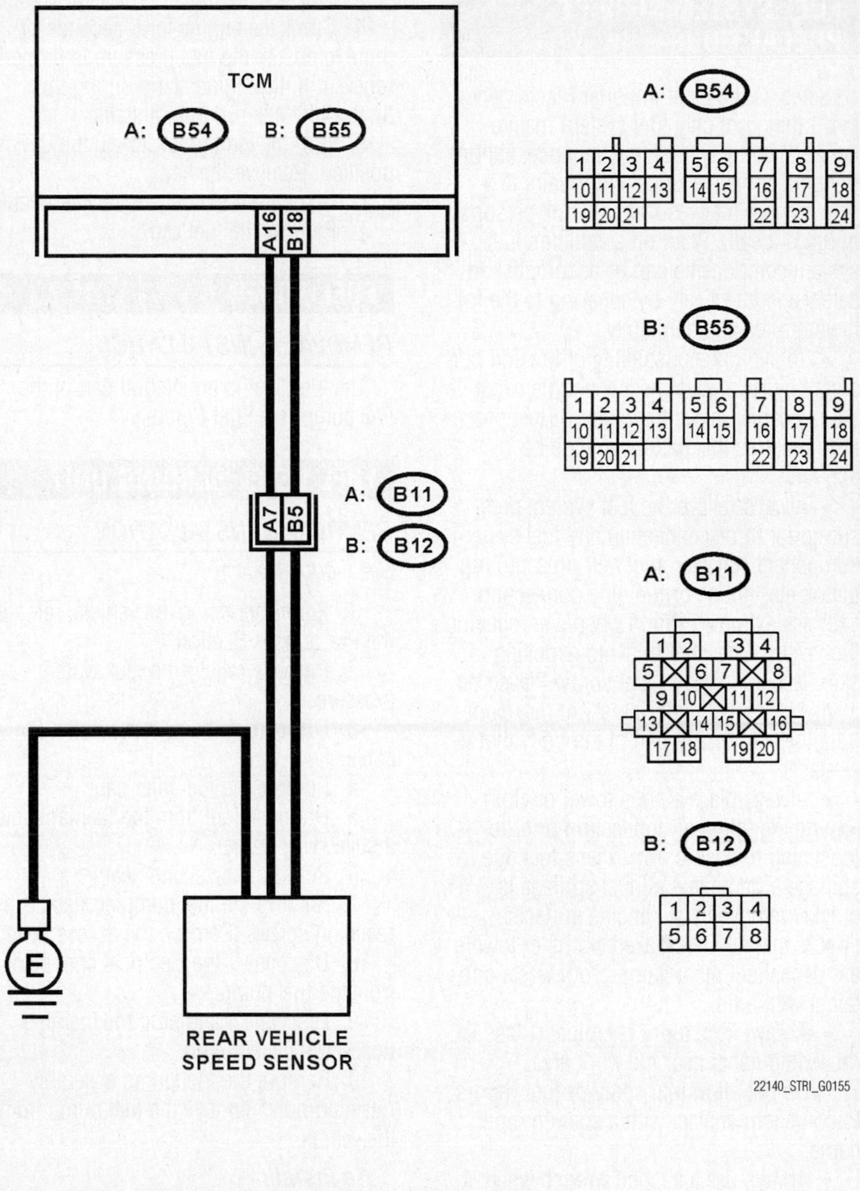

TCM
A: B54   B: B55

A16
B18

A7
B5

A: B11
B: B12

E

REAR VEHICLE
SPEED SENSOR

A: B54

1	2	3	4		5	6		7		8	9
10	11	12	13		14	15		16		17	18
19	20	21						22		23	24

B: B55

1	2	3	4		5	6		7		8	9
10	11	12	13		14	15		16		17	18
19	20	21						22		23	24

A: B11

	1	2		3	4	
5	⊠	6	7	⊠	8
9	10	⊠	11	12
13	⊠	14	15	⊠	16
17	18		19	20

B: B12

| 1 | 2 | 3 | 4 |
| 5 | 6 | 7 | 8 |

22140_STRI_G0155

Fig. 146 Rear Vehicle Speed Sensor (VSS) circuit view

REMOVAL & INSTALLATION

1. Raise and support the vehicle safely.
2. Place a drip pan below the speed sensor to catch any spilled fluid.
3. Disconnect the connector from the Vehicle Speed Sensor (VSS).

4. Remove the VSS from its mounting.

To install:

5. Installation is the reverse of the removal procedure.
6. Replace any lost fluid.

FUEL SYSTEM SERVICE PRECAUTIONS

Safety is the most important factor when performing not only fuel system maintenance, but any type of maintenance. Failure to conduct maintenance and repairs in a safe manner may result in serious personal injury or death. Work on a vehicle's fuel system components can be accomplished safely and effectively by adhering to the following rules and guidelines.

• To avoid the possibility of fire and personal injury, always disconnect the negative battery cable unless the repair or test procedure requires that battery voltage be applied.

• Always relieve the fuel system pressure prior to disconnecting any fuel system component (injector, fuel rail, pressure regulator, etc.) fitting or fuel line connection. Exercise extreme caution whenever relieving fuel system pressure to avoid exposing skin, face and eyes to fuel spray. Please be advised that fuel under pressure may penetrate the skin or any part of the body that it contacts.

• Always place a shop towel or cloth around the fitting or connection prior to loosening to absorb any excess fuel due to spillage. Ensure that all fuel spillage is quickly removed from engine surfaces. Ensure that all fuel-soaked cloths or towels are deposited into a flame-proof waste container with a lid.

• Always keep a dry chemical (Class B) fire extinguisher near the work area.

• Do not allow fuel spray or fuel vapors to come into contact with a spark or open flame.

• Always use a second wrench when loosening or tightening fuel line connection fittings. This will prevent unnecessary stress and torsion on fuel piping. Always follow the proper torque specifications.

• Always replace worn fuel fitting O-rings with new ones. Do not substitute fuel hose where rigid pipe is installed.

FUEL SYSTEM PRESSURE

RELIEVING

➡**This procedure must be performed prior to servicing any component of the fuel injection system.**

1. Before servicing the vehicle, refer to the Precautions Section.
2. Remove the fuel pump fuse from the main fuse box.

3. Start the engine and run until it stalls.
4. Crank the engine for 5 seconds or more to ensure the fuel pressure is properly relieved. If the engine starts during this time, allow it to run until it stalls.
5. Turn the ignition switch to the OFF position. Remove the key.
6. Disconnect the negative battery cable.
7. Remove the fuel cap.

FUEL FILTER

REMOVAL & INSTALLATION

The fuel filter is an integral part of the fuel pump, see Fuel Pump.

FUEL LEVEL SENDING UNIT

REMOVAL & INSTALLATION

See Figure 147.

1. Before servicing the vehicle, refer to the Precautions Section.
2. Properly relieve the fuel system pressure.
3. Disconnect the negative battery cable.
4. Remove the fuel filler cap.
5. Drain the fuel from the fuel tank into a suitable container.
6. Remove the second seat.
7. Remove the fuel pump access cover retaining screws. Remove the access cover.
8. Disconnect the electrical connector from the fuel pump.
9. Disconnect and plug the fuel line hoses at the fuel pump.
10. Remove the fuel pump assembly retaining nuts. Remove the fuel pump from the vehicle.

To install:
11. Installation is the reverse of removal procedure.

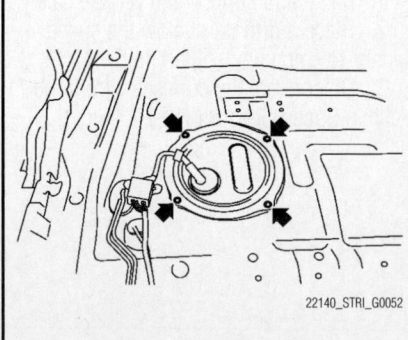

Fig. 147 Service access cover retaining screws shown

22140_STRI_G0052

12. Be sure to use a new gasket.
13. Tighten the fuel pump retaining nuts to 36 inch lbs, (4 Nm) in an alternating sequence pattern.
14. Continue the installation in the reverse order of the removal procedure.
15. Start the engine and check for leaks, correct as required.

FUEL RAIL AND INJECTOR

REMOVAL & INSTALLATION

See Figures 148 and 149.

1. Before servicing the vehicle, refer to the Precautions Section.
2. Properly relieve the fuel system pressure. Remove the fuel cap.
3. Disconnect the negative battery cable.
4. Remove the collector cover.
5. If removing the right side injectors, remove the air cleaner case.
6. If removing the left side injectors, remove the battery. Remove the alternator harness from the left side fuel line protector cover.
7. Remove the fuel line protector cover(s).
8. Disconnect the electrical connector from the fuel injector.
9. Remove the engine harness from the fuel injector line.
10. Remove the bolts that hold the fuel injector line to the intake manifold.
11. Remove the fuel injector while lifting up the fuel injector line.

To install:
12. Installation is the reverse of the removal procedure.
13. Be sure to use new O-rings and insulators.
14. Tighten the bolts that hold the fuel injector line to the intake manifold to 14 ft. lbs. (19 Nm).

Fig. 148 Fuel injector line mounting bolt shown

22140_STRI_G0053

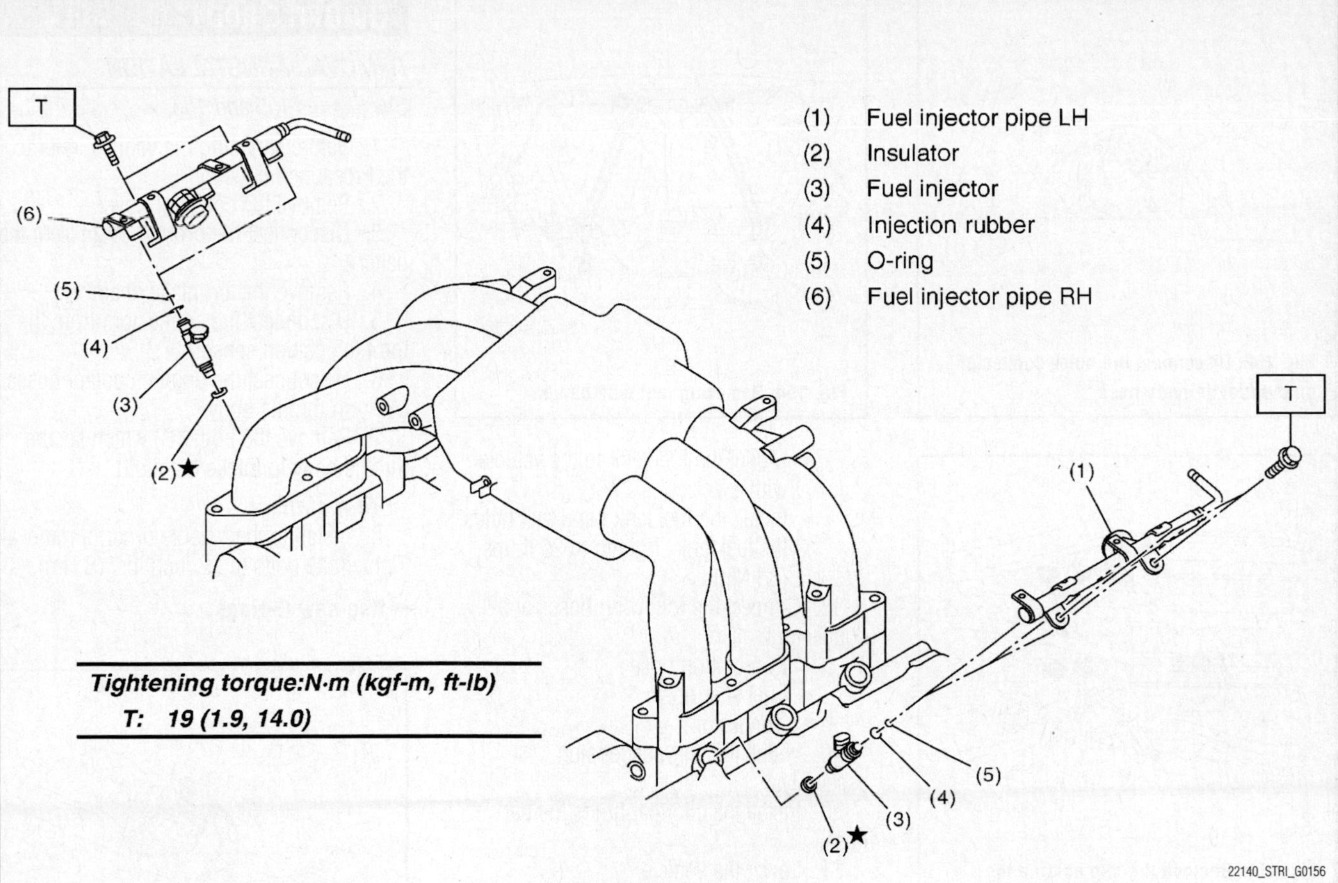

(1) Fuel injector pipe LH
(2) Insulator
(3) Fuel injector
(4) Injection rubber
(5) O-ring
(6) Fuel injector pipe RH

Tightening torque:N·m (kgf-m, ft-lb)
T: 19 (1.9, 14.0)

22140_STRI_G0156

Fig. 149 Fuel rail and injector view

15. Tighten the fuel line protector cover bolts to 14 ft. lbs. (19 Nm).

16. Start the engine and check for leaks, correct as required.

FUEL TANK

REMOVAL & INSTALLATION

See Figures 150 through 154.

1. Before servicing the vehicle, refer to the Precautions Section.

2. Set the vehicle on a lift.

3. Release the fuel pressure by performing the following:
   • Remove the fuse of fuel pump from main fuse box
   • Start the engine and run until it stalls
   • After the engine stalls, crank it for five more seconds
   • Turn the ignition switch to OFF

4. Disconnect the ground cable from battery.

5. Disconnect the ground cable from battery.

6. Drain fuel from fuel tank.

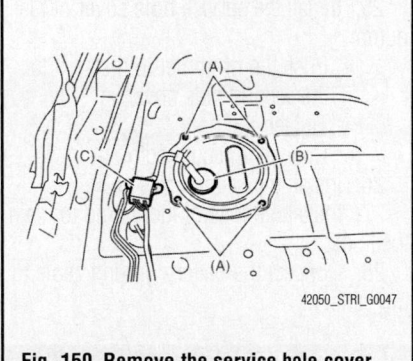

42050_STRI_G0047

Fig. 150 Remove the service hole cover from the fuel pump

7. Remove the second row seats.

8. Remove the service hole cover of fuel pump.
   • Disconnect the connector
   • Remove the bolt
   • Push the grommet down and remove service hole cover

9. Remove the service hole cover of fuel sub level sensor.

10. Disconnect the quick connector on the fuel delivery hose.

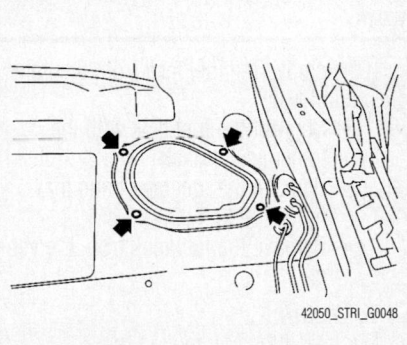

42050_STRI_G0048

Fig. 151 Remove the service hole cover of the fuel sub level sensor

11. Remove the rear wheels.

12. Lift-up the vehicle.

13. Remove the clip holding the harness.

14. Remove the rear suspension assembly.

15. Disconnect the following hoses and connectors:
   • Evaporation hose (A)
   • Fuel filler hose (C)
   • Connector (B)

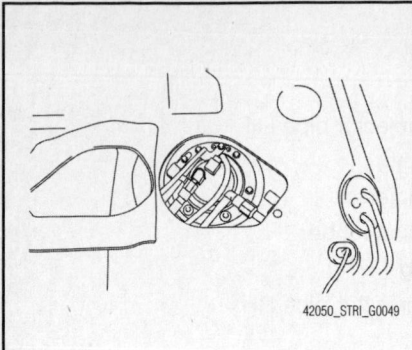

Fig. 152 Disconnect the quick connector on the fuel delivery hose

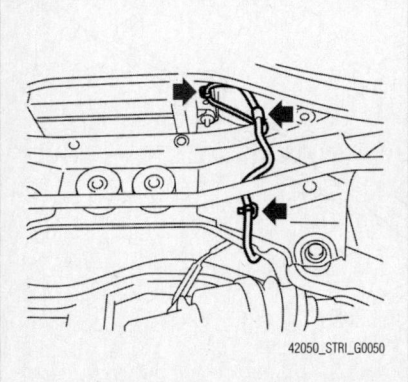

Fig. 153 Remove the clip holding the harness

❋❋ WARNING

A helper is required to perform this work.

16. Remove the fuel tank from the vehicle, as follows:
- Support the fuel tank with the transmission jack
- Remove the fuel tank band that holds the fuel tank
- Remove the fuel tank from the vehicle

To install:

17. Install the fuel tank to the vehicle.

❋❋ WARNING

A helper is required to perform this work.

- Lift the fuel tank with the transmission jack

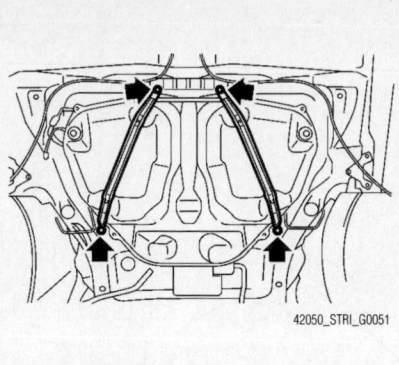

Fig. 154 Removing fuel tank bands

- Mount the fuel tank to the vehicle with a jack
- Install the fuel tank band that holds the fuel tank. Tighten to 25 ft. lbs. (33 Nm)

18. Connect the following hoses and connectors:
- Evaporation hose
- Fuel filler hose
- Connector

19. Install the rear suspension assembly.

20. Install the clip that holds the harness.

21. Lower the vehicle.

22. Install the rear wheels.

23. Connect the quick connector on fuel delivery hose.

24. Install the service hole cover of fuel sub level sensor.

25. Install the service hole cover of fuel pump:
- Push the grommet to install it to the second hole cover
- Tighten the bolt
- Connect the connector

26. Install the second row seats.

27. Install the fuse of fuel pump to main fuse box.

28. Connect the battery ground cable to the battery.

IDLE SPEED

ADJUSTMENT

Idle speed is maintained by the Powertrain Control Module (PCM). No adjustment is necessary or possible.

THROTTLE BODY

REMOVAL & INSTALLATION

See Figures 155 and 156.

1. Before servicing the vehicle, refer to the Precautions Section.

2. Remove the collector cover.

3. Disconnect the ground cable from the battery.

4. Remove the air intake chamber.

5. Disconnect the connectors from the throttle position sensor.

6. Disconnect the engine coolant hoses (A) from throttle body.

7. Remove the bolts (B) which secure throttle body to intake manifold.

To install:

8. Install in the reverse order of removal. Tighten the bolts to 72 inch lbs. (8 Nm).

➥Use new O-rings.

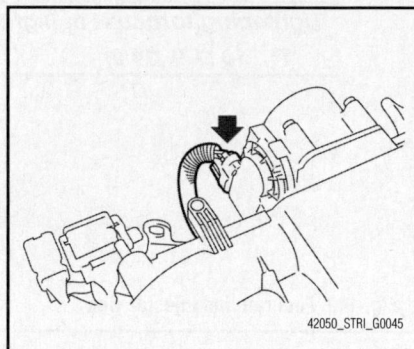

Fig. 155 Disconnect the connectors from the throttle position sensor

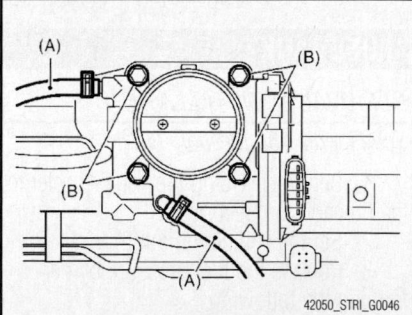

Fig. 156 Disconnecting engine coolant hoses from throttle body and removing bolts securing throttle body to the intake manifold

HEATING & AIR CONDITIONING SYSTEM

BLOWER MOTOR

REMOVAL & INSTALLATION

See Figure 157.

1. Before servicing the vehicle, refer to the Precautions Section.
2. Disconnect the ground cable from battery.
3. Remove the glove box lower cover.
4. Disconnect the blower motor electrical connector.
5. Loosen the screws to remove the blower motor.

To install:

6. Install the blower motor and mounting screws.
7. Reconnect the blower motor electrical connector
8. Install the glove box lower cover.
9. Connect the negative battery cable.

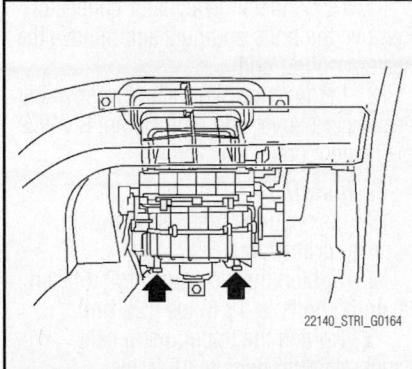

Fig. 157 Blower motor mounting screws shown

HEATER CORE

REMOVAL & INSTALLATION

See Figures 158 through 165.

➡ **Position the front wheels in the straight ahead position.**

1. Before servicing the vehicle, refer to the Precautions Section.
2. Disarm the SRS system. Wait at least 20 seconds before starting any repair work.

➡ **Air bag connectors are colored yellow. Do not use electrical equipment on these circuits. Be careful not to damage the airbag system harness when servicing.**

3. Disconnect the negative battery cable.

4. Drain the engine coolant. Discharge the air conditioning system.
5. Remove the bolts securing the expansion valve and pipe.
6. Disconnect and plug the heater hoses.
7. To remove the instrument panel upper:

- Remove the front door scuff plate and pillar lower trim
- Remove the left side instrument panel side cover
- Remove the clips, and remove the left side console lower panel
- Remove the clips, and remove the instrument panel lower undercover
- Remove the clips and hooks, disconnect the connectors, and then remove the instrument panel lower cover
- Remove the glove box
- Remove the front console panel
- Remove the instrument panel ornament and the driver's and passenger's side inner panels
- Remove the screws, and pull out the control panel. Disconnect the electrical harness and remove the control panel.
- Remove the instrument panel lower cover
- Remove the console ring indicator and front console panel cover. The ring indicator can be removed by inserting the clip remover tool or equivalent into the positions indicated by the arrows in the illustration. Remove the screws. Remove the bolts inside the upper pocket. Pull out the console lower pocket and remove the console box.
- Remove the screws and clips and remove the console side upper panel
- Remove the screws and pull out the front control panel
- Disconnect the harness connectors and remove the front control panel
- Remove the audio system retaining screws and partially pull the assembly forward. Disconnect the electrical connectors and antenna feed cable. Remove the radio assembly.
- Position the tilt steering in the lowest detent
- Remove the instrument cluster hood. Remove the instrument cluster retaining screws and pull the

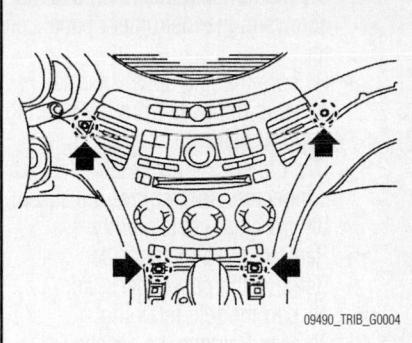

Fig. 158 Front control panel retaining clip locations

cluster forward. Disconnect the electrical connector and remove the assembly
- Lift the upper grille air vent and remove the catch on the hazard side switch
- Hold the side of the monitor and remove the catch on the upper right and left sides
- Disconnect the connectors and remove the upper air vent grille
- Remove the multifunction display or navigation monitor and warning box
- Remove the front upper pillar trim
- Remove the passenger's side air bag module retaining bolts
- Remove the screw in the center of the instrument cluster assembly

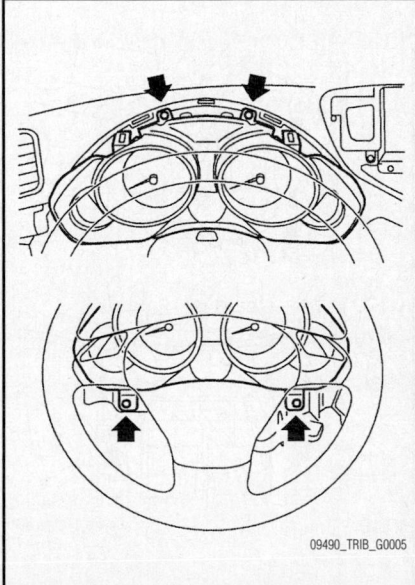

Fig. 159 Instrument cluster retaining screw locations

housing. Remove the retaining screws on the right and left side of the instrument pancl. Remove the screws in the instrument panel center
- Be sure that the upper instrument panel is removed from the steering support beam
- Disconnect the necessary electrical connectors and remove the assembly from the vehicle body
- Remove the air vent grille
- Remove the passenger's side air bag module retaining screws. Remove the air bag module.

8. To remove the steering support beam:

- Remove the knee guard plate retaining bolts, remove the knee guard plate

➡**The steering column may have to be removed from the vehicle for this procedure. Be sure the front wheels are in the straight ahead position.**

- Remove the screws on the side of the steering wheel. Disconnect the

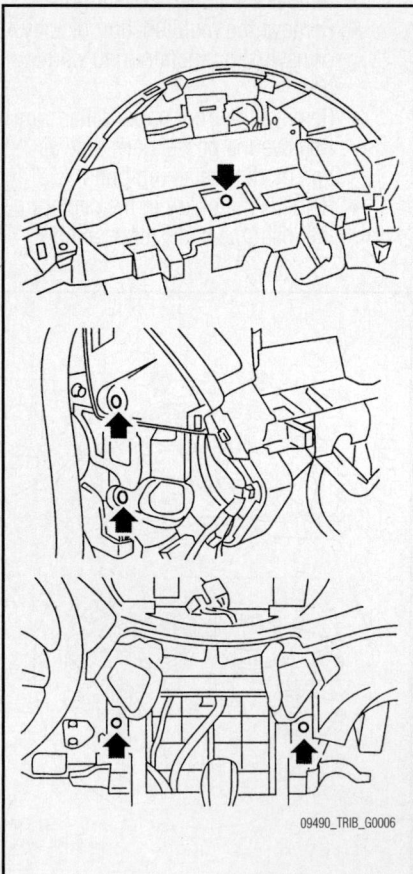

Fig. 160 Upper instrument panel retaining screw locations

horn from the harness. Disconnect the air bag connector and remove the air bag module.
- Place alignment marks on the steering wheel and the shaft. Remove the retaining nut. Using the proper tool, remove the steering wheel.
- Place alignment marks on the universal joint. Remove the joint bolt. Remove the joint.
- Disconnect the steering column electrical connectors. Remove the steering column retaining bolts. Pull the steering shaft assembly from the hole on the toe board.

➡**Be sure to remove the universal joint before removing the steering shaft assembly installing bolts when removing the steering shaft assembly or when lowering it for servicing other components.**

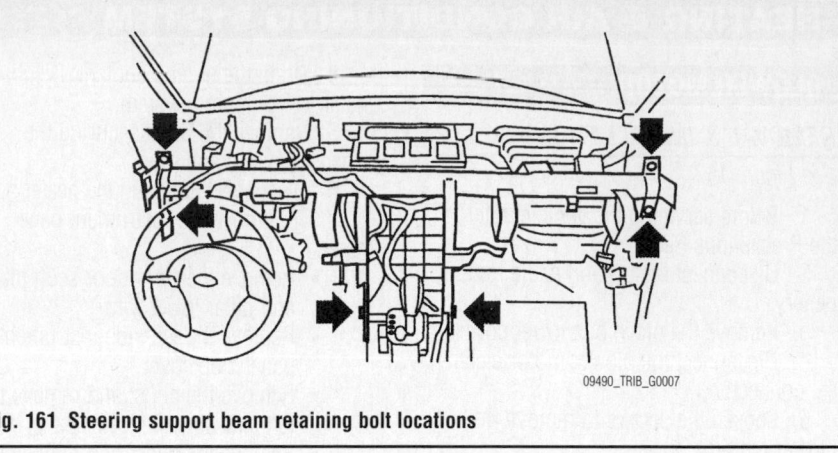

Fig. 161 Steering support beam retaining bolt locations

- Remove the bolts and remove the steering support beam

9. Disconnect the electrical connectors from the air conditioning control module, intake door actuator, blower motor, power transistor and blower resistor.

10. Lift up the floor mat. Loosen the bolt and nut and remove the blower motor assembly.

11. Disconnect the actuator connector. Remove the bolts and nuts and remove the heater/cooling unit.

12. Remove the screws and remove the heater core cover and pipe clamp. Remove the heater core.

To install:

13. Installation is the reverse of the removal procedure.

14. Tighten the steering support beam retaining bolts to 18 ft. lbs. (25 Nm).

15. Tighten the upper instrument panel retaining bolts to 18 ft. lbs. (25 Nm).

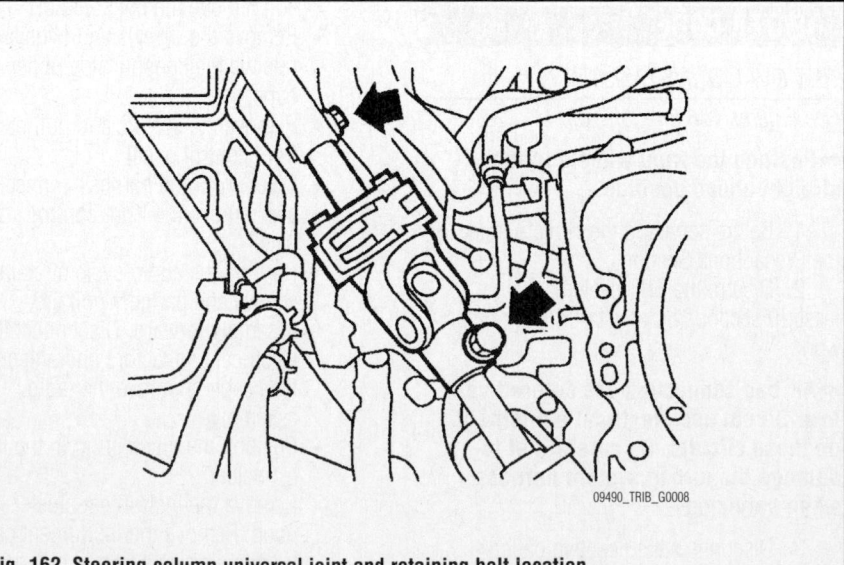

Fig. 162 Steering column universal joint and retaining bolt location

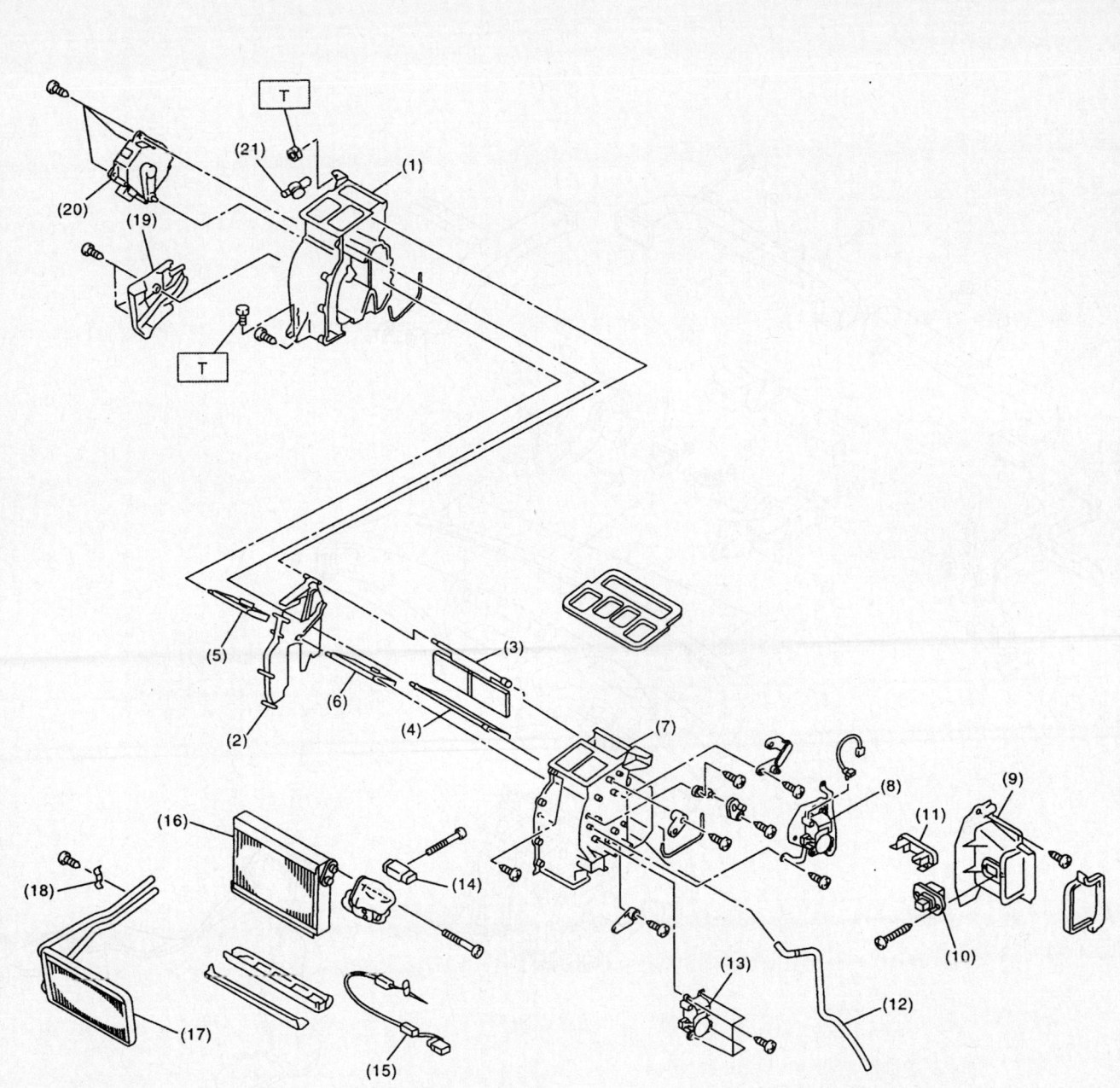

(1)	Heater unit case LH	(9)	Evaporator cover	(17)	Heater Core
(2)	Separator	(10)	Power transistor	(18)	Heater pipe clamp
(3)	Mode door RR	(11)	Pipe cover	(19)	Heater core cover
(4)	Mode door FR	(12)	Drain hose	(20)	Air mix door actuator LH
(5)	Air mix door LH	(13)	Air mix door actuator RH	(21)	Aspirator
(6)	Air mix door RH	(14)	Expansion valve		
(7)	Heater unit case RH	(15)	Evaporator sensor		
(8)	Mode door actuator	(16)	Evaporator		

Tightening torque: N·m (kgf-m, ft-lb)
T: 7.5 (0.76, 5.5)

09490_TRIB_G0009

Fig. 163 Heating and cooling unit and related components

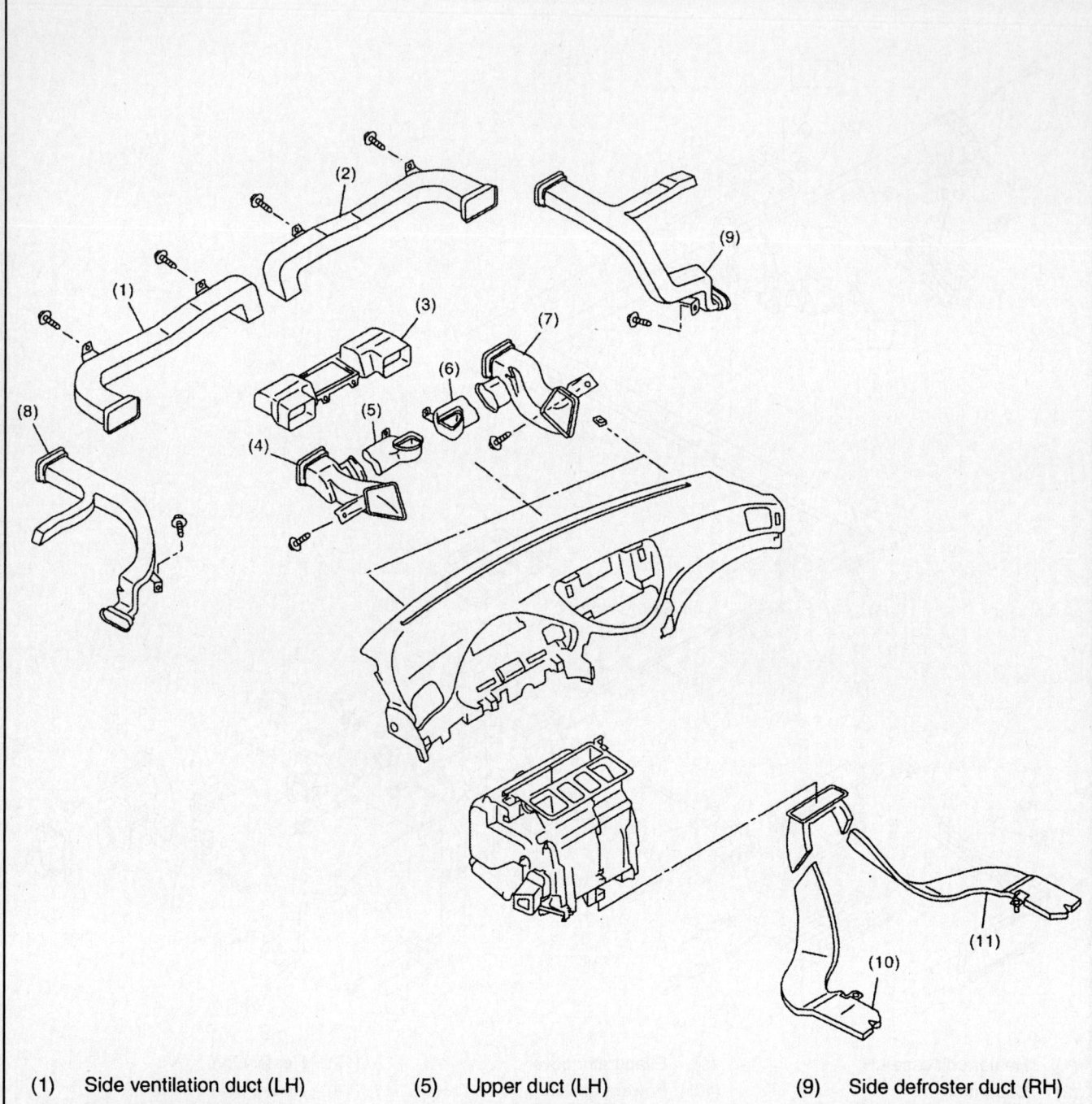

(1)	Side ventilation duct (LH)	(5)	Upper duct (LH)	(9)	Side defroster duct (RH)
(2)	Side ventilation duct (RH)	(6)	Upper duct (RH)	(10)	Rear heater duct (LH)
(3)	Center ventilation duct	(7)	Center duct (RH)	(11)	Rear heater duct (RH)
(4)	Center duct (LH)	(8)	Side defroster duct (LH)		

09490_TRIB_G0010

Fig. 164 Heater duct routing

16. When installing the steering column be sure to align the cutout portion at the serrated section on the column shaft and yoke, and then install the universal joint into the column shaft. Torque the bolt to 17.4 ft. lbs. (24 Nm).

17. Tighten the steering shaft to instrument panel retaining bolts to 18 ft. lbs. (25 Nm).

18. After installing the roll connector, be sure to check for proper adjustment:
- Check that the front wheels are in the straight ahead position
- Turn the roll connector pin (A) clockwise until it stops

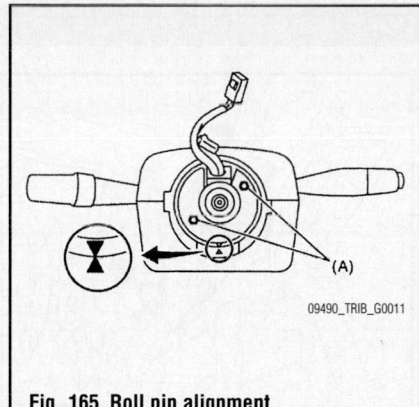

09490_TRIB_G0011

Fig. 165 Roll pin alignment

- Turn the roll connector pins (A) approximately 3.25 turns until the triangle marks are aligned

19. Tighten the steering wheel retaining nut to 33 ft. lbs. (45 Nm).

20. Refill the cooling system with the proper grade and type coolant.

21. Vacuum, leak test and recharge A/C system. For single A/C models refrigerant amount is (1.26 lbs–1.39 lbs.) For dual A/C models (1.85 lbs–1.98 lbs.)

22. Start the engine and check for leaks, correct as required.

AUXILIARY HEATING & AIR CONDITIONING SYSTEM

BLOWER MOTOR

REMOVAL & INSTALLATION

See Figure 166.

1. Before servicing the vehicle, refer to the Precautions Section.

2. Using the refrigerant recovery system, discharge the refrigerant.

3. Disconnect the ground cable from the battery.

4. Remove the left rear quarter trim.

5. Remove the bolts, then disconnect the rear tube.

6. Disconnect the harness connector.

7. Remove the nuts.

8. Remove the pipe bracket bolts and remove the bracket.

9. Remove the screws, and then remove the cooler unit.

10. Remove the blower motor mounting screws.

11. Remove blower motor.

12. Install in the reverse order of removal.

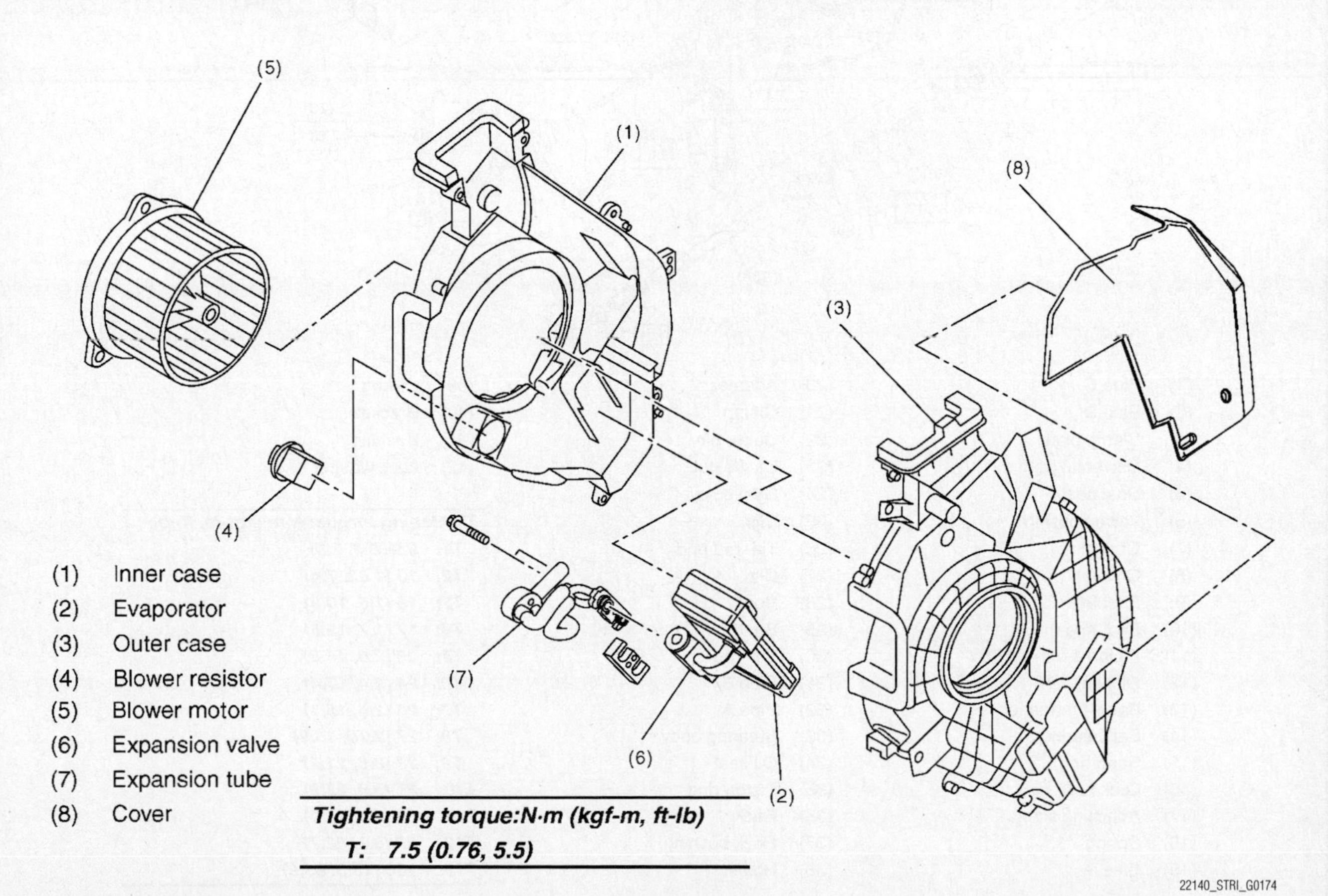

(1) Inner case
(2) Evaporator
(3) Outer case
(4) Blower resistor
(5) Blower motor
(6) Expansion valve
(7) Expansion tube
(8) Cover

Tightening torque:N·m (kgf-m, ft-lb)
   T: 7.5 (0.76, 5.5)

22140_STRI_G0174

Fig. 166 Exploded view of rear cooler unit

STEERING

POWER RACK & PINION STEERING GEAR

REMOVAL & INSTALLATION

See Figures 167 and 168.

1. Before servicing the vehicle, refer to the Precautions Section.
2. Disconnect the negative battery cable.
3. Loosen the front wheel nut.
4. Raise and support the vehicle safely.
5. Remove the tires and wheels.
6. Remove the undercover.

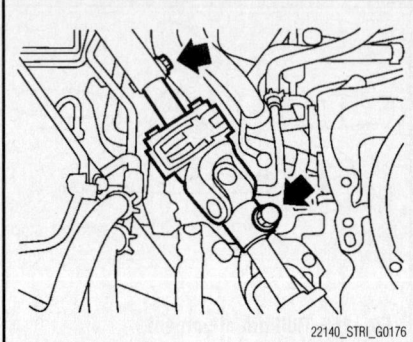

Fig. 167 Universal joint and mounting bolts view

7. Remove the front exhaust pipe assembly.
8. Remove the cotter pin and castle nut. Using a puller, remove the tie rod end from the knuckle arm.
9. Remove the front crossmember support plate. Remove the jack up plate. Remove the front stabilizer.
10. Disconnect the power steering fluid pipe at the center of the gearbox and attach a vinyl hose. Discharge the fluid into a suitable container by turning the steering wheel fully clockwise and counterclockwise. Disconnect the other fluid lines, and repeat the discharge procedure.

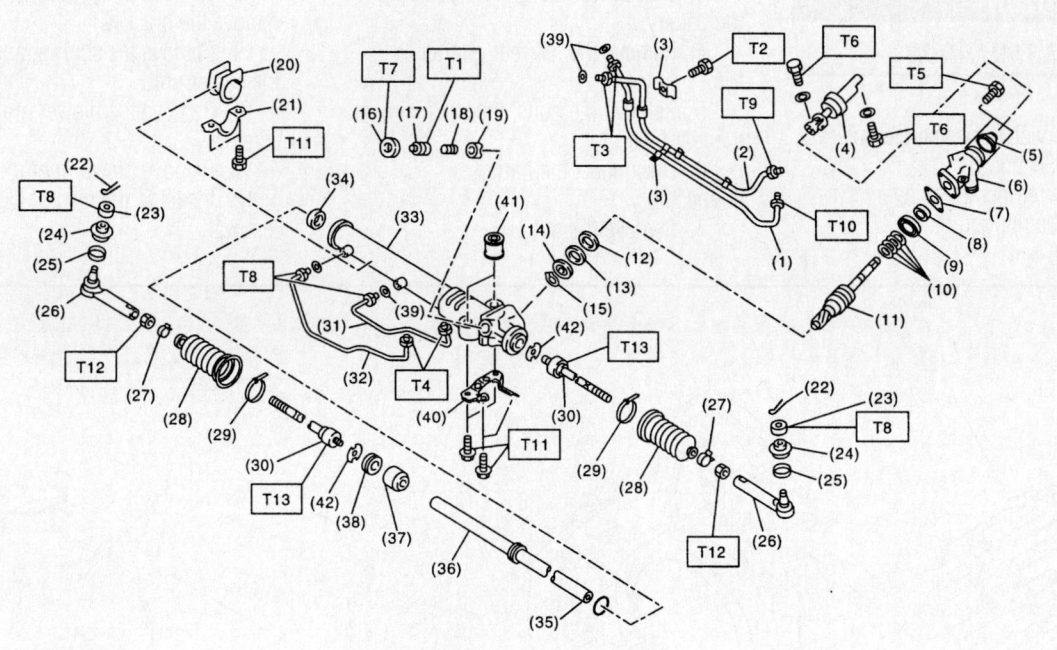

(1)	Pipe C	(20)	Adapter	(39) O–ring
(2)	Pipe D	(21)	Clamp	(40) Bracket
(3)	Clamp plate	(22)	Cotter pin	(41) Bushing
(4)	Universal joint	(23)	Castle nut	(42) Lock washer
(5)	Dust seal	(24)	Dust cover	
(6)	Valve housing	(25)	Clip	
(7)	Gasket	(26)	Tie–rod end	
(8)	Oil seal	(27)	Clip	
(9)	Bushing	(28)	Boot	
(10)	Seal ring	(29)	Band	
(11)	Pinion & valve ASSY	(30)	Tie–rod	
(12)	Oil seal	(31)	Pipe B	
(13)	Back–up washer	(32)	Pipe A	
(14)	Ball bearing	(33)	Steering body	
(15)	Snap ring	(34)	Oil seal	
(16)	Lock nut	(35)	Piston ring	
(17)	Adjusting screw	(36)	Rack	
(18)	Spring	(37)	Rack bushing	
(19)	Sleeve	(38)	Holder	

Tightening torque: N·m (kgf-m, ft-lb)

T1:	3.9 (0.4, 2.9)
T2:	10 (1.02, 7.4)
T3:	15 (1.5, 10.8)
T4:	17 (1.7, 12.5)
T5:	20 (2.0, 14.8)
T6:	24 (2.4, 17.4)
T7:	25 (2.5, 18.1)
T8:	27 (2.75, 19.9)
T9:	29 (3.0, 21.4)
T10:	37 (3.8, 27.3)
T11:	60 (6.1, 44.1)
T12:	85 (8.7, 62.7)
T13:	130 (13.3, 95.9)

Fig. 168 Power steering gear and related components

11. Remove the steering wheel. Make a matchmark on the universal joint. Remove the universal joint bolts and remove the joint from the vehicle.

12. Disconnect the fluid lines from the steering gear, pressure hose first.

13. Remove the steering gear clamp bolts and bracket securing the steering gear.

14. Remove the steering gear from the vehicle.

To install:

15. Insert the steering gear into the crossmember. Be careful not to damage the gearbox boot.

16. Tighten the steering gear to the crossmember bracket to 44 ft. lbs. (60 Nm).

17. Connect the fluid lines.

18. Align the cutout at the serrated section of the column shaft and yoke. Insert the universal joint into the column shaft.

19. Align the mating marks and insert the universal joint to serrated section of the steering gear assembly. Tighten the bolt to 18 ft. lbs. (25 Nm).

20. Continue the installation in the reverse order of the removal procedure.

21. Fill the power steering system with the proper grade and type fluid.

22. Start the engine and check for leaks. Correct as required.

POWER STEERING PUMP

REMOVAL & INSTALLATION

See Figure 169.

1. Before servicing the vehicle, refer to the Precautions Section.

2. Disconnect the ground cable from the battery.

3. Remove the air intake duct.

4. Remove the pulley belt cover.

5. Remove the drive belt.

6. Disconnect the connector from power steering pump switch.

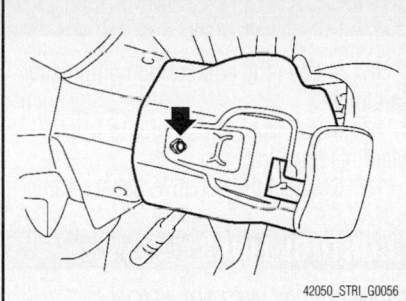

42050_STRI_G0056

Fig. 169 Disconnecting the pressure and suction hoses from the power steering pump

7. Disconnect the (2) pressure hose and (1) suction hose from power steering pump.

✴✴ CAUTION

Do not spill power steering fluid.

✴✴ CAUTION

To prevent foreign matter from entering the hose and pipe, cover the open ends with clean cloth.

8. Remove the installation bolt of the power steering pump bracket.

9. Place the oil pump bracket in a vise, and remove the two bolts from the front side of the oil pump.

✴✴ CAUTION

When securing the oil pump bracket in a vice, hold the oil pump bracket with the least possible force between two pieces of wood.

10. Remove the bolt from the rear side of oil pump.

11. Disassemble the oil pump and bracket by inserting
a flat tip screwdriver

To install

12. Install in the reverse order of removal.

13. Tighten oil pump to bracket bolts to 12 ft. lbs. (15.7 Nm).

14. Tighten the bolt from the rear side of oil pump to 27 ft. lbs. (36.8 Nm).

15. Tighten steering hose eye bolts to 30 ft. lbs. (40 Nm).

16. Replace the fluid and perform air purge.

✴✴ CAUTION

Never start the engine before filling with fluid, as doing so may cause the vane pump to seize.

BLEEDING

1. Before servicing the vehicle, refer to the Precautions Section.

2. Lift up the vehicle.

3. Add the specified fluid to reservoir tank at "MAX" level.

4. Turn the steering wheel slowly from lock to lock until the bubbles stop appearing on oil surface while keeping the fluid at the "MAX" level

➡**Normally bubbles will stop appearing after turning the steering wheel from lock to lock three times.**

5. In case bubbles do not stop appearing in the tank, or a grinding noise is generated from the oil pump leave it about half an hour and then do the step all over again.

6. Lower the vehicle, and then idle the engine.

7. Continue to turn the steering wheel from lock to lock until the bubbles stop appearing and change of the fluid level is within 0.12 inch (3 mm).

COIL SPRING

REMOVAL & INSTALLATION

See Figure 170.

1. Before servicing the vehicle, refer to the Precautions Section.
2. Disconnect the negative battery cable.
3. Raise and support the vehicle safely.
4. Remove front strut, Refer to front MacPherson Strut.
5. Using a coil spring compressor, compress the coil spring.
6. Using a strut mount socket remove the self-locking nut.
7. Remove the strut mount, spacer and upper spring seat from strut.
8. Gradually decrease the compression force of compressor, and remove the coil spring.
9. Remove the dust cover and helper spring.

To install:

10. Using a coil spring compressor, compress the coil spring.
11. Set the coil spring correctly so that its end face seats well in the spring seat as shown in the figure.
12. Install the helper and dust cover to the piston rod.
13. Pull the piston rod fully upward, and install the spring seat.
14. Install spacer and the strut mount to piston rod, and tighten a new self-locking nut temporarily.
15. Using a hexagon wrench to prevent strut rod from turning, tighten the new self-locking nut to 41 ft. lbs. (55 Nm), with a strut mount socket.

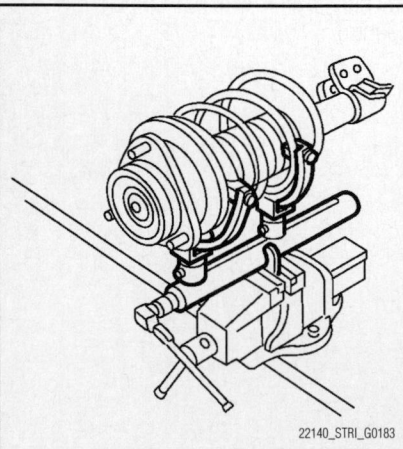

Fig. 170 Coil spring being compressed by spring compressor tool

16. Loosen the coil spring compressor carefully.
17. Install front strut, Refer to front MacPherson Strut.
18. Connect the negative battery cable.

CONTROL LINKS

REMOVAL & INSTALLATION

See Figure 171.

1. Before servicing the vehicle, refer to the Precautions Section.
2. Lift up the vehicle, and then remove the front wheels.
3. Remove the front undercover.
4. Remove the stabilizer control link upper and lower mounting nuts.

To install:

5. Install the stabilizer control link.
6. Tighten the stabilizer link upper and lower mounting nuts to 44 ft. lbs. (60 Nm).
7. Install front wheels, and lower the vehicle.

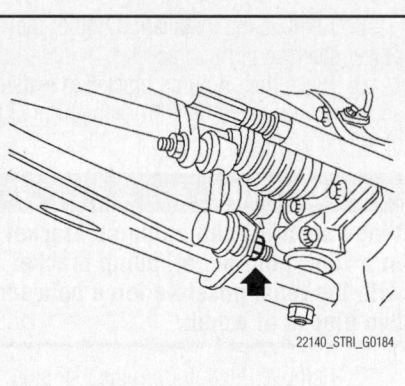

Fig. 171 Stabilizer control link and mounting nut shown

CROSSMEMBER

REMOVAL & INSTALLATION

See Figure 172.

✳✳ CAUTION

Disarm the SRS system. Wait at least 20 seconds before starting any repair work.

1. Before servicing the vehicle, refer to the Precautions Section.
2. Lift up the vehicle, and then remove the front wheels.
3. Remove the front undercover.
4. Remove the front crossmember support plate.

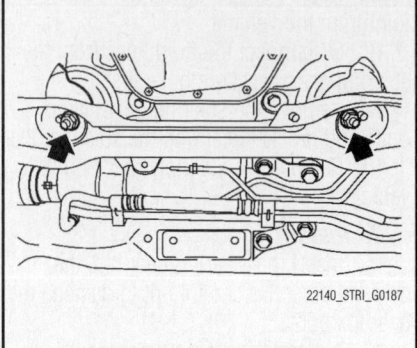

Fig. 172 Front crossmember engine mount nuts shown

5. Remove the front stabilizer.
6. Disconnect the tie-rod end from housing.
7. Remove the front arm.
8. Remove the nuts attaching the engine mount cushion rubber to crossmember.
9. Remove the steering universal joint.
10. Disconnect the power steering hose from steering gearbox.
11. Lift the engine approx. 0.39 inch (10 mm) using the chain block.
12. Support the crossmember with a jack, remove the bolts securing crossmember to body, and then gradually lower the crossmember with steering gearbox as a unit.

✳✳ CAUTION

When removing the crossmember downward, be careful that the tie-rod end does not interfere with drive shaft boot.

To install:

13. Install in the reverse order of removal.
14. Adhere to the following components specifications.
- Crossmember to body bolts to 70 ft. lbs. (95 Nm).
- Engine mounting to crossmember nuts to 62.7 ft. lbs. (85 Nm).
- Front arm to crossmember bolts to 70 ft. lbs. (95 Nm).
- Front arm to support plate bolts to 65 ft. lbs. (88 Nm).
- Support plate to body bolts to 111 ft. lbs. (150 Nm).
- Tie-rod end to housing to 20 ft. lbs. (27 Nm).
- Crossmember to body bolts to 70 ft. lbs. (95 Nm). After tightening to the specified torque, tighten the

castle nut further but within 60° until the hole in the ball stud is aligned with a slot in castle nut.

- Universal joint to 17 ft. lbs. (24 Nm).
- Stabilizer link to 44 ft. lbs. (60 Nm).
- Stabilizer clamp to 18 ft. lbs. (25 Nm).
- Power steering hose to Steering gearbox 11 ft. lbs. (15 Nm).

15. Purge air from the power steering system.

16. Inspect the wheel alignment and adjust if necessary.

LOWER BALL JOINT

REMOVAL & INSTALLATION

See Figure 173.

1. Before servicing the vehicle, refer to the Precautions Section.
2. Raise and support the vehicle safely.
3. Remove the tire and wheel.
4. Remove the stabilizer bar brackets and bushings (both sides).
5. Remove the cotter pin from the ball stud. Remove the castle nut. Extract the ball stud from the front arm.
6. Remove the bolt securing the ball joint to the housing. Extract the ball joint from the housing.

To install:

7. Installation is the reverse of the removal procedure.
8. Install the ball joint to the front arm and tighten the castle nut to 33.2 ft. lbs. (45 Nm). Tighten the castle nut an additional 60 degrees until the slot in the castle nut is aligned with the cotter pin hole in the ball joint.
9. Install the stabilizer bracket. Tighten to 18 ft. lbs. (25 Nm).
10. Always fully tighten the rubber bushings when the wheels are in full contact

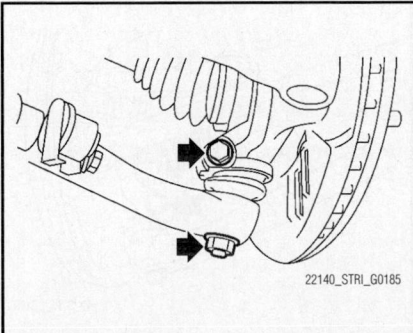

Fig. 173 Ball joint mounting nut and bolt shown

with the ground and the vehicle is at curb height.

11. Check and adjust alignment, as required.

LOWER CONTROL ARM

REMOVAL & INSTALLATION

1. Before servicing the vehicle, refer to the Precautions Section.
2. Raise and support the vehicle safely.
3. Remove the tire and wheel.
4. Remove the front crossmember support plate.
5. Remove the stabilizer bar.
6. Remove the ball joint from the front arm.
7. Remove the nut securing the front arm to the crossmember.

➡ **Do not remove the bolt.**

8. Remove the front arm support plate.
9. Remove the bolt securing the front arm to the crossmember and pull the front arm out of the crossmember.
10. To remove the stud bolt, use tool ST-20299AG020.

➡**Do not remove the stud bolt unless it is necessary. Always replace the removed parts with new ones.**

To install:

11. Installation is the reverse of the removal procedure.
12. Tighten the stud bolt to 81 ft. lbs. (110 Nm), if removed.
13. Tighten the support plate to front arm bolts to 107 ft. lbs. (145 Nm).
14. Tighten the support plate to body bolts to 111 ft. lbs. (150 Nm).
15. Always fully tighten the rubber bushings when the wheels are in full contact with the ground and the vehicle is at curb height.
16. Check and adjust alignment, as required.

CONTROL ARM BUSHING REPLACEMENT

See Figure 174.

1. Remove the control arm from the vehicle.
2. Mount the control arm in a soft jawed vise.
3. Use either a press or a control arm bushing fixture (C-clamp like tool) along with a slotted washer and a piece of pipe (slightly larger than the bushing) and press out the old bushing.
4. Clean the inside bushing contact surfaces of rust and old rubber.

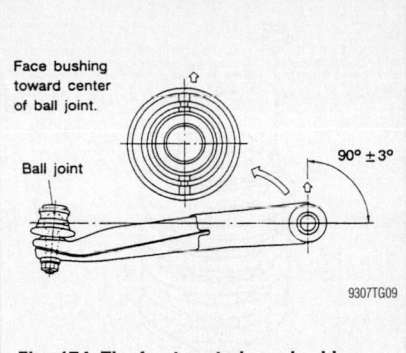

Fig. 174 The front control arm bushing must be installed in the proper direction

To install:

5. Apply a light coating of grease to both the replacement busing and bushing contact surfaces on the control arm.
6. Align the bushing.
7. Install the bushing using the press tool. A bushing install clamp can also be used to compress the bushing into the control arm.
8. Install the control arm on the vehicle.

MACPHERSON STRUT

REMOVAL & INSTALLATION

See Figure 175.

1. Before servicing the vehicle, refer to the Precautions Section.
2. Disconnect the negative battery cable.
3. Raise and support the vehicle safely.
4. Remove the tire and wheel.
5. Remove the bolt retaining the brake hose to the strut.
6. Make and alignment mark on the camber adjusting bolt which secures the strut to the housing.
7. Remove the clip retaining the ABS wheel speed sensor harness.
8. Remove the two bolts retaining the strut to the housing.

➡**While holding the head of the adjusting bolt, loosen the self locking nut.**

9. Remove the three upper strut retaining nuts.
10. Remove the strut from the vehicle.

To install:

11. Installation is the reverse of the removal procedure.
12. Tighten the upper retaining nuts to 15 ft. lbs. (20 Nm). Tighten the lower retaining bolts to 129 ft. lbs. (174.9 Nm)
13. Position the alignment mark on the camber adjusting bolt with the alignment mark on the lower side of the strut. Install using a new self locking nut.

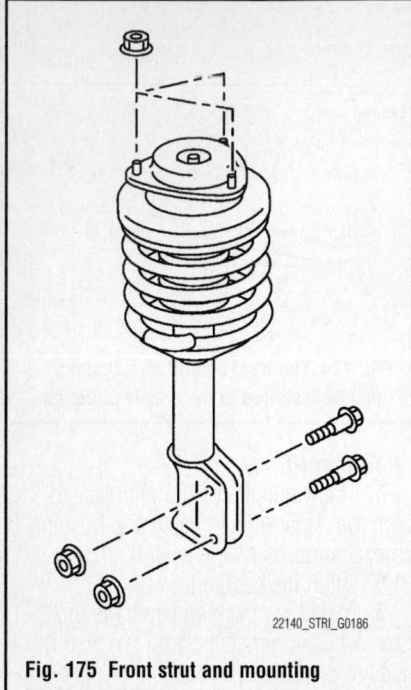

Fig. 175 Front strut and mounting hardware shown

➡ **While holding the head of the adjusting bolt, tighten the self locking nut.**

14. Check and adjust wheel alignment, as necessary.

OVERHAUL

See Figure 176.

1. Before servicing the vehicle, refer to the Precautions Section.

2. Remove the strut from the vehicle.

3. Using a coil spring compressor tool, carefully compress the spring. Remove the self locking nut.

4. Remove the strut mount, upper spring and rubber seat from the strut.

5. Gradually decrease the compression force of the spring compressor tool. Remove the coil spring.

6. Remove the dust cover and helper spring.

7. Check for the presence of air in the damping force generating mechanism.

8. Using the spring compression tool, compress the coil spring.

➡ **Be sure to properly install the coil spring.**

9. Position the coil spring so that its end face fits good into the spring seat.

10. Install the helper spring and dust cover to the piston rod.

11. Pull the piston rod fully upward, and install the rubber seat and spring seat.

12. Install the strut mount to the piston rod, and then tighten the self locking nut,

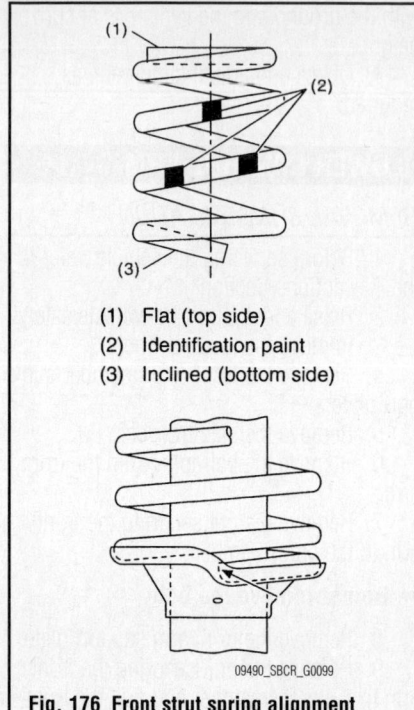

(1) Flat (top side)
(2) Identification paint
(3) Inclined (bottom side)

Fig. 176 Front strut spring alignment

temporarily. Be sure to use a new self locking nut.

13. Use a hexagon wrench to prevent the strut rod from turning. Tighten the self locking nut to 41 ft. lbs. (55 Nm).

14. Carefully loosen the coil spring.

STEERING KNUCKLE

REMOVAL & INSTALLATION

See Figures 177 through 180.

1. Before servicing the vehicle, refer to the Precautions Section.

2. Raise and support the vehicle safely.

3. Remove the tire and wheels.

4. Remove the front undercover.

5. Remove the stabilizer link.

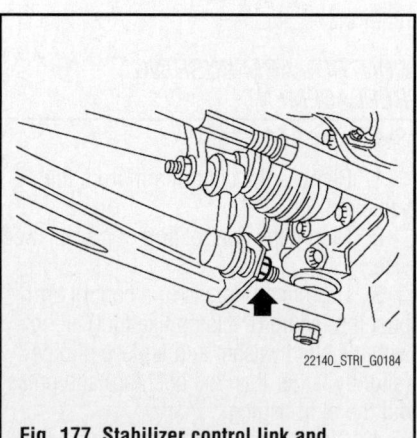

Fig. 177 Stabilizer control link and mounting nut shown

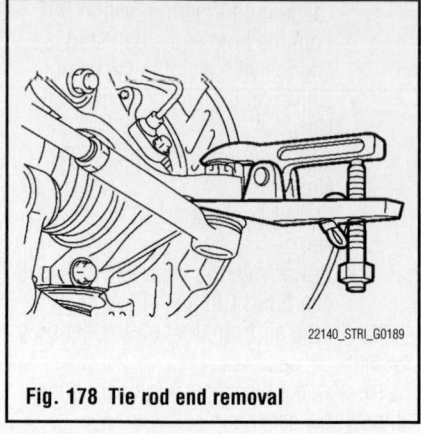

Fig. 178 Tie rod end removal

6. Remove the disc brake caliper from the front housing, and suspend it from strut using a piece of wire.

7. Remove the disc rotor from the hub.

8. Remove the cotter pin and castle nut securing the tie—rod end to the front housing knuckle arm.

9. Using a puller, remove the tie rod ball joint from knuckle arm.

10. Remove the ABS wheel speed sensor assembly and harness.

11. Remove the lower strut mounting bolts.

12. Remove the front arm ball joint from the front housing.

13. Remove the front drive shaft assembly from the steering knuckle hub.

14. After scribing an alignment mark on the camber adjusting bolt head, remove the bolts which connect the front housing and strut, and disconnect the steering knuckle from the strut.

15. Remove the front steering knuckle hub.

To install:

16. Install the front steering knuckle hub.

17. Align the alignment mark on the camber adjusting bolt head, and affix the

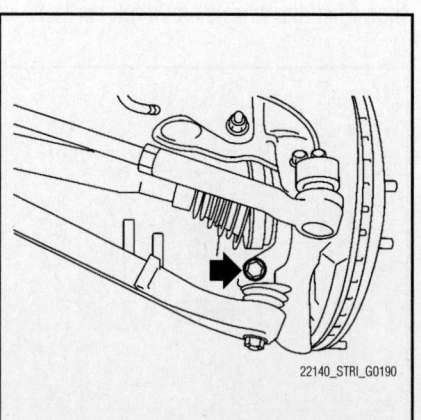

Fig. 179 Ball joint knuckle retaining bolt shown

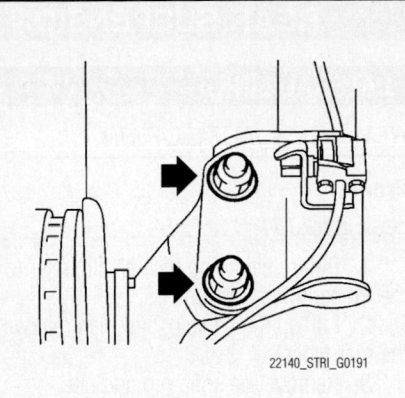

Fig. 180 Strut to knuckle retaining bolts shown

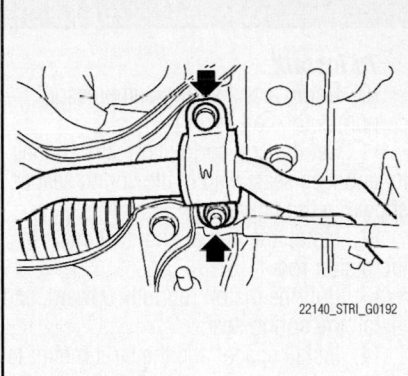

Fig. 181 Stabilizer bar bracket bolts shown

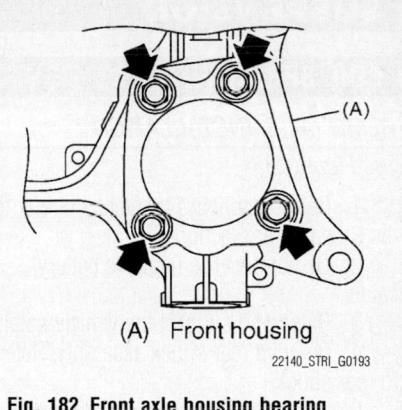

(A) Front housing

Fig. 182 Front axle housing bearing assembly retaining bolts

front housing and strut together using a new self—locking nut. Tighten to 129 ft. lbs. (175 Nm).

18. Install the front drive shaft.

19. Install the front arm ball joint to the front housing. Tighten mounting bolt to 37 ft. lbs. (50 Nm).

20. Install the ABS wheel speed sensor on the front housing. Tighten to 5.5 ft. lbs. (7.5 Nm).

21. Install the disc rotor to hub.

22. Install the disc brake caliper to the front housing. Tighten to 88 ft. lbs. (120 Nm).

23. Install the stabilizer link.

24. Connect the tie-rod end ball joint to the knuckle arm with a castle nut. Tighten to 20 ft. lbs. (27 Nm).

25. Tighten the castle nut to specified torque and tighten further within 60° until the pin hole is aligned with the slot in nut. Bend the cotter pin to lock.

26. While depressing the brake pedal, tighten a new axle nut to 177 ft. lbs. (240 Nm). and lock it securely.

27. Install the tire and wheels.

28. Lower the vehicle.

29. Inspect the wheel alignment and adjust if necessary.

STABILIZER BAR

REMOVAL & INSTALLATION

See Figures 177 and 181.

1. Before servicing the vehicle, refer to the Precautions Section.

2. Raise and support the vehicle safely.

3. Remove the tire and wheels.

4. Remove the front undercover.

5. Remove the front crossmember support plate.

6. Remove the stabilizer bar link.

7. Remove the stabilizer bar bracket bolts and bushings.

8. Remove the stabilizer bar from the vehicle.

To install:

9. Installation is the reverse of the removal procedure.

10. Be sure to use new self locking nuts, as required.

11. Install the rubber bushing, so that the paint mark on the stabilizer bar is on the left side of the vehicle.

12. Install the stabilizer bushing (front crossmember side), while aligning it with the paint mark on the stabilizer bar.

13. Tighten the stabilizer link bolts to 44 ft. lbs. (60 Nm). Tighten the stabilizer bar clamp to 18 ft. lbs. (25 Nm).

14. Always fully tighten the rubber bushings when the wheels are in full contact with the ground and the vehicle is at curb height.

WHEEL BEARINGS

REMOVAL & INSTALLATION

See Figure 182.

1. Before servicing the vehicle, refer to the Precautions Section.

2. Disconnect the ground cable from the battery.

3. Lift up the vehicle, and remove the front wheels.

4. Lift the crimped section of axle nut.

5. Remove the axle nut using a socket wrench while depressing the brake pedal.

6. Remove the disc brake caliper from the front housing, and suspend it from strut using a piece of wire.

7. Remove the disc rotor from the hub.

8. Remove four bolts from the front housing.

9. Remove the front hub unit bearing. If it is hard to remove, use a puller.

To install:

10. Place the disc cover between front housing and front hub unit, and tighten the four bolts to 48 ft. lbs. (65 Nm).

11. Tighten the axle nut temporarily.

12. Install the disc rotor to hub.

13. Install the disc brake caliper to the front housing and tighten mounting bolts to 88 ft. lbs. (120 Nm).

14. While depressing the brake pedal, tighten a new axle nut to 177 ft. lbs. (240 Nm). and lock it securely.

15. Install the tire and wheels.

16. Lower the vehicle.

17. Inspect the wheel alignment and adjust if necessary.

ADJUSTMENT

The wheel bearing is of a sealed design, and no adjustment is possible.

SUSPENSION

REAR SUSPENSION

COIL SPRING

REMOVAL & INSTALLATION

See Figure 183.

1. Before servicing the vehicle, refer to the Precautions Section.
2. Disconnect the negative battery cable.
3. Raise and support the vehicle safely.
4. Remove rear shock assembly, Refer to rear shock.
5. Using a coil spring compressor, compress the coil spring.
6. Using a strut mount socket remove the self-locking nut.
7. Remove the strut mount, spacer and upper spring seat from strut.
8. Gradually decrease the compression force of compressor, and remove the coil spring.
9. Remove the dust cover and helper spring.

To install:

10. Using a coil spring compressor, compress the coil spring.
11. Set the coil spring correctly so that its end face seats well in the spring seat as shown in the figure.
12. Install the helper and dust cover to the piston rod.
13. Pull the piston rod fully upward, and install the spring seat.
14. Install spacer and the strut mount to piston rod, and tighten a new self—locking nut temporarily.
15. Using a hexagon wrench to prevent strut rod from turning, tighten the new self-locking nut to 26 ft. lbs. (35 Nm), with a strut mount socket.
16. Loosen the coil spring compressor carefully.
17. Install rear shock assembly, Refer to rear shock.
18. Connect the negative battery cable.

LATERAL LINK

REMOVAL & INSTALLATION

Front

See Figure 184.

1. Before servicing the vehicle, refer to the Precautions Section.
2. Lift-up the vehicle, and then remove the rear wheels.
3. Remove the snap pin and nut.
4. Use a puller to detach the ball joint.
5. Scribe an alignment mark on the front lateral link adjustment bolt and the rear sub frame.

> ❋❋ **CAUTION**
>
> **When removing the adjusting bolt, loosen the nut with the bolt head secured.**

6. Remove the adjusting bolt, and then remove the front lateral link.

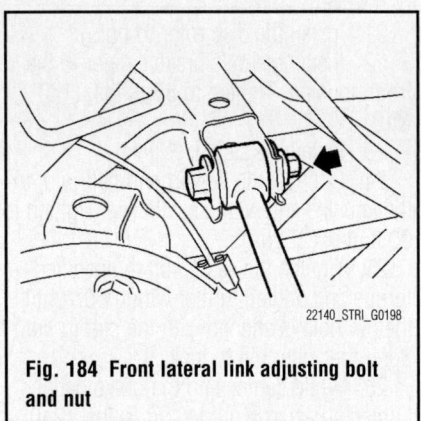

22140_STRI_G0198

Fig. 184 Front lateral link adjusting bolt and nut

To install:

7. Installation is the reverse of the removal procedure.
8. Be sure to use new bolts and nuts, as required.
9. Always fully tighten the rubber bushings when the wheels are in full contact with the ground and the vehicle is at curb height.
10. Tighten the front lateral link to sub frame to 89 ft. lbs. (120 Nm).
11. Tighten the front lateral link to rear axle housing to 20 ft. lbs. (27 Nm).
12. Check and adjust the wheel alignment, as necessary.

Rear

See Figure 185.

1. Before servicing the vehicle, refer to the Precautions Section.

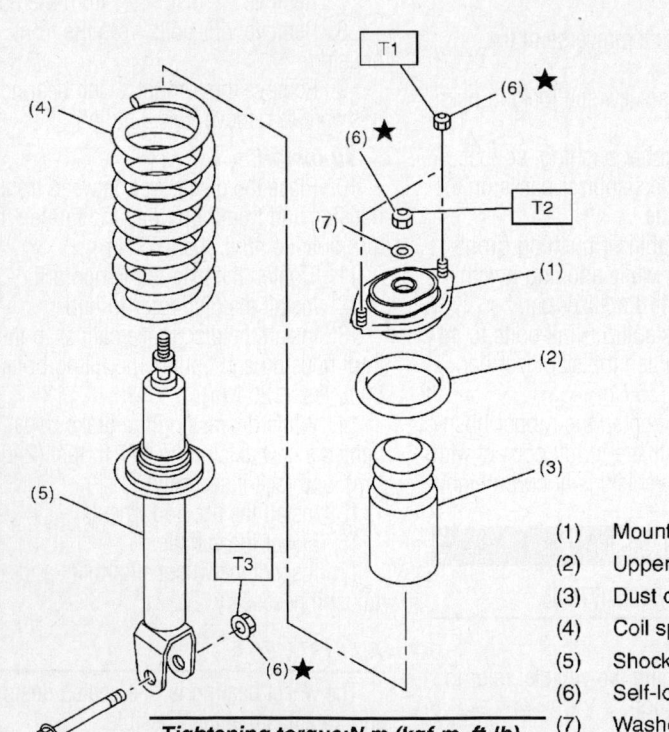

(1)	Mount
(2)	Upper rubber sheet
(3)	Dust cover
(4)	Coil spring
(5)	Shock absorber
(6)	Self-locking nut
(7)	Washer

Tightening torque:N·m (kgf-m, ft-lb)
T1: 30 (3.1, 22.4)
T2: 35 (3.6, 26)
T3: 120 (12.2, 89)

22140_STRI_G0194

Fig. 183 Exploded view of rear shock assembly

Fig. 185 Rear lateral link removal points shown

2. Lift-up the vehicle, and then remove the rear wheels.

3. Remove the nut and detach the stabilizer link.

4. Remove the bolts on the bottom side of the shock absorber.

5. Remove the bolts and remove the rear lateral link.

To install:

6. Installation is the reverse of the removal procedure.

7. Be sure to use new bolts and nuts, as required.

8. Always fully tighten the rubber bushings when the wheels are in full contact with the ground and the vehicle is at curb height.

9. Tighten the rear lateral link to 89 ft. lbs. (120 Nm).

10. Tighten the shock absorber to 89 ft. lbs. (120 Nm).

11. Tighten the stabilizer link to 44 ft. lbs. (60 Nm).

12. Check and adjust the wheel alignment, as necessary.

REAR TRAILING LINK

REMOVAL & INSTALLATION

See Figure 186.

1. Before servicing the vehicle, refer to the Precautions Section.

2. Lift-up the vehicle, and then remove the rear wheels.

3. Remove the bracket, and remove the parking brake cable from the guide.

4. Remove the ABS wheel speed sensor harness from the trailing link

5. Remove the trailing link

To install:

6. Installation is the reverse of the removal procedure, noting the following:

   a. Be sure to use new bolts and nuts, as required.

   b. Always fully tighten the rubber bushings when the wheels are in full

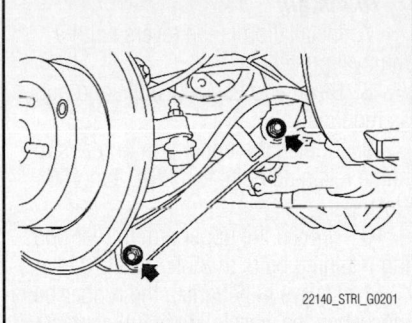

Fig. 186 Rear trailing link removal points shown

contact with the ground and the vehicle is at curb height.

   c. Tighten the trailing link mounting bolts to 59 ft. lbs. (80 Nm).

   d. Tighten the parking brake cable bracket bolt to 24 ft. lbs. (33 Nm).

   e. Check and adjust the wheel alignment, as necessary.

SHOCK ABSORBER

REMOVAL & INSTALLATION

See Figure 187.

1. Before servicing the vehicle, refer to the Precautions Section.

2. Remove the strut cap (of the quarter trim).

3. Loosen the rear wheel lug nuts.

4. Raise and support the vehicle safely.

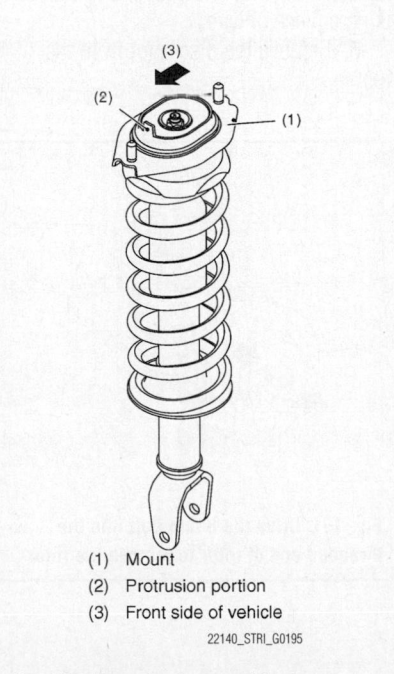

(1) Mount
(2) Protrusion portion
(3) Front side of vehicle

22140_STRI_G0195

Fig. 187 Rear shock assembly view

5. Remove the tire and wheel.

6. Remove the nut and detach the rear stabilizer link.

7. Using a jack, support the shock absorber.

8. Remove the bolts on the bottom of the shock absorber.

9. Detach the rear lateral link.

10. Remove the nuts that retain the shock absorber mount to the vehicle.

11. Remove the shock absorber from the vehicle.

To install:

12. Installation is the reverse of the removal procedure.

13. Be sure to use new bolts and nuts, as required.

14. Always fully tighten the rubber bushings when the wheels are in full contact with the ground and the vehicle is at curb height.

15. Check and adjust the wheel alignment, as necessary.

STABILIZER BAR

REMOVAL & INSTALLATION

See Figures 188 and 189.

1. Before servicing the vehicle, refer to the Precautions Section.

2. Loosen the rear wheel lug nuts.

3. Raise and support the vehicle safely.

4. Remove the tire and wheel.

5. Remove the stabilizer link.

6. Remove the clamp and bushing bolts which secure the stabilizer bar. Remove the clamps and bushings.

7. Remove the stabilizer bar from the vehicle.

To install:

8. Installation is the reverse of the removal procedure.

9. Be sure that the stabilizer bar and the bushings have the same markings and color.

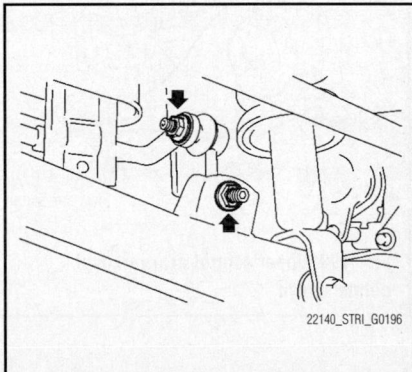

Fig. 188 Stabilizer link retaining nuts shown

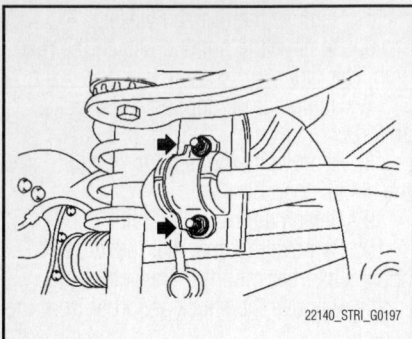

Fig. 189 Stabilizer bar retaining nuts shown

10. Be sure to use new bolts and nuts, as required.

11. Tighten the stabilizer link retaining bolts to 44 ft. lbs (60 Nm).

12. Tighten the stabilizer clamp retaining bolts to 28 ft. lbs (38 Nm).

13. Always fully tighten the rubber bushings when the wheels are in full contact with the ground and the vehicle is at curb height.

14. Check and adjust the wheel alignment, as necessary.

UPPER CONTROL ARM

REMOVAL & INSTALLATION

See Figure 190.

1. Before servicing the vehicle, refer to the Precautions Section.

2. Loosen the wheel nuts.

3. Raise and support the vehicle safely.

4. Remove the tire and wheel.

5. As required, support the assembly before removing the upper control arm.

6. Remove the bolts. Remove the upper control arm from the vehicle.

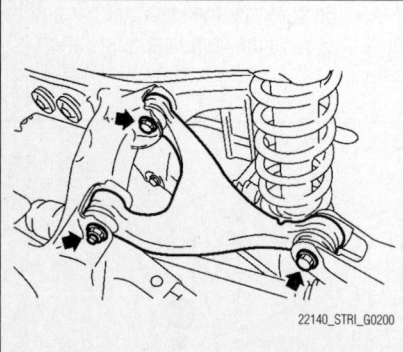

Fig. 190 Upper control arm removal points shown

To install:

7. Installation is the reverse of the removal procedure.

8. Be sure to use new bolts and nuts, as required.

9. Tighten the upper arm to rear sub frame retaining bolts to 111 ft. lbs (150 Nm).

10. Tighten the upper arm to rear housing retaining bolts to 89 ft. lbs (120 Nm).

11. Always fully tighten the rubber bushings when the wheels are in full contact with the ground and the vehicle is at curb height.

12. Check and adjust the wheel alignment, as necessary.

WHEEL HUB & BEARING

REMOVAL & INSTALLATION

See Figure 191.

1. Disconnect the ground cable from the battery.

2. Lift-up the vehicle, and remove the front wheels.

3. Lift the crimped section of axle nut.

4. Remove the axle nut using a socket wrench while depressing the brake pedal.

✳✳ CAUTION

Remove the wheel before loosening the axle nut. Failure to follow this rule may damage the wheel bearings.

5. Remove the disc brake caliper from the front housing, and suspend it from strut using a piece of wire.

6. Remove the disc rotor from the hub.

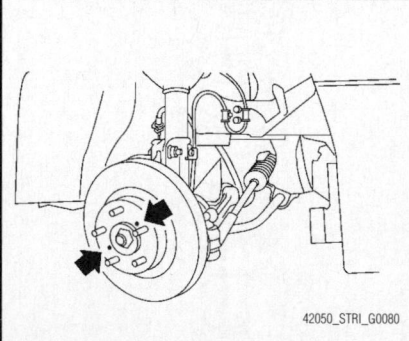

Fig. 191 Drive the 8 mm bolt into the threaded end of rotor to remove the rotor

✳✳ CAUTION

If it is difficult to remove the disc rotor from the hub, drive the 8 mm of bolt into the threaded end of rotor, then remove the rotor.

7. Remove four bolts from the front housing.

✳✳ CAUTION

Do not get close to the tool which charged magnetism to magnetic encoder.

✳✳ CAUTION

Be careful not to damage the magnetic encoder.

8. Remove the front hub unit bearing. If it is hard to remove, use the ST.

To install:

9. Place the disc cover between front housing and front hub unit, and tighten the four bolts. Tighten to 47.9 ft. lbs. (65 Nm).

10. Install the front drive shaft.

11. Tighten the axle nut temporarily.

12. Install the disc rotor to hub.

13. Install the disc brake caliper to the front housing. Tighten to 89 ft. lbs. (120 Nm).

14. While depressing the brake pedal, tighten a new axle nut (olive color) to the specified torque and lock it securely. Tighten to 177 ft. lbs. (240 Nm).

✳✳ CAUTION

Install the wheel after installation of axle nut. Failure to follow this rule may damage the wheel bearing.

✳✳ CAUTION

Be sure to tighten the axle nut to specified torque. Do not over tighten it as this may damage the wheel bearing.

15. After tightening the axle nut, lock it securely.

16. Install the wheel and tighten the wheel nuts to specified torque. Tighten to 81 ft. lbs. (110 Nm).

ADJUSTMENT

The wheel bearing is of a sealed design, and no adjustment is possible.

DIAGNOSTIC TROUBLE CODES

OBD II VEHICLE APPLICATIONS

SUBARU

Forester
2009-2010
- 2.5L I4 MFI (SOHC) . . . . VIN 6
- 2.5L I4 MFI (DOHC) . . . . VIN 6

Impreza
2009-2010
- 2.5L I4 MFI (SOHC) . . . . VIN 7
- 2.5L I4 MFI (DOHC) . . . . VIN 6

Legacy
2009-2010
- 2.5L I4 MFI (SOHC) . . . . VIN 6
- 2.5L I4 MFI (DOHC) . . . . VIN 6
- 2.5L I4 MFI (DOHC) . . . . VIN 7
- 3.0L V6 MFI . . . . VIN 8
- 3.0L V6 MFI . . . . VIN 9

Outback
2009-2010
- 2.5L I4 MFI (SOHC) . . . . VIN 6
- 2.5L I4 MFI (DOHC) . . . . VIN 6

- 2.5L I4 MFI (DOHC) . . . . VIN 7
- 3.0L V6 MFI . . . . VIN 8
- 3.0L V6 MFI . . . . VIN 9

Tribeca
2009-2010
- 3.6L V6 MFI . . . . VIN 9

OBD II Trouble Code List (P0XXX Codes)

DTC	Trouble Code Title, Conditions & Possible Causes

DTC: P0011
2T ECM, MIL: Yes
Year: 2009, 2010
Model: Forester, Impreza, Legacy, Outback, Tribeca
Engine: 2.5L H4, 3.0L H6, 3.6L H6
Transmission: All
	Intake Camshaft Position - Timing Over-Advanced (Bank 1):
	Time of establishing all secondary parameter conditions $\geq$ 3000 ms
	Battery voltage $\geq$ 10.9 V
	Engine speed $\geq$ 1300 rpm
	Engine coolant temperature $\geq$ 60 °C (140 °F)
	AVCS control - Operates
	Target timing advance change amount (per 64 ms) - < 1.07 °CA
	The amount of AVCS actual timing advance does not approach to the amount of AVCS target timing advance.
	Possible Causes:
	   • Intake camshaft damaged
	   • Timing chain problem
	   • Oil pipe clog
	   • Clogged oil flow control solenoid valve

DTC: P0014
2T ECM, MIL: Yes
Year: 2009, 2010
Model: Outback, Tribeca
Engine: 3.6L H6
Transmission: All
	Exhaust AVCS System 1 (Range/Performance):
	Engine stalls.
	Improper idling.
	The amount of exhaust AVCS actual timing advance does not approach to the amount of exhaust AVCS target timing advance
	Possible Causes:
	   • Oil pipe clog
	   • Oil flow control solenoid valve clogged
	   • Exhaust camshaft dirty or damaged

DTC: P0016
2T ECM, MIL: Yes
Year: 2009, 2010
Model: Forester, Impreza, Legacy, Outback
Engine: 2.5L H4, 3.0L H6
Transmission: All
	Crankshaft Position - Camshaft Position Correlation (Bank 1):
	Battery voltage $\geq$ 10.9 V
	Engine speed $\geq$ 500 rpm
	Engine coolant temperature $\geq$ 60 °C (140 °F)
	AVCS control - Not in operation
	Target timing advance - 0°CA
	Engine stalls, Improper idling
	Possible Causes:
	   • Oil pipe clog
	   • Oil flow control solenoid valve clogged
	   • Intake camshaft dirty or damaged
	   • Timing belt problem

DTC: P0017
2T ECM, MIL: Yes
Year: 2009, 2010
Model: Outback, Tribeca
Engine: 3.6L H6
Transmission: All
	Crank and Cam Timing B System Failure (Bank 1):
	Engine stalls.
	Improper idling.
	Standard timing advance amount is far from learning angle.
	Possible Causes:
	   • Oil pipe clog
	   • Oil flow control solenoid valve clogged
	   • Exhaust camshaft dirty or damaged
	   • Timing belt problem

DTC: P0018
2T ECM, MIL: Yes
Year: 2009, 2010
Model: Forester, Impreza, Legacy, Outback, Tribeca
Engine: 2.5L H4, 3.0L H6, 3.6L H6
Transmission: All
	Crankshaft Position - Camshaft Position Correlation (Bank 2):
	Battery voltage $\geq$ 10.9 V
	Engine speed $\geq$ 500 rpm
	Engine coolant temperature $\geq$ 60 °C (140 °F)
	AVCS control - Not in operation
	Target timing advance - 0°CA
	Engine stalls, Improper idling
	Possible Causes:
	   • Oil pipe clog
	   • Oil flow control solenoid valve clogged
	   • Intake camshaft dirty or damaged
	   • Timing belt problem

DTC	Trouble Code Title, Conditions & Possible Causes
DTC: P0019 **2T ECM, MIL: Yes** **Year:** 2009, 2010 **Model:** Outback, Tribeca **Engine:** 3.6L H6 **Transmission:** All	**Crank and Cam Timing B System Failure (Bank 2):** Engine stalls. Improper idling. Standard timing advance amount is far from learning angle. **Possible Causes:** • Oil pipe clog • Oil flow control solenoid valve clogged • Exhaust camshaft dirty or damaged • Timing belt problem
DTC: P0021 **2T ECM, MIL: Yes** **Year:** 2009, 2010 **Model:** Forester, Impreza, Legacy, Outback, Tribeca **Engine:** 2.5L H4, 3.0L H6, 3.6L H6 **Transmission:** All	**Intake Camshaft Position - Timing Over-Advanced (Bank 2):** Time of establishing all secondary parameter conditions $\geq 3000\,ms$ Battery voltage $\geq 10.9\,V$ Engine speed $\geq 1300\,rpm$ Engine coolant temperature $\geq 60\,°C$ (140 °F) AVCS control – In operation Target timing advance change amount (per 64 ms) $< 1.07\,°CA$ Engine stalls or improper idling. **Possible Causes:** • Oil pipe clog • Oil flow control solenoid valve clogged • Intake camshaft dirty or damaged • Timing belt problem
DTC: P0024 **2T ECM, MIL: Yes** **Year:** 2009, 2010 **Model:** Outback, Tribeca **Engine:** 3.6L H6 **Transmission:** All	**Exhaust AVCS System 2 (Range/Performance):** Engine stalls, Improper idling. Amount of exhaust AVCS actual timing advance does not approach to the amount of exhaust AVCS target timing advance. **Possible Causes:** • Oil pipe clog • Oil flow control solenoid valve clogged • Exhaust camshaft dirty or damaged
DTC: P0026 **1T ECM, MIL: Yes** **Year:** 2009 **Model:** Legacy, Outback **Engine:** 3.0L H6 **Transmission:** All	**Intake Valve Control Solenoid Circuit Range/Performance (Bank 1):** Improper idling. Battery voltage $\geq 10.9\,V.$ After engine starting $\geq 6000\,ms.$ Engine oil temperature $\geq 0\,°C$ (32 °F). **Possible Causes:** • Wiring harness between the ECM and variable valve lift diagnosis oil pressure switch needs repair. • Defective oil switching solenoid valve.
DTC: P0026 **1T ECM, MIL: Yes** **Year:** 2009 **Model:** Forester **Engine:** 2.5L H4 **Transmission:** All	**Intake Valve Control Solenoid Circuit Range (Bank 1):** Battery voltage $\geq 10.9\,V$ After engine starting $\geq 6000\,ms$ Engine oil temperature $\geq 15\,°C$ (59 °F) Variable valve lift control - In operation Improper idling **Possible Causes:** • Bad wiring harness • Open or short circuit • Damaged oil switching solenoid valve
DTC: P0028 **1T ECM, MIL: Yes** **Year:** 2009 **Model:** Forester **Engine:** 2.5L H4 **Transmission:** All	**Intake Valve Control Solenoid Circuit (Bank 2):** Battery voltage $\geq 10.9\,V$ After engine starting $\geq 6000\,ms$ Engine oil temperature $\geq 15\,°C$ (59 °F) Variable valve lift control - In operation Improper idling **Possible Causes:** • Bad wiring harness • Open or short circuit • Damaged oil switching solenoid valve.

DTC	Trouble Code Title, Conditions & Possible Causes
DTC: P0028 **1T ECM, MIL:** Yes **Year:** 2009 **Model:** Legacy, Outback **Engine:** 3.0L H6 **Transmission:** All	**Intake Valve Control Solenoid Circuit Range/Performance (Bank 2):** Improper idling. Battery voltage ≥ 10.9 V. After engine starting ≥ 6000 ms. Engine oil temperature ≥ 0 °C (32 °F). **Possible Causes:** • Wiring harness between the ECM and variable valve lift diagnosis oil pressure switch needs repair. • Defective oil switching solenoid valve.
DTC: P0030 **2T ECM, MIL:** Yes **Year:** 2009, 2010 **Model:** Forester, Impreza, Legacy, Outback, Tribeca **Engine:** 2.5L H4, 3.0L H6, 3.6L H6 **Transmission:** All	**HO2S Heater Control Circuit (Bank 1 Sensor 1):** Poor drivability Condition established time ≥ 42000 ms Battery voltage ≥ 10.9 V Heater current - Permitted Control duty ≥ 35 % - Experienced After fuel cut - ≥ 20000 ms **Possible Causes:** • Bad wiring harness between ECM and front oxygen sensor connector • Defective front oxygen sensor • Poor contact in ECM or front oxygen sensor connector
DTC: P0031 **1T ECM, MIL:** Yes **Year:** 2009, 2010 **Model:** Forester, Impreza, Legacy, Outback, Tribeca **Engine:** 2.5L H4, 3.0L H6, 3.6L H6 **Transmission:** All	**HO2S Heater Control Circuit Low (Bank 1 Sensor 1):** Low voltage condition on the Front Oxygen (Air Fuel) sensor heater control circuit. Battery Voltage ≥ 10.9 V **Possible Causes:** • Power supply line or main relay needs repair or replacement. • Wiring harness between ECM and front oxygen sensor needs repair. • Wiring harness and connector between ECM and chassis ground needs repair. • Front oxygen sensor needs replacement.
DTC: P0032 **1T ECM, MIL:** Yes **Year:** 2009, 2010 **Model:** Forester, Impreza, Legacy, Outback, Tribeca **Engine:** 2.5L H4, 3.0L H6, 3.6L H6 **Transmission:** All	**HO2S Heater Control Circuit High (Bank 1 Sensor 1):** High voltage condition on front oxygen (Air Fuel) sensor heater control circuit. Battery voltage ≥ 10.9V **Possible Causes:** • Wiring harness between ECM and front oxygen sensor needs repair. • Poor contact between ECM connector and ground circuit.
DTC: P0037 **2T ECM, MIL:** Yes **Year:** 2009, 2010 **Model:** Forester, Impreza, Legacy, Outback, Tribeca **Engine:** 2.5L H4, 3.0L H6, 3.6L H6 **Transmission:** All	**HO2S Heater Control Circuit Low (Bank 1 Sensor 2):** Low voltage condition on rear oxygen sensor Battery voltage ≥ 10.9V Engine started Engine Speed < 8000 RPM **Possible Causes:** • Power supply or main relay to rear oxygen sensor needs repair or replacement. • Wiring harness between ECM and rear oxygen sensor needs repair. • Wiring harness or connector between ECM and chassis ground needs repair. • Defective rear oxygen sensor.
DTC: P0038 **2T ECM, MIL:** Yes **Year:** 2009, 2010 **Model:** Forester, Impreza, Legacy, Outback, Tribeca **Engine:** 2.5L H4, 3.0L H6, 3.6L H6 **Transmission:** All	**HO2S Heater Control Circuit High:** High voltage condition detected on rear oxygen sensor Engine started Battery voltage ≥ 10.9V Engine speed < 8000 RPM **Possible Causes:** • Wiring harness between ECM and rear oxygen sensor needs repair. • Wiring harness or connector between ECM and chassis ground needs repair.
DTC: P0050 **2T CCM, MIL:** Yes **Year:** 2009, 2010 **Model:** Legacy, Outback, Tribeca **Engine:** 3.0L H6, 3.6L H6 **Transmission:** All	**HO2S Heater Control Circuit Bank 2 Sensor 1:** Poor drivability. **Possible Causes:** • Check harness between ECM and front oxygen sensor connector • Check harness between main relay and front oxygen sensor connector • Check front oxygen sensor • Check poor contact

DTC	Trouble Code Title, Conditions & Possible Causes
DTC: P0051 **2T CCM, MIL: Yes** **Year:** 2009, 2010 **Model:** Legacy, Outback, Tribeca **Engine:** 3.0L H6, 3.6L H6 **Transmission:** All	**HO2S Heater Control Circuit Low Bank 2 Sensor 1:** Engine started, system voltage from 11-16 volts, and the PCM detected an unexpected "low" voltage condition on the HO2S-21 heater control circuit during the CCM test. **Possible Causes:** • Check power supply to front oxygen sensor • Check ground circuit for ECM • Using a scan tool check current data • Check output signal of ECM • Check front oxygen sensor
DTC: P0052 **2T CCM, MIL: Yes** **Year:** 2009, 2010 **Model:** Legacy, Outback, Tribeca **Engine:** 3.0L H6, 3.6L H6 **Transmission:** All	**HO2S Heater Control Circuit High Bank 2 Sensor 1:** Engine started, system voltage from 11-16 volts, and the PCM detected an unexpected "high" voltage condition on the HO2S-21 heater control circuit during the CCM test. **Possible Causes:** • Check output signal of ECM • Check front oxygen sensor heater current • Check output signal of ECM
DTC: P0057 **2T CCM, MIL: Yes** **Year:** 2009, 2010 **Model:** Legacy, Outback, Tribeca **Engine:** 3.0L H6, 3.6L H6 **Transmission:** All	**HO2S Heater Control Circuit Low Bank 2 Sensor 2:** Poor drivability. **Possible Causes:** • Check power supply to rear oxygen sensor • Check ground circuit for ECM • Using a scan tool check current data • Check output signal of ECM • Check rear oxygen sensor
DTC: P0058 **2T CCM, MIL: Yes** **Year:** 2009, 2010 **Model:** Legacy, Outback, Tribeca **Engine:** 3.0L H6, 3.6L H6 **Transmission:** All	**HO2S Heater Control Circuit High Bank 2 Sensor 2:** Poor drivability. **Possible Causes:** • Check input signal of ECM • Using a scan tool check current data • Check poor contact
DTC: P0068 **2T ECM, MIL: Yes** **Year:** 2009, 2010 **Model:** Forester, Impreza, Legacy, Outback, Tribeca **Engine:** 2.5L H4, 3.0L H6, 3.6L H6 **Transmission:** All	**MAP/MAF - Throttle Position Correlation:** Rough idling High idle Engine coolant temperature ≥ 167 degrees F **Possible Causes:** • Loose mounting bolts or holes in the air intake system. • Manifold absolute pressure sensor needs replacement • Throttle opening angle incorrect, electronic throttle control needs replacement.
DTC: P0076 **1T ECM, MIL: Yes** **Year:** 2009 **Model:** Legacy, Outback **Engine:** 3.0L H6 **Transmission:** All	**Intake Valve Control Solenoid Circuit Low (Bank 1):** Improper idling. Duty ratio ≥ 30 %. Intake Valve Control Solenoid circuit current < 0.026 A. **Possible Causes:** • Wiring harness between ECM and oil switching solenoid valve needs repair. • Poor contact of ECM and oil switching solenoid valve connector. • Defective oil switching solenoid valve.
DTC: P0076 **1T ECM, MIL: Yes** **Year:** 2009 **Model:** Forester **Engine:** 2.5L H4 **Transmission:** All	**Intake Valve Control Solenoid Circuit Low (Bank 1):** Improper idling **Possible Causes:** • Wiring harness between ECM and oil switching solenoid valve needs repair. • Oil switching solenoid valve needs replacement.

DTC	Trouble Code Title, Conditions & Possible Causes

DTC: P0077
1T ECM, MIL: Yes
Year: 2009
Model: Legacy, Outback
Engine: 3.0L H6
Transmission: All
	Intake Valve Control Solenoid Circuit High (Bank 1):
	Improper idling.
	Duty ratio $\geq$ 30 %.
	Intake Valve Control Solenoid circuit current $\geq$ 0.465 A.
	Possible Causes:
	    • Wiring harness between ECM and oil switching solenoid valve needs repair.
	    • Poor contact of ECM and oil switching solenoid valve connector.
	    • Defective oil switching solenoid valve.

DTC: P0077
1T ECM, MIL: Yes
Year: 2009
Model: Forester
Engine: 2.5L H4
Transmission: All
	Intake Valve Control Solenoid Circuit High (Bank 1):
	Improper idling.
	Battery voltage $\geq$ 10.9V
	Possible Causes:
	    • Wiring harness between ECM and oil switching solenoid valve needs repair.
	    • Oil switching solenoid valve needs replacement.

DTC: P0082
1T ECM, MIL: Yes
Year: 2009
Model: Forester
Engine: 2.5L H4
Transmission: All
	Intake Valve Control Solenoid Circuit Low (Bank 2):
	Improper idling.
	Battery voltage $\geq$ 10.9V
	Possible Causes:
	    • Wiring harness between ECM and oil switching solenoid valve needs repair.
	    • Oil switching solenoid valve needs replacement.

DTC: P0082
1T ECM, MIL: Yes
Year: 2009
Model: Legacy, Outback
Engine: 3.0L H6
Transmission: All
	Intake Valve Control Solenoid Circuit Low (Bank 2):
	Improper idling.
	Duty ratio $\geq$ 30 %.
	Intake Valve Control Solenoid circuit current < 0.026 A.
	Possible Causes:
	    • Wiring harness between ECM and oil switching solenoid valve needs repair.
	    • Poor contact of ECM and oil switching solenoid valve connector.
	    • Defective oil switching solenoid valve.

DTC: P0083
1T ECM, MIL: Yes
Year: 2009
Model: Forester
Engine: 2.5L H4
Transmission: All
	Intake Valve Control Solenoid Circuit High (Bank 2):
	Improper idling.
	Battery voltage $\geq$ 10.9V
	Possible Causes:
	    • Wiring harness between ECM and oil switching solenoid valve needs repair.
	    • Oil switching solenoid valve needs replacement.

DTC: P0083
1T ECM, MIL: Yes
Year: 2009
Model: Legacy, Outback
Engine: 3.0L H6
Transmission: All
	Intake Valve Control Solenoid Circuit High (Bank 2):
	Improper idling.
	Duty ratio $\geq$ 30 %.
	Intake Valve Control Solenoid circuit current $\geq$ 0.465 A.
	Possible Causes:
	    • Wiring harness between ECM and oil switching solenoid valve needs repair.
	    • Poor contact of ECM and oil switching solenoid valve connector.
	    • Defective oil switching solenoid valve.

DTC: P0101 **2T ECM, MIL: Yes** **Year:** 2009, 2010 **Model:** Forester, Impreza, Legacy, Outback, Tribeca **Engine:** 2.5L H4, 3.0L H6, 3.6L H6 **Transmission:** All	**Mass or Volume Air Flow Circuit Range/Performance:** Improper idling. Engine stalls. Poor driving performance. Engine coolant temperature > 167 degrees F **Possible Causes:** • If no other DTC are displayed, mass air flow and intake air temperature sensor needs replacement.

DTC	Trouble Code Title, Conditions & Possible Causes
DTC: P0102 **1T ECM, MIL: Yes** **Year:** 2009, 2010 **Model:** Forester, Impreza, Legacy, Outback, Tribeca **Engine:** 2.5L H4, 3.0L H6, 3.6L H6 **Transmission:** All	**Mass or Volume Air Flow Circuit Low Input:** Improper idling. Engine stalls. Poor driving performance. **Possible Causes:** • Temporary poor contact of the air flow sensor connector. • Wiring harness between power supply and mass air flow sensor or intake air temperature sensor needs repair. • Wiring harness between ECM, mass air flow sensor, intake air temperature sensor and chassis ground needs repair. • Poor contact in the ECM, mass air flow, or intake air temperature sensor connector. • Defective mass air flow and intake air temperature sensor.
DTC: P0103 **1T ECM, MIL: Yes** **Year:** 2009, 2010 **Model:** Forester, Impreza, Legacy, Outback, Tribeca **Engine:** 2.5L H4, 3.0L H6, 3.6L H6 **Transmission:** All	**Mass or Volume Air Flow Circuit High Input:** Improper idling. Engine stalls. Poor driving performance. **Possible Causes:** • Temporary poor contact of the air flow sensor connector. • Wiring harness between ECM, mass air flow sensor, intake air temperature sensor and chassis ground needs repair. • Poor contact between the mass air flow and intake air temperature sensor connectors. • Defective mass air flow and intake air temperature sensor.
DTC: P0107 **1T ECM, MIL: Yes** **Year:** 2009, 2010 **Model:** Forester, Impreza, Legacy, Outback, Tribeca **Engine:** 2.5L H4, 3.0L H6, 3.6L H6 **Transmission:** All	**Manifold Absolute Pressure/Barometric Pressure Circuit Low Input:** Engine running. Output voltage of MAP is $\leq$ 0.573 V **Possible Causes:** • Temporary poor contact of the intake manifold absolute pressure sensor connector. • Wiring harness between power supply and manifold absolute pressure sensor needs repair. • Wiring harness between ECM and manifold absolute pressure sensor needs repair. • Wiring harness between ECM and chassis ground needs repair. • Poor contact between ECM and manifold absolute pressure sensor connectors.
DTC: P0108 **1T ECM, MIL: Yes** **Year:** 2009, 2010 **Model:** Forester, Impreza, Legacy, Outback, Tribeca **Engine:** 2.5L H4, 3.0L H6, 3.6L H6 **Transmission:** All	**Manifold Absolute Pressure/Barometric Pressure Circuit High Input:** Engine running. Output voltage of MAP sensor is $\geq$ 4.38757221 V **Possible Causes:** • Temporary poor contact of the intake manifold absolute pressure sensor connector. • Wiring harness between ECM and manifold absolute pressure sensor connector needs repair. • Poor contact of manifold absolute pressure sensor connector.
DTC: P0111 **2T ECM, MIL: Yes** **Year:** 2009, 2010 **Model:** Forester, Impreza, Legacy, Outback, Tribeca **Engine:** 2.5L H4, 3.0L H6, 3.6L H6 **Transmission:** All	**Intake Air Temperature Sensor 1 Circuit Range/Performance:** Improper idling. Poor driving performance. Engine coolant temperature at engine starting < 30 °C (86 °F). Engine coolant temperature $\geq$ 100 °C (212 °F). Battery voltage $\geq$ 10.9 V. Continuous time when the vehicle speed is less than 60 km/h (37.3 MPH) $\geq$ 600 s. Output voltage difference between Max. and Min < 0.02 V **Possible Causes:** • Mass air flow and intake air temperature sensor needs to be replaced. • Insufficient coolant temperature.
DTC: P0112 **1T ECM, MIL: Yes** **Year:** 2009, 2010 **Model:** Forester, Impreza, Legacy, Outback, Tribeca **Engine:** 2.5L H4, 3.0L H6, 3.6L H6 **Transmission:** All	**Intake Air Temperature Sensor 1 Circuit Low:** Improper idling. Poor driving performance. Output voltage of intake air temperature sensor 1 < 0.230975449 V. **Possible Causes:** • Temporary poor contact of the intake air temperature sensor. • Wiring harness between ECM, mass air flow and intake air temperature sensor connectors needs repair.

DTC	Trouble Code Title, Conditions & Possible Causes

DTC: P0113
1T ECM, MIL: Yes
Year: 2009, 2010
Model: Forester, Impreza, Legacy, Outback, Tribeca
Engine: 2.5L H4, 3.0L H6, 3.6L H6
Transmission: All
	Intake Air Temperature Sensor 1 Circuit High:
	Improper idling
	Poor driving performance
	Intake Air Temperature sensor output voltage ≥ 4.716 V
	Possible Causes:
	  • Poor contact in the ECM or the mass air flow and intake air temperature sensor connectors.
	  • Wiring harness between ECM, mass air flow and intake air temperature sensor connectors needs to be repaired.
	  • Mass air flow and intake air temperature sensors need to be replaced.

DTC: P0117
1T ECM, MIL: Yes
Year: 2009, 2010
Model: Forester, Impreza, Legacy, Outback, Tribeca
Engine: 2.5L H4, 3.0L H6, 3.6L H6
Transmission: All
	Engine Coolant Temperature Circuit Low:
	Hard to start.
	Improper idling.
	Poor driving performance.
	Engine coolant temperature sensor output voltage < 0.264738528 V.
	Possible Causes:
	  • Temporary connector contact failure.
	  • Wiring harness between ECM and engine coolant temperature sensor needs repair.

DTC: P0118
1T ECM, MIL: Yes
Year: 2009, 2010
Model: Forester, Impreza, Legacy, Outback, Tribeca
Engine: 2.5L H4, 3.0L H6, 3.6L H6
Transmission: All
	Engine Coolant Temperature Circuit High:
	Hard to start.
	Improper idling.
	Poor driving performance.
	Engine coolant temperature sensor output ≥ 4.716 V.
	Possible Causes:
	  • Temporary connector contact failure.
	  • Wiring harness between ECM and engine coolant temperature sensor needs repair.
	  • Defective engine coolant temperature sensor.

DTC: P0122
1T ECM, MIL: Yes
Year: 2009, 2010
Model: Forester, Impreza, Legacy, Outback, Tribeca
Engine: 2.5L H4, 3.0L H6, 3.6L H6
Transmission: All
	Throttle/Pedal Position Sensor/Switch "A" Circuit Low:
	Improper idling.
	Engine stalls.
	Poor driving performance.
	Ignition switch - ON
	Battery voltage ≥ 6 V
	Throttle position sensor ≤ 0.217 V
	Possible Causes:
	  • Wiring harness between ECM and electronic throttle control needs repair.
	  • Defective electronic throttle control.
	  • Defective ECM.

DTC: P0123
1T ECM, MIL: Yes
Year: 2009, 2010
Model: Forester, Impreza, Legacy, Outback, Tribeca
Engine: 2.5L H4, 3.0L H6, 3.6L H6
Transmission: All
	Throttle/Pedal Position Sensor/Switch "A" Circuit High:
	Improper idling.
	Engine stalls.
	Poor driving performance.
	Ignition switch - ON
	Battery voltage ≥ 6 V
	Throttle position sensor ≥ 4.858 V
	Possible Causes:
	  • Wiring harness between ECM and electronic throttle control needs repair.
	  • Poor contact between ECM and electronic throttle control.

DTC: P0125
2T CCM, MIL: Yes
Year: 2009
Model: Legacy, Outback
Engine: 3.0L H6
Transmission: All
	Insufficient Coolant Temperature For Closed loop Fuel Control:
	DTC P0115, P0116 and P0117 not set, Engine started, vehicle driven to a speed of 25-40 at over 1500 rpm, and the PCM detected the coolant temperature did not reach closed loop status.
	Possible Causes:
	  • Check any other DTC on display
	  • Check tire size
	  • Check engine coolant
	  • Check thermostat

DTC	Trouble Code Title, Conditions & Possible Causes

DTC: P0125
2T ECM, MIL: Yes
Year: 2009, 2010
Model: Forester, Impreza, Outback, Tribeca
Engine: 2.5L H4, 3.6L H6
Transmission: All
	Insufficient Coolant Temperature for Closed Loop Fuel Control:
	Engine does not return to idle.
	Battery voltage $\geq$ 10.9 V.
	Engine Speed $\geq$ 500 RPM
	Possible Causes:
	  • Incorrect tire size.
	  • Incorrect coolant level
	  • Defective engine coolant temperature sensor.

DTC: P0126
2T ECM, MIL: Yes
Year: 2009, 2010
Model: Forester, Impreza, Legacy, Outback, Tribeca
Engine: 2.5L H4, 3.0L H6, 3.6L H6
Transmission: All
	Insufficient Engine Coolant Temperature for Stable Operation:
	Hard to start.
	Improper idling.
	Poor driving performance.
	Engine coolant temperature does not decrease when it should.
	Battery voltage $\geq$ 10.9 V.
	Refueling from the last engine stop till the current engine start - None.
	Fuel level $\geq$ 15 L (3.96 US gal, 3.3 Imp gal).
	Engine coolant temperature at the last engine stop $\geq$ 75 °C (167 °F).
	Possible Causes:
	  • Poor contact of the ECM connector.
	  • Defective engine coolant temperature sensor.

DTC: P0128
2T ECM, MIL: Yes
Year: 2009, 2010
Model: Forester, Impreza, Legacy, Outback, Tribeca
Engine: 2.5L H4, 3.0L H6, 3.6L H6
Transmission: All
	Coolant Thermostat (Engine Coolant Temperature Below Thermostat Regulating Temperature):
	Thermostat remains open.
	Engine coolant temperature at engine starting < 55 °C (131 °F).
	Engine coolant temperature is lower than the estimated engine coolant temperature and the difference between them is greater than 86 degrees F.
	Possible Causes:
	  • Incorrect mixture ratio of coolant to anti-freeze solution.
	  • Radiator ran circuit needs repair.
	  • Defective thermostat.

DTC: P0131
1T ECM, MIL: Yes
Year: 2009, 2010
Model: Forester, Impreza, Legacy, Outback, Tribeca
Engine: 2.5L H4, 3.0L H6, 3.6L H6
Transmission: All
	O2 Sensor Circuit Low Voltage (Bank 1 Sensor 1):
	The input voltage of the O2 sensor meets one of the following conditions for longer than 1000ms:
	Input voltage (+) < 1.128 V
	Input voltage (-) < 0.23 V
	or
	(Input voltage (+) minus the Input voltage (-)) < 0.644 V
	Possible Causes:
	  • Water in the front oxygen sensor connector or coupling connector.
	  • Wiring harness between ECM and front oxygen sensor connector needs repair.
	  • Defective front oxygen sensor.

DTC: P0132
1T ECM, MIL: Yes
Year: 2009, 2010
Model: Forester, Impreza, Legacy, Outback, Tribeca
Engine: 2.5L H4, 3.0L H6, 3.6L H6
Transmission: All
	O2 Sensor Circuit High Voltage (Bank 1 Sensor 1):
	The input voltage of the O2 sensor meets one of the following conditions for longer than 1000ms:
	Input voltage (+) > 3.589 V
	or
	Input voltage (-) > 3.541 V
	Possible Causes:
	  • Water in the front oxygen sensor connector or coupling connector.
	  • Wiring harness between ECM and front oxygen sensor connector needs repair.
	  • Defective front oxygen sensor.

DTC	Trouble Code Title, Conditions & Possible Causes
DTC: P0133 **2T ECM, MIL: Yes** **Year:** 2009, 2010 **Model:** Forester, Impreza, Legacy, Outback, Tribeca **Engine:** 2.5L H4, 3.0L H6, 3.6L H6 **Transmission:** All	**O2 Sensor Circuit Slow Response (Bank 1 Sensor 1):** Time needed for all secondary parameters to be in enable conditions $\geq$ 1024 ms: Battery voltage $\geq$ 10.9 V. Barometric pressure > 75 kPa (563 mmHg, 22.2 in. Hg). Closed loop control with main feedback - In Operation. Front oxygen (A/F) sensor impedance = 0 Ω — 50 Ω. Elapsed time after starting the engine $\geq$ 120000 ms. Engine coolant temperature $\geq$ 75 °C (167 °F). Engine speed = 1000 rpm — 3200 rpm. Vehicle speed 10 km/h — 120 km/h (6.2 MPH — 74.6 MPH). Amount of intake air = 10 g/s — 40 g/s (0.35 oz/s — 1.41 oz/s). Engine load < 0.02 g/rev (0 oz/rev). Learning value of EVAP conc. during purge < 0.2. Total time of operating canister purge $\geq$ 19.9 s. **Possible Causes:** • Loose exhaust system connections. • Damaged exhaust system components. • Defective front oxygen sensor.
DTC: P0134 **1T ECM, MIL: Yes** **Year:** 2009, 2010 **Model:** Forester, Impreza, Legacy, Outback, Tribeca **Engine:** 2.5L H4, 3.0L H6, 3.6L H6 **Transmission:** All	**O2 Sensor Circuit No Activity Detected (Bank 1 Sensor 1):** Battery voltage $\geq$ 10.9 V. Time of heater control duty at 70 % or more $\geq$ 36000 ms. Front oxygen (A/F) sensor impedance. > 500 ohms. **Possible Causes:** • Wiring harness ECM and front oxygen sensor needs repair. • Poor contact of ECM and front oxygen sensor. • Defective front oxygen sensor.
DTC: P0137 **2T ECM, MIL: Yes** **Year:** 2009, 2010 **Model:** Forester, Impreza, Legacy, Outback, Tribeca **Engine:** 2.5L H4, 3.0L H6, 3.6L H6 **Transmission:** All	**O2 Sensor Circuit Low Voltage (Bank 1 Sensor 2):** Oxygen sensor output voltage < 0.03 V during normal operation conditions. **Possible Causes:** • Water in the front oxygen sensor connector or coupling connector. • Wiring harness between ECM and rear oxygen sensor connector needs repair. • Loose or damaged exhaust system parts. • Defective rear oxygen sensor.
DTC: P0138 **2T ECM, MIL: Yes** **Year:** 2009, 2010 **Model:** Forester, Impreza, Legacy, Outback, Tribeca **Engine:** 2.5L H4, 3.0L H6, 3.6L H6 **Transmission:** All	**O2 Sensor Circuit High Voltage (Bank 1 Sensor 2):** Rear oxygen sensor output voltage > 1.2 V during normal operating conditions. **Possible Causes:** • Water in the front oxygen sensor connector or coupling connector. • Wiring harness between ECM and rear oxygen sensor connector needs repair. • Loose or damaged exhaust system parts. • Defective rear oxygen sensor.
DTC: P0139 **2T ECM, MIL: Yes** **Year:** 2009, 2010 **Model:** Forester, Impreza, Legacy, Outback, Tribeca **Engine:** 2.5L H4, 3.0L H6, 3.6L H6 **Transmission:** All	**O2 Sensor Circuit Slow Response (Bank 1 Sensor 2):** The measured response time for oxygen sensor output, when the air/fuel ratio changes from rich to lean > 491 ms. The measured response time for oxygen sensor output, when the air/fuel ratio changes from lean to rich > 2000 ms. **Possible Causes:** • Wiring harness between ECM and rear oxygen sensor needs repair. • Defective rear oxygen sensor.
DTC: P0140 **2T ECM, MIL: Yes** **Year:** 2009, 2010 **Model:** Forester, Impreza, Legacy, Outback, Tribeca **Engine:** 2.5L H4, 3.0L H6, 3.6L H6 **Transmission:** All	**O2 Sensor Circuit No Activity Detected (Bank 1 Sensor 2):** The output voltage of the rear O2 sensor meets one of the following conditions for longer than 0.2 seconds: Minimum output voltage > 0.15 V. Maximum output voltage < 0.55 V. **Possible Causes:** • Water in the rear oxygen sensor connector or coupling connector. • Wiring harness between ECM and rear oxygen sensor connector needs repair. • Loose or damaged exhaust system parts. • Defective rear oxygen sensor.

DTC	Trouble Code Title, Conditions & Possible Causes
DTC: P0141 **2T ECM, MIL: Yes** **Year:** 2009, 2010 **Model:** Outback, Tribeca **Engine:** 3.6L H6 **Transmission:** All	**O2 Sensor Heater Circuit (Bank 1 Sensor 2):** Battery voltage ≥ 10.9 V. Elapsed time after starting the engine ≥ 1 second. Engine speed < 4500 rpm. Rear oxygen sensor heater control duty output voltage level < 75 %. **Possible Causes:** • Water in the front oxygen sensor connector or coupling connector. • Wiring harness between ECM and rear oxygen sensor connector needs repair. • Loose or damaged exhaust system parts. • Defective rear oxygen sensor.
DTC: P0151 **1T CCM, MIL: Yes** **Year:** 2009, 2010 **Model:** Legacy, Outback, Tribeca **Engine:** 3.0L H6, 3.6L H6 **Transmission:** All	**O2 Sensor Circuit Low Voltage Bank 2 Sensor 1:** Engine started, engine running in closed loop at hot idle speed, and the PCM detected an unexpected "low" voltage condition on the front A/F sensor signal circuit for over 6 seconds during the CCM test. **Possible Causes:** • Check the harness between the ECM and the front sensor connector
DTC: P0152 **1T CCM, MIL: Yes** **Year:** 2009, 2010 **Model:** Legacy, Outback, Tribeca **Engine:** 3.0L H6, 3.6L H6 **Transmission:** All	**O2 Sensor Circuit High Voltage Bank 2 Sensor 1:** Engine started, engine running in closed loop at hot idle speed, and the PCM detected an unexpected "high" voltage condition on the front A/F sensor signal circuit for over 6 seconds in the CCM test. **Possible Causes:** • Check the harness between the ECM and the front sensor connector
DTC: P0152 **1T ECM, MIL: Yes** **Year:** 2009 **Model:** Legacy, Outback **Engine:** 3.0L H6 **Transmission:** All	**O2 Sensor Circuit High Voltage (Bank 2 Sensor 1):** The input voltage of the O2 sensor meets one of the following conditions for longer than 1000ms: Input voltage (+) > 3.589 V or Input voltage (-) > 3.541 V **Possible Causes:** • Water in the front oxygen sensor connector or coupling connector. • Wiring harness between ECM and front oxygen sensor connector needs repair. • Defective front oxygen sensor.
DTC: P0153 **2T ECM, MIL: Yes** **Year:** 2009 **Model:** Legacy, Outback **Engine:** 3.0L H6 **Transmission:** All	**O2 Sensor Circuit Slow Response (Bank 2 Sensor 1):** Time needed for all secondary parameters to be in enable conditions ≥ 1024 ms: Battery voltage ≥ 10.9 V. Barometric pressure > 75 kPa (563 mmHg, 22.2 in. Hg). Closed loop control with main feedback - In Operation. Front oxygen (A/F) sensor impedance = 0 Ω — 50 Ω. Elapsed time after starting the engine ≥ 120000 ms. Engine coolant temperature ≥ 75 °C (167 °F). Engine speed = 1000 rpm — 3200 rpm. Vehicle speed 10 km/h — 120 km/h (6.2 MPH — 74.6 MPH). Amount of intake air = 10 g/s — 40 g/s (0.35 oz/s — 1.41 oz/s). Engine load < 0.02 g/rev (0 oz/rev). Learning value of EVAP conc. during purge < 0.2. Total time of operating canister purge ≥ 19.9 s. **Possible Causes:** • Loose exhaust system connections. • Damaged exhaust system components. • Defective front oxygen sensor.
DTC: P0153 **2T CCM, MIL: Yes** **Year:** 2009, 2010 **Model:** Outback, Tribeca **Engine:** 3.6L H6 **Transmission:** All	**O2 Sensor Circuit Slow Response Bank 2 Sensor 1:** Engine started, engine running in closed loop, ECT sensor more than 170°F, and the PCM detected the number of A/FS-21 rich-to-lean or lean-to-rich switches was less than a calibrated amount. **Possible Causes:** • Check any other DTC on display • Check exhaust system

DTC	Trouble Code Title, Conditions & Possible Causes
DTC: P0154 **1T CCM, MIL: Yes** **Year:** 2009, 2010 **Model:** Outback, Tribeca **Engine:** 3.6L H6 **Transmission:** All	**O2 Sensor Circuit No Activity Detected Bank 2 Sensor 1:** Poor drivability. **Possible Causes:** • Check the harness between the ECM and the front sensor • Check poor contact
DTC: P0154 **1T ECM, MIL: Yes** **Year:** 2009 **Model:** Legacy, Outback **Engine:** 3.0L H6 **Transmission:** All	**O2 Sensor Circuit No Activity Detected (Bank 2 Sensor 1):** Battery voltage ≥ 10.9 V. Time of heater control duty at 70 % or more ≥ 36000 ms. Front oxygen (A/F) sensor impedance. > 500 ohms. **Possible Causes:** • Wiring harness ECM and front oxygen sensor needs repair. • Poor contact of ECM and front oxygen sensor. • Defective front oxygen sensor.
DTC: P0157 **2T CCM, MIL: Yes** **Year:** 2009, 2010 **Model:** Legacy, Outback, Tribeca **Engine:** 3.0L H6, 3.6L H6 **Transmission:** All	**O2 Sensor Circuit Low Voltage Bank 2 Sensor 2:** Poor drivability. **Possible Causes:** • Check any other DTC on display • Using a scan tool, check data • Check harness between ECM and rear sensor connector • Check harness between rear sensor and ECM connector • Check exhaust system
DTC: P0158 **2T CCM, MIL: Yes** **Year:** 2009, 2010 **Model:** Legacy, Outback, Tribeca **Engine:** 3.0L H6, 3.6L H6 **Transmission:** All	**O2 Sensor Circuit High Voltage Bank 2 Sensor 2:** Poor drivability. **Possible Causes:** • Check any other DTC on display • Using a scan tool, check data • Check harness between ECM and rear sensor connector • Check harness between rear sensor and ECM connector • Check exhaust system
DTC: P0159 **2T CCM, MIL: Yes** **Year:** 2009, 2010 **Model:** Legacy, Outback, Tribeca **Engine:** 3.0L H6, 3.6L H6 **Transmission:** All	**O2 Sensor Circuit Slow Response Bank 2 Sensor 2:** Poor drivability. **Possible Causes:** • Check any other DTC on display • Check harness between ECM and rear sensor connector • Check sensor
DTC: P0160 **2T CCM, MIL: Yes** **Year:** 2009, 2010 **Model:** Legacy, Outback, Tribeca **Engine:** 3.0L H6, 3.6L H6 **Transmission:** All	**O2 Sensor Circuit No Activity Detected Bank 2 Sensor 2:** Poor drivability. **Possible Causes:** • Check any other DTC on display • Using a scan tool, check data • Check harness between ECM and rear sensor connector • Check harness between rear sensor and ECM connector • Check exhaust system
DTC: P0171 **2T ECM, MIL: Yes** **Year:** 2009, 2010 **Model:** Forester, Impreza, Legacy, Outback, Tribeca **Engine:** 2.5L H4, 3.0L H6, 3.6L H6 **Transmission:** All	**System Too Lean (Bank 1):** Improper idling. Engine stalls. Poor driving performance. **Possible Causes:** • Loose or damaged exhaust system parts. • Loose or damaged air intake system. • Clogged fuel line or bent hose • Improper fuel pump discharge • Defective engine coolant temperature sensor • Defective mass air flow and intake air temperature sensor.

DTC	Trouble Code Title, Conditions & Possible Causes
DTC: P0172 **2T ECM, MIL: Yes** **Year:** 2009, 2010 **Model:** Forester, Impreza, Legacy, Outback, Tribeca **Engine:** 2.5L H4, 3.0L H6, 3.6L H6 **Transmission:** All	**System too Rich (Bank 1):** Improper idling. Engine stalls. Poor driving performance. **Possible Causes:** • Loose or damaged exhaust system parts. • Loose or damaged air intake system. • Clogged fuel line or bent hose • Improper fuel pump discharge • Defective engine coolant temperature sensor • Defective mass air flow and intake air temperature sensor.
DTC: P0174 **2T ECM, MIL: Yes** **Year:** 2009 **Model:** Legacy, Outback **Engine:** 3.0L H6 **Transmission:** All	**System Too Lean (Bank 2):** Improper idling. Engine stalls. Poor driving performance. **Possible Causes:** • Loose or damaged exhaust system parts. • Loose or damaged air intake system. • Clogged fuel line or bent hose • Improper fuel pump discharge • Defective engine coolant temperature sensor • Defective mass air flow and intake air temperature sensor.
DTC: P0174 **2T ECM, MIL: Yes** **Year:** 2009, 2010 **Model:** Outback, Tribeca **Engine:** 3.6L H6 **Transmission:** All	**System Too Lean (Bank 2):** Improper idling. Engine stalls. Poor driving performance. **Possible Causes:** • Loose or damaged exhaust system parts. • Loose or damaged air intake system. • Clogged fuel line or bent hose • Improper fuel pump discharge • Defective engine coolant temperature sensor • Defective mass air flow and intake air temperature sensor.
DTC: P0175 **2T ECM, MIL: Yes** **Year:** 2009 **Model:** Legacy, Outback **Engine:** 3.0L H6 **Transmission:** All	**System Too Rich (Bank 2):** Improper idling. Engine stalls. Poor driving performance. **Possible Causes:** • Loose or damaged exhaust system parts. • Loose or damaged alr Intake system. • Clogged fuel line or bent hose • Improper fuel pump discharge • Defective engine coolant temperature sensor • Defective mass air flow and intake air temperature sensor.
DTC: P0175 **2T ECM, MIL: Yes** **Year:** 2009, 2010 **Model:** Outback, Tribeca **Engine:** 3.6L H6 **Transmission:** All	**System Too Rich (Bank 2):** Improper idling. Engine stalls. Poor driving performance. **Possible Causes:** • Loose or damaged exhaust system parts. • Loose or damaged air intake system. • Clogged fuel line or bent hose • Improper fuel pump discharge • Defective engine coolant temperature sensor • Defective mass air flow and intake air temperature sensor.

DTC	Trouble Code Title, Conditions & Possible Causes
DTC: P0181 **2T ECM, MIL: Yes** **Year:** 2009, 2010 **Model:** Forester, Impreza, Legacy, Outback, Tribeca **Engine:** 2.5L H4, 3.0L H6, 3.6L H6 **Transmission:** All	**Fuel Temperature Sensor "A" Circuit Range/Performance:** Fuel temperature detected as higher than the engine coolant temperature. Fuel temperature does not rise. **Possible Causes:** • Defective fuel temperature sensor.
DTC: P0182 **1T ECM, MIL: Yes** **Year:** 2009, 2010 **Model:** Forester, Impreza, Legacy, Outback, Tribeca **Engine:** 2.5L H4, 3.0L H6, 3.6L H6 **Transmission:** All	**Fuel Temperature Sensor "A" Circuit Low Input:** Fuel temperature sensor output voltage < 0.343951474 V. Battery voltage $\geq$ 10.9 V **Possible Causes:** • Defective fuel temperature sensor • Wiring harness between ECM and fuel pump connector needs repair.
DTC: P0183 **1T ECM, MIL: Yes** **Year:** 2009, 2010 **Model:** Forester, Impreza, Legacy, Outback, Tribeca **Engine:** 2.5L H4, 3.0L H6, 3.6L H6 **Transmission:** All	**Fuel Temperature Sensor "A" Circuit High Input:** Fuel temperature sensor output voltage $\geq$ 4.716 V Battery voltage $\geq$ 10.9 V **Possible Causes:** • Temporary connector contact failure • Poor contact between the ECM and fuel temperature sensor connectors. • Wiring harness between ECM and fuel temperature sensor needs repair. • Defective fuel temperature sensor.
DTC: P0196 **2T ECM, MIL: Yes** **Year:** 2009, 2010 **Model:** Forester, Outback, Tribeca **Engine:** 2.5L H4, 3.6L H6 **Transmission:** All	**Engine Oil Temperature Sensor Circuit Range/Performance:** Hard to start. Improper idling. Poor driving performance. Battery voltage $\geq$ 10.9 V. Engine Speed $\geq$ 500 RPM. The judgment value for after engine start oil temperature sensor characteristic timer falls out of range. **Possible Causes:** • Defective oil temperature sensor.
DTC: P0196 **2T CCM, MIL: Yes** **Year:** 2009 **Model:** Legacy, Outback **Engine:** 3.0L H6 **Transmission:** All	**Engine Oil Temperature Sensor Circuit Range/Performance:** Engine stalls. Erroneous idling. Poor driving performance. **Possible Causes:** • Check any other DTC on display • Defective oil temperature sensor.
DTC: P0197 **1T ECM, MIL: Yes** **Year:** 2009, 2010 **Model:** Forester, Outback, Tribeca **Engine:** 2.5L H4, 3.6L H6 **Transmission:** All	**Engine Oil Temperature Sensor Low:** Hard to start. Improper idling. Poor driving performance. Oil temperature sensor output voltage < 0.166 V. **Possible Causes:** • Temporary connector contact failure. • Wiring harness between ECM and oil temperature sensor connector needs repair. • Defective oil temperature sensor.
DTC: P0197 **1T CCM, MIL: Yes** **Year:** 2009 **Model:** Legacy, Outback **Engine:** 3.0L H6 **Transmission:** All	**Engine Oil Temperature Sensor Circuit Low:** Engine stalls. Erroneous idling. Poor driving performance. **Possible Causes:** • Check harness between sensor and ECM • Check poor contact • Defective oil temperature sensor.
DTC: P0198 **1T CCM, MIL: Yes** **Year:** 2009 **Model:** Legacy, Outback **Engine:** 3.0L H6 **Transmission:** All	**Engine Oil Temperature Sensor Circuit High:** Engine stalls. Erroneous idling. Poor driving performance. **Possible Causes:** • Check harness between sensor and ECM • Defective oil temperature sensor.

DTC	Trouble Code Title, Conditions & Possible Causes
DTC: P0198 **1T ECM, MIL:** Yes **Year:** 2009, 2010 **Model:** Forester, Outback, Tribeca **Engine:** 2.5L H4, 3.6L H6 **Transmission:** All	**Engine Oil Temperature Sensor High:** Hard to start. Improper idling. Poor driving performance. Oil temperature sensor output voltage $\geq$ 4.716 V. **Possible Causes:** • Temporary connector contact failure. • Wiring harness between ECM and oil temperature sensor needs repair. • Defective oil temperature sensor.
DTC: P0222 **1T ECM, MIL:** Yes **Year:** 2009, 2010 **Model:** Forester, Impreza, Legacy, Outback, Tribeca **Engine:** 2.5L H4, 3.0L H6, 3.6L H6 **Transmission:** All	**Throttle/Pedal Position Sensor/Switch "B" Circuit Low:** Improper idling. Poor driving performance. Engine stalls. Ignition switch - ON. Battery voltage $\geq$ 6 V. Throttle position sensor 2 input voltage $\leq$ 0.926256 V. **Possible Causes:** • Wiring harness between ECM and electronic throttle control connector needs repair. • Defective electronic throttle control. • Defective ECM.
DTC: P0223 **1T ECM, MIL:** Yes **Year:** 2009, 2010 **Model:** Forester, Impreza, Legacy, Outback, Tribeca **Engine:** 2.5L H4, 3.0L H6, 3.6L H6 **Transmission:** All	**Throttle/Pedal Position Sensor/Switch "B" Circuit High:** Improper idling. Poor driving performance. Engine stalls. Ignition switch - ON. Battery voltage $\geq$ 6 V. Throttle position sensor 2 input voltage $\geq$ 4.858 V **Possible Causes:** • Wiring harness between ECM and electronic throttle control connector needs repair. • Defective electronic throttle control.
DTC: P0230 **2T CCM, MIL:** Yes **Year:** 2009, 2010 **Model:** Forester, Impreza, Legacy, Outback **Engine:** 2.5L H4 **Transmission:** All	**Fuel Pump Primary Circuit:** Poor driveability. **Possible Causes:** • Check power supply to fuel pump control unit • Check ground circuit of fuel pump control unit • Check harness between the fuel pump control unit and fuel pump connector • Check poor contact • Check if vehicle has previously ran out of fuel
DTC: P0230 **2T CCM, MIL:** Yes **Year:** 2009, 2010 **Model:** Legacy, Outback, Tribeca **Engine:** 3.0L H6, 3.6L H6 **Transmission:** All	**Fuel Pump Primary Circuit:** Poor drivability. **Possible Causes:** • Check power supply to fuel pump control unit • Check ground circuit of fuel pump control unit • Check harness between the fuel pump control unit and fuel pump connector • Check poor contact • Check if vehicle has previously ran out of fuel • Defective fuel pump control unit.
DTC: P0244 **1T CCM, MIL:** Yes **Year:** 2009, 2010 **Model:** Forester, Impreza, Legacy, Outback **Engine:** 2.5L H4 **Transmission:** All	**Turbocharger/Supercharger Wastegate Solenoid "A" Range/Performance:** Poor driveability. **Possible Causes:** • Check any other DTC on display

DTC	Trouble Code Title, Conditions & Possible Causes
DTC: P0245 **1T CCM, MIL: Yes** **Year:** 2009, 2010 **Model:** Forester, Impreza, Legacy, Outback **Engine:** 2.5L H4 **Transmission:** All	**Turbocharger/Supercharger Wastegate Solenoid "A" Low:** Poor driveability. **Possible Causes:** • Check output signal from ECM • Check harness between wastegate control solenoid valve and ECM connector • Check poor contact • Check wastegate control solenoid valve • Check power supply to wastegate control solenoid valve • Check poor contact
DTC: P0246 **1T CCM, MIL: Yes** **Year:** 2009, 2010 **Model:** Forester, Impreza, Legacy, Outback **Engine:** 2.5L H4 **Transmission:** All	**Turbocharger/Supercharger Wastegate Solenoid "A" High:** Poor drivability. **Possible Causes:** • Check output signal from ECM • Check harness between wastegate control solenoid valve and ECM connector • Check poor contact • Check wastegate control solenoid valve
DTC: P0301 **2T ECM, MIL: Yes** **Year:** 2009, 2010 **Model:** Forester, Impreza, Legacy, Outback, Tribeca **Engine:** 2.5L H4, 3.0L H6, 3.6L H6 **Transmission:** All	**Cylinder 1 Misfire Detected:** Engine stalls. Improper idling. Rough driving. Engine started. Vehicle driven to a speed of over 3 mph. ECM detected a misfire in Cylinder 1 in the 200 (Catalyst) or 1000 rpm (High Emissions) revolution range. **NOTE: If the misfire is severe, the MIL will flash on/off on the 1st trip.** **Possible Causes:** • Wiring harness between ECM and fuel injector connector needs repair. • Faulty fuel injector. • Poor contact of ignition coil connector. • Poor contact of fuel injector connector. • Poor contact in ECM connector • Poor contact of coupling connector • Loose camshaft position sensor. • Damaged crank sprocket. • Timing belt out of alignment. • Loose or damaged air intake system parts. • Damaged spark plug or spark plug wire.
DTC: P0302 **2T ECM, MIL: Yes** **Year:** 2009, 2010 **Model:** Forester, Impreza, Legacy, Outback, Tribeca **Engine:** 2.5L H4, 3.0L H6, 3.6L H6 **Transmission:** All	**Cylinder 2 Misfire Detected:** Engine stalls. Improper idling. Rough driving. Engine started. Vehicle driven to a speed of over 3 mph. ECM detected a misfire in Cylinder 2 in the 200 (Catalyst) or 1000 rpm (High Emissions) revolution range. **NOTE: If the misfire is severe, the MIL will flash on/off on the 1st trip.** **Possible Causes:** • Wiring harness between ECM and fuel injector connector needs repair. • Faulty fuel injector. • Poor contact of ignition coil connector. • Poor contact of fuel injector connector. • Poor contact in ECM connector • Poor contact of coupling connector • Loose camshaft position sensor. • Damaged crank sprocket. • Timing belt out of alignment. • Loose or damaged air intake system parts. • Damaged spark plug or spark plug wire.

DTC	Trouble Code Title, Conditions & Possible Causes
DTC: P0303 **2T ECM, MIL: Yes** **Year:** 2009, 2010 **Model:** Forester, Impreza, Legacy, Outback, Tribeca **Engine:** 2.5L H4, 3.0L H6, 3.6L H6 **Transmission:** All	**Cylinder 3 Misfire Detected:** Engine stalls. Improper idling. Rough driving. Engine started. Vehicle driven to a speed of over 3 mph. ECM detected a misfire in Cylinder 3 in the 200 (Catalyst) or 1000 rpm (High Emissions) revolution range. **NOTE: If the misfire is severe, the MIL will flash on/off on the 1st trip.** **Possible Causes:** • Wiring harness between ECM and fuel injector connector needs repair. • Faulty fuel injector. • Poor contact of ignition coil connector. • Poor contact of fuel injector connector. • Poor contact in ECM connector • Poor contact of coupling connector • Loose camshaft position sensor. • Damaged crank sprocket. • Timing belt out of alignment. • Loose or damaged air intake system parts. • Damaged spark plug or spark plug wire.
DTC: P0304 **2T ECM, MIL: Yes** **Year:** 2009, 2010 **Model:** Forester, Impreza, Legacy, Outback, Tribeca **Engine:** 2.5L H4, 3.0L H6, 3.6L H6 **Transmission:** All	**Cylinder 4 Misfire Detected:** Engine stalls. Improper idling. Rough driving. Engine started. Vehicle driven to a speed of over 3 mph. ECM detected a misfire in Cylinder 4 in the 200 (Catalyst) or 1000 rpm (High Emissions) revolution range. **NOTE: If the misfire is severe, the MIL will flash on/off on the 1st trip.** **Possible Causes:** • Wiring harness between ECM and fuel injector connector needs repair. • Faulty fuel injector. • Poor contact of ignition coil connector. • Poor contact of fuel injector connector. • Poor contact in ECM connector • Poor contact of coupling connector • Loose camshaft position sensor. • Damaged crank sprocket. • Timing belt out of alignment. • Loose or damaged air intake system parts. • Damaged spark plug or spark plug wire.
DTC: P0305 **2T ECM, MIL: Yes** **Year:** 2009, 2010 **Model:** Legacy, Outback, Tribeca **Engine:** 3.0L H6, 3.6L H6 **Transmission:** All	**Cylinder 5 Misfire Detected:** Engine stalls. Improper idling. Rough driving. Engine started. Vehicle driven to a speed of over 3 mph. ECM detected a misfire in Cylinder 5 in the 200 (Catalyst) or 1000 rpm (High Emissions) revolution range. **NOTE: If the misfire is severe, the MIL will flash on/off on the 1st trip.** **Possible Causes:** • Wiring harness between ECM and fuel injector connector needs repair. • Faulty fuel injector. • Poor contact of ignition coil connector. • Poor contact of fuel injector connector. • Poor contact in ECM connector • Poor contact of coupling connector • Loose camshaft position sensor. • Damaged crank sprocket. • Timing belt out of alignment. • Loose or damaged air intake system parts. • Damaged spark plug or spark plug wire.

DTC	Trouble Code Title, Conditions & Possible Causes
DTC: P0306 **2T ECM, MIL: Yes** **Year:** 2009, 2010 **Model:** Legacy, Outback, Tribeca **Engine:** 3.0L H6, 3.6L H6 **Transmission:** All	**Cylinder 6 Misfire Detected:** Engine stalls. Improper idling. Rough driving. Engine started. Vehicle driven to a speed of over 3 mph. ECM detected a misfire in Cylinder 6 in the 200 (Catalyst) or 1000 rpm (High Emissions) revolution range. **NOTE: If the misfire is severe, the MIL will flash on/off on the 1st trip.** **Possible Causes:** • Wiring harness between ECM and fuel injector connector needs repair. • Faulty fuel injector. • Poor contact of ignition coil connector. • Poor contact of fuel injector connector. • Poor contact in ECM connector • Poor contact of coupling connector • Loose camshaft position sensor. • Damaged crank sprocket. • Timing belt out of alignment. • Loose or damaged air intake system parts. • Damaged spark plug or spark plug wire.
DTC: P0327 **1T ECM, MIL: Yes** **Year:** 2009, 2010 **Model:** Forester, Impreza, Legacy, Outback, Tribeca **Engine:** 2.5L H4, 3.0L H6, 3.6L H6 **Transmission:** All	**Knock Sensor 1 Circuit Low Input (Bank 1 or Single Sensor):** Poor driving performance. Knocking occurs. Knock sensor output voltage < 0.243 V. **Possible Causes:** • Wiring harness between ECM and knock sensor needs repair. • Knock sensor wiring harness or connector needs repair. • Defective knock sensor.
DTC: P0328 **1T ECM, MIL: Yes** **Year:** 2009, 2010 **Model:** Forester, Impreza, Legacy, Outback, Tribeca **Engine:** 2.5L H4, 3.0L H6, 3.6L H6 **Transmission:** All	**Knock Sensor 1 Circuit High Input (Bank 1 or Single Sensor):** Poor driving performance. Knocking occurs. Knock sensor output voltage $\geq$ 4.709 V. **Possible Causes:** • Wiring harness between ECM and knock sensor needs repair. • Knock sensor wiring harness or connector needs repair. • Defective knock sensor. • Poor contact of the ECM connector needs repair.
DTC: P0332 **1T ECM, MIL: Yes** **Year:** 2009, 2010 **Model:** Outback, Tribeca **Engine:** 3.6L H6 **Transmission:** All	**Knock Sensor 2 Circuit Low Input (Bank 2):** Poor driving performance. Knocking occurs. Knock sensor output voltage < 0.243 V. **Possible Causes:** • Wiring harness between ECM and knock sensor needs repair. • Knock sensor wiring harness or connector needs repair. • Defective knock sensor.
DTC: P0332 **1T CCM, MIL: Yes** **Year:** 2009 **Model:** Legacy, Outback **Engine:** 3.0L H6 **Transmission:** All	**Knock Sensor 2 Circuit Low Input Bank 2:** Engine started, engine speed over 1200 RPM, system voltage from 11-16 volts, and the PCM detected an unexpected "low" voltage condition on the Knock Sensor 2 circuit during the CCM test. **Possible Causes:** • Check harness between the sensor and the ECM connector • Check the sensor • Check sensor installation
DTC: P0333 **1T CCM, MIL: Yes** **Year:** 2009 **Model:** Legacy, Outback **Engine:** 3.0L H6 **Transmission:** All	**Knock Sensor 2 Circuit High Input Bank 2:** Engine started, engine speed over 1200 RPM, system voltage from 11-16 volts, and the PCM detected an unexpected "high" voltage condition on the Knock Sensor 2 circuit during the CCM test. **Possible Causes:** • Check harness between the sensor and the ECM connector • Check the sensor • Check input signal of ECM

DTC	Trouble Code Title, Conditions & Possible Causes
DTC: P0333 **1T ECM, MIL: Yes** **Year:** 2009, 2010 **Model:** Outback, Tribeca **Engine:** 3.6L H6 **Transmission:** All	**Knock Sensor 2 Circuit High (Bank 2):** Poor driving performance. Knocking occurs. Knock sensor output voltage $\geq$ 4.709 V. **Possible Causes:** • Wiring harness between ECM and knock sensor needs repair. • Knock sensor wiring harness or connector needs repair. • Defective knock sensor. • Poor contact of the ECM connector needs repair.
DTC: P0335 **1T ECM, MIL: Yes** **Year:** 2009, 2010 **Model:** Forester, Impreza, Legacy, Outback, Tribeca **Engine:** 2.5L H4, 3.0L H6, 3.6L H6 **Transmission:** All	**Crankshaft Position Sensor "A" Circuit:** Engine stalls. Failure of engine to start. Starter switch - ON Battery voltage $\geq$ 8 V No crankshaft position sensor signal detected. **Possible Causes:** • Crankshaft position sensor bolt not tightened. • Wiring harness between ECM and crankshaft position sensor needs repair. • Defective crankshaft position sensor.
DTC: P0336 **2T ECM, MIL: Yes** **Year:** 2009, 2010 **Model:** Forester, Impreza, Legacy, Outback, Tribeca **Engine:** 2.5L H4, 3.0L H6, 3.6L H6 **Transmission:** All	**Crankshaft Position Sensor "A" Circuit Range/Performance:** Engine stalls. Failure of engine to start. Battery voltage $\geq$ 8 V. Engine speed < 4000 rpm. Amount of crank sensor signal during 1 rev. Not = 30. There is a problem in the number of crankshaft signals for every revolution. **Possible Causes:** • Crankshaft position sensor bolt not tightened. • Crank sprocket teeth cracked or damaged. • Timing belt out of alignment. • Defective crankshaft position sensor.
DTC: P0340 **1T ECM, MIL: Yes** **Year:** 2009, 2010 **Model:** Forester, Impreza, Legacy, Outback, Tribeca **Engine:** 2.5L H4, 3.0L H6, 3.6L H6 **Transmission:** All	**Camshaft Position Sensor "A" Circuit (Bank 1 or Single Sensor):** Engine stalls. Failure of engine to start. Battery voltage $\geq$ 8 V. Amount of camshaft sensor signal during 2 revs. not = 2. **Possible Causes:** • Wiring harness between ECM and camshaft position sensor needs repair. • Camshaft position sensor bolt needs tightening. • Defective camshaft position sensor.
DTC: P0341 **2T ECM, MIL: Yes** **Year:** 2009 **Model:** Forester **Engine:** 2.5L H4 **Transmission:** All	**Camshaft Position Sensor "A" Circuit Bank 1 or Single Sensor:** Engine stalls. Failure of engine to start. Cylinder number identification completed. Battery voltage $\geq$ 8 V. Engine speed - 550 rpm — 1000 rpm. Engine operation - Idling. Misfire not detected. Engine load change during 4 milliseconds $\leq$ 12799.8 rpm. Position of camshaft position sensor signal not between BTDC 10°CA and BTDC 80°CA. **Possible Causes:** • Wiring harness between ECM and camshaft position sensor needs repair. • Wiring harness between camshaft position sensor and chassis ground needs repair. • Camshaft position sensor bolt needs tightening. • Defective camshaft position sensor. • Camshaft sprocket teeth damaged or cracked. • Timing belt dislocated from proper position.

DTC	Trouble Code Title, Conditions & Possible Causes
DTC: P0345 **1T ECM, MIL: Yes** **Year:** 2009, 2010 **Model:** Legacy, Outback, Tribeca **Engine:** 3.0L H6, 3.6L H6 **Transmission:** All	**Camshaft Position Sensor "A" Circuit (Bank 2):** Engine stalls. Failure of engine to start. Battery voltage $\geq$ 8 V. Amount of camshaft sensor signal during 2 revs. not = 2. **Possible Causes:** • Wiring harness between ECM and camshaft position sensor needs repair. • Camshaft position sensor bolt needs tightening. • Defective camshaft position sensor.
DTC: P0345 **1T CCM, MIL: Yes** **Year:** 2009, 2010 **Model:** Forester, Impreza, Legacy, Outback **Engine:** 2.5L H4 **Transmission:** All	**Camshaft Position Sensor "A" Circuit Bank 2:** Engine stalls. Failure of engine to start. **Possible Causes:** • Check any other DTC on display • Check power supply • Check harness between sensor connector and ECM • Check the sensor • Check poor contact • Check the installation and condition of the sensor
DTC: P0365 **1T ECM, MIL: Yes** **Year:** 2009, 2010 **Model:** Outback, Tribeca **Engine:** 3.6L H6 **Transmission:** All	**Camshaft Position Sensor "B" Circuit (Bank 1):** Engine stalls. Failure of engine to start. Battery voltage $\geq$ 8 V. Amount of camshaft sensor signal during 2 revs < 3 **Possible Causes:** • Wiring harness between ECM and camshaft position sensor needs repair. • Camshaft position sensor bolt needs tightening. • Defective camshaft position sensor.
DTC: P0390 **1T ECM, MIL: Yes** **Year:** 2009, 2010 **Model:** Outback, Tribeca **Engine:** 3.6L H6 **Transmission:** All	**Camshaft Position Sensor "B" Circuit (Bank 2):** Engine stalls. Failure of engine to start. Battery voltage $\geq$ 8 V. Amount of camshaft sensor signal during 2 revs < 3 **Possible Causes:** • Wiring harness between ECM and camshaft position sensor needs repair. • Camshaft position sensor bolt needs tightening. • Defective camshaft position sensor.
DTC: P0400 **2T ECM, MIL: Yes** **Year:** 2009, 2010 **Model:** Forester, Outback, Tribeca **Engine:** 2.5L H4, 3.6L H6 **Transmission:** All	**Exhaust Gas Recirculation Flow:** Movement performance problem when engine is low speed. Improper idling. Movement performance problem. Change in intake manifold pressure detected by ECM is out of range. **Possible Causes:** • EGR valve, manifold absolute pressure sensor and throttle body not installed properly. • Holes, plugged piping or foreign objects caught in the EGR system.
DTC: P0400 **2T CCM, MIL: Yes** **Year:** 2009 **Model:** Legacy, Outback **Engine:** 3.0L H6 **Transmission:** All	**Exhaust Gas Recirculation Flow Malfunction:** Engine started, vehicle driven at a steady speed of 52-58 mph for 1-3 minutes in closed loop, and the PCM detected little or no change in Pressure sensor signal after the EGR solenoid was cycled on/off. **Possible Causes:** • Check any other DTC on display • Check EGR control solenoid valve circuit • Check vacuum line • Check EGR system

DTC	Trouble Code Title, Conditions & Possible Causes
DTC: P0410 **2T CCM, MIL: Yes** **Year:** 2009, 2010 **Model:** Forester, Impreza, Legacy, Outback **Engine:** 2.5L H4 **Transmission:** All	**Secondary Air Injection System:** Poor drivability. **Possible Causes:** • Check air pump operation • Check duct between pump and combination valve • Check power supply to pump • Check harness between pump relay and pump connector • Check pump relay • Check pump relay power source • Check harness between ECM and pump relay connector
DTC: P0411 **2T CCM, MIL: Yes** **Year:** 2009, 2010 **Model:** Forester, Impreza, Legacy, Outback **Engine:** 2.5L H4 **Transmission:** All	**Secondary Air Injection System Incorrect Flow Detected:** Poor drivability. **Possible Causes:** • Check secondary air combination valve
DTC: P0413 **1T CCM, MIL: Yes** **Year:** 2009, 2010 **Model:** Forester, Impreza, Legacy, Outback **Engine:** 2.5L H4 **Transmission:** All	**Secondary Air Injection System Switching Valve "A" Circuit Open:** Poor drivability. **Possible Causes:** • Check harness between ECM and valve relay #1
DTC: P0414 **1T CCM, MIL: Yes** **Year:** 2009, 2010 **Model:** Forester, Impreza, Legacy, Outback **Engine:** 2.5L H4 **Transmission:** All	**Secondary Air Injection System Switching Valve "A" Circuit Shorted:** Poor drivability. **Possible Causes:** • Check harness between ECM and valve relay #1
DTC: P0416 **1T CCM, MIL: Yes** **Year:** 2009, 2010 **Model:** Forester, Impreza, Legacy, Outback **Engine:** 2.5L H4 **Transmission:** All	**Secondary Air Injection System Switching Valve "B" Circuit Open:** Poor drivability. **Possible Causes:** • Check harness between ECM and valve relay #2
DTC: P0417 **1T CCM, MIL: Yes** **Year:** 2009, 2010 **Model:** Forester, Impreza, Legacy, Outback **Engine:** 2.5L H4 **Transmission:** All	**Secondary Air Injection System Switching Valve "B" Circuit Shorted:** Poor drivability. **Possible Causes:** • Check harness between ECM and valve relay #2
DTC: P0418 **1T CCM, MIL: Yes** **Year:** 2009, 2010 **Model:** Forester, Impreza, Legacy, Outback **Engine:** 2.5L H4 **Transmission:** All	**Secondary Air Injection System Switching Valve "A" Circuit Open:** Poor drivability. **Possible Causes:** • Check harness between ECM and pump relay

DTC	Trouble Code Title, Conditions & Possible Causes
DTC: P0420 **2T ECM, MIL: Yes** **Year:** 2009, 2010 **Model:** Forester, Impreza, Legacy, Outback, Tribeca **Engine:** 2.5L H4, 3.0L H6, 3.6L H6 **Transmission:** All	**Catalyst System Efficiency Below Threshold (Bank 1):** Engine stalls. Idle mixture is out of specifications. Vehicle is driven > 43.5 mph and ECM detected amplitudes signals that are too similar during the test. **Possible Causes:** • Loose or damaged exhaust system parts. • Defective catalytic converter. • Water in the rear oxygen sensor connector or coupling connector. • Wiring harness between ECM and rear oxygen sensor needs repair. • Defective rear oxygen sensor.
DTC: P0442 **2T ECM, MIL: Yes** **Year:** 2009, 2010 **Model:** Forester, Impreza, Legacy, Outback, Tribeca **Engine:** 2.5L H4, 3.0L H6, 3.6L H6 **Transmission:** All	**Evaporative Emission Control System Leak Detected (Small Leak):** Fuel odor. There is a hole of more than 1.0 mm (0.04 in) dia. in evaporation system or fuel tank. ECT and IAT sensors more than 90°F and within 12°F at startup, system voltage at 11-16 volts, throttle angle over 75%, MAP sensor more than 75 kPa, vehicle driven at a steady cruise speed of greater than 31 mph, fuel from 15-85%, and the PCM detected a vacuum decaying condition existed during the test. **Possible Causes:** • Fuel filler cap not tightened. • Fuel filler cap not genuine. • Damage to the seal between fuel filler cap and fuel filler pipe. • Drain valve does not operate. • Defective purge control solenoid valve. • Evaporative emission control system line needs repair. • Damaged canister. • Damaged fuel tank. • Hole, crack, clogging, or disconnections, bend, misconnection of hoses or pipes in evaporative emission control system.
DTC: P0447 **1T ECM, MIL: Yes** **Year:** 2009, 2010 **Model:** Forester, Impreza, Legacy, Outback, Tribeca **Engine:** 2.5L H4, 3.0L H6, 3.6L H6 **Transmission:** All	**Evaporative Emission Control System Vent Control Circuit Open:** Battery voltage ≥ 10.9 V. Elapsed time after starting the engine ≥ 1 second. Vent control terminal output voltage when ECM outputs OFF signal - Low. **Possible Causes:** • Power supply circuit needs repair. • Wiring harness between ECM and drain valve needs repair. • Poor contact of the ECM connector. • Defective drain valve.
DTC: P0448 **1T ECM, MIL: Yes** **Year:** 2009, 2010 **Model:** Forester, Impreza, Legacy, Outback, Tribeca **Engine:** 2.5L H4, 3.0L H6, 3.6L H6 **Transmission:** All	**Evaporative Emission Control System Vent Control Circuit Shorted:** Battery voltage ≥ 10.9 V. Elapsed time after starting the engine ≥ 1 second. Vent control solenoid terminal output voltage when ECM outputs ON signal - High. **Possible Causes:** • Wiring harness between ECM and drain valve connector needs repair. • Defective drain valve. • Poor contact in ECM connector.
DTC: P0451 **2T ECM, MIL: Yes** **Year:** 2009, 2010 **Model:** Forester, Impreza, Legacy, Outback, Tribeca **Engine:** 2.5L H4, 3.0L H6, 3.6L H6 **Transmission:** All	**Evaporative Emission Control System Pressure Sensor:** Elapsed time after starting the engine ≥ 60 s. Fuel level ≥ 9.6 L (2.54 US gal, 2.11 Imp gal). Fuel temperature < 35 °C (95 °F). Battery voltage ≥ 10.9 V. Barometric pressure ≥ 75 kPa (563 mmHg, 22.2 in. Hg). Maximum - Minimum tank pressure < 0 kPa (0.375 mmHg, 0 in. Hg). **Possible Causes:** • Fuel filler cap not secure. • Disconnection, leakage and clogging of the vacuum hoses and pipes between fuel tank pressure sensor and fuel tank. • Disconnection, leakage and clogging of air ventilation hoses and pipes between fuel filler pipe and fuel tank. • Defective fuel tank pressure sensor.

DTC	Trouble Code Title, Conditions & Possible Causes
DTC: P0452 **1T ECM, MIL: Yes** **Year:** 2009, 2010 **Model:** Forester, Impreza, Legacy, Outback, Tribeca **Engine:** 2.5L H4, 3.0L H6, 3.6L H6 **Transmission:** All	**Evaporative Emission Control System Pressure Sensor Low Input:** Fuel tank pressure < -7.5 kPa (-55.9 mmHg, -2.2 in. Hg) Battery voltage ≥ 10.9 V **Possible Causes:** • Wiring harness between fuel tank pressure sensor and chassis ground needs repair. • Wiring harness between fuel tank pressure sensor and ECM needs repair. • Wiring harness between chassis ground and ECM needs repair. • Poor contact in ECM or fuel tank pressure sensor connector. • Defective fuel tank pressure sensor.
DTC: P0453 **1T ECM, MIL: Yes** **Year:** 2009, 2010 **Model:** Forester, Impreza, Legacy, Outback, Tribeca **Engine:** 2.5L H4, 3.0L H6, 3.6L H6 **Transmission:** All	**Evaporative Emission Control System Pressure Sensor High Input:** Time needed for all secondary parameters to be in enable conditions ≥ 5000 ms. Vehicle speed ≥ 2 km/h (1.2 MPH). All conditions of EVAP canister purge - Completed. Learning value of evaporation gas concentration (left and right) < 0.08. Main feedback compensation coefficient (left and right) ≥ 0.9. Battery voltage ≥ 10.9 V. Fuel tank pressure ≥ 7.9 kPa (59.6 mmHg, 2.3 in. Hg). Fuel temperature < 35 °C (95 °F). Barometric pressure ≥ 75 kPa (563 mmHg, 22.2 in. Hg). **Possible Causes:** • Wiring harness between ECM and fuel tank pressure sensor needs repair. • Wiring harness between fuel tank pressure sensor and chassis ground needs repair. • Poor contact in the fuel tank pressure sensor connector. • Defective fuel tank pressure sensor.
DTC: P0456 **2T ECM, MIL: Yes** **Year:** 2009, 2010 **Model:** Forester, Impreza, Legacy, Outback, Tribeca **Engine:** 2.5L H4, 3.0L H6, 3.6L H6 **Transmission:** All	**Evaporative Emission Control System Leak Detected (Very Small Leak):** Fuel odor. There is a hole of more than 1.0 mm (0.04 in) dia. in evaporation system or fuel tank. ECT and IAT sensors more than 90°F and within 12°F at startup, system voltage at 11-16 volts, throttle angle over 75%, MAP sensor more than 75 kPa, vehicle driven at a steady cruise speed of greater than 31 mph, fuel from 15-85%, and the PCM detected a vacuum decaying condition existed during the test. **Possible Causes:** • Fuel filler cap not tightened. • Fuel filler cap not genuine. • Damage to the seal between fuel filler cap and fuel filler pipe. • Drain valve does not operate. • Defective purge control solenoid valve. • Evaporative emission control system line needs repair. • Damaged canister. • Damaged fuel tank. • Hole, crack, clogging, or disconnections, bend, misconnection of hoses or pipes in evaporative emission control system.
DTC: P0457 **2T ECM, MIL: Yes** **Year:** 2009, 2010 **Model:** Forester, Impreza, Legacy, Outback, Tribeca **Engine:** 2.5L H4, 3.0L H6, 3.6L H6 **Transmission:** All	**Evaporative Emission Control System Leak Detected (Fuel Cap Loose/Off):** Fuel odor. Fuel filler cap loose or lost. There is a hole of more than 1.0 mm (0.04 in) dia. in evaporation system or fuel tank. ECT and IAT sensors more than 90°F and within 12°F at startup, system voltage at 11-16 volts, throttle angle over 75%, MAP sensor more than 75 kPa, vehicle driven at a steady cruise speed of greater than 31 mph, fuel from 15-85%, and the PCM detected a vacuum decaying condition existed during the test. **Possible Causes:** • Fuel filler cap not tightened. • Fuel filler cap not genuine. • Damage to the seal between fuel filler cap and fuel filler pipe. • Drain valve does not operate. • Defective purge control solenoid valve. • Evaporative emission control system line needs repair. • Damaged canister. • Damaged fuel tank. • Hole, crack, clogging, or disconnections, bend, misconnection of hoses or pipes in evaporative emission control system.

DTC	Trouble Code Title, Conditions & Possible Causes
DTC: P0458 **21 ECM, MIL: Yes** **Year:** 2009, 2010 **Model:** Forester, Impreza, Legacy, Outback, Tribeca **Engine:** 2.5L H4, 3.0L H6, 3.6L H6 **Transmission:** All	**Evaporative Emission System Purge Control Valve Circuit Low:** Improper idling. Battery voltage $\geq$ 10.9 V. Elapsed time after starting the engine $\geq$ 1 second. Low purge control solenoid terminal output voltage. **Possible Causes:** • Power supply circuit needs repair. • Wiring harness between ECM and purge control solenoid valve needs repair. • Poor contact of purge control solenoid valve connector. • Defective purge control solenoid valve.
DTC: P0459 **2T ECM, MIL: Yes** **Year:** 2009, 2010 **Model:** Forester, Impreza, Legacy, Outback, Tribeca **Engine:** 2.5L H4, 3.0L H6, 3.6L H6 **Transmission:** All	**Evaporative Emission System Purge Control Valve Circuit High:** Improper idling. Battery voltage $\geq$ 10.9 V. Elapsed time after starting the engine $\geq$ 1 second. High purge control solenoid terminal output voltage. **Possible Causes:** • Wiring harness between ECM and purge control solenoid valve needs repair. • Poor contact of the ECM connector. • Defective purge control solenoid valve.
DTC: P0461 **2T ECM, MIL: Yes** **Year:** 2009, 2010 **Model:** Forester, Impreza, Legacy, Outback, Tribeca **Engine:** 2.5L H4, 3.0L H6, 3.6L H6 **Transmission:** All	**Fuel Level Sensor "A" Circuit Range/Performance:** Fuel level does not vary in a particular driving condition / engine condition where it should. Accumulated amount of intake air $\geq$ 330957 g (11672.85 oz). Max. - Min. values of fuel level output < 2.6 L (0.69 US gal, 0.57 Imp gal). Battery voltage $\geq$ 10.9 V. Engine speed < 6000 rpm. Elapsed time after starting the engine $\geq$ 5000 ms. **Possible Causes:** • Defective fuel level sensor and fuel sub level sensor.
DTC: P0462 **2T ECM, MIL: Yes** **Year:** 2009, 2010 **Model:** Forester, Impreza, Legacy, Outback, Tribeca **Engine:** 2.5L H4, 3.0L H6, 3.6L H6 **Transmission:** All	**Fuel Level Sensor "A" Circuit Low:** Battery voltage $\geq$ 10.9 V. Elapsed time after starting the engine $\geq$ 3000 ms. Low output voltage for fuel level sensor. **Possible Causes:** • Wiring harness between body integrated unit and chassis ground needs repair. • Defective fuel sub level sensor. • Defective fuel level sensor. • Intermittent poor contact in fuel level sensor terminals.
DTC: P0463 **2T ECM, MIL: Yes** **Year:** 2009, 2010 **Model:** Forester, Impreza, Legacy, Outback, Tribeca **Engine:** 2.5L H4, 3.0L H6, 3.6L H6 **Transmission:** All	**Fuel Level Sensor "A" Circuit High:** Battery voltage $\geq$ 10.9 V. Elapsed time after starting the engine $\geq$ 3000 ms. Fuel level sensor output voltage $\geq$ 7.212 V. **Possible Causes:** • Wiring harness between body integrated unit and chassis ground needs repair. • Defective fuel sub level sensor. • Defective fuel level sensor. • Intermittent poor contact in fuel level sensor terminals.
DTC: P0464 **2T ECM, MIL: Yes** **Year:** 2009, 2010 **Model:** Forester, Impreza, Legacy, Outback, Tribeca **Engine:** 2.5L H4, 3.0L H6, 3.6L H6 **Transmission:** All	**Fuel Level Sensor Circuit Intermittent:** Elapsed time after starting the engine $\geq$ 1 second. Battery voltage $\geq$ 10.9 V. Idle switch = ON. Fuel level $\geq$ 9.6 L (2.54 US gal, 2.11 Imp gal) and Fuel level < 54.4 L (14.37 US gal, 11.97 Imp gal. Vehicle not moving. Output voltage variation of the fuel level sensor is larger than the threshold value. **Possible Causes:** • Wiring harness between body integrated unit and chassis ground needs repair. • Defective fuel sub level sensor. • Defective fuel level sensor. • Intermittent poor contact in fuel level sensor terminals.

DTC	Trouble Code Title, Conditions & Possible Causes
DTC: P0500 **1T CCM, MIL: Yes** **Year:** 2009, 2010 **Model:** Forester, Impreza, Legacy, Outback, Tribeca **Engine:** 2.5L H4, 3.0L H6, 3.6L H6 **Transmission:** All	**Vehicle Speed Sensor:** Engine started, vehicle driven to a speed of over 30 mph at medium engine load for 30 seconds, and the PCM did not detect any VSS signals during the CCM Rationality test. **Possible Causes:** • Check DTC of VDC
DTC: P0506 **2T ECM, MIL: Yes** **Year:** 2009, 2010 **Model:** Forester, Legacy, Outback, Tribeca **Engine:** 2.5L H4, 3.0L H6, 3.6L H6 **Transmission:** All	**Idle Control System RPM Lower Than Expected:** Hard to start the engine. Engine does not start. Improper idling. Engine stalls. Actual engine speed more than 100 rpm less than target idle speed for over 10.5 seconds. **Possible Causes:** • Clogged air cleaner element. • Foreign matter inside electronic throttle control. • Defective electronic throttle control relay. • Electronic throttle control power supply circuit needs repair. • Wiring harness between ECM and Electronic throttle control needs repair. • Defective electronic throttle control.
DTC: P0507 **2T ECM, MIL: Yes** **Year:** 2009, 2010 **Model:** Forester, Legacy, Outback, Tribeca **Engine:** 2.5L H4, 3.0L H6, 3.6L H6 **Transmission:** All	**Idle Control System RPM Higher Than Expected:** Actual idle speed was more than 200 rpm than the target idle speed for more than 10.5 seconds. **Possible Causes:** • Air suction and leaks in the air intake system. • Foreign matter inside the electronic throttle control. • Defective electronic throttle control relay. • Electronic throttle control power supply circuit needs repair. • Wiring harness between ECM and Electronic throttle control needs repair. • Defective electronic throttle control.
DTC: P0512 **1T ECM, MIL: Yes** **Year:** 2009, 2010 **Model:** Forester, Impreza, Legacy, Outback, Tribeca **Engine:** 2.5L H4, 3.0L H6, 3.6L H6 **Transmission:** All	**Starter Request Circuit:** Failure of engine to start. Starter OFF signal - Not detected. Battery voltage $\geq$ 8 V. **Possible Causes:** • Wiring harness between ECM and ignition switch needs repair. • Poor contact of the ECM connector.
DTC: P0600 **2T CCM** **Year:** 2009 **Model:** Forester, Impreza **Engine:** 2.5L H4 **Transmission:** All	**Serial Communication Link:** Key on or engine running; and the PCM detected an unexpected low or high voltage condition on the Serial Communications Link circuit. **Possible Causes:** • Check engine start failure (does engine run) • Check illumination of MIL light • Check indication of DTC on display • Using the a scan tool, perform diagnosis
DTC: P0600 **1T CCM** **Year:** 2009 **Model:** Legacy, Outback **Engine:** 3.0L H6 **Transmission:** All	**Serial Communication Link:** Key on or engine running; and the PCM detected an unexpected low or high voltage condition on the Serial Communications Link circuit. **Possible Causes:** • Check engine start failure (does engine run) • Check illumination of MIL light • Check indication of DTC on display • Using the a scan tool, perform diagnosis
DTC: P0604 **2T CCM, MIL: Yes** **Year:** 2009, 2010 **Model:** Forester, Impreza, Legacy, Outback, Tribeca **Engine:** 2.5L H4, 3.0L H6, 3.6L H6 **Transmission:** All	**Internal Control Module Random Access Memory (RAM) Error:** Engine stalls. Engine does not start. **Possible Causes:** • Check any other DTC on display

DTC	Trouble Code Title, Conditions & Possible Causes
DTC: P0605 **1T CCM, MIL: Yes** **Year:** 2009, 2010 **Model:** Forester, Impreza, Legacy, Outback, Tribeca **Engine:** 2.5L H4, 3.0L H6, 3.6L H6 **Transmission:** All	**Internal Control Module Read Only Memory (ROM) Frror:** Erroneous idling. Poor driving performance. **Possible Causes:** • Check input voltage from ECM • Check ground harness
DTC: P0607 **1T ECM, MIL: Yes** **Year:** 2009, 2010 **Model:** Forester, Impreza, Legacy, Outback, Tribeca **Engine:** 2.5L H4, 3.0L H6, 3.6L H6 **Transmission:** All	**Throttle Control System Circuit Range/Performance:** Improper idling. Poor driving performance. **Possible Causes:** • Wiring harness between ECM and chassis ground needs repair. • Wiring harness between ECM and electronic throttle control needs repair. • Poor contact of the ECM connector. • Engine ground terminals need tightening.
DTC: P0638 **1T CCM, MIL: Yes** **Year:** 2009, 2010 **Model:** Forester, Impreza, Legacy, Outback, Tribeca **Engine:** 2.5L H4, 3.0L H6, 3.6L H6 **Transmission:** All	**Throttle Actuator Control range/Performance Bank 1:** Erroneous idling. Engine stalls. Poor driving performance. **Possible Causes:** • Check electronic throttle control relay • Check power supply of electronic control relay • Check harness between ECM and electronic throttle control relay • Check sensor output • Check poor contact • Check harness between ECM and electronic throttle control • Check sensor power supply • Check for short in ECM • Check sensor output • Check harness between ECM and electronic throttle control motor • Check electronic throttle control motor harness • Check ground circuit • Check electronic throttle control
DTC: P0700 **T** **Year:** 2009, 2010 **Model:** Forester, Impreza, Legacy, Outback, Tribeca **Engine:** 2.5L H4, 3.0L H6, 3.6L H6 **Transmission:** All	**Transaxle Control System (MIL Request):** Vehicle performance issue. **Possible Causes:** • Check engine start failure (does engine run) • Check illumination of MIL light • Check indication of DTC on display • Using the a scan tool, perform diagnosis
DTC: P0851 **2T ECM, MIL: Yes** **Year:** 2009, 2010 **Model:** Forester, Impreza, Legacy, Outback **Engine:** 2.5L H4 **Transmission:** All	**Park/Neutral Switch Input Circuit Low (AT Model):** Improper idling. Battery voltage ≥ 10.9 V. Starter relay is OFF. Neutral switch signal in ECM when "P"/"N" range in TCM are "OFF" and when the other switches are "ON". **Possible Causes:** • Selector cable needs repair or adjustment. • Poor contact in the ECM connector. • Wiring harness between ECM and transmission harness needs repair. • Wiring harness between transmission harness connector and chassis ground needs repair. • Defective inhibitor switch.
DTC: P0851 **2T ECM, MIL: Yes** **Year:** 2009, 2010 **Model:** Forester, Impreza, Legacy, Outback **Engine:** 2.5L H4 **Transmission:** All	**Neutral Switch Input Circuit Low (MT Model):** Improper idling. Battery voltage ≥ 10.9 V. Starter relay is OFF. Neutral switch signal stays ON when vehicle speed changes from 0 mph to >0 mph. **Possible Causes:** • Poor contact of the ECM connector. • Wiring harness between ECM and neutral position switch needs repair. • Defective neutral position switch.

DTC	Trouble Code Title, Conditions & Possible Causes
DTC: P0852 **2T ECM, MIL: Yes** **Year:** 2009 **Model:** Forester **Engine:** 2.5L H4 **Transmission:** All	**Park/Neutral Switch Input Circuit High (AT Model):** Battery voltage $\geq$ 10.9 V. Starter relay is OFF. Neutral switch signal in ECM when "P"/"N" range in TCM are "ON" and when the other switches are "OFF". **Possible Causes:** • Selector cable needs repair or adjustment. • Poor contact of the ECM connector. • Wiring harness between ECM and inhibitor switch needs repair. • Wiring harness between inhibitor and chassis ground needs repair.
DTC: P0852 **2T ECM, MIL: Yes** **Year:** 2009, 2010 **Model:** Forester, Impreza, Legacy, Outback **Engine:** 2.5L H4 **Transmission:** All	**Neutral Switch Input Circuit High (MT Model):** Improper idling. Battery voltage $\geq$ 10.9 V. Starter relay is OFF. Neutral switch signal stays OFF when vehicle speed changes from 0 mph to >40 mph. **Possible Causes:** • Poor contact of the ECM connector. • Wiring harness between ECM and neutral position switch needs repair. • Defective neutral position switch

OBD II Trouble Code List (P1XXX Codes)

DTC	Trouble Code Title, Conditions & Possible Causes
DTC: P1152 **2T ECM, MIL: Yes** **Year:** 2009, 2010 **Model:** Forester, Impreza, Legacy, Outback, Tribeca **Engine:** 2.5L H4, 3.0L H6, 3.6L H6 **Transmission:** All	**O2 Sensor Range/Performance (Low) (Bank 1 Sensor 1):** Poor drivability. While driving at a constant speed of 12.4 mph or more: Actual air fuel ratio/Theoretical air fuel ratio < 0.85 **Possible Causes:** • Water in the front oxygen sensor connector or coupling connector. • Wiring harness between ECM and front oxygen sensor needs repair. • Poor contact of front oxygen sensor. • Defective front oxygen sensor.
DTC: P1153 **2T ECM, MIL: Yes** **Year:** 2009, 2010 **Model:** Forester, Impreza, Legacy, Outback, Tribeca **Engine:** 2.5L H4, 3.0L H6, 3.6L H6 **Transmission:** All	**O2 Sensor Circuit Range/Performance (High) (Bank 1 Sensor 1):** Poor drivability. While driving at a constant speed of 12.4 mph or more: Actual air fuel ratio/Theoretical air fuel ratio > 1.15 **Possible Causes:** • Water in the front oxygen sensor connector or coupling connector. • Wiring harness between ECM and front oxygen sensor needs repair. • Poor contact of front oxygen sensor. • Defective front oxygen sensor.
DTC: P1154 **2T ECM, MIL: Yes** **Year:** 2009, 2010 **Model:** Legacy, Outback, Tribeca **Engine:** 3.0L H6, 3.6L H6 **Transmission:** All	**O2 Sensor Circuit Range/Performance (Low) (Bank 2 Sensor 1):** Poor drivability. While driving at a constant speed of 12.4 mph or more: Actual air fuel ratio/Theoretical air fuel ratio < 0.85 **Possible Causes:** • Water in the front oxygen sensor connector or coupling connector. • Wiring harness between ECM and front oxygen sensor needs repair. • Poor contact of front oxygen sensor. • Defective front oxygen sensor.
DTC: P1155 **2T ECM, MIL: Yes** **Year:** 2009, 2010 **Model:** Legacy, Outback, Tribeca **Engine:** 3.0L H6, 3.6L H6 **Transmission:** All	**O2 Sensor Circuit Range/Performance (High) (Bank 2 Sensor 1):** Poor drivability. While driving at a constant speed of 12.4 mph or more: Actual air fuel ratio/Theoretical air fuel ratio > 1.15 **Possible Causes:** • Water in the front oxygen sensor connector or coupling connector. • Wiring harness between ECM and front oxygen sensor needs repair. • Poor contact of front oxygen sensor. • Defective front oxygen sensor.

DTC	Trouble Code Title, Conditions & Possible Causes
DTC: P1160 **1T ECM, MIL: Yes** **Year:** 2009, 2010 **Model:** Forester, Impreza, Legacy, Outback, Tribeca **Engine:** 2.5L H4, 3.0L H6, 3.6L H6 **Transmission:** All	**Return Spring Failure:** Improper idling. Poor driving performance. Engine stalls. **Possible Causes:** • Defective electronic throttle control relay. • Wiring harness between electronic throttle control relay and chassis ground needs repair. • Wiring harness between ECM and electronic throttle control relay needs repair. • Wiring harness between ECM and chassis ground needs repair. • Defective ECM
DTC: P1400 **2T ECM, MIL: Yes** **Year:** 2009, 2010 **Model:** Forester, Impreza, Legacy, Outback **Engine:** 2.5L H4, 3.0L H6 **Transmission:** All	**Fuel Tank Pressure Control Solenoid Valve Circuit Low:** Battery voltage $\geq$ 10.9 V. Elapsed time after starting the engine $\geq$ 1 second. Low fuel tank pressure solenoid terminal output voltage when ECM outputs OFF signal. **Possible Causes:** • Wiring harness between pressure control solenoid valve and chassis ground needs repair. • Wiring harness between ECM and pressure control solenoid needs repair. • Poor contact of coupling connector. • Defective pressure control solenoid valve.
DTC: P1410 **1T CCM, MIL: Yes** **Year:** 2009, 2010 **Model:** Forester, Impreza, Legacy, Outback **Engine:** 2.5L H4 **Transmission:** All	**Secondary Air Injection System Switching Valve Stuck Open:** Poor drivability. **Possible Causes:** • Using the a scan tool, check current data • Check harness between mass air flow and intake air temperature sensor and ECM
DTC: P1418 **1T CCM, MIL: Yes** **Year:** 2009, 2010 **Model:** Forester, Impreza, Legacy, Outback **Engine:** 2.5L H4 **Transmission:** All	**Secondary Air Injection System Control "A" Circuit Shorted:** Poor drivability. **Possible Causes:** • Using a scan tool, check current data • Check harness between mass air flow and intake air temperature sensor and ECM
DTC: P1420 **2T ECM, MIL: Yes** **Year:** 2009, 2010 **Model:** Forester, Impreza, Legacy, Outback **Engine:** 2.5L H4, 3.0L H6 **Transmission:** All	**Fuel Tank Pressure Control Solenoid Valve Circuit High:** Battery voltage $\geq$ 10.9 V. Elapsed time after starting the engine $\geq$ 1 second. High pressure control solenoid valve terminal output voltage when ECM outputs ON signal. **Possible Causes:** • Wiring harness between ECM and pressure control solenoid needs repair. • Poor contact of ECM connector. • Defective pressure control solenoid valve.
DTC: P1443 **1T ECM, MIL: Yes** **Year:** 2009, 2010 **Model:** Forester, Impreza, Legacy, Outback, Tribeca **Engine:** 2.5L H4, 3.0L H6, 3.6L H6 **Transmission:** All	**Vent Control Solenoid Valve Function Problem:** Improper fuel supply. Battery voltage $\geq$ 10.9 V. Barometric pressure $\geq$ 75 kPa (563 mmHg, 22.2 in. Hg). Tank pressure when starter is turned from off to on is between -0.4 kPa (-3.2 mmHg, -0.1 in. Hg) and 1.4 kPa (10.7 mmHg, 0.4 in. Hg). **Possible Causes:** • Clogging in the drain hose. • Drain valve not operating. • Poor contact of the ECM connector.
DTC: P1491 **1T ECM, MIL: Yes** **Year:** 2009, 2010 **Model:** Forester, Impreza, Legacy, Outback **Engine:** 2.5L H4 **Transmission:** All	**Positive Crankcase Ventilation (Blow-by) Function Problem:** Improper idling. With the engine started, the positive crankcase ventilation diagnosis voltage is High. **Possible Causes:** • Cracked or disconnected blow-by hose. • Wiring harness between ECM and PCV hose assembly needs repair. • Wiring harness between PCV hose assembly and ground chassis needs repair. • Poor contact in ECM and PCV hose assembly connector. • PCV hose assembly needs replacement.

DTC	Trouble Code Title, Conditions & Possible Causes
DTC: P1492 **1T ECM, MIL: Yes** **Year:** 2009, 2010 **Model:** Forester, Legacy, Outback, Tribeca **Engine:** 2.5L H4, 3.0L H6, 3.6L H6 **Transmission:** All	**EGR Solenoid Valve Signal #1 Circuit Malfunction (Low Input):** Improper idling. Poor driving performance. Engine breathing. **Possible Causes:** • Wiring harness between EGR valve and power supply needs repair. • Wiring harness between ECM and EGR valve connector needs repair. • Poor contact in the ECM and EGR valve connector. • Defective EGR valve.
DTC: P1493 **1T ECM, MIL: Yes** **Year:** 2009, 2010 **Model:** Forester, Impreza, Legacy, Outback, Tribeca **Engine:** 2.5L H4, 3.0L H6, 3.6L H6 **Transmission:** All	**EGR Solenoid Valve Signal #1 Circuit Malfunction (High Input):** Improper idling. Poor driving performance. Engine breathing. **Possible Causes:** • Wiring harness between ECM and EGR valve connector needs repair. • Wiring harness between ECM and chassis ground needs repair. • Poor contact in the ECM and EGR valve connector.
DTC: P1494 **1T ECM, MIL: Yes** **Year:** 2009, 2010 **Model:** Forester, Impreza, Legacy, Outback, Tribeca **Engine:** 2.5L H4, 3.0L H6, 3.6L H6 **Transmission:** All	**EGR Solenoid Valve Signal #2 Circuit Malfunction (Low Input):** Improper idling. Poor driving performance. Engine breathing. **Possible Causes:** • Wiring harness between EGR valve and power supply needs repair. • Wiring harness between ECM and EGR valve connector needs repair. • Poor contact in the ECM and EGR valve connector. • Defective EGR valve.
DTC: P1495 **1T ECM, MIL: Yes** **Year:** 2009, 2010 **Model:** Forester, Impreza, Legacy, Outback, Tribeca **Engine:** 2.5L H4, 3.0L H6, 3.6L H6 **Transmission:** All	**EGR Solenoid Valve Signal #2 Circuit Malfunction (High Input):** Improper idling. Poor driving performance. Engine breathing. **Possible Causes:** • Wiring harness between ECM and EGR valve connector needs repair. • Wiring harness between ECM and chassis ground needs repair. • Poor contact in the ECM and EGR valve connector.
DTC: P1496 **1T ECM, MIL: Yes** **Year:** 2009, 2010 **Model:** Forester, Legacy, Outback, Tribeca **Engine:** 2.5L H4, 3.0L H6, 3.6L H6 **Transmission:** All	**EGR Solenoid Valve Signal #3 Circuit Malfunction (Low Input):** Improper idling. Poor driving performance. Engine breathing. **Possible Causes:** • Wiring harness between EGR valve and power supply needs repair. • Wiring harness between ECM and EGR valve connector needs repair. • Poor contact in the ECM and EGR valve connector. • Defective EGR valve.
DTC: P1497 **1T ECM, MIL: Yes** **Year:** 2009, 2010 **Model:** Forester, Impreza, Legacy, Outback, Tribeca **Engine:** 2.5L H4, 3.0L H6, 3.6L H6 **Transmission:** All	**EGR Solenoid Valve Signal #3 Circuit Malfunction (High Input):** Improper idling. Poor driving performance. Engine breathing. **Possible Causes:** • Wiring harness between ECM and EGR valve connector needs repair. • Wiring harness between ECM and chassis ground needs repair. • Poor contact in the ECM and EGR valve connector.

DTC	Trouble Code Title, Conditions & Possible Causes
DTC: P1498 **1T ECM, MIL: Yes** **Year:** 2009, 2010 **Model:** Forester, Impreza, Legacy, Outback, Tribeca **Engine:** 2.5L H4, 3.0L H6, 3.6L H6 **Transmission:** All	**EGR Solenoid Valve Signal #4 Circuit Malfunction (Low Input):** Improper idling. Poor driving performance. Engine breathing. **Possible Causes:** • Wiring harness between EGR valve and power supply needs repair. • Wiring harness between ECM and EGR valve connector needs repair. • Poor contact in the ECM and EGR valve connector. • Defective EGR valve.
DTC: P1499 **1T ECM, MIL: Yes** **Year:** 2009, 2010 **Model:** Forester, Impreza, Legacy, Outback, Tribeca **Engine:** 2.5L H4, 3.0L H6, 3.6L H6 **Transmission:** All	**EGR Solenoid Valve Signal #4 Circuit Malfunction (High Input):** Improper idling. Poor driving performance. Engine breathing. **Possible Causes:** • Wiring harness between ECM and EGR valve connector needs repair. • Wiring harness between ECM and chassis ground needs repair. • Poor contact in the ECM and EGR valve connector.
DTC: P1518 **1T CCM** **Year:** 2009, 2010 **Model:** Outback, Tribeca **Engine:** 3.6L H6 **Transmission:** All	**Starter Switch Circuit Low Input:** Ignition key in crank position, and the PCM detected an unexpected "low" voltage condition on the Starter Switch circuit during the test. **NOTE: The engine will not start with this condition present.** **Possible Causes:** • Check operation of starter motor
DTC: P1560 **1T ECM, MIL: Yes** **Year:** 2009, 2010 **Model:** Forester, Impreza, Legacy, Outback, Tribeca **Engine:** 2.5L H4, 3.0L H6, 3.6L H6 **Transmission:** All	**Back-up Voltage Circuit Malfunction:** Battery voltage ≥ 10.9 V. Engine condition - After engine starting. Low voltage condition of back-up power supply. **Possible Causes:** • Wiring harness between ECM and ground chassis needs repair. • Poor contact of the ECM connector. • Defective fuse. • Wiring harness between ECM and negative battery terminal needs repair.
DTC: P1602 **2T ECM, MIL: Yes** **Year:** 2009, 2010 **Model:** Forester, Impreza, Legacy, Outback, Tribeca **Engine:** 2.5L H4, 3.0L H6, 3.6L H6 **Transmission:** All	**Control Module Programming Error:** Engine keeps running at higher speed than specified idle speed. Engine keeps running at a lower speed than the specified idle speed. Engine stalls. **Possible Causes:** • Incorrect amount of engine oil. • Holes or loose bolts on the exhaust system. • Loose mounting bolts, holes, or disconnected hoses on the air intake system. • Clogged or bent fuel lines. • Improper fuel pump discharge. • Defective engine coolant temperature sensor. • Defective mass air flow and intake air temperature sensor. • Wiring harness between ECM and fuel injector needs repair. • Faulty fuel injector. • Fuel injector wiring needs repair. • Defective electronic throttle control.

OBD II Trouble Code List (P2XXX Codes)

DTC	Trouble Code Title, Conditions & Possible Causes
DTC: P2004 **1T ECM, MIL: Yes** **Year:** 2009, 2010 **Model:** Forester, Impreza, Legacy, Outback **Engine:** 2.5L H4 **Transmission:** All	**Intake Manifold Runner Control Stuck Open (Bank 1):** Improper idling. Poor driving performance. **Possible Causes:** • Dirt or foreign objects clogging the tumble generator valve. • Defective tube generator valve assembly.

DTC	Trouble Code Title, Conditions & Possible Causes
DTC: P2005 **1T ECM, MIL: Yes** **Year:** 2009, 2010 **Model:** Forester, Impreza, Legacy, Outback **Engine:** 2.5L H4 **Transmission:** All	**Intake Manifold Runner Control Stuck Open (Bank 2):** Improper idling. Poor driving performance. **Possible Causes:** • Dirt or foreign objects clogging the tumble generator valve. • Defective tumble generator valve assembly.
DTC: P2006 **1T ECM, MIL: Yes** **Year:** 2009, 2010 **Model:** Forester, Impreza, Legacy, Outback **Engine:** 2.5L H4 **Transmission:** All	**Intake Manifold Runner Control Stuck Closed (Bank 1):** Improper idling. Poor driving performance. **Possible Causes:** • Dirt or foreign objects clogging the tumble generator valve. • Defective tumble generator valve assembly.
DTC: P2007 **1T ECM, MIL: Yes** **Year:** 2009, 2010 **Model:** Forester, Impreza, Legacy, Outback **Engine:** 2.5L H4 **Transmission:** All	**Intake Manifold Runner Control Stuck Closed (Bank 2):** Improper idling. Poor driving performance. **Possible Causes:** • Dirt or foreign objects clogging the tumble generator valve. • Defective tumble generator valve assembly.
DTC: P2008 **1T ECM, MIL: Yes** **Year:** 2009, 2010 **Model:** Forester, Impreza, Legacy, Outback **Engine:** 2.5L H4 **Transmission:** All	**Intake Manifold Runner Control Circuit / Open (Bank 1):** Improper idling. Poor driving performance. **Possible Causes:** • Wiring harness between ECM and tumble generator valve assembly needs repair. • Dirt or foreign objects clogging the tumble generator valve. • Defective tumble generator valve assembly.
DTC: P2009 **1T ECM, MIL: Yes** **Year:** 2009, 2010 **Model:** Forester, Impreza, Legacy, Outback **Engine:** 2.5L H4 **Transmission:** All	**Intake Manifold Runner Control Circuit Low (Bank 1):** Improper idling. Poor driving performance. **Possible Causes:** • Wiring harness between ECM and tumble generator valve assembly needs repair. • Defective tumble generator valve assembly.
DTC: P2011 **1T ECM, MIL: Yes** **Year:** 2009, 2010 **Model:** Forester, Impreza, Legacy, Outback **Engine:** 2.5L H4 **Transmission:** All	**Intake Manifold Runner Control Circuit / Open (Bank 2):** Improper idling. Poor driving performance. **Possible Causes:** • Wiring harness between ECM and tumble generator valve assembly needs repair. • Defective tumble generator valve assembly.
DTC: P2012 **1T ECM, MIL: Yes** **Year:** 2009, 2010 **Model:** Forester, Impreza, Legacy, Outback **Engine:** 2.5L H4 **Transmission:** All	**Intake Manifold Runner Control Circuit Low (Bank 2):** Improper idling. Poor driving performance. **Possible Causes:** • Wiring harness between ECM and tumble generator valve assembly needs repair. • Defective tumble generator valve assembly.
DTC: P2016 **1T ECM, MIL: Yes** **Year:** 2009, 2010 **Model:** Forester, Impreza, Legacy, Outback **Engine:** 2.5L H4 **Transmission:** All	**Intake Manifold Runner Position Sensor / Switch Circuit Low (Bank 1):** Improper idling. Engine stalls. Poor driving performance. **Possible Causes:** • Wiring harness between power supply and tumble generator valve assembly needs repair. • Wiring harness between ECM and tumble generator valve assembly needs repair. • Defective tumble generator valve assembly.

DTC	Trouble Code Title, Conditions & Possible Causes
DTC: P2017 **1T ECM, MIL: Yes** **Year:** 2009, 2010 **Model:** Forester, Impreza, Legacy, Outback **Engine:** 2.5L H4 **Transmission:** All	**Intake Manifold Runner Position Sensor / Switch Circuit High (Bank 1):** Improper idling. Engine stalls. Poor driving performance. **Possible Causes:** • Wiring harness between ECM and tumble generator valve assembly needs repair. • Defective tumble generator valve assembly.
DTC: P2021 **1T ECM, MIL: Yes** **Year:** 2009, 2010 **Model:** Forester, Impreza, Legacy, Outback **Engine:** 2.5L H4 **Transmission:** All	**Intake Manifold Runner Position Sensor / Switch Circuit Low (Bank 2):** Improper idling. Engine stalls. Poor driving performance. **Possible Causes:** • Wiring harness between power supply and tumble generator valve assembly needs repair. • Wiring harness between ECM and tumble generator valve assembly needs repair. • Defective tumble generator valve assembly.
DTC: P2022 **1T ECM, MIL: Yes** **Year:** 2009, 2010 **Model:** Forester, Impreza, Legacy, Outback **Engine:** 2.5L H4 **Transmission:** All	**Intake Manifold Runner Position Sensor / Switch Circuit High (Bank 2):** Improper idling. Engine stalls. Poor driving performance. **Possible Causes:** • Wiring harness between power supply and tumble generator valve assembly needs repair. • Wiring harness between ECM and tumble generator valve assembly needs repair. • Defective tumble generator valve assembly.
DTC: P2088 **1T ECM, MIL: Yes** **Year:** 2009, 2010 **Model:** Forester, Impreza, Legacy, Outback, Tribeca **Engine:** 2.5L H4, 3.0L H6, 3.6L H6 **Transmission:** All	**Intake Camshaft Position Actuator Control Circuit Low (Bank 1):** Improper idling. **Possible Causes:** • Wiring harness between ECM and oil flow control solenoid valve needs repair. • Poor contact of ECM and oil flow control solenoid valve connector. • Defective oil flow control solenoid valve.
DTC: P2089 **1T ECM, MIL: Yes** **Year:** 2009, 2010 **Model:** Forester, Impreza, Legacy, Outback, Tribeca **Engine:** 2.5L H4, 3.0L H6, 3.6L H6 **Transmission:** All	**Intake Camshaft Position Actuator Control Circuit High (Bank 1):** Improper idling. **Possible Causes:** • Wiring harness between ECM and oil flow control solenoid valve needs repair. • Poor contact of ECM and oil flow control solenoid valve connector. • Defective oil flow control solenoid valve.
DTC: P2092 **1T ECM, MIL: Yes** **Year:** 2009, 2010 **Model:** Forester, Impreza, Legacy, Outback, Tribeca **Engine:** 2.5L H4, 3.0L H6, 3.6L H6 **Transmission:** All	**Intake Camshaft Position Actuator Control Circuit Low (Bank 2):** Improper idling. **Possible Causes:** • Wiring harness between ECM and oil flow control solenoid valve needs repair. • Poor contact of ECM and oil flow control solenoid valve connector. • Defective oil flow control solenoid valve.
DTC: P2093 **1T ECM, MIL: Yes** **Year:** 2009, 2010 **Model:** Forester, Impreza, Legacy, Outback, Tribeca **Engine:** 2.5L H4, 3.0L H6, 3.6L H6 **Transmission:** All	**Intake Camshaft Position Actuator Control Circuit High (Bank 2):** Improper idling. **Possible Causes:** • Wiring harness between ECM and oil flow control solenoid valve needs repair. • Poor contact of ECM and oil flow control solenoid valve connector. • Defective oil flow control solenoid valve.
DTC: P2094 **1T ECM, MIL: Yes** **Year:** 2009, 2010 **Model:** Outback, Tribeca **Engine:** 3.6L H6 **Transmission:** All	**Exhaust Camshaft Position Actuator Control Circuit Low (Bank 2):** Improper idling. **Possible Causes:** • Wiring harness between ECM and oil flow control solenoid valve needs repair. • Poor contact of ECM and oil flow control solenoid valve connector. • Defective oil flow control solenoid valve.

DTC	Trouble Code Title, Conditions & Possible Causes
DTC: P2095 **1T ECM, MIL: Yes** **Year:** 2009, 2010 **Model:** Outback, Tribeca **Engine:** 3.6L H6 **Transmission:** All	**Exhaust Camshaft Position Actuator Control Circuit High (Bank 2):** Improper idling. **Possible Causes:** • Wiring harness between ECM and oil flow control solenoid valve needs repair. • Poor contact of ECM and oil flow control solenoid valve connector. • Defective oil flow control solenoid valve.
DTC: P2096 **2T ECM, MIL: Yes** **Year:** 2009, 2010 **Model:** Forester, Impreza, Legacy, Outback, Tribeca **Engine:** 2.5L H4, 3.0L H6, 3.6L H6 **Transmission:** All	**Post Catalyst Fuel Trim System Too Lean (Bank 1):** Poor drivability. **Possible Causes:** • Water in the front oxygen sensor connector or coupling connector. • Wiring harness between ECM and front oxygen sensor needs repair. • Wiring harness between front oxygen sensor and chassis ground needs repair. • Poor contact of the ECM connector. • Loose mounting bolts or holes in the exhaust system. • Loose mounting bolts or disconnected hoses on the air intake system. • Clogged or bent fuel lines. • Improper fuel pump discharge. • Defective engine coolant temperature sensor. • Defective mass air flow and intake air temperature sensor. • Defective front oxygen sensor. • Defective rear oxygen sensor.
DTC: P2097 **2T ECM, MIL: Yes** **Year:** 2009, 2010 **Model:** Forester, Impreza, Legacy, Outback, Tribeca **Engine:** 2.5L H4, 3.0L H6, 3.6L H6 **Transmission:** All	**Post Catalyst Fuel Trim System Too Rich (Bank 1):** Poor drivability. **Possible Causes:** • Water in the front oxygen sensor connector or coupling connector. • Wiring harness between ECM and front oxygen sensor needs repair. • Wiring harness between front oxygen sensor and chassis ground needs repair. • Poor contact of the ECM connector. • Loose mounting bolts or holes in the exhaust system. • Loose mounting bolts or disconnected hoses on the air intake system. • Clogged or bent fuel lines. • Improper fuel pump discharge. • Defective engine coolant temperature sensor. • Defective mass air flow and intake air temperature sensor. • Defective front oxygen sensor. • Defective rear oxygen sensor.
DTC: P2098 **2T ECM, MIL: Yes** **Year:** 2009, 2010 **Model:** Legacy, Outback, Tribeca **Engine:** 3.0L H6, 3.6L H6 **Transmission:** All	**Post Catalyst Fuel Trim System Too Lean Bank 2:** Poor drivability. **Possible Causes:** • Water in the front oxygen sensor connector or coupling connector. • Wiring harness between ECM and front oxygen sensor needs repair. • Wiring harness between front oxygen sensor and chassis ground needs repair. • Poor contact of the ECM connector. • Loose mounting bolts or holes in the exhaust system. • Loose mounting bolts or disconnected hoses on the air intake system. • Clogged or bent fuel lines. • Improper fuel pump discharge. • Defective engine coolant temperature sensor. • Defective mass air flow and intake air temperature sensor. • Defective front oxygen sensor. • Defective rear oxygen sensor.

DTC	Trouble Code Title, Conditions & Possible Causes
DTC: P2099 **2T ECM, MIL: Yes** **Year:** 2009, 2010 **Model:** Legacy, Outback, Tribeca **Engine:** 3.0L H6, 3.6L H6 **Transmission:** All	**Post Catalyst Fuel Trim System Too Rich Bank 2:** Poor drivability. **Possible Causes:** • Water in the front oxygen sensor connector or coupling connector. • Wiring harness between ECM and front oxygen sensor needs repair. • Wiring harness between front oxygen sensor and chassis ground needs repair. • Poor contact of the ECM connector. • Loose mounting bolts or holes in the exhaust system. • Loose mounting bolts or disconnected hoses on the air intake system. • Clogged or bent fuel lines. • Improper fuel pump discharge. • Defective engine coolant temperature sensor. • Defective mass air flow and intake air temperature sensor. • Defective front oxygen sensor. • Defective rear oxygen sensor.
DTC: P2101 **1T ECM, MIL: Yes** **Year:** 2009, 2010 **Model:** Forester, Impreza, Legacy, Outback, Tribeca **Engine:** 2.5L H4, 3.0L H6, 3.6L H6 **Transmission:** All	**Throttle Actuator Control Motor Circuit Range/Performance:** Improper idling. Poor driving performance. Engine stalls. **Possible Causes:** • Defective electronic throttle control relay. • Wiring harness between electronic throttle control relay and chassis ground needs repair. • Wiring harness between ECM and electronic throttle control relay needs repair. • Wiring harness between ECM and chassis ground needs repair. • Defective electronic throttle control. • Defective ECM
DTC: P2102 **1T ECM, MIL: Yes** **Year:** 2009, 2010 **Model:** Forester, Impreza, Legacy, Outback, Tribeca **Engine:** 2.5L H4, 3.0L H6, 3.6L H6 **Transmission:** All	**Throttle Actuator Control Motor Circuit Low:** Improper idling. Poor driving performance. Engine stalls. **Possible Causes:** • Defective electronic throttle control relay. • Wiring harness between electronic throttle control relay and chassis ground needs repair. • Wiring harness between ECM and electronic throttle control relay needs repair. • Poor contact in the ECM connector.
DTC: P2103 **1T ECM, MIL: Yes** **Year:** 2009, 2010 **Model:** Forester, Legacy, Outback, Tribeca **Engine:** 2.5L H4, 3.0L H6, 3.6L H6 **Transmission:** All	**Throttle Actuator Control Motor Circuit High:** Electronic throttle control relay output is OFF. Control Motor power voltage > 5V. **Possible Causes:** • Defective electronic throttle control relay. • Wiring harness between electronic throttle control relay connector and chassis ground needs repair. • Poor contact of the ECM connector. • Short circuit in harness to ground between ECM and electronic throttle control relay connector needs repair.
DTC: P2109 **1T ECM, MIL: Yes** **Year:** 2009, 2010 **Model:** Forester, Impreza, Legacy, Outback, Tribeca **Engine:** 2.5L H4, 3.0L H6, 3.6L H6 **Transmission:** All	**Throttle/Pedal Position Sensor "A" Minimum Stop Performance:** Improper idling. Poor driving performance. Engine stalls. **Possible Causes:** • Defective electronic throttle control relay. • Wiring harness between electronic throttle control relay and chassis ground needs repair. • Wiring harness between ECM and electronic throttle control relay needs repair. • Wiring harness between ECM and chassis ground needs repair. • Defective electronic throttle control. • Defective ECM

DTC	Trouble Code Title, Conditions & Possible Causes

DTC: P2122
1T ECM, MIL: Yes
Year: 2009, 2010
Model: Forester, Impreza, Legacy, Outback, Tribeca
Engine: 2.5L H4, 3.0L H6, 3.6L H6
Transmission: All
	Throttle/Pedal Position Sensor "D" Circuit Low Input:
	Improper idling.
	Poor driving performance.
	Possible Causes:
	   • Wiring harness between ECM and accelerator pedal position sensor needs repair.
	   • Defective accelerator pedal assembly.
	   • Defective ECM

DTC: P2123
1T ECM, MIL: Yes
Year: 2009, 2010
Model: Forester, Impreza, Legacy, Outback, Tribeca
Engine: 2.5L H4, 3.0L H6, 3.6L H6
Transmission: All
	Throttle/Pedal Position Sensor "D" Circuit High Input:
	Improper idling.
	Poor driving performance.
	Possible Causes:
	   • Wiring harness between ECM and accelerator pedal position sensor needs repair.
	   • Poor contact of accelerator pedal position sensor connector.
	   • Defective accelerator pedal assembly.

DTC: P2127
1T ECM, MIL: Yes
Year: 2009, 2010
Model: Forester, Impreza, Legacy, Outback, Tribeca
Engine: 2.5L H4, 3.0L H6, 3.6L H6
Transmission: All
	Throttle/Pedal Position Sensor "E" Circuit Low Input:
	Improper idling.
	Poor driving performance.
	Possible Causes:
	   • Wiring harness between ECM and chassis ground needs repair.
	   • Wiring harness between the ECM and accelerator pedal position sensor needs repair.
	   • Defective accelerator pedal assembly.
	   • Defective ECM.

DTC: P2128
1T ECM, MIL: Yes
Year: 2009, 2010
Model: Forester, Impreza, Legacy, Outback, Tribeca
Engine: 2.5L H4, 3.0L H6, 3.6L H6
Transmission: All
	Throttle/Pedal Position Sensor "E" Circuit High Input:
	Improper idling.
	Poor driving performance.
	Possible Causes:
	   • Wiring harness between ECM and accelerator pedal position sensor needs repair.
	   • Poor contact of accelerator pedal position sensor connector.
	   • Defective accelerator pedal assembly.

DTC: P2135
1T ECM, MIL: Yes
Year: 2009, 2010
Model: Forester, Impreza, Legacy, Outback, Tribeca
Engine: 2.5L H4, 3.0L H6, 3.6L H6
Transmission: All
	Throttle/Pedal Position Sensor/Switch "A"/"B" Voltage Correlation:
	Improper idling.
	Poor driving performance.
	Possible Causes:
	   • Wiring harness between ECM and electronic throttle control connector.
	   • Defective ECM.
	   • Poor contact of coupling connector.
	   • Open circuit of harness between ECM and engine ground.
	   • Poor contact in ECM connector.
	   • Poor contact of the electronic throttle control connector.
	   • Defective electronic throttle control.

DTC: P2138
1T ECM, MIL: Yes
Year: 2009, 2010
Model: Forester, Impreza, Legacy, Outback, Tribeca
Engine: 2.5L H4, 3.0L H6, 3.6L H6
Transmission: All
	Throttle/Pedal Position Sensor/Switch "D"/"E" Voltage Correlation:
	Improper idling.
	Poor driving performance.
	Possible Causes:
	   • Wiring harness between accelerator pedal position sensor and chassis ground needs repair.
	   • Wiring harness between ECM and accelerator pedal position sensor needs repair.
	   • Defective accelerator pedal assembly.
	   • Poor contact in ECM connector.

DTC: P2227 **2T ECM, MIL: Yes** **Year:** 2009, 2010 **Model:** Forester, Legacy, Outback, Tribeca **Engine:** 2.5L H4, 3.0L H6, 3.6L H6 **Transmission:** All	**Barometric Pressure Circuit Range/Performance:** Engine speed < 300 rpm. Vehicle speed < 0.6 mph. Barometric pressure - Intake manifold pressure $\geq$ 26.7 kPa (200 mmHg, 7.9 in. Hg). Intake manifold pressure at engine start - Intake manifold pressure < 1.3 kPa (9.99 mmHg, 0.4 in. Hg). **Possible Causes:** • Defective ECM. The barometric pressure sensor is built into the ECM.

DTC	Trouble Code Title, Conditions & Possible Causes
DTC: P2228 **1T ECM, MIL: Yes** **Year:** 2009, 2010 **Model:** Forester, Legacy, Outback, Tribeca **Engine:** 2.5L H4, 3.0L H6, 3.6L H6 **Transmission:** All	**Barometric Pressure Circuit Low:** Barometric pressure sensor output voltage < 1.706646812 V. **Possible Causes:** • Defective ECM. The barometric pressure sensor is built into the ECM.
DTC: P2229 **1T ECM, MIL: Yes** **Year:** 2009, 2010 **Model:** Forester, Legacy, Outback, Tribeca **Engine:** 2.5L H4, 3.0L H6, 3.6L H6 **Transmission:** All	**Barometric Pressure Circuit High:** Barometric pressure sensor output voltage ≥ 4.233789985 V. **Possible Causes:** • Defective ECM. The barometric pressure sensor is built into the ECM.
DTC: P2431 **2T CCM, MIL: Yes** **Year:** 2009, 2010 **Model:** Forester, Impreza, Legacy, Outback **Engine:** 2.5L H4 **Transmission:** All	**Secondary Air Injection System Air Flow/Pressure Sensor Circuit Range/Performance:** Poor drivability. **Possible Causes:** • Check any other DTC on display • Using a scan tool, check front oxygen sensor data
DTC: P2432 **1T CCM, MIL: Yes** **Year:** 2009, 2010 **Model:** Forester, Impreza, Legacy **Engine:** 2.5L H4 **Transmission:** All	**Secondary Air Injection System Air Flow/Pressure Sensor Circuit Low:** Poor drivability. **Possible Causes:** • Check harness between ECM and valve LH connector
DTC: P2433 **1T CCM, MIL: Yes** **Year:** 2009, 2010 **Model:** Forester, Impreza, Legacy **Engine:** 2.5L H4 **Transmission:** All	**Secondary Air Injection System Air Flow/Pressure Sensor Circuit High:** Poor drivability. **Possible Causes:** • Check harness between ECM and valve LH connector
DTC: P2440 **2T CCM, MIL: Yes** **Year:** 2009, 2010 **Model:** Forester, Impreza, Legacy **Engine:** 2.5L H4 **Transmission:** All	**Secondary Air Injection System Switching Valve Stock Open (bank 1):** Poor drivability. **Possible Causes:** • Check secondary air combination valve operation • Check duct between secondary air pump and secondary air combination valve • Check pipe between secondary air combination valve and cylinder head • Check power supply to valve • Check harness between secondary air combination valve relay and secondary air combination valve connector terminal • Check valve relay • Check harness between ECM and valve relay connector
DTC: P2441 **2T CCM, MIL: Yes** **Year:** 2009, 2010 **Model:** Forester, Impreza, Legacy **Engine:** 2.5L H4 **Transmission:** All	**Secondary Air Injection System Switching Valve Stock Closed (bank 1):** Poor drivability. **Possible Causes:** • Check secondary air combination valve operation • Check duct between secondary air pump and secondary air combination valve • Check pipe between secondary air combination valve and cylinder head • Check power supply to valve • Check harness between secondary air combination valve relay and secondary air combination valve connector terminal • Check valve relay • Check harness between ECM and valve relay connector

DTC	Trouble Code Title, Conditions & Possible Causes
DTC: P2442 **2T CCM, MIL: Yes** **Year:** 2009, 2010 **Model:** Forester, Impreza, Legacy **Engine:** 2.5L H4 **Transmission:** All	**Secondary Air Injection System Switching Valve Stock Open (bank 2):** Poor drivability. **Possible Causes:** • Check secondary air combination valve operation • Check duct between secondary air pump and secondary air combination valve • Check pipe between secondary air combination valve and cylinder head • Check power supply to valve • Check harness between secondary air combination valve relay and secondary air combination valve connector terminal • Check valve relay • Check harness between ECM and valve relay connector
DTC: P2443 **2T CCM, MIL: Yes** **Year:** 2009, 2010 **Model:** Forester, Impreza, Legacy **Engine:** 2.5L H4 **Transmission:** All	**Secondary Air Injection System Switching Valve Stock Closed (bank 2):** Poor drivability. **Possible Causes:** • Check secondary air combination valve operation • Check duct between secondary air pump and secondary air combination valve • Check pipe between secondary air combination valve and cylinder head • Check power supply to valve • Check harness between secondary air combination valve relay and secondary air combination valve connector terminal • Check valve relay • Check harness between ECM and valve relay connector
DTC: P2444 **1T CCM, MIL: Yes** **Year:** 2009, 2010 **Model:** Forester, Impreza, Legacy **Engine:** 2.5L H4 **Transmission:** All	**Secondary Air Injection System Pump Stuck ON:** Poor drivability. **Possible Causes:** • Check secondary air piping pressure • Check power supply to valve • Check valve relay

OBD II Trouble Code List (U0XXX Codes)

DTC	Trouble Code Title, Conditions & Possible Causes
DTC: U0073 **1T ECM, MIL: Yes** **Year:** 2009, 2010 **Model:** Impreza, Outback, Tribeca **Engine:** 2.5L H4, 3.6L H6 **Transmission:** All	**CAN Failure, Bus 'OFF' Detection:** Data from the TCM, VDC CM and body integrated unit are not normal. Battery voltage > 10.9 V. Starter switch - OFF. Engine running. **Possible Causes:** • No battery voltage • Defective fuse. • Wiring harness or coupling needs repair. • Incompatible setting in the body integrated unit or ECM. • Additional DTC displayed.
DTC: U0101 **1T ECM, MIL: Yes** **Year:** 2009, 2010 **Model:** Forester, Impreza, Legacy, Outback, Tribeca **Engine:** 2.5L H4, 3.0L H6, 3.6L H6 **Transmission:** All	**CAN (TCU) Data Not Loaded:** Malfunction of CAN communication. **Possible Causes:** • No battery voltage • Defective fuse. • Wiring harness or coupling needs repair. • Incompatible setting in the body integrated unit or ECM. • Additional DTC displayed.
DTC: U0122 **1T ECM, MIL: Yes** **Year:** 2009, 2010 **Model:** Forester, Impreza, Legacy, Outback, Tribeca **Engine:** 2.5L H4, 3.0L H6, 3.6L H6 **Transmission:** All	**CAN (VDC) Data not Loaded:** Malfunction of CAN communication. **Possible Causes:** • No battery voltage • Defective fuse. • Wiring harness or coupling needs repair. • Incompatible setting in the body integrated unit or ECM. • Additional DTC displayed.

DTC	Trouble Code Title, Conditions & Possible Causes
DTC: U0140 **1T ECM, MIL: Yes** **Year:** 2009, 2010 **Model:** Forester, Impreza, Legacy, Outback, Tribeca **Engine:** 2.5L H4, 3.0L H6, 3.6L H6 **Transmission:** All	**CAN (BCU) Data not Loaded:** Malfunction of CAN communication. **Possible Causes:** • No battery voltage • Defective fuse. • Wiring harness or coupling needs repair. • Incompatible setting in the body integrated unit or ECM. • Additional DTC displayed.
DTC: U0416 **1T ECM, MIL: Yes** **Year:** 2009, 2010 **Model:** Forester, Impreza, Legacy, Outback, Tribeca **Engine:** 2.5L H4, 3.0L H6, 3.6L H6 **Transmission:** All	**CAN (VDC) Data Abnormal:** Malfunction of CAN communication. **Possible Causes:** • No battery voltage • Defective fuse. • Wiring harness or coupling needs repair. • Incompatible setting in the body integrated unit or ECM. • Additional DTC displayed.
DTC: U0422 **1T ECM, MIL: Yes** **Year:** 2009, 2010 **Model:** Forester, Impreza, Legacy, Outback, Tribeca **Engine:** 2.5L H4, 3.0L H6, 3.6L H6 **Transmission:** All	**CAN (BCU) Data Abnormal:** Malfunction of CAN communication. **Possible Causes:** • No battery voltage • Defective fuse. • Wiring harness or coupling needs repair. • Incompatible setting in the body integrated unit or ECM. • Additional DTC displayed.

OBD II Trouble Code List (U1XXX Codes)

DTC	Trouble Code Title, Conditions & Possible Causes
DTC: U1201 **1T BCM, MIL: Yes** **Year:** 2009, 2010 **Model:** Forester, Impreza, Legacy, Outback, Tribeca **Engine:** 2.5L H4, 3.0L H6, 3.6L H6 **Transmission:** All	**CAN-HS Counter Abnormal:** High speed CAN communication becomes unstable. **Possible Causes:** • Bad wiring harness • Open or short circuit • Poor connector terminal on the ECM • Defective yaw rate sensor • Defective ECM
DTC: U1202 **1T BCM, MIL: Yes** **Year:** 2009, 2010 **Model:** Forester, Impreza, Legacy, Outback, Tribeca **Engine:** 2.5L H4, 3.0L H6, 3.6L H6 **Transmission:** All	**CAN-HS Bus Off:** Open or power supply-output short, GND-output short occurs in CAN line. End resistance malfunction Internal error in each control module **Possible Causes:** • Loose wiring harness connection • Bad wiring harness • Poor connector terminal on the ECM • Defective VDC/ABS Control Module • Defective body integrated unit • Poor connector terminal
DTC: U1211 **1T BCM, MIL: Yes** **Year:** 2009, 2010 **Model:** Forester, Impreza, Legacy, Outback, Tribeca **Engine:** 2.5L H4, 3.0L H6, 3.6L H6 **Transmission:** All	**CAN-HS ECM Data Abnormal:** Received error data from ECM. **Possible Causes:** • Poor contact of wiring harness for CAN communication circuit. • Defective ECM.

DTC	Trouble Code Title, Conditions & Possible Causes
DTC: U1212 **1T BCM, MIL: Yes** **Year:** 2009, 2010 **Model:** Forester, Impreza, Legacy, Outback, Tribeca **Engine:** 2.5L H4, 3.0L H6, 3.6L H6 **Transmission:** All	**CAN-HS TCM Data Abnormal:** Received error data from TCM. Transmission control error may occur. **Possible Causes:** • Poor contact in TCM connector. • CAN communication wiring harness may need repair. • Defective TCM.
DTC: U1213 **1T BCM, MIL: Yes** **Year:** 2009, 2010 **Model:** Forester, Impreza, Legacy, Outback, Tribeca **Engine:** 2.5L H4, 3.0L H6, 3.6L H6 **Transmission:** All	**CAN-HS VDC/ABS Data Abnormal:** Received error data from VDC/ABS CM. Brake control error may occur. **Possible Causes:** • Poor contact in VDC/ABS connector. • Wiring harness used by CAN communication circuit needs repair. • Defective VDC/ABS CM.
DTC: U1221 **1T ECM, MIL: Yes** **Year:** 2009, 2010 **Model:** Forester, Impreza, Legacy, Outback, Tribeca **Engine:** 2.5L H4, 3.0L H6, 3.6L H6 **Transmission:** All	**CAN-HS ECM No-Receive Data:** Not received error data from ECM. **Possible Causes:** • Wiring harness used by CAN communication line needs repair. • Defective ECM.
DTC: U1222 **1T BCM, MIL: Yes** **Year:** 2009, 2010 **Model:** Forester, Impreza, Legacy, Outback, Tribeca **Engine:** 2.5L H4, 3.0L H6, 3.6L H6 **Transmission:** All	**CAN-HS TCM No-Receive Data:** Not received error data from TCM. **Possible Causes:** • Wiring harness connected to high speed CAN communication line need repair. • Defective TCM. • Defective body integrated unit.
DTC: U1223 **1T BCM, MIL: Yes** **Year:** 2009, 2010 **Model:** Forester, Impreza, Legacy, Outback, Tribeca **Engine:** 2.5L H4, 3.0L H6, 3.6L H6 **Transmission:** All	**CAN-HS VDC/ABS No-Receive Data:** Not received error data from VDC/ABS CM. **Possible Causes:** • Wiring harness connected to high speed CAN communication line need repair. • Defective VDC/ABS CM. • Defective body integrated unit.
DTC: U1300 **1T BCM, MIL: Yes** **Year:** 2009, 2010 **Model:** Forester, Impreza, Legacy, Outback, Tribeca **Engine:** 2.5L H4, 3.0L H6, 3.6L H6 **Transmission:** All	**CAN-LS Malfunction:** Either end of low-speed CAN communication line is open or shorted, the connector is not connected properly, or the terminal has poor crimping. **Possible Causes:** • Wiring harness used by CAN communication lines need repair. • Navigation or audio unit needs repair. • Auto A/C control module needs repair. • Body integrated unit needs repair. • Defective combination meter.
DTC: U1301 **1T BCM, MIL: Yes** **Year:** 2009, 2010 **Model:** Forester, Impreza, Legacy, Outback, Tribeca **Engine:** 2.5L H4, 3.0L H6, 3.6L H6 **Transmission:** All	**CAN-LS Counter Abnormal:** Communication becomes unstable because of low speed CAN communication error. Display error may occur in fuel gauge because the CAN communication is not transmitted (sending/receiving) normally. **Possible Causes:** • Wiring harness used by CAN communication lines need repair. • Navigation or audio unit needs repair. • Auto A/C control module needs repair. • Body integrated unit needs repair. • Defective combination meter.

DTC	Trouble Code Title, Conditions & Possible Causes
DTC: U1302 **1T BCM, MIL: Yes** **Year:** 2009, 2010 **Model:** Forester, Impreza, Legacy, Outback, Tribeca **Engine:** 2.5L H4, 3.0L H6, 3.6L H6 **Transmission:** All	**CAN-LS Bus Off:** Open or power supply-output short, GND-output short occurs in both CAN line. Internal error in each control module. Display error may occur in fuel gauge because the CAN communication is not transmitted (sending/receiving) normally. **Possible Causes:** • Wiring harness used by CAN communication lines need repair. • Navigation or audio unit needs repair. • Auto A/C control module needs repair. • Body integrated unit needs repair. • Defective combination meter.
DTC: U1311 **1T BCM, MIL: Yes** **Year:** 2009, 2010 **Model:** Forester, Impreza, Legacy, Outback, Tribeca **Engine:** 2.5L H4, 3.0L H6, 3.6L H6 **Transmission:** All	**CAN-LS Meter Unit Data Abnormal:** Error data is received from combination meter. **Possible Causes:** • Defective combination meter. • Wiring harness used for low speed CAN communications needs repair. • Poor contact in connector terminals.
DTC: U1321 **1T BCM, MIL: Yes** **Year:** 2009, 2010 **Model:** Forester, Impreza, Legacy, Outback, Tribeca **Engine:** 2.5L H4, 3.0L H6, 3.6L H6 **Transmission:** All	**CAN-LS Meter No-Receive Data:** Not received error data from combination meter. Engine may not be started. DTC U1301 or U1302 not displayed. **Possible Causes:** • Defective combination meter. • Wiring harness connected to low speed CAN communication lines need repair. • Defective body integrated unit. • Poor contact of wiring harness used for low speed CAN communication line.
DTC: U1500 **1T BCM, MIL: Yes** **Year:** 2009 **Model:** Forester **Engine:** 2.5L H4 **Transmission:** All	**Keyless UART Communication Malfunction:** UART between keyless entry control module and body integrated unit is open or shorted, the connector is not connected properly, or the terminal is crimped improperly, keyless entry control module internal error. Door lock does not operate with keyless. **Possible Causes:** • Wiring harness between body integrated unit and keyless entry control module need repair. • Power supply circuit for keyless entry control module needs repair. • Defective keyless entry control module. • Defective body integrated unit.

OBD II Trouble Code List (US0XXX Codes)

DTC	Trouble Code Title, Conditions & Possible Causes
DTC: US0402 **1T ECM, MIL: Yes** **Year:** 2009, 2010 **Model:** Forester, Impreza, Legacy, Outback, Tribeca **Engine:** 2.5L H4, 3.0L H6, 3.6L H6 **Transmission:** All	**CAN (TCU) Data Abnormal:** Malfunction of CAN communication. **Possible Causes:** • No battery voltage • Defective fuse. • Wiring harness or coupling needs repair. • Incompatible setting in the body integrated unit or ECM. • Additional DTC displayed.

SUZUKI

Equator

24

SPECIFICATIONS AND MAINTENANCE CHARTS

ENGINE AND VEHICLE IDENTIFICATION

	Engine						Model Year	
ID/Code	Liters (cc)	Cu. In.	Cyl.	Fuel Sys.	Engine Type	Eng. Mfg.	Code ①	Year
QR25DE	2.5 (2,488)	152	4	MFI	DOHC	Nissan	9	2009
VQ40DE	4.0 (3,954)	241	6	MFI	DOHC	Nissan	A	2010

MFI: Multi-port Fuel Injection

DOHC: Double Overhead Camshafts

① Indicated at the 4th digit of the Vehicle Identification Number (VIN)

② 10th digit of the Vehicle Identification Number (VIN)

37671_EQUA_C0001

GENERAL ENGINE SPECIFICATIONS

All measurements are given in inches.

Year	Model	Engine Displacement Liters	Engine Series ID	Net Horsepower @ rpm	Net Torque @ rpm (ft. lbs.)	Bore x Stroke (in.)	Compression Ratio	Oil Pressure @ rpm
2009	Equator	2.5	QR25DE	152@5200	171@4400	3.50 x 3.94	9.5:1	43@2000
		4.0	VQ40DE	261@5600	281@4000	3.76 x 3.62	9.7:1	43@2000
2010	Equator	2.5	QR25DE	152@5200	171@4400	3.50 x 3.94	9.5:1	43@2000
		4.0	VQ40DE	261@5600	281@4000	3.76 x 3.62	9.7:1	43@2000

37671_EQUA_C0002

GASOLINE ENGINE TUNE-UP SPECIFICATIONS

Year	Engine Displacement Liters	Engine ID	Spark Plug Gap (in.)	Ignition Timing (deg.) MT	Ignition Timing (deg.) AT	Fuel Pump (psi)	Idle Speed (RPM) MT	Idle Speed (RPM) AT ①	Valve Clearance (in.) Intake	Valve Clearance (in.) Exhaust
2009	2.5	QR25DE	0.043	10-20B	10-20B	51 ②	575-675	650-750	③	④
	4.0	VQ40DE	0.043	10-20B	10-20B	51 ②	575-675	650-750	⑤	⑥
2010	2.5	QR25DE	0.043	10-20B	10-20B	51 ②	575-675	650-750	③	④
	4.0	VQ40DE	0.043	10-20B	10-20B	51 ②	575-675	650-750	⑤	⑥

NOTE: The Vehicle Emission Control Information label often reflects specification changes made during production.

The label figures must be used if they differ from those in this chart.

B: Before top dead center

① Automatic Transmission (AT) in Neutral

② At idle

③ 0.009-0.013 inch (cold)
0.012-0.016 inch (hot)

④ 0.010-0.013 inch (cold)
0.012-0.017 inch (hot)

⑤ 0.010-0.013 inch (cold)
0.012-0.016 inch (hot)

⑥ 0.011-0.015 inch (cold)
0.012-0.017 inch (hot)

37671_EQUA_C0003

CAPACITIES

Year	Model	Engine Displacement Liters	Engine ID	Engine Oil with Filter (qts.)	Transmission (pts.) Manual	Transmission (pts.) Auto. ①	Transfer Case (pts.)	Drive Axle Front (pts.)	Drive Axle Rear (pts.)	Fuel Tank (gal.)	Cooling System (qts.)
2009	Equator	2.5	QR25DE	4.9	6.1	21.8	NA	NA	3.4	21.1	10.0
		4.0	VQ40DE	5.4	NA	21.8	4.25	1.75	②	21.1	10.8
2010	Equator	2.5	QR25DE	4.9	6.1	21.8	NA	NA	3.4	21.1	10.0
		4.0	VQ40DE	5.4	NA	21.8	4.25	1.75	②	21.1	10.8

NA: Not Applicable

NOTE: All capacities are approximate. Add fluid gradually and check to be sure a proper fluid level is obtained.

① Drain and refill

② C200 axle: 3.38 pts.; M226 axle: 4.25 pts.

37671_EQUA_C0004

FLUID SPECIFICATIONS

Year	Model	Engine Displacement Liters	Engine ID	Engine Oil	Manual Trans.	Auto. Trans.	Transfer Case	Drive Axle Front	Drive Axle Rear	Power Steering Fluid	Brake Master Cylinder	Cooling System
2009	Equator	2.5	QR25DE	5W-30	①	②	NA	NA	③	NS	DOT 3	④
		4.0	VQ40DE	5W-30	NA	②	⑤	⑥	⑦	NS	DOT 3	④
2010	Equator	2.5	QR25DE	5W-30	①	②	NA	NA	③	NS	DOT 3	④
		4.0	VQ40DE	5W-30	NA	②	⑤	⑥	⑦	NS	DOT 3	④

NA: Not Applicable

NS: Not specified by manufacturer at date of publication

DOT: Department Of Transportation

① Manual Transmission Fluid (MTF) API GL-4, Viscosity SAE 75W-90

② Suzuki approved Matic S ATF. If Matic S ATF is not available, SUZUKI approved Matic J ATF may also be used.

③ API GL-5 synthetic gear oil, Viscosity SAE 75W-90

④ Suzuki genuine coolant, or equivalent

⑤ Suzuki approved Matic D AT

⑥ API GL-5 Viscosity SAE 80W-90. For hot climates, viscosity SAE 90 is suitable for ambient temperatures above 32 degrees F (0 degrees C).

⑦ C200 axle: API GL-5 synthetic gear oil, Viscosity SAE 75W-90
   M226 axle: API GL-5 synthetic gear oil, Viscosity SAE 75W-140

37671_EQUA_C0005

VALVE SPECIFICATIONS

Year	Engine Displacement Liters	Engine ID	Seat Angle (deg.)	Face Angle (deg.)	Spring Test Pressure (lbs. @ in.)	Spring Installed Height (in.)	Stem-to-Guide Clearance (in.) Intake	Stem-to-Guide Clearance (in.) Exhaust	Stem Diameter (in.) Intake	Stem Diameter (in.) Exhaust
2009	2.5	QR25DE	44.15-45.45	45.00	①	1.390	0.0008-0.0021	0.0012-0.0025	0.2348-0.2354	0.2344-0.2350
	4.0	VQ40DE	44.15-45.45	45.00	②	1.457	0.0008-0.0021	0.0012-0.0025	0.2348-0.2354	0.2344-0.2350
2010	2.5	QR25DE	44.15-45.45	45.00	①	1.390	0.0008-0.0021	0.0012-0.0025	0.2348-0.2354	0.2344-0.2350
	4.0	VQ40DE	44.15-45.45	45.00	②	1.457	0.0008-0.0021	0.0012-0.0025	0.2348-0.2354	0.2344-0.2350

① Intake valve open: 79-89@0.996
   Exhaust valve open: 71-81@1.054

② Installation: 37-42@1.457
   Valve open: 84-95@1.071

37671_EQUA_C0006

CAMSHAFT AND BEARING SPECIFICATIONS CHART

All measurements arc given in inches.

Year	Engine Displ. Liters	Engine ID	Journal Dia.	Brg. Oil Clearance	Shaft End-play	Runout	Journal Bore	Lobe Lift	
								Intake	Exhaust
2009	2.5	QR25DE	①	0.0018-0.0034	0.0045-0.0074	0.0008	②	1.7722-1.7797	1.7313-1.7388
	4.0	VQ40DE	③	④	0.0045-0.0074	0.0010	⑤	1.7900-1.7921	1.7746-1.7821
2010	2.5	QR25DE	①	0.0018-0.0034	0.0045-0.0074	0.0008	②	1.7722-1.7797	1.7313-1.7388
	4.0	VQ40DE	③	④	0.0045-0.0074	0.0010	⑤	1.7900-1.7921	1.7746-1.7821

① Number 1: 1.0998-1.1006 inches
  Numbers 2, 3, 4, 5: 0.9226-0.9234 inch

② Number 1: 1.1024-1.1032 inches
  Numbers 2, 3, 4, 5: 0.9252-0.9260 inch

③ Number 1: 1.0211-1.0218 inches
  Numbers 2, 3, 4: 0.9230-0.9238 inch

④ Number 1: 0.0018-0.0034 inch
  Numbers 2, 3, 4: 0.0014-0.0030 inch

⑤ Number 1: 1.0236-1.0244 inches
  Numbers 2, 3, 4: 0.9252-0.9260 inch

37671_EQUA_C0007

CRANKSHAFT AND CONNECTING ROD SPECIFICATIONS

All measurements are given in inches.

Year	Engine Displacement Liters	Engine ID	Crankshaft				Connecting Rod		
			Main Brg. Journal Dia.	Main Brg. Oil Clearance	Shaft End-play	Thrust on No.	Journal Diameter	Oil Clearance	Side Clearance
2009	2.5	QR25DE	①	②	0.0039-0.0102	3	③	0.0014-0.0018	0.0079-0.0138
	4.0	VQ40DE	④	0.0014-0.0018	0.0039-0.0098	3	2.2441-2.2446	0.0013-0.0023	0.0079-0.0138
2010	2.5	QR25DE	①	②	0.0039-0.0102	3	③	0.0014-0.0018	0.0079-0.0138
	4.0	VQ40DE	④	0.0014-0.0018	0.0039-0.0098	3	2.2441-2.2446	0.0013-0.0023	0.0079-0.0138

NS: Not specified by manufacturer at date of publication

① There are 24 different grades, ranging from A (2.1645 inches) to 7 (2.1636 inches)

② Numbers 1, 3, 5: 0.0011-0.0017 inch
  Numbers 2, 4: 0.0016-0.0022 inch

③ There are 13 different grades, ranging from 0 (1.8898 inches) to C (1.8903 inches)

④ There are 24 different grades, ranging from A (2.7549 inches) to 7 (2.7540 inches)

37671_EQUA_C0008

PISTON AND RING SPECIFICATIONS
All measurements are given in inches.

Year	Engine Displ. Liters	Engine ID	Piston Clearance	Ring Gap			Ring Side Clearance		
				Top Compression	Bottom Compression	Oil Control	Top Compression	Bottom Compression	Oil Control
2009	2.5	QR25DE	0.0004-0.0012	0.0083-0.0122	0.0126-0.0185	0.0079-0.0236	0.0018-0.0031	0.0012-0.0028	0.0026-0.0053
	4.0	VQ40DE	0.0004-0.0012	0.0091-0.0130	0.0130-0.0189	0.0079-0.0197	0.0018-0.0031	0.0012-0.0028	0.0026-0.0053
2010	2.5	QR25DE	0.0004-0.0012	0.0083-0.0122	0.0126-0.0185	0.0079-0.0236	0.0018-0.0031	0.0012-0.0028	0.0026-0.0053
	4.0	VQ40DE	0.0004-0.0012	0.0091-0.0130	0.0130-0.0189	0.0079-0.0197	0.0018-0.0031	0.0012-0.0028	0.0026-0.0053

37671_EQUA_C0009

TORQUE SPECIFICATIONS
All readings in ft. lbs.

Year	Engine Displacement Liters	Engine ID	Cylinder Head Bolts	Main Bearing Bolts	Rod Bearing Bolts	Crankshaft Damper Bolts	Flywheel Bolts	Manifold		Spark Plugs	Oil Pan Drain Plug
								Intake	Exhaust		
2009	2.5	QR25DE	①	②	③	④	80	⑤	⑥	14	25
	4.0	VQ40DE	⑦	⑧	14	⑨	65	⑩	⑥	18	25
2010	2.5	QR25DE	①	②	③	④	80	⑤	⑥	14	25
	4.0	VQ40DE	⑦	⑧	14	⑨	65	⑩	⑥	18	25

① Step 1: 37 ft. lbs.

  Step 2: Plus 60 degrees clockwise

  Step 3: Loosen completely to 0 ft. lbs.

  Step 4: 29 ft. lbs.

  Step 5: Plus 75 degrees clockwise

  Step 6: Plus 75 degrees clockwise

② Step 1: bolts 11-22 to 19 ft. lbs.

  Step 2: bolts 1-10 to 29 ft. lbs.

  Step 3: bolts 1-10, plus 60-65 degrees

③ Step 1: 20 ft. lbs.

  Step 2: Loosen to 0 ft. lbs.

  Step 3: 14 ft. lbs.

  Step 4: Plus 85-95 degrees

④ Step 1: 31 ft. lbs.

  Step 2: Plus 60 degrees

⑤ 83 inch lbs.

⑥ Stud bolts: 11 ft. lbs.

  Nuts: 22 ft. lbs.

⑦ Step 1: 72 ft. lbs.

  Step 2: Loosen completely to 0 ft. lbs.

  Step 3: 29 ft. lbs.

  Step 4: Plus 90 degrees clockwise

  Step 5: Plus 90 degrees clockwise

⑧ Bolts: 17-24 (M8) to 16 ft. lbs.

  Install rear main seal

  Bolts: 1-16 (M10) to 26 ft. lbs.

  Bolts: 1-16 (M10), plus 90 degrees clockwise

⑨ Step 1: 33 ft. lbs.

  Step 2: Plus 84-90 degrees clockwise

⑩ Intake manifold collector:

  Bolts and nuts: 96 inch lbs.

  Stud bolts: 61 inch lbs.

  Intake manifold:

  Bolts and nuts: 65 inch lbs. and then to 21 ft. lbs.

  Studs: 96 inch lbs.

37671_EQUA_C0010

WHEEL ALIGNMENT

			Caster		Camber		
Year	Model		Range (+/-Deg.)	Preferred Setting (Deg.)	Range (+/-Deg.)	Preferred Setting (Deg.)	Toe-in (Deg.)
2009	Equator	2WD	0.75	3.00	0.75	0.25	0.12+/-0.04
		4WD	0.75	2.75	0.75	0.50	0.12+/-0.04
2010	Equator	2WD	0.75	3.00	0.75	0.25	0.12+/-0.04
		4WD	0.75	2.75	0.75	0.50	0.12+/-0.04

NOTE: Measurements given for an unladen vehicle with fuel, coolant, and engine oil must be full; spare tire, jack, hand tools, and mats in designated positions.
Some vehicles may be equipped with straight (non-adjustable) lower link bolts and washers. In order to adjust camber and caster on these vehicles, first replace
  the lower link bolts and washers with adjustable (cam) bolts and washers.

37671_EQUA_C0011

TIRE, WHEEL AND BALL JOINT SPECIFICATIONS

		OEM Tires		Tire Pressures (psi)		Wheel Size	Ball Joint Inspection	Lug Nut Torque (ft. lbs.)
Year	Model	Standard	Optional	Front	Rear			
2009	XE	P235/75R15 P265/70R16	NA	①	①	15 x 7J 16 x 7J	②	98
	SE	P265/70R16	NA	①	①	16 x 7J	②	98
	Pro-4X	P265/75R16	NA	①	①	16 x 7J	②	98
	LE	P265/60R18	NA	①	①	18 x 8J	②	98
2010	XE	P235/75R15 P265/70R16	NA	①	①	15 x 7J 16 x 7J	②	98
	SE	P265/70R16	NA	①	①	16 x 7J	②	98
	Pro-4X	P265/75R16	NA	①	①	16 x 7J	②	98
	LE	P265/60R18	NA	①	①	18 x 8J	②	98

OEM: Original Equipment Manufacturer

PSI: Pounds Per Square Inch

NA: Not Applicable

① See the tire placard on the vehicle

② Swinging force: Lower (at groove) 3-33 ft. lbs.; Upper (at hole) 2-18 ft. lbs.
  Turning force: Lower 5-56 inch lbs.; Upper 5-43 inch lbs.
  Vertical end play: 0 inch

37671_EQUA_C0012

BRAKE SPECIFICATIONS
All measurements in inches unless noted

			Brake Disc			Brake Drum Diameter			Minimum Lining Thickness	Brake Caliper	
Year	Model		Original Thickness	Minimum Thickness	Maximum Runout	Original Inside Diameter	Max. Wear Limit	Maximum Machine Diameter		Bracket Bolts (ft. lbs.)	Guide Pin Bolts (ft. lbs.)
2009	Equator	F	1.102	1.024	0.002	NA	NA	NA	0.079	136	20
		R	0.709	0.630	0.002	NA	NA	NA	0.079	76	20
2010	Equator	F	1.102	1.024	0.002	NA	NA	NA	0.079	136	20
		R	0.709	0.630	0.002	NA	NA	NA	0.079	76	20

F: Front

R: Rear

NA: Not Applicable

37671_EQUA_C0013

SUZUKI EQUATOR

TO BE SERVICED	TYPE OF SERVICE	VEHICLE MILEAGE INTERVAL (x1000)							
		7.5	15	22.5	30	37.5	45	52.5	60
Brake lines and cables	S/I		✓		✓		✓		✓
Brake pads and rotors	S/I		✓		✓		✓		✓
Cabin air filter	R		✓		✓		✓		✓
Drive belt(s) ①	S/I								✓
Drive shaft boots and propeller shaft	S/I		✓		✓		✓		✓
Engine air cleaner	R				✓				✓
Engine coolant ②	R								✓
Engine oil and filter	R	✓	✓	✓	✓	✓	✓	✓	✓
EVAP vapor lines	S/I				✓				✓
Exhaust system	S/I				✓				✓
Fuel filter	S/I	Maintenance not normally required							
Fuel lines	S/I				✓				✓
Rear final drive oil	S/I		✓		✓		✓		✓
Spark plugs	R	Replace every 105,000 miles							
Steering gear, linkage, and axle	S/I				✓				✓
Suspension parts	S/I				✓				✓
Tire rotation	S/I	✓	✓	✓	✓	✓	✓	✓	✓
Transmission fluid	S/I		✓		✓		✓		✓
Valve clearance (intake and exhaust)	S/I	Only if valve noise increases, inspect valve clearance							

R: Replace

S/I: Inspect and service, if necessary

① After 60,000 miles or 48 months, inspect every 15,000 miles or 12 months. Replace if found damaged or belt tensioner reaches the maximum limit.

② After 60,000 miles or 48 months, replace engine coolant every 30,000 miles or 24 months.

FREQUENT OPERATION MAINTENANCE (SEVERE SERVICE)

If a vehicle is operated under any of the following conditions, it is considered severe service:

- Extremely dusty areas

- 50% or more of the vehicle operation is in 90°F (32°C) or higher temperatures, or constant operation in temperatures below 32°F (0°C)

- Prolonged idling (vehicle operation in stop and go traffic)

- Frequent short running periods (engine does not warm to normal operating temperatures)

- Police, taxi, delivery usage, or trailer towing usage

- Driving in hilly or mountainous terrain

Oil & oil filter: replace every 3,750 miles

Air filter element: service or inspect more frequently

Brake pads, discs, drums and linings: service or inspect every 7,500 miles

Driveshaft boots and propeller shaft: service or inspect every 7,500 miles

Exhaust system: service or inspect every 7,500 miles

Steering gear and linkage, axle and suspension parts: service or inspect every 7,500 miles

Steering linkage ball joints and front suspension ball joints: service or inspect every 7,500 miles

Transmission oil: replace every 30,000 miles

MAINTENANCE FOR OFF-ROAD DRIVING (4WD ONLY)

After driving the vehicle off-road through sand, mud, or water; more frequent maintenance may be required for the following items:

- Brake pads and rotors

- Brake lines and hoses

- Rear final drive oil, transmission fluid, and transfer fluid

- Steering linkage

- Drive shafts

- Engine air cleaner

- Cabin air filter

37671_EQUA_C0014

BRAKES   INFORMATION AND PRECAUTIONS

ANTI-LOCK SYSTEMS

- Certain components within the ABS system are not intended to be serviced or repaired individually.
- Do not use rubber hoses or other parts not specifically specified for and ABS system. When using repair kits, replace all parts included in the kit. Partial or incorrect repair may lead to functional problems and require the replacement of components.
- Lubricate rubber parts with clean, fresh brake fluid to ease assembly. Do not use shop air to clean parts; damage to rubber components may result.
- Use only DOT 3 brake fluid from an unopened container.
- If any hydraulic component or line is removed or replaced, it may be necessary to bleed the entire system.
- A clean repair area is essential. Always clean the reservoir and cap thoroughly before removing the cap.

The slightest amount of dirt in the fluid may plug an orifice and impair the system function. Perform repairs after components have been thoroughly cleaned; use only denatured alcohol to clean components. Do not allow ABS components to come into contact with any substance containing mineral oil; this includes used shop rags.

- The Anti-Lock control unit is a microprocessor similar to other computer units in the vehicle. Ensure that the ignition switch is **OFF** before removing or installing controller harnesses. Avoid static electricity discharge at or near the controller.
- If any arc welding is to be done on the vehicle, the control unit should be unplugged before welding operations begin.

DISC AND DRUM SYSTEMS

✳✳ CAUTION

Dust and dirt accumulating on brake parts during normal use may contain asbestos fibers from production or aftermarket brake linings. Breathing excessive concentrations of asbestos fibers can cause serious bodily harm. Exercise care when servicing brake parts. Do not sand or grind brake lining unless equipment used is designed to contain the dust residue. Do not clean brake parts with compressed air or by dry brushing. Cleaning should be done by dampening the brake components with a fine mist of water, then wiping the brake components clean with a dampened cloth. Dispose of cloth and all residue containing asbestos fibers in an impermeable container with the appropriate label. Follow practices prescribed by the Occupational Safety and Health Administration (OSHA) and the Environmental Protection Agency (EPA) for the handling, processing, and disposing of dust or debris that may contain asbestos fibers.

BRAKES   BLEEDING THE BRAKE SYSTEM

BLEEDING PROCEDURE

BLEEDING PROCEDURE

✳✳ CAUTION

Brake fluid contains polyglycol ethers and polyglycols. Avoid contact with the eyes and wash your hands thoroughly after handling brake fluid. If you do get brake fluid in your eyes, flush your eyes with clean, running water for 15 minutes. If eye irritation persists, or if you have taken brake fluid internally, IMMEDIATELY seek medical assistance.

✳ WARNING

Clean, high quality brake fluid is essential to the safe and proper oper- ation of the brake system. You should always buy the highest quality brake fluid that is available. If the brake fluid becomes contaminated, drain and flush the system, then refill the master cylinder with new fluid. Never reuse any brake fluid. Any brake fluid that is removed from the system should be discarded. Also, do not allow any brake fluid to come in contact with a painted surface; it will damage the paint.

1. Before servicing the vehicle, refer to the Precautions Section.
2. Clean all around the brake fluid reservoir filler cap.

➡**While bleeding the brake system, pay attention to the master cylinder fluid level.**

3. Raise and safely support the vehicle.
4. Attach a vinyl tube to the right, rear bleeder valve.
5. Depress the brake pedal fully 4 or 5 times.
6. With the brake pedal depressed, loosen the bleeder valve to let the air out, then tighten it immediately.
7. Repeat until no more air comes out.
8. Tighten the bleeder valve.
9. Perform steps 4–8 at each wheel, with the master cylinder reservoir tank filled at least half way, bleed the air from the front left, rear left, and front right bleed valve, in that order.
10. Fill the master cylinder reservoir.

BRAKES   ANTI-LOCK BRAKE SYSTEM (ABS)

SPEED SENSORS

REMOVAL & INSTALLATION

See Figure 1.

1. Before servicing the vehicle, refer to the Precautions Section.
2. Disconnect the negative battery cable.
3. Raise and safely support the vehicle.
4. Remove the tire and wheel assembly.
5. Remove the brake rotor, for front wheel speed sensor access. Refer to Brake Disc (Rotor), removal & installation.

6. Remove the wheel speed sensor mounting bolts and clip.
7. Pull the sensor out, being careful to turn it as little as possible. Do not pull on the sensor harness.

To install:

8. Before installing the sensor, be sure no foreign materials such as iron fragments adhere to the pick-up part of the sensor or to the inside of the sensor mounting hole or on the rotor mounting surface.

9. Inspect and replace the wheel sensor if damaged.
10. Clean the wheel sensor hole and mating surface with brake cleaner and a lint-free cloth. Be careful that dirt and debris do not enter the hub and bearing assembly or the rear axle.
11. Be sure the harness is not twisted when installed.
12. Continue the installation in the reverse order of the removal procedure.

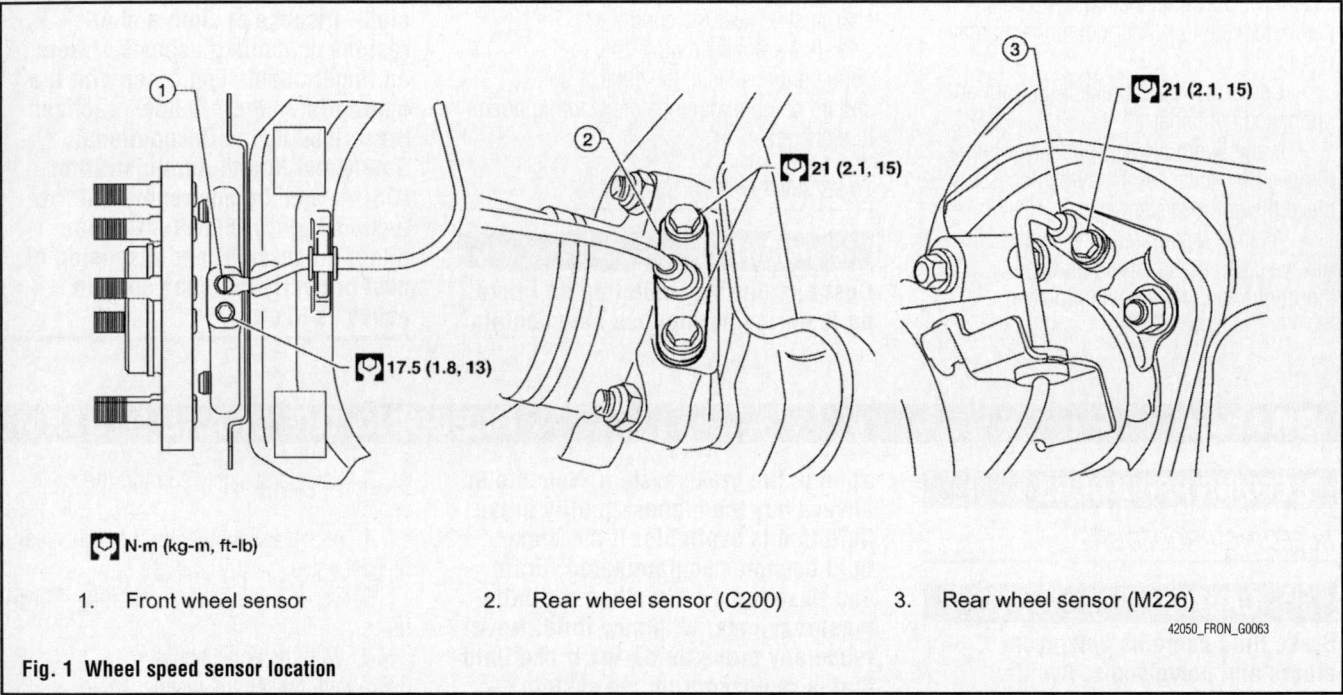

1. Front wheel sensor
2. Rear wheel sensor (C200)
3. Rear wheel sensor (M226)

42050_FRON_G0063

Fig. 1 Wheel speed sensor location

BRAKES   FRONT DISC BRAKES

BRAKE CALIPER

REMOVAL & INSTALLATION

See Figure 2.

1. Before servicing the vehicle, refer to the Precautions Section.
2. Drain the brake fluid, as necessary.
3. Raise and safely support the vehicle.
4. Remove the tire and wheel assembly.
5. Remove the bolt attaching the brake hose to the caliper. Plug the brake hose to prevent brake fluid loss.
6. Remove the caliper support mounting bolts and lift the caliper assembly from the knuckle.

To install:

7. Position the caliper assembly onto the knuckle and install the bolts. Make sure the rotor fits between the brake pads. Torque the bolts to specification.
8. Using new copper washers, connect the brake hose to the caliper. Torque the brake hose attaching bolt to specification.
9. Bleed the brake system.
10. Apply the brake pedal and inspect the system. Ensure proper operation and no leakage.
11. Install the tire and wheel assembly. Tighten the wheel nuts to 98 ft. lbs. (133 Nm).
12. Lower the vehicle and road test.

DISC BRAKE PADS

REMOVAL & INSTALLATION

See Figure 3.

1. Before servicing the vehicle, refer to the Precautions Section.
2. Remove the master cylinder reservoir cap.
3. Partially drain the brake fluid.
4. Raise and safely support the vehicle.
5. Remove the tire and wheel assembly.
6. Remove the caliper sliding pin bolts.

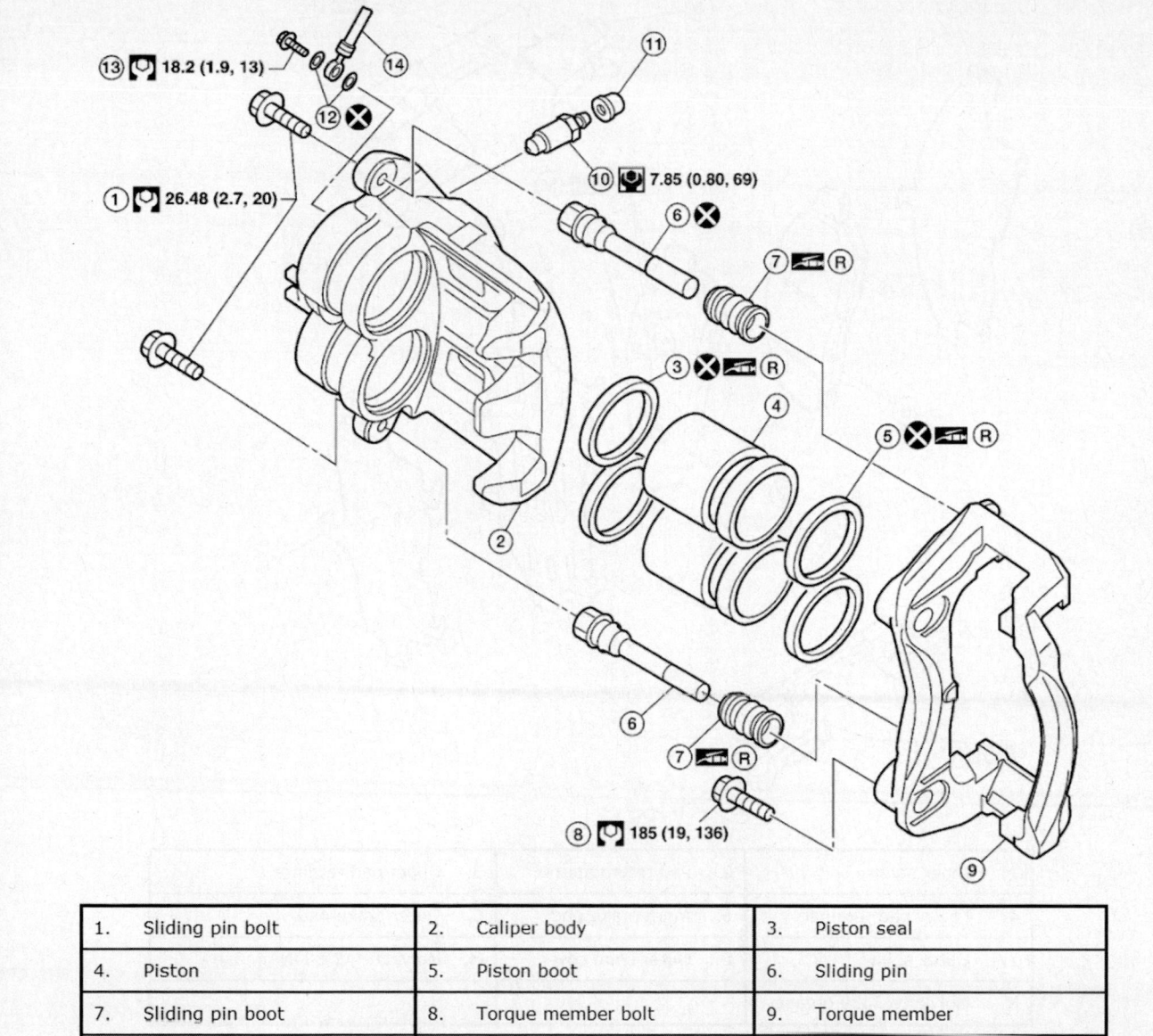

Fig. 2 Front disc brake caliper and related components

1.	Sliding pin bolt	2.	Caliper body	3.	Piston seal
4.	Piston	5.	Piston boot	6.	Sliding pin
7.	Sliding pin boot	8.	Torque member bolt	9.	Torque member
10.	Bleed valve	11.	Cap	12.	Copper washers
13.	Union bolt	14.	Front brake hose	R.	Rubber grease

37671_EQUA_G0034

7. Support the caliper body with a suitable wire to avoid pulling on the front brake hose.

8. Remove the front inner and outer brake pads, shims, shim cover, pad return spring, and retainers from the torque member.

To install:

9. Apply Molykote® AS-880N grease, or equivalent, between the outer brake pad backing and the shims, then attach the shim and shim cover to the brake pads.

10. Apply Molykote® 7439 grease to the upper pad retainer.

11. Attach the pad retainer to the torque member, then install the brake pad, shims, and shim covers to the torque member.

➡**When attaching the pad retainer, attach it firmly so that it is flush with the torque member.**

12. Using a suitable tool, push the pistons into the caliper body.

➡**When pushing in the piston, brake fluid returns to the master cylinder**

reservoir tank. Watch the level of the brake fluid in the reservoir tank.

13. Install the pad return spring to the bottom edge of the brake pads in the holes provided.

14. Install the caliper body.

15. Install the sliding pin bolts and tighten to 20 ft. lbs. (27 Nm).

16. Check the brakes for drag.

17. Inspect the brake fluid level, then install the master cylinder reservoir cap.

18. Install the front wheel and tire. Tighten the wheel nuts to 98 ft. lbs. (133 Nm).

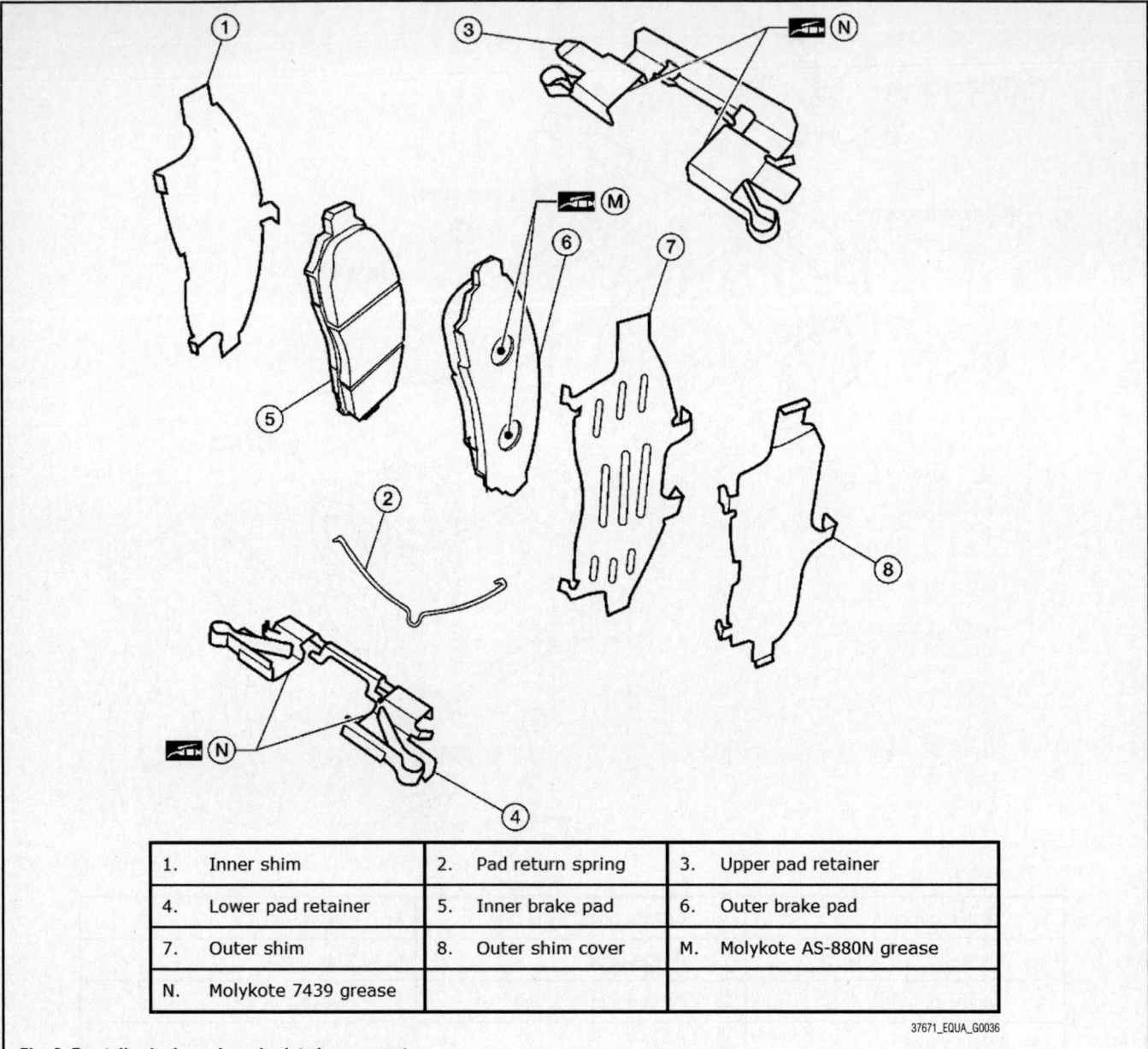

1.	Inner shim	2.	Pad return spring	3.	Upper pad retainer
4.	Lower pad retainer	5.	Inner brake pad	6.	Outer brake pad
7.	Outer shim	8.	Outer shim cover	M.	Molykote AS-880N grease
N.	Molykote 7439 grease				

37671_EQUA_G0036

Fig. 3 Front disc brake pads and related components

BRAKES

REAR DISC BRAKES

BRAKE CALIPER

REMOVAL & INSTALLATION

See Figure 4.

1. Before servicing the vehicle, refer to the Precautions Section.
2. Remove the rear wheel and tire.
3. Drain the brake fluid, as needed.
4. Remove the union bolt, then disconnect the brake hose and discard the copper washers.

➡**Discard the copper washers, do not reuse.**

5. Remove the sliding pin bolts and remove the caliper body from the torque member.

To install:

6. Install the caliper body and sliding pins, then tighten the sliding pin bolts to 20 ft. lbs. (27 Nm).

➡**When installing the caliper body to the torque member, wipe any oil off of the knuckle spindle, washers, and caliper body attachment surfaces.**

7. Install the brake hose by aligning it with the protrusion on the caliper body.

8. Install new copper washers and the union bolt. Tighten the union bolt to 13 ft. lbs. (18 Nm).

➡**Use new copper washers for installation.**

9. Refill with new brake fluid as necessary and bleed the air. Refer to Bleeding The Brake System.
10. Refill with new brake fluid. Do not reuse drained brake fluid.
11. Install the rear wheel and tire. Tighten the wheel nuts to 98 ft. lbs. (133 Nm).
12. Lower the vehicle and road test.

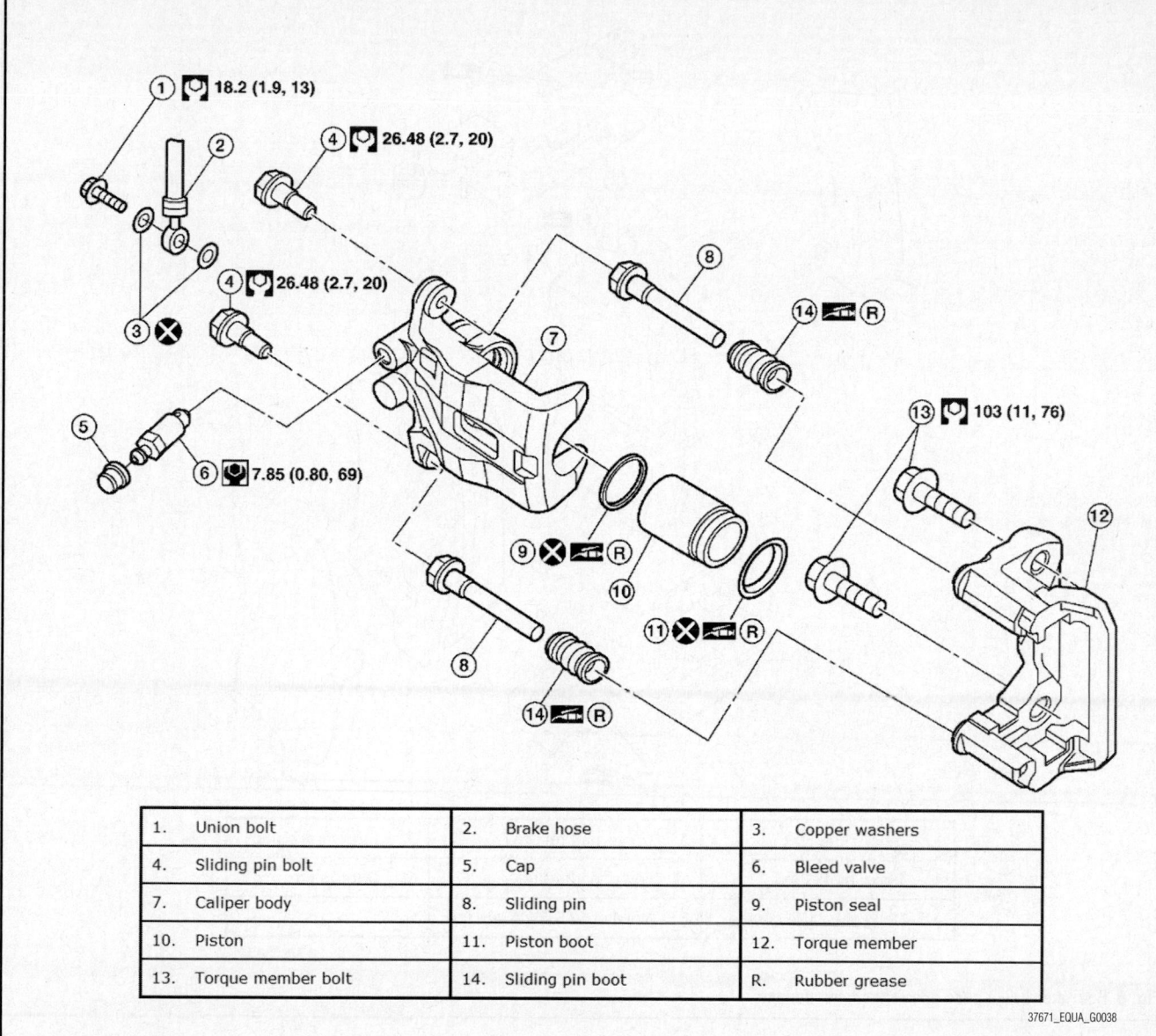

1.	Union bolt	2.	Brake hose	3.	Copper washers
4.	Sliding pin bolt	5.	Cap	6.	Bleed valve
7.	Caliper body	8.	Sliding pin	9.	Piston seal
10.	Piston	11.	Piston boot	12.	Torque member
13.	Torque member bolt	14.	Sliding pin boot	R.	Rubber grease

37671_EQUA_G0038

Fig. 4 Rear disc brake caliper and related components

DISC BRAKE PADS

REMOVAL & INSTALLATION

See Figure 5.

1. Before servicing the vehicle, refer to the Precautions Section.
2. Remove the rear wheel and tire.
3. Drain the brake fluid, as needed.
4. Remove caliper sliding pin bolts.
5. Support the caliper body with a suitable wire to avoid pulling on the rear brake hose.

6. Remove the rear inner and outer brake pads, shims, and retainers from the torque member.

To install:

7. Apply Molykote® AS-880N grease between the brake pad back plates and shims, then attach the shims to the brake pads.
8. Apply Molykote® 7439 grease to the pad retainers.
9. Attach the pad retainer to the torque member, then install the brake pad and shim assemblies.

➡**When attaching the pad retainer, attach it firmly so that it is flush with the torque member.**

10. Using a suitable tool push the piston into the caliper body.

➡**By pushing in the piston, brake fluid returns to the master cylinder reservoir tank. Watch the level of fluid in the reservoir tank.**

11. Install the caliper body.
12. Install the sliding pin bolts and tighten to 20 ft. lbs. (27 Nm).
13. Check the brakes for drag.
14. Inspect the brake fluid level, then install the master cylinder reservoir cap.
15. Install the rear wheel and tire. Tighten the wheel nuts to 98 ft. lbs. (133 Nm).

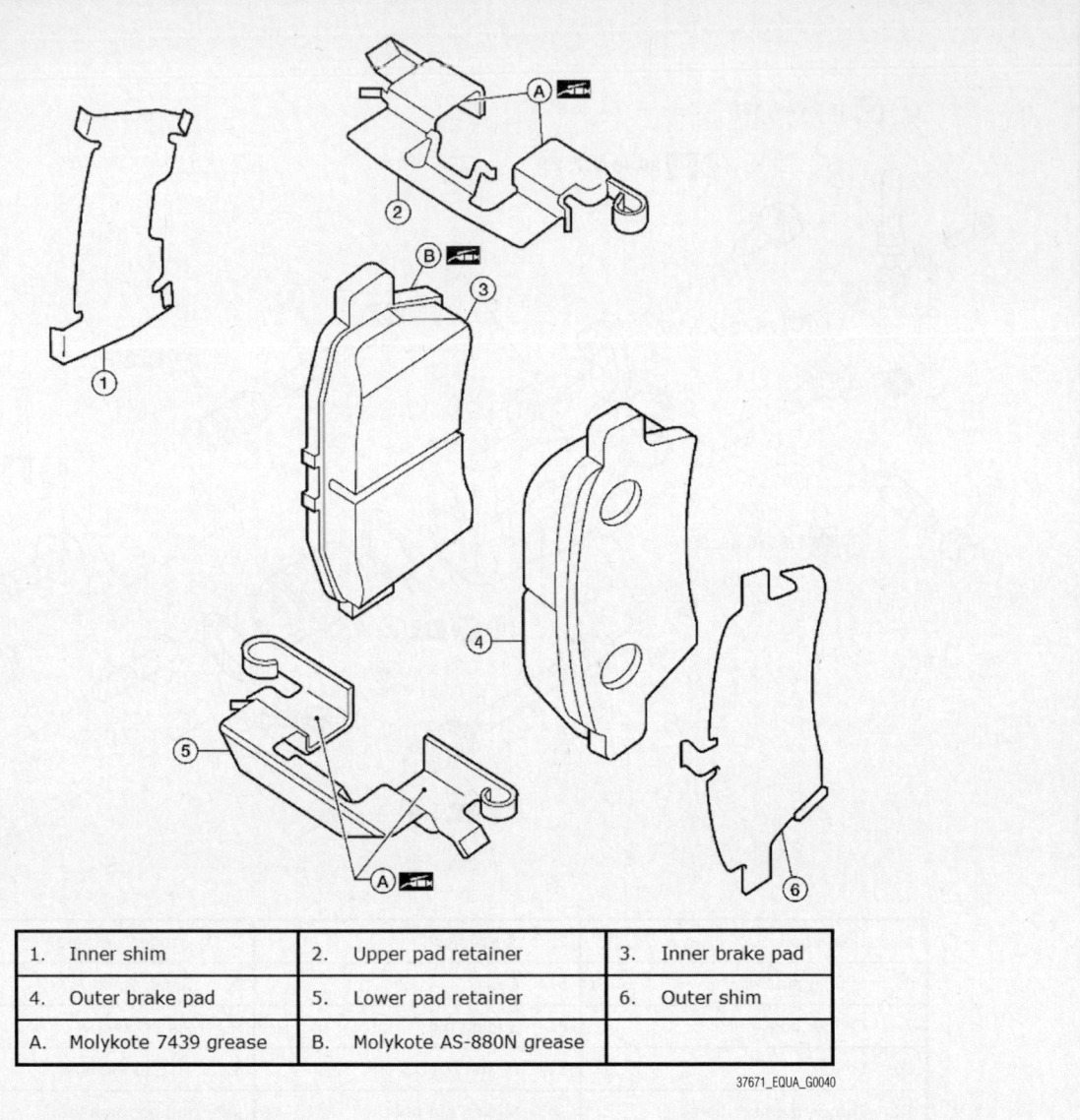

1. Inner shim	2. Upper pad retainer	3. Inner brake pad
4. Outer brake pad	5. Lower pad retainer	6. Outer shim
A. Molykote 7439 grease	B. Molykote AS-880N grease	

37671_EQUA_G0040

Fig. 5 Rear disc brake pads and related components

BRAKES   **PARKING BRAKE**

PARKING BRAKE CABLES

ADJUSTMENT

1. Before servicing the vehicle, refer to the Precautions Section.
2. Remove the rear half of the center console.
3. Rotate the adjusting nut and loosen the cable until the tension is sufficiently released.
4. Remove the wheel and tire.
5. Remove the rotor and measure the inner diameter at the widest point.
6. Transfer the measurement less 0.02 inch (0.6mm) to the parking brake shoes and adjust accordingly.

7. Using the wheel nuts, secure the disc to the hub to prevent it from tilting.
8. Rotate the disc rotor to make sure there is no drag.
9. Operate the parking brake lever 10 or more times with a force of 110 lbs. (490 N).
10. Rotate the adjusting nut to adjust the lever stroke to 6–8 notches with 44 lbs. (196 N) force.
11. With the parking brake lever completely disengaged, make sure there is no drag on the parking brake.

PARKING BRAKE SHOES

REMOVAL & INSTALLATION

See Figures 6 and 7.

1. Before servicing the vehicle, refer to the Precautions Section.

❋❋ CAUTION

Clean the brakes with a vacuum dust collector to minimize the hazard of airborne particles or other materials.

➡**Remove the disc rotor with the parking brake completely disengaged.**

2. Raise and safely support the vehicle.
3. Release the parking brake.
4. Remove the rear wheels.
5. Remove the rotor. Refer to Brake Disc (Rotor), removal & installation.

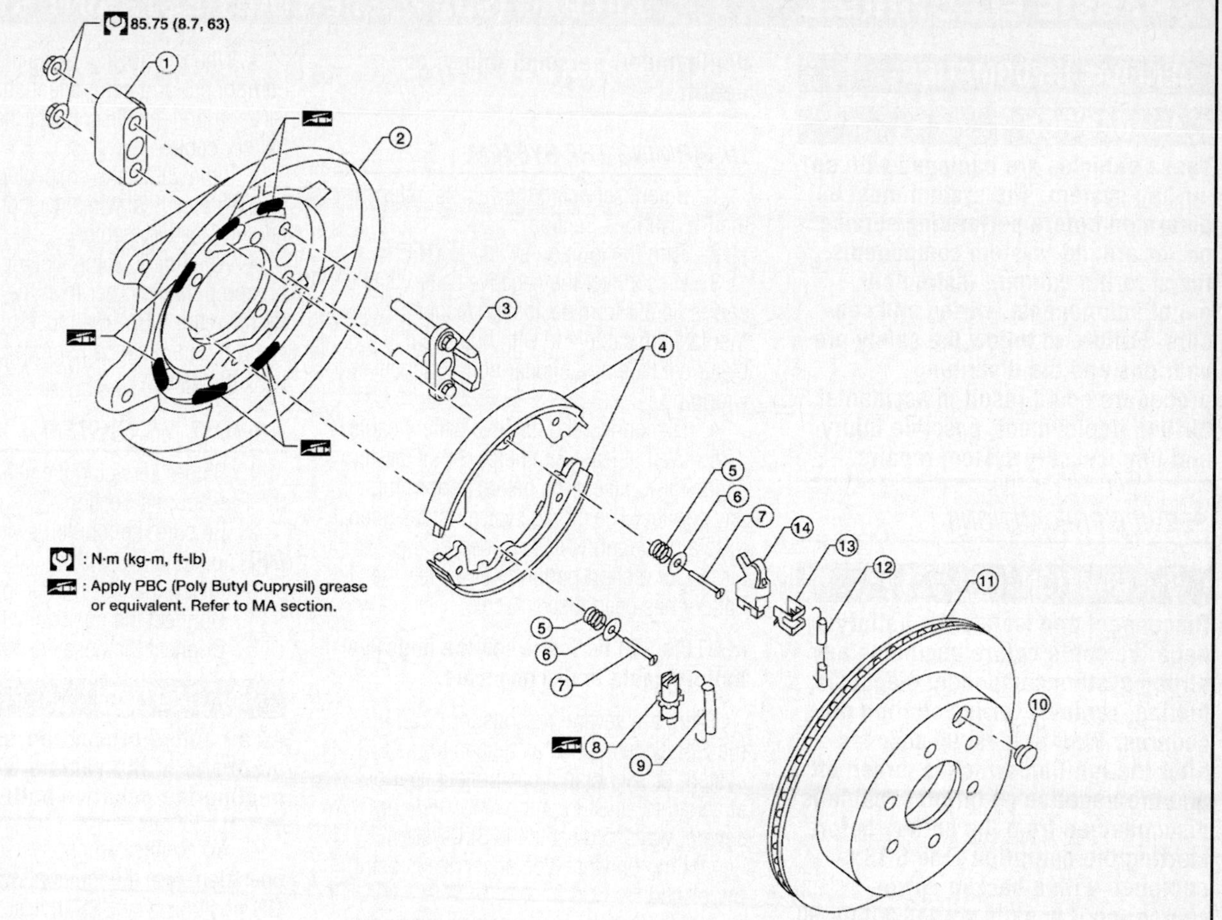

85.75 (8.7, 63)

: N·m (kg-m, ft-lb)

: Apply PBC (Poly Butyl Cuprysil) grease or equivalent. Refer to MA section.

1.	Anchor block	2.	Back plate	3.	Anchor
4.	Shoes	5.	Shoe hold-down spring	6.	Retainer
7.	Shoe hold-down pin	8.	Adjuster	9.	Rear return spring
10.	Adjuster access plug	11.	Disc rotor	12.	Front return spring
13.	Pin retainer	14.	Toggle lever		

42050_FRON_G0055

Fig. 6 Parking brake and related components

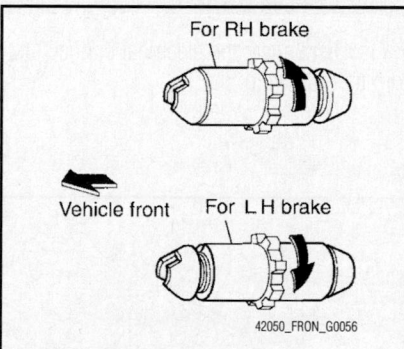

For RH brake

Vehicle front   For L H brake

42050_FRON_G0056

Fig. 7 Parking brake shoe adjuster identification

6. Remove the return springs.

7. Remove the adjuster.

8. Remove the retainers, anti-rattle pins, and shoes.

9. Remove the pin retainer. Disconnect the parking brake cable from the toggle lever.

10. Remove the back plate.

To install:

11. Installation is the reverse of the removal procedure.

12. Assemble the adjuster so that the threaded part expands when rotating it in the direction shown by the arrow. Shorten the adjuster by rotating it in the opposite direction shown by the arrow.

13. Perform the parking brake break-in operation as follows: Safely drive forward at approximately 25 mph (40 km/h) with the parking brake set with a force of approx. 45 lbs. (200 N) for about 30 seconds.

14. After the break-in operation, check the pedal stroke of the parking brake. Readjust, if necessary.

➡**To prevent the lining from getting too hot, allow a cool off period of approximately 5 minutes after every break-in operation.**

15. Check and adjust the parking brake pedal stroke. Correct as required.

GENERAL INFORMATION

✳✴ CAUTION

These vehicles are equipped with an air bag system. The system must be disarmed before performing service on, or around, system components, the steering column, instrument panel components, wiring and sensors. Failure to follow the safety precautions and the disarming procedure could result in accidental air bag deployment, possible injury and unnecessary system repairs.

SERVICE PRECAUTIONS

✳✴ CAUTION

Disconnect and isolate the battery negative cable before beginning any airbag system component diagnosis, testing, removal, or installation procedures. Wait at least 90 seconds after the ignition switch is turned off and the negative (-) terminal cable is disconnected from the battery before starting the operation. The SRS is equipped with a backup power source, so if work is started within 90 seconds after disconnecting the negative (-) terminal cable from the battery, the SRS may be deployed. Failure to disable the airbag system may result in accidental airbag deployment, personal injury, or death.

DISARMING THE SYSTEM

1. Before servicing the vehicle, refer to the Precautions Section.
2. Turn the ignition switch to **OFF**.
3. Disconnect the negative battery cable and isolate it from accidental reconnection. Insulate the cable end with high-quality electrical tape or a similar non-conductive wrapping.
4. Disconnect the positive battery cable.
5. Wait at least 3 minutes for the system capacitor to discharge before performing any service. The airbag system is designed to retain enough voltage to deploy the airbag for a short period of time after the battery has been disconnected.

➡**DTC's will be lost when the negative battery cable is disconnected.**

There are several reasons for disabling the SIR system, such as repairs to the SIR system or servicing a component near or attached to an SIR component. There are several ways to disable the SIR system depending on what type of service is being performed.

- If the vehicle was involved in an accident with an air bag deployment: Disconnect the negative battery cable
- When performing SIR diagnostics: Follow the appropriate SIR service manual diagnostic procedure(s)
- When removing or replacing an SIR component or a component attached to an SIR component: Disconnect the negative battery cable
- If the vehicle is suspected of having shorted electrical wires: Disconnect the negative battery cable
- When performing electrical diagnosis on components other than the SIR system: Remove the SIR/Airbag fuse(s) when indicated by the diagnostic procedure to disable the SIR system

ARMING THE SYSTEM

1. Before servicing the vehicle, refer to the Precautions Section.
2. Be sure the ignition switch is in the **OFF** position.
3. Install the fuses, if removed.
4. Connect the positive battery cable.
5. Connect the negative battery cable.

✳✴ CAUTION

As an added precaution, make sure no one is in the vehicle when reconnecting the negative battery cable.

6. To confirm proper system operation, turn the ignition switch to the **ON** position. The SRS indicator light should light for at least 7 seconds and then go off.
7. If the AIR BAG warning indicator does not operate as described, perform a diagnostic system check.

DRIVE TRAIN

TRANSFER CASE ASSEMBLY

REMOVAL & INSTALLATION

See Figures 8 through 10.

1. Before servicing the vehicle, refer to the Precautions Section.
2. Switch the 4WD shift switch to 2WD. Set the transfer assembly to 2WD.
3. Partially drain the transfer fluid.
4. Remove the transmission undercover.
5. Remove the center exhaust tube and main muffler.
6. Remove the front and rear propeller shafts. Refer to Driveshaft, removal & installation.

✳✴ WARNING

Do not damage the spline, sleeve yoke, or rear oil seal when removing the rear propeller shaft.

7. Insert a plug into the rear oil seal after removing the rear propeller shaft.
8. Remove transmission-to-crossmember bolts.
9. Position 2 suitable jacks under the transmission and transfer assembly.
10. Remove the transmission crossmember.

➡**Support the transmission and transfer assembly using 2 suitable jacks while removing the transmission crossmember.**

11. Disconnect the electrical connectors from the following:
- The ATP switch
- The 4LO switch

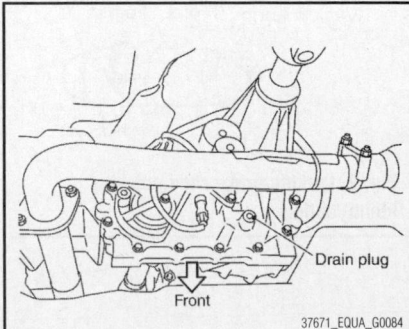

Fig. 8 Location of transfer case drain plug

37671_EQUA_G0084

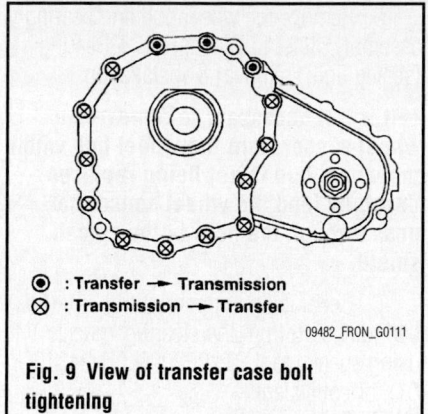

: Transfer → Transmission
: Transmission → Transfer

09482_FRON_G0111

Fig. 9 View of transfer case bolt tightening

• The wait detection switch
• The transfer control device

12. Remove the wire harness from the retainers.

13. Disconnect each air breather hose from the transfer control device and the breather tube (transfer).

14. Remove the transfer-to-transmission and transmission-to-transfer bolts.

❄ WARNING

Support the transfer assembly with a suitable jack while removing it.

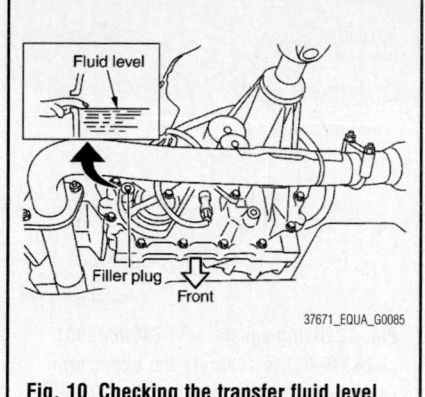

Fluid level

Filler plug

Front

37671_EQUA_G0085

Fig. 10 Checking the transfer fluid level

15. Remove the transfer assembly from the vehicle.

❄ WARNING

Do not damage automatic transmission rear oil seal.

To install:

16. Installation is the reverse of the removal procedure.

17. Tighten the transfer case-to-transmission retaining bolts to 27 ft. lbs. (36 Nm).

18. Fill the transfer with new fluid to the proper level.

19. Check the transfer fluid.

20. Start and run the engine for 1 minute. Stop the engine and recheck the transfer fluid.

21. After the installation, check the 4WD shift indicator pattern. If it is not right, adjust the position between the transfer assembly and transfer control unit.

DRIVEN DISC & PRESSURE PLATE

REMOVAL & INSTALLATION

See Figure 11.

1. Before servicing the vehicle, refer to the Precautions Section.

➡**Do not clean the clutch disc with solvent. When installing, do not get grease from the main drive shaft onto the clutch disc friction surface.**

2. Remove the manual transmission from the vehicle. Refer to Manual Transmission Assembly, removal & installation.

3. Remove the clutch cover bolts. Remove the clutch cover and clutch disc.

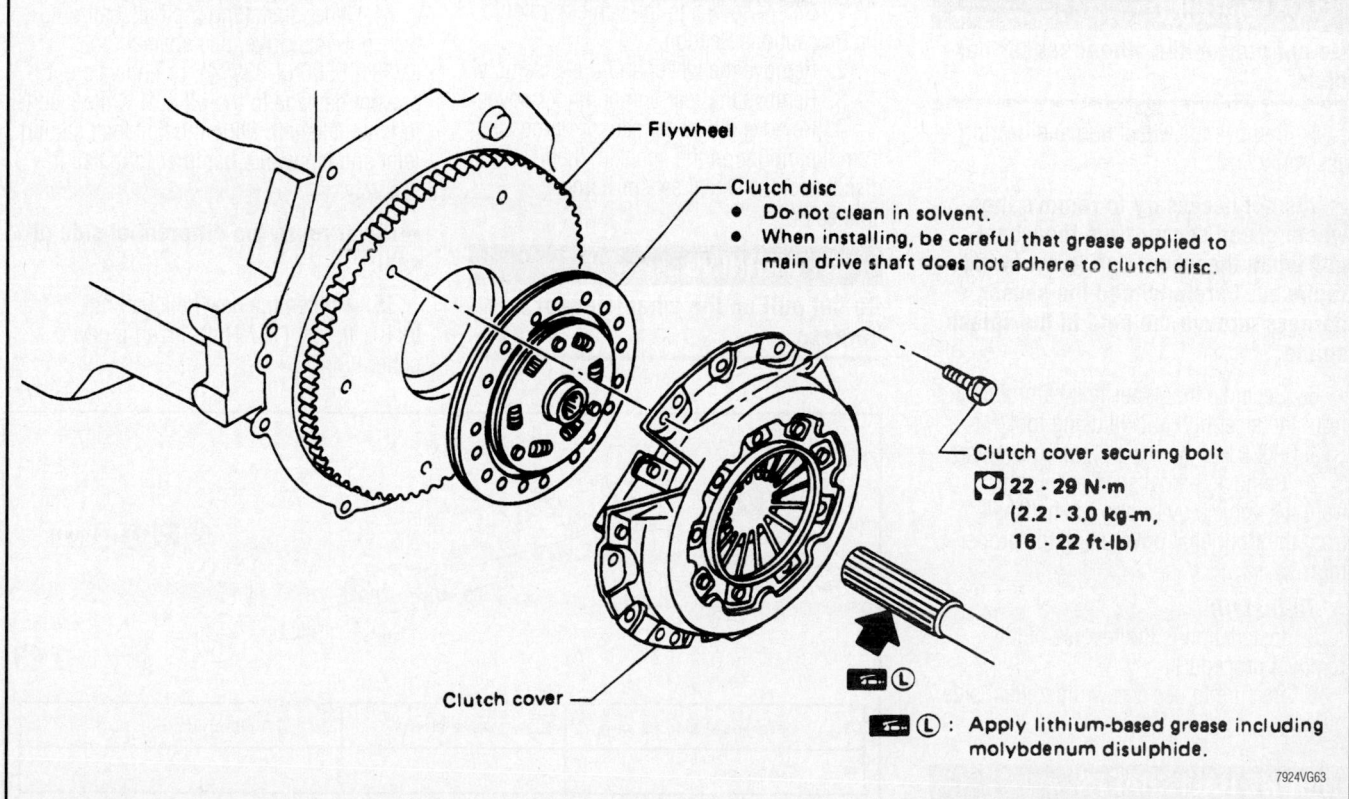

— Flywheel

— Clutch disc
• Do not clean in solvent.
• When installing, be careful that grease applied to main drive shaft does not adhere to clutch disc.

— Clutch cover securing bolt
22 - 29 N·m
(2.2 - 3.0 kg-m,
16 - 22 ft-lb)

Clutch cover —

▣ⓛ : Apply lithium-based grease including molybdenum disulphide.

7924VG63

Fig. 11 Clutch and related components

To install:

4. Apply the recommended grease to the clutch disc and main drive shaft spline.

➡ **Do not allow grease to contaminate the clutch facing.**

5. Install the clutch disc and clutch cover. Pre-tighten the bolts and install special tool ST20630000 (J-26366), or equivalent.

6. Tighten the clutch cover bolts evenly in 2 steps:
   a. Step 1: 11 ft. lbs. (15 Nm).
   b. Step 2: 19 ft. lbs. (25 Nm).

7. Install the manual transmission. Refer to Manual Transmission Assembly, removal & installation.

FRONT AXLE SHAFT, BEARING & SEAL

REMOVAL & INSTALLATION

1. Before servicing the vehicle, refer to the Precautions Section.

2. Raise and support the vehicle safely. Remove the tire and wheel assembly.

3. Remove the rear engine cover.

4. Remove the wheel sensor harness from the mount on the knuckle. Disconnect the harness connector.

❋❋ WARNING

Do not pull on the wheel sensor harness.

5. Remove the wheel hub and bearing assembly.

➡ **It is not necessary to remove the wheel speed sensor from the wheel hub when the wheel hub is not being replaced. Carefully feed the sensor harness through the hole in the splash shield.**

6. Separate the upper link ball joint stud from the steering knuckle using tool ST29020001 (J-24319-01), or equivalent.

7. Remove the halfshaft assembly from the vehicle by prying the halfshaft from the front final drive using the proper tool.

To install:

8. Installation is the reverse of the removal procedure.

9. Be sure to use a new differential side oil seal.

FRONT HALFSHAFTS

REMOVAL & INSTALLATION

See Figures 12 through 14.

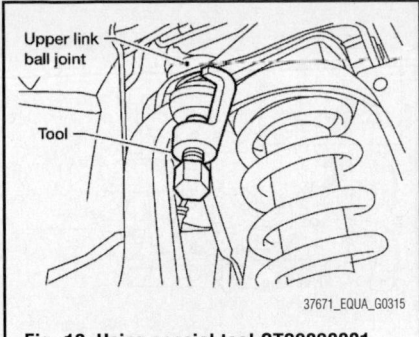

Fig. 12 Using special tool ST29020001 (J-24319-01) to separate the upper arm (link) ball joint from the steering knuckle

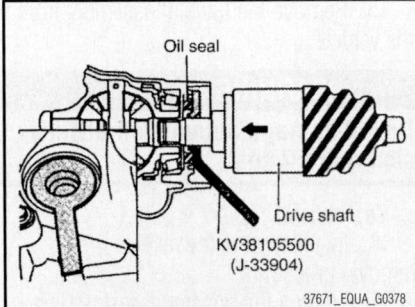

Fig. 13 Using tool KV38105500 (J-33904) to prevent damage to the oil seal while inserting the halfshaft

1. Before servicing the vehicle, refer to the Precautions Section.

2. Remove the wheel and tire assembly.

3. Remove the rear engine under cover.

4. Remove the wheel sensor harness from the mount on the knuckle, then disconnect the wheel sensor harness connector.

❋❋ WARNING

Do not pull on the wheel sensor harness.

5. Remove the wheel hub and bearing assembly. Refer to Wheel Hub & Bearing (sealed unit), removal & installation.

➡ **It is not necessary to remove the wheel sensor from the wheel hub when the wheel hub is not being replaced. Carefully feed the wheel sensor harness through the hole in the splash shield.**

6. Separate the upper control arm (link) ball joint stud from the steering knuckle using special tool ST29020001 (J-24319-01), or equivalent.

7. Support the lower control arm with a jack.

8. Pry the halfshaft front the final drive using a suitable tool.

9. Remove the differential side oil seal.

To install:

10. Move the joint up, down, left, right, and in an axial direction. Check for any rough movement or significant looseness.

11. Check the boot for cracks or other damage, and for grease leakage.

12. If damaged, disassemble the halfshaft to verify damage, and repair or replace as necessary.

13. Installation is in the reverse order of removal.

14. When installing the halfshaft onto the front final drive, use special tool KV38105500 (J-33904), or equivalent, to prevent damage to the oil seal while inserting the halfshaft. Slide the halfshaft sliding joint and tap with a hammer to install it securely.

➡ **Never reuse the differential side oil seal.**

15. Tighten the new halfshaft nut to 101 ft. lbs. (137 Nm). Insert a new cotter pin.

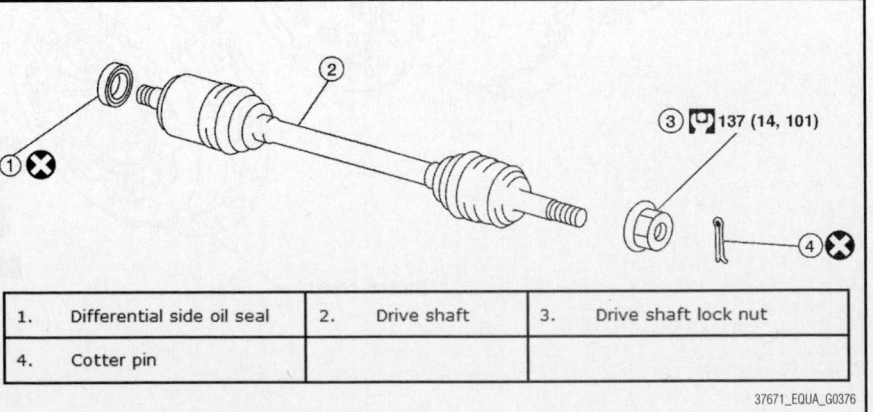

1.	Differential side oil seal	2.	Drive shaft	3.	Drive shaft lock nut
4.	Cotter pin				

Fig. 14 View of front halfshaft assembly

16. Tighten the wheel nuts to 98 ft. lbs. (133 Nm).

CV-BOOTS INSPECTION

1. Before servicing the vehicle, refer to the Precautions Section.
2. Check the halfshaft boots for damage and deterioration:
   - Raise the front of the vehicle
   - Rotate the axle and inspect for cracked or ripped CV boot material on the inner and outer CV-joints and check for missing clamps. Repeat this step on both sides of the vehicle
   - Inspect for excessive grease deposits on or around the CV boot
3. Replace the boot if it is damaged or deteriorated.

FRONT PINION SEAL

REMOVAL & INSTALLATION

See Figures 15 through 18.

1. Before servicing the vehicle, refer to the Precautions Section.
2. Remove the halfshafts from the front final drive assembly. Refer to Halfshafts, removal & installation.
3. Remove the front propeller shaft from the front final drive assembly. Refer to Driveshaft, removal & installation.
4. Measure the total preload torque. Record the total preload torque measurement.
5. Remove the drive pinion lock nut using a suitable tool.
6. Put matching marks on the companion flange and drive pinion using paint.

❉ WARNING

Use paint to make the matching marks. Do not damage the companion flange or drive pinion.

Fig. 15 Remove the drive pinion lock nut using a suitable tool

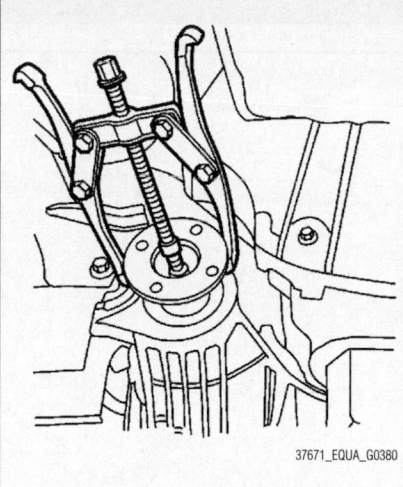

37671_EQUA_G0380

Fig. 16 Remove the companion flange using a suitable tool

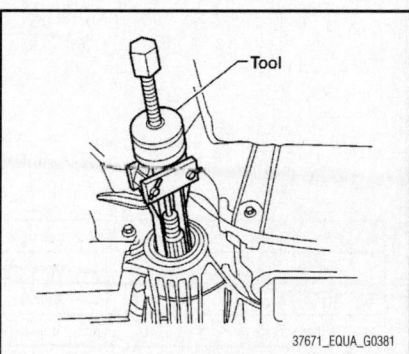

37671_EQUA_G0381

Fig. 17 Using special tool KV381054S0 (J-34286) to remove the front oil pinion seal

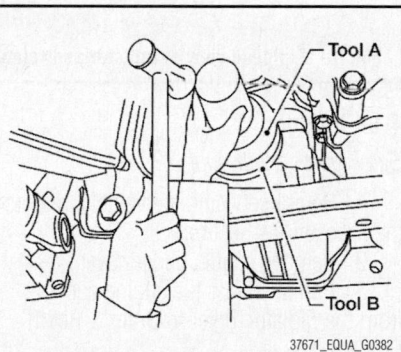

37671_EQUA_G0382

Fig. 18 Using special tools A: ST30720000 (J-25405) and B: ST27863000 to drive the new front pinion seal

7. Remove the companion flange using a suitable tool.
8. Remove the front oil seal using special tool KV381054S0 (J-34286), or equivalent.

To install:

9. Apply multi-purpose grease to the lips and differential gear oil to the circum-ference of the new front oil pinion seal. Then drive the new seal in evenly until it becomes flush with the gear carrier using special tools A: ST30720000 (J-25405) and B: ST27863000, or equivalent.

➡**Do not reuse the front oil seal. Do not incline the new front oil seal when installing. Apply multi-purpose grease to the lips and differential gear oil to the circumference of the new front oil seal.**

10. Install the companion flange to the drive pinion while aligning the matching marks.
11. Apply anti-corrosive oil to the threads of the drive pinion and the seating surface of the new drive pinion lock nut. Then adjust the drive pinion lock nut tightening torque using a suitable tool.
   a. The total preload torque should be within the total preload torque specification. When not replacing the collapsible spacer, it should also be equal to the measurement taken during removal plus an additional 5 inch lbs. (0.56 Nm).
   b. If the total preload torque is low, tighten the drive pinion lock nut in 60 inch lbs. (7 Nm) increments until the total preload torque is met.
   c. Preload torque specification: 15–24 inch lbs. (2–3 Nm).

➡**Do not reuse the drive pinion lock nut. Apply anti-corrosive oil to the threads of the drive pinion and the seating surface of the new drive pinion lock nut. Adjust the drive pinion lock nut tightening torque to the lower limit of the preload torque first. Do not exceed the drive pinion lock nut specified torque.**

➡**Do not loosen the drive pinion lock nut to adjust the total preload torque. If the drive pinion lock nut torque or the total preload torque exceeds the specifications, replace the collapsible spacer and tighten it again to adjust.**

12. After adjustment, rotate the drive pinion back and forth 2–3 times to check for unusual noise, rotation malfunction, and other malfunctions.
13. Install new side oil seals into the front final drive assembly.
14. Installation of the remaining components is in the reverse order of removal.
15. Check the differential gear oil level after installation.

ENGINE COOLING

ENGINE FAN

REMOVAL & INSTALLATION

✳✳ CAUTION

Never open, service or drain the radiator or cooling system when hot; serious burns can occur from the steam and hot coolant. Also, when draining engine coolant, keep in mind that cats and dogs are attracted to ethylene glycol antifreeze and could drink any that is left in an uncovered container or in puddles on the ground. This will prove fatal in sufficient quantities. Always drain coolant into a sealable container. Coolant should be reused unless it is contaminated or is several years old.

2.5L Engine

See Figure 19.

1. Before servicing the vehicle, refer to the Precautions Section.
2. Remove the engine front under cover.
3. Partially drain the engine coolant. Refer to Drain & Refill Procedure.
4. Remove the air duct and resonator assembly and air duct mounting brackets.
5. Remove the upper radiator hose.
6. Disconnect the reservoir tank hose from the upper shroud and radiator.
7. Remove the upper and lower fan shrouds.
8. Remove the drive belt. Refer to Accessory Drive Belts, removal & installation.
9. Remove the cooling fan.
10. Remove the fan coupling, if necessary.
11. Remove the water pump pulley, if necessary.

To install:

12. Inspect the fan coupling for oil leakage and bimetal conditions. If there are any unusual concerns, replace the fan coupling.
13. Inspect the cooling fan for cracks or unusual bends. If there are any unusual concerns, replace the cooling fan.
14. Installation is in the reverse order of removal.
15. Install the cooling fan with its front mark "F" facing the front of the engine.
16. Start and warm up the engine. Visually inspect for leaks of the engine coolant.

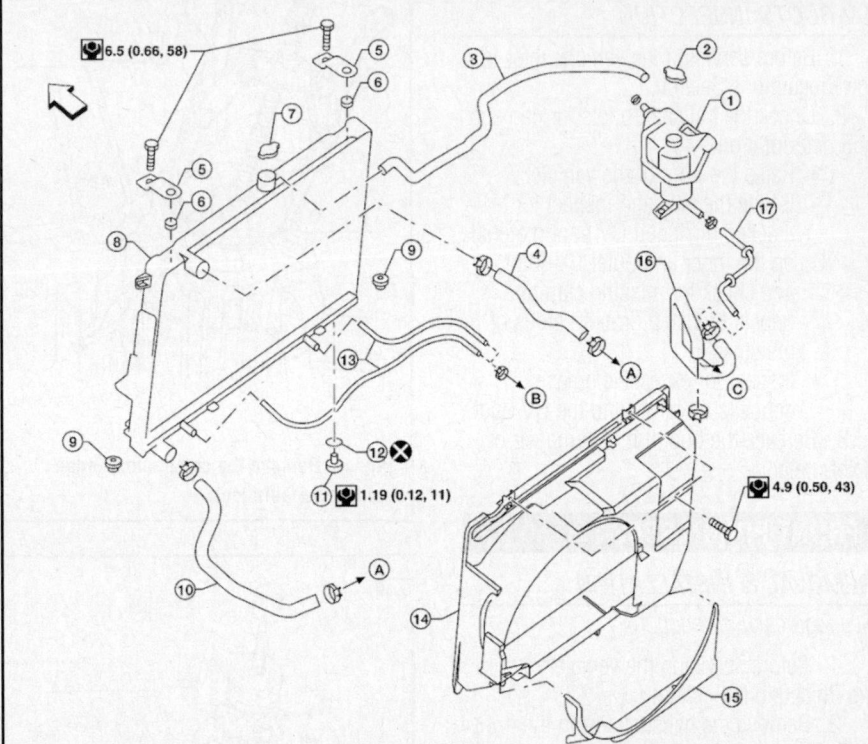

1.	Reservoir tank	2.	Reservoir tank cap	3.	Reservoir tank hose
4.	Radiator hose (upper)	5.	Upper mount bracket	6.	Mounting rubber (upper)
7.	Radiator cap	8.	Radiator	9.	Mounting rubber (lower)
10.	Radiator hose (lower)	11.	Radiator drain plug	12.	O-ring
13.	A/T fluid cooler hose (if equipped)	14.	Upper shroud	15.	Lower shroud
16.	Heater bypass hose	17.	Heater bypass tube	A.	To water inlet
B.	To A/T fluid cooler tube	C.	To heater tube	⬅	Front

37671_EQUA_G0109

Fig. 19 Exploded view of radiator and related components—2.5L engine

4.0L Engine

See Figures 20 through 22.

1. Before servicing the vehicle, refer to the Precautions Section.
2. Remove engine under cover.
3. Partially drain the engine coolant from the radiator. Refer to Drain & Refill Procedure.
   - Perform this step when engine is cold
   - Do not spill engine coolant on the drive belts
4. Remove the engine room cover.
5. Remove air duct and resonator assembly.
6. Remove reservoir tank hose from the shroud.
7. Removal the radiator hose (upper) from the radiator.
8. Release the radiator shroud (lower)

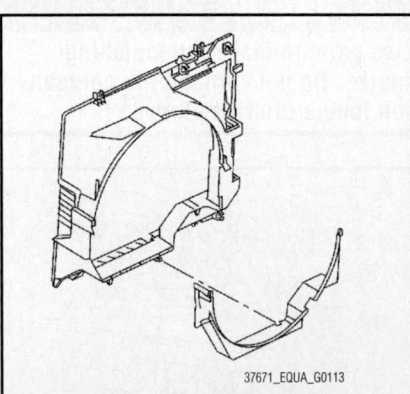

37671_EQUA_G0113

Fig. 20 Release the radiator shroud (lower) from the radiator shroud (upper)— 4.0L engine

from the radiator shroud (upper) and position aside. Release the tabs, pull the radiator shroud (lower) rearwards and down.

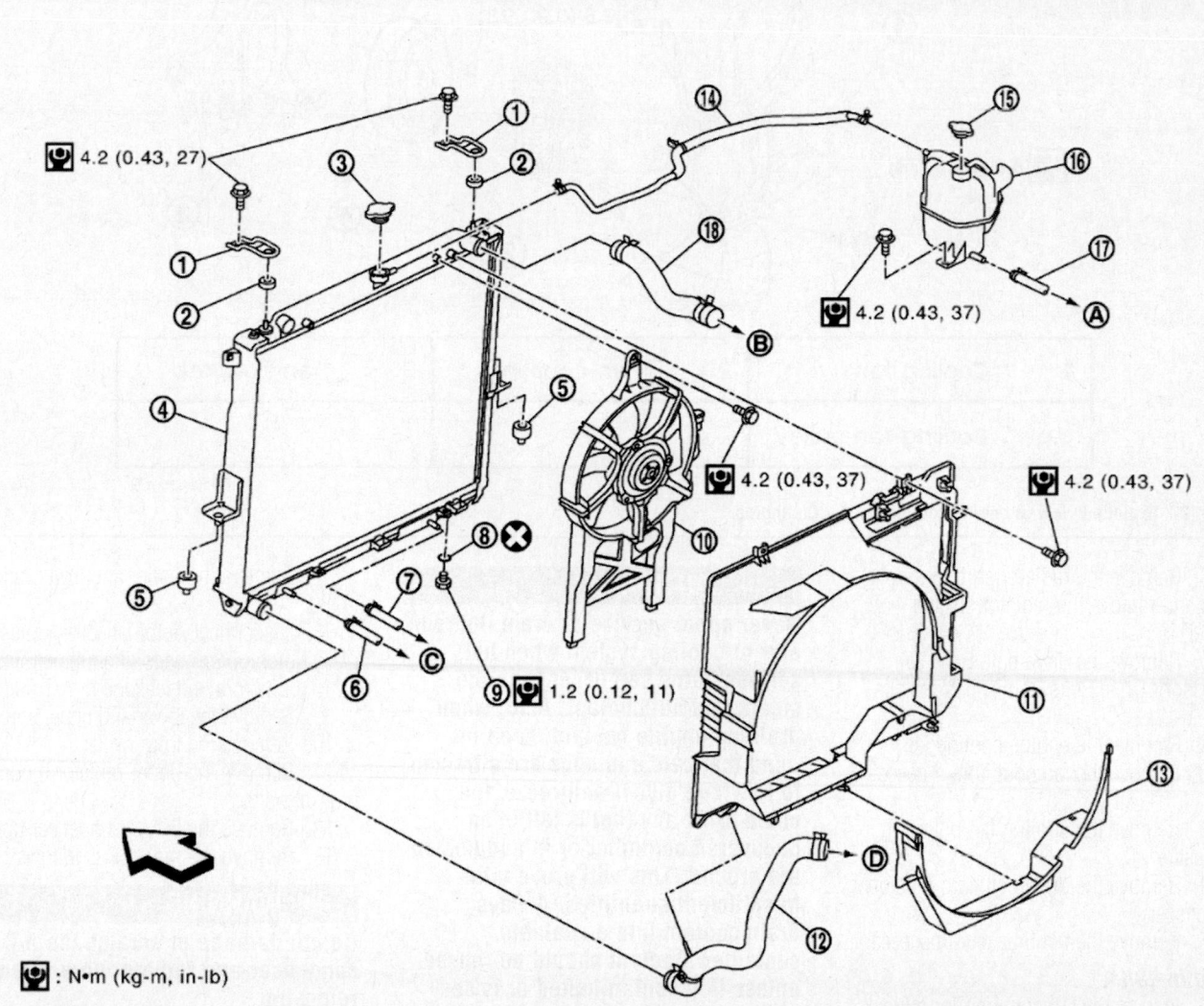

$\boxed{\text{N}}$ 4.2 (0.43, 27)

$\boxed{\text{N}}$ 4.2 (0.43, 37)

$\boxed{\text{N}}$ 4.2 (0.43, 37)

$\boxed{\text{N}}$ 4.2 (0.43, 37)

$\boxed{\text{N}}$ 1.2 (0.12, 11)

$\boxed{\text{N}}$: N•m (kg-m, in-lb)

1.	Radiator mounting bracket	2.	Mounting rubber (upper)	3.	Radiator cap
4.	Radiator	5.	Mounting rubber (lower)	6.	A/T fluid cooler hose
7.	A/T fluid cooler hose (if equipped)	8.	O-ring	9.	Drain plug
10.	Cooling fan assembly (Motor driven type)	11.	Radiator shroud (upper)	12.	Radiator hose (lower)
13.	Radiator shroud (lower)	14.	Reservoir tank hose	15.	Reservoir tank cap
16.	Reservoir tank	17.	Water hose	18.	Radiator hose (upper)
A.	To heater return tube	B.	To water pipe	C.	To A/T cooler tube
D.	To water inlet and thermostat assembly	⇦	Vehicle front		

37671_EQUA_G0114

Fig. 21 Exploded view of radiator and related components—4.0L engine

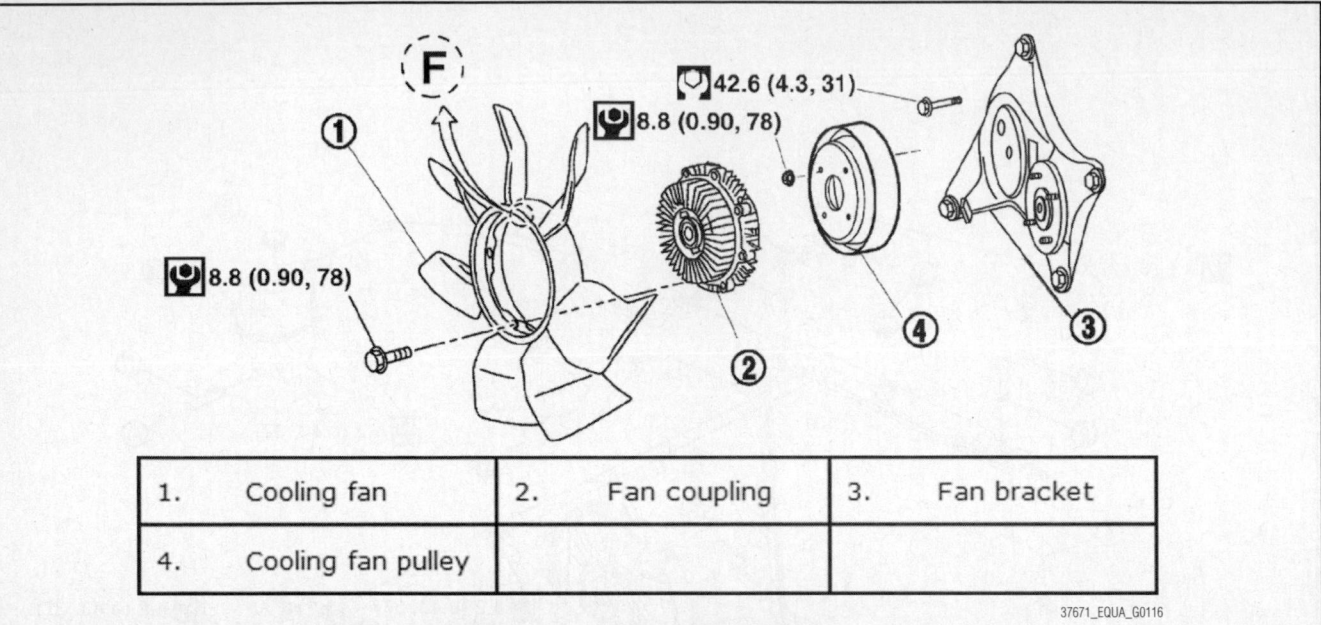

1.	Cooling fan	2.	Fan coupling	3.	Fan bracket
4.	Cooling fan pulley				

37671_EQUA_G0116

Fig. 22 Exploded view of engine cooling fan—4.0L engine

9. Remove the radiator shroud (upper) bolts and remove the radiator shroud (upper).

10. Remove the drive belt. Refer to Accessory Drive Belts, removal & installation.

11. Remove the engine cooling fan.

12. Remove the fan coupling, if necessary.

13. Remove the cooling fan pulley, if necessary.

14. Remove the drive belt auto-tensioner, if necessary.

15. Remove the fan bracket, if necessary.

To install:

16. Inspect the fan coupling for oil leakage and bimetal conditions. If there are any unusual concerns, replace the fan coupling.

17. Visually check that there is no significant looseness in the fan bracket shaft, and that it turns smoothly by hand. If there are any unusual concerns, replace the fan bracket assembly.

18. Installation is in the reverse order of removal.

19. Install the cooling fan with its front mark "F" facing the front of the engine.

20. Start and warm up the engine. Visually inspect for leaks of the engine coolant.

RADIATOR

REMOVAL & INSTALLATION

See Figures 19 and 21.

✳✳ CAUTION

Never open, service or drain the radiator or cooling system when hot; serious burns can occur from the steam and hot coolant. Also, when draining engine coolant, keep in mind that cats and dogs are attracted to ethylene glycol antifreeze and could drink any that is left in an uncovered container or in puddles on the ground. This will prove fatal in sufficient quantities. Always drain coolant into a sealable container. Coolant should be reused unless it is contaminated or is several years old.

1. Before servicing the vehicle, refer to the Precautions Section.

2. Do not remove the radiator cap when the engine is hot. Serious burns could occur from the high-pressure engine coolant escaping from the radiator. Wrap a thick cloth around the cap. Slowly turn it a quarter of a turn to release built-up pressure. Carefully remove radiator cap by turning it all the way.

3. Remove engine under cover.

4. Drain engine coolant from radiator. Refer to Drain & Refill Procedure.
- Perform this step when the engine is cold
- Do not spill engine coolant on the drive belts

5. Remove the air duct and resonator assembly and air duct brackets.

6. Remove the reservoir tank hose.

7. Remove the upper and lower radiator hoses.

8. Disconnect Automatic Transmission (A/T) fluid cooler hoses, if equipped. Install blind plugs to avoid leakage of A/T fluid.

9. Remove the lower and upper shrouds.

10. Remove the front grille.

11. Remove the upper radiator mounting bracket bolts.

12. Remove the 2 A/C condenser bolts.

13. Remove the radiator as follows:

✳✳ WARNING

Do not damage or scratch the A/C condenser and radiator core when removing.

a. While lifting and pulling the radiator in a rearward direction, disassemble the mounting rubber (lower) from the radiator core support center.

➡ **Because the A/C condenser is attached to the front-lower portion of the radiator, moving it in the rearward direction should be at a minimum.**

b. Lift the A/C condenser up and remove the radiator after disengaging the fitting at the front-bottom surface.

✳✳ WARNING

Lifting the A/C condenser should be minimum to prevent a load to the A/C piping.

c. After removing the radiator, put the A/C condenser on the radiator core support center to prevent a load to the A/C piping, and temporarily secure it with a rope or by similar means.

To install:

14. Installation is in the reverse order of removal.

15. Fill the engine coolant. Refer to Drain & Refill Procedure.

16. Start and warm up the engine. Visually inspect for leaks of the engine coolant.

THERMOSTAT

REMOVAL & INSTALLATION

2.5L Engine

See Figures 23 and 24.

✷✷ CAUTION

Never open, service or drain the radiator or cooling system when hot; serious burns can occur from the steam and hot coolant. Also, when draining engine coolant, keep in mind that cats and dogs are attracted to ethylene glycol antifreeze and could drink any that is left in an uncovered container or in puddles on the ground. This will prove fatal in sufficient quantities. Always drain coolant into a sealable container. Coolant should be reused unless it is contaminated or is several years old.

1. Before servicing the vehicle, refer to the Precautions Section.

2. Do not remove the radiator cap when the engine is hot. Serious burns could occur from the high-pressure engine coolant escaping from the radiator. Wrap a thick cloth around the cap. Slowly turn it a quarter of a turn to release built-up pressure. Carefully remove radiator cap by turning it all the way.

3. Remove engine under cover.

4. Drain engine coolant from the radiator. Refer to Drain & Refill Procedure.

- Perform this step when the engine is cold
- Do not spill engine coolant on the drive belts

5. Remove the air duct and resonator assembly and air duct brackets.

6. Disconnect the lower radiator hose at the water inlet side (engine side).

7. Remove the water inlet retaining bolts.

8. Remove the water inlet and the thermostat.

To install:

9. Installation is the reverse of the removal procedure.

10. Be sure to apply a continuous bead of the proper grade and type RTV sealant to the housing.

11. Install the thermostat with the rubber ring groove positioned to fit the thermostat flange (the whole circumference).

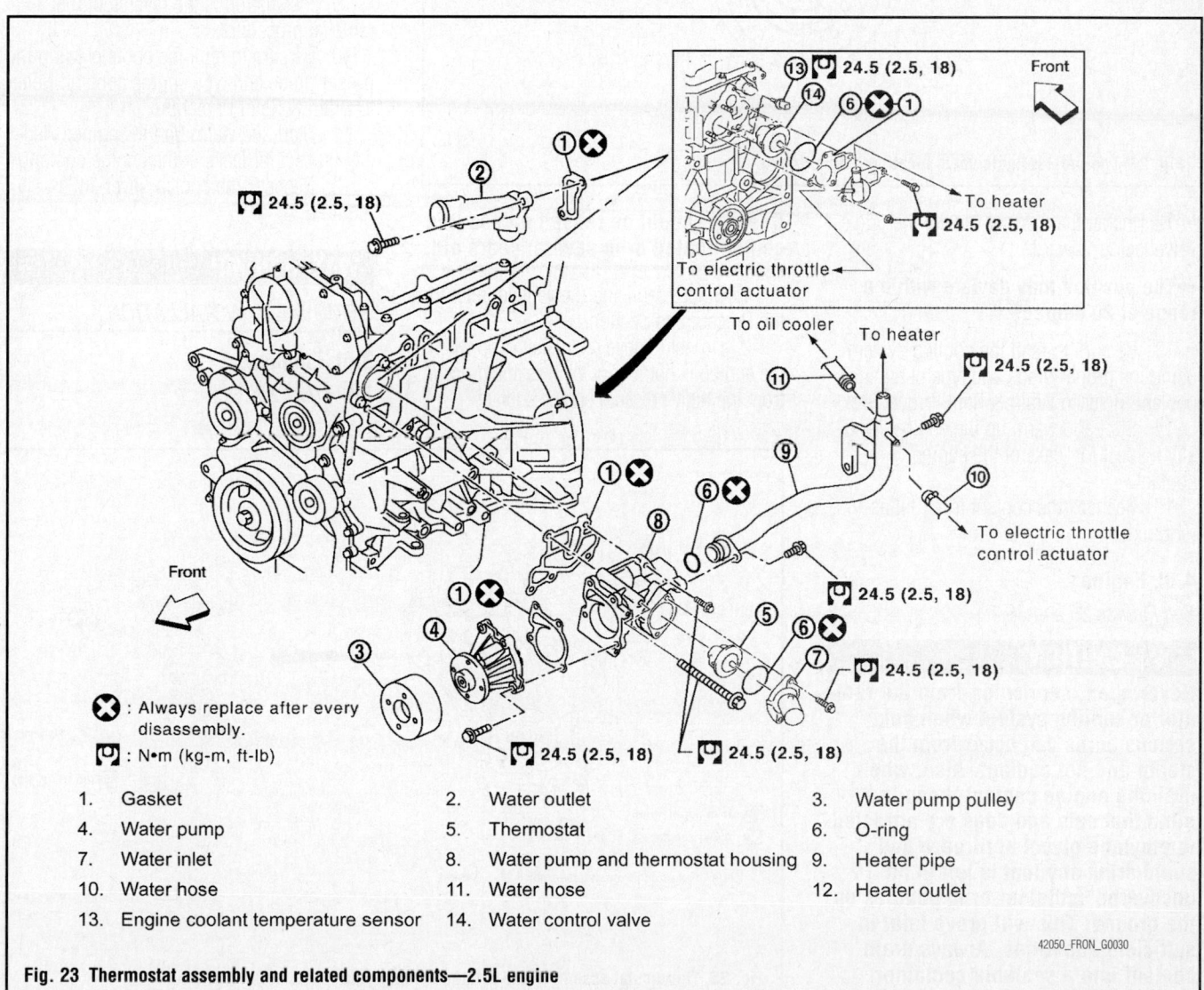

1. Gasket
2. Water outlet
3. Water pump pulley
4. Water pump
5. Thermostat
6. O-ring
7. Water inlet
8. Water pump and thermostat housing
9. Heater pipe
10. Water hose
11. Water hose
12. Heater outlet
13. Engine coolant temperature sensor
14. Water control valve

✖ : Always replace after every disassembly.
🔧 : N•m (kg-m, ft-lb)

42050_FRON_G0030

Fig. 23 Thermostat assembly and related components—2.5L engine

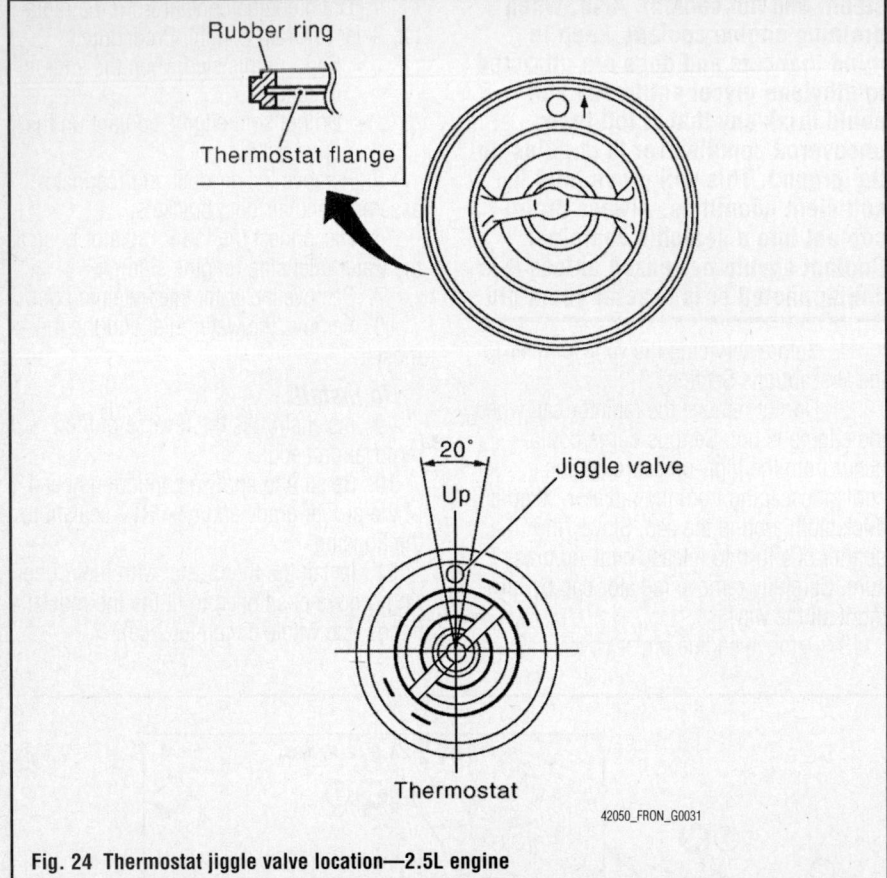

Fig. 24 Thermostat jiggle valve location—2.5L engine

12. Install the thermostat with the jiggle valve facing upward.

➡**The position may deviate within a range of 20 degrees.**

13. Be sure to refill the cooling system using the proper grade and type of engine coolant. Refer to Drain & Refill Procedure.

14. Start and warm up the engine. Visually inspect for leaks of the engine coolant.

15. Recheck the coolant level, fill as required.

4.0L Engine

See Figures 25 and 26

✷✷ CAUTION

Never open, service or drain the radiator or cooling system when hot; serious burns can occur from the steam and hot coolant. Also, when draining engine coolant, keep in mind that cats and dogs are attracted to ethylene glycol antifreeze and could drink any that is left in an uncovered container or in puddles on the ground. This will prove fatal in sufficient quantities. Always drain coolant into a sealable container.

Coolant should be reused unless it is contaminated or is several years old.

1. Before servicing the vehicle, refer to the Precautions Section.

2. Do not remove the radiator cap when the engine is hot. Serious burns could occur from the high-pressure engine coolant

escaping from the radiator. Wrap a thick cloth around the cap. Slowly turn it a quarter of a turn to release built-up pressure. Carefully remove radiator cap by turning it all the way.

3. Remove engine under cover.

4. Drain engine coolant from the radiator.
   - Perform this step when the engine is cold
   - Do not spill engine coolant on the drive belts

5. Remove the air duct and resonator assembly and air cleaner case.

6. Disconnect radiator hose (lower) and oil cooler hose from water inlet and thermostat assembly.

7. Remove the water inlet retaining bolts.

8. Remove the water inlet and the thermostat assembly.

➡**Do not disassemble the water inlet and thermostat assembly. Replace as a unit, if required.**

To install:

9. Installation is the reverse of the removal procedure.

10. Be sure to refill the cooling using the proper grade and type engine coolant. Refer to Drain & Refill Procedure.

11. Start and warm up the engine. Visually inspect for leaks of the engine coolant.

12. Recheck the coolant level, fill as required.

WATER PUMP

REMOVAL & INSTALLATION

2.5L Engine

See Figure 23.

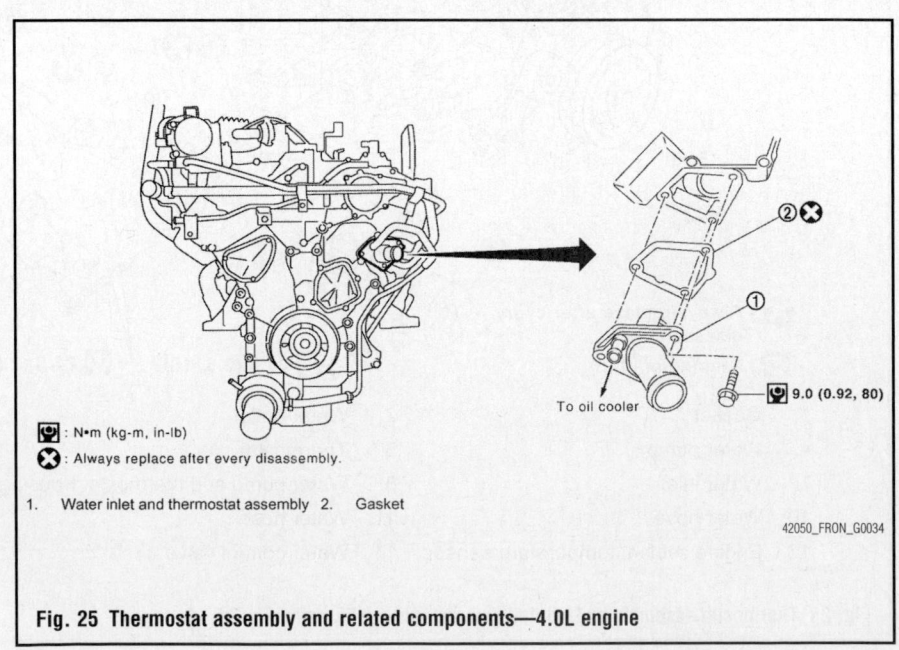

⬛ : N•m (kg-m, in-lb)
❌ : Always replace after every disassembly.

1. Water inlet and thermostat assembly 2. Gasket

42050_FRON_G0034

Fig. 25 Thermostat assembly and related components—4.0L engine

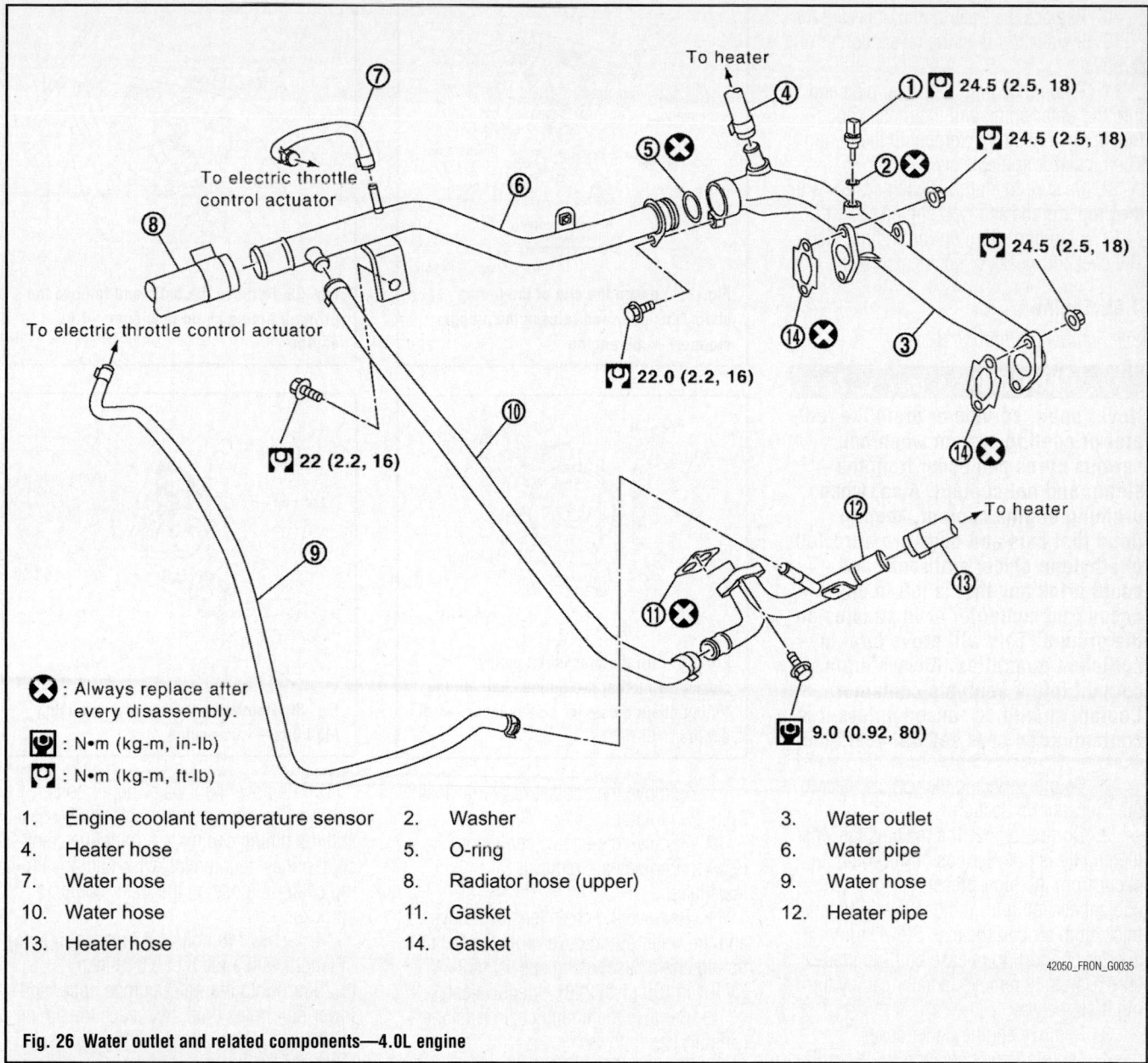

To heater

To electric throttle
control actuator

⑦

⑧

To electric throttle control actuator

① 🔧 24.5 (2.5, 18)

② ✖

🔧 24.5 (2.5, 18)

🔧 24.5 (2.5, 18)

🔧 22.0 (2.2, 16)

🔧 22 (2.2, 16)

To heater

🔧 9.0 (0.92, 80)

✖ : Always replace after every disassembly.

🔧 : N•m (kg-m, in-lb)

🔧 : N•m (kg-m, ft-lb)

1. Engine coolant temperature sensor	2. Washer	3. Water outlet
4. Heater hose	5. O-ring	6. Water pipe
7. Water hose	8. Radiator hose (upper)	9. Water hose
10. Water hose	11. Gasket	12. Heater pipe
13. Heater hose	14. Gasket	

42050_FRON_G0035

Fig. 26 Water outlet and related components—4.0L engine

❋❋ CAUTION

Never open, service or drain the radiator or cooling system when hot; serious burns can occur from the steam and hot coolant. Also, when draining engine coolant, keep in mind that cats and dogs are attracted to ethylene glycol antifreeze and could drink any that is left in an uncovered container or in puddles on the ground. This will prove fatal in sufficient quantities. Always drain coolant into a sealable container. Coolant should be reused unless it is contaminated or is several years old.

1. Before servicing the vehicle, refer to the Precautions Section.

2. Do not remove the radiator cap when the engine is hot. Serious burns could occur from the high-pressure engine coolant escaping from the radiator. Wrap a thick cloth around the cap. Slowly turn it a quarter of a turn to release built-up pressure. Carefully remove radiator cap by turning it all the way.

3. Remove engine under cover.

4. Drain engine coolant from the radiator. Refer to Drain & Refill Procedure.
   • Perform this step when the engine is cold
   • Do not spill engine coolant on the drive belts

5. Remove the cooling fan and water pump pulley. Refer to Engine Fan, removal & installation.

6. Remove the water pump bolts retaining bolts and remove the water pump from the engine.
   • Handle the water pump vane so that it does not contact any other parts
   • The water pump cannot be disassembled and should be replaced as a unit

➡Engine coolant will leak from the cylinder block; have a receptacle ready.

To install:

7. Visually check for dirt or rusting on the water pump body and vane.

8. Ensure that the vane shaft is not loose and that it turns smoothly when rotated by hand.

9. Replace the water pump, if necessary.

10. Installation is in the reverse order of removal.

11. When inserting the heater pipe end into the water pump and thermostat housing, apply a neutral detergent to the O-ring. Then insert it immediately.

12. Be sure to fill the cooling system with the proper grade and type engine coolant.

13. Start and warm up the engine. Visually check for leaks of engine coolant.

4.0L Engine

See Figures 27 through 32.

✳✳ CAUTION

Never open, service or drain the radiator or cooling system when hot; serious burns can occur from the steam and hot coolant. Also, when draining engine coolant, keep in mind that cats and dogs are attracted to ethylene glycol antifreeze and could drink any that is left in an uncovered container or in puddles on the ground. This will prove fatal in sufficient quantities. Always drain coolant into a sealable container. Coolant should be reused unless it is contaminated or is several years old.

1. Before servicing the vehicle, refer to the Precautions Section.

2. Do not remove the radiator cap when the engine is hot. Serious burns could occur from the high-pressure engine coolant escaping from the radiator. Wrap a thick cloth around the cap. Slowly turn it a quarter of a turn to release built-up pressure. Carefully remove radiator cap by turning it all the way.

3. Remove engine under cover.

4. Drain engine coolant from the radiator. Refer to Drain & Refill Procedure.

- Perform this step when the engine is cold
- Do not spill engine coolant on the drive belts

➡**When removing the water pump assembly, be careful not to get engine coolant on the timing chain. The water pump cannot be disassembled and should be replaced as a unit.**

5. Remove the engine room cover.

6. Remove the air duct and resonator assembly.

7. Remove the drive belt. Refer to Accessory Drive Belts, removal & installation.

8. Remove the radiator hose (upper).

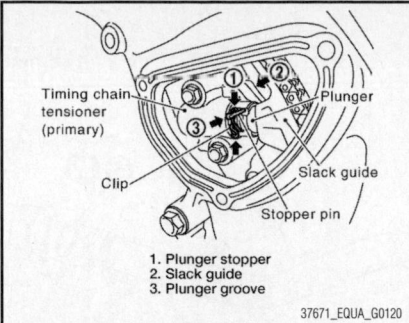

Fig. 27 Loosen the clip of the timing chain tensioner and release the plunger stopper—4.0L engine

1. Plunger stopper
2. Slack guide
3. Plunger groove

37671_EQUA_G0120

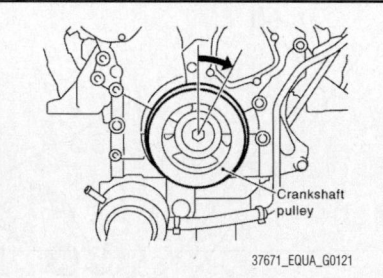

Fig. 28 Turn the crankshaft pulley clockwise so that the timing chain on the timing chain tensioner side is loose—4.0L engine

37671_EQUA_G0121

9. Remove the coolant reservoir hose from the radiator.

10. Remove the engine cooling fan. Refer to Engine Fan, removal & installation.

11. Remove the chain tensioner cover and the water pump cover from the front timing chain case, using special tool KV10111100 (J-37228), or equivalent.

12. Remove the timing chain tensioner (primary) as follows:

a. Loosen the clip of the timing chain tensioner and release the plunger stopper.

b. Insert the plunger into the tensioner body by pressing the slack guide.

c. Keep the slack guide pressed and hold the plunger in by pushing the stopper pin through the tensioner body hole and the plunger groove.

d. Turn the crankshaft pulley clockwise so that the timing chain on the timing chain tensioner side is loose.

e. Remove the bolts and remove the timing chain tensioner (primary).

✳✳ WARNING

Be careful not to drop the bolts inside the timing chain case.

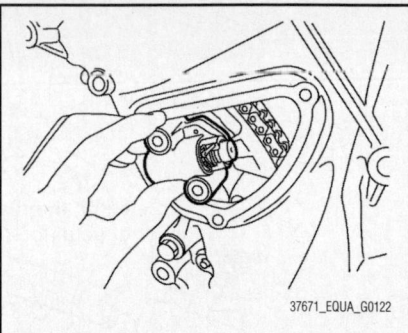

Fig. 29 Remove the bolts and remove the primary timing chain tensioner—4.0L engine

37671_EQUA_G0122

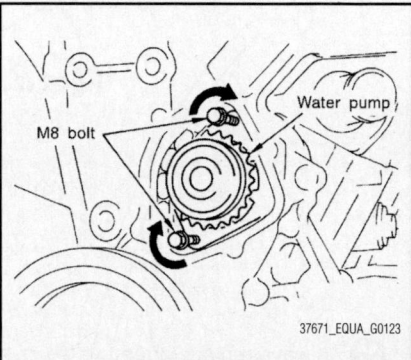

Fig. 30 Removing the water pump using M8 bolts—4.0L engine

37671_EQUA_G0123

13. Remove the 3 water pump bolts. Secure a gap between the water pump gear and the timing chain, by turning the crankshaft pulley counterclockwise until the timing chain is loose on the water pump sprocket.

14. Screw M8 bolts 1.97 inches (50mm) in length with a pitch of 0.049 inch (1.25mm) into the water pumps upper and lower bolt holes until they reach the timing chain case. Then, alternately tighten each bolt for a half turn, and pull out the water pump.

➡**Pull the water pump straight out while preventing the vane from contacting the socket in the installation area. Remove the water pump without causing the sprocket to contact the timing chain.**

15. Remove the M8 bolts. Remove and discard the O-rings.

To install:

16. Check that the water pump is not badly rusted or corroded.

17. Check for rough operation due to excessive end play.

18. Replace the water pump, if necessary.

19. Install new O-rings to the water pump.

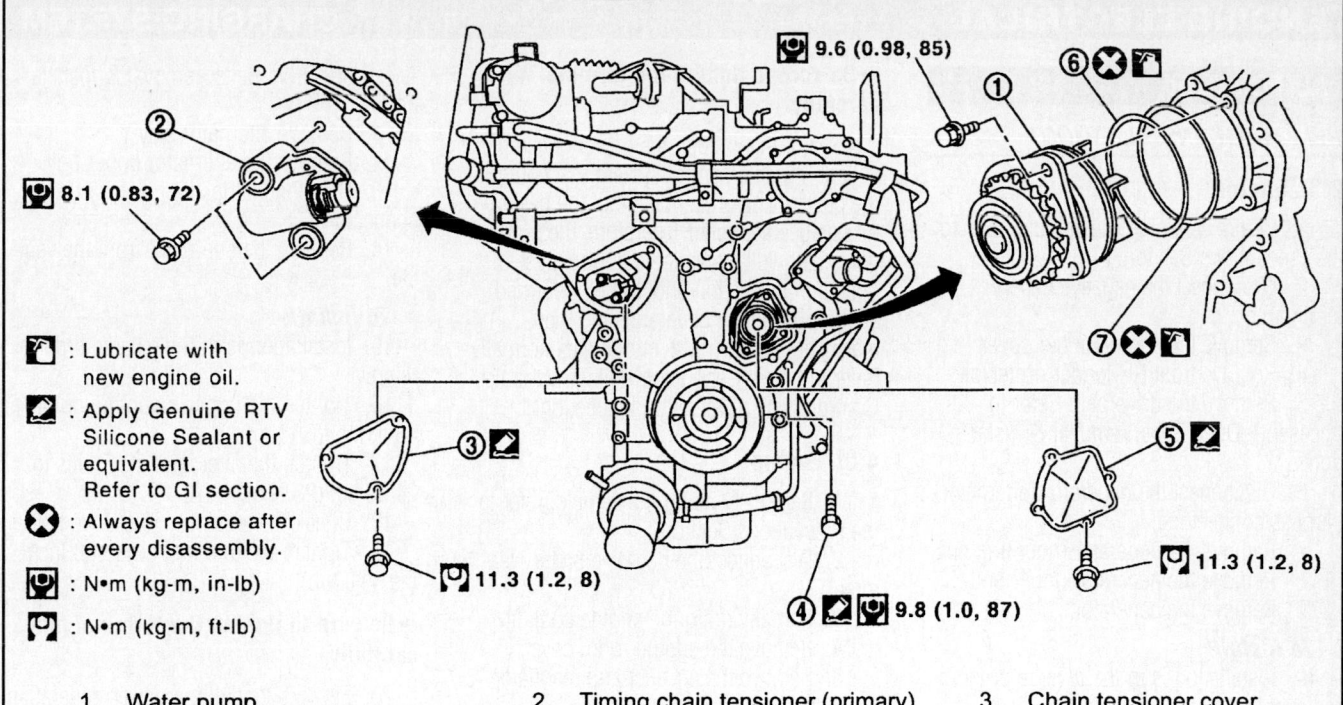

🔧 : Lubricate with new engine oil.

✍ : Apply Genuine RTV Silicone Sealant or equivalent. Refer to GI section.

✖ : Always replace after every disassembly.

🔧 : N•m (kg-m, in-lb)

🔧 : N•m (kg-m, ft-lb)

1. Water pump
2. Timing chain tensioner (primary)
3. Chain tensioner cover
4. Water drain plug (front)
5. Water pump cover
6. O-ring
7. O-ring

09482_FRON_G0016

Fig. 31 Water pump and related components—4.0L engine

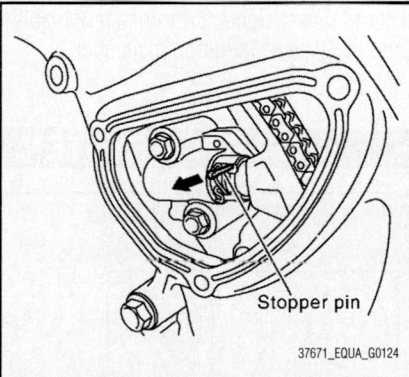

Stopper pin

37671_EQUA_G0124

Fig. 32 Remove the stopper pin from the primary chain tensioner once installed into position—4.0L engine

➡ Apply engine oil to the O-rings. Locate the O-ring with a white paint mark toward the engine front side.

20. Install the water pump.

✴✴ WARNING

Do not allow the timing chain case to pinch the O-rings when installing the water pump.

21. Make sure that the timing chain and the water pump sprocket are engaged.

22. Insert the water pump by tightening the bolts alternately and evenly.

23. Install the timing chain tensioner (primary) as follows:

a. Remove any dust and foreign material completely from the backside of the timing chain tensioner (primary) and from the installation area of the rear timing chain case.

b. Turn the crankshaft pulley clockwise so that the timing chain on the timing chain tensioner (primary) side is loose.

c. Install the timing chain tensioner (primary) with its stopper pin attached.

✴✴ WARNING

Be careful not to drop the bolts inside the timing chain case.

d. Remove the stopper pin.

e. Make sure again that timing chain and water pump sprocket are engaged.

24. Install the chain tensioner cover and the water pump cover as follows:

a. Remove all traces of the old liquid gasket from the mating surface of the

water pump cover and the chain tensioner cover using a scraper. Also, remove the traces of the old liquid gasket from the mating surface of the front timing chain case.

b. Apply a continuous bead of liquid gasket, to the mating surface of the chain tensioner and the water pump cover. Use Genuine RTV Silicone Sealant, or equivalent.

c. Attach the mating surfaces within 5 minutes after applying liquid gasket.

d. Tighten the bolts to the specified torque. Refer to illustration.

25. Installation of the remaining components is in the reverse order of removal.

26. Be sure to fill the cooling system with the proper grade and type engine coolant.

27. Start the engine and check for leaks.

28. Let the engine idle for about 3 minutes, then rev it up to 3,000 RPM under a no load condition to purge air from the high pressure chamber of the chain tensioner. The engine may produce a rattling noise. This indicates that air still remains in the chamber and is not a matter of concern.

ENGINE ELECTRICAL   CHARGING SYSTEM

ALTERNATOR

REMOVAL & INSTALLATION

2.5L Engine

1. Before servicing the vehicle, refer to the Precautions Section.

2. Disconnect the negative battery terminal.

3. Remove the engine under cover.

4. Remove front RH fender protector.

5. Remove the drive belt. Refer to Accessory Drive Belts, removal & installation.

6. Disconnect the generator (alternator) harness connectors.

7. Remove the generator mounting nut.

8. Remove the generator upper bolt.

9. Remove the generator.

To install:

10. Installation is in the reverse order of removal.

11. Be sure the generator spacer is in place on the lower stud.

12. Tighten the lower mounting nut and the upper mounting bolt to 48 ft. lbs. (65 Nm).

13. Check the tension of the drive belt.

14. Tighten the terminal nut to 96 inch lbs. (11 Nm).

➡ Be sure to tighten the terminal nut carefully.

15. For this model, the power generation voltage variable control system that controls the power generation voltage of the generator has been adopted. Therefore, the power generation voltage variable control system operation inspection should be performed after replacing the generator, and then make sure that the system operates normally. Refer to Power Generation Voltage Variable Control System Operation Inspection.

4.0L Engine

1. Before servicing the vehicle, refer to the Precautions Section.

2. Disconnect the negative battery terminal.

3. Partially drain the engine coolant.

4. Remove the engine room cover.

5. Remove the air duct and resonator assembly.

6. Remove the upper radiator hose.

7. Disconnect the coolant reservoir hose from the radiator.

8. Remove the fan shroud.

9. Remove the engine cooling fan (motor driven type). Refer to Engine Fan, removal & installation.

10. Remove the drive belt. Refer to Accessory Drive Belts, removal & installation.

11. Remove alternator stay.

12. Remove the alternator upper bolt.

13. Disconnect the alternator harness connectors.

14. Remove the alternator from the vehicle.

To install:

15. Installation is in the reverse order of removal.

16. Tighten the upper mounting bolt to 48 ft. lbs. (65 Nm).

17. Tighten the alternator stay bolts to 21 ft. lbs. (28 Nm).

18. Check the tension of the drive belt.

19. Tighten the terminal nut to 96 inch lbs. (11 Nm).

➡ Be sure to tighten the terminal nut carefully.

20. For this model, the power generation voltage variable control system that controls the power generation voltage of the generator has been adopted. Therefore, the power generation voltage variable control system operation inspection should be performed after replacing the generator, and then make sure that the system operates normally. Refer to Power Generation Voltage Variable Control System Operation Inspection.

ENGINE ELECTRICAL   IGNITION SYSTEM

FIRING ORDER

See Figures 33 and 34.

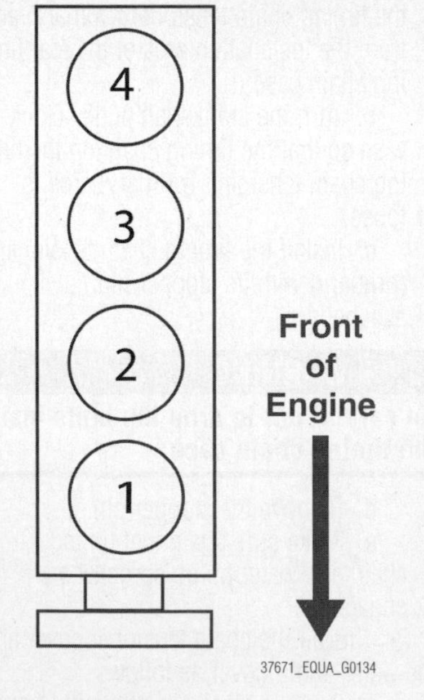

IGNITION COIL

REMOVAL & INSTALLATION

2.5L Engine

See Figures 35.

1. Before servicing the vehicle, refer to the Precautions Section.

2. Remove the intake manifold. Refer to Intake Manifold, removal & installation.

3. Disconnect the harness connector from the ignition coil.

4. Remove the ignition coil retaining bolt.

5. Remove the ignition coil from the vehicle.

To install:

6. Installation is the reverse of the removal procedure.

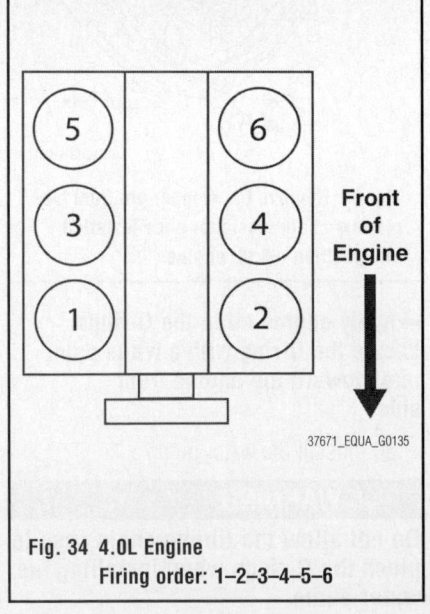

37671_EQUA_G0134

Fig. 33 2.5L Engine
Firing order: 1–3–4–2

37671_EQUA_G0135

Fig. 34 4.0L Engine
Firing order: 1–2–3–4–5–6

7. Tighten the ignition coil retaining bolt to 62 inch lbs. (7 Nm).

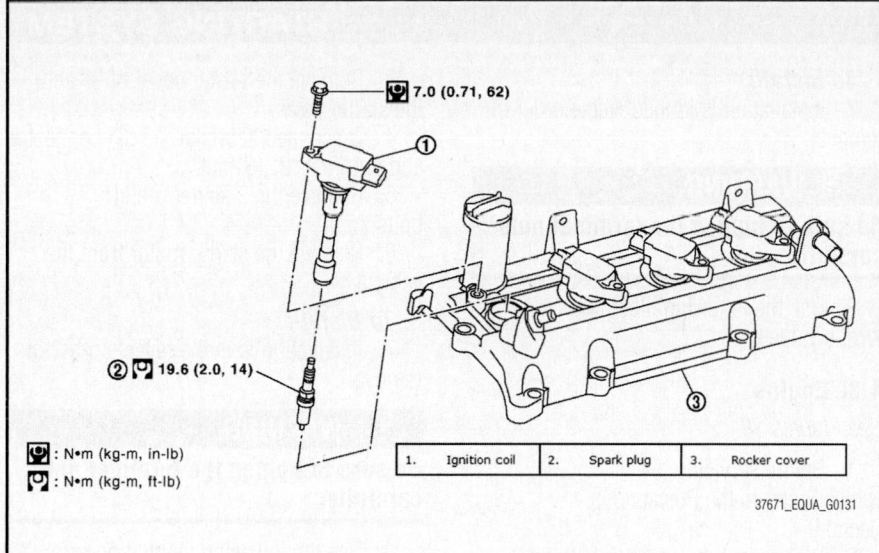

Fig. 35 View of ignition coil and spark plug removal—2.5L engine

4.0L Engine

Left Bank

See Figure 36.

1. Before servicing the vehicle, refer to the Precautions Section.
2. Remove engine room cover.
3. Move aside the harness, harness bracket, and hoses located above the ignition coil.
4. Disconnect the harness connector from the ignition coil.
5. Remove the ignition coil from the vehicle.

To install:

6. Installation is the reverse of the removal procedure.
7. Tighten the ignition coil retaining bolt to 62 inch lbs. (7 Nm).

Right Bank

See Figure 36.

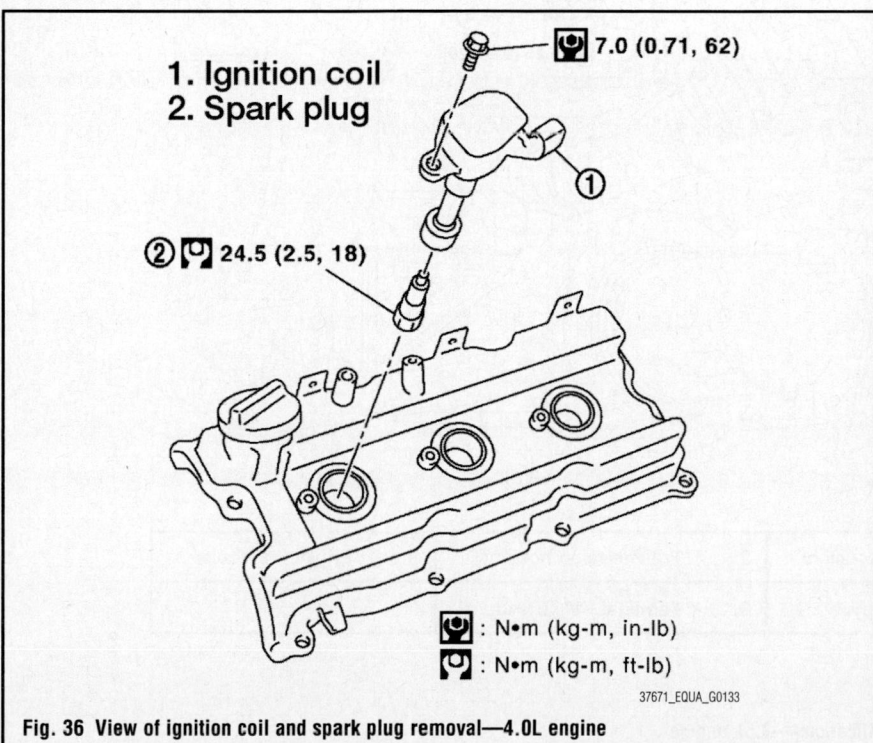

Fig. 36 View of ignition coil and spark plug removal—4.0L engine

1. Before servicing the vehicle, refer to the Precautions Section.
2. Remove intake manifold collector.
3. Move aside the harness, harness bracket, and hoses located above the ignition coil.
4. Disconnect the harness connector from the ignition coil.
5. Remove the ignition coil from the vehicle.

To install:

6. Installation is the reverse of the removal procedure.
7. Tighten the ignition coil retaining bolt to 62 inch lbs. (7 Nm).

SPARK PLUGS

REMOVAL & INSTALLATION

2.5L Engine

See Figures 35.

1. Before servicing the vehicle, refer to the Precautions Section.
2. Remove the intake manifold. Refer to Intake Manifold, removal & installation.

➡ **If removing the number one spark plug only, it is not necessary to remove the intake manifold.**

3. Remove the ignition coil. Refer to Ignition Coil, removal & installation.
4. Remove the spark plug using a spark plug socket and wrench.

To install:

5. Installation is the reverse of the removal procedure.
6. Ensure the spark plug gap is to specification: 0.043 inch (1.1mm).
7. Tighten the spark plug to 14 ft. lbs. (20 Nm).

4.0L Engine

See Figure 36.

1. Before servicing the vehicle, refer to the Precautions Section.
2. Remove the engine room cover.
3. Remove the ignition coil. Refer to Ignition Coil, removal & installation.
4. Remove the spark plug using a spark plug socket and wrench.

To install:

5. Installation is the reverse of the removal procedure.
6. Ensure the spark plug gap is to specification: 0.043 inch (1.1mm).
7. Tighten the spark plug to 18 ft. lbs. (25 Nm).

STARTER

REMOVAL & INSTALLATION

2.5L Engine

See Figures 37.

1. Before servicing the vehicle, refer to the Precautions Section.
2. Disconnect the negative battery cable.
3. Remove the air cleaner cover and the air cleaner to intake manifold collector duct.
4. Remove the "S" terminal and terminal "B" nuts.
5. Remove the 2 starter motor bolts.
6. Remove the starter motor from the vehicle.

To install:

7. Installation is in the reverse order of removal.

> **☀☀ WARNING**
>
> **Be sure to tighten the terminal nuts carefully.**

8. Use the following illustration for torque values.

4.0L Engine

See Figures 38.

1. Before servicing the vehicle, refer to the Precautions Section.
2. Disconnect the negative battery cable.
3. Remove the starter cover bolts and the starter cover.
4. Disconnect terminal "1" (S) connector and terminal "2" (B) nut.
5. Remove the 2 starter motor bolts.
6. Remove the starter motor from the vehicle.

To install:

7. Installation is in the reverse order of removal.

> **☀☀ WARNING**
>
> **Be sure to tighten the terminal nuts carefully.**

8. Use the following illustration for torque values.

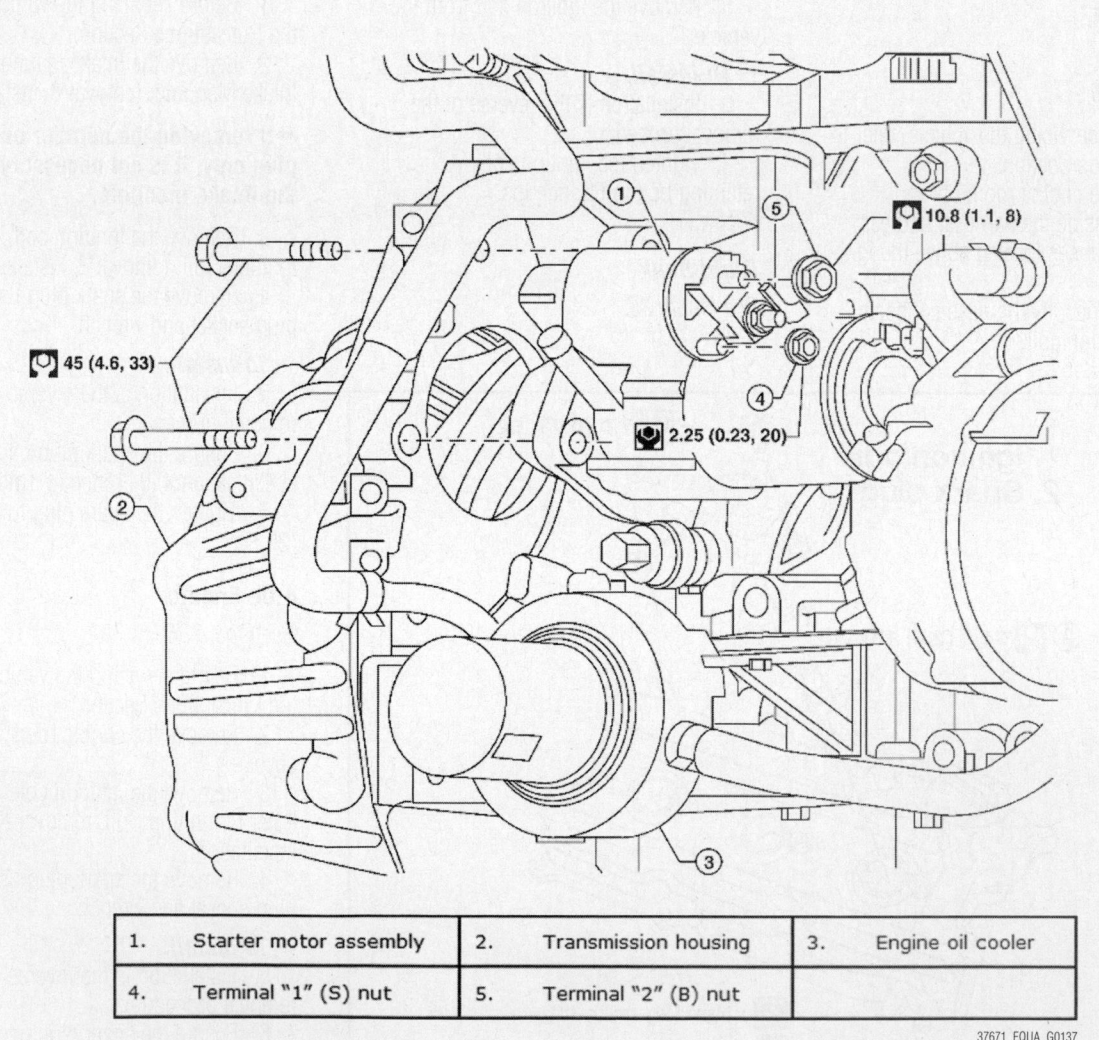

1.	Starter motor assembly	2.	Transmission housing	3.	Engine oil cooler
4.	Terminal "1" (S) nut	5.	Terminal "2" (B) nut		

37671_EQUA_G0137

Fig. 37 View of starter installation and torque specifications—2.5L engine

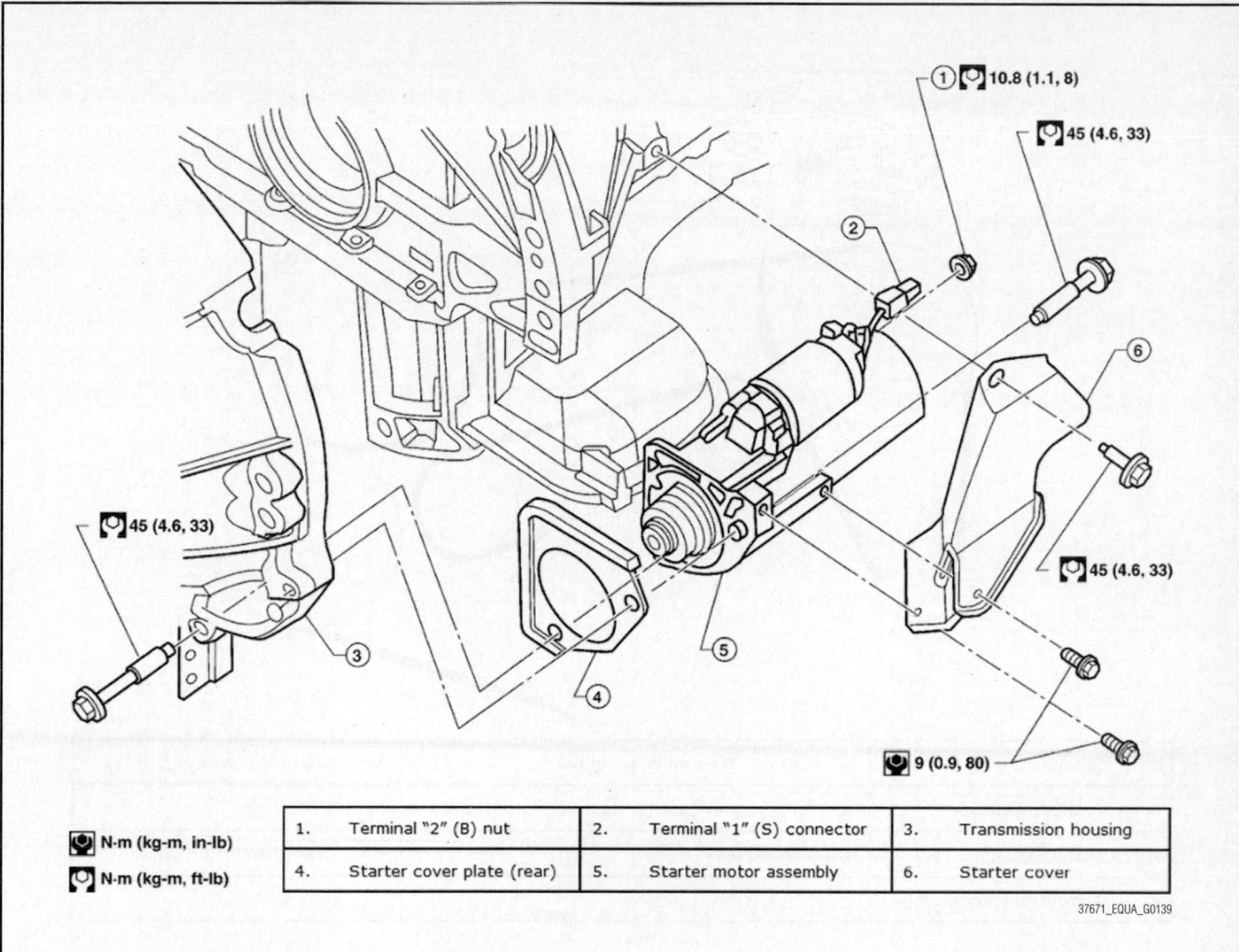

1.	Terminal "2" (B) nut	2.	Terminal "1" (S) connector	3.	Transmission housing
4.	Starter cover plate (rear)	5.	Starter motor assembly	6.	Starter cover

37671_EQUA_G0139

Fig. 38 View of starter installation and torque specifications—4.0L engine

ENGINE MECHANICAL

➡**Disconnecting the negative battery cable may interfere with the functions of the on board computer systems and may require the computer to undergo a relearning process, once the negative battery cable is reconnected.**

ACCESSORY DRIVE BELTS

ACCESSORY BELT ROUTING

2.5L Engine

See Figures 39.

Refer to the accompanying illustrations.

4.0L Engine

See Figures 40.

Refer to the accompanying illustrations.

INSPECTION

Inspect the drive belt for signs of glazing or cracking. A glazed belt will be perfectly smooth from slippage, while a good belt will have a slight texture of fabric visible. Cracks will usually start at the inner edge of the belt and run outward. All worn or damaged drive belts should be replaced immediately.

ADJUSTMENT

There is no manual drive belt tension adjustment. The drive belt tension is automatically adjusted by the drive belt auto tensioner.

REMOVAL & INSTALLATION

2.5L Engine

See Figures 41 and 42.

1. Before servicing the vehicle, refer to the Precautions Section.

2. Install special tool J-46535, or equivalent, on the auto tensioner pulley bolt and move in a clockwise direction (loosening direction of tensioner).

✷✷ CAUTION

Avoid placing your hand in a location where pinching may occur if the holding tool accidentally comes off.

➡**Do not loosen the auto-tensioner pulley bolt. (Do not turn it counterclockwise.) If it is turned counterclockwise, the complete auto-tensioner must be replaced as a unit, including the pulley.**

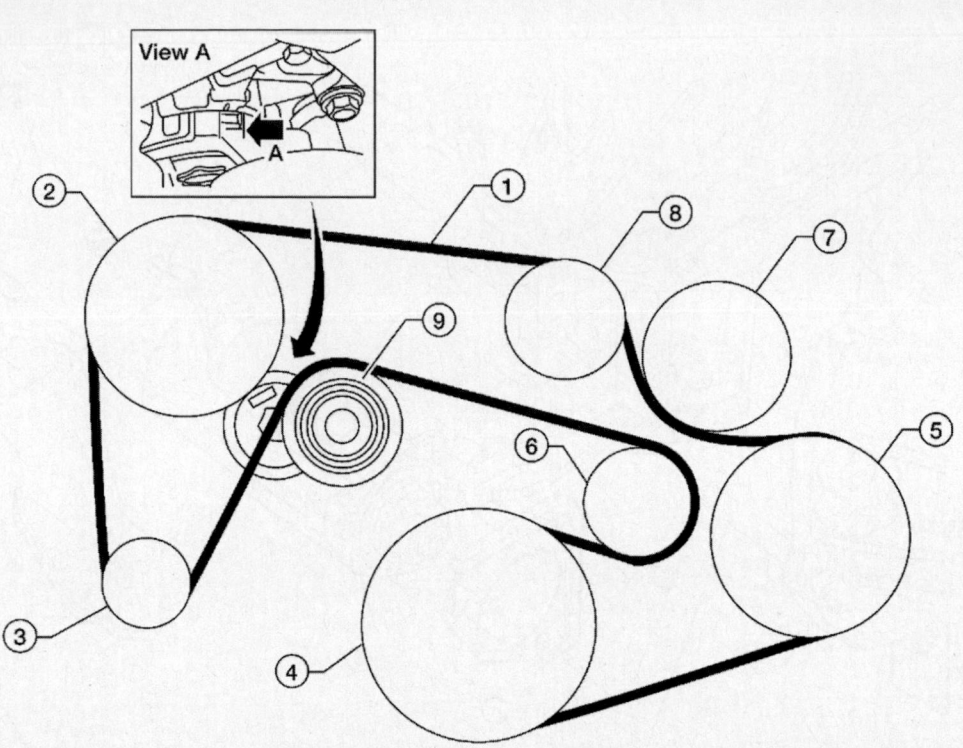

1.	Drive belt	2.	Power steering oil pump pulley	3.	Generator pulley
4.	Crankshaft pulley	5.	A/C compressor (if equipped) or idler pulley	6.	Idler pulley
7.	Water pump	8.	Idler pulley	9.	Drive belt auto- tensioner
A.	Allowable working range				

37671_EQUA_G0141

Fig. 39 Accessory drive belt routing—2.5L engine

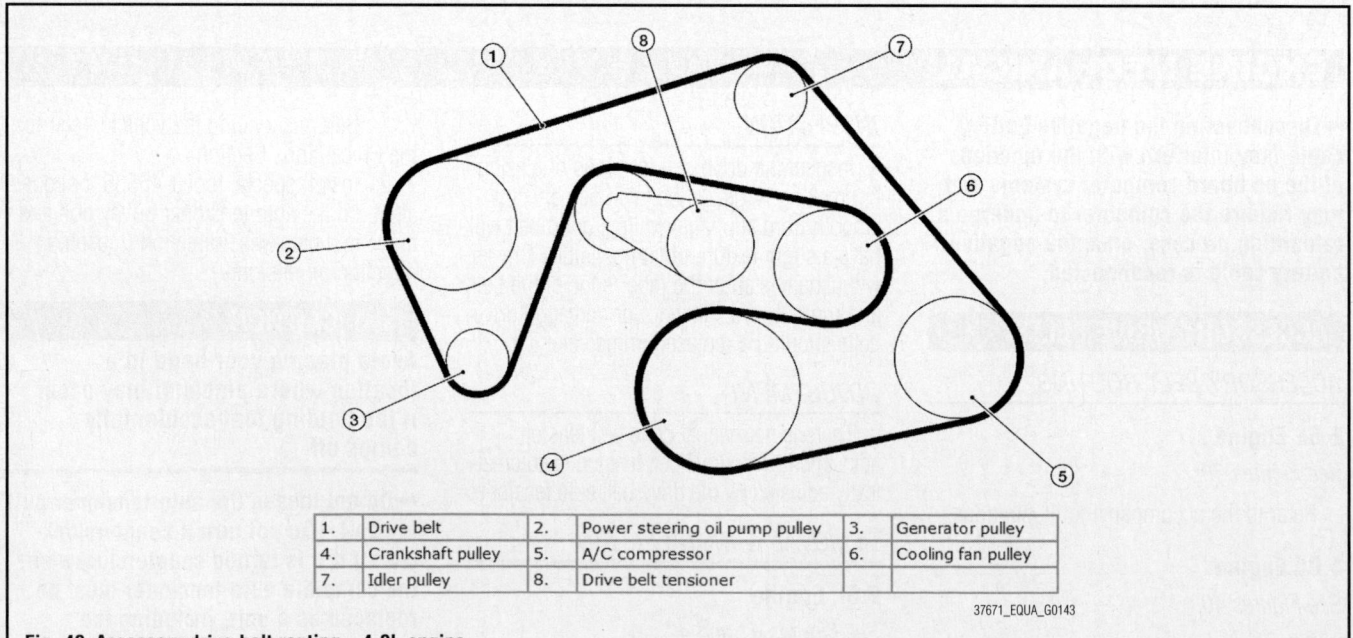

1.	Drive belt	2.	Power steering oil pump pulley	3.	Generator pulley
4.	Crankshaft pulley	5.	A/C compressor	6.	Cooling fan pulley
7.	Idler pulley	8.	Drive belt tensioner		

37671_EQUA_G0143

Fig. 40 Accessory drive belt routing—4.0L engine

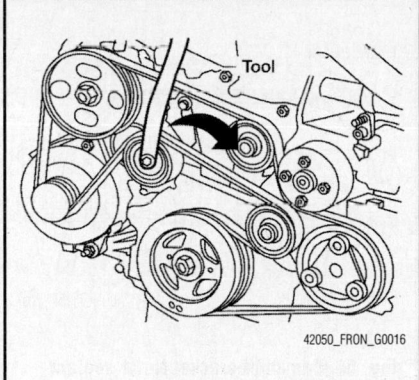

Fig. 41 Drive belt tension tool installation and removal direction—2.5L engine

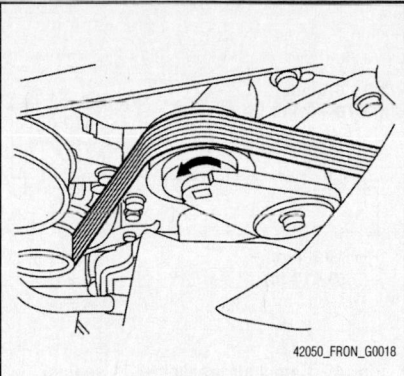

Fig. 43 Drive belt tension tool installation and removal direction—4.0L engine

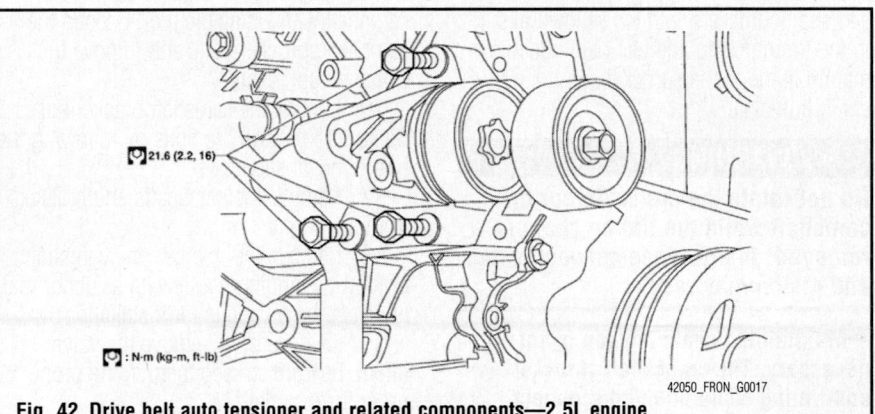

Fig. 42 Drive belt auto tensioner and related components—2.5L engine

3. Remove the drive belt.

To install:

4. Installation is the reverse of the removal procedure.

5. Confirm the belts are completely set on the pulleys.

6. Ensure that there is no engine oil or engine coolant on the drive belt or pulley grooves.

7. Turn the crankshaft pulley clockwise several times to equalize the tension between each pulley.

8. Confirm the tension of the drive belt indicator (fixed side) is within the allowable working range. Refer to the Accessory Belt Routing illustration.

4.0L Engine

See Figures 43 and 44.

1. Before servicing the vehicle, refer to the Precautions Section.

2. Remove the engine room cover.

3. Remove the air duct and resonator assembly.

4. Rotate the drive belt auto tensioner in a counterclockwise direction (loosening direction of tensioner).

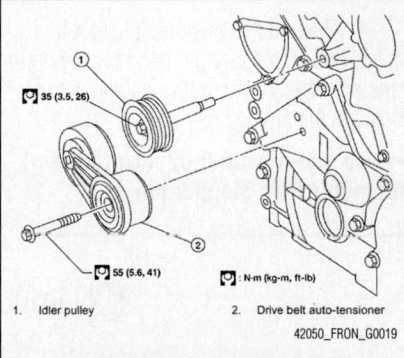

1. Idler pulley   2. Drive belt auto-tensioner

Fig. 44 Drive belt auto tensioner and related components—4.0L engine

✳✳ CAUTION

Avoid placing your hand in a location where pinching may occur if the tool accidentally comes off.

5. Remove the drive belt.

To install:

6. Installation is the reverse of the removal procedure.

7. Make sure the drive belt is securely installed around all the pulleys.

8. Ensure that there is no engine oil or

engine coolant on the drive belt or pulley grooves.

CAMSHAFT AND VALVE LIFTERS

REMOVAL & INSTALLATION

2.5L Engine

See Figures 45 through 56.

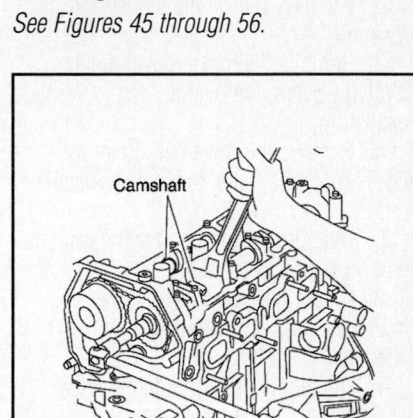

Fig. 45 Secure the camshaft with a suitable tool and loosen the camshaft sprocket bolts—2.5L engine

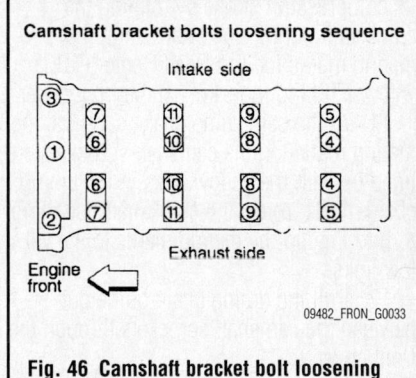

Fig. 46 Camshaft bracket bolt loosening sequence—2.5L engine

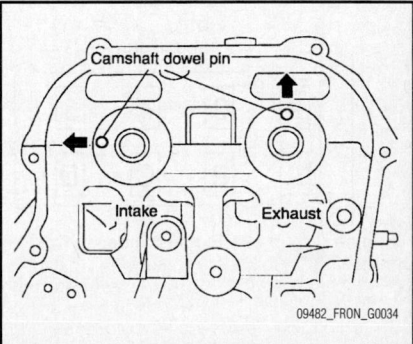

Fig. 47 Camshaft positioning—2.5L engine

→The procedure below describes removal and installation of the camshaft without removing the front cover. If the front cover is removed, refer to Timing Chain & Sprockets, removal & installation.

1. Before servicing the vehicle, refer to the Precautions Section.

2. Properly relieve the fuel system pressure.

3. Remove the rocker cover. Refer to Valve Covers, removal & installation.

4. Remove the drive belt. Refer to Accessory Drive Belts, removal & installation.

5. Disconnect and remove the camshaft position sensor (PHASE).

6. Disconnect the Intake Valve Timing (IVT) control solenoid electrical connector.

7. Disconnect the ground electrical connectors from the front cover.

8. Remove the IVT control solenoid retaining bolts. Be sure to remove the bolts by reversing the order of the tightening torque sequence.

9. Remove the cover by cutting the sealant using tool KV10111100 (J-37228), or equivalent.

10. Position the number one cylinder on its compression stroke by rotating the crankshaft pulley clockwise. Align the mating marks for Top Dead Center (TDC) with the timing indicator on the front cover.

11. At the same time, make sure that the mating marks on the camshaft sprockets are lined up with the yellow links in the timing chain. If not, rotate the crankshaft one more turn to line up the mating marks to the yellow links.

12. Pull the timing chain guide out between the camshaft sprockets through the front cover.

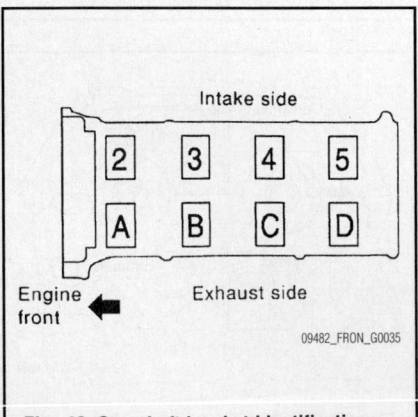

Fig. 48 Camshaft bracket identification—2.5L engine

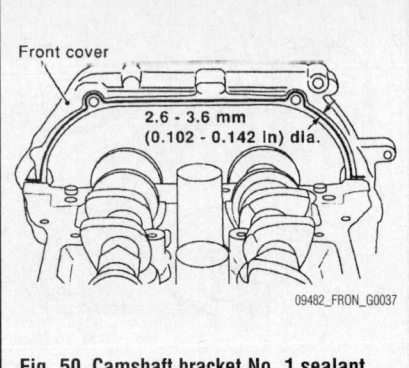

Fig. 49 Camshaft bracket No. 1 sealant application point "A"—2.5L engine

13. Line up the mating marks on the camshaft sprockets with the yellow links in the timing chain and paint an indelible mating mark on the sprocket and timing chain link plate.

✳✳ WARNING

Do not rotate the crankshaft or the camshaft while the timing chain is removed. Interference between valve and piston may result.

→Maintaining chain tension is not necessary. The crankshaft sprocket and timing chain do not disconnect structurally while the front cover is attached.

14. Push in the tensioner plunger and hold. Insert a stopper pin into the hole on the tensioner body to hold the chain tensioner. Remove the timing chain tensioner.

→Use a wire with 0.02 inch (0.5mm) diameter for a stopper pin.

15. Secure the hexagonal part of the camshaft with a suitable tool. Loosen the camshaft sprocket bolts and remove the camshaft sprockets.

16. Loosen the camshaft bracket bolts. Be sure to remove the bolts by following the bolt removal sequence.

17. Remove the camshafts and brackets from the engine.

18. Remove the number one camshaft bracket by tapping lightly with a rubber mallet. Note the positions for installation.

19. As necessary, remove the valve lifters. Be sure to keep them in the proper order for installation.

To install:

20. Inspect the camshafts, replace as required. Refer to Inspection procedure.

21. Install the camshafts so that the camshaft dowel pins on the front side are positioned as indicated in the illustration.

Fig. 50 Camshaft bracket No. 1 sealant application outside bolt hole—2.5L engine

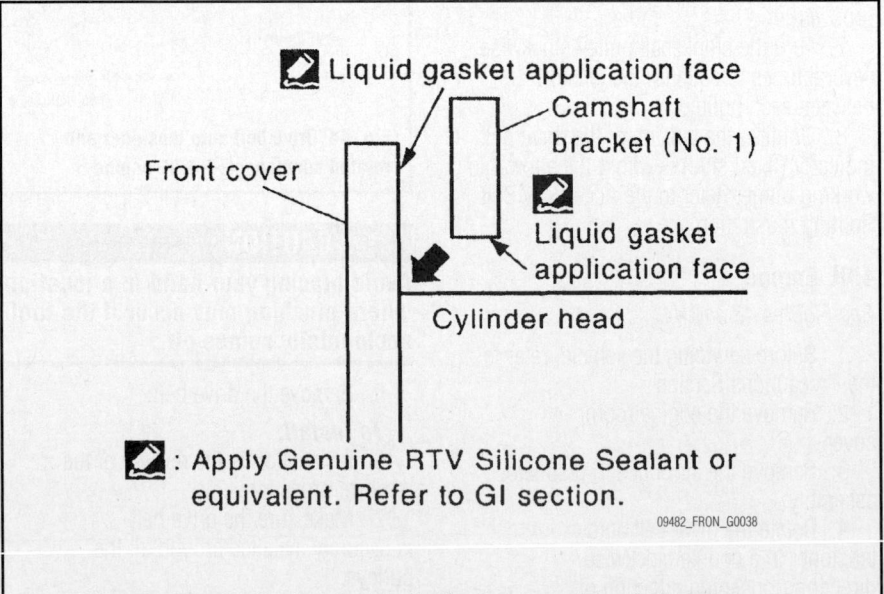

Fig. 51 Camshaft bracket No. 1 sealant application locating points—2.5L engine

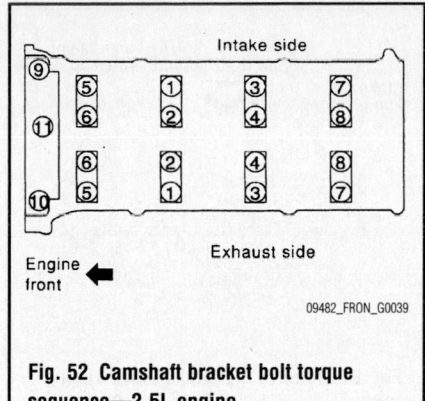

Fig. 52 Camshaft bracket bolt torque sequence—2.5L engine

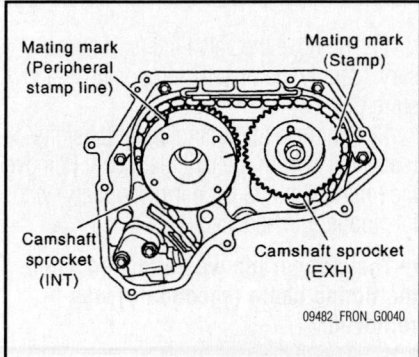

Fig. 53 Camshaft alignment mating marks—2.5L engine

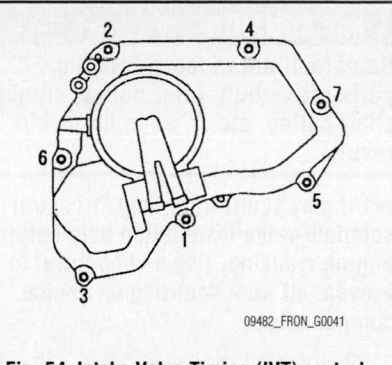

Fig. 54 Intake Valve Timing (IVT) control valve cover bolt torque sequence—2.5L engine

22. Remove any foreign material from the camshaft bracket backside and from the cylinder head face.

23. Install the camshaft brackets (No. 2–No. 5) aligning the identification marks on the upper surface as indicated in the illustration.

➡**Install so that the identification mark can be correctly read when viewed from the exhaust side.**

24. To install camshaft bracket No. 1, apply liquid gasket to the bracket.

➡**After installation, be sure to wipe excessive gasket material from part "A", as indicated in the illustration. Be sure to use genuine RTV silicone sealant, or equivalent.**

25. Apply liquid gasket to camshaft bracket No. 1 contact surface on the front cover backside. Apply liquid gasket to the outside bolt hole on the front cover. Be sure to use genuine RTV silicone sealant or equivalent.

26. Locate camshaft bracket No. 1 near installation position and install it without disturbing the liquid gasket applied to the surfaces. Be sure to use genuine RTV silicone sealant, or equivalent.

27. Tighten the camshaft bracket bolts in the proper sequence and to specification:
   a. Bolts 9–11 to 17 inch lbs. (2 Nm).
   b. Bolts 1–8 to 17 inch lbs. (2 Nm).
   c. Bolts 1–11 to 52 inch lbs. (6 Nm).
   d. Bolts 1–11 to 92 inch lbs. (10 Nm).

➡**After tightening the bolts be sure to wipe off any excessive liquid gasket. Be sure to use genuine RTV silicone sealant, or equivalent.**

28. Install the camshaft position sensor. Install the camshaft sprockets by aligning the mating marks on each camshaft sprocket with the paint marks on the timing chain link plates, which were made during removal.

➡**Aligned mating marks could slip. Therefore, after matching them, hold the timing chain in place by hand. Before and after installing the chain tensioner, make sure that the mating marks have not slipped.**

29. Install the chain tensioner. After installation, pull the stopper pin off completely, and make sure that the chain tensioner plunger is released.

➡**Before installation of the chain tensioner, it is possible to rematch the marks on the timing chain with new ones on each sprocket.**

30. Install the chain guide. Install the oil rings to the camshaft sprocket (INT) insertion points on the backside of the IVT control cover. Install the O-ring to the front cover.

31. Apply a 0.083–0.122 inch (2.1–3.1mm) diameter bead of liquid

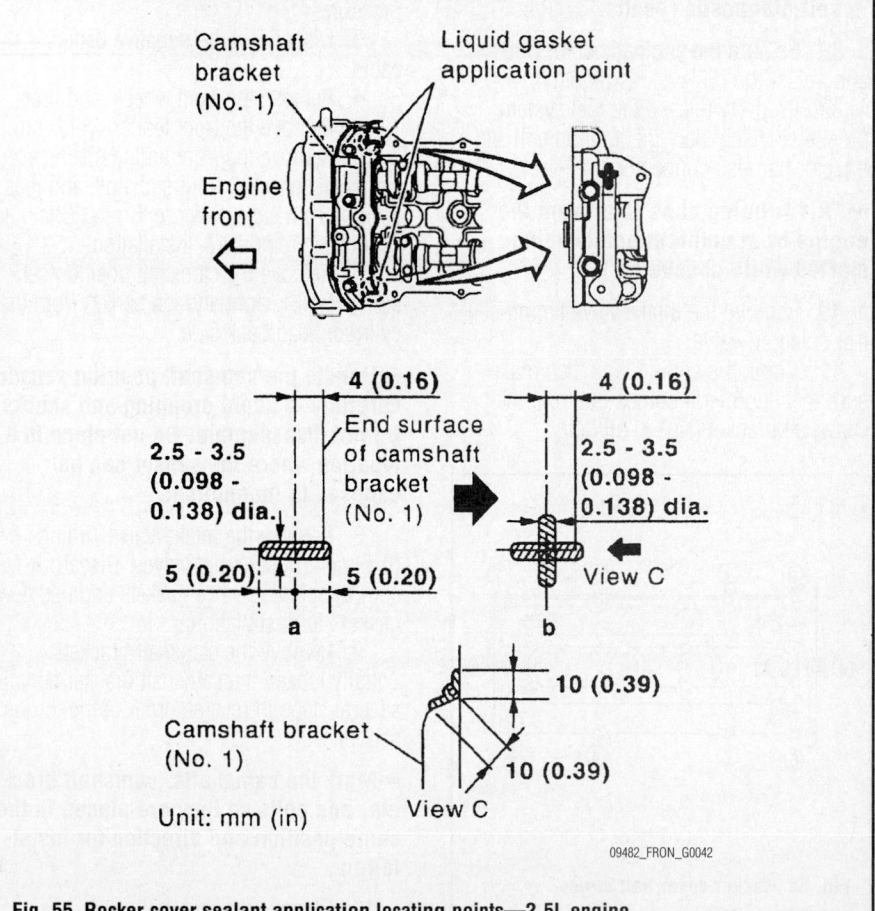

Fig. 55 Rocker cover sealant application locating points—2.5L engine

gasket to the IVT control cover. Be sure to use genuine RTV silicone sealant, or equivalent.

32. Install the cover. Tighten the bolts in the proper sequence. Connect the ground cables and install the harness clip.

33. Check and adjust the valve clearance, as required.

➡If hydraulic pressure inside the timing chain tensioner drops after removal/installation, slack in the guide may generate a pounding noise during and just after engine start. This is normal. The noise will stop after the hydraulic pressure rises.

34. Continue the installation in the reverse order of the removal procedure.

35. Apply liquid gasket, be sure to use genuine RTV silicone sealant, or equivalent, to the positions shown in the illustration.

36. Install the rocker cover. Torque the retaining bolts to 18 inch lbs. (2 Nm) and then to 73 inch lbs. (8 Nm), in the proper sequence.

37. Inspect the camshaft sprocket (INT) oil groove.

➡Perform this inspection only when DTC P0011 or DTC P0021 are detected in self-diagnostic results.

38. Be sure the engine is cold. Check and adjust the oil level, as required.

39. Properly release the fuel system pressure. Disconnect the ignition coil and injector harness connectors.

➡This is being done to prevent the engine from unintentionally being started while checking.

40. Remove the intake valve timing control solenoid valve.

41. Crank the engine, and then make sure that engine oil comes out from the camshaft bracket (No. 1) oil hole.

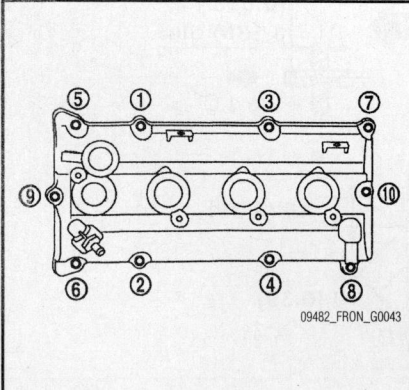

Fig. 56 Rocker cover bolt torque sequence—2.5L engine

❊❊ CAUTION

Be careful not to touch rotating parts (drive belt, idler pulley, crankshaft pulley, etc.,), as injury could result.

➡Oil may squirt from the IVT control solenoid valve installation hole during engine cranking. Use a shop towel to prevent oil from squirting on engine components.

42. Clean the oil groove between the oil strainer and the intake timing control solenoid valve if engine oil does not come out from the camshaft bracket (No. 1) oil hole.

43. Remove the components between the IVT control solenoid valve and the camshaft sprocket (INT). Check each oil groove for clogging.

44. After inspection, install any removed components.

4.0L Engine

See Figures 57 through 66.

1. Before servicing the vehicle, refer to the Precautions Section.

2. Properly relieve the fuel system pressure.

3. Disconnect the negative battery cable.

4. Remove the front wheels and tires.

5. Remove the front fender protectors.

6. Remove the front timing chain case, camshaft sprocket, timing chains, and rear timing chain case. Refer to Timing Chain & Sprockets, removal & installation.

7. Remove the camshaft position sensor (PHASE) (right and left banks) from the cylinder head back side.

➡Handle the camshaft position sensor carefully to avoid dropping and shocks. Do not disassemble. Do not place in a location where the sensor can be exposed to magnetism.

8. Remove the Intake Valve Timing (IVT) control solenoid valves. Discard IVT control solenoid valve gaskets and use new gaskets for installation.

9. Remove the camshaft brackets. Equally loosen the camshaft bracket bolts in several steps in reverse order of the torque sequence.

➡Mark the camshafts, camshaft brackets, and bolts so they are placed in the same position and direction for installation.

10. Remove the camshafts.

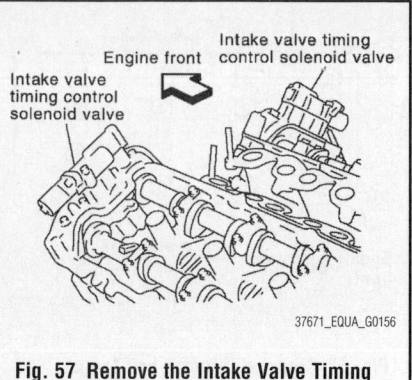

Fig. 57 Remove the Intake Valve Timing (IVT) control solenoid valves—4.0L engine

11. Remove the valve lifters, if necessary. Identify the installation positions and store them.

12. Remove the timing chain tensioner (secondary) from the cylinder head. Remove the timing chain tensioner (secondary) with its stopper pin attached.

➡The stopper pin was attached when the timing chain (secondary) was removed.

To install:

13. Inspect the camshafts, replace as required. Refer to Inspection procedure.

14. Install the timing chain tensioners (secondary) on both sides of the cylinder head. Be sure to use new O-rings.

➡Install the tensioner with its stopper pin attached. Install the tensioner with the sliding part facing downward on the right cylinder head and with the sliding part facing upward on the left cylinder head.

15. Install the valve lifters, in their original bores.

16. Install the camshafts, with the dowel pin attached to its front end face on the exhaust side.

➡Follow the identification marks for proper placement and direction.

17. Install the camshaft so that the dowel pin hole and dowel pin on the front end face are positioned as shown in the illustration (No. 1 piston at TDC on its compression stroke).

➡Large and small pin holes are located on the front end face of the camshaft (INT), at intervals of 180 degrees. Face the small diameter side pin hole upward (in the cylinder head upper face direction).

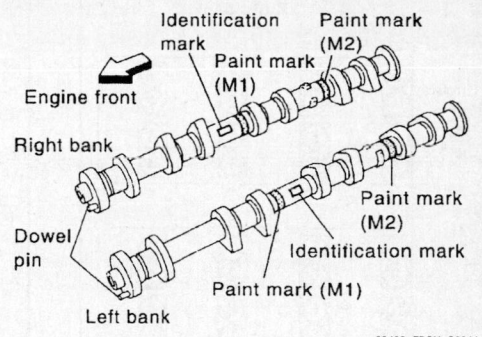

| Bank | INT/EXH | Dowel pin | Paint marks | | Identification mark |
			M1	M2	
RH	INT	No	Green	No	RE
	EXH	Yes	No	White	RE
LH	INT	No	Green	No	LH
	EXH	Yes	No	White	LH

Fig. 58 Camshaft identification—4.0L engine

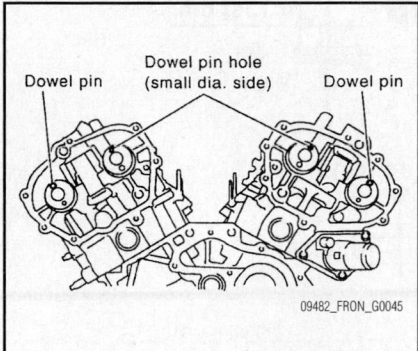

Fig. 59 Camshaft dowel pin positioning—4.0L engine

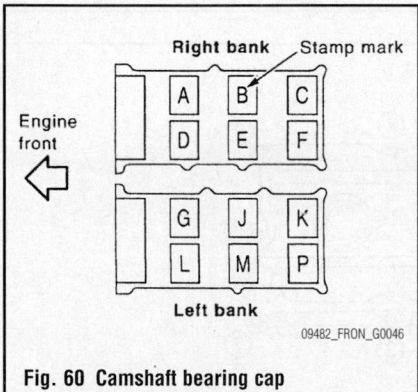

Fig. 60 Camshaft bearing cap identification—4.0L engine

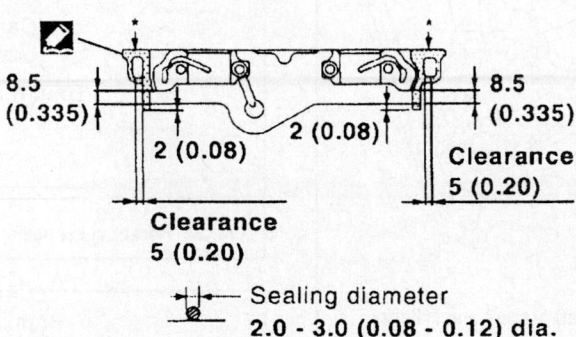

Camshaft bracket (No. 1)

8.5 (0.335)   8.5 (0.335)

2 (0.08)   2 (0.08)

Clearance 5 (0.20)

Clearance 5 (0.20)

Sealing diameter 2.0 - 3.0 (0.08 - 0.12) dia.

* : Remove the protruding liquid gasket from front face. (Remove the hardened liquid gasket from surface only.)

◢ : Apply Genuine RTV Silicone Sealant or equivalent. Refer to GI section.

Unit: mm (in)

Fig. 61 Camshaft sealant application and location—4.0L engine

➡**Though the camshaft does not stop at the portion as shown, for placement of the cam nose, it is generally accepted that the camshaft is placed for the same direction as shown.**

18. Install the camshaft brackets in the same position that they were removed. Install brackets No. 2–No. 4 aligning the stamp marks as indicated in the illustration.

➡**There are no identification marks indicating left or right for camshaft bracket No. 1.**

19. Apply liquid gasket to the mating surfaces of camshaft bracket No. 1 as shown in the illustration on both the left and right cylinder heads. Be sure to use genuine RTV sealant, or equivalent.

20. Tighten the camshaft bracket bolts in the proper sequence and to specification:
   a. Bolts 7–10 to 17 inch lbs. (2 Nm).
   b. Bolts 1–6 to 17 inch lbs. (2 Nm).
   c. All bolts to 52 inch lbs. (6 Nm).
   d. All bolts to 92 inch lbs. (10 Nm).

21. Measure the difference in levels between the front end faces of the camshaft bracket No. 1 and the cylinder head. Specification should be -0.0055–0.0055 inch (-0.14–0.14mm). If not within specification, reinstall camshaft bracket No. 1.

➡**Measure two positions (both intake and exhaust side) for a single bank.**

22. Check and adjust valve clearance, as required.

23. Apply liquid gasket, be sure to use genuine RTV silicone sealant, or equivalent, to the positions shown in the illustration.

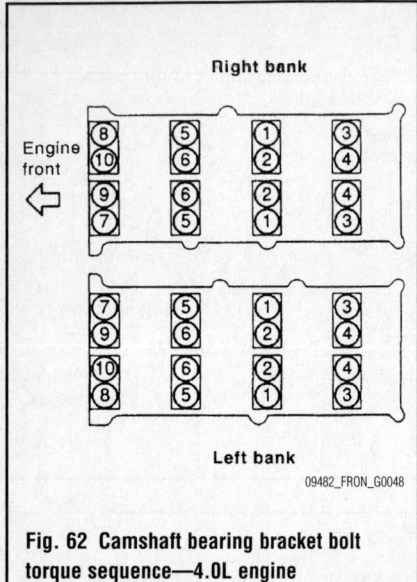

Fig. 62 Camshaft bearing bracket bolt torque sequence—4.0L engine

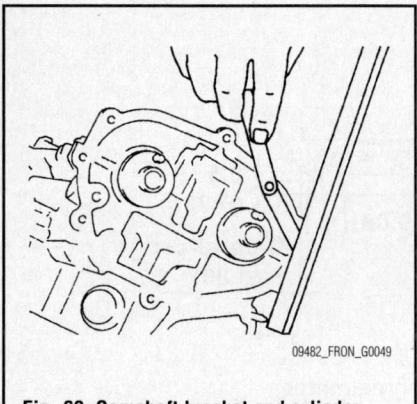

Fig. 63 Camshaft bracket and cylinder head measurement—4.0L engine

24. Install the rocker cover. Refer to Valve Covers, removal & installation.

25. Continue the installation in the reverse order of the removal procedure.

26. Inspect the camshaft sprocket (INT) oil groove.

➡**Perform this inspection only when DTC P0011 or DTC P0021 is detected in self-diagnostic results.**

27. Be sure the engine is cold. Check and adjust oil level, as required.

28. Properly release the fuel system pressure. Disconnect the ignition coil and injector harness connectors.

➡**This is being done to prevent the engine from unintentionally being started while checking.**

29. Remove the intake valve timing control solenoid valve.

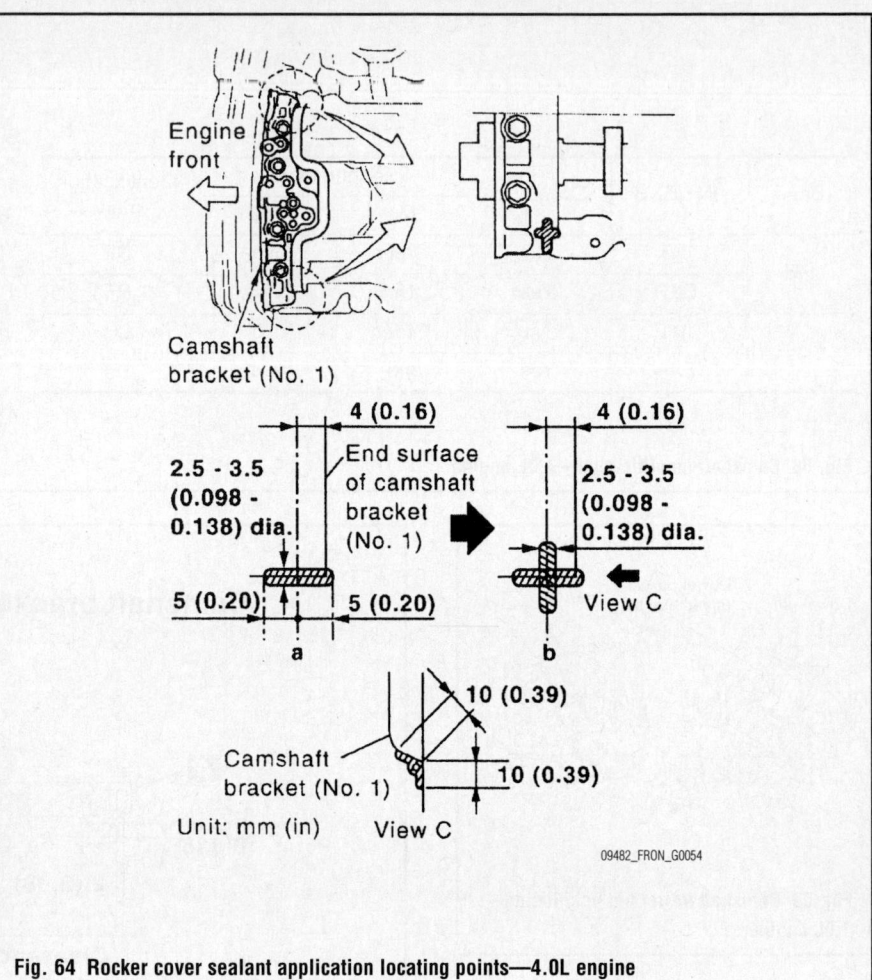

Fig. 64 Rocker cover sealant application locating points—4.0L engine

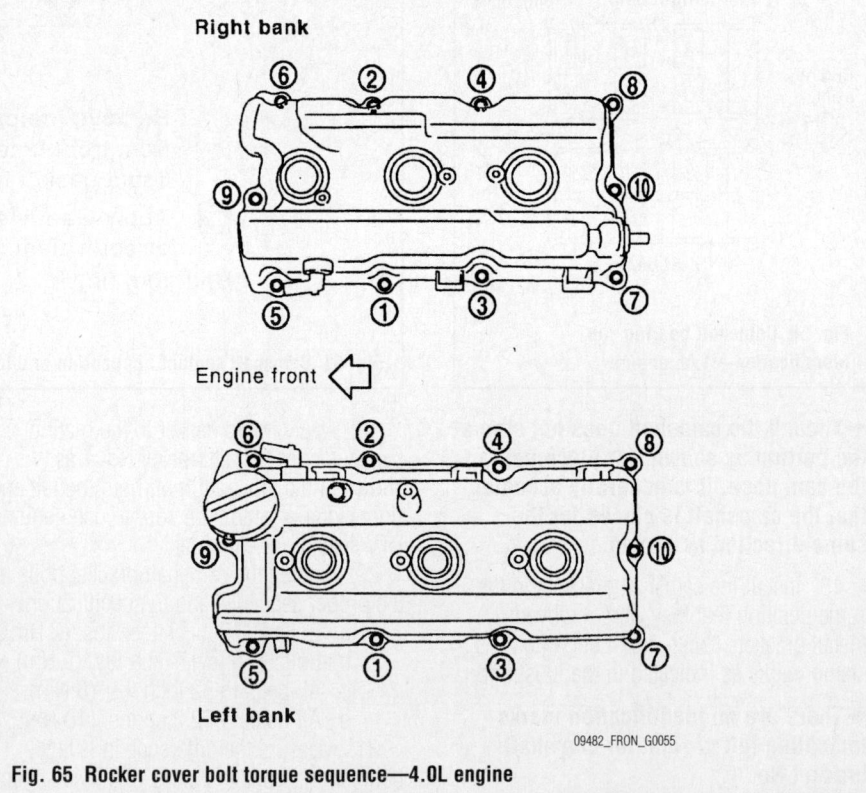

Fig. 65 Rocker cover bolt torque sequence—4.0L engine

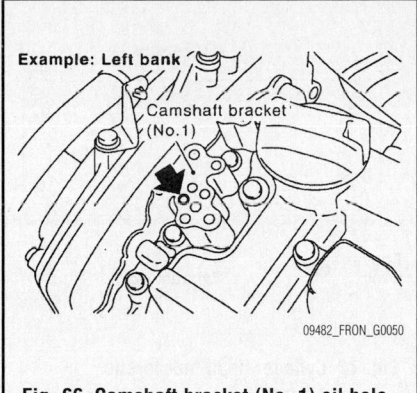

Fig. 66 Camshaft bracket (No. 1) oil hole location

30. Crank the engine, and then make sure that engine oil comes out from the camshaft bracket (No. 1) oil hole.

❋❋ CAUTION

Be careful not to touch rotating parts, (drive belt, idler pulley, crankshaft pulley, etc.) as injury could result.

➡**Oil may squirt from the intake valve timing control solenoid valve installation hole during engine cranking. Use a shop towel to prevent oil from squirting on engine components.**

31. Clean the oil groove between the oil strainer and the intake timing control solenoid valve if engine oil does not come out from the camshaft bracket (No. 1) oil hole.
32. Remove the components between the intake valve timing control solenoid valve and the camshaft sprocket (INT). Check each oil groove for clogging.
33. After inspection, install any removed components.

CRANKSHAFT FRONT SEAL

REMOVAL & INSTALLATION

2.5L Engine

See Figure 67.

1. Before servicing the vehicle, refer to the Precautions Section.
2. Remove the engine undercover.
3. Remove the fan shroud. Remove the cooling fan. Refer to Engine Fan, removal & installation.
4. Remove the drive belt. Refer to Accessory Drive Belts, removal & installation.

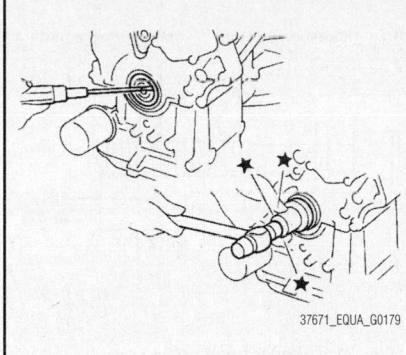

Fig. 67 View of removal and installation of the crankshaft front seal—2.5L engine

5. Remove the crankshaft pulley. Refer to Crankshaft Damper, removal & installation.
6. Using a seal removal tool, remove the oil seal from its mounting.

➡**Be careful not to damage the front cover and/or the crankshaft.**

To install:

7. Installation is the reverse order of the removal procedure.
8. Press fit the seal until it is flush with the front end surface of the front cover, using the proper tools.

4.0L Engine

See Figures 68 through 70.

1. Before servicing the vehicle, refer to the Precautions Section.
2. Remove the engine undercover.
3. Remove the drive belts. Refer to Accessory Drive Belts, removal & installation.
4. Remove the cooling fan. Refer to Engine Fan, removal & installation.

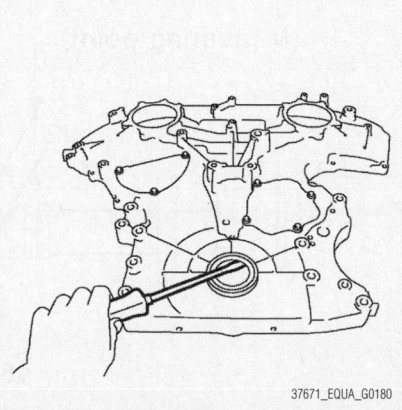

Fig. 68 View of crankshaft front seal removal—4.0L engine

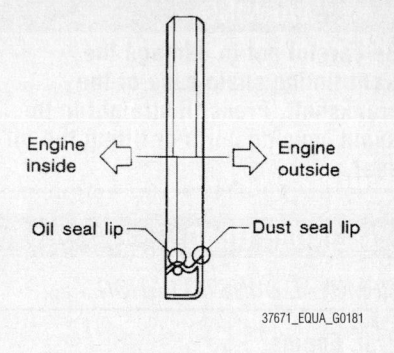

Fig. 69 Install the front oil seal so that each seal lip is oriented as shown—4.0L engine

5. Remove the crankshaft pulley. Refer to Crankshaft Damper, removal & installation.
6. Using a seal removal tool, remove the oil seal from its mounting.

➡**Be careful not to damage the front cover and/or the crankshaft.**

To install:

7. Installation is the reverse order of the removal procedure.
8. Install a new front oil seal on the front timing chain case.
9. Apply new engine oil to both the oil seal lip and the dust seal lip.
10. Install the front oil seal so that each seal lip is oriented properly.
11. Press-fit the oil seal until it becomes flush with the front timing chain case end face using a suitable drift with an outer diameter of 2.36 inches (60mm).
12. Make sure the garter spring is in position and seal lip is not inverted.

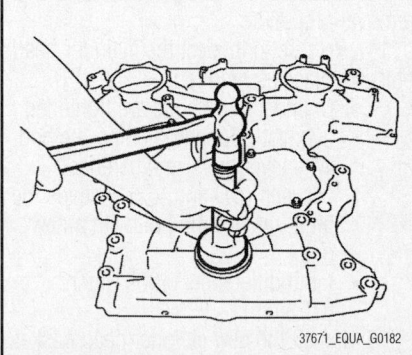

Fig. 70 Press-fit the oil seal until it becomes flush with the front timing chain case end face using a suitable drift—4.0L engine

> ※※ **WARNING**
>
> Be careful not to damage the front timing chain case or the crankshaft. Press-fit straight in to avoid causing burrs or tilting the oil seal.

CYLINDER HEAD

REMOVAL & INSTALLATION

2.5L Engine

See Figures 71 and 72.

1. Before servicing the vehicle, refer to the Precautions Section.
2. Properly relieve the fuel system pressure.
3. Drain the cooling system.

➡**Ensure the engine is cold before performing work. Do not spill engine coolant or oil on the drive belt.**

4. Drain the engine oil.
5. Remove the intake manifold and fuel tube assembly.
6. Remove the fuel injector and fuel tube assembly.
7. Remove the exhaust manifold and the three way catalyst. Refer to Exhaust Manifold, removal & installation.
8. Remove the water outlet. Remove the heater outlet.
9. Remove the front cover and the timing chain.
10. Remove the camshafts.
11. Remove the cylinder head retaining bolts. Be sure to remove the bolts by reversing the order of the tightening torque sequence.
12. Remove the cylinder head from the engine. Discard the gasket.

To install:

13. Installation is the reverse of the removal procedure.
14. Be sure to inspect the cylinder head bolts. Replace as required.
   - Cylinder head bolts are tightened by plastic zone tightening method. Whenever the size difference between (d1) and (d2) exceeds the limit, replace the bolt with a new one
   - Limit difference: 0.0091 inch (0.23mm)
15. Install the new cylinder head gasket. Apply engine oil to the cylinder head bolt threads and seating surfaces. Torque the cylinder head bolts to specification and in the proper sequence.
   a. Step 1: 37 ft. lbs. (50 Nm).

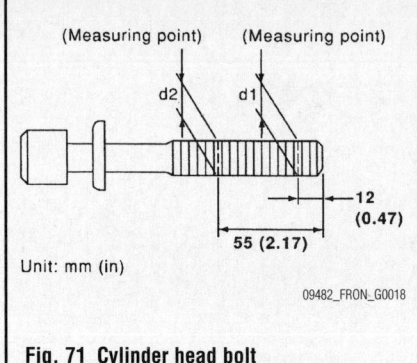

Fig. 71 Cylinder head bolt measurement—2.5L engine

   b. Step 2: 60° clockwise.
   c. Step 3: Loosen to 0 ft. lbs. (0 Nm) in the reverse order of tightening.
   d. Step 4: 29 ft. lbs. (39 Nm).
   e. Step 5: 75° clockwise.
   f. Step 6: 75° clockwise.
16. Be sure to fill the cooling system with the proper grade and type engine coolant.
17. Be sure to fill the engine with the proper grade and type motor oil.
18. Start the engine and check for leaks.

4.0L Engine

See Figures 73 through 76.

1. Before servicing the vehicle, refer to the Precautions Section.
2. Properly relieve the fuel system pressure.
3. Drain the cooling system.

➡**Ensure the engine is cold before performing work. Do not spill engine coolant or oil on the drive belt.**

4. Remove the camshaft. Refer to Camshaft and Valve Lifters, removal & installation.

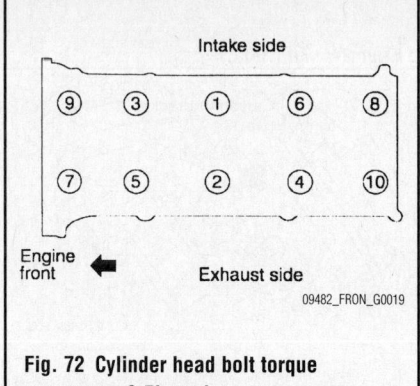

Fig. 72 Cylinder head bolt torque sequence—2.5L engine

5. Remove the intake manifold. Refer to Intake Manifold, removal & installation.
6. Remove the exhaust manifold. Refer to Exhaust Manifold, removal & installation.
7. Remove the water inlet and thermostat assembly.
8. Remove the water outlet, water pipe and heater pipe.
9. Remove the cylinder head retaining bolts. Be sure to remove the bolts by reversing the order of the tightening torque sequence.
10. Remove the cylinder head from the engine. Discard the gasket.

To install:

11. Installation is the reverse of the removal procedure.
12. Be sure to inspect the cylinder head bolts. Replace as required.
   - Cylinder head bolts are tightened by plastic zone tightening method. Whenever the size difference between (d1) and (d2) exceeds the limit, replace the bolt with a new one

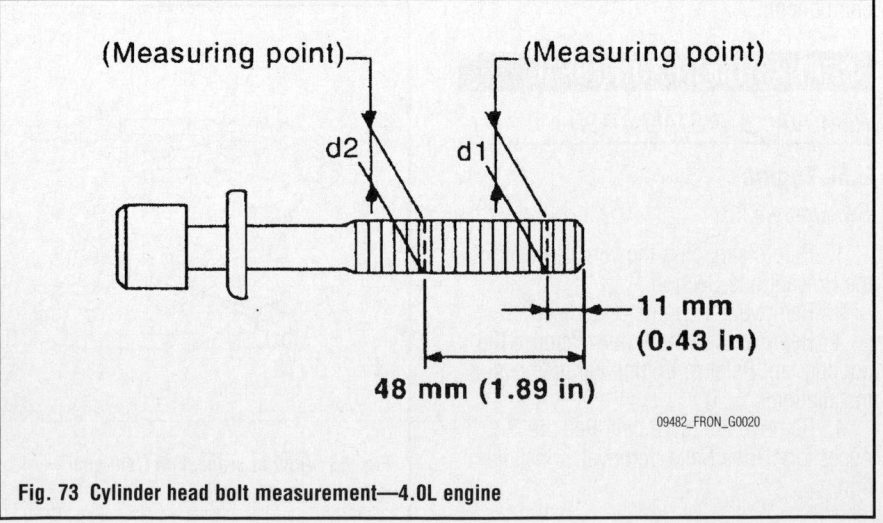

Fig. 73 Cylinder head bolt measurement—4.0L engine

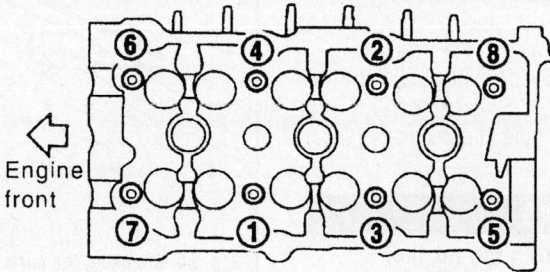

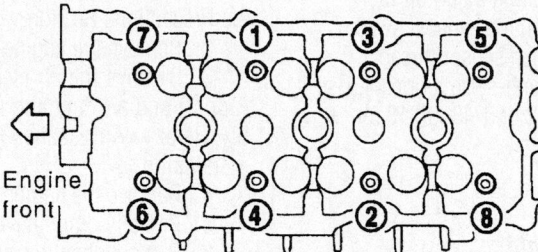

Fig. 74 Cylinder head bolt torque sequence—4.0L engine

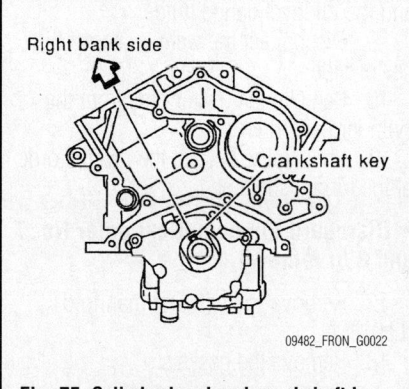

Fig. 75 Cylinder head and crankshaft key alignment—4.0L engine

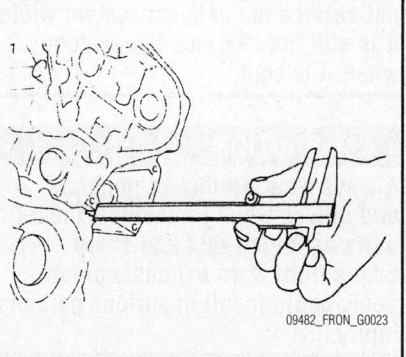

Fig. 76 Cylinder head to cylinder block installation measurement—4.0L engine

- Limit difference: 0.0043 inch (0.11mm)

13. Install the new cylinder head gasket. Turn the crankshaft until the number one piston is at TDC.

➡**The crankshaft key should line up with the right bank center line, see illustration.**

14. Torque the cylinder head bolts to specification and in the proper sequence.
    a. Step 1: 72 ft. lbs. (98 Nm).
    b. Step 2: Loosen to 0 ft. lbs. (0 Nm) in the reverse order of tightening.

    c. Step 3: 29 ft. lbs. (39 Nm).
    d. Step 4: 90° clockwise.
    e. Step 5: 90° clockwise.

15. Measure the distance between the front end faces of the cylinder block and the cylinder head on both the left and right banks. If the measured value is not within specification reinstall the cylinder head. Specification is 0.555–0.587 inch (14.1–14.9mm).

16. Be sure to fill the cooling system with the proper grade and type engine coolant.

17. Start the engine and check for leaks.

➡**If the hydraulic pressure inside the**

timing chain tensioner drops after removal and installation, slack in the guide may generate a pounding noise during and just after engine start. This is normal. The noise will stop after the hydraulic pressure rises.

EXHAUST MANIFOLD

REMOVAL & INSTALLATION

2.5L Engine

See Figures 77 through 78.

Catalytic converter precautions:
- If a large amount of unburned fuel flows into the catalyst, the catalyst temperature will be excessively high.
- Use unleaded gasoline only. Leaded gasoline will seriously damage the three way catalyst.
- When checking for ignition spark or measuring engine compression, make tests quickly and only when necessary.
- Do not run the engine when the fuel tank level is low, otherwise the engine may misfire, causing damage to the catalyst.
- Do not place the vehicle over flammable material. Keep flammable material off the exhaust pipe and the three way catalyst.

✳✳ CAUTION

In order to avoid being burned, do not service the exhaust system while it is still hot. Service the system when it is cool.

✳✳ CAUTION

Always wear protective goggles and gloves when removing exhaust parts as falling rust and sharp edges from worn exhaust components could result in serious personal injury.

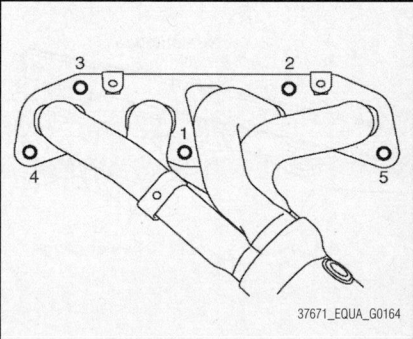

Fig. 77 Exhaust manifold torque sequence—2.5L engine

The exhaust manifold is removed with the three way catalyst as an assembly.

1. Before servicing the vehicle, refer to the Precautions Section.

2. Disconnect the harness connector of the air fuel ratio sensor 1 and the harness from the bracket and the middle clamp.

3. Remove the air fuel ratio sensor 1 using special tool J-44626, or equivalent.

✳✳ WARNING

Be careful not to damage the air fuel ratio sensor 1. Discard any air fuel ratio sensor 1 which has been dropped from a height of more than 20 inches (0.5m) onto a hard surface such as a concrete floor. Replace as needed.

4. Remove the exhaust front tube.

5. Remove the exhaust manifold cover.

6. Remove the bracket between the exhaust manifold/three way catalyst assembly and the transmission assembly.

7. Loosen the nuts in the reverse order of the torque sequence to remove the exhaust manifold/three way catalyst assembly.

8. Remove the exhaust manifold gasket.

✳✳ WARNING

Cover the engine openings to avoid entry of foreign materials.

To install:

9. Check the surface distortion of the exhaust manifold/three way catalyst assembly mating surface with a straightedge and a feeler gauge.
   • Limit: 0.012 inch (0.3mm)
   • If the limit is exceeded, replace the exhaust manifold/three way catalyst assembly.

10. Installation is in the reverse order of removal.

11. If the stud bolts were removed, install them and tighten to 11 ft. lbs. (15 Nm).

12. Tighten the nuts in the torque sequence according to the illustrations shown.

13. Tighten the nuts in numerical order as shown again.

✳✳ WARNING

Do not over-tighten the air fuel ratio sensor 1. Doing so may cause damage to the air fuel ratio sensor 1, resulting in the MIL coming on.

14. Before installing a new air fuel ratio sensor 1, clean the exhaust system threads using oxygen sensor thread cleaner and apply anti-seize lubricant. Use cleaner J-43897-12 or J-43897 18, or equivalent.

4.0L Engine

Left Side Manifold

See Figures 79 through 81.

✳✳ CAUTION

In order to avoid being burned, do not service the exhaust system while it is still hot. Service the system when it is cool.

✳✳ CAUTION

Always wear protective goggles and gloves when removing exhaust parts as falling rust and sharp edges from worn exhaust components could result in serious personal injury.

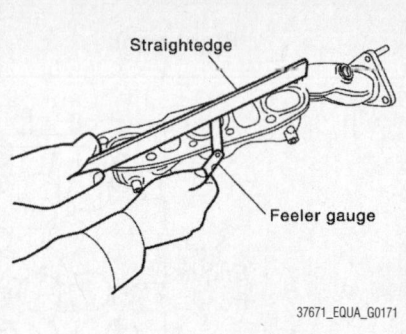

Fig. 80 Checking the surface distortion of the exhaust manifold mating surface with a straightedge and feeler gauge—4.0L engine

1. Before servicing the vehicle, refer to the Precautions Section.

2. Remove the engine room cover.

3. Remove the air cleaner case (upper) and air duct and resonator assembly. Refer to Air Cleaner, removal & installation.

4. Remove the engine under cover.

5. Partially drain the engine coolant.

6. Remove the three way catalyst (LH).

7. Remove the exhaust manifold cover (LH).

8. Remove the oil level gauge (dipstick) and the oil level gauge guide.

9. Disconnect the water hoses at the heater pipe.

10. Remove the heater pipe from the cylinder head (LH).

11. Loosen the nuts in the reverse order of the torque sequence.

➡ **Disregard the numerical order No. 7 and 8 in removal.**

12. Remove the exhaust manifold (LH).

13. Remove the gaskets.

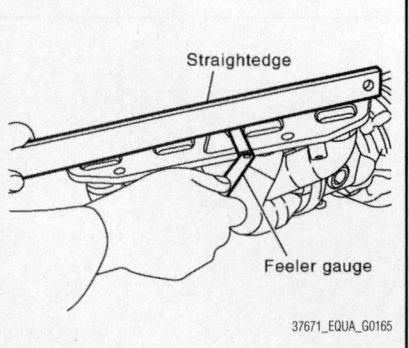

Fig. 78 Checking the surface distortion of the exhaust manifold/three way catalyst assembly mating surface—2.5L engine

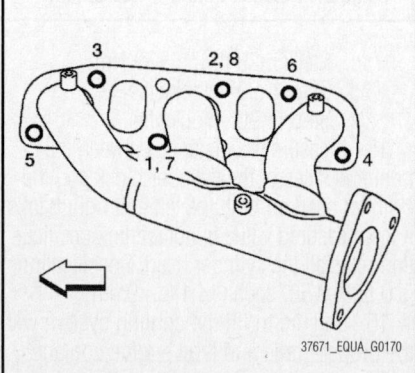

Fig. 79 Exhaust manifold torque sequence (left side)—4.0L engine

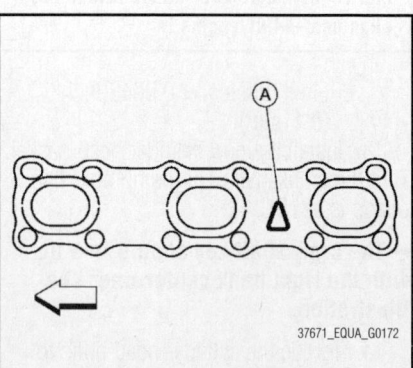

Fig. 81 Install the exhaust manifold gasket with the identification hole (A) pointing as shown—4.0L engine

※※ WARNING

Cover the engine openings to avoid entry of foreign materials.

To install:

14. Check the surface distortion of the exhaust manifold mating surface with a straightedge and feeler gauge.
   • Limit: 0.012 inch (0.3mm)
   • If the limit is exceeded, replace the exhaust manifold.

15. Installation is in the reverse order of removal.

16. Install the exhaust manifold gasket with the identification hole as shown.

17. If the exhaust manifold studs were removed, install and tighten to 11 ft. lbs. (15 Nm).

18. Install the exhaust manifold and tighten the nuts in numerical order shown.

➡**Use new exhaust manifold nuts for installation.**

19. Tighten nuts No. 1 and 2 in two steps. The numerical order No. 7 and 8 show the second step.

Right Side Manifold

See Figures 80, 82 and 83.

※※ CAUTION

In order to avoid being burned, do not service the exhaust system while it is still hot. Service the system when it is cool.

※※ CAUTION

Always wear protective goggles and gloves when removing exhaust parts as falling rust and sharp edges from worn exhaust components could result in serious personal injury.

1. Before servicing the vehicle, refer to the Precautions Section.

2. Remove the three way catalyst (RH).

3. Remove the heat shield from the lower dash panel.

4. Remove the support bolts from the A/T fluid charging pipe (automatic transmission models).

5. Loosen the nuts in the reverse order of the torque sequence.

➡**Disregard the numerical order No. 7 and 8 in removal.**

6. Remove the exhaust manifold (RH) and the exhaust manifold cover (RH) together.

7. Remove the gaskets.

※※ WARNING

Cover the engine openings to avoid entry of foreign materials.

To install:

8. Check the surface distortion of the exhaust manifold mating surface with a straightedge and feeler gauge.
   • Limit: 0.012 inch (0.3mm)
   • If the limit is exceeded, replace the exhaust manifold.

9. Installation is in the reverse order of removal.

➡**If necessary, a crowfoot may be used to tighten the exhaust manifold nuts.**

10. Install the exhaust manifold gasket with the identification hole as shown.

11. If the exhaust manifold studs were removed, install and tighten to 11 ft. lbs. (15 Nm).

12. Install the exhaust manifold and tighten the nuts in numerical order shown.

➡**Use new exhaust manifold nuts for installation.**

13. Tighten nuts No. 1 and 2 in two steps. The numerical order No. 7 and 8 show the second step.

INTAKE MANIFOLD

REMOVAL & INSTALLATION

2.5L Engine

See Figures 84 and 85.

1. Before servicing the vehicle, refer to the Precautions Section.

2. Properly relieve the fuel system pressure.

3. Disconnect the negative battery cable. Drain the cooling system.

4. Remove the air cleaner case, air cleaner and air duct.

5. Disconnect and plug the water hoses from the electric throttle control actuator.

6. Remove the mass air flow sensor from the intake manifold. Remove the quick connector cap and disconnect the quick connector at the engine side.

7. Remove the electric throttle control actuator harness connector and retaining bolts. Be sure to remove the bolts by reversing the order of the tightening torque sequence.

8. Remove the electric throttle control actuator and gasket.

9. Disconnect the harness, vacuum hoses, and PCV hoses from the intake manifold and position them to the side.

10. Remove intake manifold support.

11. Remove the intake manifold retaining bolts. Be sure to remove the bolts by reversing the order of the tightening torque sequence.

12. Remove the intake manifold, fuel tube protector, and gasket from the engine.

13. As necessary, remove the EVAP canister purge volume solenoid valve and vacuum hose adapter from the intake manifold.

14. Disconnect the sub frame harness from the fuel injectors. Remove the fuel tube and fuel injector assembly from the intake manifold, if required.

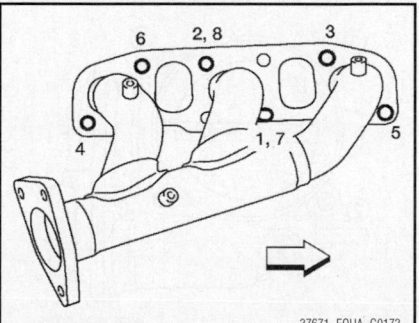

Fig. 82 Exhaust manifold torque sequence (right side)—4.0L engine

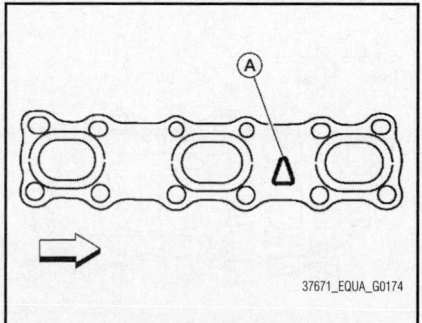

Fig. 83 Install the exhaust manifold gasket with the identification hole (A) pointing as shown—4.0L engine

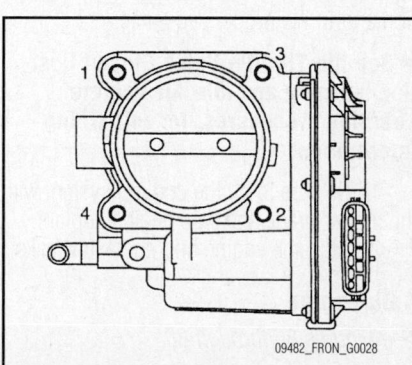

Fig. 84 Intake manifold bolt torque sequence—2.5L engine

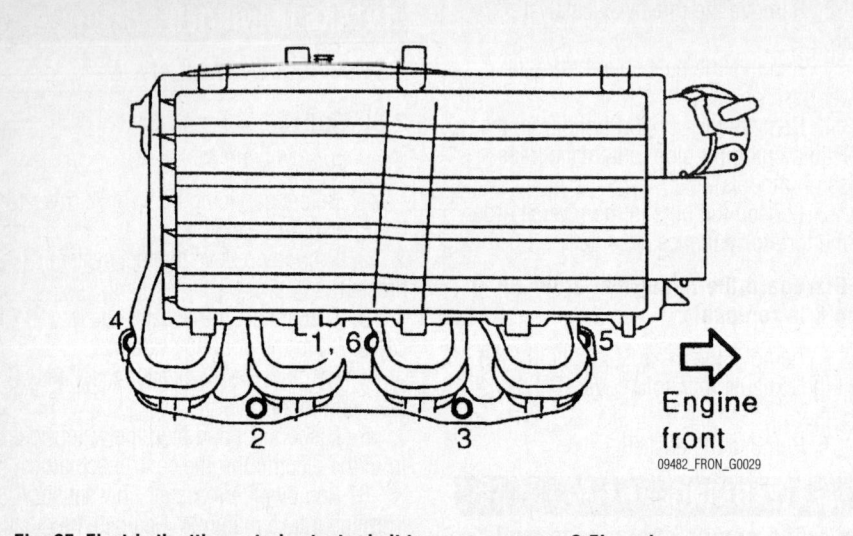

Fig. 85 Electric throttle control actuator bolt torque sequence—2.5L engine

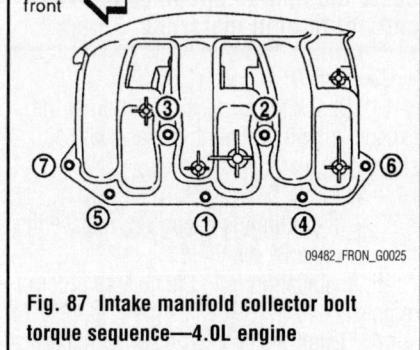

Fig. 87 Intake manifold collector bolt torque sequence—4.0L engine

⁂ WARNING

Cover engine openings to avoid entry of foreign materials. Do not disassemble the intake manifold.

To install:

15. Installation is the reverse of the removal procedure.

16. Be sure to use new gaskets.

17. If the stud bolts were removed, install and tighten to 83 inch lbs. (9 Nm).

18. Be sure to tighten the intake manifold retaining bolts to 83 inch lbs. (9 Nm) in the proper sequence.

➡️ **Refer to the torque sequence illustration, No. 6 means double tightening of bolt No. 1.**

19. Use the following for locating bolts and nuts:
  • M8 x 1.50 inches (38mm) are green in color: No. 1, 6
  • M8 x 1.38 inches (35mm): No. 2, 3
  • Nuts: No. 4, 5

20. Be sure to tighten the electric throttle control actuator retaining bolts to specification and in the proper sequence.

➡️ **See the Throttle Valve Closed Position learning and Idle Air Volume Learning procedures, for relearning information.**

21. Be sure to fill the cooling system with the proper grade and type engine coolant.

22. Start the engine and check for leaks.

4.0L Engine

See Figures 86 through 88.

➡️ **The upper intake manifold is also referred to as the intake manifold collector.**

1. Before servicing the vehicle, refer to the Precautions Section.

2. Properly relieve the fuel system pressure.

3. Disconnect the negative battery cable.

4. Drain the cooling system.

5. Remove the engine cover. Remove the air cleaner case (upper) with the mass air flow sensor and air duct assembly.

6. Disconnect the water hoses from the electric throttle control actuator. Disconnect the harness connector.

7. Remove the electric throttle control actuator retaining bolts. Be sure to remove the bolts by reversing the order of the tightening torque sequence.

8. Remove the electric throttle control actuator.

9. Remove the brake booster vacuum hose and the PCV hose. Remove the intake manifold collector support.

10. Disconnect the EVAP hoses and harness connector from the EVAP

canister purge volume control solenoid valve. Remove the EVAP canister purge volume control solenoid valve.

11. Remove the VIAS control solenoid valve and vacuum tank.

12. Remove the intake manifold collector retaining bolts. Be sure to remove the bolts by reversing the order of the tightening torque sequence.

13. Remove the intake manifold collector from the engine.

14. Remove the fuel tube and fuel injector assembly.

15. Remove the intake manifold retaining bolts. Be sure to remove the bolts by reversing the order of the tightening torque sequence.

16. Remove the intake manifold from the engine.

To install:

17. Installation is the reverse of the removal procedure.

18. Be sure to use new gaskets.

19. Be sure to tighten the intake manifold retaining bolts to specification and in the proper sequence in two or more steps.

20. Be sure to tighten the intake manifold collector retaining bolts to specification and in the proper sequence.

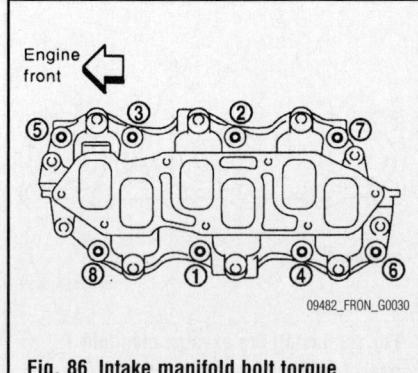

Fig. 86 Intake manifold bolt torque sequence—4.0L engine

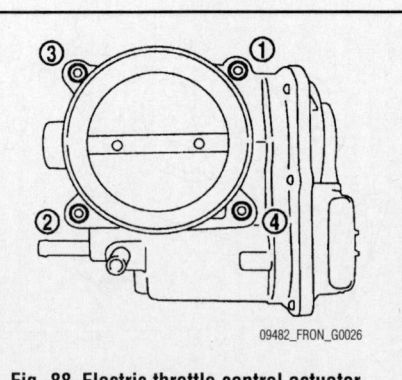

Fig. 88 Electric throttle control actuator bolt torque sequence—4.0L engine

21. Be sure to tighten the electric throttle control actuator retaining bolts to specification and in the proper sequence.

➡**See throttle valve closed position learning and idle air volume learning procedures, for relearning information.**

22. Be sure to fill the cooling system with the proper grade and type engine coolant.

23. Start the engine and check for leaks.

OIL PAN

REMOVAL & INSTALLATION

2.5L Engine

See Figure 89.

1. Before servicing the vehicle, refer to the Precautions Section.

2. Disconnect the negative battery cable.

3. Remove the engine under cover. Drain the engine oil.

4. Remove the third crossmember assembly.

5. Remove the lower joint shaft pinch bolt at the steering gear.

6. Remove the steering gear bolts.

7. If equipped with an automatic transmission, remove the fluid cooler tube.

8. Loosen the oil pan retaining bolts, in the reverse order of the installation sequence.

9. Insert a seal cutter tool between the oil pan and the cylinder block, and slide it by tapping on the side of the tool with a hammer.

10. Remove the oil pan from the engine.

To install:

11. Be sure to clean all the oil gasket material from both the oil pan and the cylinder block surfaces, using the proper tools.

❊❊ WARNING

Do not scratch or damage the mating surfaces when cleaning off the old liquid gasket.

12. Apply a continuous bead of sealant 0.138–0.177 inches (3.5–4.5mm) to the oil pan mating surface.

13. Install the oil pan to the cylinder block. This must be done within 5 minutes after applying the liquid gasket.

14. Torque the oil pan bolts to 61 inch lbs. (7 Nm) in the proper sequence.

15. Continue the installation in the reverse order of the removal procedure.

➡**Wait 30 minutes after installation of the oil pan to allow the sealant to cure before adding oil.**

16. Fill the crankcase with the proper grade and amount of oil.

17. Start the engine and check for leaks.

4.0L Engine

Lower

See Figure 90.

1. Before servicing the vehicle, refer to the Precautions Section.

2. Disconnect the negative battery cable.

3. Remove the engine under cover. Drain the engine oil.

4. Loosen the oil pan retaining bolts, in the reverse order of the installation sequence.

5. Insert a seal cutter tool between the oil pan and the cylinder block, and slide it by tapping on the side of the tool with a hammer.

6. Remove the oil pan from the engine.

To install:

7. Be sure to clean all the oil gasket material from both the oil pan and the cylinder block surfaces, using the proper tools.

8. Apply a continuous bead of sealant 0.138–0.177 inches (3.5–4.5mm) to the oil pan mating surface. Be sure to use genuine RTV sealant, or equivalent.

9. Install the oil pan to the cylinder block. This must be done within 5 minutes after applying the liquid gasket.

10. Torque the bolts to 80 inch lbs. (9 Nm) in the proper sequence.

11. Continue the installation in the reverse order of the removal procedure.

➡**Wait 30 minutes after installation of the oil pan to allow the sealant to cure before adding oil.**

12. Fill the crankcase with the proper grade and amount of oil.

13. Start the engine and check for leaks.

Upper

See Figures 91 and 92.

1. Before servicing the vehicle, refer to the Precautions Section.

2. Disconnect the negative battery cable.

3. Remove the air duct. Remove the engine under cover.

4. Drain the engine oil.

5. Drain the engine coolant.

6. Remove the final drive, if equipped with 4WD.

7. Disconnect the steering gear lower shaft joint bolt and steering gear nuts and bolts, position the assembly out of the way.

8. Remove the starter.

9. Disconnect the automatic transmission fluid cooler brackets, if equipped and position them out of the way.

10. Remove the oil filter, as necessary. Remove the oil cooler.

11. Remove the lower oil pan. Remove the oil strainer.

12. Remove the transmission joint bolts which pierce the oil pan.

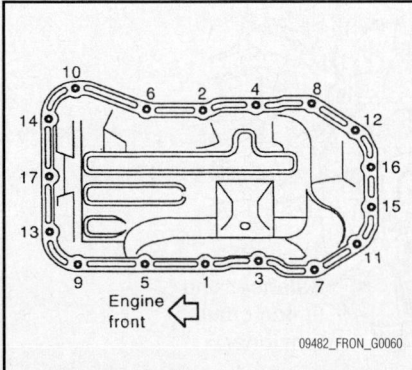

Fig. 89 Oil pan bolt torque sequence— 2.5L engine

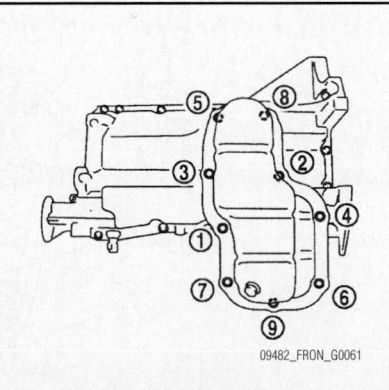

Fig. 90 Lower oil pan bolt torque sequence—4.0L engine

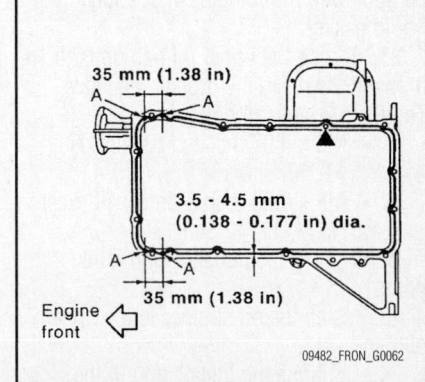

Fig. 91 Upper oil pan sealant application—4.0L engine

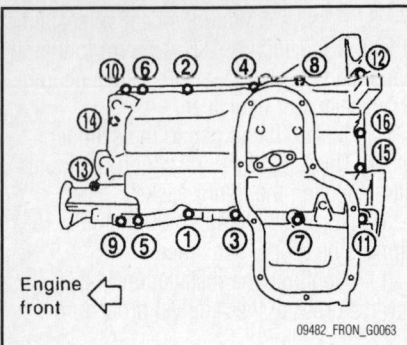

Fig. 92 Upper oil pan bolt torque sequence—4.0L engine

13. Remove the rear cover plate.

14. Loosen the upper oil pan retaining bolts, in the reverse order of the installation sequence.

15. Insert a seal cutter tool between the oil pan and the cylinder block, and slide it by tapping on the side of the tool with a hammer.

16. Remove the oil pan from the engine. Remove the O-rings from the bottom lower cylinder block and oil pump.

To install:

17. Be sure to clean all the oil gasket material from both the oil pan and the cylinder block surfaces, using the proper tools.

18. Install new O-rings on the bottom lower cylinder block and oil pump.

19. Apply a continuous bead of sealant 0.138–0.177 inches (3.5–4.5mm) to the lower cylinder block mating surfaces of the upper oil pan. Be sure to use genuine RTV sealant, or equivalent.

➡**For bolt holes marked with a solid black triangle, apply liquid gasket outside the hole. Apply a bead of sealant 0.177–0.217 inch (4.5–5.5mm) in diameter to the "A" section.**

20. Install the upper oil pan. This must be done within 5 minutes after applying the liquid gasket.

21. Torque the upper oil pan bolts to 16 ft. lbs. (22 Nm) in the proper sequence. There are 2 types of bolts:
- M8 x 3.97 inches (100mm) No. 7, 11, 12, and 13
- M8 x 0.98 inch (25mm) all except No. 7, 11, 12 and 13

22. Tighten the transmission joint bolts.

23. Install the oil strainer to the upper oil pan.

24. Continue the installation in the reverse order of the removal procedure.

➡**Wait 30 minutes after installation of the oil pan to allow the sealant to cure before adding oil.**

25. Fill the crankcase with the proper grade and amount of oil.

26. Start the engine and check for leaks.

OIL PUMP

REMOVAL & INSTALLATION

2.5L Engine

See Figures 93 through 98.

The oil pump is integral to the balancer unit and is serviced as an assembly. Do not disassemble the balancer unit.

1. Before servicing the vehicle, refer to the Precautions Section.

2. Remove the timing chain. Refer to Timing Chain & Sprockets, removal & installation.

3. Remove the balancer unit timing chain tensioner with the following procedure:

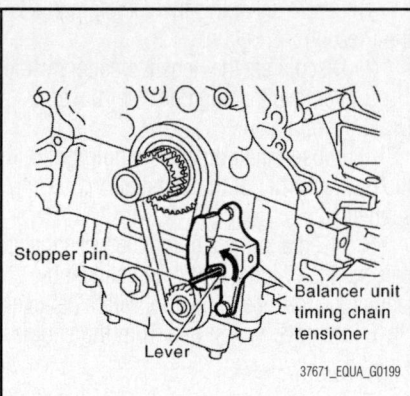

Fig. 93 Balancer unit timing chain tensioner release—2.5L engine

a. Lift the lever up and release the ratchet claw for return proof.

b. Push the tensioner sleeve in and hold it.

c. Matching the hole on the lever with the one on the body, insert a stopper pin to secure the tensioner sleeve.

➡**Use a hard metal pin as a stopper pin—approximately 0.04 inch (1mm) in diameter.**

d. Remove the balancer unit timing chain tensioner.

4. Secure the hexagonal portion of the balancer shaft using a suitable tool. Loosen the balancer unit sprocket bolt.

5. Remove the balancer unit timing chain, balancer unit sprocket, and crankshaft sprocket.

➡**When removing the balancer unit timing chain, remove the crankshaft sprocket and balancer unit sprocket at the same time.**

6. Loosen the bolts of the balancer unit in reverse order of torque sequence.

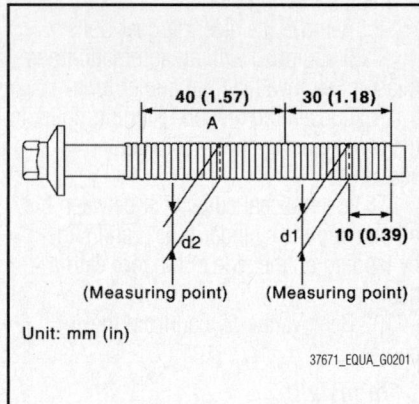

Fig. 95 Measuring balancer unit bolt diameters—2.5L engine

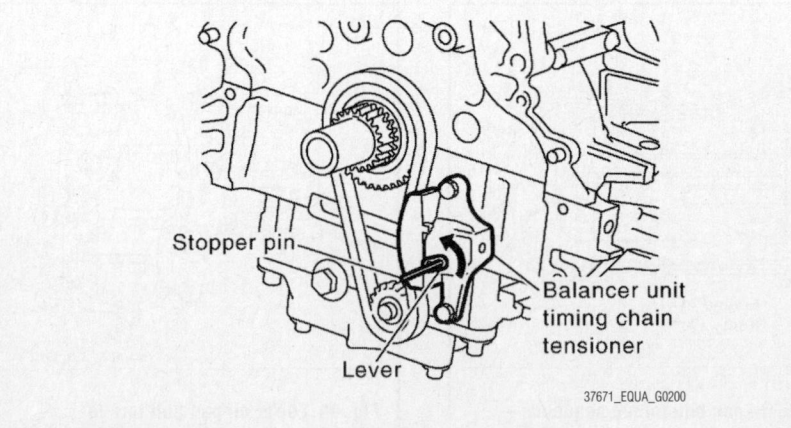

Fig. 94 Balancer unit bolt torque sequence shown—2.5L engine

✳✳ WARNING

Do not disassemble the balancer unit.

➡ **Use TORX® socket size E14 for bolts No. 1 to 4.**

To install:

7. Check the balancer unit timing chain for cracks and any excessive wear at the roller links of the chain. Replace the timing chain, if necessary.

8. Measure the balancer unit bolt outer diameters at 2 positions.
   - If a reduction appears in (A) range, regard it as (d2)
   - Limit of (d1) minus (d2): 0.0059 inch (0.15mm)
   - If the limit is exceeded, replace the balancer unit bolt with a new one.

9. Measure the balancer unit bolt length.
   - If it exceeds the limit, replace the balancer unit bolt with a new one.
   - Limit: 6.77 inches (172mm)

10. Make sure that the crankshaft key points straight up.

11. Install the O-ring to the balancer unit.

12. Tighten the bolts in numerical order with the following procedure to install the balancer unit, using special tool KV10112100 (BT8653-A), or equivalent. Apply new engine oil to the threads and seat surfaces of the bolts.

✳✳ WARNING

If the bolts are to be re-used, check their outer diameter before installation.

✳✳ WARNING

Check tightening angle using the special tool or a protractor. Do not make a judgment by a visual check alone.

   a. Step 1 — Bolts 1–4: 35 ft. lbs. (48 Nm).
   b. Step 2 — Bolts 1–4: 100° clockwise.
   c. Step 3 — Bolts 1–4: 0 ft. lbs. (0 Nm) in reverse order of the tightening sequence.
   d. Step 4 — Bolts 1–4: 35 ft. lbs. (48 Nm).
   e. Step 5 — Bolts 1–4: 100° clockwise.
   f. Step 6 — Bolts 5–6: 22 ft. lbs. (30 Nm).

13. Install the crankshaft sprocket, balancer unit sprocket, and balancer unit timing chain.

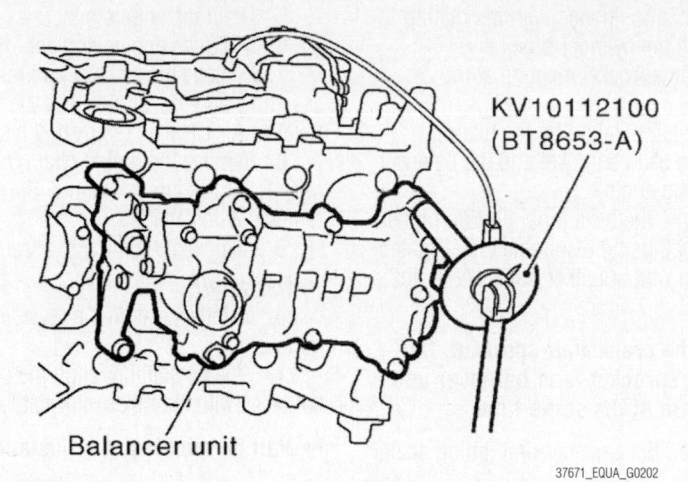

Fig. 96 Special tool KV10112100 (BT8653-A) shown tightening balancer unit bolts—2.5L engine

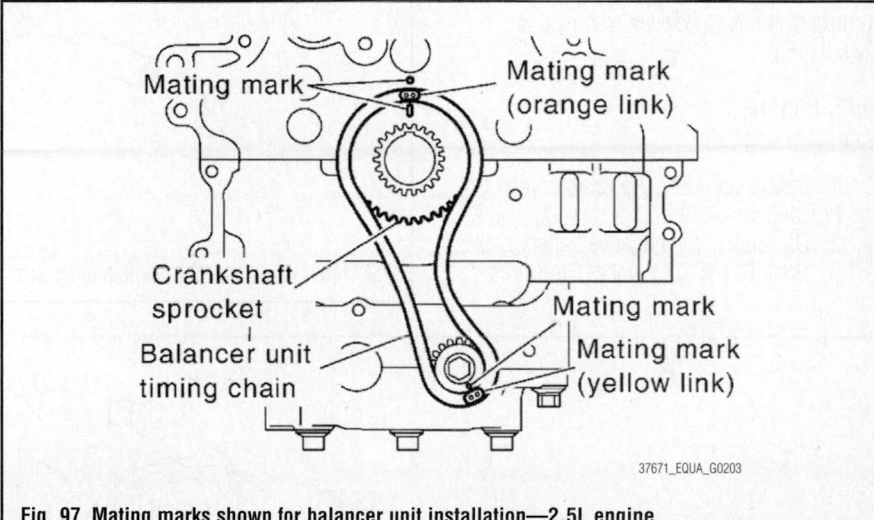

Fig. 97 Mating marks shown for balancer unit installation—2.5L engine

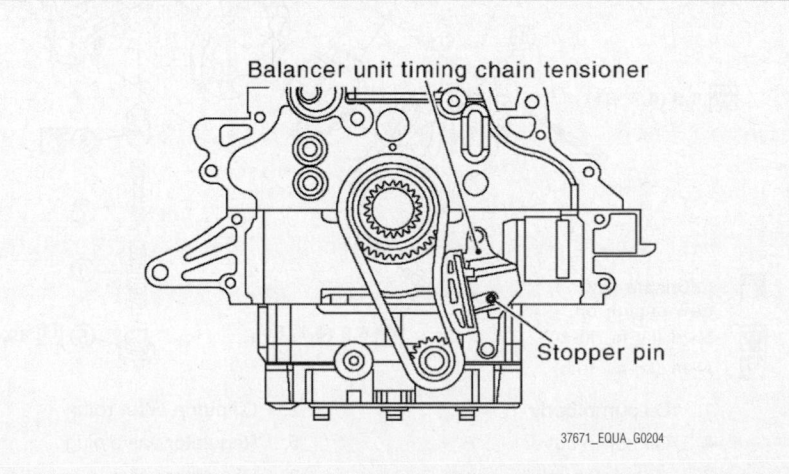

Fig. 98 Install the balancer unit timing chain tensioner and remove the stopper pin—2.5L engine

a. Make sure that the crankshaft sprocket is positioned with the mating marks on the cylinder block and crankshaft sprocket meeting at the top.

b. Install by aligning the mating marks on each sprocket and the balancer unit timing chain.

14. Secure the hexagonal portion of the balancer shaft using a suitable tool. Tighten the balancer unit sprocket bolt to 48 ft. lbs. (65 Nm).

➡**Install the crankshaft sprocket, balancer unit sprocket, and balancer unit timing chain at the same time.**

15. Install the balancer unit timing chain tensioner.

a. After installation, make sure the mating marks have not slipped.

b. Remove the stopper pin and release the tensioner sleeve.

16. Install the timing chain. Refer to Timing Chain & Sprockets, removal & installation.

4.0L Engine

See Figure 99.

1. Before servicing the vehicle, refer to the Precautions Section.
2. Disconnect the negative battery cable.
3. Remove the front wheels and tires.

4. Remove the front fender protectors.
5. Drain the engine oil.
6. Drain the engine coolant.
7. Remove the air duct and resonator assembly and the air cleaner case (upper). Refer to Air Cleaner, removal & installation.
8. Remove the timing chain (primary) only. Refer to Timing Chain & Sprockets, removal & installation.
9. Remove the oil pump assembly.

To install:

10. Installation is in the reverse order of removal.
11. When installing, align the crankshaft flat faces with the inner rotor flat faces.

➡**Wait 30 minutes after installation of**

the oil pan to allow the sealant to cure before adding oil.

12. Fill the crankcase with the proper grade and amount of oil.
13. Start the engine and check for leaks.
14. Stop the engine and wait for 10 minutes.
15. Check the engine oil level and adjust as required.

PISTON AND RING

POSITIONING

2.5L Engine

See Figures 100 and 101.

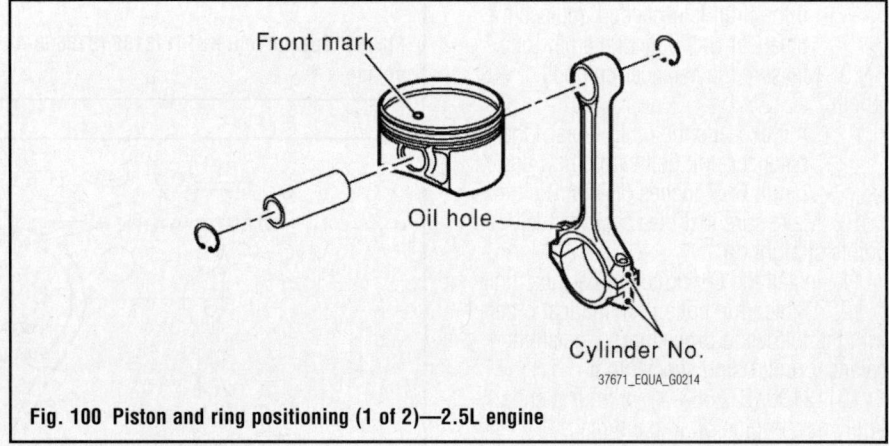

Front mark

Oil hole

Cylinder No.

37671_EQUA_G0214

Fig. 100 Piston and ring positioning (1 of 2)—2.5L engine

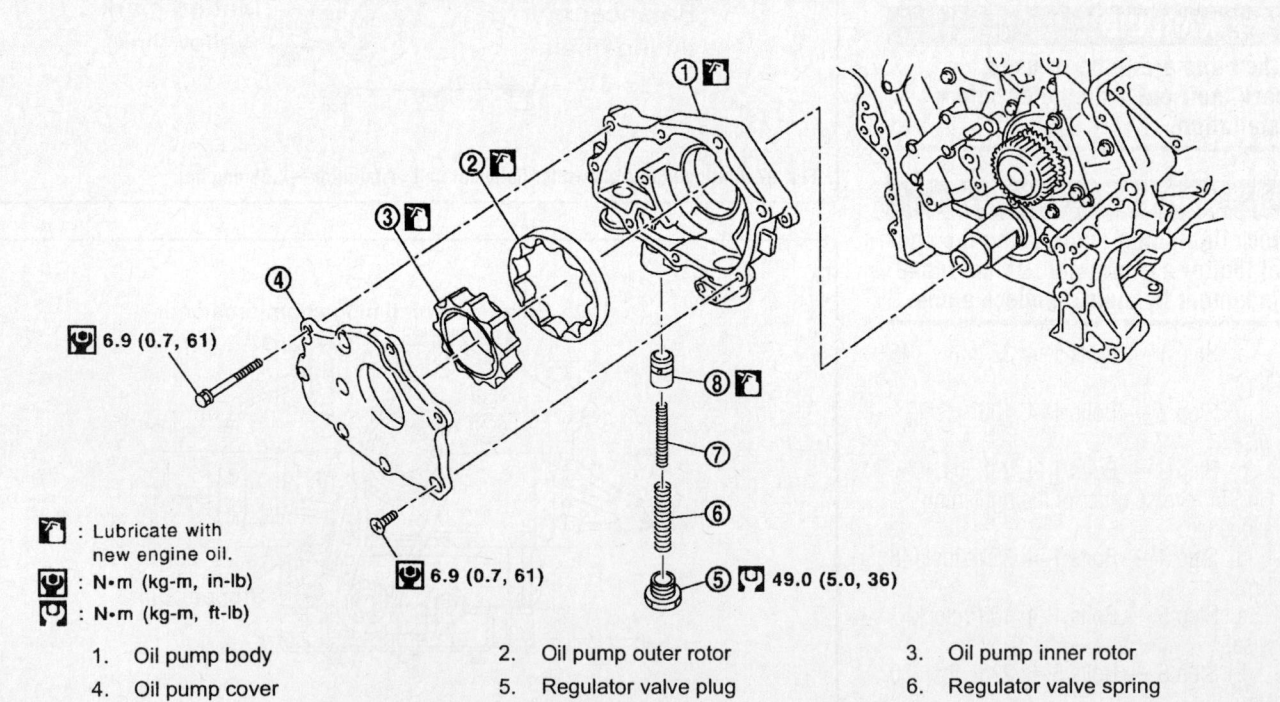

6.9 (0.7, 61)

6.9 (0.7, 61)

49.0 (5.0, 36)

: Lubricate with new engine oil.

: N·m (kg-m, in-lb)

: N·m (kg-m, ft-lb)

1. Oil pump body	2. Oil pump outer rotor	3. Oil pump inner rotor
4. Oil pump cover	5. Regulator valve plug	6. Regulator valve spring
7. Regulator valve spring	8. Regulator valve	

09482_FRON_G0065

Fig. 99 Exploded view of oil pump—4.0L engine

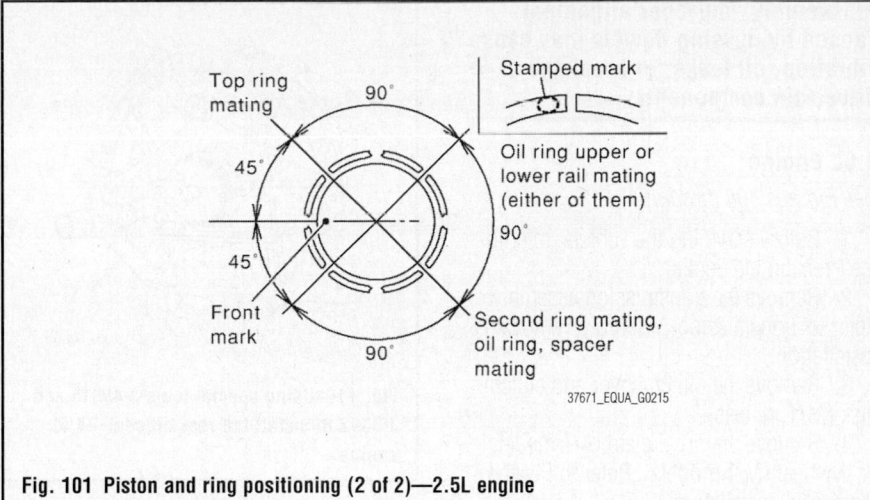

Fig. 101 Piston and ring positioning (2 of 2)—2.5L engine

4.0L Engine

See Figures 102 through 104.

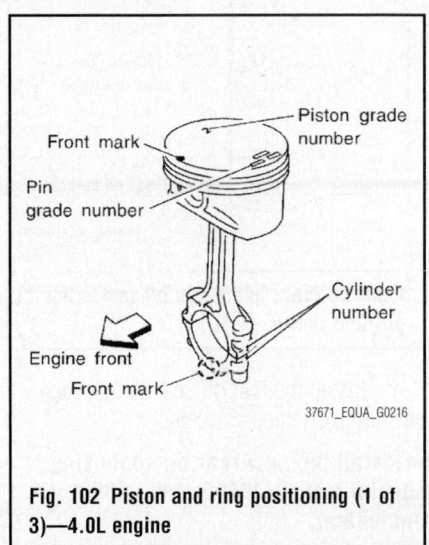

Fig. 102 Piston and ring positioning (1 of 3)—4.0L engine

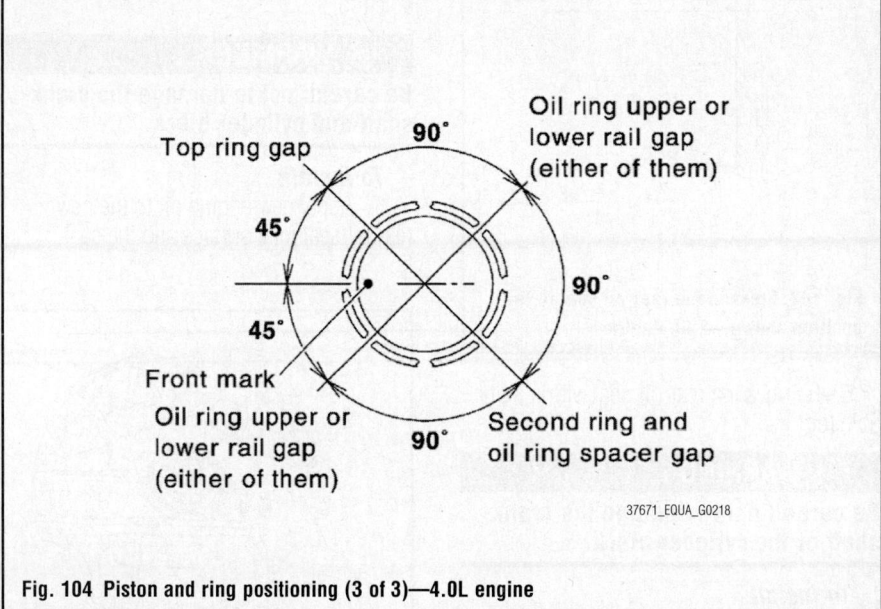

Fig. 104 Piston and ring positioning (3 of 3)—4.0L engine

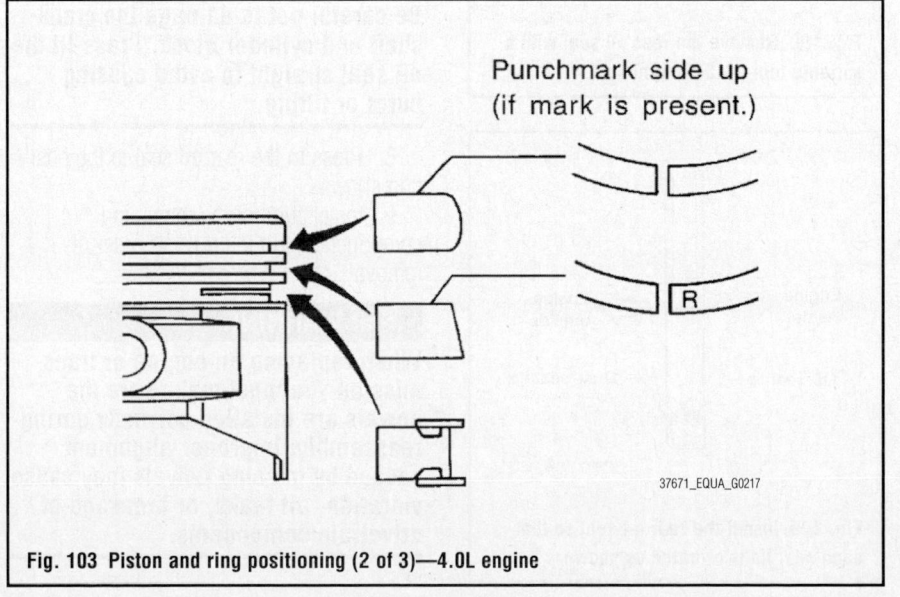

Fig. 103 Piston and ring positioning (2 of 3)—4.0L engine

REAR MAIN SEAL

REMOVAL & INSTALLATION

2.5L Engine

See Figures 105 through 107.

1. Before servicing the vehicle, refer to the Precautions Section.

2. Remove the transmission assembly. Refer to Transmission Assembly, removal & installation.

3. Remove the clutch cover and clutch disk (M/T models).

4. Remove the drive plate (A/T models) or the flywheel (M/T models). Refer to Flywheel, removal & installation.

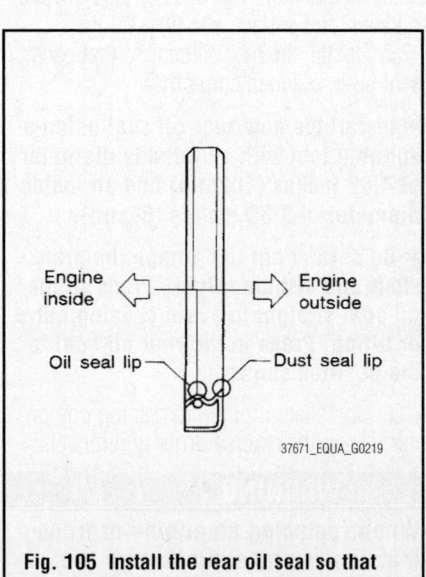

Fig. 105 Install the rear oil seal so that each seal lip is oriented as shown

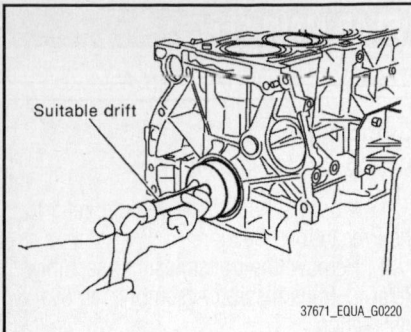

Fig. 106 Install the rear oil seal using a suitable drift—2.5L engine

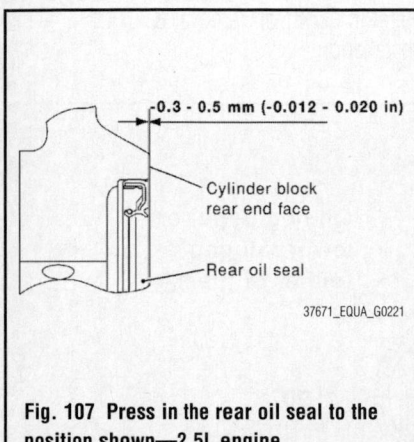

Fig. 107 Press in the rear oil seal to the position shown—2.5L engine

5. Remove the rear oil seal with a suitable tool.

⁑ WARNING

Be careful not to damage the crankshaft or the cylinder block.

To install:

6. Apply new engine oil to the new rear oil seal joint surface and the seal lip.

7. Install the rear oil seal so that each seal lip is oriented correctly.

➡Install the new rear oil seal using a suitable tool with an outside diameter of 4.02 inches (102mm) and an inside diameter of 3.39 inches (86mm).

➡Be careful not to damage the crankshaft and cylinder block. Press-fit the oil seal straight to avoid causing burrs or tilting. Press in the rear oil seal to the position shown.

8. Installation of the remaining components is in the reverse order of removal.

⁑ WARNING

When replacing an engine or transmission you must make sure the dowels are installed correctly during

reassembly. Improper alignment caused by missing dowels may cause vibration, oil leaks, or breakage of drivetrain components.

4.0L Engine

See Figures 108 through 111.

1. Before servicing the vehicle, refer to the Precautions Section.

2. Remove the transmission assembly. Refer to Transmission Assembly, removal & installation.

3. Remove the clutch cover and clutch disk (M/T models).

4. Remove the drive plate (A/T models) or flywheel (M/T models). Refer to Flywheel, removal & installation.

5. Remove the rear oil seal with a suitable tool.

⁑ WARNING

Be careful not to damage the crankshaft and cylinder block.

To install:

6. Apply new engine oil to the new rear oil seal joint surface and the seal lip.

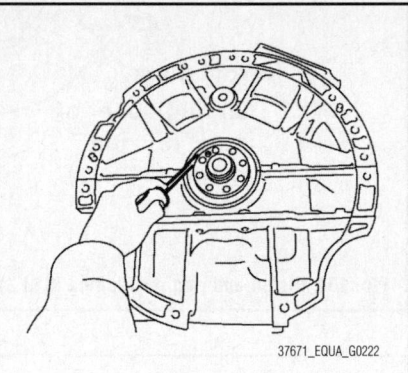

Fig. 108 Remove the rear oil seal with a suitable tool—4.0L engine

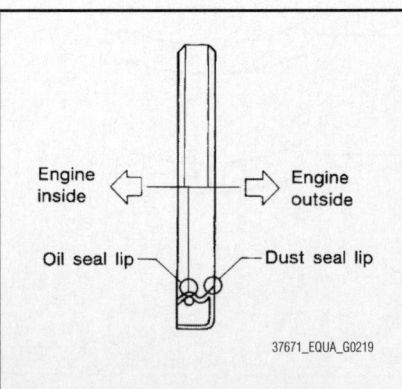

Fig. 109 Install the rear oil seal so that each seal lip is oriented as shown

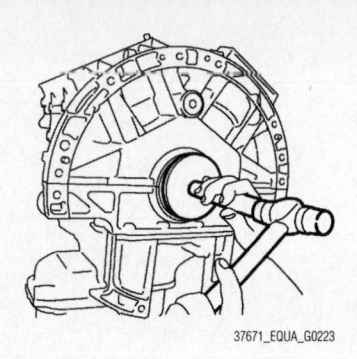

Fig. 110 Using special tools J-49815 and J-8092 to install the rear oil seal—4.0L engine

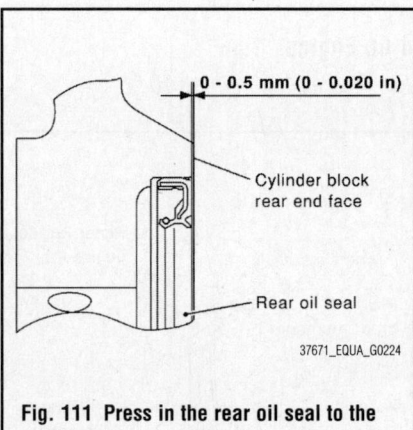

Fig. 111 Press in the rear oil seal to the position shown—4.0L engine

7. Install the rear oil seal so that each seal lip is oriented correctly.

➡Install the new rear oil seal using special tools J-49815 and J-8092, or equivalent.

⁑ WARNING

Be careful not to damage the crankshaft and cylinder block. Press-fit the oil seal straight to avoid causing burrs or tilting.

8. Press in the rear oil seal to the position shown.

9. Installation of the remaining components is in the reverse order of removal.

⁑ WARNING

When replacing an engine or transmission you must make sure the dowels are installed correctly during reassembly. Improper alignment caused by missing dowels may cause vibration, oil leaks, or breakage of drivetrain components.

TIMING CHAIN & SPROCKETS

REMOVAL & INSTALLATION

2.5L Engine

See Figures 112 through 120.

1. Before servicing the vehicle, refer to the Precautions Section.

2. Properly relieve the fuel system pressure.

3. Disconnect the negative battery cable.

4. Remove the engine undercover.

5. Remove the air cleaner and the air duct assembly. Refer to Air Cleaner, removal & installation.

6. Remove the spark plugs.

7. Disconnect the PCV hose from the rocker cover. Remove the ignition coil.

8. Remove the PCV valve and O-ring from the rocker cover, if necessary.

9. Remove the oil filler cap from the rocker cover, if necessary.

10. Remove the rocker cover retaining bolts. Be sure to remove the bolts by reversing the order of the tightening torque sequence.

11. Remove the rocker cover. Discard the gasket.

12. Remove the coolant reservoir tank. Remove the auxiliary drive belt auto-tensioner.

13. Remove the alternator. Remove the strut tower brace.

14. Remove the air conditioning compressor and position it to the side. Do not disconnect the refrigerant lines.

15. Remove the power steering pump and reservoir tank; position the assembly to the side. Do not disconnect the fluid lines.

16. Remove the upper and lower oil pan. Remove the strainer.

17. Remove the Intake Valve Timing (IVT) control cover bolts. Be sure to remove the bolts by reversing the order of the tightening sequence.

18. Remove the cover by cutting the sealant using tool KV10111100 (J-37228), or equivalent.

19. Position the number one cylinder on its compression stroke by rotating the crankshaft pulley clockwise. Align the mating marks for Top Dead Center (TDC) with the timing indicator on the front cover.

20. At the same time, make sure that the mating marks on the camshaft sprockets are lined up as indicated in the illustration. If not, rotate the crankshaft one more turn to line up the mating marks.

21. Remove the crankshaft pulley. Refer to Crankshaft Damper, removal & installation.

22. Loosen the front cover retaining bolts, in the order indicated in the bolt loosening sequence illustration. Remove the front cover. Be careful not to damage the mating surfaces.

23. Using a seal removal tool, remove the oil seal, as required.

24. To remove the timing chain, push in on the chain tensioner plunger. Insert a stopper pin into the hole on the chain tensioner body to secure the chain tensioner plunger. Remove the chain tensioner. Remove the timing chain.

➡ **Use a metal pin as a stopper pin approximately 0.02 inch (0.5mm) diameter.**

❊❊ WARNING

Do not rotate the crankshaft or camshafts with the chain removed. It causes interference between the valves and the pistons.

25. Remove the camshaft sprockets.

26. Remove the timing chain slack guide, timing chain tensioner guide and spacer.

27. Remove the balancer unit timing chain tensioner by lifting the lever up and releasing the ratchet claw for return proof.

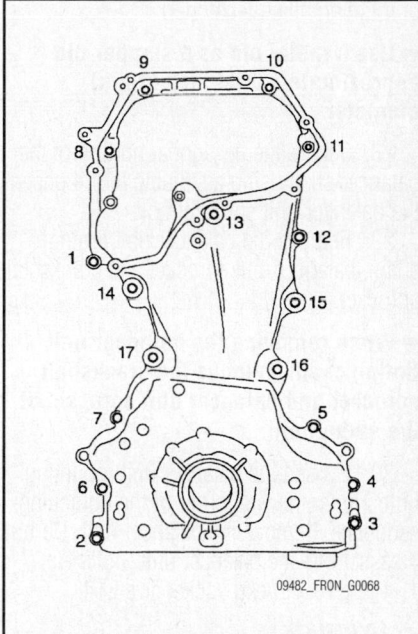

Fig. 114 Front cover bolt removal sequence—2.5L engine

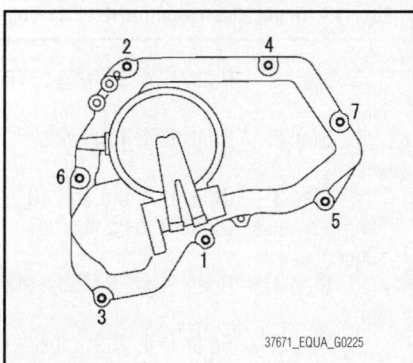

Fig. 112 Intake Valve Timing (IVT) control cover bolt tightening sequence—2.5L engine

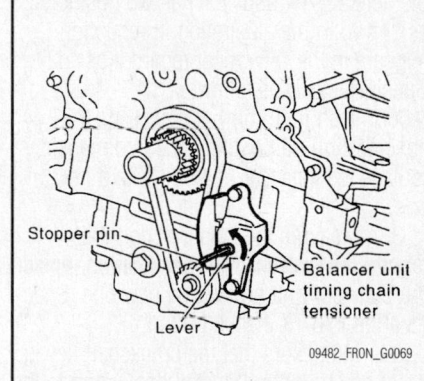

Fig. 115 Balance unit stopper pin installation—2.5L engine

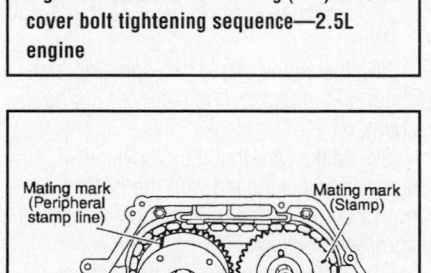

Fig. 113 Timing chain alignment marks—2.5L engine

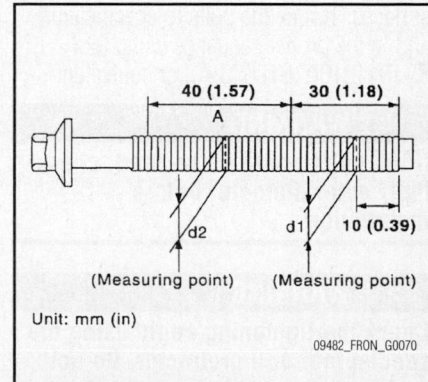

Fig. 116 Balance unit bolt measurement—2.5L engine

Push the tensioner sleeve in and hold it. Matching the hole on the lever with the one on the body, insert a stopper pin to secure the tensioner sleeve. Remove the balancer unit timing chain tensioner.

➡**Use a metal pin as a stopper pin approximately 0.04 inch (1mm) diameter.**

28. Secure the hexagonal portion of the balancer shaft using a suitable tool. Loosen the balancer unit sprocket bolt.

29. Remove the balancer unit timing chain, balancer unit sprocket and crankshaft sprocket.

➡**When removing the balancer unit timing chain, remove the crankshaft sprocket and balancer unit sprocket at the same time.**

30. Loosen the balancer unit mounting bolts, in the reverse order of the tightening sequence. Remove the balancer unit. Do not disassemble the balancer unit. Bolts No. 1–4 use Torx® head socket size E14.

To install:

31. Check the chain for cracks and excessive wear, replace as required.

32. Measure the balancer unit bolt outer diameters ("d1" and "d2") at two positions, as shown in the illustration. If reduction appears in the "A" range, regard it as "d2". Specification is as follows: ("d1" - "d2"): 0.0059 inch (0.15mm). If it exceeds the specification (large difference in dimensions) replace the balancer unit bolt with a new one.

33. Measure the balancer bolt unit length. If it exceeds the specification replace the balancer unit bolt with a new one. Specification is 6.77 inches (172mm).

34. Make sure that the crankshaft key is pointing straight up. Install the O-ring to the balancer unit.

35. Install the balancer unit. Apply engine oil to the bolt threads and sealing surfaces. Torque the bolts to specification and in the proper sequence using tool KV10112100 (BT8653-A), or equivalent.

✳✳ WARNING

If the bolts are to be re-used, check their outer diameter before installation.

✳✳ WARNING

Check the tightening angle using the special tool or a protractor. Do not make a judgment by a visual check alone.

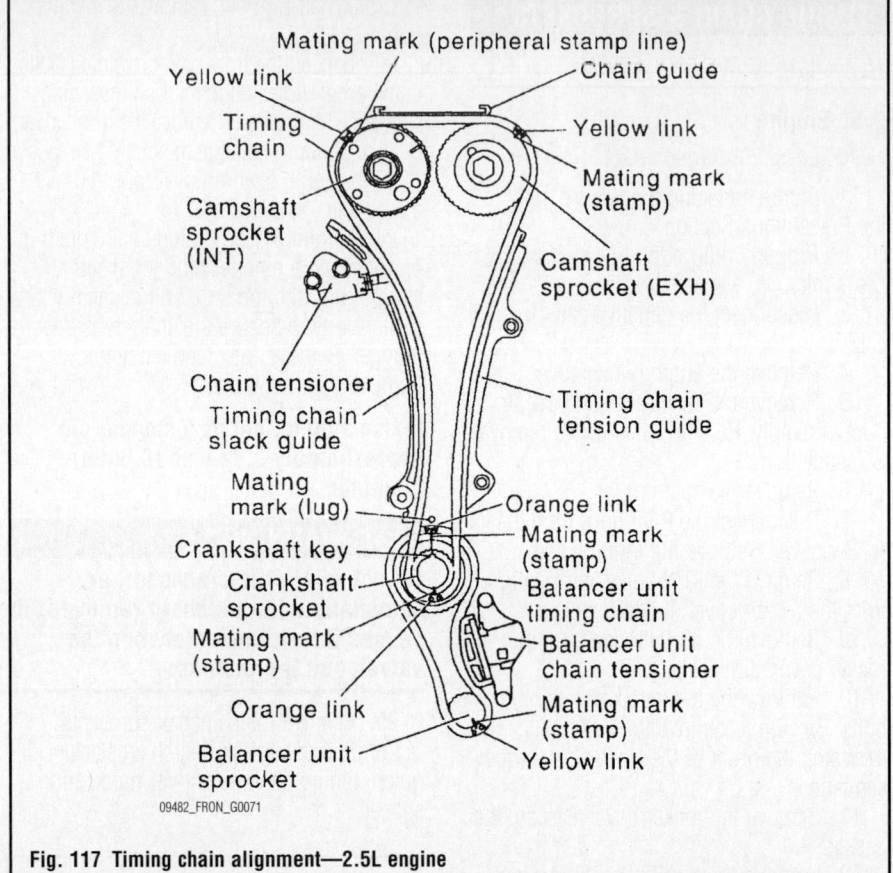

Fig. 117 Timing chain alignment—2.5L engine

a. Step 1 — Bolts 1–4: 35 ft. lbs. (48 Nm).

b. Step 2 — Bolts 1–4: 100° clockwise.

c. Step 3 — Bolts 1–4: 0 ft. lbs. (0 Nm) in reverse order of the tightening sequence.

d. Step 4 — Bolts 1–4: 35 ft. lbs. (48 Nm).

e. Step 5 — Bolts 1–4: 100° clockwise.

f. Step 6 — Bolts 5–6: 22 ft. lbs. (30 Nm).

36. Install the crankshaft sprocket, balancer unit sprocket, and balancer timing chain.

37. Make sure that the crankshaft sprocket is positioned with the mating marks on the cylinder block and crankshaft sprocket meeting at the top.

38. Install it by aligning the mating marks on each sprocket and balancer unit timing chain.

39. Secure the hexagonal portion of the balancer shaft using a suitable tool. Tighten the balancer unit sprocket bolt to specification.

➡**Install the crankshaft sprocket, balancer unit sprocket, and balancer unit timing chain at the same time.**

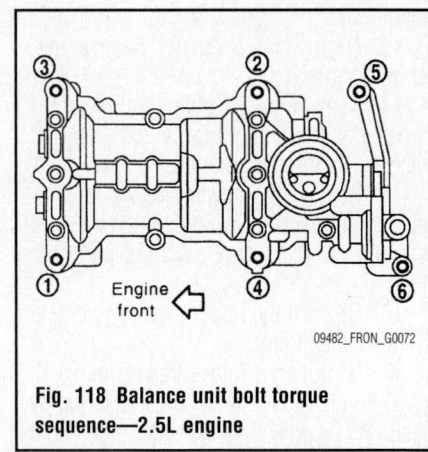

Fig. 118 Balance unit bolt torque sequence—2.5L engine

40. Install the balancer unit timing chain tensioner.

➡**After installation, make sure that the mating marks have not slipped. Remove the stopper pin and release the tensioner sleeve.**

41. Align the mating marks on each sprocket and timing chain. Install the timing chain and related parts.

42. Before and after installing the chain tensioner, check again to be sure that the mating marks have not slipped.

43. After installing the chain tensioner,

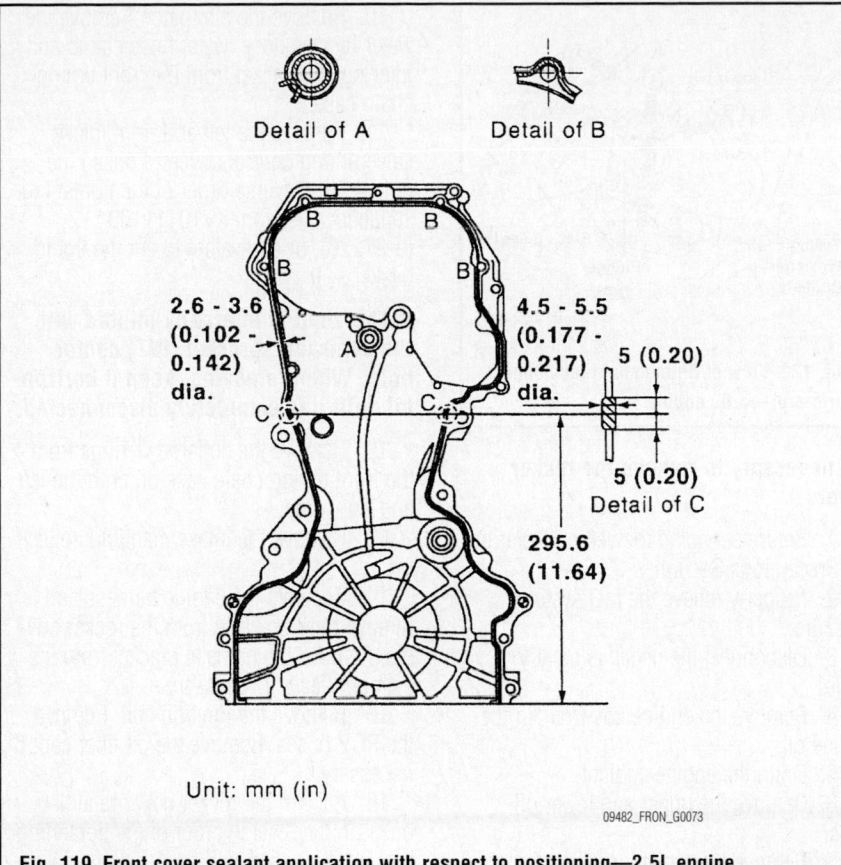

Fig. 119 Front cover sealant application with respect to positioning—2.5L engine

Detail of A   Detail of B

2.6 - 3.6 (0.102 - 0.142) dia.

4.5 - 5.5 (0.177 - 0.217) dia.

5 (0.20)

5 (0.20)
Detail of C

295.6 (11.64)

Unit: mm (in)

09482_FRON_G0073

46. Make sure that the mating marks of the chain and each sprocket are still aligned.

47. Install the front cover. Torque the retaining bolts to specification and in the proper sequence.

48. Use the following for locating the M6 bolts:

- Bolt position 5, 10, 14 and 17 — Bolt length: 1.77 inches (45mm)
- Bolts except the above (except 1 to 4) — Bolt length: 0.79 inch (20mm)

49. Tighten the bolts to the specified torque:

- Bolt position 5–17: 9 ft. lbs. (13 Nm)
- Bolt position 1–4: 36 ft. lbs. (49 Nm)
- After all bolts are tightened, retighten them to specification in the numerical order shown

➡ **Be sure to wipe off any excess liquid gasket leaking to the surface from installing the oil pan.**

50. Install the chain guide between the camshaft sprockets.

51. Install the oil rings to the camshaft sprocket (INT) insertion points on the back-

remove the stopper pin. Make sure that the tensioner moves freely.

➡ **After the mating marks are aligned, keep them aligned by holding them with your hand. To avoid skipped teeth, do not rotate the crankshaft and camshaft until the cover is installed.**

➡ **Before installing the chain tensioner, it is possible to change the position of the mating mark on the timing chain for that on each sprocket for alignment.**

44. Install the front cover oil seal. Install O-rings to the cylinder head and the cylinder block.

45. Apply a continuous bead of liquid gasket to the front cover. Be sure to use genuine RTV sealant, or equivalent.

➡ **Sealant application instructions differ depending on position, refer to the illustration for positioning. Detail "A", cross over the start of the application and the end. Detail "B", apply liquid gasket outside of the bolt holes. For all bolt holes other than "B", apply to the inside. Detail "C", between here only, apply a bead of sealant 0.177–0.217 inch (4.5–5.5mm) diameter.**

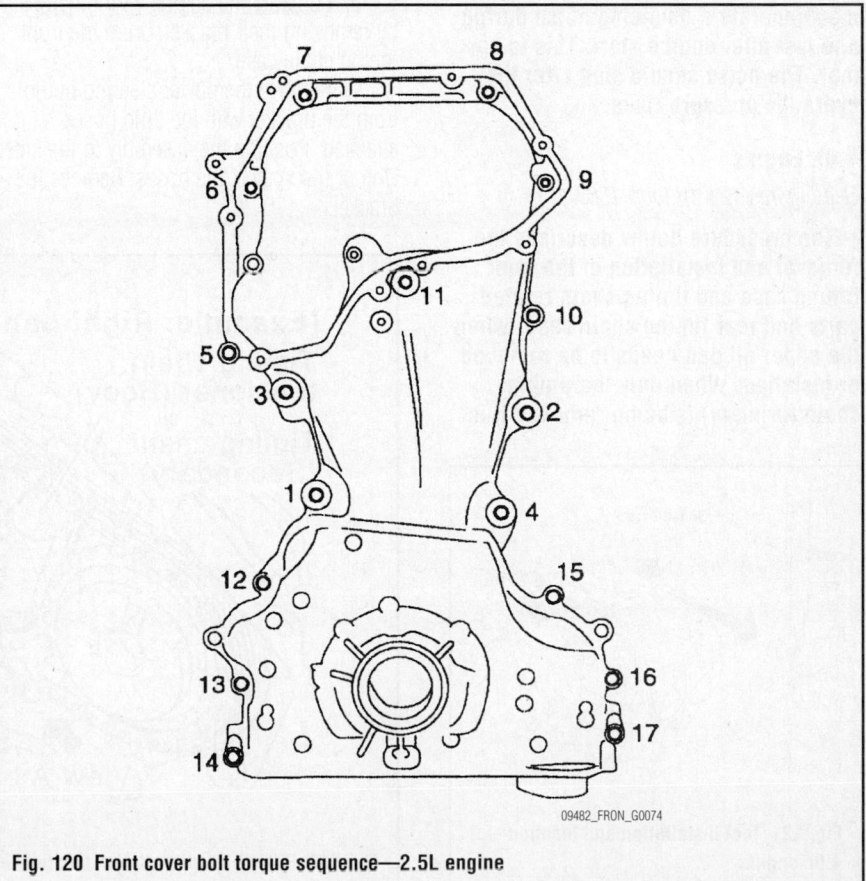

09482_FRON_G0074

Fig. 120 Front cover bolt torque sequence—2.5L engine

side of the intake valve timing control cover. Install the O-ring to the front cover.

52. Apply a continuous bead of liquid gasket, 0.083–0.122 inch (2.1–3.1mm) in diameter, to the front cover. Be sure to use genuine RTV sealant, or equivalent.

53. Install the intake valve timing control cover. Tighten the bolts in the proper sequence to specification.

54. Install the intake valve timing control solenoid valve to the intake valve timing control cover, if removed.

55. Connect the ground cables, and install the harness clip.

56. Install the crankshaft pulley. Refer to Crankshaft Damper, removal & installation.

57. When installing the rocker cover, apply liquid gasket, be sure to use genuine RTV silicone sealant, or equivalent, to the positions shown in the illustration. Refer to figure "a" to apply liquid gasket to joint part of camshaft bracket No. 1 and cylinder head. Refer to figure "b" to apply liquid gasket in 90 degrees to figure "b".

58. Install the rocker cover. Refer to Valve Cover, removal & installation.

59. Continue the installation in the reverse order of the removal procedure.

➡**If hydraulic pressure inside the timing chain tensioner drops after removal/installation, slack in the guide may generate a pounding noise during and just after engine start. This is normal. The noise should stop after the hydraulic pressure rises.**

4.0L Engine

See Figures 121 through 136.

➡**The procedure below describes the removal and installation of the front timing case and timing chain related parts and rear timing chain case, when the upper oil pan needs to be removed or installed. When only the timing chain (primary) is being**

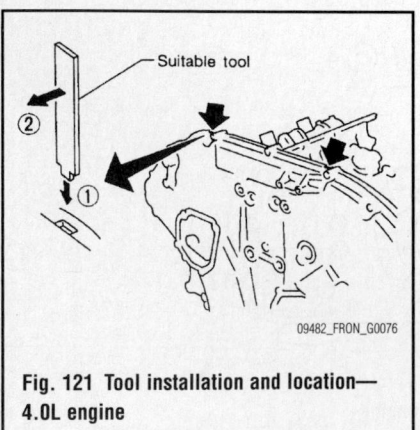

Fig. 121 Tool installation and location—4.0L engine

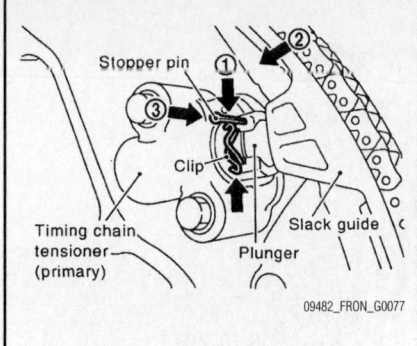

Fig. 122 View of timing chain tensioner (primary)—4.0L engine

not necessary to remove the rocker covers.

1. Before servicing the vehicle, refer to the Precautions Section.

2. Properly relieve the fuel system pressure.

3. Disconnect the negative battery cable.

4. Remove the engine cover. Drain the engine oil.

5. Drain the engine coolant.

6. Remove the upper and lower oil pans.

7. Remove the radiator cooling fan assembly. Remove the drive belts.

8. Separate the engine wiring harnesses by removing their brackets from the front timing chain case.

9. Remove the power steering pump from the bracket with the fluid hoses attached. Position the assembly to the side. Do not disconnect the hoses. Remove the bracket.

10. Remove the alternator. Remove the water bypass hose, water hose clamp and idler pulley bracket from the front timing chain case.

11. Remove the left and right intake valve timing control covers. Loosen the bolts in the reverse order of the tightening sequence. Use tool KV10111100 (J-37228), or equivalent to cut the liquid gasket seal.

➡**The shaft is internally jointed with the camshaft sprocket (INT) center hole. When removing, keep it horizontal until it is completely disconnected.**

12. Remove the collared O-rings from the front timing chain case on both the left and right side.

13. Remove the intake manifold collector.

14. Separate the engine harness and remove their brackets from the rocker covers. Remove the harness bracket from the cylinder head, if necessary.

15. Remove the ignition coil. Remove the PCV hoses. Remove the oil filler cap, if necessary.

16. Loosen the rocker cover retaining bolts, in the reverse order of the tightening sequence.

17. Remove the rocker covers from the engine.

➡**When only the timing chain (primary) is being removed it is not necessary to remove the rocker covers.**

18. Set the No. 1 cylinder at Top Dead Center (TDC) of its compression stroke by rotating the crankshaft pulley clockwise to

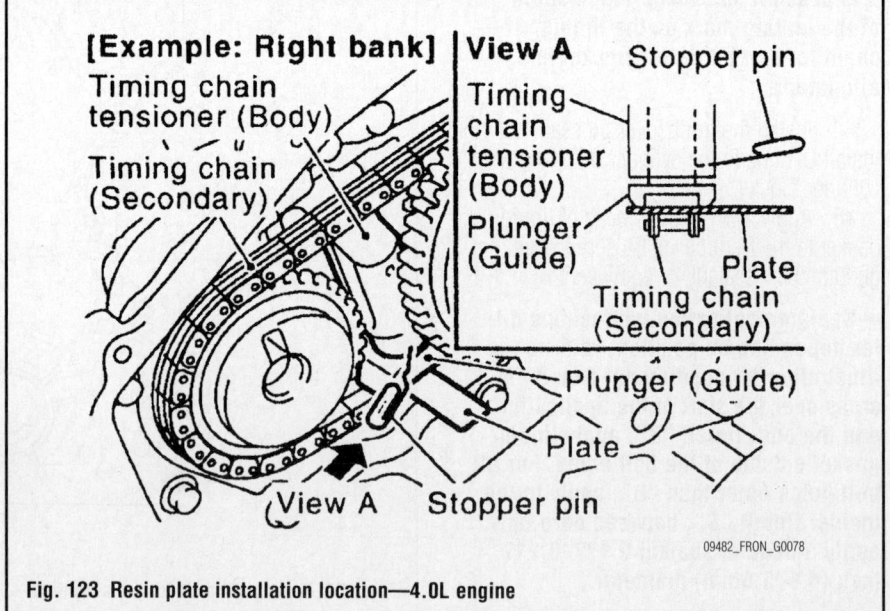

Fig. 123 Resin plate installation location—4.0L engine

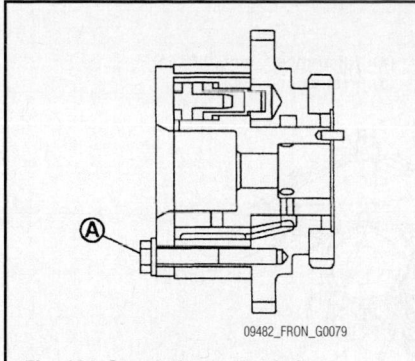

Fig. 124 Camshaft sprocket bolt location—4.0L engine

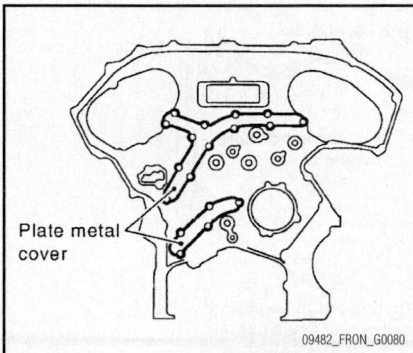

Plate metal cover

Fig. 125 Metal cover plate location on rear timing case cover—4.0L engine

align the timing mark (grooved line without color) with the timing indicator. Make sure that the intake and exhaust cam noses on No. 1 cylinder (engine front side on right bank) are in alignment as shown in the illustration. If not, rotate the crankshaft in the clockwise direction 360 degrees.

➡**When only the timing chain (primary) is removed, the rocker cover does not need to be removed. To be sure that the No. 1 cylinder is set at TDC on the compression stroke, remove the front timing chain case cover first, then check the mating marks on the camshaft sprockets.**

19. Remove the starter. Position tool KV10117700 (J-44716), or equivalent.

20. Loosen the crankshaft pulley retaining bolt and locate the bolt seating surface, which is about 0.39 inch (10mm) from its original position.

➡**Do not remove the crankshaft pulley bolt. Keep the loosened pulley bolt in place to protect the removed crankshaft pulley from dropping.**

21. Pull the pulley with both hands and remove it from its mounting.

Remove the bolt and pulley from the engine.

22. Loosen and remove the 2 bolts of the upper oil pan.

23. Loosen the front timing chain cover retaining bolts in the reverse order of the tightening sequence.

24. Insert a suitable tool in the notch at the top of the front timing chain case and pry off the case by moving the tool as shown in the illustration. Use tool KV10111100 (J-37228), or equivalent to cut the liquid gasket seal.

➡**Do not use a screwdriver or similar tool. After removal, handle the front timing chain cover case carefully so it does not tilt or warp under a load.**

25. Remove the O-rings from the rear timing chain case.

26. Remove the water pump cover and chain tensioner cover from the front timing chain case cover, as required.

27. Remove the oil seal from the front timing chain case cover, as required.

28. Remove the timing chain tensioner (primary) by loosening the clip of the timing chain tensioner (primary) and release the plunger stopper. Insert the

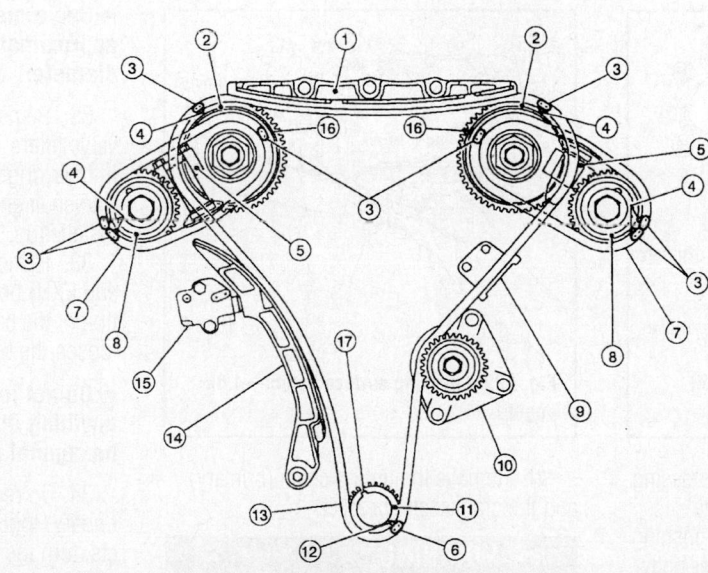

1. Internal chain guide	2. Camshaft sprocket (intake)	3. Mating mark (copper link)
4. Mating mark (punched)	5. Secondary timing chain tensioner	6. Mating mark (yellow link)
7. Secondary timing chain	8. Camshaft sprocket (exhaust)	9. Tensioner guide
10. Water pump	11. Crankshaft sprocket	12. Mating mark (notched)
13. Primary timing chain	14. Slack guide	15. Primary timing chain tensioner
16. Mating mark (back side)	17. Crankshaft key	

Fig. 126 Timing chain alignment—4.0L engine

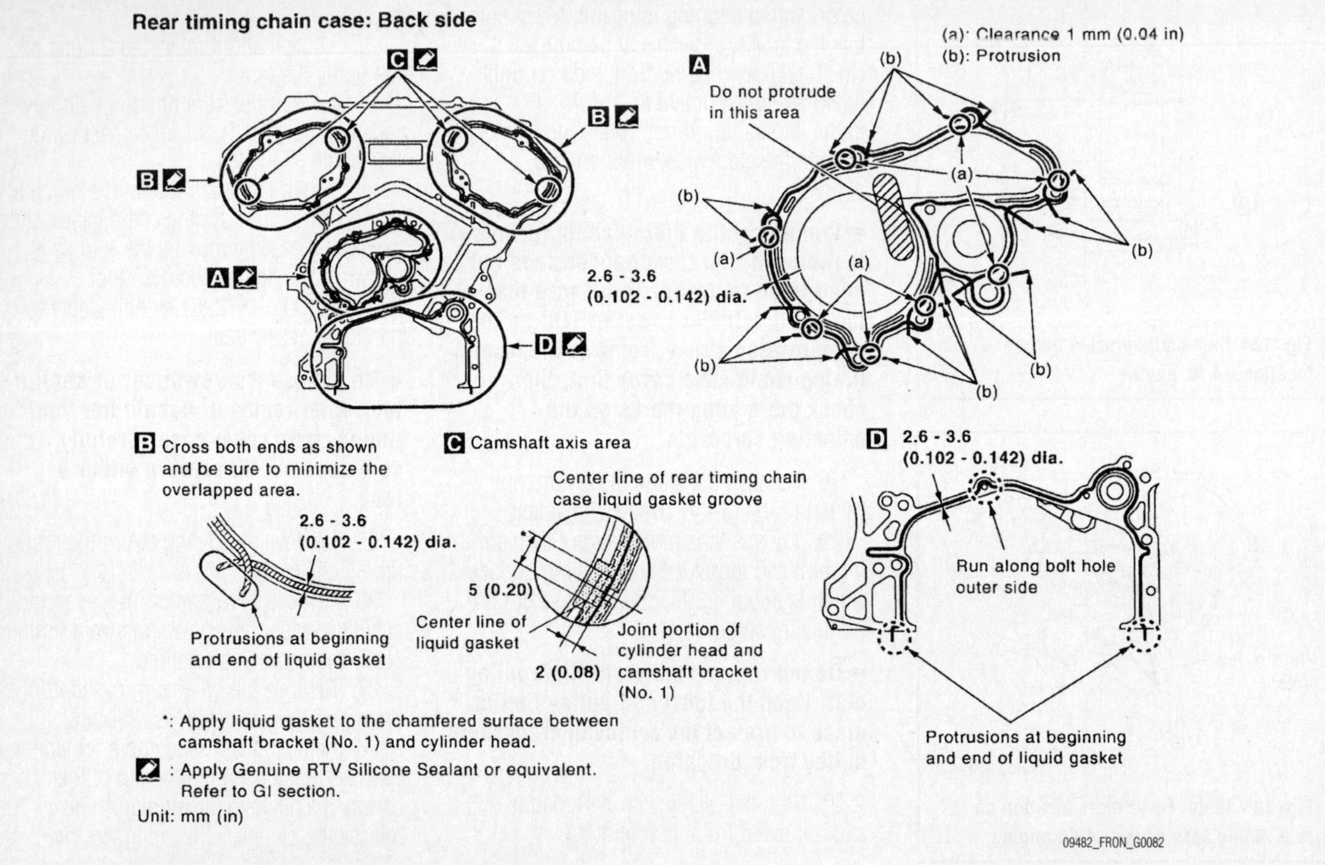

Rear timing chain case: Back side

(a): Clearance 1 mm (0.04 in)
(b): Protrusion

A Do not protrude in this area

(b) (b)

(a)

(a)

(b)

(a) (a)

2.6 - 3.6 (0.102 - 0.142) dia.

(b) (b)

(b)

(b)

B Cross both ends as shown and be sure to minimize the overlapped area.

2.6 - 3.6 (0.102 - 0.142) dia.

Protrusions at beginning and end of liquid gasket

C Camshaft axis area

Center line of rear timing chain case liquid gasket groove

5 (0.20)

Center line of liquid gasket

Joint portion of cylinder head and camshaft bracket (No. 1)

2 (0.08)

D 2.6 - 3.6 (0.102 - 0.142) dia.

Run along bolt hole outer side

Protrusions at beginning and end of liquid gasket

*: Apply liquid gasket to the chamfered surface between camshaft bracket (No. 1) and cylinder head.

🖊 : Apply Genuine RTV Silicone Sealant or equivalent. Refer to GI section.

Unit: mm (in)

09482_FRON_G0082

Fig. 127 Rear timing chain cover sealant application—4.0L engine

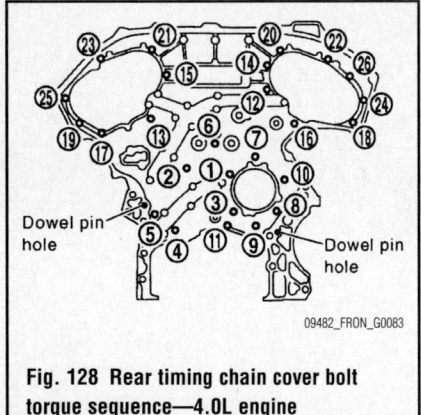

09482_FRON_G0083

Fig. 128 Rear timing chain cover bolt torque sequence—4.0L engine

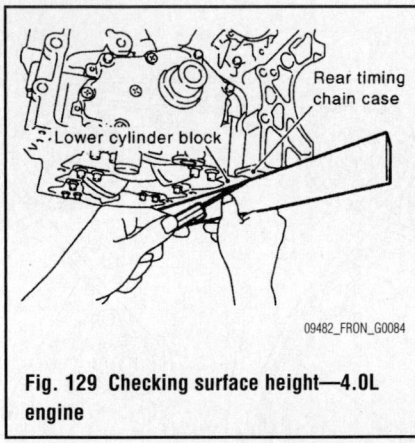

09482_FRON_G0084

Fig. 129 Checking surface height—4.0L engine

plunger into the tensioner body by pressing the slack guide. Keep the slack guide pressed and hold the plunger in by pushing the stopper pin through the tensioner body hole and the plunger groove. Remove the bolts and remove the timing chain tensioner (primary).

29. Remove the internal chain guide, tension guide and slack guide.

➡**The tension guide can be removed after removing the timing chain (primary).**

30. Remove the timing chain (primary) and the crankshaft sprocket.

※※ WARNING

After removing the timing chain (primary), do not turn the crankshaft and camshaft separately or the valves will strike the piston heads.

31. To remove the timing chain (secondary) and camshaft sprockets, attach a suitable stopper pin to the right and left timing chain tensioner (secondary).

➡**Use a metal pin as a stopper pin approximately 0.02 inch (0.5mm) in diameter.**

32. Remove the camshafts. Remove the valve lifters. Identify them for reinstallation in their original locations. Refer to Camshaft and Valve Lifters, removal & installation.

33. Remove the camshaft sprocket (INT and EXH) bolts. Secure the hexagonal portion of the camshaft using a wrench to loosen the bolts.

➡**Do not loosen the bolts by securing anything other than the camshaft hexagonal portion.**

34. To remove the timing chain (secondary) together with the camshaft sprockets, turn the crankshaft slightly to secure slackness of the timing chain on the timing chain tensioner (secondary) side.

35. Insert a 0.02 inch (0.5mm) thick metal or resin plate between the timing chain and timing chain plunger (guide). Remove the timing chain (secondary) together with the camshaft sprockets with the timing chain loose from the guide groove.

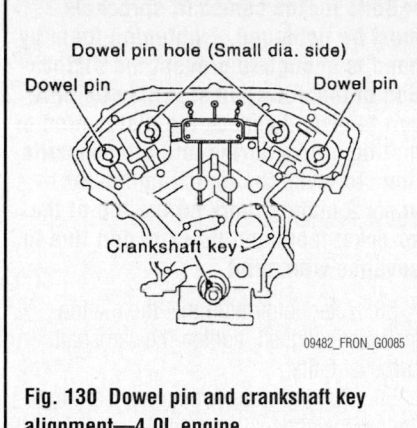

Fig. 130 Dowel pin and crankshaft key alignment—4.0L engine

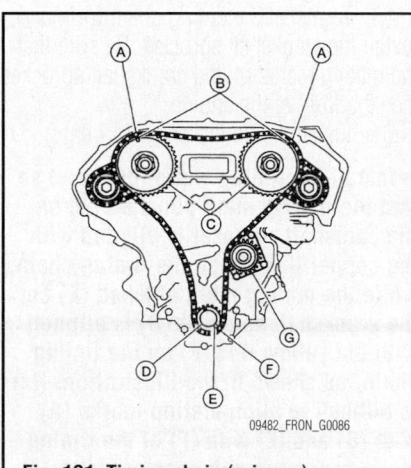

Fig. 131 Timing chain (primary) alignment—4.0L engine

✳✳ CAUTION

Be careful of the plunger coming off when removing the timing chain (secondary). This is because the plunger of the timing chain tensioner (secondary) moves during operation, leading it to coming off its fixed stopper pin.

➡**The camshaft sprocket (INT) is a one piece integrated design sprocket for the timing chain (primary) and for the timing chain (secondary). When handling the sprocket, avoid shock to the sprocket. Do not disassemble or loosen bolt "A", as shown in the illustration.**

36. Remove the water pump.

37. Remove the rear timing chain case cover bolts, in the reverse order of the tightening sequence. Using the proper tool, cut the liquid gasket sealant seal. Remove the cover.

➡**Do not remove the metal cover of the oil passage. After removal, handle the**

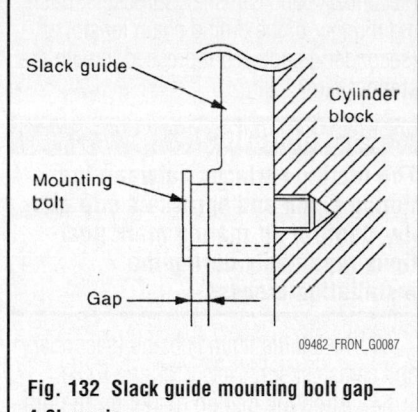

Fig. 132 Slack guide mounting bolt gap—4.0L engine

case carefully so it does not tilt or warp under a load.

38. Remove the O-rings from the cylinder head and No. 1 camshaft bracket. Remove the O-rings from the cylinder block.

39. If necessary, remove the timing chain tensioners (secondary) from the cylinder head by first removing the No. 1 camshaft bracket. Remove the timing chain tensioners (secondary) with the stopper pin attached.

To install:

40. Check the chain for cracks and excessive wear, replace as required.

41. Be sure to remove all old gasket material from bolts and bolt holes.

42. If removed, install the timing chain tensioners (secondary) to the cylinder head.

43. Install camshaft bracket No. 1.

44. To install the rear timing chain case cover, first install new O-rings to the cylinder block. Install new O-rings to the cylinder head and camshaft bracket No. 1.

45. Apply liquid gasket sealant to the rear timing chain case back side, as shown in the illustration. Be sure to use genuine RTV sealant, or equivalent.

➡**For "A" in the figure, completely wipe out excessive liquid gasket extended on a portion touching at engine coolant. Apply liquid gasket on the installation position of the water pump and cylinder head very completely.**

46. Align the rear timing case with dowel pins (right and left) on the cylinder block. Install the rear timing chain case. Make sure that the O-rings stay in place during installation to the cylinder block, cylinder head and camshaft bracket No. 1.

47. Tighten the bolts to specification and in the proper sequence.

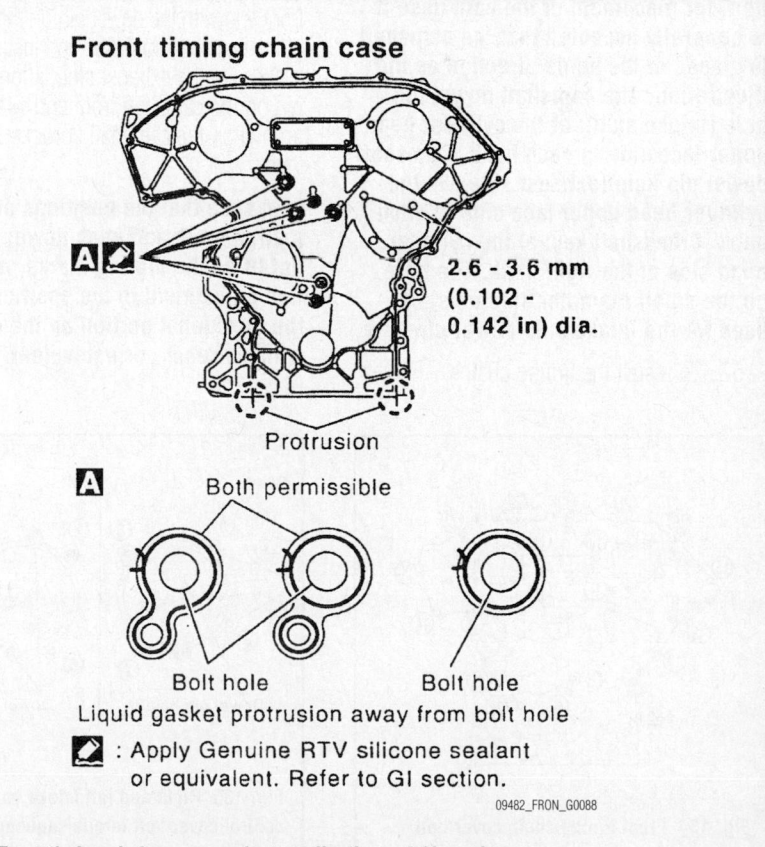

Fig. 133 Front timing chain cover sealant application—4.0L engine

- Bolt length 0.79 inch (20mm) — Bolt position 1, 2, 3, 6, 7, 8, 9, and 10: Tighten to 9 ft. lbs. (13 Nm)
- Bolt length 0.63 inch (16mm) — Bolt position 4, 5, and 11: Tighten to 9 ft. lbs. (13 Nm)
- Bolt length 0.63 inch (16mm) — Bolt position 12–26: Tighten to 11 ft. lbs. (15 Nm)

48. After all bolts are tightened, retighten them to specification and in the proper sequence.

➡ **Be sure to wipe off any excess liquid gasket leaking to the surface for installing the oil pan.**

49. After installing the rear timing case, check the surface height deference between the rear timing chain case and the lower cylinder block. Specification should be -0.0094–0.0055 inch (-0.24–0.14mm). If not within specification, repeat the installation procedure.

50. Install the water pump, using new O-rings.

51. Make sure that the dowel pin hole, dowel pin of the camshaft, and the crankshaft key are located with the number one piston at TDC on the compression stroke.

➡ **Though the camshaft does not stop at the position, as shown in the illustration, for placement of the cam nose it is generally accepted that the camshaft is placed in the same direction as the illustration. The camshaft dowel pin hole (intake side): at the cylinder head upper face side in each bank. Camshaft dowel pin hole (exhaust side): at the cylinder head upper face side in each bank. Crankshaft key: at the cylinder head side of the right bank. The hole on the small diameter side must be used for the intake side dowel pin hole.**

52. To install the timing chains

(secondary) and camshaft sprockets, push the plunger of the timing chain tensioner (secondary) and keep it pressed in with the stopper pin.

✳ WARNING

The mating surfaces between the timing chain and sprockets slip easily. Confirm all mating mark positions repeatedly during the installation process.

53. Install the timing chains (secondary) and camshaft sprockets (INT and EXH).

54. Align the mating marks on the timing chain (secondary) cooper color link, with the ones on the camshaft sprockets (INT and EXH) punched and install them.

➡ **Mating marks for the camshaft sprocket (INT) are on the back side of the camshaft sprocket (secondary). There are 2 types of mating marks, circle and oval. They should be used for the right and the left banks, respectively. Right bank: circle type. Left bank: oval type.**

55. Align the dowel pin and pin hole on the camshafts with the groove and the dowel pin on the sprockets, and install them.

56. On the exhaust side, align the pin hole on the small diameter side of the camshaft front end with the dowel pin on the back side of the camshaft sprocket, and install them.

57. On the exhaust side, align the dowel pin on the camshaft front end with the pin groove on the camshaft sprocket, and install them.

➡ **In case that the positions of each mating mark and each dowel pin will not fit on the mating marks, make a fine adjustment to the position holding the hexagonal portion on the camshaft with a wrench, or equivalent.**

➡ **Bolts for the camshaft sprockets must be tightened. Tightening them by hand is enough to prevent the dislocation of the dowel pins. It may be difficult to visually check the dislocation of mating marks during and after installation. To make the matching easier, make a mating mark on the top of the sprocket teeth and its extended line in advance with paint.**

58. After confirming that the mating marks are aligned, tighten the camshaft sprocket bolts.

59. Pull the stopper pins out from the timing chain tensioners (secondary). Install the tension guide.

60. To install the timing chain (primary), install the crankshaft sprocket. Be sure that the mating marks on the crankshaft sprocket face the front of the engine.

61. Install the timing chain (primary).

➡ **Install the timing chain (primary) so that the mating mark punched (B) on the camshaft sprocket is aligned with the copper link (A) on the timing chain, while the mating mark notched (E) on the crankshaft sprocket (D) is aligned with the yellow link (F) on the timing chain, as shown in the illustration. If it is difficult to align mating marks (A) with (B) and (E) with (F) of the timing chain (primary) with each sprocket, gradually turn the camshaft using a wrench on the hexagonal portion to align it with the timing marks. During alignment be careful to prevent dislocation of the mating marks alignments of the timing chains (secondary). Note (G) indicates the water pump.**

Fig. 200 Timing chain (primary) alignment—4.0L engine

09482_FRON_G0086

62. Install the internal chain guide, slack guide and timing chain tensioner (primary).

➡ **Do not over tighten the slack guide bolts. It is normal for a gap to exist under the bolt seats when the bolts are tightened to specification.**

Fig. 201 Slack guide mounting bolt gap—4.0L engine

09482_FRON_G0087

63. When installing the timing chain tensioner (primary), push in the plunger and keep it pressed in with the stopper pin. Remove any dirt on the surfaces. After installation, pull out the stopper pin by pressing the slack guide.

64. Make sure, again, that the mating marks on the camshaft sprockets and timing chain have not slipped out of alignment.

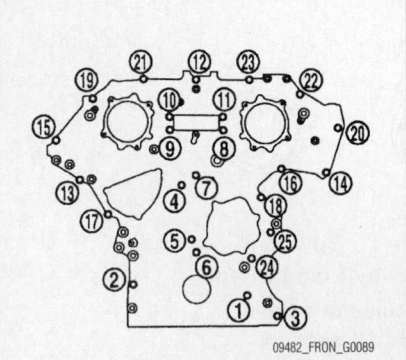

Fig. 134 Front timing chain cover bolt torque sequence—4.0L engine

09482_FRON_G0089

Right | Left

Dowel pin hole | Dowel pin hole

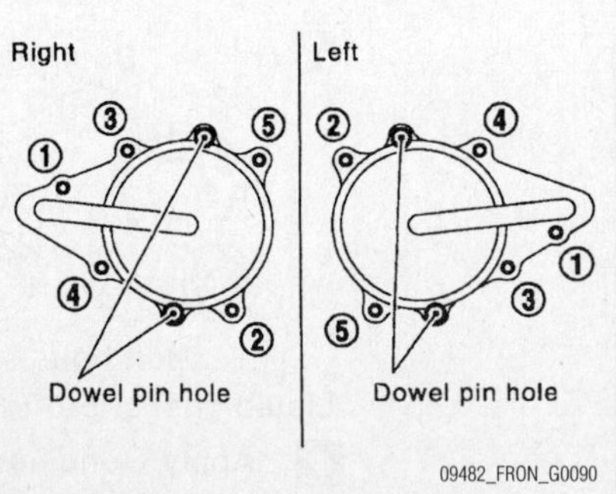

09482_FRON_G0090

Fig. 135 Right and left intake valve timing control cover bolt torque sequence—4.0L engine

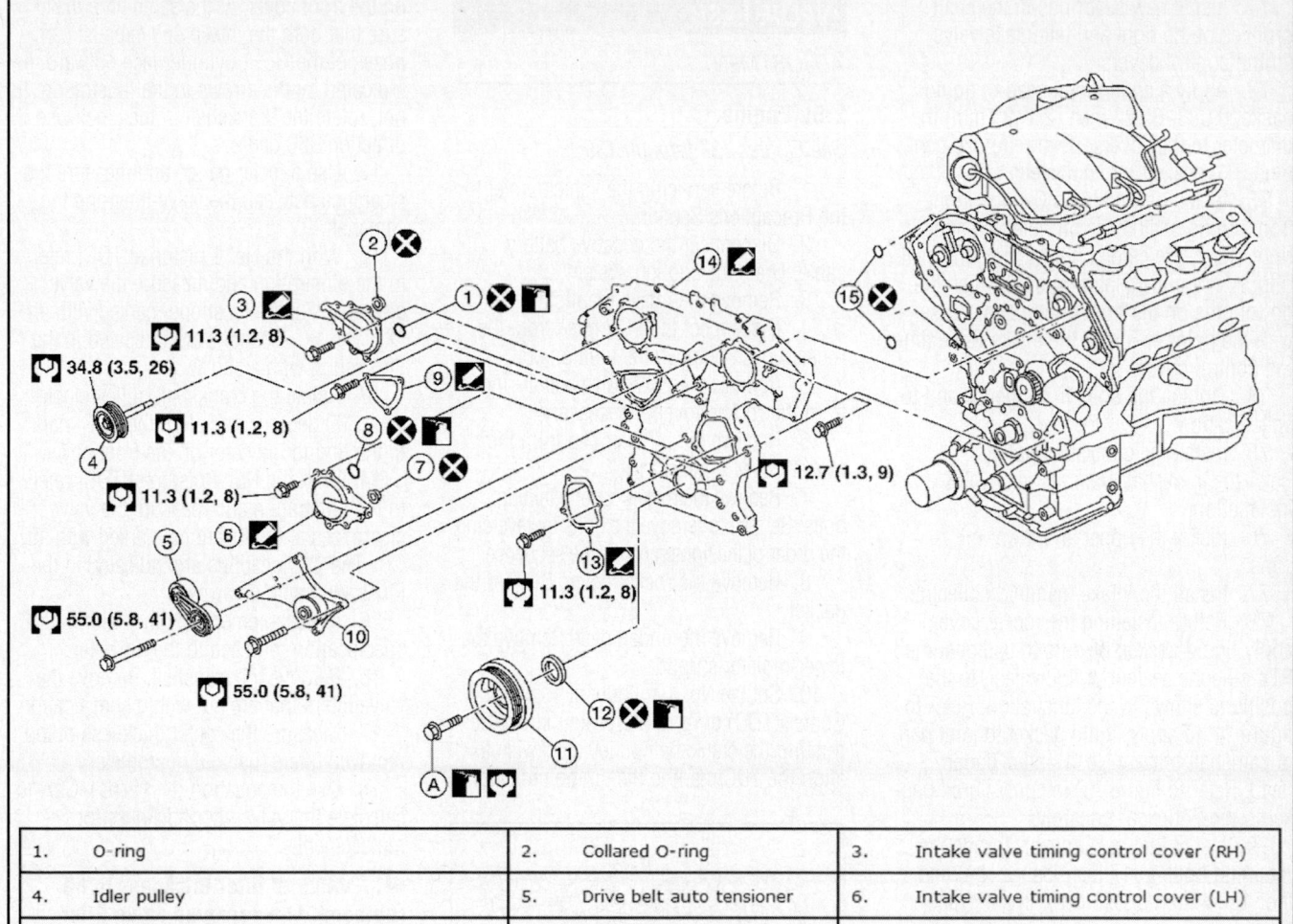

1.	O-ring	2.	Collared O-ring	3.	Intake valve timing control cover (RH)
4.	Idler pulley	5.	Drive belt auto tensioner	6.	Intake valve timing control cover (LH)
7.	Collared O-ring	8.	O-ring	9.	Chain tensioner cover
10.	Cooling fan bracket	11.	Crankshaft pulley	12.	Front oil seal
13.	Water pump cover	14.	Front timing chain case	15.	O-ring
A.	Refer to Crankshaft Damper, removal & instalation				

37671_FOUA_G0267

Fig. 136 Exploded view of front timing chain case—4.0L engine

Install new O-rings on the rear timing chain case.

66. Install a new front seal in the front timing chain case cover.

66. Install the water pump cover and chain tensioner cover to the front timing chain case cover. Apply a continuous bead of liquid gasket 0.091–0.130 inch (2.31–3.30mm) in diameter to the front timing chain case cover before installing the water pump cover and chain tensioner cover. Be sure to use genuine RTV sealant, or equivalent.

67. Before installing the front timing chain case cover apply a continuous bead of liquid gasket 0.102–0.142 inch (2.6–3.6mm) in diameter to the front timing

chain case back side, as shown in the illustration. Be sure to use genuine RTV sealant, or equivalent.

68. Install new O-rings on the rear timing chain case. To assemble the front timing chain case cover, fit the lower end of the front timing chain case tightly onto the top face of the oil pan (upper). From the fitting point, make the entire front timing chain case contact the rear timing chain case completely.

➡**Since the front timing chain case cover is offset for the difference of bolt holes, tighten the bolts temporarily while holding the front timing chain case cover from the front and the top. Then, insert a dowel pin while holding**

the front timing chain case cover from the front and the top.

69. Once the cover is installed, torque the retaining bolts to specification and in the proper sequence. There are 4 different types of bolts:

- Bolt diameter 0.39 inch (10mm) — Bolt position 1–5: Tighten to 41 ft. lbs. (55 Nm)
- Bolt diameter 0.24 inch (6mm) — Bolt position 6–25: Tighten to 9 ft. lbs. (13 Nm)
- After all bolts are tightened, retighten them to specification and in the proper sequence

70. Install the 2 bolts in the oil pan (upper). Torque to 16 ft. lbs. (22 Nm).

71. Install new seal rings in the shaft grooves of the right and left intake valve timing control covers.

72. Apply a continuous bead of liquid gasket 0.083–0.122 inch (2.1–3.1mm) in diameter to the covers. Be sure to use genuine RTV sealant, or equivalent.

73. Install new collared O-rings in the front timing chain case oil hole (left and right sides). Be careful not to move the seal ring from the installation groove, align the dowel pins on the front timing chain case with the holes to install the intake valve timing control covers.

74. Tighten the bolts in sequence and to specification.

75. Install the crankshaft pulley. Refer to Crankshaft Damper, removal & installation.

76. Install the upper and lower oil pans.

77. Install the intake manifold collector.

78. Before installing the rocker cover, apply liquid gasket, be sure to use genuine RTV silicone sealant or equivalent, to the positions shown in the illustration. Refer to figure "a" to apply liquid gasket to joint part of camshaft bracket No. 1 and cylinder head. Refer to figure "b" to apply liquid gasket to the figure "a" squarely.

79. Install the rocker cover. Torque the retaining bolts to 17 inch lbs. (2 Nm) and then to 74 inch lbs. (8 Nm), in the proper sequence.

80. Continue the installation in the reverse order of the removal procedure.

➡ **If hydraulic pressure inside the timing chain tensioner drops after removal/installation, slack in the guide may generate a pounding noise during and just after engine start. This is normal. The noise should stop after the hydraulic pressure rises.**

VALVE LASH

ADJUSTMENT

2.5L Engine

See Figures 137 through 139.

1. Before servicing the vehicle, refer to the Precautions Section.
2. Disconnect the negative battery cable. Drain the cooling system.
3. Remove the intake manifold.
4. Disconnect the PCV hose from the rocker cover. Remove the ignition coil.
5. Remove the PCV valve and O-ring from the rocker cover, if necessary.
6. Remove the oil filler cap from the rocker cover, if necessary.
7. Remove the rocker cover retaining bolts. Be sure to remove the bolts by reversing the order of the tightening torque sequence.
8. Remove the rocker cover. Discard the gasket.
9. Remove the undercover. Remove the lower radiator shroud.
10. Set the No. 1 cylinder at Top Dead Center (TDC) of its compression stroke by rotating the crankshaft pulley clockwise to align the TDC mark to the timing indicator

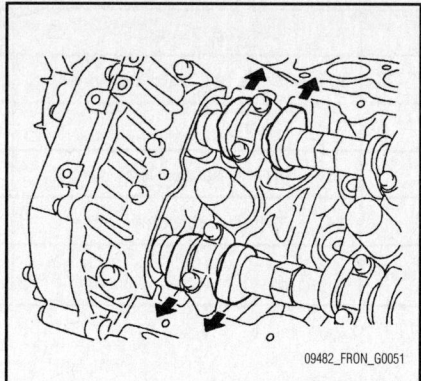

09482_FRON_G0051

Fig. 137 View of No. 1 cylinder at TDC (compression stroke)—2.5L engine

on the front cover. At the same time make sure that both the intake and exhaust cam noses of the No. 1 cylinder face outward, as indicated by the arrows in the illustration. If not, rotate the crankshaft in the clockwise direction 360 degrees.

11. Use a feeler gauge and measure the clearance between the valve lifter and the camshaft.

12. With the No. 1 piston at TDC, refer to the illustration and measure the valve clearances at the locations marked with an "X". The "X" locations are indicated in the illustration with an arrow.

13. Rotate the crankshaft pulley clockwise 360 degrees and align the TDC mark to the timing indicator on the front cover.

14. With the No. 4 piston at TDC, refer to the illustration and measure the valve clearances at the locations marked with an "X". The "X" locations are indicated in the illustration with an arrow.

15. If measurements are not within specification, proceed to the next step.

16. Remove the camshaft. Remove the valve lifters that are not within specification.

17. Measure the center thickness of the removed lifters, using a micrometer.

18. Use the equation $(t = t1+(C1-C2))$ to calculate the valve lifter thickness for replacement.

➡ **t = valve of lifter thickness to be replaced. t1 = removed valve lifter thickness. C1 = measured valve clearance. C2 = standard valve clearance.**

19. The thickness of the new valve lifter can be identified by the stamp mark on the reverse side (inside the cylinder). The stamp mark "696" indicates 0.2740 inch (6.96mm) thickness.

➡ **Available thickness of a valve lifter ranges from 0.2740–0.2937 inch (6.96–7.46mm) in steps of 0.0008 inch (0.02mm). There are 26 different sizes.**

Measuring position		No. 1 CYL.	No. 2 CYL.	No. 3 CYL.	No. 4 CYL.
No. 1 cylinder at compression TDC	INT	×	×		
	EXH	×		×	

Fig. 138 Valve adjustment measurement No. 1 cylinder at TDC (compression stroke)—2.5L engine

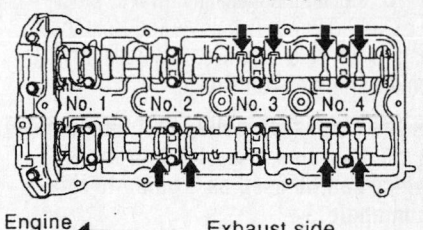

No. 4 cylinder compression TDC

Measuring position		No. 1 CYL.	No. 2 CYL.	No. 3 CYL.	No. 4 CYL.
No. 4 cylinder at compression TDC	INT			×	×
	EXH		×		×

Fig. 139 Valve adjustment measurement No. 4 cylinder at TDC (compression stroke)—2.5L engine

20. Install the selected valve lifters.
21. Install the camshaft.
22. Manually rotate the crankshaft pulley in the clockwise direction a few rotations.
23. Check the valve clearance and be sure it is within specification.
24. When installing the rocker cover, apply liquid gasket, be sure to use genuine RTV silicone sealant, or equivalent.
25. Install the rocker cover. Refer to Valve Cover, removal & installation.
26. Continue the installation in the reverse of the removal procedure.

4.0L Engine

See Figures 140 through 143.

1. Before servicing the vehicle, refer to the Precautions Section.
2. Disconnect the negative battery cable. Remove the engine under cover.

3. Remove the intake manifold collector.
4. Separate the engine harness and remove the brackets from the rocker covers. Remove the harness bracket from the cylinder head, if necessary.
5. Remove the ignition coil. Remove the PCV hoses. Remove the oil filler cap, if necessary.
6. Remove the rocker covers from the engine. Refer to Valve Covers, removal & installation.
7. Set the No. 1 cylinder at Top Dead Center (TDC) of its compression stroke by rotating the crankshaft pulley clockwise to align the timing mark (grooved line without color) with the timing indicator. Make sure that the intake and exhaust cam noses on No. 1 cylinder (engine front side on right bank) are in alignment as shown in the illustration. If not, rotate the crankshaft in the clockwise direction 360 degrees.

8. Use a feeler gauge and measure the clearance between the valve lifter and the camshaft.
9. With the No. 1 piston at TDC, refer to the illustration and measure the valve

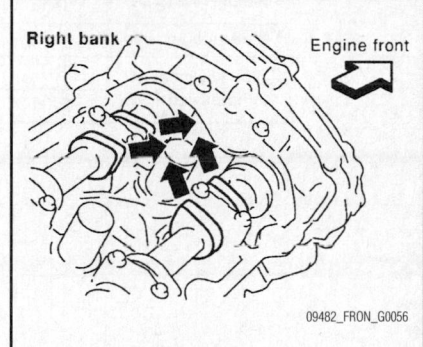

Fig. 140 No. 1 cylinder at TDC (compression stroke)—4.0L engine

Right bank

Measuring position (right bank)		No. 1 CYL.	No. 3 CYL.	No. 5 CYL.
No. 1 cylinder at compression TDC	EXH		×	
	INT	×		
Measuring position (left bank)		No. 2 CYL.	No. 4 CYL.	No. 6 CYL.
No. 1 cylinder at compression TDC	INT			×
	EXH	×		

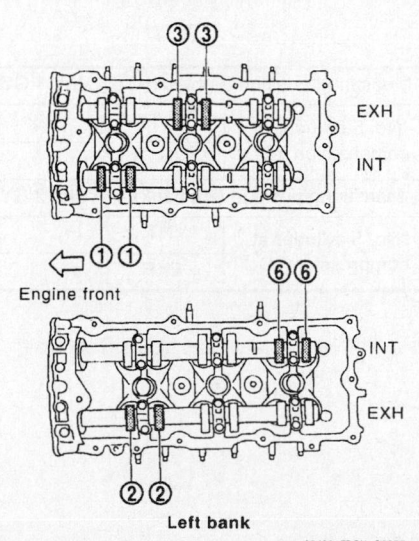

Fig. 141 Valve adjustment measurement No. 1 cylinder at TDC (compression stroke)—4.0L engine

clearances at the locations marked with an "X". The "X" locations are indicated in the illustration with an arrow

10. Rotate the crankshaft pulley clockwise 240 degrees (when viewed from the engine front) to align No. 3 cylinder at TDC on the compression stroke.

➡**The crankshaft pulley bolt flange has a stamped line every 60 degrees, which can be used as a guide to rotation angle.**

11. With the No. 3 piston at TDC, refer to the illustration and measure the valve

clearances at the locations marked with an "X". The "X" locations are indicated in the illustration with an arrow.

12. Rotate the crankshaft pulley clockwise 240 degrees (when viewed from the engine front) to align No. 5 cylinder at TDC on the compression stroke.

➡**The crankshaft pulley bolt flange has a stamped line every 60 degrees, which can be used as a guide to rotation angle.**

13. With the No. 5 piston at TDC, refer to the illustration and measure the valve

clearances at the locations marked with an "X". The "X" locations are indicated in the illustration with an arrow.

14. If measurements are not within specification, proceed to the next step.

15. Remove the camshaft. Remove the valve lifters that are not within specification.

16. Measure the center thickness of the removed lifters, using a micrometer.

17. Use the equation ($t = t1+(C1-C2)$) to calculate the valve lifter thickness for replacement.

➡**t = valve lifter thickness to be replaced. t1 = removed valve lifter**

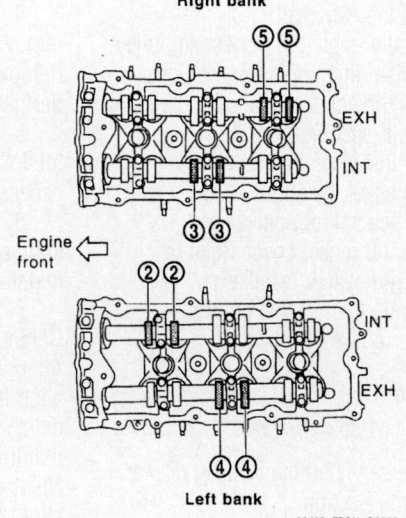

Measuring position (right bank)		No. 1 CYL.	No. 3 CYL.	No. 5 CYL.
No. 3 cylinder at compression TDC	EXH			×
	INT		×	
Measuring position (left bank)		No. 2 CYL.	No. 4 CYL.	No. 6 CYL.
No. 3 cylinder at compression TDC	INT	×		
	EXH		×	

09482_FRON_G0058

Fig. 142 Valve adjustment measurement No. 3 cylinder at TDC (compression stroke)—4.0L engine

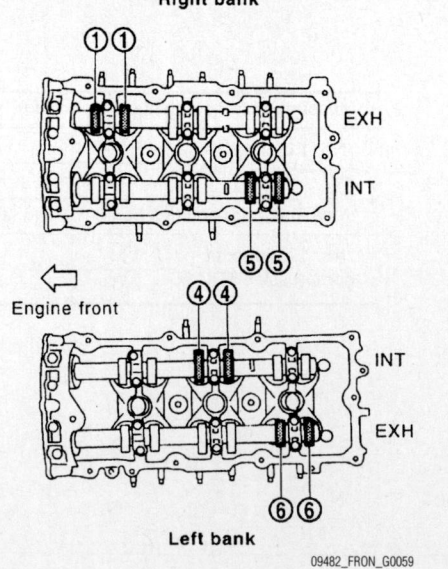

Measuring position (right bank)		No. 1 CYL.	No. 3 CYL.	No. 5 CYL.
No. 5 cylinder at compression TDC	EXH	×		
	INT			×
Measuring position (left bank)		No. 2 CYL.	No. 4 CYL.	No. 6 CYL.
No. 5 cylinder at compression TDC	INT		×	
	EXH			×

09482_FRON_G0059

Fig. 143 Valve adjustment measurement No. 5 cylinder at TDC (compression stroke)—4.0L engine

thickness. **C1 = measured valve clearance. C2 = standard valve clearance.**

18. The intake valve lifter thickness of the new valve lifter can be identified by the stamp mark on the reverse side (inside the cylinder). The stamp mark "788U" indicates 0.3102 inch (7.88mm) thickness.

➡**Available thickness of a valve lifter ranges from 0.3102–0.3307 inch (7.88–8.40mm) in steps of 0.0008 inch (0.02mm). There are 27 different sizes.**

19. Exhaust valve lifter thickness of the new valve lifter can be identified by the stamp mark on the reverse side (inside the cylinder). The stamp mark "N788" indicates 0.3102 inch (7.88mm) thickness.

➡**Available thickness of a valve lifter ranges from 0.3102–0.3291 inch (7.88–8.36mm) in steps of 0.0008 inch (0.02mm). There are 25 different sizes.**

20. Install the selected valve lifters.
21. Install the camshaft.

22. Manually rotate the crankshaft pulley in the clockwise direction a few rotations.
23. Check the valve clearance and be sure it is within specification.
24. When installing the rocker cover, apply liquid gasket, be sure to use genuine RTV silicone sealant, or equivalent.
25. Install the rocker cover. Refer to Valve Covers, removal & installation.
26. Continue the installation in the reverse of the removal procedure.

ENGINE PERFORMANCE & EMISSION CONTROLS

CAMSHAFT POSITION (CMP) SENSOR

LOCATION

2.5L Engine

See Figure 144.

Refer to the accompanying illustration.

4.0L Engine

See Figure 145.

Refer to the accompanying illustration.

REMOVAL & INSTALLATION

2.5L Engine

1. Before servicing the vehicle, refer to the Precautions Section.
2. Loosen the fixing bolt of the sensor.
3. Disconnect the CMP sensor (PHASE) harness connector.
4. Remove the CMP sensor.

To install:

5. Installation is the reverse of the removal procedure.
6. Visually check the sensor for chipping.
7. Tighten the CMP sensor (PHASE) fixing bolt to 62 inch lbs. (7 Nm).

4.0L Engine

1. Before servicing the vehicle, refer to the Precautions Section.
2. Loosen the fixing bolt of the sensor.
3. Disconnect the CMP sensor (PHASE) harness connector.
4. Remove the CMP sensor.

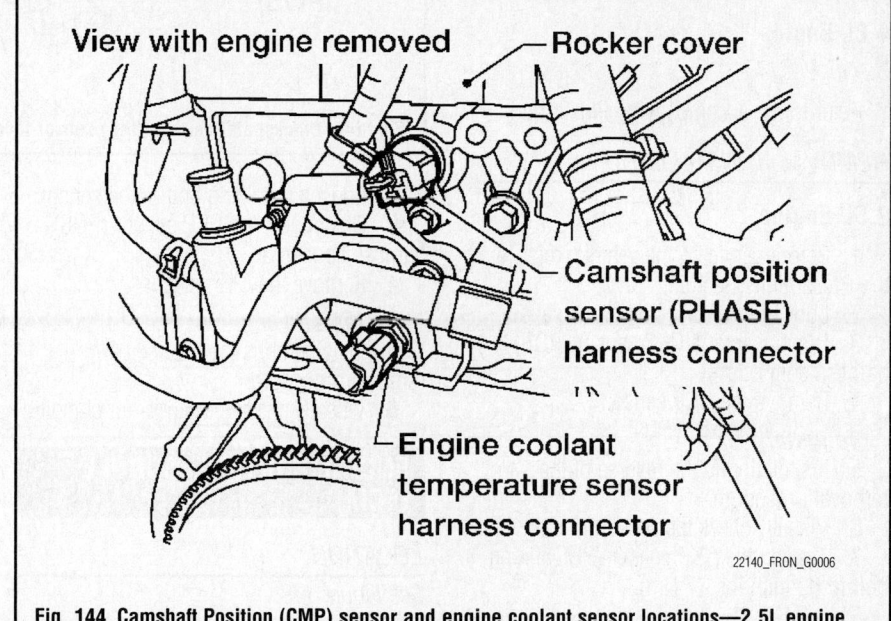

Fig. 144 Camshaft Position (CMP) sensor and engine coolant sensor locations—2.5L engine

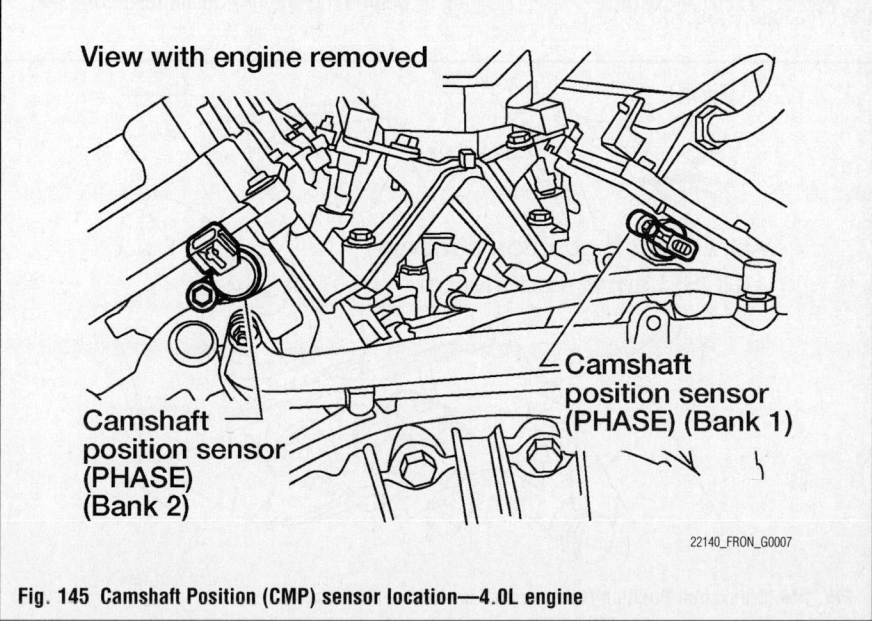

Fig. 145 Camshaft Position (CMP) sensor location—4.0L engine

To install:

5. Installation is the reverse of the removal procedure.

6. Visually check the sensor for chipping.

7. Tighten the CMP sensor (PHASE) fixing bolt to 86 inch lbs. (10 Nm).

CRANKSHAFT POSITION (CKP) SENSOR

LOCATION

2.5L Engine

See Figure 146.

Refer to the accompanying illustration.

4.0L Engine

See Figure 147.

Refer to the accompanying illustration.

REMOVAL & INSTALLATION

2.5L Engine

1. Before servicing the vehicle, refer to the Precautions Section.

2. Loosen the fixing bolt of the sensor.

3. Disconnect the CKP sensor (POS) harness connector.

4. Remove the CKP sensor.

To install:

5. Installation is the reverse of the removal procedure.

6. Visually check the sensor for chipping.

7. Tighten the CKP sensor (POS) fixing bolt to 62 inch lbs. (7 Nm).

4.0L Engine

1. Before servicing the vehicle, refer to the Precautions Section.

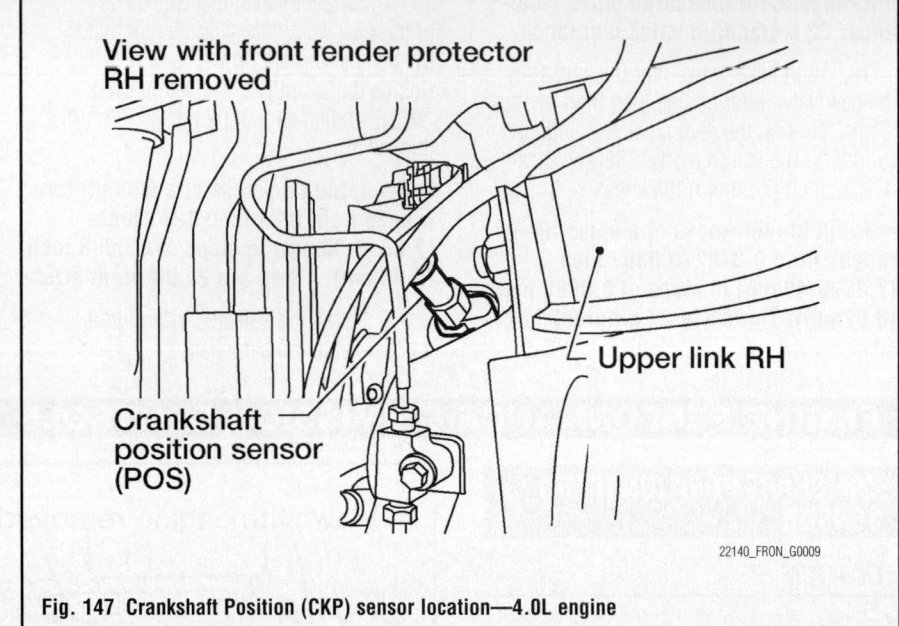

Fig. 147 Crankshaft Position (CKP) sensor location—4.0L engine

22140_FRON_G0009

2. Loosen the fixing bolt of the sensor.

3. Disconnect the CKP sensor (POS) harness connector.

4. Remove the CKP sensor.

To install:

5. Installation is the reverse of the removal procedure.

6. Visually check the sensor for chipping.

ENGINE CONTROL MODULE (ECM)

LOCATION

See Figure 148.

The Engine Control Module (ECM) is located in the engine room passenger side behind the engine coolant reservoir tank.

RESET

When replacing the ECM, the following procedures must be performed in the order listed.

- Initialize The Immobilizer System
- VIN Registration
- Accelerator Pedal Released Position Learning
- Throttle Valve Closed Position Learning
- Idle Air Volume Learning

Accelerator Pedal Released Position Learning

Accelerator Pedal Released Position Learning is an operation to learn the fully released position of the accelerator pedal by monitoring the accelerator pedal position sensor output signal. It must be performed each time the harness connector of the accelerator pedal position sensor or ECM is disconnected.

1. Before servicing the vehicle, refer to the Precautions Section.

2. Check that the accelerator pedal is fully released.

3. Turn the ignition switch ON and wait at least 2 seconds.

4. Turn the ignition switch OFF and wait at least 10 seconds.

5. Turn the ignition switch ON and wait at least 2 seconds.

6. Turn the ignition switch OFF and wait at least 10 seconds.

Idle Air Volume Learning

Idle Air Volume Learning is an operation to learn the idle air volume that keeps each engine within the specific range. It must be performed under the following conditions:

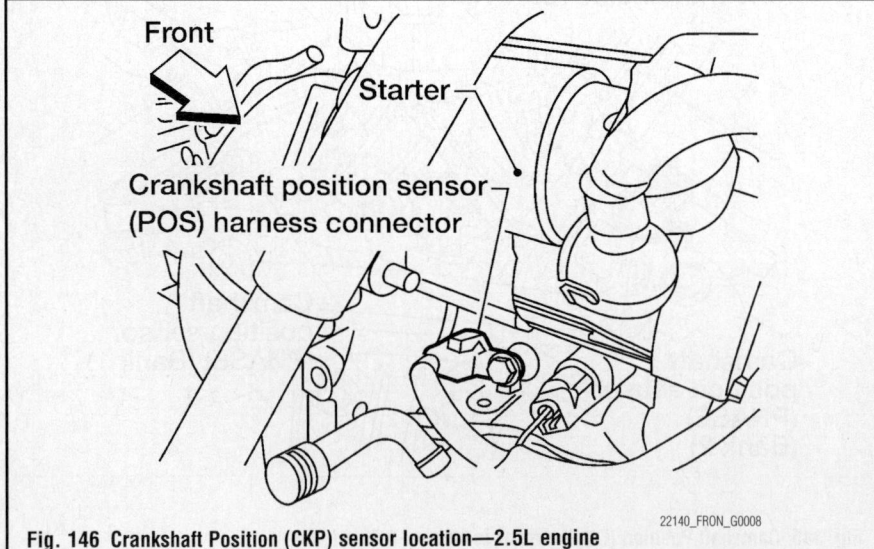

Fig. 146 Crankshaft Position (CKP) sensor location—2.5L engine

22140_FRON_G0008

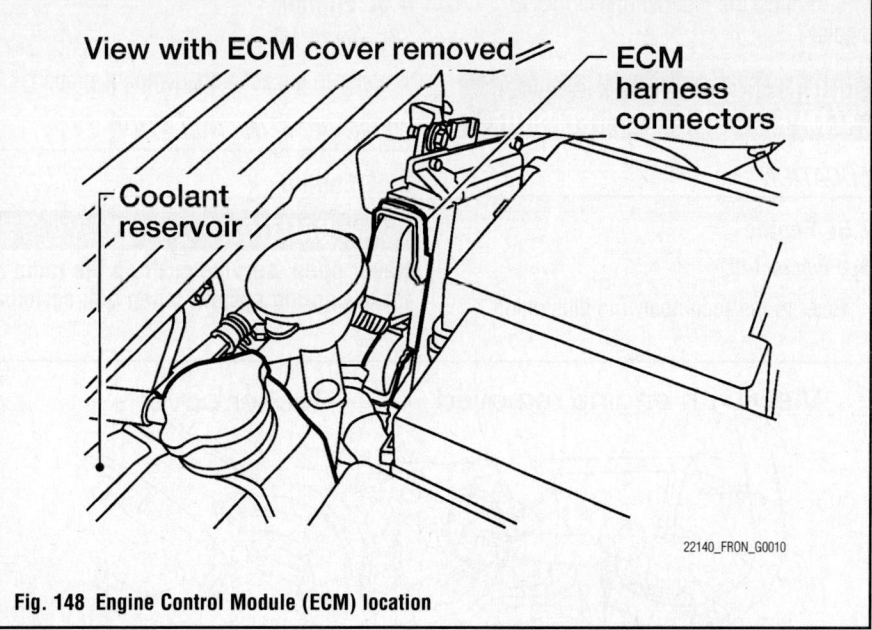

View with ECM cover removed

ECM harness connectors

Coolant reservoir

22140_FRON_G0010

Fig. 148 Engine Control Module (ECM) location

- Each time the electric throttle control actuator or ECM is replaced
- The idle speed or ignition timing is out of specification

1. Before servicing the vehicle, refer to the Precautions Section.

Preparation

2. Before performing the Idle Air Volume Learning procedure, check that all of the following conditions are satisfied. Learning will be cancelled if any of the following conditions are missed for even a moment:

- Battery voltage: More than 12.9 V (at idle)
- Engine coolant temperature: 158–212°F (70–100°C)
- PNP switch (M/T): ON
- Selector lever (A/T): P or N
- Electric load switch: OFF (air conditioner, headlamp, rear window defogger). On vehicles equipped with daytime light systems, if the parking brake is applied before the engine is started, the headlamp will not illuminate
- Steering wheel: Neutral (straight-ahead position)
- Vehicle speed: Stopped
- Transmission: Warmed-up

➡ **With SDT: Drive the vehicle until "ATF TEMP SE" in "Data List" mode of the "A/T" system indicates less than 0.9 V. Without SDT: Drive the vehicle for 10 minutes.**

Operations Procedure—with SDT

3. Perform the Accelerator Pedal Released Position Learning procedure.

4. Perform the Throttle Valve Closed Position Learning procedure.

5. Start the engine and warm it up to a normal operating temperature.

6. Check that all items listed under the topic PREPARATION are in good order.

7. Select "IDLE AIR VOL LEARN" in "Utility" mode.

8. Touch "START" and wait 20 seconds.

9. Check that "CMPLT" is displayed on the Smart Diagnostic Tool (SDT) screen. If "CMPLT" is not displayed, Idle Air Volume Learning will not be carried out successfully. In this case, find the cause of the incident by referring to the Diagnostic Procedure.

10. Rev up the engine 2–3 times and check that the idle speed and the ignition timing are within specifications.

   a. Idle speed M/T: 625 plus or minus 50 RPM (in Neutral position).
   b. Idle speed A/T: 625 plus or minus 50 RPM (in P or N position).
   c. Ignition timing M/T: 15 plus or minus 5° BTDC (in Neutral position).
   d. Ignition timing A/T: 15 plus or minus 5° BTDC (in P or N position).

Operations Procedure—without SDT

➡ **It is better to count the time accurately with a clock. It is impossible to switch the diagnostic mode when an accelerator pedal position sensor circuit has a malfunction.**

11. Perform the Accelerator Pedal Released Position Learning procedure.

12. Perform the Throttle Valve Closed Position Learning procedure.

13. Start the engine and warm it up to a normal operating temperature.

14. Check that all items listed under the topic PREPARATION are in good order.

15. Turn ignition switch OFF and wait at least 10 seconds.

16. Confirm that accelerator pedal is fully released, then turn the ignition switch ON and wait 3 seconds.

17. Repeat the following procedure quickly 5 times within 5 seconds.
   a. Fully depress the accelerator pedal.
   b. Fully release the accelerator pedal.

18. Wait 7 seconds, fully depress the accelerator pedal for approx. 20 seconds until the MIL stops blinking and turns ON.

19. Fully release the accelerator pedal within 3 seconds after the MIL turns ON.

20. Start the engine and let it idle.

21. Wait 20 seconds.

22. Rev up the engine 2–3 times and check that idle speed and ignition timing are within specifications.

   a. Idle speed M/T: 625 plus or minus 50 RPM (in Neutral position).
   b. Idle speed A/T: 625 plus or minus 50 RPM (in P or N position).
   c. Ignition timing M/T: 15 plus or minus 5° BTDC (in Neutral position).
   d. Ignition timing A/T: 15 plus or minus 5° BTDC (in P or N position).

23. If the idle speed and the ignition timing are not within specification, Idle Air Volume Learning will not be carried out successfully. In this case, find the cause of the incident by referring to the DIAGNOSTIC PROCEDURE.

Diagnostic Procedure

If the Idle Air Volume Learning cannot be performed successfully, proceed as follows:

24. Check that the throttle valve is fully closed.

25. Check the PCV valve operation.

26. Check that the downstream of throttle valve is free from air leakage.

27. If the above 3 items check out OK, engine component parts and their installation condition are questionable. Check and eliminate the cause of the incident.

28. If the engine stalls or has an

incorrect idle after the engine has started, eliminate the cause of the incident and perform the Idle Air Volume Learning again.

Initialize the Immobilizer System

1. Before servicing the vehicle, refer to the Precautions Section.
2. Perform an initialization of the immobilizer system and registration of all immobilizer system ignition key IDs.
3. Install the ECM into the vehicle.
4. Using a registered key, turn the ignition switch to ON.

➡**Use the key that has been used before performing the ECM replacement.**

5. Maintain the ignition switch in the ON position for at least 5 seconds.
6. Turn the ignition switch to OFF.
7. Start the engine.
   • If the engine starts, the procedure is completed.
   • If the engine fails to start, initialize the control unit using the Smart Diagnostic Tool (SDT) and follow the operation manual.

Throttle Valve Closed Position Learning

Throttle Valve Closed Position Learning is an operation to learn the fully closed position of the throttle valve by monitoring the throttle position sensor output signal. It must be performed each time the harness connector of electric throttle control actuator or ECM is disconnected.

1. Before servicing the vehicle, refer to the Precautions Section.
2. Check that the accelerator pedal is fully released.
3. Turn the ignition switch ON.
4. Turn the ignition switch OFF and wait at least 10 seconds.
5. Check that throttle valve moves during the above 10 seconds by confirming the operating sound.

VIN Registration

VIN Registration is an operation registering the VIN in the ECM. It must be performed each time the ECM is replaced.

1. Perform the VIN Registration procedure.
2. Before servicing the vehicle, refer to the Precautions Section.
3. Check the VIN of the vehicle and note it.
4. Turn ignition switch ON and engine stopped.
5. Select "VIN REGISTRATION" on the Smart Diagnostic Tool (SDT) in "Utility" mode.

6. Follow the instructions of the SDT display.

ENGINE COOLANT TEMPERATURE (ECT) SENSOR

LOCATION

2.5L Engine

See Figure 149.

Refer to the accompanying illustration.

4.0L Engine

See Figure 150.

Refer to the accompanying illustration.

REMOVAL & INSTALLATION

2.5L Engine

✷✷ CAUTION

Never open, service or drain the radiator or cooling system when hot; serious

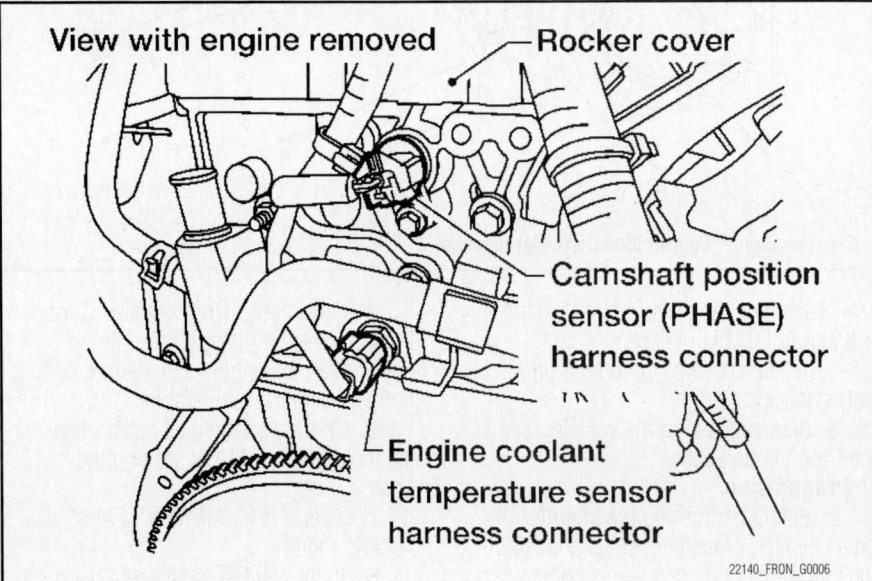

Fig. 149 Camshaft Position (CMP) sensor and Engine Coolant Temperature (ECT) sensor locations—2.5L engine

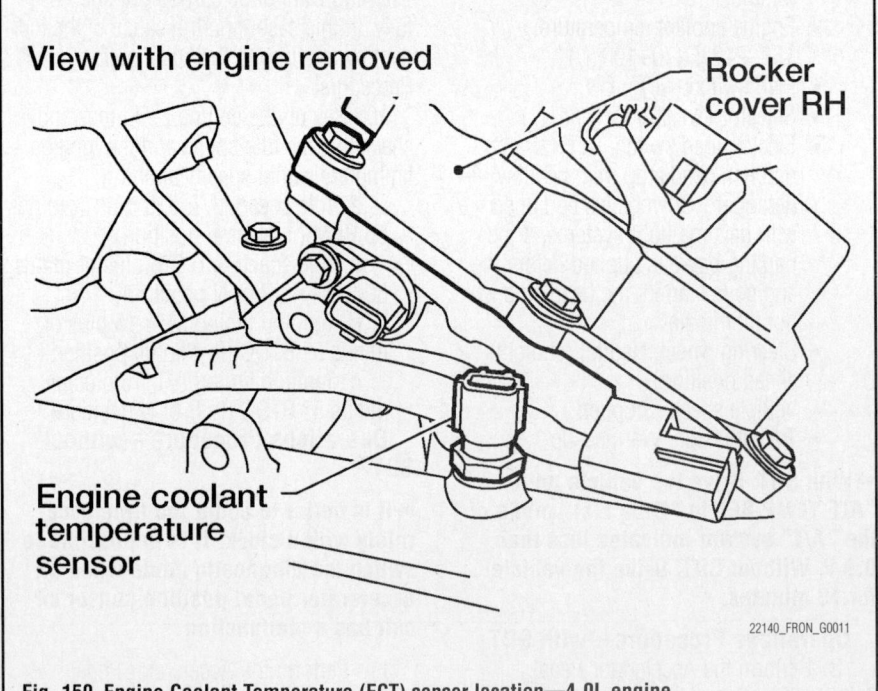

Fig. 150 Engine Coolant Temperature (ECT) sensor location—4.0L engine

burns can occur from the steam and hot coolant. Also, when draining engine coolant, keep in mind that cats and dogs are attracted to ethylene glycol antifreeze and could drink any that is left in an uncovered container or in puddles on the ground. This will prove fatal in sufficient quantities. Always drain coolant into a sealable container. Coolant should be reused unless it is contaminated or is several years old.

1. Before servicing the vehicle, refer to the Precautions Section.
2. Loosen the fixing bolt of the Engine Coolant Temperature (ECT) sensor.
3. Disconnect the ECT sensor harness connector.
4. Remove the ECT sensor.

To install:
5. Installation is the reverse of the removal procedure.
6. Tighten the ECT sensor fixing bolt to 18 ft. lbs. (25 Nm).
7. Fill the cooling system and check for leaks.

4.0L Engine

✳✳ CAUTION

Never open, service or drain the radiator or cooling system when hot; serious burns can occur from the steam and hot coolant. Also, when draining engine coolant, keep in mind that cats and dogs are attracted to ethylene glycol antifreeze and could drink any that is left in an uncovered container or in puddles on the ground. This will prove fatal in sufficient quantities. Always drain coolant into a sealable container. Coolant should be reused unless it is contaminated or is several years old.

1. Before servicing the vehicle, refer to the Precautions Section.
2. Loosen the fixing bolt of the Engine Coolant Temperature (ECT) sensor.
3. Disconnect the ECT sensor harness connector.
4. Remove the ECT sensor.

To install:
5. Installation is the reverse of the removal procedure.
6. Tighten the ECT sensor fixing bolt to 18 ft. lbs. (25 Nm).
7. Fill the cooling system and check for leaks.

HEATED OXYGEN (HO2S) SENSOR

LOCATION

2.5L Engine

See Figures 151 and 152.

Refer to the accompanying illustrations.

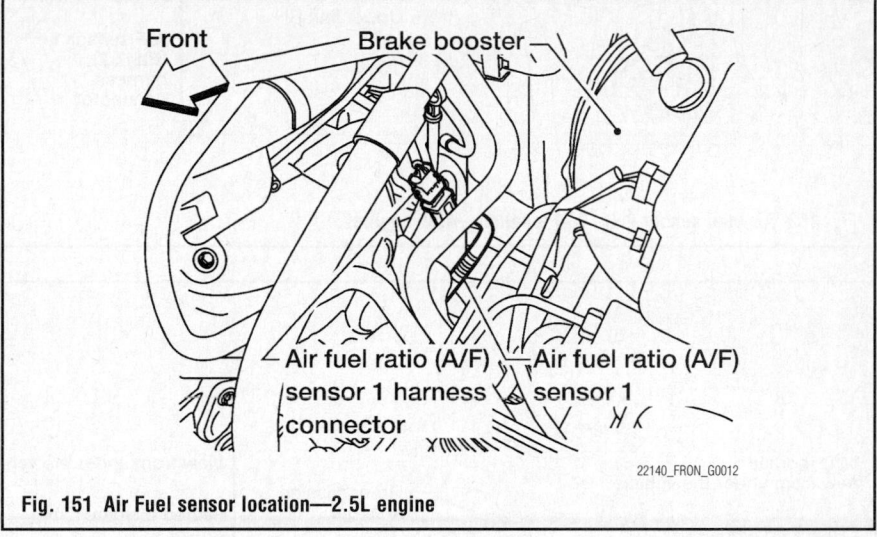

Fig. 151 Air Fuel sensor location—2.5L engine

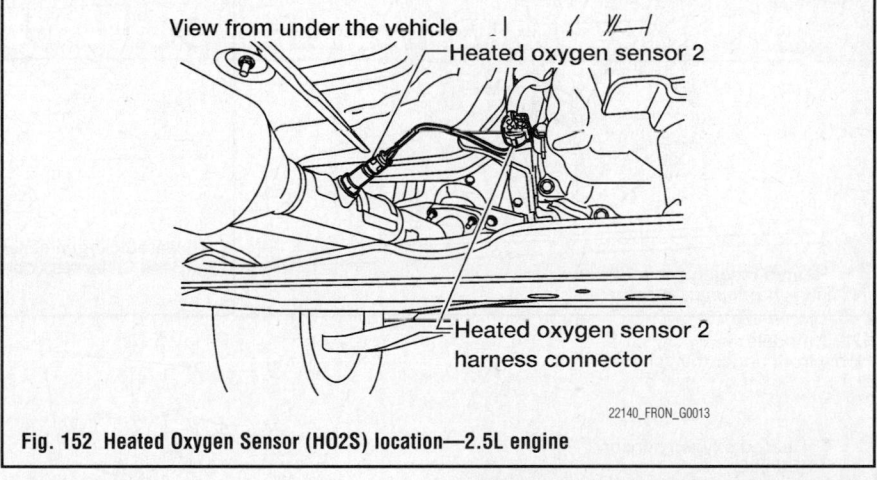

Fig. 152 Heated Oxygen Sensor (HO2S) location—2.5L engine

4.0L Engine

See Figures 153 and 154.

Refer to the accompanying illustrations.

REMOVAL & INSTALLATION

2.5L Engine

✳✳ CAUTION

Perform the operation with the exhaust system fully cooled. The system will be hot just after the engine stops.

1. Before servicing the vehicle, refer to the Precautions Section.
2. Disconnect the sensor harness connector.
3. Remove the sensor using the heated oxygen sensor wrench KV10114400 (J-38365), or equivalent.

✳✳ WARNING

Be careful not to damage the sensor. Discard any sensor which has been dropped from a height of more than 19.7 inches (0.5m) onto a hard surface such as a concrete floor. Replace with a new one.

To install:

✳✳ WARNING

Do not over-tighten the sensor. Doing so may cause damage to the sensor, resulting in the MIL coming on.

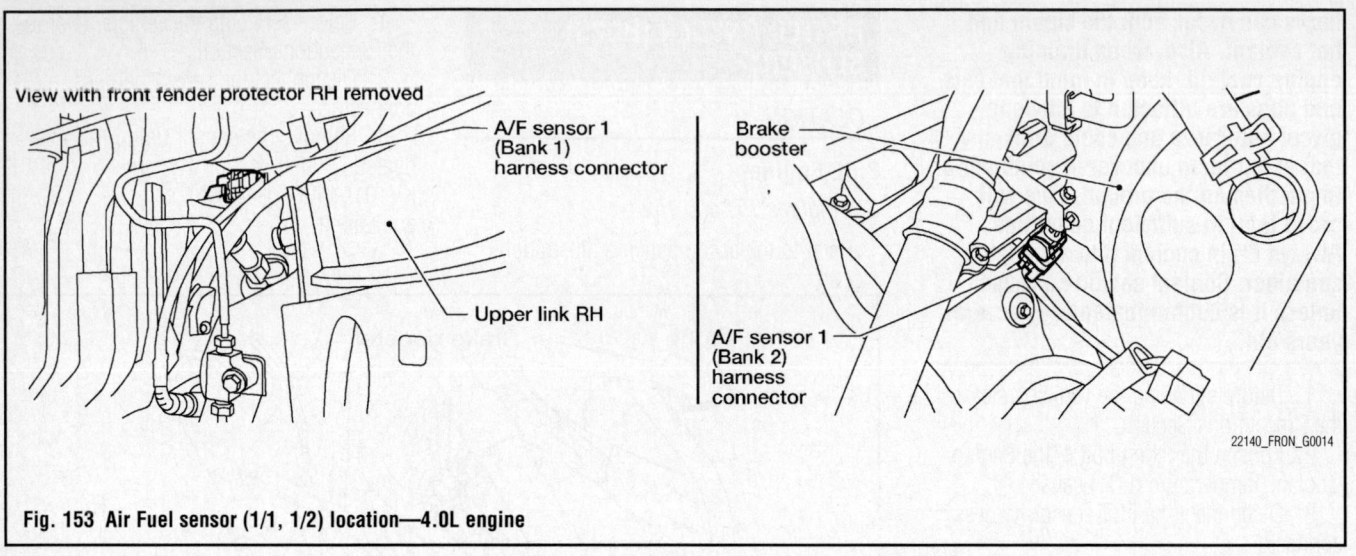

View with front fender protector RH removed

A/F sensor 1 (Bank 1) harness connector

Upper link RH

Brake booster

A/F sensor 1 (Bank 2) harness connector

22140_FRON_G0014

Fig. 153 Air Fuel sensor (1/1, 1/2) location—4.0L engine

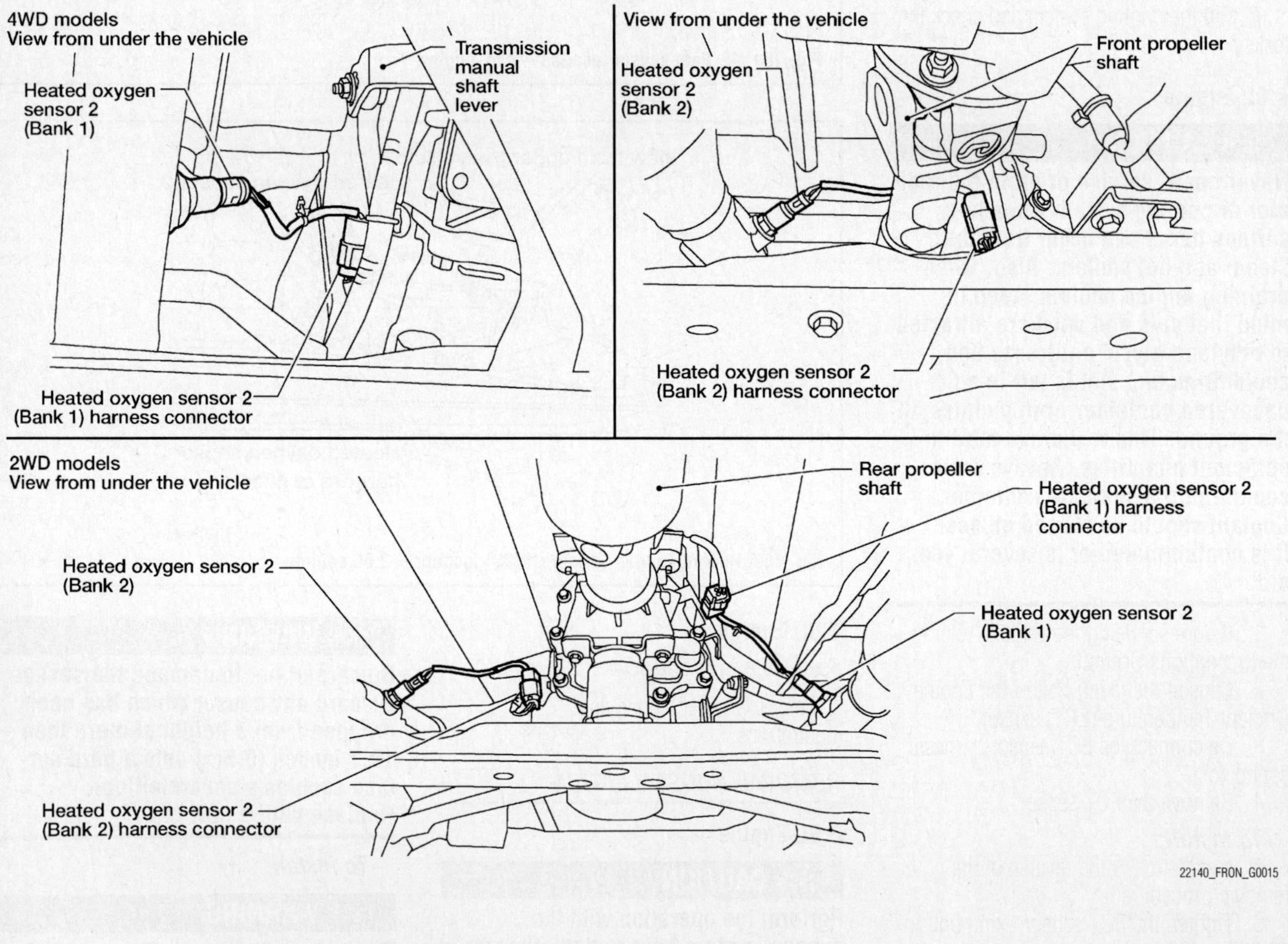

4WD models
View from under the vehicle

Heated oxygen sensor 2 (Bank 1)

Transmission manual shaft lever

Heated oxygen sensor 2 (Bank 1) harness connector

View from under the vehicle

Heated oxygen sensor 2 (Bank 2)

Front propeller shaft

Heated oxygen sensor 2 (Bank 2) harness connector

2WD models
View from under the vehicle

Rear propeller shaft

Heated oxygen sensor 2 (Bank 1) harness connector

Heated oxygen sensor 2 (Bank 2)

Heated oxygen sensor 2 (Bank 1)

Heated oxygen sensor 2 (Bank 2) harness connector

22140_FRON_G0015

Fig. 154 Heated Oxygen Sensor (HO2S) (2/1, 2/2) location—4.0L engine

4. Before installing a new sensor, clean the exhaust system threads using a suitable tool and apply antiseize lubricant. Use oxygen sensor thread cleaner J-43897-12 or J-43897-18, or equivalent.

5. Install the sensor using special tool KV10114400 (J-38365), or equivalent. Tighten to 37 ft. lbs. (50 Nm).

4.0L Engine

※※ CAUTION

Perform the operation with the exhaust system fully cooled. The system will be hot just after the engine stops.

1. Before servicing the vehicle, refer to the Precautions Section.
2. Disconnect the sensor harness connector.
3. Remove the sensor using the heated oxygen sensor wrench KV10114400 (J-38365), or equivalent.

※※ WARNING

Be careful not to damage the sensor. Discard any sensor which has been dropped from a height of more than 19.7 inches (0.5m) onto a hard surface such as a concrete floor. Replace with a new one.

To install:

※※ WARNING

Do not over-tighten the sensor. Doing so may cause damage to the sensor, resulting in the MIL coming on.

4. Before installing a new sensor, clean the exhaust system threads using a suitable tool and apply antiseize lubricant. Use oxygen sensor thread cleaner J-43897-12 or J-43897-18, or equivalent.

5. Install the sensor using special tool KV10114400 (J-38365), or equivalent. Tighten to 37 ft. lbs. (50 Nm).

INTAKE AIR TEMPERATURE (IAT) SENSOR

LOCATION

2.5L Engine

See Figure 155.

The Intake Air Temperature (IAT) sensor is built into the Mass Air Flow (MAF) sensor.

4.0L Engine

See Figure 156.

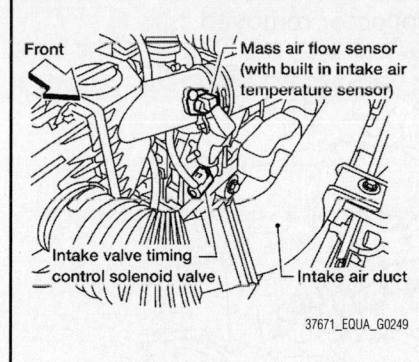

Fig. 155 Mass Air Flow (MAF)/Idle Air Control (IAC) valve location—2.5L engine

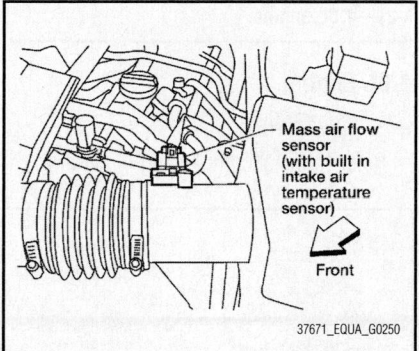

Fig. 156 Mass Air Flow (MAF)/Idle Air Control (IAC) valve location—4.0L engine

The Intake Air Temperature (IAT) sensor is built into the Mass Air Flow (MAF) sensor.

REMOVAL & INSTALLATION

To service the Intake Air Temperature (IAT) sensor, refer to Mass Air Flow (MAF) Sensor (Hot Wire), removal & installation.

KNOCK SENSOR (KS)

LOCATION

2.5L Engine

See Figure 157.

Refer to the accompanying illustration.

4.0L Engine

See Figure 158.

Refer to the accompanying illustration.

REMOVAL & INSTALLATION

2.5L Engine

1. Before servicing the vehicle, refer to the Precautions Section.
2. Remove the sensor harness. Remove the knock sensor.

To install:

3. Installation is the reverse of the removal procedure.

※※ WARNING

Use care when installing the sensor. Do not use any knock sensors that have been dropped or physically damaged. Replace these with new ones.

4. Tighten the knock sensor bolt to 16 ft. lbs. (21 Nm).

4.0L Engine

1. Before servicing the vehicle, refer to the Precautions Section.
2. Remove the intake collector to gain access to the knock sensors.
3. Remove the sensor harness. Remove the sensor.

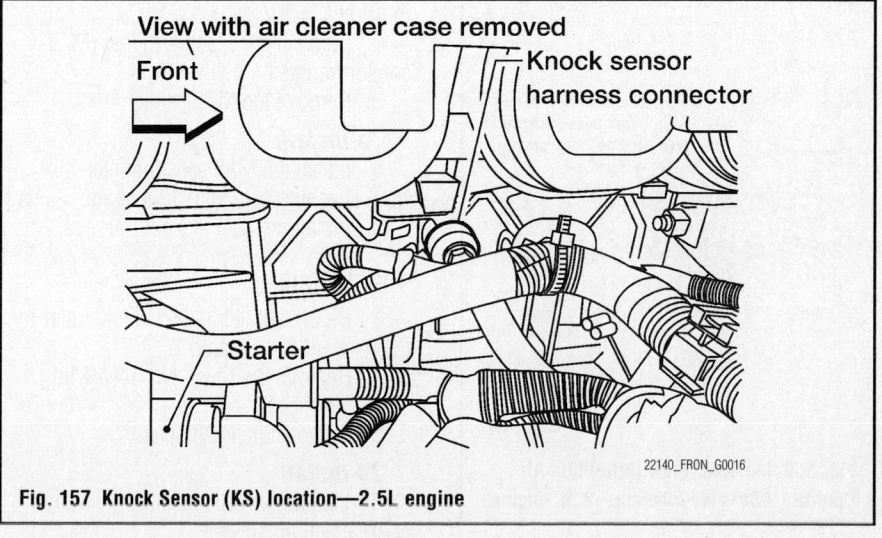

Fig. 157 Knock Sensor (KS) location—2.5L engine

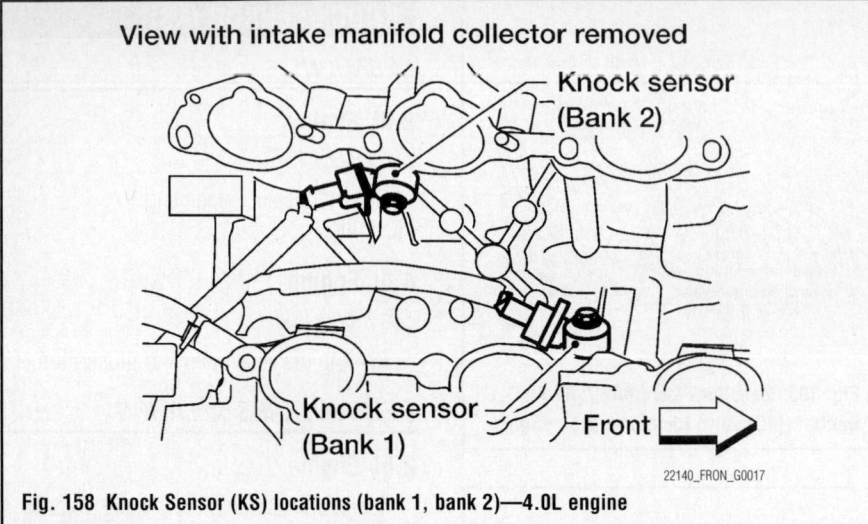

Fig. 158 Knock Sensor (KS) locations (bank 1, bank 2)—4.0L engine

To install:

4. Installation is the reverse of the removal procedure.

5. Tighten the knock sensor bolt to 13 ft. lbs. (18 Nm).

MASS AIR FLOW (MAF) SENSOR

LOCATION

2.5L Engine

See Figure 159.

Refer to the accompanying illustration.

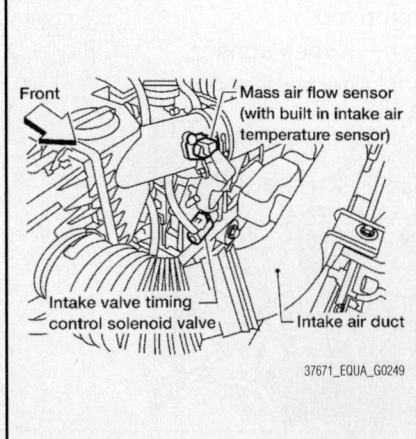

Fig. 159 Mass Air Flow (MAF)/Idle Air Control (IAC) valve location—2.5L engine

4.0L Engine

See Figure 160.

Refer to the accompanying illustration.

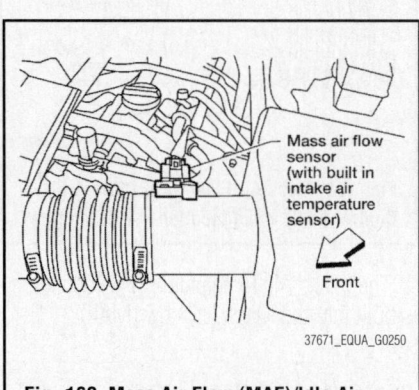

Fig. 160 Mass Air Flow (MAF)/Idle Air Control (IAC) valve location—4.0L engine

REMOVAL & INSTALLATION

2.5L Engine

1. Before servicing the vehicle, refer to the Precautions Section.
2. Remove the Mass Air Flow (MAF) sensor harness.
3. Remove the MAF sensor.

To install:

4. Install the MAF sensor tighten the MAF attaching screws to 13 inch lbs. (2 Nm).
5. Install the MAF sensor harness.

4.0L Engine

1. Before servicing the vehicle, refer to the Precautions Section.
2. Remove the Mass Air Flow (MAF) sensor harness.
3. Remove the MAF sensor.

To install:

4. Install the MAF sensor tighten the MAF attaching screws.
5. Install the MAF sensor harness.

THROTTLE POSITION SENSOR (TPS)

LOCATION

2.5L Engine

See Figure 161.

The electric throttle control actuator includes a throttle control motor and Throttle Position Sensor (TPS). The TPS responds to the throttle valve movement. The TPS has two sensors.

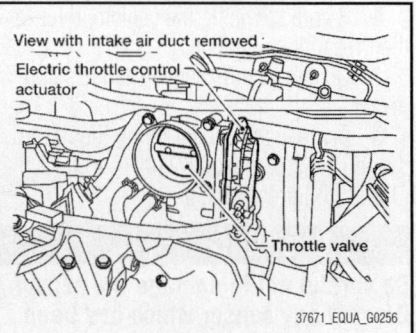

Fig. 161 Throttle Control Actuator (TAC) location—2.5L engine

4.0L Engine

See Figure 162.

The electric throttle control actuator includes a throttle control motor and Throttle Position Sensor (TPS). The TPS responds to the throttle valve movement. The TPS has two sensors.

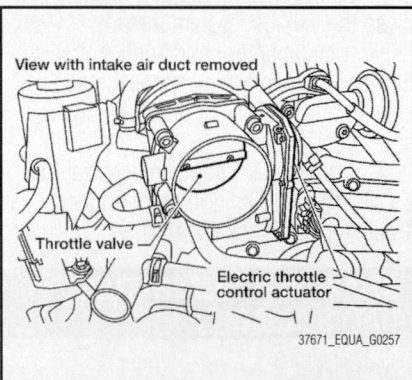

Fig. 162 Throttle Control Actuator (TAC) location—4.0L engine

REMOVAL & INSTALLATION

The Throttle Position Sensor (TPS) is replaced with the electric throttle control actuator. Refer to Throttle Control Actuator (TAC), removal & installation.

FUEL SYSTEM SERVICE PRECAUTIONS

Safety is the most important factor when performing not only fuel system maintenance but any type of maintenance. Failure to conduct maintenance and repairs in a safe manner may result in serious personal injury or death. Maintenance and testing of the vehicle's fuel system components can be accomplished safely and effectively by adhering to the following rules and guidelines.

• To avoid the possibility of fire and personal injury, always disconnect the negative battery cable unless the repair or test procedure requires that battery voltage be applied.

• Always relieve the fuel system pressure prior to disconnecting any fuel system component (injector, fuel rail, pressure regulator, etc.), fitting or fuel line connection. Exercise extreme caution whenever relieving fuel system pressure to avoid exposing skin, face and eyes to fuel spray. Please be advised that fuel under pressure may penetrate the skin or any part of the body that it contacts.

• Always place a shop towel or cloth around the fitting or connection prior to loosening to absorb any excess fuel due to spillage. Ensure that all fuel spillage (should it occur) is quickly removed from engine surfaces. Ensure that all fuel soaked cloths or towels are deposited into a suitable waste container.

• Always keep a dry chemical (Class B) fire extinguisher near the work area.

• Do not allow fuel spray or fuel vapors to come into contact with a spark or open flame.

• Always use a back-up wrench when loosening and tightening fuel line connection fittings. This will prevent unnecessary stress and torsion to fuel line piping.

• Always replace worn fuel fitting O-rings with new Do not substitute fuel hose or equivalent where fuel pipe is installed.

Before servicing the vehicle, make sure to also refer to the precautions in the beginning of this section as well.

RELIEVING FUEL SYSTEM PRESSURE

With Smart Diagnostic Tool (SDT)

1. Before servicing the vehicle, refer to the Precautions Section.

2. Turn the ignition switch ON.
3. Perform "FUEL PRESSURE RELEASE" in "Utility" mode with the SDT.
4. Start the engine.
5. After the engine stalls, crank it 2—3 times to release all the fuel pressure.
6. Turn the ignition switch OFF.

Without Smart Diagnostic Tool (SDT)

See Figure 163.

1. Before servicing the vehicle, refer to the Precautions Section.
2. Remove the fuel pump fuse located in the IPDM E/R.
3. Start the engine.
4. After the engine stalls, crank it 2—3 times to release all the fuel pressure.
5. Turn the ignition switch OFF.
6. When repairs are complete, replace the fuel pump fuse and connect the negative battery cable.

FUEL FILTER

REMOVAL & INSTALLATION

The fuel filter is part of the fuel level sensor unit, fuel filter, and fuel pump assembly. Refer to Fuel Pump Module, removal & installation.

FUEL PUMP MODULE

REMOVAL & INSTALLATION

See Figures 164.

➡**Be sure to check the fuel gauge indicator. Make sure that it reads less than FULL. If not, drain some fuel until the gauge reads less than FULL.**

1. Before servicing the vehicle, refer to the Precautions Section.
2. Properly relieve the fuel system pressure.

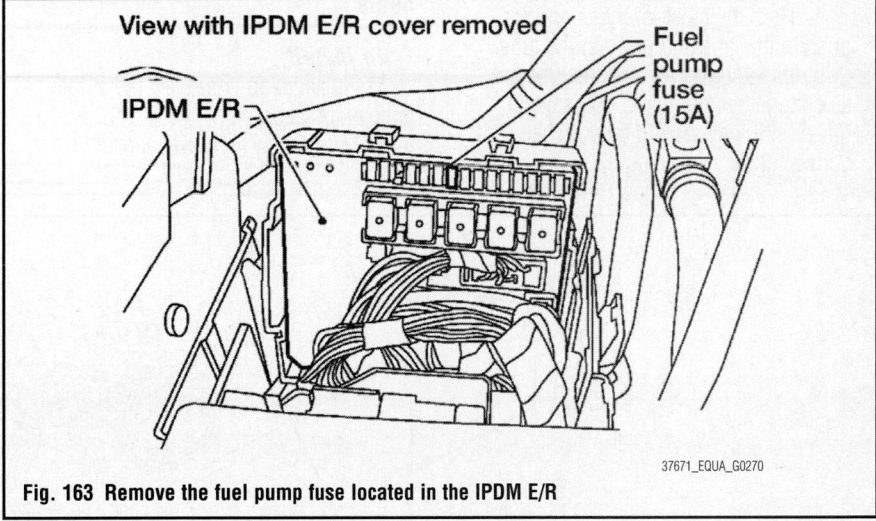

Fig. 163 Remove the fuel pump fuse located in the IPDM E/R

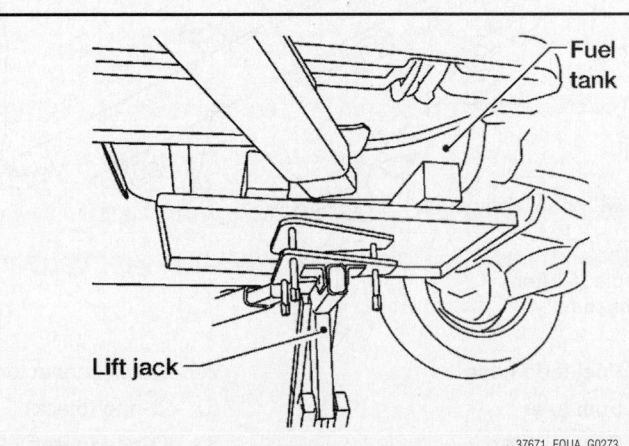

Fig. 164 Remove the fuel tank strap bolts while supporting the fuel tank with a suitable lift jack

3. Remove the fuel filler cap to release the pressure from inside the fuel tank.

4. Remove the left rear wheel and tire.

→**Fuel will be spilled when removing the fuel pump module for the fuel level is above the fuel level sensor, fuel filter, and fuel pump assembly fuel tank opening.**

5. Disconnect the battery negative terminal.

6. Remove the fuel tank shield.

7. Remove the fuel tank strap bolts while supporting the fuel tank with a suitable lift jack.

8. Lower the fuel tank using a suitable lift jack to access the top of the fuel pump module.

9. Disconnect the fuel pump module electrical connector, EVAP hose, and the fuel feed hose. Disconnect the fuel feed hose from the molded clip in the side of the fuel tank.

10. Disconnect the quick connector as follows:

a. Hold the sides of the connector, push in the tabs and pull out the tube.

b. If the connector and the tube are stuck together, push and pull several times until they start to move. Then disconnect them by pulling.

→**The quick connector can be disconnected when the tabs are completely depressed. Do not twist the quick connector more than necessary. Do not use any tools to disconnect the quick connector.**

11. Lower the fuel tank using a suitable lift jack and remove it from the vehicle to access the fuel pump module assembly.

12. Remove the lock ring using special tool J-45722, or equivalent.

13. Disconnect the EVAP hose from the molded clip in the top of the fuel tank.

14. Remove the fuel pump module assembly. Remove and discard the fuel level sensor, fuel filter, and fuel pump assembly O-ring.

⁂ WARNING

Do not bend the float arm during removal. Avoid impacts such as dropping when handling the components.

To install:

15. Installation is the reverse of the removal procedure.

16. Use a new fuel level sensor,

fuel filter, and fuel pump assembly O-ring.

17. Connect the quick connector.

a. Check the connection for any damage or foreign materials.

b. Align the connector with the pipe, then insert the connector straight into the pipe until a click is heard.

c. After connecting the quick connector, make sure that the connection is secure by pulling the tube and the connector to make sure they are securely connected. Visually inspect the connector to make sure the two retainer tabs are securely connected.

18. Turn the ignition switch ON, but do not start the engine. Check the fuel pipes and hose connections for leaks while applying fuel pressure to the system.

19. Start the engine and rev it above idle speed, then check that there are no fuel leaks at any of the fuel pipe and hose connections.

FUEL RAIL AND INJECTOR

REMOVAL & INSTALLATION

2.5L Engine

See Figures 165 and 166.

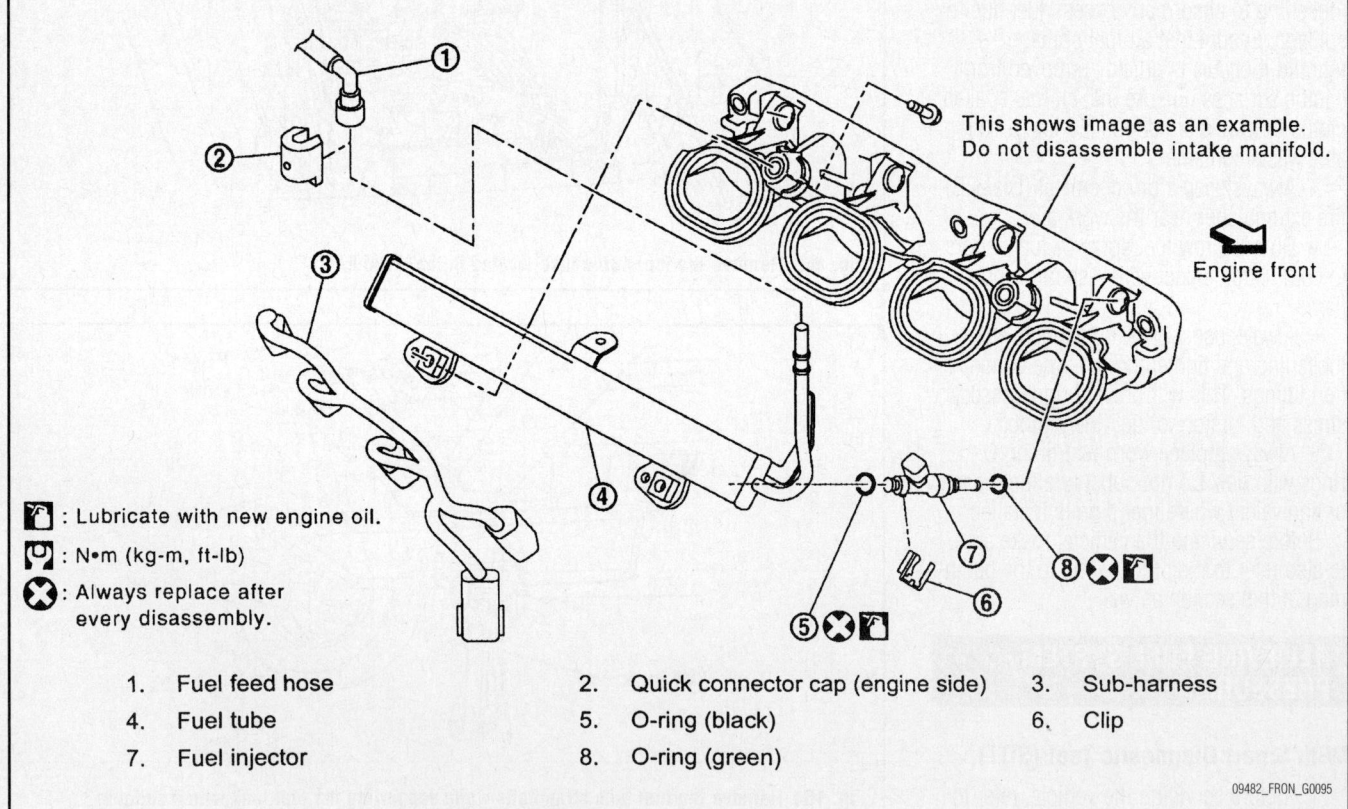

🛢 : Lubricate with new engine oil.

🔧 : N•m (kg-m, ft-lb)

✖ : Always replace after every disassembly.

1.	Fuel feed hose	2.	Quick connector cap (engine side)	3.	Sub-harness
4.	Fuel tube	5.	O-ring (black)	6.	Clip
7.	Fuel injector	8.	O-ring (green)		

This shows image as an example. Do not disassemble intake manifold.

Engine front

09482_FRON_G0095

Fig. 165 Fuel injector tube and related components—2.5L engine

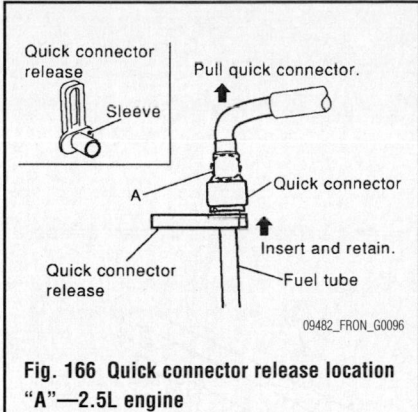

Fig. 166 Quick connector release location "A"—2.5L engine

1. Before servicing the vehicle, refer to the Precautions Section.

2. Properly relieve the fuel system pressure.

3. Disconnect the negative battery cable.

4. Remove the fuel filler cap.

5. Remove the quick connector cap (engine side). With the sleeve side of the quick connector release facing the quick connector, install the quick connector release on to the tube. Insert the quick connector release into the quick connector until the sleeve contacts and goes no further. Hold the quick connector release in that position.

➡**Disconnect the quick connector using tool J-45488, or equivalent, not by picking out the retainer tabs. Inserting the quick connector hard will not disconnect the quick connector. Hold the quick connector release where it contacts and goes no further.**

6. Draw and pull out the quick connector straight from the fuel tube. Grasp the quick connector holding "A" in the illustration. Do not pull with lateral force applied and the O-ring inside the quick connector could be damaged.

➡**Have a cloth ready, as fuel will leak out. Avoid fire and sparks. Keep parts away from heat. Do not bend or twist the connection between the quick connector and the fuel feed hose. Cover the openings with a plastic bag.**

7. Remove the intake manifold.

8. Disconnect the sub harness for the fuel injector.

9. Loosen the retaining bolts. Remove the fuel tube and fuel injector assembly.

10. To remove the fuel injectors from the fuel tube, open and remove the clip. Remove the injector by pulling it straight out.

To install:

➡**Use new O-ring seals for assembly. Note that the upper and lower O-rings are different. Do not confuse them. Fuel tube side: Black. Nozzle side: Green.**

11. Installation is the reverse of the removal procedure.

12. Handle the O-ring with bare hands. Do not wear gloves. Lubricate the O-ring with new engine oil. Do not clean the O-ring with solvent.

13. Make sure that O-ring and its mating part are free of foreign material. When installing the O-ring, be careful not to scratch it with a tool or fingernails. Do not twist or stretch the O-ring. If the O-ring was stretched while it was being attached, allow it to retract before inserting it into the fuel tube. Insert a new O-ring straight into the fuel tube. Do not angle or twist it.

14. When installing the fuel feed tube be sure to torque the retaining bolts to 9 ft. lbs. (12 Nm) and then to 21 ft. lbs. (28 Nm) in an alternating order.

15. Turn the ignition switch ON, but do not start the engine. Check the fuel lines and hose connections for leaks while applying fuel pressure to the system.

16. Start the engine and check for fuel leaks, correct as required.

4.0L Engine

See Figure 167.

1. Before servicing the vehicle, refer to the Precautions Section.

2. Properly relieve the fuel system pressure.

3. Disconnect the negative battery cable.

4. Remove the fuel filler cap.

5. Remove the intake manifold collector.

6. Remove the quick connector cap (engine side). With the sleeve side of the quick connector release facing the quick connector, install the quick connector release on to the tube. Insert the quick connector release into the quick connector until the sleeve contacts and goes no further. Hold the quick connector release in that position.

➡**Disconnect the quick connector using tool J-45488, or equivalent, not by picking out the retainer tabs. Inserting the quick connector hard will not dis- connect the quick connector. Hold the quick connector release where it con- tacts and goes no further.**

7. Draw and pull out the quick connector straight from the fuel tube. Grasp the quick connector holding "A" in the illustration. Do not pull with lateral force applied and the O-ring inside the quick connector could be damaged.

➡**Have a cloth ready, as fuel will leak out. Avoid fire and sparks. Keep parts away from heat. Do not bend or twist the connection between the quick con- nector and the fuel feed hose. Cover the openings with a plastic bag.**

8. Remove the PCV hose between the rocker covers.

9. Disconnect the harness for the fuel injector.

10. Loosen the retaining bolts. Remove the fuel tube and fuel injector assembly. Remove the bolts which connect the left and right fuel tubes.

11. To remove the fuel injectors from the fuel tube, open and remove the clip. Remove the injector by pulling it straight out.

12. Disconnect the right fuel tube from the left fuel tube. Loosen the bolts, to remove the fuel damper cap and fuel damper, if necessary.

To install:

➡**Use new O-ring seals for assembly. Note that the upper and lower O-rings are different. Do not confuse them. Fuel tube side: Blue. Nozzle side: Brown.**

13. Installation is the reverse of the removal procedure.

14. Handle the O-ring with bare hands. Do not wear gloves. Lubricate the O-ring with new engine oil. Do not clean the O-ring with solvent.

15. Make sure that O-ring and its mating part are free of foreign material. When installing the O-ring, be careful not to scratch it with a tool or fingernails. Do not twist or stretch the O-ring. If the O-ring was stretched while it was being attached, allow it to retract before insert- ing it into the fuel tube. Insert a new O-ring straight into the fuel tube. Do not angle or twist it.

16. When installing the fuel feed tube be sure to torque the retaining bolts to 7 ft. lbs. (10 Nm) and then to 16 ft. lbs. (22 Nm) in an alternating order.

17. Turn the ignition switch ON, but do not start the engine. Check the

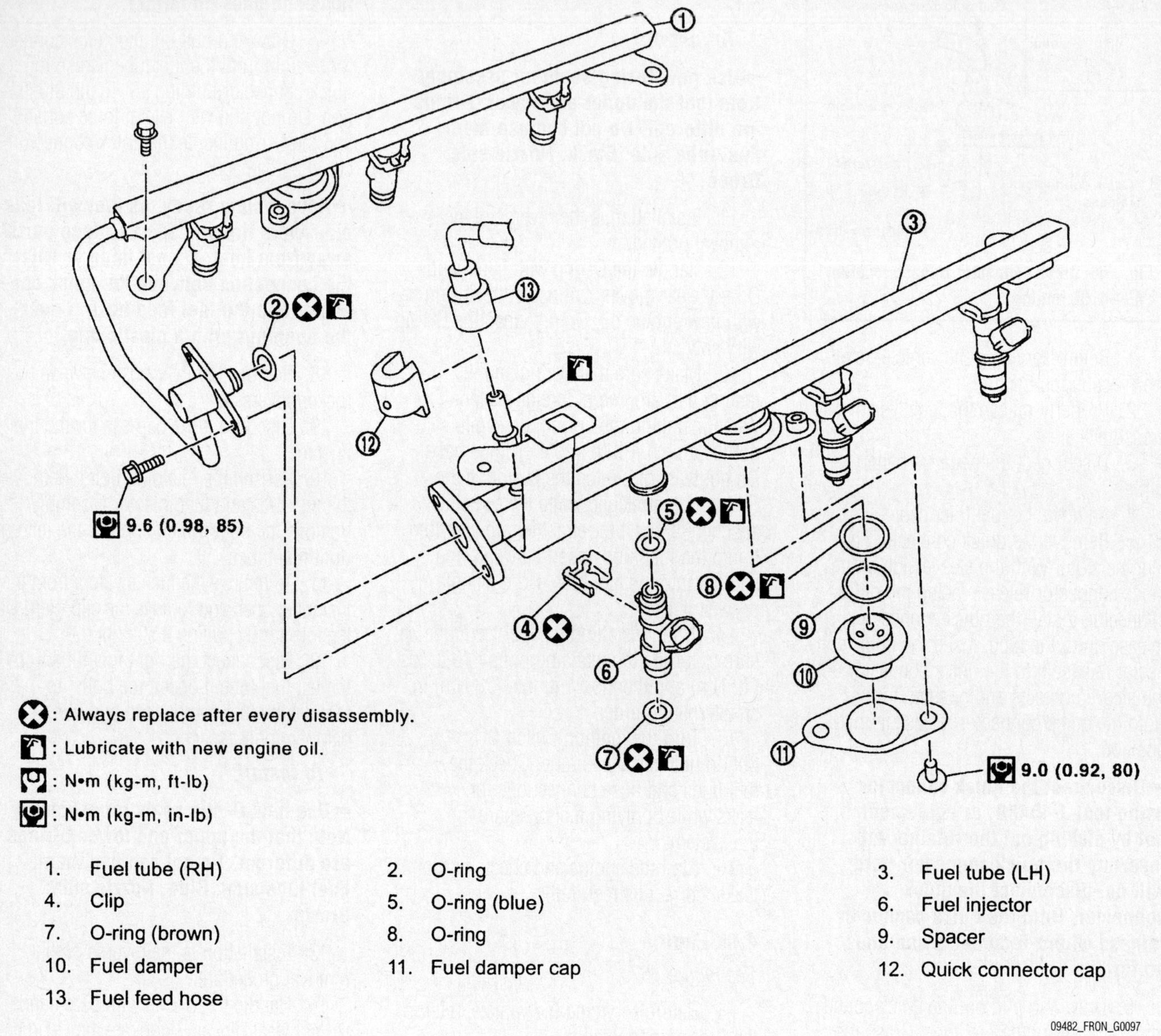

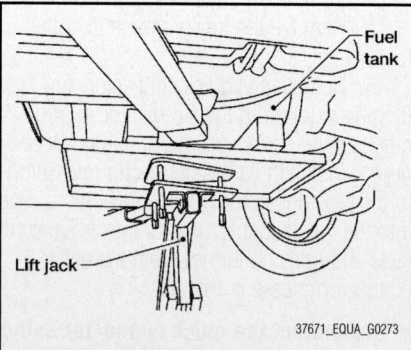

⊗ : Always replace after every disassembly.

🛢 : Lubricate with new engine oil.

🔧 : N•m (kg-m, ft-lb)

🔧 : N•m (kg-m, in-lb)

1. Fuel tube (RH)
2. O-ring
3. Fuel tube (LH)
4. Clip
5. O-ring (blue)
6. Fuel injector
7. O-ring (brown)
8. O-ring
9. Spacer
10. Fuel damper
11. Fuel damper cap
12. Quick connector cap
13. Fuel feed hose

09482_FRON_G0097

Fig. 167 Fuel injector tube and related components—4.0L engine

fuel lines and hose connections for leaks while applying fuel pressure to the system.

18. Start the engine and check for fuel leaks, correct as required.

FUEL TANK

REMOVAL & INSTALLATION

See Figures 168.

➡Be sure to check the fuel gauge indicator. Make sure that it reads less than FULL. If not, drain some fuel until the gauge reads less than FULL.

Fuel tank

Lift jack

37671_EQUA_G0273

Fig. 168 Remove the fuel tank strap bolts while supporting the fuel tank with a suitable lift jack

1. Before servicing the vehicle, refer to the Precautions Section.

2. Properly relieve the fuel system pressure.

3. Remove the fuel filler cap to release the pressure from inside the fuel tank.

4. Remove the left rear wheel and tire.

➡Fuel will be spilled when removing the fuel pump module for the fuel level is above the fuel level sensor, fuel filter, and fuel pump assembly fuel tank opening.

5. Disconnect the battery negative terminal.

6. Remove the fuel tank shield.

7. Remove the fuel tank strap bolts while supporting the fuel tank with a suitable lift jack.

8. Lower the fuel tank using a suitable lift jack to access the top of the fuel pump module.

9. Disconnect the fuel pump module electrical connector, EVAP hose, and the fuel feed hose. Disconnect the fuel feed hose from the molded clip in the side of the fuel tank.

10. Disconnect the quick connector as follows:

a. Hold the sides of the connector, push in the tabs and pull out the tube.

b. If the connector and the tube are stuck together, push and pull several times until they start to move. Then disconnect them by pulling.

➡ **The quick connector can be disconnected when the tabs are completely depressed. Do not twist the quick connector more than necessary. Do not use any tools to disconnect the quick connector.**

11. Lower the fuel tank using a suitable lift jack and remove it from the vehicle to access the fuel pump module assembly.

12. Remove the lock ring using special tool J-45722, or equivalent.

13. Disconnect the EVAP hose from the molded clip in the top of the fuel tank.

14. Remove the fuel pump module assembly, as necessary. Remove and discard the fuel level sensor, fuel filter, and fuel pump assembly O-ring.

❊❊ WARNING

Do not bend the float arm during removal. Avoid impacts such as dropping when handling the components.

15. Remove the fuel tank from the vehicle.

To install:

16. Installation is the reverse of the removal procedure.

17. Use a new fuel level sensor, fuel filter, and fuel pump assembly O-ring.

18. Connect the quick connector.

a. Check the connection for any damage or foreign materials.

b. Align the connector with the pipe, then insert the connector straight into the pipe until a click is heard.

c. After connecting the quick connector, make sure that the connection is secure by pulling the tube and the connector to make sure they are securely connected. Visually inspect the connector to make sure the two retainer tabs are securely connected.

19. Turn the ignition switch ON, but do not start the engine. Check the fuel pipes and hose connections for leaks while applying fuel pressure to the system.

20. Start the engine and rev it above idle speed, then check that there are no fuel leaks at any of the fuel pipe and hose connections.

HEATING & AIR CONDITIONING SYSTEM

BLOWER MOTOR

REMOVAL & INSTALLATION

See Figure 169.

➡ **Before servicing or working around the SRS system, turn the ignition switch OFF, disconnect both battery cables and wait at least 3 minutes. When servicing or working around the SRS system, do not work directly in front of the air bag module.**

1. Before servicing the vehicle, refer to the Precautions Section.

2. Disconnect the negative battery cable. Disconnect the positive battery cable.

3. Remove the lower glove box assembly.

4. Disconnect the blower motor electrical connector.

5. Remove the blower motor retaining screws.

6. Remove the blower motor from its mounting.

To install:

7. Installation is the reverse of the removal procedure.

8. Connect the battery positive cable and then the negative terminal.

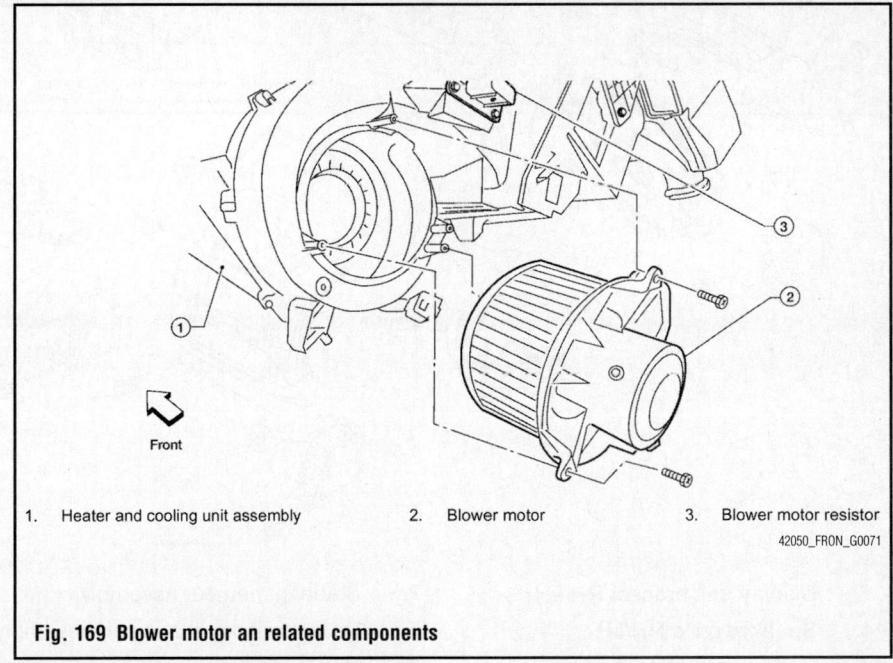

| 1. Heater and cooling unit assembly | 2. Blower motor | 3. Blower motor resistor |

42050_FRON_G0071

Fig. 169 Blower motor an related components

HEATER CORE

REMOVAL & INSTALLATION

See Figures 170 through 174.

➡ **Be sure to disarm the SRS system, prior to working on the vehicle. Turn the ignition switch OFF, disconnect both battery cables and wait at least 3 minutes before starting any work.**

1. Before servicing the vehicle, refer to the Precautions Section.

2. Position the front wheels in the straight ahead direction.

3. Disconnect the negative battery cable. Disconnect the positive battery cable.

4. Drain the engine coolant.

5. Properly discharge the air conditioning system.

6. If equipped with the 4.0L engine, remove the right side heater core pipe nuts.

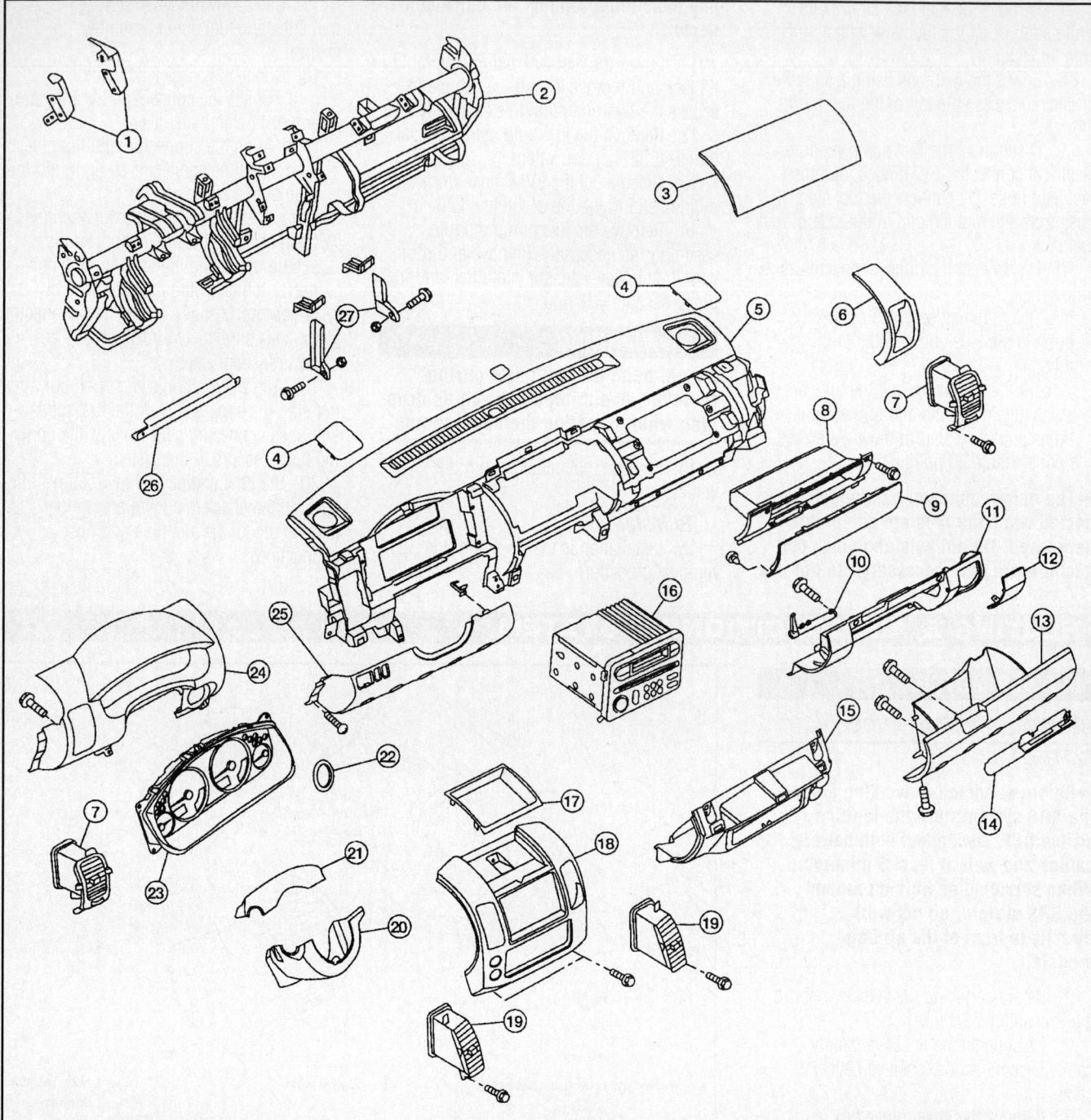

1.	Display unit bracket RH/LH	2.	Steering member assembly	3.	Passenger air bag module cover
4.	Speaker grille RH/LH	5.	Instrument panel and pad assembly	6.	Instrument side finisher
7.	Side ventilator assembly RH/LH	8.	Upper glove box bin	9.	Upper glove box door
10.	Lower glove box damper assembly	11.	Lower instrument panel RH	12.	Fuse block cover
13.	Lower glove box assembly	14.	Lower glove box latch assembly	15.	Cluster lid D
16.	Audio unit	17.	Storage tray	18.	Cluster lid C
19.	Center ventilator assembly RH/LH	20.	Steering column cover lower	21.	Steering column cover upper
22.	Steering lock escutcheon	23.	Combination meter	24.	Cluster lid A
25.	Lower instrument panel LH	26.	Knee protector brace	27.	Instrument stay RH/LH

09482_FRON_G0009

Fig. 170 Instrument panel and related components

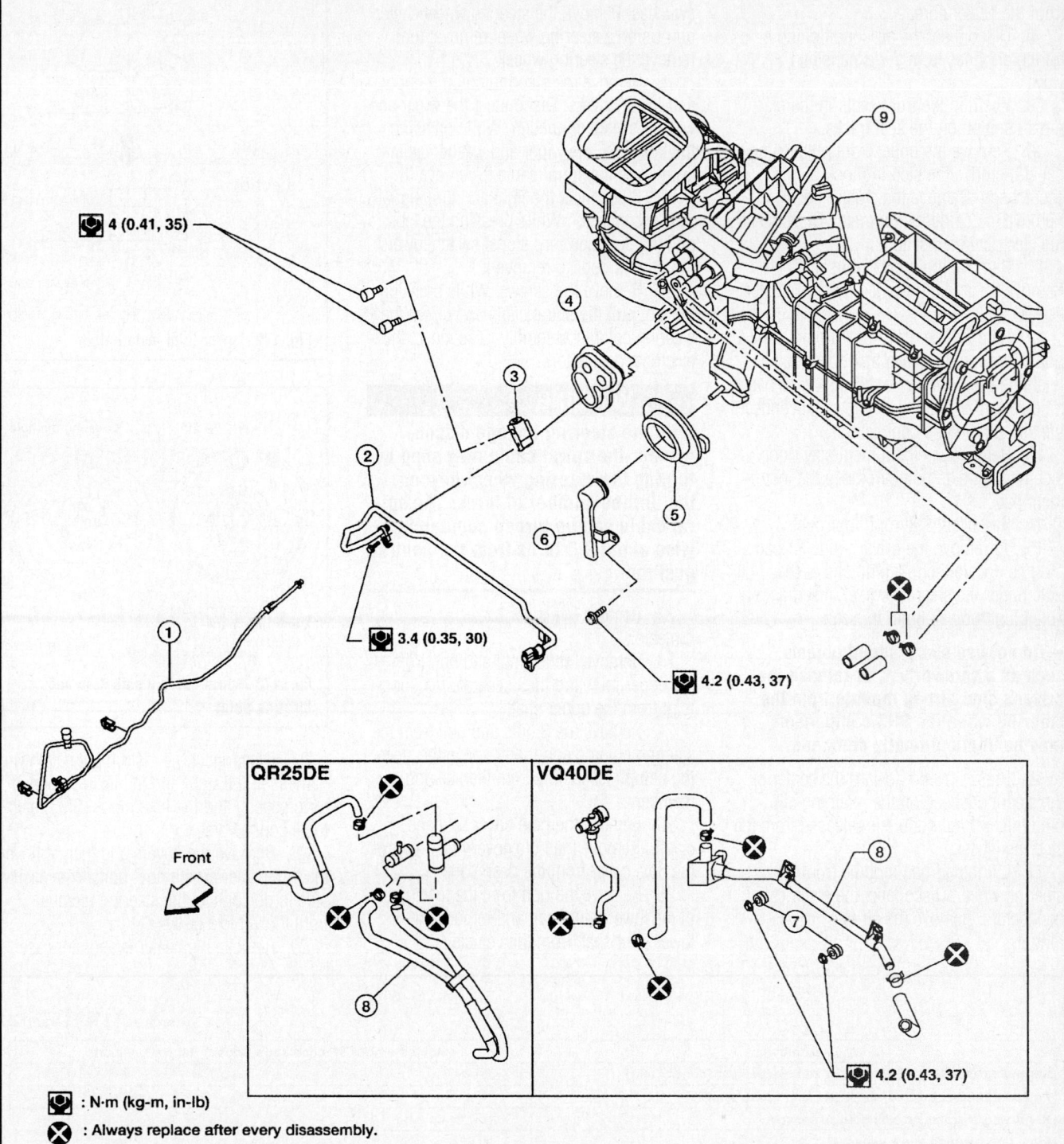

4 (0.41, 35)

3.4 (0.35, 30)

4.2 (0.43, 37)

QR25DE

VQ40DE

Front

4.2 (0.43, 37)

: N·m (kg-m, in-lb)

: Always replace after every disassembly.

1. High-pressure A/C pipe
2. Low-pressure A/C pipe
3. Expansion valve
4. Heater core and evaporator pipes grommet
5. A/C drain hose grommet
6. A/C drain hose
7. Heater core pipe mounts
8. Heater core pipes
9. Heater and cooling unit assembly

09482_FRON_G0012

Fig. 171 Heater/evaporator core and related components

7. Disconnect the heater core hoses from the heater core.

8. Disconnect the air conditioning refrigerant lines from the expansion valve.

9. Position the front seats in the rear-most position on the seat tracks.

10. Remove the upper front pillar trim panel. Remove the steering lock escutcheon. Remove the cluster lid "A". Remove the combination meter. Disconnect the electrical connections.

11. Remove the optical sensor. Remove the audio unit. Remove the cluster lid "D".

12. Remove the glove box. Remove the 2 bolts, through the glove box opening, retaining the front passenger's side air bag module to the steering member. Disconnect the air bag module connectors.

13. Remove the instrument stay right side and left side bolts. Remove the instrument panel.

14. Remove the 2 front floor ducts.

15. To remove the driver's side air bag module, locate the retaining clip access hole under the steering wheel. Insert a suitable blunt tool (4–6mm in size).

➡ Do not use sharp edged objects, such as a screwdriver, to release the driver's side airbag module from the steering wheel as SRS components may be unintentionally damaged.

16. Press upward, toward the center of the steering wheel, on the retaining clip until the air bag module is released from the steering wheel.

17. Lift the air bag module from the steering wheel. Disconnect the electrical connectors. Remove the air bag module.

18. Disconnect the steering wheel switches. Remove the steering wheel center nut. Using a steering wheel removal tool, remove the steering wheel.

19. Remove the steering column upper and lower covers. Disconnect the wiper and washer switch connector. While pressing the tabs, pull the wiper and washer switch away from the spiral cable to remove it.

20. Disconnect the light and turn signal switch connector. While pressing the tabs, pull the light and turn signal switch toward the driver's door to remove it.

21. Remove the screws. While pressing the tab, pull the spiral cable away from the steering column assembly. Disconnect the electrical connectors.

❋❋ WARNING

With the steering linkage disconnected, the spiral cable may snap by turning the steering wheel beyond the limited number of turns. The spiral cable can be turned counterclockwise about 2.5 turns from the neutral position.

22. Remove the lower knee protector.

23. Remove the locknut and bolt from the upper joint and then separate the upper joint from the upper shaft.

24. Remove the 3 nuts and bolt from the steering column and then remove the steering column assembly from the steering member.

25. Remove the hole cover seal and clamp. Remove the hole cover nuts, remove the hole cover from the dash panel.

26. Remove the bolt from the lower joint of the lower joint shaft and remove the lower joint shaft from the vehicle.

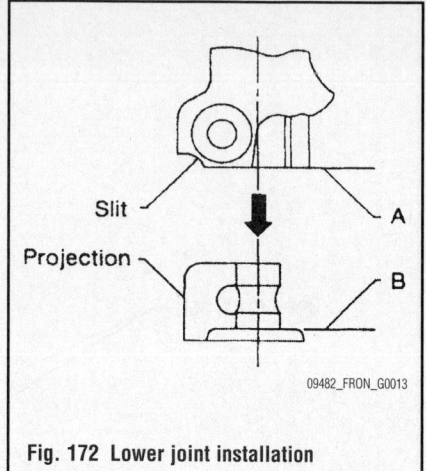

Fig. 172 Lower joint installation

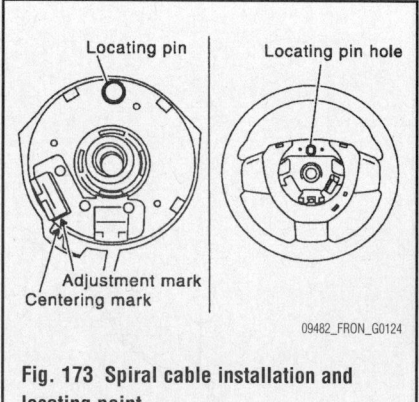

Fig. 173 Spiral cable installation and locating point

27. Disconnect the instrument panel wire harness at the right and left in-line connector brackets, and the fuse block (SMJ) electrical connectors.

28. Remove the covers and then remove the three steering member bolts from each side to disconnect the steering member from the vehicle body.

29. Remove the heater/evaporator case

Situation	Adjustment of steering angle sensor neutral position
Removing/Installing ABS actuator and electric unit (control unit)	−
Replacing ABS actuator and electric unit (control unit)	×
Removing/Installing steering angle sensor	×
Replacing steering angle sensor	×
Removing/Installing steering components	×
Replacing steering components	×
Removing/Installing suspension components	×
Replacing suspension components	×
Change tires to new ones	−
Tire rotation	−
Adjusting wheel alignment	×
Battery disconnection	×

×: Required −: Not required

Fig. 174 Steering angle sensor neutral position adjustment requirement table

assembly with it attached to the steering member from the vehicle.

30. Separate the steering member from the heater/evaporator unit.

31. Remove the heater cover retaining screws. Remove the cover.

32. Remove the heater core and the evaporator pipe bracket. Remove the heater core.

To install:

33. Installation is the reverse of the removal procedure.

➡️**If the in-cabin microfilters are contaminated with coolant, replace them.**

34. Be sure to use new steering column retaining bolts and pinch bolt, as required.

➡️**When installing the steering column, finger tighten all of the lower bracket and joint bolts and then tighten them to specification. Do not apply undue stress to the steering column.**

35. With the wheels in the straight ahead position align the slit of the lower joint with the projection on the dust cover. Insert the joint until surface "A" contacts surface "B".

36. Be sure to align the spiral cable correctly when installing the steering wheel. Make sure that the cable is in the neutral position. The neutral position is detected by turning left 2.6 revolutions from the right end position and ending with the locating pin at the top.

37. Refer to the table below to determine if adjustment of the steering angle sensor neutral position is required.

➡️**To adjust the neutral position of the steering angle sensor, make sure to use the Smart Diagnostic Tool (SDT). Adjustment cannot be done without SDT.**

38. Stop the vehicle with the front wheels in the straight-ahead position.

39. On the SDT screen, touch UTILITY and ST ANG SEN ADJUSTMENT in order.

40. Touch START.

➡️**Do not touch the steering wheel while adjusting the steering angle sensor.**

41. After approximately 10 seconds, touch END.

➡️**After approximately 60 seconds, the program ends automatically.**

42. Turn the ignition switch OFF, then turn it ON again.

43. Run the vehicle with the front wheels in the straight-ahead position, then stop.

44. Select DATA LIST. Then make sure STR ANGLE SIG is within 0 plus or minus 2.5°.

45. Erase the self-diagnosis memory of the ABS actuator and electric unit (control unit) and ECM.

46. After erasing DTC memory, start the engine and drive the vehicle at 19 MPH (30 km/h) or more for approximately 1 minute as the final inspection, and make sure that the ABS warning lamp, VDC OFF indicator lamp, SLIP indicator lamp and brake warning lamp turn OFF.

47. Be sure to fill the cooling system with the proper grade and type coolant.

48. Recharge the air conditioning system. Check for leaks.

49. Check and adjust the front end alignment, as necessary.

STEERING

POWER RACK & PINION STEERING GEAR

REMOVAL & INSTALLATION

See Figures 175 and 176.

❊❊ WARNING

The spiral cable may snap due to steering operation if the steering column is separated from the steering gear assembly. Be sure to secure the steering wheel to avoid turning.

1. Before servicing the vehicle, refer to the Precautions Section.

2. Position the front wheels in the straight ahead position.

3. Disarm the SRS system. Refer to Air Bag (Supplemental Restraint System), Disarming The System.

4. Disconnect the negative battery cable.

5. Drain the power steering fluid.

6. Raise and support the vehicle safely. Remove the tire and wheel assemblies.

7. Remove the undercover.

8. If equipped with 4WD, remove the final drive, then support the halfshafts, using wire.

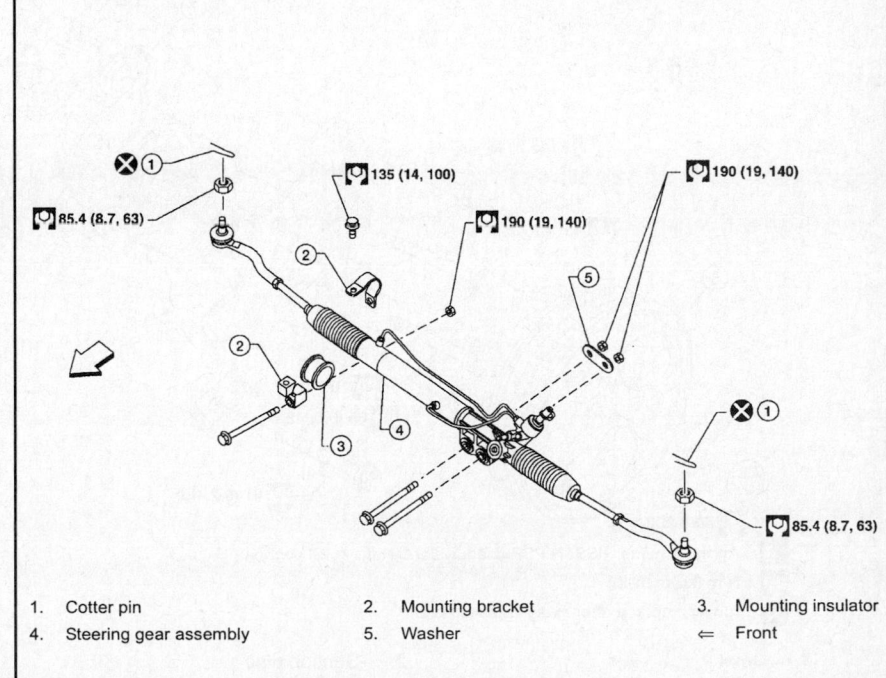

| 1. | Cotter pin | 2. | Mounting bracket | 3. | Mounting insulator |
| 4. | Steering gear assembly | 5. | Washer | ⇐ | Front |

09482_FRON_G0113

Fig. 175 Power steering gear and related component locations

Slit
A
Projection
B
09482_FRON_G0114

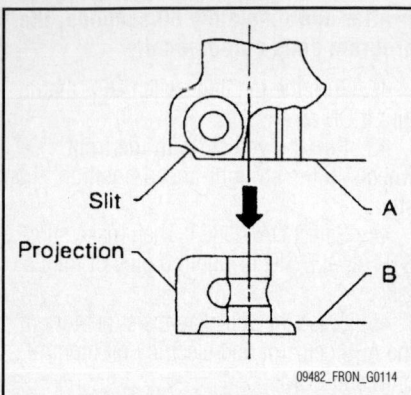

Fig. 176 Power steering gear lower joint installation alignment

9. Remove the stabilizer bar brackets, and position the stabilizer bar aside.
10. Remove and discard the cotter pins at the steering outer sockets. Loosen the outer socket locknuts.
11. Remove the steering gear outer sockets from the steering knuckles, using tool HT72520000 (J-25730-A), or equivalent.
12. Disconnect and plug the power steering fluid lines at the steering gear.
13. Remove the bolt from the lower joint of the lower joint assembly. Separate the lower joint from the steering gear assembly. Be careful not to damage the lower joint.
14. Remove the steering gear retaining nuts and bolts. Remove the steering gear from the vehicle.

To install:

15. With the steering wheel in the straight ahead position, align the slit of the lower joint with the projection on the dust cover. Insert the joint until both surfaces contact each other.
16. Continue the installation in the reverse order of the removal procedure.
17. Check and adjust the front alignment, as required.
18. Bleed the power steering system.
19. Fill the power steering pump with the proper grade and type of fluid.
20. Perform the steering angle neutral position adjustment. Refer to Steering Angle Position Adjustment.

POWER STEERING PUMP

REMOVAL & INSTALLATION

See Figures 177 and 178.

1. Before servicing the vehicle, refer to the Precautions Section.
2. Disconnect the negative battery cable.

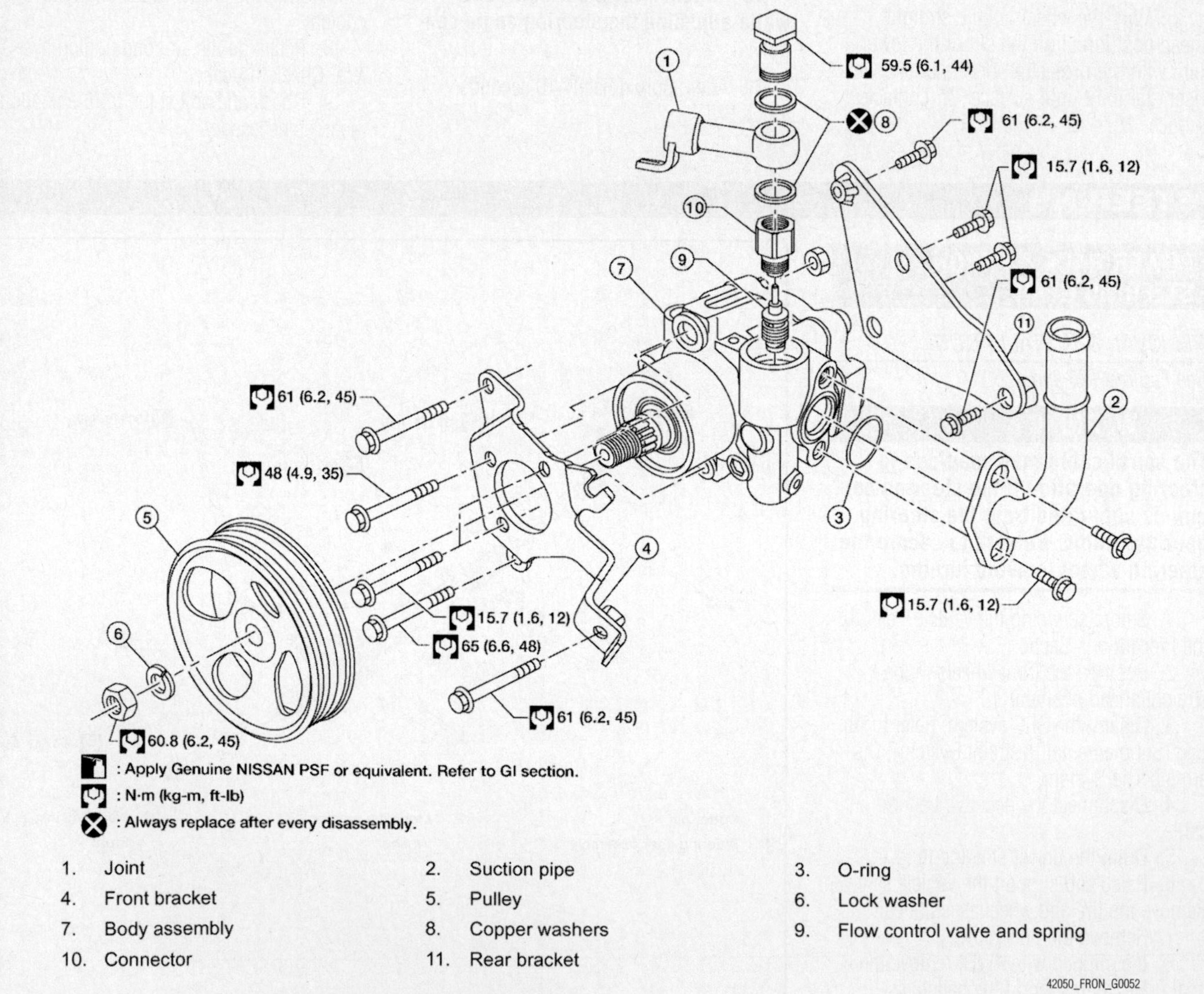

59.5 (6.1, 44)
8
61 (6.2, 45)
15.7 (1.6, 12)
61 (6.2, 45)
61 (6.2, 45)
48 (4.9, 35)
15.7 (1.6, 12)
65 (6.6, 48)
61 (6.2, 45)
60.8 (6.2, 45)
15.7 (1.6, 12)
: Apply Genuine NISSAN PSF or equivalent. Refer to GI section.
: N·m (kg-m, ft-lb)
: Always replace after every disassembly.
1. Joint
2. Suction pipe
3. O-ring
4. Front bracket
5. Pulley
6. Lock washer
7. Body assembly
8. Copper washers
9. Flow control valve and spring
10. Connector
11. Rear bracket
42050_FRON_G0052

Fig. 177 Power steering pump and related components—2.5L engine

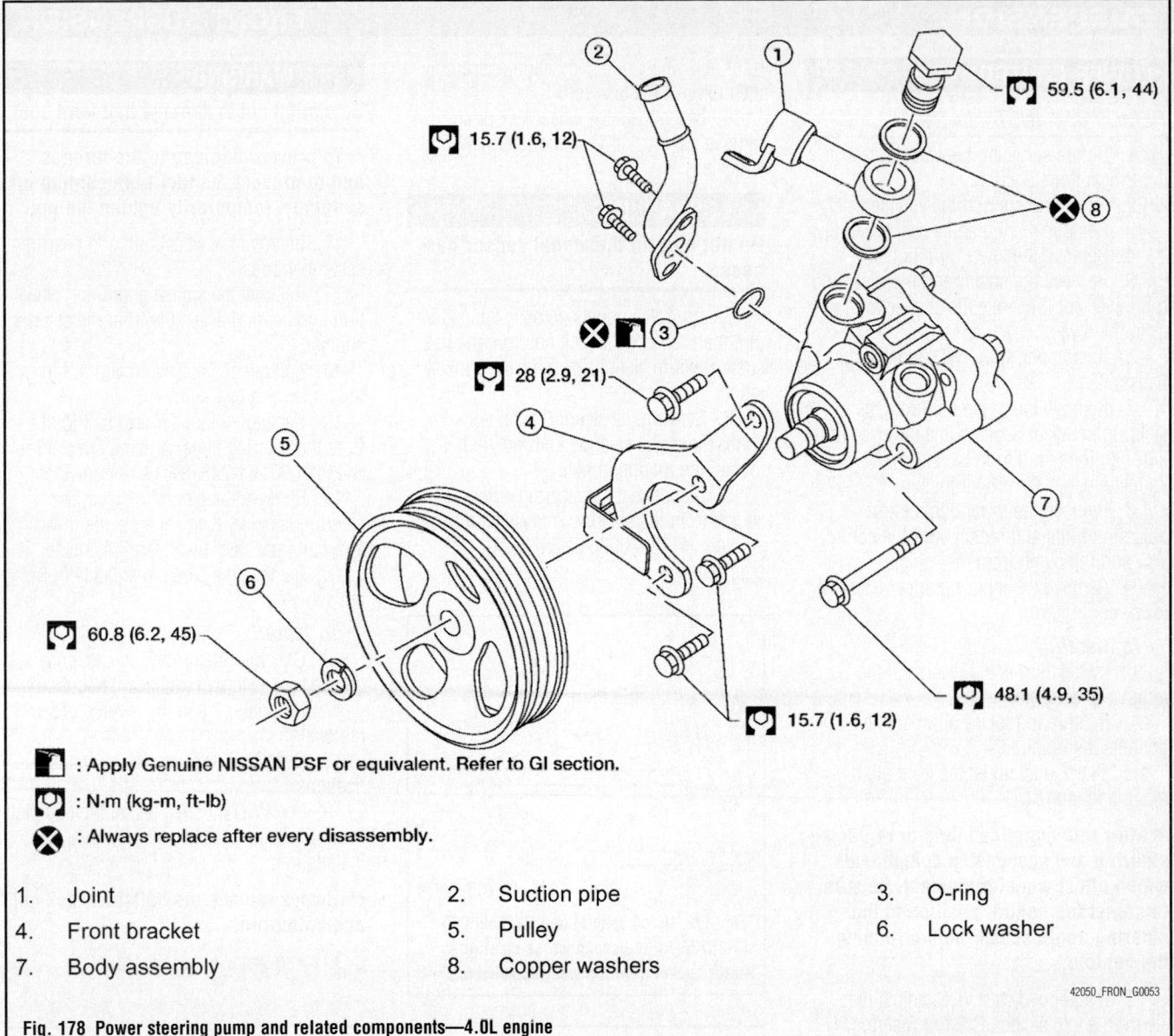

15.7 (1.6, 12)

59.5 (6.1, 44)

28 (2.9, 21)

60.8 (6.2, 45)

48.1 (4.9, 35)

15.7 (1.6, 12)

: Apply Genuine NISSAN PSF or equivalent. Refer to GI section.

: N·m (kg-m, ft-lb)

: Always replace after every disassembly.

1. Joint	2. Suction pipe	3. O-ring
4. Front bracket	5. Pulley	6. Lock washer
7. Body assembly	8. Copper washers	

42050_FRON_G0053

Fig. 178 Power steering pump and related components—4.0L engine

3. Drain the power steering fluid from the reservoir tank. Properly dispose of used fluid.

4. On the 4.0L engine, remove the engine cover.

5. Remove the air duct assembly. Refer to Air Cleaner, removal & installation.

6. Remove the drive belt. Refer to Accessory Drive Belts, removal & installation.

7. Disconnect the pressure sensor electrical connector.

8. Disconnect and plug the fluid lines.

9. Remove the pump retaining bolts.

10. Remove the pump from the vehicle.

To install:

11. Installation is the reverse of the removal procedure.

12. Bleed the power steering system. Refer to Bleeding.

BLEEDING

1. Before servicing the vehicle, refer to the Precautions Section.

2. Fill the power steering system with the proper grade and type of steering fluid.

➡**Do not allow the fluid level in the reservoir tank to go below the MIN level line. Check and add fluid as needed.**

3. Raise and safely support the vehicle.

4. Quickly turn the steering wheel to the full right and left detents and lightly touch the steering stoppers.

➡**Do not hold the steering wheel in the locked position for more than 10 seconds.**

5. Repeat this operation until the fluid level no longer decreases.

6. Start the engine.

7. Quickly turn the steering wheel to the full right and left detents and lightly touch the steering stoppers.

➡**Do not hold the steering wheel in the locked position for more than 10 seconds.**

8. Check for air bubbles or cloudy fluid. If found, repeat the bleeding procedure.

9. Stop the engine and check the fluid level. Correct as required.

LOWER CONTROL ARM

REMOVAL & INSTALLATION

1. Before servicing the vehicle, refer to the Precautions Section.
2. Raise and support the vehicle safely.
3. Remove the tire and wheel assembly.
4. Remove the lower strut bolt.
5. Remove the stabilizer bar connecting rod lower nut. Separate the connecting rod from the lower control arm.
6. If equipped with 4WD, remove the halfshaft.
7. Remove the pinch bolt from the steering knuckle. Separate the lower control arm ball joint stud from the steering knuckle, using the proper tool.
8. Remove the lower control arm adjusting bolts and nuts. Lower the control arm and remove it from the vehicle.
9. Remove the jounce bumper from the lower control arm.

To install:

10. Installation is the reverse of the removal procedure.
11. Be sure to replace all wearable components, as required.
12. Check and adjust the front alignment, as required.

➡**After removing/installing or replacing steering and suspension components which effect wheel alignment, be sure to adjust the neutral position of the steering angle sensor before running the vehicle.**

13. If equipped with VDC, adjust the steering angle sensor. Refer to Steering Column, Steering Angle Position Adjustment.

STEERING KNUCKLE

REMOVAL & INSTALLATION

See Figures 179 through 181.

1. Before servicing the vehicle, refer to the Precautions Section.
2. Raise and support the vehicle safely.
3. Remove the wheel and tire assembly.
4. Without disassembling the hydraulic lines, remove the brake caliper. Reposition it aside with a wire. Refer to Brake Caliper, removal & installation.

➡**Avoid depressing the brake pedal while the brake caliper is removed.**

5. Put alignment marks on the disc rotor and wheel hub and bearing assembly, then remove the disc rotor.
6. Disconnect the wheel sensor and remove the bracket from the steering knuckle.

✳✳ WARNING

Do not pull on the wheel sensor harness.

7. On 4WD models, remove the cotter pin, then remove the lock nut from the half-shaft. Refer to Halfshafts, removal & installation.
8. Remove the steering outer socket cotter pin at the steering knuckle, then loosen the mounting nut.
9. Disconnect the steering outer socket from the steering knuckle using special tool HT72520000 (J-25730-A), or equivalent.

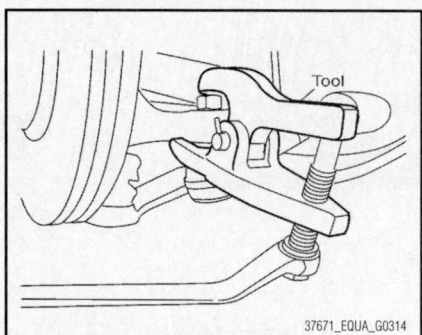

37671_EQUA_G0314

Fig. 179 Using special tool HT72520000 (J-25730-A) to disconnect the steering outer socket from the steering knuckle

✳✳ WARNING

Be careful not to damage ball joint boot.

➡**To prevent damage to the threads and to prevent the tool from coming off suddenly, temporarily tighten the nut.**

10. Remove the wheel hub and bearing assembly bolts.
11. Remove the splash guard and wheel hub and bearing assembly from the steering knuckle.
12. Remove the cotter pin and nut from the upper arm ball joint.
13. Separate the upper arm ball joint from the steering knuckle using special tool ST29020001 (J-24319-01), or equivalent.
14. Remove the pinch bolt from the steering knuckle, then separate the lower arm ball joint from the steering knuckle.
15. Remove the steering knuckle from the vehicle.

To install:

16. Check for deformity, cracks, and damage on each part, replace if necessary.
17. Installation is in the reverse order of removal.
18. Refer to illustration for torque specifications.
19. For 4WD models, install the halfshaft lock nut. Refer to Halfshafts, removal & installation.

➡**Always replace the halfshaft lock nut and cotter pin.**

20. Perform a wheel alignment.

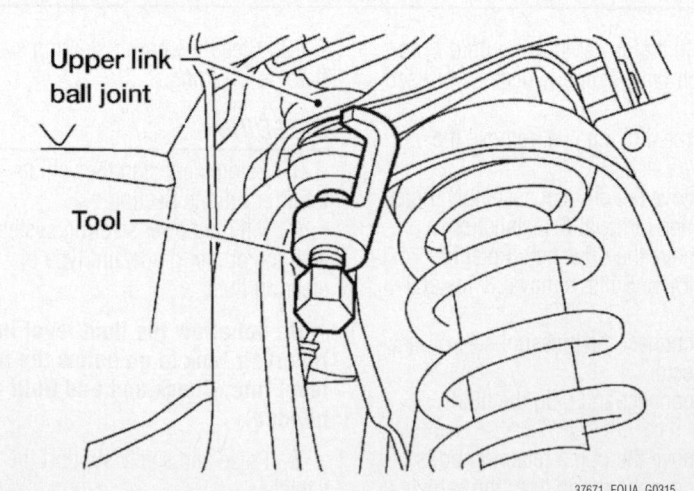

37671_EQUA_G0315

Fig. 180 Using special tool ST29020001 (J-24319-01) to separate the upper arm (link) ball joint from the steering knuckle

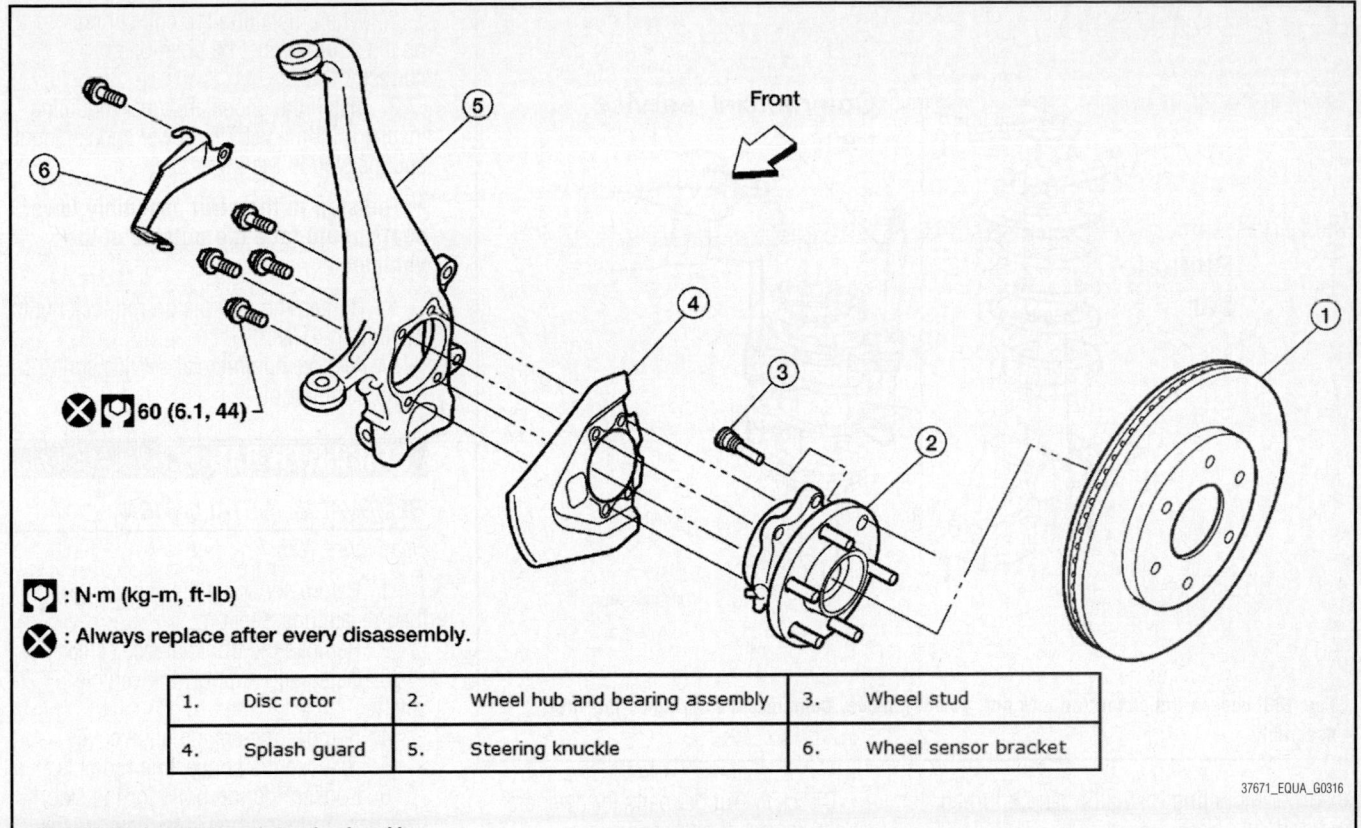

: N·m (kg-m, ft-lb)

: Always replace after every disassembly.

1.	Disc rotor	2.	Wheel hub and bearing assembly	3.	Wheel stud
4.	Splash guard	5.	Steering knuckle	6.	Wheel sensor bracket

37671_EQUA_G0316

Fig. 181 Exploded view of steering knuckle

STRUT

REMOVAL & INSTALLATION

See Figure 182.

1. Before servicing the vehicle, refer to the Precautions Section.

2. Raise and support the vehicle safely.

3. Remove the wheel and tire assembly.

4. Support the lower control arm (link) using a suitable jack.

5. Remove the connecting rod upper joints from the stabilizer bar. Swing the stabilizer bar down, repositioning it out of the way, to access the shock absorber lower mount.

6. Remove the shock absorber lower bolt and nut.

7. Remove the 3 shock absorber upper mounting nuts.

8. Remove the coil spring and shock absorber assembly. Turn the steering knuckle out to gain enough clearance for removal.

To install:

9. Installation is the reverse of the removal procedure.

10. The step in the strut assembly lower seat should face the outside of the vehicle.

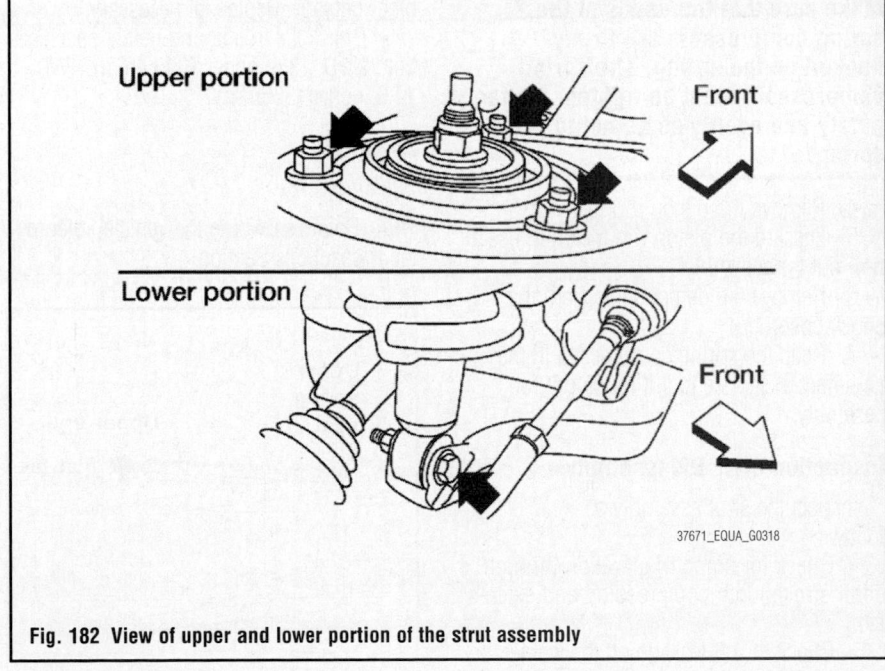

37671_EQUA_G0318

Fig. 182 View of upper and lower portion of the strut assembly

OVERHAUL

Disassembly

See Figure 183.

1. Before servicing the vehicle, refer to the Precautions Section.

2. Set the shock absorber in a vise, then loosen (without removing) the piston rod lock nut.

➡**Do not remove the piston rod lock nut at this time.**

3. Compress the spring using a commercial service tool until the shock

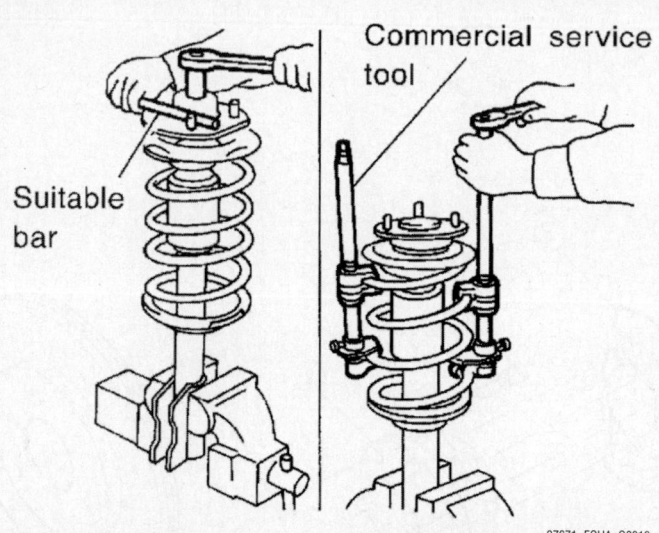

Fig. 183 Loosen the piston rod lock nut, do not remove. Compress the spring on the strut assembly

absorber mounting insulator can be turned by hand.

✳✳ CAUTION

Make sure that the pawls of the 2 spring compressors are firmly hooked on the spring. The spring compressors must be tightened alternately and evenly so as not to tilt the spring.

4. Remove the piston rod lock nut.
5. Discard the piston rod lock nut, use a new nut for assembly.
6. Remove the components from the shock absorber.
7. Keep the spring compressed in the commercial service tool if reusing it for assembly.

Inspection after Disassembly

Inspect the strut assembly as follows:
• Check for smooth operation through a full stroke, both compression and extension
• Check for oil leakage on the welded or gland packing portions
• Check the piston rod for cracks, deformation, or other damage. Replace if necessary
Inspect the mounting insulator and rubber parts:
• Check the cemented rubber-to-metal portion for separation or cracks

• Check the rubber parts for deterioration and replace if necessary
Inspect the coil spring:
• Check for cracks, deformation, or other damage. Replace if necessary
• Check the free spring height specification. 2WD: 13.6 inches (345.4mm), 4WD: 14.0 inches (356mm)

Assembly

See Figure 184.

1. Before servicing the vehicle, refer to the Precautions Section.

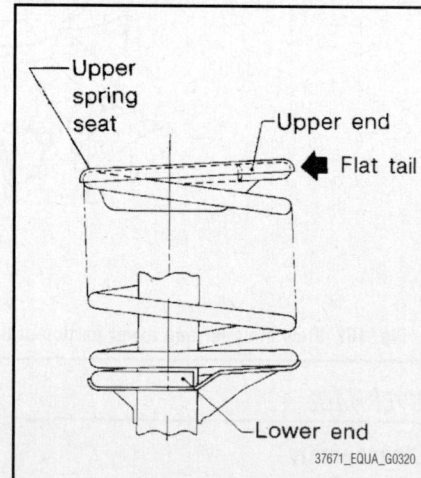

Fig. 184 When installing the coil spring on the strut, it must be positioned as shown

2. When installing the coil spring on the strut, it must be positioned correctly.
3. Install the shock absorber mounting insulator in line with the lower shock mount and the step in the lower seat.

➡ **The step in the strut assembly lower seat should face the outside of the vehicle.**

4. Tighten the new piston rod lock nut to 30 ft. lbs. (41 Nm).
5. Remove commercial service spring compression tool.

STABILIZER BAR

REMOVAL & INSTALLATION

See Figure 185.

1. Before servicing the vehicle, refer to the Precautions Section.
2. Remove the front valance center.
3. Raise and support the vehicle safely.
4. Remove the engine undercover.
5. Remove the connecting rod nuts.
6. Loosen the top bolts for the stabilizer bar mounting brackets. Remove the lower bolts from the mounting brackets.
7. Remove the stabilizer bar from the vehicle.
8. Remove the bushings from the stabilizer bar.

To install:

9. Installation is the reverse of the removal procedure.
10. Check the stabilizer bar for twist and deformation. Replace if necessary.
11. Check rubber bushing for cracks, wear, and deterioration. Replace if necessary.
12. Tighten all nuts and bolts to specification. Refer to front suspension component locations illustration.

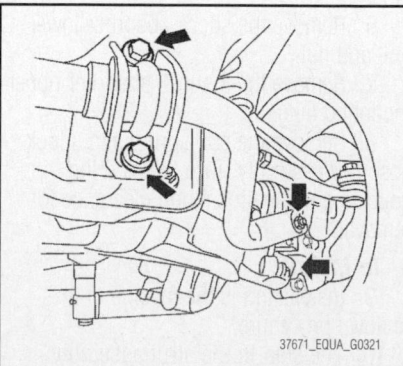

Fig. 185 Remove the stabilizer mounting brackets

UPPER CONTROL ARM

REMOVAL & INSTALLATION

See Figures 186 and 187.

1. Before servicing the vehicle, refer to the Precautions Section.

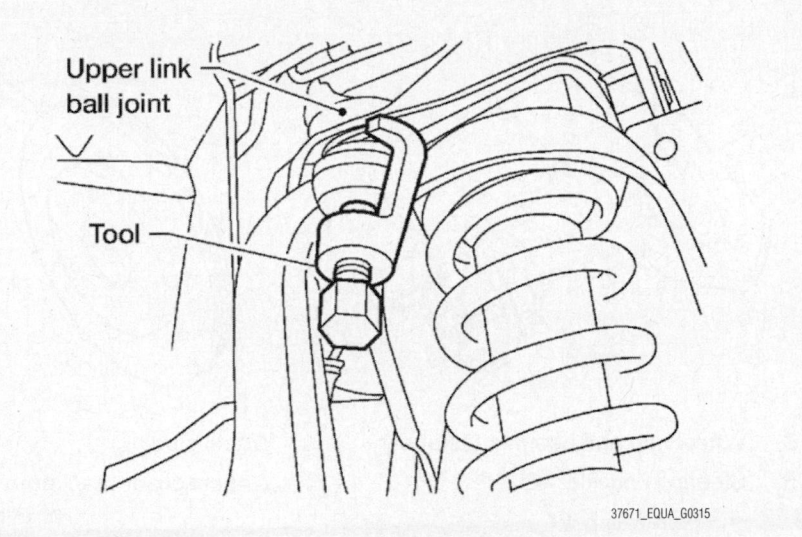

Fig. 186 Using special tool ST29020001 (J-24319-01) to separate the upper arm (link) ball joint from the steering knuckle

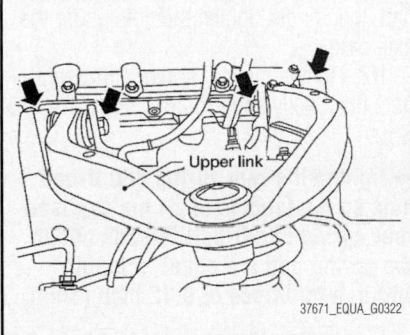

Fig. 187 Remove the upper control arm (link) retaining bolts and nuts

2. Raise and support the vehicle safely.
3. Remove the tire and wheel assembly.
4. Using a suitable jack, support the lower control arm.
5. If working on the left side, remove the bolt from the lower joint of the lower joint shaft, then reposition the lower joint shaft out of the way. Do not damage the lower joint.
6. Remove the cotter pin and nut from the upper control arm ball joint.
7. Separate the upper control arm ball joint stud from the steering knuckle, using special tool ST29020001 (J-24319-01), or equivalent.
8. Remove the upper control arm retaining bolts and nuts.
9. Remove the upper control arm from the vehicle.

To install:

10. Installation is the reverse of the removal procedure.
11. Be sure to replace all wearable components, as required.
12. Check and adjust the front alignment, as required.

WHEEL HUB & BEARING

REMOVAL & INSTALLATION

See Figure 188.

1. Before servicing the vehicle, refer to the Precautions Section.
2. Raise and support the vehicle safely.
3. Remove the tire and wheel assembly.
4. Without disassembling the hydraulic lines, remove caliper torque member bolts. Then, reposition the brake caliper aside with wire. Refer to Brake Caliper, removal & installation.

➡**Do not press the brake pedal while the brake caliper is removed.**

5. Matchmark the disc rotor and wheel hub and bearing assembly, then remove the disc rotor.
6. On models with 4WD, remove the cotter pin, then remove the lock nut from the halfshaft.
7. Remove the wheel sensor from the wheel hub and bearing assembly.
- Inspect the wheel sensor O-ring, replace the wheel sensor assembly if damaged
- Clean the wheel sensor hole and mounting surface with a suitable brake cleaner and clean lint-free shop rag. Be careful that dirt and debris does not enter the axle bearing area
- Apply a coat of suitable grease to the wheel sensor O-ring and mounting hole

✳✳ WARNING

Do not pull on the wheel sensor harness.

8. On models with 4WD, separate the halfshaft from the wheel hub and bearing assembly.
9. Remove the wheel hub and bearing assembly bolts.
10. Remove the splash guard and wheel hub and bearing assembly from the steering knuckle.
11. Carefully remove the wheel sensor and harness through the hole in the splash guard.

To install:

12. Check for deformity, cracks, and damage on each part and replace if necessary.
13. Installation is in the reverse order of removal.
14. Use new bolts when installing the wheel hub and bearing assembly.
15. When installing the disc rotor on the wheel hub and bearing assembly, position the disc rotor according to the alignment matchmark.

ADJUSTMENT

The front wheel bearings are not adjustable. If the lateral run-out on the hub with the disc removed exceeds specification, the hub must be replaced.

1. Move the wheel hub in the axial direction by hand. Make sure there is no looseness of the wheel bearing. Axial end play limit: 0.002 inch (0.05mm) or less. If out of specification, replace the wheel hub and bearing assembly.
2. Rotate the wheel hub and make sure there is no unusual noise or other irregular conditions. If there are any irregular conditions, replace the wheel hub and bearing assembly.

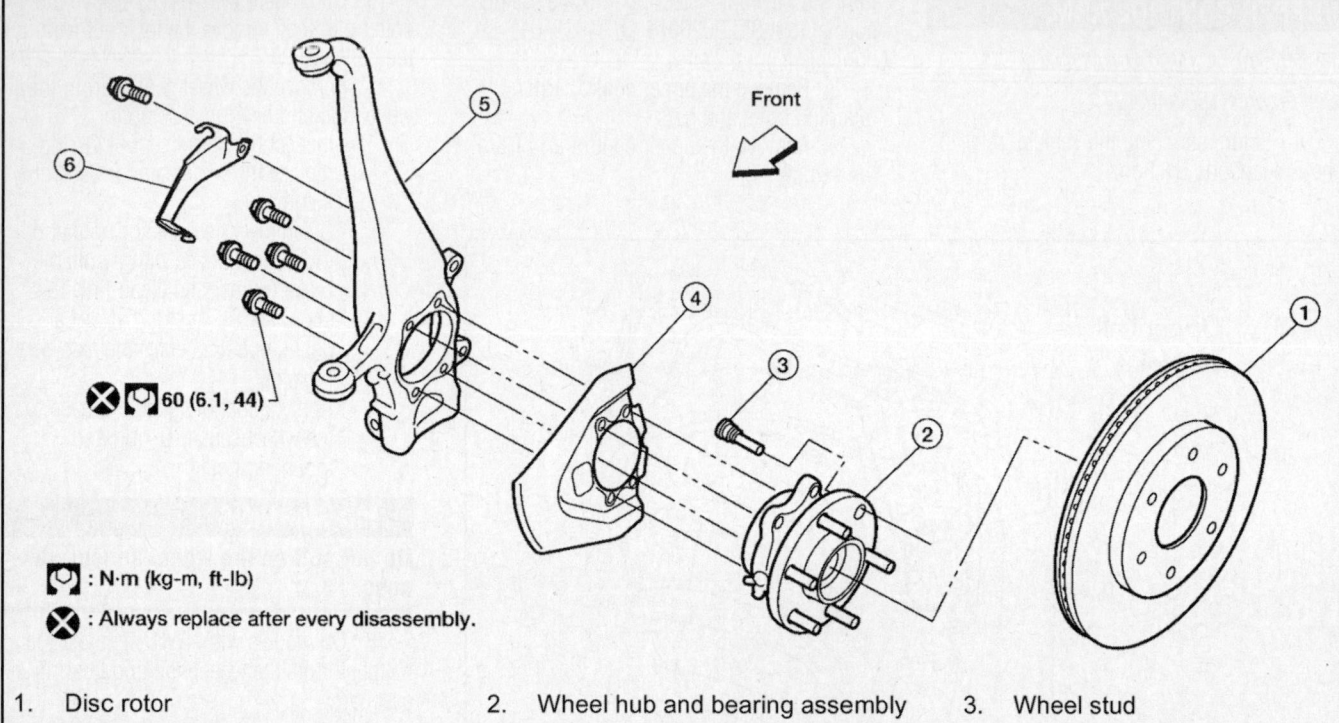

$\bigotimes$ $\boxed{\square}$ 60 (6.1, 44)

$\boxed{\square}$: N·m (kg-m, ft-lb)

$\bigotimes$: Always replace after every disassembly.

1. Disc rotor
4. Splash guard
2. Wheel hub and bearing assembly
5. Steering knuckle
3. Wheel stud
6. Wheel sensor bracket

42050_FRON_G0104

Fig. 188 Front wheel hub and bearing assembly

SUSPENSION   REAR SUSPENSION

LEAF SPRING

REMOVAL & INSTALLATION

See Figures 189.

➡**When installing the components with rubber bushings, the final tightening of the nuts and bolts must be done with the vehicle in an unladen condition (the fuel, engine coolant, and engine oil full; the spare tire, jack, hand tools, and mats in their designated positions) with the tires on the ground.**

1. Before servicing the vehicle, refer to the Precautions Section.
2. Raise and safely support the vehicle.
3. Remove the tire and wheel assembly.
4. Support the rear final drive assembly with a suitable jack to relieve the tension from the rear leaf spring.

➡**The axle weight should be supported, but there should be no compression in the rear leaf spring.**

5. Remove the 4 rear spring clip U-bolt nuts, then remove the rear spring pad and bumper.
6. Remove the rear spring shackle and bushings.
7. Remove the rear leaf spring front nut and bolt.
8. Remove the rear leaf spring.

To install:

9. Check the rear leaf spring for any cracks or damage. Replace the rear leaf spring if necessary.
10. Check the rear spring shackle, rear spring clip U-bolts, bumper, and rear spring pad for excessive wear, cracks, straightness, and damage. Replace any components if necessary.
11. Check all bushings for deformation and cracks. Replace any bushings if necessary.
12. Apply soapsuds to all of the rubber bushings.
13. Install the rear spring shackle and rear leaf spring front nut and bolt. Finger-tighten the nuts.
14. Install the rear spring clip U-bolts and bumper on top of the rear leaf spring.
15. Install the bumper and rear spring

pad, then finger-tighten the nuts under the axle case.
16. Tighten the rear spring clip U-bolt nuts diagonally and evenly to 54 ft. lbs. (73 Nm).

➡**Tighten the rear spring clip U-bolt nuts so the lengths of all the exposed rear spring clip U-bolt threads under the spring pad are equal in length within a tolerance of 0.12 inch (3mm).**

17. Remove the jack supporting the rear final drive assembly and bounce the rear of the vehicle to stabilize the suspension.
18. Tighten the rear spring shackle nuts, rear leaf spring front nut, and shock absorber nuts to specification.
   a. Lower shock absorber bolt/nut to 148 ft. lbs. (200 Nm).
   b. Front spring nut and bolt to 165 ft. lbs. (190 Nm).
   c. Rear spring shackle nuts to 78 ft. lbs. (106 Nm).

➡**When installing the components with rubber bushings, the final nut tightening must be carried out under unladen conditions with the tires on level ground.**

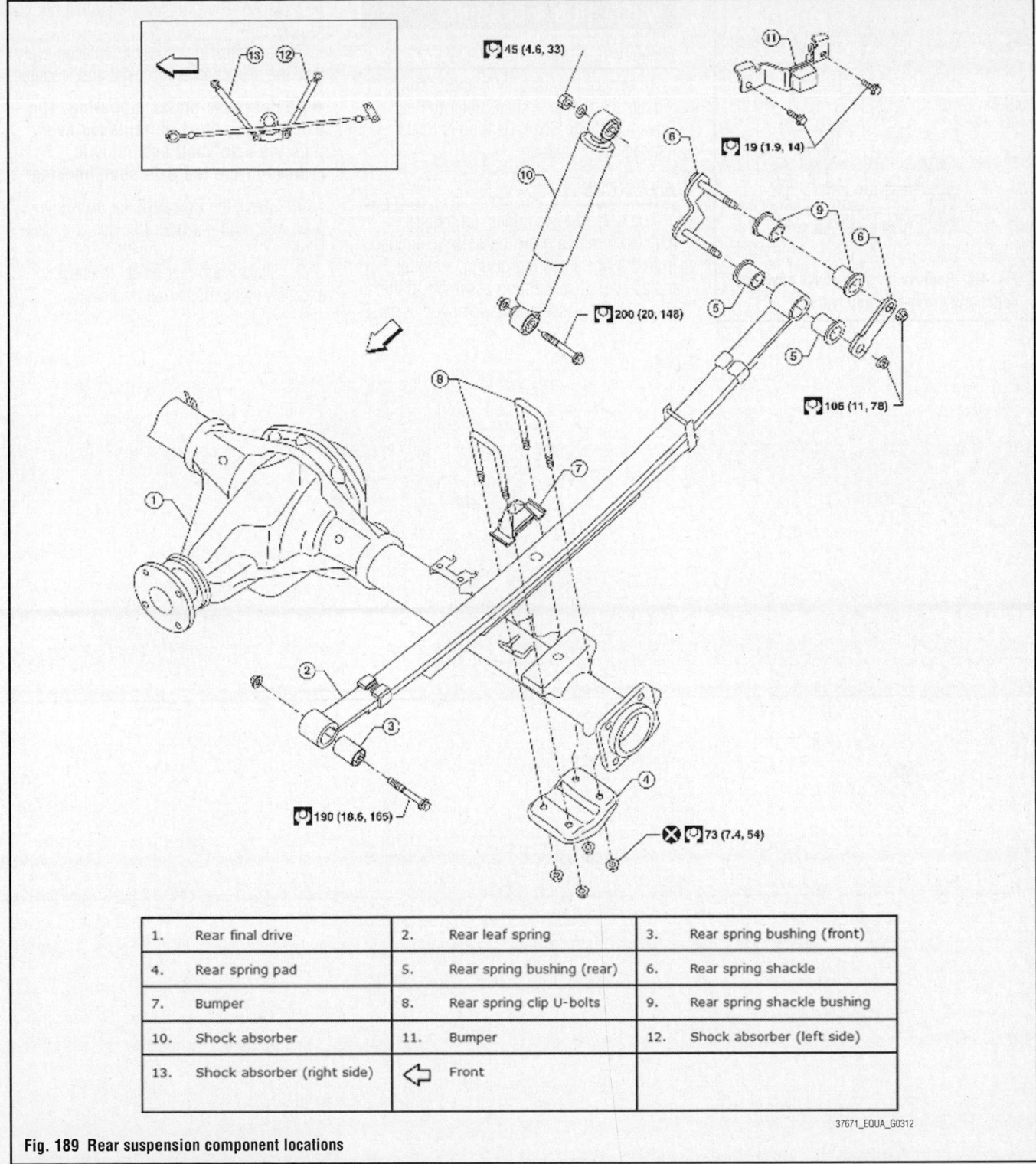

1.	Rear final drive	2.	Rear leaf spring	3.	Rear spring bushing (front)
4.	Rear spring pad	5.	Rear spring bushing (rear)	6.	Rear spring shackle
7.	Bumper	8.	Rear spring clip U-bolts	9.	Rear spring shackle bushing
10.	Shock absorber	11.	Bumper	12.	Shock absorber (left side)
13.	Shock absorber (right side)	⇦	Front		

37671_EQUA_G0312

Fig. 189 Rear suspension component locations

SHOCK ABSORBER

REMOVAL & INSTALLATION

See Figures 189 and 190.

1. Before servicing the vehicle, refer to the Precautions Section.

2. Raise and safely support the vehicle.

3. Support the rear final drive and suspension assembly using a suitable jack.

4. Remove the shock absorber upper and lower nuts and bolts.

5. Remove the shock absorber from the vehicle.

To install:

6. Inspect the shock absorber for any oil leaks, cracks, or deformations. Replace the shock absorber as necessary.

7. Installation is in the reverse order of removal.

8. Tighten the upper shock absorber nut to 33 ft. lbs. (45 Nm).

9. Tighten the lower shock absorber bolt/nut to 148 ft. lbs. (200 Nm).

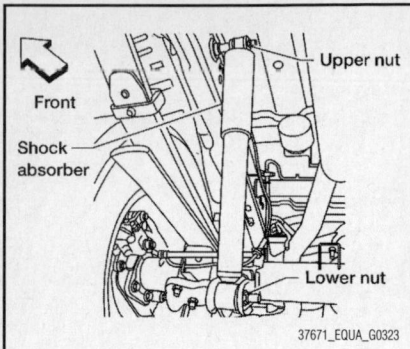

Fig. 190 Remove the rear shock absorber upper and lower nuts and bolts

WHEEL HUB & BEARING

REMOVAL & INSTALLATION

To service the wheel hub and bearing, remove the rear axle shaft. Refer to Rear Drive Axle, Axle Shaft, Bearing & Seal, removal & installation.

ADJUSTMENT

The wheel bearing does not require maintenance. If a growling noise is emitted from the wheel bearing during operation, replace the wheel bearing assembly. If the wheel bearing assembly is removed, it must be replaced. The old assembly must not be re-used.

The axle bearing must be replaced every time the axle shaft is removed and installed.

➡**Do not reuse the axle bearing. The axle bearing must be replaced every time the axle shaft assembly is removed from the axle shaft housing.**

1. Check the axle parts for excessive play, wear, and damage. Replace as necessary.

2. Shake each rear wheel to check for excessive play. Repair as necessary.

SUZUKI

Grand Vitara

25

SPECIFICATIONS AND MAINTENANCE CHARTS

ENGINE AND VEHICLE IDENTIFICATION

ID/Code	Liters (cc)	Cu. In.	Cyl.	Fuel Sys.	Engine Type	Eng. Mfg.	Code ①	Year
		Engine					Model Year	
J24B	2.4 (2,393)	146	4	MFI	DOHC	Suzuki	9	2009
N32A	3.2 (3,196)	195	6	MFI	DOHC	Suzuki	A	2010

MFI: Multi-port Fuel Injection

DOHC: Double Overhead Camshafts

① Indicated at the 4th digit of the Vehicle Identification Number (VIN)

37671_GVIT_C0001

GENERAL ENGINE SPECIFICATIONS
All measurements are given in inches.

Year	Model	Engine Displacement Liters	Engine Series ID	Net Horsepower @ rpm	Net Torque @ rpm (ft. lbs.)	Bore x Stroke (in.)	Compression Ratio	Oil Pressure @ rpm
2009	Grand Vitara	2.4	J24B	166@6,000	162@4,000	3.62 x 3.54	10.0:1	44-73@4000
		3.2	N32A	230@6,200	213@3,500	3.50 x 3.37	10.0:1	44-73@4000
2010	Grand Vitara	2.4	J24B	166@6,000	162@4,000	3.62 x 3.54	10.0:1	44-73@4000
		3.2	N32A	230@6,200	213@3,500	3.50 x 3.37	10.0:1	44-73@4000

37671_GVIT_C0002

GASOLINE ENGINE TUNE-UP SPECIFICATIONS

Year	Engine Displacement Liters	Engine ID	Spark Plug Gap (in.)	Ignition Timing (deg.) MT	Ignition Timing (deg.) AT	Fuel Pump (psi)	Idle Speed (RPM) MT	Idle Speed (RPM) AT	Valve Clearance (in.) Intake	Valve Clearance (in.) Exhaust
2009	2.4	J24B	0.039-0.043	①	①	54-56 ②	630-730	630-730	③	④
	3.2	N32A	0.043-0.047	NA	①	54-56 ②	600-700	600-700	HYD	HYD
2010	2.4	J24B	0.039-0.043	①	①	54-56 ②	630-730	630-730	③	④
	3.2	N32A	0.043-0.047	NA	①	54-56 ②	600-700	600-700	HYD	HYD

NOTE: The Vehicle Emission Control Information label often reflects specification changes made during production.

The label figures must be used if they differ from those in this chart.

NA: Not Applicable   HYD: Hydraulic

① Ignition timing is preset and cannot be adjusted

② With the fuel pump operating and engine stopped

③ 0.0063-0.0094 inch (cold)

④ 0.0123-0.0153 inch (cold)

37671_GVIT_C0003

CAPACITIES

Year	Model	Engine Displacement Liters	Engine ID	Engine Oil with Filter (qts.)	Transmission (pts.) Manual	Transmission (pts.) Auto. ①	Transfer Case (pts.)	Drive Axle Front (pts.)	Drive Axle Rear (pts.)	Fuel Tank (gal.)	Cooling System (qts.)
2009	Grand Vitara	2.4	J24B	5.5	4.0	②	③	1.9-2.1	④	17.4	⑤
		3.2	N32A	6.6	NA	②	③	1.9-2.1	④	17.4	10.0
2010	Grand Vitara	2.4	J24B	5.5	4.0	②	③	1.9-2.1	④	17.4	⑤
		3.2	N32A	6.6	NA	②	③	1.9-2.1	④	17.4	10.0

NA: Not Applicable

NOTE: All capacities are approximate. Add fluid gradually and check to be sure a proper fluid level is obtained.

① Drain and refill
② With 4-speed (4AT): 5.3 pints (2.5 liters)
   With 5-speed (5AT): 6.3 pints (3.0 liters)
③ Motor-shift type: 3.4 pints (1.6 liters)
   Non-shift type: 3.6 pints (1.7 liters)

④ 4WD models and 2WD models with 4AT: 1.7 pints (0.8 liters)
   2WD models with 5AT or 5MT: 2.3 pints (1.1 liters)
⑤ With 4-speed automatic (4AT): 7.6 quarts (7.2 liters)
   With 5-speed manual transmission (5MT): 7.7 quarts (7.3 liters)

37671_GVIT_C0004

FLUID SPECIFICATIONS

Year	Model	Engine Displacement Liters	Engine ID	Engine Oil	Manual Trans.	Auto. Trans.	Transfer Case	Drive Axle Front	Drive Axle Rear	Power Steering Fluid	Brake Master Cylinder	Cooling System
2009	Grand Vitara	2.4	J24B	5W-30	①	②	③	③	③	④	⑤	⑥
		3.2	N32A	5W-30	①	②	③	③	③	④	⑤	⑥
2010	Grand Vitara	2.4	J24B	5W-30	①	②	③	③	③	④	⑤	⑥
		3.2	N32A	5W-30	①	②	③	③	③	④	⑤	⑥

① API GL-4 gear oil, Viscosity SAE 75W-90
② SUZUKI ATF 3317 or Mobil ATF 3309
③ API GL-5 gear oil, Viscosity SAE 80W-90

④ ATF Dexron-II, or equivalent
⑤ Refer to fluid specification indicated on the brake reservoir cap of the vehicle
⑥ Ethylene glycol base antifreeze

37671_GVIT_C0005

VALVE SPECIFICATIONS

Year	Engine Displacement Liters	Engine ID	Seat Angle (deg.)	Face Angle (deg.)	Spring Test Pressure (lbs. @ in.)	Spring Installed Height (in.)	Stem-to-Guide Clearance (in.) Intake	Stem-to-Guide Clearance (in.) Exhaust	Stem Diameter (in.) Intake	Stem Diameter (in.) Exhaust
2009	2.4	J24B	①	45	②	③	0.0008-0.0018	0.0018-0.0028	0.2152-0.2157	0.2142-0.2147
	3.2	N32A	④	45	⑤	⑥	0.0010-0.0026	0.0014-0.0030	0.2344-0.2352	0.2341-0.2348
2010	2.4	J24B	①	45	②	③	0.0008-0.0018	0.0018-0.0028	0.2152-0.2157	0.2142-0.2147
	3.2	N32A	④	45	⑤	⑥	0.0010-0.0026	0.0014-0.0030	0.2344-0.2352	0.2341-0.2348

① Intake valve seat: First cut at 22 degrees, 2nd cut at 60 degrees, 3rd cut at 45 degrees. The 3rd cut must produce the desired seat width (0.0414-0.0531 inch).
   Exhaust valve seat: First cut at 22 degrees, 2nd cut at 65 degrees, 3rd cut at 45 degrees. The 3rd cut must produce the desired seat width (0.0441-0.0559 inch).

② Valve spring preload when compressed to 1.61 inches Standard: 38.3-44.0 lbs., Limit: 37.1 lbs.
   Valve spring preload when compressed to 1.23 inches Standard: 85.2-98.2 lbs., Limit: 84.3 lbs.

③ Valve spring free length Standard: 2.021 inches, Limit: 1.981 inches

④ Intake valve seat: First cut at 30 degrees, 2nd cut at 60 degrees, 3rd cut at 45 degrees. The 3rd cut must produce the desired seat width (0.0394-0.0551 inch).
   Exhaust valve seat: First cut at 30 degrees, 2nd cut at 60 degrees, 3rd cut at 45 degrees. The 3rd cut must produce the desired seat width (0.0551-0.0709 inch).

⑤ Valve spring preload when compressed to 1.378 inches Standard: 55.5-61.4 lbs., Limit: 54.0 lbs.

⑥ Valve spring free length Standard: 1.6732-1.7913 inches, Limit: 1.6535 inches

37671_GVIT_C0006

CAMSHAFT AND BEARING SPECIFICATIONS CHART

All measurements are given in inches.

Year	Engine Displ. Liters	Engine ID	Journal Dia.	Brg. Oil Clearance	Shaft End-play	Runout	Journal Bore	Cam Height Intake	Exhaust
2009	2.4	J24B	①	②	0.0040-0.0138	0.0005	③	1.8513-1.8576	1.8324-1.8387
	3.2	N32A	⑤	0.0016-0.0033	0.0018-0.0085	④	⑥	1.6687-1.6805	1.6703-1.6821
2010	2.4	J24B	①	②	0.0040-0.0138	0.0005	③	1.8513-1.8576	1.8324-1.8387
	3.2	N32A	⑤	0.0016-0.0033	0.0018-0.0085	④	⑥	1.6687-1.6805	1.6703-1.6821

① Number 1 intake side: 1.1402-1.1409 inches
Number 1 exhaust side: 1.0614-1.0622 inches
All others: 1.0220-1.0228 inches

② Number 1 intake side Standard: 0.00079-0.00283 inch, Limit: 0.0039 inch
All others Standard: 0.00079-0.00244 inch, Limit: 0.00374 inch

③ Number 1 intake side: 1.2599-1.2608 inches
Number 1 exhaust side: 1.0630-1.0638 inches
All others: 1.0237-1.0244 inches

④ Outer journals: 0.0010 inch; inner journals: 0.0020 inch

⑤ Number 1 intake and exhaust side Standard: 1.3754-1.3764 inches
All others Standard: 1.0605-1.0614 inches

⑥ Number 1 intake and exhaust side Standard: 1.3780-1.3787 inches
All others Standard: 1.0630-1.06384 inches

37671_GVIT_C0007

CRANKSHAFT AND CONNECTING ROD SPECIFICATIONS

All measurements are given in inches.

Year	Engine Displacement Liters	Engine ID	Crankshaft Main Brg. Journal Dia.	Main Brg. Oil Clearance	Shaft End-play	Thrust on No.	Connecting Rod Journal Diameter	Oil Clearance	Side Clearance
2009	2.4	J24B	2.0466-2.0472	0.00087-0.00133	0.0040-0.0137	3	1.9678-1.9685	0.0018-0.0024	0.0099-0.0157
	3.2	N32A	2.6769-2.6775	0.0013-0.0033	0.0039-0.0130	3	2.2044-2.2050	0.0011-0.0032	0.0037-0.0140
2010	2.4	J24B	2.0466-2.0472	0.00087-0.00133	0.0040-0.0137	3	1.9678-1.9685	0.0018-0.0024	0.0099-0.0157
	3.2	N32A	2.6769-2.6775	0.0013-0.0033	0.0039-0.0130	3	2.2044-2.2050	0.0011-0.0032	0.0037-0.0140

37671_GVIT_C0008

PISTON AND RING SPECIFICATIONS

All measurements are given in inches.

Year	Engine Displ. Liters	Engine ID	Piston Clearance	Ring Gap Top Compression	Bottom Compression	Oil Control	Ring Side Clearance Top Compression	Bottom Compression	Oil Control
2009	2.4	J24B	①	0.0079-0.0129	0.0126-0.0188	0.0079-0.0196	0.0095-0.0110	0.0012-0.0027	0.0016-0.0070
	3.2	N32A	0.0012-0.0023	0.0079-0.0126	0.0130-0.0177	0.0039-0.0197	0.0015-0.0031	0.0012-0.0028	0.0016-0.0047
2010	2.4	J24B	①	0.0079-0.0129	0.0126-0.0188	0.0079-0.0196	0.0095-0.0110	0.0012-0.0027	0.0016-0.0070
	3.2	N32A	0.0012-0.0023	0.0079-0.0126	0.0130-0.0177	0.0039-0.0197	0.0015-0.0031	0.0012-0.0028	0.0016-0.0047

① Piston clearance limit: 0.0047 inch

37671_GVIT_C0009

TORQUE SPECIFICATIONS
All readings in ft. lbs.

Year	Engine Displacement Liters	Engine ID	Cylinder Head Bolts	Main Bearing Bolts	Rod Bearing Bolts	Crankshaft Damper Bolts	Flywheel Bolts	Manifold Intake	Manifold Exhaust	Spark Plugs	Oil Pan Drain Plug
2009	2.4	J24B	①	②	③	111	52	NS	NS	18	26
	3.2	N32A	④	⑤	⑥	⑦	⑧	⑨	15	13	26
2010	2.4	J24B	①	②	③	111	52	NS	NS	18	26
	3.2	N32A	④	⑤	⑥	⑦	⑧	⑨	15	13	26

NS: Not Specified

① Apply engine oil to bolts, refer to procedure for tightening sequence

Step 1: Bolts 1-10 to 15 ft. lbs.

Step 2: Bolts 1-10 to 30 ft. lbs.

Step 3: Bolts 1-10 plus 60 degrees

Step 4: Bolts 1-10 plus 80 degrees

② Apply engine oil to bolts, refer to procedure for tightening sequence

Step 1: Bolts 1-10 to 37 ft. lbs.

Step 2: Bolts 1-10 to 0 ft. lbs.

Step 3: Bolts 1-10 to 15 ft. lbs.

Step 4: Bolts 1-10 to 26 ft. lbs.

Step 5: Bolts 1-10 plus 40 degrees

Step 6: Bolts 1-10 plus 40 degrees, again

③ Apply engine oil to bolt threads and seats of cap bolts

Step 1: 11 ft. lbs.

Step 2: Plus 45 degrees

Step 3: Plus 45 degrees

④ Refer to procedure for tightening sequence; do not apply oil to bolts

Step 1: Bolts 1-8 to 22 ft. lbs.

Step 2: Bolts 1-8 plus 150 degrees

Step 3: Bolts 9-10 to 11 ft. lbs.

Step 4: Bolts 9-10 plus 75 degrees

⑤ Refer to procedure for tightening sequence; do not apply oil to bolts

Step 1: Bolts 1-8 to 15 ft. lbs.

Step 2: Bolts 1-8 plus 80 degrees

Step 3: Bolts 9-16 to 11 ft. lbs.

Step 4: Bolts 9-16 plus 110 degrees

Step 5: Bolts 17-24 to 22 ft. lbs.

Step 6: Bolts 17-24 plus 60 degrees

⑥ Do not apply oil to connecting rod bolts

Step 1: 22 ft. lbs.

Step 2: 0 ft. lbs.

Step 3: 18 ft. lbs.

Step 4: Plus 110 degrees

⑦ Step 1: 89 ft. lbs.

Step 2: 0 ft. lbs.

Step 3: 74 ft. lbs.

Step 4: Plus 150 degrees

⑧ Step 1: 22 ft. lbs.

Step 2: Plus 45 degrees

⑨ Refer to procedure for tightening sequence

Step 1: Bolts 1-6 to 89 inch lbs.

Step 2: Bolts 1-8 to 18 ft. lbs.

37671_GVIT_C0010

WHEEL ALIGNMENT

Year	Model		Caster Range (+/-Deg.)	Caster Preferred Setting (Deg.)	Camber Range (+/-Deg.)	Camber Preferred Setting (Deg.)	Toe-in (Deg.)
2009	Grand Vitara	F	①	①	②	②	③
		R	NA	NA	④	④	⑤
2010	Grand Vitara	F	①	①	②	②	③
		R	NA	NA	④	④	⑤

NOTE: Measurements given for an unladen vehicle with fuel, coolant, and engine oil full; spare tire, jack, hand tools, and mats in designated positions.

F: Front   NA: Not Applicable

R: Rear

① Front caster specification: 2 degrees 30'

② Front camber specification: 0 degrees 00' +/- 1 degree

③ Front toe specification: 0 degrees +/- 5'

④ Rear camber specification: -1 degree 15' +/- 40'

⑤ Rear toe specification: 0 degrees 14' +/- 5'

37671_GVIT_C0011

TIRE, WHEEL AND BALL JOINT SPECIFICATIONS

Year	Model	OEM Tires Standard	OEM Tires Optional	Tire Pressures (psi) Front	Tire Pressures (psi) Rear	Wheel Size	Ball Joint Inspection	Lug Nut Torque (ft. lbs.)
2009	Grand Vitara Base	P225/70R16 101H	NS	①	①	16 x 6.5J	②	74
	Grand Vitara Premium	P225/70R16 101H	NS	①	①	16 x 6.5J	②	74
	Grand Vitara Special Edition	P225/70R16 101H	NS	①	①	16 x 6.5J	②	74
	Grand Vitara XSport	P225/70R16 101H	P225/65R17 100H	①	①	16 x 6.5J or 17 x 6.5J	②	74
	Grand Vitara Limited	P225/65R17 100H	P225/60R18 100H	①	①	17 x 6.5J or 18 x 7J	②	74
2010	Grand Vitara Base	P225/70R16 101H	NS	①	①	16 x 6.5J	②	74
	Grand Vitara Premium	P225/70R16 101H	NS	①	①	16 x 6.5J	②	74
	Grand Vitara Special Edition	P225/70R16 101H	NS	①	①	16 x 6.5J	②	74
	Grand Vitara XSport	P225/70R16 101H	P225/65R17 100H	①	①	16 x 6.5J or 17 x 6.5J	②	74
	Grand Vitara Limited	P225/65R17 100H	P225/60R18 100H	①	①	17 x 6.5J or 18 x 7J	②	74

OEM: Original Equipment Manufacturer

PSI: Pounds Per Square Inch

NS: Not Specified

① See the tire placard on the vehicle

② Check for smooth rotation of the ball stud and for any damage to the ball stud or dust cover. The suspension control arm and ball joint cannot be separated.

If there is any damage to either part, the suspension control arm must be replaced as a complete unit.

37671_GVIT_C0012

BRAKE SPECIFICATIONS
All measurements in inches unless noted

Year	Model		Brake Disc Original Thickness	Brake Disc Minimum Thickness	Brake Disc Maximum Runout	Brake Drum Diameter Original Inside Diameter	Brake Drum Diameter Max. Wear Limit	Brake Drum Diameter Maximum Machine Diameter	Minimum Lining Thickness	Brake Caliper Carrier Bolts (ft. lbs.)	Brake Caliper Guide Pin Bolts (ft. lbs.)
2009	Grand Vitara	F	0.984	0.905	0.004	NA	NA	NA	0.079	62	26
		R	0.630	0.551	0.004	NA	NA	NA	0.059	63	32
2010	Grand Vitara	F	0.984	0.905	0.004	NA	NA	NA	0.079	62	26
		R	0.630	0.551	0.004	NA	NA	NA	0.059	63	32

F: Front

R: Rear

NA: Not Applicable

37671_GVIT_C0013

MAINTENANCE SERVICES
SUZUKI GRAND VITARA

TO BE SERVICED	TYPE OF SERVICE	7.5	15	22.5	30	37.5	45	52.5	60	67.5	75	82.5	90	97.5	105	112.5	120	127.5	135	142.5	150
Engine and Emission Control																					
Accessory drive belt (tension, damage)	S/I				✓				✓				✓				✓				✓
Valve lash (clearance) J24B engine	S/I								✓				✓								
Engine oil and filter	R	✓	✓	✓	✓	✓	✓	✓	✓	✓	✓	✓	✓	✓	✓	✓	✓	✓	✓	✓	✓
Cooling system, hoses and connections (leakage, damage)	S/I				✓				✓				✓				✓				✓
Engine coolant (Suzuki long life standard: green color)	R				✓				✓				✓				✓				✓
Engine coolant (Suzuki super long life: blue color)	R	First time only: Replace at 90,000 miles or 90 months. Second time and after: Replace every 45,000 miles or 45 months.																			
Exhaust system (Replace mounting rubber every 60,000 miles)	S/I				✓				✓								✓				✓
EVAP canister (EVAP leak check model)	R	Replace every 150,000 or 180 months																			
Emission-related hoses and tubes	S/I								✓								✓				✓
Ignition System																					
Spark Plugs (Nickel type)	R				✓				✓				✓				✓				✓
Spark Plugs (Iridium or platinum type)	R								✓								✓				
Ignition coil (plug cap)	S/I				✓				✓				✓				✓				✓
Fuel System																					
Engine air cleaner (paved road usage)	R				✓				✓				✓				✓				✓
Fuel tank and cap (leakage, damage)	S/I				✓				✓				✓				✓				✓
Fuel filter	R																				✓
Fuel lines and connections	S/I				✓				✓								✓				✓
EVAP canister air suction filter (EVAP leak check model)	R	Replace every 150,000 or 180 months																			
Brake System																					
Brake pads and rotors (wear, damage)	S/I		✓		✓		✓		✓		✓		✓		✓		✓		✓		✓
Brake hoses and pipes (leakage, damage)	S/I		✓		✓		✓		✓		✓		✓		✓		✓		✓		✓
Brake fluid	S/I		✓		✓		✓				✓		✓		✓				✓		
Brake fluid (replace)	R								✓								✓				
Parking brake lever and cable (damage, stroke, operation)	S/I		✓		✓		✓		✓		✓		✓		✓		✓		✓		✓
Brake pedal	S/I		✓		✓		✓		✓		✓		✓		✓		✓		✓		✓
Chassis and Body																					
Clutch pedal / fluid (M/T)	S/I		✓		✓		✓		✓		✓		✓		✓		✓		✓		✓
Tires / wheel discs (wear, damage, rotation)	S/I	✓	✓	✓	✓	✓	✓	✓	✓	✓	✓	✓	✓	✓	✓	✓	✓	✓	✓	✓	✓
Wheel Bearings	S/I		✓		✓		✓		✓		✓		✓		✓		✓		✓		✓
Drive shaft boots and propeller shaft	S/I		✓		✓		✓		✓		✓		✓		✓		✓		✓		✓
Suspension system (tightness, damage, rattle, breakage)	S/I		✓		✓		✓		✓		✓		✓		✓		✓		✓		✓
Steering system (tightness, damage, breakage, rattle)	S/I	✓	✓	✓	✓	✓	✓	✓	✓	✓	✓	✓	✓	✓	✓	✓	✓	✓	✓	✓	✓
Manual transmission oil (leakage, level)	S/I	✓				✓				✓				✓				✓			
Manual transmission oil (replace)	R				✓				✓				✓				✓				✓
Automatic Transmission (A/T) fluid level	S/I		✓		✓		✓		✓		✓		✓				✓		✓		✓
A/T fluid change	R	Replace every 105,000 miles																			
A/T fluid hose	S/I					✓							✓						✓		
Transfer oil (4WD) (leakage, level)	S/I	✓			✓		✓				✓		✓		✓				✓		✓
Transfer oil (4WD) (replace)	R								✓								✓				
Differential oil and extension case oil (2WD) (leakage, level)	S/I												✓		✓		✓		✓		✓
Differential oil and extension case oil (2WD)	R	✓																			
All latches, hinges and locks	S/I	✓	✓	✓	✓	✓	✓	✓	✓	✓	✓	✓	✓	✓	✓	✓	✓	✓	✓	✓	✓
Power steering system	S/I	✓	✓	✓	✓	✓	✓	✓	✓	✓	✓	✓	✓	✓	✓	✓	✓	✓	✓	✓	✓
Cabin air filter	S/I		✓			✓					✓					✓					
Cabin air filter (replace)	R				✓				✓				✓				✓				✓

R: Replace
S/I: Inspect and service, if necessary

FREQUENT OPERATION MAINTENANCE (SEVERE SERVICE)

If a vehicle is operated under any of the following conditions, it is considered severe service:
- Extremely dusty areas
- 50% or more of the vehicle operation is in 90°F (32°C) or higher temperatures, or constant operation in temperatures below 32°F (0°C)
- Prolonged idling (vehicle operation in stop and go traffic)
- Frequent short running periods (engine does not warm to normal operating temperatures)
- Police, taxi, delivery usage, or trailer towing usage
- Driving in hilly or mountainous terrain

Accessory drive belt: service or inspect every 15,000 miles or 15 months
Oil & oil filter: replace every 3,000 miles or 3 months
Exhaust system: service or inspect every 6,000 miles or 6 months
Engine air cleaner: service or inspect every 3,000 miles or 3 months
Engine air cleaner: replace every 15,000 miles or 15 months
EVAP canister air suction filter (EVAP leak check model): replace every 60,000 miles or 60 months
Brake pads and rotors (wear, damage): service or inspect every 6,000 miles or 6 months
Drive shaft boots and propeller shaft: service or inspect every 6,000 miles or 6 months
Suspension system (tightness, damage, rattle, breakage): service or inspect every 6,000 miles or 6 months
Manual transmission oil: replace every 15,000 miles or 15 months
Automatic transmission fluid: replace every 15,000 miles or 15 months
Transfer oil (4WD), differential oil and extension case oil (2WD): replace every 15,000 miles or 15 months
Cabin air filter: service or inspect every 6,000 miles or 6 months; replace every 30,000 miles or 30 months

BRAKES — INFORMATION AND PRECAUTIONS

ANTI-LOCK SYSTEMS

• Certain components within the ABS system are not intended to be serviced or repaired individually.

• Do not use rubber hoses or other parts not specifically specified for and ABS system. When using repair kits, replace all parts included in the kit. Partial or incorrect repair may lead to functional problems and require the replacement of components.

• Lubricate rubber parts with clean, fresh brake fluid to ease assembly. Do not use shop air to clean parts; damage to rubber components may result.

• Use only DOT 3 brake fluid from an unopened container.

• If any hydraulic component or line is removed or replaced, it may be necessary to bleed the entire system.

• A clean repair area is essential. Always clean the reservoir and cap thoroughly before removing the cap. The slightest amount of dirt in the fluid may plug an orifice and impair the system function. Perform repairs after components have been thoroughly cleaned; use only denatured alcohol to clean components. Do not allow ABS components to come into contact with any substance containing mineral oil; this includes used shop rags.

• The Anti-Lock control unit is a microprocessor similar to other computer units in the vehicle. Ensure that the ignition switch is **OFF** before removing or installing controller harnesses. Avoid static electricity discharge at or near the controller.

• If any arc welding is to be done on the vehicle, the control unit should be unplugged before welding operations begin.

DISC AND DRUM SYSTEMS

> ✳✳ **CAUTION**
>
> Dust and dirt accumulating on brake parts during normal use may contain asbestos fibers from production or aftermarket brake linings. Breathing excessive concentrations of asbestos fibers can cause serious bodily harm. Exercise care when servicing brake parts. Do not sand or grind brake lining unless equipment used is designed to contain the dust residue. Do not clean brake parts with compressed air or by dry brushing. Cleaning should be done by dampening the brake components with a fine mist of water, then wiping the brake components clean with a dampened cloth. Dispose of cloth and all residue containing asbestos fibers in an impermeable container with the appropriate label. Follow practices prescribed by the Occupational Safety and Health Administration (OSHA) and the Environmental Protection Agency (EPA) for the handling, processing, and disposing of dust or debris that may contain asbestos fibers.

BRAKES — BLEEDING THE BRAKE SYSTEM

BLEEDING PROCEDURE

BLEEDING PROCEDURE

See Figure 1.

> ✳✳ **CAUTION**
>
> Brake fluid contains polyglycol ethers and polyglycols. Avoid contact with the eyes and wash your hands thoroughly after handling brake fluid. If you do get brake fluid in your eyes, flush your eyes with clean, running water for 15 minutes. If eye irritation persists, or if you have taken brake fluid internally, IMMEDIATELY seek medical assistance.

> ✳ **WARNING**
>
> Clean, high quality brake fluid is essential to the safe and proper operation of the brake system. You should always buy the highest quality brake fluid that is available. If the brake fluid becomes contaminated, drain and flush the system, then refill the master cylinder with new fluid. Never reuse any brake fluid. Any brake fluid that is removed from the system should be discarded. Also, do not allow any brake fluid to come in contact with a painted surface; it will damage the paint.

➡ The bleeding operation is necessary to remove air whenever it enters the hydraulic brake system. The hydraulic lines of the brake system are based on the diagonal split system. When a brake pipe or hose was disconnected at the wheel, a bleeding operation must be performed at both ends of the line of the removed pipe or hose. When any joint part of the master cylinder or other joint part between the master cylinder and each brake (wheel) was removed, the hydraulic brake system must be bled at all 4 wheel brakes.

➡ Perform the bleeding operation starting with the rear brake caliper or the wheel cylinder farthest from the master cylinder and then at the front caliper of the same brake line. Do the same on the other brake line.

1. Before servicing the vehicle, refer to the Precautions Section.
2. Clean all around the brake fluid reservoir filler cap.
3. Fill the master cylinder reservoir with the specified brake fluid and keep it at least ½ full of fluid during the bleeding operation.
4. Raise and safely support the vehicle.
5. Remove the bleeder plug cap. Attach a vinyl tube to the bleeder plug, and insert the other end into a container.
6. Depress the brake pedal several times, and then while holding it depressed, loosen the bleeder plug about one-third to one-half turn.
7. When the fluid pressure in the cylinder is almost depleted, retighten the bleeder plug.
8. Repeat this operation until there are no more air bubbles in the hydraulic line.
9. When the bubbles stop, depress and hold the brake pedal and tighten the bleeder plug.

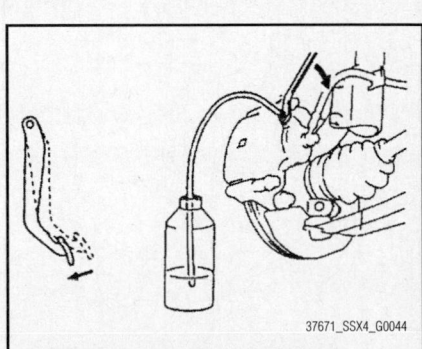

37671_SSX4_G0044

Fig. 1 Remove the bleeder plug cap, attach a vinyl tube to the bleeder plug, and insert the other end into a container

- Tighten the front brake caliper bleeder plug to 62 inch lbs. (7 Nm)
- Tighten the rear brake caliper bleeder plug to 71 inch lbs. (8 Nm)

10. Attach the bleeder plug cap when finished at the bleeder plug.

11. After completing the bleeding operation, apply fluid pressure to the pipe line and check for leakage.

12. Replenish the fluid in the reservoir to the specified level. Refer to Fluid Fill Procedure.

13. Check the brake pedal for sponginess. If found spongy, repeat the entire bleeding procedure.

✳✳ CAUTION

Do not attempt to drive a vehicle until the brakes are functioning properly.

BRAKES   ANTI-LOCK BRAKE SYSTEM (ABS)

SPEED SENSORS

REMOVAL & INSTALLATION

Front Speed Sensor

See Figure 2.

✳✳ WARNING

Do not pull on wire harness when removing and installing the front wheel speed sensor. Do not cause damage to the surface of the front wheel speed sensor and do not allow dust or other foreign material to enter the installation hole.

1. Before servicing the vehicle, refer to the Precautions Section.

2. Disconnect the negative battery cable.

3. Disconnect the front wheel speed sensor coupler.

4. Raise the vehicle and remove the wheel.

5. Remove the harness clamp, clamp bolts, and grommet.

6. Remove the front wheel speed sensor from the knuckle.

To install:

7. Installation is the reverse of the removal procedure.

8. Check that no foreign material is attached to the sensor and mating encoder.

9. Tighten the wheel speed sensor and speed sensor harness clamp bolts to 97 inch lbs. (11 Nm).

10. Reconnect the speed sensor harness coupler.

11. Check that there is no clearance between the sensor and the knuckle.

12. Connect the negative battery cable.

Rear Speed Sensor

See Figure 3.

✳✳ WARNING

Do not pull on the wire harness when removing and installing the rear wheel speed sensor. Do not cause damage to the surface of the rear wheel speed sensor and do not allow dust or other foreign material to enter the installation hole.

1. Before servicing the vehicle, refer to the Precautions Section.

2. Disconnect the negative battery cable.

3. Disconnect the rear wheel speed sensor coupler.

4. Raise the vehicle and remove the rear wheel.

5. Remove the harness clamp and the clamp bolts.

6. Remove the rear wheel speed sensor from the knuckle.

To install:

7. Installation is the reverse of the removal procedure.

8. Check that no foreign material is attached to the sensor and mating encoder.

9. Be sure to install the wheel speed sensor and its bolt at the correct position.

10. Tighten the sensor bolt and harness clamp bolts to 97 inch lbs. (11 Nm).

11. Check that there is no clearance between the sensor and the brake back plate.

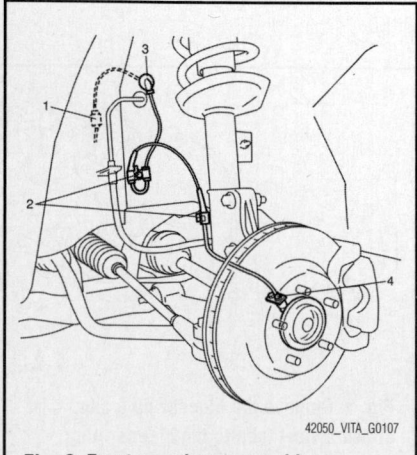

42050_VITA_G0107

Fig. 2 Front speed sensor and harness

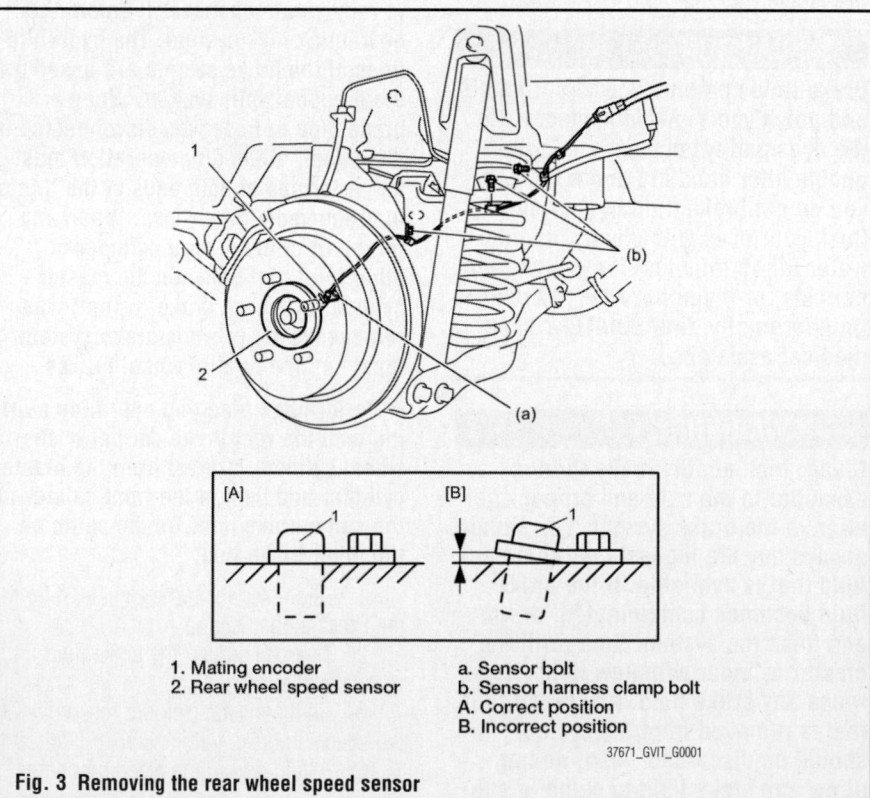

1. Mating encoder
2. Rear wheel speed sensor

a. Sensor bolt
b. Sensor harness clamp bolt
A. Correct position
B. Incorrect position

37671_GVIT_G0001

Fig. 3 Removing the rear wheel speed sensor

BRAKE CALIPER

REMOVAL & INSTALLATION

See Figure 4.

✻✻ WARNING

Do not allow brake fluid to get on painted surfaces. Painted surfaces will be damaged by brake fluid, flush all affected areas with water immediately if any fluid is spilled.

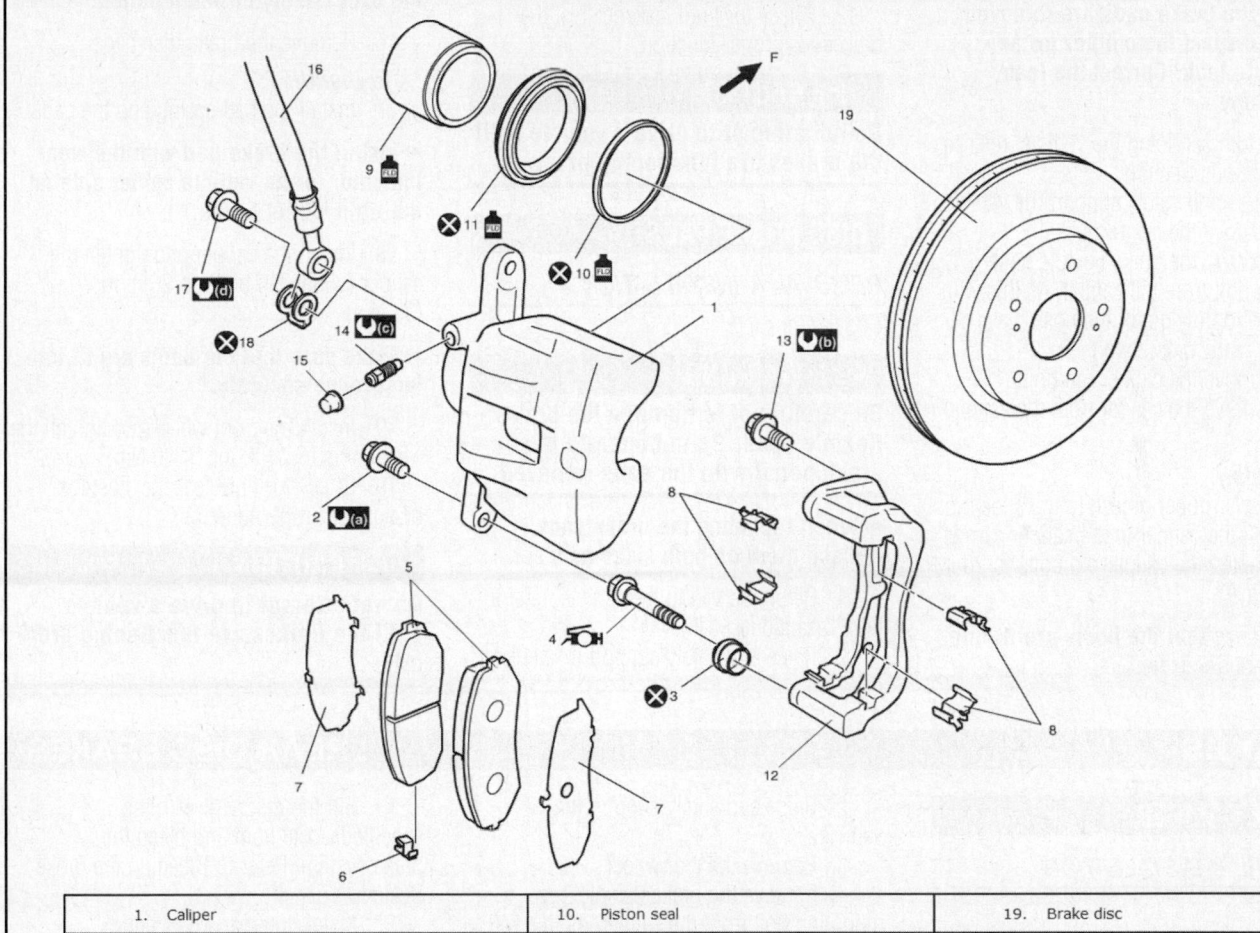

1.	Caliper	10.	Piston seal	19.	Brake disc	
2.	Caliper pin bolt	11.	Piston boot	F:	Forward	
3.	Slide pin boot	12.	Brake caliper carrier	(a)	36 N·m (3.6 kgf-m, 26.0 lbf-ft)	
4.	Slide pin : Use included grease for pin boot.	13.	Caliper carrier bolt	(b)	85 N·m (8.5 kgf-m, 61.5 lbf-ft)	
5.	Brake pad	14.	Front caliper bleeder plug	(c)	7 N·m (0.7 kgf-m, 5.0 lbf-ft)	
6.	Wear indicator (vehicle center side on right wheel brake only)	15.	Bleeder plug cap	(d)	23 N·m (2.3 kgf-m, 17.0 lbf-ft)	
7.	Shim	16.	Brake flexible hose	✕	Do not reuse.	
8.	Pad spring	17.	Flexible hose joint bolt	FLD	Apply brake fluid	
9.	Disc brake piston	18.	Hose washer			

37671_GVIT_G0003

Fig. 4 Exploded view of front disc brake components

✳ WARNING

During removal, be careful not to damage the brake flexible hose.

→Do not reuse the brake flexible hose washers. Otherwise, the brake fluid may leak.

→When the brake pads are removed, visually inspect the caliper for any brake fluid leak. Correct the leaky point, if any.

1. Before servicing the vehicle, refer to the Precautions Section.
2. Raise and safely support the vehicle.
3. Remove the front wheel.
4. Remove the brake flexible hose mounting bolt from the caliper. As this will allow fluid to flow out of the hose, have a container ready beforehand.
5. Remove the caliper pin bolts.
6. Remove the caliper from the caliper carrier.

To install:

7. Apply rubber grease to the slide pin, then install the caliper to the caliper carrier.
8. Tighten the caliper pin bolts to 26 ft. lbs. (36 Nm).

→Make sure that the boots are fit into the groove securely.

9. Install the brake flexible hose and tighten the flexible hose joint bolt to 17 ft. lbs. (23 Nm).
10. Fill the reservoir with the specified brake fluid and bleed the brake system. Refer to Bleeding the Brake System.
11. Install the front wheel and tighten the wheel nuts to 74 ft. lbs. (100 Nm).
12. Check for fluid leakage. Test the brakes for proper function.

✳ CAUTION

Do not attempt to drive a vehicle until the brakes are functioning properly.

DISC BRAKE PADS

REMOVAL & INSTALLATION
See Figure 4.

✳ WARNING

Be careful not to damage the brake flexible hose. Do not operate the brake pedal with the pads removed.

→When replacing the brake pads, replace them on both sides as a set.

1. Before servicing the vehicle, refer to the Precautions Section.
2. Raise and safely support the vehicle.

3. Remove the front wheels.
4. Remove the caliper pin bolts.
5. Remove the caliper from the caliper carrier.

→Hang the removed caliper with a wire hook, or the like, so as to prevent the brake hose from bending and twisting excessively or being pulled.

6. Remove the brake disc pads.

To install:

7. Install the pad spring and the pads.

→Install the brake pad with the wear indicator to the vehicle center side on the right wheel brake.

8. Install the caliper and tighten the caliper (slide) pin bolts to 26 ft. lbs. (36 Nm).

→Make sure that the boots are fit into the groove securely.

9. Install the front wheel and tighten the wheel nuts to 74 ft. lbs. (100 Nm).
10. Check for fluid leakage. Test the brakes for proper function.

✳ CAUTION

Do not attempt to drive a vehicle until the brakes are functioning properly.

BRAKES

BRAKE CALIPER

REMOVAL & INSTALLATION
See Figure 5.

✳ WARNING

Do not allow brake fluid to get on any painted surfaces. The painted surfaces may be damaged by brake fluid. Flush the surface with water immediately if any fluid is spilled.

✳ WARNING

Be careful not to damage the brake flexible hose and do not depress the brake pedal.

→Do not reuse the brake flexible hose washers. Otherwise, brake fluid may leak.

→When the pads are removed, visually inspect the caliper for a brake fluid leak. Correct leaky point, if any.

1. Before servicing the vehicle, refer to the Precautions Section.

2. Raise and safely support the vehicle.
3. Remove the rear wheel.
4. Remove the brake flexible hose mounting bolt from the caliper. As this will allow fluid to flow out of the hose, have a container ready beforehand.

✳ WARNING

Be careful not to twist the flexible hose while loosening the bolt.

5. Remove the caliper pin bolts.
6. Remove the caliper from the caliper carrier.

To install:

7. Install the caliper to the caliper carrier.
8. Tighten the caliper pin bolts to 32 ft. lbs. (43 Nm).

→Make sure that the boots are fit into the groove securely.

9. Connect the flexible hose with new washers to the caliper and tighten the flexible hose joint bolt to 17 ft. lbs. (23 Nm).

REAR DISC BRAKES

10. Fill the reservoir with the specified brake fluid and bleed the brake system. Refer to Bleeding the Brake System.
11. Install the rear wheel and tighten the wheel nuts to 74 ft. lbs. (100 Nm).
12. Check for fluid leakage. Test the brakes for proper function.

✳ CAUTION

Do not attempt to drive a vehicle until the brakes are functioning properly.

DISC BRAKE PADS

REMOVAL & INSTALLATION
See Figure 5.

✳ WARNING

Be careful not to damage the brake flexible hose during this procedure. Do not operate the brake pedal with the pads removed.

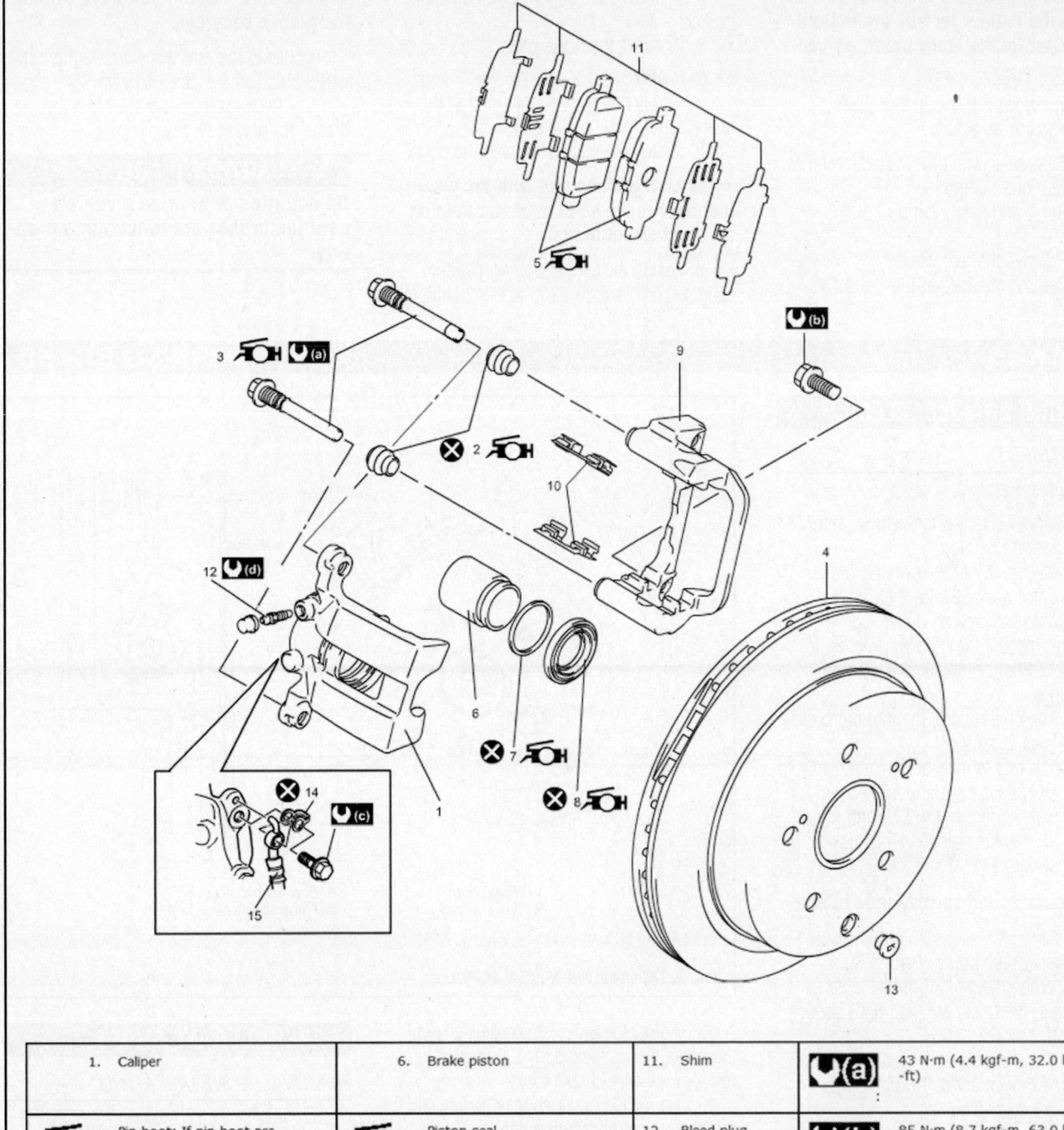

1.	Caliper	6.	Brake piston	11.	Shim	(a)	43 N·m (4.4 kgf-m, 32.0 lbf-ft)
2.	Pin boot: If pin boot are replaced, apply rubber grease included in pin boot set of supply parts.	7.	Piston seal : Apply rubber grease include in piston seal set of supply parts.	12.	Bleed plug	(b)	85 N·m (8.7 kgf-m, 63.0 lbf-ft)
3.	Slide bolt: if slide bolt are replaced, apply rubber grease included in pin boot set of supply parts.	8.	Cylinder boot: Apply rubber grease include in piston seal set of supply parts.	13.	Service plug	(c)	23 N·m (2.3 kgf-m, 17.0 lbf-ft)
4.	Brake disc	9.	Caliper carrier	14.	Washer	(d)	8 N·m (0.8 kgf-m, 6.0 lbf-ft)
5.	Brake pad: If brake pad are replaced, apply rubber grease included in brake pad set of supply parts.	10.	Pad spring	15.	Brake flexible hose	⊗	Do not reuse.

37671_GVIT_G0005

Fig. 5 Exploded view of rear disc brake components

➡When the pads are removed, visually inspect the caliper for any brake fluid leaks. Correct the leaky point, as necessary.

➡When replacing the brake pads, replace them as a set.

1. Before servicing the vehicle, refer to the Precautions Section.
2. Raise and safely support the vehicle.
3. Remove the rear wheel.
4. Remove the caliper pin bolts.

5. Remove the caliper from the caliper carrier.
6. Remove the brake pads.

To install:

7. Apply pad grease (included in the spare parts) to the brake pad.
8. Install the pad spring and the pads.

➡Install the brake pad with the wear indicator to the vehicle center side on the right wheel brake.

9. Install the caliper and tighten the caliper (slide) bolt to 32 ft. lbs. (43 Nm).

➡Make sure that the boots are fit into the groove securely.

10. Install the rear wheel and tighten the wheel nuts to 74 ft. lbs. (100 Nm).
11. Check for fluid leakage. Test the brakes for proper function.

✱✱ **CAUTION**

Do not attempt to drive a vehicle until the brakes are functioning properly.

BRAKES   **PARKING BRAKE**

PARKING BRAKE CABLES

ADJUSTMENT

See Figures 6 and 7.

1. Before servicing the vehicle, refer to the Precautions Section.
2. Make sure that the parking brake shoes are not worn beyond their limit.
3. When the parking brake lever is pulled up, inspect the following items. If defective, adjust or replace the parking brake cable.

- The operation of the parking brake lever is smooth
- The rear wheels are locked firmly
- Operate the parking brake lever with 45 ft. lbs. (200 N) and count the ratchet notches. The count should be 5–7 notches.

➡One click sound corresponds to one notch.

4. Inspect the brake warning light.
   a. When the parking brake lever is released, the brake warning light should turn off.
   b. The warning light should turn on with the movement of 1 notch of the parking brake lever.
5. Before adjusting the parking brake cables, make sure the parking brake shoes are not worn beyond their limit and that the self-adjusting mechanism operates properly.
6. Remove the console box.
7. Raise and safely support the vehicle.
8. Remove the rear wheel.
9. Release the parking brake lever.
10. Remove the service plug from the rear brake disc.
11. Align the service hole and parking brake shoe adjuster by rotating the rear brake disc.

➡Use wheel nuts to hold the disc securely against the hub.

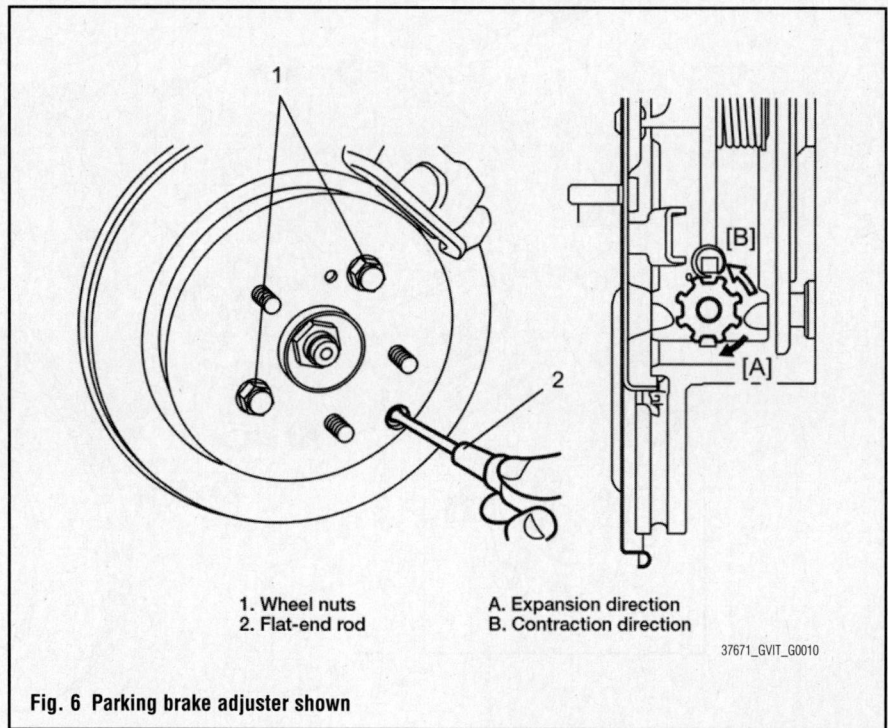

1. Wheel nuts
2. Flat-end rod

A. Expansion direction
B. Contraction direction

37671_GVIT_G0010

Fig. 6 Parking brake adjuster shown

12. Adjust the brake shoe adjuster until the brake disc is locked, by a using flat-end rod, or equivalent brake tool.
13. Adjust the brake shoe adjuster until the brake disc rotates smoothly.
14. Operate the parking brake lever more than 3 times.
15. Release the parking brake, tighten or loosen the adjusting nut so as to obtain 5–7 notches in the parking brake stroke with an operating force of 44 ft. lbs. (200 Nm).
16. After completing the adjustment, install the service plug on the rear brake disc.
17. Install the rear wheel and tighten the wheel nuts to 74 ft. lbs. (100 Nm).
18. Lower the vehicle and make sure that the brake disc is not dragging.
19. Install the console box.

PARKING BRAKE SHOES

REMOVAL & INSTALLATION

See Figure 8.

✱✱ **CAUTION**

Use special care when installing the brake shoe return spring. Failure to install it properly may allow it to spring back and cause personal injury.

1. Before servicing the vehicle, refer to the Precautions Section.
2. Remove the rear brake disc. Refer to Brake Disc (Rotor), removal & installation.
3. Push the shoe hold pin and rotate it 90°.

4. Remove the shoe hold pin, washer cup, and shoe hold spring.

5. Remove the upper and lower shoe return spring.

6. Remove the brake shoe adjuster, shoe strut, and parking brake shoe from the operating lever.

7. Remove the push nut and the operating lever from the parking brake shoe.

To install:

8. Installation is the reverse of the removal procedure.

9. Before installing the brake shoe to the brake back plate, clean the brake back plate and apply heat-resistant brake grease.

10. Install the inner cable to the brake shoe lever securely.

11. Adjust the parking brake cable. Refer to Parking Brake Cables, Adjustment.

12. Check to ensure that the rear brake is free from dragging and that the brake system operates properly.

ADJUSTMENT

For adjustment procedures, refer to Parking Brake Cables, Adjustment.

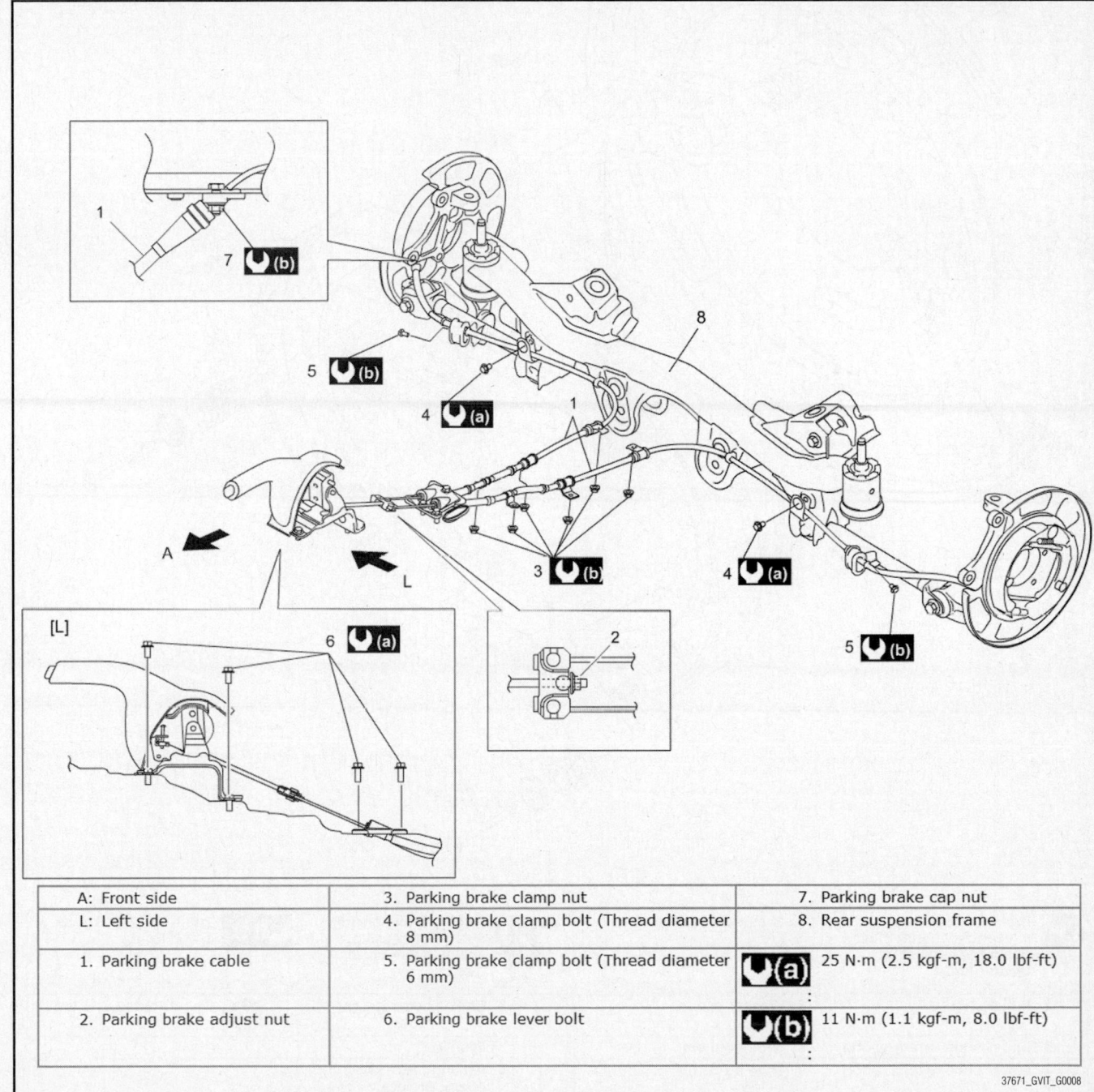

A: Front side	3. Parking brake clamp nut	7. Parking brake cap nut
L: Left side	4. Parking brake clamp bolt (Thread diameter 8 mm)	8. Rear suspension frame
1. Parking brake cable	5. Parking brake clamp bolt (Thread diameter 6 mm)	(a) 25 N·m (2.5 kgf-m, 18.0 lbf-ft)
2. Parking brake adjust nut	6. Parking brake lever bolt	(b) 11 N·m (1.1 kgf-m, 8.0 lbf-ft)

Fig. 7 Parking brake cable location

37671_GVIT_G0008

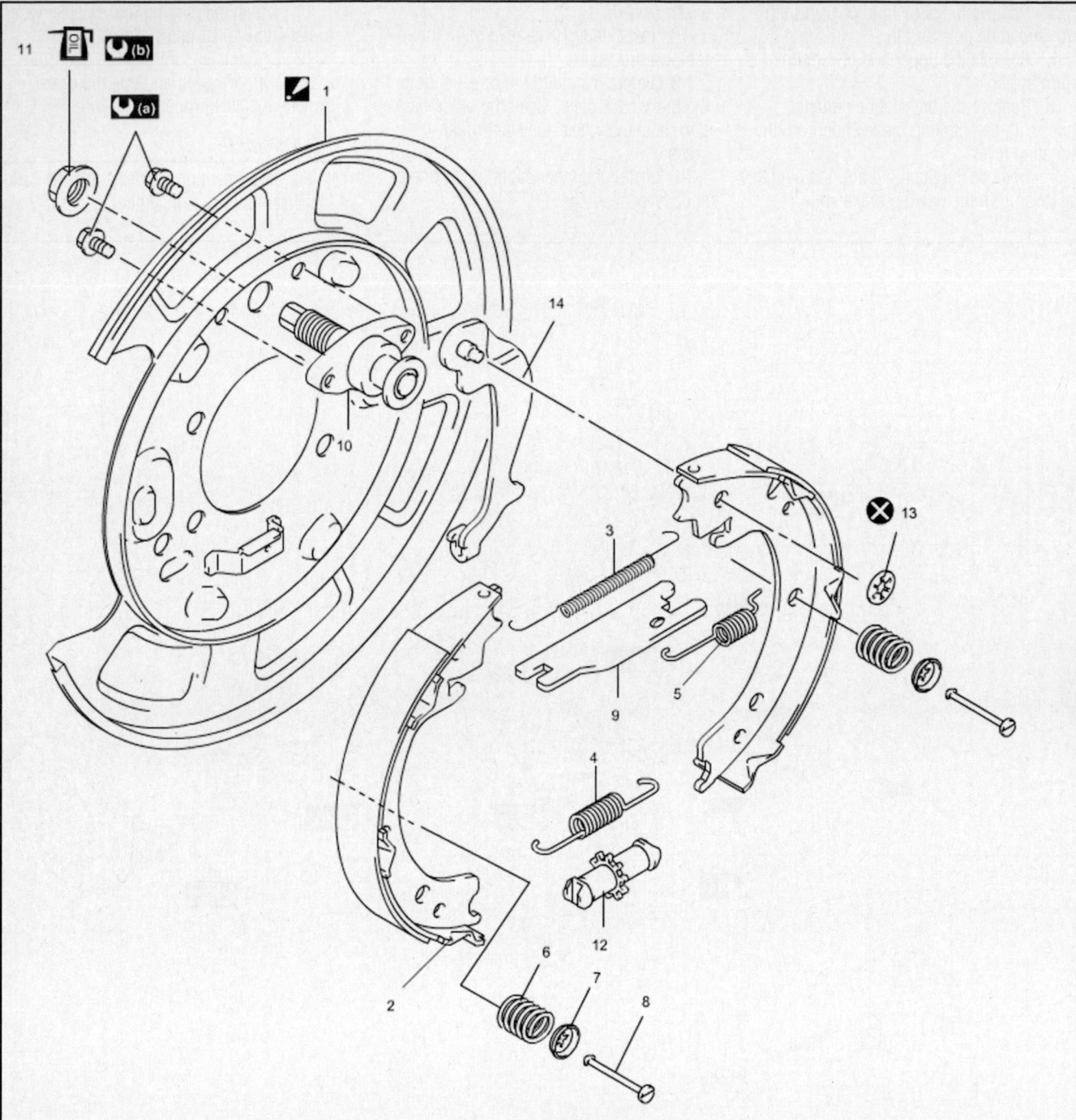

▟	Brake back plate : After clean back plate, apply heat-resistant brake grease.	6. Shoe hold spring	Parking brake anchor nut: when reusing, apply engine oil to thread. 11.	(b) 200 N·m (20.4 kgf-m, 147.5 lbf-ft) :
1.				
2.	Brake shoe	7. Washer cup	12. Brake shoe adjuster	✕ : Do not reuse
3.	Upper shoe return spring	8. Shoe hold pin	13. Push nut	
4.	Lower shoe return spring	9. Shoe strut	14. Parking brake shoe operating lever	
5.	Antirattle spring	10. Parking brake anchor assembly	(a) 12 N·m (1.2 kgf-m, 9.0 lbf-ft) :	

37671_GVIT_G0011

Fig. 8 Exploded view of parking brake components

CHASSIS ELECTRICAL

AIR BAG (SUPPLEMENTAL RESTRAINT SYSTEM)

GENERAL INFORMATION

✳✳ CAUTION

These vehicles are equipped with an air bag system. The system must be disarmed before performing service on, or around, system components, the steering column, instrument panel components, wiring and sensors. Failure to follow the safety precautions and the disarming procedure could result in accidental air bag deployment, possible injury and unnecessary system repairs.

SERVICE PRECAUTIONS

✳✳ CAUTION

Disconnect and isolate the battery negative cable before beginning any airbag system component diagnosis, testing, removal, or installation procedures. Wait at least 90 seconds after the ignition switch is turned off and the negative (-) terminal cable is disconnected from the battery before starting the operation. The SRS is equipped with a backup power source, so if work is started within 90 seconds after disconnecting the neg-

ative (-) terminal cable from the battery, the SRS may be deployed. Failure to disable the airbag system may result in accidental airbag deployment, personal injury, or death.

DISARMING THE SYSTEM

1. Before servicing the vehicle, refer to the Precautions Section.
2. Turn the steering wheel so that front tires are pointing straight ahead.
3. Disconnect the negative battery cable and isolate it from accidental reconnection. Insulate the cable end with high-quality electrical tape or a similar non-conductive wrapping.

➡**DTC's will be lost when the negative battery cable is disconnected.**

4. Turn the ignition switch to the LOCK position and remove the key.
5. Remove the A/B fuse from the fuse box.
6. Wait at least 3 minutes for the system capacitor to discharge before performing any service. The airbag system is designed to retain enough voltage to deploy the airbag for a short period of time after the battery has been disconnected.

➡**With the A/B fuse removed and the ignition switch ON, the AIR BAG warn-**

ing light will be ON. This is considered normal and does not necessarily indicate that the air bag system is malfunctioning.

7. Disconnect the air bag module connectors as necessary.

ARMING THE SYSTEM

1. Before servicing the vehicle, refer to the Precautions Section.
2. Confirm that the battery negative (-) cable is disconnected.
3. Turn the ignition switch to the LOCK position and remove the key.
4. Connect any air bag module connectors that were disconnected.
5. Install the A/B fuse to the fuse box.

✳✳ CAUTION

As an added precaution, make sure no one is in the vehicle when reconnecting the negative battery cable.

6. Connect the negative (-) cable at the battery.
7. Turn the ignition switch to the ON position and verify that AIR BAG warning light flashes 6 times and then turns OFF. If the AIR BAG warning indicator does not operate as described, perform a diagnostic system check.

DRIVE TRAIN

TRANSFER CASE ASSEMBLY

REMOVAL & INSTALLATION

See Figure 9.

1. Before servicing the vehicle, refer to the Precautions Section.
2. Shift the transfer to the 4H position operating the transfer switch.

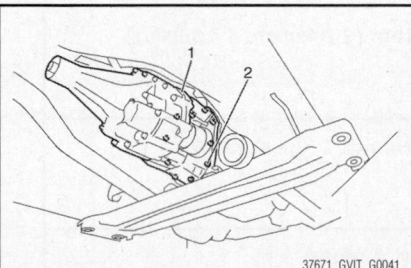

37671_GVIT_G0041

Fig. 9 Support the transfer assembly (1) and remove the transfer-to-transmission bolts (2)

3. Disconnect the negative cable at the battery.
4. Remove the gear shift control lever, Manuel Transmission (M/T model).
5. Drain the transfer oil.
6. Remove the front propeller shaft and the rear propeller shaft. Refer to Driveshaft, removal & installation.
7. Remove the exhaust pipe No. 2.
8. Disconnect the transfer actuator connector, center differential lock switch connector, and 4L/N switch connector.
9. Support the engine rear mounting member with a transmission jack.
10. Remove the engine rear mounting bolts, and then slant the transmission with the transfer.
11. Remove the gear shift control lever rear case from the transfer (for M/T model).
12. Remove the transfer-to-transmission bolts (upper side), and then install the engine rear mounting member with the transmission and transfer.

13. Support the transfer assembly with a transmission jack.
14. Remove the transfer-to-transmission bolts (lower side), and then lower the transfer assembly.

To install:

15. Installation is the reverse of the removal procedure.
16. Fill the transfer with the proper type and amount of oil.
17. Connect the battery and check all components for function.

DRIVEN DISC & PRESSURE PLATE

REMOVAL & INSTALLATION

See Figures 10 and 11.

1. Before servicing the vehicle, refer to the Precautions Section.
2. Remove the manual transmission from the vehicle. Refer to Manual Transaxle Assembly, removal & installation.

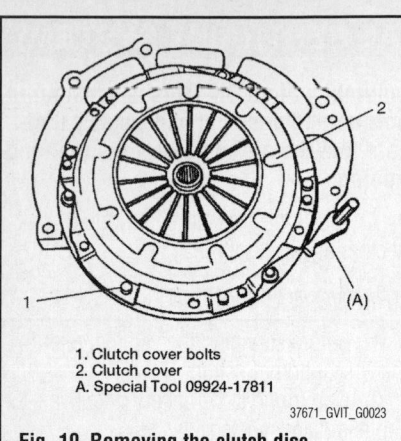

1. Clutch cover bolts
2. Clutch cover
A. Special Tool 09924-17811

37671_GVIT_G0023

Fig. 10 Removing the clutch disc

3. Hold the flywheel stationary with Special Tool 09924-17811, or equivalent.

4. Remove the clutch cover bolts, clutch cover, and clutch disc.

5. Pull out the input shaft bearing using Special Tools 09921-26020 and 09930-30104, or equivalent.

6. Remove the flywheel from the crankshaft.

To install:

→**Before installation, make sure that the flywheel surface and the pressure plate surface have been cleaned and dried thoroughly.**

7. Install the flywheel to the crankshaft and tighten the new bolts to 52 ft. lbs. (70 Nm).

8. Using Special Tool 09925-98210, or equivalent, install the input shaft bearing to the flywheel.

9. Align the clutch disc to the flywheel center using Special Tool 09923-36320, or equivalent.

10. Install the clutch cover and bolts. Tighten the bolts to 17 ft. lbs. (23 Nm).

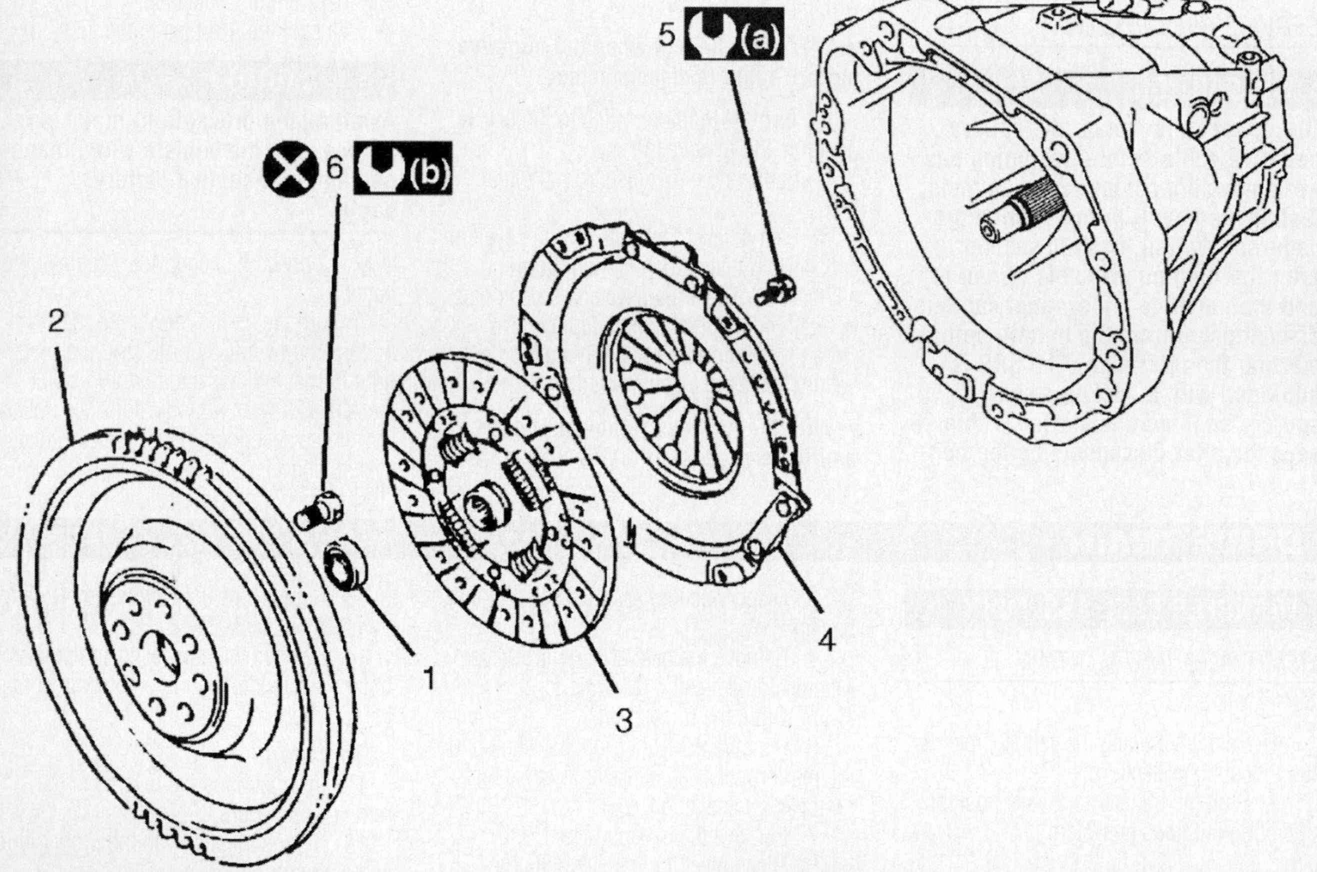

1.	Input shaft bearing	4.	Clutch cover	⊌(a) :	23 N·m (2.3 kgf-m, 17.0 lbf-ft)
2.	Flywheel	5.	Clutch cover bolt	⊌(b) :	70 N·m (7.1 kgf-m, 52.0 lbf-ft)
3.	Clutch disc	6.	Flywheel bolt	⊗ :	Do not reuse.

37671_GVIT_G0021

Fig. 11 Clutch cover, disc, and flywheel component view

➥While tightening the clutch cover bolts, compress the clutch disc with the special tool by hand so that the disc is centered. Tighten the cover bolts little by little evenly in a diagonal pattern.

11. Apply a slight amount of grease to the input shaft. Use Grease 99000-25210 (SUZUKI Super Grease I), or equivalent.

12. Join the transmission assembly to the engine. Refer to Manual Transaxle Assembly, removal & installation.

➥Turn the crankshaft with a wrench from the front while inserting the transmission input shaft to the clutch disc until the splines mesh.

FRONT DRIVESHAFT

REMOVAL & INSTALLATION

See Figures 12 through 14.

✳ WARNING

For the 4WD model, when removing the rear propeller shaft (driveshaft), never pull the shaft or the cover part. Hold the ball joint part securely, and then remove the rear propeller shaft. If the front side joint is disassembled, it is not reusable.

1. Before servicing the vehicle, refer to the Precautions Section.
2. Raise and safely support the vehicle.
3. Make matchmarks on the joint flange and the propeller shaft (driveshaft).
4. Remove the rear propeller shaft.
5. Remove the transfer rear output flange from the rear output shaft using Special Tools 09922-66010 and 09942-15511, or equivalent (4WD model only).

To install:

6. Installation is the reverse of the removal procedure.

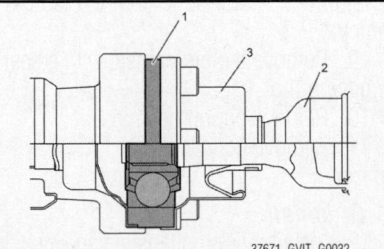

Fig. 12 When removing the rear propeller shaft (driveshaft), never pull the shaft (2) or the cover part (3). Hold the ball joint part (1) securely, and then remove the rear propeller shaft—4WD model

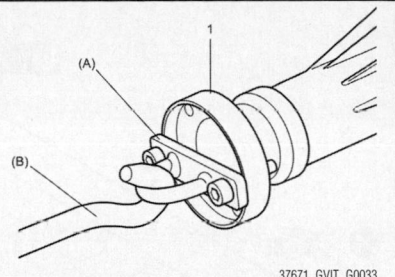

Fig. 13 Using Special Tools 09922-66010 (A) and 09942-15511 (B) to remove the transfer rear output flange (1) from the rear output shaft—4WD model

7. Clean and inspect the sliding portion of the propeller shaft end (where the oil seal contacts) and if even a small dent or scratch exists, correct and clean it again.
8. Apply grease inside the splines of the propeller shaft. Use Grease 99000-25011 (SUZUKI Super Grease A), or equivalent.
9. Install the propeller shaft aligning the matchmarks. Otherwise, vibration may occur during driving.
10. Tighten the front propeller shaft flange bolt to 22 ft. lbs. (30 Nm).
11. Tighten the rear propeller shaft flange nut to 37 ft. lbs. (50 Nm).
12. Fill the transfer oil, if the rear propeller shaft (2WD model), or transfer rear output flange (4WD model) is removed.
13. Use a new transfer rear output shaft, if the transfer rear output flange (4WD model only) is removed.

FRONT HALFSHAFTS

REMOVAL & INSTALLATION

See Figures 15 and 16.

1. Before servicing the vehicle, refer to the Precautions Section.
2. Undo the caulking and remove the halfshaft nut.
3. Raise and safely support the vehicle.
4. Remove the tire and wheel assembly.
5. Drain the front differential oil.
6. Disconnect the tie-rod end from the steering knuckle.
7. Remove the stabilizer joint from the stabilizer bar.
8. Remove the brake hose mounting bolt.
9. Remove the suspension control arm.
10. Remove the front halfshaft flange nuts.
11. Remove the halfshaft assembly from the front differential.

✳ WARNING

To prevent damage to the halfshaft boots (wheel side and differential side), do not contact them with other parts when removing the halfshaft assembly.

To install:

✳ WARNING

Be careful not to damage the oil seals and the boots when installing the halfshaft. Do not hit the joint boot with a hammer. Insert the joint only by hand.

12. Installation is the reverse of the removal procedure.
13. Make sure that the differential side joint is inserted fully and its snap ring is seated properly.
14. Use a new differential side joint circlip.
15. Tighten the front inboard halfshaft flange nuts to 58 ft. lbs. (80 Nm).
16. Tighten the new front halfshaft nut to 160 ft. lbs. (200 Nm). Caulk the nut securely.
17. Fill the front differential with the proper type and amount of oil.

CV-BOOTS INSPECTION

1. Before servicing the vehicle, refer to the Precautions Section.
2. Check the halfshaft boots for damage and deterioration:
   - Raise the front of the vehicle
   - Rotate the axle and inspect for cracked or ripped CV boot material on the inner and outer CV-joints and check for missing clamps. Repeat this step on both sides of the vehicle
   - Inspect for excessive grease deposits on or around the CV boot
3. Replace the boot if it is damaged or deteriorated.

FRONT PINION SEAL

REMOVAL & INSTALLATION

1. Before servicing the vehicle, refer to the Precautions Section.
2. Raise the vehicle, remove the wheels and brake pads and calipers.
3. Matchmark the propeller shaft (driveshaft) and pinion flange.

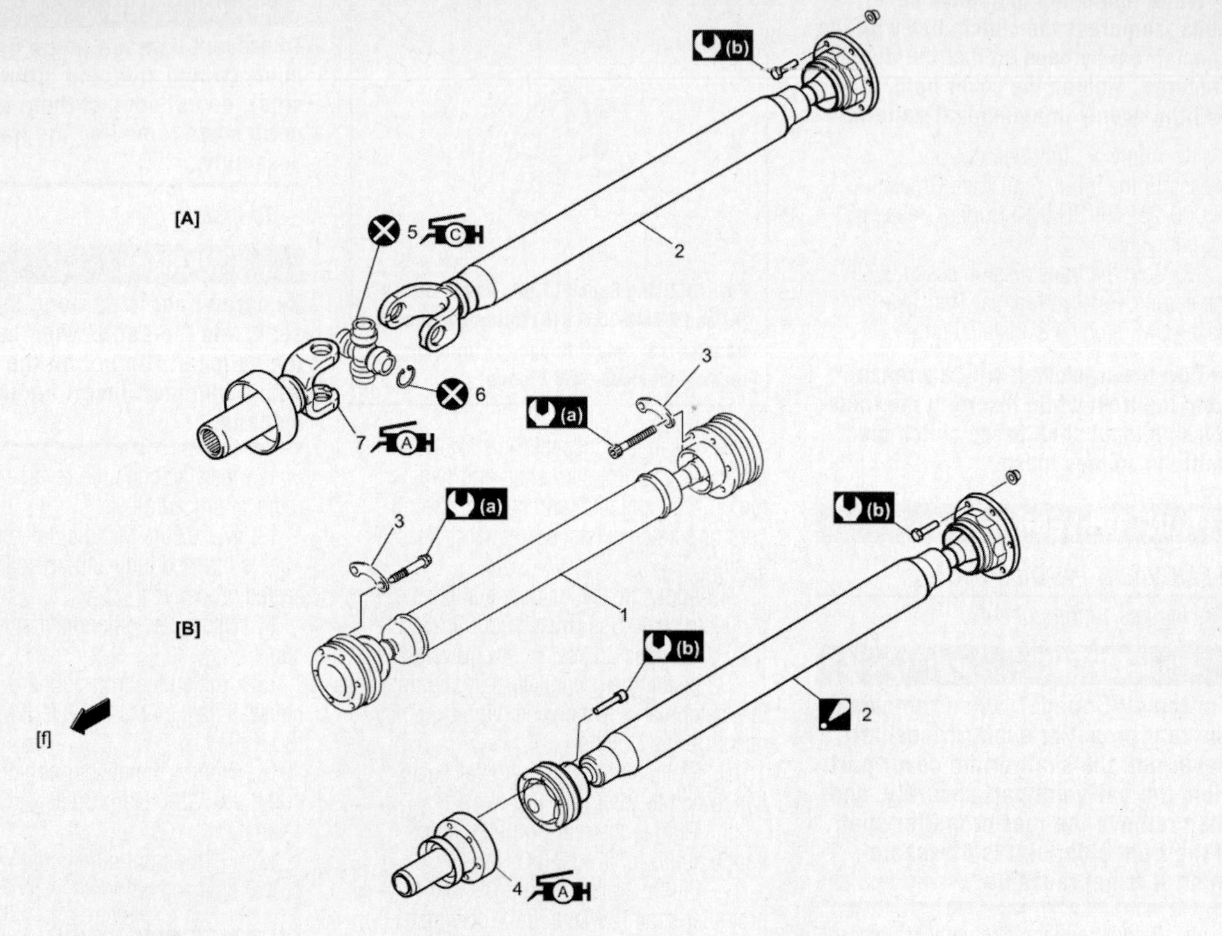

[A]

5

6

7

[B]

[f]

3

3

1

2

2

4

	[A]:	2WD model		Transfer rear output flange : Apply grease 99000-25011 to spline. 4.		30 N·m (3.0 kgf-m, 22.0 lbf-ft) (a):
	[B]:	4WD model		Spider joint assembly : Apply grease 99000-25030 to spider bearing race. 5.		50 N·m (5.1 kgf-m, 37.0 lbf-ft) (b):
	1.	Front propeller shaft		6. Circlip		[f]: Forward
	2.	Rear propeller shaft : Never disassemble.		Sliding yoke : Apply grease 99000-25011 to spline. 7.		
	3.	Support washer		Do not reuse.		

Fig. 14 Propeller shaft (driveshaft) component view

37671_GVIT_G0030

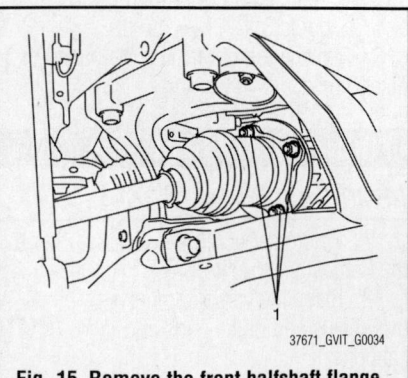

37671_GVIT_G0034

Fig. 15 Remove the front halfshaft flange nuts (1)

4. Remove the front propeller shaft. Refer to Driveshaft, removal & installation.

5. Drain the front axle, or install a catch pan under the work area.

➡**The brake calipers and pads must be removed so that there is no additional drag when measuring the pinion bearing preload.**

6. Remove the pinion nut and record the bearing preload.

7. Make mating marks on the drive bevel pinion and companion flange.

8. Remove the flange from the drive bevel pinion. Use a puller tool if it is hard to remove.

9. Remove the pinion seal with a seal removing tool.

10. Remove pinion bearing.

11. Remove collapsible spacer.

To install:

12. Install a new collapsible spacer.

13. Install the pinion bearing.

14. Lube the pinion seal lip and carefully install.

15. Install the pinion flange according to the mating marks.

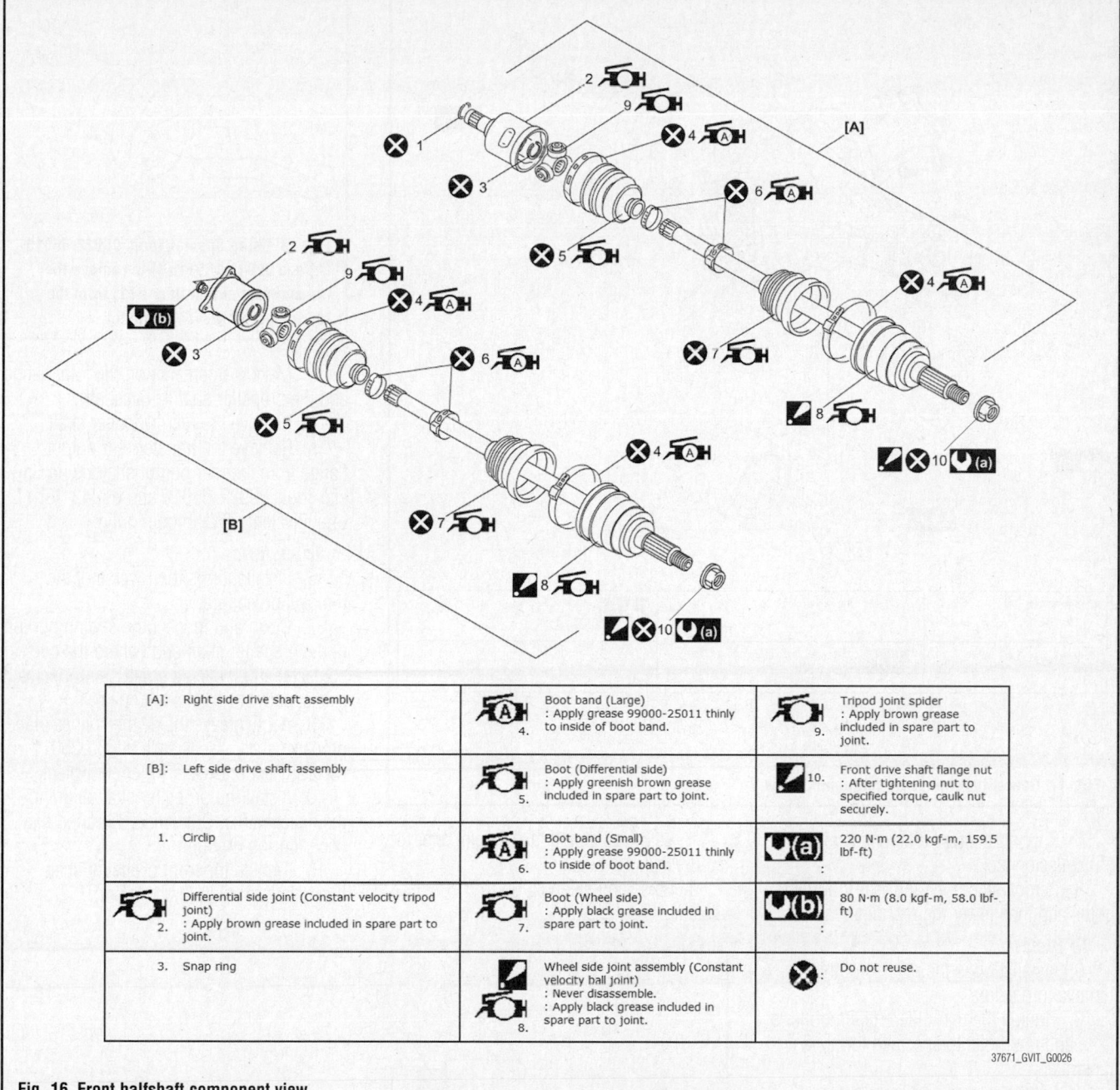

[A]:	Right side drive shaft assembly		Boot band (Large) : Apply grease 99000-25011 thinly to inside of boot band. 4.		Tripod joint spider : Apply brown grease included in spare part to joint. 9.
[B]:	Left side drive shaft assembly		Boot (Differential side) : Apply greenish brown grease included in spare part to joint. 5.		Front drive shaft flange nut : After tightening nut to specified torque, caulk nut securely. 10.
1.	Circlip		Boot band (Small) : Apply grease 99000-25011 thinly to inside of boot band. 6.		220 N·m (22.0 kgf-m, 159.5 lbf-ft) (a)
2.	Differential side joint (Constant velocity tripod joint) : Apply brown grease included in spare part to joint.		Boot (Wheel side) : Apply black grease included in spare part to joint. 7.		80 N·m (8.0 kgf-m, 58.0 lbf-ft) (b)
3.	Snap ring		Wheel side joint assembly (Constant velocity ball joint) : Never disassemble. : Apply black grease included in spare part to joint. 8.		Do not reuse.

37671_GVIT_G0026

Fig. 16 Front halfshaft component view

16. Check the pinion flange nut and install thread locking liquid.

17. Tighten the pinion flange nut gradually then set it to the recorded preload. Standard preload is 8–15 inch. lbs. (0.9–1.7 Nm).

18. Install the front propeller shaft, align the mating marks.

19. Tighten the propeller shaft mounting bolts to 62 ft. lbs. (85 Nm).

20. Install the wheels, brake pads, and calipers.

21. Refill the front drive axle housing with the proper type and amount of oil.

REAR AXLE HOUSING

REMOVAL & INSTALLATION

See Figure 17.

1. Before servicing the vehicle, refer to the Precautions Section.

❊❊ WARNING

Do not the damage spline, companion flange, or front oil seal when removing the propeller shaft. Before removing the final drive assembly or rear axle assembly, disconnect the ABS sensor harness connector from the assembly and move it away from the final drive/rear axle assembly area. Failure to do so may result in the sensor wires being damaged and the sensor becoming inoperative.

2. Raise and safely support the vehicle.

3. Drain the differential gear oil.

4. Remove the rear propeller shaft. Refer to Driveshaft, removal & installation.

5. Remove the exhaust center pipe.

6. Remove the rear halfshafts. Refer to Halfshafts, removal & installation.

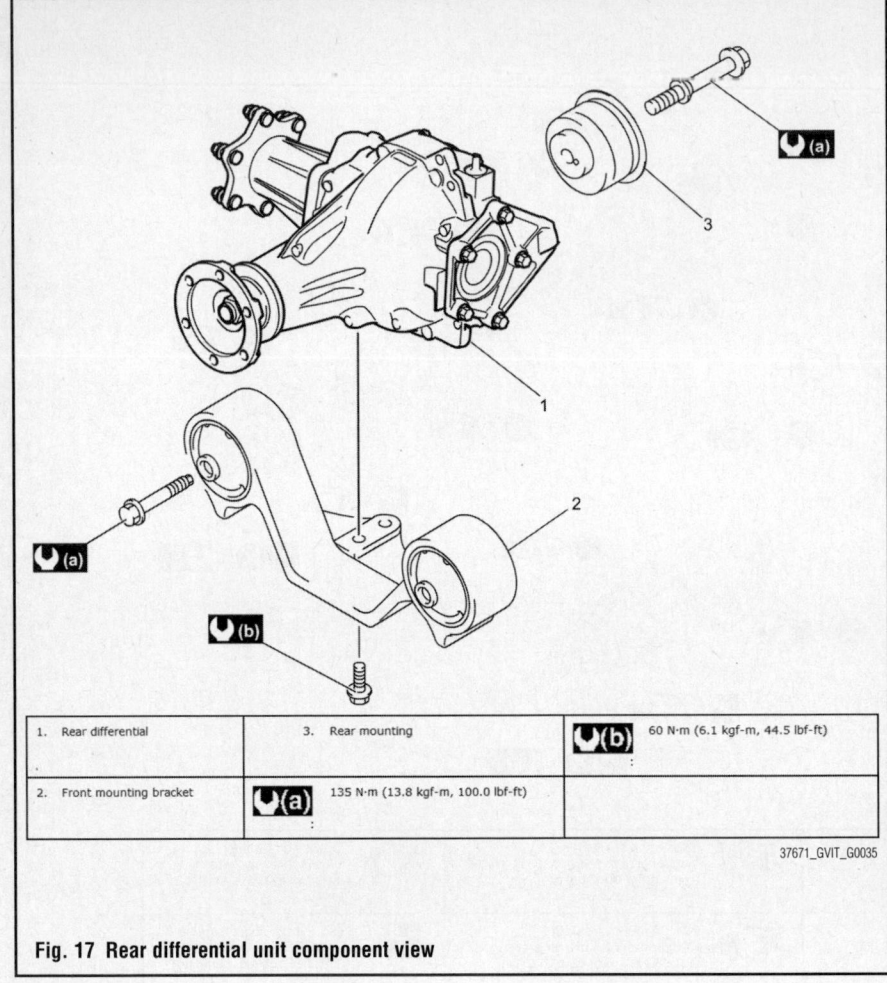

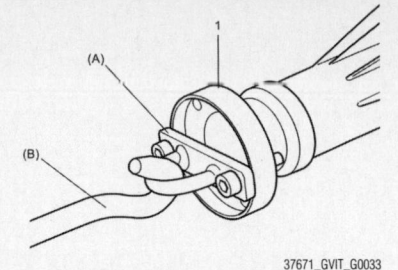

Fig. 19 Using Special Tools 09922-66010 (A) and 09942-15511 (B) to remove the transfer rear output flange (1) from the rear output shaft—4WD model

| 1. | Rear differential | 3. | Rear mounting | 🔧(b) | 60 N·m (6.1 kgf-m, 44.5 lbf-ft) |
| 2. | Front mounting bracket | 🔧(a) | 135 N·m (13.8 kgf-m, 100.0 lbf-ft) | | |

37671_GVIT_G0035

Fig. 17 Rear differential unit component view

7. Support the rear differential with a transmission jack.

8. Remove the front and rear mounting bolts, and then lower the rear differential.

To install:

9. Installation is the reverse of the removal procedure.

10. Tighten the rear differential mounting front and rear bolts to specified torque as illustrated.

11. Fill the rear differential with the proper type and amount of oil.

REAR DRIVESHAFT

REMOVAL & INSTALLATION

See Figures 18 through 20.

✳✳ WARNING

For the 4WD model, when removing the rear propeller shaft (driveshaft), never pull the shaft or the cover part. Hold the ball joint part securely, and then remove the rear propeller shaft. If the front side joint is disassembled, it is not reusable.

1. Before servicing the vehicle, refer to the Precautions Section.

2. Raise and safely support the vehicle.

3. Make matchmarks on the joint flange and the propeller shaft (driveshaft).

4. Remove the rear propeller shaft.

5. Remove the transfer rear output flange from the rear output shaft using Special Tools 09922-66010 and 09942-15511, or equivalent (4WD model only).

To install:

6. Installation is the reverse of the removal procedure.

7. Clean and inspect the sliding portion of the propeller shaft end (where the oil seal contacts) and if even a small dent or scratch exists, correct and clean it again.

8. Apply grease inside the splines of the propeller shaft. Use Grease 99000-25011 (SUZUKI Super Grease A), or equivalent.

9. Install the propeller shaft aligning the matchmarks. Otherwise, vibration may occur during driving.

10. Tighten the front propeller shaft flange bolt to 22 ft. lbs. (30 Nm).

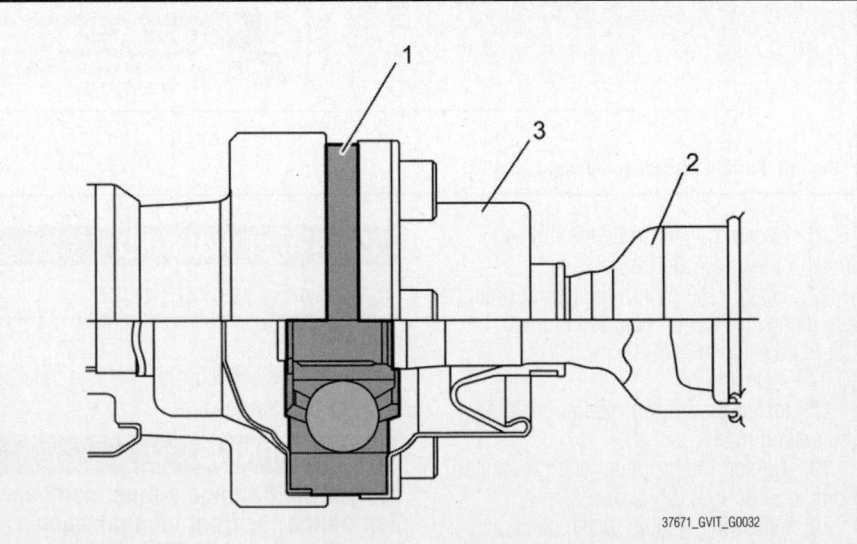

Fig. 18 When removing the rear propeller shaft (driveshaft), never pull the shaft (2) or the cover part (3). Hold the ball joint part (1) securely, and then remove the rear propeller shaft—4WD model

37671_GVIT_G0032

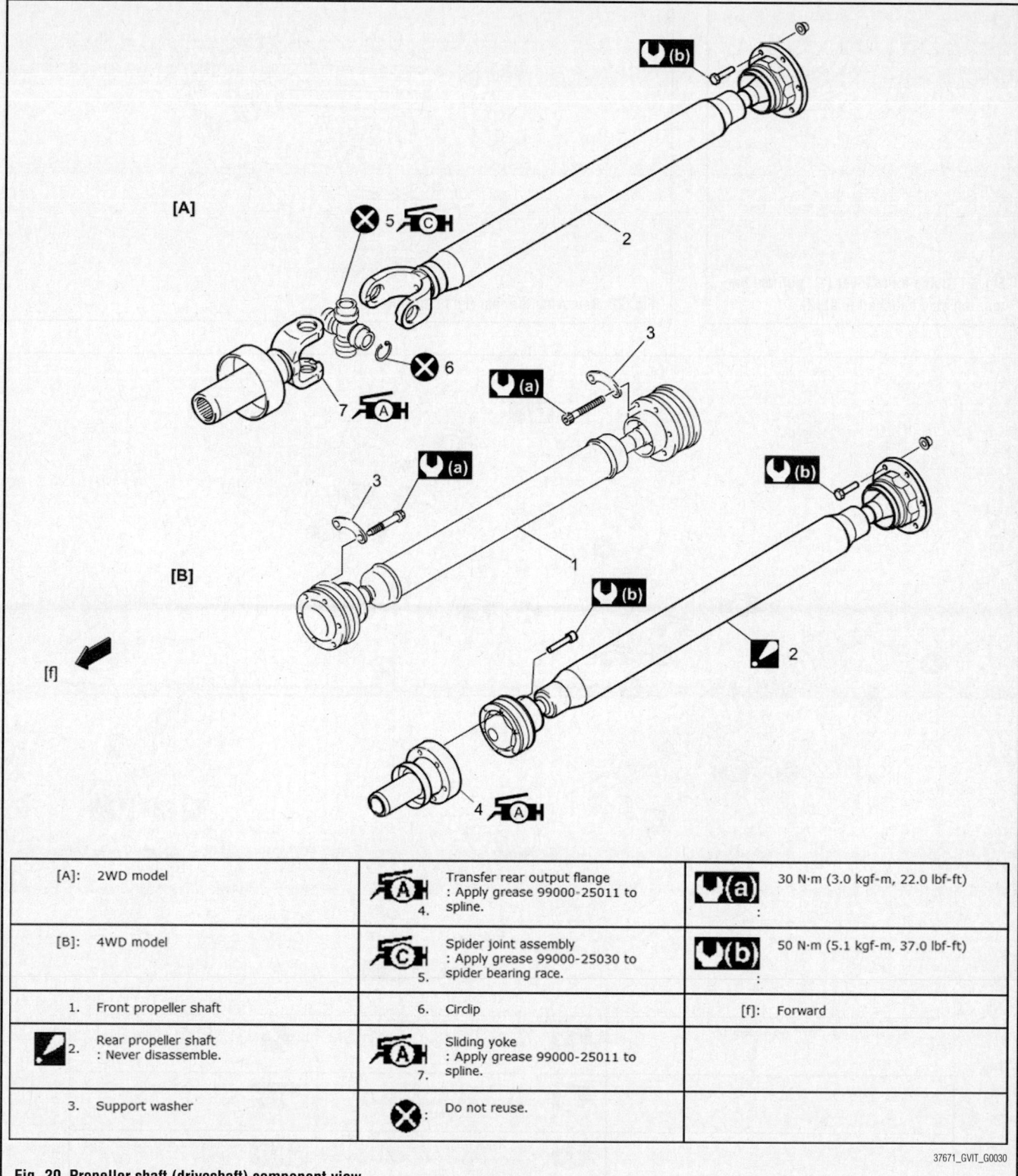

[A]:	2WD model		Transfer rear output flange : Apply grease 99000-25011 to 4. spline.		30 N·m (3.0 kgf-m, 22.0 lbf-ft)
[B]:	4WD model		Spider joint assembly : Apply grease 99000-25030 to 5. spider bearing race.		50 N·m (5.1 kgf-m, 37.0 lbf-ft)
1.	Front propeller shaft		Circlip	[f]:	Forward
2.	Rear propeller shaft : Never disassemble.		Sliding yoke : Apply grease 99000-25011 to 7. spline.		
3.	Support washer		Do not reuse.		

37671_GVIT_G0030

Fig. 20 Propeller shaft (driveshaft) component view

11. Tighten the rear propeller shaft flange nut to 37 ft. lbs. (50 Nm).

12. Fill the transfer oil, if the rear propeller shaft (2WD model), or transfer rear output flange (4WD model) is removed.

13. Use a new transfer rear output shaft, if the transfer rear output flange (4WD model only) is removed.

REAR HALFSHAFTS

REMOVAL & INSTALLATION

See Figures 21 through 23.

1. Before servicing the vehicle, refer to the Precautions Section.

2. Undo the caulking of the halfshaft nut and then remove the halfshaft nut.

3. Raise and safely support the vehicle.

4. Remove the wheel.

5. Drain the rear differential oil.

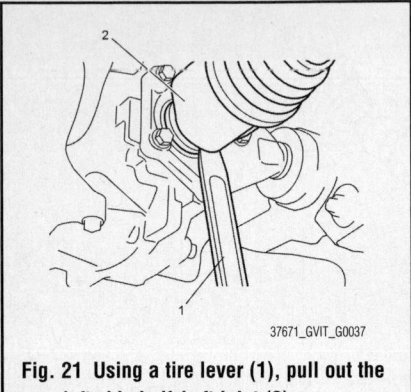

Fig. 21 Using a tire lever (1), pull out the rear left side halfshaft joint (2)

37671_GVIT_G0037

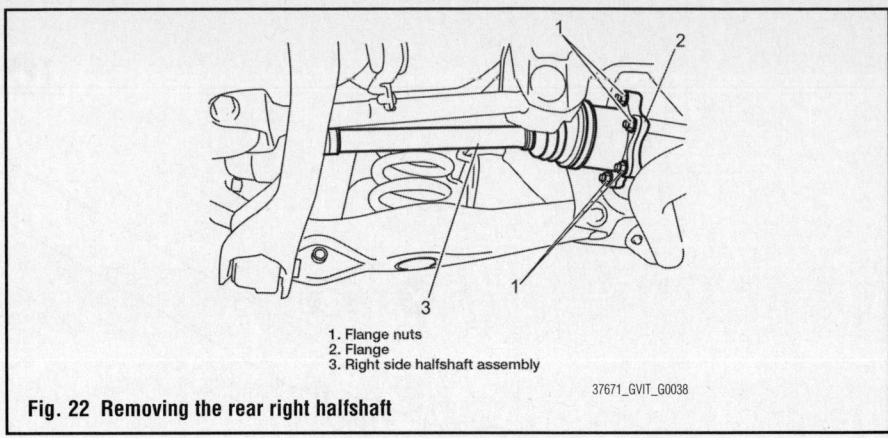

1. Flange nuts
2. Flange
3. Right side halfshaft assembly

Fig. 22 Removing the rear right halfshaft

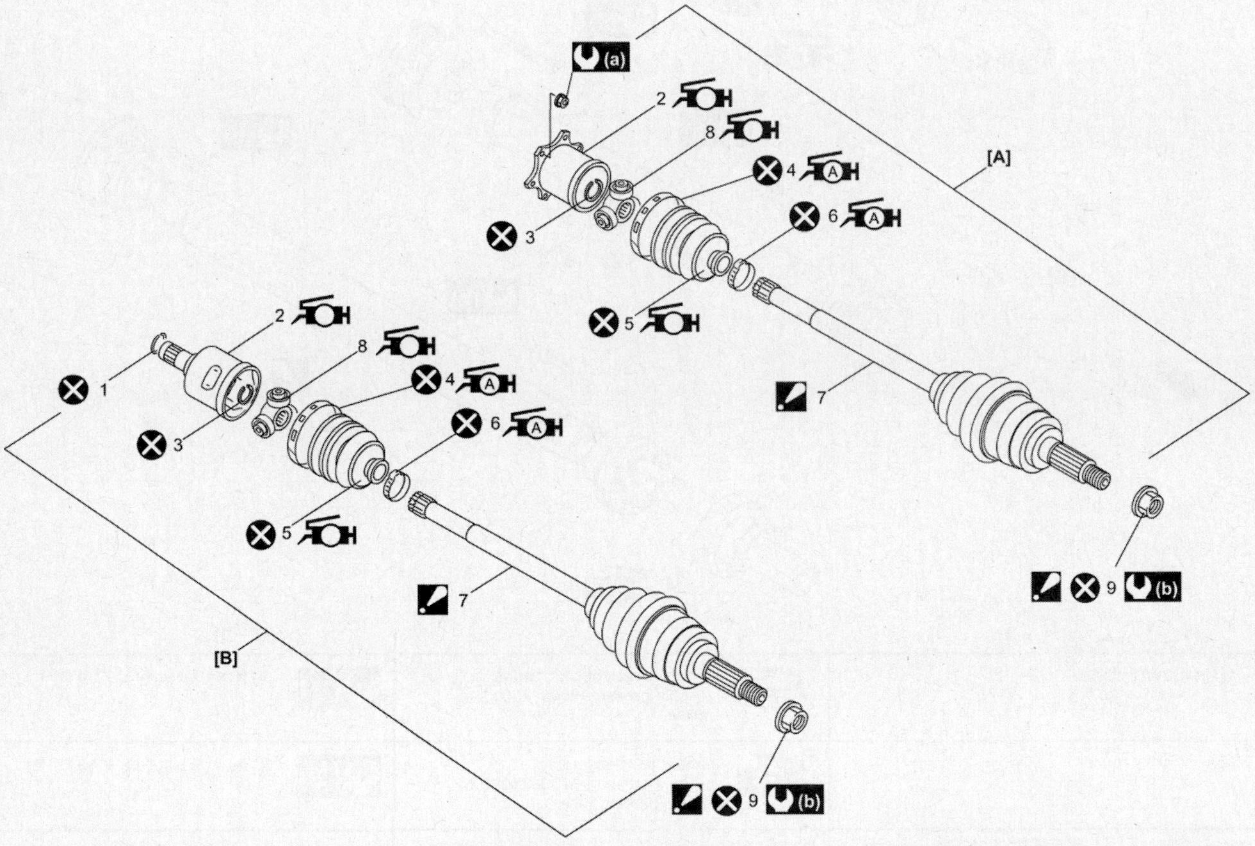

37671_GVIT_G0038

[A]:	Right side drive shaft assembly	4.	Boot band (Large) : Apply grease 99000-25011 thinly to inside of boot band.	9.	Drive shaft nut : After tightening nut, caulk nut securely.
[B]:	Left side drive shaft assembly	5.	Boot (Differential side) : Apply brown grease included in spare part to joint.	(a)	80 N·m (8.0 kgf-m, 58.0 lbf-ft)
1.	Circlip	6.	Boot band (Small) : Apply grease 99000-25011 thinly to inside of boot band.	(b)	220 N·m (22.0 kgf-m, 159.5 lbf-ft)
2.	Differential side joint (Constant velocity tripod joint) : Apply greenish brown grease included in spare part to joint.	7.	Wheel side joint assembly (Constant velocity ball joint) : Never disassemble.		Do not reuse.
3.	Snap ring	8.	Tripod joint spider : Apply brown grease included in spare part to joint.		

37671_GVIT_G0039

Fig. 23 Rear halfshaft component view

6. Remove the left rear halfshaft assembly as follows:

a. Support the lower arm with a jack.

b. Remove the left side rear wheel speed sensor.

c. Remove the left side control rod outer bolt, trailing rod rear bolt, and lower arm outer bolt.

d. Using a tire lever, pull out the left side halfshaft joint while turning the left side halfshaft joint in parallel with the rear differential gear case so as to release the snap ring fitting of the joint spline at the differential side.

e. Remove the left side halfshaft assembly.

7. Remove the right side halfshaft assembly as follows:

a. Remove the rear halfshaft flange nuts.

b. Remove the right side halfshaft assembly.

To install:

8. Install the halfshaft assembly by reversing the removal procedure.

9. Protect the oil seals and boots from any damage, preventing them from unnecessary contact while installing the halfshaft.

10. Do not hit the joint boot with a hammer. Inserting the joint only by hand.

11. Tighten the rear halfshaft nuts to the specified torque, as illustrated.

12. Fill the rear differential with the proper type and amount of oil.

CV-BOOTS INSPECTION

1. Before servicing the vehicle, refer to the Precautions Section.

2. Check the halfshaft boots for damage and deterioration:

- Raise the front of the vehicle
- Rotate the axle and inspect for cracked or ripped CV boot material on the inner and outer CV-joints and check for missing clamps. Repeat this step on both sides of the vehicle
- Inspect for excessive grease deposits on or around the CV boot

3. Replace the boot if it is damaged or deteriorated.

REAR PINION SEAL

REMOVAL & INSTALLATION

1. Before servicing the vehicle, refer to the Precautions Section.

2. Raise and safely support the vehicle.

3. Remove the wheels.

4. Mark the propeller shaft and pinion flange for installation.

5. Remove the rear propeller shaft. Refer to Driveshaft, removal & installation.

6. Drain the rear axle, or install a catch pan under the work area.

7. Remove the pinion nut and record the bearing preload.

➡The brake calipers and pads must be removed so that there is no additional drag when measuring the pinion bearing preload.

8. Make mating marks on the drive bevel pinion and companion flange.

9. Remove the flange from the drive bevel pinion. Use a puller tool if it is hard to remove.

10. Remove the pinion seal with a seal removing tool.

11. Remove the pinion bearing.

12. Remove the collapsible spacer.

To install:

13. Install a new collapsible spacer.

14. Install the pinion bearing.

15. Lube the pinion seal lip and carefully install it.

16. Install the pinion flange using the alignment marks.

17. Check the pinion flange nut and install thread locking liquid.

18. Tighten the pinion flange nut gradually to match the recorded pinion bearing preload. Standard preload is 8–15 inch. lbs. (0.9–1.7 Nm).

19. Install the front propeller shaft aligning the mating marks.

20. Tighten the rear propeller shaft mounting bolts to 62 ft. lbs. (85 Nm).

21. Fill the rear differential with the proper type and amount of oil.

22. Install the brake pads, calipers, and wheels.

23. Check the pinion seal for leaks.

ENGINE COOLING

ENGINE FAN

REMOVAL & INSTALLATION

2.4L Engine

1. Before servicing the vehicle, refer to the Precautions Section.

2. Disconnect the negative cable from the battery.

3. Disconnect the main fan motor connector (connector color: black) and the sub fan motor connector (connector color: gray).

4. Remove the air cleaner assembly and air cleaner outlet hose.

5. Detach the Power Steering (P/S) fluid reservoir with the reservoir bracket and hose connected.

➡There is no need to the drain P/S fluid.

6. Disconnect the ground terminal and remove the reservoir from the radiator.

7. Remove the cooling fan assembly from the vehicle.

To install:

8. Installation is the reverse of the removal procedure.

9. Fill the cooling system with the proper type and amount of fluid.

10. Verify there is no coolant leakage at each connection.

3.2L Engine

1. Before servicing the vehicle, refer to the Precautions Section.

2. Disconnect the negative cable from the battery.

3. Disconnect the radiator cooling fan motor connectors.

4. With the hose connected, detach the Power Steering reservoir tank from its bracket.

5. Disconnect the ground cable.

6. Remove the reservoir tank from the radiator.

7. Remove the air cleaner suction pipe.

8. Remove the radiator cooling fan assembly from the radiator.

9. Installation is the reverse of the removal procedure.

RADIATOR

REMOVAL & INSTALLATION

2.4L Engine

1. Before servicing the vehicle, refer to the Precautions Section.

2. Disconnect the negative (-) cable from the battery.

3. Drain the transmission fluid (4-speed automatic model).

4. Drain the engine coolant.

5. Remove the cooling fan assembly. Refer to Engine Fan, removal & installation.

6. Remove the transmission fluid cooler hoses (4-speed automatic model).

7. Remove the radiator inlet hose and outlet hose from the radiator.

8. Remove the front grill.

9. Remove the condenser upper brackets and lower brackets from the radiator.

10. Remove the radiator from the vehicle.

To install:

11. Installation is the reverse of the removal procedure.

12. Fill the cooling system with the proper type and amount of fluid.

13. Verify that there is no coolant leakage at each connection.

14. Fill the transmission fluid with the proper type and amount of fluid (4-speed automatic model).

3.2L Engine

1. Before servicing the vehicle, refer to the Precautions Section.

2. Disconnect the negative cable from the battery.

3. Drain the engine coolant.

4. Drain the transmission fluid (5-speed automatic model).

5. Remove the radiator inlet hose and outlet hose from the radiator.

6. Remove the oil hoses No. 3 and No. 4 from the radiator (5-speed automatic model).

7. Remove the radiator cooling fan assembly. Refer to Engine Fan, removal & installation.

8. Remove the front bumper.

9. Remove the condenser bracket bolt from the condenser.

10. Remove the radiator bracket.

11. Remove the radiator from the vehicle.

To install:

12. Installation is the reverse of the removal procedure.

13. Fill the cooling system with the proper type and amount of fluid.

14. Verify that there is no coolant leakage at each connection.

15. Fill the transmission fluid with the proper type and amount of fluid (5-speed automatic model).

THERMOSTAT

REMOVAL & INSTALLATION

2.4L Engine

See Figure 24.

1. Before servicing the vehicle, refer to the Precautions Section.

2. Drain the engine coolant.

3. Disconnect the radiator outlet hose from the thermostat cap.

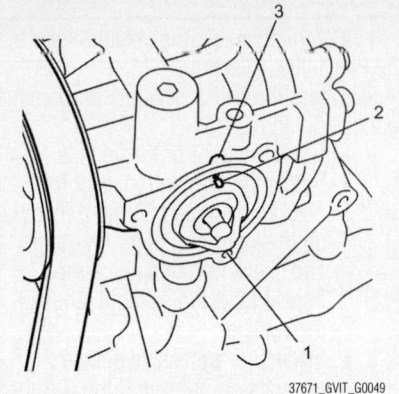

Fig. 24 When installing the thermostat (1), be sure to position it so that the air bleed valve (2) is aligned with the match mark (3) on the water pump case

4. Remove the thermostat cap from the water pump.

5. Remove the thermostat from the water pump.

To install:

6. Installation is the reverse of the removal procedure.

7. When installing the thermostat on the water pump case, be sure to position it so that the air bleed valve is aligned with the match mark.

8. Use a new gasket when installing the thermostat.

9. Fill the cooling system with the proper type and amount of fluid.

10. Verify that there is no coolant leakage at each connection.

3.2L Engine

See Figures 25 and 26.

1. Before servicing the vehicle, refer to the Precautions Section.

2. Drain the engine coolant.

3. Remove the intake upper manifold.

4. Remove the wiring harness of the HO2S-1 (bank 1) from the water inlet pipe.

5. Remove the water inlet pipe.

6. Remove the heater inlet hose, heater outlet hose, water inlet hose, and water outlet hose from the thermostat assembly.

7. Remove the thermostat assembly from the cylinder block.

To install:

8. Installation is the reverse of the removal procedure.

9. Use a new gasket when installing the thermostat assembly.

10. Tighten the water inlet pipe bolts in sequence.

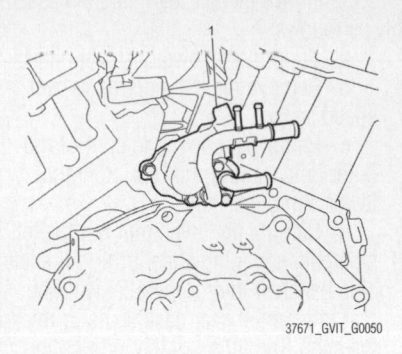

Fig. 25 Remove the thermostat assembly (1) from the cylinder block—3.2L engine

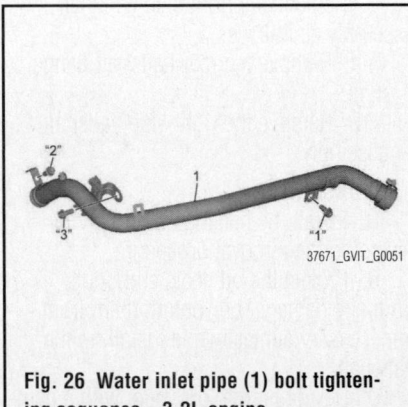

Fig. 26 Water inlet pipe (1) bolt tightening sequence—3.2L engine

11. Fill the cooling system with the proper type and amount of fluid.

12. Verify that there is no coolant leakage at each connection.

WATER PUMP

REMOVAL & INSTALLATION

2.4L Engine

See Figure 27.

1. Before servicing the vehicle, refer to the Precautions Section.

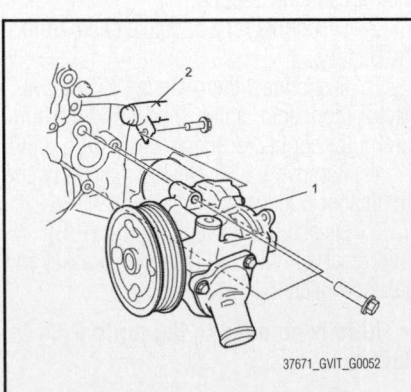

Fig. 27 Remove the water inlet pipe (2) from the water pump (1)—2.4L engine

2. Disconnect the negative (-) cable from the battery.

3. Drain the engine coolant.

4. Remove the accessory drive belt.

5. Remove the exhaust manifold cover.

6. Disconnect the radiator outlet hose from the water pump.

7. Remove the water inlet pipe from the water pump.

8. Remove the water pump from the cylinder block.

To install:

9. Installation is the reverse of the removal procedure.

10. Use a new gasket and O-ring.

11. Tighten the water pump bolts to 19 ft. lbs. (25 Nm).

12. Fill the cooling system with the proper type and amount of fluid.

13. Verify that there is no coolant leakage.

3.2L Engine

See Figures 28 and 29.

1. Before servicing the vehicle, refer to the Precautions Section.

2. Disconnect the negative cable from the battery.

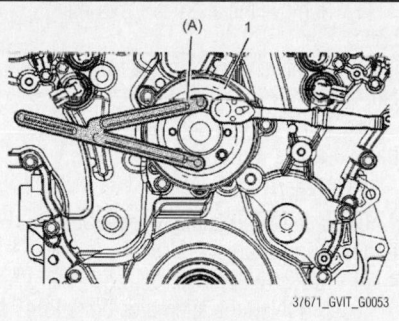

Fig. 28 Using Special Tool (EN-46104) to hold the water pump pulley (1)—3.2L engine

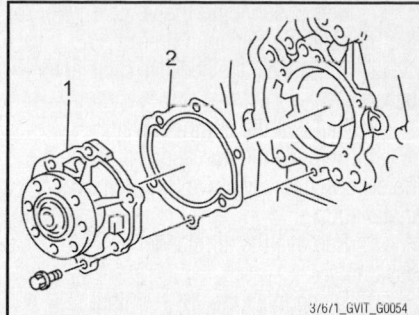

Fig. 29 Removing the water pump assembly (1) and gasket (2) from the cylinder block—3.2L engine

3. Drain the engine coolant.

4. Remove the coolant reservoir.

5. Remove the accessory drive belt.

6. Remove the water pump pulley using the water pump pulley holding tool, Special Tool (EN-46104), or equivalent.

7. Remove the water pump assembly and gasket from the cylinder block.

To install:

8. Clean the mating surfaces of the water pump assembly and timing chain cover.

9. Install a new gasket to the timing chain cover.

10. Install the water pump assembly to the timing chain cover and tighten the bolts to 89 inch lbs. (10 Nm).

11. Install the water pump pulley and tighten the bolts using Special Tool (EN-46104), or equivalent. Tighten the bolts to 106 inch lbs. (12 Nm).

12. Install the accessory drive belt.

13. Install the coolant reservoir.

14. Fill the cooling system with the proper type and amount of fluid.

15. Verify that there is no coolant leakage.

16. Connect the negative cable to the battery.

ENGINE ELECTRICAL

CHARGING SYSTEM

ALTERNATOR

REMOVAL & INSTALLATION

See Figures 30 and 31.

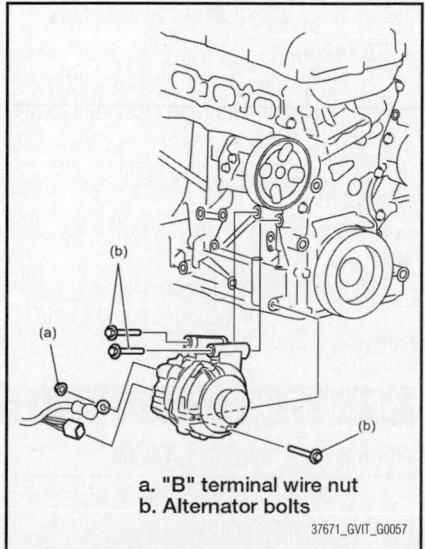

a. "B" terminal wire nut
b. Alternator bolts

Fig. 30 View of alternator installation—2.4L engine

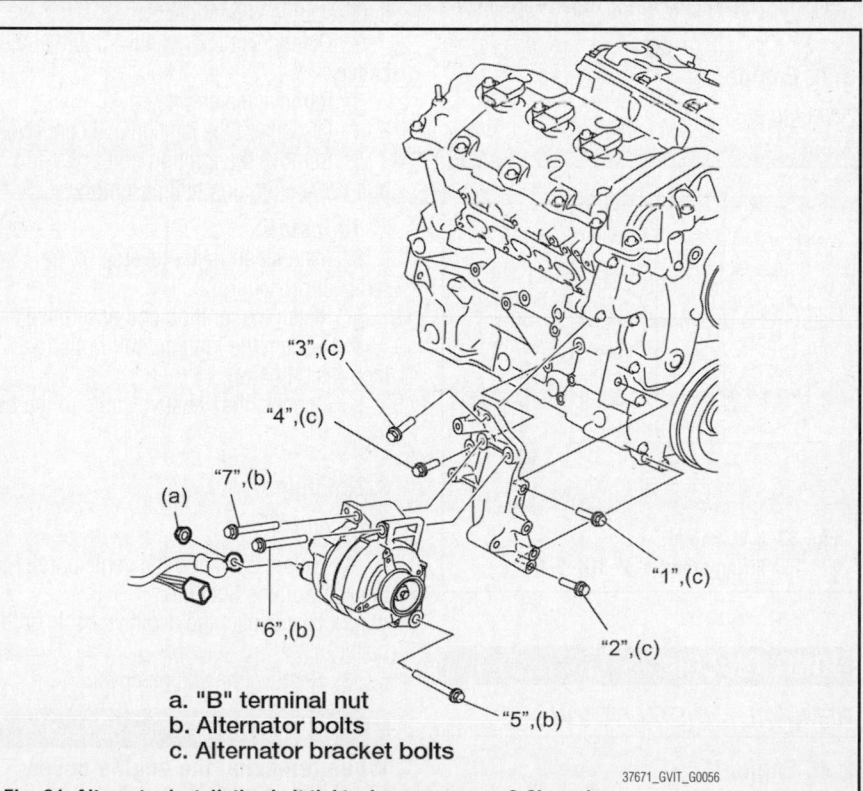

a. "B" terminal nut
b. Alternator bolts
c. Alternator bracket bolts

Fig. 31 Alternator installation bolt tightening sequence—3.2L engine

1. Before servicing the vehicle, refer to the Precautions Section.

2. Disconnect the negative cable at the battery.

3. Remove the engine cover.

4. Remove the accessory drive belt. Refer to Accessory Drive Belt, removal & installation.

5. Remove the air cleaner assembly.

6. Disconnect the "B" terminal wire and connector from the alternator.

7. Remove the alternator mounting bolts.

8. Remove the alternator from the vehicle.

9. For 3.2L engines, remove the alternator bracket, if necessary.

To install:

10. Installation is the reverse of the removal procedure.

11. For 2.4L engines:

   a. Tighten the alternator bolts to 41 ft. lbs. (55 Nm).

   b. Tighten the "B" terminal wire nut to 97 inch lbs. (11 Nm).

12. For 3.2L engines:

   a. If the alternator bracket was removed, tighten the alternator bracket bolts evenly and gradually in the proper tightening sequence. Tighten the bracket bolts to 41 ft. lbs. (55 Nm).

   b. Tighten the alternator bolts, in sequence, to 41 ft. lbs. (55 Nm).

   c. Tighten the "B" terminal wire nut to 97 inch lbs. (11 Nm).

ENGINE ELECTRICAL

FIRING ORDER

2.4L Engine
See Figure 32.

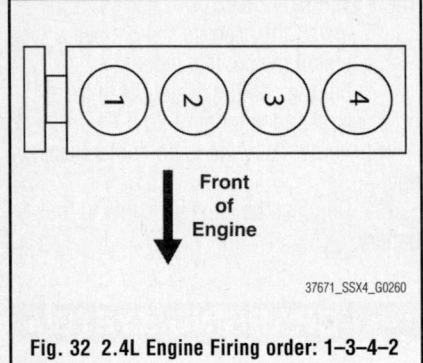

Fig. 32 2.4L Engine Firing order: 1–3–4–2

3.2L Engine
See Figure 33.

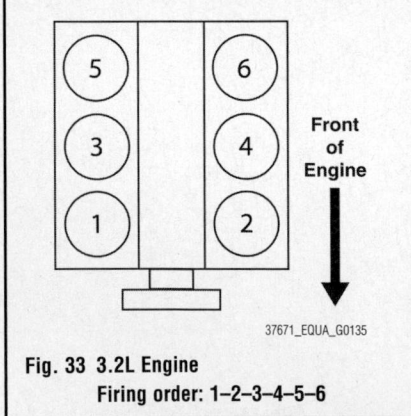

Fig. 33 3.2L Engine
Firing order: 1–2–3–4–5–6

IGNITION COIL

REMOVAL & INSTALLATION

2.4L Engine
See Figure 34.

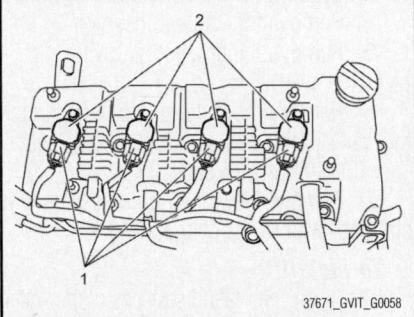

Fig. 34 Disconnect the ignition coil connector (1), remove the ignition coil bolt, and pull out the ignition coil assembly (2)—2.4L engine

1. Before servicing the vehicle, refer to the Precautions Section.

2. Disconnect the negative cable at the battery.

3. Remove the engine cover.

4. Disconnect the ignition coil connectors.

5. Remove the ignition coil bolts and pull out the ignition coil assembly.

To install:

6. Installation is the reverse of the removal procedure.

7. Install the ignition coil assembly.

8. Tighten the ignition coil bolts to 97 inch lbs. (11 Nm).

9. Connect the negative cable to the battery.

3.2L Engine
See Figure 35.

1. Before servicing the vehicle, refer to the Precautions Section.

2. Disconnect the negative cable at the battery.

3. Remove the engine cover.

❀❀ WARNING

When removing the engine cover, be sure to lift up its front end to

IGNITION SYSTEM

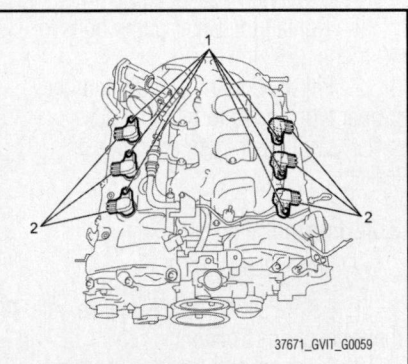

Fig. 35 Remove the ignition coil bolts (1) and pull out the ignition coil assembly (2)—3.2L engine

prevent breakage of the engine cover claws.

4. Remove the air cleaner outlet hose.

5. Disconnect the ignition coil connectors.

6. Remove the ignition coil bolts and then pull out the ignition coil assembly.

To install:

7. Install the ignition coil assembly.

8. Tighten the ignition coil bolts to 89 inch lbs. (10 Nm).

9. Connect the ignition coil connectors.

10. Install the air cleaner outlet hose.

11. Install the engine cover.

12. Connect the negative cable to the battery.

SPARK PLUGS

REMOVAL & INSTALLATION

1. Before servicing the vehicle, refer to the Precautions Section.

2. Remove the ignition coil assemblies. Refer to Ignition Coil, removal & installation.

3. Remove the spark plug using a spark plug socket and wrench.

To install:

4. Installation is the reverse of the removal procedure.

5. For 2.4L engine, tighten the spark plugs to 18 ft. lbs. (25 Nm).

6. For 3.2L engine, tighten the spark plugs to 13 ft. lbs. (18 Nm).

7. Install the ignition coil assemblies. Refer to Ignition Coil, removal & installation.

ENGINE ELECTRICAL

STARTER

REMOVAL & INSTALLATION

2.4L Engine

See Figure 36.

1. Before servicing the vehicle, refer to the Precautions Section.

2. Disconnect negative cable from battery.

3. Remove the intake manifold. Refer to Intake Manifold, removal & installation.

4. Disconnect the magnetic switch lead wire and the starting motor cable from the starting motor.

5. Remove the starting motor mounting bolts.

6. Remove the starting motor.

To install:

7. Installation is the reverse of the removal procedure.

8. Tighten the starting motor mounting bolts to 34 ft. lbs. (45 Nm).

9. Tighten the cable nut to 97 inch lbs. (11 Nm).

3.2L Engine

See Figure 37.

1. Before servicing the vehicle, refer to the Precautions Section.

2. Remove the bank-2 exhaust manifold.

STARTING SYSTEM

Refer to Exhaust Manifold, removal & installation.

3. Remove the knock sensor.

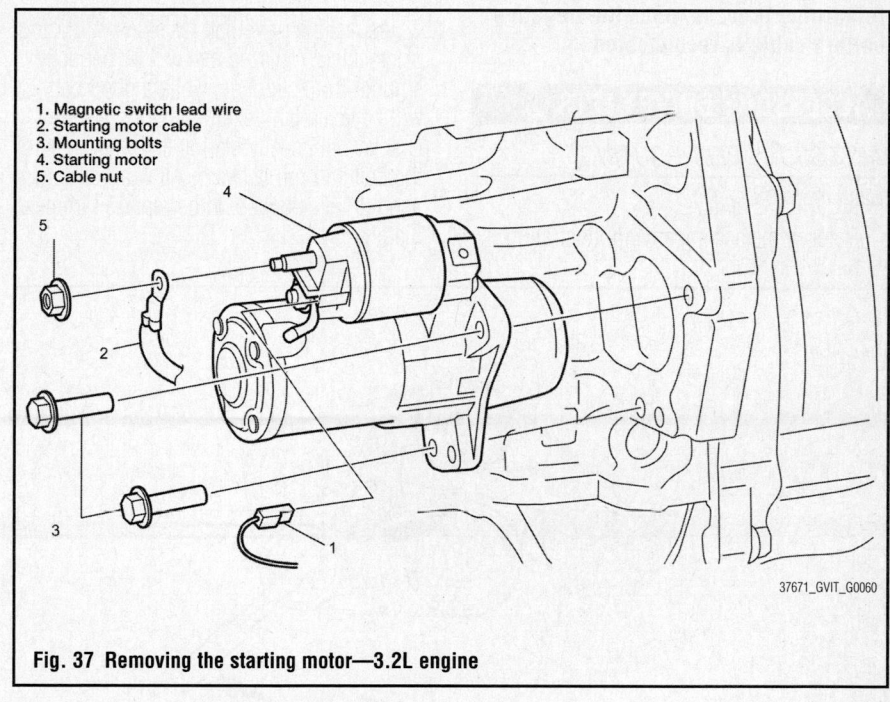

1. Magnetic switch lead wire
2. Starting motor cable
3. Mounting bolts
4. Starting motor
5. Cable nut

Fig. 37 Removing the starting motor—3.2L engine

37671_GVIT_G0060

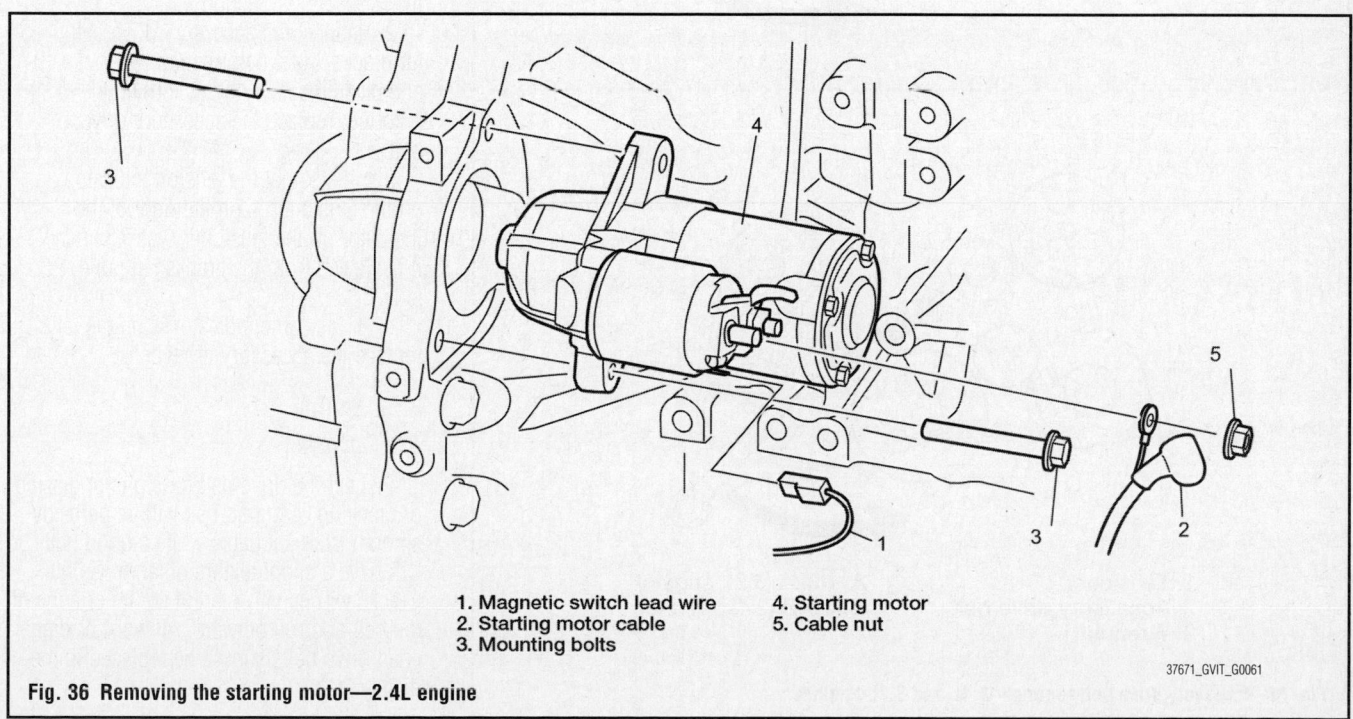

1. Magnetic switch lead wire
2. Starting motor cable
3. Mounting bolts
4. Starting motor
5. Cable nut

Fig. 36 Removing the starting motor—2.4L engine

37671_GVIT_G0061

4. Disconnect the magnetic switch lead wire and the starting motor cable from the starting motor.

5. Remove the starting motor mounting bolts.

6. Remove the starting motor from the vehicle.

To install:

7. Installation is the reverse of the removal procedure.

8. Tighten the starting motor mounting bolts to 34 ft. lbs. (45 Nm).

9. Tighten the starting motor cable nut to 97 inch lbs. (11 Nm).

ENGINE MECHANICAL

➡**Disconnecting the negative battery cable may interfere with the functions of the on board computer systems and may require the computer to undergo a relearning process, once the negative battery cable is reconnected.**

ACCESSORY DRIVE BELTS

ACCESSORY BELT ROUTING

See Figure 38.

Refer to the accompanying illustration.

INSPECTION

2.4L Engine

See Figure 39.

Inspect the drive belt for signs of glazing or cracking. A glazed belt will be perfectly smooth from slippage, while a good belt will have a slight texture of fabric visible. Cracks will usually start at the inner edge of the belt and run outward. All worn or damaged drive belts should be replaced immediately.

1. Tension indicator
2. Tension indicator

37671_SSX4_G0268

Fig. 39 Accessory drive belt inspection—2.4L engine

1. Before servicing the vehicle, refer to the Precautions Section.

2. Inspect the accessory drive belt for cracks, cuts, deformation, wear, and cleanliness. If any abnormality is found, replace the accessory drive belt.

3. Check that the tension indicator on the tensioner is on the right of the indicator on the tensioner bracket when it is viewed from the crankshaft pulley side.

4. If the same indicator is on the left, replace the accessory drive belt.

3.2L Engine

See Figure 40.

Inspect the drive belt for signs of glazing or cracking. A glazed belt will be perfectly smooth from slippage, while a good belt will have a slight texture of fabric visible. Cracks will usually start at the inner edge of the belt and run outward. All worn or damaged drive belts should be replaced immediately.

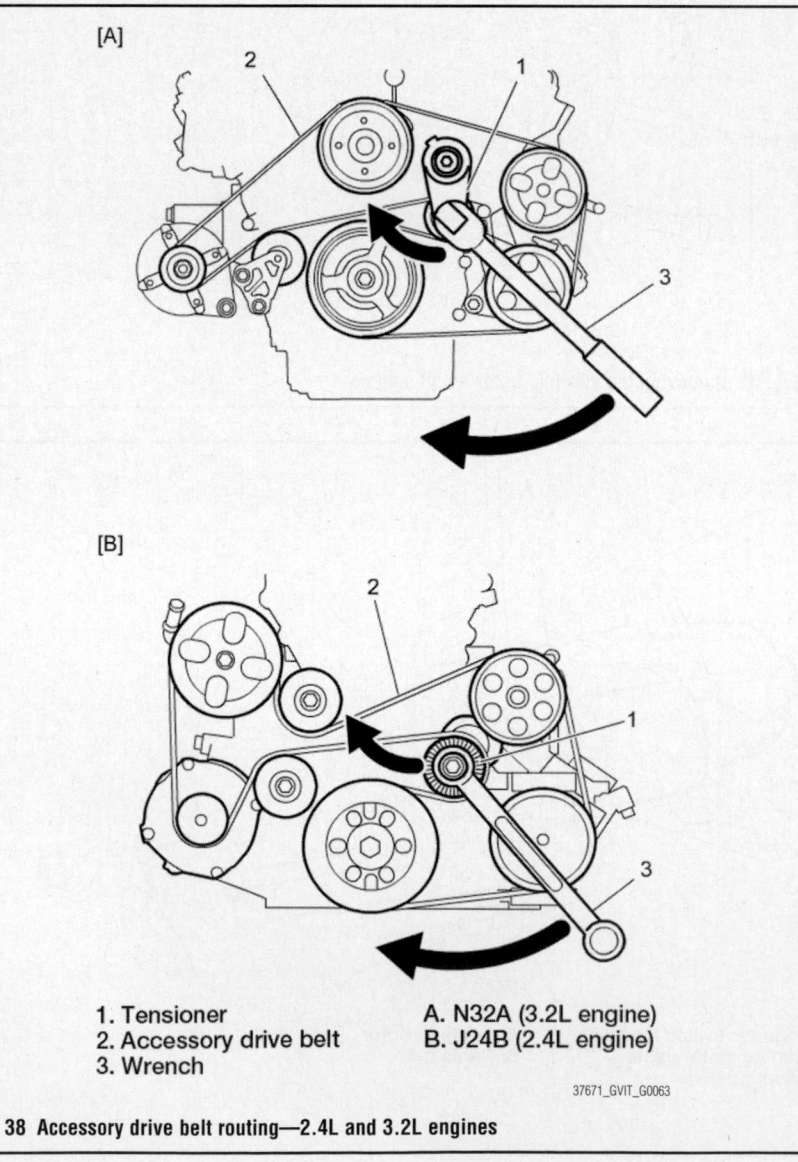

1. Tensioner
2. Accessory drive belt
3. Wrench

A. N32A (3.2L engine)
B. J24B (2.4L engine)

37671_GVIT_G0063

Fig. 38 Accessory drive belt routing—2.4L and 3.2L engines

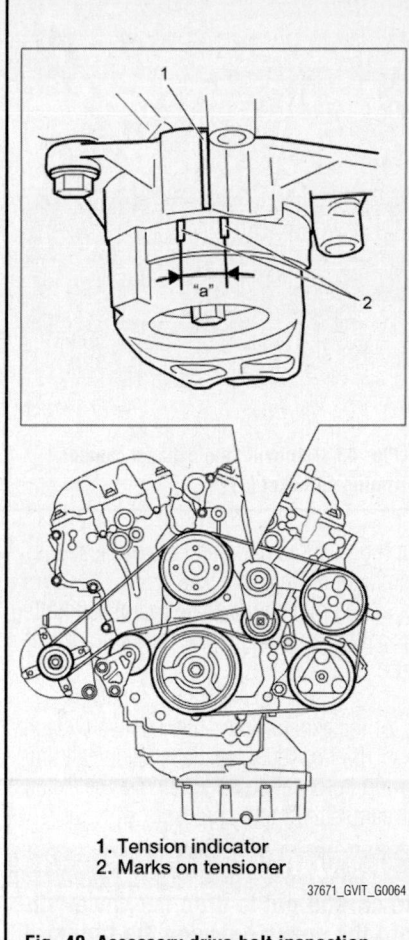

1. Tension indicator
2. Marks on tensioner

37671_GVIT_G0064

Fig. 40 Accessory drive belt inspection—3.2L engine

1. Before servicing the vehicle, refer to the Precautions Section.

2. Disconnect the negative battery cable.

3. Inspect the accessory drive belt for cracks, cuts, deformation, wear, and cleanliness. If any abnormality is found, replace the accessory drive belt.

4. Check that the tension indicator is between the marks.

5. If the indicator is not between the marks, replace the accessory drive belt.

ADJUSTMENT

There is no manual drive belt tension adjustment. The drive belt tension is automatically adjusted by the drive belt auto tensioner.

REMOVAL & INSTALLATION

See Figure 38.

Inspect the drive belt for signs of glazing or cracking. A glazed belt will be perfectly smooth from slippage, while a good belt will have a slight texture of fabric visible.

Cracks will usually start at the inner edge of the belt and run outward. All worn or damaged drive belts should be replaced immediately.

1. Before servicing the vehicle, refer to the Precautions Section.

2. Disconnect the negative battery cable.

3. Loosen the belt tensioner by turning the tensioner clockwise.

  a. On 2.4L engines, do not turn the tensioner pulley with a torque higher than 44 ft. lbs. (59 Nm).

  b. On 3.2L engines, do not turn the tensioner pulley with a torque higher than 59 ft. lbs. (80 Nm).

4. While keeping the tensioner in the loose position, remove the accessory drive belt.

To install:

5. Loosen the belt tension by turning the tensioner clockwise.

6. While keeping the tensioner in the loose position, install the accessory drive belt.

7. Connect the negative battery cable.

CAMSHAFT AND VALVE LIFTERS

REMOVAL & INSTALLATION

2.4L Engine

See Figures 41 through 49.

1. Before servicing the vehicle, refer to the Precautions Section.

2. Remove the cylinder head cover. Refer to Cylinder Head, removal & installation.

3. Remove the accessory drive belt. Refer to Accessory Drive Belts, removal & installation.

4. Turn the crankshaft clockwise with a 19mm wrench, and position piston No. 1 at TDC on the compression stroke according to the following procedure.

  a. Align the timing mark on the CMP actuator and the timing mark on the exhaust camshaft timing sprocket with the match marks on the camshaft housing No. 1.

  b. Align the notch on the crankshaft pulley with "0" on the timing chain cover.

5. Fix the timing chain tensioner according to the following procedure.

  a. Remove the timing chain cover plug and gasket.

  b. Insert the fore-end of Special Tool 09917-16710, or equivalent, into the timing chain tensioner hole through the timing chain cover plug hole.

➡**Make sure that the special tool is inserted in the timing chain tensioner hole. Use a mirror if necessary.**

  c. Fix the special tool by tightening the fixing bolt to 71 inch lbs. (8 Nm).

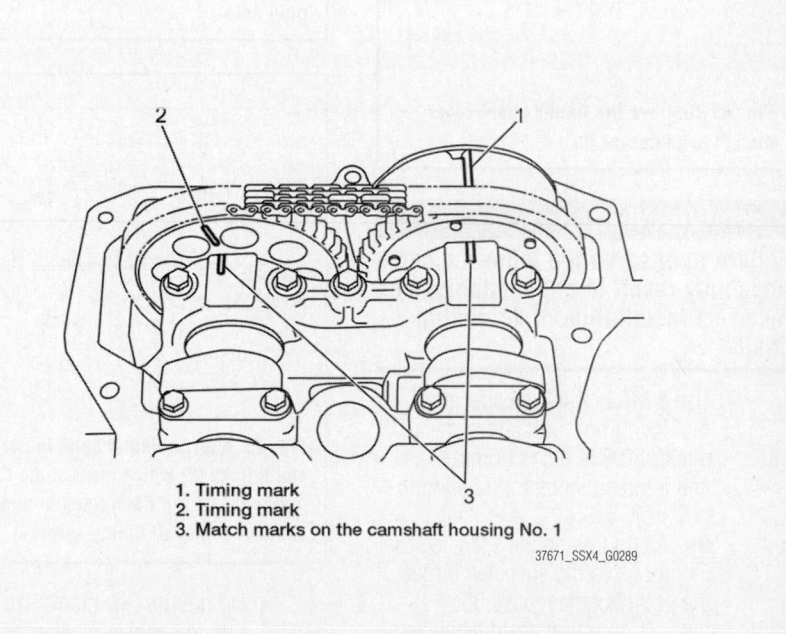

1. Timing mark
2. Timing mark
3. Match marks on the camshaft housing No. 1

37671_SSX4_G0289

Fig. 41 Align the timing mark on the CMP actuator and the timing mark on the exhaust camshaft timing sprocket with the match marks on the camshaft housing No. 1

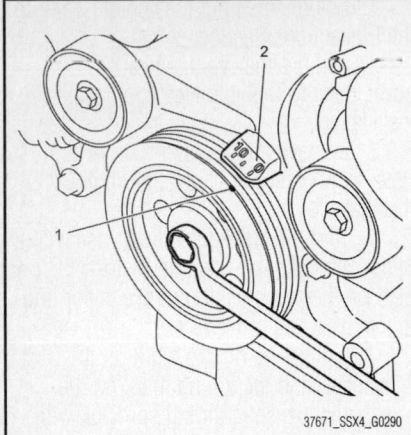

Fig. 42 Align the notch (1) on the crankshaft pulley with "0" (2) on the timing chain cover

Fig. 43 Remove the timing chain cover plug (1) and gasket (2)

✳✳ WARNING

Failure to observe the following cautions may result in engine damage or incorrect installation of the timing chain.

- Use a M6 bolt for the special tool fixing bolt with a thread pitch of 0.039 inch (1.0mm) and a thread length of 0.59 inch (15mm)
- Be sure to tighten the specified tool fixing bolt to the specified torque. If the bolt becomes loose, the plunger of timing chain tensioner adjuster may extend, making it difficult to reinstall the timing chain properly

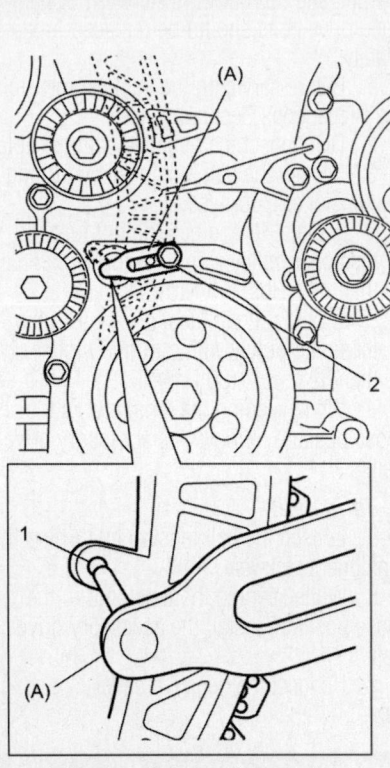

1. Timing chain tensioner hole
2. Fixing bolt
A. Special Tool 09917-16710

Fig. 44 Insert the fore-end of Special Tool 09917-16710 into the timing chain tensioner hole through the timing chain cover plug hole

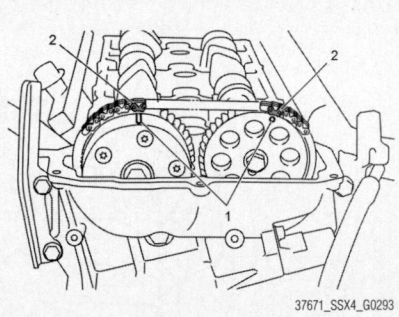

Fig. 45 Apply a dab of paint to two timing chain links (2) which meet at the timing marks (1) on the CMP actuator and the exhaust camshaft timing sprocket

- Do not turn the crankshaft after the timing chain is fixed with the special tool
6. Apply a dab of paint to two timing chain links which meet at the timing marks

Fig. 46 Removing the exhaust camshaft timing sprocket (1)

on the CMP actuator and the exhaust camshaft timing sprocket.

7. Remove the exhaust camshaft timing sprocket according to the following procedure.

   a. Hold the hexagonal section of the exhaust camshaft using a wrench.

   b. Loosen the exhaust camshaft timing sprocket bolt and remove the exhaust camshaft timing sprocket.

✳✳ WARNING

Be careful not to drop the dowel pin into the space between the timing chain cover and the cylinder block. Do not remove the CMP actuator during this procedure.

8. Remove the dowel pin from the exhaust camshaft.
9. Loosen the camshaft housing bolts, evenly and gradually, in the numerical order illustrated.
10. Remove the intake camshaft and the exhaust camshaft.

➡ **Be careful not to drop the timing chain into the timing chain cover.**

11. Remove the camshaft bearings and tappets as necessary.
12. Remove the CMP actuator from the intake camshaft according to the following procedure, if necessary.

   a. Hold the hexagonal section of the intake camshaft using a vice.

   b. Loosen the CMP actuator bolt and remove the CMP actuator.

➡ **Do not disassemble the CMP actuator.**

To install:

13. Install the CMP actuator to the intake camshaft. Tighten the CMP actuator bolt to 45 ft. lbs. (60 Nm).

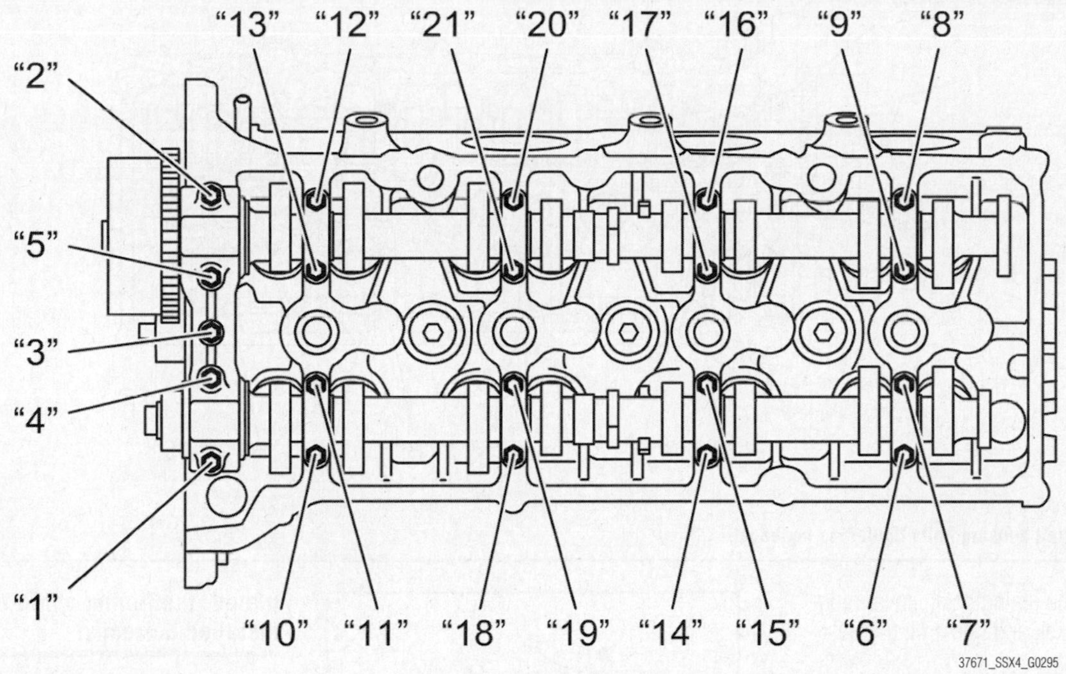

"2" "13" "12" "21" "20" "17" "16" "9" "8"

"5"

"3"

"4"

"1"

"10" "11" "18" "19" "14" "15" "6" "7"

37671_SSX4_G0295

Fig. 47 Camshaft housing bolts loosening sequence

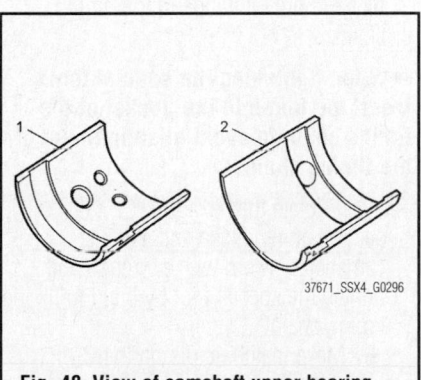

37671_SSX4_G0296

Fig. 48 View of camshaft upper bearing half (1) and lower bearing half (2)

14. Apply engine oil to the contact surface of the tappets and install the tappets to the cylinder head.

15. Install the camshaft bearings according to the following procedure.

➡**The camshaft upper bearing half is different in shape from the lower bearing half. Distinguish the upper and lower bearing halves when installing.**

   a. Fit the tab of the camshaft bearing to the groove of the cylinder head or camshaft housing.

   b. Press the camshaft bearing end until it fully seats in the cylinder head or camshaft housing.

   c. Apply engine oil to the sliding surface of the camshaft bearing halves.

➡**Do not apply engine oil between the bearing halves and the camshaft housing or cylinder head.**

16. Apply engine oil to the sliding surfaces of the camshafts, and then install the camshafts to the cylinder head according to the following procedure.

17. Align the timing chain link, which was painted in the removal procedure, with the timing mark on the CMP actuator and install the intake camshaft.

   a. Install the exhaust camshaft and position the dowel pin hole of the exhaust camshaft downward.

18. Install the camshaft housings to the proper places distinguished by the character, number, and arrow direction on each camshaft housing.

19. Tighten the camshaft housing bolts according to the following procedure.

   a. Apply engine oil to the bolt thread and seat of the camshaft housing bolts.

   b. Install the camshaft housing No. 1.

   c. Install the camshaft housing bolts, and tighten them by hand.

   d. Tighten the camshaft housing bolts, evenly and gradually, in numerical order to 97 inch lbs. (11 Nm).

20. Install the dowel pin to the exhaust camshaft.

➡**Be careful not to the drop dowel pin into the space between the timing chain cover and the cylinder block.**

21. Install the exhaust camshaft timing sprocket to the exhaust camshaft according to the following procedure.

22. Align the timing chain link, which was painted during removal, with the timing mark on the exhaust camshaft timing sprocket, and loop the timing chain up over the exhaust camshaft timing sprocket.

   a. Install the exhaust camshaft timing sprocket with the timing chain to the exhaust camshaft.

23. Tighten the exhaust camshaft timing sprocket bolt using Special Tool 09911-05120, or equivalent torque wrench. Tighten to 45 ft. lbs. (60 Nm).

24. Confirm that the painted timing chain links are aligned with the timing marks on the CMP actuator and the exhaust camshaft timing sprocket.

25. Remove the special tool from the timing chain cover.

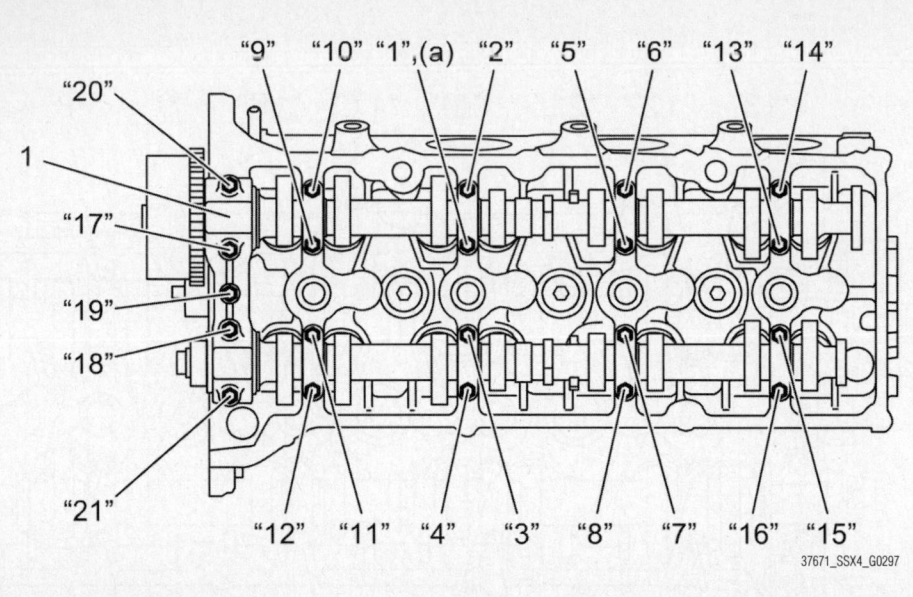

Fig. 49 Camshaft housing bolts tightening sequence

26. Install the timing chain cover plug with a new gasket, and tighten it to 20 ft. lbs. (27 Nm).

27. Check that the camshaft and the timing chain are installed properly as follows.

a. Check that the timing mark on the CMP actuator and the timing mark on the exhaust camshaft timing sprocket are aligned with the match marks on the camshaft housing No. 1.

b. Check that the notch on the crankshaft pulley is aligned with the "0" on the timing chain cover.

c. Turn the crankshaft clockwise twice and repeat the timing mark check.

➡Be sure to turn the crankshaft fully twice. If it is once, the timing marks on the CMP actuator and the exhaust camshaft timing sprockets do not meet the match marks on the camshaft housing No. 1. After turning the crankshaft twice, the painted links of the timing chain are not aligned with the timing marks on the CMP actuator and the exhaust timing sprocket, but this is normal.

28. Check the valve clearance. Refer to Valve Lash, Inspection.

29. Install the cylinder head cover. Refer to Cylinder Head, removal & installation.

3.2L Engine

See Figures 50 through 55.

1. Before servicing the vehicle, refer to the Precautions Section.

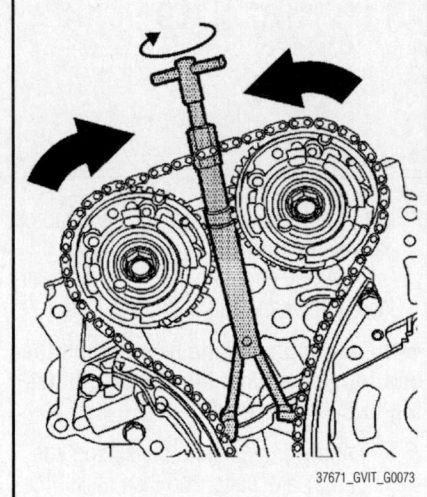

Fig. 50 Hold the timing chain using Special Tool 09918-57800—3.2L engine

2. Remove the cylinder head cover. Refer to Cylinder Head, removal & installation.

3. Remove the Camshaft Position (CMP) sensor.

4. Make sure that the flat sections of the camshafts are parallel to the cylinder head.

5. Hold the timing chain using Special Tool 09918-57800, or equivalent.

❊❊ WARNING

Do not remove the special tool until the camshaft and timing chain are installed. If the special tool is removed before operations are com-

pleted, the timing chain may not be installed correctly.

a. Insert the special tool, align the groove of the special tool and the top surface of the cylinder head.

b. Tighten the special tool by hand until it hangs in the timing chain.

➡When tightening the special tool, insert the feet into the pocket of the timing chain to avoid misalignment of the timing chain.

c. Rotate the hexagonal section of the intake and exhaust camshafts inward with a wrench and tighten the special tool by hand to fix the timing chain.

6. Make match marks on the CMP actuator and the timing chain.

7. Hold the hexagonal section of the camshaft stationary with a wrench and remove the CMP actuator bolt.

8. Loosen the camshaft housing bolts in numerical order, and remove the intake and exhaust camshafts.

9. Remove the intake and exhaust CMP actuators.

10. Remove the valve lash adjusters and valve rocker arms.

To install:

11. Apply engine oil to the sliding part of the valve lash adjusters and valve rocker arms, and install them to the cylinder head.

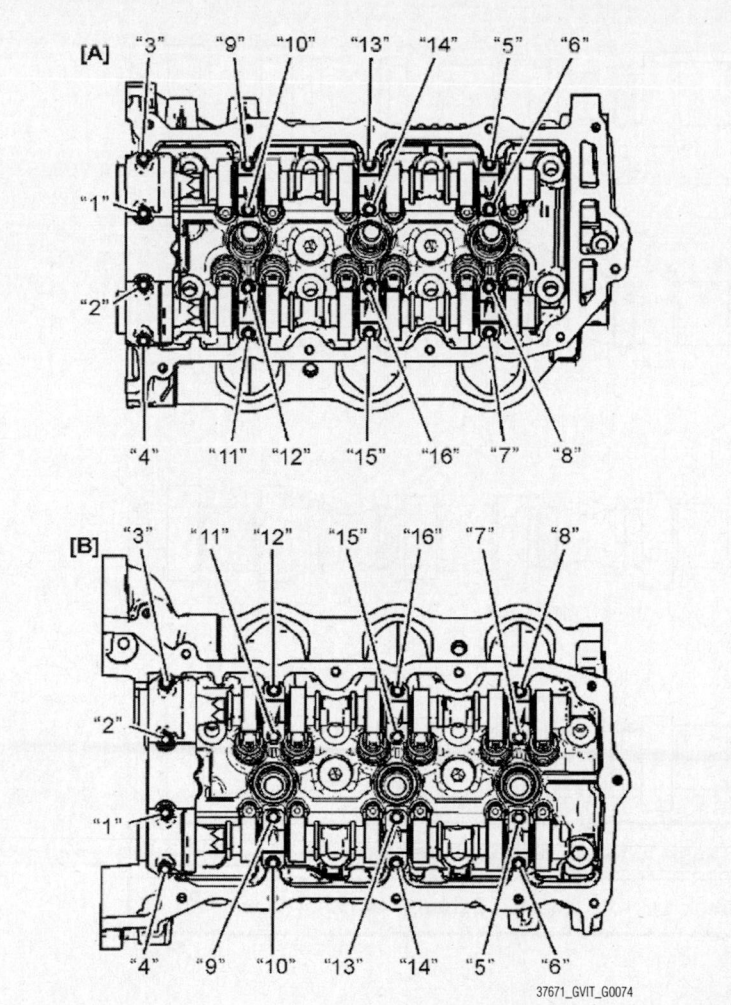

Fig. 51 Loosen the camshaft housing bolts of bank 1 (A) and bank 2 (B) in numerical order—3.2L engine

Do not expose the valve lash adjuster to excessive shock so as to prevent oil leakage.

12. Install new seal rings to the camshafts.

→ **Position the end gap of the seal ring in the proper range.**

13. Confirm the identification mark of the camshafts, to install them to the correct positions in the cylinder heads.

14. Apply engine oil to the sliding part of the camshafts. Install the camshafts to the cylinder head as follows.

    a. Align the match marks on the CMP actuators and the timing chain, and install the intake and exhaust CMP actuators.

    b. Align the lock pin of the CMP actuator with the groove of the camshaft, and install the CMP actuator bolt by hand.

    c. Ensure that the flat sections of the camshafts are parallel to the cylinder head.

    d. Install the camshaft housings according to the embossed mark.

    e. Tighten the new camshaft housing bolts evenly and gradually to 89 inch lbs. (10 Nm) in the proper tightening sequence.

Do not apply engine oil to the camshaft housing bolt.

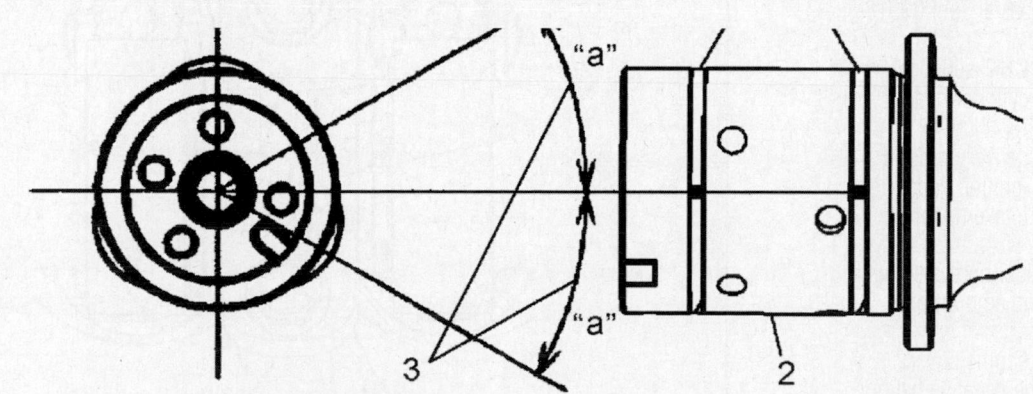

1. Seal rings
2. Camshafts
3. End gap of seal ring installation position
a. 30 degrees

Fig. 52 Install new seal rings to the camshafts. Position the end gap of the seal ring in the proper range

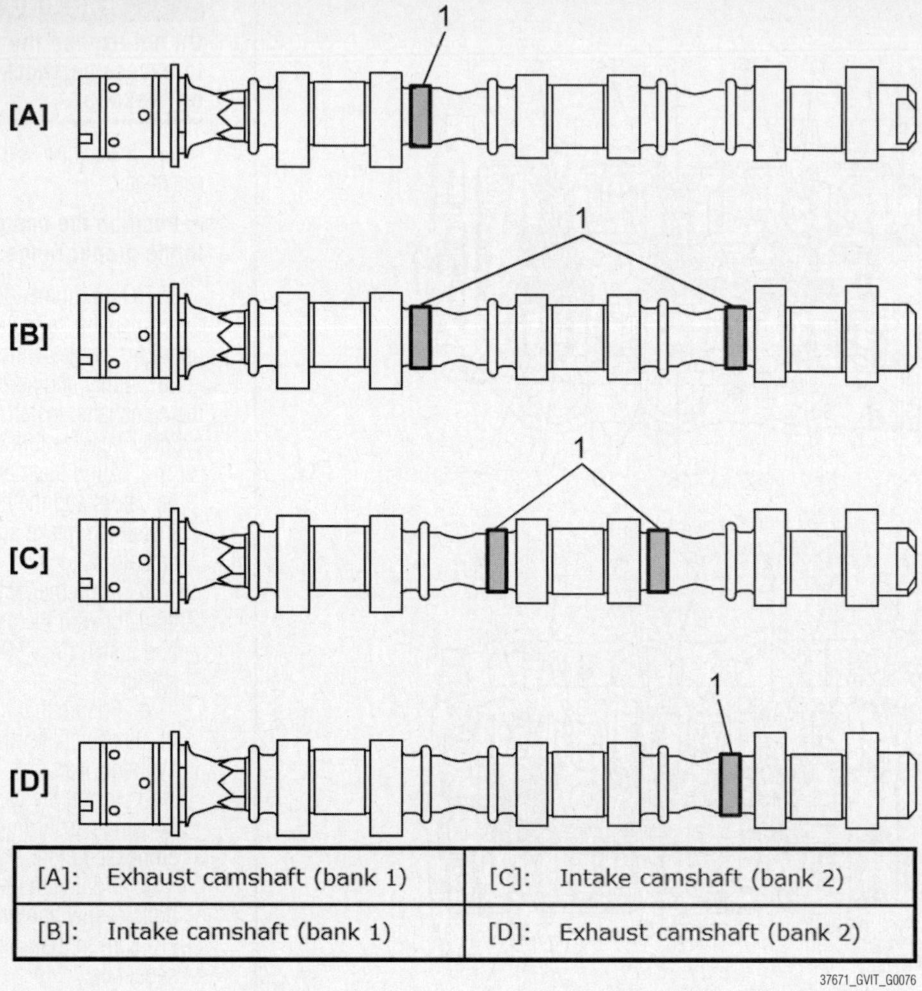

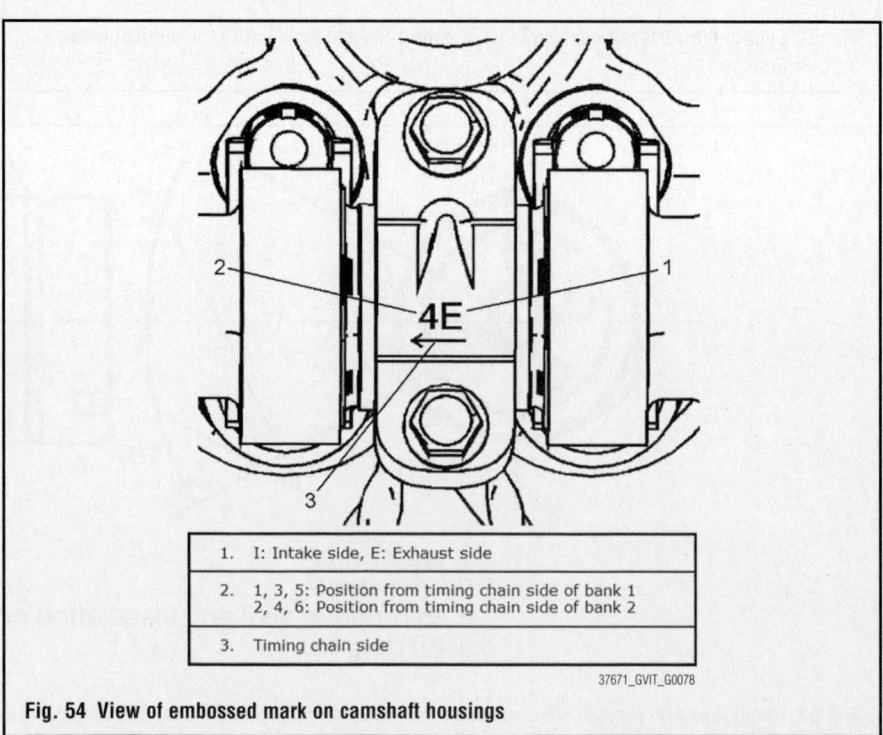

[A]:	Exhaust camshaft (bank 1)	[C]:	Intake camshaft (bank 2)
[B]:	Intake camshaft (bank 1)	[D]:	Exhaust camshaft (bank 2)

37671_GVIT_G0076

Fig. 53 View of identification marks (1) on the camshafts

f. With the hexagonal section of the camshaft held with a wrench, install the CMP actuator bolt and tighten to 43 ft. lbs. (58 Nm).

15. Remove Special Tool 09918-57800.

16. Install the CMP sensors.

17. Install the cylinder head covers.

18. Install the delivery pipe.

19. Install the ignition coils.

20. Install the intake upper manifold.

21. Refill the cooling system with the proper type and amount of coolant.

22. Install the engine cover.

23. Connect the negative battery cable.

24. Turn the ignition switch to the ON position without starting the engine and check for fuel leaks.

25. Start the engine and check for engine coolant leaks.

1.	I: Intake side, E: Exhaust side
2.	1, 3, 5: Position from timing chain side of bank 1 2, 4, 6: Position from timing chain side of bank 2
3.	Timing chain side

37671_GVIT_G0078

Fig. 54 View of embossed mark on camshaft housings

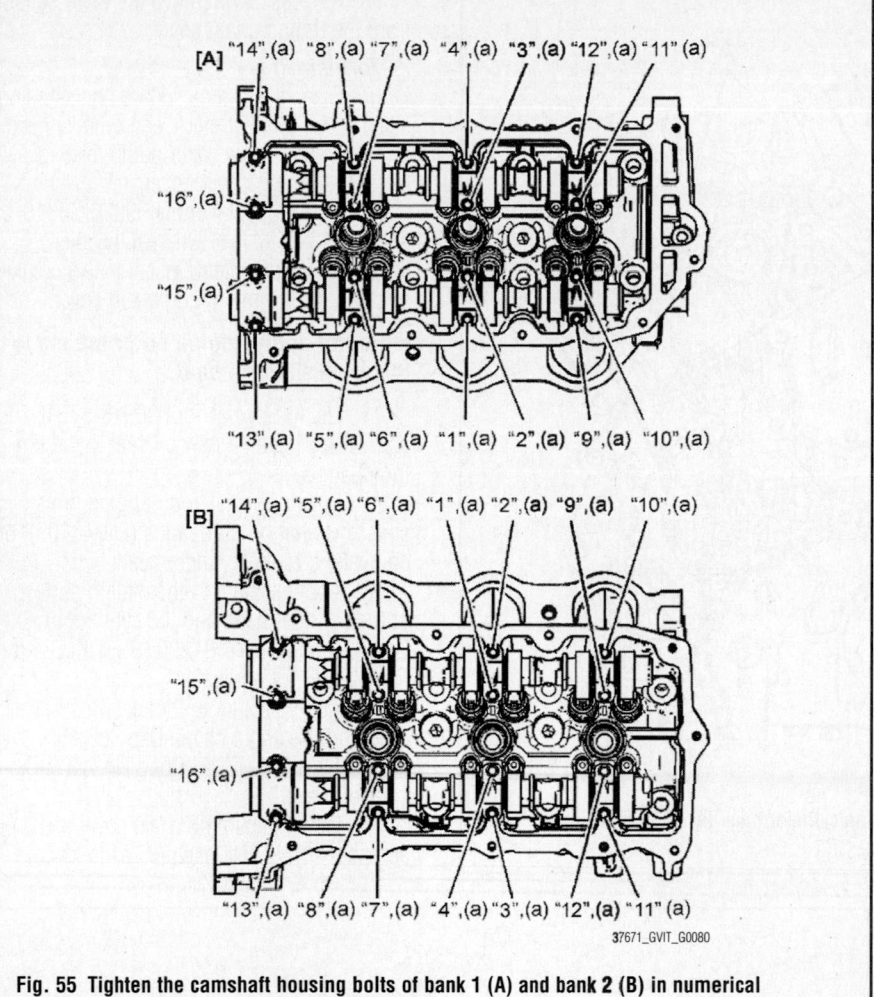

Fig. 55 Tighten the camshaft housing bolts of bank 1 (A) and bank 2 (B) in numerical order—3.2L engine

CRANKSHAFT FRONT SEAL

REMOVAL & INSTALLATION

2.4L Engine

See Figure 56.

1. Before servicing the vehicle, refer to the Precautions Section.

2. Remove the drive belt. Refer to Accessory Drive Belts, removal & installation.

3. Remove the crankshaft pulley. Refer to Crankshaft Damper, removal & installation.

4. Remove the crankshaft oil seal from the timing chain cover using a flat-head screwdriver, or equivalent.

➡**Be careful not to damage the front cover and/or the crankshaft.**

To install:

5. Installation is the reverse order of the removal procedure.

6. Press fit the seal until it is flush with the front end surface of the front cover, using the proper tools.

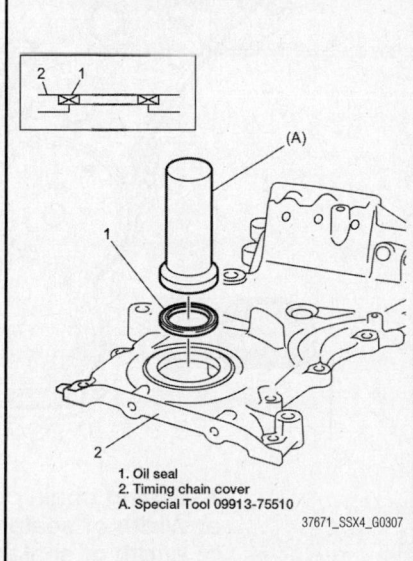

1. Oil seal
2. Timing chain cover
A. Special Tool 09913-75510

37671_SSX4_G0307

Fig. 56 Using Special Tool 09913-75510 to install the crankshaft front seal—2.4L engine

3.2L Engine

See Figures 57 through 64.

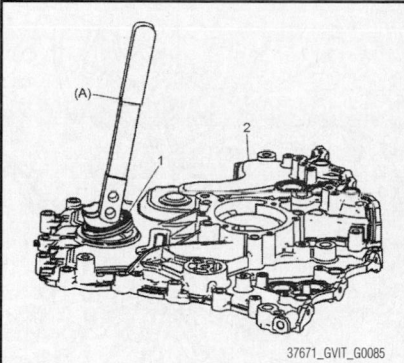

37671_GVIT_G0085

Fig. 57 Using Special Tool 09913-57830 to remove the crankshaft oil seal (1) from the timing chain cover (2)

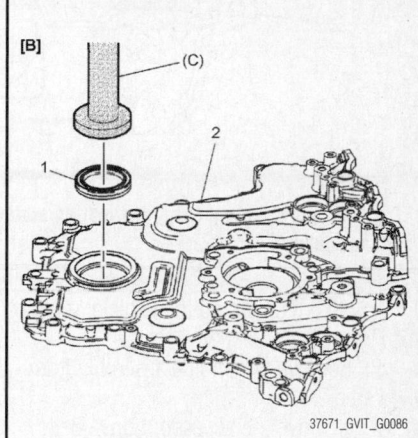

37671_GVIT_G0086

Fig. 58 Using Special Tool J-29184 to install a new crankshaft oil seal (1) to the timing chain cover (2)

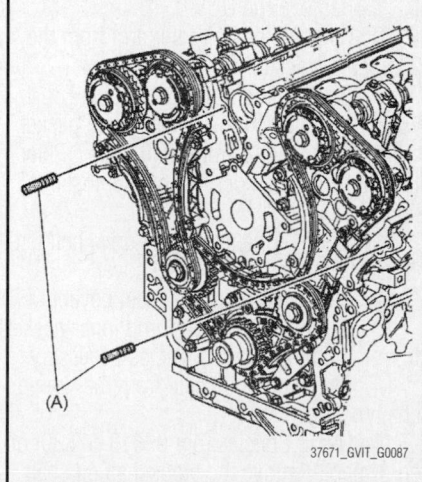

37671_GVIT_G0087

Fig. 59 Install Special Tool: Engine front cover installation guide pins (EN-46109) to the cylinder head

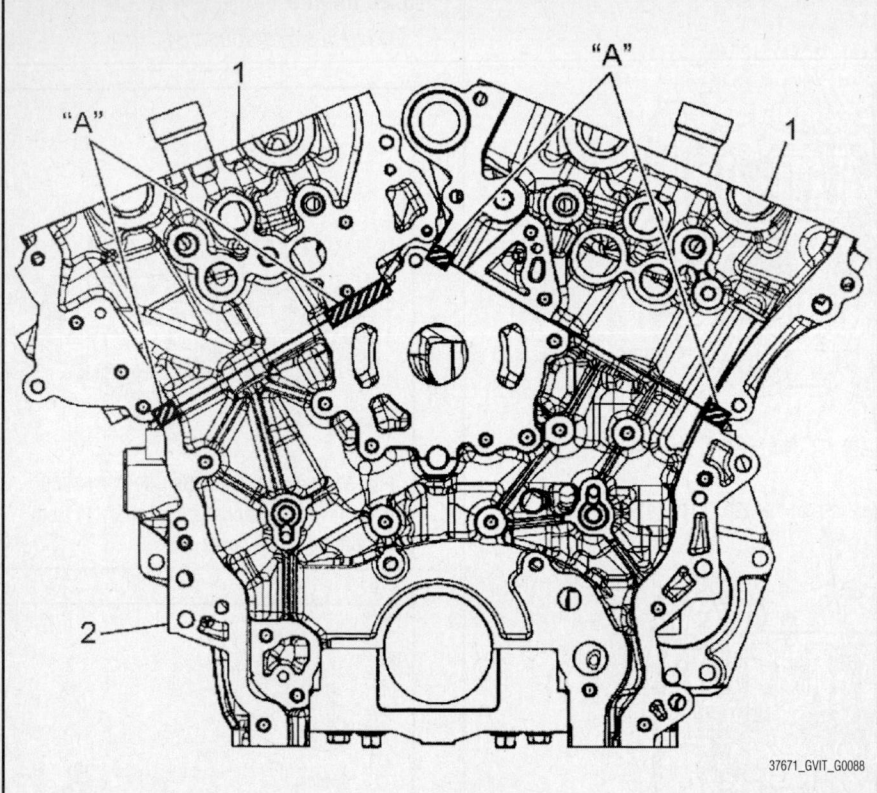

Fig. 60 Apply sealant (A) to the mating surface of the cylinder head (1) and the cylinder block (2) as shown

17. If necessary, remove the CMP sensor from the timing chain cover.

To install:

18. Clean the sealing surface on the timing chain cover, cylinder block, and cylinder head.

19. Install a new water pump inner gasket to the timing chain cover.

20. Using Special Tools 09926-68310 and 09913-75821, or Oil seal installer J-29184, or equivalent, install a new crankshaft oil seal to the timing chain cover.

➡ **Do not apply engine oil or grease to the crankshaft oil seal.**

21. Using Special Tool 09913-57810, or equivalent, install a new oil seal to the timing chain cover.

22. Install Special Tool: Engine front cover installation guide pins (EN-46109), or equivalent, to the cylinder head.

23. Apply sealant to the mating surface of the cylinder head and the cylinder block. Use Sealant 99000- 31290 (SUZUKI Bond No. 1217F), or equivalent.

24. Apply sealant to the mating surface of the timing chain cover. Use Sealant 99000-31290 (SUZUKI Bond No. 1217F), or equivalent.

25. Install the timing chain cover and the accessory drive belt tensioner as follows.

---

1. Before servicing the vehicle, refer to the Precautions Section.

2. Remove the engine assembly from the vehicle.

3. Remove the accessory drive belt. Refer to Accessory Drive Belt, removal & installation.

4. Remove the cylinder head cover.

5. Remove the water pump pulley.

6. Remove the P/S pump bracket.

7. Remove the OCV.

8. Remove the belt idler arm from the alternator bracket.

9. Remove the water outlet pipe No. 1, water outlet pipe No. 2, O-ring, and gasket.

10. Remove the crankshaft pulley. Refer to Crankshaft Damper, removal & installation.

11. Remove the accessory drive belt tensioner from the timing chain cover.

12. Remove the timing chain cover.

13. Remove the water pump inner gasket from the timing chain cover, as necessary.

14. Remove the oil seal from the timing chain cover.

15. Using Special Tool 09913-57830, or equivalent, remove the crankshaft oil seal from the timing chain cover.

16. If necessary, remove the water pump from the timing chain cover.

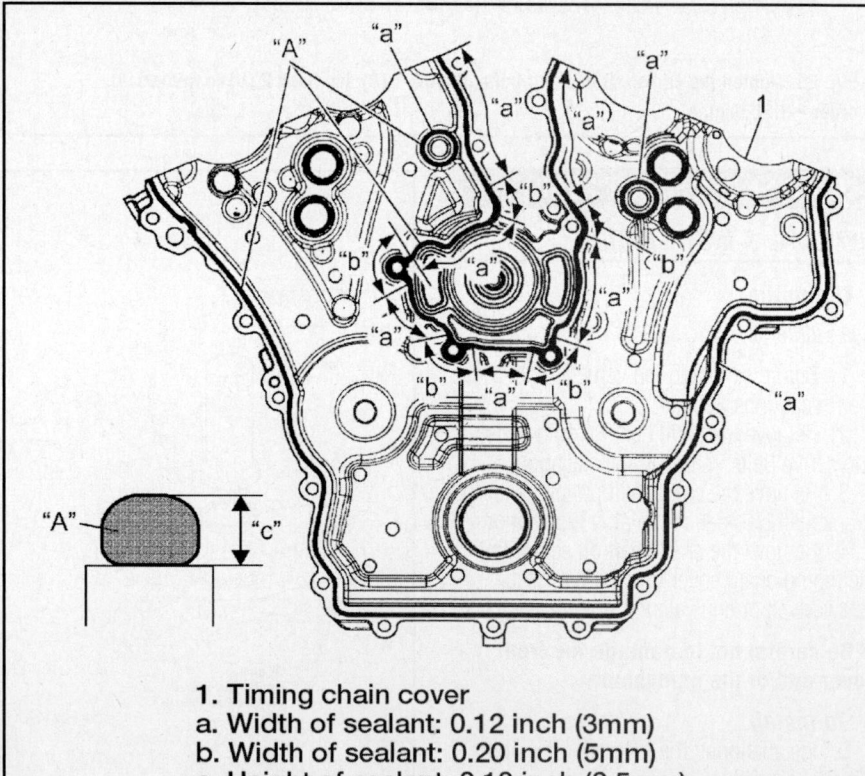

1. Timing chain cover
a. Width of sealant: 0.12 inch (3mm)
b. Width of sealant: 0.20 inch (5mm)
c. Height of sealant: 0.10 inch (2.5mm)
A. Sealant

Fig. 61 Apply sealant to the mating surface of the timing chain cover

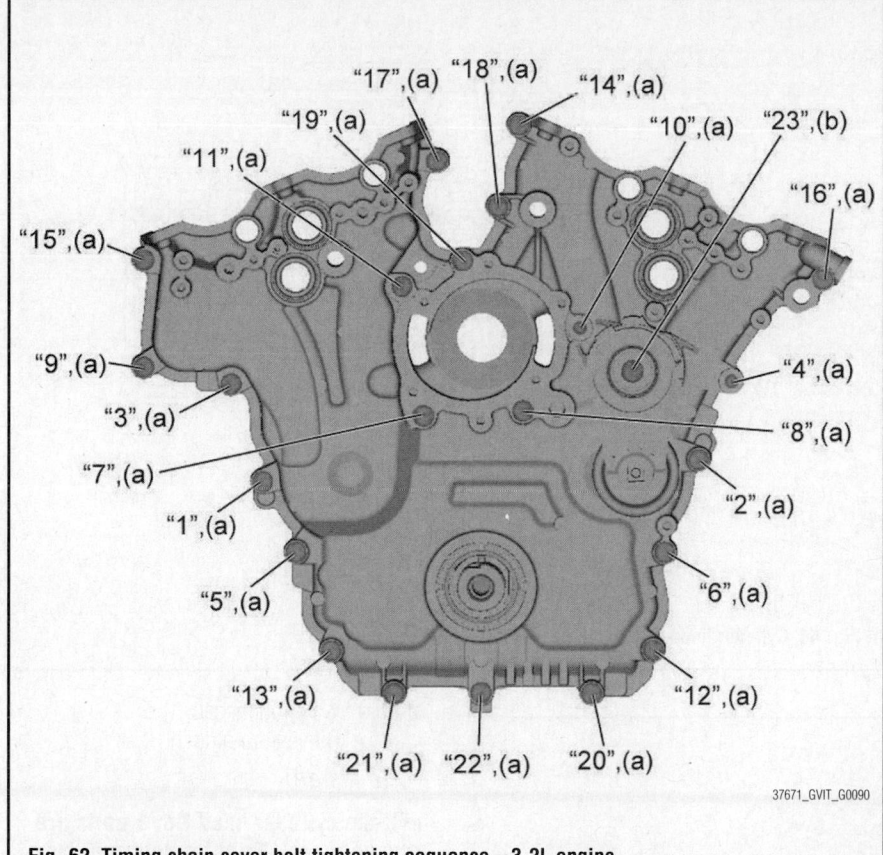

Fig. 62 Timing chain cover bolt tightening sequence—3.2L engine

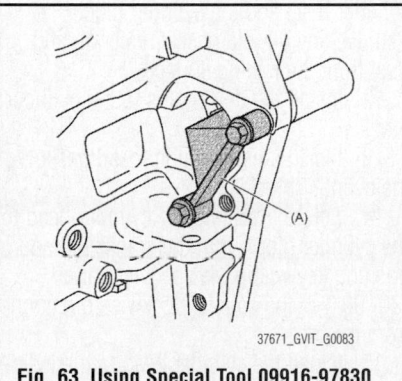

Fig. 63 Using Special Tool 09916-97830 (A) to lock the flywheel

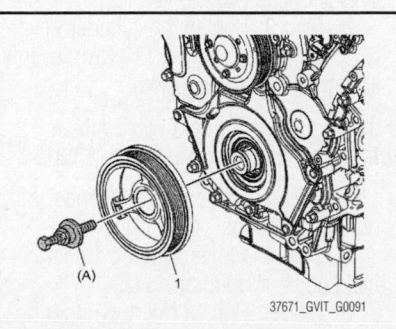

Fig. 64 Using Special Tool 09912-37820 (A), install the crankshaft pulley (1) to the crankshaft

➡**Before installing the timing chain cover, check that the dowel pin is securely fitted.**

    a. Tighten the timing chain cover bolts evenly and gradually to 15 ft. lbs. (20 Nm) according to the tightening sequence.

    b. In the same manner, tighten the cover bolts an additional 60°.

    c. Tighten the accessory drive belt tensioner bolt to 18 ft. lbs. (25 Nm).

26. Remove Special Tool: Engine front cover installation guide pins (EN-46109).

27. Install a new O-ring and new gasket for the water outlet pipe.

28. Install the water outlet pipe No. 1 and water outlet pipe No. 2.

29. Lock the flywheel with Special Tool 09916-97830, or equivalent.

30. Apply engine oil between the crankshaft pulley and the crankshaft.

➡**Do not apply engine oil between the crankshaft oil seal and the crankshaft pulley.**

31. Using Special Tool 09912-37820, or equivalent, install the crankshaft pulley to the crankshaft.

32. Remove the special tool and tighten the crankshaft pulley bolt.

    a. Step 1: 89 ft. lbs. (120 Nm).
    b. Step 2: 0 ft. lbs. (0 Nm).
    c. Step 3: 74 ft. lbs. (100 Nm).
    d. Step 4: Plus 150°.

33. Install the belt idler arm to the alternator bracket.

34. Install the OCV.

35. Install the P/S pump bracket.

36. Install the water pump pulley.

37. Install the cylinder head cover.

38. Install the accessory drive belt.

39. Install the engine assembly to the vehicle.

CYLINDER HEAD

REMOVAL & INSTALLATION

2.4L Engine
See Figures 65 through 68.

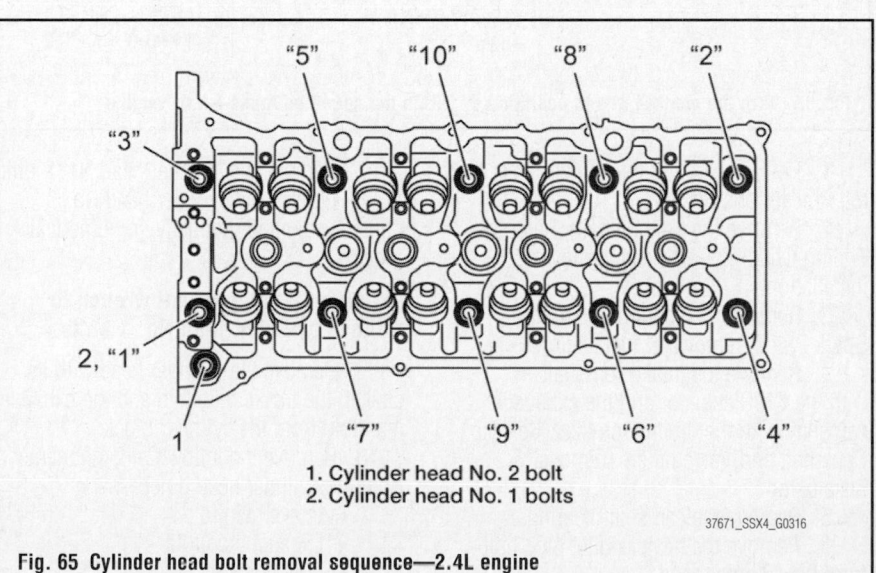

1. Cylinder head No. 2 bolt
2. Cylinder head No. 1 bolts

Fig. 65 Cylinder head bolt removal sequence—2.4L engine

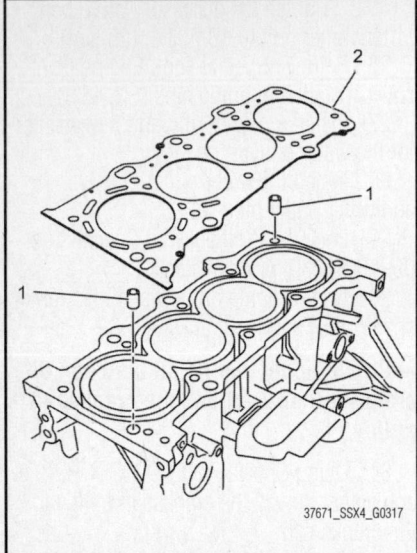

Fig. 66 Install the dowel pins (1) and a new cylinder head gasket (2) to the cylinder block—2.4L engine

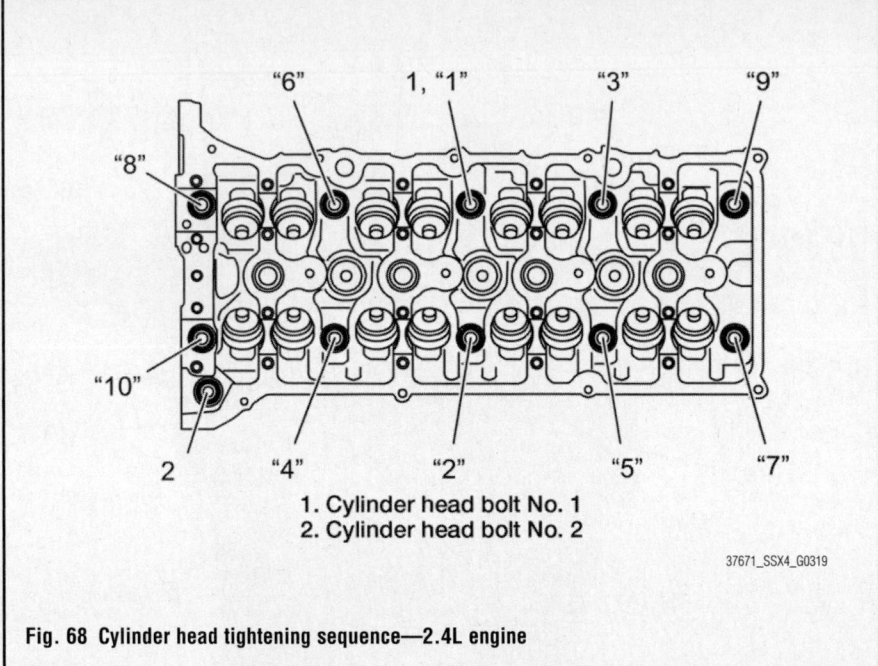

1. Cylinder head bolt No. 1
2. Cylinder head bolt No. 2

37671_SSX4_G0319

Fig. 68 Cylinder head tightening sequence—2.4L engine

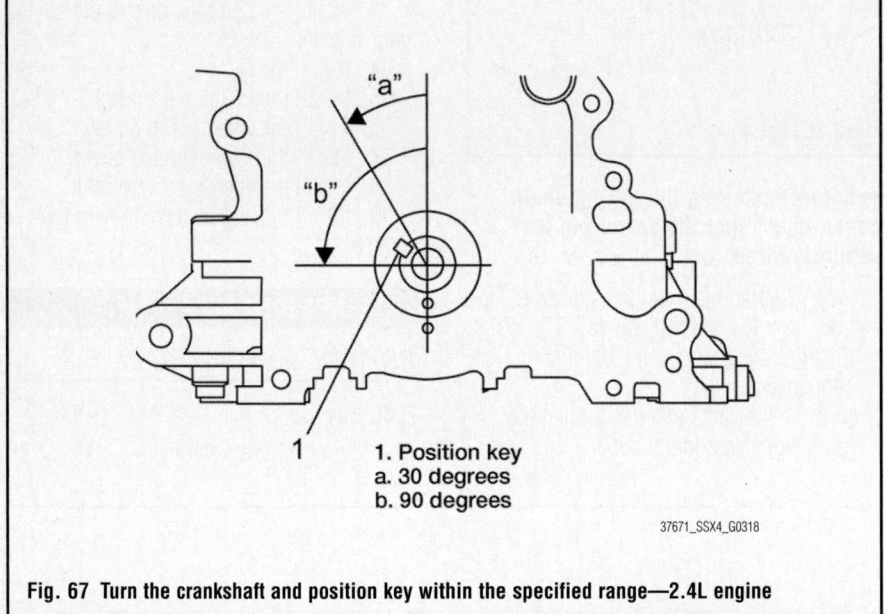

1. Position key
a. 30 degrees
b. 90 degrees

37671_SSX4_G0318

Fig. 67 Turn the crankshaft and position key within the specified range—2.4L engine

1. Before servicing the vehicle, refer to the Precautions Section.

2. Remove the timing chain. Refer to Timing Chain & Sprockets, removal & installation.

3. Remove the spark plugs. Refer to Spark Plugs, removal & installation.

4. Remove the intake camshaft with the CMP actuator and the exhaust camshaft with the timing sprocket. Refer to Camshaft and Valve Lifters, removal & installation.

5. Remove the camshaft bearings.

6. Remove the heater outlet pipe bolt from the cylinder head.

7. Remove the cylinder head No. 2 bolt.

8. Loosen the cylinder head No. 1 bolts, evenly and gradually, in numerical order.

➡Use a 12-point socket wrench to remove cylinder head No. 1 bolts.

9. Remove the cylinder head with its gasket, the intake manifold and the exhaust manifold from the cylinder block.

10. Remove the following components from the cylinder head if necessary.
- Intake manifold
- Exhaust manifold
- Fuel injector
- Water outlet cap
- Oil venture plug

To install:

➡If the cylinder head No. 1 bolts are reused, check them for deformation.

11. Clean the mating surface of the cylinder head and the cylinder block. Remove any oil, old gasket material, and dust from the mating surface.

12. Install the dowel pins to the cylinder block.

13. Install a new cylinder head gasket to the cylinder block.

14. Before installing the cylinder head to the cylinder block, turn the crankshaft and position key within the specified range (30–90°) in the counterclockwise direction from the top.

15. Install the cylinder head to the cylinder block.

  a. Step 1: Apply engine oil to the cylinder head No. 1 bolt threads.

  b. Step 2: Tighten the cylinder head No. 1 bolts to 15 ft. lbs. (20 Nm), evenly and gradually, in sequence.

➡Use a 12-point socket wrench to tighten the cylinder head No. 1 bolts.

  c. Step 3: Tighten No. 1 cylinder head bolts to 30 ft. lbs. (40 Nm).

  d. Step 4: Tighten No. 1 cylinder head bolts an additional 60°.

  e. Step 5: Tighten No. 1 cylinder head bolts an additional 80°.

  f. Step 6: Tighten No. 2 cylinder head bolt to 19 ft. lbs. (25 Nm).

16. Install the intake camshaft and the exhaust camshaft. Refer to Camshaft and Valve Lifters, removal & installation.

17. Install the timing chain. Refer to Timing Chain & Sprockets, removal & installation.

18. Install the timing chain cover. Refer to Timing Chain Front Cover, removal & installation.

19. Install the cylinder head cover.

20. Install the oil pan. Refer to Oil Pan, removal & installation.

21. Install the water outlet cap if removed.

22. Install the heater outlet pipe bolt to the cylinder head.

23. Install the spark plugs. Refer to Spark Plugs, removal & installation.

24. Install the exhaust manifold if removed. Refer to Exhaust Manifold, removal & installation.

25. Install the intake manifold if removed. Refer to Intake Manifold, removal & installation.

26. Install the fuel injectors if removed.

27. Install the engine assembly.

3.2L Engine

See Figures 69 through 71.

1. Before servicing the vehicle, refer to the Precautions Section.

2. Remove the engine assembly from vehicle.

3. Remove the cylinder head cover.

4. Remove the alternator bracket.

5. Remove the timing chain cover. Refer to Timing Chain Front Cover, removal & installation.

6. Remove the 2nd timing chain (bank 1), timing chain tensioner adjuster No. 1, timing chain tensioner No. 1, and timing chain guide No. 1.

7. Remove the 2nd timing chain (bank 2), timing chain tensioner adjuster No. 3, timing chain tensioner No. 2, and timing chain guide No. 3.

8. Remove the cylinder head bolts No. 1 and No. 2 in the proper loosening order.

9. If necessary, remove the exhaust manifold.

10. If necessary, remove the valve lash adjusters, valve rocker arms, intake and exhaust camshafts.

11. Remove the cylinder head.

To install:

12. Clean the mating surface of the cylinder head and the cylinder block.

13. Install dowel pins to the cylinder block.

14. Install new cylinder head gaskets to the cylinder block.

✳✳ WARNING

Do not apply engine oil to the cylinder head bolts No. 1 and No. 2.

15. Tighten the cylinder head bolts evenly and gradually in the proper tightening sequence:

a. Step 1: Tighten the No. 1 bolts to 23 ft. lbs. (30 Nm).

b. Step 2: Tighten the No. 1 bolts an additional 150°.

c. Step 3: Tighten the No. 2 bolts to 11 ft. lbs. (15 Nm).

d. Step 4: Tighten the No. 2 bolts an additional 75°.

16. Install the 2nd timing chain (bank 2), timing chain tensioner adjuster No. 3, timing chain tensioner No. 2, and timing chain guide No. 3.

17. Install the 2nd timing chain (bank 1), timing chain tensioner adjuster No. 1, timing chain tensioner No. 1, and timing chain guide No. 1.

18. Install the timing chain cover.

19. Install the alternator bracket.

20. Install the cylinder head cover.

21. Install the engine assembly to the vehicle.

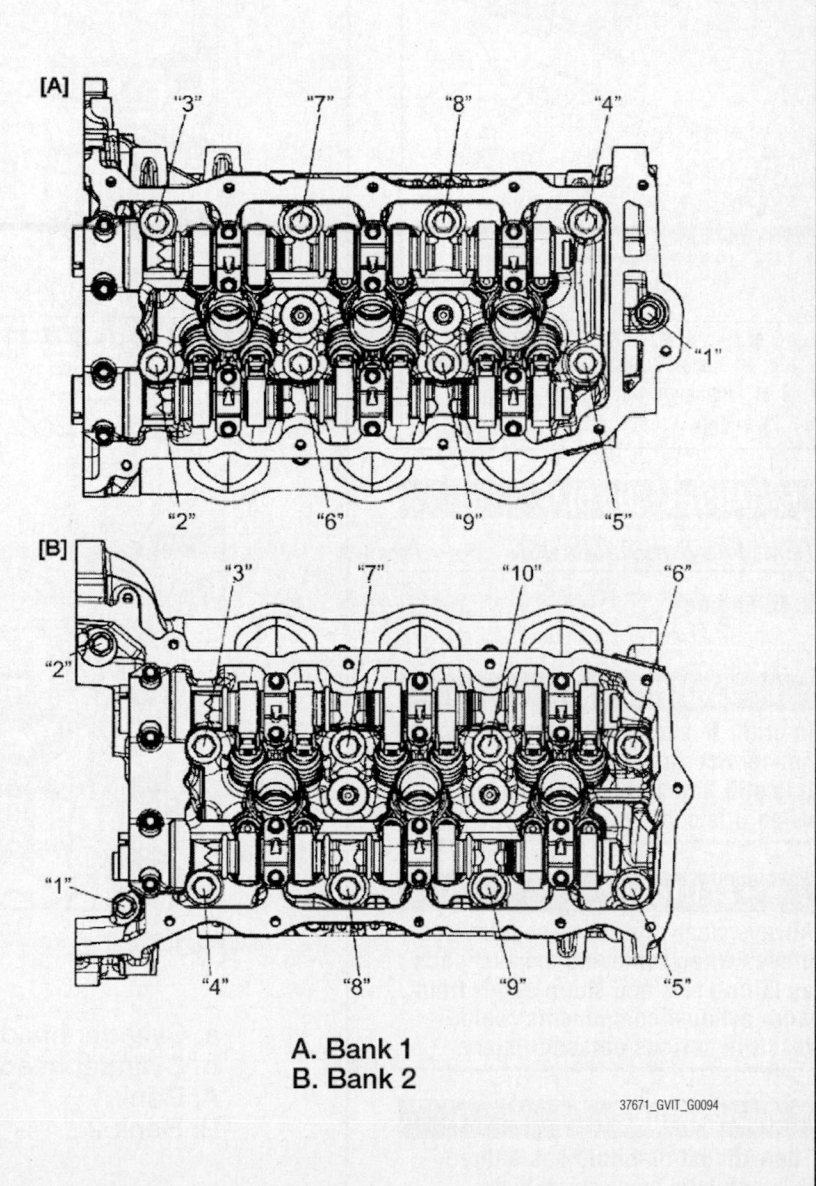

[A]
[B]

A. Bank 1
B. Bank 2

37671_GVIT_G0094

Fig. 69 Cylinder head bolt loosening order—3.2L engine

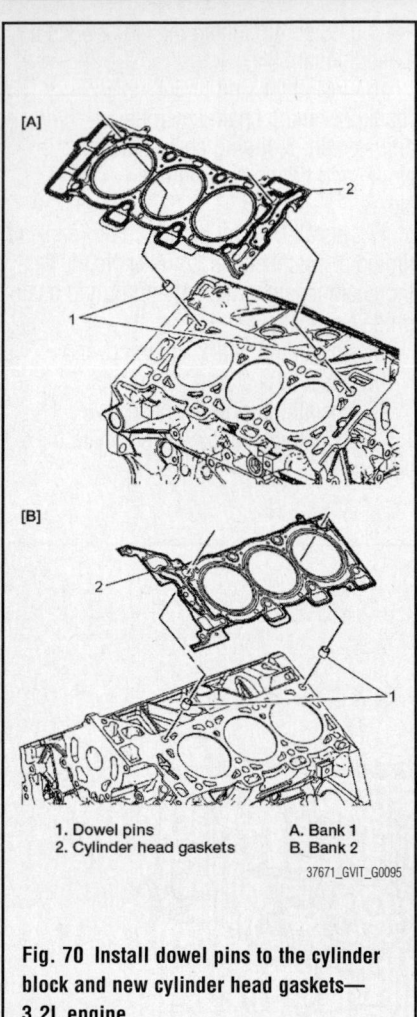

1. Dowel pins   A. Bank 1
2. Cylinder head gaskets B. Bank 2

37671_GVIT_G0095

Fig. 70 Install dowel pins to the cylinder block and new cylinder head gaskets— 3.2L engine

EXHAUST MANIFOLD

REMOVAL & INSTALLATION

2.4L Engine

See Figures 72 and 73.

❊❊ CAUTION

In order to avoid being burned, do not service the exhaust system while it is still hot. Service the system when it is cool.

❊❊ CAUTION

Always wear protective goggles and gloves when removing exhaust parts as falling rust and sharp edges from worn exhaust components could result in serious personal injury.

❊❊ WARNING

The exhaust manifold has a three-way catalytic converter which reduces harmful exhaust emissions.

Be careful not to drop the exhaust manifold. If it is dropped, the inside of the catalytic converter may be damaged on impact.

1. Before servicing the vehicle, refer to the Precautions Section.
2. Disconnect the negative battery cable.
3. Remove the exhaust manifold cover.
4. Disconnect the A/F sensor connector and the HO2S connector, and then detach them from their brackets.
5. Remove exhaust pipe No. 1.
6. Remove the oil level gauge and oil level gauge guide.
7. Loosen the exhaust manifold bolts and nuts, then remove them.

8. Remove the exhaust manifold and its gasket from the cylinder head.
9. Remove the A/F sensor and HO2S, if necessary.

To install:
10. Installation is the reverse of the removal procedure.
11. Use new gaskets and new seal rings.
12. Install the A/F sensor and HO2S.
13. Use a new O-ring for the oil level gauge guide. Apply engine oil to aid in the installation to the oil pan.
14. Tighten the oil level gauge guide bolt to 97 inch lbs. (11 Nm).
15. Run the engine and check for exhaust leaks.

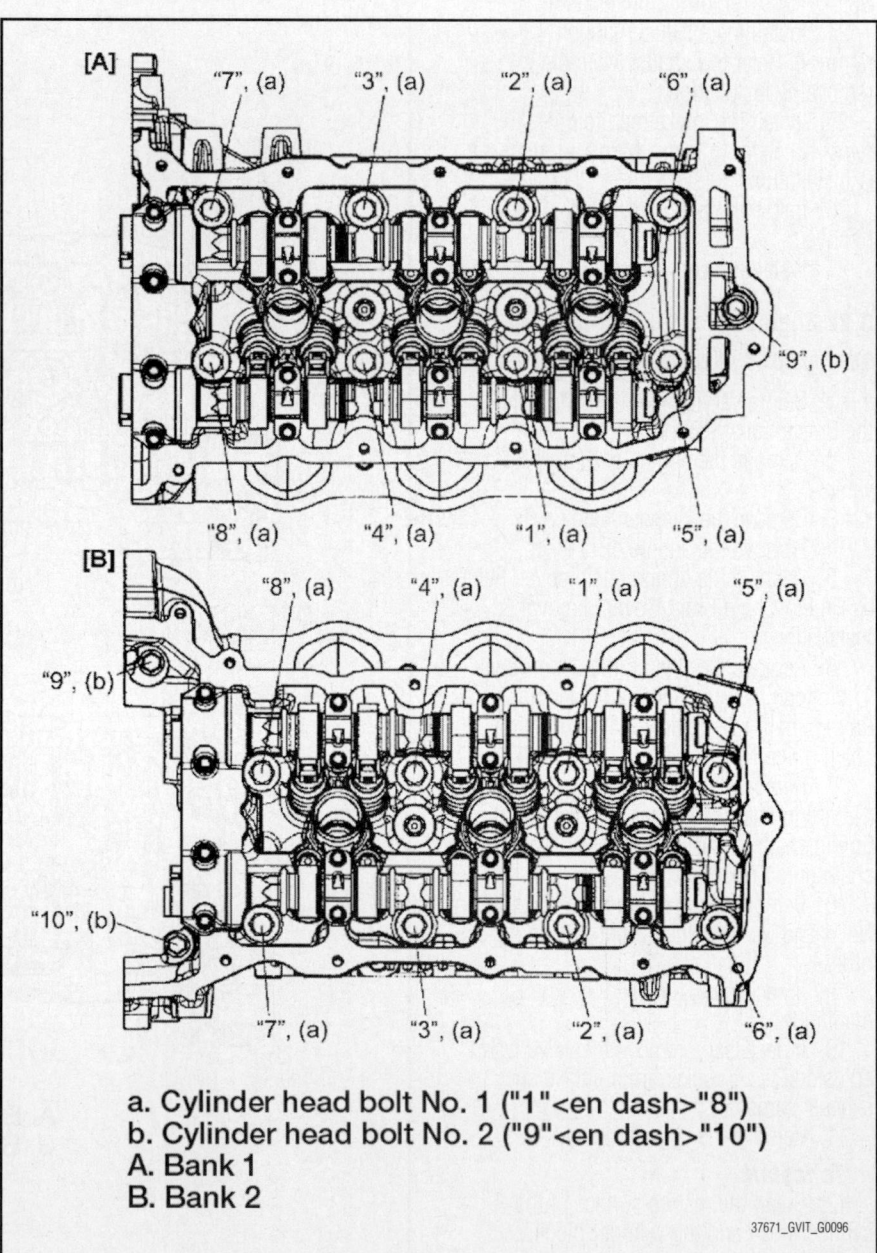

a. Cylinder head bolt No. 1 ("1"<en dash>"8")
b. Cylinder head bolt No. 2 ("9"<en dash>"10")
A. Bank 1
B. Bank 2

37671_GVIT_G0096

Fig. 71 Cylinder head bolt tightening sequence—3.2L engine

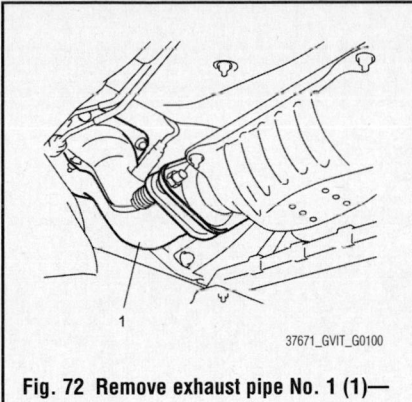

Fig. 72 Remove exhaust pipe No. 1 (1)—2.4L engine

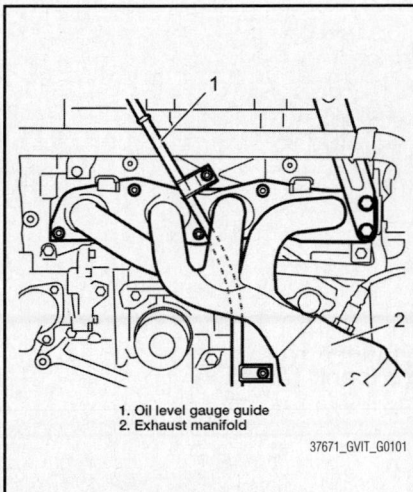

1. Oil level gauge guide
2. Exhaust manifold

37671_GVIT_G0101

Fig. 73 Exhaust manifold and oil level gauge guide shown—2.4L engine

3.2L Engine

Bank 1

See Figures 74 through 78.

> ✳✳ **CAUTION**
>
> In order to avoid being burned, do not service the exhaust system while it is still hot. Service the system when it is cool.

> ✳✳ **CAUTION**
>
> Always wear protective goggles and gloves when removing exhaust parts as falling rust and sharp edges from worn exhaust components could result in serious personal injury.

> ✳✳ **WARNING**
>
> The exhaust manifold has a three-way catalytic converter which reduces harmful exhaust emissions. Be careful not to drop the exhaust

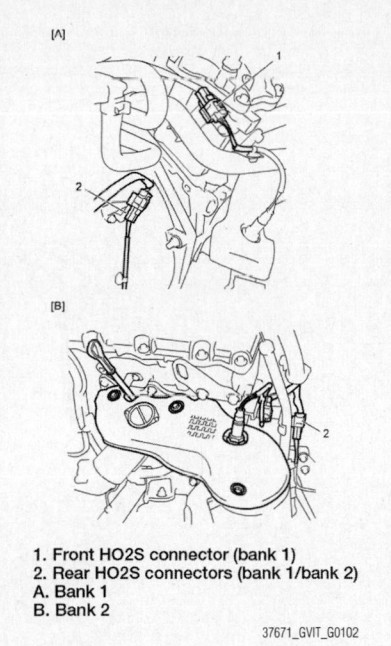

1. Front HO2S connector (bank 1)
2. Rear HO2S connectors (bank 1/bank 2)
A. Bank 1
B. Bank 2

37671_GVIT_G0102

Fig. 74 Location of HO2S connectors—3.2L engine

manifold. If it is dropped, the inside of the catalytic converter may be damaged on impact.

1. Before servicing the vehicle, refer to the Precautions Section.
2. Disconnect the battery cables and remove the battery.
3. Remove the battery tray and bracket.
4. Remove the air cleaner assembly.
5. Disconnect the front HO2S connector of bank 1, the rear HO2S connectors of both bank 1 and bank 2. Detach the disconnected connectors from their brackets.

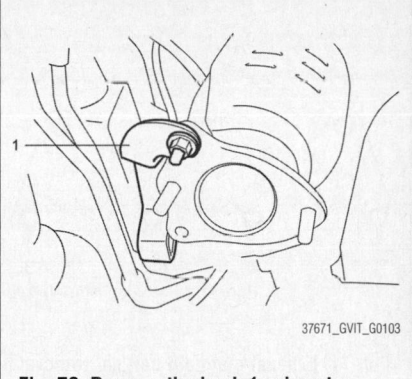

Fig. 76 Remove the bank 1 exhaust manifold stiffener (1)—3.2L engine

6. Remove the exhaust manifold cover.
7. Remove exhaust pipe No. 1.
   a. Remove the engine rear mounting.
   b. Remove exhaust pipe No. 1.
   c. Install the engine rear mounting without installing exhaust pipe No. 1. Tighten the engine rear mounting bolt to 41 ft. lbs. (55 Nm).
8. Remove the exhaust manifold stiffener.
9. Remove the exhaust manifold bolts and nuts evenly and gradually in sequence.
10. Remove the bank 1 exhaust manifold and its gasket from the cylinder head.
11. Remove the front HO2S and the rear HO2S, if necessary.

To install:
12. Installation is the reverse of the removal procedure.
13. Use new gaskets and new seal rings.
14. Use a new O-ring for the oil level gauge guide and apply engine oil to the O-ring when installing oil level gauge guide to the cylinder block.

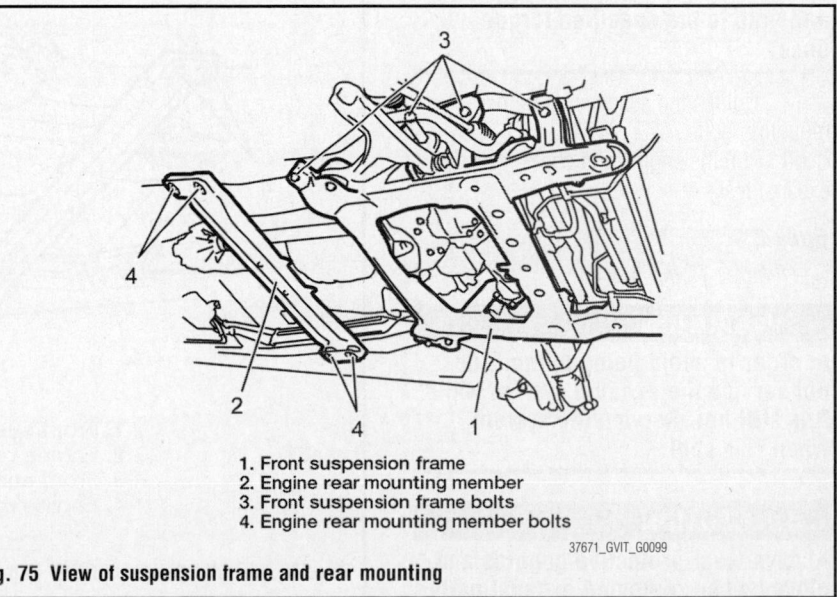

1. Front suspension frame
2. Engine rear mounting member
3. Front suspension frame bolts
4. Engine rear mounting member bolts

37671_GVIT_G0099

Fig. 75 View of suspension frame and rear mounting

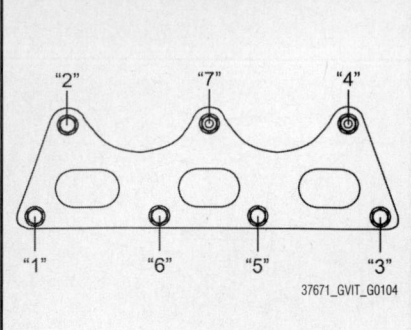

Fig. 77 Exhaust manifold bolt/nut removal sequence—3.2L engine

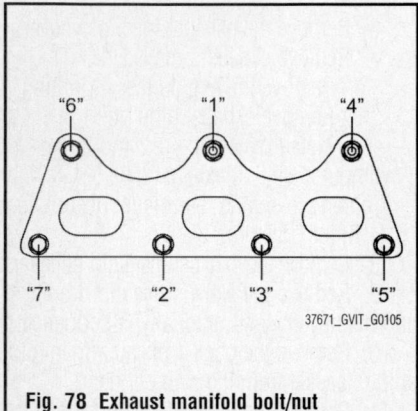

Fig. 78 Exhaust manifold bolt/nut tightening sequence—3.2L engine

15. Tighten the oil level gauge guide bolt to 89 inch lbs. (10 Nm).

16. Use new exhaust manifold bolts. Tighten the bolts and nuts evenly and gradually in sequence to 15 ft. lbs. (20 Nm).

⚠ WARNING

Retighten the exhaust manifold bolts and nuts to the specified torque once.

17. Tighten the suspension frame mounting bolts.

18. Run the engine and check for exhaust leaks.

Bank 2

See Figures 77 through 81.

❊❊ CAUTION

In order to avoid being burned, do not service the exhaust system while it is still hot. Service the system when it is cool.

⚠ CAUTION

Always wear protective goggles and gloves when removing exhaust parts

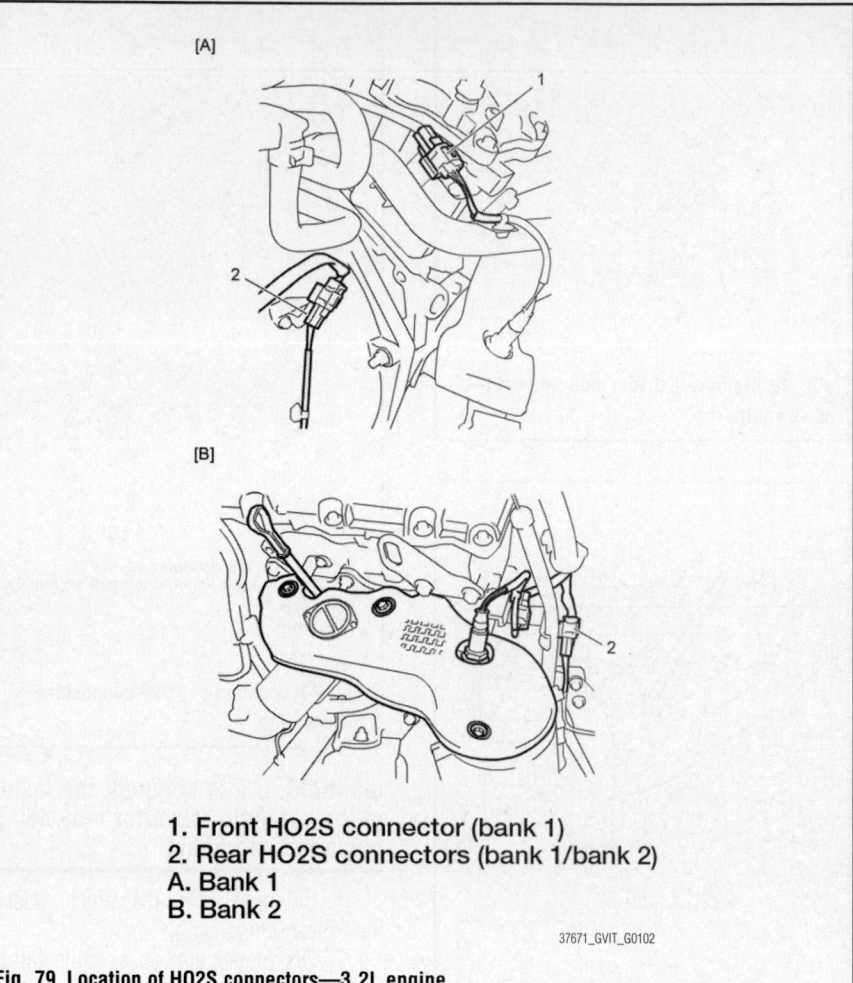

1. Front HO2S connector (bank 1)
2. Rear HO2S connectors (bank 1/bank 2)
A. Bank 1
B. Bank 2

Fig. 79 Location of HO2S connectors—3.2L engine

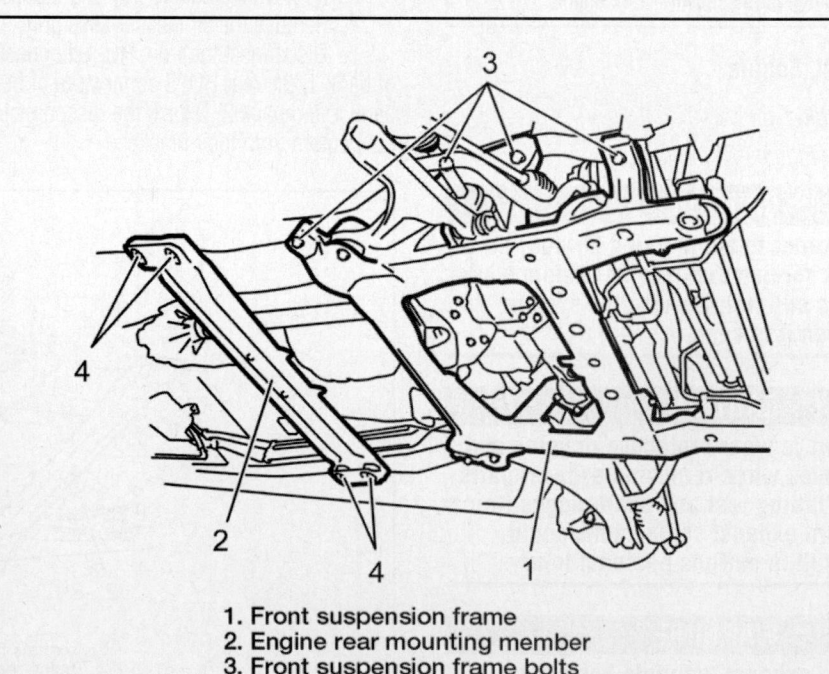

1. Front suspension frame
2. Engine rear mounting member
3. Front suspension frame bolts
4. Engine rear mounting member bolts

Fig. 80 View of suspension frame and rear mounting

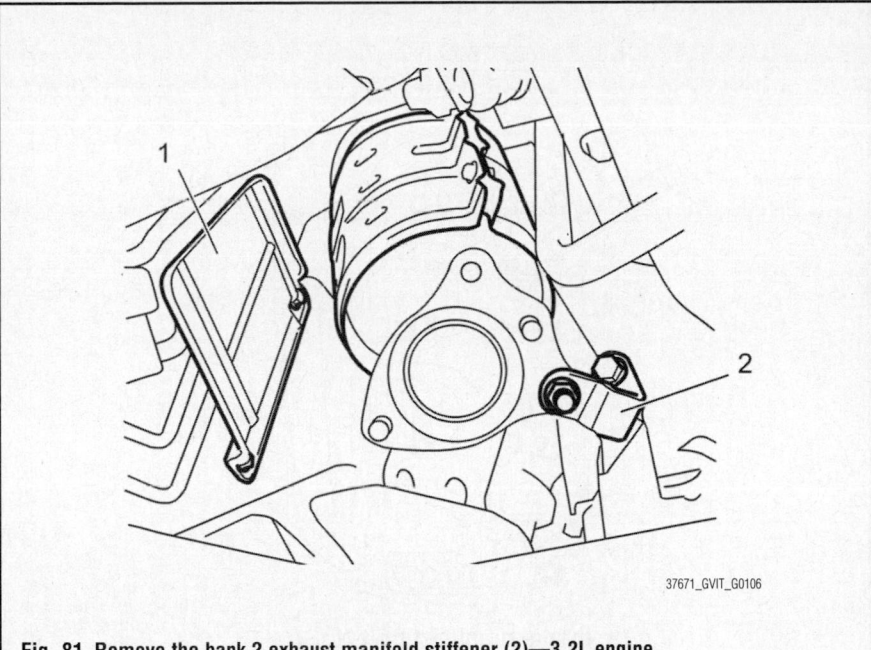

Fig. 81 Remove the bank 2 exhaust manifold stiffener (2)—3.2L engine

as falling rust and sharp edges from worn exhaust components could result in serious personal injury.

✳✳ WARNING

The exhaust manifold has a three-way catalytic converter which reduces harmful exhaust emissions. Be careful not to drop the exhaust manifold. If it is dropped, the inside of the catalytic converter may be damaged on impact.

1. Before servicing the vehicle, refer to the Precautions Section.
2. Disconnect the negative battery cable from the battery.
3. Remove the front HO2S from the bank 2 exhaust manifold.
4. Remove the exhaust manifold cover.
5. Disconnect the rear HO2S connectors, and then detach them from their bracket.
6. Remove the oil level gauge and the oil level gauge guide.
7. Remove exhaust pipe No. 1.
   a. Remove the engine rear mounting.
   b. Remove exhaust pipe No. 1.
   c. Install the engine rear mounting without installing exhaust pipe No. 1. Tighten the engine rear mounting bolt to 41 ft. lbs. (55 Nm).
8. Remove the fuel pipe cover.
9. Remove the exhaust manifold stiffener.

10. Remove the exhaust manifold bolts and nuts evenly and gradually in sequence.
11. Remove the heat protector panel.
12. Support the front suspension frame with a transmission jack, and loosen the suspension frame bolts.
13. Lower the suspension frame a little to make a space above it, then remove the bank 2 exhaust manifold and its gasket from the cylinder head.
14. Remove the rear HO2S, if necessary.

To install:

15. Installation is the reverse of the removal procedure.
16. Use new gaskets and new seal rings.
17. Use a new O-ring for the oil level gauge guide and apply engine oil to the O-ring when installing oil level gauge guide to the cylinder block.
18. Tighten the oil level gauge guide bolt to 89 inch lbs. (10 Nm).
19. Use new exhaust manifold bolts. Tighten the bolts and nuts evenly and gradually in sequence to 15 ft. lbs. (20 Nm).

✳✳ WARNING

Retighten the exhaust manifold bolts and nuts to the specified torque once.

20. Tighten the suspension frame mounting bolts.
21. Run the engine and check for exhaust leaks.

INTAKE MANIFOLD

REMOVAL & INSTALLATION

2.4L Engine

See Figures 82 and 83.

1. Before servicing the vehicle, refer to the Precautions Section.
2. Disconnect the negative battery cable.
3. Remove the engine cover.
4. Remove the air cleaner outlet hose and the air cleaner upper case.

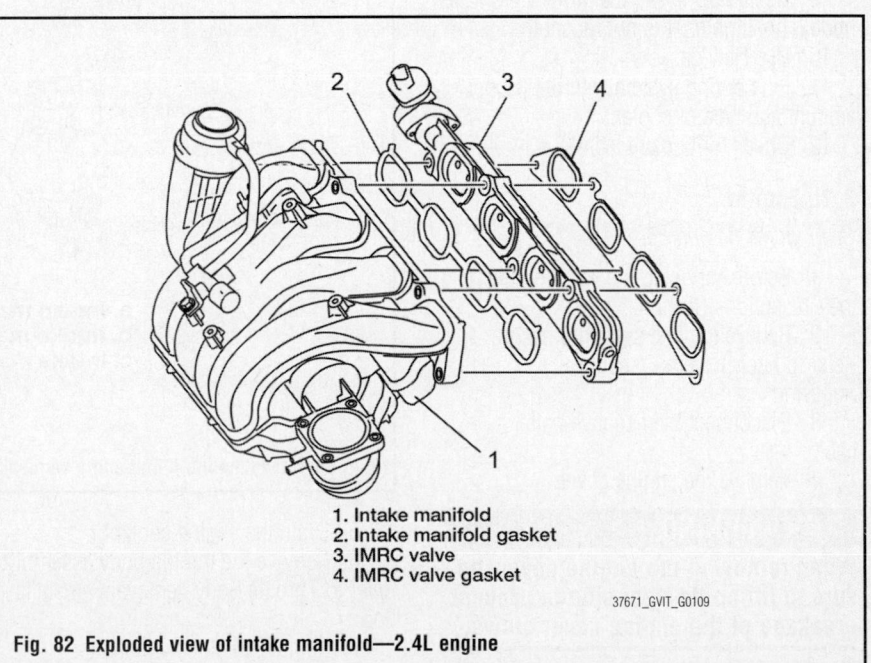

1. Intake manifold
2. Intake manifold gasket
3. IMRC valve
4. IMRC valve gasket

Fig. 82 Exploded view of intake manifold—2.4L engine

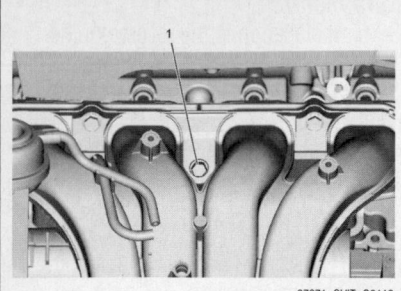

Fig. 83 Use a new intake manifold bolt in the middle position (1) as it is not reusable

5. Remove the throttle body assembly.
6. Disconnect the following connectors and hoses:
- EVAP canister purge valve connector
- IMT vacuum solenoid valve connector
- Brake booster hose
- PCV valve hose
- Breather hose
- Engine coolant hoses
- Purge hose
- IMRC valve actuator connector
- A/T oil filler tube (4 A/T model)
- Engine harness clamps
- Brake vacuum pipe

7. Remove the intake manifold, intake manifold gasket, IMRC valve, and IMRC valve gasket from the cylinder head.

To install:

8. Installation is the reverse of the removal procedure.
9. Use a new intake manifold bolt in the middle position as it is not reusable.
10. Use New gaskets.
11. Fill cooling system with the proper amount and type of coolant.
12. Check for coolant leaks.

3.2L Engine

See Figures 84 and 85.

1. Before servicing the vehicle, refer to the Precautions Section.
2. Relieve the fuel system pressure. Refer to Fuel System Pressure, Relieving.
3. Disconnect the negative battery cable.
4. Remove the engine cover.

✳✳ WARNING

When removing the engine cover, be sure to lift up its front side to prevent breakage of the engine cover claws.

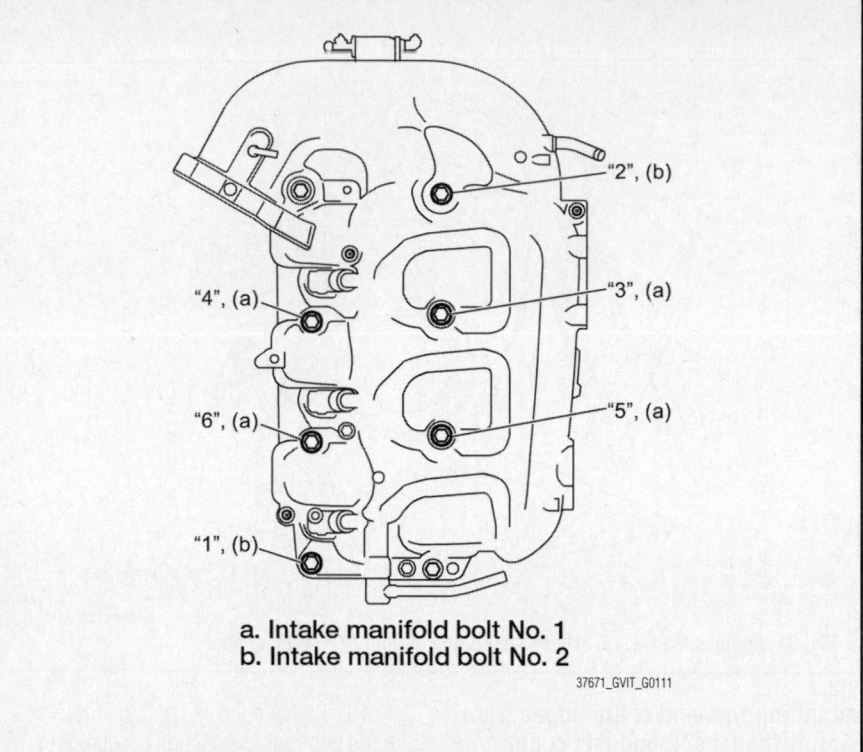

a. Intake manifold bolt No. 1
b. Intake manifold bolt No. 2

Fig. 84 Intake manifold tightening sequence ("1"–"6")

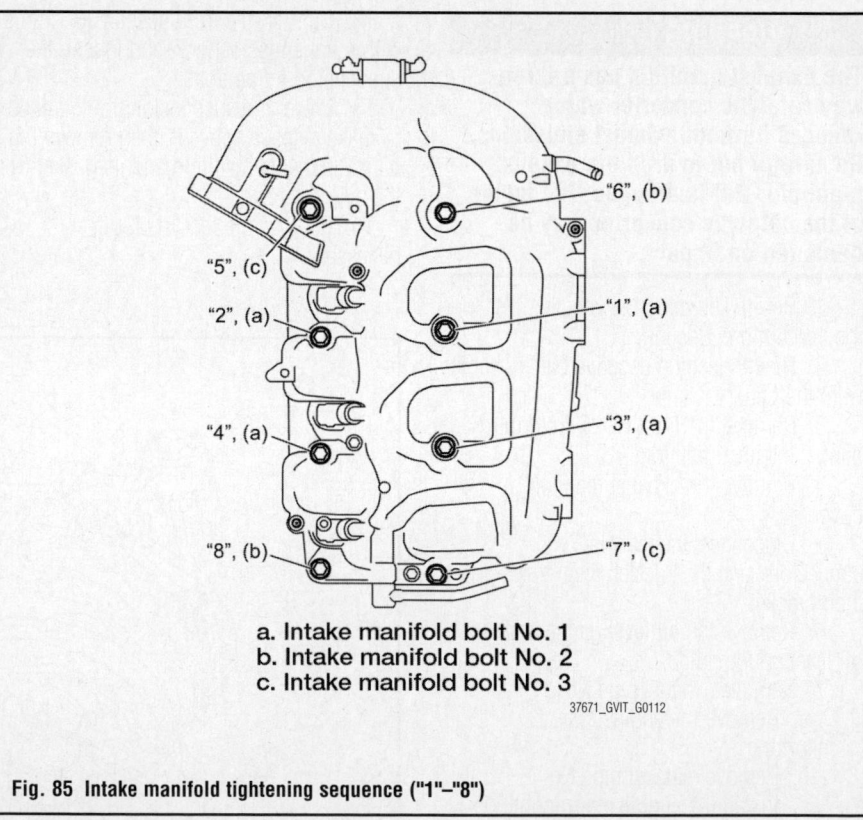

a. Intake manifold bolt No. 1
b. Intake manifold bolt No. 2
c. Intake manifold bolt No. 3

Fig. 85 Intake manifold tightening sequence ("1"–"8")

5. Drain the engine coolant.
6. Remove the throttle body assembly. Refer to Throttle Body, removal & installation.
7. Remove the following parts:

- Purge pipe bolt
- Purge valve bracket bolt
- Electric wire bracket bolt
- Water bypass hose
- PCV hose

- Brake booster hose
- Purge hose

8. Remove the intake upper manifold and the intake manifold upper gasket.

9. Remove the fuel injectors.

10. Remove the intake lower manifold and the intake manifold lower gasket.

To install:

11. Installation is the reverse of the removal procedure.

12. Install new gaskets.

13. Tighten the intake manifold bolts evenly and gradually in sequence.

    a. Step 1: Tighten intake manifold bolts (No. 1 and No. 2) to 89 inch lbs. (10 Nm).

    b. Step 2: Tighten intake manifold bolts (No. 1, No. 2, and No. 3) to 18 ft. lbs. (25 Nm).

14. Tighten the purge pipe bolt and the purge valve bracket bolt to 89 inch lbs. (10 Nm).

15. Check to ensure that all the removed parts are back in place.

16. Turn the ignition switch to the ON position without starting the engine and check for fuel leaks.

17. Start the engine and check for engine coolant leaks.

OIL PAN

REMOVAL & INSTALLATION

2.4L Engine

See Figures 86 through 89.

1. Before servicing the vehicle, refer to the Precautions Section.

2. Remove the front suspension frame.

3. Drain the engine oil.

4. Remove the oil level gauge and the oil level gauge guide.

5. Remove the transmission bolts, then remove the driveplate cover for 4 A/T model, or the clutch housing lower plate for 5 M/T model.

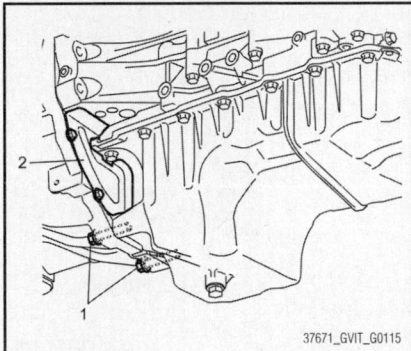

Fig. 87 Remove the transmission bolts (1), then remove the driveplate cover (2)

6. Remove the oil pan bolts evenly and gradually in the removal sequence.

7. Remove the oil pan from the lower crank case.

To install:

8. Install dowel pins to the oil pan.

9. Apply sealant to the oil pan mating surface. Use Sealant 99000-31260 (SUZUKI Bond No. 1217G), or equivalent.

➡**Apply sealant to the oil pan so that the width is 0.12 inch (3mm) and the height is 0.08 inch (2mm).**

10. Fit the oil pan to the lower crank case.

11. Install the oil pan bolts and tighten evenly and gradually in sequence.

    a. Step 1: 11 ft. lbs. (15 Nm).

    b. Step 2: 18 ft. lbs. (25 Nm).

12. Install the transmission bolts and then the driveplate cover for 4 A/T model, or the clutch housing lower plate for 5 M/T model.

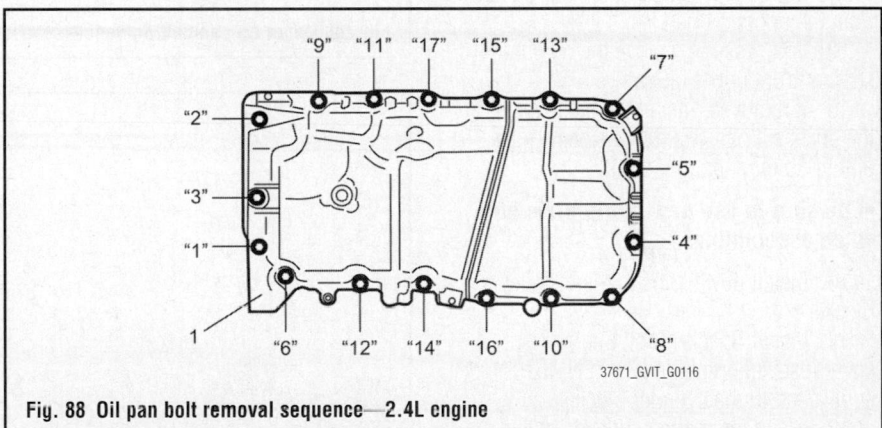

Fig. 88 Oil pan bolt removal sequence—2.4L engine

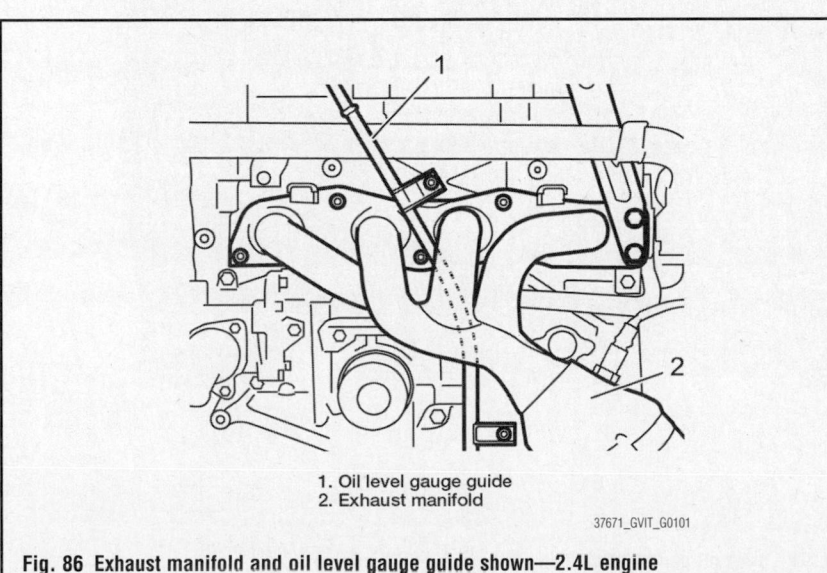

1. Oil level gauge guide
2. Exhaust manifold

Fig. 86 Exhaust manifold and oil level gauge guide shown—2.4L engine

Fig. 89 Oil pan bolt tightening sequence—2.4L engine

13. Install the oil level gauge and the oil level gauge guide. Tighten the oil level gauge guide bolt to 97 inch lbs. (11 Nm).

14. Install the drain plug and fill the engine with the proper type and amount of engine oil.

15. Install the front suspension frame.

16. Run the engine and check for engine oil leaks.

3.2L Engine

See Figures 90 and 91.

1. Before servicing the vehicle, refer to the Precautions Section.

2. Remove the engine assembly from the vehicle.

3. Remove the cylinder head cover.

4. Remove the alternator bracket.

5. Remove the timing chain cover.

6. Remove the oil level gauge.

7. Remove the lower oil pan bolts.

8. Remove the lower oil pan from the upper oil pan using a plastic hammer.

9. Remove the upper oil pan bolts No. 1, upper oil pan bolts No. 2, and upper oil pan bolt No. 3.

10. Remove the upper oil pan from the cylinder block using a plastic hammer.

11. If necessary, remove the baffle plate from the upper oil pan.

12. If necessary, remove the oil pump strainer from upper oil pan.

15. If removed, install the baffle plate to the upper oil pan. Tighten the baffle plate bolt to 89 inch lbs. (10 Nm).

➡ **Be sure to use new baffle plate bolt when assembling.**

16. Install dowel pins to the cylinder block.

17. Install Special Tool: Engine front cover installation guide pins (EN-46109), or equivalent, to the cylinder block.

18. Clean the mating surfaces of the upper oil pan and the cylinder block.

19. Apply sealant to the oil pan mating surface. Use Sealant 99000–31290 (SUZUKI Bond No. 1217F), or equivalent. The sealant bead size width should be 0.12 inch (3mm) and the bead height of 0.08 inch (2mm).

20. Install the upper oil pan to the cylinder block and tighten the upper oil pan bolts No. 1, upper oil pan bolts No. 2, and upper oil pan bolt No. 3 by hand.

21. Tighten the upper oil pan bolts No. 1, upper oil pan bolts No. 2, and upper oil pan bolt No. 3 in proper sequence.

   a. Upper oil pan bolts (No. 1): 18 ft. lbs. (25 Nm).

   b. Upper oil pan bolts (No. 2 and No. 3): 89 inch lbs. (10 Nm).

22. Clean the mating surfaces of the upper oil pan and the cylinder block.

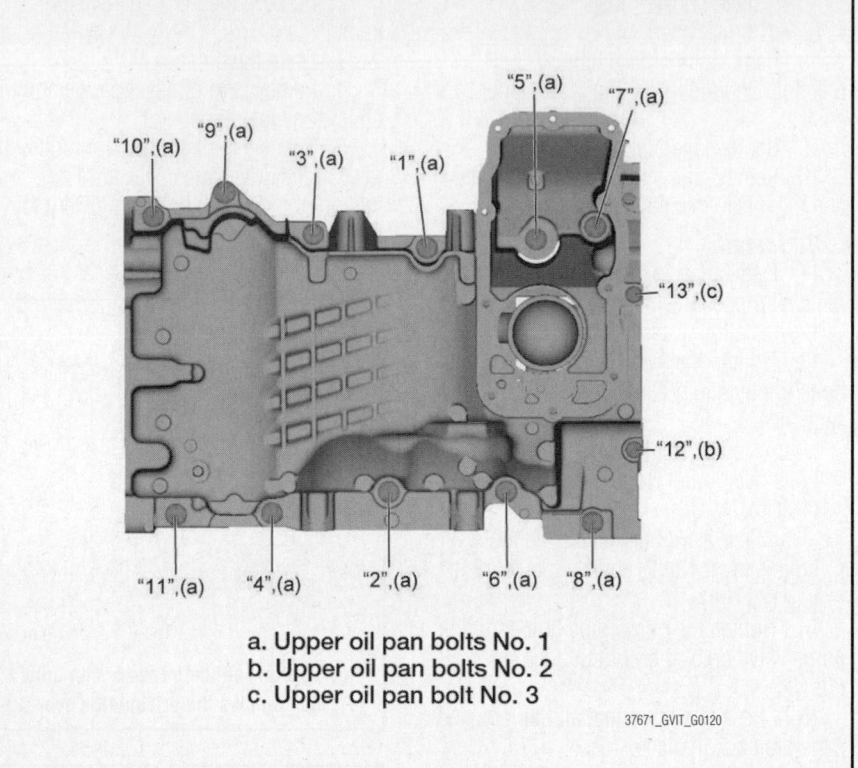

a. Upper oil pan bolts No. 1
b. Upper oil pan bolts No. 2
c. Upper oil pan bolt No. 3

37671_GVIT_G0120

Fig. 90 Upper oil pan bolt tightening sequence—3.2L engine

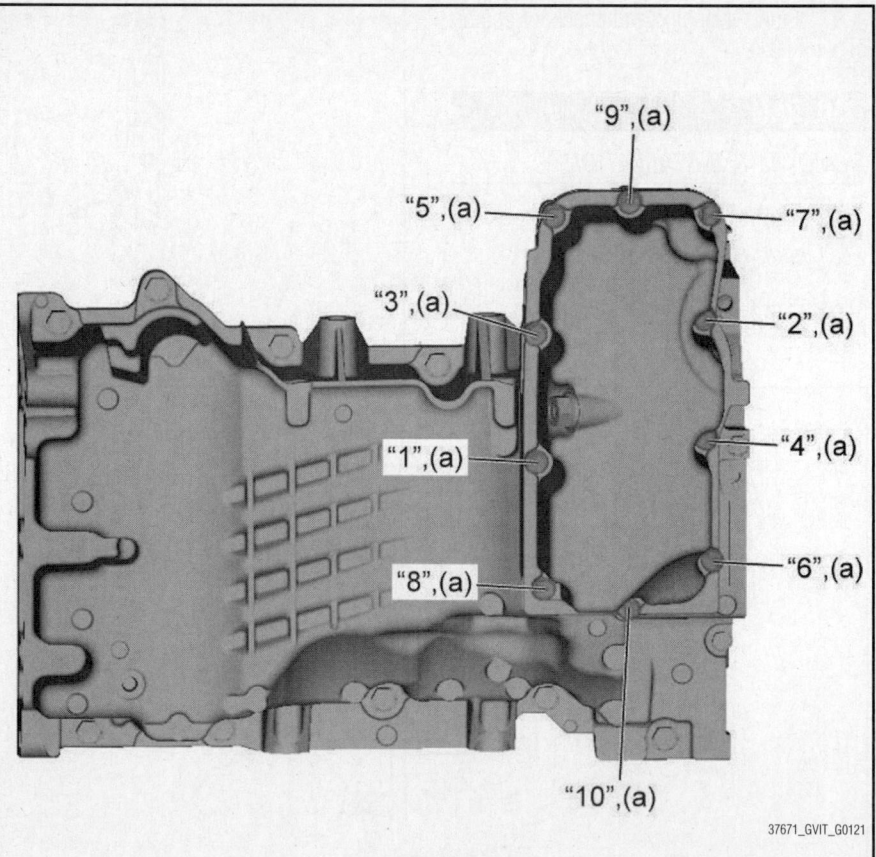

37671_GVIT_G0121

Fig. 91 Lower oil pan bolt tightening sequence—3.2L engine

23. Apply sealant to the oil pan mating surface. Use Sealant 99000-31290 (SUZUKI Bond No. 1217F), or equivalent. The sealant bead size width should be 0.12 inch (3mm) and the bead height of 0.08 inch (2mm).

24. Install the lower oil pan to the upper oil pan and tighten the lower oil pan bolts by hand.

25. Tighten the lower oil pan bolts in sequence to 89 inch lbs. (10 Nm).

26. Install the timing chain cover.

27. Install the alternator bracket.

28. Install the belt idler arm to the alternator bracket.

29. Install the OCV.

30. Install the P/S pump bracket.

31. Install the cylinder head cover.

32. Install accessory drive belt.

33. Install the engine assembly to the vehicle.

OIL PUMP

REMOVAL & INSTALLATION

2.4L Engine

See Figures 92 through 94.

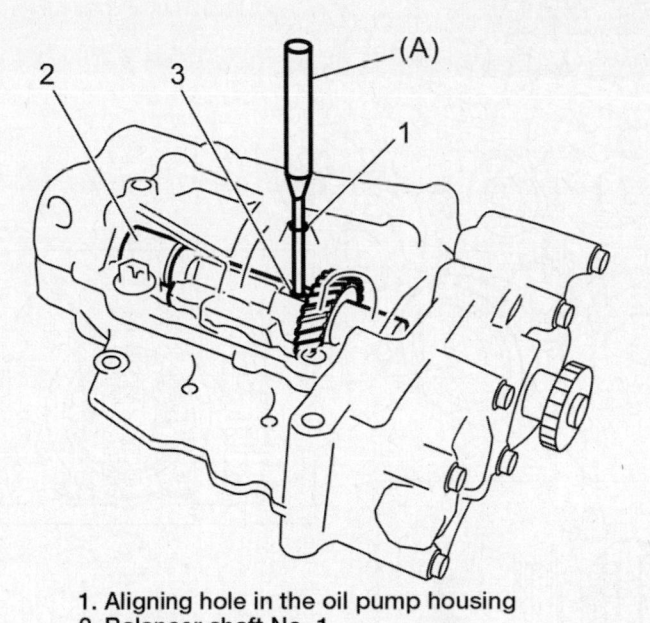

1. Aligning hole in the oil pump housing
2. Balancer shaft No. 1
3. Aligning hole in the balancer shaft
A. Special Tool 09922-85811

37671_GVIT_G0125

Fig. 93 Using Special Tool 09922-85811 to align balancer shaft No. 1—2.4L engine

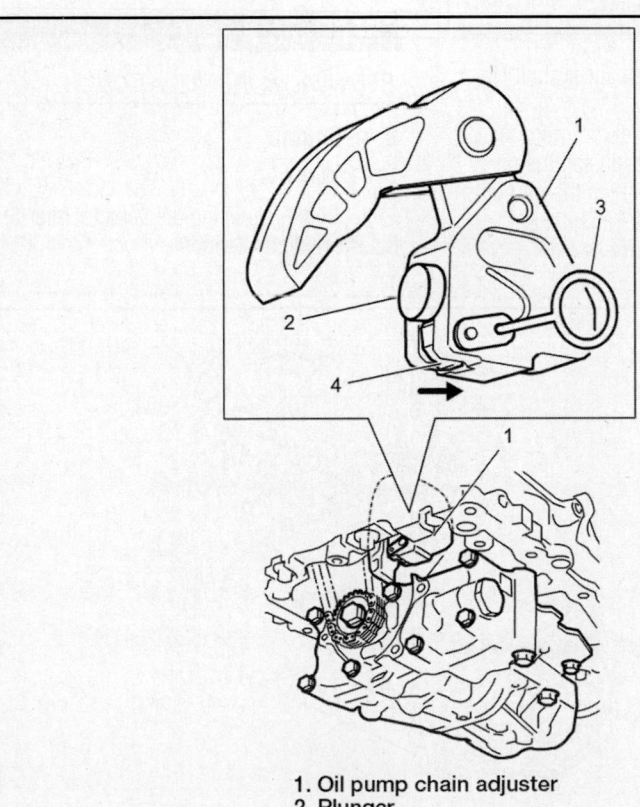

1. Oil pump chain adjuster
2. Plunger
3. Retainer

37671_GVIT_G0124

Fig. 92 Loosening the oil pump chain adjuster—2.4L engine

1. Before servicing the vehicle, refer to the Precautions Section.

2. Remove the oil pan, oil strainer, and oil pan baffle plate.

3. Remove the oil pump sprocket cover.

4. Slacken the tension of the oil pump chain according to the following procedure.

   a. Raise the latch of the oil pump chain adjuster to disengage the latch from the plunger teeth.

   b. Push the plunger into the oil pump chain adjuster and install a retainer (3mm wire) to hold the plunger in place.

5. Remove the oil pump bolts.

6. Remove the oil pump chain from the oil pump sprocket.

7. Remove the oil pump assembly from the lower crank case.

To install:

8. Make sure the dowel pins are installed.

9. Align balancer shaft No. 1 at the specified position.

   a. Insert Special Tool 09922-85811, or equivalent, into the aligning hole in the oil pump housing.

   b. Turn balancer shaft No. 1 by hand and insert the special tool into the aligning hole in the balancer shaft.

10. Rotate the crankshaft so that it is correctly positioned.

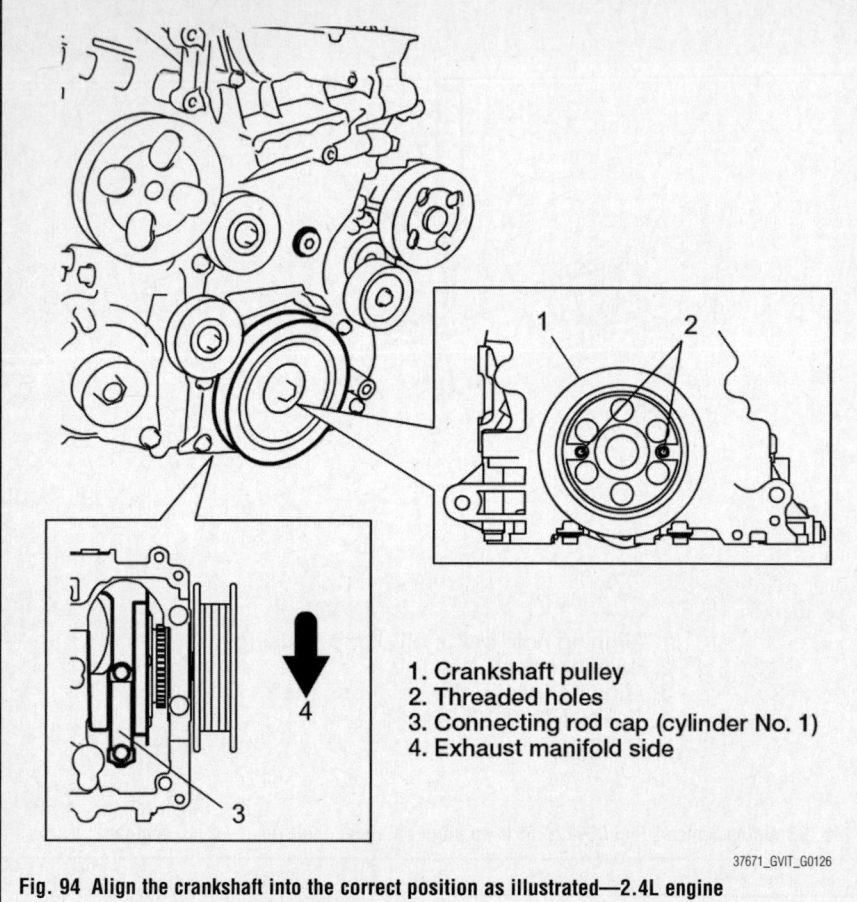

1. Crankshaft pulley
2. Threaded holes
3. Connecting rod cap (cylinder No. 1)
4. Exhaust manifold side

37671_GVIT_G0126

Fig. 94 Align the crankshaft into the correct position as illustrated—2.4L engine

a. The threaded holes on the crankshaft pulley and the lower crank case contact surface are set in parallel.

b. The connecting rod cap in cylinder No. 1 is positioned close to the exhaust manifold side.

11. Install the oil pump chain to the oil pump sprocket.

12. Install the oil pump assembly to the lower crank case.

a. Tighten the M10 oil pump bolts to 41 ft. lbs. (55 Nm).

b. Tighten the M8 oil pump bolts to 18 ft. lbs. (25 Nm).

13. Remove the retainer (3mm wire) from the oil pump chain adjuster.

14. Remove the special tool from the oil pump housing and rotate the crankshaft to the correct position.

a. Ensure the threaded holes on the crankshaft pulley and the lower crank case contact surface are aligned in parallel.

b. Check that connecting rod cap in cylinder No. 1 is positioned close to the exhaust manifold side.

15. Insert Special Tool 09922-85811, or equivalent, to the aligning hole on the oil pump housing and make sure the position

of the aligning hole on balancer shaft No. 1 is in proper condition.

16. Remove the special tool from the oil pump assembly, and then install the oil pump sprocket cover. Tighten the oil pump sprocket cover bolts to 97 inch lbs. (11 Nm).

17. Install the oil pan, oil strainer, and oil pan baffle plate.

3.2L Engine

See Figure 95.

1. Before servicing the vehicle, refer to the Precautions Section.

2. Remove the engine assembly from vehicle.

3. Remove the cylinder head cover.

4. Remove the alternator bracket.

5. Remove the timing chain cover.

6. Remove the 1st timing chain, timing chain lower guide, and crankshaft timing sprocket.

7. Remove the oil pump bolts and the oil pump.

To install:

8. Install the oil pump. Tighten the oil pump bolts to 17 ft. lbs. (23 Nm).

9. Install the 1st timing chain.

10. Install the timing chain cover.

11. Install the alternator bracket.

12. Install the OCV.

13. Install the P/S pump bracket.

14. Install the cylinder head cover.

15. Install the accessory drive belt.

16. Install the engine assembly to the vehicle.

REAR MAIN SEAL

REMOVAL & INSTALLATION

2.4L Engine

See Figure 96.

1. Before servicing the vehicle, refer to the Precautions Section.

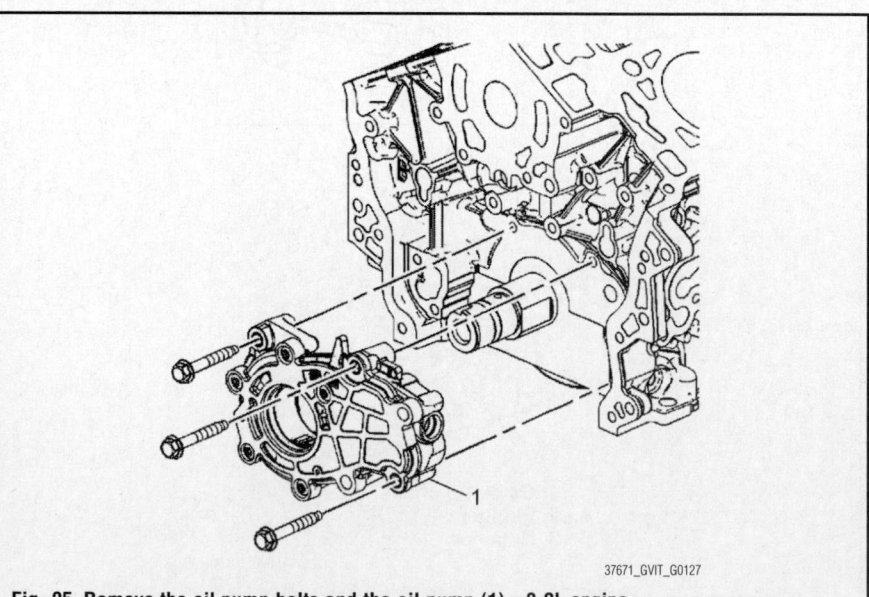

37671_GVIT_G0127

Fig. 95 Remove the oil pump bolts and the oil pump (1)—3.2L engine

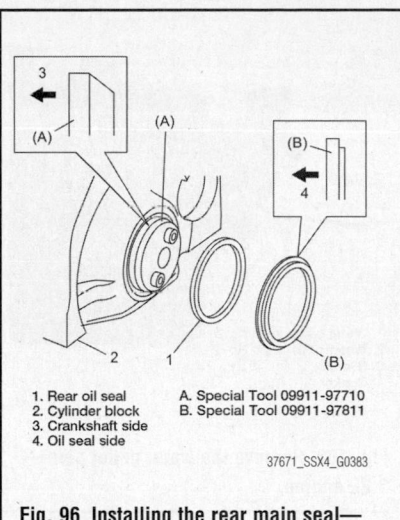

1. Rear oil seal
2. Cylinder block
3. Crankshaft side
4. Oil seal side

A. Special Tool 09911-97710
B. Special Tool 09911-97811

37671_SSX4_G0383

Fig. 96 Installing the rear main seal—2.4L engine

2. Remove the transaxle assembly. Refer to Automatic or Manual Transaxle Assembly, removal & installation.

3. Remove the clutch cover and clutch disk (M/T models).

4. Remove the drive plate (A/T models) or the flywheel (M/T models). Refer to Flywheel/Driveplate, removal & installation.

5. Remove the rear oil seal with a suitable tool.

✲✲ WARNING

Be careful not to damage the crankshaft or the cylinder block.

To install:

6. Apply engine oil to the new rear oil seal and install it to the cylinder block using Special Tools 09911-97710 and 09911-97811, or equivalent.

7. Install the rear oil seal so that each seal lip is oriented correctly and it becomes flush with the cylinder block.

3.2L Engine

See Figure 97.

1. Before servicing the vehicle, refer to the Precautions Section.

2. Remove the transaxle assembly.

3. Remove the driveplate. Refer to Flywheel/Driveplate, removal & installation.

4. Remove the oil pan. Refer to Oil Pan, removal & installation.

5. Remove the bolts and remove the rear oil seal housing.

To install:

6. Installation is the reverse of the removal procedure.

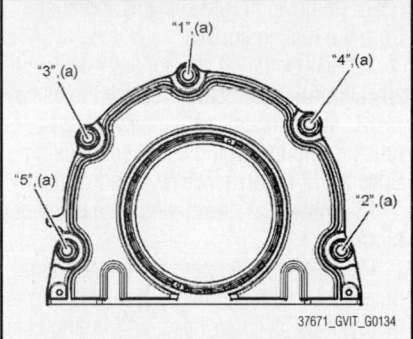

37671_GVIT_G0134

Fig. 97 Rear main seal bolt tightening sequence—3.2L engine

7. Clean the sealing surface between the oil seal housing and the cylinder block.

8. Apply sealant to the mating surface of the new rear oil seal housing. Use Sealant 99000-31290 (SUZUKI Bond No. 1217F), or equivalent.

- Sealant bead size for rear oil seal housing: Width 0.12 inch (3mm); Height 0.10 inch (2.5mm)

9. Install the oil seal housing to the cylinder block.

➡**Do not apply engine oil to the oil seal housing.**

10. Tighten the oil seal housing bolts evenly and gradually, to 89 inch lbs. (10 Nm), according to the tightening sequence.

➡**Confirm that the engine rear crankshaft oil seal lip is not turned over after the engine rear crankshaft oil seal housing is installed.**

TIMING CHAIN FRONT COVER

REMOVAL & INSTALLATION

2.4L Engine

See Figures 98 through 102.

➡**Keep the working table, tools, and hands clean while working. Use special care to handle the aluminum parts so as not to damage them. Do not expose removed parts to dust. Keep them clean.**

1. Before servicing the vehicle, refer to the Precautions Section.

2. Remove the engine assembly from the vehicle.

3. Remove crankshaft pulley. Refer to Crankshaft Damper, removal & installation.

✲✲ WARNING

Avoid applying unnecessary load to the timing chain, sprockets, and any other related parts.

4. Remove the cylinder head cover.

5. Remove the oil pan. Refer to Oil Pan, removal & installation.

6. Remove the tensioner pulley and the idler pulley.

7. Remove the timing chain cover by loosening the timing chain cover bolts evenly and gradually according to the loosening order.

8. Remove the crankshaft oil seal from the timing chain cover using a flat-head screwdriver or the equivalent, if necessary.

9. Remove the oil seal from the timing chain cover, if necessary.

To install:

10. Clean the mating surface of the timing chain cover, cylinder block, and cylinder

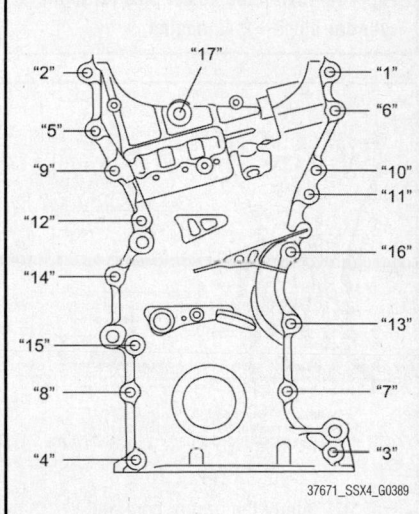

37671_SSX4_G0389

Fig. 98 Timing chain cover bolt loosening sequence—2.4L engine

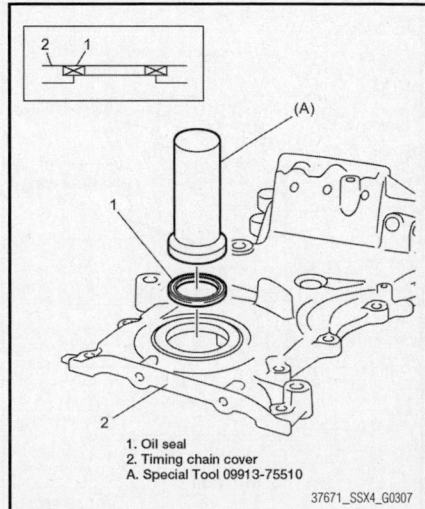

1. Oil seal
2. Timing chain cover
A. Special Tool 09913-75510

37671_SSX4_G0307

Fig. 99 Using Special Tool 09913-75510 to install the crankshaft front seal—2.4L engine

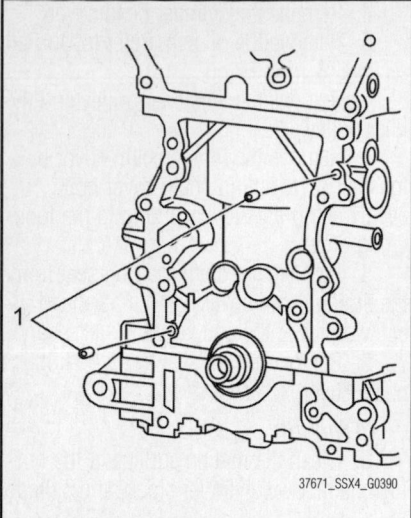

Fig. 100 Install the dowel pins (1) to the cylinder block—2.4L engine

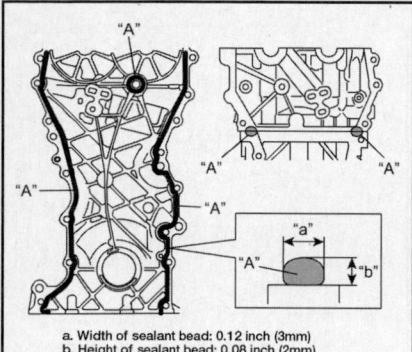

a. Width of sealant bead: 0.12 inch (3mm)
b. Height of sealant bead: 0.08 inch (2mm)
A. Sealant 99000-31260 (SUZUKI Bond No. 1217G)

37671_SSX4_G0391

Fig. 101 Apply the proper type and amount of sealant to the mating surface of the cylinder, cylinder head, and timing chain cover—2.4L engine

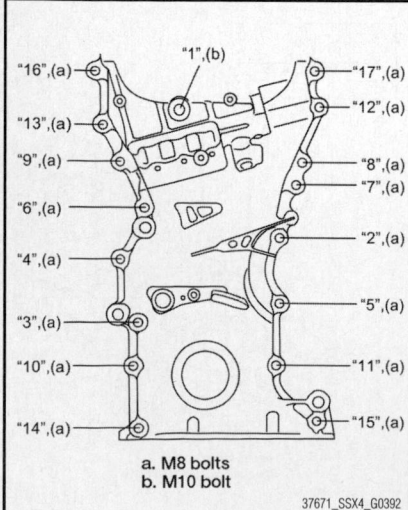

a. M8 bolts
b. M10 bolt

37671_SSX4_G0392

Fig. 102 Timing chain cover bolt tightening sequence—2.4L engine

head. Remove oil, old sealant, and dust from the mating surface.

11. Check the oil seal for any damage. If any abnormality is found, replace the oil seal.

12. When installing the new oil seal, press it into the timing chain cover using Special Tool 09913-75510, or equivalent.

13. Install the dowel pins to the cylinder block.

14. Apply the proper type and amount of sealant to the mating surface of the cylinder, cylinder head, and timing chain cover.

➡**Use Sealant 99000-31260 (SUZUKI Bond No. 1217G), or equivalent.**

15. Apply engine oil to the oil seal lip, then install the timing chain cover.

16. Tighten the timing chain cover bolts evenly and gradually to the specified torque in numerical order.

➡**Before installing the timing chain cover, check that the dowel pins are securely fitted.**

a. Tighten the M8 cover bolts to 19 ft. lbs. (25 Nm).
b. Tighten the M10 cover bolt to 41 ft. lbs. (55 Nm).

17. Install the tensioner pulley and idler pulley.

18. Install the cylinder head cover.

19. Install the oil pan. Refer to Oil Pan, removal & installation.

20. Install the crankshaft pulley. Refer to Crankshaft Damper, removal & installation.

21. Install the engine assembly to the vehicle.

3.2L Engine

See Figures 103 through 111.

✳✳ WARNING

Use special care when handling aluminum parts so as not to damage them.

1. Before servicing the vehicle, refer to the Precautions Section.

2. Remove the engine assembly from the vehicle. Refer Engine Assembly, removal & installation.

3. Remove the accessory drive belt. Refer to Accessory Drive Belt, removal & installation.

4. Remove the cylinder head cover.

5. Remove the water pump pulley.

6. Remove the Power Steering (P/S) pump bracket.

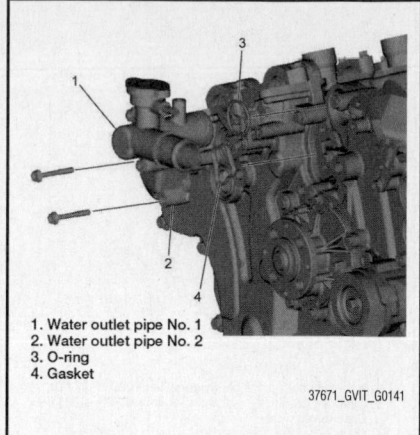

1. Water outlet pipe No. 1
2. Water outlet pipe No. 2
3. O-ring
4. Gasket

37671_GVIT_G0141

Fig. 103 Remove the water outlet pipe—3.2L engine

7. Remove the OCV.

8. Remove the belt idler arm from the alternator bracket.

9. Remove the water outlet pipe No. 1, water outlet pipe No. 2, O-ring, and gasket.

10. Remove the crankshaft pulley. Refer to Crankshaft Damper, removal & installation.

11. Remove the accessory drive belt tensioner from the timing chain cover.

12. Remove the timing chain cover.

13. Remove the water pump inner gasket from the timing chain cover.

14. Remove the oil seal from the timing chain cover.

15. Using Special Tool 09913-57830, or equivalent, remove the crankshaft oil seal from the timing chain cover.

16. If necessary, remove the water pump from the timing chain cover.

17. If necessary, remove the CMP sensor from the timing chain cover.

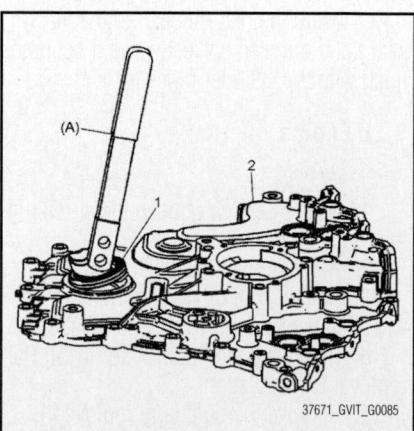

37671_GVIT_G0085

Fig. 104 Using Special Tool 09913-57830 to remove the crankshaft oil seal (1) from the timing chain cover (2)

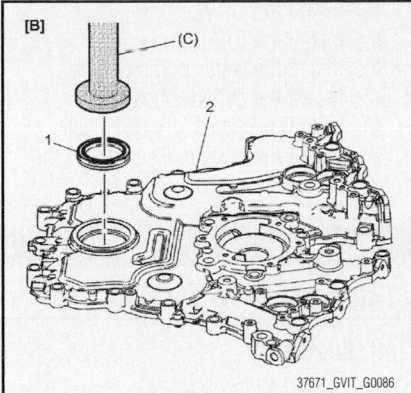

Fig. 105 Using Special Tool J-29184 to install a new crankshaft oil seal (1) to the timing chain cover (2)

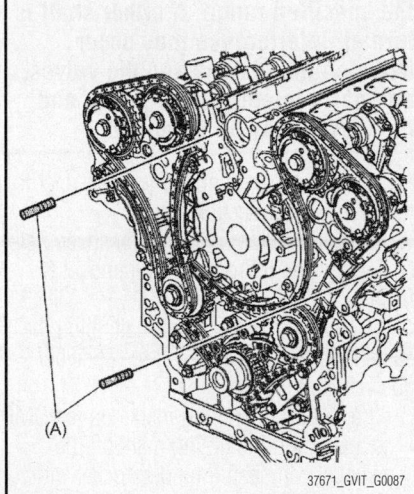

Fig. 106 Install Special Tool: Engine front cover installation guide pins (EN-46109) to the cylinder head

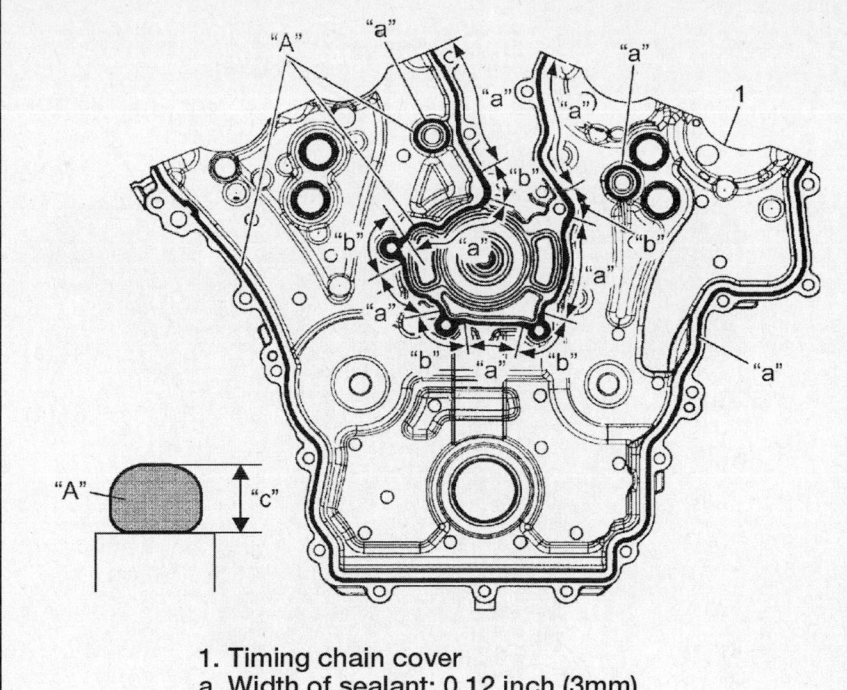

1. Timing chain cover
a. Width of sealant: 0.12 inch (3mm)
b. Width of sealant: 0.20 inch (5mm)
c. Height of sealant: 0.10 inch (2.5mm)
A. Sealant

Fig. 108 Apply sealant to the mating surface of the timing chain cover

To install:

18. Clean the sealing surface on the timing chain cover, cylinder block, and cylinder head.

19. Install a new water pump inner gasket to the timing chain cover.

20. Using Special Tools 09926-68310 and 09913-75821, or Oil seal installer J-29184, or equivalent, install a new crankshaft oil seal to the timing chain cover.

➡**Do not apply engine oil or grease to the crankshaft oil seal.**

21. Using Special Tool 09913 57810, or equivalent, install a new oil seal to the timing chain cover.

22. Install Special Tool: Engine front cover installation guide pins (EN-46109), or equivalent, to the cylinder head.

23. Apply sealant to the mating surface of the cylinder head and the cylinder block. Use Sealant 99000- 31290 (SUZUKI Bond No. 1217F), or equivalent.

24. Apply sealant to the mating surface of the timing chain cover. Use Sealant 99000-31290 (SUZUKI Bond No. 1217F), or equivalent.

25. Install the timing chain cover and the accessory drive belt tensioner as follows.

➡**Before installing the timing chain cover, check that the dowel pin is securely fitted.**

a. Tighten the timing chain cover bolts evenly and gradually to 15 ft. lbs. (20 Nm) according to the tightening sequence.

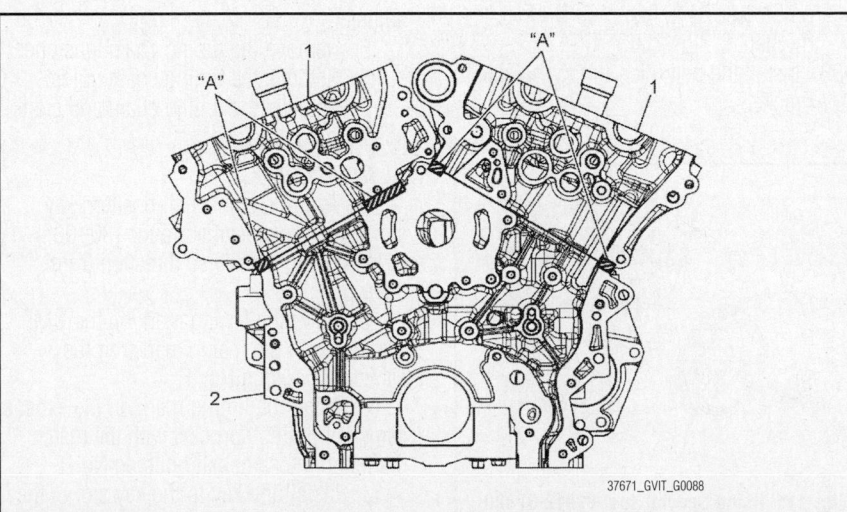

Fig. 107 Apply sealant (A) to the mating surface of the cylinder head (1) and the cylinder block (2) as shown

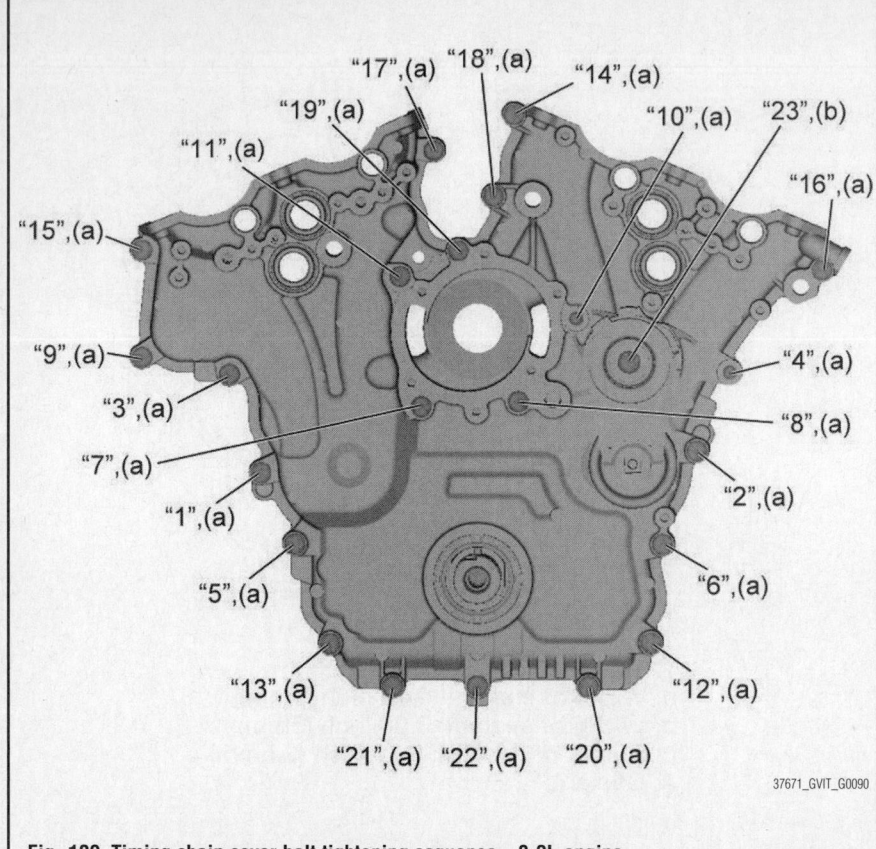

Fig. 109 Timing chain cover bolt tightening sequence—3.2L engine

b. In the same manner, tighten the cover bolts an additional 60°.

c. Tighten the accessory drive belt tensioner bolt to 18 ft. lbs. (25 Nm).

26. Remove Special Tool: Engine front cover installation guide pins (EN-46109).

27. Install a new O-ring and new gasket for the water outlet pipe.

28. Install the water outlet pipe No. 1 and water outlet pipe No. 2.

29. Lock the flywheel with Special Tool 09916-97830, or equivalent.

30. Apply engine oil between the crankshaft pulley and the crankshaft.

➡**Do not apply engine oil between the crankshaft oil seal and the crankshaft pulley.**

31. Using Special Tool 09912-37820, or equivalent, install the crankshaft pulley to the crankshaft.

32. Remove the special tool and tighten the crankshaft pulley bolt.

a. Step 1: 89 ft. lbs. (120 Nm).

b. Step 2: 0 ft. lbs. (0 Nm).

c. Step 3: 74 ft. lbs. (100 Nm).

d. Step 4: Plus 150°.

33. Install the belt idler arm to the alternator bracket.

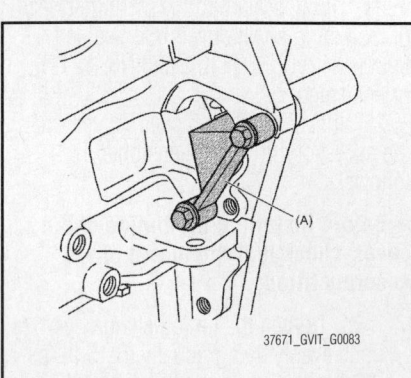

Fig. 110 Using Special Tool 09916-97830 (A) to lock the flywheel

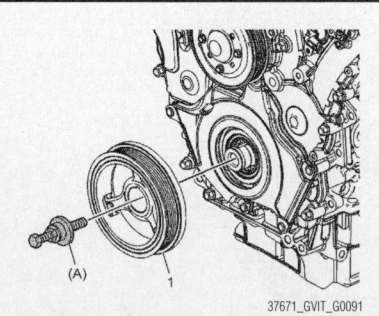

Fig. 111 Using Special Tool 09912-37820 (A), install the crankshaft pulley (1) to the crankshaft

34. Install the OCV.

35. Install the P/S pump bracket.

36. Install the water pump pulley.

37. Install the cylinder head cover.

38. Install the accessory drive belt.

39. Install the engine assembly to the vehicle.

TIMING CHAIN & SPROCKETS

REMOVAL & INSTALLATION

2.4L Engine

See Figures 112 through 114.

✳✳ WARNING

After the timing chain is removed, never turn the crankshaft or camshafts independently more than the specified range. If either shaft is turned, interference may occur between the pistons and the valves, and parts related to the piston and the valves may be damaged.

1. Before servicing the vehicle, refer to the Precautions Section.

2. Remove the timing chain cover. Refer to Timing Chain Front Cover, removal & installation.

3. By turning the crankshaft, align the camshafts and crankshaft at the specific position as follows.

a. Align the timing marks on the CMP actuator and the timing mark on the exhaust camshaft timing sprocket with the match marks on the camshaft housing No. 1.

b. Align the timing mark on the crankshaft timing sprocket with the match mark on the lower crank case.

4. Remove the timing chain tensioner adjuster.

5. Remove the timing chain tensioner.

6. Remove the timing chain guide.

7. Remove the timing chain and crankshaft timing sprocket.

To install:

8. Turn the crankshaft position key slot within the specified range (30–90°) in the counterclockwise direction from the top.

9. Align the timing mark on the CMP actuator with the match marks on the camshaft housing No. 1.

10. Align the timing mark on the exhaust camshaft timing sprocket with the match marks on the camshaft housing No. 1.

11. Install the key to the key slot of the crankshaft.

a. Align the key slot of the crankshaft

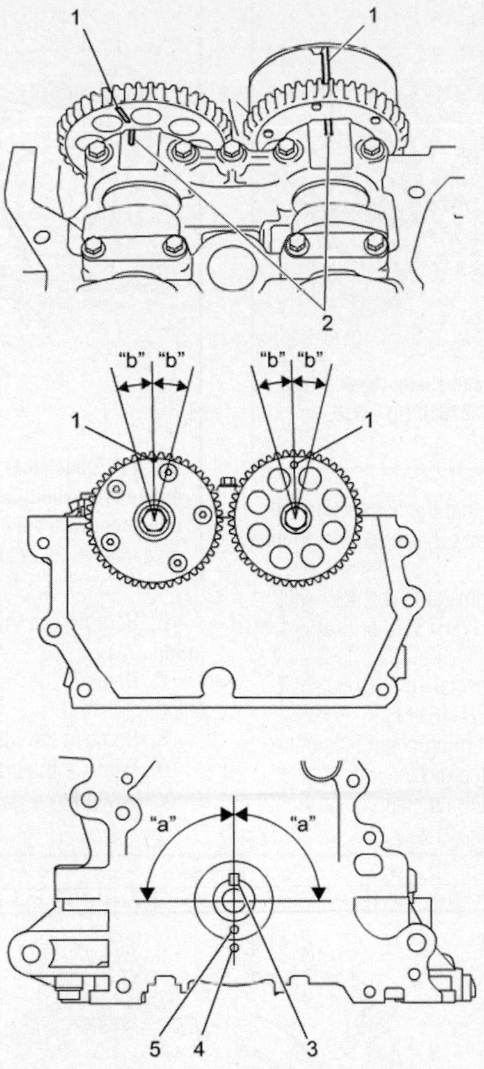

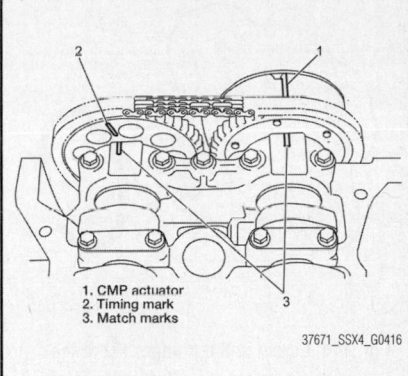

Fig. 113 Align the timing marks on the
CMP actuator and the timing mark on the
exhaust camshaft timing sprocket with
the match marks on the camshaft housing
No. 1—2.4L engine

1.	Timing marks on camshaft timing sprockets
2.	Match marks on camshaft housing No.1
3.	Key
4.	Match mark on lower crank case
5.	Timing mark on crankshaft timing sprocket
"a":	90°
"b":	15°

37671_SSX4_G0414

Fig. 112 Interference engine: allowable movement of components shown with timing chain removed—2.4L engine

timing sprocket with the key, and then install the crankshaft timing sprocket to the crankshaft.

b. Align the timing mark on the crankshaft timing sprocket with the match mark on the lower crank case.

12. Install the timing chain while align-

ing the 2 blue plates with the timing marks on the CMP actuator and the exhaust camshaft timing sprockets.

13. Install the timing chain while aligning the blue plate with the timing mark on the crankshaft timing sprocket.

14. Install the timing chain guide and

tighten the timing chain guide bolts to 80 inch lbs. (9 Nm).

15. Apply engine oil to the sliding surface of the timing chain guide.

16. Attach a spacer to the timing chain tensioner.

17. Install the timing chain tensioner and tighten the timing chain tensioner bolt to 19 ft. lbs. (25 Nm).

18. Apply engine oil to the sliding surface of the timing chain tensioner.

19. Make sure that all blue plated plates are aligned with the timing marks on the corresponding timing sprockets.

20. Screw in the plunger of the timing chain tensioner adjuster clockwise and install a retainer. Use a 0.06 inch (1.4mm) diameter wire, or equivalent, to hold the plunger in place.

21. Install the timing chain tensioner adjuster with the retainer. Tighten the timing chain tensioner adjuster bolts to 97 inch lbs. (11 Nm).

22. Remove the retainer from the timing chain tensioner adjuster.

23. Apply engine oil to the timing chain, and then turn the crankshaft clockwise fully twice, and make sure the timing marks are aligned.

➡ Be sure to turn the crankshaft fully twice. If it is rotated only once, the timing marks on the CMP actuator and the exhaust camshaft timing sprocket will not meet the match marks on the camshaft housing No. 1. After turning the crankshaft twice, the discrimination links of the timing chain are not aligned with the timing marks on the CMP actuator and the exhaust timing sprocket, but this is normal.

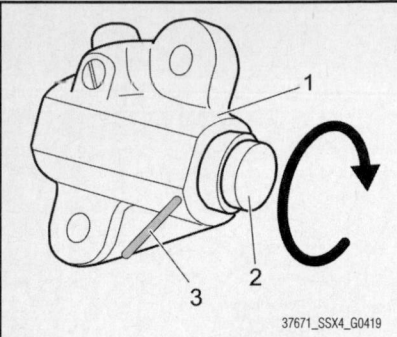

Fig. 114 Screw in the plunger (2) of the timing chain tensioner adjuster (1) clockwise and install a retainer (3)—2.4L engine

24. Install the timing chain cover.

25. Install the cylinder head cover.

26. Install the oil pan. Refer to Oil Pan, removal & installation.

27. Install the engine assembly.

3.2L Engine

First Timing Chain

See Figures 115 through 121.

1. Before servicing the vehicle, refer to the Precautions Section.

2. Remove the 2nd timing chain (bank 1). Refer to Second Timing Chain, removal & installation.

3. By turning the crankshaft, align the camshaft (bank 2) and the crankshaft at the proper position.

  a. Using Special Tool: Crankshaft rotation socket (EN-46111), or equivalent, align the circle mark on the crankshaft timing sprocket with the circle mark on the oil pump.

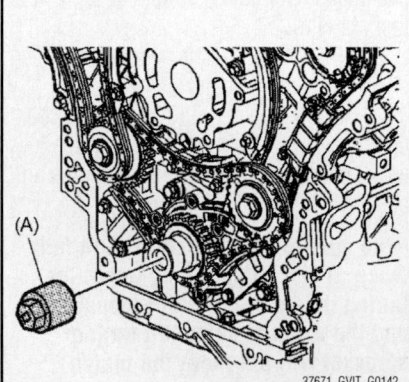

Fig. 115 Using Special Tool: Crankshaft rotation socket (EN-46111) (A) to align the crankshaft timing sprocket with the oil pump—3.2L engine

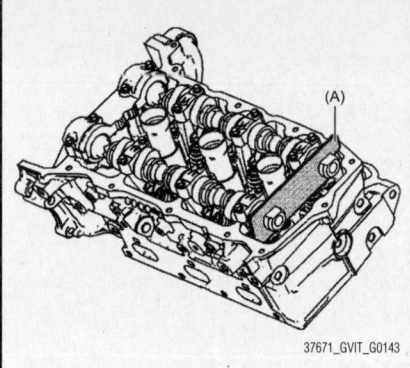

Fig. 116 Hold the camshafts (bank 2) with Special Tool 09917-67810 (A)—3.2L engine

  b. Make sure that the flat sections of the camshafts (bank 2) are parallel to the cylinder head.

  c. Hold the camshafts (bank 2) with Special Tool 09917-67810, or equivalent.

  d. Remove Special Tool: Crankshaft rotation socket (EN-46111).

4. Remove the timing chain tensioner adjuster No. 2 and the gasket.

5. Remove the timing chain guide No. 2.

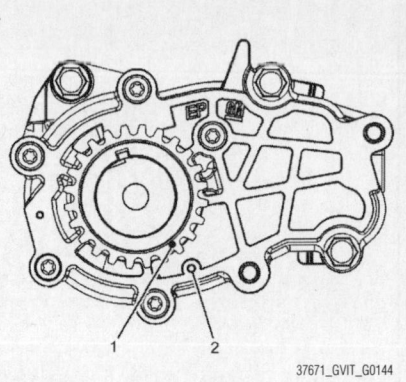

Fig. 117 Check that the circle mark (1) on the crankshaft timing sprocket is in alignment with the circle mark (2) on the oil pump—3.2L engine

6. Remove the timing chain lower guide.

7. Remove the 1st timing chain.

8. Remove the idler sprocket No. 1.

9. Remove the crankshaft timing sprocket and key.

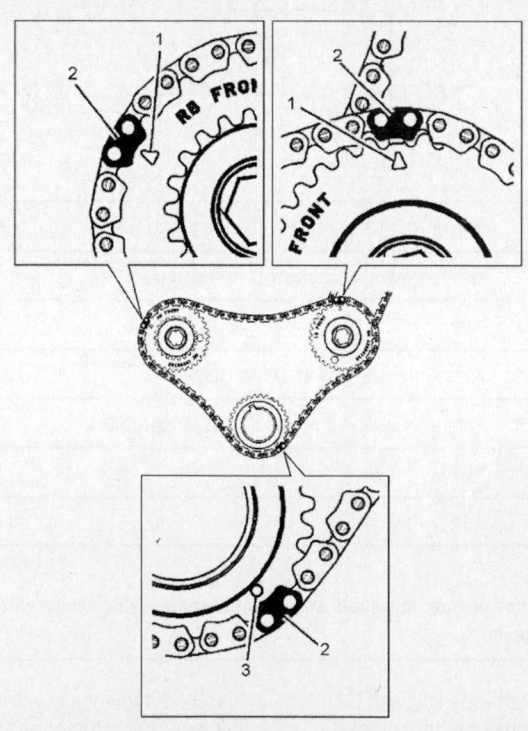

1. Triangle mark on the idler sprocket No. 1 and No. 2
2. Timing chain discrimination plates
3. Circle mark on the crankshaft timing sprocket

Fig. 118 First timing chain alignment marks—3.2L engine

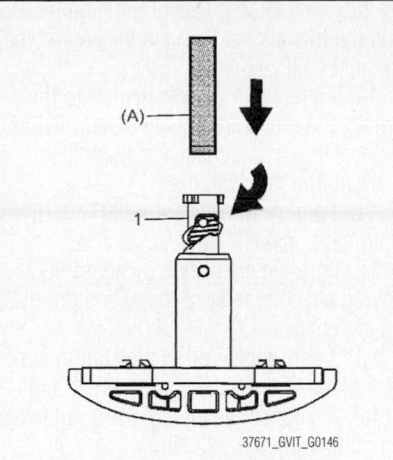

Fig. 119 Using Special Tool 09918-57810 (A), turn the plunger (1) in the direction of the arrow to contract it—3.2L engine

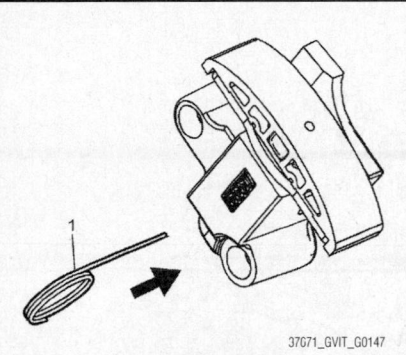

Fig. 120 Hold the plunger using a lock pin (1)—3.2L engine

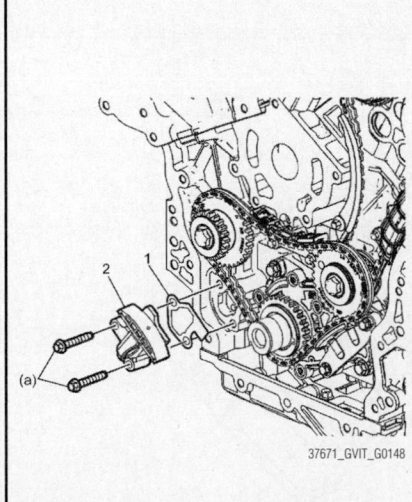

Fig. 121 Install a new gasket (1) and the timing chain tensioner adjuster No. 2 (2) and tighten the timing chain tensioner adjuster No. 2 bolts (a)—3.2L engine

To install:

10. Install the idler sprocket No. 1 and tighten the bolt to 43 ft. lbs. (58 Nm).

11. Install the key to the crankshaft.

12. Install the crankshaft timing sprocket to the crankshaft.

➡ **Be sure to face the circle mark on the crankshaft timing sprocket to the timing chain cover side.**

13. Check that the circle mark on the crankshaft timing sprocket is in alignment with the circle mark on the oil pump.

14. Install the timing chain by aligning the discrimination plates of the timing chain and the triangle mark on the idler sprocket No. 1 and No. 2.

15. Fit the crankshaft timing sprocket to the timing chain by aligning the discrimination plate of the timing chain and the circle mark on the crankshaft timing sprocket.

16. Apply engine oil to the sliding surface of the timing chain lower guide and install it. Tighten the timing chain lower guide bolt to 17 ft. lbs. (23 Nm).

17. Apply engine oil to the sliding surface of the timing chain guide No. 2 and install it. Tighten the timing chain guide No. 2 bolt to 17 ft. lbs. (23 Nm).

18. Fix the plunger in the timing chain tensioner adjuster No. 2 as follows.

   a. Remove the plunger from the timing chain tensioner adjuster No. 2.

   b. Using Special Tool 09918-57810, or equivalent, turn the plunger in to contract it and install it to the timing chain tensioner adjuster No. 2.

   c. Hold the plunger using a lock pin.

19. Install a new gasket and the timing chain tensioner adjuster No. 2 and tighten the timing chain tensioner adjuster No. 2 bolt to 17 ft. lbs. (23 Nm).

20. Check that the triangle mark on the idler sprocket No. 1, triangle mark on the idler sprocket No. 2, and the circle mark on the crankshaft timing sprocket are in alignment with the discrimination plates of the timing chain.

21. Remove the lock pin from the timing chain tensioner adjuster No. 2.

22. Turn the crankshaft clockwise by 2 revolutions and check that the circle mark on the crankshaft timing sprocket are in alignment with the circle mark on the oil pump.

23. Confirm that the special tool is installed to the camshaft (bank 2).

24. Remove Special Tool 09917-67810 from the camshaft (bank 2).

25. Install the 2nd timing chain (bank 1).

26. Install the timing chain cover.

27. Install the belt idler arm to the alternator bracket.

28. Install the OCV.

29. Install the P/S pump bracket.

30. Install the water pump pulley.

31. Install the cylinder head cover.

32. Install the accessory drive belt.

33. Install the engine assembly to the vehicle.

Second Timing Chain—Bank 1

See Figures 115, 122 through 127.

1. Before servicing the vehicle, refer to the Precautions Section.

2. Remove the timing chain cover. Refer to Timing Chain Front Cover, removal & installation.

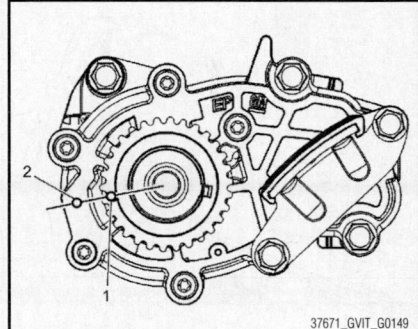

Fig. 122 Align the circle mark (1) on the crankshaft timing sprocket with the circle mark (2) on the oil pump—3.2L engine

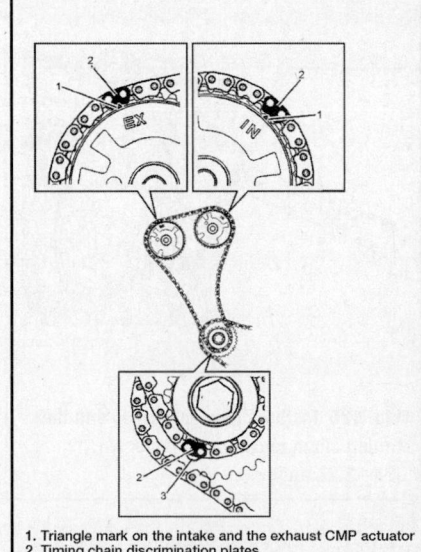

1. Triangle mark on the intake and the exhaust CMP actuator
2. Timing chain discrimination plates
3. Match mark on the idler sprocket No. 1

Fig. 123 Second timing chain alignment marks (bank 1)—3.2L engine

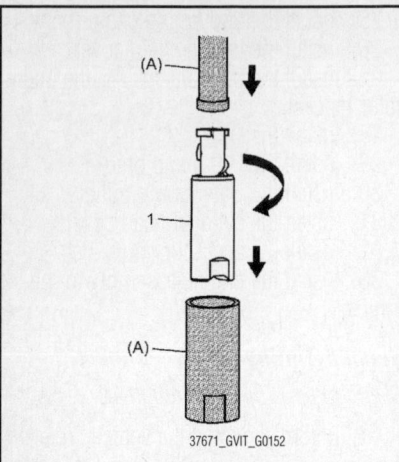

Fig. 124 Using Special Tool 09917-67810 (A) to turn the plunger to contract it—3.2L engine

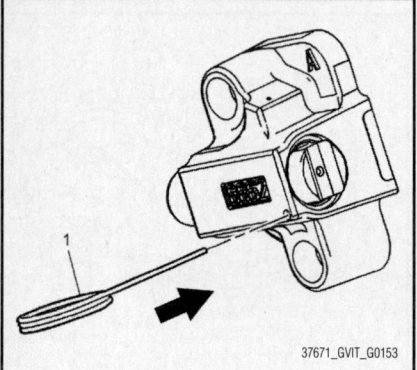

Fig. 125 Hold the plunger using a lock pin (1)—3.2L engine

Fig. 126 Install a new gasket (1) and the timing chain tensioner adjuster No. 1 (2)—3.2L engine

3. Turn the crankshaft and align the camshaft (bank 1) and the crankshaft.

a. Using Special Tool Crankshaft rotation socket (EN-46111), or equivalent, align the circle mark on the crank-

shaft timing sprocket with the circle mark on the oil pump.

b. Make sure that the flat sections of the camshafts are parallel to the cylinder head.

c. Hold the camshafts (bank 1) with Special Tool 09917-67810, or equivalent.

4. Remove the timing chain tensioner adjuster No. 1 and the gasket.

5. Remove the timing chain tensioner No. 1.

6. Remove the timing chain guide No. 1.

7. Remove the 2nd timing chain (Bank 1).

To install:

8. Check that the circle mark on the crankshaft timing sprocket is in alignment with the circle mark on the oil pump.

9. Install the timing chain by aligning

the discrimination plates of the timing chain and the triangle mark on the intake and the exhaust CMP actuator.

10. Fit the idler sprocket No. 1 to the timing chain by aligning the discrimination plate of the timing chain and the match mark on the idler sprocket No. 1.

11. Remove Special Tool 09917-67810 from the camshaft (bank 1).

12. Turn the crankshaft by about 90° paying attention to keep the alignment of the timing chain.

13. Apply engine oil to the sliding surface of the timing chain guide No. 1 and install it. Tighten the timing chain guide No. 1 bolt to 17 ft. lbs. (23 Nm).

14. Apply engine oil to the sliding surface of the timing chain tensioner No. 1 and install it. Tighten the timing chain tensioner No. 1 bolt to 17 ft. lbs. (23 Nm).

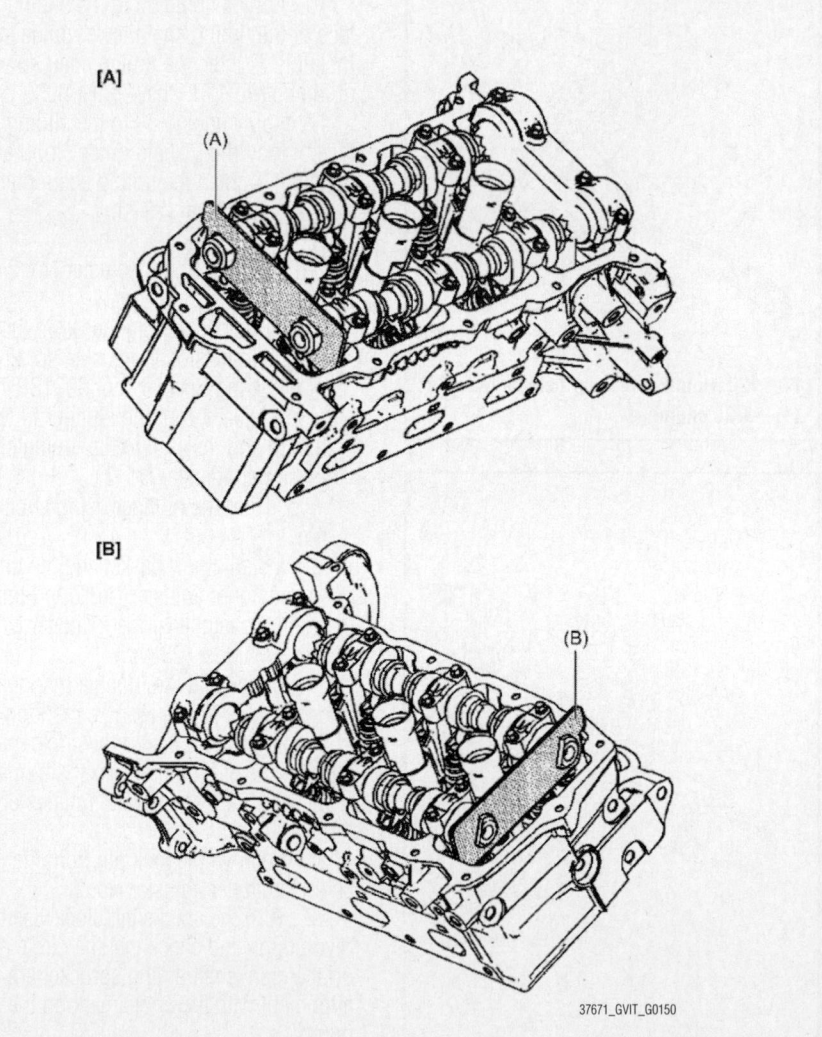

Fig. 127 Using Special Tools 09917-67830 (A) and 09917-67810 (B) installed to the camshaft (bank 1) [A] and camshaft (bank 2) [B]—3.2L engine

15. Fix the plunger in the timing chain tensioner adjuster No. 1 as follows.

a. Remove the plunger from the timing chain tensioner adjuster No. 1.

b. Using Special Tool 09917-67810, or equivalent, turn the plunger to contract it and install it to the timing chain tensioner adjuster No. 1.

c. Hold the plunger using a lock pin.

16. Install a new gasket and the timing chain tensioner adjuster No. 1 and tighten the timing chain tensioner adjuster No. 1 bolt to 17 ft. lbs. (23 Nm).

17. Check that the triangle mark on the intake CMP actuator, the triangle mark on the exhaust CMP actuator, and the match mark on the idler sprocket No. 1 are in alignment with the discrimination plates of the timing chain.

18. Remove the lock pin from the timing chain tensioner adjuster No. 1.

19. Turn the crankshaft clockwise by 2 revolutions and check that the circle mark on the crankshaft timing sprocket are in alignment with the circle mark on the oil pump.

20. Confirm that Special Tools 09917-67830 and 09917-67810, or equivalent, are installed to the camshaft (bank 1) and camshaft (bank 2).

21. Remove Special Tool: Crankshaft rotation socket (EN-46111).

22. Install the timing chain cover.

23. Install the belt idler arm to the alternator bracket.

24. Install the OCV.

25. Install the P/S pump bracket.

26. Install the water pump pulley.

27. Install the cylinder head cover.

28. Install the accessory drive belt.

29. Install the engine assembly to the vehicle.

Second Timing Chain—Bank 2

See Figures 128 through 131.

1. Before servicing the vehicle, refer to the Precautions Section.

2. Remove the 1st timing chain. Refer to First Timing Chain, removal & installation.

3. Remove the timing chain tensioner adjuster No. 3 and gasket.

4. Remove the timing chain tensioner No. 2.

5. Remove the timing chain guide No. 3.

6. Remove the 2nd timing chain (bank 2).

7. Remove the idler sprocket No. 2.

To install:

8. Install the idler sprocket No. 2 and

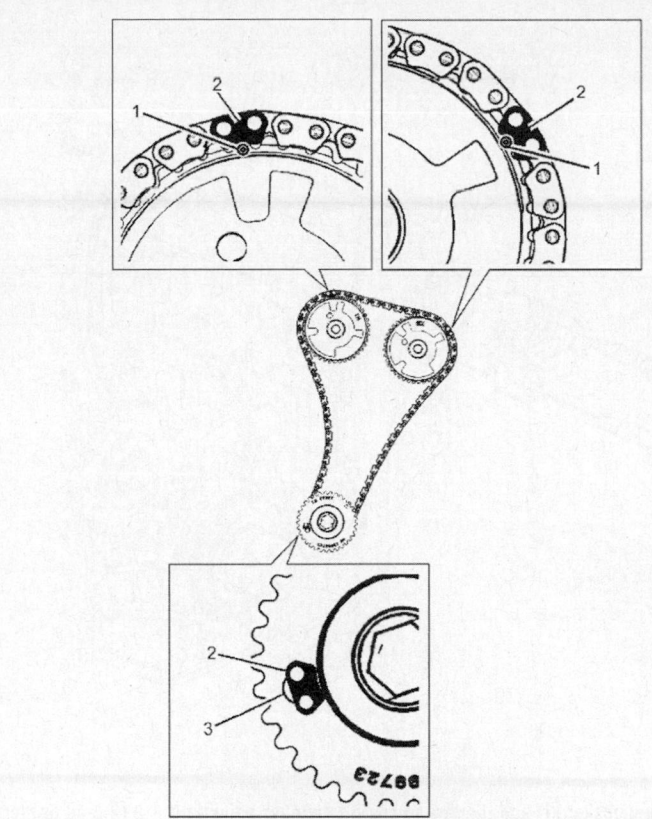

1. Circle mark on the intake and the exhaust CMP actuator
2. Timing chain aligning discrimination plates
3. Match mark on the idler sprocket No. 2

37671_GVIT_G0155

Fig. 128 Second timing chain alignment marks (bank 2)—3.2L engine

tighten the idler sprocket No. 2 bolt to 43 ft. lbs. (58 Nm).

9. Install the timing chain by aligning the discrimination plates of the timing chain and the circle mark on the intake and the exhaust CMP actuator.

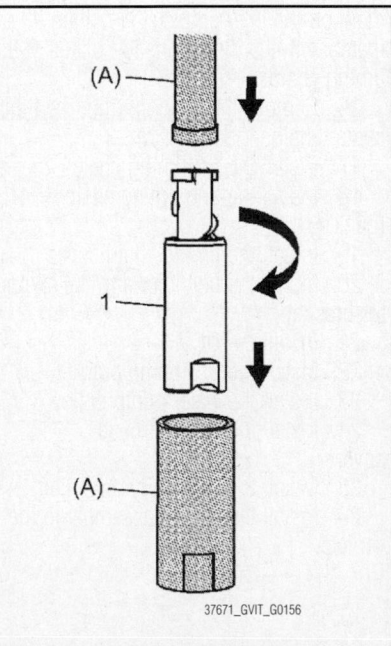

37671_GVIT_G0156

Fig. 129 Using Special Tool 09918-57810 (A) to turn the plunger (1) in to contract it—3.2L engine

37671_GVIT_G0157

Fig. 130 Hold the plunger using a lock pin (1)—3.2L engine

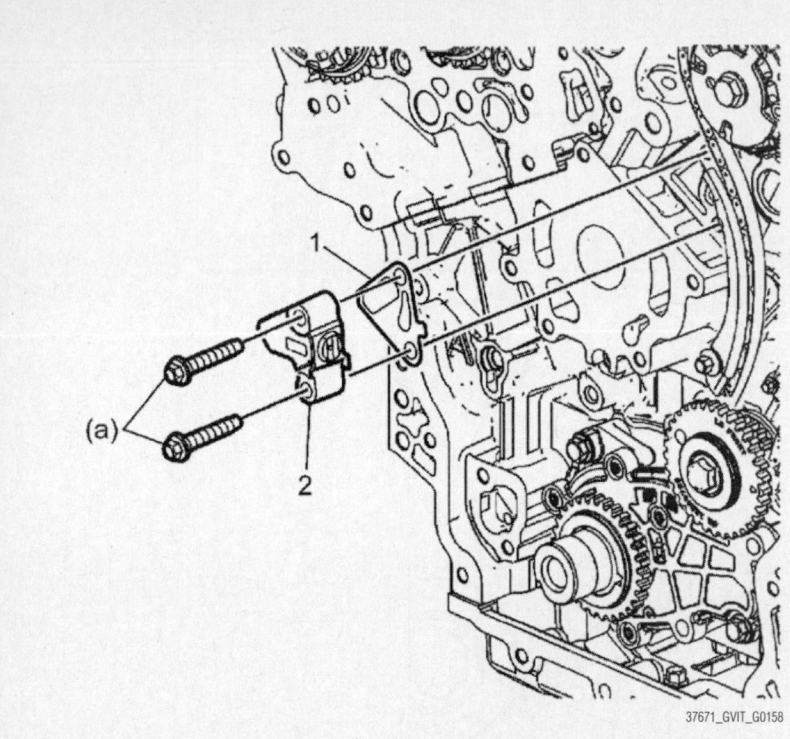

37671_GVIT_G0158

Fig. 131 Install a new gasket (1) and the timing chain tensioner adjuster No. 3 (2) and tighten the timing chain tensioner adjuster No. 3 bolts (a)—3.2L engine

10. Fit the idler sprocket No. 2 to the timing chain by aligning the discrimination plate of the timing chain and the match mark on the idler sprocket No. 2.

11. Apply engine oil to the sliding surface of the timing chain guide No. 3 and install it. Tighten the timing chain guide No. 3 bolt to 17 ft. lbs. (23 Nm).

12. Apply engine oil to the sliding surface of the timing chain tensioner No. 2 and install it. Tighten the timing chain tensioner No. 2 bolt to 17 ft. lbs. (23 Nm).

13. Fix the plunger in the timing chain tensioner adjuster No. 3 as follows.

   a. Remove the plunger from the timing chain tensioner adjuster No. 3.

   b. Using Special Tool 09918-57810, or equivalent, turn the plunger in to contract it and install it to the timing chain tensioner adjuster No. 3.

   c. Hold the plunger using a lock pin.

14. Install a new gasket and the timing chain tensioner adjuster No. 3 and tighten

the timing chain tensioner adjuster No. 3 bolt to 17 ft. lbs. (23 Nm).

15. Check that the circle mark on the intake CMP actuator, the circle mark on the exhaust CMP actuator, and the alignment mark on the idler sprocket No. 2 are in match with the discrimination plates of the timing chain.

16. Remove the lock pin from the timing chain tensioner adjuster No. 3.

17. Install the 1st timing chain.

18. Install the 2nd timing chain (bank 1).

19. Install the timing chain cover.

20. Install the belt idler arm the alternator bracket.

21. Install the OCV.

22. Install the P/S pump bracket.

23. Install the water pump pulley.

24. Install the cylinder head cover.

25. Install the accessory drive belt.

26. Install the engine assembly to the vehicle.

VALVE LASH

ADJUSTMENT

2.4L Engine

See Figures 132 and 133.

1. Before servicing the vehicle, refer to the Precautions Section.

2. Disconnect the negative battery cable.

3. Remove the tappet to be replaced.

4. Select the proper size of the tappet as follows.

   a. Using a micrometer, measure the thickness of the removed tappet.

   b. Calculate the thickness of the new tappet by the formula below.

   c. Select a new tappet closest to the calculated value from the available sizes.

5. Install the tappets and the camshafts.

6. Recheck the valve clearance.

Intake side:
A = B + C − 0.20 mm (0.0079 in.)
Exhaust side:
A = B + C − 0.35 mm (0.0138 in.)
A: Thickness "a" of new tappet
B: Thickness "a" of removed tappet
C: Measured valve clearance

37671_SSX4_G0438

Fig. 132 Formula used to determine new tappet—2.4L engine

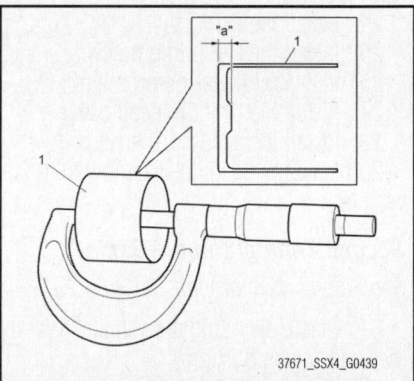

37671_SSX4_G0439

Fig. 133 Measuring the tappet (1) thickness (a) with a micrometer—2.4L engine

3.2L Engine

This engine uses hydraulic valve lash adjusters. No adjustment is necessary.

ENGINE PERFORMANCE & EMISSION CONTROLS

CAMSHAFT POSITION (CMP) SENSOR

LOCATION

2.4L Engine

See Figure 134.

Refer to the accompanying illustration.

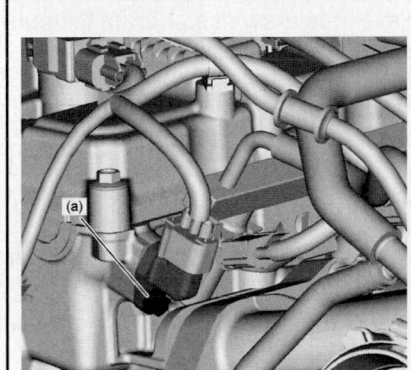

Fig. 134 Camshaft Position (CMP) sensor (a) location—2.4L engine

3.2L Engine

See Figures 135 and 136.

Refer to the accompanying illustrations.

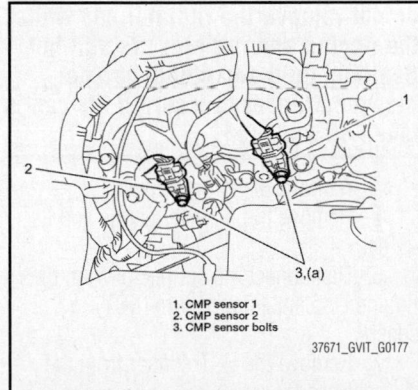

1. CMP sensor 1
2. CMP sensor 2
3. CMP sensor bolts

37671_GVIT_G0177

Fig. 135 Camshaft Position (CMP) sensor location (bank 1)—3.2L engine

REMOVAL & INSTALLATION

2.4L Engine

1. Before servicing the vehicle, refer to the Precautions Section.
2. Disconnect the negative battery cable.

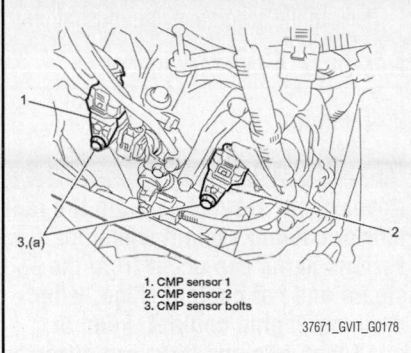

1. CMP sensor 1
2. CMP sensor 2
3. CMP sensor bolts

37671_GVIT_G0178

Fig. 136 Camshaft Position (CMP) sensor location (bank 2)—3.2L engine

3. Disconnect the connector from the CMP sensor.
4. Remove the CMP sensor from the cylinder head.

To install:

5. Installation is the reverse of the removal procedure.
6. Apply engine oil to the O-ring of the CMP sensor.
7. Install the CMP sensor to the cylinder head.
8. Tighten the CMP sensor bolt to 97 inch lbs. (11 Nm).
9. Connect the connector to the CMP sensor securely.
10. Connect the negative battery cable.

3.2L Engine

1. Before servicing the vehicle, refer to the Precautions Section.
2. Disconnect the negative battery cable.
3. Disconnect the CMP sensor connector.
4. Remove the CMP sensor from the timing chain cover.

To install:

5. Installation is the reverse of the removal procedure.
6. Apply engine oil to the O-ring of the sensor.
7. Tighten the new CMP sensor bolt to 89 inch lbs. (10 Nm).

CRANKSHAFT POSITION (CKP) SENSOR

LOCATION

2.4L Engine

See Figure 137.

Refer to the accompanying illustration.

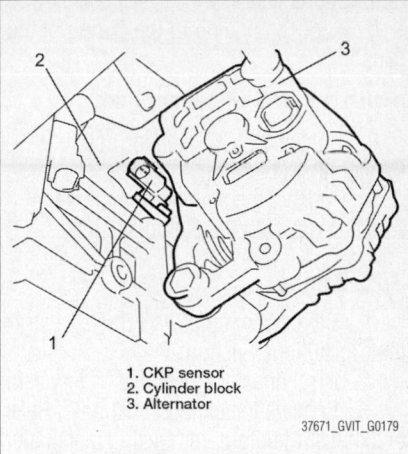

1. CKP sensor
2. Cylinder block
3. Alternator

37671_GVIT_G0179

Fig. 137 Crankshaft Position (CKP) sensor location—2.4L engine

3.2L Engine

See Figure 138.

Refer to the accompanying illustration.

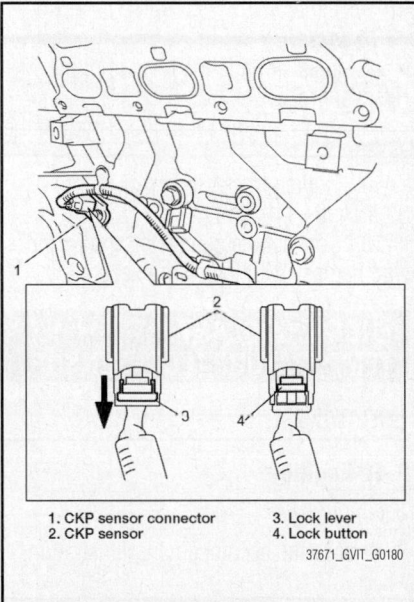

1. CKP sensor connector
2. CKP sensor
3. Lock lever
4. Lock button

37671_GVIT_G0180

Fig. 138 Crankshaft Position (CKP) sensor location—3.2L engine

REMOVAL & INSTALLATION

2.4L Engine

1. Before servicing the vehicle, refer to the Precautions Section.
2. Disconnect the negative battery cable.
3. Remove the alternator.
4. Disconnect the connector from the CKP sensor.
5. Remove the CKP sensor from the cylinder block.

To install:

6. Installation is the reverse of the removal procedure.

7. Apply engine oil to the O-ring of the sensor.

➡**Use a new CKP sensor bolt.**

8. Tighten the CKP sensor bolt to 97 inch lbs. (11 Nm).

9. Connect the connector and fix the wire harness with the clamp securely.

3.2L Engine

1. Before servicing the vehicle, refer to the Precautions Section.

2. Disconnect the negative battery cable.

3. Remove the exhaust manifold. Refer to Exhaust Manifold, removal & installation.

4. Remove the bank-1 side engine mounting and bracket.

5. Disconnect the connector from the CKP sensor as follows.

   a. Pull out the lock lever.

   b. Disconnect the connector from the CKP sensor while pushing the lock button.

6. Remove the CKP sensor from the cylinder block.

To install:

7. Installation is the reverse of the removal procedure.

8. Apply engine oil to the O-ring of the sensor.

9. Tighten the CKP sensor bolt to 89 inch lbs. (10 Nm).

10. Connect the connector and fix the wire harness with the clamp securely.

ENGINE COOLANT TEMPERATURE (ECT) SENSOR

LOCATION

2.4L Engine

See Figure 139.

Refer to the accompanying illustration.

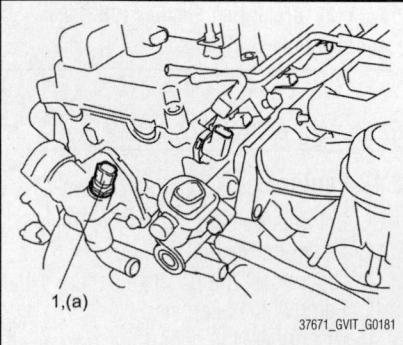

1,(a)

37671_GVIT_G0181

Fig. 139 Engine Coolant Temperature (ECT) sensor (1) location—2.4L engine

3.2L Engine

See Figure 140.

Refer to the accompanying illustration.

REMOVAL & INSTALLATION

2.4L Engine

❊❊ **CAUTION**

Never open, service or drain the radiator or cooling system when hot; serious burns can occur from the steam and hot coolant. Also, when draining engine coolant, keep in mind that cats and dogs are attracted to ethylene glycol antifreeze and could drink any that is left in an uncovered container or in puddles on the ground. This will prove fatal in sufficient quantities. Always drain coolant into a sealable container. Coolant should be reused unless it is contaminated or is several years old.

1. Before servicing the vehicle, refer to the Precautions Section.

2. Disconnect the negative battery cable.

3. Drain the engine coolant.

❊❊ **CAUTION**

To avoid the danger of being burned, do not remove the radiator cap while the engine and radiator are still hot. Scalding fluid and steam can be blown out under pressure if the cap is taken off too soon.

4. Disconnect the connector from the Engine Coolant Temperature (ECT) sensor.

5. Remove the ECT sensor.

To install:

6. Installation is the reverse of the removal procedure.

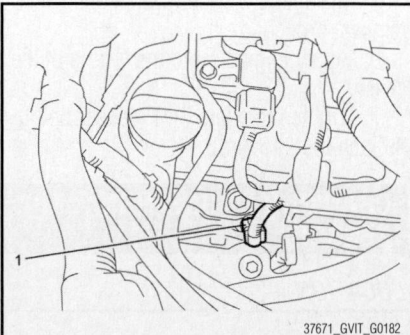

1

37671_GVIT_G0182

Fig. 140 Engine Coolant Temperature (ECT) sensor (1) location—3.2L engine

7. Check the O-ring for damage and replace if necessary.

8. Tighten the ECT sensor to 106 inch lbs. (12 Nm).

9. Connect the connector to the ECT sensor.

10. Refill the cooling system with the proper amount and type of fluid.

11. Check for coolant leakage.

3.2L Engine

❊❊ **CAUTION**

Never open, service or drain the radiator or cooling system when hot; serious burns can occur from the steam and hot coolant. Also, when draining engine coolant, keep in mind that cats and dogs are attracted to ethylene glycol antifreeze and could drink any that is left in an uncovered container or in puddles on the ground. This will prove fatal in sufficient quantities. Always drain coolant into a sealable container. Coolant should be reused unless it is contaminated or is several years old.

1. Before servicing the vehicle, refer to the Precautions Section.

2. Disconnect the negative battery cable.

3. Drain the engine coolant.

❊❊ **CAUTION**

To avoid the danger of being burned, do not remove the radiator cap while the engine and radiator are still hot. Scalding fluid and steam can be blown out under pressure if the cap is taken off too soon.

4. Remove the engine cover.

5. Remove the oil level gauge and its guide.

6. Disconnect the connector from the Engine Coolant Temperature (ECT) sensor.

7. Remove the ECT sensor from the cylinder head.

To install:

8. Installation is the reverse of the removal procedure.

9. Tighten the new ECT sensor to 17 ft. lbs. (22 Nm).

10. Refill the cooling system with the proper amount and type of fluid.

11. Check for coolant leakage from the ECT sensor with the engine running.

HEATED OXYGEN (HO2S) SENSOR

LOCATION

2.4L Engine

See Figure 141.

Refer to the accompanying illustration.

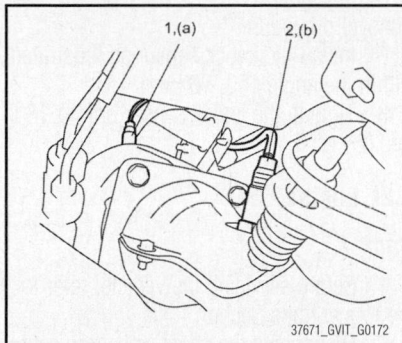

Fig. 141 Air-Fuel Ratio Sensor (AFR) (1) and Heated Oxygen (HO2S) Sensor (2) locations—2.4L engine

3.2L Engine

See Figures 142 through 145.

Refer to the accompanying illustrations.

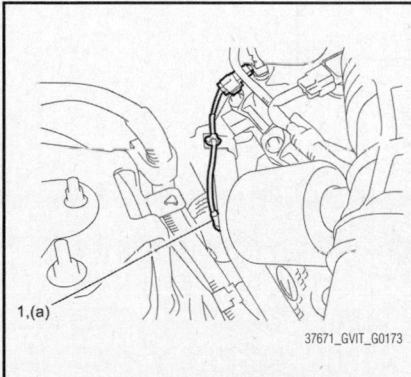

Fig. 142 Heated Oxygen (HO2S) Sensor location (sensor 1, bank 1)—3.2L engine

REMOVAL & INSTALLATION

2.4L Engine

⁂ **CAUTION**

To avoid the danger of being burned, do not touch the exhaust system when the system is hot. The Heated Oxygen Sensor (HO2S) removal should be performed when the system is cool.

1. Before servicing the vehicle, refer to the Precautions Section.

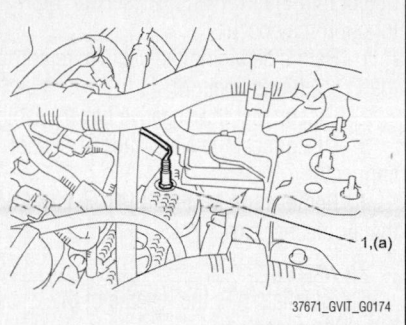

Fig. 143 Heated Oxygen (HO2S) Sensor location (sensor 1, bank 2)—3.2L engine

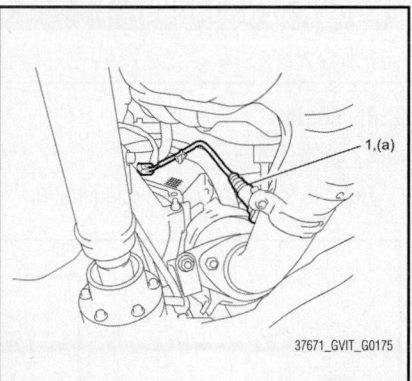

Fig. 144 Heated Oxygen (HO2S) Sensor location (sensor 2, bank 1)—3.2L engine

2. Disconnect the negative battery cable.
3. Disconnect the connector of the AFR sensor and/or HO2S.
4. Remove the exhaust manifold cover.
5. Remove the HO2S sensor from the exhaust manifold.

To install:
6. Installation is the reverse of the removal procedure.
7. Tighten the HO2S sensor to 33 ft. lbs. (45 Nm).

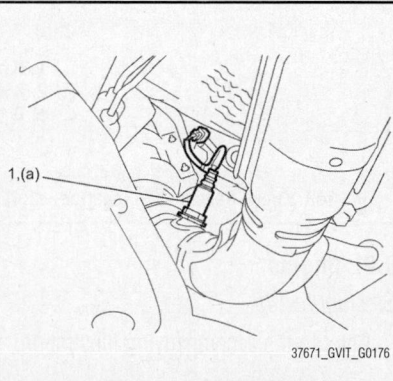

Fig. 145 Heated Oxygen (HO2S) Sensor location (sensor 2, bank 2)—3.2L engine

8. Check for exhaust leakage from the sensor with the engine running.

3.2L Engine

⁂ **CAUTION**

To avoid the danger of being burned, do not touch the exhaust system when the system is hot. The Heated Oxygen Sensor (HO2S) removal should be performed when the system is cool.

1. Before servicing the vehicle, refer to the Precautions Section.
2. Disconnect the negative battery cable.
3. Disconnect the connectors of the HO2S.
4. Remove the HO2S from the exhaust manifold.

To install:
5. Installation is the reverse of the removal procedure.
6. Tighten the HO2S to 34 ft. lbs. (45 Nm).
7. Check exhaust leakage from the sensor with the engine running.

INTAKE AIR TEMPERATURE (IAT) SENSOR

LOCATION

2.4L Engine

See Figure 146.

The Intake Air Temperature (IAT) sensor is built into the Mass Air Flow (MAF) sensor.

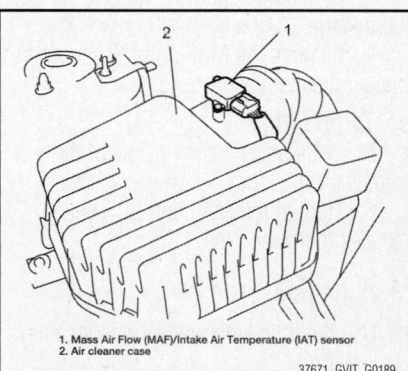

1. Mass Air Flow (MAF)/Intake Air Temperature (IAT) sensor
2. Air cleaner case

Fig. 146 Mass Air Flow (MAF)/Intake Air Temperature (IAT) sensor locations—2.4L engine

3.2L Engine

See Figure 147.

The Intake Air Temperature (IAT) sensor is built into the Mass Air Flow (MAF) sensor.

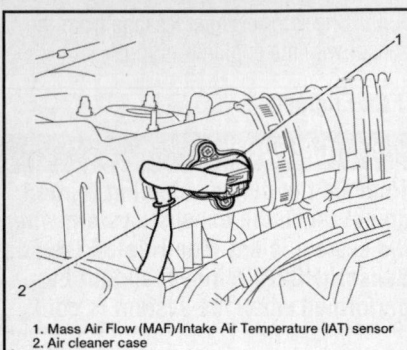

1. Mass Air Flow (MAF)/Intake Air Temperature (IAT) sensor
2. Air cleaner case

37671_GVIT_G0190

Fig. 147 Mass Air Flow (MAF)/Intake Air Temperature (IAT) sensor locations—3.2L engine

REMOVAL & INSTALLATION

2.4L Engine

Use the following precautions for this procedure.
- Do not disassemble the MAF and IAT sensor
- Do not expose the MAF and IAT sensor to any shock
- Do not clean the MAF and IAT sensor
- If the MAF and IAT sensor has been dropped, it should be replaced
- Do not blow compressed air through the MAF and IAT sensor
- Do not place a finger or any other object into the MAF and IAT sensor. Malfunction may occur

1. Before servicing the vehicle, refer to the Precautions Section.
2. Disconnect the negative battery cable.
3. Disconnect the MAF and IAT sensor connector.
4. Remove the MAF and IAT sensor from the air cleaner case.

To install:

5. Installation is the reverse of the removal procedure.
6. Tighten the MAF and IAT sensor screws to 10 inch lbs. (1 Nm).

3.2L Engine

Use the following precautions for this procedure.
- Do not disassemble the MAF and IAT sensor
- Do not expose the MAF and IAT sensor to any shock
- Do not clean the MAF and IAT sensor
- If the MAF and IAT sensor has been dropped, it should be replaced
- Do not blow compressed air through the MAF and IAT sensor
- Do not place a finger or any other

object into the MAF and IAT sensor. Malfunction may occur

1. Before servicing the vehicle, refer to the Precautions Section.
2. Disconnect the negative battery cable.
3. Disconnect the MAF and IAT sensor connector.
4. Remove the MAF and IAT sensor from the air cleaner case.

To install:

5. Installation is the reverse of the removal procedure.
6. Tighten the MAF and IAT sensor screws to 27 inch lbs. (3 Nm).

KNOCK SENSOR (KS)

LOCATION

2.4L Engine

See Figure 148.

Refer to the accompanying illustration.

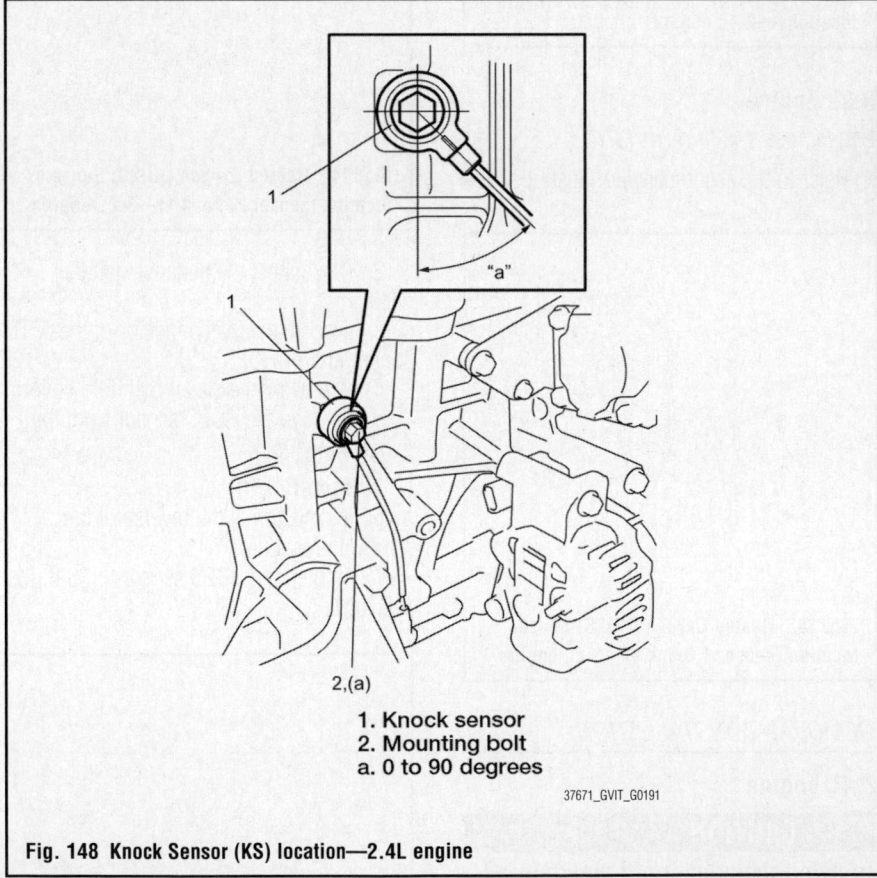

2,(a)

1. Knock sensor
2. Mounting bolt
a. 0 to 90 degrees

37671_GVIT_G0191

Fig. 148 Knock Sensor (KS) location—2.4L engine

3.2L Engine

See Figure 149.

Refer to the accompanying illustration.

REMOVAL & INSTALLATION

2.4L Engine

1. Before servicing the vehicle, refer to the Precautions Section.
2. Disconnect the negative battery cable.
3. Remove the intake manifold.
4. Disconnect the knock sensor connector.
5. Remove the knock sensor from the cylinder block.

To install:

6. Installation is the reverse of the removal procedure.
7. Install the knock sensor as illustrated within the angle of 0–90°.
8. Tighten the knock sensor bolt to 18 ft. lbs. (25 Nm).

3.2L Engine

Bank 1

1. Before servicing the vehicle, refer to the Precautions Section.
2. Disconnect the negative battery cable.

3. Remove the exhaust manifold. Refer to Exhaust Manifold, removal & installation.
4. Remove the engine mounting and bracket.
5. Disconnect the knock sensor connector.

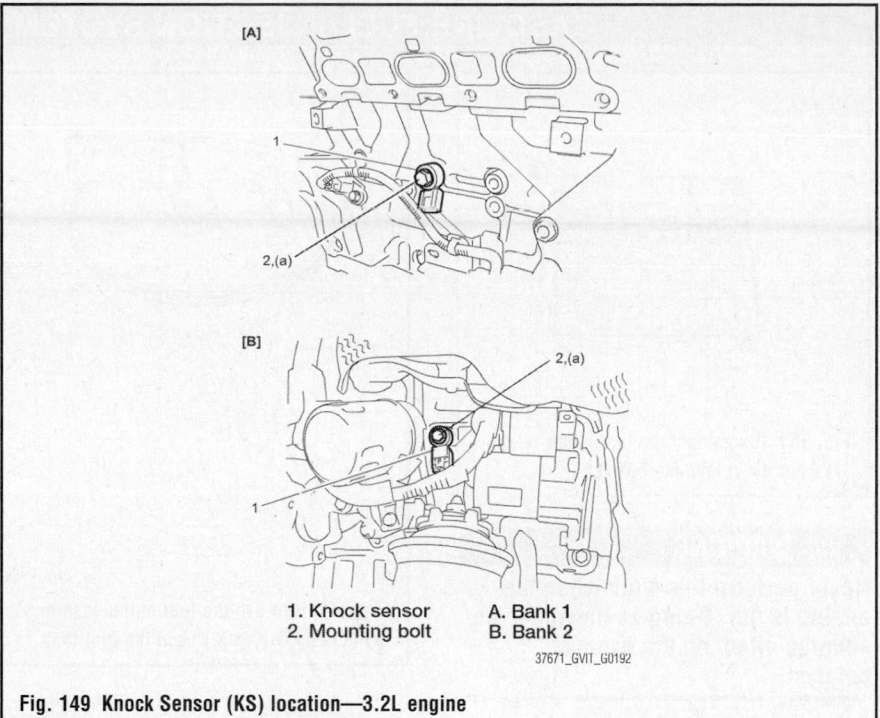

1. Knock sensor   A. Bank 1
2. Mounting bolt   B. Bank 2

37671_GVIT_G0192

Fig. 149 Knock Sensor (KS) location—3.2L engine

6. Remove the knock sensor from the cylinder block.

To install:

7. Installation is the reverse of the removal procedure.

8. Tighten the knock sensor bolt to 17 ft. lbs. (23 Nm).

Bank 2

1. Before servicing the vehicle, refer to the Precautions Section.

2. Disconnect the negative battery cable.

3. Disconnect the knock sensor connector.

4. Remove the knock sensor from the cylinder.

To install:

5. Installation is the reverse of the removal procedure.

6. Tighten the knock sensor bolt to 17 ft. lbs. (23 Nm).

MASS AIR FLOW (MAF) SENSOR

LOCATION

2.4L Engine

See Figure 150.

The Intake Air Temperature (IAT) sensor is built into the Mass Air Flow (MAF) sensor.

3.2L Engine

See Figure 147.

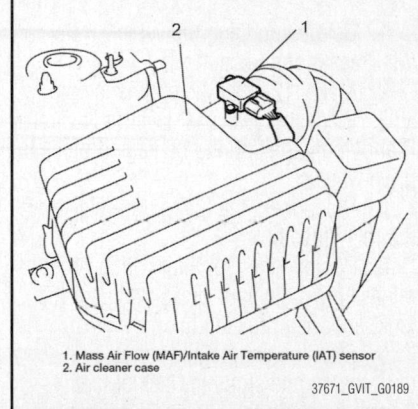

1. Mass Air Flow (MAF)/Intake Air Temperature (IAT) sensor
2. Air cleaner case

37671_GVIT_G0189

Fig. 150 Mass Air Flow (MAF)/Intake Air Temperature (IAT) sensor locations—2.4L engine

The Intake Air Temperature (IAT) sensor is built into the Mass Air Flow (MAF) sensor.

REMOVAL & INSTALLATION

2.4L Engine

Use the following precautions for this procedure.

• Do not disassemble the MAF and IAT sensor

• Do not expose the MAF and IAT sensor to any shock

• Do not clean the MAF and IAT sensor

• If the MAF and IAT sensor has been dropped, it should be replaced

• Do not blow compressed air through the MAF and IAT sensor

• Do not place a finger or any other object into the MAF and IAT sensor. Malfunction may occur

1. Before servicing the vehicle, refer to the Precautions Section.

2. Disconnect the negative battery cable.

3. Disconnect the MAF and IAT sensor connector.

4. Remove the MAF and IAT sensor from the air cleaner case.

To install:

5. Installation is the reverse of the removal procedure.

6. Tighten the MAF and IAT sensor screws to 10 inch lbs. (1 Nm).

3.2L Engine

Use the following precautions for this procedure.

• Do not disassemble the MAF and IAT sensor

• Do not expose the MAF and IAT sensor to any shock

• Do not clean the MAF and IAT sensor

• If the MAF and IAT sensor has been dropped, it should be replaced

• Do not blow compressed air through the MAF and IAT sensor

• Do not place a finger or any other object into the MAF and IAT sensor. Malfunction may occur

1. Before servicing the vehicle, refer to the Precautions Section.

2. Disconnect the negative battery cable.

3. Disconnect the MAF and IAT sensor connector.

4. Remove the MAF and IAT sensor from the air cleaner case.

To install:

5. Installation is the reverse of the removal procedure.

6. Tighten the MAF and IAT sensor screws to 27 inch lbs. (3 Nm).

THROTTLE POSITION SENSOR (TPS)

LOCATION

The electric throttle body includes a throttle control motor and Throttle Position Sensors (TPS). The TPS responds to the throttle valve movement. The TPS has two sensors.

REMOVAL & INSTALLATION

The Throttle Position Sensors (TPS) are integral to the electronic throttle body. Refer to Throttle Body, removal & installation.

FUEL   **GASOLINE FUEL INJECTION SYSTEM**

FUEL SYSTEM SERVICE PRECAUTIONS

Safety is the most important factor when performing not only fuel system maintenance but any type of maintenance. Failure to conduct maintenance and repairs in a safe manner may result in serious personal injury or death. Maintenance and testing of the vehicle's fuel system components can be accomplished safely and effectively by adhering to the following rules and guidelines.

• To avoid the possibility of fire and personal injury, always disconnect the negative battery cable unless the repair or test procedure requires that battery voltage be applied.

• Always relieve the fuel system pressure prior to disconnecting any fuel system component (injector, fuel rail, pressure regulator, etc.), fitting or fuel line connection. Exercise extreme caution whenever relieving fuel system pressure to avoid exposing skin, face and eyes to fuel spray. Please be advised that fuel under pressure may penetrate the skin or any part of the body that it contacts.

• Always place a shop towel or cloth around the fitting or connection prior to loosening to absorb any excess fuel due to spillage. Ensure that all fuel spillage (should it occur) is quickly removed from engine surfaces. Ensure that all fuel soaked cloths or towels are deposited into a suitable waste container.

• Always keep a dry chemical (Class B) fire extinguisher near the work area.

• Do not allow fuel spray or fuel vapors to come into contact with a spark or open flame.

• Always use a back-up wrench when loosening and tightening fuel line connection fittings. This will prevent unnecessary stress and torsion to fuel line piping.

• Always replace worn fuel fitting O-rings with new Do not substitute fuel hose or equivalent where fuel pipe is installed.

Before servicing the vehicle, make sure to also refer to the precautions in the beginning of this section as well.

RELIEVING FUEL SYSTEM PRESSURE

See Figure 151.

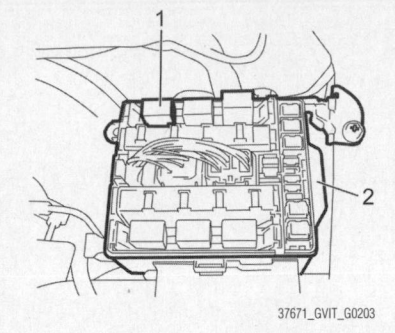

Fig. 151 Disconnect the fuel pump relay (1) from the relay/fuse box (2)

✳✳ WARNING

Never perform this work when the engine is hot. Doing so may have an adverse effect on the exhaust catalyst.

➡**If the ECM detects DTC(s) after servicing, clear the DTC(s).**

1. Before servicing the vehicle, refer to the Precautions Section.
2. Make sure that the engine is cold.
3. Place the gear shift lever in Neutral or P and apply the parking brake and block the drive wheels.
4. Disconnect the fuel pump relay from the relay/fuse box.
5. Remove the fuel filler cap in order to release the fuel vapor pressure in the fuel tank, and then reinstall it.
6. Start the engine and run it until the engine stops for lack of fuel. Repeat cranking the engine 2–3 times for about 3 seconds each time to dissipate the fuel pressure in the lines. The fuel connections are now safe for servicing.
7. After servicing, connect the fuel pump relay to the relay/fuse box and install the relay/fuse box cover.

FUEL FILTER

REMOVAL & INSTALLATION

The fuel filter is part of the fuel level sensor unit, fuel filter, and fuel pump assembly. Refer to Fuel Pump Module, removal & installation.

FUEL PUMP MODULE

REMOVAL & INSTALLATION

See Figure 152.

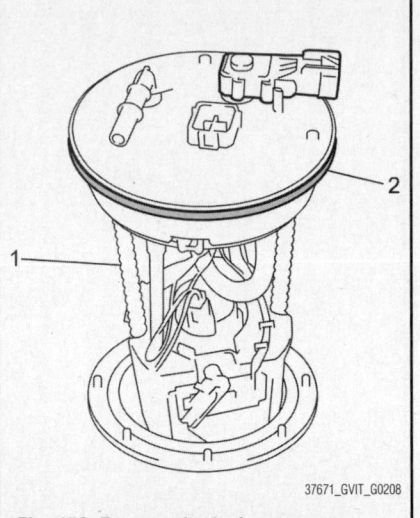

Fig. 152 Remove the fuel pump assembly (1) and O-ring (2) from the fuel tank

1. Before servicing the vehicle, refer to the Precautions Section.
2. Properly relieve the fuel system pressure. Refer to Fuel System Pressure, Relieving.
3. Remove the fuel filler cap to release the pressure from inside the fuel tank.
4. Remove the fuel tank from the vehicle. Refer to Fuel Tank, removal & installation.
5. Disconnect the fuel feed pipe from the fuel pump assembly.
6. Disconnect the fuel pump connector from the fuel pump.
7. Remove the fuel suction pipe from the fuel pump.
8. Remove the fuel pump assembly and O-ring from the fuel tank.

To install:

➡**When connecting a joint, clean the outside surface of the pipe where the joint is to be connected, push the joint into the pipe until the joint lock clicks and check to ensure that the pipes are connected securely, or a fuel leak may occur.**

9. Connect the suction pipe to the fuel pump.
10. Put the plate on the fuel pump assembly by matching the protrusion of the fuel pump assembly to the plate hole.
11. Clean the mating surfaces of the fuel pump assembly and fuel tank.
12. Install a new O-ring and the fuel pump assembly to the fuel tank. Tighten the fuel pump nut to 97 inch lbs. (11 Nm).

13. Connect the fuel pipe and the fuel pump connector to the fuel pump.

14. Install the fuel tank.

FUEL RAIL AND INJECTOR

REMOVAL & INSTALLATION

2.4L Engine

See Figure 153.

1. Before servicing the vehicle, refer to the Precautions Section.

2. Remove the fuel filler cap to release the pressure from inside the fuel tank.

3. Properly relieve the fuel system pressure. Refer to Fuel System Pressure, Relieving.

4. Disconnect the negative battery cable.

5. Disconnect the fuel injector connectors.

6. Disconnect the fuel feed hose from the fuel delivery pipe.

7. Remove the fuel delivery pipe.

8. Remove the fuel injector(s).

To install:

9. Installation is the reverse of the removal procedure.

10. Replace the injector O-rings and cushions with new ones using care not to damage them.

11. Apply a thin coat of fuel to the O-rings, and then install the injectors into the delivery pipe and the cylinder head.

12. Tighten the delivery pipe nuts to 18 ft. lbs. (25 Nm).

13. Make sure that the injectors rotate smoothly. If not, the probable cause is an incorrectly installed O-ring. Replace the O-ring with new one.

14. With the engine stationary and the ignition switch in the ON position, check for fuel leaks around the fuel line connection.

3.2L Engine

See Figure 154.

1. Before servicing the vehicle, refer to the Precautions Section.

2. Remove the fuel filler cap to release the pressure from inside the fuel tank.

3. Properly relieve the fuel system pressure. Refer to Fuel System Pressure, Relieving.

4. Disconnect the negative battery cable.

5. Remove the intake upper manifold. Refer to Intake Manifold, removal & installation.

6. Disconnect the fuel injector connectors.

7. Disconnect the fuel feed hose from the fuel delivery pipe.

8. Remove the fuel delivery pipe bolts.

9. Remove the clip(s).

10. Remove the fuel injector(s).

To install:

11. Installation is the reverse of the removal procedure.

12. Replace the injector O-rings with new ones using care not to damage them.

13. Apply a thin coat of fuel to the O-rings, and then install the injectors to the delivery pipe and the intake lower manifold.

14. Make sure that the injectors rotate smoothly. If not, the probable cause is an incorrectly installed of O-ring. Replace the O-ring with a new one.

15. Tighten the delivery pipe bolts to 89 inch lbs. (10 Nm).

16. Make sure that the injectors rotate smoothly.

17. With the engine stationary and the ignition switch in the ON position, check for fuel leaks around the fuel line connection.

FUEL TANK

REMOVAL & INSTALLATION

See Figures 155 through 157.

• When servicing the fuel tank, it should be treated with respect. Make sure that no contact with sharp edges or hot surfaces is made. In addition, the fuel tank should not be dropped since the fuel tank, fuel pump, and other components can be damaged by the impact. If dropped, all components should be replaced because of the risk of damage

1. Before servicing the vehicle, refer to the Precautions Section.

2. Remove the fuel filler cap to release the pressure from inside the fuel tank.

3. Properly relieve the fuel system pressure. Refer to Fuel System Pressure, Relieving.

4. Disconnect the negative battery cable. Refer to Battery, removal & installation.

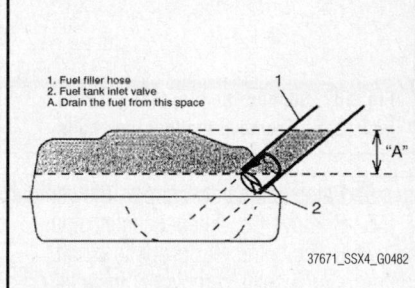

1. Fuel filler hose
2. Fuel tank inlet valve
A. Drain the fuel from this space

37671_SSX4_G0482

Fig. 155 Insert a hose into the fuel filler hose and drain the fuel in the fuel filler hose

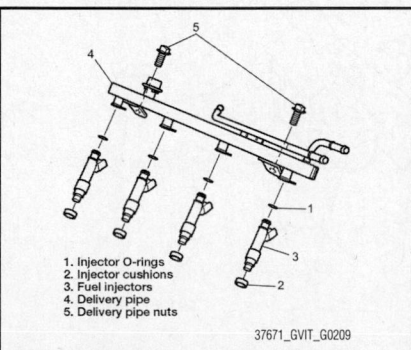

1. Injector O-rings
2. Injector cushions
3. Fuel injectors
4. Delivery pipe
5. Delivery pipe nuts

37671_GVIT_G0209

Fig. 153 View of fuel rail and injectors—2.4L engine

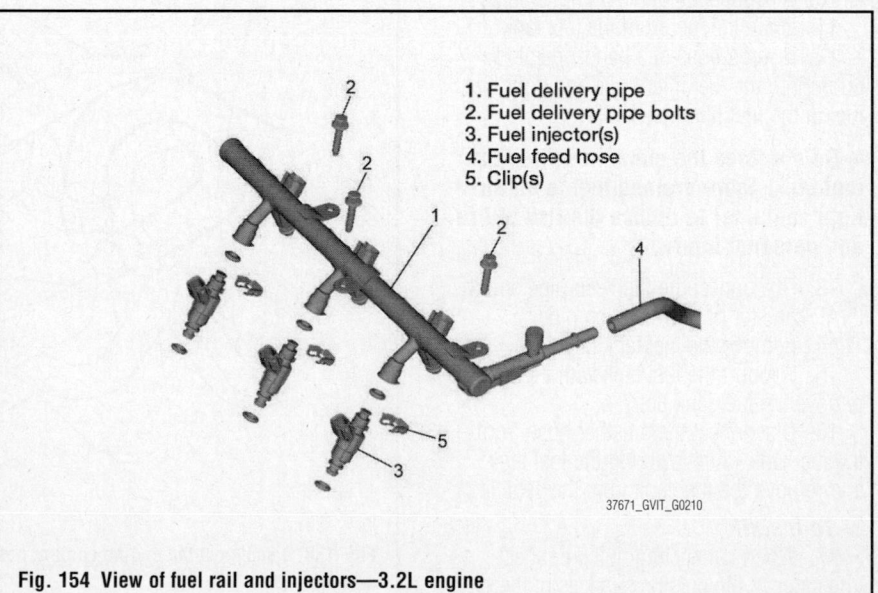

1. Fuel delivery pipe
2. Fuel delivery pipe bolts
3. Fuel injector(s)
4. Fuel feed hose
5. Clip(s)

37671_GVIT_G0210

Fig. 154 View of fuel rail and injectors—3.2L engine

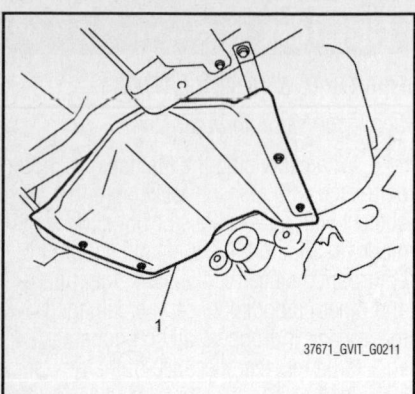

Fig. 156 Remove the fuel tank cover (1)

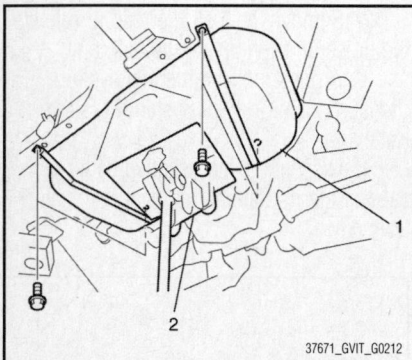

Fig. 157 Support the fuel tank (1) with a jack (2) and remove the fuel tank bolts

5. Raise and safely support the vehicle.
6. Remove the exhaust center pipe.
7. Remove the rear propeller shaft. Refer to Driveshaft, removal & installation.
8. With the cable connected, detach the parking brake cable clamp from the fuel tank cover.
9. Disconnect the fuel filler hose from fuel filler neck.
10. Disconnect the fuel pump connector.
11. Drain the fuel from the fuel tank.
12. Insert a hose of a hand-operated pump into the fuel filler hose and drain the fuel in the fuel filler hose.

➡ Do not force the pump hose into the fuel tank. Store drained fuel in an airtight container to reduce the risk of fire and personal injury.

13. Disconnect the fuel feed pipe and the EVAP canister hose from each pipe.
14. Remove the fuel tank cover.
15. Support the fuel tank with a jack and remove the fuel tank bolts.
16. Disconnect the breather hose from the fuel tank while lowering the fuel tank, and remove the fuel tank from the vehicle.

To install:

17. When connecting quick-connect joints, clean the outside surfaces of the pipe where the joint is to be installed, push the joint into the pipe until the joint lock clicks. Check to ensure that pipes are connected securely, or a fuel leak may occur.

18. If parts have been removed from the fuel tank, install them before installing the fuel tank to the vehicle.
19. Raise the fuel tank with a jack, and connect the breather hose to the fuel tank.
20. Install the fuel tank to the vehicle. Tighten the fuel tank bolt to 34 ft. lbs. (45 Nm).
21. Install the fuel tank cover.
22. Connect the fuel filler hose to the fuel tank, and clamp securely. Tighten the fuel filler hose clamp to 31 inch lbs. (4 Nm).
23. Connect the fuel pump connector.
24. Connect the fuel feed pipe and the EVAP canister hose to each pipe, and clamp them securely.
25. Install the parking brake cable clamp to the fuel tank cover.
26. Install the rear propeller shaft.
27. Install the exhaust center pipe.
28. Connect the negative battery cable. Refer to Battery, removal & installation.
29. With the engine stationary, turn the ignition switch to the ON position and check for fuel leaks.

IDLE SPEED

ADJUSTMENT

The idle speed is controlled by the Engine Control Module (ECM). No adjustment is necessary.

THROTTLE BODY

REMOVAL & INSTALLATION

2.4L Engine

See Figures 158 and 159.

✳✳ WARNING

Never disassemble the throttle body. Disassembly will spoil its original performance. If a faulty condition is found, replace the throttle body with new one as an assembly.

1. Before servicing the vehicle, refer to the Precautions Section.
2. Disconnect the negative battery cable.
3. Drain the engine coolant.
4. Remove the water bypass pipe bolt.
5. Remove the air cleaner outlet hose.
6. Disconnect the engine coolant hoses and the connector from the throttle body assembly.
7. Remove the throttle body assembly and its gasket from the intake manifold.

To install:

8. Clean the mating surfaces and install a new throttle body gasket to the intake manifold.
9. Install the throttle body assembly to the intake manifold.
10. Connect the connector and the coolant hoses to the throttle body assembly.

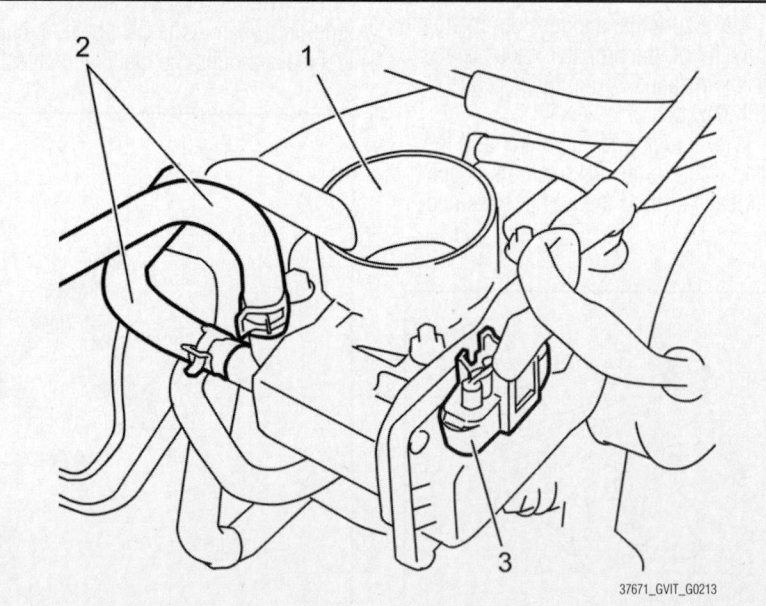

Fig. 158 Disconnect the engine coolant hoses (2) and the connector (3) from the throttle body assembly (1)

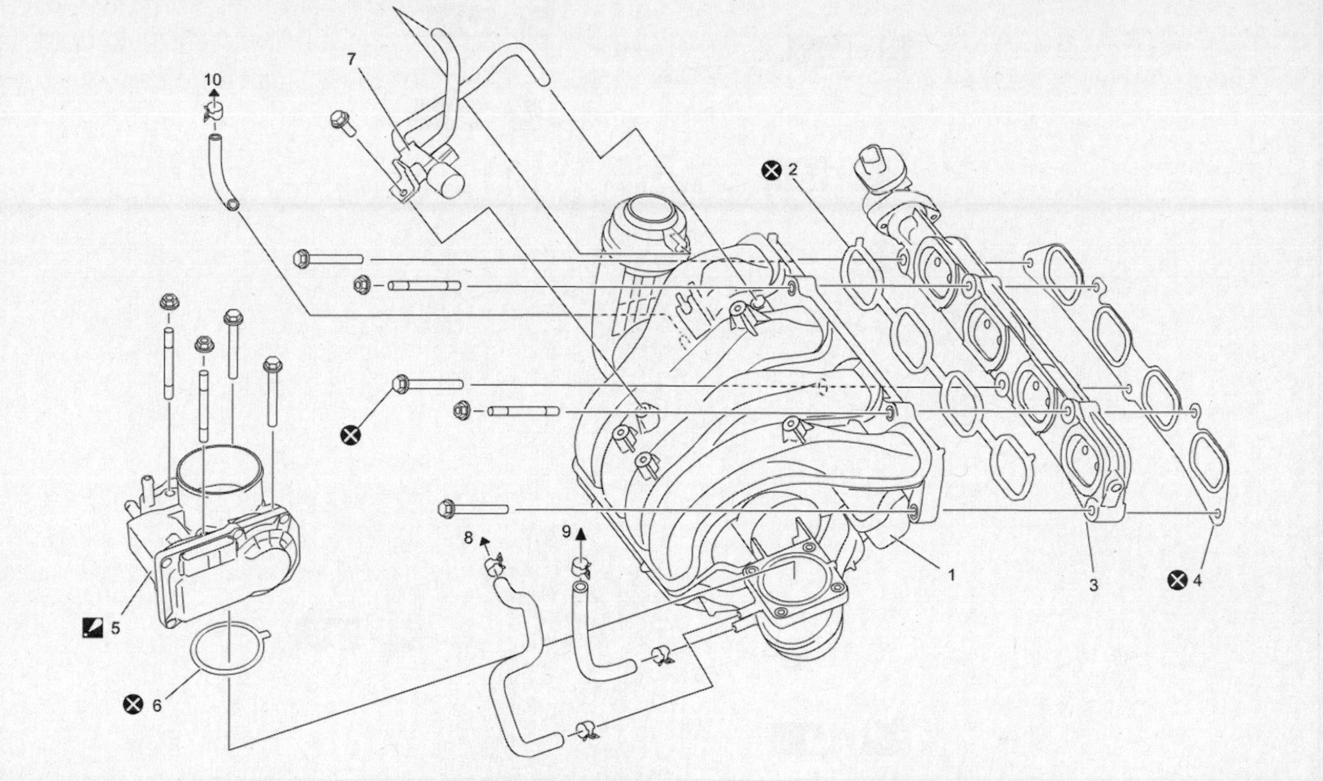

Fig. 159 Exploded view of throttle body and intake manifold—2.4L engine

1.	Intake manifold	7.	IMT vacuum solenoid valve
2.	Intake manifold gasket	8.	To EVAP canister purge valve
3.	IMRC valve	9.	To PCV valve
4.	IMRC valve gasket	10.	To brake booster
5.	Throttle body : Do not disassemble.		Do not reuse.
6.	Throttle body gasket		

37671_GVIT_G0214

11. Install the air cleaner outlet hose.

12. Refill the cooling system with the proper type and amount of fluid.

13. Connect the negative battery cable.

14. Check for coolant leaks.

3.2L Engine

See Figures 160 and 161.

> ❉❉ **WARNING**
>
> **Never disassemble the throttle body. Disassembly will spoil its original performance. If a faulty condition is found, replace the throttle body with new one as an assembly.**

1. Before servicing the vehicle, refer to the Precautions Section.

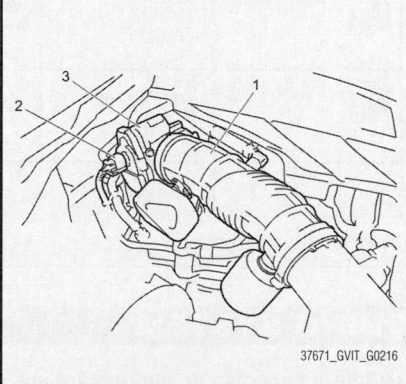

Fig. 160 Disconnect the air cleaner outlet hose (1) and the throttle body connector (2) from the throttle body (3)—3.2L engine

37671_GVIT_G0216

2. Disconnect the negative battery cable.

3. Disconnect the air cleaner outlet hose and the throttle body connector from the throttle body.

4. Remove the throttle body from the intake manifold.

To install:

5. Clean the mating surfaces of the intake manifold and the throttle body.

6. Install a new gasket and the throttle body to the intake manifold.

7. Connect the air cleaner outlet hose and the throttle body connector to the throttle body securely.

8. Connect the negative battery cable.

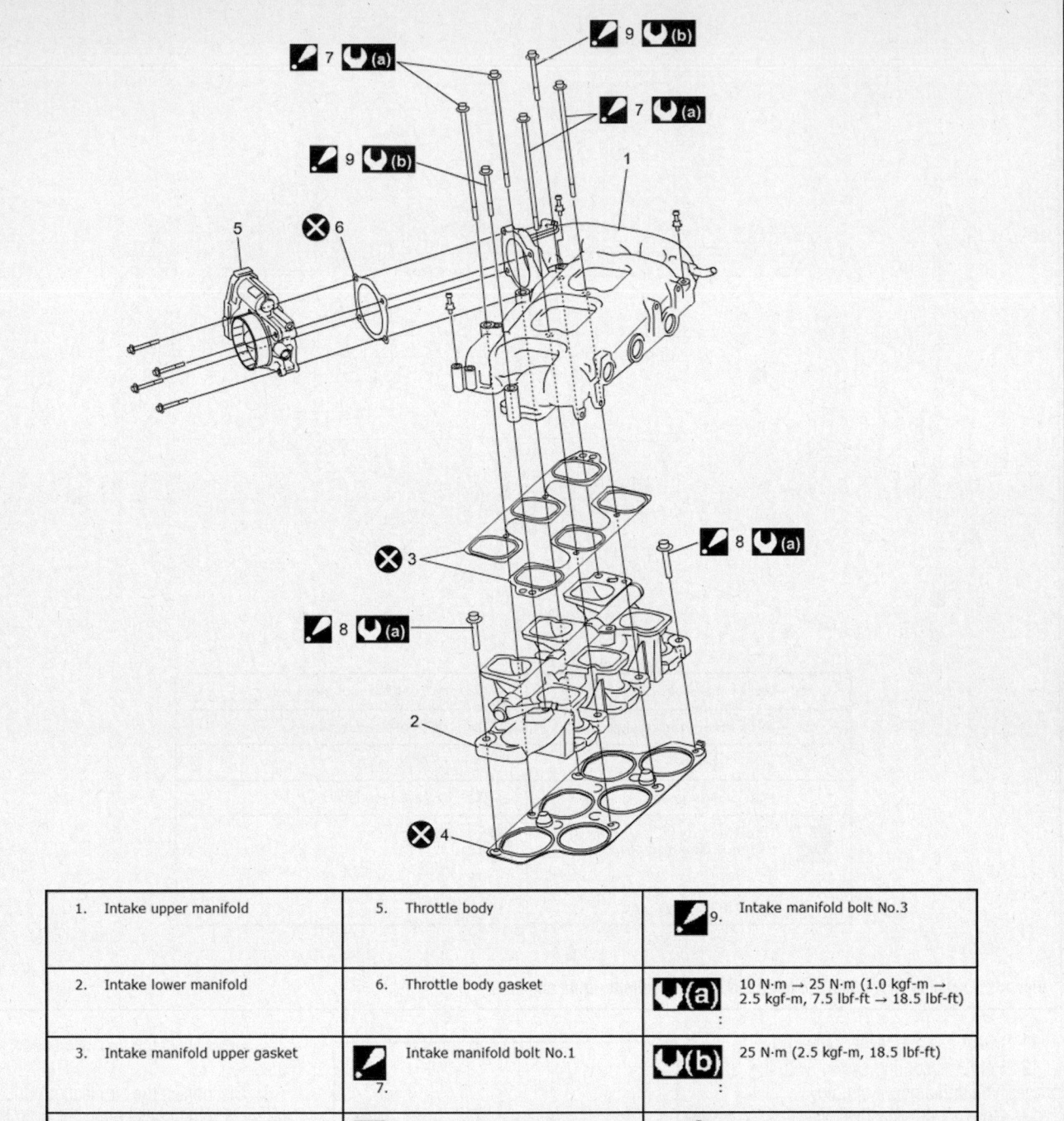

1.	Intake upper manifold	5.	Throttle body		9.	Intake manifold bolt No.3
2.	Intake lower manifold	6.	Throttle body gasket	(a):		10 N·m → 25 N·m (1.0 kgf-m → 2.5 kgf-m, 7.5 lbf-ft → 18.5 lbf-ft)
3.	Intake manifold upper gasket	7.	Intake manifold bolt No.1	(b):		25 N·m (2.5 kgf-m, 18.5 lbf-ft)
4.	Intake manifold lower gasket	8.	Intake manifold bolt No.2	⊗:		Do not reuse.

37671_GVIT_G0217

Fig. 161 Exploded view of throttle body and intake manifold—3.2L engine

HEATING & AIR CONDITIONING SYSTEM

BLOWER MOTOR

REMOVAL & INSTALLATION

See Figure 162.

➡Before servicing or working around the SRS system, turn the ignition switch OFF, disconnect both battery cables and wait at least 3 minutes. When servicing or working around the SRS system, do not work directly in front of the air bag module.

1. Before servicing the vehicle, refer to the Precautions Section.

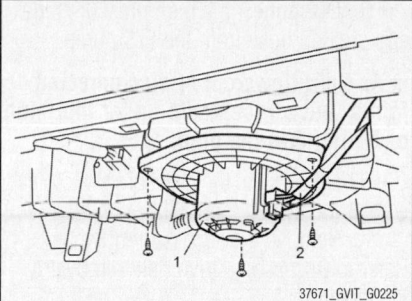

Fig. 162 Disconnect the blower motor lead wire (2) at the coupler and remove the blower motor (1) from the HVAC unit

2. Disconnect the negative battery cable. Disconnect the positive battery cable.

3. Disable the air bag system. Refer to Air Bag (Supplemental Restraint System), disarming the system.

4. Disconnect the blower motor lead wire at the coupler.

5. Remove the blower motor from the HVAC unit.

To install:

6. Installation is the reverse of the removal procedure.

7. Enable air bag system. Refer to Air Bag (Supplemental Restraint System), arming the system.

HEATER CORE

REMOVAL & INSTALLATION

See Figure 163.

1. Before servicing the vehicle, refer to the Precautions Section.

2. Remove the instrument panel.

3. Remove the HVAC unit.

4. Remove the heater core pipe clamp.

5. Pull out the heater core from the HVAC unit.

To install:

6. Installation is the reverse of the removal procedure.

7. When installing the heater core, be careful not to damage the fins of the heater core.

8. When installing each part, be careful not to catch or pinch any wiring harness components.

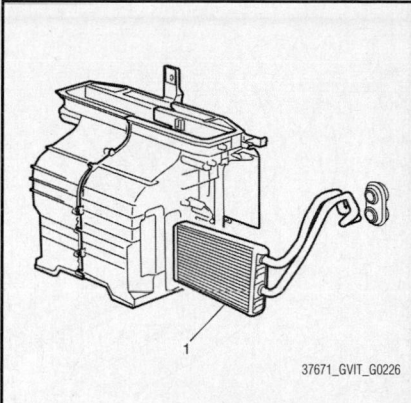

Fig. 163 Pull out the heater core (1) from the HVAC unit

STEERING

POWER RACK & PINION STEERING GEAR

REMOVAL & INSTALLATION

See Figures 164 and 165.

✷✷ WARNING

Never disassemble the Power Steering (P/S) gear case assembly. Disassembling may adversely affect the original performance of the P/S gear case assembly.

✷✷ WARNING

The spiral cable may snap due to steering operation if the steering column is separated from the steering gear assembly. Be sure to secure the steering wheel to avoid turning.

1. Before servicing the vehicle, refer to the Precautions Section.

2. Position the front wheels in the straight ahead position.

3. Disarm the air bag system. Refer to Air Bag (Supplemental Restraint System), Disarming The System.

4. Disconnect the negative battery cable.

5. Drain out the fluid in the Power Steering (P/S) fluid reservoir with a syringe.

6. Raise and safely support the vehicle.

7. Remove both right and left wheels.

8. Disconnect both right and left tie-rod ends from the knuckle.

9. Disconnect the steering lower shaft assembly from the pinion shaft of the P/S gear case assembly.

10. Disconnect the high pressure pipe from the P/S gear case assembly.

11. Disconnect the low pressure pipe from the P/S gear case assembly.

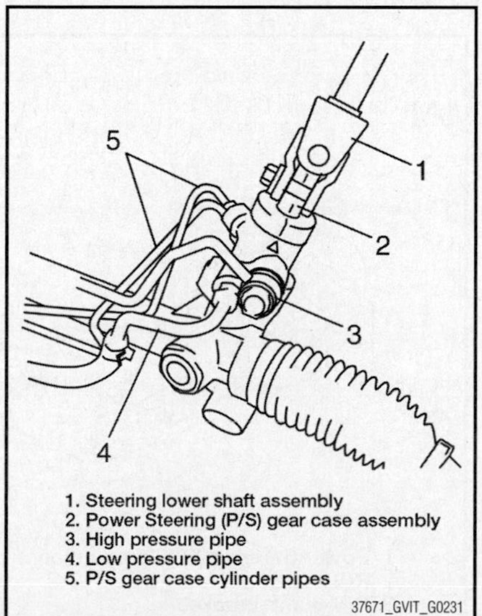

Fig. 164 Remove the power steering gear case cylinder pipes

1. Steering lower shaft assembly
2. Power Steering (P/S) gear case assembly
3. High pressure pipe
4. Low pressure pipe
5. P/S gear case cylinder pipes

12. Remove the P/S gear case cylinder pipes from the P/S gear case assembly.

13. Remove the stabilizer bar mount bolt and the stabilizer joint.

➡**Do not remove the stabilizer bar from the vehicle.**

14. Remove the stabilizer bar mount bracket from the left side of the front suspension frame.

15. Remove the bolts and then take off the P/S gear case assembly from the left side of the vehicle.

➡**The P/S gear case assembly cannot be removed from the right side of the vehicle.**

To install:

16. Installation is the reverse of the removal procedure.

17. After confirming that the front tire is in the straight position, install the P/S gear case to the body temporarily.

18. With the tie-rod end installed to the knuckle, set the rack in position close to neutral. Obtain the neutral state by aligning the match marks on the pinion shaft and the steering gear case and insert the steering lower joint into the pinion shaft.

➡**Be sure to confirm that the steering wheel and front tires (wheels) are in the straight ahead position when inserting the steering lower joint into the steering pinion shaft.**

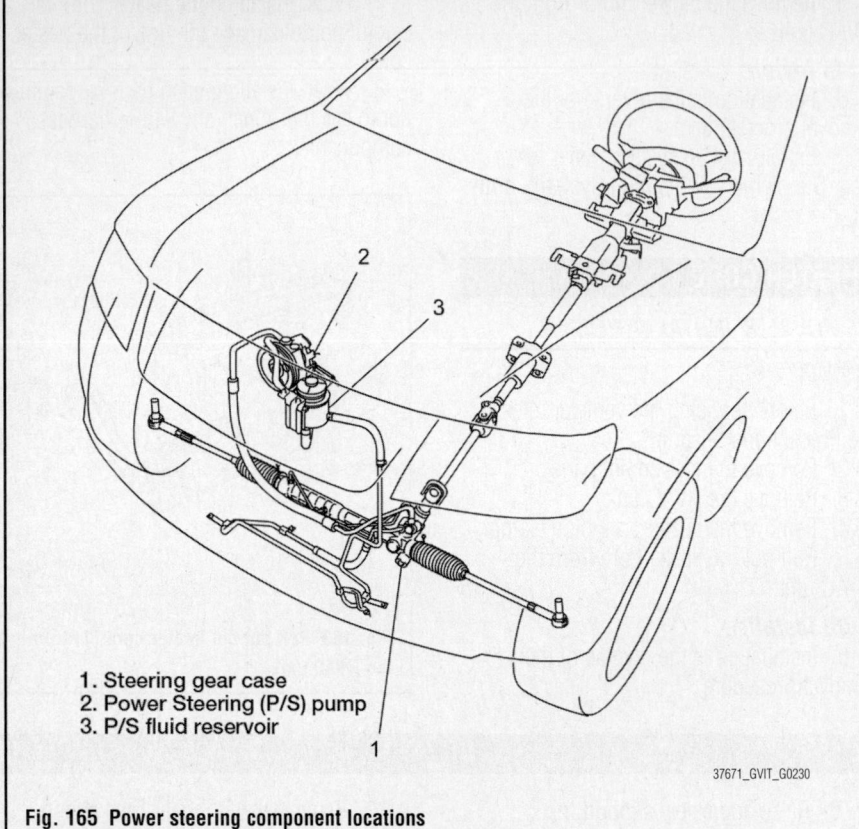

1. Steering gear case
2. Power Steering (P/S) pump
3. P/S fluid reservoir

37671_GVIT_G0230

Fig. 165 Power steering component locations

19. Tighten the steering lower shaft bolt to 18 ft. lbs. (25 Nm).

20. Tighten the gear case high pressure pipe union bolt to 26 ft. lbs. (35 Nm).

21. Tighten the gear case cylinder pipe flare nut to 18 ft. lbs. (25 Nm).

22. Tighten the gear case mounting bolt to 78 ft. lbs. (105 Nm).

23. Tighten the gear case low pressure pipe flare nut to 26 ft. lbs. (35 Nm).

24. Tighten the stabilizer bar mount bracket mount bolt to 45 ft. lbs. (60 Nm).

25. Fill the P/S fluid and bleed the air.

26. Check the toe setting. Adjust as required.

POWER STEERING PUMP

REMOVAL & INSTALLATION

See Figure 166.

❈❈ WARNING

Never disassemble the Power Steering (P/S) pump. Disassembly may spoil its original performance. If a faulty condition is found, replace the P/S pump with a new one.

➡ Be sure to clean each joint of the suction and discharge sides thoroughly before removal.

1. Before servicing the vehicle, refer to the Precautions Section.

2. Disconnect the negative battery cable.

3. Drain out the fluid in the Power Steering (P/S) fluid reservoir with a syringe.

4. For 2.4L engine (J24B model), remove the intake manifold. Refer to Intake Manifold, removal & installation.

5. Disconnect the high pressure pipe and suction hose from the P/S pump.

➡ **As fluid flows out of disconnected joints, put a receptacle under the joints or a plug into the pipe.**

6. Disconnect the pressure switch lead wire at the switch terminal.

7. Remove the accessory drive belt. Refer to Accessory Drive Belt, removal & installation.

8. Remove the P/S pump mounting bolts.

9. Remove the P/S pump.

➡ **Plug each port of the removed pump to prevent dust or any other foreign matter from entering.**

To install:

10. Installation is the reverse of the removal procedure.

11. Fill the power steering fluid and bleed the air.

12. Tighten the P/S pump mounting bolts to 18 ft. lbs. (25 Nm).

13. Tighten the P/S pump high pressure pipe union bolt to 45 ft. lbs. (60 Nm).

BLEEDING

1. Before servicing the vehicle, refer to the Precautions Section.

2. Fill the power steering system with the proper grade and type of steering fluid.

➡ **Do not allow the fluid level in the reservoir tank to go below the MIN level line. Check and add fluid as needed.**

3. Raise and safely support the vehicle.

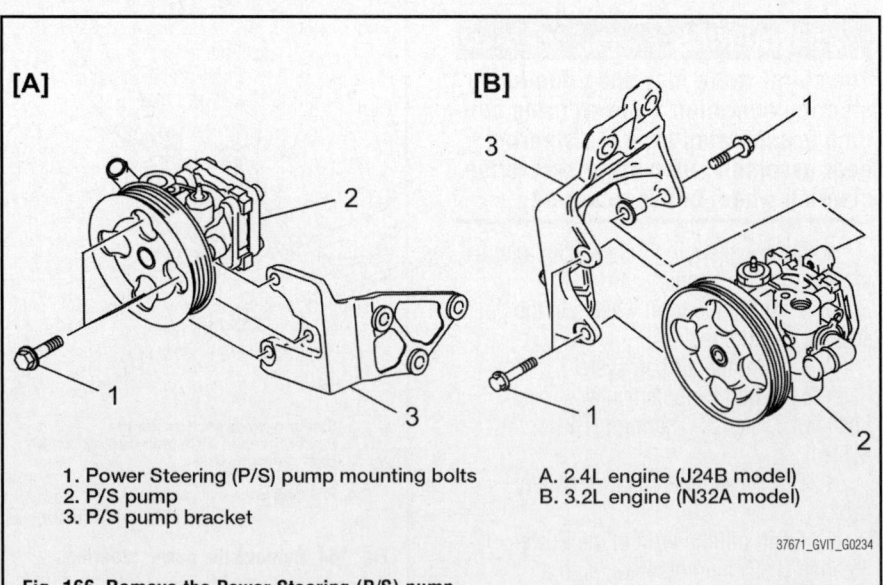

[A]   [B]

1. Power Steering (P/S) pump mounting bolts
2. P/S pump
3. P/S pump bracket

A. 2.4L engine (J24B model)
B. 3.2L engine (N32A model)

37671_GVIT_G0234

Fig. 166 Remove the Power Steering (P/S) pump

4. Quickly turn the steering wheel to the full right and left detents and lightly touch the steering stoppers.

➡**Do not hold the steering wheel in the locked position for more than 10 seconds.**

5. Repeat this operation until the fluid level no longer decreases.
6. Start the engine.
7. Quickly turn the steering wheel to the full right and left detents and lightly touch the steering stoppers.

➡**Do not hold the steering wheel in the locked position for more than 10 seconds.**

8. Check for air bubbles or cloudy fluid. If found, repeat the bleeding procedure.
9. Stop the engine and check the fluid level. Correct as required.

SUSPENSION

CONTROL LINKS

REMOVAL & INSTALLATION

See Figure 167.

1. Before servicing the vehicle, refer to the Precautions Section.
2. Raise and safely support the vehicle.
3. Remove the front wheels.
4. Remove the engine under cover.
5. Remove the suspension control arm.
6. Remove the right side and left side front halfshaft assembly, if equipped.
7. Remove the stabilizer control links (joints). When loosening the joint nut, hold the stud with hexagon wrench.

To install:

8. Installation is the reverse of the removal procedure.

9. Install the stabilizer control links (joints), and tighten the nuts to 44 ft. lbs. (60 Nm).

➡**When tightening, hold the stud with a hexagon wrench.**

LOWER CONTROL ARM

REMOVAL & INSTALLATION

See Figures 168 and 169.

1. Before servicing the vehicle, refer to the Precautions Section.
2. Raise and safely support the vehicle.
3. Remove the wheel.
4. Uncaulk the halfshaft nut.
5. Depress the foot brake pedal and hold it. Remove the halfshaft nut.
6. Disconnect the tie-rod end from the steering knuckle.
7. For headlight auto leveling model, remove the front height sensor.
8. Loosen the ball joint nut.
9. Remove the suspension control arm bolts.
10. Disconnect the halfshaft from the steering knuckle.
11. Remove the steering knuckle from the control arm using a puller and then remove the control arm.

To install:

12. Install the suspension control arm, and then install the suspension control arm bolts. Tighten the

FRONT SUSPENSION

suspension control arm nuts temporarily by hand.

➡**If the suspension control arm bolt and nut are reused, apply engine oil to the threads, bearing, and trunk surface.**

13. For headlight auto leveling model, install the front height sensor.
14. Connect the suspension control arm to the steering knuckle and then tighten a new suspension control arm ball joint nut to 40 ft. lbs. (55 Nm).

✳ **WARNING**

Never reuse the removed suspension control arm ball joint nut.

15. Connect the tie-rod end to the steering knuckle.
16. Depress the foot brake pedal and hold it there. Tighten a new halfshaft nut to 160 ft. lbs. (220 Nm).

✳ **WARNING**

Never reuse the halfshaft nut.

17. Caulk the halfshaft nut.
18. Install the wheel and lower the vehicle. Tighten the wheel nuts to 74 ft. lbs. (100 Nm).
19. Tighten the suspension control arm nuts to 98 ft. lbs. (135 Nm) with the vehicle weight on the suspension.

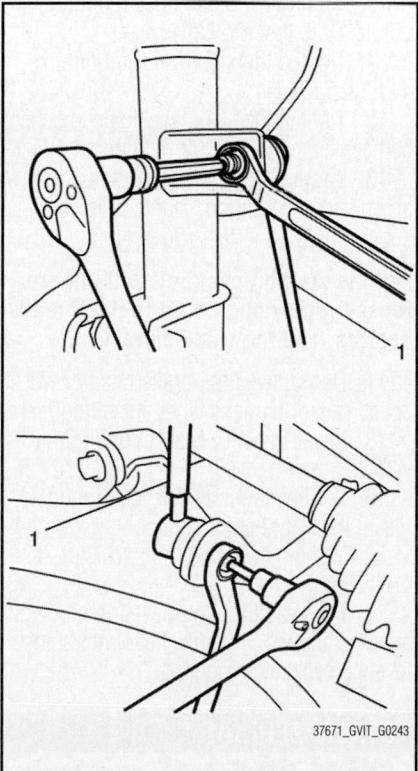

37671_GVIT_G0243

Fig. 167 Removing the stabilizer control links (1)

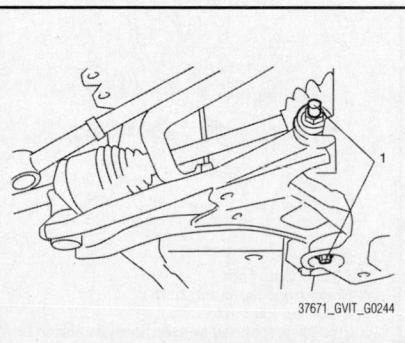

37671_GVIT_G0244

Fig. 168 Remove the suspension control arm bolts (1)

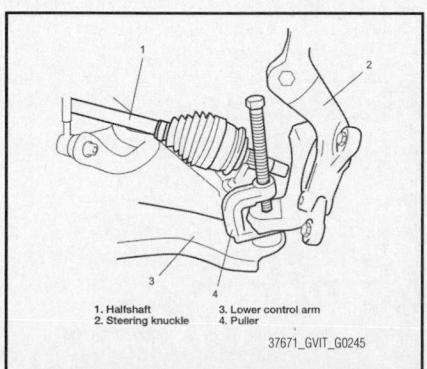

1. Halfshaft
2. Steering knuckle
3. Lower control arm
4. Puller

37671_GVIT_G0245

Fig. 169 Remove the steering knuckle from the lower control arm

➡It is the most desirable to have vehicle off hoist and in non-loaded condition when tightening them.

Tightening torque

Suspension control arm nut:

20. For headlight auto leveling model, initialize the headlight auto leveling system, as needed.

STEERING KNUCKLE

REMOVAL & INSTALLATION

See Figure 170.

1. Before servicing the vehicle, refer to the Precautions Section.
2. Raise and safely support the vehicle.
3. Remove the wheel.
4. Remove the front wheel hub assembly.
5. Disconnect the tie-rod end from the steering knuckle.
6. Remove the front wheel speed sensor from the knuckle.
7. Loosen the ball joint nut.
8. Remove the strut bracket bolts and nuts.
9. Disconnect the halfshaft from the steering knuckle.
10. Remove the steering knuckle from the control arm using a puller.

To install:

11. Installation is the reverse of the removal procedure.
12. Connect the steering knuckle to the suspension control arm.
13. Install the strut bracket bolts and nuts.
14. Tighten the strut bracket nuts to 98 ft. lbs. (135 Nm).
15. Tighten a new suspension control arm ball joint nut to 40 ft. lbs. (55 Nm).

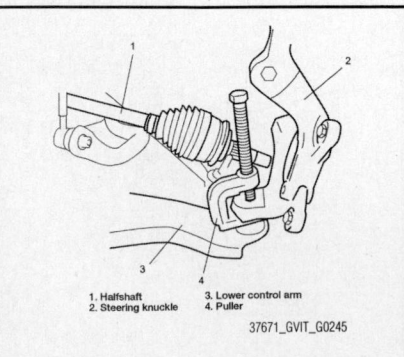

1. Halfshaft
2. Steering knuckle
3. Lower control arm
4. Puller

37671_GVIT_G0245

Fig. 170 Remove the steering knuckle from the lower control arm

❄❄ WARNING
Never reuse the removed suspension control arm ball joint nut.

16. Install the front wheel speed sensor and tighten the front wheel speed sensor bolt to 97 inch lbs. (11 Nm).
17. Connect the tie-rod end to the steering knuckle.
18. Install the front wheel hub assembly and the dust cover to the steering knuckle.
19. Install the front wheel and tighten wheel nuts to 74 ft. lbs. (100 Nm).
20. Check the front wheel alignment and adjust as necessary.

STRUT

REMOVAL & INSTALLATION
See Figure 171.

➡When servicing component parts of the strut assembly, remove the strut rod cap and loosen the strut nut a little before removing the strut assembly. This will make service work easier.

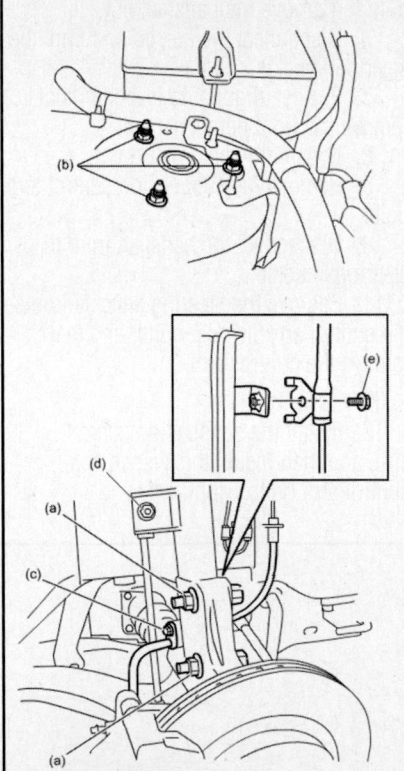

a. Strut bracket nut
b. Strut support nut
c. Brake hose mounting bolt
d. Stabilizer joint nut
e. Front wheel speed sensor harness clamp bolt

37671_GVIT_G0246

Fig. 171 Front strut installation

Note that the nut must not be removed at this point.

1. Before servicing the vehicle, refer to the Precautions Section.
2. Raise the vehicle, allowing the front suspension to hang free.
3. Remove the wheel and disconnect the stabilizer joint from the strut bracket. When loosening the joint nut, hold the stud with a hexagon wrench.
4. For headlight auto leveling model, remove the front height sensor.
5. Remove the front wheel speed sensor harness clamp bolt and then detach the front speed sensor harness from the strut bracket.
6. Remove the brake hose mounting bolt and remove the brake hose from the bracket.
7. Remove the strut bracket nut and the bolt.
8. Remove the strut support nuts.

➡Hold the strut by hand so that it will not fall off.

9. Remove the strut assembly from the vehicle.

To install:

10. Installation is the reverse of the removal procedure.
11. Insert the bolts in the correct direction for the strut installation.
12. Tighten the strut bracket nut to 98 ft. lbs. (135 Nm).
13. Tighten the brake hose mounting bolt to 18 ft. lbs. (25 Nm).
14. Tighten the stabilizer joint nut to 44 ft. lbs. (60 Nm).
15. Tighten the front wheel speed sensor harness clamp bolt to 97 inch lbs. (11 Nm).
16. Lower the vehicle in so that it is in an unloaded condition. Tighten the strut support nuts 37 ft. lbs. (50 Nm).

➡If the strut bracket bolt and nut are reused, apply engine oil to the threads, bearing, and the trunk surface.

17. Do not twist the brake hose or wheel speed sensor harness when installing them.
18. Insert the strut bracket bolt to the vehicle.
19. For headlight auto leveling model, install the front height sensor.
20. Tighten the wheel nuts to 74 ft. lbs. (100 Nm).
21. For headlight auto leveling model, after installation, initialize headlight auto leveling system, as needed.

STABILIZER BAR

REMOVAL & INSTALLATION
See Figures 172 and 173.

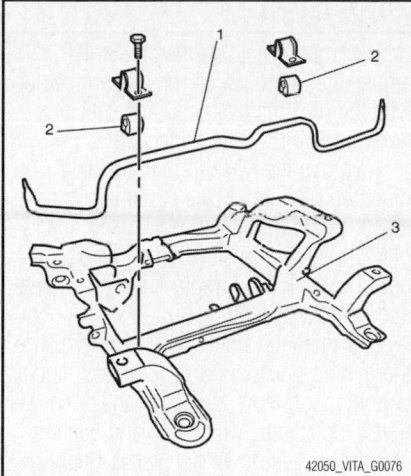

**Fig. 172 Stabilizer bar removal from
sub-frame**

1. Before servicing the vehicle, refer to
the Precautions Section.
2. Raise the vehicle and remove the
wheels.
3. Remove the suspension control arm.
4. Remove the right side and left side
front halfshafts.
5. Remove the control links. When
loosening the joint nut, hold the stud with a
hexagon wrench.
6. Disconnect the steering lower shaft
from the pinion shaft.
7. Detach the low pressure return hose
from the low pressure return pipe and then
disconnect the pipe bracket.
8. Remove the gear box union bolt.
9. Remove the front propeller shaft (if
equipped).
10. Fix the radiator to the body
with a rope to avoid having the radiator fall
when the front suspension trame is lowered.
11. Remove the hood.
12. Support the engine assembly using a
chain hoist or engine support system.
13. Support the suspension frame.
14. Remove the engine front body side
mounting nuts.

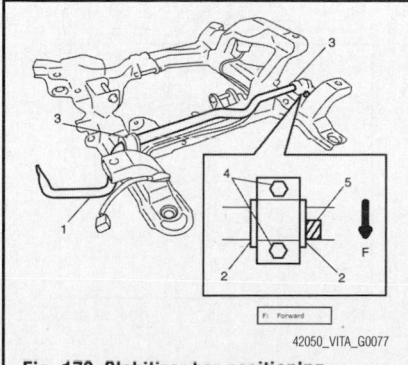

Fig. 173 Stabilizer bar positioning

15. Remove the suspension frame
mounting bolts, and then the lower suspension frame with the stabilizer bar, P/S gear
box assembly, and the front differential
assembly (if equipped).
16. Remove the P/S gear box assembly
and the front differential assembly (if
equipped).
17. Remove the stabilizer bar and bushing from the suspension frame.

To install:

➡**Install the stabilizer bar with the
marking is to the front.**

18. When installing the stabilizer, loosely
assemble all the components while insuring
that the stabilizer is centered, side-to-side.
19. Install the stabilizer bar, stabilizer
bushing, and the stabilizer mounting
bracket to the suspension frame.
20. Tighten the stabilizer bar
mounting bracket bolts to 37 ft. lbs.
(50 Nm).
21. Install the P/S gear box assembly
and the front differential assembly (if
equipped).
22. Install the suspension frame.
23. Tighten the suspension frame
mounting bolts to 98 ft. lbs. (135 Nm) and
the engine front body side mounting nuts to
40 ft. lbs. (55 Nm).
24. Remove the chain hoist or engine
support system from the engine.
25. Install the hood.
26. Install the front propeller shaft (if
equipped).
27. Tighten the pipe bracket bolts to 97
inch lbs. (11 Nm) and then insert the low
pressure return hose to the low pressure
return pipe.
28. Tighten the union gear box bolt to 26
ft. lbs. (35 Nm).
29. Connect the steering lower shaft to
the pinion shaft.
30. Install the control link and tighten
the nuts to 44 ft. lbs. (60 Nm).
31. Install the right and left front halfshaft assemblies (if equipped).
32. Install the suspension control arm
33. Install the engine under cover.
34. Install the wheels and lower vehicle.
Tighten the wheel nuts to 74 ft. lbs. (100 Nm).
35. Fill the power steering fluid and
bleed the system.
36. Check the wheel alignment and
adjust as needed.

WHEEL HUB & BEARING

REMOVAL & INSTALLATION

See Figure 174.

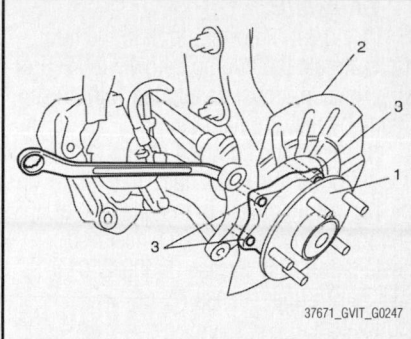

**Fig. 174 Remove the wheel hub housing
bolts (3), wheel hub assembly (1), and
dust cover (2)**

✳✳ WARNING

**Never disassemble the front wheel
hub assembly. Disassembly may
spoil its original function. If a faulty
condition is found, replace the wheel
hub assembly with a new one.**

➡**The end face of one bolt among the
hub bolts is painted as a match mark at
the brake disc installation. Be sure to
paint the end face of the new bolt for
the painted bolt when replacing the hub
bolts with new ones.**

1. Before servicing the vehicle, refer to
the Precautions Section.
2. Raise and safely support the
vehicle.
3. Remove the wheel.
4. Uncaulk the halfshaft nut.
5. Depress the foot brake pedal and
hold it. Remove the halfshaft nut.
6. For 2WD model, remove the front
wheel spindle.
7. Remove the caliper with the
carrier.

➡**Hang the removed caliper with a
wire hook, or equivalent, so as to prevent the brake hose from bending or
twisting or receiving excess tension.
Do not operate brake pedal with the
pads removed.**

8. Pull the brake disc off by using two
8mm bolts.
9. Remove the wheel hub housing
bolts, and then remove the wheel hub
assembly and the dust cover.
10. Remove the hub bolts with a copper
hammer or hydraulic press after putting a
mark by the side of a painted bolt on the
wheel hub.

➡**Never remove a hub bolt unless
replacement is necessary. Be sure to
use a new bolt for replacement.**

To install:

11. Insert a new hub bolt in the hub hole. Rotate the hub bolt slowly to assure that the serrations are aligned with those made by the original bolt.

12. Paint the end face of one bolt beside the mark on the wheel hub.

13. Apply grease to the end face of the inner ring before front wheel hub installation. Use Grease 99000-25011 (SUZUKI Super Grease A), or equivalent.

❈❈ WARNING

Do not apply grease to the encoder section or the encoder may malfunction.

14. Install the wheel hub and dust cover to the steering knuckle.

15. Tighten the wheel hub housing bolts to 37 ft. lbs. (50 Nm).

16. Install the brake disc and brake caliper.

17. For 2WD model, install the front wheel spindle.

18. Depress the foot brake pedal and hold it there. Tighten a new halfshaft nut to 160 ft. lbs. (220 Nm).

❈❈ WARNING

Never reuse the halfshaft nut.

19. Caulk halfshaft nut.

20. Install the front wheel and tighten wheel nuts to 74 ft. lbs. (100 Nm).

ADJUSTMENT

The front wheel bearings are not adjustable. If the lateral run-out on the hub with the disc removed exceeds specification, the hub must be replaced.

1. Move the wheel hub in the axial direction by hand. Make sure there is no looseness of the wheel bearing. Axial end play limit: 0.004 inch (0.1mm) or less. If out of specification, replace the wheel hub and bearing assembly.

2. Rotate the wheel hub and make sure there is no unusual noise or other irregular conditions. If there are any irregular conditions, replace the wheel hub and bearing assembly. Refer to Wheel Hub & Bearing (sealed unit), removal & installation.

SUSPENSION

COIL SPRING

REMOVAL & INSTALLATION

See Figures 175 and 176.

1. Before servicing the vehicle, refer to the Precautions Section.

2. Raise the vehicle, allowing the rear suspension to hang free.

3. Remove the rear wheels.

4. Remove the rear shock absorber

5. Put match marks on the lower arm washer and on the suspension frame to install the bolts correctly in position.

6. Loosen the lower arm mount nut.

7. Remove the lower arm outer bolt.

8. Lower the jack and then remove the rear coil spring and coil spring rubber seat.

To install:

9. Install the coil spring on the lower arm and place the coil spring end onto the lower arm.

➡**The flat end of the coil spring should be upward. The upper end of the coil spring has to be firmly mated to the coil spring rubber seat. The lower end of the coil spring has to be match to the marking. The end of the coil spring**

REAR SUSPENSION

must not interfere with the step of the spring lower seat.

10. Support the lower arm with a jack.

11. Hoist the jack and install the lower arm outer bolt and tighten the bolt temporarily by hand.

12. Align the marks on the lower arm washer and the rear suspension frame, tighten the lower arm mount nut temporarily by hand.

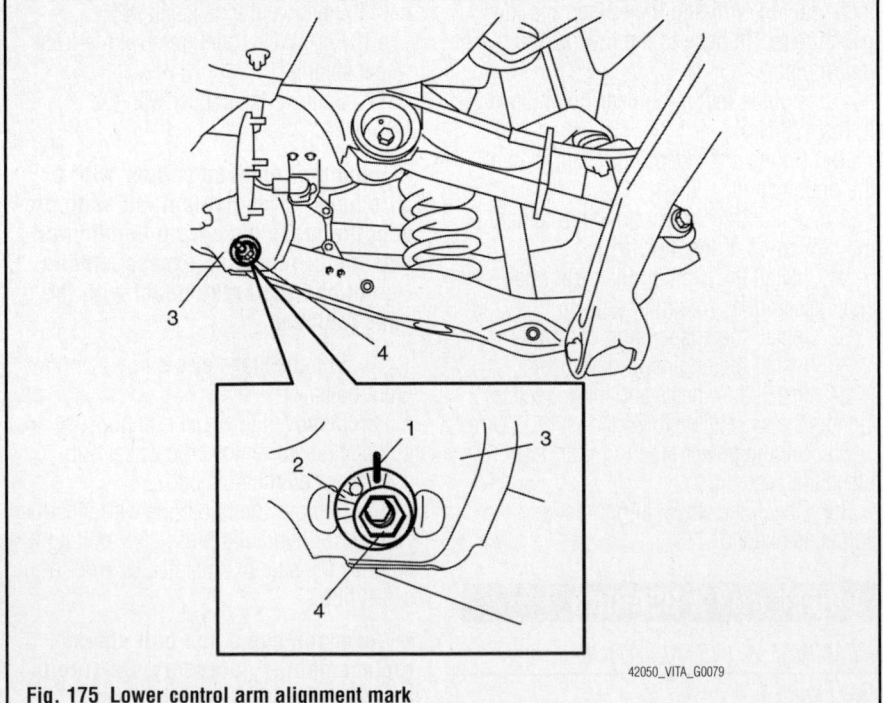

Fig. 175 Lower control arm alignment mark

42050_VITA_G0079

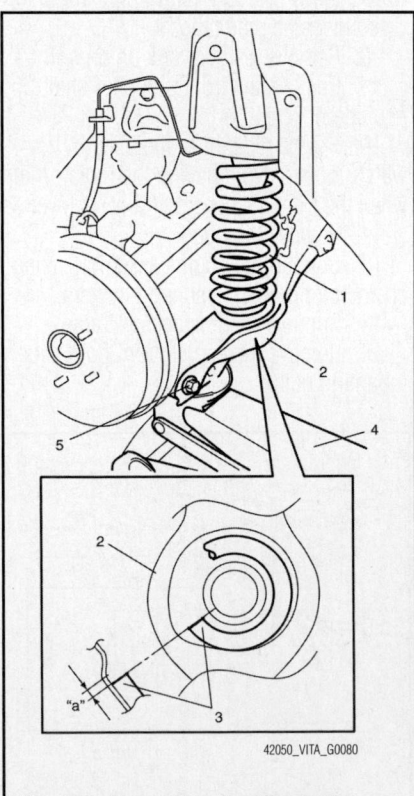

Fig. 176 Rear coil spring alignment

42050_VITA_G0080

13. Install the rear shock absorber
14. Install the wheel with nuts and lower the vehicle.
15. Tighten the wheel nuts to 74 ft. lbs. (100 Nm).
16. Tighten the lower arm outer bolts and the lower arm mount nut to 98 ft. lbs. (135 Nm).

➡**Tighten the lower arm nut and bolt in a non-loaded condition.**

17. Tighten the shock absorber upper bolt to 44 ft. lbs. (60 Nm).
18. Tighten the lower shock absorber bolt to 65 ft. lbs. (90 Nm) with the vehicle weight on the suspension.
19. Check the rear wheel alignment and adjust as necessary

CONTROL ARMS/LINKS

REMOVAL & INSTALLATION

Lower Control Arm
See Figures 177 and 178.

1. Before servicing the vehicle, refer to the Precautions Section.
2. Raise the vehicle and remove the rear wheels.
3. Put match marks on the lower arm washer and on the suspension frame to install the bolts correctly in position.
4. Loosen the lower arm mount nut.
5. Remove the rear coil spring.
6. Remove the suspension rod mount bolt and then the lower arm.

To install:
7. Install the lower arm.
8. Install the lower arm to the rear suspension frame.
9. Insert the suspension lower arm inner bolt to the vehicle rearward.
10. Install the lower arm washer with its graduated part facing up.
11. The marks on the lower arm washer and the rear suspension frame

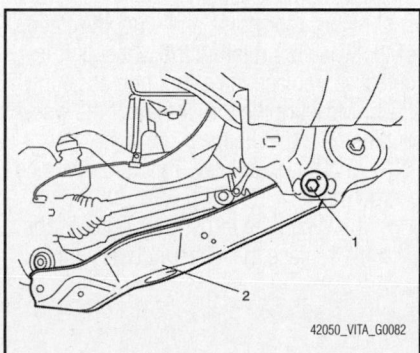

Fig. 177 View of lower control arm

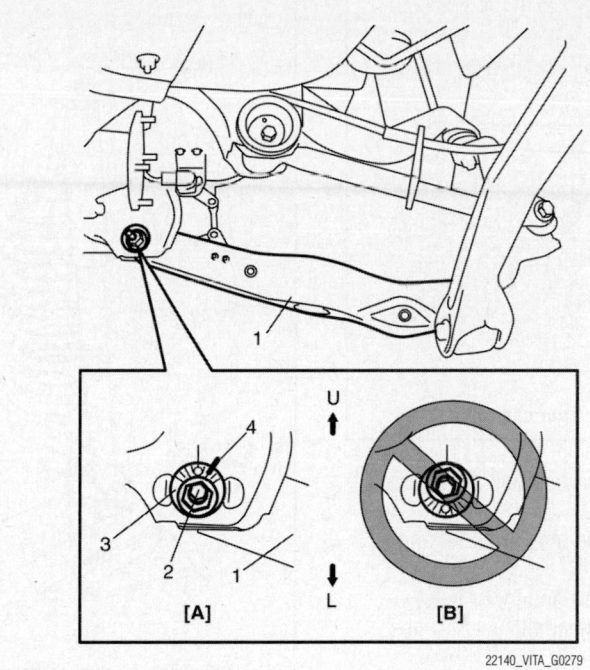

Fig. 178 Lower control arm washer with its graduated part facing up

marked before removal must be aligned. Tighten the bolt and nut temporarily by hand.
12. Install the rear coil spring.
13. Install the wheel with nuts and lower the vehicle.
14. Tighten the lower arm outer bolt, lower arm mount nut, and shock absorber bolts to specified torque with the vehicle weight on the suspension.
- Lower arm outer bolt 98 ft. lbs. (135 Nm).
- Lower arm mount nut 98 ft. lbs. (135 Nm).
- Shock absorber upper bolt 44 ft. lbs. (60 Nm).
- Shock absorber lower bolt 65 ft. lbs. (90 Nm).
15. Check the rear alignment and adjust as necessary.

Upper Control Arm
See Figure 179.

1. Before servicing the vehicle, refer to the Precautions Section.
2. Raise the vehicle and remove the rear wheels.
3. Remove the control rod.
4. Remove the trailing rod.
5. Remove the lower arm.
6. Remove the rear suspension knuckle.
7. Remove the wheel sensor bolts from the upper arm.

8. Remove the upper arm bolts and then the upper arm.

To install:
9. Install the upper arm to the rear suspension frame.
10. Insert the upper arm bolt to the upper arm inside.
11. Tighten the upper arm mount nuts temporarily by hand.
12. Install the rear suspension knuckle
13. Install the trailing rod.
14. Install the control rod.
15. Install the lower arm
16. Install the rear wheels and lower the vehicle.
17. Tighten the bolts and nuts to the specified torque with the vehicle weight on the suspension.
- Wheel sensor bolt: 97 inch lbs. (11 Nm)
- Upper arm mount nut: 98 ft. lbs. (135 Nm)
- Shock absorber upper bolt: 44 ft. lbs. (60 Nm)
- Shock absorber lower bolt: 65 ft. lbs. (90 Nm)
- Lower arm outer bolt: 98 ft. lbs. (135 Nm)
- Lower arm mount nut 98 ft. lbs. (135 Nm).
- Control rod outer bolt 98 ft. lbs. (135 Nm).
- Control rod mount nut 98 ft. lbs. (135 Nm).

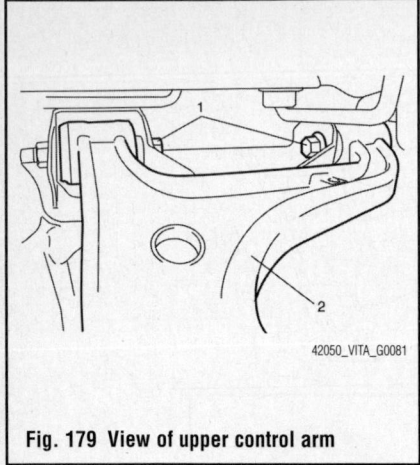

Fig. 179 View of upper control arm

- Trailing rod rear bolt 98 ft. lbs. (135 Nm).
- Trailing rod mount nut 98 ft. lbs. (135 Nm).
- Rear wheels and lower hoist
- Tighten wheel nuts to 74 ft. lbs. (100 Nm).

18. Check rear toe and camber adjust it as necessary.

SHOCK ABSORBER

REMOVAL & INSTALLATION

See Figure 180.

1. Before servicing the vehicle, refer to the Precautions Section.
2. Raise the vehicle and remove the rear wheels.
3. Support the lower arm with a jack and remove the shock absorber bolts.
4. Compress the shock absorber enough to remove it from body.

To install:

5. Installation is the reverse of the removal procedure.
6. Tighten the shock absorber upper bolt to 44 ft. lbs. (60 Nm).
7. Tighten the shock absorber lower bolt to 65 ft. lbs. (90 Nm).

➡**If the shock absorber bolt is reused, apply engine oil to the threads, bearing, and trunk surface. It is the most desirable to have the vehicle off the hoist and in a non-loaded condition when tightening the shock absorber bolts.**

8. Install the rear wheels and tighten the wheel nuts to 74 ft. lbs. (100 Nm).

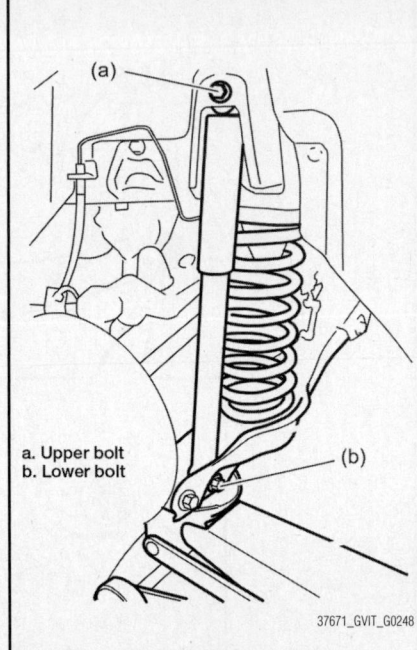

a. Upper bolt
b. Lower bolt

Fig. 180 Rear shock absorber installation

WHEEL HUB & BEARING

REMOVAL & INSTALLATION

See Figure 181.

1. Before servicing the vehicle, refer to the Precautions Section.
2. Raise the vehicle and remove the rear wheel.
3. Uncaulk the rear halfshaft nut.
4. Pull up the parking brake lever fully and remove the rear halfshaft nut.
5. Remove the rear brake disc and brake caliper.
6. Remove the rear wheel hub housing bolts and then remove the rear wheel hub assembly and back plate.

To install:

7. Apply grease to the end face of the inner ring before the rear wheel hub installation. Use Grease 99000-25011 (SUZUKI Super Grease A), or equivalent.

✷✷ WARNING

Do not apply grease to the encoder section in order to avoid an encoder malfunction.

8. Install the rear wheel hub assembly and tighten the rear wheel hub housing bolts to 37 ft. lbs. (50 Nm).

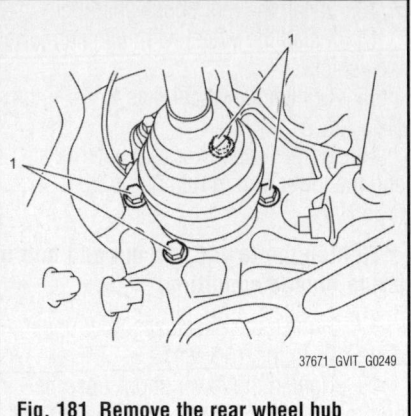

Fig. 181 Remove the rear wheel hub housing bolts (1)

9. Install the rear brake disc and brake caliper.
10. Pull up the parking brake lever fully and tighten a new rear halfshaft nut to 160 ft. lbs. (220 Nm).
11. Caulk the rear halfshaft nut.

✷✷ WARNING

Be careful while caulking the nut so that no crack will occur in the calked part of the nut. A cracked nut must be replaced with a new one.

12. Adjust the parking brake cable, as needed.
13. Install the rear wheel and tighten the wheel nuts to 74 ft. lbs. (100 Nm).
14. Check to ensure that brake drum is free from dragging and proper braking is obtained.
15. Perform a brake test (foot brake and parking brake).

ADJUSTMENT

The wheel bearing does not require maintenance or adjustment. If a growling noise is emitted from the wheel bearing during operation, replace the wheel hub and bearing assembly.

1. Before servicing the vehicle, refer to the Precautions Section.
2. Check for noise and smooth rotation of the wheel by rotating the wheel. If it is defective, replace the wheel hub assembly.
3. Measure the thrust play of the wheel hub using a dial gauge.
   a. The thrust play limit is 0.004 inch (0.1mm).
   b. When the thrust play exceeds the limit, replace the wheel hub assembly.

SPECIFICATIONS AND MAINTENANCE CHARTS

ENGINE AND VEHICLE IDENTIFICATION

			Engine				Model Year	
ID/Code	Liters (cc)	Cu. In.	Cyl.	Fuel Sys.	Engine Type	Eng. Mfg.	Code ①	Year
J	2.0 (1,995)	122	4	MFI	DOHC	Suzuki	9	2009
							A	2010

MFI: Multi-port Fuel Injection

DOHC: Double Overhead Camshafts

① Indicated at the 4th digit of the Vehicle Identification Number (VIN)

37671_SSX4_C0001

GENERAL ENGINE SPECIFICATIONS

All measurements are given in inches.

Year	Model	Engine Displacement Liters	Engine Series ID	Net Horsepower @ rpm	Net Torque @ rpm (ft. lbs.)	Bore x Stroke (in.)	Compression Ratio	Oil Pressure @ rpm
2009	SX4	2.0	J	143@5800	136@3500	3.31 x 3.54	10.5:1	56@4000
2010	SX4	2.0	J	①	140@3500	3.31 x 3.54	10.2:1	44-73@4000

① 150@5800 with manual transmission
148@6000 with CVT transmission

37671_SSX4_C0002

GASOLINE ENGINE TUNE-UP SPECIFICATIONS

Year	Engine Displacement Liters	Engine ID	Spark Plug Gap (in.)	Ignition Timing (deg.) MT	AT	Fuel Pump (psi)	Idle Speed (RPM) MT	AT ①	Valve Clearance (in.) Intake	Exhaust
2009	2.0	J	0.040-0.043	2-8B	2-8B	51-57 ②	680-780	680-780	③	④
2010	2.0	J	0.040-0.043	2-8B	2-8B	51-57 ②	600-700	600-700	⑤	⑥

NOTE: The Vehicle Emission Control Information label often reflects specification changes made during production.

The label figures must be used if they differ from those in this chart.

B: Before top dead center

① Automatic Transmission (AT) in Neutral

② At idle

③ 0.007-0.009 inch (cold)
0.008-0.011 inch (hot)

④ 0.011-0.013 inch (cold)
0.012-0.014 inch (hot)

⑤ 0.0063-0.0094 inch (cold)

⑥ 0.0123-0.0153 inch (cold)

37671_SSX4_C0003

CAPACITIES

Year	Model	Engine Displacement Liters	Engine ID	Engine Oil with Filter (qts.)	Transmission (pts.) Manual	Transmission (pts.) Auto. ①	Transfer Case (pts.)	Drive Axle Front (pts.)	Drive Axle Rear (pts.)	Fuel Tank (gal.)	Cooling System (qts.)
2009	SX4	2.0	J	5.1	5.3	9.6	1.2	NA	1.2-1.9	②	③
2010	SX4	2.0	J	5.3	5.3	17.6	1.2	NA	1.2-1.9	②	④

NA: Not Applicable

NOTE: All capacities are approximate. Add fluid gradually and check to be sure a proper fluid level is obtained.

① Drain and refill

② FWD: 13.2 gallons; AWD: 11.9 gallons

③ With manual transaxle: 6.3 qts. (including coolant reservoir)
  With automatic transaxle: 6.7 qts. (including coolant reservoir)

④ With manual transaxle: 7.5 qts. (including coolant reservoir)
  With CVT transaxle: 7.4 qts. (including coolant reservoir)

37671_SSX4_C0004

FLUID SPECIFICATIONS

Year	Model	Engine Displacement Liters	Engine ID	Engine Oil	Manual Trans.	Auto. Trans.	Transfer Case	Drive Axle Front	Drive Axle Rear	Power Steering Fluid	Brake Master Cylinder	Cooling System
2009	SX4	2.0	J	5W-30	①	②	③	NA	④	⑤	⑥	⑦
2010	SX4	2.0	J	5W-30	①	⑧	③	NA	④	⑤	⑥	⑦

NA: Not Applicable

① Suzuki Gear Oil 99000-22B21-036, API GL-4, Viscosity SAE 75W-80

② Dexron-III or equivalent

③ API GL-5 gear oil, Viscosity SAE 80W-90

④ Hypoid gear oil, API GL-5 gear oil, Viscosity SAE 80W-90

⑤ ATF Dexron-II (ESSO JWS2326) or ATF Dexron-III or equivalent

⑥ Refer to fluid specification indicated on the brake reservoir cap of the vehicle

⑦ Suzuki super long life coolant, or equivalent

⑧ Suzuki CVT FLUID GREEN 1 or Shell GREEN-1 V

37671_SSX4_C0005

VALVE SPECIFICATIONS

Year	Engine Displacement Liters	Engine ID	Seat Angle (deg.)	Face Angle (deg.)	Spring Test Pressure (lbs. @ in.)	Spring Installed Height (in.)	Stem-to-Guide Clearance (in.) Intake	Stem-to-Guide Clearance (in.) Exhaust	Stem Diameter (in.) Intake	Stem Diameter (in.) Exhaust
2009	2.0	J	①	45	46.1-53.1 @1.480	②	0.0008-0.0018	0.0017-0.0028	0.2348-0.2354	0.2339-0.2344
2010	2.0	J	③	45	④	⑤	0.0008-0.0018	0.0018-0.0028	0.2152-0.2157	0.2142-0.2147

① Intake valve seat: First cut at 25 degrees, 2nd cut at 60 degrees, 3rd cut at 45 degrees. The 3rd cut must produce the desired seat width (0.0389-0.0551 inch).
  Exhaust valve seat: First cut at 15 degrees, 2nd cut at 45 degrees. The 2nd cut must produce the desired seat width (0.0389-0.0551 inch).

② Valve spring free length Standard: 2.013 inches, Limit: 1.974 inches

③ Intake valve seat: First cut at 22 degrees, 2nd cut at 60 degrees, 3rd cut at 45 degrees. The 3rd cut must produce the desired seat width (0.0414-0.0531 inch).
  Exhaust valve seat: First cut at 22 degrees, 2nd cut at 65 degrees, 3rd cut at 45 degrees. The 3rd cut must produce the desired seat width (0.0441-0.0559 inch).

④ Valve spring preload when compressed to 1.61 inches Standard: 38.3-44.0 lbs., Limit: 37.1 lbs.
  Valve spring preload when compressed to 1.23 inches Standard: 85.2-98.2 lbs., Limit: 84.3 lbs.

⑤ Valve spring free length Standard: 2.021 inches, Limit: 1.981 inches

37671_SSX4_C0006

CAMSHAFT AND BEARING SPECIFICATIONS CHART
All measurements are given in inches.

Year	Engine Displ. Liters	Engine ID	Journal Dia.	Brg. Oil Clearance	Shaft End-play	Runout	Journal Bore	Lobe Lift Intake	Lobe Lift Exhaust
2009	2.0	J	1.0221-1.0228	0.0008-0.0029	NS	0.0012	1.0236-1.0249	1.7980-1.8043	1.7933-1.7996
2010	2.0	J	①	②	NS	0.0005	③	1.8513-1.8576	1.8324-1.8387

NS: Not Specified

① Number 1 intake side: 1.1402-1.1409 inches
　Number 1 exhaust side: 1.0614-1.0622 inches
　All others: 1.0220-1.0228 inches

② Number 1 intake side Standard: 0.00079-0.00283 inch, Limit: 0.0039 inch
　All others Standard: 0.00079-0.00244 inch, Limit: 0.00374 inch

③ Number 1 intake side: 1.2599-1.2608 inches
　Number 1 exhaust side: 1.0630-1.0638 inches
　All others: 1.0237-1.0244 inches

37671_SSX4_C0007

CRANKSHAFT AND CONNECTING ROD SPECIFICATIONS
All measurements are given in inches.

Year	Engine Displacement Liters	Engine ID	Crankshaft Main Brg. Journal Dia.	Crankshaft Main Brg. Oil Clearance	Crankshaft Shaft End-play	Crankshaft Thrust on No.	Connecting Rod Journal Diameter	Connecting Rod Oil Clearance	Connecting Rod Side Clearance
2009	2.0	J	①	0.00079-0.00149	0.0039-0.0137	3	②	0.0018-0.0025	0.0099-0.0150
2010	2.0	J	③	0.00087-0.00133	0.0040-0.0137	3	④	0.0018-0.0024	0.0099-0.0157

① There are 6 different grades, ranging from 4 (2.28358-2.28369 inches) to 9 (2.28299-2.28310 inches)

② There are 3 different grades, ranging from A (1.9683-1.9685 inches) to C (1.9677-1.9680 inches)

③ There are 6 different grades:
　Number 4 (2.04713-2.04724 inches), Number 5 (2.04701-2.04712 inches)
　Number 6 (2.04689-2.04700 inches), Number 7 (2.04678-2.04688 inches)
　Number 8 (2.04666-2.04677 inches), Number 9 (2.04654-2.04665 inches)

④ There are 3 different grades, ranging from A (1.96827-1.96850 inches) to C (1.96780-1.96802 inches)

37671_SSX4_C0008

PISTON AND RING SPECIFICATIONS
All measurements are given in inches.

Year	Engine Displ. Liters	Engine ID	Piston Clearance	Ring Gap Top Compression	Ring Gap Bottom Compression	Ring Gap Oil Control	Ring Side Clearance Top Compression	Ring Side Clearance Bottom Compression	Ring Side Clearance Oil Control
2009	2.0	J	0.0008-0.0015	0.0079-0.0129	0.0129-0.0188	0.0079-0.0196	0.0012-0.0027	0.0008-0.0023	0.0024-0.0059
2010	2.0	J	①	0.0079-0.0129	0.0126-0.0188	0.0079-0.0196	0.0012-0.0027	0.0008-0.0023	0.0012-0.0066

① Piston clearance limit: 0.0047 inch

37671_SSX4_C0009

TORQUE SPECIFICATIONS
All readings in ft. lbs.

Year	Engine Displacement Liters	Engine ID	Cylinder Head Bolts	Main Bearing Bolts	Rod Bearing Bolts	Crankshaft Damper Bolts	Flywheel Bolts	Manifold Intake	Manifold Exhaust	Spark Plugs	Oil Pan Drain Plug
2009	2.0	J	①	②	③	109	51	NS	37	18	26
2010	2.0	J	④	⑤	③	111	52	NS	37	18	26

NS: Not Specified

① Apply engine oil to bolts, refer to procedure for tightening sequence

    Step 1: M10 bolts to 38 ft. lbs.

    Step 2: M10 bolts to 60 ft. lbs.

    Step 3: M10 bolts to 0 ft. lbs.

    Step 4: M10 bolts to 38 ft. lbs.

    Step 5: M10 bolts to 75 ft. lbs.

    Step 6: M6 bolts to 97 inch lbs.

② Apply engine oil to bolts, refer to procedure for tightening sequence

    Step 1: Bolts 1-10 to 29 ft. lbs.

    Step 2: Bolts 1-10 to 0 ft. lbs.

    Step 3: Bolts 1-10 to 29 ft. lbs.

    Step 4: Bolts 1-10 to 42 ft. lbs.

    Step 5: Bolts 11-22 to 19 ft. lbs.

③ Apply engine oil to bolts, refer to procedure for tightening sequence

    Step 1: 11 ft. lbs.

    Step 2: Plus 45 degrees

    Step 3: Plus 45 degrees

④ Refer to procedure for tightening sequence

    Step 1: Engine oil on bolts 1-10 (No. 1's), tighten to 15 ft. lbs.

    Step 2: Bolts 1-10 to 30 ft. lbs.

    Step 3: Bolts 1-10 plus 60 degrees

    Step 4: Bolts 1-10 plus 80 degrees

    Step 5: No. 2 bolt to 19 ft. lbs.

⑤ Refer to procedure for tightening sequence

    Step 1: Engine oil on bolts 1-10 (No. 1's), tighten to 37 ft. lbs.

    Step 2: Bolts 1-10 to 0 ft. lbs.

    Step 3: Bolts 1-10 to 15 ft. lbs.

    Step 4: Bolts 1-10 to 26 ft. lbs.

    Step 5: Bolts 1-10 plus 40 degrees

    Step 6: Bolts 1-10 plus 40 degrees

    Step 7: Bolts 1-12 (No. 2's) to 18 ft. lbs.

37671_SSX4_C0010

WHEEL ALIGNMENT

Year	Model		Caster Range (+/-Deg.)	Caster Preferred Setting (Deg.)	Camber Range (+/-Deg.)	Camber Preferred Setting (Deg.)	Toe-in (Deg.)
2009	SX4	F	①	①	②	②	0.04+/-0.04
		R	NA	NA	1.00	-1.00	0.12+/-0.12
2010	SX4	F	③	③	④	④	0.04+/-0.04
		R	NA	NA	1.00	-1.00	0.20+/-0.20

NOTE: Measurements given for an unladen vehicle with fuel, coolant, and engine oil full; spare tire, jack, hand tools, and mats in designated positions.

F: Front   NA: Not Applicable

R: Rear

① Sedan model (15 inch wheels) and hatchback model: 3 degrees 40' +/- 2 degrees

    Sedan model (17 inch wheels): 3 degrees 46' +/- 2 degrees

② Sedan model (15 inch wheels) and hatchback model: -10' +/- 1 degree

    Sedan model (17 inch wheels): -17' +/- 1 degree

③ Other than 17 inch wheel model: 3 degrees 40' +/- 2 degrees

    17 inch wheel model: 3 degrees 46' +/- 2 degrees

④ Other than 17 inch wheel model: -10' +/- 1 degree

    17 inch wheel model: -17' +/- 1 degree

37671_SSX4_C0011

TIRE, WHEEL AND BALL JOINT SPECIFICATIONS

Year	Model	OEM Tires Standard	Optional	Tire Pressures (psi) Front	Rear	Wheel Size	Ball Joint Inspection	Lug Nut Torque (ft. lbs.)
2009	SX4 Hatchback	P195/65R15 89H	NS	①	①	15 x 6J	②	62
		P205/60R16 91H		①	①	16 x 6J		
		P205/60R16 92H		①	①	16 x 6J		
	SX4 Sedan	P195/65R15 89H	NS	①	①	15 x 6J	②	62
		P195/65R15 91H		①	①	15 x 6J		
		P205/50R17 88V		①	①	17 x 6.5J		
2010	SX4 Hatchback	P195/65R15 89H	NS	①	①	15 x 6J	②	62
		P205/60R16 91H		①	①	16 x 6J		
		P205/60R16 92H		①	①	16 x 6J		
		P205/50R17 88V		①	①	17 x 6.5J		
	SX4 Sedan	P195/65R15 89H	NS	①	①	15 x 6J	②	62
		P205/50R17 88V		①	①	17 x 6.5J		
		P205/50R17 89V		①	①	17 x 6.5J		

OEM: Original Equipment Manufacturer

PSI: Pounds Per Square Inch

NS: Not Specified

① See the tire placard on the vehicle

② Check for smooth rotation of the ball stud and for any damage to the ball stud or dust cover. The suspension arm and ball joint cannot be separated.

   If there is any damage to either part, the suspension arm must be replaced as a complete unit.

37671_SSX4_C0012

BRAKE SPECIFICATIONS
All measurements in inches unless noted

Year	Model		Brake Disc Original Thickness	Minimum Thickness	Maximum Runout	Brake Drum Diameter Original Inside Diameter	Max. Wear Limit	Maximum Machine Diameter	Minimum Lining Thickness	Brake Caliper Carrier Bolts (ft. lbs.)	Guide Pin Bolts (ft. lbs.)
2009	SX4	F	0.870	0.790	0.004	NA	NA	NA	0.080	62	19
		R	0.354	0.315	0.004	8.66	8.74	NS	①	44	17
2010	SX4	F	0.870	0.790	0.004	NA	NA	NA	0.080	62	19
		R	0.354	0.315	0.004	8.66	8.74	NS	①	44	17

F: Front   NA: Not Applicable

R: Rear   NS: Not Specified

① Rear drum brake shoe lining thickness Standard: 0.157 inch, Limit: 0.040 inch

   Rear disc brake pad lining thickness Standard: 0.354 inch, Limit: 0.039 inch

37671_SSX4_C0013

MAINTENANCE SERVICES
2009-10 SUZUKI SX4

TO BE SERVICED	TYPE OF SERVICE	7.5	15	22.5	30	37.5	45	52.5	60	67.5	75	82.5	90	97.5	105	112.5	120	127.5	135	142.5	150
Engine and Emission Control																					
Drive belt (tension, damage)	S/I				✓				✓				✓				✓				✓
Valve lash (clearance) 2009 models	S/I				✓				✓				✓				✓				✓
Valve lash (clearance) 2010 models	S/I				✓				✓				✓				✓				
Engine oil and filter	R	✓	✓	✓	✓	✓	✓	✓	✓	✓	✓	✓	✓	✓	✓	✓	✓	✓	✓	✓	✓
Cooling system, hoses and connections (leakage, damage)	S/I				✓				✓				✓				✓				✓
Engine coolant (Suzuki long life standard: green color)	R				✓				✓				✓				✓				✓
Engine coolant (Suzuki super long life: blue color)	R	First time only: Replace at 90,000 miles or 90 months. Second time and after: Replace every 45,000 miles or 45 months.																			
Exhaust system	S/I				✓				✓				✓				✓				✓
EVAP canister (EVAP leak check model)	R	Replace every 150,000 or 180 months																			
Emission-related hoses and tubes	S/I																✓				✓
Ignition System																					
Spark Plugs (Nickel type)	R				✓				✓				✓				✓				✓
Spark Plugs (Iridium type)	R								✓								✓				
Ignition coil (plug cap)	S/I				✓				✓				✓				✓				✓
Fuel System																					
Engine air cleaner (paved road usage)	R				✓				✓				✓				✓				✓
Fuel tank and cap (leakage, damage)	S/I		✓		✓		✓		✓		✓		✓		✓		✓		✓		✓
Fuel filter	R																				✓
Fuel lines and connections	S/I				✓				✓				✓				✓				✓
EVAP canister air suction filter (EVAP leak check model)	R	Replace every 150,000 or 180 months																			
Brake System																					
Brake pads and rotors (wear, damage)	S/I	✓		✓		✓		✓		✓		✓		✓		✓		✓		✓	
Brake drums and shoes (wear, damage)	S/I	✓		✓		✓		✓		✓		✓		✓		✓		✓		✓	
Brake hoses and pipes (leakage, damage)	S/I	✓		✓		✓		✓		✓		✓		✓		✓		✓		✓	
Brake fluid	S/I		✓		✓		✓		✓		✓		✓		✓		✓		✓		✓
Brake fluid (replace)	R								✓								✓				
Parking brake lever and cable (damage, stroke, operation)	S/I	✓		✓		✓		✓		✓		✓		✓		✓		✓		✓	
Brake pedal	S/I		✓		✓		✓		✓		✓		✓		✓		✓		✓		✓
Chassis and Body																					
Clutch pedal / fluid (M/T)	S/I		✓		✓		✓		✓		✓		✓		✓		✓		✓		✓
Tires / wheel discs (wear, damage, rotation)	S/I	✓	✓	✓	✓	✓	✓	✓	✓	✓	✓	✓	✓	✓	✓	✓	✓	✓	✓	✓	✓
Drive shaft boots and propeller shaft (4WD)	S/I	✓	✓	✓	✓	✓	✓	✓	✓	✓	✓	✓	✓	✓	✓	✓	✓	✓	✓	✓	✓
Suspension system (tightness, damage, rattle, breakage)	S/I	✓	✓	✓	✓	✓	✓	✓	✓	✓	✓	✓	✓	✓	✓	✓	✓	✓	✓	✓	✓
Steering system (tightness, damage, breakage, rattle)	S/I	✓	✓	✓	✓	✓	✓	✓	✓	✓	✓	✓	✓	✓	✓	✓	✓	✓	✓	✓	✓
Manual transaxle oil (leakage, level) 2009 models	S/I	✓						✓						✓						✓	
Manual transaxle oil (replace) 2009 models	R				✓				✓				✓				✓				✓
Manual transaxle oil (leakage, level) 2010 models	S/I	✓			✓		✓		✓		✓		✓		✓		✓		✓		
Manual transaxle oil (replace) 2010 models	R												✓								
Automatic Transmission (A/T) fluid level (2009 models)	S/I		✓		✓		✓		✓		✓		✓		✓		✓		✓		✓
A/T fluid change (2009 models)	R	Replace every 105,000 miles																			
A/T fluid hose (2009 models)	S/I					✓							✓					✓			
Continuously Variable Transaxle (CVT) fluid level and deterioration (2010 models)	S/I		✓		✓		✓		✓		✓		✓		✓		✓		✓		✓
CVT fluid hose (2010 models)	S/I					✓							✓					✓			
Transfer oil (4WD) (leakage, level)	S/I	✓					✓				✓				✓				✓		
Transfer oil (4WD) (replace)	R								✓								✓				
Rear differential oil (4WD) (leakage, level)	S/I				✓				✓				✓				✓				✓
Rear differential oil (4WD) (replace)	R	✓																			
All latches, hinges and locks	S/I	✓	✓	✓	✓	✓	✓	✓	✓	✓	✓	✓	✓	✓	✓	✓	✓	✓	✓	✓	✓
Power steering system (Hydraulic type)	S/I	✓	✓	✓	✓	✓	✓	✓	✓	✓	✓	✓	✓	✓	✓	✓	✓	✓	✓	✓	✓
Cabin air filter	S/I		✓			✓							✓					✓			
Cabin air filter (replace)	R				✓				✓				✓				✓				✓

R: Replace

S/I: Inspect and service, if necessary

FREQUENT OPERATION MAINTENANCE (SEVERE SERVICE)

If a vehicle is operated under any of the following conditions, it is considered severe service:

- Extremely dusty areas
- 50% or more of the vehicle operation is in 90°F (32°C) or higher temperatures, or constant operation in temperatures below 32°F (0°C)
- Prolonged idling (vehicle operation in stop and go traffic)
- Frequent short running periods (engine does not warm to normal operating temperatures)
- Police, taxi, delivery usage, or trailer towing usage
- Driving in hilly or mountainous terrain

Oil & oil filter: replace every 3,000 miles

Cooling system, hoses and connections (leakage, damage): service or inspect every 15,000 miles

Exhaust system: service or inspect every 15,000 miles

Engine air cleaner: service or inspect every 3,000 miles

Fuel filter: replace every 150,000 miles or 150 months

EVAP canister air suction filter (EVAP leak check model): replace every 75,000 miles

Brake pads and rotors (wear, damage): service or inspect initially at 6,000 miles and 12,000 miles, then every 12,000 miles thereafter

Brake drums and shoes (wear, damage): service or inspect initially at 6,000 miles and 12,000 miles, then every 12,000 miles thereafter

Brake hoses and pipes (leakage, damage): service or inspect initially at 6,000 miles and 12,000 miles, then every 12,000 miles thereafter

Parking brake lever and cable (damage, stroke, operation): service or inspect initially at 6,000 miles and 15,000 miles, then every 15,000 miles thereafter

Tires / wheel discs (wear, damage, rotation): service or inspect every 6,000 miles

Wheel bearings (looseness, wear, noise, damage): service or inspect every 12,000 miles

Drive shaft boots and propeller shaft (4WD): service or inspect every 6,000 miles

Suspension system (tightness, damage, rattle, breakage): service or inspect every 6,000 miles

Steering system (tightness, damage, breakage, rattle): service or inspect every 6,000 miles

Manual transaxle oil (2009 models): replace every 15,000 miles

Manual transaxle oil (2010 models): service or inspect every 15,000 miles, replace every 60,000 miles

Automatic transaxle fluid (2009 models): replace every 48,000 miles

Transfer oil (4WD): replace every 15,000 miles

Rear differential oil (4WD): replace every 15,000 miles

All latches, hinges and locks: service or inspect every 6,000 miles

Power steering system (Hydraulic type): service or inspect every 6,000 miles

Cabin air filter: service or inspect every 6,000 miles

BRAKES   INFORMATION AND PRECAUTIONS

ANTI-LOCK SYSTEMS

• Certain components within the ABS system are not intended to be serviced or repaired individually.

• Do not use rubber hoses or other parts not specifically specified for and ABS system. When using repair kits, replace all parts included in the kit. Partial or incorrect repair may lead to functional problems and require the replacement of components.

• Lubricate rubber parts with clean, fresh brake fluid to ease assembly. Do not use shop air to clean parts; damage to rubber components may result.

• Use only DOT 3 brake fluid from an unopened container.

• If any hydraulic component or line is removed or replaced, it may be necessary to bleed the entire system.

• A clean repair area is essential. Always clean the reservoir and cap thoroughly before removing the cap. The slightest amount of dirt in the fluid may plug an orifice and impair the system function. Perform repairs after components have been thoroughly cleaned; use only denatured alcohol to clean components. Do not allow ABS components to come into contact with any substance containing mineral oil; this includes used shop rags.

• The Anti-Lock control unit is a microprocessor similar to other computer units in the vehicle. Ensure that the ignition switch is **OFF** before removing or installing controller harnesses. Avoid static electricity discharge at or near the controller.

• If any arc welding is to be done on the vehicle, the control unit should be unplugged before welding operations begin.

DISC AND DRUM SYSTEMS

> ✳✳ **CAUTION**
>
> Dust and dirt accumulating on brake parts during normal use may contain asbestos fibers from production or aftermarket brake linings. Breathing excessive concentrations of asbestos fibers can cause serious bodily harm. Exercise care when servicing brake parts. Do not sand or grind brake lining unless equipment used is designed to contain the dust residue. Do not clean brake parts with compressed air or by dry brushing. Cleaning should be done by dampening the brake components with a fine mist of water, then wiping the brake components clean with a dampened cloth. Dispose of cloth and all residue containing asbestos fibers in an impermeable container with the appropriate label. Follow practices prescribed by the Occupational Safety and Health Administration (OSHA) and the Environmental Protection Agency (EPA) for the handling, processing, and disposing of dust or debris that may contain asbestos fibers.

BRAKES   BLEEDING THE BRAKE SYSTEM

BLEEDING PROCEDURE

BLEEDING PROCEDURE

See Figure 1.

> ✳✳ **CAUTION**
>
> Brake fluid contains polyglycol ethers and polyglycols. Avoid contact with the eyes and wash your hands thoroughly after handling brake fluid. If you do get brake fluid in your eyes, flush your eyes with clean, running water for 15 minutes. If eye irritation persists, or if you have taken brake fluid internally, IMMEDIATELY seek medical assistance.

> ✳ **WARNING**
>
> Clean, high quality brake fluid is essential to the safe and proper operation of the brake system. You should always buy the highest quality brake fluid that is available. If the brake fluid becomes contaminated, drain and flush the system, then refill the master cylinder with new fluid. Never reuse any brake fluid. Any brake fluid that is removed from the system should be discarded. Also, do not allow any brake fluid to come in contact with a painted surface; it will damage the paint.

➡ The bleeding operation is necessary to remove air whenever it enters the hydraulic brake system. The hydraulic lines of the brake system are based on the diagonal split system. When a brake pipe or hose was disconnected at the wheel, a bleeding operation must be performed at both ends of the line of the removed pipe or hose. When any joint part of the master cylinder or other joint part between the master cylinder and each brake (wheel) was removed, the hydraulic brake system must be bled at all 4 wheel brakes.

➡ Perform the bleeding operation starting with the rear brake caliper or the wheel cylinder farthest from the master cylinder and then at the front caliper of the same brake line. Do the same on the other brake line.

1. Before servicing the vehicle, refer to the Precautions Section.
2. Clean all around the brake fluid reservoir filler cap.
3. Fill the master cylinder reservoir with the specified brake fluid and keep it at least ½ full of fluid during the bleeding operation.
4. Raise and safely support the vehicle.

5. Remove the bleeder plug cap. Attach a vinyl tube to the bleeder plug, and insert the other end into a container.
6. Depress the brake pedal several times, and then while holding it depressed, loosen the bleeder plug about one-third to one-half turn.
7. When the fluid pressure in the cylinder is almost depleted, retighten the bleeder plug.
8. Repeat this operation until there are no more air bubbles in the hydraulic line.
9. When the bubbles stop, depress and hold the brake pedal and tighten the bleeder plug.

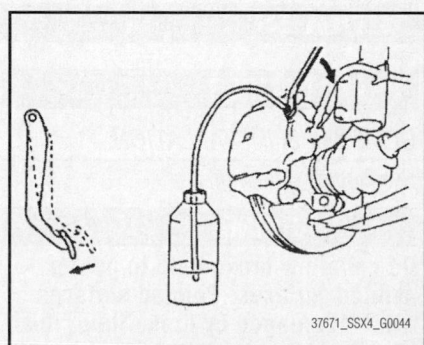

37671_SSX4_G0044

Fig. 1 Remove the bleeder plug cap, attach a vinyl tube to the bleeder plug, and insert the other end into a container

- Tighten the front brake caliper bleeder plug to 80 inch lbs. (9 Nm)
- Tighten the rear brake caliper bleeder plug (disc brake model) to 80 inch lbs. (9 Nm)
- Tighten the rear wheel cylinder bleeder plug (drum brake model) to 71 inch lbs. (8 Nm)

10. Attach the bleeder plug cap when finished at the bleeder plug.

11. After completing the bleeding operation, apply fluid pressure to the pipe line and check for leakage.

12. Replenish the fluid in the reservoir to the specified level. Refer to Fluid Fill Procedure.

13. Check the brake pedal for sponginess. If found spongy, repeat the entire bleeding procedure.

❈❈ CAUTION

Do not attempt to drive a vehicle until the brakes are functioning properly.

BRAKES   ANTI-LOCK BRAKE SYSTEM (ABS)

SPEED SENSORS

REMOVAL & INSTALLATION

Front Speed Sensor
See Figure 2.

❈❈ WARNING

Do not pull on wire harness when removing and installing the front wheel speed sensor. Do not cause damage to the surface of the front wheel speed sensor and do not allow dust or other foreign material to enter the installation hole.

1. Before servicing the vehicle, refer to the Precautions Section.
2. Disconnect the negative battery cable.
3. Disconnect the front wheel speed sensor connector.
4. Hoist the vehicle and remove the wheel.
5. Remove the harness clamp and clamp bolt.
6. Remove the front wheel speed sensor from the knuckle.

To install:
7. Installation is the reverse of the removal procedure.
8. Check that no foreign material is attached to the sensor or the wheel speed

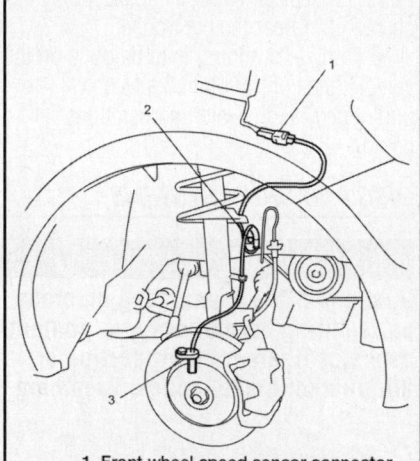

1. Front wheel speed sensor connector
2. Harness clamp and clamp bolt
3. Front wheel speed sensor

37671_SSX4_G0043

Fig. 2 Removing the front wheel speed sensor

sensor encoder (included in the wheel hub assembly).

9. Before fitting the wheel speed sensor, be sure to apply silicon grease to the O-ring.

10. Tighten the front wheel speed sensor bolt and clamp to 96 inch lbs. (11 Nm).

11. Check that there is no clearance between the sensor and the knuckle.

Rear Speed Sensor

❈❈ WARNING

Do not pull on the wire harness when removing and installing the rear wheel speed sensor. Do not cause damage to the surface of the rear wheel speed sensor and do not allow dust or other foreign material to enter the installation hole.

1. Before servicing the vehicle, refer to the Precautions Section.
2. Disconnect the negative battery cable.
3. Remove the rear wheel.
4. Disconnect the rear wheel speed sensor connector.
5. Remove the harness clamp.
6. Remove the rear wheel speed sensor from the wheel hub.

To install:
7. Installation is the reverse of the removal procedure.
8. Check that no foreign material is attached to the sensor or encoder (included in the wheel hub assembly).
9. Before fitting the wheel speed sensor, be sure to apply silicon grease to the O-ring.
10. Tighten the rear wheel speed sensor bolt to 96 inch lbs. (11 Nm).
11. Check that there is no clearance between the sensor and the wheel hub.

BRAKES   FRONT DISC BRAKES

BRAKE CALIPER

REMOVAL & INSTALLATION

See Figures 3 through 5.

❈❈ WARNING

Do not allow brake fluid to get on painted surfaces. Painted surfaces will be damaged by brake fluid, flush all affected areas with water immediately if any fluid is spilled.

❈❈ WARNING

During removal, be careful not to damage the brake flexible hose.

➡ Do not reuse the brake flexible hose washers. Otherwise, the brake fluid may leak.

➡ When the brake pads are removed, visually inspect the caliper for any brake fluid leak. Correct the leaky point, if any.

1. Before servicing the vehicle, refer to the Precautions Section.
2. Raise and safely support the vehicle.
3. Remove the front wheel.
4. Remove the brake flexible hose joint bolt from the caliper. As this will allow fluid to flow out of hose, have a container ready beforehand.
5. Remove the caliper pin bolts.
6. Remove the caliper from the caliper carrier.

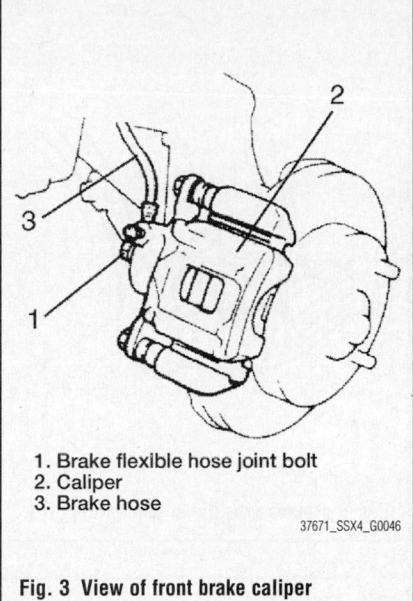

1. Brake flexible hose joint bolt
2. Caliper
3. Brake hose

37671_SSX4_G0046

Fig. 3 View of front brake caliper

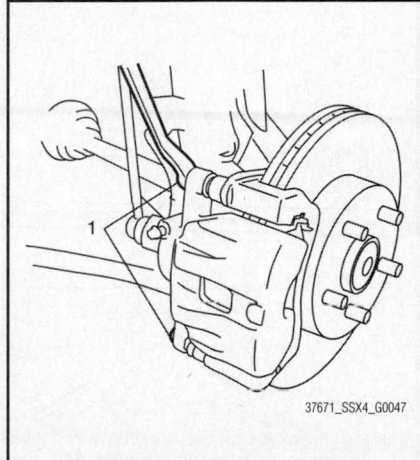

37671_SSX4_G0047

Fig. 4 Remove the caliper pin bolts (1)

To install:

7. Install the caliper to the caliper carrier. Tighten the caliper pin bolts to 19 ft. lbs. (26 Nm).

➡**Make sure that the boots are fit into the groove securely.**

8. Connect the flexible hose to the caliper.

9. Tighten the flexible hose joint bolt to 17 ft. lbs. (23 Nm).

❊❊ **WARNING**

Make sure that the flexible hose is not twisted when tightening the joint bolt. If it is twisted, reconnect it using care not to twist it.

➡**Be sure the directing mark on the brake flexible hose points to the inside of the vehicle.**

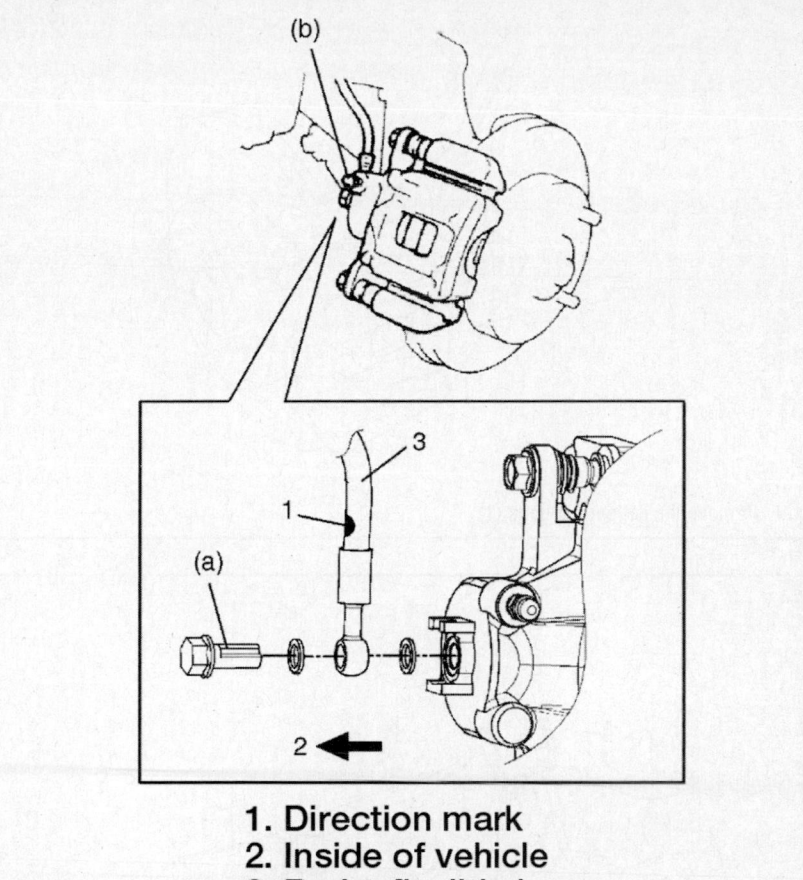

1. Direction mark
2. Inside of vehicle
3. Brake flexible hose

37671_SSX4_G0048

Fig. 5 Connect the flexible hose to the caliper

10. Tighten the bleeder plug to 80 inch lbs. (9 Nm).

11. Install the front wheel and tighten the lug nuts to 62 ft. lbs. (85 Nm).

12. Fill the reservoir with brake fluid and bleed any air from the brake system. Refer to Bleeding the Brake System.

13. Check for fluid leakage. Test the brakes for proper function.

❊❊ **CAUTION**

Do not attempt to drive a vehicle until the brakes are functioning properly.

DISC BRAKE PADS

REMOVAL & INSTALLATION
See Figures 6 through 9.

❊❊ **WARNING**

Be careful not to damage the brake flexible hose. Do not operate the brake pedal with the pads removed.

➡**When replacing the brake pads, replace them on both sides as a set.**

1. Before servicing the vehicle, refer to the Precautions Section.

2. Raise and safely support the vehicle.

3. Remove the front wheels.

4. Remove the caliper pin bolts.

5. Remove the caliper from the caliper carrier.

➡**Hang the removed caliper with a wire hook or the like, so as to prevent the brake hose from bending and twisting excessively or being pulled.**

6. Remove the brake pads.

7. Remove the brake pad spring.

To install:

8. Apply a small amount of pad grease to the pad spring.

9. Set the brake pad springs, and

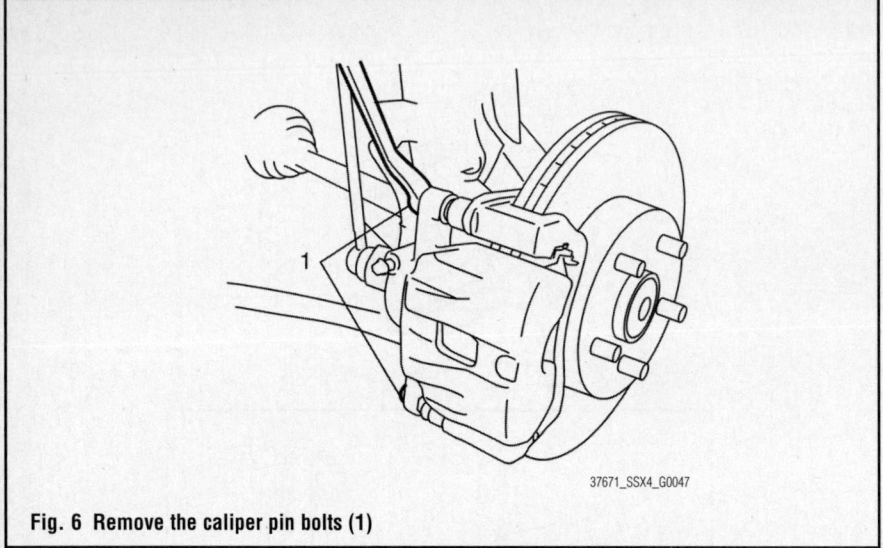

Fig. 6 Remove the caliper pin bolts (1)

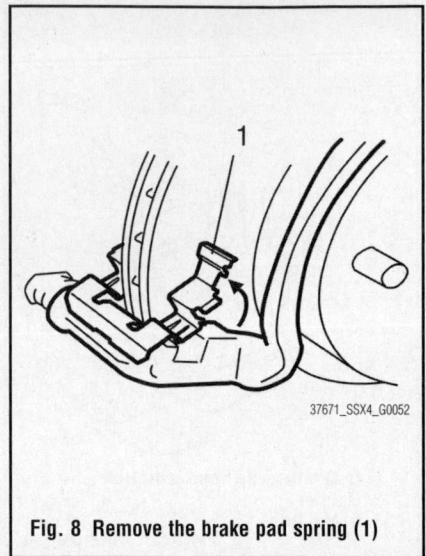

Fig. 8 Remove the brake pad spring (1)

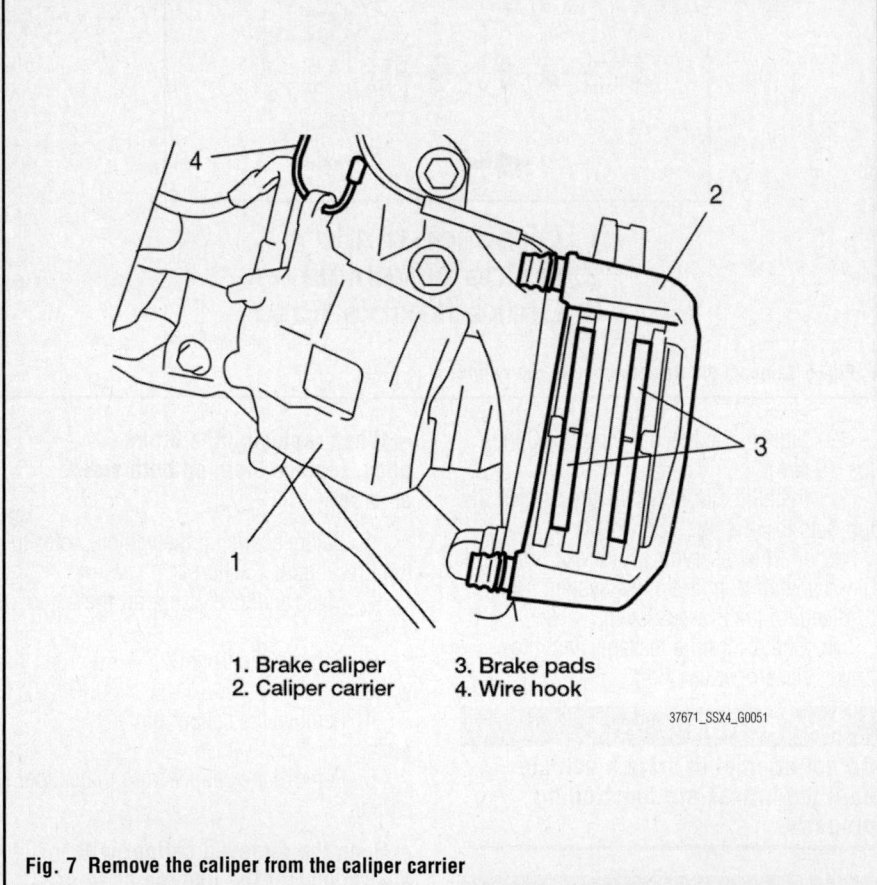

1. Brake caliper
2. Caliper carrier
3. Brake pads
4. Wire hook

Fig. 7 Remove the caliper from the caliper carrier

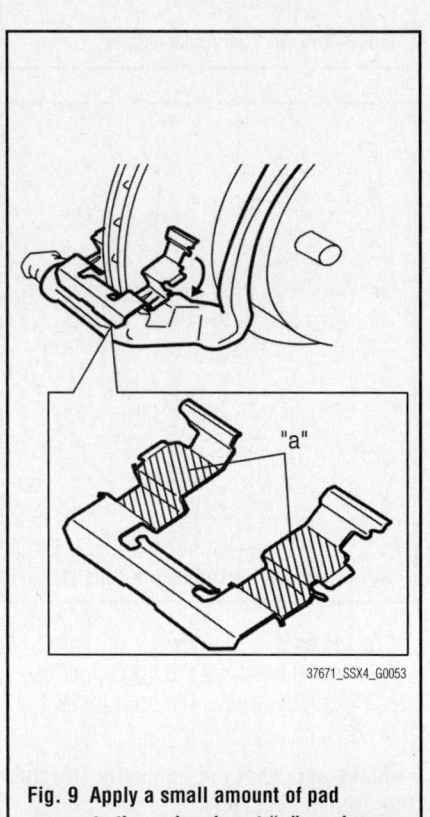

Fig. 9 Apply a small amount of pad grease to the pad spring at "a" as shown

install the brake pads, shims, and shim cover.

➡**Install the brake pad with the wear indicator to the vehicle center side of the front right brake pad.**

10. Install the caliper and tighten the caliper pin bolts to 19 ft. lbs. (26 Nm).

11. Install the front wheel and tighten the lug nuts to 62 ft. lbs. (85 Nm).

12. Check for proper brake function.

✳✳ CAUTION

Do not attempt to drive a vehicle until the brakes are functioning properly.

BRAKE CALIPER

REMOVAL & INSTALLATION

See Figures 10 through 15.

**** WARNING**

Do not allow brake fluid to get on any painted surfaces. The painted surfaces may be damaged by brake fluid. Flush the surface with water immediately if any fluid is spilled.

**** WARNING**

Be careful not to damage the brake flexible hose or parking brake cable.

➡Do not reuse the brake flexible hose washers. Otherwise, brake fluid may leak.

➡When the pads are removed, visually inspect the caliper for a brake fluid leak. Correct leaky point, if any.

1. Before servicing the vehicle, refer to the Precautions Section.
2. Raise and safely support the vehicle.
3. Remove the rear wheel.
4. Release the parking brake lever.
5. Remove the clip attaching the parking brake cable.
6. Loosen the brake pipe flare nut using a flare nut wrench, then remove the clip.
7. Remove the flexible hose joint bolt, and disconnect the brake flexible hose from the brake caliper.
8. Remove the caliper pin bolts.

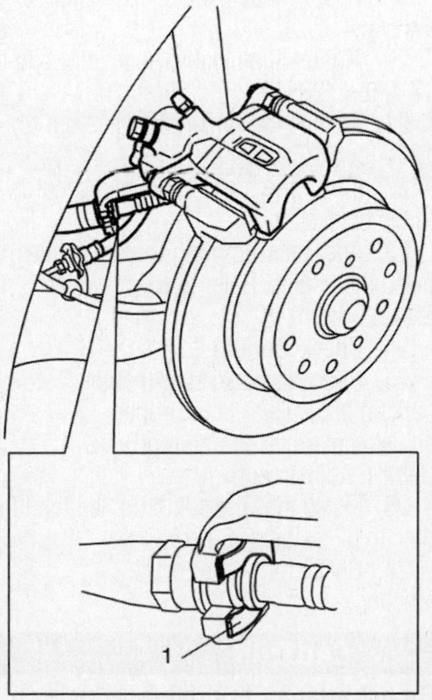
1
37671_SSX4_G0058

Fig. 10 Remove the clip (1) attaching the parking brake cable

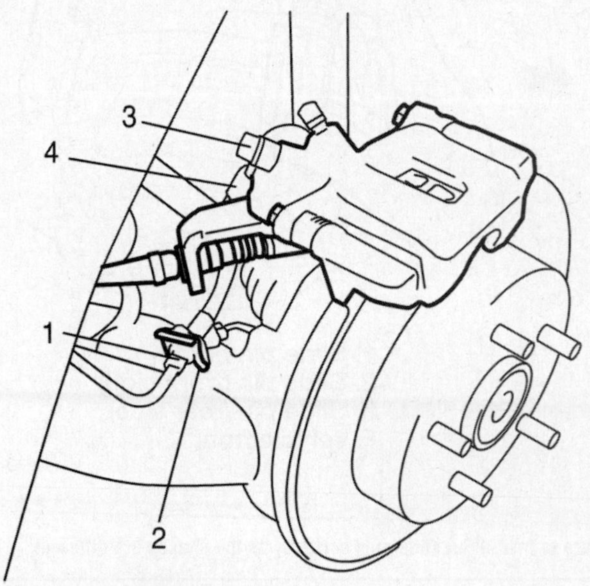
3
4
1
2
1. Brake pipe flare nut
2. Clip
3. Flexible hose joint bolt
4. Brake flexible hose
37671_SSX4_G0059

Fig. 11 Remove the flexible hose joint bolt and disconnect the brake flexible hose from the brake caliper

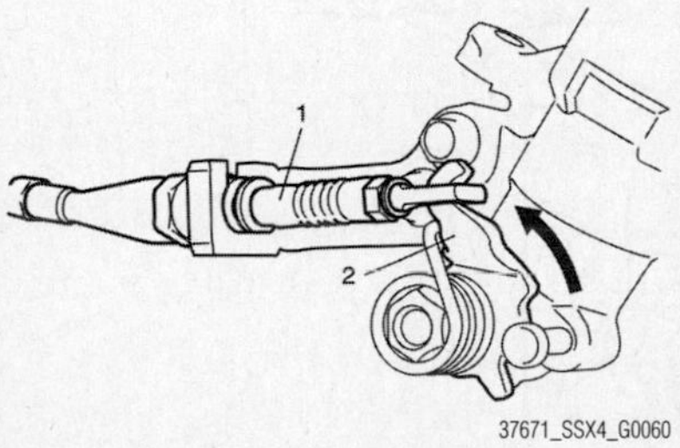
1
2
37671_SSX4_G0060

Fig. 12 Disconnect the parking brake cable (1) from the lever (2) while rotating the lever

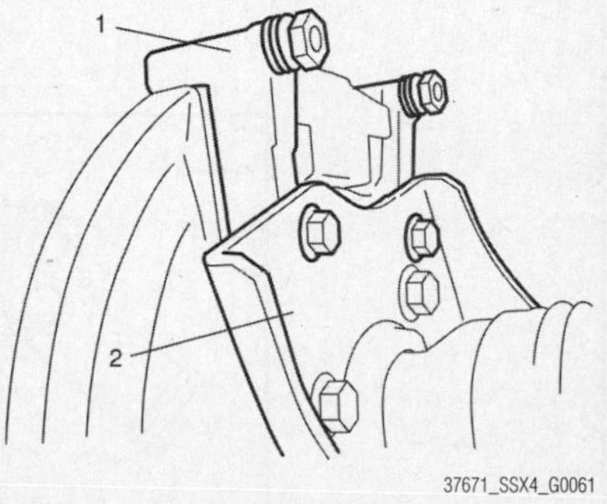
1
2
37671_SSX4_G0061

Fig. 13 Remove the brake caliper carrier (1) from the rear axle (2)

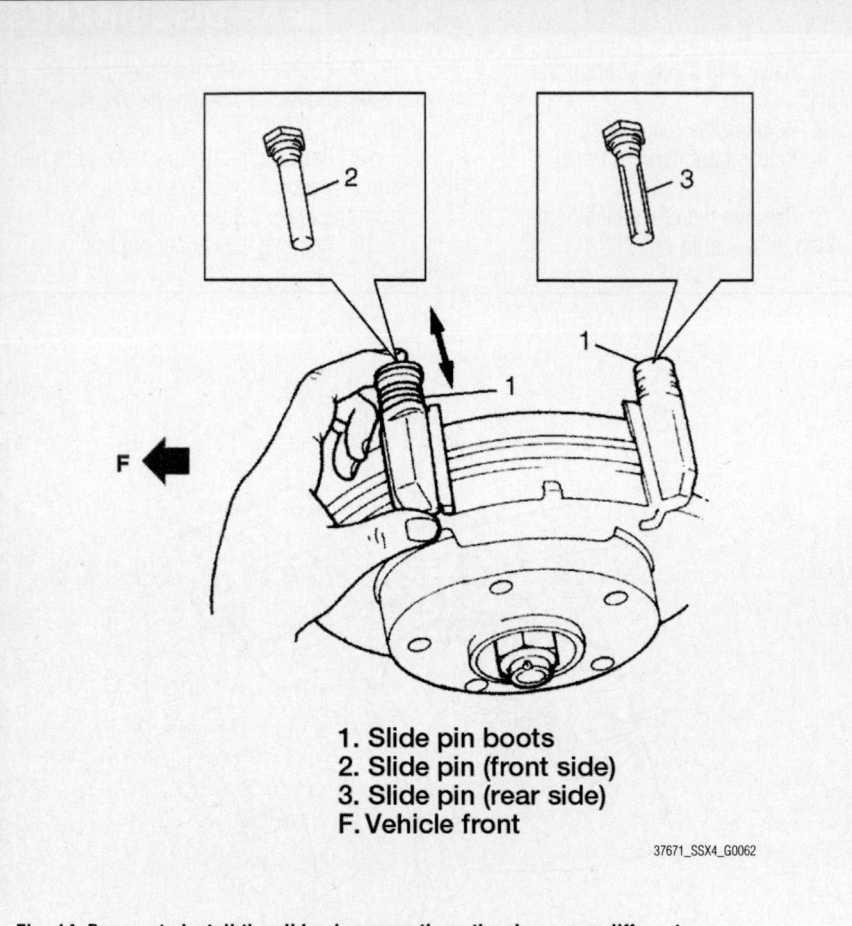

1. Slide pin boots
2. Slide pin (front side)
3. Slide pin (rear side)
F. Vehicle front

37671_SSX4_G0062

Fig. 14 Be sure to install the slide pins correctly as the shapes are different

1. Parking brake cable
2. Lever
3. Bracket of the caliper
4. Lever
5. Pin

37671_SSX4_G0063

Fig. 15 Installing the parking brake cable

9. Remove the caliper from the brake caliper carrier.

10. Disconnect the parking brake cable from the lever while rotating the lever.

11. Remove the brake pads, slide pins, and slide pin boots from the brake caliper carrier.

12. Remove the brake caliper carrier from the rear axle.

To install:

13. Install the brake caliper carrier to the rear axle. Tighten the caliper carrier bolt to 44 ft. lbs. (60 Nm).

14. Install new slide pin boots to the brake caliper carrier.

15. Apply rubber grease to the slide pins, then install slide pin to the front side and the slide pin to the rear side of the brake caliper carrier.

➡**Be sure to install the slide pins correctly as the shapes are different.**

16. Pass the parking brake cable through the bracket of the caliper.

17. Hang the parking brake cable on the lever. Make sure that the lever contacts the pin.

18. Install the brake pads and caliper.

19. Connect the flexible hose with new washers to the caliper and brake pipe temporarily by hand. Then install the clip.

20. Tighten the flexible hose joint bolt to 17 ft. lbs. (23 Nm).

21. Tighten the brake pipe flare nut to 12 ft. lbs. (16 Nm).

22. Install the brake cable clip securely to the bracket.

23. After reassembling the brake lines, bleed the brakes. Refer to Bleeding the Brake System.

24. Check for fluid leakage.

25. Check to make sure that the parking brake lever is in proper adjustment. Refer to Parking Brake Cables, Adjustment.

26. Install the wheel and tighten the lug nuts to 62 ft. lbs. (85 Nm).

27. Check for proper brake function.

✳✳ CAUTION

Do not attempt to drive a vehicle until the brakes are functioning properly.

DISC BRAKE PADS

REMOVAL & INSTALLATION

See Figures 16 through 19.

⁂ WARNING

Be careful not to damage the brake flexible hose during this procedure. Do not operate the brake pedal with the pads removed.

➡ **When the pads are removed, visually inspect the caliper for any brake fluid leaks. Correct the leaky point, as necessary.**

➡ **When replacing the brake pads, replace them as a set.**

1. Before servicing the vehicle, refer to the Precautions Section.
2. Raise and safely support the vehicle.
3. Remove the rear wheel.
4. Release the parking brake lever.

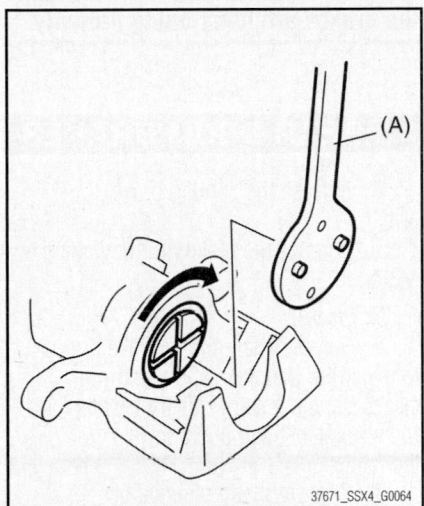

Fig. 16 Using special tool 09945-16060 to turn the brake caliper piston clockwise to obtain clearance between the brake disc and pads

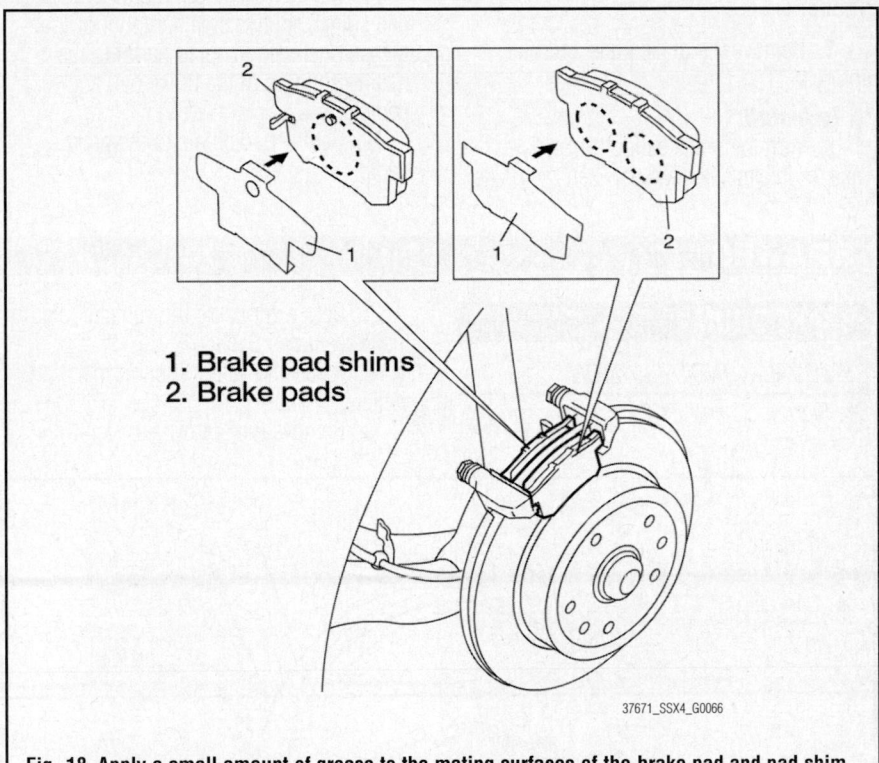

1. Brake pad shims
2. Brake pads

Fig. 18 Apply a small amount of grease to the mating surfaces of the brake pad and pad shim

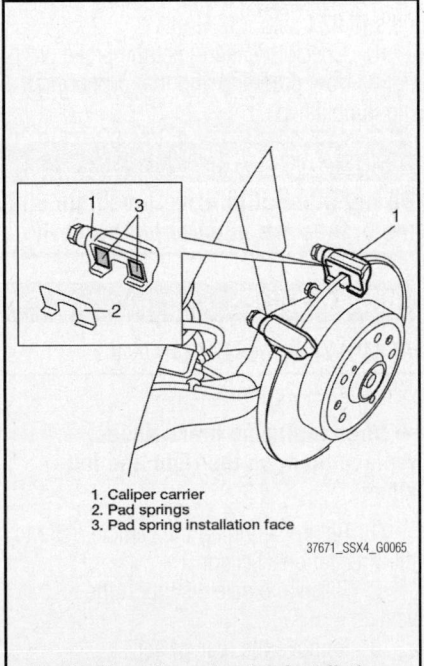

1. Caliper carrier
2. Pad springs
3. Pad spring installation face

Fig. 17 Clean the pad spring installation face of the caliper carrier. Attach the pad springs to the caliper carrier

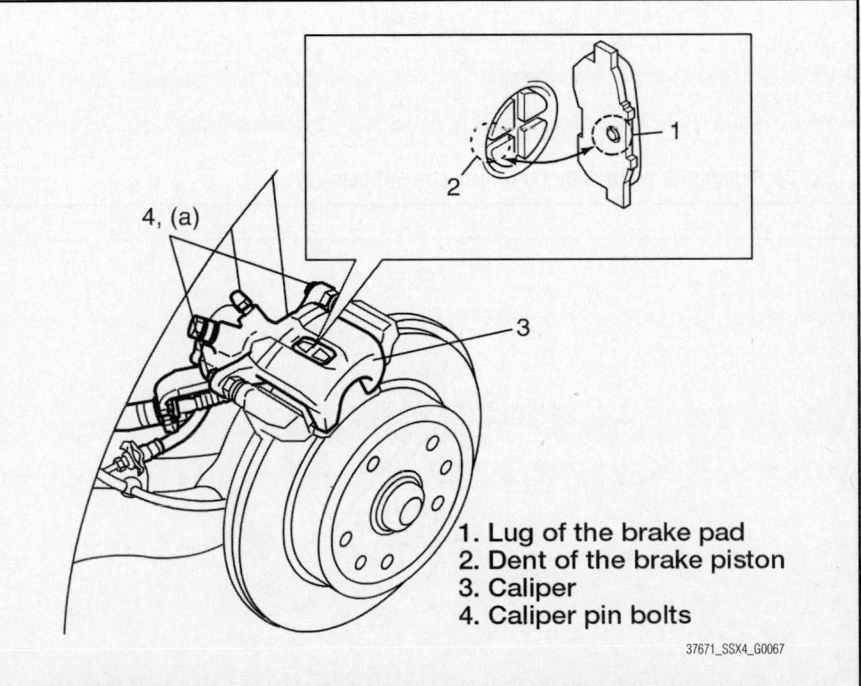

1. Lug of the brake pad
2. Dent of the brake piston
3. Caliper
4. Caliper pin bolts

Fig. 19 With the lug of the brake pad matched with the dent of the brake piston, install the caliper to the caliper carrier

5. Remove the caliper pin bolts.

6. Remove the caliper from the caliper carrier.

➡ **Hang the removed caliper with a wire hook, or the like, to prevent the brake hose from bending and twisting excessively or being pulled.**

7. Remove the brake pads and pad springs.

To install:

8. Turn the brake caliper piston clockwise to obtain clearance between the brake disc and pads. Use special tool 09945-16060, or equivalent.

9. Clean (and degrease) the pad spring installation face of the caliper carrier.

10. Attach the pad springs to the caliper carrier.

11. Before installing the brake pad shims, apply a small amount of grease (included in spare parts) to the mating surfaces of the brake pad and pad shim.

12. Set the pad shims to the brake pads.

13. With the lug of the brake pad matched with the dent of the brake piston, install the caliper to the caliper carrier.

14. Tighten the caliper pin bolts to 17 ft. lbs. (23 Nm).

15. Install the wheel and tighten the lug nuts to 62 ft. lbs. (85 Nm).

16. Check for proper brake function.

✳✳ CAUTION

Do not attempt to drive a vehicle until the brakes are functioning properly.

BRAKES

BRAKE DRUM

REMOVAL & INSTALLATION

See Figures 20 and 21.

1. Before servicing the vehicle, refer to the Precautions Section.

2. Raise and safely support the vehicle.

3. Remove the rear wheel.

REAR DRUM BRAKES

4. Release the parking brake lever.

5. Remove the brake drum by using two M8 bolts.

To install:

6. Before installing the brake drum, to maximize the brake shoe-to-drum clearance, put a screwdriver between the rod and ratchet and push down the ratchet.

7. Confirm that the inside of the brake drum and the brake shoes are free from dirt and oil, then install the brake drum.

8. Adjust the parking brake cable, as needed. Refer to Parking Brake Cables, Adjustment.

9. Install the wheel and tighten the lug nuts to 62 ft. lbs. (85 Nm).

10. Check to ensure that the brake drum is free from dragging and that proper braking is obtained.

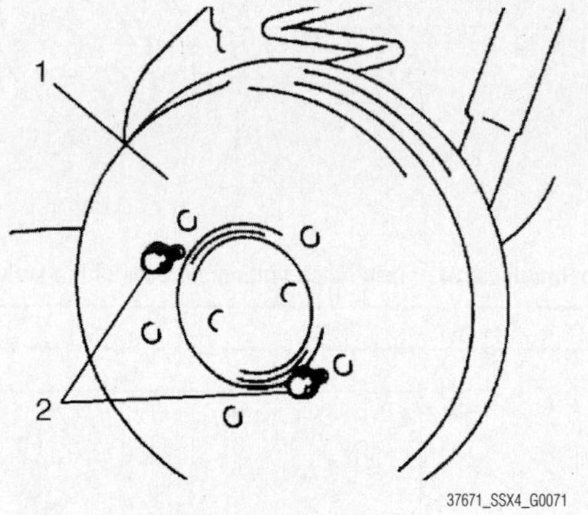

✳✳ CAUTION

Do not attempt to drive a vehicle until the brakes are functioning properly.

BRAKE SHOES

REMOVAL & INSTALLATION

See Figures 22 through 25.

➡ **When replacing brake shoes, replace them on the right and left sides.**

1. Before servicing the vehicle, refer to the Precautions Section.

2. Raise and safely support the vehicle.

3. Remove the rear wheel.

4. Remove the brake drum. Refer to Brake Drums, removal & installation.

5. Push and rotate the hold down pin

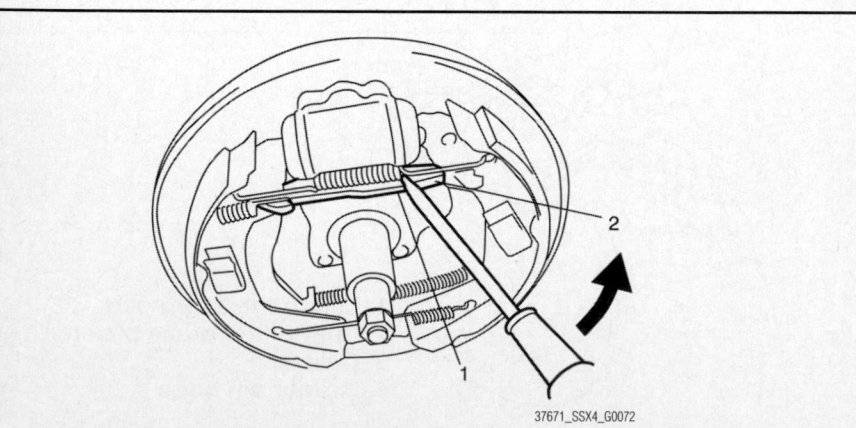

37671_SSX4_G0071

Fig. 20 Remove the brake drum (1) by using two M8 bolts (2)

37671_SSX4_G0072

Fig. 21 Use a screwdriver between the rod (1) and the ratchet (2) and push down the ratchet to maximize the brake shoe-to-drum clearance

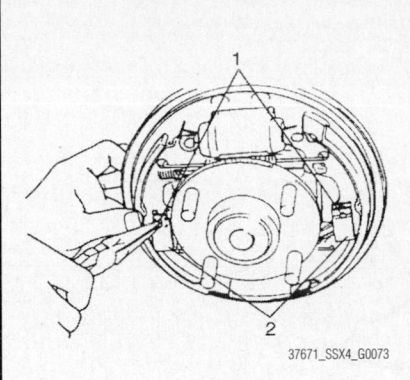

Fig. 22 Push and rotate the hold down pin (1) 90 degrees. Remove the hold down pin and the hold down spring (2)

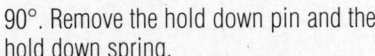

90°. Remove the hold down pin and the hold down spring.

    6. Remove the return springs, brake shoes, and adjuster.

    7. Disconnect the parking brake shoe lever from the parking brake cable.

    8. Remove the push nut.

    9. Remove the parking brake shoe lever from the shoe rim.

To install:

    10. Installation is the reverse of the removal procedure.

    11. Before installing the rear brake shoe to the brake back plate, clean the brake back plate and apply heat-resistance brake grease to the position on which the shoe rims rest.

    12. Push and rotate the hold down pin 90° and install the hold down pin and the hold down spring.

    13. Install the brake drum. Refer to Brake Drums, removal & installation.

ADJUSTMENT

See Figure 26.

    The brake shoes utilize automatic adjusters.

    1. Before servicing the vehicle, refer to the Precautions Section.

    2. Check the ratchet of the brake adjuster assembly for wear or damage.

    3. Check the shoe return spring, anti-rattle spring, quadrant spring, and hold down spring for damage, corrosion and weakening.

    4. Check for smooth movement of the brake shoe lever along the shoe rim.

    5. If any defective or malfunction is found, repair or replace.

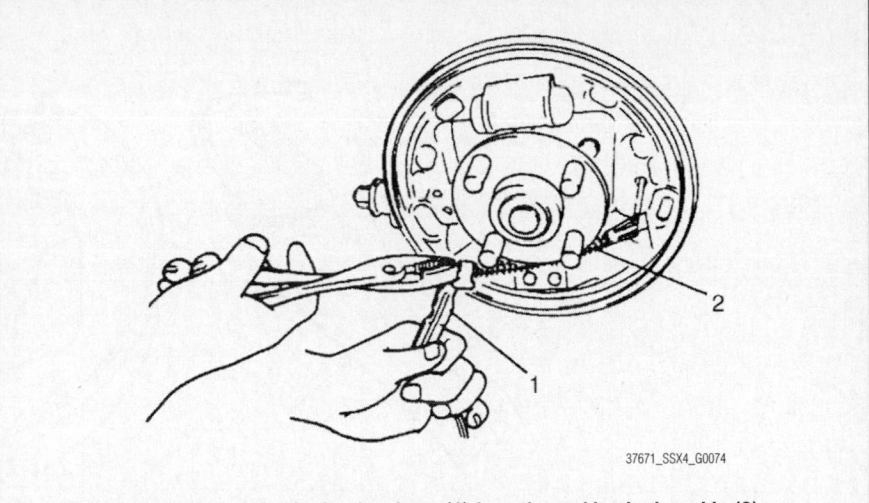

Fig. 23 Disconnect the parking brake shoe lever (1) from the parking brake cable (2)

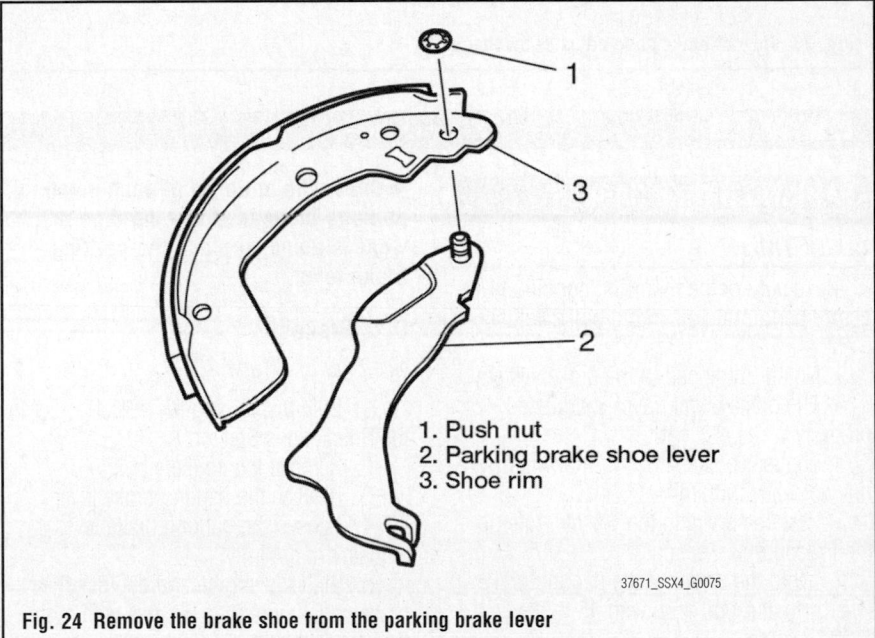

1. Push nut
2. Parking brake shoe lever
3. Shoe rim

Fig. 24 Remove the brake shoe from the parking brake lever

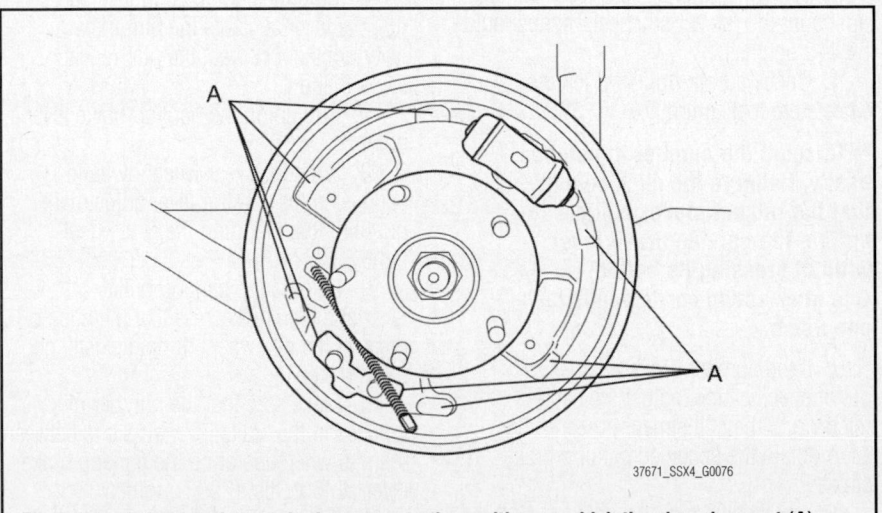

Fig. 25 Apply heat-resistance brake grease to the position on which the shoe rims rest (A)

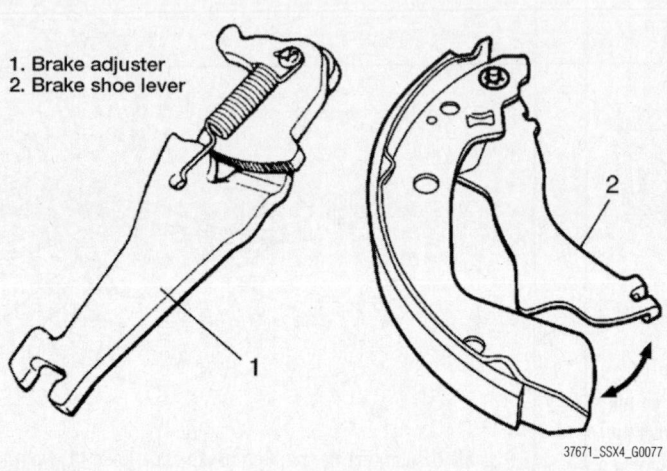

1. Brake adjuster
2. Brake shoe lever

37671_SSX4_G0077

Fig. 26 View of brake shoe adjuster mechanism

BRAKES

PARKING BRAKE

PARKING BRAKE CABLES

ADJUSTMENT

Make sure of the following conditions before performing a parking brake adjustment:

- No air is trapped in the brake system
- The brake pedal travel is adjusted properly
- The rear brake shoes or brake pads are not worn beyond limit

1. Before servicing the vehicle, refer to the Precautions Section.

2. Hold the center of the parking brake lever grip and pull it up with 45 ft. lbs. (200 N) force.

3. With the parking brake lever pulled up, count the ratchet notches. There should be 4–9 notches.

4. Check if both right and left rear wheels are locked firmly.

➡**To count the number of notches easily, listen to the click sounds that the ratchet makes while pulling the parking brake lever without pressing its button. One click sound corresponds to one notch.**

5. If the number of the notches is out of specification, adjust the cable referring to the adjustment procedure so as to obtain the specified parking brake stroke.

➡**Check the tooth tip of each notch for damage or wear. If any damage or wear is found, replace the parking brake lever.**

Disc Brakes

See Figures 27 and 28.

1. Before servicing the vehicle, refer to the Precautions Section.

2. Remove the console box.

3. Release the parking brake lever.

4. Loosen the parking brake lever adjustment nut fully.

5. Start the engine and depress the brake pedal more than 3 times with more than 34 ft. lbs. (150 N) force.

6. Stop the engine and make sure that the lever contacts the pin. If the lever does not contact the pin, repeat steps 3 and 4.

7. Pull up on the parking brake lever 1 notch.

8. Rotate the rear wheel by hand and fasten the parking lever adjustment nut until it is dragging the rear wheel lightly.

9. Release the parking brake lever and then make sure that there is no drag on the rear wheel. If drag exists, go back to step 3.

10. Make sure that the number of notches in the brake lever stroke is between 4 and 9 when operating the parking brake lever at 45 ft. lbs. (200 N) force.

11. If the number of notches

is not between 4 and 9, replace the parking brake cable and/or inspect the rear brake caliper.

Drum Brakes

See Figure 27.

1. Before servicing the vehicle, refer to the Precautions Section.

2. Remove the console box.

3. Release the parking brake lever.

4. Loosen the adjusting nut until the parking brake cable has no tension.

5. Start the engine and depress the brake pedal more than 3 times with more than 34 ft. lbs. (150 N) so as to obtain the proper drum-to-shoe clearance.

6. Stop the engine and adjust the parking brake lever stroke to the specified value with the adjusting nut.

➡**The parking brake lever must be released while adjusting.**

7. Hold the center of the parking brake lever grip and pull it up with 45 ft. lbs. (200 N) force.

8. With the parking brake lever pulled up, count the ratchet notches. There should be 4–9 notches.

9. After completion of adjustment, make sure that the brake drum is not dragging.

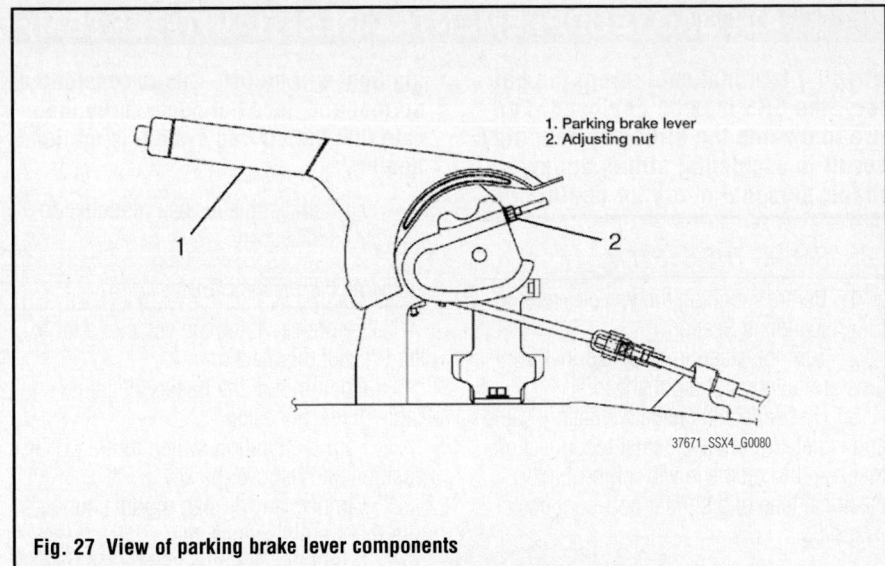

1. Parking brake lever
2. Adjusting nut

37671_SSX4_G0080

Fig. 27 View of parking brake lever components

PARKING BRAKE SHOES

REMOVAL & INSTALLATION

Disc Brakes

On disc brakes, the parking brakes utilize the regular service brakes. Refer to Rear Disc Brakes, Disc Brake Pads, removal & installation.

Drum Brakes

On drum brakes, the parking brakes utilize the regular service brakes. Refer to Rear Drum Brakes, Brake Shoes, removal & installation.

ADJUSTMENT

For adjustment procedures, refer to Parking Brake Cables, Adjustment.

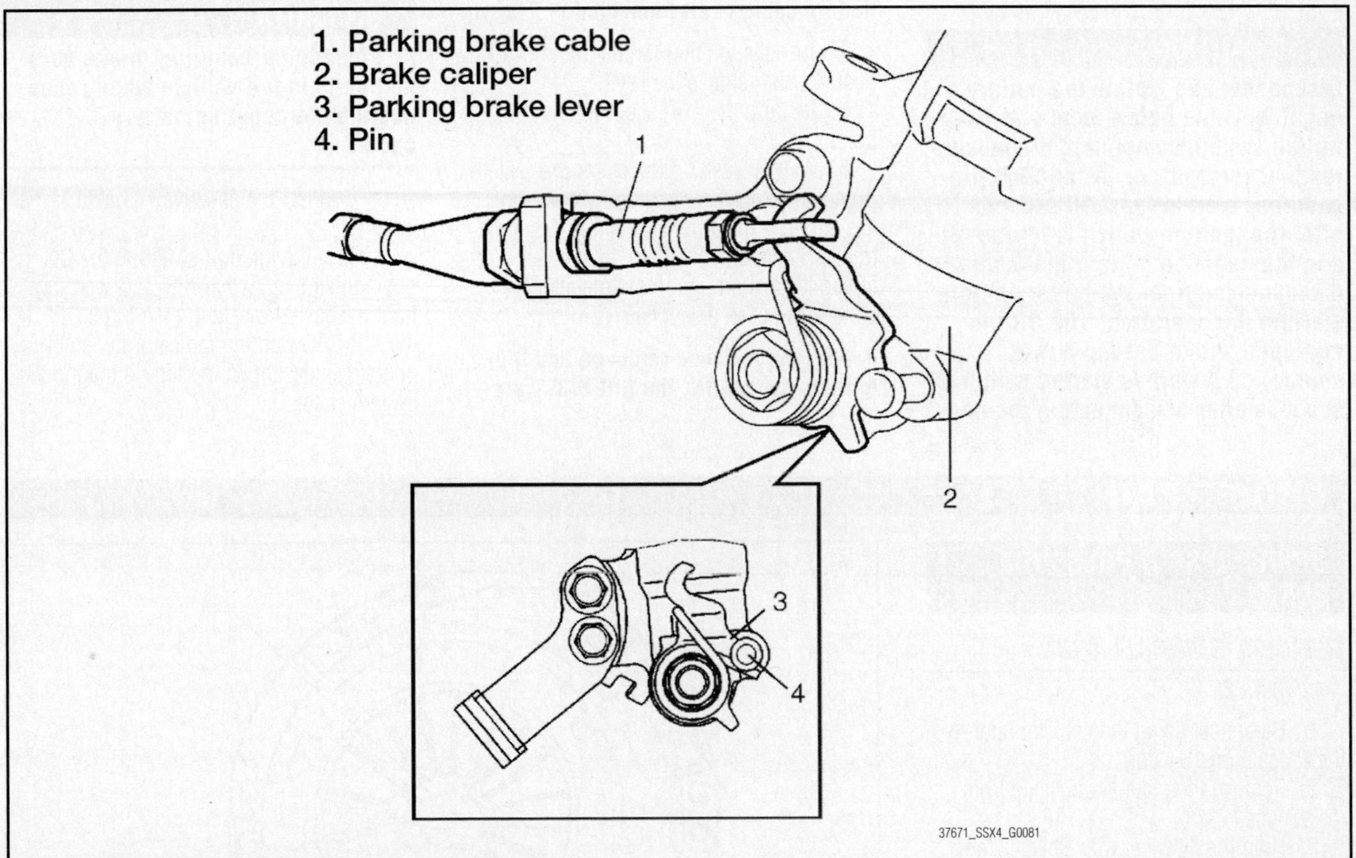

1. Parking brake cable
2. Brake caliper
3. Parking brake lever
4. Pin

37671_SSX4_G0081

Fig. 28 Make sure the parking brake lever contacts the pin

GENERAL INFORMATION

✳✳ CAUTION

These vehicles are equipped wi th an air bag system. The system must be disarmed before performing service on, or around, system components, the steering column, instrument panel components, wiring and sensors. Failure to follow the safety precautions and the disarming procedure could result in accidental air bag deployment, possible injury and unnecessary system repairs.

SERVICE PRECAUTIONS

✳✳ CAUTION

Disconnect and isolate the battery negative cable before beginning any airbag system component diagnosis, testing, removal, or installation procedures. Wait at least 90 seconds after the ignition switch is turned off and the negative (-) terminal cable is disconnected from the battery before starting the operation. The SRS is equipped with a backup power source, so if work is started within 90 seconds after disconnecting the neg-

ative (-) terminal cable from the battery, the SRS may be deployed. Failure to disable the airbag system may result in accidental airbag deployment, personal injury, or death.

DISARMING THE SYSTEM

1. Before servicing the vehicle, refer to the Precautions Section.
2. Turn the steering wheel so that front tires are pointing straight ahead.
3. Disconnect the negative battery cable and isolate it from accidental reconnection. Insulate the cable end with high-quality electrical tape or a similar non-conductive wrapping.

➡ **DTC's will be lost when the negative battery cable is disconnected.**

4. Turn the ignition switch to the LOCK position and remove the key.
5. Remove the A/B fuse from the fuse box.
6. Wait at least 3 minutes for the system capacitor to discharge before performing any service. The airbag system is designed to retain enough voltage to deploy the airbag for a short period of time after the battery has been disconnected.

➡ **With the A/B fuse removed and the ignition switch ON, the AIR BAG warn-**

ing light will be ON. This is considered normal and does not necessarily indicate that the air bag system is malfunctioning.

7. Disconnect the air bag module connectors as necessary.

ARMING THE SYSTEM

1. Before servicing the vehicle, refer to the Precautions Section.
2. Confirm that the battery negative (-) cable is disconnected.
3. Turn the ignition switch to the LOCK position and remove the key.
4. Connect any air bag module connectors that were disconnected.
5. Install the A/B fuse to the fuse box.

✳✳ CAUTION

As an added precaution, make sure no one is in the vehicle when reconnecting the negative battery cable.

6. Connect the negative (-) cable at the battery.
7. Turn the ignition switch to the ON position and verify that AIR BAG warning light flashes 6 times and then turns OFF. If the AIR BAG warning indicator does not operate as described, perform a diagnostic system check.

DRIVE TRAIN

DRIVEN DISC & PRESSURE PLATE

REMOVAL & INSTALLATION
See Figure 29.

1. Before servicing the vehicle, refer to the Precautions Section.
2. Remove the manual transaxle from the vehicle.
3. Hold the flywheel with Special Tool 09924-17811, or equivalent, and remove the clutch cover bolts, clutch cover, and clutch disc.

To install:

➡ Before assembling, make sure that the flywheel surface and the pressure plate surface have been cleaned and dried thoroughly.

4. Align the clutch disc to the flywheel center using Special Tools 09924-17811 and 09923-36320, or equivalent.

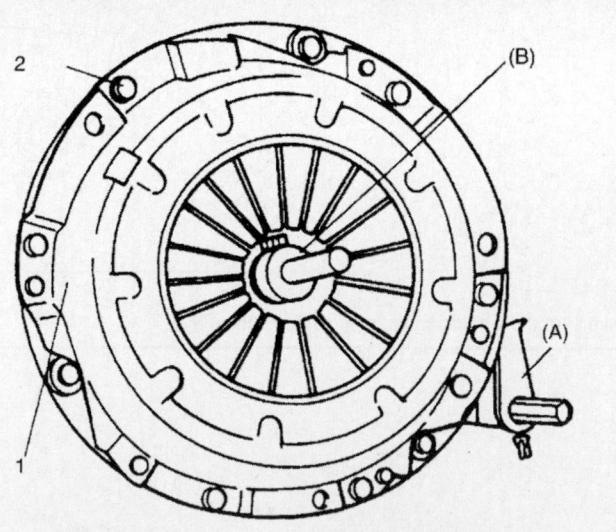

1. Clutch cover bolts   A. Special Tool 09924-17811
2. Clutch cover   B. Special Tool 09923-36320

37671_SSX4_G0144

Fig. 29 View of flywheel and special tools

5. Install the clutch cover and bolts. Tighten the bolts to 17 ft. lbs. (23 Nm).

➡**While tightening the clutch cover bolts, compress the clutch disc with special tool (clutch center guide) by hand so that the disc is centered. Tighten the cover bolts little by little and evenly in a diagonal order.**

6. Apply a small amount of grease to the input shaft, then join the manual transaxle assembly with the engine. Use grease 99000-25210 (SUZUKI Super Grease I), or equivalent.

➡**When inserting the transaxle input shaft to the clutch disc, turn the crankshaft little by little to match the splines.**

TRANSFER CASE ASSEMBLY

REMOVAL & INSTALLATION

2009 Model

See Figures 30 through 32.

1. Before servicing the vehicle, refer to the Precautions Section.

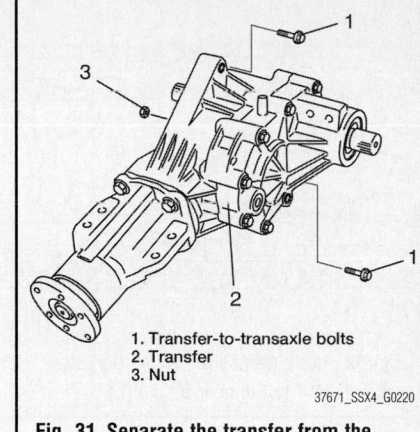

1. Transfer-to-transaxle bolts
2. Transfer
3. Nut

37671_SSX4_G0220

Fig. 31 Separate the transfer from the transaxle

2. Disconnect the negative cable at the battery.
3. Drain the transaxle oil and transfer oil.
4. Remove the front halfshaft assembly. Refer to Halfshafts, removal & installation.
5. Remove the exhaust center pipes.
6. Remove the propeller shaft. Refer to Driveshaft, removal & installation.
7. Remove the engine rear mounting upper nut, lower nut, and stiffeners.
8. Remove the stabilizer bar. Refer to Stabilizer Bar, removal & installation.
9. Remove the front suspension frame front mounting bolt and loosen the front suspension frame rear mounting bolt a little.
10. Remove the transfer-to-transaxle bolts and nut (A/T model), and then separate the transfer with the engine rear mounting from the transaxle.

➡**When separating the transfer and the engine rear mounting from the transaxle, check the front suspension frame for dents. If any dent is found on it, apply rust retardant to the dent.**

11. Remove the front suspension frame rear bolt.

❊❊ WARNING

When removing the transfer, be careful not to drop it. Otherwise, transfer damage and personal injury may result.

12. Lower the front suspension frame with the steering gear case and transfer, and then remove the transfer.

To install:

13. Reverse the removal procedure for remounting of the transfer.
14. Apply grease to the left case and a new O-ring. Apply grease to the O-ring.
  • Use Grease 99000-25010 (SUZUKI Super Grease A), or equivalent

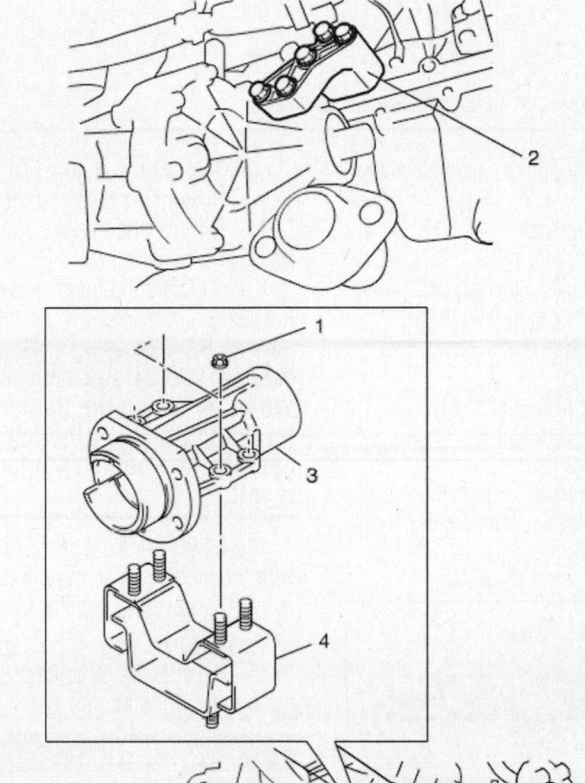

1. Mounting nuts
2. Stiffeners
3. Transfer
4. Engine rear mounting
5. Front suspension frame

37671_SSX4_G0219

Fig. 30 Remove the engine rear mounting nuts and stiffeners

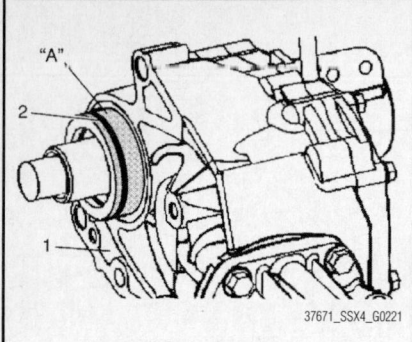

Fig. 32 Apply grease at "A" on the left case (1) and to the new O-ring (2)

15. Tighten the stiffener No. 1 bolt and No. 2 bolt to 40 ft. lbs. (55 Nm).

16. Tighten the transfer-to-transaxle bolt to 71 ft. lbs. (98 Nm).

17. Tighten the engine rear mounting upper nut to 18 ft. lbs. (25 Nm).

18. Tighten the engine rear mounting lower nut to 40 ft. lbs. (55 Nm).

19. Fill the transaxle oil.

20. Fill the transfer oil.

21. Fill the power steering fluid and bleed the air.

22. Check the front wheel alignment and adjust as needed.

2010 Model

See Figures 33 through 35.

1. Before servicing the vehicle, refer to the Precautions Section.

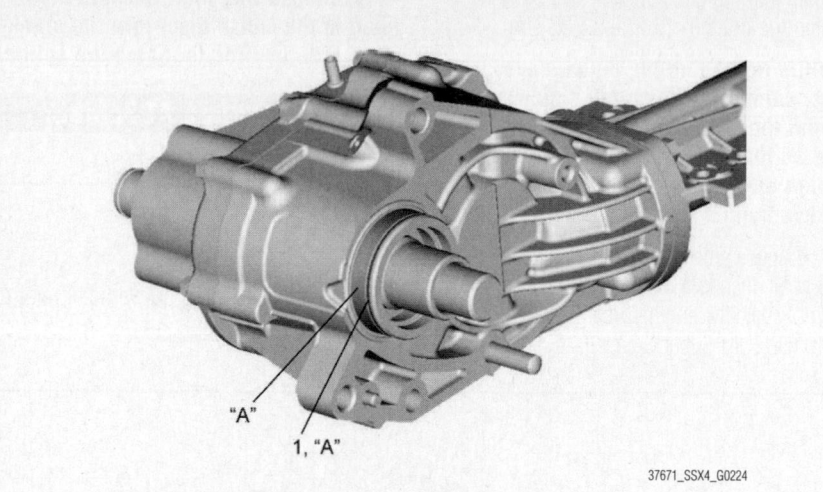

Fig. 34 Apply grease "A" to the left case and O-ring (1)

2. Disconnect the negative cable at the battery.

3. Drain the transaxle.

4. Drain the transfer oil.

5. Drain the Power Steering (P/S) fluid.

6. Remove the right side halfshaft assembly. Refer to Halfshafts, removal & installation.

7. Remove the propeller shaft. Refer to Driveshaft, removal & installation.

8. Remove the front suspension frame (2WD models). Refer to Stabilizer Bar, removal & installation.

9. Remove the engine rear mounting.

10. Disconnect the breather hose.

11. Remove the transfer stiffener, transfer bolts, and transfer nut.

12. Separate the transfer from the transaxle.

✳✳ WARNING

When removing the transfer, be careful not to drop it. Otherwise, transfer damage and personal injury may result.

13. Lower the front suspension frame with the steering gear case and the transfer.

14. Remove the transfer from the vehicle.

To install:

15. Reverse the removal procedure for remounting the transfer.

✳✳ WARNING

Never reuse the suspension arm mounting bolt or front suspension frame mounting bolt. Use new bolts whenever loosening these bolts or they may loosen while under service.

16. Remove the O-ring and dowel pin.

17. Use a new O-ring, halfshaft nut, and a new cotter pin.

18. Apply grease to the left case and O-ring.

- Use Grease 99000-25011 (SUZUKI Super Grease A), or equivalent

19. Tighten the transfer bolts and transfer nut to 65 ft. lbs. (88 Nm).

20. Tighten the transfer stiffener bolts in numerical order evenly and gradually to 41 ft. lbs. (55 Nm).

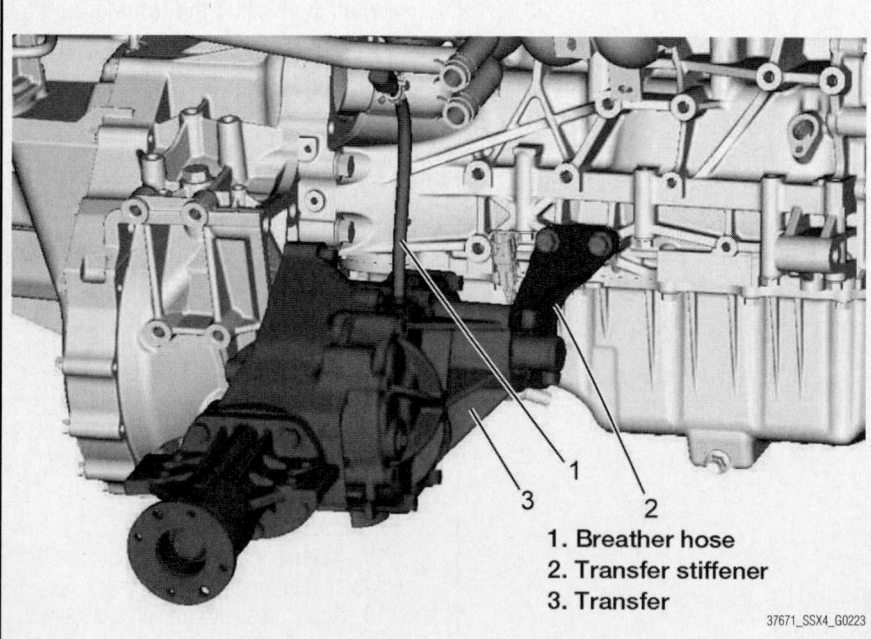

1. Breather hose
2. Transfer stiffener
3. Transfer

Fig. 33 View of transfer stiffener

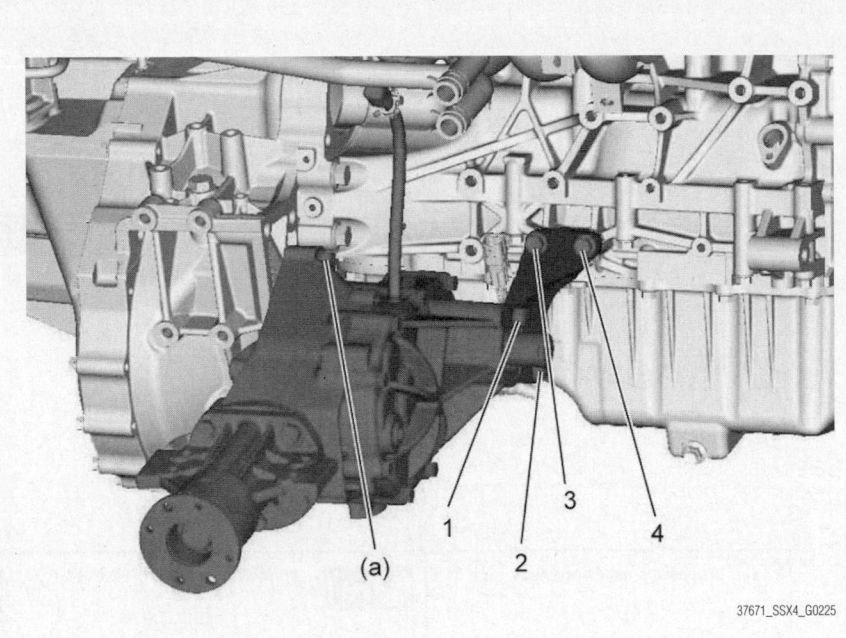

Fig. 35 Transfer stiffener bolt tightening sequence and transfer bolt (a)

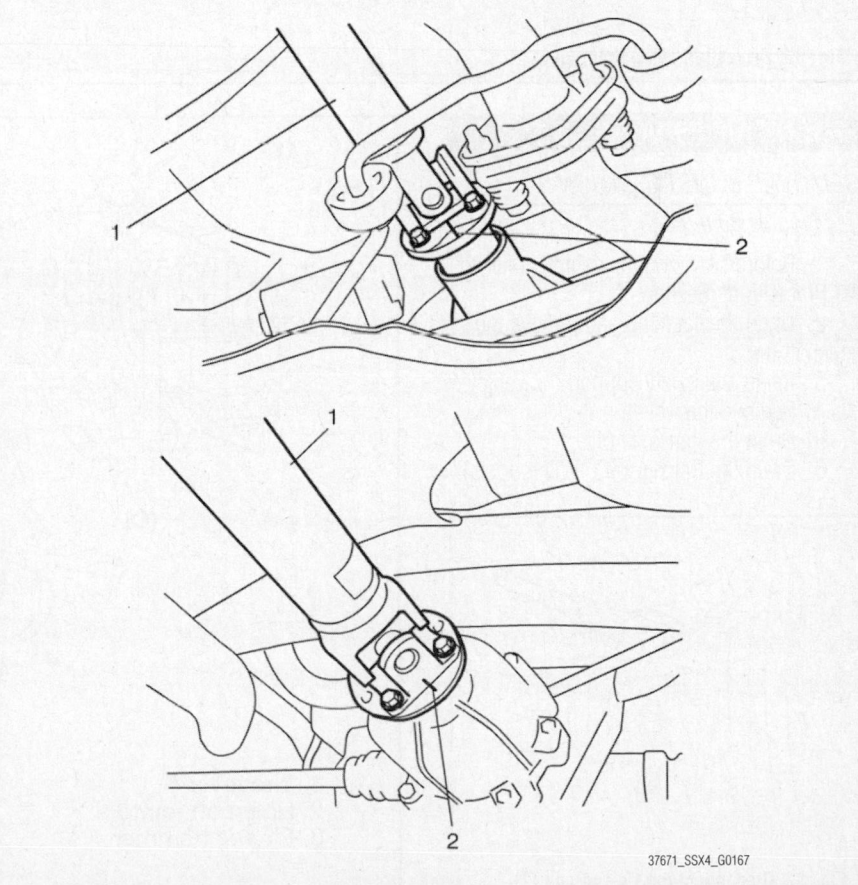

Fig. 37 Remove the propeller shaft by removing the center support nuts (1)

7. When installing the propeller shaft, align the match marks. Otherwise, vibration may occur during driving.

8. Tighten the propeller shaft bolt to 17 ft. lbs. (23 Nm).

9. Tighten the center support nut to 37 ft. lbs. (50 Nm).

21. Fill the transaxle oil.

22. Fill the transfer oil.

23. Fill the power steering fluid and bleed the air.

24. Check the front wheel alignment and adjust as needed.

FRONT DRIVESHAFT

REMOVAL & INSTALLATION

4WD Models

See Figures 36 through 38.

1. Installation is the reverse of the removal procedure.

2. Raise and safely support the vehicle.

3. Before removing the propeller shaft (driveshaft) assembly, make match marks on the joint flange and propeller shaft.

4. Separate the propeller shaft assembly from the transfer output flange and rear differential flange.

5. Remove the propeller shaft by removing the center support nuts.

✳ WARNING

Use care not to drop the propeller shaft. Otherwise, vibration may occur during driving.

To install:

6. Installation is the reverse of the removal procedure.

Fig. 36 Make match marks (2) on the joint flange and propeller shaft. Separate the propeller shaft assembly (1) from the transfer output flange and rear differential flange

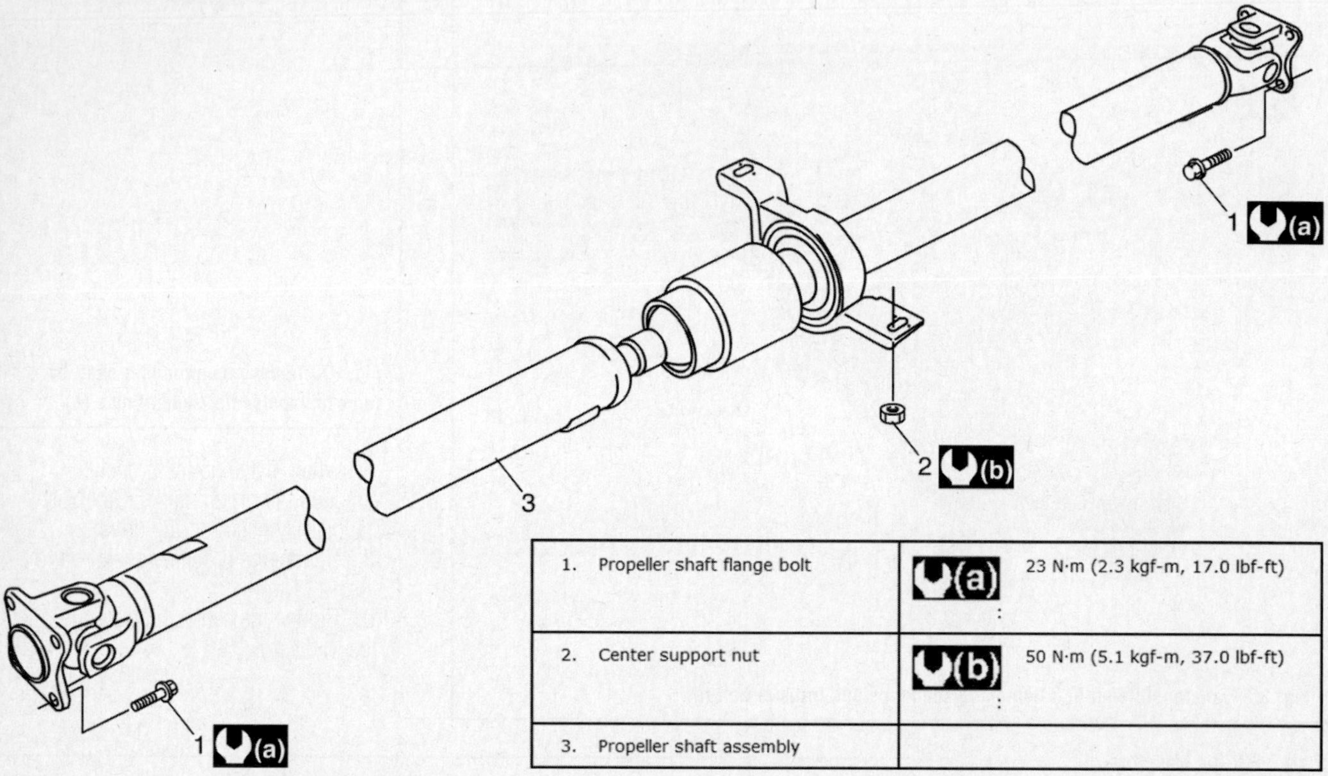

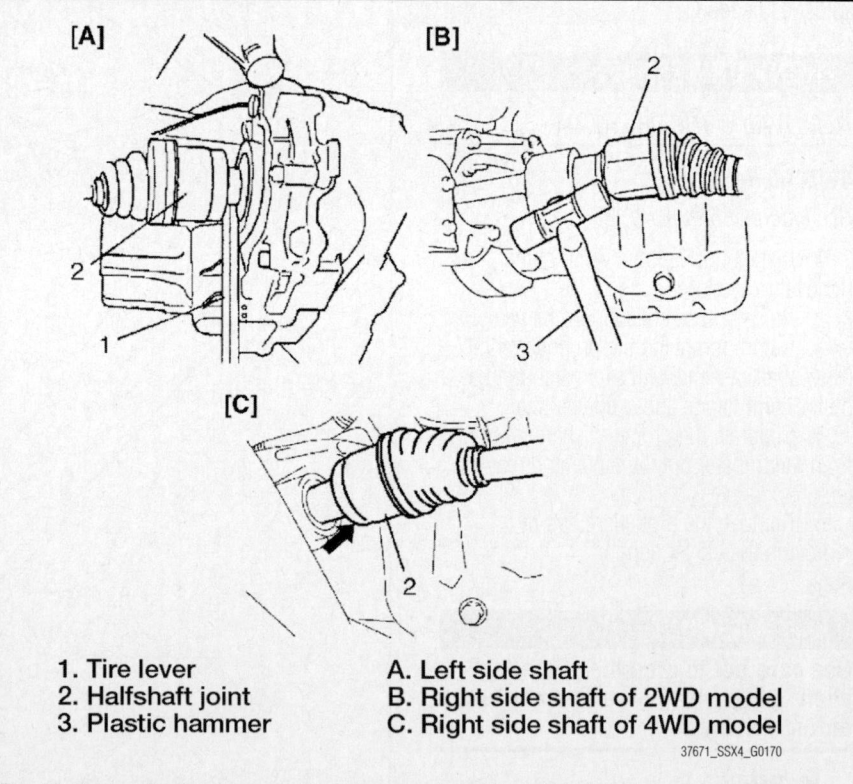

1.	Propeller shaft flange bolt	(a) :	23 N·m (2.3 kgf-m, 17.0 lbf-ft)
2.	Center support nut	(b) :	50 N·m (5.1 kgf-m, 37.0 lbf-ft)
3.	Propeller shaft assembly		

37671_SSX4_G0165

Fig. 38 Driveshaft component view

FRONT HALFSHAFTS

REMOVAL & INSTALLATION

See Figures 39 through 41.

1. Before servicing the vehicle, refer to the Precautions Section.
2. Undo the caulking and remove the halfshaft nut.
3. Raise and safely support the vehicle.
4. Remove the wheel.
5. Drain the transaxle oil.
6. Drain the transfer oil (4WD models).

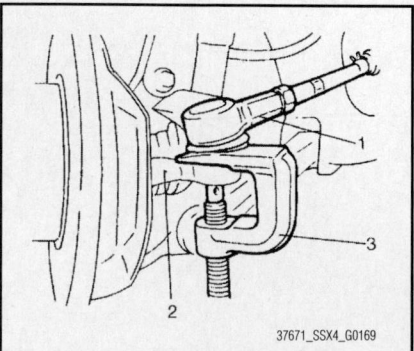

37671_SSX4_G0169

Fig. 39 Disconnect the tie-rod end (1) from the steering knuckle (2) using a puller (3)

1. Tire lever
2. Halfshaft joint
3. Plastic hammer

A. Left side shaft
B. Right side shaft of 2WD model
C. Right side shaft of 4WD model

37671_SSX4_G0170

Fig. 40 Using a tire lever or a plastic hammer, pull out the halfshaft joint so as to release the snap ring fitting of the joint spline

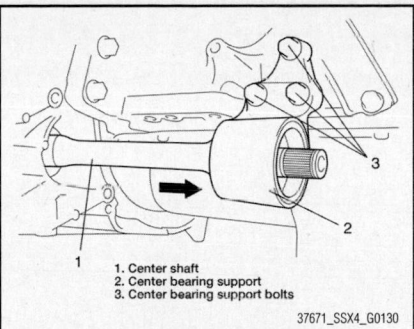

Fig. 41 Remove the center bearing support bolts and center bearing support with the center shaft—2WD model

1. Center shaft
2. Center bearing support
3. Center bearing support bolts

37671_SSX4_G0130

7. Disconnect the tie-rod end from the steering knuckle using a puller.

8. Remove the stabilizer joint.

9. Remove the brake hose mounting bolt.

10. Remove the wheel speed sensor and the suspension control arm ball joint bolt.

11. Disconnect the front suspension control arm ball joint stud from the steering knuckle.

12. Using a tire lever or a plastic hammer, pull out the halfshaft joint so as to release the snap ring fitting of the joint spline at the differential side or at the center shaft.

13. Remove the halfshaft assembly.

14. Remove the center bearing support bolts and remove the center bearing support with the center shaft from the differential side gear, if equipped.

To install:

✷✷ WARNING

Be careful not to damage the oil seals and boots when installing the halfshaft. Do not hit the joint boot with a hammer. Inserting the joint by hand is allowed. Make sure that the differential side joint is inserted fully and its snap ring is seated properly.

15. Install the halfshaft assembly by reversing the removal procedure.

16. Tighten each bolt and nut to the specified torque referring to illustrations and related procedures.

17. Tighten a new halfshaft nut to 145 ft. lbs. (200 Nm). Caulk the halfshaft nut.

18. Tighten the brake hose mounting bolt to 18 ft. lbs. (25 Nm).

19. Fill the transaxle with the proper type and amount of oil.

20. Check the toe setting and adjust as required.

CV-BOOTS INSPECTION

1. Before servicing the vehicle, refer to the Precautions Section.

2. Check the halfshaft boots for damage and deterioration:
- Raise the front of the vehicle
- Rotate the axle and inspect for cracked or ripped CV boot material on the inner and outer CV-joints and check for missing clamps. Repeat this step on both sides of the vehicle
- Inspect for excessive grease deposits on or around the CV boot

3. Replace the boot if it is damaged or deteriorated.

FRONT PINION SEAL

REMOVAL & INSTALLATION

See Figures 42 through 53.

1. Before servicing the vehicle, refer to the Precautions Section.

2. Remove the transfer case. Refer to Transfer Case Assembly, removal & installation.

3. Remove the transfer output flange nut while holding the flange using Special Tool 09930-40113, or equivalent.

4. Remove the transfer output flange using a bearing puller.

5. Drive out the bevel pinion from the transfer output retainer by tapping it with a plastic hammer.

6. Drive out the front taper roller bearing from the bevel pinion by using a bearing puller and hydraulic press.

7. Remove the oil seal using Special Tool 09913-50121, or equivalent.

8. Remove the rear taper roller bearing, pump seal, and spacer, as needed.

9. Drive out the outer races (front and rear) by using a brass bar, if necessary.

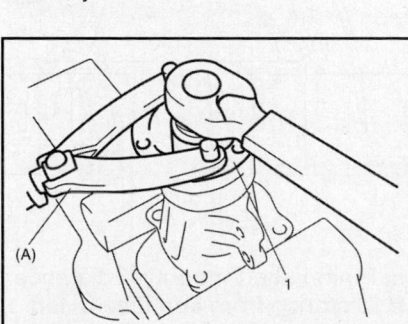

Fig. 42 Using Special Tool 09930-40113 to remove the transfer output flange nut while holding the flange (1)

37671_SSX4_G0174

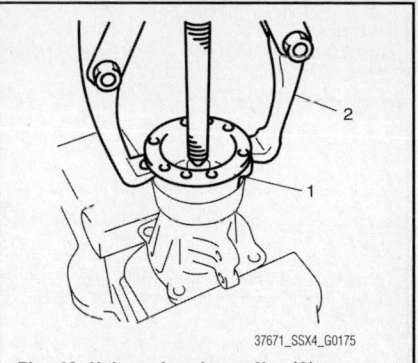

Fig. 43 Using a bearing puller (2) to remove the transfer output flange (1)

37671_SSX4_G0175

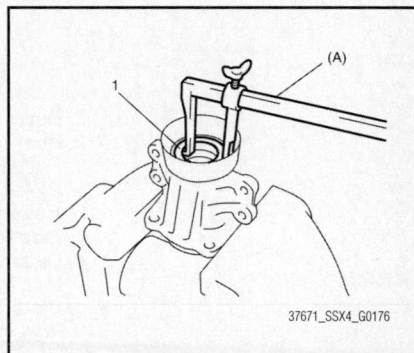

Fig. 44 Using Special Tool 09913-50121 to remove the oil seal (1)

37671_SSX4_G0176

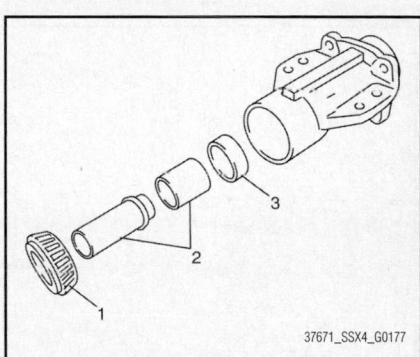

Fig. 45 Removing the rear taper roller bearing (1), pump seal (3), and spacer (2)

37671_SSX4_G0177

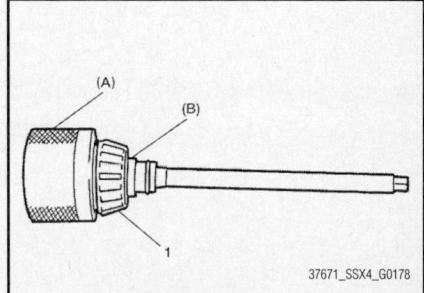

Fig. 46 Install the front taper roller bearing (1) to the bevel pinion dummy Special Tools 09922-76140 and 09922-76430

37671_SSX4_G0178

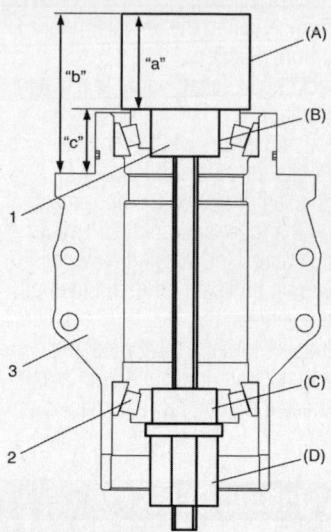

1. Bevel pinion dummy
2. Rear taper roller bearing
3. Transfer output retainer
a. Pinion dummy height
b. Height from retainer installation
   face to top surface of pinion dummy
c. Distance from retainer installation
   face to end face of bearing race
c. Distance from retainer installation
   face to end face of bearing race

A. Special Tool 09922-76140
B. Special Tool 09922-76430
C. Special Tool 09922-76340
D. Special Tool 09922-76150

37671_SSX4_G0179

Fig. 47 Using Special Tools 09922-76140, 09922-76420, 09922-76340, and 09922-76150 for installation

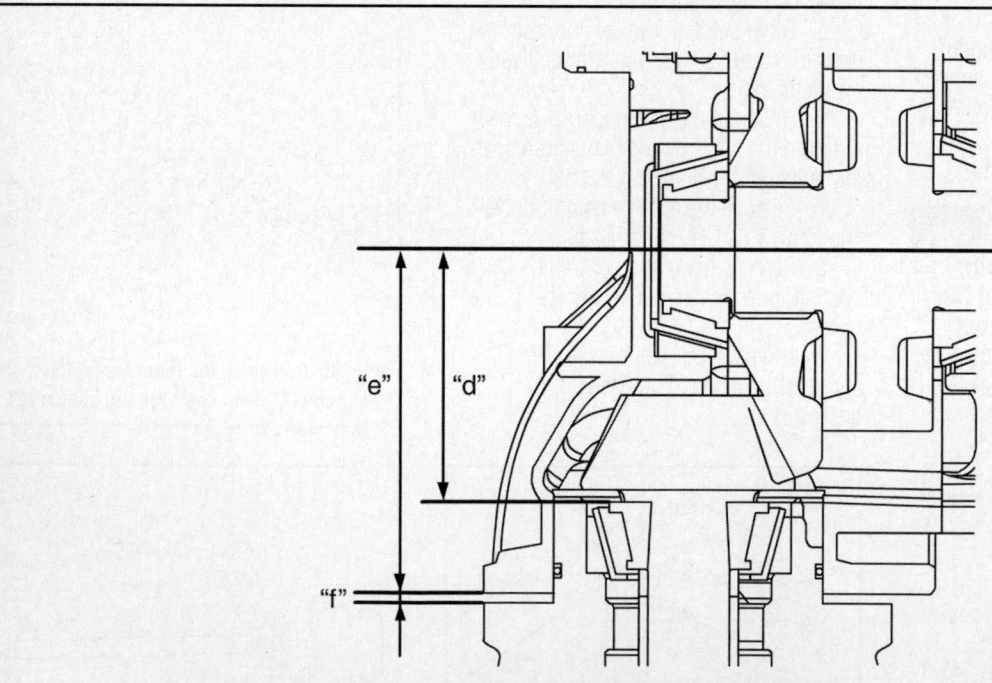

a. Pinion shaft mounting distance
b. Distance from end face of left
   case to axis of reduction driven gear
c. Necessary shim thickness

37671_SSX4_G0180

Fig. 48 Measuring to find the appropriate adjusting shim thickness

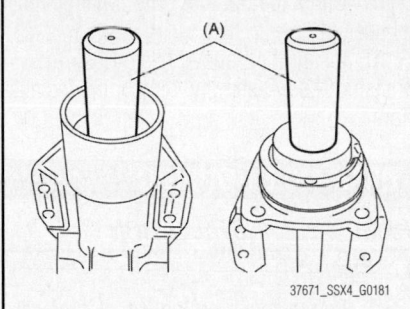

Fig. 49 Using Special Tool 09913-75520 (A) to press-fit the outer races (front and rear)

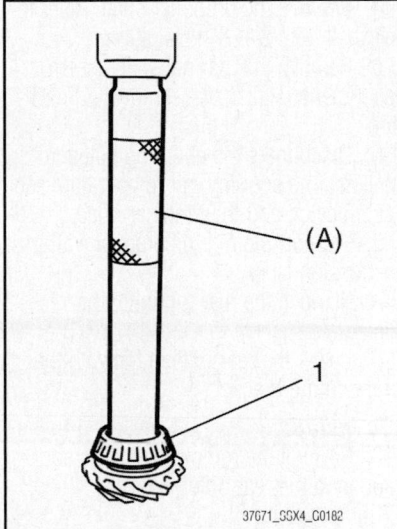

Fig. 50 Using Special Tool 09925-18011 (A) to press-fit the front taper roller bearing (1)

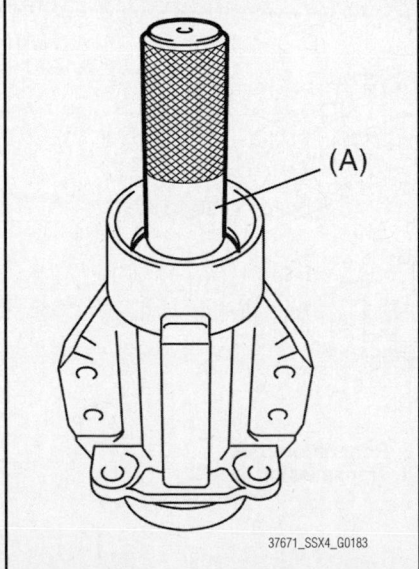

Fig. 51 Using Special Tool 09913-84510 to drive in the rear taper roller bearing

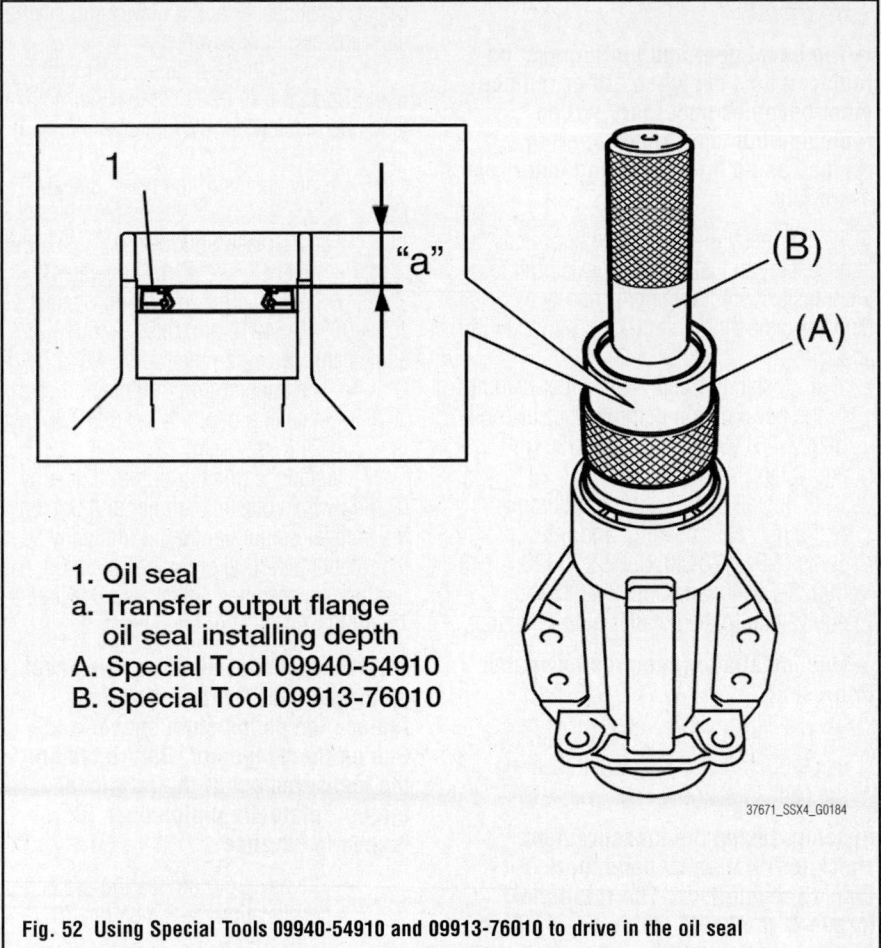

1. Oil seal
a. Transfer output flange oil seal installing depth
A. Special Tool 09940-54910
B. Special Tool 09913-76010

Fig. 52 Using Special Tools 09940-54910 and 09913-76010 to drive in the oil seal

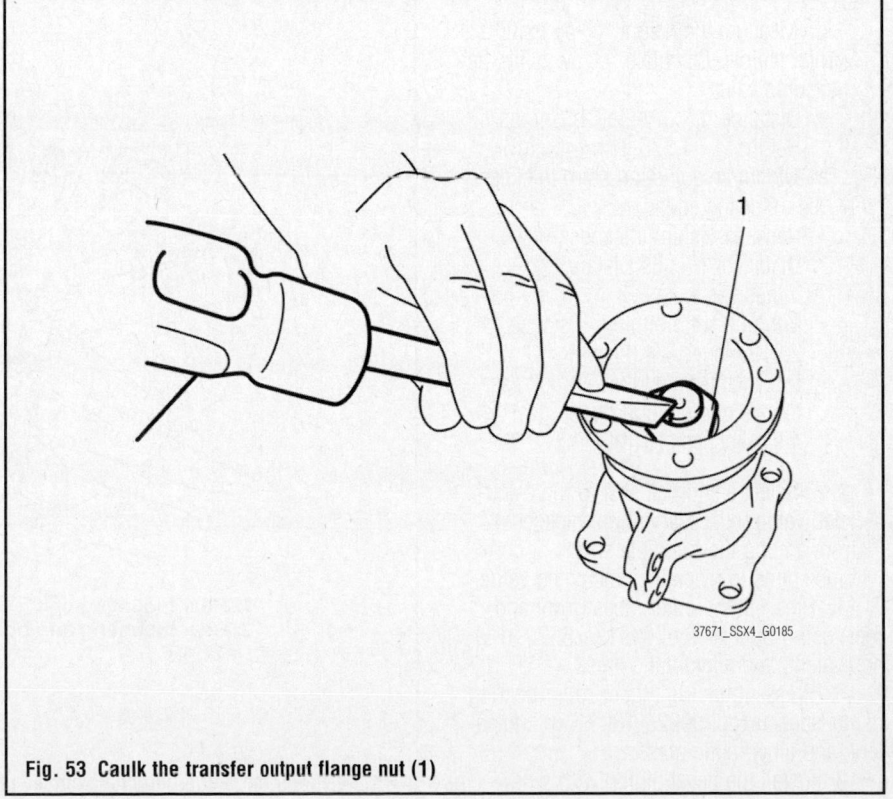

Fig. 53 Caulk the transfer output flange nut (1)

To install:

➡The bevel gear and pinion must be replaced as a set when either replacement becomes necessary. When replacing the taper roller bearing, replace as an inner race and outer race assembly.

10. To mesh the bevel gears correctly, it is necessary to install the bevel pinion to the transfer output retainer properly by using an adjusting shim (bevel pinion shim) as follows.

 a. Install the front taper roller bearing to the bevel pinion dummy Special Tools 09922-76140 and 09922-76430, or equivalent.

 b. Install the bevel pinion dummy, rear taper roller bearing, and Special Tools 09922-76140, 09922-76420, 09922-76340, and 09922-76150, or equivalent, to the transfer output retainer.

➡This installation requires no spacer or oil seal.

 c. Tighten the bevel pinion nut (special tool) so that the specified bearing preload is obtained.

➡Before taking the measurement, check for rotation by hand more than 15 revolutions. The rotational torque of the bevel pinion (bearing preload) should be 3–8 inch lbs. (0.5–1.3 Nm).

 d. Measure the height "b" by using a vernier caliper. Calculate "c" by using the measured value.
 - Distance "c" = Height "b" minus Height "a" (1.575 inch (40mm))

 e. Obtain an adjusting shim thickness by the following equation.
 - Necessary shim thickness "f" = Distance "c" plus Distance "d" minus Distance "e"
 - Pinion shaft mounting distance "d": 2.913 inches (74.0mm)
 - Distance from end face of left case to axis of reduction driven gear "e": 4.014 inches (101.95mm)

37671_ssx4_g0180

 f. Select a shim closest to the calculated value (necessary shim thickness) from among the available shims or combine shims to reach the calculated value.

11. Press-fit the outer races (front and rear) using Special Tool 09913-75520, or equivalent, and a hydraulic press.

12. Press-fit the front taper roller bearing using Special Tool 09925-18011, or equivalent, and a hydraulic press.

13. Install the bevel pinion with a new pinion shaft spacer and a new pump seal to the transfer output retainer.

14. Drive in the rear taper roller bearing using Special Tool 09913-84510, or equivalent, and tapping lightly with a plastic hammer.

15. Apply grease to the new oil seal lip.
 - Use Grease 99000-25011 (SUZUKI Super Grease A), or equivalent

16. Drive in the oil seal using Special Tools 09940-54910 and 09913-76010, or equivalent, and a hammer.
 - The transfer output flange oil seal installing depth "a": 0.82–0.86 inch (21.0–22.0mm)

17. Install the transfer output flange by tapping with a plastic hammer and tighten the transfer output flange nut gradually so the rotational torque of the bevel pinion reaches the specified value. Use Special Tool 09930-40113, or equivalent.

➡If the rotational torque of the bevel pinion exceeds the specification, replace the pinion shaft spacer and tighten the flange nut. Before taking the measurement of the rotational torque, rotate the pinion over 10 rounds in advance.

 - The bevel pinion bearing preload specification: 3–8 inch lbs. (0.5–1.3 Nm).

18. Caulk the transfer output flange nut, as needed.

19. Install the transfer case. Refer to Transfer Case Assembly, removal & installation.

REAR AXLE HOUSING

REMOVAL & INSTALLATION

See Figure 54.

1. Before servicing the vehicle, refer to the Precautions Section.
2. Raise and safely support the vehicle.
3. Drain the oil from the rear differential.
4. Remove the propeller shaft. Refer to Driveshaft, removal & installation.
5. Remove the rear halfshaft assemblies. Refer to Halfshafts, removal & installation.
6. Disconnect the coupling solenoid connector, the coupling air temperature sensor connector, and the breather hose.
7. Support the rear differential with a transmission jack.
8. Remove the rear mounting arm bolts.
9. Lower the rear differential with the rear mounting bracket.

To install:

10. Reverse the removal procedure for remounting the rear differential.

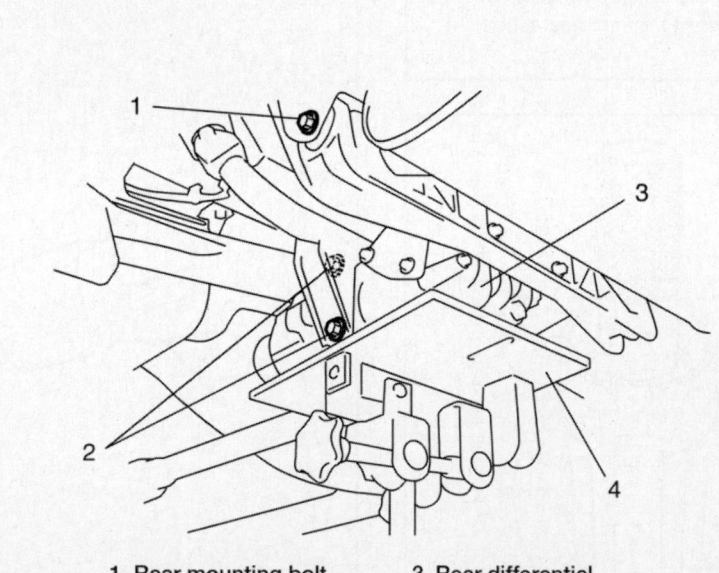

1. Rear mounting bolt
2. Rear mounting arm bolts
3. Rear differential
4. Transmission jack

37671_SSX4_G0207

Fig. 54 Lower the rear differential with the rear mounting bracket

11. Tighten each bolt to the specified torque. Refer to illustrations.

12. Fill the gear oil in the rear differential.

REAR DRIVESHAFT

REMOVAL & INSTALLATION

4WD Models

See Figures 55 through 57.

1. Installation is the reverse of the removal procedure.

2. Raise and safely support the vehicle.

3. Before removing the propeller shaft (driveshaft) assembly, make match marks on the joint flange and propeller shaft.

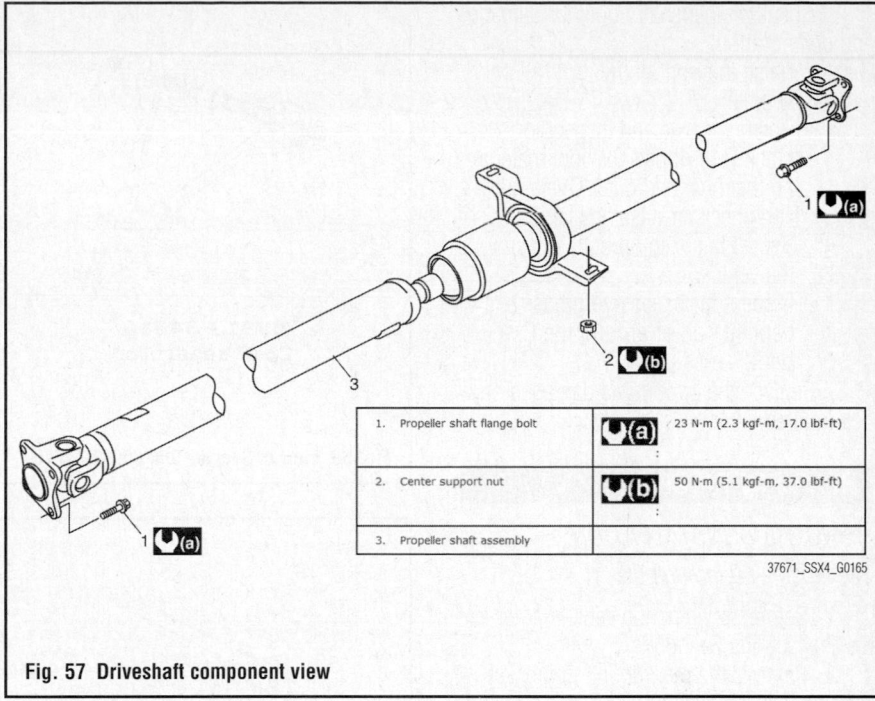

1.	Propeller shaft flange bolt	(a)	23 N·m (2.3 kgf-m, 17.0 lbf-ft)
2.	Center support nut	(b)	50 N·m (5.1 kgf-m, 37.0 lbf-ft)
3.	Propeller shaft assembly		

37671_SSX4_G0165

Fig. 57 Driveshaft component view

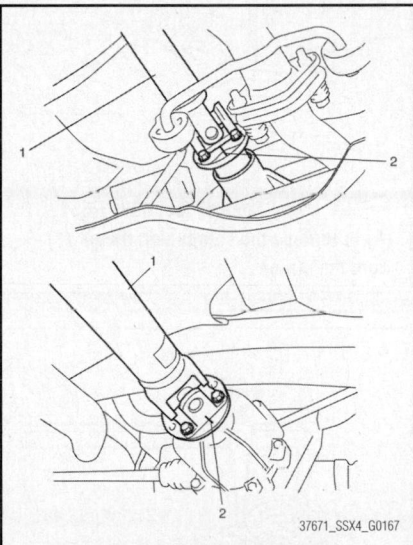

Fig. 55 Make match marks (2) on the joint flange and propeller shaft. Separate the propeller shaft assembly (1) from the transfer output flange and rear differential flange

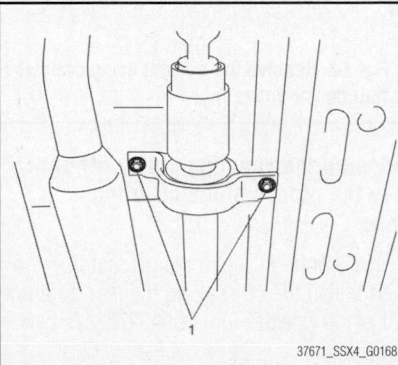

Fig. 56 Remove the propeller shaft by removing the center support nuts (1)

4. Separate the propeller shaft assembly from the transfer output flange and rear differential flange.

5. Remove the propeller shaft by removing the center support nuts.

✶✶ WARNING

Use care not to drop the propeller shaft. Otherwise, vibration may occur during driving.

To install:

6. Installation is the reverse of the removal procedure.

7. When installing the propeller shaft, align the match marks. Otherwise, vibration may occur during driving.

8. Tighten the propeller shaft bolt to 17 ft. lbs. (23 Nm).

9. Tighten the center support nut to 37 ft. lbs. (50 Nm).

REAR HALFSHAFTS

REMOVAL & INSTALLATION

See Figure 58.

1. Before servicing the vehicle, refer to the Precautions Section.

2. Undo the caulking of the halfshaft nut and then remove the halfshaft nut.

3. Raise and safely support the vehicle.

4. Remove the wheel.

5. Remove the rear halfshaft flange

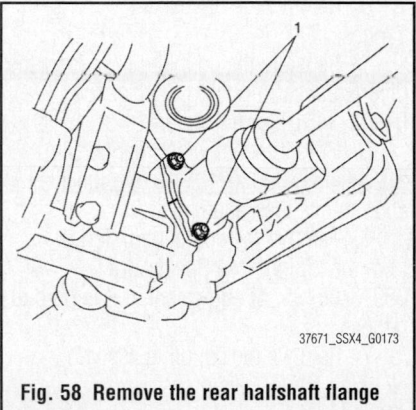

37671_SSX4_G0173

Fig. 58 Remove the rear halfshaft flange nuts (1)

nuts, and then remove the rear halfshaft.

To install:

6. Install the halfshaft assembly by reversing the removal procedure.

✶✶ WARNING

Protect the boots from any damage, preventing them from unnecessary contact while installing the halfshaft. Do not hit the joint boot with a hammer. Inserting the joint by hand is allowed.

7. Tighten a new halfshaft nut to 127 ft. lbs. (175 Nm).

CV-BOOTS INSPECTION

1. Before servicing the vehicle, refer to the Precautions Section.

2. Check the halfshaft boots for damage and deterioration:

- Raise the front of the vehicle
- Rotate the axle and inspect for cracked or ripped CV boot material on the inner and outer CV-joints and check for missing clamps. Repeat this step on both sides of the vehicle
- Inspect for excessive grease deposits on or around the CV boot

3. Replace the boot if it is damaged or deteriorated.

REAR PINION SEAL

REMOVAL & INSTALLATION

See Figures 59 through 66.

1. Before servicing the vehicle, refer to the Precautions Section.

2. Remove the rear mounting bracket from the rear differential.

3. Remove the driveshaft flanges.

4. Separate the coupling case from the differential carrier using Special Tool 09912-34510, or equivalent.

5. Hold the companion flange with Special Tool 09930-40113, or equivalent, and remove the flange nut.

6. Remove the companion flange from the pinion. Use Special Tool 09913-65135, or equivalent, if it is hard to remove.

7. Remove the coupling assembly by using a hydraulic press, as needed.

✳✳ WARNING

Do not drop the coupling assembly. If it is dropped, replace it with a new one.

8. Remove the oil seal using Special Tool 09913-50121, or equivalent.

To install:

9. Apply grease to the oil seal lip, install the oil seal to the coupling case using Special Tool 09913-75810, or equivalent.

➡**Install the oil seal horizontally to the surface of the coupling case. Use Grease 99000-25011 (SUZUKI Super Grease A), or equivalent. The distance between the case and the oil seal should be 0.06–0.10 inch (1.5–2.5mm).**

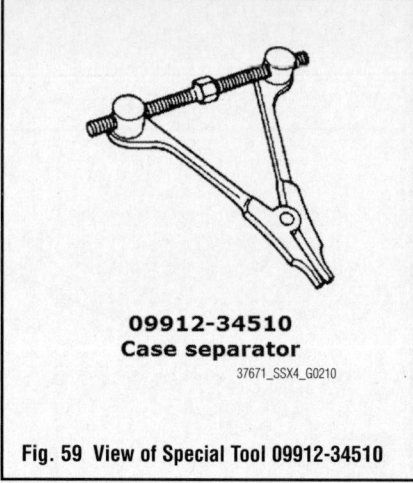

**09912-34510
Case separator**
37671_SSX4_G0210

Fig. 59 View of Special Tool 09912-34510

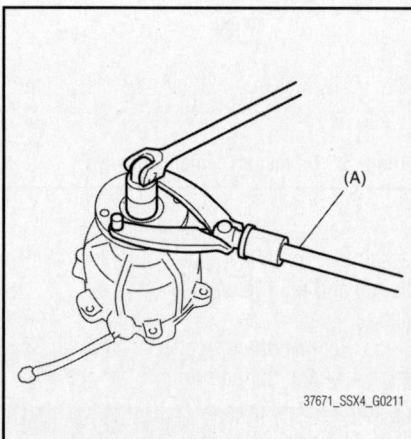

37671_SSX4_G0211

Fig. 60 Hold the companion flange with Special Tool 09930-40113 (A) and remove the flange nut

10. Install the spacer to the coupling assembly.

11. Install the grommet of the coupling harness into the groove of the coupling case and then install the coupling assembly by using a hydraulic press.

12. Install the companion flange to the pinion, and then hold the companion flange with Special Tool 09930-40113, or equivalent, and tighten the flange nut to 80 ft. lbs. (110 Nm.)

13. Clean the mating surface of the differential carrier and the coupling case.

14. Apply sealant to the carrier in a 0.059 inch (1.5mm) bead diameter and mate the coupling case with the differential carrier. Tighten the bolts to 17 ft. lbs. (23 Nm).

- Use Sealant 99000-31260 (SUZUKI Bond No. 1217G), or equivalent.

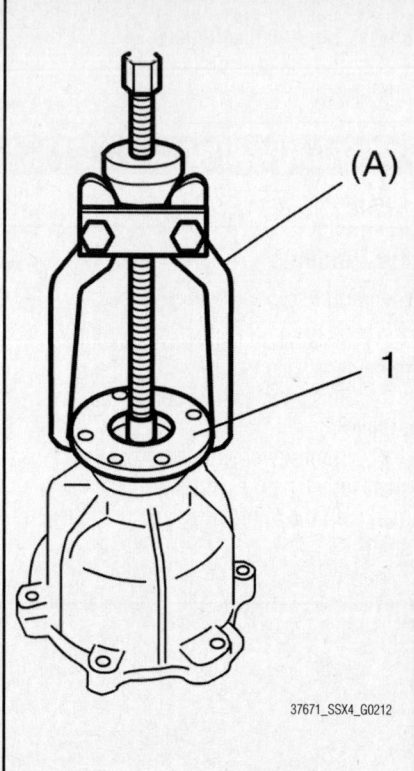

37671_SSX4_G0212

Fig. 61 Using Special Tool 09913-65135 (A) to remove the companion flange (1) from the pinion

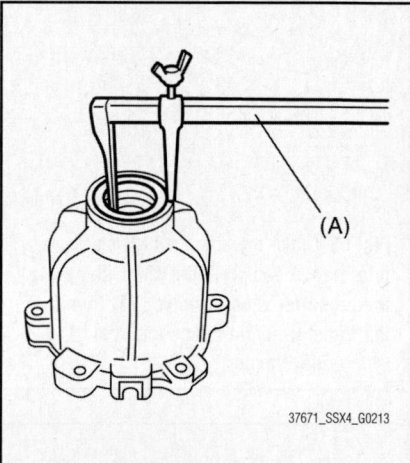

37671_SSX4_G0213

Fig. 62 Remove the oil seal using Special Tool 09913-50121 (A)

➡**Install the coupling pin by fitting it to the groove of the coupling case.**

15. Apply grease to the oil seal lip, and then install the oil seals to the rear differential using Special Tool 09913-75810, or equivalent.

➡**Install the oil seal horizontally to the surface of the rear differential**

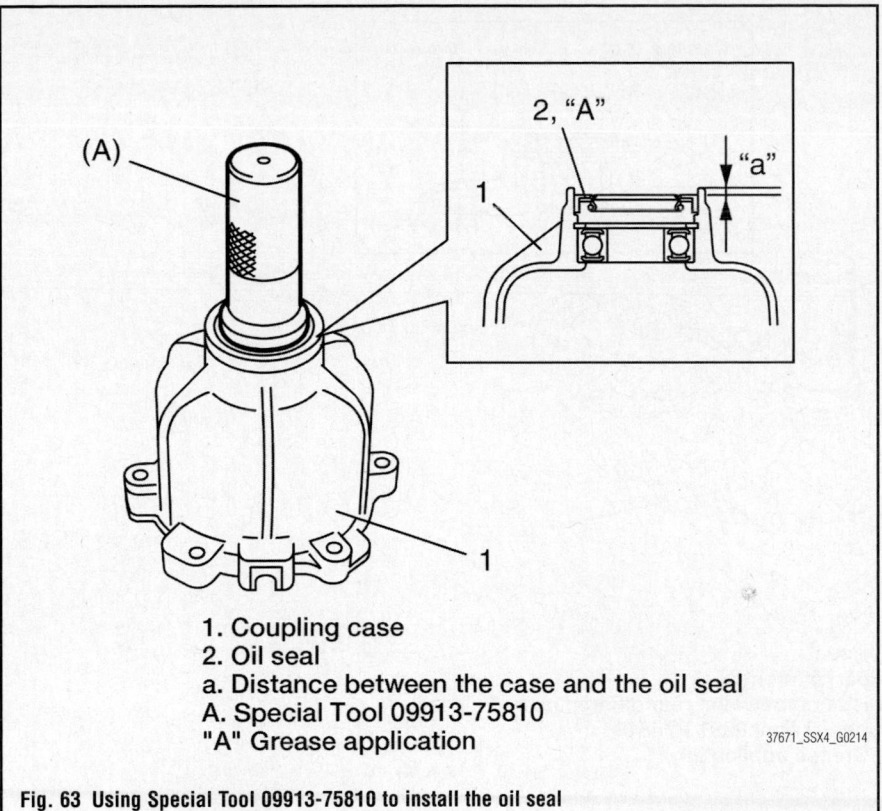

1. Coupling case
2. Oil seal
a. Distance between the case and the oil seal
A. Special Tool 09913-75810
"A" Grease application

37671_SSX4_G0214

Fig. 63 Using Special Tool 09913-75810 to install the oil seal

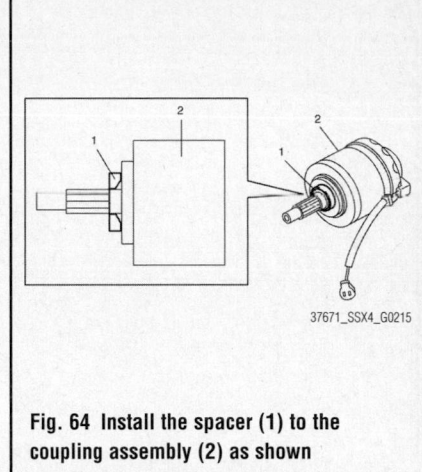

37671_SSX4_G0215

Fig. 64 Install the spacer (1) to the coupling assembly (2) as shown

case. Use Grease 99000-25011 (SUZUKI Super Grease A), or equivalent. The distance between the rear differential and the oil seal should be 0.04–0.08 inch (1.0–2.0mm).

16. Install the drive shaft flange.

17. Install the rear mounting bracket to the rear differential.

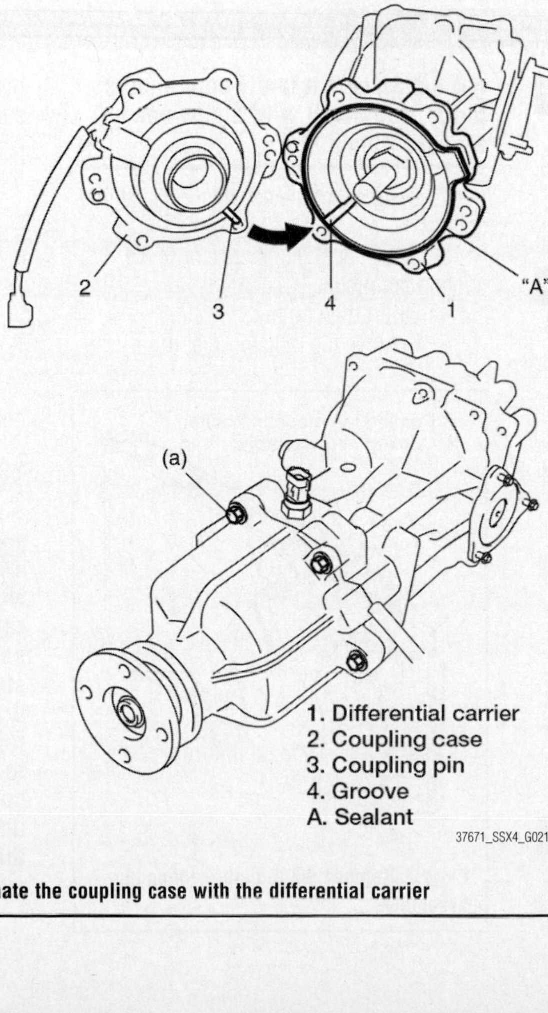

1. Differential carrier
2. Coupling case
3. Coupling pin
4. Groove
A. Sealant

37671_SSX4_G0216

Fig. 65 Apply sealant to the carrier and mate the coupling case with the differential carrier

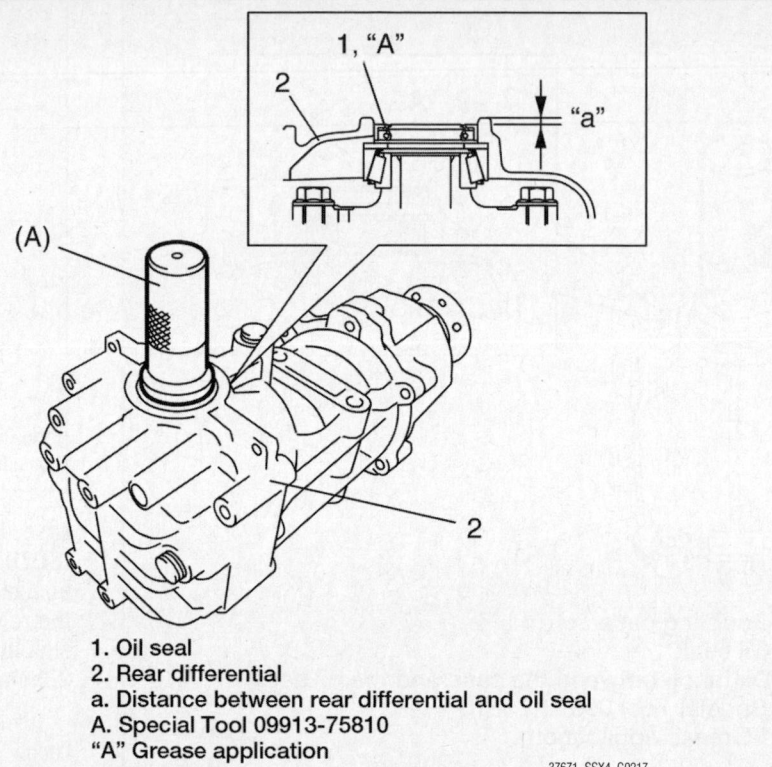

1. Oil seal
2. Rear differential
a. Distance between rear differential and oil seal
A. Special Tool 09913-75810
"A" Grease application

37671_SSX4_G0217

Fig. 66 Using Special Tool 09913-75810 to install the oil seal

ENGINE COOLING

ENGINE FAN

REMOVAL & INSTALLATION

2009 Model

See Figure 67.

> ✳ **CAUTION**
>
> Never open, service or drain the radiator or cooling system when hot; serious burns can occur from the steam and hot coolant. Also, when draining engine coolant, keep in mind that cats and dogs are attracted to ethylene glycol antifreeze and could drink any that is left in an uncovered container or in puddles on the ground. This will prove fatal in sufficient quantities. Always drain coolant into a sealable container. Coolant should be reused unless it is contaminated or is several years old.

1. Before servicing the vehicle, refer to the Precautions Section.

> ✳ **WARNING**
>
> Never disassemble radiator cooling fan. Disassembly will spoil its origi-

nal function. If a faulty condition is found, replace it with a new one as an assembly.

2. Disconnect the negative cable at the battery.
3. Disconnect the connector of the cooling fan motor.
4. Drain the coolant.
5. Remove the radiator inlet hose,

1. Cooling fan mounting bolts
2. Cooling fan assembly

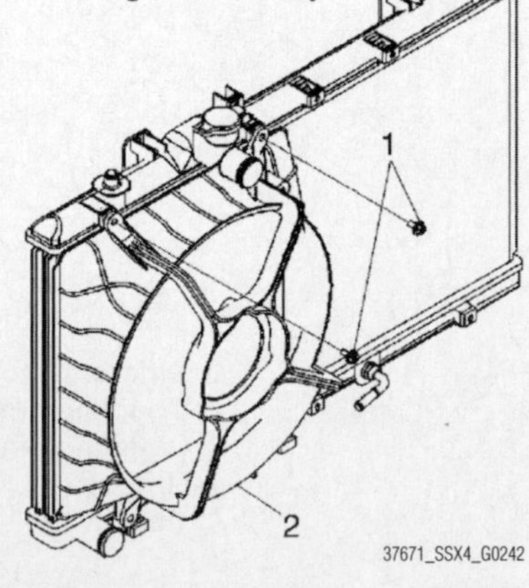

37671_SSX4_G0242

Fig. 67 Remove the radiator cooling fan assembly

reservoir, air suction hose, and air cleaner suction pipe.
6. Remove the cooling fan mounting bolts.
7. Remove the radiator cooling fan assembly.

   To install:
8. Installation is the reverse of the removal procedure.
9. Fill the cooling system.
10. Check coolant leakage at each connection.

2010 Model

See Figures 67 and 68.

> ✳ **CAUTION**
>
> Never open, service or drain the radiator or cooling system when hot; serious burns can occur from the steam and hot coolant. Also, when draining engine coolant, keep in mind that cats and dogs are attracted to ethylene glycol antifreeze and could drink any that is left in an uncovered container or in puddles on the ground. This will prove fatal in sufficient quantities. Always drain

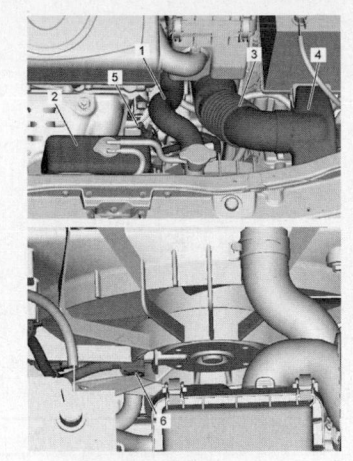

1. Radiator inlet hose
2. Reservoir
3. Air suction hose
4. Air cleaner suction pipe
5. Fluid level gauge guide (CVT model)
6. Cooling fan motor connector 37671_SSX4_G0243

Fig. 68 Remove the hoses and pipes above the cooling fan motor

coolant into a sealable container. **Coolant should be reused unless it is contaminated or is several years old.**

1. Before servicing the vehicle, refer to the Precautions Section.
2. Disconnect the negative cable at the battery.
3. Drain the coolant.
4. Remove the radiator inlet hose, reservoir, air suction hose, air cleaner suction pipe, fluid level gauge, and fluid level gauge guide (CVT model).
5. Disconnect the connector of the cooling fan motor.
6. Remove the cooling fan mounting bolts.
7. Remove the radiator cooling fan assembly.

To install:

8. Installation is the reverse of the removal procedure.
9. Fill the cooling system.

10. Check coolant leakage at each connection.

RADIATOR

REMOVAL & INSTALLATION

See Figures 69 and 70.

✳✳ CAUTION

Never open, service or drain the radiator or cooling system when hot; serious burns can occur from the steam and hot coolant. Also, when draining engine coolant, keep in mind that cats and dogs are attracted to ethylene glycol antifreeze and could drink any that is left in an uncovered container or in puddles on the ground. This will prove fatal in sufficient quantities. Always drain coolant into a sealable container. Coolant should be reused unless it is contaminated or is several years old.

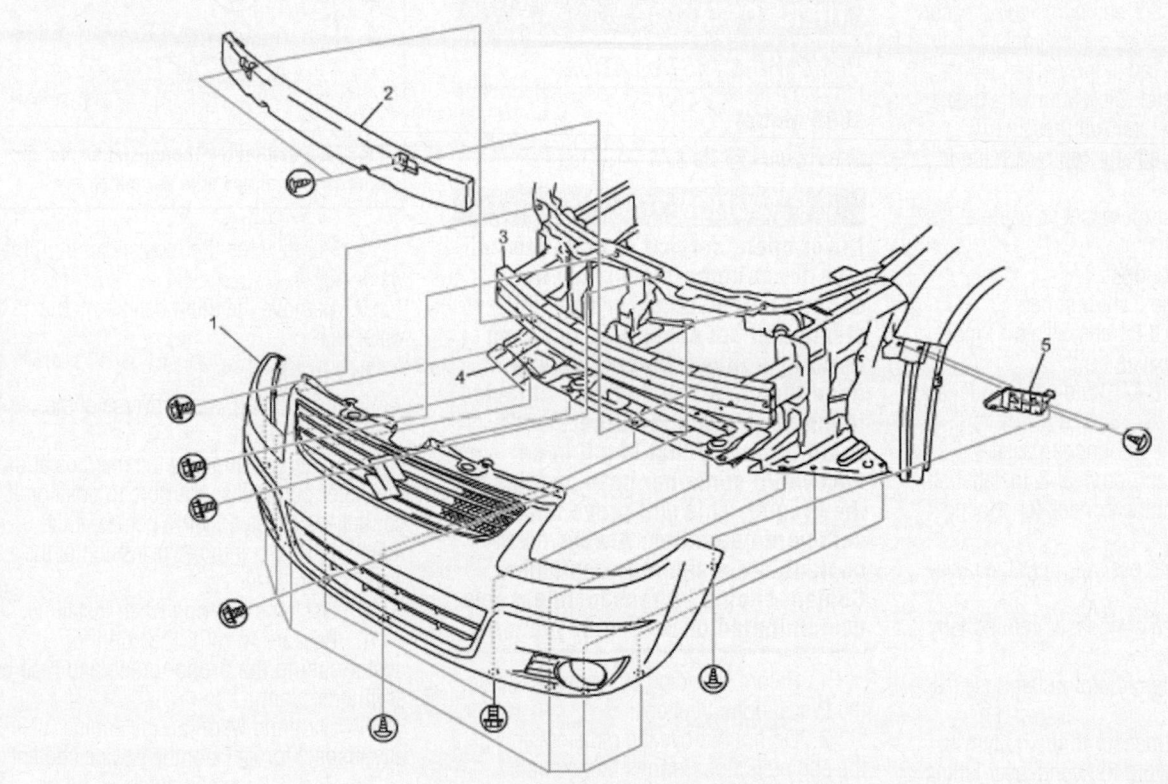

1. Front bumper
2. Front bumper absorber (if equipped)
3. Front bumper upper member
4. Front bumper lower member
5. Front bumper holder

37671_SSX4_G0116

Fig. 69 Exploded view of the front bumper

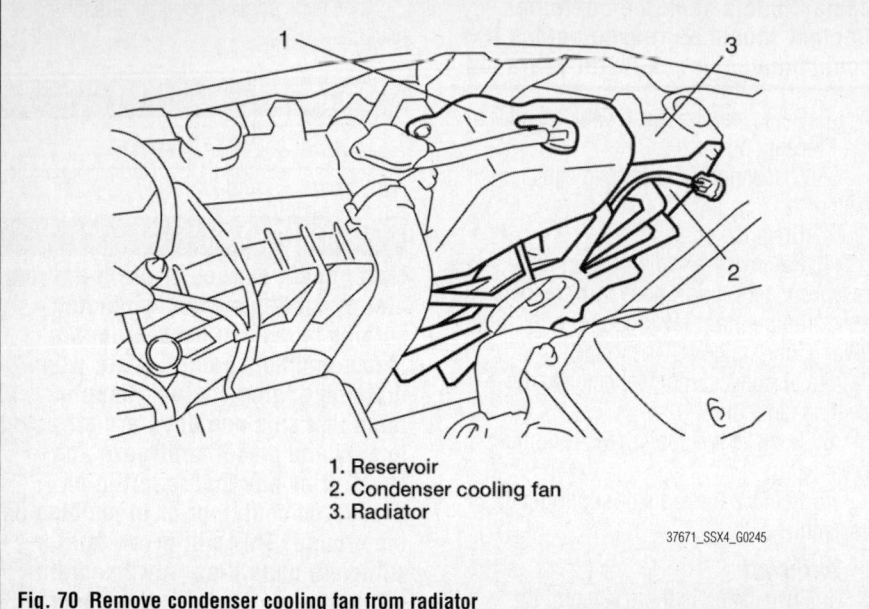

1. Reservoir
2. Condenser cooling fan
3. Radiator

37671_SSX4_G0245

Fig. 70 Remove condenser cooling fan from radiator

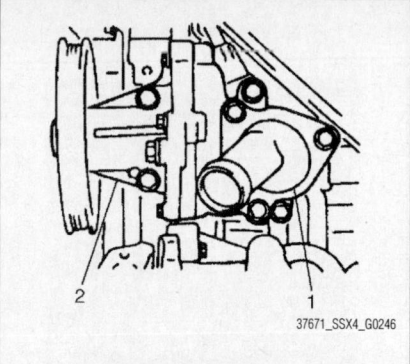

37671_SSX4_G0246

Fig. 71 Remove the thermostat cap from the water pump

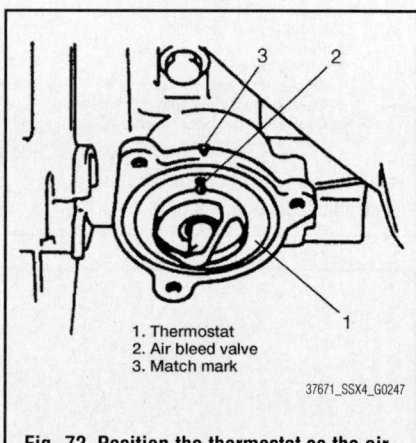

1. Thermostat
2. Air bleed valve
3. Match mark

37671_SSX4_G0247

Fig. 72 Position the thermostat so the air bleed valve aligns with the match mark

1. Before servicing the vehicle, refer to the Precautions Section.

2. Do not remove the radiator cap when the engine is hot. Serious burns could occur from the high-pressure engine coolant escaping from the radiator. Wrap a thick cloth around the cap. Slowly turn it a quarter of a turn to release built-up pressure. Carefully remove radiator cap by turning it all the way.

3. Disconnect the negative cable at the battery.

4. Drain the coolant.

5. Remove the front bumper.

6. For automatic transaxle (A/T) models, drain the transaxle fluid.

7. Remove the A/T fluid cooler inlet and outlet hoses.

8. Remove the cooling fan assembly. Refer to Engine Fan, removal & installation.

9. Disconnect the condenser cooling fan motor coupler.

10. Remove the reservoir from the radiator.

11. Remove the condenser cooling fan from the radiator.

12. Remove the radiator outlet hose from the radiator.

13. Fix the condenser to the body with rope in order to avoid the condenser falling off when removing the radiator.

14. Remove the radiator from the vehicle.

To install:

15. Installation is the reverse of the removal procedure.

16. Fill the engine coolant.

17. For A/T models, fill the fluid.

18. Start and warm up the engine.

Visually inspect for leaks of the engine coolant.

THERMOSTAT

REMOVAL & INSTALLATION

2009 Model

See Figures 71 and 72.

✳✳ CAUTION

Never open, service or drain the radiator or cooling system when hot; serious burns can occur from the steam and hot coolant. Also, when draining engine coolant, keep in mind that cats and dogs are attracted to ethylene glycol antifreeze and could drink any that is left in an uncovered container or in puddles on the ground. This will prove fatal in sufficient quantities. Always drain coolant into a sealable container. Coolant should be reused unless it is contaminated or is several years old.

1. Before servicing the vehicle, refer to the Precautions Section.

2. Do not remove the radiator cap when the engine is hot. Serious burns could occur from the high-pressure engine coolant escaping from the radiator. Wrap a thick cloth around the cap. Slowly turn it a quarter of a turn to release built-up pressure. Carefully remove radiator cap by turning it all the way.

3. Drain the coolant.

4. Disconnect the radiator outlet hose from the thermostat cap.

5. Remove the thermostat cap from the water pump.

6. Remove the thermostat from the water pump.

To install:

7. Installation is the reverse of the removal procedure.

8. When positioning the thermostat on the water pump case, be sure to position it so that the air bleed valve comes at the match mark and into the recession of the water pump case.

9. Use a new O-ring when installing.

10. Be sure to refill the cooling system using the proper grade and type of engine coolant.

11. Start and warm up the engine. Visually inspect for leaks of the engine coolant.

12. Recheck the coolant level, fill as required.

2010 Model

See Figures 69, 73 through 75.

✳✳ CAUTION

Never open, service or drain the radiator or cooling system when hot;

serious burns can occur from the steam and hot coolant. Also, when draining engine coolant, keep in mind that cats and dogs are attracted to ethylene glycol antifreeze and could drink any that is left in an uncovered container or in puddles on the ground. This will prove fatal in sufficient quantities. Always drain coolant into a sealable container. Coolant should be reused unless it is contaminated or is several years old.

1. Before servicing the vehicle, refer to the Precautions Section.
2. Do not remove the radiator cap when the engine is hot. Serious burns could occur from the high-pressure engine coolant escaping from the radiator. Wrap a thick cloth around the cap. Slowly turn it a quarter of a turn to release built-up pressure. Carefully remove radiator cap by turning it all the way.
3. Drain the coolant.
4. Remove the front bumper.

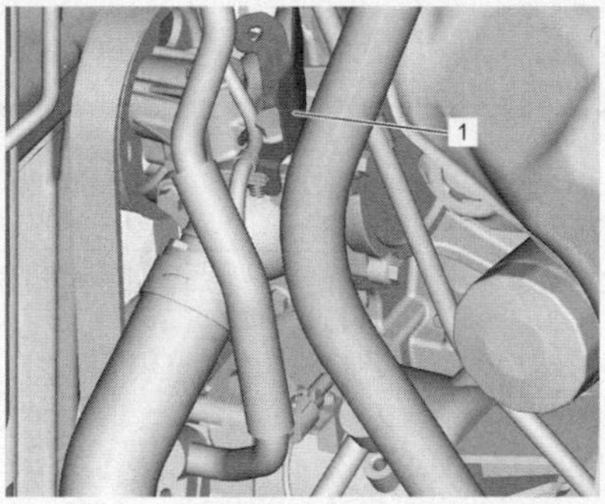
1
37671_SSX4_G0248

Fig. 73 Remove the engine wire harness bracket (1)

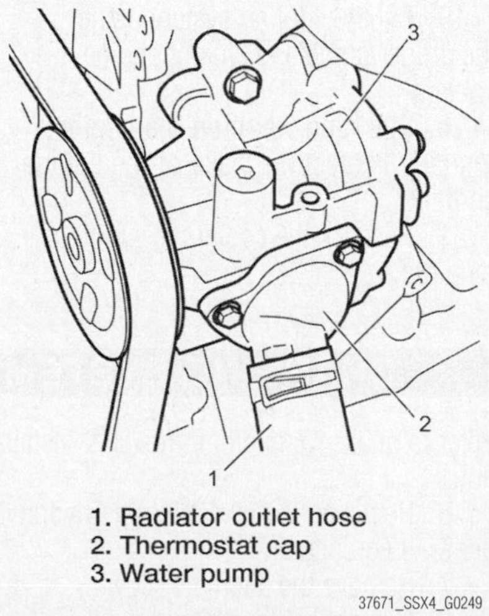
3
2
1
1. Radiator outlet hose
2. Thermostat cap
3. Water pump
37671_SSX4_G0249

Fig. 74 Remove the thermostat cap from the water pump

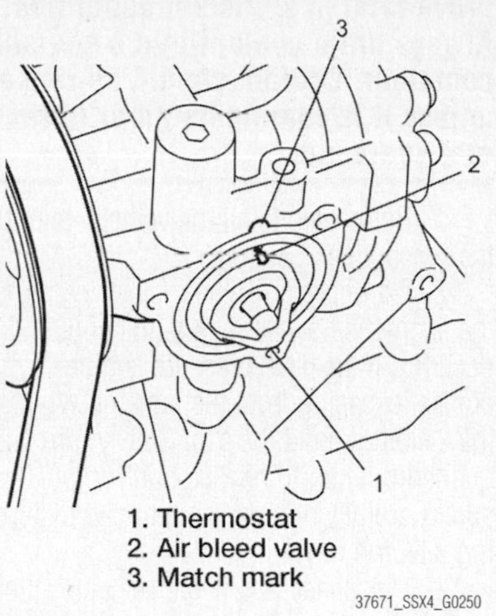
3
2
1
1. Thermostat
2. Air bleed valve
3. Match mark
37671_SSX4_G0250

Fig. 75 Position the thermostat so the air bleed valve aligns with the match mark

5. Remove the engine wire harness bracket.
6. Disconnect the radiator outlet hose from the thermostat cap.
7. Remove the thermostat cap from the water pump.
8. Remove the thermostat from the water pump.

To install:

9. Installation is the reverse of the removal procedure.
10. When positioning the thermostat on the water pump case, be sure to position it so that the air bleed valve is aligned with the match mark.
11. Use a new gasket when installing the thermostat.
12. Fill the cooling system using the proper grade and type of engine coolant.
13. Start and warm up the engine. Visually inspect for leaks of the engine coolant.
14. Recheck the coolant level, fill as required.

WATER PUMP

REMOVAL & INSTALLATION

2009 Model

See Figures 69 and 76.

⁂ CAUTION

Never open, service or drain the radiator or cooling system when hot; serious burns can occur from the steam and hot coolant. Also, when draining engine coolant, keep in mind that cats and dogs are attracted to ethylene glycol antifreeze and could drink any that is left in an uncovered container or in puddles on the ground. This will prove fatal in sufficient quantities. Always drain coolant into a sealable container. Coolant should be reused unless it is contaminated or is several years old.

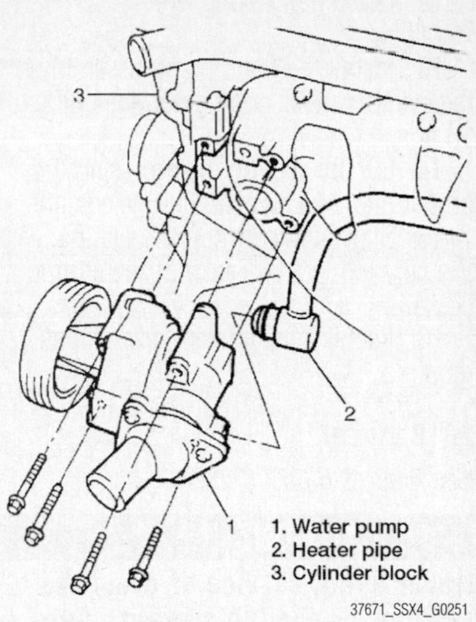
3
2
1
1. Water pump
2. Heater pipe
3. Cylinder block
37671_SSX4_G0251

Fig. 76 Removing the water pump—2009 model

1. Before servicing the vehicle, refer to the Precautions Section.
2. Do not remove the radiator cap when the engine is hot. Serious burns could occur from the high-pressure engine coolant escaping from the radiator. Wrap a thick cloth around the cap. Slowly turn it a quarter of a turn to release built-up pressure. Carefully remove radiator cap by turning it all the way.
3. Disconnect the negative cable at the battery.
4. Drain the coolant.
5. Remove the water pump and alternator drive belt. Refer to Accessory Drive Belts, removal & installation.
6. Remove the front bumper.
7. Remove the exhaust manifold. Refer to Exhaust Manifold, removal & installation.
8. Remove the heater pipe from the water pump and cylinder block.
9. Remove the thermostat. Refer to Thermostat, removal & installation.
10. Remove the water pump from the cylinder block.

To install:

11. Installation is the reverse of the removal procedure.

12. Use a new O-ring when installing.

13. Install the water pump assembly to the cylinder block and tighten the bolts to 18 ft. lbs. (25 Nm).

14. Fill the cooling system using the proper grade and type of engine coolant.

15. Start and warm up the engine. Visually inspect for leaks of the engine coolant.

16. Recheck the coolant level, fill as required.

2010 Model

See Figures 69, 77 and 78.

✳✳ CAUTION

Never open, service or drain the radiator or cooling system when hot; serious burns can occur from the steam and hot coolant. Also, when draining engine coolant, keep in mind that cats and dogs are attracted to ethylene glycol antifreeze and could drink any that is left in an uncovered container or in puddles on the ground. This will

prove fatal in sufficient quantities. Always drain coolant into a sealable container. Coolant should be reused unless it is contaminated or is several years old.

1. Before servicing the vehicle, refer to the Precautions Section.

2. Do not remove the radiator cap when the engine is hot. Serious burns could occur from the high-pressure engine coolant escaping from the radiator. Wrap a thick cloth around the cap. Slowly turn it a quarter of a turn to release built-up pressure. Carefully remove radiator cap by turning it all the way.

3. Disconnect the negative cable at the battery.

4. Drain the coolant.

5. Remove the accessory drive belt. Refer to Accessory Drive Belts, removal & installation.

6. Remove the front bumper.

7. Disconnect the condenser cooling fan motor coupler.

8. Remove the reservoir from the radiator.

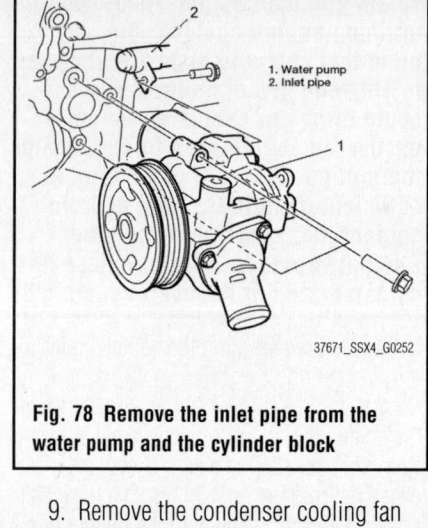

1. Water pump
2. Inlet pipe

37671_SSX4_G0252

Fig. 78 Remove the inlet pipe from the water pump and the cylinder block

9. Remove the condenser cooling fan from the radiator.

10. Remove the oil level gauge and the oil level gauge guide.

11. Remove the exhaust manifold. Refer to Exhaust Manifold, removal & installation.

12. With the hose connected, detach the A/C compressor from the cylinder block.

13. Disconnect the radiator outlet hose from the water pump.

14. Remove the inlet pipe from the water pump and the cylinder block.

15. Remove the water pump from the cylinder block.

To install:

16. Installation is the reverse of the removal procedure.

17. Use a new gasket and O-ring.

18. Install the water pump assembly to the cylinder block and tighten the bolts to 19 ft. lbs. (25 Nm).

19. Fill the cooling system using the proper grade and type of engine coolant.

20. Start and warm up the engine. Visually inspect for leaks of the engine coolant.

21. Recheck the coolant level, fill as required.

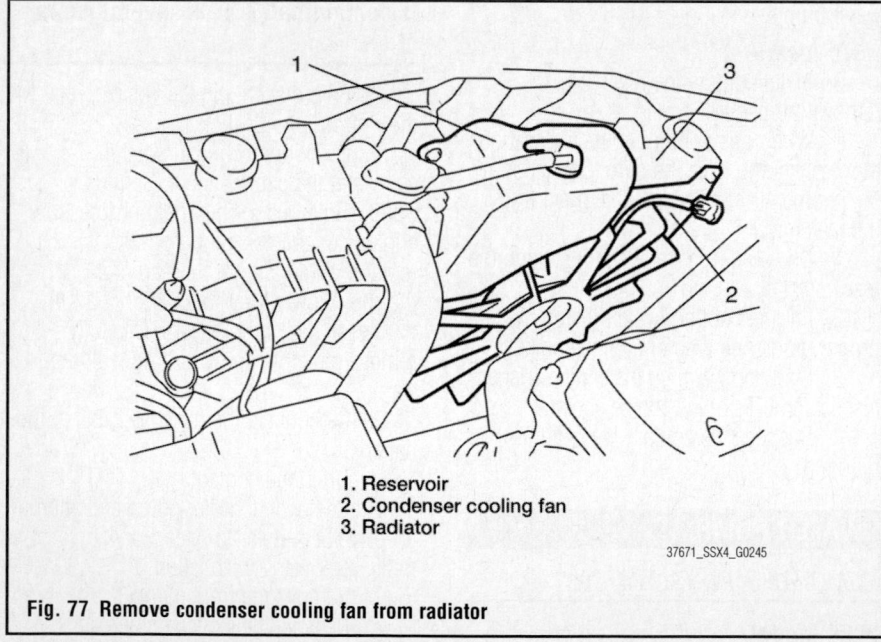

1. Reservoir
2. Condenser cooling fan
3. Radiator

37671_SSX4_G0245

Fig. 77 Remove condenser cooling fan from radiator

ENGINE ELECTRICAL

ALTERNATOR

REMOVAL & INSTALLATION

2009 Model

See Figure 79.

1. Before servicing the vehicle, refer to the Precautions Section.

2. Disconnect the negative cable at the battery.

3. Remove the water pump and alternator drive belt. Refer to Accessory Drive Belts, removal & installation.

4. Remove the air cleaner case.

5. Remove the intake manifold.

CHARGING SYSTEM

Refer to Intake Manifold, removal & installation.

6. Remove the power steering hydraulic pump, if equipped.

7. Remove the alternator cover.

8. Disconnect the "B" terminal wire and the coupler from the alternator.

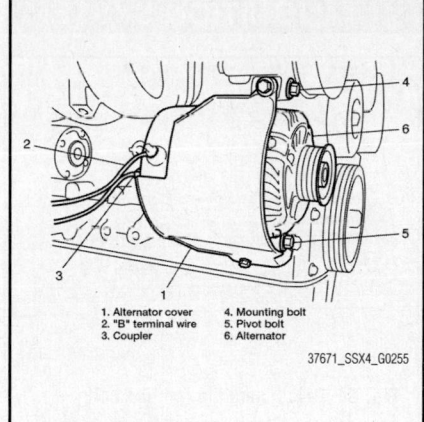

1. Alternator cover
2. "B" terminal wire
3. Coupler
4. Mounting bolt
5. Pivot bolt
6. Alternator

37671_SSX4_G0255

Fig. 79 View of alternator removal—2009 model

9. Remove the alternator mounting bolt and the alternator pivot bolt from the alternator.

10. Remove the alternator from the engine.

To install:

11. Installation is the reverse of the removal procedure.

12. Tighten the "B" terminal nut to 44 inch lbs. (5 Nm).

13. Tighten the alternator mounting bolt to 19 ft. lbs. (25 Nm).

14. Tighten the alternator pivot bolt to 38 ft. lbs. (53 Nm).

2010 Model

See Figure 80.

1. Before servicing the vehicle, refer to the Precautions Section.

2. Disconnect the negative cable at the battery.

3. Remove the intake manifold. Refer to Intake Manifold, removal & installation.

4. Remove the accessory drive belt. Refer to Accessory Drive Belts, removal & installation.

5. Disconnect the "B" terminal wire and the connector from the alternator.

6. Remove the alternator.

To install:

7. Installation is the reverse of the removal procedure.

8. Tighten the "B" terminal wire nut to 97 inch lbs. (11 Nm).

[A]

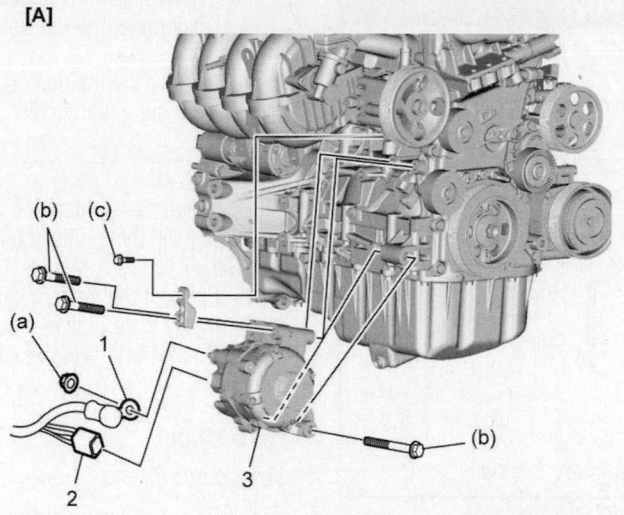

[B]

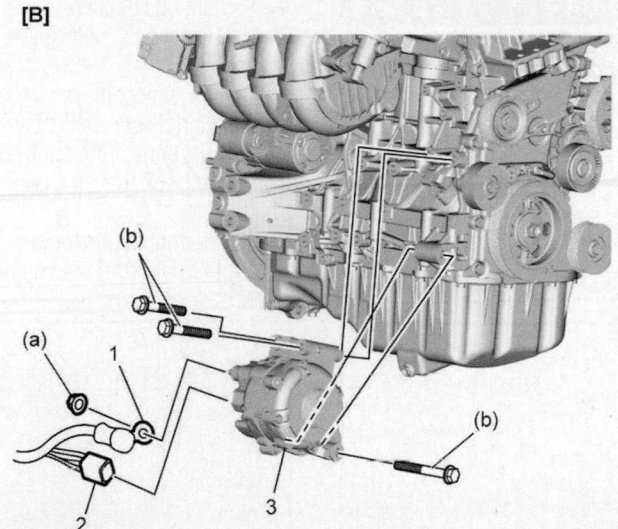

1. "B" terminal wire
2. Connector
3. Alternator
a. "B" terminal wire nut
b. Alternator bolts
c. Power steering pump stay bolt

A. Hydraulic P/S model
B. EPS model

37671_SSX4_G0256

Fig. 80 View of alternator removal—2010 model

9. Tighten the alternator bolt to 34 ft. lbs. (45 Nm).

10. Tighten the power steering pump stay bolt to 19 ft. lbs. (25 Nm).

FIRING ORDER

See Figure 81.

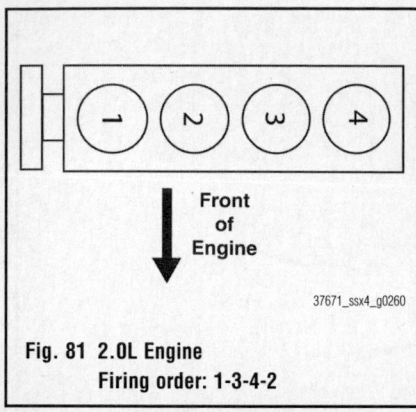

**Fig. 81 2.0L Engine
Firing order: 1-3-4-2**

IGNITION COIL

REMOVAL & INSTALLATION

2009 Model

See Figure 82.

1. Before servicing the vehicle, refer to the Precautions Section.
2. Disconnect the negative cable at the battery.

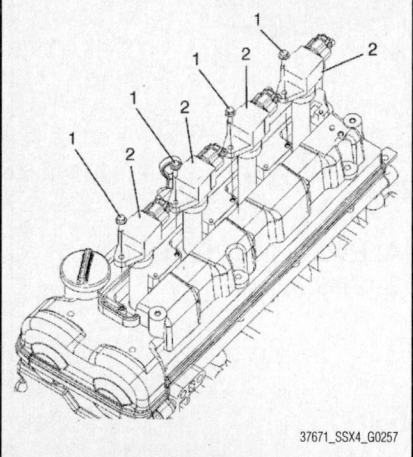

37671_SSX4_G0257

Fig. 82 Remove the ignition coil bolts (1) and pull out the ignition coil assembly (2)—2009 model

3. Remove the air cleaner case.
4. Disconnect the ignition coil coupler.
5. Remove the ignition coil bolts and pull out the ignition coil assembly.

To install:

6. Install the ignition coil assembly.
7. Tighten the ignition coil bolts to 58 inch lbs. (7 Nm).
8. Connect the ignition coil coupler.
9. Install the air cleaner case.
10. Connect the negative cable to the battery.

2010 Model

See Figures 83 and 84.

1. Before servicing the vehicle, refer to the Precautions Section.
2. Disconnect the negative cable at the battery.
3. Disconnect the engine cover hose from the air cleaner assembly.
4. Disconnect the engine cover front side hook from the engine cover pin.
5. Disconnect the engine cover rear side hook from the engine cover pin.

1. Engine cover hose
2. Engine cover front side hook
3. Engine cover rear side hook

37671_SSX4_G0259

Fig. 83 Remove the engine cover

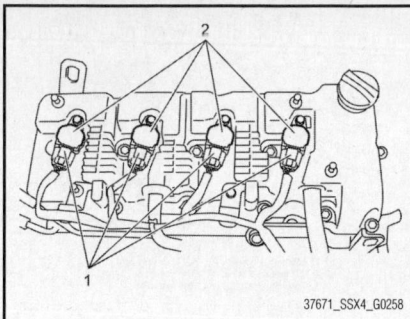

37671_SSX4_G0258

Fig. 84 Disconnect the ignition coil connectors (1) and remove the ignition coil assemblies (2)—2010 model

6. Remove the engine cover from the vehicle.

✴✴ WARNING

When removing the engine cover, be careful not to hit it against the cowling front panel.

7. Disconnect the ignition coil connectors.
8. Remove the ignition coil assembly.

To install:

9. Installation is the reverse of the removal procedure.
10. Tighten the ignition coil bolts to 75 inch lbs. (11 Nm).

SPARK PLUGS

REMOVAL & INSTALLATION

1. Before servicing the vehicle, refer to the Precautions Section.
2. Remove the ignition coil assemblies. Refer to Ignition Coil, removal & installation.
3. Remove the spark plug using a spark plug socket and wrench.

To install:

4. Installation is the reverse of the removal procedure.
5. Tighten the spark plugs to 18 ft. lbs. (25 Nm).
6. Install the ignition coil assemblies. Refer to Ignition Coil, removal & installation.

STARTER

REMOVAL & INSTALLATION

2009 Model

See Figure 85.

1. Before servicing the vehicle, refer to the Precautions Section.
2. Remove the battery and the battery tray with the ECM.
3. Remove the magnetic switch lead wire and the battery cable.

4. Remove the starting motor mount bolt and then the starting motor and bracket.

To install:

5. Installation is the reverse of the removal procedure.

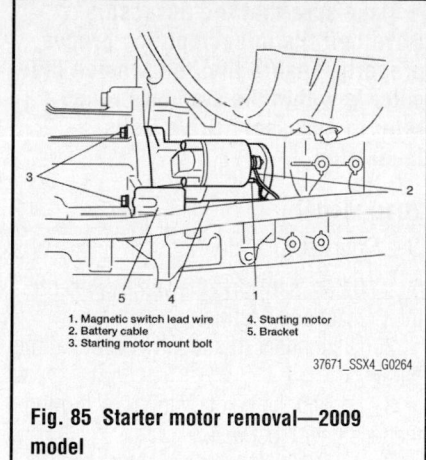

Fig. 85 Starter motor removal—2009 model

1. Magnetic switch lead wire
2. Battery cable
3. Starting motor mount bolt
4. Starting motor
5. Bracket

37671_SSX4_G0264

6. Tighten the battery cable nut to 71 inch lbs. (11 Nm).

7. Tighten the starting motor mount bolt to 18 ft. lbs. (25 Nm).

2010 Model

See Figure 86.

1. Before servicing the vehicle, refer to the Precautions Section.

2. Disconnect negative cable from battery.

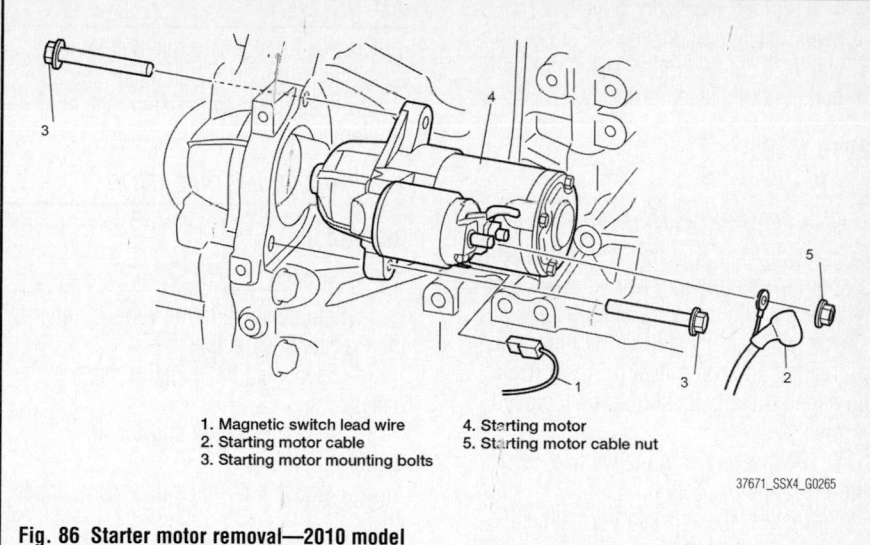

1. Magnetic switch lead wire
2. Starting motor cable
3. Starting motor mounting bolts
4. Starting motor
5. Starting motor cable nut

37671_SSX4_G0265

Fig. 86 Starter motor removal—2010 model

3. Remove the air cleaner assembly.

4. Remove the air cleaner bracket.

5. Disconnect the magnetic switch lead wire and the starting motor cable from the starting motor.

6. Remove the starting motor mounting bolts.

7. Remove the starting motor.

To install:

8. Installation is the reverse of the removal procedure.

9. Tighten the starting motor mounting bolts to 34 ft. lbs. (45 Nm).

10. Tighten the starting motor cable nut 71 inch lbs. (11 Nm).

ENGINE MECHANICAL

➡ **Disconnecting the negative battery cable may interfere with the functions of the on board computer systems and may require the computer to undergo a relearning process, once the negative battery cable is reconnected.**

ACCESSORY DRIVE BELTS

ACCESSORY BELT ROUTING

2009 Model

See Figure 87.

2010 Model

See Figure 88.

INSPECTION

2009 Model

See Figure 87.

Inspect the drive belt for signs of glazing or cracking. A glazed belt will be perfectly smooth from slippage, while a good belt will have a slight texture of fabric visible. Cracks will usually start at the inner edge of the belt and run outward. All worn or damaged drive belts should be replaced immediately.

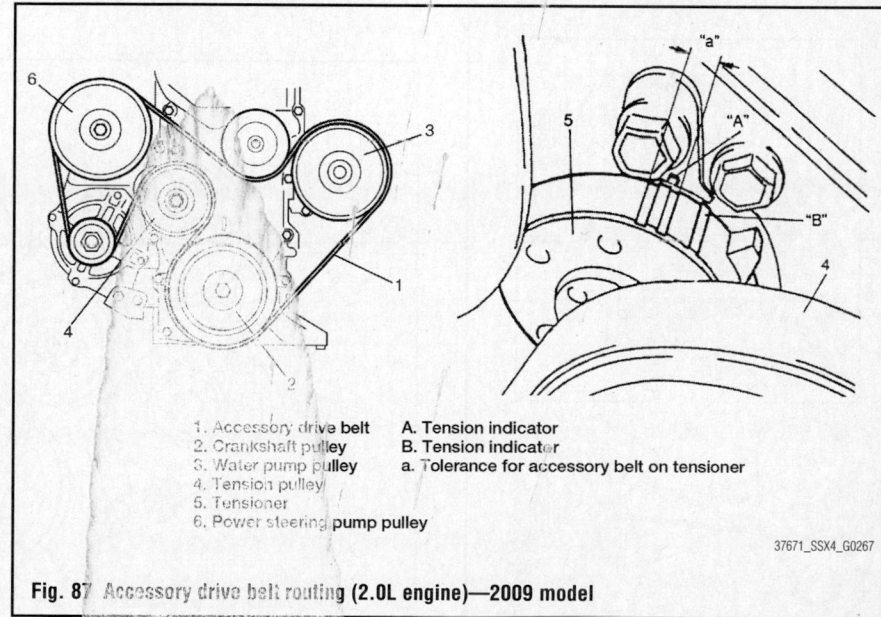

1. Accessory drive belt
2. Crankshaft pulley
3. Water pump pulley
4. Tension pulley
5. Tensioner
6. Power steering pump pulley

A. Tension indicator
B. Tension indicator
a. Tolerance for accessory belt on tensioner

37671_SSX4_G0267

Fig. 87 Accessory drive belt routing (2.0L engine)—2009 model

1. Before servicing the vehicle, refer to the Precautions Section.

2. Inspect the belt for cracks, cuts, deformation, wear, and cleanliness. If any of these conditions are found, replace the belt. Refer to Accessory Drive Belts, removal & installation.

3. Using a mirror, check to make sure that the tension indicators are as indicated in the figure.

a. If the tension indicator on the tensioner is found to the left of the indicator on the tensioner bracket, replace the accessory drive belt.

b. If new accessory drive belt has been installed, indicator "A" should be within "a" of the figure. If it is not, the belt is not installed properly.

2010 Model

See Figure 88.

Inspect the drive belt for signs of glazing or cracking. A glazed belt will be perfectly smooth from slippage, while a good belt will have a slight texture of fabric visible. Cracks will usually start at the inner edge of the belt and run outward. All worn or damaged drive belts should be replaced immediately.

1. Before servicing the vehicle, refer to the Precautions Section.
2. Inspect the accessory drive belt for cracks, cuts, deformation, wear, and cleanliness. If any abnormality is found, replace the accessory drive belt.
3. Check that the tension indicator on the tensioner is on the right of the indicator on the tensioner bracket when it is viewed from the crankshaft pulley side.
4. If the same indicator is on the left, replace the accessory drive belt.

1. Tension indicator
2. Tension indicator

37671_SSX4_G0268

Fig. 88 Accessory drive belt routing (2.0L engine)—2010 model

ADJUSTMENT

There is no manual drive belt tension adjustment. The drive belt tension is automatically adjusted by the drive belt auto tensioner.

REMOVAL & INSTALLATION

2009 Model

See Figure 89.

1. Before servicing the vehicle, refer to the Precautions Section.
2. Disconnect the negative cable at the battery.
3. Raise and safely support the vehicle.
4. Remove the right side engine under cover.
5. Loosen the tensioner by turning the tensioner pulley clockwise.
6. While holding the tensioner and belt loose, remove the accessory drive belt.

To install:

7. Loosen the tensioner by turning the tensioner pulley clockwise.
8. While holding the tensioner, install the accessory drive belt.
9. Install the right side engine under cover.

➡ Make sure that the accessory drive belt fits in each pulley groove properly. Ensure that the tension indicator is within the standard range, refer to Accessory Drive Belts, Inspection.

2010 Model

See Figure 90.

1. Before servicing the vehicle, refer to the Precautions Section.
2. Disconnect the negative cable at the battery.
3. Loosen the belt tensioner by turning the tensioner clockwise.

> ※※ **WARNING**
>
> **Do not turn the tensioner pulley with a torque higher than 44 ft. lbs. (59 Nm).**

4. While keeping the tensioner in the loose position, remove the accessory drive belt.

To install:

5. Loosen the belt tension by turning the tensioner clockwise.
6. While keeping the tensioner in the loose position, install the accessory drive belt.
7. Connect the negative cable to the battery.

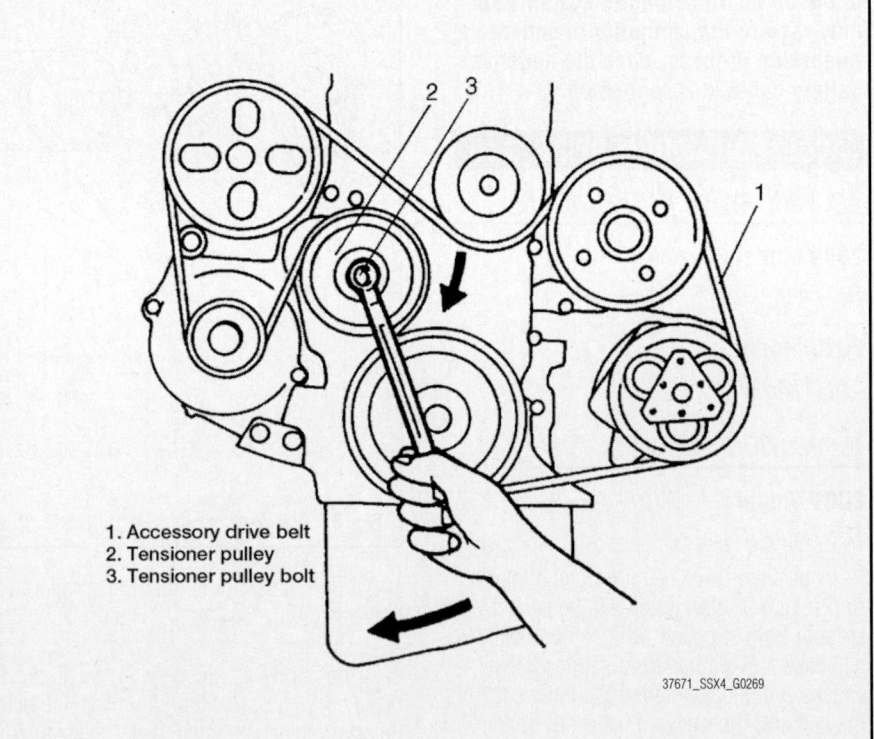

1. Accessory drive belt
2. Tensioner pulley
3. Tensioner pulley bolt

37671_SSX4_G0269

Fig. 89 Accessory drive belt removal—2009 model

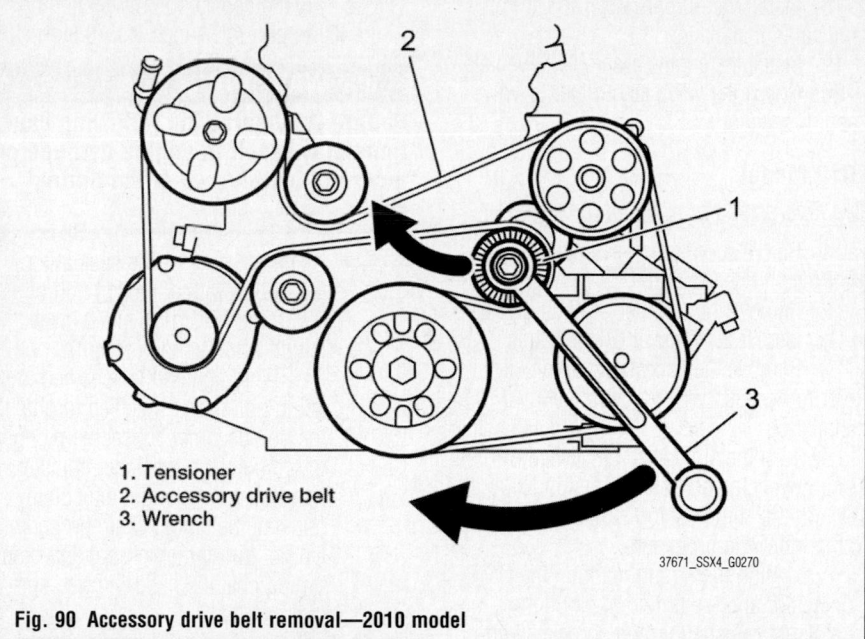

1. Tensioner
2. Accessory drive belt
3. Wrench

37671_SSX4_G0270

Fig. 90 Accessory drive belt removal—2010 model

CAMSHAFT AND VALVE LIFTERS

REMOVAL & INSTALLATION

2009 Model

See Figures 91 through 95.

➡ **Keep the working table, tools, and hands clean while overhauling. Use special care when handling aluminum parts in order not to damage them. Do not expose the removed parts to dust. Always keep them clean.**

1. Before servicing the vehicle, refer to the Precautions Section.
2. Remove the 2nd timing chain. Refer to Timing Chain & Sprockets, removal & installation.
3. Loosen the camshaft housing bolts in numerical order and remove them.
4. Remove the camshaft housings.
5. Remove the camshafts.
6. Remove the tappets with the shims.

To install:

7. Apply engine oil around the tappet and shim, and then install the tappets with the shims to the cylinder head.

➡ **When installing the shim, make sure to direct the shim No. side toward the tappet.**

8. Align the match mark on the crank timing sprocket and the mating surface of the cylinder block and lower crankcase.
9. Make sure that the arrow mark on the idler sprocket is at the correct position.

10. Install the camshafts. Apply oil to the sliding surface of each camshaft and camshaft journal then install them by aligning the match marks on the cylinder head and the camshafts.
11. Install the camshaft housing pins as shown in figure.
12. Check the position of the camshaft housings. Embossed marks are provided on each camshaft housing, indicating the position and the direction for installation. Install the housings as indicated by these marks.
13. After applying oil to the housing bolts, tighten them temporarily at first. Then, tighten in the sequence illustrated. Tighten a little at a time and evenly among all the bolts and repeat the tightening sequence 2–3 times before they are tight-

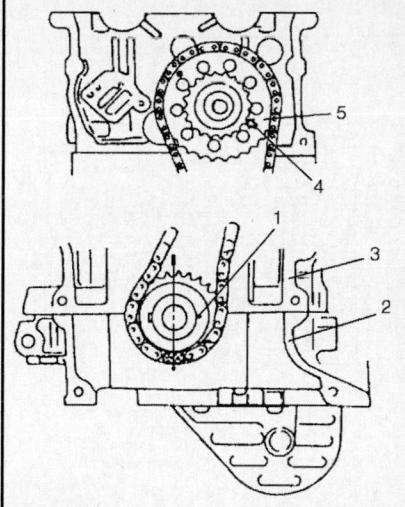

1. Match mark on the crank timing sprocket
2. Lower crankcase
3. Mating surface of the cylinder block
4. Arrow mark
5. Idler sprocket

37671_SSX4_G0286

Fig. 92 Align the match mark on the crank timing sprocket and the mating surface of the cylinder block and lower crankcase

ened the specified torque of 97 inch lbs. (11 Nm).
14. Turn the crankshaft clockwise then align the crankshaft timing sprocket key with the timing mark.
15. Install the 2nd timing chain. Refer to Timing Chain & Sprockets, removal & installation.
16. Install the timing chain cover. Refer to Timing Chain Front Cover, removal & installation.
17. Install the cylinder head cover. Refer to Cylinder Head, removal & installation.

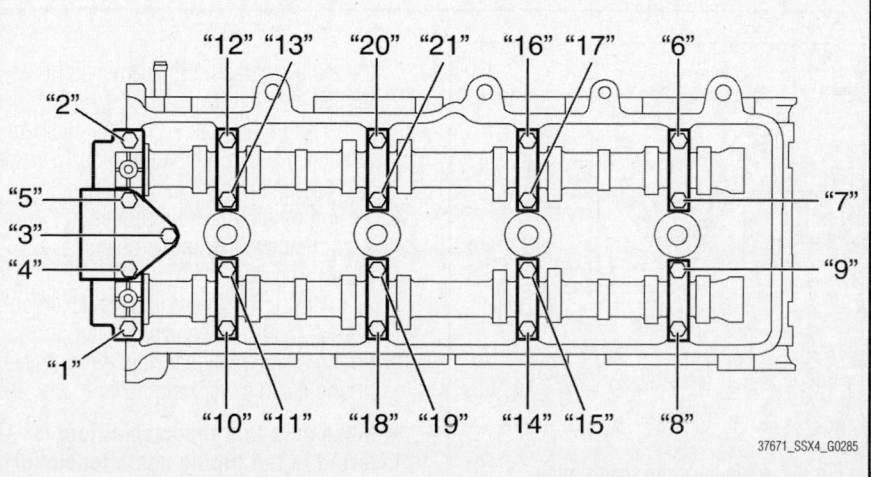

37671_SSX4_G0285

Fig. 91 Camshaft housing bolts removal order

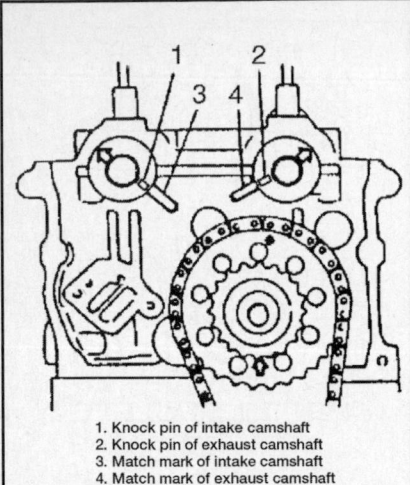

1. Knock pin of intake camshaft
2. Knock pin of exhaust camshaft
3. Match mark of intake camshaft
4. Match mark of exhaust camshaft

37671_SSX4_G0287

Fig. 93 Install the camshafts aligning the match marks on the cylinder head and the camshafts

18. Install the oil pan. Refer to Oil Pan, removal & installation.

19. Install the engine assembly to vehicle.

20. Check the valve lash. Refer to Valve Lash, Inspection.

2010 Model

See Figures 96 through 104.

1. Before servicing the vehicle, refer to the Precautions Section.

2. Remove the cylinder head cover. Refer to Cylinder Head, removal & installation.

3. Remove the accessory drive belt. Refer to Accessory Drive Belts, removal & installation.

4. Turn the crankshaft clockwise with a 19mm wrench, and position piston No. 1 at TDC on the compression stroke according to the following procedure.

a. Align the timing mark on the CMP actuator and the timing mark on the exhaust camshaft timing sprocket with

c. Fix the special tool by tightening the fixing bolt to 71 inch lbs. (8 Nm).

✳✳ WARNING

Failure to observe the following cautions may result in engine damage or incorrect installation of the timing chain.

- Use a M6 bolt for the special tool fixing bolt with a thread pitch of 0.039 inch (1.0mm) and a thread length of 0.59 inch (15mm)
- Be sure to tighten the specified tool fixing bolt to the specified torque. If the bolt becomes loose, the plunger of timing chain tensioner adjuster may extend, making it difficult to reinstall the timing chain properly
- Do not turn the crankshaft after the timing chain is fixed with the special tool

6. Apply a dab of paint to two timing chain links which meet at the timing marks on the CMP actuator and the exhaust camshaft timing sprocket.

7. Remove the exhaust camshaft timing sprocket according to the following procedure.

a. Hold the hexagonal section of the exhaust camshaft using a wrench.

b. Loosen the exhaust camshaft timing sprocket bolt and remove the exhaust camshaft timing sprocket (1).

✳✳ WARNING

Be careful not to drop the dowel pin into the space between the timing chain cover and the cylinder block. Do not remove the CMP actuator during this procedure.

8. Remove the dowel pin from the exhaust camshaft.

9. Loosen the camshaft housing bolts, evenly and gradually, in the numerical order illustrated.

10. Remove the intake camshaft and the exhaust camshaft.

➡**Be careful not to drop the timing chain into the timing chain cover.**

11. Remove the camshaft bearings and tappets as necessary.

12. Remove the CMP actuator from the intake camshaft according to the following procedure, if necessary.

a. Hold the hexagonal section of the intake camshaft using a vice.

b. Loosen the CMP actuator bolt and remove the CMP actuator.

"10", (a)   "2", (a)   "6", (a)   "16", (a)
"9", (a)   "1", (a)   "5", (a)
"20", (a)
"17", (a)   "15", (a)
"19", (a)
"18", (a)   "13", (a)
"21", (a)
"11", (a)   "3", (a)   "7", (a)
"12", (a)   "4", (a)   "8", (a)   "14", (a)

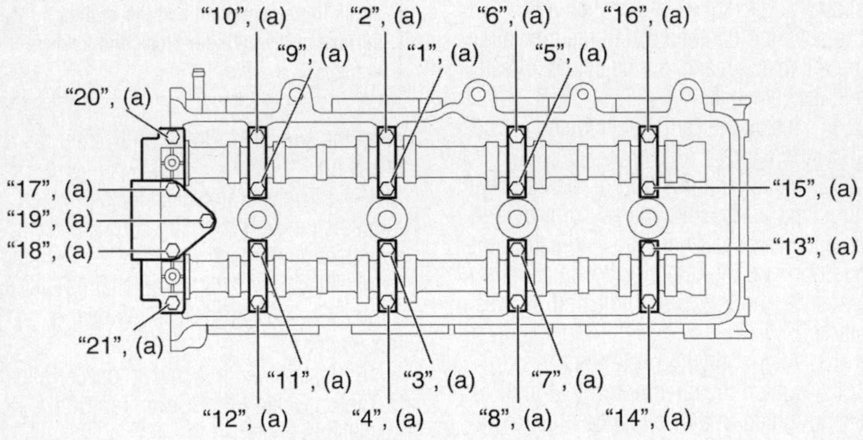

a. Camshaft housing bolt torque: 97 inch lbs. (11 Nm)

37671_SSX4_G0280

Fig. 94 Camshaft housing bolt tightening sequence

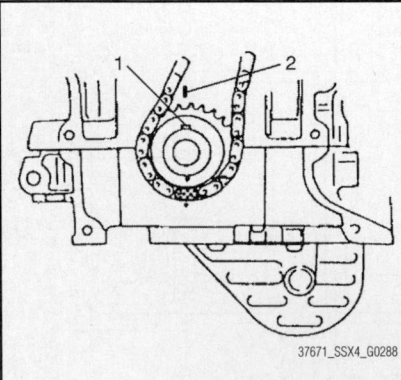

37671_SSX4_G0288

Fig. 95 Align the crankshaft timing sprocket key (1) with the timing mark (2)

the match marks on the camshaft housing No. 1.

b. Align the notch on the crankshaft pulley with "0" on the timing chain cover.

5. Fix the timing chain tensioner according to the following procedure.

a. Remove the timing chain cover plug and gasket.

b. Insert the fore-end of Special Tool 09917-16710, or equivalent, into the timing chain tensioner hole through the timing chain cover plug hole.

➡**Make sure that the special tool is inserted in the timing chain tensioner hole. Use a mirror if necessary.**

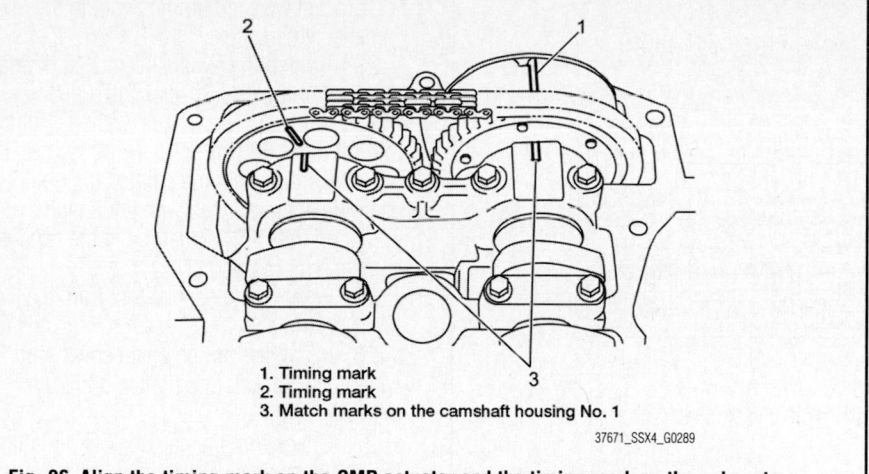

1. Timing mark
2. Timing mark
3. Match marks on the camshaft housing No. 1

37671_SSX4_G0289

Fig. 96 Align the timing mark on the CMP actuator and the timing mark on the exhaust camshaft timing sprocket with the match marks on the camshaft housing No. 1

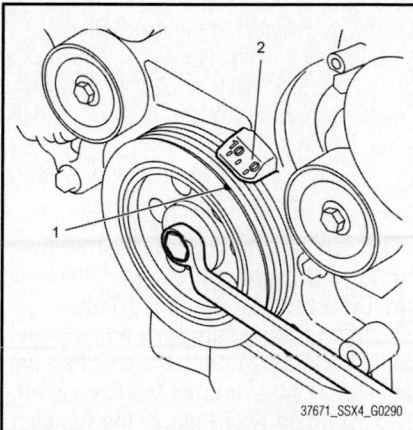

37671_SSX4_G0290

Fig. 97 Align the notch (1) on the crankshaft pulley with "0" (2) on the timing chain cover

37671_SSX4_G0291

Fig. 98 Remove the timing chain cover plug (1) and gasket (2)

➡**Do not disassemble the CMP actuator.**

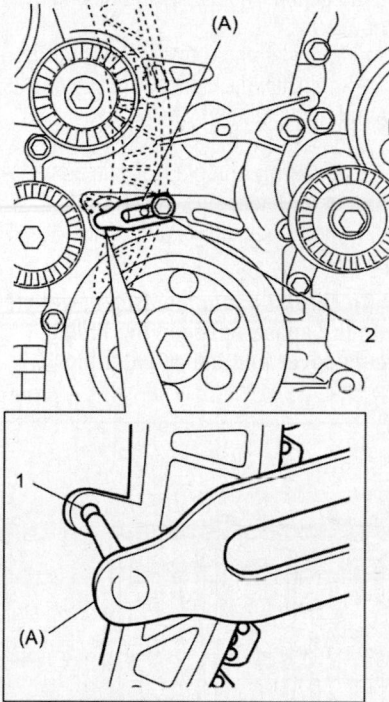

1. Timing chain tensioner hole
2. Fixing bolt
A. Special Tool 09917-16710

37671_SSX4_G0292

Fig. 99 Insert the fore-end of Special Tool 09917-16710 into the timing chain tensioner hole through the timing chain cover plug hole

To install:

13. Install the CMP actuator to the intake camshaft. Tighten the CMP actuator bolt to 45 ft. lbs. (60 Nm).

14. Apply engine oil to the contact surface of the tappets and install the tappets to the cylinder head.

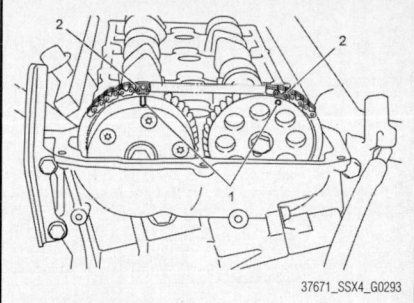

37671_SSX4_G0293

Fig. 100 Apply a dab of paint to two timing chain links (2) which meet at the timing marks (1) on the CMP actuator and the exhaust camshaft timing sprocket

37671_SSX4_G0294

Fig. 101 Removing the exhaust camshaft timing sprocket (1)

15. Install the camshaft bearings according to the following procedure.

➡**The camshaft upper bearing half (1) is different in shape from the lower bearing half (2). Distinguish the upper and lower bearing halves when installing.**

a. Fit the tab of the camshaft bearing to the groove of the cylinder head or camshaft housing.

b. Press the camshaft bearing end until it fully seats in the cylinder head or camshaft housing.

c. Apply engine oil to the sliding surface of the camshaft bearing halves.

➡**Do not apply engine oil between the bearing halves and the camshaft housing or cylinder head.**

16. Apply engine oil to the sliding surfaces of the camshafts, and then install the camshafts to the cylinder head according to the following procedure.

17. Align the timing chain link, which was painted in the removal procedure, with the timing mark on the CMP actuator and install the intake camshaft.

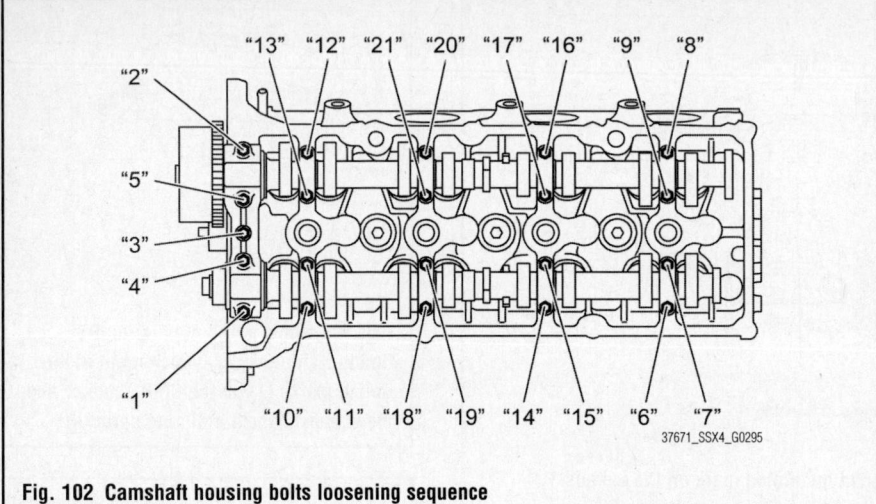

Fig. 102 Camshaft housing bolts loosening sequence

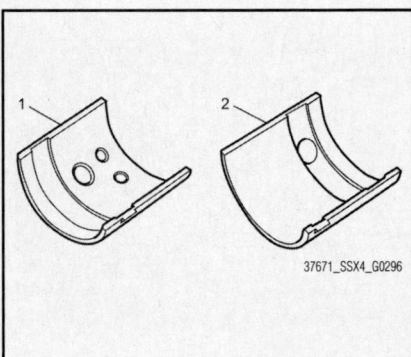

Fig. 103 View of camshaft upper bearing half (1) and lower bearing half (2)

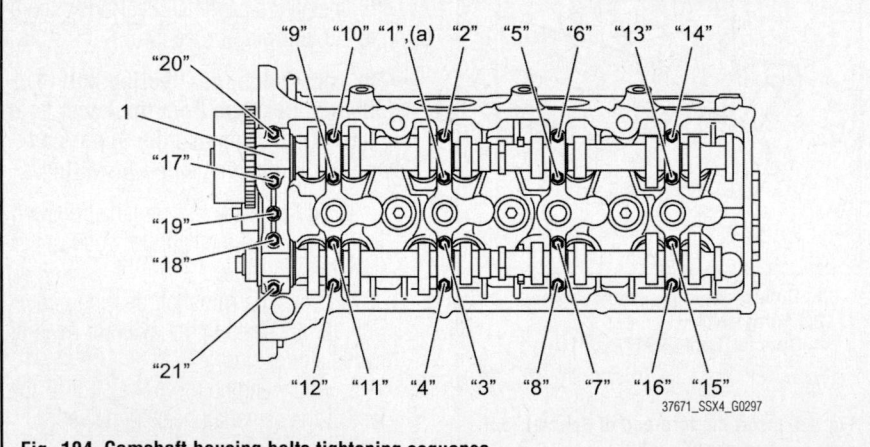

Fig. 104 Camshaft housing bolts tightening sequence

a. Install the exhaust camshaft and position the dowel pin hole of the exhaust camshaft downward.

18. Install the camshaft housings to the proper places distinguished by the character, number, and arrow direction on each camshaft housing.

19. Tighten the camshaft housing bolts according to the following procedure.

a. Apply engine oil to the bolt thread and seat of the camshaft housing bolts.

b. Install the camshaft housing No. 1.

c. Install the camshaft housing bolts, and tighten them by hand.

d. Tighten the camshaft housing bolts, evenly and gradually, in numerical order to 97 inch lbs. (11 Nm).

20. Install the dowel pin to the exhaust camshaft.

➡Be careful not to the drop dowel pin into the space between the timing chain cover and the cylinder block.

21. Install the exhaust camshaft timing sprocket to the exhaust camshaft according to the following procedure.

22. Align the timing chain link, which was painted during removal, with the timing mark on the exhaust camshaft timing sprocket, and loop the timing chain up over the exhaust camshaft timing sprocket.

a. Install the exhaust camshaft timing sprocket with the timing chain to the exhaust camshaft.

23. Tighten the exhaust camshaft timing sprocket bolt using Special Tool 09911-05120, or equivalent torque wrench. Tighten to 45 ft. lbs. (60 Nm).

24. Confirm that the painted timing chain links are aligned with the timing marks on the CMP actuator and the exhaust camshaft timing sprocket.

25. Remove the special tool from the timing chain cover.

26. Install the timing chain cover plug with a new gasket, and tighten it to 20 ft. lbs. (27 Nm).

27. Check that the camshaft and the timing chain are installed properly as follows.

a. Check that the timing mark on the CMP actuator and the timing mark on the exhaust camshaft timing sprocket are aligned with the match marks on the camshaft housing No. 1.

b. Check that the notch on the crankshaft pulley is aligned with the "0" on the timing chain cover.

c. Turn the crankshaft clockwise twice and repeat the timing mark check.

➡Be sure to turn the crankshaft fully twice. If it is once, the timing marks on the CMP actuator and the exhaust camshaft timing sprockets do not meet the match marks on the camshaft housing No. 1. After turning the crankshaft twice, the painted links of the timing chain are not aligned with the timing marks on the CMP actuator and the exhaust timing sprocket, but this is normal.

28. Check the valve clearance. Refer to Valve Lash, Inspection.

29. Install the cylinder head cover. Refer to Cylinder Head, removal & installation.

CRANKSHAFT FRONT SEAL

REMOVAL & INSTALLATION

2009 Model

See Figure 105.

1. Before servicing the vehicle, refer to the Precautions Section.

2. Remove the drive belt. Refer to Accessory Drive Belts, removal & installation.

3. Remove the crankshaft pulley. Refer to Crankshaft Damper, removal & installation.

4. Using a seal removal tool, remove the oil seal from its mounting.

➡Be careful not to damage the front cover and/or the crankshaft.

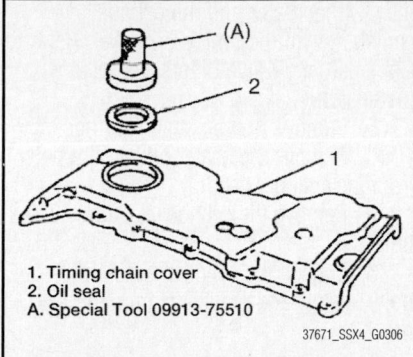

1. Timing chain cover
2. Oil seal
A. Special Tool 09913-75510

37671_SSX4_G0306

Fig. 105 Using Special Tool 09913-75510 to install the crankshaft front seal—2009 model

To install:

5. Installation is the reverse order of the removal procedure.

6. Press fit the seal until it is flush with the front end surface of the front cover, using the proper tools.

2010 Model

See Figure 106.

1. Before servicing the vehicle, refer to the Precautions Section.

2. Remove the drive belt. Refer to Accessory Drive Belts, removal & installation.

3. Remove the crankshaft pulley. Refer to Crankshaft Damper, removal & installation.

4. Remove the crankshaft oil seal from the timing chain cover using a flat-head screwdriver, or equivalent.

➡**Be careful not to damage the front cover and/or the crankshaft.**

To install:

5. Installation is the reverse order of the removal procedure.

6. Press fit the seal until it is flush with the front end surface of the front cover, using the proper tools.

CYLINDER HEAD

REMOVAL & INSTALLATION

2009 Model

See Figures 107 through 110.

1. Before servicing the vehicle, refer to the Precautions Section.

2. Remove the camshafts, tappets, and shims. Refer to Camshaft and Valve Lifters, removal & installation.

3. Loosen the cylinder head bolts in numerical order and remove the bolts.

➡**Be sure to remove the (M6) cylinder head bolt.**

4. Check all around the cylinder head for any other parts required to be removed or disconnected and remove or disconnect whatever is necessary.

5. Remove the cylinder head with the intake manifold, exhaust manifold, and water outlet cap. Use a lifting device, if necessary.

6. Remove the intake manifold. Refer to Intake Manifold, removal & installation.

7. Remove the exhaust manifold. Refer to Exhaust Manifold, removal & installation.

To install:

8. Align the match mark on the crank timing sprocket and mating surface of the cylinder block and lower crankcase.

9. Clean the mating surface of the cylinder head and cylinder block. Remove any oil, old gasket material, and dust from the mating surface.

10. Install the knock pins to the cylinder block.

11. Install a new cylinder head gasket to the cylinder block. The identification number provided on the gasket goes toward the crankshaft pulley side, facing up (toward the cylinder head side).

12. Install the cylinder head to the cylinder block.

13. Apply engine oil to the cylinder head bolts and tighten gradually as follows.

   a. Step 1: Tighten the cylinder head bolts (M10) to 38 ft. lbs. (52 Nm) in sequence.

   b. Step 2: Tighten the cylinder head bolts (M10) to 60 ft. lbs. (82 Nm) in sequence.

   c. Step 3: Loosen the cylinder head bolts (M10) to 0 ft. lbs. (0 Nm) in numerical order for loosening.

   d. Step 4: Tighten the cylinder head bolts (M10) to 38 ft. lbs. (52 Nm) in sequence.

   e. Step 5: Tighten the cylinder head bolts (M10) to 75 ft. lbs. (103 Nm).

   f. Step 6: Tighten the cylinder head bolt (M6) to 97 inch lbs. (11 Nm).

14. Install the camshafts, tappets, and shims. Refer to Camshaft and Valve Lifters, removal & installation.

15. Install the 1st and 2nd timing chain. Refer to Timing Chain & Sprockets, removal & installation.

16. Install the timing chain cover. Refer to Timing Chain Front Cover, removal & installation.

17. Check the intake and exhaust valve lash. Refer to Valve Lash, Inspection.

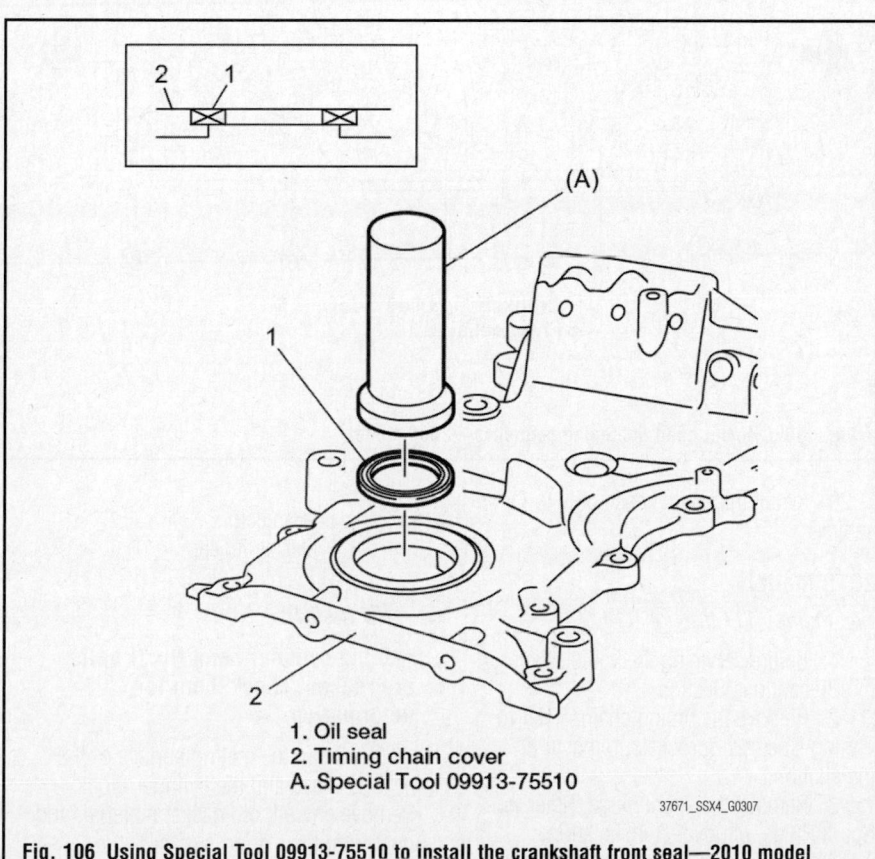

1. Oil seal
2. Timing chain cover
A. Special Tool 09913-75510

37671_SSX4_G0307

Fig. 106 Using Special Tool 09913-75510 to install the crankshaft front seal—2010 model

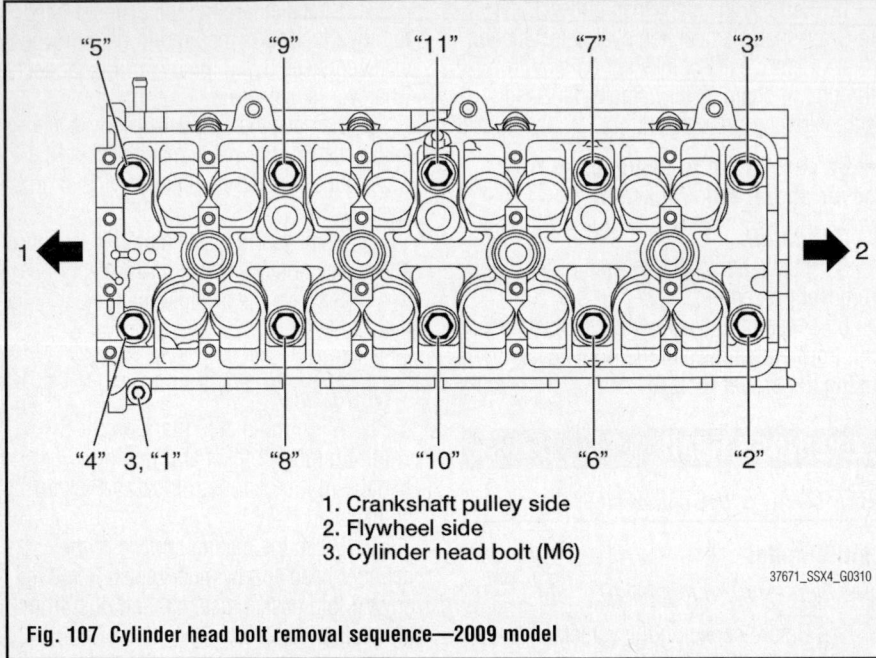

1. Crankshaft pulley side
2. Flywheel side
3. Cylinder head bolt (M6)

37671_SSX4_G0310

Fig. 107 Cylinder head bolt removal sequence—2009 model

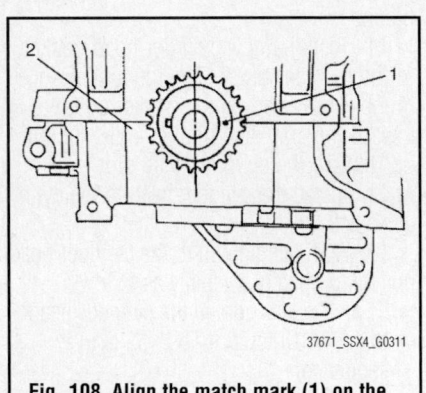

37671_SSX4_G0311

Fig. 108 Align the match mark (1) on the crank timing sprocket and mating surface (2) of the cylinder block and lower crankcase—2009 model

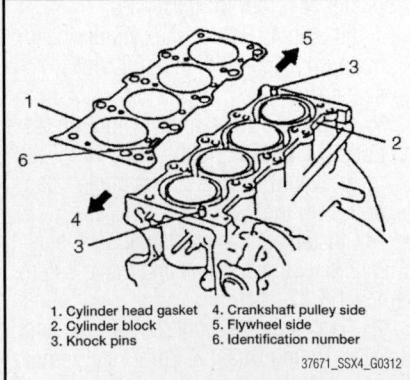

1. Cylinder head gasket
2. Cylinder block
3. Knock pins
4. Crankshaft pulley side
5. Flywheel side
6. Identification number

37671_SSX4_G0312

Fig. 109 Install a new cylinder head gasket—2009 model

18. Install the cylinder head cover.
19. Install the oil pan. Refer to Oil Pan, removal & installation.

the CMP actuator and the exhaust camshaft with the timing sprocket. Refer to Camshaft and Valve Lifters, removal & installation.

5. Remove the camshaft bearings.
6. Remove the heater outlet pipe bolt from the cylinder head.
7. Remove the cylinder head No. 2 bolt.
8. Loosen the cylinder head No. 1 bolts, evenly and gradually, in numerical order.

➡**Use a 12-point socket wrench to remove cylinder head No. 1 bolts.**

9. Remove the cylinder head with its gasket, the intake manifold and the exhaust manifold from the cylinder block.
10. Remove the following components from the cylinder head if necessary.

* Intake manifold
* Exhaust manifold

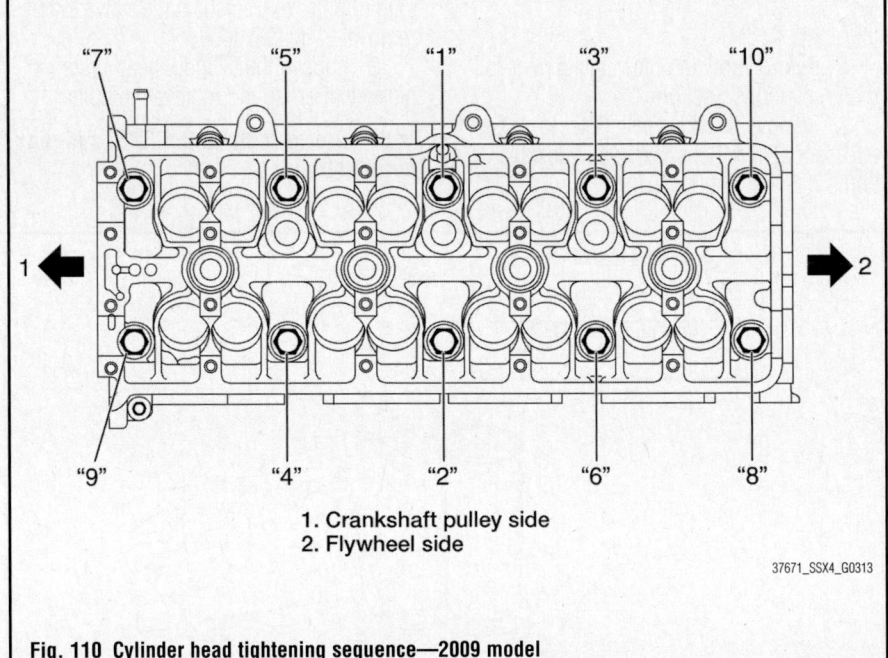

1. Crankshaft pulley side
2. Flywheel side

37671_SSX4_G0313

Fig. 110 Cylinder head tightening sequence—2009 model

20. Install the engine assembly to the vehicle.

2010 Model

See Figures 111 through 114.

1. Before servicing the vehicle, refer to the Precautions Section.
2. Remove the timing chain. Refer to Timing Chain & Sprockets, removal & installation.
3. Remove the spark plugs. Refer to Spark Plugs, removal & installation.
4. Remove the intake camshaft with

* Fuel injector
* Water outlet cap
* Oil venture plug

To install:

➡**If the cylinder head No. 1 bolts are reused, check them for deformation.**

11. Clean the mating surface of the cylinder head and the cylinder block. Remove any oil, old gasket material, and dust from the mating surface.

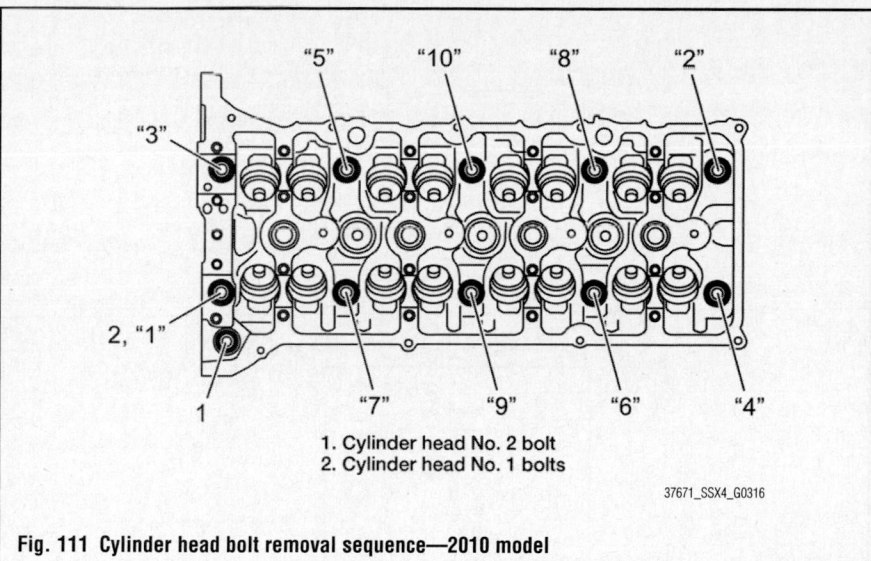

Fig. 111 Cylinder head bolt removal sequence—2010 model

1. Cylinder head No. 2 bolt
2. Cylinder head No. 1 bolts

37671_SSX4_G0316

c. Step 3: Tighten No. 1 cylinder head bolts to 30 ft. lbs. (40 Nm).

d. Step 4: Tighten No. 1 cylinder head bolts an additional 60°.

e. Step 5: Tighten No. 1 cylinder head bolts an additional 80°.

f. Step 6: Tighten No. 2 cylinder head bolt to 19 ft. lbs. (25 Nm).

16. Install the intake camshaft and the exhaust camshaft. Refer to Camshaft and Valve Lifters, removal & installation.

17. Install the timing chain. Refer to Timing Chain & Sprockets, removal & installation.

18. Install the timing chain cover. Refer to Timing Chain Front Cover, removal & installation.

19. Install the cylinder head cover.

12. Install the dowel pins to the cylinder block.

13. Install a new cylinder head gasket to the cylinder block.

14. Before installing the cylinder head to the cylinder block, turn the crankshaft and position key within the specified range (30–90°) in the counterclockwise direction from the top.

15. Install the cylinder head to the cylinder block.

a. Step 1: Apply engine oil to the cylinder head No. 1 bolt threads.

b. Step 2: Tighten the cylinder head No. 1 bolts to 15 ft. lbs. (20 Nm), evenly and gradually, in sequence.

➡**Use a 12-point socket wrench to tighten the cylinder head No. 1 bolts.**

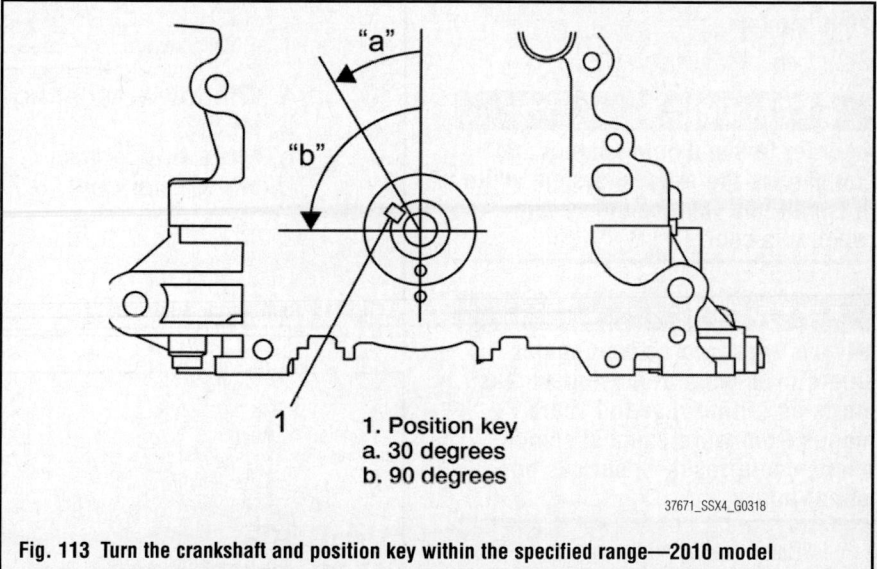

1. Position key
a. 30 degrees
b. 90 degrees

37671_SSX4_G0318

Fig. 113 Turn the crankshaft and position key within the specified range—2010 model

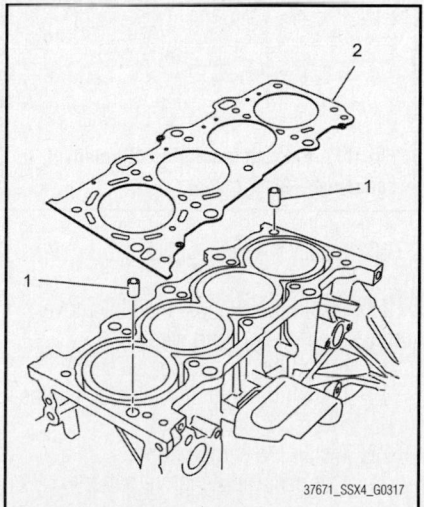

Fig. 112 Install the dowel pins (1) and a new cylinder head gasket (2) to the cylinder block—2010 model

37671_SSX4_G0317

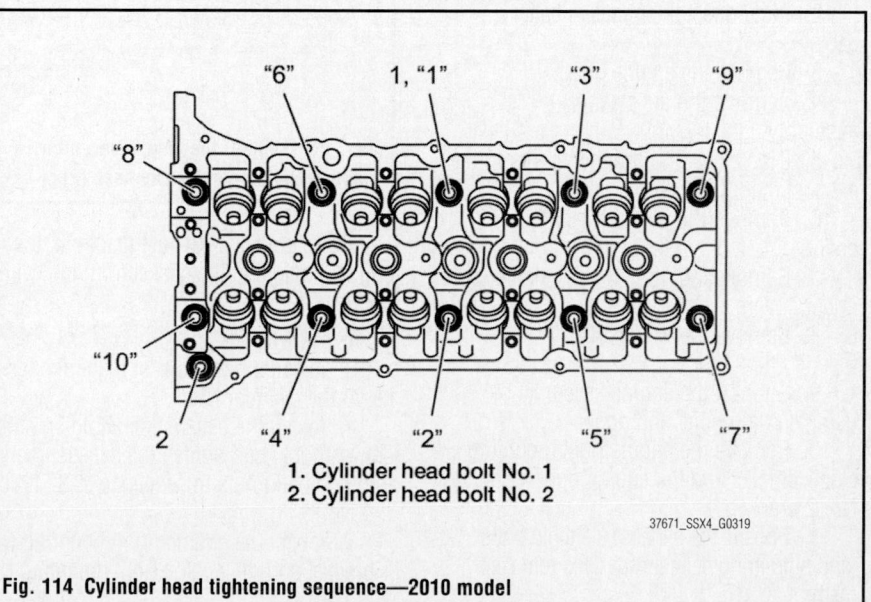

1. Cylinder head bolt No. 1
2. Cylinder head bolt No. 2

37671_SSX4_G0319

Fig. 114 Cylinder head tightening sequence—2010 model

20. Install the oil pan. Refer to Oil Pan, removal & installation.

21. Install the water outlet cap if removed.

22. Install the heater outlet pipe bolt to the cylinder head.

23. Install the spark plugs. Refer to Spark Plugs, removal & installation.

24. Install the exhaust manifold if removed. Refer to Exhaust Manifold, removal & installation.

25. Install the intake manifold if removed. Refer to Intake Manifold, removal & installation.

26. Install the fuel injectors if removed.

27. Install the engine assembly.

EXHAUST MANIFOLD

REMOVAL & INSTALLATION

2009 Model

See Figures 115 through 118.

> ※※ **CAUTION**
>
> **In order to avoid being burned, do not service the exhaust system while it is still hot. Service the system when it is cool.**

> ※※ **CAUTION**
>
> **Always wear protective goggles and gloves when removing exhaust parts as falling rust and sharp edges from worn exhaust components could result in serious personal injury.**

1. Before servicing the vehicle, refer to the Precautions Section.

2. Disconnect the negative battery cable.

3. Remove the engine hood.

4. Remove the air cleaner assembly.

5. Disconnect the ground wire from the exhaust manifold cover.

6. Remove the exhaust manifold cover.

7. Remove the right and left side engine under covers.

8. Remove the exhaust No. 1 pipe.

9. Support the engine assembly by using a chain hoist and hook.

10. Remove the engine right mounting No. 1 bracket and the engine right mounting No. 2 bracket.

11. Loosen the exhaust manifold bolts and nuts in numerical order and remove them.

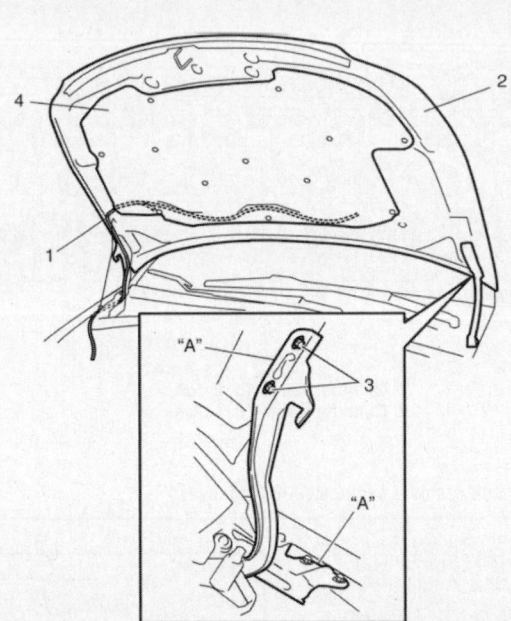

a. Window washer hose A. Sealant contact face
b. Hood
c. Mounting bolts
d. Hood silencer

37671_SSX4_G0337

Fig. 115 Remove the engine hood

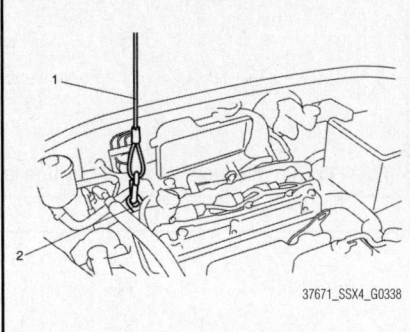

37671_SSX4_G0338

Fig. 116 Support the engine assembly by using a chain hoist (1) and hook (2)

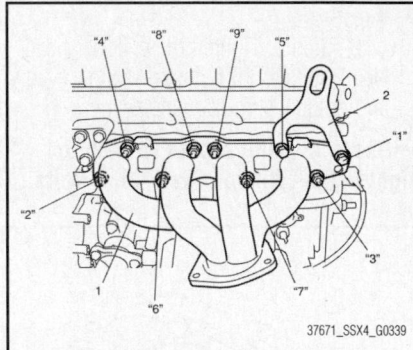

37671_SSX4_G0339

Fig. 117 Exhaust manifold bolt removal sequence—2009 model

12. Remove the exhaust manifold, the engine hook, and the gasket from the cylinder head.

To install:

13. Install a new exhaust manifold gasket to the cylinder head.

14. Install the exhaust manifold and the engine hook, and tighten the exhaust manifold bolts and nuts in sequence to 37 ft. lbs. (50 Nm).

15. Install the engine right mounting No. 1 bracket and the engine right mounting No. 2 bracket.

16. Install the exhaust No. 1 pipe.

17. Install the exhaust manifold cover and connect the ground wire.

18. Install the air cleaner assembly.

19. Install the right and left side engine under covers.

20. Install the engine hood.

21. Connect the negative battery cable.

22. Check the exhaust system for exhaust gas leakage.

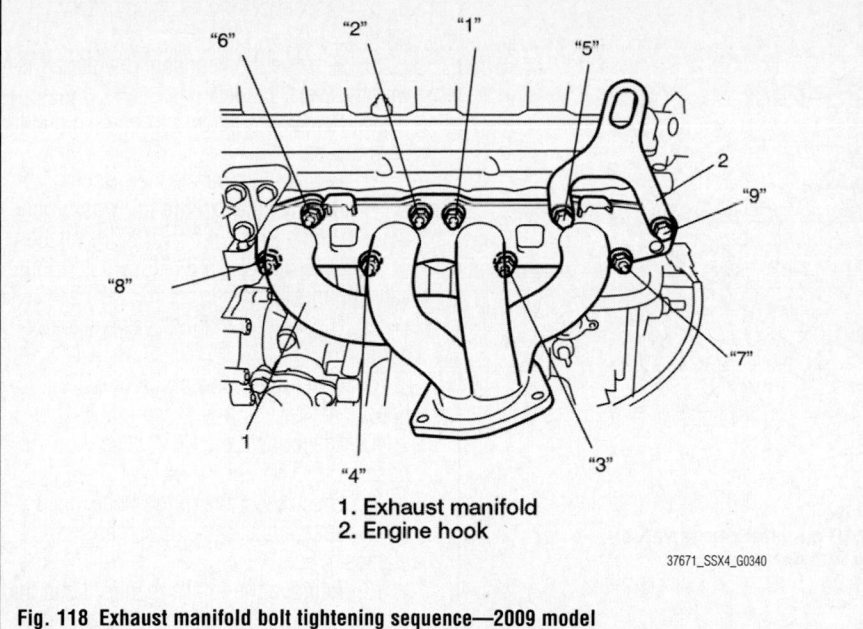

1. Exhaust manifold
2. Engine hook

37671_SSX4_G0340

Fig. 118 Exhaust manifold bolt tightening sequence—2009 model

2010 Model

See Figures 119 and 120.

※ CAUTION

In order to avoid being burned, do not service the exhaust system while it is still hot. Service the system when it is cool.

※ CAUTION

Always wear protective goggles and gloves when removing exhaust parts as falling rust and sharp edges from worn exhaust components could result in serious personal injury.

※ WARNING

The exhaust manifold has a three-way catalytic converter which reduces harmful exhaust emissions. Be careful not to drop the exhaust manifold. If it is dropped, the inside of the catalytic converter may be damaged on impact.

1. Before servicing the vehicle, refer to the Precautions Section.
2. Disconnect the negative battery cable.
3. Remove the right and left side engine under covers.
4. Remove the air cleaner assembly.
5. Remove the condenser cooling fan.
6. Disconnect the ground wire from the exhaust manifold cover.

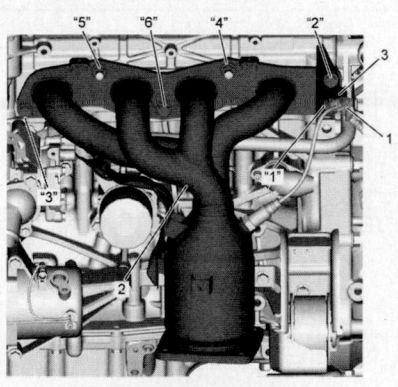

1. A/F sensor connector
2. Exhaust manifold
3. Engine hook

37671_SSX4_G0341

Fig. 119 Exhaust manifold bolt removal sequence—2010 model

7. Remove the exhaust manifold cover.
8. Remove the exhaust No. 1 pipe.
9. Disconnect the A/F sensor connector.
10. Loosen the exhaust manifold bolts and nuts in sequence, then remove them.
11. Remove the exhaust manifold, the engine hook, and the gasket from the cylinder head.

To install:
12. Install a new exhaust manifold gasket to the cylinder head.
13. Install the exhaust manifold and engine hook, and tighten the exhaust manifold bolts and nuts in sequence to 37 ft. lbs. (50 Nm).

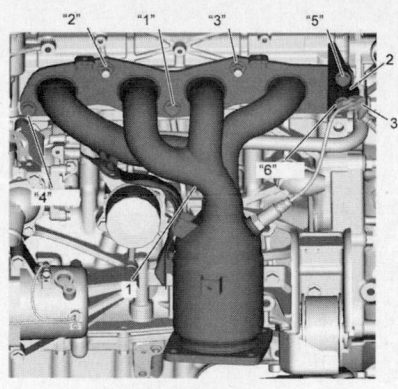

1. Exhaust manifold
2. Engine hook
3. A/F sensor connector

37671_SSX4_G0342

Fig. 120 Exhaust manifold bolt tightening sequence—2010 model

14. Connect the A/F sensor connector.
15. Install the exhaust No. 1 pipe.
16. Install the exhaust manifold cover and connect the ground wire.
17. Install the condenser cooling fan.
18. Install the air cleaner assembly.
19. Install the right and left side engine under covers.
20. Connect the negative battery cable.
21. Check the exhaust system for exhaust gas leakage.

INTAKE MANIFOLD

REMOVAL & INSTALLATION

2009 Model

See Figures 121 and 122.

※ WARNING

Never disassemble the intake manifold. Disassembly may spoil its original performance. If a faulty condition is found, replace it with a new one.

1. Before servicing the vehicle, refer to the Precautions Section.
2. Disconnect the negative battery cable.
3. Drain the engine coolant.
4. Remove the electric throttle body assembly.
5. Disconnect the electric lead wires to the IMT connector.
6. Disconnect the brake booster hose from the intake manifold.
7. Disconnect the PCV hose from the PCV valve.

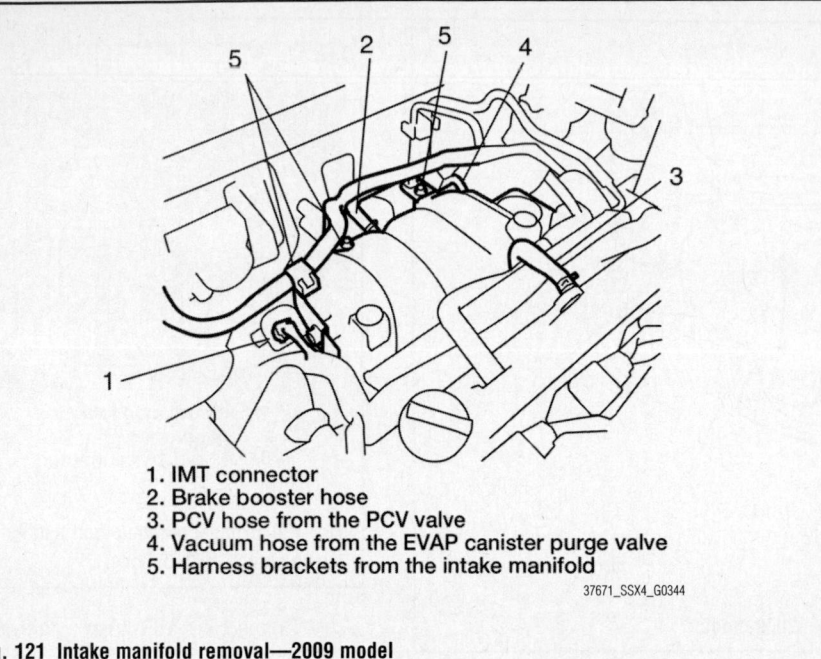

1. IMT connector
2. Brake booster hose
3. PCV hose from the PCV valve
4. Vacuum hose from the EVAP canister purge valve
5. Harness brackets from the intake manifold

37671_SSX4_G0344

Fig. 121 Intake manifold removal—2009 model

2. Disconnect the negative battery cable.

3. Remove the windshield wiper arms with the wiper blades.

4. Remove the cowl top garnish and the cowl top panel.

5. Remove the air cleaner assembly.

6. Remove the throttle body assembly. Refer Throttle Body, removal & installation.

7. Disconnect the EVAP canister purge valve connector.

8. Disconnect the IMT valve actuator connector.

9. Disconnect the brake booster hose.

10. Disconnect the PCV valve hose.

11. Disconnect the purge hose.

12. Disconnect the engine harness clamps.

13. Remove the intake manifold and the intake manifold gaskets from the cylinder head.

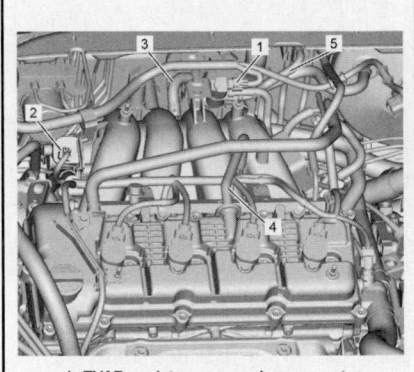

1. EVAP canister purge valve connector
2. IMT valve actuator connector
3. Brake booster hose
4. PCV valve hose
5. Purge hose

37671_SSX4_G0346

Fig. 123 Intake manifold removal—2010 model

1. Intake manifold bolts
2. Intake manifold nuts
A. Apply thread lock cement to the intake manifold bolt

37671_SSX4_G0345

Fig. 122 Intake manifold installation—2009 model

8. Disconnect the vacuum hose from the EVAP canister purge valve.

9. Disconnect the breather hose from the intake manifold.

10. Remove the harness brackets from the intake manifold.

11. Remove the intake manifold and the O-ring from the cylinder head.

To install:

12. Installation is the reverse of the removal procedure.

13. Use a new intake manifold O-ring.

14. Install the intake manifold bolts and nuts.

15. Apply thread lock cement to the intake manifold bolt.

➡ **Use thread lock cement 99000-32110 (Thread Lock Cement Super 1322), or equivalent.**

16. Check to ensure that all the removed parts are back in place. Install any necessary parts which have not been reinstalled.

17. Fill the cooling system.

18. Turn the ignition switch ON with the engine OFF and check for fuel leaks.

19. Start the engine and check for engine coolant leaks.

2010 Model

See Figures 123 through 125.

1. Before servicing the vehicle, refer to the Precautions Section.

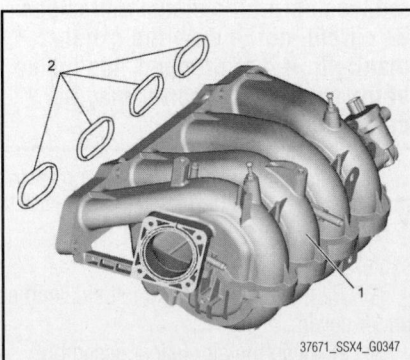

37671_SSX4_G0347

Fig. 124 Remove the intake manifold (1) and the intake manifold gaskets (2) from the cylinder head—2010 model

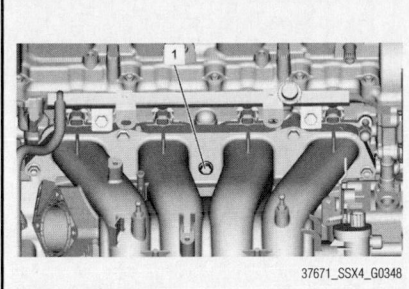

Fig. 125 Use a new intake manifold bolt in the specified position (1)—2010 model

To install:

14. Installation is the reverse of the removal procedure.

15. Use a new intake manifold bolt in the specified position, as it is not reusable.

16. Use new gaskets.

17. Fill the cooling system.

18. Start the engine and check for engine coolant leaks.

OIL PAN

REMOVAL & INSTALLATION

2009 Model

See Figures 126 and 127.

1. Before servicing the vehicle, refer to the Precautions Section.

2. Remove the oil level gauge.

3. Raise and safely support the vehicle.

4. Remove the engine under cover.

5. Remove the exhaust No. 1 pipe.

6. Remove the transfer (4WD models). Refer to Transfer Case, removal & installation.

7. Remove the clutch housing plate.

8. Drain the engine oil by removing the drain plug.

9. Cut the sealant using Special Tool 09921-96510, or equivalent, and a hammer.

➡**Be careful not to damage the stud bolt between the oil pan and the crankcase when cutting the sealant.**

10. Remove the oil pan.

11. Remove the oil pump strainer, and O-rings from the lower crankcase, as necessary.

To install:

12. Install new O-rings to the oil pump strainer, if removed. Tighten the strainer bolts to 97 inch lbs. (11 Nm).

13. Apply sealant to the oil pan mating

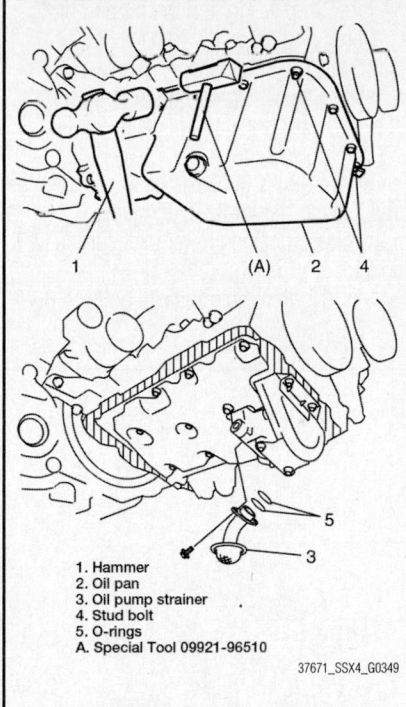

1. Hammer
2. Oil pan
3. Oil pump strainer
4. Stud bolt
5. O-rings
A. Special Tool 09921-96510

Fig. 126 Cut the sealant from the oil pan using Special Tool 09921-96510 and a hammer

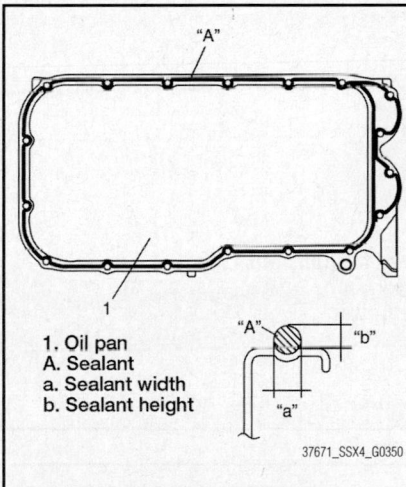

1. Oil pan
A. Sealant
a. Sealant width
b. Sealant height

Fig. 127 Apply sealant to the oil pan

surface in a continuous bead 0.08 inch (2mm) high and wide.

➡**Use Sealant 99000-31260 (SUZUKI Bond No. 1217G), or equivalent.**

14. After fitting the oil pan to the cylinder block, install the securing bolts and start tightening at the center and move outward, tightening one bolt at a time.

a. Tighten the oil pan bolts and nut to 97 inch lbs. (11 Nm).

b. Tighten the transaxle case No. 1 bolts to 40 ft. lbs. (55 Nm).

15. Install a new gasket and drain plug to

the oil pan after applying engine oil. Tighten the drain plug to 26 ft. lbs. (35 Nm).

16. Install the clutch housing plate.

➡**Tighten the upper bolt first, then tighten lower bolt.**

a. Tighten the clutch housing plate upper bolt to 22 ft. lbs. (30 Nm).

b. Tighten the clutch housing plate lower bolt to 97 inch lbs. (11 Nm).

17. Install the transfer. Refer to Transfer Case, removal & installation.

18. Install the exhaust No. 1 pipe.

19. Install the engine under cover.

20. Install the oil level gauge.

21. Fill the engine with engine oil.

2010 Model

See Figures 128 through 134.

1. Before servicing the vehicle, refer to the Precautions Section.

2. Drain the engine oil.

3. Remove the right and left side engine under covers.

4. Remove the front bumper.

5. For the CVT model, detach the CVT fluid cooler from the engine front mounting member, with the CVT fluid cooler hose connected.

6. Fix the radiator, A/C condenser, and the CVT fluid cooler (CVT model) to the vehicle body with a rope so that they do not fall off while lowering the engine front mounting member.

7. Remove the front suspension frame.

8. Remove the engine front mounting member and mounting member.

9. Remove the oil level gauge and the oil level gauge guide.

10. Remove the transaxle bolts, then remove the drive plate cover for the CVT model, or the clutch housing lower plate for the M/T model.

11. Loosen the oil pan bolts, evenly and gradually, in sequence then remove the oil pan from the lower crank case.

12. Remove the oil pump strainer from the oil pump assembly, as necessary.

To install:

13. Apply engine oil to new O-rings and install them to the oil strainer, if removed.

14. Install the oil strainer to the oil pump. Tighten the oil strainer bolt to 97 inch lbs. (11 Nm).

15. Install the dowel pins to the oil pan.

16. Apply sealant to the oil pan mating surface in a bead 0.12 inch (3mm) wide and 0.08 inch (2mm) high.

➡**Use Sealant 99000-31260 (SUZUKI Bond No. 1217G), or equivalent.**

17. Install the oil pan according to the following procedure.

   a. Fit the oil pan to the lower crank case.

   b. Step 1: Tighten the oil pan bolts evenly and gradually to 11 ft. lbs. (15 Nm) in sequence.

   c. Step 2: Tighten the oil pan bolts evenly and gradually to 19 ft. lbs. (25 Nm) in sequence.

18. Install the transaxle bolts and then the drive plate cover for the CVT model, or the clutch housing lower plate for the M/T model.

   • Tighten the transaxle bolts to 63 ft. lbs. (85 Nm)

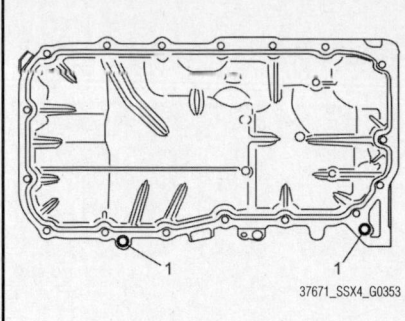

Fig. 131 Install the dowel pins (1) to the oil pan

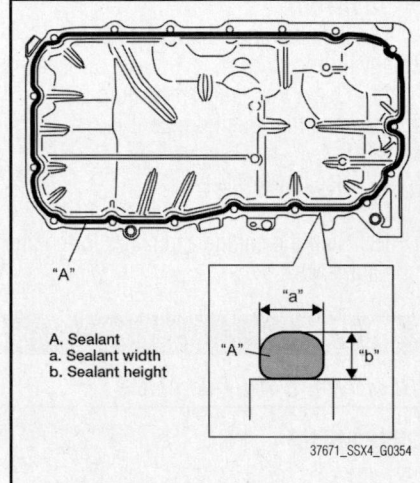

A. Sealant
a. Sealant width
b. Sealant height

Fig. 132 Apply sealant to the oil pan mating surface

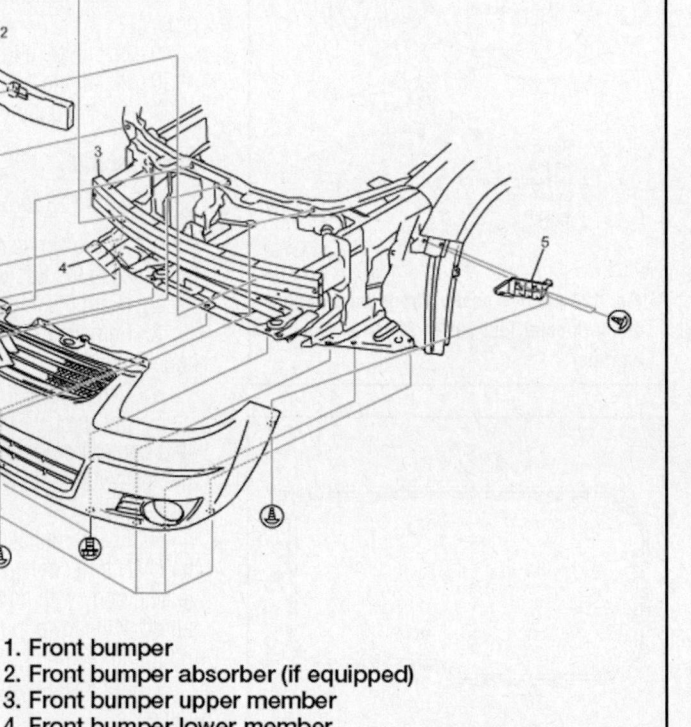

1. Front bumper
2. Front bumper absorber (if equipped)
3. Front bumper upper member
4. Front bumper lower member
5. Front bumper holder

Fig. 128 Exploded view of the front bumper

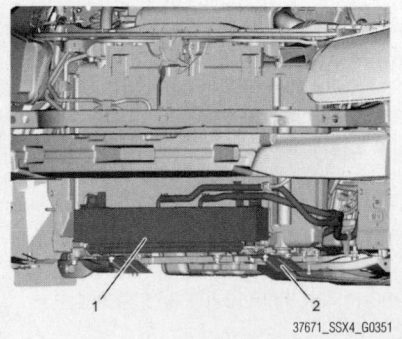

Fig. 129 Detach the CVT fluid cooler (1) from the engine front mounting member (2) with the CVT fluid cooler hose connected

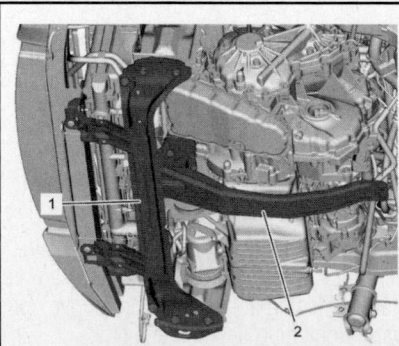

1. Engine front mounting member
2. Mounting member

Fig. 130 Remove the engine front mounting member

   • Tighten the drive plate cover bolts to 97 inch lbs. (11 Nm) (CVT model)
   • Tighten the clutch housing lower plate bolts to 97 inch lbs. (11 Nm) (M/T model)

19. Install the oil level gauge and the oil level gauge guide. Tightening the oil level gauge guide bolt to 97 inch lbs. (11 Nm).

20. Install the engine front mounting members and tighten the bolts.

21. Install the front suspension frame.

22. For CVT model, attach the CVT fluid cooler to the engine front mounting member.

23. Install the front bumper.

24. Install the drain plug and refill the engine with engine oil.

25. Verify that there is no oil leakage or exhaust gas leakage at each connection.

26. Install the right and left side engine under covers.

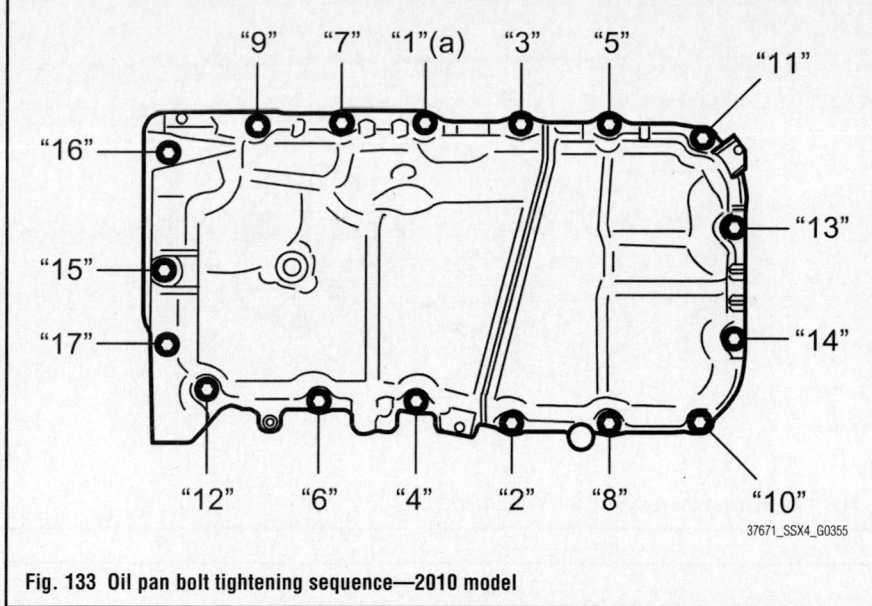

Fig. 133 Oil pan bolt tightening sequence—2010 model

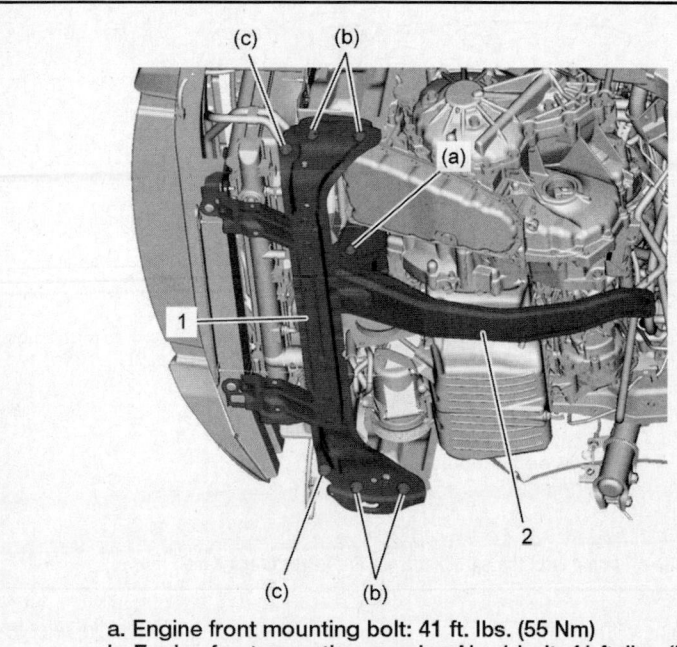

a. Engine front mounting bolt: 41 ft. lbs. (55 Nm)
b. Engine front mounting member No. 1 bolt: 41 ft. lbs. (55 Nm)
c. Engine front mounting member No. 2 bolt: 44 ft. lbs. (59 Nm)

37671_SSX4_G0356

Fig. 134 Install the engine front mounting members and tighten the bolts

OIL PUMP

REMOVAL & INSTALLATION

2009 Model

See Figures 135 through 137.

※※ WARNING

Do not remove the sprocket and the inner rotor from the oil pump, otherwise damage of the oil pump

center shaft and abnormal operation of the oil pump could result.

1. Before servicing the vehicle, refer to the Precautions Section.
2. Disconnect the negative battery cable.
3. Drain the engine oil.
4. Remove the oil pan and oil pump strainer. Refer to Oil Pan, removal & installation.
5. Remove the baffle plate from the lower crank case.

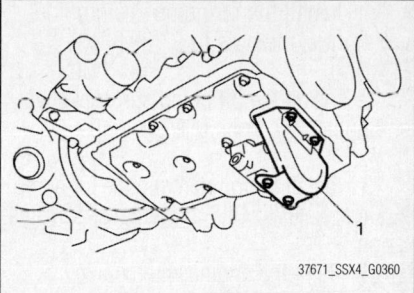

Fig. 135 Remove the oil pump sprocket cover (1)

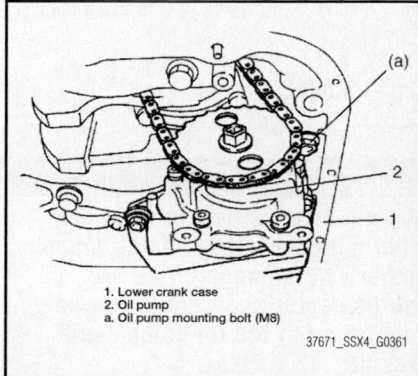

1. Lower crank case
2. Oil pump
a. Oil pump mounting bolt (M8)

37671_SSX4_G0361

Fig. 136 Remove the oil pump with the sprocket from the lower crank case

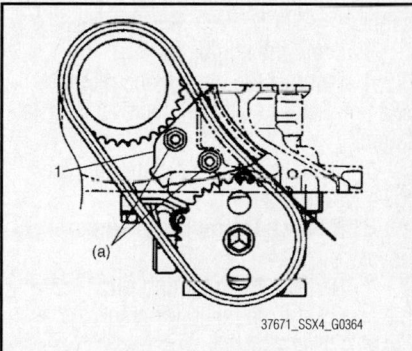

37671_SSX4_G0364

Fig. 137 Install the oil pump chain guide (1), and tighten the nuts (a)

6. Remove the oil pump chain guide.
7. Remove the oil pump sprocket cover.
8. Remove the oil pump with the sprocket from the lower crank case.

To install:

9. Install the oil pump and tighten the M8 mounting bolt to 18 ft. lbs. (25 Nm).
10. Install the baffle plate to the lower crank case and tighten the M6 baffle plate bolt to 97 inch lbs. (11 Nm).

➡When installing the oil pump, be careful not to allow the pins to fall off.

11. Install the oil pump chain guide, and tighten the nuts to 97 inch lbs. (11 Nm).

12. Install the oil pump sprocket cover, and tighten the bolts to 97 inch lbs. (11 Nm).

13. Install the oil pan and the oil pump strainer. Refer to Oil Pan, removal & installation.

14. Refill the engine with engine oil.

15. Connect the negative battery cable.

16. After completing the installation, check the oil pressure by running the engine.

2010 Model

See Figures 138 and 139.

❋❋ WARNING

Do not remove the sprocket and the inner rotor from the oil pump, otherwise damage of the oil pump center shaft and abnormal operation of the oil pump could result.

1. Before servicing the vehicle, refer to the Precautions Section.

2. Disconnect the negative battery cable.

3. Drain the engine oil.

4. Remove the oil pan and oil pump strainer. Refer to Oil Pan, removal & installation.

5. Remove the baffle plate from the lower crank case.

6. Remove the oil pump sprocket cover.

7. Remove the oil pump with the sprocket from the lower crank case.

To install:

8. Install the oil pump and tighten the mounting bolt to 18 ft. lbs. (25 Nm).

9. Install the baffle plate to the lower crank case and tighten the M6 baffle plate bolt to 97 inch lbs. (11 Nm). Tighten the M10 baffle plate bolt to 41 ft. lbs. (55 Nm).

➡**When installing the oil pump, be careful not to allow the pins to fall off.**

10. Install the oil pump sprocket cover, and tighten bolts to 97 inch lbs. (11 Nm).

11. Install the oil pan and oil pump strainer. Refer to Oil Pan, removal & installation.

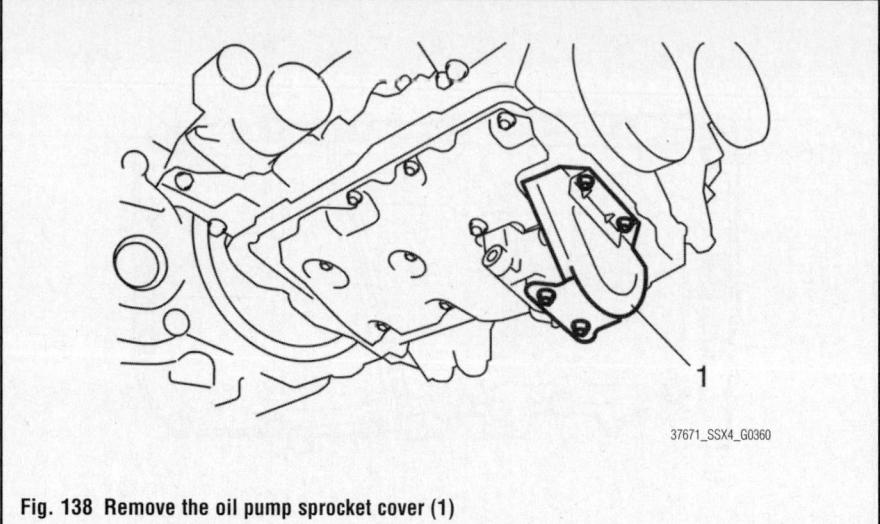

37671_SSX4_G0360

Fig. 138 Remove the oil pump sprocket cover (1)

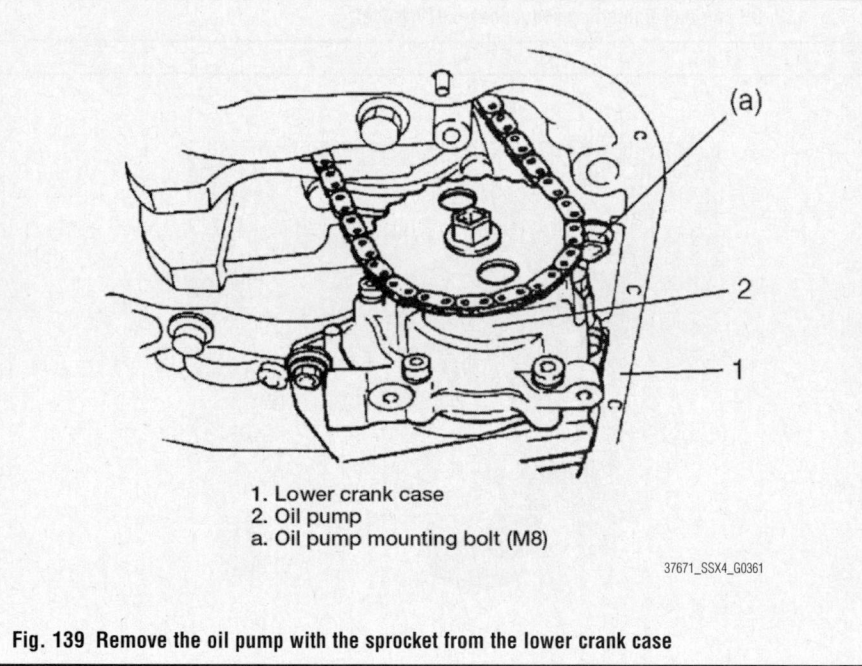

1. Lower crank case
2. Oil pump
a. Oil pump mounting bolt (M8)

37671_SSX4_G0361

Fig. 139 Remove the oil pump with the sprocket from the lower crank case

12. Refill the engine with engine oil. Refer to Engine Oil & Filter, Replacement.

13. Connect the negative battery cable.

14. After completing the installation, check the oil pressure by running the engine.

15. Verify that there is no oil leakage or exhaust gas leakage at each connection.

REAR MAIN SEAL

REMOVAL & INSTALLATION

2009 Model

See Figure 140.

1. Before servicing the vehicle, refer to the Precautions Section.

2. Remove the transaxle assembly.

3. Remove the clutch cover and clutch disk (M/T models).

4. Remove the drive plate (A/T models) or the flywheel (M/T models). Refer to Flywheel/Driveplate, removal & installation.

5. Remove the rear oil seal with a suitable tool.

❋❋ WARNING

Be careful not to damage the crankshaft or the cylinder block.

To install:

6. Pull out the dowel pin from the crank-

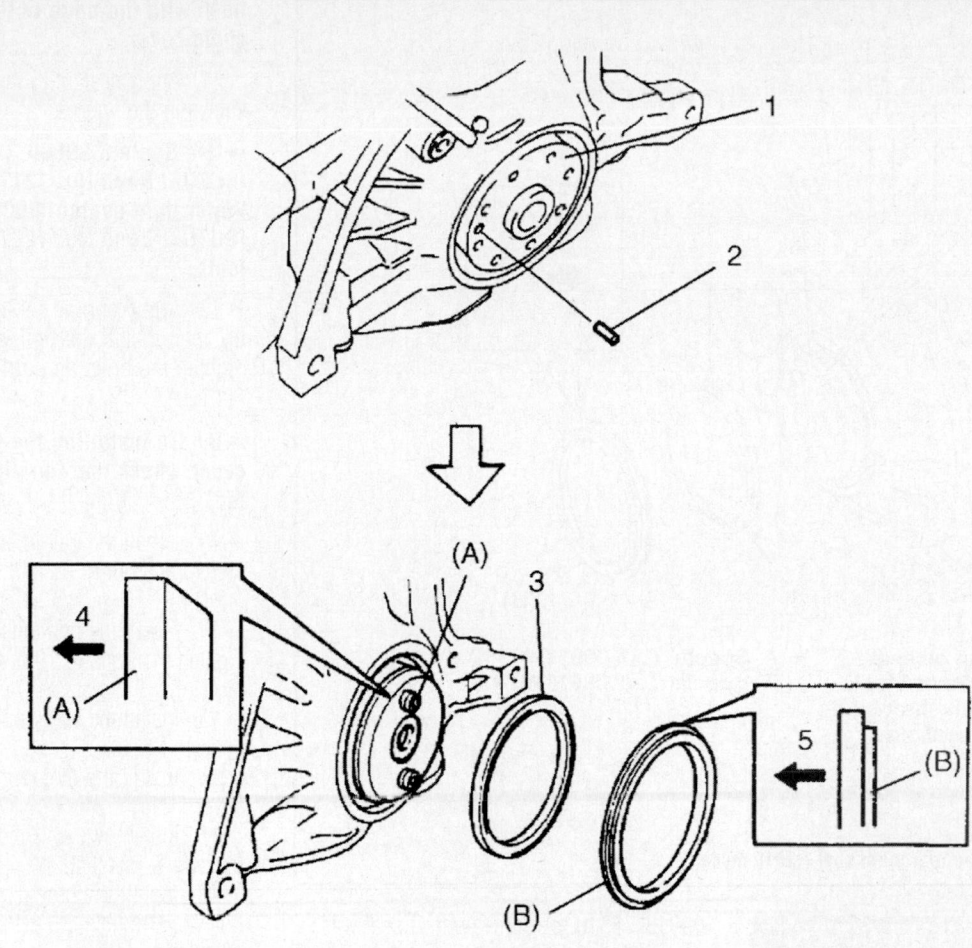

1. Crankshaft
2. Dowel pin
3. Rear oil seal
4. Crankshaft side
5. Oil seal side
A. Special Tool 09911-97710
B. Special Tool 09911-97811

37671_SSX4_G0380

Fig. 140 Installing the rear main seal—2009 model

shaft and then install the rear oil seal until it becomes flush with the cylinder block surface using Special Tools 09911-97710 and 09911-97811 (or equivalent) and a plastic hammer.

➡**Install the rear oil seal so that each seal lip is oriented correctly.**

7. Install the dowel pin.
8. Install the CKP sensor and fix its wire harness with the bracket. Tightening the CKP sensor bolt to 97 inch lbs. (11 Nm).
9. Install the flywheel (drive plate for A/T). Refer to Flywheel/Driveplate, removal & installation.

➡**Use new flywheel or drive plate bolts.**

2010 Model

See Figure 141.

1. Before servicing the vehicle, refer to the Precautions Section.
2. Remove the transaxle assembly.
3. Remove the clutch cover and clutch disk (M/T models).
4. Remove the drive plate (A/T models) or the flywheel (M/T models). Refer to Flywheel/Driveplate, removal & installation.
5. Remove the rear oil seal with a suitable tool.

⁑ **WARNING**

Be careful not to damage the crankshaft or the cylinder block.

To install:

6. Apply engine oil to the new rear oil seal and install it to the cylinder block using Special Tools 09911-97710 and 09911-97811, or equivalent.
7. Install the rear oil seal so that each seal lip is oriented correctly and it becomes flush with the cylinder block.

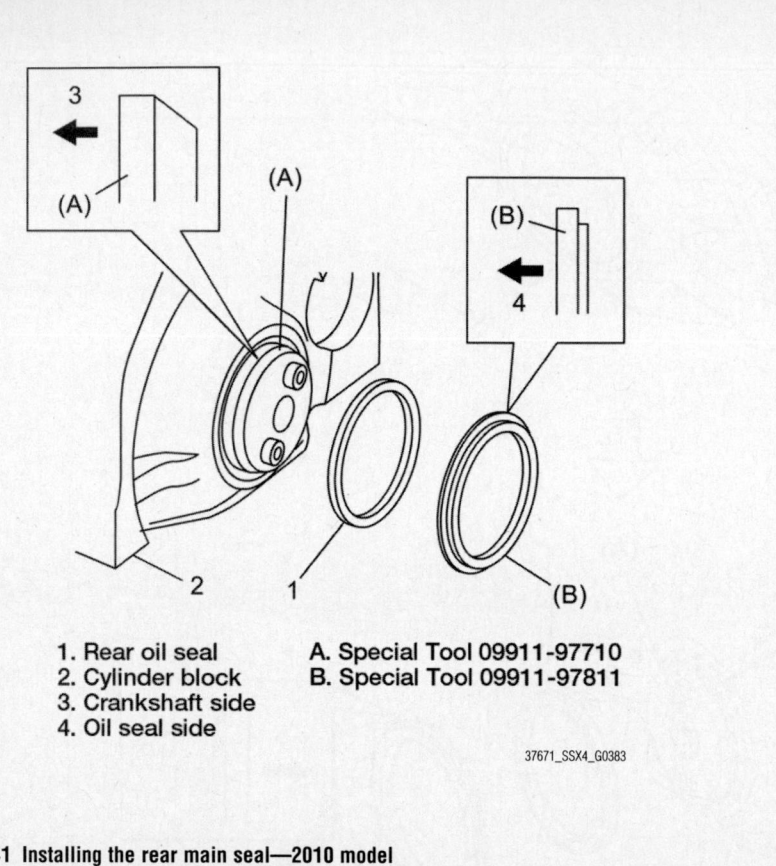

1. Rear oil seal
2. Cylinder block
3. Crankshaft side
4. Oil seal side

A. Special Tool 09911-97710
B. Special Tool 09911-97811

37671_SSX4_G0383

Fig. 141 Installing the rear main seal—2010 model

TIMING CHAIN FRONT COVER

REMOVAL & INSTALLATION

2009 Model

See Figures 142 and 143.

1. Before servicing the vehicle, refer to the Precautions Section.
2. Remove the engine assembly from the vehicle.
3. Remove the oil pan. Refer to Oil Pan, removal & installation.
4. Remove the cylinder head cover.
5. Remove the crankshaft pulley. Refer to Crankshaft Damper, removal & installation.
6. Remove the idler pulley, water pump pulley, and belt tensioner.
7. Remove the timing chain cover bolts and nut.
8. Remove the timing chain cover.

To install:

9. Clean the sealing surfaces on the timing chain cover, cylinder block, and cylinder head. Remove oil, old sealant, and dust from the sealing surface.
10. Install a new oil seal to the timing chain cover using Special Tool 09913-75510, or equivalent, if seal was removed.

➡ When installing the new oil seal, drive it until its surface is flush with the edge of the timing chain cover.

11. Apply sealant to the specific areas shown in the figure.

➡ **Use Sealant 99000-31260 (SUZUKI Bond No. 1217G) and Water tight sealant 99000-31140 (SUZUKI Bond No. 1207B) or equivalents.**

12. Apply engine oil to the oil seal lip, then install the timing chain cover. Tighten the bolts and nut to 97 inch lbs. (11 Nm).

➡ **Before installing the timing chain cover, check that the pin is securely fitted.**

13. Install the belt idler pulley. Tighten the nut to 31 ft. lbs. (42 Nm).
14. Install the belt tensioner. Tighten the bolts to 19 ft. lbs. (25 Nm).
15. Install the water pump pulley.
16. Install the cylinder head cover.
17. Install oil pan. Refer to Oil Pan, removal & installation.
18. Install the crankshaft pulley. Refer to Crankshaft Damper, removal & installation.
19. Install the engine assembly to the vehicle.

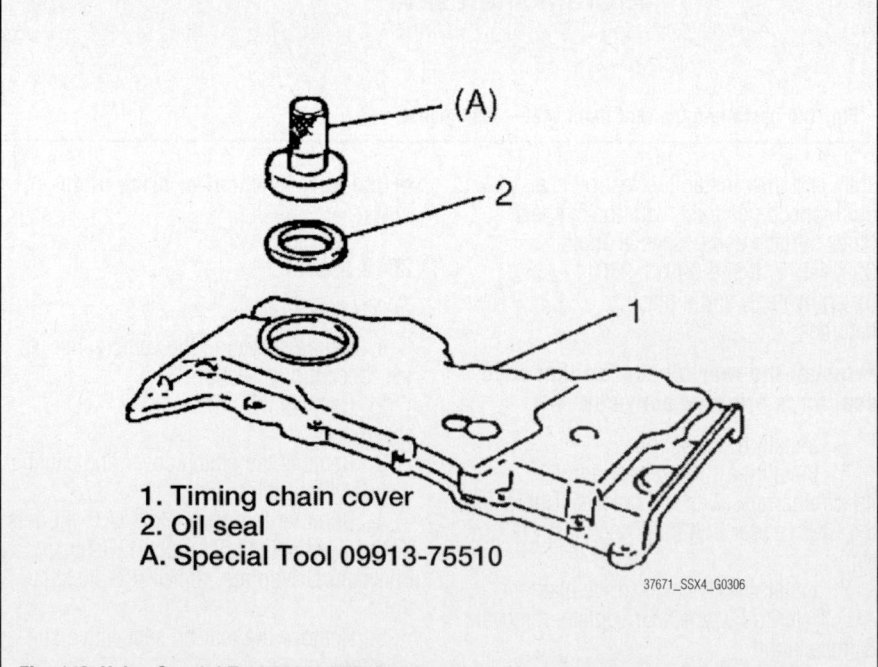

1. Timing chain cover
2. Oil seal
A. Special Tool 09913-75510

37671_SSX4_G0306

Fig. 142 Using Special Tool 09913-75510 to install the crankshaft front seal—2009 model

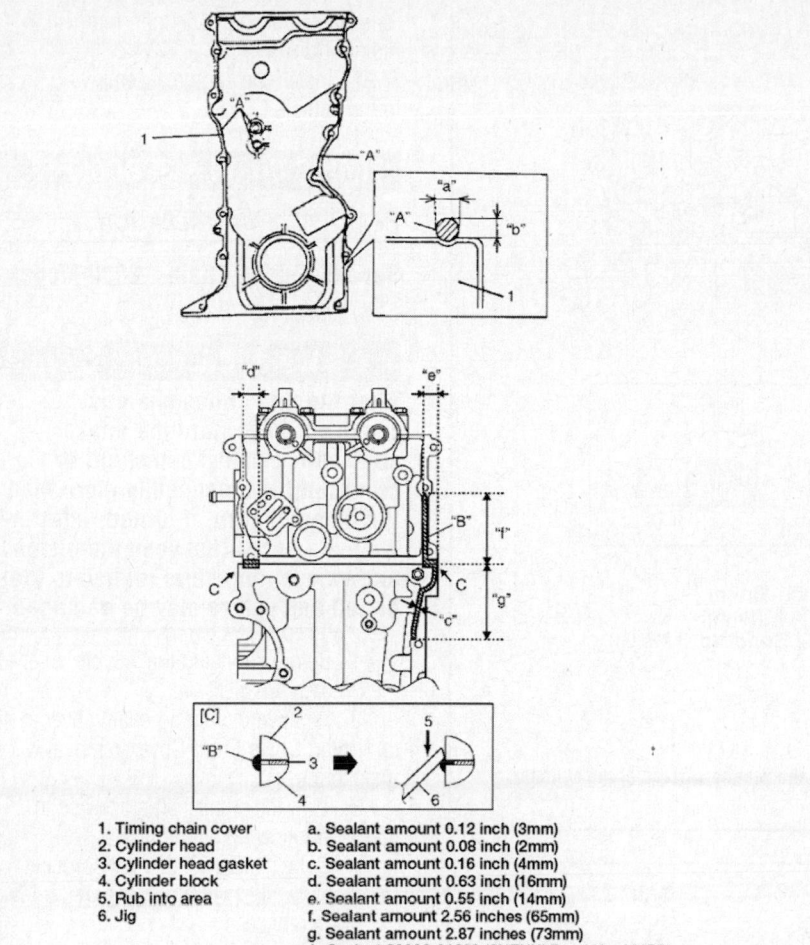

1. Timing chain cover
2. Cylinder head
3. Cylinder head gasket
4. Cylinder block
5. Rub into area
6. Jig

a. Sealant amount 0.12 inch (3mm)
b. Sealant amount 0.08 inch (2mm)
c. Sealant amount 0.16 inch (4mm)
d. Sealant amount 0.63 inch (16mm)
e. Sealant amount 0.55 inch (14mm)
f. Sealant amount 2.56 inches (65mm)
g. Sealant amount 2.87 inches (73mm)
A. Sealant 99000-31260 (SUZUKI Bond No. 1217G)
B. Water tight sealant 99000-31140 (SUZUKI Bond No. 1207B)
C. View C

37671_SSX4_G0386

Fig. 143 Apply the proper type and amount of sealant to the timing cover in the areas shown—2009 model

2010 Model

See Figures 144 through 148.

➡ Keep the working table, tools, and hands clean while working. Use special care to handle the aluminum parts so as not to damage them. Do not expose removed parts to dust. Keep them clean.

1. Before servicing the vehicle, refer to the Precautions Section.
2. Remove the engine assembly from the vehicle.
3. Remove crankshaft pulley. Refer to Crankshaft Damper, removal & installation.

✷✷ WARNING

Avoid applying unnecessary load to the timing chain, sprockets, and any other related parts.

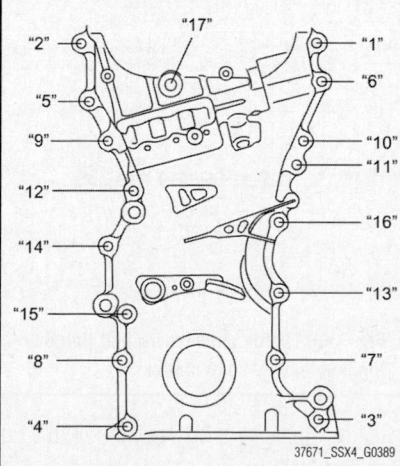

37671_SSX4_G0389

Fig. 144 Timing chain cover bolt loosening sequence—2010 model

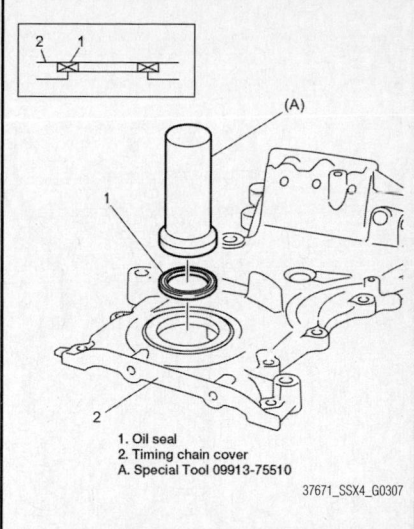

1. Oil seal
2. Timing chain cover
A. Special Tool 09913-75510

37671_SSX4_G0307

Fig. 145 Using Special Tool 09913-75510 to install the crankshaft front seal—2010 model

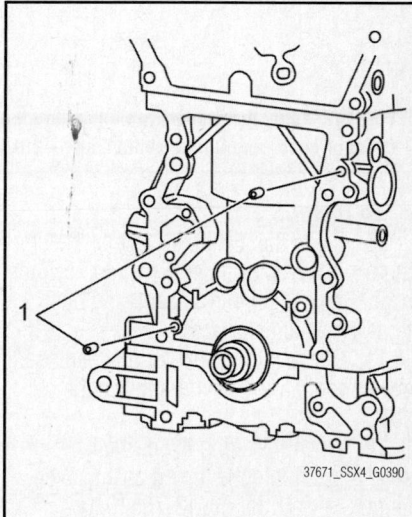

37671_SSX4_G0390

Fig. 146 Install the dowel pins (1) to the cylinder block—2010 model

4. Remove the cylinder head cover.
5. Remove the oil pan. Refer to Oil Pan, removal & installation.
6. Remove the tensioner pulley and the idler pulley.
7. Remove the timing chain cover by loosening the timing chain cover bolts evenly and gradually according to the loosening order.
8. Remove the crankshaft oil seal from the timing chain cover using a flathead screwdriver or the equivalent, if necessary.
9. Remove the oil seal from the timing chain cover, if necessary.

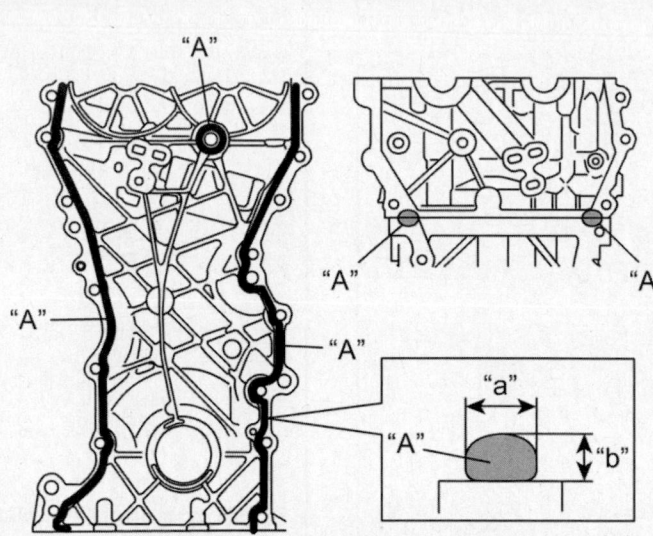

a. Width of sealant bead: 0.12 inch (3mm)
b. Height of sealant bead: 0.08 inch (2mm)
A. Sealant 99000-31260 (SUZUKI Bond No. 1217G)

37671_SSX4_G0391

Fig. 147 Apply the proper type and amount of sealant to the mating surface of the cylinder, cylinder head, and timing chain cover—2010 model

To install:

10. Clean the mating surface of the timing chain cover, cylinder block, and cylinder head. Remove oil, old sealant, and dust from the mating surface.

11. Check the oil seal for any damage. If any abnormality is found, replace the oil seal.

12. When installing the new oil seal, press it into the timing chain cover using Special Tool 09913-75510, or equivalent.

13. Install the dowel pins to the cylinder block.

14. Apply the proper type and amount of sealant to the mating surface of the cylinder, cylinder head, and timing chain cover.

➡ **Use Sealant 99000-31260 (SUZUKI Bond No. 1217G), or equivalent.**

15. Apply engine oil to the oil seal lip, then install the timing chain cover.

16. Tighten the timing chain cover bolts evenly and gradually to the specified torque in numerical order.

➡ **Before installing the timing chain cover, check that the dowel pins are securely fitted.**

a. Tighten the M8 cover bolts to 19 ft. lbs. (25 Nm).

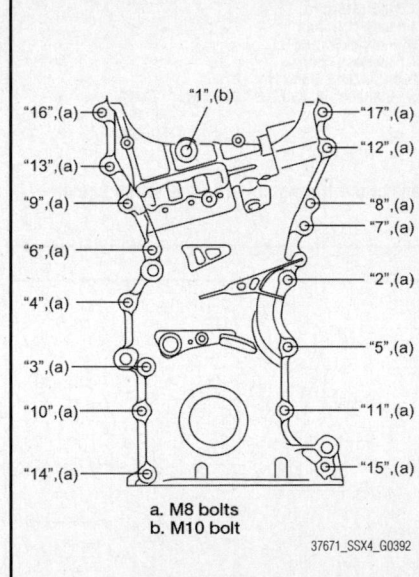

a. M8 bolts
b. M10 bolt

37671_SSX4_G0392

Fig. 148 Timing chain cover bolt tightening sequence—2010 model

b. Tighten the M10 cover bolt to 41 ft. lbs. (55 Nm).

17. Install the tensioner pulley and idler pulley.

18. Install the cylinder head cover.

19. Install the oil pan. Refer to Oil Pan, removal & installation.

20. Install the crankshaft pulley. Refer to Crankshaft Damper, removal & installation

21. Install the engine assembly to the vehicle.

TIMING CHAIN & SPROCKETS

REMOVAL & INSTALLATION

Second Timing Chain—2009 Model

See Figures 149 through 156.

※ **WARNING**

After the 2nd timing chain is removed, never turn the intake camshaft, exhaust camshaft, or crankshaft independently more than the extent shown. If turned, interference may occur between the piston and valves, and parts related to the piston and valves may be damaged.

1. Before servicing the vehicle, refer to the Precautions Section.

2. Remove the timing chain cover. Refer to Timing Chain Front Cover, removal & installation.

3. Turn the crankshaft clockwise to meet the following conditions.
   - The mark on the crank sprocket matches with the mark on the lower crankcase
   - The arrow mark on the idler sprocket points upward
   - The marks on the cam sprockets match with the marks on the cylinder head

4. Remove the timing chain tensioner adjuster No. 2 and the gasket. To remove them, slacken the 2nd timing chain by turning the intake camshaft counterclockwise a little while pushing back the pad.

5. Remove the intake and exhaust camshaft timing sprocket bolts. To remove them, fit a spanner to the hexagonal part at the center of the camshaft to hold it stationary.

6. Remove the camshaft timing sprockets and the 2nd timing chain.

To install:

7. Check that the match mark on the crank timing sprocket matches with the timing mark on the lower crankcase.

8. Check that the arrow mark on the idler sprocket faces upward.

9. Check that the knock pins of the intake and exhaust camshafts are aligned with the timing marks on the cylinder head.

10. Install the 2nd timing chain by align-

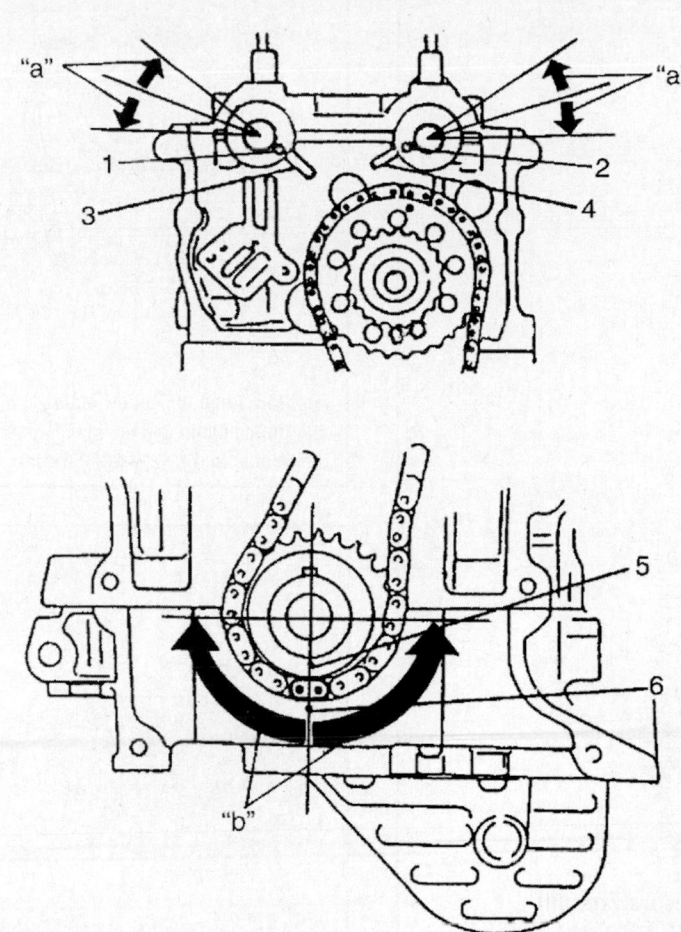

1.	Knock pin of intake camshaft
2.	Knock pin of exhaust camshaft
3.	Timing mark of intake side
4.	Timing mark of exhaust side
5.	Match mark on crank timing sprocket
6.	Timing mark on lower crankcase
"a":	Camshafts (IN & EX) allowable turning range..... Within 20° on both right and left
"b":	Crankshaft allowable turning range..... Within 90° on both right and left

37671_SSX4_G0395

Fig. 149 Interference engine: allowable movement of components shown with 2nd timing chain removed—2009 model

ing the yellow plate of the 2nd timing chain and the match marks on the idler sprocket.

11. Install the sprockets to the intake and exhaust camshafts by aligning the dark blue plate of the 2nd timing chain, match marks on the intake sprocket, and exhaust sprocket, respectively.

❋❋ WARNING

Do not turn the components more than the allowable turning range. If turned excessively, the valves and pistons may be damaged.

➡**As an arrow mark is provided on both sides, the camshaft timing sprocket has no specific installation direction.**

12. Tighten the intake and exhaust camshaft timing sprocket bolts to the specified torque.

　a. Fit a spanner to the hexagonal part at the center of the camshaft to hold it stationary.

　b. Tighten the camshaft timing sprocket bolts to 58 ft. lbs. (80 Nm).

13. Push back the plunger into the tensioner body, and hold it at the position by inserting a stopper into the body.

14. Install the timing chain tensioner adjuster No. 2 with a new gasket.

　a. Tighten the timing chain tensioner adjuster No. 2 bolt to 97 inch lbs. (11 Nm).

　b. Tighten the timing chain tensioner adjuster No. 2 nut to 33 ft. lbs. (45 Nm).

15. Pull out the stopper from the timing chain tensioner adjuster No. 2.

16. Turn the crankshaft 2 rotations clockwise, and then align the timing mark on the crankshaft and the timing mark on the cylinder block.

17. At this time, check the timing marks of the sprockets matching with the timing marks of the cylinder head, the cylinder block, and the lower crank case.

18. Apply oil to the timing chains, tensioner, tensioner adjusters, sprockets, and guides.

19. Install the timing chain cover. Refer to Timing Chain Front Cover, removal & installation.

20. Install the cylinder head cover.

21. Install the oil pan. Refer to Oil Pan, removal & installation.

22. Install the engine assembly to the vehicle..

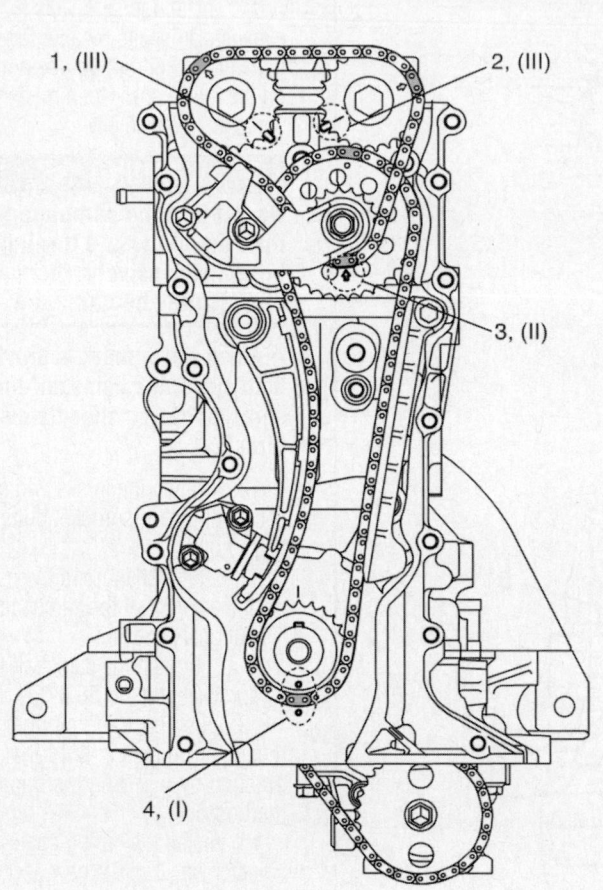

1. Timing marks of intake camshaft timing sprocket (III)
2. Timing marks of exhaust camshaft timing sprocket (III)
3. Arrow mark on idler sprocket (II)
4. Timing mark of crankshaft timing sprocket (I)

37671_SSX4_G0397

Fig. 150 Timing mark alignment shown—2009 model

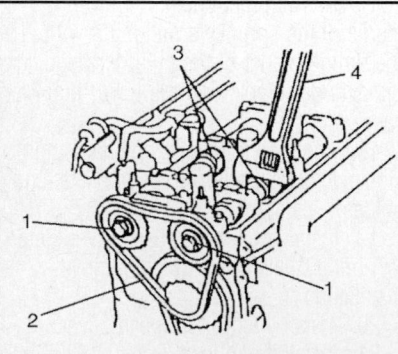

1. Intake and exhaust camshaft timing sprocket bolts
2. 2nd timing chain
3. Hexagonal part (center of camshaft)
4. Spanner

37671_SSX4_G0398

Fig. 151 Removing the 2nd timing chain and sprockets—2009 model

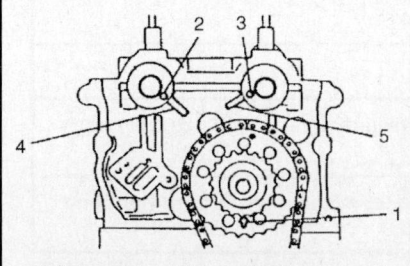

1. Arrow mark on idler sprocket facing upward
2. Knock pin of intake camshaft
3. Knock pin of exhaust camshaft
4. Timing mark of intake side
5. Timing mark of exhaust side

37671_SSX4_G0399

Fig. 152 Check that the arrow mark on the idler sprocket faces upward and the knock pins of the intake and exhaust camshafts are aligned with the timing marks on the cylinder head—2009 model

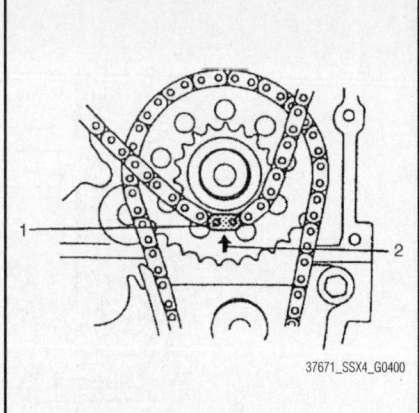

37671_SSX4_G0400

Fig. 153 Align the yellow plate (1) of the 2nd timing chain and the match mark on the idler sprocket (2)—2009 model

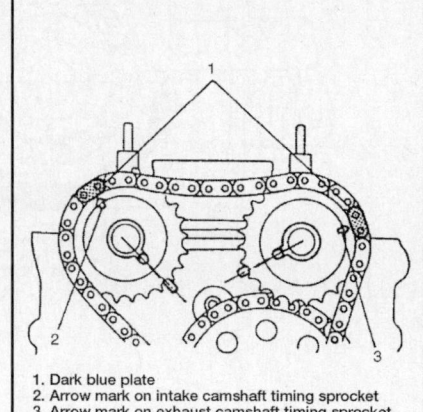

1. Dark blue plate
2. Arrow mark on intake camshaft timing sprocket
3. Arrow mark on exhaust camshaft timing sprocket

37671_SSX4_G0401

Fig. 154 Match the marks on the intake and exhaust camshafts with the 2nd timing chain—2009 model

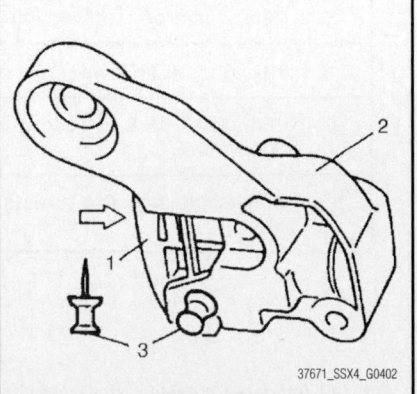

37671_SSX4_G0402

Fig. 155 Push back the plunger (1) into the tensioner body (2), and hold it at the position by inserting a stopper (3) into the body—2009 model

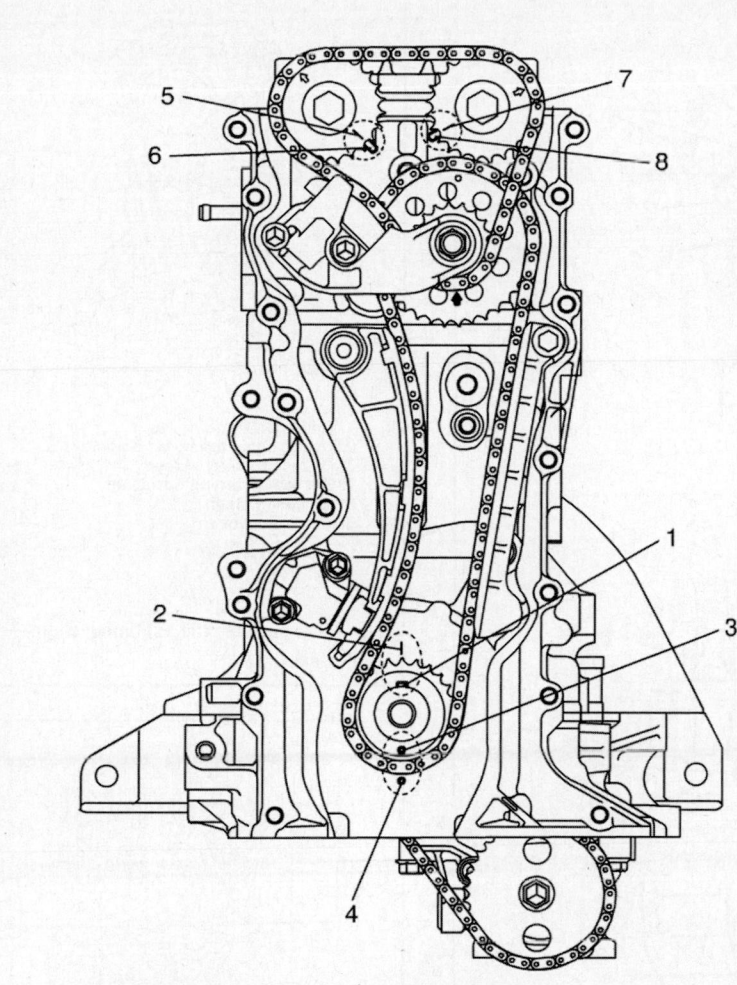

1. Timing mark on the crankshaft
2. Timing mark on the cylinder block
3. Timing mark on crank timing sprocket
4. Timing mark on lower crankcase
5. Timing mark on intake camshaft timing sprocket
6. Timing mark of intake camshaft timing sprocket
7. Timing mark on exhaust camshaft timing sprocket
8. Timing mark of exhaust camshaft timing sprocket

37671_SSX4_G0403

Fig. 156 Make sure all timing marks are aligned—2009 model

First Timing Chain—2009 Model

See Figures 157 through 163.

> ※※ **WARNING**
>
> **After the 2nd timing chain is removed, never turn the intake camshaft, exhaust camshaft, or crankshaft independently more than the extent shown. If turned, interference may occur between the piston and valves, and parts related to the piston and valves may be damaged.**

1. Before servicing the vehicle, refer to the Precautions Section.
2. Remove the timing chain cover. Refer to Timing Chain Front Cover, removal & installation.
3. Remove the 2nd timing chain. Refer to Second Timing Chain—2009 Model, removal & installation.
4. Remove the timing chain guide No. 1.
5. Remove the timing chain tensioner adjuster No. 1.
6. Remove the timing chain tensioner.
7. Remove the idler sprocket and the 1st timing chain.
8. Remove the crankshaft timing sprocket.

To install:

9. Install the crankshaft timing sprocket.
10. Check that the match mark on the crankshaft timing sprocket matches with the timing mark on the lower crankcase.
11. Apply engine oil to the idler sprocket shaft and idler sprocket.
12. Install the idler sprocket and the idler sprocket shaft.
13. Install the 1st timing chain by aligning the dark blue plate of the 1st timing chain and the match mark on the idler sprocket.
14. Bring the gold plate of the 1st timing chain into match with the match mark on the crankshaft timing sprocket.
15. Apply engine oil to the sliding surface of the timing chain tensioner and then install it. Tighten the tensioner nut to 18 ft. lbs. (25 Nm).
16. With the latch of the tensioner adjuster No. 1 returned and the plunger pushed back into the body, insert the stopper into the latch and the body. After inserting it, check to make sure that the plunger will not come out.
17. Install the timing chain tensioner adjuster No. 1. Tighten the chain tensioner adjuster bolt to 97 inch lbs. (11 Nm).
18. Pull out the stopper from the timing chain tensioner adjuster No. 1.
19. Apply engine oil to the sliding surface of the timing chain guide No. 1 and then install it.
Tighten the guide bolts to 80 inch lbs. (9 Nm).
20. Check that the dark blue and gold plates of the 1st timing chain are in match with the match marks on the sprockets respectively.
21. Install the 2nd timing chain. Second Timing Chain—2009 Model, removal & installation.
22. Install the timing chain cover. Refer to Timing Chain Front Cover, removal & installation.
23. Install the cylinder head cover.

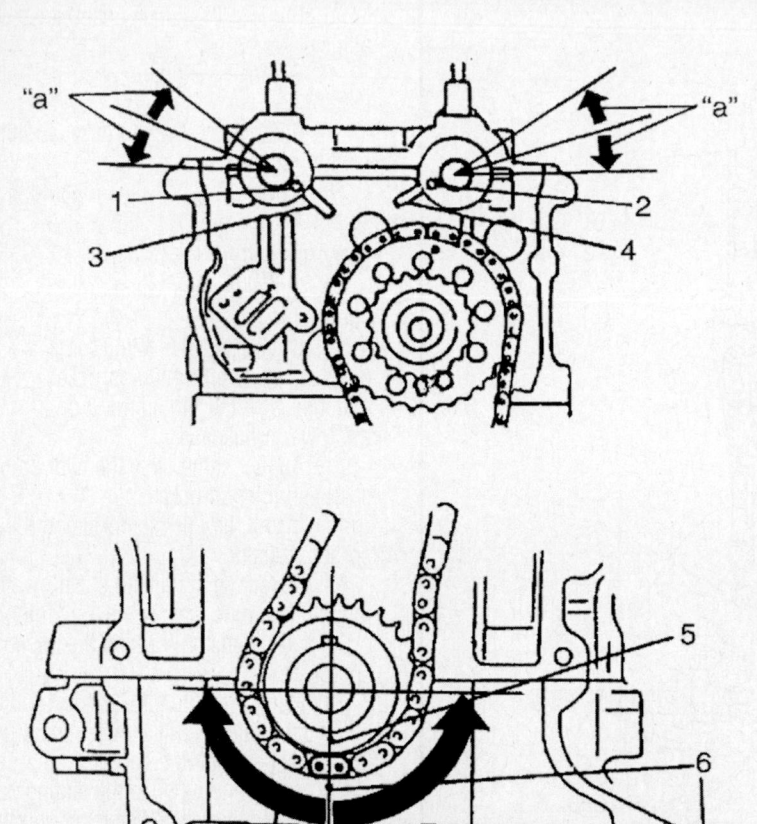

1.	Knock pin of intake camshaft
2.	Knock pin of exhaust camshaft
3.	Timing mark of intake side
4.	Timing mark of exhaust side
5.	Match mark on crank timing sprocket
6.	Timing mark on lower crankcase
"a":	Camshafts (IN & EX) allowable turning range..... Within 20° on both right and left
"b":	Crankshaft allowable turning range..... Within 90° on both right and left

37671_SSX4_G0395

Fig. 157 Interference engine: allowable movement of components shown with 2nd timing chain removed—2009 model

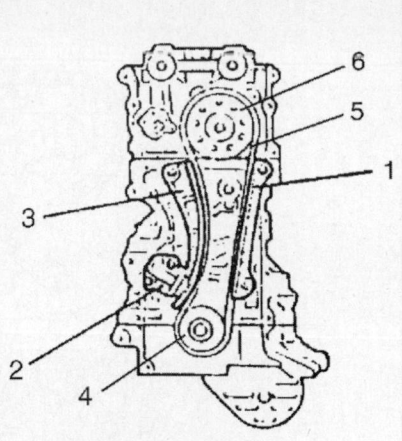

1. Timing chain guide No. 1
2. Timing chain tensioner adjuster No. 1
3. Timing chain tensioner
4. Crankshaft timing sprocket
5. 1st timing chain
6. Idler sprocket

37671_SSX4_G0406

Fig. 158 Removing the 1st timing chain— 2009 model

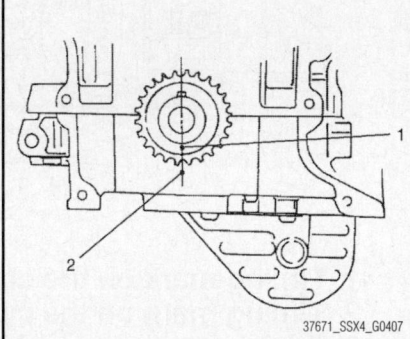

37671_SSX4_G0407

Fig. 159 Check that the match mark (1) on the crankshaft timing sprocket matches with the timing mark (2) on the lower crankcase—2009 model

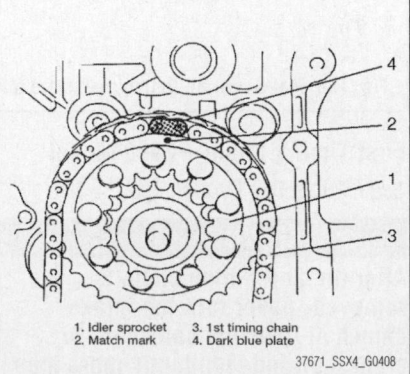

1. Idler sprocket   3. 1st timing chain
2. Match mark   4. Dark blue plate

37671_SSX4_G0408

Fig. 160 Installing the 1st timing chain on the idler sprocket—2009 model

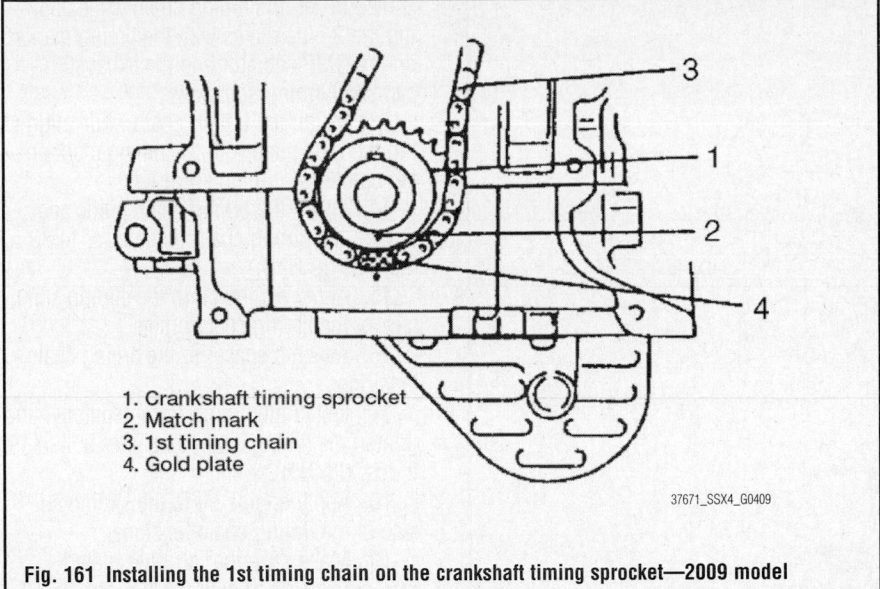

1. Crankshaft timing sprocket
2. Match mark
3. 1st timing chain
4. Gold plate

37671_SSX4_G0409

Fig. 161 Installing the 1st timing chain on the crankshaft timing sprocket—2009 model

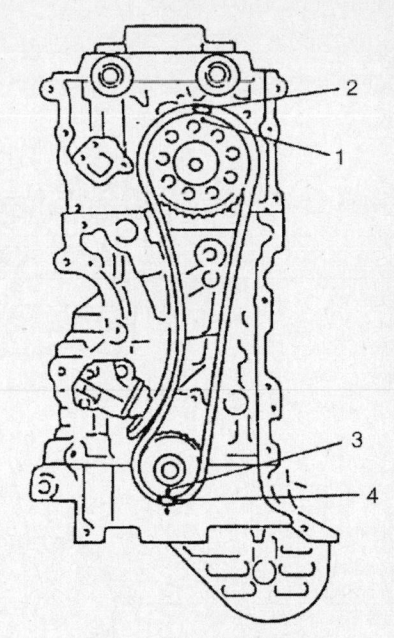

1. Match mark on idler sprocket
2. Dark blue plate
3. Match mark on crankshaft timing sprocket
4. Gold plate

37671_SSX4_G0411

Fig. 163 Check the alignment of all timing marks—2009 model

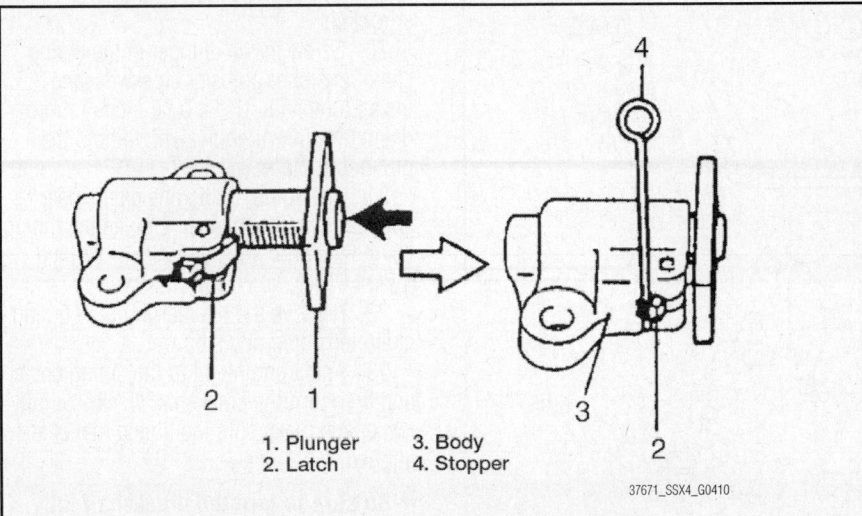

1. Plunger
2. Latch
3. Body
4. Stopper

37671_SSX4_G0410

Fig. 162 With the latch of the tensioner adjuster No. 1 returned and the plunger pushed back into the body, insert the stopper into the latch and the body—2009 model

24. Install the oil pan. Refer to Oil Pan, removal & installation.

25. Install the engine assembly to the vehicle.

2010 Model

See Figures 164 through 167.

⁑ **WARNING**

After the timing chain is removed, never turn the crankshaft or camshafts independently more than the specified range. If either shaft is turned, interference may occur between the pistons and the valves, and parts related to the piston and the valves may be damaged.

1. Before servicing the vehicle, refer to the Precautions Section.

2. Remove the timing chain cover. Refer to Timing Chain Front Cover, removal & installation.

3. By turning the crankshaft, align the camshafts and crankshaft at the specific position as follows.

   a. Align the timing marks on the CMP actuator and the timing mark on the exhaust camshaft timing sprocket with the match marks on the camshaft housing No. 1.

   b. Align the timing mark on the crankshaft timing sprocket with the match mark on the lower crank case.

4. Remove the timing chain tensioner adjuster.

5. Remove the timing chain tensioner.

6. Remove the timing chain guide.

7. Remove the timing chain and crankshaft timing sprocket.

To install:

8. Turn the crankshaft position key slot within the specified range (30–90°) in the counterclockwise direction from the top.

9. Align the timing mark on the CMP actuator with the match marks on the camshaft housing No. 1.

10. Align the timing mark on the exhaust camshaft timing sprocket with the match marks on the camshaft housing No. 1.

11. Install the key to the key slot of the crankshaft.

   a. Align the key slot of the crankshaft timing sprocket with the key, and then install the crankshaft timing sprocket to the crankshaft.

   b. Align the timing mark on the crankshaft timing sprocket with the match mark on the lower crank case.

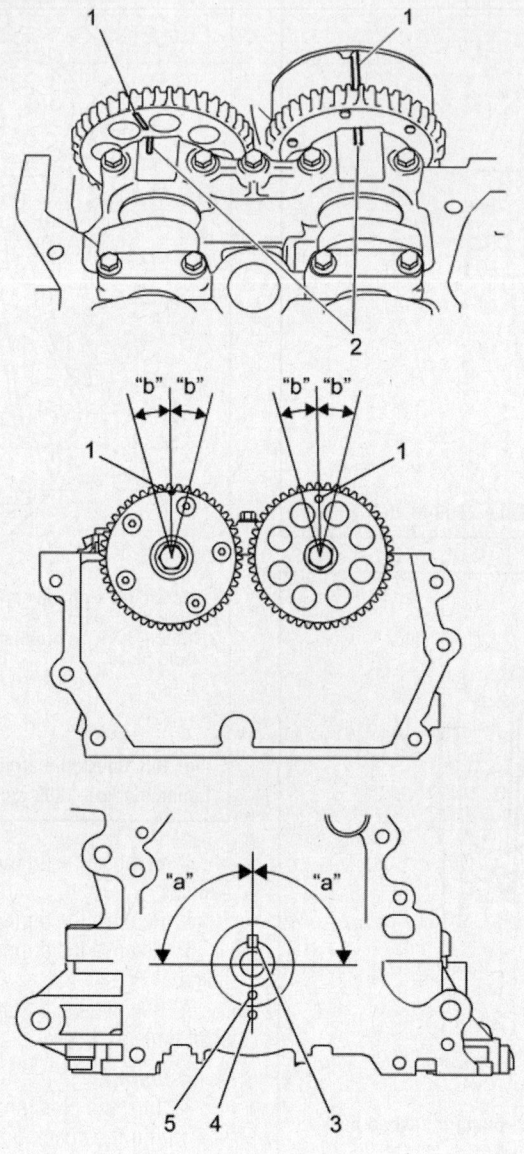

1.	Timing marks on camshaft timing sprockets
2.	Match marks on camshaft housing No.1
3.	Key
4.	Match mark on lower crank case
5.	Timing mark on crankshaft timing sprocket
"a":	90°
"b":	15°

37671_SSX4_G0414

Fig. 164 Interference engine: allowable movement of components shown with timing chain removed—2010 model

12. Install the timing chain while aligning the 2 blue plates with the timing marks on the CMP actuator and the exhaust camshaft timing sprockets.

13. Install the timing chain while aligning the blue plate with the timing mark on the crankshaft timing sprocket.

14. Install the timing chain guide and tighten the timing chain guide bolts to 80 inch lbs. (9 Nm).

15. Apply engine oil to the sliding surface of the timing chain guide.

16. Attach a spacer to the timing chain tensioner.

17. Install the timing chain tensioner and tighten the timing chain tensioner bolt to 19 ft. lbs. (25 Nm).

18. Apply engine oil to the sliding surface of the timing chain tensioner.

19. Make sure that all blue plated plates are aligned with the timing marks on the corresponding timing sprockets.

20. Screw in the plunger of the timing chain tensioner adjuster clockwise and install a retainer. Use a 0.06 inch (1.4mm) diameter wire, or equivalent, to hold the plunger in place.

21. Install the timing chain tensioner adjuster with the retainer. Tighten the timing chain tensioner adjuster bolts to 97 inch lbs. (11 Nm).

22. Remove the retainer from the timing chain tensioner adjuster.

23. Apply engine oil to the timing chain, and then turn the crankshaft clockwise fully twice, and make sure the timing marks are aligned.

➡**Be sure to turn the crankshaft fully twice. If it is rotated only once, the timing marks on the CMP actuator and the exhaust camshaft timing sprocket will not meet the match marks on the camshaft housing No. 1. After turning the crankshaft twice, the discrimination links of the timing chain are not aligned with the timing marks on the CMP actuator and the exhaust timing sprocket, but this is normal.**

24. Install the timing chain cover. Refer to Timing Chain Front Cover, removal & installation.

25. Install the cylinder head cover.

26. Install the oil pan. Refer to Oil Pan, removal & installation.

27. Install the engine assembly.

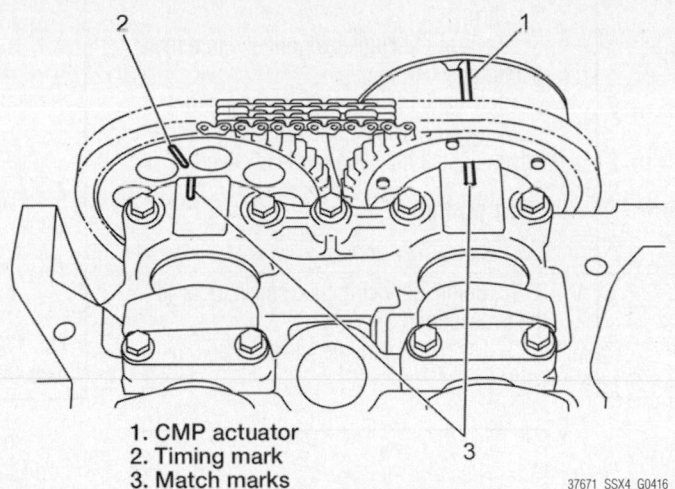

1. CMP actuator
2. Timing mark
3. Match marks

37671_SSX4_G0416

Fig. 165 Align the timing marks on the CMP actuator and the timing mark on the exhaust camshaft timing sprocket with the match marks on the camshaft housing No. 1—2010 model

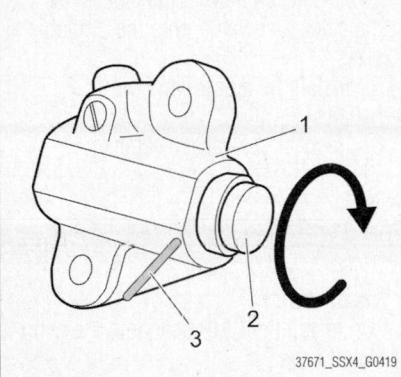

37671_SSX4_G0419

Fig. 166 Screw in the plunger (2) of the timing chain tensioner adjuster (1) clockwise and install a retainer (3)—2010 model

VALVE LASH

ADJUSTMENT

2009 Model

See Figures 167 through 170.

1. Before servicing the vehicle, refer to the Precautions Section.

2. Disconnect the negative battery cable.

3. Close the valve whose shim is to be replaced by turning the crankshaft, then turn the tappet until the cut section faces inside.

4. Turn the crankshaft to 360°.

5. Hold the tappet at that position using a special tool as follows.

   a. Remove the housing bolts.

   b. Check the housing No. and select a

special tool corresponding to the housing No., referring to the Special tool selection table.

   c. Hold down the tappet so as not to contact the shim by installing Special Tool 09916-66510, or equivalent, on the camshaft housing with the housing bolt. Tighten the housing bolts by hand.

6. Turn the camshaft approximately 90° clockwise and remove the shim.

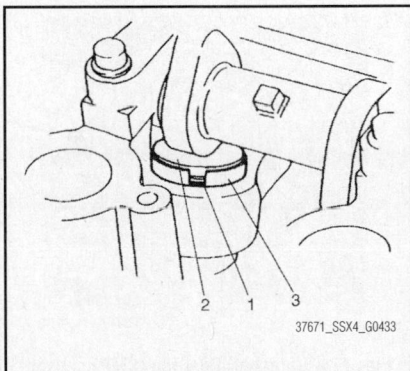

37671_SSX4_G0433

Fig. 167 Close the valve whose shim (2) is to be replaced by turning the crankshaft, then turn the tappet (3) until the cut section (1) faces inside—2009 model

✳✳ CAUTION

Never put your hand between the camshaft and the tappet.

7. Using a micrometer, measure the

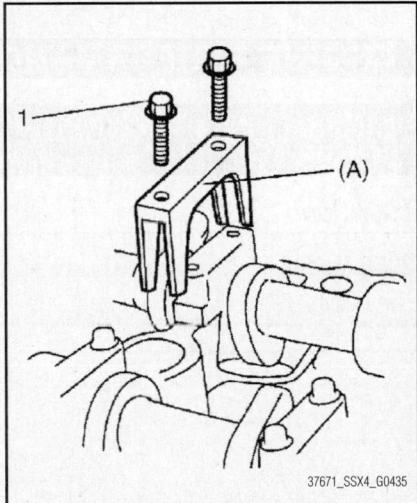

37671_SSX4_G0435

Fig. 169 Using Special Tool 09916-66510 (A) on the camshaft housing with the housing bolt (1)—2009 model

No. on camshaft housing	Embossed mark on special tool
I2, I3, I4, I5	IN
E2, E3, E4, E5	EX

37671_SSX4_G0434

Fig. 168 Special Tool Selection Table—2009 model

Shim thickness specification

Intake side:
A = B + C − 0.20 mm (0.008 in.)
Exhaust side:
A = B + C − 0.30 mm (0.012 in.)
A: Thickness of new shim
B: Thickness of removed shim
C: Measured valve clearance

37671_SSX4_G0436

Fig. 170 Shim thickness specification table—2009 model

thickness of the removed shim, and determine the replacement shim by calculating the thickness of the new shim with the following formula and table.

8. Select a new shim with a thickness as close as possible to the calculated value.

9. Install the new shim facing the shim No. side with the tappet.

10. Lift the valve by turning the crankshaft counterclockwise and remove Special Tool 09916-66510.

11. Install the camshaft housing

and tighten the bolts to 97 inch lbs. (11 Nm).

12. Turn the crankshaft pulley more than 4 rotations.

13. Check the valve clearance again.

14. After checking and adjusting all the valves, install the cylinder head cover.

2010 Model

See Figures 171 and 172.

1. Before servicing the vehicle, refer to the Precautions Section.

2. Disconnect the negative battery cable.

Intake side:
A = B + C − 0.20 mm (0.0079 in.)
Exhaust side:
A = B + C − 0.35 mm (0.0138 in.)
A: Thickness "a" of new tappet
B: Thickness "a" of removed tappet
C: Measured valve clearance

37671_SSX4_G0438

Fig. 171 Formula used to determine new tappet—2010 model

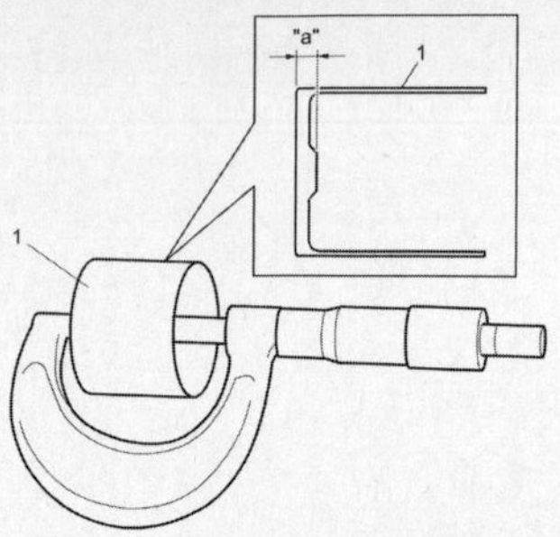

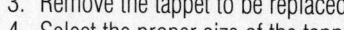

37671_SSX4_G0439

Fig. 172 Measuring the tappet (1) thickness (a) with a micrometer—2010 model

3. Remove the tappet to be replaced.

4. Select the proper size of the tappet as follows.

   a. Using a micrometer, measure the thickness of the removed tappet.

   b. Calculate the thickness of the new tappet by the formula below.

   c. Select a new tappet closest to the calculated value from the available sizes.

5. Install the tappets and the camshafts.

6. Recheck the valve clearance.

ENGINE PERFORMANCE & EMISSION CONTROLS

CAMSHAFT POSITION (CMP) SENSOR

LOCATION

2009 Model

See Figure 173.

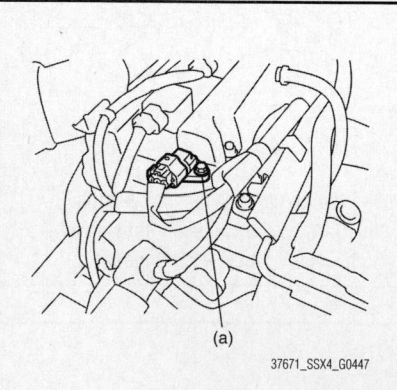

37671_SSX4_G0447

Fig. 173 Camshaft Position (CMP) sensor (a) location—2009 model

2010 Model

See Figure 174.

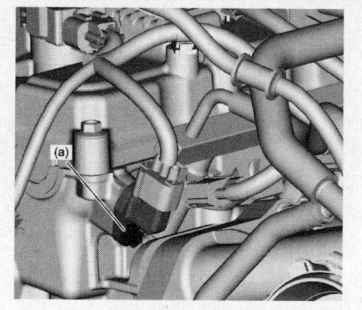

37671_SSX4_G0448

Fig. 174 Camshaft Position (CMP) sensor (a) location—2010 model

REMOVAL & INSTALLATION

2009 Model

1. Before servicing the vehicle, refer to the Precautions Section.

2. Disconnect the negative battery cable.

3. Remove the air cleaner case.

4. Disconnect the connector from CMP sensor.

5. Remove the CMP sensor from the cylinder head cover.

To install:

6. Install the CMP sensor to the cylinder head cover.

7. Tighten the CMP sensor bolt to 97 inch lbs. (11 Nm).

8. Connect the connector to the CMP sensor securely.

9. Install the air cleaner case.

10. Connect the negative battery cable.

2010 Model

1. Before servicing the vehicle, refer to the Precautions Section.

2. Disconnect the negative battery cable.

3. Remove the air cleaner case.

4. Disconnect the connector from the CMP sensor.

5. Remove the CMP sensor from the cylinder head.

To install:

6. Install the CMP sensor to the cylinder head.

7. Tighten the CMP sensor bolt to 97 inch lbs. (11 Nm).

8. Connect the connector to the CMP sensor securely.

9. Install the air cleaner case.
10. Connect the negative battery cable.

CRANKSHAFT POSITION (CKP) SENSOR

LOCATION

2009 Model

See Figure 175.

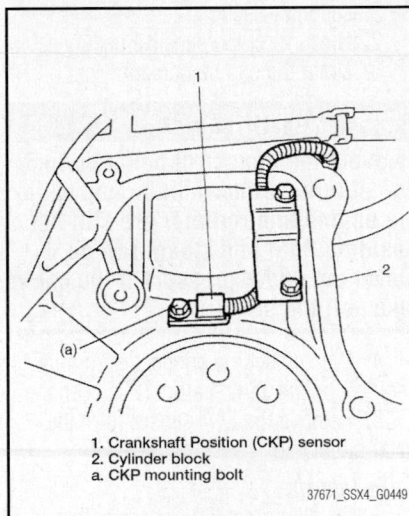

1. Crankshaft Position (CKP) sensor
2. Cylinder block
a. CKP mounting bolt

37671_SSX4_G0449

Fig. 175 Crankshaft Position (CKP) sensor location—2009 model

2010 Model

See Figure 176.

REMOVAL & INSTALLATION

2009 Model

1. Before servicing the vehicle, refer to the Precautions Section.
2. Disconnect the negative battery cable.

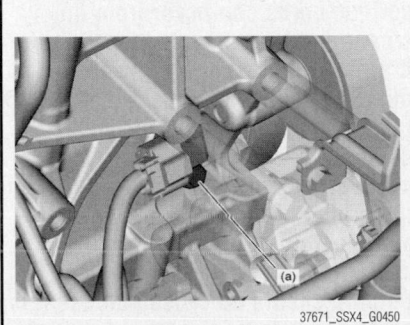

37671_SSX4_G0450

Fig. 176 Crankshaft Position (CKP) sensor (a) location—2010 model

3. Remove the transmission assembly from the vehicle.
4. Remove the drive plate or flywheel from the crankshaft.
5. Disconnect the connector from the Crankshaft Position (CKP) sensor.
6. Remove the CKP sensor from the cylinder block.

To install:

7. Installation is the reverse of the removal procedure.
8. Apply engine oil to the O-ring of the sensor.
9. Tighten the CKP sensor bolt to 97 inch lbs. (11 Nm).
10. Connect the connector and securely fix the wire harness with a clamp.

2010 Model

1. Before servicing the vehicle, refer to the Precautions Section.
2. Disconnect the negative battery cable.
3. Remove the alternator. Refer to Alternator, removal & installation.
4. Disconnect the Crankshaft Position (CKP) sensor connector.
5. Remove the CKP sensor (1) from the cylinder block.

To install:

6. Installation is the reverse of the removal procedure.
7. Apply engine oil to the O-ring of the sensor.
8. Tighten the CKP sensor bolt to 97 inch lbs. (11 Nm).
9. Connect the connector and securely fix the wire harness with a clamp.

ELECTRONIC CONTROL MODULE (ECM)

LOCATION

See Figure 177.

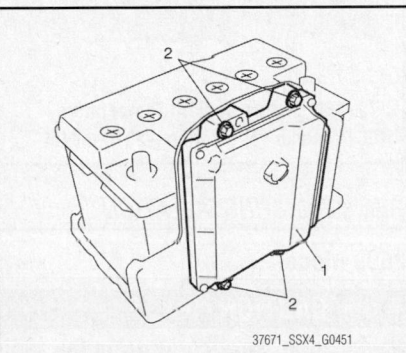

37671_SSX4_G0451

Fig. 177 Engine Control Module (ECM) (1) and mounting nuts (2) location

REMOVAL & INSTALLATION

See Figure 178.

✳✳ WARNING

As the Engine Control Module (ECM) consists of precision parts, be careful not to expose it to excessive shock.

1. Before servicing the vehicle, refer to the Precautions Section.
2. Remove the battery.
3. Disconnect the connectors from the ECM as follows.
   a. Push the lock to release the locking of the lock lever.
   b. Turn the lock lever until it stops.
4. Remove the battery tray with the ECM.
5. Remove the ECM from the battery tray by removing its mounting nuts.

To install:

6. Installation is the reverse of the removal procedure.
7. Connect the connectors to the ECM as follows.
   a. Make sure that the lock lever of the ECM connector is in the unlock position.
   b. Insert the ECM connectors to the ECM until it stops with the unlocked lock lever.
   c. Lock the ECM connectors securely by pulling its lock lever up.
8. For vehicles equipped with an EVAP leak check system:
   • If the ECM is replaced with a new one (VIN data is still not registered) or with a used one (VIN data of other vehicle has been registered previously), make sure to re-register the VIN data correctly. Otherwise, the MIL may remain on even after starting the engine. It may cause inconsistency between the registered VIN data in the ECM and the actual VIN.
   • Refer to Reset, VIN Registration (EVAP Leak Check Model).
9. For vehicles equipped with an immobilizer control system:
   • If the ECM is replaced with a new one or with another one, make sure to register the immobilizer transponder code to the ECM.
   • Refer to Reset, Immobilizer Key Registration.

RESET

VIN Registration (EVAP Leak Check Model)

1. Before servicing the vehicle, refer to the Precautions Section.
2. Connect the SUZUKI scan tool to the

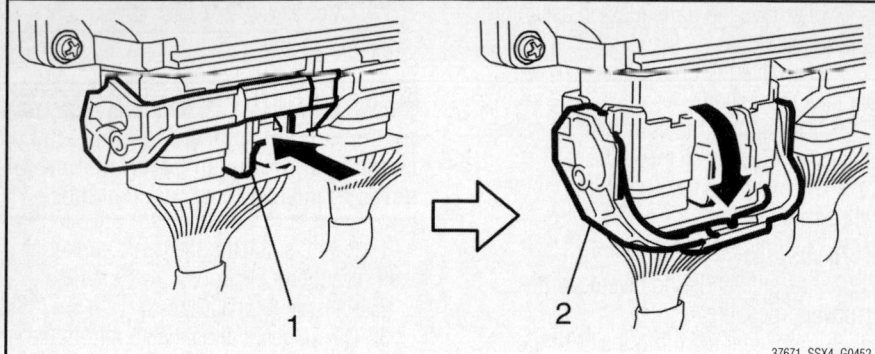

Fig. 178 Disconnect the connectors from the ECM by pushing the lock (1) to release the locking of the lock lever (2)

Data Link Connector (DLC) located on the underside of the instrument panel at the driver's seat side. Use Special Tool SUZUKI scan tool (SUZUKI-SDT).

3. Check the VIN shown on the left side of the instrument panel.

4. Turn the ignition switch to the ON position.

5. Select "VIN registration" under the "Utility" menu of the SUZUKI scan tool.

6. Register the VIN according to the instructions indicated on the SUZUKI scan tool.

➡**For details, refer to the operator's manual for the SUZUKI scan tool.**

7. Before completing the VIN registration, check that the VIN data indicated on the SUZUKI scan tool is correct. If not, re-register the VIN data into the ECM correctly.

Immobilizer Key Registration

1. Before servicing the vehicle, refer to the Precautions Section.

2. After the ECM is replaced with a new one or used one, the transponder code in the transponder built into the ignition key must be registered with the ECM.

3. To register the transponder code in the ignition key with ECM, perform the "Immobilizer Key Registration" mode of the SUZUKI scan tool referring to the SUZUKI scan tool Operator's Manual.

➡**A maximum of 4 transponder codes can be registered with the ECM.**

ENGINE COOLANT TEMPERATURE (ECT) SENSOR

LOCATION

2009 Model
See Figure 179.

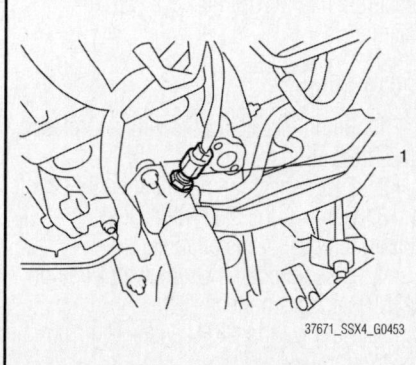

Fig. 179 Engine Coolant Temperature (ECT) sensor (1) location—2009 model

2010 Model
See Figure 180.

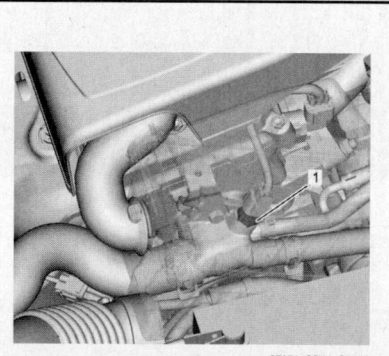

Fig. 180 Engine Coolant Temperature (ECT) sensor (1) location—2010 model

REMOVAL & INSTALLATION

2009 Model

✳✳ CAUTION

Never open, service or drain the radiator or cooling system when hot; serious burns can occur from the steam and hot coolant. Also, when

draining engine coolant, keep in mind that cats and dogs are attracted to ethylene glycol antifreeze and could drink any that is left in an uncovered container or in puddles on the ground. This will prove fatal in sufficient quantities. Always drain coolant into a sealable container. Coolant should be reused unless it is contaminated or is several years old.

1. Before servicing the vehicle, refer to the Precautions Section.

2. Disconnect the negative battery cable.

3. Drain the engine coolant.

✳✳ CAUTION

To avoid the danger of being burned, do not remove the radiator cap while the engine and radiator are still hot. Scalding fluid and steam can be blown out under pressure if the cap is taken off too soon.

4. Disconnect the connector from the Engine Coolant Temperature (ECT) sensor.

5. Remove the ECT sensor from the water outlet cap.

To install:

6. Installation is the reverse of the removal procedure.

7. Clean the mating surfaces of the ECT sensor and the water outlet cap.

8. Check the O-ring for damage and replace, if necessary.

9. Tighten the ECT sensor to 111 inch lbs. (13 Nm).

10. Connect the connector to the ECT sensor securely.

11. Fill the engine coolant with the proper type and amount of fluid.

2010 Model

✳✳ CAUTION

Never open, service or drain the radiator or cooling system when hot; serious burns can occur from the steam and hot coolant. Also, when draining engine coolant, keep in mind that cats and dogs are attracted to ethylene glycol antifreeze and could drink any that is left in an uncovered container or in puddles on the ground. This will prove fatal in sufficient quantities. Always drain coolant into a sealable container. Coolant should be reused unless it is contaminated or is several years old.

1. Before servicing the vehicle, refer to the Precautions Section.

2. Remove the air cleaner assembly.
3. Disconnect the negative battery cable.
4. Drain the engine coolant.

✻✻ CAUTION

To avoid the danger of being burned, do not remove the radiator cap while the engine and radiator are still hot. Scalding fluid and steam can be blown out under pressure if the cap is taken off too soon.

5. Disconnect the connector from the Engine Coolant Temperature (ECT) sensor.
6. Remove the ECT sensor from the water outlet cap.

To install:

7. Installation is the reverse of the removal procedure.
8. Clean the mating surfaces of the ECT sensor and the water outlet cap.
9. Check the O-ring for damage and replace, if necessary.
10. Tighten the ECT sensor to 111 inch lbs. (13 Nm).
11. Connect the connector to the ECT sensor securely.
12. Fill the engine coolant with the proper type and amount of fluid.

HEATED OXYGEN (HO2S) SENSOR

LOCATION

2009 Model

See Figure 181.

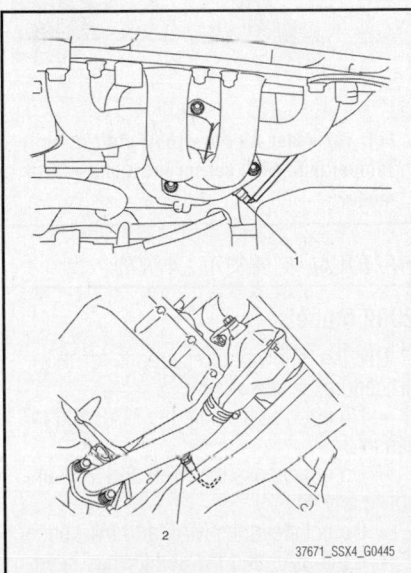

Fig. 181 Air-Fuel Ratio Sensor (AFR) (1) and Heated Oxygen (HO2S) Sensor (2) locations—2009 model

2010 Model

See Figure 182.

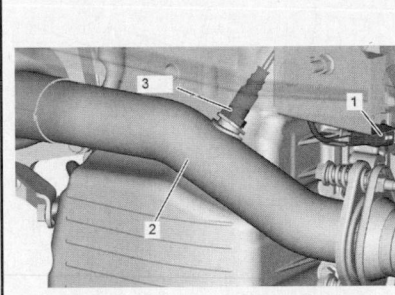

1. Heated Oxygen Sensor (HO2S) connector
2. Exhaust No. 1 pipe
3. HO2S

37671_SSX4_G0457

Fig. 182 Heated Oxygen Sensor (HO2S) location—2010 model

REMOVAL & INSTALLATION

2009 Model

✻✻ CAUTION

To avoid the danger of being burned, do not touch the exhaust system when the system is hot. The Heated Oxygen Sensor (HO2S) removal should be performed when the system is cool.

1. Before servicing the vehicle, refer to the Precautions Section.
2. Disconnect the negative battery cable.
3. Disconnect the connector from the HO2S.
4. Remove the HO2S from the exhaust No. 1 pipe.

To install:

5. Installation is the reverse of the removal procedure.
6. Tighten the HO2S to 33 ft. lbs. (45 Nm).

2010 Model

✻✻ CAUTION

To avoid the danger of being burned, do not touch exhaust system when system is hot. The Heated Oxygen Sensor (HO2S) removal should be performed when the system is cool.

1. Before servicing the vehicle, refer to the Precautions Section.
2. Disconnect the negative battery cable.
3. Raise and safely support the vehicle.

4. Disconnect the HO2S connector and remove its bracket.
5. Remove the exhaust No. 1 pipe.
6. Remove the HO2S from the exhaust No. 1 pipe.

To install:

7. Installation is the reverse of the removal procedure.
8. Tighten the HO2S to 33 ft. lbs. (45 Nm).

INTAKE AIR TEMPERATURE (IAT) SENSOR

LOCATION

2009 Model

See Figure 183.

The Intake Air Temperature (IAT) sensor is built into the Mass Air Flow (MAF) sensor.

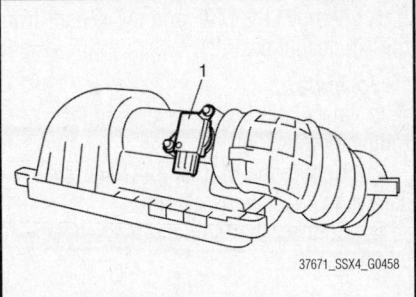

37671_SSX4_G0458

Fig. 183 Mass Air Flow (MAF)/Intake Air Temperature (IAT) sensor locations—2009 model

2010 Model

See Figure 184.

The Intake Air Temperature (IAT) sensor is built into the Mass Air Flow (MAF) sensor.

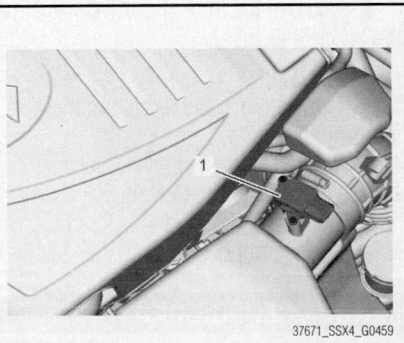

37671_SSX4_G0459

Fig. 184 Mass Air Flow (MAF)/Intake Air Temperature (IAT) sensor locations—2010 model

REMOVAL & INSTALLATION

2009 Model

Use the following precautions for this procedure.
- Do not disassemble the MAF and IAT sensor
- Do not expose the MAF and IAT sensor to any shock
- Do not clean the MAF and IAT sensor
- If the MAF and IAT sensor has been dropped, it should be replaced
- Do not blow compressed air through the MAF and IAT sensor
- Do not place a finger or any other object into the MAF and IAT sensor. Malfunction may occur

1. Before servicing the vehicle, refer to the Precautions Section.
2. Disconnect the negative battery cable.
3. Disconnect the MAF and IAT sensor connector.
4. Remove the air cleaner case.
5. Remove the MAF and IAT sensor from the air cleaner case.

To install:

6. Installation is the reverse of the removal procedure.
7. Tighten the MAF and IAT sensor screws to 11 inch lbs. (1 Nm).
8. Connect the MAF and IAT sensor connector securely.

2010 Model

> **⚛ WARNING**
>
> **Do not expose the MAF and IAT sensor to any shock. If it has been dropped, it should be replaced.**

1. Before servicing the vehicle, refer to the Precautions Section.
2. Disconnect the negative battery cable.
3. Disconnect the MAF and IAT sensor connector.
4. Remove the MAF and IAT sensor from the air cleaner case.

To install:

5. Installation is the reverse of the removal procedure.
6. Tighten the MAF and IAT sensor screws to 11 inch lbs. (1 Nm).
7. Connect the MAF and IAT sensor connector securely.

KNOCK SENSOR (KS)

LOCATION

2009 Model

See Figure 185.

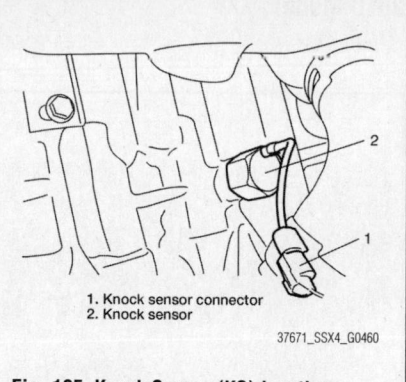

1. Knock sensor connector
2. Knock sensor

37671_SSX4_G0460

Fig. 185 Knock Sensor (KS) location—2009 model

2010 Model

See Figure 186.

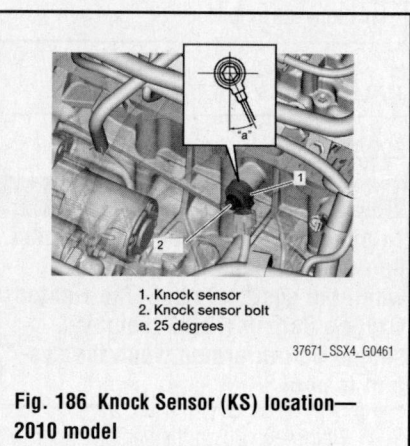

1. Knock sensor
2. Knock sensor bolt
a. 25 degrees

37671_SSX4_G0461

Fig. 186 Knock Sensor (KS) location—2010 model

REMOVAL & INSTALLATION

2009 Model

1. Before servicing the vehicle, refer to the Precautions Section.
2. Disconnect the negative battery cable.
3. Raise and safely support the vehicle.
4. Remove the right side halfshaft. Refer to Halfshafts, removal & installation.
5. Disconnect the knock sensor connector.
6. Remove the knock sensor from the cylinder block.

To install:

7. Installation is the reverse of the removal procedure.
8. Tighten the knock sensor to 16 ft. lbs. (22 Nm).

2010 Model

1. Before servicing the vehicle, refer to the Precautions Section.
2. Disconnect the negative battery cable.
3. Raise and safely support the vehicle.
4. Disconnect the knock sensor connector.
5. Remove the knock sensor from the cylinder block.

To install:

6. Installation is the reverse of the removal procedure.
7. Tighten the knock sensor bolt to 19 ft. lbs. (25 Nm).

MASS AIR FLOW (MAF) SENSOR

LOCATION

2009 Model

See Figure 183.

The Intake Air Temperature (IAT) sensor is built into the Mass Air Flow (MAF) sensor.

2010 Model

See Figure 187.

The Intake Air Temperature (IAT) sensor is built into the Mass Air Flow (MAF) sensor.

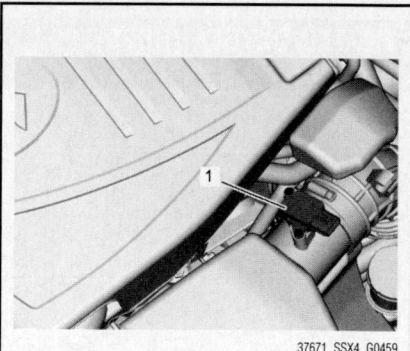

37671_SSX4_G0459

Fig. 187 Mass Air Flow (MAF)/Intake Air Temperature (IAT) sensor locations—2010 model

REMOVAL & INSTALLATION

2009 Model

Use the following precautions for this procedure.
- Do not disassemble the MAF and IAT sensor
- Do not expose the MAF and IAT sensor to any shock
- Do not clean the MAF and IAT sensor
- If the MAF and IAT sensor has been dropped, it should be replaced
- Do not blow compressed air through the MAF and IAT sensor

- Do not place a finger or any other object into the MAF and IAT sensor. Malfunction may occur
1. Before servicing the vehicle, refer to the Precautions Section.
2. Disconnect the negative battery cable.
3. Disconnect the MAF and IAT sensor connector.
4. Remove the air cleaner case.
5. Remove the MAF and IAT sensor from the air cleaner case.

To install:
6. Installation is the reverse of the removal procedure.

7. Tighten the MAF and IAT sensor screws to 11 inch lbs. (1 Nm).
8. Connect the MAF and IAT sensor connector securely.

2010 Model

✷✷ WARNING

Do not expose the MAF and IAT sensor to any shock. If it has been dropped, it should be replaced.

1. Before servicing the vehicle, refer to the Precautions Section.

2. Disconnect the negative battery cable.
3. Disconnect the MAF and IAT sensor connector.
4. Remove the MAF and IAT sensor from the air cleaner case.

To install:
5. Installation is the reverse of the removal procedure.
6. Tighten the MAF and IAT sensor screws to 11 inch lbs. (1 Nm).
7. Connect the MAF and IAT sensor connector securely.

FUEL   GASOLINE FUEL INJECTION SYSTEM

FUEL SYSTEM SERVICE PRECAUTIONS

Safety is the most important factor when performing not only fuel system maintenance but any type of maintenance. Failure to conduct maintenance and repairs in a safe manner may result in serious personal injury or death. Maintenance and testing of the vehicle's fuel system components can be accomplished safely and effectively by adhering to the following rules and guidelines.

- To avoid the possibility of fire and personal injury, always disconnect the negative battery cable unless the repair or test procedure requires that battery voltage be applied.
- Always relieve the fuel system pressure prior to disconnecting any fuel system component (injector, fuel rail, pressure regulator, etc.), fitting or fuel line connection. Exercise extreme caution whenever relieving fuel system pressure to avoid exposing skin, face and eyes to fuel spray. Please be advised that fuel under pressure may penetrate the skin or any part of the body that it contacts.
- Always place a shop towel or cloth around the fitting or connection prior to loosening to absorb any excess fuel due to spillage. Ensure that all fuel spillage (should it occur) is quickly removed from engine surfaces. Ensure that all fuel soaked cloths or towels are deposited into a suitable waste container.
- Always keep a dry chemical (Class B) fire extinguisher near the work area.
- Do not allow fuel spray or fuel vapors to come into contact with a spark or open flame.
- Always use a back-up wrench when loosening and tightening fuel line connection fittings. This will prevent unnecessary stress and torsion to fuel line piping.
- Always replace worn fuel fitting O-

rings with new Do not substitute fuel hose or equivalent where fuel pipe is installed.
Before servicing the vehicle, make sure to also refer to the precautions in the beginning of this section as well.

RELIEVING FUEL SYSTEM PRESSURE

See Figure 188.

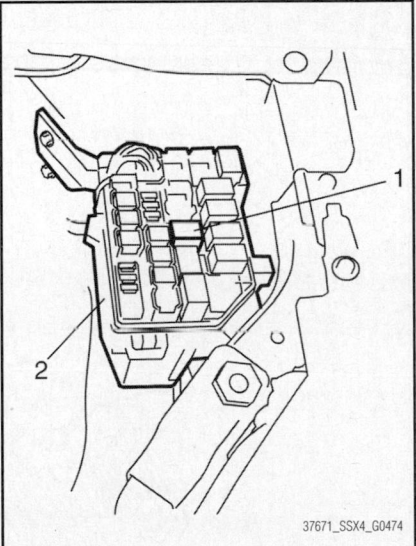

37671_SSX4_G0474

Fig. 188 Disconnect the fuel pump relay (1) from the relay/fuse box No. 1 (2)

✷✷ WARNING

Never perform this work when the engine is hot. Doing so may have an adverse effect on the exhaust catalyst.

➡**If ECM detects DTC(s) after servicing, clear the DTC(s).**

1. Before servicing the vehicle, refer to the Precautions Section.

2. Make sure that the engine is cold.
3. Place the gear shift lever in Neutral or P and apply the parking brake and block the drive wheels.
4. Disconnect the fuel pump relay from the relay/fuse box No. 1.
5. Remove the fuel filler cap in order to release the fuel vapor pressure in the fuel tank, and then reinstall it.
6. Start the engine and run it until the engine stops for lack of fuel. Repeat cranking the engine 2–3 times for about 3 seconds each time to dissipate the fuel pressure in the lines. The fuel connections are now safe for servicing.
7. After servicing, connect the fuel pump relay to the relay/fuse box No. 1 and install the relay/fuse box cover.

FUEL FILTER

REMOVAL & INSTALLATION

The fuel filter is part of the fuel level sensor unit, fuel filter, and fuel pump assembly. Refer to Fuel Pump Module, removal & installation.

FUEL PUMP MODULE

REMOVAL & INSTALLATION

See Figures 189 and 190.

1. Before servicing the vehicle, refer to the Precautions Section.
2. Properly relieve the fuel system pressure. Refer to Fuel System Pressure, Relieving.
3. Remove the fuel filler cap to release the pressure from inside the fuel tank.
4. Remove the fuel tank from the vehicle. Refer to Fuel Tank, removal & installation.
5. Disconnect the fuel feed pipe from the fuel pump assembly.

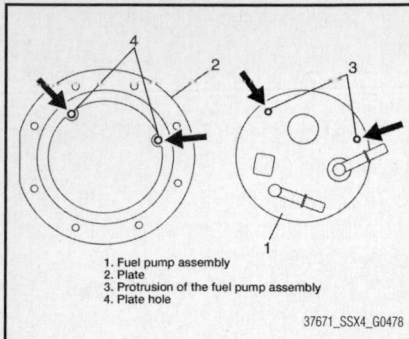

1. Fuel pump assembly
2. Plate
3. Protrusion of the fuel pump assembly
4. Plate hole

37671_SSX4_G0478

Fig. 189 Put the plate on fuel pump assembly by matching the protrusion of the fuel pump assembly to the plate hole

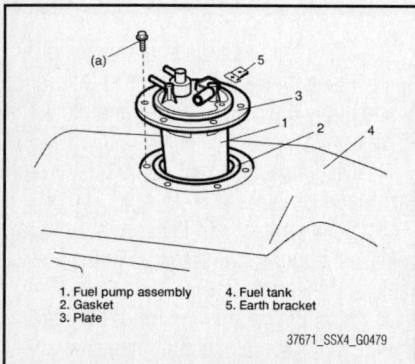

1. Fuel pump assembly
2. Gasket
3. Plate
4. Fuel tank
5. Earth bracket

37671_SSX4_G0479

Fig. 190 Install a new gasket, the fuel pump assembly, and the earth bracket with the plate to the fuel tank

6. Remove the fuel pump assembly and earth bracket from the fuel tank.

To install:

➡ **When connecting joints, clean the outside surface of the pipe where the joint is to be inserted, push the joint into the pipe until the joint lock clicks and check to ensure that the pipes are connected securely, or a fuel leak may occur.**

7. Clean the mating surfaces of the fuel pump assembly and the fuel tank.

8. Put the plate on fuel pump assembly by matching the protrusion of the fuel pump assembly to the plate hole.

9. Install a new gasket, the fuel pump assembly, and the earth bracket with the plate to the fuel tank.

10. Tighten the fuel pump assembly bolt.
- On 2009 model, tighten to 97 inch lbs. (11 Nm)
- On 2010 model, tighten to 89 inch lbs. (10 Nm)

11. Connect the fuel feed pipe to the fuel pump assembly.

12. Install the fuel tank to the vehicle. Refer to Fuel Tank, removal & installation.

FUEL RAIL AND INJECTOR

REMOVAL & INSTALLATION

2009 Model

See Figure 191.

1. Before servicing the vehicle, refer to the Precautions Section.

2. Remove the fuel filler cap to release the pressure from inside the fuel tank.

3. Properly relieve the fuel system pressure. Refer to Fuel System Pressure, Relieving.

4. Disconnect the negative battery cable.

5. Remove the air cleaner case and air suction hose.

6. Disconnect the fuel injector couplers.

7. Disconnect the fuel feed hose from the fuel delivery pipe.

8. Remove the fuel delivery pipe bolts.

9. Remove the fuel injector(s).

To install:

10. Installation is the reverse of the removal procedure.

11. Replace the injector O-ring with a new one using care not to damage it.

12. Check if the cushion is scored or damaged. If it is, replace it with a new one.

13. Apply a thin coat of fuel to the O-rings, and then install the injectors into the delivery pipe and the cylinder head.

a. Make sure that the injectors rotate smoothly.

b. If the injectors do not rotate

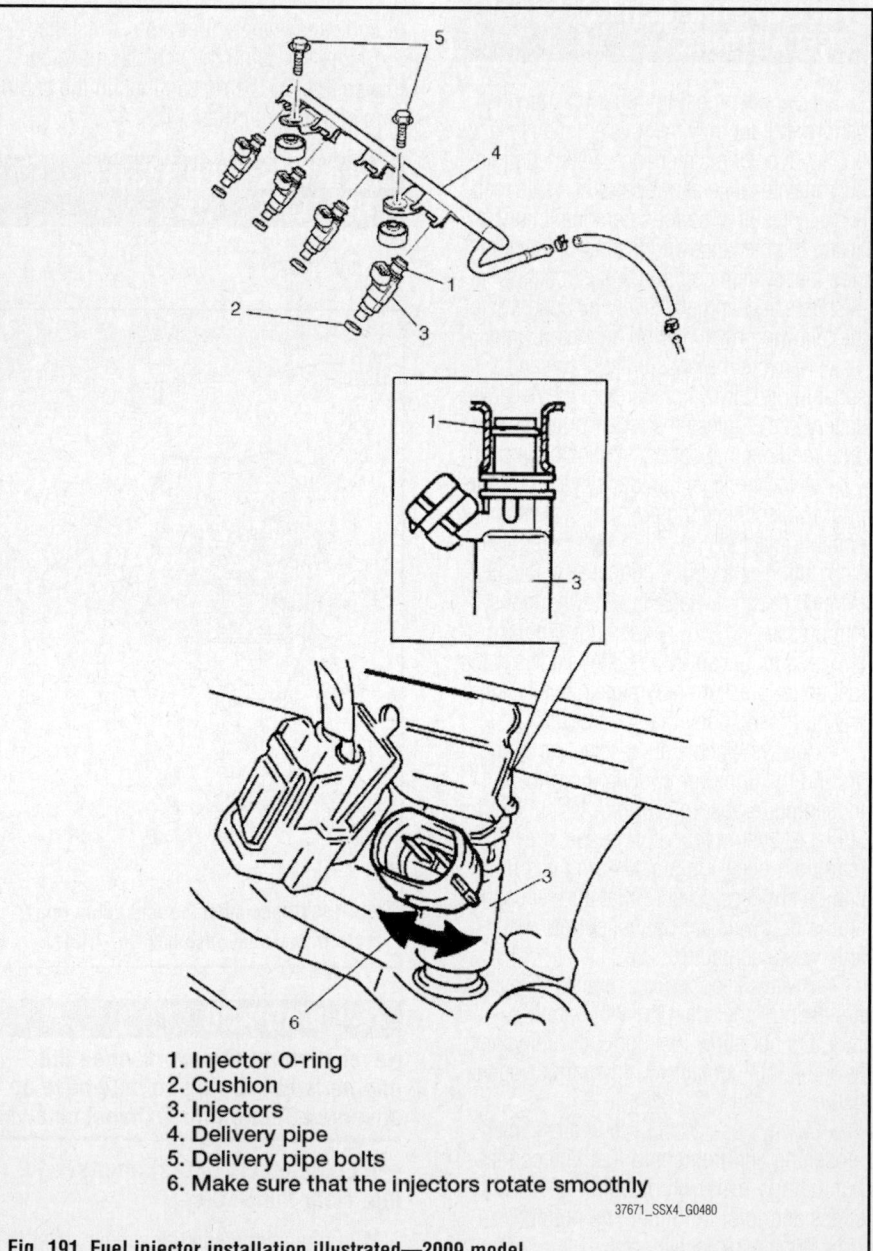

1. Injector O-ring
2. Cushion
3. Injectors
4. Delivery pipe
5. Delivery pipe bolts
6. Make sure that the injectors rotate smoothly

37671_SSX4_G0480

Fig. 191 Fuel injector installation illustrated—2009 model

smoothly, the probable cause is incorrect installation of the O-ring. Replace the O-ring with a new one.

14. Tighten the delivery pipe bolts to the specified torque and make sure that the injectors rotate smoothly. Tighten the fuel delivery pipe bolt to 18 ft. lbs. (25 Nm).

15. After installation, with the engine OFF and the ignition switch ON, check for fuel leaks around the fuel line connection.

2010 Model

See Figures 192 and 193.

1. Before servicing the vehicle, refer to the Precautions Section.

2. Remove the fuel filler cap to release the pressure from inside the fuel tank.

3. Properly relieve the fuel system pressure. Refer to Fuel System Pressure, Relieving.

4. Disconnect the negative battery cable.

5. Disconnect the engine cover hose from the air cleaner assembly.

6. Disconnect the engine cover front side hook from the engine cover pin.

7. Disconnect the engine cover rear side hook from the engine cover pin.

8. Remove the engine cover from the vehicle.

✷✷ WARNING

When removing the engine cover, be careful not to hit it against the cowling front panel.

9. Disconnect the fuel injector connectors.

10. Disconnect the fuel feed hose from the fuel delivery pipe.

11. Remove the fuel delivery pipe bolts.

12. Remove the fuel injector(s).

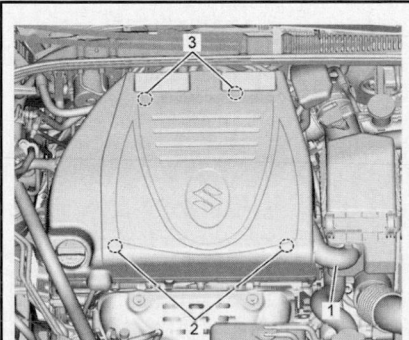

1. Engine cover hose
2. Engine cover front side hook
3. Engine cover rear side hook

37671_SSX4_G0259

Fig. 192 Remove the engine cover

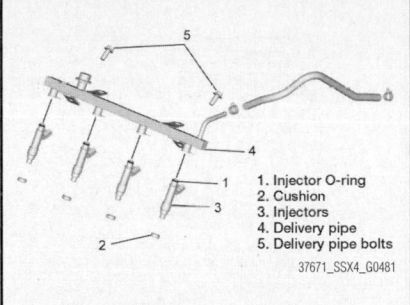

1. Injector O-ring
2. Cushion
3. Injectors
4. Delivery pipe
5. Delivery pipe bolts

37671_SSX4_G0481

Fig. 193 Fuel injector installation illustrated—2010 model

To install:

13. Installation is the reverse of the removal procedure.

14. Replace the injector O-ring with a new one using care not to damage it.

15. Check if the cushion is scored or damaged. If it is, replace with a new one.

16. Apply a thin coat of fuel to the O-rings, and then install the injectors into the delivery pipe and the cylinder head.

17. Tighten the delivery pipe bolts to 19 ft. lbs. (25 Nm).

18. Make sure that the injectors rotate smoothly. If not, the probable cause is incorrect installation of the O-ring. Replace the O-ring with a new one.

19. After installation, with the engine stationary and the ignition switch in the ON position, check for fuel leaks around the fuel line connection.

FUEL TANK

REMOVAL & INSTALLATION

See Figures 194 and 195.

• When servicing the fuel tank, it should be treated with respect. Make sure that no contact with sharp edges or hot surfaces is made. In addition, the fuel tank should not be dropped since the fuel tank, fuel pump, and other components can be damaged by the impact. If dropped, all components should be replaced because of the risk of damage

1. Before servicing the vehicle, refer to the Precautions Section.

2. Remove the fuel filler cap to release the pressure from inside the fuel tank.

3. Properly relieve the fuel system pressure. Refer to Fuel System Pressure, Relieving.

4. Disconnect the negative battery cable.

5. Insert a hose of a hand operated pump into the fuel filler hose and drain the fuel in the fuel filler hose.

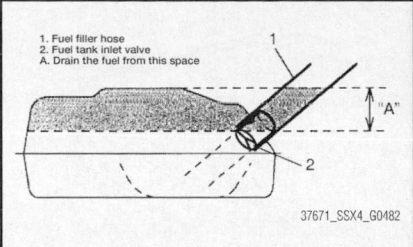

1. Fuel filler hose
2. Fuel tank inlet valve
A. Drain the fuel from this space

37671_SSX4_G0482

Fig. 194 Insert a hose into the fuel filler hose and drain the fuel in the fuel filler hose

➡**Do not force the pump hose into the fuel tank.**

6. Raise and safely support the vehicle.

7. Remove the clamps, the fuel filler hose, and the breather hose from the fuel filler neck.

8. Remove the EVAP canister.

9. Remove the exhaust center pipe.

10. Remove the propeller shaft (4WD model). Refer to Driveshaft, removal & installation.

11. Due to the absence of a fuel tank drain plug, drain the fuel tank by pumping the fuel out through the fuel tank filler. Use a hand operated pump device to drain the fuel tank.

✷✷ WARNING

Do not force the pump hose into the fuel tank. Never store fuel in an open container due to the possibility of fire or explosion.

12. Disconnect the quick joints from the fuel pipes.

13. Support the fuel tank with a jack and remove the fuel tank mounting bolts.

14. Lower the fuel tank a little in order to disconnect the wire harness at the connector and the ground wire.

15. Remove the fuel tank from the vehicle.

To install:

➡**When connecting joints, clean the outside surfaces of the pipe where the joint is to be inserted, push the joint into the pipe until the joint lock clicks and check to ensure that the pipes are connected securely, or a fuel leak may occur.**

➡**Never let the fuel hoses touch the wheel speed sensor harness (if equipped).**

16. If parts have been removed from the fuel tank, install them before installing the fuel tank to the vehicle.

17. Raise the fuel tank with a jack and connect the fuel pump connector, ground wire, and clamp wire harness.

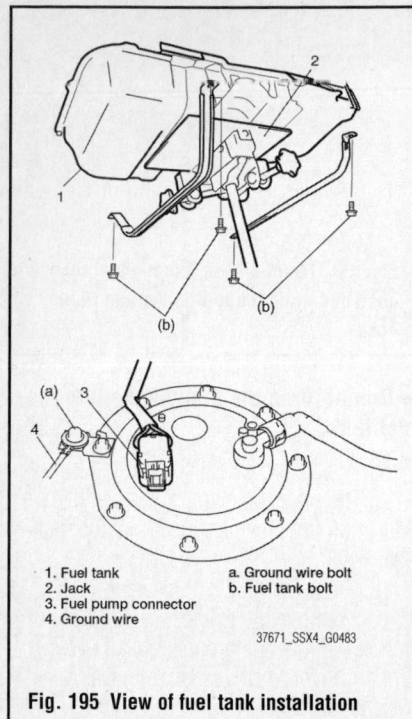

1. Fuel tank
2. Jack
3. Fuel pump connector
4. Ground wire

a. Ground wire bolt
b. Fuel tank bolt

37671_SSX4_G0483

Fig. 195 View of fuel tank installation

18. Tighten the ground wire bolt to 97 inch lbs. (11 Nm).

19. Install the fuel tank to the vehicle, and tighten new fuel tank bolts to 34 ft. lbs. (45 Nm).

20. Connect the fuel filler hose, breather hose, and clamps to the fuel filler neck, and clamp them securely.

a. Tighten the fuel tank hose clamp (Non-EVAP-leak-check model) to 13 inch lbs. (2 Nm).

b. Tighten the fuel tank hose clamp (EVAP leak check model) to 31 inch lbs. (4 Nm).

21. Connect the quick joints to each pipe.

22. Install the exhaust center pipe.

23. Install the propeller shaft (4WD model). Refer to Driveshaft, removal & installation.

24. Install the EVAP canister.

25. Connect the negative battery cable.

26. With the engine stationary, turn the ignition switch to the ON position and check for fuel leaks. Correct as needed.

THROTTLE BODY

REMOVAL & INSTALLATION

2009 Model

See Figures 196 and 197.

1. Before servicing the vehicle, refer to the Precautions Section.

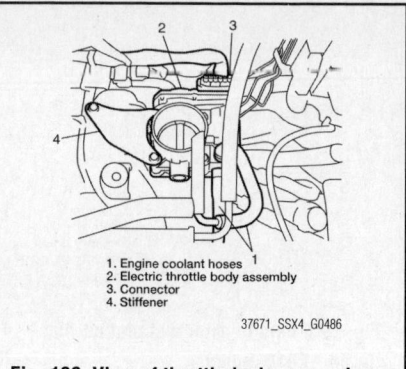

1. Engine coolant hoses
2. Electric throttle body assembly
3. Connector
4. Stiffener

37671_SSX4_G0486

Fig. 196 View of throttle body removal— 2009 model

2. Disconnect the negative battery cable.

3. Drain the engine coolant.

4. Remove the air cleaner assembly.

5. Disconnect the engine coolant hoses from the electric throttle body assembly.

6. Disconnect the connector from the electric throttle body assembly.

7. Remove the stiffener.

8. Remove the electric throttle body assembly from the intake manifold.

To install:

9. Clean the mating surfaces, and Install a new throttle body gasket to the intake manifold.

10. Install the electric throttle body assembly with the stiffener to the intake manifold.

11. Connect the connector to the electric throttle body assembly securely.

12. Connect the engine coolant hoses to the electric throttle body assembly.

13. Install the air cleaner assembly.

14. Fill the engine coolant.

15. Connect the negative battery cable.

16. If the throttle body assembly was replaced, perform a calibration of the electric throttle body assembly.

a. Disconnect the negative cable at the battery for 20 seconds, or more, for the purpose of clearing the calibration data of the closed throttle position from the memory in the ECM.

b. Connect the negative battery cable.

c. Keep the ignition switch at the ON position for 5 seconds, or more, without running the engine.

d. Calibration should be completed.

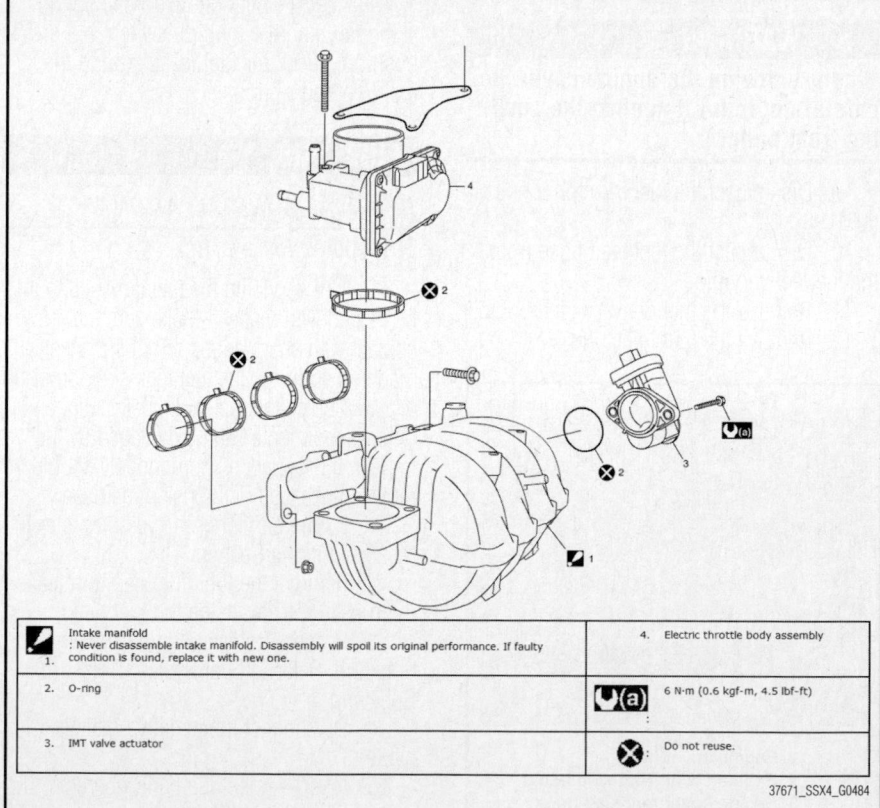

1.	Intake manifold : Never disassemble intake manifold. Disassembly will spoil its original performance. If faulty condition is found, replace it with new one.	4.	Electric throttle body assembly
2.	O-ring	(a)	6 N·m (0.6 kgf-m, 4.5 lbf-ft)
3.	IMT valve actuator	✕	Do not reuse.

37671_SSX4_G0484

Fig. 197 View of throttle body and intake manifold components—2009 model

2010 Model

See Figures 198 and 199.

> ❄❄ **WARNING**
>
> **Never disassemble the throttle body. Disassembly will spoil its original performance. If a faulty condition is found, replace the throttle body with new one as an assembly.**

1. Before servicing the vehicle, refer to the Precautions Section.
2. Disconnect the negative battery cable.
3. Drain the engine coolant.
4. Remove the air cleaner assembly.
5. Disconnect the engine coolant hoses and the electric connector from the throttle body assembly.
6. Remove the throttle body assembly and its gasket from the intake manifold.

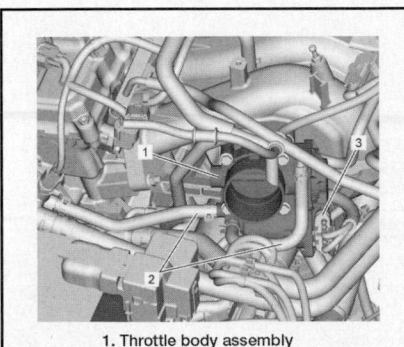

1. Throttle body assembly
2. Engine coolant hoses
3. Electric connector

37671_SSX4_G0489

Fig. 198 View of throttle body removal—2010 model

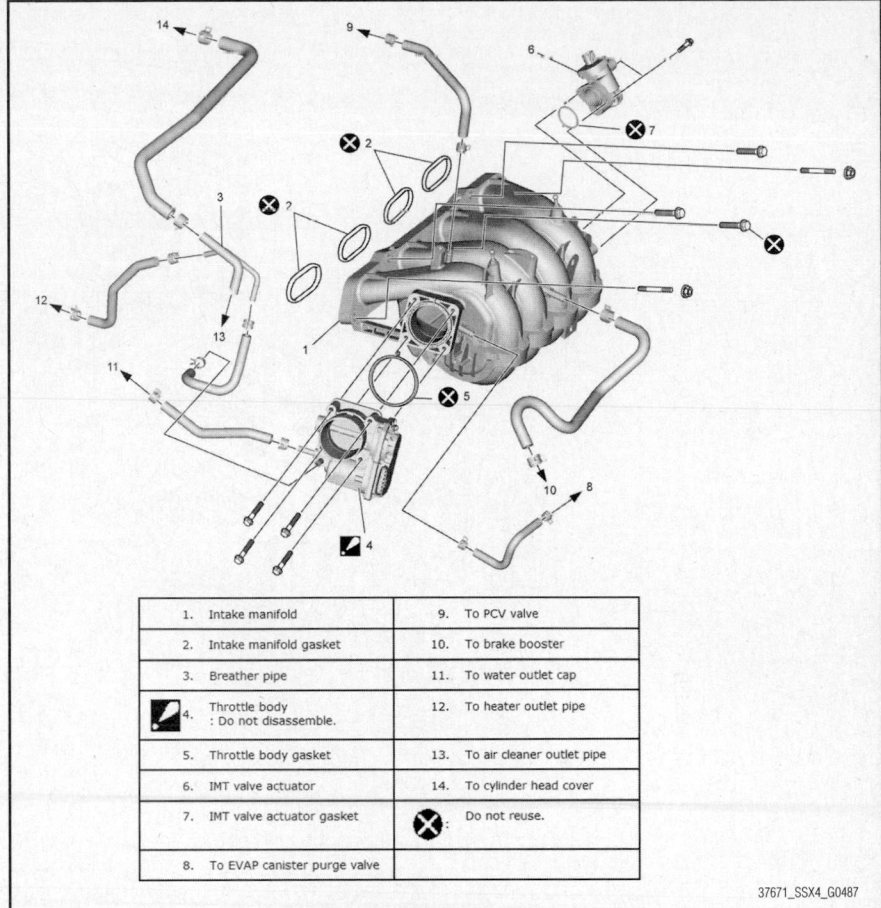

1.	Intake manifold	9.	To PCV valve
2.	Intake manifold gasket	10.	To brake booster
3.	Breather pipe	11.	To water outlet cap
4.	Throttle body : Do not disassemble.	12.	To heater outlet pipe
5.	Throttle body gasket	13.	To air cleaner outlet pipe
6.	IMT valve actuator	14.	To cylinder head cover
7.	IMT valve actuator gasket	❌	Do not reuse.
8.	To EVAP canister purge valve		

37671_SSX4_G0487

Fig. 199 View of throttle body and intake manifold components—2010 model

To install:

7. Clean the mating surfaces and install a new throttle body gasket to the intake manifold.
8. Install the throttle body assembly to the intake manifold.
9. Connect the electric connector and coolant hoses to the throttle body assembly.
10. Install the air cleaner assembly.
11. Fill the engine coolant.
12. Connect the negative battery cable.
13. Check for coolant leaks.

HEATING & AIR CONDITIONING SYSTEM

BLOWER MOTOR

REMOVAL & INSTALLATION

See Figure 200.

➡**Before servicing or working around the SRS system, turn the ignition switch OFF, disconnect both battery cables and wait at least 3 minutes. When servicing or working around the SRS system, do not work directly in front of the air bag module.**

1. Before servicing the vehicle, refer to the Precautions Section.
2. Disconnect the negative battery cable. Disconnect the positive battery cable.

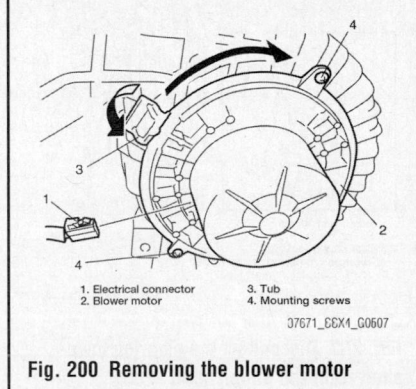

1. Electrical connector
2. Blower motor
3. Tub
4. Mounting screws

37671_SSX4_G0607

Fig. 200 Removing the blower motor

3. Disconnect the electrical connector from the blower motor.
4. Remove the blower motor from the HVAC unit as follows:

a. Remove the screws.
b. Release the tub by turning.
c. Hold the tub and turn the blower motor to remove.
5. Installation is the reverse of the removal procedure.

HEATER CORE

REMOVAL AND INSTALLATION

See Figure 201.

Before servicing the vehicle, refer to the Precautions Section.
The manufacturer does not provide a specific removal and installation procedure for this component. Use the illustration as a guide when servicing this component.

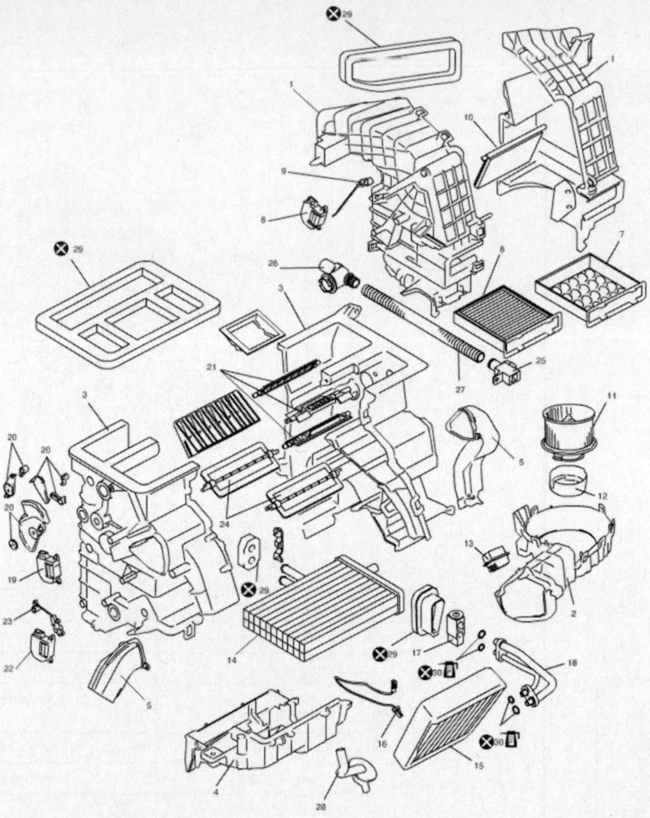

Fig. 201 Exploded view of HVAC unit component locations

1. Blower upper case	9. Air intake control link	17. Expansion valve	25. Inside air temperature sensor
2. Blower lower case	10. Air intake control door	18. Expansion pipe	26. Aspirator
3. Heater unit upper case	11. Blower motor	19. Air flow control actuator	27. Aspirator hose
4. Heater unit lower case	12. Blower motor cap	20. Air flow control links	28. Drain hose
5. Foot duct	13. Blower motor controller	21. Air flow control door assembly	29. Packing
6. HVAC air filter (if equipped)	14. Heater core	22. Temperature control actuator	O-ring : Apply compressor oil.
7. Cover (without HVAC air filter)	15. Evaporator	23. Temperature control link	30.
8. Air intake control actuator	16. Evaporator temperature sensor	24. Temperature control door assembly	Do not reuse.

37671_SSX4_G0499

STEERING

POWER RACK & PINION STEERING GEAR

REMOVAL & INSTALLATION

Electric Type

See Figures 202 and 203.

✱✱ WARNING

The spiral cable may snap due to steering operation if the steering column is separated from the steering gear assembly. Be sure to secure the steering wheel to avoid turning.

1. Before servicing the vehicle, refer to the Precautions Section.
2. Position the front wheels in the straight ahead position.

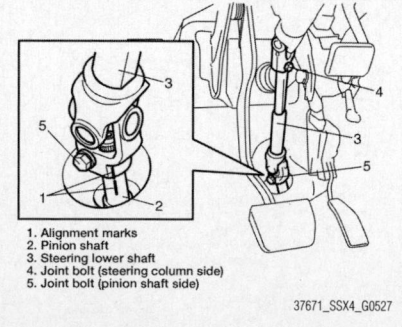

1. Alignment marks
2. Pinion shaft
3. Steering lower shaft
4. Joint bolt (steering column side)
5. Joint bolt (pinion shaft side)

37671_SSX4_G0527

Fig. 202 Disconnect the steering lower shaft from the pinion shaft

3. Disarm the SRS system. Refer to Air Bag (Supplemental Restraint System), Disarming The System.

4. Disconnect the negative battery cable.
5. Remove the steering joint cover.
6. Make alignment marks on the pinion shaft and the joint of the steering lower shaft as a guide for installation.
7. Loosen the joint bolt (steering column side) and remove the joint bolt (pinion shaft side) and disconnect the steering lower shaft from the pinion shaft.
8. Disconnect the torque sensor connector and the Power Steering (P/S) motor connector from the steering gear case.
9. Remove the front suspension frame with the steering gear case from the vehicle.
10. Remove the steering gear case from the front suspension frame.

To install:

11. Install the grommet to the steering gear case.

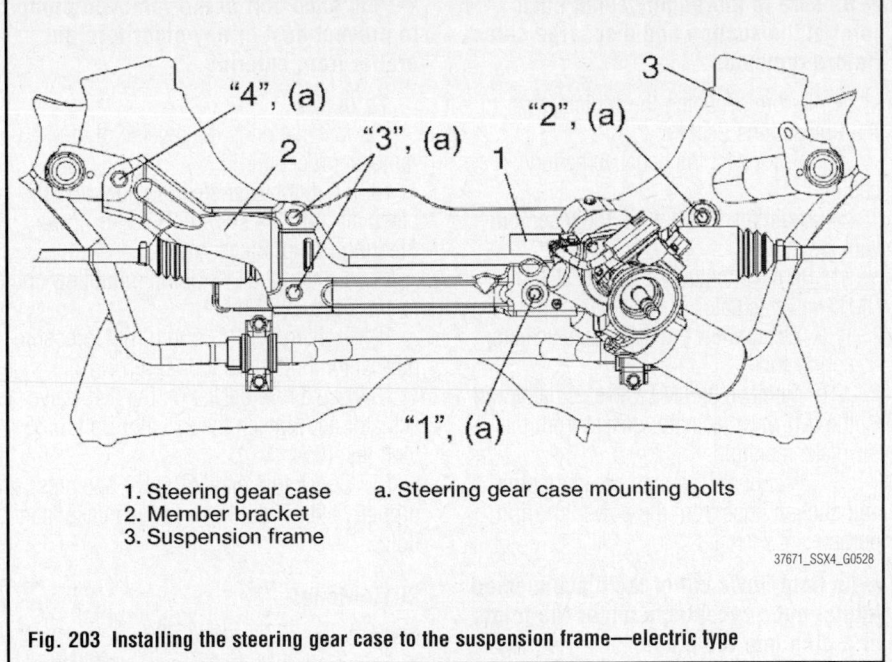

1. Steering gear case
2. Member bracket
3. Suspension frame

a. Steering gear case mounting bolts

37671_SSX4_G0528

Fig. 203 Installing the steering gear case to the suspension frame—electric type

12. Install the steering gear case to the suspension frame as follows.

    a. Set the member bracket to the steering gear case.

    b. Install all the steering gear case mounting bolts by hand.

    c. Tighten the steering mounting bolts in numerical order and tighten to 51 ft. lbs. (70 Nm).

13. Install thc front suspension frame with the steering gear case to the vehicle.

14. Install the steering gear case to the vehicle.

15. Connect the torque sensor connector and the P/S motor connector to the steering gear case.

16. Be sure that steering wheel and brake discs (right and left) are all in the straight-ahead position and then insert the steering lower shaft into the steering pinion shaft with the matching marks.

17. Tighten the steering shaft joint lower bolt and then the upper bolt to 19 ft. lbs. (25 Nm).

18. Install the steering joint cover.

Hydraulic Type

See Figure 202 and 204.

✳✳ WARNING

The spiral cable may snap due to steerIng operation if the steering column is separated from the steering gear assembly. Be sure to secure the steering wheel to avoid turning.

    1. Before servicing the vehicle, refer to the Precautions Section.

    2. Position the front wheels in the straight ahead position.

    3. Disarm the SRS system. Refer to Air Bag (Supplemental Restraint System), Disarming The System.

    4. Disconnect the negative battery cable.

    5. Drain out the fluid in the Power Steering (P/S) fluid reservoir with a syringe.

    6. Remove the steering joint cover.

    7. Make alignment marks on the pinion shaft and the joint of the steering lower shaft for a guide during installation.

    8. Loosen the joint bolt (steering column side) and remove the joint bolt (pinion shaft side), and then disconnect the steering lower shaft from the pinion shaft.

    9. Disconnect the high pressure pipe and the low pressure return hose.

    10. Remove the front suspension frame with steering gear case from the vehicle.

    11. Remove the cylinder pipes from the steering gear case using a flare nut wrench.

    12. Disconnect the high pressure pipe and the low pressure return hose from the steering gear case using a flare nut wrench.

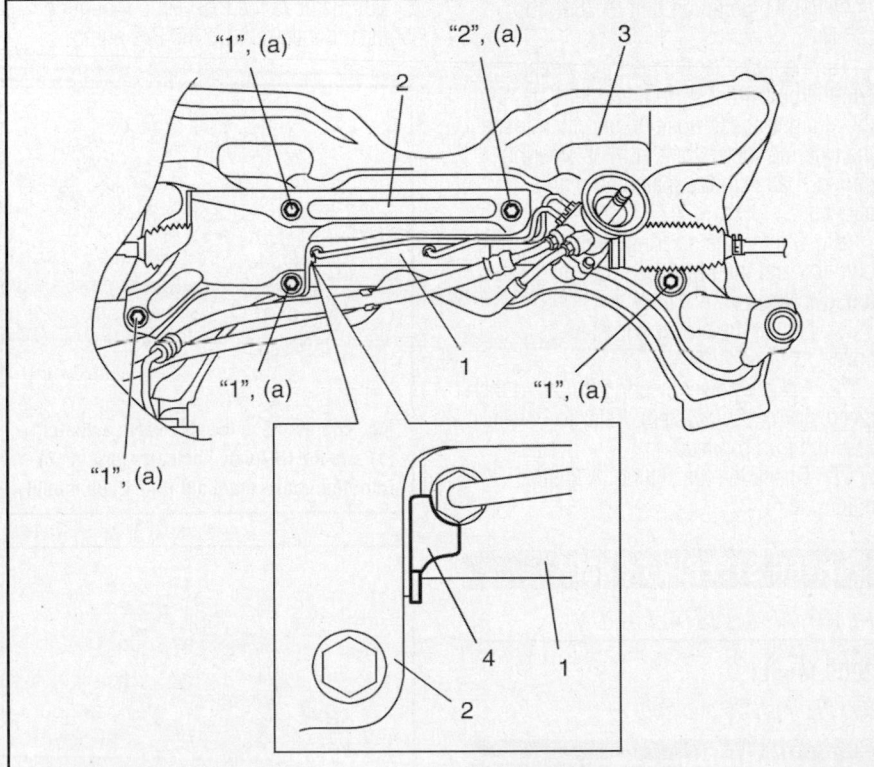

1. Steering gear case
2. Member bracket
3. Suspension frame
4. Mounting rubber protrusion

a. Steering gear case mounting bolts

37671_SSX4_G0529

FIg. 204 Installing the steering gear case to the suspension frame—hydraulic type

➡️As fluid flows out of the disconnected joints, put a receptacle under the joints or a plug into the pipe.

To install:

13. Install the cylinder pipes to the steering gear case using a flare nut wrench. Tighten the cylinder pipes to 10 ft. lbs. (13 Nm).

14. Install the high pressure pipe and the low pressure return pipe to the steering gear case and then connect the low pressure return hose to the low pressure return pipe. Tighten the pressure pipes to 26 ft. lbs. (35 Nm).

15. Install the grommet to the steering.

16. Install the steering gear case to the suspension frame as follows.

a. Set the member bracket to the steering gear case.

b. Install all the steering gear case mounting bolts by hand.

c. Tighten the steering mounting bolts in numerical order and Tighten to 51 ft. lbs. (70 Nm).

17. Install the front suspension frame with the steering gear case to the vehicle.

18. Connect the high pressure pipe and the low pressure return hose and tighten the pipe union bolt to 48 ft. lbs. (65 Nm).

19. Be sure that the steering wheel and the brake discs (right and left) are all in the straight-ahead position and then insert the steering lower shaft into the steering pinion shaft with the matching marks aligned.

20. Tighten the steering shaft joint lower bolt first and then the upper bolt to 19 ft. lbs. (25 Nm).

21. Install the steering joint cover.

22. Fill the power steering fluid and the bleed air from the system. Refer to Power Steering Pump, Bleeding.

23. Check the toe setting. Adjust as required.

POWER STEERING PUMP

REMOVAL & INSTALLATION

2009 Model

See Figures 205 and 206.

✳️✳️ WARNING

Never disassemble the Power Steering (P/S) pump. Disassembly may spoil its original performance. If a faulty condition is found, replace the P/S pump with a new one.

➡️Be sure to thoroughly clean each joint of the suction and discharge sides before removal.

1. Before servicing the vehicle, refer to the Precautions Section.

2. Disconnect the negative battery cable.

3. Drain the P/S fluid in the reservoir with a syringe.

4. Remove the Intake Manifold Tuning (IMT) valve actuator.

a. Disconnect the IMT valve actuator connector.

b. Remove the IMT valve actuator and the IMT valve actuator gasket from the intake manifold.

5. Disconnect the high pressure pipe and suction hose from the power steering pump.

➡️As fluid flows out of the disconnected joints, put a receptacle under the joints or a plug into the pipe.

6. Disconnect the pressure switch lead wire at the switch terminal.

7. Remove the water pump and alternator drive belt. Refer to Accessory Drive Belts, removal & installation.

8. Remove the P/S pump mounting bolts and then remove the P/S pump.

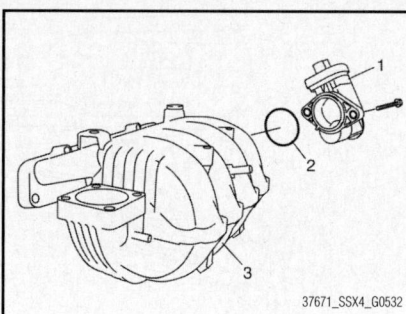

37671_SSX4_G0532

Fig. 205 Remove the IMT valve actuator (1) and the IMT valve actuator gasket (2) from the intake manifold (3)—2009 model

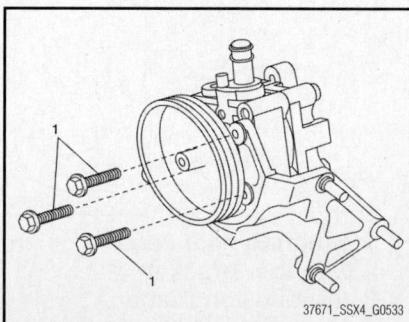

37671_SSX4_G0533

Fig. 206 Remove the P/S pump mounting bolts (1)

➡️Plug each port of the removed pump to prevent dust or any other foreign matter from entering.

To install:

9. Installation is the reverse of the removal procedure.

10. Fill the power steering fluid and the bleed air from the system. Refer to Power Steering Pump, Bleeding.

11. Tighten the P/S pump mounting bolt to 18 ft. lbs. (25 Nm).

12. Tighten the P/S pump high pressure pipe union bolt to 48 ft. lbs. (65 Nm).

13. Use a new gasket on the IMT valve actuator and tighten the actuator bolt to 53 inch lbs. (6 Nm).

14. Check that the IMT valve operates properly after installing it to the intake manifold.

2010 Model

See Figures 207 through 209.

✳️✳️ WARNING

Never disassemble the Power Steering (P/S) pump. Disassembly may spoil its original performance. If a faulty condition is found, replace the P/S pump with a new one.

➡️Be sure to thoroughly clean each joint of the suction and discharge sides before removal.

1. Before servicing the vehicle, refer to the Precautions Section.

2. Disconnect the negative battery cable.

3. Drain the Power Steering (P/S) fluid in the reservoir with a syringe.

4. Disconnect the engine cover hose from the air cleaner assembly.

5. Disconnect the engine cover front side hook from the engine cover pin.

6. Disconnect the engine cover rear side hook from the engine cover pin.

7. Remove the engine cover from the vehicle.

✳️✳️ WARNING

When removing the engine cover, be careful not to hit it against the cowling front panel.

8. Disconnect the Intake Manifold Tuning (IMT) valve actuator connector.

9. Disconnect the fuel injector connector for cylinder No. 1.

10. Disconnect the high pressure pipe and suction hose from the power steering pump.

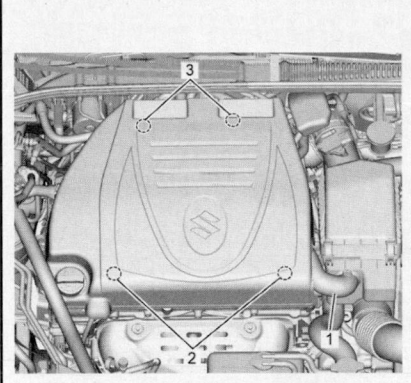

1. Engine cover hose
2. Engine cover front side hook
3. Engine cover rear side hook

37671_SSX4_G0259

Fig. 207 Remove the engine cover

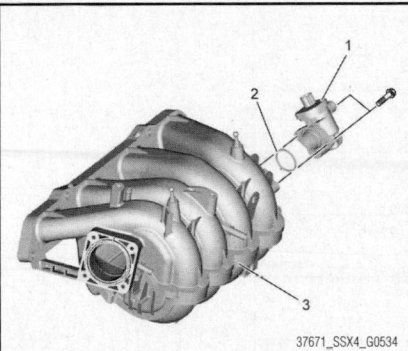

37671_SSX4_G0534

Fig. 208 Remove the IMT valve actuator (1) and the IMT valve actuator gasket (2) from the intake manifold (3)—2010 model

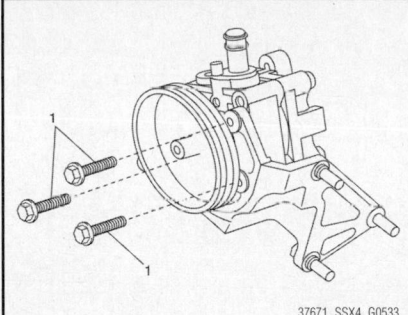

37671_SSX4_G0533

Fig. 209 Remove the P/S pump mounting bolts (1)

➡ **As fluid flows out of the disconnected joints, put a receptacle under the joints or a plug into the pipe.**

11. Disconnect the pressure switch lead wire at the switch terminal.

12. Remove the accessory drive belt. Refer to Accessory Drive Belts, removal & installation.

13. Remove the P/S pump mounting bolts and then remove the P/S pump.

➡ **Plug each port of the removed pump to prevent dust or any other foreign matter from entering.**

14. If necessary, detach the engine harness clamps at the engine.

To install:

15. Installation is the reverse of the removal procedure.

16. When installing the P/S pump high pressure pipe union bolt to the P/S pump, be sure to use a new washer.

17. Fill the power steering fluid and the

bleed air from the system. Refer to Power Steering Pump, Bleeding.

18. Tighten the P/S pump mounting bolt to 18 ft. lbs. (25 Nm).

19. Tighten the P/S pump high pressure pipe union bolt to 48 ft. lbs. (65 Nm).

BLEEDING

1. Before servicing the vehicle, refer to the Precautions Section.

2. Fill the power steering system with the proper grade and type of steering fluid.

➡ **Do not allow the fluid level in the reservoir tank to go below the MIN level line. Check and add fluid as needed.**

3. Raise and safely support the vehicle.

4. Quickly turn the steering wheel to the full right and left detents and lightly touch the steering stoppers.

➡ **Do not hold the steering wheel in the locked position for more than 10 seconds.**

5. Repeat this operation until the fluid level no longer decreases.

6. Start the engine.

7. Quickly turn the steering wheel to the full right and left detents and lightly touch the steering stoppers.

➡ **Do not hold the steering wheel in the locked position for more than 10 seconds.**

8. Check for air bubbles or cloudy fluid. If found, repeat the bleeding procedure.

9. Stop the engine and check the fluid level. Correct as required.

SUSPENSION

CONTROL LINKS

REMOVAL & INSTALLATION

See Figure 210.

Before servicing the vehicle, refer to the Precautions Section.

The manufacturer does not provide a specific removal and installation procedure for this component. Use the illustration as a guide when servicing this component.

LOWER BALL JOINT

REMOVAL & INSTALLATION

The lower ball joint is serviced with the lower control arm as an assembly. Refer to Lower Control Arm, removal & installation.

LOWER CONTROL ARM

REMOVAL AND & INSTALLATION

See Figure 211.

✻✻ WARNING

When removing and installing the steering knuckle assembly, be careful not to damage the dust boots of the suspension arm joint.

1. Before servicing the vehicle, refer to the Precautions Section.

2. Raise and safely support the vehicle.

3. Remove the steering knuckle with the front wheel hub. Refer to Steering Knuckle, removal & installation.

FRONT SUSPENSION

4. Remove the suspension arm (lower control arm) bolts.

5. Remove the suspension arm.

To install:

➡ **The suspension arm bolt is pre-coated with friction stabilizer. Never reuse the removed bolt. Otherwise, the bolt may loosen.**

6. Install the suspension arm and tighten the suspension arm bolts temporarily.

7. Install the steering knuckle with front wheel hub. Refer to Steering Knuckle, removal & installation.

8. Install the wheel and tighten the lug nuts to 62 ft. lbs. (85 Nm).

9. Lower the vehicle in an unloaded

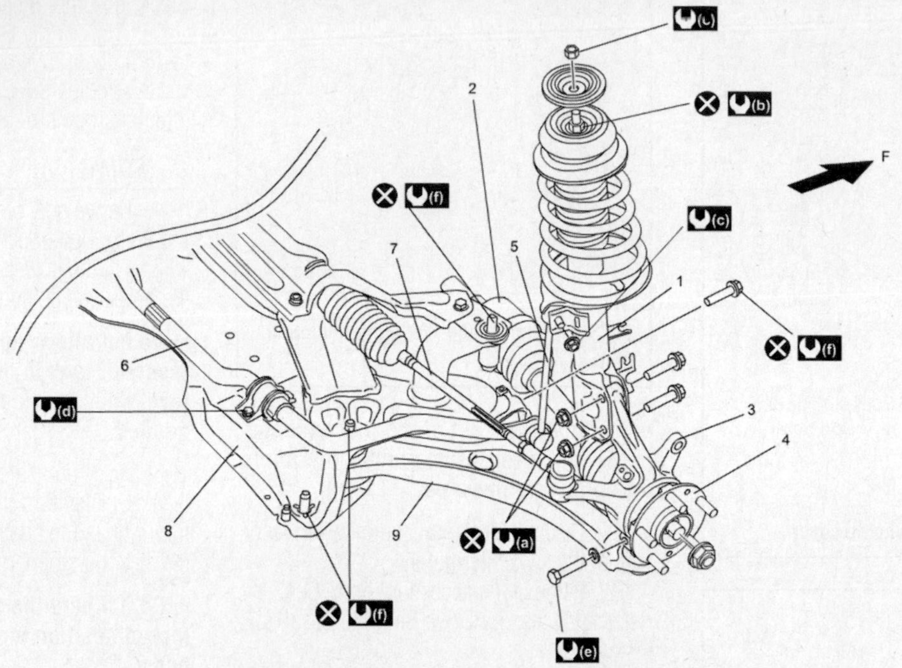

Fig. 210 Front suspension component locations

1.	Front strut assembly	7.	Tie-rod	(d) 23 N·m (2.3 kgf-m, 17.0 lbf-ft)
2.	Front drive shaft	8.	Front suspension frame	(e) 60 N·m (6.0 kgf-m, 43.5 lbf-ft)
3.	Steering knuckle	9.	Suspension arm	(f) 95 N·m (9.5 kgf-m, 69.0 lbf-ft)
4.	Front wheel hub	(a) 140 N·m (14.0 kgf-m, 101.5 lbf-ft)		⊗ Do not reuse.
5.	Stabilizer joint	(b) 55 N·m (5.5 kgf-m, 40.0 lbf-ft)		F: Forward
6.	Stabilizer bar	(c) 50 N·m (5.0 kgf-m, 36.5 lbf-ft)		

37671_SSX4_G0548

condition. Tighten the suspension arm bolts to 69 ft. lbs. (95 Nm).

10. Check the front wheel alignment and adjust as necessary.

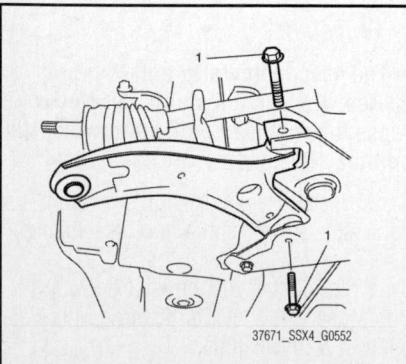

Fig. 211 Install the suspension arm and tighten the suspension arm bolts (1)

STEERING KNUCKLE

REMOVAL & INSTALLATION

See Figures 212 and 213.

❋❋ WARNING

When removing and installing the steering knuckle assembly, be careful not to damage the dust boots of the suspension arm (lower control arm) joint.

1. Before servicing the vehicle, refer to the Precautions Section.
2. Raise and safely support the vehicle.
3. Remove the wheel.
4. Uncaulk the halfshaft nut.
5. Depress the foot brake pedal and hold it. Remove the halfshaft nut.

6. Remove the caliper carrier bolts and then remove the caliper with the carrier.

➥Hang removed the caliper with a wire hook so as to prevent the brake hose from the bending, twisting, or receiving increased tension. Do not depress the brake pedal during caliper removal. Do not operate the brake pedal with the caliper removed.

7. Pull the brake disc off by using two M8 bolts.
8. Pull out the wheel hub with Special Tools 09943-17912 and 09942-15511, or equivalent.

➥If the wheel hub is removed, replace the wheel bearing with a new one.

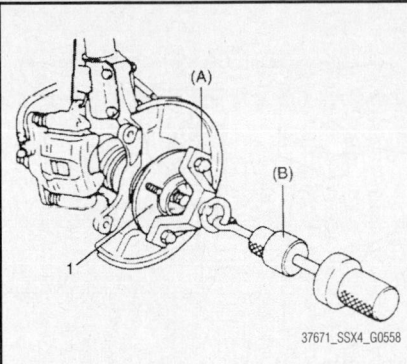

**Fig. 212 Using Special Tools 09943-17912
(A) and 09942-15511 (B) to pull out the
wheel hub (1)**

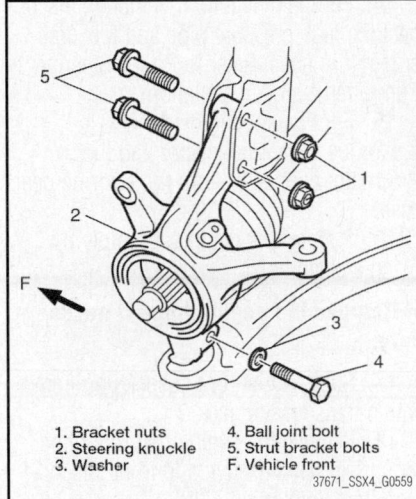

1. Bracket nuts
2. Steering knuckle
3. Washer
4. Ball joint bolt
5. Strut bracket bolts
F. Vehicle front
37671_SSX4_G0559

**Fig. 213 Remove the steering knuckle
from the vehicle**

9. Disconnect the tie-rod end from the steering knuckle.

10. Remove the wheel speed sensor from the knuckle.

11. Loosen the strut bracket nuts.

12. Remove the ball joint bolt and washer.

13. Remove the strut bracket bolts from the strut bracket.

14. Remove the steering knuckle from the vehicle.

To install:

15. Install the steering knuckle with the wheel hub and bearing so that foreign material does not enter the wheel speed sensing point.

16. Install the ball joint bolt, washer, strut bracket bolts, and new nuts to the steering knuckle.

17. Tighten the suspension arm ball joint bolt to 44 ft. lbs. (60 Nm).

18. Tighten new strut bracket nuts to 102 ft. lbs. (140 Nm).

19. Install the wheel speed sensor and tightening the mounting bolt to 97 inch lbs. (11 Nm).

20. Connect the tie-rod end to the steering knuckle.

21. Install the brake disc and the brake caliper.

22. Tighten the caliper carrier bolt to the specified torque.

23. Depress foot brake pedal and hold it there. Tighten a new halfshaft nut to the specified torque.

✳✳ WARNING

Never reuse the halfshaft nut.

24. Caulk the halfshaft nut.

✳✳ WARNING

Be careful not to damage the halfshaft nut while caulking it. If it is damaged, replace it with a new one.

25. Install the wheel and tighten the lug nuts to 62 ft. lbs. (85 Nm).

26. Check the front wheel alignment and adjust as necessary.

STRUT

REMOVAL & INSTALLATION

See Figures 214 and 215.

➡**When the rebound stopper and strut assembly are removed, check the strut support lower nut for the specified torque before installing the strut assembly.**

1. Before servicing the vehicle, refer to the Precautions Section.

2. Remove the hood rear seal, and then remove the cowl top garnish from the vehicle.

➡**When servicing the component parts of the strut assembly, loosen the strut nut a little before removing the strut assembly. This will make the service work easier. The nut must not be removed at this point.**

3. Raise and safely support the vehicle allowing the front suspension to hang free.

4. Remove the wheel and disconnect the stabilizer joint from the strut bracket. When loosening the joint nut, hold the stud with a hexagon wrench.

5. Remove the brake hose mounting

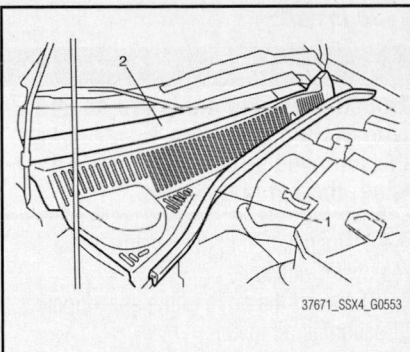

**Fig. 214 Remove the hood rear seal (1) and
the cowl top garnish (2) from the vehicle**

bolt. Remove the brake hose from the bracket and the wheel speed sensor harness from the strut bracket.

6. Remove the strut bracket bolts and nuts.

7. Remove the strut nut, and remove the rebound stopper.

➡**Hold the strut by hand so that it will not fall off.**

8. Remove the strut assembly from the vehicle.

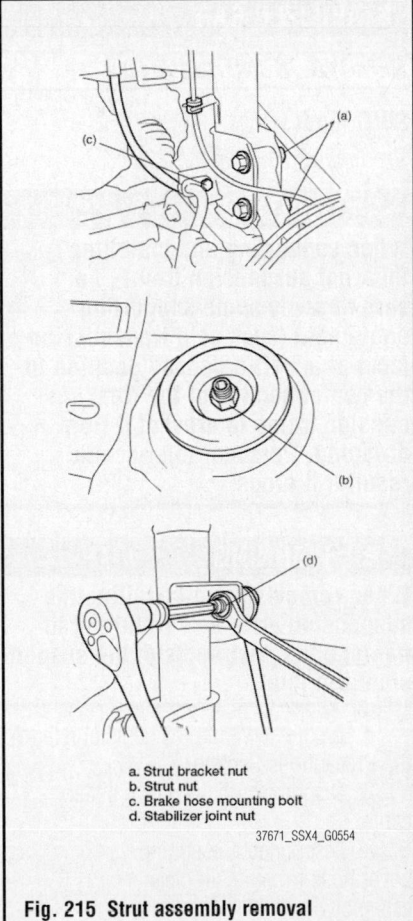

a. Strut bracket nut
b. Strut nut
c. Brake hose mounting bolt
d. Stabilizer joint nut
37671_SSX4_G0554

Fig. 215 Strut assembly removal

To install:

❋❋ WARNING

The strut bracket nut is pre-coated with friction stabilizer. Never reuse the removed strut bracket nut. Otherwise, the nut may loosen.

9. Installation is the reverse of the removal procedure.

10. Insert the bolts in the appropriate direction as removed.

11. Tighten the strut bracket nut to 102 ft. lbs. (140 Nm).

12. Tighten the brake hose mounting bolt to 19 ft. lbs. (26 Nm).

13. Tighten the stabilizer joint nut to 37 ft. lbs. (50 Nm).

14. Lower the vehicle in an unloaded condition, and tighten the strut nut to 37 ft. lbs. (50 Nm).

➡**Do not twist the brake hose and the wheel speed sensor harness when installing them.**

15. Install the wheel and tighten the lug nuts to 62 ft. lbs. (85 Nm).

16. Check the front wheel alignment and adjust as necessary.

STABILIZER BAR

REMOVAL & INSTALLATION

2WD Model

See Figures 216 through 220.

❋❋ CAUTION

When supporting and installing the front suspension frame, be sure to apply some supporting equipment (such as a transmission jack) at a well-balanced position in the center section of the front suspension frame to prevent it from dropping. Personal injury could result if it drops.

❋ WARNING

When removing and installing the suspension arm, be careful not to damage the dust boots of the suspension arm joint.

1. Before servicing the vehicle, refer to the Precautions Section.

2. Disconnect the negative battery cable.

3. Disconnect the engine cover hose from the air cleaner assembly.

1. Engine cover hose
2. Engine cover front side hook
3. Engine cover rear side hook

37671_SSX4_G0259

Fig. 216 Remove the engine cover

4. Disconnect the engine cover front side hook from the engine cover pin.

5. Disconnect the engine cover rear side hook from the engine cover pin.

6. Remove the engine cover from the vehicle.

❋❋ WARNING

When removing the engine cover, be careful not to hit it against the cowling front panel.

7. For the hydraulic P/S model, take out the fluid in the P/S fluid reservoir with a syringe.

8. Disconnect the steering lower shaft from the pinion shaft.

9. Raise and safely support the vehicle.

10. Remove the front wheels.

11. Remove the exhaust No. 1, No. 2, and center pipes.

12. Remove the suspension arms and disconnect the stabilizer joints.

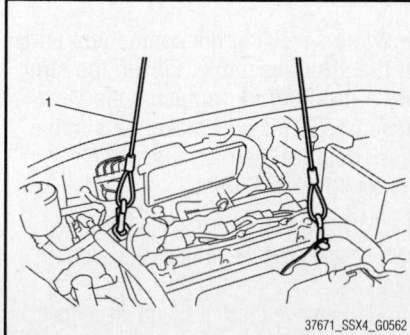

37671_SSX4_G0562

Fig. 217 Support the engine assembly by using a chain hoist (1)

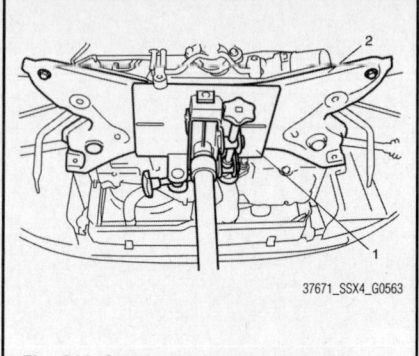

37671_SSX4_G0563

Fig. 218 Support the front suspension frame (2) with a transmission jack (1)

13. Disconnect the tie-rod end from the steering knuckle.

14. For the hydraulic P/S model, disconnect the high pressure pipe and low pressure return hose. Refer to Power Steering Pump, removal & installation.

15. For EPS model, disconnect the torque sensor connector and the P/S motor connector from the steering gear case.

16. Support the engine assembly by using a chain hoist.

➡**Remove the engine hood if necessary.**

17. Support the front suspension frame with a transmission jack.

18. Remove the engine rear mounting, mounting member bolt, and front suspension frame mounting bolts.

19. Lower the front suspension frame and the steering gear case.

20. Remove the steering gear

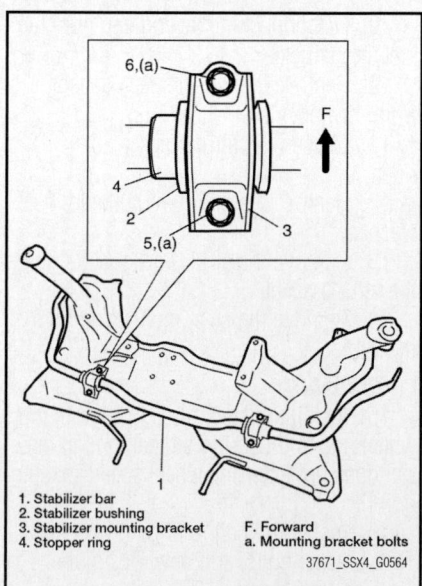

1. Stabilizer bar
2. Stabilizer bushing
3. Stabilizer mounting bracket
4. Stopper ring

F. Forward
a. Mounting bracket bolts

37671_SSX4_G0564

Fig. 219 Installing the stabilizer bar to the front suspension frame

case assembly from the front suspension frame.

21. Remove the stabilizer bar with the bushing from the suspension frame.

To install:

➡**Never reuse the suspension arm bolt or the front suspension frame mounting bolt in order to prevent it from loosening. Be sure to replace the suspension arm bolt and front suspension frame mounting bolt with a new one whenever loosening them once.**

22. Install the stabilizer bar, stabilizer bushing, and stabilizer mounting bracket to the front suspension frame.

➡**For the proper installing direction of the stabilizer mounting bracket, place the oblong hole to the rear and the circular hole to the front. For the proper installation of the stabilizer bar, align the stabilizer-bush-end to the stopper ring of the stabilizer bar.**

23. Tighten the stabilizer bar mounting bracket bolts (rear bolt before the front bolt) to 17 ft. lbs. (23 Nm).

24. Install the steering gear case assembly to the front suspension frame. Refer to Power Rack & Pinion Steering Gear, removal & installation.

25. Support the front suspension frame with a transmission jack and jack it up.

26. Align the lugs (right and left) of the front suspension frame with the hole in the vehicle body.

27. Tighten the front suspension frame mounting bolts to 69 ft. lbs. (95 Nm).

28. Remove the transmission jack from the front suspension frame.

29. Install the engine rear mounting and tighten the mounting member bolts.

30. Remove the chain hoist from the engine assembly.

31. For the hydraulic P/S model, install the high pressure pipe and the low pressure return hose.

32. For the EPS model, connect the torque sensor connector and the P/S motor connector to the steering gear case.

33. Install the suspension arms. Refer to Lower Control Arm, removal & installation.

34. Install the stabilizer joint and tighten the nut to the specified torque.

35. Connect the tie-rod end to the steering knuckle.

36. Install the exhaust No. 1, No. 2, and center pipes.

37. Install the wheel and tighten the lug nuts to 62 ft. lbs. (85 Nm).

38. Install the engine cover.

39. Install the hood.

40. Lower the hoist and vehicle in an unloaded condition. Tighten the suspension arm bolts to 69 ft. lbs. (95 Nm).

41. Connect the negative battery cable.

42. For the hydraulic P/S model, fill the P/S fluid and bleed the air.

43. Check the front wheel alignment and adjust as necessary.

4WD Model

See Figures 221 and 222.

✳✳ CAUTION

When supporting and installing the front suspension frame, be sure to apply some supporting equipment (such as a transmission jack) at a well-balanced position in the center section of the front suspension frame to prevent it from dropping. Personal injury could result if it drops.

✳✳ WARNING

When removing and installing the suspension arm, be careful not to damage the dust boots of the suspension arm joint.

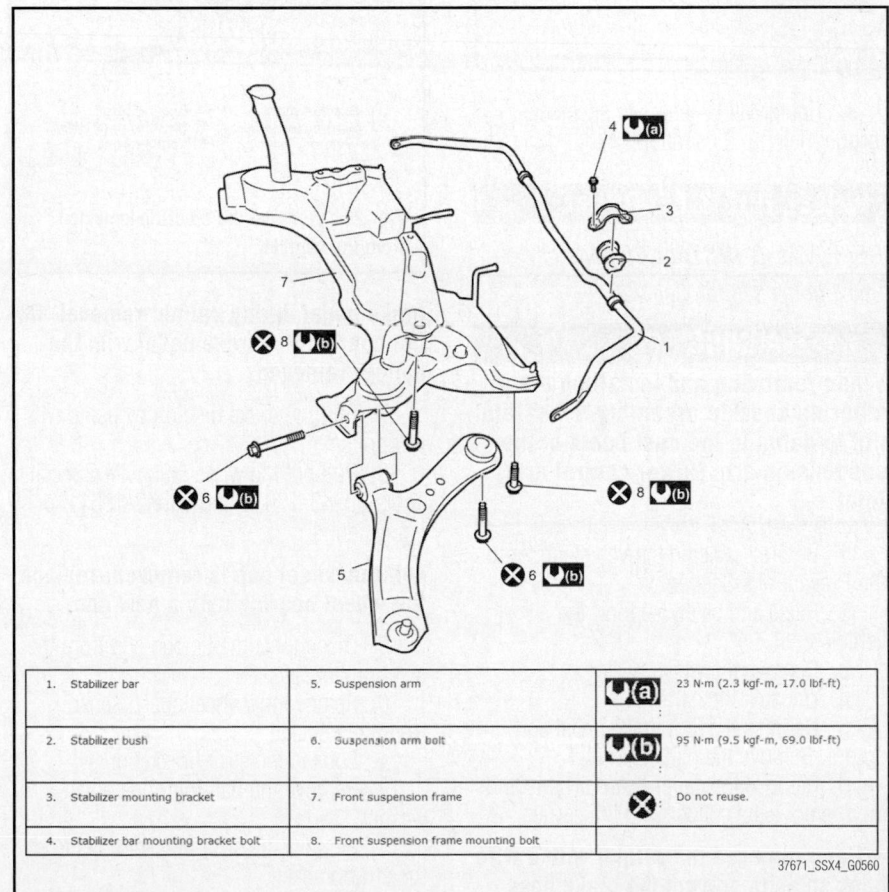

1. Stabilizer bar	5. Suspension arm	☑(a)	23 N·m (2.3 kgf-m, 17.0 lbf-ft)
2. Stabilizer bush	6. Suspension arm bolt	☑(b)	95 N·m (9.5 kgf-m, 69.0 lbf-ft)
3. Stabilizer mounting bracket	7. Front suspension frame	✖	Do not reuse.
4. Stabilizer bar mounting bracket bolt	8. Front suspension frame mounting bolt		

37671_SSX4_G0560

Fig. 220 Stabilizer bar and related components

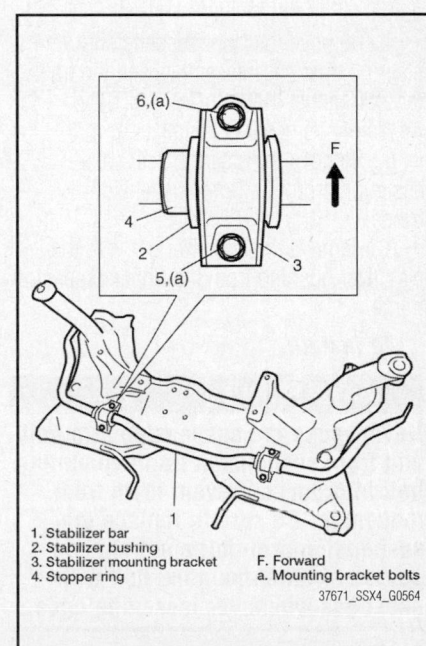

1. Stabilizer bar
2. Stabilizer bushing
3. Stabilizer mounting bracket
4. Stopper ring
F. Forward
a. Mounting bracket bolts

37671_SSX4_G0564

Fig. 221 Installing the stabilizer bar to the front suspension frame

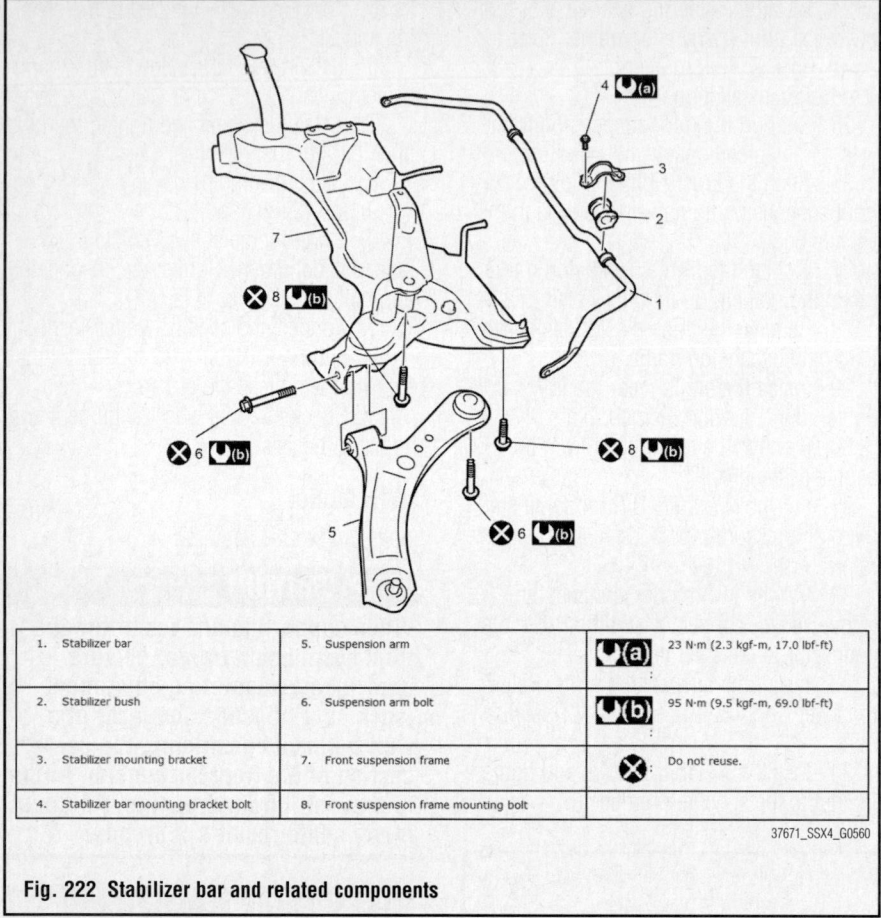

1. Stabilizer bar	5. Suspension arm		⚙(a)	23 N·m (2.3 kgf-m, 17.0 lbf-ft)
2. Stabilizer bush	6. Suspension arm bolt		⚙(b)	95 N·m (9.5 kgf-m, 69.0 lbf-ft)
3. Stabilizer mounting bracket	7. Front suspension frame		✕	Do not reuse.
4. Stabilizer bar mounting bracket bolt	8. Front suspension frame mounting bolt			

37671_SSX4_G0560

Fig. 222 Stabilizer bar and related components

1. Before servicing the vehicle, refer to the Precautions Section.
2. Disconnect the negative battery cable.
3. Remove the suspension arms. Refer to Lower Control Arm, removal & installation.
4. Remove the front suspension frame, steering gear case assembly, and the transfer all together. Refer to Transfer Case Assembly, removal & installation.
5. Remove the steering gear case assembly from the front suspension frame.
6. Remove the stabilizer bar with the stabilizer bushing from the front suspension frame.

To install:

※ WARNING

Never reuse the suspension arm bolt and front suspension frame mounting bolt in order to prevent them from loosening. Be sure to replace the suspension arm bolt and front suspension frame mounting bolt with new ones whenever loosening once.

7. Installation is the reverse of the removal procedure.

8. Tighten all fasteners to specified torque. Refer to illustration.

WHEEL HUB & BEARING

REMOVAL & INSTALLATION

See Figures 223 through 229.

※ WARNING

When removing and installing the steering knuckle assembly, be careful not to damage the dust boots of the suspension arm (lower control arm) joint.

1. Before servicing the vehicle, refer to the Precautions Section.
2. Raise and safely support the vehicle.
3. Remove the wheel.
4. Uncaulk the halfshaft nut.
5. Depress the foot brake pedal and hold it. Remove the halfshaft nut.
6. Remove the caliper carrier bolts and then remove the caliper with the carrier.

➡ **Hang removed the caliper with a wire hook so as to prevent the brake hose from the bending, twisting, or receiving increased tension. Do not depress the**

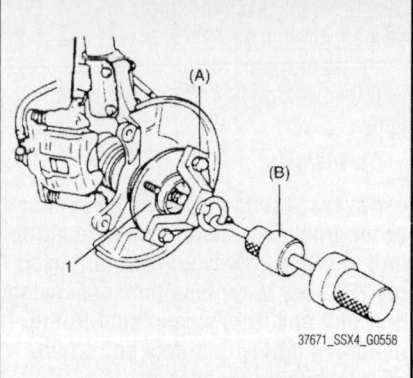

37671_SSX4_G0558

Fig. 223 Using Special Tools 09943-17912 (A) and 09942-15511 (B) to pull out the wheel hub (1)

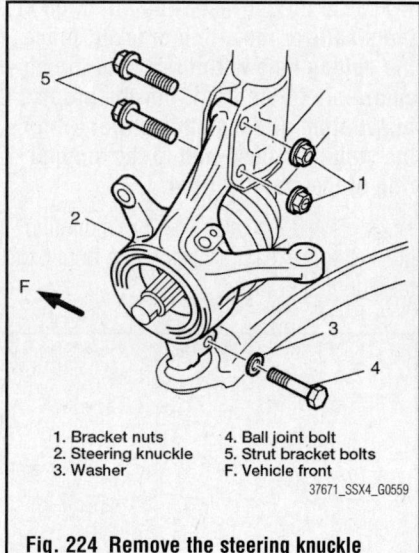

1. Bracket nuts
2. Steering knuckle
3. Washer
4. Ball joint bolt
5. Strut bracket bolts
F. Vehicle front

37671_SSX4_G0559

Fig. 224 Remove the steering knuckle from the vehicle

brake pedal during caliper removal. Do not operate the brake pedal with the caliper removed.

7. Pull the brake disc off by using two M8 bolts.
8. Pull out the wheel hub with Special Tools 09943-17912 and 09942-15511, or equivalent.

➡ **If the wheel hub is removed, replace the wheel bearing with a new one.**

9. Disconnect the tie-rod end from the steering knuckle.
10. Remove the wheel speed sensor from the knuckle.
11. Loosen the strut bracket nuts.
12. Remove the ball joint bolt and washer.
13. Remove the strut bracket bolts from the strut bracket.
14. Remove the steering knuckle from the vehicle.

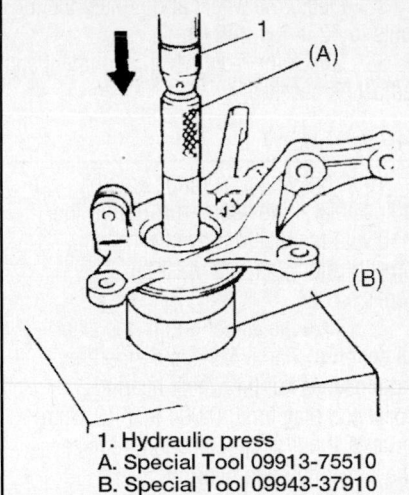

1. Hydraulic press
A. Special Tool 09913-75510
B. Special Tool 09943-37910

37671_SSX4_G0565

Fig. 225 Using Special Tools 09913-75510 and 09943-37910 with hydraulic press to remove the wheel bearing

15. Uncaulk and remove the dust cover.
16. Remove the circlip from the knuckle.
17. Using an hydraulic press and Special Tools 09913-75510 and 09943-37910, or equivalent, remove the wheel bearing.

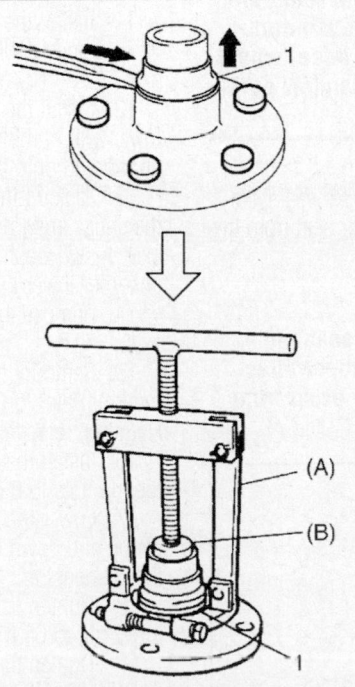

1. Wheel bearing outside inner race
A. Special Tool 09913-65810
B. Special Tool 09913-85230

37671_SSX4_G0566

Fig. 226 Using Special Tools 09913-65810 and 09913-85230 to remove the wheel bearing outside inner race

18. Using Special Tools 09913-65810 and 09913-85230, or equivalent, remove the wheel bearing outside inner race.
19. Remove the hub bolts with a copper hammer or hydraulic press.

➡ **Never remove the bolts unless replacement is necessary. Use new bolt(s) for replacement.**

To install:

❊❊ **WARNING**

Never reuse the wheel bearing and circlip. Otherwise, abnormal noise or vibration may occur.

❊❊ **WARNING**

The strut bracket nut is pre-coated with friction stabilizer. Never reuse the removed strut bracket nut. Otherwise, the nut may loosen.

20. Face the grooved rubber seal side of a new wheel bearing upward and press-fit it into the knuckle using Special Tools 09913-75510, 09944-78220, and 09925-14520, or equivalent tools.
21. Install a new circlip.
22. Insert a new hub bolt in the hub hole. Rotate the hub bolt slowly to assure

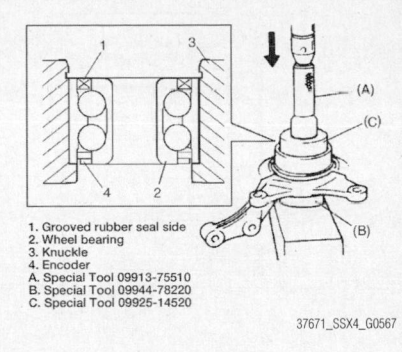

1. Grooved rubber seal side
2. Wheel bearing
3. Knuckle
4. Encoder
A. Special Tool 09913-75510
B. Special Tool 09944-78220
C. Special Tool 09925-14520

37671_SSX4_G0567

Fig. 227 Face the grooved rubber seal side of a new wheel bearing upward and press-fit it into the knuckle using Special Tools 09913-75510, 09944-78220, and 09925-14520

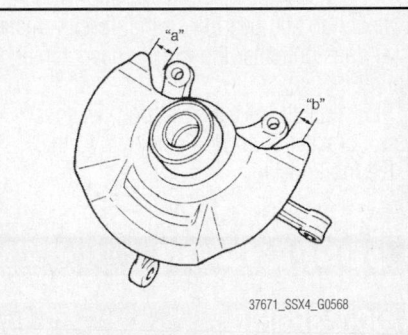

37671_SSX4_G0568

Fig. 228 Drive in the dust cover so that the dimensions from the steering knuckle to the winged portion of the dust cover are equal ("a" equals "b")

that the serrations are aligned with those made by the original bolt.

23. Drive in the dust cover so that the dimensions from the steering knuckle to the winged portion of the dust cover are equal.

❊❊ **WARNING**

When driving in the dust cover, be careful not to deform it.

24. Caulk more than 6 places on the dust cover with a punch.
25. Using Special Tool 09913-75510, or equivalent, and an hydraulic press, press fit the wheel hub into the wheel bearing. Face the grooved rubber seal side to the wheel hub.
26. Apply grease lightly to the end face of the inner ring.

➡ **Use Grease 99000-25121 (SUZUKI Super Grease H), or equivalent. Do not apply the grease to the encoder section to avoid encoder malfunction.**

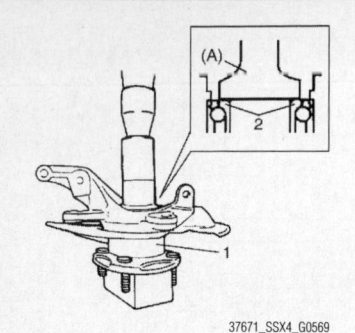

Fig. 229 Using Special Tool 09913-75510 (A) and an hydraulic press, press fit the wheel hub (1) into the wheel bearing (2). Face the grooved rubber seal side to the wheel hub

27. Install the steering knuckle with the wheel hub and bearing so that foreign material does not enter the wheel speed sensing point.

28. Install the ball joint bolt, washer, strut bracket bolts, and new nuts to the steering knuckle.

29. Tighten the suspension arm ball joint bolt to 44 ft. lbs. (60 Nm).

30. Tighten new strut bracket nuts to 102 ft. lbs. (140 Nm).

31. Install the wheel speed sensor and tightening the mounting bolt to 97 inch lbs. (11 Nm).

32. Connect the tie-rod end to the steering knuckle.

33. Install the brake disc and the brake caliper.

34. Tighten the caliper carrier bolt to the specified torque.

35. Depress foot brake pedal and hold it there. Tighten a new halfshaft nut to the specified torque.

✳✳ WARNING

Never reuse the halfshaft nut.

36. Caulk the halfshaft nut.

✳✳ WARNING

Be careful not to damage the halfshaft nut while caulking it. If it is damaged, replace it with a new one.

37. Install the wheel and tighten the lug nuts to 62 ft. lbs. (85 Nm).

38. Check the front wheel alignment and adjust as necessary.

ADJUSTMENT

The front wheel bearings are not adjustable. If the lateral run-out on the hub with the disc removed exceeds specification, the hub must be replaced.

1. Move the wheel hub in the axial direction by hand. Make sure there is no looseness of the wheel bearing. Axial end play limit: 0.004 inch (0.1mm) or less. If out of specification, replace the wheel hub and bearing assembly.

2. Rotate the wheel hub and make sure there is no unusual noise or other irregular conditions. If there are any irregular conditions, replace the wheel hub and bearing assembly. Refer to Wheel Hub & Bearing (sealed unit), removal & installation.

SUSPENSION

REAR SUSPENSION

COIL SPRING

REMOVAL & INSTALLATION

See Figures 230 and 231.

➡**Remove and install both coil springs (right and left) at the same time in order to avoid rear axle twisting and other damage.**

1. Before servicing the vehicle, refer to the Precautions Section.

2. Raise and safely support the vehicle.

3. Remove the rear wheels.

4. Dismount the rear differential (4WD model). Refer to Rear Drive Axle, Axle Housing, removal & installation.

5. Remove the rear fender lining and then loosen the rear axle bolt a little.

✳ WARNING

Do not reuse the rear axle bolt. Otherwise, the bolt may loosen.

6. Support both ends of the rear axle by using 2 floor jacks, or suitable equipment.

7. Detach each lower end of the shock absorbers (right and left) from the rear axle.

8. Lower the rear axle gradually as far down as needed for the coil spring to be removed.

✳✳ WARNING

Be careful not to lower the rear axle down too much. It may cause damage to the brake flexible hose, wheel speed sensor lead wire, and/or parking brake cable.

9. Remove the coil spring.

10. Remove the spring upper seat from the vehicle body and the lower seat from the rear axle.

To install:

➡**Install the spring upper seat and spring lower seat in the proper direction. Otherwise noise may occur from**

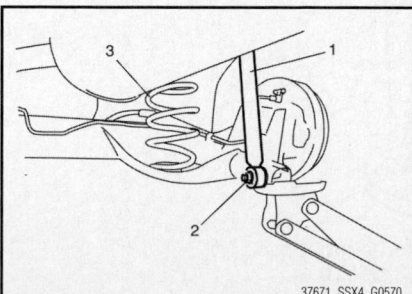

Fig. 230 Detach each lower end (2) of the shock absorber (1) from the rear axle to remove the coil spring (3)

the coil spring while the vehicle is moving.

11. Install the spring upper seat to the vehicle body and lower the seat to the rear axle.

12. Install the coil spring on the spring lower seat of the rear axle and place the coil spring end onto the spring lower seat.

13. Jack up the rear axle and then install the shock absorber lower ends to the rear axle. Install washers and tighten the shock absorber lower nuts temporarily at this step.

14. Remove the floor jacks from the rear axle.

15. Remount the rear differential (4WD model). Refer to Rear Drive Axle, Axle Housing, removal & installation.

16. Install the rear wheels and tighten the lug nuts to 62 ft. lbs. (85 Nm).

17. Lower the hoist and bounce the vehicle up and down several times to stabilize the suspension.

18. Tighten the rear shock absorber lower nut to 65 ft. lbs. (90 Nm).

19. Tighten the rear axle bolt to 53 ft. lbs. (73 Nm).

➡**When tightening these nuts and bolts, be sure that the vehicle is not on a hoist and that it is in an unloaded condition.**

20. Install the rear fender lining.

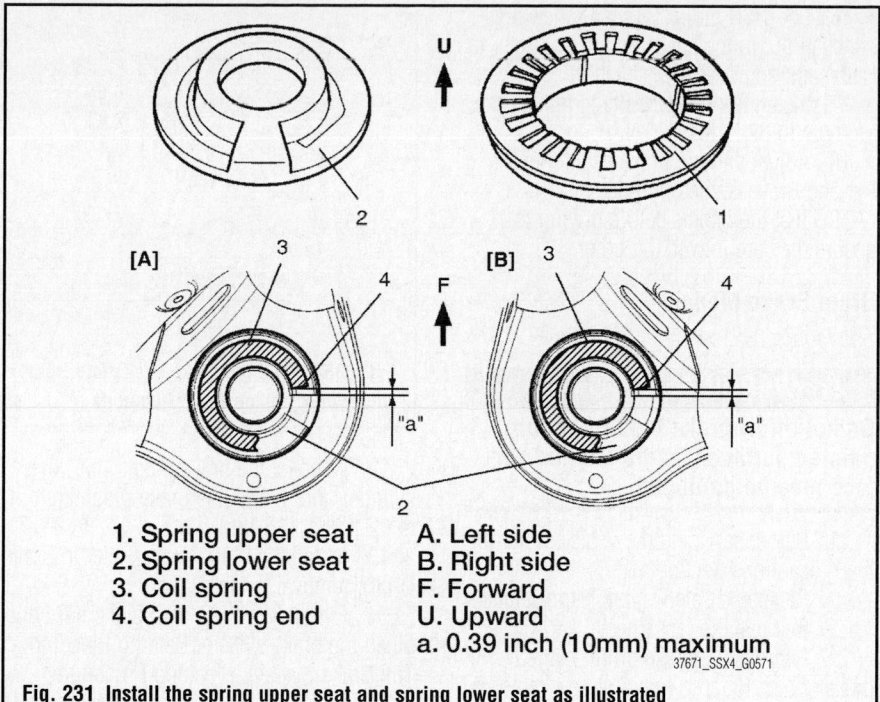

1. Spring upper seat
2. Spring lower seat
3. Coil spring
4. Coil spring end

A. Left side
B. Right side
F. Forward
U. Upward
a. 0.39 inch (10mm) maximum

37671_SSX4_G0571

Fig. 231 Install the spring upper seat and spring lower seat as illustrated

SHOCK ABSORBER

REMOVAL & INSTALLATION

See Figures 232 and 233.

1. Before servicing the vehicle, refer to the Precautions Section.
2. Raise and safely support the vehicle.
3. Remove the rear wheels.
4. For hatchback model, remove the access hole cover in the quarter inner trim.
5. For sedan model, remove the quarter inner trim.

6. Support the rear axle by using a floor jack to prevent it from lowering.
7. Remove the absorber lower nut and lower washer.
8. Remove the absorber upper nut. Then remove the shock absorber, upper washers, and absorber upper bushes.

To install:

9. Install the absorber upper bushes.
10. Install the shock absorber and upper

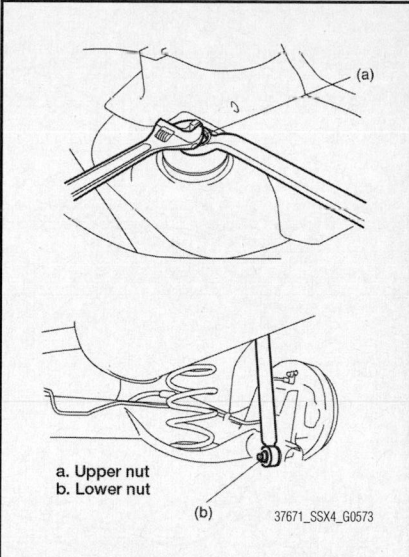

a. Upper nut
b. Lower nut

37671_SSX4_G0573

Fig. 233 Tighten the rear shock absorber nuts as illustrated

washers. Tighten a new rear shock absorber upper nut and lower nut temporarily at this step.

➡**Use a new rear shock absorber upper nut. Otherwise, the nut may loosen.**

11. Remove the floor jack from the rear axle.
12. Install the rear wheels and tighten the lug nuts to 62 ft. lbs. (85 Nm).
13. Lower the hoist and bounce the vehicle up and down several times to stabilize the suspension.
14. Tighten the rear shock absorber upper nut to 22 ft. lbs. (30 Nm).
15. Tighten the rear shock absorber lower nut to 65 ft. lbs. (90 Nm).
16. For hatchback model, install the access hole cover in the quarter inner trim.
17. For sedan model, install the quarter inner trim.

WHEEL HUB & BEARING

REMOVAL & INSTALLATION

Disc Brake Model

See Figure 234.

1. Before servicing the vehicle, refer to the Precautions Section.
2. Raise and safely support the vehicle.
3. Remove the rear wheel.
4. Set the parking brake lever.
5. Uncaulk the rear halfshaft nut and remove the rear halfshaft nut (4WD model).
6. Remove the rear brake caliper assembly and rear brake disc.
7. Disconnect the wheel speed sensor connector (4WD model).

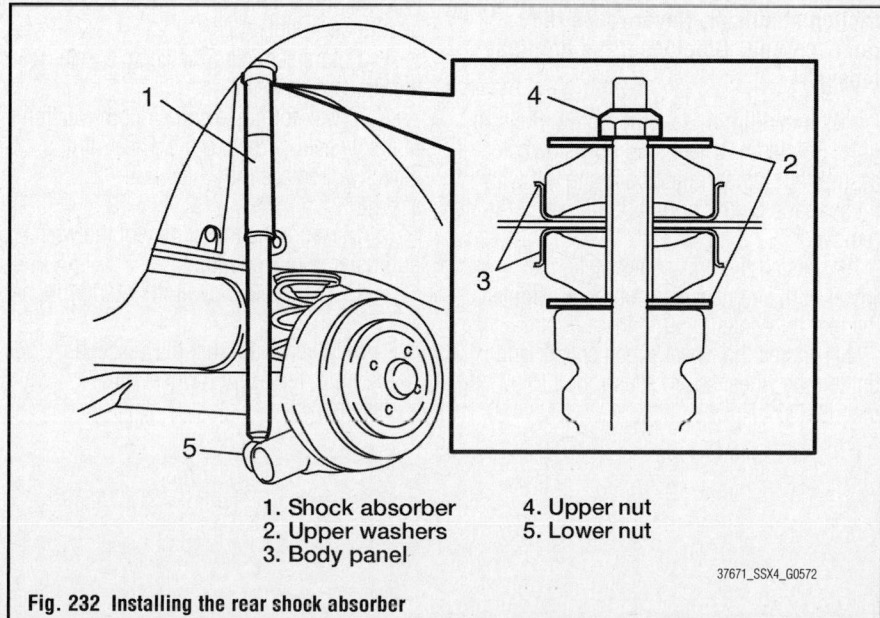

1. Shock absorber
2. Upper washers
3. Body panel

4. Upper nut
5. Lower nut

37671_SSX4_G0572

Fig. 232 Installing the rear shock absorber

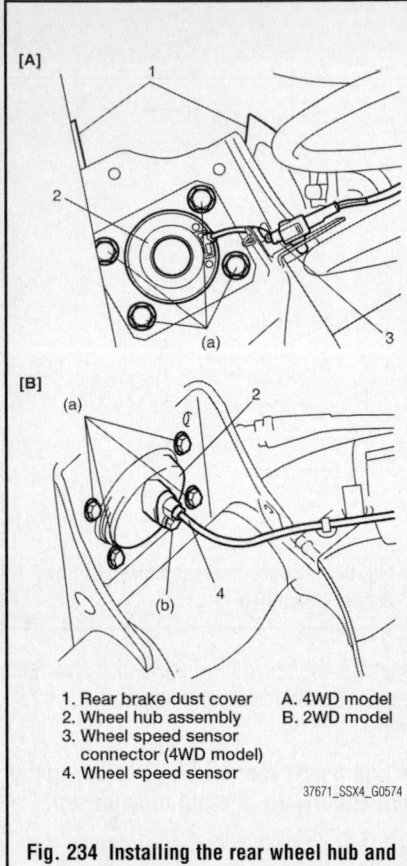

1. Rear brake dust cover
2. Wheel hub assembly
3. Wheel speed sensor connector (4WD model)
4. Wheel speed sensor

A. 4WD model
B. 2WD model

37671_SSX4_G0574

Fig. 234 Installing the rear wheel hub and bearing assembly

8. Remove the wheel speed sensor (2WD model).

9. Remove the brake dust cover and the rear wheel hub assembly from the rear axle.

To install:

➡**The rear hub bolt is pre-coated with friction stabilizer. Never reuse the removed bolt. Otherwise, the bolt may loosen.**

10. Install the stud bolts to the rear wheel hub assembly, if removed.

11. Install the rear brake dust cover and the rear wheel hub assembly and then tighten the new rear hub bolts to 53 ft. lbs. (73 Nm).

12. Connect the wheel speed sensor connector (4WD model).

13. Install the wheel speed sensor and tighten the wheel speed sensor bolt to 97 inch lbs. (11 Nm).

14. Install the rear brake disc and rear brake caliper assembly.

15. Pull up the parking brake lever.

16. For 4WD model, tighten a new rear halfshaft nut to the specified torque. Refer to Halfshafts, removal & installation.

17. Install the rear wheel and tighten the lug nuts to 62 ft. lbs. (85 Nm).

18. Adjust parking brake cable. Refer to Parking Brake Cables, Adjustment.

19. Test the brakes before driving to ensure they are in working order.

Drum Brake Model

See Figure 235.

✳✳ WARNING

Do not allow brake fluid to get on painted surfaces as the painted surface may be damaged.

1. Before servicing the vehicle, refer to the Precautions Section.

2. Raise and safely support the vehicle.

3. Remove the rear wheel.

4. Remove the brake drum and rear brake shoes. Refer to Brake Shoes, removal & installation.

5. Disconnect the brake pipe flare nut from the wheel cylinder and put the bleeder plug cap onto the pipe to prevent fluid from spilling.

6. Remove the wheel speed sensor.

7. Remove the brake back plate and the rear wheel hub from the rear axle.

8. Remove the wheel stud bolts with a copper hammer or hydraulic press.

➡**Never remove the stud bolts unless replacement is necessary. Be sure to use a new bolt(s) for replacement.**

To install:

➡**The rear hub bolt is pre-coated with friction stabilizer. Never reuse the removed bolt. Otherwise, the bolt may loosen.**

9. Insert the wheel stud bolt in the hub hole. Rotate the wheel stud bolt slowly to assure that the serrations are aligned with those made by the original bolt, if removed.

10. Install the brake back plate, rear wheel hub, and new rear hub bolts. Tighten the rear hub bolts to 53 ft. lbs. (73 Nm).

11. Install the wheel speed sensor and tighten the wheel speed sensor bolt to 97 inch lbs. (11 Nm).

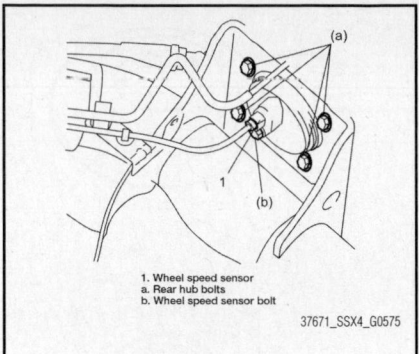

1. Wheel speed sensor
a. Rear hub bolts
b. Wheel speed sensor bolt

37671_SSX4_G0575

Fig. 235 Install the brake back plate, rear wheel hub, and new rear hub bolts

12. Connect the brake pipe to the wheel cylinder and tighten the brake pipe flare nut to 12 ft. lbs. (16 Nm).

13. Install the brake drum. Refer to Brake Drum, removal & installation.

14. Fill the reservoir with brake fluid and bleed the brake system. Refer to Bleeding The Brake System, Bleeding Procedure.

15. Install the rear wheel and tighten the lug nuts to 62 ft. lbs. (85 Nm).

16. Adjust parking brake cable. Refer to Parking Brake Cables, Adjustment.

17. Check to ensure that the brake drum is free from dragging and that proper braking is obtained.

18. Test the foot and parking brakes before driving to ensure they are in working order.

19. Check each installed part for fluid leakage.

ADJUSTMENT

The wheel bearing does not require maintenance or adjustment. If a growling noise is emitted from the wheel bearing during operation, replace the wheel hub and bearing assembly.

1. Before servicing the vehicle, refer to the Precautions Section.

2. Check for noise and smooth rotation of the wheel by rotating the wheel. If it is defective, replace the wheel hub assembly.

3. Measure the thrust play of the wheel hub using a dial gauge.

　a. The thrust play limit is 0.004 inch (0.1mm).

　b. When the thrust play exceeds the limit, replace the wheel hub assembly.

SUZUKI

XL-7

SPECIFICATIONS AND MAINTENANCE CHARTS

ENGINE AND VEHICLE IDENTIFICATION

	Engine						Model Year	
ID/Code	Liters (cc)	Cu. In.	Cyl.	Fuel Sys.	Engine Type	Eng. Mfg.	Code ①	Year
LY7	3.6 (3,556)	217	6	SFI	DOHC	Suzuki	9	2009

SFI: Sequential Multiport Fuel Injection

DOHC: Double Overhead Camshafts

① Indicated at the 4th digit of the Vehicle Identification Number (VIN)

37671_SXL7_C0001

GENERAL ENGINE SPECIFICATIONS
All measurements are given in inches.

Year	Model	Engine Displacement Liters	Engine Series ID	Net Horsepower @ rpm	Net Torque @ rpm (ft. lbs.)	Bore x Stroke (in.)	Com-pression Ratio	Oil Pressure @ rpm
2009	XL7	3.6	LY7	252@6,400	243@2,300	3.70 x 3.37	10.2:1	20@2,000

37671_SXL7_C0002

GASOLINE ENGINE TUNE-UP SPECIFICATIONS

Year	Engine Displacement Liters	Engine ID	Spark Plug Gap (in.)	Ignition Timing (deg.) MT	Ignition Timing (deg.) AT	Fuel Pump (psi)	Idle Speed (RPM) MT	Idle Speed (RPM) AT	Valve Clearance (in.) Intake	Valve Clearance (in.) Exhaust
2009	3.6	LY7	0.043-	NA	①	55-60 ②	NA	600-700	HYD	HYD

NOTE: The Vehicle Emission Control Information label often reflects specification changes made during production.

The label figures must be used if they differ from those in this chart.

NA: Not Applicable   HYD: Hydraulic

① Ignition timing is preset and cannot be adjusted

② At idle

37671_SXL7_C0003

CAPACITIES

Year	Model	Engine Displacement Liters	Engine ID	Engine Oil with Filter (qts.)	Transmission (pts.)		Transfer Case (pts.)	Drive Axle		Fuel Tank (gal.)	Cooling System (qts.)
					Manual	Auto. ①		Front (pts.)	Rear (pts.)		
2009	XL7	3.6	LY7	5.5	NA	8.4-12.6	1.7	NA	1.1	18.5	11-12

NA: Not Applicable

NOTE: All capacities are approximate. Add fluid gradually and check to be sure a proper fluid level is obtained.

① Drain and refill

37671_SXL7_C0004

FLUID SPECIFICATIONS

Year	Model	Engine Displacement Liters	Engine ID	Engine Oil	Manual Trans.	Auto. Trans.	Transfer Case	Drive Axle		Power Steering Fluid	Brake Master Cylinder	Cooling System
								Front	Rear			
2009	XL7	3.6	LY7	5W-30	NA	Dexron VI	①	NA	①	②	DOT 3	Dex-Cool

NA: Not Applicable

DOT: Department Of Transportation

① API GL-5 synthetic gear oil, Viscosity SAE 75W-90

② Power steering fluid (GM P/N 89021184), or equivalent

37671_SXL7_C0005

VALVE SPECIFICATIONS

Year	Engine Displacement Liters	Engine ID	Seat Angle (deg.)	Face Angle (deg.)	Spring Test Pressure (lbs. @ in.)	Spring Installed Height (in.)	Stem-to-Guide Clearance (in.)		Stem Diameter (in.)	
							Intake	Exhaust	Intake	Exhaust
2009	3.6	LY7	①	44.25	56-61@1.3779	1.3779	0.0010-0.0026	0.0014-0.0030	0.2344-0.2352	0.2341-0.2348

① Seating surface: 45 degrees

Relief surface: 30 degrees

Undercut surface: 60 degrees

37671_SXL7_C0006

CAMSHAFT AND BEARING SPECIFICATIONS CHART

All measurements are given in inches.

Year	Engine Displ. Liters	Engine ID	Journal Dia.	Brg. Oil Clearance	Shaft End-play	Runout	Journal Bore	Lobe Lift	
								Intake	Exhaust
2009	3.6	LY7	①	0.0016-0.0033	0.0018-0.0085	②	③	1.6687-1.6805	1.6703-1.6821

① Front Number 1: 1.3754-1.3764 inches
   Middle and Rear number 2-4: 1.0605-1.0614 inches

② Front and Rear number 1 and 4: 0.0010 inch
   Middle 2 and 3: 0.0020 inch

③ Front number 1: 1.3779-1.3787 inches
   Middle and Rear number 2-4: 1.0630-1.0638 inches

37671_SXL7_C0007

CRANKSHAFT AND CONNECTING ROD SPECIFICATIONS

All measurements are given in inches.

Year	Engine Displacement Liters	Engine ID	Crankshaft				Connecting Rod		
			Main Brg. Journal Dia.	Main Brg. Oil Clearance	Shaft End-play	Thrust on No.	Journal Diameter	Oil Clearance	Side Clearance
2009	3.6	LY7	2.6769-2.6775	0.0004-0.0024	0.0039-0.0130	3	2.2044-2.2050	0.0004-0.0028	0.0037-0.0140

37671_SXL7_C0008

PISTON AND RING SPECIFICATIONS

All measurements are given in inches.

Year	Engine Displ. Liters	Engine ID	Piston Clearance	Ring Gap			Ring Side Clearance		
				Top Compression	Bottom Compression	Oil Control	Top Compression	Bottom Compression	Oil Control
2009	3.6	LY7	0.0010-0.0021	0.0059-0.0118	0.0110-0.0189	0.0059-0.0236	0.0012-0.0026	0.0006-0.0024	0.0012-0.0067

37671_SXL7_C0009

TORQUE SPECIFICATIONS
All readings in ft. lbs.

Year	Engine Displacement Liters	Engine ID	Cylinder Head Bolts	Main Bearing Bolts	Rod Bearing Bolts	Crankshaft Damper Bolts	Flywheel Bolts	Manifold		Spark Plugs	Oil Pan Drain Plug
								Intake	Exhaust		
2009	3.6	LY7	①	②	③	④	⑤	17	18	15	15

① Refer to procedure for tightening sequence; do not apply oil to bolts

Step 1: Bolts (M11) 1-8 to 22 ft. lbs.

Step 2: Bolts (M11) 1-8 plus 150 degrees

Step 3: Bolts (M8) 9-10 to 11 ft. lbs.

Step 4: Bolts (M8) 9-10 plus 75 degrees

② Refer to procedure for tightening sequence; do not apply oil to bolts

Step 1: Bolts 1-8 to 15 ft. lbs.

Step 2: Bolts 1-8 plus 80 degrees

Step 3: Bolts 9-16 to 11 ft. lbs.

Step 4: Bolts 9-16 plus 110 degrees

Step 5: Bolts 17-20 to 22 ft. lbs.

Step 6: Bolts 17-20 plus 60 degrees

Step 7: Bolts 21-24 to 22 ft. lbs.

Step 8: Bolts 21-24 plus 60 degrees

③ Do not apply oil to connecting rod bolts

Step 1: 22 ft. lbs.

Step 2: 0 ft. lbs.

Step 3: 18 ft. lbs.

Step 4: Plus 110 degrees

④ Step 1: 74 ft. lbs.

Step 2: Plus 150 degrees

⑤ Step 1: 22 ft. lbs.

Step 2: Plus 45 degrees

37671_SXL7_C0010

WHEEL ALIGNMENT

Year	Model		Caster		Camber		Toe-in (Deg.)
			Range (+/-Deg.)	Preferred Setting (Deg.)	Range (+/-Deg.)	Preferred Setting (Deg.)	
2009	XL7	F	0.75	3.00	0.75	-0.60	+0.15+/-0.20
		R	NA	NA	0.75	-0.50	+0.20+/-0.20

NOTE: Measurements given for an unladen vehicle with fuel, coolant, and engine oil full; spare tire, jack, hand tools, and mats in designated positions.

NA: Not Applicable

F: Front

R: Rear

37671_SXL7_C0011

TIRE, WHEEL AND BALL JOINT SPECIFICATIONS

Year	Model	OEM Tires		Tire Pressures (psi)		Wheel Size	Ball Joint Inspection	Lug Nut Torque (ft. lbs.)
		Standard	Optional	Front	Rear			
2009	XL7	P225/65R17	NS	①	①	17 x 6.5J	②	100

OEM: Original Equipment Manufacturer

PSI: Pounds Per Square Inch

NS: Not Specified

① See the tire placard on the vehicle

② Gently lift or pry the suspension to induce ball joint movement. If the dial indicator indicates a reading greater than 0.02 inch (0.5mm), replace the ball joint.

BRAKE SPECIFICATIONS

All measurements in inches unless noted

Year	Model		Brake Disc			Brake Drum Diameter			Minimum Lining Thickness	Brake Caliper	
			Original Thickness	Minimum Thickness	Maximum Runout	Original Inside Diameter	Max. Wear Limit	Maximum Machine Diameter		Bracket Bolts (ft. lbs.)	Guide Pin Bolts (ft. lbs.)
2009	XL7	F	NA	1.079	0.002	NA	NA	NA	0.080	136	20
		R	NA	0.724	0.002	NA	NA	NA	0.080	89	20

F: Front

R: Rear

NA: Not Applicable

MAINTENANCE I AND II SERVICE SCHEDULES
Suzuki XL7

When the CHANGE ENGINE OIL light appears, certain services and inspections are required.
Required services are described as Maintenance I and Maintenance II.
The first service of a vehicle should be Maintenance I, and the second service should be Maintenance II.
Alternate between the 2 services thereafter. However, in some cases, Maintenance II may be required more often.
Maintenance I: Use Maintenance I if the CHANGE ENGINE OIL light comes on within 10 months since the vehicle was purchased or, if Maintenance II was performed.
Maintenance II: Use Maintenance II if the previous service performed was Maintenance I. Always use Maintenance II whenever the CHANGE ENGINE OIL light comes on 10 months or more since the last service, or, if the CHANGE ENGINE OIL light has not come on at all for one year.

Service	Maintenance I	Maintenance II
Change the engine oil and filter. Reset the oil life system.①	✓	✓
Visually inspect the vehicle for leaks or damage. A fluid loss in the vehicle system could indicate a problem. Inspect, repair, and add fluid to the system as necessary.	✓	✓
Inspect the engine air cleaner filter. If necessary, replace the filter. If driving in dusty conditions, inspect the filter at each engine oil change.		✓
Rotate the tires. Inspect the tire inflation pressures and the tire wear.	✓	✓
Visually inspect the brake lines and hoses for proper hook-up, binding, leaks, cracks, chafing, etc. Inspect the disc brake pads for wear and the rotors for surface condition. Inspect other brake parts, including wheel cylinders, calipers, parking brake, etc. Inspect the parking brake adjustment.	✓	✓
Check the engine coolant and windshield washer fluid levels and add fluid as needed.	✓	✓
Inspect the suspension and steering components. Inspect the front and rear suspension and the steering system for damaged, loose or missing parts, or signs of wear. Inspect the power steering lines and the hoses for proper hook-up, binding, leaks, cracks, chafing, etc.		✓
Inspect the coolant hoses and replace the hoses if they are cracked, swollen, or deteriorated. Inspect all pipes, fittings, and clamps; replace with GM parts as needed. To help ensure proper operation, a pressure test of the cooling system and pressure cap. Cleaning the outside of the radiator and air conditioning condenser is recommended at least once a year.		✓
Inspect the wiper blades for wear or cracking.		✓
Inspect the restraint system components. Ensure that the safety belt reminder light and all the belts, buckles, latch plates, retractors, and anchorages are working properly. Look for any other loose or damaged safety belt system parts. If you see anything that might keep a safety belt system from working correctly, repair or replace the damaged part. Replace torn or frayed safety belts. Inspect for any opened or broken air bag coverings and repair or replace, as needed. The air bag system requires regular maintenance.		✓
Lubricate the body components. Lubricate all key lock cylinders, hood latch assemblies, secondary latches, pivots, spring anchor and release pawl, hood and door hinges, rear folding seats, and liftgate hinges. Frequent lubrication may be required when exposed to a corrosive environment. Apply dielectric silicone grease GM P/N 12345579 (Canadian P/N 1974984), or equivalent, on the weatherstrips with a clean cloth.		✓
Inspect the transmission fluid level and add fluid as needed.		✓
Inspect the throttle system for interference or binding and for damaged or missing parts. Replace the parts as needed. Replace any components that have high effort or excessive wear.		✓
Replace the passenger compartment air filter, if necessary.		✓

① If the system is reset accidentally, perform the maintenance service within 3,000 miles since the last service.

To reset the CHANGE ENGINE OIL light:
Vehicles with Driver Information Center (DIC)
NOTE: When the system has calculated that oil life has been diminished, it will indicate that an oil change is necessary. A CHANGE ENGINE OIL light will come on. Change the engine oil as soon as possible within the next 600 miles. It is possible that, if driving under the best conditions, the oil life system may not indicate that an oil change is necessary for over a year. However, the engine oil and filter must be changed at least once a year and at this time the system must be reset.
NOTE: The Engine Oil Life System calculates when to change the engine oil and filter based on vehicle use. Anytime the oil is changed, reset the system so it can calculate when the next oil change is required. If a situation occurs where the oil is changed prior to a CHANGE ENGINE OIL SOON message being turned on, reset the system.

1. Press the vehicle information button on the Driver Information Center (DIC) until OIL LIFE REMAINING displays.
2. Press and hold the Set/Reset button on the DIC for more than 5 seconds. The oil life will change to 100 percent.

NOTE: If the light or message comes back on when starting the vehicle, the oil life system has not reset. Repeat the procedure.

ADDITIONAL MAINTENANCE SERVICES
Suzuki XL7

TO BE SERVICED	TYPE OF SERVICE	VEHICLE MILEAGE INTERVAL (x1000)					
		25	50	75	100	125	150
Air cleaner filter (engine)	R		✓		✓		✓
Accessory drive belt	I						✓
Automatic transmission fluid ①	R				✓		
Engine coolant ②	R	Every 5 years or 150,000 miles					
Fuel system (for damage or leaks)	I	✓	✓	✓	✓	✓	✓
Exhaust system	S/I	✓	✓	✓	✓	✓	✓
Spark plugs	R				✓		

R: Replace

S/I: Inspect and service, if necessary

FREQUENT OPERATION MAINTENANCE (SEVERE SERVICE)

If a vehicle is operated under any of the following conditions, it is considered severe service:

- **Extremely dusty areas**

- **50% or more of the vehicle operation is in 90°F (32°C) or higher temperatures, or constant operation in temperatures below 32°F (0°C)**

- **Prolonged idling (vehicle operation in stop and go traffic)**

- **Frequent short running periods (engine does not warm to normal operating temperatures)**

- **Police, taxi, delivery usage, or trailer towing usage**

- **Driving in hilly or mountainous terrain**

① Replace the fluid every 50,000 miles under Severe Service

② Drain, flush, and refill the cooling system. Inspect hoses. Clean the radiator, condenser, pressure cap, and filler neck. Pressure test the cooling system and pressure cap.

37671_SXL7_C0015

BRAKES   INFORMATION AND PRECAUTIONS

ANTI-LOCK SYSTEMS

• Certain components within the ABS system are not intended to be serviced or repaired individually.

• Do not use rubber hoses or other parts not specifically specified for and ABS system. When using repair kits, replace all parts included in the kit. Partial or incorrect repair may lead to functional problems and require the replacement of components.

• Lubricate rubber parts with clean, fresh brake fluid to ease assembly. Do not use shop air to clean parts; damage to rubber components may result.

• Use only DOT 3 brake fluid from an unopened container.

• If any hydraulic component or line is removed or replaced, it may be necessary to bleed the entire system.

• A clean repair area is essential. Always clean the reservoir and cap thoroughly before removing the cap. The slightest amount of dirt in the fluid may plug an ori-

fice and impair the system function. Perform repairs after components have been thoroughly cleaned; use only denatured alcohol to clean components. Do not allow ABS components to come into contact with any substance containing mineral oil; this includes used shop rags.

• The Anti-Lock control unit is a microprocessor similar to other computer units in the vehicle. Ensure that the ignition switch is **OFF** before removing or installing controller harnesses. Avoid static electricity discharge at or near the controller.

• If any arc welding is to be done on the vehicle, the control unit should be unplugged before welding operations begin.

DISC AND DRUM SYSTEMS

※※ CAUTION

Dust and dirt accumulating on brake parts during normal use may contain asbestos fibers from production or

aftermarket brake linings. Breathing excessive concentrations of asbestos fibers can cause serious bodily harm. Exercise care when servicing brake parts. Do not sand or grind brake lining unless equipment used is designed to contain the dust residue. Do not clean brake parts with compressed air or by dry brushing. Cleaning should be done by dampening the brake components with a fine mist of water, then wiping the brake components clean with a dampened cloth. Dispose of cloth and all residue containing asbestos fibers in an impermeable container with the appropriate label. Follow practices prescribed by the Occupational Safety and Health Administration (OSHA) and the Environmental Protection Agency (EPA) for the handling, processing, and disposing of dust or debris that may contain asbestos fibers.

BRAKES   BLEEDING THE BRAKE SYSTEM

BLEEDING PROCEDURE

BLEEDING PROCEDURE

Manual Bleeding

➡**When adding fluid to the brake master cylinder reservoir, use only GM approved or equivalent DOT-3 brake fluid from a clean, sealed brake fluid container. The use of any type of fluid other than the recommended type of brake fluid may cause contamination which could result in damage to the internal rubber seals and/or rubber linings of hydraulic brake system components.**

➡**Avoid spilling brake fluid onto painted surfaces, electrical connections, wiring, or cables. Brake fluid will damage painted surfaces and cause corrosion to electrical components. If any brake fluid comes in contact with painted surfaces, immediately flush the area with water. If any brake fluid comes in contact with electrical connections, wiring, or cables, use a clean shop cloth to wipe away the fluid.**

1. Place a clean shop cloth beneath the brake master cylinder to catch brake fluid spills.

2. With the ignition OFF and the brakes cool, apply the brakes 3–5 times, or until the brake pedal effort increases significantly, in order to deplete the brake booster power reserve.

3. If you have performed a brake master cylinder bench bleeding on this vehicle, or if you disconnected the brake pipes from the master cylinder, or if you have disconnected the brake pipes from the proportioning valve assembly or the brake modulator assembly, you must perform the following steps to bleed air at the ports of the hydraulic component:

  a. If removal of the reservoir cap and diaphragm is necessary, clean the outside of the reservoir on and around the cap prior to removal.

  b. With the brake pipes installed securely to the master cylinder, proportioning valve assembly, or brake modulator assembly, loosen and separate one of the brake pipes from the port of the component.

4. For the proportioning valve assembly or the brake modulator assembly, perform these steps in the sequence of system flow; begin with the fluid feed pipes from the master cylinder.

  a. Allow a small amount of brake fluid to gravity bleed from the open port of the component.

  b. Reconnect the brake pipe to the component and tighten securely.

  c. Have an assistant slowly depress the brake pedal fully and maintain steady pressure on the pedal.

  d. Loosen the same brake pipe to purge air from the open port of the component.

  e. Tighten the brake pipe, then have the assistant slowly release the brake pedal.

  f. Wait 15 seconds, then repeat steps until all air is purged from the same port of the component.

  g. With the brake pipe installed securely to the master cylinder, proportioning valve assembly, or brake modulator assembly, and after all air has been purged from the first port of the component that was bled, loosen and separate the next brake pipe from the component, then repeat steps until each of the ports on the component has been bled.

  h. After completing the final component port bleeding procedure, ensure that each of the brake pipe-to-component fittings is properly tightened.

5. Ensure the brake master cylinder reservoir remains at least half-full during this bleeding procedure. Add fluid as needed to maintain the proper level.

➡Clean the outside of the reservoir on and around the reservoir cap prior to removing the cap and diaphragm.

6. Install a proper box-end wrench onto the RIGHT REAR wheel hydraulic circuit bleeder valve.

7. Install a transparent hose over the end of the bleeder valve.

8. Have an assistant slowly depress the brake pedal fully and maintain steady pressure on the pedal.

9. Loosen the bleeder valve to purge air from the wheel hydraulic circuit.

10. Tighten the bleeder valve, then have the assistant slowly release the brake pedal.

11. Wait 15 seconds, then repeat steps 8–10 until all air is purged from the same wheel hydraulic circuit.

12. With the right rear wheel hydraulic circuit bleeder valve tightened securely, and after all air has been purged from the right rear hydraulic circuit, install a proper box-end wrench onto the LEFT FRONT wheel hydraulic circuit bleeder valve.

13. Install a transparent hose over the end of the bleeder valve, then repeat steps 8–11.

14. With the left front wheel hydraulic circuit bleeder valve tightened securely, and after all air has been purged from the left front hydraulic circuit, install a proper box-end wrench onto the LEFT REAR wheel hydraulic circuit bleeder valve.

15. Install a transparent hose over the end of the bleeder valve, then repeat steps 8–11.

16. With the left rear wheel hydraulic circuit bleeder valve tightened securely, and after all air has been purged from the left rear hydraulic circuit, install a proper box-end wrench onto the RIGHT FRONT wheel hydraulic circuit bleeder valve.

17. Install a transparent hose over the end of the bleeder valve, then repeat steps 8–11.

18. After completing the final wheel hydraulic circuit bleeding procedure, ensure that each of the 4 wheel hydraulic circuit bleeder valves is properly tightened.

19. Slowly depress and release the brake pedal. Observe the feel of the brake pedal.

20. If the brake pedal feels spongy, repeat the bleeding procedure again. If the brake pedal still feels spongy after repeating the bleeding procedure, perform the following steps:

a. Inspect the brake system for external leaks.

b. Pressure bleed the hydraulic brake system in order to purge any air that may still be trapped in the system.

21. Turn the ignition key ON, with the engine OFF. Check to see if the brake system warning lamp remains illuminated.

✳✳ CAUTION

If the brake system warning lamp remains illuminated, DO NOT allow the vehicle to be driven until it is diagnosed and repaired.

Pressure Bleeding

➡When adding fluid to the brake master cylinder reservoir, use only GM approved or equivalent DOT-3 brake fluid from a clean, sealed brake fluid container. The use of any type of fluid other than the recommended type of brake fluid may cause contamination which could result in damage to the internal rubber seals and/or rubber linings of hydraulic brake system components.

➡Avoid spilling brake fluid onto painted surfaces, electrical connections, wiring, or cables. Brake fluid will damage painted surfaces and cause corrosion to electrical components. If any brake fluid comes in contact with painted surfaces, immediately flush the area with water. If any brake fluid comes in contact with electrical connections, wiring, or cables, use a clean shop cloth to wipe away the fluid.

1. Place a clean shop cloth beneath the brake master cylinder to catch brake fluid spills.

2. With the ignition OFF and the brakes cool, apply the brakes 3–5 times, or until the brake pedal becomes firm, in order to deplete the brake booster power reserve.

3. If you have performed a brake master cylinder bench bleeding on this vehicle, or if you disconnected the brake pipes from the master cylinder, or if you have disconnected the brake pipes from the proportioning valve assembly or the brake modulator assembly, you must perform the following steps to bleed air at the ports of the hydraulic component:

a. If removal of the reservoir cap and diaphragm is necessary, clean the outside of the reservoir on and around the cap prior to removal.

b. With the brake pipes installed securely to the master cylinder, proportioning valve assembly, or brake modulator assembly, loosen and separate one of the brake pipes from the port of the component.

4. For the proportioning valve assembly or the brake modulator assembly, perform these steps in the sequence of system flow; begin with the fluid feed pipes from the master cylinder.

a. Allow a small amount of brake fluid to gravity bleed from the open port of the component.

b. Reconnect the brake pipe to the component and tighten securely.

c. Have an assistant slowly depress the brake pedal fully and maintain steady pressure on the pedal.

d. Loosen the same brake pipe to purge air from the open port of the component.

e. Tighten the brake pipe, then have the assistant slowly release the brake pedal.

f. Wait 15 seconds, then repeat steps until all air is purged from the same port of the component.

g. With the brake pipe installed securely to the master cylinder, proportioning valve assembly, or brake modulator assembly, and after all air has been purged from the first port of the component that was bled, loosen and separate the next brake pipe from the component, then repeat steps until each of the ports on the component has been bled.

h. After completing the final component port bleeding procedure, ensure that each of the brake pipe-to-component fittings is properly tightened.

5. Clean the outside of the reservoir on and around the reservoir cap prior to removing the cap and diaphragm.

6. Install the J 44894-A to the brake master cylinder reservoir.

7. Connect the J 29532, or equivalent, to the J 44894-A.

8. Charge the J 29532, or equivalent, air tank to 25–30 psi (175–205 kPa).

9. Open the J 29532, or equivalent, fluid tank valve to allow pressurized brake fluid to enter the brake system.

10. Wait approximately 30 seconds, then inspect the entire hydraulic brake system in order to ensure that there are no existing external brake fluid leaks.

➡Any brake fluid leaks identified require repair prior to completing this procedure.

11. Install a proper box-end wrench onto the RIGHT REAR wheel hydraulic circuit bleeder valve.

12. Install a transparent hose over the end of the bleeder valve.

13. Loosen the bleeder valve to purge air from the wheel hydraulic circuit. Allow fluid

to flow until air bubbles stop flowing from the bleeder, then tighten the bleeder valve.

14. With the right rear wheel hydraulic circuit bleeder valve tightened securely, and after all air has been purged from the right rear hydraulic circuit, install a proper box-end wrench onto the LEFT FRONT wheel hydraulic circuit bleeder valve.

15. Install a transparent hose over the end of the bleeder valve, then repeat steps 13–14.

16. With the left front wheel hydraulic circuit bleeder valve tightened securely, and after all air has been purged from the left front hydraulic circuit, install a proper box-end wrench onto the LEFT REAR wheel hydraulic circuit bleeder valve.

17. Install a transparent hose over the end of the bleeder valve, then repeat steps 13–14.

With the left rear wheel hydraulic circuit bleeder valve tightened securely, and after all air has been purged from the left rear hydraulic circuit, install a proper box-end wrench onto the RIGHT FRONT wheel hydraulic circuit bleeder valve.

18. Install a transparent hose over the end of the bleeder valve, then repeat steps 13–14.

19. After completing the final wheel hydraulic circuit bleeding procedure, ensure that each of the 4 wheel hydraulic circuit bleeder valves is properly tightened.

20. Close the J 29532, or equivalent, fluid tank valve, then disconnect the J 29532, or equivalent, from the J 44894-A.

21. Remove the J 44894-A from the brake master cylinder reservoir.

22. Slowly depress and release the brake pedal. Observe the feel of the brake pedal.

23. If the brake pedal feels spongy perform the following steps:

    a. Inspect the brake system for external leaks.

    b. If equipped with antilock brakes, using a scan tool, perform the antilock brake system automated bleeding procedure to remove any air that may have been trapped in the brake pressure modulator valve (BPMV).

24. Turn the ignition key ON, with the engine OFF. Check to see if the brake system warning lamp remains illuminated.

✴✴ CAUTION

If the brake system warning lamp remains illuminated, DO NOT allow the vehicle to be driven until it is diagnosed and repaired.

BLEEDING THE ABS SYSTEM

✴✴ WARNING

When adding fluid to the brake master cylinder reservoir, use only DOT-3 brake fluid from a clean, sealed brake fluid container. The use of any type of fluid other than the recommended type of brake fluid, may cause contamination which could result in damage to the internal rubber seals and/or rubber linings of hydraulic brake system components.

✴✴ WARNING

Avoid spilling brake fluid onto painted surfaces, electrical connections, wiring, or cables. Brake fluid will damage painted surfaces and cause corrosion to electrical components. If any brake fluid comes in contact with painted surfaces, immediately flush the area with water. If any brake fluid comes in contact with electrical connections, wiring, or cables, use a clean shop cloth to wipe away the fluid.

➡The base hydraulic brake system must be bled before performing this automated bleeding procedure. Refer to Bleeding The Brake System.

The automated bleed procedure is recommended when one of the following conditions exist:

• The base brake system bleeding does not achieve the desired pedal height or feel

• An extreme loss of brake fluid has occurred

• Air ingestion is suspected in the secondary circuits of the brake modulator assembly

The ABS Automated Bleed Procedure uses a scan tool to cycle the system solenoid valves and run the pump in order to purge any air from the secondary circuits. These circuits are normally closed off, and are only opened during system initialization at vehicle start up and during ABS operation. The automated bleed procedure opens these secondary circuits and allows any air trapped in these circuits to flow out toward the brake corners.

➡The Auto Bleed Procedure may be terminated at any time during the process by pressing the EXIT button. No further Scan Tool prompts pertaining to the Auto Bleed procedure will be given. After exiting the bleed procedure, relieve bleed pressure and disconnect bleed equipment per manufacturer's instructions. Failure to properly relieve pressure may result in spilled brake fluid causing damage to components and painted surfaces.

1. Before servicing the vehicle, refer to the Precautions Section.

2. Raise and support the vehicle.

3. Remove all 4 tire and wheel assemblies.

4. Inspect the brake system for leaks and visual damage. Repair or replace components as needed.

5. Lower the vehicle.

6. Inspect the battery state of charge.

7. Install a scan tool.

8. Turn the ignition ON, with the engine OFF.

9. With the scan tool, establish communications with the ABS system. Select Special Functions. Select Automated Bleed from the Special Functions menu.

10. Raise and support the vehicle.

11. Following the directions given on the scan tool, pressure bleed the base brake system.

12. Follow the scan tool directions until the desired brake pedal height is achieved.

13. If the bleed procedure is aborted, a malfunction exists. Perform the following steps before resuming the bleed procedure:

    a. If a DTC is detected, diagnose the appropriate DTC.

    b. If the brake pedal feels spongy, perform the conventional brake bleed procedure again.

14. When the desired pedal height is achieved, press the brake pedal to inspect for firmness.

15. Lower the vehicle.

16. Remove the scan tool.

17. Install the tire and wheel assemblies. Tighten the wheel nuts to 100 ft. lbs. (140 Nm).

18. Inspect the brake fluid level.

19. Road test the vehicle while inspecting that the pedal remains high and firm.

20. If the brake pedal feels spongy, repeat the automated bleeding procedure until a firm brake pedal is obtained.

BRAKES

SPEED SENSORS

REMOVAL & INSTALLATION

Front Sensor

See Figure 1.

1. Before servicing the vehicle, refer to the Precautions Section.
2. Raise and support the vehicle.
3. Remove the tire and wheel assembly.
4. Remove the brake rotor.
5. Disconnect the wheel speed sensor electrical connector.
6. Remove the wheel speed sensor bolt.
7. Remove the wheel speed sensor.

To install:

8. Install the wheel speed sensor to the wheel bearing/hub assembly.
9. Install the wheel speed sensor mounting bolt and tighten to 71 inch lbs. (8 Nm).

10. Connect the wheel speed sensor electrical connector.
11. Install the brake rotor.
12. Install the tire and wheel assembly.
13. Lower the vehicle.

Rear Sensor

See Figure 2.

1. Before servicing the vehicle, refer to the Precautions Section.
2. Raise and support the vehicle.
3. Remove the tire and wheel assembly.
4. Remove the parking brake shoes.
5. Disconnect the wheel speed sensor electrical connector.
6. Remove the wheel speed sensor bolt.

7. Remove the wheel speed sensor through the backing plate.

To install:

8. Install the wheel speed sensor through the drum brake backing plate to the wheel bearing/hub assembly.
9. Seat the wheel speed sensor harness grommet into the backing plate.
10. Install the wheel speed sensor mounting bolt and tighten to 71 inch lbs. (8 Nm).
11. Connect the wheel speed sensor electrical connector.
12. Install the parking brake shoes.
13. Install the tire and wheel assembly.
14. Lower the vehicle.

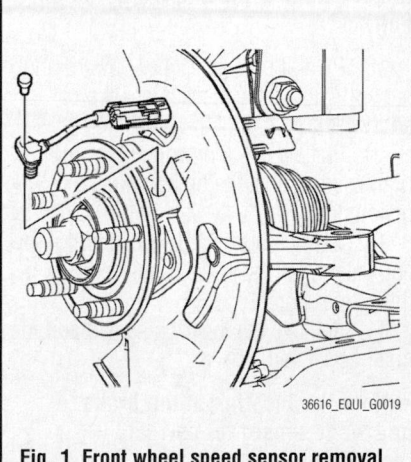

36616_EQUI_G0019

Fig. 1 Front wheel speed sensor removal shown

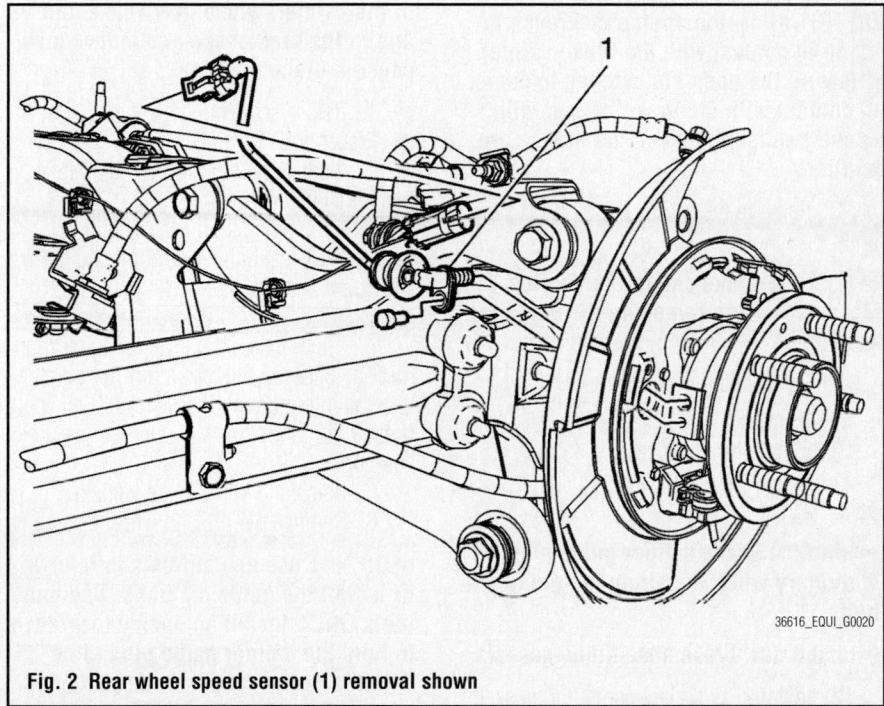

36616_EQUI_G0020

Fig. 2 Rear wheel speed sensor (1) removal shown

BRAKES

BRAKE CALIPER

REMOVAL & INSTALLATION

See Figures 3 and 4.

✳✳ WARNING

Do not allow brake fluid to get on painted surfaces. Painted surfaces will be damaged by brake fluid, flush all affected areas with water immediately if any fluid is spilled.

✳✳ WARNING

During removal, be careful not to damage the brake flexible hose.

➡**Do not reuse the brake flexible hose washers. Otherwise, the brake fluid may leak.**

➡**When the brake pads are removed, visually inspect the caliper for any brake fluid leak. Correct the leaky point, if any.**

1. Before servicing the vehicle, refer to the Precautions Section.

2. Raise and safely support the vehicle.
3. Remove the tire and wheel assembly.
4. Remove the brake hose fitting bolt.
5. Remove the brake hose fitting from the brake caliper.

➡**Do not reuse the brake hose fitting gaskets.**

6. Remove and discard the brake hose fitting gaskets.
7. Cap the brake hose fitting to prevent brake fluid loss and contamination.

➡**DO NOT use any air tools to remove or install the guide pin bolts. Use hand**

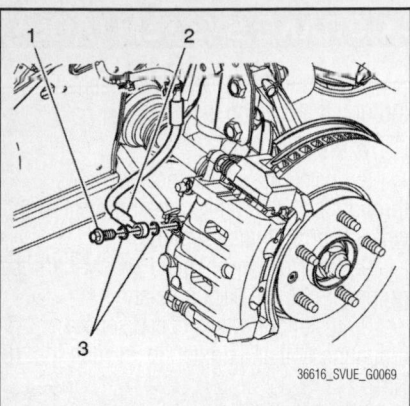

Fig. 3 Remove the brake hose fitting bolt (1)

tools ONLY. Install an open end wrench to hold the caliper guide pin in line with the brake caliper while removing or installing the caliper guide pin bolt. DO NOT allow the open end wrench to come in contact with the brake caliper. Allowing the open end wrench to come in contact with the brake caliper will cause a pulsation when the brakes are applied.

8. Remove the brake caliper guide pin bolts.

➡**Hold the brake caliper guide pins stationary when removing the guide pin bolts.**

9. Remove the brake caliper.

To install:

10. Install the brake caliper.

11. Install the brake caliper guide pin bolts. Tighten the bolts to 20 ft. lbs. (27 Nm).

➡**Hold the brake caliper guide pins stationary when installing the guide pin bolts.**

➡**Install new brake hose fitting gaskets.**

12. Install new brake hose fitting gaskets to the brake hose fitting.

13. Install the brake hose fitting to the brake caliper.

14. Install the brake hose fitting bolt. Tighten the bolt to 38 ft. lbs. (52 Nm).

15. Bleed the hydraulic brake system.

16. Install the tire and wheel assembly.

17. Lower the vehicle.

DISC BRAKE PADS

REMOVAL & INSTALLATION

See Figures 4 through 7.

➡**Support the brake caliper with heavy mechanic wire, or equivalent, whenever it is separated from its mount and the hydraulic flexible brake hose is still connected. Failure to support the caliper in this manner will cause the flexible brake hose to bear the weight of the caliper, which may cause damage to the brake hose and in turn may cause a brake fluid leak.**

1. Before servicing the vehicle, refer to the Precautions Section.

2. Inspect the fluid level in the brake master cylinder reservoir.

3. If the brake fluid level is midway between the maximum-full point and the minimum allowable level, no brake fluid needs to be removed before proceeding.

4. If the brake fluid level is higher than midway between the maximum-full point and the minimum allowable level, remove brake fluid to the midway point before proceeding.

5. Raise and support the vehicle.

6. Remove the tire and wheel assembly.

➡**DO NOT use any air tools to remove or install the guide pin bolts. Use hand tools ONLY. Install an open end wrench to hold the caliper guide pin in line with the brake caliper while removing or installing the caliper guide pin bolt. DO NOT allow the open end wrench to**

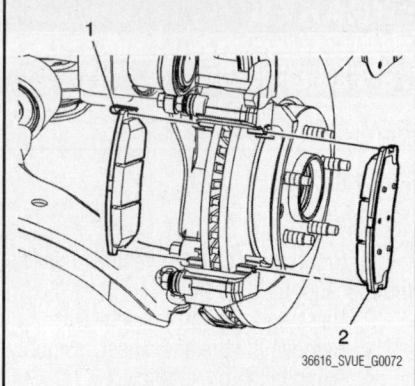

Fig. 6 Remove the inner brake pad (1) and the outer brake pad (2)

come in contact with the brake caliper. Allowing the open end wrench to come in contact with the brake caliper will cause a pulsation when the brakes are applied.

7. Remove the lower brake caliper guide pin bolt.

➡**Hold the brake caliper guide pin stationary when removing the guide pin bolt.**

8. Pivot the brake caliper upward and support with heavy mechanics wire or equivalent.

9. Place a block of wood or an old brake pad against the brake caliper pistons.

10. Using a brake pad spreader tool, or equivalent, fully seat the caliper pistons in the caliper bores.

11. Remove the inner brake pad and the outer brake pad.

➡**Note the location of the brake pad wear sensor for correct installation.**

12. Remove the upper and lower brake pad shims.

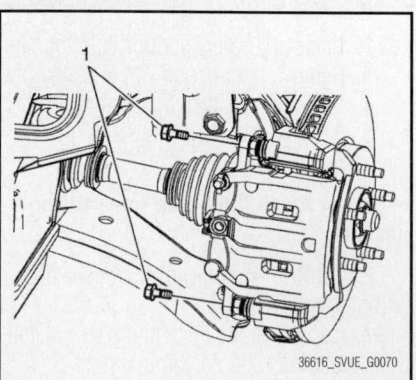

Fig. 4 Remove the brake caliper guide pin bolts (1)

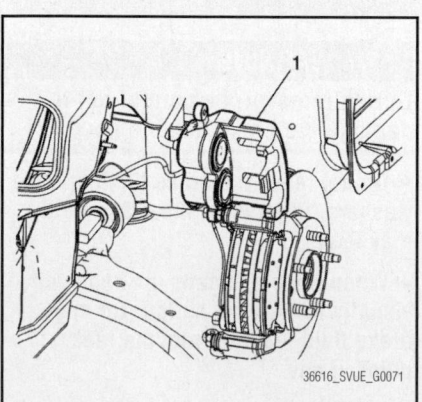

Fig. 5 Pivot the brake caliper (1) upward

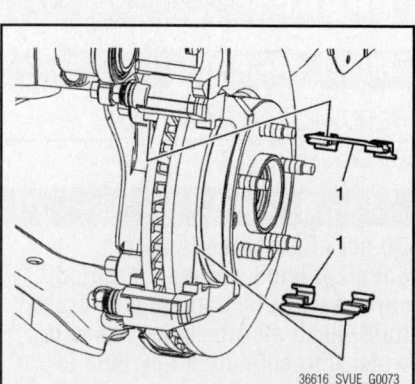

Fig. 7 Remove the upper and lower brake pad shims (1)

➡️**If installing new brake pads, discard the shims.**

To install:

13. Install the upper and lower brake pad shims.

➡️**If installing new brake pads, install new shims.**

14. Install the inner brake pad and the outer brake pad.

➡️**Note the location of the brake pad wear sensor for correct installation.**

15. Pivot the brake caliper into position and install the lower brake caliper guide pin bolt. Tighten the bolt to 20 ft. lbs. (27 Nm).

➡️**Hold the brake caliper guide pin stationary when installing the guide pin bolt.**

16. Install the tire and wheel assembly.
17. Lower the vehicle.
18. With the engine OFF, gradually apply the brake pedal to approximately ⅔ of its travel distance.
19. Slowly release the brake pedal.
20. Wait 15 seconds, then repeat steps until a firm brake pedal is obtained. This will properly seat the brake caliper pistons and brake pads.
21. Fill the master cylinder reservoir to the proper level.
22. Burnish the pads and rotors.

BRAKES

BRAKE CALIPER

REMOVAL & INSTALLATION
See Figure 8.

❋❋ WARNING

Do not allow brake fluid to get on any painted surfaces. The painted surfaces may be damaged by brake fluid. Flush the surface with water immediately if any fluid is spilled.

❋❋ WARNING

Be careful not to damage the brake flexible hose and do not depress the brake pedal.

➡️**Do not reuse the brake flexible hose washers. Otherwise, brake fluid may leak.**

➡️**When the pads are removed, visually inspect the caliper for a brake fluid leak. Correct leaky point, if any.**

1. Before servicing the vehicle, refer to the Precautions Section.
2. Raise and safely support the vehicle.
3. Remove the rear wheel.
4. Remove the brake hose fitting bolt and discard the gasket.

❋❋ WARNING

Cap the brake hose to prevent any contamination of the brake fluid.

5. Remove the brake caliper guide pin bolts.
6. Remove the brake caliper from the caliper bracket.

To install:

7. Thoroughly clean the thread locker residue from the guide pin bolt threads with

REAR DISC BRAKES

denatured alcohol or equivalent and allow to dry.

8. Apply high-temperature, high-strength thread locker to ⅔ of the threaded length of the brake caliper guide pin bolt.
9. Install the brake caliper guide pin bolts to the caliper mounting bracket. Tighten the guide pin bolts to 20 ft. lbs. (27 Nm).
10. Install the brake hoses with NEW gaskets. Tighten the brake hose fitting bolt to 38 ft. lbs. (52 Nm).
11. Install the rear wheel.
12. Bleed the brake system.
13. Lower the vehicle.
14. Fill the master cylinder reservoir to the proper level.

DISC BRAKE PADS

REMOVAL & INSTALLATION
See Figure 9.

1. Before servicing the vehicle, refer to the Precautions Section.
2. Inspect the fluid level in the brake master cylinder auxiliary reservoir.
3. If the brake fluid level is midway between the maximum-full point and the minimum allowable level, no brake fluid needs to be removed from the reservoir before proceeding.
4. If the brake fluid level is higher than midway between the maximum-full point and the minimum allowable level, remove brake fluid to the midway point before proceeding.
5. Raise and safely support the vehicle.
6. Remove the rear wheel.
7. Remove one caliper guide pin bolt.
8. Rotate the caliper up and to the rear until it rests on the mounting bracket and support with heavy mechanics wire or equivalent.

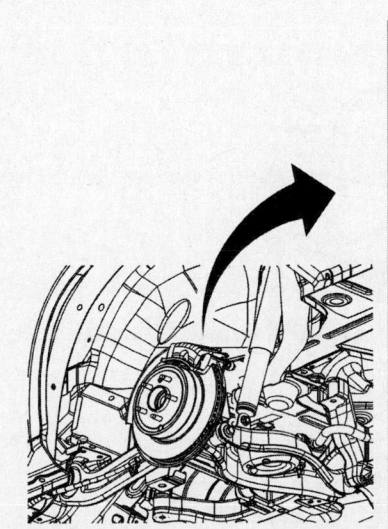
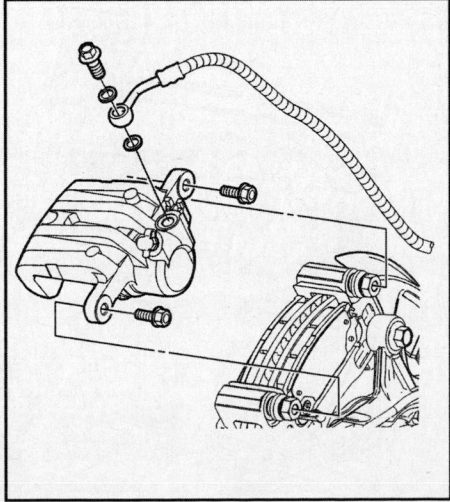

36616_EQUI_G0023

Fig. 8 Rear brake caliper removal shown

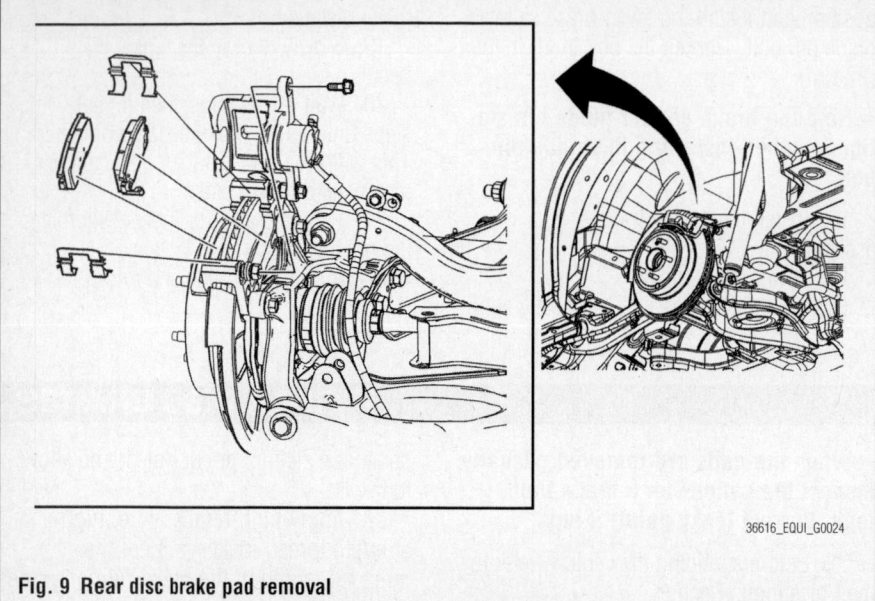

36616_EQUI_G0024

Fig. 9 Rear disc brake pad removal

9. Place a block of wood or an old brake pad against the brake caliper piston.

10. Using a brake pad spreader tool, or equivalent, fully seat the caliper piston in the caliper bore.

11. Remove the brake pads from the caliper. If the original brake pads are being reused, mark the position of the pads and springs. If the brake pads are being replaced, discard the springs.

To install:

12. Install the brake pads with new springs to the caliper. **DO NOT reuse the springs.**

13. Rotate the caliper down and tighten the guide pin bolt to 20 ft. lbs. (27 Nm).

a. If reusing the guide pin bolt, the threads of the guide pin bolt and the threads of the guide pin must be free of thread locker residue and debris prior to the application of a thread locker to ensure proper adhesion and fastener retention.

b. If reusing the guide pin bolt, prepare the guide pin bolt and the guide pin as follows:

- Thoroughly clean the thread locker residue from the guide pin bolt threads with denatured alcohol, or equivalent, and allow to dry
- Thoroughly clean the thread locker residue from the guide pin threads with denatured alcohol, or equivalent, and allow to dry
- Apply high temperature, high strength thread locker to ⅔ of the threaded length of the guide pin bolt. Ensure there are no gaps along the length of the filled area of the guide pin bolt
- Allow the thread locker to cure approximately 10 minutes before installing the guide pin bolt.

14. Install the tire and wheel assembly. Tighten the wheel nuts to 100 ft. lbs. (140 Nm).

15. Lower the vehicle.

16. With the engine OFF, gradually apply the brake pedal approximately ⅔ of its travel distance.

17. Slowly release the brake pedal.

18. Wait 15 seconds, then gradually apply the brake pedal approximately ⅔ of its travel distance again until a firm brake pedal apply is obtained. This will properly seat the brake caliper pistons and brake pads.

19. Fill the master cylinder reservoir to the proper level.

20. Burnish the pads and rotors.

BRAKES   PARKING BRAKE

PARKING BRAKE CABLES

ADJUSTMENT

See Figures 10 and 11.

1. Before servicing the vehicle, refer to the Precautions Section.

2. Remove the front floor console.

3. With the park brake lever in the fully released position, using **ONLY hand tools, loosen the adjusting nut completely to the end of the front cable threaded rod.**

4. Raise the park brake lever 1 detent position.

5. Using **ONLY hand tools, tighten the park brake cable adjusting nut until light to moderate drag is exhibited while rotating the rear wheels.**

6. Attempt to rotate the rear wheels. There should be no rotation forward or rearward.

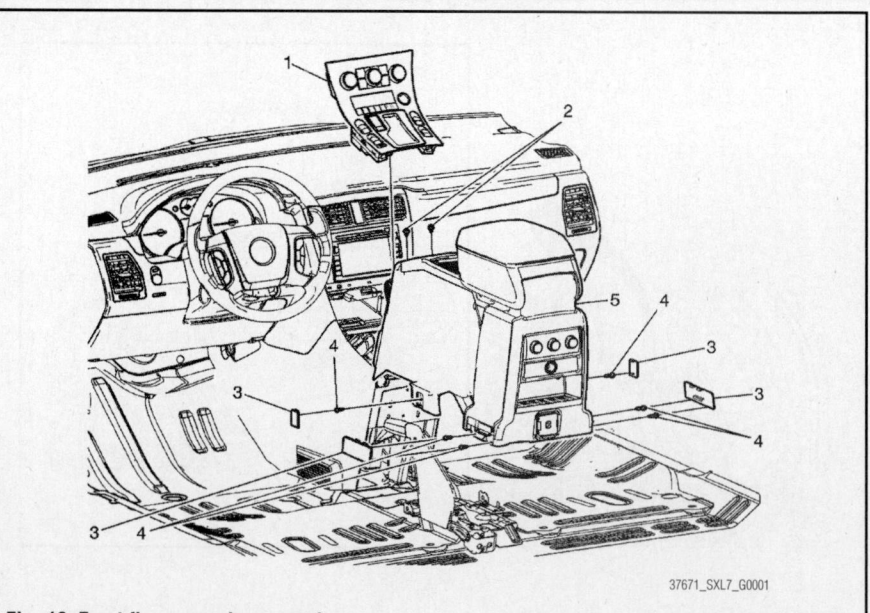

37671_SXL7_G0001

Fig. 10 Front floor console removal

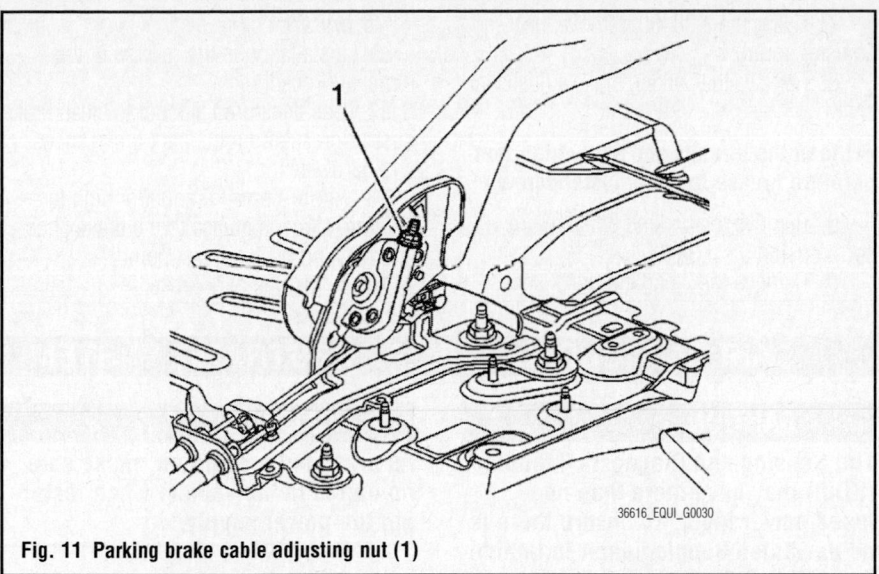

36616_EQUI_G0030

Fig. 11 Parking brake cable adjusting nut (1)

7. Fully release the park brake lever.

8. Verify the park brake is released by rotating the rear wheels. The wheels should rotate freely and exhibit no park brake shoe drag.

9. If the wheels do not rotate freely, repeat the park brake cable adjustment procedure.

10. Raise the park brake lever 3 detent positions and attempt to rotate the rear wheels as follows:
- One of the wheels should not rotate forward or rearward.
- The other wheel should not rotate forward or rearward, or should require substantial effort to rotate.

11. Install the front floor console.

12. Release the park brake lever.

PARKING BRAKE SHOES

REMOVAL & INSTALLATION

See Figure 12.

1. Before servicing the vehicle, refer to the Precautions Section.

2. Raise and safely support the vehicle.

3. Remove the tire and wheel.

4. Remove the brake rotor.

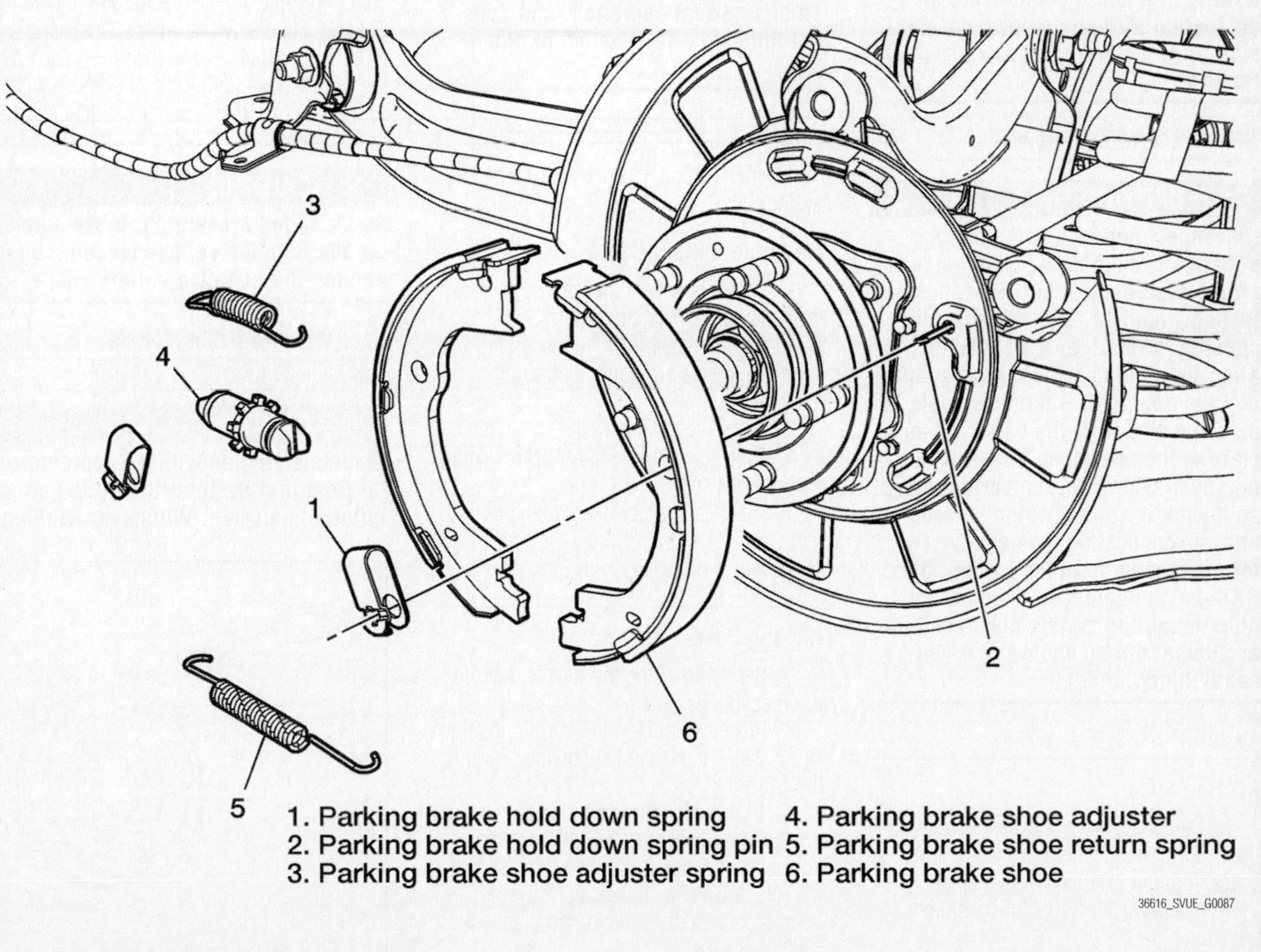

1. Parking brake hold down spring
2. Parking brake hold down spring pin
3. Parking brake shoe adjuster spring
4. Parking brake shoe adjuster
5. Parking brake shoe return spring
6. Parking brake shoe

36616_SVUE_G0087

Fig. 12 Exploded view or parking brake assembly

→ **If reinstalling the park brake shoes, note the location of the park brake shoes for Installation.**

5. Remove the two parking brake hold down springs.

   a. Compress the spring and rotate ¼ turn to release.

6. Remove the two parking brake hold down spring pins.

7. Using the J 38400, remove the adjuster spring.

8. Remove the parking brake adjuster screw.

→ **Clean the threads and apply high temperature grease to the adjuster screw.**

9. Remove the parking brake return spring using the J 38400.

10. Remove the 2 parking brake shoes.

To install:

11. Installation is the reverse of the removal procedure.

12. Use denatured alcohol to clean brake dust or grease from the park brake shoes and hardware.

13. Apply a small amount of high temperature silicone grease to the brake shoe and backing plate contact points.

14. Adjust the park brake.

CHASSIS ELECTRICAL

AIR BAG (SUPPLEMENTAL RESTRAINT SYSTEM)

GENERAL INFORMATION

✳ CAUTION

These vehicles are equipped with an air bag system. The system must be disarmed before performing service on, or around, system components, the steering column, instrument panel components, wiring and sensors. Failure to follow the safety precautions and the disarming procedure could result in accidental air bag deployment, possible injury and unnecessary system repairs.

SERVICE PRECAUTIONS

✳ CAUTION

Disconnect and isolate the battery negative cable before beginning any airbag system component diagnosis, testing, removal, or installation procedures. Wait at least 90 seconds after the ignition switch is turned off and the negative (-) terminal cable is disconnected from the battery before starting the operation. The SRS is equipped with a backup power source, so if work is started within 90 seconds after disconnecting the negative (-) terminal cable from the battery, the SRS may be deployed. Failure to disable the airbag system may result in accidental airbag deployment, personal injury, or death.

DISARMING THE SYSTEM

Air Bag Fuse

1. Before servicing the vehicle, refer to the Precautions Section.

2. Turn the steering wheel so that the vehicles wheels are pointing straight ahead.

3. Place the ignition in the OFF position.

✳✳ CAUTION

The Sensing and Diagnostic Module (SDM) may have more than one fused power input. To ensure there is no unwanted Supplemental Inflatable Restraint (SIR) deployment, personal injury, or unnecessary SIR system repairs, remove all fuses supplying power to the SDM. With all SDM fuses removed and the ignition switch in the ON position, the AIR BAG warning indicator illuminates. This is normal operation, and does not indicate a SIR system malfunction.

4. Locate and remove the fuse(s) supplying power to the SDM.

5. Wait 1 minute before working on the system

Negative Battery Cable

1. Before servicing the vehicle, refer to the Precautions Section.

2. Turn the steering wheel so that the vehicles wheels are pointing straight ahead.

3. Place the ignition in the OFF position.

4. Disconnect the negative battery cable from the battery.

5. Wait 1 minute before working on system.

ARMING THE SYSTEM

Air Bag Fuse

1. Before servicing the vehicle, refer to the Precautions Section.

2. Place the ignition in the OFF position.

3. Install the fuse(s) supplying power to the SDM.

4. Turn the ignition switch to the ON position and verify that AIR BAG warning light flashes 6 times and then turns OFF. If the AIR BAG warning indicator does not operate as described, perform a diagnostic system check.

✳✳ CAUTION

As an added precaution, make sure no one is in the vehicle when restoring the power supply.

Negative Battery Cable

1. Before servicing the vehicle, refer to the Precautions Section.

2. Place the ignition in the OFF position.

3. Connect the negative battery cable to the battery.

4. Turn the ignition switch to the ON position and verify that AIR BAG warning light flashes 6 times and then turns OFF. If the AIR BAG warning indicator does not operate as described, perform a diagnostic system check.

✳✳ CAUTION

As an added precaution, make sure no one is in the vehicle when reconnecting the negative battery cable.

CLOCKSPRING CENTERING
See Figure 13.

✳✳ CAUTION

Models equipped with a Supplemental Restraint System (SRS), use an inflatable air bag. Whenever working

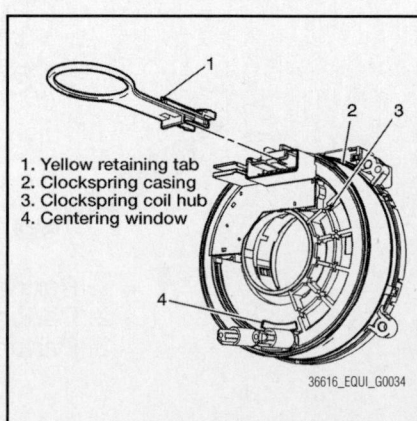

1. Yellow retaining tab
2. Clockspring casing
3. Clockspring coil hub
4. Centering window

36616_EQUI_G0034

Fig. 13 Clockspring components shown

near any of the SRS components, such as the impact sensors, the air bag module, steering column, and instrument panel, disable the SRS. Refer to Disarming The System.

✳✳ WARNING

The contact cable may snap due to steering operation if the cable is not installed in the correct position. With the steering linkage disconnected, the cable may snap by turning the steering wheel beyond the limited number of turns. The contact cable can be turned counterclockwise about 2.5 turns from the neutral position.

✳✳ WARNING

The new Supplemental Inflatable Restraint (SIR) coil assembly will be centered. Improper alignment of the SIR coil assembly may damage the unit, causing an inflatable restraint malfunction.

1. Before servicing the vehicle, refer to the Precautions.
2. Check that the front tires are set at the straight-ahead position.
3. Check that the block tooth and the centering mark of the steering shaft is in the 12 o'clock position.
4. If available, remove the yellow retaining tab from the SIR steering wheel module coil and save the tab for reassembly.

5. Hold the SIR steering wheel module coil face up by the casing.
6. Slowly turn the SIR steering wheel module coil hub clockwise until the coil ribbon stops.
7. Slowly rotate the SIR steering wheel module coil hub counterclockwise 2.5 revolutions until the centering window turns yellow. This indicates the CENTER position.

➡**If the retaining tab is not available, the use of tape to secure the SIR steering wheel module coil is recommended for installation to the steering column.**

8. Install the yellow retaining tab to the SIR steering wheel module coil.
9. Slide the centered SIR steering wheel module coil onto the steering shaft.

DRIVE TRAIN

TRANSFER CASE ASSEMBLY

REMOVAL & INSTALLATION

See Figures 14 and 15.

1. Before servicing the vehicle, refer to the Precautions Section.
2. Raise and support the vehicle.
3. Drain the transfer case fluid.
4. Remove the driveshaft from the vehicle. Refer to Driveshaft, removal & installation.
5. Remove the right halfshaft. Refer to Halfshafts, removal & installation.
6. Remove both catalytic converters.
7. Remove the transfer case mounting bracket bolts.
8. Remove the transfer case mounting bracket.

9. Support the transaxle with a jack stand.
10. Remove the rear transmission mount and bracket.
11. Remove the bolts securing the transfer case to the transaxle.
12. Remove the transfer case from the transaxle.
13. If replacing the transfer case, complete the following steps:
   • Remove the transfer case heat shield bolts
   • Remove the transfer case heat shield

To install:

14. If the transfer case heat shield was previously removed, complete the following steps:
   • Install the transfer case heat shield
   • Install the transfer case heat shield bolts and tighten to 89 inch lbs. (10 Nm).
15. Install the transfer case to the transaxle. Install a jack stand for support.
16. Install the bolts securing the transfer case to the transaxle and tighten the bolts to 37 ft. lbs. (50 Nm).

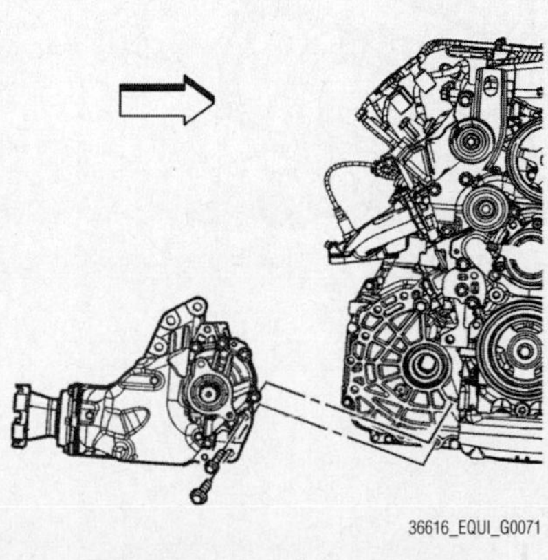

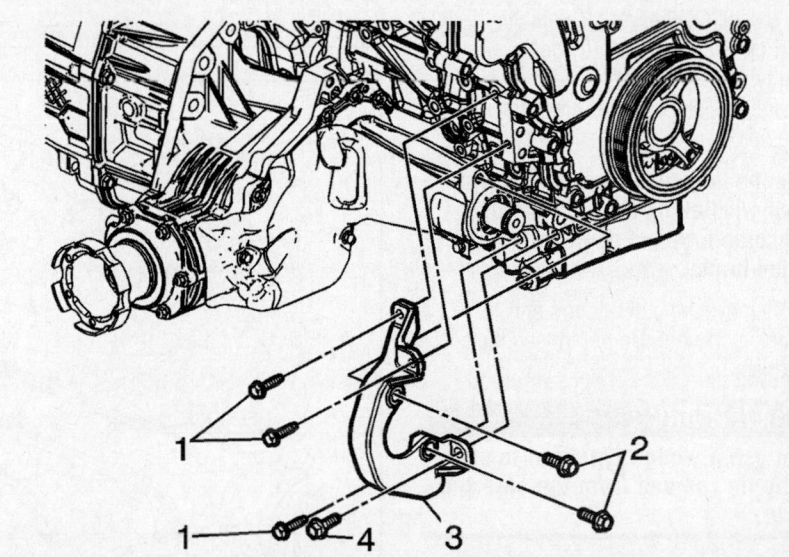

1. Bracket bolts—17 ft. lbs. (23 Nm)
2. Bracket bolts—37 ft. lbs. (50 Nm)
3. Transfer case mounting bracket
4. Bracket bolt—37 ft. lbs. (50 Nm)

36616_EQUI_G0072

Fig. 15 Transfer case mounting bracket bolts tightening sequence

36616_EQUI_G0071

Fig. 14 Transfer case removal—Getrag 760 unit

17. Install the rear transmission mount and bracket.

18. Remove the jack stand supporting the transaxle.

➡**Tighten bolts in the specified sequential order.**

19. Install the transfer case mounting bracket.

20. Install the transfer case mounting bracket bolts and tighten to 17 ft. lbs. (23 Nm) or 37 ft. lbs. (50 Nm) in sequence, as illustrated.

21. Install both catalytic converters.

22. Install the right wheel halfshaft. Refer to Halfshafts, removal & installation.

23. Install the driveshaft to the vehicle. Refer to Driveshaft, removal & installation.

24. Fill the transfer case with the proper type and amount of fluid.

25. Lower the vehicle.

FRONT HALFSHAFTS

REMOVAL & INSTALLATION

See Figures 16 and 17.

Special Tools needed:
- J-2619-01 Slide Hammer
- J-24319-B Steering Linkage and Tie Rod Puller
- J-43828 Ball Joint Separator
- J-44015 Steering Linkage Installer
- J-44394 Seal Protector
- J-45341 Rear Wheel Drive Shaft Removal Tool

1. Before servicing the vehicle, refer to the Precautions Section.

2. Raise and support the vehicle.

3. Remove the tire and wheel assembly.

4. Remove and discard the halfshaft spindle nut.

➡**Hold the ball stud from turning when removing/installing the nut. The boot can become torn and damaged if the ball stud turns.**

5. Remove the outer tie rod end-to-steering knuckle nut. Do not loosen the tie rod end jam nut.

✳✳ WARNING

Do not use a wedge type tool to separate the tie rod end from the steering knuckle.

6. Using a 2-jawed puller, separate the tie rod end from the steering knuckle.

7. Remove and discard the cotter pin from the lower ball joint stud.

8. Remove the ball joint stud nut.

9. Using a ball joint separator, separate the lower ball joint stud from the steering knuckle.

10. Using a backup wrench on the stud, remove the nut securing the lower stabilizer bar link and disengage the link.

11. Disengage the halfshaft spindle from the wheel hub assembly. If necessary, place a wood block against the end of the halfshaft spindle and tap with a hammer to aid removal.

✳✳ WARNING

Use care not to damage the joint seal when removing the halfshaft.

12. Assemble tools J 45341 and J 2619-01, or equivalent to the halfshaft inner tripot joint.

✳✳ WARNING

On vehicles equipped with AWD, the stub shaft may disengage from the Power Takeoff Unit (PTU). If necessary, cap the opening in the PTU to prevent fluid loss.

13. Disengage the halfshaft from the transaxle.

14. Remove the halfshaft from the vehicle.

To install:
15. Install a new halfshaft retaining ring to the output shaft.

16. Install tool J 44394, or equivalent to the halfshaft oil seal.

17. Install the halfshaft to the output shaft as follows:

- Guide the halfshaft tripot joint squarely onto the output shaft.
- After the splined end of the halfshaft passes the oil seal, remove the tool from the oil seal.
- Firmly engage the halfshaft to the output shaft.
- Ensure that the tripot joint is fully seated on the output shaft by grasping the tripot joint and attempting to pull free of the output shaft.

18. Insert the Constant Velocity (CV) joint spindle to the wheel hub/bearing assembly of the steering knuckle.

19. Hand install a new halfshaft spindle nut.

20. Install the lower ball joint stud to the steering knuckle.

21. Install the lower ball joint castle nut to the stud. Tighten the nut to 89 inch lbs. (10 Nm). Tighten the nut an additional 150 degrees.

22. Install the cotter pin to the ball joint stud.

23. If necessary, tighten the nut one additional flat at a time until the castle nut aligns with the hole in the ball joint stud.

24. Secure the cotter pin to the ball joint stud by folding one tine over the end of the ball joint stud. Cut off any excess length of the cotter pin tines.

25. Install the lower link to the stabilizer bar.

26. Install a new nut to the stabilizer bar link stud.

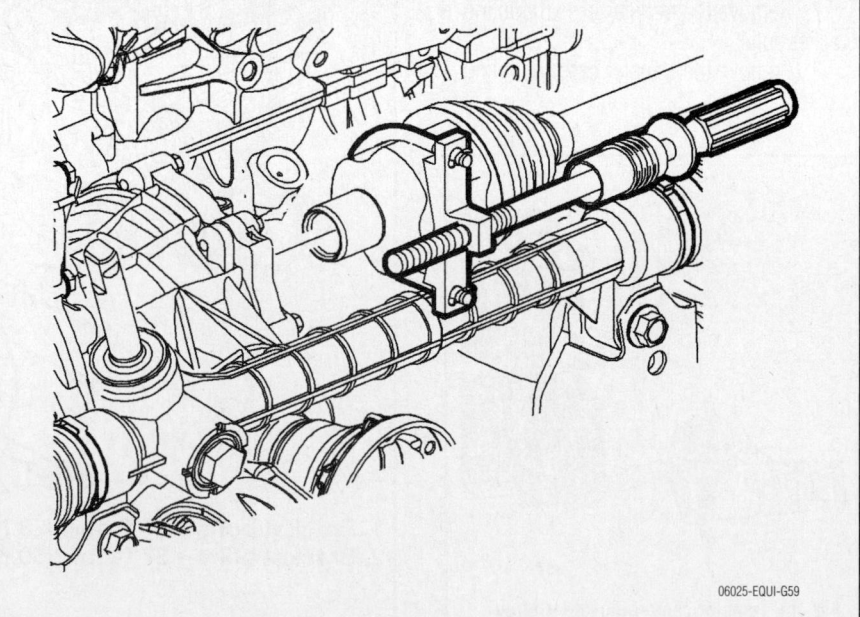

06025-EQUI-G59

Fig. 16 J 45341 and J 2619-01 assembled on the halfshaft

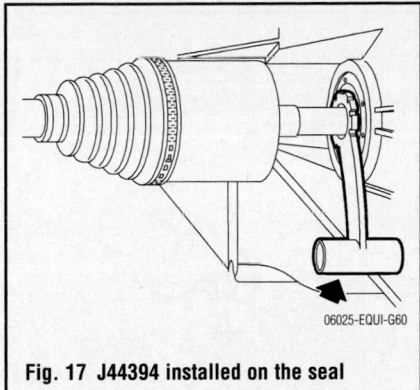

Fig. 17 J44394 installed on the seal

✳✳ WARNING

In order to prevent damaging the stabilizer bar link stud seal, do not allow the stud to rotate while tightening the nut.

27. Use a backup wrench on the stud and tighten the nut. Tighten the nut to 48 ft. lbs. (65 Nm).

28. Install the tie rod end to the steering knuckle. Install a new nut to the tie rod end stud. Tighten the nut to 37 ft. lbs. (50 Nm).

29. Tighten the halfshaft spindle nut to 151 ft. lbs. (205 Nm).

30. Install the tire and wheel assembly. Tighten the wheel nuts to 100 ft. lbs. (140 Nm).

31. Lower the vehicle.

32. Inspect the transaxle fluid level.

CV-BOOTS INSPECTION

1. Before servicing the vehicle, refer to the Precautions Section.

2. Check the halfshaft boots for damage and deterioration:
   - Raise the front of the vehicle
   - Rotate the axle and inspect for cracked or ripped CV-boot material on the inner and outer CV-joints and check for missing clamps. Repeat this step on both sides of the vehicle
   - Inspect for excessive grease deposits on or around the CV-boot

3. Replace the boot if it is damaged or deteriorated.

REAR AXLE SHAFT, BEARING & SEAL

REMOVAL & INSTALLATION

See Figure 18.

1. Before servicing the vehicle, refer to the Precautions Section.

2. Raise and support the vehicle.

3. Drain the rear differential.

4. Remove the exhaust system.

5. Remove the spare tire.

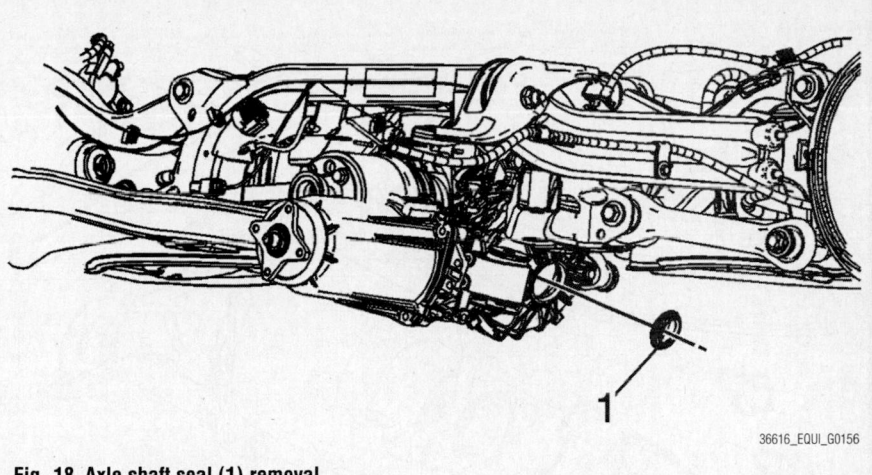

Fig. 18 Axle shaft seal (1) removal

➥In the following service procedure, it is not necessary to completely remove the driveshaft. Relocate the driveshaft to the side and secure with mechanics wire or equivalent.

6. Remove the driveshaft from the rear differential.

7. Remove the rear halfshaft.

8. Support rear differential with a transaxle jack stand.

9. Remove the rear differential support bushing bolt.

10. Remove the differential mount.

11. Remove the differential support bushing nut.

12. Lower the differential to gain access to the axle shaft seal.

13. Using a suitable pry tool, remove the axle shaft seal.

To install:

14. Install the new output shaft seal using a seal installer.

15. Raise and position the differential in the rear cradle.

16. Install the rear support bolt.

17. Install the differential support bushing nut.

18. Install the differential mount.

19. The transaxle jack stand may be removed at this time.

20. Install the rear halfshaft.

21. Install the drive shaft.

22. Install the exhaust system.

23. Install the spare tire.

24. Inspect the rear differential fluid level.

25. Lower the vehicle.

REAR DRIVESHAFT

REMOVAL & INSTALLATION

See Figures 19 through 21.

1. Before servicing the vehicle, refer to the Precautions Section.

2. Place the transaxle in neutral.

3. Raise and support the vehicle.

4. Index mark the relationship of the driveshaft to the rear drive module flange.

5. Support the driveshaft at the rear differential.

6. Remove the mounting bolts for the driveshaft at the rear differential drive flange.

7. Support driveshaft at the transfer case.

8. Remove the mounting bolts for the driveshaft at the transfer case.

9. Place a support under the driveshaft at the center support bearing.

10. Remove the bolts securing the driveshaft support bearing to the vehicle underbody.

11. With the aid of an assistant, remove the driveshaft from the vehicle.

To install:

12. Thoroughly clean the driveshaft flange mounting bolts and apply thread locker, GM P/N 89021297 (Canadian P/N 10953488), or equivalent, to the bolt threads.

13. With the aid of an assistant, position the driveshaft on the supports.

14. Align the reference marks on the front and rear of the driveshaft to the transfer case and rear differential.

15. Position the driveshaft on the transfer case output flange.

16. Finger tighten the mounting bolts for the driveshaft at the transfer case output flange.

17. Position the center support bearing of the driveshaft on the vehicle.

18. Finger tighten the mounting bolts for the center support bearing.

19. Position the driveshaft on the rear differential drive flange.

20. Finger tighten the rear mounting bolts for the driveshaft.

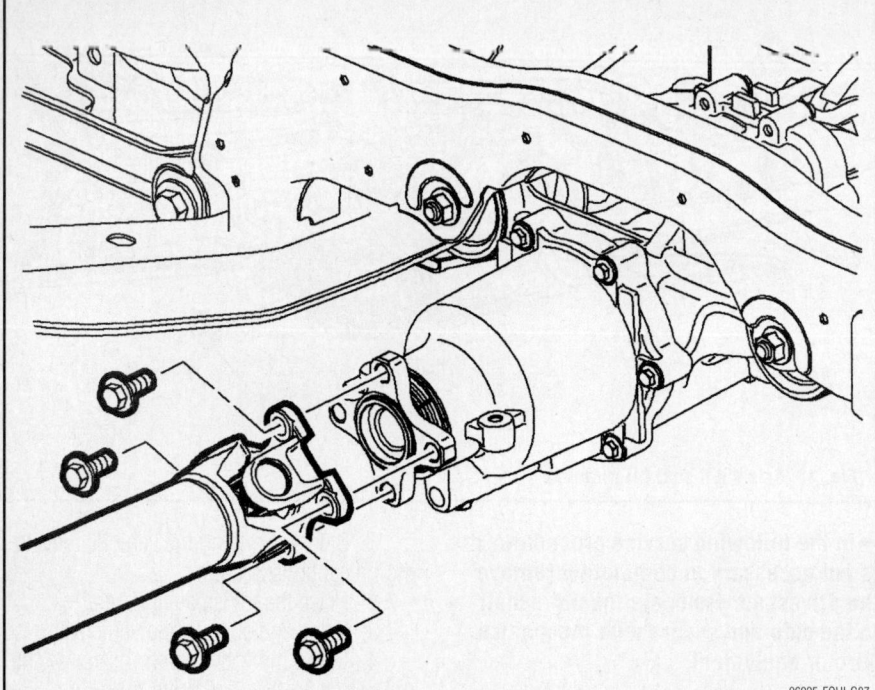

Fig. 19 View of bolts securing the driveshaft yoke flange to the rear drive module flange

06025-EQUI-G87

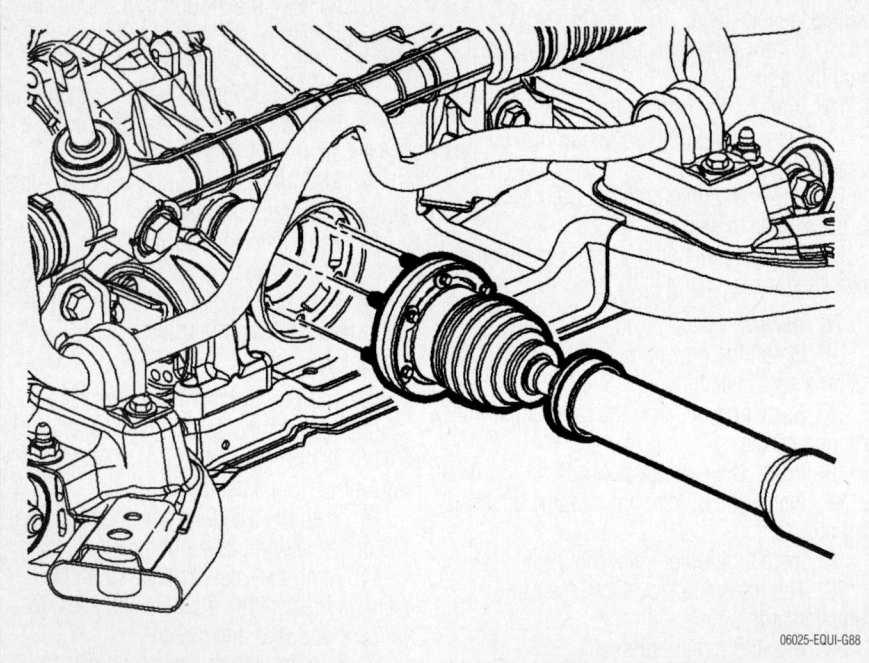

Fig. 20 View of bolts securing the driveshaft to the transfer case

06025-EQUI-G88

21. Tighten the front mounting bolts for the driveshaft to 25 ft. lbs. (35 Nm).

22. Tighten the mounting bolts for the center support bearing to 18 ft. lbs. (25 Nm).

23. Tighten the rear mounting bolts for the driveshaft yoke to rear drive axle to 37 ft. lbs. (50 Nm).

24. Lower the vehicle.

REAR HALFSHAFTS

REMOVAL & INSTALLATION

See Figures 22 through 24.

1. Before servicing the vehicle, refer to the Precautions Section.

2. Raise and safely support the vehicle.

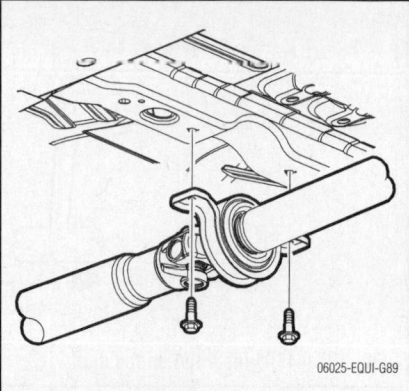

Fig. 21 View of bolts securing the drive-shaft support bearing to the vehicle under-body

06025-EQUI-G89

3. Remove the tire and wheel assembly.

4. Insert a drift or punch into the rotor, against the brake caliper bracket. Using a suitable tool, loosen the wheel halfshaft spindle nut and discard.

5. Using Special Tool J-42129, or suitable wheel hub removal tool, disengage the halfshaft from the wheel hub.

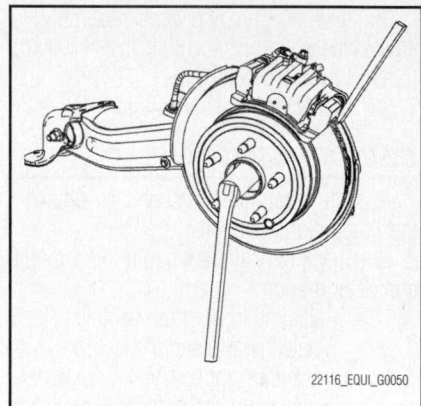

Fig. 22 Insert a drift into the rotor to prevent the wheel from turning while removing the halfshaft spindle nut

22116_EQUI_G0050

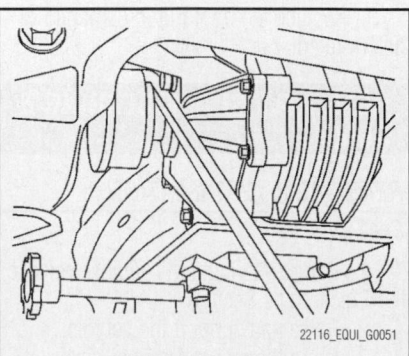

Fig. 23 Carefully release the halfshaft from the RDM using a pry tool

22116_EQUI_G0051

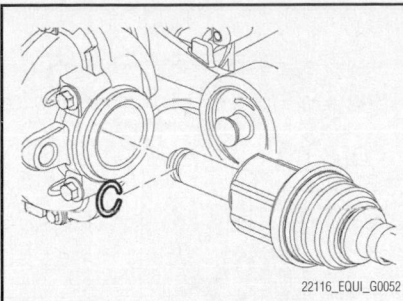

Fig. 24 Before reinstalling the halfshaft, the tripod joint retaining clip must be replaced

6. Remove the rear brake caliper. Refer to Brake Caliper, removal & installation.

7. Remove the wheel bearing/hub assembly.

8. Remove the control arm-to-knuckle mounting bolts.

9. Remove the toe link-to-knuckle bolt.

10. Remove the 3 trailing arm-to-knuckle bolts.

11. Remove the rear suspension knuckle.

12. Using a suitable pry tool, carefully release the halfshaft from the Rear Drive Module (RDM).

➡**Because of the design of the half-shaft seal, the output seal will come out with the halfshaft when removed. Do not re-use the inner halfshaft seal. The seal must be replaced.**

To install:

13. Replace the retaining clip on the tri-pod joint.

14. Install a new halfshaft seal.

➡**When installing the wheel drive shaft, you will notice a slight resistance. This is the wheel drive shaft seal. A snap or click should be heard when the wheel drive shaft is fully seated.**

15. Install the halfshaft.

16. Install the rear suspension knuckle. Install all of the bolts loosely at first. Then tighten the bolts in sequence as follows:
- Tighten the knuckle-to-lower control arm bolt and nut to 118 ft. lbs. (160 Nm).
- Tighten the knuckle-to-upper control arm bolt and nut to 118 ft. lbs. (160 Nm).
- Tighten the knuckle-to-toe link bolt and nut to 118 ft. lbs. (160 Nm).
- Tighten the 3 trailing arm-to-knuckle bolts to 81 ft. lbs. (110 Nm).

17. Install the bearing/hub assembly.

18. Install the brake caliper assembly.

19. Install the a new halfshaft spindle nut. Hand tighten at this time.

20. Insert a drift or punch into the rotor, against the brake caliper bracket. Tighten the wheel halfshaft spindle nut to 151 ft. lbs. (205 Nm).

21. Install the tire/wheel assembly. Tighten the wheel nuts to 100 ft. lbs. (140 Nm).

22. Lower the vehicle.

CV-BOOTS INSPECTION

1. Before servicing the vehicle, refer to the Precautions Section.

2. Check the halfshaft boots for damage and deterioration:
- Raise the front of the vehicle
- Rotate the axle and inspect for cracked or ripped CV-boot material on the inner and outer CV-joints and check for missing clamps. Repeat this step on both sides of the vehicle
- Inspect for excessive grease deposits on or around the CV-boot

3. Replace the boot if it is damaged or deteriorated.

REAR PINION SEAL

REMOVAL & INSTALLATION

See Figures 25 and 26.

1. Before servicing the vehicle, refer to the Precautions Section.

2. Raise and support the vehicle.

3. Remove the differential clutch drum assembly as follows:

a. Remove the driveshaft from the differential clutch drum assembly. Relocate the rear portion of the driveshaft to the side and support it with mechanics wire.

b. Remove the electrical clutch control connector and wire retaining clips.

c. Remove the 4 differential clutch drum bolts.

d. Remove the differential clutch drum.

➡**The pinion flange, dust cover and pinion nut are serviced with the differential clutch drum assembly.**

4. Install a sheet metal screw into the seal.

5. Attach a pair of pliers or a slider hammer to the screw and remove the seal.

6. Using the slide hammer or pliers, remove the seal.

To install:

7. Using the seal installation tool J 44636-1, install the pinion seal.

8. Install the differential clutch drum assembly as follows:

a. Install the differential clutch drum gasket.

b. Install the clutch drum assembly and hand tighten the 4 mounting bolts.

c. Tighten the bolts to 21 ft. lbs. (29 Nm) in a crisscross pattern.

d. Connect the electrical clutch control connector and wire retaining clips.

e. Install the driveshaft to the differential clutch drum assembly, tighten the mounting bolts to 37 ft. lbs. (50 Nm).

f. Inspect the fluid level.

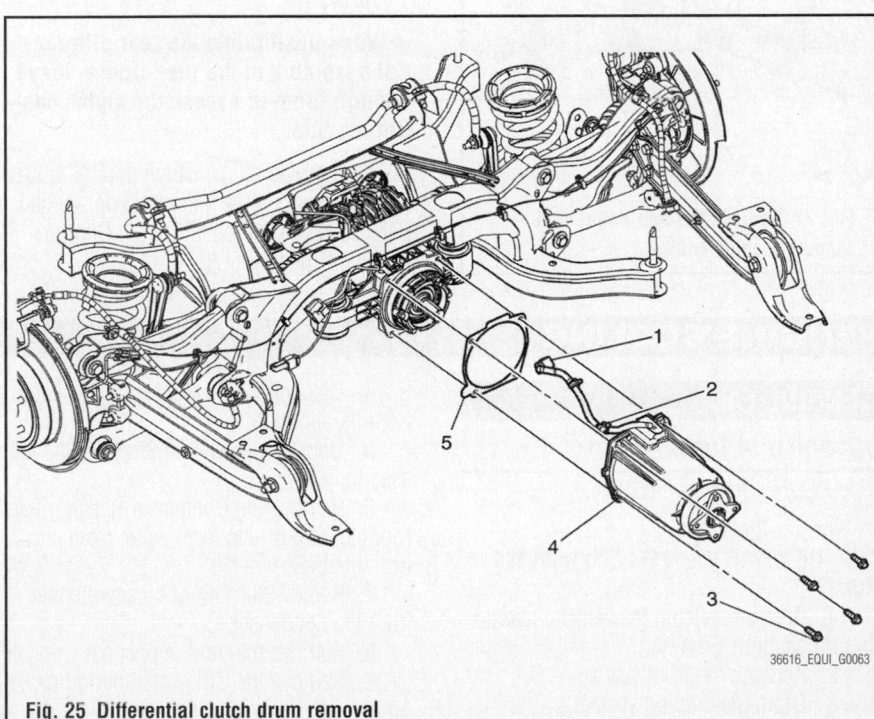

Fig. 25 Differential clutch drum removal

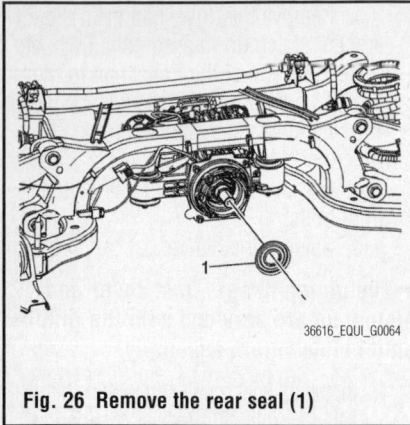

Fig. 26 Remove the rear seal (1)

REAR DIFFERENTIAL

REMOVAL & INSTALLATION

See Figures 27 through 29.

1. Before servicing the vehicle, refer to the Precautions Section.
2. Raise and support the vehicle.
3. Drain the rear differential, if needed.
4. Remove the spare tire.
5. Remove the rear exhaust assembly.
6. Remove the rear halfshafts. Refer to Halfshafts, removal & installation.
7. Remove the driveshaft from the

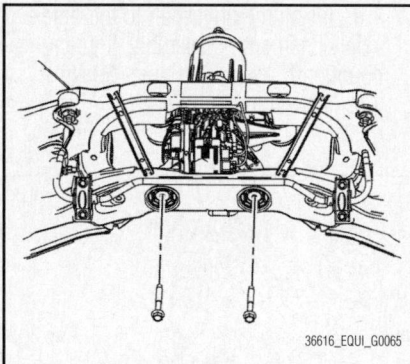

Fig. 27 Remove the rear differential support bushing bolts

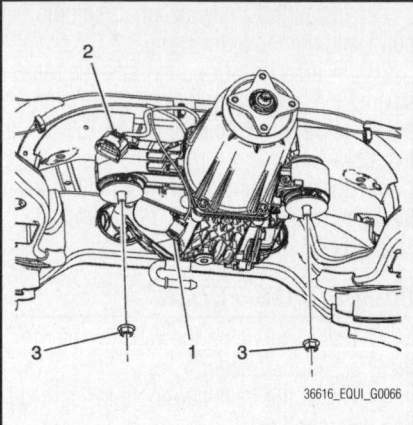

Fig. 28 Remove the electrical connectors (1, 2) and mounting nuts (3)

vehicle. Refer to Driveshaft, removal & installation.

8. Support the rear differential with a transmission jack stand.
9. Remove the rear differential support bushing bolts.
10. Remove the front mounting nuts for the rear differential.
11. Lower the rear differential to gain access to the rear differential clutch control module.
12. Remove the small electrical connector from the control module.
13. Remove the large electrical connector from the control module.
14. Remove the rear differential assembly from the vehicle.

To install:

➡**When positioning the rear differential assembly in the rear cradle, leave enough room to access the clutch control module.**

15. If replacing the clutch control module, the new module must be programmed. Refer to Rear Differential Clutch Control Module Programming.

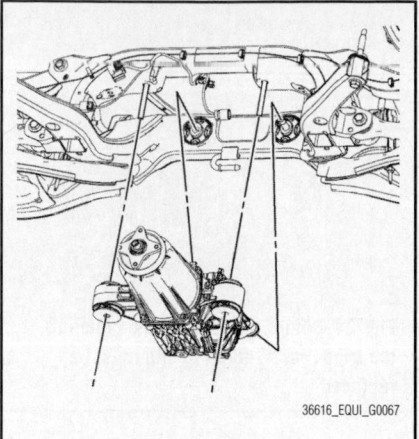

Fig. 29 Rear differential assembly removal

16. Position the rear differential assembly in the rear cradle.
17. Install the small and large electrical connectors.
18. Install the front mounting nuts.
19. Install the rear differential support bushing bolts.
20. Raise the rear differential assembly into place.
21. Tighten the front mounting nuts to 90 ft. lbs. (122 Nm).
22. Tighten the rear differential support bushing bolts to 139 ft. lbs. (188 Nm).
23. Remove the transmission jack stand.
24. Install the driveshaft to the vehicle. Refer to Driveshaft, removal & installation.
25. Install the rear halfshafts. Refer to Halfshafts, removal & installation.
26. Install the rear exhaust assembly. Install a new gasket and tighten the nuts to 18 ft. lbs. (25 Nm).
27. Install the spare tire.
28. Fill the rear differential assembly with fluid, if drained.
29. Lower the vehicle.

ENGINE COOLING

ENGINE FAN

REMOVAL & INSTALLATION

See Figures 30 through 32.

Special Tools:
• GE-47716 Vac N Fill Coolant Refill Tool

1. Before servicing the vehicle, refer to the Precautions Section.
2. Remove the front fascia.
3. Drain the cooling system.

4. Disconnect the electrical connectors from the fan motors.
5. Unclip the wire harness from the fan assembly.
6. Remove the Condenser Radiator Fan Module (CRFM) closeout panel retainers from the condenser.
7. Remove the CRFM closeout panel from the condenser.
8. Remove the front impact bar.
9. Remove the CRFM mounting bracket bolts from the radiator support.

10. Remove the CRFM mounting brackets from the radiator support.
11. Remove the radiator inlet hose clamp from the radiator.
12. Remove the radiator inlet hose from the radiator.
13. Disconnect the upper transaxle cooler line from the radiator.
14. Unclip the transaxle cooler lines from the fan assembly.
15. Lift the CRFM assembly from the lower mounts and carefully move the bot-

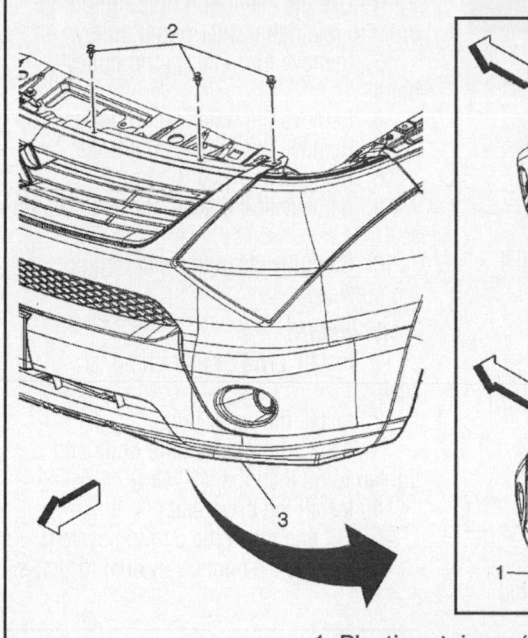

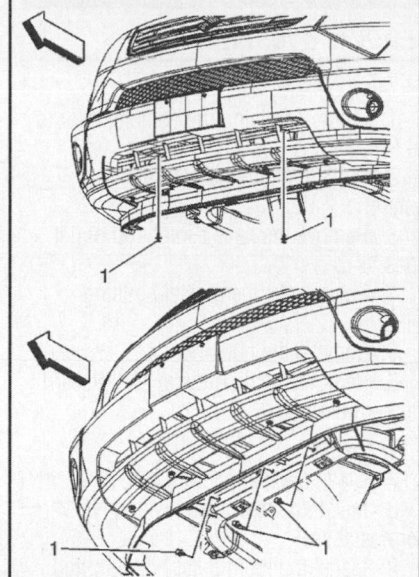

1. Plastic retainers (Qty: 5)
2. Plastic retainers (Qty: 6)
3. Front Bumper Fascia

37671_SXL7_G0007

Fig. 30 View of front fascia

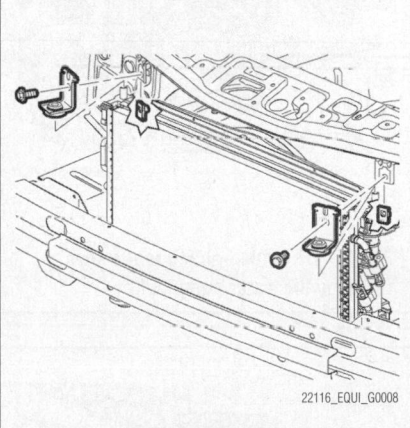

22116_EQUI_G0008

Fig. 31 Remove the CRFM mounting bracket bolts from the radiator support

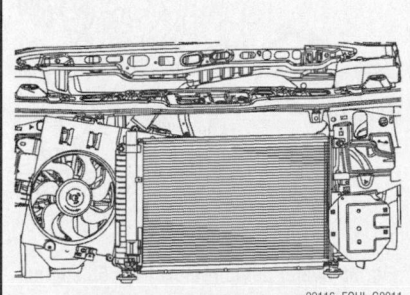

22116_EQUI_G0011

Fig. 32 Removing the fan assembly from the radiator

tom of the assembly rearward while tilting the top forward.

16. Remove the fan assembly bolts from the radiator.

17. Remove the fan assembly from the radiator.

To install:

18. Install the fan to the motor.

19. Align the scribe marks previously made on the fan hub and the motor shaft.

20. Install a new fan retaining clip to the motor shaft. Ensure the retaining clip is fully seated.

21. Install the fan assembly to the radiator by guiding the lower tabs into the corresponding hooks on the radiator.

22. Install the fan assembly bolts to the radiator and tighten to 80 inch lbs. (9 Nm).

23. Position the CRFM assembly onto the lower mounts.

24. Install the radiator inlet hose to the radiator.

25. Install the radiator inlet hose clamp to the radiator.

26. Clip the transaxle cooler lines to the fan assembly.

27. Connect the upper transaxle cooler line to the radiator.

28. Install the CRFM mounting brackets to the radiator support.

29. Install the CRFM mounting bracket bolts to the radiator support and tighten to 15 ft. lbs. (20 Nm).

30. Install the front impact bar.

31. Install the CRFM closeout panel to the condenser.

32. Install the CRFM closeout panel retainers to the condenser.

33. Clip the transaxle cooler lines to the fan assembly.

34. Clip the engine wire harness to fan assembly.

35. Install the electrical connectors to the fan motors.

36. Install the front fascia.

37. Refill and bleed the cooling system.

RADIATOR

REMOVAL & INSTALLATION

See Figures 33 through 35.

1. Before servicing the vehicle, refer to the Precautions Section.

2. Drain the cooling system.

3. Raise and support the vehicle.

4. Remove the front bumper air deflector, if equipped.

5. Remove the front fascia.

6. Remove the front energy absorber.

7. Remove the front bumper impact bar.

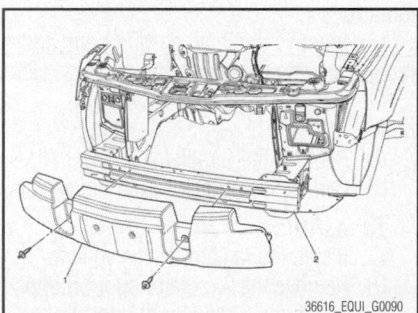

36616_EQUI_G0090

Fig. 33 Remove the energy absorber (1) and impact bar (2)

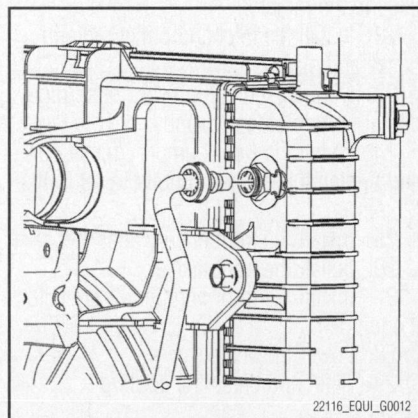

22116_EQUI_G0012

Fig. 34 Disconnect the transaxle cooler lines from the radiator.

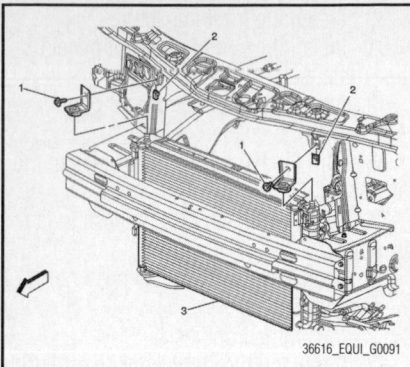

Fig. 35 Remove the support bolts (1) clips (2) and radiator (3)

8. Remove radiator closeout panel.
9. Disconnect the compressor hose/pipe clip at bottom of fan shroud.
10. Remove mounting bolts from condenser using care with the upper left bolt to clear the compressor hose/pipe, reposition and support the condenser.
11. Remove the inlet and outlet radiator hoses.
12. Disconnect the transmission oil cooler lines from the attaching clips on radiator.
13. Remove the 2 fan shroud bolts and reposition the fan shroud.
14. Remove the 2 radiator support bolts.
15. Remove the 2 radiator support clips.
16. Remove the radiator from the vehicle.

To install:
17. Install the radiator to the vehicle.
18. Remove the 2 radiator support clips.
19. Install the 2 radiator support bolts and tighten to 15 ft. lbs. (20 Nm).
20. Reposition the fan shroud, install the bolts and tighten to 80 inch lbs. (9 Nm).
21. Connect the transmission oil cooler lines to the radiator.
22. Install the inlet and outlet radiator hoses and clamps
23. Connect the compressor hose/pipe clip at bottom of fan shroud.
24. Install the front bumper impact bar and tighten the mounting bolts to 18 ft. lbs. (25 Nm).
25. Install the front energy absorber.
26. Install the front fascia.
27. Install the front bumper air deflector, if equipped.
28. Lower the vehicle.
29. Refill and bleed the cooling system.

THERMOSTAT

REMOVAL & INSTALLATION
See Figure 36.

1. Before servicing the vehicle, refer to the Precautions Section.
2. Partially drain the cooling system.
3. Remove the heater inlet and outlet pipes.
4. Remove the thermostat housing bolts.
5. Remove the housing.
6. Remove the thermostat and discard the thermostat seal.

To install:
7. Clean the mating surfaces.
8. Install the thermostat with a NEW thermostat seal.
9. Install the thermostat housing and bolts, tighten to 89 inch lbs. (10 Nm).
10. Install the heater inlet and outlet pipes.
11. Refill and bleed the cooling system.

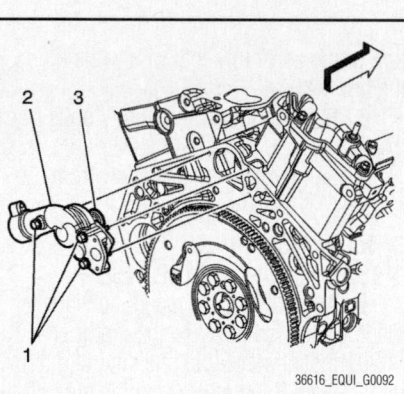

Fig. 36 Remove the housing bolts (1) housing (2) and the thermostat (3)

WATER PUMP

REMOVAL & INSTALLATION
See Figures 37 and 38.

Special Tool:
• EN 46104 Water Pump Pulley Holding Tool

1. Before servicing the vehicle, refer to the Precautions Section.
2. Drain the cooling system until the coolant is below the level of the water pump.
3. Remove the drive belt. Refer to Accessory Drive Belt, removal & installation.

4. Use the EN 46104 holding tool in order to retain the water pump pulley.
5. Remove the water pump pulley bolts.
6. Remove the water pump pulley.
7. Remove the water pump bolts.
8. Remove the water pump.
9. Remove and discard the water pump seal.
10. Carefully clean the water pump sealing surfaces.

To install:
11. Install a NEW water pump seal.
12. Install the water pump.
13. Install the water pump bolts and tighten to 89 inch lbs. (10 Nm)
14. Install the drive belt
15. Fill and bleed the cooling system.
16. Inspect the cooling system for leaks.

Fig. 37 EN 46104 holding tool shown retaining the water pump pulley

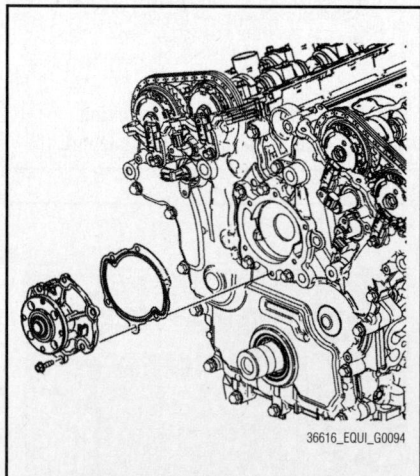

Fig. 38 Water pump removal—3.6L engine

ENGINE ELECTRICAL   **CHARGING SYSTEM**

ALTERNATOR

REMOVAL & INSTALLATION

See Figure 39.

1. Before servicing the vehicle, refer to the Precautions Section.
2. Disconnect the negative battery cable.
3. Remove the accessory drive belt.
4. Reposition the engine wiring harness boot.
5. Remove the alternator terminal nut.
6. Remove the engine wiring harness terminal lead from the alternator.
7. Disconnect the engine wiring harness electrical connector from the alternator.
8. Remove the idler pulley mounting bolt.
9. Remove the idler pulley.
10. Remove the alternator mounting bolts.
11. Remove the alternator from the vehicle.

To install:

12. Position the alternator to the engine.
13. Loosely install the engine bolts.
14. Install the idler pulley and tighten the mounting bolt to 37 ft. lbs. (50 Nm).
15. Tighten the alternating mounting bolts to 37 ft. lbs. (50 Nm).
16. Connect the engine wiring harness electrical connector to the generator.
17. Install the engine wiring harness terminal lead to the generator.

18. Install the generator terminal nut and tighten to 15 ft. lbs. (20 Nm).
19. Reposition the engine wiring harness boot.

20. Install the accessory drive belt. Refer to Accessory Drive Belt, removal & installation.
21. Connect the negative battery cable.

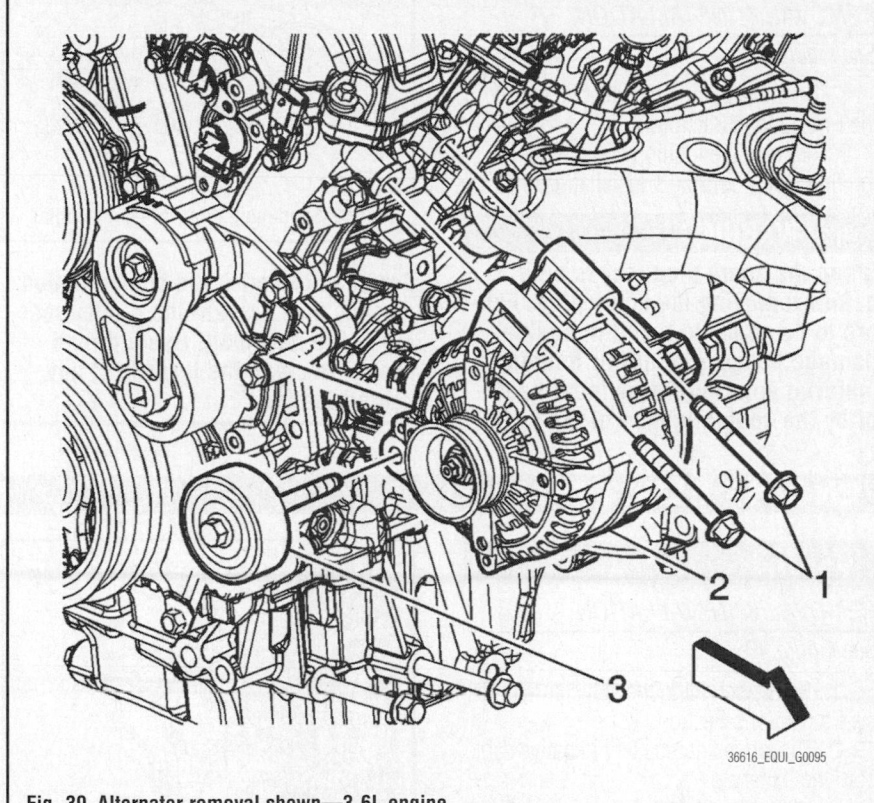

36616_EQUI_G0095

Fig. 39 Alternator removal shown—3.6L engine

ENGINE ELECTRICAL   **IGNITION SYSTEM**

FIRING ORDER

See Figure 40.

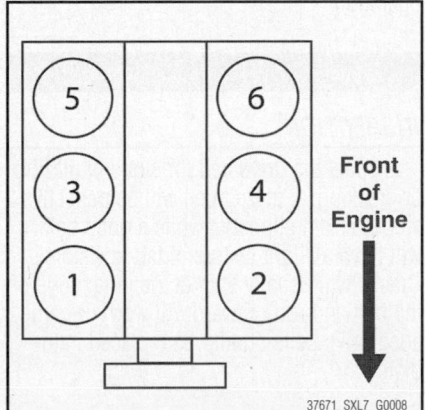

37671_SXL7_G0008

Front of Engine

Fig. 40 3.6L Engine Firing order: 1–2–3–4–5–6

IGNITION COIL

REMOVAL & INSTALLATION

See Figure 41.

1. Before servicing the vehicle, refer to the Precautions Section.
2. Remove the fuel injector sight shield.
3. Remove the air cleaner outlet duct for coil removal (Bank 2).
4. Disconnect the negative battery cable.
5. Disconnect the engine wiring harness electrical connectors from the ignition coils.
6. Remove the ignition coil bolts.
7. Remove the ignition coils.

To install:

8. Install the ignition coils and tighten the bolts to 89 inch lbs. (10 Nm).
9. Connect the engine wiring harness electrical connectors to the ignition coils.

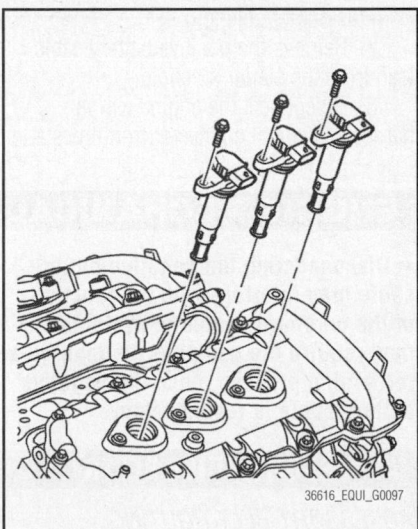

36616_EQUI_G0097

Fig. 41 Bank 1 ignition coil removal shown—3.6L engine

10. Install the fuel injector sight shield.

11. Install the air cleaner outlet duct for coils (Bank 2).

12. Connect the negative battery cable.

SPARK PLUGS

REMOVAL & INSTALLATION

See Figure 42.

1. Before servicing the vehicle, refer to the Precautions Section.

2. Remove the ignition coils. Refer to Ignition Coil, removal & installation.

✳✳ WARNING

Clean the spark plug recess area before removing the spark plug. Failure to do so could result in engine damage because of dirt or foreign material entering the cylinder head, or by the contamination of the cylin-

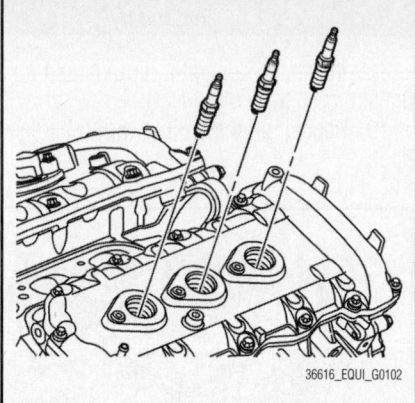

Fig. 42 Spark plug removal—3.6L engine

der head threads. The contaminated threads may prevent the proper seating of the new plug. Use a thread chaser to clean the threads of any contamination.

3. Use compressed air in order to remove debris from the spark plug cavity.

4. Remove the spark plugs.

To install:

✳✳ WARNING

Use only the spark plugs specified for use in the vehicle. Do not install spark plugs that are either hotter or colder than those specified for the vehicle. Installing spark plugs of another type can severely damage the engine.

5. Ensure that the spark plug gap is equivalent to the spark plug gap specification: 0.044 inch.

6. Install the spark plug and tighten to 15 ft. lbs. (20 Nm).

7. Install the ignition coil and tighten the retaining bolt to 89 inch lbs. (10 Nm).

ENGINE ELECTRICAL

STARTER

REMOVAL & INSTALLATION

See Figure 43.

1. Before servicing the vehicle, refer to the Precautions Section.

2. Disconnect the battery negative cable from the battery.

3. Raise and suitably support the vehicle.

4. Remove the front catalytic converter.

5. Remove the positive battery cable terminal-to-starter solenoid nut.

6. Remove the alternator wiring lead from the starter solenoid.

7. Remove the positive battery cable lead from the starter solenoid.

8. Disconnect the engine wiring harness electrical connector from the starter.

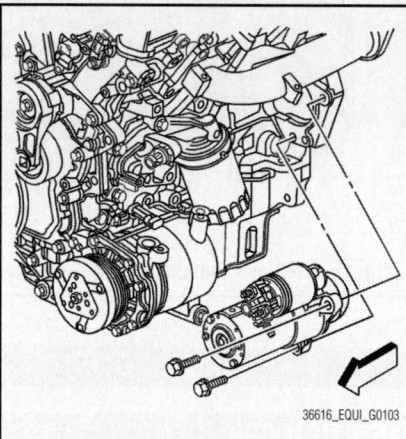

Fig. 43 Starter removal—3.6L engine

9. Remove the starter bolts and the starter.

STARTING SYSTEM

To install:

10. Position the starter motor in the engine block.

11. Install the starter bolts and tighten to 37 ft. lbs. (50 Nm).

12. Connect the engine wiring harness electrical connector to the starter.

13. Install the positive battery cable lead to the starter solenoid.

14. Install the generator wiring lead to the starter solenoid.

15. Install the positive battery cable terminal to starter solenoid nut and tighten to 89 inch lbs. (10 Nm).

16. Install the front catalytic converter.

17. Lower the vehicle.

18. Connect the negative battery cable.

19. Check the starter for the correct operation.

ENGINE MECHANICAL

➡Disconnecting the negative battery cable may interfere with the functions of the on board computer systems and may require the computer to undergo a relearning process, once the negative battery cable is reconnected.

ACCESSORY DRIVE BELTS

ACCESSORY BELT ROUTING

See Figure 44.

Refer to the accompanying illustration.

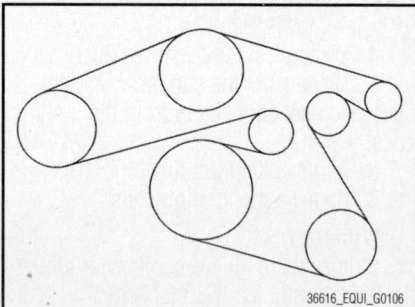

Fig. 44 Accessory drive belt routing— 3.6L engine

INSPECTION

Inspect the drive belt for signs of glazing or cracking. A glazed belt will be perfectly smooth from slippage, while a good belt will have a slight texture of fabric visible. Cracks will usually start at the inner edge of the belt and run outward. All worn or damaged drive belts should be replaced immediately.

ADJUSTMENT

The engine accessory drive belt uses an auto-tensioner. No adjustment is necessary.

REMOVAL & INSTALLATION

See Figures 45 through 48.

1. Before servicing the vehicle, refer to the Precautions Section.

2. Remove the air cleaner assembly.

3. Install the engine support fixture.

4. Remove the engine mount strut bracket.

   a. Remove the air cleaner assembly.

   b. Remove the engine mount strut-to-bracket bolts.

   c. Rotate the strut vertical.

   d. Remove the engine mount strut bracket-to-cylinder head bolts.

5. Raise and support the vehicle.

6. Remove the engine splash shield.

7. Remove the right side engine mount bracket.

8. Lower the vehicle.

9. Rotate the drive belt tensioner clockwise to release the drive belt tension.

10. Slide the drive belt off of the belt idler pulley.

11. Slowly release the drive belt tensioner.

12. Remove the drive belt from the accessory drive pulleys.

To install:

13. Install the drive belt to the crankshaft pulley, the tensioner, and the alternator.

14. Rotate the drive belt tensioner clockwise.

15. Install the drive belt to the idler pulley.

➡**Ensure the drive belt is properly aligned and seated into the grooves of the accessory drive pulleys.**

16. Slowly release the drive belt tensioner.

17. Raise the vehicle.

18. Partially thread the lower right side engine mount bracket bolt in place.

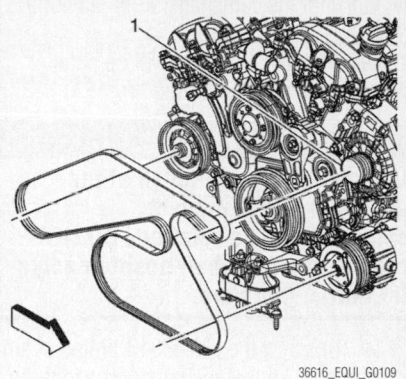

Fig. 48 Accessory drive belt removal—3.6L engine

19. Install the right side engine mount bracket as follows:

- Install the engine mount and the lower bracket to the vehicle. Tighten the lower bracket bolts to 37 ft. lbs. (50 Nm).
- Install the upper engine mount bracket nuts and tighten to 37 ft. lbs. (50 Nm).
- Install the engine mount upper bracket and bolt to the engine. Do not tighten the bolt.
- Tighten the right upper engine mount bracket bolt to 37 ft. lbs. (50 Nm).
- Install the upper engine mount bracket bolts and tighten to 81 ft. lbs. (110 Nm).

20. Install the engine splash shield.

21. Lower the vehicle.

22. Remove the engine support.

23. Install the air cleaner assembly.

CAMSHAFT AND VALVE LIFTERS

REMOVAL & INSTALLATION

Left Camshaft & Lifters

See Figures 49 through 55.

1. Before servicing the vehicle, refer to the Precautions Section.

2. Disconnect the negative battery cable.

3. Remove the lower intake manifold.

4. Remove the left bank camshaft cover.

5. Remove the camshaft sensors.

6. Remove the camshaft position actuator solenoid.

7. Remove the crankshaft balancer. Refer to Crankshaft Damper, removal & installation.

8. Rotate the crankshaft with the EN

Fig. 45 View the engine support fixture (1) installed

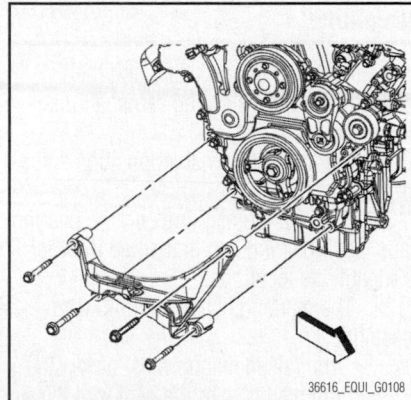

Fig. 47 Right side engine mount bracket removal—3.6L engine

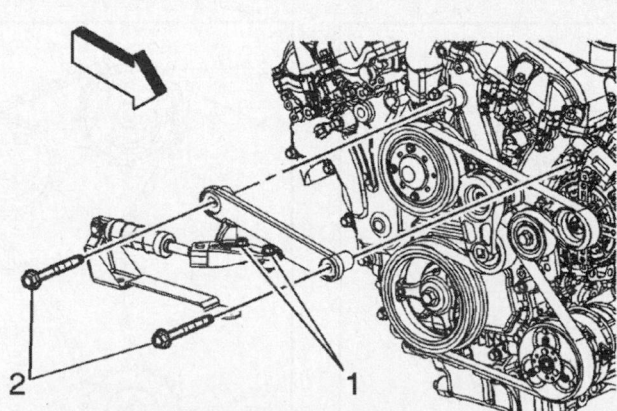

1. Engine mount strut-to-bracket bolts
2. Engine mount strut bracket-to-cylinder head bolts

Fig. 46 Removing the engine mount strut bracket

46111 until the camshafts are in a neutral (low tension) position.

9. The camshaft flats will be parallel with the camshaft cover rail.

✵ WARNING

Use an open-end wrench at the camshaft hex to prevent camshaft/engine rotation. DO NOT remove the camshaft position actuator bolt at this time.

10. Unscrew the EN-48313 timing chain holding tool so that the legs of the tool are retracted.

11. Insert the EN-48313 between the camshaft actuators, rearward of the timing chain until the bottom line that is scribed in the body of the tool is adjacent to the top surface of the cylinder head. This is the approximate installed position.

12. Ensure that the feet on the legs of the tool are facing the front of the engine.

13. Partially expand the legs of the EN-48313 by turning the T-shaped handle clockwise.

14. Insert the leg of the tool behind the timing chain guide.

15. Continue expanding the EN-48313 until the legs contact the timing chain. Do not tighten at this time.

➡**Ensure that the foot of the EN-48313 is engaged into one of the link pockets to prevent tool slippage during tightening of the EN-48313.**

16. Hand tighten the EN-48313.

17. Use an open end wrench on the hex cast into the left intake and exhaust camshafts and rotate the camshafts toward each other in order to create slack in the chain between the actuators.

18. The EN-48313 is now properly installed to hold the timing chain in position.

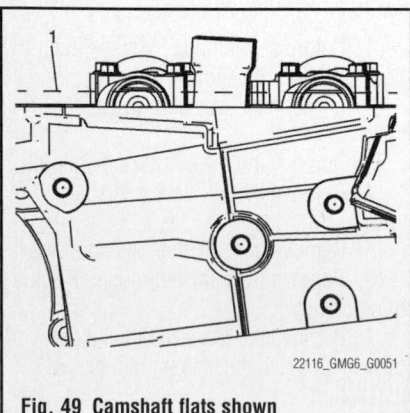

Fig. 49 Camshaft flats shown

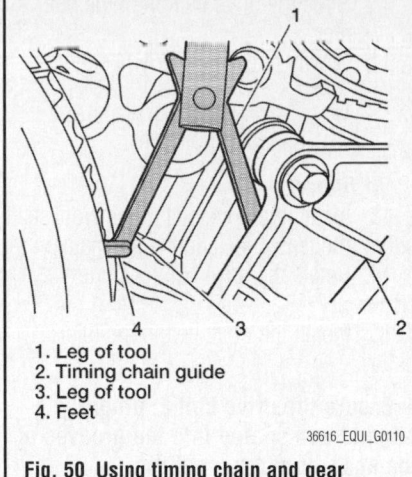

1. Leg of tool
2. Timing chain guide
3. Leg of tool
4. Feet

Fig. 50 Using timing chain and gear holding tool EN-48313

✵ WARNING

Ensure that the camshaft timing chain and the camshaft position actuators are marked for proper assembly.

19. Mark the timing chain and the respective locations on the camshaft position actuators.

20. Remove the camshaft position actuator bolt.

21. Observe the markings on the bearing caps. Each bearing cap is marked in order to identify its location.

22. The markings have the following meanings:

- The raised feature must always be oriented toward the center of the cylinder head.
- The I indicates the intake camshaft.

Fig. 51 Another view of the timing chain and gear holding tool EN-48313 properly installed

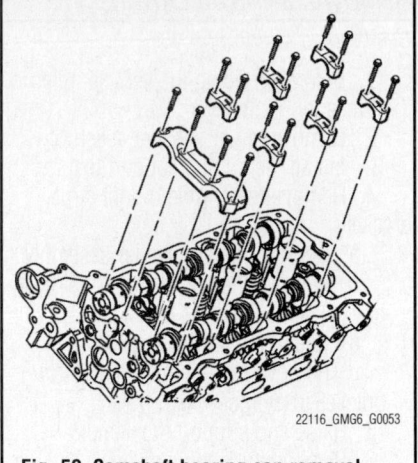

Fig. 52 Camshaft bearing cap removal— 3.6L engine

- The E indicates the exhaust camshaft
- The number indicates the journal position from the front of the engine

23. Remove the camshaft bearing cap bolts.

24. Remove the camshaft bearing caps.

25. Remove the camshafts.

26. Remove the rocker arms.

27. Remove the lifters.

28. Replace the camshaft bearing caps and bolts.

To install:

✵ WARNING

Ensure that the marks on the camshaft position actuator and the timing chain are aligned. DO NOT tighten the camshaft position actuator bolt at this time.

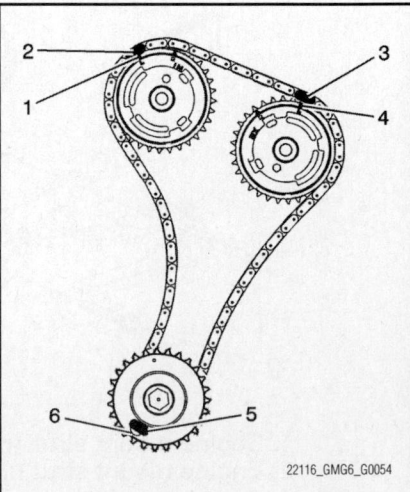

Fig. 53 Timing chain alignment marks— left bank

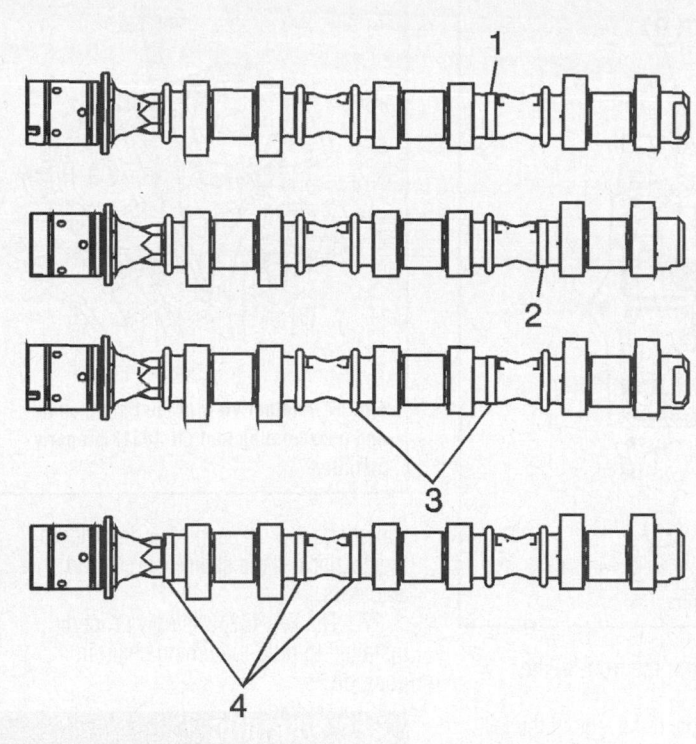

1. The number 4 identification ring for the left intake camshaft is machined off (1) - Third Design, Camshaft Timing Drive System.

2. The number 5 identification ring for the left exhaust camshaft is machined off (2) - Third Design and Fourth, except High Output, Camshaft Timing Drive System.

3. The number 3 and 4 identification rings for the left intake camshaft is machined off (3) - Fourth Design, Camshaft Timing Drive System.

4. The number 1, 2 and 3 identification rings for the left exhaust camshaft is machined off (4) - Fourth Design High Output, Camshaft Timing Drive System.

22116_GMG6_G0055

Fig. 54 Camshaft design identification—3.6L engine

29. Locate the camshafts to the cylinder head and assemble the camshaft actuators to the camshafts.

30. Ensure that the crankshaft is in the stage one timing drive assembly position using the EN 46111.

31. Ensure that the camshaft sealing rings are in place in the camshaft grooves. Camshaft sealing rings must be in place below the surface of the camshaft journal in order to avoid being pinched between the cylinder head and the camshaft caps.

➡**Ensure each valve lifter is filled with clean engine oil and the valve lifter does not tip over (plunger down) before the installation of the valve lifters. The loss of oil in the valve lifter lower pressure chamber or the dry stroking/cycling of the valve lifter plunger will allow air to travel into the high pressure chamber of the valve lifter. Air in the high pressure chamber of the valve lifter may not be purged causing extensive engine component damage.**

32. Install valve lifters.
33. Install rocker arms.
34. Apply a liberal amount of lubricant to the camshaft journals and the left cylinder head camshaft carriers.

35. Place the left intake and left exhaust camshafts in position in the left cylinder head.

36. Position the camshaft lobes in a neutral position with the flats on the back of the camshafts up and parallel with the left cylinder head camshaft cover rail.

37. Observe the markings on the left cylinder head camshaft bearing caps. Each bearing cap is marked in order to identify its location.

38. The markings have the following meanings:
   • The raised feature must always be oriented toward the center of the cylinder head.
   • The I indicates the intake camshaft.
   • The E indicates the exhaust camshaft
   • The number indicates the journal position from the front of the engine

39. Apply a liberal amount of lubricant to the camshaft bearing caps.

40. Install the camshaft bearing thrust cap in the first journal of the left cylinder head.

41. Install the remaining bearing caps with their orientation mark toward the center of the cylinder head.

42. Hand start all the camshaft bearing cap bolts.

43. Tighten the bearing cap bolts by following the next few steps:
   • Tighten the camshaft bearing cap bolts in sequence to 89 inch lbs.(10 Nm).
   • Loosen the center intake camshaft bearing cap bolts 1, 2 and the center exhaust camshaft bearing cap bolts.
   • Retighten the center camshaft bearing cap bolts 1, 2, 3, and 4.
   • Retighten the camshaft bearing cap bolts to 89 inch lbs.(10 Nm).

44. Remove the EN-48313 timing chain holding tool

45. Install and tighten the camshaft position actuators.

46. Install the intake camshaft position actuator solenoid.

47. Install the camshaft sensors.
48. Install the crankshaft balancer.
49. Install the camshaft cover.
50. Install the lower intake manifold.
51. Connect the negative battery cable.
52. Drain crankcase and install recommended motor oil.
53. Start the vehicle, check for leaks and repair if necessary.

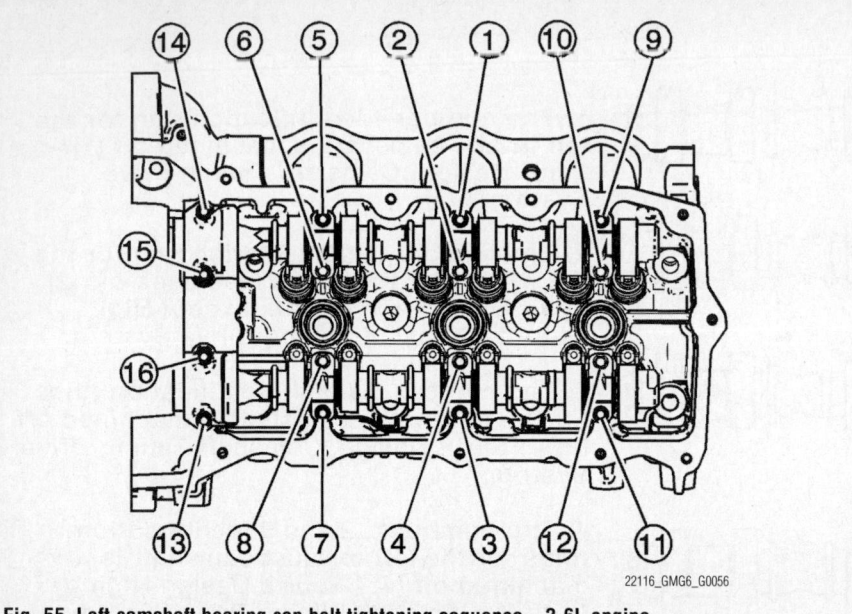

Fig. 55 Left camshaft bearing cap bolt tightening sequence—3.6L engine

Right Camshaft & Lifters

See Figures 56 through 61.

1. Before servicing the vehicle, refer to the Precautions Section.

2. Disconnect the negative battery cable.

3. Remove the lower intake manifold.

4. Remove the camshaft cover.

5. Remove the camshaft sensors.

6. Remove the intake camshaft position actuator solenoid.

7. Rotate the crankshaft with the EN 46111 until the camshafts are in a neutral (low tension) position. The camshaft flats will be parallel with the camshaft cover rail.

✳✳ WARNING

Use an open-end wrench at the camshaft hex to prevent camshaft/engine rotation. DO NOT remove the camshaft position actuator bolt at this time.

8. Loosen the camshaft position actuator bolt.

9. Unscrew the EN-48313 timing chain holding tool so that the legs of the tool are retracted.

10. Insert the EN-48313 between the camshaft actuators, rearward of the timing chain until the bottom line that is scribed in the body of the tool is adjacent to the top surface of the cylinder head. This is the approximate installed position.

11. Ensure that the feet on the legs

of the tool are facing the front of the engine.

12. Partially expand the legs of the EN-48313 by turning the T-shaped handle clockwise.

13. Insert the leg of the tool behind the timing chain guide.

14. Continue expanding the EN-48313 until the legs contact the timing chain. Do not tighten at this time.

➡**Ensure that the foot of the EN-48313 is engaged into one of the link pockets to prevent tool slippage during tightening of the EN-48313.**

15. Hand tighten the EN-48313.

16. Use an open end wrench on the hex cast into the left intake and exhaust camshafts and rotate the

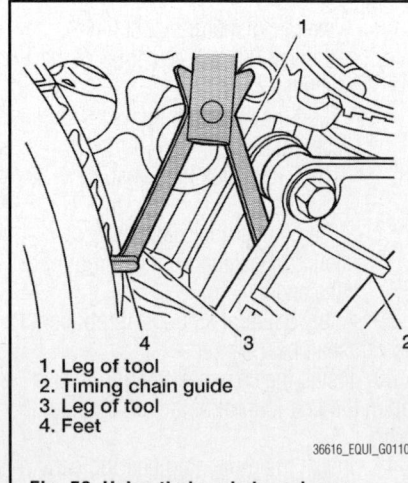

1. Leg of tool
2. Timing chain guide
3. Leg of tool
4. Feet

36616_EQUI_G0110

Fig. 56 Using timing chain and gear holding tool EN-48313

36616_EQUI_G0111

Fig. 57 Another view of the timing chain and gear holding tool EN-48313 properly installed

camshafts toward each other in order to create slack in the chain between the actuators.

17. The EN-48313 is now properly installed to hold the timing chain in position.

✳✳ WARNING

Ensure that the camshaft timing chain and the camshaft position actuators are marked for proper assembly.

18. Mark the timing chain and the respective locations on camshaft position actuators.

19. Remove the camshaft position actuator bolt.

20. Observe the markings on the right cylinder head camshaft bearing caps. Each bearing cap is marked in order to identify its location.

21. The markings have the following meanings:
- The raised feature must always be oriented toward the center of the cylinder head.
- The I indicates the intake camshaft.
- The E indicates the exhaust camshaft
- The number indicates the journal position from the front of the engine

22. Remove the camshaft bearing cap bolts.

23. Remove the camshaft bearing caps.

24. Remove the camshafts.

25. Remove the rocker arms.

26. Remove lifters.

27. Replace the camshaft bearing caps and bolts.

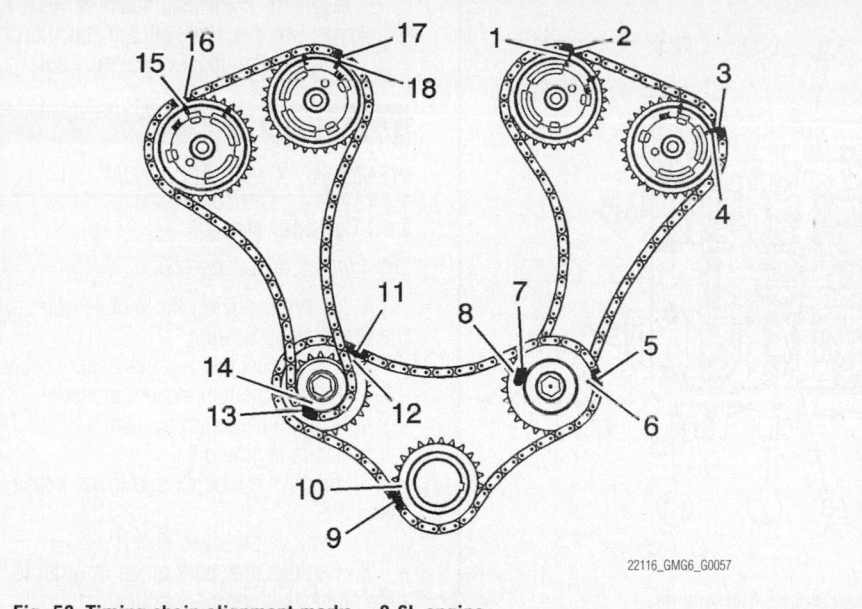

Fig. 58 Timing chain alignment marks —3.6L engine

To install:

✷✷ WARNING

Ensure that the marks on the camshaft position actuator and the timing chain are aligned. DO NOT tighten the camshaft position actuator bolt at this time.

28. Locate the camshafts to the cylinder head and assemble the camshaft actuators to the camshafts.

29. Ensure that the crankshaft is in the stage one timing drive assembly position using the EN 46111.

30. Ensure that the camshaft sealing rings are in place in the camshaft grooves. Camshaft sealing rings must be in place below the surface of the camshaft journal in order to avoid being pinched between the cylinder head and the camshaft caps.

➡**Ensure each valve lifter is filled with clean engine oil and the valve lifter does not tip over (plunger down) before the installation of the valve lifters. The loss of oil in the valve lifter lower pressure chamber or the dry stroking/cycling of the valve lifter plunger will allow air to travel into the high pressure chamber of the valve lifter. Air in the high pressure chamber of the valve lifter may not be purged causing extensive engine component damage.**

31. Install the valve lifters.
32. Install the rocker arms.
33. Apply a liberal amount of lubricant to the camshaft journals and the right cylinder head camshaft carriers.

34. Place the right intake and right exhaust camshafts in position in the right cylinder head.

35. Position the camshaft lobes in a neutral position with the flats on the back of the camshafts up and parallel with the right cylinder head camshaft cover rail.

36. Observe the markings on the right cylinder head camshaft bearing caps. Each bearing cap is marked in order to identify its location.

37. The markings have the following meanings:
- The raised feature must always be oriented toward the center of the cylinder head.
- The I indicates the intake camshaft.
- The E indicates the exhaust camshaft
- The number indicates the journal position from the front of the engine

38. Apply a liberal amount of lubricant to the camshaft bearing caps.

39. Install the camshaft bearing thrust cap in the first journal of the right cylinder head.

40. Install the remaining bearing caps with their orientation mark toward the center of the cylinder head.

41. Hand start all the camshaft bearing cap bolts.

42. Tighten the bearing cap bolts:
   a. Tighten the camshaft bearing cap bolts in sequence to 89 inch lbs. (10 Nm).
   b. Loosen the center intake camshaft bearing cap bolts (1, 2) and the center exhaust camshaft bearing cap bolts (3, 4).
   c. Retighten the center camshaft bearing cap bolts 1, 2, 3, and 4.
   d. Retighten the camshaft bearing cap bolts to 89 inch lbs. (10 Nm).

43. Install and tighten the camshaft position actuators.

Fig. 59 Bearing cap markings view

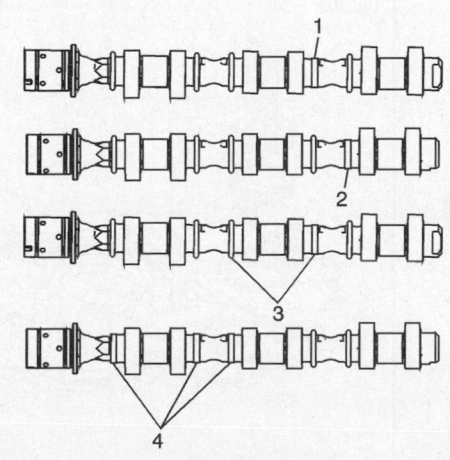

1. The number 4 identification ring for the left intake camshaft is machined off (1) - Third Design, Camshaft Timing Drive System.

2. The number 5 identification ring for the left exhaust camshaft is machined off (2) - Third Design and Fourth, except High Output, Camshaft Timing Drive System.

3. The number 3 and 4 identification rings for the left intake camshaft is machined off (3) - Fourth Design, Camshaft Timing Drive System.

4. The number 1, 2 and 3 identification rings for the left exhaust camshaft is machined off (4) - Fourth Design High Output, Camshaft Timing Drive System.

Fig. 60 Camshaft design identification—3.6L engine

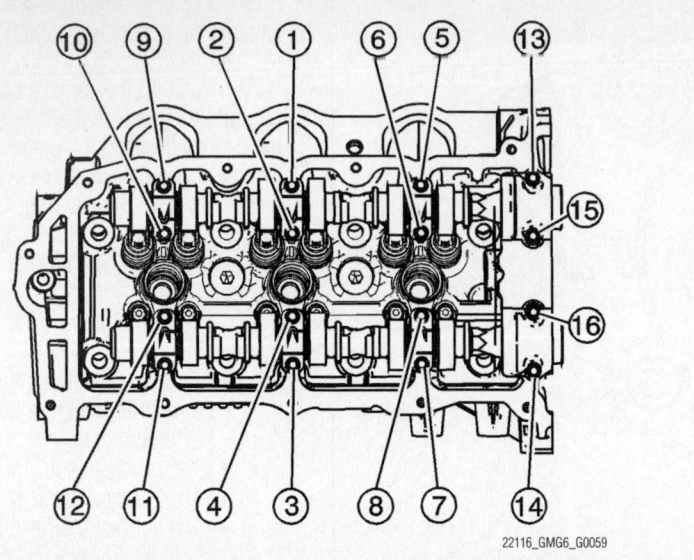

Fig. 61 Right camshaft bearing cap bolt tightening sequence—3.6L engine

44. Install the intake camshaft position actuator solenoid.

45. Install the camshaft sensors.

46. Install the crankshaft balancer. Refer to Crankshaft Damper, removal & installation.

47. Install the camshaft cover.

48. Install the lower intake manifold.

49. Connect the negative battery cable.

50. Drain crankcase and install recommended motor oil.

51. Start the vehicle, check for leaks and repair if necessary.

CRANKSHAFT FRONT SEAL

REMOVAL & INSTALLATION

See Figures 62 and 63.

1. Before servicing the vehicle, refer to the Precautions Section.

2. Disconnect the negative battery cable.

3. Remove the crankshaft damper. Refer to Crankshaft Damper, removal & installation.

4. Use a flat-bladed tool in order to remove the crankshaft oil seal.

☀ WARNING

Use care not to damage the engine front cover or the crankshaft.

To install:

➡**Do not lubricate the crankshaft front oil seal or the crankshaft balancer sealing surfaces.**

5. Use the J-29184 installation tool or equivalent to install the crankshaft front oil seal.

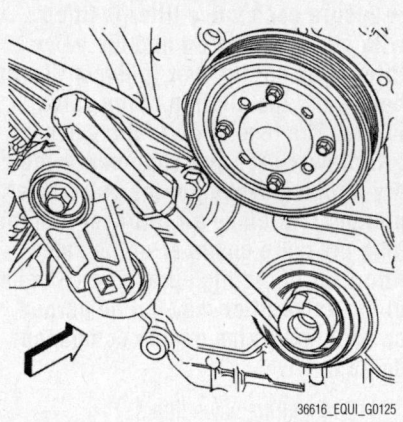

Fig. 62 Carefully pry out the crankshaft front seal with a flat-bladed tool—3.6L engine

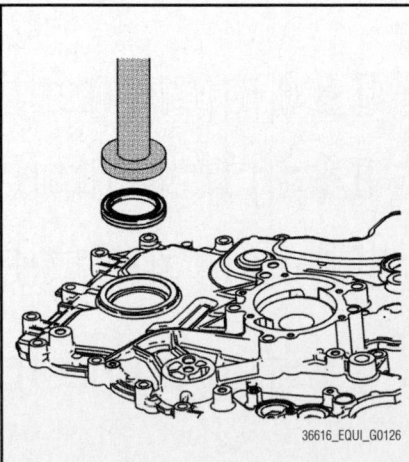

Fig. 63 Install the front seal using Special Tool J-29184 installation tool—3.6L engine

6. Install the crankshaft balancer. Refer to Crankshaft Damper, removal & installation.

7. Connect the negative battery cable.

CYLINDER HEAD

REMOVAL & INSTALLATION

Left Cylinder Head

See Figures 64 through 66.

1. Before servicing the vehicle, refer to the Precautions Section.

2. Disconnect the negative battery cable.

3. Relieve the fuel system pressure.

4. Drain the cooling system.

5. Drain engine oil.

6. Remove the upper and lower intake manifolds.

7. Remove the valve covers.

8. Remove the spark plugs in order to ease crankshaft/engine rotation.

9. Remove the engine front cover.

10. Remove the right bank secondary camshaft drive chain.

11. Remove the primary camshaft drive chain.

12. Remove the left bank secondary camshaft drive chain tensioner.

13. Remove the left bank secondary camshaft drive chain shoe.

14. Remove the left bank secondary camshaft drive chain guide.

15. Remove the left bank camshaft intermediate drive chain idler.

16. Remove the left bank secondary camshaft drive chain.

17. Remove the oil level indicator.

18. Disconnect the coolant temperature sensor electrical connector.

19. Remove the wiring harness ground from the cylinder head.

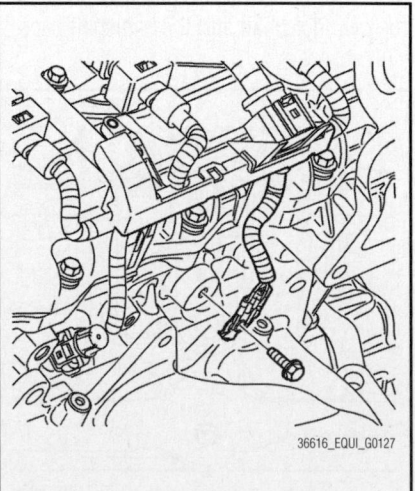

Fig. 64 Cylinder head ground removal—3.6L engine

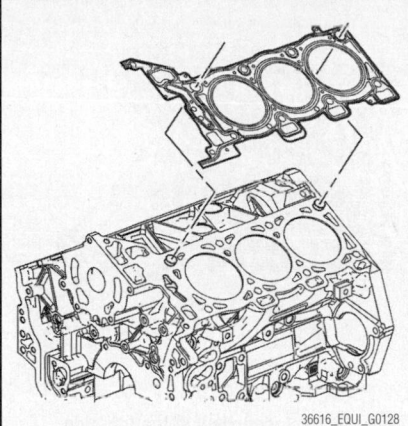

Fig. 65 Remove and discard the cylinder head gasket

20. Remove the catalytic converter.
21. Remove the two front M8 left cylinder head bolts.
22. Remove the left cylinder head bolts.
23. Remove the cylinder head with the exhaust manifold.
24. Remove and discard the cylinder head gasket.
25. Clean and inspect the cylinder head and the engine block sealing surfaces.

To install:

26. Ensure the cylinder head locating pins are securely mounted in the cylinder block deck face.
27. Install a new left cylinder head gasket using the deck face locating pins for retention.
28. Carefully install and align the left cylinder head with the deck face locating pins.

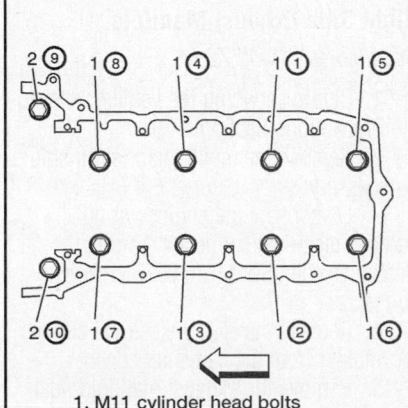

1. M11 cylinder head bolts
2. M8 cylinder head bolts

Fig. 66 Left cylinder head bolt tightening sequence—3.6L engine

☀ WARNING

DO NOT allow oil on the cylinder head bolt bosses or reuse the old M11 cylinder head bolts.

29. Install new M11 cylinder head bolts.
30. Tighten the M11 cylinder head bolts a first pass in sequence to 22 ft. lbs. (30 Nm).
31. Tighten the M11 cylinder head bolts a second pass in sequence an additional 150 degrees using an angle meter.
32. Install the 2 front M8 left cylinder head bolts.
33. Tighten the M8 cylinder head bolts a first pass to 11 ft. lbs. (15 Nm).
34. Tighten the M8 cylinder head bolts a second pass in sequence an additional 75 degrees using an angle meter.
35. Install the catalytic converter to the exhaust manifold.
36. Connect the wiring harness electrical connector located at the side of the cylinder head.
37. Install the wiring harness ground to the cylinder head and tighten mounting bolt to 89 inch lbs. (10 Nm).
38. Install the coolant temperature sensor electrical connector.
39. Install the oil level indicator with new seal.
40. Install the left bank secondary camshaft drive chain
41. Install the left bank camshaft intermediate drive chain idler.
42. Install the left bank secondary camshaft drive chain guide.
43. Install the left bank secondary camshaft drive chain shoe.
44. Install the left bank secondary camshaft drive chain tensioner.
45. Install the primary camshaft drive chain.
46. Install the right bank secondary camshaft drive chain.
47. Install the engine front cover.
48. Install the spark plugs.
49. Install the valve covers.
50. Install the upper and lower intake manifolds.
51. Fill the engine with clean oil.
52. Fill and bleed the cooling system.
53. Connect the negative battery cable.
54. Start the engine and check for leaks; repair if necessary.

Right Cylinder Head

See Figures 67 and 68.

1. Before servicing the vehicle, refer to the Precautions Section.
2. Disconnect the negative battery cable.
3. Remove the hood.

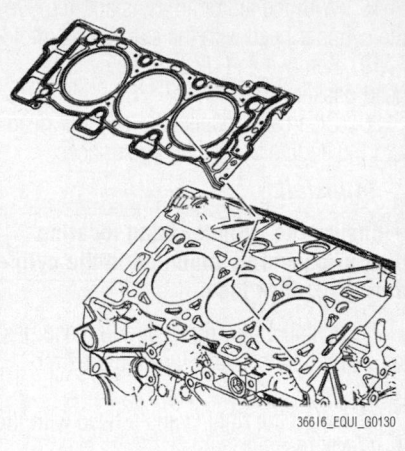

Fig. 67 Remove and discard the cylinder head gasket

4. Relieve the fuel system pressure.
5. Drain the cooling system.
6. Drain engine oil.
7. Remove the upper and lower intake manifolds.
8. Remove the valve covers.
9. Remove the spark plugs in order to ease crankshaft/engine rotation.
10. Remove the ground wires from the cylinder head.
11. Remove the catalytic converter.
12. Remove the engine front cover.
13. Remove the right bank secondary camshaft drive chain tensioner.
14. Remove the right bank secondary camshaft drive chain shoe
15. Remove the right bank secondary camshaft drive chain guide.
16. Remove the right bank secondary camshaft drive chain.
17. Remove the right cylinder head bolts.

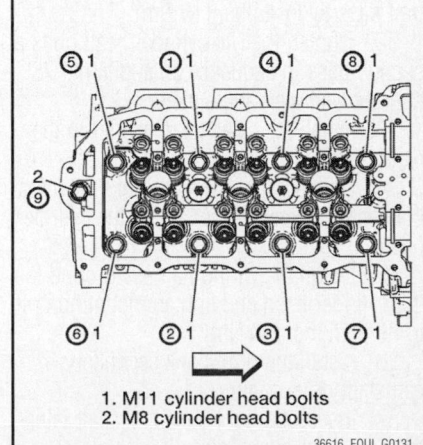

1. M11 cylinder head bolts
2. M8 cylinder head bolts

Fig. 68 Right cylinder head bolt tightening sequence—3.6L engine

18. With the aid of an assistant, remove the cylinder head with the exhaust manifold.

19. Remove and discard the cylinder head gasket.

20. Clean and inspect the cylinder head and the engine block sealing surfaces.

To install:

➡ **Ensure the cylinder head locating pins are securely mounted in the cylinder block deck face.**

21. Install a new right cylinder head gasket using the deck face locating pins for retention.

22. Align the right cylinder head with the deck face locating pins.

23. With the aid of an assistant, carefully install the cylinder head with the exhaust manifold to the engine.

24. Ensure the cylinder head locating pins are securely mounted in the cylinder block deck face.

25. Install a new left cylinder head gasket using the deck face locating pins for retention.

26. Carefully install and align the left cylinder head with the deck face locating pins.

✳✳ WARNING

DO NOT allow oil on the cylinder head bolt bosses or reuse the old M11 cylinder head bolts.

27. Install new M11 cylinder head bolts.

28. Tighten the M11 cylinder head bolts a first pass in sequence to 22 ft. lbs. (30 Nm).

29. Tighten the M11 cylinder head bolts a second pass in sequence an additional 150 degrees using an angle meter.

30. Install the 2 front M8 left cylinder head bolts.

31. Tighten the M8 cylinder head bolts a first pass to 11 ft. lbs. (15 Nm).

32. Tighten the M8 cylinder head bolts a second pass in sequence an additional 75 degrees using an angle meter.

33. Install the catalytic converter to the exhaust manifold.

34. Connect the wiring harness electrical connector located at the side of the cylinder head.

35. Install the wiring harness ground to the cylinder head and tighten mounting bolt to 89 inch lbs. (10 Nm).

36. Install the right bank secondary camshaft drive chain

37. Install the right bank camshaft intermediate drive chain idler.

38. Install the right bank secondary camshaft drive chain guide.

39. Install the right bank secondary camshaft drive chain shoe.

40. Install the right bank secondary camshaft drive chain tensioner.

41. Install the engine front cover.

42. Install the spark plugs.

43. Install the valve covers.

44. Install the upper and lower intake manifolds.

45. Fill the engine with clean oil.

46. Fill and bleed the cooling system.

47. Connect the negative battery cable.

48. Start the engine and check for leaks; repair if necessary.

EXHAUST MANIFOLD

REMOVAL & INSTALLATION

Left Side Exhaust Manifold

See Figures 69 and 70.

1. Before servicing the vehicle, refer to the Precautions Section.

2. Remove the fuel injector sight shield, if necessary.

3. Disconnect the engine wiring harness electrical connector from the Heated Oxygen Sensor (HO2S) electrical connector.

4. Remove the HO2S electrical connector retainer from the retainer clip.

5. Remove the exhaust manifold heat shield bolts.

6. Remove the exhaust manifold heat shield, sliding the shield up over the HO2S pigtail.

7. Remove the oil level indicator.

8. Remove the catalytic converter-to-exhaust manifold nuts.

9. Remove the exhaust manifold bolts.

10. Remove the exhaust manifold and gasket. Discard the gasket.

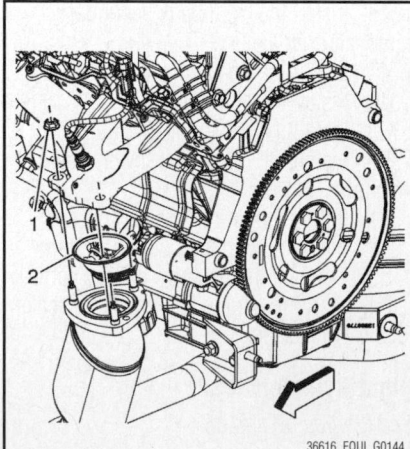

Fig. 69 Catalytic converter removal from the exhaust manifold—left bank

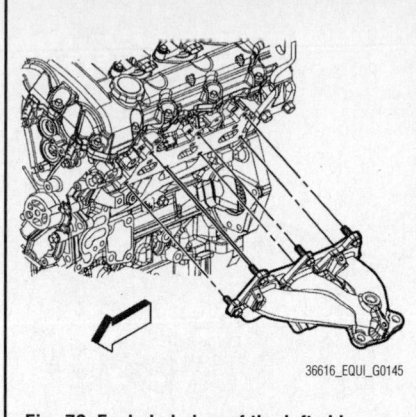

Fig. 70 Exploded view of the left side exhaust manifold—3.6L engine

To install:

11. Install one exhaust manifold bolt to the exhaust manifold.

12. Install the **NEW** exhaust manifold gasket onto the cylinder head and bolt.

13. Install the exhaust manifold (with gasket) to the catalytic converter and the cylinder head.

14. Install the remaining exhaust manifold bolts and tighten to 15 ft. lbs. (20 Nm).

15. Install the oil level indicator.

16. Install the exhaust manifold heat shield, sliding the shield down over the HO2S pigtail.

17. Install the exhaust manifold heat shield bolts and tighten to 89 inch lbs. (10 Nm).

18. Connect the engine wiring harness electrical connector to the HO2S electrical connector.

19. Install the HO2S electrical connector retainer to the retainer clip.

20. Install the fuel injector sight shield, if necessary.

Right Side Exhaust Manifold

See Figures 71 and 72.

1. Before servicing the vehicle, refer to the Precautions Section.

2. Remove the fuel injector sight shield, if necessary.

3. Disconnect the engine wiring harness electrical connector from the Heated Oxygen Sensor (HO2S) electrical connector.

4. Remove the HO2S electrical connector retainer from the camshaft cover.

5. Remove the exhaust manifold heat shield bolts.

6. Remove the exhaust manifold heat shield, sliding the shield up over the HO2S pigtail.

7. Remove the catalytic converter to exhaust manifold nuts.

8. Remove the exhaust manifold bolts.

9. Remove the exhaust manifold and gasket. Discard the gasket.

10. Remove the right catalytic converter. Refer to Catalytic Converter, removal & installation.

11. Remove the exhaust manifold bolts.

12. Remove the exhaust manifold and gasket out from the vehicle. Discard the gasket.

To install:

13. Install one exhaust manifold bolt to the exhaust manifold.

14. Install the NEW exhaust manifold gasket onto the cylinder head and bolt.

15. Install the exhaust manifold (with gasket) to the catalytic converter and the cylinder head.

16. Install the exhaust manifold bolts and tighten to 15 ft. lbs. (20 Nm).

17. Install the right catalytic converter.

18. Install the catalytic converter nuts and tighten to 37 ft. lbs. (50 Nm).

19. Install the exhaust manifold heat shield, sliding the shield down over the HO2S pigtail.

20. Install the exhaust manifold heat shield bolts and tighten to 89 inch lbs. (10 Nm).

21. Connect the engine wiring harness electrical connector to the HO2S electrical connector.

22. Install the HO2S electrical connector retainer to the camshaft cover.

23. Install the fuel injector sight shield, if necessary

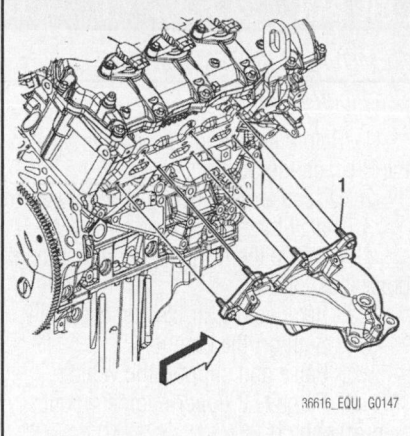

Fig. 72 Exploded view of the right side exhaust manifold—3.6L engine

INTAKE MANIFOLD

REMOVAL & INSTALLATION

Upper Intake Manifold

See Figure 73.

1. Before servicing the vehicle, refer to the Precautions Section.

2. Disconnect the negative battery cable.

3. Remove the engine cover.

4. Remove the air cleaner outlet duct.

5. Reposition the fresh air Positive

Crankcase Ventilation (PCV) line from the air cleaner inlet tube.

6. Disconnect the Electronic Throttle Control (ETC) electrical connector.

7. Disconnect the PCV line from the top of the intake manifold and reposition aside.

8. Disconnect the Evaporative Emissions (EVAP) canister purge line and reposition aside.

9. Remove the bleed pipe bolts.

10. Remove the bleed pipe hose clamp.

11. Reposition the bleed pipe.

12. Remove the brake booster vacuum hose from the intake manifold.

13. Remove the engine harness retaining clips.

14. Remove the upper intake retaining bolts.

15. Remove the upper intake manifold and gasket. Discard gasket.

To install:

16. Clean and inspect the intake manifold and the sealing surfaces.

17. Install a **NEW** upper intake manifold gaskets.

18. Carefully install the intake manifold.

19. Install the intake manifold mounting bolts and tighten in sequence to 17 ft. (23 Nm).

20. Install the engine harness retaining clips.

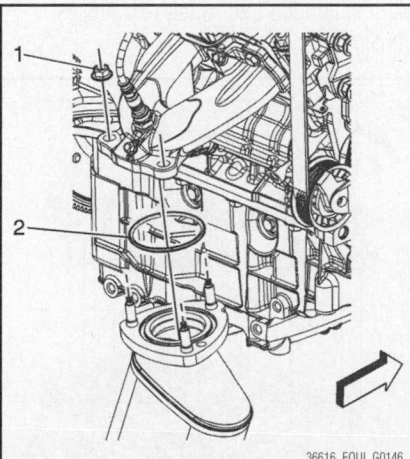

Fig. 71 Catalytic converter removal from the exhaust manifold—right bank

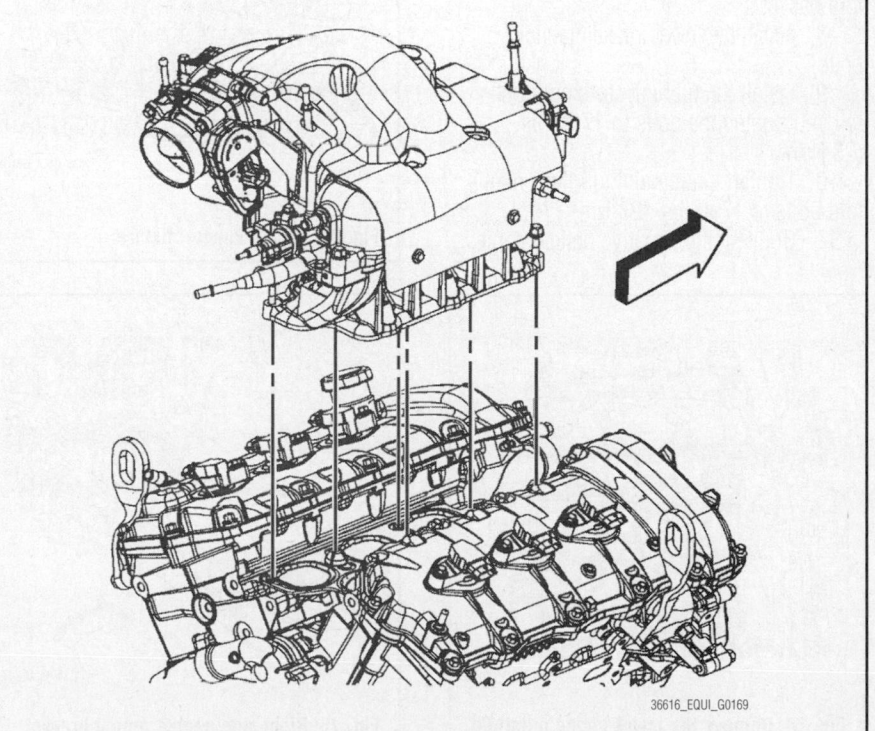

Fig. 73 Upper intake manifold removal—3.6L engine

21. Install the brake booster vacuum hose to the intake manifold.

22. Position the bleed pipe.

23. Install the bleed pipe hose clamp.

24. Install the bleed pipe bolts and tighten to 89 inch lbs. (10 Nm).

25. Connect the coolant hose bleed pipe.

26. Connect the EVAP canister purge line.

27. Connect the PCV line to the top of the intake manifold.

28. Connect the ETC electrical connector.

29. Install the air cleaner outlet duct.

30. Install the fresh air PCV line to the air cleaner inlet duct.

31. Install the engine cover.

32. Connect the negative battery cable.

Lower Intake Manifold

See Figure 74.

1. Before servicing the vehicle, refer to the Precautions Section.

2. Disconnect the negative battery cable.

3. Remove the fuel injectors and fuel rail. Refer to Fuel Rail & Injectors, removal & installation.

4. Remove the lower intake manifold bolts.

5. Remove the lower intake manifold and gasket from engine. Discard the gasket.

6. Clean and inspect the intake manifold and the sealing surfaces.

To install:

7. Install the **NEW** lower intake manifold gasket.

8. Install the lower intake manifold bolts

9. Install the fuel injectors and fuel rail and tighten the bolts to 17 ft. lbs. (23 Nm).

10. Tighten the remaining intake manifold bolts to 17 ft. lbs. (23 Nm).

11. Connect the negative battery cable.

OIL PAN

REMOVAL & INSTALLATION

See Figures 75 through 80.

1. Before servicing the vehicle, refer to the Precautions Section.

2. Disconnect the negative battery cable.

3. Install the engine support fixture.

4. Remove the right side engine mount bracket as follows:

   a. Remove the air cleaner assembly.

   b. Support the engine.

   c. Raise and support the vehicle.

   d. Remove 3 upper engine mount bracket bolts.

   e. Loosen right upper front bolt.

   f. Remove the ABS connector clip from the frame.

   g. Remove the upper engine mount bracket nuts.

   h. Remove the lower bracket bolts.

   i. Remove the upper bracket bolt from the bracket.

   j. Remove the mount and the lower bracket from the vehicle.

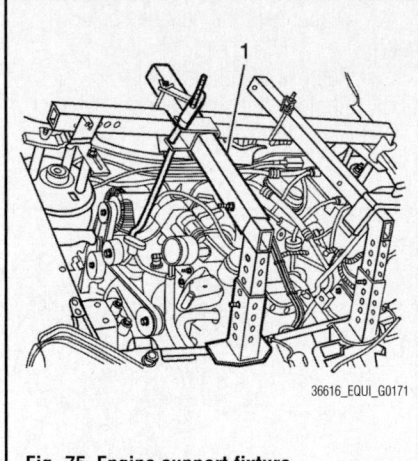

Fig. 75 Engine support fixture

k. Remove the upper bracket.

l. Remove the engine mount.

5. Raise and support the vehicle.

6. Drain the engine oil and remove the oil filter.

7. Remove the catalytic converter. Refer to Catalytic Converter, removal & installation.

8. Remove the Air Conditioning (A/C) compressor.

9. Remove the oil pan bolts.

10. Remove the oil pan.

To install:

11. Clean the oil pan and the engine block gasket surface.

12. Install the 0.315 inch (8mm) oil pan guides EN 46109 into the center oil pan rail bolt hole on each side of the engine block.

13. Place a 0.118 inch (3mm) bead of RTV sealant, GM P/N 12378521 (Canadian P/N 88901148), or equivalent, on the block pan rail and the crankshaft rear oil seal housing.

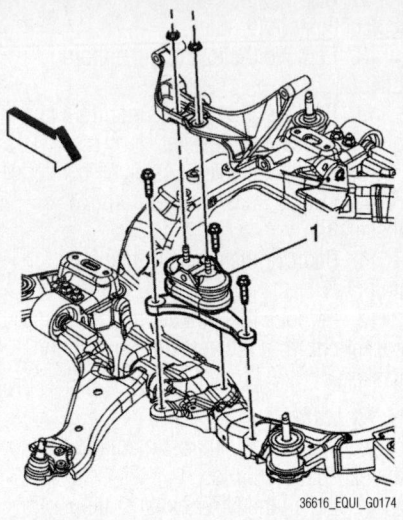

Fig. 77 Remove the engine mount—3.6L engine

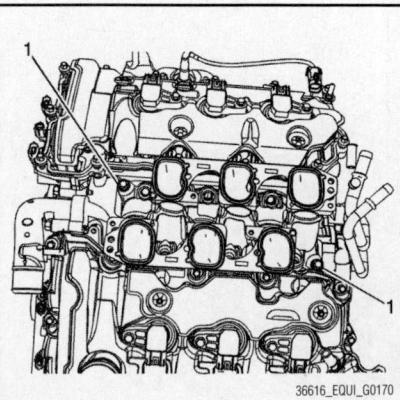

Fig. 74 Remove the lower intake manifold bolts (1)

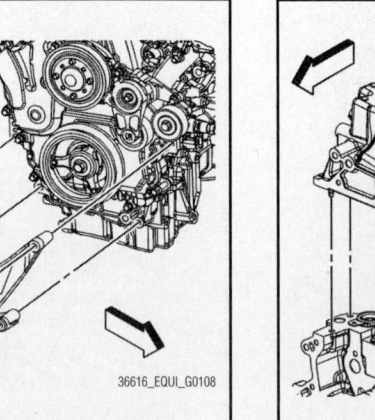

Fig. 76 Right side engine mount bracket removal—3.6L engine

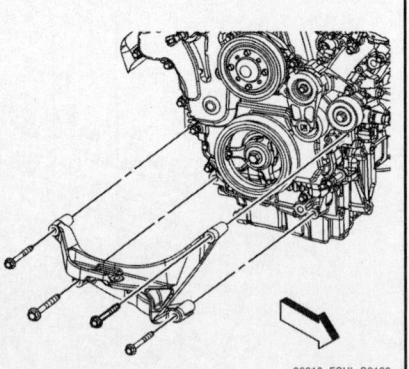

Fig. 78 Oil pan removal—3.6L engine

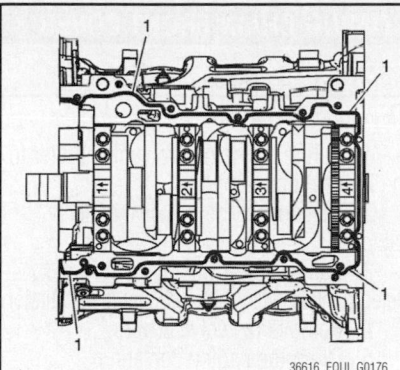

Fig. 79 Place a 0.118 inch (3mm) bead (1) of RTV sealant on the block pan rail and the crankshaft rear oil seal housing

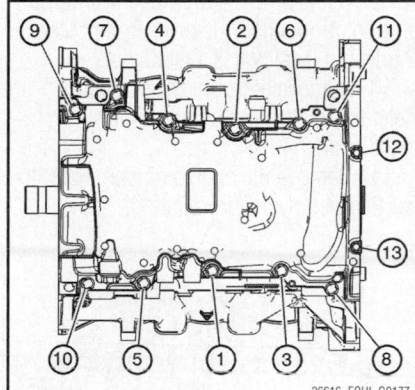

Fig. 80 Oil pan tightening sequence—3.6L engine

14. Position the oil pan onto the block.

15. Remove the EN 46109 guides from the engine block.

16. Loosely install the oil pan bolts.

17. Tighten the oil pan bolts in sequence:

   a. Tighten the 8mm bolts (1–11) to 17 ft. lbs. (23 Nm).

   b. Tighten the 6mm bolts (12, 13) to 89 inch lbs. (10 Nm).

18. Install the A/C compressor.

19. Install the catalytic converter.

20. Install a new oil filter.

21. Lower the vehicle.

22. Refill the engine with oil.

23. Install the right side engine mount bracket as follows:

   a. Install the engine mount and the lower bracket to the vehicle. Tighten the lower bracket bolts to 37 ft. lbs. (50 Nm).

   b. Install the ABS connector clip to the frame.

   c. Install the upper engine mount bracket nuts and tighten to 37 ft. lbs. (50 Nm).

   d. Install the engine mount upper bracket and bolt to the engine. Do not tighten the bolt.

   e. Tighten the right upper engine mount bracket bolt to 37 ft. lbs. (50 Nm).

   f. Install the upper engine mount bracket bolts and tighten to 81 ft. lbs. (110 Nm).

24. Remove the engine support fixture.

25. Connect the negative battery cable.

26. Start the engine and check for leaks.

OIL PUMP

REMOVAL & INSTALLATION
See Figure 81.

1. Before servicing the vehicle, refer to the Precautions Section.

2. Remove the front timing cover. Refer to Front Timing Cover and Seal, removal & installation.

3. Remove the primary timing chain. Refer to Timing Chain and Sprockets, removal & installation.

4. Remove the oil pump bolts and the oil pump.

To install:

5. Align the oil pump generator with the crankshaft flats and install the oil pump to the engine block.

6. Align the pump body with the mounting holes in the cylinder block.

7. Install the oil pump bolts and tighten to 17 ft. lbs. (23 Nm).

8. Install the primary timing chain.

9. Install the front timing cover.

10. Connect the negative battery cable.

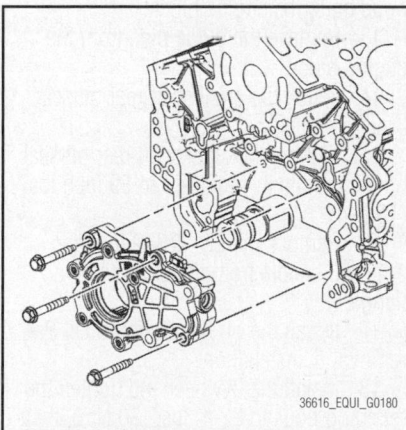

Fig. 81 Oil pump mounting view—3.6L engine

REAR MAIN SEAL

REMOVAL & INSTALLATION
See Figures 82 through 88.

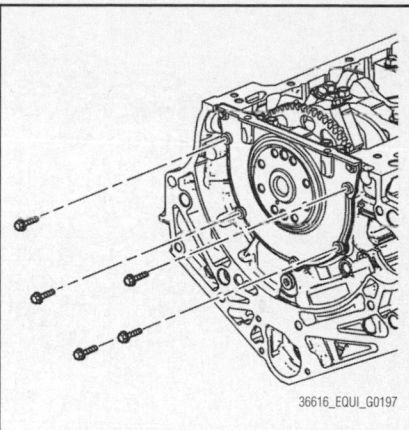

Fig. 82 Crankshaft rear oil seal housing bolt removal—3.6L engine

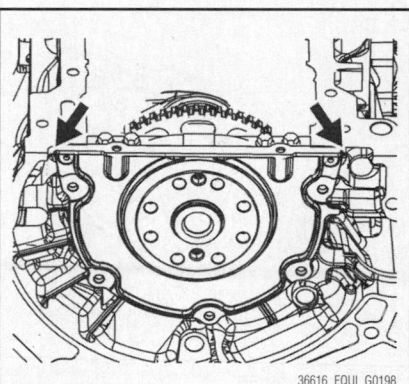

Fig. 83 Crankshaft rear oil seal housing pry points—3.6L engine

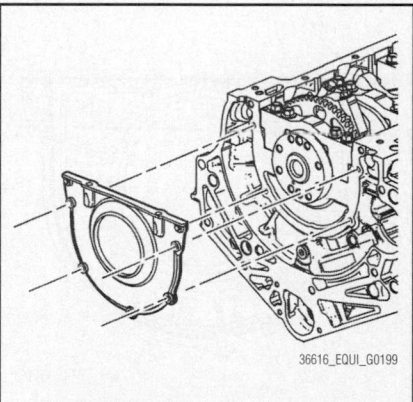

Fig. 84 Remove the crankshaft rear oil seal housing—3.6L engine

1. Before servicing the vehicle, refer to the Precautions Section.

2. Disconnect the negative battery cable.

3. Remove the transaxle assembly. Refer to Automatic Transaxle Assembly, removal & installation.

4. Remove the oil pan. Refer to Oil Pan, removal & installation.

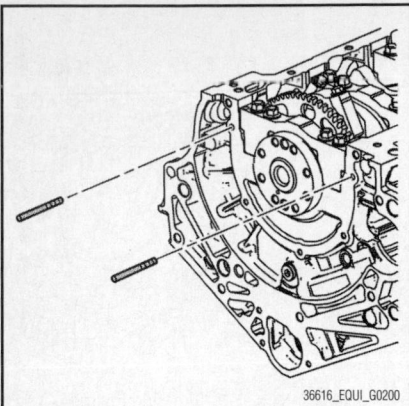

Fig. 85 Install the guide pins into the 2 crankshaft rear oil seal housing corner bolt holes of the engine block

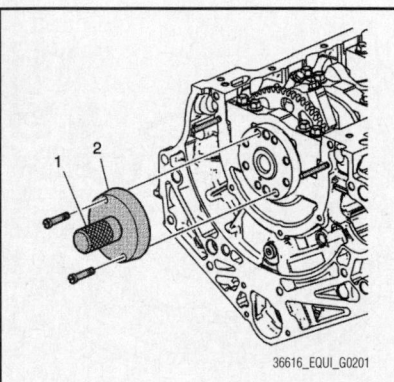

Fig. 86 Install the EN-47839 tool (1) with the J-42183 handle (2) onto the rear of the crankshaft flange

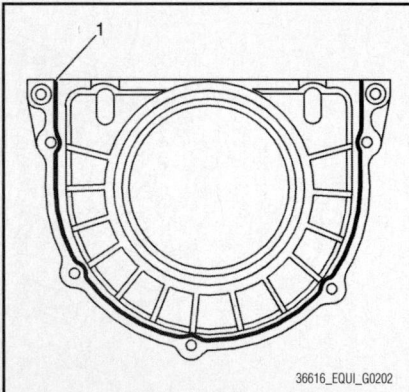

Fig. 87 Place a 0.118 inch (3mm) bead of RTV sealant on seal housing

5. Remove the engine flywheel bolts and flywheel. Refer to Flywheel, removal & installation.

6. Remove the crankshaft rear oil seal housing bolts.

7. Use the pry points located at the edge of the crankshaft rear oil seal housing to separate the RTV sealant.

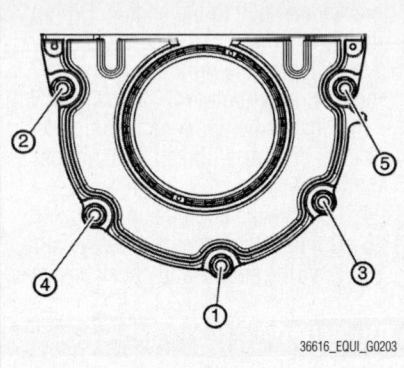

Fig. 88 Rear oil seal housing tightening sequence

8. Remove and discard the crankshaft rear oil seal housing.

To install:

9. Install the 0.236 inch (6mm) guides from the EN-46109 pin set into the 2 crankshaft rear oil seal housing corner bolt holes of the engine block.

10. Install the EN-47839 tool with the J-42183 handle onto the rear of the crankshaft flange.

11. Place a 0.118 inch (3mm) bead of RTV sealant, GM P/N 12378521 (Canadian P/N 88901148), or equivalent, to the NEW crankshaft rear oil seal housing.

➡ **DO NOT allow any engine oil on the area where the crankshaft rear oil seal housing is to be installed.**

12. Install the crankshaft rear oil seal housing to the engine block.

13. Remove the guide pins from the engine block.

14. Install the crankshaft rear oil seal housing bolts.

15. Tighten the crankshaft rear oil seal housing bolts in sequence to 89 inch lbs. (10 Nm).

16. Remove the EN-47839 tool and J-42183 handle from the crankshaft flange.

17. Install the oil pan. Refer to Oil Pan, removal & installation.

18. Install the flywheel and tighten the retaining bolts to 22 ft. lbs. (30 Nm), then an additional 45 degrees. Refer to Flywheel, removal & installation.

19. Install the transaxle assembly. Refer to Automatic Transaxle Assembly, removal & installation.

20. Connect the negative battery cable.

TIMING CHAIN FRONT COVER

REMOVAL & INSTALLATION

See Figures 89 through 96.

1. Before servicing the vehicle, refer to the Precautions Section.

2. Disconnect the negative battery cable.

3. Remove the lower intake manifold. Refer to Intake Manifold, removal & installation.

4. Remove the valve covers.

5. Drain the engine coolant.

6. Remove the water outlet housing assembly.

7. Remove the drive belt tensioner.

8. Remove the water pump. Refer to Water Pump, removal & installation.

9. Remove the power steering pump and position aside. Refer to Power Steering Pump, removal & installation.

10. Remove the crankshaft balancer. Refer to Crankshaft Damper, removal & installation.

11. Remove the engine oil pan. Refer to Oil Pan, removal & installation.

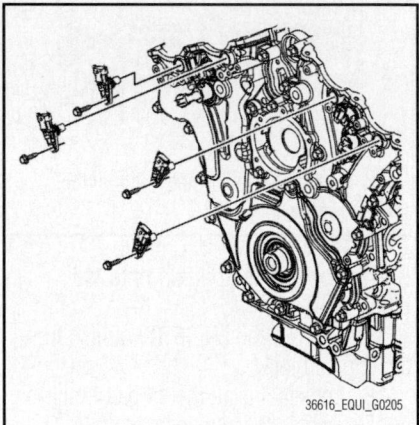

Fig. 89 Camshaft position sensor removal—3.6L engine

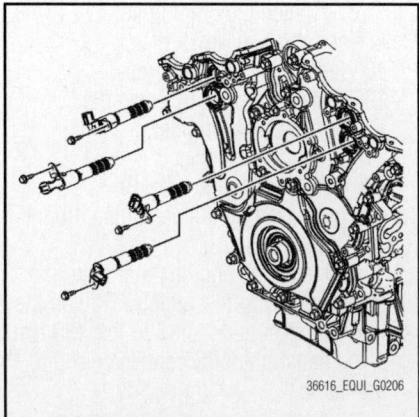

Fig. 90 Camshaft position actuator removal—3.6L engine

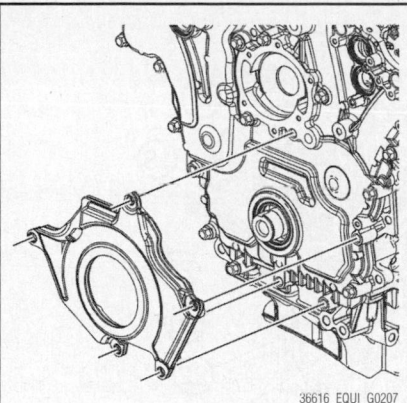

Fig. 91 Engine front cover deadener removal—3.6L engine

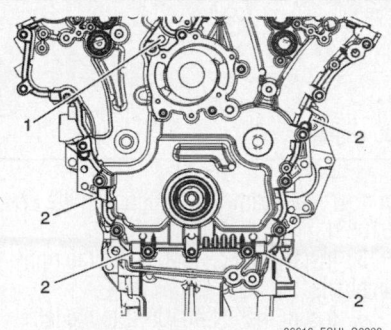

Fig. 92 Using the pry points (2) located at the edge of the front cover and the jackscrew, separate the RTV sealant—3.6L engine

12. Remove the camshaft position sensor bolts.

13. Remove the camshaft position sensors.

14. Remove the camshaft position actuator valve bolts.

15. Remove the camshaft position actuator valves from the front cover.

✳✳ WARNING

The camshaft position actuator valves must be removed from the front cover prior to front cover removal or damage to the valves may occur.

16. Remove the engine front cover bolts that hold the engine front cover deadener into position.

17. Remove the engine front cover deadener.

➡**There are a total of 22 M8 bolts that must be removed and 3 optional M12 bolts that may need to be removed before the front cover will separate from the engine block.**

18. Remove the remaining engine front cover bolts.

19. Loosely install a 10 x 1.5mm bolt in the jackscrew hole.

✳✳ WARNING

Do not use the jackscrew hole without first removing all engine front cover bolts. Failure to remove all engine front cover bolts before using the jackscrew hole could result in damage to components.

✳✳ WARNING

Do not pry between the engine front cover and the camshaft position sensors or the camshaft position actuators in order to separate the RTV. Use the pry points and a bolt in the jackscrew hole in order to remove the engine front cover. Damage to the camshaft position sensors or the camshaft position actuators may occur if the camshaft position sensors or the camshaft position actuators are used to pry against in order to remove the engine front cover.

20. Using the pry points located at the edge of the front cover and the jackscrew, separate the RTV sealant.

21. Remove the engine front cover.

To install:

22. Install the 0.315 inch (8mm) guide from the EN-46109 pins into the cylinder block positions.

23. Install the NEW engine front cover-to-cylinder block seal.

24. Place a 0.118 inch (3mm) bead of RTV sealant, GM P/N 12378521 (Canadian P/N 88901148), or equivalent, on the engine front cover.

25. Install the engine front cover onto the EN-46109 pins and slide into position.

26. Remove the pins from the cylinder block.

27. Install the engine front cover deadener.

28. Loosely install the engine front cover bolts to hold the engine front cover deadener into position.

29. Loosely install the remaining engine front cover bolts.

30. Tighten the engine front cover bolts in sequence as follows:
  - Step 1: Tighten the bolts (1–22) to 14 ft. lbs. (20 Nm)
  - Step 2: Tighten the bolts (1–22) an additional 60°
  - Step 3: Tighten bolt (23) to 48 ft. lbs. (65 Nm)

➡**Engine front cover bolts in the number 23 location are model dependent and may not apply.**

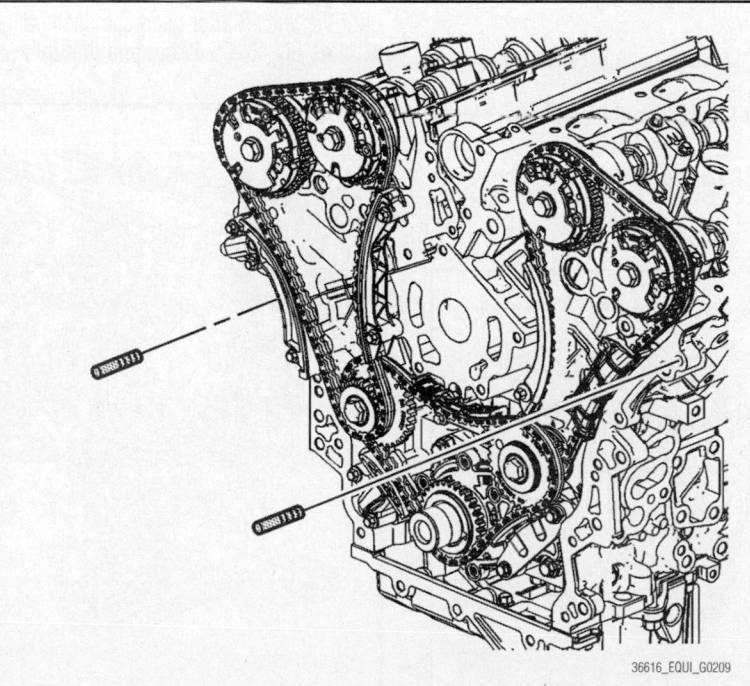

Fig. 93 Install the guide from the EN-46109 pins into the cylinder block positions—3.6L engine

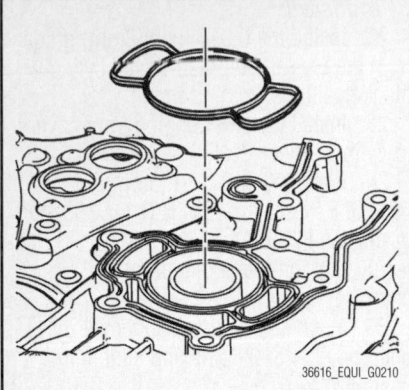

Fig. 94 Engine front cover-to-cylinder block seal

31. Install the camshaft position actuator valves and tighten the bolts to 89 inch lbs. (10 Nm).

32. Install NEW O-rings on the camshaft position sensor.

33. Install the camshaft position sensors and tighten the mounting bolts to 89 inch lbs. (10 Nm).

34. Install the engine oil pan.

35. Install the crankshaft balancer.

36. Install the power steering pump.

37. Install the water pump.

38. Install the water outlet housing assembly.

39. Install the drive belt tensioner.

40. Install the valve covers.

41. Install the lower intake manifold.

42. Connect the negative battery cable.

43. Refill and bleed the cooling system.

44. Start the engine and check for leaks.

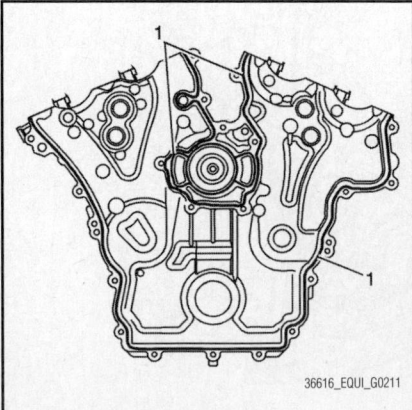

Fig. 95 Place a 0.118 inch (3mm) bead of RTV sealant on the engine front cover (1)

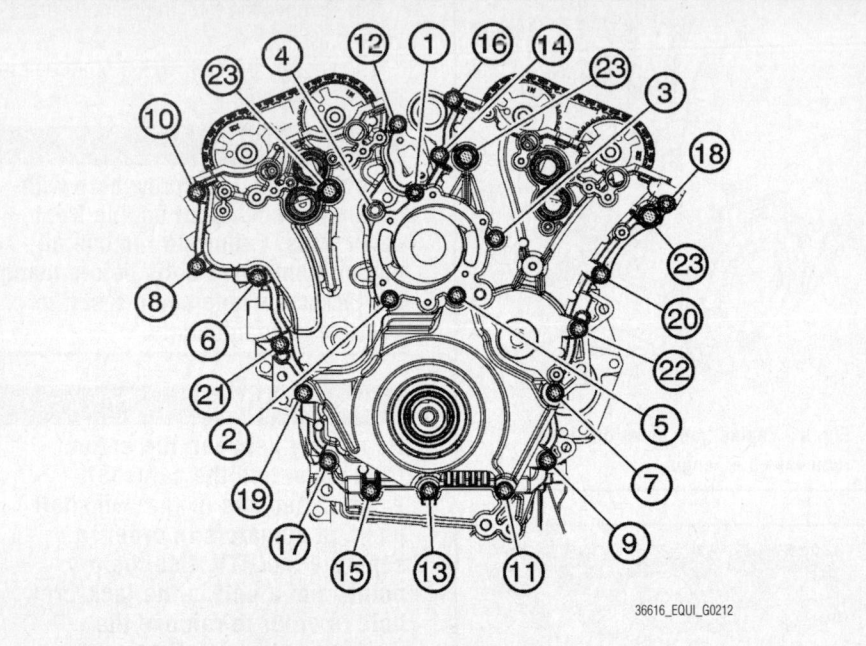

Fig. 96 Engine front cover tightening sequence—3.6L engine

TIMING CHAIN & SPROCKETS

REMOVAL & INSTALLATION

See Figures 97 through 108.

1. Before servicing the vehicle, refer to the Precautions Section.

2. Disconnect the negative battery cable.

3. Remove the timing chain front cover.

4. Using the special camshaft locking tool EN 46111, rotate the crankshaft until the left cylinder head camshafts align with the EN 46105-2 tool and the right

cylinder head camshafts align with the EN 46105-1.

5. Install the EN 46105-1 to the right camshafts.

6. Install the EN 46105-2 to the left camshafts.

7. Remove the right secondary camshaft drive chain from the right camshaft position actuators and the right camshaft intermediate drive chain idler sprocket.

8. Remove the primary camshaft drive chain.

9. Remove the left secondary camshaft drive chain tensioner bolts.

10. Remove the left secondary camshaft drive chain tensioner.

11. Remove and discard the left secondary camshaft drive chain tensioner gasket.

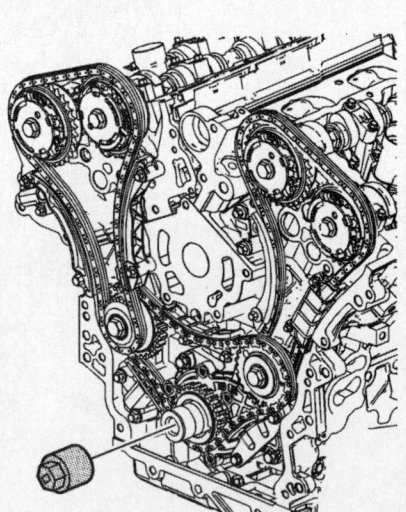

Fig. 97 Rotating crankshaft with tool EN 46111 for camshaft alignment—3.6L engine

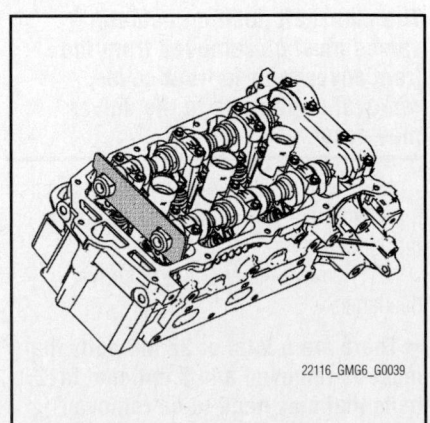

Fig. 98 Right camshaft alignment shown with tool EN 46105-1 (left similar)

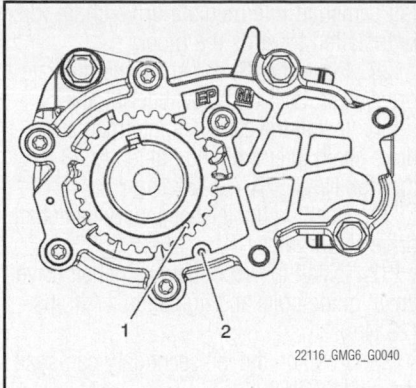

Fig. 99 Crankshaft sprocket (1) and oil pump (2) alignment marks

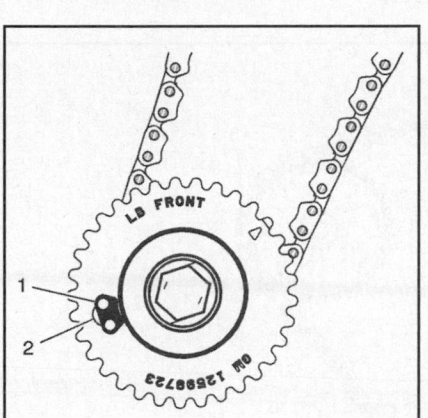

Fig. 100 Left camshaft intermediate drive chain idler outer sprocket and chain alignment.

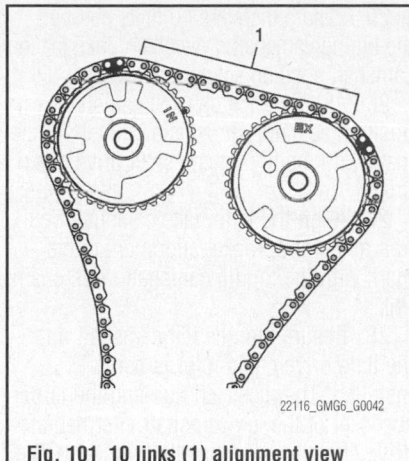

Fig. 101 10 links (1) alignment view

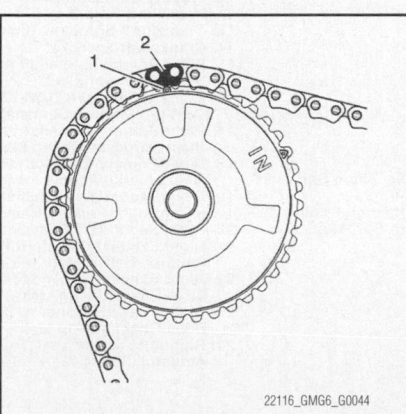

Fig. 102 Exhaust camshaft alignment marks

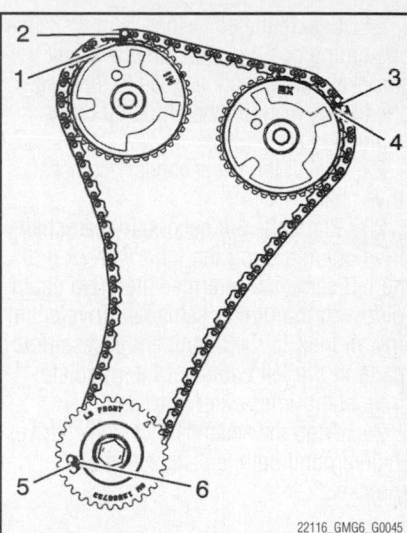

Fig. 104 Left secondary camshaft drive chain timing mark alignments (1—6)

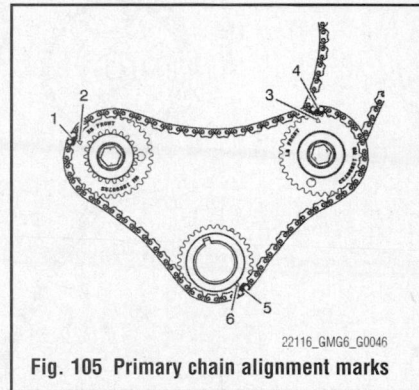

Fig. 105 Primary chain alignment marks

➡Inspect the left secondary camshaft drive chain tensioner mounting surface on the left cylinder head for burrs or any defects that would degrade the sealing of the new left secondary camshaft drive chain tensioner gasket.

12. Remove the left secondary camshaft drive chain shoe bolt.

13. Remove the left secondary camshaft drive chain shoe.

14. Remove the left secondary camshaft drive chain guide bolts.

15. Remove the left secondary camshaft drive chain guide.

16. Remove the left camshaft intermediate drive chain idler bolt

17. Remove the left camshaft intermediate drive chain idler.

18. Remove the left secondary camshaft drive chain from the left camshaft position actuators and the left camshaft intermediate drive chain idler sprocket.

19. Clean and inspect all of the camshaft timing drive components.

To install:

❊❊ WARNING

All camshafts must be locked in place before installation of any camshaft drive chains.

20. Ensure that the EN 46105-1 is fully seated onto the camshafts.

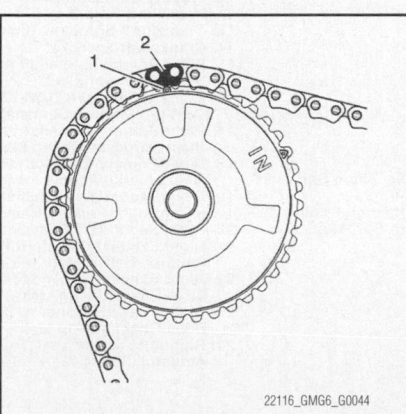

Fig. 103 Intake camshaft alignment marks

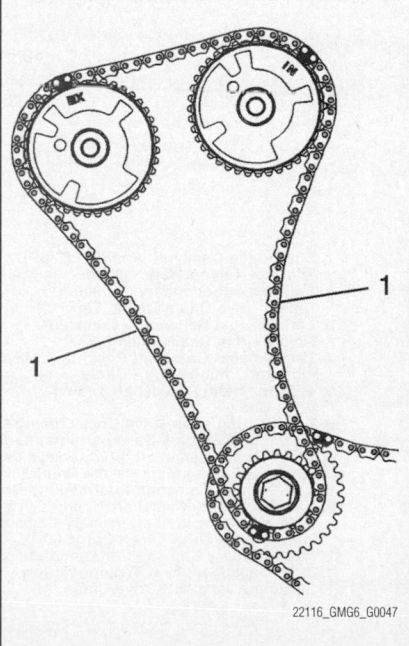

Fig. 106 Secondary drive chain alignment marks

21. Ensure the crankshaft is in the stage one timing position with the crankshaft sprocket timing mark aligned to the stage one timing mark on the oil pump cover using the EN-48589.

22. Install the left secondary camshaft drive chain.

23. Place the left secondary camshaft drive chain around the inner sprocket of the left camshaft intermediate drive chain idler with the timing camshaft drive chain link aligned to the alignment access hole made in the left camshaft intermediate drive chain idler outer sprocket.

24. Wrap the secondary camshaft drive chain around both left actuator drive sprockets.

25. Ensure there are 10 links between the timing camshaft drive chain links for the camshaft position actuator sprockets.

26. Align the left exhaust camshaft position actuator sprocket alignment circle mark with the timing camshaft drive chain link.

27. Align the left intake camshaft position actuator sprocket alignment circle mark with the timing camshaft drive chain link.

28. Ensure that the left camshaft intermediate drive chain idler is being installed. The recessed hub and the larger sprocket of the left camshaft intermediate drive chain idler is installed outward. The raised hub and the smaller sprocket of the left camshaft intermediate drive chain idler is installed towards the block.

29. Place the left camshaft intermediate drive chain idler to the cylinder block.

30. Install the camshaft intermediate drive chain idler bolt and tighten to 43 ft. lbs. (58 Nm).

31. Position the left secondary camshaft drive chain guide.

32. Install the secondary camshaft drive chain guide bolts and tighten to 17 ft. lbs. (23 Nm).

33. Position the left secondary camshaft drive chain shoe.

34. Install the secondary camshaft drive chain shoe bolt and tighten to 17 ft. lbs. (23 Nm).

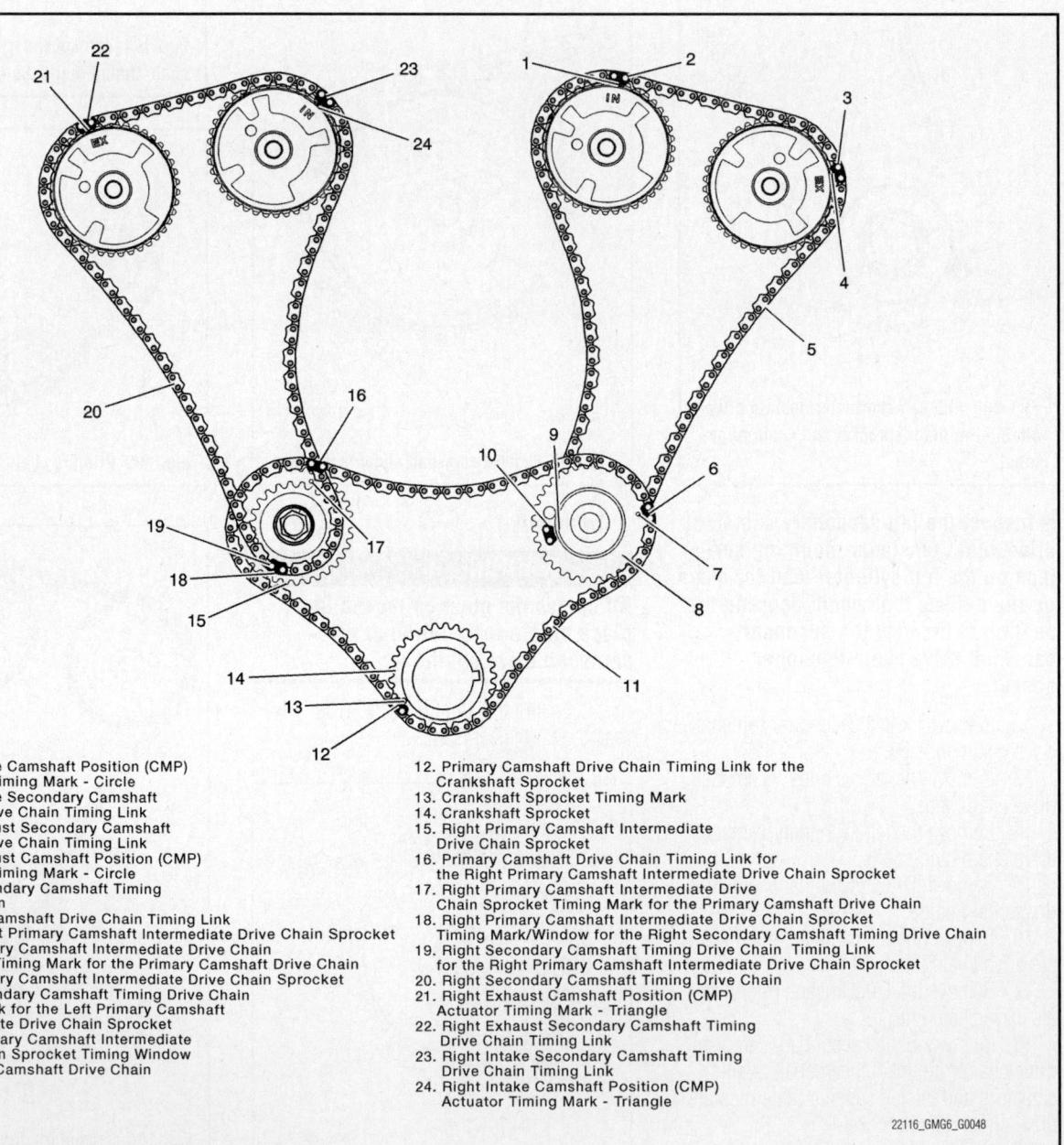

1. Left Intake Camshaft Position (CMP) Actuator Timing Mark - Circle
2. Left Intake Secondary Camshaft Timing Drive Chain Timing Link
3. Left Exhaust Secondary Camshaft Timing Drive Chain Timing Link
4. Left Exhaust Camshaft Position (CMP) Actuator Timing Mark - Circle
5. Left Secondary Camshaft Timing Drive Chain
6. Primary Camshaft Drive Chain Timing Link for the Left Primary Camshaft Intermediate Drive Chain Sprocket
7. Left Primary Camshaft Intermediate Drive Chain Sprocket Timing Mark for the Primary Camshaft Drive Chain
8. Left Primary Camshaft Intermediate Drive Chain Sprocket
9. Left Secondary Camshaft Timing Drive Chain Timing Link for the Left Primary Camshaft Intermediate Drive Chain Sprocket
10. Left Primary Camshaft Intermediate Drive Chain Sprocket Timing Window
11. Primary Camshaft Drive Chain
12. Primary Camshaft Drive Chain Timing Link for the Crankshaft Sprocket
13. Crankshaft Sprocket Timing Mark
14. Crankshaft Sprocket
15. Right Primary Camshaft Intermediate Drive Chain Sprocket
16. Primary Camshaft Drive Chain Timing Link for the Right Primary Camshaft Intermediate Drive Chain Sprocket
17. Right Primary Camshaft Intermediate Drive Chain Sprocket Timing Mark for the Primary Camshaft Drive Chain
18. Right Primary Camshaft Intermediate Drive Chain Sprocket Timing Mark/Window for the Right Secondary Camshaft Timing Drive Chain
19. Right Secondary Camshaft Timing Drive Chain Timing Link for the Right Primary Camshaft Intermediate Drive Chain Sprocket
20. Right Secondary Camshaft Timing Drive Chain
21. Right Exhaust Camshaft Position (CMP) Actuator Timing Mark - Triangle
22. Right Exhaust Secondary Camshaft Timing Drive Chain Timing Link
23. Right Intake Secondary Camshaft Timing Drive Chain Timing Link
24. Right Intake Camshaft Position (CMP) Actuator Timing Mark - Triangle

22116_GMG6_G0048

Fig. 107 Timing drive chain alignment diagram (Fourth Design)—3.6L engine

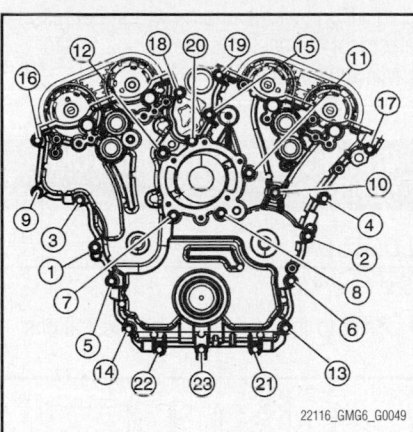

Fig. 108 Engine front cover tightening sequence—3.6L engine

35. Using the J 45027, reset the left secondary camshaft drive chain tensioner plunger.

36. Install the plunger into the left secondary camshaft drive chain tensioner body.

37. Compress the plunger into the body and lock the left secondary camshaft drive chain tensioner by inserting the EN 46112 into the access hole in the side of the left secondary camshaft drive chain tensioner body.

38. Slowly release pressure on the left secondary camshaft drive chain tensioner. The left secondary camshaft drive chain tensioner should remain compressed.

39. Install a NEW left secondary camshaft drive chain tensioner gasket to the left secondary camshaft drive chain tensioner.

40. Install the left secondary camshaft drive chain tensioner bolts through the left secondary camshaft drive chain tensioner and gasket.

41. Ensure the left secondary camshaft drive chain tensioner mounting surface on the left cylinder head does not have any burrs or defects that would degrade the sealing of the NEW left secondary camshaft drive chain tensioner gasket.

42. Place the left secondary camshaft drive chain tensioner into position and loosely install the bolts to the block.

43. Verify the proper placement of the left secondary camshaft drive chain tensioner gasket tab.

44. Tighten the left secondary camshaft drive chain tensioner bolts in two steps:

   a. Tighten tensioner bolts to 44 inch lbs. (5 Nm).

   b. Tighten tensioner bots an additional 17 ft. lbs. (23 Nm).

45. Release the left secondary camshaft drive chain tensioner by pulling out the EN 46112 pin and unlocking the tensioner plunger.

46. Verify the left secondary camshaft drive chain timing mark alignments.

⁂ WARNING

Ensure that the crankshaft is in the stage one timing drive assembly position.

47. Install the primary camshaft drive chain.

48. Wrap the primary camshaft drive chain around the large sprockets of each camshaft intermediate drive chain idler and the crankshaft sprocket.

49. The left camshaft intermediate drive chain idler timing mark will align with a timing camshaft drive chain link.

50. The right camshaft intermediate drive chain idler timing mark will align with a timing camshaft drive chain link.

51. The crankshaft sprocket timing mark will align with a timing camshaft drive chain link.

52. Ensure all the timing marks are properly aligned with the timing camshaft drive chain links.

53. Ensure that the crankshaft is in the stage 2 timing drive assembly position.

54. Install the right secondary camshaft drive chain.

55. Place the secondary camshaft drive chain around the right camshaft intermediate drive chain idler outer sprocket, aligning the timing camshaft drive chain link with the alignment access hole made in the right camshaft intermediate drive chain idler inner sprocket.

56. Wrap the secondary camshaft drive chain around both right actuator drive sprockets.

57. Ensure there are 10 links between the timing camshaft drive chain links for the camshaft position actuator sprockets

58. Align the right exhaust camshaft position actuator sprocket alignment triangle mark with the timing camshaft drive chain link.

59. Align the right intake camshaft position actuator sprocket alignment triangle mark with the timing camshaft drive chain link.

60. There will be 22 links between the right camshaft intermediate drive chain idler timing camshaft drive chain link and each right camshaft position actuator sprocket timing camshaft drive chain link.

61. Install the 0.315 inch. (8mm) guide pins from the EN 46109 into the cylinder block.

62. Install the engine front cover to cylinder block seal.

63. Place a 0.118 inch (3mm) bead of RTV sealant, or equivalent, on the engine front cover.

64. Place the engine front cover onto the guide pins EN 46109 and slide into position.

65. Remove the EN 46109 guide pins.

66. Hand start all the front cover bolts.

67. Tighten the engine front cover bolts in sequence to 17 ft. lbs. (23 Nm).

68. Fill and bleed the cooling system.

69. Start the engine and check for leaks.

ENGINE PERFORMANCE & EMISSION CONTROLS

CAMSHAFT POSITION (CMP) SENSOR

LOCATION

See Figures 109 through 112.

Refer to the accompanying illustrations.

REMOVAL & INSTALLATION

Camshaft Position (CMP) Sensor— Bank 1, Exhaust

1. Before servicing the vehicle, refer to the Precautions Section.
2. Remove the air cleaner assembly.
3. Disconnect the engine wiring harness electrical connector from the bank 1 exhaust Camshaft Position (CMP) sensor.
4. Remove the CMP sensor bolt.
5. Remove the bank 1 exhaust CMP sensor.

To install:

6. Install the CMP sensor.

7. Install the CMP sensor bolt and tighten to 89 inch lbs. (10 Nm.).
8. Connect the engine wiring harness electrical connector to the bank 1 exhaust CMP sensor.
9. Install the air cleaner assembly.

Camshaft Position (CMP) Sensor— Bank 1, Intake

1. Before servicing the vehicle, refer to the Precautions Section.
2. Remove the air cleaner assembly.

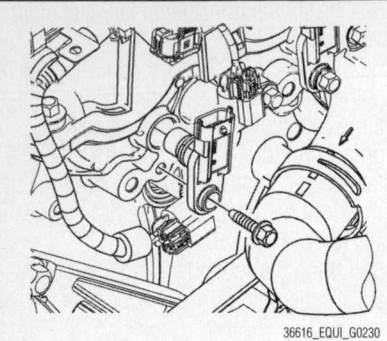

Fig. 109 Camshaft Position (CMP) sensor location—bank 1, exhaust

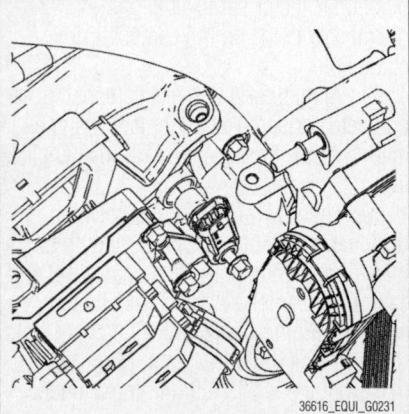

Fig. 110 Camshaft Position (CMP) sensor location—bank 1, intake

Fig. 111 Camshaft Position (CMP) sensor location—bank 2, exhaust

3. Disconnect the engine wiring harness electrical connector from the bank 1 intake Camshaft Position (CMP) sensor.

4. Remove the CMP sensor bolt.

5. Remove the bank 1 intake CMP sensor.

To install:

6. Install the CMP sensor.

7. Install the CMP sensor

bolt and tighten to 89 inch lbs. (10 Nm.).

8. Connect the engine wiring harness electrical connector to the bank 1 intake CMP sensor.

9. Install the air cleaner assembly.

Camshaft Position (CMP) Sensor— Bank 2, Exhaust

1. Before servicing the vehicle, refer to the Precautions Section.

2. Remove the air cleaner assembly.

3. Disconnect the engine wiring harness electrical connector from the bank 2 exhaust Camshaft Position (CMP) sensor.

4. Remove the CMP sensor bolt.

5. Remove the bank 2 exhaust CMP sensor.

To install:

6. Install the CMP sensor.

7. Install the CMP sensor bolt and tighten to 89 inch lbs. (10 Nm.).

8. Connect the engine wiring harness electrical connector to the bank 2 exhaust CMP sensor.

9. Install the air cleaner assembly.

Camshaft Position (CMP) Sensor— Bank 2, Intake

1. Before servicing the vehicle, refer to the Precautions Section.

2. Remove the air cleaner assembly.

3. Disconnect the engine wiring harness electrical connector from the bank 2 intake Camshaft Position (CMP) sensor.

4. Remove the CMP sensor bolt.

5. Remove the bank 2 intake CMP sensor.

To install:

6. Install the CMP sensor.

7. Install the CMP sensor bolt and tighten to 89 inch lbs. (10 Nm.).

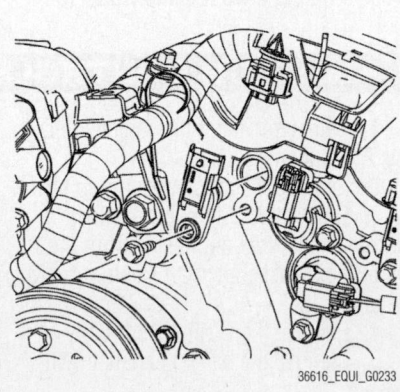

Fig. 112 Camshaft Position (CMP) sensor location—bank 2, intake

8. Connect the engine wiring harness electrical connector to the bank 2 intake CMP sensor.

9. Install the air cleaner assembly.

CRANKSHAFT POSITION (CKP) SENSOR

LOCATION

See Figure 113.

Refer to the accompanying illustrations.

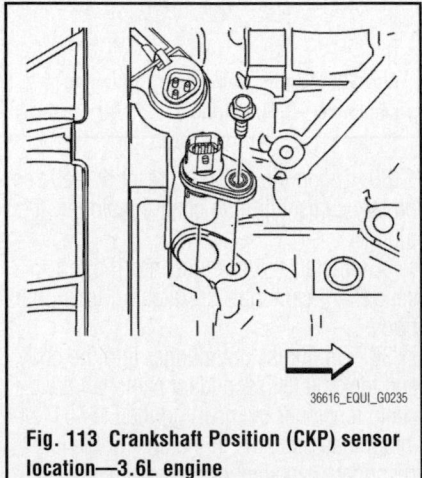

Fig. 113 Crankshaft Position (CKP) sensor location—3.6L engine

REMOVAL & INSTALLATION

1. Before servicing the vehicle, refer to the Precautions Section.

2. Raise and support the vehicle.

3. If equipped with All Wheel Drive (AWD), remove the transfer case. Refer to Transfer Case Assembly, removal & installation.

4. Disconnect the engine wiring harness electrical connector from the Crankshaft Position (CKP) sensor.

5. Remove the CKP sensor bolt.

6. Remove the CKP sensor.

To install:

7. Install the CKP sensor.

8. Install the CKP sensor bolt and tighten to 89 inch lbs. (10 Nm).

9. Connect the engine wiring harness electrical connector to the CKP sensor.

10. If equipped with all AWD, install the transfer case.

11. Lower the vehicle.

ELECTRONIC CONTROL MODULE (ECM)

LOCATION

The Electronic Control Module (ECM), also called the Engine Control Module

(ECM), is located at the left front corner of the engine compartment, mounted on top of the battery cover.

REMOVAL & INSTALLATION

See Figure 114.

※※ WARNING

Turn the ignition OFF when installing or removing the control module connectors and disconnecting or reconnecting the power to the control module (battery cable, Engine Control Module (ECM)/Transaxle Control Module (TCM) pigtail, control module fuse, jumper cables, etc.) in order to prevent internal control module damage.

※※ WARNING

Control module damage may result when the metal case contacts battery voltage. DO NOT contact the control module metal case with battery voltage when servicing a control module, using battery booster cables, or when charging the vehicle battery. In order to prevent any possible electrostatic discharge damage to the control module, do not touch the connector pins or the soldered components on the circuit board.

※※ WARNING

Remove any debris from around the control module connector surfaces before servicing the control module. Inspect the control module connector

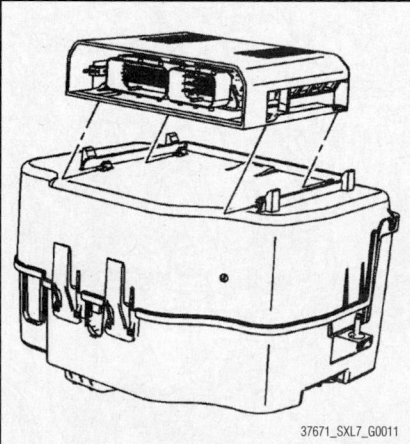

Fig. 114 Grasp the ECM retainer and slide the retainer rearward until the retainer disengages from the clips on the battery cover

gaskets when diagnosing or replacing the control module. Ensure that the gaskets are installed correctly. The gaskets prevent contaminant intrusion into the control module.

➡**The replacement control module must be programmed.**

It is necessary to record the remaining engine oil life. If the replacement module is not programmed with the remaining engine oil life, the engine oil life will default to 100 percent. If the replacement module is not programmed with the remaining engine oil life, the engine oil will need to be changed at 3,000 miles (5,000 km) from the last engine oil change.

It is necessary to record the remaining automatic transmission fluid life. If the replacement module is not programmed with the remaining automatic transmission fluid life, the automatic transmission fluid life will default to 100 percent. If the replacement module is not programmed with the remaining automatic transmission fluid life, the automatic transmission fluid will need to be changed at 50,000 miles (83,000 km).

1. Before servicing the vehicle, refer to the Precautions Section.

2. Turn the ignition OFF.

3. Using a scan tool, retrieve the percentage of remaining engine oil and the remaining automatic transmission fluid life. Record the remaining engine oil and the remaining automatic transmission fluid life.

4. Disconnect the negative battery cable.

5. Disconnect the engine wiring harness electrical connector from the ECM.

6. Grasp the ECM retainer and slide the retainer rearward until the retainer disengages from the 4 clips on the battery cover.

7. Disengage the clips on the ECM retainer from the ECM, and slide the ECM out of the retainer.

To install:

8. Slide the ECM into the retainer until the clips engage the ECM.

9. Grasp the ECM retainer and slide the retainer forward until the retainer locks against the 4 clips on the battery cover.

10. Connect the engine wiring harness electrical connector to the ECM.

11. Connect the negative battery cable.

12. If replacing the ECM, program the ECM. Specialized equipment is required to reprogram the ECM.

RESET

1. Before servicing the vehicle, refer to the Precautions Section.

2. For step-by-step programming instructions, please refer to the Techline Information System (TIS) terminal.

3. Review the information below to ensure proper programming protocol.

- **DO NOT** program a control module unless you are directed by a service procedure or you are directed by a service bulletin. Programming a control module at any other time will not permanently correct a concern.

- It is essential that the Tech 2, MDI and the TIS terminal are all equipped with the latest software before performing service programming.

- Due to the time requirements of programming a controller, install a battery charger to maintain system voltage. Stable battery voltage is critical during programming. Any fluctuation, spiking, over voltage or loss of voltage will interrupt programming. To ensure trouble-free programming, SUZUKI recommends using one of the following external power sources: A Midtronics PSC charger or a fully charged 12V jumper or booster pack disconnected from the AC voltage supply

- Some modules will require additional programming/setup events to be performed before or after programming

- Some vehicles may require the use of a CANDi or MDI module for programming.

- Review the appropriate service information for these procedures.

- DTC's may set during programming. Clear DTC's after programming is complete.

- Clearing powertrain DTC's will set the Inspection/Maintenance (I/M) system status indicators to NO.

4. Ensure the following conditions are met before programming a control module:

- There is not a charging system concern. All charging system concerns must be repaired before programming a control module

- Battery voltage is greater than 12 volts, but less than 16 volts. The battery must be fully charged before programming the control module

5. Turn **OFF** or disable any system that may put a load on the vehicle battery, such as the following components:
- Twilight sentinel
- Interior lights
- Daytime Running Lights (DRL)—Applying the parking brake, on most vehicles, disables the DRL system
- Heating, Ventilation, and Air Conditioning (HVAC) systems
- Engine cooling fans, radio, etc.

6. The ignition switch must be in the proper position. SPS prompts you to turn ON the ignition, with the engine OFF. **DO NOT** change the position of the ignition switch during the programming procedure, unless instructed to do so.

7. Make certain all tool connections are secure, including the following components and circuits:
   a. Tech 2
- The RS-232 communication cable port
- The connection at the Data Link Connector (DLC)
- The voltage supply circuits
   b. MDI
- The USB, Ethernet or Wireless communication port
- The connection at the Data Link Connector (DLC)

8. **DO NOT** disturb the tool harnesses while programming. If an interruption occurs during the programming procedure, programming failure or control module damage may occur.

9. **DO NOT** turn OFF the ignition if the programming procedure is interrupted or unsuccessful. Ensure that all control module and DLC connections are secure and the TIS terminal operating software is up to date. Attempt to reprogram the control module. If the control module cannot be programmed, replace the control module.

ENGINE COOLANT TEMPERATURE (ECT) SENSOR

LOCATION

See Figure 115.

Refer to the accompanying illustration.

REMOVAL & INSTALLATION

1. Before servicing the vehicle, refer to the Precautions Section.
2. Partially drain the cooling system.
3. Disconnect the engine wiring harness electrical connector from the Engine Coolant Temperature (ECT) sensor.
4. Remove the ECT sensor.

Fig. 115 Engine Coolant Temperature (ECT) sensor location—3.6L engine

To install:

5. Install the ECT sensor. Tighten the sensor to 16 ft. lbs. (22 Nm).
6. Connect the engine wiring harness electrical connector to the ECT sensor.
7. Fill the cooling system with the proper type and amount of fluid.

HEATED OXYGEN (HO2S) SENSOR

LOCATION

See Figures 116 through 119.

The Heated Oxygen Sensor (HO2S) (Bank 1 Sensor 1) is attached to the rear exhaust bank, above the catalytic converter.

The Heated Oxygen Sensor (HO2S) (Bank 1 Sensor 2) is attached to the rear exhaust bank, below the catalytic converter.

The Heated Oxygen Sensor (HO2S) (Bank 2 Sensor 1) is attached to the front exhaust bank, above the catalytic converter.

The Heated Oxygen Sensor (HO2S) (Bank 2 Sensor 2) is attached to the front exhaust bank, below the catalytic converter.

REMOVAL & INSTALLATION

Bank 1, Sensor 1

> **✳✳ CAUTION**
>
> **To avoid the danger of being burned, do not touch the exhaust system when the system is hot. The Heated Oxygen Sensor (HO2S) removal should be performed when the system is cool.**

1. Before servicing the vehicle, refer to the Precautions Section.
2. Remove the fuel injector sight shield, if necessary.
3. Disconnect the engine wiring harness electrical connector from the Heated Oxygen Sensor (HO2S) electrical connector. Remove

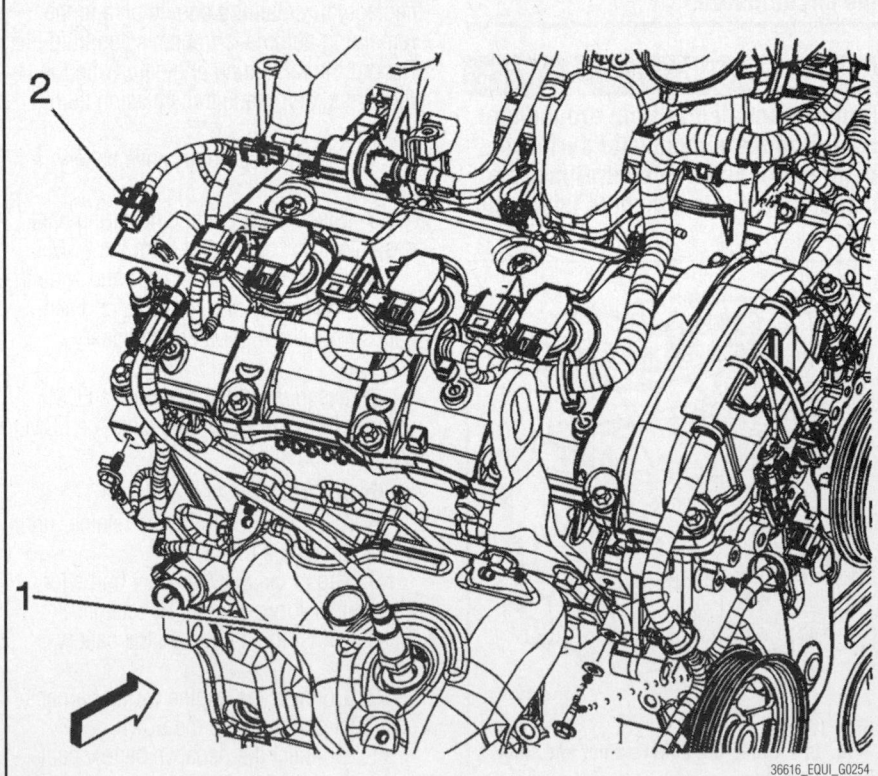

Fig. 116 Heated Oxygen (HO2S) Sensor location (bank 1, sensor 1)

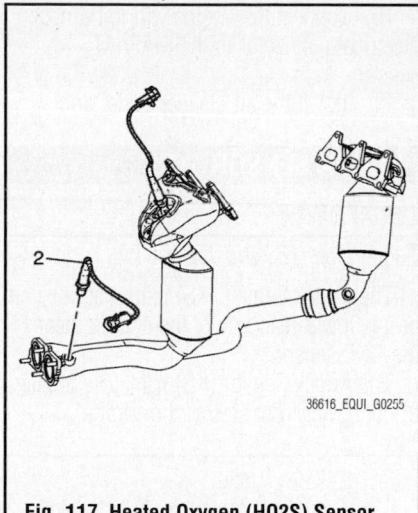

Fig. 117 Heated Oxygen (HO2S) Sensor location (bank 1, sensor 2)

the HO2S electrical connector retainer from the camshaft cover.

4. Remove the HO2S from the exhaust manifold.

To install:

→A special anti-seize compound is used in the HO2S threads. The compound consists of liquid graphite and glass beads. The graphite tends to burn away, but the glass beads remain, making the sensor easier to remove. New, or service replacement sensors already have the compound applied to the threads. If the sensor is removed from an exhaust component and if for any reason the sensor is to reinstalled, the threads must have anti-seize compound applied before the reinstallation.

5. Apply a small amount ot anti-seize compound to the threads of the HO2S.

6. Install the HO2S to the exhaust manifold and tighten to 31 ft. lbs. (42 Nm).

7. Connect the engine wiring harness electrical connector to the HO2S electrical connector.

8. Install the HO2S electrical connector retainer to the camshaft cover.

9. Install the fuel injector sight shield, if necessary.

Bank 1, Sensor 2

✷✷ CAUTION

To avoid the danger of being burned, do not touch the exhaust system when the system is hot. The Heated Oxygen Sensor (HO2S) removal should be performed when the system is cool.

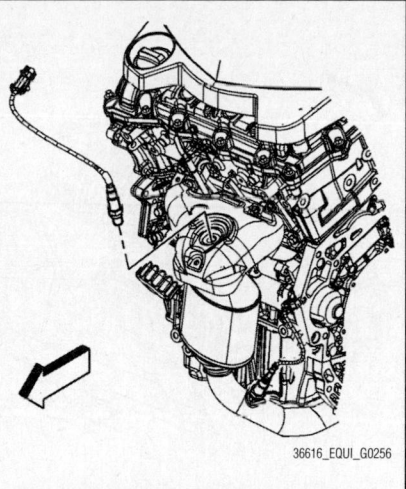

Fig. 118 Heated Oxygen (HO2S) Sensor location (bank 2, sensor 1)

1. Before servicing the vehicle, refer to the Precautions Section.

2. Remove the fuel injector sight shield, if necessary.

3. Disconnect the engine wiring harness electrical connector from the heated Oxygen Sensor (HO2S) electrical connector.

4. Remove the bank 1 sensor 2 HO2S from the catalytic converter.

To install:

→A special anti-seize compound is used in the HO2S threads. The compound consists of liquid graphite and glass beads. The graphite tends to burn away, but the glass beads remain, making the sensor easier to remove. New, or service replacement sensors already have the compound applied to the threads. If the sensor is removed from an exhaust component and if for any reason the sensor is to reinstalled, the threads must have anti-seize compound applied before the reinstallation.

5. Apply a small amount of anti-seize compound to the threads of the HO2S.

6. Install the HO2S to the exhaust manifold and tighten to 31 ft. lbs. (42 Nm).

7. Connect the engine wiring harness electrical connector to the HO2S electrical connector.

8. Install the fuel injector sight shield, if necessary.

Bank 2, Sensor 1

✷✷ CAUTION

To avoid the danger of being burned, do not touch the exhaust system

when the system is hot. The Heated Oxygen Sensor (HO2S) removal should be performed when the system is cool.

1. Before servicing the vehicle, refer to the Precautions Section.

2. Remove the fuel injector sight shield, if necessary.

3. Disconnect the engine wiring harness electrical connector from the heated Oxygen Sensor (HO2S) electrical connector.

4. Remove the HO2S electrical connector retainer from the retainer clip.

5. Remove the bank 2 sensor 1 HO2S from the exhaust manifold.

To install:

→A special anti-seize compound is used in the HO2S threads. The compound consists of liquid graphite and glass beads. The graphite tends to burn away, but the glass beads remain, making the sensor easier to remove. New, or service replacement sensors already have the compound applied to the threads. If the sensor is removed from an exhaust component and if for any reason the sensor is to reinstalled, the threads must have anti-seize compound applied before the reinstallation.

6. Apply a small amount of anti-seize compound to the threads of the HO2S.

7. Install the HO2S to the exhaust manifold and tighten to 31 ft. lbs. (42 Nm).

8. Connect the engine wiring harness electrical connector to the HO2S electrical connector.

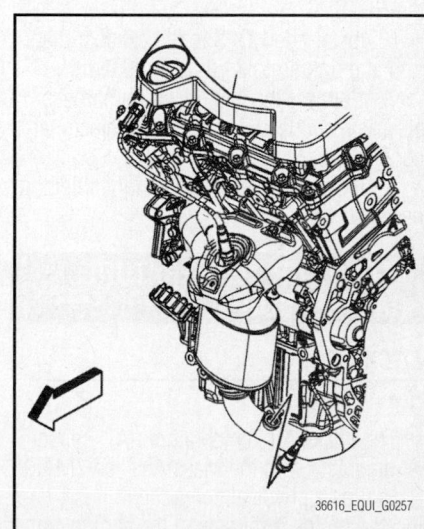

Fig. 119 Heated Oxygen (HO2S) Sensor location (bank 2, sensor 2)

9. Install the HO2S electrical connector retainer to the retainer clip.

10. Install the fuel injector sight shield, if necessary.

Bank 2, Sensor 2

✳✳ CAUTION

To avoid the danger of being burned, do not touch the exhaust system when the system is hot. The Heated Oxygen Sensor (HO2S) removal should be performed when the system is cool.

1. Before servicing the vehicle, refer to the Precautions Section.

2. Remove the fuel injector sight shield, if necessary.

3. Disconnect the engine wiring harness electrical connector from the heated Oxygen Sensor (HO2S) electrical connector.

4. Remove the bank 2 sensor 2 HO2S from the exhaust manifold.

To install:

➡A special anti-seize compound is used in the HO2S threads. The compound consists of liquid graphite and glass beads. The graphite tends to burn away, but the glass beads remain, making the sensor easier to remove. New, or service replacement sensors already have the compound applied to the threads. If the sensor is removed from an exhaust component and if for any reason the sensor is to reinstalled, the threads must have anti-seize compound applied before the reinstallation.

5. Apply a small amount of anti-seize compound to the threads of the HO2S.

6. Install the HO2S to the exhaust manifold and tighten to 31 ft. lbs. (42 Nm).

7. Connect the engine wiring harness electrical connector to the HO2S electrical connector.

8. Install the fuel injector sight shield, if necessary.

INTAKE AIR TEMPERATURE (IAT) SENSOR

LOCATION

See Figure 120.

The Intake Air Temperature (IAT) sensor is integrated with the Mass Air Flow (MAF) Sensor as a physical component. It is located on the right side of the engine compartment, to the right side of the throttle body, in the cold air duct.

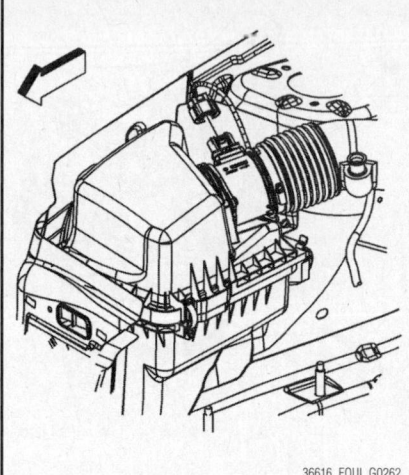

Fig. 120 Mass Air Flow (MAF)/Intake Air Temperature (IAT) sensor location

REMOVAL & INSTALLATION

See Figure 121.

1. Before servicing the vehicle, refer to the Precautions Section.

2. Remove the air cleaner outlet duct.

3. Disconnect the engine wiring harness electrical connector from the Mass Air Flow (MAF)/Intake Air Temperature (IAT) sensor.

4. Remove the MAF/IAT sensor screws.

5. Remove the MAF/IAT sensor from the air cleaner assembly.

6. Remove the MAF/IAT sensor seal.

To install:

7. Install the MAF/IAT sensor seal to the MAF/IAT sensor.

8. Install the MAF/IAT sensor to the air cleaner assembly.

9. Install the MAF/IAT sensor screws and tighten to 35 inch lbs. (4 Nm).

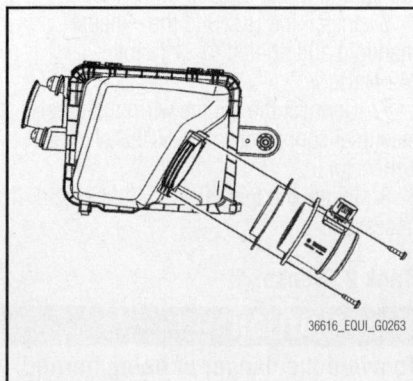

Fig. 121 MAF/IAT sensor removal

10. Connect the engine wiring harness electrical connector to the MAF/IAT sensor.

11. Install the air cleaner outlet duct.

KNOCK SENSOR (KS)

LOCATION

See Figures 122 and 123.

The Knock Sensor (KS) bank 1 is located on the lower rear side of the engine, near the CKP sensor.

The Knock Sensor (KS) bank 2 is located on the lower front side of the engine.

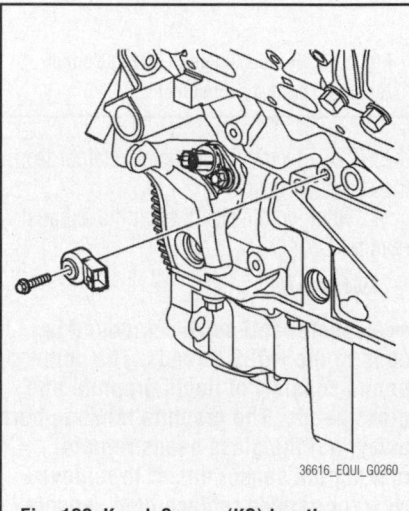

Fig. 122 Knock Sensor (KS) location—bank 1

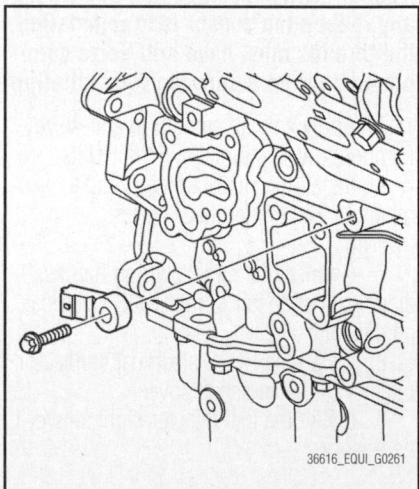

Fig. 123 Knock Sensor (KS) location—bank 2

REMOVAL & INSTALLATION

Knock Sensor (Bank 1)

1. Before servicing the vehicle, refer to the Precautions Section.

2. Remove the right catalytic converter. Refer to Catalytic Converter, removal & installation.

3. Remove the exhaust manifold heat shield.

4. Disconnect the engine wiring harness electrical connector from the bank 1 knock sensor.

5. Loosen the knock sensor bolt and remove the knock sensor.

To install:

6. Position the knock sensor to the engine and tighten the bolt to 17 ft. lbs. (23 Nm).

7. Connect the engine wiring harness electrical connector to the bank 1 knock sensor.

8. Install the exhaust manifold heat shield.

9. Install the right catalytic converter.

Knock Sensor (Bank 2)

1. Before servicing the vehicle, refer to the Precautions Section.

2. Raise and suitably support the vehicle.

3. Disconnect the engine wiring harness electrical connector from the bank 2 knock sensor.

4. Loosen the knock sensor bolt and remove the knock sensor.

To install:

5. Position the knock sensor to the engine and tighten the bolt to 17 ft. lbs. (23 Nm).

6. Connect the engine wiring harness electrical connector to the bank 2 knock sensor.

MASS AIR FLOW (MAF) SENSOR

LOCATION

See Figure 120.

The Mass Air Flow (MAF) sensor is located on the right side of the engine compartment, to the right side of the throttle body, in the cold air duct.

REMOVAL & INSTALLATION

See Figure 124.

1. Before servicing the vehicle, refer to the Precautions Section.

2. Remove the air cleaner outlet duct.

3. Disconnect the engine wiring harness electrical connector from the Mass Air Flow (MAF)/Intake Air Temperature (IAT) sensor.

4. Remove the MAF/IAT sensor screws.

5. Remove the MAF/IAT sensor from the air cleaner assembly.

6. Remove the MAF/IAT sensor seal.

To install:

7. Install the MAF/IAT sensor seal to the MAF/IAT sensor.

8. Install the MAF/IAT sensor to the air cleaner assembly.

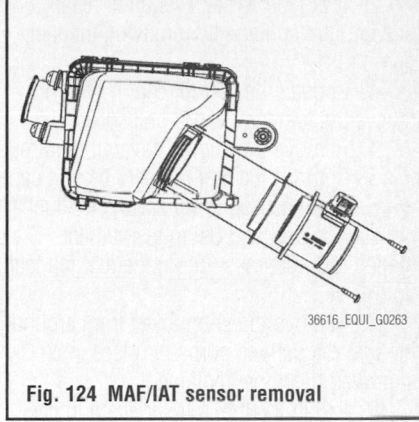

Fig. 124 MAF/IAT sensor removal

36616_EQUI_G0263

9. Install the MAF/IAT sensor screws and tighten to 35 inch lbs. (4 Nm).

10. Connect the engine wiring harness electrical connector to the MAF/IAT sensor.

11. Install the air cleaner outlet duct.

THROTTLE POSITION SENSOR (TPS)

LOCATION

The Throttle Position Sensor (TPS) is integral to the Throttle Control Actuator (TAC).

REMOVAL & INSTALLATION

The Throttle Position Sensors (TPS) are integral to the electronic throttle body. Refer to Throttle Body, removal & installation.

FUEL   **GASOLINE FUEL INJECTION SYSTEM**

FUEL SYSTEM SERVICE PRECAUTIONS

Safety is the most important factor when performing not only fuel system maintenance but any type of maintenance. Failure to conduct maintenance and repairs in a safe manner may result in serious personal injury or death. Maintenance and testing of the vehicle's fuel system components can be accomplished safely and effectively by adhering to the following rules and guidelines.

• To avoid the possibility of fire and personal injury, always disconnect the negative battery cable unless the repair or test procedure requires that battery voltage be applied.

• Always relieve the fuel system pressure prior to disconnecting any fuel system component (injector, fuel rail, pressure regulator,

etc.), fitting or fuel line connection. Exercise extreme caution whenever relieving fuel system pressure to avoid exposing skin, face and eyes to fuel spray. Please be advised that fuel under pressure may penetrate the skin or any part of the body that it contacts.

• Always place a shop towel or cloth around the fitting or connection prior to loosening to absorb any excess fuel due to spillage. Ensure that all fuel spillage (should it occur) is quickly removed from engine surfaces. Ensure that all fuel soaked cloths or towels are deposited into a suitable waste container.

• Always keep a dry chemical (Class B) fire extinguisher near the work area.

• Do not allow fuel spray or fuel vapors to come into contact with a spark or open flame.

• Always use a back-up wrench when loosening and tightening fuel line connec-

tion fittings. This will prevent unnecessary stress and torsion to fuel line piping.

• Always replace worn fuel fitting O-rings with new Do not substitute fuel hose or equivalent where fuel pipe is installed.

Before servicing the vehicle, make sure to also refer to the precautions in the beginning of this section as well.

RELIEVING FUEL SYSTEM PRESSURE

Fuel Pressure Relief—Without Fuel Pressure Gauge

✷✷ CAUTION

Always keep a dry chemical (Class B) fire extinguisher near the work area.

1. Before servicing the vehicle, refer to the Precautions Section.

2. If the fuel system requires repair, prevent fuel spillage by removing the fuel pump fuse.

3. Loosen the fuel fill cap in order to relieve the fuel tank vapor pressure.

4. Remove the engine cover, if required.

5. Remove the fuel rail service port cap.

6. Wrap a shop towel around the fuel rail service port and using a small flat-bladed tool, depress (open) the fuel rail test port valve.

7. Remove the shop towel from around the fuel rail service port, and place in an approved gasoline container.

8. Install the fuel rail service port cap.

9. Install the engine cover, if required.

10. Tighten the fuel fill cap.

Fuel Pressure Relief—With Fuel Pressure Gauge

✳✳ CAUTION

Always keep a dry chemical (Class B) fire extinguisher near the work area.

1. Before servicing the vehicle, refer to the Precautions Section.

2. If the fuel system requires repair, prevent fuel spillage by removing the fuel pump fuse.

3. Remove the engine cover, if required.

4. Loosen the fuel fill cap in order to relieve the fuel tank vapor pressure.

5. Remove the fuel rail service port cap.

✳✳ CAUTION °

Wrap a shop towel around the fuel pressure connection in order to reduce the risk of fire and personal injury. The towel will absorb any fuel leakage that occurs during the connection of the fuel pressure gage. Place the towel in an approved container when the connection of the fuel pressure gage is complete.

6. Wrap a shop towel around the fuel rail service port.

7. Connect the fuel pressure gauge to the fuel rail service port.

8. Place the hose on the fuel pressure gauge into an approved gasoline container.

9. Open the valve on the fuel pressure gauge in order to bleed any fuel from the fuel rail.

10. Close the valve on the fuel pressure gauge

11. Remove the hose on the fuel pressure gauge from the approved gasoline container.

12. Disconnect the fuel pressure gauge from the fuel rail service port.

13. Remove the shop towel from around the fuel rail service port, and place in an approved gasoline container.

14. Install the fuel rail service port cap.

15. Install the engine cover, if required.

16. Tighten the fuel fill cap.

FUEL FILTER

REMOVAL & INSTALLATION

See Figure 125.

The fuel filter is part of the fuel level sensor unit, fuel filter, and fuel pump assembly. Refer to Fuel Pump Module, removal & installation.

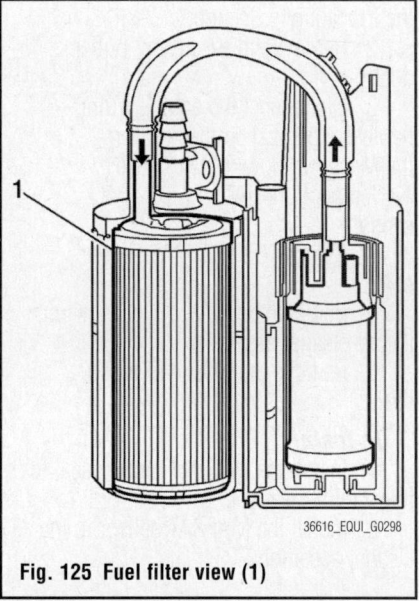

36616_EQUI_G0298

Fig. 125 Fuel filter view (1)

FUEL PUMP MODULE

REMOVAL & INSTALLATION

Primary Module
See Figures 126 and 127.

✳✳ WARNING

NEW fuel tank module seals are necessary each time the fuel tank module is serviced. Obtain NEW seals for both the primary and secondary modules prior to beginning this service procedure.

✳✳ CAUTION

Whenever fuel line fittings are loosened or removed, wrap a shop cloth around the fitting and have an approved container available to collect any fuel.

➡Clean all fuel pipe and hose connections and surrounding areas before disassembling to avoid possible contamination of the fuel system. Spray the fuel pump module cam-lock ring tang with penetrating oil prior to attempting removal.

✳✳ CAUTION

Always keep a dry chemical (Class B) fire extinguisher near the work area.

1. Before servicing the vehicle, refer to the precautions section.

2. Relieve the fuel system pressure. Refer to Fuel System Pressure, Relieving.

3. Ensure that the fuel level in the tank is less than ¼ full. If necessary, drain the fuel tank to at least this level.

4. Remove the fuel tank. Refer to Fuel Tank, removal & installation.

5. Remove the secondary fuel pump module. Refer to Fuel Pump Module, Secondary Module, removal & installation.

6. Disconnect the electrical connectors from the primary fuel pump module and fuel tank pressure sensor.

➡Avoid damaging the lock ring. Use only tool J-45722 to prevent damage to the lock ring.

➡Do not handle the fuel sender assembly by the fuel pipes. The amount of leverage generated by handling the fuel pipes could damage the joints.

✳✳ WARNING

Do NOT use impact tools. Significant force will be required to release the lock ring. The use of a hammer and screwdriver is not recommended. Secure the fuel tank in order to prevent fuel tank rotation.

7. Use tool J 45722, or equivalent, and a long breaker-bar in order to unlock the fuel sender lock ring. Turn the fuel sender lock ring in a counterclockwise direction.

8. Disconnect the fuel feed and vent lines from the fuel tank.

➡To prevent bending of the sending unit float arm during removal, lift the pump module up slightly to disengage the orientation tabs in the tank and rotate the module 45 degrees.

9. Remove the primary fuel pump module assembly.

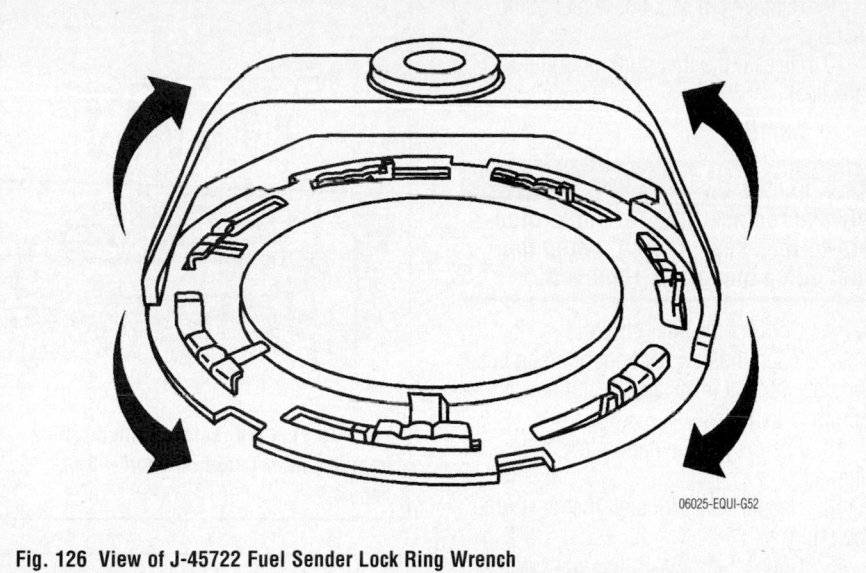

Fig. 126 View of J-45722 Fuel Sender Lock Ring Wrench

To install:

※※ WARNING

Always replace the fuel pump module-to-tank seal, O-ring, when the fuel pump module is removed.

➡ **Some lock rings were manufactured with DO NOT REUSE stamped into them. These lock rings may be reused if they are not damaged or warped. Inspect the lock ring for damage due to improper removal or installation procedures. If damage is found, install a NEW lock ring.**

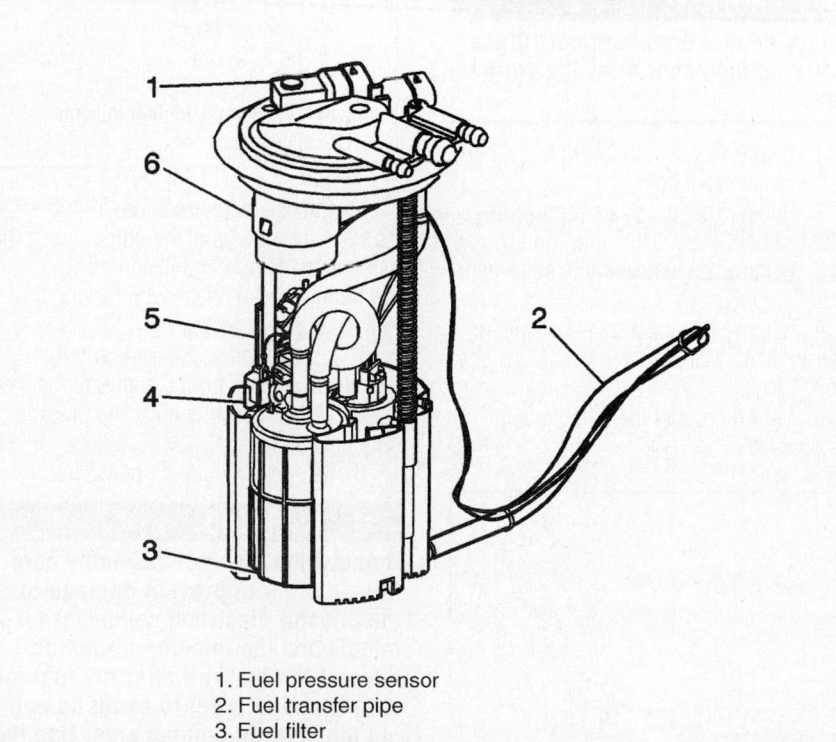

1. Fuel pressure sensor
2. Fuel transfer pipe
3. Fuel filter
4. Fuel level sensor
5. Fuel pressure regulator
6. Fill limiter vent valve

Fig. 127 Primary fuel tank module removal

※※ WARNING

Check the lock ring for flatness.

10. Discard the fuel pump module-to-tank seal.

11. Place the lock ring on a flat surface. Measure the clearance between to lock ring and the flat surface using a feeler gage at 7 points. If the warpage is less than 0.016 inch (0.41mm), the lock ring does not require replacement. If the warpage is greater than 0.016 inch (0.41 mm), the lock ring must be replaced.

12. Insert the new primary fuel pump module assembly with the level sender and the new fuel pump-to-tank seal. Ensure the orientation tabs are aligned.

※※ WARNING

Always replace the fuel sender seal when installing the fuel sender assembly. Replace the lock ring if necessary. Do not apply any type of lubrication in the seal groove.

13. Ensure the lock ring is installed with the correct side facing upward. A correctly installed lock ring will only turn in a clockwise direction.

14. Use tool J 45722 in order to install the fuel sender lock ring. Turn the fuel sender lock ring in a clockwise direction.

15. Connect the wiring harness to the primary fuel pump module and fuel tank pressure sensor.

16. Install the secondary fuel pump module.

17. Install the fuel tank.

18. Install any fuel that was removed and check for leaks.

Secondary Module

See Figure 128.

※※ WARNING

A NEW fuel tank module seal is necessary each time the fuel tank module is serviced. Obtain a NEW seal prior to beginning this service procedure.

※※ CAUTION

Whenever fuel line fittings are loosened or removed, wrap a shop cloth around the fitting and have an approved container available to collect any fuel.

➡ **Clean all fuel pipe and hose connections and surrounding areas before disassembling to avoid possible**

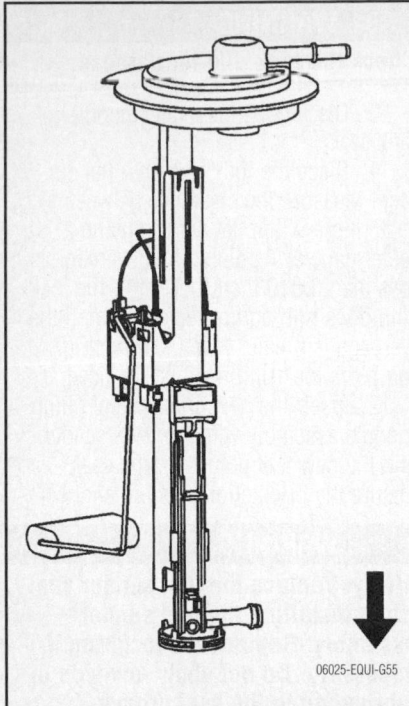

Fig. 128 Secondary fuel tank module

contamination of the fuel system. Spray the fuel pump module cam-lock ring tang with penetrating oil prior to attempting removal.

❋❋ CAUTION

Always keep a dry chemical (Class B) fire extinguisher near the work area.

1. Before servicing the vehicle, refer to the Precautions Section.
2. Relieve the fuel system pressure. Refer to Fuel System Pressure, Relieving.
3. Ensure that the fuel level in the tank is less than ¼ full. If necessary, drain the fuel tank to at least this level.
4. Remove the fuel tank. Refer to Fuel Tank, removal & installation.
5. Disconnect the EVAP vent line quick connect.

➥**To prevent retainer damage, do not attempt to remove the retainer with a 12 inch or shorter ratchet/breaker bar.**

6. Use tool J39765-A and remove the fuel pump module retaining ring.
7. Disconnect the secondary level sensor electrical connector.
8. Disconnect the suction port attaching tube by pressing down on the tab.

➥**To prevent bending of the sending unit float arm during removal, lift the pump module up slightly to disengage the orientation tabs in the tank and rotate the module 45 degrees.**

9. Remove the secondary fuel pump module.
10. Discard the fuel pump module-to-tank seal.

To install:

❋❋ WARNING

Always replace the fuel pump module-to-tank seal, O-ring, when the fuel pump module is removed.

11. Connect the suction port
12. Insert the new secondary fuel pump module with the level sender and new fuel pump-to-tank seal.
13. Ensure the orientation tabs are aligned.
14. Use the tool to install the fuel pump lock ring.
15. Connect the EVAP line quick connect.
16. Install the fuel tank.
17. Install any fuel that was removed and check for leaks.

FUEL RAIL AND INJECTOR

REMOVAL & INSTALLATION

See Figures 129 through 133.

❋❋ CAUTION

Always keep a dry chemical (Class B) fire extinguisher near the work area.

1. Before servicing the vehicle, refer to the Precautions Section.
2. Relieve the fuel system pressure. Refer to Fuel System Pressure, Relieving.
3. Remove the fuel feed line quick connect fitting retainer.
4. Remove the upper intake manifold. Refer to Intake Manifold, removal & installation.
5. Blow dirt out of the fitting using compressed air.

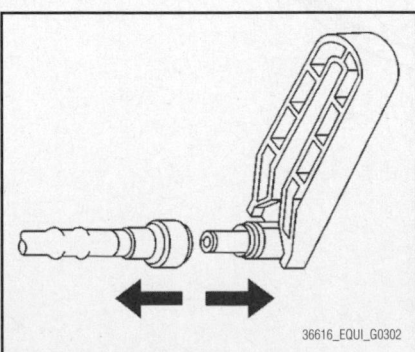

Fig. 129 Fuel feed line removal with quick connect tool J 37088-A

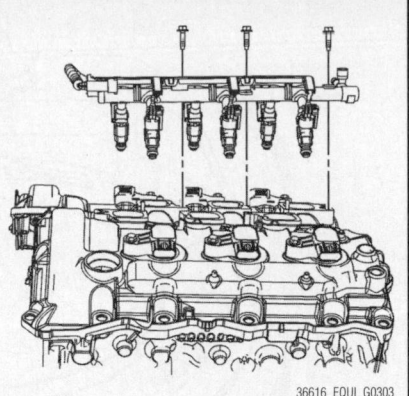

Fig. 130 Fuel rail removal with injectors from the lower intake manifold—3.6L engine

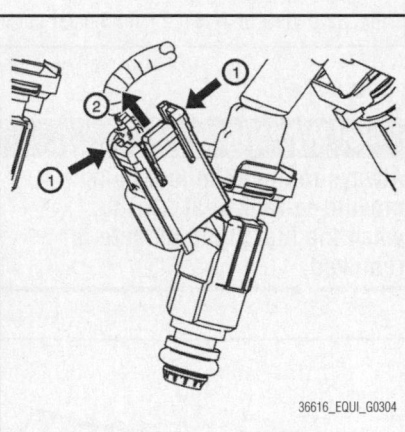

Fig. 131 Disengage the fuel injector electrical connector lock

6. Choose the correct tool from the J 37088-A for the size of the fitting. Insert the tool into the female connector, then push inward in order to release the locking tabs. Pull the connection apart.
7. Use compressed air in order to remove any debris from the around the area where the fuel injectors enter the lower intake manifold.
8. Remove the fuel rail bolts.

❋❋ WARNING

Remove the fuel rail assembly carefully in order to prevent damage to the injector electrical connector terminals and the injector spray tips. Support the fuel rail after the fuel rail is removed in order to avoid damaging the fuel rail components. Cap the fittings and plug the holes when servicing the fuel system in order to prevent dirt and other contaminants from entering open pipes and passages.

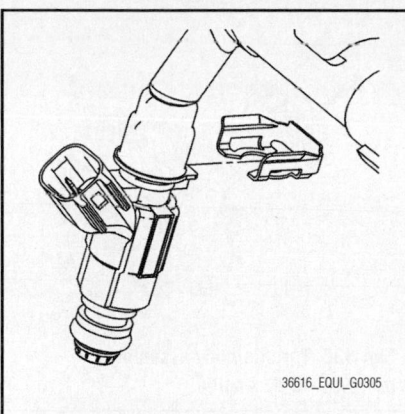

Fig. 132 Fuel injector retainer clip removal

9. Remove the fuel rail with fuel injectors from the lower intake manifold.

10. Disengage the fuel injector electrical connector lock.

11. Remove the fuel injector electrical connector.

12. Remove the fuel injector retainer clip.

13. Remove the fuel injector.

14. Remove and discard the fuel injector seals.

To install:

15. Install new fuel injector seals.

16. Install the fuel injector.

17. Install the fuel injector retainer clip.

18. Install the fuel injector electrical connector.

19. Engage the fuel injector electrical connector lock.

20. Carefully Install the fuel rail with fuel injectors to the lower intake manifold.

➡️**Apply a little petroleum jelly to the lower injector O-rings to help with the installation.**

21. Install the fuel rail bolts and tighten to 89 inch lbs. (10 Nm).

22. Before installing the fuel feed line, use a clean shop towel in order to wipe off the male pipe end.

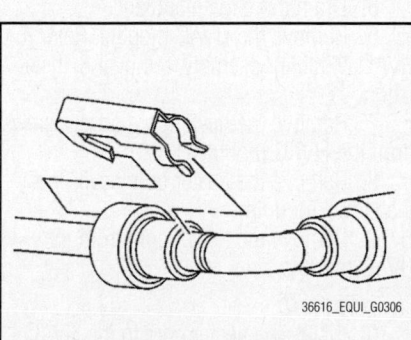

Fig. 133 Install the safety retainer to the quick-connect fitting

23. Inspect both ends of the fitting for dirt and burrs. Clean or replace the components as required.

24. Apply a few drops of clean engine oil to the male pipe end.

25. Push both sides of the fitting together in order to snap the retaining tabs into place.

26. Once installed, pull on both sides of the fitting in order to make sure the connection is secure.

27. Install the retainer to the quick-connect fitting.

28. Install the upper intake manifold.

29. Pressurize the fuel system and check for leaks as follows:

- Turn ON the ignition for 2 seconds
- Turn OFF the ignition for 10 seconds
- Turn ON the ignition
- Inspect for fuel leaks

FUEL TANK

REMOVAL & INSTALLATION

See Figures 134 and 135.

> ✳️✳️ **CAUTION**
>
> **Always keep a dry chemical (Class B) fire extinguisher near the work area.**

> ✳️✳️ **CAUTION**
>
> **Do not allow smoking or the use of open flames in the area where work on the fuel or EVAP system is taking place. Anytime work is being done on the fuel system, disconnect the negative battery cable, except for those tests where battery voltage is required.**

1. Before servicing the vehicle, refer to the Precautions Section.

2. Ensure that the fuel level in the tank is less than ¼ full. If necessary, drain the fuel tank to at least this level. Refer to Fuel Tank, Draining.

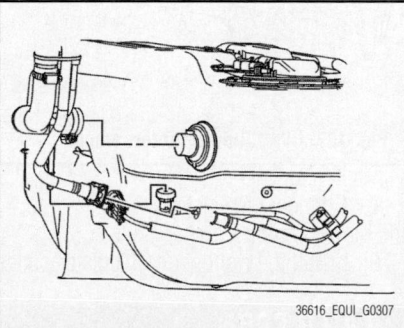

Fig. 134 Fuel supply and EVAP hose removal from the fuel tank

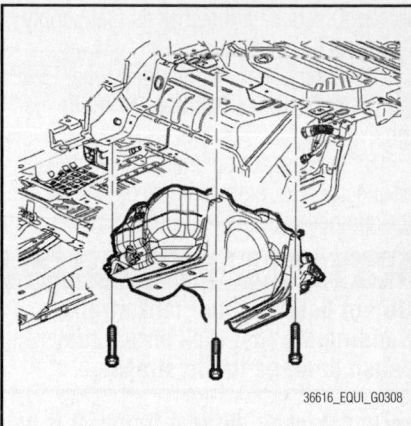

Fig. 135 Lower the fuel tank from the underbody of the vehicle

> ✳️✳️ **CAUTION**
>
> **Fuel supply lines will remain pressurized for long periods of time after the engine is shutdown. This pressure must be relieved before servicing the fuel system.**

3. Relieve the fuel system pressure. Refer to Fuel System Pressure, Relieving.

4. Disconnect the negative battery cable.

5. Raise and safely support the vehicle.

6. Remove the catalytic converter pipe flange-to-exhaust system pipe flange nuts.

7. Separate the pipes and discard the gasket.

8. Separate the rubber isolators from the hangers.

9. Remove the exhaust system.

10. If equipped with All Wheel Drive (AWD), remove the driveshaft. Refer to Driveshaft, removal & installation.

11. If equipped with Front Wheel Drive (FWD), disconnect the fuel tank vent pipe from the evaporative emission canister.

12. If equipped with AWD, remove the Evaporative Emission (EVAP) canister. Refer to Evaporative Emissions (EVAP) Canister, removal & installation.

> ✳️✳️ **WARNING**
>
> **Clean all fuel pipe connections and surrounding areas before disconnecting the fuel pipes to avoid contamination of the fuel system.**

> ✳️✳️ **CAUTION**
>
> **Whenever fuel lines are removed, catch fuel in an approved container. Container opening must be a minimum of 12 inches (300mm) diameter to adequately catch the fluid.**

13. Disconnect the chassis fuel supply line from the fuel tank.

14. Disconnect the fuel filler tube, EVAP vent hose, and fresh air hose from the fuel tank.

15. Disconnect the fuel tank electrical connector and remove the electrical connector retainer from the rear frame.

✳✳ WARNING

Do not bend the fuel tank straps. Bending the fuel tank straps may cause damage to the straps.

➡ **Do not lower the rear frame. It is not necessary to lower the rear frame for fuel tank removal.**

16. Support the fuel tank.

17. Remove the fuel tank strap bolts and fuel tank straps.

18. Lower the fuel tank from the underbody of the vehicle.

To install:

19. Install the fuel tank heat shield and fuel tank assembly to the vehicle.

20. Install the fuel tank straps and the fuel tank strap-to-body bolts. Tighten the bolts to 18 ft. lbs. (25 Nm).

21. Connect the fuel tank electrical connector and install the electrical connector retainer to the rear frame.

22. Connect the EVAP vent, and fresh air hoses to the fuel tank.

23. Connect the fuel filler tube to the fuel tank. Tighten the fuel filler tube clamp to 44 inch lbs. (5 Nm).

24. Connect the chassis fuel supply line to the fuel tank.

25. If equipped with FWD, connect the fuel tank vent pipe from the evaporative emission canister.

26. If equipped with AWD, install the Evaporative Emission (EVAP) canister.

27. If equipped with AWD, install the driveshaft.

28. Lower the vehicle.

29. Fill the fuel tank with gasoline.

30. Connect the negative battery cable.

31. Prime the fuel system:
- Cycle the ignition ON for 5 seconds and then OFF for 10 seconds
- Repeat the previous step twice
- Crank the engine until it starts. The maximum starter motor cranking time is 20 seconds
- If the engine does not start, repeat the priming procedure

32. Check the fuel system for leakage.

THROTTLE BODY

REMOVAL & INSTALLATION

See Figure 136.

➡ **The Throttle Actuator Control (TAC) module is not replaceable separate of the throttle body assembly. If the TAC module requires replacement, the entire throttle body assembly must be replaced.**

1. Before servicing the vehicle, refer to the Precautions Section.

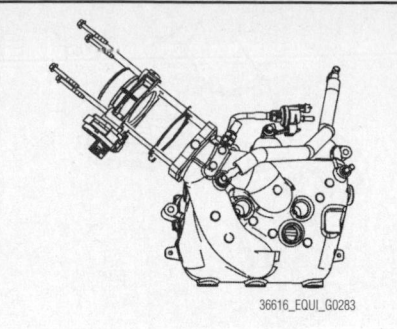

Fig. 136 Throttle body assembly removal—3.6L engine

2. Remove the air cleaner outlet duct.

3. Disconnect the engine wiring harness electrical connector from the Throttle Actuator Control (TAC) module.

4. Remove the throttle body bolts.

5. Remove the throttle body and gasket. Discard the gasket.

To install:

6. Position a NEW throttle body gasket to the upper intake manifold.

7. Position the throttle body to the upper intake manifold.

8. Install the throttle body bolts and tighten to 89 inch lbs. (10 Nm).

9. Connect the engine wiring harness electrical connector to the TAC module.

10. Install the air cleaner outlet duct.

HEATING & AIR CONDITIONING SYSTEM

BLOWER MOTOR

REMOVAL & INSTALLATION

See Figure 137.

1. Before servicing the vehicle, refer to the Precautions Section.

2. Remove the right sound insulator panel.

3. Disconnect the electrical connector from the blower motor.

4. Remove the blower motor screws from the HVAC module.

5. Remove the blower motor from the HVAC module.

To install:

6. Install the blower motor to the HVAC module.

7. Install the blower motor screws to the HVAC module. Tighten the screws to 13 inch lbs. (2 Nm).

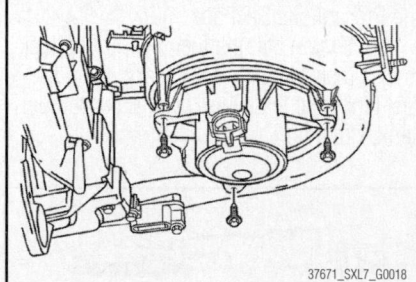

Fig. 137 HVAC Blower motor removal

8. Connect the electrical connector to the blower motor.

9. Install the right sound insulator panel.

HEATER CORE

REMOVAL & INSTALLATION

See Figures 138 and 139.

1. Before servicing the vehicle, refer to the Precautions Section.

2. Disable the frontal and curtain air bags.

3. Disconnect the negative battery cable.

4. Recover the refrigerant.

5. Drain the engine coolant.

6. Remove the HVAC module. Refer to HVAC Module Assembly, removal & installation.

7. Remove the heater core cover screws from the HVAC module.

8. Remove the heater core cover from the HVAC module.

9. Remove the heater core from the HVAC module.

To install:

10. Install the heater core to the HVAC module.

11. Install the heater core cover to the HVAC module.

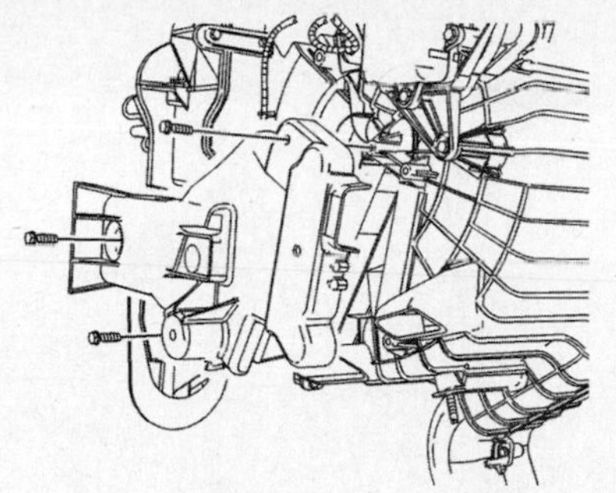

Fig. 138 Remove the heater core cover screws from the HVAC module

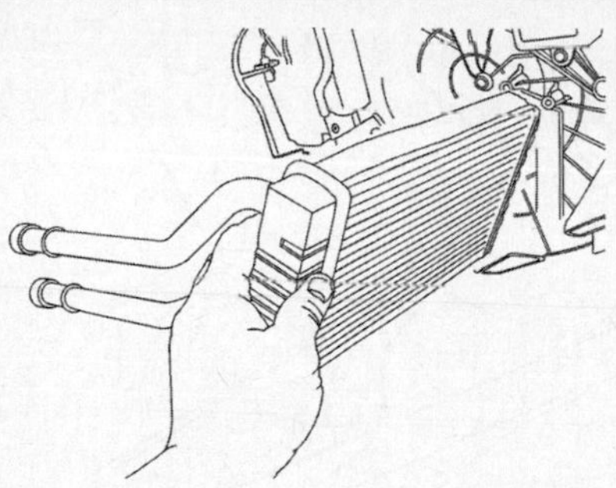

Fig. 139 Remove the heater core from the HVAC module

12. Install the heater core cover screws to the HVAC module. Tighten the screw to 13 inch lbs. (2 Nm)
13. Install the HVAC module to the vehicle.
14. Enable the frontal and curtain air bags.
15. Connect the negative battery cable.
16. Fill and bleed the cooling system.
17. Evacuate and charge the A/C system.
18. Test the affected A/C joints for leaks using a halogen leak detector.

STEERING

POWER RACK & PINION STEERING GEAR

REMOVAL & INSTALLATION

See Figures 140 through 145.

WARNING

With the wheels of the vehicle facing straight ahead, secure the steering wheel utilizing steering column anti-rotation pin, steering column lock, or a strap to prevent rotation. Locking of the steering column will prevent damage and a possible malfunction of the SRS system.

1. Before servicing the vehicle, refer to the Precautions Section.
2. Disable the SRS air bag system.
3. Disconnect the negative battery cable.
4. Raise and safely support the vehicle.
5. Remove the front wheels.
6. Remove both outer tie rod-to-steering knuckle nuts. Discard the nuts.

WARNING

Do not free the ball stud by using a pickle fork or a wedge-type tool. Damage to the seal or bushing may result.

➡Hold the ball stud to prevent turning during removal of the nut.

7. Rotate the intermediate steering shaft in order to gain access to the intermediate shaft pinch bolt.
8. Remove the intermediate to steering gear pinch bolt. Discard the bolt.
9. Disconnect the intermediate shaft from the steering gear.
10. Place drain pans under the vehicle as needed.

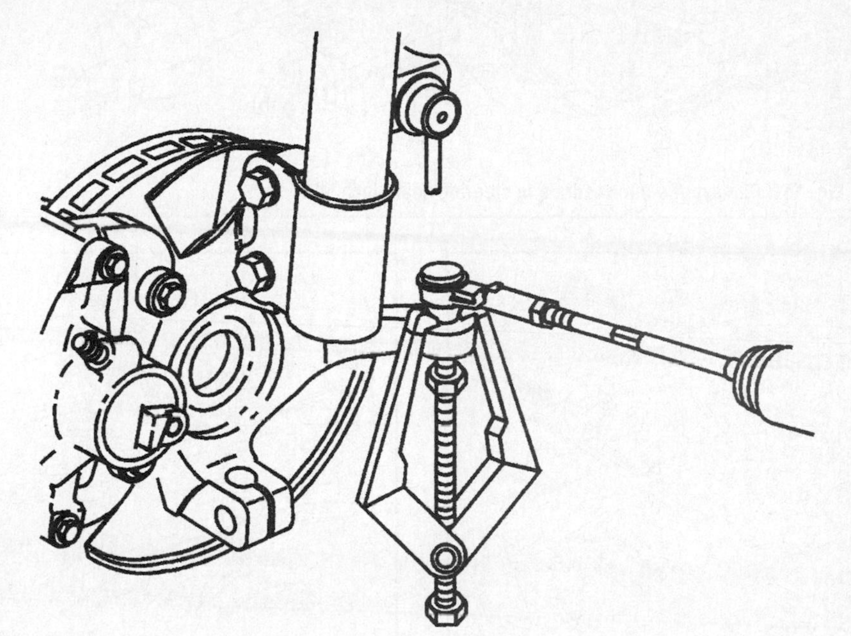

Fig. 140 Separate the tie rods from the steering knuckles

11. Remove the transaxle mount through bolt.
12. Remove the transaxle mount bolts and position the transaxle mount aside.
13. Disconnect the power steering gear inlet hose and the power steering cooler hose from the power steering gear.
14. If equipped, remove the power steering gear hose bracket.
15. Remove the power steering gear bolts.
16. Remove the power steering gear through the left wheel house area.
17. Transfer any parts as needed.

To install:

18. Position the power steering gear into the vehicle through the left wheel house area.
19. Install the power steering gear mounting bolts and tighten to 81 ft. lbs. (110 Nm).
20. If equipped, install the power steering gear hoses bracket and tighten the mounting bolt to 80 inch lbs. (9 Nm).
21. Install the power steering hoses and tighten to 18 ft. lbs. (25 Nm).
22. Install the transaxle mount and transaxle mount bolts, tighten the bolts to 37 ft. lbs. (50 Nm).
23. Install the transaxle mount bolt through and tighten to 81 ft. lbs. (110 Nm).
24. Clean any excess fluid from the vehicle and remove the drain pans.

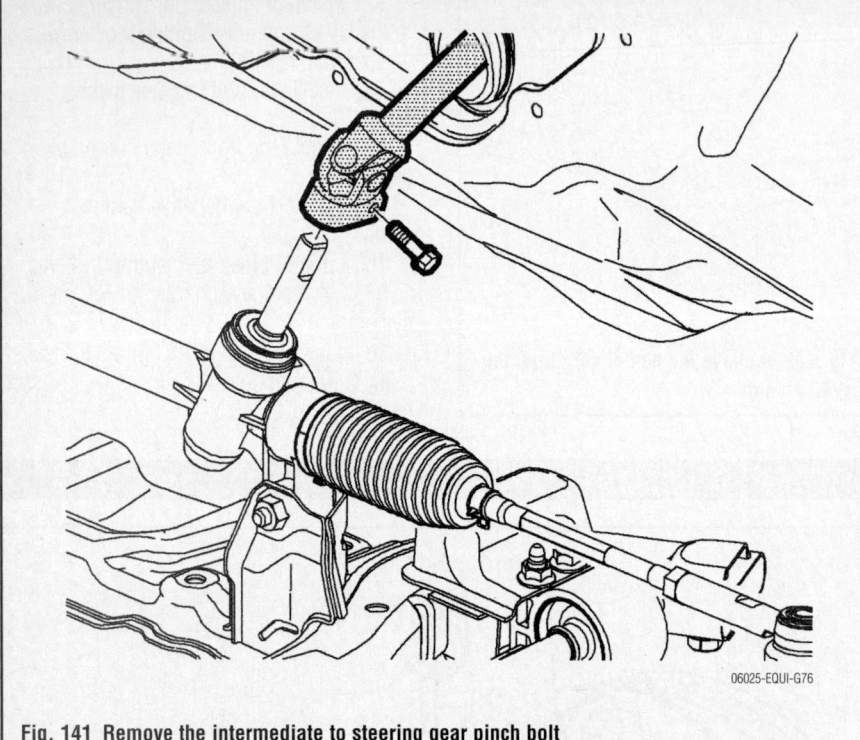

Fig. 141 Remove the intermediate to steering gear pinch bolt

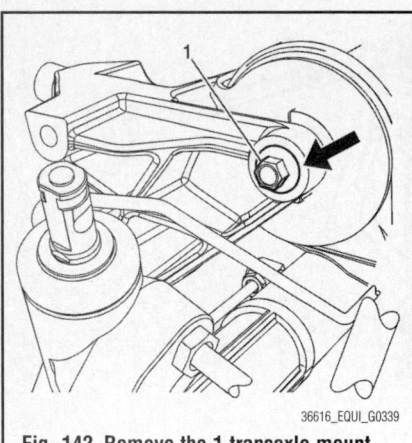

Fig. 142 Remove the 1 transaxle mount bolt (1)

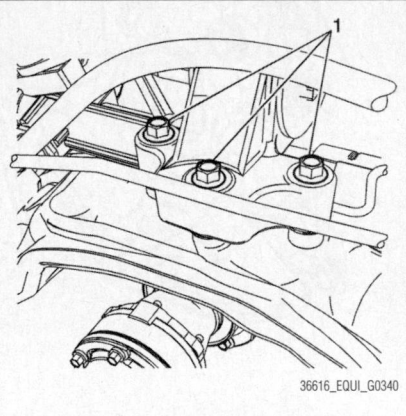

Fig. 143 Remove the 3 transaxle mount bolts (1)

25. Connect the intermediate shaft to the steering gear and install a new pinch bolt. Tighten the intermediate pinch bolt to 25 ft. lbs. (34 Nm).

26. Connect the tie rod to the knuckle and install a new nut. Tighten the nut to 44 ft. lbs. (60 Nm).

27. Install the front wheels.

28. Enable the SRS air bag system.

29. Connect the negative battery cable.

30. Check the front wheel alignment and align as necessary.

31. Lower the vehicle.

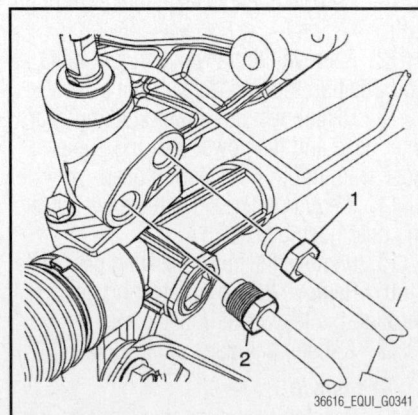

Fig. 144 Disconnect the power steering gear hoses (1 and 2)

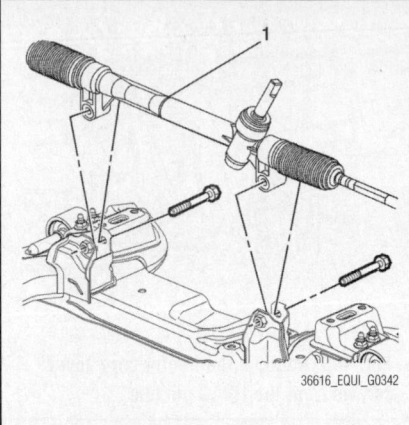

Fig. 145 Hydraulic Power Steering (HPS) gear removal (1)

POWER STEERING PUMP

REMOVAL & INSTALLATION

See Figures 146 and 147.

1. Before servicing the vehicle, refer to the Precautions Section.

2. Remove the drive belt. Refer to Accessory Drive Belt, removal & installation.

3. Remove as much power steering fluid from the remote power steering fluid reservoir as possible.

4. Place drain pans under the vehicle, as needed.

5. Remove the power steering reservoir return hose clamp and disconnect the return hose from the power steering pump.

6. Disconnect the power steering pressure hose from the power steering pump.

7. Remove the power steering pump bolts.

8. Remove the power steering pump through the right wheelhouse area.

9. Transfer the power steering pump pulley if needed.

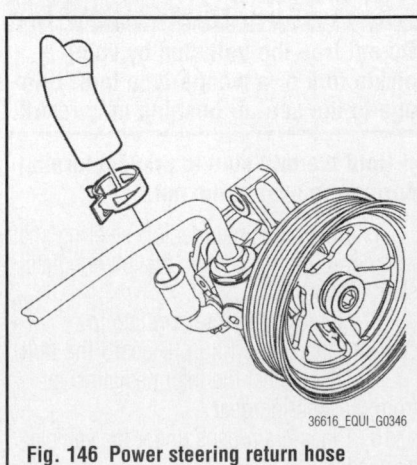

Fig. 146 Power steering return hose removal

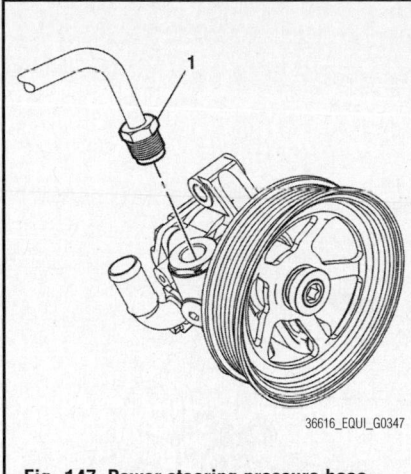

Fig. 147 Power steering pressure hose removal (1)

To install:

10. Position the power steering pump to the vehicle through the right wheelhouse area.

11. Install the power steering pump bolts and tighten to 37 ft. lbs. (50 Nm).

12. Connect the power steering gear pressure hose to the power steering pump. Tighten the hose to 18 ft. lbs. (25 Nm).

13. Install the power steering return hose and reposition the clamp.

14. Install the drive belt.

15. Fill and bleed the power steering system.

16. Remove the drain pans and clean off any excess fluid.

BLEEDING

1. Before servicing the vehicle, refer to the Precautions Section.

2. Fill pump reservoir with fluid to minimum system level, FULL COLD level, or middle of hash mark on cap stick fluid level indicator.

3. If equipped with hydro-boost, fully charge the hydro-boost accumulator using the following procedure:

- Start the engine
- Firmly apply the brake pedal 10–15 times
- Turn the engine OFF

4. Raise the vehicle until the front wheels are off the ground.

5. Key ON engine OFF, turn the steering wheel from stop to stop 12 times.

➡**Vehicles equipped with hydro-boost systems or longer length power steering hoses may require turns up to 15–20 stop to stops.**

6. Verify power steering fluid level per operating specification.

7. Start the engine. Rotate the steering wheel from left to right. Check for sign of cavitation or fluid aeration (pump noise/whining).

8. Verify the fluid level. Repeat the bleed procedure, if necessary.

SUSPENSION

CONTROL LINKS

REMOVAL & INSTALLATION

See Figure 148.

1. Before servicing the vehicle, refer to the Precautions Section.
2. Raise and safely support the vehicle.
3. Remove the front wheels.
4. Remove the upper and lower stabilizer link retaining nuts.

➡**Use the proper size Allen wrench to keep thc stabilizer link ball stud from rotate while removing or installing the nut.**

To install:

5. Install the stabilizer link and retaining nuts.

➡**Use the proper size Allen wrench to keep the stabilizer link ball stud from rotate while removing or installing the nut.**

6. Tighten the retaining nuts as follows:

- Tighten the upper retaining nut to 55 ft. lbs. (75 Nm).
- Tighten the lower retaining nut to 63 ft. lbs. (85 Nm).

7. Install the front wheels. Tighten the wheel nuts to 100 ft. lbs. (140 Nm).

8. Lower the vehicle.

LOWER BALL JOINT

REMOVAL & INSTALLATION

See Figures 149 and 150.

1. Before servicing the vehicle, refer to the Precautions Section.

2. Remove the lower control arm. Refer to Lower Control Arm, removal & installation.

3. Place the control arm in a vise or suitable holding device.

4. Remove the ball joint rivets using the following procedure:

- Drill through the rivets using a 5/16 inch (8mm) drill bit
- Enlarge the hole using a 31/64 inch (12mm) drill bit

FRONT SUSPENSION

Fig. 149 Drill out the ball joint rivets

- Remove any remaining burs from the control arm

5. Remove the ball joint from the control arm. Note the position of the ball joint for reassembly.

To install:

➡**The control arm must be clean and free of debris.**

6. Install the ball joint to the control arm as previously noted.

✳✳ WARNING

Only use hardware provided with the new ball joint. The bolts must be installed with the bolt head on top of the ball joint.

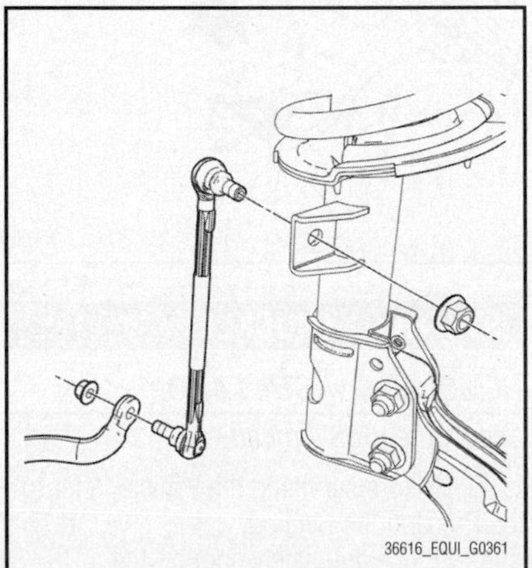

Fig. 148 Stabilizer link removal

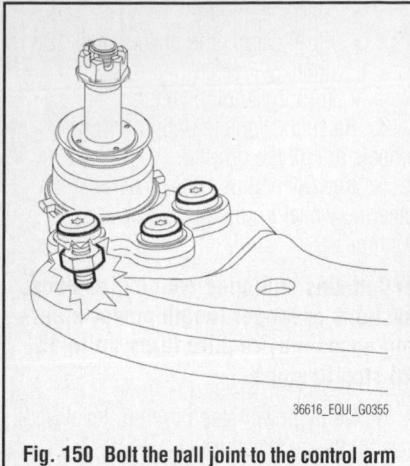

Fig. 150 Bolt the ball joint to the control arm

7. Install the ball joint to control arm bolts. Tighten the bolts and nuts to 50 ft. lbs. (68 Nm).

8. Install the lower control arm.

9. Perform a wheel alignment.

LOWER CONTROL ARM

REMOVAL & INSTALLATION

See Figures 151 and 152.

1. Before servicing the vehicle, refer to the Precautions Section.

2. Raise and support the vehicle.

3. Remove the wheel and tire assembly.

4. Remove the control arm ball stud cotter pin. Discard the cotter pin.

5. Loosen the ball stud nut until the nut is level with the top of the ball stud.

6. Separate the lower control arm from the steering knuckle.

7. Remove the ball stud nut.

8. Remove the control arm-to-frame front bolt and nut. Discard the bolt and nut.

9. Remove the control arm-to-frame rear bolts and nuts. Discard the bolts and nuts.

10. Remove the control arm.

To install:

11. Install the control arm to the frame.

12. Install new control arm-to-frame rear bolts and nuts. Tighten the nuts to 52 ft. lbs. (70 Nm).

13. Install a new arm-to-frame front bolt and nut. Tighten the bolt to 74 ft. lbs. (100 Nm).

14. Position the control arm ball stud into the steering knuckle, tighten the nut to 30 ft. lbs. (40 Nm).

✴✴ WARNING

Do not loosen the castle nut, only tighten to align the ball stud slot. Ensure that the cotter pin ends do not

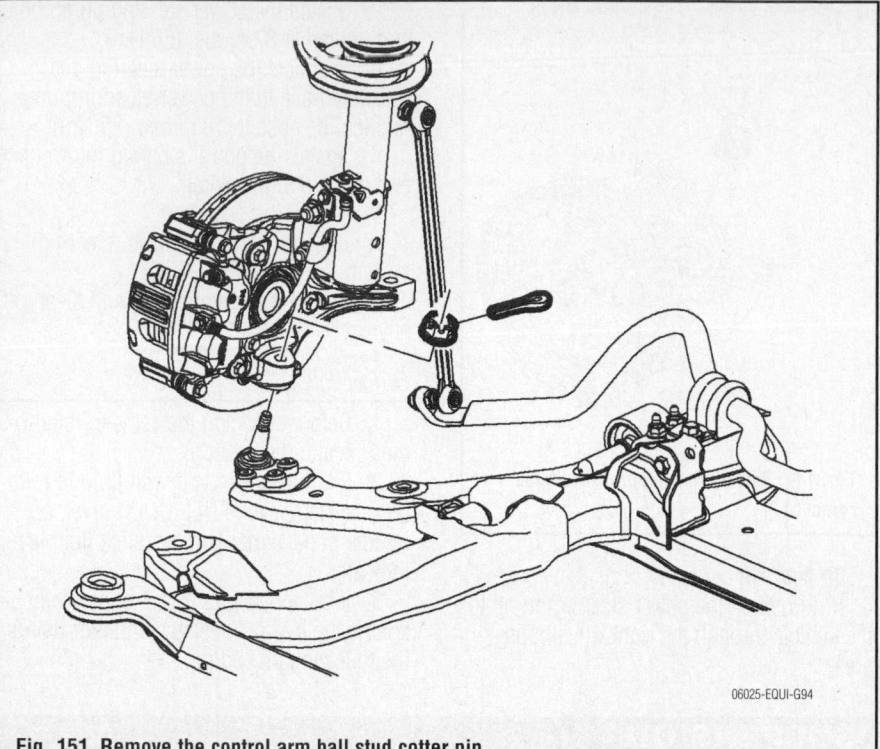

Fig. 151 Remove the control arm ball stud cotter pin

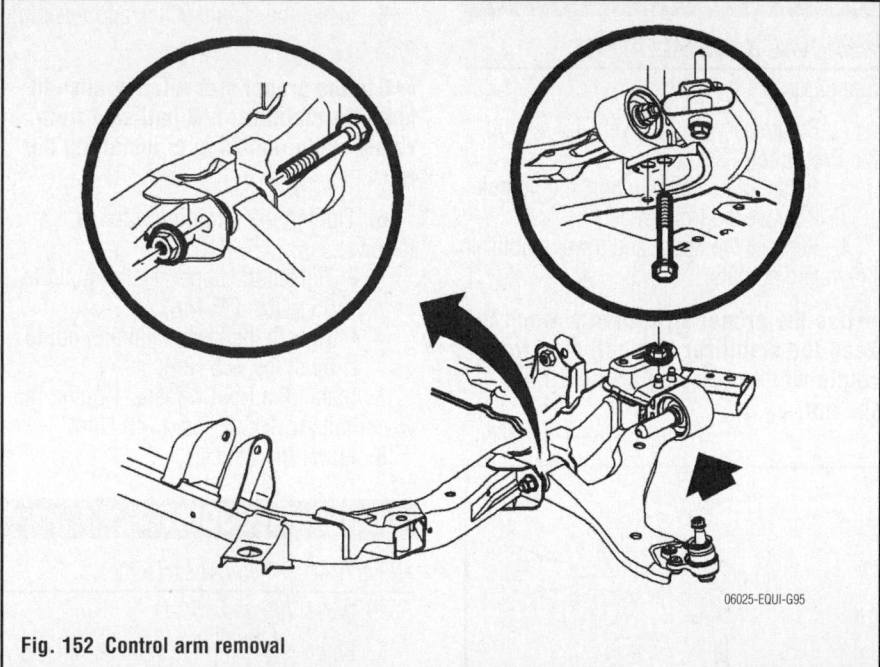

Fig. 152 Control arm removal

contact the Antilock Brake System (ABS) sensor harness or drive axle.

15. Continue to tighten the nut only enough to align the castle nut slots with the ball stud, install the cotter pin.

16. Install the wheel and tire assembly. Tighten the wheel nuts to 100 ft. lbs. (140 Nm).

17. Lower the vehicle.

18. Perform a wheel alignment.

STEERING KNUCKLE

REMOVAL & INSTALLATION

See Figures 153 through 156.

1. Before servicing the vehicle, refer to the Precautions Section.

2. Raise and safely support the vehicle.

3. Remove the tire and wheel.

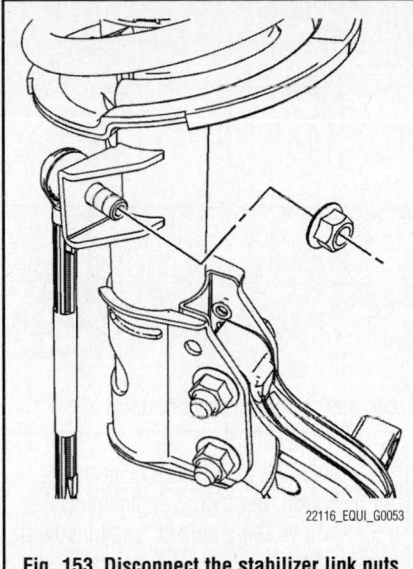

Fig. 153 Disconnect the stabilizer link nuts

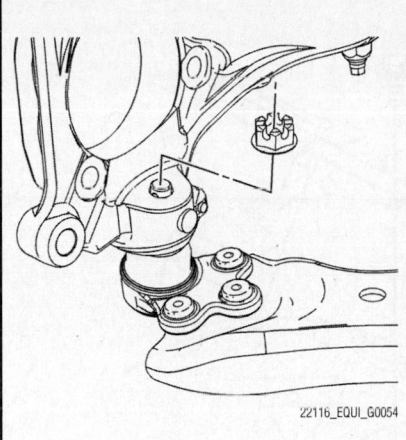

Fig. 155 Remove the lower ball joint nut to separate the lower control arm from the knuckle

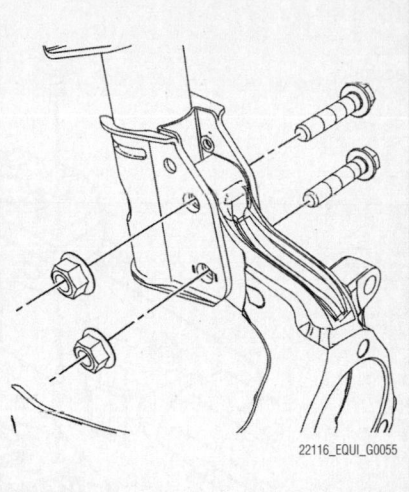

Fig. 156 Remove the steering knuckle-to-strut bolts

➡**Do not allow the stabilizer link ball stud to rotate while removing the link nut.**

4. Disconnect the stabilizer link from the strut assembly.

5. Loosen the steering knuckle to strut bolts and nuts.

6. Remove the wheel bearing/hub assembly. Refer to Wheel Hub & Bearing (sealed unit), removal & installation.

7. Remove and discard the lower ball joint cotter pin.

8. Loosen the ball stud nut, until level with the top of the ball stud.

9. Separate the lower control arm from the steering knuckle.

10. Remove the lower control arm and nut from the steering knuckle.

11. Remove the outer tie rod end-to-knuckle nut.

12. Separate the outer tie rod from the steering knuckle.

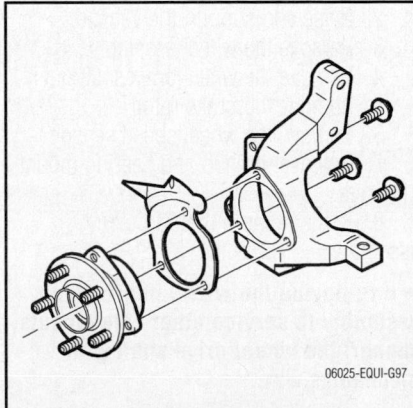

Fig. 154 Hub and bearing assembly removed from the steering knuckle

13. Remove the steering knuckle-to-strut bolts and nuts. Discard the bolts and nuts.

14. Remove the steering knuckle from the vehicle.

To install:

15. Install the steering knuckle to strut assembly.

16. Loosely install the strut-to-steering knuckle bolts and nuts.

17. Install the control arm ball stud into the steering knuckle.

18. Install the ball stud and tighten the nut to 30 ft. lbs. (40 Nm).

19. Tighten the strut-to-steering knuckle bolts and nuts. Tighten the bolts and nuts to 148 ft. lbs. (200 Nm).

➡**Do not loosen the castle nut for cotter pin installation.**

20. Tighten the castle nut enough to allow for cotter pin installation.

➡**The cotter pin must not contact the wheel speed sensor or drive axle.**

21. Install the cotter pin.

22. Install the wheel bearing/hub assembly.

23. Connect the outer tie rod end to the steering knuckle.

24. Seat the ball stud taper. Tighten to 30 ft. lbs. (40 Nm).

25. Install a new tie rod retention nut. Tighten the nut to 37 ft. lbs. (50 Nm).

26. Connect the stabilizer link to the strut assembly. Tighten the nut to 48 ft. lbs. (65 Nm).

27. Install the tire and wheel. Tighten the wheel nuts to 100 ft. lbs. (140 Nm).

28. Lower the vehicle.

29. Perform a wheel alignment.

STRUT

REMOVAL & INSTALLATION

See Figures 157 and 158.

1. Before servicing the vehicle, refer to the Precautions Section.

2. Remove the 3 upper strut mount bolts.

3. Raise and safely support the vehicle.

4. Remove the wheel assembly.

5. Remove the brake hose bracket from the strut assembly.

6. Disconnect the stabilizer link from the strut assembly.

7. Remove the strut to knuckle bolts and nuts.

8. Remove the strut assembly from the vehicle.

To install:

9. Position the strut assembly to the vehicle.

10. Install the 3 upper strut mount bolts and tighten to 18 ft. lbs. (25 Nm).

11. Attach the strut to the steering knuckle and install the nuts and bolts. Tighten the bolts and nuts to 148 ft. lbs. (200 Nm).

➡**Inspect the stabilizer link seals for damage and replace the link as necessary. Do not allow the stabilizer link ball stud to rotate while installing the link nut.**

12. Connect the stabilizer link to the strut. Tighten the nut to 48 ft. lbs. (65 Nm).

13. Install the brake hose bracket to the strut assembly. Tighten the brake bracket bolt to 11 ft. lbs. (15 Nm).

14. Install the wheel assembly. Tighten the wheel nuts to 100 ft. lbs. (140 Nm).

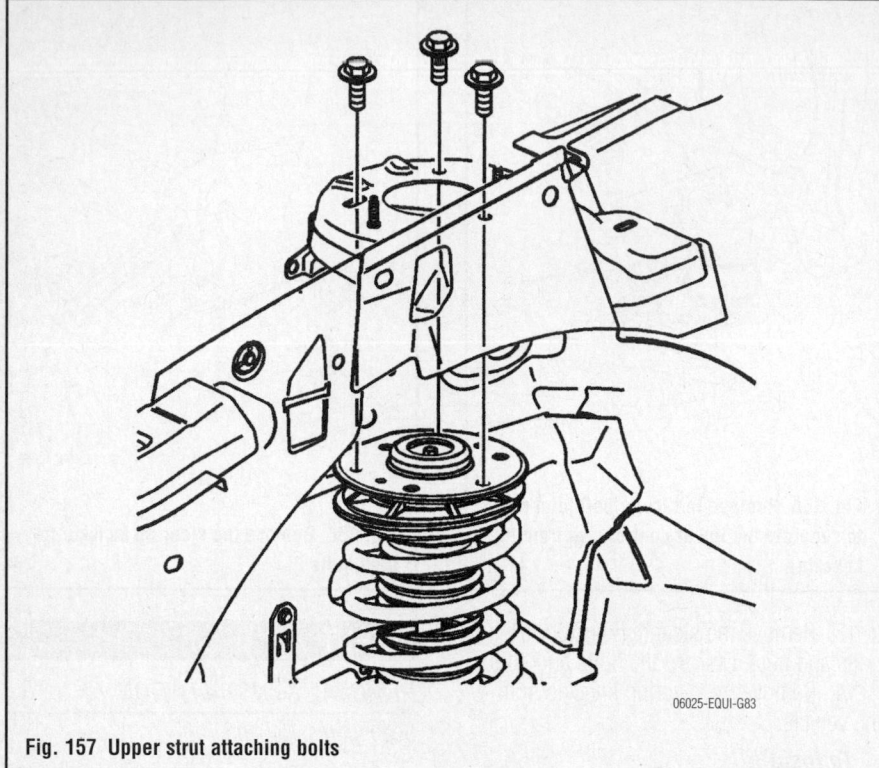

Fig. 157 Upper strut attaching bolts

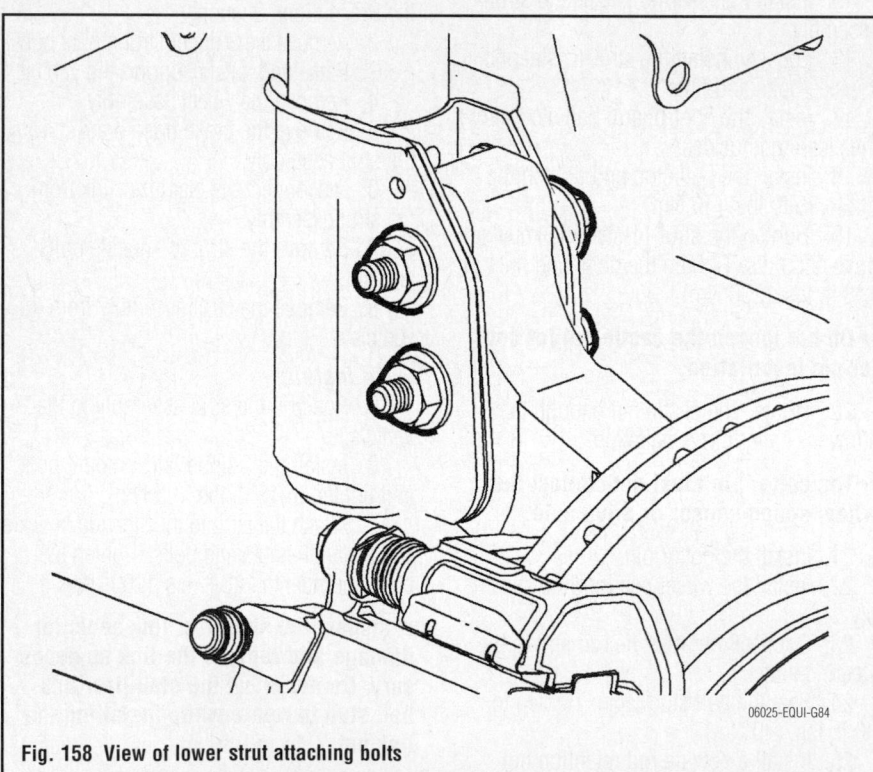

Fig. 158 View of lower strut attaching bolts

15. Lower the vehicle.
16. Perform a wheel alignment.

STABILIZER BAR

REMOVAL & INSTALLATION

See Figure 159.

1. Before servicing the vehicle, refer to the Precautions Section.
2. Position the front wheels in the straight ahead position.
3. Raise and safely support the vehicle.
4. Remove the front wheels.

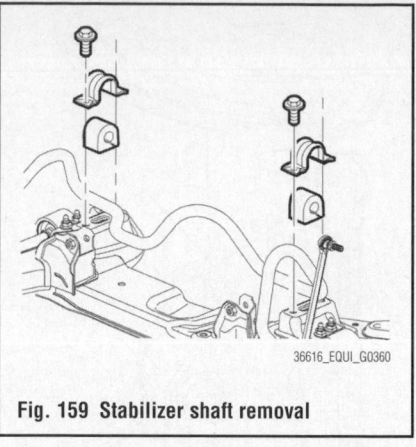

Fig. 159 Stabilizer shaft removal

5. Lower the front suspension frame. Refer to Frame, removal & installation.
6. Remove the stabilizer shaft insulator bracket bolts.
7. Remove the stabilizer shaft assembly from the frame.
8. Remove the stabilizer shaft insulator brackets.
9. Remove the stabilizer shaft insulators.

To install:

10. Install the stabilizer shaft insulators to the stabilizer shaft.
11. Install the stabilizer shaft insulator brackets.
12. Install the stabilizer shaft assembly to the frame.
13. Install the stabilizer shaft insulator bracket bolts to 37 ft. lbs. (50 Nm).
14. Install the front suspension frame.
15. Lower the vehicle.
16. Perform a wheel alignment.

WHEEL HUB & BEARING

REMOVAL & INSTALLATION

See Figure 160.

1. Before servicing the vehicle, refer to the Precautions Section.
2. Raise and support the vehicle.
3. Remove the wheel assembly.
4. Remove the wheel drive shaft nut.
5. Remove the brake rotor.
6. Remove the wheel speed sensor.
7. Remove the hub and bearing mounting bolts.
8. Remove the hub and bearing assembly.

➡**If removing the wheel bearing/hub assembly to service other components, support the wheel drive shaft with mechanics wire.**

To install:

9. Install the hub and bearing assembly.

10. Install the hub and bearing mounting bolts and tighten to 96 ft. lbs. (130 Nm).

11. Install the wheel speed sensor.

12. Install the brake rotor.

13. Install the wheel drive shaft nut and tighten to 151 ft. lbs. (205 Nm).

14. Install the wheel assembly.

15. Lower the vehicle.

ADJUSTMENT

The front wheel bearings are not adjustable. If the lateral run-out on the hub with the disc removed exceeds specification, the hub must be replaced.

1. Move the wheel hub in the axial direction by hand. Make sure there is no looseness of the wheel bearing. Axial end play limit: 0.004 inch (0.1mm) or less. If out of specification, replace the wheel hub and bearing assembly.

2. Rotate the wheel hub and make sure there is no unusual noise or other irregular conditions. If there are any irregular conditions, replace the wheel hub and bearing assembly. Refer to Wheel Hub & Bearing (sealed unit), removal & installation.

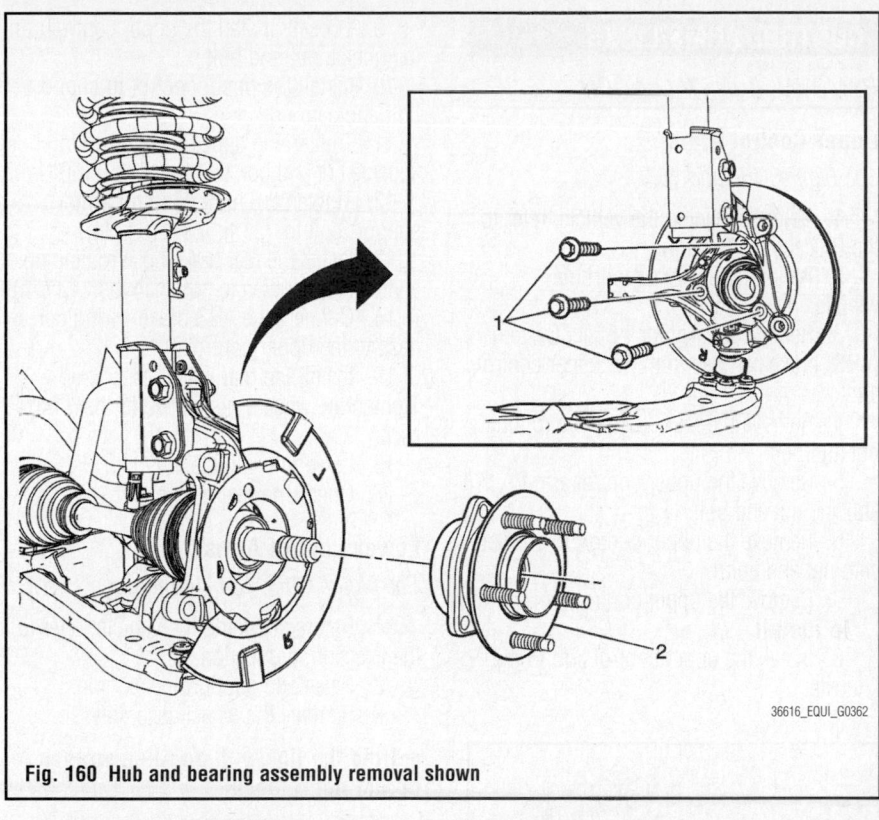

Fig. 160 Hub and bearing assembly removal shown

36616_EQUI_G0362

SUSPENSION

COIL SPRING

REMOVAL & INSTALLATION

See Figures 161 through 165.

1. Before servicing the vehicle, refer to the Precautions Section.

2. Raise and support the vehicle.

3. Remove the wheel assembly.

➡**Hold the link with an Allen wrench during nut removal.**

4. Remove the stabilizer link to lower control arm nut.

5. Position a jack stand underneath the lower control arm.

6. Raise the jack stand slightly to compress the coil spring.

7. Remove the lower shock bolt and nut.

8. Loosen the lower control arm to support frame bolt and nut.

9. Remove the lower control arm to knuckle nut and bolt.

10. Slowly lower the control arm in order to unload the coil spring.

11. Remove the coil spring and insulators.

To install:

12. Fully seat the top and bottom coil spring insulators to the spring.

➡**Spray silicon lubricant on the insulators to aid in installation. Ensure that part number identification tape located on the coil spring is oriented outboard of the vehicle and at the top of the spring.**

13. Position the spring with the rubber insulators into the vehicle.

14. Raise the jack stand to compress the spring.

15. Install the knuckle to the lower control arm. Tighten the lower control arm to knuckle bolt to 118 ft. lbs. (160 Nm).

16. Tighten the lower control arm to support nut and bolt. Tighten the bolt to 81 ft. lbs. (110 Nm).

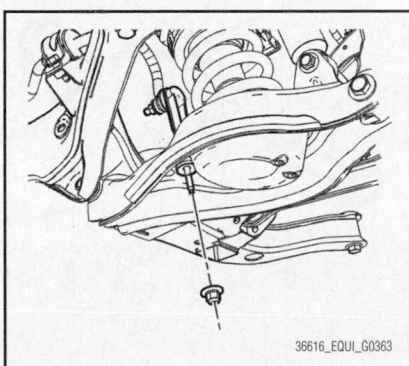

Fig. 161 Stabilizer link to lower control arm nut removal

36616_EQUI_G0363

REAR SUSPENSION

17. Install the shock to the lower control arm. Tighten the lower shock bolt to 81 ft. lbs. (110 Nm).

18. Remove the jack stand from under the vehicle.

➡**Hold the link with a wrench during nut installation.**

19. Install the stabilizer link to the lower control arm. Tighten the nut to 11 ft. lbs. (15 Nm).

20. Push the trailing arm upward to align the front bracket to body bolt.

21. Use a drift to aid in bracket alignment and install the remaining bolts.

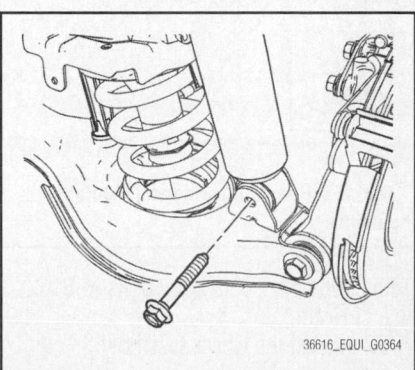

Fig. 162 Lower shock mounting bolt removal

36616_EQUI_G0364

Fig. 163 Lower control arm to support frame bolt (1)

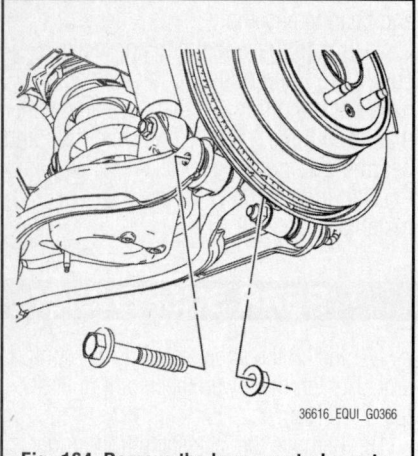

Fig. 164 Remove the lower control arm to knuckle nut and bolt

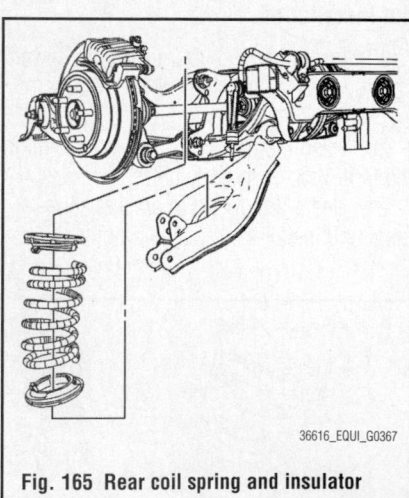

Fig. 165 Rear coil spring and insulator removal

Tighten the bracket to body bolts to 81 ft. lbs. (110 Nm).

22. Install the wheel assembly.
23. Lower the vehicle.
24. Check the rear alignment.

CONTROL ARMS/LINKS

REMOVAL & INSTALLATION

Upper Control Arms

See Figures 166 and 167.

1. Before servicing the vehicle, refer to the Precautions Section.
2. Raise and safely support the vehicle.
3. Remove the Antilock Brake System (ABS) brake harness from the upper control arm.
4. Remove the rear brake hose routing nut and bolt.
5. Remove the upper control arm to knuckle nut and bolt.
6. Remove the upper control to support cam nut and bolt.
7. Remove the upper control arm.

To install

8. Install the upper control arm to the knuckle.

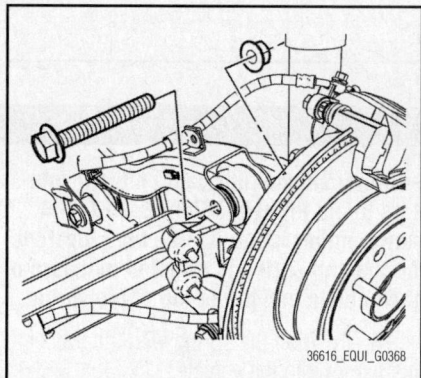

Fig. 166 View of upper control arm to knuckle nut and bolt

9. Loosely install the upper control arm to knuckle nut and bolt.
10. Install the upper control to support bolt and cam nut.
11. Tighten the upper control arm to knuckle nut and bolt to 118 ft. lbs. (160 Nm).
12. Tighten the upper control arm to support bolt to 121 ft. lbs. (164 Nm).
13. Install the rear brake hose routing nut and bolt and tighten to 106 inch lbs. (12 Nm).
14. Connect the ABS brake wiring harness to the upper control arm.
15. Install the rear wheel assembly. Tighten the wheel nuts to 100 ft. lbs. (140 Nm).
16. Lower the vehicle.
17. Check the rear alignment.

Lower Control Arms

See Figures 163, 164, 168 through 170.

1. Before servicing the vehicle, refer to the Precautions Section.
2. Raise and support the vehicle.
3. Remove the wheel assembly.

➡**Hold the link with an Allen wrench during nut removal.**

4. Remove the stabilizer link to lower control arm nut.
5. Position a jack stand underneath the lower control arm.
6. Raise the jack stand slightly to compress the coil spring.
7. Remove the lower shock bolt and nut.
8. Loosen the lower control arm to support frame bolt and nut.
9. Remove the lower control arm to knuckle nut and bolt.
10. Slowly lower the control arm in order to unload the coil spring.

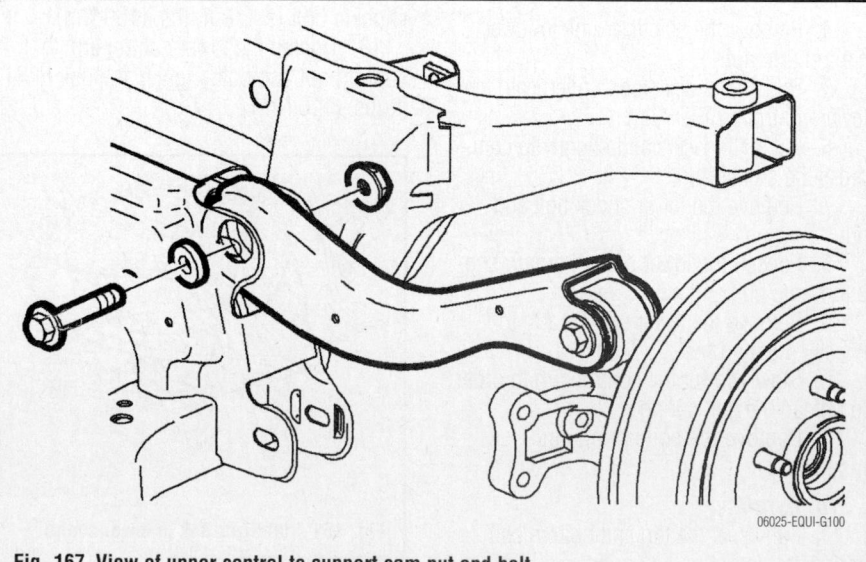

Fig. 167 View of upper control to support cam nut and bolt

11. Remove the coil spring.
12. Remove the jack stand.
13. Remove the lower control arm support nut and bolt.
14. Remove the lower control arm.

To install:

15. Inspect the coil spring upper and lower insulators, if damage exists replace the insulators.
16. Position the lower control to the support and hand tighten the bolt and nut.

➡ **Spray silicon lubricant on the insulators to aid in installation. Ensure the spring is properly seated.**

17. Position the spring with the rubber insulators into the vehicle.
18. Use a screw type jack stand to compress the spring.
19. Install the knuckle to the lower control arm. Tighten the lower control arm to knuckle nut and bolt to 118 ft. lbs. (160 Nm).
20. Tighten the lower control arm to support nut and bolt to 118 ft. lbs. (160 Nm).

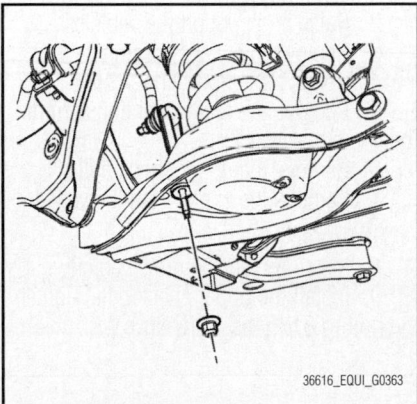

Fig. 168 Stabilizer link to lower control arm nut removal

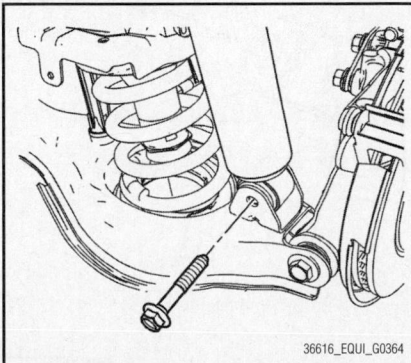

Fig. 169 Lower shock mounting bolt removal

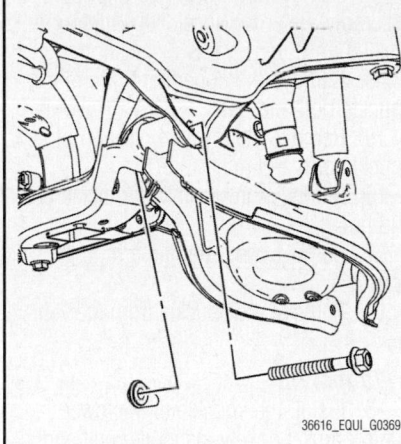

Fig. 170 Remove the control arm support nut and bolt

21. Install the shock to the lower control arm. Tighten the lower shock bolt to 81 ft. lbs. (110 Nm).
22. Remove the jack stand.
23. Install the stabilizer link to the lower control arm. Tighten the nut to 11 ft. lbs. (15 Nm).

➡ **Hold the link with a wrench during nut installation.**

24. Install the wheel assembly. Tighten the wheel nuts to 100 ft. lbs. (140 Nm).
25. Lower the vehicle.
26. Check the rear alignment.

Trailing Arm

See Figures 171 through 173.

1. Before servicing the vehicle, refer to the Precautions Section.
2. Raise and safely support the vehicle.
3. Remove the wheel assembly.
4. Remove the park brake cable bolt from the trailing arm and from the frame.
5. Remove the trailing arm bracket to body bolts.
6. Remove the trailing arm bushing to bracket nut and bolt.
7. Remove the trailing arm to knuckle bolts.
8. Remove the trailing arm.

To install:

9. Position the trailing arm to the vehicle.
10. Install the trailing arm to knuckle bolts and tighten to 81 ft. lbs. (110 Nm).
11. Position the trailing arm bracket to the trailing arm.
12. Loosely install the trailing arm bushing to bracket nut and bolt.

13. Install the trail arm bracket.
14. Tighten the trailing arm bushing to bracket nut and bolt to 118 ft. lbs. (160 Nm).
15. Install the park brake cable bolt to trailing arm and to the frame.
16. Install the wheel assembly. Tighten the wheel nuts to 100 ft. lbs. (140 Nm).
17. Lower the vehicle.

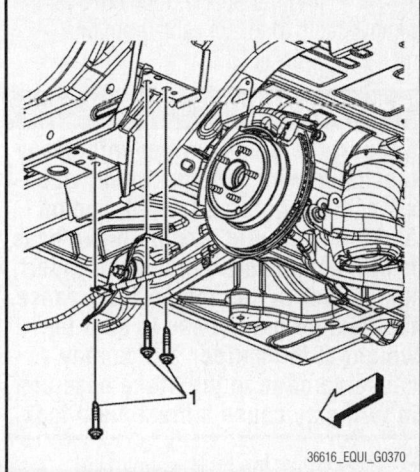

Fig. 171 Remove the trailing arm bracket to body bolts (1)

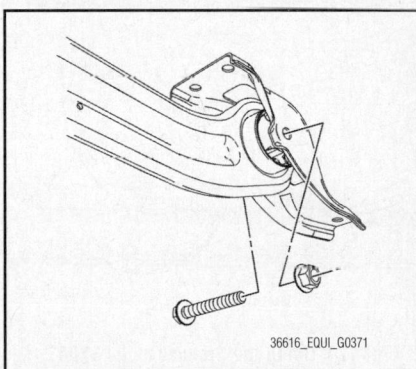

Fig. 172 Remove the trailing arm bushing to bracket nut and bolt

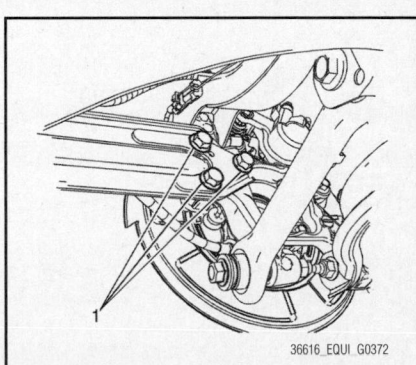

Fig. 173 Remove the trailing arm to knuckle bolts

KNUCKLE

REMOVAL & INSTALLATION

See Figures 174 through 177.

1. Before servicing the vehicle, refer to the Precautions Section.
2. Raise and safely support the vehicle.
3. Remove the wheel assembly.
4. Disconnect the rear park brake cable from the park brake actuator.
5. Using the Special tool J 37043, remove the park brake cable from the mounting bracket.

✳✳ WARNING

Support the brake caliper with heavy mechanic wire, or equivalent, whenever it is separated from its mount and the hydraulic flexible brake hose is still connected. Failure to support the caliper in this manner will cause the flexible brake hose to bear the weight of the caliper, which may cause damage to the brake hose and in turn may cause a brake fluid leak.

6. Remove the brake caliper and bracket as an assembly and support it with heavy mechanics wire or equivalent.

7. Remove the wheel bearing/hub assembly. Refer to Wheel Hub & Bearing (sealed unit), removal & installation.
8. Remove the upper control arm to knuckle bolt and nut.
9. Remove the lower control arm to knuckle bolt and nut.
10. Remove the toe link to knuckle bolt and nut.
11. Remove the 3 trailing arm to knuckle bolts.
12. Remove the knuckle from the vehicle.

To install:
13. Install the knuckle to the lower control arm. Loosely install the bolt and nut.
14. Install the knuckle to the upper control arm. Loosely install the bolt and nut.
15. Install the knuckle to the toe link. Loosely install the bolt and nut.
16. Install the 3 trailing arm to knuckle bolts. Loosely install the bolt and nut.
17. Tighten the bolts and nuts in the following sequence:
   - Tighten the knuckle to lower control arm bolt and nut to 118 ft. lbs. (160 Nm)
   - Tighten the knuckle to upper control arm bolt and nut to 118 ft. lbs. (160 Nm)
   - Tighten the knuckle to toe link bolt and nut to 118 ft. lbs. (160 Nm)
   - Tighten the 3 trailing arm to knuckle bolts to 81 ft. lbs. (110 Nm)
18. Install the wheel bearing/hub assembly.
19. Remove the supporting wire and position the brake caliper and bracket assemblies back onto the knuckles.
20. Connect the rear park brake cable through the mounting bracket and onto the park brake actuator.
21. Install the tire and wheel. Tighten the wheel nuts to 100 ft. lbs. (140 Nm).
22. Lower the vehicle.
23. Perform a vehicle wheel alignment.

SHOCK ABSORBER

REMOVAL & INSTALLATION

See Figures 178 and 179.

1. Before servicing the vehicle, refer to the Precautions Section.
2. Raise and support the vehicle.
3. Remove the wheel assembly.
4. Remove the lower shock bolt.
5. Remove the rear wheel house from the vehicle.
6. Remove the upper shock bolt.
7. Remove the shock from the vehicle.

To install:
8. Install the shock to the vehicle.
9. Install the upper shock bolt. Tighten the bolt to 81 ft. lbs. (110 Nm).

Fig. 174 Using the Special tool J 37043, remove the park brake cable from the mounting bracket

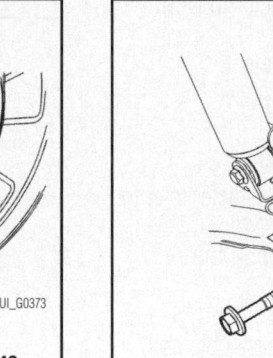

Fig. 176 Remove the lower control arm to knuckle bolt and nut.

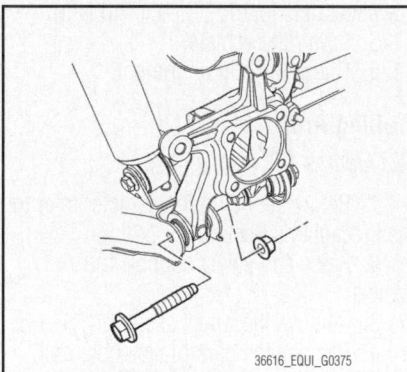

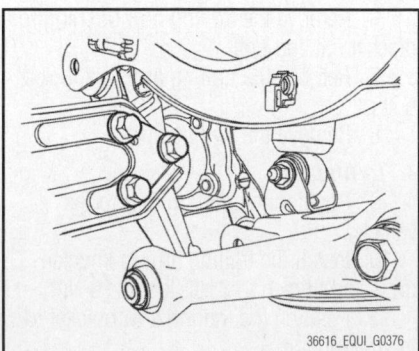

Fig. 175 Remove the upper control arm to knuckle bolt and nut

Fig. 177 Remove the 3 trailing arm to knuckle bolts

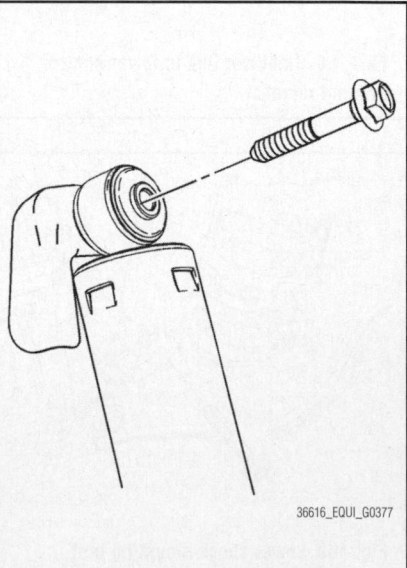

Fig. 178 Shock absorber upper bolt

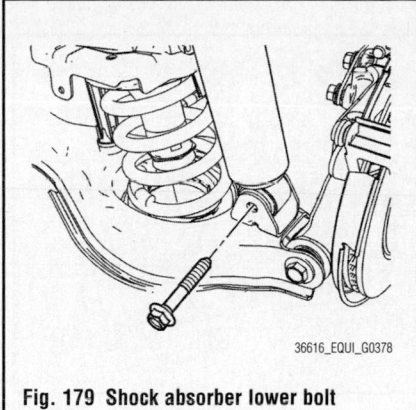

Fig. 179 Shock absorber lower bolt

10. Install the lower shock bolt. Tighten the bolt to 81 ft. lbs. (110 Nm).

11. Install the rear wheel house in the vehicle.

12. Install the wheel assembly. Tighten the wheel nuts to 100 ft. lbs. (140 Nm).

13. Lower the vehicle.

STABILIZER BAR

REMOVAL & INSTALLATION

See Figures 180 and 181.

1. Before servicing the vehicle, refer to the Precautions Section.

2. Raise and support the vehicle.

3. Remove the rear wheel assemblies.

4. Lower the rear suspension support.

➡**Hold the ball shaft secure with a TORX® bit, when installing the nut.**

5. Remove the stabilizer link to stabilizer shaft nut.

6. Remove the stabilizer shaft clamp bolts.

7. Disengage the stabilizer shaft from the stabilizer link ball studs, while removing the stabilizer bar from the vehicle.

To install

8. Position the stabilizer shaft in the

Fig. 180 Remove the stabilizer link to stabilizer shaft nut

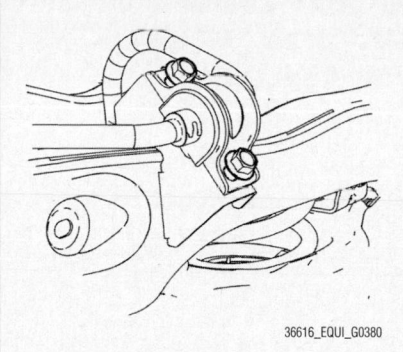

Fig. 181 Remove the stabilizer shaft clamp bolts

vehicle, while positioning the links to the stabilizer bar.

9. Install the stabilizer shaft clamps and bushings to the stabilizer shaft.

10. Install the stabilizer shaft clamp bolts. Tighten the bolts to 52 ft. lbs. (70 Nm).

11. Install the stabilizer link to stabilizer bar nut. Tighten the nut to 37 ft. lbs. (50 Nm).

12. Install the rear wheel assemblies.

13. Lower the vehicle.

STABILIZER LINKS

REMOVAL & INSTALLATION

See Figures 181 and 182.

1. Before servicing the vehicle, refer to the Precautions Section.

2. Raise and support the vehicle.

3. Loosen the stabilizer bar clamp bolts.

➡**Use a 90 degree bend TORX® bit to hold the ball stud when loosening or tightening the nut.**

4. Remove the stabilizer link to stabilizer bar nut.

➡**When disconnecting the stabilizer link, hold the link with a wrench to prevent turning.**

5. Remove the stabilizer link to control arm nut.

6. Remove the stabilizer link from the vehicle.

To install:

7. Position the stabilizer link through the control arm.

8. Install the stabilizer link to control arm nut. Tighten the nut to 11 ft. lbs. (15 Nm).

9. Install the stabilizer link to stabilizer bar nut. Tighten the nut to 37 ft. lbs. (50 Nm).

10. Tighten the loose stabilizer bar clamp bolts to 52 ft. lbs. (70 Nm).

11. Lower the vehicle.

Fig. 182 Remove the stabilizer link to stabilizer bar nut

WHEEL HUB & BEARING

REMOVAL & INSTALLATION

See Figure 183.

1. Before servicing the vehicle, refer to the Precautions Section.

2. Raise and support the vehicle.

3. Remove the rear wheel assembly.

4. Remove the brake caliper and bracket.

5. Remove the rear brake rotor.

6. Remove the parking brake shoes. Refer to Parking Brake Shoes, removal & installation.

7. On vehicles with AWD, remove the halfshaft spindle nut.

8. Disconnect the wheel speed sensor electrical connector.

✳✳ WARNING

Do not damage the halfshaft joint seal.

9. Support the halfshaft with heavy mechanics wire, or equivalent.

10. Remove the wheel hub and bearing mounting bolts.

11. Remove the wheel hub and bearing assembly from the suspension knuckle.

To install:

12. Install the wheel hub and bearing assembly to the steering knuckle.

13. Install the wheel hub and bearing mounting bolts. Tighten the bolts to 55 ft. lbs. (75 Nm)

14. On AWD vehicles, install the halfshaft spindle nut. Tighten the nut to 151 ft. lbs. (205 Nm).

15. Route the wheel speed sensor electrical harness through the backing plate and seat the grommet.

16. Connect the wheel speed sensor electrical connector.

17. Install the parking brake shoes.

18. Install the rear brake rotor.

19. Install the brake caliper and bracket.

20. Install the rear wheel assembly. Tighten the wheel nuts to 100 ft. lbs. (140 Nm).

21. Lower the vehicle

ADJUSTMENT

The wheel bearing does not require maintenance or adjustment. If a growling noise is emitted from the wheel bearing during operation, replace the wheel hub and bearing assembly.

1. Before servicing the vehicle, refer to the Precautions Section.

2. Check for noise and smooth rotation of the wheel by rotating the wheel. If it is defective, replace the wheel hub assembly.

3. Measure the thrust play of the wheel hub using a dial gauge.

   a. The thrust play limit is 0.004 inch (0.1mm).

   b. When the thrust play exceeds the limit, replace the wheel hub assembly.

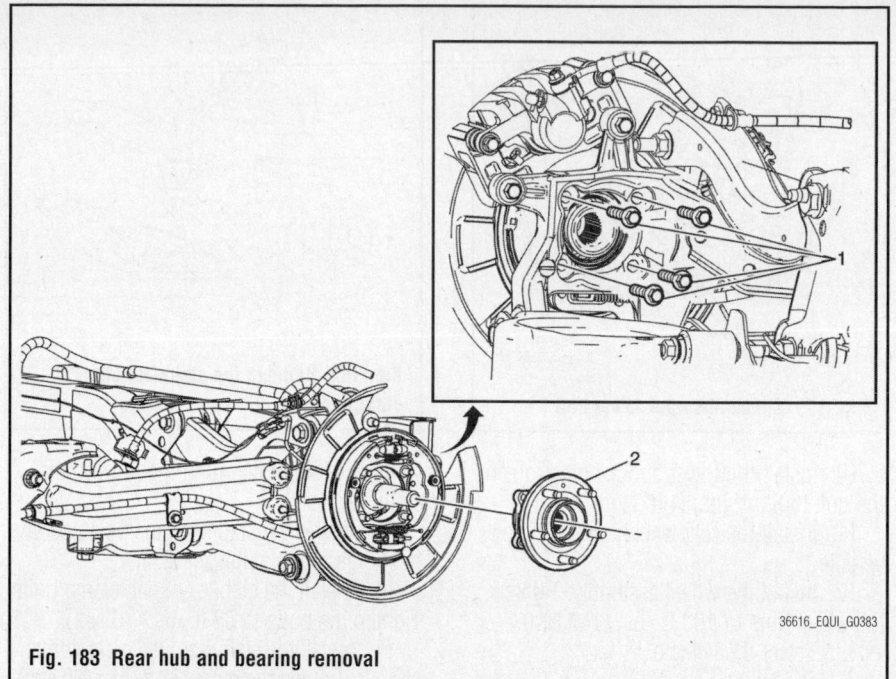

36616_EQUI_G0383

Fig. 183 Rear hub and bearing removal

SUZUKI

Diagnostic Trouble Codes

28

DIAGNOSTIC TROUBLE CODES

OBD II VEHICLE APPLICATIONS

SUZUKI

Equator
2009-2010
- 2.5L I4 .... VIN QR25DE
- 4.0L I4 .... VIN VQ40DE

Grand Vitara
2009-2010
- 2.4L V6................... VIN 0
- 3.2L V6................... VIN 1

SX4
2009-2010
- 2.0L I4 .... VIN J

XL-7
2009-2010
- 3.6L V6................... VIN 7

OBD II Trouble Code List (P0XXX Codes)

DTC	Trouble Code Title, Conditions & Possible Causes
DTC: P0008 **2T, MIL: Yes** **Year:** 2009, 2010 **Model:** Grand Vitara, XL-7 **Engine:** 3.2L V6 VIN 1, 3.6L V6 **Transmission:** All	**Engine Position System Performance Bank 1:** Difference between intake side CMP and exhaust side CMP on bank 1 is out of specified range. **Possible Causes:** • Timing chain • Timing chain tensioner • Idler sprocket • CMP actuator • ECM
DTC: P0009 **2T, MIL: Yes** **Year:** 2009, 2010 **Model:** Grand Vitara, XL-7 **Engine:** 3.2L V6 VIN 1, 3.6L V6 **Transmission:** All	**Engine Position System Performance Bank 2:** Difference between intake side CMP and exhaust side CMP on bank 2 is out of specified range. **Possible Causes:** • Timing chain • Timing chain tensioner • Idler sprocket • CMP actuator • ECM
DTC: P000A **T, MIL: Yes** **Year:** 2009 **Model:** XL-7 **Engine:** 3.6L V6 **Transmission:** All	**Intake Camshaft Position (CMP) System Slow Response Bank 1:** • The ECM detects the difference between the desired camshaft position angle and the actual camshaft position angle is greater than 6–11 degrees. OR • The ECM detects a slow response, a deviation greater than 1.5 degrees in time greater than 2.5 seconds, for the actual camshaft position angle to match the desired position angle during the test. • Either condition exists for greater than 1 second or a cumulative of 5 seconds. **Possible Causes:** • CMP actuator solenoid and/or circuit • ECM • Engine oil old, low or leaking
DTC: P000B **1T, MIL: Yes** **Year:** 2009 **Model:** XL-7 **Engine:** 3.6L V6 **Transmission:** All	**Exhaust Camshaft Position (CMP) System Slow Response Bank 1:** • The ECM detects the difference between the desired camshaft position angle and the actual camshaft position angle is greater than 6–11 degrees. OR • The ECM detects a slow response, a deviation greater than 1.5 degrees in time greater than 2.5 seconds, for the actual camshaft position angle to match the desired position angle during the test. • Either condition exists for greater than 1 second or a cumulative of 5 seconds. **Possible Causes:** • CMP actuator solenoid and/or circuit • ECM • Engine oil old, low or leaking
DTC: P000C **1T, MIL: Yes** **Year:** 2009 **Model:** XL-7 **Engine:** 3.6L V6 **Transmission:** All	**Intake Camshaft Position (CMP) System Slow Response Bank 2:** • The ECM detects the difference between the desired camshaft position angle and the actual camshaft position angle is greater than 6–11 degrees. OR • The ECM detects a slow response, a deviation greater than 1.5 degrees in time greater than 2.5 seconds, for the actual camshaft position angle to match the desired position angle during the test. • Either condition exists for greater than 1 second or a cumulative of 5 seconds. **Possible Causes:** • CMP actuator solenoid and/or circuit • ECM • Engine oil old, low or leaking

DTC	Trouble Code Title, Conditions & Possible Causes
DTC: P000D **1T, MIL: Yes** **Year:** 2009 **Model:** XL-7 **Engine:** 3.6L V6 **Transmission:** All	**Exhaust Camshaft Position (CMP) System Slow Response Bank 2:** • The ECM detects the difference between the desired camshaft position angle and the actual camshaft position angle is greater than 6–11 degrees. OR • The ECM detects a slow response, a deviation greater than 1.5 degrees in time greater than 2.5 seconds, for the actual camshaft position angle to match the desired position angle during the test. • Either condition exists for greater than 1 second or a cumulative of 5 seconds. **Possible Causes:** • CMP actuator solenoid and/or circuit • ECM • Engine oil old, low or leaking
DTC: P0010 **2T, MIL: Yes** **Year:** 2009, 2010 **Model:** Grand Vitara, SX4 **Engine:** 2.0L L4, 2.0L L4 VIN 5, 2.4L L4 VIN 9, 3.2L V6 VIN 1 **Transmission:** All	**"A" Camshaft Position Actuator Circuit Bank 1:** OCV (bank 1 / intake side) control circuit is open circuit. **Possible Causes:** • OCV and its circuit • ECM
DTC: P0010 **1T, MIL: Yes** **Year:** 2009 **Model:** XL-7 **Engine:** 3.6L V6 **Transmission:** All	**Intake Camshaft Position (CMP) Actuator Solenoid Control Circuit Bank 1:** The ECM detects an open in the CMP actuator solenoid circuits for greater than 1 second or a cumulative of 5 seconds, when the solenoid is commanded OFF. **Possible Causes:** • CMP actuator solenoid and/or circuit • ECM
DTC: P0010 **2T, MIL: Yes** **Year:** 2009, 2010 **Model:** Grand Vitara **Engine:** 2.4L L4 VIN 0 **Transmission:** All	**Camshaft Position Actuator Circuit:** • Electric current of OCV drive circuit is lower than 0.2 A for 2 sec. even if control duty of OCV is higher than 45%. • Electric current of OCV drive circuit is higher than 0.8 A for 2 sec. even if control duty of OCV is lower than 15%. **Possible Causes:** • OCV and/or its circuit • ECM
DTC: P0011 **2T, MIL: Yes** **Year:** 2009, 2010 **Model:** Equator, Grand Vitara **Engine:** 2.4L L4 VIN 0, 2.5L L4, 3.2L V6 VIN 1, 4.0L V6 **Transmission:** All	**Intake valve timing control solenoid valve:** Intake valve timing control solenoid valve **Possible Causes:** • Intake valve timing control solenoid valve
DTC: P0011 **1T, MIL: Yes** **Year:** 2009 **Model:** XL-7 **Engine:** 3.6L V6 **Transmission:** All	**Intake Camshaft Position (CMP) System Performance Bank 1:** • The ECM detects the difference between the desired camshaft position angle and the actual camshaft position angle is greater than 6–11 degrees. OR • The ECM detects a slow response, a deviation greater than 1.5 degrees in time greater than 2.5 seconds, for the actual camshaft position angle to match the desired position angle during the test. • Either condition exists for greater than 1 second or a cumulative of 5 seconds. **Possible Causes:** • CMP actuator solenoid and/or circuit • ECM • Engine oil old, low or leaking
DTC: P0011 **2T, MIL: Yes** **Year:** 2009, 2010 **Model:** SX4 **Engine:** 2.0L L4, 2.0L L4 VIN 5, 2.4L L4 VIN 9 **Transmission:** All	**Camshaft Position-Timing Over-Advanced or System Performance:** Measured CMP is advanced more than 12° from targeted CMP for 4 sec. when CMP control system is operated. **Possible Causes:** • OCV and/or its circuit • Lubrication system • VVT oil passenger • CMP actuator • Timing chain • Timing chain tensioner • ECM

DTC	Trouble Code Title, Conditions & Possible Causes
DTC: P0012 **2T, MIL: Yes** **Year:** 2009, 2010 **Model:** Grand Vitara, SX4 **Engine:** 2.0L L4, 2.0L L4 VIN 5, 2.4L L4 VIN 0, 2.4L L4 VIN 9 **Transmission:** All	**Camshaft position-timing over-retarded:** Measured CMP is retarded 15° from targeted CMP for specified time when CMP control system is operated. **Possible Causes:** • OCV and/or its circuit • Oil pressure • Oil control circuit • CKP sensor • Sensor plate • CMP sensor • Signal rotor • CMP actuator • Valve timing • ECM
DTC: P0013 **2T, MIL: Yes** **Year:** 2009, 2010 **Model:** Grand Vitara **Engine:** 3.2L V6 VIN 1 **Transmission:** All	**"B" Camshaft Position - Actuator Circuit Bank 1:** OCV (bank 1 / exhaust side) control circuit is open circuit. **Possible Causes:** • OCV and its circuit • ECM
DTC: P0013 **1T, MIL: Yes** **Year:** 2009 **Model:** XL-7 **Engine:** 3.6L V6 **Transmission:** All	**Exhaust Camshaft Position (CMP) Actuator Solenoid Control Circuit Bank 1:** The ECM detects an open in the CMP actuator solenoid circuits for greater than 1 second or a cumulative of 5 seconds, when the solenoid is commanded OFF. **Possible Causes:** • CMP actuator solenoid and/or circuit • ECM
DTC: P0014 **1T, MIL: Yes** **Year:** 2009 **Model:** XL-7 **Engine:** 3.6L V6 **Transmission:** All	**Exhaust Camshaft Position (CMP) System Performance Bank 1:** • The ECM detects the difference between the desired camshaft position angle and the actual camshaft position angle is greater than 6–11 degrees. OR • The ECM detects a slow response, a deviation greater than 1.5 degrees in time greater than 2.5 seconds, for the actual camshaft position angle to match the desired position angle during the test. • Either condition exists for greater than 1 second or a cumulative of 5 seconds. **Possible Causes:** • CMP actuator solenoid and/or circuit • ECM • Engine oil old, low or leaking
DTC: P0014 **2T, MIL: Yes** **Year:** 2009, 2010 **Model:** Grand Vitara **Engine:** 3.2L V6 VIN 1 **Transmission:** All	**"B" Camshaft Position - Timing Over-Advanced or System Performance Bank 1:** This DTC is detected if any one of the following conditions is met at camshaft (bank 1 / exhaust side). • Difference between measured CMP and targeted CMP is more than specified value for specified time. • Measured CMP does not reach targeted CMP for specified time. • Measured camshaft position is unchanged for specified time. **Possible Causes:** • OCV and/or its circuit • Lubrication system • VVT oil passage • CMP actuator • ECM
DTC: P0016 **2T, MIL: Yes** **Year:** 2009, 2010 **Model:** Grand Vitara, SX4 **Engine:** 2.0L L4, 2.0L L4 VIN 5, 2.4L L4 VIN 0, 2.4L L4 VIN 9 **Transmission:** All	**Crankshaft Position-Camshaft Position Correlation:** • Difference between CKP minus CMP at the end of previous driving cycle and CKP minus CMP is out of specified range (−10° through 10°) for 1 sec. • Difference between CKP and CMP is out of specified range (1° through 31°) for 1 sec. **Possible Causes:** • CKP sensor and/or its circuit • CMP sensor and/or its circuit • Sensor plate • Signal rotor • Valve timing • ECM

DTC	Trouble Code Title, Conditions & Possible Causes
DTC: P0016 **2T, MIL: Yes** **Year:** 2009, 2010 **Model:** Grand Vitara **Engine:** 3.2L V6 VIN 1 **Transmission:** All	**Crankshaft Position – Camshaft Position Correlation Bank 1 Sensor "A":** Difference between camshaft (bank 1 / intake side) position and crankshaft position is out of specified range. **Possible Causes:** • CMP sensor • CKP sensor • Signal rotor • Timing chain • Timing chain tensioner • CMP actuator • ECM
DTC: P0017 **2T, MIL: Yes** **Year:** 2009, 2010 **Model:** Grand Vitara **Engine:** 3.2L V6 VIN 1 **Transmission:** All	**Crankshaft Position – Camshaft Position Correlation Bank 1 Sensor "B":** Difference between camshaft (bank 1 / exhaust side) position and crankshaft position is out of specified range. **Possible Causes:** • CMP sensor • CKP sensor • Signal rotor • Timing chain • Timing chain tensioner • CMP actuator • ECM
DTC: P0018 **2T, MIL: Yes** **Year:** 2009, 2010 **Model:** Grand Vitara **Engine:** 3.2L V6 VIN 1 **Transmission:** All	**Crankshaft Position – Camshaft Position Correlation Bank 2 Sensor "A":** Difference between camshaft (bank 2 / intake side) position and crankshaft position is out of specified range. **Possible Causes:** • CMP sensor • CKP sensor • Signal rotor • Timing chain • Timing chain tensioner • CMP actuator • ECM
DTC: P0019 **2T, MIL: Yes** **Year:** 2009, 2010 **Model:** Grand Vitara **Engine:** 3.2L V6 VIN 1 **Transmission:** All	**Crankshaft Position – Camshaft Position Correlation Bank 2 Sensor "B":** Difference between camshaft (bank 2 / exhaust side) position and crankshaft position is out of specified range. **Possible Causes:** • CMP sensor • CKP sensor • Signal rotor • Timing chain • Timing chain tensioner • CMP actuator • ECM
DTC: P0020 **2T, MIL: Yes** **Year:** 2009, 2010 **Model:** Grand Vitara **Engine:** 3.2L V6 VIN 1 **Transmission:** All	**"A" Camshaft Position Actuator Circuit Bank 2:** OCV (bank 2 / intake side) control circuit is open circuit. **Possible Causes:** • OCV and its circuit • ECM
DTC: P0020 **1T, MIL: Yes** **Year:** 2009 **Model:** XL-7 **Engine:** 3.6L V6 **Transmission:** All	**Intake Camshaft Position (CMP) Actuator Solenoid Control Circuit Bank 2:** The ECM detects an open in the CMP actuator solenoid circuits for greater than 1 second or a cumulative of 5 seconds, when the solenoid is commanded OFF. **Possible Causes:** • CMP actuator solenoid and/or circuit • ECM

DTC	Trouble Code Title, Conditions & Possible Causes
DTC: P0021 **1T, MIL: Yes** **Year:** 2009 **Model:** XL-7 **Engine:** 3.6L V6 **Transmission:** All	**Intake Camshaft Position (CMP) System Performance Bank 2:** • The ECM detects the difference between the desired camshaft position angle and the actual camshaft position angle is greater than 6–11 degrees. OR • The ECM detects a slow response, a deviation greater than 1.5 degrees in time greater than 2.5 seconds, for the actual camshaft position angle to match the desired position angle during the test. • Either condition exists for greater than 1 second or a cumulative of 5 seconds. **Possible Causes:** • CMP actuator solenoid and/or circuit • ECM • Engine oil old, low or leaking
DTC: P0021 **2T, MIL: Yes** **Year:** 2009, 2010 **Model:** Equator, Grand Vitara **Engine:** 3.2L V6 VIN 1, 4.0L V6 **Transmission:** All	**Intake valve timing control performance (bank 2):** There is a gap between angle of target and phase-control angle degree. **Possible Causes:** • Crankshaft position sensor (POS) • Camshaft position sensor (PHASE) • Intake valve timing control solenoid valve • Accumulation of debris to the signal pick-up portion of the camshaft • Timing chain installation • Foreign matter caught in the oil groove for intake valve timing control
DTC: P0023 **2T, MIL: Yes** **Year:** 2009, 2010 **Model:** Grand Vitara **Engine:** 3.2L V6 VIN 1 **Transmission:** All	**"B" Camshaft Position - Actuator Circuit Bank 2:** OCV (bank 2 / exhaust side) control circuit is open circuit. **Possible Causes:** • OCV and its circuit • ECM
DTC: P0023 **1T, MIL: Yes** **Year:** 2009 **Model:** XL-7 **Engine:** 3.6L V6 **Transmission:** All	**Exhaust Camshaft Position (CMP) Actuator Solenoid Control Circuit Bank 2:** The ECM detects an open in the CMP actuator solenoid circuits for greater than 1 second or a cumulative of 5 seconds, when the solenoid is commanded OFF. **Possible Causes:** • CMP actuator solenoid and/or circuit • ECM
DTC: P0024 **2T, MIL: Yes** **Year:** 2009, 2010 **Model:** Grand Vitara **Engine:** 3.2L V6 VIN 1 **Transmission:** All	**"B" Camshaft Position - Timing Over-Advanced or System Performance Bank 2:** This DTC is detected if any one of the following conditions is met at camshaft (bank 2 / exhaust side). • Difference between measured CMP and targeted CMP is more than specified value for specified time. • Measured CMP does not reach targeted CMP for specified time. • Measured camshaft position is unchanged for specified time. **Possible Causes:** • OCV and/or its circuit • Lubrication system • VVT oil passage • CMP actuator • ECM
DTC: P0024 **1T, MIL: Yes** **Year:** 2009 **Model:** XL-7 **Engine:** 3.6L V6 **Transmission:** All	**Exhaust Camshaft Position (CMP) System Performance Bank 2:** • The ECM detects the difference between the desired camshaft position angle and the actual camshaft position angle is greater than 6–11 degrees. OR • The ECM detects a slow response, a deviation greater than 1.5 degrees in time greater than 2.5 seconds, for the actual camshaft position angle to match the desired position angle during the test. • Either condition exists for greater than 1 second or a cumulative of 5 seconds. **Possible Causes:** • CMP actuator solenoid and/or circuit • ECM • Engine oil old, low or leaking
DTC: P0030 **2T, MIL: Yes** **Year:** 2009, 2010 **Model:** Grand Vitara, XL-7 **Engine:** 3.2L V6 VIN 1, 3.6L V6 **Transmission:** All	**HO2S Heater Control Circuit Bank 1 Sensor 1:** HO2S-1 heater (bank 1) control circuit is open circuit. **Possible Causes:** • HO2S heater and/or its circuit • ECM

DTC	Trouble Code Title, Conditions & Possible Causes
DTC: P0031 **2T CCM, MIL: Yes** **Year:** 2009, 2010 **Model:** Equator, Grand Vitara, SX4, XL-7 **Engine:** 2.0L L4, 2.0L L4 VIN 5, 2.4L L4 VIN 0, 2.4L L4 VIN 9, 2.5L L4, 3.2L V6 VIN 1, 3.6L V6, 4.0L V6 **Transmission:** All	**HO2S-11 (Bank 1 Sensor 1) Heater Circuit Low Input:** Engine started, vehicle driven to a speed over 45 mph for 6 minutes at an engine speed from 2500-3000 rpm, followed by a deceleration period of 3 seconds, fuel level from 15-85%, BARO sensor more than 75 kPa, IAT sensor more than 6.8°F, and the PCM detected the maximum HO2S-11 voltage was less than 300 mv during the test. **Possible Causes:** • HO2S signal ground circuit is open • HO2S is contaminated or has failed • Vehicle driven until it ran out of fuel, or while very low on fuel • PCM has failed
DTC: P0032 **2T CCM, MIL: Yes** **Year:** 2009, 2010 **Model:** Equator, Grand Vitara, SX4, XL-7 **Engine:** 2.0L L4, 2.0L L4 VIN 5, 2.4L L4 VIN 0, 2.4L L4 VIN 9, 2.5L L4, 3.2L V6 VIN 1, 3.6L V6, 4.0L V6 **Transmission:** All	**HO2S-11 (Bank 1 Sensor 1) Heater Circuit High Input:** Engine started, vehicle driven to a speed over 45 mph for 6 minutes at an engine speed from 2500-3000 rpm, followed by a deceleration period of 3 seconds, fuel level from 15-85%, BARO sensor more than 75 kPa, IAT sensor more than 6.8°F, and the PCM detected the HO2S-11 signal did not go below 600 mv during the CCM test. **Possible Causes:** • HO2S signal circuit is open or shorted to system power • Fuel supply system is too rich (fuel injector or regulator leaking) • HO2S is contaminated, damaged or has failed • PCM has failed • TSB TS015 (12/95) contains information related to this code
DTC: P0036 **2T, MIL: Yes** **Year:** 2009, 2010 **Model:** Grand Vitara, XL-7 **Engine:** 3.2L V6 VIN 1, 3.6L V6 **Transmission:** All	**HO2S Heater Control Circuit Bank 1 Sensor 2:** HO2S-2 heater (bank 1) control circuit is open circuit. **Possible Causes:** • HO2S heater and/or its circuit • ECM
DTC: P0037 **2T CCM, MIL: Yes** **Year:** 2009, 2010 **Model:** Equator, Grand Vitara, SX4, XL-7 **Engine:** 2.0L L4, 2.0L L4 VIN 5, 2.4L L4 VIN 0, 2.4L L4 VIN 9, 2.5L L4, 3.2L V6 VIN 1, 3.6L V6, 4.0L V6 **Transmission:** All	**HO2S-12 (Bank 1 Sensor 2) Heater Circuit Low Input:** Engine started, vehicle driven to a speed over 45 mph for 6 minutes at an engine speed from 2500-3000 rpm, followed by a deceleration period of 3 seconds, fuel level from 15-85%, BARO sensor more than 75 kPa, IAT sensor more than 6.8°F, and the PCM detected the maximum HO2S-12 voltage was less than 300 mv during the test. **Possible Causes:** • HO2S signal ground circuit is open • HO2S is contaminated or has failed • Vehicle driven until it ran out of fuel, or while very low on fuel • PCM has failed
DTC: P0038 **2T CCM, MIL: Yes** **Year:** 2009, 2010 **Model:** Equator, Grand Vitara, SX4, XL-7 **Engine:** 2.0L L4, 2.0L L4 VIN 5, 2.4L L4 VIN 0, 2.4L L4 VIN 9, 2.5L L4, 3.2L V6 VIN 1, 3.6L V6, 4.0L V6 **Transmission:** All	**HO2S-12 (Bank 1 Sensor 2) Heater Circuit High Input:** Engine started, vehicle driven to a speed over 45 mph for 6 minutes at an engine speed from 2500-3000 rpm, followed by a deceleration period of 3 seconds, fuel level from 15-85%, BARO sensor more than 75 kPa, IAT sensor more than 6.8°F, and the PCM detected the HO2S-12 signal did not go below 600 mv during the CCM test. **Possible Causes:** • HO2S signal circuit is open or shorted to system power • Fuel supply system is too rich (fuel injector or regulator leaking) • HO2S is contaminated, damaged or has failed • PCM has failed • TSB TS015 (12/95) contains information related to this code
DTC: P0050 **2T, MIL: Yes** **Year:** 2009, 2010 **Model:** Grand Vitara, XL-7 **Engine:** 3.2L V6 VIN 1, 3.6L V6 **Transmission:** All	**HO2S Heater Control Circuit Bank 2 Sensor 1:** HO2S-1 heater (bank 2) control circuit is open circuit. **Possible Causes:** • HO2S heater and/or its circuit • ECM
DTC: P0051 **2T CCM, MIL: Yes** **Year:** 2009, 2010 **Model:** Equator, Grand Vitara, XL-7 **Engine:** 3.2L V6 VIN 1, 3.6L V6, 4.0L V6 **Transmission:** All	**HO2S-12 (Bank 2 Sensor 1) Heater Circuit Low Input:** Engine started, vehicle driven to a speed over 45 mph for 6 minutes at an engine speed from 2500-3000 rpm, followed by a deceleration period of 3 seconds, fuel level from 15-85%, BARO sensor more than 75 kPa, IAT sensor more than 6.8°F, and the PCM detected the maximum HO2S-21 voltage was less than 300 mv during the test. **Possible Causes:** • HO2S signal ground circuit is open • HO2S is contaminated or has failed • Vehicle driven until it ran out of fuel, or while very low on fuel • PCM has failed

DTC	Trouble Code Title, Conditions & Possible Causes
DTC: P0052 **2T CCM, MIL: Yes** **Year:** 2009, 2010 **Model:** Equator, Grand Vitara, XL-7 **Engine:** 3.2L V6 VIN 1, 3.6L V6, 4.0L V6 **Transmission:** All	**HO2S-12 (Bank 2 Sensor 1) Heater Circuit High Input:** Engine started, vehicle driven to a speed over 45 mph for 6 minutes at an engine speed from 2500-3000 rpm, followed by a deceleration period of 3 seconds, fuel level from 15-85%, BARO sensor more than 75 kPa, IAT sensor more than 6.8°F, and the PCM detected the HO2S-21 signal did not go below 600 mv during the CCM test. **Possible Causes:** • HO2S signal circuit is open or shorted to system power • Fuel supply system is too rich (fuel injector or regulator leaking) • HO2S is contaminated, damaged or has failed • PCM has failed • TSB TS015 (12/95) contains information related to this code
DTC: P0056 **2T, MIL: Yes** **Year:** 2009, 2010 **Model:** Grand Vitara, XL-7 **Engine:** 3.2L V6 VIN 1, 3.6L V6 **Transmission:** All	**HO2S Heater Control Circuit Bank 2 Sensor 2:** HO2S-2 heater (bank 2) control circuit is open circuit. **Possible Causes:** • HO2S heater and/or its circuit • ECM
DTC: P0057 **2T CCM, MIL: Yes** **Year:** 2009, 2010 **Model:** Equator, Grand Vitara, XL-7 **Engine:** 3.2L V6 VIN 1, 3.6L V6, 4.0L V6 **Transmission:** All	**HO2S-12 (Bank 2 Sensor 2) Heater Circuit Low Input:** Engine started, vehicle driven to a speed over 45 mph for 6 minutes at an engine speed from 2500-3000 rpm, followed by a deceleration period of 3 seconds, fuel level from 15-85%, BARO sensor more than 75 kPa, IAT sensor more than 6.8°F, and the PCM detected the maximum HO2S-22 voltage was less than 300 mv during the test. **Possible Causes:** • HO2S signal ground circuit is open • HO2S is contaminated or has failed • Vehicle driven until it ran out of fuel, or while very low on fuel • PCM has failed
DTC: P0058 **2T CCM, MIL: Yes** **Year:** 2009, 2010 **Model:** Equator, Grand Vitara, XL-7 **Engine:** 3.2L V6 VIN 1, 3.6L V6, 4.0L V6 **Transmission:** All	**HO2S-12 (Bank 2 Sensor 2) Heater Circuit High Input:** Engine started, vehicle driven to a speed over 45 mph for 6 minutes at an engine speed from 2500-3000 rpm, followed by a deceleration period of 3 seconds, fuel level from 15-85%, BARO sensor more than 75 kPa, IAT sensor more than 6.8°F, and the PCM detected the HO2S-22 signal did not go below 600 mv during the CCM test. **Possible Causes:** • HO2S signal circuit is open or shorted to system power • Fuel supply system is too rich (fuel injector or regulator leaking) • HO2S is contaminated, damaged or has failed • PCM has failed • TSB TS015 (12/95) contains information related to this code
DTC: P0075 **2T, MIL: Yes** **Year:** 2009, 2010 **Model:** Equator **Engine:** 2.5L L4, 4.0L V6 **Transmission:** All	**Intake valve timing control solenoid valve circuit:** An improper voltage is sent to the ECM through intake valve timing control solenoid valve. **Possible Causes:** • Harness or connectors • (Intake valve timing control solenoid valve circuit is open or shorted.) • Intake valve timing control solenoid valve
DTC: P0081 **2T, MIL: Yes** **Year:** 2009, 2010 **Model:** Equator **Engine:** 4.0L V6 **Transmission:** All	**Intake valve timing control solenoid valve circuit (bank 2):** An improper voltage is sent to the ECM through intake valve timing control solenoid valve. **Possible Causes:** • Harness or connectors • (Intake valve timing control solenoid valve circuit is open or shorted.) • Intake valve timing control solenoid valve
DTC: P0100 **2T, MIL: Yes** **Year:** 2009, 2010 **Model:** Grand Vitara, XL-7 **Engine:** 3.2L V6 VIN 1, 3.6L V6 **Transmission:** All	**Mass or Volume Air Flow Circuit:** Frequency of MAF sensor signal is fixed at 0 (zero) for 5 sec. **Possible Causes:** • MAF sensor and its circuit • ECM

DTC	Trouble Code Title, Conditions & Possible Causes
DTC: P0101 **2T CCM, MIL: Yes** **Year:** 2009, 2010 **Model:** Equator, Grand Vitara, SX4, XL-7 **Engine:** 2.0L L4, 2.0L L4 VIN 5, 2.4L L4 VIN 0, 2.4L L4 VIN 9, 2.5L L4, 3.2L V6 VIN 1, 3.6L V6, 4.0L V6 **Transmission:** All	**Mass Airflow Sensor Range/Performance:** Engine started, vehicle driven at a speed of over 35 mph in closed loop, then back to idle speed, altitude less than 8000 ft, ECT sensor from 18-230°F, IAT sensor more than 6.8°F, and the PCM detected the MAF signal was less than 0.82-1.08 lbs/min at 2500 rpm, or that it was less than 0.22-0.44 lbs/min at idle speed during the CCM test. **Possible Causes:** • Intake air system is restricted, or the air filter is clogged • MAF sensor signal or ground circuit has high resistance • MAF sensor is contaminated, damaged or has failed • PCM has failed
DTC: P0102 **1T CCM, MIL: Yes** **Year:** 2009, 2010 **Model:** Equator, Grand Vitara, SX4, XL-7 **Engine:** 2.0L L4, 2.0L L4 VIN 5, 2.4L L4 VIN 0, 2.4L L4 VIN 9, 2.5L L4, 3.2L V6 VIN 1, 3.6L V6, 4.0L V6 **Transmission:** All	**Mass Airflow Sensor Circuit Low Input:** Engine started, engine running in closed loop, altitude less than 8000 ft, ECT signal from 18-230°F, IAT signal from 18-122°F, and the PCM detected the MAF signal indicated less than (Scan Tool reads close to '0' lbs/minute) for 1 second during the CCM test. **Possible Causes:** • MAF sensor signal circuit is open or shorted to ground • MAF sensor power circuit is open • MAF sensor is damaged or has failed • PCM has failed
DTC: P0103 **1T CCM, MIL: Yes** **Year:** 2009, 2010 **Model:** Equator, Grand Vitara, SX4, XL-7 **Engine:** 2.0L L4, 2.0L L4 VIN 5, 2.4L L4 VIN 0, 2.4L L4 VIN 9, 2.5L L4, 3.2L V6 VIN 1, 3.6L V6, 4.0L V6 **Transmission:** All	**Mass Airflow Sensor Circuit High Input:** Engine started engine running at idle speed in closed loop, and the PCM detected the MAF sensor indicated more than 4.90v (Scan Tool reads 28 lbs/minute) for 1 second during the CCM test. **Possible Causes:** • MAF sensor ground circuit is open • MAF sensor signal is shorted to VREF or system power (B+) • MAF sensor is damaged or has failed • PCM has failed
DTC: P0111 **2T, MIL: Yes** **Year:** 2009, 2010 **Model:** Grand Vitara, SX4, XL-7 **Engine:** 2.0L L4, 2.0L L4 VIN 5, 2.4L L4 VIN 0, 2.4L L4 VIN 9, 3.2L V6 VIN 1, 3.6L V6 **Transmission:** All	**Intake Air Temperature Sensor 1 Circuit Range/Performance:** Difference between measured max. IAT value and measured min. IAT value is lower than 1.5 °C (2.7 °F). **Possible Causes:** • IAT sensor and/or its circuit • ECM
DTC: P0112 **1T CCM, MIL: Yes** **Year:** 2009, 2010 **Model:** Equator, Grand Vitara, SX4, XL-7 **Engine:** 2.0L L4, 2.0L L4 VIN 5, 2.4L L4 VIN 0, 2.4L L4 VIN 9, 2.5L L4, 3.2L V6 VIN 1, 3.6L V6, 4.0L V6 **Transmission:** All	**Intake Air Temperature Sensor Circuit Low Input:** Key on or engine running, and the PCM detected an unexpected "low" voltage condition on the IAT sensor circuit (Scan Tool reads 246°F or higher) during the CCM test. **Possible Causes:** • IAT sensor signal circuit is shorted to sensor ground • IAT sensor signal circuit is shorted to chassis ground • IAT sensor is damaged or has failed (it may be shorted) • PCM has failed
DTC: P0113 **1T CCM, MIL: Yes** **Year:** 2009, 2010 **Model:** Equator, Grand Vitara, SX4, XL-7 **Engine:** 2.0L L4, 2.0L L4 VIN 5, 2.4L L4 VIN 0, 2.4L L4 VIN 9, 2.5L L4, 3.2L V6 VIN 1, 3.6L V6, 4.0L V6 **Transmission:** All	**Intake Air Temperature Sensor Circuit High Input:** Key on or engine running, and the PCM detected the IAT sensor was more than 4.90v (Scan Tool reads -40°F or lower) in the test. **Possible Causes:** • IAT sensor signal circuit or ground circuit is open • IAT sensor signal circuit is shorted to VREF or system power • IAT sensor is damaged or has failed • PCM has failed

DTC	Trouble Code Title, Conditions & Possible Causes
DTC: P0116 **2T, MIL: Yes** **Year:** 2009, 2010 **Model:** Equator, Grand Vitara, SX4, XL-7 **Engine:** 2.0L L4, 2.0L L4 VIN 5, 2.4L L4 VIN 0, 2.4L L4 VIN 9, 2.5L L4, 3.2L V6 VIN 1, 3.6L V6, 4.0L V6 **Transmission:** All	**Engine coolant temperature sensor circuit range/performance:** Engine coolant temperature signal from engine coolant temperature sensor does not fluctuate, even when some time has passed after starting the engine with pre-warming up condition. **Possible Causes:** • Harness or connectors • (High or low resistance in the circuit) • Engine coolant temperature sensor
DTC: P0117 **1T CCM, MIL: Yes** **Year:** 2009, 2010 **Model:** Equator, Grand Vitara, SX4, XL-7 **Engine:** 2.0L L4, 2.0L L4 VIN 5, 2.4L L4 VIN 0, 2.4L L4 VIN 9, 2.5L L4, 3.2L V6 VIN 1, 3.6L V6, 4.0L V6 **Transmission:** All	**Engine Coolant Temperature Sensor Circuit Low Input:** Key on and engine running, and the PCM detected the ECT signal indicated more than 0.20v (Scan Tool reads 246°F) during the test. **Possible Causes:** • ECT sensor signal circuit is shorted to sensor ground • ECT sensor signal circuit is shorted to chassis ground • ECT sensor is damaged or has failed (it may be shorted) • PCM has failed
DTC: P0118 **1T CCM, MIL: Yes** **Year:** 2009, 2010 **Model:** Equator, Grand Vitara, SX4, XL-7 **Engine:** 2.0L L4, 2.0L L4 VIN 5, 2.4L L4 VIN 0, 2.4L L4 VIN 9, 2.5L L4, 3.2L V6 VIN 1, 3.6L V6, 4.0L V6 **Transmission:** All	**Engine Coolant Temperature Sensor Circuit High Input:** Key on or engine running, and the PCM detected the ECT sensor was more than 4.90v (Scan Tool reads -40°F or lower) in the test. **Possible Causes:** • ECT sensor signal circuit is open between sensor and the PCM • ECT sensor signal circuit is shorted to VREF or system power • ECT sensor ground circuit is open between sensor and PCM • ECT sensor is damaged or has failed • PCM has failed
DTC: P0119 **2T, MIL: Yes** **Year:** 2009 **Model:** XL-7 **Engine:** 3.6L V6 **Transmission:** All	**Engine Coolant Temperature (ECT) Sensor Circuit Intermittent:** • The ECM detects that the ECT changed +/-5°C (+/-9°F) greater than the expected value during a calibrated amount of time. • The condition exists for greater than 1 second or a cumulative of 10 seconds. **Possible Causes:** • ECT sensor and/or circuit • ECM
DTC: P011B **2T, MIL: Yes** **Year:** 2009, 2010 **Model:** SX4 **Engine:** 2.0L L4, 2.0L L4 VIN 5, 2.4L L4 VIN 9 **Transmission:** All	**Engine Coolant Temperature / Intake Air Temperature Correlation:** Difference between ECT and IAT is more than 30 °C (86 °F) for 5 sec. after engine has stopped for 5 hours. **Possible Causes:** • ECT sensor and/or its circuit • IAT sensor and/or its circuit • ECM
DTC: P011B **2T, MIL: Yes** **Year:** 2009, 2010 **Model:** Grand Vitara **Engine:** 2.4L L4 VIN 0, 3.2L V6 VIN 1 **Transmission:** All	**Engine Coolant Temperature (ECT) Sensor Performance:** Difference between measured ECT and calculated ECT is out of specified range (−20.3 through 20.3 °C (−36.5 through 36.5 °F)) for 8 sec. when ignition switch is turned to ON position after 5 hours from previous driving cycle. **Possible Causes:** • Coolant • ECT sensor and its circuit • Thermostat • ECM
DTC: P0121 **T, MIL: Yes** **Year:** 2009 **Model:** XL-7 **Engine:** 3.6L V6 **Transmission:** All	**Throttle Position (TP) Sensor 1 Performance:** • The ignition voltage is greater than 7 volts. • The TP sensor 1 voltage is between 0.18–4.62 volts. **Possible Causes:** • TP sensor and/or circuit • ECM

DTC	Trouble Code Title, Conditions & Possible Causes

DTC: P0122
1T, MIL: Yes
Year: 2009, 2010
Model: SX4
Engine: 2.0L L4, 2.0L L4 VIN 5, 2.4L L4 VIN 9
Transmission: All
	Throttle/Pedal Position Sensor/Switch "A" Circuit Low:
	Output voltage of TP sensor (main) signal circuit is lower than 0.3 V.
	Possible Causes:
	   • TP sensor (main) and/or its circuit
	   • ECM

DTC: P0122
2T, MIL: Yes
Year: 2009
Model: XL-7
Engine: 3.6L V6
Transmission: All
	Throttle Position (TP) Sensor 1 Circuit Low Voltage:
	   • The ignition is ON, with the engine OFF, or the engine is operating.
	   • The ignition voltage is greater than 7 volts.
	Possible Causes:
	   • TP sensor and/or circuit
	   • ECM

DTC: P0122
1T, MIL: Yes
Year: 2009, 2010
Model: Equator, Grand Vitara
Engine: 2.4L L4 VIN 0, 2.5L L4, 3.2L V6 VIN 1, 4.0L V6
Transmission: All
	Throttle position sensor 2 circuit low input:
	An excessively low voltage from the TP sensor 2 is sent to ECM.
	Possible Causes:
	   • Harness or connectors
	   • (The TP sensor 2 circuit is open or shorted.)
	   • (APP sensor 2 circuit is shorted.)
	   • Electric throttle control actuator
	   • (TP sensor 2)
	   • Accelerator pedal position sensor
	   • (APP sensor 2)

DTC: P0123
1T, MIL: Yes
Year: 2009, 2010
Model: Equator, Grand Vitara
Engine: 2.4L L4 VIN 0, 2.5L L4, 3.2L V6 VIN 1, 4.0L V6
Transmission: All
	Throttle position sensor 2 circuit high input:
	An excessively high voltage from the TP sensor 2 is sent to ECM.
	Possible Causes:
	   • Harness or connectors
	   • (The TP sensor 2 circuit is open or shorted.)
	   • (APP sensor 2 circuit is shorted.)
	   • Electric throttle control actuator
	   • (TP sensor 2)
	   • Accelerator pedal position sensor
	   • (APP sensor 2)

DTC: P0123
2T, MIL: Yes
Year: 2009
Model: XL-7
Engine: 3.6L V6
Transmission: All
	Throttle Position (TP) Sensor 1 Circuit High Voltage:
	   • The ignition is ON, with the engine OFF, or the engine is operating.
	   • The ignition voltage is greater than 7 volts.
	Possible Causes:
	   • TP sensor and/or circuit
	   • ECM

DTC: P0123
1T, MIL: Yes
Year: 2009, 2010
Model: SX4
Engine: 2.0L L4, 2.0L L4 VIN 5, 2.4L L4 VIN 9
Transmission: All
	Throttle/Pedal Position Sensor/Switch "A" Circuit High:
	Output voltage of TP sensor (main) signal circuit is higher than 4.65 V.
	Possible Causes:
	   • TP sensor (main) and/or its circuit
	   • ECM

DTC: P0125
2T, MIL: Yes
Year: 2009, 2010
Model: Equator, Grand Vitara, SX4
Engine: 2.0L L4, 2.0L L4 VIN 5, 2.4L L4 VIN 0, 2.4L L4 VIN 9, 2.5L L4, 4.0L V6
Transmission: All
	Insufficient Coolant Temperature For Closed Loop:
	Engine started, ECT sensor from 18-230°F, IAT sensor over 18°F, BARO sensor more than 75 kPa, vehicle driven at over 35 mph for 5 minutes under normal stop and go or cruise conditions, then back to idle speed for 5 minutes, and the PCM detected the engine coolant temperature did not reach a value that allows closed loop operation.
	Possible Causes:
	   • Inspect for low coolant level or an incorrect coolant mixture
	   • Check the operation of the thermostat (it may be stuck open)
	   • ECT sensor signal circuit has high resistance
	   • ECT sensor has failed

DTC	Trouble Code Title, Conditions & Possible Causes
DTC: P0127 **2T, MIL: Yes** **Year:** 2009, 2010 **Model:** Equator **Engine:** 2.5L L4, 4.0L V6 **Transmission:** All	**Intake air temperature too high:** Rationally incorrect voltage from the sensor is sent to ECM, compared with the voltage signal from engine coolant temperature sensor. **Possible Causes:** • Harness or connectors • (The sensor circuit is open or shorted) • Intake air temperature sensor
DTC: P0128 **2T, MIL: Yes** **Year:** 2009, 2010 **Model:** Equator, Grand Vitara, SX4, XL-7 **Engine:** 2.0L L4, 2.0L L4 VIN 5, 2.4L L4 VIN 0, 2.4L L4 VIN 9, 2.5L L4, 3.2L V6 VIN 1, 3.6L V6, 4.0L V6 **Transmission:** All	**Coolant Temperature Below Thermostat Regulating Temperature:** Engine started, ECT sensor from 18-230°F, IAT sensor over 18°F, BARO sensor more than 75 kPa, vehicle driven at over 35 mph for 5 minutes under normal stop and go or cruise conditions, then back to idle speed for 5 minutes, and the PCM detected the engine coolant temperature did not reach a value over that of the thermostat regulating temperature. **Possible Causes:** • Inspect for low coolant level or an incorrect coolant mixture • Check the operation of the thermostat (it may be stuck open) • ECT sensor signal circuit has high resistance • ECT sensor has failed
DTC: P012E **2T, MIL: Yes** **Year:** 2009, 2010 **Model:** Grand Vitara **Engine:** 3.2L V6 VIN 1 **Transmission:** All	**O2 Sensor Delayed Response – Rich to Lean Bank 1 Sensor 2:** Elapsed time until HO2S-2 (bank 1) signal voltage is reached specified value is longer than specified time at fuel cut. **Possible Causes:** • Oil level gauge • Oil filler cap • Engine oil • Coolant • Air intake system • Exhaust system • EVAP canister • EVAP canister purge valve • HO2S and/or its circuit • HO2S heater and/or its circuit • Fuel system • ECM
DTC: P012E **2T, MIL: Yes** **Year:** 2009, 2010 **Model:** Grand Vitara **Engine:** 2.4L L4 VIN 0 **Transmission:** All	**O2 Sensor Delayed Response – Rich to Lean (Bank-1 Sensor-2):** Shifting time which HO2S signal circuit voltage change from 0.65 V to 0.20 V is longer than 1.6 sec. at fuel cut. **Possible Causes:** • HO2S and/or its circuit • ECM
DTC: P0130 **2T, MIL: Yes** **Year:** 2009, 2010 **Model:** Equator, Grand Vitara **Engine:** 2.4L L4 VIN 0, 2.5L L4, 3.2L V6 VIN 1, 4.0L V6 **Transmission:** All	**Air fuel ratio (A/F) sensor 1 circuit:** A) The A/F signal computed by ECM from the A/F sensor 1 signal is constantly in a range other than approx. 1.5V. B) The A/F signal computed by ECM from the A/F sensor 1 signal is constantly approx. 1.5V. **Possible Causes:** • Harness or connectors • (A/F sensor 1 circuit is open or shorted.) • Air fuel ratio (A/F) sensor 1
DTC: P0130 **T, MIL: Yes** **Year:** 2009 **Model:** XL-7 **Engine:** 3.6L V6 **Transmission:** All	**HO2S Signal Circuit Shorted to Heater Circuit Bank 1 Sensor 1:** Condition 1 The ECM detects that the HO2S 1 signal voltage increases greater than 2 volts in 4 out of 6 HO2S heater switch OFF samples. Condition 2 The HO2S 1 voltage is between 600–1,150 mV and the corresponding HO2S 2 voltage is less than 100 mV. Condition 3 The HO2S 1 voltage is between 60–400 mV and the corresponding HO2S 2 voltage is greater than 500 mV. **Possible Causes:** • O2 sensor and/or circuit • ECM

DTC	Trouble Code Title, Conditions & Possible Causes
DTC: P0131 **2T, MIL: Yes** **Year:** 2009 **Model:** XL-7 **Engine:** 3.6L V6 **Transmission:** All	**HO2S Circuit Low Voltage Bank 1 Sensor 1:** • The ECM detects that the HO2S 1 voltage is less than 60mV while the corresponding HO2S 2 voltage is greater than 500mV. • The ECM detects that the HO2S 1 voltage is less than 60mV after a cold start with the engine coolant temperature (ECT) less than 40°C (104°F). **Possible Causes:** • O2 sensor and/or circuit • ECM
DTC: P0131 **2T CCM, MIL: Yes** **Year:** 2009, 2010 **Model:** Equator, Grand Vitara, SX4 **Engine:** 2.0L L4, 2.0L L4 VIN 5, 2.4L L4 VIN 0, 2.4L L4 VIN 9, 2.5L L4, 3.2L V6 VIN 1, 4.0L V6 **Transmission:** All	**HO2S-11 (Bank 1 Sensor 1) Circuit Low Input:** Engine started, vehicle driven to a speed over 45 mph for 6 minutes at an engine speed from 2500-3000 rpm, followed by a deceleration period of 3 seconds, fuel level from 15-85%, BARO sensor more than 75 kPa, IAT sensor more than 6.8°F, and the PCM detected the maximum HO2S-11 voltage was less than 300 mv during the test. **Possible Causes:** • HO2S signal ground circuit is open • HO2S is contaminated or has failed • Vehicle driven until it ran out of fuel, or while very low on fuel • PCM has failed
DTC: P0132 **2T CCM, MIL: Yes** **Year:** 2009, 2010 **Model:** Equator, Grand Vitara, SX4 **Engine:** 2.0L L4, 2.0L L4 VIN 5, 2.4L L4 VIN 0, 2.4L L4 VIN 9, 2.5L L4, 3.2L V6 VIN 1, 4.0L V6 **Transmission:** All	**HO2S-11 (Bank 1 Sensor 1) Circuit High Input:** Engine started, vehicle driven to a speed over 45 mph for 6 minutes at an engine speed from 2500-3000 rpm, followed by a deceleration period of 3 seconds, fuel level from 15-85%, BARO sensor more than 75 kPa, IAT sensor more than 6.8°F, and the PCM detected the HO2S-11 signal did not go below 600 mv during the CCM test. **Possible Causes:** • HO2S signal circuit is open or shorted to system power • Fuel supply system is too rich (fuel injector or regulator leaking) • HO2S is contaminated, damaged or has failed • PCM has failed • TSB TS015 (12/95) contains information related to this code
DTC: P0133 **2T, MIL: Yes** **Year:** 2009, 2010 **Model:** Equator, Grand Vitara, SX4 **Engine:** 2.0L L4, 2.0L L4 VIN 5, 2.4L L4 VIN 0, 2.4L L4 VIN 9, 2.5L L4, 3.2L V6 VIN 1, 4.0L V6 **Transmission:** All	**HO2S-11 (Bank 1 Sensor 1) Slow Response:** DTC P0131 and P0132 not set, vehicle driven at a speed of 35 mph for 2 minutes in closed loop, then back to idle speed for 2 minutes, ambient air temperature over 14°F, fuel level from 15-85%, BARO sensor more than 75 kPa, IAT sensor more than 14°F, and the PCM detected the HO2S-11 signal response time to change from rich-to-lean or lean-to-rich was more than 1 second during the CCM test. **Possible Causes:** • Air leaks in the intake manifold or exhaust manifold or pipes • IAT sensor or MAF sensor has deteriorated (out of calibration) • HO2S signal circuit is open or shorted to ground (intermittent) • HO2S contaminated with wrong fuel, has deteriorated or failed • PCM has failed • TSB TS015 (12/95) contains information related to this code
DTC: P0134 **2T, MIL: Yes** **Year:** 2009, 2010 **Model:** Grand Vitara, SX4 **Engine:** 2.0L L4, 2.0L L4 VIN 5, 2.4L L4 VIN 0, 2.4L L4 VIN 9, 3.2L V6 VIN 1 **Transmission:** All	**HO2S-11 (Bank 1 Sensor 1) No Activity Detected:** DTC P0131 and P0132 not set, vehicle driven at a speed of 35 mph for 2 minutes in closed loop, then back to idle speed for 2 minutes, ambient air temperature over 14°F, fuel level from 15-85%, BARO sensor more than 75 kPa, IAT sensor more than 14°F, and the PCM detected the HO2S-11 signal remained "high" or "low" in the test. **Possible Causes:** • Air leaks in the intake manifold or exhaust manifold or pipes • HO2S is contaminated (wrong fuel), has deteriorated or failed • HO2S heater element is damaged or has failed • PCM has failed • TSB TS015 (12/95) contains information related to this code
DTC: P0135 **2T, MIL: Yes** **Year:** 2009, 2010 **Model:** Grand Vitara **Engine:** 3.2L V6 VIN 1 **Transmission:** All	**HO2S-11 (Bank 1 Sensor 1) Heater Circuit Malfunction:** Engine started, engine runtime 1 minute, and the PCM detected the HO2S-11 heater signal was "low" with the heater commanded "off", or that it was "high: with the heater commanded "on" during the test. **Possible Causes:** • HO2S heater control circuit is open or shorted to ground • HO2S heater power circuit is open (check the IG fuse) • HO2S heater control circuit is shorted to power • HO2S heater is damaged or has failed • PCM has failed

DTC	Trouble Code Title, Conditions & Possible Causes
DTC: P0137 **2T CCM, MIL: Yes** **Year:** 2009, 2010 **Model:** Equator, Grand Vitara, SX4, XL-7 **Engine:** 2.0L L4, 2.0L L4 VIN 5, 2.4L L4 VIN 0, 2.4L L4 VIN 9, 2.5L L4, 3.2L V6 VIN 1, 3.6L V6, 4.0L V6 **Transmission:** All	**HO2S-12 (Bank 1 Sensor 2) Circuit Low Input:** Engine started, vehicle driven to a speed over 45 mph for 6 minutes at an engine speed from 2500-3000 rpm, followed by a deceleration period of 3 seconds, fuel level from 15-85%, BARO sensor more than 75 kPa, IAT sensor more than 6.8°F, and the PCM detected the maximum HO2S-12 voltage was less than 600 mv, or the minimum voltage was more than 400 mv, or the maximum HO2S-12 voltage was more than 4.5v during the CCM test. **Possible Causes:** • HO2S signal circuit open (sensor signal reads over 4.5v) • HO2S is shorted to ground • HO2S signal ground circuit is open • HO2S is contaminated or has failed • Vehicle driven until it ran out of fuel, or while very low on fuel • PCM has failed
DTC: P0138 **2T CCM, MIL: Yes** **Year:** 2009, 2010 **Model:** Equator, Grand Vitara, SX4, XL-7 **Engine:** 2.0L L4, 2.0L L4 VIN 5, 2.4L L4 VIN 0, 2.4L L4 VIN 9, 2.5L L4, 3.2L V6 VIN 1, 3.6L V6, 4.0L V6 **Transmission:** All	**HO2S-12 (Bank 1 Sensor 2) Circuit High Input:** Engine started, vehicle driven to a speed over 45 mph for 6 minutes at an engine speed from 2500-3000 rpm, followed by a deceleration period of 3 seconds, fuel level from 15-85%, BARO sensor more than 75 kPa, IAT sensor more than 6.8°F, and the PCM detected the maximum HO2S-12 voltage was less than 600 mv, or the minimum voltage was more than 400 mv, or the maximum HO2S-12 voltage was more than 4.5v during the CCM test. **Possible Causes:** • HO2S signal circuit open (sensor signal reads over 4.5v) • HO2S is shorted to ground • HO2S signal ground circuit is open • HO2S is contaminated or has failed • Vehicle driven until it ran out of fuel, or while very low on fuel • PCM has failed
DTC: P0139 **2T, MIL: Yes** **Year:** 2009, 2010 **Model:** Equator, Grand Vitara, SX4 **Engine:** 2.0L L4, 2.0L L4 VIN 5, 2.4L L4 VIN 0, 2.4L L4 VIN 9, 2.5L L4, 4.0L V6 **Transmission:** All	**Heated oxygen sensor 2 circuit slow response:** It takes more time for the sensor to respond between rich and lean than the specified time. **Possible Causes:** • Harness or connectors • (The sensor circuit is open or shorted) • Heated oxygen sensor 2 • Fuel pressure • Fuel injector • Intake air leaks
DTC: P013A **1T, MIL: Yes** **Year:** 2009, 2010 **Model:** Grand Vitara, XL-7 **Engine:** 3.2L V6 VIN 1, 3.6L V6 **Transmission:** All	**O2 Sensor Slow Response - Rich to Lean Bank 1 Sensor 2:** Average time which shifts HO2S-2 (bank 1) signal voltage from 0.4 V to 0.2 V is longer than 0.35 sec. at fuel cut. **Possible Causes:** • Oil level gauge • Oil filler cap • Engine oil • Coolant • Air intake system • Exhaust system • EVAP canister • EVAP canister purge solenoid valve • HO2S-2 and/or its circuit • HO2S-2 heater and/or its circuit • Fuel system • ECM

DTC	Trouble Code Title, Conditions & Possible Causes
DTC: P013C **1T, MIL: Yes** **Year:** 2009, 2010 **Model:** Grand Vitara, XL-7 **Engine:** 3.2L V6 VIN 1, 3.6L V6 **Transmission:** All	**O2 Sensor Slow Response - Rich to Lean Bank 2 Sensor 2:** Average time which shifts HO2S-2 (bank 2) signal voltage from 0.4 V to 0.2 V is longer than 0.35 sec. at fuel cut. **Possible Causes:** • Oil level gauge • Oil filler cap • Engine oil • Coolant • Air intake system • Exhaust system • EVAP canister • EVAP canister purge solenoid valve • HO2S-2 and/or its circuit • HO2S-2 heater and/or its circuit • Fuel system • ECM
DTC: P013E **2T, MIL: Yes** **Year:** 2009, 2010 **Model:** SX4, XL-7 **Engine:** 2.0L L4, 2.0L L4 VIN 5, 2.4L L4 VIN 9, 3.6L V6 **Transmission:** All	**O2 Sensor Delayed Response Rich to Lean (Sensor 2):** Shifting time which HO2S signal circuit voltage change from 0.65 V to 0.2 V is longer than 1.6 sec. at fuel cut. **Possible Causes:** • HO2S and/or its circuit • ECM
DTC: P0140 **2T, MIL: Yes** **Year:** 2009 **Model:** XL-7 **Engine:** 3.6L V6 **Transmission:** All	**HO2S Circuit Insufficient Activity Bank 1 Sensor 2:** • The ECM detects that the HO2S voltage is between 401–519 mV, when the calculated exhaust temperature is colder than 800°C (1472°F). • The ECM detects that the HO2S voltage is between 401–548 mV, when the calculated exhaust temperature is 800°C (1472°F) or warmer. • The ECM detects that the measured internal resistance of the HO2S is greater than 40,000 ohms, when the calculated exhaust temperature is warmer than 450°C (802°F). • The ECM detects any of the above conditions for greater than 1 second or for a cumulative of 10 seconds. **Possible Causes:** • O2 sensor and/or circuit • ECM
DTC: P0140 **2T CCM, MIL: Yes** **Year:** 2009, 2010 **Model:** Grand Vitara **Engine:** 3.2L V6 VIN 1 **Transmission:** All	**HO2S-12 (Bank 1 Sensor 2) Circuit, No Activity Detected:** DTC P0131 and P0132 not set, vehicle driven at a speed of 35 mph for 2 minutes in closed loop, then back to idle speed for 2 minutes, ambient air temperature over 14°F, fuel level from 15-85%, BARO sensor more than 75 kPa, IAT sensor more than 14°F, and the PCM detected the HO2S-11 signal remained "high" or "low" in the test. **Possible Causes:** • Air leaks in the intake manifold or exhaust manifold or pipes • HO2S is contaminated (wrong fuel), has deteriorated or failed • HO2S heater element is damaged or has failed • PCM has failed • TSB TS015 (12/95) contains information related to this code
DTC: P0141 **2T, MIL: Yes** **Year:** 2009, 2010 **Model:** Grand Vitara **Engine:** 3.2L V6 VIN 1 **Transmission:** All	**HO2S-12 (Bank 1 Sensor2) Heater Circuit Malfunction:** Engine started, engine runtime over 1 minute, IAT sensor more than 14°F, ECT sensor from 18-230°F, and the PCM detected the HO2S-12 heater current level was more than 5.3A or less than 0.09A. **Possible Causes:** • HO2S heater control circuit is open or shorted to ground • HO2S heater power circuit is open (check the IG fuse) • HO2S heater control circuit is shorted to power • HO2S heater is damaged or has failed • PCM has failed

DTC	Trouble Code Title, Conditions & Possible Causes
DTC: P014A **2T, MIL:** Yes **Year:** 2009, 2010 **Model:** Grand Vitara, XL-7 **Engine:** 3.2L V6 VIN 1, 3.6L V6 **Transmission:** All	**O2 Sensor Delayed Response -Rich to Lean Bank 2 Sensor 2:** Elapsed time until HO2S-2 (bank 2) signal voltage is reached specified value is longer than specified time at fuel cut. **Possible Causes:** • Oil level gauge • Oil filler cap • Engine oil • Coolant • Air intake system • Exhaust system • EVAP canister • EVAP canister purge valve • HO2S and/or its circuit • HO2S heater and/or its circuit • Fuel system • ECM
DTC: P0150 **2T, MIL:** Yes **Year:** 2009, 2010 **Model:** Equator, Grand Vitara **Engine:** 3.2L V6 VIN 1, 4.0L V6 **Transmission:** All	**Air fuel ratio (A/F) sensor 1 circuit (bank 2):** A) The A/F signal computed by ECM from the A/F sensor 1 signal is constantly in a range other than approx. 2.2V. B) The A/F signal computed by ECM from the A/F sensor 1 signal is constantly approx. 2.2V. **Possible Causes:** • Harness or connectors • (The A/F sensor 1 circuit is open or shorted.) • Air fuel ratio (A/F) sensor 1
DTC: P0150 **2T, MIL:** Yes **Year:** 2009 **Model:** XL-7 **Engine:** 3.6L V6 **Transmission:** All	**HO2S Signal Circuit Shorted to Heater Circuit Bank 2 Sensor 1:** Condition 1 The ECM detects that the HO2S 1 signal voltage increases greater than 2 volts in 4 out of 6 HO2S heater switch OFF samples. Condition 2 The HO2S 1 voltage is between 600–1,150 mV and the corresponding HO2S 2 voltage is less than 100 mV. Condition 3 The HO2S 1 voltage is between 60–400 mV and the corresponding HO2S 2 voltage is greater than 500 mV. **Possible Causes:** • O2 sensor and/or circuit • ECM
DTC: P0151 **2T CCM, MIL:** Yes **Year:** 2009, 2010 **Model:** Equator, Grand Vitara **Engine:** 3.2L V6 VIN 1, 4.0L V6 **Transmission:** All	**HO2S-21 (Bank 2 Sensor 1) Circuit Low Input:** Engine started, vehicle driven at a speed of over 35 mph for 2 minutes, then back to idle speed for 1 minute, BARO sensor more than 75 kPa, ECT sensor from 18-230°F, IAT sensor more than 18°F, fuel level from 15-85%, and the PCM detected the maximum HO2S-12 voltage was less than 300 mv during the CCM test. **Possible Causes:** • HO2S is shorted to ground • HO2S is contaminated, damaged or it has failed • Vehicle driven until it ran out of fuel, or while very low on fuel • PCM has failed
DTC: P0151 **2T, MIL:** Yes **Year:** 2009 **Model:** XL-7 **Engine:** 3.6L V6 **Transmission:** All	**HO2S Circuit Low Voltage Bank 2 Sensor 1:** • The ECM detects that the HO2S 1 voltage is less than 60 mV while the corresponding HO2S 2 voltage is greater than 500 mV. • The ECM detects that the HO2S 1 voltage is less than 60 mV after a cold start with the engine coolant temperature (ECT) less than 40°C (104°F). **Possible Causes:** • O2 sensor and/or circuit • ECM
DTC: P0152 **2T CCM, MIL:** Yes **Year:** 2009, 2010 **Model:** Equator, Grand Vitara **Engine:** 3.2L V6 VIN 1, 4.0L V6 **Transmission:** All	**HO2S-21 (Bank 2 Sensor 1) Circuit High Input:** Engine started, vehicle driven at a speed of over 35 mph for 2 minutes, then back to idle speed for 1 minute, BARO sensor more than 75 kPa, ECT sensor from 18-230°F, IAT sensor more than 18°F, fuel level from 15-85%, and the PCM detected the maximum HO2S-12 voltage was less than 600 mv during the CCM test. **Possible Causes:** • HO2S signal circuit is open or shorted to system power • Fuel supply system is too rich (fuel injector or regulator leaking) • HO2S is contaminated, damaged or has failed • PCM has failed

DTC	Trouble Code Title, Conditions & Possible Causes
DTC: P0153 **2T CCM, MIL: Yes** **Year:** 2009, 2010 **Model:** Equator, Grand Vitara **Engine:** 3.2L V6 VIN 1, 4.0L V6 **Transmission:** All	**HO2S-21 (Bank 2 Sensor 1) Slow Response:** Engine started, vehicle driven at a constant speed of over 35 mph for 2 minutes in closed loop, then back to idle speed for 1 minute, ambient air temperature over 14°F, fuel level from 15-85%, BARO sensor more than 75 kPa, IAI sensor from 18-230°F, and the PCM detected the response rate of the HO2S-21 signal was too long during the CCM test. **Possible Causes:** • Air leaks in the intake manifold or exhaust manifold or pipes • IAT sensor or MAF sensor has deteriorated (out of calibration) • HO2S signal circuit is open or shorted to ground (intermittent) • HO2S contaminated with wrong fuel, has deteriorated or failed • PCM has failed
DTC: P0154 **2T CCM, MIL: Yes** **Year:** 2009, 2010 **Model:** Grand Vitara **Engine:** 3.2L V6 VIN 1 **Transmission:** All	**HO2S-21 (Bank 2 Sensor 1) No Activity Detected:** Engine started, vehicle driven at a constant speed of over 35 mph for 2 minutes in closed loop, then back to idle speed for 1 minute, ambient air temperature over 14°F, fuel level from 15-85%, BARO sensor more than 75 kPa, IAT sensor more than 18°F, and the PCM detected the HO2S-21 signal did not exceed 0.45v during the test. **Possible Causes:** • Air leaks in the intake manifold or exhaust manifold or pipes • HO2S is contaminated (wrong fuel), has deteriorated or failed • HO2S heater element is damaged or has failed • PCM has failed
DTC: P0155 **2T, MIL: Yes** **Year:** 2009, 2010 **Model:** Grand Vitara **Engine:** 3.2L V6 VIN 1 **Transmission:** All	**HO2S-21 (Bank 2 Sensor 1) Heater Circuit Malfunction:** Engine started, engine runtime over 1 minute, IAT sensor more than 14°F, ECT sensor from 18-230°F, and the PCM detected the HO2S-21 heater current level was more than 5.3A or less than 0.09A. **Possible Causes:** • HO2S heater control circuit is open or shorted to ground • HO2S heater power circuit is open (check the IG fuse) • HO2S heater control circuit is shorted to power • HO2S heater is damaged or has failed • PCM has failed
DTC: P0157 **2T CCM, MIL: Yes** **Year:** 2009, 2010 **Model:** Equator, Grand Vitara, XL-7 **Engine:** 3.2L V6 VIN 1, 3.6L V6, 4.0L V6 **Transmission:** All	**HO2S-22 (Bank 2 Sensor 2) Circuit Low Input:** Engine started, vehicle driven at a speed of over 35 mph for 2 minutes, then back to idle speed for 1 minute, BARO sensor more than 75 kPa, ECT sensor from 18-230°F, IAT sensor more than 18°F, fuel level from 15-85%, and the PCM detected the maximum HO2S 12 voltage was less than 600 mv during the CCM test. **Possible Causes:** • HO2S signal circuit is open or shorted to system power • Fuel supply system is too rich (fuel injector or regulator leaking) • HO2S is contaminated, damaged or has failed • PCM has failed
DTC: P0158 **2T CCM, MIL: Yes** **Year:** 2009, 2010 **Model:** Equator, Grand Vitara, XL-7 **Engine:** 3.2L V6 VIN 1, 3.6L V6, 4.0L V6 **Transmission:** All	**HO2S-21 (Bank 2 Sensor 2) Circuit High Input:** Engine started, vehicle driven at a speed of over 35 mph for 2 minutes, then back to idle speed for 1 minute, BARO sensor more than 75 kPa, ECT sensor from 18-230°F, IAT sensor more than 18°F, fuel level from 15-85%, and the PCM detected the maximum HO2S-12 voltage was less than 600 mv during the CCM test. **Possible Causes:** • HO2S signal circuit is open or shorted to system power • Fuel supply system is too rich (fuel injector or regulator leaking) • HO2S is contaminated, damaged or has failed • PCM has failed
DTC: P0159 **2T, MIL: Yes** **Year:** 2009, 2010 **Model:** Equator **Engine:** 4.0L V6 **Transmission:** All	**Heated oxygen sensor 2 (bank 2) circuit slow response:** The switching time between rich and lean of a heated oxygen sensor 2 signal delays more than the specified time computed by ECM. **Possible Causes:** • Harness or connectors • (The sensor circuit is open or shorted) • Heated oxygen sensor 2 • Fuel system • EVAP system • Intake air system

DTC	Trouble Code Title, Conditions & Possible Causes
DTC: P0160 **2T CCM, MIL: Yes** **Year:** 2009, 2010 **Model:** Grand Vitara **Engine:** 3.2L V6 VIN 1 **Transmission:** All	**HO2S-12 (Bank 2 Sensor 2) Circuit, No Activity Detected:** DTC P0131 and P0132 not set, vehicle driven at a speed of 35 mph for 2 minutes in closed loop, then back to idle speed for 2 minutes, ambient air temperature over 14°F, fuel level from 15-85%, BARO sensor more than 75 kPa, IAT sensor more than 14°F, and the PCM detected the HO2S-11 signal remained "high" or "low" in the test. **Possible Causes:** • Air leaks in the intake manifold or exhaust manifold or pipes • HO2S is contaminated (wrong fuel), has deteriorated or failed • HO2S heater element is damaged or has failed • PCM has failed • TSB TS015 (12/95) contains information related to this code
DTC: P0160 **2T, MIL: Yes** **Year:** 2009 **Model:** XL-7 **Engine:** 3.6L V6 **Transmission:** All	**HO2S Circuit Insufficient Activity Bank 2 Sensor 2:** • The ECM detects that the HO2S voltage is between 401–519 mV, when the calculated exhaust temperature is colder than 800°C (1472°F). • The ECM detects that the HO2S voltage is between 401–548 mV, when the calculated exhaust temperature is 800°C (1472°F) or warmer. • The ECM detects that the measured internal resistance of the HO2S is greater than 40,000 ohms, when the calculated exhaust temperature is warmer than 450°C (802°F). • The ECM detects any of the above conditions for greater than 1 second or for a cumulative of 10 seconds. **Possible Causes:** • O2 sensor and/or circuit • ECM
DTC: P0161 **2T, MIL: Yes** **Year:** 2009, 2010 **Model:** Grand Vitara **Engine:** 3.2L V6 VIN 1 **Transmission:** All	**HO2S-22 (Bank 2 Sensor 2) Heater Circuit Malfunction:** Engine started, engine runtime over 1 minute, IAT sensor more than 14°F, ECT sensor from 18-230°F, and the PCM detected the HO2S-22 heater current level was more than 11 mA or less than 0.32 mA. **Possible Causes:** • HO2S heater control circuit is open or shorted to ground • HO2S heater power circuit is open (check the IG fuse) • HO2S heater control circuit is shorted to power • HO2S heater is damaged or has failed • PCM has failed
DTC: P0171 **2T, MIL: Yes** **Year:** 2009, 2010 **Model:** Equator, Grand Vitara, SX4 **Engine:** 2.0L L4, 2.0L L4 VIN 5, 2.4L L4 VIN 0, 2.4L L4 VIN 9, 2.5L L4, 4.0L V6 **Transmission:** All	**Fuel System Too Lean (Bank 1):** DTC P0101, P0102, P0103, P0106, P0107, P0108, P0112, P0113, P0117, P0118, P0121, P0131, P0132, P0133, P0135, P0136, P0141, P0301-306, P0400, P0400 and P0601 not set, engine started, vehicle driven at a speed of 30-40 mph for 5 minutes at an engine speed from 2500-3000 rpm, BARO sensor more than 75 kPa, IAT sensor from 14-158°F, ambient temperature over 14°F, fuel level from 15-85%, and the PCM detected the total fuel trim (Short and Long Term fuel trim total) exceed +31% in the Fuel System test. **Possible Causes:** • Air leaks after the MAF sensor, or in the EGR or PCV system • Base engine "mechanical" fault affecting one or more cylinders • Exhaust leaks before or near where the front HO2S is mounted • Fuel control sensor is out of calibration (i.e., ECT, IAT or MAP) • Fuel delivery system supplying too little fuel during cruise or idle periods (e.g., faulty fuel pump or dirty, restricted fuel filter) • Fuel injector (one or more) dirty or pressure regulator has failed • HO2S is contaminated, deteriorated or it has failed • Vehicle driven low on fuel or until it ran out of fuel
DTC: P0172 **2T, MIL: Yes** **Year:** 2009, 2010 **Model:** Equator, Grand Vitara, SX4 **Engine:** 2.0L L4, 2.0L L4 VIN 5, 2.4L L4 VIN 0, 2.4L L4 VIN 9, 2.5L L4, 4.0L V6 **Transmission:** All	**Fuel System Too Rich (Bank 1):** DTC P0101, P0102, P0103, P0106, P0107, P0108, P0112, P0113, P0117, P0118, P0121, P0131, P0132, P0133, P0135, P0136, P0141, P0301-306, P0400, P0400 and P0601 not set, engine started, vehicle driven at a speed of 30-40 mph for 5 minutes at an engine speed from 2500-3000 rpm, BARO sensor more than 75 kPa, IAT sensor from 14-158°F, ambient temperature over 14°F, fuel level from 15-85%, and the PCM detected the total fuel trim exceeded -31 (Short and Long Term fuel trim total) during the test. **Possible Causes:** • Base engine "mechanical" fault affecting one or more cylinders • EVAP system component has failed or canister fuel saturated • Fuel control sensor is out of calibration (i.e., ECT, IAT or MAP) • Fuel delivery system supplying too much fuel during cruise or idle periods (e.g., faulty fuel pump, or faulty pressure regulator) • Fuel injector(s) is leaking or stuck partially open (one or more) • HO2S is contaminated, deteriorated or it has failed

DTC	Trouble Code Title, Conditions & Possible Causes
DTC: P0174 **2T, MIL:** Yes **Year:** 2009, 2010 **Model:** Equator **Engine:** 4.0L V6 **Transmission:** All	**Fuel System Too Lean (Bank 2):** DTC P0101, P0102, P0103, P0106, P0107, P0108, P0112, P0113, P0117, P0118, P0121, P0131, P0132, P0133, P0135, P0136, P0141, P0301-306, P0400, P0400 and P0601 not set, engine started, vehicle driven at a speed of 30-40 mph for 5 minutes at an engine speed from 2500-3000 rpm, BARO sensor more than 75 kPa, IAT sensor from 14-158°F, ambient temperature over 14°F, fuel level from 15-85%, and the PCM detected the total fuel trim (Short and Long Term fuel trim total) exceed +31% in the Fuel System test. **Possible Causes:** • Air leaks after the MAF sensor, or in the EGR or PCV system • Base engine "mechanical" fault affecting one or more cylinders • Exhaust leaks before or near where the front HO2S is mounted • Fuel control sensor is out of calibration (i.e., ECT, IAT or MAP) • Fuel delivery system supplying too little fuel during cruise or idle periods (e.g., faulty fuel pump or dirty, restricted fuel filter) • Fuel injector (one or more) dirty or pressure regulator has failed • HO2S is contaminated, deteriorated or it has failed • Vehicle driven low on fuel or until it ran out of fuel
DTC: P0175 **2T, MIL:** Yes **Year:** 2009, 2010 **Model:** Equator **Engine:** 4.0L V6 **Transmission:** All	**Fuel System Too Rich (Bank 2):** DTC P0101, P0102, P0103, P0106, P0107, P0108, P0112, P0113, P0117, P0118, P0121, P0131, P0132, P0133, P0135, P0136, P0141, P0301-306, P0400, P0400 and P0601 not set, engine started, vehicle driven at a speed of 30-40 rpm for 5 minutes at an engine speed from 2500-3000 rpm, BARO sensor more than 75 kPa, IAT sensor from 14-158°F, ambient temperature over 14°F, fuel level from 15-85%, and the PCM detected the total fuel trim (Short and Long Term fuel trim total) exceed -31% in the Fuel System test. **Possible Causes:** • Base engine "mechanical" fault affecting one or more cylinders • EVAP system component has failed or canister fuel saturated • Fuel control sensor is out of calibration (i.e., ECT, IAT or MAP) • Fuel delivery system supplying too much fuel during cruise or idle periods (e.g., faulty fuel pump, or faulty pressure regulator) • Fuel injector(s) is leaking or stuck partially open (one or more) • HO2S is contaminated, deteriorated or it has failed
DTC: P0181 **2T, MIL:** Yes **Year:** 2009, 2010 **Model:** Equator **Engine:** 2.5L L4, 4.0L V6 **Transmission:** All	**Fuel tank temperature sensor circuit range/performance:** Rationally incorrect voltage from the sensor is sent to ECM, compared with the voltage signals from engine coolant temperature sensor and intake air temperature sensor. **Possible Causes:** • Harness or connectors • (The sensor circuit is open or shorted) • Fuel tank temperature sensor
DTC: P0182 **2T, MIL:** Yes **Year:** 2009, 2010 **Model:** Equator **Engine:** 2.5L L4, 4.0L V6 **Transmission:** All	**Fuel tank temperature sensor circuit low input:** An excessively low voltage from the sensor is sent to ECM. **Possible Causes:** • Harness or connectors • (The sensor circuit is open or shorted.) • Fuel tank temperature sensor
DTC: P0183 **2T, MIL:** Yes **Year:** 2009, 2010 **Model:** Equator **Engine:** 2.5L L4, 4.0L V6 **Transmission:** All	**Fuel tank temperature sensor circuit high input:** An excessively high voltage from the sensor is sent to ECM. **Possible Causes:** • Harness or connectors • (The sensor circuit is open or shorted.) • Fuel tank temperature sensor
DTC: P0201 **1T, MIL:** Yes **Year:** 2009 **Model:** XL-7 **Engine:** 3.6L V6 **Transmission:** All	**Fuel Injector 1 Circuit Malfunction:** Engine started, engine speed less than 1000 rpm, TP sensor less than 0.7v, Actuator Tests all "off", and the PCM detected the injector coil surge voltage was too low (i.e., it did not reach system voltage +2v) within 2 seconds after the injector turned "off" during the test. **Possible Causes:** • Injector control circuit is open or shorted to ground • Injector power circuit is open (check power from the MFI relay) • Fuel injector is damaged or has failed

DTC	Trouble Code Title, Conditions & Possible Causes
DTC: P0202 **1I, MIL: Yes** **Year:** 2009 **Model:** XL-7 **Engine:** 3.6L V6 **Transmission:** All	**Fuel Injector 2 Circuit Malfunction:** Engine started, engine speed less than 1000 rpm, TP sensor less than 0.7v, Actuator Tests all "off", and the PCM detected the injector coil surge voltage was too low (i.e., it did not reach system voltage +2v) within 2 seconds after the injector turned "off" during the test. **Possible Causes:** • Injector control circuit is open or shorted to ground • Injector power circuit is open (check power from the MFI relay) • Fuel injector is damaged or has failed
DTC: P0203 **1T, MIL: Yes** **Year:** 2009 **Model:** XL-7 **Engine:** 3.6L V6 **Transmission:** All	**Fuel Injector 3 Circuit Malfunction:** Engine started, engine speed less than 1000 rpm, TP sensor less than 0.7v, Actuator Tests all "off", and the PCM detected the injector coil surge voltage was too low (i.e., it did not reach system voltage +2v) within 2 seconds after the injector turned "off" during the test. **Possible Causes:** • Injector control circuit is open or shorted to ground • Injector power circuit is open (check power from the MFI relay) • Fuel injector is damaged or has failed
DTC: P0204 **1T, MIL: Yes** **Year:** 2009 **Model:** XL-7 **Engine:** 3.6L V6 **Transmission:** All	**Fuel Injector 4 Circuit Malfunction:** Engine started, engine speed less than 1000 rpm, TP sensor less than 0.7v, Actuator Tests all "off", and the PCM detected the injector coil surge voltage was too low (i.e., it did not reach system voltage +2v) within 2 seconds after the injector turned "off" during the test. **Possible Causes:** • Injector control circuit is open or shorted to ground • Injector power circuit is open (check power from the MFI relay) • Fuel injector is damaged or has failed
DTC: P0205 **2T, MIL: Yes** **Year:** 2009 **Model:** XL-7 **Engine:** 3.6L V6 **Transmission:** All	**Injector 5 Control Circuit:** The ECM detects the injector control circuit is open for greater than 4 seconds or for a cumulative of 30 seconds. **Possible Causes:** • Injector and/or circuit • ECM
DTC: P0206 **2T, MIL: Yes** **Year:** 2009 **Model:** XL-7 **Engine:** 3.6L V6 **Transmission:** All	**Injector 6 Control Circuit:** The ECM detects the injector control circuit is open for greater than 4 seconds or for a cumulative of 30 seconds. **Possible Causes:** • Injector and/or circuit • ECM
DTC: P0221 **2T, MIL: Yes** **Year:** 2009 **Model:** XL-7 **Engine:** 3.6L V6 **Transmission:** All	**Throttle Position (TP) Sensor 2 Performance:** • The ignition is ON, with the engine OFF, or the engine is operating. • The ignition voltage is greater than 7 volts. **Possible Causes:** • TP sensor and/or circuit • ECM
DTC: P0222 **2T, MIL: Yes** **Year:** 2009 **Model:** XL-7 **Engine:** 3.6L V6 **Transmission:** All	**Throttle Position (TP) Sensor 2 Circuit Low Voltage:** • The ignition is ON, with the engine OFF, or the engine is operating. • The ignition voltage is greater than 7 volts. **Possible Causes:** • TP sensor and/or circuit • ECM
DTC: P0222 **1T, MIL: Yes** **Year:** 2009, 2010 **Model:** Equator, Grand Vitara **Engine:** 2.4L L4 VIN 0, 2.5L L4, 3.2L V6 VIN 1, 4.0L V6 **Transmission:** All	**Throttle position sensor 1 circuit low input:** An excessively low voltage from the TP sensor 1 is sent to ECM. **Possible Causes:** • Harness or connectors • (The TP sensor 1 circuit is open or shorted.) • (APP sensor 2 circuit is shorted.) • Electric throttle control actuator • (TP sensor 1) • Accelerator pedal position sensor • (APP sensor 2)

DTC	Trouble Code Title, Conditions & Possible Causes
DTC: P0222 **1T, MIL: Yes** **Year:** 2009, 2010 **Model:** SX4 **Engine:** 2.0L L4, 2.0L L4 VIN 5, 2.4L L4 VIN 9 **Transmission:** All	**Throttle/Pedal Position Sensor/Switch "B" Circuit Low:** Output voltage of TP sensor (sub) signal circuit is lower than 0.74 V. **Possible Causes:** • TP sensor (sub) and/or its circuit • ECM
DTC: P0223 **1T, MIL: Yes** **Year:** 2009, 2010 **Model:** SX4 **Engine:** 2.0L L4, 2.0L L4 VIN 5, 2.4L L4 VIN 9 **Transmission:** All	**Throttle/Pedal Position Sensor/Switch "B" Circuit High:** Output voltage of TP sensor (sub) signal circuit is higher than 4.75 V. **Possible Causes:** • TP sensor (sub) and/or its circuit • ECM
DTC: P0223 **2T, MIL: Yes** **Year:** 2009 **Model:** XL-7 **Engine:** 3.6L V6 **Transmission:** All	**Throttle Position (TP) Sensor 2 Circuit High Voltage:** • The ignition is ON, with the engine OFF, or the engine is operating. • The ignition voltage is greater than 7 volts. **Possible Causes:** • TP sensor and/or circuit • ECM
DTC: P0223 **1T, MIL: Yes** **Year:** 2009, 2010 **Model:** Equator, Grand Vitara **Engine:** 2.4L L4 VIN 0, 2.5L L4, 3.2L V6 VIN 1, 4.0L V6 **Transmission:** All	**Throttle position sensor 1 circuit high input:** An excessively high voltage from the TP sensor 1 is sent to ECM. **Possible Causes:** • Harness or connectors • (The TP sensor 1 circuit is open or shorted.) • (APP sensor 2 circuit is shorted.) • Electric throttle control actuator • (TP sensor 1) • Accelerator pedal position sensor • (APP sensor 2)
DTC: P0261 **2T, MIL: Yes** **Year:** 2009 **Model:** XL-7 **Engine:** 3.6L V6 **Transmission:** All	**Injector 1 Control Circuit Low Voltage:** The ECM detects the injector control circuit is shorted to ground for greater than 4 seconds or for a cumulative of 30 seconds. **Possible Causes:** • Injector and/or circuit • ECM
DTC: P0262 **2T, MIL: Yes** **Year:** 2009 **Model:** XL-7 **Engine:** 3.6L V6 **Transmission:** All	**Injector 1 Control Circuit High Voltage:** The ECM detects the injector control circuit is shorted to voltage for greater than 4 seconds or for a cumulative of 30 seconds. **Possible Causes:** • Injector and/or circuit • ECM
DTC: P0264 **2T, MIL: Yes** **Year:** 2009 **Model:** XL-7 **Engine:** 3.6L V6 **Transmission:** All	**Injector 2 Control Circuit Low Voltage:** The ECM detects the injector control circuit is shorted to ground for greater than 4 seconds or for a cumulative of 30 seconds. **Possible Causes:** • Injector and/or circuit • ECM
DTC: P0265 **2T, MIL: Yes** **Year:** 2009 **Model:** XL-7 **Engine:** 3.6L V6 **Transmission:** All	**Injector 2 Control Circuit High Voltage:** The ECM detects the injector control circuit is shorted to voltage for greater than 4 seconds or for a cumulative of 30 seconds. **Possible Causes:** • Injector and/or circuit • ECM
DTC: P0267 **2T, MIL: Yes** **Year:** 2009 **Model:** XL-7 **Engine:** 3.6L V6 **Transmission:** All	**Injector 3 Control Circuit Low Voltage:** The ECM detects the injector control circuit is shorted to ground for greater than 4 seconds or for a cumulative of 30 seconds. **Possible Causes:** • Injector and/or circuit • ECM

DTC	Trouble Code Title, Conditions & Possible Causes
DTC: P0268 **2T, MIL: Yes** **Year:** 2009 **Model:** XL-7 **Engine:** 3.6L V6 **Transmission:** All	**Injector 3 Control Circuit High Voltage:** The ECM detects the injector control circuit is shorted to voltage for greater than 4 seconds or for a cumulative of 30 seconds. **Possible Causes:** • Injector and/or circuit • ECM
DTC: P0270 **2T, MIL: Yes** **Year:** 2009 **Model:** XL-7 **Engine:** 3.6L V6 **Transmission:** All	**Injector 4 Control Circuit Low Voltage:** The ECM detects the injector control circuit is shorted to ground for greater than 4 seconds or for a cumulative of 30 seconds. **Possible Causes:** • Injector and/or circuit • ECM
DTC: P0271 **2T, MIL: Yes** **Year:** 2009 **Model:** XL-7 **Engine:** 3.6L V6 **Transmission:** All	**Injector 4 Control Circuit High Voltage:** The ECM detects the injector control circuit is shorted to voltage for greater than 4 seconds or for a cumulative of 30 seconds. **Possible Causes:** • Injector and/or circuit • ECM
DTC: P0273 **2T, MIL: Yes** **Year:** 2009 **Model:** XL-7 **Engine:** 3.6L V6 **Transmission:** All	**Injector 5 Control Circuit Low Voltage:** The ECM detects the injector control circuit is shorted to ground for greater than 4 seconds or for a cumulative of 30 seconds. **Possible Causes:** • Injector and/or circuit • ECM
DTC: P0274 **2T, MIL: Yes** **Year:** 2009 **Model:** XL-7 **Engine:** 3.6L V6 **Transmission:** All	**Injector 5 Control Circuit High Voltage:** The ECM detects the injector control circuit is shorted to voltage for greater than 4 seconds or for a cumulative of 30 seconds. **Possible Causes:** • Injector and/or circuit • ECM
DTC: P0276 **2T, MIL: Yes** **Year:** 2009 **Model:** XL-7 **Engine:** 3.6L V6 **Transmission:** All	**Injector 6 Control Circuit Low Voltage:** The ECM detects the injector control circuit is shorted to ground for greater than 4 seconds or for a cumulative of 30 seconds. **Possible Causes:** • Injector and/or circuit • ECM
DTC: P0277 **2T, MIL: Yes** **Year:** 2009 **Model:** XL-7 **Engine:** 3.6L V6 **Transmission:** All	**Injector 6 Control Circuit High Voltage:** The ECM detects the injector control circuit is shorted to voltage for greater than 4 seconds or for a cumulative of 30 seconds. **Possible Causes:** • Injector and/or circuit • ECM
DTC: P0300 **2T CCM, MIL: Yes** **Year:** 2009, 2010 **Model:** Equator, Grand Vitara, SX4, XL-7 **Engine:** 2.0L L4, 2.0L L4 VIN 5, 2.4L L4 VIN 0, 2.4L L4 VIN 9, 2.5L L4, 3.2L V6 VIN 1, 3.6L V6, 4.0L V6 **Transmission:** All	**Multiple Cylinder Misfire Detected:** DTC P0335 and P0340 not set, altitude less than 9150 feet, fuel level from 15-85%, IAT signal from 6.8-158°F, ambient temperature over 14°F, ECT sensor from 14-230°F, engine runtime over 1 minute, and the PCM detected a misfire in more than 1 cylinders in the 200 (Catalyst) or 1000-rpm (High Emissions) revolution range. **NOTE: If the misfire is severe, the MIL will flash on/off on the 1st trip!** **Possible Causes:** • Air leak in the intake manifold, or in the EGR or PCM system • Base engine mechanical fault that affects one or more cylinders • Fuel delivery component fault that affects one or more cylinders (i.e., one or more contaminated, dirty or sticking fuel injectors) • Ignition system problem (coil or plug) in one or more cylinders

DTC	Trouble Code Title, Conditions & Possible Causes
DTC: P0301 **2T CCM, MIL: Yes** **Year:** 2009, 2010 **Model:** Equator, Grand Vitara, SX4, XL-7 **Engine:** 2.0L L4, 2.0L L4 VIN 5, 2.4L L4 VIN 0, 2.4L L4 VIN 9, 2.5L L4, 3.2L V6 VIN 1, 3.6L V6, 4.0L V6 **Transmission:** All	**Cylinder 1 Misfire Detected:** DTC P0335 and P0340 not set, BARO sensor more than 75 kPa, fuel level from 15-85%, IAT signal more than 18°F, ECT sensor from 18-230°F, engine running, and the PCM detected a misfire in one cylinder in the 200 (Catalyst) or 1000-rpm (High Emissions) range. **NOTE: If the misfire is severe, the MIL will flash on/off on the 1st trip!** **Possible Causes:** • Air leak in the intake manifold, or in the EGR or PCM system • Base engine mechanical fault that affects only one cylinder • Fuel delivery component fault that affects only one cylinder (i.e., a contaminated, dirty or sticking fuel injector) • Ignition system problem (coil or plug) that affects one cylinder
DTC: P0302 **2T CCM, MIL: Yes** **Year:** 2009, 2010 **Model:** Equator, Grand Vitara, SX4, XL-7 **Engine:** 2.0L L4, 2.0L L4 VIN 5, 2.4L L4 VIN 0, 2.4L L4 VIN 9, 2.5L L4, 3.2L V6 VIN 1, 3.6L V6, 4.0L V6 **Transmission:** All	**Cylinder 2 Misfire Detected:** DTC P0335 and P0340 not set, BARO sensor more than 75 kPa, fuel level from 15-85%, IAT signal more than 18°F, ECT sensor from 18-230°F, engine running, and the PCM detected a misfire in one cylinder in the 200 (Catalyst) or 1000-rpm (High Emissions) range. **NOTE: If the misfire is severe, the MIL will flash on/off on the 1st trip!** **Possible Causes:** • Air leak in the intake manifold, or in the EGR or PCM system • Base engine mechanical fault that affects only one cylinder • Fuel delivery component fault that affects only one cylinder (i.e., a contaminated, dirty or sticking fuel injector) • Ignition system problem (coil or plug) that affects one cylinder
DTC: P0303 **2T CCM, MIL: Yes** **Year:** 2009, 2010 **Model:** Equator, Grand Vitara, SX4, XL-7 **Engine:** 2.0L L4, 2.0L L4 VIN 5, 2.4L L4 VIN 0, 2.4L L4 VIN 9, 2.5L L4, 3.2L V6 VIN 1, 3.6L V6, 4.0L V6 **Transmission:** All	**Cylinder 3 Misfire Detected:** DTC P0335 and P0340 not set, BARO sensor more than 75 kPa, fuel level from 15-85%, IAT signal more than 18°F, ECT sensor from 18-230°F, engine running, and the PCM detected a misfire in one cylinder in the 200 (Catalyst) or 1000-rpm (High Emissions) range. **NOTE: If the misfire is severe, the MIL will flash on/off on the 1st trip!** **Possible Causes:** • Air leak in the intake manifold, or in the EGR or PCM system • Base engine mechanical fault that affects only one cylinder • Fuel delivery component fault that affects only one cylinder (i.e., a contaminated, dirty or sticking fuel injector) • Ignition system problem (coil or plug) that affects one cylinder
DTC: P0304 **2T CCM, MIL: Yes** **Year:** 2009, 2010 **Model:** Equator, Grand Vitara, SX4, XL-7 **Engine:** 2.0L L4, 2.0L L4 VIN 5, 2.4L L4 VIN 0, 2.4L L4 VIN 9, 2.5L L4, 3.2L V6 VIN 1, 3.6L V6, 4.0L V6 **Transmission:** All	**Cylinder 4 Misfire Detected:** DTC P0335 and P0340 not set, BARO sensor more than 75 kPa, fuel level from 15-85%, IAT signal more than 18°F, ECT sensor from 18-230°F, engine running, and the PCM detected a misfire in one cylinder in the 200 (Catalyst) or 1000-rpm (High Emissions) range. **NOTE: If the misfire is severe, the MIL will flash on/off on the 1st trip!** **Possible Causes:** • Air leak in the intake manifold, or in the EGR or PCM system • Base engine mechanical fault that affects only one cylinder • Fuel delivery component fault that affects only one cylinder (i.e., a contaminated, dirty or sticking fuel injector) • Ignition system problem (coil or plug) that affects one cylinder
DTC: P0305 **2T CCM, MIL: Yes** **Year:** 2009, 2010 **Model:** Equator, Grand Vitara, XL-7 **Engine:** 3.2L V6 VIN 1, 3.6L V6, 4.0L V6 **Transmission:** All	**Cylinder 5 Misfire Detected:** DTC P0335 and P0340 not set, BARO sensor more than 75 kPa, fuel level from 15-85%, IAT signal more than 18°F, ECT sensor from 18-230°F, engine running, and the PCM detected a misfire in one cylinder in the 200 (Catalyst) or 1000-rpm (High Emissions) range. **NOTE: If the misfire is severe, the MIL will flash on/off on the 1st trip!** **Possible Causes:** • Air leak in the intake manifold, or in the EGR or PCM system • Base engine mechanical fault that affects only one cylinder • Fuel delivery component fault that affects only one cylinder (i.e., a contaminated, dirty or sticking fuel injector) • Ignition system problem (coil or plug) that affects one cylinder
DTC: P0306 **2T CCM, MIL: Yes** **Year:** 2009, 2010 **Model:** Equator, Grand Vitara, XL-7 **Engine:** 3.2L V6 VIN 1, 3.6L V6, 4.0L V6 **Transmission:** All	**Cylinder 6 Misfire Detected:** DTC P0335 and P0340 not set, BARO sensor more than 75 kPa, fuel level from 15-85%, IAT signal more than 18°F, ECT sensor from 18-230°F, engine running, and the PCM detected a misfire in one cylinder in the 200 (Catalyst) or 1000-rpm (High Emissions) range. **NOTE: If the misfire is severe, the MIL will flash on/off on the 1st trip!** **Possible Causes:** • Air leak in the intake manifold, or in the EGR or PCM system • Base engine mechanical fault that affects only one cylinder • Fuel delivery component fault that affects only one cylinder (i.e., a contaminated, dirty or sticking fuel injector) • Ignition system problem (coil or plug) that affects one cylinder
DTC: P0324 **2T, MIL: Yes** **Year:** 2009, 2010 **Model:** Grand Vitara, XL-7 **Engine:** 3.2L V6 VIN 1, 3.6L V6 **Transmission:** All	**Knock Control System Error:** Failure of knock sensor signal processing IC in ECM is detected. **Possible Causes:** • Knock sensor and/or its circuit • ECM

DTC	Trouble Code Title, Conditions & Possible Causes
DTC: P0326 **2T, MIL: Yes** **Year:** 2009, 2010 **Model:** Grand Vitara, XL-7 **Engine:** 3.2L V6 VIN 1, 3.6L V6 **Transmission:** All	**Knock Sensor 1 Circuit Range/Performance Bank 1 or Single Sensor:** Knock sensor (bank 1) signal is not detected for specified time within specified ignition times. **Possible Causes:** • Knock sensor and/or its circuit • ECM
DTC: P0327 **2T, MIL: Yes** **Year:** 2009, 2010 **Model:** Equator, XL-7 **Engine:** 2.5L L4, 3.6L V6, 4.0L V6 **Transmission:** All	**Knock sensor circuit low input (bank 1):** An excessively low voltage from the sensor is sent to ECM. **Possible Causes:** • Harness or connectors • (The sensor circuit is open or shorted.) • Knock sensor
DTC: P0327 **1T CCM, MIL: Yes** **Year:** 2009, 2010 **Model:** Equator, Grand Vitara, SX4 **Engine:** 2.0L L4, 2.0L L4 VIN 5, 2.4L L4 VIN 0, 2.4L L4 VIN 9, 2.5L L4, 3.2L V6 VIN 1, 4.0L V6 **Transmission:** All	**Knock Sensor Circuit Low Input:** Key on or engine running, and the PCM detected an unexpected "high" voltage condition (0.90v or less) on the Knock Sensor circuit. **Possible Causes:** • Knock sensor signal circuit is shorted to sensor ground • Knock sensor signal circuit is shorted to chassis ground • Knock sensor is damaged or has failed • PCM has failed
DTC: P0328 **2T, MIL: Yes** **Year:** 2009, 2010 **Model:** Equator, Grand Vitara, SX4, XL-7 **Engine:** 2.0L L4, 2.0L L4 VIN 5, 2.4L L4 VIN 0, 2.4L L4 VIN 9, 3.2L V6 VIN 1, 3.6L V6, 4.0L V6 **Transmission:** All	**Knock sensor circuit high input (bank 1):** An excessively high voltage from the sensor is sent to ECM. **Possible Causes:** • Harness or connectors • (The sensor circuit is open or shorted.) • Knock sensor
DTC: P0331 **2T, MIL: Yes** **Year:** 2009, 2010 **Model:** Grand Vitara, XL-7 **Engine:** 3.2L V6 VIN 1, 3.6L V6 **Transmission:** All	**Knock Sensor 2 Circuit Range/Performance Bank 2:** Knock sensor (bank 2) signal is not detected for specified time within specified ignition times. **Possible Causes:** • Knock sensor and/or its circuit • ECM
DTC: P0332 **2T, MIL: Yes** **Year:** 2009, 2010 **Model:** Equator, Grand Vitara, XL-7 **Engine:** 3.2L V6 VIN 1, 3.6L V6, 4.0L V6 **Transmission:** All	**Knock sensor circuit low input (bank 2):** An excessively low voltage from the sensor is sent to ECM. **Possible Causes:** • Harness or connectors • (The sensor circuit is open or shorted.) • Knock sensor
DTC: P0333 **2T, MIL: Yes** **Year:** 2009, 2010 **Model:** Equator, Grand Vitara, XL-7 **Engine:** 3.2L V6 VIN 1, 3.6L V6, 4.0L V6 **Transmission:** All	**Knock sensor circuit high input (bank 2):** An excessively high voltage from the sensor is sent to ECM. **Possible Causes:** • Harness or connectors • (The sensor circuit is open or shorted.) • Knock sensor
DTC: P0335 **1T CCM, MIL: Yes** **Year:** 2009, 2010 **Model:** Equator, Grand Vitara, SX4, XL-7 **Engine:** 2.0L L4, 2.0L L4 VIN 5, 2.4L L4 VIN 0, 2.4L L4 VIN 9, 2.5L L4, 3.2L V6 VIN 1, 3.6L V6, 4.0L V6 **Transmission:** All	**Crankshaft Position Sensor Circuit Malfunction:** Engine started, and the PCM did not detect any CKP signals for 3 seconds or after 100 CMP pulses were received in the CCM test. **Possible Causes:** • CKP sensor positive (+) circuit is open or shorted to ground • CKP sensor negative (-) circuit is open or shorted to ground • CKP sensor is damaged or has failed • PCM has failed

DTC	Trouble Code Title, Conditions & Possible Causes

DTC: P0336
2T, MIL: Yes
Year: 2009, 2010
Model: Grand Vitara, XL-7
Engine: 3.2L V6 VIN 1, 3.6L V6
Transmission: All
	Crankshaft Position Sensor "A" Circuit Range/Performance:
	This DTC is detected if any one of the following conditions is met.
	   • Even if CMP sensor signal is inputting, CKP sensor signal is not input.
	   • Space of CKP sensor signal rotor is not detected for 3 times consecutively.
	Possible Causes:
	   • CKP sensor and/or its circuit
	   • Signal rotor
	   • ECM

DTC: P0338
T, MIL: Yes
Year: 2009
Model: XL-7
Engine: 3.6L V6
Transmission: All
	Crankshaft Position (CKP) Sensor Circuit High Duty Cycle:
	The ECM detects a difference of greater than 8 teeth between reference gap position pulses for greater than 1 second or a cumulative of 10 seconds.
	Possible Causes:
	   • CKP sensor and/or circuit
	   • ECM

DTC: P0340
1T CCM, MIL: Yes
Year: 2009, 2010
Model: Equator, Grand Vitara, SX4
Engine: 2.0L L4, 2.0L L4 VIN 5, 2.4L L4 VIN 0, 2.4L L4 VIN 9, 2.5L L4, 4.0L V6
Transmission: All
	Camshaft Position Sensor Circuit Malfunction:
	Engine cranking, engine start signal received (from starter circuit), and the PCM did not detect any CMP sensor signals for 5 seconds.
	Possible Causes:
	   • CMP sensor signal circuit is open between sensor and PCM
	   • CMP sensor signal circuit is shorted to ground
	   • CMP sensor power circuit is open (check power from the relay)
	   • CMP sensor is damaged or has failed
	   • PCM has failed

DTC: P0341
1T CCM, MIL: Yes
Year: 2009, 2010
Model: Grand Vitara
Engine: 3.2L V6 VIN 1
Transmission: All
	Camshaft Position Sensor Circuit Rationality:
	Engine cranking, engine start signal received (from starter circuit), and the PCM did not detect any CMP sensor signals within the expected parameters for 5 seconds.
	Possible Causes:
	   • CMP sensor signal circuit is open between sensor and PCM
	   • CMP sensor signal circuit is shorted to ground
	   • CMP sensor power circuit is open (check power from the relay)
	   • CMP sensor is damaged or has failed
	   • PCM has failed

DTC: P0341
2T, MIL: Yes
Year: 2009
Model: XL-7
Engine: 3.6L V6
Transmission: All
	Intake Camshaft Position (CMP) Sensor Performance Bank 1:
	   • The ECM detects a signal from the CMP sensor, but the number of pulses are less than or greater than what is expected for one crankshaft revolution.
	OR
	   • The CMP sensor does NOT correlate to the crankshaft position.
	   • Either condition must exist for greater than 1 second or cumulative of 10 seconds.
	Possible Causes:
	   • CMP sensor and/or circuit
	   • CMP sensor and/or reluctor wheel installation or damage
	   • ECM

DTC: P0342
1T CCM, MIL: Yes
Year: 2009, 2010
Model: Grand Vitara
Engine: 3.2L V6 VIN 1
Transmission: All
	Camshaft Position Sensor Circuit No Signal:
	Engine cranking, engine start signal received (from starter circuit), and the PCM did not detect any CMP sensor signals for 5 seconds.
	Possible Causes:
	   • CMP sensor signal circuit is open between sensor and PCM
	   • CMP sensor signal circuit is shorted to ground
	   • CMP sensor power circuit is open (check power from the relay)
	   • CMP sensor is damaged or has failed
	   • PCM has failed

DTC: P0342
2T, MIL: Yes
Year: 2009
Model: XL-7
Engine: 3.6L V6
Transmission: All
	Intake Camshaft Position (CMP) Sensor Circuit Low Voltage Bank 1:
	The CMP sensor signal voltage is always low and the ECM detects no pulses from the CMP sensor for greater than 1 second, or cumulative of 10 seconds.
	Possible Causes:
	   • CMP sensor and/or circuit
	   • CMP sensor and/or reluctor wheel installation or damage
	   • ECM

DTC	Trouble Code Title, Conditions & Possible Causes
DTC: P0343 **2T, MIL:** Yes **Year:** 2009 **Model:** XL-7 **Engine:** 3.6L V6 **Transmission:** All	**Intake Camshaft Position (CMP) Sensor Circuit High Voltage Bank 1:** The CMP sensor signal voltage is always high and the ECM detects no pulses from the CMP sensor for greater than 1 second or cumulative of 10 seconds. **Possible Causes:** • CMP sensor and/or circuit • CMP sensor and/or reluctor wheel installation or damage • ECM
DTC: P0343 **2T, MIL:** Yes **Year:** 2009, 2010 **Model:** Grand Vitara **Engine:** 3.2L V6 VIN 1 **Transmission:** All	**Camshaft Position Sensor "A" Circuit High Bank 1 or Single Sensor:** CMP sensor (bank 1 / intake side) signal voltage is unchanged from high level for specified crankshaft revolutions. **Possible Causes:** • CMP sensor and/or its circuit • Signal rotor • ECM
DTC: P0345 **2T, MIL:** Yes **Year:** 2009, 2010 **Model:** Equator **Engine:** 4.0L V6 **Transmission:** All	**Camshaft position sensor (PHASE) circuit (bank 2):** • The cylinder No. signal is not sent to ECM for the first few seconds during engine cranking. • The cylinder No. signal is not sent to ECM during engine running. • The cylinder No. signal is not in the normal pattern during engine running. **Possible Causes:** • Harness or connectors • (The sensor circuit is open or shorted) • Camshaft position sensor (PHASE) • Camshaft (Intake) • Starter motor • Starting system circuit • Dead (Weak) battery
DTC: P0346 **2T, MIL:** Yes **Year:** 2009, 2010 **Model:** Grand Vitara **Engine:** 3.2L V6 VIN 1 **Transmission:** All	**Camshaft Position Sensor "A" Circuit Range/Performance Bank 2:** Abnormal CMP sensor signal (bank 2 / intake side) is detected more than specified times. **Possible Causes:** • CMP sensor and/or its circuit • CKP sensor and/or its circuit • Signal rotor • ECM
DTC: P0346 **2T, MIL:** Yes **Year:** 2009 **Model:** XL-7 **Engine:** 3.6L V6 **Transmission:** All	**Intake Camshaft Position (CMP) Sensor Performance Bank 2:** • The ECM detects a signal from the CMP sensor, but the number of pulses are less than or greater than what is expected for one crankshaft revolution. OR • The CMP sensor does NOT correlate to the crankshaft position. • Either condition must exist for greater than 1 second or cumulative of 10 seconds. **Possible Causes:** • CMP sensor and/or circuit • CMP sensor and/or reluctor wheel installation or damage • ECM
DTC: P0347 **2T, MIL:** Yes **Year:** 2009 **Model:** XL-7 **Engine:** 3.6L V6 **Transmission:** All	**Intake Camshaft Position (CMP) Sensor Circuit Low Voltage Bank 2:** The CMP sensor signal voltage is always low and the ECM detects no pulses from the CMP sensor for greater than 1 second, or cumulative of 10 seconds. **Possible Causes:** • CMP sensor and/or circuit • CMP sensor and/or reluctor wheel installation or damage • ECM
DTC: P0347 **2T, MIL:** Yes **Year:** 2009, 2010 **Model:** Grand Vitara **Engine:** 3.2L V6 VIN 1 **Transmission:** All	**Camshaft Position Sensor "A" Circuit Low Bank 2:** CMP sensor (bank 2 / intake side) signal voltage is unchanged from low level for specified crankshaft revolutions. **Possible Causes:** • CMP sensor and/or its circuit • Signal rotor • ECM

DTC	Trouble Code Title, Conditions & Possible Causes

DTC: P0348
2T, MIL: Yes
Year: 2009, 2010
Model: Grand Vitara
Engine: 3.2L V6 VIN 1
Transmission: All
	Camshaft Position Sensor "A" Circuit High Bank 2:
	CMP sensor (bank 2 / intake side) signal voltage is unchanged from high level for specified crankshaft revolutions.
	Possible Causes:
	   • CMP sensor and/or its circuit
	   • Signal rotor
	   • ECM

DTC: P0348
2T, MIL: Yes
Year: 2009
Model: XL-7
Engine: 3.6L V6
Transmission: All
	Intake Camshaft Position (CMP) Sensor Circuit High Voltage Bank 2:
	The CMP sensor signal voltage is always high and the ECM detects no pulses from the CMP sensor for greater than 1 second or cumulative of 10 seconds.
	Possible Causes:
	   • CMP sensor and/or circuit
	   • CMP sensor and/or reluctor wheel installation or damage
	   • ECM

DTC: P0351
2T
Year: 2009, 2010
Model: Grand Vitara, XL-7
Engine: 3.2L V6 VIN 1, 3.6L V6
Transmission: All
	Ignition Coil "A" Primary/Secondary Circuit:
	Ignition coil No.1 control circuit is open circuit.
	Possible Causes:
	   • Ignition coil and/or its circuit
	   • CMP sensor and/or its circuit
	   • CKP sensor and/or its circuit
	   • Signal rotor
	   • ECM

DTC: P0352
2T
Year: 2009, 2010
Model: Grand Vitara, XL-7
Engine: 3.2L V6 VIN 1, 3.6L V6
Transmission: All
	Ignition Coil "B" Primary/Secondary Circuit:
	Ignition coil No.2 control circuit is open circuit.
	Possible Causes:
	   • (2 D/C detection logic but MIL does not light up) • Ignition coil and/or its circuit
	   • CMP sensor and/or its circuit
	   • CKP sensor and/or its circuit
	   • Signal rotor
	   • ECM

DTC: P0353
2T
Year: 2009, 2010
Model: Grand Vitara, XL-7
Engine: 3.2L V6 VIN 1, 3.6L V6
Transmission: All
	Ignition Coil "C" Primary/Secondary Circuit:
	Ignition coil No.3 control circuit is open circuit.
	Possible Causes:
	   • Ignition coil and/or its circuit
	   • CMP sensor and/or its circuit
	   • CKP sensor and/or its circuit
	   • Signal rotor
	   • ECM

DTC: P0354
2T
Year: 2009, 2010
Model: Grand Vitara, XL-7
Engine: 3.2L V6 VIN 1, 3.6L V6
Transmission: All
	Ignition Coil "D" Primary/Secondary Circuit:
	Ignition coil No.4 control circuit is open circuit.
	Possible Causes:
	   • Ignition coil and/or its circuit
	   • CMP sensor and/or its circuit
	   • CKP sensor and/or its circuit
	   • Signal rotor
	   • ECM

DTC: P0355
2T
Year: 2009, 2010
Model: Grand Vitara, XL-7
Engine: 3.2L V6 VIN 1, 3.6L V6
Transmission: All
	Ignition Coil "E" Primary/Secondary Circuit:
	Ignition coil No.5 control circuit is open circuit.
	Possible Causes:
	   • Ignition coil and/or its circuit
	   • CMP sensor and/or its circuit
	   • CKP sensor and/or its circuit
	   • Signal rotor
	   • ECM

DTC	Trouble Code Title, Conditions & Possible Causes
DTC: P0356 **2T** **Year:** 2009, 2010 **Model:** Grand Vitara, XL-7 **Engine:** 3.2L V6 VIN 1, 3.6L V6 **Transmission:** All	**Ignition Coil "F" Primary/Secondary Circuit:** Ignition coil No.6 control circuit is open circuit. **Possible Causes:** • Ignition coil and/or its circuit • CMP sensor and/or its circuit • CKP sensor and/or its circuit • Signal rotor • ECM
DTC: P0366 **2T, MIL: Yes** **Year:** 2009 **Model:** XL-7 **Engine:** 3.6L V6 **Transmission:** All	**Exhaust Camshaft Position (CMP) Sensor Performance Bank 1:** • The ECM detects a signal from the CMP sensor, but the number of pulses are less than or greater than what is expected for one crankshaft revolution. OR • The CMP sensor does NOT correlate to the crankshaft position. • Either condition must exist for greater than 1 second or cumulative of 10 seconds. **Possible Causes:** • CMP sensor and/or circuit • CMP sensor and/or reluctor wheel installation or damage • ECM
DTC: P0366 **2T, MIL: Yes** **Year:** 2009, 2010 **Model:** Grand Vitara **Engine:** 3.2L V6 VIN 1 **Transmission:** All	**Camshaft Position Sensor "B" Circuit Range/Performance Bank 1:** Abnormal CMP sensor signal (bank 1 / exhaust side) is detected more than specified times. **Possible Causes:** • CMP sensor and/or its circuit • CKP sensor and/or its circuit • Signal rotor • ECM
DTC: P0367 **2T, MIL: Yes** **Year:** 2009, 2010 **Model:** Grand Vitara **Engine:** 3.2L V6 VIN 1 **Transmission:** All	**Camshaft Position Sensor "B" Circuit Low Bank 1:** CMP sensor (bank 1 / exhaust side) signal voltage is unchanged from low level for specified crankshaft revolutions. **Possible Causes:** • CMP sensor and/or its circuit • Signal rotor • ECM
DTC: P0367 **2T, MIL: Yes** **Year:** 2009 **Model:** XL-7 **Engine:** 3.6L V6 **Transmission:** All	**Exhaust Camshaft Position (CMP) Sensor Circuit Low Voltage Bank 1:** The CMP sensor signal voltage is always low and the ECM detects no pulses from the CMP sensor for greater than 1 second, or cumulative of 10 seconds. **Possible Causes:** • CMP sensor and/or circuit • CMP sensor and/or reluctor wheel installation or damage • ECM
DTC: P0368 **2T, MIL: Yes** **Year:** 2009 **Model:** XL-7 **Engine:** 3.6L V6 **Transmission:** All	**Exhaust Camshaft Position (CMP) Sensor Circuit High Voltage Bank 1:** The CMP sensor signal voltage is always high and the ECM detects no pulses from the CMP sensor for greater than 1 second or cumulative of 10 seconds. **Possible Causes:** • CMP sensor and/or circuit • CMP sensor and/or reluctor wheel installation or damage • ECM
DTC: P0368 **2T, MIL: Yes** **Year:** 2009, 2010 **Model:** Grand Vitara **Engine:** 3.2L V6 VIN 1 **Transmission:** All	**Camshaft Position Sensor "B" Circuit High Bank 1:** CMP sensor (bank 1 / exhaust side) signal voltage is unchanged from high level for specified crankshaft revolutions. **Possible Causes:** • CMP sensor and/or its circuit • Signal rotor • ECM
DTC: P0391 **2T, MIL: Yes** **Year:** 2009, 2010 **Model:** Grand Vitara **Engine:** 3.2L V6 VIN 1 **Transmission:** All	**Camshaft Position Sensor "B" Circuit Range/Performance Bank 2:** Abnormal CMP sensor signal (bank 2 / exhaust side) is detected more than specified times. **Possible Causes:** • CMP sensor and/or its circuit • CKP sensor and/or its circuit • Signal rotor • ECM

DTC	Trouble Code Title, Conditions & Possible Causes

DTC: P0391
2T, MIL: Yes
Year: 2009
Model: XL-7
Engine: 3.6L V6
Transmission: All
	Exhaust Camshaft Position (CMP) Sensor Performance Bank 2:
	    • The ECM detects a signal from the CMP sensor, but the number of pulses are less than or greater than what is expected for one crankshaft revolution.
	OR
	    • The CMP sensor does NOT correlate to the crankshaft position.
	    • Either condition must exist for greater than 1 second or cumulative of 10 seconds.
	Possible Causes:
	    • CMP sensor and/or circuit
	    • CMP sensor and/or reluctor wheel installation or damage
	    • ECM

DTC: P0392
2T, MIL: Yes
Year: 2009
Model: XL-7
Engine: 3.6L V6
Transmission: All
	Exhaust Camshaft Position (CMP) Sensor Circuit Low Voltage Bank 2:
	The CMP sensor signal voltage is always low and the ECM detects no pulses from the CMP sensor for greater than 1 second, or cumulative of 10 seconds.
	Possible Causes:
	    • CMP sensor and/or circuit
	    • CMP sensor and/or reluctor wheel installation or damage
	    • ECM

DTC: P0392
2T, MIL: Yes
Year: 2009, 2010
Model: Grand Vitara
Engine: 3.2L V6 VIN 1
Transmission: All
	Camshaft Position Sensor "B" Circuit Low Bank 2:
	CMP sensor (bank 2 / exhaust side) signal voltage is unchanged from low level for specified crankshaft revolutions.
	Possible Causes:
	    • CMP sensor and/or its circuit
	    • Signal rotor
	    • ECM

DTC: P0393
2T, MIL: Yes
Year: 2009
Model: XL-7
Engine: 3.6L V6
Transmission: All
	Exhaust Camshaft Position (CMP) Sensor Circuit High Voltage Bank 2:
	The CMP sensor signal voltage is always high and the ECM detects no pulses from the CMP sensor for greater than 1 second or cumulative of 10 seconds.
	Possible Causes:
	    • CMP sensor and/or circuit
	    • CMP sensor and/or reluctor wheel installation or damage
	    • ECM

DTC: P0393
2T, MIL: Yes
Year: 2009, 2010
Model: Grand Vitara
Engine: 3.2L V6 VIN 1
Transmission: All
	Camshaft Position Sensor "B" Circuit High Bank 2:
	CMP sensor (bank 2 / exhaust side) signal voltage is unchanged from high level for specified crankshaft revolutions.
	Possible Causes:
	    • CMP sensor and/or its circuit
	    • Signal rotor
	    • ECM

DTC: P0420
2T CCM, MIL: Yes
Year: 2009, 2010
Model: Equator, Grand Vitara, SX4, XL-7
Engine: 2.0L L4, 2.0L L4 VIN 5, 2.4L L4 VIN 0, 2.4L L4 VIN 9, 2.5L L4, 3.2L V6 VIN 1, 3.6L V6, 4.0L V6
Transmission: All
	Catalyst Efficiency Below Normal (Bank 1):
	DTC P0101, P0102, P0103, P0111-P0113, P0116-P0118, P0131, P0132, P0133, P0134, P0135, P0136, P0141, P0151, P0152, P0153, P0154, P0155, P0156, P0161, P0335, P0461, P0463, P0500, P1450 and P1451 not set, IAT signal more than 18°F, ECT sensor from 1858-230°F, BARO sensor over 75 kPa, vehicle driven at over 35 mph, then at 55-60 mph for 5 minutes; and the PCM detected the HO2S-12 and HO2S-11 signals were too similar.
	Possible Causes:
	    • Air leaks at the exhaust manifold or in the exhaust pipes
	    • Catalytic converter is damaged, contaminated or has failed
	    • Front HO2S or rear HO2S is contaminated with fuel or moisture
	    • Front HO2S and/or the rear HO2S is loose in the mounting hole
	    • Front HO2S is older (aged) than the rear HO2S (HO2S is lazy)

DTC: P0430
2T CCM, MIL: Yes
Year: 2009, 2010
Model: Equator, Grand Vitara, XL-7
Engine: 3.2L V6 VIN 1, 3.6L V6, 4.0L V6
Transmission: All
	Catalyst Efficiency Below Normal (Bank 2):
	DTC P0101, P0102, P0103, P0111-P0113, P0116-P0118, P0131, P0132, P0133, P0134, P0135, P0136, P0141, P0151, P0152, P0153, P0154, P0155, P0156, P0161, P0335, P0461, P0463, P0500, P1450 and P1451 not set, IAT signal more than 18°F, ECT sensor from 158°F-230°F, BARO sensor over 75 kPa, vehicle driven at over 35 mph, then at 55-60 mph for 5 minutes; and the PCM detected the HO2S-22 and HO2S-21 signals were too similar.
	Possible Causes:
	    • Air leaks at the exhaust manifold or in the exhaust pipes
	    • Catalytic converter is damaged, contaminated or has failed
	    • Front HO2S or rear HO2S is contaminated with fuel or moisture
	    • Front HO2S and/or the rear HO2S is loose in the mounting hole
	    • Front HO2S is older (aged) than the rear HO2S (HO2S is lazy)

DTC	Trouble Code Title, Conditions & Possible Causes
DTC: P0441 **2T, MIL: Yes** **Year:** 2009, 2010 **Model:** SX4 **Engine:** 2.0L L4, 2.0L L4 VIN 5, 2.4L L4 VIN 9 **Transmission:** All	**Evaporative Emission System Incorrect Purge Flow:** Variation of fuel tank pressure is less than 1.26 kPa for 3 sec. when EVAP vent valve is closed and EVAP canister purge valve is ON (CLOSE). **Possible Causes:** • EVAP canister purge valve and/or its circuit • Purge line of EVAP system • ECM
DTC: P0441 **2T CCM, MIL: Yes** **Year:** 2009, 2010 **Model:** Equator, Grand Vitara **Engine:** 2.4L L4 VIN 0, 2.5L L4, 3.2L V6 VIN 1, 4.0L V6 **Transmission:** All	**EVAP System Leak Detected (very small leak):** Engine started, vehicle driven to a speed of 35-40 for 20 minutes under conditions that do not include high-load, high engine speed, rapid acceleration or deceleration events, then at a constant speed of 30-40 mph for 3 minutes, fuel level from 25-75%, BARO sensor over 75 kPa, ambient temperature over 14°F, IAT sensor less than 158°F, ECT sensor from 158°F-230°F, then after the pressure in the EVAP system reached a predetermined value, and with the Purge valve command closed, the PCM detected a leak in the system. **Possible Causes:** • Canister air valve (solenoid) is damaged or has failed • Charcoal canister is clogged, loaded with fuel or moisture • Fuel filler cap loose, cross-threaded, incorrect part or damaged • Fuel tank pressure sensor is damaged or has failed • Fuel tank or fuel tank sender assembly 'O' ring is leaking • Fuel tank vapor line(s) blocked, damaged or disconnected • Tank pressure control solenoid is contaminated or damaged • Purge solenoid control circuit is open or shorted to ground • Purge solenoid power circuit is open (test FI fuse in relay box) • Purge solenoid is contaminated, damaged or has failed • PCM has failed
DTC: P0442 **2T CCM, MIL: Yes** **Year:** 2009, 2010 **Model:** Equator, Grand Vitara, XL-7 **Engine:** 2.5L L4, 3.2L V6 VIN 1, 3.6L V6, 4.0L V6 **Transmission:** All	**EVAP System Leak Detected (small leak):** Engine started, vehicle driven to a speed of 35-40 for 20 minutes under conditions that do not include high-load, high engine speed, rapid acceleration or deceleration events, then at a constant speed of 30-40 mph for 3 minutes, fuel level from 25-75%, BARO sensor over 75 kPa, ambient temperature over 14°F, IAT sensor less than 158°F, ECT sensor from 158°F-230°F, then after the pressure in the EVAP system reached a predetermined value, and with the Purge valve command closed, the PCM detected a leak in the system. **Possible Causes:** • Canister air valve (solenoid) is damaged or has failed • Charcoal canister is clogged, loaded with fuel or moisture • Fuel filler cap loose, cross-threaded, incorrect part or damaged • Fuel tank pressure sensor is damaged or has failed • Fuel tank or fuel tank sender assembly 'O' ring is leaking • Fuel tank vapor line(s) blocked, damaged or disconnected • Tank pressure control solenoid is contaminated or damaged • Purge solenoid control circuit is open or shorted to ground • Purge solenoid power circuit is open (test FI fuse in relay box) • Purge solenoid is contaminated, damaged or has failed • PCM has failed
DTC: P0443 **2T CCM, MIL: Yes** **Year:** 2009, 2010 **Model:** Equator, Grand Vitara, SX4, XL-7 **Engine:** 2.0L L4, 2.0L L4 VIN 5, 2.4L L4 VIN 0, 2.4L L4 VIN 9, 2.5L L4, 3.2L V6 VIN 1, 3.6L V6, 4.0L V6 **Transmission:** All	**EVAP Canister Purge Solenoid Circuit Malfunction:** Engine started, IAT signal less than 122°F, ambient temperature over 14°F, engine running at a stable idle speed, fuel level from 25-75%, all accessory loads off, and the PCM detected an unexpected voltage condition on the EVAP Canister Purge solenoid circuit. **Possible Causes:** • Purge solenoid control circuit is open or shorted to ground • Purge solenoid power circuit is open (check IG fuse in the J/B) • Purge solenoid is damaged or has failed • PCM has failed
DTC: P0444 **2T, MIL: Yes** **Year:** 2009, 2010 **Model:** Equator **Engine:** 2.5L L4, 4.0L V6 **Transmission:** All	**EVAP canister purge volume control solenoid valve circuit open:** An excessively low voltage signal is sent to ECM through the valve **Possible Causes:** • Harness or connectors • (The solenoid valve circuit is open or shorted.) • EVAP canister purge volume control solenoid valve

DTC	Trouble Code Title, Conditions & Possible Causes
DTC: P0445 **2T, MIL: Yes** **Year:** 2009, 2010 **Model:** Equator **Engine:** 2.5L L4, 4.0L V6 **Transmission:** All	**EVAP canister purge volume control solenoid valve circuit shorted:** An excessively high voltage signal is sent to ECM through the valve **Possible Causes:** • Harness or connectors • (The solenoid valve circuit is shorted.) • EVAP canister purge volume control solenoid valve
DTC: P0446 **2T, MIL: Yes** **Year:** 2009 **Model:** XL-7 **Engine:** 3.6L V6 **Transmission:** All	**Evaporative Emissions (EVAP) Vent System Performance:** • The FTP is less than -7.5 mm Hg (-4.0 in. H2O). • The condition is present for greater than 4 seconds or a cumulative of 30 seconds. **Possible Causes:** • Leak in EVAP system
DTC: P0447 **2T, MIL: Yes** **Year:** 2009, 2010 **Model:** Equator **Engine:** 2.5L L4, 4.0L V6 **Transmission:** All	**EVAP canister vent control valve circuit open:** An improper voltage signal is sent to ECM through EVAP canister vent control valve. **Possible Causes:** • Harness or connectors • (The valve circuit is open or shorted.) • EVAP canister vent control valve
DTC: P0448 **2T, MIL: Yes** **Year:** 2009, 2010 **Model:** Equator **Engine:** 2.5L L4, 4.0L V6 **Transmission:** All	**EVAP canister vent control valve close:** EVAP canister vent control valve remains closed under specified driving conditions. **Possible Causes:** • EVAP canister vent control valve • EVAP control system pressure sensor and the circuit • Blocked rubber tube to EVAP canister vent control valve • EVAP canister is saturated with water
DTC: P0449 **2T CCM, MIL: Yes** **Year:** 2009, 2010 **Model:** Grand Vitara, SX4, XL-7 **Engine:** 2.0L L4, 2.0L L4 VIN 5, 2.4L L4 VIN 0, 2.4L L4 VIN 9, 3.2L V6 VIN 1, 3.6L V6 **Transmission:** All	**EVAP Canister Control System Vent Control Valve/Solenoid Circuit Malfunction:** Engine started, IAT signal less than 122°F, ambient temperature over 14°F, engine running at a stable idle speed, fuel level from 25-75%, all accessory loads off, and the PCM detected an unexpected voltage condition on the EVAP Canister Purge solenoid circuit. **Possible Causes:** • Purge solenoid control circuit is open or shorted to ground • Purge solenoid power circuit is open (check IG fuse in the J/B) • Purge solenoid is damaged or has failed • PCM has failed
DTC: P0450 **2T, MIL: Yes** **Year:** 2009 **Model:** XL-7 **Engine:** 3.6L V6 **Transmission:** All	**Fuel Tank Pressure (FTP) Sensor Circuit:** The ECM detects that the FTP sensor signal oscillates greater than 6.09 mm/Hg (3.26 in. H2O) for 4 seconds, or for a cumulative of 30 seconds. **Possible Causes:** • FTP sensor and/or circuit • ECM
DTC: P0451 **2T, MIL: Yes** **Year:** 2009, 2010 **Model:** Equator, Grand Vitara **Engine:** 2.5L L4, 3.2L V6 VIN 1, 4.0L V6 **Transmission:** All	**EVAP control system pressure sensor performance:** ECM detects a sloshing signal from the EVAP control system pressure sensor **Possible Causes:** • Harness or connectors • EVAP control system pressure sensor
DTC: P0451 **2T, MIL: Yes** **Year:** 2009 **Model:** XL-7 **Engine:** 3.6L V6 **Transmission:** All	**Fuel Tank Pressure (FTP) Sensor Performance:** • The ECM detects that the FTP is less than -26.2 mm/Hg (-14.1 in. H2O) or greater than 11.0 mm/Hg (5.9 in. H2O) for 10 seconds or for a cumulative of 30 seconds. OR • The ECM detects a change in the zero point of FTP sensor signal greater than +/-5.16 mm/Hg (2.76 in. H2O) from the zero point at start up for 10 seconds, or a cumulative of 30 seconds. **Possible Causes:** • FTP sensor and/or circuit • ECM

DTC	Trouble Code Title, Conditions & Possible Causes
DTC: P0452 **2T CCM, MIL: Yes** **Year:** 2009, 2010 **Model:** Equator, Grand Vitara **Engine:** 2.5L L4, 3.2L V6 VIN 1, 4.0L V6 **Transmission:** All	**EVAP Pressure Sensor Low Input:** Engine started, vehicle driven to a speed over 45 mph for 6 minutes at an engine speed from 2500-3000 rpm, followed by a deceleration period of 3 seconds, fuel level from 15-85%, BARO sensor more than 75 kPa, IAT sensor more than 158°F, ambient temperature over 14°F, ECT sensor from 158-230°F, and the PCM detected an unexpected voltage change under these test conditions. **Possible Causes:** • Pressure sensor signal circuit is open or shorted to ground • Pressure sensor signal is shorted to VREF or system power • Pressure sensor is damaged or has failed • PCM has failed
DTC: P0452 **2T, MIL: Yes** **Year:** 2009 **Model:** XL-7 **Engine:** 3.6L V6 **Transmission:** All	**Fuel Tank Pressure (FTP) Sensor Circuit Low Voltage:** The ECM detects the FTP sensor signal voltage is less than 0.20 volt for 4 seconds, or for a cumulative of 30 seconds. **Possible Causes:** • FTP sensor and/or circuit • ECM
DTC: P0453 **2T CCM, MIL: Yes** **Year:** 2009, 2010 **Model:** Equator **Engine:** 2.5L L4, 4.0L V6 **Transmission:** All	**EVAP Pressure Sensor High Input:** Engine started, vehicle driven to a speed over 45 mph for 6 minutes at an engine speed from 2500-3000 rpm, followed by a deceleration period of 3 seconds, fuel level from 15-85%, BARO sensor more than 75 kPa, IAT sensor more than 158°F, ambient temperature over 14°F, ECT sensor from 158-230°F, and the PCM detected an unexpected voltage change under these test conditions. **Possible Causes:** • Pressure sensor signal circuit is open or shorted to ground • Pressure sensor signal is shorted to VREF or system power • Pressure sensor is damaged or has failed • PCM has failed
DTC: P0453 **2T, MIL: Yes** **Year:** 2009 **Model:** XL-7 **Engine:** 3.6L V6 **Transmission:** All	**Fuel Tank Pressure (FTP) Sensor Circuit High Voltage:** The ECM detects the FTP sensor signal voltage is greater than 4.85 volts for 4 seconds, or for a cumulative of 30 seconds. **Possible Causes:** • FTP sensor and/or circuit • ECM
DTC: P0455 **2T CCM, MIL: Yes** **Year:** 2009, 2010 **Model:** Equator, Grand Vitara, XL-7 **Engine:** 2.5L L4, 3.2L V6 VIN 1, 3.6L V6, 4.0L V6 **Transmission:** All	**EVAP System Large Leak (0.080") Detected:** Cold engine startup, vehicle driven at 35-55 mph for 2 minutes, then returned to idle speed, ambient temperature over 14°F, ECT signal from 18-230°F, and the PCM detected a large change in the FTP sensor indicating a large leak (0.080") present during the leak test. **Possible Causes:** • Canister vent (CV) solenoid is stuck open • EVAP canister tube, EVAP canister purge outlet tube or EVAP return tube disconnected or cracked, or canister is damaged • EVAP canister purge valve stuck closed, or canister damaged • Fuel filler cap missing, loose (not tightened) or the wrong part • Fuel vapor hoses/tubes blocked or restricted, or fuel vapor control valve tube or fuel vapor vent valve assembly blocked • Fuel tank pressure (FTP) sensor has failed (mechanical fault) • Fuel tank control valve is contaminated, damaged or has failed
DTC: P0456 **2T CCM, MIL: Yes** **Year:** 2009, 2010 **Model:** Equator, Grand Vitara, SX4 **Engine:** 2.0L L4, 2.0L L4 VIN 5, 2.4L L4 VIN 0, 2.4L L4 VIN 9, 2.5L L4, 3.2L V6 VIN 1, 4.0L V6 **Transmission:** All	**EVAP System Small Leak (0.080") Detected:** Cold engine startup, vehicle driven at 35-55 mph for 2 minutes, then returned to idle speed, ambient temperature over 14°F, ECT signal from 18-230°F, and the PCM detected a large change in the FTP sensor indicating a small leak (0.040") present during the leak test. **Possible Causes:** • Canister vent (CV) solenoid is stuck open • EVAP canister tube, EVAP canister purge outlet tube or EVAP return tube disconnected or cracked, or canister is damaged • EVAP canister purge valve stuck closed, or canister damaged • Fuel filler cap missing, loose (not tightened) or the wrong part • Fuel vapor hoses/tubes blocked or restricted, or fuel vapor control valve tube or fuel vapor vent valve assembly blocked • Fuel tank pressure (FTP) sensor has failed (mechanical fault) • Fuel tank control valve is contaminated, damaged or has failed

DTC	Trouble Code Title, Conditions & Possible Causes

DTC: P0458
2T, MIL: Yes
Year: 2009, 2010
Model: Grand Vitara, XL-7
Engine: 3.2L V6 VIN 1, 3.6L V6
Transmission: All
	Evaporative Emission System Purge Control Valve Circuit Low:
	EVAP canister purge valve control circuit is shorted to ground.
	Possible Causes:
	    • EVAP canister purge valve and/or its circuit
	    • ECM

DTC: P0459
2T, MIL: Yes
Year: 2009, 2010
Model: Grand Vitara, XL-7
Engine: 3.2L V6 VIN 1, 3.6L V6
Transmission: All
	Evaporative Emission System Purge Control Valve Circuit High:
	EVAP canister purge valve control circuit is shorted to power supply circuit.
	Possible Causes:
	    • EVAP canister purge valve and/or its circuit
	    • ECM

DTC: P0460
2T, MIL: Yes
Year: 2009, 2010
Model: Equator
Engine: 2.5L L4, 4.0L V6
Transmission: All
	Fuel level sensor circuit noise:
	Even though the vehicle is parked, a signal being varied is sent from the fuel level sensor to ECM.
	Possible Causes:
	    • Harness or connectors
	    • (The CAN communication line is open or shorted)
	    • Harness or connectors
	    • (The sensor circuit is open or shorted)
	    • Combination meter
	    • Fuel level sensor

DTC: P0461
2T CCM, MIL: Yes
Year: 2009, 2010
Model: Equator, Grand Vitara, SX4
Engine: 2.0L L4, 2.0L L4 VIN 5, 2.4L L4 VIN 0, 2.4L L4 VIN 9, 2.5L L4, 3.2L V6 VIN 1, 4.0L V6
Transmission: All
	Fuel Level Sensor Circuit Range / Performance :
	DTC P0461 not set, engine started, then after the vehicle was driven a predetermined amount of miles, the PCM detected the fuel level signal value changed less than 0.14v, or the PCM detected a signal from the Fuel Level sensor that was too "low" or too "high" during the CCM Rationality test.
	Possible Causes:
	    • Fuel tank empty or overfull (fuel sender is stuck mechanically)
	    • Wrong fuel gauge is installed, or instrument panel is damaged
	    • Fuel gauge sender unit is damaged or has failed
	    • PCM has failed

DTC: P0462
2T CCM, MIL: Yes
Year: 2009, 2010
Model: Equator, Grand Vitara, SX4
Engine: 2.0L L4, 2.0L L4 VIN 5, 2.4L L4 VIN 0, 2.4L L4 VIN 9, 2.5L L4, 3.2L V6 VIN 1, 4.0L V6
Transmission: All
	Fuel Level Sensor Circuit Low Input:
	Key on or engine running, and the PCM detected the Fuel Level Sensor signal indicated an unexpected voltage during the CCM test.
	Possible Causes:
	    • Fuel level signal circuit is open between the splice and the I/P
	    • Wrong fuel gauge is installed, or instrument panel is damaged
	    • Fuel gauge sender unit is damaged or has failed
	    • PCM has failed

DTC: P0463
2T CCM, MIL: Yes
Year: 2009, 2010
Model: Equator, Grand Vitara, SX4
Engine: 2.0L L4, 2.0L L4 VIN 5, 2.4L L4 VIN 0, 2.4L L4 VIN 9, 2.5L L4, 3.2L V6 VIN 1, 4.0L V6
Transmission: All
	Fuel Level Sensor Circuit High Input:
	Key on or engine running, and the PCM detected the Fuel Level Sensor signal indicated more than 4.60v during the CCM test.
	Possible Causes:
	    • Fuel level signal circuit is open between the splice and the I/P
	    • Wrong fuel gauge is installed, or instrument panel is damaged
	    • Fuel gauge sender unit is damaged or has failed
	    • PCM has failed

DTC: P0480
1T
Year: 2009, 2010
Model: Grand Vitara, SX4
Engine: 2.0L L4, 2.0L L4 VIN 5, 2.4L L4 VIN 0, 2.4L L4 VIN 9, 3.2L V6 VIN 1
Transmission: All
	Fan 1 Control Circuit:
	    • Monitor signal of radiator cooling fan relay No.1 drive circuit is high level (OFF) for 5 sec. even if command signal of radiator cooling fan is at low speed.
	    • Monitor signal of radiator cooling fan relay No.1 drive circuit is low level (ON) for 5 sec. even if command signal of radiator cooling fan is NOT at low speed.
	Possible Causes:
	    • Radiator cooling fan relay No.1 (coil side) and/or its circuit
	    • ECM

DTC	Trouble Code Title, Conditions & Possible Causes
DTC: P0481 **1T** **Year:** 2009, 2010 **Model:** Grand Vitara, SX4 **Engine:** 2.0L L4, 2.0L L4 VIN 5, 2.4L L4 VIN 0, 2.4L L4 VIN 9, 3.2L V6 VIN 1 **Transmission:** All	**Fan 2 Control Circuit:** • Monitor signal of radiator cooling fan relay No.2 drive circuit is high level (OFF) for 5 sec. even if command signal of radiator cooling fan is at high speed. • Monitor signal of radiator cooling fan relay No.2 drive circuit is low level (ON) for 5 sec. even if command signal of radiator cooling fan is NOT at high speed. **Possible Causes:** • Radiator cooling fan relay No.2 (coil side) and/or its circuit • ECM
DTC: P0482 **2T, MIL: Yes** **Year:** 2009, 2010 **Model:** SX4 **Engine:** 2.0L L4, 2.0L L4 VIN 5, 2.4L L4 VIN 9 **Transmission:** All	**Fan 3 Control Circuit:** • Monitor signal of A/C condenser cooling fan relay drive circuit is high level (OFF) for 5 sec. even if command signal of radiator cooling fan is at high speed. • Monitor signal of A/C condenser cooling fan relay drive circuit is low level (ON) for 5 sec. even if command signal of radiator cooling fan is NOT at high speed. **Possible Causes:** • A/C condenser fan relay (coil side) and/or its circuit • ECM
DTC: P0483 **1T** **Year:** 2009, 2010 **Model:** Grand Vitara **Engine:** 2.4L L4 VIN 0, 3.2L V6 VIN 1 **Transmission:** All	**Fan 3 Control Circuit:** • Monitor signal of radiator cooling fan relay No.3 drive circuit is high level (OFF) for 5 sec. even if command signal of radiator cooling fan is at high speed. • Monitor signal of radiator cooling fan relay No.3 drive circuit is low level (ON) for 5 sec. even if command signal of radiator cooling fan is NOT at high speed. **Possible Causes:** • Radiator cooling fan relay No.3 (coil side) and/or its circuit • ECM
DTC: P0496 **2T, MIL: Yes** **Year:** 2009 **Model:** XL-7 **Engine:** 3.6L V6 **Transmission:** All	**Evaporative Emissions (EVAP) System Continuous Purge Flow:** • The control module detects vacuum during a non-purge condition. • The condition exists for greater than 4 seconds, or for a cumulative of 30 seconds. **Possible Causes:** • FTP sensor • EVAP purge solenoid valve
DTC: P0497 **2T, MIL: Yes** **Year:** 2009 **Model:** XL-7 **Engine:** 3.6L V6 **Transmission:** All	**Evaporative Emission (EVAP) System No Flow During Purge:** • The ECM detects the EVAP system is not able to achieve or maintain vacuum during the diagnostic test. • The condition exists for greater than 4 seconds or for a cumulative of 30 seconds. **Possible Causes:** • A loose, missing, or damaged fuel fill cap • A stuck closed, blocked, or restricted EVAP canister purge solenoid valve • A blockage or restriction in the EVAP canister purge solenoid valve vacuum supply hose, EVAP canister purge solenoid valve purge pipe, EVAP canister, or vapor pipe • A temporary blockage in the EVAP canister purge solenoid valve, purge pipe or EVAP canister
DTC: P0498 **2T, MIL: Yes** **Year:** 2009, 2010 **Model:** Grand Vitara, XL-7 **Engine:** 3.2L V6 VIN 1, 3.6L V6 **Transmission:** All	**Evaporative Emission System Vent Valve Control Circuit Low:** EVAP canister vent solenoid valve control circuit is shorted to ground circuit. **Possible Causes:** • EVAP canister vent solenoid valve and/or its circuit • ECM
DTC: P0499 **2T, MIL: Yes** **Year:** 2009, 2010 **Model:** Grand Vitara, XL-7 **Engine:** 3.2L V6 VIN 1, 3.6L V6 **Transmission:** All	**Evaporative Emission System Vent Valve Control Circuit High:** EVAP canister vent solenoid valve control circuit is shorted to power supply circuit. **Possible Causes:** • EVAP canister vent solenoid valve and/or its circuit • ECM

DTC	Trouble Code Title, Conditions & Possible Causes

DTC: P0500
2T, MIL: Yes
Year: 2009, 2010
Model: Equator, Grand Vitara, SX4
Engine: 2.0L L4, 2.0L L4 VIN 5, 2.4L L4 VIN 0, 2.4L L4 VIN 9, 2.5L L4, 3.2L V6 VIN 1, 4.0L V6
Transmission: All
	Vehicle speed sensor:
	NOTE:
	  • If DTC P0500 is displayed with DTC UXXXX, first perform the trouble diagnosis for DTC UXXXX.
	  • If DTC P0500 is displayed with DTC P0607, first perform the trouble diagnosis for DTC P0607.
	The almost 0 km/h (0 MPH) signal from vehicle speed sensor is sent to ECM even when vehicle is being driven.
	Possible Causes:
	  • Harness or connectors
	  • (The CAN communication line is open or shorted)
	  • Harness or connectors
	  • (The vehicle speed signal circuit is open or shorted)
	  • Wheel sensor
	  • Combination meter
	  • ABS actuator and electric unit (control unit)

DTC: P0501
2T, MIL: Yes
Year: 2009, 2010
Model: Grand Vitara
Engine: 3.2L V6 VIN 1
Transmission: All
	Vehicle Speed Sensor "A" Range/Performance:
	Difference between rear-right wheel speed and rear-left wheel speed is out of specified value at driving condition.
	Possible Causes:
	  • Wheel speed sensor and/or its circuit
	  • ESP® control module
	  • ECM

DTC: P0504
T
Year: 2009, 2010
Model: Grand Vitara, SX4
Engine: 2.0L L4, 2.0L L4 VIN 5, 2.4L L4 VIN 0, 2.4L L4 VIN 9, 3.2L V6 VIN 1
Transmission: All
	Brake Switch "A"/"B" Correlation:
	4 cylinder model: Both brake light switch signal voltage and brake pedal position switch signal voltage remain at low level for specified time continuously
	V6 model: Both brake light switch signal voltage and brake pedal position switch signal voltage remain at low level for specified time continuously when brake pedal is depressed
	Possible Causes:
	  • Brake light switch and/or circuit
	  • brake pedal position switch and/or circuit
	  • ECM

DTC: P0506
2T CCM, MIL: Yes
Year: 2009, 2010
Model: Equator, Grand Vitara, SX4, XL-7
Engine: 2.0L L4, 2.0L L4 VIN 5, 2.4L L4 VIN 0, 2.4L L4 VIN 9, 2.5L L4, 3.2L V6 VIN 1, 3.6L V6, 4.0L V6
Transmission: All
	Idle Speed Control System Lower Than Expected:
	Engine started, engine running at idle speed, throttle switch is "on", IAT signal less than 158°F, ambient temperature over 14°F, ECT sensor from 158-230°F, and the PCM detected the Actual idle speed was over 100 rpm less than the Desired idle speed in the test.
	Possible Causes:
	  • IAC valve control circuit is open or shorted to ground
	  • IAC valve power circuit is open (intermittent fault)
	  • IAC valve ground circuit is open (intermittent fault)
	  • Throttle plate is carbon fouled (it may need to be cleaned)
	  • PCM has failed

DTC: P0507
2T CCM, MIL: Yes
Year: 2009, 2010
Model: Equator, Grand Vitara, SX4, XL-7
Engine: 2.0L L4, 2.0L L4 VIN 5, 2.4L L4 VIN 0, 2.4L L4 VIN 9, 2.5L L4, 3.2L V6 VIN 1, 3.6L V6, 4.0L V6
Transmission: All
	Idle Speed Control System Higher Than Expected:
	Engine started, engine running at idle speed, throttle switch is "on", IAT signal less than 122°F, ambient temperature over 14°F, ECT sensor from 158-230°F, and the PCM detected the Actual idle speed was over 100 rpm more than the Desired idle speed in the test.
	Possible Causes:
	  • IAC valve control circuit is open (intermittent fault)
	  • IAC valve power circuit is open (intermittent fault)
	  • IAC valve ground circuit is open (intermittent fault)
	  • Throttle plate is carbon fouled (it may need to be cleaned)
	  • PCM has failed

DTC	Trouble Code Title, Conditions & Possible Causes
DTC: P050A **2T, MIL: Yes** **Year:** 2009, 2010 **Model:** Grand Vitara, SX4, XL-7 **Engine:** 2.0L L4, 2.0L L4 VIN 5, 2.4L L4 VIN 0, 2.4L L4 VIN 9, 3.6L V6 **Transmission:** All	**Cold Start Idle Air Control System Performance:** • Difference between command duty of IAC and reference duty of IAC is out of range (−20% through +48%) for 10 sec. at cold starting. • Command duty of IAC is 1% or 99% at cold starting. **Possible Causes:** • Air intake system • PCV valve and/or hose • Exhaust system • Throttle body assembly and/or its circuit • ECT sensor and/or its circuit • Ignition system • EVAP system • Engine compression • Engine vacuum • Valve clearance • Valve timing • ECM
DTC: P050A **2T, MIL: Yes** **Year:** 2009, 2010 **Model:** Grand Vitara **Engine:** 3.2L V6 VIN 1 **Transmission:** All	**Idle Air Control System RPM out of range during catalyst heating:** • Measured engine idle speed is lower than targeted engine idle speed by 100 rpm for 10 sec. at cold start. • Measured engine idle speed is higher than targeted engine idle speed by 200 rpm for 10 sec. at cold start. **Possible Causes:** • Air intake system • Exhaust system • Ignition system • Throttle body • APP sensor • Engine mechanical system • Fuel system • ECM
DTC: P050B **2T, MIL: Yes** **Year:** 2009, 2010 **Model:** Grand Vitara, SX4 **Engine:** 2.0L L4, 2.0L L4 VIN 5, 2.4L L4 VIN 0, 2.4L L4 VIN 9, 3.2L V6 VIN 1 **Transmission:** All	**Cold Start Ignition Timing Performance:** Difference between commanded ignition timing and reference ignition timing is out of specified range (1° through 20°) for 10 sec. at idle speed and cold starting. **Possible Causes:** • Air intake system • PCV valve and/or hose • Exhaust system • Throttle body assembly and/or its circuit • ECT sensor and/or its circuit • Ignition system • EVAP system • Engine compression • Engine vacuum • Valve clearance • Valve timing • ECM
DTC: P050E **2T** **Year:** 2009, 2010 **Model:** Equator **Engine:** 4.0L V6 **Transmission:** All	**Cold start engine exhaust temperature too low:** The temperature of the catalyst inlet does not rise to the proper temperature when the engine is started with pre-warming up condition. **Possible Causes:** • Lack of intake air volume • Fuel injection system • ECM
DTC: P052A **2T, MIL: Yes** **Year:** 2009, 2010 **Model:** Grand Vitara **Engine:** 3.2L V6 VIN 1 **Transmission:** All	**Cold Start "A" Camshaft Position Timing Over-Advanced Bank 1:** Difference between intake side measured CMP and target CMP on bank 1 is more than specified value for 12 sec. at cold start. **Possible Causes:** • OCV and/or its circuit • Lubrication system • VVT oil passage • CMP actuator • ECM

DTC	Trouble Code Title, Conditions & Possible Causes
DTC: P052C **2T, MIL: Yes** **Year:** 2009, 2010 **Model:** Grand Vitara **Engine:** 3.2L V6 VIN 1 **Transmission:** All	**Cold Start "A" Camshaft Position Timing Over-Advanced Bank 2:** Difference between intake side measured CMP and target CMP on bank 2 is more than specified value for 12 sec. at cold start. **Possible Causes:** • OCV and/or its circuit • Lubrication system • VVT oil passage • CMP actuator • ECM
DTC: P0532 **2T CCM, MIL: Yes** **Year:** 2009, 2010 **Model:** Grand Vitara, SX4 **Engine:** 2.0L L4, 2.0L L4 VIN 5, 2.4L L4 VIN 0, 2.4L L4 VIN 9, 3.2L V6 VIN 1 **Transmission:** All	**A/C Pressure Sensor Circuit Low Voltage:** Engine started, vehicle driven, and the PCM detected an unexpected voltage on the A/C pressure sensor circuit **Possible Causes:** • Sensor signal circuit is open or shorted to ground • Sensor ground circuit is open • Sensor is damaged or has failed (drive gear may be damaged) • PCM has failed
DTC: P0533 **2T CCM, MIL: Yes** **Year:** 2009, 2010 **Model:** Grand Vitara, SX4 **Engine:** 2.0L L4, 2.0L L4 VIN 5, 2.4L L4 VIN 0, 2.4L L4 VIN 9, 3.2L V6 VIN 1 **Transmission:** All	**A/C Pressure Sensor Circuit High Voltage:** Engine started, vehicle driven, and the PCM detected an unexpected voltage on the A/C pressure sensor circuit **Possible Causes:** • Sensor signal circuit is open or shorted to ground • Sensor ground circuit is open • Sensor is damaged or has failed (drive gear may be damaged) • PCM has failed
DTC: P054A **2T, MIL: Yes** **Year:** 2009, 2010 **Model:** Grand Vitara **Engine:** 3.2L V6 VIN 1 **Transmission:** All	**Cold Start "B" Camshaft Position Timing Over-Advanced Bank 1:** Difference between exhaust side measured CMP and target CMP on bank 1 is more than specified value for 12 sec. at cold start. **Possible Causes:** • OCV and/or its circuit • Lubrication system • VVT oil passage • CMP actuator • ECM
DTC: P054C **2T, MIL: Yes** **Year:** 2009, 2010 **Model:** Grand Vitara **Engine:** 3.2L V6 VIN 1 **Transmission:** All	**Cold Start "B" Camshaft Position Timing Over-Advanced Bank 2:** Difference between exhaust side measured CMP and target CMP on bank 2 is more than specified value for 12 sec. at cold start. **Possible Causes:** • OCV and/or its circuit • Lubrication system • VVT oil passage • CMP actuator • ECM
DTC: P0550 **2T** **Year:** 2009, 2010 **Model:** Equator **Engine:** 2.5L L4, 4.0L V6 **Transmission:** All	**Power steering pressure sensor circuit:** **NOTE:** If DTC P0550 is displayed with DTC P0643, first perform the trouble diagnosis for DTC P0643. An excessively low or high voltage from the sensor is sent to ECM. **Possible Causes:** • Harness or connectors • (The sensor circuit is open or shorted.) • Power steering pressure sensor
DTC: P0560 **1T, MIL: Yes** **Year:** 2009, 2010 **Model:** Grand Vitara, SX4 **Engine:** 2.0L L4, 2.0L L4 VIN 5, 2.4L L4 VIN 0, 2.4L L4 VIN 9, 3.2L V6 VIN 1 **Transmission:** All	**ECM Back-Up Power Supply Malfunction:** Monitor voltage of backup power supply circuit is lower than specified value for 5 sec. **Possible Causes:** • Backup power supply circuit • Charging system • ECM

DTC	Trouble Code Title, Conditions & Possible Causes
DTC: P0562 **2T CCM, MIL: Yes** **Year:** 2009, 2010 **Model:** Grand Vitara, XL-7 **Engine:** 3.2L V6 VIN 1, 3.6L V6 **Transmission:** All	**System Voltage Too Low (Engine Side):** Engine started, vehicle driven, and the PCM detected an unexpected voltage on the system. **Possible Causes:** • There is a signal circuit open or shorted to ground • There is a ground circuit is open • Check the battery voltage • PCM has failed
DTC: P0563 **2T CCM, MIL: Yes** **Year:** 2009, 2010 **Model:** Grand Vitara, XL-7 **Engine:** 3.2L V6 VIN 1, 3.6L V6 **Transmission:** All	**System Voltage Too High (Engine Side):** Engine started, vehicle driven, and the PCM detected an unexpected voltage on the system. **Possible Causes:** • There is a signal circuit open or shorted to ground • There is a ground circuit is open • Check the battery voltage • PCM has failed
DTC: P0575 **1T** **Year:** 2009, 2010 **Model:** Grand Vitara **Engine:** 2.4L L4 VIN 0, 2.4L L4 VIN 9, 3.2L V6 VIN 1 **Transmission:** All	**Cruise Control Input Circuit:** • MAIN (ON/OFF) switch signal voltage remains at high level for specified time continuously • Cruise control command switch (CANCEL switch, RES/ACC switch and/or SET/COAST switch) signal voltage remains at low level for specified time continuously **Possible Causes:** • Cruise control switch and/or circuit • ECM
DTC: P0601 **1T CCM, MIL: Yes** **Year:** 2009, 2010 **Model:** Grand Vitara, SX4 **Engine:** 2.0L L4, 2.0L L4 VIN 5, 2.4L L4 VIN 0, 2.4L L4 VIN 9 **Transmission:** All	**PCM Memory Check Sum Error:** Key on, and the PCM detected a Memory Check Sum Error in the initial check during the CCM test. **NOTE: The engine may not start when this trouble code is set.** **Possible Causes:** • Clear the trouble code, and recheck for the same code. • If this trouble code resets, the PCM must be replaced
DTC: P0601 **5T, MIL: Yes** **Year:** 2009 **Model:** XL-7 **Engine:** 3.6L V6 **Transmission:** All	**Control Module Read Only Memory (ROM):** The ECM detects a read only memory (ROM) checksum error at ECM power up. The error is detected 5 times, then a 5 second delay for MIL ON after the fifth detection. **Possible Causes:** • ECM
DTC: P0602 **1T, MIL: Yes** **Year:** 2009 **Model:** XL-7 **Engine:** 3.6L V6 **Transmission:** All	**Control Module Not Programmed:** The ECM detects that programming is incomplete. The condition exists, then a 5 second delay for MIL ON. **Possible Causes:** • ECM
DTC: P0602 **1T, MIL: Yes** **Year:** 2009, 2010 **Model:** Grand Vitara, SX4 **Engine:** 2.0L L4, 2.0L L4 VIN 5, 2.4L L4 VIN 0, 2.4L L4 VIN 9, 3.2L V6 VIN 1 **Transmission:** All	**ECM Memory Error:** Key on, and the ECM detected a Memory Error. **Possible Causes:** • The contents of the ECM have changed • Clear the trouble code, and recheck for the same code. • If this trouble code resets, the ECM must be replaced
DTC: P0603 **2T, MIL: Yes** **Year:** 2009, 2010 **Model:** Equator **Engine:** 2.5L L4, 4.0L V6 **Transmission:** All	**ECM power supply circuit:** ECM back up RAM system does not function properly. **Possible Causes:** • Harness or connectors • [ECM power supply (back up) circuit is open or shorted.] • ECM
DTC: P0603 **1T, MIL: Yes** **Year:** 2009 **Model:** XL-7 **Engine:** 3.6L V6 **Transmission:** All	**Control Module Long Term Memory Reset:** The ECM detects a reset error of the internal software that operates the throttle actuator control (TAC) system, during the shutdown path test. The condition exists, then a 5 second delay for MIL ON. **Possible Causes:** • ECM

DTC	Trouble Code Title, Conditions & Possible Causes
DTC: P0604 **1T, MIL: Yes** **Year:** 2009 **Model:** XL-7 **Engine:** 3.6L V6 **Transmission:** All	**Control Module Random Access Memory (RAM):** The ECM detects random access memory (RAM) errors. The condition exists, then a 5 second delay for MIL ON. **Possible Causes:** • ECM
DTC: P0605 **1T, MIL: Yes** **Year:** 2009, 2010 **Model:** Equator **Engine:** 2.5L L4, 4.0L V6 **Transmission:** All	**Engine control module:** A) ECM calculation function is malfunctioning. B) ECM EEP-ROM system is malfunctioning. C) ECM self shut-off function is malfunctioning. **Possible Causes:** • ECM
DTC: P0606 **1T, MIL: Yes** **Year:** 2009 **Model:** XL-7 **Engine:** 3.6L V6 **Transmission:** All	**Control Module Internal Performance:** The ECM detects an internal error of the monitoring software for the throttle actuator control (TAC) system. The condition exists, then a 5 second delay for MIL ON. **Possible Causes:** • ECM
DTC: P0607 **1T, MIL: Yes** **Year:** 2009, 2010 **Model:** Equator, Grand Vitara **Engine:** 2.4L L4 VIN 0, 2.5L L4, 3.2L V6 VIN 1, 4.0L V6 **Transmission:** All	**CAN communication bus:** **NOTE** This self-diagnosis has the one trip detection logic (A/T models). The MIL will not light up for this diagnosis (M/T models). When detecting error during the initial diagnosis for CAN controller of each control unit. **Possible Causes:** • ECM
DTC: P0607 **1T, MIL: Yes** **Year:** 2009, 2010 **Model:** SX4 **Engine:** 2.0L L4, 2.0L L4 VIN 5, 2.4L L4 VIN 9 **Transmission:** All	**Control Module Performance:** ECM internal failure related to throttle control system. **Possible Causes:** • ECM power supply circuit and/or ground circuit • ECM
DTC: P0616 **2T, MIL: Yes** **Year:** 2009, 2010 **Model:** Grand Vitara, SX4 **Engine:** 2.0L L4, 2.0L L4 VIN 5, 2.4L L4 VIN 0, 2.4L L4 VIN 9 **Transmission:** All	**Starter Relay Circuit Low:** Starter switch signal is undetectable even if engine has been started. **Possible Causes:** • Engine start signal circuit • ECM
DTC: P0617 **2T, MIL: Yes** **Year:** 2009, 2010 **Model:** Grand Vitara, SX4 **Engine:** 2.0L L4, 2.0L L4 VIN 5, 2.4L L4 VIN 0, 2.4L L4 VIN 9 **Transmission:** All	**Starter Relay Circuit High:** Monitor signal of starting motor relay is high level (ON) for 180 sec. after starting engine. **Possible Causes:** • Engine start signal circuit • ECM
DTC: P0620 **2T, MIL: Yes** **Year:** 2009, 2010 **Model:** SX4 **Engine:** 2.0L L4, 2.0L L4 VIN 5, 2.4L L4 VIN 9 **Transmission:** All	**Generator Control Circuit:** • Duty control signal of generator control is maximum (100%) for 10 sec. even if battery voltage is higher than 13.8 V. • Duty control signal of generator control is minimum (0%) for 10 sec. even if battery voltage is lower than 12.6 V and electric load is less than 16 A. **Possible Causes:** • Generator and/or its circuit • ECM

DTC	Trouble Code Title, Conditions & Possible Causes

DTC: P0625
1T
Year: 2009, 2010
Model: Grand Vitara, SX4
Engine: 2.0L L4, 2.0L L4 VIN 5, 2.4L L4 VIN 0, 2.4L L4 VIN 9, 3.2L V6 VIN 1
Transmission: All
	Generator Field Terminal Circuit Low:
	Duty signal of generator field coil is higher than 99.9% (low voltage) for 10 sec. even if duty control signal of generator control is maximum (100%).
	Generator field coil signal circuit is shorted to ground.
	Possible Causes:
	   • Generator and/or its circuit
	   • ECM

DTC: P0626
1T
Year: 2009, 2010
Model: Grand Vitara, SX4
Engine: 2.0L L4, 2.0L L4 VIN 5, 2.4L L4 VIN 0, 2.4L L4 VIN 9, 3.2L V6 VIN 1
Transmission: All
	Generator Field Terminal Circuit High:
	Duty control signal of generator field coil is lower than 1% (high voltage) for 10 sec. even if duty control signal of generator control is minimum (0%).
	Generator field coil signal circuit is shorted to power supply circuit.
	Possible Causes:
	   • Generator and/or its circuit
	   • ECM

DTC: P0627
2T
Year: 2009, 2010
Model: Grand Vitara, XL-7
Engine: 3.2L V6 VIN 1, 3.6L V6
Transmission: All
	Fuel Pump "A" Control Circuit /Open:
	Fuel pump relay drive circuit is open circuit.
	Possible Causes:
	   • Fuel pump relay (coil side) and/or its circuit
	   • Fuel pump relay
	   • ECM

DTC: P0628
2T
Year: 2009, 2010
Model: Grand Vitara, XL-7
Engine: 3.2L V6 VIN 1, 3.6L V6
Transmission: All
	Fuel Pump "A" Control Circuit Low:
	Fuel pump relay drive circuit is shorted to ground.
	Possible Causes:
	   • Fuel pump relay (coil side) and/or its circuit
	   • Fuel pump relay
	   • ECM

DTC: P0629
2T
Year: 2009, 2010
Model: Grand Vitara, XL-7
Engine: 3.2L V6 VIN 1, 3.6L V6
Transmission: All
	Fuel Pump "A" Control Circuit High:
	Fuel pump relay drive circuit is shorted to power supply circuit.
	Possible Causes:
	   • Fuel pump relay (coil side) and/or its circuit
	   • Fuel pump relay
	   • ECM

DTC: P062F
1T, MIL: Yes
Year: 2009, 2010
Model: Grand Vitara, SX4, XL-7
Engine: 2.0L L4, 2.0L L4 VIN 5, 2.4L L4 VIN 0, 2.4L L4 VIN 9, 3.2L V6 VIN 1, 3.6L V6
Transmission: All
	Internal Control Module EEPROM Error:
	ECM internal failure related to check sum error.
	Possible Causes:
	   • ECM power supply circuit and/or ground circuit
	   • ECM

DTC: P0630
1T
Year: 2009, 2010
Model: Grand Vitara
Engine: 2.4L L4 VIN 0
Transmission: All
	Generator Control Circuit:
	   • Duty control signal of generator control is maximum (100%) for 10 sec. even if battery voltage is higher than 13.8 V.
	   • Duty control signal of generator control is minimum (0%) for 10 sec. even if battery voltage is lower than 12.6 V and electric load is less than 16 A.
	Possible Causes:
	   • Generator and/or its circuit
	   • ECM

DTC: P0630
1T, MIL: Yes
Year: 2009, 2010
Model: Grand Vitara, SX4, XL-7
Engine: 2.0L L4, 2.0L L4 VIN 5, 2.4L L4 VIN 0, 2.4L L4 VIN 9, 3.2L V6 VIN 1, 3.6L V6
Transmission: All
	VIN Not Programmed or Incompatible-ECM/PCM:
	Key on, and the PCM detected incompatibility error in regards to the VIN.
	Possible Causes:
	   • The contents of the ECM have changed
	   • Check the validity of the VIN
	   • PCM is faulty

DTC	Trouble Code Title, Conditions & Possible Causes

DTC: P0638
2T, MIL: Yes
Year: 2009
Model: XL-7
Engine: 3.6L V6
Transmission: All
	Throttle Actuator Control (TAC) Command Performance:
	    • The ECM detects that the commanded duty cycle for the range test high is greater than 80 percent.
	OR
	    • The ECM detects that the commanded duty cycle for the range test low is greater than 80 percent.
	    • Either condition exists for greater than 1 second, or a cumulative of 10 seconds.
	Possible Causes:
	    • Throttle Actuator Control
	    • Throttle body
	    • ECM

DTC: P0641
2T, MIL: Yes
Year: 2009, 2010
Model: Grand Vitara, XL-7
Engine: 2.4L L4 VIN 9, 3.2L V6 VIN 1, 3.6L V6
Transmission: All
	Sensor Reference Voltage "A" Circuit/Open:
	5 V power supply circuit of A/C refrigerant pressure sensor, CKP sensor, APP sensor (main) and fuel tank pressure sensor is open circuit.
	Possible Causes:
	    • A/C refrigerant pressure sensor and/or its circuit
	    • APP sensor (main) and/or its circuit
	    • CKP sensor and/or its circuit
	    • Fuel tank pressure sensor
	    • ECM

DTC: P0642
2T, MIL: Yes
Year: 2009, 2010
Model: Grand Vitara, XL-7
Engine: 3.2L V6 VIN 1, 3.6L V6
Transmission: All
	Sensor Reference Voltage "A" Circuit Low:
	5 V power supply circuit of A/C refrigerant pressure sensor, CKP sensor, APP sensor (main) and fuel tank pressure sensor is shorted to ground.
	Possible Causes:
	    • A/C refrigerant pressure sensor and/or its circuit
	    • APP sensor (main) and/or its circuit
	    • CKP sensor and/or its circuit
	    • Fuel tank pressure sensor
	    • ECM

DTC: P0643
2T, MIL: Yes
Year: 2009, 2010
Model: Grand Vitara, XL-7
Engine: 3.2L V6 VIN 1, 3.6L V6
Transmission: All
	Sensor Reference Voltage "A" Circuit High:
	5 V power supply circuit of A/C refrigerant pressure sensor, CKP sensor, APP sensor (main) and fuel tank pressure sensor is shorted to power supply circuit.
	Possible Causes:
	    • A/C refrigerant pressure sensor and/or its circuit
	    • APP sensor (main) and/or its circuit
	    • CKP sensor and/or its circuit
	    • Fuel tank pressure sensor
	    • ECM

DTC: P0643
1T, MIL: Yes
Year: 2009, 2010
Model: Equator
Engine: 2.5L L4, 4.0L V6
Transmission: All
	Sensor power supply circuit short:
	ECM detects that the voltage of power source for sensor is excessively low or high.
	Possible Causes:
	    • Harness or connectors
	    • (APP sensor 1 circuit is shorted.)
	    • (PSP sensor circuit is shorted.)
	    • (Refrigerant pressure sensor circuit is shorted.)
	    • (Battery current sensor circuit is shorted.)
	    • (EVAP control system pressure sensor circuit is shorted.)
	    • Accelerator pedal position sensor
	    • (APP sensor 1)
	    • Power steering pressure sensor
	    • Refrigerant pressure sensor
	    • Battery current sensor
	    • EVAP control system pressure sensor

DTC: P0645
2T
Year: 2009, 2010
Model: Grand Vitara
Engine: 3.2L V6 VIN 1
Transmission: All
	A/C Clutch Relay Control Circuit:
	A/C compressor relay drive circuit is open circuit.
	Possible Causes:
	    • A/C compressor relay (coil side) and/or its circuit
	    • ECM

DTC	Trouble Code Title, Conditions & Possible Causes

DTC: P0646
2T
Year: 2009, 2010
Model: Grand Vitara
Engine: 3.2L V6 VIN 1
Transmission: All
	A/C Clutch Relay Control Circuit Low;
	A/C compressor relay drive circuit is shorted to ground.
	Possible Causes:
	   • A/C compressor relay (coil side) and/or its circuit
	   • ECM

DTC: P0647
2T
Year: 2009, 2010
Model: Grand Vitara
Engine: 3.2L V6 VIN 1
Transmission: All
	A/C Clutch Relay Control Circuit High:
	A/C compressor relay drive circuit is shorted to power supply circuit.
	Possible Causes:
	   • A/C compressor relay (coil side) and/or its circuit
	   • ECM

DTC: P0649
1T
Year: 2009, 2010
Model: Grand Vitara
Engine: 2.4L L4 VIN 0, 2.4L L4 VIN 9, 3.2L V6 VIN 1
Transmission: All
	Speed Control Lamp Control Circuit:
	"CRUISE / SET indicator light lighting signal" from combination meter remains unchanged even if ECM transmits request to turn on or turn off "CRUISE / SET indicator light control signal" to combination meter through CAN communication.
	Possible Causes:
	   • ECM
	   • combination meter
	   • CAN communication

DTC: P0650
2T
Year: 2009
Model: XL-7
Engine: 3.6L V6
Transmission: All
	Malfunction Indicator Lamp (MIL) Control Circuit:
	The ECM detects an open, a short to ground, or a short to voltage on the circuit that controls the MIL.
	Possible Causes:
	   • MIL bulb and/or circuit
	   • ECM

DTC: P0651
2T, MIL: Yes
Year: 2009, 2010
Model: Grand Vitara, XL-7
Engine: 3.2L V6 VIN 1, 3.6L V6
Transmission: All
	Sensor Reference Voltage "B" Circuit/Open:
	5 V power supply circuit of CMP sensor is open circuit.
	Possible Causes:
	   • CMP sensor and/or its circuit
	   • ECM

DTC: P0652
2T, MIL: Yes
Year: 2009, 2010
Model: Grand Vitara, XL-7
Engine: 3.2L V6 VIN 1, 3.6L V6
Transmission: All
	Sensor Reference Voltage "B" Circuit Low:
	5 V power supply circuit of CMP sensor is shorted to ground.
	Possible Causes:
	   • CMP sensor and/or its circuit
	   • ECM

DTC: P0653
2T, MIL: Yes
Year: 2009, 2010
Model: Grand Vitara, XL-7
Engine: 3.2L V6 VIN 1, 3.6L V6
Transmission: All
	Sensor Reference Voltage "B" High Input:
	5 V power supply circuit of CMP sensor is shorted to power supply circuit.
	Possible Causes:
	   • CMP sensor and/or its circuit
	   • ECM

DTC: P0660
2T, MIL: Yes
Year: 2009, 2010
Model: Grand Vitara, SX4
Engine: 2.0L L4, 2.0L L4 VIN 5, 2.4L L4 VIN 0, 2.4L L4 VIN 9
Transmission: All
	Intake Manifold Tuning Valve Control Circuit Open:
	The PCM detected an unexpected voltage across the intake manifold tuning valve control circuit
	Possible Causes:
	   • Faulty valve
	   • There is a short or an open circuit
	   • PCM is faulty

DTC: P0685
2T, MIL: Yes
Year: 2009
Model: XL-7
Engine: 3.6L V6
Transmission: All
	Engine Controls Ignition Relay Control Circuit:
	The commanded state of the ODM and the actual state of the control circuit do not match.
	Possible Causes:
	   • Fuse
	   • Powertrain relay and/or circuit
	   • ECM

DTC	Trouble Code Title, Conditions & Possible Causes
DTC: P0686 **1T** **Year:** 2009 **Model:** XL-7 **Engine:** 3.6L V6 **Transmission:** All	**Engine Controls Ignition Relay Control Circuit Low Voltage:** • The ignition voltage is between 10–18 volts. • The engine speed is greater than 80 RPM. • The powertrain relay has been commanded ON and OFF. • The DTCs run continuously when the above conditions are met. **Possible Causes:** • Fuse • Powertrain relay and/or circuit • ECM
DTC: P0687 **1T** **Year:** 2009 **Model:** XL-7 **Engine:** 3.6L V6 **Transmission:** All	**Engine Controls Ignition Relay Control Circuit High Voltage:** • The ignition voltage is between 10–18 volts. • The engine speed is greater than 80 RPM. • The powertrain relay has been commanded ON and OFF. • The DTCs run continuously when the above conditions are met. **Possible Causes:** • Fuse • Powertrain relay and/or circuit • ECM
DTC: P0689 **1T** **Year:** 2009 **Model:** XL-7 **Engine:** 3.6L V6 **Transmission:** All	**Engine Controls Ignition Relay Feedback Circuit Low Voltage:** • The ignition voltage is between 10–18 volts. • The ignition is ON. • The DTCs runs continuously when the above conditions are met. **Possible Causes:** • Fuse • Powertrain relay and/or circuit • ECM
DTC: P0690 **1T, MIL: Yes** **Year:** 2009 **Model:** XL-7 **Engine:** 3.6L V6 **Transmission:** All	**Engine Controls Ignition Relay Feedback Circuit High Voltage:** • The ignition voltage is between 10–18 volts. • The ignition is ON. • The DTCs runs continuously when the above conditions are met. **Possible Causes:** • Fuse • Powertrain relay and/or circuit • ECM
DTC: P0697 **2T, MIL: Yes** **Year:** 2009, 2010 **Model:** Grand Vitara, XL-7 **Engine:** 3.2L V6 VIN 1, 3.6L V6 **Transmission:** All	**Sensor Reference Voltage "C" Circuit/Open:** 5 V power supply circuit and TP sensor or APP sensor (sub) is open circuit. **Possible Causes:** • TP sensor and/or its circuit • APP sensor (sub) and/or its circuit • ECM
DTC: P0698 **2T, MIL: Yes** **Year:** 2009, 2010 **Model:** Grand Vitara, XL-7 **Engine:** 3.2L V6 VIN 1, 3.6L V6 **Transmission:** All	**Sensor Reference Voltage "C" Circuit Low:** 5 V power supply circuit of TP sensor and APP sensor (sub) is shorted to ground. **Possible Causes:** • TP sensor and/or its circuit • APP sensor (sub) and/or its circuit • ECM
DTC: P0699 **2T, MIL: Yes** **Year:** 2009, 2010 **Model:** Grand Vitara, XL-7 **Engine:** 3.2L V6 VIN 1, 3.6L V6 **Transmission:** All	**Sensor Reference Voltage "C" Circuit High:** 5 V power supply circuit of TP sensor and APP sensor (sub) is shorted to power supply circuit. **Possible Causes:** • TP sensor and/or its circuit • APP sensor (sub) and/or its circuit • ECM
DTC: P0700 **1T, MIL: Yes** **Year:** 2009, 2010 **Model:** Equator, XL-7 **Engine:** 2.5L L4, 3.6L V6, 4.0L V6 **Transmission:** All	**Transmission Control:** • This is an OBD-II self-diagnostic item. • Diagnostic trouble code "P0700" with SDT is detected when the TCM is malfunctioning. **Possible Causes:** • TCM.

DTC	Trouble Code Title, Conditions & Possible Causes
DTC: P0705 **2T, MIL: Yes** **Year:** 2009, 2010 **Model:** Equator, Grand Vitara **Engine:** 2.4L L4 VIN 9, 2.5L L4, 3.2L V6 VIN 1, 4.0L V6 **Transmission:** All	**Transmission Range Switch A:** — When TCM does not receive the correct voltage signal from the transmission range switch 1, 2, 3, 4 based on the gear position. — When no other position but "P" position is detected from "N" positions. **Possible Causes:** • Harness or connectors • (The transmission range switch 1, 2, 3, 4 and TCM circuit is open or shorted.) • Transmission range switch 1, 2, 3, 4
DTC: P0707 **2T, MIL: Yes** **Year:** 2009, 2010 **Model:** Grand Vitara **Engine:** 2.4L L4 VIN 9, 3.2L V6 VIN 1 **Transmission:** All	**Transmission Range Sensor Circuit Low:** Transmission range sensor signal (P, R, N, D, 3, L) is not inputted for more than 2 seconds in the following condition. • Vehicle speed is more than 30 km/h (19 mile/h). • Engine speed is more than 1500 rpm. **Possible Causes:** • Select cable maladjusted. • Transmission range sensor maladjusted. • Transmission range sensor or its circuit malfunction. • TCM
DTC: P0710 **2T, MIL: Yes** **Year:** 2009, 2010 **Model:** Equator **Engine:** 2.5L L4, 4.0L V6 **Transmission:** All	**Transmission Fluid Temperature Sensor:** TCM receives an excessively low or high voltage from the sensor. **Possible Causes:** • Harness or connectors • (The sensor circuit is open or shorted.) • A/T fluid temperature sensor 1
DTC: P0711 **2T, MIL: Yes** **Year:** 2009, 2010 **Model:** Grand Vitara, XL-7 **Engine:** 2.4L L4 VIN 9, 3.2L V6 VIN 1, 3.6L V6 **Transmission:** All	**Transmission Fluid Temperature Sensor "A" Circuit Range / Performance:** Transmission fluid temperature sensor variation is less than specified value even though engine was running for specified time after engine start. **Possible Causes:** • Transmission fluid temperature sensor or its circuit. • TCM
DTC: P0712 **2T, MIL: Yes** **Year:** 2009, 2010 **Model:** Grand Vitara, XL-7 **Engine:** 2.4L L4 VIN 9, 3.2L V6 VIN 1, 3.6L V6 **Transmission:** All	**Transmission Fluid Temperature Sensor "A" Circuit Low:** Transmission fluid temperature sensor terminal voltage is less than 0.05 V for 5 seconds or more with engine running. **Possible Causes:** • Transmission fluid temperature sensor or its circuit. • TCM
DTC: P0713 **2T, MIL: Yes** **Year:** 2009, 2010 **Model:** Grand Vitara, XL-7 **Engine:** 2.4L L4 VIN 9, 3.2L V6 VIN 1, 3.6L V6 **Transmission:** All	**Transmission Fluid Temperature Sensor "A" Circuit High:** Transmission fluid temperature sensor terminal voltage is more than 4.89 V for 5 seconds or more with engine running. **Possible Causes:** • Transmission fluid temperature sensor or its circuit. • TCM
DTC: P0717 **2T, MIL: Yes** **Year:** 2009, 2010 **Model:** Equator **Engine:** 2.5L L4, 4.0L V6 **Transmission:** All	**Input Speed Sensor A:** — When TCM does not receive the proper voltage signal from the sensor. — When TCM detects an irregularity only for input speed sensor 2. **Possible Causes:** • Harness or connectors • (The sensor circuit is open or shorted.) • Input speed sensor 1, 2
DTC: P0717 **1T, MIL: Yes** **Year:** 2009, 2010 **Model:** Grand Vitara **Engine:** 2.4L L4 VIN 9, 3.2L V6 VIN 1 **Transmission:** All	**Input / Turbine Speed Sensor "A" Circuit No Signal:** No pulse signal of input shaft speed sensor is inputted for 5 pulses period of output shaft speed sensor through it is detected more than 600 rpm. **Possible Causes:** • Input shaft speed sensor or its circuit malfunction. • Improper input shaft speed sensor installation. • Damaged clutch drum. • Foreign material attachment to sensor or drum. • TCM

DTC	Trouble Code Title, Conditions & Possible Causes
DTC: P0720 **2T, MIL:** Yes **Year:** 2009, 2010 **Model:** Equator **Engine:** 2.5L L4, 4.0L V6 **Transmission:** All	**Output Speed Sensor:** — When TCM does not receive the proper voltage signal from the sensor. — After ignition switch is turned "ON", irregular signal input from vehicle speed signal before the vehicle starts moving. **Possible Causes:** • Harness or connectors • (The sensor circuit is open or shorted.) • Output speed sensor • Vehicle speed signal
DTC: P0722 **1T, MIL:** Yes **Year:** 2009, 2010 **Model:** Grand Vitara **Engine:** 2.4l L4 VIN 9, 3.2L V6 VIN 1 **Transmission:** All	**Output Speed Sensor Circuit No Signal:** No pulse signal of output shaft speed sensor is inputted for 18 pulses period of input shaft speed sensor. **Possible Causes:** • Output shaft speed sensor or its circuit malfunction. • Improper output shaft speed sensor installation. • Damaged sensor rotor. • Foreign material attachment to sensor or rotor. • TCM
DTC: P0731 **2T, MIL:** Yes **Year:** 2009, 2010 **Model:** Equator **Engine:** 2.5L L4, 4.0L V6 **Transmission:** All	**1Gr Incorrect Ratio:** TCM detects any inconsistency in the actual gear ratio. **Possible Causes:** • Input clutch solenoid valve • Front brake solenoid valve • Direct clutch solenoid valve • High and low reverse clutch solenoid valve • Each clutch • Hydraulic control circuit
DTC: P0732 **2T, MIL:** Yes **Year:** 2009, 2010 **Model:** Equator **Engine:** 2.5L L4, 4.0L V6 **Transmission:** All	**2Gr Incorrect Ratio:** TCM detects any inconsistency in the actual gear ratio. **Possible Causes:** • Input clutch solenoid valve • Front brake solenoid valve • Direct clutch solenoid valve • High and low reverse clutch solenoid valve • Each clutch • Hydraulic control circuit
DTC: P0733 **1T, MIL:** Yes **Year:** 2009, 2010 **Model:** Equator **Engine:** 2.5L L4, 4.0L V6 **Transmission:** All	**3Gr Incorrect Ratio:** TCM detects any inconsistency in the actual gear ratio. **Possible Causes:** • Input clutch solenoid valve • Front brake solenoid valve • Direct clutch solenoid valve • High and low reverse clutch solenoid valve • Each clutch • Hydraulic control circuit
DTC: P0734 **1T, MIL:** Yes **Year:** 2009, 2010 **Model:** Equator **Engine:** 2.5L L4, 4.0L V6 **Transmission:** All	**4Gr Incorrect Ratio:** TCM detects any inconsistency in the actual gear ratio. **Possible Causes:** • Input clutch solenoid valve • Front brake solenoid valve • Direct clutch solenoid valve • High and low reverse clutch solenoid valve • Each clutch • Hydraulic control circuit

DTC	Trouble Code Title, Conditions & Possible Causes

DTC: P0735
1T, MIL: Yes
Year: 2009, 2010
Model: Equator
Engine: 2.5L L4, 4.0L V6
Transmission: All
	5Gr Incorrect Ratio:
	TCM detects any inconsistency in the actual gear ratio.
	Possible Causes:
	    • Input clutch solenoid valve
	    • Front brake solenoid valve
	    • Direct clutch solenoid valve
	    • High and low reverse clutch solenoid valve
	    • Each clutch
	    • Hydraulic control circuit

DTC: P0740
2T, MIL: Yes
Year: 2009, 2010
Model: Equator
Engine: 2.5L L4, 4.0L V6
Transmission: All
	Torque converter clutch solenoid valve:
	— When TCM detects an improper voltage drop when it tries to operate the solenoid valve.
	— When TCM detects as irregular by comparing target value with monitor value.
	Possible Causes:
	    • Torque converter clutch solenoid valve
	    • Harness or connectors
	    • (The solenoid circuit is open or shorted.)

DTC: P0741
2T, MIL: Yes
Year: 2009, 2010
Model: Grand Vitara, XL-7
Engine: 2.4L L4 VIN 9, 3.2L V6
VIN 1, 3.6L V6
Transmission: All
	Torque Converter Clutch Circuit Performance or Stuck OFF:
	When driving vehicle with 4th or 5th gear in "D" range, difference in revolution between engine and A/T input (input shaft speed) is larger than specification although TCM commanded TCC pressure control solenoid to turn ON.
	Possible Causes:
	    • Mechanical malfunction of TCC pressure control solenoid valve.
	    • Malfunction of valve body assembly.
	    • Fluid passage clogged or leaking.
	    • Torque converter clutch malfunction.

DTC: P0742
2T, MIL: Yes
Year: 2009, 2010
Model: Grand Vitara, XL-7
Engine: 3.2L V6 VIN 1, 3.6L V6
Transmission: All
	Torque Converter Clutch Circuit Performance or Stuck ON:
	When driving vehicle with 2nd, 3rd, 4th or 5th gear in "D" range, difference in revolution between engine and A/T input (input shaft speed) is smaller than specification although TCM commanded TCC pressure control solenoid to turn OFF.
	Possible Causes:
	    • Mechanical malfunction of TCC pressure control solenoid valve.
	    • Malfunction of valve body assembly.
	    • Fluid passage clogged or leaking.
	    • Torque converter clutch malfunction.

DTC: P0744
2T, MIL: Yes
Year: 2009, 2010
Model: Equator
Engine: 2.5L L4, 4.0L V6
Transmission: All
	A/T Torque Converter does not lock-up:
	— When A/T cannot perform lock-up even if electrical circuit is good.
	— When TCM detects as irregular by comparing difference value with slip rotation.
	Possible Causes:
	    • Harness or connectors
	    • (The solenoid circuit is open or shorted.)
	    • Torque converter clutch solenoid valve
	    • Hydraulic control circuit

DTC: P0745
2T, MIL: Yes
Year: 2009, 2010
Model: Equator
Engine: 2.5L L4, 4.0L V6
Transmission: All
	Pressure Control Solenoid A:
	— When TCM detects an improper voltage drop when it tries to operate the solenoid valve.
	— When TCM detects as irregular by comparing target value with monitor value.
	Possible Causes:
	    • Harness or connectors
	    • (The solenoid circuit is open or shorted.)
	    • Line pressure solenoid valve

DTC: P0751
2T, MIL: Yes
Year: 2009, 2010
Model: Grand Vitara
Engine: 3.2L V6 VIN 1
Transmission: All
	Shift Solenoid "A" Performance or Stuck OFF:
	4th gear ratio is detected although TCM command is for 1st gear and 3rd gear ratio is detected although TCM command is for 2nd gear while vehicle running at 2 km/h (1.5 mile/h) or more in "D" range after engine being warmed up.
	Possible Causes:
	    • Mechanical malfunction of shift solenoid valve-A.
	    • Malfunction of valve body assembly.
	    • Fluid passage clogged or leaking.
	    • Mechanical malfunction of automatic transmission (clutch, brake or gear etc.).

DTC	Trouble Code Title, Conditions & Possible Causes

DTC: P0752
T, MIL: Yes
Year: 2009, 2010
Model: Grand Vitara
Engine: 3.2L V6 VIN 1
Transmission: All
	Shift Solenoid "A" Performance or Stuck ON:
	When one of the following condition was detected while vehicle running at 2 km/h (1.5 mile/h) or more in "D" range after engine being warmed up.
	    • 2nd gear ratio is detected although TCM command is for 3rd gear and 1st gear is detected although TCM command is for 4th gear.
	Or
	    • 2nd gear ratio is detected although TCM command is for 3rd gear and Neutral gear is detected although TCM command is for 5th gear.
	Or
	    • 1st gear ratio is detected although TCM command is for 4th gear and Neutral gear is detected although TCM command is for 5th gear.
	Possible Causes:
	    • Mechanical malfunction of shift solenoid valve-A.
	    • Malfunction of valve body assembly.
	    • Fluid passage clogged or leaking.
	    • Mechanical malfunction of automatic transmission (clutch, brake or gear etc.).

DTC: P0756
2T, MIL: Yes
Year: 2009, 2010
Model: Grand Vitara
Engine: 3.2L V6 VIN 1
Transmission: All
	Shift Solenoid "B" Performance or Stuck OFF:
	Gear ratio is detected although TCM command is for 2nd gear and 4th gear ratio is detected although TCM command is for 3rd gear while vehicle running at 2 km/h (1.5 mile/h) or more in "D" range after engine being warmed up.
	Possible Causes:
	    • Mechanical malfunction of shift solenoid valve-B.
	    • Malfunction of valve body assembly.
	    • Fluid passage clogged or leaking.
	    • Mechanical malfunction of automatic transmission (clutch, brake or gear etc.).

DTC: P0757
2T, MIL: Yes
Year: 2009, 2010
Model: Grand Vitara
Engine: 3.2L V6 VIN 1
Transmission: All
	Shift Solenoid "B" Performance or Stuck ON:
	When one of the following condition was detected while vehicle running at 2 km/h (1.5 mile/h) or more in "D" range after engine being warmed up.
	    • 2nd gear ratio is detected although TCM command is for 1st gear and Neutral gear is detected although TCM command is for 5th gear.
	Or
	    • 3rd gear ratio is detected although TCM command is for 4th gear and Neutral gear is detected although TCM command is for 5th gear.
	Or
	    • 2nd gear ratio is detected although TCM command is 1st gear and 3rd gear ratio is detected although TCM command is 4th gear.
	Possible Causes:
	    • Mechanical malfunction of shift solenoid valve-B.
	    • Malfunction of valve body assembly.
	    • Fluid passage clogged or leaking.
	    • Mechanical malfunction of automatic transmission (clutch, brake or gear etc.).

DTC: P0771
2T, MIL: Yes
Year: 2009, 2010
Model: Grand Vitara
Engine: 3.2L V6 VIN 1
Transmission: All
	Shift Solenoid "E" Performance or Stuck OFF:
	4th gear ratio is detected although TCM command is for 5th gear while vehicle running at 2 km/h (1.5 mile/h) or more in "D" range after engine being warmed up.
	Possible Causes:
	    • Mechanical malfunction of shift solenoid valve-E.
	    • Malfunction of valve body assembly.
	    • Fluid passage clogged or leaking.
	    • Mechanical malfunction of automatic transmission (clutch, brake or gear etc.).

DTC: P0772
2T, MIL: Yes
Year: 2009, 2010
Model: Grand Vitara
Engine: 3.2L V6 VIN 1
Transmission: All
	Shift Solenoid "E" Performance or Stuck ON:
	TCM detected the turn rise of input speed over the specified value when upshifting 1st to 2nd gear.
	Possible Causes:
	    • Mechanical malfunction of shift solenoid valve-E.
	    • Malfunction of valve body assembly.
	    • Fluid passage clogged or leaking.
	    • Mechanical malfunction of automatic transmission (clutch, brake or gear etc.).

DTC	Trouble Code Title, Conditions & Possible Causes

DTC: P0781
2T, MIL: Yes
Year: 2009, 2010
Model: Grand Vitara
Engine: 3.2L V6 VIN 1
Transmission: All
	1-2 Shift:
	When one of the following condition was detected while vehicle running at 2 km/h (1.5 mile/h) or more in "D" range after engine being warmed up.
	    • 1st gear ratio is detected although TCM command is for 2nd gear and 3rd gear is detected although TCM command is for 4th gear.
	Or
	    • 1st gear ratio is detected although TCM command is for 2nd gear and Neutral gear is detected although TCM command is for 5th gear.
	Possible Causes:
	• Mechanical malfunction of 1-2 shift valve.
	• Malfunction of valve body assembly.
	• Mechanical malfunction of automatic transmission (clutch, brake or gear etc.).

DTC: P0797 **2T, MIL: Yes** **Year:** 2009, 2010 **Model:** Grand Vitara **Engine:** 3.2L V6 VIN 1 **Transmission:** All	**Pressure Control Solenoid "C" Stuck ON:** 4th gear ratio is detected although TCM command is for 4th gear and Neutral gear is detected although TCM command is for 5th gear while vehicle running at 2 km/h (1.5 mile/h) or more in "D" range after engine being warmed up. **Possible Causes:** • Mechanical malfunction of pressure control solenoid valve-C. • Malfunction of valve body assembly. • Fluid passage clogged or leaking. • Mechanical malfunction of automatic transmission.
DTC: P0850 **2T, MIL: Yes** **Year:** 2009, 2010 **Model:** Equator **Engine:** 2.5L L4, 4.0L V6 **Transmission:** All	**PNP Switch:** The signal of the park/neutral position (PNP) does not change during driving after the engine is started. **Possible Causes:** • Harness or connectors • [The park/neutral position (PNP) signal circuit is open or shorted.] • Park/neutral position (PNP) switch (M/T) • Transmission range switch (A/T) • Combination meter • TCM (A/T)
DTC: P0856 **1T** **Year:** 2009 **Model:** XL-7 **Engine:** 3.6L V6 **Transmission:** All	**Traction Control Torque Request Circuit:** The ECM detects an invalid signal from the EBCM. **Possible Causes:** • EBCM • GMLAN serial data circuit • ECM
DTC: P0961 **1T, MIL: Yes** **Year:** 2009, 2010 **Model:** Grand Vitara **Engine:** 2.4L L4 VIN 9, 3.2L V6 VIN 1 **Transmission:** All	**Pressure Control Solenoid "A" Control Circuit Range / Performance:** Difference between target current of control solenoid circuit and monitor current of control solenoid circuit is more than specified. **Possible Causes:** • Malfunction of pressure control solenoid valve or its circuit • TCM
DTC: P0962 **1T, MIL: Yes** **Year:** 2009, 2010 **Model:** Grand Vitara **Engine:** 2.4L L4 VIN 9, 3.2L V6 VIN 1 **Transmission:** All	**Pressure Control Solenoid "A" Control Circuit Low:** Pressure control solenoid valve output voltage is too low comparing with TCM command value. **Possible Causes:** • Pressure control solenoid valve circuit open or shorted to ground. • Malfunction of pressure control solenoid valve. • TCM
DTC: P0963 **1T, MIL: Yes** **Year:** 2009, 2010 **Model:** Grand Vitara **Engine:** 2.4L L4 VIN 9, 3.2L V6 VIN 1 **Transmission:** All	**Pressure Control Solenoid "A" Control Circuit High:** Pressure control solenoid valve output voltage is too high comparing with TCM command value. **Possible Causes:** • Pressure control solenoid valve circuit shorted to power circuit. • Pressure control solenoid valve malfunction. • TCM

DTC	Trouble Code Title, Conditions & Possible Causes
DTC: P0965 **1T, MIL: Yes** **Year:** 2009, 2010 **Model:** Grand Vitara **Engine:** 3.2L V6 VIN 1 **Transmission:** All	**Pressure Control Solenoid "B" Control Circuit Range / Performance:** Difference between target current of control solenoid circuit and monitor current of control solenoid circuit is more than specified. **Possible Causes:** • Malfunction of pressure control solenoid valve or its circuit • TCM
DTC: P0966 **1T, MIL: Yes** **Year:** 2009, 2010 **Model:** Grand Vitara **Engine:** 2.4L L4 VIN 9, 3.2L V6 VIN 1 **Transmission:** All	**Pressure Control Solenoid "B" Control Circuit Low:** Pressure control solenoid valve output voltage is too low comparing with TCM command value. **Possible Causes:** • Pressure control solenoid valve circuit open or shorted to ground. • Malfunction of pressure control solenoid valve. • TCM
DTC: P0967 **1T, MIL: Yes** **Year:** 2009, 2010 **Model:** Grand Vitara **Engine:** 2.4L L4 VIN 9, 3.2L V6 VIN 1 **Transmission:** All	**Pressure Control Solenoid "B" Control Circuit High:** Pressure control solenoid valve output voltage is too high comparing with TCM command value. **Possible Causes:** • Pressure control solenoid valve circuit shorted to power circuit. • Pressure control solenoid valve malfunction. • TCM
DTC: P0969 **1T, MIL: Yes** **Year:** 2009, 2010 **Model:** Grand Vitara **Engine:** 3.2L V6 VIN 1 **Transmission:** All	**Pressure Control Solenoid "C" Control Circuit Range / Performance:** Difference between target current of control solenoid circuit and monitor current of control solenoid circuit is more than specified. **Possible Causes:** • Malfunction of pressure control solenoid valve or its circuit • TCM
DTC: P0970 **1T, MIL: Yes** **Year:** 2009, 2010 **Model:** Grand Vitara **Engine:** 3.2L V6 VIN 1 **Transmission:** All	**Pressure Control Solenoid "C" Control Circuit Low:** Pressure control solenoid valve output voltage is too low comparing with TCM command value. **Possible Causes:** • Pressure control solenoid valve circuit open or shorted to ground. • Malfunction of pressure control solenoid valve. • TCM
DTC: P0971 **1T, MIL: Yes** **Year:** 2009, 2010 **Model:** Grand Vitara **Engine:** 3.2L V6 VIN 1 **Transmission:** All	**Pressure Control Solenoid "C" Control Circuit High:** Pressure control solenoid valve output voltage is too high comparing with TCM command value. **Possible Causes:** • Pressure control solenoid valve circuit shorted to power circuit. • Pressure control solenoid valve malfunction. • TCM
DTC: P0973 **1T, MIL: Yes** **Year:** 2009, 2010 **Model:** Grand Vitara **Engine:** 3.2L V6 VIN 1 **Transmission:** All	**Shift Solenoid "A" Control Circuit Low:** Voltage of shift solenoid valve TCM terminal is low although TCM is commanding shift solenoid to turn ON. **Possible Causes:** • Shift solenoid valve circuit shorted to ground. • Malfunction of shift solenoid valve. • TCM
DTC: P0974 **1T, MIL: Yes** **Year:** 2009, 2010 **Model:** Grand Vitara **Engine:** 3.2L V6 VIN 1 **Transmission:** All	**Shift Solenoid "A" Control Circuit High:** Voltage of shift solenoid valve TCM terminal is high although TCM is commanding shift solenoid to turn OFF. **Possible Causes:** • Shift solenoid valve circuit open or shorted to power circuit. • Malfunction of shift solenoid valve. • TCM
DTC: P0976 **1T, MIL: Yes** **Year:** 2009, 2010 **Model:** Grand Vitara **Engine:** 3.2L V6 VIN 1 **Transmission:** All	**Shift Solenoid "B" Control Circuit Low:** • Shift solenoid valve circuit shorted to ground. • Malfunction of shift solenoid valve. • TCM **Possible Causes:** • Shift solenoid valve circuit shorted to ground. • Malfunction of shift solenoid valve. • TCM

DTC	Trouble Code Title, Conditions & Possible Causes
DTC: P0977 **1T, MIL:** Yes **Year:** 2009, 2010 **Model:** Grand Vitara **Engine:** 3.2L V6 VIN 1 **Transmission:** All	**Shift Solenoid "B" Control Circuit High:** Voltage of shift solenoid valve TCM terminal is high although TCM is commanding shift solenoid to turn OFF. **Possible Causes:** • Shift solenoid valve circuit open or shorted to power circuit. • Malfunction of shift solenoid valve. • TCM
DTC: P0985 **1T, MIL:** Yes **Year:** 2009, 2010 **Model:** Grand Vitara **Engine:** 3.2L V6 VIN 1 **Transmission:** All	**Shift Solenoid "E" Control Circuit Low:** Voltage of shift solenoid valve TCM terminal is low although TCM is commanding shift solenoid to turn ON. **Possible Causes:** • Shift solenoid valve circuit shorted to ground. • Malfunction of shift solenoid valve. • TCM
DTC: P0986 **1T, MIL:** Yes **Year:** 2009, 2010 **Model:** Grand Vitara **Engine:** 3.2L V6 VIN 1 **Transmission:** All	**Shift Solenoid "E" Control Circuit High:** Voltage of shift solenoid valve TCM terminal is high although TCM is commanding shift solenoid to turn OFF. **Possible Causes:** • Shift solenoid valve circuit open or shorted to power circuit. • Malfunction of shift solenoid valve. • TCM

OBD II Trouble Code List (P1XXX Codes)

DTC	Trouble Code Title, Conditions & Possible Causes
DTC: P1011 **2T** **Year:** 2009, 2010 **Model:** Grand Vitara **Engine:** 3.2L V6 VIN 1 **Transmission:** All	**Cams Locking Pin Intake Bank 1 Low:** CMP actuator (bank 1 / intake side) is not in parked position (the most retarded position) at engine starting. **Possible Causes:** • CMP actuator • ECM
DTC: P1011 **1T** **Year:** 2009 **Model:** XL-7 **Engine:** 3.6L V6 **Transmission:** All	**Intake Camshaft Position (CMP) Actuator Park Position Bank 1:** The ECM detects that average of the actual camshaft angle measurement is greater than 10 degrees of the parked position angle at engine start-up. **Possible Causes:** • Oil level and/or condition • CMP actuator
DTC: P1012 **1T** **Year:** 2009 **Model:** XL-7 **Engine:** 3.6L V6 **Transmission:** All	**Exhaust Camshaft Position (CMP) Actuator Park Position Bank 1:** The ECM detects that average of the actual camshaft angle measurement is greater than 10 degrees of the parked position angle at engine start-up. **Possible Causes:** • Oil level and/or condition • CMP actuator
DTC: P1012 **2T** **Year:** 2009, 2010 **Model:** Grand Vitara **Engine:** 3.2L V6 VIN 1 **Transmission:** All	**Engine Position System Performance Bank 2:** CMP actuator (bank 1 / exhaust side) is not in parked position (the most advanced position) at engine starting. **Possible Causes:** • CMP actuator • ECM
DTC: P1013 **1T** **Year:** 2009 **Model:** XL-7 **Engine:** 3.6L V6 **Transmission:** All	**Intake Camshaft Position (CMP) Actuator Park Position Bank 2:** The ECM detects that average of the actual camshaft angle measurement is greater than 10 degrees of the parked position angle at engine start-up. **Possible Causes:** • Oil level and/or condition • CMP actuator
DTC: P1013 **2T** **Year:** 2009, 2010 **Model:** Grand Vitara **Engine:** 3.2L V6 VIN 1 **Transmission:** All	**Cams Locking Pin Intake Bank 2 Low:** CMP actuator (bank 2 / intake side) is not in parked position (the most retarded position) at engine starting. **Possible Causes:** • CMP actuator • ECM

DTC	Trouble Code Title, Conditions & Possible Causes
DTC: P1014 **1T** **Year:** 2009 **Model:** XL-7 **Engine:** 3.6L V6 **Transmission:** All	**Exhaust Camshaft Position (CMP) Actuator Park Position Bank 2:** The ECM detects that average of the actual camshaft angle measurement is greater than 10 degrees of the parked position angle at engine start-up. **Possible Causes:** • Oil level and/or condition • CMP actuator
DTC: P1014 **2T** **Year:** 2009, 2010 **Model:** Grand Vitara **Engine:** 3.2L V6 VIN 1 **Transmission:** All	**Cams Locking Pin Exhaust Bank 2 Low:** CMP actuator (bank 2 / exhaust side) is not in parked position (the most advanced position) at engine starting. **Possible Causes:** • CMP actuator • ECM
DTC: P1148 **1T, MIL: Yes** **Year:** 2009, 2010 **Model:** Equator **Engine:** 2.5L L4, 4.0L V6 **Transmission:** All	**Closed loop control function:** The closed loop control function does not operate even when vehicle is being driven in the specified condition. **Possible Causes:** • Harness or connectors • [The air fuel ratio (A/F) sensor 1 circuit is open or shorted.] • Air fuel ratio (A/F) sensor 1 • Air fuel ratio (A/F) sensor 1 heater
DTC: P1168 **1T, MIL: Yes** **Year:** 2009, 2010 **Model:** Equator **Engine:** 4.0L V6 **Transmission:** All	**Closed loop control function (Bank 2) :** The closed loop control function for bank 2 does not operate even when vehicle is being driven in the specified condition. **Possible Causes:** • Harness or connectors • [The air fuel ratio (A/F) sensor 1 circuit is open or shorted.] • Air fuel ratio (A/F) sensor 1 • Air fuel ratio (A/F) sensor 1 heater
DTC: P1211 **2T** **Year:** 2009, 2010 **Model:** Equator **Engine:** 4.0L V6 **Transmission:** All	**TCS control unit:** ECM receives malfunction information from "ABS actuator and electric unit (Control unit)". **Possible Causes:** • ABS actuator and electric unit (control unit) • TCS related parts
DTC: P1212 **2T** **Year:** 2009, 2010 **Model:** Equator **Engine:** 4.0L V6 **Transmission:** All	**TCS communication line:** ECM cannot receive the information from "ABS actuator and electric unit (control unit)". **Possible Causes:** • Harness or connectors • (The CAN communication line is open or shorted.) • ABS actuator and electric unit (control unit) • Dead (Weak) battery
DTC: P1217 **1T, MIL: Yes** **Year:** 2009, 2010 **Model:** Equator **Engine:** 2.5L L4, 4.0L V6 **Transmission:** All	**Engine Over Temperature:** CAUTION: When a malfunction is indicated, always replace the coolant. Also, replace the engine oil. • Cooling fan does not operate properly (Overheat). • Cooling fan system does not operate properly (Overheat). • Engine coolant was not added to the system using the proper filling method. • Engine coolant is not within the specified range. **Possible Causes:** • Cooling fan (crankshaft driven) • Radiator hose • Radiator • Radiator cap • Water pump • Thermostat • Engine coolant temperature sensor
DTC: P1225 **2T, MIL: Yes** **Year:** 2009, 2010 **Model:** Equator **Engine:** 2.5L L4, 4.0L V6 **Transmission:** All	**Closed throttle position learning performance problem:** Closed throttle position learning value is excessively low. **Possible Causes:** • Electric throttle control actuator • (TP sensor 1 and 2)

DTC	Trouble Code Title, Conditions & Possible Causes
DTC: P1226 **2T, MIL: Yes** **Year:** 2009, 2010 **Model:** Equator **Engine:** 2.5L L4, 4.0L V6 **Transmission:** All	**Closed throttle position learning performance problem:** Closed throttle position learning is not performed successfully, repeatedly. **Possible Causes:** • Electric throttle control actuator • (TP sensor 1 and 2)
DTC: P1441 **2T, MIL: Yes** **Year:** 2009, 2010 **Model:** Equator **Engine:** 2.5L L4 **Transmission:** All	**Cold start emission reduction strategy monitoring:** ECM does not control ignition timing and engine idle speed properly when engine is started with prewarming up condition. **Possible Causes:** • Lack of intake air volume • Fuel injection system • ECM
DTC: P1550 **2T** **Year:** 2009, 2010 **Model:** Equator **Engine:** 2.5L L4, 4.0L V6 **Transmission:** All	**Battery current sensor circuit range/performance:** The output voltage of the battery current sensor remains within the specified range while engine is running. **Possible Causes:** • Harness or connectors • (The sensor circuit is open or shorted.) • Battery current sensor
DTC: P1551 **1T, MIL: Yes** **Year:** 2009 **Model:** XL-7 **Engine:** 3.6L V6 **Transmission:** All	**Throttle Valve Rest Position Not Reached During Learn:** • The ECM detects the TP sensor angle is less than 10 percent or greater than 40 percent when the throttle actuator control motor is deactivated. **Possible Causes:** • Throttle body
DTC: P1551 **2T** **Year:** 2009, 2010 **Model:** Equator **Engine:** 2.5L L4, 4.0L V6 **Transmission:** All	**Battery current sensor circuit low input:** An excessively low voltage from the sensor is sent to ECM. **Possible Causes:** • Harness or connectors • (The sensor circuit is open or shorted.) • Battery current sensor
DTC: P1552 **2T** **Year:** 2009, 2010 **Model:** Equator **Engine:** 2.5L L4, 4.0L V6 **Transmission:** All	**Battery Current Sensor High Input:** An excessively high voltage from the sensor is sent to ECM. **Possible Causes:** • Harness or connectors • (The sensor circuit is open or shorted.) • Battery current sensor
DTC: P1553 **2T** **Year:** 2009, 2010 **Model:** Equator **Engine:** 2.5L L4, 4.0L V6 **Transmission:** All	**Battery current sensor performance:** The signal voltage transmitted from the sensor to ECM is higher than the amount of the maximum power generation. **Possible Causes:** • Harness or connectors • (The sensor circuit is open or shorted.) • Battery current sensor
DTC: P1554 **2T** **Year:** 2009, 2010 **Model:** Equator **Engine:** 2.5L L4, 4.0L V6 **Transmission:** All	**Battery current sensor performance:** The output voltage of the battery current sensor is lower than the specified value while the battery voltage is high enough. **Possible Causes:** • Harness or connectors • (The sensor circuit is open or shorted.) • Battery current sensor
DTC: P1564 **1T** **Year:** 2009, 2010 **Model:** Equator **Engine:** 2.5L L4, 4.0L V6 **Transmission:** All	**ASCD steering switch:** • An excessively high voltage signal from the ASCD steering switch is sent to ECM. • ECM detects that input signal from the ASCD steering switch is out of the specified range. • ECM detects that the ASCD steering switch is stuck ON. **Possible Causes:** • Harness or connectors • (The switch circuit is open or shorted.) • ASCD steering switch • ECM

DTC	Trouble Code Title, Conditions & Possible Causes
DTC: P1572 **1T** **Year:** 2009, 2010 **Model:** Equator **Engine:** 2.5L L4, 4.0L V6 **Transmission:** All	**ASCD Brake Switch:** A) • When the vehicle speed is above 30km/h (19 MPH), ON signals from the stop lamp switch and the ASCD brake switch are sent to ECM at the same time. B) • ASCD brake switch signal is not sent to ECM for extremely long time while the vehicle is driving **Possible Causes:** • Harness or connectors • (The stop lamp switch circuit is shorted.) • Harness or connectors • (The ASCD brake switch circuit is shorted.) • Harness or connectors • (The ASCD clutch switch circuit is shorted.) (M/T models) • Stop lamp switch • ASCD brake switch • ASCD clutch switch (M/T models) • Incorrect stop lamp switch installation • Incorrect ASCD brake switch installation • Incorrect ASCD clutch switch installation • (M/T models) • ECM
DTC: P1572 **1T, MIL: Yes** **Year:** 2009, 2010 **Model:** Equator **Engine:** 2.5L L4 **Transmission:** All	**Input Clutch Solenoid:** — When TCM detects an improper voltage drop when it tries to operate the solenoid valve. — When TCM detects as irregular by comparing target value with monitor value. **Possible Causes:** • Harness or connectors • (The solenoid circuit is open or shorted.) • Input clutch solenoid valve
DTC: P1574 **1T** **Year:** 2009, 2010 **Model:** Equator **Engine:** 2.5L L4, 4.0L V6 **Transmission:** All	**ASCD Vehicle Speed Sensor:** ECM detects a difference between two vehicle speed signals is out of the specified range. **Possible Causes:** • Harness or connectors • (The CAN communication line is open or shorted.) • Harness or connectors • (The combination meter circuit is open or shorted.) • Combination meter • Wheel sensor • ABS actuator and electric unit (control unit) • TCM (A/T models) • ECM
DTC: P1610 **2T** **Year:** 2009, 2010 **Model:** Equator **Engine:** 2.5L L4, 4.0L V6 **Transmission:** All	**Lock Mode:** When the starting operation is carried out five or more times consecutively under the following conditions. • Unregistered mechanical key • BCM or ECM's malfunctioning. **Possible Causes:** • -
DTC: P1611 **2T** **Year:** 2009, 2010 **Model:** Equator **Engine:** 2.5L L4, 4.0L V6 **Transmission:** All	**ID Discord, IMMU-ECM:** The ID verification results between BCM and ECM are NG. The registration is necessary. **Possible Causes:** • BCM • ECM
DTC: P1612 **2T** **Year:** 2009, 2010 **Model:** Equator **Engine:** 2.5L L4, 4.0L V6 **Transmission:** All	**Chain of ECM-IMMU:** Inactive communication between ECM and BCM **Possible Causes:** • Harness or connectors • (The CAN communication line is open or short) • BCM • ECM

DTC	Trouble Code Title, Conditions & Possible Causes

DTC: P1614
2T
Year: 2009, 2010
Model: Equator
Engine: 2.5L L4, 4.0L V6
Transmission: All
	Immobilizer System Antenna Amp.:
	   • Inactive communication between immobilizer system antenna amp. and BCM.
	   • Ignition key is malfunctioning.
	Possible Causes:
	   • Harness or connectors
	   • (The immobilizer system antenna amp. circuit is open or shorted)
	   • Ignition key
	   • Immobilizer system antenna amp.
	   • BCM

DTC: P1614 **1T, PATS: Yes** **Year:** 2009, 2010 **Model:** Grand Vitara, SX4 **Engine:** 2.0L L4, 2.0L L4 VIN 5, 2.4L L4 VIN 0, 3.2L V6 VIN 1 **Transmission:** All	**Transponder response error:** Transponder code in the transponder built in the ignition key cannot be read through ICM. **Possible Causes:** • -

DTC: P1615
1T, PATS: Yes
Year: 2009, 2010
Model: Grand Vitara, SX4
Engine: 2.0L L4, 2.0L L4 VIN 5, 2.4L L4 VIN 0, 3.2L V6 VIN 1
Transmission: All
	Steering lock unit communication error (for vehicle with keyless start system):
	   • While registering the transponder code in the transponder built in the ignition key in ECM, the keyless start control module sent a signal to ECM indicating that the remote controller ID code could not be registered.
	   • The remote controller ID code could not be registered in the keyless start control module or ECM. And, the registration procedure of the transponder code in the transponder built in the ignition key was terminated forcibly.
	Possible Causes:
	   • -

DTC: P1615 **2T** **Year:** 2009, 2010 **Model:** Equator **Engine:** 2.5L L4, 4.0L V6 **Transmission:** All	**Difference of Key:** The ID verification results between BCM and mechanical key are NG. The registration is necessary. **Possible Causes:** • Mechanical key
DTC: P1616 **1T, PATS: Yes** **Year:** 2009, 2010 **Model:** Grand Vitara, SX4 **Engine:** 2.0L L4, 2.0L L4 VIN 5, 2.4L L4 VIN 0, 3.2L V6 VIN 1 **Transmission:** All	**Unregistered keyless start control module (for vehicle with keyless start system):** ECM detects different ID codes registered in ECM and keyless start system. **Possible Causes:** • -
DTC: P1618 **1T, PATS: Yes** **Year:** 2009, 2010 **Model:** Grand Vitara, SX4 **Engine:** 2.0L L4, 2.0L L4 VIN 5, 2.4L L4 VIN 0, 3.2L V6 VIN 1 **Transmission:** All	**Keyless start control module CAN communication error (for vehicle with keyless start system):** Reception error of communication data for keyless start control module is detected for longer than specified time continuously. **Possible Causes:** • -
DTC: P1621 **1T, PATS: Yes** **Year:** 2009, 2010 **Model:** Grand Vitara, SX4 **Engine:** 2.0L L4, 2.0L L4 VIN 5, 2.4L L4 VIN 0, 3.2L V6 VIN 1 **Transmission:** All	**Immobilizer communication line error:** Communication error between ICM and ECM is detected by ECM. **Possible Causes:** • -
DTC: P1622 **1T, PATS: Yes** **Year:** 2009, 2010 **Model:** Grand Vitara, SX4 **Engine:** 2.0L L4, 2.0L L4 VIN 5, 2.4L L4 VIN 0, 3.2L V6 VIN 1 **Transmission:** All	**EEPROM reading/writing error:** EEPROM in ECM is corrupted. **Possible Causes:** • -

DTC	Trouble Code Title, Conditions & Possible Causes
DTC: P1623 **1T, PATS: Yes** **Year:** 2009, 2010 **Model:** Grand Vitara, SX4 **Engine:** 2.0L L4, 2.0L L4 VIN 5, 2.4L L4 VIN 0, 3.2L V6 VIN 1 **Transmission:** All	**Unregistered transponder:** Transponder code in the transponder built in the ignition key is invalid. **Possible Causes:** • -
DTC: P1625 **1T, PATS: Yes** **Year:** 2009, 2010 **Model:** Grand Vitara, SX4 **Engine:** 2.0L L4, 2.0L L4 VIN 5, 2.4L L4 VIN 0, 3.2L V6 VIN 1 **Transmission:** All	**Immobilizer antenna error:** ICM is faulty. **Possible Causes:** • -
DTC: P1636 **1T** **Year:** 2009, 2010 **Model:** Grand Vitara, SX4 **Engine:** 2.0L L4, 2.0L L4 VIN 5, 2.4L L4 VIN 0, 3.2L V6 VIN 1 **Transmission:** All	**Immobilizer information registration failure:** Communication error between ECM and BCM is detected by ECM. **Possible Causes:** • -
DTC: P1638 **1T** **Year:** 2009, 2010 **Model:** Grand Vitara, SX4 **Engine:** 2.0L L4, 2.0L L4 VIN 5, 2.4L L4 VIN 0, 3.2L V6 VIN 1 **Transmission:** All	**Immobilizer information mismatched:** • The immobilizer control system information in ECM and the one in BCM does not match. • The registration of the immobilizer control system information in ECM is failed. **Possible Causes:** • -
DTC: P1642 **T** **Year:** 2010 **Model:** Kizashi **Engine:** 2.4L L4 VIN 9 **Transmission:** All	**Immobilizer Communication Line Error:** Communication error with keyless start control module. **Possible Causes:** • CAN communication circuit / connector • Control module connected with CAN communication line • ECM • Keyless start control module
DTC: P1645 **1T, PATS: Yes** **Year:** 2009, 2010 **Model:** Grand Vitara, SX4 **Engine:** 2.0L L4, 2.0L L4 VIN 5, 2.4L L4 VIN 0, 3.2L V6 VIN 1 **Transmission:** All	**ID code communication error:** No response from ID controller. **Possible Causes:** • -
DTC: P1646 **1T, PATS: Yes** **Year:** 2009, 2010 **Model:** Grand Vitara, SX4 **Engine:** 2.0L L4, 2.0L L4 VIN 5, 2.4L L4 VIN 0, 2.4L L4 VIN 9, 3.2L V6 VIN 1 **Transmission:** All	**ID code incorrectness:** • Different ID codes registered in ECM and ID controller. • Communication error of ECM. • Transponder code is not registered in ECM. **Possible Causes:** • -
DTC: P167D **1T, MIL: Yes** **Year:** 2009 **Model:** XL-7 **Engine:** 3.6L V6 **Transmission:** All	**Control Module Ignition Coil Internal Circuit:** The ECM detects a condition with the integrated circuits of the ignition driver module for greater than 4 seconds, or for a cumulative time of 30 seconds. **Possible Causes:** • ECM • ignition driver module

DTC	Trouble Code Title, Conditions & Possible Causes
DTC: P1702 **1T, MIL:** Yes **Year:** 2009, 2010 **Model:** Grand Vitara **Engine:** 3.2L V6 VIN 1 **Transmission:** All	**Internal Control Module Memory Check Sum Error:** Calculation of current data stored in TCM is not correct comparing with pre-stored checking data in TCM **Possible Causes:** • -
DTC: P1706 **1T, MIL:** Yes **Year:** 2009, 2010 **Model:** Grand Vitara **Engine:** 2.4L L4 VIN 9, 3.2L V6 VIN 1 **Transmission:** All	**Torque Request Communication Error from TCM:** ECM detects abnormality of CAN communication data which TCM transmits. **Possible Causes:** • TCM • ECM
DTC: P1715 **2T** **Year:** 2009, 2010 **Model:** Equator **Engine:** 2.5L L4, 4.0L V6 **Transmission:** All	**Input speed sensor (TCM output):** Input speed sensor signal is different from the theoretical value calculated by ECM from output speed sensor signal and engine rpm signal. **Possible Causes:** • Harness or connectors • (The CAN communication line is open or shorted) • Harness or connectors • (Input speed sensor circuit is open or shorted) • TCM
DTC: P1723 **1T, MIL:** Yes **Year:** 2009, 2010 **Model:** Grand Vitara **Engine:** 3.2L V6 VIN 1 **Transmission:** All	**Range Select Switch Malfunction:** "4" position switch signal is inputted out of specified value. **Possible Causes:** • "4" position switch or its circuit malfunction • TCM
DTC: P1730 **1T, MIL:** Yes **Year:** 2009, 2010 **Model:** Equator **Engine:** 2.5L L4, 4.0L V6 **Transmission:** All	**Interlock:** • Diagnostic trouble code is detected when TCM does not receive the proper voltage signal from the sensor and switch. • TCM monitors and compares gear position and conditions of each ATF pressure switch when gear is steady. **Possible Causes:** • Harness or connectors • (The solenoid and switch circuit is open or shorted.) • Low coast brake solenoid valve • ATF pressure switch 2
DTC: P1752 **1T, MIL:** Yes **Year:** 2009, 2010 **Model:** Equator **Engine:** 2.5L L4, 4.0L V6 **Transmission:** All	**Input Clutch Solenoid:** — When TCM detects an improper voltage drop when it tries to operate the solenoid valve. — When TCM detects as irregular by comparing target value with monitor value. **Possible Causes:** • Harness or connectors • (The solenoid circuit is open or shorted.) • Input clutch solenoid valve
DTC: P1754 **1T, MIL:** Yes **Year:** 2009, 2010 **Model:** Equator **Engine:** 2.5L L4 **Transmission:** All	**Input Clutch Solenoid:** — When TCM detects an improper voltage drop when it tries to operate the solenoid valve. — When TCM detects as irregular by comparing target value with monitor value. **Possible Causes:** • Harness or connectors • (The solenoid circuit is open or shorted.) • Input clutch solenoid valve
DTC: P1757 **1T, MIL:** Yes **Year:** 2009, 2010 **Model:** Equator **Engine:** 2.5L L4, 4.0L V6 **Transmission:** All	**Front Brake Solenoid:** — When TCM detects an improper voltage drop when it tries to operate the solenoid valve. — When TCM detects as irregular by comparing target value with monitor value. **Possible Causes:** • Harness or connectors • (The solenoid circuit is open or shorted.) • Front brake solenoid valve

DTC	Trouble Code Title, Conditions & Possible Causes

DTC: P1759
1T, MIL: Yes
Year: 2009, 2010
Model: Equator
Engine: 2.5L L4
Transmission: All
	Front Brake Solenoid:
	— When TCM detects an improper voltage drop when it tries to operate the solenoid valve.
	— When TCM detects as irregular by comparing target value with monitor value.
	Possible Causes:
	  • Harness or connectors
	  • (The solenoid circuit is open or shorted.)
	  • Front brake solenoid valve

DTC: P1762
1T, MIL: Yes
Year: 2009, 2010
Model: Equator
Engine: 2.5L L4, 4.0L V6
Transmission: All
	Direct Clutch Solenoid:
	— When TCM detects an improper voltage drop when it tries to operate the solenoid valve.
	— When TCM detects as irregular by comparing target value with monitor value.
	Possible Causes:
	  • Harness or connectors
	  • (The solenoid circuit is open or shorted.)
	  • Direct clutch solenoid valve

DTC: P1767
1T, MIL: Yes
Year: 2009, 2010
Model: Equator
Engine: 2.5L L4, 4.0L V6
Transmission: All
	High and Low Reverse Clutch Solenoid:
	— When TCM detects an improper voltage drop when it tries to operate the solenoid valve.
	— When TCM detects as irregular by comparing target value with monitor value.
	Possible Causes:
	  • Harness or connectors
	  • (The solenoid circuit is open or shorted.)
	  • High and low reverse clutch solenoid valve

DTC: P1772
1T, MIL: Yes
Year: 2009, 2010
Model: Equator
Engine: 2.5L L4, 4.0L V6
Transmission: All
	Low Coast Brake Solenoid:
	TCM detects an improper voltage drop when it tries to operate the solenoid valve.
	Possible Causes:
	  • Harness or connectors
	  • (The solenoid circuit is open or shorted.)
	  • Low coast brake solenoid valve

DTC: P1774
1T
Year: 2009, 2010
Model: Equator
Engine: 4.0L V6
Transmission: All
	Low Coast Brake Solenoid:
	— When TCM detects that actual gear ratio is irregular, and relation between gear position and condition of ATF pressure switch 2 is irregular during depressing accelerator pedal. (Other than during shift change)
	— When TCM detects that relation between gear position and condition of ATF pressure switch 2 is irregular during releasing accelerator pedal. (Other than during shift change)
	Possible Causes:
	  • Harness or connectors
	  • (The solenoid and switch circuits are open or shorted.)
	  • Low coast brake solenoid valve
	  • ATF pressure switch 2

DTC: P1800
T
Year: 2009, 2010
Model: Equator
Engine: 4.0L V6
Transmission: All
	VIAS control solenoid valve circuit :
	An excessively low or high voltage signal is sent to ECM through the valve
	Possible Causes:
	  • Harness or connectors
	  • (The solenoid valve circuit is open or shorted.)
	  • VIAS control solenoid valve

DTC: P1805
2T
Year: 2009, 2010
Model: Equator
Engine: 2.5L L4, 4.0L V6
Transmission: All
	Brake Switch:
	A brake switch signal is not sent to ECM for extremely long time while the vehicle is driving.
	Possible Causes:
	  • Harness or connectors
	  • (Stop lamp switch circuit is open or shorted.)
	  • Stop lamp switch

DTC: P1874
2T CCM, MIL: Yes
Year: 2009, 2010
Model: Grand Vitara
Engine: 3.2L V6 VIN 1
Transmission: All
	4WD Low Switch Circuit Malfunction (Shorted):
	Engine started, engine running, and the PCM detected an unexpected "low" voltage condition on the PCM Torque Reduction circuit (5-volt) during the CCM test.
	Possible Causes:
	  • Engine Torque control signal circuit is open
	  • Engine Torque control signal is shorted to ground
	  • 4WD switch is damaged or has failed
	  • TCM or PCM has failed

DTC	Trouble Code Title, Conditions & Possible Causes
DTC: P1875 **1T, MIL: Yes** **Year:** 2009, 2010 **Model:** Grand Vitara **Engine:** 3.2L V6 VIN 1 **Transmission:** All	**4WD Low Switch Circuit Malfunction (Open):** Actual transfer position is 4L although TCM detected low switch is turned OFF with vehicle speed between 29 km/h (18 mile/h) and 88 km/h (55 mile/h). (if equipped with 4L/N switch) **Possible Causes:** • 4L/N switch or its circuit • TCM
DTC: P1878 **20T** **Year:** 2009, 2010 **Model:** Grand Vitara **Engine:** 3.2L V6 VIN 1 **Transmission:** All	**Torque Converter Clutch Shudder:** The acceleration slip control function stops when the variation in the output revolution speed of the specified amplitude and specified cycle is detected within a specified period of time. When the specified variation is not detected after the acceleration slip control stops. **Possible Causes:** • Mismatching ATF • Torque converter clutch malfunction • TCM

OBD II Trouble Code List (P2XXX Codes)

DTC	Trouble Code Title, Conditions & Possible Causes
DTC: P2004 **2T, MIL: Yes** **Year:** 2009, 2010 **Model:** Grand Vitara **Engine:** 2.4L L4 VIN 0 **Transmission:** All	**Intake Manifold Runner Control Stuck Open:** EVAP-Leak-Check Model Opening angle of IMRC valve is higher than 9.7° even if command signal of IMRC valve is fully closed signal for 5 sec. Non-EVAP-Leak-Check Model Opening angle of IMRC valve is higher than 45° even if command signal of IMRC valve is fully closed signal for 5 sec. **Possible Causes:** • IMRC valve motor and/or its circuit • IMRC valve • ECM
DTC: P2006 **2T, MIL: Yes** **Year:** 2009, 2010 **Model:** Grand Vitara **Engine:** 2.4L L4 VIN 0 **Transmission:** All	**Intake Manifold Runner Control Stuck Closed:** EVAP-Leak-Check Model Opening angle of IMRC valve is higher than 9.7° even if command signal of IMRC valve is fully closed signal for 5 sec. Non-EVAP-Leak-Check Model Opening angle of IMRC valve is higher than 45° even if command signal of IMRC valve is fully closed signal for 5 sec. **Possible Causes:** • IMRC valve motor and/or its circuit • IMRC valve • ECM
DTC: P2008 **2T, MIL: Yes** **Year:** 2009, 2010 **Model:** Grand Vitara **Engine:** 2.4L L4 VIN 0 **Transmission:** All	**Intake Manifold Runner Control Circuit/Open:** • Monitor voltage of IMRC valve motor drive circuit is lower than specified value for 5 times. • Electric current of IMRC valve motor drive circuit is higher than 6.5 A for 5 times. **Possible Causes:** • IMRC valve motor and/or its circuit • ECM
DTC: P2016 **2T, MIL: Yes** **Year:** 2009, 2010 **Model:** Grand Vitara **Engine:** 2.4L L4 VIN 0 **Transmission:** All	**Intake Manifold Runner Position Sensor/Switch Circuit Low:** Output voltage of IMRC valve position sensor signal circuit is lower than 0.22 V. **Possible Causes:** • IMRC valve position sensor and/or its circuit • ECM
DTC: P2017 **2T, MIL: Yes** **Year:** 2009, 2010 **Model:** Grand Vitara **Engine:** 2.4L L4 VIN 0 **Transmission:** All	**Intake Manifold Runner Position Sensor/Switch Circuit High:** Output voltage of IMRC valve position sensor signal circuit is higher than 4.3 V. **Possible Causes:** • IMRC valve position sensor and/or its circuit • ECM

DTC	Trouble Code Title, Conditions & Possible Causes
DTC: P2088 **2T, MIL: Yes** **Year:** 2009 **Model:** XL-7 **Engine:** 3.6L V6 **Transmission:** All	**Intake Camshaft Position (CMP) Actuator Solenoid Control Circuit Low Voltage Bank 1:** The ECM detects a short to ground in the CMP actuator solenoid circuits for greater than 1 second or a cumulative of 5 seconds, when the solenoid is commanded OFF. **Possible Causes:** • CMP actuator solenoid and/or circuit • ECM
DTC: P2088 **2T, MIL: Yes** **Year:** 2009, 2010 **Model:** Grand Vitara **Engine:** 3.2L V6 VIN 1 **Transmission:** All	**"A" Camshaft Position Actuator Control Circuit Low Bank 1:** OCV (bank 1 / intake side) control circuit is shorted to ground. **Possible Causes:** • OCV and its circuit • ECM
DTC: P2088 **1T, MIL: Yes** **Year:** 2009 **Model:** XL-7 **Engine:** 3.6L V6 **Transmission:** All	**Intake Camshaft Position (CMP) Actuator Solenoid Control Circuit Low Voltage Bank 1:** The ECM detects a short to ground in the CMP actuator solenoid circuits for greater than 1 second or a cumulative of 5 seconds, when the solenoid is commanded OFF. **Possible Causes:** • CMP actuator solenoid and/or circuit • ECM
DTC: P2089 **1T, MIL: Yes** **Year:** 2009 **Model:** XL-7 **Engine:** 3.6L V6 **Transmission:** All	**Intake Camshaft Position (CMP) Actuator Solenoid Control Circuit High Voltage Bank 1:** The ECM detects a short to voltage in the CMP actuator solenoid circuits for greater than 1 second or a cumulative of 5 seconds, when the solenoid is commanded ON. **Possible Causes:** • CMP actuator solenoid and/or circuit • ECM
DTC: P2089 **2T, MIL: Yes** **Year:** 2009, 2010 **Model:** Grand Vitara **Engine:** 3.2L V6 VIN 1 **Transmission:** All	**"A" Camshaft Position Actuator Control Circuit High Bank 1:** OCV (bank 1 / intake side) control circuit is shorted to power supply circuit. **Possible Causes:** • OCV and its circuit • ECM
DTC: P2089 **2T, MIL: Yes** **Year:** 2009 **Model:** XL-7 **Engine:** 3.6L V6 **Transmission:** All	**Intake Camshaft Position (CMP) Actuator Solenoid Control Circuit High Voltage Bank 1:** The ECM detects a short to voltage in the CMP actuator solenoid circuits for greater than 1 second or a cumulative of 5 seconds, when the solenoid is commanded ON. **Possible Causes:** • CMP actuator solenoid and/or circuit • ECM
DTC: P2090 **1T, MIL: Yes** **Year:** 2009 **Model:** XL-7 **Engine:** 3.6L V6 **Transmission:** All	**Exhaust Camshaft Position (CMP) Actuator Solenoid Control Circuit Low Voltage Bank 1:** The ECM detects a short to ground in the CMP actuator solenoid circuits for greater than 1 second or a cumulative of 5 seconds, when the solenoid is commanded OFF. **Possible Causes:** • CMP actuator solenoid and/or circuit • ECM
DTC: P2090 **2T, MIL: Yes** **Year:** 2009 **Model:** XL-7 **Engine:** 3.6L V6 **Transmission:** All	**Exhaust Camshaft Position (CMP) Actuator Solenoid Control Circuit Low Voltage Bank 1:** The ECM detects a short to ground in the CMP actuator solenoid circuits for greater than 1 second or a cumulative of 5 seconds, when the solenoid is commanded OFF. **Possible Causes:** • CMP actuator solenoid and/or circuit • ECM
DTC: P2090 **2T, MIL: Yes** **Year:** 2009, 2010 **Model:** Grand Vitara **Engine:** 3.2L V6 VIN 1 **Transmission:** All	**"B" Camshaft Position Actuator Control Circuit Low Bank 1:** OCV (bank 1 / exhaust side) control circuit voltage is shorted to ground. **Possible Causes:** • OCV and its circuit • ECM

DTC	Trouble Code Title, Conditions & Possible Causes
DTC: P2091 **2T, MIL:** Yes **Year:** 2009 **Model:** XL-7 **Engine:** 3.6L V6 **Transmission:** All	**Exhaust Camshaft Position (CMP) Actuator Solenoid Control Circuit High Voltage Bank 1:** The ECM detects a short to voltage in the CMP actuator solenoid circuits for greater than 1 second or a cumulative of 5 seconds, when the solenoid is commanded ON. **Possible Causes:** • CMP actuator solenoid and/or circuit • ECM
DTC: P2091 **2T, MIL:** Yes **Year:** 2009, 2010 **Model:** Grand Vitara **Engine:** 3.2L V6 VIN 1 **Transmission:** All	**"B" Camshaft Position Actuator Control Circuit High Bank 1:** OCV (bank 1 / exhaust side) control circuit is shorted to power supply circuit. **Possible Causes:** • OCV and its circuit • ECM
DTC: P2091 **1T, MIL:** Yes **Year:** 2009 **Model:** XL-7 **Engine:** 3.6L V6 **Transmission:** All	**Exhaust Camshaft Position (CMP) Actuator Solenoid Control Circuit High Voltage Bank 1:** The ECM detects a short to voltage in the CMP actuator solenoid circuits for greater than 1 second or a cumulative of 5 seconds, when the solenoid is commanded ON. **Possible Causes:** • CMP actuator solenoid and/or circuit • ECM
DTC: P2092 **1T, MIL:** Yes **Year:** 2009 **Model:** XL-7 **Engine:** 3.6L V6 **Transmission:** All	**Intake Camshaft Position (CMP) Actuator Solenoid Control Circuit Low Voltage Bank 2:** The ECM detects a short to ground in the CMP actuator solenoid circuits for greater than 1 second or a cumulative of 5 seconds, when the solenoid is commanded OFF. **Possible Causes:** • CMP actuator solenoid and/or circuit • ECM
DTC: P2092 **2T, MIL:** Yes **Year:** 2009, 2010 **Model:** Grand Vitara **Engine:** 3.2L V6 VIN 1 **Transmission:** All	**"A" Camshaft Position Actuator Control Circuit Low Bank 2:** OCV (bank 2 / intake side) control circuit is shorted to ground. **Possible Causes:** • OCV and its circuit • ECM
DTC: P2092 **2T, MIL:** Yes **Year:** 2009 **Model:** XL-7 **Engine:** 3.6L V6 **Transmission:** All	**Intake Camshaft Position (CMP) Actuator Solenoid Control Circuit Low Voltage Bank 2:** The ECM detects a short to ground in the CMP actuator solenoid circuits for greater than 1 second or a cumulative of 5 seconds, when the solenoid is commanded OFF. **Possible Causes:** • CMP actuator solenoid and/or circuit • ECM
DTC: P2093 **2T, MIL:** Yes **Year:** 2009, 2010 **Model:** Grand Vitara **Engine:** 3.2L V6 VIN 1 **Transmission:** All	**"A" Camshaft Position Actuator Control Circuit High Bank 2:** OCV (bank 2 / intake side) control circuit is shorted to power supply circuit. **Possible Causes:** • OCV and its circuit • ECM
DTC: P2093 **1T, MIL:** Yes **Year:** 2009 **Model:** XL-7 **Engine:** 3.6L V6 **Transmission:** All	**Intake Camshaft Position (CMP) Actuator Solenoid Control Circuit High Voltage Bank 2:** The ECM detects a short to voltage in the CMP actuator solenoid circuits for greater than 1 second or a cumulative of 5 seconds, when the solenoid is commanded ON. **Possible Causes:** • CMP actuator solenoid and/or circuit • ECM
DTC: P2094 **2T, MIL:** Yes **Year:** 2009, 2010 **Model:** Grand Vitara **Engine:** 3.2L V6 VIN 1 **Transmission:** All	**"B" Camshaft Position Actuator Control Circuit Low Bank 2:** OCV (bank 2 / exhaust side) control circuit is shorted to ground. **Possible Causes:** • OCV and its circuit • ECM

DTC	Trouble Code Title, Conditions & Possible Causes
DTC: P2094 **2T, MIL: Yes** **Year:** 2009 **Model:** XL-7 **Engine:** 3.6L V6 **Transmission:** All	**Exhaust Camshaft Position (CMP) Actuator Solenoid Control Circuit Low Voltage Bank 2:** The ECM detects a short to ground in the CMP actuator solenoid circuits for greater than 1 second or a cumulative of 5 seconds, when the solenoid is commanded OFF. **Possible Causes:** • CMP actuator solenoid and/or circuit • ECM
DTC: P2094 **1T, MIL: Yes** **Year:** 2009 **Model:** XL-7 **Engine:** 3.6L V6 **Transmission:** All	**Exhaust Camshaft Position (CMP) Actuator Solenoid Control Circuit Low Voltage Bank 2:** The ECM detects a short to ground in the CMP actuator solenoid circuits for greater than 1 second or a cumulative of 5 seconds, when the solenoid is commanded OFF. **Possible Causes:** • CMP actuator solenoid and/or circuit • ECM
DTC: P2095 **2T, MIL: Yes** **Year:** 2009 **Model:** XL-7 **Engine:** 3.6L V6 **Transmission:** All	**Exhaust Camshaft Position (CMP) Actuator Solenoid Control Circuit High Voltage Bank 2:** The ECM detects a short to voltage in the CMP actuator solenoid circuits for greater than 1 second or a cumulative of 5 seconds, when the solenoid is commanded ON. **Possible Causes:** • CMP actuator solenoid and/or circuit • ECM
DTC: P2095 **2T, MIL: Yes** **Year:** 2009, 2010 **Model:** Grand Vitara **Engine:** 3.2L V6 VIN 1 **Transmission:** All	**"B" Camshaft Position Actuator Control Circuit High Bank 2:** OCV (bank 2 / exhaust side) control circuit is shorted to power supply circuit. **Possible Causes:** • OCV and its circuit • ECM
DTC: P2095 **1T, MIL: Yes** **Year:** 2009 **Model:** XL-7 **Engine:** 3.6L V6 **Transmission:** All	**Exhaust Camshaft Position (CMP) Actuator Solenoid Control Circuit High Voltage Bank 2:** The ECM detects a short to voltage in the CMP actuator solenoid circuits for greater than 1 second or a cumulative of 5 seconds, when the solenoid is commanded ON. **Possible Causes:** • CMP actuator solenoid and/or circuit • ECM
DTC: P2096 **2T, MIL: Yes** **Year:** 2009, 2010 **Model:** Grand Vitara, SX4, XL-7 **Engine:** 2.0L L4, 2.0L L4 VIN 5, 2.4L L4 VIN 0, 2.4L L4 VIN 9, 3.2L V6 VIN 1, 3.6L V6 **Transmission:** All	**Post Catalyst Fuel Trim System Too Lean Bank 1:** Average of short term fuel trim based on HO2S is lower than -1.95% for 5 sec. **Possible Causes:** • Vacuum leakage • Exhaust gas leakage • Fuel injector malfunction • HO2S and/or its circuit • A/F sensor and/or its circuit • Engine compression • Valve clearance • Intake and exhaust valve • Valve timing • ECM
DTC: P2097 **2T, MIL: Yes** **Year:** 2009, 2010 **Model:** Grand Vitara, SX4, XL-7 **Engine:** 2.0L L4, 2.0L L4 VIN 5, 2.4L L4 VIN 0, 2.4L L4 VIN 9, 3.2L V6 VIN 1, 3.6L V6 **Transmission:** All	**Post Catalyst Fuel Trim System Too Rich Bank 1:** Average of short term fuel trim based on HO2S is higher than 2.20% for 5 sec. **Possible Causes:** • Vacuum leakage • Exhaust gas leakage • Fuel injector malfunction • HO2S and/or its circuit • A/F sensor and/or its circuit • Engine compression • Valve clearance • Intake and exhaust valve • Valve timing • ECM

DTC	Trouble Code Title, Conditions & Possible Causes
DTC: P2098 **2T, MIL: Yes** **Year:** 2009, 2010 **Model:** Grand Vitara, XL-7 **Engine:** 3.2L V6 VIN 1, 3.6L V6 **Transmission:** All	**Post Catalyst Fuel Trim System Too Lean Bank 2:** HO2S-2 (bank 2) fuel trim correction (rich command) is higher than specified value at specified driving condition. **Possible Causes:** • Oil level gauge • Oil filler cap • Engine oil • Coolant • Air intake system • Exhaust system • Ignition system • HO2S-1 and its circuit • HO2S-1 heater and its circuit • HO2S-2 and its circuit • HO2S-2 heater and its circuit • Engine mechanical system • Fuel system • TWC • ECM
DTC: P2099 **2T, MIL: Yes** **Year:** 2009, 2010 **Model:** Grand Vitara, XL-7 **Engine:** 3.2L V6 VIN 1, 3.6L V6 **Transmission:** All	**Post Catalyst Fuel Trim System Too Rich Bank 2:** HO2S-2 (bank 2) fuel trim correction (rich command) is higher than specified value at specified driving condition. **Possible Causes:** • Oil level gauge • Oil filler cap • Engine oil • Coolant • Air intake system • Exhaust system • Ignition system • HO2S-1 and its circuit • HO2S-1 heater and its circuit • HO2S-2 and its circuit • HO2S-2 heater and its circuit • Engine mechanical system • Fuel system • TWC • ECM
DTC: P2100 **1T, MIL: Yes** **Year:** 2009, 2010 **Model:** Equator, Grand Vitara **Engine:** 2.5L L4, 3.2L V6 VIN 1, 4.0L V6 **Transmission:** All	**Throttle control motor relay circuit open:** ECM detects a voltage of power source for throttle control motor is excessively low **Possible Causes:** • Harness or connectors • (Throttle control motor relay circuit is open) • Throttle control motor relay
DTC: P2100 **1T, MIL: Yes** **Year:** 2009 **Model:** XL-7 **Engine:** 3.6L V6 **Transmission:** All	**Throttle Actuator Control (TAC) Motor Control Circuit:** The ECM detects the output circuit for the TAC motor is open, shorted to ground, or shorted to a voltage. The condition exists, then a 5 second delay for MIL ON. **Possible Causes:** • TAC motor and/or circuit • Throttle body • ECM

DTC	Trouble Code Title, Conditions & Possible Causes

DTC: P2101
1T, MIL: Yes
Year: 2009, 2010
Model: Grand Vitara, SX4
Engine: 2.0L L4, 2.0L L4 VIN 5, 2.4L L4 VIN 0, 2.4L L4 VIN 9, 3.2L V6 VIN 1
Transmission: All
	Throttle Motor Control Motor Circuit Range / Performance:
	4 cylinder models:
	Drive circuit of throttle motor is in malfunction condition (open or excess temperature of device).
	V6 models:
	   • Duty ratio of throttle actuator control signal is higher than specified valve for specified time.
	   • Difference between targeted TP and measured TP is out of specified range.
	Possible Causes:
	   • Throttle motor and/or its circuit
	   • ECM

DTC: P2101
1T, MIL: Yes
Year: 2009
Model: XL-7
Engine: 3.6L V6
Transmission: All
	Control Module Throttle Actuator Position Performance:
	The ECM detects a 4-50 percent difference between the commanded and the actual throttle plate position, dependent upon the rate of commanded throttle movement. The condition exists, then a 5 second delay for MIL ON.
	Possible Causes:
	   • TAC motor and/or circuit
	   • Throttle body
	   • ECM

DTC: P2101
1T, MIL: Yes
Year: 2009, 2010
Model: Equator
Engine: 2.5L L4, 4.0L V6
Transmission: All
	Electric throttle control performance problem:
	Electric throttle control function does not operate properly.
	Possible Causes:
	   • Harness or connectors
	   • (Throttle control motor circuit is open or shorted)
	   • Electric throttle control actuator

DTC: P2102
1T, MIL: Yes
Year: 2009, 2010
Model: Grand Vitara, SX4
Engine: 2.0L L4, 2.0L L4 VIN 5, 2.4L L4 VIN 0, 2.4L L4 VIN 9
Transmission: All
	Throttle Actuator Control Motor Circuit Low:
	Monitor voltage of throttle motor power supply circuit is lower than 5 V even if throttle motor control relay is ON (ignition switch is ON).
	Possible Causes:
	   • Throttle motor control relay and/or its circuit
	   • "THR MOT" fuse
	   • ECM

DTC: P2103
1T, MIL: Yes
Year: 2009, 2010
Model: Equator
Engine: 2.5L L4, 4.0L V6
Transmission: All
	Throttle control motor relay circuit short:
	ECM detects the throttle control motor relay is stuck ON.
	Possible Causes:
	   • Harness or connectors
	   • (Throttle control motor relay circuit is shorted)
	   • Throttle control motor relay

DTC: P2103
1T, MIL: Yes
Year: 2009, 2010
Model: Grand Vitara, SX4
Engine: 2.0L L4, 2.0L L4 VIN 5, 2.4L L4 VIN 0, 2.4L L4 VIN 9
Transmission: All
	Throttle Actuator Control Motor Circuit High:
	Output voltage of throttle motor power supply circuit is higher than 5 V even if throttle motor control relay is OFF (ignition switch is OFF).
	Possible Causes:
	   • Throttle motor control relay and/or its circuit
	   • ECM

DTC: P2105
1T, MIL: Yes
Year: 2009
Model: XL-7
Engine: 3.6L V6
Transmission: All
	Throttle Actuator Control (TAC) System – Forced Engine Shutdown:
	   • The ECM detects an incorrect voltage level at the ignition voltage supply circuits.
	OR
	   • The ECM detects an internal communication error.
	   • A condition exists, then a 5 second delay for MIL ON.
	Possible Causes:
	   • Fuse
	   • ECM circuits
	   • ECM

DTC	Trouble Code Title, Conditions & Possible Causes
DTC: P2111 **1T, MIL: Yes** **Year:** 2009, 2010 **Model:** Grand Vitara, SX4 **Engine:** 2.0L L4, 2.0L L4 VIN 5, 2.4L L4 VIN 0, 2.4L L4 VIN 9 **Transmission:** All	**Throttle Actuator Control System-Stuck Open·** In diagnosing throttle valve at ignition switch turned OFF, throttle valve does NOT close even if throttle valve is opened at specified value. **Possible Causes:** • Throttle valve • TP sensor (main) • ECM
DTC: P2118 **1T, MIL: Yes** **Year:** 2009, 2010 **Model:** Equator **Engine:** 2.5L L4, 4.0L V6 **Transmission:** All	**Throttle control motor circuit short:** ECM detects short in both circuits between ECM and throttle control motor. **Possible Causes:** • Harness or connectors • (Throttle control motor circuit is shorted.) • Electric throttle control actuator • (Throttle control motor)
DTC: P2119 **1T, MIL: Yes** **Year:** 2009, 2010 **Model:** Grand Vitara, SX4 **Engine:** 2.0L L4, 2.0L L4 VIN 5, 2.4L L4 VIN 0, 2.4L L4 VIN 9, 3.2L V6 VIN 1 **Transmission:** All	**Throttle Actuator Control Throttle Body Range / Performance:** 4 cylinder models: Difference between targeted TP and measured TP is more than specified value. V6 models: • Measured default TP is out of specified range. • Return spring malfunction of throttle valve is detected. **Possible Causes:** • Throttle motor and/or its circuit • TP sensor (main) circuit • Throttle body assembly • ECM
DTC: P2119 **1T, MIL: Yes** **Year:** 2009, 2010 **Model:** Equator **Engine:** 2.5L L4, 4.0L V6 **Transmission:** All	**Electric Throttle Control Actuator:** A) Electric throttle control actuator does not function properly due to the return spring malfunction. B) Throttle valve opening angle in fail-safe mode is not in specified range. C) ECM detects the throttle valve is stuck open. **Possible Causes:** • Electric throttle control actuator
DTC: P2119 **1T, MIL: Yes** **Year:** 2009 **Model:** XL-7 **Engine:** 3.6L V6 **Transmission:** All	**Throttle Closed Position Performance:** The ECM determines that the throttle valve did not return to the rest position within 500 mS. The condition exists, then a 5 second delay for MIL ON. **Possible Causes:** • TAC motor and/or circuit • Throttle body • ECM
DTC: P2122 **1T, MIL: Yes** **Year:** 2009, 2010 **Model:** Grand Vitara, SX4 **Engine:** 2.0L L4, 2.0L L4 VIN 5, 2.4L L4 VIN 0, 2.4L L4 VIN 9, 3.2L V6 VIN 1 **Transmission:** All	**Throttle/Pedal Position Sensor/Switch "D" Circuit Low Input:** Output voltage of APP sensor (main) signal circuit is lower than 0.45 V. **Possible Causes:** • APP sensor (main) and/or its circuit • ECM
DTC: P2122 **1T, MIL: Yes** **Year:** 2009, 2010 **Model:** Equator **Engine:** 2.5L L4, 4.0L V6 **Transmission:** All	**Accelerator pedal position sensor 1 circuit low input:** An excessively low voltage from the APP sensor 1 is sent to ECM. **Possible Causes:** • Harness or connectors • (The APP sensor 1 circuit is open or shorted.) • Accelerator pedal position sensor • (Accelerator pedal position sensor 1)
DTC: P2122 **1T, MIL: Yes** **Year:** 2009 **Model:** XL-7 **Engine:** 3.6L V6 **Transmission:** All	**Accelerator Pedal Position (APP) Sensor 1 Circuit Low Voltage:** The APP sensor 1 voltage is less than 0.86 volt, then a 5 second delay for MIL ON. **Possible Causes:** • APP sensor and/or circuit • ECM

DTC	Trouble Code Title, Conditions & Possible Causes

DTC: P2123
1T, MIL: Yes
Year: 2009
Model: XL-7
Engine: 3.6L V6
Transmission: All
	Accelerator Pedal Position (APP) Sensor 1 Circuit High Voltage:
	The APP sensor 1 voltage is greater than 4.82 volts, then a 5 second delay for MIL ON.
	Possible Causes:
	  • APP sensor and/or circuit
	  • ECM

DTC: P2123
1T, MIL: Yes
Year: 2009, 2010
Model: Grand Vitara, SX4
Engine: 2.0L L4, 2.0L L4 VIN 5, 2.4L L4 VIN 0, 2.4L L4 VIN 9, 3.2L V6 VIN 1
Transmission: All
	Throttle/Pedal Position Sensor/Switch "D" Circuit High Input:
	Output voltage of APP sensor (main) signal circuit is higher than 4.8 V.
	Possible Causes:
	  • APP sensor (main) and/or its circuit
	  • ECM

DTC: P2123
1T, MIL: Yes
Year: 2009, 2010
Model: Equator
Engine: 2.5L L4, 4.0L V6
Transmission: All
	Accelerator pedal position sensor 1 circuit high input:
	An excessively high voltage from the APP sensor 1 is sent to ECM.
	Possible Causes:
	  • Harness or connectors
	  • (The APP sensor 1 circuit is open or shorted.)
	  • Accelerator pedal position sensor
	  • (Accelerator pedal position sensor 1)

DTC: P2127
1T, MIL: Yes
Year: 2009
Model: XL-7
Engine: 3.6L V6
Transmission: All
	Accelerator Pedal Position (APP) Sensor 2 Circuit Low Voltage:
	The APP sensor 2 voltage is less than 0.63 volt, then a 5 second delay for MIL ON.
	Possible Causes:
	  • APP sensor and/or circuit
	  • ECM

DTC: P2127
1T, MIL: Yes
Year: 2009, 2010
Model: Equator
Engine: 2.5L L4, 4.0L V6
Transmission: All
	Accelerator pedal position sensor 2 circuit low input:
	An excessively low voltage from the APP sensor 2 is sent to ECM.
	Possible Causes:
	  • Harness or connectors
	  • (The APP sensor 2 circuit is open or shorted.)
	  • (TP sensor circuit is shorted.)
	  • Accelerator pedal position sensor
	  • (Accelerator pedal position sensor 2)
	  • Electric throttle control actuator
	  • (TP sensor 1 and 2)

DTC: P2127
1T, MIL: Yes
Year: 2009, 2010
Model: Grand Vitara, SX4
Engine: 2.0L L4, 2.0L L4 VIN 5, 2.4L L4 VIN 0, 2.4L L4 VIN 9, 3.2L V6 VIN 1
Transmission: All
	Throttle/Pedal Position Sensor/Switch "E" Circuit Low Input:
	Output voltage of APP sensor (sub) signal circuit is lower than 0.23 V.
	Possible Causes:
	  • APP sensor (sub) and/or its circuit
	  • ECM

DTC: P2128
1T, MIL: Yes
Year: 2009, 2010
Model: Grand Vitara, SX4
Engine: 2.0L L4, 2.0L L4 VIN 5, 2.4L L4 VIN 0, 2.4L L4 VIN 9, 3.2L V6 VIN 1
Transmission: All
	Throttle/Pedal Position Sensor/Switch "E" Circuit High Input:
	Output voltage of APP sensor (sub) signal circuit is higher than 2.4 V.
	Possible Causes:
	  • APP sensor (sub) and/or its circuit
	  • ECM

DTC: P2128
1T, MIL: Yes
Year: 2009
Model: XL-7
Engine: 3.6L V6
Transmission: All
	Accelerator Pedal Position (APP) Sensor 2 Circuit High Voltage:
	The APP sensor 2 voltage is greater than 4.82 volts, then a 5 second delay for MIL ON.
	Possible Causes:
	  • APP sensor and/or circuit
	  • ECM

DTC	Trouble Code Title, Conditions & Possible Causes
DTC: P2128 **1T, MIL:** Yes **Year:** 2009, 2010 **Model:** Equator **Engine:** 2.5L L4 **Transmission:** All	**Accelerator pedal position sensor 2 circuit low input:** An excessively low voltage from the APP sensor 2 is sent to ECM. **Possible Causes:** • Harness or connectors • (The APP sensor 2 circuit is open or shorted.) • (TP sensor circuit is shorted.) • Accelerator pedal position sensor • (Accelerator pedal position sensor 2) • Electric throttle control actuator • (TP sensor 1 and 2)
DTC: P2135 **1T, MIL:** Yes **Year:** 2009, 2010 **Model:** Grand Vitara, SX4 **Engine:** 2.0L L4, 2.0L L4 VIN 5, 2.4L L4 VIN 0, 2.4L L4 VIN 9, 3.2L V6 VIN 1 **Transmission:** All	**Throttle/Pedal Position Sensor/Switch "A" / "B" Voltage Correlation:** Difference between TP (main) and TP (sub) is higher than specified value. **Possible Causes:** • TP sensor and/or its circuit • Throttle body assembly • ECM
DTC: P2135 **1T, MIL:** Yes **Year:** 2009, 2010 **Model:** Equator **Engine:** 2.5L L4, 4.0L V6 **Transmission:** All	**Throttle position sensor circuit range/performance problem:** Rationally incorrect voltage is sent to ECM compared with the signals from TP sensor 1 and TP sensor 2. **Possible Causes:** • Harness or connector • (The TP sensor 1 and 2 circuit is open or shorted.) • (APP sensor 2 circuit is shorted.) • Electric throttle control actuator • (TP sensor 1 and 2) • Accelerator pedal position sensor • (APP sensor 2)
DTC: P2138 **1T, MIL:** Yes **Year:** 2009, 2010 **Model:** Equator **Engine:** 2.5L L4, 4.0L V6 **Transmission:** All	**Accelerator pedal position sensor circuit range/performance problem:** Rationally incorrect voltage is sent to ECM compared with the signals from APP sensor 1 and APP sensor 2. **Possible Causes:** • Harness or connector • (The APP sensor 1 and 2 circuit is open or shorted.) • (TP sensor circuit is shorted.) • Accelerator pedal position sensor • (APP sensor 1 and 2) • Electric throttle control actuator • (TP sensor 1 and 2)
DTC: P2138 **1T, MIL:** Yes **Year:** 2009, 2010 **Model:** Grand Vitara, SX4 **Engine:** 2.0L L4, 2.0L L4 VIN 5, 2.4L L4 VIN 0, 2.4L L4 VIN 9, 3.2L V6 VIN 1 **Transmission:** All	**Throttle/Pedal Position Sensor/Switch "D" / "E" Voltage Correlation:** Difference between APP (main) and APP (sub) is more than specified value. **Possible Causes:** • APP sensor and/or its circuit • ECM
DTC: P2138 **1T, MIL:** Yes **Year:** 2009 **Model:** XL-7 **Engine:** 3.6L V6 **Transmission:** All	**Accelerator Pedal Position (APP) Sensor 1-2 Correlation:** • The ECM detects that the voltage difference between APP sensor 1 and 2 is greater than 0.23 volt. • The ECM detects that the voltage difference between APP sensor 1 and 2 is greater than 0.29 volt with a partially pressed pedal. • The ECM detects that the voltage difference between APP sensor 1 and 2 is greater than 1.66 volts with a fully pressed pedal. • Any of the above conditions exist, then a 5 second delay for MIL ON. **Possible Causes:** • APP sensor and/or circuit • ECM

DTC	Trouble Code Title, Conditions & Possible Causes
DTC: P2176 **1T, MIL: Yes** **Year:** 2009, 2010 **Model:** Grand Vitara, XL-7 **Engine:** 3.2L V6 VIN 1, 3.6L V6 **Transmission:** All	**Throttle Actuator Control System – Idle Position Not Learned:** TP sensor (main) voltage or TP sensor (sub) voltage is out of specified range when accelerator pedal is released. **Possible Causes:** • TP sensor (main and sub) and/or its circuit • ECM
DTC: P2177 **2T, MIL: Yes** **Year:** 2009, 2010 **Model:** Grand Vitara, XL-7 **Engine:** 3.2L V6 VIN 1, 3.6L V6 **Transmission:** All	**System Too Lean Off Idle Bank 1:** Air / fuel ratio on bank 1 is unchanged from lean state for specified time at specified driving condition. **Possible Causes:** • Oil level gauge • Oil filler cap • Engine oil • Coolant • Air intake system • Exhaust system • MAF & IAT sensor and/or its circuit • EVAP canister • EVAP canister purge solenoid valve • HO2S and/or its circuit • HO2S heater and/or its circuit • Engine mechanical system • Fuel system • ECM
DTC: P2178 **1T, MIL: Yes** **Year:** 2009, 2010 **Model:** Grand Vitara, XL-7 **Engine:** 3.2L V6 VIN 1, 3.6L V6 **Transmission:** All	**System Too Rich Off Idle Bank 1:** Air / fuel ratio on bank 1 is unchanged from rich state for specified time at specified driving condition. **Possible Causes:** • Oll level gauge • Oil filler cap • Engine oil • Coolant • Air intake system • Exhaust system • MAF & IAT sensor and/or its circuit • EVAP canister • EVAP canister purge solenoid valve • HO2S and/or its circuit • HO2S heater and/or its circuit • Engine mechanical system • Fuel system • ECM
DTC: P2179 **2T, MIL: Yes** **Year:** 2009, 2010 **Model:** Grand Vitara, XL-7 **Engine:** 3.2L V6 VIN 1, 3.6L V6 **Transmission:** All	**System Too Lean Off Idle Bank 2:** Air / fuel ratio on bank 2 is unchanged from lean state for specified time at specified driving condition. **Possible Causes:** • Oil level gauge • Oil filler cap • Engine oil • Coolant • Air intake system • Exhaust system • MAF & IAT sensor and/or its circuit • EVAP canister • EVAP canister purge solenoid valve • HO2S and/or its circuit • HO2S heater and/or its circuit • Engine mechanical system • Fuel system • ECM

DTC	Trouble Code Title, Conditions & Possible Causes
DTC: P2180 **2T, MIL:** Yes **Year:** 2009, 2010 **Model:** Grand Vitara, XL-7 **Engine:** 3.2L V6 VIN 1, 3.6L V6 **Transmission:** All	**System Too Rich Off Idle Bank 2:** Air / fuel ratio on bank 2 is unchanged from rich state for specified time at specified driving condition. **Possible Causes:** • Oil level gauge • Oil filler cap • Engine oil • Coolant • Air intake system • Exhaust system • MAF & IAT sensor and/or its circuit • EVAP canister • EVAP canister purge solenoid valve • HO2S and/or its circuit • HO2S heater and/or its circuit • Engine mechanical system • Fuel system • ECM
DTC: P2187 **2T, MIL:** Yes **Year:** 2009, 2010 **Model:** Grand Vitara, XL-7 **Engine:** 3.2L V6 VIN 1, 3.6L V6 **Transmission:** All	**System Too Lean at Idle Bank 1:** Long term fuel trim on bank 1 based on engine torque and engine speed is higher than 5.5% at idling condition. **Possible Causes:** • Oil level gauge • Oil filler cap • Engine oil • Coolant • Air intake system • Exhaust system • MAF & IAT sensor and/or its circuit • EVAP canister • EVAP canister purge solenoid valve • HO2S and/or its circuit • HO2S heater and/or its circuit • Engine mechanical system • Fuel system • ECM
DTC: P2188 **2T, MIL:** Yes **Year:** 2009, 2010 **Model:** Grand Vitara, XL-7 **Engine:** 3.2L V6 VIN 1, 3.6L V6 **Transmission:** All	**System Too Rich at Idle Bank 1:** Long term fuel trim on bank 1 based on engine torque and engine speed is lower than -5.5% at idling condition. **Possible Causes:** • Oil level gauge • Oil filler cap • Engine oil • Coolant • Air intake system • Exhaust system • MAF & IAT sensor and/or its circuit • EVAP canister • EVAP canister purge solenoid valve • HO2S and/or its circuit • HO2S heater and/or its circuit • Engine mechanical system • Fuel system • ECM

DTC	Trouble Code Title, Conditions & Possible Causes
DTC: P2189 **2T, MIL: Yes** **Year:** 2009, 2010 **Model:** Grand Vitara, XL-7 **Engine:** 3.2L V6 VIN 1, 3.6L V6 **Transmission:** All	**System Too Lean at Idle Bank 2:** Long term fuel trim on bank 2 based on engine torque and engine speed is higher than 5.5% at idling condition. **Possible Causes:** • Oil level gauge • Oil filler cap • Engine oil • Coolant • Air intake system • Exhaust system • MAF & IAT sensor and/or its circuit • EVAP canister • EVAP canister purge solenoid valve • HO2S and/or its circuit • HO2S heater and/or its circuit • Engine mechanical system • Fuel system • ECM
DTC: P2190 **2T, MIL: Yes** **Year:** 2009, 2010 **Model:** Grand Vitara, XL-7 **Engine:** 3.2L V6 VIN 1, 3.6L V6 **Transmission:** All	**System Too Rich at Idle Bank 2:** Long term fuel trim on bank 2 based on engine torque and engine speed is lower than -5.5% at idling condition. **Possible Causes:** • Oil level gauge • Oil filler cap • Engine oil • Coolant • Air intake system • Exhaust system • MAF & IAT sensor and/or its circuit • EVAP canister • EVAP canister purge solenoid valve • HO2S and/or its circuit • HO2S heater and/or its circuit • Engine mechanical system • Fuel system • ECM
DTC: P2195 **2T, MIL: Yes** **Year:** 2009, 2010 **Model:** Grand Vitara, SX4 **Engine:** 2.0L L4, 2.0L L4 VIN 5, 2.4L L4 VIN 0, 2.4L L4 VIN 9 **Transmission:** All	**O2 Sensor Signal Stuck Lean (Bank-1 Sensor-1):** A/F mixture output from A/F sensor is lean state for 10 sec. even if HO2S signal circuit is higher than 0.2 V (rich state). **Possible Causes:** • A/F sensor and/or its circuit • ECM
DTC: P2196 **2T, MIL: Yes** **Year:** 2009, 2010 **Model:** Grand Vitara, SX4 **Engine:** 2.0L L4, 2.0L L4 VIN 5, 2.4L L4 VIN 0, 2.4L L4 VIN 9 **Transmission:** All	**O2 Sensor Signal Stuck Rich (Bank-1 Sensor-1):** A/F mixture output from A/F sensor is rich state for 10 sec. even if HO2S signal circuit is lower than 0.7 V (lean state). **Possible Causes:** • A/F sensor and/or its circuit • ECM
DTC: P2227 **2T, MIL: Yes** **Year:** 2009, 2010 **Model:** Grand Vitara, SX4, XL-7 **Engine:** 2.0L L4, 2.0L L4 VIN 5, 2.4L L4 VIN 0, 2.4L L4 VIN 9, 3.2L V6 VIN 1, 3.6L V6 **Transmission:** All	**Barometric Pressure Circuit Performance:** • BARO sensor signal is out of specified range for 2 sec. • Variation of BARO sensor signal is out of specified range for 20 sec. at engine running condition. • Variation of BARO sensor signal between last driving cycle and actual driving cycle is out of specified range and difference between measured BARO and standard BARO is more than specified valve. **Possible Causes:** • EVAP leak check pressure sensor (in EVAP leak check module) and its circuit • ECM

DTC	Trouble Code Title, Conditions & Possible Causes
DTC: P2228 **2T, MIL: Yes** **Year:** 2009, 2010 **Model:** Grand Vitara, SX4, XL-7 **Engine:** 2.0L L4, 2.0L L4 VIN 5, 2.4L L4 VIN 0, 2.4L L4 VIN 9, 3.2L V6 VIN 1, 3.6L V6 **Transmission:** All	**Barometric Pressure Circuit Low:** Output voltage of EVAP leak check pressure sensor signal circuit is excessively low for 5 sec. **Possible Causes:** • EVAP leak check pressure sensor (in EVAP leak check module) and/or its circuit • ECM
DTC: P2229 **2T, MIL: Yes** **Year:** 2009, 2010 **Model:** Grand Vitara, SX4, XL-7 **Engine:** 2.0L L4, 2.0L L4 VIN 5, 2.4L L4 VIN 0, 2.4L L4 VIN 9, 3.2L V6 VIN 1, 3.6L V6 **Transmission:** All	**Barometric Pressure Circuit High:** Output voltage of EVAP leak check pressure sensor signal circuit is excessively high for 5 sec. **Possible Causes:** • EVAP leak check pressure sensor (in EVAP leak check module) and/or its circuit • ECM
DTC: P2232 **2T, MIL: Yes** **Year:** 2009, 2010 **Model:** Grand Vitara, XL-7 **Engine:** 3.2L V6 VIN 1, 3.6L V6 **Transmission:** All	**O2 Sensor Signal Circuit Shorted to Heater Circuit Bank 1 Sensor 2:** Variation of HO2S-2 (bank 1) signal voltage is higher than 2 V 4 times in 6 times when HO2S heater is turned off. **Possible Causes:** • HO2S-2 and/or its circuit • HO2S-2 heater and/or its circuit • ECM
DTC: P2235 **2T, MIL: Yes** **Year:** 2009, 2010 **Model:** Grand Vitara, XL-7 **Engine:** 3.2L V6 VIN 1, 3.6L V6 **Transmission:** All	**O2 Sensor Signal Circuit Shorted to Heater Circuit Bank 2 Sensor 2:** Variation of HO2S-2 (bank 2) signal voltage is higher than 2 V 4 times in 6 times when HO2S heater is turned off. **Possible Causes:** • HO2S-2 and/or its circuit • HO2S-2 heater and/or its circuit • ECM
DTC: P2237 **2T, MIL: Yes** **Year:** 2009, 2010 **Model:** Grand Vitara, SX4 **Engine:** 2.0L L4, 2.0L L4 VIN 5, 2.4L L4 VIN 0, 2.4L L4 VIN 9 **Transmission:** All	**O2 Sensor Positive Current Control Circuit / Open:** Impedance of A/F sensor element is higher than 500 W for 15 sec. even if A/F sensor heater is turned ON for 20 sec. **Possible Causes:** • A/F sensor and/or its circuit • ECM
DTC: P2270 **2T, MIL: Yes** **Year:** 2009, 2010 **Model:** Grand Vitara, XL-7 **Engine:** 3.2L V6 VIN 1, 3.6L V6 **Transmission:** All	**O2 Sensor Signal Stuck Lean Bank 1 Sensor 2:** HO2S-2 signal (bank 1) voltage is lower than specified value for 120 sec. **Possible Causes:** • Oil level gauge • Oil filler cap • Engine oil • Coolant • Air intake system • Exhaust system • Ignition system • HO2S-2 and its circuit • HO2S-2 heater and its circuit • Engine mechanical system • Fuel system • ECM

DTC	Trouble Code Title, Conditions & Possible Causes
DTC: P2271 **2T, MIL: Yes** **Year:** 2009, 2010 **Model:** Grand Vitara, XL-7 **Engine:** 3.2L V6 VIN 1, 3.6L V6 **Transmission:** All	**O2 Sensor Signal Stuck Rich Bank 1 Sensor 2:** HO2S-2 signal (bank 1) voltage is higher than specified value for 120 sec. **Possible Causes:** • Oil level gauge • Oil filler cap • Engine oil • Coolant • Air intake system • Exhaust system • Ignition system • HO2S-2 and its circuit • HO2S-2 heater and its circuit • Engine mechanical system • Fuel system • ECM
DTC: P2272 **2T, MIL: Yes** **Year:** 2009, 2010 **Model:** Grand Vitara, XL-7 **Engine:** 3.2L V6 VIN 1, 3.6L V6 **Transmission:** All	**O2 Sensor Signal Stuck Lean Bank 2 Sensor 2:** HO2S-2 signal (bank 2) voltage is lower than specified value for 120 sec. **Possible Causes:** • Oil level gauge • Oil filler cap • Engine oil • Coolant • Air intake system • Exhaust system • Ignition system • HO2S-2 and its circuit • HO2S-2 heater and its circuit • Engine mechanical system • Fuel system • ECM
DTC: P2273 **2T, MIL: Yes** **Year:** 2009, 2010 **Model:** Grand Vitara, XL-7 **Engine:** 3.2L V6 VIN 1, 3.6L V6 **Transmission:** All	**O2 Sensor Signal Stuck Rich Bank 2 Sensor 2:** HO2S-2 signal (bank 2) voltage is higher than specified value for 120 sec. **Possible Causes:** • Oil level gauge • Oil filler cap • Engine oil • Coolant • Air intake system • Exhaust system • Ignition system • HO2S-2 and its circuit • HO2S-2 heater and its circuit • Engine mechanical system • Fuel system • ECM
DTC: P2300 **2T, MIL: Yes** **Year:** 2009 **Model:** XL-7 **Engine:** 3.6L V6 **Transmission:** All	**Ignition Coil 1 Control Circuit Low Voltage:** The ECM detects the ignition control circuit is shorted to ground for more than 4 seconds or a cumulative of 30 seconds. **Possible Causes:** • Ignition coil and/or circuit • ECM
DTC: P2300 **2T** **Year:** 2009, 2010 **Model:** Grand Vitara **Engine:** 3.2L V6 VIN 1 **Transmission:** All	**Ignition Coil "A" Primary Control Circuit Low:** Ignition coil No.1 control circuit is shorted to ground. **Possible Causes:** • Ignition coil and/or its circuit • CMP sensor and/or its circuit • CKP sensor and/or its circuit • Signal rotor • ECM

DTC	Trouble Code Title, Conditions & Possible Causes
DTC: P2301 **2T** **Year:** 2009, 2010 **Model:** Grand Vitara **Engine:** 3.2L V6 VIN 1 **Transmission:** All	**Ignition Coil "A" Primary Control Circuit High:** Ignition coil No.1 control circuit is shorted to power circuit. **Possible Causes:** • Ignition coil and/or its circuit • CMP sensor and/or its circuit • CKP sensor and/or its circuit • Signal rotor • ECM
DTC: P2301 **2T, MIL: Yes** **Year:** 2009 **Model:** XL-7 **Engine:** 3.6L V6 **Transmission:** All	**Ignition Coil 1 Control Circuit High Voltage:** The ECM detects the ignition control circuit is shorted to voltage for more than 4 seconds or a cumulative of 30 seconds. **Possible Causes:** • Ignition coil and/or circuit • ECM
DTC: P2303 **2T** **Year:** 2009, 2010 **Model:** Grand Vitara **Engine:** 3.2L V6 VIN 1 **Transmission:** All	**Ignition Coil "B" Primary Control Circuit Low:** Ignition coil No.2 control circuit is shorted to ground. **Possible Causes:** • Ignition coil and/or its circuit • CMP sensor and/or its circuit • CKP sensor and/or its circuit • Signal rotor • ECM
DTC: P2303 **2T, MIL: Yes** **Year:** 2009 **Model:** XL-7 **Engine:** 3.6L V6 **Transmission:** All	**Ignition Coil 2 Control Circuit Low Voltage:** The ECM detects the ignition control circuit is shorted to ground for more than 4 seconds or a cumulative of 30 seconds. **Possible Causes:** • Ignition coil and/or circuit • ECM
DTC: P2304 **2T** **Year:** 2009, 2010 **Model:** Grand Vitara **Engine:** 3.2L V6 VIN 1 **Transmission:** All	**Ignition Coil "B" Primary Control Circuit High:** Ignition coil No.2 control circuit is shorted to power circuit. **Possible Causes:** • Ignition coil and/or its circuit • CMP sensor and/or its circuit • CKP sensor and/or its circuit • Signal rotor • ECM
DTC: P2304 **2T, MIL: Yes** **Year:** 2009 **Model:** XL-7 **Engine:** 3.6L V6 **Transmission:** All	**Ignition Coil 2 Control Circuit High Voltage:** The ECM detects the ignition control circuit is shorted to voltage for more than 4 seconds or a cumulative of 30 seconds. **Possible Causes:** • Ignition coil and/or circuit • ECM
DTC: P2306 **2T, MIL: Yes** **Year:** 2009 **Model:** XL-7 **Engine:** 3.6L V6 **Transmission:** All	**Ignition Coil 3 Control Circuit Low Voltage:** The ECM detects the ignition control circuit is shorted to ground for more than 4 seconds or a cumulative of 30 seconds. **Possible Causes:** • Ignition coil and/or circuit • ECM
DTC: P2306 **2T** **Year:** 2009, 2010 **Model:** Grand Vitara **Engine:** 3.2L V6 VIN 1 **Transmission:** All	**Ignition Coil "C" Primary Control Circuit Low:** Ignition coil No.3 control circuit is shorted to ground. **Possible Causes:** • Ignition coil and/or its circuit • CMP sensor and/or its circuit • CKP sensor and/or its circuit • Signal rotor • ECM

DTC	Trouble Code Title, Conditions & Possible Causes
DTC: P2307 **2T, MIL: Yes** **Year:** 2009 **Model:** XL-7 **Engine:** 3.6L V6 **Transmission:** All	**Ignition Coil 3 Control Circuit High Voltage:** The ECM detects the ignition control circuit is shorted to voltage for more than 4 seconds or a cumulative of 30 seconds. **Possible Causes:** • Ignition coil and/or circuit • ECM
DTC: P2307 **T** **Year:** 2009, 2010 **Model:** Grand Vitara **Engine:** 3.2L V6 VIN 1 **Transmission:** All	**Ignition Coil "C" Primary Control Circuit High:** Ignition coil No.3 control circuit is shorted to power circuit. **Possible Causes:** • Ignition coil and/or its circuit • CMP sensor and/or its circuit • CKP sensor and/or its circuit • Signal rotor • ECM
DTC: P2309 **2T** **Year:** 2009, 2010 **Model:** Grand Vitara **Engine:** 3.2L V6 VIN 1 **Transmission:** All	**Ignition Coil "D" Primary Control Circuit Low:** Ignition coil No.4 control circuit is shorted to ground. **Possible Causes:** • Ignition coil and/or its circuit • CMP sensor and/or its circuit • CKP sensor and/or its circuit • Signal rotor • ECM
DTC: P2309 **2T, MIL: Yes** **Year:** 2009 **Model:** XL-7 **Engine:** 3.6L V6 **Transmission:** All	**Ignition Coil 4 Control Circuit Low Voltage:** The ECM detects the ignition control circuit is shorted to ground for more than 4 seconds or a cumulative of 30 seconds. **Possible Causes:** • Ignition coil and/or circuit • ECM
DTC: P2310 **2T, MIL: Yes** **Year:** 2009 **Model:** XL-7 **Engine:** 3.6L V6 **Transmission:** All	**Ignition Coil 4 Control Circuit High Voltage:** The ECM detects the ignition control circuit is shorted to voltage for more than 4 seconds or a cumulative of 30 seconds. **Possible Causes:** • Ignition coil and/or circuit • ECM
DTC: P2310 **2T** **Year:** 2009, 2010 **Model:** Grand Vitara **Engine:** 3.2L V6 VIN 1 **Transmission:** All	**Ignition Coil "D" Primary Control Circuit High:** Ignition coil No.4 control circuit is shorted to power circuit. **Possible Causes:** • Ignition coil and/or its circuit • CMP sensor and/or its circuit • CKP sensor and/or its circuit • Signal rotor • ECM
DTC: P2312 **2T** **Year:** 2009, 2010 **Model:** Grand Vitara **Engine:** 3.2L V6 VIN 1 **Transmission:** All	**Ignition Coil "E" Primary Control Circuit Low:** Ignition coil No.5 control circuit is shorted to ground. **Possible Causes:** • Ignition coil and/or its circuit • CMP sensor and/or its circuit • CKP sensor and/or its circuit • Signal rotor • ECM
DTC: P2312 **2T, MIL: Yes** **Year:** 2009 **Model:** XL-7 **Engine:** 3.6L V6 **Transmission:** All	**Ignition Coil 5 Control Circuit Low Voltage:** The ECM detects the ignition control circuit is shorted to ground for more than 4 seconds or a cumulative of 30 seconds. **Possible Causes:** • Ignition coil and/or circuit • ECM

DTC	Trouble Code Title, Conditions & Possible Causes
DTC: P2313 **2T, MIL:** Yes **Year:** 2009 **Model:** XL-7 **Engine:** 3.6L V6 **Transmission:** All	**Ignition Coil 5 Control Circuit High Voltage:** The ECM detects the ignition control circuit is shorted to voltage for more than 4 seconds or a cumulative of 30 seconds. **Possible Causes:** • Ignition coil and/or circuit • ECM
DTC: P2313 **2T** **Year:** 2009, 2010 **Model:** Grand Vitara **Engine:** 3.2L V6 VIN 1 **Transmission:** All	**Ignition Coil "E" Primary Control Circuit High:** Ignition coil No.5 control circuit is shorted to power circuit. **Possible Causes:** • Ignition coil and/or its circuit • CMP sensor and/or its circuit • CKP sensor and/or its circuit • Signal rotor • ECM
DTC: P2315 **2T, MIL:** Yes **Year:** 2009 **Model:** XL-7 **Engine:** 3.6L V6 **Transmission:** All	**Ignition Coil 6 Control Circuit Low Voltage:** The ECM detects the ignition control circuit is shorted to ground for more than 4 seconds or a cumulative of 30 seconds. **Possible Causes:** • Ignition coil and/or circuit • ECM
DTC: P2315 **2T** **Year:** 2009, 2010 **Model:** Grand Vitara **Engine:** 3.2L V6 VIN 1 **Transmission:** All	**Ignition Coil "F" Primary Control Circuit Low:** Ignition coil No.6 control circuit is shorted to ground. **Possible Causes:** • Ignition coil and/or its circuit • CMP sensor and/or its circuit • CKP sensor and/or its circuit • Signal rotor • ECM
DTC: P2316 **2T** **Year:** 2009, 2010 **Model:** Grand Vitara **Engine:** 3.2L V6 VIN 1 **Transmission:** All	**Ignition Coil "F" Primary Control Circuit High:** Ignition coil No.6 control circuit is shorted to power circuit. **Possible Causes:** • Ignition coil and/or its circuit • CMP sensor and/or its circuit • CKP sensor and/or its circuit • Signal rotor • ECM
DTC: P2316 **2T, MIL:** Yes **Year:** 2009 **Model:** XL-7 **Engine:** 3.6L V6 **Transmission:** All	**Ignition Coil 6 Control Circuit High Voltage:** The ECM detects the ignition control circuit is shorted to voltage for more than 4 seconds or a cumulative of 30 seconds. **Possible Causes:** • Ignition coil and/or circuit • ECM
DTC: P2400 **2T, MIL:** Yes **Year:** 2009, 2010 **Model:** Grand Vitara, SX4 **Engine:** 2.0L L4, 2.0L L4 VIN 5, 2.4L L4 VIN 0, 2.4L L4 VIN 9 **Transmission:** All	**Evaporative Emission System Leak Detection Pump Control Circuit / Open:** Output voltage of EVAP leak detection pump drive circuit is higher than 4 V even if EVAP leak detection pump is OFF. **Possible Causes:** • EVAP leak check module and/or its circuit • ECM

DTC	Trouble Code Title, Conditions & Possible Causes
DTC: P2404 **1T, MIL: Yes** **Year:** 2009, 2010 **Model:** Grand Vitara, SX4 **Engine:** 2.0L L4, 2.0L L4 VIN 5, 2.4L L4 VIN 0, 2.4L L4 VIN 9 **Transmission:** All	**Evaporative Emission System Leak Detection Pump Sense Circuit Range / Performance:** EVAP vent valve stuck open: EVAP leak check pressure is lower than -0.14 kPa even if EVAP vent valve is operated from OFF (OPEN) to ON (CLOSE) with EVAP leak detection pump ON. EVAP vent valve stuck closed: EVAP leak check pressure is unstable (difference between maximum and minimum is 0.36 kPa or more) when EVAP vent valve is OFF (OPEN) with EVAP leak detection pump ON. EVAP leak detection pump stuck OFF: EVAP leak check pressure is higher than -0.14 kPa even if EVAP leak detection pump is ON. EVAP leak detection pump flow high: Reference pressure of EVAP leak check pressure is lower than specified value (depending on BARO) EVAP leak detection pump flow Low: Reference pressure of EVAP leak check pressure is higher than specified value (depending on BARO). **Possible Causes:** • EVAP leak detection pump (in EVAP leak check module) and/or its circuit • EVAP vent valve (in EVAP leak check module) and/or its circuit • EVAP leak check module and/or its circuit • ECM
DTC: P2422 **2T, MIL: Yes** **Year:** 2009, 2010 **Model:** Grand Vitara **Engine:** 3.2L V6 VIN 1 **Transmission:** All	**Evaporative Emission System Vent Valve Stuck Closed:** Fuel tank pressure is lower than -2.2 kPa at phase A in "Large leak, small leak and stuck open/close test". **Possible Causes:** • EVAP canister vent solenoid valve and/or its circuit • EVAP canister • EVAP canister air suction filter • EVAP pipes and hoses • Fuel tank pressure sensor and/or its circuit • ECM
DTC: P2610 **2T, MIL: Yes** **Year:** 2009, 2010 **Model:** Grand Vitara, SX4 **Engine:** 2.0L L4, 2.0L L4 VIN 5, 2.4L L4 VIN 0, 2.4L L4 VIN 9, 3.2L V6 VIN 1 **Transmission:** All	**ECM / PCM Internal Engine Off Timer Performance:** ECM internal failure related to engine off timer. **Possible Causes:** • ECM power supply circuit and/or ground circuit • ECM
DTC: P2610 **1T, MIL: Yes** **Year:** 2009 **Model:** XL-7 **Engine:** 3.6L V6 **Transmission:** All	**Control Module Ignition Off Timer Performance:** The ECM detects the engine off timer signal is missing or not valid for greater than 1 second, or a cumulative of 10 seconds. **Possible Causes:** • ECM
DTC: P2636 **1T** **Year:** 2009 **Model:** XL-7 **Engine:** 3.6L V6 **Transmission:** All	**Fuel Transfer Pump Flow Insufficient:** The ECM detects that the left side fuel level is less than 11 percent, and the right side fuel level is greater than 22 percent. **Possible Causes:** • Transfer pump and/or circuit • Fuel level sensors and/or circuits
DTC: P2762 **1T** **Year:** 2009, 2010 **Model:** Grand Vitara **Engine:** 3.2L V6 VIN 1 **Transmission:** All	**Torque Converter Clutch Pressure Control Solenoid Control Circuit Range / Performance:** Difference between target current of TCC solenoid valve circuit and monitor current of TCC solenoid valve circuit is more than specified value. **Possible Causes:** • TCC pressure control solenoid valve or its circuit • TCM
DTC: P2763 **1T, MIL: Yes** **Year:** 2009, 2010 **Model:** Grand Vitara **Engine:** 3.2L V6 VIN 1 **Transmission:** All	**Torque Converter Clutch Pressure Control Solenoid Control Circuit High:** Voltage of TCC pressure control solenoid valve TCM terminal is high although TCM is commanding TCC pressure control solenoid to turn OFF. **Possible Causes:** • TCC pressure control solenoid valve circuit shorted to power circuit. • Malfunction of TCC pressure control solenoid valve. • TCM

DTC	Trouble Code Title, Conditions & Possible Causes
DTC: P2763 **1T, MIL: Yes** **Year:** 2010 **Model:** Kizashi **Engine:** 2.4L L4 VIN 9 **Transmission:** All	**Torque Converter Clutch Pressure Control Solenoid Control Circuit High:** Monitored circuit current of TCC (Torque Converter Clutch) solenoid valve is lower than limit value for 5 sec. even if TCM outputs power to its control circuit. **Possible Causes:** • Solenoid and/or circuit • ECM
DTC: P2764 **1T, MIL: Yes** **Year:** 2009, 2010 **Model:** Grand Vitara **Engine:** 3.2L V6 VIN 1 **Transmission:** All	**Torque Converter Clutch Pressure Control Solenoid Control Circuit Low:** Voltage of TCC pressure control solenoid valve TCM terminal is low although TCM is commanding TCC pressure control solenoid to turn ON. **Possible Causes:** • TCC pressure control solenoid valve circuit open or shorted to ground. • Malfunction of TCC pressure control solenoid valve • TCM
DTC: P2764 **1T, MIL: Yes** **Year:** 2010 **Model:** Kizashi **Engine:** 2.4L L4 VIN 9 **Transmission:** All	**P2764 Torque Converter Clutch Pressure Control Solenoid Control Circuit Low:** Monitored circuit current of TCC (Torque Converter Clutch) solenoid valve is higher than limit value even if TCM outputs power to its control circuit. **Possible Causes:** • Solenoid and/or circuit • ECM
DTC: P2A00 **2T, MIL: Yes** **Year:** 2009, 2010 **Model:** Equator **Engine:** 4.0L V6 **Transmission:** All	**Air fuel ratio (A/F) sensor 1 circuit range/performance (bank 1):** • The output voltage computed by ECM from the A/F sensor 1 signal shift to the lean side for a specified period. • The A/F signal computed by ECM from the A/F sensor 1 signal shift to the rich side for a specified period. **Possible Causes:** • A/F sensor 1 • A/F sensor 1 heater • Fuel pressure • Fuel injector • Intake air leaks
DTC: P2A01 **2T, MIL: Yes** **Year:** 2009, 2010 **Model:** Grand Vitara, SX4 **Engine:** 2.0L L4, 2.0L L4 VIN 5, 2.4L L4 VIN 0, 2.4L L4 VIN 9 **Transmission:** All	**O2 Sensor Circuit Range / Performance (Bank 1, Sensor 2):** EVAP-Leak-Check Model: Stuck lean: Output voltage of HO2S signal circuit is lower than 0.4 V (A/F mixture: lean state) for 40 sec. Stuck rich: Output voltage of HO2S signal circuit is higher than 0.85 V (A/F mixture: rich state) for 40 sec. High output performance malfunction: Output voltage of HO2S signal circuit is lower than 0.64 V (A/F mixture: lean state) for 15 sec. even if A/F mixture based on A/F sensor has been rich state after fuel cut. Low output performance malfunction: Output voltage of HO2S signal circuit is higher than 0.1 V (A/F mixture: rich state) even if fuel cut is activated for 8 sec. Non-EVAP-Leak-Check Model: Stuck lean: Output voltage of HO2S signal circuit is lower than 0.4 V (A/F mixture: lean state) for 40 sec. Stuck rich: Output voltage of HO2S signal circuit is higher than 0.85 V (A/F mixture: rich state) for 40 sec. **Possible Causes:** • Exhaust system leakage • HO2S and/or its circuit • ECM
DTC: P2A03 **2T, MIL: Yes** **Year:** 2009, 2010 **Model:** Equator **Engine:** 4.0L V6 **Transmission:** All	**Air fuel ratio (A/F) sensor 1 circuit range/performance (bank 2):** • The output voltage computed by ECM from the A/F sensor 1 signal shift to the lean side for a specified period. • The A/F signal computed by ECM from the A/F sensor 1 signal shift to the rich side for a specified period. **Possible Causes:** • A/F sensor 1 • A/F sensor 1 heater • Fuel pressure • Fuel injector • Intake air leaks

OBD II Trouble Code List (U0XXX Codes)

DTC	Trouble Code Title, Conditions & Possible Causes

DTC: U0073
T, MIL: Yes
Year: 2010
Model: Kizashi
Engine: 2.4L L4 VIN 9
Transmission: All
	Control Module Communication Bus Off:
	TCM fails to transmit and receive the data via CAN for specified time continuously.
	Possible Causes:
	   • TCM
	   • circuits
	   • ECM

DTC: U0073 **Year:** 2009, 2010 **Model:** Grand Vitara **Engine:** 3.2L V6 VIN 1 **Transmission:** All	**Control Module Communication Bus Off:** Transmitting and receiving error of ECM for specified time continuously.
DTC: U0073 **1T, MIL:** Yes **Year:** 2009, 2010 **Model:** Grand Vitara, SX4 **Engine:** 2.0L L4, 2.0L L4 VIN 5, 2.4L L4 VIN 0, 2.4L L4 VIN 9, 3.2L V6 VIN 1 **Transmission:** All	**Control Module Communication Bus Off:** Transmitting and receiving error of ECM for specified time continuously.
DTC: U0100 **Year:** 2009, 2010 **Model:** Grand Vitara **Engine:** 2.4L L4 VIN 9, 3.2L V6 VIN 1 **Transmission:** All	**Lost Communication With ECM:** Receiving error of TCM from ECM for specified time continuously.

DTC: U0101
1T, MIL: Yes
Year: 2009, 2010
Model: Equator, Grand Vitara, SX4
Engine: 2.0L L4, 2.0L L4 VIN 5, 2.4L L4 VIN 0, 2.4L L4 VIN 9, 2.5L L4, 3.2L V6 VIN 1
Transmission: All
	CAN COMM Circuit - Lost communication with TCM:
	When ECM is not transmitting or receiving CAN communication signal of OBD (emission related diagnosis) with TCM for 2 seconds or more.
	Possible Causes:
	   • CAN communication line between TCM and ECM
	   • CAN communication line open or shorted

DTC: U0121 **1T, MIL:** Yes **Year:** 2009, 2010 **Model:** Grand Vitara, SX4 **Engine:** 2.0L L4, 2.0L L4 VIN 5, 2.4L L4 VIN 0, 2.4L L4 VIN 9, 3.2L V6 VIN 1 **Transmission:** All	**Lost Communication With ABS / ESP® Control Module:** Receiving error of ECM from ABS / ESP® control module for specified time continuously.

DTC: U0131
Year: 2010
Model: Kizashi
Engine: 2.4L L4 VIN 9
Transmission: All
	Lost Communication With Power Steering Control Module:
	Receiving error of the data from P/S control module via CAN for specified time continuously.
	Possible Causes:
	   • P/S control module
	   • CAN
	   • ECM

DTC: U0140
1T, MIL: Yes
Year: 2009, 2010
Model: Equator, Grand Vitara, SX4
Engine: 2.0L L4, 2.0L L4 VIN 5, 2.4L L4 VIN 0, 2.4L L4 VIN 9, 2.5L L4, 3.2L V6 VIN 1
Transmission: All
	CAN COMM Circuit - Lost communication with BCM:
	When ECM is not transmitting or receiving CAN communication signal of OBD (emission related diagnosis) with BCM for 2 seconds or more.
	Possible Causes:
	   • CAN communication line between BCM and ECM
	   • CAN communication line open or shorted

DTC	Trouble Code Title, Conditions & Possible Causes

DTC: U0141
Year: 2010
Model: Kizashi
Engine: 2.4L L4 VIN 9
Transmission: All
	Lost Communication With Body Control Module "keyless start":
	Receiving error of the data from keyless start control module via CAN for specified time continuously.
	Possible Causes:
	   • keyless start control module
	   • CAN
	   • ECM

DTC: U0155 **1T, MIL: Yes** **Year:** 2009, 2010 **Model:** Grand Vitara, SX4 **Engine:** 2.0L L4, 2.0L L4 VIN 5, 2.4L L4 VIN 0, 2.4L L4 VIN 9, 3.2L V6 VIN 1 **Transmission:** All	**Lost Communication With Instrument Panel Cluster (IPC) Control Module:** Receiving error of ECM from combination meter for specified time continuously.

OBD II Trouble Code List (U1XXX Codes)

DTC	Trouble Code Title, Conditions & Possible Causes

DTC: U1001
2T
Year: 2009, 2010
Model: Equator
Engine: 2.5L L4
Transmission: All
	CAN communication line:
	When ECM is not transmitting or receiving CAN communication signal other than OBD (emission related diagnosis) for 2 seconds or more.
	Possible Causes:
	   • Harness or connectors
	   • (CAN communication line is open or shorted)

GLOSSARY

ABS: Anti-lock braking system. An electro-mechanical braking system which is designed to minimize or prevent wheel lock-up during braking.

ABSOLUTE PRESSURE: Atmospheric (barometric) pressure plus the pressure gauge reading.

ACCELERATOR PUMP: A small pump located in the carburetor that feeds fuel into the air/fuel mixture during acceleration.

ACCUMULATOR: A device that controls shift quality by cushioning the shock of hydraulic oil pressure being applied to a clutch or band.

ACTUATING MECHANISM: The mechanical output devices of a hydraulic system, for example, clutch pistons and band servos.

ACTUATOR: The output component of a hydraulic or electronic system.

ADVANCE: Setting the ignition timing so that spark occurs earlier before the piston reaches top dead center (TDC).

ADAPTIVE MEMORY (ADAPTIVE STRATEGY): The learning ability of the TCM or PCM to redefine its decision-making process to provide optimum shift quality.

AFTER TOP DEAD CENTER (ATDC): The point after the piston reaches the top of its travel on the compression stroke.

AIR BAG: Device on the inside of the car designed to inflate on impact of crash, protecting the occupants of the car.

AIR CHARGE TEMPERATURE (ACT) SENSOR: The temperature of the airflow into the engine is measured by an ACT sensor, usually located in the lower intake manifold or air cleaner.

AIR CLEANER: An assembly consisting of a housing, filter and any connecting ductwork. The filter element is made up of a porous paper, sometimes with a wire mesh screening, and is designed to prevent airborne particles from entering the engine through the carburetor or throttle body.

AIR INJECTION: One method of reducing harmful exhaust emissions by injecting air into each of the exhaust ports of an engine. The fresh air entering the hot exhaust manifold causes any remaining fuel to be burned before it can exit the tailpipe.

AIR PUMP: An emission control device that supplies fresh air to the exhaust manifold to aid in more completely burning exhaust gases.

AIR/FUEL RATIO: The ratio of air-to-gasoline by weight in the fuel mixture drawn into the engine.

ALDL (assembly line diagnostic link): Electrical connector for scanning ECM/PCM/TCM input and output devices.

ALIGNMENT RACK: A special drive-on vehicle lift apparatus/measuring device used to adjust a vehicle's toe, caster and camber angles.

ALL WHEEL DRIVE: Term used to describe a full time four wheel drive system or any other vehicle drive system that continuously delivers power to all four wheels. This system is found primarily on station wagon vehicles and SUVs not utilized for significant off road use.

ALTERNATING CURRENT (AC): Electric current that flows first in one direction, then in the opposite direction, continually reversing flow.

ALTERNATOR: A device which produces AC (alternating current) which is converted to DC (direct current) to charge the car battery.

AMMETER: An instrument, calibrated in amperes, used to measure the flow of an electrical current in a circuit. Ammeters are always connected in series with the circuit being tested.

AMPERAGE: The total amount of current (amperes) flowing in a circuit.

AMPLIFIER: A device used in an electrical circuit to increase the voltage of an output signal.

AMP/HR. RATING (BATTERY): Measurement of the ability of a battery to deliver a stated amount of current for a stated period of time. The higher the amp/hr. rating, the better the battery.

AMPERE: The rate of flow of electrical current present when one volt of electrical pressure is applied against one ohm of electrical resistance.

ANALOG COMPUTER: Any microprocessor that uses similar (analogous) electrical signals to make its calculations.

ANODIZED: A special coating applied to the surface of aluminum valves for extended service life.

ANTIFREEZE: A substance (ethylene or propylene glycol) added to the coolant to prevent freezing in cold weather.

ANTI-FOAM AGENTS: Minimize fluid foaming from the whipping action encountered in the converter and planetary action.

ANTI-WEAR AGENTS: Zinc agents that control wear on the gears, bushings, and thrust washers.

ANTI-LOCK BRAKING SYSTEM: A supplementary system to the base hydraulic system that prevents sustained lock-up of the wheels during braking as well as automatically controlling wheel slip.

ANTI-ROLL BAR: See stabilizer bar.

ARC: A flow of electricity through the air between two electrodes or contact points that produces a spark.

ARMATURE: A laminated, soft iron core wrapped by a wire that converts electrical energy to mechanical energy as in a motor or relay. When rotated in a magnetic field, it changes mechanical energy into electrical energy as in a generator.

ATDC: After Top Dead Center.

ATF: Automatic transmission fluid.

ATMOSPHERIC PRESSURE: The pressure on the Earth's surface caused by the weight of the air in the atmosphere. At sea level, this pressure is 14.7 psi at 32°F (101 kPa at 0°C).

ATOMIZATION: The breaking down of a liquid into a fine mist that can be suspended in air.

AUXILIARY ADD-ON COOLER: A supplemental transmission fluid cooling device that is installed in series with the heat exchanger (cooler), located inside the radiator, to provide additional support to cool the hot fluid leaving the torque converter.

AUXILIARY PRESSURE: An added fluid pressure that is introduced into a regulator or balanced valve system to control valve movement. The auxiliary pressure itself can be either a fixed or a variable value. (See balanced valve; regulator valve.)

AWD: All wheel drive.

AXIAL FORCE: A side or end thrust force acting in or along the same plane as the power flow.

AXIAL PLAY: Movement parallel to a shaft or bearing bore.

AXLE CAPACITY: The maximum load-carrying capacity of the axle itself, as specified by the manufacturer. This is usually a higher number than the GAWR.

AXLE RATIO: This is a number (3.07:1, 4.56:1, for example) expressing the ratio between driveshaft revolutions and wheel revolutions. A low numerical ratio allows the engine to work easier because it doesn't have to turn as fast. A high numerical ratio means that the engine has to turn more rpm's to move the wheels through the same number of turns.

BACKFIRE: The sudden combustion of gases in the intake or exhaust system that results in a loud explosion.

BACKLASH: The clearance or play between two parts, such as meshed gears.

BACKPRESSURE: Restrictions in the exhaust system that slow the exit of exhaust gases from the combustion chamber.

BAKELITE®: A heat resistant, plastic insulator material commonly used in printed circuit boards and transistorized components.

BALANCED VALVE: A valve that is positioned by opposing auxiliary hydraulic pressures and/or spring force. Examples include mainline regulator, throttle, and governor valves. (See regulator valve.)

BAND: A flexible ring of steel with an inner lining of friction material. When tightened around the outside of a drum, a planetary member is held stationary to the transmission/transaxle case.

BALL BEARING: A bearing made up of hardened inner and outer races between which hardened steel balls roll.

BALL JOINT: A ball and matching socket connecting suspension components (steering knuckle to lower control arms). It permits rotating movement in any direction between the components that are joined.

BARO (BAROMETRIC PRESSURE SENSOR): Measures the change in the intake manifold pressure caused by changes in altitude.

BAROMETRIC MANIFOLD ABSOLUTE PRESSURE (BMAP) SENSOR: Operates similarly to a conventional MAP sensor; reads intake manifold pressure and is also responsible for determining altitude and barometric pressure prior to engine operation.

BAROMETRIC PRESSURE: (See atmospheric pressure.)

BALLAST RESISTOR: A resistor in the primary ignition circuit that lowers voltage after the engine is started to reduce wear on ignition components.

BATTERY: A direct current electrical storage unit, consisting of the basic active materials of lead and sulfuric acid, which converts chemical energy into electrical energy. Used to provide current for the operation of the starter as well as other equipment, such as the radio, lighting, etc.

BEAD: The portion of a tire that holds it on the rim.

BEARING: A friction reducing, supportive device usually located between a stationary part and a moving part.

BEFORE TOP DEAD CENTER (BTDC): The point just before the piston reaches the top of its travel on the compression stroke.

BELTED TIRE: Tire construction similar to bias-ply tires, but using two or more layers of reinforced belts between body plies and the tread.

BEZEL: Piece of metal surrounding radio, headlights, gauges or similar components; sometimes used to hold the glass face of a gauge in the dash.

BIAS-PLY TIRE: Tire construction, using body ply reinforcing cords which run at alternating angles to the center line of the tread.

BI-METAL TEMPERATURE SENSOR: Any sensor or switch made of two dissimilar types of metal that bend when heated or cooled due to the different expansion rates of the alloys. These types of sensors usually function as an on/off switch.

BLOCK: See Engine Block.

BLOW-BY: Combustion gases, composed of water vapor and unburned fuel, that leak past the piston rings into the crankcase during normal engine operation. These gases are removed by the PCV system to prevent the buildup of harmful acids in the crankcase.

BOOK TIME: See Labor Time.

BOOK VALUE: The average value of a car, widely used to determine trade-in and resale value.

BOOST VALVE: Used at the base of the regulator valve to increase mainline pressure.

BORE: Diameter of a cylinder.

BRAKE CALIPER: The housing that fits over the brake disc. The caliper holds the brake pads, which are pressed against the discs by the caliper pistons when the brake pedal is depressed.

BRAKE HORSEPOWER (BHP): The actual horsepower available at the engine flywheel as measured by a dynamometer.

BRAKE FADE: Loss of braking power, usually caused by excessive heat after repeated brake applications.

BRAKE HORSEPOWER: Usable horsepower of an engine measured at the crankshaft.

BRAKE PAD: A brake shoe and lining assembly used with disc brakes.

BRAKE PROPORTIONING VALVE: A valve on the master cylinder which restricts hydraulic brake pressure to the wheels to a specified amount, preventing wheel lock-up.

BREAKAWAY: Often used by Chrysler to identify first-gear operation in D and 2 ranges. In these ranges, first-gear operation depends on a one-way roller clutch that holds on acceleration and releases (breaks away) on deceleration, resulting in a freewheeling coast-down condition.

BRAKE SHOE: The backing for the brake lining. The term is, however, usually applied to the assembly of the brake backing and lining.

BREAKER POINTS: A set of points inside the distributor, operated by a cam, which make and break the ignition circuit.

BRINNELLING: A wear pattern identified by a series of indentations at regular intervals. This condition is caused by a lack of lube, overload situations, and/or vibrations.

BTDC: Before Top Dead Center.

BUMP: Sudden and forceful apply of a clutch or band.

BUSHING: A liner, usually removable, for a bearing; an anti-friction liner used in place of a bearing.

CALIFORNIA ENGINE: An engine certified by the EPA for use in California only; conforms to more stringent emission regulations than Federal engine.

CALIPER: A hydraulically activated device in a disc brake system,

which is mounted straddling the brake rotor (disc). The caliper contains at least one piston and two brake pads. Hydraulic pressure on the piston(s) forces the pads against the rotor.

CAPACITY: The quantity of electricity that can be delivered from a unit, as from a battery in ampere-hours, or output, as from a generator.

CAMBER: One of the factors of wheel alignment. Viewed from the front of the car, it is the inward or outward tilt of the wheel. The top of the tire will lean outward (positive camber) or inward (negative camber).

CAMSHAFT: A shaft in the engine on which are the lobes (cams) which operate the valves. The camshaft is driven by the crankshaft, via a belt, chain or gears, at one half the crankshaft speed.

CAPACITOR: A device which stores an electrical charge.

CARBON MONOXIDE (CO): A colorless, odorless gas given off as a normal byproduct of combustion. It is poisonous and extremely dangerous in confined areas, building up slowly to toxic levels without warning if adequate ventilation is not available.

CARBURETOR: A device, usually mounted on the intake manifold of an engine, which mixes the air and fuel in the proper proportion to allow even combustion.

CASTER: The forward or rearward tilt of an imaginary line drawn through the upper ball joint and the center of the wheel. Viewed from the sides, positive caster (forward tilt) lends directional stability, while negative caster (rearward tilt) produces instability.

CATALYTIC CONVERTER: A device installed in the exhaust system, like a muffler, that converts harmful byproducts of combustion into carbon dioxide and water vapor by means of a heat-producing chemical reaction.

CENTRIFUGAL ADVANCE: A mechanical method of advancing the spark timing by using flyweights in the distributor that react to centrifugal force generated by the distributor shaft rotation.

CENTRIFUGAL FORCE: The outward pull of a revolving object, away from the center of revolution. Centrifugal force increases with the speed of rotation.

CETANE RATING: A measure of the ignition value of diesel fuel. The higher the cetane rating, the better the fuel. Diesel fuel cetane rating is roughly comparable to gasoline octane rating.

CHECK VALVE: Any one-way valve installed to permit the flow of air, fuel or vacuum in one direction only.

CHOKE: The valve/plate that restricts the amount of air entering an engine on the induction stroke, thereby enriching the air/fuel ratio.

CHUGGLE: Bucking or jerking condition that may be engine related and may be most noticeable when converter clutch is engaged; similar to the feel of towing a trailer.

CIRCLIP: A split steel snapring that fits into a groove to hold various parts in place.

CIRCUIT BREAKER: A switch which protects an electrical circuit from overload by opening the circuit when the current flow exceeds a pre-determined level. Some circuit breakers must be reset manually, while most reset automatically.

CIRCUIT: Any unbroken path through which an electrical current can flow. Also used to describe fuel flow in some instances.

CIRCUIT, BYPASS: Another circuit in parallel with the major circuit through which power is diverted.

CIRCUIT, CLOSED: An electrical circuit in which there is no interruption of current flow.

CIRCUIT, GROUND: The non-insulated portion of a complete circuit used as a common potential point. In automotive circuits, the ground is composed of metal parts, such as the engine, body sheet metal, and frame and is usually a negative potential.

CIRCUIT, HOT: That portion of a circuit not at ground potential. The hot circuit is usually insulated and is connected to the positive side of the battery.

CIRCUIT, OPEN: A break or lack of contact in an electrical circuit, either intentional (switch) or unintentional (bad connection or broken wire).

CIRCUIT, PARALLEL: A circuit having two or more paths for current flow with common positive and negative tie points. The same voltage is applied to each load device or parallel branch.

CIRCUIT, SERIES: An electrical system in which separate parts are connected end to end, using one wire, to form a single path for current to flow.

CIRCUIT, SHORT: A circuit that is accidentally completed in an electrical path for which it was not intended.

CLAMPING (ISOLATION) DIODES: Diodes positioned in a circuit to prevent self-induction from damaging electronic components.

CLEARCOAT: A transparent layer which, when sprayed over a vehicle's paint job, adds gloss and depth as well as an additional protective coating to the finish.

CLUTCH: Part of the power train used to connect/disconnect power to the rear wheels.

CLUTCH, FLUID: The same as a fluid coupling. A fluid clutch or coupling performs the same function as a friction clutch by utilizing fluid friction and inertia as opposed to solid friction used by a friction clutch. (See fluid coupling.)

CLUTCH, FRICTION: A coupling device that provides a means of smooth and positive engagement and disengagement of engine torque to the vehicle powertrain. Transmission of power through the clutch is accomplished by bringing one or more rotating drive members into contact with complementing driven members.

COAST: Vehicle deceleration caused by engine braking conditions.

COEFFICIENT OF FRICTION: The amount of surface tension between two contacting surfaces; identified by a scientifically calculated number.

COIL: Part of the ignition system that boosts the relatively low voltage supplied by the car's electrical system to the high voltage required to fire the spark plugs.

COMBINATION MANIFOLD: An assembly which includes both the intake and exhaust manifolds in one casting.

COMBINATION VALVE: A device used in some fuel systems that routes fuel vapors to a charcoal storage canister instead of venting them into the atmosphere. The valve relieves fuel tank pressure and allows fresh air into the tank as the fuel level drops to prevent a vapor lock situation.

COMBUSTION CHAMBER: The part of the engine in the cylinder head where combustion takes place.

COMPOUND GEAR: A gear consisting of two or more simple gears with a common shaft.

COMPOUND PLANETARY: A gearset that has more than the three elements found in a simple gearset and is constructed by combining members of two planetary gearsets to create additional gear ratio possibilities.

COMPRESSION CHECK: A test involving removing each spark plug and inserting a gauge. When the engine is cranked, the gauge will record a pressure reading in the individual cylinder. General operating condition can be determined from a compression check.

COMPRESSION RATIO: The ratio of the volume between the piston and cylinder head when the piston is at the bottom of its stroke (bottom dead center) and when the piston is at the top of its stroke (top dead center).

COMPUTER: An electronic control module that correlates input data according to prearranged engineered instructions; used for the management of an actuator system or systems.

CONDENSER: An electrical device which acts to store an electrical charge, preventing voltage surges.

2. A radiator-like device in the air conditioning system in which refrigerant gas condenses into a liquid, giving off heat.

CONDUCTOR: Any material through which an electrical current can be transmitted easily.

CONNECTING ROD: The connecting link between the crankshaft and piston.

CONSTANT VELOCITY JOINT: Type of universal joint in a halfshaft assembly in which the output shaft turns at a constant angular velocity without variation, provided that the speed of the input shaft is constant.

CONTINUITY: Continuous or complete circuit. Can be checked with an ohmmeter.

CONTROL ARM: The upper or lower suspension components which are mounted on the frame and support the ball joints and steering knuckles.

CONVENTIONAL IGNITION: Ignition system which uses breaker points.

CONVERTER: (See torque converter.)

CONVERTER LOCKUP: The switching from hydrodynamic to direct mechanical drive, usually through the application of a friction element called the converter clutch.

COOLANT: Mixture of water and anti-freeze circulated through the engine to carry off heat produced by the engine.

CORROSION INHIBITOR: An inhibitor in ATF that prevents corrosion of bushings, thrust washers, and oil cooler brazed joints.

COUNTERSHAFT: An intermediate shaft which is rotated by a mainshaft and transmits, in turn, that rotation to a working part.

COUPLING PHASE: Occurs when the torque converter is operating at its greatest hydraulic efficiency. The speed differential between the impeller and the turbine is at its minimum. At this point, the stator freewheels, and there is no torque multiplication.

CRANKCASE: The lower part of an engine in which the crankshaft and related parts operate.

CRANKSHAFT: Engine component (connected to pistons by connecting rods) which converts the reciprocating (up and down) motion of pistons to rotary motion used to turn the driveshaft.

CURB WEIGHT: The weight of a vehicle without passengers or payload, but including all fluids (oil, gas, coolant, etc.) and other equipment specified as standard.

CURRENT: The flow (or rate) of electrons moving through a circuit. Current is measured in amperes (amp).

CURRENT FLOW CONVENTIONAL: Current flows through a circuit from the positive terminal of the source to the negative terminal (plus to minus).

CURRENT FLOW, ELECTRON: Current or electrons flow from the negative terminal of the source, through the circuit, to the positive terminal (minus to plus).

CV-JOINT: Constant velocity joint.

CYCLIC VIBRATIONS: The off-center movement of a rotating object that is affected by its initial balance, speed of rotation, and working angles.

CYLINDER BLOCK: See engine block.

CYLINDER HEAD: The detachable portion of the engine, usually fastened to the top of the cylinder block and containing all or most of the combustion chambers. On overhead valve engines, it contains the valves and their operating parts. On overhead cam engines, it contains the camshaft as well.

CYLINDER: In an engine, the round hole in the engine block in which the piston(s) ride.

DATA LINK CONNECTOR (DLC): Current acronym/term applied to the federally mandated, diagnostic junction connector that is used to monitor ECM/PC/TCM inputs, processing strategies, and outputs including diagnostic trouble codes (DTCs).

DEAD CENTER: The extreme top or bottom of the piston stroke.

DECELERATION BUMP: When referring to a torque converter clutch in the applied position, a sudden release of the accelerator pedal causes a forceful reversal of power through the drivetrain (engine braking), just prior to the apply plate actually being released.

DELAYED (LATE OR EXTENDED): Condition where shift is expected but does not occur for a period of time, for example, where clutch or band engagement does not occur as quickly as expected during part throttle or wide open throttle apply of accelerator or when manually downshifting to a lower range.

DETENT: A spring-loaded plunger, pin, ball, or pawl used as a holding device on a ratchet wheel or shaft. In automatic transmissions, a detent mechanism is used for locking the manual valve in place.

DETENT DOWNSHIFT: (See kickdown.)

DETERGENT: An additive in engine oil to improve its operating characteristics.

DETONATION: An unwanted explosion of the air/fuel mixture in the combustion chamber caused by excess heat and compression, advanced timing, or an overly lean mixture. Also referred to as "ping".

DEXRON®: A brand of automatic transmission fluid.

DIAGNOSTIC TROUBLE CODES (DTCs): A digital display from the control module memory that identifies the input, processor, or output device circuit that is related to the powertrain emission/driveability malfunction detected. Diagnostic trouble codes can be read by the MIL to flash any codes or by using a handheld scanner.

DIAPHRAGM: A thin, flexible wall separating two cavities, such as in a vacuum advance unit.

DIESELING: The engine continues to run after the car is shut off; caused by fuel continuing to be burned in the combustion chamber.

DIFFERENTIAL: A geared assembly which allows the transmission of motion between drive axles, giving one axle the ability to rotate faster than the other, as in cornering.

DIFFERENTIAL AREAS: When opposing faces of a spool valve are acted upon by the same pressure but their areas differ in size, the face with the larger area produces the differential force and valve movement. (See spool valve.)

DIFFERENTIAL FORCE: (See differential areas)

DIGITAL READOUT: A display of numbers or a combination of numbers and letters.

DIGITAL VOLT OHMMETER: An electronic diagnostic tool used to measure voltage, ohms and amps as well as several other functions, with the readings displayed on a digital screen in tenths, hundredths and thousandths.

DIODE: An electrical device that will allow current to flow in one direction only.

DIRECT CURRENT (DC): Electrical current that flows in one direction only.

DIRECT DRIVE: The gear ratio is 1:1, with no change occurring in the torque and speed input/output relationship.

DISC BRAKE: A hydraulic braking assembly consisting of a brake disc, or rotor, mounted on an axle shaft, and a caliper assembly containing, usually two brake pads which are activated by hydraulic pressure. The pads are forced against the sides of the disc, creating friction which slows the vehicle.

DISPERSANTS: Suspend dirt and prevent sludge buildup in a liquid, such as engine oil.

DOUBLE BUMP (DOUBLE FEEL): Two sudden and forceful applies of a clutch or band.

DISPLACEMENT: The total volume of air that is displaced by all pistons as the engine turns through one complete revolution.

DISTRIBUTOR: A mechanically driven device on an engine which is responsible for electrically firing the spark plug at a pre-determined point of the piston stroke.

DOHC: Double overhead camshaft.

DOUBLE OVERHEAD CAMSHAFT: The engine utilizes two camshafts mounted in one cylinder head. One camshaft operates the exhaust valves, while the other operates the intake valves.

DOWEL PIN: A pin, inserted in mating holes in two different parts allowing those parts to maintain a fixed relationship.

DRIVELINE: The drive connection between the transmission and the drive wheels.

DRIVE TRAIN: The components that transmit the flow of power from the engine to the wheels. The components include the clutch, transmission, driveshafts (or axle shafts in front wheel drive), U-joints and differential.

DRUM BRAKE: A braking system which consists of two brake shoes and one or two wheel cylinders, mounted on a fixed backing plate, and a brake drum, mounted on an axle, which revolves around the assembly.

DRY CHARGED BATTERY: Battery to which electrolyte is added when the battery is placed in service.

DVOM: Digital volt ohmmeter

DWELL: The rate, measured in degrees of shaft rotation, at which an electrical circuit cycles on and off.

DYNAMIC: An application in which there is rotating or reciprocating motion between the parts.

EARLY: Condition where shift occurs before vehicle has reached proper speed, which tends to labor engine after upshift.

EBCM: See Electronic Control Unit (ECU).

ECM: See Electronic Control Unit (ECU).

ECU: Electronic control unit.

ELECTRODE: Conductor (positive or negative) of electric current.

ELECTROLYSIS: A surface etching or bonding of current conducting transmission/transaxle components that may occur when grounding straps are missing or in poor condition.

ELECTROLYTE: A solution of water and sulfuric acid used to activate the battery. Electrolyte is extremely corrosive.

ELECTROMAGNET: A coil that produces a magnetic field when current flows through its windings.

ELECTROMAGNETIC INDUCTION: A method to create (generate) current flow through the use of magnetism.

ELECTROMAGNETISM: The effects surrounding the relationship between electricity and magnetism.

ELECTROMOTIVE FORCE (EMF): The force or pressure (voltage) that causes current movement in an electrical circuit.

ELECTRONIC CONTROL UNIT: A digital computer that controls engine (and sometimes transmission, brake or other vehicle system) functions based on data received from various sensors. Examples used by some manufacturers include Electronic Brake Control Module (EBCM), Engine Control Module (ECM), Powertrain Control Module (PCM) or Vehicle Control Module (VCM).

ELECTRONIC IGNITION: A system in which the timing and firing of the spark plugs is controlled by an electronic control unit, usually called a module. These systems have no points or condenser.

ELECTRONIC PRESSURE CONTROL (EPC) SOLENOID: A specially designed solenoid containing a spool valve and spring assembly to control fluid mainline pressure. A variable current flow, controlled by the ECM/PCM, varies the internal force of the solenoid on the spool valve and resulting mainline pressure. (See variable force solenoid.)

ELECTRONICS: Miniaturized electrical circuits utilizing semiconductors, solid-state devices, and printed circuits. Electronic circuits utilize small amounts of power.

ELECTRONIFICATION: The application of electronic circuitry to a mechanical device. Regarding automatic transmissions, electrification is incorporated into converter clutch lockup, shift scheduling, and line pressure control systems.

ELECTROSTATIC DISCHARGE (ESD): An unwanted, high-voltage electrical current released by an individual who has taken on a static charge of electricity. Electronic components can be easily damaged by ESD.

ELEMENT: A device within a hydrodynamic drive unit designed with a set of blades to direct fluid flow.

ENAMEL: Type of paint that dries to a smooth, glossy finish.

END BUMP (END FEEL OR SLIP BUMP): Firmer feel at end of shift when compared with feel at start of shift.

END-PLAY: The clearance/gap between two components that allows for expansion of the parts as they warm up, to prevent binding and to allow space for lubrication.

ENERGY: The ability or capacity to do work.

ENGINE: The primary motor or power apparatus of a vehicle, which converts liquid or gas fuel into mechanical energy.

ENGINE BLOCK: The basic engine casting containing the cylinders, the crankshaft main bearings, as well as machined surfaces for the mounting of other components such as the cylinder head, oil pan, transmission, etc.

ENGINE BRAKING: Use of engine to slow vehicle by manually downshifting during zero-throttle coast down.

ENGINE CONTROL MODULE (ECM): Manages the engine and incorporates output control over the torque converter clutch solenoid. (Note: Current designation for the ECM in late model vehicles is PCM.)

ENGINE COOLANT TEMPERATURE (ECT) SENSOR: Prevents converter clutch engagement with a cold engine; also used for shift timing and shift quality.

EP LUBRICANT: EP (extreme pressure) lubricants are specially formulated for use with gears involving heavy loads (transmissions, differentials, etc.).

ETHYL: A substance added to gasoline to improve its resistance to knock, by slowing down the rate of combustion.

ETHYLENE GLYCOL: The base substance of antifreeze.

EXHAUST MANIFOLD: A set of cast passages or pipes which conduct exhaust gases from the engine.

FAIL-SAFE (BACKUP) CONTROL: A substitute value used by the PCM/TCM to replace a faulty signal from an input sensor. The temporary value allows the vehicle to continue to be operated.

FAST IDLE: The speed of the engine when the choke is on. Fast idle speeds engine warm-up.

FEDERAL ENGINE: An engine certified by the EPA for use in any of the 49 states (except California).

FEEDBACK: A circuit malfunction whereby current can find another path to feed load devices.

FEELER GAUGE: A blade, usually metal, of precisely predetermined thickness, used to measure the clearance between two parts.

FILAMENT: The part of a bulb that glows; the filament creates high resistance to current flow and actually glows from the resulting heat.

FINAL DRIVE: An essential part of the axle drive assembly where final gear reduction takes place in the powertrain. In RWD applications and north-south FWD applications, it must also change the power flow direction to the axle shaft by ninety degrees. (Also see axle ratio).

FIRING ORDER: The order in which combustion occurs in the cylinders of an engine. Also the order in which spark is distributed to the plugs by the distributor.

FIRM: A noticeable quick apply of a clutch or band that is considered normal with medium to heavy throttle shift; should not be confused with harsh or rough.

FLAME FRONT: The term used to describe certain aspects of the fuel explosion in the cylinders. The flame front should move in a controlled pattern across the cylinder, rather than simply exploding immediately.

FLARE (SLIPPING): A quick increase in engine rpm accompanied by momentary loss of torque; generally occurs during shift.

FLAT ENGINE: Engine design in which the pistons are horizontally opposed. Porsche, Subaru and some old VW are common examples of flat engines.

FLAT RATE: A dealership term referring to the amount of money paid to a technician for a repair or diagnostic service based on that particular service versus dealership's labor time (NOT based on the actual time the technician spent on the job).

FLAT SPOT: A point during acceleration when the engine seems to lose power for an instant.

FLOODING: The presence of too much fuel in the intake manifold and combustion chamber which prevents the air/fuel mixture from firing, thereby causing a no-start situation.

FLUID: A fluid can be either liquid or gas. In hydraulics, a liquid is used for transmitting force or motion.

FLUID COUPLING: The simplest form of hydrodynamic drive, the fluid coupling consists of two look-alike members with straight radial varies referred to as the impeller (pump) and the turbine. Input torque is always equal to the output torque.

FLUID DRIVE: Either a fluid coupling or a fluid torque converter. (See hydrodynamic drive units.)

FLUID TORQUE CONVERTER: A hydrodynamic drive that has the ability to act both as a torque multiplier and fluid coupling. (See hydrodynamic drive units; torque converter.)

FLUID VISCOSITY: The resistance of a liquid to flow. A cold fluid (oil) has greater viscosity and flows more slowly than a hot fluid (oil).

FLYWHEEL: A heavy disc of metal attached to the rear of the crankshaft. It smoothes the firing impulses of the engine and keeps the crankshaft turning during periods when no firing takes place. The starter also engages the flywheel to start the engine.

FOOT POUND (ft. lbs., lbs. ft. or sometimes, ft. lb.): The amount of energy or work needed to raise an item weighing one pound, a distance of one foot.

FREEZE PLUG: A plug in the engine block which will be pushed out if the coolant freezes. Sometimes called expansion plugs, they protect the block from cracking should the coolant freeze.

FRICTION: The resistance that occurs between contacting surfaces. This relationship is expressed by a ratio called the coefficient of friction (CL).

FRICTION, COEFFICIENT OF: The amount of surface tension between two contacting surfaces; expressed by a scientifically calculated number.

FRONT END ALIGNMENT: A service to set caster, camber and toe-in to the correct specifications. This will ensure that the car steers and handles properly and that the tires wear properly.

FRICTION MODIFIER: Changes the coefficient of friction of the fluid between the mating steel and composition clutch/band surfaces during the engagement process and allows for a certain amount of intentional slipping for a good "shift-feel".

FRONTAL AREA: The total frontal area of a vehicle exposed to air flow.

FUEL FILTER: A component of the fuel system containing a porous paper element used to prevent any impurities from entering the engine through the fuel system. It usually takes the form of a canister-like housing, mounted in-line with the fuel hose, located anywhere on a vehicle between the fuel tank and engine.

FUEL INJECTION: A system replacing the carburetor that sprays fuel into the cylinder through nozzles. The amount of fuel can be more precisely controlled with fuel injection.

FULL FLOATING AXLE: An axle in which the axle housing extends through the wheel giving bearing support on the outside of the housing. The front axle of a four-wheel drive vehicle is usually a full floating axle, as are the rear axles of many larger (1 ton and over) pick-ups and vans.

FULL-TIME FOUR-WHEEL DRIVE: A four-wheel drive system that continuously delivers power to all four wheels. A differential between the front and rear driveshafts permits variations in axle speeds to control gear wind-up without damage.

FULL THROTTLE DETENT DOWNSHIFT: A quick apply of accelerator pedal to its full travel, forcing a downshift.

FUSE: A protective device in a circuit which prevents circuit overload by breaking the circuit when a specific amperage is present. The device is constructed around a strip or wire of a lower amperage rating than the circuit it is designed to protect. When an amperage higher than that stamped on the fuse is present in the circuit, the strip or wire melts, opening the circuit.

FUSIBLE LINK: A piece of wire in a wiring harness that performs the same job as a fuse. If overloaded, the fusible link will melt and interrupt the circuit.

FWD: Front wheel drive.

GAWR: (Gross axle weight rating) the total maximum weight an axle is designed to carry.

GCW: (Gross combined weight) total combined weight of a tow vehicle and trailer.

GARAGE SHIFT: initial engagement feel of transmission, neutral to reverse or neutral to a forward drive.

GARAGE SHIFT FEEL: A quick check of the engagement quality and responsiveness of reverse and forward gears. This test is done with the vehicle stationary.

GEAR: A toothed mechanical device that acts as a rotating lever to transmit power or turning effort from one shaft to another. (See gear ratio.)

GEAR RATIO: A ratio expressing the number of turns a smaller gear will make to turn a larger gear through one revolution. The ratio is found by dividing the number of teeth on the smaller gear into the number of teeth on the larger gear.

GEARBOX: Transmission

GEAR REDUCTION: Torque is multiplied and speed decreased by the factor of the gear ratio. For example, a 3:1 gear ratio changes an input torque of 180 ft. lbs. and an input speed of 2700 rpm to 540 Ft. lbs. and 900 rpm, respectively. (No account is taken of frictional losses, which are always present.)

GEARTRAIN: A succession of intermeshing gears that form an assembly and provide for one or more torque changes as the power input is transmitted to the power output.

GEL COAT: A thin coat of plastic resin covering fiberglass body panels.

GENERATOR: A device which produces direct current (DC) necessary to charge the battery.

GOVERNOR: A device that senses vehicle speed and generates a hydraulic oil pressure. As vehicle speed increases, governor oil pressure rises.

GROUND CIRCUIT: (See circuit, ground.)

GROUND SIDE SWITCHING: The electrical/electronic circuit control switch is located after the circuit load.

GVWR: (Gross vehicle weight rating) total maximum weight a vehicle is designed to carry including the weight of the vehicle, passengers, equipment, gas, oil, etc.

HALOGEN: A special type of lamp known for its quality of brilliant white light. Originally used for fog lights and driving lights.

HARD CODES: DTCs that are present at the time of testing; also called continuous or current codes.

HARSH(ROUGH): An apply of a clutch or band that is more noticeable than a firm one; considered undesirable at any throttle position.

HEADER TANK: An expansion tank for the radiator coolant. It can be located remotely or built into the radiator.

HEAT RANGE: A term used to describe the ability of a spark plug to carry away heat. Plugs with longer nosed insulators take longer to carry heat off effectively.

HEAT RISER: A flapper in the exhaust manifold that is closed when the engine is cold, causing hot exhaust gases to heat the intake manifold providing better cold engine operation. A thermostatic spring opens the flapper when the engine warms up.

HEAVY THROTTLE: Approximately three-fourths of accelerator pedal travel.

HEMI: A name given an engine using hemispherical combustion chambers.

HERTZ (HZ): The international unit of frequency equal to one cycle per second (10,000 Hertz equals 10,000 cycles per second).

HIGH-IMPEDANCE DVOM (DIGITAL VOLT-OHMMETER): This styled device provides a built-in resistance value and is capable of limiting circuit current flow to safe milliamp levels.

HIGH RESISTANCE: Often refers to a circuit where there is an excessive amount of opposition to normal current flow.

HORSEPOWER: A measurement of the amount of work; one horsepower is the amount of work necessary to lift 33,000 lbs. one foot in one minute. Brake horsepower (bhp) is the horsepower delivered by an engine on a dynamometer. Net horsepower is the power remaining (measured at the flywheel of the engine) that can be used to turn the wheels after power is consumed through friction and running the engine accessories (water pump, alternator, air pump, fan etc.)

HOT CIRCUIT: (See circuit, hot; hot lead.)

HOT LEAD: A wire or conductor in the power side of the circuit. (See circuit, hot.)

HOT SIDE SWITCHING: The electrical/electronic circuit control switch is located before the circuit load.

HUB: The center part of a wheel or gear.

HUNTING (BUSYNESS): Repeating quick series of up-shifts and downshifts that causes noticeable change in engine rpm, for example, as in a 4-3-4 shift pattern.

HYDRAULICS: The use of liquid under pressure to transfer force of motion.

HYDROCARBON (HC): Any chemical compound made up of hydrogen and carbon. A major pollutant formed by the engine as a by-product of combustion.

HYDRODYNAMIC DRIVE UNITS: Devices that transmit power solely by the action of a kinetic fluid flow in a closed recirculating path. An impeller energizes the fluid and discharges the high-speed jet stream into the turbine for power output.

HYDROMETER: An instrument used to measure the specific gravity of a solution.

HYDROPLANING: A phenomenon of driving when water builds up under the tire tread, causing it to lose contact with the road. Slowing down will usually restore normal tire contact with the road.

HYPOID GEARSET: The drive pinion gear may be placed below or above the centerline of the driven gear; often used as a final drive gearset.

IDLE MIXTURE: The mixture of air and fuel (usually about 14:1) being fed to the cylinders. The idle mixture screw(s) are sometimes adjusted as part of a tune-up.

IDLER ARM: Component of the steering linkage which is a geometric duplicate of the steering gear arm. It supports the right side of the center steering link.

IMPELLER: Often called a pump, the impeller is the power input (drive) member of a hydrodynamic drive. As part of the torque converter cover, it acts as a centrifugal pump and puts the fluid in motion.

INCH POUND (inch lbs.; sometimes in. lb. or in. lbs.): One twelfth of a foot pound.

INDUCTANCE: The force that produces voltage when a conductor is passed through a magnetic field.

INDUCTION: A means of transferring electrical energy in the form of a magnetic field. Principle used in the ignition coil to increase voltage.

INITIAL FEEL: A distinct firmer feel at start of shift when compared with feel at finish of shift.

INJECTOR: A device which receives metered fuel under relatively low pressure and is activated to inject the fuel into the engine under relatively high pressure at a predetermined time.

INPUT: In an automatic transmission, the source of power from the engine is absorbed by the torque converter, which provides the power input into the transmission. The turbine drives the input(turbine)shaft.

INPUT SHAFT: The shaft to which torque is applied, usually carrying the driving gear or gears.

INTAKE MANIFOLD: A casting of passages or pipes used to conduct air or a fuel/air mixture to the cylinders.

INTERNAL GEAR: The ring-like outer gear of a planetary gearset with the gear teeth cut on the inside of the ring to provide a mesh with the planet pinions.

ISOLATION (CLAMPING) DIODES: Diodes positioned in a circuit to prevent self-induction from damaging electronic components.

IX ROTARY GEAR PUMP: Contains two rotating members, one shaped with internal gear teeth and the other with external gear teeth. As the gears separate, the fluid fills the gaps between gear teeth, is pulled across a crescent-shaped divider, and then is forced to flow through the outlet as the gears mesh.

IX ROTARY LOBE PUMP: Sometimes referred to as a gerotor type pump. Two rotating members, one shaped with internal lobes and the other with external lobes, separate and then mesh to cause fluid to flow.

JOURNAL: The bearing surface within which a shaft operates.

JUMPER CABLES: Two heavy duty wires with large alligator clips used to provide power from a charged battery to a discharged battery mounted in a vehicle.

JUMPSTART: Utilizing the sufficiently charged battery of one vehicle to start the engine of another vehicle with a discharged battery by the use of jumper cables.

KEY: A small block usually fitted in a notch between a shaft and a hub to prevent slippage of the two parts.

KICKDOWN: Detent downshift system; either linkage, cable, or electrically controlled.

KILO: A prefix used in the metric system to indicate one thousand.

KNOCK: Noise which results from the spontaneous ignition of a portion of the air-fuel mixture in the engine cylinder caused by overly advanced ignition timing or use of incorrectly low octane fuel for that engine.

KNOCK SENSOR: An input device that responds to spark knock, caused by over advanced ignition timing.

LABOR TIME: A specific amount of time required to perform a certain repair or diagnostic service as defined by a vehicle or after-market manufacturer .

LACQUER: A quick-drying automotive paint.

LATE: Shift that occurs when engine is at higher than normal rpm for given amount of throttle.

LIGHT-EMITTING DIODE (LED): A semiconductor diode that emits light as electrical current flows through it; used in some electronic display devices to emit a red or other color light.

LIGHT THROTTLE: Approximately one-fourth of accelerator pedal travel.

LIMITED SLIP: A type of differential which transfers driving force to the wheel with the best traction.

LIMP-IN MODE: Electrical shutdown of the transmission/ transaxle output solenoids, allowing only forward and reverse gears that are hydraulically energized by the manual valve. This permits the vehicle to be driven to a service facility for repair.

LIP SEAL: Molded synthetic rubber seal designed with an outer sealing edge (lip) that points into the fluid containing area to be sealed. This type of seal is used where rotational and axial forces are present.

LITHIUM-BASE GREASE: Chassis and wheel bearing grease using lithium as a base. Not compatible with sodium-base grease.

LOAD DEVICE: A circuit's resistance that converts the electrical energy into light, sound, heat, or mechanical movement.

LOAD RANGE: Indicates the number of plies at which a tire is rated. Load range B equals four-ply rating; C equals six-ply rating; and, D equals an eight-ply rating.

LOAD TORQUE: The amount of output torque needed from the transmission/transaxle to overcome the vehicle load.

LOCKING HUBS: Accessories used on part-time four-wheel drive systems that allow the front wheels to be disengaged from the drive train when four-wheel drive is not being used. When four-wheel drive is desired, the hubs are engaged, locking the wheels to the drive train.

LOCKUP CONVERTER: A torque converter that operates hydraulically and mechanically. When an internal apply plate (lockup plate) clamps to the torque converter cover, hydraulic slippage is eliminated.

LOCK RING: See Circlip or Snapring

MAGNET: Any body with the property of attracting iron or steel.

MAGNETIC FIELD: The area surrounding the poles of a magnet that is affected by its attraction or repulsion forces.

MAIN LINE PRESSURE: Often called control pressure or line pressure, it refers to the pressure of the oil leaving the pump and is controlled by the pressure regulator valve.

MALFUNCTION INDICATOR LAMP (MIL): Previously known as a check engine light, the dash-mounted MIL illuminates and signals the driver that an emission or driveability problem with the powertrain has been detected by the ECM/PCM. When this occurs, at least one diagnostic trouble code (DTC) has been stored into the control module memory.

MANIFOLD ABSOLUTE PRESSURE (MAP) SENSOR: Reads the amount of air pressure (vacuum) in the engine's intake manifold system; its signal is used to analyze engine load conditions.

MANIFOLD VACUUM: Low pressure in an engine intake manifold formed just below the throttle plates. Manifold vacuum is highest at idle and drops under acceleration.

MANIFOLD: A casting of passages or set of pipes which connect the cylinders to an inlet or outlet source.

MANUAL LEVER POSITION SWITCH (MLPS): A mechanical switching unit that is typically mounted externally to the transmission/transaxle to inform the PCM/ECM which gear range the driver has selected.

MANUAL VALVE: Located inside the transmission/transaxle, it is directly connected to the driver's shift lever. The position of the manual valve determines which hydraulic circuits will be charged with oil pressure and the operating mode of the transmission.

MANUAL VALVE LEVER POSITION SENSOR (MVLPS): The input from this device tells the TCM what gear range was selected.

MASS AIR FLOW (MAF) SENSOR: Measures the airflow into the engine.

MASTER CYLINDER: The primary fluid pressurizing device in a hydraulic system. In automotive use, it is found in brake and hydraulic clutch systems and is pedal activated, either directly or, in a power brake system, through the power booster.

MacPherson STRUT: A suspension component combining a shock absorber and spring in one unit.

MEDIUM THROTTLE: Approximately one-half of accelerator pedal travel.

MEGA: A metric prefix indicating one million.

MEMBER: An independent component of a hydrodynamic unit such as an impeller, a stator, or a turbine. It may have one or more elements.

MERCON: A fluid developed by Ford Motor Company in 1988. It contains a friction modifier and closely resembles operating characteristics of Dexron.

METAL SEALING RINGS: Made from cast iron or aluminum, their primary application is with dynamic components involving pressure sealing circuits of rotating members. These rings are designed with either butt or hook lock end joints.

METER (ANALOG): A linear-style meter representing data as lengths; a needle-style instrument interfacing with logical numerical increments. This style of electrical meter uses relatively low impedance internal resistance and cannot be used for testing electronic circuitry.

METER (DIGITAL): Uses numbers as a direct readout to show values. Most meters of this style use high impedance internal resistance and must be used for testing low current electronic circuitry.

MICRO: A metric prefix indicating one-millionth (0.000001).

MILLI: A metric prefix indicating one-thousandth (0.001).

MINIMUM THROTTLE: The least amount of throttle opening required for upshift; normally close to zero throttle.

MISFIRE: Condition occurring when the fuel mixture in a cylinder fails to ignite, causing the engine to run roughly.

MODULE: Electronic control unit, amplifier or igniter of solid state or integrated design which controls the current flow in the ignition primary circuit based on input from the pick-up coil. When the module opens the primary circuit, high secondary voltage is induced in the coil.

MODULATED: In an electronic-hydraulic converter clutch system (or shift valve system), the term modulated refers to the pulsing of a solenoid, at a variable rate. This action controls the buildup of oil pressure in the hydraulic circuit to allow a controlled amount of clutch slippage.

MODULATED CONVERTER CLUTCH CONTROL (MCCC): A pulse width duty cycle valve that controls the converter lockup apply pressure and maximizes smoother transitions between lock and unlock conditions.

MODULATOR PRESSURE (THROTTLE PRESSURE): A hydraulic signal oil pressure relating to the amount of engine load, based on either the amount of throttle plate opening or engine vacuum.

MODULATOR VALVE: A regulator valve that is controlled by engine vacuum, providing a hydraulic pressure that varies in relation to engine torque. The hydraulic torque signal functions to delay the shift pattern and provide a line pressure boost. (See throttle valve.)

MOTOR: An electromagnetic device used to convert electrical energy into mechanical energy.

MULTIPLE-DISC CLUTCH: A grouping of steel and friction lined plates that, when compressed together by hydraulic pressure acting upon a piston, lock or unlock a planetary member.

MULTI-WEIGHT: Type of oil that provides adequate lubrication at both high and low temperatures.

needed to move one amp through a resistance of one ohm.

MUSHY: Same as soft; slow and drawn out clutch apply with very little shift feel.

MUTUAL INDUCTION: The generation of current from one wire circuit to another by movement of the magnetic field surrounding a current-carrying circuit as its ampere flow increases or decreases.

NEEDLE BEARING: A bearing which consists of a number (usually a large number) of long, thin rollers.

NITROGEN OXIDE (NOx): One of the three basic pollutants found in the exhaust emission of an internal combustion engine. The amount of NOx usually varies in an inverse proportion to the amount of HC and CO.

NONPOSITIVE SEALING: A sealing method that allows some minor leakage, which normally assists in lubrication.

O2 SENSOR: Located in the engine's exhaust system, it is an input device to the ECM/PCM for managing the fuel delivery and ignition system. A scanner can be used to observe the fluctuating voltage readings produced by an O2 sensor as the oxygen content of the exhaust is analyzed.

O-RING SEAL: Molded synthetic rubber seal designed with a circular cross-section. This type of seal is used primarily in static applications.

OBD II (ON-BOARD DIAGNOSTICS, SECOND GENERATION): Refers to the federal law mandating tighter control of 1996 and newer vehicle emissions, active monitoring of related devices, and standardization of terminology, data link connectors, and other technician concerns.

OCTANE RATING: A number, indicating the quality of gasoline based on its ability to resist knock. The higher the number, the better the quality. Higher compression engines require higher octane gas.

OEM: Original Equipment Manufactured. OEM equipment is that furnished standard by the manufacturer.

OFFSET: The distance between the vertical center of the wheel and the mounting surface at the lugs. Offset is positive if the center is outside the lug circle; negative offset puts the center line inside the lug circle.

OHM'S LAW: A law of electricity that states the relationship between voltage, current, and resistance. Volts = amperes x ohms

OHM: The unit used to measure the resistance of conductor-to-electrical flow. One ohm is the amount of resistance that limits current flow to one ampere in a circuit with one volt of pressure.

OHMMETER: An instrument used for measuring the resistance, in ohms, in an electrical circuit.

ONE-WAY CLUTCH: A mechanical clutch of roller or sprag design that resists torque or transmits power in one direction only. It is used to either hold or drive a planetary member.

ONE-WAY ROLLER CLUTCH: A mechanical device that transmits or holds torque in one direction only.

OPEN CIRCUIT: A break or lack of contact in an electrical circuit, either intentional (switch) or unintentional (bad connection or broken wire).

ORIFICE: Located in hydraulic oil circuits, it acts as a restriction. It slows down fluid flow to either create back pressure or delay pressure buildup downstream.

OSCILLOSCOPE: A piece of test equipment that shows electric impulses as a pattern on a screen. Engine performance can be analyzed by interpreting these patterns.

OUTPUT SHAFT: The shaft which transmits torque from a device, such as a transmission.

OUTPUT SPEED SENSOR (OSS): Identifies transmission/transaxle output shaft speed for shift timing and may be used to calculate TCC slip; often functions as the VSS (vehicle speed sensor).

OVERDRIVE: (1.) A device attached to or incorporated in a transmission/transaxle that allows the engine to turn less than one full revolution for every complete revolution of the wheels. The net effect is to reduce engine rpm, thereby using less fuel. A typical overdrive gear ratio would be .87:1, instead of the normal 1:1 in high gear. (2.) A gear assembly which produces more shaft revolutions than that transmitted to it.

OVERDRIVE PLANETARY GEARSET: A single planetary gearset designed to provide a direct drive and overdrive ratio. When coupled to a three-speed transmission/transaxle configuration, a four-speed/overdrive unit is present.

OVERHEAD CAMSHAFT (OHC): An engine configuration in which the camshaft is mounted on top of the cylinder head and operates the valve either directly or by means of rocker arms.

OVERHEAD VALVE (OHV): An engine configuration in which all of the valves are located in the cylinder head and the camshaft is located in the cylinder block. The camshaft operates the valves via lifters and pushrods.

OVERRUNCLUTCH: Another name for a one-way mechanical clutch. Applies to both roller and sprag designs.

OVERSTEER: The tendency of some vehicles, when steering into a turn, to over-respond or steer more than required, which could result in excessive slip of the rear wheels. Opposite of under-steer.

OXIDATION STABILIZERS: Absorb and dissipate heat. Automatic transmission fluid has high resistance to varnish and sludge buildup that occurs from excessive heat that is generated primarily in the torque converter. Local temperatures as high as 6000F (3150C) can occur at the clutch plates during engagement, and this heat must be absorbed and dissipated. If the fluid cannot withstand the heat, it burns or oxidizes, resulting in an almost immediate destruction of friction materials, clogged filter screen and hydraulic passages, and sticky valves.

OXIDES OF NITROGEN: See nitrogen oxide (NOx).

OXYGEN SENSOR: Used with a feedback system to sense the presence of oxygen in the exhaust gas and signal the computer which can use the voltage signal to determine engine operating efficiency and adjust the air/fuel ratio.

PARALLEL CIRCUIT: (See circuit, parallel.)

PARTS WASHER: A basin or tub, usually with a built-in pump mechanism and hose used for circulating chemical solvent for the purpose of cleaning greasy, oily and dirty components.

PART-TIME FOUR WHEEL DRIVE: A system that is normally in the two wheel drive mode and only runs in four-wheel drive when the system is manually engaged because more traction is desired. Two or four wheel drive is normally selected by a lever to engage the front axle, but if locking hubs are used, these must also be manually engaged in the Lock position. Otherwise, the front axle will not drive the front wheels.

PASSIVE RESTRAINT: Safety systems such as air bags or automatic seat belts which operate with no action required on the part of the driver or passenger. Mandated by Federal regulations on all vehicles sold in the U.S. after 1990.

PAYLOAD: The weight the vehicle is capable of carrying in addition to its own weight. Payload includes weight of the driver, passengers and cargo, but not coolant, fuel, lubricant, spare tire, etc.

PCM: Powertrain control module.

PCV VALVE: A valve usually located in the rocker cover that vents crankcase vapors back into the engine to be reburned.

PERCOLATION: A condition in which the fuel actually "boils," due to excessive heat. Percolation prevents proper atomization of the fuel causing rough running.

PICK-UP COIL: The coil in which voltage is induced in an electronic ignition.

PING: A metallic rattling sound produced by the engine during acceleration. It is usually due to incorrect ignition timing or a poor grade of gasoline.

PINION: The smaller of two gears. The rear axle pinion drives the ring gear which transmits motion to the axle shafts.

PINION GEAR: The smallest gear in a drive gear assembly.

PISTON: A disc or cup that fits in a cylinder bore and is free to move. In hydraulics, it provides the means of converting hydraulic pressure into a usable force. Examples of piston applications are found in servo, clutch, and accumulator units.

PISTON RING: An open-ended ring which fits into a groove on the outer diameter of the piston. Its chief function is to form a seal between the piston and cylinder wall. Most automotive pistons have three rings: two for compression sealing; one for oil sealing.

PITMAN ARM: A lever which transmits steering force from the steering gear to the steering linkage.

PLANET CARRIER: A basic member of a planetary gear assembly that carries the pinion gears.

PLANET PINIONS: Gears housed in a planet carrier that are in constant mesh with the sun gear and internal gear. Because they have their own independent rotating centers, the pinions are capable of rotating around the sun gear or the inside of the internal gear.

PLANETARY GEAR RATIO: The reduction or overdrive ratio developed by a planetary gearset.

PLANETARY GEARSET: In its simplest form, it is made up of a basic assembly group containing a sun gear, internal gear, and planet carrier. The gears are always in constant mesh and offer a wide range of gear ratio possibilities.

PLANETARY GEARSET (COMPOUND): Two planetary gearsets combined together.

PLANETARY GEARSET (SIMPLE): An assembly of gears in constant mesh consisting of a sun gear, several pinion gears mounted in a carrier, and a ring gear. It provides gear ratio and direction changes, in addition to a direct drive and a neutral.

PLY RATING: A. rating given a tire which indicates strength (but not necessarily actual plies). A two-ply/four-ply rating has only two plies, but the strength of a four-ply tire.

POLARITY: Indication (positive or negative) of the two poles of a battery.

PORT: An opening for fluid intake or exhaust.

POSITIVE SEALING: A sealing method that completely prevents leakage.

POTENTIAL: Electrical force measured in volts; sometimes used interchangeably with voltage.

POWER: The ability to do work per unit of time, as expressed in horsepower; one horsepower equals 33,000 ft. lbs. of work per minute, or 550 ft. lbs. of work per second.

POWER FLOW: The systematic flow or transmission of power through the gears, from the input shaft to the output shaft.

POWER-TO-WEIGHT RATIO: Ratio of horsepower to weight of car.

POWERTRAIN: See Drivetrain.

POWERTRAIN CONTROL MODULE (PCM): Current designation for the engine control module (ECM). In many cases, late model vehicle control units manage the engine as well as the transmission. In other settings, the PCM controls the engine and is interfaced with a TCM to control transmission functions.

Ppm: Parts per million; unit used to measure exhaust emissions.

PREIGNITION: Early ignition of fuel in the cylinder, sometimes due to glowing carbon deposits in the combustion chamber. Preignition can be damaging since combustion takes place prematurely.

PRELOAD: A predetermined load placed on a bearing during assembly or by adjustment.

PRESS FIT: The mating of two parts under pressure, due to the inner diameter of one being smaller than the outer diameter of the other, or vice versa; an interference fit.

PRESSURE: The amount of force exerted upon a surface area.

PRESSURE CONTROL SOLENOID (PCS): An output device that provides a boost oil pressure to the mainline regulator valve to control line pressure. Its operation is determined by the amount of current sent from the PCM.

PRESSURE GAUGE: An instrument used for measuring the fluid pressure in a hydraulic circuit.

PRESSURE REGULATOR VALVE: In automatic transmissions, its purpose is to regulate the pressure of the pump output and supply the basic fluid pressure necessary to operate the transmission. The regulated fluid pressure may be referred to as mainline pressure, line pressure, or control pressure.

PRESSURE SWITCH ASSEMBLY (PSA): Mounted inside the transmission, it is a grouping of oil pressure switches that inputs to the PCM when certain hydraulic passages are charged with oil pressure.

PRESSURE PLATE: A spring-loaded plate (part of the clutch) that transmits power to the driven (friction) plate when the clutch is engaged.

PRIMARY CIRCUIT: The low voltage side of the ignition system which consists of the ignition switch, ballast resistor or resistance wire, bypass, coil, electronic control unit and pick-up coil as well as the connecting wires and harnesses.

PROFILE: Term used for tire measurement (tire series), which is the ratio of tire height to tread width.

PROM (PROGRAMMABLE READ-ONLY MEMORY): The heart of the computer that compares input data and makes the engineered program or strategy decisions about when to trigger the appropriate output based on stored computer instructions.

PULSE GENERATOR: A two-wire pickup sensor used to produce a fluctuating electrical signal. This changing signal is read by the controller to determine the speed of the object and can be used to measure transmission/transaxle input speed, output speed, and vehicle speed.

PSI: Pounds per square inch; a measurement of pressure.

PULSE WIDTH DUTY CYCLE SOLENOID (PULSE WIDTH MODULATED SOLENOID): A computer-controlled solenoid that turns on and off at a variable rate producing a modulated oil pressure; often referred to as a pulse width modulated (PWM) solenoid. Employed in many electronic automatic transmissions and transaxles, these solenoids are used to manage shift control and converter clutch hydraulic circuits.

PUSHROD: A steel rod between the hydraulic valve lifter and the valve rocker arm in overhead valve (OHV) engines.

PUMP: A mechanical device designed to create fluid flow and pressure buildup in a hydraulic system.

QUARTER PANEL: General term used to refer to a rear fender. Quarter panel is the area from the rear door opening to the tail light area and from rear wheel well to the base of the trunk and roof-line.

RACE: The surface on the inner or outer ring of a bearing on which the balls, needles or rollers move.

RACK AND PINION: A type of automotive steering system using a pinion gear attached to the end of the steering shaft. The pinion meshes with a long rack attached to the steering linkage.

RADIAL TIRE: Tire design which uses body cords running at right angles to the center line of the tire. Two or more belts are used to give tread strength. Radials can be identified by their characteristic sidewall bulge.

RADIATOR: Part of the cooling system for a water-cooled engine, mounted in the front of the vehicle and connected to the engine with rubber hoses. Through the radiator, excess combustion heat is dissipated into the atmosphere through forced convection using a water and glycol based mixture that circulates through, and cools, the engine.

RANGE REFERENCE AND CLUTCH/BAND APPLY CHART: A guide that shows the application of clutches and bands for each gear, within the selector range positions. These charts are extremely useful for understanding how the unit operates and for diagnosing malfunctions.

RAVIGNEAUX GEARSET: A compound planetary gearset that features matched dual planetary pinions (sets of two) mounted in a single planet carrier. Two sun gears and one ring mesh with the carrier pinions.

REACTION MEMBER: The stationary planetary member, in a planetary gearset, that is grounded to the transmission/transaxle case through the use of friction and wedging devices known as bands, disc clutches, and one-way clutches.

REACTION PRESSURE: The fluid pressure that moves a spool valve against an opposing force or forces; the area on which the opposing force acts. The opposing force can be a spring or a combination of spring force and auxiliary hydraulic force.

REACTOR, TORQUE CONVERTER: The reaction member of a fluid torque converter, more commonly called a stator. (See stator.)

REAR MAIN OIL SEAL: A synthetic or rope-type seal that prevents oil from leaking out of the engine past the rear main crankshaft bearing.

RECIRCULATING BALL: Type of steering system in which recirculating steel balls occupy the area between the nut and worm wheel, causing a reduction in friction.

RECTIFIER: A device (used primarily in alternators) that permits electrical current to flow in one direction only.

REDUCTION: (See gear reduction.)

REGULATOR VALVE: A valve that changes the pressure of the oil in a hydraulic circuit as the oil passes through the valve by bleeding off (or exhausting) some of the volume of oil supplied to the valve.

REFRIGERANT 12 (R-12) or 134 (R-134): The generic name of the refrigerant used in automotive air conditioning systems.

REGULATOR: A device which maintains the amperage and/or voltage levels of a circuit at predetermined values.

RELAY: A switch which automatically opens and/or closes a circuit.

RELAY VALVE: A valve that directs flow and pressure. Relay valves simply connect or disconnect interrelated passages without restricting the fluid flow or changing the pressure.

RELIEF VALVE: A spring-loaded, pressure-operated valve that limits oil pressure buildup in a hydraulic circuit to a predetermined maximum value.

RELUCTOR: A wheel that rotates inside the distributor and triggers the release of voltage in an electronic ignition.

RESERVOIR: The storage area for fluid in a hydraulic system; often called a sump.

RESIN: A liquid plastic used in body work.

RESIDUAL MAGNETISM: The magnetic strength stored in a material after a magnetizing field has been removed.

RESISTANCE: The opposition to the flow of current through a circuit or electrical device, and is measured in ohms. Resistance is equal to the voltage divided by the amperage.

RESISTOR SPARK PLUG: A spark plug using a resistor to shorten the spark duration. This suppresses radio interference and lengthens plug life.

RESISTOR: A device, usually made of wire, which offers a preset amount of resistance in an electrical circuit.

RESULTANT FORCE: The single effective directional thrust of the fluid force on the turbine produced by the vortex and rotary forces acting in different planes.

RETARD: Set the ignition timing so that spark occurs later (fewer degrees before TDC).

RHEOSTAT: A device for regulating a current by means of a variable resistance.

RING GEAR: The name given to a ring-shaped gear attached to a differential case, or affixed to a flywheel or as part of a planetary gear set.

ROADLOAD: grade.

ROCKER ARM: A lever which rotates around a shaft pushing down (opening) the valve with an end when the other end is pushed up by the pushrod. Spring pressure will later close the valve.

ROCKER PANEL: The body panel below the doors between the wheel opening.

ROLLER BEARING: A bearing made up of hardened inner and outer races between which hardened steel rollers move.

ROLLER CLUTCH: A type of one-way clutch design using rollers and springs mounted within an inner and outer cam race assembly.

ROTARY FLOW: The path of the fluid trapped between the blades of the members as they revolve with the rotation of the torque converter cover (rotational inertia).

ROTOR: (1.) The disc-shaped part of a disc brake assembly, upon which the brake pads bear; also called, brake disc. (2.) The device mounted atop the distributor shaft, which passes current to the distributor cap tower contacts.

ROTARY ENGINE: See Wankel engine.

RPM: Revolutions per minute (usually indicates engine speed).

RTV: A gasket making compound that cures as it is exposed to the atmosphere. It is used between surfaces that are not perfectly machined to one another, leaving a slight gap that the RTV fills and in which it hardens. The letters RTV represent room temperature vulcanizing.

RUN-ON: Condition when the engine continues to run, even when the key is turned off. See dieseling.

SEALED BEAM: A automotive headlight. The lens, reflector and filament from a single unit.

SEATBELT INTERLOCK: A system whereby the car cannot be started unless the seatbelt is buckled.

SECONDARY CIRCUIT: The high voltage side of the ignition system, usually above 20,000 volts. The secondary includes the ignition coil, coil wire, distributor cap and rotor, spark plug wires and spark plugs.

SELF-INDUCTION: The generation of voltage in a current-carrying wire by changing the amount of current flowing within that wire.

SEMI-CONDUCTOR: A material (silicon or germanium) that is neither a good conductor nor an insulator; used in diodes and transistors.

SEMI-FLOATING AXLE: In this design, a wheel is attached to the axle shaft, which takes both drive and cornering loads. Almost all solid axle passenger cars and light trucks use this design.

SENDING UNIT: A mechanical, electrical, hydraulic or electromagnetic device which transmits information to a gauge.

SENSOR: Any device designed to measure engine operating conditions or ambient pressures and temperatures. Usually electronic in nature and designed to send a voltage signal to an on-board computer, some sensors may operate as a simple on/off switch or they may provide a variable voltage signal (like a potentiometer) as conditions or measured parameters change.

SERIES CIRCUIT: (See circuit, series.)

SERPENTINE BELT: An accessory drive belt, with small multiple v-ribs, routed around most or all of the engine-powered accessories such as the alternator and power steering pump. Usually both the front and the back side of the belt comes into contact with various pulleys.

SERVO: In an automatic transmission, it is a piston in a cylinder assembly that converts hydraulic pressure into mechanical force and movement; used for the application of the bands and clutches.

SHIFT BUSYNESS: When referring to a torque converter clutch, it is the frequent apply and release of the clutch plate due to uncommon driving conditions.

SHIFT VALVE: Classified as a relay valve, it triggers the automatic shift in response to a governor and a throttle signal by directing fluid to the appropriate band and clutch apply combination to cause the shift to occur.

SHIM: Spacers of precise, predetermined thickness used between parts to establish a proper working relationship.

SHIMMY: Vibration (sometimes violent) in the front end caused by misaligned front end, out of balance tires or worn suspension components.

SHORT CIRCUIT: An electrical malfunction where current takes the path of least resistance to ground (usually through damaged insulation). Current flow is excessive from low resistance resulting in a blown fuse.

SHUDDER: Repeated jerking or stick-slip sensation, similar to chuggle but more severe and rapid in nature, that may be most noticeable during certain ranges of vehicle speed; also used to define condition after converter clutch engagement.

SIMPSON GEARSET: A compound planetary gear train that integrates two simple planetary gearsets referred to as the front planetary and the rear planetary.

SINGLE OVERHEAD CAMSHAFT: See overhead camshaft.

SKIDPLATE: A metal plate attached to the underside of the body to protect the fuel tank, transfer case or other vulnerable parts from damage.

SLAVE CYLINDER: In automotive use, a device in the hydraulic clutch system which is activated by hydraulic force, disengaging the clutch.

SLIPPING: Noticeable increase in engine rpm without vehicle speed increase; usually occurs during or after initial clutch or band engagement.

SLUDGE: Thick, black deposits in engine formed from dirt, oil, water, etc. It is usually formed in engines when oil changes are neglected.

SNAP RING: A circular retaining clip used inside or outside a shaft or part to secure a shaft, such as a floating wrist pin.

SOFT: Slow, almost unnoticeable clutch apply with very little shift feel.

SOFTCODES: DTCs that have been set into the PCM memory but are not present at the time of testing; often referred to as history or intermittent codes.

SOHC: Single overhead camshaft.

SOLENOID: An electrically operated, magnetic switching device.

SPALLING: A wear pattern identified by metal chips flaking off the hardened surface. This condition is caused by foreign particles, overloading situations, and/or normal wear.

SPARK PLUG: A device screwed into the combustion chamber of a spark ignition engine. The basic construction is a conductive core inside of a ceramic insulator, mounted in an outer conductive base. An electrical charge from the spark plug wire travels along the conductive core and jumps a preset air gap to a grounding point or points at the end of the conductive base. The resultant spark ignites the fuel/air mixture in the combustion chamber.

SPECIFIC GRAVITY (BATTERY): The relative weight of liquid (battery electrolyte) as compared to the weight of an equal volume of water.

SPLINES: Ridges machined or cast onto the outer diameter of a shaft or inner diameter of a bore to enable parts to mate without rotation.

SPLIT TORQUE DRIVE: In a torque converter, it refers to parallel paths of torque transmission, one of which is mechanical and the other hydraulic.

SPONGY PEDAL: A soft or spongy feeling when the brake pedal is depressed. It is usually due to air in the brake lines.

SPOOLVALVE: A precision-machined, cylindrically shaped valve made up of lands and grooves. Depending on its position in the valve bore, various interconnecting hydraulic circuit passages are either opened or closed.

SPRAG CLUTCH: A type of one-way clutch design using cams or contoured-shaped sprags between inner and outer races. (See one-way clutch.)

SPRUNG WEIGHT: The weight of a car supported by the springs.

SQUARE-CUT SEAL: Molded synthetic rubber seal designed with a square- or rectangular-shaped cross-section. This type of seal is used for both dynamic and static applications.

SRS: Supplemental restraint system

STABILIZER (SWAY) BAR: A bar linking both sides of the suspension. It resists sway on turns by taking some of added load from one wheel and putting it on the other.

STAGE: The number of turbine sets separated by a stator. A turbine set may be made up of one or more turbine members. A three-element converter is classified as a single stage.

STALL: In fluid drive transmission/transaxle applications, stall refers to engine rpm with the transmission/transaxle engaged and the vehicle stationary; throttle valve can be in any position between closed and wide open.

STALL SPEED: In fluid drive transmission/transaxle applications, stall speed refers to the maximum engine rpm with the transmission/transaxle engaged and vehicle stationary, when the throttle valve is wide open. (See stall; stall test.)

STALL TEST: A procedure recommended by many manufacturers to help determine the integrity of an engine, the torque converter stator, and certain clutch and band combinations. With the shift lever in each of the forward and reverse positions and with the brakes firmly applied, the accelerator pedal is momentarily pressed to the wide open throttle (WOT) position. The engine rpm reading at full throttle can provide clues for diagnosing the condition of the items listed above.

STALL TORQUE: The maximum design or engineered torque ratio of a fluid torque converter, produced under stall speed conditions. (See stall speed.)

STARTER: A high-torque electric motor used for the purpose of starting the engine, typically through a high ratio geared drive connected to the flywheel ring gear.

STATIC: A sealing application in which the parts being sealed do not move in relation to each other.

STATOR (REACTOR): The reaction member of a fluid torque converter that changes the direction of the fluid as it leaves the turbine to enter the impeller vanes. During the torque multiplication phase, this action assists the impeller's rotary force and results in an increase in torque.

STEERING GEOMETRY: Combination of various angles of suspension components (caster, camber, toe-in); roughly equivalent to front end alignment.

STRAIGHT WEIGHT: Term designating motor oil as suitable for use within a narrow range of temperatures. Outside the narrow temperature range its flow characteristics will not adequately lubricate.

STROKE: The distance the piston travels from bottom dead center to top dead center.

SUBSTITUTION: Replacing one part suspected of a defect with a like part of known quality.

SUMP: The storage vessel or reservoir that provides a ready source of fluid to the pump. In an automatic transmission, the sump is the oil pan. All fluid eventually returns to the sump for recycling into the hydraulic system.

SUN GEAR: In a planetary gearset, it is the center gear that meshes with a cluster of planet pinions.

SUPERCHARGER: An air pump driven mechanically by the engine through belts, chains, shafts or gears from the crankshaft. Two general types of supercharger are the positive displacement and centrifugal type, which pump air in direct relationship to the speed of the engine.

SUPPLEMENTAL RESTRAINT SYSTEM: See air bag.

SURGE: Repeating engine-related feeling of acceleration and deceleration that is less intense than chuggle.

SWITCH: A device used to open, close, or redirect the current in an electrical circuit.

SYNCHROMESH: A manual transmission/transaxle that is equipped with devices (synchronizers) that match the gear speeds so that the transmission/transaxle can be downshifted without clashing gears.

SYNTHETIC OIL: Non-petroleum based oil.

TACHOMETER: A device used to measure the rotary speed of an engine, shaft, gear, etc., usually in rotations per minute.

TDC: Top dead center. The exact top of the piston's stroke.

TEFLON SEALING RINGS: Teflon is a soft, durable, plastic-like material that is resistant to heat and provides excellent sealing. These rings are designed with either scarf-cut joints or as one-piece rings. Teflon sealing rings have replaced many metal ring applications.

TERMINAL: A device attached to the end of a wire or cable to make an electrical connection.

TEST LIGHT, CIRCUIT-POWERED: Uses available circuit voltage to test circuit continuity.

TEST LIGHT, SELF-POWERED: Uses its own battery source to test circuit continuity.

THERMISTOR: A special resistor used to measure fluid temperature; it decreases its resistance with increases in temperature.

THERMOSTAT: A valve, located in the cooling system of an engine, which is closed when cold and opens gradually in response to engine heating, controlling the temperature of the coolant and rate of coolant flow.

THERMOSTATIC ELEMENT: A heat-sensitive, spring-type device that controls a drain port from the upper sump area to the lower sump. When the transaxle fluid reaches operating temperature, the port is closed and the upper sump fills, thus reducing the fluid level in the lower sump.

THROTTLE POSITION (TP) SENSOR: Reads the degree of throttle opening; its signal is used to analyze engine load conditions. The ECM/PCM decides to apply the TCC, or to disengage it for coast or load conditions that need a converter torque boost.

THROTTLE PRESSURE/MODULATOR PRESSURE: A hydraulic signal oil pressure relating to the amount of engine load, based on either the amount of throttle plate opening or engine vacuum.

THROTTLE VALVE: A regulating or balanced valve that is controlled mechanically by throttle linkage or engine vacuum. It sends a hydraulic signal to the shift valve body to control shift timing and shift quality. (See balanced valve; modulator valve.)

THROW-OUT BEARING: As the clutch pedal is depressed, the throwout bearing moves against the spring fingers of the pressure plate, forcing the pressure plate to disengage from the driven disc.

TIE ROD: A rod connecting the steering arms. Tie rods have threaded ends that are used to adjust toe-in.

TIE-UP: Condition where two opposing clutches are attempting to apply at same time, causing engine to labor with noticeable loss of engine rpm.

TIMING BELT: A square-toothed, reinforced rubber belt that is driven by the crankshaft and operates the camshaft.

TIMING CHAIN: A roller chain that is driven by the crankshaft and operates the camshaft.

TIRE ROTATION: Moving the tires from one position to another to make the tires wear evenly.

TOE-IN (OUT): A term comparing the extreme front and rear of the front tires. Closer together at the front is toe-in; farther apart at the front is toe-out.

TOP DEAD CENTER (TDC): The point at which the piston reaches the top of its travel on the compression stroke.

TORQUE: Measurement of turning or twisting force, expressed as foot-pounds or inch-pounds.

TORQUE CONVERTER: A turbine used to transmit power from a driving member to a driven member via hydraulic action, providing changes in drive ratio and torque. In automotive use, it links the driveplate at the rear of the engine to the automatic transmission.

TORQUE CONVERTER CLUTCH: The apply plate (lockup plate) assembly used for mechanical power flow through the converter.

TORQUE PHASE: Sometimes referred to as slip phase or stall phase, torque multiplication occurs when the turbine is turning at a slower speed than the impeller, and the stator is reactionary (stationary). This sequence generates a boost in output torque.

TORQUE RATING (STALL TORQUE): The maximum torque multiplication that occurs during stall conditions, with the engine at wide open throttle (WOT) and zero turbine speed.

TORQUE RATIO: An expression of the gear ratio factor on torque effect. A 3:1 gear ratio or 3:1 torque ratio increases the torque input by the ratio factor of 3. Input torque (100 ft. lbs.) x 3 = output torque (300 ft. lbs.)

TRACTION: The amount of usable tractive effort before the drive wheels slip on the road contact surface.

TORSION BAR SUSPENSION: Long rods of spring steel which take the place of springs. One end of the bar is anchored and the other arm (attached to the suspension) is free to twist. The bars' resistance to twisting causes springing action.

TRACK: Distance between the centers of the tires where they contact the ground.

TRACTION CONTROL: A control system that prevents the spinning of a vehicle's drive wheels when excess power is applied.

TRACTIVE EFFORT: The amount of force available to the drive wheels, to move the vehicle.

TRANSAXLE: A single housing containing the transmission and differential. Transaxles are usually found on front engine/front wheel drive or rear engine/rear wheel drive cars.

TRANSDUCER: A device that changes energy from one form to another. For example, a transducer in a microphone changes sound energy to electrical energy. In automotive air-conditioning controls used in automatic temperature systems, a transducer changes an electrical signal to a vacuum signal, which operates mechanical doors.

TRANSMISSION: A powertrain component designed to modify torque and speed developed by the engine; also provides direct drive, reverse, and neutral.

TRANSMISSION CONTROL MODULE (TCM): Manages transmission functions. These vary according to the manufacturer's product design but may include converter clutch operation, electronic shift scheduling, and mainline pressure.

TRANSMISSION FLUID TEMPERATURE (TFT) SENSOR: Originally called a transmission oil temperature (TOT) sensor, this input device to the ECM/PCM senses the fluid temperature and provides a resistance value. It operates on the thermistor principle.

TRANSMISSION INPUT SPEED (TIS) SENSOR: Measures turbine shaft (input shaft) rpm's and compares to engine rpm's to determine torque converter slip. When compared to the transmission output speed sensor or VSS, gear ratio and clutch engagement timing can be determined.

TRANSMISSION OIL TEMPERATURE (TOT) SENSOR: (See transmission fluid temperature (TFT) sensor.)

TRANSMISSION RANGE SELECTOR (TRS) SWITCH: Tells the module which gear shift position the driver has chosen.

TRANSFER CASE: A gearbox driven from the transmission that delivers power to both front and rear driveshafts in a four-wheel drive system. Transfer cases usually have a high and low range set of gears, used depending on how much pulling power is needed.

TRANSISTOR: A semi-conductor component which can be actuated by a small voltage to perform an electrical switching function.

TREAD WEAR INDICATOR: Bars molded into the tire at right angles to the tread that appear as horizontal bars when 1/16 in. of tread remains.

TREAD WEAR PATTERN: The pattern of wear on tires which can be "read" to diagnose problems in the front suspension.

TUNE-UP: A regular maintenance function, usually associated with the replacement and adjustment of parts and components in the electrical and fuel systems of a vehicle for the purpose of attaining optimum performance.

TURBINE: The output (driven) member of a fluid coupling or fluid torque converter. It is splined to the input (turbine) shaft of the transmission.

TURBOCHARGER: An exhaust driven pump which compresses intake air and forces it into the combustion chambers at higher than atmospheric pressures. The increased air pressure allows more fuel to be burned and results in increased horsepower being produced.

TURBULENCE: The interference of molecules of a fluid (or vapor) with each other in a fluid flow.

TYPE F: Transmission fluid developed and used by Ford Motor Company up to 1982. This fluid type provides a high coefficient of friction.

TYPE 7176: The preferred choice of transmission fluid for Chrysler automatic transmissions and transaxles. Developed in 1986, it closely resembles Dexron and Mercon. Type 7176 is the recommended service fill fluid for all Chrysler products utilizing a lockup torque converter dating back to 1978.

U-JOINT (UNIVERSAL JOINT): A flexible coupling in the drive train that allows the driveshafts or axle shafts to operate at different angles and still transmit rotary power.

UNDERSTEER: The tendency of a car to continue straight ahead while negotiating a turn.

UNIT BODY: Design in which the car body acts as the frame.

UNLEADED FUEL: Fuel which contains no lead (a common gasoline additive). The presence of lead in fuel will destroy the functioning elements of a catalytic converter, making it useless.

UNSPRUNG WEIGHT: The weight of car components not supported by the springs (wheels, tires, brakes, rear axle, control arms, etc.).

UPSHIFT: A shift that results in a decrease in torque ratio and an increase in speed.

VACUUM: A negative pressure; any pressure less than atmospheric pressure.

VACUUM ADVANCE: A device which advances the ignition timing in response to increased engine vacuum.

VACUUM GAUGE: An instrument used for measuring the existing vacuum in a vacuum circuit or chamber. The unit of measure is inches (of mercury in a barometer).

VACUUM MODULATOR: Generates a hydraulic oil pressure in response to the amount of engine vacuum.

VALVES: Devices that can open or close fluid passages in a hydraulic system and are used for directing fluid flow and controlling pressure.

VALVE BODY ASSEMBLY: The main hydraulic control assembly of the transmission/transaxle that contains numerous valves, check balls, and other components to control the distribution of pressurized oil throughout the transmission.

VALVE CLEARANCE: The measured gap between the end of the valve stem and the rocker arm, cam lobe or follower that activates the valve.

VALVE GUIDES: The guide through which the stem of the valve passes.

The guide is designed to keep the valve in proper alignment.

VALVE LASH (clearance): The operating clearance in the valve train.

VALVE TRAIN: The system that operates intake and exhaust valves, consisting of camshaft, valves and springs, lifters, pushrods and rocker arms.

VAPOR LOCK: Boiling of the fuel in the fuel lines due to excess heat. This will interfere with the flow of fuel in the lines and can completely stop the flow. Vapor lock normally only occurs in hot weather.

VARIABLE DISPLACEMENT (VARIABLE CAPACITY) VANE PUMP: Slipper-type vanes, mounted in a revolving rotor and contained within the bore of a movable slide, capture and then force fluid to flow. Movement of the slide to various positions changes the size of the vane chambers and the amount of fluid flow. **Note:** GM refers to this pump design as variable displacement, and Ford terms it variable capacity.

VARIABLE FORCE SOLENOID (VFS): Commonly referred to as the electronic pressure control (EPC) solenoid, it replaces the cable/linkage style of TV system control and is integrated with a spool valve and spring assembly to control pressure. A variable computer-controlled current flow varies the internal force of the solenoid on the spool valve and resulting control pressure.

VARIABLE ORIFICE THERMAL VALVE: Temperature-sensitive hydraulic oil control device that adjusts the size of a circuit path opening. By altering the size of the opening, the oil flow rate is adapted for cold to hot oil viscosity changes.

VARNISH: Term applied to the residue formed when gasoline gets old and stale.

VCM: See Electronic Control Unit (ECU).

VEHICLE SPEED SENSOR (VSS): Provides an electrical signal to the computer module, measuring vehicle speed, and affects the torque converter clutch engagement and release.

VESPEL SEALING RINGS: Hard plastic material that produces excellent sealing in dynamic settings. These rings are found in late versions of the 4T60 and in all 4T60-E and 4T80-E transaxles.

VISCOSITY: The ability of a fluid to flow. The lower the viscosity rating, the easier the fluid will flow. 10 weight motor oil will flow much easier than 40 weight motor oil.

VISCOSITY INDEX IMPROVERS: Keeps the viscosity nearly constant with changes in temperature. This is especially important at low temperatures, when the oil needs to be thin to aid in shifting and for cold-weather starting. Yet it must not be so thin that at high temperatures it will cause excessive hydraulic leakage so that pumps are unable to maintain the proper pressures.

VISCOUS CLUTCH: A specially designed torque converter clutch apply plate that, through the use of a silicon fluid, clamps smoothly and absorbs torsional vibrations.

VOLT: Unit used to measure the force or pressure of electricity. It is defined as the pressure needed to move one amp through the resistance of one ohm.

VOLTAGE: The electrical pressure that causes current to flow. Voltage is measured in volts (V).

VOLTAGE, APPLIED: The actual voltage read at a given point in a circuit. It equals the available voltage of the power supply minus the losses in the circuit up to that point.

VOLTAGE DROP: The voltage lost or used in a circuit by normal loads such as a motor or lamp or by abnormal loads such as a poor (high-resistance) lead or terminal connection.

VOLTAGE REGULATOR: A device that controls the current output of the alternator or generator.

VOLTMETER: An instrument used for measuring electrical force in units called volts. Voltmeters are always connected parallel with the circuit being tested.

VORTEX FLOW: The crosswise or circulatory flow of oil between the blades of the members caused by the centrifugal pumping action of the impeller.

WANKEL ENGINE: An engine which uses no pistons. In place of pistons, triangular-shaped rotors revolve in specially shaped housings.

WATER PUMP: A belt driven component of the cooling system that mounts on the engine, circulating the coolant under pressure.

WATT: The unit for measuring electrical power. One watt is the product of one ampere and one volt (watts equals amps times volts). Wattage is the horsepower of electricity (746 watts equal one horsepower).

WHEEL ALIGNMENT: Inclusive term to describe the front end geometry (caster, camber, toe-in/out).

WHEEL CYLINDER: Found in the automotive drum brake assembly, it is a device, actuated by hydraulic pressure, which, through internal pistons, pushes the brake shoes outward against the drums.

WHEEL WEIGHT: Small weights attached to the wheel to balance the wheel and tire assembly. Out-of-balance tires quickly wear out and also give erratic handling when installed on the front.

WHEELBASE: Distance between the center of front wheels and the center of rear wheels.

WIDE OPEN THROTTLE (WOT): Full travel of accelerator pedal.

WORK: The force exerted to move a mass or object. Work involves motion; if a force is exerted and no motion takes place, no work is done. Work per unit of time is called power. Work = force x distance = ft. lbs. 33,000 ft. lbs. in one minute = 1 horsepower

ZERO-THROTTLE COAST DOWN: A full release of accelerator pedal while vehicle is in motion and in drive range.

Commonly Used Abbreviations

2
2WD	Two Wheel Drive

4
4WD	Four Wheel Drive

A
A/C	Air Conditioning
ABDC	After Bottom Dead Center
ABS	Anti-lock Brakes
AC	Alternating Current
ACL	Air cleaner
ACT	Air Charge Temperature
AIR	Secondary Air Injection
ALCL	Assembly Line Communications Link
ALDL	Assembly Line Diagnostic Link
AT	Automatic Transaxle/Transmission
ATDC	After Top Dead Center
ATF	Automatic Transmission Fluid
ATS	Air Temperature Sensor
AWD	All Wheel Drive

B
BAP	Barometric Absolute Pressure
BARO	Barometric Pressure
BBDC	Before Bottom Dead Center
BCM	Body Control Module
BDC	Bottom Dead Center
BPT	Backpressure Transducer
BTDC	Before Top Dead Center
BVSV	Bimetallic Vacuum Switching Valve

C
CAC	Charge Air Cooler
CARB	California Air Resources Board
CAT	Catalytic Converter
CCC	Computer Command Control
CCCC	Computer Controlled Catalytic Converter
CCCI	Computer Controlled Coil Ignition
CCD	Computer Controlled Dwell
CDI	Capacitor Discharge Ignition
CEC	Computerized Engine Control
CFI	Continuous Fuel Injection
CIS	Continuous Injection System
CIS-E	Continuous Injection System - Electronic
CKP	Crankshaft Position
CL	Closed Loop
CMP	Camshaft Position
CPP	Clutch Pedal Position
CTOX	Continuous Trap Oxidizer System
CTP	Closed Throttle Position
CVC	Constant Vacuum Control
CYL	Cylinder

D
DBC	Dual Bed Catalyst
DC	Direct Current
DFI	Direct Fuel Injection
DIS	Distributorless Ignition System
DLC	Data Link Connector
DMM	Digital Multimeter
DOHC	Double Overhead Camshaft
DRB	Diagnostic Readout Box
DTC	Diagnostic Trouble Code
DTM	Diagnostic Test Mode
DVOM	Digital Volt/Ohmmeter

E
EBCM	Electronic Brake Control Module
ECM	Engine Control Module
ECT	Engine Coolant Temperature
ECU	Engine Control Unit or Electronic Control Unit
EDIS	Electronic Distributorless Ignition System
EEC	Electronic Engine Control
EEPROM	Electrically Erasable Programmable Read Only Memory
EFE	Early Fuel Evaporation
EGR	Exhaust Gas Recirculation
EGRT	Exhaust Gas Recirculation Temperature
EGRVC	EGR Valve Control
EPROM	Erasable Programmable Read Only Memory
EVAP	Evaporative Emissions
EVP	EGR Valve Position

F
FBC	Feedback Carburetor
FEEPROM	Flash Electrically Erasable Programmable Read Only Memory
FF	Flexible Fuel
FI	Fuel Injection
FT	Fuel Trim
FWD	Front Wheel Drive

G
GND	Ground

H
HAC	High Altitude Compensation
HEGO	Heated Exhaust Gas Oxygen sensor
HEI	High Energy Ignition
HO2 Sensor	Heated Oxygen Sensor

I
IAC	Idle Air Control
IAT	Intake Air Temperature
ICM	Ignition Control Module
IFI	Indirect Fuel Injection
IFS	Inertia Fuel Shutoff
ISC	Idle Speed Control
IVSV	Idle Vacuum Switching Valve

Commonly Used Abbreviations

K

KOEO	Key On, Engine Off
KOER	Key ON, Engine Running
KS	Knock Sensor

M

MAF	Mass Air Flow
MAP	Manifold Absolute Pressure
MAT	Manifold Air Temperature
MC	Mixture Control
MDP	Manifold Differential Pressure
MFI	Multiport Fuel Injection
MIL	Malfunction Indicator Lamp or Maintenance
MST	Manifold Surface Temperature
MVZ	Manifold Vacuum Zone

N

NVRAM	Nonvolatile Random Access Memory

O

O2 Sensor	Oxygen Sensor
OBD	On-Board Diagnostic
OC	Oxidation Catalyst
OHC	Overhead Camshaft
OL	Open Loop

P

P/S	Power Steering
PAIR	Pulsed Secondary Air Injection
PCM	Powertrain Control Module
PCS	Purge Control Solenoid
PCV	Positive Crankcase Ventilation
PIP	Profile Ignition Pick-up
PNP	Park/Neutral Position
PROM	Programmable Read Only Memory
PSP	Power Steering Pressure
PTO	Power Take-Off
PTOX	Periodic Trap Oxidizer System

R

RABS	Rear Anti-lock Brake System
RAM	Random Access Memory
ROM	Read Only Memory
RPM	Revolutions Per Minute
RWAL	Rear Wheel Anti-lock Brakes
RWD	Rear Wheel Drive

S

SBC	Single Bed Converter
SBEC	Single Board Engine Controller
SC	Supercharger
SCB	Supercharger Bypass
SFI	Sequential Multiport Fuel Injection
SIR	Supplemental Inflatible Restraint
SOHC	Single Overhead Camshaft
SPL	Smoke Puff Limiter
SPOUT	Spark Output
SRI	Service Reminder Indicator
SRS	Supplemental Restraint System
SRT	System Readiness Test
SSI	Solid State Ignition
ST	Scan Tool
STO	Self-Test Output

T

TAC	Thermostatic Air Clearner
TBI	Throttle Body Fuel Injection
TC	Turbocharger
TCC	Torque Converter Clutch
TCM	Transmission Control Module
TDC	Top Dead Center
TFI	Thick Film Ignition
TP	Throttle Position
TR Sensor	Transaxle/Transmission Range Sensor
TVV	Thermal Vacuum Valve
TWC	Three-way Catalytic Converter

V

VAF	Volume Air Flow, or Vane Air Flow
VAPS	Variable Assist Power Steering
VRV	Vacuum Regulator Valve
VSS	Vehicle Speed Sensor
VSV	Vacuum Switching Valve

W

WOT	Wide Open Throttle
WU-TWC	Warm Up Three-way Catalytic Converter

CHILTON LABOR GUIDE

Whether you are looking for labor times in print, or on CD-ROM, Chilton is your source! Chilton's editors have carefully crafted the latest edition of the famous Chilton Labor Guide to bring you the most accurate repair information available. Chilton's editors consider warranty times, component locations, component type, the environment in which technicians work, the training they receive, and the tools they use when calculating a labor time. To allow for vehicle age, operating conditions, and type of service, the Chilton Labor Guide provides standard and severe service times, plus OEM warranty times. Vehicle makes and models conform to current Automotive Aftermarket Industry Association (AAIA) standards.

978-1-1115-4291-7 Chilton 2011 Labor Guide Manual Set (Domestic & Import)
978-1-1115-4294-8 Chilton 2011 Labor Guide CD-ROM (Domestic & Import)

CD-ROM FEATURES

- access labor times for 1981-2011 import and domestic vehicle models
- save time with automatically calculated labor charges, taxes, & parts as total job is estimated
- create professional estimates for your customer and worksheets for your technicians, printing them whenever needed
- keep track of customers, prior estimates, and your own parts or package jobs with less paper
- choose part names for estimates from an industry standard database to reduce typing
- estimate and track your work status with improved forms
- communicate easily with customers using re-designed printouts which show all labor and parts in an easy-to-read format.
- simplify adding parts to your estimate or work order with a helpful parts list
- locate information quick with a keyword search engine
- quickly locate work requests by day, week and month using the calendar feature

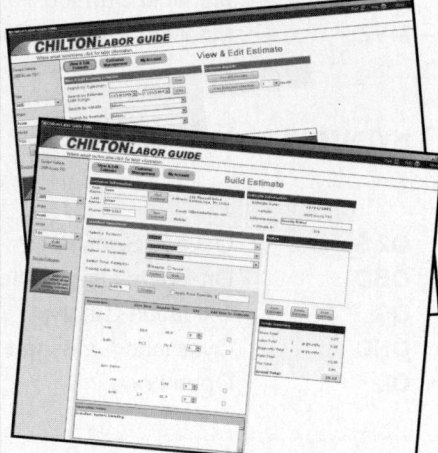

Manual FEATURES

- more than 2,500 pages of updated Chilton labor times split into two volumes includes vehicle information from 1981 to 2011
- trusted by more service professionals than any other labor guide
- less flipping though pages with separate domestic and imported vehicle manuals
- convenient tabs display contents by manufacturer and model
- easy-to-find manufacturers are arranged alphabetically within each volume
- search using two-indexes - labor operations and systems - in each model group
- page numbers include manufacturer code so you know where you are in the book

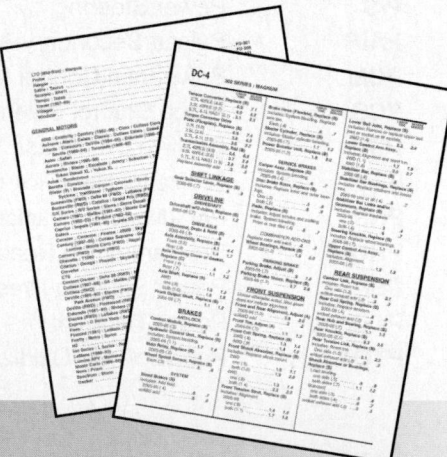

Chilton's labor times are so trusted, even a competing publisher uses them!

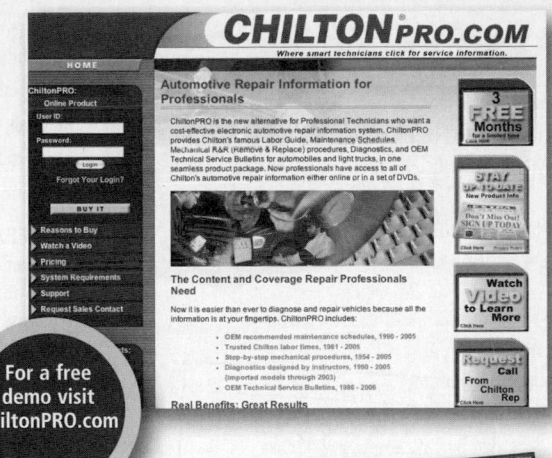
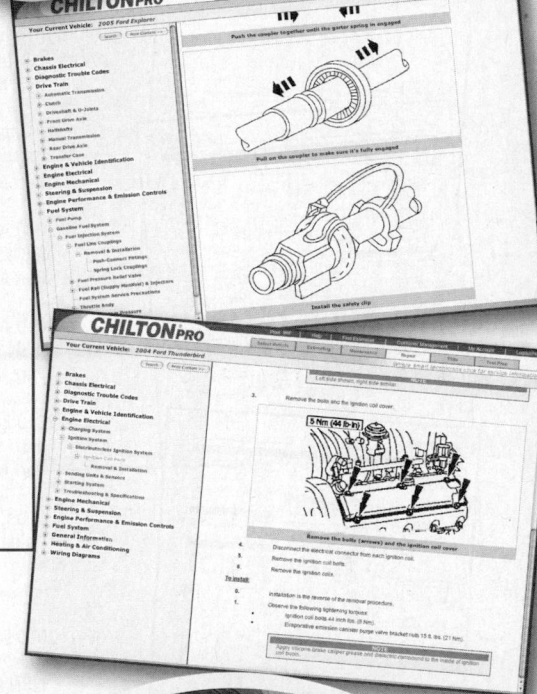
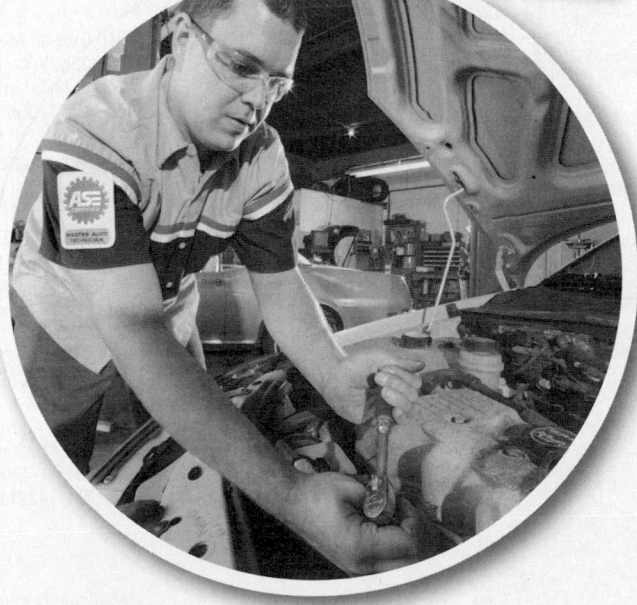
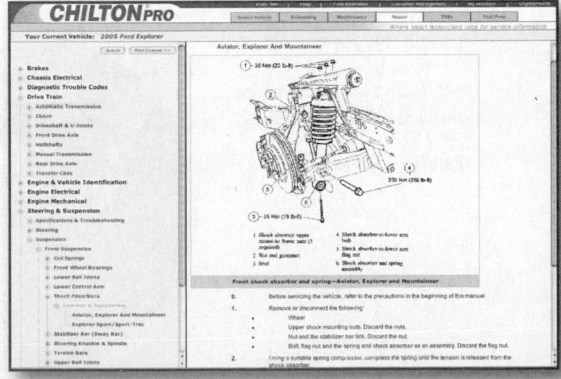

Chilton® 2010 Service Manuals

The Chilton 2010 Service Manuals now include even better graphics and expanded procedures! Chilton's editors have put together the most current automotive repair information available to assist users during daily repairs. These new manuals allow users to accurately and efficiently diagnose and repair late-model cars and trucks. Trust the step-by-step procedures and helpful illustrations that only Chilton can provide. The 2010 Service Manuals cover 2008 and 2009 models plus available 2010 models.

KEY FEATURES

- organized by vehicle manufacturer
- provides thousands of pages of expertly written content
- access new year, make, and model information without repeating previous edition's content
- comprehensive, technically detailed content, including exploded view illustrations, diagnostics and specification charts, arranged alphabetically by model group for quick, easy access

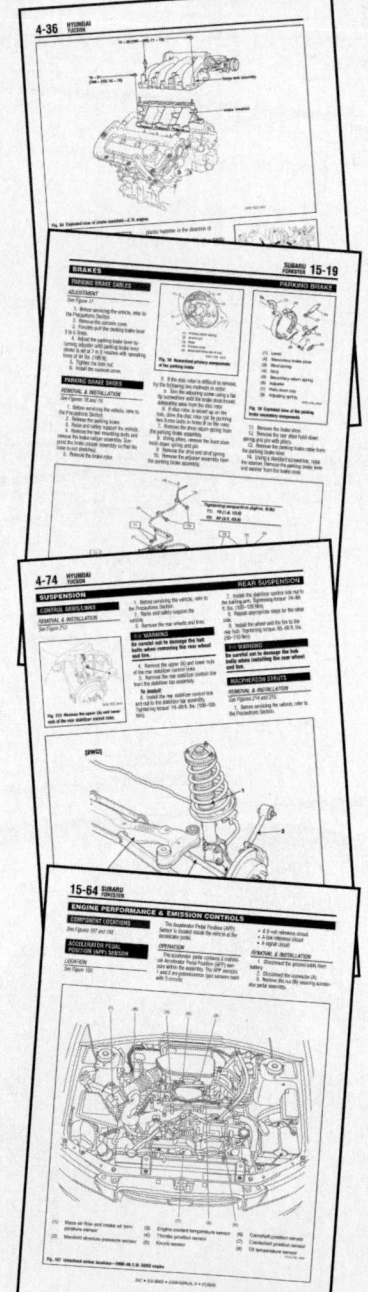

2010 EDITIONS

2010 Asian Service Manual Vol. 1*
ISBN 978-1-1110-3764-2
Part No. 163764

2010 Asian Service Manual Vol. 2*
ISBN 978-1-1110-3765-9
Part No. 163765

2010 Asian Service Manual Vol. 3*
ISBN 978-1-1110-3766-6
Part No. 163766

2010 Asian Service Manual Vol. 4*
ISBN 978-1-1110-3767-3
Part No. 163767

2010 Asian Service Manual Vol. 5*
ISBN 978-1-1110-3768-0
Part No. 163768

2010 European Service Manual*
ISBN 978-1-1110-3769-7
Part No. 163769

2010 Chrysler Service Manual,
Volumes 1 & 2
ISBN 978-1-1110-3654-6
Part No. 163654

2010 Ford Service Manual,
Vols. 1 & 2
ISBN 978-1-1110-3657-7
Part No. 163657

2010 General Motors Service
Manuals, Vols. 1, 2, & 3
ISBN 978-1-111-03661-4
Part No. 163661

2008 EDITIONS

2008 Chrysler Service Manual,
Vols. 1 & 2
ISBN 978-1-4283-2204-2
Part No. 142204

2008 Ford Service Manuals,
Vols. 1 & 2
ISBN 978-1-4283-2208-0
Part No. 142208

2008 Edition General Motors
Service Manuals, Vols. 1 & 2
ISBN 978-1-4283-2211-0
Part No. 142211

2008 Asian Service Manuals,
Vols. 1-4
ISBN 978-1-4283-2214-1
Part No. 142214

2008 Asian Service Manual, Vol. 1
ISBN 978-1-4283-2215-8
Part No. 142215

2008 Asian Service Manual, Vol. 2
ISBN 978-1-4283-2216-5
Part No. 142216

2008 Asian Service Manual, Vol. 3
ISBN 978-1-4283-2217-2
Part No. 142217

2008 Asian Service Manual, Vol. 4
ISBN 978-1-4283-2218-9
Part No. 142218

2008 European Service Manual
ISBN 978-1-4283-2220-2
Part No. 142220

2006 EDITIONS

2006 DaimlerChrysler Diagnostic
Service Manual
ISBN 978-1-4180-2118-4
Part No. 132118

2006 General Motors Diagnostic
Service Manual
ISBN 978-1-4180-2120-7
Part No. 132120

2006 Asian Diagnostic Service
Manual, Vol. 1
ISBN 978-1-4180-2913-5
Part No. 132913

2006 Asian Diagnostic Service
Manual, Vol. 2
ISBN 978-1-4180-2914-2
Part No. 132914

2006 Asian Diagnostic Service
Manual, Vol. 3
ISBN 978-1-4180-2915-9
Part No. 132915

2006 European Diagnostic Service
Manual
ISBN 978-1-4180-2924-1
Part No. 132924

2006 DaimlerChrysler Mechanical
Service Manual
ISBN 978-1-4180-0600-6
Part No. 130600

2006 Asian Mechanical Service
Manual, Vol. 1
ISBN 978-1-4180-0947-2
Part No. 130947

2006 Asian Mechanical Service
Manual, Vol. 2
ISBN 978-1-4180-0948-9
Part No. 130948

2006 Asian Mechanical Service
Manual, Vol. 3
ISBN 978-1-4180-0949-6
Part No. 130949

2006 Asian Mechanical Service
Manual, 3 Vol. Set
ISBN 978-1-4180-0603-7
Part No. 130603

2006 European Mechanical Service
Manual
ISBN 978-1-4180-0604-4
Part No. 130604

*Available December 2010

Chilton® Mechanical Service Manuals–Perennial Editions

These manuals contain repair and maintenance information for all major systems. Included are repair and overhaul procedures using thousands of illustrations.

CHILTON AUTO REPAIR MANUALS

1998-2002
ISBN 978-0-8019-9362-6/Part No. 9362
Covers all popular American and Canadian cars. An added feature includes scheduled maintenance interval charts.

1993-97
ISBN 978-0-8019-7919-4/Part No. 7919
Covers all popular American and Canadian cars.

1980-87
ISBN 978-0-8019-7670-4/Part No. 7670
Covers all popular American and Canadian cars.

CHILTON IMPORT AUTO REPAIR MANUALS

1998-2002
ISBN 978-0-8019-9363-3/Part No. 9363
Covers all popular Import cars. An added feature includes scheduled maintenance intervals charts.

1993-97
ISBN 978-0-8019-7920-0/Part No. 7920
Covers all popular Import cars.

1988-92
ISBN 978-0-8019-7907-1/Part No. 7907
Covers all popular Import cars.

1980-87
ISBN 978-0-8019-7672-8/Part No. 7672
Covers all popular Import cars.

CHILTON TRUCK AND VAN REPAIR MANUALS

1998-2002
ISBN 978-0-8019-9364-0/Part No. 9364
Covers popular U.S., Canadian, and Import Pick-Ups, Vans, and 4WDs. An added feature includes scheduled maintenance interval charts.

1993-97
ISBN 978-0-8019-7921-7/Part No. 7921
Covers popular U.S., Canadian, and Import Pick-Ups, Sport-Utilities, Vans, RVs and 4 wheel drives.

1991-95
ISBN 978-0-8019-7911-8/Part No. 7911
Covers popular U.S., Canadian, and Import Pick-Ups, Vans, RVs and 4 wheel drives.

1986-90
ISBN 978-08019-7902-6/Part No. 7902
Covers popular U.S., Canadian, and Import Pick-Us, Vans, RVs and 4 wheel drives.

1979-86
ISBN 978-08019-7655-1/Part No. 7655
Covers popular U.S., Canadian, and Import Pick-Ups, Vans, RVs and 4 wheel drives.

CHILTON SUV REPAIR MANUAL

1998-2002
ISBN 978-08019-9365-7/Part No. 9365
Covers popular U.S., Canadian, and import SUVs. An added feature includes scheduled maintenance intervals charts.

COLLECTOR'S SERIES

CHILTON AUTO REPAIR MANUAL 1964-1971
ISBN 978-08019-5974-5/Part No. 5974
1971-1978
ISBN 978-08019-7012-2/Part No. 7012

Chilton Timing Belts, 1985-2005

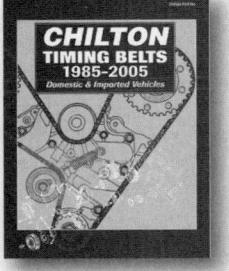

ISBN 978-1-4018-9880-9
Part No. 129880
544 pp, 8" x 11", SC, ©2006

Timing belt procedures can represent increased profits for automotive repair shops and service stations, and this manual contains all the information automotive technicians need to properly service timing belts on domestic and imported cars, vans, and light trucks through 2005 models. Clear, straightforward procedures, illustrations, and specifications help to communicate 20 years of vehicle applications for fast, accurate inspection, replacement, and tensioning of timing belts. Users will learn how to perform key procedures quickly and safely, while learning the correct labor time to charge for the service.

ALSO AVAILABLE:
Quick-Reference Manuals

The Chilton Professional Series offers *Quick-Reference Manuals* for the automotive professional, providing complete coverage on repair and maintenance, adjustments, and diagnostic procedures for specific systems and components.

KEY FEATURES

- step-by-step procedures
- detailed illustrations and exploded views
- easy-to-use manufacturer and model indexing
- handy specifications or data charts

Heater Core Service 1990-2000,
ISBN 978-0-8019-9311-4
Part No. 9311

Brake Specifications and Service 1990-2000
ISBN 978-0-8019-9312-1
Part No. 9312

Electric Cooling Fans, Accessory Drive Belts & Water Pumps, 1995-1999,
ISBN 978-0-8019-9126-4
Part No. 9126

Powertrain Codes & Oxygen Sensors, 1990-1999,
ISBN 978-0-8019-9127-1
Part No. 9127